Handbook of
INDUSTRIAL SURFACTANTS

Handbook of INDUSTRIAL SURFACTANTS

An International Guide to More Than 16,000 Products by Tradename, Application, Composition & Manufacturer

Compiled by

Michael and Irene Ash

LONDON AND NEW YORK

First published 1993 by Gower Publishing

Reissued 2019 by Routledge
2 Park Square, Milton Park, Abingdon, Oxon, OX14 4RN
52 Vanderbilt Avenue, New York, NY 10017

Routledge is an imprint of the Taylor & Francis Group, an informa business

A Library of Congress record exists under LC control number:

ISBN 13: 978-0-367-17851-2 (hbk)
ISBN 13: 978-0-367-17854-3 (pbk)
ISBN 13: 978-0-429-05806-6 (ebk)

Contents

Preface

Over 16,000 tradename surface-active agents for industrial applications, manufactured worldwide, are contained in this edition. General-use surfactants, such as emulsifiers, wetting agents, foaming agents, detergents, dispersants, and solubilizers are included, as well as detergent raw materials, defoamers, and antifoaming agents.

The applications for these products are diverse and an attempt has been made to focus on surfactant products used in industry as opposed to personal care and cosmetic preparations. There is , however, an overlap because of similarity in chemical composition. Many times, a product's application is determined by the manufacturer's marketing focus.

The demand for surfactant products is growing by approximately 6% each year. Much of this growth can be attributed to: development of new areas of applications , environmental demands for new products that satisfy increasingly stringent ecological requirements, and research and development of new products for improved performance in traditional applications.

The types and quantities of surfactants available commercially are numerous and the difficulty in making choices between products may become overwhelming. It is the purpose of this book to guide those who are involved in the selection of these materials through the process of identifying, classifying, and selecting the most appropriate products for their requirements. Therefore, this reference is organized so that the user can search for and locate products based on a variety of essential distinguishing attributes:

tradename
manufacturer
chemical composition
application area
CAS (Chemical Abstract Service) Registry number
EINECS (European Inventory of Existing Commercial Chemical Substances) number
ionic classification
HLB (hydrophile-lipophile balance) value

Identifying surfactant chemicals through these cross references, permits the user to make an informed decision on what to purchase based on relevant criteria for end-product requirements.

Another important feature of this reference is the inclusion of historical tracing for both products and their manufacturers. Products that have been discontinued by companies have been noted; products with name changes are cross referenced to the new tradename; and a list of companies that have been acquired by other chemical manufacturers is included in the Appendix.

The book is divided into four sections:

Part I—*Tradename Reference* contains an alphabetical arrangement of surfactant-type tradename product entries that have been gathered from approximately 500 manufacturer sources. Each entry references its manufacturer, chemical composition, associated CAS number, general properties such as specific gravity, HLB, flash point, surface tension, Draves wetting, and Ross-Miles foam height, functions, applications, toxicology, and compliance and regulatory information as provided by the manufacturer.

Part II—*Tradename Application Cross Reference* lists tradename products under seventeen broad-based application areas. Some of the categories included are: agricultural chemicals ; industrial, institutional, and consumer cleaning; textiles and fibers; mining and mineral processing; paints, coatings, lacquers, inks, and adhesives; metalworking, cleaning, processing, corrosion inhibiting, cutting and drilling oils, pickling and plating; etc.

Part III— *Chemical Component Cross Reference* contains an alphabetical listing of surfactant chemical compounds. Each chemical entry lists the tradename products that is equivalent to the chemical compound or contains it as the tradename's major chemical constituent. Wherever possible, CAS numbers, EINECS numbers, and CTFA (Cosmetic, Toiletry, and Fragrance Association) adopted names for the chemical entry are included. Synonyms for the chemical compounds are grouped with the entry and major synonyms are thoroughly cross referenced back to the main entry.

Part IV— *Manufacturers Directory* contains detailed contact information for the manufacturers of the more than 16,000 tradename products referenced in this handbook. Approximately 500 companies, divisions, subsidiaries, and branch offices worldwide are included. Wherever possible, telephone numbers, toll-free 800 numbers, fax numbers, telex numbers and complete mailing addresses are listed for each manufacturer .

The Appendix contains several additional cross references:

CAS Number-to-Tradename Cross Reference orders many tradenames found in Part I by identifying CAS numbers; it should be noted that tradenames often contain more than one chemical component and the CAS numbers listed in this section refer to each tradename's primary component.

CAS Number-to-Chemical Compound Cross Reference orders chemicals found in Part III by CAS numbers.

EINECS-to-CAS Number Cross Reference relates EINECS numbers to CAS identifiers and includes the specific chemical referenced.

Ionic Classification categorizes the tradename products by four major classes of surfactants: anionic, cationic, nonionic, and amphoteric. The ionic nature of products is often a determining factor for the designated application.

HLB Classification orders the tradename products by the HLB value. It is an indicator for selection of surfactants based on functional requirements, e.g., W/O emulsions, wetting, O/W emulsions, detergency, or solubilization. However, products with similar HLB values may still have disparate performance because of differences in chemical structure or physical chemistry. These values are a guide but not a definitive indicator.

This book is the culmination of many months of research, investigation of product sources, and sorting through a variety of technical data sheets and brochures acquired through trade journals, personal contacts, and correspondence with major chemical manufacturers worldwide. We are especially grateful to Roberta Dakan for her skill and dedication in the development and maintenance of this chemical tradename database resource. Her talent and persistence have been instrumental in the production of this reference.

M. & I. Ash

Abbreviations

act.	active
agric.	agricultural
agrochem.	agrochemical
a.i.	active ingredient
anhyd.	anhydrous
APHA	American Public Health Association
applic(s)	application(s)
aq.	aqueous
ASTM	American Society for Testing and Materials
aux.	auxiliary
avail.	available
avg.	average
b.p.	boiling point
B/S	butadiene/styrene
BGA	Federal Republic of Germany Health Dept. certification
BHA	butylated hydroxyanisole
BHT	butylated hydroxytoluene
biodeg.	biodegradable
blk, blk.	black
BP	British Pharmacopeia
B&R	Ball & Ring
br., brn.	brown
brnsh.	brownish
C	degrees Centigrade
cap	capillary
CAS	Chemical Abstracts Service
CC	closed cup
cc	cubic centimeter(s)
CCl_4	carbon tetrachloride
CFR	Code of Federal Regulations
char.	characteristic
chem(s)	chemical(s)
cm	centimeter(s)
cm^3	cubic centimeter(s)
CMC	critical Micelle concentration
COC	Cleveland Open Cup
compd.	compound
conc.	concentrated, concentration
cosolv.	cosolvent
CP	Canadian Pharmacopeia
cps	centipoise(s)
cryst.	crystalline, crystallization
cs or cSt	centistoke(s)
CTFA	Cosmetic, Toiletry and Fragrance Association
DEA	diethanolamide, diethanolamine
dec.	decomposes
DEG	diethylene glycol
dens.	density
deriv.	derivative(s)
dg	decigram(s)
diam.	diameter
disp.	dispersible, dispersion
dist.	distilled

dk.	dark
DOT	Department of Transportation
EDTA	ethylene diamine tetraacetic acid
elec.	electrical
EO	ethylene oxide
EP	extreme pressure
EPA	Environmental Protection Agency
EPDM	ethylene-propylene-diene rubbers
EPM	ethylene-propylene rubbers
EPR	ethylene-propylene rubber
equip.	equipment
esp.	especially
EVA	ethylene vinyl acetate
exc	excellent
F	degrees Fahrenheit
f.p.	freezing point
FD&C	Foods, Drugs, and Cosmetics
FDA	Food and Drug Administration
FFA	free fatty acid
flam(m).	flammable, flammability
ft	foot, feet
F-T	Fischer-Tropsch
G	giga
g	gram(s)
G-H	Gardner-Holdt
gal	gallon(s)
gr.	gravity
gran.	granules, granular
grn.	green
h.	hour(s)
HC	hydrocarbon
HCl	hydrochloric acid
Hg	mercury
HLB	hydrophilic lipophilic balance
hyd	hydroxyl, hydrogenated
hydrog	hydrogenated
i.b.p.	initial boiling point
IIR	isobutylene-isoprene rubber
in.	inch(es)
incl	including
ingred	ingredient(s)
inorg.	inorganic
insol.	insoluble
IPA	isopropyl alcohol, isopropanol
IPM	isopropyl myristate
IPP	isopropyl palmitate
IR	isoprene rubber (synthetic)
k	kilo
kg	kilogram(s)
KU	Krebs units
l or L	liter(s)
lb	pound(s)
LDPE	low-density polyethylene
liq.	liquid
lt.	light
M	mega
m	milli or meter(s)

m^3	cubic meters
max.	maximum
MEA	monoethanolamine, monoethanolamide
med.	medium
MEK	methyl ethyl ketone
mfg.	manufacture
mg	milligram(s)
MIBK	methyl isobutyl ketone
min	minute(s), mineral, minimum
MIPA	monoisopropanolamine, monoisopropanolamide
misc	miscible
mixt	mixture(s)
ml	milliliter(s)
mm	millimeter(s)
mN	millinewton(s)
mod.	modulus, moderately
m.p.	melting point
m.w.	molecular weight
NaCl	sodium chloride
nat	natural
NBR	nitrile-butadiene rubber
NC	nitrocellulose
NF	National Formulary
no.	number
nonflamm.	nonflammable
nonyel.	nonyellowing
NR	isoprene rubber (natural)
NV	nonvolatiles
o/w	oil-in-water
OC	open crucible
org.	organic
Pa	Pascal
PE	polyethylene
PEG	polyethylene glycol
petrol	petroleum
pH	hydrogen-ion concentration
pkg	packaging
P-M	Pensky-Martens
PMCC	Pensky-Martens closed cup
PO	propylene oxide
POE	polyoxyethylene, polyoxyethylated
POP	polyoxypropylene, polyoxypropylated
powd	powder
PP	polypropylene
PPG	polypropylene glycol
pract	practically
prep	preparation(s)
prod	product(s), production
props.	properties
pt.	point
PU	polyurethane
PVAc	polyvinyl acetate
PVC	polyvinyl chloride
PVP	polyvinylpyrrolidone
quat.	quaternary
R&B	Ring & Ball
R.T.	room temperature

rdsh.	reddish
ref.	refractive
resist.	resistance, resistant, resistivity
resp.	respectively
s	second(s)
S/B	styrene/butadiene
SAN	styrene-acrylonitrile
sapon	saponification
sat	saturated
SBR	styrene/butadiene rubber
SDA	specially denatured alcohol
SE	self-emulsifying
sec	secondary
sm.	small
SMA	styrene maleic anhydride
soften	softening
sol	soluble, solubility
sol'n.	solution
solid	solidification
solv(s)	solvent(s)
sp	specific
std	standard
Stod	Stoddard solvent
str	strength
surf	surface
SUS	Saybolt Universal Seconds
syn	synthetic
t	tertiary
TCC	Tagliabue closed cup
TEA	triethanolamine, triethanolamide
tech	technical
temp	temperature
tens.	tensile or tension
tert	tertiary
thru	through
typ	typical
USP	Unites States Pharmacopeia
uv	ultraviolet
VA	vinyl acetate
veg	vegetable
visc	viscous, viscosity
w/o	water-in-oil
wh	white
wt.	weight
yel	yellow
ylsh.	yellowish
#	number
%	percent
<	less than
>	greater than
@	at
μ	mu (10^{-6})
≈	approximately

Part I
Tradename Reference

AA#2 Lime Additive. [Crown Tech.] Surfactants/antifoams blend; for lime or borax coating operations; clear to milky wh. liq., pleasant odor; dens. 8.5 lb/gal; pH 7-8.

AA Standard. [CasChem] Castor oil; emollient for industrial applics. where lt. color, high purity, and low acidity are desirable; plasticizer, wetting agent, lubricant for rubber compding.; FDA approval; Gardner 1+; sol. in alcohols, esters, ethers, ketone, and aromatic solvs.; sp.gr. 0.959; visc. 7.5 stokes; pour pt. -10 F; acid no. 2; iodine no. 86; sapon. no. 180; hyd. no. 164.

AB®. [Angus] 2-Amino-1-butanol; pigment dispersant, neutralizing/emulsifying amine, corrosion inhibitor, acid-salt catalyst, pH buffer, chemical and pharmaceutical intermediate, solubilizer; m.w. 89.1; water-sol.; m.p. –2 C; b.p. 178 C; flash pt. 193 F (TCC); pH 11.1 (0.1M aq. sol'n.); 99% conc.

Abex® 12S. [Rhone-Poulenc Surf.] Surfactant; anionic; emulsifier for vinyl acrylics and acrylic polymerization; liq.; 30% conc.

Abex® 18S. [Rhone-Poulenc Surf.; Rhone-Poulenc France] Proprietary surfactant; anionic; detergent, low foaming emulsifier for polymerization of vinyl acetate and acrylics; mineral oil emulsifier; rewetting agent; FDA compliance; liq.; surf. tens. 48 dynes/cm (@ CMC); 35% conc.

Abex®22S. [Rhone-Poulenc Surf.] Proprietary surfactant; anionic; detergent, low foaming emulsifier for use in polymerization of vinyl acetate and acrylics; mineral oil emulsifier; rewetting agent; FDA compliance; liq.; surf. tens. 55 dynes/cm (@ CMC); 25% conc.

Abex®23S. [Rhone-Poulenc Surf.] Proprietary surfactant; anionic; emulsifier for emulsion polymerization; liq.; surf. tens. 40 dynes/cm (@ CMC); 60% conc.

Abex® 26S. [Rhone-Poulenc Surf.; Rhone-Poulenc France] Proprietary surfactant; anionic; emulsifier for emulsion polymerization of vinyl acetate and acrylics; FDA compliance; liq.; surf. tens. 55 dynes/cm (@ CMC); 33% conc.

Abex®33S. [Rhone-Poulenc Surf.] Proprietary surfactant; anionic; emulsifier for vinyl acrylics and acrylic polymerization; liq.; surf. tens. 37 dynes/cm (@ CMC); 27% conc.

Abex® 1404. [Rhone-Poulenc Surf.] Proprietary surfactant; anionic; surfactant for emulsion polymerization; liq.; 42% act. DISCONTINUED.

Abex® AAE-301. [Rhone-Poulenc Surf.] Surfactant; anionic; polymerization emulsifier; liq.; 20% conc.

Abex® EP-110 (formerly Alipal® EP-110). [Rhone-Poulenc Surf.; Rhone-Poulenc France] Ammonium nonoxynol-9 sulfate; anionic; primary emulsifier and stabilizing agent for the preparation of vinyl acetate, vinyl acetate/acrylic, all acrylic, styrene/acrylic, and S/B emulsion copolymers; wetting agent, dispersant for agric. formulations; FDA, EPA compliance; pale yel. liq.; sp.gr. 1.04; visc. 91.0 cks; pour pt. 0 C; surf. tens. 38.3 dynes/cm (1%); 30% act.

Abex® EP-115 (formerly Alipal® EP-115). [Rhone-Poulenc Surf.] Ammonium nonoxynol-20 sulfate; anionic; emulsifier for agric. formulations and polymerization of most monomer systems; for use in latexes where extended shear stability and freeze/thaw stability are desired; liq.; surf. tens. 40 dynes/cm (@ CMC); 30% conc.

Abex® EP-120 (formerly Alipal® EP-120). [Rhone-Poulenc Surf.; Rhone-Poulenc France] Ammonium nonoxynol-30 sulfate; anionic; emulsifier for emulsion polymerization of vinyl acetate; provides mech. and freeze/thaw stability; wetting agent, dispersant for agric. formulations; FDA compliance; pale yel. liq.; sp.gr. 1.06; visc. 111 cks; pour pt. 8 C; surf. tens. 40 dynes/cm (1%); 30% solids.

Abex® EP-227 (formerly Alipal® EP-227). [Rhone-Poulenc Surf.] Ammonium nonoxynol-77 sulfate; anionic; emulsifier, stabilizer for prep. of vinyl acetate, vinyl acetate/acrylic, acrylic, styrene/acrylic, and styrene/butadiene emulsion copolymers; emulsifier for agric. formulation; liq.; surf. tens. 44 dynes/cm (@ CMC); 30% act.

Abex® JKB. [Rhone-Poulenc Surf.; Rhone-Poulenc France] Proprietary surfactant; anionic; emulsifier for high-acid polymerization systems; FDA compliance; liq.; surf. tens. 40 dynes/cm (@ CMC); 30% conc.

Abex® LIV/30 (formerly Geropon® LIV/30). [Rhone-Poulenc Geronazzo] Ammonium alkylaryl ether sulfate; anionic; emulsifier for emulsion polymerization of acrylic, styrene-acrylic, vinyl acetate; detergent, emulsifier, foam stabilizer, wetting agent in household and industrial detergents, shampoos, bubble baths; liq.; 30% conc.

Abex® LIV/2330 (formerly Geropon® LIV/2330). [Rhone-Poulenc Geronazzo] Ammonium alkylaryl ether sulfate; anionic; primary emulsifier for acrylic, vinyl acetate, vinyl acrylic, styrene acrylic, styrene-butadiene; liq.; 30% conc.

Abex® VA 50. [Rhone-Poulenc Surf.; Rhone-Poulenc France] Octoxynol-33, sodium laureth sulfate; anionic; emulsifier for high solids vinyl acetate emulsions; FDA compliance; liq.; surf. tens. 44 dynes/cm (@ CMC); 46% conc.

Abil® B 8843, B 8847. [Goldschmidt; Goldschmidt AG] Dimethicone copolyol; CAS 68937-55-3; nonionic; surfactant, conditioner used in personal care prods.; emollient for hair and skin care prods., aerosol shaving lather, deodorants, antiperspirants, creams and lotions, perfumes and colognes; liq.; water-sol.; 100% conc.

Abil® B 9806, B 9808. [Goldschmidt] Cetyl dimethicone copolyol; surfactant, emollient for creams and lotions; emulsifier for cyclomethicone.

Abil® B 88184. [Goldschmidt; Goldschmidt AG] Dimethicone copolyol; CAS 68937-55-3; nonionic; surfactant for hair and skin care prods.; anticracking agent for soap bars; liq.; 100% conc.

Ablufoam HT. [Taiwan Surf.] Organic compd.; antifoamer for high temp. jet dyeing; water-dilutable.

Ablufoam SAE. [Taiwan Surf.] Silicone emulsion; textile antifoamer effective over wide pH range; water-dilutable.

Abluhide DS. [Taiwan Surf.] Blend; anionic/nonionic; soaking, degreasing and rewetting agent for leather mfg.

Abluhide F Series. [Taiwan Surf.] Surfactant; fatliquoring agent for leather mfg.

Abluhide I. [Taiwan Surf.] Blend; anionic/nonionic; emulsifier, degreaser for leather mfg.

Abluhide NNO. [Taiwan Surf.] Aromatic compd.; leather dyeing aux., leveling agent, penetrant; for bleaching and neutralizing chrome leathers.

Ablumide CDE. [Taiwan Surf.] Cocamide DEA (1:1); nonionic; foam stabilizer, thickener for shampoos, bubble baths, liq. detergents, toiletries; liq.; 90% conc.

Ablumide CME. [Taiwan Surf.] Cocamide MEA (1:1); CAS 68140-00-1; nonionic; foam stabilizer, thickener for shampoos, bubble baths, liq. detergents, toiletries; solid; 95% conc.

Ablumide LDE. [Taiwan Surf.] Lauramide DEA (1:1); CAS 120-40-1; nonionic; foam stabilizer, thickener for shampoos, bubble baths, liq. detergents, toiletries; solid; 92% conc.

Ablumide LME. [Taiwan Surf.] Lauramide MEA (1:1); CAS 142-78-9; nonionic; foam stabilizer, thickener for shampoos, bubble baths, liq. detergents, toiletries; solid; 95% conc.

Ablumide SDE. [Taiwan Surf.] Stearamide DEA (1:1); nonionic; thickener, emulsifier for min. and veg. oils, microcrystalline wax; solid; 90% conc.

Ablumine 08. [Taiwan Surf.] Alkyl (99% C8) benzyl dimethyl quat.; cationic; leveling agent for acrylic fiber; liq.; 50% act.

Ablumine 10. [Taiwan Surf.] Alkyl (99% C10) benzyl dimethyl quat.; cationic; leveling agent for acrylic fiber; liq.; 50% act.

Ablumine 12. [Taiwan Surf.] Alkyl (97% C12, 2% C14) benzyl dimethyl quat.; cationic; retarder for acrylic fiber; disinfectant, sanitizer, germicide, algicide for swimming pools, water cooling systems, cleaners for hospitals, institutions, breweries, and food plants; liq.; 50 or 80% act.

Ablumine 18. [Taiwan Surf.] Alkyl (4% C16, 92% C18) benzyl dimethyl quat.; cationic; conditioner, softener, antistat for human hair, wool, cotton, other cellulosic fibers; paste; 25% act.

Ablumine 230. [Taiwan Surf.] N-alkyl dimethyl ammonium chloride; cationic; algicide for industrial cooling towers and swimming pools; liq.; 50% act.

Ablumine 1214. [Taiwan Surf.] Alkyl(63% C12, 30% C14, 7% C16) benzyl dimethyl quat.; cationic; retarder for acrylic fiber; disinfectant, sanitizer, germicide, algicide for swimming pools, water cooling systems, cleaners for hospitals, institutions, breweries, and food plants; liq.; 50 or 80% act.

Ablumine 3500. [Taiwan Surf.] Alkyl (40% C12, 50% C14, 10% C16) benzyl dimethyl quat.; cationic; retarder for acrylic fiber; disinfectant, sanitizer, germicide, algicide for swimming pools, water cooling systems, cleaners for hospitals, institutions, breweries, and food plants; liq.; 50 or 80% act.

Ablumine AN. [Taiwan Surf.] Quat. ammonium salt; cationic; leveling agent for dyeing acrylic fibers; high migrating power.

Ablumine D10. [Taiwan Surf.] Didecyl dimethyl ammonium methosulfate; cationic; disinfectant, sanitizer, germicide; liq.; 50% act.

Ablumine DHT75. [Taiwan Surf.] Dihydrog. tallow dimethyl ammonium methosulfate; cationic; antistat, fabric softener suitable for dryer sheets; soft paste; 75% act.

Ablumine DHT90. [Taiwan Surf.] Dihydrog. tallow dimethyl ammonium methosulfate; cationic; antistat, fabric softener suitable for dryer sheets; almost wh. solid; 90% act.

Ablumine DT. [Taiwan Surf.] Ditallow dimethyl ammonium methosulfate; cationic; household softener with wide dispersion stability; paste; 90% act.

Ablumine PN. [Taiwan Surf.] Complex alkyl dimethy benzyl chloride; cationic; retarder for acrylic fiber; prevents unlevel dyeing.

Ablumox C-7. [Taiwan Surf.] PEG-7 cocamine; CAS 61791-14-8; nonionic; corrosion inhibitor for acid cleaners; liq.

Ablumox CAPO. [Taiwan Surf.] Cocamidopropyl amine oxide; nonionic; foamer, wetting agent, foam stabilizer, antistat, detergent, emollient; liq.

Ablumox LO. [Taiwan Surf.] Lauramine oxide; nonionic; foamer, wetting agent, foam stabilizer, antistat, detergent, emollient; liq.

Ablumox T-15. [Taiwan Surf.] PEG-15 tallow amine; nonionic; leveling agent for dyeing of nylon with anionic dyes; controls rate of strike and dye migration; liq.; 100% conc.

Ablumox T-20. [Taiwan Surf.] PEG-20 tallowamine; nonionic; leveling agent for dyeing of nylon with anionic dyes; controls rate of strike and dye migration; liq.; 100% conc.

Ablumul AG-306. [Taiwan Surf.] Blend; anionic/nonionic; emulsifier for Acephate (agric.); liq.; sp.gr. 1.02.

Ablumul AG-420. [Taiwan Surf.] Blend; anionic/nonionic; emulsifier for DDVP (agric.); liq.; sp.gr. 1.04; HLB 15.2.

Ablumul AG-900. [Taiwan Surf.] Blend; cationic/nonionic; spreading and wetting agent for Paraquat (agric.); liq.; sp.gr. 1.02; 70% conc.

Ablumul AG-909. [Taiwan Surf.] Blend; nonionic; spreading and wetting agent for agric. chemicals; liq.; sp.gr. 1.04.

Ablumul AG-910. [Taiwan Surf.] Blend; anionic/nonionic; spreading and wetting agent for Tamarone, Azodrin (agric.); liq.; sp.gr. 1.05.

Ablumul AG-1214. [Taiwan Surf.] Blend; cationic/nonionic; emulsifier for BLA-S (agric.); liq.; sp.gr. 0.96; 50% conc.

Ablumul AG-AH. [Taiwan Surf.] Blend; nonionic; emulsifier for Dimethoate, etc. (agric.); liq.; sp.gr. 1.03; HLB 13.5.

Ablumul AG-GL. [Taiwan Surf.] Blend; anionic/nonionic; spreading and wetting agent for Glyphosate (agric.); liq.

Ablumul AG-H. [Taiwan Surf.] Blend; anionic/nonionic; emulsifier for Methyl Parathion (agric.); liq.;

sp.gr. 1.05; HLB 13.9.

Ablumul AG-KP3. [Taiwan Surf.] Blend; anionic/nonionic; emulsifier for Kitazin-P (agric.); liq.; sp.gr. 1.04; HLB 13.9.

Ablumul AG-KTM. [Taiwan Surf.] Blend; anionic/nonionic; emulsifier for Mon, Butachlor (agric.); liq.; sp.gr. 1.04; HLB 13.5.

Ablumul AG-L. [Taiwan Surf.] Blend; anionic/nonionic; matched pair emulsifier with Ablumul AG-MBX for agric. toxicants; liq.; sp.gr. 1.05; HLB 13.7.

Ablumul AG-MBX. [Taiwan Surf.] Blend; anionic/nonionic; matched pair emulsifier with Ablumul AG-L for agric. toxicants; liq.; sp.gr. 1.04; HLB 14.0.

Ablumul AG-SB. [Taiwan Surf.] Blend; anionic/nonionic; spreading and wetting agent for agric. chemicals; liq.; sp.gr. 1.02; 35% conc.

Ablumul AG-WP. [Taiwan Surf.] Blend; anionic/nonionic; emulsifiers for agric. wettable powds.; powd.; sp.gr. 0.15-0.30.

Ablumul AG-WPS. [Taiwan Surf.] Anionic/nonionic; agric. surfactant for wettable powds.; liq.; sp.gr. 1.03.

Ablumul EP. [Taiwan Surf.] Surfactant; nonionic; emulsifier, thickener for textile printing.

Ablumul M. [Taiwan Surf.] Surfactant; anionic/nonionic; emulsifier for methyl naphthalene carriers.

Ablumul MI. [Taiwan Surf.] Blend; anionic/nonionic; emulsifier for mfg. of MN type carrier; liq.

Ablumul OCE. [Taiwan Surf.] Surfactant; anionic; emulsifier for prep. of biocides; liq.

Ablumul OP. [Taiwan Surf.] Surfactant; anionic/nonionic; emulsifier fororthodichlorobenzene carriers.

Ablumul S Series. [Taiwan Surf.] Surfactant; anionic/nonionic; emulsifier for spindle oil, machine oil, agric. applics.; liq.; sp.gr. 1.03; 97% conc.

Ablumul S2S. [Taiwan Surf.] Blend; anionic/nonionic; emulsifier for summer oil, machine oil, spindle oil; liq.; 3% conc.

Ablumul T. [Taiwan Surf.] Surfactant; anionic/nonionic; emulsifier for trichlorobenzene carriers.

Ablumul T2. [Taiwan Surf.] Blend; anionic/nonionic; emulsifier for mfg. of trichlorobenzene carrier; liq.; 3% conc.

Ablumul TN. [Taiwan Surf.] Surfactant; nonionic; emulsifier for chlorobenzene carriers used in textile dyeing.

Ablunol 200ML. [Taiwan Surf.] PEG 200 laurate; CAS 9004-81-3; nonionic; emulsifier, lubricant, dispersing and leveling agent, defoamer used in cosmetic, textile, paint, dyestuffs, and other industrial uses; liq.; HLB 9.5; 100% act.

Ablunol 200MO. [Taiwan Surf.] PEG 200 oleate; CAS 9004-96-0; nonionic; emulsifier, lubricant, dispersing and leveling agents used in cosmetic, textile, leather, paint and other industrial uses; liq.; HLB 7.9; 100% act.

Ablunol 200MS. [Taiwan Surf.] PEG 200 stearate; CAS 9004-99-3; nonionic; emulsifier, thickener, lubricant, softener, defoamer, dispersing and leveling agent used in cosmetic, textile, paint and other industrial uses; solid; HLB 8.0; 100% act.

Ablunol 400ML. [Taiwan Surf.] PEG 400 laurate; CAS 9004-81-3; nonionic; emulsifier, lubricant, dispersing and leveling agent, defoamer used in cosmetic, textile, paint, dyestuffs, and other industrial uses; liq.; HLB 13.1; 100% act.

Ablunol 400MO. [Taiwan Surf.] PEG 400 oleate; CAS 9004-96-0; nonionic; emulsifiers, lubricants, dispersing and leveling agents used in cosmetic, textile, leather, paint and other industrial uses; liq.; HLB 11.5; 100% act.

Ablunol 400MS. [Taiwan Surf.] PEG 400 stearate; CAS 9004-99-3; nonionic; emulsifier, thickener, lubricant, softener, defoamer, dispersing and leveling agent used in cosmetic, textile, paint and other industrial uses; solid; HLB 11.6; 100% act.

Ablunol 600ML. [Taiwan Surf.] PEG 600 laurate; CAS 9004-81-3; nonionic; emulsifier, lubricant, dispersing and leveling agent, defoamer used in cosmetic, textile, paint, dyestuffs, and other industrial uses; liq.; HLB 15.0; 100% act.

Ablunol 600MO. [Taiwan Surf.] PEG 600 oleate; CAS 9004-96-0; nonionic; emulsifiers, lubricants, dispersing and leveling agents used in cosmetic, textile, leather, paint and other industrial uses; liq.; HLB 13.5; 100% act.

Ablunol 600MS. [Taiwan Surf.] PEG 600 stearate; CAS 9004-99-3; nonionic; emulsifier, thickener, lubricant, softener, defoamer, dispersing and leveling agent used in cosmetic, textile, paint and other industrial uses; solid; HLB 13.6; 100% act.

Ablunol 1000MO. [Taiwan Surf.] PEG 1000 oleate; CAS 9004-96-0; nonionic; emulsifiers, lubricants, dispersing and leveling agents used in cosmetic, textile, paint and other industrial uses; solid; HLB 15.2; 100% conc.

Ablunol 1000MS. [Taiwan Surf.] PEG 1000 stearate; CAS 9004-99-3; nonionic; emulsifiers, lubricants, dispersing and leveling agents used in cosmetic, textile, paint and other industrial uses; solid; HLB 15.2; 100% conc.

Ablunol 6000DS. [Taiwan Surf.] PEG 6000 distearate; CAS 9005-08-7; nonionic; emulsifier, thickener for cosmetic, pigment preparations, textile printing; flake; HLB 19.0; 100% act.

Ablunol CO 5. [Taiwan Surf.] PEG-5 castor oil; nonionic; emulsifier, lubricant, antistat; liq.; HLB 4.0.

Ablunol CO 10. [Taiwan Surf.] PEG-10 castor oil; nonionic; emulsifier for oils, solvs. and waxes, lubricant, antistat; liq.; HLB 6.6; cloud pt. 44-50 C (10% in 25% butyl Carbitol).

Ablunol CO 15. [Taiwan Surf.] PEG-15 castor oil; nonionic; emulsifier, lubricant, antistat; liq.; HLB 8.5.

Ablunol CO 30. [Taiwan Surf.] PEG-30 castor oil; nonionic; emulsifier, lubricant, antistat; liq.; HLB 11.8.

Ablunol CO 45. [Taiwan Surf.] PEG-45 castor oil; nonionic; emulsifier for oils, solvs. and waxes, lubricant, antistat; liq.; HLB 13.8; cloud pt. 50-56 C (1% in 5% sodium sulfate).

Ablunol DEGMS. [Taiwan Surf.] Diethylene glycol stearate; CAS 9004-99-3; nonionic; opacifier, pearlescent for cosmetics, detergents; solid; HLB 3.7.

Ablunol EGMS. [Taiwan Surf.] Glycol stearate; nonionic; opacifier, pearlescent for cosmetics, detergents; solid; HLB 2.9.

Ablunol GML. [Taiwan Surf.] Glyceryl laurate; nonionic; component in mold release agents; amber liq./paste; 100% act.

Ablunol GMO. [Taiwan Surf.] Glyceryl oleate; CAS 111-03-5; nonionic; internal lubricant, antistat, antifogging agent for PVC film; mold release agent and rust prevention for compounded oils; amber liq. to paste.

Ablunol GMS. [Taiwan Surf.] Glyceryl stearate; nonionic; emulsifier for hand creams, lotions, cosmetics; textile lubricant-softener; solid; 100% act.

Ablunol LA-3. [Taiwan Surf.] Laureth-3; nonionic; emulsifier, dispersant, detergent, wetting agent used in textile processing, cosmetics, metalworking compds., agric., industrial cleaners; liq.; HLB 7.9; 100% act.

Ablunol LA-5. [Taiwan Surf.] Laureth-5; nonionic; emulsifier, dispersant, detergent, wetting agent used in textile processing, cosmetics, metalworking compds., agric., industrial cleaners; liq.; HLB 10.5; 100% act.

Ablunol LA-7. [Taiwan Surf.] Laureth-7; nonionic; emulsifier, dispersant, detergent, wetting agent used in textile processing, cosmetics, metalworking compds., agric., industrial cleaners; liq.; HLB 12.1; cloud pt. 28-38 C; 100% act.

Ablunol LA-9. [Taiwan Surf.] Laureth-9; nonionic; emulsifier, dispersant, detergent, wetting agent used in textile processing, cosmetics, metalworking compds., agric., industrial cleaners; liq.; HLB 13.3; cloud pt. 51-61 C; 100% act.

Ablunol LA-12. [Taiwan Surf.] Laureth-12; nonionic; emulsifier, dispersant, detergent, wetting agent used in textile processing, cosmetics, metalworking compds., agric., industrial cleaners; liq.; HLB 14.5; cloud pt. 78-88 C; 100% act.

Ablunol LA-16. [Taiwan Surf.] Laureth-16; nonionic; emulsifier, dispersant, detergent, wetting agent used in textile processing, cosmetics, metalworking compds., agric., industrial cleaners; semisolid; HLB 15.2; 100% act.

Ablunol LA-40. [Taiwan Surf.] Laureth-40; nonionic; emulsifier, dispersant, detergent, wetting agent used in textile processing, cosmetics, metalworking compds., agric., industrial cleaners; solid; HLB 18.0; 100% act.

Ablunol LMO. [Taiwan Surf.] Surfactant; nonionic; leveling agent for dyeing polyester fibers with disperse dyes.

Ablunol LN. [Taiwan Surf.] Surfactant; nonionic; leveling agent for dyeing nylon fibers with acid dyes.

Ablunol NP4. [Taiwan Surf.] Nonoxynol-4; CAS 9016-45-9; nonionic; detergent and dispersant for use in petrol. oils; intermediate for mfg. of surfactants and antistatic agents; co-emulsifier for fats, oils and waxes; liq.; oil-sol.; sp.gr. 1.02; HLB 8.9; cloud pt. 56-61 C (10% in 25% butyl Carbitol); 100% act.

Ablunol NP6. [Taiwan Surf.] Nonoxynol-6; CAS 9016-45-9; nonionic; emulsifier or coemulsifier; coupling agent; intermediate for mfg. of surfactants; emulsifier for silicone and min. oil; liq.; oil-sol.; sp.gr. 1.03; HLB 10.9; cloud pt. 68-72 C (1% in 10% NaCl); 100% act.

Ablunol NP8. [Taiwan Surf.] Nonoxynol-8; CAS 9016-45-9; nonionic; detergent, wetting agent, emulsifier for textile scouring, warp sizing, carbonizing and bleaching, household/industrial cleaners; liq.; water-sol.; sp.gr. 1.05; HLB 12.3; cloud pt. 37-42 C; 100% act.

Ablunol NP9. [Taiwan Surf.] Nonoxynol-9; CAS 9016-45-9; nonionic; detergent, wetting agent, emulsifier for textile scouring, warp sizing, carbonizing and bleaching, household/industrial cleaners, paper, and leather industries; metal processing as a wetting agent with min. acid; liq.; water-sol.; sp.gr. 1.05; HLB 12.9; cloud pt. 50-56 C; 100% act.

Ablunol NP10. [Taiwan Surf.] Nonoxynol-10; CAS 9016-45-9; nonionic; detergent, wetting agent, emulsifier for textile scouring, warp sizing, carbonizing and bleaching, household/industrial cleaners; liq.; water-sol.; sp.gr. 1.06; HLB 13.3; cloud pt. 60-66 C.

Ablunol NP12. [Taiwan Surf.] Nonoxynol-12; CAS 9016-45-9; nonionic; detergent, wetting agent, emulsifier for textile scouring, warp sizing, carbonizing and bleaching, household/industrial cleaners; liq.; water-sol.; sp.gr. 1.07; HLB 14.1; cloud pt. 80-86 C.

Ablunol NP15. [Taiwan Surf.] Nonoxynol-15; CAS 9016-45-9; nonionic; surfactant, wetting agent, penetrant used in heavy-duty alkaline cleaners, metal cleaners and bottle washing formulations; high temp. textile scouring agent; liq.; sp.gr. 1.07 (40 C); HLB 15.0; cloud pt. 92-96 C; 100% act.

Ablunol NP16. [Taiwan Surf.] Nonoxynol-16; CAS 9016-45-9; nonionic; surfactant, wetting agent, penetrant used in heavy-duty alkaline cleaners, metal cleaners and bottle washing formulations; high temp. textile scouring agent; liq.; sp.gr. 1.07 (40 C); HLB 15.2; cloud pt. 97-100 C.

Ablunol NP20. [Taiwan Surf.] Nonoxynol-20; CAS 9016-45-9; nonionic; surfactant, wetting agent, penetrant used in heavy-duty alkaline cleaners, metal cleaners and bottle washing formulations; high temp. textile scouring agent; semisolid; sp.gr. 1.07 (50 C); HLB 16.0; cloud pt. 71 C (1% in 10% NaCl); 100% act.

Ablunol NP30. [Taiwan Surf.] Nonoxynol-30; CAS 9016-45-9; nonionic; emulsifier for vinyl acetate and acrylate emulsion polymerization; stabilizer; used in latex paints, floor finishes, paper coatings, textiles; solid; sp.gr. 1.08 (50 C); HLB 17.1; cloud pt. 74 C (1% in 10% NaCl); 100% act.

Ablunol NP30 70%. [Taiwan Surf.] Nonoxynol-30; CAS 9016-45-9; nonionic; emulsifier for vinyl acetate and acrylate emulsion polymerizations; stabilizer; used in latex paints, floor finishes, paper coatings, textiles; liq.; HLB 17.1; cloud pt. 74 C (1% in 10% NaCl); 70% act.

Ablunol NP40. [Taiwan Surf.] Nonoxynol-40; CAS 9016-45-9; nonionic; emulsifier for vinyl acetate and acrylate emulsion polymerizations; stabilizer; used in latex paints, floor finishes, paper coatings, textiles; solid; sp.gr. 1.08 (50 C); HLB 17.8; cloud pt. 76 C (1% in 10% NaCl); 100% act.

Ablunol NP40 70%. [Taiwan Surf.] Nonoxynol-40; CAS 9016-45-9; nonionic; emulsifier for vinyl acetate and acrylate emulsion polymerizations; stabilizer; used in latex paints, floor finishes, paper coatings, textiles; liq.; HLB 17.8; cloud pt. 76 C (1% in 10% NaCl); 70% act.

Ablunol NP50. [Taiwan Surf.] Nonoxynol-50; CAS 9016-45-9; nonionic; emulsifier for vinyl acetate and acrylate emulsion polymerizations; stabilizer; used in latex paints, floor finishes, paper coatings, textiles; solid; sp.gr. 1.08 (60 C); HLB 18.2; cloud pt. 76 C (1% in 10% NaCl); 100% act.

Ablunol NP50 70%. [Taiwan Surf.] Nonoxynol-50; CAS 9016-45-9; nonionic; emulsifier for vinyl acetate and acrylate emulsion polymerizations; stabilizer; used in latex paints, floor finishes, paper coatings, textiles; liq.; HLB 18.2; cloud pt. 76 C (1% in 10% NaCl); 70% act.

Ablunol OA-6. [Taiwan Surf.] Oleth-6; CAS 9004-98-2; nonionic; emulsifier for min. oil and cosmetics; liq.; HLB 10.0; 100% act.

Ablunol OA-7. [Taiwan Surf.] Oleth-7; CAS 9004-98-2; nonionic; emulsifier for min. oil and cosmetics; liq.; HLB 10.7; 100% act.

Ablunol S-20. [Taiwan Surf.] Sorbitan laurate; CAS 1338-39-2; nonionic; emulsifier, emulsion stabilizer, thickener for cosmetic, pharmaceutical, food applics.; textile fiber lubricant, softener; antifog agent; liq.; oil-sol.; water-disp.; HLB 8.6; 100% act.

Ablunol S-40. [Taiwan Surf.] Sorbitan palmitate; CAS 26266-57-9; nonionic; emulsifier, emulsion stabilizer, thickener for cosmetic, pharmaceutical, food applics.; textile fiber lubricant, softener; antifog agent; waxy solid; oil-sol.; HLB 6.7.

Ablunol S-60. [Taiwan Surf.] Sorbitan stearate; CAS 1338-41-6; nonionic; emulsifier, emulsion stabilizer, thickener for cosmetic, pharmaceutical, food applics.; textile fiber lubricant, softener; antifog agent; silicone defoamer emulsions; waxy solid; HLB 4.7; 100% act.

Ablunol S-80. [Taiwan Surf.] Sorbitan oleate; CAS 1338-43-8; nonionic; emulsifier, emulsion stabilizer, thickener for cosmetic, pharmaceutical, food applics.; textile fiber lubricant, softener; antifog agent; wet processing of syn. PU leather; oily liq.; HLB 4.3; 100% act.

Ablunol S-85. [Taiwan Surf.] Sorbitan trioleate; CAS 26266-58-0; nonionic; emulsifier, emulsion stabilizer, thickener for cosmetic, pharmaceutical, food applics.; textile fiber lubricant, softener; antifog agent; oily liq.; HLB 1.8; 100% act.

Ablunol SA-7. [Taiwan Surf.] Steareth-7; CAS 9005-00-9; nonionic; emulsifier for wax and cosmetics; solid; HLB 10.7; 100% act.

Abluphat AP Series. [Taiwan Surf.] Complex org. phosphate ester, free acid; anionic; antistat, penetrant, wetting agent, solubilizer, detergent; for cotton and synthetics processing; high alkaline and acid tolerance; liq.; 100% conc.

Abluphat LP Series. [Taiwan Surf.] Phosphate ester sodium salt; antistat for syn. fibers; penetrant, wetting agent, solubilizer, detergent; liq.; 100% conc.

Abluphat OP Series. [Taiwan Surf.] Complex org. phosphate ester, free acid; anionic; antistat, penetrant, wetting agent, solubilizer, detergent; liq.; 100% conc.

Ablupol AF. [Taiwan Surf.] Mixt.; antifoaming agent for coating color formulation of paper.

Ablupol SAE. [Taiwan Surf.] Silicone compd.; antifoamer; liq.; water-disp.

Ablusoft A. [Taiwan Surf.] Polyamide type surfactant; cationic; softener for acrylic fibers, cotton and synthetics; flake; 100% conc.

Ablusoft ES. [Taiwan Surf.] Epoxy-modified silicone; finishing agent imparting durable softness, wrinkle resistance to cotton and syn. fiber blends.

Ablusoft ND. [Taiwan Surf.] Surfactant; weakly cationic; general purpose softener, antistatic properties in resin finishing of cotton and synthetics; flake; 100% conc.

Ablusoft PE. [Taiwan Surf.] Polyethylene emulsion; nonionic; finishing agent improving tear str., abrasion resistance and handle of fabrics; liq.

Ablusoft SF. [Taiwan Surf.] Silicone emulsion; finishing agent imparting silky hand to yarn and fabrics; improves sewability.

Ablusoft SN. [Taiwan Surf.] Amino-modified silicone; cationic; softener for fabrics; imparts soft, silky hand to cotton and syn. fiber blends; liq.

Ablusoft SNC. [Taiwan Surf.] Amino-modified silicone; cationic; softener for fabrics; liq.; 100% conc.

Ablusol BX Series. [Taiwan Surf.] Anionic; dispersant, penetrant, wetting agent.

Ablusol C-70. [Taiwan Surf.] Dioctyl sulfosuccinate; anionic; wetting agent for industrial applics.; penetrant/wetting agent for bleaching of cotton, agric. applics.; liq.; sp.gr. 1.05; 70% act.

Ablusol C-78. [Taiwan Surf.] Sodium dioctyl sulfosuccinate; anionic; wetting agent for textile, PU leather; emulsion polymerization, agric. emulsion, and other industrial applics.; liq.; 70% conc.

Ablusol CDE. [Taiwan Surf.] Cocamide DEA sulfosuccinate monoester; anionic; detergent, foam booster/stabilizer for low irritation shampoos, bubble baths, liq. detergents; liq.; 40% act.

Ablusol DA. [Taiwan Surf.] Ethoxylated decyl alcohol sulfosuccinate monoester; anionic; emulsifier for polyacrylate emulsion polymerization; liq.; water-sol.; 30% act.

Ablusol DBC. [Taiwan Surf.] Calcium dodecylbenzene sulfonate; anionic; emulsifier for agric. chem., oils, solvs.; liq.; oil-sol.; 70% act.

Ablusol DBD. [Taiwan Surf.] DEA dodecylbenzene sulfonate; anionic; surfactant, emulsifier, wetting agent for bubble bath, shampoos, detergents; liq.; water-sol.; 100% act.

Ablusol DBM. [Taiwan Surf.] Ammonium dodecylbenzene sulfonate; anionic; surfactant, emulsifier, wetting agent for bubble bath, shampoos, detergents; liq.; water-sol.; 100% act.

Ablusol DBT. [Taiwan Surf.] TEA dodecylbenzene sulfonate; anionic; surfactant, emulsifier, wetting agent for bubble bath, shampoos, detergents; liq.; water-sol.; 100% act.

Ablusol LDE. [Taiwan Surf.] Lauramide DEA sulfosuccinate monoester; anionic; detergent, foam booster/stabilizer for low irritation shampoos, bubble baths, liq. detergents; liq.; water-sol.; 40% act.

Ablusol LME. [Taiwan Surf.] Lauramide MEA sulfosuccinate monoester; anionic; detergent producing copious lather and extra dry residue for high foaming rug shampoos; paste; 40% act.

Ablusol M-75. [Taiwan Surf.] Dioctyl sulfosuccinate; anionic; wetting agent for industrial applics.; suitable for adding to PU resin; liq.; 75% act.

Ablusol ML. [Taiwan Surf.] Sodium naphthalene sulfonate formaldehyde condensate; anionic; plasticizer and water reducing agent used in pourable and high strength concrete; liq.; 40% act.

Ablusol N. [Taiwan Surf.] Naphthalene sulfonate condensate; anionic; dispersant for dyestuffs, pigments, agrochem.; powd.; 95% act.

Ablusol NL. [Taiwan Surf.] Sodium naphthalene sulfonate formaldehyde condensate; anionic; dispersant and antiagglomerant for dyestuffs, pigments, agrochem.; used in the process of polyester and acetate fabrics; liq.; 40% act.

Ablusol OA. [Taiwan Surf.] Oleyl sulfosuccinamate; anionic; foaming agent for noncarboxylated SBR latexes where stable uniform cell structure is necessary; for foam carpet underlay, foam carpet backing; liq.; 35% act.

Ablusol OK. [Taiwan Surf.] Surfactant blend; anionic; low foam penetrating and wetting agent for continous scouring-bleaching systems in textile industry; stable to high alkali.

Ablusol P Series. [Taiwan Surf.] Anionic; penetrant, wetting agent, foaming agent.

Ablusol PAC. [Taiwan Surf.] Surfactant; cationic; accelerant for caustic treatment of polyester fibers.

Ablusol PM. [Taiwan Surf.] Alkyl sulfate; anionic; penetrant, scouring agent for mercerizing of cotton.
Ablusol SF Series. [Taiwan Surf.] Anionic; emulsifier for emulsion polymerization; foaming agent, scouring and dyeing assistant for textiles.
Ablusol SN Series. [Taiwan Surf.] Anionic; emulsifier for emulsion polymerization; foaming agent, scouring and dyeing assistant for textiles.
Ablusol TA. [Taiwan Surf.] Octadecyl sulfosuccinamate; anionic; foaming agent for noncarboxylated SBR latexes where stable uniform cell structure is necessary; for foam carpet underlay, foam carpet backing; paste; 35% act.
Abluter BE. [Taiwan Surf.] Cocamidopropyl betaine; amphoteric; detergent for mild cleansing prods.; antistat, softener, germicide, spreading/wetting agent; liq.; 30% conc.
Abluter GL Series. [Taiwan Surf.] Glycine derivs.; amphoteric; antistat, softener, germicide, spreading/wetting agent.
Abluton 700. [Taiwan Surf.] Mixt.; anionic/nonionic; dispersing and leveling agent for dyeing polyester and its blends.
Abluton A. [Taiwan Surf.] Surfactant blend; anionic/nonionic; penetrant, scouring agent for cotton and synthetics.
Abluton BN. [Taiwan Surf.] Mixt.; cationic/nonionic; leveling and retarding agent for dyeing nylon with acid dyes; prevents barré.
Abluton BT. [Taiwan Surf.] Organic; anionic/nonionic; carrier for high temp. dyeing of polyester fibers; leveling agent; prevents barré.
Abluton CDL. [Taiwan Surf.] Blend; anionic; imparts smoothness to fabrics; used in carding processes; liq.; 100% conc.
Abluton CMN. [Taiwan Surf.] Methyl naphthalene; anionic/nonionic; dye carrier for polyester fiber and its blends; liq.; 100% conc.
Abluton CTP. [Taiwan Surf.] Trichlorobenzene; anionic/nonionic; carrier for bleaching polyester fiber and its blends; emulsifiable; liq.; 100% conc.
Abluton EP. [Taiwan Surf.] Blend; nonionic; thickener for textile pigment printing; liq.
Abluton LMO. [Taiwan Surf.] Surfactant; nonionic; dye dispersant and leveling agent for polyester; liq.; 100% conc.
Abluton LN. [Taiwan Surf.] Surfactant; nonionic; leveling agent for acid and metal complex dyes; liq.; 100% conc.
Abluton N. [Taiwan Surf.] Org. chemical/surfactant; nonionic; dyeing auxiliary for polyamide fiber and blends; liq.; 100% conc.
Abluton RT430. [Taiwan Surf.] Mixt.; anionic/nonionic; low foam dispersing and leveling agent for dyeing polyester and its blends.
Abluton T. [Taiwan Surf.] Org. chemical/surfactant; nonionic; leveling agent for polyester dyes; aids barré reduction; liq.; 100% conc.
A.B.S. 87%. [Triantaphyllou] Alkylbenzene sulfonic acid; detergent intermediate; liq.; 87% conc.
Accobetaine CL. [Karlshamns] Complex coco betaine; detergent, wetting agent, emulsifier, high foaming agent, solubilizer, household and cosmetic uses; yel. liq.; pleasant odor; water-sol.; f.p. 0 C; 35% act.
Accomeen C2. [Karlshamns] PEG-2 cocamine; CAS 61791-14-8; nonionic; emulsifier, antistat, surfactant; Gardner 4–6 liq., amine odor; sol. in org. solv.; dens. 7.25 lb/gal; sp.gr. 0.87; 99% act.
Accomeen C5. [Karlshamns] PEG-5 cocamine; CAS 61791-14-8; nonionic; emulsifier, antistat, surfactant; Gardner 4–6 liq., amine odor; sol. in org. solv., water; dens. 8.15 lb/gal; sp.gr. 0.98; 99% act.
Accomeen C10. [Karlshamns] PEG-10 cocamine; CAS 61791-14-8; nonionic; emulsifier, antistat, surfactant; Gardner 4–6 liq., amine odor; sol. in org. solv., water; dens. 8.3 lb/gal; sp.gr. 1.0; 99% act.
Accomeen C15. [Karlshamns] PEG-15 cocamine; CAS 61791-14-8; nonionic; emulsifier, antistat, surfactant; Gardner 4–6 liq., amine odor; sol. in org. solv.; dens. 8.7 lb/gal; sp.gr. 1.04; 99% act.
Accomeen S2. [Karlshamns] PEG-2 soyamine; CAS 61791-24-0; nonionic; emulsifier, antistat, surfactant; Gardner 5–10 liq.; amine odor; sol. in org. solv.; dens. 7.6 lb/gal; sp.gr. 0.91; surf. tens. 31.3 (0.1%); 99% act.
Accomeen S5. [Karlshamns] PEG-5 soyamine; CAS 61791-24-0; nonionic; emulsifier, antistat, surfactant; Gardner 5–10 liq.; amine odor; sol. in org. solv.; water-disp.; dens. 7.9 lb/gal; sp.gr. 0.95; 99% act.
Accomeen S10. [Karlshamns] PEG-10 soyamine; CAS 61791-24-0; nonionic; emulsifier, antistat, surfactant; Gardner 5–10 liq.; amine odor; sol. in org. solv., water; dens. 8.5 lb/gal; sp.gr. 1.02; 99% act.
Accomeen S15. [Karlshamns] PEG-15 soyamine; CAS 61791-24-0; surfactant.
Accomeen T2. [Karlshamns] PEG-2 tallow amine; nonionic; emulsifier, antistat, surfactant, dispersant; Gardner 3–5; amine odor; sol. in org. solv., insol. in water; dens. 7.7 lb/gal; sp.gr. 0.92; 99% act.
Accomeen T5. [Karlshamns] PEG-5 tallow amine; nonionic; emulsifier, antistat, surfactant, dispersant; Gardner 3–5; amine odor; sol. in org. solv., insol. in water; 99% act.
Accomeen T15. [Karlshamns] PEG-15 tallow amine; nonionic; emulsifier, antistat, surfactant, dispersant; Gardner 3–5; amine odor; sol. in org. solv., insol. in water; dens. 8.1 lb/gal; sp.gr. 0.97); 99% act.
Accomid 50. [Karlshamns] Palm kernelamide DEA.
Accomid C. [Karlshamns] Cocamide DEA; nonionic; detergent, stabilizer, visc. improver, foam booster for shampoos and dishwashes; biodeg.; Gardner 4 liq., mild odor; water-sol.; sp.gr. 0.99; dens. 8.3 lb/gal; pH 9-10.5; 98% act.
Accomid PK. [Karlshamns] Palm kernelamide DEA (1:1); nonionic; visc. builder, foam booster/stabilizer, emulsifier for shampoos, liq. soaps, dish detergents, bubble bath prods.; Gardner 5 liq.; pH 9–10.6 (10% aq.).
Acconon 200-DL. [Karlshamns] PEG-4 dilaurate; CAS 9005-02-1; nonionic; surfactant used as emulsifier, dispersant, solubilizer, visc. control agent for cosmetics, pharmaceuticals, and industrial applics.
Acconon 200-MS. [Karlshamns] PEG-4 stearate; CAS 9004-99-3; nonionic; surfactant used as emulsifier, dispersant, solubilizer, visc. control agent for cosmetics, pharmaceuticals, and industrial applics.; Gardner 4 max. solid; HLB 8; pH 5.5–6.5.
Acconon 300-MO. [Karlshamns] PEG 300 oleate; CAS 9004-96-0; nonionic; emulsifier, lubricant, chemical intermediate; for cosmetics, food, agriculture, plastics; Gardner 6 liq.; sol. in org. solv.; dens. 8.3 lb/gal resp.; sp.gr. 0.99; m.p. < –5 C; 99% act.
Acconon 400-DO. [Karlshamns] PEG-8 dioleate; CAS 9005-07-6; nonionic; surfactant used as emulsifier, dispersant, solubilizer, visc. control agent for cosmetics, pharmaceuticals, and industrial applics.
Acconon 400-ML. [Karlshamns] PEG-8 laurate; CAS

9004-81-3; nonionic; surfactant used as emulsifier, dispersant, solubilizer, visc. control agent for cosmetics, pharmaceuticals, and industrial applics.

Acconon 400-MO. [Karlshamns] PEG-8 oleate; CAS 9004-96-0; nonionic; emulsifier, dispersant, lubricant, chemical intermediate, solubilizer, visc. control agent; for cosmetics, pharmaceuticals, food, agric., plastics; Gardner 4 max. liq.; sol. in org. solv.; water-disp.; dens. 8.4 lb/gal; sp.gr. 1.01; HLB 12; m.p. < 10 C; pH 5.5–6.5; 99% act.

Acconon 400-MS. [Karlshamns] PEG-8 stearate; CAS 9004-99-3; nonionic; surfactant used as emulsifier, dispersant, solubilizer, visc. control agent for cosmetics, pharmaceuticals, and industrial applics.; Gardner 4 max. solid; HLB 12; pH 5.5–6.5.

Acconon 1300. [Karlshamns] PPG-3-laureth-9; nonionic; surfactant used as emulsifier, dispersant, solubilizer, visc. control agent for cosmetics, pharmaceuticals, and industrial applics.; Gardner 3 max. liq.; sp.gr. 1.016–1.019; dens. 8.35–8.45 lb/gal; HLB 11; cloud pt. 135–145 F; pH 6.0–7.0.

Acconon CA-5. [Karlshamns] PEG-5 castor oil; nonionic; surfactant used as emulsifier, dispersant, solubilizer, visc. control agent for cosmetics, pharmaceuticals, and industrial applics.; Gardner 3 max. liq.; HLB 8.0; pH 6.0–7.0.

Acconon CA-8. [Karlshamns] PEG-8 castor oil; nonionic; surfactant used as emulsifier, dispersant, solubilizer, visc. control agent for cosmetics, pharmaceuticals, and industrial applics.; Gardner 3 max. liq.; HLB 8; pH 6.0–7.0.

Acconon CA-9. [Karlshamns] PEG-9 castor oil; nonionic; surfactant used as emulsifier, lubricant, dispersant, solubilizer, visc. control agent for cosmetics, pharmaceuticals, and industrial applics.; Gardner 3 max. liq.; water-disp.; HLB 12.0; pH 6-7.

Acconon CA-15. [Karlshamns] PEG-15 castor oil; nonionic; surfactant used as emulsifier, lubricant, dispersant, solubilizer, visc. control agent for cosmetics, pharmaceuticals, and industrial applics.; Gardner 3 max. liq.; water-disp.; HLB 16; pH 6.0–7.0.

Acconon CON. [Karlshamns] PEG-10 propylene glycol glyceryl laurate; nonionic; surfactant used as emulsifier, dispersant, solubilizer, visc. control agent for cosmetics, pharmaceuticals, and industrial applics.; Gardner 2 max. liq.; HLB 10; pH 6.0–7.0.

Acconon E. [Karlshamns] PPG-15 stearyl ether; nonionic; surfactant used as emulsifier, dispersant, solubilizer, visc. control agent for cosmetics, pharmaceuticals, and industrial applics.; Gardner 3 max. liq.; HLB 16; pH 6.0–7.0.

Acconon ETG. [Karlshamns] Glycereth-26; nonionic; humectant; lubricant for skin care prods., creams, lotions, industrial applics.; Gardner 3 max. liq.; sol. in water, alcohol, ketones, esters; HLB 15; pH 5.5–6.5.

Acconon TGH. [Karlshamns] PEG-20-PPG-10 glyceryl stearate; nonionic; surfactant used as emulsifier, dispersant, solubilizer, visc. control agent, wetting and foaming agent for cosmetics, pharmaceuticals, and industrial applics.; biodeg.; Gardner 3 max. liq., mild odor; HLB 16; pH 6.0–7.0; 100% conc.

Acconon W230. [Karlshamns] Ceteareth-20; CAS 68439-49-6; nonionic; surfactant used as emulsifier, dispersant, solubilizer, visc. control agent for cosmetics, pharmaceuticals, and industrial applics.; Gardner 3 max. solid; HLB 15; pH 6.5–7.5.

Accoquat 2C-75. [Karlshamns] Dicocodimonium chloride; cationic; emulsifier, coupling agent; used for car spray waxes, dust control oil, spot removal; amber liq., mild odor; dens. 7.2 lb/gal; sp.gr. 0.86; flamm.; 75% act. in IPA.

Accoquat 2C-75H. [Karlshamns] Dicocodimonium chloride; cationic; emulsifier, coupling agent; amber liq., mild odor; 75% act. in hexylene glycol.

Accosoft 440-75. [Stepan; Stepan Canada] Methyl bis (hydrog. tallowamidoethyl) 2-hydroxyethyl ammonium methyl sulfate; cationic; fabric softener quat. for textile industry, household and commercial use; good lubricity and scorch resistance; nonyel.; cream colored semisolid; IPA odor; visc. 200 cps (125 F); pour pt. 113 F; pH 6.0 (10% IPA/water); 75% solids.

Accosoft 540. [Stepan; Stepan Canada] Modified tallow diamidoamine quat. ammonium methylsulfate; cationic; quat. for use in rinse cycle fabric softeners; gives high visc. dispersions at low solids; Gardner 6 paste; dens. 8.12 lb/gal (50 C); visc. 1500 cps (95 F); pour pt. 86 F; pH 6 (10% water/IPA); 90% solids.

Accosoft 540 HC. [Stepan; Stepan Canada] Methyl bis (modified tallowamidoethyl) 2-hydroxyethyl ammonium methyl sulfate; fabric softener quat. for household prods. and textile processing; yel. opaque paste; mild odor; dens. 8.35 lb/gal (50 C); visc. 1000 cps (100 F); pour pt. 88 F; pH 6 (10% IPA/water); 90% solids.

Accosoft 550-75. [Stepan; Stepan Canada] Tallow diamidoamine quat. ammonium methylsulfate; cationic; nonyel. quat. with good rewetting props. for fabric softening in industrial, household and textile industries; Gardner 6 liq.; dens. 7.94 lb/gal; visc. 700 cps (65 F); pour pt. 54 F; pH 6 (10% water/IPA); 90% solids.

Accosoft 550-90 HHV. [Stepan; Stepan Canada] Methyl bis (tallowamidoethyl) 2-hydroxyethyl ammonium methyl sulfate; cationic; fabric softener and antistat with good rewet props. for laundry prods., industrial textile processing; yel. opaque paste; mild odor; readily dispersible in water; dens. 8.14 lb/gal (50 C); visc. 800 cps (100 F); pour pt. 86 F; pH 6.0 (10% IPA/water); 90% solids.

Accosoft 550 HC. [Stepan; Stepan Canada] Tallow diamidoamine quat. ammonium methyl sulfate; cationic; surfactant used in household and industrial fabric softener, fabric antistat, textile processing; Gardner 6 liq.; dens. 8.43 lb/gal; visc. 1500 cps (75 F); pour pt. 65 F; pH 6 (10% aq. IPA); 88% solids.

Accosoft 550L-90. [Stepan; Stepan Canada] Methyl bis (tallowamidoethyl) 2-hydroxyethyl ammonium methyl sulfate; cationic; fabric softener quat. for household laundry prods., textile processing; yel. visc. liq.; mild odor; dens. 8.31 lb/gal; visc. 1500 cps (75 F); pour pt. 65 F; pH 6.0 (10% IPA/water); 88% solids.

Accosoft 580. [Stepan; Stepan Canada] Modified tallow diamidoamine quat. ammonium methylsulfate; cationic; softener without greasy feel for rinse cycle fabric softening; Gardner 6 solid; dens. 8.07 lb/gal (50 C); visc. 1500 cps (110 F); pour pt. 100 F; pH 6 (10% water/IPA); 90% solids.

Accosoft 580 HC. [Stepan; Stepan Canada] Modified tallow diamidoamine quat. ammonium methylsulfate; cationic; quat. for household fabric softeners, textile processing; Gardner 6 solid; dens. 8.28 lb/gal (50 C); visc. 1500 cps (105 F); pour pt. 95 F; pH 6 (10% water/IPA); 90% solids.

Accosoft 620-90. [Stepan; Stepan Canada] Methyl bis (tallowamidoethyl) 2-hydroxypropyl ammonium methyl sulfate; cationic; fabric softener and antistat for industrial and household prods.; exc. rewet props.; nonyel.; U.S. patent #3,933,871; yel. visc. liq.; mild odor; readily dispersible in water; dens. 8.23 lb/gal; visc. 3000 cps (75 F); pour pt. 65 F; pH 6.0 (10% IPA/water); 89% solids.

Accosoft 707. [Karlshamns] Dimethyl dihydrog. tallow ammonium chloride; CAS 61789-80-8; cationic; surfactant used in household and industrial fabric softener, fabric antistat; Gardner 3 paste; sol. in IPA; flam.; dens. 7.3; sp.gr. 8.7; 75% act. DISCONTINUED.

Accosoft 748. [Karlshamns] Dimethyl dihydrog. tallow ammonium methyl sulfate; cationic; surfactant used in household and industrial fabric softener or antistat esp. suited to dryer use; Gardner 4 paste, solid; m.p. 125–135 C; biodeg.; flam.; 90% act. in IPA. DISCONTINUED.

Accosoft 750. [Stepan; Stepan Canada] Methyl bis (oleylamidoethyl) 2-hydroxyethyl ammonium methyl sulfate; cationic; fabric softener quat. for heavy duty liq. laundry detergents; amber liq.; IPA odor; dens. 8.36 lb/gal; visc. 3400 cps (45 F); pour pt. < 40 F; pH 6.0 (10% IPA/water); 90% solids.

Accosoft 808. [Stepan; Stepan Canada] Tallow imidazoline quat.; cationic; fabric softener and antistat for household and commerical laundries; biodeg.; Gardner 6 liq.; dens. 7.96 lb/gal; visc. 500 cps (60 F); pour pt. 50 F; pH 6 (10% water/IPA); 80% solids.

Accosoft 808HT. [Stepan; Stepan Canada] Methyl-1-hydrog. tallowamidoethyl-2-hydrog. tallow imidazolinium-methyl sulfate; cationic; fabric softener and antistat for household and industrial applics., dryer prods.; Gardner 6 solid; visc. 2500 cps (155 F); pour pt. 144 F; pH 6.0 (10% IPA/water); 91% solids.

Accosoft 870. [Stepan; Stepan Canada] Hydrog. tallow imidazolinium quat. ammonium compd./ethoxylated nonionic blend; cationic; softener, antistat for dyer sheet fabric softeners; Gardner 6 solid; visc. 2500 cps (145 F); pour pt. 135 F; pH 6.0 (10% IPA/water); 93% solids.

Accosoft A-155. [Karlshamns] Tallow amido amine quat. compd.; cationic; surfactant, household fabric softener base, dispersion, antistat agent; Gardner 7 gel; biodeg.; 75 or 95% act. DISCONTINUED.

Accosoft M 1154. [Karlshamns] Tallow amine quat. deriv.; cationic; fabric softener for detergent softener systems; liq.; 98% conc. DISCONTINUED.

Accosperse 20. [Karlshamns] PEG-20 sorbitan laurate; CAS 9005-64-5; nonionic; emulsifier, solubilizer; biodeg.; liq.; 100% conc.

Accosperse 60. [Karlshamns] PEG-20 sorbitan stearate; CAS 9005-67-8; nonionic; emulsifier, solubilizer; biodeg.; amber liq.; sp.gr. 1.1; visc. 500–600 cps; HLB 14.9–15.0; sapon. no. 45–55; 98% act.

Accosperse 80. [Karlshamns] PEG-20 sorbitan oleate; CAS 9005-65-6; nonionic; emulsifier, solubilizer; biodeg.; amber liq.; sp.gr. 1.1; visc. 400–450 cps; HLB 15.0; sapon. no. 45–55; 98% act.

Accosperse TGH. [Karlshamns] Alkoxylated mono-, diglyceride; nonionic; emulsifier, foaming and wetting agent; yel. liq.; mild odor; f.p. 30 F; cloud pt. 140–146 F (1% aq.); biodeg.; 98–100% act.

Acetamin 24. [Kao Corp. SA] Cocamine acetate; CAS 2016-56-0; cationic; surface coating agent for pigments, anticaking agent for fertilizer; emulsifier, dispersant, and softening agent for textiles; min. flotation reagent; solid; 100% conc.

Acetamin 86. [Kao Corp. SA] Stearamine acetate; CAS 2190-04-7; surface coating agent for pigments, anticaking agent for fertilizer; emulsifier, dispersant, and softening agent for textiles; min. flotation reagent; flake; 100% conc..

Acetamin C. [Kao Corp. SA] n-Cocamine acetate; cationic; flotation of mins., anticaking agents, emulsifier bactericide; paste, solid; 100% conc.

Acetamin HT. [Kao Corp. SA] n-Hyd. tallow amine acetate; cationic; flotation of mins., anticaking agents, emulsifier bactericide; solid; 100% conc.

Acetamin T. [Kao Corp. SA] n-Tallow amine acetate; cationic; flotation of mins., anticaking agents, emulsifier bactericide; solid; 100% conc.

Acetoquat CPB. [Aceto] Cetyl pyridinium bromide; CAS 140-72-7; cationic; germicide, sanitizing agent; powd.; 95% act.

Acetoquat CPC. [Aceto] Cetyl pyridinium chloride; cationic; germicide, sanitizing agent; powd.; 100% act.

Acetoquat CTAB. [Aceto] Cetrimonium bromide; cationic; germicide, sanitizing agent; powd.; 95% act.

Acid Aid X. [Crown Tech.] Accelerator and extender for hydrochloric and sulfuric acids; detergency props.; for metal cleaning, pickling; liq., pleasant aromatic odor; sol. in water; dens. 8.6 lb/gal.

Acid Felt Scour. [Hart Chem. Ltd.] Specially formulated prod.; nonionic; pulp and paper felt cleaner for batch and continuous processing; liq.; 40% act.

Acid Foamer. [Exxon/Tomah] Quat. ammonium chloride; cationic; surfactant; foaming agent for strong acids; used in aluminum trailer cleaner, brightener, acid inhibitors and cleaners, chrome plating baths; dk. amber iq.; dens. 8.8 lb/gal; pH 6-9 (5%); 100% conc.

Acid Thickener. [Exxon/Tomah] Surfactant; cationic; visc. builder, wetting agent, corrosion inhibitor for acid-based cleaners, e.g., acid bowl cleaners, truck cleaners, building restoration cleaners; perfume solubilizer; amber paste; sp.gr. 0.91; dens. 7.6 lb/gal; visc. 5000 cps (9.5% HCl with 3% Acid Thickener); pour pt. 80 F; flash pt. > 200 F.

Acintol® 736. [Arizona] Tall oil acid; CAS 61790-12-3; emulsifier for SBR polymerization; Gardner 7; sp.gr. 0.900; dens. 7.48 lb/gal; flash pt. (CC) > 200 F; acid no. 189; 87% fatty acids.

Acintol® 746. [Arizona] Tall oil acid; CAS 61790-12-3; emulsifier for SBR polymerization; Gardner 4; sp.gr. 0.905; dens. 7.52 lb/gal; flash pt. (CC) > 200 F; acid no. 198; 97.5% fatty acids.

Acintol® 2122. [Arizona] Tall oil heads; CAS 61790-12-3; surfactant; Gardner 16 solid; sp.gr. 0.92; dens. 7.60 lb/gal; visc. 15 cps (50 C); flash pt. (CC) > 200 F; acid no. 135-155; sapon. no. 105-155; 76.2-68.2% fatty acids.

Acintol® 7002. [Arizona] Tall oil acid; CAS 61790-12-3; surfactant; Gardner > 18; sp.gr. 0.97; dens. 7.89 lb/gal; visc. 1600-2000 cps; flash pt. (CC) > 200 F; acid no. 140-142; sapon. no. 179; 76.2-68.2% fatty acids.

Acintol® D25LR. [Arizona] Tall oil acid, distilled; CAS 61790-12-3; surfactant for asphalt emulsifiers, concrete form release and air entraining agents, metalworking fluids, varnishes, printing inks, soaps, cleaners, degreasers; Gardner 5+; sp.gr. 0.94; dens. 7.83 lb/gal; visc. 95 cps; acid no. 188; sapon. no. 192; flash pt. (OC) 204 F; 72.7% fatty acids.

Acintol® D30E. [Arizona] Dist. tall oil fatty acid; CAS 61790-12-3; surfactant for asphalt emulsifiers, concrete form release and air entraining agents, metalworking fluids, varnishes, printing inks, soaps, cleaners, degreasers; Gardner 12+; sp.gr. 0.94; visc. 350 cps; acid no. 150-185; sapon. no. 167-187; flash pt. (OC) 204 C.

Acintol® D30LR. [Arizona] Tall oil acid, distilled; CAS 61790-12-3; surfactant for asphalt emulsifiers, concrete form release and air entraining agents, metalworking fluids, varnishes, printing inks, soaps, cleaners, degreasers; Gardner 6-; sp.gr. 0.94; dens. 7.83 lb/gal; visc. 110 cps; acid no. 186; sapon. no. 190; flash pt. (OC) 204 C; 67.8% fatty acids.

Acintol® D40LR. [Arizona] Tall oil acid, distilled; CAS 61790-12-3; surfactant for asphalt emulsifiers, concrete form release and air entraining agents, metalworking fluids, varnishes, printing inks, soaps, cleaners, degreasers; Gardner 5+; sp.gr. 0.95; dens. 7.89 lb/gal; visc. 180 cps; acid no. 183; sapon. no. 188; flash pt. (OC) 204 C; 59.1 fatty acids.

Acintol® D40T. [Arizona] Dist. tall oil fatty acid; CAS 61790-12-3; surfactant for asphalt emulsifiers, concrete form release and air entraining agents, metalworking fluids, varnishes, printing inks, soaps, cleaners, degreasers; Gardner 13-14; sp.gr. 0.90; visc. 1290 cps; acid no. 170; sapon. no. 184; flash pt. (CC) > 94 C.

Acintol® D60LR. [Arizona] Dist. tall oil fatty acid; CAS 61790-12-3; surfactant for asphalt emulsifiers, concrete form release and air entraining agents, metalworking fluids, varnishes, printing inks, soaps, cleaners, degreasers; Gardner 6+; sp.gr. 0.90; acid no. 175-182; flash pt. (OC) 204 C.

Acintol® DFA. [Arizona] Tall oil acid; CAS 61790-12-3; for metalworking fluids, lubricant additives, oilfield chems., asphalt emulsifiers, alkyd resins, industrial/household cleaners, plasticizers, textile drawing lubricants, surf. coatings, rubber prods.; Gardner 8; sp.gr. 0.91; visc. 20 cps; acid no. 190; iodine no. 135; sapon. no. 197; flash pt. (OC) 204 C.

Acintol® EPG. [Arizona] Tall oil acid; CAS 61790-12-3; surfactant for metalworking fluids, lubricant additives, oilfield chems., asphalt emulsifiers, alkyd resins, industrial/household cleaners, plasticizers, textile drawing lubricants, surf. coatings, rubber prods.; epoxy grade; Gardner 1+; sp.gr. 0.897; dens. 7.45 lb/gal; visc. 20 cps; acid no. 199; iodine no. 130; sapon. no. 200; flash pt. (OC) 204 C; 99.0% fatty acids.

Acintol® FA-1. [Arizona] Tall oil acid; CAS 61790-12-3; surfactant for metalworking fluids, lubricant additives, oilfield chems., asphalt emulsifiers, alkyd resins, industrial/household cleaners, plasticizers, textile drawing lubricants, surf. coatings, rubber prods.; Gardner 5; sp.gr. 0.906; dens. 7.53 lb/gal; visc. 20 cps; acid no. 194; iodine no. 131; sapon. no. 197; flash pt. (OC) 204 C; 92.8% fatty acids.

Acintol® FA-1 Special. [Arizona] Tall oil acid; CAS 61790-12-3; surfactant for metalworking fluids, lubricant additives, oilfield chems., asphalt emulsifiers, alkyd resins, industrial/household cleaners, plasticizers, textile drawing lubricants, surf. coatings, rubber prods.; Gardner 4+; sp.gr. 0.91; dens. 7.50 lb/gal; visc. 20 cps; acid no. 195; iodine no. 131; sapon. no. 198; flash pt. (OC) 204 C; 95.2% fatty acids.

Acintol® FA-2. [Arizona] Tall oil acid; CAS 61790-12-3; surfactant for metalworking fluids, lubricant additives, oilfield chems., asphalt emulsifiers, alkyd resins, industrial/household cleaners, plasticizers, textile drawing lubricants, surf. coatings, rubber prods.; Gardner 3+; sp.gr. 0.898; dens. 7.47 lb/gal; visc. 20 cps; acid no. 197; iodine no. 130; sapon. no. 199; flash pt. (OC) 204 C; 97.8% fatty acids.

Acintol® FA-3. [Arizona] Tall oil acid; CAS 61790-12-3; surfactant for metalworking fluids, lubricant additives, oilfield chems., asphalt emulsifiers, alkyd resins, industrial/household cleaners, plasticizers, textile drawing lubricants, surf. coatings, rubber prods.; Gardner 2+; sp.gr. 0.897; dens. 7.45 lb/gal; visc. 20 cps; acid no. 198; iodine no. 130; sapon. no. 200; flash pt. (OC) 204 C; 98.8% fatty acids.

Acintol® R Type 3A. [Arizona] Tall oil rosin; printing ink binder as resin or salt, paper sizing agent, emulsifier for SBR polymerization as soap, tackifier resin in adhesives, imidazoline modifier in corrosion inhibitors, elastomer modifier in emulsion polymerization, dust control additive; film former/plasticizer in lacquers and varnishes; Gardner 4+ color; sp.gr. 1.03; soften. pt. (R&B) 72 C; acid no. 168; flash pt. (OC) 226 C.

Acintol® R Type S. [Arizona] Tall oil rosin; printing ink binder as resin or salt, paper sizing agent, emulsifier for SBR polymerization as soap, tackifier resin in adhesives, imidazoline modifier in corrosion inhibitors, elastomer modifier in emulsion polymerization, dust control additive; film former/plasticizer in lacquers and varnishes; Gardner 7 color; sp.gr. 1.01; soften. pt. (R&B) 76 C; acid no. 175-178; flash pt. (OC) 226 C.

Acintol® R Type SB. [Arizona] Tall oil rosin; printing ink binder as resin or salt, paper sizing agent, emulsifier for SBR polymerization as soap, tackifier resin in adhesives, imidazoline modifier in corrosion inhibitors, elastomer modifier in emulsion polymerization, dust control additive; film former/plasticizer in lacquers and varnishes; Gardner 8 color; sp.gr. 1.01; soften. pt. (R&B) 75 C; acid no. 165-166; flash pt. (OC) 226 C.

Aconol X6. [Hart Chem. Ltd.] PEG-6 tallate; CAS 61791-00-2; nonionic; emulsifier for min. oil; liq.; HLB 10.5; 100% act.

Aconol X10. [Hart Chem. Ltd.] PEG-10 tallate; CAS 61791-00-2; nonionic; emulsifier, low-foaming surfactant for built detergent systems; stable in moderate concs. of acid; clear liq.; clear sol.; sp.gr. 1.039; visc. 140 cps; HLB 13.0; 95% act.

Acra-500. [Exxon/Tomah] Fatty amine complex; cationic; asphalt wetting and antistripping agent; paste; 100% conc.

Acrilev ADK Special. [Finetex] Potassium salt of phosphate ester; caustic stable wetting agent, leveling agent, scouring agent; liq.; 85% conc.

Acrilev AM, AM-Special. [Finetex] Phosphate ester, potassium salt; anionic; detergent, wetter, dye leveler used in textiles; liq.; 35 and 65% conc. resp.

Acrilev OJP-25N. [Finetex] Alkylalkoxylated phosphate ester, sodium salt; anionic; emulsifier, detergent, wetting agent, dispersant; liq.; 30% conc.

Acrylic Resin AS. [ICI Surf. UK] Acrylic copolymer aq. disp.; anionic/nonionic; finishing agent giving full supple hand on all fibers; liq.

Acto 450. [Exxon] Alkylaryl sodium sulfonate; anionic; detergent, wetting agent, emulsifier, rust preventative; reddish-amber liq.; alcohol odor; m.w. 470; oil sol., slight sol. in water but readily disp.; dens. 8.24 lb/gal; sp.gr. 0.99; flam.; 44% act.

Acto 500. [Exxon] Alkylaryl sodium sulfonate; anionic; detergent, wetting agent, emulsifier, rust preventative; amber, visc. liq.; petrol. odor; m.w. 460; oil sol., slight sol. in water but readily disp.; dens.

8.33 lb/gal; sp.gr. 1.00; 50% act.

Acto 630. [Exxon] Alkylaryl sodium sulfonate; anionic; detergent, wetting agent, emulsifier, rust preventative; amber visc. liq.; bland, petrol. odor; flam.; m.w. 470; oil sol., slight sol. in water but readily disp.; dens. 8.41 lb/gal; sp.gr. 1.01; 63% act.

Acto 632. [Exxon] Alkylaryl sodium sulfonate; anionic; detergent, wetting agent, emulsifier, rust preventative; reddish-amber liq.; alcohol odor; flam.; m.w. 430; oil sol., slight sol. in water but readily disp.; dens. 8.41 lb/gal; sp.gr. 1.01; 63% act.

Acto 636. [Exxon] Alkylaryl sodium sulfonate; anionic; detergent, wetting agent, emulsifier, rust preventative; orange, visc. liq.; alcohol odor; flam.; m.w. 470; oil sol., slight sol. in water but readily disp.; dens. 8.5 lb/gal; sp.gr. 1.02; 63% act.

Acto 639. [Exxon] Alkylaryl sodium sulfonate; anionic; detergent, wetting agent, emulsifier, rust preventative; reddish-amber liq.; alcohol odor; flam.; m.w. 520; oil sol., slight sol. in water but readily disp.; dens. 8.5 lb/gal; sp.gr. 1.02; 63% act.

ACtone® 1. [Allied-Signal/A-C® Perf. Addit.] Low m.w. proprietary ionomer; pigment wetting agent, dispersion aid to increase color strength; used in thermoplastic and thermoset resins; free-flowing powd.; visc. 5500 cps (190 C); bulk dens. 410 kg/m^3; soften. pt. 101 C; acid no. 37; hardness 1.5 dmm.

ACtone® 2000V. [Allied-Signal/A-C® Perf. Addit.] color enhancer, dispersion aid for PVC colorants, esp. difficult-to-disperse organics; visc. 9400-14,500 cps (190 C); bulk dens. 26.8 lb/ft^3; hardness 3.8 ± 0.7 dmm.

ACtone® 2010, 2010P. [Allied-Signal/A-C® Perf. Addit.] Low m.w. proprietary ionomer resins; pigment wetting agent, dispersion aid for color concs. and masterbatches for polyester and styrenics; gran. and powd. resp.; visc. 500,000 cps (190 C); melt flow 1.0 g/10 min.

ACtone® 2461. [Allied-Signal/A-C® Perf. Addit.] color enhancer, dispersion aid for PE colorants, esp. difficult-to-disperse organics.

ACtone® N. [Allied-Signal/A-C® Perf. Addit.] Low m.w. proprietary ionomer; pigment wetting agent, dispersion aid, color enhancer for color concs. for nylon and polyester; gran.; visc. 30,000 cps (190 C); bulk dens. 30.21 lb/ft^3; acid no. 5; soften. pt. 100 C.

Actrabase 31-A. [Climax Performance] Soap sulfonate; anionic; primary emulsifier for oil systems, metalworking fluids; base for naphthenic oils; liq.; 100% conc.

Actrabase 163. [Climax Performance] Soap sulfonate; anionic; emulsifier base for sol. oils; liq.; 100% conc.

Actrabase 215. [Climax Performance] emulsifier for metalworking fluids; low use level base for paraffinic oils.

Actrabase 264. [Climax Performance] emulsifier for metalworking fluids; low use level base for naphthenic oils.

Actrabase PS-470. [Climax Performance] Med. m.w. petrol. sulfonate, sodium salt; anionic; emulsifier and rust inhibitor for cutting and lube oils; dispersant for sol. oil and semi-syns. for metalworking fluids; visc. liq.; oil-sol.; 100% conc.

Actrabase SS-503. [Climax Performance] Semisyn. conc. base, emulsifier for metalworking fluids.

Actrabase SS-523. [Climax Performance] SS-503 with no oil or water in conc.; emulsifier designed to add naphthenic oil and water for metalworking fluids.

Actrafoam A, B, C, S. [Climax Performance] Blend of glycols, fatty acids, and nonionic surfactants in a hydrocarbon base; general purpose defoamer (A, B); defoamer for water sewage applics. (C, S).

Actrafos 110, 110A. [Climax Performance] Complex aliphatic hydroxyl compd. phosphate ester; anionic; pressure additive for cutting and rolling oils; hydrotrope for cleaning compds.; lubricant emulsifier and rust inhibitor; exc. for aluminum; 110A has higher m.p.; liq.; water-sol.; 100% conc.

Actrafos 152A. [Climax Performance] Org. phosphate ester; anionic; extreme pressure lubricant for cutting oils; high phosphorus content; liq.; water-sol.; 100% conc.

Actrafos 161, 315. [Climax Performance] Phosphate ester; anionic; lubricant and emulsifier for use in cutting oils; liq.; sol. in water and oil; 100% conc.

Actrafos 186. [Climax Performance] Aliphatic alcohol phosphate ester; anionic; lubricant and emulsifier for use in cutting oils; liq.; oil and water sol.; 100% conc.

Actrafos 208. [Climax Performance] Phosphate ester; coupler for nonionics in cleaner formulations; lubricant and emulsifier for cutting fluids; stable in electrolytes; sol. in water and oil.

Actrafos 216. [Climax Performance] Phosphate ester; coupler for sulfated oils in cleaner formulations; lubricant, emulsifier for syn, semi-syn. and water-based cutting, grinding, and drawing fluids; liq.; water-sol.

Actrafos 306. [Climax Performance] Phosphate ester; coupler for cleaner formulations; lubricant, emulsifier for cutting, grinding, and drawing fluids; liq.; water-sol.

Actrafos 314. [Climax Performance] Phosphate ester; coupler for nonionics in cleaner formulations; lubricant and emulsifier for cutting fluids; stable in electrolytes; water-sol.

Actrafos 800. [Climax Performance] Phosphate ester; defoamer in metalworking fluids; liq.

Actrafos 822. [Climax Performance] Complex aliphatic alcohol phosphate ester; anionic; lubricant and emulsifier for use in cutting oils; liq.; oil and water sol.; 100% conc.

Actrafos SA-208. [Climax Performance] Linear alcohol phosphate ester; anionic; lubricant and emulsifier in cutting oils; liq.; oil and water sol.; 100% conc.

Actrafos SA-216. [Climax Performance] Phosphate ester; anionic; lubricant and emulsifier for cutting oils; liq.; sol. in oil and water; 100% conc.

Actrafos SN-306. [Climax Performance] Phosphate ester; anionic; used in liq. cleaners; liq.; water-sol.; 100% conc.

Actrafos SN-314. [Climax Performance] Phosphate ester; anionic; used in waterless hand cleaners and solv. cleaners; liq.; 100% conc.

Actrafos SN-315. [Climax Performance] Phosphate ester; anionic; lubricant, emulsifier for metal working; aluminum corrosion inhibitor; liq.; oil-sol.; 100% conc.

Actrafos T. [Climax Performance] Tridecyl alcohol phosphate ester; anionic; extreme pressure lubricant and release agent for cutting oils; liq.; oil-sol.; 100% conc.

Actralube SOS. [Climax Performance] Ester; nonionic; emulsifier, lubricant, metalworking additive; substitute for sperm oil; liq.; oil-sol.; 100% conc.

Actralube Syn-147. [Climax Performance] Complex diester; nonionic; lubricity additive for syn. cutting fluids; liq.; water-sol.; 90% conc.

Actralube Syn-153. [Climax Performance] Soap; anionic; lt. duty syn. lubricant and rust inhibitor for syn. cutting fluids; liq.; water-sol.; 95% conc.

Actramide 202. [Climax Performance] 2:1 Tall oil fatty acid alkanolamide; nonionic; emulsifier for sol. oils, metalworking fluids and emulsion cleaners; corrosion inhibitor; liq.; water-sol.

Actramide 410. [Climax Performance] Alkanolamide; nonionic; sec. emulsifier with lubricating properties; liq.; oil-sol.; 100% conc.

Actramide 5264. [Climax Performance] Modified 2:1 tall oil fatty acid alkanolamide; emulsifier, lubricant, rust inhibitor; water-sol.

Actrasol 6092. [Climax Performance] Sulfated rapeseed oil; lubricant, emulsifier in pigment flushing, cleaners, textiles, paper processing.

Actrasol C-50, C-75, C-85. [Climax Performance] Sulfated castor oil; anionic; pigment wetting and disp.; lubricant and emulsifier for metalworking fluids; for pigment flushing, cleaners, textile, and paper processing; liq.; 50, 70, and 75% conc. resp.

Actrasol CS-75. [Climax Performance] Sulfonated soyabean oil, sodium neutralized; lubricant, emulsifier in pigment flushing, cleaners, textiles, paper processing.

Actrasol EO. [Climax Performance] Sulfated glyceryl trioleate, sodium neutralized; anionic; surfactant for shampoos, metalworking; liq.; 75% conc.

Actrasol MY-75. [Climax Performance] Sulfated methyl ester of soya fatty acid, sodium neutralized; anionic; lubricant and emulsifier in metalworking fluids, water-based drilling muds; oil field defoamer; biodeg.; liq.; water-disp.; 75% conc.

Actrasol OY-75. [Climax Performance] Sulfated soyabean oil, sodium neutralized; lubricant, emulsifier in pigment flushing, cleaners, textiles, paper processing; biodeg.; liq.; water-disp.; 75% conc.

Actrasol PSR. [Climax Performance] Sulfated ricinoleic acid, potassium neutralized; anionic; pigment wetting and disp.; aluminum lubricant; lubricant, emulsifier for pigment flushing, cleaners, textiles, paper processing; liq.; 75% conc.

Actrasol SBO. [Climax Performance] Sulfated butyl oleate, sodium neutralized; lubricant, emulsifier in pigment flushing, cleaners, textiles, paper processing.

Actrasol SP. [Climax Performance] Sulfonated tall oil fatty acid, sodium neutralized; anionic; wet process phosphoric acid defoamer; lubricant, emulsifier for pigment flushing, cleaners, textiles, paper processing; liq.; 50% conc.

Actrasol SP 175K. [Climax Performance] Sulfated tall oil, potassium neutralized; anionic; lubricant, emulsifier for pigment flushing, cleaners, textiles, paper processing; biodeg.; liq.; water-disp.; 75% conc.

Actrasol SR 75. [Climax Performance] Sulfated oleic acid, ammonium neutralized; anionic; mold release agent; lubricant, emulsifier for pigment flushing, cleaners, textiles, paper processing; biodeg.; liq.; water-disp.; 75% conc.

Actrasol SRK 75. [Climax Performance] Sulfated oleic acid, potassium neutralized; anionic; mold release agent; lubricant, emulsifier for pigment flushing, cleaners, textiles, paper processing; biodeg.; liq.; water-disp.; 75% conc.

Actrasol SS. [Climax Performance] Sulfated tall oil; anionic; rust preventative, lubricant, metal polish; solid; 75% conc.

Actrol 4DP. [Climax Performance] PEG diester; nonionic; emulsifier and lubricity additive for metalworking fluids; liq.; HLB < 5; 95% conc.

Actrol 4MP, 628. [Climax Performance] PEG ester; nonionic; emulsifier and lubricity additive for metalworking fluids; liq.; HLB > 10; 95% conc.

Acusol® 410N. [Rohm & Haas] Sodium polyacrylate; detergent polymer, dispersant for cleaners, water treatment, min. processing; m.w. 10,000; visc. 500-1500 cps; pH 6.5-8.0; 40% solids.

Acusol® 445. [Rohm & Haas] Polyacrylic acid; detergent polymer for detergents and cleaners, water treatment, min. processing, other industrial markets; m.w. 4500; pH 3; 48% solids.

Acusol® 445N. [Rohm & Haas] Sodium polyacrylate; detergent polymer for detergents and cleaners, water treatment, min. processing, other industrial markets; m.w. 4500; pH 7; 45% solids.

Acusol® 445ND. [Rohm & Haas] Sodium polyacrylate; detergent polymer for detergents and cleaners, water treatment, min. processing, other industrial markets; dry form; m.w. 4500; 92% solids.

Acusol® 460ND. [Rohm & Haas] Maleic acid/olefin copolymer, sodium salt; detergent polymer for detergents and cleaners, water treatment, min. processing, other industrial markets; 92% solids.

Acusol® 479N. [Rohm & Haas] Sodium acrylic acid/maleic acid copolymer; detergent polymer for detergents and cleaners, water treatment, min. processing, other industrial markets; m.w. 70,000; pH 7; 40% solids.

Acusol® 479ND. [Rohm & Haas] Sodium acrylic acid/maleic acid copolymer; detergent polymer for detergents and cleaners, water treatment, min. processing, other industrial markets; dry form; m.w. 70,000; 92% solids.

Acusol® 480N. [Rohm & Haas] Modified polyacrylic acid, sodium salt; detergent polymer for detergents and cleaners, water treatment, min. processing, other industrial markets; m.w. 3500; pH 7; 46% solids.

Acusol® 810. [Rohm & Haas] Acrylic crosslinked copolymer; detergent polymer, processing aid, thickener for detergents and cleaners, water treatment, min. processing, other industrial markets; pH 3; 18% solids.

Acusol® 820. [Rohm & Haas] Acrylic copolymer; detergent polymer, processing aid, thickener for detergents and cleaners, water treatment, min. processing, other industrial markets; m.w. 500,000; pH 3; 30% solids.

Acusol® 830. [Rohm & Haas] Acrylates copolymer; detergent polymer, thickener for detergents, water treatment, min. processing, other industrial markets; sp.gr. 1.054; visc. 3000-5000 cps; pH 2.1-4.0.

Acusol® 840. [Rohm & Haas] Acrylic copolymer; detergent polymer, processing aid, thickener for detergents and cleaners, water treatment, min. processing, other industrial markets; m.w. 2,000,000; pH 3; 18% solids.

Acusol® 860N. [Rohm & Haas] Sodium polyacrylate; detergent polymer, processing aid for detergents, water treatment, min. processing, other industrial markets; m.w. 40,000; visc. 2500 cps; pH 7.5-9.0; 34-35% solids.

Acylglutamate AS-12. [Ajinomoto] Sodium oleoyl glutamate, sodium cocoyl glutamate; good detergency for liq. dishwashing detergents, textile detergents, etc.; biodeg.; colorless or lt. yel. liq.; m.w. 414; pH 6.2 (1%, 40 C); 25% aq. sol'n.

Acylglutamate CS-11. [Ajinomoto] Sodium cocoyl

glutamate; anionic; detergent, emollient for personal care prods.; bacteriostatic effect; biodeg.; wh. powd.; m.w. 359; pH 5.5 (1%, 40 C); 100% act.

Acylglutamate CS-21. [Ajinomoto] Disodium cocoyl glutamate; detergent, emollient for personal care prods.; bacteriostatic effect; corrosion inhibitor; biodeg.; wh. powd.; m.w. 378; pH 9.0 (1%, 40 C).

Acylglutamate CT-12. [Ajinomoto] TEA N-cocoyl-L-glutamate; anionic; detergent, emollient for personal care prods.; bacteriostatic effect; biodeg.; colorless or lt. yel. liq.; m.w. 483; pH 5.4 (1%, 40 C); toxicology: LD50 (mice, acute oral) 6.5 g/kg; 30% aq. sol'n.

Acylglutamate DL-12. [Ajinomoto] Monolithium salt of N-distilled cocoyl-L-glutamic acid; foamer for carpet cleansers; easily becomes a powd. after drying; 20% aq. sol'n.

Acylglutamate GS-11. [Ajinomoto] Sodium hydrogenated tallow glutamate, sodium cocoyl glutamate; detergent, emollient for personal care prods.; bacteriostatic effect; biodeg.; wh. powd.; biodeg.; m.w. 420; pH 6.6 (1%, 40 C); 100% conc.

Acylglutamate GS-21. [Ajinomoto] Disodium cocoyl/tallowyl glutamate; basic material for heavy-duty detergents; reduces adverse reactions on human skin in toiletries; capturing agent of heavy metal ions; biodeg.; wh. powd.; m.w. 438; pH 9.0 (1%, 40 C).

Acylglutamate HS-11. [Ajinomoto] Sodium hydrogeanted tallow glutamate; detergent, emollient for personal care prods.; bacteriostatic effect; biodeg.; wh. powd.; m.w. 432; pH 6.9 (1%, 40 C); toxicology: LD50 (mice, acute oral) 4.3 g/kg; 100% conc.

Acylglutamate HS-21. [Ajinomoto] Disodium stearoyl glutamate; basic material for heavy-duty detergents; reduces adverse reactions on human skin in toiletries; capturing agent of heavy metal ions; biodeg.; wh. powd.; m.w. 452; pH 9.0 (1%, 40 C); toxicology: LD50 (mice, acute oral) 3.5 g/kg.

Acylglutamate LS-11. [Ajinomoto] Sodium lauroyl glutamate; detergent, emollient for personal care prods.; bacteriostatic effect; biodeg.; wh. powd.; m.w. 355; pH 5.3 (1%, 40 C); toxicology: LD50 (mice, acute oral) 5.5 g/kg; 100% conc.

Acylglutamate LT-12. [Ajinomoto] TEA lauroyl glutamate; detergent, emollient for personal care prods.; bacteriostatic effect; biodeg.; colorless or lt. yel. aq. sol'n.; m.w. 479; pH 5.2 (1%, 40 C); 30% aq. sol'n.

Acylglutamate MS-11. [Ajinomoto] Sodium myristoyl glutamate; anionic; detergent, emollient for personal care prods.; bacteriostatic effect; biodeg.; wh. powd.; m.w. 383; pH 6.1 (1%, 40 C); 100% conc.

AD-700. [Anedco] Surfactant/coupler/alkaline blend; biodeg. steam cleaner.

AD-709. [Anedco] Alkyl sulfonate; paraffin dispersant.

AD-710. [Anedco] Terpenes/amides/surfactants/solvent blend; paraffin dispersant to prevent or remove paraffin; 40% active.

AD-713. [Anedco] Multicomponent paraffin dispersant.

AD-713C. [Anedco] Diamines/amides/surfactant blend; conc. paraffin dispersant to remove or prevent paraffin; 100% active.

AD-716. [Anedco] Blend; anionic/nonionic; surfactant for cleaning oil rigs.

AD-742. [Anedco] Surfactant/coupler blend; heavy-duty degreaser.

AD-742C. [Anedco] Surfactant/coupler blend; biodeg. heavy-duty degreaser.

AD-747. [Anedco] Alkyl succinate full ester; surfactant; 30% active.

AD-747C. [Anedco] Alkyl succinate full ester; surfactant; 100% active.

AD-748. [Anedco] Polyphosphate; cleaner and degreaser.

AD-749. [Anedco] Ethoxylated nonyl phenol and surfactant; drilling detergent; water-sol.

AD-750, AD-750C. [Anedco] Alkanolamide/polyphosphate; drilling detergent; water-sol.

AD-752. [Anedco] Half ester polyolefin anhydride; pour pt. depressant.

AD-763. [Anedco] Pine oil, surfactants, couplers; for pine oil cleaners.

ADF-600. [Anedco] Alcohol-based; general purpose defoamer for water-based drilling muds; water-sol.

ADF-610. [Anedco] Silicone-based; defoamer; water-sol.

ADF-620. [Anedco] Silicone-based; defoamer; oil-sol.

ADF-630. [Anedco] Silicone-based; anionic; defoamer; oil-sol.

Adinol OT16. [Croda Chem. Ltd.] Sodium methyl oleoyl taurate; CAS 137-20-2; anionic; biodeg. surfactant, wetting agent, detergent, emulsifier, foamer, dispersant for cosmetic and industrial applics., esp. textile processing; liq.; 16+% conc.

ADM-407. [Anedco] Alkyl benzyl sulfonic acid; anionic; demulsifier; 62.5% active.

ADM-407C. [Anedco] Alkyl benzyl sulfonic acid; anionic; demulsifier; 100% active.

ADM-408. [Anedco] Partially neutralized sulfonate; demulsifier; 50% active.

ADM-408C. [Anedco] Partially neutralized sulfonate; demulsifier; 100% active.

ADM-409. [Anedco] Neutralized alkyl sulfonate; demulsifier; 50% active.

ADM-409C. [Anedco] Neutralized alkyl sulfonate; demulsifier; 100% active.

ADM-410. [Anedco] Alkylaryl sulfonate; demulsifier; 50% active.

ADM-410C. [Anedco] Alkylaryl sulfonate; demulsifier; 100% active.

ADM-411. [Anedco] Amine sulfonate; demulsifier; 50% active.

ADM-412. [Anedco] Naphthalene sulfonate; demulsifier; 50% active.

ADM-412C. [Anedco] Naphthalene sulfonate; demulsifier; 100% active.

ADM-456. [Anedco] Oxyalkylated phenolic resin; demulsifier; 40% active.

ADM-456C. [Anedco] Combination of phenolic resins; demulsifier; 60% active.

ADM-457. [Anedco] Mixed resin and sulfonates; demulsifier for bad tank bottoms; 35% active.

ADM-457C. [Anedco] Sulfonate/resin blend; demulsifier for bad tank bottoms; 70% active.

ADM-458. [Anedco] Polyol; demulsifier.

ADM-467. [Anedco] Diepoxide resin; demulsifier; 35.5% active.

ADM-477. [Anedco] Resin/wetting agent blend; demulsifier; 28% active.

ADM-477C. [Anedco] Resin/wetting agent blend; demulsifier; 56% active.

ADM-487. [Anedco] Resin/surfactant blend; demulsifier; 28% active.

ADM-487C. [Anedco] Resin/wetting agent blend; demulsifier; 56% active.

Adma® 8. [Ethyl] Octyldimethylamine; CAS 7378-99-6; cationic; intermediate for quat. ammonium compds., amine oxides, betaines; APHA < 10 liq., fatty amine odor; sp.gr. 0.765; f.p. -57 C; amine no. 352; flash pt. (TCC) 64 C; corrosive; 100% conc.

Adma® 10. [Ethyl] Decyldimethylamine; CAS 1120-24-7; cationic; intermediate for quat. ammonium compds., amine oxides, betaines; clear liq., fatty amine odor; sp.gr. 0.778; f.p. -35 C; amine no. 300; flash pt. (TCC) 91 C; corrosive; 100% conc.

Adma® 12. [Ethyl] Dodecyl dimethylamine; CAS 112-18-5; cationic; intermediate for quat. ammonium compds., amine oxides, betaines; clear liq., fatty amine odor; sp.gr. 0.778; f.p. -22 C; amine no. 259; flash pt. (PM) 114 C; corrosive; 100% conc.

Adma® 14. [Ethyl] Tetradecyl dimethylamine; CAS 112-75-4; cationic; intermediate for quat. ammonium compds., amine oxides, betaines; clear liq.; sp.gr. 0.794; f.p. -6 C; amine no. 229; flash pt. (PM) 132 C; corrosive; 100% conc.

Adma® 16. [Ethyl] Hexadecyl dimethylamine; CAS 112-69-6; cationic; intermediate for quat. ammonium compds., amine oxides, betaines; clear liq., fatty amine odor; sp.gr. 0.800; f.p. 8 C; amine no. 206; flash pt. (PM) 142 C; corrosive; 100% conc.

Adma® 18. [Ethyl] Octadecyl dimethylamine; CAS 124-28-7; intermediate for mfg. of quaternary ammonium compds. for biocides, textile chems., oilfield chems., amine oxides, betaines, polyurethane foam catalysts, epoxy curing agents; clear liq., fatty amine odor; sp.gr. 0.807; f.p. 21 C; amine no. 186; flash pt. (PM) 163 C; corrosive; 98% tert. amine.

Adma® 246-451. [Ethyl] Dodecyl dimethylamine (40%), tetradecyl dimethylamine (50%), hexadecyl dimethylamine (10%); cationic; intermediate for quat. ammonium compds., amine oxides, betaines; clear liq., fatty amine odor; sp.gr. 0.792; f.p. -13 C; amine no. 238; flash pt. (PM) 114 C; corrosive; 100% conc.

Adma® 246-621. [Ethyl] Dodecyl dimethylamine (65%), tetradecyl dimethylamine (25%), hexadecyl dimethylamine (10%); cationic; intermediate for quat. ammonium compds., amine oxides, betaines; clear liq., fatty amine odor; sp.gr. 0.791; f.p. -18 C; amine no. 245; flash pt. (PM) 121 C; corrosive; 100% conc.

Adma® 1214. [Ethyl] Dodecyl dimethylamine (65%), tetradecyl dimethylamine (35%); cationic; intermediate for quat. ammonium compds., amine oxides; clear liq., fatty amine odor; sp.gr. 0.791; f.p. -18 C; amine no. 249.2; flash pt. (PM) 121 C; corrosive; 100% conc.

Adma® 1416. [Ethyl] Dodecyl dimethylamine (5%), tetradecyl dimethylamine (60%), hexadecyl dimethylamine (30%), octadecyl dimethylamine (5%); cationic; intermediate for quat. ammonium compds., amine oxides, betaines; APHA < 10 liq.; sp.gr. 0.796; amine no. 220; 100% conc.

Adma® WC. [Ethyl] Octyl dimethylamine (7%), decyl dimethylamine (6%), dodecyl dimethylamine (53%), tetradecyl dimethylamine (19%), hexadecyl dimethylamine (9%), octadecyl dimethylamine (6%); cationic; intermediate for quat. ammonium compds., amine oxides, betaines; clear liq., fatty amine odor; sp.gr. 0.798; f.p. -22 C; amine no. 249; flash pt. (PM) 102 C; corrosive; 100% conc.

Admox® 1214. [Ethyl] Alkyldimethylamine oxide; nonionic; high foaming material to improve foam profile of anionic surfactants; visc. modifier, emollient; liq.; 30% conc.

Admox® 14-85. [Ethyl] Myristamine oxide; CAS 3332-27-2; for soap bars, shaving creams, fabric softeners, hard surf. cleaners, laundry detergents, oxygen bleach powds., tootphaste, agric., automatic dishwash, cellulose extraction, gasoline additives, bubble baths; solid; m.w. 257; m.p. 40-42 C; flash pt. > 93 C; 87 ± 2% amine oxide.

Admox® 18-85. [Ethyl] Stearamine oxide; CAS 2571-88-2; for soap bars, shaving creams, fabric softeners, hard surf. cleaners, laundry detergents, oxygen bleach powds., tootphaste, agric., automatic dishwash, cellulose extraction, gasoline additives, bubble baths; solid; m.w. 313; m.p. 61-62 C; flash pt. > 93 C; 87 ± 2% amine oxide.

Adogen® 444. [Sherex/Div. of Witco] Palmityl trimethyl ammonium chloride; cationic; specialty quat.; emulsifier, dispersant, cream rinse, fermentation aid; Gardner 6 max. liq.; m.w. 319; flash pt. (PM) 58 F; 49–52% quat.

Adogen® 461. [Sherex/Div. of Witco] Coco trimethyl ammonium chloride; cationic; emulsifier, dispersant; used in corrosion inhibitor formulations for oilfield brines and HCl acidizing systems; liq.; 50% conc.

Adogen® 462. [Sherex/Div. of Witco] Dicocodimonium chloride, IPA/water; cationic; antistat, emulsifier, flocculating agent, dispersant used in corrosion inhibitor formulations for oil-field chemicals; Gardner 5 max. liq.; m.w. 439; flash pt. 68 F (PM); 74–77% quat.

Adogen® 464. [Sherex/Div. of Witco] Methyl tri (C8–C10) ammonium chloride; cationic; flotation agent, emulsifier for solv. extraction of metals; in corrosion inhibitor formulations for oilfield brines and HCl acidizing systems; Gardner 5 max. liq.; m.w. 437; flash pt. 156 F (PM); 85% quat.

Adogen® 470. [Sherex/Div. of Witco] Ditallowdimonium chloride; specialty quat. for nonionic laundry detergent-softeners; Gardner 6 max. liq.; m.w. 564; flash pt. 65 F (PM); 74–77% quat.

Adogen® 471. [Sherex/Div. of Witco] Tallow trimonium chloride, IPA; cationic; dispersant, antistat, emulsifier; used in corrosion inhibitor formulations for oilfield brines and HCl acidizing systems; Gardner 6 max. liq.; m.w. 339; flash pt. 58 F (PM); 49–52% quat.

Adogen® 477. [Sherex/Div. of Witco] N-Tallow pentamethyl propane diammonium chloride; cationic; emulsifier, dispersant; liq.; 50% conc.

Adogen® 560. [Sherex/Div. of Witco] Coco amino propyl amines; cationic; corrosion inhibitors, emulsifiers; intermediates for corrosion inhibitors, emulsifiers for asphalt; liq.; 100% conc.

Adogen® 570-S. [Sherex/Div. of Witco] Tallow amino propyl amines; cationic; corrosion inhibitors, emulsifiers; intermediates for corrosion inhibitors, emulsifiers for asphalt; liq.; 100% conc.

Adogen® 572. [Sherex/Div. of Witco] Oleyl amino propyl amines; cationic; corrosion inhibitors, emulsifiers; intermediates for corrosion inhibitors, emulsifiers for asphalt; liq.; 100% conc.

Adogen® MA-112 SF. [Sherex/Div. of Witco] Dimethyl behenamine; neutralizer, conditioner, coemulsifier; paste; 99% solids.

Adol® 42. [Sherex/Div. of Witco] Tallow alcohol; nonionic; emulsifier, lubricant, foam control agent, cosolvent, plasticizer, stabilizer, emollient, intermediate; for metal rolling oils and lubricants, gas scrubbing, printing inks, auto antifreeze, textile dyeing and finishing, emulsion systems,; min. processing, oil field chemicals, fabric softeners; Lovibond 5Y/0.5R color; m.w. 258; sol. in fatty alcohols, IPA, benzene, trichlorethylene, acetone, turpentine, VMP naphtha, kerosene, lt. min. oil; sp.gr. 0.814 (60/25 C); cloud pt. 36 C max.; sapon. no. 3.5 max.; 100% conc.

Adol® 52 NF. [Sherex/Div. of Witco] Cetyl alcohol; CAS 36653-82-4; nonionic; coemulsifier, lubricant, foam control agent, cosolvent, plasticizer, stabilizer, emollient, intermediate; for metal lubricants, inks, textiles, emulsions, paper, cosmetics, mineral processing, oil field chemicals, fabric softeners; Lovibond 5Y/0.5R max.; m.w. 247; sol. in fatty alcohols, IPA, benzene, trichloroethylene, acetone, turpentine, VMP naphtha, kerosene, lt. min. oil; sp.gr. 0.815 (60/25 C); m.p. 45–50 C; acid no. 1.0 max.; sapon. no. 3.0 max.

Adol® 60. [Sherex/Div. of Witco] Behenyl alcohol; lubricant, foam control agent, cosolv., plasticizer, emollient; Lovibond 10Y/1.0R; m.w. 316; sp. gr. 0.79 (100/25 C); acid no. 1.5 max.; iodine no. 6 max.; sapon. no. 3 max.

Adol® 61 NF. [Sherex/Div. of Witco] Stearyl alcohol; CAS 112-92-5; nonionic; used in emulsifiers, surfactants, cosmetics, wax formulations; see also Adol 42; wh. cryst. solid; odorless; sol. in IPA, acetone, naphtha, lt. min. oil; m.w. 272; sp.gr. 0.817 (60/25 C); m.p. 56–60 C; acid no. 0.5; sapon. no. 1.0 max.; 100% conc.

Adol® 62 NF. [Sherex/Div. of Witco] Stearyl alcohol; CAS 112-92-5; nonionic; emollient, glass frit binders, waxes, emulsion stabilizers, esters, tertiary amines, surfactants, polymers, chemical intermediate; cosmetic formulations; see also Adol 42; Lovibond 5Y/0.54 max; odorless; m.w. 272; sol. see Adol 61NF; sp.gr. 0.817 (60/25 C); visc. 42 SSU (210 F); m.p. 56–60 C; b.p. 337–360 C (760 mm, 90%); acid no. 1.0 max.; sapon. no. 3.0 max.; 100% conc.

Adol® 63. [Sherex/Div. of Witco] Cetearyl alcohol; nonionic; emulsifiers; prime base for detergents; used in plasticizers, tert. amines, lube oil additives, textile auxiliaries, mold lubricants, polymers, org. synthesis, chemical intermediates; see also Adol 42; Lovibond 5Y/0.5R max. flake; m.w. 268; sp.gr. 0.816 (60/25 C); visc. 44 SSU (210 F); solid pt. 48–53 C; b.p. 312–344 C (760 mm, 90%); acid no. 1.0 max.; sapon. no. 3.0 max.; 100% conc.

Adol® 66. [Sherex/Div. of Witco] Isostearyl alcohol; nonionic; coemulsifier, lubricant, foam control agent, cosolvent, plasticizer, stabilizer, emollient, intermediate; for metal lubricants, inks, textiles, emulsions, paper, cosmetics, mineral processing, oil field chemicals, fabric softeners; Lovibond 5Y/0.5R max. liq.; m.w. 295; sol. in fatty alcohols, IPA, benzene, trichlorethylene, acetone, turpentine, VMP naphtha, kerosene, lt. min. oil; sp.gr. 0.861; cloud pt. 8.0 C max.; acid no. 1.0 max.; sapon. no. 2.0 max.; 100% conc.

Adol® 80. [Sherex/Div. of Witco] Oleyl alcohol; CAS 143-28-2; nonionic; coemulsifier, lubricant, foam control agent, cosolvent, plasticizer, stabilizer, emollient, intermediate; for metal rolling oils and lubricants, gas scrubbing, printing inks, auto antifreeze, textile dyeing and finishing, emulsion systems, paper pulping, cosmetics, cationic surfactants, min. processing, oil field chemicals, fabric softeners; Lovibond 5Y/0.5R liq.; m.w. 263; sol. in fatty alcohols, IPA, benzene, trichlorethylene, acetone, turpentine, VMP naphtha, kerosene, lt. min. oil; sp.gr. 0.840; cloud pt. 13 C max.; sapon. no. 3.0 max.; 100% conc.

Adol® 85 NF. [Sherex/Div. of Witco] Oleyl alcohol; CAS 143-28-2; nonionic; emulsifier, lubricant, foam control agent, cosolv., plasticizer, emollient; Lovibond 3Y/0.3R liq.; m.w. 268; sp.gr. 0.84; cloud pt. 10 C max.; acid no. 1.0; iodine no. 85–95; sapon no. 2 max.; 100% conc.

Adol® 90 NF. [Sherex/Div. of Witco] Oleyl alcohol; CAS 143-28-2; nonionic; cosmetic and pharmaceutical grade; emollient; coupling agent; plasticizer for hair sprays, lubricant for aerosols; emulsion stabilizer; used in lotions, creams, bath oils; lt. clear liq.; low odor; m.w. 268; b.p. 282–349 C (760 mm, 90%); sol. in ethanol, IPA, benzene, ethyl ether, acetone, turpentine, VM&P naphtha, kerosene, lt. min. oil; sp.gr. 0.840; acid no. 0.5; iodine no. 90; sapon. no. 1.5; hyd. no. 210; cloud pt. 5 C.; 100% conc.

Adol® 320, 330, 340. [Sherex/Div. of Witco] Nonionic; emulsfiers, emollients; liq.; 100% conc.

Adol® 520 NF. [Sherex/Div. of Witco] Cetyl alcohol; CAS 36653-82-4; nonionic; coemulsifier, lubricant, foam control agent, cosolvent, plasticizer, stabilizer, emollient, intermediate; for metal lubricants, inks, textiles, emulsions, paper, cosmetics, mineral processing, oil field chemicals, fabric softeners; APHA 50 max. flake; m.w. 246; sol. in fatty alcohols, IPA, benzene, trichlorethylene, acetone, turpentine, VMP naphtha, kerosene, lt. min. oil; sp.gr. 0.815 (60/25 C); m.p. 45–50; solid pt. 48–53 C; acid no. 1.0 max.; sapon. no. 2.0 max.; 100% conc.

Adol® 620 NF. [Sherex/Div. of Witco] Stearyl alcohol; CAS 112-92-5; nonionic; coemulsifier, lubricant, foam control agent, cosolvent, plasticizer, stabilizer, emollient, intermediate; for metal lubricants, inks, textiles, emulsions, paper, cosmetics, mineral processing, oil field chemicals, fabric softeners; APHA 50 max. flake; m.w. 267; sol. see Adol 42; sp.gr. 0.817 (60/25 C); m.p. 55–60 C; acid no. 2.0 max.; sapon. no. 2.0 max.; 100% conc.

Adol® 630. [Sherex/Div. of Witco] Cetearyl alcohol; nonionic; emulsifier; flake; 100% conc.

Adol® 640. [Sherex/Div. of Witco] Cetearyl alcohol; nonionic; emulsifier, lubricant, foam control agent, cosolv., plasticizer, emollient; APHA 40 max. flake; m.w. 253; sp.gr. 0.82 (50/50 C); m.p. 43–46 C; acid no. 1.0 max.; iodine no. 1.5 max.; sapon. no. 1.0 max.; 100% conc.

Adsee® 775. [Witco/Organics] POE ethers and special resins; nonionic; spreader sticker, wetting agent, and penetrant for agric. spray; surfactant for monosodium methane arsonate formulations; EPA clearance; liq.; 100% conc.

Adsee® 799. [Witco/Organics] Alkyl POE ether; nonionic; agric. surfactant; soil penetrant; EPA clearance; liq.; 100% conc.

Adsee® 801. [Witco/Organics] Surfactant blend; nonionic; agric. spreader-penetrant, wetting agent, spray-tank adjuvant; EPA clearance; liq.; 100% conc.

Adsee® 1080. [Witco/Organics] Surfactant blend; nonionic; agric. wetting agent, spray-tank adjuvant; EPA clearance; liq.; 100% conc.

Adsee® 2141. [Witco/Organics] Complex surfactant; anionic/nonionic; agric. spreader-penetrant, premix surfactant for monosodium methane arsonate; liq.

Adsee® AK31-73. [Witco/Organics] Surfactant blend; nonionic; veg. oil emulsifier, foaming agent for agric. applics.; EPA clearance; liq.

Advantage 5 Defoamer. [Hercules] Defoamer for acid and alkaline papermaking systems, deinking systems; FDA compliance; brick, soapy odor; disp. in water; sp.gr. 0.96; m.p. 54 C; 100% conc.

Advantage 6 Defoamer. [Hercules] Defoamer for acid and alkaline papermaking systems, deinking systems; FDA compliance; brick, soapy odor; disp. in water; sp.gr. 0.96; m.p. 54 C; 100% conc.

Advantage 7 Defoamer. [Hercules] Defoamer for acid and alkaline papermaking systems, deinking systems, kraft pulp mill screening operations; FDA compliance; brick, soapy odor; disp. in water; sp.gr. 0.96; m.p. 54 C; 100% conc.

Advantage 10 Defoamer. [Hercules] Defoamer for acid and alkaline papermaking systems, deinking systems; FDA compliance; brick, soapy odor; emulsifiable in water; sp.gr. 0.96; m.p. 54 C; 100% conc.

Advantage 52-B. [Hercules] Oil-based defoamer for pulp mill brownstock washing operations or other high-temp. surfactant-stabilized foam systems; improves drainage; FDA compliance; tannish-gray liq.; dens. 7.5 lb/gal; visc. 3500 ± 1000 cps.

Advantage 52EH Defoamer. [Hercules] Oil-based defoamer for pulp mill brownstock washing operations or other high-temp. surfactant-stabilized foam systems; improves drainage; FDA compliance; liq.; dens. 7.59 lb/gal; visc. 1500-3000 cps.

Advantage 52-JS. [Hercules] Oil-based defoamer for pulp mill brownstock washing operations or other high-temp. surfactant-stabilized foam systems; improves drainage; FDA compliance; tannish-gray liq.; dens. 7.5 lb/gal; visc. 3500 ± 1000 cps.

Advantage 70DYX. [Hercules] Silicone-based drainage aid for improved drainage in kraft pulpmill brownstock washing operations by removing entrained air and surface foam; also for cold-stock systems, pulpmill bleaching and screening; FDA compliance; wh. opaque liq.; dens. 8.32 lb/gal; visc. 1500-4000 cps; pH 6.5-8.5.

Advantage 70PHE. [Hercules] Silicone-based drainage aid for improved drainage in kraft pulpmill cold-stock systems, e.g., bleaching and screening operations, by removing entrained air and surface foam; FDA compliance; wh. opaque liq.; dens. 8.34 lb/gal; visc. 1000 ± 300 cps; pH 7.0-8.5.

Advantage 70WLH. [Hercules] Silicone-based drainage aid for improved drainage in kraft pulpmill brownstock washing operations by removing entrained air and surface foam; also for cold-stock systems, pulpmill bleaching and screening; FDA compliance; wh. opaque liq.; dens. 8.34 lb/gal; visc. 1000 ± 300 cps; pH 7.0-8.5.

Advantage 91WW. [Hercules] Silicone-based drainage aid for improved drainage in kraft pulpmill brownstock washing operations by removing entrained air and surface foam; also for cold-stock systems, pulpmill bleaching and screening; FDA compliance; off-wh. liq.; dens. 8.32 lb/gal; visc. 1500-4000 cps.

Advantage 136 Defoamer. [Hercules] Hydrocarbon oil-based defoamer for use as drainage aid and foam killer in kraft pulpmill brownstock washing operations; FDA compliance; lt. tan liq.; dens. 7.3 lb/gal; visc. 1500-2500 cps.

Advantage 136Z Defoamer. [Hercules] Hydrocarbon oil-based defoamer for use as drainage aid and foam killer in kraft pulpmill brownstock washing operations; FDA compliance; lt. tan liq.; dens. 7.3 lb/gal; visc. 1500-2500 cps.

Advantage 187 Defoamer. [Hercules] Hydrocarbon oil-based defoamer for use as drainage aid and foam killer in kraft pulpmill brownstock washing operations; FDA compliance; lt. brn. liq.; dens. 7.7 lb/gal; visc. 1500-3000 cps.

Advantage 187Z Defoamer. [Hercules] Hydrocarbon oil-based defoamer for use as drainage aid and foam killer in kraft pulpmill brownstock washing operations; FDA compliance; lt. brn. liq.; dens. 7.7 lb/gal; visc. 1500-3000 cps.

Advantage 344 Defoamer. [Hercules] All-purpose defoamer for wet-end use in acid or alkaline papermaking systems or pulpmill effluents; FDA compliance; straw liq.; dens. 7.3 lb/gal.

Advantage 357 Defoamer. [Hercules] Hydrocarbon oil-based antifoam for wet-end acid and alkaline papermaking operations, coating operations; controls entrained air and surface foam; FDA compliance; greenish-brn. oily liq.; dens. 7.62 lb/gal.

Advantage 388 Defoamer. [Hercules] Hydrocarbon oil-based all-purpose defoamer for aq. foaming systems, acid and alkaline systems, at high temps., in surfactant-stabilized foam systems, kraft pulpmill screening and bleaching operations; FDA compliance; tan oily liq.; dens. 7.2 lb/gal; 100% conc.

Advantage 470A Defoamer. [Hercules] Water-based defoamer for acid and alkaline papermaking systems, pulpmill bleaching operations, effluents, other aq. foaming systems; FDA compliance; wh. liq., hydrocarbon odor; dens. 8.25 lb/gal; visc. < 100 cps; pH 9.7.

Advantage 491A Defoamer. [Hercules] Water-based defoamer for acid and alkaline papermaking systems, calender and size press sol'ns., other aq. foaming systems; FDA compliance; wh. liq., hydrocarbon odor; dens. 7.85 lb/gal; visc. 300 cps; pH 7.4.

Advantage 831 Defoamer. [Hercules] Hydrocarbon oil-based defoamer for use in size press applics. and paper coatings; FDA compliance; lt. gray liq.; dens. 7.62 lb/gal; visc. 800-1000 cps.

Advantage 833 Defoamer. [Hercules] Hydrocarbon oil-based defoamer for use in size press applics. and paper coatings; FDA compliance; tan oily liq.; dens. 7.2 lb/gal; visc. 2000 cps.

Advantage 951 Defoamer. [Hercules] Hydrocarbon oil-based defoamer for use as drainage aid and foam killer in kraft pulpmill brownstock washing operations; FDA compliance; lt. tan liq.; dens. 7.3 lb/gal; visc. 1500-2500 cps.

Advantage 1007B Defoamer. [Hercules] Water-based all-purpose defoamer for solid and surfactant-stabilized aq. foaming systems, e.g., acid and alkaline paper machines, pulpmill screening, deinking operations, size press and calender sol'ns., mill effluent systems; FDA compliance; wh. creamy liq.; dens. 7.5 lb/gal; visc. 1000-2000 cps; pH 8.5.

Advantage 1275PD. [Hercules] Water-based high-efficiency production aid for elimination of entrained air and surface foam for pulp and paper industry, wet-end processes, size press sol'ns., calender stack, effluent systems; FDA compliance; wh. emulsion; dens. 8.0 lb/gal; visc. 300-1500 cps.

Advantage 1280PD. [Hercules] Water-based high-efficiency production aid for elimination of entrained air and surface foam for pulp and paper industry, wet-end processes, size press sol'ns., calender stack, effluent systems; FDA compliance; wh. emulsion; dens. 8.21 lb/gal; visc. 500-2500 cps.

Advantage 1512 Defoamer. [Hercules] Hydrocarbon oil-based defoamer for use as drainage aid and foam killer in kraft pulpmill brownstock washing systems; FDA compliance; gray liq.; dens. 7.7 lb/gal; visc. 600 cps.

Advantage 5271 Production Aid. [Hercules] Hydrocarbon oil-based defoamer for use in size press applics. and paper coatings; FDA compliance; lt. gray liq.; dens. 7.62 lb/gal; visc. 800-1000 cps.

Advantage DF 110. [Hercules] Hydrocarbon oil-based defoamer for papermaking systems; FDA compliance; straw liq., hydrocarbon odor; disp. in water; dens. 7.42 lb/gal; visc. 30 cps.

Advantage DF 244. [Hercules] Hydrocarbon oil-based defoamer for papermaking systems; FDA compliance; straw liq., hydrocarbon odor; disp. in water; dens. 7.67 lb/gal; visc. 35 cps.

Advantage DF 285. [Hercules] Hydrocarbon oil-based defoamer for papermaking systems; FDA compliance; straw liq., hydrocarbon odor; dens. 8.40 lb/gal; visc. 400 cps.

Advantage Eff-101 Defoamer. [Hercules] Hydrocarbon oil-based multipurpose defoamer for paper machines, pulpmills, and other industrial operations; FDA compliance; grayish-brn. oily liq.; dens. 7.62 lb/gal; visc. 500-1000 cps.

Advantage M104 Defoamer. [Hercules] Drainage aid and foam killer for kraft pulpmill brownstock washing systems; FDA compliance; lt. tan liq.; dens. 7.3 lb/gal; visc. 1500-4000 cps.

Advantage M133A Defoamer. [Hercules] Water-extended defoamer for use as drainage aid and foam killer in kraft pulpmill brownstock washing systems, screen rooms, paper machine systems, effluents; FDA compliance; wh. liq.; dens. 7.6 lb/gal; visc. 1500-4000 cps.

Advantage M201 Defoamer. [Hercules] Water-extended defoamer for use as drainage aid and foam killer in kraft paper machine systems, effluent treatment; FDA compliance; wh. liq.; dens. 7.7 lb/gal; visc. 2000-4000 cps.

Advantage M1250 Defoamer. [Hercules] Anionic; water-based defoamer for use in industrial applics., screen room operations, effluent treatment, paper machine systems; FDA compliance; wh. emulsion; dens. 7.9 lb/gal; visc. 600-1000 cps.

Advantage M1251 Production Aid. [Hercules] Production aid removing entrained air and surface foam in papermaking applics., and effluent treatment; FDA compliance; colorless liq.; dens. 8.21 lb/gal; visc. 200-600 cps; pH 6.5-7.5.

AE-1. [Procter & Gamble] Ethoxylated lauryl alcohol; intermediate in mfg. of surfactants; liq.; 100% conc.

AE-3. [Procter & Gamble] Ethoxylated lauryl alcohol; nonionic; detergent, emulsifier; surfactant intermediate; liq.; 100% conc.

AE-7. [Procter & Gamble] Ethoxylated lauryl alcohol; nonionic; textile lubricants, textile wetting; surfactant intermediate; solid; 100% conc.

AE-1214/3. [Procter & Gamble] Laureth-3; nonionic; detergent, emulsifier; liq.; 100% act.

AE-1214/6. [Procter & Gamble] Ethoxylated linear alcohols; nonionic; lubricant, wetting agent, used in textiles; solid; 100% act.

AEPD®. [Angus] 2-Amino-2-ethyl-1,3-propanediol; CAS 115-70-8; nonionic; pigment dispersant, neutralizing amine, corrosion inhibitor, acid-salt catalyst, pH buffer, chemical and pharmaceutical intermediate, solubilizer; m.w. 119.2; water-sol.; m.p. 37.5 C; b.p. 152 C; flash pt. > 200 F (TCC); pH 10.8 (0.1M aq. sol'n.); 100% conc.

Aerosol® 18. [Am. Cyanamid; Cyanamid BV] Disodium stearyl sulfosuccinamate; CAS 14481-60-8; anionic; emulsifier, dispersant, foamer, detergent, solubilizer for soaps and surfactants; alkaline cleaner formulations, brick and tile cleaners, emulsion polymerization of vinyl chloride and SBRs; emulsifying oils and waxes, household detergents, cleaning paper mill felts; foamer for foamed latexes and plastics; biodeg.; Gardner 10 max. creamy paste; m.w. 493; water-disp.; sp.gr. 1.07; dens. 8.9 lb/gal; acid no. 4.0 max.; flash pt. (Seta CC) > 200 F; surf. tens. 39 dynes/cm; toxicology: LD50 (rat, oral) 2.68 mg/kg; mild skin, mild to moderate eye irritation; 35 ± 1.5% solids.

Aerosol® 19. [Am. Cyanamid] Disodium N-alkyl sulfosuccinamate; emulsifier, foaming and wetting agent; biodeg.; Gardner 10 clear liq.; water sol.; dens. 8.9 lb/gal; sp.gr. 1.07; surf. tens. 41 dynes/cm; 35% act.

Aerosol® 22. [Am. Cyanamid; Cyanamid BV] Tetrasodium dicarboxyethyl stearyl sulfosuccinamate; CAS 38916-42-6; anionic; emulsifier, dispersant, solubilizer, surfactant; emulsion polymerization of vinyl monomers; polishing waxes; surf. tension depressant for writing and drawing inks; demulsifier for w/o emulsions; cleaning of paper mill felts; industrial, household, and metal cleaners; biodeg.; lt. tan clear to cloudy liq.; water-sol.; m.w. 653; sp.gr. 1.12; dens. 9.4 lb/gal; visc. 53 cps; m.p. > 200 C; flash pt. (Seta CC) 143 F; acid no. 2.0; pH 7-8; surf. tens. 41 dynes/cm; toxicology: LD50 (rat, acute oral) 18.7 mL/kg (sol'n.); mild skin and eye irritant; 35% act. in water/alcohol.

Aerosol® 501. [Am. Cyanamid; Cyanamid BV] Disodium alkyl sulfosuccinate; anionic; dispersant, emulsifier, wetting agent, dispersant, foaming agent; used for acrylic and vinyl acetate emulsions; self-crosslinking latexes; textile wetting and foaming applics.; biodeg.; APHA 60 max. clear liq.; water-sol.; sp.gr. 1.16; dens. 9.66 lb/gal; visc. 260 cps; f.p. –3 C; flash pt. (Seta CC) > 200 F; acid no. 7.0; pH 6.0-7.0; surf. tens. 28 dynes/cm; toxicology: nonirritating or sensitizing to skin; 50% act. in water.

Aerosol® A-102. [Am. Cyanamid; Cyanamid BV] Disodium deceth-6 sulfosuccinate; CAS 39354-45-5; anionic; emulsifier, solubilizer, foamer, dispersant, surfactant, wetting agent; used in emulsion polymerization of PVAc/acrylics, textiles, cosmetics, shampoos, wallboard, adhesives; biodeg.; stable to acid media; colorless to lt. yel. clear liq.; sol. in water, dimethyl sulfoxide; m.w. 614; sp.gr. 1.08; dens. 9.01 lb/gal; visc. 40 cps; f.p. –4 C; flash pt. (Seta CC) > 200 F; acid no. 6 max.; pH 4.5-5.5; surf. tens. 33 dynes/cm; toxicology: LD50 (rat, oral) 30.8 mL/kg, (rabbit, dermal) > 5.0 mL/kg; moderate skin, minimal eye irritation; 30% act. in water.

Aerosol® A-103. [Am. Cyanamid; Cyanamid BV] Disodium nonoxynol-10 sulfosuccinate; CAS 9040-38-4; anionic; emulsifier, solubilizer, wetting agent, surfactant, surf. tens. depressant; used in PVAc acrylic emulsions; textile emulsions, padbath additive, textile wetting; cosmetics, shampoos, wallboard, adhesives; colorless to lt. yel. clear liq.; sol. in water, MEK; partly sol. in other polar solvs.; m.w. 854; sp.gr. 1.09; dens. 9.1 lb/gal; visc. 170–190 cps; f.p. –9 C; flash pt. (Seta CC) > 200 F; acid no. 10 max.; pH 4.5-5.5; surf. tens. 34 dynes/cm; toxicology: LD50 (rat, oral) > 10 mL/kg; not appreciably irritating to skin and eyes; 35% act. in water.

Aerosol® A-196-40. [Am. Cyanamid; Cyanamid BV]

Dicyclohexyl sodium sulfosuccinate; anionic; dispersant, surfactant; sole emulsifier for modified S/B; post additive to stabilize latex and promote adhesion; biodeg.; clear to cloudy liq.; sol. warm in org. solvs.; sol. 25 g/100 ml water; flash pt. (Seta CC) > 200 F; surf. tens. 39 dynes/cm; 40% act. in water.

Aerosol® A-196-85. [Amerchol] Dicyclohexyl sodium sulfosuccinate; anionic; dispersant, surfactant; sole emulsifier for modified S/B; post additive to stabilize latex and promote adhesion; biodeg.; wh. pellets; sol. warm in org. solvs.; sol. 10 g/100 ml water; surf. tens. 39 dynes/cm; 85% conc.

Aerosol® A-268. [Am. Cyanamid; Cyanamid BV] Disodium isodecyl sulfosuccinate; anionic; surfactant, sole emulsifier for PVC latexes-vinyl, vinylidene chloride, acrylics, surf. tens. depressant, solubilizer; clear liq.; water sol.; dens. 9.96 lb/gal; sp.gr. 1.19; visc. 150 cps; f.p. –5 C; flash pt. (Seta CC) > 200 F; surf. tens. 28 dynes/cm; toxicology: LD50 (rat, oral) 5.2 mL/kg (sol'n.), (rabbit, dermal) > 10 mL/kg; mild skin, moderate eye irritation; 50% act. in water.

Aerosol® AY-65. [Am. Cyanamid; Cyanamid BV] Diamyl sodium sulfosuccinate; anionic; wetting agent, dispersant, surfactant; used in agriculture, emulsion polymerization, electroplating, ore leaching, cleaning of porcelain, tile, brick, cement; biodeg.; clear liq.; sol. in water/ethanol; sp.gr. 1.081; dens. 9.0 lb/gal; f.p. < 18 C; flash pt. (Seta CC) 77 F; surf. tens. 29 dynes/cm; 65% conc. in water/alcohol.

Aerosol® AY-100. [Am. Cyanamid; Cyanamid BV] Diamyl sodium sulfosuccinate; anionic; wetting agent, dispersant, surfactant; used in agriculture, emulsion polymerization, electroplating, ore leaching, cleaning of porcelain, tile, brick, cement; biodeg.; clear liq.; sol. in water and org. solvs.; sp.gr. 1.2; dens. 10.0 lb/gal; f.p. < 18 C; surf. tens. 30 dynes/cm; 98% conc.

Aerosol® C-61. [Am. Cyanamid] Ethoxylated alkyl guanidine-amine complex; cationic; antistat, pigment dispersant, flushing agent, wetting agent, settling agent; alkaline, cement, brick, and tile cleaner formulations for crystal growth control, emulsion breaking, alkaline metal and paint brush cleaners,; paint removers; textile softener; demulsifying agent; foaming agent; for plastics, paper, textiles, adhesive industries; lt. tan creamy paste; strong ammoniacal odor; sol. in org. solvs. in presence of alcohol; disp. in water; sp.gr. 1.00; dens. 8 lb/gal; flash pt. (Seta CC) 94 F; surf. tens. 34 dynes/cm; partially biodeg.; 70.4% act.

Aerosol® DPOS-45. [Am. Cyanamid; Cyanamid BV] Disodium mono- and didodecyl diphenyl oxide disulfonate; CAS 25167-32-2; anionic; emulsifier, dispersant, solubilizer, primary surfactant for emulsion polymerization systems; coupling agent; high electrolyte tolerance; stable in highly acidic and alkaline sol'ns. and at elevated temps.; liq.; water-sol.; insol. in org. solvs.; flash pt. (Seta CC) > 200 F; surf. tens. 34 dyne/cm; 45% act. in water.

Aerosol® GPG. [Am. Cyanamid; Cyanamid BV] Sodium dioctyl sulfosuccinate; anionic; wetting agent, surf. tens. depressant, emulsifier, dispersant; for dust control, industrial cleaners, emulsifying waxes; biodeg.; clear, slightly visc. liq.; limited sol. in water; sol. in org. solv.; visc. 200 cps; f.p. –40 C; flash pt. (Seta CC) 103 F; surf. tens. 26 dynes/cm; 70% act. in water/alcohol.

Aerosol® IB-45. [Am. Cyanamid; Cyanamid BV] Sodium diisobutyl sulfosuccinate; CAS 127-39-9; anionic; emulsifier, wetting agent; emulsion polymerization of styrene, butadiene and copolymers; dye and pigment dispersant; for leaching, electroplating; biodeg.; clear liq.; water-sol.; extremely hydrophilic; m.w. 332; sol. in water; sp.gr. 1.12; dens. 9.3 lb/gal; f.p. 20–21 C; flash pt. (Seta CC) > 200 F; acid no. 2 max.; pH 5-7 (10%); surf. tens. 49 dynes/cm; biodeg.; toxicology: LD50 (rat, oral) 6.16 g/kg (solids); LD50 (rabbit, dermal) > 5 g/kg; nontoxic by single-dose ingestion; 45% act. in water.

Aerosol® MA-80. [Am. Cyanamid; Cyanamid BV] Dihexyl sodium sulfosuccinate; anionic; dispersant, textile wetting agent, emulsifier, solubilizer, penetrant; used for emulsion polymerization, battery separators, electroplating, ore leaching; germicidal act.; not as rapidly biodeg. as Aerosol 18 and 22; APHA 50 max. clear slightly visc. liq.; sol. in water, alcohol, and org. solvs.; sp.gr. 1.13; dens. 9.4 lb/gal; f.p. –28 C; m.p. 199–292 C; flash pt. (Seta CC) 115 F; surf. tens. 28 dynes/cm; 80% act. in water/alcohol.

Aerosol® NPES 458. [Am. Cyanamid] Ammonium salt of sulfated nonylphenoxy POE ethanol; CAS 9051-57-4; anionic; high foaming surfactant for emulsion polymerization of acrylic, styrene and vinyl acetate systems, dishwashing detergents, germicides, pesticides, general purpose cleaners, cosmetics, and textile wet processing applics.; pale yel. clear liq., alcoholic odor; sol. in water; partly sol. in org. solvs.; m.w. 493; sp.gr. 1.065; dens. 8.9 lb/gal; visc. 100 cps; f.p. < 0 C; flash pt. (PMCC) 83 F; pH 6.5-7.5; surf. tens. 31 dynes.cm; toxicology: LD50 (rat, oral) > 5 g/kg; severe eye, moderate skin irritant; 58% conc. in water/alcohol.

Aerosol® NPES 930. [Am. Cyanamid] Ammonium salt of sulfated nonylphenoxy POE ethanol; CAS 9051-57-4; anionic; emulsifier for emulsion polymerization of vinyl acetate, acrylic copolymers, styrene acrylic copolymers; imparts superior water resistance in films; yel. clear liq.; water-sol.; insol. in org. solvs.; m.w. 713; sp.gr. 1.04; dens. 8.7 lb/gal; visc. 90 cps; gel pt. < 55 F; flash pt. (Seta CC) > 200 F; pH 7.0-7.5 (10%); surf. tens. 33 dynes/cm; toxicology: LD50 (rat, oral) > 10 g/kg; severe skin and eye irritation; 30% act. in water.

Aerosol® NPES 2030. [Am. Cyanamid] Ammonium salt of sulfated nonylphenoxy POE ethanol; CAS 9051-57-4; anionic; for emulsion polymerization of acrylic monomers where small particle size and water resistant props. are required; yel. clear liq.; water-sol.; insol. in org. solvs.; m.w. 1197; sp.gr. 1.05-1.07; dens. 8.8 lb/gal; visc. 100 cps; gel pt. < 55 F; flash pt. (Seta CC) > 200 F; pH 7.0-7.5 (10%); surf. tens. 43 dynes/cm; toxicology: LD50 (rat, oral) > 10 g/kg; moderate skin irritation on prolonged contact; mild eye irritation; 30% act. in water.

Aerosol® NPES 3030. [Am. Cyanamid] Ammonium salt of sulfated nonylphenoxy POE ethanol; CAS 9051-57-4; anionic; emulsifier for emulsion polymerization of acrylic, vinyl acetate and styrene-acrylic systems; forms films with superior water resist.; yel. clear liq.; water-sol.; insol. in org. solvs.; m.w. 1637; sp.gr. 1.05-1.07; dens. 8.8 lb/gal; visc. 110 cps; gel pt. < 55 F; flash pt. (Seta CC) > 200 F; pH 7.0-7.5 (10%); surf. tens. 43 dynes/cm; toxicology: LD50 (rat, oral) > 10 g/kg; minimal skin and eye irritation; 30% act. in water.

Aerosol® NS. [Am. Cyanamid] Sodium naphthalene

sulfonate; anionic; dispersant for pigments, extenders, and fillers in aq. media over broad pH range; powd.; sol. 42 g/100 ml water; partly sol. in org. solvs.; surf. tens. 72 dynes/cm; slowly biodeg.; 87% act.

Aerosol® OS. [Am. Cyanamid; Cyanamid BV] Sodium diisopropylnaphthalene sulfonate; anionic; emulsifier, dispersant and wetting agent; used in alkaline cleaning formulations; antigelling agents, automotive radiator cleaners, metal, cement, brick, and tile cleaners for crystal growth control,; electroplating, filtration, glass cleaning, household detergents, leaching ores and slags, pigment disps.; soap additive; adjuvant in agric. chem. prods.; slowly biodeg.; wh. to lt. cream powd.; partly sol. in polar org. solvs.; sol. > 20 g/100 ml in water; sp.gr. 1.43; dens. 11.93 lb/gal; m.p. 205 C; surf. tens. 37 dynes/cm; pH 9 ± 1.0 (5% aq.); 75 act., 25% sodium sulfate.

Aerosol® OT-70 PG. [Am. Cyanamid; Cyanamid BV] Sodium dioctyl sulfosuccinate, propylene glycol/water; anionic; wetting agent, surf. tens. depressant, emulsifier, surfactant; for use where high flash required; biodeg.; clear to slightly yel. liq.; limited water sol.; sol. in org. solv.; sp.gr. 1.09; visc. 200–400 cps; flash pt. (Seta CC) > 200 F; surf. tens. 26 dynes/cm; 70% act.

Aerosol® OT-75%. [Am. Cyanamid; Cyanamid BV] Dioctyl sodium sulfosuccinate; anionic; wetting agent and surf. tens. depressant used in textile, rubber, petrol., paper, metal, paint, plastic, and agric. industries; antistat for cosmetics, dry cleaning detergents, emulsion, plastic, pipelines, and suspension polymerization; emulsifier wax for polish, firefighting, germicide, metal cleaner, mold release agent; dispersant in paints and inks, paper, photography, process aid, rust preventative, soldering flux, wallpaper removal; APHA 100 max. clear visc. liq.; m.w. 444; sol. org. solv.; limited water sol.; sp.gr. 1.09; visc. 200 cps; flash pt. 85 C (OC); acid no. 2.5 max.; surf. tens. 28.7 dynes/cm (0.1% aq.); biodeg.; 75 ± 2% solids in water/alcohol.

Aerosol® OT-100%. [Am. Cyanamid; Cyanamid BV] Dioctyl sodium sulfosuccinate; anionic; emulsifier, dispersant, lubricant, wetting agent, mold release agent for emulsion and suspension polymerization, drycleaning, emulsifying waxes, industrial cleaners, paints; surfactant for water-free systems; biodeg.; APHA 100 max. waxy solid; m.w. 444; sol. in polar and nonpolar solv.; sol. in oil, fat, and wax @ 75 C; disp. in water; sp.gr. 1.1; m.p. 153–157 C; acid no. 2.5 max.; surf. tens. 28.7 dynes/cm (0.1% aq.); 100% conc.

Aerosol® OT-B. [Am. Cyanamid; Cyanamid BV] Dioctyl sodium sulfosuccinate, sodium benzoate; anionic; wetting agent, dispersant, solubilizer, adjuvant for agric. chem. wettable powds.; pigment dispersant in plastics; wh. powd.; bulk particle size 15–150 μ; m.w. 444; water-disp.; sp.gr. 1.1; m.p. < 300 C; acid no. 2.5 max.; surf. tens. 28.7 dynes/cm (0.1% aq.); 85% act., 15% sodium benzoate.

Aerosol® OT-MSO. [Am. Cyanamid] Dioctyl sodium sulfosuccinate in min. seal oil; anionic; wetting agent, lubricant, detergent for dry cleaning, corrosion resist. lubricants, agric. emulsions, org. solvent systems; used when a higher flash is required; lt. amber transparent liq.; very sol. in org. solvs.; sp.gr. 0.96-0.98; dens. 8.1 lb/gal; visc. 125-300 cps; flash pt. (PMCC) 255 F; acid no. 2.5 max.; surf. tens. 26 dyne/cm; biodeg.; toxicology: LD50 (rat, oral) > 5 mL/kg; LD50 (rabbit, dermal) > 10 mL/kg; eye irritant; 62% conc.

Aerosol® OT-S. [Am. Cyanamid; Cyanamid BV] Dioctyl sodium sulfosuccinate; anionic; wetting agent, surf. tens. depressant, emulsifier for plastics, organosols, lacquers, varnishes, all org. media, dry cleaning, corrosion resistant lubricants, agric. emulsions; biodeg.; lt. amber, transparent liq.; limited water sol., good org. solv.; sp.gr. 1.0; visc. 200–300 cps; flash pt. (Seta CC) 134 F; surf. tens. 26 dynes/cm; 70% act. in lt. petrol. distillate.

Aerosol® TR-70. [Am. Cyanamid; Cyanamid BV] Ditridecyl sodium sulfosuccinate; anionic; emulsifier, surfactant; used in emulsion polymerization of vinyl chloride and vinyl acetate, suspension polymerization of vinyl chloride; dispersant for resins, pigments, polymers, and dyes in org. systems; pigment dispersant in printing inks; rust preventative; biodeg.; clear liq.; sol. in org. media; limited water sol.; sp.gr. 0.995; dens. 8.3 lb/gal; visc. 110 cps; f.p. –40 C; surf. tens. 26 dynes/cm; 70% act. in water/alcohol.

AF 10 FG. [Harcros] Silicone antifoam; nonionic; antifoam agent used for general food, poultry, and meat processing applics.; FDA compliance; kosher; wh. liq.; disp. in water; sp.gr. 1.00; dens. 8.3 lb/gal; flash pt. > 212 F (PMCC); pH 4-5 (1% aq.); 10% act.

AF 10 IND. [Harcros] Silicone antifoam; nonionic; antifoam for agric., cutting oils, drilling muds, effluent, inks, chemicals, detergents and textiles; FDA compliance; wh. liq.; water-disp.; sp.gr. 1.00; dens. 8.3 lb/gal; flash pt. > 212 F (PMCC); pH 4–5 (1% aq.); 10% act.

AF 30 FG. [Harcros] Silicone antifoam; nonionic; antifoam for general food, meat, and poultry processing; FDA compliance; kosher; wh. liq.; water-disp.; sp.gr. 1.01; dens. 8.4 lb/gal; flash pt. > 212 F (PMCC); pH 4–5 (1% aq.); 30% act.

AF 30 IND. [Harcros] Silicone antifoam; nonionic; antifoam for effluent, agric. formulations, antifreeze, detergents; FDA compliance; wh. liq.; water-disp.; sp.gr. 1.01; dens. 8.4 lb/gal; flash pt. > 212 F (PMCC); pH 4–5 (1% aq.); 30% act.

AF 60. [GE Silicones] Dimethyl polysiloxane aq. emulsion; nonionic; defoaming agent used in adhesive, ink, latex, soap, starch, and paint mfg. and other aq. industrial systems; wh. fluid, sorbic acid odor; sol. in water; sp.gr. 1.01; dens. 8.4 lb/gal; visc. 1000 cps; 30% silicone, 44.5% solids.

AF 66. [GE Silicones] Filled polydimethylsiloxane; antifoamer in nonaq. industrial applic.; adhesive, ink, and paint mfg., resin polymerization, petrol. processing; grayish wh. fluid; odorless; sol. in aliphatic, aromatic, chlorinated solvs.; negligible sol. in water; sp.gr. 1.01; dens. 8.4 lb/gal; visc. 500 cSt max.; OC flash pt. 600 F min.; 100% act.

AF 70. [GE Silicones] 100% silicone compd.; defoamer in petrol. refining, cutting oils, chem. processing, antifoam formulating; food additive for food processing; liq.; sol. in aliphatic, aromatic, and chlorinated hydrocarbons; sp.gr. 1.01; dens. 8.4 lb/gal; visc. 1500 cps; flash pt. (OC) 315 C.

AF 72. [GE Silicones] PEG-40 stearate, sorbitan stearate, and silica; nonionic; food-grade antifoam agent, surfactant; also for industrial applics. such as textile dyeing and finishing, leather finishing, latex processing, soap and detergent mfg., adhesive mfg., and as a boiler feed water defoamer; wh. fluid; slight sorbic acid odor; sol. in water; sp.gr. 1.01; dens. 8.4 lb/gal; visc. 1000 cps; 30% silicone; 44.2% solids.

AF 75. [GE Silicones] PEG-40 stearate, sorbitan

stearate, and silica; nonionic; food-grade antifoam emulsion, surfactant; used as direct and indirect food additive, in the mfg. of food-pkg. materials,; chemical processing (adhesive mfg., water-based ink mfg., latex processing, soap mfg., starch processing, resin polymerization), textiles, paper, leather finishing, metalworking; wh. emulsion; slight sorbic acid odor; water-sol.; sp.gr. 1.02; dens. 8.4 lb/gal; visc. 3000 cps; 10% silicone; 13.75% solids.

AF 100 FG. [Harcros] Silicone antifoam; antifoam for food, edible oils, meat and poultry processing; FDA complaince; kosher; translucent liq.; insol. in water; sp.gr. 1.01; dens. 8.4 lb/gal; pour pt. < 0 F; flash pt. > 600 F (PMCC); 100% conc.

AF 100 IND. [Harcros] Silicone antifoam; antifoam for agric. formulations, drilling muds, vacuum distillations; FDA compliance; translucent liq.; insol. in water; sp.gr. 1.01; dens. 8.4 lb/gal; pour pt. < 0 F; flash pt. > 600 F (PMCC); 100% conc.

AF 112. [Harcros] Nonsilicone antifoam; used in solvs., adhesives, latex paints, inks; FDA compliance; creamy yel. liq.; water-insol.; sp.gr. 0.92; dens. 7.7 lb/gal; pour pt. 40 F; flash pt. > 200 F (PMCC); 100% act.

AF-800. [Anedco] Alcohol ether sulfate; fresh water foamer for drilling with gas.

AF-801. [Anedco] Alcohol ether sulfate; brine water foamer for drilling with gas.

AF-802. [Anedco] Hydrocarbon; foamer for drilling with gas.

AF-803. [Anedco] Sulfates and sulfonic salts; foaming agent.

AF 1025. [Harcros] Silicone antifoam; antifoam for delayed coker units; clear liq.; water-insol.; sp.gr. 0.82; dens. 6.8 lb/gal; pour pt. < 0 F; flash pt. (PMCC) 140 F; 25% act.

AF 6050. [Harcros] Silicone antifoam; for delayed coker units; sp.gr. 0.86.

AF 8805 FG. [Harcros] Silicone antifoam; nonionic; antifoam for general food, poultry, and meat processing, agric.; kosher; FDA compliance; wh. liq.; water-disp.; sp.gr. 1.00; dens. 8.3 lb/gal; flash pt. (PMCC) > 212 F; pH 4-5 (1% aq.); 5% act.

AF 8805. [Harcros] Silicone antifoam; for agric., cutting oils, drilling muds, effluent, inks, chemicals, detergents, textiles; FDA compliance; wh. liq.; water-disp.; sp.gr. 1.00; dens. 8.3 lb/gal; flash pt. (PMCC) > 212 F; pH 4-5 (1% aq.); 5% act.

AF 8810. [Harcros] Silicone antifoam; nonionic; antifoam for agric., cutting oils, drilling muds, effluent, inks, chemicals, detergents, textiles; FDA compliance; wh. liq.; sp.gr. 1.00; dens. 8.3 lb/gal; flash pt. (PMCC) > 212 F; pH 4-5 (1% aq.); 10% act.

AF 8810 FG. [Harcros] Silicone antifoam; nonionic; antifoam for general food, poultry and meat processing, agric. use; kosher; FDA compliance; wh. liq.; sp.gr. 1.00; dens. 8.3 lb/gal; flash pt. (PMCC) > 212 F; pH 4-5 (1% aq.); 10% act.

AF 8820. [Harcros] Silicone antifoam; nonionic; antifoam for effluent, agric., antifreeze, detergent applics.; dilutable; FDA compliance; wh. liq.; water-disp.; sp.gr. 1.00; dens. 8.3 lb/gal; flash pt. (PMCC) > 212 F; pH 4-5 (1% aq.); 20% act.

AF 8820 FG. [Harcros] Silicone antifoam; nonionic; antifoam for general food, meat, and poultry processing, agric.; Kosher; dilutable for spa and hot tub applics.; FDA compliance; wh. liq.; water-disp.; sp.gr. 1.00; dens. 8.3 lb/gal; flash pt. (PMCC) > 212 F; pH 4-5 (1% aq.); 20% act.

AF 8830. [Harcros] Silicone antifoam; nonionic; antifoam for effluent, agric., antifreeze, detergents; dilutable; FDA compliance; wh. liq.; water-disp.; sp.gr. 1.00; dens. 8.3 lb/gal; flash pt. (PMCC) > 212 F; pH 4-5 (1% aq.); 30% act.

AF 8830 FG. [Harcros] Silicone antifoam; nonionic; antifoam for general food, meat and poultry processing, agric. use; kosher; FDA compliance; wh. liq.; water-disp.; sp.gr. 1.00; dens. 8.3 lb/gal; flash pt. (PMCC) > 212 F; pH 4-5 (1% aq.); 30% act.

AF 9000. [GE Silicones] Silicone compd.; defoamer in many nonaq. direct and indirect food additive and industrial applics. incl. chemical processing (adhesive mfg., ink mfg., paint mfg., resin polymerization, formulating antifoam emulsions, insecticides/herbicides, wool fats); petrol. processing; sol. in aliphatic, aromatic, and chlorinated solvs.; sp.gr. 1.01; dens. 8.4 lb/gal; visc. 2500 cps; flash pt. (OC) 315 C; 100% silicone content.

AF 9020. [GE Silicones] Dimethicone aq. emulsion; nonionic; defoamer for industrial and food-processing systems incl. chemical processing (adhesive mfg., water-based ink mfg., latex processing, soap mfg., starch processing, paint additive, alcohol fermentation), waste treatment,; petrochemical (resin polymerization, glycol dehydrators, ethylene oxide prod., urea prod.); wh. emulsion; disp. in warm or cold water with mild agitation; sp.gr. 1.01; dens. 8.4 lb/gal; visc. 3500 cps; 20% silicone, 28.75% solids.

AF 9021. [GE Silicones] Polydimethylsiloxane emulsion; antifoam for industrial applics.; wh. emulsion; disp. in water; sp.gr. 1.01; dens. 8.4 lb/gal; visc. 3500 cps; 20% silicone, 28.75% solids.

AF 93. [GE Silicones] Polydimethylsiloxane emulsion; defoamer used in textiles, soap mfg., resin prod., aq. systems, for extreme acid and base pH ranges; wh. emulsion; water-disp.; sp.gr. 1.01; dens. 8.4 lb/gal; visc. 3500 cps; 30% silicone, 35.5% solids.

AF CM Conc. [Harcros] Silicone antifoam; nonionic; antifoam for carpet extraction machines; wh. visc. liq.; water-disp.; sp.gr. 0.97; dens. 8.1 lb/gal; flash pt. > 212 F (PMCC); pH 6–8 (1% aq.); 20% act.

AF GN-11-P. [Harcros] Nonsilicone antifoam; nonionic; used for fermentation, drilling muds, effluent, adhesives, gas treating; FDA compliance; clear liq.; water-disp; sp.gr. 1.01; dens. 8.4 lb/gal; pour pt. –50 F; flash pt. > 200 F (PMCC); pH 5–7 (1% aq.); 100% act.

AF GN-23. [Harcros] Nonsilicone antifoam; for pulp and paper, effluent, wastewater treatment; sp.gr. 0.94.

AF HL-21. [Harcros] Nonsilicone antifoam; nonionic; for latex paints, adhesives, inks, solvs.; FDA compliance; opaque liq.; water-insol.; sp.gr. 0.83; dens. 6.9 lb/gal; flash pt. (PMCC) 270 F; 100% act.

AF HL-23. [Harcros] Nonsilicone antifoam; for cutting oil, boiler water additives, detergents, cleaning sol'ns, syn. lubricants; FDA compliance; clear liq.; water-insol.; sp.gr. 1.00; dens. 8.3 lb/gal; pour pt. –20 F; flash pt. > 300 F (PMCC); pH 4–6 (1% in 15% IPA/water); 100% act.

AF HL-26. [Harcros] Nonsilicone antifoam; nonionic; antifoam for waste water treatment, electroplating, coatings, cleaning aids, dyes, inks; yel. clear liq.; sol. in oil; sp.gr. 1.00; dens. 8.3 lb/gal; pour pt. < 30 F; flash pt. > 220 F (PMCC); 50% act.

AF HL-27. [Harcros] Nonsilicone antifoam; for urea and phenolic resins, latex paints, inks, adhesives; FDA compliance; yel. liq.; water-insol.; sp.gr. 0.95;

dens. 7.9 lb/gal; pour pt. 32 F; flash pt. > 300 F (PMCC); 100% act.

AF HL-36. [Harcros] Nonsilicone antifoam; for fermentation, processing beet sugar and yeast, distillation; Kosher; FDA compliance; clear liq.; water-disp.; sp.gr. 1.00; dens. 8.3 lb/gal; flash pt. (PMCC) > 300 F; 100% act.

AF HL-40. [Harcros] Nonsilicone antifoam; nonionic; for latex paints, adhesives, inks, solvs., chemical processing, textiles, paper, paper coatings, electroplating; FDA compliance; creamy yel. liq.; oil-sol.; sp.gr. 0.91; dens. 7.5 lb/gal; flash pt. (PMCC) > 200 F; 100% act.

AF HL-52. [Harcros] Nonsilicone antifoam; nonionic; for solvs., latex paints, inks, chemical processing, adhesives, paper, paper coatings; FDA compliance; creamy yel. liq.; oil-sol.; sp.gr. 0.87; dens. 7.25 lb/gal; flash pt. (PMCC) > 200 F; 100% act.

Afilan EHS. [Hoechst Celanese/Colorants & Surf.] 2-Ethylhexyl stearate; surfactant for textile processing; sl. yel. clear liq.; visc. 20 cps; acid no. < 1; iodine no. < 2; sapon. no. 150; hyd. no. 3.

Afilan ICS. [Hoechst Celanese/Colorants & Surf.] Isocetyl stearate; surfactant for textile processing; sl. yel. clear liq.; visc. 30 cps; acid no. < 1; iodine no. < 2; sapon. no. 115; hyd. no. 2.

Afilan ODA. [Hoechst Celanese/Colorants & Surf.] Di-C8-10 adipate; surfactant for textile processing; sl. yel. clear liq.; visc. 15 cps; acid no. < 1; iodine no. < 2; sapon. no. 290; hyd. no. 3.

Afilan POD. [Hoechst Celanese/Colorants & Surf.] C8-10 pentaerythritol tetra ester; surfactant for textile processing; sl. yel. clear liq.; visc. 60 cps; acid no. < 1; iodine no. < 2; sapon. no. 330; hyd. no. 2.

Afilan PP. [Hoechst Celanese/Colorants & Surf.] Pentaerythrityl tetrapelargonate; surfactant for textile processing; sl. yel. clear liq.; visc. 50 cps; acid no. < 1; iodine no. < 2; sapon. no. 320; hyd. no. 1.

Afilan SME. [Hoechst Celanese/Colorants & Surf.] Blend; surfactant for textile processing; yel. liq.; visc. 60 cps; pH 7.0 (1% aq.); 80% act.

Afilan TDA. [Hoechst Celanese/Colorants & Surf.] Ditridecyl adipate; surfactant for textile processing; sl. yel. clear liq.; visc. 30 cps; acid no. < 1; iodine no. < 2; sapon. no. 271; hyd. no. 1.

Afilan TDS. [Hoechst Celanese/Colorants & Surf.] Tridecyl stearate; surfactant for textile processing; sl. yel. clear liq.; visc. 30 cps; acid no. < 1; iodine no. < 2; sapon. no. 123; hyd. no. 3.

Afilan TMOD. [Hoechst Celanese/Colorants & Surf.] C8-10 trimethylolpropane tri ester; surfactant for textile processing; sl. yel. clear liq.; visc. 34 cps; acid no. < 1; iodine no. < 2; sapon. no. 310; hyd. no. 3.

Afilan TMPP. [Hoechst Celanese/Colorants & Surf.] Trimethylolpropane tripelargonate; surfactant for textile processing; sl. yel. clear liq.; visc. 36 cps; acid no. < 1; iodine no. < 2; sapon. no. 310; hyd. no. 3.

Afilan TXE. [Hoechst Celanese/Colorants & Surf.] Blend; surfactant for textile processing; yel. liq.; visc. 50 cps; pH 7.0 (1% aq.); 95% act.

Afranil®. [BASF AG] Alcohol and fatty acid derivs.; grease and foam inhibitor, pulp deaerator for papermaking.

Ageflex FM-1. [CPS] Dimethylaminoethyl methacrylate; CAS 2867-47-2; detergent and sludge dispersant in lubricants; visc. index improver; flocculant for waste water treatment; retention aid for paper mfg.; acid scavenger in PU foams; corrosion inhibitor; resin and rubber modifier; used in acrylic polishes and paints, hair prep. copolymers, sugar clarification, adhesives, water clarification; APHA 50 clear liq.; very sol. in water; sol. in org. solvs.; m.w. 157.21; visc. 1.38 cst; b.p. 68.5 C (10 mm); f.p. < -60 C; toxicology: poison; harmful if swallowed; severe eye burns and skin irritation; irritating vapor; 99% assay.

Agent 2A-2S. [Norman, Fox] Modified amide; anionic/nonionic; air entraining admixture; liq.; 100% conc.

Agesperse 71. [CPS] Polymeric carboxylic acid, sodium salt; low foaming dispersant, emulsifier, stabilizer for paper, paints, carpet backcoating, rubber, mining, textiles, ceramic slip, detergents, boiler water compds., cooling water compds., adhesives; FDA compliance; lt. yel. clear liq.; sp.gr. 1.104; dens. 9.2 lb/gal; visc. 150 cps; 25% act.

Agesperse 80. [CPS] Polymeric carboxylic acid, sodium salt; dispersant, emulsifier, stabilizer for paper, rubber, mining, textiles, ceramic slip, detergents, boiler water compds., cooling water compds., adhesives; FDA compliance; APHA 100 liq.; sp.gr. 1.10; visc. 150 cps; 30% act.

Agesperse 81. [CPS] Polymeric carboxylic acid, sodium salt; dispersant, emulsifier, stabilizer for paper, rubber, mining, textiles, ceramic slip, detergents, boiler water compds., cooling water compds., adhesives; FDA compliance; APHA 60 liq.; sp.gr. 1.10; visc. 75 cps; 25% act.

Agesperse 82. [CPS] Polymeric carboxylic acid, ammonium salt; dispersant, emulsifier, stabilizer for paper, rubber, mining, textiles, ceramic slip, detergents, boiler water compds., cooling water compds., adhesives; FDA compliance; APHA > 250 liq.; sp.gr. 1.10; visc. 35 cps; 25% act.

Agitan 217. [Münzing Chemie GmbH] Silicone defoamer; for emulsion paints, emulsion polymers, adhesives.

Agitan 218. [Münzing Chemie GmbH] Silicone defoamer; for emulsion paints, aq. coatings, syn. renderings, adhesives.

Agitan 232. [Münzing Chemie GmbH] Silicone-free defoamer for emulsion paints, syn. renderings.

Agitan 260. [Münzing Chemie GmbH] Silicone-free defoamer for emulsion paints, gloss emulsion paints, syn. renderings, adhesives, silicate paints, aq. epoxy resin systems; BGA compliance.

Agitan 280. [Münzing Chemie GmbH] Silicone-free defoamer for emulsion paints, emulsion polymers, syn. renderings, adhesives.

Agitan 281. [Münzing Chemie GmbH] Silicone-free defoamer for emulsion paints, emulsion polymers, adhesives, aq. systems, silicate paints.

Agitan 285. [Münzing Chemie GmbH] Anionic; defoamer for strongly acid media; low foaming wetting agent; free of min. oil.

Agitan 288. [Münzing Chemie GmbH] Anionic; silicone-free defoamer and leveling agent for varnishes and other aq. systems; free of min. oil; BGA compliance.

Agitan 290. [Münzing Chemie GmbH] Nonionic; silicone-free defoamer for sol'ns. of polyvinyl alcohol, deliming and cleaning agents, acid media, printing inks; ashless.

Agitan 295. [Münzing Chemie GmbH] Silicone-free defoamer for emulsion paints, gloss emulsion paints, printing inks, emulsion polymers, adhesives, aq. systems, polymerization processes, wood preservative stains.

Agitan 296. [Münzing Chemie GmbH] Silicone-free defoamer for emulsion paints, emulsion polymers, aq. systems.

Agitan 301. [Münzing Chemie GmbH] Silicone defoamer; biodeg. defoamer for emulsion paints, emulsion polymers, syn. renderings, adhesives; BGA compliance.

Agitan 305. [Münzing Chemie GmbH] Silicone-free defoamer for emulsion paints, emulsion polymers, syn. renderings, adhesives; BGA compliance.

Agitan 315. [Münzing Chemie GmbH] Silicone-free defoamer with very low odor for odorless inner emulsion paints, syn. renderings, adhesives, gloss emulsion paints, aq. systems; BGA compliance.

Agitan 633. [Münzing Chemie GmbH] Nonionic; Silicone-free defoamer for water paints, water-dilutable resin systems, impregnations, adhesives, printing inks; easily emulsifiable, good leveling props.

Agitan 650. [Münzing Chemie GmbH] Silicone-free defoamer for emulsion polymers, water-dilutable systems, pigmented and unpigmented systems; easily emulsifiable; BGA compliance.

Agitan 655. [Münzing Chemie GmbH] Silicone-free defoamer for emulsion polymers, water-dilutable systems, pigmented and unpigmented systems; easily emulsifiable; BGA compliance.

Agitan 700. [Münzing Chemie GmbH] Silicone defoamer; for emulsion paints, hydrosols, aq. coatings, adhesives, silicate paints, impregnations.

Agitan 701. [Münzing Chemie GmbH] Silicone defoamer; for emulsion paints, gloss emulsion paints, silicate paints, hydrosols, aq. systems.

Agitan 702. [Münzing Chemie GmbH] Silicone-free defoamer for hydrosols, emulsion paints, emulsion polymers, aq. systems.

Agitan 703 N. [Münzing Chemie GmbH] Silicone-free defoamer for emulsion paints, water paints, emulsion polymers, aq. resins, printing inks; easily emulsifiable; BGA compliance.

Agitan 730 N. [Münzing Chemie GmbH] Silicone-free defoamer for aq. flexographic inks, other low visc. aq. systems; emulsifiable, stable; also suitable for post-dosage; BGA compliance.

Agitan 731. [Münzing Chemie GmbH] Silicone compd.; defoamer for gloss emulsion paints, wood preservative stains, aq. flexographic inks.

Agitan 745. [Münzing Chemie GmbH] Polysiloxane compd.; defoamer for aq. systems, varnish for woodwork, furniture lacquers, parquet floor lacquers, systems of corrosion prevention, printing inks.

Agitan E 255. [Münzing Chemie GmbH] Silicone emulsion; defoamer for emulsion paints, gloss emulsion paints, syn. renderings, adhesives, aq. systems, glazes, aq. printing inks, polymerization processes; BGA compliance.

Agitan E 256. [Münzing Chemie GmbH] Silicone emulsion; defoamer for emulsion paints, gloss emulsion paints, syn. renderings, adhesives, aq. systems, glazes, aq. printing inks, aq. systems, polymerization processes, aq. epoxy resin systems, corrosion prevention.

Agitan LA 742. [Münzing Chemie GmbH] Silicone-free defoamer for flow coat paints for furnitures based on nitrocellulose, acid curing systems.

Agitan P 800. [Münzing Chemie GmbH] Defoamer for powder systems, powder coatings, syn. renderings, plasters, fillers, mortars, cements.

Agitan P 801. [Münzing Chemie GmbH] Silicone defoamer; for powder systems, powder coatings, syn. renderings, plasters, fillers, mortars, cements.

Agitan P 803. [Münzing Chemie GmbH] Defoamer with leveling and liquefication effect for powder coatings, syn. renderings, plasters, fillers, mortars, cements.

Agitan P 813. [Münzing Chemie GmbH] Silicone-free defoamer with leveling props. for powder coatings, adhesives, plasters, fillers, mortars, cements.

Agitan VP 725. [Münzing Chemie GmbH] Silicone compd.; defoamer for lacquers, solv.-free systems, printing inks.

Agitan VP E 251. [Münzing Chemie GmbH] Silicone-free w/o emulsion; defoamer for emulsion paints, syn. renderings.

Agitan VP P 804. [Münzing Chemie GmbH] Silicone defoamer; for powder coatings, adhesives, plasters, fillers, mortars, cements.

Agrilan® A. [Harcros UK] Blend of oil-sol. sulfonates and POE ethers; anionic/nonionic; balanced pair emulsifier for agrochem. toxicants; brn. visc. liq.; faint odor; turbid water-sol.; sp.gr. 1.062; visc. 2400 cs; flash pt. (PMCC) > 150 C; pour pt. < 0 C; pH 4.5–6.5 (1% aq.); 100% act.

Agrilan® AC. [Harcros UK] Blend of oil-sol. sulfonates and POE ethers; anionic/nonionic; emulsifier for agric. formulations; amber visc. liq.; alcoholic odor; water-sol.; sp.gr. 1.052; visc. 500 cs; flash pt. > 120 F (Abel CC); pour pt. 17 C; pH 4.5–6.5 (1% aq.); 90% act.

Agrilan® AEC123. [Harcros UK] Alkoxylate; nonionic; coemulsifier, emulsion stabilizer for agric. toxicant emulsifiable concs.; EPA approved; wh. soft paste, mild odor; sp.gr. 1.040 (40 C); visc. 402 cs (40 C); pour pt. 28 C; cloud pt. 74 C (1% aq.); flash pt. (COC) > 200 C; pH 6.9 (1% aq.); 100% act.

Agrilan® AEC145. [Harcros UK] Polyaromatic ethoxylate; nonionic; coemulsifier used in agrochemical toxicant-emulsion concs. and oil flowables, herbicides and pesticides; EPA approved; off-wh. paste/liq., mild odor; sp.gr. 0.080 (40 C); visc. 141 cs (60 C); pour pt. 14 C; cloud pt. 55 C (1% aq.); flash pt. (PMCC) 190 C; pH 6.0 (1% aq.); 100% act.

Agrilan® AEC156. [Harcros UK] Fatty acid ester; nonionic; emulsifier for citrus and other agrochemical spray oil systems; liq.; HLB 9.0; 100% conc.

Agrilan® AEC167. [Harcros UK] EO/PO copolymer complex; nonionic; coemulsifier, emulsion stabilizer for agric. toxicants; EPA approved; colorless clear liq., mild odor; sp.gr. 1.040; visc. 800 cs; pour pt. 8 C; cloud pt. 44 C (1% aq.); flash pt. (PMCC) > 150 C; 100% act.

Agrilan® AEC178. [Harcros UK] EO/PO copolymer complex; anionic; coemulsifier, emulsion stabilizer for agric. toxicants; EPA approved; colorless clear liq., mild odor; sp.gr. 1.043; visc. 650 cs; pour pt. < 0 C; cloud pt. 34 C (1% aq.); flash pt. (PMCC) > 150 C; 100% act.

Agrilan® AEC189. [Harcros UK] EO/PO copolymer complex; anionic/nonionic; blended emulsifier for prod. of pesticides; EPA approved; amber visc. liq., alcoholic odor; water-sol.; visc. 978 cs (40 C); pour pt. 1 C; flash pt. (PMCC) 47 C; pH 5.0-8.0 (1% aq.); 83% act.

Agrilan® AEC200. [Harcros UK] EO/PO copolymer complex; anionic/nonionic; blended emulsifier for prod. of pesticides; EPA approved; amber visc. liq., alcoholic odor; water-sol.; visc. 468 cs; pour pt. 12 C; flash pt. (PMCC) 48 C; pH 5.0-8.0 (1% aq.); 85% act.

Agrilan® AEC211. [Harcros UK] Balanced pair

emulsifier; anionic/nonionic; emulsifier blend for agrochem. formulations, esp. esters of 2,4-D and 2,5-T; EPA approved; amber visc. liq., alcoholic odor; water-sol.; visc. 930 cs (40 C); pour pt. 6 C; flash pt. (PMCC) 47 C; pH 5.0-8.0 (1% aq.); 85% act.

Agrilan® AEC266. [Harcros UK] Sodium dioctyl sulfosuccinate, ethanol; anionic; emulsifier for prod. of insecticide emulsifiable concs. and microemulsion formulations; EPA approved; pale yel. clear liq., alcoholic odor; insol. in water; sp.gr. 0.996; visc. 43 cs; pour pt. < 0 C; flash pt. (Abel CC) 27 C; pH 5-8 (1% aq.); flamm.; 58% act.

Agrilan® AEC299. [Harcros UK] Sodium dioctyl sulfosuccinate, Solvesso 150; anionic; emulsifiable concs. for agrochem. actives; pale yel. clear liq., mild odor; insol. in water; sp.gr. 0.992; visc. 360 cs; pour pt. < 0 C; flash pt. (PMCC) 67 C; pH 6-8 (1% aq.); flamm.; 68% act.

Agrilan® AEC310. [Harcros UK] Blend; anionic/nonionic; blended emulsifier for prod. of pesticides, esp. 2,4-D esters and organo-phosphorus; EPA approved; amber liq., alcoholic odor; water-sol.; visc. 740 cs; pour pt. < 0 C; flash pt. (PMCC) 48 C; pH 5.0-8.0 (1% aq.); 88% act.

Agrilan® BA. [Harcros UK] Blend of oil-sol. sulfonates and POE ethers; anionic/nonionic; emulsifier for agric. formulations; brn. visc. liq.; alcoholic odor; turbid water-sol.; sp.gr. 1.034; visc. 3100 cs; flash pt. > 90 F (Abel CC); pour pt. < 0 C; pH 4.5–6.5 (1% aq.); 80% act.

Agrilan® BM. [Harcros UK] Blend of oil-sol. sulfonates and POE ethers; anionic/nonionic; emulsifier for agric. formulations; amber liq.; char. odor; turbid water-sol.; sp.gr. 1.022; visc. 196 cs; flash pt. 78 F (Abel CC); pour pt. < 0 C; pH 5.0–6.5 (1% aq.); 70% act.

Agrilan® C91D. [Harcros UK] Phosphate ester; anionic; compatibility agent for agrochemical toxicants; liq.; 70% conc.

Agrilan® D54. [Harcros UK] Alkyl phenol ethoxylate condensate; nonionic; emulsifier, wetting agent, dispersant for emulsifiable concs. of agrochemical toxicants; EPA approved; brn. sl. hazy liq., aromatic odor; sp.gr. 1.030; visc. 950 cs; pour pt. -17 C; cloud pt. 60 C (5%); flash pt. (PMCC) 70 C; flamm.; 80% act.

Agrilan® DG102. [Harcros UK] Alkylaryl ether sulfate; anionic; wetting agents for prod. of water-disp. gran. agrochemical toxicants; EPA approved; amber hazy visc. liq., faint ammoniacal odor; sp.gr. 1.110; visc. 6000 cs; pour pt. 7 C; flash pt. (Abel CC) 30 C; pH 7.8 (1% aq.); flamm.; 90% act, 1% water, 9% ethanol.

Agrilan® DG113. [Harcros UK] Alcohol alkoxylate; nonionic; wetting agent for prod. of water-disp. gran. agrochemical toxicants; EPA approved; colorless sl. hazy liq., mild odor; sp.gr. 1.006; visc. 66 cs; HLB 13.0; pour pt. -7 C; flash pt. (COC) > 210 C; pH 7.0 (1% aq.); 96% act. in water.

Agrilan® EA14. [Harcros UK] Surfactant, activity optimizer for agrochem. toxicant formulations; wh. hazy liq.; sp.gr. 0.928; visc. 32 cs; pour pt. < 0 C; flash pt. (PMCC) 46 C; pH 7.0 (1% aq.); flamm.; 80% act. in isobutanol.

Agrilan® EA25. [Harcros UK] Surfactant, activity optimizer for agrochem. toxicant formulations; pale amber liq.; sp.gr. 1.013; visc. 155 cs; pour pt. < 0 C; flash pt. (PMCC) > 150 C; pH 7.0 (1% aq.); 100% act.

Agrilan® EA36. [Harcros UK] Surfactant, activity optimizer for agrochem. toxicant formulations; colorless clear liq.; sp.gr. 0.973; visc. 58 cs; pour pt. < 0 C; flash pt. (PMCC) > 150 C; pH 7.0 (1% aq.); 100% act.

Agrilan® EA47. [Harcros UK] Alcohol alkoxylate; nonionic; surfactant, activity optimizers for agrochemical toxicant formulations; straw clear liq.; sp.gr. 0.980; visc. 195 cs; pour pt. < 0 C; flash pt. (PMCC) > 150 C; pH 9.6 (1% aq.); 100% act.

Agrilan® EA58. [Harcros UK] Surfactant, activity optimizer for agrochem. toxicant formulations; straw hazy liq.; sp.gr. 0.941; visc. 58 cs; pour pt. 5 C; flash pt. (PMCC) > 150 C; pH 9.6 (1% aq.); 100% act.

Agrilan® EA69. [Harcros UK] Surfactant, activity optimizer for agrochem. toxicant formulations; brn. liq.; sp.gr. 1.030; visc. 250 cs; pour pt. 0 C; flash pt. (PMCC) > 150 C; pH 9.5 (1% aq.); 100% act.

Agrilan® EA80. [Harcros UK] Surfactant, activity optimizer for agrochem. toxicant formulations; pale yel. liq.; sp.gr. 1.055; visc. 366 cs; pour pt. < 0 C; flash pt. (PMCC) > 150 C; pH 6.5 (1% aq.); 100% act.

Agrilan® F460. [Harcros UK] Alcohol alkoxylate; nonionic/anionic; adjuvant, wetting agent, dispersant for prod. of agric. suspension conc. flowables; EPA approved; pale yel. clear liq.; sp.gr. 1.035; visc. 230 cs; pour pt. -12 C; flash pt. (COC) 45 C; pH 6.7 (1% aq.); 60% act.

Agrilan® F491. [Harcros UK] Complex phosphate ester; anionic; wetter/dispersant for agric. flowables; liq.

Agrilan® F502. [Harcros UK] Alkylene oxide copolymer; nonionic; wetting and dispersing agent for agric. flowable systems; EPA approved; wh. soft solid; sp.gr. 1.061; visc. 380 cs (50 C); pour pt. 35 C; flash pt. (COC) > 200 C; pH 7.0 (1% aq.); 100% act.

Agrilan® F513. [Harcros UK] Polyaromatic ethoxylate phosphate ester; anionic; wetting and dispersing agent for agric. toxicants; pale amber visc. liq.; sp.gr. 1.130; visc. 4300 cs; pour pt. 9 C; flash pt. (PMCC) > 150 C; pH 6.4 (1% aq.); 100% act.

Agrilan® F524. [Harcros UK] Complex sulfonated carboxylic acid, neutral salt; anionic; adjuvant for prod. of suspension conc./flowables; prevents sedimentation and caking, improves solubility and freeze/thaw props.; aids visc. control; dk. amber clear liq.; sp.gr. 1.125; visc. 270 cs; pour pt. < 0 C; flash pt. (PMCC) > 95 C; pH 6.0 (1% aq.); 55% act.

Agrilan® F535. [Harcros UK] Phosphate ester; anionic; wetting agent, dispersant adjuvant for prod. of suspension conc./flowables; EPA approved; pale amber visc. liq.; sp.gr. 1.098; visc. 1600 cs (60 C); pour pt. 14 C; flash pt. (PMCC) > 150 C; pH 6.2 (1% aq.); 100% act.

Agrilan® F546. [Harcros UK] Complex phosphate ester, free acid; anionic; emulsifier for agrochem. flowable/emulsifiable conc. blends; EPA approved; pale amber visc. liq.; sp.gr. 1.180; visc. 3200 cs; pour pt. 10 C; flash pt. (PMCC) > 150 C; pH 2.6 (1% aq.); corrosive org. acid; toxicology: extremely irritating to skin and eyes; 1005 act.

Agrilan® F557. [Harcros UK] Phosphate ester; anionic; formulation of agric. suspension concs.; pale amber visc. liq.; sp.gr. 1.133; visc. 15,000 cs; pour pt. 0 C; flash pt. (PMCC) > 100 C; pH 6.5 (1% aq.); 60% act.

Agrilan® FS101. [Harcros UK] Epoxidized veg. oils;

nonionic; stabilizers for emulsifiable concs. of agric. toxicants; EPA, FDA approved; clear liq; sp.gr. 0.995; visc. 350 cs; pour pt. -13 C; flash pt. (COC) 285 C; usage level: 0.5-2.0%; 100% act.

Agrilan® FS112. [Harcros UK] Epoxidized veg. oils; nonionic; stabilizer for emulsifiable concs. of agric. toxicants; EPA, FDA approved; clear liq.; sp.gr. 1.030; visc. 800 cs; pour pt. -13 C; flash pt. (COC) 235 C; usage level: 0.5-2.0%; 100% act.

Agrilan® MC-90. [Harcros UK] Epoxilated veg. oils; nonionic/anionic; adjuvant for insecticides; liq.; 100% conc.

Agrilan® TKA103. [Harcros UK] Quat. ammonium compd.; cationic; wetting and compatibility agent for quat. herbicides; lt. brn. visc. liq., char. odor; water-sol.; sp.gr. 1.081; visc. 400 cs; pour pt. < 0 C; flash pt. (PMCC) > 150 C; pH 6.0-9.0 (1% aq.); 100% act.

Agrilan® TKA114. [Harcros UK] Surfactant; amphoteric; foaming agent for agric. formulations; straw liq., mild odor; sp.gr. 1.030; visc. 45 cs; pour pt. -5 C; flash pt. (PMCC) > 100 C; pH 4-6 (1% aq.); 35% act. in water.

Agrilan® TKA125. [Harcros UK] Complex phosphate ester, free acid; anionic; compatibility agent for formulated pesticides, hard water conditions; amber liq., alcoholic odor; sol. in water; sp.gr. 1.099; visc. 475 cs; pour pt. 15 C; flash pt. (PMCC) 41 C; pH 2.5 (1% aq.); flamm.; 70% act., 15% water.

Agrilan® TKA147. [Harcros UK] Complex phosphate ester, potassium salt; anionic; compatibility agent for agric. toxicants, hard water conditions; pale straw liq., mild odor; sol. in water; sp.gr. 1.294; visc. 180 cs; pour pt. -18 C; flash pt. (PMCC) > 100 C; pH 7.2 (1% aq.); 65% act. in water.

Agrilan® WP101. [Harcros UK] Aromatic sulfonate; anionic; wetting agent for prod. of water-disp. granules and wettable powds. of agrochem. toxicants; EPA approved; off-wh. powd./flake; sol. in water; pH 9.5 (10% aq.); 85% act.

Agrilan® WP112. [Harcros UK] Surfactant on inert silica carrier; nonionic; wetting agent for prod. of water-disp. granules of agrochem. toxicants; EPA approved; wh. powd.; disp. in water; HLB 13.0; pH 6.6 (10% aq.); 60% conc.

Agrilan® WP123. [Harcros UK] Wetting agent for formulation of wettable powds. of agric. toxicants; EPA approved; off-wh. powd.; sol. in water @ 10%; pH 9.0 (10% aq.); 65.5% act.

Agrilan® WP134. [Harcros UK] Wetting agent/ dispersant for formulation of agric. toxicant water-disp. granules; EPA approved; off-wh. powd.; sol. in water @ 10%; pH 7.8; 67% act.

Agrilan® WP145. [Harcros UK] Alkyl naphthalene sulfonate; anionic; wetting agent, dispersant for water-disp. granules, wettable powds. and suspension concs.; EPA approved; tan powd.; sol. in water @ 10%; pH 9.3 (10% aq.); 84% act.

Agrilan® WP156. [Harcros UK] Aromatic sulfonate; anionic; dispersant for water-disp. granules, wettable powds., and suspension concs. of agrochem. toxicants; EPA approved; tan powd.; sol. in water @ 10%; pH 9.5 (10% aq.); 87% act.

Agrilan® WP167. [Harcros UK] Aromatic sulfonate; amphoteric; wetting agent for prod. of water-disp. granules of agrochem. toxicants; EPA approved; off-wh. powd.; sol. in water @ 10%; pH 9.0 (10% aq.); 75.5% act.

Agrilan® WP178. [Harcros UK] Polyaromatic sulfonate; anionic; dispersant for wettable powds. and agrochemical suspension concs.; off-wh. powd.; sol. in water @ 10%; pH 6.5 (10% aq.); 96% act.

Agrilan® X98. [Harcros UK] Calcium dodecylbenzene sulfonate; anionic; emulsifier for herbicide and pesticide formulations; EPA approved; visc. liq., alcoholic odor; insol. in water; sp.gr. 0.970; visc. 1080 cs; pour pt. -13 C; flash pt. (Abel CC) 31 C; pH 5-8 (1% aq.); flamm.; 58% act.

Agrilan® X109. [Harcros UK] Calcium alkylbenzene sulfonate; anionic; coemulsifier for conc. of agrochemical toxicants; liq.

Agrimul 26-B. [Henkel/Emery; Henkel-Nopco] Ethoxylated fatty acid; nonionic; oil spray emulsifier for agric. systems; amber liq.; dens. 8.3 lb/gal; visc. 95 cps; HLB 14.0; 100% conc. DISCONTINUED.

Agrimul 70-A. [Henkel/Emery; Henkel-Nopco] Alkylaryl polyether alcohol; nonionic; agric. emulsifier for chlorinated hydrocarbons; EPA-exempt; dk. amber liq.; sol. @ 5% in xylene; disp. in water, min. oil; dens. 8.8 lb/gal; visc. 8850 cps; HLB 10.5; pour pt. 0 C; 100% conc.

Agrimul A-300. [Henkel/Emery; Henkel-Nopco] Aromatic sulfonate-oxide condensate blend; anionic; agric. emulsifier for chlorinated and phosphated insecticide concs.; EPA-exempt; amber liq.; sol. @ 5% in xylene, methyl oleate; disp. in water; dens. 8.8 lb/gal; visc. 1050 cps; pour pt. 5 C; 100% conc. DISCONTINUED.

Agrimul N-300. [Henkel/Emery; Henkel-Nopco] Aromatic sulfonate-oxide condensate blend; anionic; agric. emulsifier for chlorinated and phosphated insecticide concs.; liq.; 100% conc.

Agrimul S-300. [Henkel/Emery] Aromatic sulfonate-oxide condensate blend; anionic/nonionic; agric. emulsifier for phosphated toxicants; EPA-exempt; amber liq.; sol. @ 5% in water, xylene; disp. in methyl oleate; dens. 8.6 lb/gal; visc. 2240 cps; 100% conc. DISCONTINUED.

Agrimul VUQ. [Henkel/Emery] Formulated prod.; anionic; emulsifier for paraffinic oils; liq.; 100% conc.

Agrisol PX401. [Harcros UK] Aromatic alkoxylates; nonionic; cosolv., penetrant, flow promotor for agrochemical toxicants; clear liq., faint odor; insol. in water; sp.gr. 1.098; visc. 27 cs; pour pt. -15 C; flash pt. (COC) > 100 C; pH 7.0 (1% aq.); toxicology: damaging to eyes; 100% act.

Agrisol PX413 [Harcros UK] Aromatic alkoxylate; nonionic; cosolv., penetrant, flow promotor for formulation of agrochem. toxicants; clear liq., faint odor; sol. in water; sp.gr. 1.121; visc. 64 cs; pour pt. -6 C; flash pt. (COC) > 150 C; pH 6.4 (1% aq.); 100% act.

Agriwet 1186A. [Henkel/Emery] Dioctyl sodium sulfosuccinate-based formulated prod.; anionic; wetting agent, emulsifier, solubilizer, penetrant for agric. toxicants; EPA-exempt; clear liq.; pH 7 (1%); 70% act. DISCONTINUED.

Agriwet CA, C91, C92. [Henkel] Phosphate ester; anionic; wetting agents, detergents, emulsifiers and lubricants; compatibility and resuspension aid for pesticides; coupling agent; liq.; 100% conc.

Agriwet FOA. [Henkel] Phosphate ester; anionic; resuspension aid in pesticide formulations; liq.; 100% conc.

Agriwet T-F. [Henkel/Emery] Formulated prod.; anionic; wetting agent and dispersant for pesticide fomulations, wettable powds., dry flowables, aq. suspensions; flake; 70% conc.

Ahco 759. [ICI Am.] Sorbitan laurate; CAS 1338-39-

2; nonionic; emulsifier, solubilizer for textile use; liq.; HLB 8.6; 100% conc.

Ahco 832. [ICI Am.] Sorbitan oleate; CAS 1338-43-8; nonionic; emulsifier, solubilizer for textile use; liq.; 100% conc.

Ahco 909. [ICI Am.] Sorbitan stearate; CAS 1338-41-6; nonionic; emulsifier, solubilizer for textile use; solid; HLB 4.7; 100% conc.

Ahco 944. [ICI Am.] Sorbitan oleate; CAS 1338-43-8; nonionic; emulsifier, solubilizer for textile use; liq.; 100% conc.

Ahco 3998. [ICI Am.] Oleth-20; CAS 9004-98-2; nonionic; dye leveling agent and emulsifier; liq.; 100% conc.

Ahco 7166T. [ICI Am.] Polysorbate 65; CAS 9005-71-4; nonionic; dye leveling agent and emulsifier; solid; 100% conc.

Ahco 7596D. [ICI Am.] Polysorbate 21; CAS 9005-64-5; nonionic; dye leveling agent and emulsifier; liq.; 100% conc.

Ahco 7596T. [ICI Am.] Polysorbate 20; CAS 9005-64-5; nonionic; dye leveling agent and emulsifier; liq.; 100% conc.

Ahco A-117. [ICI Am.] Alkylaryl sulfonate; anionic; emulsifier, surfactant, used as dispersant, detergent, wetting agent; amber liq.; faint odor; sol. IPA; sp.gr. 1.0; visc. 7000 cps; 100% act.

Ahco AB 100. [ICI Am.] Formulated prod.; anionic; solvent system emulsifiers for chlorinated solvs., dye carrier formulations, scourants, etc.; amber liq.; solv. odor; water or IPA sol.

Ahco AB 118. [ICI Am.] Formulated prod.; anionic; solvent system emulsifiers for chlorinated solvs., dye carrier formulations, scourants, etc.; liq.

Ahco AB 120. [ICI Am.] Formulated prod.; anionic; solvent system emulsifiers for chlorinated solvs., dye carrier formulations, scourants, etc.; liq.

Ahco AB 135. [ICI Am.] Formulated prod.; anionic; solvent system emulsifiers for chlorinated solvs., dye carrier formulations, scourants, etc.; liq.

Ahco AB 146. [ICI Am.] Formulated prod.; anionic; solvent system emulsifiers for chlorinated solvs., dye carrier formulations, scourants, etc.; liq.

Ahco AB 160. [ICI Am.] Formulated prod.; anionic; solvent system emulsifiers for chlorinated solvs., dye carrier formulations, scourants, etc.; liq.

Ahco AB 228. [ICI Am.] Formulated prod.; anionic; solvent system emulsifiers for chlorinated solvs., dye carrier formulations, scourants, etc.; liq.

Ahco AJ-110. [ICI Am.] Sulfated castor oil; anionic; wetting agent, lubricant, penetrant, dispersant, emulsifier, detergent for textile and leather industries; dyeing assistant for textiles; amber liq.; castor oil odor; water sol.; sp.gr. 1.05; HLB 11.0; 50% act. in water.

Ahco AY 2200. [ICI Am.] Phosphate ester; anionic; low-soiling textile antistat, lubricant; liq.; 45% conc.

Ahco C330. [ICI Am.] Quat. ammonium compd.; cationic; heat-stable textile fiber antistat; liq.; 100% conc.

Ahco DFO-100. [ICI Am.] Polysorbate 81; CAS 9005-65-6; nonionic; emulsifier, solubilizer for textile use; liq.; HLB 10.0; 100% conc.

Ahco DFO-110. [ICI Am.] Polysorbate 85; CAS 9005-70-3; nonionic; emulsifier, solubilizer for textile use; liq.; HLB 11.0; 100% conc.

Ahco DFO-150. [ICI Am.] Polysorbate 80; CAS 9005-65-6; nonionic; emulsifier, solubilizer for textile use; liq.; HLB 15.0; 100% conc.

Ahco DFP-156. [ICI Am.] Polysorbate 40; CAS 9005-66-7; nonionic; emulsifier, solubilizer for textile use; liq.; 100% conc.

Ahco DFS-96. [ICI Am.] Polysorbate 61; CAS 9005-67-8; nonionic; emulsifier, solubilizer for textile use; solid; HLB 9.6; 100% conc.

Ahco DFS-149. [ICI Am.] Polysorbate 60; CAS 9005-67-8; nonionic; emulsifier, solubilizer for textile use; solid; HLB 14.9; 100% conc.

Ahco DHS-111. [ICI Am.] POE stearate; nonionic; textile lubricant and all-purpose emulsifier; solid; HLB 11.1; 100% conc.

Ahco EO-102. [ICI Am.] POE sorbitol polyoleate; nonionic; emulsifier, solubilizer for textile use; liq.; HLB 10.2; 100% conc.

Ahco EO-114. [ICI Am.] POE sorbital polyoleate; nonionic; emulsifier, solubilizer for textile use; liq.; HLB 11.4; 100% conc.

Ahco FO-18. [ICI Am.] Sorbitan trioleate; CAS 26266-58-0; nonionic; emulsifier, solubilizer for textile use; liq.; 100% conc.

Ahco FP-67. [ICI Am.] Sorbitan palmitate; CAS 26266-57-9; nonionic; emulsifier, solubilizer for textile use; solid; 100% conc.

Ahco FS-21. [ICI Am.] Sorbitan tristearate; CAS 26658-19-5; nonionic; emulsifier, solubilizer for textile use; solid; HLB 2.1; 100% conc.

Ahcovel Base 500. [ICI Am.] Fatty carbamide; cationic; textile softener; hard wax; 100% conc.

Ahcovel Base 700. [ICI Am.] Fatty amide acetate salt; cationic; textile softener; solid; 94% conc.

Ahcovel Base N-15. [ICI Am.] Glyceryl stearate, emulsifiable; nonionic; textile softener; solid; 100% conc.

Ahcovel Base N-62. [ICI Am.] Ester blend; nonionic; textile softener; liq.; 100% conc.

Ahcovel Base N-64. [ICI Am.] Ester blend; nonionic; knitting lubricant, softener, cellulosic and acrylic fiber finishes, antistat; solid; 100% conc.

Ahcovel Base OB. [ICI Am.] Fatty carbamide, modified; cationic; base for textile softeners; waxy solid; 100% act.

Ahcovel R Base. [ICI Am.] Fatty carbamide salt; anionic; textile softener, fiber lubricant; solid; 100% conc.

Ahcowet DQ-114. [ICI Am.] Trideceth-6; CAS 24938-91-8; nonionic; wetting agent, dispersant, detergent in textiles; liq.; HLB 11.4; 100% conc.

Ahcowet DQ-145. [ICI Am.] Trideceth-12; CAS 24938-91-8; nonionic; scour and dye leveling agent in textiles; liq.; HLB 14.5; 100% conc.

Ahcowet RS. [ICI Am.] Fatty acid ester, sulfated; anionic; wetting agent for textile processing; liq.; 65% conc.

Airrol CT-1. [Toho Chem. Industry] Dialkyl sulfosuccinate; anionic; wetting agent and dyeing assistant; liq.; 70% conc.

Ajicoat SPG. [Ajinomoto] Sodium polyglutamate; anionic; surface modifier, coemulsifier, codispersant; solid; 100% conc.

Ajidew A-100. [Ajinomoto] PCA; nat. humectant used in cosmetics, soaps, dentifrices, medicinal supplies, tobacco, cellulose film, paper prods., fiber prods., paints; additive to dyeing agent, softening agent, finishing agent, and antistatic agent; intermediate for synthesis; wh. cryst., odorless, sl. acidic taste, nonhygroscopic; m.w. 129.11; m.p. 181 C.

Ajidew N-50. [Ajinomoto] Sodium PCA; see Ajidew A-100; 50% aq. sol'n.

Ajidew SP-100. [Ajinomoto] Sodium PCA and PCA;

see Ajidew A-100; wh. cryst.

Akypo 1690 S. [Chem-Y GmbH] Laureth-5 carboxylic acid; additive for liq. heavy-duty detergent formulations; thickener for NaOH; liq.; 93% act.

Akypo ITD 30 N. [Chem-Y GmbH] Sodium trideceth-3 carboxylic acid; CAS 68891-17-8; anionic; emulsifier for silicone oil; biodeg.; sl. ylsh. cloudy, visc. liq.; acid no. 2 max.; pH 6-8 (10% aq.); 96.5% act.

Akypo ITD 70 BV. [Chem-Y GmbH] Sodium trideceth-8 carboxylate; anionic; biodeg. surfactant, wetting agent for textile industry; APHA < 100 paste; visc. 63,000 mPa•s; pH 7-9 (10% water/ethanol 1:1); 66% act., 30% water.

Akypo LF 1. [Chem-Y GmbH] Capryleth-6 carboxylic acid; anionic; low foaming surfactant for industrial, institutional and household cleaning; alkaline and acid stable; biodeg.; ylsh. liq.; visc. < 200 mPa•s; HLB 13.0; cloud pt. 39 C (1% aq., pH 3.5); pH 1.5-3.0 (1%); surf. tens. 28.9 mN/m (1%, pH 3.5); 91% act.

Akypo LF 2. [Chem-Y GmbH] Capryleth-9 carboxylic acid; anionic; low foaming surfactant for industrial, institutional and household cleaning, cooling tower cleaners, disinfectant cleaners, high-pressure cleaners, metalworking fluids, electroplating, PU foam for orthopedic uses; alkaline and acid stable; biodeg.; ylsh. liq.; visc. < 200 mPa•s; HLB 16.0; cloud pt. > 100 C (1% aq., pH 3.5); pH 1.5-3.0 (1%); surf. tens. 30.8 mN/m (1%, pH 3.5); 92% act.

Akypo LF 3. [Chem-Y GmbH] Hexeth-4 carboxylic acid; CAS 105391-15-9; anionic; low foaming surfactant for industrial, institutional and household cleaning; alkaline and acid stable; biodeg.; ylsh. liq.; visc. < 200 mPa•s; HLB 12.0; cloud pt. > 100 C (1% aq., pH 12); pH 1.5-3.0 (1%); surf. tens. 36.2 mN/m (1%, pH 12); 90% act.

Akypo LF 4. [Chem-Y GmbH] Capryleth-9 carboxylic acid, hexeth-4 carboxylic acid; anionic; low foaming surfactant for industrial, institutional and household cleaning, cutting and drilling oils, drawing and rolling oils, water treatment prods., high-pressure cleaners, cooling tower cleaners, electroplating, film developing; solubilizer; alkaline and acid stable; biodeg.; ylsh. clear liq.; visc. < 200 mPa•s; HLB 10.0; cloud pt. 57 C (1% aq., pH 3.5); pH 1.5-3.0 (1%); surf. tens. 30.5 mN/m (1%, pH 3.5); 93% act.

Akypo LF 4N. [Chem-Y GmbH] Sodium capryleth-9 carboxylate, sodium hexeth-4 carboxylate; anionic; low foaming surfactant for industrial cleaning; alkaline and acid stable; biodeg.; liq.; HLB 10.2; 80% act.

Akypo LF 5. [Chem-Y GmbH] Buteth-2 carboxylic acid; CAS 105391-15-9; anionic; low foaming surfactant for industrial, institutional and household cleaning; alkaline and acid stable; biodeg.; ylsh. liq.; visc. < 200 mPa•s; HLB 10.0; cloud pt. > 100 C (1% aq., pH 12); pH 1.5-3.0 (1%); surf. tens. 38.8 mN/m (1%, pH 12); 87% act.

Akypo LF 6. [Chem-Y GmbH] Buteth-2 carboxylic acid, capryleth-9 carboxylic acid; anionic; low foaming surfactant for industrial, institutional and household cleaning, metalworking fluids, disinfectant cleaners, high-pressure cleaners, engine cleaners, automatic dishwash, cooling water systems, electroplating, textile pretreatment; alkaline and acid stable; biodeg.; ylsh. liq.; visc. < 200 mPa•s; HLB 14.0; pH 1.5-3.0 (1%); cloud pt. > 100 C (1% aq., pH 12); surf. tens. 30.3 mN/m (1%, pH 3.5); 93% act.

Akypo MB 1585. [Chem-Y GmbH] Fatty alcohol polyglycol ether carboxylic acid; CAS 105391-15-9; anionic; surfactant for industrial cleaning; liq.; HLB 15.6; 92% conc.

Akypo MB 1614/1. [Chem-Y GmbH] Capryleth-4 carboxylic acid; anionic; surfactant for industrial cleaning; liq.; HLB 11.1; 90% conc.

Akypo MB 1614/2. [Chem-Y GmbH] Fatty alcohol polyglycol ether carboxylic acid; anionic; surfactant for industrial cleaning; liq.; HLB 12.5; 93% conc.

Akypo MB 2528S. [Chem-Y GmbH] Hexeth-4 carboxylic acid; anionic; surfactant for industrial cleaning; liq.; HLB 12.2.

Akypo MB 2621 S. [Chem-Y GmbH] Fatty alcohol polyglycol ether carboxylic acid; anionic; surfactant for industrial cleaning, metalworking fluids; lime soap dispersant; liq.; sol. in water; insol. in oils; acid no. 102-115; HLB 10.2; pH 2-3 (1%); surf. tens. 35 mN/m; 93% conc.

Akypo MB 2705 S. [Chem-Y GmbH] Fatty alcohol polyglycol ether carboxylic acid; anionic; surfactant for industrial cleaning; liq.; HLB 10.2.

Akypo MB 2717 S. [Chem-Y GmbH] Fatty alcohol polyglycol ether carboxylic acid; anionic; surfactant for industrial cleaning, metalworking fluids; lime soap dispersant; liq.; sol. in water; insol. in oils; acid no. 180-192; HLB 14.2; pH 2-3 (1%); surf. tens. 48 mN/m; 86% conc.

Akypo NP 70. [Chem-Y GmbH] Nonoxynol-8 carboxylic acid; CAS 3115-49-9; anionic; emulsifier; Gardner 4 liq.; visc. 1500 mPa•s; pH 2-3 (1%); surf. tens. 37 mN/m; 90% conc.

Akypo NTS. [Chem-Y GmbH] Sodium laureth-6 carboxylate; CAS 68987-89-3; anionic; detergent for carpet and upholstery cleaners esp. aerosols; leak detector spray; liq.; 22% act.

Akypo OCD 10 NV. [Chem-Y GmbH] Sodium deceth-2 carboxylate, sodium capryleth-2 carboxylate; defoamer for phosphoric acid industry (suitable for nitrate process); liq.; 20% act.

Akypo OP 80. [Chem-Y GmbH] Octoxynol-9 carboxylic acid; CAS 107628-08-0; 72160-13-5; anionic/nonionic; detergent, emulsifier; used in aq. sol'ns, metalworking fluids, emulsion polymerization, film developing baths; lime soap dispersant; moderate foam; Gardner 2 clear liq.; sol. in water; insol. in oils; sp.gr. 1.08; visc. 2000 mPa•s; acid no. 73-83; pH 2-3 (1%); surf. tens. 33 mN/m; 90% act.

Akypo OP 115. [Chem-Y GmbH] Alkylphenol polyglycol ether carboxylic acids; CAS 107628-08-0; anionic/nonionic; detergent, emulsifier; used in aq. sol'ns.; liq.; 90% conc.

Akypo OP 190. [Chem-Y GmbH] Octoxynol-20 carboxylic acid; CAS 107628-08-0; 72160-13-5; anionic/nonionic; detergent, emulsifier; used in aq. sol'ns., metalworking fluids, electroplating; lime soap diseprsant; moderate foam; ylsh. liq.; sol. in water; insol. in oils; visc. 1000 mPa•s; acid no. 50-62; pH 2-3 (1%); surf. tens. 37 mN/m; 88% act.

Akypo RCS 60. [Chem-Y GmbH] Ceteareth-7 carboxylic acid; CAS 68954-89-2; surfactant; ylsh. wh. solid; pH 2-3 (1%); surf. tens. 44 mN/m; 90% act.

Akypo RLM 25. [Chem-Y GmbH] Laureth-4 carboxylic acid; CAS 68954-89-2; nonionic; emulsifier, dispersant for emulsion and dispersion use; metalworking fluids; lime soap dispersant; moderate foam; liq.; disp. in water; sol. in oil; visc. 1000 mPa•s; acid no. 95-112; pH 2-3 (1%); surf. tens. 33 mN/m; 92% conc.

Akypo RLM 38. [Chem-Y GmbH] Laureth-5 carboxylic acid; surfactant; Gardner 1 liq.; visc. 500 mPa•s; pH 2-3 (1%); surf. tens. 33 mN/m; 90% act.

Akypo RLM 38 NV. [Chem-Y GmbH] Fatty alcohol polyglycol ether carboxylic acid, sodium salt; anionic; detergent for shampoo, liq. soap and foam bath; emulsifier and wetting agent; liq.; toxicology: LD50 (rat, oral) 4 g/kg; 22% conc.

Akypo RLM 45. [Chem-Y GmbH] Laureth-6 carboxylic acid; CAS 68954-89-2; anionic/nonionic; emulsifier, dispersant, superfatting and foam stabilizing agent for emulsion and detergent use, shampoos, bubble baths; Gardner 1 liq.; visc. 500 mPa•s; HLB 11.0; pH 2-3 (1%); surf. tens. 34 mN/m; toxicology: LD50 (rat, oral) 3.5 g/kg; 90% act.

Akypo RLM 100. [Chem-Y GmbH] Laureth-11 carboxylic acid; anionic/nonionic; emulsifier for cosmetics applics.; foam booster for cleaners, heavy-duty detergent formulations; Gardner 1 liq.; visc. 500 mPa•s; HLB 14.8; pH 2-3 (1%); surf. tens. 38 mN/m; 90% act.

Akypo RLM 100 MGV. [Chem-Y GmbH] Fatty alcohol polyglycol ether carboxylic acid, magnesium salt; anionic; detergent for shampoos, liq. soap and foam bath; emulsifier, wetting agent; liq.; 22% conc.

Akypo RLM 130. [Chem-Y GmbH] Laureth-14 carboxylic acid; CAS 68954-89-2; nonionic; emulsifier; Gardner 1 liq.; visc. 2000 mPa•s; HLB 15.9; pH 2-3 (1%); surf. tens. 40 mN/m; 90% conc.

Akypo RLM 160. [Chem-Y GmbH] Laureth-17 carboxylic acid; CAS 27306-90-7; anionic/nonionic; emulsifier; ylsh. wh. solid; pH 2-3 (1%); surf. tens. 47 mN/m; 85% conc.

Akypo RLM 160 NV. [Chem-Y GmbH] Fatty alcohol polyglycol ether carboxylate, sodium salt; anionic; detergent for shampoo, liq. soap, and foam bath; emulsifier and wetting agent; liq.; HLB 33.6; 22% conc.

Akypo RLMQ 38. [Chem-Y GmbH] Laureth-5 carboxylic acid; CAS 68954-89-2; nonionic; emulsifier, dispersant, additive for personal care prods., household and industrial formulas; primary emulsifier for syn. latex; liq.; HLB 10.2; 90% conc.

Akypo RO 20. [Chem-Y GmbH] Oleth-3 carboxylic acid; emulsifier for metalworking fluids; lime soap dispersant; very sl. foaming; liq.; insol. in water; sol. in oil; acid no. 95-106; pH 2-3 (1%); 90% act.

Akypo RO 50. [Chem-Y GmbH] Oleth-6 carboxylic acid; CAS 57635-48-0; anionic; emulsifier for cleaning agents and metal cooling liqs.; chain lubricant; lime soap dispersant; sl. foaming; brnsh. liq.; sol. in oil and water; visc. 1000 mPa•s; acid no. 84-95; pH 2-3 (1%); surf. tens. 35 mN/m; 90% act.

Akypo RO 90. [Chem-Y GmbH] Oleth-10 carboxylic acid; CAS 57635-48-0; anionic; emulsifier for metalworking fluids; lime soap dispersant; sl. foaming; stable to hard water; biodeg.; brnsh. liq.; sol. in water, oil; m.w. 698; visc. 1000 mPa•s; acid no. 67-78; pH 2-3 (1%); surf. tens. 37 mN/m; 90% act.

Akypo RS 60. [Chem-Y GmbH] Steareth-7 carboxylic acid; CAS 68954-89-2; 59559-30-7; surfactant; ylsh. wh. solid; pH 2-3 (1%); surf. tens. 44 mN/m; 90% act.

Akypo RS 100. [Chem-Y GmbH] Steareth-11 carboxylic acid; CAS 68954-89-2; surfactant; ylsh. wh. solid; pH 2-3 (1%); surf. tens. 45 mN/m; 90% act.

Akypo RT 60. [Chem-Y GmbH] Talloweth-7 carboxylic acid; CAS 68954-89-2; surfactant; ylsh. wh. solid; pH 2-3 (1%); surf. tens. 39 mN/m; 90% act.

Akypo TBP 40. [Chem-Y GmbH] Butoxynol-5 carboxylic acid; surfactant for metalworking fluids; liq.; 93% act.

Akypo TBP 180. [Chem-Y GmbH] Butoxynol-19 carboxylic acid; CAS 104909-82-2; nonionic; surfactant for zinc galvanization processes; liq.; 90% act.

Akypo TEC-AM. [Chem-Y GmbH] Alkyl ether carboxylic acid; surfactant for metalworking cooling lubricants; liq.; 90% act.

Akypo TFC-S. [Chem-Y GmbH] Laureth-5 carboxylic acid and sodium octyl sulfate; biodeg. foaming detergent, thickener, disinfectant for dairy, brewery, sanitary cleaning; lt. yel. clear liq.; visc. 500 mPa•s; pH 3-4 (1%); 58% act.

Akypo TFC-SN. [Chem-Y GmbH] Laureth-5 carboxylic acid, sodium octyl sulfate, and isostearic acid; biodeg. foaming detergent, thickener, disinfectant for dairy, brewery, sanitary cleaning, thickened toilet cleaners; chlorine stable; yel. sl. cloudy visc. liq.; visc. 4000-9000 mPa•s; pH 3-4 (1%); 61% act.

Akypo TPR. [Chem-Y GmbH] Sodium hexeth-4 carboxylate and trideceth-2; foam-suppressant surfactant for powd. cleaners for dishwash, steam carpet cleaners, metalworking industry; liq.; 93% act.

Akypogene FP 35 T. [Chem-Y GmbH] TEA cocoate; anionic; cosmetics surfactant for shower, shampoo and bath formulations, liq. hand cleaner; liq.; 35% conc.

Akypogene Jod F. [Chem-Y GmbH] Sodium laureth-11 carboxylate, iodine; antibacterial surfactant for food, dairy, beverage industries, agric., and hospitals; dk. brn. liq.; misc. with water, methanol, ethanol; sp.gr. 1.26 kg/l; visc. 400 mPa•s; pH 2.5-3.0 (1:9 with water); 20% act. iodine.

Akypogene Jod MB 1918. [Chem-Y GmbH] Iodine and fatty alcohol polyglycol-ether carboxylic acid; anionic; disinfectant; liq.; 70% act.

Akypogene KTS. [Chem-Y GmbH] Ammonium polyacrylate and sodium laureth-6 carboxylate; anionic; base for rug shampoo and upholstery cleaner with antistatic and anticorrosive properties; esp. for aerosols; wh. opaque liq.; sp.gr. 1.0 kg/l; visc. 300 mPa•s; f.p. 0 C; pH 8.5-9.5; 31% act.

Akypogene SO. [Chem-Y GmbH] MEA-PPG-6-laureth-7 carboxylate, nonoxynol-2; surfactant for metalworking cooling lubricants; emulsifier, lime soap dispersant; paste; 95% act.

Akypogene SV. [Chem-Y GmbH] Mixt. of alkyl ether carboxylic acid and nonionic; thickener for acids; liq.; 99% act.

Akypogene VSM. [Chem-Y GmbH] Blend; CAS 24938-91-8; anionic/nonionic; detergent for liq. scour; liq.; 92% conc.

Akypogene VSM-N. [Chem-Y GmbH] Trideceth-2, sodium dodecylbenzene sulfonate; anionic/nonionic; for mfg. of liq. scour creams with exc. stability; brn. cloudy visc. liq.; sp.gr. 1.1 kg/l; visc. 800-2000 mPa•s; pH 6-8; 92% act.

Akypogene WSW-W. [Chem-Y GmbH] Sodium laureth sulfate, sodium dodecylbenzene sulfonate, MEA laureth-6 carboxylate, cocamide DEA; surfactant for mild wool detergent formulations; liq.; 25% act.

Akypogene ZA 97 SP. [Chem-Y GmbH] Potassium xylene sulfonate, potassium tallate, potassium cocoate; anionic; surfactant blend for liq. soap; liq.; 25% conc.

Akypomine® AL. [Chem-Y GmbH] Surfactant for mining industry; collector for flotation of scheelite;

liq.; 90% act.

Akypomine® AT. [Chem-Y GmbH] Nonoxynol carboxylic acid; anionic; surfactant activator for min. flotation; lt. brn. visc. liq.; flash pt. > 100 C; 93% act.

Akypomine® BC 50. [Chem-Y GmbH] Fatty alcohol ether sulfate; CAS 68130-43-8; anionic; surfactant for mineral industry; flotation agent for barite (selective); collector for flotation of typical salt minerals such as fluorspar, magnesite or scheelite; wh.-yel. paste; pH 7-9 (50% aq.); 45% act.

Akypomine® BC/S. [Chem-Y GmbH] Fatty alcohol sulfate; flotation agent for barite; paste; 45% act.

Akypomine® DB. [Chem-Y GmbH] Sodium polynyl sulfonate; surfactant for mining industry; barite and monacite depressant; liq.; 40% act.

Akypomine® FA. [Chem-Y GmbH] Mixt. of alkyl ether carboxylic acid and tall oil fatty acid; surfactant for mining industry; barite activator; liq.; 97% act.

Akypomine® MW 05. [Chem-Y GmbH] Laureth-7 phosphate; surfactant for mining industry; fluorspar collector selective for barite; liq.; 30% act.

Akypomine® P 191. [Chem-Y GmbH] Oleamine hydroxypropyl bistrimonium chloride, polyacrylamide; cationic; filtration aux. with flocculant for mining industry; liq.; 20% act.

Akypopress DB. [Chem-Y GmbH] Syn. polymer; CAS 9002-97-5; anionic; depressant for metal ions; liq.; 30% conc.

Akypoquat 40. [Chem-Y GmbH] Oleoyl PG-trimonium chloride, stearoyl PG-trimonium chloride, behenoyl PG-trimonium chloride, palmitoyl PG-trimonium chloride, trideceth-2; cationic; environmentally safe laundry softener conc., textile softener; antistat, rewetting agent; fully biodeg.; lt. yel. nontransparent liq.; flash pt. 78 C; pH 3.5-4.5; 83% act.

Akypoquat 129. [Chem-Y GmbH] Isostearoyl PG-trimonium chloride and behenoyl PG-trimonium chloride; cationic; raw material for mfg. of laundry softeners; molecule is fully biodeg.; liq.; 80% act.

Akypoquat 1295. [Chem-Y GmbH] Mixt. of cationic fatty acids; cationic; environmentally safe raw material for fabric softeners and concs.; paste; 68% act.

Akyporox CO 400. [Chem-Y GmbH] PEG-40 hydrog. castor oil; CAS 61788-85-0; nonionic; perfume solubilizer, o/w emulsifier for cosmetic prods.; eliminates oil bath turbidity; paste; m.p. 25 C; cloud pt. > 100 C (1%); surf. tens. 47 mN/m; 100% act.

Akyporox CO 600. [Chem-Y GmbH] PEG-60 hydrog. castor oil; CAS 61788-85-0; nonionic; perfume solubilizer, o/w emulsifier for cosmetic prods.; eliminates oil bath turbidity; solid; m.p. 35 C; cloud pt. > 100 C (1%); surf. tens. 48 mN/m; 100% act.

Akyporox NP 15. [Chem-Y GmbH] Nonoxynol-1; CAS 27986-36-3; nonionic; emulsifier for emulsion polymerization; liq.; sp.gr. 1.0; HLB 4.7; 100% act.

Akyporox NP 30. [Chem-Y GmbH] Nonoxynol-3; CAS 9016-45-9; nonionic; for mfg. of hair dye formulations; emulsifier for emulsion polymerization; liq.; 100% act.

Akyporox NP 40. [Chem-Y GmbH] Nonoxynol-4; CAS 9016-45-9; nonionic; emulsifier for film developing baths; liq.; sp.gr. 1.01; HLB 7.9; surf. tens. 32 mN/m; 100% act.

Akyporox NP 90. [Chem-Y GmbH] Nonoxynol-9; CAS 9016-45-9; nonionic; emulsifier; liq.; sp.gr. 1.06; HLB 12.9; cloud pt. 62 C (1%); surf. tens. 33 mN/m; 100% act.

Akyporox NP 95. [Chem-Y GmbH] Nonoxynol-10; CAS 9016-45-9; nonionic; emulsifier for calcium stearate; liq.; 100% act.

Akyporox NP 105. [Chem-Y GmbH] Nonoxynol-10; CAS 9016-45-9; nonionic; emulsifier; liq.; sp.gr. 1.06; HLB 13.5; cloud pt. 68 C (1%); surf. tens. 34 mN/m; 100% act.

Akyporox NP 150. [Chem-Y GmbH] Nonoxynol-15; CAS 9016-45-9; 26027-38-3; nonionic; emulsifier; liq.; 100% conc.

Akyporox NP 200. [Chem-Y GmbH] Nonoxynol-20; CAS 9016-45-9; nonionic; emulsifier, wetting agent used in textile prods., emulsion polymerization, degreasing baths, electroplating industry; solid; sp.gr. 1.08; m.p. 33 C; HLB 16.0; cloud pt. > 100 C (1%); surf. tens. 42 mN/m; 100% act.

Akyporox NP 300. [Chem-Y GmbH] Alkylphenol polyglycol ether; nonionic; emulsifier; solid; 100% conc.

Akyporox NP 300V. [Chem-Y GmbH] Nonoxynol-30; CAS 9016-45-9; nonionic; emulsifier for calcium stearate, emulsion polymerization; liq.; 70% act.

Akyporox NP 1200V. [Chem-Y GmbH] Nonoxynol-120; CAS 9016-45-9; nonionic; emulsifier for emulsion polymerization; Gardner 2 clear liq.; sp.gr. 1.08; visc. 1000 mPa•s; HLB 19.2; cloud pt. > 100 C (1%); pH 7-8; surf. tens. 52 mN/m; 50% act.

Akyporox OP 40. [Chem-Y GmbH] Alkylphenol polyglycol ether; nonionic; emulsifier; liq.; 100% conc.

Akyporox OP 100. [Chem-Y GmbH] Octoxynol-10; CAS 9002-93-1; nonionic; emulsifier; dust suppressant for coal mining industry; liq.; sp.gr. 1.04; cloud pt. 70 C (1%); HLB 13.6; surf. tens. 30 mN/m; 100% act.

Akyporox OP 115SPC. [Chem-Y GmbH] Octoxynol-12; CAS 9002-93-1; nonionic; emulsifier, wetting agent; liq.; 70% conc.

Akyporox OP 200. [Chem-Y GmbH] Octoxynol-20.; CAS 9002-93-1; nonionic; emulsifier.

Akyporox OP 250 V. [Chem-Y GmbH] Octoxynol-25; CAS 9002-93-1; nonionic; emulsifier for emulsion polymerization; perfume solubilizer; Gardner 2 clear liq.; sp.gr. 1.07; visc. 1000 mPa•s; cloud pt. > 100 C (1%); HLB 17.3; pH 8-9 (5%); surf. tens. 40 mN/m; 70% act.

Akyporox OP 400V. [Chem-Y GmbH] Octoxynol-40; CAS 9002-93-1; nonionic; emulsifier for emulsion polymerization; Gardner 2 clear liq.; sp.gr. 1.1; visc. 1200 mPa•s; cloud pt. > 100 C (1%); HLB 17.9; pH 7-8.5 (5%); surf. tens. 45 mM/m; 70% conc.

Akyporox RLM 22. [Chem-Y GmbH] Laureth-2; CAS 3055-93-4; nonionic; emulsifier for mfg. of hair dye formulations; additive for mfg. of snow from spray cans; liq.; oil-sol.; sp.gr. 0.91; HLB 6.3; surf. tens. 29 mN/m; 100% act.

Akyporox RLM 40. [Chem-Y GmbH] Laureth-4; CAS 5274-68-0; nonionic; emulsifier for mfg. of hair dye formulations, cosmetic aerosols, oil bath formulations, window cleaners, hand cleaners, heavy-duty detergents; liq.; sp.gr. 0.95; HLB 9.4; surf. tens. 29 mN/m; 100% act.

Akyporox RLM 80. [Chem-Y GmbH] Laureth-8; CAS 9002-92-0; nonionic; emulsifier; paste; sp.gr. 1.0; m.p. 31 C; HLB 12.8; cloud pt. > 100 C (1%); surf. tens. 33 mN/m; 100% act.

Akyporox RLM 80V. [Chem-Y GmbH] Laureth-8; nonionic; emulsifier for mfg. of cosmetic aerosols, heavy-duty detergents, all-purpose cleaners; liq.; 85% act.

Akyporox RLM 160. [Chem-Y GmbH] Laureth-16; CAS 9002-92-0; nonionic; emulsifier for emulsion polymerization; liq.; sp.gr. 1.04; m.p. 35 C; cloud pt. > 100 C (1%); HLB 15.6; surf. tens. 45 mN/m; 100% act.

Akyporox RO 90. [Chem-Y GmbH] Oleth-9; CAS 9004-98-2; nonionic; emulsifier, wetting agent for textile prods., metalworking fluids, all-purpose cleaners, hand cleaners, creams and lotions; liq.; sp.gr. 0.99; HLB 12.1; cloud pt. > 60 C (1%); surf. tens. 34 mN/m; 100% act.

Akyporox RTO 70. [Chem-Y GmbH] Oleth-7; CAS 9004-98-2; nonionic; emulsifier for emulsion polymerization; liq.; 100% conc.

Akyporox RZO 30. [Chem-Y GmbH] PEG-3 castor oil; surfactant for metalworking cooling lubricants; liq.; 100% act.

Akyporox SAL SAS. [Chem-Y GmbH] Sodium lauryl sulfate; CAS 142-87-0; anionic; detergent, shampoo base, emulsifier for emulsion polymerization; liq.; 28% conc.

Akyposal 100 DE. [Chem-Y GmbH] Lauryl ether sulfate; anionic; emulsifier for pesticides; liq.; 100% conc.

Akyposal 100 DEG. [Chem-Y GmbH] Lauryl ether sulfate; anionic; emulsifier; liq.; 100% conc.

Akyposal 9278 R. [Chem-Y GmbH] Sodium laureth sulfate; CAS 9004-82-4; emulsifier for emulsion polymerization; liq.; 30% act.

Akyposal ALS 33. [Chem-Y GmbH] Ammonium lauryl sulfate; CAS 2235-54-3; anionic; detergent, emulsifier; shampoo base; used in emulsion polymerization; liq.; 33% conc.

Akyposal BA 28. [Chem-Y GmbH] Sodium trideceth sulfate; anionic; detergent for personal care, dishwashing, and textile prods.; liq.; 28% conc.

Akyposal BD. [Chem-Y GmbH] Sodium octoxynol-6 sulfate; CAS 69011-84-3; anionic; emulsifier for emulsion polymerization; Gardner 5 clear liq.; sp.gr. 1.0; visc. 100 mPa•s; pH 2-4; 32% act.

Akyposal EO 20 MW. [Chem-Y GmbH] Sodium laureth sulfate; CAS 9004-82-4; anionic; detergent, base for shampoos and bubble baths, emulsifier for emulsion polymerization; liq.; 28% conc.

Akyposal EO 20 PA. [Chem-Y GmbH] Sodium laureth sulfate; CAS 9004-82-4; anionic; detergent; liq.; sol. in cold water; 56% conc.

Akyposal EO 20 PA/TS. [Chem-Y GmbH] Sodium laureth sulfate; CAS 9004-82-4; anionic; detergent; liq.; sol. in cold water; 59% conc.

Akyposal EO 20 SF. [Chem-Y GmbH] Sodium laureth sulfate; CAS 9004-82-4; anionic; foaming agent for neutralizing liqs.; liq.; 28% conc.

Akyposal MS SPC. [Chem-Y GmbH] Sodium laureth sulfate; CAS 9004-82-4; anionic; detergent; liq.; 22% conc.

Akyposal NAF. [Chem-Y GmbH] Sodium dodecylbenzene sulfonate; emulsifier for emulsion polymerization; liq.; 24% act.

Akyposal NLS. [Chem-Y GmbH] Sodium lauryl sulfate; CAS 151-21-3; anionic; detergent; liq.; 31% conc.

Akyposal NPS 60. [Chem-Y GmbH] Sodium nonoxynol-6 sulfate; anionic; emulsifier for emulsion polymerization; Gardner 3 clear liq.; sp.gr. 1.05; visc. 15,000 mPa•s; pH 6-7 (10%); 32% act.

Akyposal NPS 100. [Chem-Y GmbH] Sodium nonoxynol-10 sulfate; CAS 9014-90-8; anionic; emulsifier for polymerization; Gardner 3 clear liq.; sp.gr. 1.05; visc. 500 mPa•s; pH 6-7; 34% conc.

Akyposal NPS 250. [Chem-Y GmbH] Sodium nonoxynol-25 sulfate; CAS 9014-90-8; anionic; emulsifier for polymerization; liq.; 34% conc.

Akyposal OP 80. [Chem-Y GmbH] Octoxynol-9 carboxylic acid; emulsifier for emulsion polymerization; liq.; 90% act.

Akyposal RLM 56 S. [Chem-Y GmbH] Fatty alcohol ether sulfate; CAS 3088-31-1; anionic; detergent for personal care prods., liq. soaps, dishwashing prods.; liq.; 56% conc.

Akyposal TLS 42. [Chem-Y GmbH] TEA-lauryl sulfate; anionic; detergent; base for personal care prods. and car shampoos; foaming agent for agrochemicals, fire extinguishers; liq.; 42% act.

Akypo®-Soft 100 MgV. [Chem-Y GmbH] Magnesium laureth-11 carboxylate; CAS 99330-44-6; anionic; detergent, emulsifier, wetting agent for shampoos, liq. soaps, foam baths, low irritation formulas; biodeg.; water-wh. clear liq.; sp.gr. 1.0 kg/l; visc. 500 mPa•s; pH 5-6; 22% act.

Akypo®-Soft 100 NV. [Chem-Y GmbH] Sodium laureth-11 carboxylate; CAS 68987-89-3; anionic; detergent, emulsifier, wetting agent for shampoos, liq. soaps, foam baths, low irritation formulas; biodeg.; water-wh. clear liq.; sp.gr. 1.0 kg/l; visc. 100 mPa•s; pH 6.5-7.5; toxicology: nonirritant; 22% act.

Akypo®-Soft 130 NV. [Chem-Y GmbH] Sodium laureth-14 carboxylate; CAS 68987-89-3; anionic; detergent, emulsifier, wetting agent for shampoos, liq. soaps, foam baths, low irritation formulas; liq.; 22% conc.

Akypo®-Soft 160 NV. [Chem-Y GmbH] Sodium laureth-17 carboxylate; CAS 68987-89-3; anionic; detergent, emulsifier, wetting agent for shampoos, liq. soaps, foam baths, low irritation formulas; biodeg.; water-wh. clear liq.; sp.gr. 1.0 kg/l; visc. 100 mPa•s; pH 6-7; 22% act.

Akypo® Soft KA 250 BVC. [Chem-Y GmbH] Sodium PEG-6 cocamide carboxylate; economical surfactant for prep. of mild shampoos, foam baths, shower baths, liq. soaps not irritating to optic mucosa, mild dishwash formulations; liq.; 52% act.

AL 2070. [ICI Am.] POE/POP block polymer; nonionic; dispersant for agric. formulations; solid; 100% conc.

Alarsol AL. [Auschem SpA] Linear dodecylbenzene sulfonic acid; CAS 27172-87-0; anionic; general detergent; liq.; 100% conc.

Albalan. [Westbrook Lanolin] Lanolin wax; nonionic; emollient, emulsifier; forms stable w/o emulsions; soft wax; HLB 5.0; 100% conc.

Albatex® FFC. [Ciba-Geigy/Dyestuffs] Hydrotropic proprietary blend; anionic; deaerator and penetrant for alkaline bleach or dye systems; liq.

Albatex® OR. [Ciba-Geigy/Dyestuffs] Low m.w. polyamide; nonionic; leveling and penetrating agent for vat dyes; liq.

Albegal® A. [Ciba-Geigy/Dyestuffs] Polyglycol ether deriv.; amphoteric; nonfoaming wool leveling agent for premetallized, acid, chrome, and selected reactive dyes; liq.

Albegal® B. [Ciba-Geigy/Dyestuffs] Polyglycol ether deriv.; amphoteric; nonfoaming wool leveling agent for 1:2 metal complex and selected reactive dyes; liq.

Albegal® BMD. [Ciba-Geigy/Dyestuffs] Hydrotropic proprietary blend; nonionic; penetrant and deaerator in beam and pkg. dyeing machines and in fabric padding; visc. liq.

Albigen® A. [BASF] Nonionic; textile stripping agent, leveling agent, washing-off agent; prevents re-exhaustion of dyes in washing of prints.

Albrite MALP. [Albright & Wilson UK] Mono aluminum orthophosphate; refractory bonding agent.

Alcocare® 6000. [Rhone-Poulenc Surf.] Formulated sanitizer conc.

Alcodet® 218 (formerly Siponic® 218). [Rhone-Poulenc Surf.; Rhone-Poulenc France] PEG-10 isolauryl thioether; nonionic; emulsifier, wetting agent, detergent, carbon soil and grease cleaners, metal cleaning specialties, textile scouring, steel processing, insecticide emulsions, cosmetics, wood and paper industries; Gardner 6 max. liq.; sol. in water and alkaline detergent builders; sp.gr. 1.05; dens. 8.6 lb/gal; HLB 13.9; cloud pt. 52 C (1% aq.); surf. tens. 28 dynes/cm (0.05%); 100% conc.

Alcodet® 260 (formerly Siponic® 260). [Rhone-Poulenc Surf.] PEG-6 isolauryl thioether; nonionic; emulsifier, wetting agent, detergent, carbon soil and grease cleaners, metal cleaning specialties, textile scouring, insecticide emulsions, cosmetics, wood pulp and paper industries; Gardner 6 max. liq.; sol. in org. solvs.; sp.gr. 1.01; dens. 8.4 lb/gal; HLB 11.0; surf. tens. 29 dynes/cm (0.05%); 100% conc.

Alcodet® HSC-1000. [Rhone-Poulenc Surf.] POE thioether; CAS 9004-83-5; nonionic; emulsifier, wetting agent, detergent, hard surf. cleaner, carbon soil and grease cleaner; low odor liq.; HLB 12.0; cloud pt. 19 C (1% aq.); 100% conc.

Alcodet® IL-3500. [Rhone-Poulenc Surf.] POE thioether; CAS 9004-83-5; nonionic; emulsifier, wetting agent, detergent, carbon soil and grease cleaner; liq.; HLB 12.0; 100% conc.

Alcodet® MC-2000. [Rhone-Poulenc Surf.] POE thioether; CAS 9004-83-5; nonionic; emulsifier, wetting agent, detergent, corrosion inhibitor, carbon soil and grease cleaner, metal cleaner; low odor liq.; oil-sol.; HLB 12.0; 98% act.

Alcodet® SK (formerly Siponic® SK). [Rhone-Poulenc Surf.] PEG-8 isolauryl thioether; nonionic; emulsifier, wetting agent, detergent, carbon soil and grease cleaners, metal cleaning specialties, textile scouring, insecticide emulsions, paints, hair prods., wood and paper industries; emulsifier for petrol oils, chlorinated solvs., silicones; Gardner 6 max. liq.; sol. in water; sp.gr. 1.03; dens. 8.5 lb/gal; HLB 12.7; cloud pt. 28 C (1% aq.); surf. tens. 31 dynes/cm (0.05%); 100% conc.

Alcodet® TX 4000. [Rhone-Poulenc Surf.] POE thioether; CAS 9004-83-5; nonionic; wool scour/degreaser, emulsifier, detergent for poly/cotton blend scouring; liq.; HLB 12.1; 100% conc.

Alcojet®. [Alconox] Sodium metasilicate, sodium carbonate, POE ester of mixed fatty and resin acids; nonionic; detergent used in mechanical washers for stain removal; for cleaning healthcare instruments, laboratory ware, electronic components, pharmaceutical apparatus, industrial parts, etc.; wh. powd. and gran., sl. acrid odor; moderate water-sol.; pH 12.0 (1%); surf. tens. 35 dyne/cm (1%); biodeg.; toxicology: moderate eye irritant if not rinsed; powd. potential irritant by inhalation; nontoxic orally; 100% conc.

Alcolec® 439-C. [Am. Lecithin] Lecithin; CAS 8002-43-5; nonionic; wetting agent, emulsifier, release agent; for water-based paints, coatings, textiles; amber liq.; bland odor; sol. in fat solvs. except acetone; water-disp.; sp.gr. 1.03; visc. 7000 cps; acid no. 30 max.; 100% act.

Alcolec® 440-WD. [Am. Lecithin] Lecithin; CAS 8002-43-5; nonionic; emulsifier, stabilizer, visc. control agent, wetting agent, pigment grinding aid and dispersant for water-based paints, coatings; Gardner 14 liq.; visc. 10,000 cP; acid no. 1.0.; 100% act.

Alcolec® 495. [Am. Lecithin] Lecithin; CAS 8002-43-5; nonionic; o/w emulsifier, wetting agent for aq. and oil-base systems; approved for food use; amber fluid; visc. < 10,000 cP; HLB 5-6; acid no. < 25; 100% conc.

Alcolec® BS. [Am. Lecithin] Single bleached lecithin; CAS 8002-43-5; nonionic; commercial lecithin emulsifier, wetting and dispersing agent, stabilizer, release and lubricating agent, foam suppressant, solubilizer for food and industrial applics.; choline source; Gardner 14 liq.; acid no. 32 max.

Alcolec® Extra A. [Am. Lecithin] Lecithin; CAS 8002-43-5; nonionic; emulsifier for aq. and oil-base systems, food industry; higher phosphatide content; liq.; 100% conc.

Alcolec® F-100. [Am. Lecithin] De-oiled lecithin; CAS 8002-43-5; emulsifier for industrial and food applics.; instantizing for milk powd., cake mixes, etc.; choline source; lt. tan/yel. powd.; acid no. 36 max.

Alcolec® FF-100. [Am. Lecithin] De-oiled lecithin; CAS 8002-43-5; emulsifier for industrial and food applics.; instantizing for milk powd., cake mixes, etc.; choline source; lt. tan/yel. fine powd.; acid no. 36 max.

Alcolec® Granules. [Am. Lecithin] Lecithin; CAS 8002-43-5; nonionic; wetting agent, release agent, emulsifier, stabilizer, diet supplement, in industrial, cosmetics, pharmaceuticals, food applics.; instantizing for milk powd., cake mixes, etc.; choline source; lt. tan/yel. gran.; bland odor and taste; sp.gr. 0.5; acid no. 36 max.; 97% act.

Alcolec® S. [Am. Lecithin] Unbleached lecithin; CAS 8002-43-5; nonionic; commercial lecithin emulsifier, wetting and dispersing agent, stabilizer, release and lubricating agent, foam suppressant, solubilizer for food and industrial applics.; choline source; Gardner 17 liq.; acid no. 32 max.

Alconate® 2CH. [Rhone-Poulenc Surf.] Dicyclohexyl sodium sulfosuccinate; anionic; emulsifier in S/B emulsion polymerization; paste; 43% act. DISCONTINUED.

Alconate® CPA. [Rhone-Poulenc Surf.] Disodium cocamido MIPA sulfosuccinate; anionic; surfactant for personal care products, rug and upholstery shampoos; Gardner 3 clear liq.; dens. 8.9 lb/gal; cloud pt. –5 C; pH 6.5 (5%); 40% solids. DISCONTINUED.

Alconate® SBDO (see Geropon® SS-O-75). [Rhone-Poulenc Surf.] Dioctyl sodium sulfosuccinate; anionic; wetting agent, surfactant, dispersant, and penetrant for industrial and mining applics.; liq.; 70% act.

Alconate® SBL 203 (redesignated Geropon® SBL 203). [Rhone-Poulenc Surf.].

Alconate® SBN-862. [Rhone-Poulenc Surf.] Disodium nonoxynol-10 sulfosuccinate; emulsifier, dispersant for emulsion polymerization of vinyl acetate and acrylates; liq.; 32% act. DISCONTINUED.

Alconate® SBTA-269. [Rhone-Poulenc Surf.] Disodium alkyl (C18) sulfosuccinamate; foam generator for latex systems; liq.; 36% act. DISCONTINUED.

Alconox®. [Alconox] Blend of alkylaryl sulfonates, lauryl alcohol sulfates, phosphates, carbonates; anionic; detergent with wetting, sequestering and synergistic agents; for manual cleaning of laboratory and hospital glassware and instruments; colorless powd.; odorless; sol. in water; flash pt. none; pH 9.5 (1%); surf. tens. 32 dyne/cm (1%); biodeg.; toxicology: mild to moderate eye irritant if not rinsed; powd. potential irritant by inhalation; nontoxic orally; 100% act.

Alcotabs®. [Alconox] Blend of alkylaryl sulfonates, lauryl alcohol sulfates, phosphates, carbonates; anionic; detergent with wetting, sequestering and synergistic agents; for cleaning pipettes and tubes in hospital, clinical, education, R&D, and industrial laboratories; effervescent tablet; flash pt. none; pH 6.5 (1%); surf. tens. 32 dyne/cm (1%); biodeg.; toxicology: mild to moderate eye irritant if not rinsed; nonirritating by inhalation; nontoxic orally; 100% active.

Aldo® DC. [Lonza] Propylene glycol dicaprylate/dicaprate; nonionic; coupling agent for mixed solv. systems; emulsifier for food, cosmetic, industrial use; Gardner 2 max. liq.; sol. in ethanol, min. and veg. oil; insol. in water; sp.gr. 0.92; HLB 2 ± 1; iodine no. 0.5 max.; sapon. no. 315–335.; 100% conc.

Aldo® HMS. [Lonza] Glyceryl stearate; CAS 123-94-4; nonionic; emulsifier for w/o systems; stabilizer, consistency builder, emollient in personal care prods.; wh. beads; sol. in ethanol, min. and veg. oil; disp. in water; HLB 2.8; m.p. 61–68 C; sapon. no. 165–175; 52–56% mono content.

Aldo® MC. [Lonza] Glyceryl cocoate; nonionic; emulsifier; cream-colored, soft solid; sol. in ethanol, ethyl acetate, toluol, naphtha, min. and veg. oils, disp. in water; sp.gr. 0.97; m.p. 20–26 C; HLB 6.8; sapon. 180–190; pH 7.5–8.5 (5% aq); 100% conc.

Aldo® ML. [Lonza] Glyceryl monodilaurate; nonionic; coupling agent for mixed solv. systems; emulsifier for food, cosmetic, industrial use; liq.; HLB 5.2; 100% conc.

Aldo® MLD. [Lonza] Glyceryl laurate, dispersible; nonionic; emulsifier for cosmetic, pharmaceutical and industrial use; cream soft solid; sol. in ethanol, ethyl acetate, toluol, naphtha, min. and veg. oils, disp. in water; sp.gr. 0.97; m.p. 21–26 C; HLB 6.8; sapon. no. 185–195; pH 7.5–8.5 (5% aq.); 100% conc.

Aldo® MO. [Lonza] Glyceryl oleate; CAS 111-03-5; nonionic; emulsifier, defoamer; yel., soft solid; sol. in ethyl acetate; sp.gr. 0.95; m.p. < 25 C; HLB 3.4; sapon. no. 170–180; pH 4.5–6.5; 100% conc.

Aldo® MOD. [Lonza] Glyceryl oleate SE; nonionic; emulsifier; avail. in animal, veg., food, and Kosher grades; yel. liq.; sol. in toluol, naphtha, min. and veg. oils, disp. in water; sp.gr. 0.95; m.p. < 0 C; HLB 5.0; sapon. no. 141–147; pH 8.5–9.5 (5% aq.); 100% conc.

Aldo® MOD FG. [Lonza] Glyceryl monodioleate, disp.; nonionic; emulsifier and antifoam for foods; emulsifier, solubilizer for cosmetic, pharmaceutical, and industrial applics.; liq.; HLB 4.0; 100% conc.

Aldo® MO FG. [Lonza] Glyceryl mono- and dioleate; nonionic; emulsifier and antifoam for foods; emulsifier, solubilizer for cosmetic, pharmaceutical, and industrial applics.; liq.; HLB 3.0; 100% conc.

Aldo® MO Tech. [Lonza] Glyceryl oleate, tech. grade; CAS 111-03-5; nonionic; emulsifier; amber soft liq.; sol. in ethyl acetate; sp.gr. 0.95; m.p. < 25 C; HLB 3.4; sapon. no. 170–180; pH 4.5–6.5; 100% conc.

Aldo® MR. [Lonza] Glyceryl ricinoleate; nonionic; emulsifier, solubilizer for cosmetic, pharmaceutical and industrial applics.; yel. liq.; sol. in methanol, ethanol, ethyl acetate, toluol, veg. oil, disp. in water; sp.gr. 1.02; m.p. < –8 C; HLB 6.0; sapon. no. 66–69; pH 8.3–9.3; 100% conc.

Aldo® MS. [Lonza] Glyceryl stearate; CAS 123-94-4; nonionic; emulsifier for cosmetic, pharmaceutical and industrial use; wh. beads; sol. hot in methanol, ethanol, toluol, naphtha, min. and veg. oils, disp. in hot water; sp.gr. 0.97; m.p. 57–61 C; HLB 4.0; sapon. no. 158–165; pH 7.6–8.6 (3% aq.); 100% conc.

Aldo® MS-20 FG. [Lonza] PEG-20 glyceryl stearate; nonionic; emulsifier, antifoam for foods; emulsifier, solubilizer for cosmetic, pharmaceutical, and industrial applics.; solid; HLB 13.0; 100% conc.

Aldo® MS Industrial. [Lonza] Glyceryl stearate, industrial grade; CAS 123-94-4; nonionic; emulsifier; tan flakes; sol. hot in methanol, ethanol, toluol, naphtha, min. and veg. oils, disp. in hot water; sp.gr. 0.97; m.p. 55–62 C; HLB 3.6; sapon. no. 157–177; pH 6.5–8.5 (5% aq.); 100% conc.

Aldo® MS LG. [Lonza] Glyceryl stearate; CAS 123-94-4; nonionic; emulsifier for w/o systems; stabilizer, consistency builder, emollient in personal care prods.; wh. beads; sol. in ethanol, min. and veg. oil; disp. in water; HLB 3.3; m.p. 58–62 C; sapon. no. 160–170; 40–45% mono content.

Aldo® MSA. [Lonza] Glyceryl stearate; CAS 123-94-4; nonionic; emulsifier for o/w personal care prods., cosmetics, pharmaceuticals, industrial use; wh. beads; sol. in ethanol, min. and veg. oil; disp. in water; HLB 11 ± 1; m.p. 56–60 C; sapon. no. 90–100; 100% conc., 17.5% min. mono content.

Aldo® MSD. [Lonza] Glyceryl stearate SE; CAS 123-94-4; nonionic; emulsifier, solubilizer for cosmetic, pharmaceutical and industrial applics.; wh. beads; sol. hot in methanol, ethanol, toluol, naphtha, min. and veg. oils, disp. in hot water; sp.gr. 0.97; m.p. 56–60 C; HLB 6.0; sapon. no. 140–150; pH 9.2–10.2 (3% aq.); 100% conc.

Aldo® PMS. [Lonza] Propylene glycol stearate; nonionic; emulsifier, stabilizer, solubilizer, emollient, lubricant, thickener, plasticizing agent, antiirritant, and conditioner in pharmaceutical, cosmetic, and toiletry preparations; foam modifier for detergent systems; wh. flakes; sol. in ethanol, min. and veg. oils; insol. in water; HLB 3 ± 1; iodine no. 1.0 max.; sapon. no. 170–185.

Aldosperse® ML 23. [Lonza] PEG-23 glyceryl laurate; nonionic; emulsifier, solubilizer, suspending and dispersing agent used in personal care prods., textiles; straw liq.; sol. in ethanol, veg. oil; sp.gr. 1.09; HLB 17 ± 1; acid no. 3; sapon. no. 42–50.

Aldosperse® MS-20. [Lonza] PEG-20 glyceryl stearate; CAS 51158-08-8; nonionic; emulsifier, solubilizer, suspending and dispersing agent used in personal care prods., food industry, textiles; wh. soft solid; sol. in ethanol; disp. in water; HLB 13 ± 1; m.p. 32–38 C; acid no. 2; sapon. no. 65–75; hyd. no. 73.

Aldosperse® O-20 FG. [Lonza] 80% Glyceryl stearate, 20% polysorbate 80; nonionic; emulsifier, solubilizer for frozen desserts, cosmetic, pharmaceutical, and industrial applics.; bead; HLB 5.0; 100% conc.

Alfol® 4. [Vista] n-Butyl alcohol; intermediate for

amines, plasticizers, fatty acid esters, detergents; colorless liq.; char. penetrating odor; sol. in water; dens. 6.76; sp.gr. 0.809; visc. 3.4; m.p. –128 C; b.p. 117.7; 99.3% act.

Alfol® 6. [Vista] Hexyl alcohol; detergent and plasticizer intermediate; colorless liq.; dens. 6.83; sp.gr. 0.823; visc. 5.5; m.p. –49 C; b.p. 315 C; 98.5% act.

Alfol® 8. [Vista] Caprylic alcohol; wetting agent, emulsifier, surfactant intermediate, raw material; colorless liq.; mild aromatic odor; dens. 6.86; sp.gr. 0.823; m.p. 5 C; b.p. 383 C; 99.2% act.

Alfol® 10. [Vista] Decyl alcohol; emulsifier, surfactant intermediate; colorless liq., wh. waxy solid; sweet, fat-like odor; sol. in alcohol, ether, benzene, glacial acetic acid; water insol.; dens. 6.90; sp.gr. 0.828; m.p. 44 C; b.p. 448 C; 99% act.

Alfol® 12. [Vista] Lauryl alcohol; CAS 112-53-8; detergent intermediate; colorless liq.; sp.gr. 0.831; m.p. 75 C; b.p. 498 C; 99.1% act.

Alfol® 14. [Vista] Myristyl alcohol; CAS 112-72-1; detergent intermediate; wh. solid, cryst.; typ. fatty alcohol odor; sp.gr. 0.824; m.p. 99 C; b.p. 552 C; 99% act.

Alfol® 16. [Vista] Cetyl alcohol; CAS 36653-82-4; detergent intermediate; emollient used in cosmetics; plastics additive; wh. waxy solid; typ. fatty alcohol odor; sol. in alcohol, acetone, ether; water-insol.; dens. 6.77; sp.gr. 0.813; visc. 6.77; m.p. 117 C; b.p. 604 C; 98.9% act.

Alfol® 16 NF. [Vista] Cetyl alcohol; CAS 36653-82-4; biodeg. detergent intermediate; waxy solid; 100% conc.

Alfol® 18. [Vista] Stearyl alcohol; CAS 112-92-5; surfactant intermediate; emollient used in cosmetics; plastics additive; wh. waxy solid; typ. fatty alcohol odor; sol. in alcohol, acetone, ether; water-insol.; dens. 6.71; sp.gr. 0.8075; visc. 13.5; m.p. 135 C; b.p. 640 C; 98.7% act.

Alfol® 18 NF. [Vista] Stearyl alcohol; CAS 112-92-5; biodeg. detergent intermediate; waxy solid; 100% conc.

Alfol® 20+. [Vista] Blend of C20 and higher even-carbon-number primary linear alcohols; detergent intermediate; lubricant, defoamer; emollient; used in fuel oil; waxes and polishers used in paper pulp defoamers; off-wh. solid; sol. in alcohol, acetone, ether; water-insol.; sol. in alcohol, acetone, ether; water-insol.; dens. 6.80; sp.gr. 0.817; m.p. 131; 100% conc.

Alfol® 22+. [Vista] C22 and higher linear alcohols; biodeg. detergent intermediate; lubricant, defoamer; emollient; used in fuel oil; waxes and polishers used in paper pulp defoamers; wh. solid; sol. in alcohol, acetone, ether; water-insol.; 100% conc.

Alfol® 610. [Vista] C6-C10 linear primary alcohol; intermediate for surfactants, plasticizers; colorless liq.; mild, sweet odor; sol. in alcohol, ether, benzene, insol. in water; dens. 6.92; sp.gr. 0.829; visc. 11.0; m.p. –8 C; b.p. 350 C; 99% act.

Alfol® 610ADE. [Vista] C6, C8, C10 alcohol blend; chemical intermediate; also for lube oil additives, plasticizers, surfactant feedstocks; clear colorless liq.; m.w. 140; 100% conc.

Alfol® 610 AFC. [Vista] C6-C10 linear primary alcohol; biodeg. detergent intermediate, plasticizer intermediate; liq.; 100% conc.

Alfol® 810. [Vista] C8–C10 linear primary alcohol; intermediate for plasticizers and biodeg. surfactants; colorless liq.; mild, sweet odor; sol. in alcohol, ether, benzene; dens. 6.93; sp.gr. 0.831; visc. 8.9; m.p. 7 C; b.p. 400 C; 99% act.

Alfol® 810FD. [Vista] C8, C10 alcohol blend; chemical intermediate; also for lube oil additives, plasticizers, surfactant feedstocks; clear colorless liq.; m.w. 141; 100% conc.

Alfol® 1012 CDC, 1012 HA. [Vista] C10–C12 linear primary alcohol; intermediate for surfactants, plasticizer; lubricant and metal rolling rolls; colorless liq.; sweet odor; sol. in alcohol, acetone, ether; water-insol.; dens. 7.0; sp.gr. 0.834; visc. 15; m.p. 35–40 C; b.p. 427 C; 99% act.

Alfol® 1214. [Vista] C12–C14 linear primary alcohol; intermediate for surfactants; lubricant and metal rolling rolls; biodeg.; colorless liq.; sweet, typ. fatty alcohol odor; sol. see Alfol 20; dens. 7.0; sp.gr. 0.838; visc. 14.3; m.p. 73–75 C; b.p. 518 C; 99% act.

Alfol® 1214 GC. [Vista] C12 and C14 alcohol blend; chemical intermediate; also for lube oil additives, plasticizers, surfactant feedstocks; biodeg.; clear colorless liq.; m.w. 195; 100% conc.

Alfol® 1216. [Vista] C12–C16 linear alcohols; intermediate for surfactants, plastics; lubricant and metal rolling rolls; clear, colorless liq.; sweet, typ. fatty alcohol odor; sol. see Alfol 20; dens. 7.0; sp.gr. 0.830; visc. 15.5; m.p. 64–70 C; b.p. 514 C; 98.5% act.

Alfol® 1216 CO. [Vista] C12 and C14 alcohol blend; chemical intermediate; also for lube oil additives, plasticizers, surfactant feedstocks; clear colorless liq.; m.w. 198; 100% conc.

Alfol® 1216 DCBA. [Vista] C12-16 linear primary alcohol; biodeg. detergent intermediate; liq.; 100% conc.

Alfol® 1412. [Vista] C12-C14 linear primary alcohol; chemical intermediate; also for lube oil additives, plasticizers, surfactant feedstocks; biodeg.; clear colorless liq.; m.w. 205; 100% conc.

Alfol® 1418 DDB. [Vista] C14, C16, C18 alcohol blend; chemical intermediate; also for lube oil additives, plasticizers, surfactant feedstocks; biodeg.; wh. solid; m.w. 243; 100% conc.

Alfol® 1618. [Vista] C16–C18 linear primary alcohol; biodeg. intermediate for surfactants; wh. waxy solid; typ. fatty alcohol odor; sol. in water, alcohol, chloroform, ether, benzene, glacial acetic acid; dens. 6.81; sp.gr. 0.820; visc. 22; m.p. 110–115 C; b.p. 609 C; 98.7% act.

Alfol® 1618 CG. [Vista] C16-C18 linear primary alcohol; biodeg. chemical intermediate; also for lube oil additives, plasticizers, surfactant feedstocks; wh. solid; m.w. 261; 100% conc.

Alfol® 1620. [Vista] C16–C20 linear primary alcohol; biodeg. intermediate for surfactants; wh. waxy solid; typ. fatty alcohol odor; dens. 6.81; sp.gr. 0.817; visc. 13.5; m.p. 110–113 C; b.p. 605 C; 98.5% act.

Alfol® C22. [Vista] Blend of C22 and higher even-carbon-number primary linear alcohols; lubricant, defoamer; off-wh. solid; dens. 6.81; sp.gr. 0.817; m.p. 132.

Alfonic® 610-50R. [Vista] C6-10 pareth-3; nonionic; surfactant intermediate, emulsifier, hard surf. cleaner, oil well applics.; clear liq.; HLB 10.0; 100% conc.

Alfonic® 810-40. [Vista] C8-10 pareth-2.2; nonionic; surfactant intermediate, emulsifier, hard surf. cleaner, oil well applics.; biodeg.; clear liq.; HLB 8.0; 100% conc.

Alfonic® 810-60. [Vista] C8-10 pareth-5; nonionic; surfactant intermediate, emulsifier, hard surf.

cleaner, oil well applics.; biodeg.; clear liq.; HLB 12.0; 100% conc.

Alfonic® 1012-40. [Vista] C10-12 pareth-2.5; nonionic; surfactant intermediate, dispersant, foam stabilizer, hard surf. cleaner; biodeg.; clear liq.; water-sol.; HLB 8.0; 100% conc.

Alfonic® 1012-60. [Vista] C10-12 pareth-5; nonionic; detergent, wetting agent, emulsifier, foaming agent, foam stabilizer, intermediate, dispersant, hard surf. cleaner; biodeg.; APHA 20 liq.; sweet odor; dens. 8.2; sp.gr. 0.98; visc. 20.5 cps; m.p. 36; HLB 12.0; 100% conc.

Alfonic® 1214-GC-30. [Vista] C12-14 pareth-2; surfactant intermediate; clear liq.; HLB 6.2.

Alfonic® 1214-GC-40. [Vista] C12-14 pareth-3; surfactant intermediate; clear liq.; HLB 7.8.

Alfonic® 1216-22. [Vista] C12-16 pareth-1.3; nonionic; surfactant intermediate, emulsifier; biodeg.; clear liq.; HLB 4.4; 100% conc.

Alfonic® 1216-30. [Vista] C12-16 pareth-2; nonionic; surfactant, emulsifier; biodeg.; liq.; HLB 6.0; 100% conc.

Alfonic® 1412-40. [Vista] C14-12 pareth-3; nonionic; detergent intermediate, emulsifier, dispersant; biodeg.; clear liq.; slight odor; oil-sol.; dens. 7.64; sp.gr. 0.91; visc. 52 cps; m.p. 41; HLB 8.0; 100% act.

Alfonic® 1412-60. [Vista] C14-12 pareth-7; nonionic; detergent, wetting agent, emulsifier, foaming agent, surfactant; biodeg.; cloudy liq.; sweet, mild odor; dens. 8.2; sp.gr. 0.98; visc. 33 cps; HLB 12.0; 100% act.

Alfonic® 1412-A Ether Sulfate. [Vista] Sulfated ethoxylated alcohol deriv., ammonium salt; anionic; detergent, wetting agent, emulsifier, foaming agent for detergent formulations, shampoos, emulsion polymerization, oil well drilling, agriculture; clear liq.; mild odor; dens. 8.4; visc. 76 cps; biodeg.; 58% act.

Alfonic® 1412-S Ether Sulfate. [Vista] Sulfated ethoxylated alcohol deriv.; anionic; detergent, wetting agent, emulsifier, foaming agent; clear liq.; mild odor; dens. 8.8; visc. 64 cps; biodeg.; 38% act.

Alfonic® 1618-65. [Vista] Ethoxylated linear alcohols (65% E.O.); nonionic; detergent, emulsifier, foaming and wetting agent; near wh. solid; sol. in water, ethanol, monobutyl glycol ether, IPA, xylene, perchlorethylene, slight in kerosene, insol. in min. oil; dens. 8.2; sp.gr. 0.98; visc. 81 cps; m.p. 88; b.p. 234; biodeg.; 100% act.

Algepon AK. [Sandoz] Fatty quat. ammonium compd.; cationic; detergent stripping compd.; paste.

Alipal® CD-128 (redesignated Rhodapex CD-128). [Rhone-Poulenc Surf.; Rhone-Poulenc France].

Alipal® CO-433 (redesignated Rhodapex® CO-433). [Rhone-Poulenc Surf.; Rhone-Poulenc France].

Alipal® CO-436 (redesignated Rhodapex® CO-436). [Rhone-Poulenc Surf.; Rhone-Poulenc France].

Alipal® EP-100. [Rhone-Poulenc Surf.] Ammonium salt of sulfated alkylphenol ethoxylate (4 EO); anionic; emulsifier for emulsion polymerization, surfactant for mild light-duty liqs.; liq.; surf. tens. 33 dynes/cm (@ CMC); 30% act. DISCONTINUED.

Alipal® EP-110 (redesignated Abex® EP-110). [Rhone-Poulenc Surf.; Rhone-Poulenc France].

Alipal® EP-115 (redesignated Abex® EP-115). [Rhone-Poulenc Surf.].

Alipal® EP-120 (redesignated Abex® EP-120). [Rhone-Poulenc Surf.; Rhone-Poulenc France].

Alipal® EP-227 (redesignated Abex EP-227). [Rhone-Poulenc Surf.].

Alkadet 15. [ICI Australia] Fatty alkyl polyglucoside; nonionic; biodeg. wetting agent, detergent for industrial hard surface cleaners and bottle washing; liq.; 70% conc.

Alkadet DCB-100 (see Rhodaterge DCB-100). [Rhone-Poulenc Surf.] Formulated prod.; anionic/nonionic; base for drycleaning charge soap yielding good soil suspension, water tolerance, cleaning efficiency, antistatic props.; liq.; 100% conc. DISCONTINUED.

Alkafoam D. [Rhone-Poulenc Surf.] Foam dyeing agent.

Alkali Surfactant NM. [Exxon/Tomah] Amphoteric; surfactant, wetting and coupling agent used in alkaline formulations for hard surf. cleaning, floor strippers, heavy-duty degreasers, steam, soak tank, and household/institutional cleaners; solubilizer for nonionics into high electrolyte cleaners; lt. amber liq.; sol. 5% in water, IPA, 5% sodium lauryl sulfate, 5% benzalkonium chloride, 10% sodium hydroxide; sp.gr. 1.04; pH 5–9 (5%); 35% act. in water.

Alkamide® 101 CG. [Rhone-Poulenc Surf.] Alkanolamide; nonionic; thickener, detergent, emulsifier for lower boiling aliphatic hydrocarbons; for cosmetic and industrial applics.; produces emulsions stable in presence of alcohols, glycols, and phenols; liq.; 100% conc. DISCONTINUED.

Alkamide® 200 CGN (formerly Cyclomide 200 CGN). [Rhone-Poulenc Surf.] Cocamide DEA; nonionic; thickener, detergent, emulsifier, foam stabilizer; for hard surf. cleaners, floor cleaners, rinsable degreasers, metal cleaners, metalworking compds.; corrosion inhibition chars.; liq.; 100% conc.

Alkamide® 206 CGN (formerly Cyclomide 206 CGN). [Rhone-Poulenc Surf.] Cocamide alkanolamide; anionic; emulsifier for fats and greases for dishwashing detergents; detergent for hard surf. cleaners; foam stabilizer, thickener; liq.; 100% conc.

Alkamide® 210 CGN (formerly Cyclomide 210 CGN). [Rhone-Poulenc Surf.] Cocamide alkanolamide; anionic/cationic; emulsifier for aliphatic hydrocarbons for rinsable degreasers; conveyor chain lubricant; liq.; 100% conc.

Alkamide® 491. [Rhone-Poulenc Surf.] Amide blend; nonionic; detergent for floor care and hard surf. cleaners; amber liq. DISCONTINUED.

Alkamide® 1002. [Rhone-Poulenc Surf.] 2:1 Cocamide DEA; nonionic-anionic; surfactant, detergent, thickener, foam stabilizer for hair shampoo, bubble bath, specialty detergents, household and industrial cleaners; straw-color liq.; water sol.; 100% act. DISCONTINUED.

Alkamide® 2104. [Rhone-Poulenc Surf.] 2:1 Cocamide DEA; nonionic-anionic; detergent, emulsifier, foam booster; base for floor and general purpose cleaners; lubricant for syn. grinding and cutting fluids; amber visc. liq.; low odor; sol. in aromatic, aliphatic, and chlorinated solvs., water; 100% act.

Alkamide® 2106. [Rhone-Poulenc Surf.] 2:1 Cocamide DEA; nonionic-anionic; detergent base for floor and hard surface cleaners; solubilizer; amber visc. liq.; sol. in water, aromatic solv., perchloroethylene; dens. 1.0 g/ml; pH 8–11; 100% act. DISCONTINUED.

Alkamide® 2110. [Rhone-Poulenc Surf.] 2:1

Cocamide DEA; nonionic; detergent base to solubilize high silicate concs. in liq. steam cleaners and wax strippers; amber liq.; sol. in water, aromatic solv., perchloroethylene; dens. 1.0 g/ml; pH 8–11 (1% DW); 100% act.

Alkamide® 2112. [Rhone-Poulenc Surf.] 2:1 Fatty acid DEA, sulfonate blend; nonionic-anionic; detergent base for floor cleaners, alkaline strippers, and degreasers; detergent, wetting, lubricant, dispersing and air entertaining agent for construction materials; brn. visc. liq.; low odor; sol. in water, aromatic solv, perchloroethylene; dens. 1.0 g/ml; cloud pt. 45 F max.; pH 8–11; 100% act.

Alkamide® 2124. [Rhone-Poulenc Surf.] 2:1 Lauramide DEA; CAS 120-40-1; nonionic-anionic; detergent, wetting agent, emulsifier, foaming agent, foam stabilizer, thickener for shampoos, bubble baths, liq. detergents; amber visc. liq.; sol. in aromatic hydrocarbons and chlorinated solvs., water; sp.gr. 1.0–1.05; 100% act. DISCONTINUED.

Alkamide® 2204. [Rhone-Poulenc Surf.] 2:1 Cocamide DEA; nonionic-anionic; detergent, rust inhibitor; base for hand soap, floor cleaners, all-purpose cleaners, wax strippers; amber visc. liq.; low odor; water sol.; 100% act.

Alkamide® 2204A. [Rhone-Poulenc Surf.] 2:1 Fatty acid alkanolamide; anionic/nonionic; detergent; more readily disp. in aq. systems than Alkamide 2204; liq.; 100% conc.

Alkamide® C-2 (formerly Alkamidox C-2). [Rhone-Poulenc Surf.] Cocamide MEA ethoxylate; nonionic; thickener, foam stabilizer, and emulsifier for formulated detergents and cosmetics; tan paste; sol. in aromatic solv., perchloroethylene; disp. in water; pH 10.0-11.5 (5% DW).

Alkamide® C-212 (formerly Cyclomide C-212). [Rhone-Poulenc Surf.] Cocamide MEA (1:1); CAS 68140-00-1; nonionic; thickener, foam builder and stabilizer for soap or syn. based washing powds.; cream-colored flakes; 100% act.; 95% amide.

Alkamide® CDE. [Rhone-Poulenc Surf.] Cocamide DEA; nonionic; detergent, emulsifier, stabilizer, thickener, foam stabilizer for personal care and detergent prods.; lt. amber liq.; low odor; sol. in detergent systems, aromatic and chlorinated aliphatic solv., petrol. solv., min. oils; water-disp.; sp.gr. 1.04; dens. 1.0 g/ml; pH 8–11 (1% DW); 90% act.

Alkamide® CDM. [Rhone-Poulenc Surf.] Cocamide DEA.; nonionic; emulsifier, thickener, foam stabilizer for liq. shampoos, bubble baths, liq. detergents; liq.; 100% conc.

Alkamide® CDO. [Rhone-Poulenc Surf.; Rhone-Poulenc France] Cocamide DEA; nonionic; detergent, emulsifier, foam stabilizer, thickener for personal care and detergent prods.; lt. amber liq.; low odor; sol. in aromatic and chlorinated aliphatic solv., petrol. solv.; min. oils, water; sp.gr. 1.04; dens. 1.0 g/ml; pH 8–11 (1% DW); 82% act.

Alkamide® CL63. [Rhone-Poulenc Surf.] Cocamide DEA; nonionic; detergent, thickener, emulsifier, foam stabilizer in toiletry and cleaning preparations; amber clear to cloudy liq.; sol. in detergent systems, aromatic and chlorinated aliphatic solv., petrol. solv., min. oils; water-disp.; sp.gr. 1.03–1.05; dens. 1.0 g/ml; pH 8–11 (1% DW); 92% act.

Alkamide® CME. [Rhone-Poulenc Surf.] Cocamide MEA; CAS 68140-00-1; nonionic; detergent, foam boosters, visc. builder, opacifier for liq. and powd. detergents, shampoos; wh. waxy solid; sol. in detergent systems, various solv., min. oils; m.p. 56 C; 95% act.

Alkamide® CMO. [Rhone-Poulenc Surf.] Cocamide MEA; CAS 68140-00-1; detergent, foam boosters, visc. builder, opacifier; cream-colored, waxy solid; sol. in detergent systems, various solvs., min. oils; m.p. 50 C; 90% act. DISCONTINUED.

Alkamide® CP-1255 (formerly Cyclomide CP-1255). [Rhone-Poulenc Surf.] Fatty alkanolamide; nonionic.

Alkamide® CP-6565 (formerly Cyclomide CP-6565). [Rhone-Poulenc Surf.] Fatty alkanolamide; nonionic.

Alkamide® DC-212 (formerly Cyclomide DC-212). [Rhone-Poulenc Surf.] Cocamide DEA (2:1), diethanolamine; nonionic; visc. booster, foaming agent, emulsifier, dispersant, detergent used in cosmetic and laundry prods.; amber liq.; 100% act.

Alkamide® DC-212/M (formerly Cyclomide DC-212/M). [Rhone-Poulenc Surf.] Modified cocamide DEA (2:1); nonionic; thickener, emulsifier, wetting agent, dispersant, base for hard surf. cleaners, solventless degreasers, and wax strippers; corrosion inhibitor for metalworking fluids; high alkalki compatibility; amber liq.; water-sol.; dens. 8.25 lb/gal; flash pt. > 200 C; 100% act.

Alkamide® DC-212/MP (formerly Cyclomide DC-212/MP). [Rhone-Poulenc Surf.] Fatty alkanolamide; nonionic.

Alkamide® DC-212/S (formerly Cyclomide DC-212/S). [Rhone-Poulenc Surf.; Rhone-Poulenc France] Cocamide DEA (1:1); nonionic; emulsifier, thickener, foam stabilizer and visc. booster with low cloud pt.; for liq. shampoos, bubble baths, liq. detergents; liq.; 100% act.; 85% amide.

Alkamide® DC-212/SE (formerly Cyclomide DC-212/SE). [Rhone-Poulenc Surf.] 1:1 Cocamide DEA; nonionic; detergent, foam stabilizer, thickener, foam booster in shampoos and detergent formulations; liq.; 100% act.; 95% amide.

Alkamide® DIN 100. [Rhone-Poulenc Surf.] Lauric/linoleic diethanolamide; nonionic; thickener, foam stabilizer for toiletries, cosmetics, and detergents; liq.; 100% conc. DISCONTINUED.

Alkamide® DIN295 (formerly Cyclomide DIN295). [Rhone-Poulenc Surf.] 2:1 Linoleamide DEA; visc. builder, grease cutting aid for cleaners; lubricant for metal treatment; liq.

Alkamide® DIN295/S (formerly Cyclomide DIN295/S). [Rhone-Poulenc Surf.; Rhone-Poulenc France] Linoleamide DEA (1:1); nonionic; foam booster, emulsifier, visc. builder, thickener for shampoos, industrial cleaners; conditioning to hair; liq.; 100% conc.

Alkamide® DL-203. [Rhone-Poulenc Surf.] 2:1 Lauric diethanolamide; CAS 120-40-1; nonionic; lather stabilizer for shampoos and liq. detergents; paste; 100% conc. DISCONTINUED.

Alkamide® DL-203/S (formerly Cyclomide DL-203/S). [Rhone-Poulenc Surf.; Rhone-Poulenc France] 1:1 Lauramide DEA; CAS 120-40-1; nonionic; foam stabilizer, visc. and detergency booster for cosmetic prods.; superfatting agent; lubricant for metalworking fluids; off-wh. cryst. solid; sol. in oil and water; pour pt. 60 C; 95% act.

Alkamide® DL-207/S (formerly Cyclomide DL-207/S). [Rhone-Poulenc Surf.; Rhone-Poulenc France] 1:1 Lauramide DEA; CAS 120-40-1; nonionic; foam and visc. booster, wetting agent, superfatting agent; off-wh. liq., wh. paste; 100% act.

Alkamide® DO-280 (formerly Cyclomide DO-280). [Rhone-Poulenc Surf.] 2:1 Oleamide DEA and diethanolamine; nonionic; emulsifier for sol. oils; corrosive inhibitor; liq.; 100% conc.

Alkamide® DS-280. [Rhone-Poulenc Surf.] 2:1 Stearamide DEA; visc. builder, grease cutting aid for cleaners; lubricant for metal treatment; solid. DISCONTINUED.

Alkamide® DS-280/S (formerly Cyclomide DS-280/S). [Rhone-Poulenc Surf.] Stearamide DEA (1:1); nonionic; visc. builder, thickener, foam booster, dispersant for nonionic and cationic systems, shampoos, bath preps., industrial cleaners; emulsifier, corrosion inhibitor, lubricant for metalworking fluids; conditioner for hair and skin care prods.; liq.; sol. in oil and water; dens. 8.33 lb/gal; pour pt. 36 C; flash pt. > 200 C; 100% act.

Alkamide® HTDE. [Rhone-Poulenc Surf.] Stearamide DEA; nonionic; detergent, thickener, visc. builder, emulsifier for kerosene, veg. and min. oil, microcryst. wax; cream-colored waxy solid; low odor; sol. in min. spirits, aromatic solvs., perchloroethylene; water-disp.; m.p. 50–55 C; pH 8–11 (1%); 90% act. DISCONTINUED.

Alkamide® HTME. [Rhone-Poulenc Surf.] Stearamide MEA; nonionic; detergent, opacifier, thickener for cosmetics, other aq. systems; cream waxy solid; low odor; insol. in water; m.p. 85–90 C; pH 8–11 (1%); 90% act. DISCONTINUED.

Alkamide® KD (formerly Cyclomide KD). [Rhone-Poulenc Surf.] Cocamide DEA (1:1); nonionic; visc. builder, foam booster/stabilizer for industrial, institutional and household cleaners; pale yel. liq.; 100% act.; 95% amide.

Alkamide® L-203 (formerly Cyclomide L-203). [Rhone-Poulenc Surf.; Rhone-Poulenc France] Lauramide MEA (1:1); CAS 142-78-9; nonionic; foam builder/stabilizer for soap and syn. washing powds.; visc. builder; off-wh. flakes; 100% act.; 95% amide.

Alkamide® L7DE. [Rhone-Poulenc Surf.] Lauric-myristic DEA; nonionic; detergent, foam booster/stabilizer, superfatting and thickening agent for toiletry and cleaning formulations; fortifier for perfumes in soaps; wh. paste, liq. > 25 C; low odor; sol. in min. spirits, aromatic solv., perchloroethylene; water-disp.; sp.gr. 1.03–1.05; m.p. 34 C; pH 8–11 (1% DW); 92% act.

Alkamide® L7DE-PG. [Rhone-Poulenc Surf.] Lauramide DEA; CAS 120-40-1; nonionic; liq. version of Alkamide L7DE; amber liq. DISCONTINUED.

Alkamide® L7ME. [Rhone-Poulenc Surf.] Lauramide MEA; CAS 142-78-9; nonionic; foam and visc. booster and for liq. and powd. detergents and shampoos; wh. waxy solid; m.p. 60 C; 100% conc. DISCONTINUED.

Alkamide® L9DE. [Rhone-Poulenc Surf.; Rhone-Poulenc France] Lauramide DEA, high purity; CAS 120-40-1; nonionic; detergent, emulsifier, foam stabilizer and booster, thickener for toiletry and cleaning applics., industrial and household detergents; wh. waxy solid, liq. > 25 C; low odor; sol. in aromatic and chlorinated solvs., min. oils; sp.gr. 1.04; f.p. 25 C; m.p. 51 C; 92% act.

Alkamide® L9ME. [Rhone-Poulenc Surf.] Lauramide MEA; CAS 142-78-9; nonionic; detergent, foam stabilizer and booster, visc. builder; wh., waxy solid; low odor; sol. in detergent systems, various solvs., min. oils; m.p. 60 C; 95% act. DISCONTINUED.

Alkamide® LIPA. [Rhone-Poulenc Surf.] Lauramide MIPA; CAS 142-54-1; nonionic; foam booster/stabilizer, detergent in detergents, laundry compds., shampoos; visc. builder in liq. formulations; mild to skin; lt. yel. solid; low odor; disp. in water; m.p. 63–68 C; pH 8–11 (1% DW); 95% amide. DISCONTINUED.

Alkamide® LIPA/C. [Rhone-Poulenc Surf.] Nonionic; high foaming lubricant for metalworking fluids; solid; sol. in oil and water; pour pt. 63-68 C; flash pt. > 200 C; 95% act.

Alkamide® OIP. [Rhone-Poulenc Surf.] Oleamide MIPA; nonionic; foam modifier for high-temp. cleaners, esp. liq. and powd. laundry detergents; lubricant for metalworking fluids; clear liq. to wh. paste; sol. in oil, water; m.p. 28 C; flash pt. > 200 C; 90% act. DISCONTINUED.

Alkamide® RODEA (formerly Cyclomide RODEA). [Rhone-Poulenc Surf.] 2:1 Ricinoleamide DEA; CAS 40716-42-5; visc. builder, grease cutting aid for cleaners; lubricant for metal treatment; liq.

Alkamide® S-280 (formerly Cyclomide S-280). [Rhone-Poulenc Surf.] Stearamide MEA (1:1); nonionic; viscosifier for industrial cleaners; skin protectant in toilet bars, creams, lotions, pastes; off-wh. flakes; 100% act.; 95% amide.

Alkamide® SDO. [Rhone-Poulenc Surf.] Soyamide DEA; nonionic; foam stabilizer, visc. builder, superfatting agent for toiletries, cutting and sol. oils, textiles, household and industrial cleaners, corrosion inhibitor; amber visc. liq.; low odor; sol. in min. oil and spirits, perchloroethylene; water-disp.; sp.gr. 1.04; dens. 1.0 g/ml; pH 8–11 (1%); 100% conc.

Alkamide® SODI (see Miramine® SODI). [Rhone-Poulenc Surf.; Rhone-Poulenc France] Stearamidopropyl dimethylamine; emulsifier, conditioner; produces cationic emulsions; off-wh. solid; 100% act.; 98% amide. DISCONTINUED.

Alkamide® STEDA. [Rhone-Poulenc Surf.] Ethylene bis-stearamide; nonionic; additive in pulp and paper defoamer formulations; lubricant, plasticizer, antistat, pigment dispersant for resins and plastics; solid; 100% conc.

Alkamide® W-197A. [Rhone-Poulenc Surf.] Cocamide DEA; nonionic; detergent for floor care and hard surf. cleaners; liq. DISCONTINUED.

Alkamide® WRS 1-66 (formerly Cyclomide WRS 1-66). [Rhone-Poulenc Surf.] Unsat. fatty acids diethanolamide; anionic/nonionic; emulsifier for highly nonpolar aliphatic hydrocarbons and chlorinated aliphatic hydrocarbons; rust inhibitor; visc. liq.; 100% conc.

Alkaminox® C-2. [Rhone-Poulenc Surf.] PEG-2 cocamine; CAS 61791-14-8; cationic; textile scouring, dyeing assistant, softener, antistatic agent; used as corrosion inhibitor in steam generating and circulating systems; coemulsifier; clear amber liq.; sol. in most org. solv., min. acid, insol. in water; sp.gr. 0.87; 100% conc. DISCONTINUED.

Alkaminox® C-5. [Rhone-Poulenc Surf.] PEG-5 cocamine; CAS 61791-14-8; textile scouring, dyeing assistant, softener, antistatic agent; used as corrosion inhibitor in steam generating and circulating systems; coemulsifier; Gardner 12 liq.; sol. in aromatic solv., perchloroethylene; water-disp.; dens. 0.98 g/ml; HLB 14; 98% min. tert. amine.

Alkaminox® T-12 (redesignated Rhodameen® T-12). [Rhone-Poulenc Surf.]

Alkaminox® T-15 (redesignated Rhodameen® T-15). [Rhone-Poulenc Surf.]

Alkaminox® T-30 (redesignated Rhodameen® T-30). [Rhone-Poulenc Surf.].

Alkaminox® T-50 (redesignated Rhodameen® T-50). [Rhone-Poulenc Surf.]

Alkamox® CAPO (redesignated Rhodamox® CAPO). [Rhone-Poulenc Surf.]

Alkamox® L20. [Rhone-Poulenc Surf.] Lauramine oxide; cationic; foamer, wetter, foam stabilizer for rug shampoos, fine laundry detergents, dishwashing liqs., shampoos, bubble baths, antistatic textile softeners; emollient; foam stabilizer in foam rubber, in electroplating, paper coatings,; as pour pt. depressant for min. oils; clear liq.; 29-31% amine oxide. DISCONTINUED.

Alkamox® LO (redesignated Rhodamox® LO). [Rhone-Poulenc Surf.]

Alkamuls® 14/R (formerly Soprophor 14/R). [Rhone-Poulenc France] PEG-60 castor oil; nonionic; surfactant.

Alkamuls® 200-DL. [Rhone-Poulenc Surf.] PEG-4 dilaurate; CAS 9005-02-1; nonionic; w/o emulsifier, dispersant, defoamer, coemulsifier, and lubricant in industrial and textile oils, cosmetics; as paper softener; yel. liq. to paste; sol. in min. oil and spirits, aromatic solv., perchloroethylene; disp. in water; dens. 0.96 g/ml; HLB 7.4; sapon. no. 190–200; 100% conc. DISCONTINUED.

Alkamuls® 200-DO. [Rhone-Poulenc Surf.] PEG-4 dioleate; CAS 9005-07-6; nonionic; w/o emulsifier, dispersant, defoamer, coemulsifier, and lubricant in industrial and textile oils, cosmetics; as paper softener; amber liq.; sol. in min. oil, min. spirits, aromatic solv., perchlorethylene; disp. in water; dens. 0.95 g/ml; HLB 6.0; sapon. no. 147–157; 100% conc. DISCONTINUED.

Alkamuls® 200-DS. [Rhone-Poulenc Surf.] PEG-4 distearate; CAS 9005-08-7; nonionic; lubricant and softener in textile applics.; opacifier and emulsifier in cosmetic preparations; cream solid; sol. in min. oil and spirits, aromatic solv., perchloroethylene; disp. in water; HLB 5.0; sapon. no. 160–170; 100% conc. DISCONTINUED.

Alkamuls® 200-ML. [Rhone-Poulenc Surf.] PEG-4 laurate; CAS 9004-81-3; nonionic; emulsifier, coupling agent, and solubilizer in metal working fluids, industrial and textile lubricants; yel. liq.; disp. in water; dens. 1.00 g/ml; HLB 9.8; sapon. no. 142–152. DISCONTINUED.

Alkamuls® 200-MO. [Rhone-Poulenc Surf.] PEG-4 oleate; CAS 9004-96-0; nonionic; w/o emulsifier for min. oils, fatty acids, and solv.; used in industrial applics. such as cleaners, degreasers and pesticide formulations; lubricant-softener in tanning and textile processes; yel. liq.; disp. in water; dens. 0.99 g/ml; HLB 8.0; sapon. no. 111–121. DISCONTINUED.

Alkamuls® 200-MS. [Rhone-Poulenc Surf.] PEG-4 stearate; CAS 9004-99-3; nonionic; emulsifier for fats and oils; softener and lubricant for textiles and leather; wh. solid; disp. in water; HLB 8.0; sapon. no. 120–130. DISCONTINUED.

Alkamuls® 400-DL. [Rhone-Poulenc Surf.] PEG-8 dilaurate; CAS 9005-02-1; nonionic; lipophilic emulsifier, solubilizer, dispersing agent used in personal care prods. and industrial applics., agric. chemical sprays, industrial and textile lubricants; yel. liq. to paste; sol. in min. oil and spirits, aromatic solv., perchloroethylene; disp. in water; dens. 1.0 g/ml; HLB 10.0; sapon. no. 132–142; 100% act. DISCONTINUED.

Alkamuls® 400-DO. [Rhone-Poulenc Surf.] PEG-8 dioleate; CAS 9005-07-6; nonionic; emulsifier, solubilizer, lubricant, wetting agent for cosmetic, textile, metalworking, and agric. uses; clear, amber liq.; sol. in min. spirits and oil, aromatic solv, perchloroethylene; water-disp.; sp.gr. 0.97–0.99; dens. 8.163 lb/gal; HLB 7.2; sapon. no. 105–115; 100% act.

Alkamuls® 400-DS. [Rhone-Poulenc Surf.] PEG-8 distearate; CAS 9005-08-7; nonionic; emulsifier and thickener for cosmetic and industrial emulsions; lubricant and softener in textile applic., opacifier and emulsifier in cosmetic preparations; wh. waxy flakes; sol. in min. oil and spirits, aromatic solv., perchloroethylene; water-disp.; HLB 7.8; sapon. no. 120–130; 100% act.

Alkamuls® 400-ML. [Rhone-Poulenc Surf.] PEG-8 laurate; CAS 9004-81-3; nonionic; defoamer, leveling agent for latex paints; dispersant for pigments, dyes; emulsifier for cosmetics and toiletries; clear amber liq.; sol. (@ 10%) in aromatic solv., perchlorethylene; disp. in water, min. oil, min. spirits; sp.gr. 1.02–1.04; cloud pt. 50 F; HLB 12.8; sapon. no. 191–201; 100% act. DISCONTINUED.

Alkamuls® 400-MO. [Rhone-Poulenc Surf.] PEG-9 oleate; CAS 9004-96-0; nonionic; emulsifier for fats, wetting agent, dispersant, lubricant used in dairy industry, cosmetic, metalworking, and industrial applics.; amber clear liquid; sol. in min. oil; disp. in water; sp.gr. 1.01–1.03; dens. 8.497 lb/gal; HLB 11.0; flash pt. > 200 C; 100% act.

Alkamuls® 400-MS. [Rhone-Poulenc Surf.] PEG-8 stearate; CAS 9004-99-3; nonionic; lipophilic emulsifier, solubilizer, dispersing agent used in personal care prods. and industrial applics., agric. chemical sprays, industrial and textile lubricants; wh. paste; sol. in aromatic solv., perchloroethylene; water-disp.; HLB 11.2; sapon. no. 80–90; 100% act. DISCONTINUED.

Alkamuls® 600-DL. [Rhone-Poulenc Surf.] PEG-12 dilaurate; CAS 9005-02-1; nonionic; lipophilic emulsifier, solubilizer, dispersing agent used in personal care prods. and industrial applics., agric. chemical sprays, industrial and textile lubricants; yel. liq., paste; sol. in water, aromatic solv., perchloroethylene; dens. 1.03 g/ml; HLB 11.5; sapon. no. 106–116; 100% conc. DISCONTINUED.

Alkamuls® 600-DO. [Rhone-Poulenc Surf.] PEG-12 dioleate; CAS 9005-07-6; nonionic; dispersant, emulsifier for o/w emulsions; for cosmetic, metalworking, and industrial use; amber liq.; sol. in min. spirits, aromatic solvs., perchloroethylene; disp. in min. oil; sp.gr. 1.01–1.03; dens. 8.497 lb/gal; HLB 10.0; sapon. no. 99–104; flash pt. > 200 C; 100% act.

Alkamuls® 600-DS. [Rhone-Poulenc Surf.] PEG-12 distearate; CAS 9005-08-7; nonionic; lubricant and softener in textile applics.; opacifier and emulsifier in cosmetic preparations; cream solid; sol. in min. oil and spirits, aromatic solv, perchloroethylene; disp. in water; HLB 10.6; sapon. no. 96–106; 100% conc. DISCONTINUED.

Alkamuls® 600-GML. [Rhone-Poulenc Surf.] PEG-12 laurate; CAS 9004-81-3; surfactant.

Alkamuls® 600-ML. [Rhone-Poulenc Surf.] PEG-12 laurate; CAS 9004-81-3; nonionic; defoamer, leveling agent for latex paints; dispersant for pigments, dyes; emulsifier for cosmetics and toiletries; yel. liq. to paste; sol. (@ 10%) in water; disp. in min. oil, min.

spirits, aromatic solv., perchlorethylene; dens. 1.05 g/ml; HLB 14.6; sapon. no. 67–77; 100% act. DISCONTINUED.

Alkamuls® 600-MO. [Rhone-Poulenc Surf.] PEG-12 oleate; CAS 9004-96-0; nonionic; dispersant, o/w emulsifier for cosmetic and industrial applics.; liq.; HLB 13.0; 100% act.

Alkamuls® 600-MS. [Rhone-Poulenc Surf.] PEG-12 stearate; CAS 9004-99-3; nonionic; defoamer, leveling agent for latex paints; dispersant and dye leveling agent for textiles; wh. solid; sol. @ 10% in water, perchlorethylene; disp. in min. oil, min. spirits; HLB 13.2; sapon. no. 60–70; 100% act. DISCONTINUED.

Alkamuls® 6000-DS. [Rhone-Poulenc Surf.] PEG-150 distearate; CAS 9005-08-7; nonionic; hydrophilic emulsifier, thickener for cosmetics, textiles, printing, pigment mfg.; cream solid; sol. in water; HLB 18.4; sapon. no. 14–20; 100% conc. DISCONTINUED.

Alkamuls® 783/P (formerly Geronol 783/P). [Rhone-Poulenc France] PEG-60 oleate; CAS 9004-96-0; nonionic; surfactant.

Alkamuls® A (formerly Rhodiasurf A). [Rhone-Poulenc Surf.; Rhone-Poulenc France] PEG-6 oleate; CAS 9004-96-0; nonionic; emulsifier, lubricant for sol. oils, most aliphatic solvs.; for lubricating and cutting oils, agric. formulations; EPA compliance; liq.; HLB 9.7; 99% conc.

Alkamuls® AG-821. [Rhone-Poulenc Surf. Canada] Ethoxylated ester blend; nonionic; agric. emulsifiers and adjuvants for crop oil/surfactant concs.; amber clear liq.; pH 7.5–8.5; 91% act.

Alkamuls® AG-900. [Rhone-Poulenc Surf. Canada] Ethoxylate blend; nonionic; spreading and wetting agents for aq. pesticide systems; clear liq.; HLB 10.0–14.0; pH 7.0–7.5 (5% DW); 92% act.

Alkamuls® AP (formerly Rhodiasurf AP). [Rhone-Poulenc France] PEG-6 oleate; CAS 9004-96-0; nonionic; emulsifier, lubricant for sol. oils, most aliphatic solvs.; for lubricating and cutting oils, agric. formulations; visc. liq.; 99% conc.

Alkamuls® B (formerly Rhodiasurf B). [Rhone-Poulenc France] PEG-33 castor oil; nonionic; emulsifier, dispersant for textiles, metallurgy, metal degreasing, personal care prods.; dye leveler, fabric softener; liq.; HLB 11.5; 96% conc.

Alkamuls® BR (formerly Rhodiasurf BR). [Rhone-Poulenc France] PEG-33 castor oil; nonionic; emulsifier, dispersant for textiles, metallurgy, metal degreasing, personal care prods.; dye leveler, fabric softener; visc. liq.; 96% conc.

Alkamuls® CO-15 (formerly Alkasurf® CO-15). [Rhone-Poulenc Surf.] PEG-15 castor oil; nonionic; emulsifier, dispersant, detergent, wetting agent, defoamer, antistat; liq.; 100% conc.

Alkamuls® CO-40 (formerly Alkasurf® CO-40). [Rhone-Poulenc Surf. Canada] PEG-40 castor oil; nonionic; low foaming emulsifier for oils and waxes; coemulsifier for dye carriers in textiles; lubricant in fat liquoring baths for leather processing and primary fiber lubricant for nylon and polyester; liq.; water-sol.; HLB 13.0; 100% conc.

Alkamuls® COH-5 (formerly Emulphor® COH-5). [Rhone-Poulenc Surf.] PEG-5 hydrogenated castor oil; CAS 61788-85-0; nonionic; lubricant, softener, antistat, emulsifier, detergent; liq.; HLB 3.9; 100% conc.

Alkamuls® D-10 (formerly Texafor D-10). [Rhone-Poulenc France] PEG-10 castor oil; nonionic; surfactant.

Alkamuls® D-20 (formerly Texafor D-20). [Rhone-Poulenc France] PEG-20 castor oil; nonionic; surfactant.

Alkamuls® D-30. [Rhone-Poulenc France] PEG-30 castor oil; nonionic; surfactant.

Alkamuls® EGMS/C (formerly Cyclochem EGMS/C). [Rhone-Poulenc Surf.] Ethylene glycol monostearate; nonionic; visc. booster, opacifying and pearlescing agent for liq. cosmetic and detergent compds.; wh. flakes; insol. in water; m.p. 56 C; HLB 2.9; sapon. no. 174–184.

Alkamuls® EL-620 (formerly Emulphor® EL-620). [Rhone-Poulenc Surf.; Rhone-Poulenc France] PEG-30 castor oil; nonionic; emulsifier, wetting agent, pigment dispersant, antistat, lubricant, solubilizer for industrial/household cleaners, cosmetics, pharmaceuticals, metalworking fluids, leather, pesticides, herbicides, paper industries; FDA, EPA compliance; lt. brn. clear liq.; sol. in water, acetone, CCl_4, alcohols, veg. oil, ethers, toluene, xylene; sp.gr. 1.04–1.05; dens. 8.705 lb/gal; visc. 600–1000 cps; HLB 12.0; cloud pt. 42 C (1% aq.); flash pt. 291–295 C; surf. tens. 41 dynes/; 100% conc.

Alkamuls® EL-620L (formerly Emulphor® EL-620L). [Rhone-Poulenc Surf.] PEG-30 castor oil; nonionic; emulsifier, dispersant, softener, rewetting agent, lubricant, emulsion stabilizer, dyeing assistant, antistat, solubilizer for textiles, wet-str. papers, fat liquoring, emulsion paints, oleoresinous binders, glass-reinforced plastics,; PU foams, perfumes, cosmetics; low dioxane; EPA compliance; liq.; HLB 12.0; 10% act.

Alkamuls® EL-719 (formerly Emulphor® EL-719). [Rhone-Poulenc Surf.; Rhone-Poulenc France] PEG-40 castor oil; nonionic; emulsifier, wetting agent for industrial/household cleaners; dispersant for pigments; for pesticides, paper, leather, plastics, paint, textile and cosmetics industries; emulsifier for vitamins and drugs; FDA, EPA compliance; liq.; sol. in water, acetone, CCl_4, alcohols, veg. oil, ether, toluene, xylene; sp.gr. 1.06-1.07; dens. 8.9-9.0 lb/gal; visc. 500-800 cps; HLB 13.6; cloud pt. 80 C (1% aq.); flash pt. 275-279 C; surf. tens. 38 dynes/cm (0.1%); 96% act.

Alkamuls® EL-719L (formerly Emulphor® EL-719L). [Rhone-Poulenc Surf.] PEG-40 castor oil; nonionic; emulsifier for agric. formulations; low dioxane; EPA compliance.

Alkamuls® EL-980 (formerly Emulphor® EL-980). [Rhone-Poulenc Surf.] PEG-200 castor oil; nonionic; emulsifier for min. oil, triglycerides, alkyl esters, textile antistat, syn. fiber lubricant, dyeing assistant, metalworking surfactant; sol. in water; dens. 8.30 lb/gal; HLB 18.5; pour pt. 50 C; flash pt. > 200 C; 100% act.

Alkamuls® EL-985 (formerly Emulphor® EL-985). [Rhone-Poulenc Surf.] PEG-200 castor oil; nonionic; emulsifier, antistat, syn. fiber lubricant for textile compounding, metalworking fluids; VSC 4 max. visc. liq.; water-sol.; dens. 8.30 lb/gal; flash pt. > 200 C; 50% act.

Alkamuls® EPS (see Rhodaterge EPS). [Rhone-Poulenc France] Modified alkylphenol ethoxylate; nonionic; emulsifier for prep. of stable emulsions of solvs. and min. oils; liq.; 99% conc. DISCONTINUED.

Alkamuls® GMO. [Rhone-Poulenc Surf.] Glyceryl oleate; CAS 111-03-5; nonionic; emulsifier, wetting agent, lubricant, antistat in cosmetic, agric., textile industries; lt. yel. liq.; m.p. 25 C; HLB 3.4; 100% act. DISCONTINUED.

Alkamuls® GMO-45LG. [Rhone-Poulenc Surf.] Glyceryl oleate; CAS 111-03-5; emulsifier, lubricant, antistat, corrosion inhibitor for cosmetic, metalworking, and industrial applics.; amber liq. to paste; sol. in aromatic solv., perchloroethylene; insol. in water; sp.gr. 0.93 g/ml; dens. 8.163 lb/gal; HLB 3.0; flash pt. > 200 C; 45% act. DISCONTINUED.

Alkamuls® GMR-55LG. [Rhone-Poulenc Surf. Canada] Glyceryl mono/dioleate; nonionic; coemulsifier, lubricant, softener, emollient, rust preventive additive for mold release agents, syn. fiber spin finishes, compounded oils; antistat, antifog for PVC film processing; lt. amber liq. to paste, very mild char. odor; sol. in most aromatic solvs., disp. in min. oil, most aliphatic solvs., insol. in water; dens. 0.96 g/ml; HLB 3.0; 42% monoglyceride.

Alkamuls® GMS/C (formerly Cyclochem GMS, Dermalcare® GMS). [Rhone-Poulenc Surf.] Glyceryl stearate; nonionic; emulsifier, wetting agent for cosmetic, agric., textile industries; coupler used to bind waxes together; emollient and thickener in cosmetic creams; flake; water-disp.; m.p. 58-63 C; HLB 3.4; 100% conc.

Alkamuls® L-9 (formerly Alkasurf® L-9). [Rhone-Poulenc Surf.] PEG-9 laurate; CAS 9004-81-3; nonionic; emulsifier, coemulsifier for cosmetic and toiletry preps.; defoamer, leveling agent for latex paints; dispersant for dyes and pigments; yel. liq. to paste; sol. in aromatic solv., perchloroethylene; disp. in water; dens. 1.03 g/ml; HLB 12.8; sapon. no. 91-101; 100% conc.

Alkamuls® M-6 (formerly Texafor M-6). [Rhone-Poulenc France] PEG-6 oleate; CAS 9004-96-0; nonionic; surfactant.

Alkamuls® MS-40. [Rhone-Poulenc Surf.] Ethoxylated fatty acids; nonionic; lubricant, softener, antistat, emulsifier for personal care and textile prods.; solid; HLB 16.9; 100% conc. DISCONTINUED.

Alkamuls® O-14 (formerly Alkasurf® O-14). [Rhone-Poulenc Surf.] PEG-14 oleate; CAS 9004-96-0; nonionic; coemulsifier for industrial applics.; yel. liq.; sol. in water, aromatic solvs., perchloroethylene; dens. 1.04 g/ml; HLB 13.5; sapon. no. 57-67; 100% conc.

Alkamuls® OR/36 (formerly Soprophor® OR/36). [Rhone-Poulenc Geronazzo] PEG-36 castor oil; nonionic; general purpose, wide range surfactant; paste; 100% conc.

Alkamuls® PE/220 (formerly Soprophor® PE/220). [Rhone-Poulenc Geronazzo] PEG 220 laurate; CAS 9004-81-3; nonionic; low foam detergent, visc. depressant for PVC emulsions; liq.; 100% conc.

Alkamuls® PE/400 (formerly Soprophor PE/400). [Rhone-Poulenc Geronazzo] PEG 400 laurate; CAS 9004-81-3; nonionic; wetting agent for emulsion paints; PVC visc. depressant; grease improver in cosmetics; liq.; 100% conc.

Alkamuls® PEG 200-DS. [Rhone-Poulenc Surf.] Nonionic; emulsifier, lubricant for metalworking fluids; sol. in oil; dens. 7.580 lb/gal; pour pt. 35 C; flash pt. > 200 C; 100% act. DISCONTINUED.

Alkamuls® PEL-9. [Rhone-Poulenc Surf.] Nonionic; emulsifier for metalworking fluids; sol. in water, oil; dens. 8.746 lb/gal; HLB 12.6; 100% act. DISCONTINUED.

Alkamuls® PSML-4. [Rhone-Poulenc Surf.] PEG-4 sorbitan laurate; CAS 9005-64-5; nonionic; emulsifier for PVC emulsion polymerization; liq.; HLB 13.3; 100% conc. DISCONTINUED.

Alkamuls® PSML-20. [Rhone-Poulenc Surf.] Polysorbate 20; CAS 9005-64-5; nonionic; emulsifier, solubilizer, antistat, visc. modifier, lubricant for textiles, cosmetics, pharmaceuticals; yel. liq.; sol. in water, aromatic solv.; dens. 1.1 g/ml; HLB 16.7; sapon. no. 40–50; 97% act.

Alkamuls® PSMO-5. [Rhone-Poulenc Surf.] Polysorbate 81; CAS 9005-65-6; nonionic; emulsifier, solubilizer, antistat, lubricant for paint, food, cosmetic, insecticides, herbicides, fungicides, textiles, cutting oils; amber liq. to paste; sp.gr. 1.0 g/ml; dens. 8.330 lb/gal; HLB 10; sapon. no. 96–104; flash pt. > 200 C; 100% act.

Alkamuls® PSMO-20. [Rhone-Poulenc Surf.] Polysorbate 80; CAS 9005-65-6; nonionic; emulsifier, wetting agent for cosmetic, food, agric. applics.; coemulsifier for aliphatic alcohols, petrol. oils, fats, solvs., waxes; yel. liq.; water-sol.; HLB 15; 97% act.

Alkamuls® PSMS-4. [Rhone-Poulenc Surf.] Polysorbate 61; CAS 9005-67-8; nonionic; emulsifier; fiber-to-metal lubricant for fibers and yarns; used in suppositories in pharmaceutical industry; tan solid; typ. odor; water-disp.; dens. 1.1 g/ml; HLB 9.6; sapon. no. 98–113; 97% act. DISCONTINUED.

Alkamuls® PSMS-20. [Rhone-Poulenc Surf.] Polysorbate 60; CAS 9005-67-8; nonionic; wetting agent, emulsifier for cosmetic and food applics., textiles, paper coatings; fiber-to-metal lubricant for fibers and yarns; yel. visc. liq., gel on standing; typ. odor; water disp.; dens. 1.1 g/ml; HLB 14.9; sapon. no. 45–55; 97% act.

Alkamuls® PSTO-20. [Rhone-Poulenc Surf.] Polysorbate 85; CAS 9005-70-3; nonionic; emulsifier for cosmetic and food applics.; textile and leather lubricant; amber liq.; typ. odor; water disp.; dens. 1.0 g/ml; HLB 11; sapon. no. 80–95; 97% act.

Alkamuls® PSTS-20. [Rhone-Poulenc Surf.] Polysorbate 65; CAS 9005-71-4; nonionic; wetting agent, emulsifier for cosmetic and food formulations; lubricant, softener for textiles; tan solid; typ. odor; sol. in min. oils, min. spirits; water disp.; dens. 1.0 g/ml; HLB 10.5; sapon. no. 88–98; 97% act. DISCONTINUED.

Alkamuls® R81 (formerly Rhodiasurf R81). [Rhone-Poulenc France] PEG-18 castor oil; nonionic; softener, antistat for textile finishing, plastics processing; liq.; 99% conc.

Alkamuls® RC (formerly Soprophor RC). [Rhone-Poulenc France] PEG-22 castor oil; nonionic; surfactant.

Alkamuls® S-6 (formerly Rhodiasurf S-6). [Rhone-Poulenc France] PEG-6 stearate; CAS 9004-99-3; nonionic; coemulsifier, softener, lubricant for textile processing; emulsifier for cosmetic, pharmaceutical and food applics.; wax; 99% conc.

Alkamuls® S-8 (formerly Alkasurf® S-8). [Rhone-Poulenc Surf.] Ethoxylated stearic acid; nonionic; emulsifier, self-emulsifying lubricant and softener for syn. fibers; solid; HLB 11.2; 100% conc.

Alkamuls® S-20 (formerly Soprofor S/20). [Rhone-Poulenc Geronazzo] Sorbitan laurate; CAS 1338-39-2; nonionic; w/o emulsifier, lubricant and softener for the textile industry; sec. suspending agent, porosity modifier in PVC suspensions; liq.; sol. in oils; 100% conc.

Alkamuls® S-60 (formerly Soprofor S/60). [Rhone-Poulenc Geronazzo] Sorbitan stearate; CAS 1338-41-6; nonionic; w/o emulsifier, lubricant and softener for the textile industry; sec. suspending agent, porosity modifier in PVC suspensions; solid; 100% conc.

Alkamuls® S-65 (formerly Soprofor S/65). [Rhone-Poulenc Geronazzo] Sorbitan tristearate; CAS 26658-19-5; nonionic; w/o emulsifier, lubricant and softener for the textile industry; sec. suspending agent, porosity modifier in PVC suspensions; solid; 100% conc.

Alkamuls® S-65-8 (formerly Alkasurf® S-65-8). [Rhone-Poulenc Surf.] PEG-8 stearate; CAS 9004-99-3; nonionic; emulsifier, self-emulsifying lubricant and softener for syn. fibers; wh. flake; HLB 11.2; 100% conc.

Alkamuls® S-65-40 (formerly Alkasurf® S-65-40). [Rhone-Poulenc Surf.] PEG-40 stearate; CAS 9004-99-3; nonionic; emulsifier, self-emulsifying lubricant and softener for syn. fibers; off-wh. flakes; water-sol.; HLB 17.0; 100% conc.

Alkamuls® S-80 (formerly Soprofor S/80). [Rhone-Poulenc Geronazzo] Sorbitan oleate; CAS 1338-43-8; nonionic; w/o emulsifier for min. and veg. oils, in metalworking; oil spill dispersant; liq.; 100% conc.

Alkamuls® S-85 (formerly Soprofor S/85). [Rhone-Poulenc Geronazzo] Sorbitan trioleate; CAS 26266-58-0; nonionic; w/o emulsifier, lubricant and softener for the textile industry; sec. suspending agent, porosity modifier in PVC suspensions; liq.; 100% conc.

Alkamuls® SEG (formerly Cyclochem® SEG). [Rhone-Poulenc Surf.] Ethylene glycol monostearate; nonionic; opacifier and pearling agent for shampoos, creams, liq. hand soaps, liq. detergents; emulsion stabilizer, visc. builder; flake; m.p. 55-60 C; 100% conc.

Alkamuls® SML. [Rhone-Poulenc Surf.] Sorbitan laurate; CAS 1338-39-2; nonionic; emulsifier for oils and fats in cosmetic, metalworking and industrial oil prods.; corrosion inhibitor; antistat for PVC; amber liq.; typ. odor; disp. in water; moderately sol. most alcohols, veg. and min. oils; sp.gr. 1.05 (60 F); dens. 8.330 lb/gal; HLB 8.6; sapon. no. 160–170; hyd. no. 320–350; flash pt. > 200 C; 100% act.

Alkamuls® SMO. [Rhone-Poulenc Surf.] Sorbitan oleate; CAS 1338-43-8; nonionic; emulsifier, coupling agent, wetting agent for medicants, petrol. oils, fats, and waxes in the industrial, textile, metalworking, and cosmetic industries; textile and leather lubricant and softener; corrosion inhibitor; amber liq.; sol. in most veg., min. oils, aromatic solv., perchloroethylene; insol. in water; sp.gr. 1.0; dens. 8.330 lb/gal; HLB 4.3; sapon. no. 145–160; hyd. no. 193–210; flash pt. > 200 C; 100% act.

Alkamuls® SMS. [Rhone-Poulenc Surf.] Sorbitan stearate; CAS 1338-41-6; nonionic; emulsifier and coupling agent; used to prepare silicone defoamer emulsions for industrial applics., paraffin wax emulsions for processing paper coatings; textile process lubricant; internal PVC film lubricant; cosmetics; foods; cream flakes; sol. (10%) in aromatic solv., perchloroethylene; dens. 1.0 g/ml; HLB 4.7; sapon. no 147–157; hyd. no. 235–260; 98.5% act.

Alkamuls® STO. [Rhone-Poulenc Surf.] Sorbitan trioleate; CAS 26266-58-0; nonionic; emulsifier and coupling agent; used to compd. textile and leather softener finishes; in metalworking fluids; amber liq.; typ. odor; sol. in min. oil and spirits, aromatic solv., perchloroethylene; insol. in water; sp.gr. 1.0 g/ml; dens. 8.330 lb/gal; HLB 1.8; sapon. no. 170–190; hyd. no. 55–70; flash pt. > 200 C; 100% act.

Alkamuls® STS. [Rhone-Poulenc Surf.] Sorbitan tristearate; CAS 26658-19-5; nonionic; hydrophobic emulsifier for use as a fiber-to-metal lubricant for syn. and cotton fibers; cosmetics, foods; cream flakes; typ. odor; sol. (10%) in aromatic solv., perchloroethylene; insol. in water; dens. 1.0 g/ml; HLB 2.1; sapon. no. 176–188; hyd. no. 66–80; 100% act.

Alkamuls® T-20 (formerly Soprofor T/20). [Rhone-Poulenc Geronazzo] Polysorbate 20; CAS 9005-64-5; nonionic; emulsifier, solubilizer, antistat and lubricant for textile industry; solubilizer for essential oils; raw material for no-tears shampoo; liq.; 100% conc.

Alkamuls® T-60 (formerly Soprofor T/60). [Rhone-Poulenc Geronazzo] PEG-20 sorbitan stearate; CAS 9005-67-8; nonionic; emulsifier, solubilizer, antistat and lubricant for textile industry; solubilizer for essential oils; raw material for no-tears shampoo; paste; 100% conc.

Alkamuls® T-65 (formerly Soprofor T/65). [Rhone-Poulenc Geronazzo] Sorbitan tristearate, ethoxylated; nonionic; emulsifier, solubilizer, antistat and lubricant for textile industry; solubilizer for essential oils; raw material for no-tears shampoo; solid; 100% conc.

Alkamuls® T-80 (formerly Soprofor T/80). [Rhone-Poulenc Geronazzo] Polysorbate 80; CAS 9005-65-6; nonionic; o/w emulsifier, solubilizer, textile fiber antistat/lubricant; used in hot and cold rolling formulations; liq.; 100% conc.

Alkamuls® T-85 (formerly Soprofor T/85). [Rhone-Poulenc Geronazzo] PEG-20 sorbitan trioleate; CAS 9005-70-3; nonionic; emulsifier, solubilizer, antistat and lubricant for textile industry; solubilizer for essential oils; raw material for no-tears shampoo; liq.; 100% conc.

Alkamuls® TD-41 (formerly Rhodiasurf TD-41). [Rhone-Poulenc France] PEG-14 tallate; CAS 61791-00-2; nonionic; low foaming emulsifier, wetting agent for laundry detergents, degreaser in textile processing, metalworking; visc. liq.; HLB 14.0; 99% conc.

Alkamuls® VN-430. [Rhone-Poulenc Surf.] Nonionic; emulsifier, lubricant for metalworking fluids; sol. in oil; dens. 7.913 lb/gal; HLB 7.7; flash pt. > 200 C; 100% act. DISCONTINUED.

Alkanol® 189-S. [DuPont] Sodium alkyl sulfonate; anionic; wetting agent, detergent, penetrant, foamer for textiles, elastomers, plastics, film, metal cleaning and pickling, hard surf. cleaning, and chemical mfg.; effective in acid and alkali media; reddish-br. liq., alcoholic odor; sol. in water; sp.gr. 1.06 g/mL; dens. 8.8 lb/gal; visc. 30 cps; cloud pt. < 0 C; flash pt. (PMCC) 21 C; pH 7.5-9.0 (1%); surf. tens. 38 dynes/cm (0.1%); flamm.; 31.5% act.; contains IPA.

Alkanol® 6112. [DuPont] Fatty alcohol ethoxylate; nonionic; nonrewetting wetting agent for textile processing; stable to acids, bases; milky wh. emulsion, mild soapy odor; misc. with water; sp.gr. 0.91 g/mL; dens. 7.6 lb/gal; flash pt. (PMCC) 113 C; pH 6.6; surf. tens. 26 dynes/cm (0.1%); 100% act.

Alkanol® A-CN. [DuPont] Amine ethoxylate; nonionic; surfactant, antiprecipitant, dyeing assistant, dye solubilizer, contrast agent for textile dyeing; lt. amber liq., alcoholic odor; 50% sol. in water; sol. in polar solvs.; sp.gr. 0.98 g/mL; dens. 8.2 lb/gal; HLB 14.4; cloud pt. > 100 C (upper, 1%); flash pt. (PMCC) 45 C; pH 7-9; surf. tens. 39 dynes/cm

(0.1%); combustible; 60% act.

Alkanol® CNR. [DuPont] Amphoteric; surfactant; dye assistant; amber-colored clear liq.; misc. with water; dens. 8.61 lb/gal; sp.gr. 1.033.

Alkanol® DW. [DuPont] Sodium alkylaryl sulfonate; anionic; emulsifier, wetting agent, detergent; yel. liq.; alcoholic odor; misc. in water; dens. 8.9 lb/ gal; cloud pt. 5 C (max); 28% act.

Alkanol® ND. [DuPont] Sodium alkyl diaryl sulfonate; anionic; foaming agent, dyeing assistant, surfactant; for textiles, chemical mfg.; leveling agent for acid dyes on nylon; clear yel. liq.; terpene odor; misc. with water; sp.gr. 1.15 g/mL; dens. 9.6 lb/gal; visc. 118 cP; cloud pt. < 0 C; flash pt. extinguishes flame; surf. tens. 34 dynes/cm (0.1% aq.); 45% act.

Alkanol® S. [DuPont] Sodium tetrahydronaphthalene sulfonate; anionic; dispersant, stabilizer, solubilizing agent; lt. cream-colored, small flakes; water sol. 20%; dens. 2.6 lb/gal; 98% act.

Alkanol® WXN. [DuPont] Sodium alkylbenzene sulfonate; anionic; wetting, rewetting agent, foaming agent, emulsifier, dyeing assistant; for textiles, paper, chemical mfg., alkaline and acid cleaners; leveling agent for acid dyes; stable to acid or alkaline media; lt. yel. liq.; alcoholic odor; misc. in water; sp.gr. 1.03 g/mL; dens. 8.6 lb/gal; cloud pt. 0 C max.; flash pt. (PMCC) 28 C; pH 7.5-9 (1% aq.); surf. tens. 36 dynes/cm (0.1%); flamm.; 30% act.; contains IPA.

Alkanol® XC. [DuPont] Sodium alkylnaphthalene sulfonate; anionic; wetting agent, dispersant, penetrant, low foaming; used in bleaching and dyeing of textiles, leather, paper, chemical mfg., photography; reduces shrinkage in ceramics mfg.; dry colors mfg.; lt. buff powd., naphthenic odor; sol. in ethyl alcohol, acetone, benzene; 8–10% in water; sp.gr. 0.41 g/mL; dens. 3.4 lb/gal; pH 9.5-10.0 (1% aq.); surf. tens. 41 dynes/cm (0.1% aq.); 90% act.

Alkapol PEG 300. [Rhone-Poulenc Surf.] PEG-6; intermediate for surfactants; binder/lubricant in pharmaceuticals; plasticizer; paper softener; humectant; solvent; antistat; for cosmetics, textile, plastics processing, dyes and inks; liq.; m.w. 285-315, water-sol.; dens. 1.13 g/ml; pH 5-8 (5% DW); 100% conc. DISCONTINUED.

Alkapol PEG 600. [Rhone-Poulenc Surf. Canada] PEG-14; plasticizer, solv.; lubricant, binder for pharmaceuticals; color stabilizer for fuel oils; intermediate for processing surfactants; wh. liq.; m.w. 570–630; water sol.; sp.gr. 1.13 (30/15.5 C); f.p. 20-25 C; pH 5–8; 100% conc. DISCONTINUED.

Alkaquat® DAET-90 (redesignated Rhodaquat® DAET-90). [Rhone-Poulenc Surf.]

Alkaquat® DMB-451-50, DMB-451-80. [Rhone-Poulenc Surf. Canada] Benzalkonium chloride; cationic; wetting agent, emulsifier, biocide, disinfectant for use in beverage industry, dairy industry, food processing, water treatment, paper industry, pest control, preservatives, antidandruff rinses,; general disinfection and sanitization for hospitals, laundries; pale-yel. liq.; sol. in water, ethanol, acetone, aliphatic solv.; sp.gr. 0.96; surf. tens. 33 dynes/cm (1%); 50 and 80% act.

Alkaquat® DMB-ST, 25%. [Rhone-Poulenc Surf.] Stearalkonium chloride; cationic; antistat, conditioner, softener for fibers, hair, paper prods.; wh. paste; sol. in water; dens. 0.95 g/ml; pH 3.0–5.0 (0.5% aq.); 24–26% act. DISCONTINUED.

Alkaquat® T (redesignated Rhodaquat® T). [Rhone-Poulenc Surf.]

Alkasil® HNM 1223-15 (70%). [Rhone-Poulenc Surf. Canada] Organo-modified polydimethylsiloxane in aromatic hydrocarbon solvs.; for surface modification, waterproofing, release props. on wood, masonry, silica, mineral granules, paper, etc.; colorless to lt. yel. clear liq.; sp.gr. 0.945 g/ml; visc. 40-160 cps; flash pt. (PMCC) 64 C; pH 6-8 (5% in 1:1 IPA/ water); flash pt. (PMCC) > 100 C; pH 6-8 (5% aq.); 70% solids in aromatic hydrocarbon solv.

Alkasil® NE 58-50. [Rhone-Poulenc Surf.; Rhone-Poulenc France] Silicone polyalkoxylate block copolymer; nonionic; nonhydrolyzable surfactant; intermediate for prod. of rigid PU foams; also used in cosmetics, toiletries, textiles, coatings, and as release agent; colorless to lt. amber liq.; 100% act.

Alkasil® NEP 73-70. [Rhone-Poulenc Surf.] Silicone polyalkoxylate block copolymer; nonionic; nonhydrolyzable surfactant; intermediate for prod. of rigid PU foams; also used in cosmetics, toiletries, textiles, coatings, and as release agent; lower visc. and f.p. than 58-50 for improved handling and convenience; colorless to lt. amber liq.; 100% act.

Alkasurf® CA (redesignated Rhodacal® CA). [Rhone-Poulenc Surf.]

Alkasurf® CO-20. [Rhone-Poulenc Surf.] PEG-20 castor oil; nonionic; emulsifier, lubricant, softener, coemulsifier, antistat, degreaser for textiles, leather processing; pigment dispersants; Gardner 4 liq.; sol. in water, aromatic solv., perchloroethylene; dens. 1.02 g/ml; HLB 10.3; cloud pt. 65-69 C; sapon.no. 91-96; 100% conc. DISCONTINUED.

Alkasurf® CO-40 (redesignated Alkamuls® CO-40). [Rhone-Poulenc Surf.]

Alkasurf® EA-60. [Rhone-Poulenc Surf.] Ammonium laureth sulfate; anionic; detergent, surfactant, wetting agent, emulsifier; clear, lt. amber-colored, slightly visc. liq.; typ. odor; water sol.; sp.gr. 1.04 @ 20 C; cloud pt. < 40 F; 58% act. DISCONTINUED.

Alkasurf® ES-60. [Rhone-Poulenc Surf.] Sodium laureth sulfate; CAS 9004-82-4; anionic; general purpose surfactant; detergent, emulsifier, wetting agent; 58% act. DISCONTINUED.

Alkasurf® IPAM (redesignated Rhodacal® IPAM). [Rhone-Poulenc Surf.]

Alkasurf® L-9 (redesignated Alkamuls® L-9). [Rhone-Poulenc Surf.]

Alkasurf® L-14. [Rhone-Poulenc Surf.] PEG-14 laurate; CAS 9004-81-3; nonionic; emulsifier for cosmetics, toiletries; defoamer, leveling agent for latex paints; dispersant for pigment and dye systems; yel. liq.to paste; sol. in water; dens. 1.05 g/ml; HLB 14.6; sapon. no. 67-77. DISCONTINUED.

Alkasurf® LA-3 (redesignated Rhodasurf® LA-3). [Rhone-Poulenc Surf.]

Alkasurf® LA-7 (redesignated Rhodasurf® LA-7). [Rhone-Poulenc Surf.]

Alkasurf® LA-12 (redesignated Rhodasurf® LA-12). [Rhone-Poulenc Surf.]

Alkasurf® LA Acid (redesignated Rhodacal® LA Acid). [Rhone-Poulenc Surf.]

Alkasurf® LAN-15. [Rhone-Poulenc Surf.] Laureth-15; nonionic; coemulsifier for styrene and styrene/ acrylic emulsion polymerization; wh. solid; sol. in water, aromatic solv., perchloroethylene; HLB 15.5; cloud pt. 85-88 C (1% in 5% NaCl); 100% conc. DISCONTINUED.

Alkasurf® LAN-23 (redesignated Rhodasurf® LAN-23). [Rhone-Poulenc Surf.]

Alkasurf® NP-1. [Rhone-Poulenc Surf. Canada] Nonoxynol-1; nonionic; emulsifier and dispersing agent for petroleum oils; coemulsifier and retardant in hair care formulations; defoamer; liq.; sol. in min. oil and spirits, aromatic solv, perchloroethylene; insol. in water; dens. 0.99; HLB 4.6; pH 5-8 (5% DW); 99.0% min. act.

Alkasurf® NP-4. [Rhone-Poulenc Surf. Canada] Nonoxynol-4; CAS 9016-45-9; nonionic; emulsifier, detergent, dispersant, intermediate, stabilizer; plasticizer, antistat for plastics, surfactants, household, industrial, and cosmetic use, fat liquoring, cutting and sol. oils; lt. liq.; low odor; water insol.; sp.gr. 1.02; HLB 9; 100% act.

Alkasurf® NP-6. [Rhone-Poulenc Surf. Canada] Nonoxynol-6; CAS 9016-45-9; nonionic; emulsifier, coemulsifier, oil-sol. dispersant; used for household and industrial cleaners; intermediate; plasticizer and antistat for plastics; emulsifier for min. oils; insecticides, fungicides, herbicides; fat liquoring; making of cutting and sol. oils; dispersing waxes, pigments, resins; printing; preparation of emulsified paint; lt. liq.; low odor; oil-sol., water insol.; sp.gr. 1.04; HLB 11; 100% act.

Alkasurf® NP-8. [Rhone-Poulenc Surf. Canada] Nonoxynol-8; CAS 9016-45-9; nonionic; detergent, wetting agent, emulsifier; lt. liq.; low odor; water sol.; sp.gr. 1.05; HLB 12; 100% act.

Alkasurf® NP-9. [Rhone-Poulenc Surf. Canada] Nonoxynol-9; CAS 9016-45-9; nonionic; detergent, wetting agent, emulsifier, dispersant for household and industrial cleaners, textile processing, laundry detergent, pesticides; lt. liq.; low odor; water sol.; sp.gr. 1.06; HLB 13.4; 100% act.

Alkasurf® NP-10. [Rhone-Poulenc Surf. Canada] Nonoxynol-10; CAS 9016-45-9; nonionic; detergent, wetting agent, emulsifier, solubilizer, dispersant for textiles, household and industrial cleaners, antimicrobials; lt. liq.; low odor; sol. in water, aromatic solv, perchloroethylene; sp.gr. 1.06; HLB 13.5; cloud pt. 62–66 C; pH 5–8; 100% act.

Alkasurf® NP-11. [Rhone-Poulenc Surf. Canada] Nonoxynol-11; CAS 9016-45-9; nonionic; detergent, wetting agent, emulsifier for household and industrial cleaners, textile processing, laundry detergents, pesticides; lt. liq.; low odor; sol. in water, aromatic solv., perchloroethylene; sp.gr. 1.06; HLB 13.8; cloud pt. 70–74 C; 100% act.

Alkasurf® NP-12. [Rhone-Poulenc Surf. Canada] Nonoxynol-12; CAS 9016-45-9; nonionic; surfactant for household and industrial cleaning formulations; liq.; HLB 13.9; 100% conc.

Alkasurf® NP-15. [Rhone-Poulenc Surf. Canada] Nonoxynol-15; CAS 9016-45-9; nonionic; detergent, wetting agent, emulsifier; lt. paste; low odor; water sol.; sp.gr. 1.07; HLB 15; 100% act.

Alkasurf® NP-15, 80%. [Rhone-Poulenc Surf. Canada] Nonoxynol-15; CAS 9016-45-9; nonionic; detergent, wetting agent, emulsifier; liq.; HLB 15.0; 80% conc.

Alkasurf® NP-20, NP-20 70%. [Rhone-Poulenc Surf.] Nonoxynol-20; CAS 9016-45-9; nonionic; wetting agent, stabilizer, dispersant, penetrant; textile scouring agent; coemulsifier for oils, fats, waxes, and solv.; demulsifier for petroleum oil emulsions; liq.; sol. in water; dens. 1.08 g/ml; HLB 16.0; cloud pt. 71 C; pH 5-8 (5% DW); 99 and 70% min. act. resp. DISCONTINUED.

Alkasurf® NP-30, 70%. [Rhone-Poulenc Surf. Canada] Nonoxynol-30; CAS 9016-45-9; nonionic; solubilizer, coemulsifier for highly polar substances; liq.; HLB 17.1; 70% conc.

Alkasurf® NP-40, 70%. [Rhone-Poulenc Surf. Canada] Nonoxynol-40; CAS 9016-45-9; nonionic; wetting agent, stabilizer, penetrant, emulsifier, dispersant; lt. liq.; low odor; water sol.; f.p. 50-60F; HLB 17.6; 70% act.

Alkasurf® NP-50 70%. [Rhone-Poulenc Surf. Canada] Nonoxynol-50; CAS 9016-45-9; nonionic; coemulsifier and vinyl/acrylic latex stabilizer; liq.; sol. in water; dens. 1.09 g/ml; HLB 18.0; cloud pt. 76 C; pH 5-8 (5% DW); 70% conc.

Alkasurf® O-9. [Rhone-Poulenc Surf.] PEG-9 oleate; CAS 9004-96-0; nonionic; surfactant used as a dyeing assistant in the textile industry and as an emulsifier for neats-foot oil fat liquors in leather processing; yel. liq.; sol. in aromatic solv., perchloroethylene; disp. in water; dens. 1.02 g/ml; HLB 11.0; sapon. no. 75-85; 100% conc. DISCONTINUED.

Alkasurf® O-14 (redesignated Alkamuls® O-14). [Rhone-Poulenc Surf.]

Alkasurf® O75-7. [Rhone-Poulenc Surf.] PEG-7 oleate; CAS 9004-96-0; nonionic; emulsifier, lubricant, antistat for textile fiber additives; dyeing assistant; emulsifier for leather fatliquors; liq.; water-disp.; HLB 10.0; 100% conc. DISCONTINUED.

Alkasurf® O75-9. [Rhone-Poulenc Surf.] PEG-9 oleate; CAS 9004-96-0; nonionic; emulsifier, lubricant, antistat for textile fiber additives; emulsifier for leather fatliquors; liq.; water-disp.; HLB 11.0; 100% conc. DISCONTINUED.

Alkasurf® OP-1. [Rhone-Poulenc Surf. Canada] Octoxynol-1; CAS 9002-93-1; nonionic; emulsifier and dispersant for petroleum oils; coemulsifier and retardant in hair color preparations; coupling agent; liq.; sol. in min. oil and spirits, aromatic solv., perchloroethylene; dens. 0.99 g/ml; HLB 3.6; pH 5-8 (5% DW); 100% act.

Alkasurf® OP-5. [Rhone-Poulenc Surf. Canada] Octoxynol-5; CAS 9002-93-1; nonionic; emulsifier, dispersant, surfactant, dry cleaning detergents, insecticides and wax emulsions; lt.-colored liq.; char. odor; sol. in min. spirits, aromatic solv., perchloroethylene; insol. in water; sp.gr. 1.05 (15/15 C); HLB 10.4; cloud pt. 63–66 C; pH 5–8; 100% act.

Alkasurf® OP-8. [Rhone-Poulenc Surf. Canada] Octoxynol-8; CAS 9002-93-1; nonionic; controlled-foam detergent, emulsifier, wetting agent for household and industrial laundry prods.; liq.; HLB 12.5; 100% conc.

Alkasurf® OP-10. [Rhone-Poulenc Surf. Canada] Octoxynol-10; CAS 9002-93-1; nonionic; detergent, emulsifier, wetting agent, dispersant for household and industrial cleaners, textile processing, wool scouring, metal cleaning; lt.-colored liq.; low odor; water sol.; sp.gr. 1.1 (15/15 C); HLB 13.5; 100% act.

Alkasurf® OP-12. [Rhone-Poulenc Surf. Canada] Octoxynol-12; CAS 9002-93-1; nonionic; wetting agent, detergent for metal cleaning, industrial and household liq. detergents; lt.-colored liq., low odor; HLB 14.5; 100% conc.

Alkasurf® OP-30, 70%. [Rhone-Poulenc Surf. Canada] Octoxynol-30; CAS 9002-93-1; nonionic; emulsifier, detergent, wetting agent, coemulsifier, vinyl/acrylic latex stabilizer; lt.-colored liq., low odor; water sol.; HLB 17.3; 70% conc.

Alkasurf® OP-40, 70%. [Rhone-Poulenc Surf. Canada] Octoxynol-40; CAS 9002-93-1; nonionic; detergent, emulsifier, wetting agent; coemulsifier and vinyl/acrylic latex stabilizer; lt.-colored liq.;

low odor; water sol.; sp.gr. 1.1; HLB 18; 70% act.

Alkasurf® PEL-9. [Rhone-Poulenc Surf.] Ethoxlyated pelargonic acid; nonionic; emulsifier for textile industry as dyeing assistant, fiber additive (lubricant and antistat); emulsifier in fatliquors for leather industry; clear to hazy liq.; sol. in water; HLB 12.6; DISCONTINUED.

Alkasurf® S65-8 (redesignated Alkamuls® S-65-8). [Rhone-Poulenc Surf.]

Alkasurf® S65-40 (redesignated Alkamuls® S-65-40). [Rhone-Poulenc Surf.]

Alkasurf® SA-20. [Rhone-Poulenc Surf.] Steareth-20; CAS 9005-00-9; nonionic; detergent and lubricant for fiber and fabric scouring; emulsifier for topical cosmetics; aids disp. and suspension in roll-on deodorants; wh. solid; sol. in water @ 10%, in aromatic solv. and perchloroethylene @ 1%; HLB 15.3; cloud pt. 73-77 C; 100% conc. DISCONTINUED.

Alkasurf® SS-L7DE (redesignated Geropon® SS-L7DE). [Rhone-Poulenc Surf.].

Alkasurf® SS-L9ME. [Rhone-Poulenc Surf.] Disodium lauramido MEA-sulfosuccinate; anionic; detergent for toiletry and carpet shampoo formulations; cream wh., visc. liq.; bland odor; sp.gr. 1.00; biodeg.; 40% act. DISCONTINUED.

Alkasurf® SS-LA-3. [Rhone-Poulenc Surf.] Disodium laureth sulfosuccinate; anionic; mild detergent, emulsifier, wetter for shampoo and bubble bath base, skin cleansers, liq. detergents; lt. clear liq.; typ. odor; water sol.; sp.gr. 1.10; 30% act. DISCONTINUED.

Alkasurf® T [Rhone-Poulenc Surf.] TEA-dodecylbenzene sulfonate, cosmetic grade; anionic; detergent, emulsifier, wetting agent for liq. detergents, bubble baths, shampoos; clear straw-colored liq.; typ. odor; complete sol. in water; visc. 2500 cps max.; 60% act. DISCONTINUED.

Alkasurf® TDA-12. [Rhone-Poulenc Surf.] Trideceth-12; CAS 24938-91-8; nonionic; emulsifier, high-temp. detergent; leveling agent, solubilizer; wh. solid; mild odor; water sol.; m.p. 27 C; 100% act. DISCONTINUED.

Alkasurf® WHC-347. [Rhone-Poulenc Surf. Canada] Waterless hand cleaner base for formulation with kerosene and water; amber clear visc. liq.; dens. 1.00 g/ml; visc. 650 cps; 97.5% act.

Alkaterge®-C. [Angus] Oxazolidine; cationic; surfactant, emulsifier, emulsion stabilizer, wetting agent, acid acceptor; pigment grinding and dispersion aid; penetrant for textile and paper industries, metal cleaners; coatings; antifoam for antibiotic fermentation; antioxidant; Gardner 15 clear prod.; misc. with min. and veg. oils, naphtha, benzene, kerosene, CCl_4, acetone, methanol, butanol, etc.; m.w. 350; dens. 7.76 lb/gal; visc. 122 cP; flash pt. (COC) 400 F; surf. tens. 42 dynes/cm (0.001% aq.); toxicology: relatively nontoxic orally; 70% active.

Alkaterge®-E. [Angus] Ethyl hydroxymethyl oleyl oxazoline; CAS 88543-32-2; amphoteric; detergent, emulsifier, wetting agent, antifoamer, antioxidant; used in salt, soap, paper, textiles, and metal cleaners; emulsion stabilizer; acid acceptor; pigment grinding and disp.; Gardner 15 max. clear liq.; sol. in most org. liq., slight sol. in water; sp.gr. 0.9; dens. 7.74 lb/gal; visc. 155 cp; f.p. –31 C; flash pt. > 200 F; surf. tens. 40 dynes/cm; HLB 4.0-5.0; 70% conc.

Alkaterge®-T. [Angus] Oxazoline-type compd.; CAS 28984-69-2; 75499-49-9; amphoteric; detergent, invert emulsifier, pigment dispersant, corrosion inhibitor; wetting agent; emulsion stabilizer; grinding aid; acid acceptor; antifoam agent; antioxidant; lubricant; buff to brn. waxy solid; high sol. in aromatic hydrocarbons; flash pt. > 200 F; surf. tens. 30.4 dynes/cm; HLB 4.0; 60% conc.

Alkaterge®-T-IV. [Angus] Oxazoline deriv., ethoxylated; CAS 95706-86-8; nonionic; acid scavenger, offers filming protection to metal surfs.; corrosion inhibitor; o/w emulsifier; dispersant in aq. and nonaq. systems; wetting agent; liq.; oil-sol.; water-disp.; HLB 8.5; 60% conc.

Alkateric® 2CIB (redesignated Miranol® 2CIB). [Rhone-Poulenc Surf.].

Alkateric® A2P-OS. [Rhone-Poulenc Surf.] Octyl propionate; amphoteric; low-foaming surfactant for acid and alkaline cleaners; pH stable; pale yel. liq.; 49-51% solids. DISCONTINUED.

Alkateric® A2P-TS (redesignated Mirataine® A2P-TS-30). [Rhone-Poulenc Surf.].

Alkateric® AP-C. [Rhone-Poulenc Surf.] Coco propionate; amphoteric; foaming agent for alkaline cleaning compd. formulations; clear liq.; 42-44% solids. DISCONTINUED.

Alkateric® CIB. [Rhone-Poulenc Surf.] Sodium cocoamphoacetate; amphoteric; detergent, foaming agent, wetting agent; amber liq.; bland odor; sp.gr. 1.1; 40% act. DISCONTINUED.

Alkateric® LAB (see Mirataine® BB). [Rhone-Poulenc Surf.].

Alkawet® CF. [Lonza] Proprietary; amphoteric; wetting agent, detergent for industrial use; electrolyte-tolerant, controlled foaming; liq.; 100% conc.

Alkawet® LF. [Lonza] Amphoteric; low foam, electrolyte-stable surfactant for industrial use; liq.; 68% solids.

Alkawet® N. [Lonza] Mixed phosphate ester; anionic/nonionic; surfactant for industrial cleaners; pH 9.2 (0.1%); 100% act.

Alkawet® NP-6. [Lonza] Mono and dialkylarylphenoxy POE acid phosphates; anionic; emulsifier; intermediate surfactant for cleaners; liq.; 100% conc.

Alkazine® C (redesignated Miramine® C). [Rhone-Poulenc Surf.].

Alkazine® O (redesignated Miramine® O). [Rhone-Poulenc Surf.].

Alkenyl Succinic Anhydrides. [Humphrey] Alkenyl succinic anhydrides from C6-C18 branched and linear; intermediate for prod. of amide and imide rust inhihbitor lube oil additives; detergent and dispersant additives in gasoline; alkali metal salts used as detergents in industrial cleaning formulations; liq./solid; 100% conc.

Alkylate 215. [Monsanto/Detergents & Phosphates] Linear dodecylbenzene; CAS 68648-87-3; anionic; detergent intermediate used in lt.-duty detergents, dishwash, laundry, industrial cleaners; water-wh. liq.; m.w. 234; sp.gr. 0.855-0.870; dens. 7.2 lb/gal; toxicology: eye and skin irritant; 100% conc.

Alkylate 225. [Monsanto/Detergents & Phosphates] Linear dodecylbenzene; CAS 68648-87-3; anionic; detergent intermediate used in liq. and powd. heavy duty detergent, dishwash, laundry, industrial cleaners; water-wh. liq.; m.w. 242; sp.gr. 0.855-0.870; dens. 7.2 lb/gal; toxicology: eye and skin irritant; 100% conc.

Alkylate 227. [Mona Industries] Linear dodecylbenzene; anionic; detergent intermediate for lt. or heavy duty liq. or powd. detergents; water-wh. liq.; m.w.

249; sp.gr. 0.855-0.870; dens. 7.2 lb/gal; toxicology: eye and skin irritant; 100% conc.

Alkylate 230. [Monsanto/Detergents & Phosphates] Linear tridecylbenzene; anionic; detergent intermediate used in heavy duty detergents; water-wh. liq.; m.w. 261; sp.gr. 0.855-0.870; dens. 7.2 lb/gal; toxicology: eye and skin irritant; 100% conc.

Alkylate H230H, H230L. [Monsanto/Detergents & Phosphates] By-prod. of alkylbenzene mfg.; anionic; surfactant intermediate; yel. liq.; m.w. 360 and 335 resp.; sp.gr. 0.890-0.910; visc. 140-220 and 100–160 SUS (100 F); I.B.P. 290 C min.

Alkylene Series. [Hart Prods. Corp.] Phosphated ester; anionic; wetting agents, emulsifiers; liq.; 100% conc.

Alkylox P1904. [Seppic] Alkylpolyethoxyether; nonionic; textile detergent; liq.; 50% conc.

All Wet. [W.A. Cleary] Ethoxylated nonylphenol and polyglycol mixture; nonionic; detergent, wetting, emulsifier used in pesticides; liq.; 80% conc.

Alox® 100. [Alox] Org. acids, oxy-acids, esters, lactones, and some unsaponifiable matter; film-forming rust preventive by conversion to metallic soaps; lubricity agent by conversion of free acid to esters; emulsifier or de-emulsifier by conversion to amine or alkanolamine soaps; corrosion inhibitor by blending with other surfactants; solid; m.w. 350; sp.gr. 0.917; dens. 7.7 lb/gal; m.p. 109 F; i.b.p. 149 F; flash pt. (COC) 305 F; fire pt. (COC) 340 F; acid no. 80; sapon. no. 148.

Alox® 100D. [Alox] Org. acids, oxy-acids, esters, lactones, and some unsaponifiable matter; film forming rust preventive by conversion to metallic soaps; oil-sol. lubricity agent by conversion of free acid to esters; emulsifier or de-emulsifier by conversion to amine or alkanolamine soaps; corrosion inhibitor by blending with other surf. ac; solid; m.w. 422; sp.gr. 0.864; dens. 7.2 lb/gal; m.p. 122 F; i.b.p. 185 F; flash pt. (COC) 343 F; fire pt. (COC) 415 F; acid no. 16; sapon. no. 32.

Alox® 102. [Alox] Org. acids, oxy-acids, esters, lactones, and some unsaponifiable matter; film forming rust preventive by conversion to metallic soaps; oil-sol. lubricity agent by conversion of free acid to esters; emulsifier or de-emulsifier by conversion to amine or alkanolamine soaps; corrosion inhibitor by blending with other surf. ac; solid; m.w. 291; sp.gr. 0.965; dens. 8.1 lb/gal; m.p. 102 F; i.b.p. 131 F; flash pt. (COC) 348 F; fire pt. (COC) 295 F; acid no. 128; sapon. no. 225.

Alox® 575. [Alox] Oxygenated hydrocarbon with barium and a sodium petrol. sulfonate; emulsifier, corrosion inhibitor for cutting and sol. oils; tan liq., gel; mild odor; oil sol.; sp.gr. 1.02 (15.6 C); dens. 8.5 lb/gal; flash pt. 400 F; pour pt. 80 F; acid no. 9–11; sapon. no. 28–31; 100% act.

Alox® 601. [Alox] Org. acids, oxy-acids, esters, lactones, and some unsaponifiable matter; film forming rust preventive by conversion to metallic soaps; oil-sol. lubricity agent by conversion of free acid to esters; emulsifier or de-emulsifier by conversion to amine or alkanolamine soaps; corrosion inhibitor by blending with other surf. ac; solid; m.w. 807; sp.gr. 0.927; dens. 7.7 lb/gal; m.p. 135 F; i.b.p. 149 F; flash pt. (COC) 375 F; fire pt. (COC) 495 F; acid no. 25; sapon. no. 60.

Alox® 1689. [Alox] Ester of an oxygenated hydrocarbon; nonionic; emulsifier, lubricity agent for cutting and sol. oils, marine engine lubricants; rolling oil additive for stainless steel sheets and titanium sheets; dk. brn. solid; mild odor; sol. in nonpolar solv.; sp.gr. 0.93; dens. 7.7 lb/gal; m.p. 115 F; flash pt. (OC) 400 F; acid no. 16 max.; sapon. no. 75 max.; 100% act.

Alox® 1843. [Alox] Org. acid amine salt; corrosion inhibitor; surfactant; pale yel. liq.; mild odor; sol. in water, alcohols, glycols; sp.gr. 1.12; dens. 9.3 lb/gal; visc. 138 cs (100 F); pour pt. 15 F; pH neutral; biodeg. 80%; 75% act. in water.

Alpha-Step® LD-200. [Stepan; Stepan Canada] Formulated detergent base; biodeg. liq. dishwash conc. for fine fabric, carwash and general-purpose cleaners; clear liq.; 50% act.

Alpha-Step® MC-48. [Stepan; Stepan Canada] Sodium alpha sulfo methyl cocoate; surfactant, foam booster/stabilizer for dishwashing liqs.; yel. clear liq.; 39% act.

Alpha-Step® ML-40. [Stepan; Stepan Canada; Stepan Europe] Sodium methyl-2 sulfolaurate and sodium ethyl-2 sulfolaurate; anionic; biodeg. surfactant, foaming agent, hydrotrope for dishwashing liqs., fine fabric washes, hard surf. cleaners and bubble baths; scouring, leveling, coupling and foaming agent for textiles; metalworking formulations; yel. clear liq.; pH 7.0 (1%); 37% act.

Alpha-Step® ML-A. [Stepan; Stepan Canada] Sodium alpha sulfo methyl laurate, lauric/myristic MEA; anionic; surfactant, foam booster/stabilizer for dishwashing liqs., fine fabric washes, hard surf. cleaners; yel. opaque liq.; 44% act.

Alphenate GA 65. [Henkel-Nopco] Alkyl polyglycol ether and esters blend; nonionic; detergent for soaping of textiles after dyeing and printing; liq.; 38% conc.

Alphenate HM. [Henkel-Nopco] Alkyl polyglycol ether blend; nonionic; detergent with high cloud pt.; liq.; 60% conc.

Alphenate PE Extra. [Henkel-Nopco] Alkyl polyglycol ester and sulfate semiester of alkylaryl polyglycol ether blend; anionic/nonionic; dyeing assistant for polyester and acrylic fibers; liq.; 50% conc.

Alphenate TFC-76. [Henkel-Nopco] Sodium alkyl sulfate, modified; anionic; dye leveling agent; liq.; 50% conc.

Alphenate TH 454. [Henkel-Nopco] Alkyl sulfosuccinate; anionic; wetting agent, emulsifier for emulsion polymerization; liq.; 50% conc.

Alphoxat MS 130. [Zschimmer & Schwarz] Blended fatty acid with 30 moles EO; dispersant for chemo-tech. prods.; wax; 100% act.

Alphoxat O 105. [Zschimmer & Schwarz] PEG-5 oleate; CAS 9004-96-0; nonionic; basic material for textile industry; liq.; 100% act.

Alphoxat O 110. [Zschimmer & Schwarz] PEG-10 oleate; CAS 9004-96-0; nonionic; dispersant, defoamer; aux. agent for textile, ceramic, and chemical tech. industries; liq.; 100% act.

Alphoxat O 115. [Zschimmer & Schwarz] PEG-15 oleate; CAS 9004-96-0; nonionic; dispersant for chemo-tech. prods.; paste; 100% act.

Alphoxat S 110. [Zschimmer & Schwarz] PEG-10 stearate; CAS 9004-99-3; nonionic; emulsifier and dispersant, greasing agent for textile and chemical technical industries; wax; 100% act.

Alphoxat S 120. [Zschimmer & Schwarz] PEG-20 stearate; CAS 9004-99-3; nonionic; emulsifier and dispersant, greasing agent for textile and chemical technical industries; wax; 100% act.

Alresat 640C. [Hoechst AG] Modified maleinic resin;

wetting agent for lustering emulsions.

Alrodyne 6104. [Ciba-Geigy] Polyoxyalkylene fatty ester polymer; nonionic; emulsifier; dk. amber visc. liq.

Alrosol B. [Ciba-Geigy] Cocamide DEA; used in liq. soaps and detergents; amber, visc. liq.

Alrosol C. [Ciba-Geigy] Capramide DEA; CAS 136-26-5; foam booster, wetting agent; lt. amber, visc. liq.

Alrosol Conc. [Ciba-Geigy] Fatty acid amide; foam stabilizer, dispersing, emulsifying, wetting agent for SR and latexes; clear yel., visc. liq., mild odor; sp.gr. 0.99.

Alrosol O. [Ciba-Geigy] Oleamide DEA; CAS 93-83-4; emulsifier for hydrocarbons, pigments, etc.; amber visc. liq.

Alrosperse 11P Flake. [Ciba-Geigy] Fatty acid amide; nonionic; emulsifier, lubricant used in hair rinses, hand modifiers for textiles, spreading agent in paste waxes and polishes; off-wh., waxy flake.

Alrosperse® 100. [M.S. Paisner] Surfactant blend; nonionic/cationic; surfactant, interfacial tens. depressant, dispersant, deflocculant, solubilizer, emulsifier, corrosion inhibitor, antistat for metal processing, petrol. prods., drycleaning, spotting compds., leather and upholstery cleaners, emulsions, paints, inks; sol. in all org. solvs., disp. in water; sp.gr. 0.96; visc. 50 cps; f.p. -10 C; flash pt. > 115 C.

Alrowet® D-65. [Ciba-Geigy] Dioctyl sodium sulfosuccinate; anionic; wetting and rewetting agent for textile applics.; dispersant, antistat, wetting agent for NR and SR latexes; clear liq., slight odor; sp.gr. 1.10; 65% conc.

Alscoap AF Series. [Toho Chem. Industry] Surfactants; foaming agent for air drilling; liq.

Alscoap LN-40, LN-90. [Toho Chem. Industry] Sodium lauryl sulfate; CAS 151-21-3; anionic; detergent, shampoo base, toothpaste; polymerization emulsifier for syn. resins and latex; liq. and powd. resp.; 40 and 90% conc.

Ambiteric D40. [Rhone-Poulenc UK] Alkyl dimethyl betaine; amphoteric; foam booster/stabilizer, solubilizer in high electrolyte sol'ns. and hypochlorite; liq.; 40% conc.

Ameenex 70 WS. [Chemron] Amine salt; conc. for oilfield down-hole corrosion inhibition; amber liq.; water-sol.; 60% act.

Ameenex 73 WS. [Chemron] Amine salt; conc. for oilfield corrosion control; amber liq.; water-sol.; 60% act.

Ameenex C-18. [Chemron] Tall oil amido-amine; film-forming corrosion inhibitor; wetting, emulsifying and antistripping agent with asphalt compds., coal tar pitches; drilling fluid additive; useful in nom-metallic min. flotation; lt. amber solid; acetic acid and hydroxyacetic salts are highly water-sol.; 100% conc.

Ameenex C-20. [Chemron] Complex aliphatic amine blend; cationic; film-former for corrosion inhibitor formulations, boiler-condensate corrosion control; drilling fluid additive; neutralizing amine; dk. amber liq.; sol. in water and brine; 100% conc.

Ameenex Polymer. [Chemron] Complex resinous polyamine; corrosion inhibitor intermediate; exc. film persistency; low emulsification tendency; high temp. stability; for oilfield applics.; dk. visc. liq.; 100% act.

Amerchol® C. [Amerchol; Amerchol Europe] Petrolatum, lanolin, lanolin alcohol; nonionic; absorp. base, aux. emulsifier for o/w systems, conditioner, emollient, moisturizer, stabilizer for cosmetics and pharmaceuticals, textile finishes; pale yel.-cream soft solid, slight, char. sterol odor; oil sol.; HLB 9.5; m.p. 40–46 C; acid no. 1.0 max.; sapon. no. 10–20; 100% conc.

Amerchol® CAB. [Amerchol; Amerchol Europe] Petrolatum, lanolin alcohol; nonionic; emollient, emulsifier, moisturizer, stabilizer, plasticizer for therapeutic ointments, burn preparations, dermatological prods., hypoallergenic preparations, cosmetics, pharmaceuticals, absorp. bases, textile finishes; pale cream soft solid, faint, char. sterol odor; oil sol.; HLB 9.0; m.p. 40–46 C; acid no. 1 max.; sapon. no. 1.0 max.; 100% conc.

Amerchol® L-101. [Amerchol] Min. oil, lanolin alcohol; nonionic; emollient, penetrant, emulsifier, moisturizer, softener, stabilizer for cosmetics, creams, makeup, hair dressing, pharmaceuticals, aerosols, baby prods., textile finishes; plasticizer for hair sprays; pale yel. oily liq., faint char. sterol odor; oil sol.; sp.gr. 0.840–0.860; visc. 20–30 cps; HLB 8; acid no. 1 max.; sapon. no. 1 max.; 100% conc.

Amergel® 100. [Drew Ind. Div.] Organics, surfactants, water blend; antifoam for aq. process systems where oils, solvs., waxes, silicas are undesirable, e.g., pulp and paper, nowovens, effluent systems; FDA compliance; wh. creamy liq.; disp. in water; sp.gr. 0.97; visc. 600-1800 cps.

Amergel® 200. [Drew Ind. Div.] Organics, surfactants, water blend; antifoam for aq. process systems where oils, solvs., waxes, silicas are undesirable, e.g., pulp and paper, nowovens, effluent systems; FDA compliance; wh. creamy liq.; disp. in water; sp.gr. 0.97; visc. 400 cps; f.p. < -3 C; flash pt. (PMCC) > 93 C.

Amergel® 500. [Drew Ind. Div.] Organics, surfactants, water blend; antifoam for aq. process systems where oils, solvs., waxes, silicas are undesirable, e.g., pulp and paper, nowovens, effluent systems; FDA compliance; wh. creamy liq.; disp. in water; sp.gr. 0.97; visc. 400 cps; f.p. < 32 F; flash pt. (PMCC) > 200 F.

Amerlate® LFA. [Amerchol; Amerchol Europe] Lanolin acid; anionic; emulsifier, stabilizer, emollient for fatty acid systems, aerosol shave creams, cream shampoos, wax systems, household prods.; pigment dispersant; increases tack and plasticity of wax films; ylsh.-tan firm, waxy solid; mild, waxy odor; m.p. 55–62 C acid no. 130–150; sapon. no. 170–190; 100% conc.

Amerlate® P. [Amerchol; Amerchol Europe] Isopropyl lanolate; nonionic; conditioner, penetrant, lubricant, moisturizer, emollient, w/o emulsifier, stabilizer, opacifier for cosmetics and pharmaceuticals; pigment dispersant; wetting agent and dispersant for solids; plasticizer for wax and pigment systems; yel. buttery solid, faint char. odor; HLB 9; acid no. 18 max.; sapon. no. 130–155; 100% act.

Amerlate® WFA. [Amerchol; Amerchol Europe] Lanolin acid; anionic; emulsifier, stabilizer for emulsions, aerosols, shampoos; stabilizer for conventional soap emulsions; wets and disperses pigments in makeups; ylsh.-tan waxy solid, mild waxy odor; misc. with warm min. oil, IPP, IPM, castor oil; m.p. 60 C; acid no. 120; sapon. no. 165; 100% conc.

Ameroxol® LE-4. [Amerchol; Amerchol Europe] Laureth-4; CAS 5274-68-0; nonionic; o/w emulsifier and solubilizer; broad pH tolerance; liq.; HLB 9.7; 100% conc.

Ameroxol® LE-23. [Amerchol; Amerchol Europe] Laureth-23; nonionic; o/w emulsifier and solubilizer; liq.; HLB 17.0; 100% conc.

Ameroxol® OE-2. [Amerchol; Amerchol Europe] Oleth-2; CAS 9004-98-2; nonionic; solubilizer, emulsifier, dispersant, stabilizer, lipophilic cosolv. for creams and lotions, shampoos, and detergents, fluid and gelled transparent emulsions, fragrance prods., and aerosols; pale straw-colored clear liq.; bland odor; sol. in min. oil, isopropyl esters, anhyd. ethanol; HLB 5.0; acid no. 0.2 max.; sapon. no. 2 max.; pH 4.5–7.0; 100% conc.

Amfotex FV-15. [Pulcra SA] Sulfonate coconut deriv.; foamer, corrosion inhibitor used in metal cleaning processes; liq.; 38% conc.

Amfotex FV-16. [Pulcra SA] Monocarboxylic lauric deriv., sodium salt; raw material for detergents, shampoos, dishwashers; textile softeners, paint emulsifiers, all-purpose washing agents; liq.; 38% conc.

Amgard Series. [Albright & Wilson UK] Phosphorus-based formulations; flame retardant for textile fabrics.

Amidex 1248. [Chemron] Complex fatty amido phosphate; detergent, emulsifier, wetting agent for aq. degreaser, all-purpose industrial, and institutional cleaners; removes oil or water-based muds prior to completion and stimulation; lt. amber liq.; 100% conc.

Amidex 1285. [Chemron] Modified coco diethanolamide (2:1); phosphate-compat. detergent, wetting agent, emulsifier for high-alkaline industrial and specialty cleaning compds., e.g., degreasers, floor strippers; compat. with high conc. of inorg. in aq. systems without need for hydrotrope; yel. visc. liq.; 100% conc.

Amidex 1351. [Chemron] Modified coco diethanolamide (2:1); detergent, wetting agent, emulsifier for industrial and specialty cleaning compds., floor strippers, degreasers, household and hard surf. cleaners; foam stabilizer and visc. builder; compat. with inorg. builders in aq. systems; amber visc. liq.; 100% conc.

Amidex C. [Chemron] Modified coco diethanolamide; detergent, thickener, emulsifier, wetting agent, foam stabilizer, visc. builder; for industrial and household cleaners; yel. liq.; 100% conc.

Amidex CA. [Chemron] Modified coco diethanolamide; detergent, wetting agent, emulsifier for household and industrial hard surf. cleaners, floor strippers, degreasers, specialty cleaning compds.; foam stabilizer, visc. builder; compat. with inorg. builders in aq. systems; yel. liq.; 100% conc.

Amidex CE. [Chemron] Cocamide DEA (1:1); nonionic; detergent, thickener, visc. builder, foam stabilizer for shampoos, cleaners, bubble baths, industrial cleaners, car shampoos, dishwashes, drycleaning detergents, waterless cleaners, solv. cleaners; yel. liq.; 100% conc.

Amidex CME. [Chemron] Cocamide MEA; CAS 68140-00-1; nonionic; visc. builder, foam enhancer for personal care prods., soap systems, syn. powd. detergents, liq. dishwashing formulations; waxy flake; 100% conc.

Amidex CO-1. [Chemron] Modified coco diethanolamide; nonionic; detergent, wetting agent, emulsifier, lubricant, dispersant, foamer for industrial and specialty cleaning compds., conveyor chain lubricants, drilling fluids; lowers surf. tens. for increased penetration; compat. with inorg. builders in aq. systems; amber visc. liq.; 100% conc.

Amidex KME. [Chemron] Cocamide MEA; CAS 68140-00-1; nonionic; foam builder, visc. booster, stabilizer for personal care prods., syn. powd. and liq. detergent systems; waxy flake; 100% conc.

Amidex L-9. [Chemron] Lauramide DEA; CAS 120-40-1; nonionic; thickener, flash foamer, visc. enhancer, foam stabilizer/builder; for liq. detergents, household, institutional and industrial cleaning compds.; yel. liq. to wh. paste; 100% conc.

Amidex LD. [Chemron] Lauramide DEA; CAS 120-40-1; nonionic; thickener, visc. builder, foam booster/stabilizer, detergent, emulsifier; for household, institutional and industrial cleaners, personal care prods.; waxy solid; 100% conc.

Amidex OE. [Chemron] Polyoxyalkylene amide ester; nonionic; emulsifier for industrial pale oils, sol. oils, cutting oils, drawing oils, water-rinsable lubricants, heavy crudes; paraffin and tank bottom dispersant; amber visc. liq.; oil-sol.; 100% conc.

Amidex TD. [Chemron] Tallowamide DEA; detergent for dry laundry compds. and specialty cleaners; amber visc. liq.; 100% conc.

Amidex WD. [Chemron] Fatty amine salt; industrial surfactant exhibiting exc. carbon removal props. formulated with ortho-dichlorobenzene, kerosene, butyl cellosolve and water; cold-dip degreasing formulation; amber liq. to soft paste; 80% conc.

Amido Betaine C. [Zohar Detergent Factory] Coconut amido alkyl betaine; amphoteric; component of personal care prods.; industrial foamer; liq.; gel; 30% conc.

Amido Betaine C-45. [Zohar Detergent Factory] Coconut amido alkyl betaine; amphoteric; component of personal care prods.; industrial foamer; liq.; 39% conc.

Amido Betaine C Conc. [Zohar Detergent Factory] Coconut amido alkyl betaine; amphoteric; component of nonirritating shampoos, conditioning shampoos, bubble baths; industrial foamer.

Amidox® C-2. [Stepan; Stepan Canada] PEG-3 cocamide; nonionic; emulsifier, detergent, wetting agent for dishwashing detergents, shampoos, emulsions; textile wetting and leveling agent; Gardner 5 pasty liq.; sol. in ethanol, xylene; disp. in water; pH 10.0–11.5 (1% aq.); 100% conc.

Amidox® C-5. [Stepan; Stepan Canada] PEG-6 cocamide; nonionic; emulsifier, detergent, wetting agent for dishwashing detergents, shampoos, emulsions; textile wetting and leveling agent; Gardner 5 liq.; sol. in ethanol, xylene, water; pH 10.0–11.5 (1% aq.); 100% conc.

Amidox® L-2. [Stepan; Stepan Canada] PEG-3 lauramide; nonionic; emulsifier, detergent, wetting agent for dishwashing detergents, shampoos, emulsions; Gardner 3 waxy solid; pH 10.0–11.5 (1% aq.); 100% conc.

Amidox® L-5. [Stepan; Stepan Canada] PEG-5 lauramide; nonionic; emulsifier, detergent, wetting agent for dishwashing detergents, shampoos, emulsions; Gardner 6 max. clear visc. liq.; sol. in water, ethanol, xylene; pH 10.0–11.5 (1% aq.); 100% conc.

Amiet 102, 105, 110, 115, 202, 205, 210, 215, 302, 305, 310, 315, 402, 405, 410. [Kao Corp. SA] Tert. amines EO condensation prods. of coco (100 series), soya (200), hydrog. tallow (300), oleyl (400), tallow (500); cationic/nonionic; antistat, textile dyeing assistant, softening and dispersing agent; liq., paste; 100% conc.

Amiet CD/14. [Kao Corp. SA] PEG-14 cocamine;

CAS 61791-14-8; nonionic/cationic; antistat, textile dyeing assistant, softener, dispersant; liq.; 100% conc.

Amiet CD/17. [Kao Corp. SA] PEG-17 cocamine; CAS 61791-14-8; nonionic/cationic; antistat, textile dyeing assistant, softener, dispersant; liq.; 100% conc.

Amiet CD/22. [Kao Corp. SA] PEG-22 cocamine; CAS 61791-14-8; nonionic/cationic; antistat, textile dyeing assistant, softener, dispersant; liq.; 100% conc.

Amiet CD/27. [Kao Corp. SA] PEG-27 cocamine; CAS 61791-14-8; nonionic/cationic; antistat, textile dyeing assistant, softener, dispersant; liq.; 100% conc.

Amiet DT/15. [Kao Corp. SA] Propylene tallow diamine ethoxylate; cationic; emulsifier, dispersant, corrosion inhibitor, wetting agent; liq./paste; 100% conc.

Amiet DT/20. [Kao Corp. SA] Propylene tallow diamine ethoxylate; cationic; emulsifier, dispersant, corrosion inhibitor, wetting agent; liq./paste; 100% conc.

Amiet DT/32. [Kao Corp. SA] Propylene tallow diamine ethoxylate; cationic; emulsifier, dispersant, corrosion inhibitor, wetting agent; liq./paste; 100% conc.

Amiet OD/14. [Kao Corp. SA] PEG-14 oleamine; nonionic/cationic; antistat, textile dyeing assistant, softener, dispersant; liq./paste; 100% conc.

Amiet TD/14. [Kao Corp. SA] PEG-14 tallowamine; nonionic/cationic; antistat, textile dyeing assistant, softener, dispersant; liq.; 100% conc.

Amiet TD/17. [Kao Corp. SA] PEG-17 tallowamine; nonionic/cationic; antistat, textile dyeing assistant, softener, dispersant; liq.; 100% conc.

Amiet TD/22. [Kao Corp. SA] PEG-22 tallowamine; nonionic/cationic; antistat, textile dyeing assistant, softener, dispersant; liq.; 100% conc.

Amiet TD/27. [Kao Corp. SA] PEG-27 tallowamine; nonionic/cationic; antistat, textile dyeing assistant, softener, dispersant; liq.; 100% conc.

Amiet THD/14. [Kao Corp. SA] PEG-14 hydrog. tallow amine; nonionic/cationic; antistat, textile dyeing assistant, softener, dispersant; solid/paste; 100% conc.

Amiet THD/17. [Kao Corp. SA] PEG-17 hydrog. tallow amine; nonionic/cationic; antistat, textile dyeing assistant, softener, dispersant; solid/paste; 100% conc.

Amiet THD/22. [Kao Corp. SA] PEG-22 hydrog. tallow amine; nonionic/cationic; antistat, textile dyeing assistant, softener, dispersant; solid/paste; 100% conc.

Amiet THD/27. [Kao Corp. SA] PEG-27 hydrog. tallow amine; nonionic/cationic; antistat, textile dyeing assistant, softener, dispersant; solid/paste; 100% conc.

Amihope LL-11. [Ajinomoto] Lauroyl lysine; CAS 52315-75-0; amphoteric; surface modifier, coemulsifier, codispersant; in cosmetics, medical, painting and other fields; filler for ink and paint; chelating agent; wh. fine powd.; insol. in almost all solvs. except strong acidic and alkaline sol'ns.; sol. in water @ pH < 1 and > 12; sp.gr. 1.2; 100% conc.

Amiladin. [Dai-ichi Kogyo Seiyaku] PEG alkyl amine ether; nonionic; leveling agent for dyestuffs; paste; 50% conc.

Amiladin C-1802. [Dai-ichi Kogyo Seiyaku] PEG alkyl amine ether; CAS 26635-92-7; nonionic; dispersant, washing agent; liq.; 100% conc.

Amine 12-98D. [Berol Nobel AB] n-Dodecylamine; CAS 124-22-1; cationic; emulsifier; solid; 98% conc.

Amine 14D. [Berol Nobel AB] n-Tetradecylamine; CAS 2016-42-4; cationic; emulsifier; solid; 98% conc.

Amine 16D. [Berol Nobel AB] n-Hexadecylamine; CAS 143-27-1; cationic; emulsifier; solid; 98% conc.

Amine 18D. [Berol Nobel AB] n-Octadecylamine; CAS 124-30-1; cationic; emulsifier; solid; 98% conc.

Amine 2HBG. [Berol Nobel AB] N,N-Di(hydrog. tallow) amine; cationic; surfactant intermediate; paste; 88% conc.

Amine 2M12D. [Berol Nobel AB] Dimethyl dodecyl amine; CAS 67700-98-5; cationic; surfactant intermediate; liq.; 96% conc.

Amine 2M14D. [Berol Nobel AB] Dimethyl tetradecylamine; CAS 68439-70-3; cationic; surfactant intermediate; liq.; 96% conc.

Amine 2M16D. [Berol Nobel AB] Dimethyl hexadecyl amine; CAS 68037-93-4; cationic; surfactant intermediate; liq.; 96% conc.

Amine 2M18D. [Berol Nobel AB] Dimethyl octadecyl amine; cationic; surfactant intermediate; liq.; 96% conc.

Amine 2MHBGD. [Berol Nobel AB] Dimethyl hydrog. tallowamine; CAS 61788-95-2; cationic; surfactant intermediate; liq.; 96% conc.

Amine 2MKKD. [Berol Nobel AB] Dimethyl coco amine; CAS 61788-93-0; cationic; surfactant intermediate; liq.; 96% conc.

Amine BG. [Berol Nobel AB] Tallow amine; CAS 61790-33-8; cationic; emulsifier, corrosion inhibitor; paste; 95% conc.

Amine BGD. [Berol Nobel AB] Distilled tallow amine; CAS 61790-33-8; cationic; emulsifier, corrosion inhibitor; paste; 97% conc.

Amine C. [Ciba-Geigy] Imidazoline deriv.; cationic; emulsifier, dispersant, detergent; used in acid cleaners; antistat for textiles, plastics; paints; corrosion inhibitor; cutting and rust preventive oils; yel. liq.; sol. in polar org. solv., hydrocarbons, relatively insol. water; 100% act.

Amine CS-1135®. [Angus] Oxazolidine; CAS 51200-87-4; emulsifying amine, corrosion inhibitor, alkaline pH stabilizer; for metalworking fluids and aq. systems; sol. with water; m.w. 101.1; dens. 8.2 lb/gal; m.p. < -20 C; b.p. 99 C; flash pt. (TCC) 120 F; pH 10.8 (0.1 M aq.); 78% aq. sol'n. of water and hydrocarbon sol. act. materials.

Amine HBG. [Berol Nobel AB] Hydrog. tallow amine; CAS 61788-45-2; cationic; emulsifier, corrosion inhibitor; solid; 95% conc.

Amine HBGD. [Berol Nobel AB] Distilled hydrog. tallow amine; CAS 61788-45-2; cationic; emulsifier, corrosion inhibitor; solid; 97% conc.

Amine KK. [Berol Nobel AB] Cocamine; CAS 61788-46-3; cationic; emulsifier, corrosion inhibitor; liq.; 95% conc.

Amine KKD. [Berol Nobel AB] Cocamine; CAS 61788-46-3; cationic; emulsifier, corrosion inhibitor; liq.; 98% conc.

Amine M218. [Berol Nobel AB] Methyl dioctadecyl amine; cationic; surfactant intermediate; paste; 96% conc.

Amine M2HBG. [Berol Nobel AB] Methyl di-(hydrogenated tallow) amine; CAS 67700-99-6; cationic;

surfactant intermediate; paste; 96% conc.

Amine O. [Ciba-Geigy] Oleyl imidazoline; CAS 21652-27-7; cationic; emulsifier, dispersant, detergent, antistat used in acid cleaners and corrosion inhibitor formulations; amber liq.; sol. in polar org. solv., hydrocarbons, relatively insol. water; 100% act.

Amine OL. [Berol Nobel AB] Oleamine; CAS 112-90-3; cationic; emulsifier, corrosion inhibitor; liq.; 95% conc.

Amine OLD. [Berol Nobel AB] Distilled oleamine; CAS 112-90-3; cationic; emulsifier, corrosion inhibitor; liq.; 97% conc.

Amine S. [Ciba-Geigy] Stearyl imidazoline; cationic; emulsifier, dispersant, detergent, antistat; tan wax; sol. in polar org. solv., hydrocarbons, relatively insol. water; 100% act.

Amine T. [Ciba-Geigy] Tall oil fatty acid imidazoline; see Amine O; dk. amber liq.; sol. in polar org. solv., hydrocarbons, relatively insol. water; 100% act.

Aminol CA-2. [Finetex] Ricinoleamide DEA; CAS 40716-42-5; nonionic; softening agent for textiles, low-foaming emulsifier, dispersant for dyes and oils, cosmetic emulsions; liq.; 100% conc.

Aminol CM, CM Flakes, CM-C Flakes, CM-D Flakes. [Finetex] Cocamide MEA; CAS 68140-00-1; nonionic; soap additive, foam booster/stabilizer, thickener for toiletry, household prods., hair shampoos; solid, flakes; 100% conc.

Aminol COR-2. [Finetex] Cocamide DEA; nonionic; surfactant used in industrial applications; biodeg.; liq.; 100% conc.

Aminol COR-2C. [Finetex] Cocamide DEA and diethanolamine; nonionic; foam booster/stabilizer, thickener, emulsifier for w/o emulsions, cosmetic applics.; liq.; 100% conc.

Aminol COR-4. [Finetex] Cocamide DEA superamide; nonionic; foam booster/thickener, detergent for drycleaning; liq.; 100% conc.

Aminol COR-4C. [Finetex] Cocamide DEA; nonionic; foam booster, stabilizer, emulsifier, thickener, detergent for cosmetic applics.; liq.; 100% conc.

Aminol HCA. [Finetex] Cocamide DEA; nonionic; foam booster/stabilizer in shampoos and household detergents; liq.; 100% conc.

Aminol KDE. [Chem-Y GmbH] Cocamide DEA; CAS 68603-42-9; nonionic; foam booster/stabilizer, superfatting agent for personal care prods.; solubilizer for perfumes, veg. oils; clear yel. liq.; sp.gr. 1.0; visc. 1200 mPa•s; cloud pt. < 7 C; pH 9.5-10.5 (1% aq.); 100% conc.

Aminol LM-30C, LM-30C Special. [Finetex] Lauramide DEA; CAS 120-40-1; nonionic; foam booster/stabilizer in personal care and household prods.; liq.; 100% conc.

Aminol N-1918. [Finetex] Stearamide DEA and diethanolamine; nonionic; softener for textile; solid; 100% conc.

Aminol TEC N. [Chem-Y GmbH] PEG-4 rapeseedamide; CAS 85536-23-8; nonionic; biodeg. emulsifier for metalworking fluids and lubricants, conveyor chain lubricant, in anticorrosive formulations; clear liq.; visc. 500 mPa•s; pH 9.0-10.5 (10% in water/ethanol 1:1); 92% act.

Aminol VR-14. [Finetex] Mixed acid alkanolamide; nonionic; thickener, foam booster/stabilizer, detergent used in detergent compositions for household and industrial use; wool scouring and fulling agent; liq.; 97% conc.

Aminoxid A 4080. [Goldschmidt; Goldschmidt AG] Cocamine oxide; CAS 61788-90-7; nonionic; surfactant for acid and alkali-stable formulations; liq.; 30% conc.

Aminoxid WS 35. [Goldschmidt; Goldschmidt AG] Cocamidopropylamine oxide; CAS 68155-09-9; nonionic; detergent, emulsifier, wetting agent, softener, foam stabilizer for detergent preparations, cosmetic and pharmaceutical emulsions; liq.; 35% conc.

Amisoft CS-11. [Ajinomoto; Ajinomoto USA] Sodium cocoyl glutamate; anionic; nonirritating high foaming detergent, emollient, bacteriostat; biodeg.; powd.; 100% conc.

Amisoft CT-12. [Ajinomoto; Ajinomoto USA] TEA cocoyl glutamate; anionic; emulsifier for cosmetics; nonirritating high foaming detergent, emollient, bacteriostat; biodeg.; liq.; 100% act.

Amisoft GS-11. [Ajinomoto; Ajinomoto USA] Sodium hydrog. tallow glutamate, sodium cocoyl glutamate; anionic; emulsifier for cosmetics; nonirritating high foaming detergent, emollient, bacteriostat; biodeg.; powd.; 100% act.

Amisoft HS-11. [Ajinomoto; Ajinomoto USA] Sodium hydrog. tallow glutamate; CAS 38517-23-6; anionic; emulsifier for cosmetics; nonirritating high foaming detergent, emollient, bacteriostat; biodeg.; powd.; 100% act.

Amisoft LS-11. [Ajinomoto; Ajinomoto USA] Sodium lauroyl glutamate; CAS 29923-31-7; anionic; emulsifier for cosmetics; nonirritating high foaming detergent, emollient, bacteriostat; biodeg.; powd.; 100% act.

Amisoft MS-11. [Ajinomoto; Ajinomoto USA] Sodium myristoyl glutamate; anionic; nonirritating high foaming detergent, emollient, bacteriostat; biodeg.; powd.; 100% conc.

Amjet A-4. [Am. Emulsions] Leveling carrier for pressure dyeing of stock yarn and piece goods.

Amlev ACY-Super. [Am. Emulsions] Complex inorg. salt of org. amine; nonretarding leveling agent for polyacrylic dyeing.

Amlev CH641. [Am. Emulsions] Surfactant blend; leveling agent for dyeing nylon fiber, yarns, fabric, carpet.

Amlev DAS. [Am. Emulsions] Quat. ammonium compd.; leveling and retarding agent for Dacro 62 with cationic dyes; also suitable for acrylics.

Amlev HBL. [Am. Emulsions] Proprietary; low foaming penetrant, dispersant for jet machine dyeing of polyester, polyester/nylon carpet.

Amlev KC-2. [Am. Emulsions] Surfactant; nonionic; leveling agent for polyester and blends.

Amlev KOF-7. [Am. Emulsions] Surfactant; nonionic; nonfoaming leveling agent for disperse dyes.

Amlev LD. [Am. Emulsions] Surfactant; mildly anionic; leveling agent for disperse dyes on polyester.

Amlev MDV. [Am. Emulsions] Surfactant; anionic; low foaming, economical acid dye leveler.

Amlev MOD. [Am. Emulsions] Surfactant; anionic; leveling agent for acid dyes on nylon; barré improvement.

Amlev MRC. [Am. Emulsions] Sulfonated surfactant; acid dye leveler for continuous and batch dye applics.

Amlev OMF. [Am. Emulsions] Surfactant blend; anionic; leveling agent for acid or disperse dyes on nylon; exc. barré coverage.

Amlev PNL. [Am. Emulsions] Blend; nonionic; low foaming leveling agent for disperse dyes on nylon and carrierless dyeable polyester.

Amlev RDC. [Am. Emulsions] Complex surfactant blend; nonionic/cationic; leveling agent for nylon carpet.

Ammonium Cumene Sulfonate 60. [Hüls Am.] Ammonium cumene sulfonate; hydrotrope; solubilizer and coupler for liq. detergents; water-sol.

Ammonium Stearate 33% Liq. [Hart Chem. Ltd.] Ammonium stearate; anionic; foam stabilizer, hand modifier for frothed latex systems; liq.

Ammonyx® CDO. [Stepan; Stepan Canada] Cocamidopropylamine oxide; nonionic; wetting, foaming agent, foam stabilizer, conditioner for bubble baths, bath oils, dishwashing, hair color systems, softeners, cleansers; liq.; sp.gr. 1.02; flash pt. > 200 F; 30% amine oxide.

Ammonyx® CO. [Stepan; Stepan Canada] Palmitamine oxide; nonionic; conditioner, detergent, foam stabilizer, visc. builder used in cosmetic, household, and janitorial prods.; wetting agent in conc. electrolyte sol'ns.; textile lubricant, emulsifier, wetter, dye dispersant; liq.; sp.gr. 0.96; flash pt. > 200 F; 29.0–31.0% amine oxide.

Ammonyx® DMCD-40. [Stepan; Stepan Canada] Lauramine oxide; nonionic; wetting, foaming agent, foam stabilizer for cosmetic, home and janitorial prods.; liq.; sp.gr. 0.91; flash pt. 86 F; 40–42% amine oxide.

Ammonyx® LO. [Stepan; Stepan Canada] Lauramine oxide; nonionic; foamer/foam stabilizer, wetting agent, visc. builder, grease emulsifier for shampoos, bath prods., fine fabric cleaners, hard surf. cleaners containing acids or bleach, dishwash, shaving creams, lotions; textile lubricant, emulsifier, dye dispersant; liq., sp.gr. 0.96; flash pt. > 200 F; 29–31% amine oxide.

Ammonyx® MCO. [Stepan; Stepan Canada] Myristamine oxide; nonionic; foamer/foam stabilizer; liq.; 30% conc.

Ammonyx® MO. [Stepan; Stepan Canada] Myristamine oxide; nonionic; wetting and foaming agent, foam stabilizer for cosmetics, home and janitorial prods.; textile lubricant, emulsifier, wetter, dye dispersant; liq.; sp.gr. 0.96; flash pt. > 200 F; 29–31% amine oxide.

Ammonyx® SO. [Stepan; Stepan Canada] Stearamine oxide; nonionic; conditioner, detergent, foam stabilizer, visc. builder, conditioner, emulsifier used in cosmetic, household, and janitorial prods.; wetting agent in conc. electrolyte sol'ns.; textile lubricant, emulsifier, wetter, dye dispersant; paste; sp.gr. 0.99; flash pt. > 200 F; 24.5–26.5% amine oxide.

Amollan®. [BASF AG] wetting agents, lubricants for leather and furs.

Amollan® A. [BASF AG] Oxyethylated fatty amine; wetting and emulsifying agent for bating and degreasing leather.

Amollan® L. [BASF AG] Org. esters and fatty acids; leveling agent for applic. of pigment finishes.

Amollan® R. [BASF AG] Ethoxylated fatty alcohol and ethoxylated amine; for improving penetration dyeing and levelness of dyeings.

Amollan® S. [BASF AG] Ethoxylated syn. fatty alcohol; wetting agent, degreaser for leather.

Amonyl BR 1244. [Seppic] Lauralkonium bromide; germicide; liq.

AMP. [Angus] 2-Amino-2-methyl-1-propanol; CAS 124-68-5; nonionic; emulsifier, catalyst; dispersant for pigments and latex paints; corrosion inhibitor; stabilizer; resin solubilizer; APHA 20 solid; m.w. 89.14; sp.gr. 0.928; dens. 7.78; visc. 102 cp (30 C); m.p. 30 C; b.p. 165 C; flash pt. 172 F; 100% act.

AMP-95. [Angus] 2-Amino-2-methyl-1-propanol; CAS 124-68-5; nonionic; emulsifier, catalyst; dispersant for pigments and latex paints; corrosion inhibitor; stabilizer; resin solubilizer; colorless liq.; m.w. 89.14; dens. 7.85; sp.gr. 0.942; visc. 147 cp; f.p. –2 C; flash pt. 182 F (TCC); surf. tens. 36–38 dynes/cm; 95% act. in water.

AMPD. [Angus] 2-Amino-2-methyl-1,3-propanediol; pigment dispersant, neutralizing amine, corrosion inhibitor, acid-salt catalyst, pH buffer, chemical and pharmaceutical intermediate; solubilizer or emulsifier system component in personal care prods.; m.w. 105.1; sol. 250 g/100 ml water; m.p. 109 C; b.p. 151 C; pH 10.8 (0.1M aq. sol'n.); 99% conc.

Amphionic 25B. [Rhone-Poulenc France] Amphoteric; sanitizer.

Amphionic SFB. [Rhone-Poulenc France] Amphoteric; sanitizer.

Amphionic XL. [Rhone-Poulenc France; Rhone-Poulenc UK] Betaine; amphoteric; sanitizer, chelating agent, solubilizer for nonionic surfactants in alkaline sol'ns.; liq.; 40% conc.

Amphisol K. [Bernel] Potassium cetyl phosphate; anionic; emulsifier; stable over wide pH range; colorless solid; sol. in water, oil; usage level: 1-3%.

Amphisol. [Bernel] DEA-cetyl phosphate; anionic; acid pH emulsifier; colorless solid; sol. in water, oil; usage level: 1-3%.

Ampho B11-34. [Karlshamns] Complex coco betaine; amphoteric; detergent, wetting and foaming agent, solubilizer for organics and inorganics; biodeg.; liq.; 30% conc.

Ampho T-35. [Karlshamns] Complex tallow ammonium carboxylate; amphoteric; detergent, wetting and foaming agent, solubilizer for organics and inorganics; biodeg.; liq.; 60% conc.

Amphocerin E. [Henkel KGaA] Fatty alcohols and wax esters; w/o emulsifier; solid.

Ampholak 7CX. [Berol Nobel AB] Cocoamphopolycarboxyglycinate; CAS 97659-53-5; amphoteric; med. to high foaming surfactant for cleaners, nonirritating toiletries, liq. soap, washing-up liqs.; liq.; 39-41% conc.

Ampholak 7TX. [Berol Nobel AB] Tallowamphopolycarboxyglycinate; CAS 97659-53-5; amphoteric; med. foaming detergent for hard surf. cleaners, nonirritating toiletries, as softening agent; liq.; 39-41% conc.

Ampholak 7TY. [Berol Nobel AB] Tallowamphopolycarboxypropionate; CAS 97488-62-5; amphoteric; low foaming surfactant for alkaline cleaners, toiletries; liq.; 30-32% conc.

Ampholak BCA-30. [Berol Nobel AB] Cocamidopropyl betaine; CAS 70851-07-9; amphoteric; foam booster, visc. regulator, mild surfactant, thickener for liq. soaps, washing-up liqs., toiletries; 34-36% conc.

Ampholak BTH-35. [Berol Nobel AB] Tallow bis(hydroxyethyl) betaine; CAS 70750-46-8; amphoteric; thickener for household acid cleaners; wide pH stability; liq.; 39-41% conc.

Ampholak XCE. [Berol Nobel AB] Cocoiminodiglycinate; CAS 97659-51-3; amphoteric; med. foaming surfactant, hydrotrope, detergent for industrial applics. in strong alkaline sol'ns.; 38-40% conc.

Ampholak XCO-30. [Berol Nobel AB] Sodium cocoamphoacetate; CAS 68608-65-1; amphoteric; med.

foaming surfactant for toiletries, nonirritating shampoos, acid hard surf. cleaners; liq.; 38-41% conc.

Ampholak XJO. [Berol Nobel AB] Disodium capryloamphodiacetate; CAS 68608-64-0; amphoteric; low foaming wetting agent, hydrotrope for high alkaline industrial hard surf. cleaners; liq.; 35-38% conc.

Ampholak XO7. [Berol Nobel AB] Oleoamphocarboxyglycinate; CAS 97659-53-5; amphoteric; med. foaming, multipurpose cleaner component; for nonirritating toiletries, conditioners, liq. soap, as softener; liq.; 39-41% conc.

Ampholak XO7-SD-55. [Berol Nobel AB] Oleoamphocarboxyglycinate; CAS 97659-53-5; amphoteric; for household detergents and hard surf. cleaners; powd.; 55% conc.

Ampholak XOO-30. [Berol Nobel AB] Oleoamphocarboxyglycinate; CAS 97659-53-5; amphoteric; med. foaming for household detergents and high visc. industrial and hard surf. cleaners; liq.; 32-34% conc.

Ampholak YCA/P. [Berol Nobel AB] Cocoiminodipropionate; CAS 91995-05-0; amphoteric; med. to high foaming surfactant for alkaline cleaners; high stability to alkali; liq.; 30% conc.

Ampholak YCE. [Berol Nobel AB] Cocoiminodipropionate; CAS 97659-50-2; amphoteric; med. foaming surfactant for industrial alkaline cleaners, cosmetic preps.; hydrotrope; liq.; 29-31% conc.

Ampholak YJH. [Berol Nobel AB] Octyliminodipropionate; CAS 52663-87-3; amphoteric; surfactant for strong alkali cleaners; liq.; 38-40% conc.

Ampholan® B-171. [Harcros UK] Betaine; amphoteric; foaming agent for shampoos, foam based cleaners, cement, and industrial applications; liq.; 30% conc.

Ampholan® E210. [Harcros UK] Cocodimethyl betaine; amphoteric; foaming agent, thickener; liq.

Ampholan® U 203. [Harcros UK] Cocoiminodipropionate; amphoteric; detergent, foaming and stabilizing agent, dispersant, hydrotrope; pale amber liq.; char. odor; water sol.; sp.gr. 1.035; visc. 132 cs; flash pt. > 200 F (COC); pour pt. < 0 C; pH 6.0–8.0 (1% aq); 30% act.

Ampholyt JA 120. [Hüls AG] C10-12 fatty acid amidoethyl (2-hydroxyethyl) glycinate; amphoteric; surfactant for cosmetics, shampoos, detergents, strongly alkaline industrial cleaners; liq.; 33% act.

Ampholyt JA 140. [Hüls Am.; Hüls AG] Sodium lauroamphoacetate; amphoteric; mild surfactant for cosmetics, shampoos, detergents; liq.; 33% act.

Ampholyt JB 130. [Hüls Am.; Hüls AG] Cocamidopropyl betaine; amphoteric; surfactant for cosmetics, shampoos, detergents; liq.; 30% act.

Ampholyte KKE. [Berol Nobel AB] Coco alkyl aminopropionic acid; CAS 84812-94-2; amphoteric; surfactant for detergents, toiletries; 70% conc.

Amphoram CB A30. [Ceca SA] Coco-alkyl betaine; amphoteric; detergent, bactericidal, emulsifier; liq.; 30% conc.

Amphoram CP1. [Ceca SA] N-Coco amino propionic acid; CAS 1462-54-0; amphoteric; detergents, foaming agents; used in cosmetics; liq.; 65% conc.

Amphoram CT 30. [Ceca SA] Coco-alkyl taurine; amphoteric; detergent, bactericidal, emulsifier; liq.; 30% conc.

Amphosol® CA. [Stepan; Stepan Canada; Stepan Europe] Cocamidopropyl betaine; amphoteric; mild conditioner, detergent, wetting agent, visc. builder, foam enhancer, base for cosmetics and household and industrial liq. detergents; straw clear liq.; pH 5.0 (10%); 30% act.

Amphosol® CB3. [Stepan Europe] C8-18 alkylamido betaine; amphoteric; detergent, foaming and wetting agent for household and industrial cleaners; water-wh. to pale yel. liq.; 30% act.

Amphosol® CG. [Stepan; Stepan Canada] Cocamidopropyl betaine; amphoteric; foam booster, visc. builder, and lime soap dispersant; used in cosmetics and liq. detergents; amber clear liq.; pH 4.5–6.5; 29–31% act.

Amphosol® DM. [Stepan Europe] Alkyl betaine; amphoteric; mild foaming, conditioning, thickening base for shampoos, foam baths, liq. soaps, household and industrial cleaners; bacteriostatic; stable in acid systems; water-wh. to pale yel. liq.; 30% act.

Amphoteen 24. [Berol Nobel AB] C12-14 alkyldimethyl betaine; CAS 66455-29-6; amphoteric; surfactant for low-irritation shampoos, washing-up liqs., hard surf. cleaners, vehicle cleaners; liq.; 36-38% conc.

Amphotensid B4. [Zschimmer & Schwarz] Cocamidopropyl betaine; amphoteric; surfactant for cosmetics, shampoos, detergents; liq.; 30% conc.

Amphotensid CT. [Zschimmer & Schwarz] Coconut imidazoline deriv.; amphoteric; detergent for alkaline or acid liq. cleaners; fluid; 38% conc.

Amphotensid D1. [Zschimmer & Schwarz] N-Alkyl amino acid ammonium salt; amphoteric; wetting agent and detergent for chemo-tech. applics.; liq.; 40% conc.

Amphoterge® J-2. [Lonza] Disodium capryloamphodiacetate; amphoteric; wetting agent and detergent for personal care and industrial applics.; liq.; visc. 236 cps; pour pt. < -10 C; pH 8.5; surf. tens. 26.4 dynes/cm (0.1%); Draves wetting 30 s; 50% conc.

Amphoterge® K. [Lonza] Sodium cocoamphopropionate; amphoteric; detergents used in shampoos, skin cleansers, dishwashing; salt-free; liq.; visc. 186 cps; pour pt. 2 C; pH 9.8; surf. tens. 31.9 dynes/cm (0.1%); Draves wetting 48 s; 40% conc.

Amphoterge® K-2. [Lonza] Disodium cocoamphodipropionate; amphoteric; detergents used in shampoos, skin cleansers, dishwashing, heavy duty liq. cleaners; liq.; visc. 76 cps; pour pt. 0 C; pH 9.6; surf. tens. 38.6 dynes/cm (0.1%); Draves wetting 180 s; 40% conc.

Amphoterge® KJ-2. [Lonza] Disodium capryloamphodipropionate; amphoteric; salt-free version of Amphoterge J-2; wetting agent, detergent for personal care and industrial applics.; liq.; visc. 50 cps; pour pt. -2 C; pH 9.6; surf. tens. 27.5 dynes/cm (0.1%); Draves wetting 83 s; 40% conc.

Amphoterge® NX. [Lonza] Coco imidazoline dicarboxylate; amphoteric; industrial detergent; liq.; 40% solids.

Amphoterge® S. [Lonza] Sodium stearoamphoacetate; amphoteric; surfactant, textile softener; used in creme rinses; biodeg.; paste; 25% conc.

Amphoterge® SB. [Lonza] Sodium cocoamphohydroxypropyl sulfonate; amphoteric; surfactant used in detergent and cosmetic applic.; liq.; visc. 26 cps; pour pt. -5 C; pH 7.5 (1%); surf. tens. 33.2 dynes/cm (0.1%); Draves wetting 17 s; 45% conc.

Amphoterge® W. [Lonza] Sodium cocoamphoacetate; amphoteric; surfactant for mild shampoos, skin cleansers, heavy duty cleaners, dishwashing preps.; liq.; visc. 564 cps; pour pt. 8 C; pH 9.8; surf.

tens. 28.5 dynes/cm (0.1%); Draves wetting 22 s; 46% conc.

Amphoterge® W-2. [Lonza] Disodium cocoamphodiacetate; amphoteric; surfactant for nonirritating shampoos and skin cleansers, heavy duty liq. cleaners; gel to visc. liq.; visc. 96,000 cps; pour pt. < -10 C; pH 8.2 (20%); surf. tens. 28.5 dynes/cm (0.1%); Draves wetting 22 s; 52% conc.

Amphoteric 300. [Exxon/Tomah] Sodium eicosyloxypropyliminodipropionate; amphoteric; general surfactant; sol. in water, oil; sp.gr. 1.03; visc. 400 cSt; flash pt. (PMCC) > 210 F; toxicology: eye and skin irritant; 30% act.

Amphoteric 400. [Exxon/Tomah] Iminopropionate, partial sodium salt; amphoteric; low foam detergent, coupler for hard surf. alkaline or acid detergents, laundry, metal, acid bowl cleaners; defoamer in latex paints; corrosion inhibitor in metalworking lubricants; leather lubricant; stable in acid, alkali and conc. electrolytes; yel. clear liq.; sol. in glycols, water, alcohols; sp.gr. 1.09; pH 6-9 (5%); 50% solids.

Amphoteric L. [Exxon/Tomah] Coco deriv.; amphoteric; detergent, foam stabilizer/booster, wetting agent, mild surfactant for liq. detergents, shampoos, hand soaps, mech. foaming systems, dishwash; stable in mildly acid and alkaline media; lt. amber liq.; sp.gr. 1.04; pour pt. 35 F; pH 5–8 (5%); surf. tens. 33 dynes/cm (0.1%); 35% min. act.

Amphoteric N. [Exxon/Tomah] Sodium C12–15 alkoxypropyl iminodipropionate; amphoteric; high foam wetting agent, coupler for shampoos, detergents; corrosion inhibitor in metalworking lubricants; visc. builder; fire fighting foams; clear lt. amber liq.; sol. in glycols, water, alcohol; sp.gr. 1.04; pH 6 (5%); surf. tens. 33 dynes/cm (0.1%); 35% solids in water.

Amphoteric SC. [Exxon/Tomah] Amphoteric; detergent, coupling, foaming, and wetting agent for alkaline and acid cleaners, transportation cleaners, household and institutional cleaners; lt. amber clear liq.; sp.gr. 1.06; pour pt. 30 F; pH 5-9 (5%); surf. tens. 30 dynes/cm (0.1%); 35-38% solids.

Amsol GMS. [Am. Emulsions] Surfactant blend; nonionic; antigelling agent for conc. dye sol'ns.

Amsperse 109. [Am. Emulsions] Dispersing assistant for disperse dyes; prevents agglomeration; compatibilizes dye systems.

Amsul 70. [Harcros] Dodecylbenzene sulfonic acid, amine salt; anionic; emulsifier used in industrial and agricultural formulations; liq.; oil-sol.; 70% conc.

Amterge TC. [Am. Emulsions] Blend of org. solvs., detergents, emulsifiers; scouring and cleaning aid.

Amwet DAD. [Am. Emulsions] Ethoxylated alcohol; detergent, nonrewetting wetting agent, penetrant.

Amwet DOSS. [Am. Emulsions] Dioctyl sulfosuccinate; fast wetting agent for synthetics, cotton and blends; exc. for space dyeing.

Amwet GWF. [Am. Emulsions] Surfactant blend; wetting agent for fourth generation nylon carpet yarn, dye assistant in continuous carpet dyeing.

Amwet MS-100. [Am. Emulsions] Solv.-based penetrant, scour, wetting agent for bleaching, alkaline scouring and dyeing operations; liq.; combustible.

Amwet STA-1. [Am. Emulsions] Sulfated fatty ester; wetting and rewetting agent for variety of uses.

Amyx A-25-S 0040. [Clough] Stearyl dimethyl benzyl ammonium chloride; cationic; conditioner, softener, and emollient for hair rinses, skin creams, and lotions; emulsifier; paste; 25% min. act.

Amyx CDO 3599. [Clough] Cocamidopropylamine oxide; nonionic; mild high foaming surfactant, foam booster/stabilizer, wetting agent, hair conditioner for personal care, household, and janitorial prods.; liq.; 30-32% conc.

Amyx CO 3764. [Clough] Cetamine oxide; cationic in acid media; mild high foaming surfactant, foam booster/stabilizer, conditioner for personal care, household, and janitorial prods.; paste; 29-31% act.

Amyx LO 3594. [Clough] Lauramine oxide; cationic in acid media; mild high foaming surfactant, foam booster/stabilizer, wetting agent, grease emulsifier for personal care, household and janitorial prods.; liq.; 29-31% conc.

Amyx SO 3734. [Clough] Stearamine oxide; cationic in acid media; mild high foaming surfactant, foam booster/stabilizer, conditioner, emulsifier for personal care, household, and janitorial prods.; paste; 24.5-26.5% act.

Anedco ADM-407. [Anedco] Long-chain benzyl sulfonate; anionic; surfactant, demulsifier for "bad tank bottoms" and slop oil in the refinery; amber liq.; sol. in water, oil, xylene, kerosene, IPA; sp.gr. 1.04 (60 F); dens. 8.7 lb/gal; flash pt. (PMCC) > 200 F; pH 3; corrosive; 62.5% act.

Anedco ADM-407C. [Anedco] Alkylbenzyl sulfonic acid; anionic; surfactant, demulsifier for "bad tank bottoms"; amber liq.; sol. in water, oil, xylene, kerosene, IPA; sp.gr. 1.06 (60 F); dens. 8.8 lb/gal; flash pt. (PMCC) > 200 F; pH 2.5-3.5; corrosive; 100% act.

Anedco ADM-408, ADM-410. [Anedco] Sulfonate; anionic; surfactant, demulsifier and desalting conc. for crude oil treatment; amber liq.; sp.gr. 1.03-1.08; dens. 8.6-9.0 lb/gal; flash pt. (PMCC) > 200 F; pH 2-9; corrosive; 50-100% act.

Anedco ADM-456, ADM-457. [Anedco] Oxyalkylated resin/sulfonate blends; anionic; surfactant, demulsifier and desalting conc. for crude oil treatment; amber liq.; sp.gr. 0.96-1.0; dens. 8.0-8.3 lb/gal; flash pt. (PMCC) > 200 F; pH 5-8; 35-100% act.

Anedco AF-800. [Anedco] Ammonium salt of an alcohol ether sulfate; fresh-water foamer for air/gas drilling and well clean-out operations; lt. clear liq.; sol. in water, xylene; disp. in IPA; insol. in oil, kerosene; sp.gr. 1.05 (4 C); visc. 100 cps; flash pt. (PMCC) 24 C; pH 8.3 (5% aq.); flamm.; usage level: 0.1-0.5%; 60% act.

Anedco AF-801. [Anedco] Alkyl sulfate salt; brine water foamer for air/gas drilling and well clean-out operations; lt. clear liq.; sp.gr. 1.06 (4 C); flash pt. (PMCC) 98 F; pH 7.5 (5% aq.); flamm.; usage level: 0.1-0.5%; 60% act.

Anedco AF-802. [Anedco] Alkyl sulfonic acid salt; surfactant, foamer for air/gas drilling and well clean-out operations where hydrocarbons are encountered; dk. clear liq.; sp.gr. 1.15 (4 C); flash pt. (PMCC) 175 F; pH 7.5 (5% aq.); usage level: 0.1-0.5%; 50% act.

Anedco AF-803. [Anedco] Sulfonic acid-alkyl ether sulfate; surfactant, foamer for air/gas drilling and well clean-out operations; effective in fresh water, brine water, hydrocarbons; amber clear liq.; sp.gr. 1.074; flash pt. (PMCC) 96 F; pH 7.5 (5% aq.); usage level: 0.1-0.5%; 50% act.

Anedco AFA-804. [Anedco] Alkyl sulfate salt with foam stabilizer; surfactant, foamer for air/gas drilling and well clean-out operations; effective in brine water; lt. clear liq.; sp.gr. 1.03; flash pt. (PMCC) 65 F; flamm.; usage level: 0.1-0.5%; 86% act.

Anedco AW-395. [Anedco] Coco alkyl dimethyl benzyl quat. amine; cationic; quat. surfactant; anticlay swelling agent; corrosion inhibitor for waterfloods, oil and gas wells, pipelines; foamer additive for surfactants, cleaners, water treating; lt. yel. clear liq.; sp.gr. 0.970-0.980; f.p. < 32 F; flash pt. (PMCC) > 200 F; 50% act.

Anedco AW-396. [Anedco] Quat. amine/nonyl phenol ethoxylate blend; cationic; surfactant flush aid; anticlay swelling agent; corrosion inhibitor for waterfloods, oil and gas wells, pipelines; foamer for wetting agents, surfactants, cleaners, water treating; lt. yel. clear liq.; sp.gr. 0.970-0.980; f.p. < 32 F; flash pt. (PMCC) 72 F; pH 5-6; flamm.; 50% act.

Anedco AW-397. [Anedco] Nonyl phenol ethoxylate; nonionic; surfactant for mfg. of corrosion inhibitors; emulsifier, detergent, wetting agent, penetrant, antistat, coupling agent; liq.; water-sol.; sp.gr. 1.06; dens. 8.83 lb/gal; pour pt. 25 F; flash pt. (TCC) 460 F; 100% act.

Anedco AW-398. [Anedco] Long-chained alkyl sulfonate; anionic; surfactant, cleaner, demulsifier, emulsifier for paraffin tank bottoms, slop oil treatment; amber liq.; sol. in water and oil; sp.gr. 1.09 (60 F); dens. 9.1 lb/gal; flash pt. (PMCC) > 200 F; pH 3; corrosive.

Anedco DF-6002. [Anedco] Nonsilicone, alcohol-based; defoamer for drilling muds; clear liq.; dens. 6.6 lb/gal; visc. 5 cps; flash pt. (PMCC) 68 F; toxicology: avoid contact with eyes, skin, clothing.

Anedco DF-6031. [Anedco] Silicone emulsion; anionic; antifoamer/defoamer for aq. systems; effective in acid and alkaline media; for cooling towers, amine scrubbers, glycol dehydrators, water-based drilling muds, cleaning compds., effluents, cutting oils, abrasive slurries; milky wh. emulsion; disp. in water; sp.gr. 0.98; dens. 8.2 lb/gal; pH 9-10 (1%); flash pt. none; 10% active.

Anedco DF-6130. [Anedco] Dimethyl silicone fluid; anionic; antifoamer for gas-oil separators; clear colorless liq.; insol. in water; dilutable with diesel, kerosene, xylene; sp.gr. 0.97; dens. 8.1 lb/gal; visc. 12,500 cst; flash pt. (OC) 600 F; 100% active.

Anedco DF-6131. [Anedco] Dimethyl silicone fluid; anionic; antifoamer for gas-oil separators; clear colorless liq.; insol. in water; dilutable with diesel, kerosene, xylene; sp.gr. 0.83; dens. 6.9 lb/gal; visc. 12,500 cst; flash pt. (OC) 600 F.

Anedco DF-6231. [Anedco] Dimethyl silicone fluid in hydrocarbon solv.; anionic; antifoamer for cokers and gas-oil separators; clear amber liq.; insol. in water; dilutable with diesel, kerosene, xylene; sp.gr. 0.89; dens. 7.45 lb/gal; visc. 10 cst; flash pt. (OC) 163 F; flamm.

Anedco DF-6233. [Anedco] Dimethyl silicone fluid in hydrocarbon solv.; anionic; antifoamer for cokers and gas-oil separators; clear amber liq.; insol. in water; dilutable with diesel, kerosene, xylene; sp.gr. 0.89; dens. 7.6 lb/gal; visc. 250 cst; flash pt. (OC) 178 F; combustible.

Anedco DF-6300. [Anedco] PEG ethoxylate/modified fatty acid deriv. in hydrocarbon solv.; defoamer for water-based muds, aq. sol'ns., crude oil; amber liq.; sol. in oil; sl. disp. in water; sp.gr. 0.86; dens. 7.14 lb/gal; flash pt. (PMCC) > 200 F; toxicology: avoid contact with eyes, skin, clothing.

Anfomul 01T. [Croda Chem. Ltd.] 2-Heptadecenyl-4,4-bishydroxymethyl oxazoline; nonionic; emulsifier for emulsion explosives; waxy solid; 100% conc.

Anfomul PL. [Croda Chem. Ltd.] Wax acid pentaerythritol ester; nonionic; emulsifier for emulsion explosives; soft solid; 100% conc.

Anfomul PLR. [Croda Chem. Ltd.] Wax and olefinic fatty acids pentaerythritol ester; nonionic; emulsifier for emulsion explosives; liq.; 100% conc.

Anfomul PO. [Croda Chem. Ltd.] Olefinic fatty acid pentaerythritol ester; nonionic; emulsifier for emulsion explosives; liq.; 100% conc.

Anfomul S4. [Croda Chem. Ltd.] Sorbitan oleate; CAS 1338-43-8; nonionic; emulsifiers for emulsion explosives; liq.; HLB 4.3; 100% conc.

Anfomul S43. [Croda Chem. Ltd.] Sorbitan sesquioleate; CAS 8007-43-0; nonionic; emulsifiers for emulsion explosives; liq.; HLB 3.7; 100% conc.

Anfomul S4L. [Croda Chem. Ltd.] Polyol ester/steroidal alcohol complex; emulsifiers for emulsion explosives; 100% conc.

Anfomul S6. [Croda Chem. Ltd.] Sorbitan isostearate; nonionic; emulsifiers for emulsion explosives; liq.; HLB 4.7; 100% conc.

Anfomul S50. [Croda Chem. Ltd.] Sorbitan oleate; CAS 1338-43-8; nonionic; emulsifiers for emulsion explosives; liq.; HLB 4.3; 100% conc.

Anionyx® 12S. [Stepan; Stepan Canada] Disodium oleamido PEG-2 sulfosuccinate; anionic; detergent for personal care prods., bubble baths, dishwashing liqs.; liq.; sp.gr. 1.05; flash pt. 138 F; 20% act.

Annonyx SO. [Stepan; Stepan Canada] Stearamine oxide; conditioner, emulsifier, visc. modifier, wetting agent, foam booster/stabilizer; paste; 24.5% conc.

Anstex AK-25. [Toho Chem. Industry] Special phosphate; antistat for syn. fibers, plastics; liq.; 55% act.

Antara® HR-719 (redesignated Lubrhophos® HR-719). [Rhone-Poulenc Surf.].

Antara® LB-400 (redesignated Lubrhophos® LB-400). [Rhone-Poulenc Surf.].

Antara® LE-500 (redesignated Lubrhophos® LE-500). [Rhone-Poulenc Surf.; Rhone-Poulenc France].

Antara® LE-700 (redesignated Lubrhophos® LE-700). [Rhone-Poulenc Surf.].

Antara® LF-200 (redesignated Lubrhophos® LF-200). [Rhone-Poulenc Surf.].

Antara® LK-500 (redesignated Lubrhophos® LK-500). [Rhone-Poulenc Surf.].

Antara® LM-400 (redesignated Lubrhophos® LM-400). [Rhone-Poulenc Surf.].

Antara® LM-600 (redesignated Lubrhophos® LM-600). [Rhone-Poulenc Surf.].

Antara® LP-700 (redesignated Lubrhophos® LP-700). [Rhone-Poulenc Surf.; Rhone-Poulenc France].

Antaron FC-34. [ISP] Monocarboxyl coco imidazoline compd.; amphoteric; detergent, wetting agent, emulsifier, dispersant, emollient, surfactant; fulling agent for woolen and worsted fabrics; emulsifier for leather processing; ingred. of bubble baths, hair,; upholstery, and rug shampoos, liq. dishwashing and hard-surf. detergents; amber clear semivisc. liq.; sol. in water and high electrolyte sol'ns.; surf. tens. 32.0 dynes/cm (0.0155%); > 38% conc.

Antaron MC-44. [ISP] Dicarboxylic coco imidazoline, sodium salt; amphoteric; emulsifier, solubilizer, coupling agent for nonirritating shampoos, skin cleaners, other cosmetics, industrial and household cleaners; amber visc. liq.; 38% act.

Antarox® 17-R-2 (formerly Pegol® 17R2). [Rhone-Poulenc Surf.] Alkoxylated glycol copolymer; nonionic; defoamer, dispersant, wetting agent, emulsifier, demulsifier, leveling agent, detergent for industrial/household cleaners, fermentation, paper processing, rinse aids, automatic dishwashing, metal cleaning; liq.; HLB 8.0; pour pt. -25 C; cloud pt. 39 C (1% aq.); 100% conc.

Antarox® 25-R-2 (formerly Pegol® 25R2). [Rhone-Poulenc Surf.] Alkoxylated glycol copolymer; nonionic; defoamer, dispersant, wetting agent, emulsifier, demulsifier, leveling agent, detergent for industrial/household cleaners rinse aids, automatic dishwashing, paper processing, metal cleaning, fermentation; liq.; HLB 6.0; pour pt. -5 C; cloud pt. 33 C (1% aq.); 100% conc.

Antarox® 31-R-1 (formerly Pegol® 31R1). [Rhone-Poulenc Surf.] Alkoxylated glycol copolymer; nonionic; defoamer, dispersant, wetting agent, emulsifier, demulsifier, leveling agent, detergent for industrial/household cleaners, rinse aids, automatic dishwashing, paper processing, metal cleaning, fermentation; liq.; HLB 4.0; pour pt. 25 C; cloud pt. 25 C (1% aq.); 100% conc.

Antarox® 461/P (formerly Soprophor® 461/P). [Rhone-Poulenc Surf.; Rhone-Poulenc Geronazzo] EO/PO alkylphenol block polymer; nonionic; emulsifier, dispersant used in agric. industry for prep. of emulsifiable concs. and toxicant flowable systems; EPA compliance; paste; 99% conc.

Antarox® 487/P (formerly Soprophor® 487/P). [Rhone-Poulenc Surf.; Rhone-Poulenc Geronazzo] EO/PO alkylphenol block polymer; nonionic; emulsifier, dispersant used in agric. industry for prep. of emulsifiable concs. and toxicant flowable systems; EPA compliance; solid; 99% conc.

Antarox® AA-60 (formerly Alkasurf AA-60). [Rhone-Poulenc Surf.] Alcohol alkoxylate; nonionic.

Antarox® B-10 (formerly Supronic B-10). [Rhone-Poulenc France] Low m.w. block copolymer; nonionic; surfactant.

Antarox® B-25 (formerly Supronic B-25). [Rhone-Poulenc France] Low m.w. block copolymer; nonionic; surfactant.

Antarox® BA-PE 70 (formerly Rhodacal® BA-PE 70, Alkasurf® BA-PE 70). [Rhone-Poulenc Surf.] Aliphatic alcohol alkoxylate; nonionic; moderate foaming surfactant, emulsifier, dispersant for insecticides, herbicides, latex polymerization, leather finishing applics.; biodeg.; soft solid; HLB 16.1; 100% conc.

Antarox® BA-PE 80 (formerly Rhodacal® BA-PE 80, Alkasurf® BA-PE 80). [Rhone-Poulenc Surf.] Aliphatic alcohol alkoxylate; nonionic; moderate foaming surfactant, emulsifier, dispersant for insecticides, herbicides, latex polymerization, leather finishing applics.; biodeg.; hard solid; HLB 26.1; 100% conc.

Antarox® BL-214. [Rhone-Poulenc Surf.; Rhone-Poulenc France] Linear aliphatic EO/PO adduct, modified; nonionic; wetting, rewetting agent, detergent for metal cleaning, textile finishing, industrial/household cleaners; liq.; sol. in water and polar solv.; cloud pt. 14 C (1% aq.); 100% act.

Antarox® BL-225. [Rhone-Poulenc Surf.; Rhone-Poulenc France] Linear aliphatic EO/PO adduct, modified; nonionic; low-foaming surfactant, detergent, wetting agent for metal cleaning, rinse aids, textiles, floor cleaners; slightly yel., clear liq.; bland odor; sol. in water and org. solv. except paraffins; sp.gr. 0.99; visc. 40 cps; cloud pt. 25–29 C; flash pt. 190 C (COC); pour pt. < -18 C; surf. tens. 30 dynes/cm (0.1%); biodeg.; 100% act.

Antarox® BL-236. [Rhone-Poulenc Surf.] Linear aliphatic EO/PO adduct, modified; nonionic; low foaming surfactant, wetting agent for high temp. spray metal cleaning; rinse aids for commercial and industrial use; liq.; cloud pt. 35 C (1% aq.); 100% conc.

Antarox® BL-240. [Rhone-Poulenc Surf.; Rhone-Poulenc France] Modified linear aliphatic EO/PO adduct; nonionic; low foaming detergent, wetting agent for spray metal cleaning, rinse aids, textiles, floor cleaners; slightly yel. liq.; bland odor; water sol.; sp.gr. 0.99; visc. 35 cps; cloud pt. 38–42 C; flash pt. 124 C; pour pt. < 21 C; surf. tens. 28 dynes/cm (0.1%); biodeg.; 100% act.

Antarox® BL-330. [Rhone-Poulenc Surf.; Rhone-Poulenc France] Linear aliphatic polyether, chlorine-capped; nonionic; detergent, wetting agent for rinse aids, metal cleaning, mechanical dishwashing, household and industrial cleaners, dairy cleaners; dedusting agent for powd. detergents; liq.; sol. in water and solvs. except for paraffinics; cloud pt. 30 C (10% aq.); 95% act.

Antarox® BL-344. [Rhone-Poulenc Surf.; Rhone-Poulenc France] Linear aliphatic polyether, chlorine-capped; nonionic; low foaming detergent, wetting agent for industrial/household cleaners; stable at high temps.; deduster for caustic powds.; liq.; cloud pt. 42 C (10% aq.); 90% act.

Antarox® BL-600. [Rhone-Poulenc Surf.] Modified aliphatic polyether; nonionic; biodeg. low-foaming surfactant; hydrotrope for use in highly alkaline built liq. conc. cleaners; hard surf. cleaner; liq. DISCONTINUED.

Antarox® BO/327 (formerly Soprophor® BO/327). [Rhone-Poulenc Geronazzo] Ethoxy propoxylated alcohol; nonionic; biodeg. surfactant for controlled foam alkaline degreasing compds.; liq.; 100% conc.

Antarox® E-100 (formerly Supronic E-100). [Rhone-Poulenc France] Poloxamer 401; surfactant.

Antarox® EGE 25-2 (formerly Alkatronic® EGE 25-2). [Rhone-Poulenc Surf. Canada] Propoxylated POE glycol; nonionic; surfactant, defoamer/foam control agent for pulp and paper industry; dye leveling and wetting agent for textiles processing; liq.; 100% conc.

Antarox® EGE 31-1. [Rhone-Poulenc Surf. Canada] Propoxylated POE glycol; nonionic; wetting, spreading agent with low foaming props.; for windshield washer, antifreeze, and rinse aid formulations; liq.; 100% conc.

Antarox® F88 (formerly Pegol® F88). [Rhone-Poulenc Surf.] EO/PO block copolymer; nonionic; defoamer, dispersant, wetting agent, emulsifier, demulsifier, leveling agent, detergent, foam booster, viscosifier for industrial/household cleaners, lt. duty liqs., syndet bars, toilet tank blocks; flake; HLB 28.0; pour pt. 54 C; cloud pt. > 100 C (1% aq.); 100% conc.

Antarox® FM 33 (formerly Rhodiasurf FM). [Rhone-Poulenc France] Alkoxylated alcohol; nonionic; low foaming detergent and wetting agent for liq. detergent formulations, textile processing, latex emulsions, cutting oils; liq.; 99% conc.

Antarox® FM 53 (formerly Rhodiasurf FM 53). [Rhone-Poulenc France] Alkoxylated branched chain alcohol; nonionic; low foaming detergent and wetting agent for automatic dish detergents, liq.

detergents, textile processing, metalworking compds.; liq.; 99% conc.

Antarox® FM 63 (formerly Rhodiasurf FM 63). [Rhone-Poulenc France] Alkoxylated fatty alcohol; nonionic; low foaming detergent and wetting agent for liq. detergents, textile processing, metalworking compds.; liq.; 95% conc.

Antarox® L-61 (formerly Pegol® L-61). [Rhone-Poulenc Surf.; Rhone-Poulenc France] EO/PO block copolymer; nonionic; defoamer, dispersant, wetting agent, emulsifier, demulsifier, leveling agent, detergent, lubricant for household/industrial cleaners, metalworking fluids, agric. formulations, rinse aids, automatic dishwashing, water treatment; EPA compliance; liq.; sol. in oils; HLB 3.0; pour pt. -29 C; cloud pt. 24 C (1% aq.); flash pt. > 200 C; 100% act.

Antarox® L-62 (formerly Pegol® L-62). [Rhone-Poulenc Surf.] EO/PO block copolymer; nonionic; defoamer, dispersant, wetting agent, emulsifier, demulsifier, leveling agent, detergent for household/industrial cleaners, hard surf. cleaners, laundry, skin care prods., emulsion polymerization;; liq.; HLB 7.0; pour pt. -4 C; cloud pt. 32 C (1% aq.); 100% conc.

Antarox® L-62 LF (formerly Pegol® L-62LF). [Rhone-Poulenc Surf.] EO/PO block copolymer; nonionic; defoamer, dispersant, wetting agent, emulsifier, demulsifier, leveling agent, detergent for industrial/household cleaners, hard surf. cleaning, laundry, skin care prods., emulsion polymerization; liq.; HLB 7.0; pour pt. -10 C; cloud pt. 28 C (1% aq.); 100% conc.

Antarox® L-64 (formerly Pegol® L-64). [Rhone-Poulenc Surf.] EO/PO block copolymer; nonionic; defoamer, dispersant, wetting agent, emulsifier, demulsifier, leveling agent, detergent for industrial/household cleaners, hard surf. cleaning, laundry, skin care, emulsion polymerization;; liq.; HLB 15.0; pour pt. 16 C; cloud pt. 59 C (1% aq.); 100% conc.

Antarox® LA-EP 15 (formerly Alkasurf® LA-EP 15). [Rhone-Poulenc Surf.] Modified oxyethylated straight chain alcohol; nonionic; detergent, dispersant, wetting agent, emulsifier; for controlled foam applics., machine dishwashing, rinse aid compositions; liq.; HLB 7.0; 100% conc.

Antarox® LA-EP 16 (formerly Alkasurf® LA-EP 16). [Rhone-Poulenc Surf.] Modified oxyethylated straight chain alcohol; nonionic; detergent, dispersant, wetting agent, emulsifier; for general, industrial and household detergent prods.; nongelling in aq. dilutions; liq.; HLB 13.1; 100% conc.

Antarox® LA-EP 25 (formerly Alkasurf® LA-EP 25). [Rhone-Poulenc Surf.] Modified oxyethylated straight chain alcohol; nonionic; biodeg. surfactant, foamer, wetting agent for machine dishwashing, rinse aids, industrial and household cleaners; liq.; water-sol.; HLB 7.0; 100% conc.

Antarox® LA-EP 25LF (formerly Alkasurf® LA-EP 25LF). [Rhone-Poulenc Surf.] Modified oxyethylated straight chain alcohol; nonionic; biodeg. surfactant, foamer, wetting agent for machine dishwashing, rinse aids, industrial and household cleaners; low foam version; liq.; water-sol.; HLB 7.0; 100% conc.

Antarox® LA-EP 45 (formerly Alkasurf® LA-EP 45). [Rhone-Poulenc Surf.] Modified oxyethylated straight chain alcohol; nonionic; biodeg. low foaming surfactant, wetting agent for machine dishwashing, rinse aids, industrial and household cleaners; liq.; water-sol.; HLB 10.0; 100% conc.

Antarox® LA-EP 59 (formerly Alkasurf® LA-EP 59). [Rhone-Poulenc Surf.] Modified oxyethylated straight chain alcohol; nonionic; biodeg. low foaming surfactant, wetting agent for machine dishwashing, rinse aids, industrial and household cleaners; liq.; water-sol.; HLB 10.0; 100% conc.

Antarox® LA-EP 65. [Rhone-Poulenc Surf.] Modified oxyethylated straight chain alcohol; nonionic; biodeg. low foaming surfactant, wetting agent for machine dishwashing, rinse aids, industrial and household cleaners; liq.; water-sol.; HLB 11.5; 100% conc. DISCONTINUED.

Antarox® LA-EP 73 (formerly Alkasurf® LA-EP 73). [Rhone-Poulenc Surf.] Modified oxyethylated straight chain alcohol; nonionic; biodeg. low foaming surfactant, wetting agent for machine dishwashing, rinse aids, industrial and household cleaners; increased water-sol. and tolerance to salt levels at higher use temps.; liq.; water-sol.; HLB 14.0; 100% conc.

Antarox® LA-EPB-17 (formerly Alkasurf LA-EPB-17). [Rhone-Poulenc Surf. Canada] Benzyl capped straight chain aliphatic alcohol alkoxylate; nonionic; low foaming biodeg. surfactant for stable alkaline enhanced sol'ns.; Gardner ≤ 2 clear liq. above 35 C, mild typ. odor; sol. in water below cloud pt., most common solvs. incl. alcohols, aromatic and chlorinated solvs.; insol. in min. oils; dens. 1.02 g/ml; pour pt. < 12 C; cloud pt. 14-19 C (1% aq.); flash pt. (PMCC) > 100 C; pH 6-8 (5% aq.); 99% min. act.

Antarox® LF-222. [Rhone-Poulenc Surf.] Aromatic polyether, modified; nonionic; wetting and rewetting agent for spray metal cleaning; stable to acids and mild alkalies; liq.; cloud pt. 23 C (1% aq.); 100% conc.

Antarox® LF-224. [Rhone-Poulenc Surf.] Modified aliphatic EO/PO adduct; nonionic; low-foaming surfactant, wetting agent for industrial/household cleaners; liq.; cloud pt. 24 C (1% aq.); 99% act.

Antarox® LF-330. [Rhone-Poulenc Surf.; Rhone-Poulenc France] Linear aliphatic polyether, chlorine-capped; nonionic; low foaming detergent, wetting agent for rinse aid concs., household and industrial detergents, laundry and dairy cleaners; dedusting agent for powd. detergents; pale yel., clear liq.; bland odor; sol. water and most org. solv. except paraffins; visc. 29 cs; cloud pt. 30 C (10% aq.); surf. tens. 29 dynes/cm (0.1%); 95% act.

Antarox® LF-344. [Rhone-Poulenc Surf.] Linear aliphatic polyether, chlorine-capped; nonionic; low foaming detergent, wetting agent for industrial/household cleaners; pale yel., clear liq.; bland odor; sol. water and most org. solv. except paraffins; visc. 61 cs; cloud pt. 43 C (10% aq.); surf. tens. 33 dynes/cm (0.1%); 90% act.

Antarox® P-84 (formerly Pegol® P-84). [Rhone-Poulenc Surf.] EO/PO block copolymer; nonionic; defoamer, dispersant, wetting agent, emulsifier, demulsifier, leveling agent, detergent, viscosifier, foam booster for industrial/household cleaners, rinse aids, automatic dishwashing, metal treatment, water treatment, asphalt systems, w/o emulsions; paste; HLB 14.0; pour pt. 34 C; cloud pt. 74 C (1% aq.); 100% conc.

Antarox® P-104 (formerly Pegol® P-104). [Rhone-Poulenc Surf.] EO/PO block copolymer; nonionic; defoamer, dispersant, wetting agent, emulsifier, demulsifier, leveling agent, detergent, foam booster, viscosifier for industrial/household cleaners, agric.

formulations; EPA compliance; paste; HLB 13.0; cloud pt. 83 C (1% aq.); 100% conc.

Antarox® PGP 18-1 (formerly Alkatronic PGP 18-1). [Rhone-Poulenc Surf.] Ethoxylated propoxylated glycol; nonionic; surfactant, detergent, emulsifier; liq.; HLB 3.0; 100% conc.

Antarox® PGP 18-2 (formerly Alkatronic PGP 18-2). [Rhone-Poulenc Surf.] Ethoxylated propoxylated glycol; nonionic; surfactant, detergent, emulsifier; liq.; HLB 7.0; 100% conc.

Antarox® PGP 18-2D (formerly Alkatronic PGP 18-2D). [Rhone-Poulenc Surf.] Ethoxylated propoxylated glycol; nonionic; surfactant, detergent, emulsifier; liq.; 100% conc.

Antarox® PGP 18-2LF (formerly Alkatronic PGP 18-2LF). [Rhone-Poulenc Surf.] Ethoxylated propoxylated glycol; nonionic; surfactant, detergent, emulsifier; liq.; 100% conc.

Antarox® PGP 18-4 (formerly Alkatronic PGP 18-4). [Rhone-Poulenc Surf.] Ethoxylated propoxylated glycol; nonionic; surfactant, detergent, emulsifier; liq.; HLB 15.0; 100% conc.

Antarox® PGP 18-8 (formerly Alkatronic PGP 18-8). [Rhone-Poulenc Surf.] Ethoxylated propoxylated glycol; nonionic; surfactant, detergent, emulsifier; liq.; HLB 29.0; 100% conc.

Antarox® PGP 23-7 (formerly Alkatronic PGP 23-7). [Rhone-Poulenc Surf. Canada] Poloxamer 237; CAS 9003-11-6; nonionic; coemulsifier for cosmetics, toiletries, pulp and paper defoamers; dispersant, visc. control agent; wh. flake; sol. in water, aromatic and chlorinated solvs.; insol. in min. oil and aliphatic solvs.; m.w. 6700-8373; dens. 1.04 g/ml; HLB 24.0; cloud pt. > 100 C (1% aq.); pH 5.0-7.5 (2.5% aq.); foam height 44 mm (0.1%); 99% min. act.

Antarox® PGP 33-8. [Rhone-Poulenc Surf.] Ethoxylated propoxylated glycol; nonionic; surfactant, detergent, emulsifier; liq.; HLB 27.0; 100% conc. DISCONTINUED.

Antarox® PL/Series (formerly Soprofor PL/Series). [Rhone-Poulenc Geronazzo] POE/POP block polymer; nonionic; detergent, emulsifier, wetting and antistat, defoamer, deduster; 100% conc.

Antarox® RA 40. [Rhone-Poulenc Surf.] Alkoxylated alcohol; nonionic; low foaming surfactant for use in rinse aids and metal cleaners; liq.; 100% conc.

Antifoam 20WB. [Stockhausen] Water-based antifoam for screen room, paper machine, bleaching applics., waste water treatment, chem. processing industry, latex, coatings, metal treating and electroplating; FDA compliance; creamy wh. emulsion, mild petrol. odor; disp. in water; sp.gr. 0.93; dens. 7.8 lb/gal; visc. 800 ± 200 cps; f.p. 0 C; b.p. 212 F; flash pt. (PMCC) 330 F; toxicology: very low toxicity; mild skin or eye irritant on prolonged contact.

AntiFoam 55. [Yorkshire Pat-Chem] Silicone defoamer; free-rinsing defoamer for garment dyeing.

Antifoam 6031. [Stockhausen] Sulfocarboxylic acid/complex fatty blend; defoamer for mfg. of phos-acid and fertilizer; yel. clear liq., char. mild odor; misc. with water; sp.gr. 0.985; visc. 100-110 mPa•s; b.p. > 212 F; flash pt. (PMCC) > 200 F; pH 9.2-9.8 (10%); toxicology: eye and skin irritant.

Antifoam 7800 New. [Miles/Organic Prods.] Higher hydrocarbons and their sulfonic acid derivs.; antifoam for aq. systems in sugar, fertilizer, phosphoric acid, dyestuffs, paper, leather, plastics, and chemical industries; resistant to acid and weak alkalis; faint yel. liq., sl. turbid @ R.T.; sp.gr. 0.87; b.p. 80 C; flash pt. > 100 C.

Antifoam Base 263. [Soluol] Silicone containing compd.; defoamer; water-sol.

Antifoam CM Conc. [Harcros] Blend of antifoaming and defoaming agents with silicone synergist, o/w emulsion, nonionic emulsifier; used in industrial processing; carpet cleaning, wood pulping, paper coating, and textile mfg.; wh. liq. lt. cream; disp. in water; sp.gr. 0.972; pH 6.5–7.5 (1% aq.).

AntiFoam D-10. [Yorkshire Pat-Chem] Silicone defoamer; defoamer for all textile processes; exc. stability over wide pH range.

Antifoam E-20. [Kao Corp. SA] Emulsified denatured silicone; defoamer for tech. applic.; liq.

Antifoam-G. [Soluol] Silicone emulsion; defoamer; water-sol.

Antifoam GEB. [Am. Emulsions] Silicone defoamer; free-rinsing defoamer for continuous and beck dyeing of carpets.

AntiFoam K-20. [Yorkshire Pat-Chem] Conc. silicone defoamer; defoamer for textile applics.; recommended for dilution.

AntiFoam NIL-100. [Yorkshire Pat-Chem] Nonsilicone defoamer; for textile wet processing, dyeing, printing, and finishing applics.

Antifoam FRS. [Am. Emulsions] Blended silicones and emulsifiers; highly effective defoamer and antifoam for ambient and high temp. operations; free rinsing on carpets.

Antifoam Q-41. [Soluol] Silicone compd.; defoamer; sol. in solvs.

Anti-foam TP. [BASF AG] Antifoam for pigment printing of textiles.

Antifoam VOL. [Am. Emulsions] Polysiloxane; defoamer for use in dyebaths, finish mixes where stability is essential.

Antil® 141 Liq. [Goldschmidt] Propylene glycol, PEG-55 propylene glycol oleate; nonionic; thickener for aq. sol'ns. of surfactants, e.g., shampoos, foam baths, shower preps., liq. soaps; solubilizes essential oils into aq. surfactant systems; liq. version developed for cold process systems.; liq.; 40% conc.

Antispumin ZU. [Stockhausen] Ethoxylated/propoxylated fatty alcohols; nonionic; defoamer for beet sugar industry for flume water, diffusion, liming, and carbonation; also for paper mill effluent; yel. sl. cloudy oily visc. liq., typ. odor; forms unstable emulsion in water; sp.gr. 0.92; visc. 100 cp; b.p. 412 F; pour pt. -30 C; flash pt. 323 F; pH 5.5 (10%); toxicology: skin irritant on prolonged contact; 100% act.

Anti-Terra®-202. [Byk-Chemie USA] Alkylammonium salt of a higher m.w. polycarboxylic acid sol'n.; wetting and dispersing additive to prevent settling and flooding of pigments; gellant for organophilic bentonites; for coating systems; sp.gr. 0.83-0.87 g/cc; dens. 6.93-7.26 lb/gal; acid no. 48-54; flash pt. (Seta) 40 C; 50-54% NV in Stod./2-butoxyethanol (9/1).

Anti-Terra®-204. [Byk-Chemie USA] Sol'n. of higher m.w. carboxylic acid salts of polyamine amides; wetting and dispersing additive to improve anti-sedimentation and sagging props.; used in alkyds, PVC copolymers, chlorinated rubber, epoxy systems; gellant for organophilic bentonites; sp.gr. 0.92-0.94 g/cc; dens. 7.67-7.84 lb/gal; acid no. 37-44; flash pt. (Seta) 31 C; ref. index 1.466-1.473; 50-54% NV in methoxypropanol/naphtha (3/2).

Anti-Terra®-207. [Byk-Chemie USA] Sol'n. of an alkylammonium salt of an unsat. fatty acid; anionic; wetting and dispersing additive to improve pigment wetting and anti-sedimentation; used in alkyd, acrylic, and polyester systems; sp.gr. 0.96-1.00 g/cc; dens. 8.01-8.35 lb/gal; acid no. 90-110; flash pt. (Seta) 80 C; 78-82% NV in dipropylene glycol methyl ether.

Anti-Terra®-P. [Byk-Chemie USA] Phosphoric acid salt of long chain carboxylic acid polyamine amides; cationic; wetting and dispersing additive to prevent settling and flooding of pigments in alkyds, alkyd-melamine, chlorinated rubber systems, PVC copolymers, acid catalyzed paints; sp.gr. 0.98-1.00 g/cc; dens. 8.18-8.35 lb/gal; acid no. 160-180; flash pt. (Seta) 26 C; 39-44% NV in isobutanol/xylene/water (3/1/1).

Anti-Terra®-U. [Byk-Chemie USA] Salt of unsat. polyamine amides and higher m.w. acidic esters sol'n.; wetting and dispersing additive; gellant for organophilic bentonites; for coating systems; sp.gr. 0.93-0.95 g/cc; dens. 7.76-7.93; acid no. 20-28; flash pt. (Seta) 25 C; 48-52% NV in xylene/isobutanol (8/1).

Anti-Terra®-U80. [Byk-Chemie USA] Salt of unsat. polyamine amides and higher m.w. acidic esters sol'n.; wetting and suspending additive for latex and water-sol. coatings; wetting/dispersing agent for colored pigments; sp.gr. 0.97-1.01 g/cc; dens. 8.09-8.43 lb/gal; acid no. 32-48; flash pt. (Seta) 66 C; ref. index 1.473-1.483; 78-82% NV in 2-butoxyethanol/xylene (8/1).

Apex Pentrapex #1923. [Apex] Wetting agent, penetrating agent; yel. liq.; misc. with water.

Apexscour #1609 [Apex] Detergent, continuous scouring agent for cotton and polyester/cotton fabrics; fast wetting, good detergency and stability in strong alkali sol'ns.; liq.; water sol.

APG® 225 Glycoside. [Henkel/Emery] C8-10 alkyl polysaccharide ether; CAS 68515-73-1; nonionic; caustic-stable wetting agent and coupler for agric. formulations; biodeg.; EPA-exempt; clear liq.; m.w. 405; HLB 13.5; pH 6 (1%); surf. tens. 29 dynes/cm (0.1%); 68% act. in water. DISCONTINUED.

APG® 300 Glycoside. [Henkel/Emery] C9-11 alkyl polysaccharide ether; CAS 113976-90-2; nonionic; caustic-stable degreaser, emulsifier, dispersing and wetting agent; for general purpose and hard surf. cleaners, agric. formulations; biodeg.; clear liq.; m.w. 390; HLB 12.5; pH 9 (1%); surf. tens. 26 dynes/cm (0.1%); 50% act. in water. DISCONTINUED.

APG® 325 Glycoside. [Henkel/Emery] C9-11 alkyl polysaccharide ether; CAS 113976-90-2; nonionic; caustic-stable degreaser, emulsifier, dispersing and wetting agent; for general purpose and hard surf. cleaners; biodeg.; liq.; m.w. 420; HLB 13.0; surf. tens. 27 dynes/cm (0.1%); 50% act. in water.

APG® 600 Glycoside. [Henkel/Emery] C12-14 alkyl polysaccharide ether; CAS 110615-47-9; nonionic; caustic-stable detergent active; visc. modifier for crutcher spray-dryer slurries; biodeg.; liq.; m.w. 420; HLB 11.5; surf. tens. 27 dynes/cm (0.1%); 50% act. in water.

APG® 625 Glycoside. [Henkel/Emery] C12-14 alkyl polysaccharide ether; CAS 110615-47-9; nonionic; caustic-stable, moderate foaming detergent active; visc. modifier for crutcher spray-dryer slurries; biodeg.; liq.; m.w. 455; HLB 12.0; surf. tens. 28 dynes/cm (0.1%); 50% act. in water.

Aphrogene 5001. [ICI Surf. UK] Silicone emulsion; nonionic; antifoam for aq. textile dyeing and finishing systems; liq.

Aphrogene Jet. [ICI Surf. UK] Polydimethylsiloxane emulsion; nonionic; for high temp. dyeing in jet, beam, and package machines; low visc. liq.

Appretan Grades. [Hoechst AG] Polymer dispersions; wash-fast finishing agents.

Aquabase. [Westbrook Lanolin] Fatty alcohols/PEG ester blend; nonionic; base for o/w emulsions; flake; HLB 10.0.

Aquafoam 9451. [Aquaness] Ethoxy ether sulfate; anionic; heavy brine foaming agent for foam drilling; liq.; 99% conc.

Aquafoam 9452. [Aquaness] Ethoxy ether sulfate; anionic; foaming agent for foam drilling; liq.; 80% conc.

Aqualose L30. [Westbrook Lanolin] PEG-30 lanolin; CAS 61790-81-6; nonionic; emollient, emulsifier, plasticizer, solubilizer; wax.; HLB 14.0; 100% conc.

Aqualose L75. [Westbrook Lanolin] PEG-75 lanolin USP; CAS 61790-81-6; nonionic; emollient, emulsifier for personal care prods.; plasticizer in aerosol hair sprays; solubilizer for perfume and germicidal agents; conditioner for shampoos; superfatting agent for soap; wax; HLB 16.0; 100% conc.

Aqualose L75/50. [Westbrook Lanolin] PEG-75 lanolin USP; CAS 61790-81-6; nonionic; emollient, emulsifier; gel; 50% conc.

Aqualose LL100. [Westbrook Lanolin] PPG-40-PEG-60 lanolin oil; nonionic; emollient, emulsifier, plasticizer, solubilizer used in aq./alcoholic preparations of alcohol content; liq.; HLB 13.0; 100% conc.

Aqualose W20. [Westbrook Lanolin] Laneth-20; nonionic; plasticizer and solubilizer for hydrophobic substances; emollient, emulsifier; wax.

Aqualose W20/50. [Westbrook Lanolin] Laneth-20; nonionic; emollient, emulsifier; gel; 50% conc.

Aqualox® 225-100, 225A-100. [Alox] Amine salt; surfactant, corrosion inhibitor, lubricant, antiwear additive effective in inhibiting the attack of ferrous metals by aq. sol'ns.; for metalworking formulations; use 225-100 for EP/antiwear use; 225A-100 grade contains no phosphorous; pale yel.; mild odor; sol. in water @ < 2% and > 50%; sp.gr. 1.06 (15.6 C); dens. 8.8 lb/gal (15.6 C); visc. < 550 cs (40 C); flash pt. 150 C; pour pt. –25.5 C; pH 9.7 ± 0.3; biodeg. (225A-100).

Aqualox® 232. [Alox] Amine salts of org. acids; corrosion inhibitor, low foaming surfactant for syn. metalworking formulations, esp. in aq. sol'n.; pale amber liq.; mild odor; sol. in water, methanol, ethanol, IPA, glycol, water/alcohol mixts.; sp.gr. 1.15 (15.6 C); dens. 9.6 lb/gal (15.6 C); visc. 75 ± 15 cs (100 F); pour pt. –29 C max.; pH 7.6 ± 0.3; biodeg.; 75% act.

Aquamul Series. [Aquaness] Modified fatty acids; anionic/cationic; emulsifiers for oil external, invert drilling fluids; liq.; oil-sol.

Aquaperle D34. [ICI Surf. UK] Paraffin wax emulsion containing aluminum and zirconium salts; cationic; waterproofing agent for natural or syn. fibers; liq.

Aquaphil K. [Westbrook Lanolin] Lanolin and lanolin alcohol; nonionic; emollient, emulsifier with enhanced w/o emulsion stability; soft solid; HLB 4.5; 100% conc.

Aquasol® AR 90. [Am. Cyanamid/Textiles] Sulfonated castor oil; anionic; wetting, penetrating, leveling agent for dyeing cotton and rayon with direct colors; dispersant, penetrant for vat and naphthol colors; emulsifier for self-scouring fiber and yarn lubricants; liq.; 79% conc.

Aquasol® W 90. [Am. Cyanamid/Textiles] Sulfonated castor oil; anionic; wetting, penetrating, leveling agent for dyeing cotton and rayon with direct colors; dispersant, penetrant for vat and naphthol colors; emulsifier for self-scouring fiber and yarn lubricants; liq.; 78% conc.

Aramide CDM-4. [Aquaness] Alkanolamine condensate; nonionic; visc. builder for all-purpose cleaners, emulsifiers, and lubricants; liq.; 100% conc.

Arbreak Series. [Aquaness] Polyol, resin ester oxyalkylates; anionic/nonionic; o/w and w/o emulsion breakers; liq.; 100% conc.

Arbyl 18/50. [Grünau] Fatty alcohol polyglycol ether; nonionic; precleaning and leveling agent; liq.; 50% conc.

Arbyl ASN. [Grünau] Sulfosuccinate; anionic; wetting agent; liq.; 60% conc.

Arbyl N. [Grünau] Fatty alcohol polyglycol ether; nonionic; washing, wetting and aftersoaping agent used in textile prods.; paste; 50% conc.

Arbyl R Conc. [Grünau] Polyglycol ether; nonionic; cold wetting agent; liq.; 100% conc.

Arbylen Conc. [Grünau] Alkylphenol polyglycol ether; nonionic; washing and wetting agent used in textile prods.; liq., paste; 50–100% conc.

Ardril DMD. [Aquaness] Detergent; anionic/nonionic; detergent for mud drilling; emulsifier for oil; aids penetration by reducing torque and minimizing bit-balling; liq.

Ardril DME. [Aquaness] Blended emulsifier; anionic/nonionic; emulsifier for surfactant muds when oil is added to the drilling mud system; liq.; 100% conc.

Ardril DMS. [Aquaness] Aryl polyglycol ether; nonionic; surfactant for drilling muds; liq.; 70% conc.

Aremul AIS. [Arol Chem. Prods.] Detergent, emulsifier; amber liq.; mild odor; sol. in various aliphatic and chlorinated solv.; sp.gr. 1.03; 90–95% act.

Aremul DCL. [Arol Chem. Prods.] Blend of anionic sulfonates; detergent, emulsifier; dark amber liquid; mild odor; miscible sol. with water and ODCB, TCB, butyl benzoate and methyl salicylate; sp.gr. 1.04; 55–60% act. DISCONTINUED.

Aremul V-91, X-91. [Arol Chem. Prods.] Blend of alkylaryl sulfonates, modified nonionics and org. coupling agents; detergent, emulsifier; amber liq., mild odor (V-91); sol. in warm water, aliphatic solv. (V-91); sp.gr. 1.03 (X-91); 90–95% act. DISCONTINUED.

Arflow 168. [Aquaness] Sodium alkyl sulfate; paraffin wash or remover; liq.

Arfoam 2213. [Aquaness] Blend; anionic; foaming agent to remove water and drilled solids during drilling operations; liq.; 100% conc.

Arfoam 2386. [Aquaness] Blend; anionic; foaming agent to remove water and drilled solids during drilling operations; liq.; 100% conc.

Argobase EU. [Westbrook Lanolin] Sterols and sterol esters lanolin extracts; nonionic; emollient, emulsifier, stabilizer; absorp. base; paste; HLB 4.0; 6% conc.

Argobase LI. [Westbrook Lanolin] Sterols and sterol esters lanolin extracts; nonionic; emollient, emulsifier, stabilizer; absorp. base; liq.; HLB 4.0; 24% conc.

Argobase MS-5. [Westbrook Lanolin] Sterols and sterol esters lanolin extracts; nonionic; w/o emulsifier, emollient, stabilizer; paste; HLB 5.0; 25% conc.

Argobase SI. [Westbrook Lanolin] Sterols and sterol esters lanolin extracts; nonionic; emollient, emulsifier, stabilizer; absorp. base; paste; HLB 4.0; 92% conc.

Argowax Cosmetic Super. [Westbrook Lanolin] Lanolin alcohol; CAS 8027-33-6; nonionic; gelling agent, emulsifier; wax; low color and odor; HLB 2.6; 100% conc.

Argowax Dist. [Westbrook Lanolin] Lanolin alcohol; nonionic; gelling agent, w/o emulsifier; lt. wax; slight odor; HLB 2.6; 100% conc.

Argowax Standard. [Westbrook Lanolin] Lanolin alcohol BP; nonionic; gelling agent, w/o emulsifier; wax; HLB 2.6; 100% conc.

Aristonate H. [Pilot] Sodium petroleum sulfonate (CAS #78330-12-8); anionic; surfactant for formulating drycleaning soaps, cutting oils, textile oils, leather oils, rust preventive and fuel oil compositions; ore floation collectors; emulsifiers for agric. sprays; liq.; 62% conc.

Aristonate L. [Pilot] Sodium petroleum sulfonate (CAS #78330-12-8); anionic; surfactant for formulating drycleaning soaps, cutting oils, textile oils, leather oils, rust preventive and fuel oil compositions; ore floation collectors; emulsifiers for agric. sprays; liq.; 62% conc.

Aristonate M. [Pilot] Sodium petroleum sulfonate (CAS #78330-12-8); anionic; surfactant for formulating drycleaning soaps, cutting oils, textile oils, leather oils, rust preventive and fuel oil compositions; ore floation collectors; emulsifiers for agric. sprays; liq.; 62% conc.

Arizona 208. [Arizona] Tall oil fatty acid ester; for plasticizers, extenders, surface-act. agents in grinding and cutting oils, specialty lubricant additives, corrosion inhibitors, specialty solvs. for printing inks, metalworking, and oil well servicing; Gardner 2+ color; sp.gr. 0.87; visc. 6-7 cps; acid no. 0.5; iodine no. 98; cloud pt. < 4 C; flash pt. (CC) > 94 C.

Arizona 258. [Arizona] Tall oil fatty acid ester; for plasticizers, extenders, surface-act. agents in grinding and cutting oils, specialty lubricant additives, corrosion inhibitors, specialty solvs. for printing inks, metalworking, and oil well servicing; Gardner < 1 color; sp.gr. 0.90; visc. 20 cps; acid no. 0.2; iodine no. 91; flash pt. (OC) 204 C.

Arizona 2154. [Arizona] Tall oil fatty acid ester; for plasticizers, extenders, surface-act. agents in grinding and cutting oils, specialty lubricant additives, corrosion inhibitors, specialty solvs. for printing inks, metalworking, and oil well servicing; Gardner 2 color; sp.gr. 0.87; visc. 6-7 cps; acid no. 0.3; iodine no. 97; cloud pt. < 4 C; flash pt. (CC) > 94 C.

Arizona DR-22. [Arizona] Disproportionated rosin; emulsifier, detergent, wetting agent; used to prepare emulsifiers for SBR polymerization, as shortstop for solv. polymerizations of rubber, as plasticizer/tackifier; used to make Arizona disproportionated tall oil rosin soaps (DRS-40, 42, 43, 44); Gardner 6+ color; sp.gr. 1.06; dens. 8.2 lb/gal (150 C); soften. pt. (R&B) 59 C; acid no. 161; sapon. no. 165; flash pt. (OC) > 204 C; 100% solids.

Arizona DR-24. [Arizona] Disproportionated rosin; emulsifier for SBR prod.; intermediate for prod. of disproportionated rosin soaps; Gardner 8-9 color; sp.gr. 1.06; dens. 8.2 lb/gal (150 C); soften. pt. (R&B) 45 C; acid no. 163; flash pt. (OC) > 204 C.

Arizona DR-25. [Arizona] Sodium soap of disproportionated rosin; for preparing emulsifiers for SBR polymerization; Gardner 8 color; sp.gr. 1.06; dens. 8.2 lb/gal; (150 C); soften. pt. (R&B) 73 C; acid no. 138; flash pt. (OC) > 204 C.

Arizona DR Mix-26. [Arizona] Emulsifier-mixed disproportionated rosin and fatty acid; emulsifier for SBR industry; paste; 100% conc.

Arizona DRS-40. [Arizona] Potassium soap of disproportionated tall oil rosin; anionic; emulsifier, detergent, wetting agent; for ABS, SBR, other syn. elastomers; tan paste; bland odor; water disp.; sp.gr. 1.1; dens. 9.08 lb/gal (80 C); visc. 1800 cps (65 C); acid no. 16; flash pt. (OC) > 204 C; 79.5% solids.

Arizona DRS-42. [Arizona] Potassium soap of disproportionated tall oil rosin; anionic; emulsifier, detergent, wetting agent; for ABS, SBR, other syn. elastomers; tan paste, bland odor; water disp.; dens. 9.08 lb/gal (80 C); visc. 1500 cps (65 C); acid no. 12.5; flash pt. (OC > 204 C; 80% solids.

Arizona DRS-43. [Arizona] Sodium salt of disproportionated tall oil rosin; anionic; emulsifier, detergent, wetting agent; emulsifier for ABS, SBR, syn. elastomers; tan paste; bland odor; water disp.; sp.gr. 1.1; dens. 8.92 lb/gal (80 C); visc. 2500 cps (65 C); acid no. 11.5; flash pt. (OC) > 204 C; 71% solids.

Arizona DRS-44. [Arizona] Sodium soap of disproportionated tall oil rosin; anionic; emulsifier for prod. of ABS, SBR and other syn. elastomers; paste; 70% conc.

Arizona DRS-50. [Arizona] Potassium soap of disproportioned tall oil rosin; anionic; polymerization emulsifier for the SBR industry; paste; water-disp.; dens. 9.09 lb/gal (80 C); visc. 1800 cps (65 C); flash pt. > 400 F; acid no. 26; 84.5% solids.

Arizona DRS-51E. [Arizona] Potassium soap of disproportionated tall oil rosin; anionic; polymerization emulsifier for SBR industry; paste; water-disp.; dens. 9.08 lb/gal (80 C); visc. 800 cps (65 C); flash pt. > 400 F; acid no. 19; 79.5% solids.

Arkofil Brands. [Hoechst AG] Syn. polymers with special additives; sizing agents for spun yarns; dustfree gran.

Arkofix Grades. [Hoechst AG] Reactant and self-crosslinking resins; for resin finishing; liq.

Arkomon A Conc. [Hoechst AG] Fatty acid sarcoside, sodium salt; anionic; basic material for detergents and cleaning agents; corrosion inhibitor; antistat/softener for fiber mfg. and processing; paste; 60% conc.

Arkomon SO. [Hoechst AG] Oleoyl sarcoside; anionic; corrosion inhibitor for metalworking fluids.; starting material for emulsifiers; liq.

Arkopal N040. [Hoechst AG] Nonoxynol-4; CAS 9016-45-9; nonionic; detergent, wetting agent, emulsifier, surface-active raw material for general industrial use; 100% conc.

Arkopal N060. [Hoechst Celanese] Nonoxynol-6; CAS 9016-45-9; nonionic; detergent, wetting agent, emulsifier, surface-active raw material for general industrial use; 100% conc.

Arkopal N080. [Hoechst Celanese] Nonoxynol-8; CAS 9016-45-9; nonionic; detergent, wetting agent, emulsifier, surface-active raw material for general industrial use; 100% conc.

Arkopal N090. [Hoechst Celanese] Nonoxynol-9; CAS 9016-45-9; nonionic; detergent, wetting agent, emulsifier, surface-active raw material for general industrial use; 100% conc.

Arkopal N100. [Hoechst Celanese] Nonoxynol-10; CAS 9016-45-9; nonionic; detergent, wetting agent, emulsifier, surface-active raw material for general industrial use; 100% conc.

Arkopal N110. [Hoechst Celanese] Nonoxynol-11; CAS 9016-45-9; nonionic; detergent, wetting agent, emulsifier, surface-active raw material for general industrial use; 100% conc.

Arkopal N130. [Hoechst Celanese] Nonoxynol-13; CAS 9016-45-9; nonionic; detergent, wetting agent, emulsifier, surface-active raw material for general industrial use; 100% conc.

Arkopal N150. [Hoechst Celanese] Nonoxynol-15; CAS 9016-45-9; nonionic; detergent, wetting agent, emulsifier, surface-active raw material for general industrial use; 100% conc.

Arkopal N230. [Hoechst Celanese] Nonoxynol-23; CAS 9016-45-9; nonionic; detergent, wetting agent, emulsifier, surface-active raw material for general industrial use; 100% conc.

Arkophob NCS. [Hoechst AG] Polysiloxane emulsion; cationic; water-repellent finishing agent; emulsion.

Arkopon Brands. [Hoechst AG] Fatty acid condensation prods.; anionic; basic material for mfg. of detergents and cleaning agents; auxiliary for the metal and rubber industries; paste, powd.

Arkopon T Grades. [Hoechst AG] Sodium oleoyl methyl taurides; anionic; surfactants, wetting agents.

Arlacel® 129. [ICI Surf. Belgium] Glyceryl stearate; nonionic; coemulsifier; powd.; HLB 3.5; 100% conc.

Arlacel® 780. [ICI Surf. Belgium] Alkoxylated glyceryl sorbitan hydroxystearate; nonionic; emulsifier for w/o milks; suitable for cold emulsification technique; liq.; HLB 4.7; 100% conc.

Arlacel® A. [ICI Spec. Chem.; ICI Surf. Belgium] Mannide oleate; nonionic; surfactant for use in emulsified vaccines of the w/o type; liq.; HLB 4.3; 100% conc.

Arlacel® C. [ICI Spec. Chem.; ICI Surf. Belgium] Sorbitan sesquioleate; CAS 8007-43-0; nonionic; surfactant, w/o emulsifier; amber, oily liq.; sol. in min. oil, cottonseed oil, ethanol, IPA; sp.gr. 1.0; visc. 1000 cps; HLB 3.7; flash pt. > 300 F; HLB 3.7; 100% act.

Arlamol® F. [ICI Australia] PPG-11 stearyl ether; nonionic; surfactant; colorless liq.; sol. in alcohol, cottonseed and min. oil; insol. in water; visc. 1200 cps.

Arlamol® ISML. [ICI Australia] Isosorbide laurate; nonionic; surfactant; wh. solid; insol. in water, min. oil; pour pt. 47 C.

Arlasolve® 200. [ICI Spec. Chem.] Isoceteth-20; nonionic; surfactant, emulsifier, solubilizer for cosmetics; wh. soft waxy solid; sol. in water, alcohol, propylene glycol; HLB 15.7; pour pt. 36 C.

Arlasolve® DMI. [ICI Spec. Chem.] Dimethyl isosorbide; nonionic; surfactant, emollient; colorless liq.; sol. in water, alcohol, cottonseed oil, propylene glycol; visc. 6 cps.

Arlatone® 285. [ICI Spec. Chem.; ICI Surf. Belgium] POE castor oil; nonionic; cosmetic grade surfactant, coupling agent, solubilizer for perfumes, fragrances; semifluid; HLB 14.4; 100% conc.

Arlatone® 289. [ICI Spec. Chem.; ICI Surf. Belgium] POE hydrog. castor oil; nonionic; coupling agent, solubilizer, emulsifier for o/w creams and lotions; waxy; HLB 14.4; 100% conc.

Arlatone® 983. [ICI Spec. Chem.] POE fatty acid ester; nonionic; emulsifier; solid; 100% conc.

Arlatone® 985. [ICI Spec. Chem.; ICI Surf. Belgium] Ethoxylated stearyl stearate; nonionic; coemulsifier and visc. stabilizer for o/w milks; solid; HLB 7.5; 100% conc.

Arlatone® G. [ICI Spec. Chem.; ICI Surf. Belgium] PEG-25 hydrog. castor oil; CAS 61788-85-0; nonionic; surfactant, solubilizer, emollient; formulates clear gels; yel. visc. liq. to soft paste; sol. water, ethanol and IPA; sp.gr. 1.0; visc. 1400 cps; HLB 10.8; flash pt. > 300 F; 100% act.

Arlatone® T. [ICI Spec. Chem.; ICI Surf. Belgium] PEG-40 sorbitan peroleate; nonionic; emulsifier, solubilizer, antistat, lubricant, spreading agent; for bath oils, textile industry; yel. liq.; sol. in veg., min. oils, IPM, IPP; water disp.; sp.gr. 1; visc. 175 cps; HLB 9; flash pt. > 300F; 100% act.

Armac® 18D. [Akzo] Octadecylamine acetate, distilled; cationic; wetting agent, emulsifier, corrosion inhibitor, flotation agent for pigment flushing, froth flotation of mins., flocculation; wh. flake; water sol.; sp.gr. 0.827 (80 C); m.p. 85 C; flash pt. 320 F (COC); 100% act.

Armac® 18D-40. [Akzo] Stearamine acetate and stearamine; cationic; emulsifier, flotation agent, corrosion inhibitor; Gardner 11 max. flake; amine no. 190-200; 40% conc.

Armac® C. [Akzo; Akzo Chem. BV] Cocamine acetate; cationic; wetting agent, emulsifier, corrosion inhibitor, flotation reagent, stripping agent; yel. solid; m.w. 260; water sol.; sp.gr. 0.852 (80 C); HLB 11.2; m.p. 50 C; flash pt. 285 F (COC); 95% conc.

Armac® HT. [Akzo] (Hydrog. tallow) amine acetate; CAS 61790-59-8; cationic; wetting agent, emulsifier, corrosion inhibitor, flotation reagent; wh. flake; water-sol.; sp.gr. 0.820 (80 C); HLB 10.7; m.p. 60 C; flash pt. 325 F (COC); 95% act.

Armac® T. [Akzo; Akzo Chem. BV] Tallowamine acetate; cationic; flotation agent, corrosion inhibitor, lubricant, emulsifier; metal processing; Gardner 10 max. solid; water-sol.; sp.gr. 0.87 (60 C); HLB 10.8; m.p. 45–61 C; flash pt. 168 C (COC); 94% min act.

Armeen® 2-10. [Akzo] Didecylamine; CAS 1120-49-6; cationic; industrial surfactant; Gardner 2 max.; sp.gr. 0.84 (20 C); m.p. -10 C; iodine no. 0.5; amine no. 181; flash pt. (PMCC) > 132 C; 93% min. sec. amine.

Armeen® 2-18 [Akzo] Dioctadecylamine; CAS 112-99-2; cationic; industrial surfactant; Gardner 2 max.; sp.gr. 0.84 (75 C); visc. 11.7 cps (70 C); m.p. 80 C; iodine no. 1; amine no. 107; flash pt. (PMCC) > 149 C; 93% min. sec. amine.

Armeen® 2C. [Akzo] Dicocamine (sec. amine); CAS 61789-76-2; cationic; emulsifier, flotation agent, corrosion inhibitor; Gardner 2 max. solid; sol. in chloroform, slightly sol. in IPA, toluene, CCl_4, kerosene; sp.gr. 0.793 (60/40 C); visc. 49.1 SSU (60 C); m.p. 104–117 F; pour pt. 80 F; iodine no. 8; amine no. 140; flash pt. (PMCC) > 149 C; 93% sec. amine.

Armeen® 2HT. [Akzo] Di(hydrog. tallow) amine (sec. amine); CAS 61789-79-5; cationic; emulsifier, flotation agent, corrosion inhibitor; Gardner 2 max. solid; sp.gr. 0.79 (70 C); visc. 10.82 cps (70 C); m.p. 62 C; iodine no. 3; amine no. 110; flash pt. (PMCC) > 149 C; 93% sec. amine.

Armeen® 2T [Akzo] Ditallowamine; CAS 68783-24-4; cationic; emulsifier, flotation agent, corrosion inhibitor; Gardner 2 max.; sp.gr. 0.79 (70 C); visc. 10.32 cps (70 C); m.p. 55 C; iodine no. 30; amine no. 110; flash pt. (PMCC) > 149 C; 93% min. sec. amine.

Armeen® 3-12 [Akzo] Tridodecylamine; CAS 102-87-4; cationic; chemical intermediate for mfg. of sol. betaines and quat. ammonium salts; carrier for mfg. of citric acid and oil; Gardner 1 max. liq.; sp.gr. 0.82; m.p. -9 C; amine no. 102; flash pt. (PMCC) 190 C; 95% tert. amine.

Armeen® 3-16 [Akzo] Trihexadecylamine; CAS 67701-00-2; cationic; chemical intermediate for mfg. of oil-sol. betaines and quat. ammonium salts; Gardner 3 max. solid; m.p. 38 C; amine no. 82; flash pt. (PMCC) > 149 C; 98% tert. amine.

Armeen® 12. [Akzo] Lauramine; CAS 124-22-1; industrial surfactant; Gardner 3 max. color; sp.gr. 0.80; visc. 7.37 cps; m.p. 24 C; iodine no. 1; amine no. 294; flash pt. (PMCC) > 149 C; 97% primary amine.

Armeen® 12D. [Akzo; Akzo Chem. BV] Lauramine (primary amine); CAS 124-22-1; cationic; emulsifier, flotation agent, corrosion inhibitor; lubricant for metal treatment; Gardner 1 max. liq.; sol. in methanol, ethanol, acetone, IPA, chloroform, toluene, carbon tetrachloride, kerosene; sp.gr. 0.801; visc. 42.2 SSU; m.p. 24 C; iodine no. 1; amine no. 297; flash pt. (PMCC) > 149 C; pour pt. 80 F; 98% primary amine.

Armeen® 16. [Akzo] Palmitamine; CAS 143-27-1; cationic; industrial surfactant; Gardner 3 max. color; sp.gr. 0.79; visc. 6.35 cps; m.p. 48 C; iodine no. 2; amine no. 226; flash pt. (PMCC) > 149 C; 97% primary amine.

Armeen® 16D. [Akzo; Akzo Chem. BV] Palmitamine (primary amine); CAS 143-27-1; cationic; emulsifier, flotation agent, corrosion inhibitor; Gardner 1 max. solid; sol. in methanol, ethanol, IPA, chloroform, toluene, carbon tetrachloride, slightly sol. in acetone, kerosene; sp.gr. 0.789 (60/4 C); visc. 37.5 SSU (55 C); m.p. 100–118 F; iodine no. 2; amine no. 228; pour pt. 100 F; flash pt. (PMCC) > 149 C; 98% primary amine.

Armeen® 18. [Akzo; Akzo Chem. BV] Stearamine (primary amine); CAS 124-30-1; cationic; emulsifier, flotation agent, corrosion inhibitor, anticaking agent; hard rubber mold release agent; Gardner 3 max. solid; sol. in ethanol, IPA, chloroform, toluene, CCl_4, slightly sol. in acetone, kerosene; sp.gr. 0.792 (60/4 C); visc. 45.6 SSU; m.p. 122–133 F; pour pt. 115 F; iodine no. 3; amine no. 202; flash pt. (PMCC) > 149 C; 97% primary amine.

Armeen® 18D. [Akzo] Stearamine, dist.; CAS 124-30-1; cationic; emulsifier, flotation agent, corrosion inhibitor, anticaking agent; rubber processing auxiliary; mold release agent for plastics and rubber; Gardner 1 max. solid; sol. in ethanol, IPA, chloroform, toluene, CCl_4, slightly sol. in methanol, kerosene; sp.gr. 0.791–0.792 (60/4 C); visc. 43.7 SSU; m.p. 122–133 F; pour pt. 110 F; iodine no. 3; amine no. 204; flash pt. (PMCC) > 149 C; 98% primary amine.

Armeen® C. [Akzo; Akzo Chem. BV] Cocamine (primary amine); CAS 61788-46-3; cationic; emulsifier, flotation agent, corrosion inhibitor, stripping agent for paints; Gardner 3 max. liq.; sol. in methanol, ethanol, acetone, IPA, chloroform, toluene, CCl_4, kerosene; sp.gr. 0.805; visc. 44.2 SSU (35 C); m.p. 54–59 F; pour pt. 45 F; iodine no. 8; amine no. 272; flash pt. (PMCC) > 149 C; 97% primary amine.

Armeen® CD. [Akzo] Cocamine (primary amine); CAS 61788-46-3; cationic; emulsifier, flotation agent, corrosion inhibitor, stripping agent for paints; Gardner 1 max. liq.; sol. in methanol, ethanol, acetone, IPA, chloroform, toluene, CCl_4, kerosene; sp.gr. 0.804; visc. 43 SSU (35 C); m.p. 57–63 F; pour pt. 55 F; iodine no. 8 C; amine no. 275; flash pt. (PMCC) > 149 C; 98% primary amine.

Armeen® DM8. [Akzo Chem. BV] Tert. amine; cationic; polyurethane catalyst; corrosion inhibitor; chemical intermediate; liq.; 98% conc.

Armeen® DM10. [Akzo Chem. BV] Tert. amine; cationic; polyurethane catalyst; corrosion inhibitor; chemical intermediate; liq.; 98% conc.

Armeen® DM12. [Akzo Chem. BV] Tert. amine; cationic; polyurethane catalyst; corrosion inhibitor; chemical intermediate; liq.; 98% conc.

Armeen® DM12D. [Akzo] Dimethyl lauramine; CAS 112-18-5; cationic; surfactant intermediate; yel. liq.; amine odor; water insol.; sp.gr. 0.78; visc. 2.6 cps; f.p. -15 C; b.p. 80–115 C (3 mm Hg); iodine no. 0.5; amine no. 250; flash pt. (PMCC) > 149 C; 98% tert. amine.

Armeen® DM14. [Akzo Chem. BV] Tert. amine; cationic; polyurethane catalyst; corrosion inhibitor; chemical intermediate; liq.; 98% conc.

Armeen® DM14D. [Akzo] Dimethyl myristamine; CAS 112-75-4; cationic; surfactant intermediate; yel. liq.; amine odor; water insol.; sp.gr. 0.79; visc. 4.7 cs; f.p. -8 C; b.p. 100–125 C (3 mm Hg); flash pt. 28 C (COC); 98% act.

Armeen® DM16. [Akzo Chem. BV] Tert. amine; cationic; polyurethane catalyst; corrosion inhibitor; chemical intermediate; liq.; 98% conc.

Armeen® DM16D. [Akzo] Dimethyl palmitamine; CAS 112-69-6; cationic; chemical intermediate, raw material for surfactants; yel. liq.; amine odor; water insol.; sp.gr. 0.80; visc. 5.4 cps; f.p. 8 C; b.p. 100–136 C (3 mm Hg); iodine no. 0.5; amine no. 198; flash pt. (PMCC) > 149 C; 95% tert. amine.

Armeen® DM18D. [Akzo] Dimethyl stearamine; CAS 124-28-7; cationic; chemical intermediate, raw material for surfactants; yel. liq., amine odor; water insol.; sp.gr. 0.79; visc. 7.5 cps; f.p. 20 C; b.p. 145–160 C (3 mm Hg); iodine no. 1; amine no. 180; flash pt. (PMCC) > 149 C; 95% tert. amine.

Armeen® DMC. [Akzo Chem. BV] Tert. amine; cationic; polyurethane catalyst; corrosion inhibitor; chemical intermediate; liq.; 98% conc.

Armeen® DMCD. [Akzo; Akzo Chem. BV] Dimethyl cocamine; CAS 61788-93-0; cationic; chemical intermediate, raw material for surfactants; yel. liq.; amine odor; water insol.; sp.gr. 0.79; visc. 3.1 cps; b.p. 42–150 C (3 mm); m.p. -22 C; iodine no. 10; amine no. 234; flash pt. (PMCC) > 149 C; 95% tert. amine.

Armeen® DMHT. [Akzo Chem. BV] Tert. amine; cationic; polyurethane catalyst; corrosion inhibitor; chemical intermediate; liq.; 98% conc.

Armeen® DMHTD. [Akzo; Akzo Chem. BV] Dimethyl hydrog. tallow amine; CAS 61788-95-2; cationic; chemical intermediate, raw material for surfactants; yel. liq.; amine odor; water insol.; sp.gr. 0.80; visc. 7.0 cps; f.p. 18 C; b.p. 100–155 C; iodine no. 1; amine no. 184; flash pt. (PMCC) > 149 C; 95% tert. amine.

Armeen® DMMCD. [Akzo Chem. BV] Tert. amine; nonionic; chemical intermediate; solid; 100% conc.

Armeen® DMO. [Akzo Chem. BV] Tert. amine; cationic; polyurethane catalyst; corrosion inhibitor; chemical intermediate; liq.; 98% conc.

Armeen® DMOD. [Akzo] Oleyl dimethylamine; CAS 28061-69-0; cationic; surfactant intermediate; Gardner 1 max. liq.; sp.gr. 0.81; visc. 3.3 cps; m.p. -10 C; iodine no. 60; amine no. 183; flash pt. (PMCC) > 149 C; 95% tert. amine.

Armeen® DMSD. [Akzo] Soyaalkyl dimethylamine; CAS 61788-91-8; cationic; surfactant intermediate; Gardner 2 max. liq.; sp.gr. 0.81; visc. 3.4 cps; m.p. -10 C; iodine no. 60; amine no. 183; flash pt. (PMCC) > 149 C; 95% tert. amine.

Armeen® DMT. [Akzo Chem. BV] Tert. amine; cationic; polyurethane catalyst; corrosion inhibitor; chemical intermediate; liq.; 98% conc.

Armeen® DMTD. [Akzo] Tallowalkyl dimethylamine; CAS 68814-69-7; cationic; surfactant intermediate; Gardner 1 max. liq.; sp.gr. 0.80 (38 C); visc. 6 cps; m.p. 5 C; iodine no. 42; amine no. 184; flash pt. (PMCC) > 149 C; 95% tert. amine.

Armeen® HT. [Akzo; Akzo Chem. BV] (Hydrog. tallow) amine (primary amine); CAS 61788-45-2; cationic; emulsifier, flotation agent, corrosion inhibitor, chemical intermediate, anticaking agent; Gardner 9 solid; sol. in methanol, ethanol, IPA, chloroform, toluene, CCl_4, sp.gr. 0.795 (60/4 C); visc. 47.5 SSU (55 C); m.p. 79–136 F; cloud pt. 115 F; pour pt. 110 F; iodine no. 5; amine no. 207; flash pt. (PMCC) > 149 C; 97% primary amine.

Armeen® HTD. [Akzo; Akzo Chem. BV] (Hydrog. tallow) amine (primary amine); CAS 61788-45-2; cationic; emulsifier, flotation agent, corrosion inhibitor, chemical intermediate, anticaking agent; Gardner 9 solid; sol. in methanol, ethanol, IPA, chloroform, toluene, CCl_4, sp.gr. 0.794 (60/4 C); visc. 44.1 SSU (55 C); m.p. 70–120 F; cloud pt. 110 F; pour pt. 100 F; iodine no. 5; amine no. 209; flash pt. (PMCC) > 149 C; 98% primary amine.

Armeen® L8D. [Akzo] 2-Ethylhexylamine, distilled; CAS 104-75-6; cationic; chemical intermediate for vapor phase corrosion inhibitors; Gardner 1 max. liq.; sp.gr. 0.79; visc. 4 cps; m.p. < -18 C; iodine no. < 1; amine no. 422; flash pt. (PMCC) 59 C; 98% primary amine.

Armeen® M2-10D. [Akzo] Didecyl methylamine; CAS 7396-58-9; cationic; chemical intermediate for water-sol. betaines; catalyst for urethane resins; Gardner 1 max. liq.; sp.gr. 0.80; m.p. -4 C; iodine no. 0.5; amine no. 175; flash pt. (PMCC) > 132 C; 97% tert. amine.

Armeen® M2C. [Akzo; Akzo Chem. BV] Dicoco methylamine; CAS 61788-62-3; cationic; chemical intermediate; surfactant; for mfg. of oil-sol. betaines and quat. ammonium salts; Gardner 2 max. liq.; sp.gr. 0.81 (30 C); visc. 26 cps (35 C); m.p. -2 C; iodine no. 8; amine no. 137; flash pt. (PMCC) 210 C; 97% tert. amine.

Armeen® M2HT [Akzo; Akzo Chem. BV] Dihydrogenated tallow methylamine; CAS 61788-63-4; cationic; chemical intermediate; for mfg. of oil-sol. betaines and quat. ammonium salts; Gardner 1 max. solid; sp.gr. 0.81 (38 C); visc. 56 cps (30 C); m.p. 38 C; iodine no. 3; amine no. 105; flash pt. (PMCC) > 149 C; 97% tert. amine.

Armeen® O. [Akzo; Akzo Chem. BV] Oleamine (primary amine); CAS 112-90-3; cationic; emulsifier, wetting agent, corrosion inhibitor, dispersant, chemical intermediate, oil additive; cosmetics; Gardner 8 paste; sol. in acetone, methanol, ethanol, IPA, chloroform; toluene, carbon tetrachloride, kerosene, wh. min. oil; sp.gr. 0.820 (38/4 C); visc. 57.0 SSU; m.p. 50–68 F; flash pt. 320 F; 100% conc.

Armeen® OD. [Akzo] Oleamine (primary amine); CAS 112-90-3; cationic; wetting agent, lube oil additive, emulsifier, corrosion inhibitor, cosmetic industry dispersant, chemical intermediate; Gardner 2 paste; water insol.; sp.gr. 0.79 (60 C); visc. 56.6

SSU; m.p. 8–18 C; flash pt. 154 C (COC); 98% act.

Armeen® OL. [Akzo] Oleamine; CAS 112-90-3; cationic; emulsifier, flotation agent, corrosion inhibitor; Gardner 4 max. paste/liq.; sp.gr. 0.82 (38 C); visc. 8.15 cps (35 C); m.p. 20 C; iodine no. 85; amine no. 202; flash pt. (PMCC) > 149 C; 95% primary amine.

Armeen® OLD. [Akzo] Oleamine, dist.; CAS 112-90-3; cationic; emulsifier, flotation reagent, corrosion inhibitor; Gardner 1 max. paste/liq.; sp.gr. 0.82 (38 C); visc. 8.15 cps (35 C); m.p. 21 C; iodine no. 85; amine no. 207; flash pt. (PMCC) > 149 C; 98% primary amine.

Armeen® S. [Akzo] Soyamine; CAS 61790-18-9; cationic; emulsifier, flotation agent, corrosion inhibitor; Gardner 4 max.; sp.gr. 0.81 (38 C); visc. 8.04 cps (35 C); m.p. 29 C; iodine no. 70; amine no. 206; flash pt. (PMCC) > 149 C; 97% primary amine.

Armeen® SD. [Akzo] Soyamine, dist.; CAS 61970-18-9; cationic; emulsifier, flotation agent, corrosion inhibitor; Gardner 3 paste; sol. in methanol, ethanol, acetone, IPA, chloroform, toluene, CCl_4; sp.gr. 0.81 (38/4 C); visc. 46.2 SSU (35 C); m.p. 81–86 F; cloud pt. 85 F; pour pt. 70 F; iodine no. 70; amine no. 208; flash pt. (PMCC) > 149 C; 98% primary amine.

Armeen® SZ. [Akzo Chem. BV] Amino acid; amphoteric; wetting agent in alkaline paint strippers, freeze/thaw stable latex emulsions, detergents; liq.; HLB 13.6; 40% conc.

Armeen® T. [Akzo; Akzo Chem. BV] Tallowamine (primary amine); CAS 61790-33-8; cationic; emulsifier, flotation reagent, corrosion inhibitor, dispersant, anticaking agent, chemical intermediate, cosmetics ingredient; Gardner 11 paste; sol. in IPA, methanol, ethanol, chloroform, toluene, CCl_4; sp.gr. 0.813 (38/4 C); visc. 47 SSU (35 C); m.p. 64–117 F; cloud pt. 100 F; pour pt. 70 F; iodine no. 46; amine no. 208; flash pt. (PMCC) > 149 C; 97% primary amine.

Armeen® TD. [Akzo] Tallowamine, dist.; CAS 61790-33-8; cationic; emulsifier, flotation agent, corrosion inhibitor; Gardner 2 paste; sol. in IPA, methanol, ethanol, chloroform, toluene, CCl_4; sp.gr. 0.812 (38/4 C); visc. 45.2 SSU (35 C); m.p. 64–118 F; cloud pt. 102 F; pour pt. 70 F; iodine no. 46; amine no. 210; flash pt. (PMCC) > 149 C; 98% primary amine.

Armeen® Z. [Akzo; Akzo Chem. BV] Cocaminobutyric acid; amphoteric; pigment softening, dispersing agent; antifogging agent, foam booster, stabilizer, wetting agent in alkaline paint strippers, latex emulsions, latex rubber reclamation, inks, plastic films, cosmetics; cooling tower corrosion inhibitor; Gardner 8 pumpable slurry; sol. in water, IPA, ethyl acetate; sp.gr. 0.98; visc. 247 SSU; flash pt. (TCC) 175 F; pour pt. 65 F; pH 6.5–7.5 (10% aq.); 100% conc.

Armid® 18. [Akzo] Stearamide, antiblock agent; CAS 124-26-5; internal lubricant and slip agent for processed plastics, coatings, and films; builder, foam visc. stabilizer, and foam booster in syn. detergent formulations; water repellent for textiles; improves dye solubility in printing inks, dyes, carbon paper coatings, and fusible coatings for glassware and ceramics; intermediate for syn. waxes; pigment dispersant; thickener for paint; Gardner 7 flake; water-insol.; sp.gr. 0.52 (100 C); m.p. 99–109 C; flash pt. 225 C; 90% act.

Armid® C. [Akzo] Cocamide; CAS 61789-19-3; see Armid 18; Gardner 10 flake; insol. in water; sp.gr. 0.845 (100 C); m.p. 85 C; flash pt. 174 C; 90% act.

Armid® E. [Akzo] Erucamide; CAS 112-84-5; mold release agent for rubber and plastics; auxiliary for processing rubber; Gardner 7 max. flakes, pellets; m.p. 81 C; iodine no. 80; 90% min. amide.

Armid® HT. [Akzo] Hydrog. tallow amide; CAS 61790-31-6; see Armid 18; also antifoam in steam generator systems, lubricant additive; auxiliary for rubber processing; Gardner 7 flakes, powd. 99% – 60 mesh; m.w. 277; insol. in water; sp.gr. 0.851 (100 C); visc. 16 cps; m.p. 98–103 C; flash pt. 225 C; 90% act.

Armid® O. [Akzo] Oleamide; CAS 301-02-0; see Armid 18; also release agent in cosmetics, penetrant in paper manufacture; Gardner 7 flake, solid; bland odor; m.w. 279; sol (g/100 ml solv. with heating) 59 g in 95% IPA; 30 g in 95% ethanol; 15 g in acetone and trichloroethylene; 11 g in ethyl acetate and MIBK; insol. in water; sp.gr. 0.830 (100 C); visc. 25 cps; m.p. 68 C; flash pt. 207 C; 90% act.

Armix 146. [Witco/Organics] Blended adjuvant; anionic; surfactant used at low concs. for MSMA formulations; also for DSMA and high caustic industrial detergent cleaners; liq.; 100% conc.

Armix 176. [Witco/Organics] Formulated spray adjuvant; nonionic; wetting, sticking, spreading and penetrating agent for fungicides, insecticides, herbicides, or defoliants; liq.; 100% conc.

Armix 180-C. [Witco/Organics] Alkyl polyoxyalkylene ether; nonionic; low foaming spray adjuvant with wetting and surf. tens. props.; enhances distribution of pesticide spray sol'ns.; liq.; 100% conc.

Armix 183. [Witco/Organics] Alkyl polyoxyethylene ether; nonionic; wetting and penetratinga gent for water absorp. and retention allowing a more complete distribution of plant foods and minerals; liq.; 100% conc.

Armix 185. [Witco/Organics] Formulated prod.; anionic; conditioner improving the compatibility of liq. and foliar fertilizers with pesticides as well as combinations of different herbicides applied as a single spray sol'n.; liq.; 100% conc.

Armix 309. [Witco/Organics] Blend; anionic/nonionic; foaming agent for field making; liq.; 100% conc.

Armoblen® S. [Akzo] Long chain fatty amine derivs.; special cationic; pigment wetting agent for paints and coatings; Gardner 14 liq.; sp.gr. 0.889; dens. 7.40 lb/gal.

Armoflote Series. [Akzo Chem. BV] Formulated compds.; cationic; min. flotation agents; 100% conc.

Armotan® ML. [Akzo Chem. BV] Sorbitan laurate; CAS 1338-39-2; nonionic; w/o emulsifier; liq.; 100% conc.

Armotan® MO. [Akzo Chem. BV] Sorbitan oleate; CAS 1338-43-8; nonionic; w/o emulsifier for cosmetic and pharmaceutical preparations, used to make cutting and sol. oils; Gardner 8 liq.; sp.gr. 1.01; visc. 9.5–11 poise; pour pt. -12 C; 100% conc.

Armotan® MP. [Akzo Chem. BV] Sorbitan palmitate; CAS 26266-57-9; nonionic; w/o emulsifier; waxy substance; 100% conc.

Armotan® MS. [Akzo Chem. BV] Sorbitan stearate; CAS 1338-41-6; nonionic; w/o emulsifier; cream needle-like; m.p. 51–58 C; > 99% act.

Armotan® PML 20. [Akzo Chem. BV] PEG-20 sorbitan laurate; CAS 9005-64-5; nonionic; o/w emulsifier, solubilizer for bath oils; liq.; 100% conc.

Armotan® PMO 20. [Akzo Chem. BV] Polysorbate 80; CAS 9005-65-6; 37200-49-0; nonionic; o/w emulsifier; liq.; 100% conc.

Armul 03. [Witco/Organics] Polyoxyalkylene ether; nonionic; emulsifier for spray oils for agriculture; liq.; HLB 9.3; 100% conc.

Armul 16. [Witco/Organics] Polyoxyalkylene ether; nonionic; emulsifier for industrial applics.; liq.; HLB 11.8; 100% conc.

Armul 17. [Witco/Organics] Nonionic surfactants blend; nonionic; emulsifier for paraffinic hydrocarbon crop oils used as spray additive system with post-emergence herbicides; liq.; HLB 12.0; 100% conc.

Armul 21. [Witco/Organics] POE sorbitol monotallate; nonionic; emulsifier as plant regulators in tobacco industry; liq.; HLB 15.4; 100% conc.

Armul 22. [Witco/Organics] Formulated emulsifier; anionic/nonionic; predominantly anionic emulsifier used as a matched pair system with Armul 33 for pesticide formulations; liq.; HLB 11.2; 100% conc.

Armul 33. [Witco/Organics] Formulated emulsifier; anionic/nonionic; predominantly nonionic emulsifier used as a matched pair system with Armul 22 for pesticide formulations; liq.; HLB 11.7; 100% conc.

Armul 34. [Witco/Organics] Formulated emulsifier; anionic/nonionic; emulsifier for Propanil herbicide formulations, emulsifiable concs.; liq.; HLB 11.2; 100% conc.

Armul 44. [Witco/Organics] Formulated emulsifier; anionic/nonionic; predominantly anionic emulsifier used singly or in combination with Armul 55, 33 or 88 for pesticide formulations; liq.; HLB 11.5; 100% conc.

Armul 55. [Witco/Organics] Formulated emulsifier; anionic/nonionic; predominantly nonionic emulsifier used in combination with Armul 22, 44, 66, or 88 for pesticide formulations; liq.; HLB 12.2; 100% conc.

Armul 66. [Witco/Organics] Formulated emulsifier; anionic/nonionic; emulsifier for malathion or cythion pesticides; liq.; HLB 9.7; 100% conc.

Armul 88. [Witco/Organics] Formulated emulsifier; anionic/nonionic; emulsifier for high active organophosphate insecticides; liq.; HLB 7.3; 100% conc.

Armul 100, 101, 102. [Witco/Organics] Blended emulsifier; anionic/nonionic; formulated emulsifier for flowable formulations; liq.; 100% conc.

Armul 214. [Witco/Organics] Blended emulsifier; anionic/nonionic; matched emulsifier pair with Armul 215 for dithiophosphate insecticides; liq.; HLB 9.3; 100% conc.

Armul 215. [Witco/Organics] Blended emulsifier; anionic/nonionic; matched emulsifier pair with Armul 214 for dithiophosphate insecticides; liq.; HLB 11.6; 100% conc.

Armul 906. [Witco/Organics] Alkylphenol PEG ether; nonionic; emulsifier, detergent for styrene-butadiene latexes, leather industry for degreasing and to increase penetration of dyes and finishes; liq.; oil-sol.; HLB 10.9; 100% conc.

Armul 908. [Witco/Organics] Alkylphenol PEG ether; nonionic; emulsifier, detergent, solubilizer, wetting agent for drycleaning formulations; liq.; oil-sol., water-disp.; HLB 12.3; 100% conc.

Armul 910. [Witco/Organics] Alkylphenol PEG ether; nonionic; wetting agent for wettable powds., clays; dust control modifier, lime soap dispersant, detergents and visc. builders; for pulp and paper and textile industries; liq.; HLB 12.9; 100% conc.

Armul 912. [Witco/Organics] Nonylphenoxy polyethoxyethanol; nonionic; surfactant, emulsifier, detergent for textile processing, cosmetics; liq.; water-sol.; HLB 14.1; 100% conc.

Armul 930. [Witco/Organics] Nonylphenoxy PEG ether; nonionic; emulsifier for flowable pesticide formulations, acrylic polymer emulsions; liq.; water-sol.; HLB 17.1; 100% conc.

Armul 940. [Witco/Organics] Nonylphenoxy PEG ether; nonionic; emulsifier for flowable pesticide formulations, acrylic polymer emulsions; for use in conc. electrolyte sol'ns.; liq.; water-sol.; HLB 17.8; 100% conc.

Armul 950. [Witco/Organics] Nonylphenoxy PEG ether; nonionic; emulsifier for flowable pesticide formulations, corrosion inhibitor sticks; solid; water-sol.; HLB 18.1; 100% conc.

Armul 1003. [Witco/Organics] PEG ether of straight chain fatty alcohol; nonionic; biodeg. emulsifier, detergent, and defoamer intermediate; liq.; oil-sol.; HLB 8.1; 100% conc.

Armul 1005. [Witco/Organics] PEG ether of straight chain fatty alcohol; nonionic; detergent for household and industrial applics.; liq.; HLB 11.5; 100% conc.

Armul 1007. [Witco/Organics] PEG ether of straight chain fatty alcohol; nonionic; deterent, intermediate wetting agent for paints, oil field, textile industries; liq.; water-sol.; HLB 13.1; 100% conc.

Armul 1009. [Witco/Organics] PEG ether of straight chain fatty alcohol; nonionic; deterent, intermediate wetting agent for paints, oil field, textile industries; liq.; water-sol.; HLB 13.7; 100% conc.

Armul 2404. [Witco/Organics] Polyalkylene ether; nonionic; surfactant for flowable pesticide formulations, o/w emulsifiable concs.; solid; 100% conc.

Armul 3260. [Witco/Organics] Blended emulsifier; anionic/nonionic; emulsifier used as single system for many pesticide and industrial emulsifiable concs.; liq.; HLB 11.5; 100% conc.

Armul 5830. [Witco/Organics] Blended emulsifier; anionic/nonionic; single emulsifier for EPN and methyl parathion combinations, Toxaphene and Toxaphene-methyl combinations; liq.; HLB 8.7; 100% conc.

Arneel® 18 D. [Akzo] Stearyl nitrile; detergent, wetting agent, rust inhibitor; metal wetting with min. oils; plasticizer for syn. rubbers and plastics; Gardner 6; f.p. 39 C; b.p. 330–360 C.

Arneel® C. [Akzo] Coco nitrile; detergent, wetting agent, rust inhibitor; Gardner 6.

Arneel® OD. [Akzo] Octadecene nitrile; detergent, wetting agent, rust inhibitor; Gardner 3 liq.; sp.gr. 0.83; f.p. 5 C; b.p. > 325 C.

Arneel® T. [Akzo] Tallow nitrile; detergent, wetting agent, rust inhibitor; f.p. 4 C; b.p. 330–360 C.

Arnox BP Series. [Witco/Organics] EO/PO blocked polymers; nonionic; emulsifiers, defoamers, wetting agents, detergents, dye levelers, demulsifiers; liqs., pastes, or solids; HLB 3.0-24.0; 100% conc.

Aroclean MC-4. [Arol Chem. Prods.] Heavy-duty industrial cleaner for metal surfaces, plastic, concrete; solubilizes trimer build-up in textile processing equip.; water-wh. clear liq., mild glycol ether odor; sol. in water; sp.gr. 1.04 g/cc; dens. 8.68 lb/gal; pH 13 ± 0.5; 10% solids.

Aroclear. [Arol Chem. Prods.] Proprietary blend with alkaline builders; one-step clearing agent for reduction clearing of disperse dyes on polyester; biodeg.; removes excess dye concs. on fabric, antiredeposition agent on fabric and equip.; stripping agent for overdyed fabric lots; wh. free-flowing powd.; sol. in water; pH 10.5 ± 0.5 (1%).

Arodet 25 S. [Arol Chem. Prods.] Blend of amine condensate and alkylaryl sulfonates; detergent, wetting agent, emulsifier; amber liq.; mild odor; water sol.; sp.gr. 0.98; 100% act. DISCONTINUED.

Arodet 60 T Soft. [Arol Chem. Prods.] TEA salt of linear alkyl sulfonate; detergent, wetting agent, emulsifier, coupler used in liq. dishwashing detergents, personal care prods., textiles, skins, laundering, foods, cleaners; pale yel. liq.; mild odor; water sol.; sp.gr. 1.09; 60% act. DISCONTINUED.

Arodet AA-350. [Arol Chem. Prods.] Alkylaryl sulfonate with glycol coupling agents; detergent, wetting agent, dyeing assistant, scouring agent, leveler, retarder, dye dispersant, finishing agent, emulsifier; pale yel. liq.; mild odor; misc. in water; sp.gr. 1.04; 35% act.

Arodet AN-100. [Arol Chem. Prods.] Modified nonionic deriv.; nonionic; detergent, wetting agent, scouring agent, dye dispersant; water sol.; sp.gr. 1.11; 68–70% act.

Arodet AN-160. [Arol Chem. Prods.] Detergent, wetting agent for natural and syn. fiber processing, textile scouring, cotton desizing, kier boiling; dye dispersant, leveling and penetrating agent; stable t6o acid and alkalies, hard water, bleaching agents; water-wh. to pale straw clear liq., mild alcoholic odor; sol. in water; pH 7.5-8.5 (1%).

Arodet APS. [Arol Chem. Prods.] Low foaming scouring and wetting agent for natural and syn. fibers incl. cotton, polyester, nylon, wool; caustic-stable; water-wh. to pale yel. sl. visc. liq., mild sl. aromatic odor; cloudy sol. in 5% tap water; sp.gr. 1.003 g/cc; dens. 8.37 lb/gal; pH 7.5; Draves wetting 15 s (0.125%); usage level: 2-3% owg.

Arodet BLN Special. [Arol Chem. Prods.] Blend of long-chain ethoxylates; nonionic; detergent, wetting agent, penetrant for natural and syn. fibers; aids dyestuff dispersion; fulling agent for wool; solv. emulsifier; water-wh. clear to sl. hazy visc. liq., mild odor; sol. in hot water; dens. 8.5 lb/gal; 100% act.

Arodet BN-100. [Arol Chem. Prods.] Ethoxylated alcohol; detergent, wetting agent, emulsifier, dyeing assistant, dispersant for dyeing, finishing, textiles, pigments, resins; clear liq.; mild odor; water sol.; 100% act.

Arodet E-15. [Arol Chem. Prods.] Blend; nonionic; one-bath scouring agent for dyebaths for woolen fabrics; imparts wetting, scouring and leveling without interfering with subsequent dye procedures; lt. yel. to amber liq., mild odor; sp.gr. 1.01 g/cc; dens. 8.43 lb/gal; pH 8.0; usage level: 2-4% owg.

Arodet F. [Arol Chem. Prods.] Fatty amide deriv.; detergent, wetting agent, scouring agent for textile industry, kier boiling, bleaching, dyeing assistant; pale straw-colored clear to sl. hazy gel, mildly perfumed odor; sol. in warm water; sp.gr. 1.07; pH 7-8 (0.5%); 14–16% act.

Arodet HCS. [Arol Chem. Prods.] Ethoxylate; nonionic; general-purpose detergent, wetting agent, scouring agent for syn. and natural fibers; high cloud pt., relatively low foaming, high temp. operating stability; water-wh. sl. visc. liq., mild char. odor; sol. in hot water; cloud pt. > 100 C (10% salt sol'n.); pH 7-8 (1%); Draves wetting 18 s (0.1%); usage level: 2-4 owg; 65+% solids.

Arodet L-70 M. [Arol Chem. Prods.] EO deriv.; nonionic; detergent, wetting agent, emulsifier, dye assistant; water wh. liq.; mildly alcoholic odor; water sol.; sp.gr. 1.03. DISCONTINUED.

Arodet MER-3. [Arol Chem. Prods.] Blend of nonionics and solvs.; nonionic; detergent for cleaning build-up and scale on dyeing equip.; stable to acids, alkalies, hard water; water-wh. clear liq., mild odor; sol. in water; pH 7-8 (1%).

Arodet MKD. [Arol Chem. Prods.] Blend of nonionic and neutralized phosphate ester surfactants; nonionic/anionic; low temp. textile scouring agent for removal of sizes, waxes, and natural or syn. oils from fabrics; water-wh. liq., mild ester odor; sp.gr. 1.08 g/cc; dens. 9.01 lb/gal; pH 6.0 ± 0.3; usage level: 3-5% owg.

Arodet N-100. [Arol Chem. Prods.] Nonylphenol PEG ester; nonionic; scouring and soaping off agent for natural and syn. fibers; water-wh. clear visc. liq.; sp.gr. 1.05 g/cc; cloud pt. 52 C; pH 8.0-8.5.

Arodet N-100 Special. [Arol Chem. Prods.] Nonylphenol PEG ester; nonionic/anionic; scouring and soaping off agent for natural and syn. fibers; high temp. stability; water-wh. clear visc. liq.; sp.gr. 1.05 g/cc; cloud pt. 85-90 C; pH 8.0-8.5.

Arodet RA. [Arol Chem. Prods.] Sulfated modified alcohol deriv.; detergent, scouring agent, penetrant, dyeing assistant; pale amber liq; mild odor; warm water sol.; 32–35% act. DISCONTINUED.

Arodet SC Special. [Arol Chem. Prods.] Mixt. of alcohol sulfates, alkylaryl sulfonates, amine condensates and sequesterants; detergent, emulsifier, wetting agent; clear liq.; mild odor; water sol.; sp.gr. 1.09. DISCONTINUED.

Arodet TA-8. [Arol Chem. Prods.] Phosphate ester; anionic; multipurpose surfactant for textile processing; for desizing, kier boiling, bleaching, wetting and dispersion in jig or beck, after-scouring; emulsifier for polar and nonpolar solvs. used as carriers (trichlorobenzene, butyl benzoate, etc.); water-wh. clear liq.; forms opalescent sol'n. in 5% tap water; sp.gr. 1.02 g/cc; pH 8.0 ± 0.5; 60% solids.

Arodet TX-7. [Arol Chem. Prods.] Modified nonionic blend; nonionic; detergent, wetting agent, emulsifier; amber liq.; mildly alcoholic odor; water sol.; sp.gr. 1.01; 50–55% act. DISCONTINUED.

Arodet UG. [Arol Chem. Prods.] Blend; nonionic; low foaming detergent, wetting agent for syn., natural or blended fibers; exc. antiredeposition props.; dispesant, penetrant for dyestuffs; solv. emulsifier; water-wh. clear liq., mild odor; water-wh. sol. with breaking foam in 5% tap water; sp.gr. 1.1 g/cc; dens. 9.18 lb/gal; cloud pt. 100 C (1% salt); pH 5.5 ± 0.5.

Arodet WIN. [Arol Chem. Prods.] Blend; anionic/nonionic; wetting agent, penetrant, leveling agent for disperse dyes; for natural, syn. and blended fibers; water-wh. clear liq., mild alcoholic odor; sp.gr. 0.995 g/cc; dens. 8.3 lb/gal; cloud pt. none to boil; pH 7.0 ± 0.3; usage level: 2-4% owg; 20% solids.

Arofoam SNI. [Arol Chem. Prods.] Ethoxylates blend; nonionic; micro-foam surfactant for foam dyeing procedures on syn. fibers; dispersant for dyestuffs; dye leveling agent; antiprecipitant; water-wh. clear liq., mild odor; sol. in water; pH 7.0 ± 0.5.

Arofos 200 Conc. [Arol Chem. Prods.] Phosphate ester blend; detergent, wetting agent, emulsifier, penetrant, dye leveling agent, dispersant; water wh. to pale yel. liq.; mild odor; water sol.; sp.gr. 1.08; 85% act.

Arofos 326. [Arol Chem. Prods.] Polyphosphorylated surfactant; detergent, emulsifier, wetting agent; pale yel. liq.; mild odor; water sol.; sp.gr. 1.09; 99+% act.

Arofos TD-19 Special. [Arol Chem. Prods.] Phosphorylated EO adduct in acid form; detergent, emulsifier, wetting agent; lt. amber gel; mild odor; sol. in aromatic and chlorinated solv., water; sp.gr. 1.10; 98% act. DISCONTINUED.

Arol Biodet. [Arol Chem. Prods.] Nonionic; biodeg. scouring agent, rapid wetting agent for textile processing incl. degreasing, desizing, bleaching, dyeing, and finishing operations; water-wh. sl. visc. liq.; water-wh. sol. with high stable foam in 5% tap water; sp.gr. 1.015 g/cc; pH 7.2; usage level: 1-3% owg; 90% solids.

Arol Defoamer JT. [Arol Chem. Prods.] Complex alkyl ethoxylate blend; defoamer; water wh. liq.; mild odor; water sol.; 100% act. DISCONTINUED.

Arol Defoamer NA-3 Special. [Arol Chem. Prods.] Silicone-stearate blend; nonionic; defoamer for atmospheric dyeing operations in textile industry; milky-wh. visc. liq.; forms opal sol'n. @ 5% in tap water; sp.gr. 0.9798 g/cc; dens. 8.176 lb/gal; pH 7.0; 10.5-11% solids.

Arol Defoamer NA2X. [Arol Chem. Prods.] Silicone-stearate blend; nonionic; defoamer for atmospheric dyeing operations in textile industry; milky-wh. visc. liq.; forms opal sol'n. @ 5% in tap water; sp.gr. 0.9798 g/cc; dens. 8.176 lb/gal; pH 7.0; 12% solids.

Arolev ADL-30. [Arol Chem. Prods.] Long-chain deriv.; mildly cationic; leveling and retarding agent for wool and acrylic fibers; stable to dilute acids and alkalies, hard water, salts; amber liq.; sol. in warm water; pH 7-8 (0.5%); usage level: 2-4% owg.

Arolev ADL-86. [Arol Chem. Prods.] Long-chain deriv.; mildly cationic; leveling and retarding agent for wool and acrylic fibers; stable to dilute acids and alkalies, hard water, salts; amber liq.; sol. in warm water; pH 6-7 (0.5%); usage level: 0.5-3.0% owg.

Arolev CDD. [Arol Chem. Prods.] Sulfated fatty ester blend; anionic; surfactant, fast wetting/rewetting agent, emulsifier, leveling agent for cellulosics dyed with direct dyes, various synthetics and blends; for scouring, solv. scouring, sizing, kier bleaching, etc.; biodeg.; pale yel. to amber liq., mild fatty odor; forms water-wh. sol. @ 5% in tap water; sp.gr. 1.012 g/cc; dens. 8.445 lb/gal; pH 6.1; 35% act.

Arolev HPM. [Arol Chem. Prods.] Surfactant/high flash solv. blend; nonionic; leveler for carrier-less dyeing on polyesters; low foaming lubricant; clear to sl. hazy liq.; forms stable blue-wh. emulsion @ 5% in tap water; sp.gr. 0.962 g/cc; dens. 8.03 lb/gal; flash pt. (COC) 130 F; pH 8.2 ± 0.3; combustible; usage level: 3-8% owg; 47% solids.

Arolev IDD. [Arol Chem. Prods.] Blend; nonionic/amphoteric; leveling agent for dyeing with disperse and direct dyes on polyester, cotton, and blends; wetting agent, emulsifier; pale yel. liq., mild odor; sol. in water; sp.gr. 1.01 g/cc; pH 7.3 ± 0.5.

Arolev MTR-7. [Arol Chem. Prods.] Nonionic; low foam leveling agent, dye dispersant, lubricant; synergistic with polyester dye carriers; for piece dyeing, yarn dyeing, atmospheric or pressure equip.; amber clear to sl. hazy visc. liq., mild odor; sol. in warm water; pH 7.5 ± 0.5.

Arolterge 100M. [Arol Chem. Prods.] Blend of fatty acid alkanolamides and phosphate esters; detergent, emulsifier, wetting agent; dk. amber liq.; mild odor; water sol.; sp.gr. 1.0; 100% act.

Arolube MIT-1. [Arol Chem. Prods.] Ethoxylates blend; anionic/nonionic; softener, lubricant, crack mark inhibitor for all fibers; for dye bath, bleach bath, scouring bath; promotes leveling; stable under high temp. and pressure; milky-wh. thin liq., mild odor; disp. in water water; pH 6.0-6.5; usage level: 0.5-2.0%.

Aromox® C/12. [Akzo] Dihydroxyethyl cocamine oxide, IPA; cationic; wetting agent, emulsifier, stabilizer, antistat, foaming agent for detergents, shampoos, cosmetics, textiles, metal plating, petrol. additives, paper, plastics, rubber; Gardner 2 clear liq.; sp.gr. 0.949; visc. 52 cp; cloud pt. 18 F; flash pt. 82 F; pour pt. 0 F; surf. tens. 33 dynes/cm; biodeg.; 50% act. in aq. IPA.

Aromox® C/12-W. [Akzo; Akzo Chem. BV] Dihydroxyethyl cocamine oxide; cationic; wetting agent, emulsifier, stabilizer, antistat, foaming agent for detergents, shampoos, cosmetics, textiles, metal plating, petrol. additives, paper, plastics, rubber; gel sensitizer for latex foam; biodeg.; Gardner 2 clear liq.; sp.gr. 0.997; visc. 2097 cp; HLB 18.4; flash pt. 212 F; pour pt. 35 F; surf. tens. 30.8 dynes/cm; 40% act. in water.

Aromox® DM14D-W. [Akzo Chem. BV] Myristamine oxide; CAS 3332-27-2; nonionic; foam stabilizer for detergent and shampoo formulations; thickener; paste; HLB 12.2; 24% conc.

Aromox® DM16. [Akzo] Palmitamine oxide, IPA; CAS 7128-91-8; cationic; suds and foam stabilizer for detergent and shampoo formulations; Gardner 1 clear liq.; sp.gr. 0.885; visc. 19 cp; cloud pt. 44 F; flash pt. 80 F; pour pt. 0 F; surf. tens. 31.6 dynes/cm; biodeg.; 40% act. in aq. IPA.

Aromox® DMB. [Akzo Chem. BV] Amine oxide; nonionic; foam stabilizer for detergent and shampoo formulations; thickener; liq.; 29% conc.

Aromox® DMC. [Akzo] Dimethylcocamine oxide; CAS 61788-90-7; cationic; suds and foam stabilizer for detergent and shampoo formulations; liq.; 40% conc.

Aromox® DMCD. [Akzo] Cocamine oxide, IPA; detergent, thickener for household and cosmetic prods.; Gardner 1 max. liq.; sp.gr. 0.89; HLB 18.6; flash pt. (APCC) 21 C; pH 6–9; 39% min. act.

Aromox® DMC-W. [Akzo] Cocamine oxide; CAS 61788-90-7; cationic; suds and foam stabilizer for detergent and shampoo formulations; Gardner 1 clear liq.; sp.gr. 0.971; visc. 17 cp; cloud pt. 34 F; flash pt. > 212 F; pour pt. 35 F; surf. tens. 32.5 dynes/cm; biodeg.; 30% act. in water.

Aromox® DMHTD. [Akzo] Hydrog. tallow dimethylamine oxide; cationic; suds and foam stabilizer for detergent and shampoo formulations; Gardner 1 max. liq.; sp.gr. 0.89; flash pt. 29 C (PMCC); 39% min act.

Aromox® DMMCD-W. [Akzo Chem. BV] Amine oxide; nonionic; foam stabilizer for detergent and shampoo formulations; thickener; liq.; HLB 18.7; 30% conc.

Aromox® DMMC-W. [Akzo] Lauramine oxide; cationic; wetting agent, emulsifier, foam stabilizer, antistat, foaming agent for detergents, shampoos; Gardner 1 clear liq.; sp.gr. 0.96; visc. 90 cp; cloud pt. 22 F; flash pt. > 212 F; pour pt. 36 F; surf. tens. 31.2 dynes/cm (0.1%); 30% act. in water.

Aromox® T/12. [Akzo; Akzo Chem. BV] Dihydroxyethyl tallow amine oxide, IPA; nonionic; suds and foam stabilizers for detergent and shampoo formualtions; Gardner 4 clear liq.; sp.gr. 0.94; visc. 77 cp; cloud pt. 60 F; flash pt. 90 F; pour pt. 55 F; surf. tens. 33.0 dynes/cm (0.1%); 50% act. in aq. IPA.

Arosoft LC-15. [Arol Chem. Prods.] Fatty amide/nonionic softener blend; cationic; non-yel. softener for nylon and other syn. fibers; off-wh. soft pasty flowing liq., mild acetic odor; sol. in hot water; pH 5-6 (2%); usage level: 2-3% owg; 15% act.

Arosolve 570-HF. [Arol Chem. Prods.] Aromatic solvs./detergent blend; anionic/nonionic; non-red label solv. scouring agent, detergent for textiles; roller cleaner, tar remover; water-wh. to pale yel. thin liq., solv.-type odor; disp. in water; sp.gr. 0.88 g/cc; dens. 7.39 lb/gal; flash pt. (COC) 55 C; pH 8.5 ± 0.5; 100% act.

Arosolve 570 Special. [Arol Chem. Prods.] Aromatic solvs./detergent blend; anionic/nonionic; solv. scouring agent for textiles; roller cleaner; tar remover; water-wh. to pale yel. thin liq., solv.-type odor; disp. in water; dens. 7.39 lb/gal; flash pt. (COC) 117 F; pH 8.5 ± 0.5; combustible; 100% act.

Arosolve B-950. [Arol Chem. Prods.] Chlorinated solvs./emulsifiers blend; anionic/nonionic; solv. scour for removal of grease, graphite and oil stains on polyester, nylon and other syn. and natural fibers; degreaser and tar remover; stable to most acids, alkalies, metal salts; lt. straw liq.; pH 7.5 (0.5%); usage level: 2-6% owg; 100% act.

Arosolve CON. [Arol Chem. Prods.] Chlorinated aromatic solvs./alcohols/emulsifiers blend; anionic/nonionic; scouring agent for natural and syn. fibers for batch processing in becks or jigs; removes oil, sizes, and waxes for subsequent dyeing; yel. to amber clear liq., chlorinated aromatic-type odor; sol. in water to wh. emulsion; sp.gr. 1.028 g/cc; dens. 8.58 lb/gal; pH 6.6-7.8; usage level: 1-3% owg.

Arosolve DBM. [Arol Chem. Prods.] Aromatic hydrocarbon/emulsifiers blend; anionic/nonionic; low-toxicity carrier auxiliary producing level dyeings on syn. and natural fibers; pale yel. to amber thin liq.; sol. in water forming stable blue-wh. emulion; sp.gr. 0.94 g/cc; dens. 7.84 lb/gal; flash pt. (COC) 78 C; pH 8.0 ± 0.5 (5%); usage level: 6-15% owg.

Arosolve DMI-F. [Arol Chem. Prods.] Solvs./detergents blend; high foaming solv. scouring agent and machine cleaner for textile industry; water-wh. to pale straw clear thin liq., mild solv.-type odor; disp. in warm water; pH 8-9; 100% act.

Arosolve FX-6. [Arol Chem. Prods.] Surfactant/solv. blend; one-bath desize, scour, and dyeing auxiliary for syn., natural and blended fabrics; antiredeposition aid for soil and sizings; yel. clear liq., mild aromatic solv.-type odor; sp.gr. 0.935 g/cc; dens. 7.8 lb/gal; flash pt. (COC) 106 F; pH 7.0 ± 0.2; usage level: 2-4% owg.

Arosolve IWS. [Arol Chem. Prods.] Nonionic/anionic; solv. scouring agent for removal of soils, oils and sizes from cellulosics and syn. fabrics and their blends; wetting agent with good high caustic stability; water-wh. sl. visc. liq., sl. aromatic odor; sol. in water; sp.gr. 0.925; dens. 7.72 lb/gal; pH 7.2 ± 0.3.

Arosolve MN-LF. [Arol Chem. Prods.] Emulsified aromatic naphthas; CAS 1321-94-4; low foaming polyester dye carrier for use in jets and other high-pressure dyeing equip.; produces level and bright shades on polyester; pale yel. to amber liq., mild aromatic odor; sp.gr. 1.002 g/cc; dens. 8.36 lb/gal; flash pt. (PMCC) 190 F; pH 8.0-8.3; 82% volatile.

Arosolve MRC-A. [Arol Chem. Prods.] Aromatic and oxygenated solvs./detergents blend; nonionic/anionic; roller cleaner; pale yel. clear thin liq., mild solv. and ammonia odor; disp. in water; 95+% act.

Arosolve MRC-HF. [Arol Chem. Prods.] Aromatic and oxygenated solvs./detergents blend; nonionic/anionic; roller cleaner; pale yel. clear thin liq., mild solv. odor; disp. in water; 95+% act.

Arosolve RCB. [Arol Chem. Prods.] Low-foaming emfulsifiers/aromatic petrol. distillate/biphenyl blend; anionic/nonionic; solv. carrier for use on syn. fabrics and their blends; low foaming for high shear jets; lt. yel. to pale amber liq., mild naphthenic odor; sol. in warm water to stable blue-wh. emulsion; dens. 8.15 lb/gal; flash pt. (PMCC) 130 F; pH 6-7; combustible.

Arosolve SH-180 LF. [Arol Chem. Prods.] Solvs./low foaming detergent blend; scouring agent and equip. cleaner for rubber rollers; tar remover; dye dispersant; stable to inorg. builders, alkalies, and aqua ammonia; water-wh. to pale yel. thin liq., mild odor; misc. with water; sp.gr. 0.93; flash pt. (COC) 204 F; pH 7.7 ± 0.3; 95+% act.

Arosolve TMB. [Arol Chem. Prods.] Chlorinated/aromatic hydrocarbons/aromatic ester blend; polyester dye carrier for pressure jet procedures; lt. amber clear thin liq., mild chlorinated odor; disp. in warm water forming stable low foaming emulsions; pH 7-8.

Arosolve VNO-LF. [Arol Chem. Prods.] Polyester dye carrier imparting levelness and depth of shade to syn. fabrics dyed with disperse dyes; low foaming; pale yel. to lt. amber thin liq., mild aromatic odor; sp.gr. 1.05 g/cc; dens. 8.76 lb/gal; flash pt. 115 F; pH 6.6; combustible.

Arosolve XNF-1. [Arol Chem. Prods.] Solvs./low foaming detergent blend; nonionic; low foaming pressure jet solv. scour for difficult grease, graphite and oil stains on polyester, nylon and other syn. and natural fibers; general degreaser and tar remover; stable to most acids and alkalies; clear liq.; disp. in warm water; sp.gr. 0.91; dens. 7.6 lb/gal; pH 7.5-8.0; usage level: 2-6% owg; 95+% act.

Arosulf SBO-65. [Arol Chem. Prods.] Sulfated fatty acid ester; wetting, rewetting agent, lubricant for textile dyeing operations; dye leveling agent; emulsifier for solv. systems; dark amber clear oil; sol. in water and many org. solvs.; sp.gr. 1.02; dens. 8.6 lb/gal; pH 7 ± 1; Draves wetting 20 s (0.125%); 64 ± 1% act.

Arosulf SCO-75%. [Arol Chem. Prods.] Castor oil deriv.; textile processing auxiliary, bleaching, leveling and dyeing assistant; emulsifier, finishing agent; also for industrial waxes, polishes, paints; clear to lt. amber oil; sp.gr. 1.045 ± 0.005; dens. 8.7 lb/gal; pH 6.8 ± 0.5; 75% act.

Arosurf® 32-E20. [Sherex/Div. of Witco] PEG-20 oleyl ether; CAS 9004-98-2; nonionic; surfactant, solubilizer, emulsifier, stabilizer; paste; HLB 11.3; 100% solids.

Arosurf® 42-E6. [Sherex/Div. of Witco] Alkoxylated tallow alcohol; nonionic; emulsifier, detergent base; liq.; 100% conc.

Arosurf® 42-PE10. [Sherex/Div. of Witco] Alkoxylated tallow alcohol; nonionic; low pour pt. detergent for heavy-duty laundry liqs.; liq.; 100% conc.

Arosurf® 66-E10. [Sherex/Div. of Witco] Isosteareth-10; nonionic; emulsifier, emollient for personal care prods.; o/w and w/o systems; coupling agent, emulsion stabilizer, detergent; Gardner 1 semisolid; HLB 12.0; m.p. 22 C; pH 7 (1% DW); 100% conc.

Arosurf® 66-E2. [Sherex/Div. of Witco] Isosteareth-2; nonionic; emulsifier, emollient for personal care prods., cutting oils; o/w and w/o systems; coupling agent, emulsion stabilizer, perfume stabilizer;

Gardner 1 liq.; HLB 4.6; m.p. –5 C; pH 7 (1% DW); 100% conc.

Arosurf® 66-E20. [Sherex/Div. of Witco] Isosteareth-20; nonionic; surfactant, emulsifier; Gardner 1 soft solid; HLB 18.0; m.p. 35 C; pH 7 (1% DW); 100% solids.

Arosurf® 66-PE12. [Sherex/Div. of Witco] PPG-3-isosteareth-9; nonionic; low cloud pt. emulsifier, emollient, dispersant, bath oil spreading agent, perfume solubilizer; Gardner 1 liq.; HLB 12.2; m.p. –10 C; pH 7 (1% DW); 100% conc.

Arowet 70 E. [Arol Chem. Prods.] Sulfonated ester; anionic; wetting/rewetting agent, penetrant for textile processing, desizing, scouring, bleaching, level dyeing and printing, finishing operations; water-wh. to faint amber liq., mildly alcoholic odor; sol. in warm water; pH 6.5 (0.5%); 70% act.

Arowet ODA. [Arol Chem. Prods.] Nonionic; biodeg. wetting agent for batch and continous operations on natural and syn. fibers and blends; relatively low foaming; water-wh. liq., mild alcoholic odor; hazy sol'n. at 5% in tap water; sp.gr. 0.980 g/cc; dens. 8.18 lb/gal; pH 6.0; 60% solids.

Arowet S. [Arol Chem. Prods.] Blend of sulfonated esters; anionic; fast wetting agent and dyeing assistant for skein and pkg. work in textile industry; stable to hard water; amber clear liq.; sol. in warm water; sp.gr. 1.03; pH 6-7 (1%).

Arowet SC-75. [Arol Chem. Prods.] Dioctyl sodium sulfosuccinate; anionic; fast wetting agent, penetrant, and dyeing assistant for mild acidic or alkaline textile processing; dye leveling agent; colorless to lt. yel. clear liq., mild alocholic odor; sp.gr. 1.085-1.095 g/cc; flash pt. (PMCC) 103 F; pH 5-7; combustible; usage level: < 2% owg; 27-31% volatile (105 C).

Arquad® 2C-70 Nitrite. [Akzo] Dicoco nitrite, methanol/isopropanol; CAS 71487-01-9; biodeg. surfactant, dispersant for protective coatings, pigments, inks, textiles, agric., acid pickling baths, marine applics, metalworking, electroplating, fuel treatment, emulsion/plastic mfg., waste water treatment, min. processing, paper; Gardner 14 max. liq.; sol. in alcohols, benzene, chloroform, CCl_4; disp. in water; pH 6-8.5; flash pt. (PMCC) 20 C; 68-72% quat. in methanol/IPA.

Arquad® 2C-75. [Akzo; Akzo Chem. BV] Dicocodimonium chloride, aq. IPA; CAS 61789-77-3; cationic; biodeg. emulsifier, foaming, wetting, dispersing agents, corrosion inhibitor, softener, dyeing aid, antistat for textiles, paper, cosmetics; industrial, agriculture, plastics, petrol. industry, acid pickling baths; bactericide, algicide; Gardner 7 semiliq.; sol. in alcohols, benzene, chloroform, CCl_4; disp. in water; m.w. 447; sp.gr. 0.89; HLB 11.4; flash pt. < 80 F; pour pt. 10 F; surf. tens. 30 dynes/cm (0.1%); pH 9; flamm.; 75% act. in aq. IPA.

Arquad® 2HT-75. [Akzo] Quaternium-18, aq. IPA; CAS 61789-80-8; cationic; biodeg. emulsifier, foaming, wetting, dispersing agents, corrosion inhibitor, antistat, bacteriostat for paper softening, household laundry, hair conditioning; soft wh. paste; sol. in alcohols, benzene, chloroform, CCl_4; disp. in water; m.w. 573; sp.gr. 0.87; dens. 7.22 lb/gal; visc. 47.5 cps (120 F); f.p. 95 F; HLB 9.7; flash pt. 112 F; pour pt. 90–100 F; surf. tens. 37 dynes/cm (0.1%); flamm.; 75% act. in aq. IPA.

Arquad® 2T-75 [Akzo] Ditallow dimonium chloride, aq. ethanol; CAS 68783-78-8; biodeg. surfactant, dispersant for protective coatings, pigments, inks, textiles, agric., acid pickling baths, marine applics, metalworking, electroplating, fuel treatment, emulsion/plastic mfg., waste water treatment, min. processing, paper; Gardner 5 max. liq. to paste; sol. in alcohols, benzene, chloroform, CCl_4; disp. in water; pH 6-9; flash pt. (PMCC) 26 C; 74-77% quat. in aq. ethanol.

Arquad® 12-33. [Akzo] Laurtrimonium chloride, IPA; cationic; emulsifier, foaming, wetting, dispersion agents, corrosion inhibitor, antistat for textiles, cosmetics, industrial, agric.; bactericide, algicide; Gardner 7 liq.; m.w. (act.) 263; sp.gr. 0.98; f.p. 5 F; HLB 17.1; flash pt. 140 F; pH 5–8 (10% aq.); biodeg.; 33% act. in aq. IPA.

Arquad® 12-37W. [Akzo] Laurtrimonium chloride; CAS 112-00-5; cationic; emulsifier, corrosion inhibitor, textile softener, antistat, hair conditioner and combing aid emulsifier; biodeg.; Gardner 2 max. liq.; sol. in water, alcohols, chloroform, CCl_4; pH 6.5-9; nonflamm.; 35-39% quat. in water.

Arquad® 12-50. [Akzo] Laurtrimonium chloride, IPA; CAS 112-00-5; cationic; biodeg. emulsifier, foaming, wetting, dispersing agents, corrosion inhibitor, softener, dyeing aid, antistat for textiles, paper, cosmetics; industrial, agriculture, plastics, petrol. industry, acid pickling baths; bactericide, algicide; gel sensitizer for latex foam; Gardner 1 liq.; sol. in water, alcohols, chloroform, CCl_4; m.w. (act.) 263; sp.gr. 0.89; f.p. 13 F; HLB 17.1; flash pt. < 80 F; surf. tens. 33 dynes/cm; pH 5–8 (10% aq.); 50% act. in aq. IPA.

Arquad® 16-29. [Akzo] Cetrimonium chloride; cationic; emulsifier, foaming, wetting, dispersion agents, corrosion inhibitor, antistat for textiles, cosmetics, industrial, agric.; bactericide, algicide; Gardner 6 liq.; m.w. (act.) 319; sp.gr. 0.96; f.p. 61 F; HLB 15.8; flash pt. > 212 F; pH 5–8 (10% aq.); biodeg.; 29% act. in water.

Arquad® 16-29W. [Akzo] Cetrimonium chloride; CAS 112-02-7; cationic; emulsifier, corrosion inhibitor, textile softener, antistatic agent, hair conditioner and combing aid emulsifier; biodeg.; Gardner 3 max. liq.; sol. in water, alcohols, chloroform, CCl_4; pH 6-9; nonflamm.; 27-30% quat. in water.

Arquad® 16-50. [Akzo; Akzo Chem. BV] Cetrimonium chloride, IPA; CAS 112-02-7; cationic; emulsifier, foaming, wetting, dispersing agents, corrosion inhibitor, softener, dyeing aid, antistat for textiles, paper, cosmetics; industrial, agriculture, plastics, petrol. industry, acid pickling baths; bactericide, algicide; rubber to textile bonding agent; biodeg.; Gardner 6 liq.; sol. in water, alcohols, chloroform, CCl_4; m.w. (act.) 319; sp.gr. 0.88; f.p. 61 F; HLB 15.8; flash pt. < 80 F; surf. tens. 34 dynes/cm; pH 5–8 (10% aq.); 49-52% quat. in aq. IPA.

Arquad® 18-50. [Akzo; Akzo Chem. BV] Steartrimonium chloride, IPA; CAS 112-03-8; cationic; emulsifier, foaming, wetting, dispersing agents, corrosion inhibitor, softener, dyeing aid, antistat for textiles, paper, cosmetics; industrial, agriculture, plastics, petrol. industry, acid pickling baths; bactericide, algicide; dye leveling agent, visc. stabilizer, in lubricant compding.; biodeg.; Gardner 7 liq.; sol. in water, alcohols, chloroform, CCl_4; m.w. (act.) 347; sp.gr. 0.88; HLB 15.7; flash pt. < 80 F; surf. tens. 34 dynes/cm; pH 5–8 (10% aq.); 50% act. in aq. IPA.

Arquad® 210-50. [Akzo; Akzo Chem. BV] Didecyl dimonium chloride, aq. ethanol; CAS 7173-51-5; cationic; biodeg. surfactant, dispersant for protective coatings, pigments, inks, textiles, agric., acid pickling baths, marine applics, metalworking, electroplating, fuel treatment, emulsion/plastic mfg., waste water treatment, min. processing, paper; APHA 180 max. liq.; sol. in alcohols, benzene, chloroform, CCl_4; disp. in water; pH 7-9; flash pt. (PMCC) 57 C; 50% quat. in aq. ethanol.

Arquad® 218-100 [Akzo] Dioctadecyl dimethyl ammonium chloride; CAS 107-64-2; biodeg. surfactant, dispersant for protective coatings, pigments, inks, textiles, agric., acid pickling baths, marine applics, metalworking, electroplating, fuel treatment, emulsion/plastic mfg., waste water treatment, min. processing, paper; powd.; sol. in alcohols, benzene, chloroform, CCl_4; disp. in water; flash pt. (PMCC) > 149 C.

Arquad® 218-75 [Akzo] Dioctadecyl dimethyl ammonium chloride, aq. IPA; CAS 107-64-2; biodeg. surfactant, dispersant for protective coatings, pigments, inks, textiles, agric., acid pickling baths, marine applics, metalworking, electroplating, fuel treatment, emulsion/plastic mfg., waste water treatment, min. processing, paper; Gardner 3 max. paste; sol. in alcohols, benzene, chloroform, CCl_4; disp. in water; pH 6-9; flash pt. (Seta) 44 C; 74-77% quat. in aq. IPA.

Arquad® 316(W). [Akzo] Trihexadecylmethyl ammonium chloride, water; CAS 71060-72-5; industrial surfactant for pigment dispersing, coatings, inks, paper processing; Gardner 4 max. solid; m.p. 46-53 C; pH 6-8; flash pt. nonflamm.; 86-90% quat.

Arquad® B-50. [Akzo Chem. BV] Quat. ammonium compd.; cationic; emulsifier, bactericide, algicide, soil stabilizer, cosmetics ingred., textile antistat, flocculant, fabric softener, decolorizing aid; demulsifier in tetracycline processing; liq.; 50% conc.

Arquad® B-90. [Akzo Chem. BV] Quat. ammonium compd.; cationic; emulsifier, bactericide, algicide, soil stabilizer, cosmetics ingred., textile antistat, flocculant, fabric softener, decolorizing aid; demulsifier in tetracycline processing; liq.; 90% conc.

Arquad® B-100. [Akzo] Benzalkonium chloride, aq. IPA; CAS 68391-01-5; cationic; antimicrobial for industrial applics., sec. oil recovery, textiles, cosmetics, pharmaceuticals, sanitizers; Gardner 2 liq.; sol. in acetone, alcohol, most polar solvs., water; m.w. 380; sp.gr. 0.967; pour pt. 0 F; flash pt. (PMCC) 32 C; pH 7–8; 50% act. in aq. IPA.

Arquad® C-33. [Akzo] Cocotrimonium chloride, IPA; cationic; emulsifier, foaming, wetting, dispersion agents, corrosion inhibitor, antistat for textiles, cosmetics, industrial, agric.; bactericide, algicide; Gardner 7 liquid; m.w. (act.) 278; water-sol.; sp.gr. 0.98; f.p. 25 F; HLB 16.5; flash pt. < 80–140 F; pH 6–9 (10% aq.); biodeg.; 33% act. in aq. IPA.

Arquad® C-33W. [Akzo] Cocotrimonium chloride; CAS 61789-18-2; cationic; emulsifier, corrosion inhibitor, textile softener, antistat; hair conditioning and combing aid emulsifier; emulsion-break retardant in cosmetics; biodeg.; Gardner 4 max. liq.; sol. in water, alcohols, chloroform, CCl_4; sp.gr. 0.96; f.p. –3 C; nonflamm.; pH 5–8; 32–35% act. in water.

Arquad® C-50. [Akzo; Akzo Chem. BV] Cocotrimonium chloride, IPA; CAS 61789-18-2; cationic; biodeg. emulsifier, foaming, wetting, dispersing agents, corrosion inhibitor, softener, dyeing aid, antistat for textiles, paper, cosmetics; industrial, agriculture, plastics, petrol. industry, acid pickling baths; bactericide, algicide; gel sensitizer for latex foam; Gardner 7 liq.; sol. in water, alcohols, chloroform, CCl_4; m.w. (active) 278; sp.gr. 0.89; f.p. 5 F; HLB 16.5; flash pt. < 80 F; surf. tens. 31 dynes/cm (0.1%); pH 5–8 (10% aq.); 50% act. in aq. IPA.

Arquad® DM14B-90. [Akzo; Akzo Chem. BV] Myristyl dimethylbenzyl ammonium chloride dihydrate; CAS 139-08-2; cationic; bactericide, fungicide, germicide, disinfectant; cosmetics, textiles, soil stabilization; wh. powd.; 90% act.

Arquad® DMCB-80. [Akzo] Cocoalkyl dimethyl benzyl ammonium chloride, aq. IPA; CAS 61789-71-7; microbicide for disinfectants, sanitizers, algicides for use in swimming pools, air conditioning cooling towers, bathroom cleaners, petrol. recovery; Gardner 4 max. liq.; sol. in water and most common org. solv.; m.w. 354; sp.gr. 0.935; flash pt. (PMCC) 27 C; pH 6-8; 79% act. in aq. IPA.

Arquad® DMHTB-75. [Akzo] Hydrog. tallow dimethylbenzyl ammonium chloride, aq. IPA; CAS 61789-72-8; bactericide, disinfectant, softening agent for textiles; Gardner 4 max. soft paste; sol. in water, alcohols, acetone; f.p. 60 C; pH 7-9; flash pt. (Seta) 23 C; 75% act. in aq. IPA.

Arquad® DMMCB-50. [Akzo; Akzo Chem. BV] Alkyl (C12,C14,C16) dimethyl benzyl ammonium chloride; cationic; antistat, flocculant, emulsifier, softener, corrosion inhibitor used in cosmetics, soil stabilization, textiles, fabric softener, fungicide, bactericide; liq.; 50% act.

Arquad® DMMCB-75. [Akzo] Alkyl (C12–C14, C16) dimethyl benzyl ammonium chloride; disinfectant, bactericide, germicide, fungicide; liq.; cloud pt. –36 F; flash pt. < 80 F; biodeg.; flam.; 50% act.

Arquad® HTL8(W) MS-85. [Akzo] 2-Ethylhexyl hydrogenated tallowalkyl methosulfate; cationic; biodeg. surfactant, dispersant for protective coatings, pigments, inks, textiles, agric., acid pickling baths, marine applics, metalworking, electroplating, fuel treatment, emulsion/plastic mfg., waste water treatment, min. processing, paper; EPA listed; Gardner 5 max. liq.; sol. in alcohols, benzene, chloroform, CCl_4; disp. in water; flash pt. nonflamm.; pH 4.5-6; 81.5-84.5% quat.

Arquad® L-11. [Akzo] Quarternized beta-amines; cationic; emulsifier; Gardner 2 liq.; sp.gr. 0.885; m.p. –90 F; flash pt. 37 F; 50% act.

Arquad® L-15. [Akzo] Quarternized beta-amines; cationic; emulsifier; Gardner 9 liq.; water sol.; sp.gr. 0.877; m.p. 46 F; cloud pt. 42 F; flash pt. 100 F; 50% act.

Arquad® M2HTB-80 [Akzo] Di(hydrogenated tallow)benzyl methyl ammonium chloride, aq. IPA; CAS 61789-73-9; industrial surfactant for prep. of organophilic clays; Gardner 3 max. solid; sol. in water, alcohols, acetone; pH 7-9; flash pt. (Seta) 26 C; 78-82% quat. in aq. IPA.

Arquad® NF-50. [Akzo] Dialkyldimethyl ammonium chloride; bactericide, algicide; liq.; 50% act.

Arquad® S. [Akzo] Alkyl trimethyl ammonium chloride; microbicide for formulating disinfectants, sanitizers, and algicides; used in mold growth control, swimming pool conditioning, air conditioning systems, industrial water, sec. oil recovery; liq.; m.w. 342 (of act.); sol. in water; sp.gr. 0.89; flash pt. (PM) 60 F; pH 5–8 (10%); 49% act.

Arquad® S-2C-50. [Akzo Chem. BV] 1:1 mixt. of oleyltrimethyl ammonium chloride and dicocodimethyl ammonium chloride; cationic; bactericide, wetting agent, corrosion inhibitor; oil recovery, cosmetics, textiles; Gardner 10 max. liq.; sp.gr. 9.87; HLB 13.5; f.p. –15 C; pH 5–8; 50–55% act.

Arquad® S-50. [Akzo; Akzo Chem. BV] Soytrimonium chloride, IPA; CAS 61790-41-8; cationic; emulsifier, corrosion inhibitor, textile softener, antistat; hair conditioning and combing aid emulsifier; bitumen emulsions; slime control agent in water systems; biodeg.; Gardner 8 max. liq.; sol. in water, alcohols, chloroform, CCl_4; m.w. 343; sp.gr. 0.89; HLB 15.6; f.p. 20 C; flash pt. (PM) < 80 F; pH 5–8 (10% aq.); 49–52% act. in IPA.

Arquad® T-2C-50. [Akzo Chem. BV] 1:1 mixt. of tallow trimonium chloride and dicoco dimonium chloride; cationic; emulsifier, foaming, wetting, dispersing agents, corrosion inhibitor, softener, dyeing aid, antistat for textiles, paper, cosmetics; industrial, agriculture, plastics, petrol. industry, acid pickling baths; bactericide, algicide; Gardner 8 liq.; m.w. (act.) 394; sp.gr. 0.87; HLB 13.0; flash pt. < 80 F; flam.; 50% act. in aq. IPA.

Arquad® T-27W. [Akzo] Tallow trimonium chloride; CAS 8030-78-2; biodeg. emulsifier, foaming, wetting, dispersing agents, corrosion inhibitor, softener, dyeing aid, antistat for textiles, paper, cosmetics; industrial, agriculture, plastics, petrol. industry, acid pickling baths; bactericide, algicide; Gardner 3 max. liq.; sol. in water, alcohols, chloroform, CCl_4; m.w. 343; HLB 14.2; pH 5–8 (10% aq.); 26–29% act. in water.

Arquad® T-50. [Akzo; Akzo Chem. BV] Tallowtrimonium chloride, aq. IPA; CAS 8030-78-2; cationic; emulsifier, corrosion inhibitor, textile softener, antistat; hair conditioning and combing aid emulsifier; also used in mfg. of antibiotics; gel sensitizer for latex foam; biodeg.; Gardner 8 max. liq.; sol. in water, alcohols, chloroform, CCl_4; m.w. 340; sp.gr. 0.881; HLB 14.2; flash pt. < 80 F (PM); pour pt. 15–48 F; pH 5–8 (10% aq.); biodeg.; 50% act. in aq. IPA.

Arstim RRC. [Aquaness] Sulfonate and surfactant; surfactant for cleaning the face of oil-producing formation of such contaminants as emulsion blocks, foreign solids, water blocks, and drilling mud; liq.

Arsul DDB. [Aquaness] Dodecylbenzene sulfonic acid; anionic; intermediate to form metal or amine salts or sulfonate derivs. used as emulsifiers, dispersants, detergents, emulsion breakers, and corrosion inhibitors; liq.; 90% conc.

Arsul LAS. [Aquaness] Linear dodecylbenzene sulfonic acid; anionic; biodeg. intermediate to form metal or amine salts or sulfonate derivs. used as emulsifiers, dispersants, detergents, emulsion breakers, and corrosion inhibitors; liq.; 98% conc.

Arsurf Series. [Aquaness] Surfactants and proprietary blends; nonionic; wetting agents, emulsifiers, detergents, dispersants, penetrants, solubilizers; liq.

Artrads 6524. [Aquaness] Dimer trimer polybasic acid; intermediate reacting with diols, diamine polyols, and polyamines to produce corrosion inhibitors; liq.; 100% conc.

Artrads 6923. [Aquaness] Dimer trimer polybasic acid; intermediate reacting with diols, diamine polyols, and polyamines to produce corrosion inhibitors; liq.; 100% conc.

Artrads 7522. [Aquaness] Dimer trimer polybasic acid; intermediate reacting with diols, diamine polyols, and polyamines to produce corrosion inhibitors; liq.; 100% conc.

Arylan® CA. [Harcros; Harcros UK] Calcium dodecylbenzene sulfonate; anionic; emulsifier for degreasers, herbicides, pesticides, waxes, hydrocarbon solvs.; EPA approved; visc. liq.; alcoholic odor; water insol.; sp.gr. 1.012; visc. 3025 cs; pour pt. < 0 C; flash pt. (Abel CC) 35 C; pH 6.5 (1% aq); contains butanol; keep from heat; 70% act.

Arylan® DA 36. [Harcros] Blended liq. detergent with nonionic foam stabilizer; anionic/nonionic; detergent; liq.; biodeg.; 36% conc.

Arylan® DT. [Harcros] Specially blended liq. detergent conc. based on alkylaryl sulfonates with foam stabilizers; detergent; amber, clear liq.; mild odor; sp.gr. 1.05; visc. 330 cs; flash pt. > 200 F (COC); pour pt. < 0 C; pH 6–7.5 (1% aq.); 30% act.

Arylan® HAL. [Harcros UK] Specially blended liq. detergent conc. based on alkylaryl sulfonates with foam stabilizers; anionic/nonionic; high-foaming detergent for specialty cleaner formulations, hard. surf. cleaners, upholstery cleaners, industrial cleaning; fully biodeg.; amber clear liq.; mild, alcoholic odor; sp.gr. 1.04; visc. 300 cs; flash pt. (Abel CC) 32 C; pour pt. < 0 C; pH 7.0 (1% aq.); 95% act.

Arylan® HE Acid. [Harcros UK] Alkylbenzene sulfonic acid, branched chain; anionic; intermediate for detergent mfg. and prod. of wetting agents and emulsifiers; liq.; 95% conc.

Arylan® LQ. [Harcros; Harcros UK] Specially blended liq. detergent conc. based on alkylaryl sulfonates with foam stabilizers; anionic; high-foaming detergent for specialty cleaner formulations, hard. surf. cleaners, upholstery cleaners, industrial cleaning; fully biodeg.; lt. amber clear liq.; mild odor; sp.gr. 1.053; visc. 350 cs; flash pt. (COC) > 95 C; pour pt. < 0 C; pH 8.0 (1% aq.); 40% act.

Arylan® PWS. [Harcros; Harcros UK] Isopropylamine dodecylbenzene sulfonate; CAS 26264-05-1; anionic; surfactant; emulsifier for min. oils, kerosene, waxes, and chlorinated solvs., herbicides and insecticides; for mfg. of emulsion degreasers and kerosene-based hand cleaning gels; EPA approved; visc. liq., mild odor; sol. hazy in water; sp.gr. 1.028; visc. 4200 cst; pour pt. < 0 C; flash pt. (COC) > 175 C; pH 4.0 (1% aq.); 90% act.

Arylan® S90 Flake. [Harcros] Sodium benzene sulfonate (branched chain); anionic; surfactant.

Arylan® S Acid. [Harcros UK] Dodecylbenzene sulfonic acid (branched chain); anionic; intermediate; liq.; 96% conc.

Arylan® S Flake. [Harcros] Sodium benzene sulfonate; wetting agent, emulsifier; cream-colored flake; 82% conc.

Arylan® SBC25. [Harcros; Harcros UK] Sodium dodecylbenzene sulfonate; anionic; biodeg. wetting agent, detergent base, emulsifier for emulsion polymerization; clear liq., mild odor; water-sol.; sp.gr. 1.036; visc. 150 cst; pour pt. 0 C; flash pt. (COC) > 95 C; pH 7.5 (1% aq.); surf. tens. 33 dynes/cm (0.1%); 25% act. in water.

Arylan® SBC Acid. [Harcros; Harcros UK] Straighter chain dodecylbenzene sulfonic acid; anionic; surfactant, detergent, foaming agent, emulsifier for phenolic materials; intermediate for liq. detergents, dishwash, emulsifiers; biodeg.; brn. visc. liq., char. odor; water-sol.; sp.gr. 1.066; visc. 4300 cst; pour pt. 0 C; flash pt. (COC) 120 C; pH 2 (1% aq.); surf. tens. 33 dynes/cm (0.1%); toxicology: corrosive; extremely irritating to skin and eyes; 95% act.

Arylan® SC15. [Harcros; Harcros UK] Sodium dodecylbenzene sulfonate; anionic; biodeg. wetting agent, detergent base, emulsifier for emulsion polymerization; clear liq., mild odor; water-sol.; sp.gr. 1.008; visc. 300 cst; pour pt. 5 C; flash pt. (COC) > 95 C; pH 7.5 (1% aq.); 15% act. in water.

Arylan® SC30. [Harcros; Harcros UK] Sodium dodecylbenzene sulfonate; anionic; biodeg. wetting agent, detergent base, emulsifier for emulsion polymerization; liq./soft paste, mild odor; water-sol.; sp.gr. 1.040; visc. 5100 cst; pour pt. 12 C; flash pt. (COC) > 95 C; pH 7.5 (1% aq.); 30% act.

Arylan® SC Acid. [Harcros; Harcros UK] Dodecylbenzene sulfonic acid; anionic; biodeg. intermediate for mfg. of anionic detergents, liq. dishwashes, emulsifiers; emulsifier for phenolic systems; brn. visc. liq., char. odor; water-sol.; sp.gr. 1.056; visc. 2000 cst; pour pt. < 0 C; flash pt. (COC) > 120 C; pH 2 (1% aq.); surf. tens. 33 dynes/cm (0.1%); toxicology: corrosive; extremely irritating to skin and eyes; > 95% act.

Arylan® SKN Acid. [Harcros] Straighter chain dodecylbenzene sulfonic acid; intermediate for mfg. of anionic detergents, liq. dishwashes, high active detergents; brn. visc. liq., char. odor; water-sol.; sp.gr. 1.068; visc. 1850 cps; flash pt. (COC) 120 C; pH 2 (1% aq.); > 95% act.

Arylan® SNS. [Harcros UK] Sodium naphthalene sulfonic acid formaldehyde conc.; dispersant; aq. suspensions; buff powd.; pH neutral; 90% act.

Arylan® SO60 Acid. [Harcros] Branched chain tridecyl benzene sulfonic acid; detergent, emulsifier, wetting agent, emulsion polymerization aids; br., visc. liq.; char. odor; water sol.; sp.gr. 1.045; visc. > 25,000 cs; flash pt. > 250 F (COC); pour pt. < 0 C; pH 2.0; > 96% act.

Arylan® SP Acid. [Harcros; Harcros UK] Straighter chain dodecylbenzene sulfonic acid; anionic; surfactant, detergent, foaming agent, emulsifier for phenolic materials; intermediate for liq. detergents, dishwash, emulsifiers; biodeg.; brn. visc. liq., char. odor; water-sol.; sp.gr. 1.056; visc. 2000 cst; pour pt. 0 C; flash pt. (COC) 120 C; pH 2 (1% aq.); toxicology: corrosive; extremely irritating to skin and eyes; 95% act.

Arylan® SX85. [Harcros; Harcros UK] Sodium dodecylbenzene sulfonate; anionic; heavy-duty industrial detergent for industrial cleaning, hard surf. cleaning, textile scouring; base surfactant for powd. detergents; wetting agent for powds.; emulsifier for emulsion polymerization; off-wh. powd., mild odor; water-sol.; bulk dens. 0.50 g/cc; pH 9.5 (1% aq.); surf. tens. 35 dynes/cm (0.1%); toxicology: avoid inhalation of dust, exposure to eyes; 85% act.

Arylan® SX Flake. [Harcros] Sodium dodecylbenzene sulfonate; anionic; detergent base, wetting agent; liquid; 75% conc.

Arylan® SY30. [Harcros UK] Sodium alkylbenzene sulfonate; anionic; biodeg. emulsifier for emulsion polymerization; yel. clear liq.; surf. tens. 37 dynes/cm (0.1%); 30% act.

Arylan® SY Acid. [Harcros UK] Alkylbenzene sulfonic acid; anionic; biodeg. emulsifier for emulsion polymerization; brn. visc. liq.; surf. tens. 36 dynes/cm (0.1%); 96% act.

Arylan® TE/C. [Harcros] Benzene sulfonate, TEA salt; anionic; detergent, emulsifier, wetting agent, emulsion polymerization aids; clear liq.; mild odor; water sol.; sp.gr. 1.060; visc. 900 cs; flash pt. > 200 F (COC); pour pt. < 0 C; pH 7–8.5; biodeg.; 40% act.

Arylene M40. [Hart Chem. Ltd.] Dioctyl sodium sulfosuccinate; anionic; wetting agent, rewetting, dewatering surfactant, filtration aids; clear liq.; dens. 1.03; visc. 50 cps; 40% act.

Arylene M60. [Hart Chem. Ltd.] Dioctyl sodium sulfosuccinate; anionic; wetting agent, rewetting agent for cotton and cotton blends; dewatering surfactant, filtration aids; pale yel. clear liq.; dens. 1.06; visc. 50 cps; 60% act.

Arylene M75. [Hart Chem. Ltd.] Sodium dioctyl sulfosuccinate; anionic; wetting aid, filtration aid; liq.; 75% act.

ASA. [Ethyl] Alkenyl succinic anhydride; intermediate for defoamers, demulsifiers, emulsifiers, foam boosters, wetting agents, detergents, dispersants; sizing agent for paper; Gardner 8 liq.; visc. 145 cSt.

Ascote 5, 9, 12, 12L, 14. [Kao Corp. SA] Fatty amine and polyamine-based; cationic; antistripping agents for road applics.; liq./paste; 45 and 100% conc.

Asfier Series. [Kao Corp. SA] Fatty amine and diamine-based; cationic; emulsifier for cationic asphalt emulsions; liq./paste; 50 and 100% conc.

Asol. [Lucas Meyer] Lecithin fraction; CAS 8002-43-5; nonionic; antispatter agent, release agent, emulsifier for food, cosmetics, pharmaceuticals; liq.; 40–100% conc.

Astromid 18. [Alco] N-octadecyl sulfosuccinamate, sodium salt; anionic; foaming, frothing, emulsifying agent; paste; 100% conc. DISCONTINUED.

Astromid 22. [Alco] Tetrasodium N-alkyl, N-(1,2 dicarboxyethyl) sulfosuccinamate; anionic; emulsifier; liq.; 35% conc. DISCONTINUED.

Astromid 25. [Alco] N-alkyl sulfosuccinamate; anionic; foaming, frothing, emulsifying agent; liq.; 35% conc. DISCONTINUED.

Astrowet 102. [Alco] Alcohol half-ester sulfosuccinate, sodium salt, ethoxylated; anionic; polymerization emulsifier; liq.; 30% conc. DISCONTINUED.

Astrowet B-45. [Alco] Sodium diisobutyl sulfosuccinate; anionic; polymerization emulsifier; liq.; 45% conc. DISCONTINUED.

Astrowet H-80. [Alco] Sodium dihexyl sulfosuccinate; anionic; polymerization emulsifier; liq.; 80% conc. DISCONTINUED.

Astrowet O-70-PG. [Alco] Sodium dioctyl sulfosuccinate, propylene glycol; CAS 577-11-7; anionic; wetting, emulsifying agent; for high flash point applics.; liq.; 70% conc. in propylene glycol/water solv.

Astrowet O-75. [Alco] Sodium dioctyl sulfosuccinate; CAS 577-11-7; anionic; wetting, emulsifying agent, liq.; 75% conc.

Atcowet C. [Yorkshire Pat-Chem] Alkyl polyphosphate; anionic; detergent, wetting for scouring and kier boiling; liq.; 78% conc.

Atlas Defoamer AFC. [Atlas Refinery] Hydrophobized silicone; defoamer for textile and leather industries; translucent gray liq.; dens. 8.1 lb/gal; 100% active.

Atlas EM-2. [Atlas Refinery] Glycol ester; nonionic; fiber lubricant, emulsifier; liq.; HLB 11.4; 100% conc.

Atlas EM-13. [Atlas Refinery] Glycol ester; nonionic; fiber lubricant, emulsifier; liq.; 100% conc.

Atlas EMJ-2. [Atlas Refinery] Nonylphenoxyl polyethoxy ethanol; nonionic; detergent, dispersant, emulsifier, wetting agent, penetrant; grease dispersant; for leather, textile sizing, bleaching operations, paper industry; liq.; HLB 13.4; 50% active.

Atlas EMJ-C. [Atlas Refinery] Nonylphenoxy polyethoxy ethanol; nonionic; detergent, dispersant, emulsifier, wetting agent, penetrant; grease dispersant; for leather, textile sizing, bleaching operations, paper industry; clear liq.; water-sol.; sp.gr. 1.06; dens. 8.8 lb/gal; visc. 510 SUS (100 F); flash pt. (COC) 540 F; cloud pt. 52-58 C (1% aq.); 100%

active.

Atlas G-711. [ICI Am.] Alkylaryl sulfonate; anionic; drycleaning detergent; lt. yel. liq.; sol. in chlorinated solv., ethylene glycol, lower alcohols, aniline and petroleum solv.; sp.gr. 1; visc. 5500 cps; HLB 11.7; flash pt. > 300 F; 100% act.

Atlas G-950. [ICI Am.] Sorbide dioleate; CAS 29116-98-1; nonionic; surfactant; orange-red liq.; water disp., sol. in aromatic and aliphatic hydrocarbons, veg. oils; sp.gr. 1; visc. 250 cps; HLB 2.7; flash pt. > 300 F.

Atlas G-1045. [ICI Am.] POE sorbitol laurate; nonionic; emulsifier; lt. yel., oily liq.; disp. with haze in water; sol. in toluene, min. spirits, acetone, dioxane, cellosolve, ethyl acetate, aniline; sp.gr. 1; visc. 175 cp; HLB 11.5; flash pt. > 300 F; 100% act.

Atlas G-1045A. [ICI Am.] POE sorbitol polyoleate-laurate; nonionic; emulsifier for textiles; liq.; HLB 13.2; 100% conc.

Atlas G-1052. [ICI Am.] POE sorbitol distearate; nonionic; surfactant; cream-colored wax; sol. in IPA; HLB 8.5; flash pt. > 300 F.

Atlas G-1069. [ICI Am.] POE sorbitol ester of mixed fatty acids; nonionic; surfactant; amber liq.; disp. in water, sol. in IPA, most veg. oils, hydrocarbon solv.; sp.gr. 1.0; visc. 200 cps; HLB 9.4; pour pt. 45.

Atlas G-1086. [ICI Surf. Belgium] PEG-40 sorbitol hexaoleate; CAS 57171-56-9; nonionic; emulsifier and coupling agent for paraffinic, naphthenic and organic ester lubricants; for textile industry; yel., oily liq.; sol. in toluene, min. spirits, many veg. oils, acetone, CCl_4, cellosolve, methanol, lower alcohols, aniline; sp.gr. 1; visc. 200 cps; HLB 10.2; flash pt. > 300 F; 100% act.

Atlas G-1087. [ICI Am.] POE sorbitol polyoleate; nonionic; emulsifier for textiles; liq.; HLB 9.2; 100% conc.

Atlas G-1096. [ICI Surf. Belgium] PEG-50 sorbitol hexaoleate; CAS 57171-56-9; nonionic; emulsifier and coupling agent for textiles; amber liq.; sol. in veg. oils, acetone, cellosolve, lower alcohols, some aromatic solv., tetrachloride; sp.gr. 1.0; visc. 220 cps; HLB 11.4; flash pt. > 300 F; 100% act.

Atlas G-1186. [ICI Am.] POE sorbitol oleate; nonionic; surfactant; amber, oily liq.; sol. in lower alcohols, veg. and min. oils; sp.gr. 1; visc. 230 cps; HLB 10.2; flash pt. > 300 F; 100% act.

Atlas G-1223. [ICI Am.] POE esters of mixed fatty and resin acids with 0.025% antifoam added; nonionic; detergent; lt. amber, oily liq.; sol. in water, dioxane, cellosolve, ethyl acetate, lower alcohols, toluene, acetone; sp.gr. 1.1; visc. 350 cps; HLB 13.8; flash pt. > 300 F.

Atlas G-1256. [ICI Surf. Belgium] POE sorbitol and tall oil ester; nonionic; emulsifier for agriculture, emulsions, and textile formulations; liq.; HLB 9.7; 100% conc.

Atlas G-1281. [ICI Surf. Belgium] POE triglyceride; nonionic; emulsifier for agric., emulsions, textile formulations; liq.; HLB 9.7; 100% conc.

Atlas G-1284. [ICI Surf. Belgium] POE triglyceride; nonionic; emulsifier for agric., emulsions, textile formulations; liq.; HLB 13.1; 100% conc.

Atlas G-1285. [ICI Surf. Belgium] POE triglyceride; nonionic; emulsifier for agric., emulsions, textile formulations; liq.; HLB 14.4; 100% conc.

Atlas G-1288. [ICI Surf. Belgium] POE triglyceride; nonionic; emulsifier for agric., emulsions, textile formulations; liq.; HLB 16.0; 100% conc.

Atlas G-1289. [ICI Surf. Belgium] POE triglyceride; nonionic; emulsifier for agric., emulsions, textile formulations; liq.; HLB 14.4; 100% conc.

Atlas G-1292. [ICI Am.; ICI Surf. Belgium] POE triglyceride; nonionic; emulsifier for agric., emulsions, textile formulations; pale yel. liq. to soft paste; sol. in water, ethanol, IPA; sp.gr. 1.0; HLB 10.8; flash pt. > 300 F; 100% conc.

Atlas G-1295. [ICI Surf. Belgium] POE triglyceride; nonionic; emulsifier for agric., emulsions, textile formulations; liq.; HLB 17.5; 100% conc.

Atlas G-1300. [ICI Surf. Belgium] POE triglyceride; nonionic; emulsifier for agric., emulsions, textile formulations; tan, hard waxy solid; sol. in water, lower alcohols, acetone, chloroform; HLB 18.1; flash pt. > 300 F; 100% conc.

Atlas G-1304. [ICI Surf. Belgium] POE triglyceride; nonionic; emulsifier for agric., emulsions, textile formulations; solid; HLB 18.7; 100% conc.

Atlas G-1530. [ICI Surf. Belgium] Blend; nonionic/cationic; lubricant, emulsifier system for textiles; liq.; 100% conc.

Atlas G-1554. [ICI Surf. Belgium] Blend; nonionic; lubricant, emulsifier system for textiles; liq.; 100% conc.

Atlas G-1556. [ICI Surf. Belgium] Blend; nonionic; lubricant, emulsifier system for textiles; liq.; HLB 11.2; 100% conc.

Atlas G-1564. [ICI Surf. Belgium] Blend; nonionic; lubricant, emulsifier system for textiles; flake; 100% conc.

Atlas G-1649. [ICI Am.] POE alkyl aryl ether; nonionic; surfactant; cream-colored, waxy solid; sol. in water, toluol, methanol, IPA, CCl_4; sp.gr. 1.1; HLB 16.0; flash pt. > 300 F.

Atlas G-2079. [ICI Am.] PEG-20 palmitate; nonionic; surfactant; wh. waxy solid; sol. in water, ethanol; HLB 15.5; flash pt. > 300 F.

Atlas G-2090. [ICI Am.] Blend of POE fatty amine and POE sorbitol oleate; emulsifier blend, detergent; red-brn. liq.; water disp., sol. in IPA, xylene, min. spirits, kerosene; sp.gr. 0.95; visc. 40 cps; HLB 12.5; flash pt. > 300 F; 100% act.

Atlas G-2109. [ICI Surf. Belgium] POE laurate; nonionic; lubricant, emulsifier for textiles; liq.; HLB 13.3; 100% conc.

Atlas G-2127. [ICI Surf. Belgium] POE laurate; CAS 9004-81-3; nonionic; surfactant, lubricant, emulsifier for textiles; lt. yel. liq.; sol. in IPA, disp. in water, many veg. oils; sp.gr. 1.0; HLB 12.8; flash pt. > 300 F; 100% conc.

Atlas G-2143. [ICI Am.] PEG-10 oleate; CAS 9004-96-0; nonionic; emulsifier; lt. yel. liq.; disp. in water, sol. in most min. oils, lower alcohols; flash pt. > 300 F; pour pt. -10 C.

Atlas G-2151. [ICI Am.] PEG-30 stearate; CAS 9004-99-3; nonionic; surfactant; ivory, waxy solid; sol. in propylene glycol, ethylene glycol, water; HLB 16; flash pt. > 300 F.

Atlas G-2162. [ICI Am.] PEG-25 propylene glycol stearate; nonionic; surfactant; cream-colored semi-solid; sol. in water, methanol, ethanol, HLB 16; flash pt. > 300 F; 100% act.

Atlas G-2198. [ICI Am.] PEG-40 stearate; CAS 9004-99-3; nonionic; surfactant; lt. tan, waxy solid; sol. in water, acetone, ether, dioxane, cellosolve, CCl_4, lower alcohols, ethyl acetate, aniline and many org. solv.; HLB 16.9; flash pt. > 300 F; pour pt. 38 C.

Atlas G-2203. [ICI Surf. Belgium] Potassium phosphate ester; anionic; antistat for textiles; liq.; 100% conc.

Atlas G-2684. [ICI Am.] Blend of sorbitan oleate and POE esters of mixed fatty and rosin acids; nonionic; detergent; amber liq.; disp. in water, sol. in acetone, toluene, dioxane, CCl_4, aniline, min. oils; sp.gr. 1; visc. 500 cps; HLB 7.8; flash pt. > 300 F; pour pt. 15 F.

Atlas G-3300. [ICI Am.] Alkylaryl sulfonate; anionic; pigment dispersant; surfactant; amber liq.; sol. in water, acetone, CCl_4, Cellosolve, diethyl ether, dioxane, ethylene glycol, methanol, aniline, min. spirits, kerosene, toluene; sp.gr. 1.0; visc. 6000 cps; HLB 11.7; flash pt. > 300 F; fire pt. > 300 F; 90% act.

Atlas G-3300B. [ICI Surf. Belgium] Alkylaryl sulfonate; nonionic; emulsifier for agric. emulsions; liq.; HLB 11.4; 100% conc.

Atlas G-3634A. [ICI Am.] Quat. ammonium deriv.; antistat and softener for textiles; lt. amber liq.; sol. in water, lower alcohols, acetone, propylene glycol, CCl_4, perchlorethylene; sp.gr. 1.05; visc. 2500 cps; HLB 18.5; flash pt. > 300 F.

Atlas G-3684. [ICI Am.] POE fatty amine; surfactant; amber liq.; sol. in acetone, lower alcohols, 8% sulfuric acid, 6% sodium hydroxide, water; sp.gr. 1.03; visc. 150 cps; flash pt. > 300 F; pour pt. 20 F.

Atlas G-3780A. [ICI Am.] PEG-20 tallow amine; nonionic; surfactant, antistat for textiles; amber liq. (may become hazy); sol. in water, lower alcohols, acetone, ethyl acetate, ethylene glycol; sp.gr. 1.04; visc. 250 cps; HLB 15.5; flash pt. > 300 F; 100% conc.

Atlas G-3801. [ICI Am.] PEG-2 alkyl alcohol; nonionic; emulsifier; wh., pourable liq.; sp.gr. 0.89; visc. 60 cps; HLB 5.3.

Atlas G-3811. [ICI Am.] PEG-10 alkyl alcohol; nonionic; emulsifier; wh. pourable liq.; sp.gr. 0.992; visc. 415 cps; HLB 12.9.

Atlas G-4809. [ICI Surf. Belgium] Alkoxylated alkylphenol; nonionic; resin emulsifier for water-based coatings; liq.; 76% conc.

Atlas G-4822. [ICI Surf. Belgium] Cetostearyl alcohol, ethoxylated, self-emulsifying; nonionic; surfactant; wax; 100% conc.

Atlas G-4884. [ICI Surf. Belgium] Sorbitan oleate, tech.; CAS 1338-43-8; nonionic; lipophilic surfactant; liq.; HLB 4.3; 100% conc.

Atlas G-4885. [ICI Surf. Belgium] Sorbitan trioleate, tech.; CAS 26266-58-0; nonionic; lipophilic surfactant; liq.; HLB 1.8; 100% conc.

Atlas G-4905. [ICI Am.] PEG-20 sorbitan oleate, tech.; CAS 9005-65-6; nonionic; hydrophilic surfactant; liq.; water-sol.; HLB 15.0; 100% conc.

Atlas G-4909. [ICI Surf. Belgium] Proprietary blend; nonionic; lanolin absorp. base substitute; liq.; 100% conc.

Atlas G-5000. [ICI Surf. Belgium] Alkoxylated alcohol; CAS 99821-01-9; nonionic; industrial emulsifier; solid; HLB 16.9; 100% conc.

Atlas G-5002. [ICI Surf. Belgium] Alkoxylated alkylphenol; nonionic; industrial emulsifier; solid; 100% conc.

Atlas G-7166P. [ICI Am.] PEG-16 sorbitan tristearate; nonionic; surfactant; tan, waxy solid; disp. in warm water, most org. solv., veg. and min. oils; sp.gr. 1.05; HLB 10.

Atlas G-7596H. [ICI Am.] POE sorbitan laurate; nonionic; emulsifier; orange, oily liq.; water sol.; sp.gr. 1.1; visc. 400 cps; HLB 14.3; flash pt. > 300 F.

Atlas G-7596-J. [ICI Am.] PEG-10 sorbitan laurate; nonionic; detergent; yel. liq.; sol. in water, acetone, dioxane, cellosolve, lower alcohols, ethyl acetate, aniline; sp.gr. 1.1; visc. 400 cps; HLB 14.9; flash pt. > 300 F.

Atlas G-8916PF. [ICI Surf. Belgium] POE sorbitan ester of mixed fatty and resin acids; nonionic; general emulsifier for agric. emulsions; liq.; HLB 14.6; 100% conc.

Atlas G-8936CJ. [ICI Am.] POE sorbitan ester of mixed fatty and resin acids; nonionic; emulsifiers for agric. emulsions; liq.; HLB 12.5; 100% conc.

Atlas G-9046T. [ICI Am.] PEG-20 mannitan laurate; nonionic; surfactant; yel., oily liq.; sol. in water, ethanol, methanol, propylene glycol; sp.gr. 1.1; HLB 16.7; flash pt. > 300 F.

Atlas L-801-LF. [Atlas Refinery] Sulfated/ethoxylated oils; anionic/nonionic; single sol'n. lubricant, dye leveling and scouring agent; liq.; 70% conc.

Atlas Sul. Neats L-2. [Atlas Refinery] Sulfated neatsfoot oil; anionic; lubricant, softener for leather processing; emulsifier for min. oil and solvs.; liq.; dens. 8.31 lb/gal; pH 5.5-6.0 (10%); 75% conc.

Atlas Sul. Oil HC. [Atlas Refinery] Sulfated fish oil; anionic; penetrant; emulsifier for min. oils and solvs.; for leather applics.; dens. 8.31 lb/gal; pH 6.0-6.5 (10%); 75% active.

Atlas WA-100. [Atlas Refinery] Dioctyl sodium sulfosuccinate; anionic; wetting agent; liq.; 70% conc.

Atlasol 103. [Atlas Refinery] Sodium decyl sulfate; anionic; emulsifier, wetting agent, dispersant, fiber lubricant, syn. fatliquor; for textile, leather, and general industrial applics.; liq.; dens. 8.0 lb/gal; pH 6-7 (10%); 75% conc.

Atlasol 155. [Atlas Refinery] Anionic/nonionic emulsifiers, solv., other additives; anionic/nonionic; scouring and degreasing compd. for wet leather processes incl. soaking, bating, pickling, and tanning; colorless gel; dens. 7.8 lb/gal; pH 8.0-8.5 (10%); 75% active.

Atlasol 160-S. [Atlas Refinery] Anionic; lubricant, emulsifier for raw oils, min. oil and solvs.; penetrant; for leather processing; replacement for sulfated sperm oil; liq.; dens. 8.25 lb/gal; pH 7-8 (10%); 75% act.

Atlasol 6920. [Atlas Refinery] Syn. fatty ester and alkenes; anionic; replacement for sulfated sperm oil; emulsifier for raw oil, min. oil, solvs.; lubricant; fatliquor for leathers; dens. 7.56 lb/gal; pH 5-6 (10%); 75% active.

Atlasol 6920-VF. [Atlas Refinery] Syn. fatty ester and alkenes; anionic; replacement for sulfated sperm oil; emulsifier for raw oil, min. oil, solvs.; fatliquor for leathers esp. upholstery leather and pale colored shoe uppers; lightfast, heat stable; dens. 8.23 lb/gal; pH 6.5-7.0 (10%); 75% active.

Atlasol Base Oil S. [Atlas Refinery] Modified syn. fatty ester; nonionic; syn. replacement for raw sperm oil, base for sulfated oils; liq.; 100% conc.

Atlasol KAD. [Atlas Refinery] Stearic imidazoline; cationic; lightfast fiber lubricant, fatliquor, and softener for textile and leather applics.; smooth wh. paste; dens. 7.13 lb/gal; pH 3.8-4.0 (10%); 25% conc.

Atlasol KMM. [Atlas Refinery] Ricinoleic acid, triethanolamine salt; cationic; leather tanning surfactant, lubricant; yel. visc. liq.; sp.gr. 0.99; dens. 8.32 lb/gal; pH 4.5-5.0 (10%); 30% active.

Atlosol Series. [Aquaness] Emulsifier conc.; emulsifier for drilling muds; liq.

Atlox 80. [ICI Am.] POE sorbitan esters of mixed fatty

and resin acids; nonionic; pesticide emulsifier and adjuvant; liq.; HLB 15.4; 99% conc.

Atlox 775. [ICI Am.] POE sorbitol ester; nonionic; emulsifier for agric. spray oils, pesticides (pyrethrin formulations); liq.; HLB 9.0; 100% act.

Atlox 804. [ICI Surf. Belgium] POE alkyl ether; nonionic; emulsifier for pesticides; liq.; HLB 14.0; 100% conc.

Atlox 847. [ICI Am.] Polyethoxylated ester, nonionic surfactant; nonionic; pesticide emulsifier; liq.; HLB 11.2; 100% conc.

Atlox 848. [ICI Am.] Polyoxylated ester; nonionic; pesticide emulsifier; liq.; HLB 10.0; 100% conc.

Atlox 849. [ICI Am.] Polyethoxylated ester; nonionic; pesticide emulsifier; liq.; HLB 9.0; 100% conc.

Atlox 1045A. [ICI Am.; ICI Surf. Belgium] POE sorbitol oleate-laurate; nonionic; emulsifier, coupling agent for agric. spray oils, pesticides (pyrethrin formulations), textile industry; yel. liq.; sol. in most org. solv. and pesticides, disp. in water; sp.gr. 1.0; visc. 200 cps; HLB 11.4; flash pt. > 300 F; 100% conc.

Atlox 1087. [ICI Am.] POE sorbitol septaoleate; nonionic; emulsifier; yel. liq.; sol. in most veg. and min. oils, disp. in water; sp.gr. 1; visc. 175 cps; HLB 9.2; flash pt. > 300 F.

Atlox 1096. [ICI Am.] POE sorbitol hexaoleate; nonionic; emulsifier and coupling agent used in textile industry; liq.; HLB 11.4; 100% conc.

Atlox 1196. [ICI Am.] POE sorbitol oleate; nonionic; emulsifier; amber, oily liq.; disp. in water, sol. in xylol, methanol, acetone, most agric. conc.; sp.gr. 1; visc. 450 cp; HLB 11.4; flash pt. 475F.

Atlox 1256. [ICI Am.] POE sorbitol esters of mixed fatty and resin acids; nonionic; surfactant; dk. amber liq.; sol. in most org. solv. and pesticide formulations, disp. in water; sp.gr. 1.0; visc. 1000 cps; HLB 9.7; flash pt. > 300 F.

Atlox 1285. [ICI Surf. Belgium] POE triglyceride; nonionic; emulsifier and coupling agent used in textile industry; semisolid; HLB 14.4; 100% conc.

Atlox 2081 B. [ICI Am.] POE sorbitan esters of fatty and resin acids-alkylaryl sulfonate blend; anionic; emulsifier, surfactant; liq.; biodeg.

Atlox 3300B. [ICI Am.] POE sorbitan esters of fatty and resin acids/alkylaryl sulfonate blend; anionic; emulsifier for pesticides; biodeg.; liq.; HLB 11.4; 100% conc.

Atlox 3335. [ICI Am.] Alkylaryl sulfonate blended with POE sorbitan ester of tall oil; agric. surfactant; amber liq.; disp. in water, sol. in most aromatic solv. and pesticide formulations; sp.gr. 1.1; visc. 2200 cps; HLB 13; flash pt 160 F.

Atlox 3335 B. [ICI Surf. Belgium] POE sorbitan esters of fatty and resin acids, alkylaryl sulfonate blend; anionic; agric. emulsifier; pesticides, herbicides; biodeg.; liq.; HLB 12.6.

Atlox 3386 B. [ICI Surf. Belgium] POE sorbitol fatty esters/alkyl aryl sulfonate blend; anionic; agric. emulsifier; pesticides, herbicides; biodeg.; liq.; HLB 9.4.

Atlox 3387. [ICI Am.] Anionic-nonionic; surfactant blend; agric. surfactant; amber, visc. liq.; sol. in most aromatic solv. and pesticide formulations; sp.gr. 1.0; visc. 4000 cps.

Atlox 3387 BM. [ICI Am.] Blend; nonionic/anionic; emulsifier for pesticides and herbicides; liq.; HLB 11.0.

Atlox 3400B. [ICI Surf. Belgium] Blend; anionic/nonionic; emulsifier for pesticides, herbicides; liq.; HLB 10.9.

Atlox 3401. [ICI Am.] Alkylaryl sulfonate blend; anionic/nonionic; general purpose emulsifier for pesticide concs. where high flash pt. is required; visc. liq.; HLB 10.0; 70% conc.

Atlox 3403F. [ICI Am.] Formulated prod.; anionic/nonionic; general purpose emulsifier for pesticides; liq.; HLB 14.0.

Atlox 3404F. [ICI Am.] Formulated prod.; anionic/nonionic; general purpose emulsifier for pesticides; liq.; HLB 10.0.

Atlox 3406F. [ICI Am.] Formulated prod.; anionic/nonionic; general purpose emulsifier for pesticides; liq.; HLB 12.0.

Atlox 3409F. [ICI Am.] Formulated prod.; anionic/nonionic; general purpose emulsifier for pesticides; liq.; HLB 12.0.

Atlox 3414F. [ICI Am.] Formulated prod.; anionic/nonionic; general purpose emulsifier for pesticides; liq.

Atlox 3422F. [ICI Am.] Formulated prod.; anionic/nonionic; emulsifier component for organo-phosphate pesticides; liq.

Atlox 3450F. [ICI Am.] Formulated prod.; anionic/nonionic; general purpose emulsifier for pesticide concs.; liq.

Atlox 3453F. [ICI Am.] Formulated prod.; anionic/nonionic; general purpose emulsifier for pesticide concs. where high flash pt. is required; liq.; HLB 14.0; 100% conc.

Atlox 3454F. [ICI Am.] Alkylaryl sulfonate blend; anionic/nonionic; general purpose emulsifier for pesticide concs. where high flash pt. is required; visc. liq.; HLB 11.0; 70% conc.

Atlox 3455F. [ICI Am.] Alkylaryl sulfonate blend; anionic/nonionic; general purpose emulsifier for pesticide concs. where high flash pt. is required; visc. liq.; HLB 12.0; 70% conc.

Atlox 4851 B. [ICI Surf. Belgium] POE triglyceride alkylaryl sulfonate blend; anionic; emulsifier for pesticides, herbicides; biodeg.; liq.; HLB 13.2.

Atlox 4853 B. [ICI Surf. Belgium; ICI Europe] Blend of anionic and nonionic surfactants; emulsifier, wetter, dispersant for pesticides, herbicides; biodeg.; liq.; HLB 11.9.

Atlox 4855 B. [ICI Surf. Belgium] POE triglyceride, alkylaryl sulfonate blend; anionic; emulsifier for pesticides, herbicides; biodeg.; liq.; HLB 9.2.

Atlox 4856 B. [ICI Surf. Belgium] POE alcohol alkyl aryl sulfonate blend; anionic; emulsifier for pesticides, herbicides; biodeg.; liq.; HLB 8.7.

Atlox 4857 B. [ICI Surf. Belgium] POE alkyl phenol, alkylaryl sulfonate blend; anionic; emulsifier for pesticides, herbicides; biodeg.; liq.; HLB 12.3.

Atlox 4858 B. [ICI Surf. Belgium] POE sorbitan esters of fatty and resin acids, alkylaryl sulfonate blend; anionic; emulsifier for pesticides, herbicides; biodeg.; liq.; HLB 11.6; 100% conc.

Atlox 4861B. [ICI Surf. Belgium] Alkylaryl sulfonate; anionic; emulsifying agent for pesticide and herbicide formulations; liq.; HLB 8.6; 70% conc.

Atlox 4862. [ICI Surf. Belgium] Disodium methylene dinaphthalene sulfonate; anionic; suspending agent; wetting/dispersing agent for pesticides; powd.; 100% conc.

Atlox 4868B. [ICI Surf. Belgium] POE triglyceride/alkylaryl sulfonate; anionic/nonionic; wetting and suspending agent for agric. wettable powds.; biodeg.; liq.; HLB 11.9.

Atlox 4870B. [ICI Europe] Polyethylene sorbitan

esters of fatty and resin acids, alkylaryl sulfonate blend; wetting/suspending agent; liq.; HLB 13.0; biodeg.

Atlox 4873. [ICI Europe] POE alkyl ether; nonionic; wetting agent for pesticides and herbicides; powd.; HLB 13.6.

Atlox 4875. [ICI Surf. Belgium] POE ester; wetting and dispersing agent for pesticides and herbicides; liq.; HLB 18.2.

Atlox 4880B. [ICI Surf. Belgium] Blend; anionic/nonionic; emulsifier for pesticides, herbicides; liq.; HLB 12.8.

Atlox 4881. [ICI Surf. Belgium] POE alkyl ether; nonionic; emulsifier for pesticides; liq.; HLB 15.9; 70% conc.

Atlox 4883. [ICI Europe] POE alkyl ether; wetting agent for pesticides and herbicides; powd.; HLB 13.5.

Atlox 4885. [ICI Surf. Belgium] Sorbitan trioleate; CAS 26266-58-0; nonionic; emulsifying agent for pesticides and herbicides; liq.; HLB 1.8; 100% conc.

Atlox 4890B. [ICI Surf. Belgium] Blend; anionic/nonionic; emulsifier for pesticides; liq.; HLB 12.4; 100% conc.

Atlox 4896. [ICI Surf. Belgium] POE alkyl ether; nonionic; emulsifier for pesticides; liq.; HLB 15.9; 100% conc.

Atlox 4898. [ICI Surf. Belgium] Alkoxylated alkylphenol; nonionic; emulsifier for Dimethoate formulations; liq.; HLB 15.0; 100% conc.

Atlox 4899B. [ICI Surf. Belgium] Blend; anionic/nonionic; emulsifier for pesticides; liq.; HLB 12.4.

Atlox 4901. [ICI Surf. Belgium] POE alkyl ether; nonionic; wetting agent for pesticides; liq.; HLB 11.6; 60% conc.

Atlox 4911. [ICI Surf. Belgium] POE alkylaryl phenol; nonionic; dispersant, wetting agent for pesticides; liq.; HLB 14.0; 100% conc.

Atlox 4912. [ICI Surf. Belgium] Polymeric surfactant; nonionic; emulsifier, dispersant, wetting agent for pesticides; liq.; HLB 5.5; 100% conc.

Atlox 4990B. [ICI Surf. Belgium] Blend; anionic/nonionic; emulsifier for pesticides; liq.; HLB 10.8; 100% conc.

Atlox 4991. [ICI Surf. Belgium] POE alkylaryl ether; nonionic; surfactant, wetting agent for pesticides, herbicides; cloudy liq.; HLB 13.6; 100% conc.

Atlox 4995. [ICI Surf. Belgium] POE alkyl ether; nonionic; wetting/suspending agent for pesticides, herbicides; powd.; HLB 13.6; 100% conc.

Atlox 4996 B. [ICI Europe] Nonionic; wetting/suspending agent; powd.

Atlox 5320. [ICI Surf. Belgium] Blend; anionic; emulsifier for pesticides; liq.; 100% conc.

Atlox 5325. [ICI Surf. Belgium] Blend; anionic; emulsifier for pesticides; paste; 100% conc.

Atlox 5330. [ICI Surf. Belgium] Blend; anionic/nonionic; emulsifier for pesticides; liq.; 100% conc.

Atlox 8916P. [ICI Am.] PEG-16 sorbitan ester of tall oil; nonionic; surfactant; amber liq.; sol. in water and most aromatic solv. and pesticide formulations; sp.gr. 1.1; visc. 1320 cps; HLB 14.6; flash pt. > 300 F; 100% conc.

Atlox 8916PF. [ICI Am.; ICI Surf. Belgium] Mixed fatty and resin acids POE sorbitan esters; nonionic; emulsifier, agric. surfactant; pesticides and herbicides; amber liq.; sol. hazy in most aromatic solv. used in pesticide formulations, water sol. after heating; dens. 9.0 lb/gal; sp.gr. 1.08; visc. 500 cp; HLB 14.6; flash pt. > 200 F; pour pt. –20 F; 100% conc.

Atlox 8916TF. [ICI Am.] Alkoxylated polyol fatty acid ester; CAS 9005-65-6; nonionic; emulsifier, agric. surfactant; amber liq.; sol. hazy in most aromatic solv. used in pesticide formulations, water sol. after heating; dens. 8.3 lb/gal; sp.gr. 1.00; visc. 500 cp; flash pt. > 200 F; pour pt. –20 F; HLB 15.4.

Atolene RW. [Dexter] Oleic acid sulfated ester; anionic; lubricant, wetting, dye leveling agent used in textile processing; liq.

Atolex LDA/40. [Standard Chem. UK] Nonionic; dyebath aux.; leveling and dispersing agent.

Atolex LS/3. [Standard Chem. UK] Sulfated fatty alcohol; nonionic/anionic; detergent, wetting agent; liq.

Atolex NI/100B. [Standard Chem. UK] Nonionic; detergent, wetting agent for textiles; liq.

Atpet 545. [ICI Surf. Belgium] Polymeric surfactant; nonionic; crude demulsifier; liq.

Atpet 787. [ICI Surf. Belgium] Proprietary blend; nonionic; oil spill dispersant; liq.

Atpet 900. [ICI Surf. Belgium] Proprietary blend; nonionic; oil spill dispersant; liq.

Atplus 1034. [ICI Am.] Aromatic sulfates and polyhydroxy compds.; agric. foaming agent, spray tank additive; amber liq.; dens. 10 lb/gal; sp.gr. 1.2; visc. 140 cps; flash pt. > 200 F; pour pt. –35 F.

Atplus 1364. [ICI Am.] Nonionic; surfactant with sticker and coupler; agric. surfactant used with post-emergent WP pesticides.

Atplus 1399, 1400, 1401, 1402. [ICI Am.] Nonionic; surfactant with coupling agents; agric. surfactant.

Atplus 1403. [ICI Am.] Anionic; surfactant with coupling agents; agric. surfactant, wetting agent.

Atplus 1992. [ICI Am.] POE polyol fatty acid esters blend; nonionic; surfactant for veg. oil systems and crop oil concs.; liq.; HLB 7.0; 96% conc.

Atplus 2380. [ICI Am.] Nonionic surfactant and free acids in aq. alcohol solvent system; nonionic; pesticide spreader activator adjuvant; liq.; 90% conc.

Atplus 300F. [ICI Am.] Nonionic surfactants with coupling agents; agric. surfactant, emulsifier for pesticides; adjuvant for methane arsonates and other water-sol. herbicides; amber liq.; sol. in most aromatic solv. used in pesticide formulations, agric. grade paraffin oil, disp. in water; dens. 8.6 lb/gal; sp.gr. 1.03; visc. 400 cp; flash pt. > 200 F; pour pt. 15 F; HLB 4.3; 92% solids.

Atplus 400. [ICI Am.] Anionic; agric. surfactant; pale yel. liq.; water sol.; dens. 9 lb/gal; sp.gr. 1.08; visc. < 100 cps.

Atplus 401. [ICI Am.] Anionic surfactant with coupling agents; anionic; agric. surfactant, wetting agent and adjuvant; colorless or lt. yel. liq.; disp. readily w/out heat; sp.gr. 1.21; visc. 31 cstk; flash pt. 118 F; pour pt. –20 F; HLB 4.3.

Atplus 411F. [ICI Am.] Oil/surfactant conc.; agric. surfactant for herbicides and harvest aid chemicals.

Atplus 522. [ICI Am.] Agric. surfactant for use in sodium chlorate sol'n; amber liq.; sol. in water; dens. 10.0 lb/gal; sp.gr. 1.2; visc. 35 cs; flash pt. 115 F; pour pt. < 32 F.

Atplus 526. [ICI Am.] Conc. blend of fatty acids, fatty acid esters and alkoxylated polyhydric alcohol fatty acid esters; nonionic; agric. surfactant, spreader-sticker; specifically developed for repackaging by formulators; liq.

Atplus 535. [ICI Am.] Anionic; agric. surfactant; stabilizer for liq. fertilizer-pesticide mixtures, for repackaging/resale; lt. amber liq.; complete sol. in liq. fertilizers; dens. 10.5 lb/gal; sp.gr. 1.25; visc. 40

cps; flash pt. > 200 F.

Atplus 540. [ICI Am.] Alkoxylated alcohol ether; agric. foaming agent, spray tank additive; amber liq.; dens. 8.1 lb/gal; sp.gr. 0.97; visc. 30 cps; flash pt. 65 F; pour pt. 20 F; 75% conc.

Atplus 555. [ICI Am.] Nonionic; surfactant and coupling agent; agric. surfactant, wetting agent.

Atplus 989. [ICI Am.] Nonionic; surfactant and coupling agent; agric. surfactant, shows activity with pre-emergent Caporal®.

Atranonic Polymer 20. [Atramax] EO nonylphenyl condensate; nonionic; water-based dispersant for pigments used in textiles; used at high temps.; liq.; HLB 12.7; 90% conc.

Atranonic Polymer N200. [Atramax] EO nonylphenyl condensate; nonionic; water-reducible dispersant for pigments used in textiles; liq.; HLB 17.7; 70% conc.

Atra Polymer 10. [Atramax] Ammonium salt of carboxylic acid, styrene copolymer; anionic; water-based film-forming dispersant for pigments; vehicle for flexographic inks; avail. as metal and amine salt; liq.; 25% conc.

Atrasein 115. [Atramax] Ammonium caseinate; anionic; dispersant, leveling agent, etc.; aq. sol'n.; 15% conc.

Atsolyn BH 100. [Atsaun] Ethoxylated alcohol; nonionic; emulsifier for min. and veg. waxes; solid; 100% conc.

Atsolyn GMO. [Atsaun] Glyceryl ester; nonionic; emulsifier; semifluid; HLB 4.3; 100% conc.

Atsolyn P. [Atsaun] Veg. oil ester; nonionic; emulsifier, wetting agent, dispersant, grinding aid for pigments; liq.; HLB 11.0; 100% conc.

Atsolyn PE 27. [Atsaun] Phosphate ester, acid form; nonionic; emulsifier, wetting agent, antistat, corrosion inhibitor for emulsion polymerization, textile scouring; liq.; 100% conc.

Atsolyn PE 36. [Atsaun] Phosphate ester; anionic; emulsifier, solubilizer, antistat, corrosion inhibitor; liq.; 100% conc.

Atsolyn SUC 100. [Atsaun] Sulfosuccinate; nonionic; foaming agent; solubilizer in fat liquors; partial replacement for ether sulfate; liq.; 30% conc.

Atsolyn T. [Atsaun] Ethoxylated alkylphenol; nonionic; wetting and dispersing agent; liq.; HLB 13.0; 100% conc.

Atsolyn TD 50. [Atsaun] Ethoxylated amine; nonionic; emulsifier, corrosion inhibitor in metal treatment; liq.; 100% conc.

Atsowet P 50. [Atsaun] Sulfosuccinate; anionic; wetting and rewetting agent; liq.; 50% conc.

Atsurf 311. [ICI Am.] Surfactant blend in aq.-alcohol solv. system; anionic/nonionic; wetting agent for agric. flowables; 55% conc.

Atsurf 1910. [ICI Am.] Ammonium salt of sulfated alkyl phenol ethoxylate, aq. alcohol sol'n.; anionic; wetting/dispersing agent for agric. formulations; liq.; 58% conc.

Atsurf 2801. [ICI Am.] Polyol oleate; nonionic; industrial grade surfactant, emulsifier, for paper coatings, floor polishes, furniture polishes, fiberboard and particle board; amber liq.; sol. in IPA, perchloroethylene, xylene, cottonseed and paraffinic oil; dens. 1.00 g/ml; visc. 379 cs; HLB 5; pour pt. –39 F; 100% act.

Atsurf 2802. [ICI Am.] Polyol trioleate; nonionic; industrial grade surfactant, emulsifier, for paper coatings, floor polishes, furniture polishes, fiberboard and particle board; amber liq.; sol. in perchloroethylene, xylene, cottonseed and paraffinic oil; dens. 0.94 g/ml; visc. 204 cs; HLB 2.5; pour pt. < –35 F; 100% act.

Atsurf 2803. [ICI Am.] Polyol stearate; nonionic; industrial grade surfactant, emulsifier, for paper coatings, floor polishes, furniture polishes, fiberboard and particle board; lt. tan solid; sol. in IPA (1%), perchloroethylene, xylene; dens. 1.06 g/ml; visc. 165 cs; HLB 5; pour pt. 125 F; 100% act.

Atsurf 2821. [ICI Am.] Ethoxylated polyol oleate; nonionic; industrial grade surfactant, emulsifier, for paper coatings, floor polishes, furniture polishes, fiberboard and particle board; amber liq.; sol. in water, IPA, perchloroethylene (10%), xylene (1%), cottonseed oil (1%); dens. 1.06 g/ml; visc. 846 cs; HLB 16; pour pt. 55 F; 100% act.

Atsurf 2822. [ICI Am.] Ethoxylated polyol trioleate; nonionic; industrial grade surfactant, emulsifier, for paper coatings, floor polishes, furniture polishes, fiberboard and particle board; amber liq.; sol. in IPA, perchloroethylene, xylene, cottonseed and paraffinic oil; dens. 1.02 g/ml; visc. 349 cs; HLB 11; pour pt. 15 F; 100% act.

Atsurf 2823. [ICI Am.] Ethoxylated polyol stearate; nonionic; industrial grade surfactant, emulsifier, for paper coatings, floor polishes, furniture polishes, fiberboard and particle board; amber semisolid; sol. in water, IPA, xylene (1%), cottonseed oil; dens. 1.08 g/ml; visc. 50 cs (160 F; HLB 16; pour pt. 90 F; 100% act.

Atsurf S-80. [ICI Am.] Sorbitan fatty ester; nonionic; surfactants for use in emulsions; nonaq. wetting props., solid dispersing props.; liq.; HLB 4.3; 100% conc.

Atsurf S-85. [ICI Am.] Sorbitan fatty ester; nonionic; surfactants for use in emulsions; nonaq. wetting props., solid dispersing props.; liq.; HLB 1.8; 100% conc.

Atsurf T-80. [ICI Am.] PEG sorbitan fatty ester; nonionic; dispersant, o/w emulsifier providing good substrate wetting of aq. systems; liq.; HLB 15.0; 100% conc.

Atsurf T-85. [ICI Am.] PEG sorbitan fatty ester; nonionic; dispersant, o/w emulsifier providing good substrate wetting of aq. systems; liq.; HLB 11.0; 100% conc.

Autopoon GK 4003. [Zschimmer & Schwarz] Blend; cationic; basic material for water repellents and washing/preserving agents for car care; liq.; 94% act.

Autopoon GK 4004. [Zschimmer & Schwarz] Surfactant blend; cationic; basic material for water repellents and washing/preserving agents for car care; liq.; 96% act.

Autopoon NI. [Zschimmer & Schwarz] Cationic surfactants, solvs., and solubilizers; cationic; water-repellent conc. for preps. for lacquered surfaces, car finishing; clear red-brn. liq.; sol. in water, IPA; sp.gr. 0.95; cloud pt. 0 C; pH 4.5–5.5 (1%); 55% act.

Autopur WK 4121. [Zschimmer & Schwarz] Surfactant blend; nonionic/cationic; basic material for car shampoo formulations with water-repellent and gloss effects; slightly opalescent, brownish liq.; visc. 500 mPa•s; cloud pt. 10 C; pH 4.5–5.5 (10%); 40% act. in water.

Avanel® S-30. [PPG/Specialty Chem.] Sodium C12-15 pareth-3 sulfonate; anionic; biodeg. mild surfactant, emulsifier for personal care, household, institutional and industrial prods.; stable in presence of hypochlorite and over entire pH range; wh. paste;

sol. in water and inorg. sol'ns.; m.w. 420; sp.gr. 1.06; visc. 360 cps (35 C); solid. pt. 25 C; flash pt. 130 F; 35% conc., 5% ethanol.

Avanel® S-35. [PPG/Specialty Chem.] Sodium alkyl ether phenol ether sulfonate; anionic; mild, stable emulsifier for creams, lotions, and liq. soaps; liq.; 28% conc.

Avanel® S-70. [PPG/Specialty Chem.] Sodium C12-15 pareth-7 sulfonate; anionic; biodeg. mild surfactant, emulsifier for personal care, household, institutional and industrial prods.; stable in presence of hypochlorite and over entire pH range; clear liq., odorless; sol. in water and various inorg. sol'ns.; m.w. 600; sp.gr. 1.07; visc. 270 cps; solid. pt. -1 C; flash pt. 200 F; 35% conc.

Avanel® S-74. [PPG/Specialty Chem.] Sodium alkyl ether sulfonate; anionic; biodeg. mild surfactant, emulsifier for personal care, household, institutional and industrial prods.; stable in presence of hypochlorite and over entire pH range; clear liq., odorless; sol. in water and various inorg. sol'ns.; m.w. 260; sp.gr. 1.10; visc. 30 cps; solid. pt. -8 C; flash pt. 200 F; 35% conc.

Avanel® S-90. [PPG/Specialty Chem.] Sodium C12-15 pareth-9 sulfonate; biodeg. mild surfactant, emulsifier for personal care, household, institutional and industrial prods.; stable in presence of hypochlorite and over entire pH range; clear liq., odorless; sol. in water and various inorg. sol'ns.; m.w. 690; sp.gr. 1.07; visc. 60 cps; solid. pt. -1 C; flash pt. 200 F.

Avanel® S-150. [PPG/Specialty Chem.] Sodium C12-15 pareth-15 sulfonate; anionic; biodeg. mild surfactant, emulsifier for personal care, household, institutional and industrial prods.; stable in presence of hypochlorite and over entire pH range; almost colorless liq., odorless; sol. in water and inorg. sol'ns.; m.w. 950; sp.gr. 1.07; visc. 70 cps; solid. pt. -1 C; flash pt. 200 F; 35% conc.

Avirol® 125 E. [Henkel/Functional Prods.] Sodium alkyl ether sulfate; anionic; emulsifier for vinyl acetate copolymers, S/B latexes, vinyl chloride copolymers, acrylate homo- and copolymers; liq./paste; 62-65% solids.

Avirol® 200. [Henkel/Functional Prods.] Ammonium lauryl sulfate; CAS 2235-54-3; foaming agent and suspension aid in rock drilling, fire fighting, dispersant for dry wall mfg., dyes and pigments; lt. amber liq.; typ. odor; visc. 1000 cps; cloud pt. 10 C max; 28% conc.

Avirol® 252 S. [Henkel Canada] Sodium laureth sulfate; CAS 9004-82-4; 1335-72-4; anionic; surfactant for lt. duty detergents, dishwashing liqs., car cleaners, mfg. of gypsum boards; liq.; 25% conc.

Avirol® 270A. [Henkel Canada] Ammonium lauryl sulfate; anionic; surfactant for lt. duty detergents, high foaming cleaners, dishwashing liqs., fire fighting foams; liq.; 27% conc.

Avirol® 280 S. [Henkel Canada] Sodium lauryl sulfate; CAS 151-21-3; anionic; surfactant for lt. and heavy duty detergents, carpet shampoos, dishwashing detergents, fire fighting foams; liq.; 28% conc.

Avirol® 300. [Henkel/Functional Prods.] TEA lauryl sulfate; CAS 139-96-8; base for cleaning and cosmetic preparations; lt. amber liq.; visc. 500 cps max; 39–41% fatty alcohol sulfate.

Avirol® 400 T. [Henkel Canada] TEA lauryl sulfate; CAS 139-96-8; anionic; surfactant for lt. and heavy duty detergents, general purpose cleaners; foams in presence of oily soil; liq.; 40% conc.

Avirol® 603 A. [Henkel Canada] Ammonium lauryl ether sulfate; anionic; surfactant for lt. and heavy duty detergents, foamer in dishwashing liqs. and fire fighting foams; liq.; 60% conc.

Avirol® 603 S. [Henkel Canada] Sodium lauryl ether sulfate; CAS 9004-82-4; anionic; surfactant for lt. and heavy duty detergents, all purpose domestic and industrial cleaners; liq.; 60% conc.

Avirol® A. [Henkel/Functional Prods.] Ammonium lauryl sulfate; anionic; emulsifier for emulsion polymerization; additive for mech. latex foaming; foaming agent for acrylate disps., carpet and upholstery cleaners; air entraining agent for mortars; liq.; 30-32% solids.

Avirol® AE 3003. [Henkel/Functional Prods.] Ammonium laureth sulfate; anionic; emulsifier for vinyl acetate copolymers, S/B latexes, vinyl chloride copolymers, acrylate homo- and copolymers; liq.; 31-33% solids.

Avirol® AOO 1080. [Henkel/Functional Prods.] Oxidized oleic acid, ammonium salt; anionic; emulsifier for vinyl chloride and acrylic polymers; liq.; 40% conc.

Avirol® FES 996. [Henkel/Functional Prods.] Sodium laureth sulfate; CAS 9004-82-4; anionic; emulsifier for vinyl acetate copolymers, S/B latexes, vinyl chloride copolymers, acrylate homo- and copolymers; liq./paste; 62-65% solids.

Avirol® SA 4106. [Henkel/Functional Prods.] Sodium 2-ethylhexyl sulfate; anionic; wetting agent, stabilizer for plastics, rubber, adhesives, food contact paper; coemulsifier for vinyl chloride, acrylics, vinyl acetate copolymers; biodeg.; amber liq.; sp.gr. 1.1; dens. 9.2 lb/gal; visc. 200 cps; cloud pt. 5 C max.; surf. tens. 38 dynes/cm; pH 7–10 (10%); 43–46% solids.

Avirol® SA 4108. [Henkel/Functional Prods.] Sodium n-octyl sulfate; anionic; emulsifier, low foaming surfactant, wetting agent; liq.; 35-37% solids.

Avirol® SA 4110. [Henkel/Functional Prods.] Sodium n-decyl sulfate; anionic; wetting and emulsifying agent for plastics; amber liq.; sp.gr. 1.065; dens. 8.8 lb/gal; visc. 100 cps. max.; pH 8–10 min. (10%); cloud pt. –5 C; surf. tens. 31 dynes/cm (@ CMC); 31–33% solids.

Avirol® SA 4113. [Henkel/Functional Prods.] Sodium tridecyl sulfate; CAS 3026-63-9; anionic; emulsifier for S/B and vinyl chloride copolymers; liq.; 29-31% solids.

Avirol® SE 3002. [Henkel/Functional Prods.] Sodium laureth sulfate; CAS 9004-82-4; 1335-72-4; anionic; emulsifier for vinyl acetate copolymers, S/B latexes, vinyl chloride copolymers, acrylate homo- and copolymers; liq.; 27-30% solids.

Avirol® SE 3003. [Henkel/Functional Prods.] Sodium laureth sulfate; CAS 9004-82-4; 1335-72-4; anionic; emulsifier for vinyl acetate copolymers, S/B latexes, vinyl chloride copolymers, acrylate homo- and copolymers; liq.; 27-30% solids.

Avirol® SL 2010. [Henkel/Functional Prods.] Sodium lauryl sulfate; CAS 151-21-3; anionic; dispersing agent and emulsifier for acrylates, styrene acrylic, vinyl chloride, vinylidene chloride and vinyl acetate copolymers; foaming agent for mech. latex foaming, carpet and upholstery cleaners; air entraining agent for mortars; liq.; 30-32% solids.

Avirol® SL 2015. [Henkel/Functional Prods.] Sodium C12-C15 sulfate; anionic; emulsifier for emulsion polymerization; additive for mech. latex foaming;

foaming agent for acrylate disps., carpet and upholstery cleaners; air entraining agent for mortars; liq.; 31-34% solids.

Avirol® SL 2020. [Henkel/Functional Prods.] Sodium lauryl sulfate; CAS 151-21-3; anionic; emulsifier for emulsion polymerization; additive for mech. latex foaming; foaming agent for acrylate disps., carpet and upholstery cleaners; air entraining agent for mortars; liq.; 30-32% solids.

Avirol® SO 70P. [Henkel/Functional Prods.] Sodium dioctyl sulfosuccinate; anionic; wetting agent, emulsifier for antifog compositions, cleaners for automobiles, dry cleaning formulations, and glass surfaces; pigment dispersion in paints and inks; latex paint stabilization; used in pesticides; APHA 100 clear liq.; sol. in polar and nonpolar solv.; sp.gr. 1.06; dens. 8.8 lb/gal; flash pt. 280 F (PMCC); cloud pt. –5 C; 70% solids.

Avirol® T 40. [Henkel/Functional Prods.] TEA lauryl sulfate; CAS 139-96-8; anionic; emulsifier for emulsion polymerization; additive for mech. latex foaming; foaming agent for acrylate disps., carpet and upholstery cleaners; air entraining agent for mortars; liq.; 43-44% solids.

Aviscour 98/70. [Albright & Wilson Australia] Amine neutralized alkylbenzene sulfonic acid; anionic/nonionic; detergent conc. for household detergents; pale yel. clear-hazy visc. liq.; 95% min. solids; visc. 80 ± 25 s (No. 4 Cup); pH 7.5 ± 0.7.

Aviscour 125. [Albright & Wilson Australia] Linear alkylbenzene sulfonate and coactive nonionic surfactant blend; anionic/nonionic; detergent conc. used in detergents, hard surface and upholstery cleaning; amber visc. liq.; sol. in water; pH 7.0±0.5 (5%).

Aviscour HP50. [Albright & Wilson Australia] Alkylbenzene sulfonic acid detergent conc.; anionic/nonionic; laundry and lt. duty detergents, hard surface cleaners, car shampoos, and foaming bath additives; orange-yel. hazy visc. liq.; dilutable in water; sp.gr. 1.05; cloud pt. 5 C max.; pH 8.2 ± 0.5; 50.0 ± 1.0% solids.

Aviscour HQ50. [Albright & Wilson Australia] Surfactant blend; anionic/nonionic; formulated liq. detergent conc.; liq.; 50% conc.

Avistin® FD. [Hüls Am.; Hüls AG] Stearamidoethyl ethanolamine; CAS 141-21-9; cationic; bases for textile auxiliary agents; paste to wax; 100% conc.

Avistin® PN. [Hüls Am.; Hüls AG] Stearamidoethyl ethanolamine; CAS 141-21-9; cationic; for prod. of cationic textile auxiliary agents; flakes; 100% act.

Avitone® A. [DuPont] Sodium alkyl sulfonate; finishing agent, softener, lubricant for improving texture and hand of textiles, leather, and paper; also for elastomers; highly stable to chemicals and oxidation; lt. tan paste, bland odor; disp. in water; sp.gr. 1.02 g/mL; dens. 8.4 lb/gal; cloud pt. none; flash pt. none; pH 7.5-9.0 (10% aq.); 75% act.

Avitone® F. [DuPont] Sodium alkyl sulfonate; anionic; dyeing assistant, penetrant, softener, napping assistant, dispersant, leveling agent, lubricant; for natural and syn. fibers and their blends; visc. reddish liq., bland odor; disp. in water; sp.gr. 1.05 g/mL; dens. 8.8 lb/gal; cloud pt. < 0 C; flash pt. (PMCC) > 104 C; pH 7-9 (10% aq.); 40% act.

Avitone® T. [DuPont] Sodium alkyl sulfonate; anionic; softener, lubricant, rewetting agent, dispersant for cotton fabrics, Dacron polyester and wool blends; tan visc. paste; disp. in water; sp.gr. 1.04 g/mL; dens. 8.7 lb/gal; flash pt. none; pH 7-8.5 (10% aq.); 76% act.

Avivan® SFC. [Ciba-Geigy/Dyestuffs] Fatty acid amide condensate; amphoteric; nonyel. softener for cellulosic and syn. fibers; liq.

AW-357. [Anedco] Neutralized sulfonic; wetting agent; 50% active.

AW-357C. [Anedco] Neutralized sulfonic; wetting agent; 100% active.

AW-395. [Anedco] Quaternary surfactant; cationic; wetting agent; 50% active.

AW-396. [Anedco] Nonionic/cationic; wetting agent; surfactant flush aid; backwashing filters.

AW-397. [Anedco] Nonyl phenol ethoxylate; nonionic; surfactant, wetting agent.

AW-398. [Anedco] Benzyl sulfonic acid; anionic; surfactant, wetting agent.

Axel P 100. [Aquatec Quimica SA] Fatty alcohol and fatty ester; antifoamer for paper machines; paste.

B

B-8880-50%. [Ethox] High m.w. polyoxyalkylene polymer; dyeing assistant for acrylics; stabilizer for aq. emulsions; liq.; cloud pt. > 100 C (1% aq.); 50% act.

Babinar 715. [Marubishi Oil Chem.] Fatty acid polyethylene polyamine condensate; cationic; textile softener for polyacrylonitrile fiber; liq.

Babinar 801C. [Marubishi Oil Chem.] Fatty acid polyethylene polyamine condensate; cationic; antistatic softener for polyacrylonitrile fiber; paste.

Bactistep® MH 80. [Stepan Europe] Dialkyl dimethyl ammonium methoxy sulfate; cationic; sanitizer; 80% conc.

Barisol Super BRM. [Dexter] Complex multicarbon alcohol phosphate potassium salt; anionic; textile wetting agent for dye leveling, pectin removal from cottons, scouring of cotton and synthetic blends; dispersant; straw-colored liq.; sol. in hot or cold water; pH 8-9 (0.1%); 30% conc.

Barium Petronate 50-S Neutral. [Witco/Sonneborn] Barium petroleum sulfonate neutral; anionic; emulsifier; fuel oil additive; rust preventive; liq.; 50% conc.

Barium Petronate Basic. [Witco/Sonneborn] Barium petroleum sulfonate (CAS #61790-48-5); anionic; emulsifier; lubricant additive for lube and industrial oils and fuels, specialty oils and greases; 40–45% act.

Barlox® 10S. [Lonza] Decylamine oxide; CAS 2605-79-0; nonionic; detergent; visc. builder, emollient; liq.; 30% conc.

Barlox® 12. [Lonza] Cocamine oxide; nonionic/cationic; detergent; visc. builder, emollient; liq.; visc. 45 cps; pour pt. 4 C; pH 7.0 (1%); surf. tens. 32.3 dynes/cm (0.1%); Draves wetting 4 s; 30% conc.

Barlox® 14. [Lonza] Myristamine oxide; nonionic/cationic; detergent; visc. builder, emollient; liq.; visc. 60 cps; pour pt. 2 C; pH 7.0 (1%); surf. tens. 31.0 dynes/cm (0.1%); Draves wetting 5 s; 30% conc.

Barlox® 16S. [Lonza] Cetamine oxide; biodeg. foam stabilizer, visc. builder, emulsifier, conditioner for personal care and industrial prods.; paste; visc. 27,000 cps; pH 7.0 (1%); surf. tens. 32.4 dynes/cm (0.1%); Draves wetting 9 s; 30% act.

Barlox® 18S. [Lonza] Stearyl dimethyl amine oxide; biodeg. foam stabilizer, visc. builder, emulsifier, conditioner for personal care and industrial prods.; paste; 25% act.

Barlox® C. [Lonza] Cocamidopropylamine oxide; nonionic/cationic; foam stabilizer and visc. builder for shampoos, industrial prods.; liq.; visc. 36 cps; pour pt. 4 C; pH 7.0 (1%); surf. tens. 35.4 dynes/cm (0.1%); Draves wetting 27 s; 30% conc.

Barre® Common Degras. [RITA] Wool grease deriv.; nonionic; emollient; leather softener; w/o emulsion for corrosion prevention; dk. paste; oil-sol.; 100% conc.

Base 75. [Ferro/Keil] Sodium sulfonate-soap deriv.; anionic; o/w emulsifier for conventional and EP sol. oils and pastes for metalworking industry; imparts stability and wetting props.; liq.; sp.gr. 1.01; dens. 8.4 lb/gal; visc. 530 cSt (40 C); pour pt. < -12 C; acid no. 30; pH 8.1 (1% aq.).

Base 76. [Ferro/Keil] Sulfonate; nonionic; emulsifier for sol. oils for cutting and grinding fluids; provides stability; liq.; sp.gr. 1.02; dens. 8.5 lb/gal; visc. 1600 cSt (40 C); pour pt. -7 C; acid no. 1; pH 8.8 (1% aq.).

Base 85. [Ferro/Keil] Modified nonionic emulsifier; nonionic; ashless o/w emulsifier for prep. of sol. oils for metalworking industry, glass shear lubricants; liq.; sp.gr. 1.01; dens. 8.4 lb/gal; visc. 88 cSt (40 C); pour pt. < -12 C; acid no. 2; pH 6.0 (1% aq.).

Base 91-6P. [Clark] Ethoxylated linear alcohol phosphate, free acid; alkali-stable surfactant for heavy-duty detergents, scours, antistats, lubricants, rust preventatives.

Base 104. [Clark] 2-Ethyl hexanol phosphate, free acid; surfactant base for making low foam wetters, penetrants, antistats, lubricants, rust preventatives.

Base 610. [Clark] Ethoxylated alcohol phosphate, free acid; alkali-stable surfactant for heavy-duty detergents, scours, antistats, lubricants, rust preventatives.

Base 865. [Ferro/Keil] Sulfur chlorinated base; extreme pressure additive, detergent, lubricant for high performance fluids, machining and grinding operations; sp.gr. 1.08; dens. 9.0 lb/gal; visc. 450 cSt (40 C); pour pt. 16 C; flash pt. (COC) 218 C; acid no. 10; 3.6% sulfur, 3.2% chlorine.

Base 7800. [Ferro/Keil] Sulfonate soap; anionic; o/w emulsifier for sol. oils and pastes for metalworking industry, emulsion cleaners, syn. and semi-syn. coolants; hard water tolerance; aids rust protection; liq.; sp.gr. 1.02; dens. 8.5 lb/gal; visc. 1100 cSt (40 C); pour pt. 4 C; acid no. 4; pH 8.9 (1% aq.).

Base 8000. [Ferro/Keil] Petroleum sulfonate/auxiliary emulsifier blend; o/w emulsifier for prep. of sol. oils and pastes for metalworking industry; liq.; 100% conc.

Base 8000P. [Ferro/Keil] Emulsifier system for use in formulating sol. oils with paraffinic base oils; dk. amber clear fluid; sp.gr. 1.035; dens. 8.62 lb/gal; visc. 1470 cSt (40 C); pour pt. -7 C; pH 8.8 (20% in paraffinic oil).

Base ML. [Ferro/Keil] Methyl lardate; wetting/oiliness/lubricity agent for metalworking, lubricating, motor, and rolling oils; antiwear additive, process aid, release additive; Gardner 1 liq.; sp.gr. 0.87; dens. 7.2 lb/gal; visc. 4.5 cSt (40 C); pour pt. 10 C; cloud pt. 11 C; flash pt. (COC) 168 C; acid no. 1; sapon. no. 195; 100% conc.

Base MO. [Ferro/Keil] Methyl ester; wetting/oiliness/lubricity agent for metalworking, lubricating, motor, and rolling oils; antiwear additive, process aid, release additive; Gardner 1; sp.gr. 0.87; dens. 7.2 lb/gal; visc. 4.5 cSt (40 C); pour pt. 10 C; cloud pt. 11 C; flash pt. (COC) 168 C; acid no. 1; sapon. no. 195.

Base MT. [Ferro/Keil] Methyl tallowate; wetting/oiliness/lubricity agent for metalworking, lubricating, motor, and rolling oils; antiwear additive, process aid, release additive; Gardner 1 liq.; sp.gr. 0.87; dens. 7.2 lb/gal; visc. 4.5 cSt (40 C); pour pt. 10 C; cloud pt. 13 C; flash pt. (COC) 168 C; acid no. 1; sapon. no. 195; 100% conc.

BASF Wax LS. [BASF] Montan acid wax; emulsifying component for paraffin wax emulsions.

Basojet® PEL 200%. [BASF] EO condensate/anionic blend; anionic; nonfoaming dispersing and leveling agent for dyeing of polyester fibers with disperse dyes under high temp. conditions; amber liq.; readily sol. in cold water.

Basol® WS. [BASF] Aromatic sulfonic acid condensate; anionic; dispersant, protective colloid, dyeing assistant esp. for exhaust dyeing of polyester; stabilizes dye dispersions; stable to acids, alkalies, hard water, electrolytes; brownish microgranules; sol. in water.

Basophen® M. [BASF/Fibers] Alkyl ester of phosphoric acid with anionic/nonionic emulsifiers; anionic/nonionic; nonfoaming wetting agent, foam suppressant for textile applics., pkg. dyeing of cotton, cotton/polyester, wool, desizing, latex penetration in carpet backing; stable in hard water and liquors containing alkali, acetic acid, or sulfuric acid; lt. yel. sl. visc. liq., weak odor becomes pungent above 160 F; insol. in water; 90% conc.

Basophen® NB-U. [BASF/Fibers] Proprietary; anionic/nonionic; synergistic wetting agent/detergent combination for rapid wetting action in pretreatment, bleaching, desizing, alkaline boil-off processes; high stability to electrolytes; ylsh. paste; readily sol. in cold and hot water; pH 7.5-8 (10g/l water); 50% conc.

Basophen® RA. [BASF; BASF/Fibers] Sulfosuccinate; anionic; wetting agent for textile desizing, pretreatment, and bleaching of cellulosic fibers; resistant to chlorine, hard water, weak acids; ylsh. liq.; sol. in hot water; pH 5.5-7 (1% aq.); toxicology: attacks skin and eye mucous membranes; avoid inhalation of vapors from hot liquors; 70% conc.

Basophen® RBD. [BASF] Anionic; wetting agent for desizing and bleaching processes.

Basophor A. [BASF/Fibers] Fatty acid ester; nonionic; emulsifier for min. and fatty oils; liq.; 100% conc.

Basopon® LN. [BASF; BASF/Fibers] Alkylphenol ethoxylate; nonionic; wetting agent, detergent with dispersing effects for wool and syn. fibers, desizing and scouring applics.; stain removing agent; stable to hard water, alkalis, acids, reducing and oxidizing agents; clear ylsh. visc. liq.; sol. in water; dens. 1.05 g/cc; pH 6-8 (0.5%); 100% conc.

Basopon® TX-110. [BASF] Proprietary; nonionic; nonfoaming scouring agent with emulsifying props. for polyester knit goods; clear ylsh. visc. liq.; readily sol. in cold and warm water.

Bayhibit. [Miles] 2-Phosphono-butane-tricarboxylic acid-1,2,4; corrosion and scale inhibitor in cooling systems; deflocculation of ceramic slips and oil drilling sludges, and stabilization of pigment suspensions; formulation of cleaning agents; pickling and cleaning agent for oxidized metal surfs.; sequestrant; dispersant; colorless to straw liq.; odorless; m.w. 270; water-misc.; dens. 1.28 g/cm3; visc. 20 mPa; pH 2 (1% aq.); 45–50 aq.

Bayhibit® AM. [Miles] 2-Phosphonobutane-1,2,4-tricarboxylic acid; CAS 37971-36-1; corrosion and scale inhibitor in cooling systems; sea water evaporation units, water used for flooding in oil drilling; deflocculation of ceramic slips and oil drilling sludges, stabilization of pigment suspensions; formulation of cleaning agents; pickling and cleaning agent for oxidized metal surfaces; sequestrant; dispersant; colorless-straw liq., almost odorless; sol. in water; m.w. 270; sp.gr. 1.28; visc. 20 mPa•s; b.p. 100 C; f.p. -15 C; flash pt. > 100 C; pH 1.501.8 (1% aq.); corrosive; toxicology: LD50 (oral, rat) > 6500 mg/kg; moderately irritating to eyes, mucous membranes; 50% aq. sol'n.

Berol 02. [Berol Nobel AB] Nonylphenol ethoxylate; CAS 68412-54-4; nonionic; emulsifier; solv. cleaner; emulsion polymerization; biodeg.; Hazen < 200 clear liq.; sol. in ethanol, xylene, trichloroethylene, lt. fuel oil; disp. in water, paraffin oil, wh. spirit; sp.gr. 1.04; visc. 350 cps; HLB 10.9; cloud pt. 60–70 C; flash pt. > 100 C; pour pt. < 0 C; surf. tens. 31 dynes/cm; pH 6–7 (1% aq.); surf. tens. 31 mN/m (0.1%); Draves wetting 1.6 g/l (25 s); Ross-Miles foam 20 mm (initial, 0.05%, 50 C); 100% conc.

Berol 07. [Berol Nobel AB] Fatty alcohol ethoxylate; CAS 69227-20-9; nonionic; detergent, equalizing and dispersing agent; biodeg.; wh. wax; sol. in water, ethanol, xylene, trichlorethylene, disp. in paraffin oil, wh. spirit, lt. fuel oil; sp.gr. 1.09; visc. 50 cps (50 C); m.p. 36–38 C; HLB 15; cloud pt. 74–75 C; flash pt. > 100 C; surf. tens. 40 dynes/cm; pH 6–7 (1% aq.); 100% conc.

Berol 08 Powd. [Berol Nobel AB] Fatty alcohol ethoxylate; CAS 69227-20-9; nonionic; detergent, dispersant for detergent applics., enzyme coating; biodeg.; wh. powd.; sol. in water, ethanol, trichlorethylene, disp. in xylene; sp.gr. 0.55–0.65; visc. 200 cps (70 C); m.p. 52–54 C; HLB 18.7; cloud pt. 76–77 C; flash pt. > 100 C; pH 6–7 (1% aq.); surf. tens. 48 dynes/cm (0.1%); Draves wetting > 10 g/l (25 s); Ross-Miles foam 80 mm (initial, 0.05%, 50 C); 100% conc.

Berol 09. [Berol Nobel AB] Nonyl phenol ethoxylate; CAS 68412-54-4; nonionic; detergent, wetting agent, emulsifier, dispersant; biodeg.; Hazen < 150 clear liq.; sol. in water, ethanol, xylene, trichlorethylene; disp. in paraffin oil, wh. spirit, lt. fuel oil; sp.gr. 1.05; visc. 350 cps; HLB 13.3; cloud pt. 52–58 C; flash pt. > 100 C; pour pt. 5 C; pH 6–7 (1% aq.); surf. tens. 33 mN/m (0.1%); Draves wetting 0.5 g/l (25 s); Ross-Miles foam 80 mm (initial, 0.05%, 50 C); 100% conc.

Berol 26. [Berol Nobel AB] Nonylphenol ethoxylate; CAS 68412-54-4; nonionic; emulsifier, detergent additive; base for sulfation; biodeg.; Hazen < 250 clear liq.; sol. in ethanol, xylene, trichloroethylene, paraffin oil, wh. spirit, lt. fuel oil, disp. in water; sp.gr. 1.02; visc. 350 cps; HLB 8.9; cloud pt. 42–46 C; flash pt. > 100 C; pour pt. < 0 C; pH 6–7 (1% aq. disp.); degreasing to skin; 100% conc.

Berol 28. [Berol Nobel AB] PEG-7 oleamine; CAS 26635-93-8; emulsifier, dispersant, wetting agent, antistat, anticorrosive for agric., leather, textiles, metalworking and plastic industries; Gardner ≤ 12 liq.; sol. @ 5% in alcohols, low aromatic solvs., propylene glycol, wh. spirit, xylene, water; sp.gr. 0.980 g/cc; visc. 160 mNs/m; HLB 10.6; pour pt. -15 C; cloud pt. 42 C (1% in 10% NaCl); pH 10 (1% aq.); surf. tens. 36.5 mN/m (0.1%); Draves wetting 10 g/l (25 s); Ross-Miles foam 50 mm (initial, 0.05%, 50 C); corrosive; 100% conc.

Berol 048. [Berol Nobel AB] Ethoxylated tridecyl fatty alcohol; CAS 9043-30-5; nonionic; wetting agent, detergent; for alkaline and acid industrial cleaners; gives high brittle foam; Hazen < 100 clear liq.; sol. in water, ethanol, disp. in propylene glycol, trichloroethylene, benzene, paraffin oil, wh. spirit; sp.gr. 1.02; visc. 170 cps; HLB 13.5; cloud pt. 70 C; flash pt. > 100 C; pour pt. 0 C; pH 6-7 (1% aq.); surf. tens. 30 mN/m (0.1%); Draves wetting 0.4 g/l (25 s); Ross-Miles foam 120 mm (initial, 0.05%, 50 C); 85% conc.

Berol 050. [Berol Nobel AB] Ethoxylated fatty alcohol; CAS 68439-50-9; nonionic; intermediate for prod. of ether sulfates; Hazen < 100 clear to cloudy liq.; sol. @ 5% in ethanol, IPA, propylene glycol, wh. spirit, xylene; disp. in water; sp.gr. 0.930; visc. 30 mPa•s; HLB 8.4; pour pt. 0 C; cloud pt. 50 C (5 g in 25 g 25% butyl diglycol); clear pt. 20 C; flash pt. > 100 C; pH 7 (1% aq.); 100% act.

Berol 79. [Berol Nobel AB] Blend; nonionic; emulsifier for stable o/w emulsions; used in solv. degreasers; Gardner 13 max. clear yel. liq; sol. in ethanol, benzene, trichloroethylene, paraffin oil, light min. oil, disp. in water; sp.gr. 0.97; visc. 330 mPa•s; pour pt. 0 C; clear pt. 0 C; flash pt. > 100 C; pH 7-8 (1% in aq. ethanol); 100% conc.

Berol 087. [Berol Nobel AB] Ethoxylated alcohol; CAS 68439-51-0; nonionic; detergent for machine dishwashing, high pressure cleaning, low foaming cleaning agents; biodeg.; Hazen 75 clear liq.; sol. in water, ethanol, trichloroethylene, disp. in xylene; sp.gr. 0.997; visc. 85 cps; HLB 11.5; cloud pt. 42 C; flash pt. > 100 C; pour pt. –1 C; pH 5–7 (1% aq.); surf. tens. 34 dynes/cm (0.1%); Draves wetting 0.5 g/l (25 s); Ross-Miles foam 15 mm (initial, 0.05%, 50 C); 100% act.

Berol 106. [Berol Nobel AB] PEG-28 castor oil; CAS 61791-12-6; nonionic; o/w emulsifier for biocides and pesticides; biodeg.; Gardner < 4 cloudy liq.; sol. in water, ethanol, propylene glycol, xylene, trichloroethylene; sp.gr. 1.055; visc. 650 cps; HLB 15; cloud pt. 54–58 C; flash pt. > 100 C; pour pt. 15 C; surf. tens. 44 dynes/cm; pH 5–7 (1% aq.); 100% conc.

Berol 108. [Berol Nobel AB] PEG-40 castor oil; CAS 61791-12-6; nonionic; surfactant, emulsifier for chemical industry; as softener, rewetting agent, pigment dipsersant, dye assistant, leveling agent for paints, textiles, leather; lubricant additive and emulsifier in lubricants for plastics, metals, textiles; Gardner ≤ 5 cloudy visc. liq.; sol. @ 5% in ethanol, IPA, water, xylene; disp. in propylene glycol; sp.gr. 1.06; visc. 700 mPa•s; HLB 13.3; sapon. no. 60-64; pour pt. 20 C; cloud pt. 73-77 C (5 g in 25 ml 25% butyl diglycol); clear pt. 34 C; flash pt. > 100 C; pH 5-7 (1% aq.); surf. tens. 45 mN/m (0.1%); Draves wetting > 10 g/l (25 s); Ross-Miles foam 70 mm (initial, 0.05%, 50 C); 100% act.

Berol 173. [Berol Nobel AB] Fatty alcohol ethoxylate; CAS 68439-50-9; nonionic; low foaming detergent, emulsifier, wetting agent in cleaning prods.; biodeg.; wh. paste; sol. in water, ethanol, xylene, trichloroethylene, light fuel oil; disp. in wh. spirit; sp.gr. 0.99; visc. 30 cps; HLB 12.5; cloud pt. 65–70 C; pour pt. 25 C; flash pt. ≥ 100 C; pH 6–7 (1% aq.); surf. tens. 33 mN/m (0.1%); Draves wetting 10 g/l (25 s); Ross-Miles foam 75 mm (initial, 0.05%, 50 C); degreasing to skin; 100% act.

Berol 185. [Berol Nobel AB] Modified fatty alcohol ethoxylate; CAS 68439-51-0; nonionic; biodeg. detergent, wetting agent, emulsifier for cleaning prods.; clear, colorless liq.; less tendency than similar products to unpleasant odor > 30 C; sol. in water, ethanol, disp. in xylene, trichloroethylene, wh. spirit, lt. fuel oil; sp.gr. 1.02; visc. 100 cps; HLB 13.5; cloud pt. 65–70 C; flash pt. ≥ 100 C; pour pt. < 0 C; pH 6-7 (1% aq.); surf. tens. 31.5 dynes/cm (0.1%); Draves wetting 0.7 g/l (25 s); Ross-Miles foam 90 mm (initial, 0.05%, 50 C); 90% act.

Berol 190. [Berol Nobel AB] PEG-75 castor oil; CAS 61791-12-6; nonionic; surfactant, emulsifier for chemical industry; as softener, rewetting agent, pigment dipsersant, dye assistant, leveling agent for paint, textile, leather; lubricant additive/emulsifier in lubricants for plastics, metals, textiles; Gardner ≤ 5 wax; sol. @ 5% in ethanol, water, xylene; partly sol. in IPA; sp.gr. 1.06; visc. 260 mPa•s; HLB 15.7; sapon. no. 37.5-42.5; pour pt. 37 C; cloud pt. 65 C (1% in 10% NaCl); flash pt. > 150 C; pH 507 (1% aq.); surf. tens. 45 mN/m (0.1%); Draves wetting > 10 g/l (25 s); Ross-Miles foam 40 mm (initial, 0.05%, 50 C); 100% act.

Berol 191. [Berol Nobel AB] PEG-200 castor oil; CAS 61791-12-6; nonionic; surfactant, emulsifier for chemical industry; as softener, rewetting agent, pigment dipsersant, dye assistant, leveling agent for paint, textile, leather; lubricant additive/emulsifier in lubricants for plastics, metals, textiles; Gardner ≤ 5 wax; sol. @ 5% in ethanol, water; sp.gr. 1.08; visc. 588 mPa•s; HLB 18.1; sapon. no. 16.2-18.4; pour pt. 45 C; cloud pt. 65 C (1% in 10% NaCl); flash pt. > 150 C; pH 5-7 (1% aq.); surf. tens. 46 mN/m (0.1%); Draves wetting > 10 g/l (25 s); Ross-Miles foam 45 mm (initial, 0.05%, 50 C); 100% act.

Berol 195. [Berol Nobel AB] PEG-32 castor oil; CAS 61791-12-6; nonionic; surfactant, emulsifier for chemical industry; liq.; HLB 12.0; 95% conc.

Berol 198. [Berol Nobel AB] PEG-160 castor oil; CAS 61791-12-6; nonionic; surfactant, emulsifier for chemical industry; as softener, rewetting agent, pigment dispersant, dye assistant, leveling agent for paint, textile, leather; lubricant additive/emulsifier in lubricants for plastics, metals, and textiles; Gardner ≤ 5 wax; sol. @ 5% in water, ethanol; disp. in propylene glycol; sp.gr. 1.08; visc. 510 mPa•s; HLB 17.7; sapon. no. 19-23; pour pt. 43 C; cloud pt. 61 C (1% in 10% NaCl); flash pt. > 150 C; pH 6-8 (1% aq.); surf. tens. 46 mN/m (0.1%); Draves wetting > 10 g/l (25 s); Ross-Miles foam 45 mm (initial, 0.05%, 50 C); 100% act.

Berol 199. [Berol Nobel AB] PEG-32 castor oil; CAS 61791-12-6; nonionic; surfactant, emulsifier for chemical industry; as softener, rewetting agent, pigment dipsersant, dye assistant, leveling agent for paint, textile, leather; lubricant additive/emulsifier in lubricants for plastics, metals, textiles; Gardner ≤ 5 liq.; sol. @ 5% in ethanol, IPA, water, xylene; disp. in propylene glycol; sp.gr. 1.05; visc. 700 mPa•s; HLB 12.3; sapon. no. 68-76; pour pt. 17 C; cloud pt. 70-74 C (5 g in 25 ml 25% butyl diglycol); flash pt. > 100 C; pH 7-8 (5% aq.); surf. tens. 45 mN/m

(0.1%); Draves wetting > 10 g/l (25 s); Ross-Miles foam 40 mm (initial, 0.05%, 50 C); 100% act.

Berol 223. [Berol Nobel AB] Blend; nonionic; emulsifier for formulating microemulsions for cleaning of vehicles and for engineering applics.; Gardner 5 max. clear/sl. opalescent liq.; sol. @ 5% in acetone, ethanol, propylene glycol, xylene; disp. in water; sp.gr. 0.960; visc. 21 mPa•s; pour pt. -6 C; cloud pt. 72 ± 3 C (5 g in 25 ml of 25% butyldiglycol); flash pt. > 100 C; pH 9 ± 1 (1% aq.); 100% conc.

Berol 225. [Berol Nobel AB] Blend; nonionic/cationic; surfactant for water-based alkaline cleaners for hard surf. cleaning, high pressure cleaners, acid cleaners; Gardner 10 max. clear liq.; sol. @ 5% in ethanol, propylene glycol, water; disp. in xylene; sp.gr. 1.055; visc. 700 mPa•s; pour pt. -5 C; clear pt. -5 C; flash pt. > 100 C; pH 7 ± 1 (1% aq.); surf. tens. 32 mN/m (0.1%); Draves wetting 0.55 g/l (25 s); Ross-Miles foam 50 mm (initial, 0.05%, 50 C); 100% act.

Berol 226. [Berol Nobel AB] Blend; nonionic/cationic; surfactant blend for water-based alkaline cleaners, hard surf. cleaning, high pressure cleaners, acid cleaners; Gardner 10 max. clear/sl. opalescent liq.; sol. @ 5% in ethanol, propylene glycol, water; disp. in xylene; sp.gr. 1.010; visc. 100 mPa•s; pour pt. -10 C; clear pt. -7 C; flash pt. > 100 C; pH 7 ± 1 (1% aq.); surf. tens. 32 mN/m (0.1%); Draves wetting 0.35 g/l (25 s); Ross-Miles foam 55 mm (initial, 0.05%, 50 C); 100% act.

Berol 259. [Berol Nobel AB] Nonylphenol ethoxylate; CAS 68412-54-4; nonionic; stabilizer, foam depressor, liq. cleaner, wetting agent; biodeg.; Hazen ≤ 500 clear liq.; sol. in ethanol, xylene, trichloroethylene, paraffin oil, wh. spirit, lt. fuel oil; sp.gr. 1.01; visc. 550 cps; HLB 5.7; cloud pt. 50–54 C; flash pt. > 100 C; pour pt. < 0 C; pH 6–7 (1% aq.); degreasing to skin; 100% act.

Berol 260. [Berol Nobel AB] C9-11 pareth-4; CAS 68439-45-2; nonionic; surfactant for alkaline systems, hard surf. cleaners, industrial and institutional cleaners, vehicle cleaners; Hazen 150 max. clear/sl. opalescent liq.; sol. @ 5% in ethanol, propylene glycol, wh. spirit, xylene; disp. in water; sp.gr. 0.931; visc. 55 mPa•s; HLB 10.5; pour pt. -15 C; cloud pt. 56-59 C (5 g in 25 ml of 25% butyl diglycol); flash pt. > 100 C; pH 6-8 (1% aq.); surf. tens. 31 mN/m (0.1%); Draves wetting 0.4 g/l (25 S); Ross-Miles foam 20 mm (0.05%, 50 C); 100% act.

Berol 267. [Berol Nobel AB] Nonylphenol ethoxylate; CAS 68412-54-4; nonionic; emulsifier, detergent, wetting agent; biodeg.; Hazen < 200 clear liq.; sol. in ethanol, xylene, trichloroethylene, paraffin oil, wh. spirit, lt. fuel oil; sp.gr. 1.04; visc. 350 cps; HLB 12.3; cloud pt. 10 C; flash pt. > 100 C; pour pt. < 0 C; pH 6–7 (1% aq.); surf. tens. 31 dynes/cm; Draves wetting 1.3 g/l (25 s); Ross-Miles foam 30 mm (initial, 0.05%, 50 C); 100% act.

Berol 269. [Berol Nobel AB] Dinonylphenol ethoxylate; CAS 68891-21-4; nonionic; emulsifier for polymerization; detergent; biodeg.; Hazen < 500 opalescent liq.; sol. in ethanol, xylene, trichloroethylene, lt. fuel oil, disp. in water, wh. spirit; sp.gr. 1.02; visc. 550 cps; HLB 10.5; cloud pt. 60 C; flash pt. > 100 C; pour pt. < 0 C; pH 6–7 (1% aq.); degreasing to skin; 100% conc.

Berol 272. [Berol Nobel AB] Dinonylphenol ethoxylate; CAS 68891-21-4; nonionic; low foaming detergent with good stability for spray drying; lt. yel. paste; sol. in water, ethanol, xylene, trichloroethylene, disp. in paraffin oil; sp.gr. 1.04; visc. 300 cps (35 C); m.p. 34 C; HLB 13.4; cloud pt. 60 C (1% aq.); flash pt. >100 C; pour pt. 23 C; pH 6-7 (1% aq.); surf. tens. 30 dynes/cm (0.1%); Draves wetting > 10 g/l (25 s); Ross-Miles foam 63 mm (initial, 0.05%, 50 C); degreasing to skin; 100% conc.

Berol 277. [Berol Nobel AB] Nonylphenol ethoxylate; CAS 68412-54-4; nonionic; emulsifier for emulsion polymerization; biodeg.; Hazen < 200 clear liq.; sol. in water, ethanol; sp.gr. 1.09; visc. 1000 cps; HLB 17; cloud pt. 77 C (1% in 10/NaCl); flash pt. >100 C; pour pt. 12 C; pH 6–7 (1% aq.); surf. tens. 41 dynes/cm (0.1%); Draves wetting > 10 g/l (25 s); Ross-Miles foam 100 mm (initial, 0.05%, 50 C); degreasing to skin; 70% act.

Berol 278. [Berol Nobel AB] Nonylphenol ethoxylate; CAS 68412-54-4; nonionic; emulsifier for emulsion polymerization; biodeg.; Hazen < 200 clear liq.; sol. in water, ethanol; sp.gr. 1.08; visc. 625 cps; HLB 15.2; cloud pt. 97–99 C (1% aq.); flash pt. > 100 C; pour pt. 0 C; pH 6–7 (1% aq.); surf. tens. 37 dynes/cm (0.1%); Draves wetting 10 g/l (25 s); Ross-Miles foam 130 mm (initial, 0.5%, 50 C); 80% act.

Berol 281. [Berol Nobel AB] Nonylphenol ethoxylate; CAS 68412-54-4; nonionic; emulsifier for emulsion polymerization; biodeg.; Hazen < 200 clear liq.; sol. in water, ethanol; sp.gr. 1.09; visc. 625 cps; HLB 16; cloud pt. 73 C (1% in 10/NaCl); flash pt. >100 C; pour pt. 7 C; pH 6–7 (1% aq.); surf. tens. 39 mN/m; Draves wetting > 10 g/l (25 s); Ross-Miles foam 130 mm (initial, 0.05%, 50 C); toxicology: LD50 (oral, rat) > 4 g/kg; degreasing to skin; 80% conc.

Berol 291. [Berol Nobel AB] Nonylphenol ethoxylate; CAS 68412-54-4; nonionic; emulsifier for emulsion polymerization; Hazen < 200 hard wax; sol. in water, ethanol, xylene, trichloroethylene; sp.gr. 1.07 (60 C); visc. 300 cps (60 C); HLB 18.2; cloud pt. 77 C (1% in 10/NaCl); flash pt. >100 C; pour pt. 51 C; pH 6–7 (1% aq.); surf. tens. 45 mN/m (0.1%); Draves wetting > 10 g/l (25 s); Ross-Miles foam 100 mm (initial, 0.05%, 50 C); toxicology: LD50 (oral, rat) > 4 g/kg; non skin sensitizing; 100% act.

Berol 292. [Berol Nobel AB] Nonylphenol ethoxylate; CAS 68412-54-4; nonionic; emulsifier for emulsion polymerization; biodeg.; Hazen < 200 soft wax; sol. in water, ethanol, xylene, trichloroethylene, disp. in wh. spirit; sp.gr. 1.07 (40 C); visc. 200 cps (40 C); HLB 16.0; cloud pt. 73 C (1% in 10/NaCl); flash pt. > 100 C; pour pt. 33 C; pH 6-7 (1% aq.); surf. tens. 39 dynes/cm; Draves wetting > 10 g/l (25 s); Ross-Miles foam 130 mm (initial, 0.05%, 50 C); toxicology: degreasing to skin; 100% act.

Berol 295. [Berol Nobel AB] Nonylphenol ethoxylate; CAS 68412-54-4; nonionic; emulsifier for emulsion polymerization; Hazen < 300 clear liq.; sol. in water, ethanol; sp.gr. 1.10; visc. 1500 cps; HLB 17.8; cloud pt. 77 C (1% in 10/NaCl); flash pt. > 100 C; pour pt. 14 C; pH 5–7 (1% aq.); surf. tens. 43 mN/m (0.1%); Draves wetting > 10 g/l (25 s); Ross-Miles foam 90 mm (initial, 0.05%, 50 C); toxicology: LD50 (oral, rat) > 4 g/kg; non skin sensitizing; 70% act.

Berol 302. [Berol Nobel AB] PEG-2 oleamine; CAS 13127-82-7; cationic; emulsifier, dispersant, wetting agent, antistat, anticorrosive for agric., leather, textiles, metalworking and plastic industries; Gardner ≤ 10 clear to cloudy liq.; sol. @ 5% in ethanol, IPA, low aromatic solvs., wh. spirit, xylene; disp. in propylene glycol; sp.gr. 0.904; visc. 140 mNs/m; HLB 4.9; pour pt. 14 C; clear pt. 24 C; pH 9 (1% aq.); 100% act.

Berol 303. [Berol Nobel AB] PEG-12 oleamine; CAS

26635-93-8; emulsifier, wetting agent, antistat, anticorrosive for agric., leather, textiles, metalworking and plastic industries; Gardner ≤ 15 clear liq.; sol. @ 5% in water, ethanol, IPA, propylene glycol, xylene; disp. in wh. spirit; sp.gr. 1.02; visc. 220 mPa•s; HLB 13.2; pour pt. -5 C; cloud pt. 75 C (1% in 10% NaCl); flash pt. > 150 C; pH 10 (1% aq.); surf. tens. 41 mN/m (0.1%); Draves wetting > 10 g/l (25 s); Ross-Miles foam 100 mm (initial, 0.05%, 50 C); 100% act.

Berol 305. [Berol Nobel AB] Amine oxide; CAS 68071-48-7; surfactant; yel. clear liq.; sol. @ 5% in acetone, ethanol, water; sp.gr. 1.005; visc. 11 mPa•s; pour pt. 0 C; clear pt. 4 C; pH 7-9 (1% aq.); surf. tens. 34 mN/m (0.1%); Draves wetting 1.4 g/l (25 s); Ross-Miles foam 145 mm (initial, 0.05%, 50 C); 28% act.

Berol 307. [Berol Nobel AB] PEG-2 cocamine; CAS 61791-14-8; 61791-31-9; emulsifier, wetting agent, antistat, anticorrosive for agric., leather, textiles, metalworking and plastic industries; Gardner ≤ 6 clear liq.; sol. @ 5% in ethanol, IPA, low aromatic solvs., propylene glycol, wh. spirit, xylene; disp. in water; sp.gr. 0.910; visc. 150 mPa•s; HLB 5.9; cloud pt. 47 C (5 g in 25 ml 25% butyl diglycol); clear pt. 11 C; surf. tens. 30 mN/m (0.1%); Draves wetting 3 g/l (25 s); Ross-Miles foam 20 mm (initial, 0.05%, 50 C); corrosive; 100% act.

Berol 370. [Berol Nobel AB] EO/PO block polymer; CAS 9003-11-6; nonionic; detergent, rinse aid additive for machine dishwashing; Hazen < 150 clear liq.; m.w. 1400; sol. in ethanol, xylene, trichloroethylene, wh. spirit, water; sp.gr. 1.05; visc. 230 cps; cloud pt. 30–34 C (1% aq.); flash pt. > 100 C; pour pt. < –10 C; pH 5–7 (1% aq.); surf. tens. 41 dynes/cm (0.1%); Draves wetting 7 g/l (25 s); Ross-Miles foam 5 mm (initial, 0.05%, 50 C); 100% conc.

Berol 374. [Berol Nobel AB] EO/PO block polymer; CAS 9003-11-6; nonionic; low-foaming surfactant, emulsifier for emulsion polymerization esp. for latex paints; foam depressant/detergent for foodstuffs industry; emollient; wh. flakes; m.w. 2200; sol. in ethanol, xylene, trichloroethylene, wh. spirit; disp. water; sp.gr. 1.05; visc. 450 cps; cloud pt. 24–26 C (1% aq.); flash pt. > 100 C; pour pt. < –10 C; pH 5–7 (1% aq.); surf. tens. 40 dynes/cm (0.1%); Draves wetting 2 g/l (25 s); Ross-Miles foam 5 mm (initial, 0.05%, 50 C); 100% act.

Berol 381. [Berol Nobel AB] PEG-15 tallowamine; CAS 61791-26-2; emulsifier, wetting agent, antistat, anticorrosive for agric., leather, textiles, metalworking and plastic industries; Gardner ≤ 10 clear to cloudy liq.; sol. @ 5% in water, ethanol, IPA, propylene glycol, xylene; sp.gr. 1.03; visc. 600 mPa•s; HLB 14.2; pour pt. 5 C; cloud pt. 77-87 C (1% in 10% NaCl); clear pt. 11 C; flash pt. > 150 C; pH 9 (1% aq.); surf. tens. 40 mN/m (0.1%); Draves wetting > 10 g/l (25 s); Ross-Miles foam 55 mm (initial, 0.05%, 50 C); 100% act.

Berol 386. [Berol Nobel AB] PEG-20 tallowamine; CAS 61791-26-2; emulsifier, wetting agent, antistat, anticorrosive for agric., leather, textiles, metalworking and plastics industries; red/brn. liq.; sol. @ 5% in water, ethanol, IPA, propylene glycol, xylene; sp.gr. 1.04; visc. 470 mPa•s; HLB 15.3; pour pt. 18 C; cloud pt. 85 C (1% in 10% NaCl); clear pt. 36 C; pH 9 (1% aq.); surf. tens. 42 mN/m (0.1%); Draves wetting > 10 g/l (25 s); Ross-Miles foam 60 mm (initial, 0.05%, 50 C); corrosive; 100% act.

Berol 387. [Berol Nobel AB] PEG-40 tallowamine; CAS 61791-26-2; emulsifier, wetting agent, antistat, anticorrosive for agric., leather, textiles, metalworking and plastics industries; Gardner ≤ 15 wax; sol. @ 5% in water, ethanol, IPA, propylene glycol, xylene; sp.gr. 1.06; visc. 144 mPa•s; HLB 17.3; pour pt. 36 C; cloud pt. 81-89 C (1% in 10% NaCl); flash pt. > 150 C; pH 9 (1% aq.); surf. tens. 49 mN/m (0.1%); Draves wetting > 10 g/l (25 s); Ross-Miles foam 70 mm (initial, 0.05%, 50 C); corrosive; 100% act.

Berol 389. [Berol Nobel AB] PEG-10 tallowamine; CAS 61791-26-2; emulsifier, wetting agent, antistat, anticorrosive for agric., leather, textiles, metalworking and plastics industries; Gardner ≤ 12 clear to cloudy liq.; sol. @ 5% in water, ethanol, IPA, propylene glycol, xylene; sp.gr. 1.00; visc. 220 mPa•s; HLB 12.4; pour pt. < 2 C; cloud pt. 65-70 C (1% in 10% NaCl); flash pt. > 150 C; pH 9-11 (1% aq.); surf. tens. 40 mN/m (0.1%); Draves wetting > 10 g/l (25 s); Ross-Miles foam 60 mm (initial, 0.05%, 50 C); corrosive; 100% act.

Berol 391. [Berol Nobel AB] PEG-5 tallowamine; CAS 61791-26-2; emulsifier, wetting agent, antistat, anticorrosive for agric., leather, textiles, metalworking and plastics industries; yel./brn. liq.; sol. @ 5% in water, ethanol, IPA, low aromatic solvs., propylene glycol, wh. spirit, xylene; sp.gr. 0.96; visc. 160 mPa•s; HLB 9.0; pour pt. 8 C; cloud pt. 24-31 C (1% aq.); flash pt. >> 150 C; pH 9 (1% aq.); surf. tens. 34 mN/m (0.1%); Draves wetting 10 g/l (25 s); Ross-Miles foam 15 mm (initial, 0.05%, 50 C); corrosive; 100% act.

Berol 392. [Berol Nobel AB] PEG-15 tallowamine; CAS 61791-26-2; emulsifier, wetting agent, antistat, anticorrosive for agric., leather, textiles, metalworking and plastics industries; red/brn. liq.; sol. @ 5% in water, ethanol, IPA, propylene glycol, xylene; sp.gr. 1.03; visc. 270 mPa•s; HLB 14.2; pour pt. 8 C; cloud pt. 77-87 C (1% aq.); flash pt. >> 150 C; pH 9 (1% aq.); surf. tens. 41 mN/m (0.1%); Draves wetting > 10 g/l (25 s); Ross-Miles foam 45 mm (initial, 0.05%, 50 C); corrosive; 100% act.

Berol 397. [Berol Nobel AB] PEG-15 cocamine; CAS 61791-14-8; emulsifier, wetting agent, antistat, anticorrosive for agric., leather, textiles, metalworking and plastic industries; Gardner ≤ 12 clear to cloudy liq.; sol. @ 5% in water, ethanol, IPA, propylene glycol, xylene; sp.gr. 1.043; visc. 193 mPa•s; HLB 15.3; cloud pt. 85 C (1% in 10% NaCl); clear pt. 10 C; surf. tens. 40 mN/m (0.1%); Draves wetting > 10 g/l (25 s); Ross-Miles foam 110 mm (0.05%, 50 C); corrosive; 100% act.

Berol 452. [Berol Nobel AB] Sodium lauryl polyglycol ether sulfate; CAS 68891-38-3; foaming agent, degreaser for hair shampoos, foam baths, manual dishwash, general detergents; emulsifier for polymerization of vinyl chloride, vinyl acetate, acrylic acids; Hazen ≤ 200 clear liq.; sol. @ 5% in water, propylene glycol; disp. in ethanol, IPA; sp.gr. 1.040; visc. 200 mPa•s; clear pt. < 12 C; flash pt. >> 100 C; pH 6.5-8.0 (3% aq.); surf. tens. 37 mN/m (0.1%); Draves wetting > 10 g/l (25 s); Ross-Miles foam 185 mm (initial, 0.05%, 50 C); corrosive; 27-29% act.

Berol 455. [Berol Nobel AB] PEG-3 tallow diamine; emulsifier, wetting agent, antistat, anticorrosive for agric., leather, textiles, metalworking and plastic industries; liq.; HLB 20.1; 100% conc.

Berol 456. [Berol Nobel AB] PEG-2 tallowamine; CAS 61791-44-4; emulsifier, wetting agent, antistat, anticorrosive for agric., leather, textiles, metalworking and plastic industries; Gardner ≤ 8 paste; sol. @ 5% in ethanol, IPA, low aromatic solvs., propylene glycol, wh. spirit, xylene; disp. in water; sp.gr. 0.90; visc. 90 mPa•s; HLB 4.9; pour pt. 24 C; clear pt. 27 C; pH 9 (1% aq.); surf. tens. 33 mN/m (0.1%); Draves wetting > 10 g/l (25 s); Ross-Miles foam 10 mm (initial, 0.5%, 50 C); 100% act.

Berol 457. [Berol Nobel AB] PEG-5 tallowamine; CAS 61791-26-2; emulsifier, wetting agent, antistat, anticorrosive for agric., leather, textiles, metalworking and plastic industries; yel./brn. liq.; sol. @ 5% in water, ethanol, IPA, low aromatic solvs., propylene glycol, wh. spirit, xylene; sp.gr. 0.95; visc. 100 mPa•s; HLB 9.0; pour pt. 6 C; cloud pt. 22 C (1% aq.); clear pt. 18 C; flash pt. >> 150 C; pH 10 (1% aq.); surf. tens. 34 mN/m (0.1%); Ross-Miles foam 20 mm (initial, 0.05%, 50 C); corrosive; 100% act.

Berol 458. [Berol Nobel AB] PEG-10 tallowamine; CAS 61791-26-2; emulsifier, wetting agent, antistat, anticorrosive for agric., leather, textiles, metalworking and plastic industries; Gardner ≤ 9 clear liq.; sol. @ 5% in water, ethanol, IPA, propylene glycol, wh. spirit, xylene; sp.gr. 1.03; visc. 190 mPa•s; HLB 12.4; pour pt. 3 C; cloud pt. 71 C (1% in 10% NaCl); clear pt. 7 C; flash pt. > 150 C; pH 10 (1% aq.); surf. tens. 39 mN/m (0.1%); Ross-Miles foam 80 mm (initial, 0.05%, 50 C); 100% act.

Berol 475. [Berol Nobel AB] Sodium alkyl ether sulfate; anionic; degreaser, dispersant for aq. pigment pastes, cleaners, textile use; biodeg.; Hazen < 300 soft paste; sol. in propylene glycol and water, disp. in ethanol, trichloroethylene, xylene; sp.gr. 0.97; flash pt. > 100 C; surf. tens. 37 dynes/cm (0.1%); pH 6.5–8.0 (1% aq.); 39-41% conc.

Berol 480. [Berol Nobel AB] TEA lauryl sulfate; CAS 139-96-8; anionic; detergent, foaming agent for hair shampoos, foam baths, hand cleaners; biodeg.; Hazen < 300 soft paste; sol. in propylene glycol and water, disp. in ethanol, trichloroethylene, xylene; sp.gr. 0.97; flash pt. > 100 C; surf. tens. 37 dynes/cm (0.1%); pH 6.5–8.0 (1% aq.); 39-41% conc.

Berol 521. [Berol Nobel AB] Potassium alkyl phosphate ester; anionic; corrosion inhibitor, solubilizer of nonionic surfactants in presence of high electrolyte conc.; for liq. alkaline hard surf. cleaners; biodeg.; Gardner < 5 clear liq.; sol. @ 5% in water, propylene glycol; sp.gr. 1.145; visc. 300 mPa•s; pour pt. –10 C; clear pt. 25 C; flash pt. > 100 C; pH 7–8 (1% aq.); surf. tens. 33.5 mN/m (0.1%); Draves wetting > 10 g/l (25 s); Ross-Miles foam (125 mm (300 ppm, initial, 0.05%, 50 C); 40% act. in water.

Berol 522. [Berol Nobel AB] Potassium alkyl phosphate ester; anionic; solubilizer of nonionic surfactants in presence of high electrolyte concs.; for liq. alkaline hard surf. cleaners, laundry detergents; biodeg.; Hazen < 300 clear liq.; sol. @ 5% in water, propylene glycol; sp.gr. 1.270; visc. 300 mPa•s; pour pt. < 0 C; clear pt. < 0 C; flash pt. >> 100 C; pH 8 ± 1 (1% aq.); surf. tens. 31 mN/m (0.1%); Draves wetting > 10 g/l (25 s); Ross-Miles foam 0 mm (300 ppm, initial, 0.05%, 50 C); 40% act. in water.

Berol 525. [Berol Nobel AB] Alkyl ether phosphate ester/straight chain primary alcohol ethoxylate blend; anionic/nonionic; detergent aux.; solubilizer for most nonionic surfactants in presecne of inorganic salts; for alkaline cleaners; Hazen < 150 paste; sol. @ 5% in water, acetone, ethanol, trichlorethylene, wh. spirit, xylene; sol. sl. hazy in propylene glycol; sp.gr. 1.060; visc. 2000 mPa•s; pour pt. 23 C; clear pt. 30 C; pH 2 (1% aq.); surf. tens. 33 mN/m (0.1%); Draves wetting > 10 g/l (25 s); Ross-Miles foam 30 mm (initial, 0.05%, 50 C); 100% act.

Berol 556. [Berol Nobel AB] Alkyl polyglycol ether ammonium methyl chloride; cationic; surfactant, hydrotrope in aq. alkaline cleaning prods. for hard surf. cleaning, acid cleaners; Gardner 14 max. clear liq.; sol. @ 5% in water, ethanol, propylene glycol; sp.gr. 1.070; visc. 350 mPa•s; pour pt. 2 C; clear pt. 2 C; flash pt. > 100 C; pH 6-8 (1% aq.); surf. tens. 38 mN/m (0.1%); Draves wetting > 10 g/l (25 s); Ross-Miles foam 70 mm (initial, 300 ppm, 0.05%, 50 C); 100% act.

Berol 563. [Berol Nobel AB] Alkyl polyglycol ether ammonium methyl sulfate; cationic; detergent, alkaline degreasing and cleaning agent; hydrotrope for aq. alkaline cleaners for hard surf. cleaning, acid cleaners; Gardner 12 max. clear liq.; sol. @ 5% in water, ethanol, propylene glycol, trichloroethylene; sp.gr. 1.110; visc. 500 mPa•s; pour pt. 10 C; cloud pt. 85 C (1% in 9% NaOH); flash pt. >> 100 C; pH 7–9 (1% aq.); surf. tens. 43 mN/m (0.1%); Draves wetting > 10 g/l (25 s); Ross-Miles foam 100 mm (initial, 300 ppm, 0.05%, 50 C); 100% act.

Berol 594. [Berol Nobel AB] Hydroxyethyl alkyl imidazoline; cationic; adhesion aid in solv.-based paints and varnishes; dispersant, wetting and foaming agent; corrosion inhibitor in min. oils and solvs., aq. systems; friction modifier in automatic transmission fluids; Gardner 15 max. clear liq.; sol. in ethanol, xylene, trichloroethylene, wh. spirit, paraffin oil, and dilute sulfuric, hydrochloric, phosphoric and acetic acids; disp. in water; m.w. 360; sp.gr. 0.935; visc. 300 mPa•s; pour pt. 3 C; clear pt. 28 C; flash pt. > 100 C; pH 10-11 (1% aq.); surf. tens. 32 mN/m (0.1%); Draves wetting 0.8 g/l (25 s); Ross-Miles foam 180 mm (initial, 0.05%, 50 C); 90% act.

Berol 716. [Berol Nobel AB] Alkylphenol ethoxylate; CAS 68891-21-4; nonionic; low foaming detergent with good pigment and grease removal; Hazen < 500 cloudy liq.; sol. in ethanol, xylene, trichloroethylene, paraffin oil (opaque); disp. in water, wh. spirit, lt. fuel oil; sp.gr. 1.04; visc. 800 cps; HLB 13; pour pt. 20 C; cloud pt. 80–85 C (in 10% n-propanol); flash pt. > 100 C; pH 6-7 (1% aq.); surf. tens. 32 mN/m (0.1%); Draves wetting > 10 g/l (25 s); Ross-Miles foam 30 mm (initial, 0.05%, 50 C); toxicology: degreasing to skin; 100% act.

Berol 733. [Berol Nobel AB] Potassium alkylphenol ethoxylate phosphate; anionic; detergent, solubilizer for nonionic surfactants in presence of inorganic salts, liq. alkaline industrial cleaners; biodeg.; Hazen ≤ 200 clear liq.; sol. in water, 6% NaOH, 35% tetrapotassium pyrophosphate, 50% sodium metasilicate, propylene glycol; sp.gr. 1.050; visc. 400 cps; pour pt. –7 C; clear pt. –5 C; flash pt. > 100 C; pH 9–10 (1% aq.); surf. tens. 43 mN/m (0.1%); Draves wetting > 10 g/l (25 s); Ross-Miles foam 100 mm (initial, 0.05%, 50 C); 33.5-36.5% act. in water.

Berol 752. [Berol Nobel AB] Blend; anionic; surfactant, foam stabilizer; emulsifier for mechanically whipped PVC foam; Gardner < 7 visc. liq.; disp. in water, propylene glycol, dioctylphthalate, tributyl phosphate; sp.gr. 1.070; visc. 10,000 cps; flash pt. > 100 C; pour pt. 25 C; pH 6–8 (1% in 1:1 water/ethanol); toxicology: degreasing to skin; 96% act.

Berol 784. [Berol Nobel AB] Alkylaryl sulfonate/fatty alcohol ethoxylate blend; anionic/nonionic; high foaming surfactant blend for neutral cleaning prods.; yel.-brn. clear liq.; sol. @ 5% in water, ethanol, propylene glycol, wh. spirit; sp.gr. 1.043; visc. 2200 mPa•s; pour pt. 0 C; clear pt. 8 C; flash pt. >> 100 C; pH 6-8 (1% aq.); surf. tens. 34 mN/m (0.1%); Draves wetting 0.5 g/l (25 s); Ross-Miles foam 140 mm (initial, 0.05%, 50 C); 100% act.

Berol 797. [Berol Nobel AB] Blend; anionic/nonionic; self-separating emulsifier in solv. degreasers; Gardner 10 max. clear liq.; sol. @ 5% in ethanol, wh. spirit, xylene; sp.gr. 0.952; visc. 38 mPa•s; pour pt. -20 C; clear pt. -10 C; flash pt. 92 C; pH 6 ± 1 (1% in aq. ethanol); 100% act.

Berol 806. [Berol Nobel AB] Emulsifier/detergent blend; anionic/nonionic/cationic; detergent with textile softening and antistatic props. for lt. duty detergents, carpet cleaning, manual car wash with hydrophobic rinsing effect; lt. yel. clear liq.; sol. @ 5% in water, acetone, ethanol, propylene glycol, wh. spirit; disp. in trichloroethylene, xylene; sp.gr. 1.020; visc. 200 mPa•s; pour pt. 10 C; clear pt. 15 C; flash pt. > 100 C; pH 6 ± 1 (1% aq.); surf. tens. 32 mN/m (0.1%); Draves wetting 0.3 g/l (25 s); Ross-Miles foam 60 mm (initial, 0.05%, 50 C); 80% act. in water.

Berol 822. [Berol Nobel AB] Calcium alkylaryl sulfonate; anionic; emulsifier for pesticides; biodeg.; Gardner < 10 clear visc. liq.; sol. in ethanol, trichloroethylene, paraffin oil, xylene, wh. spirit, lt. fuel oil; disp. in water; sp.gr. 1.03; visc. 1200 cps; flash pt. 35 C; pour pt. 5 C; pH 5–8 (1% aq.); 60% conc.

Berol 824. [Berol Nobel AB] Alkylaryl sulfonate; anionic; general purpose emulsifier for pesticides; 70% conc.

Berol 829. [Berol Nobel AB] PEG-20 castor oil; CAS 61791-12-6; nonionic; surfactant, emulsifier for chemical industry; clear liq.; sol. @ 5% in ethanol, IPA, xylene; opaquely sol. in water; disp. in propylene glycol; sp.gr. 1.03; visc. 800 mPa•s; HLB 10.2; pour pt. -5 C; cloud pt. 62-66 C (5 g in 25 ml 25% butyl diglycol); clear pt. 10 C; flash pt. > 150 C; pH 5.5-7.5 (1% aq.); surf. tens. 44 mN/m (0.1%); Draves wetting > 10 g/l (25 s); Ross-Miles foam 35 mm (initial, 0.05%, 50 C); 100% act.

Berol 930. [Berol Nobel AB] Blend; anionic/nonionic; emulsifier for insecticides and herbicides; yel. cloudy liq.; sol. @ 5% in IPA, methanol, water, xylene; disp. in kerosene; dens. 990 kg/m³; visc. 140 mPa•s; HLB 14.7; pour pt. 8 C; flash pt. 40 C; pH 5-7 (1% aq.).

Berol 938. [Berol Nobel AB] Blend; nonionic; emulsifier for insecticides and herbicides; yel.-brn. clear liq.; sol. @ 5% in IPA, 2% kerosene, methanol, xylene; disp. in water; dens. 1005 kg/m³; visc. 700 mPa•s; HLB 13.0; pour pt. 5 C; flash pt. 37 C; pH 5-7 (1% aq.).

Berol 946. [Berol Nobel AB] Blend; nonionic; emulsifier for formulating Dimethoate; lt. yel. clear liq.; sol. @ 5% in water, IPA, xylene, methanol; disp. in 2% kerosene; dens. 1025 kg/m³; visc. 240 mPa•s; HLB 16.7; pour pt. 5 C; clear pt. 15 C; flash pt. 55 C; pH 5-7 (1% aq.).

Berol 947. [Berol Nobel AB] Blend; anionic/nonionic; emulsifier for chlorinated and organophosphate pesticides; yel.-brn. clear liq.; sol. @ 5% in IPA, methanol, xylene; disp. in water, 2% kerosene; dens. 1005 kg/m³; visc. 900 mPa•s; HLB 12.7; pour pt. 10 C; clear pt. 15 C; flash pt. 37 C; pH 5-7 (1% aq.).

Berol 948. [Berol Nobel AB] Blend; CAS 37251-69-7; nonionic; emulsifier for chlorinated and organophosphates pesticides; wh. paste; sol. @ 5% in IPA, methanol, water, xylene; disp. in 2% kerosene; dens. 1060 kg/m³; visc. 150 mPa•s; HLB 16.0; pour pt. 33 C; clear pt. 34 C; flash pt. > 100 C; pH 5-7 (1% aq.).

Berol 949. [Berol Nobel AB] Blend; anionic/nonionic; emulsifier for high-conc. toxicant formulations; cloudy liq.; sol. @ 5% in IPA, methanol, water, xylene (opaque); disp. in 20% kerosene; dens. 995 kg/m³; visc. 600 mPa•s; pour pt. 5 C; clear pt. 35 C; flash pt. 42 C; pH 5-7 (1% aq.).

Berol WASC. [Berol Nobel AB] Nonylphenol ethoxylate; CAS 68412-54-4; nonionic; biodeg. household and industrial detergent, textile washing agent, dispersant for paints, varnishes; Hazen < 150 clear liq.; sol. in water, ethanol; disp. in xylene, trichloroethylene, paraffin oil, wh. spirit; sp.gr. 1.06; visc. 375 cps; HLB 14.1; pour pt. 0 C; cloud pt. 75-80 C (1% aq.); flash pt. > 100 C; pH 6-7 (1% aq.); surf. tens. 34 mN/m (0.1%); Draves wetting 0.9 g/l (25 s); Ross-Miles foam 110 mm (initial, 0.05%, 50 C); toxicology: LD50 (oral, rat) > 2 g/kg; 95% act.

Beycopon 345. [Ceca SA] Amine alkylaryl sulfonate; anionic; emulsifier; liq.

Beycopon EC. [Ceca SA] Sodium/TEA alkylaryl sulfonate; anionic; surfactant; liq.

Beycopon S 3A. [Ceca SA] Sodium alkylaryl sulfonate complex; anionic; surfactant; release agent for molding; liq.; 32% conc.

Beycopon S 50. [Ceca SA] Calcium alkylaryl sulfonate; anionic; emulsifier; liq.; 47% conc.

Beycopon TB. [Ceca SA] Sodium alkylaryl sulfonate complex; anionic; degreaser, emulsifier; liq.; 17% conc.

Beycopon TL. [Ceca SA] Sodium alkylsulfate/alkylaryl sulfonate complex; anionic; surfactant; liq.

Beycostat 148 K. [Ceca SA] Potassium phosphate; anionic; surfactant; cream.

Beycostat 211 A. [Ceca SA] Phosphate ester; anionic; base for industrial detergents; liq.; disp. in water; acid no. 120; pH 2 (5%); surf. tens. 30 dynes/cm; 100% conc.

Beycostat 231. [Ceca SA] Phosphate ester, potassium salt; anionic; antistat, wetting agent; liq.

Beycostat 256A. [Ceca SA] C8 fatty alcohol phosphate ester; anionic; surfactant, release agent, corrosion inhibitor, antifoam, intermediate; liq.; disp. in water; acid no. 280; pH 2 (5%); 100% conc.

Beycostat 273 A. [Ceca SA] PEG-3 C13 fatty alcohol phosphate ester; anionic/nonionic; antistat, release agent, wetting agent; liq.; disp. in water; acid no. 165; pH 2 (5%); surf. tens. 30 dynes/cm; 100% conc.

Beycostat 273 P. [Ceca SA] Potassium phosphate; anionic; surfactant, wetting agent, detergent; fluid paste; pH 7.0.

Beycostat 319 A. [Ceca SA] PEG-6 C13 fatty alcohol phosphate ester; anionic/nonionic; antistat, degreaser; liq.; sol. in water; acid no. 110; pH 2 (5%); surf. tens. 29 dynes/cm; 100% conc.

Beycostat 319 P. [Ceca SA] Potassium phosphate; anionic; surfactant, degreaser for textile industry; liq.; pH 7.0.

Beycostat 656 A. [Ceca SA] PEG-8 alkylphenol phosphate ester; anionic/nonionic; emulsifier for emulsion polymerization; liq.; 100% conc.

Beycostat 714 A. [Ceca SA] PEG-6 C13 fatty alcohol phosphate ester; anionic/nonionic; emulsifier; industrial detergent; liq.; disp. in water; acid no. 120; pH 2 (5%); surf. tens. 29 dynes/cm; 100% conc.

Beycostat 714 P. [Ceca SA] Potassium phosphate; anionic; surfactant; liq.

Beycostat B 070 A. [Ceca SA] Phosphate ester; anionic; for acid detergent formulations; liq.; sol. in water; acid no. 770; pH 1.15 (5%); surf. tens. 48 dynes/cm; 100% conc.

Beycostat B 080 A. [Ceca SA] Phosphate ester; anionic; for acid detergent formulations; liq.; sol. in water; acid no. 690; pH 1.2 (5%); surf. tens. 46 dynes/cm; 100% conc.

Beycostat B 089 A. [Ceca SA] Phosphate ester; anionic; for acid detergent formulations; liq.; sol. in water; acid no. 720; pH 1.2 (5%); surf. tens. 55 dynes/cm; 100% conc.

Beycostat B 151. [Ceca SA] Phosphate ester, potassium salt; anionic; base for liq. detergents; stabilizer for nonionics in aq. media; industrial degreaser; liq.; pH 8.2 (1%); 41.5% active in water.

Beycostat B 231. [Ceca SA] Phosphate ester, potassium salt; anionic; surfactant; liq.; pH 5.5.

Beycostat B 327. [Ceca SA] Phosphate ester, potassium salt; anionic; emulsifier; liq.; pH 7.0.

Beycostat B 337. [Ceca SA] Phosphate ester, potassium salt; anionic; emulsifier; liq.; pH 7.0.

Beycostat B 706. [Ceca SA] C13 fatty alcohol phosphate ester; anionic; surfactant; liq.

Beycostat B 706 A. [Ceca SA] C13 fatty alcohol phosphate ester; anionic; corrosion inhibitor, EP additive; intermediate; liq.; insol. in water; acid no. 177; pH 2 (5%); surf. tens. 36 dynes/cm; 100% conc.

Beycostat B 706 E. [Ceca SA] Phosphate ester, TEA salt; anionic; antistat for textile lubricants; liq.; pH 7.1.

Beycostat C 103. [Ceca SA] Phosphate ester; anionic; dispersant; liq.; 100% conc.

Beycostat C 213. [Ceca SA] Phosphate ester; anionic; dispersant; paste; 100% conc.

Beycostat DA. [Ceca SA] PEG-12 alkylphenol phosphate ester; anionic/nonionic; emulsifier for chlorinated solvs.; liq.; sol. in water; acid no. 62; pH 2 (5%); surf. tens. 34.6 dynes/cm; 100% conc.

Beycostat DP. [Ceca SA] Potassium phosphate; anionic; surfactant; liq.

Beycostat LA. [Ceca SA] C12 fatty alcohol phosphate ester; anionic; surfactant, corrosion inhibitor, wetting agent, antifoam, intermediate, resin dispersant; solid; disp. in water; acid no. 173; pH 2.2 (5%); surf. tens. 26 dynes/cm; 100% conc.

Beycostat LP 4 A. [Ceca SA] PEG-4 C12 fatty alcohol phosphate ester; anionic/nonionic; antistat, degreaser; additive for acid or basic detergents; liq.; insol. in water; acid no. 96; pH 2 (5%); surf. tens. 30 dynes/cm; 100% conc.

Beycostat LP 9 A. [Ceca SA] PEG-9 C12 fatty alcohol phosphate ester; anionic/nonionic; emulsifier for emulsion polymerization; paste; sol. hot in water; acid no. 95; pH 2 (5%); surf. tens. 34 dynes/cm; 100% conc.

Beycostat LP 12 A. [Ceca SA] Phosphate ester; anionic; surfactant for acid or basic detergent formulation, emulsion polymerization, min. flotation; solid; sol. in water; acid no. 93; pH 2 (5%); 100% conc.

Beycostat NA. [Ceca SA] PEG-9 alkylphenol phosphate ester; anionic/nonionic; emulsifier for emulsion polymerization; liq.; sol. in water; acid no. 86; pH 2 (5%); surf. tens. 32 dynes/cm; 100% conc.

Beycostat NE. [Ceca SA] Alkyl ether sulfate and solv.; anionic; antistatic for PVC, wetting agent; liq.; pH 7.5.

Beycostat NED. [Ceca SA] Phosphate ester, TEA salt; anionic; wetting agent; liq.; pH 7.5.

Beycostat QA. [Ceca SA] PEG-4 alkylphenol phosphate ester; anionic/nonionic; emulsifier, detergent base; liq.; disp. in water; acid no. 120; pH 2 (5%); surf. tens. 29.5 dynes/cm; 100% conc.

Biodet B/D. [Auschem SpA] Dialkyl polyglycol ether citrate; anionic; nonirritating biodeg. raw material for detergents; liq./paste; 100% conc.

Biodet B/TA. [Auschem SpA] Alkyl polyglycol ether tartrate; anionic; nonirritating biodeg. raw material for detergents; liq./paste; 100% conc.

Biodet D. [Auschem SpA] Dialkyl polyglycol ether citrate, sodium salt; anionic; nonirritating biodeg. raw material for detergents; liq.; 25% conc.

Biodet T. [Auschem SpA] Trialkyl polyglycol ether citrate; nonionic; nonirritating biodeg. raw material for detergents; liq.; 80% conc.

Biodet TA. [Auschem SpA] Alkyl polyglycol ether tartrate, sodium salt; anionic; nonirritating biodeg. raw material for detergents; liq.; 25% conc.

Biopal® LF-20. [Rhone-Poulenc Surf.] Polyethoxypolypropoxypolyethoxyethanol-iodine complex; iodophor; broad spectrum germicide for formulating low-foaming cleaners and sanitizers; EPA compliance; liq.; 20% act.

Biopal® NR-20. [Rhone-Poulenc Surf.] Nonoxynol-12 iodine; anionic; iodophor used in formulating no rinse sanitizing sol'ns.; EPA compliance; dk. reddish brn. visc. liq.; 20% conc.

Biopal® NR-20 W. [Rhone-Poulenc Surf.] Nonylphenoxypoly (ethyleneoxy) ethanol-iodine complex; anionic; iodophor for formulating "no rinse" sanitizing sol'n.; liq.; 20% conc.

Biopal® VRO-20. [Rhone-Poulenc Surf.] Nonoxynol-9 iodine; iodophor conc. complexed with nonionic surfactant for formulation of rinses and cleaners, disinfecting agents; dk. brn. pourable liq.; water-sol.; sp.gr. 1.34; visc. 1700 cps; pour pt. 10.4 C; 20% conc.

Bio-Soft® 9283. [Stepan Europe] Surfactant blend; nonionic; all-purpose detergent base for industrial/household cleaners; orange to brn. visc. liq.; 90% act.

Bio-Soft® AS-40. [Stepan; Stepan Canada] Sodium olefin sulfonate; anionic; wetting agent, emulsifier, penetrant, and dye dispersant for textiles; liq.; 40% act.

Bio-Soft® CS 50. [Stepan Europe] Fatty alcohol ethoxylate (50 EO); nonionic; emulsifier, dispersant, and detergent for deodorizing blocks, laundry cleaners; leveling agent for wool dyeing; dispersant for acrylic dyeing; wh. flakes; 100% act.

Bio-Soft® D-35X. [Stepan; Stepan Canada] Sodium linear alkylate sulfonate; anionic; liq. detergent bases; high foaming; biodeg.; liq.; 35% conc.

Bio-Soft® D-40. [Stepan; Stepan Canada] Sodium dodecylbenzene sulfonate; anionic; biodeg. detergent with good foaming, wetting, emulsifying properties; detergent base; for household, industrial and institutional cleaners; emulsifier, penetrant, and dye dispersant for textile applics.; clear liq.; bland odor; water sol.; dens. 8.7 lb/gal; cloud pt. 29 C; 40% act.

Bio-Soft® D-53. [Stepan; Stepan Canada] Sodium alkylbenzene sulfonate; anionic; high foaming detergent for household, industrial and institutional cleaners; biodeg.; slurry; 53% conc.

Bio-Soft® D-62. [Stepan; Stepan Canada] Sodium dodecylbenzene sulfonate; anionic; detergent with good foaming, wetting, emulsifying properties; detergent base; for household, industrial and institutional cleaners; emulsifier, penetrant, dye dispersant for textiles; biodeg.; straw, yel. slurry; bland odor; water sol.; dens. 9.1 lb/gal; 60% act.

Bio-Soft® D-233. [Stepan Europe] Surfactant blend; nonionic; detergent, wetting agent, thickener for hydrochloric acid-based scale removing cleaners; water-wh. to pale yel. liq.; 100% act.

Bio-Soft® E-200. [Stepan Canada] Laureth-1; nonionic; emulsifier, detergent, wetting and foam stabilizing; liq.; HLB 4.0; 100% act.

Bio-Soft® E-300. [Stepan Canada] Laureth-2; nonionic; emulsifier, detergent, wetting and foam stabilizing; liq.; HLB 6.0; 100% act.

Bio-Soft® E-400. [Stepan Canada] C12-15 pareth-3; nonionic; emulsifier, detergent, wetting and foam stabilizing; liq.; HLB 8.0; 100% act.

Bio-Soft® E-670. [Stepan Canada] Laureth-9; nonionic; surfactant; 100% act.

Bio-Soft® E-710. [Stepan Canada] C12-18 pareth-12; nonionic; emulsifier, detergent, wetting and foam stabilizing; wax; 100% act.

Bio-Soft® E-840. [Stepan Canada] Laureth-23; nonionic; surfactant; 100% act.

Bio-Soft® EA-4. [Stepan; Stepan Canada] Linear primary alkoxylate; nonionic; surfactant, emulsifier, detergent; wh. hazy liq.; sol. in org. solvs.; disp. in water; dens. 7.7 lb/gal (60 C); pour pt. 5 C; cloud pt. < 25 C; 100% act.

Bio-Soft® EA-8. [Stepan; Stepan Canada; Stepan Europe] Linear alkoxypolyoxethylene ethanols; nonionic; wetting agent, emulsifier, detergent for chemical specialties, laundry detergent, agric. emulsions, sanitizers, metal cleaning and pickling, used in textile and petrol. industries; biodeg.; hazy wh. liq.; sol. in org. solv., oil, water; dens. 8.42 lb/gal; sp.gr. 1.01; cloud pt. 46–58.5 C; 100% act.

Bio-Soft® EA-10. [Stepan; Stepan Canada; Stepan Europe] Linear alkoxypolyoxethylene ethanols; nonionic; emulsifier, detergent, wetting agent for household/industrial cleaners; water-wh. liq.; 100% act.

Bio-Soft® EC-600. [Stepan Canada] Laureth-7; nonionic; surfactant; 100% act.

Bio-Soft® EL-719. [Stepan Canada] Veg. oil ethoxylate; nonionic; emulsifier, dispersant, lubricant; liq.; 96% conc.

Bio-Soft® EN 600. [Stepan Canada] C12-15 pareth-7; nonionic; emulsifier, detergent, wetting and foam stabilizing; liq.; HLB 12.2; 100% act.

Bio-Soft® ERM. [Stepan Europe] Surfactant blend; nonionic; conc. soap for mechanics grease removers; constant visc. when diluted; orange-yel. visc. liq.; 27% act.

Bio-Soft® ET-630. [Stepan Canada] C10-12 pareth-5; nonionic; emulsifier, detergent, wetting and foam stabilizing; liq.; HLB 12.5; 100% act.

Bio-Soft® FF 400. [Stepan Canada] Deceth-2; nonionic; surfactant; 100% act.

Bio-Soft® JN. [Stepan Europe] Dodecylbenzene sulfonic acid; anionic; intermediate for prod. of liq. and powd. detergents; liq.; 97% conc.

Bio-Soft® LAS-40S. [Stepan; Stepan Canada] Sodium linear alkylbenzene sulfonate; anionic; detergent for household, industrial and institutional cleaners; liq.; 40% conc.

Bio-Soft® LD-32. [Stepan; Stepan Canada] Formulated prod.; anionic; detergent base for dishwashing, carwashing, and other lt. duty hard surface cleaners; liq.; 60% conc.

Bio-Soft® LD-47. [Stepan; Stepan Canada] Formulated prod.; anionic; high foaming blend for liq. hand dishwash formulations; liq.; 47% act.

Bio-Soft® LD-95. [Stepan; Stepan Canada] Sodium dodecylbenzene sulfonate, sodium laureth sulfate, lauramide DEA, and urea; anionic-nonionic; base for dishwashing, carwashing, and other lt. duty hard surface detergents; biodeg.; pale straw clear liq.; water sol.; dens. 9.11 lb/gal; visc. 600 cps; cloud pt. 5 C; 60% act.

Bio-Soft® LD-145. [Stepan; Stepan Canada] Formulated prod.; anionic; biodeg. detergent conc.; base for liq. dishwash, carwash, fine fabric, and general purpose cleaners; clear yel. liq.; 44-47% conc.

Bio-Soft® LD-150. [Stepan; Stepan Canada] Formulated prod.; anionic; biodeg. detergent conc.; base for liq. dishwash, carwash, fine fabric, and general purpose cleaners; clear yel. liq.; 48% act.

Bio-Soft® LD-190. [Stepan; Stepan Canada] Formulated prod.; anionic; biodeg. detergent conc. for liq. laundry and dishwash concs.; liq.; 91% act.

Bio-Soft® LDL-4. [Stepan Canada] Surfactant blend; lt.-duty liq. blend.

Bio-Soft® LF 77A. [Stepan Europe] Surfactant blend; nonionic; all-purpose conc. low foaming detergent base for alkaline formulations; yel. liq.; 77% act.

Bio-Soft® N-300. [Stepan; Stepan Canada] TEA-dodecylbenzene sulfonate; anionic; liq. dishwashing, car washing detergent, wetting and foaming agent; pale yel. clear liq.; water sol.; dens. 9.0 lb/gal; sp.gr. 1.08; visc. 3200 cps; surf. tens. 38.8 dynes/cm (0.1%); biodeg.; 60% act.

Bio-Soft® N-411. [Stepan; Stepan Canada] Isopropylamine linear alkylbenzene sulfonate; anionic; surfactant for emulsion degreasers, laundry prespotters, drycleaning solvs., paint strippers, waterless hand cleaners; liq.; 46% conc.

Bio-Soft® Ninex 21. [Stepan Europe] Amide sulfonate complex; anionic; liq. dishwash detergent with flash foam; liq.; 60% conc.

Bio-Soft® PG 4. [Stepan Europe] Polyalkoxylated compd.; nonionic; wetting agent, hydrotrope, low foaming agent for dishwasher rinse prods., conc. detergents; water-wh. liq.; 100% act.

Bio-Soft® S-100. [Stepan; Stepan Canada; Stepan Europe] Linear dodecylbenzene sulfonic acid; anionic; emulsifier, detergent intermediate for formulation of built detergents, dishwash, all-purpose cleaners, acid cleaners, degreasers, industrial cleaners; textile scouring, wetting, bleaching, dyeing assistant; high foamer when neutralized; biodeg.; brn. visc. liq.; water sol.; sp.gr. 1.04; 97% act.

Bio-Soft® S-130. [Stepan; Stepan Canada] Linear alkylbenzene sulfonic acid; anionic; emulsifier, detergent intermediate for laundry detergents; as high foaming surfactant when neutralized; dk. visc. liq.; 97% conc.

Bio-Soft® TD 400. [Stepan Canada] Trideceth-3; CAS 24938-91-8; nonionic; emulsifier; liq.; oil-sol.; HLB 8.0; 100% act.

Bio-Soft® TD 630. [Stepan Canada] Trideceth-8; CAS 24938-91-8; nonionic; emulsifier, detergent; liq.; water-sol.; HLB 12.5; 100% act.

Bio-Step Series. [Stepan Europe] Surfactant blends; anionic/nonionic; biodeg. high flash pt. emulsifier for agric. emulsifiable concs.; liq.; 44% act.

Bio-Surf DC-730. [Lonza] Quat. ammonium compd.; cationic; emulsifier and dispersant; liq.; oil-sol.; 75% conc.

Bio-Surf PBC-420. [Lonza] TEA-dodecylbenzene sulfonate; anionic; surfactant and wetting agent; liq.; 60% conc.

Bio-Surf PBC-430. [Lonza] Complex organic phosphate ester, free acid; anionic; lt. colored detergent for heavy-duty liq. all-purpose formulations, as wetting agent, emulsifier, and dedusting agent; low rewetting props.; electrolyte tolerant; liq.; 100% conc.

Bio-Surf PBC-460. [Lonza] Lauryl myristyl dimethyl amine oxides; nonionic/cationic; detergent aid in stabilizing foam; emollient, visc. builder for aq. systems; liq.; 30% conc.

Bio-Terge® AS-40. [Stepan; Stepan Canada] Sodium C14–16 olefin sulfonate; anionic; detergent, foaming agent for personal care, commercial and industrial formulations; yel. liquid; water sol.; dens. 8.9 lb/gal; biodeg.; 40% act. in water; contains no phosphates.

Bio-Terge® AS-90 Beads. [Stepan; Stepan Canada] Sodium alpha olefin sulfonate; anionic; high foaming detergent for industrial cleaners; solid beads; 90% conc.

Bio-Terge® PAS-8. [Stepan Europe] Sodium octyl sulfonate; CAS 142-31-4; anionic; solubilizing wetting agent for acid, alkaline, and oxidizing systems; rug shampoos, all-purpose cleaners; liq.; 35% conc.

Bio-Terge® PAS-8S. [Stepan; Stepan Canada; Stepan Europe] Sodium 1-octane sulfonate; anionic; hydrotrope and detergent used in acid, alkaline, high electrolyte or bleach containing cleaners for industrial, institutional and household markets, e.g., acid cleaners, carpet steam cleaners, automatic dishwash; textile penetrant, dye dispersant; metalworking formulations; water-wh. to pale yel. clear liq.; cloud pt. 15 C; pH 6 (1%); 35% act.

Biozan. [Hercules] Xanthan gum; suspending agent, thickener, emulsifier in slurry explosives, foundry coatings, acid and caustic cleaning compds., cosmetics, pharmaceuticals, oil field chemicals, aq. systems; food grade avail.; water-sol.

Blancol®. [Rhone-Poulenc Surf.] Sodium naphthalenesulfonate-formaldehyde condensate; anionic; thinning, dispersant, peptizing agent, dye-leveling agent used in paper, leather, textiles, heavy-metal cleaners, concrete; tan to brn. free-flowing coarse gran.; odorless; also avail. liq. form; sol. in warm or cold water; dens. 0.77–0.82 g/ml; 88–90% act.; DISCONTINUED.

Blancol® N (redesignated Rhodacal® N). [Rhone-Poulenc Surf.]

Blendmax Series. [Central Soya] Enzyme-modified lecithin; CAS 8002-43-5; amphoteric; emulsifier with enhanced water dispersibility; liq.; HLB 8.0; 100% conc.

Bohrmittel Hoechst. [Hoechst Celanese/Colorants & Surf.; Hoechst AG] Alkylsulfamido carboxylic acid, sodium salt; anionic; corrosion inhibitor, lubricant, wetting agent, and emulsifier for metalworking fluids; extreme pressure props.; visc. liq.; oil and water sol.; sp.gr. 0.99; pH 8-9 (1% aq.).

Borax. [U.S. Borax & Chem.] Sodium tetraborate decahydrate; dispersant, wetting agent for NR, SR latexes; mold lubricant for general dry rubber molding; wh. powd., odorless; sol. in water, glycerin; sp.gr. 1.73.

Bozemine N 60. [Hoechst AG] Polyethylene derivs.; improves abrasion resistance and tear str. in resin-finished fabrics.

Bozemine N 609. [Hoechst AG] Polyethylene derivs.; softener for cellulosic and syn. fibers.

Bozemine NSI. [Hoechst AG] Silicone-based prod.; softener and raising auxiliary for fibers; dispersion.

Brij® 35. [ICI Spec. Chem.; ICI Surf. Belgium] Laureth-23; CAS 9002-92-0; nonionic; surfactant; wh. waxy solid; sol. in water, alcohol, propylene glycol; sp.gr. 1.05; HLB 16.9; flash pt. >300 F; pour pt. 33 C.

Brij® 35 Liq. [ICI Spec. Chem.] Laureth-23; CAS 9002-92-0; nonionic; surfactant; colorless liq.; sol. in water, alcohol, propylene glycol; visc. 1200 cps; 70% act.

Brij® 35 SP. [ICI Spec. Chem.] Laureth-23 with preservatives; CAS 9002-92-0; nonionic; surfactant; wh. solid; sol. in water, alcohol, propylene glycol; HLB 16.9; pour pt. 33 C.

Brij® 35 SP Liq. [ICI Spec. Chem.] Laureth-23 with preservatives; CAS 9002-92-0; nonionic; surfactant; colorless liq.; sol. in water, alcohol, propylene glycol; visc. 1200 cps; 70% act.

Brij® 56. [ICI Spec. Chem.; ICI Surf. Belgium] Ceteth-10 (antioxidants added); CAS 9004-95-9; nonionic; surfactant, emulsifier esp. for topical cosmetic applics.; solubilizer for fragrances; wh. waxy solid; sol. in alcohol; HLB 12.9; flash pt. > 300F; pour pt. 31 C; 100% conc.

Brij® 58. [ICI Spec. Chem.; ICI Surf. Belgium] Ceteth-20 with preservatives; CAS 9004-95-9; nonionic; surfactant, emulsifier; solubilizer for fragrances; wh., waxy solid; sol. in water, alcohol; HLB 15.7; flash pt. > 300 F; pour pt. 38 C; 100% conc.

Brij® 68. [ICI Spec. Chem.] Ceteareth-20; CAS 68439-49-6; nonionic; surfactant; wh. solid; sol. in water, alcohol; disp. in cottonseed oil; HLB 15.5; pour pt. 38 C.

Brij® 72. [ICI Spec. Chem.; ICI Surf. Belgium] Steareth-2 with preservatives; CAS 9005-00-9; nonionic; surfactant, emulsifier esp. for topical cosmetic applics.; wh. waxy solid; sol. in alcohol, cottonseed oil; HLB 4.9; flash pt. >300 F; pour pt. 43 C; 100% conc.

Brij® 98G. [ICI Spec. Chem.] Oleth-20 with preservatives; CAS 9004-98-2; nonionic; surfactant; cream solid; sol. in water, alcohol; disp. in cottonseed oil, propylene glycol; HLB 15.3; pour pt. 30 C.

Brij® 99. [ICI Spec. Chem.] Oleth-20 with preservatives; CAS 9004-98-2; nonionic; surfactant with low color and odor; cream solid; sol. in water, alcohol, propylene glycol; disp. in cottonseed oil; HLB 15.3; pour pt. 33 C.

Brij® 700 S. [ICI Spec. Chem.] Steareth-100; CAS 9005-00-9; nonionic; emulsifier, solubilizer; wh. solid gran.; sol. in water, alcohol; disp. in cottonseed oil; HLB 18.8; pour pt. 55 C; 100% conc.

Brij® 721 S. [ICI Spec. Chem.] Steareth-21; CAS 9005-00-9; nonionic; emulsifier, solubilizer; wh. solid gran.; disp. in water, alcohol, cottonseed oil; HLB 15.5; pour pt. 45 C; 100% conc.

Briquest® 301-30SH. [Albright & Wilson UK] Sodium nitrilotris (methylene phosphate); sequestrant for scale inhibition and corrosion control; aq.; 30% act.

Briquest® 301-50A. [Albright & Wilson UK] Nitrilotris (methylene phosphonic acid); sequestrant for water treatment, oil-drilling muds, powd. detergents, photographic applics.; aq.; 50% act.

Briquest® 462-23K. [Albright & Wilson UK] Potassium hexamethylene diamine tetrakis (methylene phosphate); sequestrant for water treatment; aq.; 23% act.

Briquest® 543-45AS. [Albright & Wilson UK] Diethylenetriamine-pentakis (methylene phosphonic acid); CAS 22042-96-2; sequestrant for peroxide stabilization in pulp bleaching and de-inking;

in liq. detergents and oil-field chemicals; aq.; 45% act.

Briquest® ADPA-60A. [Albright & Wilson Americas] 1-Hydroxyethylidene-1, 1-diphosphonic acid; used for water treatment and oil-drilling muds, in powd. detergents and photographic applics.; sequestering agent for calcium carbonate; m.w. 206.03; 60% act. in water.

Briquest® ADPA-60AW. [Albright & Wilson UK] 1-Hydroxyethylidene-1,1-diphosphonic acid aq. sol'n.; sequestrant for water treatment, scale and corrosion inhibition, detergent formulation, metal ion control, desalination, pulp and paper, metal plating and finishing, textiles, radioactive pharmaceuticals; APHA < 80 clear liq.; slight odor; m.w. 206; sp.gr. 1.46 (20 C); visc. 100 cP (20 C); pH 1.8 (1%); 60% act.

Bromat. [Hexcel] Cetyl trimethyl ammonium bromide; cationic; surfactant, emulsifier; solid; 100% conc.

BTC® 50 USP. [Stepan; Stepan Canada] Benzalkonium chloride; cationic; antimicrobial for hard surf. disinfection, sanitization, deodorization; EPA registered; liq.; sp.gr. 0.96; flash pt. 126 F; 50% act.

BTC® 65 USP. [Stepan; Stepan Canada] Benzalkonium chloride; cationic; antimicrobial for hard surf. disinfection, sanitization, deodorization; EPA registered; liq.; sp.gr. 0.97; flash pt. 132 F; 50% act.

BTC® 99. [Stepan; Stepan Canada] Didecyl dimonium chloride; low foaming algicide and slimicide for swimming pool and industrial water treatment; EPA registered; liq.; 50% act.

BTC® 776. [Stepan; Stepan Canada] Benzalkonium chloride and dialkyl methyl benzyl ammonium chloride; cationic; algicide and slimicide for swimming pool and industrial water treatment; EPA registered; liq.; sp.gr. 0.96; flash pt. 102 F; 50% act.

BTC® 818. [Stepan; Stepan Canada] Quaternium-24; CAS 32426-11-2; cationic; disinfectant, sanitizer, and fungicide for hard surfaces; exc. sanitizer in hard water to 800 ppm as $CaCO_3$; EPA registered; liq.; water-sol.; sp.gr. 0.93; flash pt. 86 F; 50% act.

BTC® 818-80. [Stepan; Stepan Canada] Quaternium-24; CAS 32426-11-2; cationic; disinfectant, sanitizer, fungicide for hard surfs.; exc. sanitizer in hard water to 800 ppm as $CaCO_3$; EPA registered; liq.; 80% act.

BTC® 824. [Stepan; Stepan Canada] Myristalkonium chloride; cationic; antimicrobial for hard surf. disinfection and sanitization; algicide for swimming pools and industrial water treatment; EPA registered; liq.; sp.gr. 0.96; flash pt. 120 F; 50% act.

BTC® 824 P100. [Stepan; Stepan Canada] Tetradecyl dimethyl benzyl ammonium chloride monohydrate; cationic; antimicrobial for tablet mfg. of disinfectants, sanitizers, deodorizers; EPA registered; powd.; water-sol.; sp.gr. 0.40; flash pt. > 200 F; 95% act.

BTC® 835. [Stepan; Stepan Canada] Benzalkonium chloride; cationic; antimicrobial for hard surf. disinfection and sanitization; algicide in swimming pool and industrial water treatment; EPA registered; liq.; sp.gr. 0.97; flash pt. 130 F; 50% act.

BTC® 885. [Stepan; Stepan Canada] n-Alkyl (50% C14, 40% C12, 10% C16) dimethyl benzyl, octyldecyl dimethyl, dioctyl dimethyl, and didecyl dimethyl ammonium chlorides; cationic; germicide for formulation of disinfectant, sanitizer, and fungicidal prods. used in hospitals, nursing homes, and public institutions; sanitizer in hard water to 1200 ppm as $CaCO_3$; EPA registered; liq.; sp.gr. 0.95; flash pt. 116 F; 50% act.

BTC® 885 P40. [Stepan; Stepan Canada] Quaternium-24, benzalkonium chloride; cationic; germicide for formulation of disinfectant, sanitizer and fungicidal prods. for hospitals, nursing homes, public institutions; EPA registered; powd.; 40% act.

BTC® 888. [Stepan; Stepan Canada] Quaternium-24, benzalkonium chloride; cationic; germicide for formulation of disinfectant, sanitizer and fungicidal prods. for hospitals, nursing homes and public institutions; sanitizer in hard water to 1200 ppm as $CaCO_3$; EPA registered; liq.; 80% act.

BTC® 1010. [Stepan; Stepan Canada] Didecyl dimonium chloride; cationic; sanitizer in hard water to 1200 ppm as $CaCO_3$; EPA registered; liq.; water-sol.; sp.gr. 0.89; flash pt. 86 F; 50% act.

BTC® 1010-80. [Stepan; Stepan Canada] Quaternium-12; cationic; fungicide for hard-surf. disinfection and sanitization; algicide in swimming pool and industrial water treatment; deodorizer; EPA registered; liq.; water-sol.; 80% act.

BTC® 2125, 2125-80, 2125 P-40. [Stepan; Stepan Canada] Myristalkonium chloride and quaternium-14; cationic; antimicrobial, hard surf. disinfectant, sanitizer, fungicide for hospitals, public institutions; algicide in swimming pool and industrial water treatment; deodorizer; EPA registered; liqs. except 2125 P-40 (powd.); sp.gr. 0.97, 0.94, and 0.55 resp.; flash pt. > 200 F except 2125 80% (130 F); 50, 80, and 40% act.

BTC® 2125M. [Stepan; Stepan Canada] Myristalkonium chloride and quaternium-14; cationic; disinfection and sanitization quat. for hospitals, nursing homes, public insitutions; algicide in swimming pool and industrial water treatment; deodorizer; EPA registered; liq.; sp.gr. 0.97; flash pt. > 200 F; 50% act.

BTC® 2125M-80, 2125M P-40. [Stepan; Stepan Canada] Benzalkonium chloride and quaternium-14; cationic; broad-spectrum bacteriological control agent used in disinfectant and sanitizer formulations for hospitals, public institutions, industry; algicide for swimming pool and water treatment; EPA registered; liq. and powd. resp.; sp.gr. 0.94 and 0.55 resp.; flash pt. > 130 and 200 F resp.; 80 and 40% act. resp.

BTC® 2565. [Stepan; Stepan Canada] Myristalkonium chloride; cationic; algicide and slimicide for swimming pool and industrial water treatment; EPA registered; liq.; sp.gr. 0.96; flash pt. 110 F; 50% act.

BTC® 2568. [Stepan; Stepan Canada] Myristalkonium chloride; cationic; algicide and slimicide for swimming pool and industrial water treatment; EPA registered; liq.; 50% act.

BTC® 8248. [Stepan; Stepan Canada] Myristalkonium chloride; cationic; antimicrobial for hard surf. disinfection and sanitization; algicide for swimming pools and industrial water treatment; EPA registered; liq.; sp.gr. 0.95; flash pt. 110 F; 80% act.

BTC® 8249. [Stepan; Stepan Canada] Myristalkonium chloride; cationic; antimicrobial for hard surf. disinfection and sanitization; algicide for swimming pools and industrial water treatment; EPA registered; liq. to paste; sp.gr. 0.94; flash pt. 118 F; 90% act.

BTC® 8358. [Stepan; Stepan Canada] Benzalkonium chloride; cationic; antimicrobial for hard surf. disinfection and sanitization; algicide in swimming pool and industrial water treatment; EPA registered; liq.;

80% act.

Bubble Breaker® 259, 260, 613-M, 622, 730, 737, 746, 748, 900, 913, 917. [Witco/Organics] Blend of org., nonsilicone compds.; defoamer used in the mfg. of water-based paper coatings, textile processing formulations, agric. chemical prods., paints, adhesives, and inks.

Bubble Breaker® 748. [Witco/Organics] Silicone-free blend; defoamer for water-based systems, paints/coatings; opaque creamy liq.; sol. @ 5% in min. spirits, disp. in water; sp.gr. 0.87; flash pt. (PMCC) > 200 F; pH 9.5; usage level: 0.1-0.5%; 100% act.

Bubble Breaker® 776, 3017-A. [Witco/Organics] Disp. of reacted silica in hydrocarbon solv.; defoamer used in latex mfg. operations, formulation of water-based paints and adhesives.

Bubble Breaker® 900. [Witco/Organics] Silicone-free blend; defoamer for water-based systems, paints/coatings; opaque wh. liq.; sol. @ 5% in min. spirits, disp. in water; sp.gr. 0.89; flash pt. (PMCC) > 200 F; pH 9.4; usage level: 0.1-0.5%; 100% act.

Bubble Breaker® 1840X. [Witco/Organics] Sodium salt of sulfonated fatty acid; defoamer for wet process prod. of phosphoric acid; corrosion inhibitor, detergent; sol. in water, alcohol.

Bubble Breaker® 3017-A. [Witco/Organics] Disp. of reacted silica in hydrocarbons; defoamer; monomer stripping and water based flexographic inks and water based coatings; water-disp.

Bubble Breaker® 3056A. [Witco/Organics] Disp. of reacted silica in hydrocarbon solv.; defoamer used in latex mfg. operations, formulation of water-based paints and adhesives, effluent water, asphalt emulsions, PVC monomer stripping; opaque oily liq.; sol. @ 5% in min. spirits, disp. in water; sp.gr. 0.89; flash pt. (PMCC) > 200 F; pH 5.0; usage level: 0.01-0.05%; 100% act.

Bubble Breaker® 3073-7, D. [Witco/Organics] Blend of org., nonsilicone compds.; defoamer for water-based drilling for the petrol. industry.

Bubble Breaker® 3295. [Witco/Organics] Reacted silica blend; defoamer/deaerator for high solids or 100% solids liq. systems, e.g., urethane, alkyds, epoxy; used where hydrocarbon or min. oil cannot be tolerated; opaque; sol. @ 5% in min. spirits, disp. in water; sp.gr. 1.03; flash pt. (PMCC) > 200 F; pH 5.0; usage level: 0.1-0.5%; 100% act.

Bubble Breaker® DMD-1. [Witco/Organics] Complex surfactant; nonionic; oilfield surfactant, defoamer; liq.; sol. in IPA; disp. in water, kerosene, xylene; dens. 8.4 lb/gal; visc. 400 cps; pour pt. < 0 F; pH 10.5.

Bubble Breaker® DMD-2. [Witco/Organics] Complex surfactant; nonionic; oilfield surfactant, defoamer; liq.; sol. in IPA; disp. in water, kerosene, xylene; dens. 8.3 lb/gal; visc. 600 cps; pour pt. < 0 F; pH 10.5.

Bubble Breaker® PR. [Witco/Organics] Sodium salt of sulfonated fatty acid; defoamer for wet process prod. of phosphoric acid; sol. in water, IPA.

Bug Remover Conc. [Sherex/Div. of Witco] Surfactant blend; bug and tar remover conc. for autos; liq.; 100% solids.

Burco Anionic APS. [Burlington Chem.] Ethoxylated sulfonate; anionic; hypochlorite-stable surfactant, emulsifier for acid and alkaline cleaners, disinfectants, personal care prods., household cleaners, tub and tile cleaners, mildew removers, textile scours, dairy cleaners; stable over entire pH range; clear liq.; m.w. 600; sp.gr. 1.06; visc. 270 cps; pH 6.5-7.5; 35% solids.

Burco Anionic APS-LF. [Burlington Chem.] Anionic; hypochlorite-stable low-foam surfactant, emulsifier for automatic dishwash detergents, CIP cleaning systems, dairy cleaners, spray cleaners; stable in acids, bases, hard water; biodeg.; water-wh. liq.; pH 7.5; flash pt. > 200 F; 35% solids.

Burco CS-LF. [Burlington Chem.] Phosphate ester of EO/PO block polymer, halogen-capped ethoxylated polyether; anionic/nonionic; biodeg. low foaming synergistic surfactant blend, scouring agent for textile processing, dishwash rinse aid, hard surf. cleaners, liq. laundry prods., steam cleaners; stable to 10% caustic; clear lt. yel. microemulsion; sp.gr. 1.02; pH 6.2; 50% active, 34.5% solids.

Burco DFE-45. [Burlington Chem.] Polyol ester; emulsifier, defoamer for paper, textiles, water treatment, coatings, metalworking applics. and low-foaming emulsions; amber liq.; sol. in oils, solvs.; sp.gr. 0.94; pH 5 (5% disp.).

Burco FAE. [Burlington Chem.] Fatty amine ethoxylate; cationic; surfactant, wetting agent, penetrant, emulsifier, detergent, lubricant, corrosion inhibitor; moderately low foaming; for indutrial lubricants, dye carriers, polyethylene textile softeners, hard surf. cleaners; amber liq.; sol. in water, min. spirits, xylene; insol. in min. oil; dens. 8.2 lb/gal; HLB 12; amine no. 110-120; pH 9.1 (1%); 95% tert. amine.

Burco HCS-50NF. [Burlington Chem.] Alkaline-stable surfactant for formulating conc. liq. alkaline detergents; nonfoaming wetting agent, detergent; for metal cleaning, food plant cleaning, industrial laundry, bottle washing, heavy-duty applics.; clear liq.; sp.gr. 1.15; pH 6.5 (1% aq.); 50% act.

Burco LAF-6. [Burlington Chem.] Aliphatic alcohol alkoxylate; nonionic; biodeg. low foam detergent, wetting agent for rinse aid formulations, automatic dishwash, metal cleaners, textile scouring agents; stable to acids and alkalies; pale yel. clear liq.; sp.gr. 0.99; cloud pt. 37 C; flash pt. > 300 F; pH 6-8 (5%); 100% act.

Burco LAF-125. [Burlington Chem.] Linear alcohol alkoxylate; nonionic; low foam surfactant for rinse aids, high-pressure spray cleaners, metal lubricants, textile jet scours; Gardner 1 max. color; more sol. in cold than hot water; sp.gr. 1.04; acid no. 1 max.; cloud pt. 10-16 C (1% aq.); pH 6 (1% aq.); Draves wetting 13 s (0.1% aq.); 100% act.

Burco LAF-DW. [Burlington Chem.] Low foam surfactant; defoamer, wetting agent for formulating mech. dishwash detergents; clear to sl. hazy liq.; sp.gr. 1.01; cloud pt. 20 C (1% aq.); pH 7 (1%); 100% act.

Burco NCS-80. [Burlington Chem.] Surfactant blend; nonionic/cationic; detergent, moderate foamer, detergent intermediate for formulating mildly alkaline cleaners for use as vehicle cleaners, floor cleaners, car wash; amber liq.; sp.gr. 1.04; cloud pt. > 100 C (1% aq.); pH 6 (5% aq.).

Burco NF-225. [Burlington Chem.] Foam booster/stabilizer for use in combination with Burco NPS-225; produces copious amounts of stable foam.

Burco NPS-50%. [Burlington Chem.] Polyglycoside; nonionic; surfactant, grease emulsifier for hand dishwash, laundry, neutral and acidic hard surf. cleaners, personal care prods., all-purpose cleaners; stable, biodeg.; EPA approval; clear amber liq.; highly water-sol.; dens. 9.4 lb/gal; visc. 550 cps;

HLB 15.0; f.p. < 5 F; cloud pt. none; pour pt. 12 F; flash pt. (CC) > 200 F; pH 5-7 (1%); surf. tens. 26.6 dynes/cm (0.1%); 50% act.

Burco NPS-225. [Burlington Chem.] Alkyl polyglycoside; CAS 68515-73-1; nonionic; dispersant, wetting agent, coupling agent in high electrolyte sol'ns.; for hard surf. cleaners (acid or alkali), all-purpose cleaners, metal cleaners; hydrotrope in highly alkaline formulations; biodeg.; stable to acids or alkalies; USDA, EPA approvals; brn. visc. liq., mild odor; water-sol.; sp.gr. 1.17 g/ml; dens. 9.7 lb/gal; visc. 4500 cps; HLB 13-14; f.p. < 5 F; pour pt. 20 F; cloud pt. none; flash pt. > 200 F; pH 5-7 (1%); surf. tens. 28.8 dynes/cm (0.1%); 70% act.

Burco TME. [Burlington Chem.] Ethoxylated dodecylmercaptan; nonionic; detergent for aq. cleaning systems, wool scouring, hard surf. cleaners; oil splitter; replaces chlorinated hydrocarbons in metal degreasing; clear yel. liq., mild grapefruit-like odor; sp.gr. 1.07; visc. 5.7 cs; HLB 13.5; cloud pt. 160 F (1% aq.); flash pt. (TCC) 127 F; pH 4 (1% aq.); toxicology: LD50 (rat, oral) 1885-2665 mg/kg; LD50 (rabbit, dermal) 3500 mg/kg; primary eye, mild skin irritant; 85% act.

Burco TM-HF. [Burlington Chem.] Anionic; surfactant, emulsifier producing copious levels of stable foam.

Burcofac 1060. [Burlington Chem.] Organic phosphate ester; wetting agent, detergent, emulsifier, hydrotrope for other surfactants, glycol ethers; for mildly alkaline cleaner formulations, all-purpose cleaners, floor cleaners/wax strippers, carwash, textile scouring; alkaline stable; clear liq.; sp.gr. 1.07; acid no. 205; pH 2.0 (1% aq.); 100% act.

Burcofac 6660K. [Burlington Chem.] Phosphate ester; hydrotrope for coupling nonionic and anionic surfactants into alkaline detergent formulations; clear liq.; sp.gr. 1.11; pH 8 (5% aq.); 50% solids.

Burcofac 7580. [Burlington Chem.] Aromatic phosphate ester; emulsifier for solvs. and oils; clear to sl. hazy visc. liq.; sp.gr. 1.05; pH 2 (1% disp.); acid no. 75-90; 99% conc.

Burcofac 9125. [Burlington Chem.] Organic phosphate ester; anionic; wetting agent, detergent, emulsifier, hydrotrope for other surfactants, glycol ethers; used in mildly alkaline detergents, all-purpose cleaners, floor cleaners/wax strippers, carwash, textile and laundry prods.; clear liq.; sp.gr. 1.08; acid no. 285; pH 2.0 (1% aq.); 100% act.

Burcol BP-181. [Burlington Chem.] Low-foaming block polymer; defoamer for textiles, paper, metalworking, antifreeze, and other applics.; lubricant base; rinse aid formulations; clear liq.; m.w. 2000; sp.gr. 1.01; visc. 340 cps; cloud pt. 16 C (10% aq.).

Burcosol ADS-40. [Burlington Chem.] Multifunctional component, chelating agent, antiredeposition agent, dispersant.

Burcosperse AP Liq. [Burlington Chem.] Low m.w. sodium polyacrylate sol'n.; chelating agent, antiredeposition agent; m.w. 2000.

Burcoterge DG-40. [Burlington Chem.] Linear alcohol derivs.; anionic/nonionic; wetting agent, detergent, emulsifier for laundry and hard surf. cleaners, microemulsions with solvs., degreasing formulations; exc. performance in hot or cold systems; clear liq.; sp.gr. 1.0; pH 7 (1% aq.); 40% solids.

Burcotreat 900-A. [Burlington Chem.] Polyacrylic acid; chelating agent, antiredeposition agent; free acid version of Burcosperse AP Liq.

Burcowet TM-LF. [Burlington Chem.] Alkoxylated linear alcohol; defoamer at high temps.; adds wetting, detergency.

Burcowet TMW. [Burlington Chem.] Linear alcohol ethoxylate; wetting agent, detergent; can function as extender for Burco TME.

BYK®-020. [Byk-Chemie USA] Sol'n. of a modified polysiloxane copolymer; defoamer for water-reducible coating systems, e.g., alkyds, polyester, epoxy esters, acrylics; sp.gr. 0.87-0.89 g/cc; dens. 7.26-7.42 lb/gal; flash pt. 50 C; ref. index 1.420-1.430; 9-11% NV in 2-butoxyethanol/2-ethylhexanol/Stod. (6/1/1/).

BYK®-022. [Byk-Chemie USA] Mixt. of hydrophobic solids and foam-destroying polysiloxanes in polyglycol; defoamer for water-based industrial coatings and emulsion paints (pigmented acrylate/polyurethane, baking enamels, hybrid systems); sp.gr. 0.99 g/cc; dens. 8.26 lb/gal; flash pt. > 100 C; 100% NV.

BYK®-023. [Byk-Chemie USA] Emulsion of hydrophobic solids, emulsifiers, foam destroying polysiloxanes; defoamer for water-based emulsion paints (exterior, gloss, semigloss paints, hybrid systems, prod. of polymeric dispersions; sp.gr. 0.97 g/cc; dens. 7.17 lb/gal; flash pt. (Seta) > 100 C; 18% NV in water.

BYK®-024. [Byk-Chemie USA] Polysiloxanes/polymer mixt.; defoamer for water-based systems incl. polyurethane, acrylate/polyurethane paints, wood varnishes, furniture paints, pigmented dispersion paints, plastic coatings; sp.gr. 1.00 g/cc; dens. 8.34 lb/gal; flash pt. (Seta) > 100 C; ref. index 1.488; > 97% NV.

BYK®-025. [Byk-Chemie USA] Polysiloxane mixt.; defoamer for water-based systems incl. polyurethane, acrylate/polyurethane paints, wood varnishes, furniture varnishes, plastic coatings; sp.gr. 0.96 g/cc; dens. 8.01 lb/gal; flash pt. (Seta) 80 C; ref. index 1.426; 19% NV in dipropylene glycol methyl ether.

BYK®-031. [Byk-Chemie USA] Emulsion of hydrophobic components and paraffin-based min. oil; defoamer for latex paints, latex-based stucco finishes, latex-based adhesives based on polymethyl acrylate, vinyl ester, and styrene copolymers; sp.gr. 0.92-0.94 g/cc; dens. 7.68-7.84 lb/gal; flash pt. (Seta) > 110 C; 51-55% NV in water/oil.

BYK®-032. [Byk-Chemie USA] Emulsion of paraffin-based min. oils and hydrophobic components; defoamer for emulsion paints, emulsion plasters, industrial emulsions; sp.gr. 0.92-0.94 g/cc; dens. 7.67-7.84 lb/gal; flash pt. (Seta) > 110 C; 49.5-53.5% NV in water/oil.

BYK®-033. [Byk-Chemie USA] Mixt. of hydrophobic components in paraffin-based min. oil; defoamer for emulsion paints, emulsion plasters, emulsion adhesives, industrial emulsions; sp.gr. 0.85-0.88 g/cc; dens. 7.09-7.34 lb/gal; flash pt. (Seta) > 110 C; ref. index 1.474-1.481; >98% NV.

BYK®-034. [Byk-Chemie USA] Mixt. of hydrophobic components in paraffin-based min. oil; contains silicone; defoamer for emulsion paints, gloss and semigloss latex systems, emulsion plasters, emulsion adhesives, industrial emulsions (water-thinnable polyurethane resins, alkyd resin emulsions); sp.gr. 0.83-0.89 g/cc; dens. 6.92-7.42 lb/gal; flash pt. (Seta) > 110 C; ref. index 1.472-1.482; 100% NV.

BYK®-035. [Byk-Chemie USA] Mixt. of hydrophobic components in paraffin-based min. oil; defoamer for gloss and semigloss latex systems based on polymethacrylate, polyvinyl esters, styrene copolymers; sp.gr. 0.85-0.88 g/cc; dens. 7.09-7.34 lb/gal; flash pt. (Seta) > 110 C; ref. index 1.474-1.481; 100% NV.

BYK®-045. [Byk-Chemie USA] Emulsion of hydrophobic solids, emulsifiers, foam-destroying polysiloxanes; defoamer for emulsion paints, paper coatings, foil coatings, stains; sp.gr. 0.93-1.03 g/cc; dens. 7.76-8.59 lb/gal; flash pt. (Seta) > 100 C; 7-10% NV in water.

BYK®-051. [Byk-Chemie USA] Silicone-free polymeric sol'n.; defoamer for org. paint systems incl. nitrocellulose, polyurethane, alkyd-melamine baking enamels, epoxy systems, oil-free polyesters, alkyd paints, acid-curing systems; sp.gr. 0.81-0.83 g/cc; dens. 6.76-6.93 lb/gal; flash pt. (Seta) > 34 C; ref. index 1.435-1.445; 19-21% NV in Stod./butylglycolate/2-butoxyethanol (71/8/1).

BYK®-052. [Byk-Chemie USA] Silicone-free polymeric sol'n.; defoamer for org. paint systems incl. nitrocellulose, polyurethane, alkyd-melamine baking enamels, epoxy systems, oil-free polyesters, alkyd paints, acid-curing systems; sp.gr. 0.81-0.83 g/cc; dens. 6.76-6.93 lb/gal; flash pt. (Seta) > 34 C; ref. index 1.435-1.445; 19-21% NV in Stod./butylglycolate/2-butoxyethanol (71/8/1).

BYK®-053. [Byk-Chemie USA] Silicone-free polymeric sol'n.; defoamer for org. paint systems incl. nitrocellulose, polyurethane, alkyd-melamine baking enamels, epoxy systems, oil-free polyesters, alkyd paints, acid-curing systems; sp.gr. 0.81-0.83 g/cc; dens. 6.76-6.93 lb/gal; flash pt. (Seta) > 34 C; ref. index 1.435-1.445; 19-21% NV in Stod./butylglycolate/2-butoxyethanol (71/8/1).

BYK®-065. [Byk-Chemie USA] Polysiloxane sol'n.; defoamer for org. systems incl. chlorinated rubber, vinyl resin, acrylic resin, alkyds; sp.gr. 0.94-0.96 g/cc; dens. 7.76-8.09 lb/gal; flash pt. (Seta) 43 C; ref. index 1.445-1.455; <1% NV in cyclohexanone.

BYK®-066. [Byk-Chemie USA] Polysiloxane sol'n.; defoamer for org. systems incl. chlorinated rubber, epoxy, 2-part polyurethane, alkyd, alkyd/melamine, self-crosslinking acrylates, polyesters; sp.gr. 0.80-0.82 g/cc; dens. 6.76-6.93 lb/gal; flash pt. (Seta) 47 C; ref. index 1.411-1.417; < 1% NV in diisobutyl ketone.

BYK®-070. [Byk-Chemie USA] Polysiloxane and polymers mixt.; defoamer for org. coating systems incl. 2-part epoxy, polyurethane, alkyd baking, and nitrocellulose systems; sp.gr. 0.88-0.90 g/cc; dens. 7.34-7.51 lb/gal; flash pt. (Seta) 25 C; ref. index 1.470-1.480; 8-10% NV in xylene/methoxypropyl acetate/butyl acetate (10/2/1).

BYK®-075. [Byk-Chemie USA] Aralkyl-modified methylalkylpolysiloxane sol'n.; defoamer for solv.-based systems, incl. air-drying baking enamels, acid-curing systems, polyurethane and nitrocellulose lacquers, industrial paints; sp.gr. 0.89 g/cc; dens. 7.42 lb/gal; flash pt. (Seta) > 100 C; ref. index 1.44; 54% NV.

BYK®-077. [Byk-Chemie USA] Methylalkyl polysiloxane sol'n.; defoamer for org. systems; improves mar resistance and leveling in air-dry and baking enamels, acid-curing systems, polyurethane and nitrocellulose lacquers, solv.-based industrial paints; sp.gr. 0.87-0.90 g/cc; dens. 7.26-7.51 lb/gal; flash pt. (Seta) 45 C; ref. index 1.468-1.478; 50-54% NV in naphtha.

BYK®-080. [Byk-Chemie USA] Polysiloxane copolymer nonaq. emulsion; defoamer for polar org. coating systems, esp. 2-part polyurethane paints and furniture finishes, acrylic paints, chlorinated rubber paints; sp.gr. 1.02-1.05 g/cc; dens. 8.50-8.76 lb/gal; flash pt. (Seta) > 110 C; 85% NV in propylene glycol.

BYK®-085. [Byk-Chemie USA] Modified polysiloxane copolymer sol'n.; defoamer improving slip and leveling in high solids systems, solv.-based systems, printing inks, paints; sp.gr. 0.89-0.92 g/cc; dens. 7.42-7.68 lb/gal; flash pt. (Seta) > 65 C; > 98% NV.

BYK®-141. [Byk-Chemie USA] Polysiloxanes/polymer mixt.; defoamer for solv.-based coatings, esp. wood finishes based on nitrocellulose, acid curing, polyester, or polyurethane, epoxy coatings, chlorinated rubber, or vinyl resin systems; sp.gr. 0.85-0.89 g/cc; dens. 7.09-7.43 lb/gal; flash pt. (Seta) 28 C; ref. index 1.478-1.488; 2-4% NV in naphtha/isobutanol (11/2).

BYK®-151. [Byk-Chemie USA] Sol'n. of an alkylolammonium salt of a polyfunctional polymer; anionic; dispersant, wetting agent for inorg. and org. pigments, fillers in aq. systems; improves color str. development and rheological props. of pigment pastes; sp.gr. 1.08-1.11 g/cc; dens. 9.01-9.26 lb/gal; acid no. 90-120; flash pt. (Seta) > 110 C; 38-42% NV in water/butylglycol (11/1).

BYK®-307. [Byk-Chemie USA] Polyether modified dimethyl polysiloxane copolymer; additive to increase surface slip, substrate wetting, scratch and mar resistance for paints, printing inks; sp.gr. 1.01-1.05 g/cc; dens. 8.42-8.76 lb/gal; flash pt. (Seta) > 65 C; > 98% NV in xylene.

BYK®-320. [Byk-Chemie USA] Sol'n. of a polyester modified methylalkyl polysiloxane copolymer; additive for improving slip props., scratch and mar resistance, flow, substrate wetting for polar coating systems such as alkyd melamine, acrylic, polyurethane systems; foam suppressor for flow coatings, dipping coatings, roller coatings; sp.gr. 0.83-0.87 g/cc; dens. 6.93-7.26 lb/gal; flash pt. (Seta) 38 C; ref. index 1.436-1.446; 50-53% NV in Stod./methoxypropyl acetate (9/1).

BYK®-321. [Byk-Chemie USA] Sol'n. of a polyether modified dimethyl polysiloxane copolymer; additive to increase mar resistance, add slip, improve flow and substrate wetting for water-reducible, colloidal disp. or latex systems, solv.-based systems; foam suppressor for flow, dip and roller coatings; sp.gr. 0.90-0.94 g/cc; dens. 7.51-7.84 lb/gal; flash pt. (Seta) 58 C; ref. index 1.429-1.439; 49-53% NV in 2-butoxyethanol.

BYK®-331. [Byk-Chemie USA] Polyether-modified dimethylpolysiloxane copolymer; additive to improve surface slip, leveling, substrate wetting in solv.-based, air drying, baking systems, e.g., alkyd-acrylic, polyester-melamine, water-borne paint systems; sp.gr. 1.030-1.050 g/cc; dens. 8.59-8.76 lb/gal; flash pt. (Seta) > 110 C; ref. index 1.447-1.455; 98% NV.

BYK®-336. [Byk-Chemie USA] Sol'n. of polyether-modified dimethyl polysiloxane copolymer; additive to increase mar resistance, slip, surface flow and gloss, substrate wetting; for solv.-based and water-borne coatings; sp.gr. 0.96-0.98 g/cc; dens. 8.05-8.22 lb/gal; flash pt. (Seta) 39 C; 24-26% NV in xylene/methoxypropyl acetate (1/8).

BYK®-341. [Byk-Chemie USA] Sol'n. of a polyether-modified dimethylpolysiloxane copolymer; additive to improve substrate wetting in solv.-based and water-reducible systems; improves surface flow, slip, scratch and mar resistance; sp.gr. 0.95-0.99 g/cc; dens. 7.93-8.26 lb/gal; flash pt. (Seta) 64 C; ref. index 1.433-1.439; 50-54% NV in 2-butoxyethanol.

BYK®-344. [Byk-Chemie USA] Sol'n. of a polyether-modified dimethylpolysiloxane copolymer; additive to increase mar resistance, slip, flow, substrate wetting for solv. and solv.-free coatings; sp.gr. 0.93-0.95 g/cc; dens. 7.76-7.92 lb/gal; flash pt. (Seta) 23 C; ref. index 1.46-1.47; 50.5-54.0% NV in xylene/isobutanol (4/1).

BYK®-354. [Byk-Chemie USA] Polyacrylate sol'n.; additive for improving leveling and surf. flow, degassing and defoaming aid for most systems; sp.gr. 0.92-0.98 g/cc; dens. 7.68-8.18 lb/gal; flash pt. (Seta) 43 C; ref. index 1.477-1.483; 48-52% NV in naphtha/diisobutyl ketone (9/1).

BYK®-A500. [Byk-Chemie USA] Silicone-free polymeric sol'n.; defoamer, air release agent for laminating, spray-up, hand lay-up molding, gel coats, and solv.-free epoxy flooring systems; prevents air entrapment and porosity; sp.gr. 0.87-0.89 g/cc; dens. 7.43 lb/gal; flash pt. (Seta) 45 C; ref. index 1.490-1.500; 5-8% NV.

BYK®-A501. [Byk-Chemie USA] Silicone-free polymeric sol'n.; defoamer, air release agent for laminating, spray-up, hand lay-up molding, gel coats, and solv.-free epoxy flooring systems; prevents air entrapment and porosity; sp.gr. 0.87-0.89 g/cc; dens. 7.43 lb/gal; flash pt. (Seta) 46 C; ref. index 1.497-1.503; 40-48% NV.

BYK®-P104. [Byk-Chemie USA] Sol'n. of higher m.w. unsat. polycarboxylic acid; anionic; wetting, dispersing, antiflooding and antisettling additive; for coating systems; sp.gr. 0.94-0.96 g/cc; dens. 7.84-8.01 lb/gal; acid no. 160-200; flash pt. (Seta) 28 C; ref. index 1.485-1.495; 48-52% NV in xylene/diisobutyl ketone (9/1).

BYK®-P104S. [Byk-Chemie USA] Sol'n. of higher m.w. unsat. polycarboxylic acid with a polysiloxane copolymer; anionic; wetting, dispersing, antiflooding additive; for coating systems; sp.gr. 0.94-0.96 g/cc; dens. 7.84-8.01 lb/gal; acid no. 130-170; flash pt. 28 C; ref. index 1.484-1.494; 48-52% NV in xylene/diisobutyl ketone (9/1).

BYK®-P105. [Byk-Chemie USA] Sol'n. of a higher m.w. unsat. polycarboxylic acid; anionic; wetting, dispersing, antiflooding, and antisettling additive for high-solids coatings, printing inks, pigment pastes, solv.-based paint systems; sp.gr. 1.03-1.06 g/cc; dens. 8.59-8.84 lb/gal; acid no. 340-390; flash pt. > 110 C; > 97% NV.

Bykumen®. [Byk-Chemie USA] Sol'n. of a higher m.w. unsat. acidic polycarboxylic acid ester; anionic; wetting, dispersing additive to improve pigment wetting and prevent settling of pigments; for solv. and solv.-free systems, alkyd trade sales systems, acrylics, polyesters; sp.gr. 0.86-0.90 g/cc; dens. 7.18-7.51 lb/gal; acid no. 30-40; flash pt. (Seta) 26 C; ref. index 1.444-1.456; 44-48% NV in Stod./isobutanol (3/2).

Bykumen®-WS. [Byk-Chemie USA] Sol'n. of polyamine amides of unsat. polycarboxylic acids; anionic; wetting, dispersing additive in water-reducible systems, industrial emulsion paints, colloidal disp. systems, latex paints; stabilizes pigment disp. against flocculation and sedimentation; improves gloss; increases corrosion and water resistance; sp.gr. 0.98-1.02 g/cc; dens. 8.17-8.51 lb/gal; acid no. 50-70; flash pt. (Seta) 72 C; 49-53% NV in 2-butyoxyethanol.

C

C-108. [Procter & Gamble] Coconut fatty acid; CAS 67701-05-7; intermediate for mfg. of soaps, amides, esters, alcoholamides, and nonsurfactant applics.; Gardner < 1 liq.; m.w. 207; acid no. 266-274; iodine no. 5 max.; sapon. no. 273; 100% conc.

C-110. [Procter & Gamble] Coconut fatty acid; CAS 67701-05-7; intermediate for mfg. of soaps, amides, esters, alcoholamides, and nonsurfactant applics.; Gardner 3 max. liq.; m.w. 207; acid no. 266-274; iodine no. 12 max.; sapon. no. 272; 100% conc.

C-810. [Procter & Gamble] Caprylic/capric fatty acids; CAS 67762-36-1; intermediate for mfg. of soaps, amides, esters, alcoholamides, and nonsurfactant applics.; liq.; m.w. 154; acid no. 358-368; iodine no. 0.5 max.; sapon. no. 370 max.; 100% conc.

C-810L. [Procter & Gamble] Caprylic/capric acids, fractionated; CAS 67762-36-1; intermediate; m.w. 157; acid no. 345-365; iodine no. 1 max.

C-895. [Procter & Gamble] Caprylic acid, fractionated; CAS 124-07-2; intermediate for mfg. of soaps, amides, esters, alcoholamides, and nonsurfactant applics.; liq.; m.w. 145; acid no. 380-394; iodine no. 0.2 max.; sapon. no. 395 max.; 100% conc.

C-899. [Procter & Gamble] Caprylic acid; CAS 124-07-2; intermediate for mfg. of soaps, amides, esters, alcoholamides, and nonsurfactant applics.; m.w. 144; acid no. 385-389; iodine no. 0.2 max.; sapon. no. 386-290.

C-1095. [Procter & Gamble] Capric acid, fractionated; CAS 334-48-5; intermediate for mfg. of soaps, amides, esters, alcoholamides, and nonsurfactant applics.; solid; m.w. 173; acid no. 320-330; iodine no. 0.5 max.; sapon. no. 331 max.; 100% conc.

C-1214. [Procter & Gamble] C12-14 acids, fractionated; CAS 68002-90-4; intermediate; m.w. 207; acid no. 267-280; iodine no. 0.5 max.

C-1298. [Procter & Gamble] Lauric acid, fractionated; CAS 143-07-7; intermediate for mfg. of soaps, amides, esters, alcoholamides, and nonsurfactant applics.; solid; m.w. 200; acid no. 277-281; iodine no. 0.2 max.; sapon. no. 278-282; 100% conc.

C-1495. [Procter & Gamble] Myristic acid, fractionated; CAS 544-63-8; intermediate for mfg. of soaps, amides, esters, alcoholamides, and nonsurfactant applics.; solid; m.w. 228; acid no. 243-248; iodine no. 0.5 max.; sapon. no. 244-249; 100% conc.

Cachalot® L-50. [M. Michel] Lauryl alcohol; CAS 112-53-8; tech. grade; Hazen 10 color; m.w. 200-212; sp.gr. 0.830; visc. 8 cps (50 C); b.p. 260-340 C; solid. pt. 20-24 C; acid no. 0.1 max.; iodine no. 0.5 max.; sapon. no. 1 max.; hyd. no. 265-280; flash pt. 130 C; ref. index 1.4337-1.4339 (50 C).

Cachalot® L-90. [M. Michel] Lauryl alcohol; CAS 112-53-8; Hazen 10 color; m.w. 184-190; sp.gr. 0.830; visc. 8 cps (50 C); b.p. 250-265 C; solid. pt. 22-24 C; acid no. 0.1 max.; iodine no. 0.3 max.; sapon. no. 0.5 max.; hyd. no. 295-305; flash pt. 120 C; ref. index 1.4315-1.4317 (50 C).

Cachalot® O-3. [M. Michel] Oleyl alcohol NF; CAS 143-28-2; Hazen 80 color; m.w. 266-270; sp.gr. 0.825 (50 C); visc. 11.7 cps (50 C); b.p. 310-350 C; acid no. 0.2 max.; iodine no. 90-96; sapon. no. 1.0 max.; hyd. no. 200-210; cloud pt. 5-8 C; flash pt. 180 C; ref. index 1.4582-1.4820.

Cachalot® O-8. [M. Michel] Oleyl alcohol; CAS 143-28-2; Hazen 150 color; m.w. 261-274; sp.gr. 0.825 (50 C); visc. 11.5 cps (50 C); b.p. 300-350 C; acid no. 0.2 max.; iodine no. 80-85; sapon. no. 1.0 max.; hyd. no. 205-215; cloud pt. 12-18 C; flash pt. 170 C; ref. index 1.4579-1.4582.

Cachalot® O-27. [M. Michel] Oleyl alcohol; CAS 143-28-2; tech. grade; Hazen 150 color; m.w. 255-274; sp.gr. 0.825 (50 C); visc. 10.7 cps (50 C); b.p. 290-350 C; acid no. 0.2 max.; iodine no. 45-55; sapon. no. 1 max.; hyd. no. 210-220; cloud pt. 30-40 C; flash pt. 170 C; ref. index 1.4485-1.4489.

CA DBS 50 SA. [Witco SA] Branched calcium dodecylbenzene sulfonate in aromatic solv.; anionic; lipophilic emulsifier for pesticide emulsifiable concs.; liq.; 50% conc.

Cadoussant AS. [Ceca SA] Cationic; softening detergent for wool and cotton.

CAE. [Ajinomoto; Ajinomoto USA] Ethyl N-cocoyl-L-arginate PCA salt; cationic; surfactant, foamer, and antistat; preservative; antiseptic; germicide; disinfectant in cosmetics, detergents, dentifrices, medical supplies; white crystalline powd.; sol. in water, ethanol, ethyleneglycol; sl. sol. in ethyl acetate, toluene; sol. 5% in water @ 30 C; m.p. 180-185 C; pH 5.0-7.0 (1% aq., 20 C); 100% conc.

Calamide C. [Pilot] Cocamide DEA; nonionic; emollient, foam booster/stabilizer, visc. builder; liq. detergents and personal care prods.; emulsifier for degreasing; lubricant, rust preventive for metal cleaning and cooling systems; detergent and solubilizer for washing floors, walls, metals; amber clear liq.; water-sol.; pH 9.5 (1% aq.); 100% conc.

Calamide CW-100. [Pilot] Cocamide DEA (2:1); nonionic; emollient, lubricant, foam stabilizer, visc. builder, detergent, solubilizer used in bubble baths, shampoos, liq. detergents, heavy-duty cleaners, emulsion systems; yel. clear liq.; water-sol.; dens. 8.5 lb/gal; pH 10.5 (1% aq.); 100% conc.

Calamide O. [Pilot] Oleamide DEA; nonionic; emollient, lubricant, foam stabilizer, visc. builder, emulsifier; used in personal care prods., emulsions, industrial applics.; amber clear liq.; disp. in water and oil; pH 10; 100% conc.

Calamide S. [Pilot] Modified soya dialkanolamide; nonionic; liq. detergent, foam stabilizer, visc. builder, w/o emulsifier; liq.; 100% conc.

Calcium Petronate. [Witco/Sonneborn] Calcium petrol. sulfonate (CAS #61789-86-4); anionic; emulsifier, detergent and rust inhibitor in lube oil additives, industrial applics.; dk. br. visc. liq.; sol. in org. solv.; sp.gr. 0.98; 45% conc.

Calcium Stearate, Regular. [Witco] Calcium stearate; lubricant for ceramics, coated foundry sands, plastics; plasticizer/lubricant for paper coatings; aux. emulsifier; water repellent for cosmetics; flatting, pigment suspending, and thickener for solv.-based paints and varnishes; wh. powd.; sp.gr. 1.03.

Calcium Sulfonate C-50N. [Witco/Sonneborn] Neutral calcium sulfonates (CAS #61789-86-4); anionic; lube oil additive, rust preventive, industrial detergent; liq.; 50% conc.

Calcium Sulfonate EP-163. [Witco/Sonneborn] Modified overbased calcium sulfonates (CAS #61789-86-4); anionic; EP additive, rust preventive; liq.; 18% conc.

Calester. [Pilot] Alpha sulfo methyl laurate; surfactant for high-quality toilet soaps, laundry detergents, automobile cleaners, spray cleaners, foamers, emulsifiers.

Calfax 10L-45. [Pilot] Sodium linear decyl diphenyl oxide disulfonate; anionic; biodeg. surfactant, solubilizer, dispersant for dye bath leveling, pigment dispersion, heavy-duty cleaners, latex emulsification, agric. chemicals, phenolic germicides, bottle washing; lt. amber aq. sol'n.; sol. in water, org. solvs.; pH 8.5; 45% act.

Calfax 16L-35. [Pilot] Sodium alkyl diphenyl disulfonate; CAS 65143-89-7; anionic; emulsifier; high electrolyte tolerance; biodeg.; liq.; 35% conc.

Calfax DB-45. [Pilot] Sodium alkyl diphenyl oxide disulfonate; CAS 28519-02-0; anionic; detergent, solubilizer for dye bath and other strongly polar applics, e.g., dip tank cleaners, electroplating baths, heavy-duty cleaners, latex emulsifiers, agric. chemicals; tolerant of high alkalinity, high acidity, high levels of electrolyte; pale yel.-brn. liq.; 45% conc.

Calfax DBA-40. [Pilot] Alkyl diphenyl oxide disulfonic acid; CAS 30260-72-1; anionic; free sulfonic acid form of Calfax DB-45; useful in acidic formulations or for conversion to specialty salts; liq.; 40% conc.

Calfax DBA-70. [Pilot] Alkyl diphenyl oxide disulfonic acid; CAS 30260-72-1; anionic; free sulfonic acid form of Calfax DB-45; useful in acidic formulations or for conversion to specialty salts; liq.; 70% conc.

Calfoam AAL. [Pilot] Alkanolamide and surfactant blend; anionic; mild lt. duty liq. detergent for dishwashing and fine fabric laundering; liq.; dens. 9.0 lb/gal; visc. 300 cps; pH 7.0; 45% solids.

Calfoam ES-30. [Pilot] Sodium laureth sulfate; CAS 9004-82-4; anionic; detergent, foam stabilizer, flash foamer, wetter for detergent systems, personal care prods., wool washing; emulsion polymerization; yel. clear liq.; mild odor; dens. 8.8 lb/gal; pH 8.0; 30% solids.

Calfoam LLD. [Pilot] Surfactant/alkanolamide blend; anionic; lt. duty liq. detergent for dishwashing, fine hand washables, and all-purpose cleaning; liq.; 42% conc.

Calfoam NEL-60. [Pilot] Ammonium lauryl ether sulfate; anionic; flash foamer, foam stabilizer, detergent, emulsifier, wetter for liq. detergents, bubble baths, shampoos, carwashing; lime soap dispersant; clear liq.; faint alcohol odor; sol. in aq. systems; dens. 8.58 lb/gal; pH 7.5; 57.5% act.

Calfoam NLS-30. [Pilot] Ammonium lauryl sulfate; anionic; mild surfactant base for neutral pH shampoos, bubble baths, rug cleaner formulations, cosmetic emulsifiers, emulsion polymerization; liq.; 30% act.

Calfoam SEL-60. [Pilot] Sodium laureth sulfate; CAS 9004-82-4; anionic; general all-purpose cleaning and wetting for use in bubble baths, shampoos, car washing, liquid detergents; liq.; 60% act.

Calfoam SLS-30. [Pilot] Sodium lauryl sulfate; CAS 151-21-3; anionic; mild detergent, foamer for personal care prods.; rug/upholstery shampoos; emulsifier for cosmetics, emulsion polymerization of latex, SBR rubber, polyacrylates, elastomers; foaming agent for foamed rubber; wh. paste; mild odor; pH 8.0; 30% act.

Calimulse EM-30. [Pilot] Sodium alkylbenzene sulfonate (hard); emulsifier for emulsion polymers and other prods. which do not enter sewage streams; paste; 30% conc.

Calimulse EM-99. [Pilot] Alkylbenzene sulfonic acid (hard); neutralized as emulsifiers for agric., emulsion polymers, and other prods. which do not enter sewage streams; thick liq.; 97% conc.

Calimulse PRS. [Pilot] Isopropylamine dodecylbenzene sulfonate; anionic; biodeg. emulsifier, solubilizer; dry cleaning; degreasers; latex emulsifier; pigment dispersant; agric. sprays, oil slick emulsifiers; clear amber liq.; dens. 8.5 lb/gal; pH 4.8; 90% conc.

Caloxylate N-9. [Pilot] Nonoxynol-9; CAS 9016-45-9; nonionic; detergent, emulsifier for hard surf. cleaners, liq. laundry prods., general purpose cleaners; biodeg.; liq.; HLB 13.2; 100% conc.

Calsoft AOS-40. [Pilot] Sodium alpha olefin sulfonate; surfactant for hand soaps, shampoos, hard surf. cleaners, household and industrial cleaners; liq.; 40% act.

Calsoft F-90. [Pilot] Sodium dodecylbenzene sulfonate; anionic; detergent, emulsifier, wetter for all-purpose and hard surface cleaners, bubble baths, degreasers, laundry powds., textile scouring aids, emulsion polymers, sanitation, emulsion paints, wettable powds., ore flotation, metal pickling; wh. free-flowing flake; water-sol.; dens. 0.45 g/cc; pH 8.0 (1%); 90% act.

Calsoft L-40. [Pilot] Sodium dodecylbenzene sulfonate; anionic; emulsion stabilizer; wetting and foaming agent, detergent, emulsifier for household and industrial detergents, agric. emulsions, dye bath leveling, rug cleaners, bubble baths, ore flotation, and air entrainment in concrete and gypsum board; washing fruits and vegetables; biodeg.; lt. yel. visc. liq.; water-sol.; sp.gr. 1.07; pH 7.5; 42% solids.

Calsoft L-60. [Pilot] Sodium dodecylbenzene sulfonate; anionic; biodeg. emulsion stabilizer; wetting and foaming agent, detergent, emulsifier for household and industrial detergents, agric. emulsions, dye bath leveling, rug cleaners, bubble baths, ore flotation, and air entrainment in concrete and gypsum board; water-wh. pasty liq.; odorless; dens. 8.7 lb/gal; pH 7.4; 60% solids.

Calsoft LAS-99. [Pilot] Dodecylbenzene sulfonic acid, linear; anionic; biodeg. detergent, emulsifier, intermediate for liq. and dry detergents, hard surf. cleaners, stripping, wetting, foaming; Klett 50 syrupy liq.; water-sol.; sp.gr. 1.06; dens. 8.83 lb/gal; visc. 1100 cps; 97.5% act.

Calsoft T-60. [Pilot] TEA-dodecylbenzene sulfonate; anionic; biodeg. detergent, wetting agent, flash foamer; liq. detergents, wool wash compds., cosmetics and shampoos, agric. emulsifiers, industrial cleaners, textile scouring, and car wash compds.; yel. clear visc. liq.; mild odor; dens. 9.1 lb/gal; visc. 2300 cps; pH 5.7; 60% solids.

Calsolene Oil HSA. [ICI Am.] Sulfated fatty acid ester; anionic; emulsifier, low foaming wetting and penetrating agent; stable in acid and alkalies; gold-yel., clear liq. (may be turbid at room temp.); dissolves in hard and soft water and in salt sol'ns.; flash pt. 200 F (OC).; 45% conc.

Calsolene Oil HSAD. [ICI Am.] Sulfated ester; anionic; wetting agent, emulsifier, lubricant, leveling agent; liq.; 45% conc.

Calsuds 81 Conc. [Pilot] Formulated prod.; anionic/nonionic; conc. base for liq. detergents, all-purpose and hard surface cleaning, dishwash, car wash, foaming and wetting sol'ns.; liq.; 60% conc.

Calsuds A. [Pilot] Anionic detergents and amide; anionic; biodeg. cleaning, foaming and wetting conc. base for liq. detergents, lt. duty cleaners; liq.; 50% conc.

Calsuds CD-6. [Pilot] Alkylaryl sulfonate/cocamide DEA blend; anionic; conc. base, foam builder/stabilizer, wetting agent, visc. modifier, and emulsifier; used in liq. detergents, shampoos, wool-washing compds., hand, felt, and janitorial cleaners, textile scours, agric. sprays; clear visc. liq.; dens. 8.6 lb/gal; visc. 2100 cps; pH 9.5 (10%); 100% conc.

Capcure Emulsifier 37S. [Henkel/Functional Prods.] Ethoxylate; nonionic; epoxy resin emulsifier; waxy solid; HLB 18.0; 100% conc.

Capcure Emulsifier 65. [Henkel/Functional Prods.] Ethoxylate; nonionic; epoxy resin emulsifier; clear liq.; HLB 18.0; 100% conc.

Caplube 8350. [Karlshamns] Glyceryl monodioleate; nonionic; o/w emulsifier for food and industrial applics.; liq.; HLB 2.7; 100% conc.

Caplube 8385. [Karlshamns] Castor oil, ethoxylated; nonionic; emulsifier, dispersant, lubricant used in textile compding.; liq.; HLB 14.5; 100% conc.

Caplube 8410. [Karlshamns] Triglyceryl oleate; nonionic; emulsifier, lubricant for textiles; liq.; HLB 8.8; 100% conc.

Caplube 8440. [Karlshamns] Decaglyceryl tetraoleate; nonionic; emulsifier, antistat, lubricant, and coupler for textile finishes; liq. HLB 6.7; 100% conc.

Caplube 8442. [Karlshamns] Decaglyceryl decaoleate; nonionic; emulsifier, lubricant; liq.; HLB 3.2; 100% conc.

Caplube 8445. [Karlshamns] Decaglyceryl tetracocate; nonionic; emulsifier, lubricant, antistat for textile applic.; liq.; HLB 8.8; 100% conc.

Caplube 8448. [Karlshamns] Decaglyceryl decastearate; nonionic; lubricant for thread finish compding.; opacifier; solid; HLB 3.5; 100% conc.

Caplube 8508. [Karlshamns] Coconut ethoxylate; nonionic; emulsifier, lubricant for textile applics.; liq.; 100% conc.

Caplube 8540B. [Karlshamns] Ethoxylated glycerides or glyceride derivs.; nonionic; emulsifier, lubricant for textile finishes; liq.; HLB 6.5; 100% conc.

Caplube 8540C. [Karlshamns] Ethoxylated glycerides or glyceride derivs.; nonionic; emulsifier, lubricant for textile finishes; liq.; HLB 10.0; 100% conc.

Caplube 8540D. [Karlshamns] Ethoxylated glycerides or glyceride derivs.; nonionic; emulsifier, lubricant for textile finishes; liq.; HLB 13.5; 100% conc.

Capmul® MCM. [Karlshamns] Glyceryl caprate/caprylate; co-solv. and coupler for org. compds.; w/o emulsifier; sol. in oil and alcohol.

Carbon Detergent K. [Henkel/Cospha; Henkel Canada] Fatty nitrogen compd.; nonionic; detergent for removing carbon from metal surfs.; soft paste; 84% conc.

Carbonox. [Baroid] Altered lignite; anionic; thinner, emulsifier, and dispersant for oil well drilling muds; powd.; 100% conc.

Carbopol® 613, 614. [BFGoodrich] Polyacrylic acid; emulsifier for solv. cleaners, emulsion stabilizer, suspending agent, thickener for detergent formulations; fluffy wh. powd.; 2-6 μ particle size; sp.gr. 1.41; bulk dens. 13 lb/ft^3; acid no. 700-750; pH 2.5-3.0 (1%); toxicology: eye irritant.

Carbopol® 615, 616, 617. [BFGoodrich] High m.w. polyacrylic acid, crosslinked with polyalkenyl polyether; anionic; emulsifier, thickener, coupler, and insol. particle suspending agent used in liq. detergents; prevents soil redeposition; fluffy wh. powd.; 2-6 μ particle size; sp.gr. 1.41; bulk dens. 13 lb/ft^3; acid no. 700-750; pH 2.5-3.0 (1%); toxicology: eye irritant.

Carbopol® 907. [BFGoodrich] Polyacrylic acid; anionic; emulsifier, thickener, stabilizer, suspending agent; used for drilling muds, photosensitive emulsions, water treatment; solid; sol. in water, polar solvs., many nonpolar solvs. blends; 100% conc.

Carbopol® 910. [BFGoodrich] Carbomer; anionic; emulsifier, thickener, stabilizer, suspending agent; used for flocking, dip coating, textile back coating; powd.; sol. see Carbopol 907; 100% conc.

Carbopol® 934. [BFGoodrich] Carbomer; anionic; emulsifier, thickener, stabilizer, suspending agent; for lubricating, quenching and silicone emulsions; graphite, polyethylene, fiber and paper suspensions; sol. see Carbopol 907; 100% conc.

Carbopol® 940. [BFGoodrich] Carbomer; anionic; emulsifier, thickener, stabilizer, suspending agent used in cosmetic applics., for die-casting and forging lubricants, thixotropic paints; solv. thickening with or without neutralizing; powd.; sol. see Carbopol 907; 100% conc.

Carbostat 2203. [Hoechst Celanese/Colorants & Surf.] Quaternary amine; cationic; surfactant for textile processing; amber liq.; 35% act.

Carbowax® MPEG 350. [Union Carbide] PEG-6 methyl ether; surfactant intermediate, lubricant for adhesives, inks, mining, soaps and detergents; liq.; sol. in water; m.w. 335–365; sp.gr. 1.0891; dens. 9.13 lb/gal; visc. 3.9 cSt (210 F); f.p. –5 to 10 C; ref. index 1.455; pH 4.0–7.5 (5% aq.); surf. tens. 40 dynes/cm.

Carbowax® MPEG 550. [Union Carbide] PEG-10 methyl ether; surfactant intermediate, lubricant for adhesives, inks, mining, soaps and detergents; liq.; sol. in water; m.w. 525–575; sp.gr. 1.1039; dens. 8.97 lb/gal (55 C); visc. 6.6 cSt (210 F); f.p. 15–25 C; flash pt. (CCC) 360 F; flash pt. (CCC) > 350 F; ref. index 1.455 (40 C); pH 4.0–7.5 (5% aq.); surf. tens. 37.5 dynes/cm (40 C).

Carbowax® MPEG 750. [Union Carbide] PEG-16 methyl ether; chemical intermediate, lubricant for adhesives; soft solid; sol. in water; m.w. 715–785; sp.gr. 1.0760 (60 C); dens. 9.02 lb/gal (55 C); visc.

10.3 cSt (210 F); f.p. 27–32 C; flash pt. (CCC) 415 F; ref. index 1.459 (40 C); pH 4.0–7.5 (5% aq.); surf. tens. 40.7 dynes/cm (40 C).

Carbowax® MPEG 2000. [Union Carbide] PEG-40 methyl ether; surfactant intermediate, lubricant for adhesives, toilet bowl cleaners; flake; sol. 68% in water; m.w. 1900–2100; sp.gr. 1.0871 (60 C); dens. 9.18 lb/gal (55 C); visc. 45.5 cSt (210 F); m.p. 49–54 C; flash pt. (Seta CC) 355 F; pH 6.0–8.0 (5% aq.).

Carbowax® MPEG 5000. [Union Carbide] PEG-100 methyl ether; surfactant intermediate, lubricant for adhesives, toilet bowl cleaners; flake; sol. 64% in water; m.w. 4750–5250; sp.gr. 1.0907 (60 C); dens. 8.95 lb/gal (80 C); visc. 320 cSt (210 F); m.p. 57–63 C; flash pt. (CCC) 415 F; pH 6.0–8.0 (5% aq.).

Carbowax® PEG 200. [Union Carbide] PEG-4; intermediate for surfactants, lubricants, urethanes; antistat, humectant, mold release agent, plasticizer for adhesives, inks, lubricants; liq.; sol. in water; m.w. 190–210; sp.gr. 1.1239; dens. 9.38 lb/gal; visc. 4.3 cSt (210 F); flash pt. (CCC) > 300 F; ref. index 1.459; pH 4.5–7.5 (5% aq.); surf. tens. 44.5 dynes/cm.

Carbowax® PEG 300. [Union Carbide] PEG-6; antistat, intermediate for surfactants, dye carrier, humectant, lubricant, release agent, plasticizer for adhesives, capsules, creams and lotions, deodorant sticks, inks, lipsticks; also as glycerin replacement; liq.; sol. in water; m.w. 285–315; sp.gr. 1.1250; dens. 9.38 lb/gal; visc. 5.8 cSt (210 F); f.p. –15 to –8 C; flash pt. (CCC) > 350 F; ref. index 1.463; pH 4.5–7.5 (5% aq.); surf. tens. 44.5 dynes/cm.

Carbowax® PEG 400. [Union Carbide] PEG-8; antistat, surfactant intermediate, dye carrier, humectant, lubricant, release agent, plasticizer for adhesives, capsules, ceramic glazes, creams and lotions, deodorant sticks, inks, lipsticks; liq.; sol. in water, methanol, ethanol, acetone, trichloroethylene, Cellosolve®, Carbitol®, dibutyl phthalate, toluene; m.w. 380–420; sp.gr. 1.1254; dens. 9.39 lb/gal; visc. 7.3 cSt (210 F); f.p. 4–8 C; flash pt. (CCC) > 350 F; ref. index 1.465; pH 4.5–7.5 (5% aq.); surf. tens. 44.5 dynes/cm.

Carbowax® PEG 600. [Union Carbide] PEG-12; antistat, surfactant intermediate, humectant, lubricant, release agent, plasticizer for adhesives, capsules, ceramic glaze, creams and lotions, dentifrices, deodorant sticks, inks, lipsticks, wood treatment; liq.; sol. in water; m.w. 570–630; sp.gr. 1.1257; dens. 9.40 lb/gal; visc. 10.8 cSt (210 F); f.p. 20–25 C; flash pt. (CCC) > 350 F; ref. index 1.46; pH 4.5–7.5 (5% aq.); surf. tens. 44.5 dynes/cm.

Carbowax® PEG 900. [Union Carbide] PEG-20; antistat, surfactant intermediate, lubricant, release agent, ointment and suppository base, plasticizer for adhesives, ceramic glaze, creams and lotions, dentifrices, deodorant sticks, wood treatment; soft solid; sol. 86% in water; m.w. 855-900; sp.gr. 1.0927 (60 C); dens. 9.16 lb/gal (55 C); visc. 15.3 cSt (210 F); f.p. 32–36; flash pt. (CCC) > 350 F; pH 4.5–7.5 (5% aq.).

Carbowax® PEG 1000. [Union Carbide] PEG-20; antistat, surfactant intermediate, lubricant, release agent, ointment and suppository base, plasticizer for adhesives, ceramic glaze, creams and lotions, dentifrices, deodorant sticks, wood treatment; soft solid; sol. 80% in water; m.w. 950–1050; sp.gr. 1.0926 (60 C); dens. 9.16 lb/gal (55 C); visc. 17.2 cSt (210 F); f.p. 37–40; flash pt. (CCC) > 350 F; pH 4.5–7.5 (5% aq.).

Carbowax® PEG 1450. [Union Carbide] PEG-32; antistat, surfactant intermediate, lubricant, release agent for adhesives, ceramic glaze, creams and lotions, dentifrices, deodorant sticks, wood treatment; soft solid or flake; sol. 72% in water; m.w. 1300–1600; sp.gr. 1.0919 (60 C); dens. 9.17 lb/gal (55 C); visc. 26.5 cSt (210 F); f.p. 43–46; flash pt. (CCC) > 350 F; pH 4.5–7.5 (5% aq.).

Carbowax® PEG 3350. [Union Carbide] PEG-75; antistat, surfactant intermediate, dye carrier, lubricant, release agent, tablet binder for adhesives, ceramic glaze, creams and lotions, dentifrices, mining, soaps and detergents, tablet coating, toilet bowl cleaners; flake or powd.; sol. 67% in water; m.w. 3000–3700; sp.gr. 1.0926 (60 C); dens. 8.94 lb/gal (80 C); visc. 90.8 cSt (210 F); f.p. 54–58; flash pt. (CCC) > 350 F; pH 4.5–7.5 (5% aq.).

Carbowax® PEG 4600. [Union Carbide] PEG-100; antistat, surfactant intermediate, dye carrier, lubricant, release agent, tablet binder for adhesives, ceramic glaze, creams and lotions, mining, soaps and detergents, tablet coating, toilet bowl cleaners; flake or powd.; sol. 65% in water; m.w. 4400–4800; sp.gr. 1.0926 (60 C); dens. 8.95 lb/gal (80 C); visc. 184 cSt (210 F); f.p. 57–61; flash pt. (CCC) > 350 F; pH 4.5–7.5 (5% aq.).

Carbowax® PEG 8000. [Union Carbide] PEG-150; antistat, ceramic binder, surfactant intermediate, dye carrier, lubricant, release agent, tablet binder for adhesives, creams and lotions, mining, powd. metallurgy, soaps and detergents, tablet coating, toilet bowl cleaners; flake or powd.; sol. 63% in water; m.w. 7000–9000; sp.gr. 1.0845 (70 C); dens. 8.96 lb/gal (80 C); visc. 822 cSt (210 F); m.p. 60–63; flash pt. (CCC) > 350 F; pH 4.5–7.5 (5% aq.).

Carding Oil A-4 Super. [Takemoto Oil & Fat] Surfactant, neutral oil; anionic/nonionic; carding oil; liq.; 100% conc.

Cardolite® NC-507. [Cardolite] 3-(n-Pentadecyl) phenol; CAS #501-24-6 and 3158-56-3; starting raw material for surfactants, antioxidants, anticorrosives; lubricant additive; cosolv. for insecticides, germicides; resin modifier; tan waxy solid; m.w. 304; sol. in org. solvs. incl. aliphatic hydrocarbons; sp.gr. > 1.0; m.p. 44–48 C (760 mm Hg); b.p. 243–249 C (10 mm Hg); ref. index 1.494.

Cardolite® NC-510. [Cardolite] 3-(n-Pentadecyl) phenol; CAS #501-24-6; starting raw material for surfactants; coupling agent for pigments and dyes; cosolv. for insecticides, germicides; resin modifier for phenolic-aldehyde, polyester, PC polymers; also in photographic industry; slight pink to wh. waxy solid; m.w. 304; sol. in org. solvs. incl. aliphatic hydrocarbons; sp.gr. > 1.0; m.p. 49–51 C (760 mm Hg); b.p. 190–195 C (1 mm Hg); ref. index 1.4750.

Carnauba Spray 200. [Sherex/Div. of Witco] Quat. blend; car rinse with carnauba wax; liq.; 63% solids.

Carriant Series. [Toho Chem. Industry] Oil mixt.; anionic/nonionic; textile dyeing assistants; carriers for polyester fibers; liq.

Carsamide® 7644. [Lonza] Modified cocamide DEA; nonionic/anionic; emulsifier, thickening agent, controlled suds detergent, textile fulling and scouring agent; Gardner 6 liq.; sol. in water, alcohol, chlorinated and aromatic hydrocarbons; disp. in min. oil, kerosene, natural fats and oils, min. spirits; dens. 8.3 lb/gal.

Carsamide® C-3. [Lonza] Cocamide DEA; intermediate for liq. detergent formulation, emulsifier for aromatic solv., improves the foaming, detergency and hard water resistance of soaps.

Carsamide® C-7944. [Lonza] Cocamide DEA; amber liq. free of foreign matter.

Carsamide® CA. [Lonza] Cocamide DEA (1:1); nonionic; detergent, dispersant, emulsifier, wetting agent, foam booster, thickener, softener for industrial, cosmetic, and household cleaners; biodeg.; liq.; sol. in water, alcohol, chlorinated and aromatic hydrocarbons, Polysolve, Cellosolves, Carbitols, and natural fats and oils; sp.gr. 1.0; dens. 8.3 lb/gal; 100% act.

Carsamide® SAC. [Lonza] Cocamide DEA (1:1); nonionic; detergent for industrial, institutional, cosmetic, and household cleaners; dispersant, emulsifier, wetting agent, visc. builder; foam stabilizer for shampoos, bubble baths, detergents; pale amber visc. liq.; mild fatty odor; ol. in water, alcohol, chlorinated and aromatic hydrocarbons, Polysolve, Cellosolves, Carbitols, and natural fats and oils; dens. 8.2 lb/gal; 100% act.

Carsamide® SAL-7. [Lonza] Lauramide DEA (1:1); CAS 120-40-1; nonionic; detergent, emulsifier, foaming agent, foam stabilizer, thickener for shampoos, bath prods., household, institutional and industrial detergents; Gardner 2 solid; mild, fatty odor; sol. in alcohol, chlorinated and aromatic hydrocarbons; dens. 8.2 lb/gal; m.p. 104–114 F; biodeg.; 100% conc.

Carsamide® SAL-9. [Lonza] Lauramide DEA (1:1); CAS 120-40-1; nonionic; foaming and thickening agent for shampoos, bath prods., detergents; solid; dens. 8.1 lb/gal; biodeg.; 100% conc.

Carsofoam® 211. [Lonza] Blend of sulfonates, nonionics, and coconut derived alkanolamides; detergent conc. for all purpose cleaners, paint strippers, aircraft cleaners (base), metal cleaners, etc.; amber liq.; mild odor; dens. 8.8 lb/gal; clear pt. 1 C; pH 8.2 ± 0.2; biodeg.; 100% act.

Carsofoam® MS Conc. [Lonza] Detergent, shampoo conc. esp. for mild shampoos; lt. yel. hazy liq.; pH 7.3 (10%); 50% act.

Carsofoam® T-60-L. [Lonza] TEA dodecylbenzene sulfonate; anionic; high foaming detergent, wetting agent for cosmetic, industrial, and institutional usage; Gardner < 5 clear visc. liq.; bland odor; dens. 9.0 lb/gal; pH 7.0–7.6; biodeg.; 60% solids.

Carsonol® ALS-R. [Lonza] Ammonium lauryl sulfate; anionic; detergent with high foam, good wetting and emulsifying properties used for cosmetics, chemical specialties; clear visc. liq.; water sol.; dens. 8.4 lb/gal; pH 6.5–7.0 (10%); biodeg.; 29% act.

Carsonol® ALS-S. [Lonza] Ammonium lauryl sulfate; anionic; biodeg. detergent, foaming agent, wetting agent, emulsifier for personal care prods., household and industrial cleaners, emulsion polymerization; liq.; 29% act.

Carsonol® ANS. [Lonza] Sulfated nonylphenol ethoxylate, aluminum salt; anionic; surfactant used in dishwashing, soaps, and shampoos; liq.; 58% conc.

Carsonol® AOS. [Lonza] Sodium C14-C16 olefin sulfonate; surfactant for shampoos, liq. soaps, industrial cleaners; pH 8.0 (5%); 40% act.

Carsonol® BD. [Lonza] Blend of carboxybetaine, sulfate and amine groups; anionic/amphoteric; detergent, foaming agent, cosmetic uses; lt. amber, moderately visc. liq.; excellent water sol.; cloud pt. 50 F; pH 7.5–8.5; 30–32% act.

Carsonol® ILS. [Lonza] Mixed isopropanolamines lauryl sulfate; anionic; detergent with high foam, good wetting and emulsifying properties used for cosmetics, chemical specialties; visc. liq.; low odor; water sol.; dens. 8.63 lb/gal; pH 7.0–7.4 (10%); biodeg.; 75% act.

Carsonol® MLS. [Lonza] Magnesium lauryl sulfate; anionic; detergent with high foam, good wetting and emulsifying properties used for cosmetics, chemical specialties, rug and upholstery formulations; visc. liq.; low odor; water sol.; dens. 8.57 lb/gal; pH 6.5–7.5 (10%); biodeg.; 30% act.

Carsonol® SES-A. [Lonza] Ammonium laureth sulfate; anionic; surfactant with excellent foaming in hard and soft water, for cosmetic, household, and industrial uses, shampoos, bubble baths, liq. cleaners; liq.; dens. 8.4 lb/gal; pH 6.5–7.0 (10%); biodeg.; 60% act.

Carsonol® SES-S. [Lonza] Sodium laureth sulfate; CAS 9004-82-4; anionic; surfactant with excellent foaming in hard and soft water, for cosmetic, household, and industrial uses, liq. carwash, laundry detergents; liq.; dens. 8.8 lb/gal; pH 7.0-8.0 (10%); biodeg.; 60% act.

Carsonol® SHS. [Lonza] Sodium 2-ethylhexyl sulfate; anionic; low foaming detergent, wetting agent, penetrant, emulsifier used in caustic sol'ns. for peeling of fruits and vegetables; stable to high concs. of electrolytes; FDA approved; Gardner 3 clear liq.; dens. 9.6 lb/gal; sp.gr. 1.15; visc. 35 cps; pH 10.3 (10%); 40% act.

Carsonol® SLES. [Lonza] Sodium laureth sulfate; CAS 9004-82-4; anionic; surfactant with excellent foaming in hard and soft water, for cosmetic, household, and industrial uses, liq. carwash, laundry detergents; liq.; dens. 8.7 lb/gal; pH 7.5–8.5 (10%); biodeg; 30% act.

Carsonol® SLS. [Lonza] Sodium lauryl sulfate; CAS 151-21-3; anionic; detergent with high foam, good wetting and emulsifying properties used for cosmetics, chemical specialties, shampoos, bubble baths, detergents; clear, liq. low color; low odor; water sol.; dens. 8.7 lb/gal; pH 7.5–8.5 (10%); biodeg.; 30% act.

Carsonol® SLS Paste B. [Lonza] Sodium lauryl sulfate; CAS 151-21-3; anionic; detergent with high foam, good wetting and emulsifying properties used for cosmetics, chemical specialties, shampoo bases, textile scouring; paste, low color; low odor; water sol.; dens. 8.8 lb/gal; pH 8.4–8.8 (10%); biodeg.; 30% act.

Carsonol® SLS-R. [Lonza] Sodium lauryl sulfate; CAS 151-21-3; anionic; biodeg. detergent, foaming agent, wetting agent, emulsifier for personal care prods., household and industrial cleaners, emulsion polymerization; liq.; 29% act.

Carsonol® SLS-S. [Lonza] Sodium lauryl sulfate; CAS 151-21-3; anionic; biodeg. detergent, foaming agent, wetting agent, emulsifier for personal care prods., household and industrial cleaners, emulsion polymerization; liq.; 29% act.

Carsonol® SLS Special. [Lonza] Sodium lauryl sulfate; CAS 151-21-3; anionic; detergent with high foam, good wetting and emulsifying properties used for cosmetics, chemical specialties; liq., low color; low odor; water sol.; dens. 8.7 lb/gal; pH 7.5–8.5 (10%); biodeg.; 30% act.

Carsonol® TLS. [Lonza] TEA-lauryl sulfate; anionic; biodeg. detergent with high foam, good wetting and emulsifying properties used for cosmetics, mild shampoos, bubble baths, chemical specialties, emulsion polymerization; clear liq.; low color, odor; water sol.; dens. 8.7 lb/gal; pH 7.0–7.5 (10%); 40% act.

Carsonon® L-985. [Lonza] Laureth-9 (9-10 EO); nonionic; wetting agent, detergent, emulsifier, dispersant used for maintenance and institutional cleaners, in textile, paper, and paint industries; colorless liq.; sol. in acid and alkaline sol'ns.; dens. 8.32 lb/gal; cloud pt. 137 F; pour pt. 39–43 F; pH neutral (3%); biodeg.; 85% act.

Carsonon® LF-5. [Lonza] Linear alcohol alkoxylate; nonionic; low foam surfactant for pulp and paper, mechanical dishwashing, metal cleaning, dairy equipment cleaning; APHA 100 color, mild odor; sol. in aromatic solv., polar solv., water below cloud pt. and in common chlorinated solv.; sp.gr. 1.02 (60F/60 F); visc. 125 cts; cloud pt. 43–49 C; flash pt. 460 F (COC); surf. tens. 33.5 dynes/cm (0.1%); pH 5.5–7.0 (1% aq.); biodeg.

Carsonon® LF-46. [Lonza] Polyoxyalkalene glycol; nonionic; low foam surfactant for pulp and paper, mechanical dishwashing, metal cleaning, dairy equipment cleaning; Gardner 1 color; mild odor; sol. in aromatic solv., polar solv., water below cloud pt. and in the common chlorinated solv.; visc. 200 cps.; cloud pt. 44 C; flash pt. > 230 F; pH 5.5–7.0 (1% aq.); biodeg.

Carsonon® N-4. [Lonza] Nonoxynol-4; CAS 9016-45-9; nonionic; emulsifier, detergent, wetting agent, dispersant for household and industrial uses; intermediate; drycleaning detergent; pale liq.; sol. in kerosene, alcohols, aromatic and chlorinated solv., Stod.; sp.gr. 1.02; dens. 8.6 lb/gal; visc. 175–250 cps; flash pt. 500–600 F; pour pt. –5 ± 2 F; solid. pt. –20 ± 2 F; HLB 9.0; 100% act.

Carsonon® N-6. [Lonza] Nonoxynol-6; CAS 9016-45-9; nonionic; emulsifier, wetting agent for household and industrial uses; pale-colored liq; mild, pleasant odor; sol. in alcohols, aromatic, chlorinated and aliphatic solv., fatty oil, disp. in water; dens. 8.68 lb/gal; visc. 180–240 cps; cloud pt. 90–102 F; flash pt. 500 F; HLB 11.0; 100% act.

Carsonon® N-8. [Lonza] Nonoxynol-8 (8–8.5 mole); CAS 9016-45-9; nonionic; emulsifier, detergent, wetting agent, dispersant for household and industrial uses; pale-colored liq.; mild, pleasant odor; sol. in water, alcohol, aromatic, chlorinated and aliphatic solv.; dens. 8.75 lb/gal; visc. 200–300 cps; cloud pt. 82–86 F; flash pt. 435 F; pour pt. 45 F; pH neutral (1%); 100% act.

Carsonon® N-9. [Lonza] Nonoxynol-9 (9–10 mole); CAS 9016-45-9; nonionic; surfactant for water-sol. household and industrial cleaning formulations; pale-colored liq.; mild, pleasant odor; sol. in water, alcohol, aromatic, chlorinated and aliphatic solv.; dens. 8.83 lb/gal; visc. 180–250 cps; flash pt. 540 F; pour pt. 35–39 F; pH neutral (1%); HLB 13.0; 100% act.

Carsonon® N-10. [Lonza] Nonoxynol-10 (10–11 mole); CAS 9016-45-9; nonionic; emulsifier, detergent, wetting agent, dispersant for household and industrial uses; pale-colored liq.; mild, pleasant odor; sol. in water, alcohol, aromatic, chlorinated and aliphatic solv.; dens. 8.84 lb/gal; visc. 180–250 cps; cloud pt. 143–146 F; flash pt. 540 F; pour pt. 35–39 F; pH neutral (1%); HLB 13.2; 100% act.

Carsonon® N-11. [Lonza] Nonoxynol-11; CAS 9016-45-9; nonionic; household and industrial cleaning formulations; liq.; water-sol.; HLB 13.8; 100% conc.

Carsonon® N-12. [Lonza] Nonoxynol-12; CAS 9016-45-9; nonionic; surfactant, emulsifier for household and industrial cleaning formulations; APHA 75 liq.; water-sol.; HLB 14.2; hyd. no. 75; cloud pt. 190 F; 100% conc.

Carsonon® N-30. [Lonza] Nonoxynol-30; CAS 9016-45-9; nonionic; emulsifier, stabilizer for emulsion polymerization, oils, fats, waxes, essential oils; pale yel. liq. to semisolid; sp.gr. 1.09; cloud pt. 167–171 F (1%); pH 6.0–7.0 (3%); HLB 17.2; 70% act., 30% water.

Carsonon® N-40, 70%. [Lonza] Nonoxynol-40; CAS 9016-45-9; nonionic; emulsifier, stabilizer; wh. to yel. semisolid; cloud pt. 212 F (1%); pH 6.0–7.0 (3%); 70% act., 30% water.

Carsonon® N-50. [Lonza] Nonoxynol-50; CAS 9016-45-9; nonionic; emulsifier, stabilizer for household and industrial use; wh. to yel semisolid; HLB 18; cloud pt. 212 F (1%); pH 6.0–7.0 (3%); 70% act., 30% water.

Carsonon® N-100, 70%. [Lonza] Nonoxynol-100; CAS 9016-45-9; nonionic; emulsifier, stabilizer; wh. to off-wh. semisolid; water sol.; cloud pt. 212 F (1%); pH 6.0–7.0 (3%); 70% act., 30% water.

Carsonon® ND-317. [Lonza] Nonylphenol ethoxylate; nonionic; surfactant, base for solid toilet bowl cleaners; Gardner 2 opaque, visc. liq.; sp.gr. 1.11; cloud pt. 162 ± 2 F (1% in 10% NaCl); pH 7.5–8.0 (3% aq.); 100% conc.

Carsonon® TD-10. [Lonza] Trideceth-10; CAS 24938-91-8; nonionic; emulsifier, detergent, wetting agent, foam builder, solubilizer used for household, industrial cleaners, textile processing, paper industry; Gardner 1 liq.; water sol.; dens. 8.5 lb/gal; HLB 13.7; cloud pt. 162–166 F (1% aq.); flash pt. > 300 F (COC); pour pt. 67 F; pH 5.0–7.0 (3% aq.); 100% act.

Carsonon® TD-11, TD-11 70%. [Lonza] Trideceth-11; CAS 24938-91-8; nonionic; detergent, foam builder, emulsifier, wetting agent for household and industrial cleaners; Gardner 1 waxy solid; dens. 8.5 lb/gal; HLB 14.6; cloud pt. 137–141 F (1% in 10% NaCl); flash pt. > 300 F (COC); pour pt. 92 F; pH 5.0–7.0 (3% aq.); 100% conc. and 70% aq. sol'n. resp.

Carsoquat® 621. [Lonza] Benzalkonium chloride; cationic; corrosion inhibitor, surfactant; almost water-wh. to pale yel. liq.; mild odor; misc. with water, lower alcohols, most polar solv.; sp.gr. 0.96; dens. 8 lb/gal; 50% act.

Carsoquat® 621 (80%). [Lonza] Benzalkonium chloride; cationic; corrosion inhibitor, surfactant; straw-colored clear liq.; m.w. 361; pH 7–8 (1%); 80% act.

Carsoquat® 816-C. [Lonza] Cetearyl alcohol, PEG-40 castor oil, and stearalkonium chloride; cationic; formulated base, cream rinse conc.; wh. waxy flakes; mild, fatty odor; disp. in water; pH 6.3 (5% disp.); 96% min act.

Carsosoft® S-75. [Lonza] Quaternium-27, IPA; CAS 86088-85-9; cationic; fabric softener base, exc. rewet properties and nonyellowing; Gardner 4 liq.; m.w. 720; pH 5–7 (5%); 75% act. in IPA.

Carsosoft® S-90-M. [Lonza] Modified alkylamidoethyl alkyl imidazolinium methyl methosulfate; cationic; fabric softener base, exc. rewet properties and nonyellowing; Gardner 4 liq.; water-disp.; pH 5.0–7.0 (5%); 88–91% act. in IPA.

Carsosoft® T-90. [Lonza] Quaternium-53; CAS 130124-24-2; cationic; fabric softener, rewetting agent, antistat, and antilint for home and commercial laundries; yel. liq.; pH 5–7 (5%); 90% act. in IPA.

Carsosulf SXS-Liq. [Lonza] Sodium xylene sulfonate; CAS 1300-72-7; anionic; coupler, solubilizer for liq. detergent systems and cleaning compd.; liq.; sp.gr. 1.18–1.22; pH 7.5–10.5; nonflam.; biodeg.; 40% act.

Carsosulf UL-100 Acid. [Lonza] Linear dodecylbenzene sulfonic acid; m.w. 316; dens. 8.84 lb/gal; sp.gr. 1.06; biodeg.; 97 ± 1% act.

Carspray #2. [Sherex/Div. of Witco] Dicoco quat.; emulsifier plus glycol ether for car rinses; liq.; 100% solids.

Carspray 205. [Sherex/Div. of Witco] Proprietary; nonionic; car rinse emulsion aid; liq.; 100% solids.

Carspray 300. [Sherex/Div. of Witco] Dicoco quat.; emulsifier for car rinses; liq.; 75% solids.

Carspray 300HF. [Sherex/Div. of Witco] Dicoco quat.; emulsifier for car rinses; high-flash version of Carspary 300; liq.; 100% solids.

Carspray 375. [Sherex/Div. of Witco] Quat.; emulsifier for car hot waxes; liq.; 100% solids.

Carspray 400. [Sherex/Div. of Witco] Quat.; emulsifier for car rinses; liq.; 80% solids.

Carspray 500. [Sherex/Div. of Witco] Quat. blend; silicone rinse aid; liq.; 60% solids.

Carspray 650. [Sherex/Div. of Witco] Quat. blend; car rinse emulsifier with silicone and carnauba wax; liq.; 53% solids.

Carspray 700. [Sherex/Div. of Witco] Quat. blend; foaming rinse aid for car rinses; liq.; 61% solids.

Carspray CW. [Sherex/Div. of Witco] Quat./surfactant blend; wash and wax conc. for autos; liq.; 58% solids.

Cassapret SRH. [Hoechst AG] Polyester copolymer; nonionic; hydrophilizing agent; paste.

Cassurit Grades. [Hoechst AG] Reactant resins; resin finishing of cellulosic fibers; liq.

Casul® 55 HF. [Harcros] Calcium sulfonate; anionic; emulsifier for agric.; FDA compliance; dk. amber liq.; sp.gr. 1.02; dens. 8.5 lb/gal; pour pt. 70 F; flash pt. (PMCC) 110 F; pH 6-7 (10% in 20% IPA/water); 100% conc.

Casul® 70 HF. [Harcros] Calcium dodecylbenzene sulfonate; anionic; coemulsifier for o/w and w/o formulations, agric. emulsions; FDA compliance; amber liq.; dens. 8.5 lb/gal; sp.gr. 1.02; visc. 18,400 cps; HLB 10.5; flash pt. 104 F (PMCC); pour pt. 46 F; pH 6–7 (10% in 20% IPA/water); 70% act.

Catalyst NKS. [Hoechst AG] Mixt. of metal salts and organic acids; acid donor for the crosslinking of reactant resins; liq.

Catalyst PAT. [ICI Surf. UK] Alkanolamine hydrochloride; cationic; for use with acid curing (thermosetting or reactant type) finishing resins; liq.

Catamine 101. [Exxon/Tomah] Fatty amine complex; cationic; asphalt emulsifier for slurry seal and CMS emulsions; liq.; 100% conc.

Catigene® 50 USP. [Stepan Europe] n-Alkyl (50% C12, 30% C14, 17% C16, 3% C18) dimethylbenzyl ammonium chloride; cationic; germicide, algicide, fungicide, deodorizing agent, antistat; liq.; 50% act.

Catigene® 65 USP. [Stepan Europe] n-Alkyl (67% C12, 25% C14, 7% C16, 1% C8, C10, C18) dimethylbenzyl ammonium chloride; cationic; germicide, algicide, fungicide, deodorizing agent, antistat; liq.; 50% act.

Catigene® 776. [Stepan Europe] n-Alkyl dimethylbenzyl ammonium chloride, n-dialkyl methylbenzyl ammonium chloride; cationic; algicide, slimicide for swimming pool and industrial water treatment; liq.; 50% act.

Catigene® 818. [Stepan Europe] Octyl decyl dimethyl ammonium chloride, dioctyl dimethyl ammonium chloride, didecyl dimethyl ammonium chloride; cationic; germicide, algicide, fungicide, deodorizing agent, antistat; liq.; 50% act.

Catigene® 818-80. [Stepan Europe] Octyl decyl dimethyl ammonium chloride, dioctyl dimethyl ammonium chloride, didecyl dimethyl ammonium chloride; cationic; germicide, algicide, fungicide, deodorizing agent, antistat; liq.; 80% act.

Catigene® 824. [Stepan Europe] n-Alkyl (60% C14, 30% C16, 5% C12, 5% C18) dimethyl benzyl ammonium chloride; cationic; algicide, slimicide for swimming pool, industrial water treatment, hard surf. disinfection; liq.; 50% act.

Catigene® 885. [Stepan Europe] n-Alkyl (50% C14, 40% C12, 10% C16) dimethyl benzyl ammonium chloride, octyl decyl dimonium chloride, dioctyl dimonium chloride, didecyl dimonium chloride; cationic; germicide, algicide, fungicide, deodorizing agent, antistat; liq.; 50% act.

Catigene® 888. [Stepan Europe] n-Alkyl (50% C14, 40% C12, 10% C16) dimethyl benzyl ammonium chloride, octyl decyl dimonium chloride, dioctyl dimonium chloride, didecyl dimonium chloride; cationic; germicide, algicide, fungicide, deodorizing agent, antistat; liq.; 80% act.

Catigene® 1011. [Stepan Europe] Didecyl dimethyl ammonium chloride; cationic; germicide, algicide, fungicide, deodorizing agent, antistat; also for oilfield prod.; water-wh. to pale yel. liq.; 50% act.

Catigene® 2125 M. [Stepan Europe] n-Alkyl (60% C14, 30% C16, 5% C12, 5% C18) dimethyl benzyl ammonium chloride, n-alkyl (68% C12, 32% C14) dimethyl ethylbenzyl ammonium chloride; cationic; germicide, algicide, fungicide, deodorizing agent, antistat; liq.; 50% act.

Catigene® 2125M P40. [Stepan Europe] n-Alkyl (60% C14, 30% C16, 5% C12, 5% C18) dimethyl benzyl ammonium chloride, n-alkyl (68% C12, 32% C14) dimethyl ethylbenzyl ammonium chloride; cationic; germicide, algicide, fungicide, deodorizing agent, antistat; powd.; 40% act.

Catigene® 2125 P40. [Stepan Europe] n-Alkyl (60% C14, 30% C16, 5% C12, 5% C18) dimethyl benzyl ammonium chloride, n-alkyl (50% C12, 30% C14, 17% C16, 3% C18) dimethyl ethylbenzyl ammonium chloride; cationic; germicide, algicide, fungicide, deodorizing agent, antistat; powd.; 40% act.

Catigene® 2565. [Stepan Europe] n-Alkyl (60% C14, 25% C12, 15% C16) dimethyl benzyl ammonium chloride; cationic; algicide, slimicide for swimming pool, industrial water treatment; liq.; 50% act.

Catigene® 4513-50. [Stepan Europe] n-Alkyl (5% C12, 60% C14, 30% C16, 5% C18) dimethyl benzyl ammonium chloride, n-alkyl (50% C12, 30% C14, 17% C16, 3% C18) dimethyl ethylbenzyl ammonium chloride; cationic; germicide, algicide, fungicide, deodorizing agent, antistat; liq.; 50% act.

Catigene® 4513-80. [Stepan Europe] n-Alkyl (5% C12, 60% C14, 30% C16, 5% C18) dimethyl benzyl ammonium chloride, n-alkyl (50% C12, 30% C14, 17% C16, 3% C18) dimethyl ethylbenzyl ammonium chloride; cationic; germicide, algicide, fungicide, deodorizing agent, antistat; liq.; 80% act.

Catigene® 4513-80 M. [Stepan Europe] n-Alkyl (5% C12, 60% C14, 30% C16, 5% C18) dimethyl benzyl ammonium chloride, n-alkyl (68% C12, 32% C14) dimethyl ethylbenzyl ammonium chloride; cationic; germicide, algicide, fungicide, deodorizing agent, antistat; liq.; 80% act.

Catigene® 8248. [Stepan Europe] n-Alkyl (60% C14, 30% C16, 5% C12, 5% C18) dimethyl benzyl ammonium chloride; cationic; algicide, slimicide for swimming pool, industrial water treatment, hard surf. disinfection; liq.; 80% act.

Catigene® B 50. [Stepan Europe] n-Alkyl (40% C12, 50% C14, 10% C16) dimethylbenzyl ammonium chloride; cationic; germicide, algicide, fungicide, deodorizing agent, antistat; liq.; 50% act.

Catigene® B 80. [Stepan Europe] n-Alkyl (40% C12, 50% C14, 10% C16) dimethylbenzyl ammonium chloride; cationic; germicide, algicide, fungicide, deodorizing agent, antistat; liq.; 80% act.

Catigene® CA 56. [Stepan Europe] Alkyl amidopropyl trimethyl ammonium methoxysulfate; cationic; antistat additive for carpet latex and bitumen; pale yel. liq.; 56% act.

Catigene® CETAC 30. [Stepan Europe] Alkyl trimethylammonium chloride; cationic; antistat, lubricant, emulsifier, softener, germicide, algicide, fungicide, deodorizing agent for household/industrial cleaners, oilfield prod.; water-wh. to pale yel. liq.; 30% act.

Catigene® DC 100. [Stepan Europe] n-Alkyl (3% C12, 95% C14, 2% C16) dimethylbenzyl ammonium chloride; cationic; germicide, algicide, fungicide, deodorizing agent, antistat; powd.; 100% act.

Catigene® T 50. [Stepan Europe] n-Alkyl (<10% C8-10, 50% C12, 19% C14, 9% C16, 8% C18) dimethylbenzyl ammonium chloride; cationic; germicide, algicide, fungicide, deodorizing agent; antistat, dispersant, retarding agent for acrylic fiber dyeing with cationic dyes; water-wh. to pale yel. liq.; 50% act.

Catigene® T 80. [Stepan Europe] n-Alkyl (<10% C8-10, 50% C12, 19% C14, 9% C16, 8% C18) dimethylbenzyl ammonium chloride; cationic; germicide, algicide, fungicide, deodorizing agent; antistat, dispersant, retarding agent for acrylic fiber dyeing with cationic dyes; water-wh. to pale yel. liq.; 80% act.

Catinal CB-50. [Toho Chem. Industry] Lauralkonium chloride; cationic; disinfectant, germicide, antistat; liq.

Catinal HTB. [Toho Chem. Industry] Alkyl trimethyl ammonium bromide; cationic; dyeing assistant for PAN fiber dyeing; liq.; 30% conc.

Catinal MB-50A. [Toho Chem. Industry] Dodecyl dimethyl benzyl ammonium chloride; cationic; germicide, disinfectant, antistat; liq.

Catinex KB-10. [Pulcra SA] Nonylphenol, ethoxylated; nonionic; antifoaming and intermediate emulsifier; liq.; HLB 6.6; 100% conc.

Catinex KB-11. [Pulcra SA] Nonylphenol, ethoxylated; nonionic; detergent intermediate, emulsifier; liq. surface act. agent used in petrol. prods.; liq.; HLB 8.9; 100% conc.

Catinex KB-13. [Pulcra SA] Nonylphenol, ethoxylated; nonionic; drycleaning agent; emulsifier for insecticides; detergent; liq.; HLB 10; 100% conc.

Catinex KB-15. [Pulcra SA] Nonylphenol, ethoxylated; nonionic; detergent, dispersant, and wetting agent; solv. cleaner formulations; emulsifier and stabilizer for paint additives; de-icing fluid and sludge dispersing additive in petrol. prods.; liq.; HLB 10.9; 100% conc.

Catinex KB-16. [Pulcra SA] Nonylphenol, ethoxylated; nonionic; detergent, dispersant, emulsifier and wetting agent; surface act. agent; liq.; water-sol.; HLB 12.3; 100% conc.

Catinex KB-18. [Pulcra SA] Nonylphenol, ethoxylated; dispersant; solubilizer, detergent, emulsifier, wetting agent for anionic detergents; water-sol.; HLB 12.8; 100% conc.

Catinex KB-19. [Pulcra SA] Nonylphenol ethoxylate; nonionic; dispersant; solubilizer, detergent, emulsifier, wetting agent for anionic detergents; HLB 13.1; 100% conc.

Catinex KB-20. [Pulcra SA] Nonylphenol ethoxylate; nonionic; dispersant; solubilizer, detergent, emulsifier, wetting agent for anionic detergents; HLB 13.3; 100% conc.

Catinex KB-22. [Pulcra SA] Nonylphenol ethoxylate; nonionic; dispersant; solubilizer, detergent, emulsifier, wetting agent for anionic detergents; HLB 14.1; 100% conc.

Catinex KB-23. [Pulcra SA] Nonylphenol, ethoxylated; nonionic; detergent, dispersant, emulsifier and wetting agent; surface act. agent; liq., paste; HLB 14.7; 100% conc.

Catinex KB-24. [Pulcra SA] Nonylphenol ethoxylate; nonionic; detergent, dispersant, emulsifier and wetting agent; surface act. agent; HLB 16.0; 100% conc.

Catinex KB-25. [Pulcra SA] Nonylphenol ethoxylate; nonionic; detergent, dispersant, emulsifier and wetting agent; surface act. agent; HLB 17.1; 100% conc.

Catinex KB-26. [Pulcra SA] Nonylphenol ethoxylate; nonionic; detergent, dispersant, emulsifier and wetting agent; surface act. agent; HLB 16.7; 100% conc.

Catinex KB-27. [Pulcra SA] Nonylphenol, ethoxylated; nonionic; detergent, dispersant, and wetting agent; emulsifier for oils, fats, and waxes; metal cleaner and degreaser; solid; HLB 15.0; 100% conc.

Catinex KB-31. [Pulcra SA] Nonylphenol, ethoxylated; nonionic; surfactant for foam control; coemulsifier; dispersant for petrol. oils; liq.; HLB 4.6; 100% conc.

Catinex KB-32. [Pulcra SA] Nonylphenol, ethoxylated; nonionic; detergent, dispersant, and wetting agent; solv. cleaner formulations; emulsifier and stabilizer for paint additives; de-icing fluid and sludge dispersing additive in petrol. prods.; liq.; HLB 11.6; 100% conc.

Catinex KB-40. [Pulcra SA] Octylphenol, ethoxylated; nonionic; detergent, emulsifier, dispersant, surface act. agent used in washing formulations and emulsion cleaners; liq.; HLB 11.2; 100% conc.

Catinex KB-41. [Pulcra SA] Octylphenol, ethoxylated; nonionic; detergent, emulsifier, wetting agent for tank cleaners; additive in acid cleaning compds. for metals and lt. duty detergents; liq.; HLB 12.6; 100% conc.

Catinex KB-42. [Pulcra SA] Octylphenol, ethoxylated; nonionic; wetting agent, detergent, dispersant, emulsifier for household and industrial cleaners, textile processes, wool scouring; solubilizer for perfumes; emulsifier for insecticides and herbicides; liq.; water-sol.; HLB 13.3; 100% conc.

Catinex KB-43. [Pulcra SA] Octylphenol, ethoxylated; nonionic; solubilizer for anionic detergents; used in hot cleaning systems, metal pickling operations and electrolytic cleaning; paste; water-sol.; HLB 14.7; 100% conc.

Catinex KB-44. [Pulcra SA] Octylphenol, ethoxylated; nonionic; detergent, wetting agent, coemulsifier and stabilizer for vinyl-acrylic latex; emulsifier for fats and waxes; solid; HLB 17.3; 100% conc.

Catinex KB-45. [Pulcra SA] Octylphenol, ethoxylated; nonionic; primary emulsifier for vinyl acetate and acrylate polymerization; post stabilizer for syn. latex; dyeing assistant and emulsifier for fats and waxes; solid; HLB 17.9; 100% conc.

Catinex KB-48. [Pulcra SA] Octylphenol, ethoxylated; nonionic; solubilizer for anionic detergents; used in hot cleaning systems, metal pickling operations and electrolytic cleaning; solid; HLB 15.7; 100% conc.

Catinex KB-49. [Pulcra SA] Octylphenol, ethoxylated; nonionic; emulsifier, detergent, dispersant, surface act. agent used in washing formulations and emulsion cleaners; liq.; HLB 10.2; 100% conc.

Catinex KB-50. [Pulcra SA] Octylphenol, ethoxylated; nonionic; emulsifier, detergent, wetting agent; coemulsifier and stabilizer for vinyl-acrylic latex; emulsifier for fats and waxes; solid; HLB 16.2; 100% conc.

Catinex KB-51. [Pulcra SA] Octylphenol, ethoxylated; nonionic; emulsifier, detergent, dispersant, surface act. agent used in wax based washing formulations and emulsion cleaners; liq.; HLB 9.2; 100% conc.

Cation DS. [Sanyo Chem. Industries] Distearyl dimethyl ammonium chloride; cationic; fabric softener; solid; 90% act.

Cation SF-10. [Sanyo Chem. Industries] Alkyl imidazoline type quat. compd.; cationic; conc. softener; easily dilutable with water; paste; 75% act.

Catisol AO 100. [Stepan Europe] Oleylamine acetate; emulsifier, wetting agent, antistat, anticorrosive agent, lubricant for textile lubricants for min. and syn. fibers; orange to yel. wax; 92% act.

Catisol AO C. [Stepan Europe] Oleylamine acetate; cationic; emulsifier, wetting agent, antistat, corrosion inhibitor, lubricant for min. and syn. fibers; wax; 95% conc.

CD-DMAPA. [Clark] Coconut oil dimethylaminopropyl amine condensate; raw material for prod. of coconut amine oxide surfactants.

Cedemide AX. [Stepan Canada] Lauramide DEA; CAS 120-40-1; nonionic; foam stabilizer and visc. modifier in liq. detergents and shampoos; wax; 100% conc.

Cedepal CA-210. [Stepan Canada] Octoxynol-1 (1.5 EO); CAS 9002-93-1; nonionic; emulsifier for nonpolar hydrocarbon; liq.; HLB 3.5; 100% conc.

Cedepal CA-520. [Stepan Canada] Octoxynol-5; CAS 9002-93-1; nonionic; emulsifier, detergent, dispersant, surfactant; liq.; oil-sol.; HLB 10.0; 100% conc.

Cedepal CA-630. [Stepan Canada] Octoxynol-9; CAS 9002-93-1; nonionic; detergent, wetting, emulsifier, surfactant; liq.; water-sol.; HLB 13.0; 100% conc.

Cedepal CA-720. [Stepan Canada] Octoxynol-12; CAS 9002-93-1; nonionic; detergent; stable to acids and alkalies; rinseable; liq.; HLB 14.6; 100% conc.

Cedepal CA-890. [Stepan Canada] Octoxynol-40; CAS 9002-93-1; nonionic; primary emulsifier for vinyl acetate and acrylate polymerization; wax; HLB 18.0; 100% conc.

Cedepal CA-897. [Stepan Canada] Octoxynol-40; CAS 9002-93-1; nonionic; emulsifier, stabilizer; liq.; HLB 18.0; 70% conc.

Cedepal CO-210. [Stepan Canada] Nonoxynol-1 (1.5 EO); nonionic; defoaming, dispersant, detergent, emulsifier, stabilizer intermediate; surfactant and chemical intermediate; liq.; oil-sol.; HLB 4.6; 100% conc.

Cedepal CO-430. [Stepan Canada] Nonoxynol-4; CAS 9016-45-9; nonionic; detergent, emulsifier, dispersant, stabilizer intermediate; surfactant and chemical intermediate; liq.; HLB 8.8; 100% conc.

Cedepal CO-436. [Stepan Canada] Ammonium alkylaryl ether sulfate; anionic; detergent, wetting agent, lime soap dispersant; emulsifier in polymers; liq.; 60% conc.

Cedepal CO-500. [Stepan Canada] Nonylphenol ethoxylate; nonionic; emulsifier, detergent, dispersant, stabilizer, intermediate; surfactant; oil-sol.; liq.; HLB 10.0; 100% conc.

Cedepal CO-530. [Stepan Canada] Nonoxynol-6; CAS 9016-45-9; nonionic; emulsifier, detergent, dispersant; surfactant and intermediate; liq.; oil-sol.; HLB 10.8; 100% conc.

Cedepal CO-610. [Stepan Canada] Nonoxynol-8; CAS 9016-45-9; nonionic; detergent, wetting, emulsifier, surfactant for detergency; liq.; water-sol.; HLB 12.2; 100% conc.

Cedepal CO-630. [Stepan Canada] Nonoxynol-9; CAS 9016-45-9; nonionic; emulsifier, detergent, wetting agent, surfactant; liq.; water-sol.; HLB 13.0; 100% conc.

Cedepal CO-710. [Stepan Canada] Nonoxynol-10; CAS 9016-45-9; nonionic; detergent, wetting, emulsifier, surfactant; liq.; water-sol.; HLB 13.6; 100% conc.

Cedepal CO-730. [Stepan Canada] Nonylphenol ethoxylate; nonionic; emulsifier, detergent, wetting agent, surfactant; liq.; water-sol.; HLB 15.0; 100% conc.

Cedepal CO-880. [Stepan Canada] Nonylphenol ethoxylate; nonionic; detergent, wetting, emulsifier; wax; HLB 17.2; 100% conc.

Cedepal CO-887. [Stepan Canada] Nonylphenol ethoxylate; nonionic; detergent, wetting, emulsifier; liq.; HLB 17.2; 70% conc.

Cedepal CO-890. [Stepan Canada] Nonylphenol ethoxylate; nonionic; emulsifier, stabilizer; wax; HLB 17.8; 100% conc.

Cedepal CO-897. [Stepan Canada] Nonyl phenol ethoxylate; nonionic; emulsifier, stabilizer; liq.; HLB 17.8; 100% conc.

Cedepal CO-970. [Stepan Canada] Nonyl phenol ethoxylate; nonionic; emulsifier, stabilizer; wax; HLB 18.2; 100% conc.

Cedepal CO-977. [Stepan Canada] Nonylphenol ethoxylate; nonionic; emulsifier, stabilizer; liq.; HLB 18.2; 70% conc.

Cedepal CO-990. [Stepan Canada] Nonylphenol ethoxylate; nonionic; emulsifier, stabilizer, surfactant; wax; water-sol.; HLB 19.0; 100% conc.

Cedepal CO-997. [Stepan Canada] Nonylphenol ethoxylate; nonionic; emulsifier, stabilizer; liq.; HLB 19.0; 70% conc.

Cedepal FA-406. [Stepan; Stepan Canada] Ammonium ether sulfate; nonionic; foaming agent, emulsifier, detergent, wetting and foam stabilizing; for gypsum board prod.; clear liq.; HLB 12.8; 60% act.

Cedepal FS-406. [Stepan Canada] Sodium deceth sulfate; anionic; surfactant for high electrolyte applics.; liq.; 60% act.

Cedepal SD-409. [Stepan Canada] DEA laureth sulfate; used in applics. where water content is carefully controlled; liq.; 90% conc.

Cedepal SS-203, -306, -403, -406. [Stepan Canada] Sodium laureth sulfate; CAS 9004-82-4; anionic; detergent, dispersant, and wetting agent used in shampoos, dishwashers, car washers; liq.; 25, 60, 25, and 60% conc. resp.

Cedepal TD-400. [Stepan Canada] Primary alcohol ethoxylate; anionic; emulsifier; liq.; oil-sol.; HLB 8.0; 90% conc.

Cedepal TD-403 MF. [Stepan Canada] Sodium trideceth sulfate; surfactant.

Cedepal TD-630. [Stepan Canada] Primary alcohol ethoxylate; nonionic; emulsifier, detergent; liq.; HLB 12.5; 100% conc.

Cedepal TDS 484. [Stepan Canada] Sodium trideceth sulfate; surfactant.

Cedephos® CP610. [Stepan Canada] Nonoxynol phosphate; anionic; surfactant; 100% act.

Cedephos® FA600. [Stepan; Stepan Canada] Deceth-4 phosphate; anionic; detergent, emulsifier, wetting agent, hard surf. detergent for industrial cleaners, metal cleaners, janitorial prods., agric., textile wetting, emulsion polymerization, oil emulsification, lubricants, corrosion inhibitor, dedusting agent; coupling agent in highly alkaline industrial detergent systems; straw colored clear liq.; 100% act.

Cedephos® FE610. [Stepan Canada] Nonoxynol phosphate; anionic; surfactant; 100% act.

Cedephos® RA600. [Stepan; Stepan Canada] Alkyl ether phosphate, acid form; hydrotrope for nonionics in alkali cleaners; emulsifier for agric., emulsion polymerization, oils; lubricants, corrosion inhibitors, pigment dispersants; esp. for heavy-duty industrial/household alkaline cleaners, soak tank formulations; amber clear liq.; 100% act.

Cedepon A 100. [Stepan Canada] Linear alkyl benzene sulfonic acid; anionic; detergent; liq.; 96% conc.

Cedepon AM. [Stepan Canada] Linear alkyl benzene sulfonic acid, amine salt; anionic; emulsifier, detergent for solv. based systems; liq.; 93% conc.

Cedepon LT-40. [Stepan Canada] TEA lauryl sulfate; anionic; used in shampoos and detergents; liq.; 40% conc.

Cedepon SX-55. [Stepan Canada] Sodium linear alkyl benzene sulfonate; anionic; used where neutralization of acid is impractical; slurry; 55% conc.

Cedepon T. [Stepan Canada] TEA linear alkyl benzene sulfonate; anionic; detergent; liq.; 60% conc.

Cedepon TS. [Stepan Canada] TEA linear alkyl benzene sulfonate plus an alkanolamide; detergent; liq.; 60% conc.

Cenegen® 7. [Crompton & Knowles] Alkylaryl sulfonate; anionic; retarding and leveling agent for dyeing nylon with acid and neutral premetallized dyes; good barre coverage; straw liq.; misc. with water; pH 6.0 ± 0.5.

Cenegen® B. [Crompton & Knowles] Alkyl ether salts; anionic/cationic; nylon dyeing assistant for leveling, retarding, and barre coverage; emulsifier; yel.-brn. clear liq.; misc. with water; pH 8.0-8.5.

Cenegen® CJB. [Crompton & Knowles] Alkylethoxy condensate; nonionic/cationic; leveling, migrating, and retarding agent for dyeing nylon with acid dyes; yel. to lt. amber liq.; misc. with water; pH 5-6.

Cenegen® EKD. [Crompton & Knowles] Alkylethoxy condensate; cationic/nonionic; leveling, migrating, and repairing agent for acid dyes on nylon; yel. to lt. amber liq.; misc. with water; pH 10.0 ± 0.5.

Cenegen® NWA. [Crompton & Knowles] Alkylaryl polyethoxy condensate; amphoteric; dyeing assistant, leveling agent for premetallized dyes on nylon and wool; lt. brn. liq.; misc. with water; cloud pt. > 100 C; pH 6.0 ± 0.5.

Cenekol® 1141. [Crompton & Knowles] Sulfonated phenolic condensate; anionic; fixing and reserving agent for nylon in cellulosic blends and acid dye fixatives; for automotive nylon/rayon blends, nylon/wool blends; dk. brn. liq.; misc. with water; pH 10.5 ± 0.5.

Cenekol® FT Supra. [Crompton & Knowles] Aromatic condensate; anionic; surfactant for the textile industry; acid dye fixative on nylon; reserving agent in dyebaths; dk. brn. liq.; sol. in water; pH 8.0-8.8 (1%).

Cenekol® Liq., NCS Liq. [Crompton & Knowles] Sulfonated phenolic condensate; anionic; acid dye fixative for nylon; stable to electrolytes; dk. red to amber liq.; misc. with water; pH 6.0 ± 0.5.

Centrol® 2FSB, 2FUB, 3FSB, 3FUB. [Central Soya] Std. grade lecithin; CAS 8002-43-5; amphoteric; emulsifiers, dispersants; amber fluid; visc. 6000 cP; acid no. 27.

Centrol® CA. [Central Soya] Special grade lecithin; CAS 8002-43-5; amphoteric; o/w emulsifier; amber fluid; visc. 4000 cP; acid no. 20; HLB 6.0; 100% conc.

Centrolene® A, S. [Central Soya] Hydroxylated lecithin; amphoteric; o/w emulsifiers, increased hydrophilic props.; Gardner 11 and 12 resp. heavy-bodied fluid; acid no. 27; 100% conc.

Centromix® CPS. [Central Soya] Special grade lecithin; CAS 8002-43-5; amphoteric; amber fluid; visc. 8500 cP; acid no. 23.

Centromix® E. [Central Soya] Special grade lecithin; CAS 8002-43-5; amphoteric; amber fluid; visc. 6500 cP; acid no. 17.

Centrophase® 31. [Central Soya] Lecithin; CAS 8002-43-5; amphoteric; wetting agent for the magnetic media industry; fluid; sol. in THF, cyclohexanone, toluene; 62% act.

Centrophase® 152. [Central Soya] Special grade lecithin; CAS 8002-43-5; amphoteric; wetting agents; low visc., bland, sprayable; amber fluid; visc. 1200 cP; acid no. 22; 100% conc.

Centrophase® C. [Central Soya] Special grade lecithin; CAS 8002-43-5; amphoteric; wetting agents; amber fluid; visc. 1000 cP; acid no. 27.

Centrophase® HR2B, HR2U. [Central Soya] Special grade lecithin; CAS 8002-43-5; amphoteric; wetting agents; amber fluid; visc. 5000 cP; acid no. 22.

Centrophase® HR4B, HR4U. [Central Soya] Special grade lecithin; CAS 8002-43-5; amphoteric; wetting agents; heavy-bodied fluid; visc. 3300 cP; acid no. 24.

Centrophase® HR6B. [Central Soya] Special grade lecithin; CAS 8002-43-5; amphoteric; wetting agents; amber fluid; visc. 2500 cP; acid no. 23.

Centrophase® NV. [Central Soya] Special grade lecithin; CAS 8002-43-5; amphoteric; wetting agents; amber fluid; visc. 215 cP; acid no. 21.

Centrophil® K. [Central Soya] Special grade lecithin; CAS 8002-43-5; amphoteric; amber plastic; acid no. 20.

Centrophil® M, W. [Central Soya] Special grade lecithin; CAS 8002-43-5; amphoteric; amber fluid; visc. 150 cP; acid no. 12.

Ceralan®. [Amerchol] Lanolin alcohols; nonionic; emollient; w/o emulsifier; wax; 100% conc.

Cerex EL 250. [Auschem SpA] PEG-25 castor oil; CAS 61791-12-6; nonionic; solubilizer for essential oils, vitamins, other actives; liq.; HLB 10.8.

Cerex EL 300. [Auschem SpA] PEG-30 castor oil; CAS 61791-12-6; nonionic; solubilizer for essential oils, vitamins, other actives; liq.; HLB 12.5.

Cerex EL 360. [Auschem SpA] PEG-36 castor oil; CAS 61791-12-6; nonionic; solubilizer for essential oils, vitamins, other actives; liq.; HLB 12.9.

Cerex EL 400. [Auschem SpA] PEG-40 castor oil; CAS 61791-12-6; nonionic; solubilizer for essential oils, vitamins, other actives; liq.; HLB 14.4.

Cerex EL 4929. [Auschem SpA] PEG-9 ricnoleate; CAS 9004-97-1; nonionic; solubilizer for active ingreds.; liq.; HLB 11.4.

Cerfak 1400. [E.F. Houghton] Alkyl POE ether; nonionic; detergent, wetting agent used in wet processing operations of textiles; liq.; sol. in water and Stoddard solv.; 85% conc.

Cerfax N-100. [E.F. Houghton] Fatty acid polyethanolamine condensate; nonionic; detergent, wetting agent used in textile scouring and dyeing; stable to acids, alkalies, and hard water; liq.; 100% conc.

Cetalox 50. [Witco UK] Fatty alcohol, ethoxylated; nonionic; nonfoaming detergent; flakes.

Cetalox 8, 25. [Witco SA] Fatty alcohol, ethoxylated; nonionic; detergent; paste.

Cetalox AT. [Witco SA] Ethoxylated fatty alcohol; nonionic; nonfoaming detergent; powd.; HLB 17.1.

Cetrimide BP. [Aceto] Cetyl trimethyl ammonium bromide; antistat in hair conditioners; biocidal applics.; phase transfer catalysis; wh. free-flowing powd., faint char. odor, bitter soapy taste; sol. in alcohol, chloroform; m.w. 365; m.p. 100 C; toxicology: skin dermatitis; dust inhalation is irritating; 94-100% act.

Charlab Condensate-K. [Catawba-Charlab] Alkanolamide; nonionic; wetting agent, emulsifier, scouring agent; amber liq.; water sol.; sp.gr. 1.01; 100% act.

Charlab LPC. [Catawba-Charlab] Lauryl pyridinium chloride; cationic; surfactant; lt. br. solid; pyridine odor; water sol.; f.p. 45 C; 97–100% act.

Charlab Leveler AT Special. [Catawba-Charlab] Sodium hydrocarbon sulfonate; anionic; detergent, emulsifier, textile leveling agent, softener, lubricant; amber solid; sol. in water (160 F); sp.gr. 1.0; m.p. 110–125; 80% act.

Charlab Leveler DSL. [Catawba-Charlab] Sodium hydrocarbon sulfonate; anionic; detergent, emulsifier, textile leveling agent, softener, lubricant; lt. brn. paste; sol. in water (160 F); dens. 1.03 lb/gal; 40% act.

Chemal 2EH-2. [Chemax] PEG-2 2-ethylhexyl ether; detergent, wetting agent, emulsifier, dispersant, solubilizer, defoamer for textiles, metal cleaners, industrial and institutional cleaners, household cleaners, hand cleaners, specialties; liq.; cloud pt. < 25 C (1% aq.); HLB 8.0.

Chemal 2EH-5. [Chemax] PEG-5 2-ethylhexyl ether; detergent, wetting agent, emulsifier, dispersant, solubilizer, defoamer for textiles, metal cleaners, industrial and institutional cleaners, household cleaners, hand cleaners, specialties; liq.; cloud pt. < 25 C (1% aq.); HLB 12.6.

Chemal BP 261. [Chemax] Difunctional block polymers ending in primary hydroxyl groups; nonionic; defoamer, emulsifier, demulsifier, dispersant, binder, stabilizer, wetting agent, chemical intermediate; for metalworking, cosmetic, paper, textiles, dishwashing detergents, rinse aids, lubricant bases; liq.; cloud pt. 24 C (1% aq.); HLB 3.0; 100% act.

Chemal BP-262. [Chemax] Difunctional block polymer ending in primary hydroxyl groups; nonionic; see Chemal BP 261; liq.; cloud pt. 30 C (1% aq.); HLB 7.0; 100% act.

Chemal BP-262LF. [Chemax] Difunctional block polymer ending in primary hydroxyl groups; nonionic; see Chemal BP 261; liq.; cloud pt. 28 C (1% aq.); HLB 6.5; 100% act.

Chemal BP-2101. [Chemax] Difunctional block polymer ending in primary hydroxyl groups; nonionic; see Chemal BP 261; liq.; cloud pt. 16 C (1% aq.); HLB 1.0; 100% act.

Chemal DA-4. [Chemax] Deceth-4; CAS 26183-52-8; nonionic; detergent, wetting and penetrating agent for textile processing, clay soils, and fire fighting prods.; emulsifier for polyethylene emulsions; dispersant, solubilizer, defoamer; metal cleaners, industrial, institutional, household, and hand cleaners; liq.; hyd. no. 165–185; cloud pt. < 25 C (1% aq.); HLB 10.5; 100% conc.

Chemal DA-6. [Chemax] Deceth-6; nonionic; see Chemal DA-4; liq.; hyd. no. 131–145; cloud pt. 41 C (1% aq.); HLB 12.4; 100% conc.

Chemal DA-9. [Chemax] Deceth-9; nonionic; see Chemal DA-4; liq.; hyd. no. 95–110; cloud pt. 80 C (1% aq.); HLB 14.3; 100% conc.

Chemal LA-4. [Chemax] Laureth-4; nonionic; o/w emulsifier, lubricant, detergent, dispersant, solubilizer, defoamer for cosmetic, household, silicone polish, and mold release prods.; liq.; hyd. no. 150–165; cloud pt. < 25 C (1% aq.); HLB 9.2; 100% conc.

Chemal LA-9. [Chemax] Laureth-9; nonionic; see Chemal LA-4; liq.; hyd. no. 90–110; cloud pt. 76 C (1% aq.); HLB 13.3; 100% conc.

Chemal LA-12. [Chemax] Laureth-12; nonionic; see Chemal LA-4; liq.; hyd. no. 72–87; 100% conc.

Chemal LA-23. [Chemax] Laureth-23; nonionic; see Chemal LA-4; solid; hyd. no. 40–55; cloud pt. > 100 C (1% aq.); HLB 16.7; 100% conc.

Chemal LF 14B, 25B, 40B. [Chemax] Alkoxylated linear alcohol; nonionic; low foaming biodeg. surfactant, wetting agent, detergent, defoamer used as rinse aids in mechanical dishwashing, spray metal cleaning formulations and detergent prods.; liq.; cloud pt. 14, 25, 40 C resp. (1% aq.); 100% conc.

Chemal LFL-10, -17, -19, -28, -38, -47. [Chemax] Alkoxylated linear alcohol; nonionic; low foaming biodeg. surfactant, wetting agent, detergent, defoamer used in rinse aids and mechanical dishwashing detergents, spray metal cleaning formulations and detergent prods.; cloud pt. 10, 17, 19, 28, 38, and 47 C resp. (1% aq.); 100% conc.

Chemal OA-4. [Chemax] Oleth-4; CAS 9004-98-2; nonionic; dispersant, detergent; emulsifier and solubilizer for topical cosmetic applics.; stabilizer and anticoagulant for natural and syn. latices; emulsifier for waxes used in coating citrus fruit; liq.; cloud pt. < 25 C (1% aq.); HLB 7.9; 100% conc.

Chemal OA-5. [Chemax] Oleth-5; CAS 9004-98-2; nonionic; emulsifier, lubricant, and solubilizer; liq.; 100% conc.

Chemal OA-9. [Chemax] Oleth-9; CAS 9004-98-2; nonionic; see Chemal OA-4; liq.; cloud pt. 52 C (1% aq.); HLB 11.9; 100% conc.

Chemal OA-20/70CWS. [Chemax] Oleth-20; CAS 9004-98-2; nonionic; emulsifier, lubricant, solubilizer; liq.; cloud pt. > 100 C (1% aq.); HLB 15.3; 70% conc.

Chemal OA-20G. [Chemax] Oleth-20; CAS 9004-98-2; nonionic; dispersant, detergent; emulsifier and solubilizer for topical cosmetic applics.; stabilizer and anticoagulant for natural and syn. latices; emulsifier for waxes used in coating citrus fruit; lubricant; semisolid; cloud pt. > 100 C (1% aq.); HLB 15.3; 100% conc.

Chemal TDA-3. [Chemax] Trideceth-3; CAS 24938-91-8; nonionic; wetting agent, detergent, emulsifier, dispersant, foam stabilizer; solubilizer, penetrant for scouring and dye leveling in textiles, in cleaning and dishwashing compds.; liq.; cloud pt. < 25 C (1% aq.); HLB 7.9; 100% conc.

Chemal TDA-6. [Chemax] Trideceth-6; CAS 24938-91-8; nonionic; see Chemal TDA-3; liq.; cloud pt. < 25 C (1% aq.); HLB 11.4; 100% conc.

Chemal TDA-9. [Chemax] Trideceth-9; CAS 24938-91-8; nonionic; see Chemal TDA-3; liq.; cloud pt. 54 C (1% aq.); HLB 13.0; 100% conc.

Chemal TDA-12. [Chemax] Trideceth-12; CAS 24938-91-8; nonionic; see Chemal TDA-3; paste; cloud pt. 70 C (1% aq.); HLB 14.5; 100% conc.

Chemal TDA-15. [Chemax] Trideceth-15; CAS 24938-91-8; nonionic; see Chemal TDA-3; solid; 100% conc.

Chemal TDA-18. [Chemax] Trideceth-18; CAS 24938-91-8; nonionic; see Chemal TDA-3; 100% conc.

Chemax AR-497. [Chemax] PEG-15 rosin acid; nonionic; emulsifier, detergent for acid cleaners, esp. for aluminum; paste; cloud pt. 75 C (1% aq.); HLB 13.0.

Chemax CO-5. [Chemax] PEG-5 castor oil; nonionic; emulsifier, lubricant for textiles; pigment dispersant in latex paints, paper; essential oils solubilizer; liq.; oil-sol.; sapon. no. 145; HLB 3.8; 100% conc.

Chemax CO-16. [Chemax] PEG-16 castor oil; nonionic; emulsifier for fiber lubricants; cutting oils and hydraulic fluids; clay and pigment dispersant, rewetting agent, softener, dyeing assistant for paint, paper, textile, and leather industries; liq.; water-sol.; sapon. no. 100; HLB 8.6; 100% conc.

Chemax CO-25. [Chemax] PEG-25 castor oil; nonionic; emulsifier for industrial lubricants; pigment dispersant in textiles, paint, paper, leather; liq.; sapon. no. 80; HLB 10.8.

Chemax CO-28. [Chemax] PEG-28 castor oil; nonionic; see Chemax CO-25; HLB 11.1.

Chemax CO-30. [Chemax] PEG-30 castor oil; nonionic; see Chemax CO-25; liq.; sapon. no. 73; HLB 11.7; 100% conc.

Chemax CO-36. [Chemax] PEG-36 castor oil; nonionic; see Chemax CO-25; liq.; sapon. no. 68; HLB 12.6.

Chemax CO-40. [Chemax] PEG-40 castor oil; nonionic; see Chemax CO-25; liq.; sapon. no. 61; HLB 12.9; 100% conc.

Chemax CO-80. [Chemax] PEG-80 castor oil; nonionic; see Chemax CO-25; solid; sapon. no. 34; HLB 15.8; 100% conc.

Chemax CO-200/50. [Chemax] PEG-200 castor oil; nonionic; see Chemax CO-25; liq.; sapon. no. 16; HLB 18.1; 50% conc.

Chemax DF-10, DF-10A. [Chemax] Silicone defoamer; defoamer/antifoam for pulp and paper, textiles, paints, effluent treatment, commercial cleaning processes, adhesives, metalworking; emulsion; 10% silicone additives.

Chemax DF-30. [Chemax] Silicone defoamer; defoamer/antifoam for pulp and paper, textiles, paints, effluent treatment, commercial cleaning processes, adhesives, metalworking; emulsion; 30% silicone additives.

Chemax DF-100. [Chemax] Silicone defoamer; defoamer/antifoam for pulp and paper, textiles, paints, effluent treatment, commercial cleaning processes, adhesives, metalworking; liq.; 100% silicone additives.

Chemax DNP-8. [Chemax] Nonyl nonoxynol-8; CAS 9014-93-1; nonionic; emulsifier for nonpolar solv. and oils; detergent for cellulosic and syn. fibers; dispersant for hard surface cleaners and laundry compds.; solubilizer; liq.; pour pt. 32 F; cloud pt. < 25 C (1% aq.); HLB 10.4; 100% conc.

Chemax DNP-15. [Chemax] Nonyl nonoxynol-15; CAS 9014-93-1; nonionic; see Chemax DNP-8; liq.; pour pt. 64 F.

Chemax DNP-18. [Chemax] Nonyl nonoxynol-18; CAS 9014-93-1; nonionic; emulsifier, detergent, solubilizer; solid; 100% conc.

Chemax DNP-150. [Chemax] Nonyl nonoxynol-150; CAS 9014-93-1; nonionic; see Chemax DNP-8; solid; cloud pt. > 100 C (1% aq.); HLB 19.0.

Chemax DNP-150/50. [Chemax] Nonyl nonoxynol-150; CAS 9014-93-1; nonionic; emulsifier, dispersant, detergent, wetting agent, solubilizer, coupler for textile, metalworking, household, industrial, agric., paper, paint, and other industries; liq.; cloud pt. > 100 C (1% aq.); HLB 19.0; 50% conc.

Chemax DOSS/70. [Chemax] Sodium dioctyl sulfosuccinate; wetting agent for textile, agric., detergent formulations, emulsion polymerization; pigment dispersant in paints and inks; solubilizer for drycleaning solvs.; liq.; 70% act.

Chemax DOSS-75E. [Chemax] Dioctyl sodium sulfosuccinate; anionic; wetting and rewetting agent, detergent, emulsifier for emulsion polymerization, textile, agric., detergent formulations; pigment dispersant in paints and inks; solubilizer for drycleaning solvs.; liq.; 75% conc.

Chemax E-200 ML. [Chemax] PEG-4 laurate; CAS 9004-81-3; nonionic; emulsifier for min. and cutting oils; dispersant, detergent, lubricant; coemulsifier and defoamer in water-based coatings; cosmetics ingred.; liq.; sapon. no. 135; HLB 9.3; 100% conc.

Chemax E 200 MO. [Chemax] PEG-5 oleate; CAS 9004-96-0; nonionic; emulsifier for min. and fatty oils, solv.; degreaser, dispersant, detergent, lubricant; metal, textile, cosmetic, plastisol formulations; liq.; oil-sol.; sapon. no. 118; HLB 8.3; 100% conc.

Chemax E-200 MS. [Chemax] PEG-5 stearate; CAS 9004-99-3; nonionic; emulsifier for min. oils and fats used in polishes and metal buffing compds.; dye assistant, lubricant, softener, antistat; for metal lubricants, textiles, cosmetic, plastisol formulations; soft solid; sapon. no. 112; HLB 8.5; 100% conc.

Chemax E-400 ML. [Chemax] PEG-8 laurate; CAS 9004-81-3; nonionic; emulsifier, dispersant, detergent, lubricant; visc. control agent in plastisol formulations; wetting agent and defoamer in latex paint; cosmetics ingred.; liq.; sapon. no. 90; HLB 13.2; 100% conc.

Chemax E-400 MO. [Chemax] PEG-9 oleate; CAS 9004-96-0; nonionic; emulsifier and lubricant for solv. and oils in pesticides and metal cleaners; detergent, dispersant; textile, cosmetics, plastisol formulations; liq.; sapon. no. 85; HLB 11.8; 100% conc.

Chemax E-400 MS. [Chemax] PEG-9 stearate; CAS 9004-99-3; nonionic; lubricant and softener for syn. fibers; dye assistant, antistat, emulsifier; for metal lubricants, textiles, cosmetics, plastisols; soft solid;

sapon. no. 87; HLB 12.0; 100% conc.

Chemax E-600 ML. [Chemax] PEG-14 laurate; CAS 9004-81-3; nonionic; emulsifier, dispersant, detergent, lubricant; metal, textile, cosmetics, plastisol formulations; liq.; water-sol.; sapon. no. 68; HLB 14.8; 100% conc.

Chemax E-600 MO. [Chemax] PEG-14 oleate; CAS 9004-96-0; nonionic; surfactant used as coemulsifier and lubricant in industrial formulations, cosmetics, metal lubricants, textiles, plastisols; dispersant, detergent; liq.; sapon. no. 65; HLB 13.6; 100% conc.

Chemax E-600 MS. [Chemax] PEG-14 stearate; CAS 9004-99-3; nonionic; dye assistant, lubricant, softener, antistat, emulsifier for cosmetic and textile formulations; paste; sapon. no. 62; HLB 13.8; 100% conc.

Chemax E-1000 MO. [Chemax] PEG-20 oleate; CAS 9004-96-0; nonionic; emulsifier for min. and fatty oils, solv.; degreaser, dispersant, detergent, lubricant; solid; 100% conc.

Chemax E-1000 MS. [Chemax] PEG-20 stearate; CAS 9004-99-3; nonionic; emulsifier for cosmetic and textile formulations; dye assistant, lubricant, softener, antistat; soft solid; water-sol.; sapon. no. 43; HLB 15.7; 100% conc.

Chemax HCO-5. [Chemax] PEG-5 hydrog. castor oil; CAS 61788-85-0; nonionic; emulsifier, lubricant, softener, dispersant; coemulsifier for syn. esters; for plastics, metals, textiles, leather, paint, and paper indutries; liq.; oil-sol.; sapon. no. 142; HLB 3.8; 100% conc.

Chemax HCO-16. [Chemax] PEG-16 hydrog. castor oil; CAS 61788-85-0; nonionic; emulsifier, lubricant, dispersant, softener; for textiles, plastics, metalworking, paint, paper, leather; liq.; sapon. no. 100; HLB 8.6; 100% conc.

Chemax HCO-25. [Chemax] PEG-25 hydrog. castor oil; CAS 61788-85-0; nonionic; emulsifier, lubricant, and softener for plastics, metals, textiles, leather, paint, and paper industries; liq.; sapon. no. 80; HLB 10.8; 100% conc.

Chemax HCO-200/50. [Chemax] PEG-200 hydrog. castor oil; CAS 61788-85-0; nonionic; emulsifier and lubricant for plastics, metals, textiles, paint, paper, leather industries; liq.; water-sol.; sapon. no. 17; HLB 18.1; 50% act.

Chemax NP-1.5. [Chemax] Nonoxynol-1.5; CAS 9016-45-9; nonionic; emulsifier, dispersant, detergent, wetting agent, solubilizer, coupler for textile, metalworking, household, industrial, agric., paper, paint, and other industries; liq.; pour pt. –3 F; cloud pt. < 25 C (1% aq.); HLB 4.6.

Chemax NP-4. [Chemax] Nonoxynol-4; CAS 9016-45-9; nonionic; emulsifier, dispersant, detergent, wetting agent, solubilizer, coupler for textile, metalworking, household, industrial, agric., paper, paint, and other industries; liq.; water oil sol.; pour pt. –15 F; cloud pt. < 25 C (1% aq.); HLB 8.9; 100% conc.

Chemax NP-6. [Chemax] Nonoxynol-6; CAS 9016-45-9; nonionic; emulsifier, dispersant, detergent, wetting agent, solubilizer, coupler for textile, metalworking, household, industrial, agric., paper, paint, and other industries; liq.; pour pt. –26 F; cloud pt. < 25 C (1% aq.); HLB 10.9; 100% conc.

Chemax NP-9. [Chemax] Nonoxynol-9; CAS 9016-45-9; nonionic; emulsifier, dispersant, detergent, wetting agent, solubilizer, coupler for textile, metalworking, household, industrial, agric., paper, paint, and other industries; liq.; pour pt. 31 F; cloud pt. 54 C (1% aq.); HLB 13.0; 100% conc.

Chemax NP-10. [Chemax] Nonoxynol-10; CAS 9016-45-9; nonionic; emulsifier, dispersant, detergent, wetting agent, solubilizer, coupler for textile, metalworking, household, industrial, agric., paper, paint, and other industries; liq.; pour pt. 49 F; cloud pt. 72 C (1% aq.); HLB 13.5; 100% conc.

Chemax NP-15. [Chemax] Nonoxynol-15; CAS 9016-45-9; nonionic; surfactant, detergent, wetting and rewetting agent, emulsifier in textile, leather, paper, paint, and metal processing; paste; pour pt. 71 F; cloud pt. 96 C (1% aq.); HLB 15.0; 100% conc.

Chemax NP-20. [Chemax] Nonoxynol-20; CAS 9016-45-9; nonionic; see Chemax NP-15; solid; pour pt. 91 F; cloud pt. > 100 C (1% aq.); HLB 16.0.

Chemax NP-30. [Chemax] Nonoxynol-30; CAS 9016-45-9; nonionic; surfactant, detergent, wetting and rewetting agent, emulsifier in textile, leather, paper, paint, and metal processing; solid; pour pt. 109 F; cloud pt. > 100 C (1% aq.); HLB 17.1; 100% conc.

Chemax NP-30/70. [Chemax] Nonoxynol-30; CAS 9016-45-9; nonionic; see Chemax NP-15; liq.; pour pt. 34 F; cloud pt. > 100 C (1% aq.); HLB 17.1.

Chemax NP-40. [Chemax] Nonoxynol-40; CAS 9016-45-9; nonionic; polymerization emulsifier for vinyl acetate and acrylic emulsions; stabilizer for syn. latices; wetting agent in electrolyte sol'ns.; solid; pour pt. 112 F; cloud pt. > 100 C (1% aq.); HLB 17.8; 100% conc.

Chemax NP-40/70. [Chemax] Nonoxynol-40; CAS 9016-45-9; nonionic; see Chemax NP-40; liq.; pour pt. 46 F; cloud pt. > 100 C (1% aq.); HLB 17.8.

Chemax NP-50. [Chemax] Nonoxynol-50; CAS 9016-45-9; nonionic; see Chemax NP-40; solid; pour pt. 114 F; cloud pt. > 100 C (1% aq.); HLB 18.2.

Chemax NP-50/70. [Chemax] Nonoxynol-50; CAS 9016-45-9; nonionic; see Chemax NP-40; liq.; pour pt. 52 F; cloud pt. > 100 C (1% aq.); HLB 18.2.

Chemax NP-100. [Chemax] Nonoxynol-100; CAS 9016-45-9; nonionic; see Chemax NP-40; solid; pour pt. 122 F; cloud pt. > 100 C (1% aq.); HLB 19.0.

Chemax NP-100/70. [Chemax] Nonoxynol-100; CAS 9016-45-9; nonionic; see Chemax NP-40; liq.; pour pt. 68 F; cloud pt. > 100 C (1% aq.); HLB 19.0.

Chemax OP-3. [Chemax] Octoxynol-3; CAS 9002-93-1; nonionic; emulsifier, detergent, stabilizer, dispersant, wetting agent; pesticides and floor finishes; liq.; pour pt. –9 F.

Chemax OP-5. [Chemax] Octoxynol-5; CAS 9002-93-1; nonionic; see Chemax OP-3; liq.; oil sol.; pour pt. –15 F; 100% conc.

Chemax OP-7. [Chemax] Octoxynol-7; CAS 9002-93-1; nonionic; detergent comps.; industrial metal cleaning, acid and waterless hand cleaners, and floor finishes; liq.; pour pt. 15 F.

Chemax OP-9. [Chemax] Octoxynol-9; CAS 9002-93-1; nonionic; see Chemax OP-7; liq.; pour pt. 45 F; 100% conc.

Chemax OP-30/70. [Chemax] Octoxynol-30; CAS 9002-93-1; nonionic; emulsifier, dispersant, detergent, wetting agent, solubilizer, coupler for textile, metalworking, household, industrial, agric., paper, paint, and other industries; liq.; cloud pt. > 100 C (1% aq.); HLB 17.3; 70% conc.

Chemax OP-40. [Chemax] Octoxynol-40; CAS 9002-93-1; nonionic; emulsifier, detergent, stabilizer, wetting agent, dispersant; solid; 100% conc.

Chemax OP-40/70. [Chemax] Octoxynol-40; CAS 9002-93-1; nonionic; emulsifier for vinyl acetate and acrylate polymerization; liq.; pour pt. 25 F;

cloud pt. > 100 C (1% aq.); HLB 17.9; 70% conc.

Chemax PEG 200 DO. [Chemax] PEG-4 dioleate; CAS 9005-07-6; surfactant; coemulsifier for oils as mold release agent; liq.; oil-sol.; HLB 5.4.

Chemax PEG 400 DO. [Chemax] PEG-8 dioleate; CAS 9005-07-6; emulsifier and solubilizer for solv., fats, and min. oils; in lubricant, softener, and defoamer formulations for agric., cosmetic, household, leather, metalworking, and textile industries; liq.; oil-sol.; sapon. no. 116; HLB 8.8.

Chemax PEG 600 DO. [Chemax] PEG-12 dioleate; CAS 9005-07-6; emulsifier in lubricant, softener, and defoamer formulations for agric., cosmetic, household, leather, metalworking, and textile industries; semiliq.; water-sol.; sapon. no. 102; HLB 10.3.

Chemax SBO. [Chemax] Sulfated butyl oleate; softener, emulsifier, wetting agent in textile and metal working industries; liq.

Chemax SCO. [Chemax] Sulfated castor oil; softener, emulsifier, wetting agent in textile and metal working industries; liq.

Chemax TO-8. [Chemax] PEG-8 tallate; CAS 61791-00-2; nonionic; emulsifier, lubricant, dye assistant; liq.; cloud pt. < 25 C (1% aq.); HLB 10.5; 100% conc.

Chemax TO-10. [Chemax] PEG-10 tallate; CAS 61791-00-2; nonionic; emulsifier, detergent in industrial lubricants and degreasers; liq.; cloud pt. 25 C (1% aq.); HLB 11.5.

Chemax TO-16. [Chemax] PEG-16 tallate; CAS 61791-00-2; nonionic; foam detergent and emulsifier; lubricant, dye assistant; liq.; cloud pt. 70 C (1% aq.); HLB 13.4; 100% conc.

Chemazine 18, C, O, TO. [Chemax] Imidazolines; cationic; softener, antistat, emulsifier, corrosion inhibitor, filming agents; solid (18, C), liq. (O, TO).

Chembetaine BC-50. [Chemron] Surfactant; amphoteric; wetting and foaming agent for sat. brines incl. $CaCl_2$, HCl, NaCl, high-temp. acidizing; drilling fluid additive; removes oil-based muds prior to cementing or stimulation; stable to 400 F; stable to acids; water-wh. liq.; 50% act.

Chembetaine BW. [Chemron] Coco betaine; anionic/cationic; visc. builder, gelling agent, industrial surfactant; lime soap dispersant; mild to skin; tolerant to hard water; stable in high-electrolyte sol'ns.; cationic in acid and anionic in alkaline media; water-wh. liq.; 45% conc.

Chembetaine C. [Chemron] Cocamidopropyl betaine; amphoteric; high foaming mild industrial and personal care surfactant; foam and visc. builder; lime soap dispersant; foaming agent for water and acid systems; stable over wide pH range; pale yel. liq.; 35% conc.

Chembetaine CAS. [Chemron] Cocamidopropylhydroxysultaine; amphoteric; anti-irritant for other surfactants; esp. for baby shampoos and baby bath prods.; detergent for heavy-duty industrial alkaline cleaners (steam cleaners, wax remover, hard surf. cleaner); wetting agent in acid pickling of metals; lime soap dispersant; visc. builder; pale yel. liq.; sol. in soft and hard water, brine and conc. electrolyte sol'ns.; 50% conc.

Chemeen 18-2. [Chemax] PEG-2 stearamine; mild cationic; emulsifier and antistat in textiles, metal buffing, and rubber compds.; lubricant for fiber glass; solid; m.w. 365; HLB 4.8.

Chemeen 18-5. [Chemax] PEG-5 stearamine; see Chemeen 18-2; solid; m.w. 540.

Chemeen 18-50. [Chemax] PEG-50 stearamine; CAS 26635-92-7; mild cationic; see Chemeen 18-2; solid; m.w. 2400; HLB 17.8.

Chemeen C-2. [Chemax] PEG-2 cocamine; CAS 61791-14-8; mild cationic; emulsifier, antistat, dye leveler, wetting agent, lubricant, dispersant; substantive to metals, fibers, and clays; liq.; m.w. 290; HLB 6.1; 100% conc.

Chemeen C-5. [Chemax] PEG-5 cocamine; CAS 61791-14-8; mild cationic; see Chemeen C-2; liq.; m.w. 425; HLB 10.4; 100% conc.

Chemeen C-10. [Chemax] PEG-10 cocamine; CAS 61791-14-8; see Chemeen C-2; liq.; m.w. 645; 100% conc.

Chemeen C-15. [Chemax] PEG-15 cocamine; CAS 61791-14-8; mild cationic; see Chemeen C-2; liq.; m.w. 890; HLB 15.0; 100% conc.

Chemeen DT-3. [Chemax] PEG-3 tallow diamine; mild cationic; emulsifier, textile dyeing assistant, corrosion inhibitor used in preparation of asphalt and agric. chemical emulsions; liq.; m.w. 535; HLB 4.9.

Chemeen DT-15. [Chemax] PEG-15 tallow aminopropylamine; CAS 61790-85-0; mild cationic; see Chemeen DT-3; liq.; m.w. 1020; HLB 13.0.

Chemeen DT-30. [Chemax] PEG-30 tallow diamine; see Chemeen DT-3; liq.; m.w. 1665; HLB 15.9.

Chemeen HT-2. [Chemax] PEG-2 hydrog. tallow amine; emulsifier, antistat, lubricant; substantive to metals, fiber and clays; solid; 100% conc.

Chemeen HT-5. [Chemax] PEG-5 hydrog. tallow amine; mild cationic; emulsifier and antistat in textiles, metal buffing and rubber compds., lubricant for fiber glass; paste; m.w. 495; HLB 9.0.

Chemeen HT-15. [Chemax] PEG-15 hydrog. tallow amine; mild cationic; emulsifier, antistat, lubricant; liq.; m.w. 925; HLB 14.3; 100% conc.

Chemeen HT-50. [Chemax] PEG-50 hydrog. tallow amine; see Chemeen HT-5; solid; m.w. 2470; 100% conc.

Chemeen O-30, O-30/80. [Chemax] PEG-30 oleamine; mild cationic; emulsifier, antistat, lubricant, and textile dyeing assistant; antiprecipitant in cross dyeing; liq.; m.w. 1600; HLB 16.5; 100% and 80% conc.

Chemeen S-2. [Chemax] PEG-2 soya amine; CAS 61791-24-0; emulsifier, antistat, lubricant; liq.; 100% conc.

Chemeen S-5. [Chemax] PEG-5 soya amine; CAS 61791-24-0; emulsifier, antistat, lubricant; liq.; 100% conc.

Chemeen S-30. [Chemax] PEG-30 soya amine; CAS 61791-24-0; see Chemeen S-2; solid; 100% conc.

Chemeen S-30/80. [Chemax] PEG-30 soya amine; CAS 61791-24-0; see Chemeen S-2; liq.; 100% conc.

Chemeen T-2. [Chemax] PEG-2 tallow amine; mild cationic; antistat for carpet shampoos; emulsifier, lubricant, dispersant, softener, antiprecipitant, leveling and migrating agent in textile dyeing process; paste; m.w. 350; HLB 5.0; 100% conc.

Chemeen T-5. [Chemax] PEG-5 tallow amine; mild cationic; see Chemeen T-2; liq.; m.w. 490; HLB 9.0; 100% conc.

Chemeen T-10. [Chemax] PEG-10 tallow amine; see Chemeen T-2; liq.; m.w. 700; 100% conc.

Chemeen T-15. [Chemax] PEG-15 tallow amine; mild cationic; see Chemeen T-2; liq.; m.w. 930; HLB 14.3; 100% conc.

Chemeen T-20. [Chemax] PEG-20 tallow amine; mild cationic; see Chemeen T-2; liq.; m.w. 1120; HLB 15.7; 100% conc.

Chemfac 100. [Chemron] Proprietary; nonionic; defoamer for vacuum filtration fluids; esp. for processing yel. cake and waste concs.; defoamer for guar, HPG, CMHPG; surfactant for polymerization; lt. yel. liq.; 100% act.

Chemfac NC-0910. [Chemax] POE alkyl phenol phosphate; wetting agent, detergent, hydrotrope, emulsifier, rust inhibitor, EP additive for alkaline detergents, metal cleaners, hard surf. cleaners, textile scours, metal and textile lubricants, drycleaning soaps, emulsion polymerization, agric. formulations; liq.; acid no. 145.

Chemfac PA-080. [Chemax] Alcohol phosphate ester; anionic; detergent, emulsifier, wetting agent, lubricant, antistat for alkaline detergents, metal cleaners, hard surf. cleaners, textile scours, metal and textile lubricants, drycleaning soaps, emulsion polymerization, agric. formulations; liq.; acid no. 345; 100% conc.

Chemfac PB-082. [Chemax] Phosphate esters; anionic; detergent, wetting, and coupling agent, antistat, emulsifier for alkaline detergents, metal cleaners, hard surf. cleaners, textile scours, metal and textile lubricants, drycleaning soaps, emulsion polymerization, agric. formulations; liq.; acid no. 195; 99% min. act.

Chemfac PB-106. [Chemax] POE alkyl ether phosphate; anionic; wetting agent, detergent, hydrotrope, emulsifier, rust inhibitor, EP additive for alkaline detergents, metal cleaners, hard surf. cleaners, textile scours, metal and textile lubricants, drycleaning soaps, emulsion polymerization, agric. formulations; liq.; acid no. 120.

Chemfac PB-135. [Chemax] POE alkyl ether phosphate; anionic; wetting agent, detergent, hydrotrope, emulsifier, rust inhibitor, EP additive for alkaline detergents, metal cleaners, hard surf. cleaners, textile scours, metal and textile lubricants, drycleaning soaps, emulsion polymerization, agric. formulations; liq.; acid no. 115.

Chemfac PB-184. [Chemax] Oleth-4 phosphate; anionic; wetting agent, detergent, hydrotrope, emulsifier, rust inhibitor, EP additive for alkaline detergents, metal cleaners, hard surf. cleaners, textile scours, metal and textile lubricants, drycleaning soaps, emulsion polymerization, agric. formulations; liq.; acid no. 140; 100% conc.

Chemfac PB-264. [Chemax] Phosphate ester; anionic; wetting agent, detergent, hydrotrope, emulsifier, rust inhibitor, EP additive for alkaline detergents, metal cleaners, hard surf. cleaners, textile scours, metal and textile lubricants, drycleaning soaps, emulsion polymerization, agric. formulations; liq.; acid no. 165; 100% conc.

Chemfac PC-006. [Chemax] Phosphate ester; CAS 39464-70-5; anionic; wetting agent, lubricant, antistat, detergent, emulsifier; foaming hydrotrope; liq.; acid no. 180; 99% min. act.

Chemfac PC-099E. [Chemax] POE alkyl phenol phosphate; anionic; detergent, wetting agent, primary emulsifier in emulsion polymerization; liq.; acid no. 115; 99% min. act.

Chemfac PC-188. [Chemax] Phosphate ester; anionic; wetting agent, detergent, hydrotrope, emulsifier, rust inhibitor, EP additive for alkaline detergents, metal cleaners, hard surf. cleaners, textile scours, metal and textile lubricants, drycleaning soaps, emulsion polymerization, agric. formulations; liq.; acid no. 85; 100% conc.

Chemfac PD-600. [Chemax] POE alkyl ether phosphate; CAS 52019-36-0; anionic; wetting agent, detergent, hydrotrope, emulsifier, rust inhibitor, EP additive for alkaline detergents, metal cleaners, hard surf. cleaners, textile scours, metal and textile lubricants, drycleaning soaps, emulsion polymerization, agric. formulations; liq.; acid no. 210.

Chemfac PD-990. [Chemax] Phosphate ester; anionic; wetting agent, detergent, emulsifier, lubricant; 90% conc.

Chemfac PF-623. [Chemax] POE alkyl ether phosphate; anionic; wetting agent, detergent, hydrotrope, emulsifier, rust inhibitor, EP additive for alkaline detergents, metal cleaners, hard surf. cleaners, textile scours, metal and textile lubricants, drycleaning soaps, emulsion polymerization, agric. formulations; liq.; acid no. 170.

Chemfac PF-636. [Chemax] POE phosphate; anionic; wetting agent, detergent, hydrotrope, emulsifier, rust inhibitor, EP additive for alkaline detergents, metal cleaners, hard surf. cleaners, textile scours, metal and textile lubricants, drycleaning soaps, emulsion polymerization, agric. formulations; liq.; acid no. 400; 90% conc.

Chemfac PN-322. [Chemax] Phosphate ester, neutralized; anionic; hydrotrope for solubilizing nonionic surfactants in high concs. of alkali or other electrolytes; wetting agent, detergent, emulsifier, rust inhibitor, EP additive for alkaline detergents, textile scours, emulsion polymerization, lubricants, agric.; liq.; 50% conc.

Chemfac PX-322. [Chemax] Phosphate ester; anionic; hydrotrope for solubilizing surfactants in alkali or other electrolytes; liq.; 50% conc.

Chemfac RD-1200. [Chemron] Proprietary; nonionic; environmentally safe industrial defoamer for offshore drilling; lt. yel. liq.; 100% act.

Chemoxide CAW. [Chemron] Cocamidopropylamine oxide; nonionic; surfactant for mild, low irritation personal care and industrial applics., e.g., shampoos, facial cleansers, bath prods.; foam and visc. builder, emollient over broad pH range; pale yel. liq.; 30% conc.

Chemoxide L. [Chemron] Lauramidopropyl betaine; amphoteric; foam booster, visc. builder; mild surfactant; liq.; 35% conc.

Chemoxide LM-30. [Chemron] Lauramine oxide; nonionic; visc. builder, foam enhancer for household and industrial cleaners, personal care prods.; tolerant to electrolytes for improved hard water performance; water-wh. liq.; 30% conc.

Chemoxide OI. [Chemron] Oleyl dimethylamine oxide; nonionic; foamer, thickener, industrial surfactant; tolerant to electrolytes; water-wh. liq.; 55% conc.

Chemphonate 22. [Chemron] Polyalkylamino-methylene phosphonic acid; scale inhibitor for oilfield prod.; dark amber liq.; pH < 2; 48% act.

Chemphonate AMP. [Chemron] Amino tris (methylene phosphonic acid); scale inhibitor for water inj., water disposal and prod. systems; pale yel. liq.; pH < 2; 50% act.

Chemphonate AMP-S. [Chemron] Sodium amino tris(methylene phosphonate); scale inhibitor for water systems; dispersant for drilling muds; yel. liq.; pH 10-11; 24% act.

Chemphonate HEDP. [Chemron] 1-Hydroxyethyl-1-diphosphonic acid; temp.-stable scale inhibitor for oilfield applic.; sequestering agent for calcium carbonate; pale yel. liq.; pH < 2; 60% act.

Chemphonate N. [Chemron] TEA phosphoric acid ester; effective scale inhibitor for control of calcium carbonate, barium sulfate, and calcium sulfate; for oilfield prod.; pale yel. liq.; pH 1.7; 85% act.

Chemphonate NP. [Chemron] Sodium TEA phosphoric acid ester; scale inhibitor for oilfield prod.; pale yel. liq.; pH 5; 60% conc.

Chemphos TC-227. [Chemron] Aromatic phosphate ester; detergent, wetting agent, emulsifier, coupling agent, surf. tension reducer; for alkaline cleaners, heavy-duty all-purpose metalworking detergents, steam cleaning, dairy cleaners, bottle washing compds., floor strippers; lubricant and detergent for drilling fluids; oil treating chemicals; emulsion polymerization of vinyl acetate, acrylates, SBR; clear visc. liq.; sol. in water, most oxygenated solvs., aromatic solvs., chlorinated solvs.; 100% act.

Chemphos TC-227S. [Chemron] Aromatic phosphate ester, sodium salt; detergent, emulsifier, dispersant, solubilizer for use with nonionic surfactants and conc. electrolyte sol'ns.; clear visc. liq.; 85% act.

Chemphos TC-310. [Chemron] Aromatic phosphate ester; anionic; detergent, wetting agent, emulsifier, dispersant, surf. tension reducer, rust inhibitor; emulsion polymerization surfactant for vinyl acetate, acrylates, SBR; clear visc. liq.; sol. in water, alcohols, hydrocarbon solvs., chlorinated solvs.; 100% act.

Chemphos TC-310D. [Chemron] Aromatic phosphate ester, amine salt; industrial surfactant for incorporation into water-rinsable, solv.-based cleaners; clear visc. liq.; dissolves readily in most org. solvs.; 85% act.

Chemphos TC-310S. [Chemron] Aromatic phosphate ester, sodium salt; detergent, emulsifier, wetting agent for heavy-duty, all-purpose, hard-surf., and metal cleaners; emulsifier for min. oil, crude oil, pesticides; paraffin dispersant for oil treatment; prolongs life of wax and resin floor finishes; clear visc. liq.; sol. in water, many polar and nonpolar solvs.; 85% act.

Chemphos TC-337. [Chemron] Nonoxynol-20 phosphate; anionic; emulsion polymerization surfactant for vinyl acetate, acrylates, SBR; emulsifier, solubilizer, antistat, substantivity agent for hair care prods., perms, straighteners, depilatories; resistant to hydrolysis; clear visc. liq.; 100% act.

Chemphos TC-444. [Chemron] Aliphatic phosphate ester; coupling agent for nonionic surfactants with liq. alkali detergent systems; surf. tension reducer; for oil treating chemicals; compat. with high concs. of sodium hydroxide and silicate builders; clear visc. liq.; 100% conc.

Chemphos TDAP. [Chemron] Alkyl acid phosphate; anionic; oil treating surfactant, asphaltine dispersant; clear visc. liq.; oil-sol.; 100% act.

Chemphos TR-414W. [Chemron] Phosphate ester; lubricant and detergent for water-based drilling fluids; clear lt. liq.; 80% act.

Chemphos TR-421. [Chemron] Aliphatic phosphate ester; coupling agent for nonionic surfactants with liq. alkali detergent systems; detergent in hard surf. and household cleaners; moderate foamer; stable in alkali systems; clear visc. liq.; water-sol.; 98% conc.

Chemphos TR-495. [Chemron] Aliphatic phosphate ester; detergent, emulsifier, dispersant, wetting agent for waterless hand cleaners, drycleaning formulations, pesticides; used in caustic boil and peroxide bleach for cotton processing; compat. in strong electrolyte systems; clear to hazy visc. liq.; 100% conc.

Chemphos TR-513. [Chemron] Aliphatic phosphate ester; detergent, emulsifier, wetting agent, hydrotrope for hard surf. detergents, liq. industrial cleaners, soak-tank metal cleaning, steam-cleaning formulations; compat. in conc. electrolyte systems; clear visc. liq.; water-sol.; 100% conc.

Chemphos TR-517. [Chemron] Aliphatic phosphate ester; very hydrophilic surfactant for cotton processing, esp. mercerizing; clear visc. liq.; sol. and stable in caustic and alkali builders; 100% conc.

Chemphos TX-625. [Chemron] Aliphatic phosphate ester; emulsifier, penetrant, anticorrosion agent for industrial use; lubricant and detergent for drilling fluids; clear lt. visc. liq.; 100% conc.

Chemphos TX-625D. [Chemron] Aliphatic phosphate ester, amine salt; detergent, wetting agent, emulsifier, anticorrosion agent, coupling agent for inorg. phosphates; for med. or heavy-duty detergents, hard-surf. and steam cleaning; clear visc. liq.; sol. in chlorinated solvs., modified hydrocarbons; 100% conc.

Chemquat 12-33. [Chemax] Laurtrimonium chloride; cationic; surfactant, corrosion inhibitor, antistat for plastics, textile dyeing aid, gel sensitizer in latex foam prod.; visc. depressant in paper and textile softener formulations; 33% act.

Chemquat 12-50. [Chemax] Laurtrimonium chloride; cationic; see Chemquat 12-33; liq.; 50% act.

Chemquat 16-50. [Chemax] Cetrimonium chloride; cationic; see Chemquat 12-33; liq.; 50% act.

Chemquat C/33W. [Chemax] Coco trimethyl ammonium chloride; cationic; see Chemquat 12-33; liq.; 33% act.

Chemsalan NLS 30. [Chemsal] Sodium lauryl sulfate; CAS 151-21-3; anionic; surfactant; liq.; 30% conc.

Chemsalan RLM 28. [Chemsal] Sodium laureth sulfate; CAS 9004-82-4; anionic; biodeg. surfactant; clear sl. ylsh. liq.; m.w. 390; visc. 200 mPa•s; pH 6-7; 28% conc.

Chemsalan RLM 56. [Chemsal] Sodium laureth sulfate; CAS 9004-82-4; anionic; biodeg. surfactant; sl. cloudy, lt. ylsh. visc. liq.; m.w. 390; visc. 5000-15,000 mPa•s; pH 6.5-8.0 (10% aq.); 56% conc.

Chemsalan RLM 70. [Chemsal] Sodium laureth sulfate; CAS 9004-82-4; anionic; biodeg. surfactant; sl. ylsh. paste; m.w. 390; pH 6.5-8.5 (10% aq.); 70% conc.

Chemsperse GMS-SE. [Chemron] Glyceryl stearate, self-emulsifying; nonionic; emulsifier, stabilizer, emollient, opacifier for netural to sl. alkaline anionic systems; flake; 100% conc.

Chemsulf 094. [Chemax] Ammonium alkyl phenol ethoxy sulfate; anionic; emulsifier, detergent, foaming agent; liq.; 58% conc.

Chemsulf S2EH-Na. [Chemax] Sodium 2-ethylhexyl sulfate; anionic; low foaming surfactant, wetting agent, emulsifier, detergent; electrolyte tolerant; liq.; 40% conc.

Chemsulf SBO/65. [Chemax] Sulfated butyl oleate; softener, wetting agent, lubricant additive, emulsifier, solubilizer for textile, metalworking industries; liq.; 65% act.

Chemsulf SCO/75. [Chemax] Sulfated castor oil; softener, wetting agent, lubricant additive, emulsifier, solubilizer for textile, metalworking industries; liq.; 75% act.

Chemteric 2C. [Chemron] Cocoamphodiacetate; amphoteric; mild surfactant, detergent; liq.; 50% conc.

Chemteric SF. [Chemron] Cocoamphodipropionate; amphoteric; surfactant for systems where sodium chloride would impair performance of final prod. or cause corrosion problems; liq.; 39% conc.

Chemzoline 1411. [Chemron] Aminoethyl tall oil imidazoline; corrosion inhibitor base for high-temp. oilfield applics.; amber visc. liq.; 100% act.

Chemzoline C-22. [Chemron] Hydroxyethyl coco imidazoline; emulsifier for oils; corrosion inhibitor for oilfield applics.; lt. tan paste; sol. in IPA, kerosene, aromatic solvs.; forms hazy disp. in water; 100% conc.

Chemzoline T-11. [Chemron] Aminoethyl tall oil imidazoline; intermediate for prod. of film-forming corrosion inhibitors; in automatic car wash rinse aids, oilfield applics.; pigment dispersant for paints; dark amber visc. liq.; 100% conc.

Chemzoline T-33. [Chemron] Aminoethyl tall oil imidazoline; industrial surfactant; produces self-emulsifiable min. oil compositions; dispersant for min. spirits, aromatic solvs.; film-forming corrosion inhibitor; antistripping agent for asphalt, coal tar pitches; pigment dispersant in paints; dark amber visc. liq.; disp. in water; sol. in min. oil, most org. solvs.; 100% conc.

Chemzoline T-44. [Chemron] Hydroxyethyl tall oil imidazoline; emulsifier for min., veg., and animal oils; emulsifier for oil-based drilling muds; corrosion inhibitor for oilfield applics.; dark visc. liq.; sol. in min. oil, veg. and animal glycerides, most org. solvs.; forms hazy, stable disp. in water; 100% conc.

Chimin BX. [Auschem SpA] Lauryl betaine; CAS 683-10-3; amphoteric; dispersant, wetting agent, antistat; soft paste; 40% conc.

Chimin DOS 70. [Auschem SpA] Sodium di 2-ethylhexyl sulfosuccinate; CAS 577-11-7; anionic; wetting agent; liq.; 68% conc.

Chimin KF3. [Auschem SpA] Sulfated syn. oil; anionic; emulsifier for fat liquoring, leather processing; liq.; 90% conc.

Chimin KSP. [Auschem SpA] Fatty alcohol ester; nonionic; fat liquoring of leather, substitute for sperm oil, textile lubricant; liq.; 100% conc.

Chimin P1A. [Auschem SpA] Complex org. phosphate ester, free acid; anionic; surfactant for drycleaning, hand cleaners, corrosion inhibitor, heavy-duty liq. formulations, polymerization; liq.; 100% conc.

Chimin P10. [Auschem SpA] Alkyl polyphosphate; anionic; wetting and penetrating agent; liq.; 100% conc.

Chimin P40. [Auschem SpA] Sodium alkylpolyglycol ether phosphate; CAS 68071-35-2; anionic; detergent, wetting agent for low foaming prods. in textile, paper, cleaners, machine washing; alkali-stable; liq.; 75% conc.

Chimin P45. [Auschem SpA] Sodium alkylpolyglycol ether phosphate; CAS 68071-35-2; anionic; detergent, antistat, wetting agent for textile, paper, cleaners, machine washing, and personal care prods.; liq.; 30% conc.

Chimin P50. [Auschem SpA] Phosphoric acid complex org. ester; nonionic; emulsifier, antistat for use in leather, textile and cosmetic applics.; liq.; sol. in water and solv.; 100% conc.

Chimin RI. [Auschem SpA] Complex phosphate ester; anionic; low foaming alkaline liq. detergent, hydrotrope; liq.; 50% conc.

Chimipal APG 400. [Auschem SpA] Polyglycol laurate; CAS 9004-81-3; nonionic; pigment dispersant and wetter, emulsion polymerization; liq.; HLB 13.0; 100% conc.

Chimipal DCL/M. [Auschem SpA] Cocamide DEA; CAS 68603-42-9; nonionic; thickener and foam stabilizer, detergent, and shampoo additive; liq.; 100% conc.

Chimipal DE 7. [Auschem SpA] Ethoxylated alkanolamide; nonionic; detergent, foam booster; liq./paste; 100% conc.

Chimipal LDA. [Auschem SpA] Lauramide DEA; CAS 120-40-1; nonionic; thickener and foam stabilizer, detergent and shampoo additive; liq.; 100% conc.

Chimipal MC. [Auschem SpA] Cocamide MEA; CAS 68140-00-1; nonionic; thickener and foam stabilizer, detergent and shampoo additive; flakes; 100% conc.

Chimipal NH 2. [Auschem SpA] Ethoxylated oleamine; CAS 26635-93-8; cationic; leveling agent, emulsifier; liq.; 100% conc.

Chimipal NH 6. [Auschem SpA] Ethoxylated oleamine; CAS 26635-93-8; cationic; leveling agent, emulsifier; liq.; 100% conc.

Chimipal NH 10. [Auschem SpA] Ethoxylated oleamine; CAS 26635-93-8; cationic; leveling agent, emulsifier; liq./paste; 100% conc.

Chimipal NH 20. [Auschem SpA] Ethoxylated oleamine; CAS 26635-93-8; cationic; leveling agent, emulsifier; paste; 100% conc.

Chimipal NH 25. [Auschem SpA] Ethoxylated oleamine; CAS 26635-93-8; cationic; leveling agent, emulsifier; solid; 100% conc.

Chimipal NH 30. [Auschem SpA] Ethoxylated oleamine; CAS 26635-93-8; cationic; leveling agent, emulsifier; solid; 100% conc.

Chimipal OLD. [Auschem SpA] Oleamide DEA; CAS 93-83-4; nonionic; thickener for shampoos, bubble baths and liq. detergents; liq.; 100% conc.

Chimipal OS 2. [Auschem SpA] Fatty alcohol ethoxylate blend; nonionic; powd. detergent, wax emulsifier, leveling agent; solid; 100% conc.

Chimipal PE 300, PE 302. [Tessilchimica] Polypropylene glycol ethoxylate; CAS 9003-11-6; nonionic; detergent, dispersant, emulsifier, foam controller; liq.; 100% conc.

Chimipal PE 402, PE 403, PE 405. [Auschem SpA] Modified alkyl polyether; nonionic; low foam detergent; liq.; 100% conc.

Chimipal PE 520. [Auschem SpA] Terminated alkylene oxide adducts; nonionic; low foam detergent for metals, bottle, alkaline, strong acid cleaners; liq.; 100% conc.

Chimipon FC. [Auschem SpA] Detergent blend; anionic/nonionic; general purpose and hard surf. cleaner; liq.; 40% conc.

Chimipon GT. [Auschem SpA] Sodium alkylaryl polyether sulfate; anionic; detergent, emulsifier in emulsion polymerization; liq.; 40% conc.

Chimipon HD. [Auschem SpA] Detergent/foam stabilizer blend; anionic/nonionic; base for heavy-duty liq. laundry detergents; paste; 56% conc.

Chimipon LD. [Auschem SpA] Dodecylbenzene sulfonate/alkylether sulfate blend; anionic/nonionic; surfactant for formulated hand dishwashing detergents; liq./paste; 55% conc.

Chimipon LDP. [Auschem SpA] Dodecylbenzene sulfonate/alkylether sulfate blend; anionic/nonionic; formulated dishwashing detergent; liq./paste;

52% conc.

Chimipon NA. [Auschem SpA] Alkylphenol ether sulfate, ammonium salt; anionic; liq. detergent base, frothing agent, emulsifier in emulsion polymerization; paste; 60% conc.

Chimipon SK. [Auschem SpA] Potassium cocoate; CAS 61789-30-8; anionic; detergent; liq.; 30% conc.

Chimipon TSB. [Auschem SpA] Linear dodecylbenzene sulfonate, TEA-sodium salt; anionic; household and industrial liq. detergents; paste; 70% conc.

CHPTA 65%. [Chem-Y GmbH] 3 Chloro-2-hydroxypropyltrimethyl ammonium chloride; CAS 3327-22-8; cationic; cationizing reagent for natural and syn. polymers; starch modifier for textiles; liq.; 65% act.

Chromasist 6B. [Henkel/Textile] Phosphate ester, free acid form; anionic; surfactant for textile use; Gardner 1 liq.; sol. @ 5% in Stod., xylene; disp. in water, min. oil; dens. 8.7 lb/gal; visc. 108 cSt (100 F); pour pt. -12 C; cloud pt. < 25 C; flash pt. > 200 F.

Chromasist 87H. [Henkel/Textile] Sodium naphthalene sulfonate; dispersing and leveling agent for disperse dyes; helps reduce dye residue build-up in machinery; dk. amber liq.; sol. @ 5% in water; dens. 10 lb/gal; visc. 11 cSt (100 F); pour pt. -1 C; cloud pt. 1 C; flash pt. 45 F; pH 9.5 (as is).

Chromasist 1487A. [Henkel/Textile] Sodium naphthalene sulfonate; dispersing and leveling agent for disperse dyes; esp. for high-temp. dyeable polyester; high m.w.; dk. amber liq.; sol. @ 5% in water; disp. in glycerol trioleate; dens. 10 lb/gal; visc. 11 cSt (100 F); pour pt. -1 C; cloud pt. 4 C; flash pt. 30 F; pH 4.5 (as is).

Chromasist KDL-1. [Henkel/Textile] Leveling agent for high temp. dyeing of polyester; non-solv.; yel. liq.; sol. forms clear sol'ns. in water; pH 7.5 (as is).

Chromasist LW. [Henkel/Textile] Leveling agent for polyester/cotton blends with disperse/direct dyes; dk. amber liq.; sol. forms clear sol'ns. in water; pH 9.0 (as is).

Chromosol SS. [Nippon Senka] Silicon complex; fixing agent for acid dyes on wool; liq.

CHT Felosan TAK-NO. [Catawba-Charlab] Low foaming stain remover and detergent.

CHT Heptol NWS. [Catawba-Charlab] Polyphosphate; for textile washing; liq.

CHT Lavotan DS. [Catawba-Charlab] Wetting, washing and cleaning surfactant.

CHT Subitol HLF Conc. [Catawba-Charlab] Low foaming alkali-stable surfactant for continuous bleaching.

CHT Subitol LS-N. [Catawba-Charlab] Low foaming alkali-stable wetting agent and detergent.

CHT Subitol SAN. [Catawba-Charlab] Low foaming rapid wetting agent and detergent.

Chupol C. [Takemoto Oil & Fat] Polyglycol ether; nonionic; reducing agent for concrete; liq.

Chupol EX. [Takemoto Oil & Fat] Hydroxylated carboxylic acid; anionic; water reducing agent for concrete; liq.

Cibaphasol® AS. [Ciba-Geigy/Dyestuffs] Sulfuric acid ester; anionic; leveling and penetrating agent for continuous dyeing and printing of nylon; liq.

Cirrasol® 185A. [ICI Am.] Acid solubilized fatty acid amide; CAS 26392-63-2; for fiberglass; liq.; 100% conc.

Cirrasol® 185AE. [ICI Am.] Acid solubilized fatty acid amide; for fiberglass; liq. to soft paste; 100% conc.

Cirrasol® 185AN. [ICI Am.] Acid solubilized fatty acid amide; for fiberglass; liq.; 100% conc.

Cirrasol® AEN-XB. [ICI Surf. Belgium] Fatty alcohol ethoxylate; nonionic; fiber processing aid; antistat for PVC belting; emulsifier for low m.p. paraffin waxes; liq.; HLB 10.2; 100% conc.

Cirrasol® AEN-XF. [ICI Surf. Belgium] Fatty alcohol ethoxylate; nonionic; fiber processing aid; emulsifier for oleins, castor oil, etc. into water; semisolid; HLB 13.8; 100% conc.

Cirrasol® AEN-XZ. [ICI Surf. Belgium] Alkylphenol ethoxylate; nonionic; fiber processing aid; detergent base; emulsifier, wetting agent, dispersant, dedusting and antistatic agent; liq.; HLB 12.3; 100% conc.

Cirrasol® ALN-FP. [ICI Surf. Belgium] Fatty acid ethoxylate; nonionic; fiber processing aid; emulsifier, antistat, gelling agent; semisolid; HLB 11.6; 100% conc.

Cirrasol® ALN-GM. [ICI Surf. Belgium] Polyester ethoxylate; nonionic; fiber processing aid; antistat for plastics and rubber; wetting agent; liq.; HLB 15.7; 100% conc.

Cirrasol® ALN-TF. [ICI Surf. Belgium] Fatty acid ethoxylate; nonionic; fiber processing aid; general wetting agent; liq.; HLB 13.6; 100% conc.

Cirrasol® ALN-TS. [ICI Surf. Belgium] Fatty acid ethoxylate; nonionic; fiber processing aid; general wetting agent; liq.; HLB 13.6; 82.4% conc.

Cirrasol® ALN-TV. [ICI Surf. Belgium] Antistatic lubricant/solubilizer blend; nonionic; fiber processing aid; antistatic lubricant, solubilizer; liq.; 100% conc.

Cirrasol® ALN-WF. [ICI Surf. Belgium] Cetyl fatty alcohol ethoxylate; CAS 9004-95-9; nonionic; fiber processing aid; antistatic lubricant for syn. fibers; emulsifier for waxes; flakes; HLB 14.9; 100% conc.

Cirrasol® ALN-WY. [ICI Surf. Belgium] Ethoxylated triglyceride; nonionic; fiber processing aid; wool lubricant; flakes; HLB 11.0; 90% conc.

Cirrasol® EN-MB. [ICI Surf. Belgium] Fatty alcohol ethoxylate; nonionic; fiber processing aid; emulsifier/stabilizer for textile lubricant compositions; liq.; HLB 6.0; 100% conc.

Cirrasol® EN-MP. [ICI Surf. Belgium] Fatty alcohol ethoxylate; nonionic; fiber processing aid; emulsifier for mfg. of self-emulsifiable min. oils; liq.; HLB 6.0; 100% conc.

Cirrasol® GM. [ICI Am.] POE fatty ester; nonionic; fiber lubricant, emulsifier and antistatic agent; liq.; HLB 16.5; 100% conc.

Cirrasol® LAN-SF. [ICI Surf. Belgium] Fatty alcohol ethoxylate; nonionic; fiber processing aid; emulsifier for min. and veg. oils; liq.; HLB 7.5; 100% conc.

Cirrasol® LC-HK. [ICI Surf. Belgium] Substituted high m.w. amine alkyl deriv.; cationic; fiber processing aid; lubricant and softener; paste; 33% conc.

Cirrasol® LC-PQ. [ICI Surf. Belgium] Fatty acid deriv. of substituted diamine; cationic; fiber processing aid; lubricant and softener; paste; 25% conc.

Cirrasol® LN-GS. [ICI Surf. Belgium] Tridecyl stearate; CAS 31556-45-3; fiber processing aid; heat-stable lubricant; liq.; 100% conc.

Cithrol 2DL. [Croda Chem. Ltd.] PEG-4 dilaurate; CAS 9005-02-1; nonionic; dispersant, w/o emulsifier, wetting agent and co-solv.; used in cosmetic and industrial applics.; paste; HLB 6.0; 97% conc.

Cithrol 2DO. [Croda Chem. Ltd.] PEG-4 dioleate; CAS 9005-07-6; nonionic; dispersant, w/o emulsifier, wetting agent and co-solv.; used in cosmetic and industrial applics.; liq.; 97% conc.

Cithrol 2DS. [Croda Chem. Ltd.] PEG-4 distearate; CAS 9005-08-7; nonionic; antistat in textile finishing; paste; HLB 5.2; 97% conc.

Cithrol 2ML. [Croda Chem. Ltd.] PEG-4 laurate; CAS 9004-81-3; nonionic; wetting agent, emulsifier, detergent, thickener, solubilizer, dispersant, softener, lubricant, antistat, dye asistant, penetrant for cosmetics, textiles, glass fiber, metal treatment; liq.; m.w. 200; HLB 8.8; sapon. no. 158–168; 100% conc.

Cithrol 2MO. [Croda Chem. Ltd.] PEG-4 oleate; CAS 9004-96-0; nonionic; wetting agent, penetrant, detergent, emulsifier, solubilizer, thickening agent, dispersant, textile aux., softener, lubricant for textiles, cosmetics, metalworking; m.w. 200; HLB 6.2; sapon. no. 110–112; 100% conc.

Cithrol 2MS. [Croda Chem. Ltd.] PEG-4 stearate; CAS 9004-99-3; nonionic; emulsifier for insecticides and cosmetics, detergent, wetting agent, solubilizer and thickening agent for perfumery, antifrothing agent, foaming agent; m.w. 200; m.p. 39–41 C; HLB 6.3; sapon. no. 110–120; 100% conc.

Cithrol 3DS. [Croda Chem. Ltd.] PEG-6 distearate; CAS 9005-08-7; nonionic; antistat in textile finishing; solid; HLB 7.3; 97% conc.

Cithrol 3MS. [Croda Chem. Ltd.] PEG-6 stearate; CAS 9004-99-3; emulsifier for insecticides and cosmetics, detergent, wetting agent, solubilizer and thickening agent for perfumery, antifrothing agent, foaming agent; paste; m.w. 300; m.p. 30–33 C; HLB 10.7; sapon. no. 95–105; 100% conc.

Cithrol 4DL. [Croda Chem. Ltd.] PEG-8 dilaurate; CAS 9005-02-1; nonionic; dispersant, w/o emulsifier, wetting agent and co-solv.; used in cosmetic and industrial applics.; liq.; HLB 10.2; 97% conc.

Cithrol 4DO. [Croda Chem. Ltd.] PEG-8 dioleate; CAS 9005-07-6; nonionic; dispersant, w/o emulsifier, wetting agent and co-solv.; used in cosmetic and industrial applics.; liq.; HLB 8.3; 97% conc.

Cithrol 4DS. [Croda Chem. Ltd.] PEG-8 distearate; CAS 9005-08-7; nonionic; antistat in textile finishing; solid; HLB 9.3; 97% conc.

Cithrol 4ML. [Croda Chem. Ltd.] PEG-8 laurate; CAS 9004-81-3; nonionic; wetting agent, emulsifier, detergent, thickener, solubilizer, dispersant, softener, lubricant, antistat, dye asistant, penetrant for cosmetics, textiles, glass fiber, metal treatment; m.w. 400; HLB 13.1; sapon. no. 92–98; 100% conc.

Cithrol 4MO. [Croda Chem. Ltd.] PEG-8 oleate; CAS 9004-96-0; nonionic; wetting agent, penetrant, detergent, emulsifier, solubilizer, thickening agent, dispersant, textile aux., softener, lubricant for textiles, cosmetics, metalworking; liq.; m.w. 400; HLB 11.4; sapon. no. 85–93; 100% conc.

Cithrol 4MS. [Croda Chem. Ltd.] PEG-8 stearate; CAS 9004-99-3; nonionic; emulsifier for insecticides and cosmetics, detergent, wetting agent, solubilizer and thickening agent for perfumery, antifrothing agent, foaming agent; paste; m.w. 400; m.p. 31–34 C; HLB 11; sapon. no. 95–105; 100% conc.

Cithrol 6DL. [Croda Chem. Ltd.] PEG-12 dilaurate; CAS 9005-02-1; nonionic; dispersant, w/o emulsifier, wetting agent and co-solv.; used in cosmetic and industrial applics.; liq.; HLB 12.0; 97% conc.

Cithrol 6DO. [Croda Chem. Ltd.] PEG-12 dioleate; CAS 9005-07-6; nonionic; dispersant, w/o emulsifier, wetting agent and co-solv.; used in cosmetic and industrial applics.; liq.; HLB 10.4; 97% conc.

Cithrol 6DS. [Croda Chem. Ltd.] PEG-12 distearate; CAS 9005-08-7; nonionic; antistat in textile finishing; solid; HLB 10.8; 97% conc.

Cithrol 6ML. [Croda Chem. Ltd.] PEG-12 laurate; CAS 9004-81-3; nonionic; wetting agent, emulsifier, detergent, thickener, solubilizer, dispersant, softener, lubricant, antistat, dye asistant, penetrant for cosmetics, textiles, glass fiber, metal treatment, leather treatment; m.w. 600; HLB 16.8; sapon. no. 64–74; 100% conc.

Cithrol 6MO. [Croda Chem. Ltd.] PEG-12 oleate; CAS 9004-96-0; nonionic; wetting agent, penetrant, detergent, emulsifier, solubilizer, thickening agent, dispersant, textile aux., softener, lubricant for textiles, cosmetics, metalworking; liq.; m.w. 600; HLB 13.1; sapon. no. 65–75; 100% conc.

Cithrol 6MS. [Croda Chem. Ltd.] PEG-12 stearate; CAS 9004-99-3; nonionic; emulsifier for insecticides and cosmetics, detergent, wetting agent, solubilizer and thickening agent for perfumery, antifrothing agent, foaming agent; paste; m.w. 600; m.p. 33–35 C; HLB 14; sapon. no. 68–76; 100% conc.

Cithrol 10DL. [Croda Chem. Ltd.] PEG-20 dilaurate; CAS 9005-02-1; nonionic; dispersant, w/o emulsifier, wetting agent and co-solv.; used in cosmetic and industrial applics.; paste; HLB 14.7; 97% conc.

Cithrol 10DO. [Croda Chem. Ltd.] PEG-20 dioleate; CAS 9005-07-6; nonionic; dispersant, w/o emulsifier, wetting agent and co-solv.; used in cosmetic and industrial applics.; liq.; HLB 13.2; 97% conc.

Cithrol 10DS. [Croda Chem. Ltd.] PEG-20 distearate; CAS 9005-08-7; nonionic; antistat in textile finishing; solid; HLB 13.7; 97% conc.

Cithrol 10ML. [Croda Chem. Ltd.] PEG-20 laurate; CAS 9004-81-3; nonionic; wetting agent, emulsifier, detergent, thickener, solubilizer, dispersant, softener, lubricant, antistat, dye asistant, penetrant for cosmetics, textiles, glass fiber, metal treatment; sol. in water, alcohol, polar solv.; m.w. 1000; HLB 18; sapon. no. 51–55; 100% conc.

Cithrol 10MO. [Croda Chem. Ltd.] PEG-20 oleate; CAS 9004-96-0; nonionic; wetting agent, penetrant, detergent, emulsifier, solubilizer, thickening agent, dispersant, textile aux., softener, lubricant for textiles, cosmetics, metalworking; liq.; m.w. 1000; HLB 16.2; sapon. no. 35–40; 100% conc.

Cithrol 10MS. [Croda Chem. Ltd.] PEG-20 stearate; CAS 9004-99-3; nonionic; emulsifier for insecticides and cosmetics, detergent, wetting agent, solubilizer and thickening agent for perfumery, antifrothing agent, foaming agent; solid; m.w. 1000; m.p. 34–40 C; HLB 16; sapon. no. 36–50; 100% conc.

Cithrol 15MS. [Croda Chem. Ltd.] PEG-1500 stearate; nonionic; emulsifier for insecticides and cosmetics, detergent, wetting agent, solubilizer and thickening agent for perfumery, antifrothing agent, foaming agent; solid; m.w. 1500; m.p. 30-41 C; HLB 17; sapon. no. 30–55; 100% conc.

Cithrol 40MO. [Croda Chem. Ltd.] PEG-75 oleate; CAS 9004-96-0; wetting agent, penetrant, detergent, emulsifier, solubilizer, thickening agent, dispersant, textile aux., softener, lubricant for textiles, cosmetics, metalworking; m.w. 4000; HLB 18.7; sapon. no. 10–15.

Cithrol 40MS. [Croda Chem. Ltd.] PEG-75 stearate; CAS 9004-99-3; nonionic; emulsifier for insecticides and cosmetics, detergent, wetting agent, solubilizer and thickening agent for perfumery, antifrothing agent, foaming agent; solid; m.w. 4000;

m.p. 30–41 C; HLB 18.8; sapon. no. 12–17; 100% conc.

Cithrol 60ML. [Croda Chem. Ltd.] PEG 6000 laurate; CAS 9004-81-3; wetting agent, emulsifier, detergent, thickener, solubilizer, dispersant, softener, lubricant, antistat, dye asistant, penetrant for cosmetics, textiles, glass fiber, metal treatment; m.w. 6000; HLB 19.2; sapon. no. 8–13.

Cithrol 60MO. [Croda Chem. Ltd.] PEG 6000 oleate; CAS 9004-96-0; wetting agent, penetrant, detergent, emulsifier, solubilizer, thickening agent, dispersant, textile aux., softener, lubricant for textiles, cosmetics, metalworking; m.w. 6000; HLB 19.0; sapon. no. 5–10.

Cithrol A. [Croda Chem. Ltd.] PEG-8 oleate; CAS 9004-96-0; nonionic; w/o emulsifier, dispersant, antistat for textile, paper processing, cutting oils, polishes, emulsion cleaners, rubber latex, wool lubricants; amber liq.; slight agreeable odor; sol. in wh. spirit and liq. paraffin; completely sol. in xylol, ethyl acetate, methylated spirits, oleic acid; disp. in water; sp.gr. 1.013; HLB 11.4; flash pt. 215 C; sapon. no. 85–93; pH 6.5–7.5 (10%); 97% conc.

Cithrol DGDL N/E. [Croda Chem. Ltd.] Diethylene glycol dilaurate; CAS 9005-02-1; nonionic; w/o emulsifier, dispersant, antistat for textile, paper processing, cutting oils, polishes, emulsion cleaners, rubber latex, wool lubricants; paste; HLB 1.5; 100% conc.

Cithrol DGDL S/E. [Croda Chem. Ltd.] Diethylene glycol dilaurate SE; CAS 9005-02-1; anionic; w/o emulsifier, dispersant, antistat for textile, paper processing, cutting oils, polishes, emulsion cleaners, rubber latex, wool lubricants; paste; HLB 2.2; 100% conc.

Cithrol DGDO N/E. [Croda Chem. Ltd.] Diethylene glycol dioleate; CAS 9005-07-6; nonionic; w/o emulsifier, dispersant, antistat for textile, paper processing, cutting oils, polishes, emulsion cleaners, rubber latex, wool lubricants; liq.; HLB 1.8; 100% conc.

Cithrol DGDO S/E. [Croda Chem. Ltd.] Diethylene glycol dioleate SE; CAS 9005-07-6; anionic; w/o emulsifier, dispersant, antistat for textile, paper processing, cutting oils, polishes, emulsion cleaners, rubber latex, wool lubricants; liq.; HLB 2.7; 100% conc.

Cithrol DGDS N/E. [Croda Chem. Ltd.] Diethylene glycol distearate; CAS 9005-08-7; nonionic; w/o emulsifier, dispersant, antistat for textile, paper processing, cutting oils, polishes, emulsion cleaners, rubber latex, wool lubricants; solid; HLB 3.0; 100% conc.

Cithrol DGDS S/E. [Croda Chem. Ltd.] Diethylene glycol distearate SE; CAS 9005-08-7; anionic; w/o emulsifier, dispersant, antistat for textile, paper processing, cutting oils, polishes, emulsion cleaners, rubber latex, wool lubricants; solid; HLB 5.5; 100% conc.

Cithrol DGML N/E. [Croda Chem. Ltd.] Diethylene glycol laurate; CAS 9004-81-3; 141-20-8; nonionic; w/o emulsifier, dispersant, antistat for textile, paper processing, cutting oils, polishes, emulsion cleaners, rubber latex, wool lubricants; paste; HLB 6.1; 100% conc.

Cithrol DGML S/E. [Croda Chem. Ltd.] Diethylene glycol laurate SE; CAS 9004-81-3; anionic; w/o emulsifier, dispersant, antistat for textile, paper processing, cutting oils, polishes, emulsion cleaners, rubber latex, wool lubricants; paste; HLB 6.7; 100% conc.

Cithrol DGMO N/E. [Croda Chem. Ltd.] Diethylene glycol oleate; CAS 9004-96-0; nonionic; w/o emulsifier, dispersant, antistat for textile, paper processing, cutting oils, polishes, emulsion cleaners, rubber latex, wool lubricants; liq.; HLB 5.0; 100% conc.

Cithrol DGMO S/E. [Croda Chem. Ltd.] Diethylene glycol oleate SE; CAS 9004-96-0; anionic; w/o emulsifier, dispersant, antistat for textile, paper processing, cutting oils, polishes, emulsion cleaners, rubber latex, wool lubricants; liq.; HLB 5.7; 100% conc.

Cithrol DGMS N/E. [Croda Chem. Ltd.] Diethylene glycol stearate; CAS 9004-99-3; nonionic; w/o emulsifier, dispersant, antistat for textile, paper processing, cutting oils, polishes, emulsion cleaners, rubber latex, wool lubricants; solid; HLB 4.4; 100% conc.

Cithrol DGMS S/E. [Croda Chem. Ltd.] Diethylene glycol stearate SE; CAS 9004-99-3; anionic; w/o emulsifier, dispersant, antistat for textile, paper processing, cutting oils, polishes, emulsion cleaners, rubber latex, wool lubricants; solid; HLB 5.0; 100% conc.

Cithrol DPGML N/E. [Croda Chem. Ltd.] Dipropylene glycol laurate; nonionic; w/o emulsifier, dispersant, antistat for textile, paper processing, cutting oils, polishes, emulsion cleaners, rubber latex, wool lubricants; liq.; 100% conc.

Cithrol DPGML S/E. [Croda Chem. Ltd.] Dipropylene glycol laurate SE; anionic; w/o emulsifier, dispersant, antistat for textile, paper processing, cutting oils, polishes, emulsion cleaners, rubber latex, wool lubricants; liq.; 100% conc.

Cithrol DPGMO N/E. [Croda Chem. Ltd.] Dipropylene glycol oleate; nonionic; w/o emulsifier, dispersant, antistat for textile, paper processing, cutting oils, polishes, emulsion cleaners, rubber latex, wool lubricants; liq.; 100% conc.

Cithrol DPGMO S/E. [Croda Chem. Ltd.] Dipropylene glycol oleate SE; anionic; w/o emulsifier, dispersant, antistat for textile, paper processing, cutting oils, polishes, emulsion cleaners, rubber latex, wool lubricants; liq.; 100% conc.

Cithrol DPGMS N/E. [Croda Chem. Ltd.] Dipropylene glycol stearate; nonionic; w/o emulsifier, dispersant, antistat for textile, paper processing, cutting oils, polishes, emulsion cleaners, rubber latex, wool lubricants; solid; 100% conc.

Cithrol DPGMS S/E. [Croda Chem. Ltd.] Dipropylene glycol stearate SE; anionic; w/o emulsifier, dispersant, antistat for textile, paper processing, cutting oils, polishes, emulsion cleaners, rubber latex, wool lubricants; solid; 100% conc.

Cithrol EGDL N/E. [Croda Chem. Ltd.] Ethylene glycol dilaurate; CAS 624-04-4; nonionic; w/o emulsifier, dispersant, antistat for textile, paper processing, cutting oils, polishes, emulsion cleaners, rubber latex, wool lubricants; solid; HLB 1.0; 100% conc.

Cithrol EGDL S/E. [Croda Chem. Ltd.] Ethylene glycol dilaurate SE; anionic; w/o emulsifier, dispersant, antistat for textile, paper processing, cutting oils, polishes, emulsion cleaners, rubber latex, wool lubricants; solid; HLB 2.0; 100% conc.

Cithrol EGDO N/E. [Croda Chem. Ltd.] Ethylene glycol dioleate; CAS 928-24-5; nonionic; w/o emulsifier, dispersant, antistat for textile, paper processing, cutting oils, polishes, emulsion cleaners, rubber latex, wool lubricants; liq.; HLB 1.0; 100% conc.

Cithrol EGDO S/E. [Croda Chem. Ltd.] Ethylene glycol dioleate SE; anionic; w/o emulsifier, dispersant, antistat for textile, paper processing, cutting oils, polishes, emulsion cleaners, rubber latex, wool lubricants; liq.; HLB 2.0; 100% conc.

Cithrol EGDS N/E. [Croda Chem. Ltd.] Ethylene glycol distearate; CAS 627-83-8; nonionic; w/o emulsifier, dispersant, antistat for textile, paper processing, cutting oils, polishes, emulsion cleaners, rubber latex, wool lubricants; solid; HLB 1.5; 100% conc.

Cithrol EGDS S/E. [Croda Chem. Ltd.] Ethylene glycol distearate SE; anionic; w/o emulsifier, dispersant, antistat for textile, paper processing, cutting oils, polishes, emulsion cleaners, rubber latex, wool lubricants; solid; HLB 2.4; 100% conc.

Cithrol EGML N/E. [Croda Chem. Ltd.] Ethylene glycol laurate; CAS 4219-48-1; nonionic; w/o emulsifier, dispersant, antistat for textile, paper processing, cutting oils, polishes, emulsion cleaners, rubber latex, wool lubricants; liq.; HLB 2.7; 100% conc.

Cithrol EGML S/E. [Croda Chem. Ltd.] Ethylene glycol laurate SE; anionic; w/o emulsifier, dispersant, antistat for textile, paper processing, cutting oils, polishes, emulsion cleaners, rubber latex, wool lubricants; liq.; HLB 3.6; 100% conc.

Cithrol EGMO N/E. [Croda Chem. Ltd.] Ethylene glycol oleate; CAS 4500-01-0; nonionic; w/o emulsifier, dispersant, antistat for textile, paper processing, cutting oils, polishes, emulsion cleaners, rubber latex, wool lubricants; liq.; HLB 1.8; 100% conc.

Cithrol EGMO S/E. [Croda Chem. Ltd.] Ethylene glycol oleate SE; anionic; w/o emulsifier, dispersant, antistat for textile, paper processing, cutting oils, polishes, emulsion cleaners, rubber latex, wool lubricants; liq.; HLB 2.7; 100% conc.

Cithrol EGMR N/E. [Croda Chem. Ltd.] Ethylene glycol ricinoleate; CAS 106-17-2; nonionic; w/o emulsifier, dispersant, antistat for textile, paper processing, cutting oils, polishes, emulsion cleaners, rubber latex, wool lubricants; liq.; HLB 1.1; 100% conc.

Cithrol EGMR S/E. [Croda Chem. Ltd.] Ethylene glycol ricinoleate SE; anionic; w/o emulsifier, dispersant, antistat for textile, paper processing, cutting oils, polishes, emulsion cleaners, rubber latex, wool lubricants; liq.; HLB 2.0; 100% conc.

Cithrol EGMS N/E. [Croda Chem. Ltd.] Ethylene glycol stearate; CAS 111-60-4; nonionic; w/o emulsifier, dispersant, antistat for textile, paper processing, cutting oils, polishes, emulsion cleaners, rubber latex, wool lubricants; solid; HLB 2.0; 100% conc.

Cithrol EGMS S/E. [Croda Chem. Ltd.] Ethylene glycol stearate SE; anionic; w/o emulsifier, dispersant, antistat for textile, paper processing, cutting oils, polishes, emulsion cleaners, rubber latex, wool lubricants; solid; HLB 2.9; 100% conc.

Cithrol GDL N/E. [Croda Chem. Ltd.] Glyceryl dilaurate; CAS 27638-00-2; nonionic; w/o emulsifier, dispersant, antistat for textile, paper processing, cutting oils, polishes, emulsion cleaners, rubber latex, wool lubricants, cosmetics, pharmaceuticals; liq.; HLB 2.0; 100% conc.

Cithrol GDL S/E. [Croda Chem. Ltd.] Glyceryl dilaurate SE; anionic; w/o emulsifier, dispersant, antistat for textile, paper processing, cutting oils, polishes, emulsion cleaners, rubber latex, wool lubricants, cosmetics, pharmaceuticals; liq.; 100% conc.

Cithrol GDO N/E. [Croda Chem. Ltd.] Glyceryl dioleate; CAS 25637-84-7; nonionic; emulsifier, coemulsifier, stabilizer, wetting agent, lubricant, and antistat; used in cosmetic, pharmaceutical, industrial, food applics.; liq.; HLB 2.0; 100% conc.

Cithrol GDO S/E. [Croda Chem. Ltd.] Glyceryl dioleate SE; anionic; emulsifier, coemulsifier, stabilizer, wetting agent, lubricant, and antistat; used in cosmetic, pharmaceutical, industrial, food applics.; liq.; HLB 2.9; 100% conc.

Cithrol GDS N/E. [Croda Chem. Ltd.] Glyceryl distearate; CAS 1323-83-7; nonionic; emulsifier, coemulsifier, stabilizer, wetting agent, lubricant, and antistat; used in cosmetic, pharmaceutical, industrial, food applics.; solid; HLB 3.4; 100% conc.

Cithrol GDS S/E. [Croda Chem. Ltd.] Glyceryl distearate SE; anionic; emulsifier, coemulsifier, stabilizer, wetting agent, lubricant, and antistat; used in cosmetic, pharmaceutical, industrial, food applics.; solid; HLB 4.2; 100% conc.

Cithrol GML N/E. [Croda Chem. Ltd.] Glyceryl laurate; nonionic; emulsifier, coemulsifier, stabilizer, wetting agent, lubricant, and antistat; used in cosmetic, pharmaceutical, industrial, food applics.; liq.; HLB 4.9; 100% conc.

Cithrol GML S/E. [Croda Chem. Ltd.] Glyceryl laurate SE; anionic; emulsifier, coemulsifier, stabilizer, wetting agent, lubricant, and antistat; used in cosmetic, pharmaceutical, industrial, food applics.; liq.; HLB 5.6; 100% conc.

Cithrol GMO N/E. [Croda Chem. Ltd.] Glyceryl oleate; CAS 111-03-5; nonionic; emulsifier, coemulsifier, stabilizer, wetting agent, lubricant, and antistat; used in cosmetic, pharmaceutical, industrial, food applics.; liq.; HLB 3.3; 100% conc.

Cithrol GMO S/E. [Croda Chem. Ltd.] Glyceryl oleate SE; anionic; emulsifier, coemulsifier, stabilizer, wetting agent, lubricant, and antistat; used in cosmetic, pharmaceutical, industrial, food applics.; liq.; HLB 4.1; 100% conc.

Cithrol GMR N/E. [Croda Chem. Ltd.] Glyceryl ricinoleate; nonionic; emulsifier, coemulsifier, stabilizer, wetting agent, lubricant, and antistat; used in cosmetic, pharmaceutical, industrial, food applics.; liq.; HLB 2.7; 100% conc.

Cithrol GMR S/E. [Croda Chem. Ltd.] Glyceryl ricinoleate SE; anionic; emulsifier, coemulsifier, stabilizer, wetting agent, lubricant, and antistat; used in cosmetic, pharmaceutical, industrial, food applics.; liq.; HLB 3.6; 100% conc.

Cithrol GMS Acid Stable. [Croda Chem. Ltd.] Glyceryl stearate SE; CAS 31566-31-1; nonionic; emulsifier, coemulsifier, stabilizer, wetting agent, lubricant, and antistat; used in cosmetic, pharmaceutical, industrial, food applics.; solid; HLB 10.9; 100% conc.

Cithrol GMS N/E. [Croda Chem. Ltd.] Glyceryl stearate; CAS 31566-31-1; nonionic; emulsifier, coemulsifier, stabilizer, wetting agent, lubricant, and antistat; used in cosmetic, pharmaceutical, industrial, food applics.; solid; HLB 3.4; 100% conc.

Cithrol GMS S/E. [Croda Chem. Ltd.] Glyceryl stearate SE; CAS 31566-31-1; anionic; emulsifier, coemulsifier, stabilizer, wetting agent, lubricant, and antistat; used in cosmetic, pharmaceutical, industrial, food applics.; solid; HLB 4.4; 100% conc.

Cithrol L, O Range. [Croda Chem. Ltd.] Polyglycol esters; dispersant for cosmetic and industrial hydrophilic systems; liq., solid; water-sol.

Cithrol PGML N/E. [Croda Chem. Ltd.] Propylene glycol laurate; nonionic; emulsifier, coemulsifier, stabilizer, wetting agent, lubricant, and antistat; used in cosmetic, pharmaceutical, industrial, food applics.; liq.; HLB 2.7; 100% conc.

Cithrol PGML S/E. [Croda Chem. Ltd.] Propylene glycol laurate SE; anionic; emulsifier, coemulsifier, stabilizer, wetting agent, lubricant, and antistat; used in cosmetic, pharmaceutical, industrial, food applics.; liq.; HLB 3.6; 100% conc.

Cithrol PGMO N/E. [Croda Chem. Ltd.] Propylene glycol oleate; CAS 1330-80-9; nonionic; emulsifier, coemulsifier, stabilizer, wetting agent, lubricant, and antistat; used in cosmetic, pharmaceutical, industrial, food applics.; liq.; HLB 3.1; 100% conc.

Cithrol PGMO S/E. [Croda Chem. Ltd.] Propylene glycol oleate SE; anionic; emulsifier, coemulsifier, stabilizer, wetting agent, lubricant, and antistat; used in cosmetic, pharmaceutical, industrial, food applics.; liq.; HLB 3.9; 100% conc.

Cithrol PGMR N/E. [Croda Chem. Ltd.] Propylene glycol ricinoleate; nonionic; emulsifier, coemulsifier, stabilizer, wetting agent, lubricant, and antistat; used in cosmetic, pharmaceutical, industrial, food applics.; liq.; HLB 2.7; 100% conc.

Cithrol PGMR S/E. [Croda Chem. Ltd.] Propylene glycol ricinoleate SE; anionic; emulsifier, coemulsifier, stabilizer, wetting agent, lubricant, and antistat; used in cosmetic, pharmaceutical, industrial, food applics.; liq.; HLB 3.6; 100% conc.

Cithrol PGMS N/E. [Croda Chem. Ltd.] Propylene glycol stearate; CAS 1323-39-3; nonionic; emulsifier, coemulsifier, stabilizer, wetting agent, lubricant, and antistat; used in cosmetic, pharmaceutical, industrial, food applics.; solid; HLB 2.4; 100% conc.

Cithrol PGMS S/E. [Croda Chem. Ltd.] Propylene glycol stearate SE; anionic; emulsifier, coemulsifier, stabilizer, wetting agent, lubricant, and antistat; used in cosmetic, pharmaceutical, industrial, food applics.; solid; 3.2; 100% conc.

Cithrol S Range. [Croda Chem. Ltd.] Polyglycol esters; dispersant for cosmetic and industrial hydrophilic systems; liq./solid; water-sol.

Citranox®. [Alconox] Blend of organic acids, anionic and nonionic surfactants, alkanolamines; anionic/nonionic; phosphate-free biodeg. acid detergent, wetting agent for cleaning dairy equip., laboratory ware, clean rooms, optical and electronic parts, pharmaceutical apparatus, industrial parts, etc.; pale yel. liq., nearly odorless; sol. in water; sp.gr. 1.120; dens. 9.15 lb/gal; b.p. 217 F; flash pt. none; pH 2.5 (1%); surf. tens. 32 dyne/cm (1%); toxicology: nontoxic orally; nonirritating by inhalation; serious eye irritation potential if not rinsed.

Clearate B-60. [W.A. Cleary] Soya lecithin; CAS 8002-43-5; nonionic; w/o emulsifier; liq.; 65% conc.

Clearate LV. [W.A. Cleary] Soya lecithin deriv.; CAS 8002-43-5; nonionic; pigment dispersant, wetting agent, and antiflocculant in paint systems; liq.; 56% conc.

Clearate Special Extra. [W.A. Cleary] Lecithin; CAS 8002-43-5; nonionic; dispersant in inks, paints, foods; wetting agent; liq.

Clearate WD. [W.A. Cleary] Soya lecithin derivs.; CAS 8002-43-5; nonionic; pigment dispersant, wetting agent, antiflocculant for water-based paint systems; liq.; water-disp.; 56% conc.

Clearbreak TEB. [Chemron] Polymeric amine; cationic; reverse emulsion breaker for oil treating; may be combined with metal salts or anionic polymers for enhanced performance; dk. amber liq.; 50% act.

Cleary's Waterless Hand Cleaner. [W.A. Cleary] Conc. heavy-duty but gentle cleaning agent for removal of pesticide residues, grease, oil, tar, ink, paint, carbon, adhesives from hands, work clothes, cars.

Clink A-26. [Yoshimura Oil Chem.] Surfactant/builders blend; nonionic; cleaning agent for metal surfaces, esp. aluminum; liq.; 18% conc.

Clink A-70. [Yoshimura Oil Chem.] Surfactant/alkali builders blend; amphoteric; cleaning agent for dyeing machine and other heavy duty purposes; liq.; 25% conc.

Closyl 30 2089. [Clough] Sodium-N-cocoyl sarcosinate; anionic; wetting, foaming detergent used in personal care and household prods.; corrosion inhibitor for mild steel; liq.; pH 7.5-9.0; 30–31% conc.

Closyl LA 3584. [Clough] Sodium lauroyl sarcosinate; anionic; soap-like detergent providing wetting and foaming; for personal care and household prods.; corrosion inhibitor for mild steel; liq.; pH 7.5-9.0; 30-31% conc.

CM-88®. [Georgia-Pacific] Amine-lignosulfonate; cationic; modifier and extender for cationic emulsions; replaces 40–60% of a primary emulsifier; solid; dens. 1.25 g/ml; 46% solid.

CNC Antifoam 1-A, 1-AP. [CNC Int'l.] Emulsified defoaming agents for aq. systems in textile industry.

CNC Antifoam 10-FG. [CNC Int'l.] Silicone; antifoam/defoamer; also avail. in food grade versions; 10% act.

CNC Antifoam 30-FG. [CNC Int'l.] Silicone emulsion; nonionic; antifoam/defoam for paper, textile, water-phase paints and other industrial uses; avail. in food grade versions; wh. visc. emulsion; easily disp. with water; dens. 8.3 lb/gal; pH 4-5; 27-28% act. silicone.

CNC Antifoam 77. [CNC Int'l.] Silicone; defoamer for aq. systems, effluents, resin systems, coatings, adhesives; liq.

CNC Antifoam 100. [CNC Int'l.] Silicone; defoamer for aq. systems in textile industry.

CNC Antifoam 495, 495-M. [CNC Int'l.] Silicone emulsions; nonionic/anionic; foam control agent for aq. systems, textile dyeing, paper, water-phase paints, industrial municipal effluents; wh. emulsion; visc. 1200 cps; pH 7.0.

CNC Antifoam A-107. [CNC Int'l.] Filled silicone fluid; defoamer for aq. and nonaq. systems; effective over very broad pH and temp. range; 100% act.

CNC Antifoam SS. [CNC Int'l.] Compounded silicone fluid; defoamer for solv. phase systems by itself or as base to make emulsion defoamers for water-phase systems.

CNC Antifoam SSD. [CNC Int'l.] Compounded silicone fluid with dispersant; defoamer for aq. and nonaq. systems.

CNC Defoamer 12, 34, 407. [CNC Int'l.] Defoamer for paper applics. for difficult, foamy high solids, or high resin content systems; liq.

CNC Defoamer 44 Series. [CNC Int'l.] Defoamer for coated or heavily surface sized sheet such as wallpaper; liq.

CNC Defoamer 69, 97. [CNC Int'l.] Nonsilicone; water-based defoamer for paper applics.

CNC Defoamer 229. [CNC Int'l.] Nonsilicone; defoamer for aq. systems in industrial applics. (effluent, pulp and paper, and municipal waste treatment, coatings, adhesives, textile scouring, dyeing, and finishing; FDA compliance; amber opaque liq.;

dens. 7.5 lb/gal; visc. 200-400 cps; pour pt. 35 F; flash pt. (COC) > 300 F.

CNC Defoamer 544-C. [CNC Int'l.] Defoamer for jet and pressure dye machines in textile industry.

CNC Detergent E. [CNC Int'l.] Coconut fatty acid amine condensate; anionic; detergent, wetting agent, dyeing assistant; leveling agent for acid dyestuffs on wool; rewetting and softening agent for sanforizing; scouring agent for synthetics; also for laundry processing of garments; amber visc. liq.; readily sol. in warm water; pH 9.8 (1%); 98-100% act.

CNC Dispersant WB Series. [CNC Int'l.] Anionic; surfactants, leveling agents, dispersants, dyeing assistants for textiles.

CNC Dispersion PE. [CNC Int'l.] Complex phosphate ester of nonionic ethoxylates; forms self-emulsfiable concs. when mixed with many solvs.; base for textile wetting agents.

CNC Foam Assist AA, AA-100. [CNC Int'l.] High solids foaming agents for textile foam finishing.

CNC Gel Series. [CNC Int'l.] Anionic; detergents for textile scouring, wetting, dispersing.

CNC Leveler JH. [CNC Int'l.] Sulfated PEG ester; low-foaming surfactant, wetting agent, dispersant for textile applics.; high compat. with high concs. of acids.

CNComerse IMP. [CNC Int'l.] Alkyl phospho-sulfate; nonfoaming biodeg. mercerizing agent, wetting and penetrating agent.

CNC PAL 100, 200, 300. [CNC Int'l.] Surfactants; anionic, nonionic, and cationic low-foaming biodeg. surfactants for rewetting towels, tissue or corrugating medium, deinking printed wastes, magazine stocks; wetting and penetrating agent, detergent.

CNC PAL 210 T. [CNC Int'l.] Surfactant; nonionic; low foaming surfactant, wetting, dispersing, rewetting and emulsifying agent; emulsifier for oils, waxes, petrol. hydrocarbons; wetter for towels, tissue, corrugated medium; deinking and repulping aid; clear semivisc. liq.; dissolves in warm or cold water; pH 7.5.

CNC PAL 500. [CNC Int'l.] Leveling and scouring surfactant for direct, acid, vat, and sulfur dyes.

CNC PAL 1000. [CNC Int'l.] Leveling and scouring surfactant for acid dyes on nylon, esp. for continuous dyeing of nylon carpet.

CNC PAL AN. [CNC Int'l.] Coconut fatty acid amide condensate; nonionic/anionic; detergent, wetting agent, penetrant, emulsifier for textile applics.

CNC PAL DS. [CNC Int'l.] Nonionic/sulfate blend; scouring, leveling, wetting, retarding agent for textiles; prevents barre; compat. with sodium sulfate, sodium chloride at almost any pH or temp.

CNC PAL NRW. [CNC Int'l.] Surfactant; nonionic; enhances stability of water repellent and fluorochemical baths; wetting agent for textiles which will not rewet after curing.

CNC PAL V-8 Supra. [CNC Int'l.] Detergent, emulsifier for textile processing.

CNC Product PW. [CNC Int'l.] Print wash used in conjunction with CNC Fix Conc. for nylon printed with acid dyes; amber liq.

CNC Product ST. [CNC Int'l.] Surfactant; anionic; print wash for use where high concs. of residual color and print paste are to be removed from cotton, rayon, or synthetics; wetting and scouring agent.

CNC Sol 72-N Series. [CNC Int'l.] Anionic; wetting and rewetting agents for textiles; stable to dilute acids and alkalies; not for use with water repellent or fluorochemical finishes.

CNC Sol BD. [CNC Int'l.] Solv./emulsifier blend; for removal of waxes, greases, oils, and greige-mill dirt from cotton, rayon, and blends.

CNC Sol UE. [CNC Int'l.] Aliphatic and aromatic solvs., terpenes, low titer soaps, and nonionic biodeg. detergents; detergent/solv. for removal of oil, dirt, waxes, graphite, for textile scouring; amber liq.; dissolves in hot or cold water.

CNC Sol XN. [CNC Int'l.] Solv./emulsifier with alkaline buidlers, nonionic detergents; for removal of waxes, oils, and greige mill dirt from cotton, rayon, and blends.

CNC Sol XNN #11. [CNC Int'l.] Solv./emulsifier blend; for removal of waxes, greases, oils, and greige-mill dirt from cotton, rayon, and blends.

CNC Solv 809. [CNC Int'l.] Solv./surfactant blend; for deinking and repulping resin coated mixed waste furnishes; prevents redeposition of agglomerated inks and binders.

CNC Wet CP. [CNC Int'l.] Neutralized phosphate ester; general wetting and dispersing aid for clays and pigments, detergent for cleaners; for pulp and paper industry.

CNC Wet CP Conc. [CNC Int'l.] Phosphated long chain alcohol; wetting and dispersing agent, leveler for pigments and dyes, in peroxide bleach baths; dyeing assistant; suitable for cotton and synthetics processing; acid and alkali stable; amber liq.; sol. in cold water; dens. 8.6 lb/gal; pH 8.1 (1%).

CNC Wet CP-X. [CNC Int'l.] Moderate foaming wetting and dispersing aid for minerals, clay, or pigment systems; for high solids fluid, industrial, and paper coatings.

CNC Wet SS-80. [CNC Int'l.] Long chain alcohol; anionic; detergent, dispersant, wetting agent for textiles; acid and alkali stable.

CO-810. [Procter & Gamble] C8-10 alcohols; CAS 68603-15-5; intermediate for mfg. of alkyl sulfates and ethoxylates, alkyl halides, esters, etc.; water-wh. mobile liq.; sp.gr. 0.823; m.p. -15 C; acid no. 0.1 max.; iodine no. 0.5 max.; sapon. no. 0.6 max.; hyd. no. 380-406; 100% conc.

CO-1214. [Procter & Gamble] C12-14 alcohols; CAS 67762-41-8; intermediate for mfg. of alkyl sulfates and ethoxylates, alkyl halides, esters, etc.; water-wh. mobile liq.; sp.gr. 0.823; m.p. 22 C; acid no. 0.1 max.; iodine no. 0.3 max.; sapon. no. 0.5 max.; hyd. no. 280-290; 100% conc.

CO-1218. [Procter & Gamble] C12-18 alcohols; CAS 67762-25-8; intermediate for mfg. of alkyl sulfates and ethoxylates, alkyl halides, esters, etc.; wh. semisolid; sp.gr. 0.826; m.p. 28 C; acid no. 1 max.; iodine no. 0.7 max.; hyd. no. 256-270; 100% conc.

CO-1270. [Procter & Gamble] C12-14 alcohols; CAS 67762-41-8; intermediate for mfg. of alkyl sulfates and ethoxylates, alkyl halides, esters, etc.; water-wh. mobile liq.; sp.gr. 0.831; acid no. 0.1 max.; iodine no. 0.3 max.; sapon. no. 0.5 max.; hyd. no. 285-295; 100% conc.

CO-1695. [Procter & Gamble] Cetyl alcohol; CAS 36653-82-4; emollient, intermediate for mfg. of alkyl sulfates and ethoxylates, alkyl halides, esters, etc.; wh. waxy solid; sp.gr. 0.814 (55 C); m.p. 47-50 C; acid no. 0.5 max.; iodine no. 2 max.; sapon. no. 1 max.; hyd. no. 220-235; 100% conc.

CO-1895. [Procter & Gamble] Stearyl alcohol; CAS 112-92-5; emollient, intermediate for mfg. of alkyl sulfates and ethoxylates, alkyl halides, esters, etc.;

wh. waxy solid; sp.gr. 0.811 (65 C); m.p. 56-60 C; acid no. 0.5 max.; iodine no. 2 max.; sapon. no. 2 max.; hyd. nno. 200-215; 100% conc.

CO-1897. [Procter & Gamble] Stearyl alcohol; CAS 112-92-5; intermediate for mfg. of alkyl sulfates and ethoxylates, alkyl halides, esters, etc.; wh. waxy solid; sp.gr. 0.811 (65 C); m.p. 56-60 C; acid no. 0.5 max.; iodine no. 2 max.; sapon. no. 1 max.; hyd. no. 200-215; 100% conc.

CO-1898. [Procter & Gamble] Stearyl alcohol; CAS 112-92-5; intermediate for mfg. of alkyl sulfates and ethoxylates, alkyl halides, esters, etc.; wh. waxy solid; sp.gr. 0.811 (65 C); m.p. 56-60 C; acid no. 0.5 max.; iodine no. 2 max.; sapon. no. 1 max.; hyd. no. 200-215.

Code 8059. [Hart Prods. Corp.] Fatty acid condensate; cationic; base for car water beading agent; liq.; 85% conc.

Colloid 111-D (formerly Geropon® 111-D). [Rhone-Poulenc Surf.] Polymeric; anionic; surfactant, dispersant for agric. formulations; EPA compliance; spray-dried; 95% act.

Collone AC. [Rhone-Poulenc UK] Alcohol ethoxylate; nonionic; emulsifying wax; waxy solid; HLB 10.4; 100% conc.

Collone HV. [Rhone-Poulenc UK] Fatty ester; anionic; emulsifying wax; waxy solid; 100% conc.

Collone NI. [Rhone-Poulenc UK] Alcohol ethoxylate blend; nonionic; Cetomacrogrol emulsifying wax BP; solid; 100% conc.

Colonial A-18. [Colonial Chem.] N-Tallow sulfosuccinamate; surfactant; 70% conc.

Colonial ALS. [Colonial Chem.] Ammonium lauryl sulfate; surfactant.

Colonial LKP. [Colonial Chem.] Sodium lauryl sulfate; CAS 151-21-3; low cloud pt. surfactant.

Colonial SLE(2)S. [Colonial Chem.] Sodium laureth sulfate; CAS 9004-82-4; surfactant; 30% conc.

Colonial SLE(3)S. [Colonial Chem.] Sodium laureth sulfate; CAS 9004-82-4; surfactant; 30% conc.

Colonial SLES-70. [Colonial Chem.] Sodium laureth sulfate; CAS 9004-82-4; surfactant; 70% conc.

Colonial SLS. [Colonial Chem.] Sodium lauryl sulfate; CAS 151-21-3; surfactant.

Colonial STDS. [Colonial Chem.] Sodium tridecyl sulfate; surfactant.

Colonial TLS. [Colonial Chem.] TEA lauryl sulfate; surfactant.

Colorin 301, 302. [Sanyo Chem. Industries] Polyether polyol; antifoamer used in acrylonitrile refining process; liq.

Colorol 20. [Lucas Meyer] High m.w. phosphoamino compds. dissolved in solvs. compat. with NC; amphoteric; wetting and dispersing agent, binder for NC and NC-modified lacquers; brn. visc. liq.; sp.gr. 1.02; visc. 2000 mPas; flash pt. 30 C; pH 6–7; 90% solids; contains xylene.

Colorol 46. [Lucas Meyer] Phosphoamino and surfactants; wetting agent for treated silicates; flatting agent; used for flat varnishes based on NC, NC-alkyds, polyester, acid curing resins; lt. yel. liq.; sp.gr. 0.85; flash pt. 15 C; pH 6.0; 28% solids; contains ethanol.

Colorol 70. [Lucas Meyer] Phosphoamino compds.; amphoteric; wetting and dispersing agent, binder for 2-part PU varnishes, epoxy-varnishes in systems based on PVC, PVA; ylsh. brn. liq.; sp.gr. 0.97; usage level: 0.5-2.0%; 50% solids; contains isopropyl acetate.

Colorol Aquasorb. [Lucas Meyer] Phosphoamino compds.; wetting agent for damp surfaces; procures better adhesion, improves anticorrosive props.; brn. liq.; dens. 0.89; flash pt. 1 C; 25% solids in xylene and alcohols.

Colorol E. [Lucas Meyer] Phosphoamino compds.; amphoteric; wetting and dispersing agent for aq. systems; used for water-thinnable paints, leather finishes, alkyd systems, electrodeposition, emulsion paints; brn. visc. liq.; sp.gr. 1.038; visc. 1500 cp; pH 5.5–6.5; usage level: 1-3%; 90% solids.

Colorol F. [Lucas Meyer] Phosphoamino compds. modified with film-formers, dissolved in aromatic hydrocarbon; amphoteric; wetting and dispersing agent, binder for pigmented air-dry and stoving finishes, anticorrosive paints, alkyd and acrylic paints, zinc dust primers, marine paints, epoxy and bitumen paints, two-part systems, putty, inks; brn. visc. liq.; sp.gr. 1.02; visc. 1300–1600 cp; flash pt. 74 C; pH 5.5–6.5; usage level: 0.5-2.0%; 90% solids; contains Shellsol AB.

Colorol Standard. [Lucas Meyer] Phosphoamino compds. modified with film-formers; amphoteric; wetting and dispersing agent, binder for pigmented air-dry and stoving finishes, anticorrosive paints, alkyd and marine paints, zinc dust paints, bituminous paints, epoxy resins, putty, solventless floor sealers, two-part finishes, inks; visc. to plastic brn. mass; sp.gr. 1.01; pH 6.5; usage level: 0.3-1.5%; 99–100% solids.

Colsol. [Scher] Fatty amide; cationic; cotton and wool softener and lubricant for the textile industry; finishing agent; paste; 25% act.

Comperlan 100. [Henkel Canada; Henkel/Emery/Cospha; Henkel KGaA] Cocamide MEA; CAS 68140-00-1; nonionic; foamer, booster for detergents, foam stabilizer, thickener; for bowl cleaners, shampoos; solid beads; 100% conc.

Comperlan CD. [Henkel Canada] Capramide DEA; nonionic; surfactant providing flash foam, no visc. buildup; amber visc. liq.; 98% conc.

Comperlan KDO. [Henkel Canada] 1:1 Cocamide DEA; nonionic; thickening and foaming agent for personal care prods., cleansers; amber visc. liq.; 100% conc.

Comperlan KM. [Henkel KGaA] Cocamide MEA; CAS 68140-00-1; booster, foam stabilizer, pearly sheen improving agent for surfactant preparations; paste; 30% conc.

Comperlan LD. [Henkel; Henkel KGaA; Pulcra SA] Lauramide DEA superamide; CAS 120-40-1; nonionic; foam booster, stabilizer, detergency and visc. builder, emulsifier for personal care prods., household detergents; waxy solid; pH 9.5-10.5 (1%); 100% conc.

Comperlan LM. [Henkel KGaA] Lauramide MEA; CAS 142-78-9; nonionic; foam booster/stabilizer and visc. builder for personal care prods. and detergents; solid; 100% conc.

Comperlan LMD. [Henkel KGaA] Lauramide DEA; CAS 52725-64-1; nonionic; foam booster/stabilizer and thickener for personal care prods.; fortifier for perfumes in soaps; liq.; 100% conc.

Comperlan LP. [Henkel Canada; Henkel KGaA] Lauramide MIPA; CAS 142-54-1; nonionic; foam stabilizer, thickener for shampoos and detergents; solid; 100% conc.

Comperlan LS. [Henkel KGaA] Cocamide DEA and laureth-12; emulsifier, thickener, and foam stabilizer for personal care prods.; dissolving intermediates; liq.; 65–75% conc.

Comperlan OD. [Henkel Canada; Henkel KGaA] Oleamide DEA; CAS 93-83-4; nonionic; foam stabilizer, thickener and superfatting agent for shampoos and bubble bath prods., body cleansers, metal spot removers, engine cleaners; liq.; sol. in oils and solvs.; 100% conc.

Comperlan P 100. [Pulcra SA] Cocamide MEA; CAS 68140-00-1; nonionic; foamer, thickener, visc. builder for powd. detergents, soaps; dispersant and blending agent for many cosmetic prods.; flakes; 100% conc.

Comperlan PD. [Henkel/Cospha; Henkel KGaA] 2:1 Cocamide DEA; nonionic; thickener, foam builder/stabilizer for personal care prods., dishwashing agents; coemulsifier for cleaning and metalworking agents; amber visc. liq.; 98% conc.

Comperlan PKDA. [Pulcra SA] Cocamide DEA (2:1); nonionic; detergent base, foam booster/stabilizer, emulsifier, wetting agent for household detergents; liq.; 100% conc.

Comperlan PVD. [Henkel KGaA] Veg. oil polydiethanolamide; nonionic; active for dishwashing agents, coemulsifier for cleaning and metalworking agents; liq.

Comperlan SD. [Henkel Canada] Cocamide DEA; nonionic; foamer; liq.; 100% conc.

Compound 170. [Henkel/Cospha; Henkel Canada] Organic phosphates blend; anionic; emulsifier for metalworking fluids; imparts lubricity, anticorrosion and mild EP props.; grease additive; liq.; 98% conc.

Compound 535. [A. Harrison] Coconut acid and amine; anionic; wetting and fulling agent; liq.; 60% conc.

Compound S.A. [A. Harrison] Cocamide DEA modified; anionic; wetting agent, fulling and scouring agent for wool and worsted fabrics with soda ash; liq.; 65% conc.

Conco X-200. [Continental Chem.] Sodium alkylaryl polyether sulfonate; anionic; detergent, wetting agent, emulsifier, dye leveler for acid dyestuffs, latex post-stabilizer; for pickling and plating baths; liq.; 28% conc.

Condensate PC. [Nat'l. Starch] Coconut-DEA condensation product; wetting and rewetting agent, detergent, high foamer, emulsifier, thickener, used in the textile industry, dishwashing; amber liq.; 100% act.

Coning Oil C Special. [Hoechst AG] Surfactants/paraffin hydrocarbon blend; nonionic; coning oil imparting antistatic props. and high smoothness to textured PA and PES yarns for processing; liq.

Consamine 15. [Consos] Alkyl POE ether; nonionic; detergent, scouring, wetting agent for textiles; liq.; 100% conc.

Consamine CA. [Consos] Fatty alkanolamide; nonionic; detergent, emulsifier, intermediate, detergent base, thickener, fulling agent used in textiles; liq.

Consamine DS Powd. [Consos] Built detergent powd.; anionic; detergent for textile scouring before and after dyeing and bleaching; biodeg.; powd.; 100% conc.

Consamine K-Gel. [Consos] Sodium methyl oleyl taurate; anionic; wetting, scouring, and detergent for textile applics.; gel.

Consamine OM. [Consos] Blend of detergents, emulsifiers, solvs.; nonionic; leveling agent and stripper for use in dyeing polyester yarn and fabric with disperse and basic dyestuffs; liq.; 100% conc.

Consamine P. [Consos] Aromatic solv. mixture with emulsifier and surfactants; anionic; solv.-based cleaner for textile printing and dyeing equip.; solv. scour; liq.; 100% conc.

Consamine PA. [Consos] Complex alcohol ester; nonionic; wetting agent, and penetrant for textile processes, kier bleaching; stable in caustic and silicate; liq.

Consamine X. [Consos] Blend of anionic and nonionic surfactants; anionic; wetting agent, scouring compd. for textile fibers; biodeg.; liq.

Consoft CP-50. [Consos] Quaternized imidazoline; cationic; softener for acrylics and nylon, napping lubricant and leveling agent; liq.

Consolevel JBT. [Consos] Blend of anionic surfactants; anionic; leveling agent for dyeing cotton and rayon yarn and knit goods; low foam; biodeg.; liq.

Consolevel N. [Consos] Surfactant blend; anionic/nonionic; leveling agent for eliminating barre in dyeing nylon with acid and disperse dyes; biodeg.; liq.

Consoluble 71. [Consos] Fatty deriv.; anionic; lubricant, dipsersant, leveling agent for dyeing syn. knit goods; prevents crease goods; liq.

Consonyl FIX. [Consos] Alkyl naphthalene sulfonate; anionic; fixative for improving wash fastness of acid dyestuffs on nylon; resist for direct and naphthol dyes on nylon; liq.

Consos Castor Oil. [Consos] Sulfated castor oil; anionic; retarder and leveling agent for dyeing applic. in textiles; dispersant and lubricant; liq.

Consoscour 47. [Consos] Alkylaryl sulfonate; anionic; detergent, scouring, and leveling agent for dyeing applics.; liq.; 47% conc.

Consoscour M. [Consos] Emulsifiable solv.; nonionic; solvent scour for removing oils, wax and grease from goods prior to dyeing; liq.; 100% conc.

Consoscour TEK. [Consos] Solvent/emulsifier blend; nonionic; low temp. solvent scour for polyester and nylon knit goods; removes heavy oil and wax deposits; low foam; biodeg.; liq.

Coptal WA OSN. [ICI Surf. UK] Sodium dioctyl sulfosuccinate in high flash solv.; anionic; wetting agent, penetrant for continuous processes; liq.

Coralon F. [Hoechst AG] Neutral salt of an aromatic condensation prod.; leveling agent for chrome leather.

Coralon GP. [Hoechst AG] Neutral salt of an aromatic condensation prod.; dyeing auxiliary.

Coralon Grades. [Hoechst AG] Neutral salt of an aromatic condensation prod.; leveling agent for suede dyeing.

Cordex DJ. [Finetex] Fatty amine deriv.; cationic; leveling agent for cationic dyes on syn. fibers; antistat for acrylics, polyamides, and acetate fibers; liq.; water-sol.

Cordon AES-65. [Finetex] Sulfonated syn. sperm oil; anionic; detergent, wetting agent; lubricant, dye leveling for direct and disperse dyes; prevents chafe marks in textiles; liq.; 65% conc.

Cordon COT. [Finetex] Organic sulfonate; anionic; direct dye leveler and stripper; liq.

Cordon DA-B. [Finetex] Sulfated fatty acid ester; anionic; wetting agent, lubricant; liq.; 65% conc.

Cordon N-400. [Finetex] Complex alkyl-naphthalene sulfonate; anionic; disperse dye leveling and dispersing agent; liq.; 22% conc.

Cordon NU 890/75. [Finetex] Sulfated castor oil; anionic; emulsifier for pine oil; plasticizer, lubricant, textile dyeing and kier bleaching assistant; dispersant for vinyl pigments; liq.; 70% conc.

Cordon PB-870. [Finetex] Sulfated tall oil; anionic; lubricant, emulsifier; liq.; 42% conc.

Corexit CL578. [Exxon] Alkanolamide and ethoxylated alkylphenol; nonionic; degreaser and multipurpose cleaner; liq.; 18% conc.

Corexit CL8500. [Exxon] Polyol; nonionic; high pH degreaser and cleaner; liq.; 15% conc.

Corexit CL8569. [Exxon] Ethoxylated alcohols; nonionic; water-based high pH cleaner and degreaser for removing oil, fuel, and grease deposits; liq.; 19% conc.

Corexit CL8594. [Exxon] Ethoxylated alkylphenol and ethoxylated alcohol; nonionic; heavy-duty hydrocarbon solv.-based cleaner and degreaser that solubilizes heavy oils, asphaltic materials and grease deposits to permit their removal by a water or solv. flush; liq.; 21% conc.

Corexit CL8662. [Exxon] Ethoxylated alkylphenol and ethoxylated alcohol; nonionic; multipurpose cleaner conc.; liq.; 45% conc.

Corexit CL8685. [Exxon] Ethoxylated alkylphenol and ethoxylated alcohol; nonionic; surfactant conc. for cleaner and degreaser formulations; liq.; 60% conc.

Cosmopon 35. [Auschem SpA] Sodium lauryl sulfate; CAS 151-21-3; anionic; spray-dried wool detergent powd., hand cleaner; paste; 48% conc.

Cosmopon BL. [Auschem SpA] Fatty alkanolamide sulfosuccinate; CAS 68784-08-7; anionic; rug shampoo, bubble bath, conditioner; paste; 50% conc.

Cosmopon BN. [Auschem SpA] Sodium oleyl sulfosuccinamate; anionic; emulsifier for emulsion polymerization; foaming agent for latex emulsions; antigelling and cleaning agents for paper mill felts; paste; 35% conc.

Cosmopon LE 50. [Auschem SpA] Sodium laureth sulfate; CAS 9004-82-4; 15826-16-1; anionic; personal care prods. and liq. detergents; paste; sol. in cold water; 50% conc.

Cosmopon MO. [Auschem SpA] MEA lauryl sulfate; CAS 4722-98-9; anionic; raw material for liq. detergent; liq.; 32% conc.

Cosmopon SES. [Auschem SpA] Sodium 2-ethylhexylsulfate; CAS 126-92-1; anionic; detergent, wetting agent; liq.; 30% conc.

Cosmopon TR. [Auschem SpA] TEA lauryl sulfate; CAS 139-96-8; anionic; hair shampoo, bubble bath, car wash, liq. hand cleaner; liq.; 40% conc.

Coursemin SWG-7. [Sanyo Chem. Industries] Anionic and nonionic surfactants; anionic; low foaming dyebath lubricant to prevent crease and wrinkle formation; liq.

Coursemin SWG-10. [Sanyo Chem. Industries] Anionic and nonionic surfactants; anionic; low foaming dyebath lubricant for dyeing, scouring, and bleaching; stable to acid, alkali, and other chemicals; paste.

CPA-Alpha®. [Dixo] Detergent blend; anionic/nonionic; detergent, wetting agent, emulsifier for water and drycleaning solv.; very lt. yel. amber liq.; slight odor; complete sol. in drycleaning solv.; 100% act. DISCONTINUED.

CPH-27-N. [C.P. Hall] PEG-4 laurate; CAS 9004-81-3; nonionic; solubilizer and emulsifier; Gardner 3 clear oily liq.; sol. in toluene, kerosene, ethanol, acetone; disp. in water; sp.gr. 0.985; f.p. 5 C; flash pt. 199 C; acid no. 4.0; sapon. no. 139; ref. index 1.454; 100% conc.

CPH-30-N. [C.P. Hall] PEG-8 laurate; CAS 9004-81-3; nonionic; solubilizer and emulsifier; Gardner 2 clear oily liq.; sol. in toluene, kerosene, ethanol, acetone, water; sp.gr. 1.030; HLB 12.9; f.p. 15 C; acid no. 4.0; sapon. no. 91; ref. index 1.458.

CPH-39-N. [C.P. Hall] PEG-4 oleate; CAS 9004-96-0; nonionic; spreading agent, defoamer; Gardner 2 clear oily liq.; m.w. 457; sol. in toluene, kerosene, ethanol, acetone; disp. water; sp.gr. 0.969; HLB 7.6; f.p. –5 C; acid no. 3.0; sapon. no. 125; ref. index 1.462; 100% conc.

CPH-41-N. [C.P. Hall] PEG-12 oleate; CAS 9004-96-0; nonionic; solubilizer and emulsifier; Gardner 2 clear oily liq.; m.w. 857; sol. in toluene, kerosene, ethanol, acetone, water; sp.gr. 1.035; HLB 13.6; f.p. 25 C; acid no. 4.0; sapon. no. 67; ref. index 1.466; 100% conc.

CPH-53-N. [C.P. Hall] Glyceryl stearate; nonionic; thickener and emulsifier; wh. flake; m.w. 330; sol. in toluene, ethanol, acetone; HLB 3.7; m.p. 56 C; acid no. 2.0; sapon. no. 170; 100% conc.

CPH-79-N. [C.P. Hall] PEG-8 dilaurate; CAS 9005-02-1; nonionic; wetting agent, pigment dispersant; Gardner 2 clear oily liq.; sol. in hexane, toluene, kerosene, ethanol, acetone; disp. in water; sp.gr. 0.990; HLB 9.7; f.p. 17 C; acid no. 9.0; sapon. no. 128; ref. index 1.457; 100% conc.

CPH-250-SE. [C.P. Hall] Glyceryl stearate; anionic; emulsifier; wh. flake; m.w. 300; sol. in toluene, ethanol, acetone; disp. in water; m.p. 56 C; acid no. 17.0; sapon. no. 155; 100% conc.

CPH-376-N. [C.P. Hall] POE laurate; nonionic; solubilizer and emulsifier; PVC antistat; Gardner 2 clear oily liq.; sol. in toluene, kerosene, ethanol, acetone, propylene glycol, min. oil; disp. water; sp.gr. 1.002; acid no. 4.0; sapon. no. 120; ref. index 1.455; 100% conc.

Crafol AP-11. [Pulcra SA] Oleth-3 phosphate; CAS 39464-69-2; anionic; foam stabilizer, antistat, conditioner for cosmetic formulations; detergent and antistat for textile finishing processes; liq.; 100% conc.

Crafol AP-16. [Pulcra SA] Ethoxylated C16 phosphate ester; anionic; surfactant, textile lubricant; paste; sp.gr. 1.006; pH 7.2-7.8 (1%); 49-51% act.

Crafol AP-20. [Pulcra SA] Phosphate ester, acid form; anionic; surfactant, antistat; liq.; sp.gr. 1.039; acid no. 370-395; pour pt. < 0 C; pH 1.5-2.5 (1%); > 99% act.

Crafol AP-21. [Pulcra SA] C8 phosphate ester; anionic; surfactant, antistat; paste; sp.gr. 1.053; pH 7.2-7.8 (1%); 77-79% act.

Crafol AP-35. [Pulcra SA] C8-10 phosphate ester, acid form; anionic; surfactant, antistat; liq.; sp.gr. 1.003; acid no. 290-310; pour pt. < 0 C; pH 1.0-2.5 (1%); > 99% act.

Crafol AP-36. [Pulcra SA] DEA C8-10 phoshpate ester; anionic; surfactant, antistat for textile industry; paste; sp.gr. 1.044-1.046 (70 C); drop pt. 40 C; acid no. 187-213; pH 6.5-7.5 (1%); > 99% act.

Crafol AP-53. [Pulcra SA] Nonylphenol ether phosphate, free acid; CAS 66197-78-2; anionic; lubricant, emulsifier, corrosion inhibitor for oil and water-based cutting fluids, hydraulic fluids, rolling oils; liq.; sp.gr. 1.113; acid no. 119-128; pour pt. < 0 C; > 99% conc.

Crafol AP-55. [Pulcra SA] Alkylphenol ether phosphate, potassium salt; anionic; wetting and dispersing agent in aq. conc. dispersions, triazine-based pesticide formulations; humectant; liq.; sp.gr. 1.079; pour pt. < 0 C; pH 7.0-8.0 (1%); 85-87% act.

Crafol AP-60. [Pulcra SA] Ethoxylated C8-10 phosphate ester, free acid; CAS 9004-80-2; detergent, emulsifier, wetting agent, humectant, disperant, hydrotrope, compatabilizer and coupler for alkaline formulations; lubricant compat. with conc. electrolyte sol'ns.; detergent for hard surfaces; liq.; water-sol.; sp.gr. 1.100; acid no. 175-200; pour pt. < 0 C; pH < 2.5 (10%); > 99% conc.

Crafol AP-63. [Pulcra SA] Alcohol ether phosphate, free acid; CAS 39464-66-9; anionic; emulsifier for min. oils, waterless hand cleaners, fiber antistat lubricant; liq./paste; sp.gr. 1.014; acid no. 165-190; pour pt. 13 C; pH < 2.5 (10%); > 99% conc.

Crafol AP-64. [Pulcra SA] Ethoxylated C12-18 phosphate (8 EO), sodium salt; anionic; wetting agent, detergent, humectant; liq./paste; sp.gr. 1.042-1.053; acid no. 8-15; pour pt. < 0 C; pH 6-7 (1%); 28-32% act.

Crafol AP-65. [Pulcra SA] Ethoxylated C12-14 phosphate (6 EO), MEA salt; anionic; surfactant, lubricant, antistat for textiles; thick liq.; sp.gr. 1.030-1.033; acid no. 85-90; pour pt. < 13 C; pH 8-9 (1%); > 99% act.

Crafol AP-67. [Pulcra SA] Ethoxylated phosphate ester; anionic; surfactant, detergent, humecatnt, dispersant for textile industry; liq.; sp.gr. 1.100; pour pt. < 0 C; pH 6.0-6.5 (5%); 85% act.

Crafol AP-68. [Pulcra SA] Ethoxylated phosphate ester; anionic; surfactant, antistat; liq.; sp.gr. 1.050; pour pt. < 0 C; pH 6.0-7.0 (5%); 39-41% act.

Crafol AP-69. [Pulcra SA] Ethoxylated C10 phosphate ester; anionic; surfactant, dispersant, humectant, antistat; paste; sp.gr. 1.129; pH 7.0-7.8 (1%); 83-85% act.

Crafol AP-70. [Pulcra SA] Ethoxylated C10 phosphate ester; anionic; surfactant, dispersant, humectant, antistat; paste; sp.gr. 1.026; pH 7.0-7.8 (1%); 83-85% act.

Crafol AP-85. [Pulcra SA] Ethoxylated phosphate ester, free acid; anionic; buffering agent for pesticide emulsions; liq.; sp.gr. 1.152; acid no. 303-325; pour pt. < 0 C; pH 1.0-2.0 (5%); > 99% act.

Crafol AP-201. [Henkel KGaA; Pulcra SA] Lauryl phosphate, free acid; anionic; lubricant, emulsifier for leather and textile auxiliaries; solid; sp.gr. 0.963-0.970 (70 C); m.p. 45 C; acid no. 220-280; pH 1.5-3.0 (1%); > 99% act.

Crafol AP-202. [Henkel KGaA; Pulcra SA] Lauryl phosphate, free acid; anionic; emulsifier, wetting agent for textile auxiliaries; solid; sp.gr. 0.920-0.926 (70 C); m.p. 46 C; acid no. 200-240; pH 2-3 (1%); > 99% act.

Crafol AP-203. [Henkel KGaA] Lauryl phosphate, free acid; anionic; wetting agent for textile auxiliaries; liq.; sp.gr. 0.978-0.991; acid no. 230-280; pour pt. < -15 C; pH 1.0-2.5 (1%); > 99% act.

Crafol AP-240. [Henkel KGaA] Ethoxylated C12-18 phosphate (10 EO), free acid; anionic; additive for leather auxiliaries; solid; sp.gr. 1.035-1.039 (70 C); drop pt. 30 C; acid no. 90-130; pH 1.5-3.0 (1%); > 99% act.

Crafol AP-241. [Henkel KGaA] Ethoxylated C12-18 phosphate (10 EO), free acid; anionic; antistat for textile auxiliaries; paste; sp.gr. 1.013 (70 C); drop pt. 33 C; acid no. 85 C; pH 2-3 (1%); > 99% act.

Crafol AP-260. [Henkel KGaA; Pulcra SA] Sodium laureth (10) phosphate; anionic; emulsifier, wetting agent, antistat for leather, textile auxiliaries, hot baths containing cyanides; liq.; sp.gr. 1.019-1.021; acid no. < 10; pour pt. < 2 C; pH 6.5-7.5 (10%); 10-14% act.

Crafol AP-261. [Henkel KGaA; Pulcra SA] Sodium laureth (10) phosphate; anionic; emulsifier, wetting agent, antistat for leather, textile auxiliaries, hot baths containing cyanides; visc. paste; sp.gr. 1.050-1.060; acid no. 5-10; pour pt. < 2 C; pH 6.5-7.5 (10%); 28-32% act.

Crafol AP-262. [Henkel KGaA] MEA laureth (6) phosphate; anionic; lubricant, antistat for textile auxiliaries; paste; sp.gr. 1.040-1.043; acid no. 60-90; pour pt. 7-14 C; pH 8-10 (1%); > 99% act.

Cralane AT-17. [Pulcra SA] Alkylaryl polyethoxyether, modified; nonionic; low foaming detergent for mechanical dishwashing and rinse aids, automatic laundering, metal cleaning, dairy equip. cleaning, nylon dyeing assistant; liq./paste; HLB 14.0; 100% conc.

Cralane AU-10. [Pulcra SA] Polymer; cationic; fixing agent for direct dyestuffs; liq.; 36% conc.

Cralane KR-13, -14. [Pulcra SA] Ethoxylated lanolin; nonionic; emollient, wetting, cleaning, and superfatting agents, o/w emulsifier, conditioner, solubilizer, and plasticizer for personal care prods. and germicides; solid; sol. in water and ethanol; 100% conc.

Cralane LR-10. [Pulcra SA] Ethoxylated lanolin (75 EO); emollient, wetting, cleaning, and superfatting agents, o/w emulsifier, conditioner, solubilizer, and plasticizer for personal care prods. and germicides; visc. liq.; 50% act.

Cralane LR-11. [Pulcra SA] Ethoxylated lanolin; nonionic; emollient, wetting, cleaning, and superfatting agents, o/w emulsifier, conditioner, solubilizer, and plasticizer for personal care prods. and germicides; visc. liq.; 50% conc.

Crapol AU-40. [Pulcra SA] Alkylamine polyglycol ether methosulfate plus additives; cationic; antistat and surface act. agent used in mfg. of heavy duty detergents; liq.; 46 ± 2% conc.

Crapol AV-10. [Pulcra SA] Coco hydroxyethyl imidazoline; cationic; corrosive inhibitor, dispersant and fluidizing agent for pigments, emulsions, and agric. prods.; acid detergent for food and dairy prods.; liq., paste; sol. in most hydrocarbon polar and chlorinated solvs., acids; water-disp.; 100% conc.

Crapol AV-11. [Pulcra SA] Lauryl hydroxyethyl imidazoline; cationic; corrosive inhibitor, dispersant and fluidizing agent for pigments, emulsions, and agric. prods.; acid detergent for food and dairy prods.; liq.; 100% conc.

Cremophor® GS-32. [BASF] Polyglyceryl-3 distearate; w/o emulsifier for nonpolar oils; flake.

Crestex REM 55. [Reilly-Whiteman] Additive for cleaning of pressure equip. in textile industry.

Crestolan NF. [Reilly-Whiteman] Detergent for textile scouring.

Crestopene 5X. [Reilly-Whiteman] Wetting agent for textile processing of cellulosics and blends.

Crestosolve 630. [Reilly-Whiteman] Detergent for textile scouring.

Crill 1. [Croda Inc.; Croda Surf. Ltd.] Sorbitan laurate; CAS 1338-39-2; nonionic; emulsifier, pigment dispersant, cosolv., wetting agent, antifoam, visc. reducer, mold release, antiblock agent, corrosion inhibitor, lubricant, antistat; used for cosmetics, food and food pkg., insecticides and herbicides, leather treatment,; metalworking fluids, oil slick dispersing, paints and inks, pharmaceuticals, plastics, polishes, textiles; pale yel. clear visc. liq.; sol. in ethanol, oleyl alcohol, min. oil; HLB 8.6; sapon. no. 158–170; 98% conc.

Crill 2. [Croda Inc.; Croda Surf. Ltd.] Sorbitan palmitate; CAS 26266-57-9; nonionic; see Crill 1; pale tan hard waxy solid; partially sol. in propylene glycol, ethyl and oleyl alcohols, olive oil, oleic acid; m.p. 46 C; HLB 6.7; sapon. no. 140–150; 98% conc.

Crill 3. [Croda Inc.; Croda Surf. Ltd.] Sorbitan stearate; CAS 1338-41-6; nonionic; emulsifier, lubricant, antistat; o/w emulsions; cosmetic and pharmaceutical creams and lotions; polishes; insecticides, herbicides; metal cleaners; buffing compds.; textile lubricants; food applics.; pale tan hard waxy solid; low odor; partially sol. in oleyl alcohol, olive oil, oleic acid; m.p. 51–54 C; HLB 4.7; sapon. no. 147–157; usage level: 0.5-5%; 98% conc.

Crill 4. [Croda Inc.; Croda Surf. Ltd.] Sorbitan oleate; CAS 1338-43-8; nonionic; w/o emulsifier, wetting agent, pigment dispersant, coupler, antifoam; cosmetic and pharmaceutical applic.; aerosol polishes; insecticidal sprays; inks and surf. coatings; metal working lubricants; cutting oils; textile lubricants; dry cleaning operations; food process antifoam; oil slick dispersant; amber visc. liq.; sol. in ethyl and oleyl alcohols, min. oil, IPM, olive oil, oleic acid; HLB 4.3; sapon. no. 147–160; usage level: 0.5-5%; 98% conc.

Crill 6. [Croda Inc.; Croda Surf. Ltd.] Sorbitan isostearate; CAS 71902-01-7; nonionic; w/o emulsifier, wetting agent, pigment dispersant for creams, lotions, aerosols; pale yel. clear visc. liq.; sol. in min. oil, olive oil, partly sol. in oleyl alcohol, IPM; HLB 4.6; sapon. no. 143–153; usage level: 0.5-5%; 98% conc.

Crill 35. [Croda Inc.; Croda Surf. Ltd.] Sorbitan tristearate; CAS 26658-19-5; nonionic; emulsifier, lubricant, antistat for cosmetic, pharmaceutical, food, and industrial applics.; pale tan hard waxy solid; partly sol. in oleyl alcohol, min. and olive oil, IPM, oleic acid; m.p. 48 C; HLB 2.1; sapon. no. 176–188; 98% conc.

Crill 43. [Croda Inc.; Croda Surf. Ltd.] Sorbitan sesquioleate; CAS 8007-43-0; nonionic; w/o emulsifier, wetting agent, pigment dispersant; for cosmetic, pharmaceutical, food, and industrial applics.; amber visc. liq.; sol. in oleyl alcohol, min. and olive oil, oleic acid; HLB 3.7; sapon. no. 149–160; 98% conc.

Crill 45. [Croda Inc.; Croda Surf. Ltd.] Sorbitan trioleate; CAS 26266-58-0; nonionic; w/o emulsifier, wetting agent, pigment dispersant; for cosmetic, pharmaceutical, food, and industrial applics.; amber visc. liq.; sol. in oleyl alcohol, min. and olive oil, IPM, oleic acid; HLB 4.3; sapon. no. 172–186; 98% conc.

Crill 50. [Croda Surf. Ltd.] Sorbitan oleate, tech.; CAS 1338-43-8; nonionic; emulsifier, dispersant, wetting agent for cosmetic, pharmaceutical, food, and industrial applics.; liq.; 98% conc.

Crillet 1. [Croda Inc.; Croda Chem. Ltd.] Polysorbate 20; CAS 9005-64-5; nonionic; solubilizer, emulsifier, dispersant, wetting agent; often combined with a member of the Crill range in emulsification systems; used in cosmetics, food and food pkg., household prods., insecticides, herbicides,; metalworking fluids, paints and inks, pharmaceuticals, textiles; clear, yel. clear liq.; low odor; sol. in water, ethyl and oleyl alcohol, oleic acid; HLB 16.7; sapon. no. 40–51; surf. tens. 38.5 dynes/cm (0.1%); 97% conc.

Crillet 2. [Croda Chem. Ltd.] PEG-20 sorbitan palmitate; CAS 9005-66-7; nonionic; emulsifier, solubilizer, wetting agent for cosmetic, pharmaceutical, food, and industrial applics.; paste; HLB 15.6; 97% conc.

Crillet 3. [Croda Inc.; Croda Chem. Ltd.] Polysorbate 60 NF; CAS 9005-67-8; nonionic; o/w emulsifier for cosmetics and pharmaceuticals, dispersant for insecticides, herbicides, cattle dyes, penetrant, leveling agent, lubricant, antistat; yel. liq. gels to soft solid on cooling; sol. in ethyl and oleyl alcohol, oleic acid; partly sol. in water; HLB 14.9; sapon. no. 45–55; surf. tens. 42.5 dynes/cm (0.1%); usage level: 0.5-5%; 97% conc.

Crillet 4. [Croda Inc.; Croda Chem. Ltd.] Polysorbate 80 NF; CAS 9005-65-6; nonionic; emulsifier, dispersant, solubilizer, lubricant, detergent, antistat, wetting agent for cosmetics, pharmaceuticals, polishes, insecticides, leather degreasing, veterinary prods.; clear yel. amber liq.; faint char. odor; sol. in water, ethyl and oleyl alcohols, oleic acid; HLB 15.0; sapon. no. 45–55; surf. tens. 42.5 dynes/cm (0.1%); usage level: 0.5-5%; 97% conc.

Crillet 6. [Croda Inc.; Croda Chem. Ltd.] PEG-20 sorbitan isostearate; CAS 66794-58-9; nonionic; o/w emulsifier, solubilizer for fragrances and perfumes; for creams, lotions, ointments; improved resistance to oxidation; clear yel. liq.; sol. in water, ethyl and oleyl alcohols, oleic acid, xylene, trichlorethylene; HLB 14.9; sapon. no. 40–50; surf. tens. 38.6 dynes/cm (0.1%); usage level: 0.5-5%; 97% conc.

Crillet 11. [Croda Chem. Ltd.] PEG-4 sorbitan laurate; CAS 9005-64-5; nonionic; emulsifier, solubilizer, wetting agent for cosmetic, pharmaceutical, food, and industrial applics.; liq.; 97% conc.

Crillet 31. [Croda Chem. Ltd.] PEG-4 sorbitan stearate; CAS 9005-67-8; nonionic; emulsifier, solubilizer, wetting agent for cosmetic, pharmaceutical, food, and industrial applics.; solid; 97% conc.

Crillet 35. [Croda Chem. Ltd.] Polysorbate 65; CAS 9005-71-4; nonionic; emulsifier, solubilizer, wetting agent for cosmetic, pharmaceutical, food, and industrial applics.; cream/buff waxy solid; sol. in ethyl and oleyl alcohols, oleic acid, trichlorethylene, partly sol. in water; HLB 10.5; sapon. no. 88–98; surf. tens. 42.5 dynes/cm (0.1%); 97% conc.

Crillet 41. [Croda Chem. Ltd.] PEG-5 sorbitan oleate; CAS 9005-65-6; nonionic; emulsifier, solubilizer, wetting agent for cosmetic, pharmaceutical, food, and industrial applics.; liq.; HLB 10.0; 97% conc.

Crillet 45. [Croda Chem. Ltd.] Polysorbate 85; CAS 9005-70-3; nonionic; emulsifier, solubilizer, wetting agent for cosmetic, pharmaceutical, food, and industrial applics.; clear amber visc. liq.; sol. in ethyl and oleyl alcohols, IPM, oleic acid, kerosene, trichlorethylene, butyl stearate; HLB 11.0; sapon. no. 82–95; surf. tens. 41 dynes/cm (0.1%); 97% conc.

Crillon CDY. [Croda Chem. Ltd.] Cocamide DEA; nonionic; sol. and cutting oils; liq.

Crillon LDE. [Croda Surf. Ltd.] Lauramide DEA; CAS 120-40-1; nonionic; detergent and foam stabilizer; o/w emulsifiers, antistat; anticorrosive; liq.

Crillon ODE. [Croda Surf. Ltd.] Oleamide DEA; CAS 93-83-4; nonionic; emulsifier, stabilizer, skin protectant, lubricant, anti-irritant used in personal care prods.; additive for cutting fluids and sol. cutting oils; liq.

Crodafos CAP. [Croda Inc.] PPG-10 cetyl ether phosphate; CAS 111019-03-5; anionic; w/o emulsifier; antistat used in personal care prods.; modifies pH, thickening, emulsifying and suspending props.; enhances hair conditioning; useful for microemulsion systems; yel. clear liq.; sol. in oil, water; usage level: 0.5-3%; 100% conc.

Crodafos CDP. [Croda Chem. Ltd.] Cetyl diethanolamine phosphate; CAS 90388-14-0; anionic; emulsifier and stabilizer for o/w emulsions; powd.; water-disp.; 100% conc.

Crodafos CS2 Acid. [Croda Chem. Ltd.] Ceteareth-2 phosphate; anionic; emulsifier and stabilizer for o/w emulsions; solid; 99% conc.

Crodafos CS5 Acid. [Croda Chem. Ltd.] Ceteareth-5 phosphate; anionic; emulsifier and stabilizer for o/w emulsions; solid; 99% conc.

Crodafos CS10 Acid. [Croda Chem. Ltd.] Ceteareth-10 phosphate; anionic; emulsifier and stabilizer for o/w emulsions; solid; 99% conc.

Crodafos N3 Acid. [Croda Inc.; Croda Chem. Ltd.] Oleth-3 phosphate; CAS 39464-69-2; anionic; surfactant, conditioner, antistat, o/w emulsifier, gelling agent for surfactants, cosmetics, pharmaceuticals, and toiletries, microemulsion gels; corrosion inhibitor and anti-gelling agent in aerosol antiperspirant systems; amber visc. liq.; sol. in oil, water; acid no. 120–135; iodine no. 45-58; sapon. no. 125–145; pH 1–3 (2% aq.); usage level: 0.5-5%; 100% act.

Crodafos N3 Neutral. [Croda Inc.] DEA-oleth-3 phosphate; CAS 58855-63-3; anionic; o/w emulsifier, gelling agent for surfactants and for prep. of clear min. oil gels and microemulsion gels; used in hair relaxers; amber visc. liq.; sol. in oil, water; acid no. 90–100; iodine no. 35-50; pH 6–7 (2% aq.); usage level: 0.5-5%; 100% conc.

Crodafos N5 Acid. [Croda Chem. Ltd.] Oleth-5 phosphate; CAS 39464-69-2; anionic; emulsifier and stabilizer for o/w emulsions; solid; 99% conc.

Crodafos N10 Acid. [Croda Inc.] Oleth-10 phosphate; CAS 39464-69-2; anionic; o/w emulsifier, gelling agent for surfactants, prep. of clear min. oil gels, skin cleansers, clear microemulsion gels; yel. visc. liq.; sol. in oil, water; acid no. 70–100; iodine no. 25-35; sapon. no. 75–115; pH 1–3 (2% aq.); usage level: 0.5-10%; 100% act.

Crodafos N10 Neutral. [Croda Inc.] DEA-oleth-10 phosphate; CAS 58855-63-3; anionic; o/w emulsifier, gelling agent for surfactants, prep. of clear min. oil gels, skin cleansers, clear microemulsion gels; yel. visc. liq.; sol. in oil, water; acid no. 65–85; iodine no. 22-32; pH 5.5–7.0 (2% aq.); usage level: 0.5-10%; 100% conc.

Crodalan AWS. [Croda Inc.; Croda Chem. Ltd.] Polysorbate 80, cetyl acetate, acetylated lanolin alcohol; nonionic; emollient, superfatting agent, conditioner, o/w emulsifier, dispersant, wetting agent, plasticizer, solubilizer used in cosmetics, pharmaceuticals, detergent systems; golden liq.; faint fatty odor; sol. in alcohol, water; sp.gr. 1.02–1.08; acid no. 3 max.; hyd. no. 55–67; pH 5–7 (10% aq.); usage level: 1-5%; 100% act.

Crodamet Series. [Croda Chem. Ltd.] Ethoxylated primary amines; cationic; emulsifier, antistat for plastics and fibers; liq., paste; 100% conc.

Crodamine 1.16D. [Croda Universal Ltd.] Primary cetyl amine; CAS 143-27-1; cationic; emulsifier for herbicides, ore flotation, pigment dispersion; aux. for textiles, leather, rubber, plastics, and metal industries; liq.; 100% conc.

Crodamine 1.18D. [Croda Universal Ltd.] Stearylamine; CAS 124-30-1; cationic; emulsifier for herbicides, ore flotation, pigment dispersion; aux. for textiles, leather, rubber, plastics, and metal industries; solid; 100% conc.

Crodamine 1.HT. [Croda Universal Ltd.] Hydrog. tallow amine; CAS 61788-45-2; cationic; anticaking agent for fertilizers; emulsifier for herbicides, ore flotation, pigment dispersion; aux. for textiles, leather, rubber, plastics, and metal industries; waxy solid; 100% conc.

Crodamine 1.O, 1.OD. [Croda Universal Ltd.] Oleyl amine; CAS 112-90-3; cationic; emulsifier for herbicides, ore flotation, pigment dispersion; aux. for textiles, leather, rubber, plastics, and metal industries; solid and liq. resp.; 100% conc.

Crodamine 1.T. [Croda Universal Ltd.] Tallow amine; CAS 61790-33-8; cationic; emulsifier for herbicides, ore flotation, pigment dispersion; aux. for textiles, leather, rubber, plastics, and metal industries; liq.; 100% conc.

Crodamine 3.A16D. [Croda Universal Ltd.] Palmityl dimethylamine; CAS 112-69-6; cationic; emulsifier for herbicides, ore flotation, pigment dispersion; aux. for textiles, leather, rubber, plastics, and metal industries; liq.; 100% conc.

Crodamine 3.A18D. [Croda Universal Ltd.] Stearyl dimethylamine, dist.; CAS 124-28-7; cationic; emulsifier for herbicides, ore flotation, pigment dispersion; aux. for textiles, leather, rubber, plastics, and metal industries; paste; 100% conc.

Crodamine 3.ABD. [Croda Universal Ltd.] Dimethyl behenylamine; CAS 21542-96-1; cationic; emulsifier for herbicides, ore flotation, pigment dispersion; aux. for textiles, leather, rubber, plastics, and metal industries; solid; 100% conc.

Crodamine 3.AED. [Croda Universal Ltd.] Dimethyl erucylamine; cationic; emulsifier for herbicides, ore flotation, pigment dispersion; aux. for textiles, leather, rubber, plastics, and metal industries; intermediate; liq.; 100% conc.

Crodamine 3.AOD. [Croda Universal Ltd.] Dimethyl oleylamine; CAS 28061-69-0; cationic; emulsifier for herbicides, ore flotation, pigment dispersion; aux. for textiles, leather, rubber, plastics, and metal industries; liq.; 100% conc.

Crodapearl Liq. [Croda Inc.] Sodium laureth sulfate, glycol MIPA stearate; pearling agent for shampoos, bubble baths, dishwashing liqs.; creamy pearly paste, mild char. odor; pH 7.5-8.5; usage level: 1-5%; 35-40% conc. in water.

Crodapearl NI Liquid. [Croda Inc.] Hydroxyethyl stearamide-MIPA, PPG-5 ceteth-20; nonionic; pearlescent for detergent systems, lotions, gels, clear rinses, bath prods.; wh. soft paste; usage level: 1-5%.

Crodasinic L. [Croda Chem. Ltd.] Lauroyl sarcosine; CAS 97-78-9; nonionic; salts with detergent props.; lubricants and metal working fluids; solid; oil-sol.; 94% conc.

Crodasinic LS30. [Croda Chem. Ltd.] Sodium N-lauroyl sarcosinate; CAS 137-16-6; anionic; foaming, wetting agent and detergent for acidic conditions; corrosion inhibitor; bacteriastat and inhibitor; used in dental care preps., pharmaceuticals, personal care prods., household and industrial applics.; clear liq.; water-sol.; 30% act.

Crodasinic LS35. [Croda Chem. Ltd.] Sodium N-lauroyl sarcosinate; CAS 137-16-6; anionic; foaming agent, wetting agent, detergent, lubricant, antistat, corrosion inhibitor, bacteriostat, penetrant used in dental, pharmaceutical, shampoos, depilatories, and shaving preparations, food pkg., household and industrial uses; biodeg.; clear liq.; water sol.; 35% act.

Crodasinic LT40. [Croda Chem. Ltd.] TEA lauroyl sarcosinate; CAS 16693-53-1; anionic; detergent, foaming agent, wetting agent, dispersant, emulsifier, anticorrosive, foam stabilizing synergist for carpet shampoos, textile and cosmetic detergent systems; liq.; 40% conc.

Crodasinic O. [Croda Chem. Ltd.] N-Oleoyl sarcosine; CAS 110-25-8; nonionic; corrosion inhibitor in oils, fuels, lubricants, greases, surface coatings; as antifog agent for food pkg. polyolefin films; liq.; 93% conc.

Crodasinic OS35. [Croda Chem. Ltd.] Sodium N-oleoyl sarcosinate; anionic; foaming agent, wetting agent, detergent, lubricant, antistat, corrosion inhibitor, bacteriostat, penetrant for dental, pharmaceutical, shampoos, depilatories, shaving preps., household and industrial uses; liq.; 35% conc.

Crodateric C. [Croda Universal Ltd.] Derived from coconut fatty acid; amphoteric; surfactant; m.w. 360; 50% act. aq. sol'n.

Crodateric Cy. [Croda Universal Ltd.] Derived from caprylic acid; CAS 63451-23-0; amphoteric; surfactant; m.w. 284; 50% act. aq. sol'n.

Crodateric L. [Croda Universal Ltd.] Derived from lauric acid; amphoteric; surfactant; m.w. 340; 50% act. aq. sol'n.

Crodateric O, O.100. [Croda Universal Ltd.] Derived from oleic acid; CAS 32456-28-6; amphoteric; surfactant; m.w. 417; 50% act. aq. sol'n. and 100% act. resp.

Crodateric S. [Croda Universal Ltd.] Derived from stearic acid; CAS 30342-62-2; amphoteric; surfactant; m.w. 410; 50% act. aq. sol'n.

Crodazoline O. [Croda Universal Ltd.] Oleic imidazoline; CAS 95-38-5; cationic; surfactant, wetting agent, emulsifier, softener, for textiles, asphalt, tar emulsion breaker, paint additive, corrosion inhibitor, lubricating metal processing aids; Gardner 6–7 liq.; m.w. 345; oil-sol.; m.p. 40–57 C; surf. tens. 26.5 (1%/0.36%); pH 11.1 (10% aq. disp.); 90% min. imidazoline.

Crodazoline S. [Croda Universal Ltd.] Stearic imidazoline; CAS 95-19-2; cationic; surfactant, wetting agent, emulsifier, softener, for textiles, asphalt, tar emulsion breaker, paint additive, corrosion inhibitor; Gardner 4–5 solid; m.w. 338; oil-sol.; m.p. 52 C; surf. tens. 43.2 (1%/0.36%); pH 11.0 (10% aq. disp.); 90% min. imidazoline.

Crodesta F-110. [Croda Inc.] Sucrose distearate and sucrose stearate; nonionic; dispersant, emulsifier, wetting agent, solubilizer, detergent in cosmetics, toiletries, pharmaceuticals; thickener and suspending agent; wh. powd.; water-sol.; HLB 12.0; m.p. 72–78 C; acid no. 5 max.; sapon. no. 85–145; usage level: 3-6%; 100% act.

Crodet C10. [Croda Chem. Ltd.] PEG-10 coconut fatty acid; CAS 61791-29-5; nonionic; surfactant for cosmetic and industrial applics.; liq.; 97+% conc.

Crodet L4. [Croda Chem. Ltd.] PEG-4 laurate; CAS 9004-81-3; nonionic; o/w emulsifier for cosmetics and pharmaceutical creams, lotions and ointments, industrial applics., wetting agent, solubilizer for perfumes or aq. alcoholic preparations; dispersant; plasticizer for hair setting sprays; pale straw liq.; sol. in ethyl, oleyl, and cetearyl alcohols, oleic acid; HLB 9.8; sapon. no. 138–150; 97% conc.

Crodet L8. [Croda Chem. Ltd.] PEG-8 laurate; CAS 9004-81-3; nonionic; o/w emulsifier for cosmetics and pharmaceutical creams, lotions and ointments, industrial applics., wetting agent, solubilizer for perfumes or aq. alcoholic preparations; dispersant; plasticizer for hair setting sprays; pale straw liq.; sol. see Crodet L4; HLB 12.7; sapon. no. 95–106; 97% conc.

Crodet L12. [Croda Chem. Ltd.] PEG-12 laurate; CAS 9004-81-3; nonionic; o/w emulsifier for cosmetics and pharmaceutical creams, lotions and ointments, industrial applics., wetting agent, solubilizer for perfumes or aq. alcoholic preparations; dispersant; plasticizer for hair setting sprays; pale straw liq.; sol. in water, ethyl, oleyl, and cetearyl alcohols, oleic acid; HLB 14.5; sapon. no. 72–82; 97% conc.

Crodet L24. [Croda Chem. Ltd.] PEG-24 laurate; CAS 9004-81-3; nonionic; o/w emulsifier for cosmetics and pharmaceutical creams, lotions and ointments, industrial applics., wetting agent, solubilizer for perfumes or aq. alcoholic preparations; dispersant; plasticizer for hair setting sprays; off-wh. soft paste; sol. in water, ethyl and cetostearyl alcohol; HLB 16.8; sapon. no. 42–48; 97% conc.

Crodet L40. [Croda Chem. Ltd.] PEG-40 laurate; CAS 9004-81-3; nonionic; o/w emulsifier for cosmetics and pharmaceutical creams, lotions and ointments, industrial applics., wetting agent, solubilizer for perfumes or aq. alcoholic preparations; dispersant; plasticizer for hair setting sprays; off-wh. waxy solid; sol. in water, ethyl and cetostearyl alcohol; HLB 17.9; sapon. no. 26–31; 97% conc.

Crodet L100. [Croda Chem. Ltd.] PEG-100 laurate; CAS 9004-81-3; nonionic; o/w emulsifier for cosmetics and pharmaceutical creams, lotions and ointments, industrial applics., wetting agent, solubilizer for perfumes or aq. alcoholic preparations; dispersant; plasticizer for hair setting sprays; pale yel. waxy solid; sol. in water, ethyl and cetostearyl alcohol; HLB 19.1; sapon. no. 11–15; 97% conc.

Crodet O4. [Croda Chem. Ltd.] PEG-4 oleate; CAS 9004-96-0; nonionic; surfactant for cosmetic and industrial applics.; liq.; 97% conc.

Crodet O8. [Croda Chem. Ltd.] PEG-8 oleate; CAS 9004-96-0; nonionic; surfactant for cosmetic and industrial applics.; liq.; HLB 10.8; 97% conc.

Crodet O12. [Croda Chem. Ltd.] PEG-12 oleate; CAS 9004-96-0; nonionic; surfactant for cosmetic and industrial applics.; liq.; HLB 13.4; 97% conc.

Crodet O24. [Croda Chem. Ltd.] PEG-24 oleate; CAS 9004-96-0; nonionic; surfactant for cosmetic and industrial applics.; paste; HLB 15.8; 97% conc.

Crodet O40. [Croda Chem. Ltd.] PEG-40 oleate; CAS 9004-96-0; nonionic; surfactant for cosmetic and industrial applics.; solid; 97% conc.

Crodet O100. [Croda Chem. Ltd.] PEG-100 oleate; CAS 9004-96-0; nonionic; surfactant for cosmetic and industrial applics.; solid; HLB 18.8; 97% conc.

Crodet S4. [Croda Chem. Ltd.] PEG-4 stearate; CAS 9004-99-3; nonionic; o/w emulsifier for cosmetics and pharmaceutical creams, lotions and ointments, industrial applics.; wetting agent, solubilizer for perfumes or aq. alcoholic preparations, dispersant; off-wh. soft paste; sol. in ethyl and oleyl alcohols, oleic acid, ceto stearyl alcohol, arachis oil and isoparaffinic solv.; HLB 7.7; sapon. no. 117–129; 97% conc.

Crodet S8. [Croda Chem. Ltd.] PEG-8 stearate; CAS 9004-99-3; nonionic; o/w emulsifier for cosmetics and pharmaceutical creams, lotions and ointments, industrial applics.; wetting agent, solubilizer for perfumes or aq. alcoholic preparations, dispersant;

off-wh. soft paste; sol. see Croduret S4; HLB 10.8; sapon. no. 84–94; 97% conc.

Crodet S12. [Croda Chem. Ltd.] PEG-12 stearate; CAS 9004-99-3; nonionic; o/w emulsifier for cosmetics and pharmaceutical creams, lotions and ointments, industrial applics.; wetting agent, solubilizer for perfumes or aq. alcoholic preparations, dispersant; off-wh. waxy solid; sol. in ethyl, oleyl, and cetostearyl alcohol, water, oleic acide; HLB 13.4; sapon. no. 65–75; 97% conc.

Crodet S24. [Croda Chem. Ltd.] PEG-24 stearate; CAS 9004-99-3; nonionic; o/w emulsifier for cosmetics and pharmaceutical creams, lotions and ointments, industrial applics.; wetting agent, solubilizer for perfumes or aq. alcoholic preparations, dispersant; off-wh. waxy solid; sol. in ethyl and cetostearyl alcohol, water; HLB 15.8; sapon. no. 38–47; 97% conc.

Crodet S40. [Croda Chem. Ltd.] PEG-40 stearate; CAS 9004-99-3; nonionic; o/w emulsifier for cosmetics and pharmaceutical creams, lotions and ointments, industrial applics.; wetting agent, solubilizer for perfumes or aq. alcoholic preparations, dispersant; off-wh. waxy solid; sol. in ethyl and cetostearyl alcohol, water; HLB 16.7; sapon. no. 23–30; 97% conc.

Crodet S100. [Croda Chem. Ltd.] PEG-100 stearate; CAS 9004-99-3; nonionic; o/w emulsifier for cosmetics and pharmaceutical creams, lotions and ointments, industrial applics.; wetting agent, solubilizer for perfumes or aq. alcoholic preparations, dispersant; off-wh. waxy solid; sol. in ethyl and cetostearyl alcohol, water; HLB 18.8; sapon. no. 10–14; 97% conc.

Crodex C. [Croda Chem. Ltd.] Cetostearyl alcohol and Cetrimide BP; cationic; emulsifying wax BPC, bactericides, for pharmaceuticals and cosmetics, hair conditioning rinses; almost wh. waxy solid; faint char. odor; water-disp.; 100% conc.

Crodex N. [Croda Chem. Ltd.] Cetostearyl alcohol and ceteth-20; nonionic; emulsifying wax BP, wetting agent, penetrant, emulsifier for most emollient materials in cosmetics and pharmaceuticals; almost wh. waxy solid; faint char. odor; water-disp.; 100% conc.

Croduret 10. [Croda Chem. Ltd.] PEG-10 hydrog. castor oil; CAS 61788-85-0; nonionic; emulsifier, solubilizer, emollient, superfatting agent, detergent used for cosmetics, textiles, metalworking fluids, emulsion polymerization, insecticides, herbicides, household detergents; straw-colored liq.; sol. in oleyl alcohol, naphtha, MEK, oleic acid, trichloroethylene; HLB 6.3; cloud pt. < 20 C; sapon. no. 120–130; surf. tens. 37 dynes/cm (0.1%); 100% conc.

Croduret 30. [Croda Chem. Ltd.] PEG-30 hydrog. castor oil; CAS 61788-85-0; emulsifier, solubilizer, emollient, superfatting agent, detergent for cosmetics, textiles, metalworking fluids, emulsion polymerization, agric., household detergents; straw colored liq.; sol. in water, ethanol, oleyl alcohol, naphtha, MEK, oleic acid, trichloroethylene; HLB 11.6; cloud pt. 48 C; sapon. no. 70–80; surf. tens. 46 dynes/cm (0.1%).

Croduret 40. [Croda Chem. Ltd.] PEG-40 hydrog. castor oil; CAS 61788-85-0; nonionic; see Croduret 10; off-wh. visc. paste; sol. in water, ethanol, naphtha, MEK, oleic acid, trichloroethylene; HLB 12.9; cloud pt. 62 C; sapon. no. 60–65; surf. tens. 46 dynes/cm (0.1%); 100% conc.

Croduret 60. [Croda Chem. Ltd.] PEG-60 hydrog. castor oil; CAS 61788-85-0; nonionic; see Croduret 10; off-wh. stiff paste; sol. in water, ethanol, MEK, oleic acid, trichloroethylene; HLB 14.7; cloud pt. 71 C; sapon. no.45–50; surf. tens. 47.5 dynes/cm (0.1%); 100% conc.

Croduret 100. [Croda Chem. Ltd.] PEG-100 hydrog. castor oil; CAS 61788-85-0; nonionic; see Croduret 10; off-wh. waxy solid; sol. in water, ethanol, MEK, trichloroethylene; HLB 16.5; cloud pt. 65 C; sapon. no. 25–35; surf. tens. 46.2 dynes/cm (0.1%); 100% conc.

Croduret 200. [Croda Chem. Ltd.] PEG-200 hydrogenated castor oil; CAS 61788-85-0; nonionic; emulsifier, wetting agent, solubilizer, lubricant, antistat; solid; water-sol.; HLB 18.0; 100% conc.

Cromeen. [Croda Chem. Ltd.] Substituted alkylamine deriv. lanolin acids; anionic; mild multifunctional surfactant with high foaming, detergency, and emulsifying props.; for shampoos, detergents, and hand cleansing preps., aerosol skin and shaving foams; soft gel; 45% min. conc.

Cromul 1540. [Croda Chem. Ltd.] Blend of anionic and nonionic emulsifiers; anionic/nonionic; emulsifier for printing ink; in emulsion mold release agents, oil well drilling emulsions, siesal rope batching lubricant; liq.; 100% conc.

Cropol. [Croda Surf. Ltd.] Sulfosuccinate; anionic; dewatering protective, emulsification aid; liq.

Crosultaine C-50. [Croda Inc.] Cocamidopropyl hydroxysultaine; amphoteric; foam booster/stabilizer effective over wide range of pH and water hardness; for shampoos, baby shampoos; lime soap dispersant; yel. visc. liq., char. odor; sol. @ 10% in water, ethanol, ethanol/water mixts., propylene glycol, glycerin; pH 6.5-8.5; usage level: 1-10%; 50% act.

Crosultaine E-30. [Croda Inc.] Erucamidopropyl hydroxysultaine; amphoteric; visc. booster and conditioner giving silky feel, comb and static control at levels less than 2% active; improves creaminess and lubricity of lather; lime soap dispersant; yel. to amber gel, char. odor; sol. @ 10% in water, water/ethanol mixts.; disp. in glycerin, propylene glycol; pH 6.5-8.5 (10% in IPA/water 60:40); usage level: 1-10%; 30% act.

Crosultaine T-30. [Croda Inc.] Tallowamidopropyl hydroxysultaine; amphoteric; foam and visc. booster; improves wet combing and hair condition; lime soap dispersant; yel. gel, char. odor; sol. @ 10% in water, ethanol, ethanol/water mixts., propylene glycol, glycerin; pH 6.5-8.5 (10% in IPA/water 60:40); usage level: 1-10%; 30% conc.

Crothix. [Croda Inc.] Polyol alkoxy ester; nonionic; mild thickener for aq. systems, shampoos, aux. emulsifier and bodying agent for creams and lotions; wh. to off-wh. solid; sol. in aq. surfactant systems; usage level: 0.5-2%.

Crovol A70. [Croda Inc.; Croda Chem. Ltd.] PEG-60 almond glycerides; nonionic; emulsifier reducing irritation potential of anionic/amphoteric surfactant systems; emollients for hydroalcoholic systems; wetting agent, solubilizer; yel. soft paste; sol. in water, ethanol, oleyl alcohol, maize oil; HLB 15.0; acid no. 2 max.; sapon. no. 45-55; hyd. no. 70-90; usage level: 1-10%.

Crown Anti-Foam. [Crown Tech.] Silicone emulsion; nonionic; antifoam for dehydrating, evaporating, fermentation processes, abrasive slurries, commercial cleaners, adhesives, latex emulsions, cutting oils, insecticides and pesticides; effective at concs.

as low as 5 ppm; stable to high shear and high pH; sp.gr. 1.00; dens. 8.34 lb/gal; pH 7.0; 10% act.

Crown Foamer 20. [Crown Tech.] Foamer, wetting agent producing a detergent-type foam for pickling sol'ns.; effective in sulfuric and hydrochloric acid baths; amber liq., pleasant odor; sol. in acid; dens. 8.5 lb/gal; pH 7-8; usage level: 1 pt/1000 gal.

Crown Foamer 20X. [Crown Tech.] Foamer producing a dense foam blanket for sulfuric, hydrochloric, and nitric-hydrofluoric pickling baths; amber liq., pleasant odor; sol. in acid; dens. 8.5 lb/gal; pH 7-8; usage level: 1 pt/1000 gal.

Crown Foamer 50. [Crown Tech.] Foamer for use in nitric-hydrofluoric acid pickling sol'ns.; amber liq., pleasant odor; sol. in acid; dens. 8.5 lb/gal; pH 7-8; usage level: 1 pt/1000 gal.

Crystal Inhibitor #5. [Harcros] Blend; nonionic; retards formation of crystals in 2,4-D amine formulation dilutions; FDA compliance; amber liq.; sp.gr. 1.05; dens. 8.7 lb/gal; pour pt. 60 F; flash pt. (PMCC) > 300 F; pH 6-7 (1% aq.); 100% conc.

CWT AF-200 Antifoam. [Drew Ind. Div.] Alcohols, silica derivs., surfactants blend; antifoam for recirculating cooling water systems; for continuous or intermittent applic.; yel. opaque liq.; dens. 7.8 lb/gal; pour pt. 32 C; pH 7.0.

Cyclochem® NI (redesignated Dermalcare® NI). [Rhone-Poulenc Surf.]

Cyclochem® PEG 200DS. [Rhone-Poulenc Surf.] PEG-4 distearate; CAS 9005-08-7; emollient, detergent, emulsifier, visc. builder for shampoos, liq. soaps, creams and lotions, facial cleansers, bath and toiletries; 100% conc. DISCONTINUED.

Cyclochem® PEG 6000DS. [Rhone-Poulenc Surf.] PEG-150 distearate; CAS 9005-08-7; emollient, detergent, emulsifier, visc. builder for shampoos, liq. soaps, creams and lotions, facial cleansers, bath and toiletries; 100% conc. DISCONTINUED.

Cyclochem® PEG 600DS. [Rhone-Poulenc Surf.] PEG-12 distearate; CAS 9005-08-7; emollient, detergent, emulsifier, visc. builder for shampoos, liq. soaps, creams and lotions, facial cleansers, bath and toiletries; 100% conc. DISCONTINUED.

Cyclomatic Dur. [Ceca SA] Anionic; wetting agent for use in hydrosulfite bleaching sol'ns. for textiles.

Cyclomox® C. [Rhone-Poulenc Surf.] Cocamine oxide; nonionic; flash foamer and viscosifier in hard water; liq.; 30% conc. DISCONTINUED.

Cyclomox® CO (see Rhodamox® CAPO). [Rhone-Poulenc Surf.] DISCONTINUED.

Cyclomox® L (redesignated Rhodamox® LO). [Rhone-Poulenc Surf.]

Cyclomox® LO. [Rhone-Poulenc Surf.] Lauramine oxide; nonionic; top foam prod.; liq.; 30% conc. DISCONTINUED.

Cyclomox® SO. [Rhone-Poulenc Surf.] Stearamidopropylamine oxide; nonionic; foamer, viscosifier, emollient, conditioner; hydrogen peroxide stabilizer in bleaching formulations, cold waves; paste; 30% conc. DISCONTINUED.

Cycloryl DCA (redesignated Rhodaterge® DCA). [Rhone-Poulenc Surf.]

Cycloryl LDC. [Rhone-Poulenc Surf.] Ammonium laureth sulfate plus sulfonate and amide; anionic; base for lt. duty detergent and liq. dishwashing compds., car washing applic.; amber liq.; 60–62% act. DISCONTINUED.

Cycloteric BET-CB. [Rhone-Poulenc Surf.] Cocamidopropyl betaine, cosmetic grade, glycerin-free; foam booster, foaming agent, thickener, conditioner for shampoos, cosmetic, and industrial applics.; irritation mollifying agent for baby shampoos; 30% act. DISCONTINUED.

Cycloteric BET-OD40. [Rhone-Poulenc Surf.] Capric/caprylic amidopropyl betaine; amphoteric; surfactant; liq.; 40% conc. DISCONTINUED.

Cycloteric BET-T2 40 (see Miranol® TM). [Rhone-Poulenc Surf.] DISCONTINUED.

Cycloteric CAPA. [Rhone-Poulenc Surf.] Cocaminopropionic acid; high foaming conditioning agent for shampoos, skin cleansers, foam baths; surfactant for hard surf. cleaners, car washes, industrial foamers; 40% act. DISCONTINUED.

Cycloteric SLIP. [Rhone-Poulenc Surf.] Sodium lauriminodipropionate; high foaming conditioning agent for shampoos, skin cleaners, foam baths; surfactant for hard surf. cleaners, car washes, industrial foamers; 30% act. DISCONTINUED.

Cycloton® D261C/70. [Rhone-Poulenc Surf.] Ditallowalkonium chloride (Quaternium 18); CAS 61789-80-8; conc. base for prep. of fabric softeners, antistatic treatments, hair conditioners; soft paste; 70% act. DISCONTINUED.

Cyncal® 80%. [Hilton-Davis] Myristalkonium chloride; cationic; antimicrobial, antistat, disinfectant, sanitizer for use in food, beverage processing, for industrial and farm use, fabrics; lt. yel. liq.; mild, pleasant odor; m.w. 359; sol. in water, lower alcohols, ketones, glycols; dens. 7.8 lb/gal; sp.gr. 0.94; 80% act. DISCONTINUED.

D

DA-14. [Exxon/Tomah] N-Isodecyloxypropyl-1,3-diaminopropane; intermediate for textile foaming agents, surfactants, ethoxylates, agric. chemicals; corrosion inhibitor for metalworking fluids; additive for fuels, lubricants, petrol. refining; crosslinking agent for epoxy resins; sp.gr. 0.86; sol. 1.8 g/100 g in water; pour pt. -50 F; flash pt. (COC) 108 C.

DA-16. [Exxon/Tomah] N-Isododecyloxypropyl-1,3-diaminopropane; intermediate for textile foaming agents, surfactants, ethoxylates, agric. chemicals; corrosion inhibitor for metalworking fluids; additive for fuels, lubricants, petrol. refining; crosslinking agent for epoxy resins; sp.gr. 0.87; pour pt. -30 F; flash pt. (COC) 160 C.

DA-17. [Exxon/Tomah] N-Isotridecyloxypropyl-1,3-diaminopropane; intermediate for textile foaming agents, surfactants, ethoxylates, agric. chemicals; corrosion inhibitor for metalworking fluids; additive for fuels, lubricants, petrol. refining; crosslinking agent for epoxy resins; sp.gr. 0.87; sol. 2.1 g/100 g in water; pour pt. -30 F; flash pt. (COC) 160 C.

Dabco® DC193. [Air Prods./Polyurethanes] Silicone glycol copolymer; surfactant used for prod. of rigid polyurethane foam; suitable for board, laminates, foam-in-place, spray foam, and shoe soles; nonhydrolyzable; Gardner 2 clear visc. liq.; sp.gr. 1.07; visc. 425 cs; f.p. 15 C; pour pt 11 C; flash pt. (CC) 92 C; ref. index 1.45; hyd. no. 75.

Dabco® DC197. [Air Prods./Polyurethanes] Silicone glycol copolymer; surfactant for prod. of high and low-density and sprayed rigid polyurethane foams; exc. emulsifying ability, good dimensional stability; nonhydrolyzable; Gardner 4 clear to sl. hazy liq.; sp.gr. 1.50; visc. 350 cs; f.p. -6.7 C; flash pt. 71 C.

Dabco® DC198. [Air Prods./Polyurethanes] Silicone glycol copolymer; high-potency surfactant used in prod. of flexible slabstock polyurethane foam; Gardner 3 clear to hazy liq.; sol. in polyol, fluorocarbon, water-amine streams; sp.gr. 1.039; visc. 1600 cs; flash pt. (OC) 84 C; pH 5.3 (30% aq.).

Dabco® DC1315. [Air Prods./Polyurethanes] Silicone glycol copolymer; wide processing-latitude surfactant for use in flame-retarded foam systems such as PU slabstock and hot molded foam; Gardner 2-4 clear to sl. hazy liq.; sol. in polyol, fluorocarbon, water-amine streams; sp.gr. 1.03; visc. 1000 cs; flash pt. (CC) 60.6 C.

Dabco® DC1630. [Air Prods./Polyurethanes] Specially formulated dimethyl polysiloxane; surfactant for use in high-resiliency, flexible polyurethane foam based on high m.w. polyols and polymeric isocyanates; clear liq.; sp.gr. 0.919; visc. 5.0 cs; flash pt. (CC) 138 C; ref. index 1.3966.

Dabco® DC5043. [Air Prods./Polyurethanes] Silicone glycol copolymer; surfactant for prod. of molded and slabstock high-resiliency polyurethane foam; nonhydrolyzable; broad processing latitude; clear to lt. straw liq.; sp.gr. 1.0; visc. 300 cs; flash pt. (CC) 60 C.

Dabco® DC5098. [Air Prods./Polyurethanes] Silicone glycol copolymer; nonhydrolyzable surfactant for prod. of polyisocyanurate and conventional rigid PU foam; lt. straw clear to sl. hazy liq.; sp.gr. 1.08; visc. 250 cs; f.p. 12 C; flash pt. (CC) 61 C; hyd. no. < 10.

Dabco® DC5103. [Air Prods./Polyurethanes] Silicone glycol copolymer; nonhydrolyzable surfactant for prod. of all types of rigid PU foam; low freeze pt.; clear to lt. straw liq.; sp.gr. 1.05; visc. 250 cs; f.p. -7 C; pour pt. -5 C; flash pt. (CC) 61 C; hyd. no. 104.

Dabco® DC5125. [Air Prods./Polyurethanes] Silicone glycol copolymer; wide-processing-latitude surfactant for prod. of conventional and flame-retarded flexible polyurethane slabstock foam systems; Gardner 2-4 clear to sl. hazy liq.; sol. in polyol, fluorocarbon, and water-amine streams; sp.gr. 1.03; visc. 1000 cs; flash pt. (CC) 60.6 C.

Dabco® DC5160. [Air Prods./Polyurethanes] Silicone glycol copolymer; surfactant used in prod. of conventional and flame-retarded slabstock polyurethane foam; liq.; sol. in polyol, fluorocarbon, water-amine streams; sp.gr. 1.03; visc. 950 cs; flash pt. (CC) 61 C.

Dabco® DC5164. [Air Prods./Polyurethanes] Silicone prod.; surfactant for prod. of difficult-to-stabilize high-resiliency molded PU foam formulations; clear to sl. hazy liq.; sp.gr. 1.0; visc. 225-375 cs; flash pt. (CC) 64 C.

Dabco® DC5169. [Air Prods./Polyurethanes] Silicone prod.; surfactant for prod. of high-resiliency MDI-based PU molded foams; clear to sl. hazy liq.; sp.gr. 1.0; visc. 29-35 cs; flash pt. (CC) 190 C.

Dabco® DC5244. [Air Prods./Polyurethanes] Silicone prod.; cell-opening surfactant for prod. of all types of high-resiliency foams; clear liq.; sp.gr. 1.05; visc. 800-1200 cs; flash pt. (CC) 87 C.

Dabco® DC5270. [Air Prods./Polyurethanes] Silicone prod.; surfactant for low-density, flexible slabstock PU foams.

Dabco® DC5418. [Air Prods./Polyurethanes] Silicone glycol copolymer; surfactant for MDI-based molded polyurethane foam systems, providing efficient cell opening and bulk stabilization in all-water-blown foams; nonhydrolyzable; clear to straw low-visc. liq.; sp.gr. 0.9658 (21 C); visc. 28.9 cps; b.p. > 149 C; flash pt. (CC) 113 C.

Dabco® X2-5357. [Air Prods./Polyurethanes] Surfactants for reduced CFC and non-CFC blown rigid polyurethane foam systems.; Gardner 3 color; sp.gr. 1.04; visc. 400 cps; flash pt. (CC) > 61 C; hyd. no. 57.

Dabco® X2-5367. [Air Prods./Polyurethanes] Surfactants for reduced CFC and non-CFC blown rigid polyurethane foam systems.; Gardner 4 color; sp.gr. 1.04; visc. 280 cps; flash pt. (CC) > 61 C; hyd. no. 31.

Dacospin 12-R. [Henkel] Ethoxylated hydrog. castor oil; emulsifier and solubilizer for esters and glycerides; water-sol.

Dacospin 869. [Henkel/Textile] PEG-9 hydrog. coconut fatty acid; nonionic; surfactant for textile use; Gardner 1 liq.; sol. @ 5% in water, min. oil, glycerol trioleate, xylene; dens. 8.6 lb/gal; visc. 46 cSt (100 F); HLB 13.2; pour pt. 6 C; cloud pt. 58 C; flash pt. 495 F.

Dacospin 1735-A. [Henkel/Textile] PEG-30 castor oil glycerides; nonionic; surfactant, lubricant for PP carpet backing giving excellent fiber-to-fiber lubricity and plasticizing properties; lt. yel. liq.; sol. @ 5% in water, xylene; dens. 8.7 lb/gal; visc. 266 cSt (100 F); HLB 11.8; pour pt. 9 C; cloud pt. 59 C; flash pt. 266 C (COC); pH 7.3 (2%); 100% act.

Dacospin 9212. [Henkel] Alkylolamide, high-purity; antistat and wetting agent for PP yarn; yel. clear liq.; dens. 1.00 g/ml; visc. 365 cP; pH 9.5 (2%); flash pt. 185 C (COC); 100% act.

Dacospin HT-118. [Henkel] Ethoxylated (9) alkyl phenol; thermally stable emulsifier; milky emulsion in water.

Dacospin HT-165. [Henkel] Ethoxylated (9) aryl phenol; thermally stable emulsifier; water-sol.

Dacospin LA-704. [Henkel/Textile] PEG-7/PPG-4 tridecyl alcohol; nonionic; surfactant for textile use; Gardner 2 liq.; sol. @ 5% in water, butyl stearate, glycerol trioleate, Stod., xylene; disp. in min. oil; dens. 7.5 lb/gal; visc. 42 cSt (100 F); HLB 8.3; pour pt. -12 C; cloud pt. 37 C; flash pt. > 200 F.

Dacospin LD-605. [Henkel/Textile] PEG-6/PPG-5 2-ethyl hexanol; nonionic; surfactant for textile use; Gardner 1 liq.; sol. @ 5% in butyl stearate, glycerol trioleate, Stod., xylene; disp. in water, min. oil; dens. 8.1 lb/gal; visc. 44 cSt (100 F); HLB 7.7; pour pt. 5 C; cloud pt. < 25 C; flash pt. > 200 F.

Dacospin PE-47. [Henkel/Textile] Phosphate ester, potassium salt; anionic; surfactant for textile use; Gardner 1 liq.; sol. @ 5% in water, xylene; disp. in min. oil; dens. 9.4 lb/gal; visc. 102 cSt (100 F); pour pt. -12 C; cloud pt. 55 C (5% saline); flash pt. > 200 F.

Dacospin PE-146. [Henkel/Textile] Phosphated aliphatic alcohol, free acid; anionic; antistat for syn. fibers; Gardner 1 liq.; sol. @ 5% in min. oil, butyl stearate, glycerol trioleate, Stod., xylene; disp. in water; dens. 8.3 lb/gal; visc. 115 cSt (100 F); pour pt. 5 C; cloud pt. < 25 C; flash pt. > 200 F.

Dacospin POE(25)HRG. [Henkel/Textile] PEG-25 hydrog. castor oil, bleached; CAS 61788-85-0; nonionic; surfactant, lubricant for PP carpet backing giving excellent fiber-to-fiber lubricity and plasticizing properties; as spin finish component with good emulsifying properties, high thermal stability, and good color; pale yel. liq.; sol. @ 5% in water; dens. 8.8 lb/gal; visc. 500 cSt (100 F); HLB 10.8; pour pt. 5 C; cloud pt. 80 C (5% saline); flash pt. (COC) 266 C; pH 6.4 (5%); 99% act.

Daisurf C. [Dai-ichi Kogyo Seiyaku] Specialty; nonionic; detergent, wetting agent, penetrant for textiles; aux. in bleaching cotton; stable in alkali; liq.; 25% conc.

Dama® 810. [Ethyl] Dioctyl/octyldecyl/didecyl methyl amines; cationic; intermediate for mfg. of quaternary ammonium compds. for biocides, textile chemicals, oil field chemicals, amine oxides, betaines, polyurethane foam catalysis, epoxy curing agent; in fabric softeners, disinfectants, laundry detergents; clear liq., fatty amine odor; sp.gr. 0.801; f.p. -38.6 C; amine no. 197; flash pt. (PM) 166 C; corrosive.

Dama® 1010. [Ethyl] Didecyl methylamine; CAS 7396-58-9; cationic; intermediate for mfg. of quaternary ammonium compds. for biocides, textile chemicals, oil field chemicals, amine oxides, betaines, polyurethane foam catalysis, epoxy curing agent; in fabric softeners, disinfectants, laundry detergents; clear liq., fatty amine odor; sp.gr. 0.807; f.p. -6.3 C; amine no. 177; flash pt. (PM) > 93 C; corrosive; 100% conc.

Damox® 1010. [Ethyl] Didecyl dimethylamine oxide; nonionic; wetting agent, emulsifier, dispersant, visc. modifier, hair conditioner with low foam; liq.; sol. in oil, disp. in water; 80% conc.

Dapral®. [Akzo Chem. BV] Phosphate derivs.; detergent, wetting agent, emulsifier for making of cutting and sol. oils, for household, cosmetic, industrial uses; liq., paste.

Dapral® AS. [Akzo Chem. BV] Phosphoric acid ester, potassium salt; anionic; polyester yarn spinning; paste; 75% conc.

Darvan® ME. [R.T. Vanderbilt] Sodium alkyl sulfates; anionic; latex stabilizer; NR, SR mold and stock lubricant; foam modifier, wetting agent, and emulsifier for latexes; wh. to cream powd.; 99.9% min. thru 10 mesh; dens. 1.19 ± 0.03 mg/m3; pH 9.0–11.0 (10%).

Darvan® No. 2. [R.T. Vanderbilt] Sodium lignosulfonate; dispersing and emulsifying agent for rubber industry, esp. for zinc oxide, clays, and sulfur; dk. brn. powd.; water-sol.; dens. 1.25 ± 0.03 mg/m3; pH 7.0–8.5 (1%); 84.5% min. act.

Darvan® NS. [R.T. Vanderbilt] c-Cetyl betaine and c-decyl betaine aq. sol'n.; nonionic; stabilizer for low alkaline latex systems; latex foam modifier and wetting agent; clear to lt. amber liq.; dens. 1.05 ± 0.02 mg/m3; pH 8.5–11.5 (10%); 32.0–36.0% total solids.

Darvan® WAQ. [R.T. Vanderbilt] Sodium alkyl sulfates; anionic; latex stabilizer for NR and syn. latexes; wetting agent and emulsifier for latexes; mold and internal lubricant; clear water-wh. paste (> 30 C); dens. 1.04 ± 0.03 mg/m3; pH 7.0–9.0 (10%).

Daxad® 11. [W.R. Grace] Low m.w. naphthalene sulfonate formaldehyde condensate, sodium salt; dispersant for pigments in aq. media; used in agric. chemicals, mastics, caulks, sealants, pigment slurries and disps.; buff powd.; dens. 35 lb/ft3; pH 8.0–10.5 (1%); surf. tens. 70 dynes/cm (1%); 87% min. act.

Daxad® 11G. [W.R. Grace] Low m.w. naphthalene sulfonate formaldehyde condensate, sodium salt; see Daxad 11; also for inks, syn. polymers, paper coating disps., paper mill slime control, pitch control, pulp digestion, and tall oil separation; buff fine gran.; dens. 42 lb/ft3; pH 8.0–10.5 (1%); surf. tens. 70–71 dynes/cm (1%); 87% min. act.

Daxad® 11KLS. [W.R. Grace] Low m.w. naphthalene sulfonate formaldehyde condensate, potassium salt; dispersant for dyes, dyestuffs, inks, latex paints, wax emulsions, wallboard coating, ore flotation; buff powd.; dens. 35 lb/ft3; pH 7.0–8.5 (1%); 85% min. act.

Daxad® 13. [W.R. Grace] Polymerized alkyl naphthalene sulfonic acid sodium salt; dispersant for use in paper making operations, pigment slurries and disp., paper coating disp., paper mill slime control, pitch control, pulp digestion, tall oil separation; amber powd.; dens. 47 lb/ft3; pH 9.5 (1%).

Daxad® 14B. [W.R. Grace] Low m.w. naphthalene sulfonate formaldehyde condensate, sodium salt sol'n.; dispersant for emulsion polymerization, dyestuffs, tanning, herbicides, pesticides, and pitch; dk. brn. liq.; 45% total solids.

Daxad® 14C. [W.R. Grace] Low m.w. naphthalene sulfonate formaldehyde condensate, sodium and potassium salts in sol'n.; dispersant for emulsion polymerization, dyestuffs, tanning, herbicides, pesticides, and pitch; designed for cold weather stability; dk. brn. liq.; 45% total solids.

Daxad® 15. [W.R. Grace] Low m.w. naphthalene sulfonate formaldehyde condensate, sodium salt; industrial grade, general purpose dispersant for emulsion polymerization, dyestuffs, tanning, herbicides, pesticides, and pitch; amber powd.; dens. 35 lb/ft3; pH 9.5 (1%); surf. tens. 70 dynes/cm (1%); 85% total solids.

Daxad® 16. [W.R. Grace] Low m.w. naphthalene sulfonate formaldehyde condensate, sodium salt; dispersant for emulsion polymerization, concrete, dyestuffs, tanning, herbicides, pesticides, and pitch; dk. brn. liq.; dens. 10.4 lb/gal; pH 9.5 (1%); 47.5% total solids.

Daxad® 17. [W.R. Grace] Low m.w. naphthalene sulfonate formaldehyde condensate, sodium salt; industrial grade dispersant for emulsion polymerization, dyestuffs, tanning, herbicides, pesticides, and pitch; amber gran.; ≤ 3% passes through 100-mesh screen; dens. 42 lb/ft3; pH 7.8–10.4 (1%); surf. tens. 70 dynes/cm (1%); 85% act.

Daxad® 19. [W.R. Grace] High m.w. naphthalene sulfonate formaldehyde condensate, sodium salt; dispersant and fluidifier for high-solids, aq. disps. and slurries; used in cement, gypsum, lime, coal, and other slurry systems; brn. powd.; very sol. in water; dens. 38 lb/ft3; pH 9.5 (1%); surf. tens. 71 dynes/cm (1%).

Daxad® 19K. [W.R. Grace] High m.w. naphthalene sulfonate formaldehyde condensate, potassium salt; version of Daxad 19 for use where the sodium salt is undesirable; brn. powd.; very sol. in water; dens. 42 lb/ft3; pH 9.5 (1%).

Daxad® 19L-33. [W.R. Grace] High m.w. naphthalene sulfonate formaldehyde condensate, sodium salt, in sol'n.; see Daxad 19; dk. brn. liq.; dens. 9.8 lb/gal; pH 9.5 (1%); surf. tens. 71 dynes/cm (1%); 33% total solids.

Daxad® 19L-40. [W.R. Grace] High m.w. naphthalene sulfonate formaldehyde condensate, sodium/potassium salt, in sol'n.; cold weather-stable dispersant, water-reducing agent, and visc.-reducer for high-solids slurries like concrete, cement, gypsum, lime, coal; dk. brn. liq.; dens. 10 lb/gal; pH 9.5 (1%); 40% total solids.

Daxad® 21. [W.R. Grace] Calcium salt of polymerized aryl alkyl sulfonic acids; dispersant for finely divided insol. particles in water; used for wettable powds., agric. chemicals, concrete admixtures, dyes, gypsum wallboard, high-strength cements, pigment slurries and disps., and water treatment chemicals; brn. powd.; dens. 30–35 lb/ft3; pH 5.0–8.0 (1%); surf. tens. 60–65 dynes/cm (1%); 92% act.

Daxad® 23. [W.R. Grace] Sodium salts of polymerized substituted benzoid alkyl sulfonic acids; dispersant for agric. chemicals, concrete admixtures, dyes, high-strength cements, linoleum pastes, pigment slurries and disps., and water treatment chemicals; dk. brn. powd.; dens. 35–40 lb/ft3; pH 7.5 (1%); surf. tens. 56–57 dynes/cm (1%); 84.5% act.

Daxad® 27. [W.R. Grace] Polymerized aryl and substituted benzoid alkyl sulfonic acid sodium salt; dispersant and suspending aid; used in agric. chemicals and pigment slurries and disp.; gray brn. powd.; dens. 48 lb/ft3; pH 8 (1%).

Daxad® 30. [W.R. Grace] Sodium polymethacrylate sol'n.; dispersant esp. for pigments in aq. sol'ns.; used in paint formulations, emulsion polymerization, water treatment, agriculture, cosmetics, industrial cleaners, in large particle suspensions; water-wh. clear liq.; sol. in water systems; sp.gr. 1.15; dens. 9.6 lb/gal; visc. 75 cps max.; pH 10.0; surf. tens. 70 dynes/cm (1%); 25% solids.

Daxad® 30-30. [W.R. Grace] Sodium polymethacrylate sol'n.; dispersant esp. for pigments in aq. sol'ns.; used in paint formulations, emulsion polymerization, water treatment (as a scale control agent for boiler systems), in large particle suspensions; water-wh. clear liq.; sol. in water systems; sp.gr. 1.21; dens. 10.1 lb/gal; visc. 150 cps max.; pH 10.0; surf. tens. 70 dynes/cm (1%); 30% solids.

Daxad® 30S. [W.R. Grace] Sodium polymethacrylate polymer; dispersant for pigments in aq. sol'ns.; used in water treatment, trade sales flat paints, dry mixes, agriculture; wh. fine powd.; sol. in water systems; dens. 32–35 lb/ft3 (tamped); pH 10.0 (1%); 90% act.

Daxad® 31. [W.R. Grace] Sodium polyisobutylene maleic anhydride copolymer sol'n.; dispersant for aq. systems; used in latex paints and coatings, enamels, polymerization, leather tanning, and water treatment; pale amber clear liq.; sol. in water systems; sp.gr. 1.11; dens. 9.2 lb/gal; visc. 30 cps; pH 10.0; surf. tens. 50 dynes/cm (1%); 25% total solids.

Daxad® 31S. [W.R. Grace] Sodium polyisobutylene maleic anhydride copolymer; see Daxad 31; wh. fine powd.; sol. in water systems; dens. 32–35 lb/ft3 (tamped); pH 10.0 (1%); 90% act.

Daxad® 32. [W.R. Grace] Ammonium polymethacrylate sol'n.; dispersant for the ceramics industry and pigment disps.; pale amber clear liq.; very sol. in water systems; dens. 9.1–9.4 lb/gal; visc. 75 cps max.; pH 8.0; 25% total solids.

Daxad® 32S. [W.R. Grace] Polymethacrylic acid buffered with ammonia; dispersant for the ceramics industry and where sodium salts are undesirable; wh. fine powd.; sol. in water systems; dens. 32–35 lb/ft3 (tamped); pH 6.3 (1%); 90% act.

Daxad® 34. [W.R. Grace] Polymethacrylic acid sol'n.; dispersant for pigments and fillers in ceramics, polymerization; pale amber clear liq.; sol. in water systems; sp.gr. 1.09; dens. 9.1 lb/gal; visc. 400 cps max.; pH 3; surf. tens. 70 dynes/cm (1%); 25% total solids.

Daxad® 34A9. [W.R. Grace] Ammonium polymethacrylate sol'n.; dispersant for pigments and fillers commonly used in latex paint systems; also for polymerization and clay coating; clear liq.; sol. in water systems; sp.gr. 1.11; dens. 9.2 lb/gal; visc. 25 cps max.; pH 9; surf. tens. 76 dynes/cm (1%); 24% total solids.

Daxad® 34N10. [W.R. Grace] Sodium polymethacrylate sol'n.; dispersant for latex polymerization, industrial cleaners, and for clay in paper coating colors; pale amber clear liq.; sol. in water systems;

sp.gr. 1.21; dens. 10.0 lb/gal; visc. 400 cps max.; pH 10.0; surf. tens. 70 dynes/cm (1%); 30% total solids.

Daxad® 34S. [W.R. Grace] Polymethacrylic acid; dispersant for preparations that start with acidic components and are further compded. to an alkaline pH; wh. fine powd.; sol. in water systems; dens. 32–35 lb/ft3 (tamped); pH 3.2 (1%); 90% act.

Daxad® 35. [W.R. Grace] Sodium polymethacrylate sol'n.; primary dispersant for clays, zinc oxide, and other paint or coating pigments and fillers; for latex paint formulations, water treatment; pale amber clear liq.; sol. in water systems; sp.gr. 1.28; dens. 10.6 lb/gal; visc. 350 cps; pH 7.0; surf. tens. 70 dynes/cm (1%); 40% total solids.

Daxad® 37LA7. [W.R. Grace] Ammonium polyacrylate sol'n.; anionic; dispersant in aq. systems when used at very low levels; esp. for latex paints and coatings, dispersing org. and inorg. pigments; slight amber clear, slightly hazy liq.; sol. in water and glycols; sp.gr. 1.17; dens. 9.8 lb/gal; visc. 80 cps; pH 7.0; surf. tens. 66 dynes/cm (1%); 40% total solids.

Daxad® 37L Acid. [W.R. Grace] Aq. sol'n. of low m.w. polyacrylic acid; for applics. requiring an acid-stable dispersant; also for use in water treatment as an antiscale agent and in latex flat paints; straw liq.; visc. 250 cps; pH 3.0; 50% total solids.

Daxad® 37LK9. [W.R. Grace] Potassium polyacrylate sol'n.; dispersant in aq. systems when used at very low levels; esp. for latex paints and coatings, dispersing org. and inorg. pigments; slight amber clear, slightly hazy liq.; sol. in water and glycols; sp.gr. 1.24; dens. 10.3 lb/gal; visc. 17 cps; pH 9.0; surf. tens. 71 dynes/cm (1%); 30% total solids.

Daxad® 37LN7. [W.R. Grace] Sodium polyacrylate sol'n.; dispersant, esp. for clays; deflocculant and dispersant for clay slurries; used in paper coating industry, in flat latex paints; as a fluidifier in the formulation of oil well drilling muds; as an antiredeposition agent in cleaning formulas; pale amber clear, slightly hazy liq.; infinite sol. in water; sp.gr. 1.30; dens. 10.9 lb/gal; visc. 500 cps; pH 7.0; surf. tens. 75 dynes/cm (1%); 45% total solids.

Daxad® 37LN10. [W.R. Grace] Sodium polyacrylate sol'n.; see Daxad 37LN7; pale amber clear, slightly hazy liq.; infinite sol. in water; sp.gr. 1.30; dens. 10.9 lb/gal; visc. 500 cps; pH 10.0; surf. tens. 71 dynes/cm (1%); 45% total solids.

Daxad® 37LN10-35. [W.R. Grace/Organics] Sodium polyacrylate sol'n.; dispersant for rapid dispersion of solid materials (esp. clays) in aq. systems; dispersant, deflocculant for pigment mfg., paper coating, latex paints, oil well drilling muds; antiredeposition agent in cleaning formulas; stable over wide pH range; pale amber sl. hazy liq.; sol. in water; sp.gr. 1.25; dens. 10.4 lb/gal; visc. 50 cps; pH 10; surf. tens. 65 dynes/cm (1%).

Daxad® 37NS. [W.R. Grace] Sodium polyacrylic acid; dispersant for solids in aq. systems; esp. for clays; as a deflocculant in the prod. of high-solids, low-visc. slurries in the pigments industry; wh. fine powd.; sol. in water systems; dens. 32–35 lb/ft3 (tamped); pH 2.7 (1%); 90% act.

Daxad® 41. [W.R. Grace] Sodium polymethacrylate polymer; dispersant designed for scale inhibition and conditioning of suspended matter in water treatment; inhibits deposition of new scale and sludge; wh. fine powd.; sol. in water systems; pH 7.5 (1%); surf. tens. 70 dynes/cm (1%); 90% act.

Daxad® CP-2. [W.R. Grace] Cationic polyelectrolyte; dispersant for fillers, pigments, other additives; precipitates wetting agents, detergent soaps, emulsifiers, and dispersants out of industrial waters; coagulates latexes; flocculates kaolin clay disps.; breaks o/w and w/o emulsions; liq.; sol. in water; dens. 9.6 lb/gal; visc. 700 cps; pH 6.0; 50% total solids.

DB-1 Defoamer. [Genesee Polymers] Organo-modified silicone polymer, silica, and waxes; defoamer for emulsifiable oils (metalworking lubricant sol. oils), paints, water treatment, coolants, agric. sprays, rolling oils, coatings; thick tan liq.; sp.gr. 0.90; dens. 7.0 lb/gal; flash pt. > 200 F; 100% act.

DB-12 Defoamer. [Genesee Polymers] Organo-modified silicone polymer, silica, and waxes; defoamer for emulsifiable oils (metalworking lubricant sol. oils), paints, water treatment, coolants, agric. sprays, rolling oils, coatings; tan opaque fluid; sp.gr. 0.90; dens. 7.5 lb/gal; flash pt. > 200 F; 100% act.

DB-19 Antifoam Compd. [Genesee Polymers] Silicone compd. with water carrier; nonionic; antifoam for aq. systems, hot or cold foaming systems, alkaline systems, industrial processing, chemical processing, cleaning prods., paints, paper and latex processing, metalworking; wh. opaque liq.; sp.gr. 1.0; dens. 8.0 lb/gal; visc. thixotropic; 10% silicone.

DBM. [Tiarco] Dibutyl maleate; intermediate for mfg. of surfactants; in copolymerization of PVC and vinyl acetates; plasticizer for vinyl resins; also in latex paints; liq.; 100% conc.

DDBS 100. [Zohar Detergent Factory] Alkylbenzene sulfonic acid, branched; anionic; detergent intermediate; liq.; 100% conc.

DDBS-100 SP. [Zohar Detergent Factory] Alkylbenzene and toluene cosulfonic acid; anionic; detergent intermediate used in spray drying; liq.; 100% conc.

DDBS Special. [Zohar Detergent Factory] Dodecylbenzene sulfonic acid; anionic; intermediate for mfg. of calcium dodecylbenzene sulfonates and similar anionics; visc. liq.; 92% conc.

DEAS Base. [Clark] Fatty alkanolamide; base for textile softeners; finished prods. yield a dry, hard, sl. waxy hand.

Deceresol Surfactant NI Conc. [Am. Cyanamid/Textiles] Polyether alcohol; nonionic; detergent, emulsifier, wetting agent, scouring agent, penetrant for textile industry; clear liq.; dissolves readily with stirring in cold water; dens. 8.8 lb/gal; sp.gr. 1.05; pH 6–7; 100% act.

Deceresol Surfactant OT 75%. [Am. Cyanamid/Textiles] Sulfonated ester; anionic; wetting and rewetting agent for textile industry; water-wh. visc. liq.; slight hexanol odor; completely sol. as dilute sol'n.; dens. 8.9 lb/gal; sp.gr. 1.082; 75% act.

Deceresol Surfactant OT Special. [Am. Cyanamid/Textiles] Sulfonated ester; anionic; wetting and rewetting agent for textile industry; colorless thin liq.; slight odor; dissolves with agitation in warm water; dens. 8.8 lb/gal; sp.gr. 1.05; pH 5.7; 70% act.

Defoamer 9. [CNC Int'l. L.P.] Oil-based liq. defoamer for pulp and paper mill effluent foam control and municpal sewage treatment plants.

Defoamer 9-W. [CNC Int'l. L.P.] Liq. defoamer for control of foam in pulp and paper mill effluent; clear straw-colored liq.; sp.gr. 0.89; dens. 7.4 lb/gal; flash pt. (COC) 120 F; pH 5.5; cloud pt. 35-40 F; 100% act.

Defoamer 10B. [CNC Int'l. L.P.] Liq. oil-base defoamers for Kraft screen room and bleach plant applic.

Defoamer 12. [CNC Int'l. L.P.] Nonsilicone oil-based;

defoamer for difficult, foamy high solids or high resin content coating systems, resin or latex mfg./ monomer stripping.; liq.

Defoamer 28. [CNC Int'l. L.P.] Liq. defoamer for paper machines, sulfate or sulfite pulp systems, brown stock washing and screening; for heavy weight grades, liner board, roofing or flooring stock, corrugated or min.-filled grades; promotes drainage.

Defoamer 28C. [CNC Int'l. L.P.] Liq. defoamer for aq. systems, resin or latex formulations, textiles and leather treatment.

Defoamer 30 Series. [CNC Int'l. L.P.] Water-based pulp mill screen defoamers for sulfate or sulfite pulps; paste or liqs.

Defoamer 34. [CNC Int'l. L.P.] Liq. defoamer for difficult, foamy high solids or high resin content coating systems.

Defoamer 44 Series. [CNC Int'l. L.P.] Liq. defoamers for coated or heavily surface-sized sheet such as wallpaper.

Defoamer 67. [CNC Int'l. L.P.] Water-based liq. defoamer for paper and paperboard.

Defoamer 69. [CNC Int'l. L.P.] Water-based liq. defoamer for paper and paperboard, min. or pigment-filled paper coatings, food pkg. adhesives, coatings and liners for food containers.

Defoamer 71. [CNC Int'l. L.P.] Liq. defoamer for starch or resin size sol'ns.

Defoamer 97. [CNC Int'l. L.P.] Water-based nonsilicone; defoamer for clay or pigment-filled paper coatings with protein or latex binders, print pastes for wallpaper; esp. for systems which cannot tolerate silicone or oils.

Defoamer 124. [CNC Int'l. L.P.] Liq. water-oil defoamer for resin or food pkg. adhesives and coatings; for nonwoven saturant or binder formulations, specialty coatings with low tolerance for oils or silicones.

Defoamer 127. [CNC Int'l. L.P.] Liq. defoamer for paper machine and specialty coating applics.; low cloud pt.

Defoamer 141. [CNC Int'l. L.P.] Water-based liq. defoamer for pulp and paper mill effluents.

Defoamer 167. [A. Harrison] Silicone emulsion; defoamer used in dyebath and finishing; water-sol.

Defoamer 177. [CNC Int'l. L.P.] Liq. defoamer for groundwood pulping.

Defoamer 224W. [CNC Int'l. L.P.] Water-extended effluent defoamer for difficult foamy systems.

Defoamer 226. [CNC Int'l. L.P.] Liq. defoamer for sulfate or sulfite pulp systems, brown stock washing and screening.

Defoamer 226W. [CNC Int'l. L.P.] Water-extended effluent defoamer for difficult foamy systems.

Defoamer 229. [CNC Int'l. L.P.] Defoamer for control of foam in aq. systems, effluents, pulp and paper, municipal waste treatment, textile scouring and finishing; amber opaque liq.; dens. 7.5 lb/gal; visc. 200-400 cps; flash pt. (COC) > 300 F; pour pt. 35 F.

Defoamer 297. [CNC Int'l. L.P.] Liq. water-extended defoamer for paper machines; for furnishes with high percentages of ground wood or recycled fibers.

Defoamer 309. [CNC Int'l. L.P.] Liq. oil-base defoamers for Kraft screen room and bleach plant applic.

Defoamer 357. [Hercules] Hydrocarbon oil; defoamer for mfg. and end-use foam control of water-based systems; water-disp.

Defoamer 388. [Hercules] Hydrocarbon oil; defoamer for chemical processing; liq.

Defoamer 403. [CNC Int'l. L.P.] Conc. defoamer for paper machines; intended for premixing with water; effective for filled, starch-sized, or resin-sized grades.

Defoamer 404. [CNC Int'l. L.P.] Liq. antifoam/ defoamer for water and water/oil-based acrylic or vinyl paints or industrial coatings.

Defoamer 407. [CNC Int'l. L.P.] Liq. defoamer for difficult, foamy high solids or high resin content coating systems.

Defoamer 407A. [CNC Int'l. L.P.] Nonsilicone prod.; defoamer for pigments, inks, textile processing; effective in jet dyeing.

Defoamer 440. [CNC Int'l. L.P.] Liq. defoamer for starch or synthetic size sol'ns.

Defoamer 477. [CNC Int'l. L.P.] Conc. defoamer for paper machines; intended for premixing with water; effective for filled, starch-sized, or resin-sized grades.

Defoamer 477L. [CNC Int'l. L.P.] Conc. general purpose antifoam/defoamer.

Defoamer 505. [CNC Int'l. L.P.] Liq. defoamer for paper machine and specialty coating applics.; low cloud pt.

Defoamer 546. [CNC Int'l. L.P.] Liq. defoamer for paper machine and specialty coating applics.; low cloud pt.

Defoamer 831. [Hercules] Hydrocarbon oil; liq. pH defoamer; low visc. for metering and rapid foam control; water-disp.

Defoamer 1030. [CNC Int'l. L.P.] Conc. liq. defoamer for heavy weight paper grades, linear board requiring good drainage.

Defoamer 1033. [CNC Int'l. L.P.] Conc. oil-based liq. defoamer for continuous or batch processing; for mining or drilling operations, latex or resin mfg., glues, adhesives.

Defoamer 1037. [CNC Int'l. L.P.] Liq. brownstock washer drainage defoamer for paper industry.

Defoamer 1063. [CNC Int'l. L.P.] Water-extended economical defoamers for washing and screening.

Defoamer 1066. [CNC Int'l. L.P.] Liq. brownstock washer drainage defoamer for paper industry.

Defoamer 1090. [CNC Int'l. L.P.] Liq. brownstock washer drainage defoamer for paper industry.

Defoamer 1191. [CNC Int'l. L.P.] Liq. oil-based defoamer for aq. systems, paper machines; for heavy weight grades, liner board, roofing or flooring stock, corrugated or min.-filled grades, for glues, adhesive, or binder formulations; promotes drainage.

Defoamer 1194. [CNC Int'l. L.P.] Oil-based defoamer for aq. systems, glues, adhesive, or binder formulations.

Defoamer 1713. [Hart Chem. Ltd.] Proprietary blend; paper machine defoamer for fine paper; visc. liq.

Defoamer A 50. [Chem-Y GmbH] Tall oil acid, octoxynol-200; cationic; defoamer for phosphate mineral processing for the sulfuric acid process in mfg. of artificial fertilizer; biodeg.; wh. opaque liq.; sp.gr. 0.95 kg/l; visc. 4000 mPa•s; pH 4-6; toxicology: LD50 (rat, oral) > 30 ml/kg; 50% act.

Defoamer B 90. [Chem-Y GmbH] Octeth-3 carboxylic acid; defoamer for phosphate mineral processing for the sulfuric acid process in mfg. of artificial fertilizer and for the regeneration of silver in electroplating baths; liq.; 90% act.

Defoamer C5B. [Hart Chem. Ltd.] Nonsilicone min. oil-based; defoamer for effluents in pulp and paper industry; effective over wide pH range; lt. brn. liq.;

100% act.

Defoamer CK-35. [Crompton & Knowles] Ethoxylated fatty acids in petrol. hydrocarbon; defoamer; eliminates entrained gas and improves drainage on paper machines; disp.

Defoamer CK-55. [Crompton & Knowles] Fatty amide derivs., hydrophobic silica, and nonionic surfactants in petrol. oil; defoamer for standard paper machines and hydropulpers; disp.

Defoamer CK-75. [Crompton & Knowles] Ethoxylated fatty acids and petrol. oil; defoamer for paper machines and hydropulpers; disp.

Defoamer DF-160-L. [Henkel/Textile] Defoamer for syn. latex coating systems, esp. acrylic emulsions and binders; also controls foaming of sodium lauryl sulfate sol'ns.; lt. yel. liq.; sol. forms stable wh. emulsions in water; pH 7.0 (10%); 100% act.

Defoamer KCE/S. [Hart Chem. Ltd.] Proprietary blend; water-based defoamer for screen room, bleachery, and effluent applics. in pulp and paper industry; liq.

Defoamer NXZ. [Henkel/Textile] Defoamer for use with latex coating and latex printing; also in atmospheric dyeing, printing and textile finishing processes; amber liq.; sol. forms stable wh. emulsions in water; pH 7.0 (10%); 100% act.

Defoamer PM. [Hart Chem. Ltd.] Proprietary blend; paper machine defoamer for tissues and towels; liq.

Defoamer S. [Hart Chem. Ltd.] Silicone-based defoamer; multipurpose defoaming agent; wh. cream; 11% act.

Defoamer SAS. [Catawba-Charlab] Silicone-based; defoamer for batch and continous dyeing systems.

Defoamer SF. [Hoechst Celanese/Colorants & Surf.] Perfluoro alkyl phosphinate/phosphonate; anionic; surfactant for agric. formulations.

Defoamer TIP. [Hoechst Celanese] Phosphoric acid esters; defoamer for waste water; clear liq.; 50% act.

Defoamer WB Series. [Hart Chem. Ltd.] Proprietary blend; water-based defoamers for screen room and linerboard applics. in pulp/paper industry; liq.

Defoamer WE Series. [Hart Chem. Ltd.] Proprietary blend; water-extended brownstock washer defoamers for enhancing deposit control and improving washer operation in pulp/paper industry; liq.

Defoamer XD2-22. [CNC Int'l. L.P.] Liq. foam control agent for anionic or nonionic surfactant systems; suggested for applics. where soaps or detergents are the major cause of foam problems.

Defoamer XD2-123D. [CNC Int'l. L.P.] Conc. liq. defoamer for acid phosphate ore processing.

Defoamer XD3-1C. [CNC Int'l. L.P.] Liq. foam control agent for anionic or nonionic surfactant systems; suggested for applics. where soaps or detergents are the major cause of foam problems.

Defoamer/Drainage Aid CK-25. [Crompton & Knowles] Fatty acids, fatty acid derivs., and surfactants in petrol. blending oil; defoamer and drainage aid for paper industry; disp.

Defomax. [Toho Chem. Industry] Nonionic surfactant/wax emulsions; nonionic; antifoaming agents for rubbers and plastics; liq./paste.

Degreez. [Alzo] Methyl oleate, methyl stearate, methyl palmitate, methyl laurate, methyl myristate; degreasing agent for removing oil stains from metal, textiles, etc.; biodeg.; yel. liq., mild typ. almost fruity odor; disp. in water; sp.gr. 0.87; b.p. > 400 F (760 mm); f.p. 18 F; cloud pt. 27 F; flash pt. (COC) 259 F; toxicology: nontoxic, nonirritating; LD50 (rat, acute oral) > 20 g/kg; mild transient eye irritation.

Degressal® SD 20. [BASF AG] Polypropoxylate; defoamer for cleaners and detergents; liq.; 100% act.

Degressal® SD 21. [BASF AG] Fatty alcohol alkoxylate; see Degressal SD 20; liq.; 100% act.

Degressal® SD 22. [BASF AG] Fatty alcohol alkoxylate; foam suppressor for sugar and chemical industries.

Degressal® SD 23. [BASF AG] Alcohol alkoxylate; defoamer for chemical and tech. industries; liq.; 100% act.

Degressal® SD 30. [BASF AG] Carboxylic acid ester; defoamer for cleaners and detergents; liq.; 100% act.

Degressal® SD 40. [BASF AG] Phosphoric acid ester; defoamer for cleaners; defoamer and plasticizer for dry-bright emulsions; liq.; 100% act.

Degressal® SNC. [BASF AG] Modified phosphoric acid monoester; defoamer for detergents and cleaners; liq.; 100% act.

Dehscofix 904. [Albright & Wilson Am.] Substituted phenol ethoxylated phosphate ester, TEA salt; anionic; emulsifier, dispersant for agric. chemicals, leather processing; liq.; 96% conc.

Dehscofix 905. [Albright & Wilson Am.] Substituted phenol ethoxylated phosphate ester, TEA salt; anionic; emulsifier, dispersant for agric. chemicals, leather processing; liq.; 96% conc.

Dehscofix 906. [Albright & Wilson Am.] Substituted phenol ethoxylate; nonionic; emulsifier, dispersant for agric. chemicals, leather processing; soft paste; 98% conc.

Dehscofix 907. [Albright & Wilson Am.] Substituted phenol ethoxylate phosphate ester, acid form; anionic; emulsifier, dispersant for agric. chemicals, leather processing; liq.; 96% conc.

Dehscofix 908. [Albright & Wilson Am.] Substituted phenol ethoxylate; nonionic; emulsifier, dispersant for agric. chemicals, leather processing; soft paste; 98% conc.

Dehscofix 909. [Albright & Wilson Am.] Substituted phenol ethoxylate; nonionic; emulsifier, dispersant for agric. chemicals, leather processing; paste; 98% conc.

Dehscofix 911. [Albright & Wilson Am.] Naphthalene sulfonic acid, formaldehyde condensate; anionic; surfactant for leather processing; paste; 40% conc.

Dehscofix 912. [Albright & Wilson Am.] Sodium naphthalene-formaldehyde sulfonate; anionic; surfactant for leather processing; liq.; 40% conc.

Dehscofix 914. [Albright & Wilson Am.] Sodium naphthalene-formaldehyde sulfonate; anionic; surfactant, pigment dispersant for leather processing; concrete plasticizer; liq.; 45% conc.

Dehscofix 914/AS. [Albright & Wilson Am.] Sodium naphthalene-formaldehyde sulfonate; anionic; surfactant for leather processing; liq.; 40% conc.

Dehscofix 914/ASL. [Albright & Wilson Am.] Sodium naphthalene-formaldehyde sulfonate; anionic; surfactant for leather processing; liq.; 40% conc.

Dehscofix 915. [Albright & Wilson Am.] Sodium naphthalene-formaldehyde sulfonate; anionic; surfactant for leather processing; concrete plasticizer; powd.; 92% conc.

Dehscofix 915/AS. [Albright & Wilson Am.] Sodium naphthalene-formaldehyde sulfonate; anionic; surfactant for leather processing; powd.; 92% conc.

Dehscofix 916. [Albright & Wilson Am.] Sodium diisopropyl naphthalene sulfonate; anionic; wetting agents for agric. chemicals, leather processing; powd.; 92% conc.

Dehscofix 916S. [Albright & Wilson Am.] Sodium diisopropyl naphthalene sulfonate; anionic; surfactant for leather processing; powd.; 92% conc.

Dehscofix 917. [Albright & Wilson Am.] Sodium di-n-butyl naphthalene sulfonate; anionic; wetting agents for agric. chemicals, leather processing; powd.; 92% conc.

Dehscofix 918. [Albright & Wilson Am.] beta-Naphthalene sulfonic acid; anionic; surfactant for leather processing; liq.; 45% conc.

Dehscofix 920. [Albright & Wilson Am.] Sodium naphthalene-formaldehyde sulfonate; anionic; wetting agents for agric. chemicals; pigment dispersant for leather processing; dispersant/emulsifier for emulsion polymerization, rubber; powd.; 92% conc.

Dehscofix 923. [Albright & Wilson Am.] Sodium dimethyl naphthalene-formaldehyde sulfonate; anionic; wetting agents for agric. chemicals, leather processing; powd.; 92% conc.

Dehscofix 926. [Albright & Wilson Am.] Sodium dimethyl naphthalene-formaldehyde sulfonate; anionic; surfactant for leather processing; concrete plasticizer; powd.; 92% conc.

Dehscofix 929. [Albright & Wilson Am.] Ammonium naphthalene-formaldehyde sulfonate; anionic; surfactant for leather processing; liq.; 40% conc.

Dehscofix 930. [Albright & Wilson Am.] Ammonium naphthalene-formaldehyde sulfonate; anionic; pigment dispersant for leather processing; dispersant/emulsifier for emulsion polymerization, rubber; powd.; 92% conc.

Dehscofix CO Series. [Albright & Wilson Am.] Castor oil ethoxylates; emulsifier for agric. chemicals.

Dehscotex. [Albright & Wilson Am.] Formulated auxiliaries for textiles and leather.

Dehscotex BA Series. [Albright & Wilson Am.] Formulated surfactant blends; bleaching agent for textile processing.

Dehscotex DT 809. [Albright & Wilson Am.] Formulated surfactant; anionic; detergent for leather processing.

Dehscotex DT Series. [Albright & Wilson Am.] Formulated surfactant blends; washing/scouring detergent for textile processing.

Dehscotex DY Series. [Albright & Wilson Am.] Formulated surfactant blends; dye bath dispersant and leveling agent for textile processing.

Dehscotex FW Series. [Albright & Wilson Am.] Formulated surfactant blends; felting auxiliary for textile processing.

Dehscotex MC Series. [Albright & Wilson Am.] Formulated surfactant blends; wetting agent for textile mercerizing and causticizing.

Dehscotex SN Series. [Albright & Wilson Am.] Formulated surfactants; nonionic/cationic; softener for textile processing.

Dehscotex VP-PF. [Albright & Wilson Am.] Nonsilicone surfactant blends; fiber antistat for textile processing.

Dehscotex WA Series. [Albright & Wilson Am.] Formulated surfactant blends; wetting agent for textile processing.

Dehscoxid 700 Series. [Albright & Wilson Am.] Synthetic alcohol ethoxylates; nonionic; wetting agents for agric. chemicals, textiles, leather; liq./paste.

Dehscoxid 730/740 Series. [Albright & Wilson UK] C13 alcohol ethoxylates; nonionic; surfactant; liq./paste.

Dehydag Wax E. [Henkel] Sodium cetearyl sulfate; anionic; o/w emulsifier used in personal care prods. and powd. cleaners; wh. powd.; water-sol.; dens. 0.180 g/ml; pH 6–8; 87% min. act.

Dehydol 25. [Henkel Canada] C12-18 fatty alcohol ethoxylate; nonionic; ingred. in household and industrial cleaners; adjuvant in concrete prod.; liq.; HLB 13.0; 25% conc.

Dehydol 100. [Henkel Canada; Henkel KGaA] Laureth-10; CAS 9004-98-2; nonionic; detergent, wetting; used in insecticides, low-foaming cleansing prods., household and industrial cleaners; paste; HLB 13.0; 100% conc.

Dehydol 737. [Henkel KGaA] Fatty alcohol polyglycol ether; nonionic; raw material for laundry detergents; paste; 100% conc.

Dehydol 980. [Henkel KGaA] Ethoxylated fatty alcohol (6 EO); nonionic; surfactant for detergents, industrial cleaners; liq.; 100% conc.

Dehydol D 3. [Henkel KGaA] Fatty alcohol polyglycol ether; nonionic; emulsifier, solubilizer for oils; liq.; HLB 9.0; 100% conc.

Dehydol G 202. [Henkel KGaA] Ethoxylated Guerbet alcohol; nonionic; surfactant, antifoam; liq.; 100% conc.

Dehydol G 205. [Henkel KGaA] Ethoxylated Guerbet alcohol; nonionic; surfactant, antifoam; dispersant, suspending agent for solid powds.; liq.; 100% conc.

Dehydol HD-FC 1, HD-FC 2. [Henkel KGaA] Fatty alcohol polyglycol ether blend; nonionic; surfactant conc. for laundry detergents; paste; 100% conc.

Dehydol HD-FC 4, HD-FC 6. [Henkel KGaA] Fatty alcohol polyglycol ether and fatty acid blend; nonionic; surfactant conc. for laundry detergents, heavy-duty liqs.; liq.; 90% conc.

Dehydol HD-L 1. [Henkel KGaA] Nonionics/fatty acid blend; nonionic; surfactant conc. for heavy-duty liqs.; liq.; 93% conc.

Dehydol LS 2. [Henkel Canada; Henkel KGaA; Pulcra SA] Laureth-2; nonionic; emulsifier, solubilizer for solvs., oils, bases for prod. of sulfates; raw material for dishwashing, cleansing agent and cold cleaners; in bath oils, waterless hand cleaners; liq.; HLB 6.2; hyd. no. 196-204; pH 6.0-7.5 (1%); 99–100% conc.

Dehydol LS 3. [Henkel; Henkel Canada; Henkel KGaA] Laureth-3; nonionic; emulsifier, solubilizer for solvs., oils, bases for prod. of sulfates; raw material for dishwashing, cleansing agent and cold cleaners; liq.; 99–100% conc.

Dehydol LS 4. [Henkel KGaA] Laureth-4; nonionic; emulsifier, solubilizer for solvs., oils, bases for prod. of sulfates; raw material for dishwashing, cleansing agent and cold cleaners; liq.; 99–100% conc.

Dehydol LT 2. [Henkel KGaA] Fatty alcohol polyglycol ether; nonionic; emulsifier, solubilizer for solvs., oils; base for prod. of ether sulfates; raw material for dishwashing, cleansing agents, cold cleaners; liq.; 99+% conc.

Dehydol LT 3. [Henkel KGaA; Pulcra SA] Coceth-27; CAS 61791-13-7; nonionic; emulsifier, solubilizer for solvs., oils; base for prod. of ether sulfates; raw material for low-foaming detergents, dishwashing, cleansing agents, cold cleaners; superfatting agent; liq.; sol. in oil; HLB 7.3; hyd. no. 163-174; pH 6.0-7.5 (1%); 100% conc.

Dehydol LT 5. [Henkel KGaA] Fatty alcohol polyglycol ether; nonionic; emulsifier, solubilizer for solvs., oils; base for prod. of ether sulfates; raw material for dishwashing, cleansing agents, cold cleaners; liq.; 99+% conc.

Dehydol LT 6. [Henkel/Cospha; Henkel Canada] Ethoxylated lauryl alcohol; nonionic; emulsifier for metalworking fluids and tech. applics.; also in low-foam detergents incl. laundry; liq.; 100% conc.

Dehydol LT 7. [Henkel KGaA; Pulcra SA] Coceth-7; CAS 61791-13-7; nonionic; emulsifier, solubilizer for solvs., oils; base for prod. of ether sulfates; surfactant for heavy-duty liq. detergents, dishwashing, cleansing agents, cold cleaners; paste; HLB 11.9; hyd. no. 106-112; pH 6.0-7.5 (1%); 100% conc.

Dehydol LT 7 L. [Henkel KGaA] Coceth-7; CAS 61791-13-7; nonionic; emulsifier for solvs., oils; liq.; 100% conc.

Dehydol O4. [Henkel; Henkel Canada; Henkel KGaA] Octeth-4; nonionic; raw material for mfg. of surfactants and cleaning agents; detergent, degreaser; butoxyethanol replacement; liq.; HLB 11.5; 100% conc.

Dehydol O4 DEO. [Henkel KGaA] Octeth-4; nonionic; surfactant for cold cleaners, degreasers; odorless version of Dehydol O4; liq.; 100% conc.

Dehydol O4 Special. [Henkel KGaA] Octeth-4; nonionic; surfactant for cold cleaners, degreasers; odorless version of Dehydol O4; liq.; 100% conc.

Dehydol PCS 6. [Pulcra SA] Ceteareth-6; CAS 68439-49-6; nonionic; emulsifier; intermediate raw material for mfg. of ethoxysulfates for use in detergents and industrial specialties; paste/solid; HLB 10.2; hyd. no. 106-110; 100% conc.

Dehydol PCS 14. [Pulcra SA] Ceteareth-14; CAS 68439-49-6; nonionic; surfactant for low-foaming heavy-duty detergents; dispersant; soap additive; solid; HLB 14.2; hyd. no. 63-67; pH 6.0-7.5 (1%); 100% conc.

Dehydol PD 253. [Pulcra SA] Ethoxylated C12-15 alcohol (3 EO); nonionic; surfactant; liq.; HLB 7.8; hyd. no. 160-164; 100% conc.

Dehydol PEH 2. [Pulcra SA] Ethoxylated C8 alcohol (2 EO); nonionic; surfactant; liq.; HLB 8.0; hyd. no. 256-260; pH 6.0-7.5 (1%); 100% conc.

Dehydol PID 6. [Pulcra SA] Laureth-6; CAS 34938-91-8; nonionic; emulsifier, wetting agent for industrial, cosmetic, pharmaceutical applics., in high electrolyte concs.; liq.; HLB 11.8; hyd. no. 178-182; pH 6.0-7.5 (1%); 100% conc.

Dehydol PIT 6. [Pulcra SA] Ethoxylated C13 alcohol (6.5 EO); nonionic; surfactant; liq.; HLB 11.8; hyd. no. 114-121; pH 6.0-7.5 (1%); 100% conc.

Dehydol PLS 1. [Pulcra SA] Ethoxylated C12-14 fatty alcohol (1 EO); nonionic; surfactant; liq.; HLB 3.7; hyd. no. 232-236; 100% conc.

Dehydol PLS 3. [Pulcra SA] Ethoxylated C12-14 fatty alcohol (3 EO); CAS 3055-94-5; nonionic; emulsifier for min. oils, aliphatic solvs.; visc. depressant for PVC plastisols; intermediate for sulfation; liq.; HLB 8.0; hyd. no. 169-173; 100% conc.

Dehydol PLS 4. [Pulcra SA] Ethoxylated C12-14 fatty alcohol (4 EO); nonionic; surfactant; liq.; HLB 9.5; hyd. no. 149-153; 100% conc.

Dehydol PLS 6. [Pulcra SA] Ethoxylated C12-14 fatty alcohol (6 EO); nonionic; surfactant; liq./paste; HLB 11.5; hyd. no. 120-124; pH 6.0-7.5 (1%); 100% conc.

Dehydol PLS 8. [Pulcra SA] Ethoxylated C12-14 fatty alcohol (8 EO); nonionic; surfactant; liq./paste; HLB 12.8; hyd. no. 100-104; pH 6.0-7.5 (1%); 100% conc.

Dehydol PLS 11/80. [Pulcra SA] Ethoxylated C12-14 fatty alcohol (11 EO); nonionic; surfactant; liq.; pH 5.5-7.5 (5%); 80% conc.

Dehydol PLS 12. [Pulcra SA] Ethoxylated C12-14 fatty alcohol (12 EO); nonionic; surfactant; solid; HLB 14.5; hyd. no. 73-81; pH 6.0-7.5 (1%); 100% conc.

Dehydol PLS 15. [Pulcra SA] Ethoxylated C12-14 fatty alcohol (1.5 EO); nonionic; surfactant; liq.; HLB 5.0; hyd. no. 205-215; pH 9-10 (1%); 100% conc.

Dehydol PLS 21. [Pulcra SA] Ethoxylated C12-14 fatty alcohol (21 EO); nonionic; surfactant; solid; HLB 16.5; hyd. no. 40-60; pH 6.0-7.5 (1%); 100% conc.

Dehydol PLS 21E. [Pulcra SA] Ethoxylated C12-14 fatty alcohol (21 EO); nonionic; surfactant; flakes; HLB 16.5; hyd. no. 40-60; pH 6.0-7.5 (1%); 100% conc.

Dehydol PLS 235. [Pulcra SA] Ethoxylated C12-14 fatty alcohol (2.35 EO); nonionic; surfactant; liq.; HLB 6.9; hyd. no. 185-189; 100% conc.

Dehydol PLT 5. [Pulcra SA] Ethoxylated C12-18 fatty alcohol (5 EO); nonionic; surfactant; paste; HLB 9.6; hyd. no. 138-144; pH 6.0-7.5 (1%); 100% conc.

Dehydol PLT 6. [Pulcra SA] Ethoxylated isotridecanol; CAS 24938-91-8; nonionic; emulsifier, wetting agent for industrial, cosmetic, pharmaceutical applics., in high electrolyte concs.; liq.; HLB 11.7; 100% conc.

Dehydol PLT 8. [Pulcra SA] Coceth-8; CAS 61791-13-7; nonionic; surfactant for heavy-duty liq. detergents; paste/solid; HLB 12.6; 100% conc.

Dehydol PO 5. [Pulcra SA] Ethoxylated C8-10 alcohol (5 EO); CAS 68439-46-3; nonionic; w/o emulsifier, coemulsifier for min. oils and waxes, paraffinic hydrocarbons; liq.; HLB 11.6; hyd. no. 145-160; pH 5.0-7.0 (1%); 100% conc.

Dehydol PTA 7. [Pulcra SA] Talloweth-7; CAS 61791-28-4; nonionic; surfactant; solid; HLB 10.8; hyd. no. 96-104; pH 6.0-7.5 (5%); 100% conc.

Dehydol PTA 23. [Pulcra SA] Talloweth-23; CAS 61791-28-4; nonionic; surfactant; solid; HLB 16.0; hyd. no. 42-50; pH 6.0-7.5 (5%); 100% conc.

Dehydol PTA 23/E. [Pulcra SA] Talloweth-23; CAS 61791-28-4; 68439-49-6; nonionic; surfactant for low-foaming heavy-duty detergents; dispersant; for soap additives; flakes; HLB 16.0; hyd. no. 42-50; pH 6.0-7.5 (1%); 100% conc.

Dehydol PTA 40. [Pulcra SA] Talloweth-40; CAS 61791-28-4; nonionic; surfactant; solid; HLB 17.4; hyd. no. 26-30; pH 6.0-7.5 (20%); 100% conc.

Dehydol PTA 80. [Pulcra SA] Talloweth-80; CAS 68439-49-6; nonionic; surfactant for powd. detergent mixtures; flakes; HLB 18.7; hyd. no. 18-22; pH 6.0-7.5 (1%); 100% conc.

Dehydol PTA 114. [Pulcra SA] Talloweth-11; CAS 61791-28-4; nonionic; surfactant for low-foaming heavy-duty detergents; dispersant; soap additive; solid; 100% conc.

Dehydol TA 5. [Henkel KGaA] Talloweth-5; CAS 61791-28-4; nonionic; raw material for low-foaming detergents, toilet freshening cubes; emulsifier; solid; 100% conc.

Dehydol TA 6. [Henkel KGaA] Fatty alcohol polyglycol ether; nonionic; emulsifier for min. oils and paraffin; solid; HLB 10.0; 100% conc.

Dehydol TA 11. [Henkel KGaA; Pulcra SA] Talloweth-11; CAS 61791-28-4; nonionic; raw material for detergents, toilet cubes; emulsifier for tech. applics.; solid; HLB 13.0; hyd. no. 75-80; pH 6.0-7.5 (1%); 100% conc.

Dehydol TA 12. [Henkel KGaA] Talloweth-12; CAS 61791-28-4; nonionic; emulsifier for solv., waxes, oils; solid; HLB 13.4; 100% conc.

Dehydol TA 14. [Henkel KGaA] Talloweth-14; CAS 61791-28-4; nonionic; raw material for low-foaming detergents, toilet cubes; emulsifier; solid; 100% conc.

Dehydol TA 20. [Henkel KGaA; Pulcra SA] Talloweth-20; CAS 61791-28-4; nonionic; emulsifier for paraffins, waxes; raw material for foam-controlled detergents, toilet cubes; solid; HLB 15.4; hyd. no. 49-53; pH 6.0-7.5 (1%); 100% conc.

Dehydol TA 30. [Henkel KGaA] Talloweth-30; CAS 61791-28-4; nonionic; emulsifier for paraffins, waxes; raw material for foam-controlled detergents, toilet cubes; solid; HLB 16.5; 100% conc.

Dehydol WM. [Henkel KGaA] Fatty alcohol polyglycol ether; nonionic; raw material for laundry detergents; paste; 100% conc.

Dehydol WM 90. [Henkel KGaA] Fatty alcohol polyglycol ether; nonionic; see Dehydol WM; liq.; 90% conc.

Dehydran 150. [Henkel/Functional Prods.] Aq. silicone compd.; defoamer for systems containing surfactants, emulsifiers, and cleaning suds; liq.

Dehydran 240. [Henkel/Functional Prods.] Modified polyalkylene glycols; defoamer for industrial cleaners, automatic dishwashing agents; liq.

Dehydran 241. [Henkel/Functional Prods.] Modified polyalkylene glycols; defoamer for industrial cleaners, automatic dishwashing agents; liq.

Dehydran 420. [Henkel/Functional Prods.] Fatty acid ester; defoamer used in inorg. acid prod.; liq.

Dehydran 520. [Henkel/Functional Prods.] Fatty acid ester; defoamer for polymerization, esp. during monomer stripping in mfg. of suspension PVC; biodeg.; liq.; 100% act.

Dehydran 610. [Henkel/Functional Prods.] Natural fatty derivs. in emulsion form; defoamer used in waste water treatment; liq.

Dehydran 630. [Henkel/Functional Prods.] Mixt. of special fatty acid esters with aliphatic hydrocarbons; defoamer used in waste water treatment; liq.

Dehydran 1019. [Henkel/Functional Prods.] Nonsilicone defoamer; defoamer for prod. and processing of polymer dispersions; liq.; 100% act.

Dehydran 1227. [Henkel/Functional Prods.] Trace silicone defoamer; defoamer and foam inhibitor for semi- and high-gloss emulsion paints; liq.; 100% act.

Dehydran 1293. [Henkel/Functional Prods.] Modified polysiloxane; defoamer for water-based industrial coatings, printing inks, wood lacquers; liq.; 100% act.

Dehydran C. [Henkel/Functional Prods.] Nonsilicone defoamer; defoamer for emulsion paints, water-based adhesives and glues; liq.; 100% act.

Dehydran F. [Henkel/Functional Prods.] Nonsilicone defoamer; defoamer for emulsion paints, water-based adhesives and glues; liq.; 100% act.

Dehydran G. [Henkel/Functional Prods.] Trace silicone defoamer; defoamer for emulsion paints, water-based adhesives and glues; liq.; 100% act.

Dehydran P 4. [Henkel/Functional Prods.] Nonsilicone defoamer; defoamer for polymerization esp. during removal of remaining monomers in prod. of suspension, microsuspension, and emulsion PVC; liq.; 100% act.

Dehydran P 11. [Henkel/Functional Prods.] Trace silicone defoamer; defoamer for polymerization esp. during removal of remaining monomers in prod. of suspension, microsuspension, and emulsion PVC; liq.; 100% act.

Dehydran P 12. [Henkel/Functional Prods.] Nonsilicone defoamer; for monomer recovery in prod. of syn. rubber; antifoaming agent for syn. polymer disps., water-based paints, adhesives, glues; liq.; 100% act.

Dehydrophen 65. [Henkel KGaA] Alkylaryl polyglycol ether; nonionic; detergent, wetting, dishwashing agent; liq.; HLB 11.3; 100% conc.

Dehydrophen 100. [Henkel KGaA] Alkylaryl polyglycol ether; nonionic; detergent, wetting, emulsifier; liq. dishwashing agent; liq.; HLB 13.3; 100% conc.

Dehydrophen 150. [Henkel KGaA] Alkylaryl polyglycol ether; nonionic; emulsifier for various hydrocarbons; paste; HLB 15.0; 100% conc.

Dehydrophen PNP 4. [Pulcra SA] Nonoxynol-6; CAS 9016-45-9; 2311-27-5; nonionic; detergent, dispersant, emulsifier, wetting agent for solv. cleaners; emulsifier/stabilizer for paint additives; as deicing fluid; sludge dispersant in petrol. prods.; liq.; HLB 8.9; pH 5.5-7.5 (5%); 100% conc.

Dehydrophen PNP 6. [Pulcra SA] Nonoxynol-6; CAS 9016-45-9; nonionic; detergent, dispersant, emulsifier, wetting agent for solv. cleaners, paint additives, deicing fluids, sludge dispesant in petrol. prods.; liq.; HLB 10.9; pH 5.5-7.5 (5%); 100% conc.

Dehydrophen PNP 8. [Pulcra SA] Nonoxynol-8; CAS 9016-45-9; nonionic; detergent, dispersant, emulsifier, wetting agent; lower foaming, greater oil affinity; liq.; HLB 12.3; pH 5.5-7.5 (5%); 100% conc.

Dehydrophen PNP 10. [Pulcra SA] Nonoxynol-10 (9.5-10 EO); CAS 9016-45-9; nonionic; surfactant; liq.; HLB 13.3; pH 5.5-7.5 (5%); 100% conc.

Dehydrophen PNP 12. [Pulcra SA] Nonoxynol-12; CAS 9016-45-9; nonionic; detergent, dispersant, emulsifier, wetting agent; effective in high temp. and high electrolyte media; liq./paste; HLB 14.1; pH 5.5-7.5 (5%); 100% conc.

Dehydrophen PNP 15. [Pulcra SA] Nonoxynol-15; CAS 9016-45-9; nonionic; detergent, dispersant, emulsifier, wetting agent; high-temp. detergent; emulsifier for oils, fats, waxes; metal cleaning and degreasing; paste; HLB 15.0; pH 5.5-7.5 (5%); 100% conc.

Dehydrophen PNP 20. [Pulcra SA] Nonoxynol-20; CAS 9016-45-9; nonionic; all-purpose dispersant, emulsifier, wetting agent; latex stabilizer; solid; HLB 16.0; pH 5.5-7.5 (5%); 100% conc.

Dehydrophen PNP 30. [Pulcra SA] Nonoxynol-30; CAS 9016-45-9; nonionic; all-purpose high-temp. wetting agent, dispersant, emulsifier; emulsion polymerization; solid; HLB 17.1; pH 5.5-7.5 (5%); 100% conc.

Dehydrophen PNP 40. [Pulcra SA] Nonoxynol-40; CAS 9016-45-9; nonionic; emulsifier for fats, oils, waxes, org. solvs., emulsion polymerization; latex stabilizer; flakes; HLB 17.8; 100% conc.

Dehydrophen POP 4. [Pulcra SA] Octoxynol-4; CAS 9002-93-1; nonionic; emulsifier, detergent, dispersant; used in wax-based washing formulations, emulsion cleaners; liq.; oil-sol.; HLB 9.2; pH 5.5-7.5 (5%); 100% conc.

Dehydrophen POP 8. [Pulcra SA] Octoxynol-8; CAS 9002-93-1; nonionic; emulsifier, detergent, wetting agent for low-temp. tank cleaners; additive in acid compds. for metals, lt.-duty cleaners; liq.; HLB 12.6; pH 5.5-7.5 (5%); 100% conc.

Dehydrophen POP 10. [Pulcra SA] Octoxynol-10; CAS 9002-93-1; nonionic; wetting agent, detergent, dispersant, emulsifier for household and industrial cleaners, textile processes, wool scouring; emulsifier for pesticides; liq.; HLB 13.2; pH 5.5-7.5 (5%); 100% conc.

Dehydrophen POP 17. [Pulcra SA] Octoxynol-17; CAS 9002-93-1; nonionic; surfactant for hot cleaning systems, metal pickling, electrolytic cleaning; solid; HLB 15.7; pH 5.5-7.5 (5%); 100% conc.

Dehydrophen POP 17/80. [Pulcra SA] Octoxynol-17; CAS 9002-93-1; nonionic; surfactant for hot cleaning systems, metal pickling, electrolytic cleaning; liq.; pH 5.5-7.5 (5%); 80% conc. in water.

Dehymuls FCE. [Henkel KGaA] Dicoco distearyl pentaerythrityl citrate; nonionic; emulsifier for w/o emulsions; flakes; 100% conc.

Dehymuls HRE 7. [Henkel KGaA] PEG-7 hydrogenated castor oil; CAS 61788-85-0; nonionic; emulsifier for w/o emulsions for personal care prods.; pale yel. visc. liq.; 100% conc.

Dehymuls SSO. [Henkel; Henkel KGaA] Sorbitan sesquioleate; CAS 8007-43-0; nonionic; w/o emulsifier and coemulsifier for waxes and oils; liq.

Dehypon Conc. [Henkel; Henkel Canada; Henkel KGaA] Ethoxylated, propoxylated fatty alcohols and anionics; anionic; wetting agent, emulsifier and dispersant; for carpet cleaners; high detergency, low foam; paste; 99% conc.

Dehypon G 2084. [Henkel KGaA] Guerbet alcohol polyglycol ether; nonionic; surfactant with foam inhibiting props. for industrial cleaners, bottle cleaners; liq.; 100% conc.

Dehypon LS-24. [Henkel Canada; Henkel KGaA] Ethoxylated, propoxylated lauryl alcohol; nonionic; detergent and wetting agent; low foaming; liq.; 100% conc.

Dehypon LS-36. [Henkel Canada; Henkel KGaA] Ethoxylated, propoxylated lauryl alcohol; nonionic; detergent, wetting agent, low foaming; liq.; cloud pt. 11 C; 100% conc.

Dehypon LS-45. [Henkel; Henkel KGaA] Fatty alcohol EO/PO adduct; nonionic; surfactant, dispersant, wetting agent for industrial cleaners, liq. laundry detergents, dishwashing agents; liq.; 100% conc.

Dehypon LS-54. [Henkel Canada; Henkel KGaA] Ethoxylated, propoxylated lauryl alcohol; nonionic; detergent, wetting agent, low foaming; as rinse aid, acid cleaning agent, glass cleaner, alkaline metal degreaser; liq.; cloud pt. 30 C; 100% conc.

Dehypon LT 24. [Henkel KGaA] Fatty alcohol EO/PO adduct; nonionic; low-foaming surfactant for industrial cleaners, liq. laundry detergents; liq.; 100% conc.

Dehypon LT 054. [Henkel KGaA] Fatty alcohol polyglycol ether; nonionic; surfactant for bottle washing; paste.

Dehypon LT 104. [Henkel Canada; Henkel KGaA] Fatty alcohol polyglycol ether; nonionic; low-foaming surfactant, antifoaming agent; stable in alkaline and acidic media; paste; 100% conc.

Dehypon OD 044. [Henkel KGaA] Fatty alcohol polyglycol ether; nonionic; surfactant with antifoaming props.; for high-pressure jet cleaners; liq.; 100% conc.

Dehyquart A. [Henkel/Cospha; Henkel/Functional Prods.] Cetrimonium chloride; CAS 112-02-7; cationic; emulsifier for emulsion polymerization; softener, conditioner, bactericide, fungicide, and odor inhibitor in personal care prods.; antistat for hair and fibers; pale yel. clear liq.; 24–26% act.

Dehyquart AU-36. [Pulcra SA] Bis (acyloxyethyl) hydroxyethyl methylammonium methosulfate, 15% IPA; cationic; highly biodeg. surfactant; raw material for prod. of softeners; good rewetting; semisolid paste @ 20 C; ylsh. liq. @ 45 C; m.w. 800; drop pt. 33-37 C; flash pt. 26-30 C; pH 2-3 (5%); 85 ± 2% act.

Dehyquart AU-56. [Pulcra SA] Bis (acyloxyethyl) hydroxyethyl methylammonium methosulfate, 10% IPA; cationic; highly biodeg. surfactant; raw material for prod. of softeners; good rewetting, antistatic props.; semisolid paste @ 20 C; ylsh. liq. @ 45 C; m.w. 800; drop pt. 37-42 C; flash pt. 26-30 C; pH 2-3 (5%); 90 ± 2% act.

Dehyquart C. [Henkel/Functional Prods.] Lauryl pyridinium chloride; CAS 104-74-5; cationic; surfactant, wetting agent, fungicide, bactericide, and disinfectant used in cleaning formulations and personal care prods.; sequestrant for min. oil industry; wh. paste, powd.; 80–82% act.

Dehyquart C Crystals. [Henkel/Cospha; Henkel/Functional Prods.] Lauryl pyridinium chloride; CAS 104-74-5; cationic; emulsifier in creams and lotions; hair conditioners, skin creams; antistat for hair and fiber; bactericide, fungicide, corrosion inhibitor, sequestrant; conditioner used in personal care prods.; recrystallized powd.; 90–94% act.

Dehyquart D. [Henkel/Cospha; Henkel/Functional Prods.] Lauryl pyridinium bisulfate; CAS 17342-21-1; cationic; germicide, wetting agent with anticorrosive effect, emulsion breaker; acid stable; liq.; 52-55% act.

Dehyquart DAM. [Henkel/Functional Prods.] Distearyl dimethyl ammonium chloride; cationic; emulsifier for plastics industry; conditioning component for hair care preparations; antistat; paste; 70–80% act.

Dehyquart LDB. [Henkel/Functional Prods.] Lauralkonium chloride; CAS 139-07-1; cationic; bactericide and fungicide for disinfectants; emulsifier; external antistat for plastics; liq.; 34–36% conc.

Dehyquart LT. [Henkel/Functional Prods.] Laurtrimonium chloride; CAS 112-00-5; cationic; emulsifier for plastics industry, wetting agent, antistat, bactericide, demulsifier, deodorant, conditioning component for hair care prods.; liq.; 34–36% conc.

Dehyquart SP. [Henkel/Cospha; Henkel Canada] Quaternium-52; CAS 58069-11-7; cationic; emulsifier, conditioning, softening and antistatic agent used in personal care prods.; metal corrosion inhibitor; lt. yel. clear visc. liq.; pH 6.8–7.2 (10%); 49.0–51.0% solids.

Dehysan Z 4904. [Henkel] Mixt. of fatty alcohol polyglycol ethers and soaps; defoamer for sugar prod. (cane and beet); liq.

Dehysan Z 7225. [Henkel] Mixt. of natural fats with hydrophilic fat derivs.; defoamer for beet sugar prod.; liq.

Dehysol R. [Henkel KGaA] Thickening and wetting agent for alkyd paints; liq.

Dehyton® G-SF. [Henkel KGaA] Cocoamphodipropionate; amphoteric; mild surfactant for cosmetic preps., esp. baby cleansers; emulsifier for o/w emulsions; liq.; 40% conc.

Dehyton® K. [Henkel KGaA] Cocamidopropyl betaine; amphoteric; raw material for mfg. of surfactants, conditioners; liq.; 30% conc.

Dehyton® KE. [Pulcra SA] Cocamidopropyl betaine; amphoteric; surfactant base; liq.; pH 4-6 (5%); 29-

31% act.

Dehyton® PMG. [Pulcra SA] Sodium lauroamphoacetate; CAS 14350-96-0; amphoteric; raw material for detergents, shampoos, dishwashers, textile softeners, paint emulsifiers, all-purpose washing agents; liq.; pH 9.0-9.5 (20%); 37-41% act.

Delestat P35. [Seppic] Quat.; cationic; resistivity control for painting; liq.; 80% conc.

Delion 342. [Takemoto Oil & Fat] Alkyl phosphate POE alkyl ether; anionic/nonionic; finishing oil for acrylic fibers; liq.; 60% conc.

Delion 624. [Takemoto Oil & Fat] Phosphate surfactant; anionic; antistat and lubricant for fibers; paste; 60% conc.

Delion 662. [Takemoto Oil & Fat] Phosphate surfactant; anionic; antistat for syn. fibers; paste.

Delion 964. [Takemoto Oil & Fat] Alkyl phosphate; anionic/nonionic; spin finish oil for polyester staple; paste; 80% conc.

Delion 966. [Takemoto Oil & Fat] Surfactant/neutral oil blend; nonionic/cationic; finish oil for polyester staple; liq.; 90% conc.

Delion 6067. [Takemoto Oil & Fat] Alkyl phosphate; anionic; staple finish oil for polyester staple; paste; 60% conc.

Delion A-016. [Takemoto Oil & Fat] Sulfonated surfactant; anionic; antistat for syn. fibers; wetting agent and detergent; liq.; 40% conc.

Delion A-160. [Takemoto Oil & Fat] Phosphated surfactant; anionic; antistat agent for wool and syn. fibers; liq.

Delion F-200 Series. [Takemoto Oil & Fat] Syn. lubricant/surfactant blend; anionic/nonionic; spin finish oil for polyester filament; liq.; 80-100% conc.

Delion F-400 Series. [Takemoto Oil & Fat] Syn. lubricant/surfactant blend; nonionic; spin finish oil for nylon filament industrial yarn; liq.; 80-100% conc.

Delion F-500 Series. [Takemoto Oil & Fat] Syn. lubricant/surfactant blend; anionic/nonionic; spin finish oil for nylon filament; liq.; 80-100% conc.

Delion F-4000 Series. [Takemoto Oil & Fat] Syn. lubricant/surfactant blend; nonionic; spin finish oil for polyester filament industrial yarn; liq.; 80-100% conc.

Delvet 68, 70. [Henkel] Nonionic; surfactant used as chlorinated paraffin flame retardant additive; fluid; 68% act.

Demelan AU-40. [Pulcra SA] Alkylamine polyglycol ether methosulfate, additives; cationic; surfactant with synergistic effect for use in mfg. of heavy-duty detergents; coupling agent for nonionics in liq. alkaline cleaners; antistat; liq.; 46% conc.

Demelan CB-10. [Pulcra SA] Alkylbenzene sulfonate blend; anionic/nonionic; detergent for liq. dishwashing formulations with foaming capacity; liq.; sol. in water; 65% conc.

Demelan CB-28. [Pulcra SA] Alkylbenzene sulfonate blend; detergent prod. for prod. of hand cleaners; liq.; 79 ± 1% conc.

Demelan CB-42. [Pulcra SA] Alkylbenzene sulfonate blend; scouring agent for textile industry; liq.; 30% conc.

Demelan CB-57. [Pulcra SA] Alkylbenzene sulfonate blend; detergent for liq. dishwashing formulations with foaming capacity; gel; 50 ± 1% conc.

Demelan CB-60. [Pulcra SA] Blend; anionic; conc. base for liq. dishwashing with foaming capacity; gel; sol. in water at all temps. and concs.; 60% conc.

Demelan CB-70. [Pulcra SA] Blend; anionic/nonionic; conc. for heavy-duty liq. detergents; liq.; 70% conc.

Demelan CM-33. [Pulcra SA] EO/PO block polymers and additives; nonionic; conc. base for home mech. dishwashing compds.; liq.; 67% conc.

Demelan CM-95. [Pulcra SA] Blend; anionic/nonionic; conc. for hard surf. liq. detergents; liq.; 50% conc.

Demelan FB-12. [Pulcra SA] Alkylbenzene sulfonate blend; detergent prod. for prod. of hand cleaners; paste; 65 ± 1% conc.

Demelan FB-16. [Pulcra SA] Alkylbenzene sulfonate blend; scouring agent for textile industry; paste; 65 ± 1% conc.

Demix® 7730. [Arizona] Catalytically disproportionated tall oil prod.; emulsifier for SBR prod.; Gardner 11; sp.gr. 1.0; flash pt. > 400 F; acid no. 180.

Demix® 7740. [Arizona] Catalytically disproportionated tall oil prod.; emulsifier for SBR prod.; Gardner 7-8; sp.gr. 1.0; flash pt. > 400 F; acid no. 177.

Demix® 7750. [Arizona] Catalytically disproportionated tall oil prod.; emulsifier for SBR prod.; Gardner 13; sp.gr. 0.9; flash pt. > 400 F; acid no. 174.

Demulfer Series. [Toho Chem. Industry] Anionic/nonionic; demulsifier for crude oil prod.; desalting agent for oil refinery; liq.

Demulsifier 3837. [Hoechst Celanese/Colorants & Surf.] Polymer; cationic; demulsifier for metalworking applics.; for use at alkaline pH and without the use of any other chemical; visc. liq.; sol. in water; dens. 1.0 g/cc; pH 9.5 (1% aq.).

Denphos 623. [Graden] Phosphate ester, free acid; anionic; emulsifier for emulsion polymerization; wetting agent, detergent for heavy-duty all-purpose cleaners; dedusting agent for alkaline powds.; emulsifier for pesticides; clear lt.-colored liq.; sol. in water; > 99% conc.

Denphos P-610. [Graden] Phosphate ester; anionic; detergent, emulsifier, wetting agent, hydrotrope for household and industrial cleaners; water-wh. to pale yel. liq.; mild odor; sol. in aq. sol'ns. with high conc. of alkaline builders; stable in hard water, acids, alkalies; sp.gr. 1.155; pH 1.5–2.5 (2%); 100% act.

Densol 284. [Graden] Highly sulfated fatty acid; anionic; stabilizer for natural and syn. latex; protects from coagulation; dk. amber liq.; fatty odor; 70% conc.

Densol 6920. [Graden] Syn. replacement for sulfated castor oil; anionic; emulsifier for min. oil, dispersant for colors in oil systems; antisettling agent for pigments during storage; liq.; sp.gr. 0.972 (60 F); pH 5.4 (10%); 75% act.

Densol 6930. [Graden] Syn. replacement for sulfated sperm oil; emulsifier; pH 5.4 (10%).

Densol BP-61, -62. [Graden] Block polymer; nonionic; defoamer, emulsifier, dispersant, wetting agent, and stabilizer; liq.; 100% conc.

Densol P-82. [Graden] Sulfated oleic ester; anionic; wetting agent, rewetting agent, emulsifier for paper towels, latex impregnation and dipping, textile dyeing; amber liq.; dens. 8.57 lb/gal; sp.gr. 1.0288; pH 5.5 (2%); 65% act.

Densulf TA-75. [Graden] Sulfonated tallow; softener for textiles; fatliquoring agent for leathers; plasticizer for cotton finishing; defoamer for glues; ivory soft paste, mild fatty odor; sol. in water; sp.gr. 1.09; 25% moisture.

Denwet CM. [Graden] Sodium dioctyl sulfosuccinate; anionic; wetting agent, emulsifier for emulsion polymerization, battery separators, glass cleaners; aids in dispersing, emulsifying, penetrating, and solubilizing; clear liq.; dens. 9.0 lb/gal; pH 7.0 ± 0.5 (2%); 75% act.

Denwet RG-7. [Graden] Sulfonated alkyl ester; anionic; wetting agent, coupling agent, solubilizer used in drycleaning, herbicides, pesticides; aids in dispersing, emulsifying, penetrating; clear liq.; dens. 9.0 lb/gal; pH 7.0 ± 0.5 (2%); 75% act.

Depasol AS-27. [Pulcra SA] Sodium dioctyl sulfosuccinate; anionic; dispersant for pigments; wetting and penetrating agent; used in drycleaning detergents and emulsion polymerization; rewetting agent for textile and paper industries; liq.; sol. in water, alcohol; 70 ± 2% conc.

Depasol CM-41. [Pulcra SA] Dialkyl sodium sulfosuccinate; anionic; dispersant for pigments; wetting and penetrating agent; used in drycleaning detergents and emulsion polymerization; rewetting agent for textile and paper industries; liq.; 50 ± 2% conc.

Deplastol. [Pulcra SA] Ethoxylated lauric acid; nonionic; visc. reducer for organosols; liq.; HLB 9.2; 100% conc.

Depsodye AR. [ICI Surf. UK] Anionic; inorganic dyebath stabilizer; improves shade reproducibility and yield of specific disperse dyes sensitive to reducing conditions; liq.

Depsodye HN. [ICI Surf. UK] Leveling agents/emulsifier/nonchlorinated solv. blend; anionic/cationic; scour dye assistant for polyamide and polyamide/lycra goods; liq.

Depsodye TCA. [ICI Surf. UK] Carrier/leveling agent/emulsifier blend; nonionic; very low foaming carrier/leveling agent for cellulose triacetate and cellulose triacetate/polyamide or polyester blends; liq.

Depsolube ACA. [ICI Surf. UK] Phosphate ester blend; anionic; anticrease lubricant for wet processing of textiles; liq.

Depsolube CP. [ICI Surf. UK] PEG fatty acid esters blend; anionic/nonionic; anticrease lubricant for polyester dyeing; liq.

Depsolube LNP. [ICI Surf. UK] Polymer; nonionic; surfactant, scour-dye agent for polyester fabrics; liq.

Depuma®. [Ciba-Geigy/Textiles] Silicone; nonionic; defoamer for atmospheric dyeing, printing, effluent systems; suitable for all textile fibers.

Depuma® C-306. [Ciba-Geigy/Textiles] Nonsilicone; nonionic; defoamer for atmospheric dyeing and finishing; suitable for nylon, cellulosic, and polyester textile fibers.

Depuma® OB New. [Ciba-Geigy/Textiles] Nonsilicone; nonionic; shear-stable defoamer for high temp. dyeing.

Deriphat 130-C. [Henkel] Sodium N-octyl-beta-iminodipropionic acid; amphoteric; surfactant used in liq. lt.-duty, transportation, alkali, or acid cleaners; wetting agent; pale to colorless liq.; sol. in water, ethanol, IPA, butyl Cellosolve; dens. 8.6 lb/gal; pH 8.6; 49% solids.

Deriphat 151. [Henkel] Sodium-N-coco beta-aminopropionate; amphoteric; wetting agent and emulsifier; wh. powd.; sol. in strong acids, alkali, and ionic systems; dens. 2 lb/gal; pH 12; 98% solids.

Deriphat 151C. [Henkel/Cospha; Henkel Canada] Lauraminopropionic acid; amphoteric; wetting agent, detergent, emulsifier, corrosion inhibitor; personal care and hard surface cleaners; general-purpose surfactant; high foaming, substantive; Gardner 5 clear liq.; sol. in strong acids, alkalis, ionic systems; sp.gr. 1.03; pH 5.5; 40% solids.

Deriphat 154. [Henkel/Cospha; Henkel Canada] Disodium N-tallow-beta iminodipropionate; amphoteric; detergent, solubilizer for personal care prods., hard surface cleaning, textiles, emulsion polymerization; good substantivity; wh. powd.; sol. in strong acids, alkalies, and ionic systems; dens. 2 lb/gal; pH 11; 98% solids.

Deriphat 154L. [Henkel/Cospha; Henkel Canada] Disodium N-tallow beta iminodipropionate; amphoteric; liq. form of Deriphat 154; liq.; 30% conc.

Deriphat 160. [Henkel/Cospha; Henkel/Functional Prods.; Henkel Canada] Disodium N-lauryl beta-iminodipropionate; CAS 3655-00-3; amphoteric; detergent, solubilizer, primary emulsifier used in org. and inorg. compds.; emulsion polymerization and stabilization; wetting agent; mild surfactant for hair and skin prods.; wh. powd.; dens. 2.0 lb/gal; 98% solids.

Deriphat 160C. [Henkel/Cospha; Henkel/Functional Prods.; Henkel Canada] Sodium-N-lauryl beta-iminodipropionate; CAS 26256-79-1; amphoteric; detergent, solubilizer, stabilizer; used in petrol. processing, emulsion polymerization, foaming cleaners, personal care prods.; amber clear liq.; sol. in strong acid, alkali, and ionic systems; sp.gr. 1.04; dens. 8.6 lb/gal; pH 7.5; 30% solids.

Deriphat 170C. [Henkel] N-Lauraminopropionic acid; CAS 3614-12-8; amphoteric; wetting agent; Gardner 3 clear liq.; sol. in alcohols; sp.gr. 1.03; pH 6; 45% solids.

Deriphat BAW. [Henkel Canada] Cocamidopropyl betaine; amphoteric; emulsifier, foaming and wetting agent in acid, alkali, or electrolyte sol'ns., and household and industrial detergent formulations; liq.; 35% conc.

Deriphat BC, BCW. [Henkel] Coco dimethyl ammonium carboxylic acid betaine; amphoteric; foam stabilizer, wetting agent, frothing agent for liq. detergents; lime soap dispersant; liq.; 59 and 43% act. resp.

Dermalcare® NI (formerly Cyclochem® NI). [Rhone-Poulenc Surf.] Cetomacrogol emulsifying wax; nonionic; broad tolerance emulsifier for o/w systems, visc. builder for cosmetic creams, lotions, ointments; flakes; m.p. 48-51 C; 100% conc.

Dermalcare® POL (formerly Cyclochem® POL). [Rhone-Poulenc Surf.] Cetearyl alcohol, ceteth-20, and glycol stearate; nonionic; lubricant SE wax, emulsifier for lotions and creams; effective over broad pH range; off-wh. wax; m.p. 48-52 C; 100% conc.

Derminol Fur Liquors HSP. [Hoechst AG] Alkane sulfonates, neutral oil, stabilizer blend; fatliquoring agent for furs.

Derminol Fur Liquors SL. [Hoechst AG] Alkane sulfonates, neutral oil, stabilizer blend; fatliquoring agents for furs.

Derminol Fur Liquors W. [Hoechst AG] Alkane sulfonates, neutral oil, stabilizer blend; fatliquoring agents for furs.

Derminol Grades. [Hoechst AG] Sulfoester derivs. of a natural oil; fatliquoring agent for leather.

Desmuldo DS 3. [Henkel-Nopco] Compounded formula; nonionic; emulsion breaker for crude petroleum; liq.; 100% conc.

Desomeen® TA-2. [Witco/Organics] PEG-2 tallow amine; cationic; emulsifier, dispersant, textile scouring, dyeing assistant, desizing assistant, softener, antistat; paste; 100% act.

Desomeen® TA-5. [Witco/Organics] PEG-5 tallow amine; cationic; emulsifier and dispersant; used as textile scouring agents, dyeing assistants, desizing agents, softening agents, antistats, etc.; liq.; HLB 5.3; cloud pt. 68–74 F; 100% act.

Desomeen® TA-15. [Witco/Organics] PEG-15 tallow amine; cationic; emulsifier, dispersant, textile scouring, dyeing assistant, desizing assistant, softener, antistat; liq.; cloud pt. 172–179 F; 100% act.

Desomeen® TA-20. [Witco/Organics] PEG-20 tallow amine; emulsifier, dispersant, textile scouring, dyeing assistant, desizing assistant, softener, antistat; liq.; cloud pt. 179–181 F (10% NaCl); 100% act.

DeSonate 50-S. [Witco/Organics] Sodium dodecylbenzene sulfonate; anionic; detergent component; liq.; 53% act.

DeSonate 60-S. [Witco/Organics] Sodium dodecylbenzene sulfonate; anionic; detergent component; liq.; 60% act.

DeSonate AOS. [Witco/Organics] Sodium alpha olefin sulfonate; anionic; detergent component for shampoos, bubble bath, detergents, liquid soaps; liq.; 40% act.

DeSonate AUS. [Witco/Organics] Sodium C14-C16 alpha olefin sulfonate; anionic; surfactant for use in shampoos and detergent prods.; liq.; 37% conc.

DeSonate SA. [Witco/Organics] Linear dodecylbenzene sulfonic acid; detergent intermediate, emulsifier, base for liquid and powd. detergents; liq.; 97% act.

DeSonate SA-H. [Witco/Organics] Branched dodecylbenzene sulfonic acid; chemical intermediate, emulsifier; liq.; 97% act.

Desonic® 1.5N. [Witco/Organics] Nonoxynol-1 (1.5 EO); nonionic; defoamer, detergent, emulsifer; liq.; oil-sol.; 100% act.

Desonic® 3K. [Witco/Organics] Linear alcohol ethoxylate (3 EO); nonionic; detergent, emulsifier, wetting agent, defoamer for industrial detergents, textile scouring; liq.; oil-sol.; 100% act.

Desonic® 4N. [Witco/Organics] Nonoxynol-4; CAS 9016-45-9; nonionic; detergent, emulsifier, defoamer for pesticide, paint, paper, and textile industries; liq.; oil-sol.; HLB 8.8; 100% act.

Desonic® 5D. [Witco/Organics] Dodoxynol-5; nonionic; emulsifier, wetter, and surfactant for agric., pesticide, industrial cleaner, and detergent formulations; liq.; HLB 9.1; 100% conc.

Desonic® 5K. [Witco/Organics] Linear alcohol ethoxylate (5 EO); nonionic; detergent, emulsifier, wetting agent for household/industrial cleaners; liq.; cloud pt. 75–79 F; 100% act.

Desonic® 5N. [Witco/Organics] Nonoxynol-5; CAS 9016-45-9; nonionic; detergent, surfactant; liq.; oil-sol.; 100% act.

Desonic® 6C. [Witco/Organics] PEG-6 castor oil; nonionic; emulsifier, lubricant, dye leveler, antistat, and dispersant for textile applics.; mfg. of PU foams; softening and rewetting agents for paper; liq.; HLB 4.4; 100% conc.

Desonic® 6D. [Witco/Organics] Dodoxynol-6; nonionic; emulsifier for solv. and emulsion cleaners; liq.; oil-sol.; HLB 10.0; 100% act.

Desonic® 6N. [Witco/Organics] Nonoxynol-6; CAS 9016-45-9; nonionic; emulsifier, detergent; liq.; HLB 10.8; 100% act.

Desonic® 6T. [Witco/Organics] Trideceth-6; CAS 24938-91-8; nonionic; low-foam wetting agent, detergent, foamer for mechanical and spray cleaning; pulp, paper, textile industries; liq.; HLB 11.4; 100% act.

Desonic® 7K. [Witco/Organics] Linear alcohol ethoxylate (7 EO); nonionic; detergent, emulsifier, wetting agent for household/industrial cleaners; liq.; cloud pt. 118–122 F; 100% act.

Desonic® 7N. [Witco/Organics] Nonoxynol-7; CAS 9016-45-9; nonionic; low-foaming surfactant, detergent, emulsifier; liq.; HLB 11.7; 100% act.

Desonic® 9D. [Witco/Organics] Dodoxynol-9; nonionic; emulsifier, detergent for emulsion cleaners; liq.; cloud pt. 61–64 F; 100% act.

Desonic® 9K. [Witco/Organics] Linear alcohol ethoxylate (9 EO); nonionic; detergent, emulsifier, wetting agent for household/industrial cleaners; liq.; cloud pt. 131–138 F; 100% act.

Desonic® 9N. [Witco/Organics] Nonoxynol-9; CAS 9016-45-9; nonionic; detergent, wetting agent, emulsifier for textile, paper, metal cleaning; liq.; HLB 12.8; cloud pt. 127–133 F; 100% act.

Desonic® 9T. [Witco/Organics] Trideceth-9; CAS 24938-91-8; nonionic; surfactant for lt. and heavy-duty detergents, textile leveling and scouring; liq.; cloud pt. 154–170 F; HLB 13.3; 100% act.

Desonic® 10D. [Witco/Organics] Dodoxynol-10; nonionic; detergent/wetting agent for industrial and heavy-duty detergents; liq.; HLB 13.2; cloud pt. 145–153 F; 100% act.

Desonic® 10N. [Witco/Organics] Nonoxynol-10; CAS 9016-45-9; nonionic; detergent, wetting agent, emulsifier for textile, paper, metal cleaning; liq.; cloud pt. 140–149 F; 100% act.

Desonic® 10T. [Witco/Organics] Trideceth-10; CAS 24938-91-8; nonionic; foam builder and detergent for cleaning formulations; scouring agent in textile industry; liq.; HLB 13.8; 100% conc.

Desonic® 11N. [Witco/Organics] Nonoxynol-11; CAS 9016-45-9; nonionic; detergent, wetting agent, emulsifier for textile, paper, metal cleaning; liq.; HLB 13.6; cloud pt. 158–162 F; 100% act.

Desonic® 12-3. [Witco/Organics] PEG-3 linear alcohol; emulsifier, dispersant, surfactant; detergent intermediate for mfg. of ethoxysulfates for specialty industrial and dishwashing applics.; liq.; oil-sol.; HLB 8.0; 100% act.

Desonic® 12D. [Witco/Organics] Dodoxynol-12; nonionic; detergent/wetting agent for industrial and heavy-duty detergents at high temps.; liq.; cloud pt. 165–174 F; 100% act.

Desonic® 12K. [Witco/Organics] Linear alcohol ethoxylate (12 EO); nonionic; detergent, wetting agent for household/industrial cleaners, high-temp. applics.; liq.; cloud pt. 187–192 F; 100% act.

Desonic® 12N. [Witco/Organics] Nonoxynol-12; CAS 9016-45-9; nonionic; wetting agent, detergent used with high temp. and electrolytes; liq.; cloud pt. 176–181 F; 100% act.

Desonic® 12T. [Witco/Organics] Trideceth-12; CAS 24938-91-8; nonionic; surfactant for lt.- and heavy-duty detergents; leveling and scouring agent; liq.; cloud pt. 187–201 F; 100% act.

Desonic® 13N. [Witco/Organics] Nonoxynol-13; CAS 9016-45-9; nonionic; detergent, emulsifier, defoamer for pesticide, paint, paper, and textile industries; liq.; HLB 14.3; 100% conc.

Desonic® 14D. [Witco/Organics] Dodoxynol-14; nonionic; emulsifier, wetter, and surfactant for agric., pesticide, industrial cleaner, and detergent formulations; solid; HLB 14.0; 100% conc.

Desonic® 15N. [Witco/Organics] Nonoxynol-15; CAS 9016-45-9; nonionic; detergent, wetting agent at high temps. and electrolyte; metal cleaning; liq.; cloud pt. 143–149 F (10% NaCl); 100% act.

Desonic® 15T. [Witco/Organics] Trideceth-15; CAS 24938-91-8; nonionic; surfactant for lt.- and heavy-duty detergents at high temps.; leveling and scouring agent; liq.; cloud pt. 156–167 F (10% NaCl); 100% act.

Desonic® 20N. [Witco/Organics] Nonoxynol-20; CAS 9016-45-9; nonionic; detergent, wetting agent at high temps. and electrolyte; solid; cloud pt. 154–162 F (10% NaCl); 100% act.

Desonic® 30C. [Witco/Organics] PEG-30 castor oil; nonionic; emulsifier, lubricant, dye leveler, antistat, dispersant for textiles; emulsifer for rigid PU foams; softener/rewetter for wet-strength paper; liq.; HLB 11.7; 100% act.

Desonic® 30N. [Witco/Organics] Nonoxynol-30; CAS 9016-45-9; nonionic; emulsifier for fats, oils, waxes; solid; HLB 17.1; cloud pt. 159–163 F (10% NaCl); 100% act.

Desonic® 30N70. [Witco/Organics] Nonoxynol-30; CAS 9016-45-9; nonionic; emulsifier for fats, oils, waxes; liq.; HLB 17.1; cloud pt. 159–163 F (10% NaCl); 70% act.

Desonic® 36C. [Witco/Organics] PEG-36 castor oil; nonionic; emulsifier, lubricant, dye leveler, antistat, dispersant for textiles; emulsifer for rigid PU foams; softener/rewetter for wet-strength paper; liq.; HLB 12.6; cloud pt. 122–140 F; 100% act.

Desonic® 40C. [Witco/Organics] PEG-40 castor oil; nonionic; emulsifier, lubricant, dye leveler, antistat, dispersant for textiles; emulsifer for rigid PU foams; softener/rewetter for wet-strength paper; liq.; HLB 13.1; cloud pt. 173–179 F; 100% act.

Desonic® 40N. [Witco/Organics] Nonoxynol-40; CAS 9016-45-9; nonionic; surfactant for high temps. and electrolytes; emulsifier; solid; cloud pt. 165–176 F (10% NaCl); 100% act.

Desonic® 40N70. [Witco/Organics] Nonoxynol-40; CAS 9016-45-9; nonionic; emulsifier; surfactant for high temps. and electrolytes; liq.; cloud pt. 165–176 F (10% NaCl); 70% act.

Desonic® 50N. [Witco/Organics] Nonoxynol-50; CAS 9016-45-9; nonionic; emulsifier for waxes, polishes; surfactant for high temps. and electrolytes; solid; 100% act.

Desonic® 50N70. [Witco/Organics] Nonoxynol-50; CAS 9016-45-9; nonionic; emulsifier for waxes, polishes; surfactant for high temps. and electrolytes; liq.; 70% act.

Desonic® 54C. [Witco/Organics] PEG-54 castor oil; nonionic; emulsifier, lubricant, dye leveler, antistat, dispersant for textiles; emulsifer for rigid PU foams; softener/rewetter for wet-strength paper; liq.; HLB 14.4; cloud pt. 136–142 F (10% NaCl); 100% act.

Desonic® 81-2. [Witco/Organics] PEG-2 linear alcohol; detergent intermediate, emulsifier, dispersant, defoamer, oil-sol. surfactant; liq.; oil-sol.; HLB 8.0; 100% act.

Desonic® 100N. [Witco/Organics] Nonoxynol-100; CAS 9016-45-9; nonionic; emulsifier for waxes, polishes; surfactant for high temps. and electrolytes; solid; 100% act.

Desonic® 100N70. [Witco/Organics] Nonoxynol-100; CAS 9016-45-9; nonionic; emulsifier for waxes, polishes; surfactant for high temps. and electrolytes; liq.; 70% act.

Desonic® 315-3. [Witco/Organics] PEG-3 linear alcohol; detergent, emulsifier, wetting, surfactant, defoamer, intermediate; liq.; oil-sol.; HLB 8.0; 100% act.

Desonic® AJ-85. [Witco/Organics] Alcohol ethoxylate; nonionic; detergent and wetting agent for household and industrial cleaners; wetting of carpet fibers for dyeing process; liq.; 85% conc.

Desonic® AJ-100. [Witco/Organics] Alcohol ethoxylate; nonionic; anhyd. version of Flo Mo AJ85; paste; 100% conc.

Desonic® DA-4. [Witco/Organics] Deceth-4; nonionic; rapid wetter/rewetter with low foaming properties for textile uses; liq.; water-disp.; HLB 10.5; 100% act.

Desonic® DA-6. [Witco/Organics] Deceth-6; nonionic; wetting agent for built scour systems; liq.; water-sol.; HLB 12.5; cloud pt. 115–120 F; 100% act.

Desonic® LFA 144, LFA 198. [Witco/Organics] Alkoxylated surfactant; nonionic; general-purpose surfactant providing detergency, emulsification, wetting; for use at high temps. and high electrolyte systems; liq.; 100% conc.

Desonic® LFB 65. [Witco/Organics] Nonionic; low-foaming surfactant with detergency and wetting for industrial and home mech. dishwashing, metal-working applics. where foam control is important; liq.; 100% conc.

Desonic® LFC 50. [Witco/Organics] Nonionic; low-foaming surfactant with detergency and wetting for industrial and home mech. dishwashing, metal-working applics. where foam control is important; liq.; 100% conc.

Desonic® LFD 97. [Witco/Organics] Nonionic; low foam surfactant, wetting agent, detergent for industrial and home mech. dishwashing, metalworking; liq.; 100% conc.

Desonic® LFE 89. [Witco/Organics] Nonionic; low-foaming surfactant with detergency and wetting for metal and hard surf. cleaners; liq.; 100% conc.

Desonic® LFO 97. [Witco/Organics] Nonionic; low-foaming surfactant with detergency and wetting for industrial and home mech. dishwashing, metal-working applics.; liq.; 100% conc.

Desonic® S-45. [Witco/Organics] Octoxynol-4 (4.5 EO); CAS 9002-93-1; nonionic; emulsifier, surfactant, emulsion cleaner, dry dishwashing detergent, polish emulsifier; liq.; 100% act.

Desonic® S-100. [Witco/Organics] Octoxynol-9 (9–10 EO); CAS 9002-93-1; nonionic; metal and textile processing surfactant; household and industrial cleaners; emulsifier for vinyl and acrylic polymerization; liq.; cloud pt. 140–158 F; 100% act.

Desonic® S-102. [Witco/Organics] Octoxynol-12 (12–13 EO); CAS 9002-93-1; nonionic; surfactant for hot cleaning and pickling systems, hard surface cleaners, household and industrial cleaners; liq.; cloud pt. 187–194 F; 100% act.

Desonic® S-114. [Witco/Organics] Octoxynol-7 (7–8 EO); CAS 9002-93-1; nonionic; surfactant for metal and acid cleaners, tank-soak, household, and industrial cleaners; liq.; cloud pt. 72–79 F; 100% act.

Desonic® S-405. [Witco/Organics] Octoxynol-40; CAS 9002-93-1; nonionic; coemulsifier for vinyl and acrylic polymerization; dye assistant; liq.; cloud pt. 165–176 F (10% NaCl); 70% act.

Desonic® SMO. [Witco/Organics] Sorbitan oleate; CAS 1338-43-8; nonionic; emulsifier, fiber lubricant and softener; liq.; sol. in oils and org. solvs.; insol. in water; HLB 4.3; 100% conc.

Desonic® SMO-20. [Witco/Organics] Polysorbate 80; CAS 9005-65-6; nonionic; emulsifier, wetting agent, surfactant exhibiting antistat and lubricating properties; liq.; HLB 15.0; 100% conc.

Desonic® SMT. [Witco/Organics] Sorbitan tallate; nonionic; emulsifier, fiber lubricant and softener; liq.; insol. in water; sol. in oils and org. solvs.; HLB 4.1; 100% conc.

Desonic® SMT-20. [Witco/Organics] PEG-20 sorbitan tallate; nonionic; emulsifier, wetting agent, surfactant exhibiting antistat and lubricating properties; liq.; HLB 15.0; 100% conc.

Desonic® TA-2, -15. [Witco/Organics] Tallow amine, ethoxylated; cationic; emulsifier, dispersant, textile scouring agent, desizing assistant, softener, and antistat; 80-100% conc.

Desonic® TA-25CWS. [Witco/Organics] Tallow amine, ethoxylated; cationic; emulsifier, dispersant, textile scouring agent, desizing assistant, softener, and antistat; liq.; 80% conc.

Desonic® TDA-9. [Witco/Organics] Ethoxylated alcohol; nonionic; wetting agent, degreaser, detergent, and emulsifier; liq.; 100% conc.

DeSonol A. (redesignated Witcolate 6431) [Witco/Organics]

DeSonol AE (redesignated Witcolate AE). [Witco/Organics]

DeSonol S (redesignated Witcolate S). [Witco/Organics]

DeSonol SE (redesignated Witcolate SE). [Witco/Organics]

DeSonol SE-2. [Witco/Organics] Sodium laureth-2 sulfate; CAS 9004-82-4; anionic; surfactant for shampoos, bubble bath, liquid detergents; liq.; 27–29% act.; DISCONTINUED.

DeSonol T (redesignated Witcolate 6434). [Witco/Organics] TEA lauryl sulfate; anionic; surfactant for shampoos, liquid detergents; liq.; 39–40% act.; DISCONTINUED.

Desophos® 3 DP. [Witco/Organics] Phosphate ester; textile wetting agent; alkaline scouring; as potassium or sodium salts in dry cleaning detergents; liq.; 100% act.

Desophos® 3 OMP. [Witco/Organics] Acid ester of aliphatic hydrophobic base; emulsifier for paraffinic oils; liq.; 100% act.

Desophos® 3 OMPDEA. [Witco/Organics] Partly neutralized DEA salt of DeSophos 3 OMP; emulsifier for paraffinic oils; liq.; 98% act.

Desophos® 4 CP. [Witco/Organics] Phosphate ester; softener, antistat for textile finishing; in lubricants for filament yarns, syn. fibers, wool; emulsifier for cosmetic oils and creams, polymerization of latexes; liq.; 95% act.

Desophos® 4 NP. [Witco/Organics] Phosphate ester; nonionic; emulsifier for emulsion polymerization; liq.; most oil-sol. of series; water-sol. when neutralized; 100% act.

Desophos® 5 AP. [Witco/Organics] Phosphate ester, free acid; anionic; coupling agent for nonionics in liq. alkaline detergents; moderate foamer; liq.; 100% act.

Desophos® 5 BMP. [Witco/Organics] Phosphate ester, free acid; anionic; coupling agent for industrial uses; liq.; high water sol.; 100% act.

Desophos® 6 DNP. [Witco/Organics] Phosphate ester; paraffinic oil emulsifier as TEA salt; rust inhibitor; most oil-sol. of DeSophos DNP series; liq.; 100% act.

Desophos® 6 DP. [Witco/Organics] Phosphate ester; dry cleaning detergent, penetrant, antistat; liq.; 100% act.

Desophos® 6 MPNa. [Witco/Organics] Phosphate ester; nonionic; imparts hard surf. detergency, corrosion retardation, and moderate foaming to detergents; dry cleaning detergent; antistat for textiles; liq.; 88% act.

Desophos® 6 NP. [Witco/Organics] Phosphate ester, free acid; anionic; emulsifier, detergent; for emulsion polymerization; liq.; 100% act.

Desophos® 6 NP4. [Witco/Organics] Phosphate ester; polymerization emulsifier for PVAc and acrylic films; also in waterless hand cleaners and laundry detergents; liq.; 100% act.

Desophos® 6 NPNa. [Witco/Organics] Phosphate ester; nonionic; hard surf. detergent, corrosion retarder, moderate foamer for detergent concs., drycleaning; antistat for textiles; liq.; 88% conc.

Desophos® 6 OMP. [Witco/Organics] Acid ester; emulsifier for paraffinic oils; more hydrophilic version of DeSophos 3 OMP; liq.; 100% act.

Desophos® 6 OMPDEA. [Witco/Organics] Partly neutralized DEA salt of DeSophos 6 OMP; liq.; 98% act.

Desophos® 7 DP. [Witco/Organics] Phosphate ester; alkali-stable emulsifier, detergent, wetting agent, dispersant; rewetter in textile applics.; liq.; 100% act.

Desophos® 7 OPNa. [Witco/Organics] Phosphate ester; dispersible surfactant in emulsification of min. oils; textile softener and antistat; in lubricants for syn. and wool fibers; liq.; sol. in aromatic solv.; 98% act.

Desophos® 8 DNP. [Witco/Organics] Phosphate ester; emulsifier for phosphated and chlorinated pesticide concs., polyethylene; in industrial cleaners and dry cleaning; liq.; sol. in aromatic solv., kerosene; disp. in water; 100% act.

Desophos® 9 NP. [Witco/Organics] Phosphate ester; nonionic; dedusting agent for dry cleaning detergent, alkaline powders, water-repellent fabric finishes; emulsion polymerization of PVAc and acrylic films; liq.; 100% act.

Desophos® 10TP. [Witco/Organics] Phosphate ester; nonionic; wetting agent in textiles, alkaline scouring; liq.; 100% conc.

Desophos® 14 DNP. [Witco/Organics] Phosphate ester, free acid; anionic; surfactant for industrial cleaners, drycleaning; emulsifier for phosphated and chlorinated pesticide concs. and polyethylene; liq.; water-sol.; 100% act.

Desophos® 30 NP. [Witco/Organics] Phosphate ester; nonionic; emulsifer, stabilizer; in preparation of PVAc and acrylic copolymers; liq.; 100% act.

Desotan® SMO. [Witco/Organics] Sorbitan oleate; CAS 1338-43-8; nonionic; lipophilic emulsifier, fiber lubricant, softener; liq.; sol. in oils, org. solvs.; insol. water; HLB 4.3; 100% act.

Desotan® SMO-20. [Witco/Organics] Polysorbate 80; CAS 9005-65-6; nonionic; hydrophilic emulsifier, wetting agent; general-purpose surfactant with antistatic and lubricating properties; liq.; HLB 15.0; cloud pt. 149–158 F (10% NaCl); 100% act.

Desotan® SMT. [Witco/Organics] Sorbitan tallate; nonionic; lipophilic emulsifier, fiber lubricant, softener; liq.; sol. oils, org. solv.; insol. water; HLB 4.1; 100% act.

Desotan® SMT-20. [Witco/Organics] PEG-20 sorbitan tallate; nonionic; hydrophilic emulsifier, wetting agent; general-purpose surfactant with antistatic and lubricating properties; liq.; HLB 15.0; cloud pt. 149–158 F (10% NaCl); 100% act.

DET. [Swastic] Alkylaryl sulfonate sodium salt with S.T.P.P. and other alkaline builders; detergent suitable for hand washing fabrics; wh. powd.; pH 10.5 (1%).

DET Liq. [Swastic] Alkylaryl sulfonate, POE alcohol ether sulfate plus additives; household cleanser, detergent; clear liq.; pH 8.0–9.5 (1%).

Deterflo A 210. [Ceca SA] Modified ethoxylated alcohol; nonionic; low foaming degreasing sol'ns. for surfaces requiring strongly acid or alkaline detergents, at low temp. and under pressure; stable to acids and bases; biodeg.; liq.; pH 3 (5%); > 95% act.

Deterflo A 215. [Ceca SA] Ethoxylated alcohol; nonionic; degreaser for surfaces requiring strongly acid or alkaline detergents, for treatment at low temp. and under pressure; liq.; 100% conc.

Deterflo A 233. [Ceca SA] Ethoxylated alcohol; nonionic; degreaser for surfaces requiring strongly acid or alkaline detergents, for treatment at low temp. and under pressure; liq.; 100% conc.

Detergent 8®. [Alconox] Blend of an alkanolamine, glycol ethers, and an alkoxylated fatty alcohol; nonionic; biodeg. detergent, wetting agent for cleaning circuit boards, electronic parts, phosphate-sensitive labware, delicate industrial parts, nuclear reactor cavities; clear liq., sl. ammonia odor; sol. in water; sp.gr. 0.994; dens. 8.25 lb/gal; b.p. 235 F; flash pt. (COC) 191 F; pH 11 (1%); surf. tens. 32 dyne/cm (1%); toxicology: nontoxic orally; potential serious eye irritant.

Detergent ADC. [ICI Surf. UK] Sodium dodecylbenzene sulfonate; anionic; general-purpose detergent; biodeg.; liq.

Detergent Concentrate 840. [Mona Industries] Amido sulfonate complex; nonionic/anionic; high-foaming detergent at low concs.; used in car wash, dishwash, bubble baths, all purpose cleaners; amber clear liq.; sol. in cold or warm water; sp.gr. 1.05; dens. 8.75 lb/gal; acid no. 90; alkali no. 23; pH 8.7 (10%); 82% conc.

Detergent CR. [Arol Chem. Prods.] Fatty ethanolamine condensate; detergent, wetting agent, emulsifier, thickener, penetrating and leveling agent, for use in textiles, personal care, food, and household prods.; pigment dispersant; pale amber liq.; mild odor; readily sol. in water; sp.gr. 1.01; 100% act.

Detergent E. [CNC Int'l. L.P.] Coconut fatty acid amine condensate; surfactant with high detergency, wetting and rewetting props. for textile scouring and dyeing operations; sanforizing assistant; amber visc. liq.; pH 9.8 (1%); 98-100% act.

Deterpal 832. [Ceca SA] Alkyl ether sulfate/solvent blend; anionic; degreaser, emulsifier; liq.

Deterpal 843. [Ceca SA] Polyethoxy ether; nonionic; detergent base; liq.; pH 5.5 ± 1.5 (10%); 99% active.

Deterpal LC. [Ceca SA] Alkylsulfate/solvent blend; anionic; dispersant and antisettling agent for alkyd paints; liq.; 46% conc.

Detersol HF. [Finetex] Hydrocarbon blend, emulsified; nonionic; high flash pt. solvent scour for textiles; low foam; liq.; 100% conc.

Detersol MC-10. [Finetex] Solvent blend; nonionic; detergent for wax and size removal; machine cleaning; caustic-free; liq.

Detersol WR. [Finetex] Phosphated blend; anionic; wetting and rewetting agent; stable in caustic and high temps.; compat. with enzymes; liq.; 65% act.

Detersol X-66. [Finetex] Solvent/surfactant blend; nonionic; general-purpose solvent scour; liq.; 100% conc.

Deteryl 955 FS. [ICI Surf. UK] Alkylaryl sulfonate plus nonionic emulsifiers; anionic; detergent with high foaming and soil-suspending props.; for scouring cotton/elastomeric and polyamide/elastomeric blends; liq.

Deteryl AG. [ICI Surf. UK] Lauramide DEA blend; CAS 120-40-1; anionic/nonionic; antifrosting agent in printing and continuous dyeing of tufted carpets; liq.

Det-O-Jet®. [Alconox] Highly alkaline detergent containing potassium hydroxide, silicate of soda, and sodium hypochlorite; nonionic; low sudsing biodeg. detergent with good wetting, emulsifying, and penetrating properties, for ultrasonic and mechanical washers, hospital and lab ware, optical and electronic components, pharmaceutical apparatus, industrial parts; clear liq., pract. odorless; sol. in water; sp.gr. 1.282; dens. 10.8 lb/gal; b.p. 212 F; flash pt. none; pH 13 (1%); surf. tens. 55 dyne/cm (1%); toxicology: DOT corrosive liq.; serious eye irritant.

DET-Washmatic. [Swastic] Sodium alkylaryl sulfonate; detergent for use in washing machines; blue, spray-dried powd.; pH 10.5 (1%).

Dexopal 555. [Dexter] Polyether; nonionic; detergent, penetrant, wetting agent; low foaming in acid, neutral and alkaline media; does not increase color transference of dispersed dyes; liq.

Dextrol OC-15. [Dexter] Complex org. phosphate ester free acid; anionic; detergent, wetting agent, emulsifier; for pesticides, PVA and acrylic polymerization; corrosion inhibitor; dispersant for magnetic oxide in aromatic solvs.; lt.-colored liq.; 100% act.

Dextrol OC-20. [Dexter] Complex org. phosphate ester free acid; CAS 51811-79-1; anionic; detergent, wetting agent, emulsifier; for pesticides, emulsion polymerization; corrosion inhibitor; dedusting agent for alkaline powd.; liq.; water-sol.; 100% act.

Dextrol OC-40. [Dexter] Phosphate ester free acid; CAS 9046-01-9; anionic; wetting agent, penetrant for nonwoven fabrics; stabilizer for resins; liq.; 100% conc.

Dextrol OC-50. [Dexter] Alkyl phenoxy ethoxylate, partially sodium neutralized; anionic; wetting and dispersing agent for latex coatings, colorant systems, detergent formulations; Gardner 2 max. clear visc. liq.; sol. in water and solvs.; sp.gr. 1.11; dens. 9.25 lb/gal; visc. 6000 ± 1500 cps; flash pt. (TCC) > 200 F; pH 5.55 ± 0.5; surf. tens. 35.2 dynes/cm (0.1%); 92 ± 1% act.

Dextrol OC-70. [Dexter] Phosphated aliphatic ethoxylate; anionic; detergent, wetting agent, dispersant, emulsifier; for alkyd paints, solv.-based coatings, colorant systems, pesticides, PVA and acrylic polymerization; corrosion inhibitor; dispersant for magnetic oxide in aromatic solvs.; Gardner 6 max. clear to hazy liq.; sp.gr. 1.06; dens. 8.83 lb/gal; visc. 800 ± 200 cps; flash pt. (TCC) > 300 F; pH 1.5 ± 0.3; surf. tens. 32.4 dynes/cm (0.1%); 100% act.

Diable AS. [Kao Corp. SA] Amine-based; cationic; slurry seal and emulsifier; liq.; 50% conc.

Diadol 11. [Mitsubishi Kasei] Higher alcohol; plasticizer, surfactant; colorless clear liq.; acid no. 0.1 max.; iodine no. 0.1 max.; hyd. no. 315 min.

Diadol 13. [Mitsubishi Kasei] Higher alcohol; surfactant for shampoo, lt. duty liq. detergents; colorless clear liq.; acid no. 0.1 max.; iodine no. 0.1 max.; hyd. no. 267 min.

Diadol 135. [Mitsubishi Kasei] Higher alcohol; surfactant for heavy duty detergent powds.; colorless clear liq.; sp.gr. 0.836; acid no. 0.1 max.; iodine no. 0.1

max.; hyd. no. 250 min.

Diamiet 503, 508, 520, AB, C. [Kao Corp. SA] n-Propylene alkyl fatty diamine EO condensation product; cationic; emulsifier, dispersant, corrosion inhibitor, wetting agent; liq., paste, solid; 100% conc.

Diamin DO, DT. [Kao Corp. SA] Diamine dioleate salts; cationic; dispersant for oil flushing agents for pigments, oil/wax emulsifier; liq./paste; 100% conc.

Diamin HT. [Kao Corp. SA] n-Propylene hydrog. tallow fatty diamine; cationic; asphalt emulsifier, corrosion inhibitor, antistripping agent; solid; 100% conc.

Diamin O. [Kao Corp. SA] n-Propylene oleyl fatty diamine; cationic; asphalt emulsifier, corrosion inhibitor; antistripping agent; liq. or paste; 100% conc.

Diamin S. [Kao Corp. SA] n-Propylene soya fatty diamine; cationic; asphalt emulsifier, corrosion inhibitor; antistripping agent; liq., paste; 100% conc.

Diamin T. [Kao Corp. SA] n-Propylene tallow fatty diamine; cationic; asphalt emulsifier, corrosion inhibitor; antistripping agent; paste; 100% conc.

Diamine B11. [Berol Nobel AB] n-Alkyl propylene diamine; CAS 15268-40-3; cationic; emulsifier, corrosion inhibitor; solid; 85% conc.

Diamine BG. [Berol Nobel AB] n-Tallow-propylene diamine; cationic; emulsifier, corrosion inhibitor; paste; 85% conc.

Diamine HBG. [Berol Nobel AB] n-Tallow propylene diamine; cationic; emulsifier, corrosion inhibitor; solid; 85% conc.

Diamine KKP. [Berol Nobel AB] n-Coco propylene diamine; cationic; emulsifier, corrosion inhibitor; liq.; 85% conc.

Diamine OL. [Berol Nobel AB] n-Oleyl propylene diamine; cationic; emulsifier, corrosion inhibitor; liq.; 85% conc.

Diamonine B. [ICI Surf. UK] Stabilized urea-formaldehyde resin; nonionic; thermosetting resin producing durable press finishes on cellulose fibers and stiff finishes on syn. fibers; liq.

Dianol. [Dai-ichi Kogyo Seiyaku] Fatty acid DEA; nonionic; emulsifier, dispersant for pigments, wax, solvs.; detergent, solubilizer; liq.; 100% conc.

Dianol 300. [Dai-ichi Kogyo Seiyaku] Fatty acid DEA; nonionic; detergent, solubilizer, emulsifier, and dispersant solvs.; liq.; 100% conc.

Diazital O Extra Conc. [Henkel; Henkel-Nopco] Fatty alcohol polyglycol ethers; nonionic; dyeing assistant; solid; 100% conc.

DIBM. [Tiarco] Diisobutyl maleate; CAS 14234-82-3; intermediate for mfg. of surfactants; in copolymerization of PVC and vinyl acetates; plasticizer for vinyl resins; also in latex paints; liq.; 100% conc.

Dikssol 201, 202. [Dai-ichi Kogyo Seiyaku] Anionic-nonionic activators; anionic; emulsifier for org. chlorine agric. formulations; liq.

Dikssol 301, 302. [Dai-ichi Kogyo Seiyaku] Anionic-nonionic activators; anionic; emulsifier for org. phosphorous agric. formulations; liq.

Dinoram C. [Ceca SA] n-Coco propylene diamine; CAS 61791-63-7; cationic; chemical intermediate; paste.

Dinoram O. [Ceca SA] n-Oleyl propylene diamine; CAS 7173-62-8; cationic; chemical intermediate; corrosion inhibitor; paste.

Dinoram S. [Ceca SA] n-Tallow propylene diamine; CAS 61791-55-7; cationic; emulsifier for bitumen; antistripping agent for road making; paste.

Dinoram SH. [Ceca SA] n-Hydrog. tallow propylene diamine; CAS 68603-64-5; cationic; chemical intermediate; solid, flakes.

Dinoramac C. [Ceca SA] n-Coco propylene diamine acetate; cationic; flotation, bactericide, emulsifier, anticaking, soil stabilization, flocculation, corrosion inhibitor; paste.

Dinoramac O. [Ceca SA] n-Oleyl propylene diamine acetate; cationic; see Dinoramac C; paste.

Dinoramac S. [Ceca SA] n-Tallow propylene diamine acetate; cationic; see Dinoramac C; paste.

Dinoramac SH. [Ceca SA] Hydrog. tallowamine acetate; cationic.

Dinoramox S3. [Ceca SA] PEG-3 tallow propylene diamine; CAS 61790-85-0; nonionic/cationic; dispersant, wetting agent, emulsifier, corrosion inhibitor, industrial detergent used in paint, agric., chemical, and textile industries; liq.; 100% conc.

Dinoramox S7. [Ceca SA] PEG-7 tallow propylene diamine; cationic; dispersant, wetting agent, emulsifier, corrosion inhibitor, industrial detergent used in paint, agric., chemical, and textile industries.

Dinoramox S12. [Ceca SA] PEG-12 tallow propylene diamine; cationic; dispersant, wetting agent, emulsifier, corrosion inhibitor, industrial detergent used in paint, agric., chemical, and textile industries.

Dinoramox SH 3. [Ceca SA] PEG-3 N-hydrog. tallow propylene diamine; cationic; detergent, emulsifier, dispersant, corrosion inhibitor for use in textile and detergent industries; solid; 100% conc.

DIOM. [Tiarco] Diisooctyl maleate; CAS 1330-76-3; intermediate for mfg. of surfactants; in copolymerization of PVC and vinyl acetates; plasticizer for vinyl resins; also in latex paints; liq.; 100% conc.

Dionil® OC. [Hüls Am.; Hüls AG] PEG-3 oleamide; CAS 26027-37-2; 31799-71-0; nonionic; detergent for lt. and heavy-duty detergents, dishwashing agents, cosmetic preps.; component in textile auxliaries; refatting agent; liq.; cloud pt. 67 (10% in 25% BDG); 100% act.

Dionil® RS. [Hüls Am.; Hüls AG] Fatty acid amide polyglycol ether and ethoxylated alcohols; nonionic; superfatting and preparation agent; liq.; 80% act.

Dionil® S 37. [Hüls Am.; Hüls AG] Fatty acid amide polyglycol ether; nonionic; detergent for lt. and heavy-duty detergents, dishwashing agents, cosmetic preps.; component for textile auxiliaries; wax; 100% conc.

Dionil® SD. [Hüls Am.; Hüls AG] Fatty acid amide polyglycol ether; nonionic; superfatting and preparation agent; wax; 100% conc.

Dionil® SH 100. [Hüls Am.; Hüls AG] PEG-6 oleamide; CAS 26027-37-2; nonionic; detergent for lt. and heavy-duty detergents, dishwashing agents, cosmetic preps., car shampoos; component for textile auxiliaries; good leveling power; liq./paste; 100% act.

Dionil® W 100. [Hüls Am.; Hüls AG] PEG-14 oleamide; CAS 26027-37-2; nonionic; detergent for lt. and heavy-duty detergents, dishwashing agents, cosmetic preps., car shampoos; component for textile auxiliaries; good leveling power; solid; cloud pt. 60 (2% in 10% NaOH); 100% act.

Diprosin A-100. [Toho Chem. Industry] Disproportionated rosin; polymerization emulsifier for syn. rubbers and plastics.; solid.

Diprosin K-80. [Toho Chem. Industry] Disproportionated rosin potassium soap; polymerization emulsifier for syn. rubbers and plastics.; paste.

Diprosin N-70. [Toho Chem. Industry] Disproportionated rosin sodium soap; polymerization emulsifier for syn. rubbers and plastics.; paste.

Dismulgan Brands. [Hoechst AG] Polyamine containing condensation prod.; cationic; reverse emulsion breaker for o/w emulsions; liq.

Disperbyk®. [Byk-Chemie USA] Sol'n. of an alkylolammonium salt of a higher m.w. polycarboxylic acid; anionic; wetting, dispersing additive to prevent settling and flooding of pigments; for solv. and aq. systems, stains, wood preservatives, anticorrosive primers, wash primers, nitrocellulose primers, fillers, antifouling paints, emulsion paints; emulsifier; sp.gr. 1.07-1.09 g/cc; dens. 8.93-9.09 lb/gal; acid no. 75-95; flash pt. (Seta) > 110 C; ref. index 1.420-1.430; 48-52% NV in water.

Disperbyk®-101. [Byk-Chemie USA] Sol'n. of a salt of long chain polyamine amides and a polar acidic ester; wetting and dispersing additive for improving wetting of pigments; for solv. and solv.-free systems, transparent wood stains; gellant for organophilic bentonites; sp.gr. 0.88-0.92 g/cc; dens. 7.34-7.68 lb/gal; acid no. 27-33; flash pt. (Seta) 35 C; ref. index 1.462-1.472; 50-54% NV in Stod./2-butoxyethanol/xylene (26/3/1).

Disperbyk®-110. [Byk-Chemie USA] Sol'n. of a copolymer with acidic groups; wetting and dispersing additive for coil coatings, OEM, general industrial, med.-high solid one-pack systems, acid-catalyzed systems; sp.gr. 1.02 g/cc; dens. 8.51 lb/gal; acid no. 51; flash pt. (Seta) 44 C; 52% NV in methoxypropyl acetate/alkylbenzene (1/1).

Disperbyk®-130. [Byk-Chemie USA] Sol'n. of polyamine amides of unsat. polycarboxylic acids; cationic; wetting and dispersing additive for carbon blks. and oxide pigments in solv.-based systems, high solids systems, alkyd melamine, acrylic; improves gloss; reduces flooding and floating of pigments; antiflocculant; sp.gr. 0.92-0.94 g/cc; dens. 7.68-7384 lb/gal; flash pt. (Seta) 45 C; ref. index 1.498-4.508; 48-54% NV in naphtha/2-butoxyethanol (5/1).

Disperbyk®-160. [Byk-Chemie USA] Sol'n. of higher m.w. block copolymers; cationic; patented wetting and dispersing additive for org. and inorganic pigments in industrial coatings, automotive coatings, automotive refinishing paints, coil coatings; prevents reflocculation of pigments; stabilizes color str.; sp.gr. 0.93-0.97 g/cc; dens. 7.76-8.09 lb/gal; flash pt. (Seta) 25 C; 28-30% NV in xylene/butyl acetate (6/1).

Disperbyk®-161. [Byk-Chemie USA] Sol'n. of higher m.w. block copolymers; cationic; patented wetting and dispersing additive for org. and inorganic pigments in industrial coatings, automotive coatings, automotive refinishing paints, coil coatings; prevents reflocculation of pigments; stabilizes color str.; sp.gr. 1.00-1.03 g/cc; dens. 8.35-8.60 lb/gal; flash pt. (Seta) 39 C; 29-31% NV in PMA/butyl acetate (6/1).

Disperbyk®-162. [Byk-Chemie USA] Sol'n. of higher m.w. block copolymers; cationic; patented wetting/dispersing additive for org. and inorganic pigments in industrial coatings, automotive coatings, automotive refinishing paints, coil coatings, tinting pastes for solv.-based systems; prevents reflocculation of pigments; stabilizes color str.; sp.gr. 0.99-1.02 g/cc; dens. 8.26-8.51 lb/gal; flash pt. (Seta) 28 C; 37-39% NV in xylene/butyl acetate/PMA (4/2/5).

Disperbyk®-163. [Byk-Chemie USA] Sol'n. of higher m.w. block copolymers; cationic; wetting and dispersing additive for org. and inorganic pigments in industrial coatings, automotive coatings, automotive refinishing paints, coil coatings, tinting pastes for solv.-based systems; prevents reflocculation of pigments; stabilizes color str.; sp.gr. 0.985-1.00 g/cc; dens. 8.21-8.34 lb/gal; flash pt. (Seta) 28 C; 44-46% NV in xylene/butyl acetate/PMA (3/1/1).

Disperbyk®-166. [Byk-Chemie USA] Sol'n. of a block copolymer; wetting and dispersing additive for org. and inorg. pigments; for high quality industrial paints, automotive paints, car refinishing paints, pigment pastes, wood and furniture coatings; prevents reflocculation of pigments; stabilizes color str.; sp.gr. 0.95-0.99 g/cc; dens. 7.92-8.26 lb/gal; flash pt. (Seta) 27 C; 29-31% NV in methoxypropyl acetate/butyl acetate (1/4).

Disperbyk®-181. [Byk-Chemie USA] Sol'n. of an alkanolammonium salt of a polyfunctional polymer; anionic/nonionic; wetting additive for emulsion paints based on polymethacrylates and copolymers, vinyl esters, styrene copolymers, water-reducible paint systems, solv.-based paint systems; improves color development, gloss; sp.gr. 1.03-1.05 g/cc; dens. 8.60-8.76 lb/gal; acid no. 25-41; flash pt. (Seta) 46 C; 60-70% NV in methoxypropylacetate/propylene glycol/methoxy propanol (5/3/2).

Disperbyk®-182. [Byk-Chemie USA] Sol'n. of higher m.w. blocked copolymer; cationic; wetting and dispersing additive for org. and inorg. pigments; for water-reducible paint systems, pigment pastes, solv.-based and high-solids coatings systems; sp.gr. 1.01-1.05 g/cc; dens. 8.42-8.76 lb/gal; flash pt. (Seta) 38 C; 41-45% NV in methoxypropyl acetate/methoxypropoxy propanol/butyl acetate (7/4/4).

Dispersing Agent SS Dry. [Hoechst AG] Polymeric org. sulfonic acid; anionic; dispersant for formulating biocidal wettable powds.; powd.

Dispersogen A. [Hoechst Celanese/Colorants & Surf.; Hoechst AG] Sodium naphthalene formaldehyde sulfonate; anionic; dispersing and dyeing aux.; emulsifier, wetting agent, adjuvant for agric. formulations; powd.

Dispersogen ASN. [Hoechst Celanese/Colorants & Surf.; Hoechst AG] Higher alcohol, ethoxylated; nonionic; dispersant for polyester dyeing; dyeing aux. for leather industry; liq.; 26% conc.

Dispersogen SI. [Hoechst Celanese/Colorants & Surf.] Sodium polynaphthalene sulfonate and sodium C12-14 alkyl sulfate; anionic; surfactant for agric. formulations.

Displasol DP. [Am. Emulsions] Quat. ammonium compd. blend; displaces acid dyes on nylon producing novel styling effects.

Disponil AAP 307. [Henkel/Functional Prods.] Alkylaryl polyglycol ether; nonionic; coemulsifier for polyacrylates, acrylate-vinyl acetate copolymers, other applics.; dispersant for emulsion paints; liq.; HLB 17.3; 70% act.

Disponil AAP 436. [Henkel/Functional Prods.] Alkylaryl polyglycol ether; nonionic; coemulsifier for polyacrylates, acrylate vinyl acetate copolymers; dispersant for emulsion paints; liq.; HLB 17.9; 60% act.

Disponil AEP 5300. [Henkel/Functional Prods.] Ether phosphate, acid ester; anionic; emulsifier for rosin, vinyl acetate and acrylate systems; liq.; 100% conc.

Disponil AEP 8100. [Henkel/Functional Prods.] Ether phosphate, acid ester; anionic; emulsifier for rosin, vinyl acetate and acrylate systems; liq.; 100% solids.

Disponil AEP 9525. [Henkel/Functional Prods.] Ether phosphate, sodium salt; anionic; emulsifier for rosin, vinyl acetate and acrylate systems; liq.; 25% solids.

Disponil AES 13. [Henkel/Functional Prods.] Sodium alkylaryl ether sulfate; anionic; emulsifier for vinyl acetate homopolymers, acrylate homo and copolymers, styrene acrylate copolymers, vinyl acetate-acrylate copolymers, VAE copolymers, PVDC latexes, vinyl chloride homo and copolymer latexes; liq.; 31-34% conc.

Disponil AES 21. [Henkel/Functional Prods.] Sodium alkylaryl ether sulfate; anionic; emulsifier for vinyl acetate homo and copolymers, acrylate homo and copolymers, styrene-acrylate, S/B latexes, vinyl propionate copolymers, PVDC, vinyl chloride homo and copolymer latexes; liq.; 30-35% solids.

Disponil AES 42. [Henkel/Functional Prods.] Sodium alkylaryl ether sulfate; anionic; emulsifier for vinyl acetate homo and copolymers, acrylate homo and copolymers, styrene-acrylate, S/B latexes, vinyl propionate copolymers, PVDC, vinyl chloride homo and copolymer latexes; liq.; 40% solids.

Disponil AES 48. [Henkel/Functional Prods.] Ammonium alkylaryl ether sulfate; anionic; emulsifier for emulsion polymerization, natural fats; dispersant for chrome and lime soaps; liq./paste; 68-72% solids.

Disponil AES 60. [Henkel/Functional Prods.] Sodium alkylaryl ether sulfate; anionic; emulsifier for vinyl acetate homo and copolymers, acrylate homo and copolymers, styrene-acrylate, S/B latexes, vinyl propionate copolymers, PVDC, vinyl chloride homo and copolymer latexes; liq.; 34-36% solids.

Disponil AES 60 E. [Pulcra SA] Sodium nonoxynol-8 sulfate; anionic; emulsifier for emulsion polymerization; liq.; 30% conc.

Disponil AES 72. [Henkel/Functional Prods.] Sodium alkylaryl ether sulfate; anionic; low foaming emulsifier for vinyl acetate homo and copolymers, acrylate homo and copolymers, styrene-acrylate, S/B latexes, vinyl propionate copolymers, PVDC, vinyl chloride homo and copolymer latexes; liq.; 34-36% solids.

Disponil APG 110. [Henkel/Functional Prods.] Linear alkyl diol polyglycol ether; nonionic; coemulsifier; latex stabilizer; liq.; HLB 14.1; 100% act.

Disponil FES 32. [Henkel/Functional Prods.] Sodium laureth sulfate; CAS 9004-82-4; anionic; emulsifier for vinyl acetate copolymers, S/B latexes, vinyl chloride copolymers, acrylate homo- and copolymers; liq.; 29-32% solids.

Disponil FES 61. [Henkel/Functional Prods.] Sodium laureth sulfate; CAS 9004-82-4; anionic; emulsifier for vinyl acetate copolymers, S/B latexes, vinyl chloride copolymers, acrylate homo- and copolymers; liq.; 29-32% solids.

Disponil FES 77. [Henkel/Functional Prods.] Sodium laureth sulfate; CAS 9004-82-4; anionic; emulsifier for polymerization of vinyl acetate copolymers, S/B latexes, vinyl chloride copolymers, acrylate homo- and copolymers; liq.; 30-34% conc.

Disponil FES 92E. [Pulcra SA] Sodium laureth-12 sulfate; anionic; surfactant for low-irritation shampoos, emulsion polymerization; liq.; m.w. 830; pH 8.0-9.0 (10%); 29-31% act.

Disponil G 200. [Henkel/Functional Prods.] Isoeicosanol; surfactant; liq. fatty alcohol; liq.; 100% act.

Disponil LS 4. [Henkel/Functional Prods.] Sat. fatty alcohol polyglycol ether; nonionic; wetting agent, detergent, emulsifier, raw material for mfg. of surfactants; liq.; HLB 9.5; 100% act.

Disponil LS 12. [Henkel/Functional Prods.] Sat. fatty alcohol polyglycol ether; nonionic; wetting agent, detergent, emulsifier, raw material for mfg. of surfactants; solid; HLB 14.7; 100% act.

Disponil LS 30. [Henkel/Functional Prods.] Sat. fatty alcohol polyglycol ether; nonionic; wetting agent, detergent, emulsifier, raw material for mfg. of surfactants; flakes; HLB 17.5; 100% act.

Disponil MGS 65. [Henkel/Functional Prods.] Surfactant blend; anionic; emulsifier for vinyl acetate homo and copolymers; liq.; 45% solids.

Disponil MGS 777. [Henkel/Functional Prods.] Surfactant blend; rheology modifier for PVC-plastisol resins; liq.; 100% act.

Disponil MGS 935. [Henkel/Functional Prods.] Surfactant blend; anionic; emulsifier for vinyl acetate homo and copolymers; foaming and wetting agent for highly water absorbent foam resins; liq.; 50% solids.

Disponil O 5. [Henkel/Functional Prods.] Cetoleth-5; nonionic; emulsifier for emulsion polymerization; wetting agent; liq.; HLB 9.5; 100% act.

Disponil O 10. [Henkel/Functional Prods.] Cetoleth-10; nonionic; emulsifier for emulsion polymerization; wetting agent; paste; HLB 12.5; 100% act.

Disponil O 20. [Henkel/Functional Prods.] Cetoleth-20; nonionic; emulsifier for emulsion polymerization; solubilizer, dispersant, latex stabilizer; solid; HLB 16.5; 100% act.

Disponil O 250. [Henkel/Functional Prods.] Cetoleth-25; nonionic; emulsifier for emulsion polymerization; solubilizer, dispersant, latex stabilizer; flakes; HLB 18.4; 100% act.

Disponil RO 40. [Henkel/Functional Prods.] POE fatty glyceride; nonionic; hydrotrope, solubilizer; emulsifier for aq. nitrocellulose and alkyd-resin emulsions; liq.; HLB 10.8; 100% act.

Disponil SML 100 F1. [Henkel/Functional Prods.] Sorbitan laurate; CAS 1338-39-2; nonionic; surfactant for polymerization; liq.; HLB 8.6; 100% act.

Disponil SML 104 F1. [Henkel/Functional Prods.] Polysorbate 21; CAS 9005-64-5; nonionic; surfactant for polymerization; liq.; HLB 13.3; 100% act.

Disponil SML 120 F1. [Henkel/Functional Prods.] Polysorbate 20; CAS 9005-64-5; nonionic; surfactant for polymerization; liq.; HLB 16.7; 100% act.

Disponil SMO 100 F1. [Henkel/Functional Prods.] Sorbitan oleate; CAS 1338-43-8; nonionic; surfactant for polymerization; liq.; HLB 4.3; 100% act.

Disponil SMO 120 F1. [Henkel/Functional Prods.] Polysorbate 80; CAS 9005-65-6; nonionic; surfactant for polymerization; liq.; HLB 15.0; 100% act.

Disponil SMP 100 F1. [Henkel/Functional Prods.] Sorbitan palmitate; CAS 26266-57-9; nonionic; surfactant for polymerization; flakes; HLB 6.7; 100% act.

Disponil SMP 120 F1. [Henkel/Functional Prods.] Polysorbate 40; CAS 9005-66-7; nonionic; surfactant for polymerization; liq.; HLB 15.6; 100% act.

Disponil SMS 100 F1. [Henkel/Functional Prods.] Sorbitan stearate; CAS 1338-41-6; nonionic; surfactant for polymerization; flakes; HLB 4.7; 100% act.

Disponil SMS 120 F1. [Henkel/Functional Prods.] Polysorbate 60; CAS 9005-67-8; nonionic; surfactant for polymerization; liq.; HLB 14.9; 100% act.

Disponil SSO 100 F1. [Henkel/Functional Prods.] Sorbitan sesquioleate; CAS 8007-43-0; nonionic; surfactant for polymerization; liq.; HLB 3.7; 100% act.

Disponil STO 100 F1. [Henkel/Functional Prods.] Sorbitan trioleate; CAS 26266-58-0; nonionic; surfactant for polymerization; liq.; HLB 1.8; 100% act.

Disponil STO 120 F1. [Henkel/Functional Prods.] Polysorbate 85; CAS 9005-70-3; nonionic; surfactant for polymerization; liq.; HLB 11.0; 100% act.

Disponil STS 100 F1. [Henkel/Functional Prods.] Sorbitan tristearate; CAS 26658-19-5; nonionic; surfactant for polymerization; flakes; HLB 2.1; 100% act.

Disponil STS 120 F1. [Henkel/Functional Prods.] Polysorbate 65; CAS 9005-71-4; nonionic; surfactant for polymerization; solid; HLB 10.5; 100% act.

Disponil SUS 29 L. [Henkel/Functional Prods.] Sodium sulfosuccinamate; anionic; emulsifier for vinyl acetate copolymer, acrylate homo and copolymer, S/B latexes, vinyl chloride copolymers; liq.; 35% solids.

Disponil SUS 65. [Henkel/Functional Prods.] Fatty alcohol ethoxylate sulfosuccinate; anionic; emulsifier for vinyl acetate copolymer, acrylate homo and copolymer, S/B latexes, vinyl chloride copolymers; liq.; 40% solids.

Disponil SUS 87 Special. [Henkel/Functional Prods.] Disodium sulfosuccinate; anionic; emulsifier in polymerization of vinyl acetate copolymers, acrylate homo and copolymers, S/B latexes, and vinyl chloride copolymers; liq.; 30-32% conc.

Disponil SUS 90. [Henkel/Functional Prods.] Sodium alkylaryl EO sulfosuccinate; anionic; emulsifier for vinyl acetate copolymer, acrylate homo and copolymer, S/B latexes, vinyl chloride copolymers; liq.; 32-35% solids.

Disponil SUS IC 8. [Henkel/Functional Prods.] Dioctyl sodium sulfosuccinate; wetting agent, coemulsifier for plastics industry; liq.; 70-75% conc.

Disponil TA 5. [Henkel/Functional Prods.] Linear sat. fatty alcohol polyglycol ether; nonionic; wetting agent, polymerization emulsifier, dispersant; liq.; HLB 9.2; 100% act.

Disponil TA 25. [Henkel/Functional Prods.] Linear sat. fatty alcohol polyglycol ether; nonionic; wetting agent, polymerization emulsifier, dispersant; solid; HLB 16.2; 100% act.

Disponil TA 430. [Henkel/Functional Prods.] Linear sat. fatty alcohol polyglycol ether; nonionic; wetting agent, polymerization emulsifier, dispersant; flake; HLB 17.4; 100% act.

Disrol SH. [Nippon Nyukazai] Sodium bis (naphthalene) sulfonate; dispersant and suspending agent; flake.

Dissolvan Brands. [Hoechst AG] PO/EO block polymers and/or oxyalkylated resins; nonionic; emulsion breaker for w/o emulsions; dehydration and desalting agents for crude oil; liq.

Dissolver GX Grades. [Hoechst AG] Alkylaryl polyglycol ethers; surfactant.

DK-100 Nonsilicone Defoamer. [Genesee Polymers] Nonsilicone; defoamer for aq. systems, paints, coatings, paperboard, paper, metalworking lubricants, waste water treatment, cleaner formulations; FDA approved; tan liq.; sp.gr. 0.85; dens. 7.0 lb/gal; flash pt. (COC) 265 F; 100% act.

DK-230 Nonsilicone Defoamer. [Genesee Polymers] Nonsilicone; defoamer for aq. systems and where air entrapment is a problem; for paints, coatings, paperboard, paper, metalworking lubricants, cutting oils, coolants, adhesives; alkaline resistant; FDA approved; straw-colored liq.; sp.gr. 0.82; dens. 7.0 lb/gal; flash pt. (PMCC) > 200 F; 100% act.

DLS Base. [Clark] Pre-neutralized emulsifiable amidoamine condensate; softener base imparting a dry, soft hand to cotton and other fabrics; disp. in hot water.

DMAMP-80. [Angus] 2-Dimethylamino-2-methyl-1-propanol; amine solubilizer for resins in aq. coatings; emulsifier for waxes; vapor-phase corrosion inhibitor; urethane catalyst; titanate solubilizer; raw material for synthesis; misc. in water; m.p. -20 C; 80% act.

DMS-33. [Hefti Ltd.] PEG-2 stearate; CAS 9004-99-3; nonionic; emulsifier for creams and ointments, shoe creams, paraffin wax, pigment suspensions; overcreasing agent in soaps; consistency regulator in soft o/w emulsions; flakes; HLB 5.0; 100% conc.

DO-33-F. [Hefti Ltd.] Sorbitan dioleate; CAS 29116-98-1; nonionic; emulsifier for creams and ointments; pigment stabilizer for paint emulsions, min. oil emulsions, anticorrosion oils; liq.; HLB 3.5; 100% conc.

Dobanol 23. [Shell UK] C12–C13 linear primary alcohol; nonionic; base material for preparation of alcohol ethoxylates, sulfates, ethoxysulfates; liq.; mild odor; dens. 0.831 kg/l; visc. 14 cs (40 C); m.p. 17–19 C; flash pt. 132 C; > 95% biodeg.; 100% act.

Dobanol 23-2. [Shell UK] C12–C13 linear primary alcohol ethoxylate; nonionic; detergent, intermediate for prod. of laundry powds.; clear liq.; dens. 0.91 kg/l; visc. 15 cs (40 C); m.p. 4–5 C; HLB 6.2; flash pt. 148 C; 100% act.

Dobanol 23-3. [Shell UK] C12-C13 linear primary alcohol ethoxylate; nonionic; detergent intermediate for cosmetic, specialty and dishwashing formulations; liq.; HLB 8.1; 100% conc.

Dobanol 23-6.5. [Shell UK] C12–C13 linear primary alcohol ethoxylate (6.5 EO); nonionic; detergent, intermediate for prod. of laundry powds.; wh. slurry; dens. 0.96 kg/l; visc. 26 cs (40 C); m.p. 17–19 C; HLB 11.9; flash pt. 172 C; 100% act.

Dobanol 25. [Shell UK] C12–C15 linear primary alcohol; base material for preparation of alcohol ethoxylates, sulfates, and ethoxysulfates; liq.; mild odor; dens. 0.831 kg/l; visc. 14 cs (40 C); m.p. 20–22 C; biodeg.; 100% act.

Dobanol 25-3. [Shell UK] Pareth-25-3; nonionic; detergent, intermediate for mfg. of cosmetic, specialty industrial, and dishwashing formulations; clear liq.; dens. 0.93 kg/l; visc. 18 cs (40 C); m.p. 5–6 C; HLB 7.8; flash pt. 162 C; > 95% biodeg.; 100% act.

Dobanol 25-7. [Shell UK] Pareth-25-7; nonionic; detergent, wetting agent in laundry powds.; wh. paste/solid; misc. in all proportions; dens. 0.97 kg/l (40 C); visc. 31 cs (40 C); HLB 12.0; cloud pt. 48–54 C; flash pt. 180 C; pH 6.8 (0.5% aq.); 100% act.

Dobanol 25-9. [Shell UK] Pareth-25-9; nonionic; detergent, wetting in laundry powds.; wh. solid; misc. in all proportions; dens. 0.98 kg/l (40 C); visc. 40 cs (40 C); HLB 13.2; cloud pt. 75–83 C; flash pt. 190 C; pH 6.8 (0.5% aq.); 100% act.

Dobanol 45. [Shell] C14–C15 linear primary alcohol; detergent, intermediate, base material; liq.; mild odor; dens. 0.824 kg/l (40 C); visc. 17 cs (40 C); m.p. 29–31 C; flash pt. 152 C; biodeg.; 100% act.

Dobanol 45-7. [Shell UK] Pareth-45-7; nonionic;

detergent, wetting agent in high performance laundry powds.; wh. solid; misc. in all proportions; dens. 0.96 kg/l (50 C); visc. 33 cs (40 C); HLB 11.6; cloud pt. 43–49 C; flash pt. 185 C; pH 6.8 (0.5% aq.); 100% act.

Dobanol 91. [Shell UK] C9–C11 linear primary alcohol; base material for preparation of alcohol ethoxylates, sulfates, ethoxysulfates; dens. 0.832 kg/l; visc. 9 cs (40 C); m.p. –10 to –8 C; flash pt. 107 C (PMCC); biodeg.; 100% act.

Dobanol 91-2.5. [Shell UK] Pareth-91-3; nonionic; detergent intermediate; clear liq.; dens. 0.93 kg/l; visc. 11 cs (40 C); HLB 8.1; flash pt. 120 C (PMCC); pH 6.8 (0.5% aq.); 95% biodeg.; 100% act.

Dobanol 91-5. [Shell UK] Pareth-91-5; nonionic; wetting agent, emulsifier; clear liq.; misc. in all proportions; dens. 0.98 kg/l; visc. 18 cs (40 C); HLB 11.6; flash pt. 146 C (PMCC); pH 6.8 (0.5% aq.); 100% act.

Dobanol 91-6. [Shell UK] Pareth-91-6; nonionic; detergent, wetting agent; clear liq.; misc. in all proportions; dens. 0.99 kg/l; visc. 21 cs (40 C); HLB 12.5; flash pt. 150 C (PMCC); pH 6.8 (0.5% aq.); 100% act.

Dobanol 91-8. [Shell UK] Pareth-91-8; nonionic; detergent; liq./slurry; misc. in all proportions; dens. 1.00 kg/l; visc. 28 cs (40 C); HLB 13.7; flash pt. 160 C (PMCC); pH 6.8 (0.5% aq.); 100% act.

Dodecenyl Succinic Anhydride. [Humphrey] Dodecenyl succinic anhydride; CAS 26544-38-7; anionic; intermediate for amide and imide rust inhibitors and sludge dispersant for lube oils and greases; epoxy hardener; mercurial fungicide; sodium salt as industrial cleaner, wetting and dispersing agent; liq.; 100% conc.

Dodicor 2565. [Hoechst Celanese/Colorants & Surf.] Quat. arylammonium chloride; cationic; metalworking surfactant; corrosion inhibitor for oil and gas industry; max. protection for zinc in acid cleaners; visc. liq.; sol. in water; dens. 1.11 g/cc; pH 3-5 (1% aq.).

Dodiflood Brands. [Hoechst AG] Ether sulfonates, ether carboxylates; anionic; surfactant for enhanced oil recovery and microemulsion flooding; liq.

Dodiflow Brands. [Hoechst AG] Ethylene copolymers and wetting agents; paraffin inhibitors for prod. of crude oil, flow improvers for middle distillates; visc. liq.

Dodifoam Brands. [Hoechst AG] Surfactant blends; anionic/nonionic; foaming agents for drilling; liq.

Dodigen 1490. [Hoechst AG] Dicoco dimethylammonium chloride; cationic; surfactant.

Doittol 14. [Henkel-Nopco] Sulfated fatty acid ester; anionic; wetting agent and leveling assistant for dyeing and bleaching; liq.; 45% conc.

Doittol 891. [Henkel-Nopco] Sodium fatty acid soap, modified; anionic; surfactant; rubber antiblocking agent; liq.; 20% act.

Doittol APS Conc. [Henkel-Nopco] Alkylaryl polyglycol ether semiester; anionic; dyeing assistant for dyeing polyester and blends; liq.; 60% conc.

Doittol FL. [Henkel-Nopco] Alkyl polyglycol ether blend; nonionic; detergent for soaping after dyeing; semiliq.; 50% conc.

DOM. [Tiarco] Dioctyl maleate; CAS 2915-53-9; intermediate for surfactant mfg.; in copolymerization of PVC and vinyl acetates; plasticizer for vinyl resins; also in latex paints; liq.; 100% conc.

Dovanox 23H, 23M, 25N, 231. [Lion] POE syn. alcohol; nonionic; detergent, penetrant, emulsifier, dispersant; liq., wax; 100% conc.

Dow Corning® 190 Surfactant. [Dow Corning] Dimethicone copolyol; nonionic; silicone surfactant, surf. tens. depressant, wetting agent, emulsifier, foam builder, humectant, softener; used for producing flexible slab stock urethane foam; ingredient in personal care prods.; plasticizer for hair resins; Gardner 2 low-visc. liq.; sol. in water and ethanol; sp.gr. 1.035; visc. 1500 cst; HLB 14.4; flash pt. (COC) 121 C;; nontoxic; 100% conc.

Dow Corning® 193 Surfactant. [Dow Corning] Dimethicone copolyol; nonionic; silicone surfactant, surf. tens. depressant, wetting agent, emulsifier, foam builder, humectant, softener; used for producing flexible slab stock urethane foam; ingredient in personal care prods.; plasticizer for hair resins; Gardner 2 clear visc. liq.; sol. in water, alcohol, hydroalcoholic systems; sp.gr. 1.07; visc. 465 cs; HLB 13.6; f.p. 50 F; flash pt. (COC) 149 C; pour pt. 52 F; usage level: 0.1-5.0%; nontoxic; 100% conc.

Dow Corning® 197 Surfactant. [Dow] Silicone surfactant; for polyurethane foam industry; sp.gr. 1.05; visc. 230 cSt; flash pt. (CC) > 60.6 C.

Dow Corning® 198 Surfactant. [Dow] Silicone surfactant; for polyurethane foam industry; sp.gr. 1.04; visc. 1600 cSt; flash pt. (CC) > 60.6 C.

Dow Corning® 200 Fluid, Food Grade. [Dow Corning] Dimethyl silicone fluid; foam control agent for nonaq. systems, food industry, meat, poultry and seafood processing, rendering, inks; FDA, EPA, USDA, kosher approved; thin clear fluid; visc. 350 cSt; 100% act.

Dow Corning® 200 Fluid. [Dow Corning] Dimethicone; foam control agent for nonaq. systems, distillation, resin mfg., asphalt, oil refining, gas-oil separation; clear fluid; dilutable in aliphatic, aromatic or chlorinated solvs.; visc. 1000–10,000 cSt and 12,500–60,000 cSt grades; 100% act.

Dow Corning® 544 Compd. [Dow Corning] Silicone compd.; foam control agent for aq. or nonaq. systems, distillation, glycol scrubbing, detergents, textile jet dyeing, cutting oils; med. off-wh. liq.; dilutable in cool water; 100% act.

Dow Corning® 1248 Fluid. [Dow Corning] Silicone polycarbinols (graft copolymer type); nonionic; surfactant for org. systems and urethanes; clear liq.; sol. in hydrocarbons; visc. 160 cs; HLB 1.5; flash pt. 250 F; pour pt. –130 F; 100% conc.

Dow Corning® 1250. [Dow Corning] Silicone in solv. (xylene); for polyurethane foam industry; profoamer for mechanical frothing; allows foam fusing at low temp.; very lt. straw-colored thin liq.; sp.gr. 1.00; visc. 6 cst; flash pt. 85 F (OC); 50% silicone.

Dow Corning® 1252. [Dow Corning] Silicone in solv. (texanol isobutyrate); for polyurethane foam industry; profoamer for mechanical frothing; allows foam fusing at low temp.; very lt. straw-colored thin liq.; sp.gr. 1.06; visc. 250 cs; flash pt. 135 F (OC); 50% silicone.

Dow Corning® 1315 Surfactant. [Dow Corning] Silicone glycol copolymer; nonionic; surfactant for PU foams; wetting agent; detackifier and plasticizer for hair fixative resins; liq.; sol. in water, ethanol; sp.gr. 1.03; visc. 1000 cSt; HLB 15.3; flash pt. (CC) > 60.6 C; 100% conc.

Dow Corning® 1400 Compd. [Dow] Silicone compd.; foam control agent for aq. or nonaq. systems, distillation, glycol scrubbing, resin mfg., detergents, oil refining, asphalt, solvs., metal cleaning/degreasing, cutting oils; med. off-wh., wh. liq.;

dilutable in aliphatic, aromatic or chlorinated solvs.; 100% act.

Dow Corning® 1500 Compd. [Dow Corning] Silica filled polydimethyl siloxane; defoamer for aq. or nonaq. systems, food processing, rendering, glycol scrubbing, cutting oils; FDA, EPA, USDA, kosher approved; med., off-wh. liq.; dilutable in food grade glycols; 100% act.

Dow Corning® 1520 Silicone Antifoam. [Dow Corning] Silica filled polydimethyl siloxane emulsion; defoamer which controls foam in aq. systems; inert, long lasting, does not interact with other system components; liq.

Dow Corning® 1920 Powdered Antifoam. [Dow] Silicone; nonionic; antifoam for food processing, ultrafiltration, waste water treatment, biotechnology; FDA and USDA approved; wh. free-flowing powd.; 20% act.

Dow Corning® 5043 Surfactant. [Dow] Silicone surfactant; for polyurethane foam industry; sp.gr. 1.00; visc. 265 cSt; flash pt. (CC) > 60.6 C.

Dow Corning® 5098. [Dow] Silicone surfactant; for polyurethane foam industry; sp.gr. 1.07; visc. 195 cSt; flash pt. (CC) > 60.6 C.

Dow Corning® 5103 Surfactant. [Dow Corning] Silicone glycol copolymer; nonionic; surfactant for PU foams; wetting agent; detackifier and plasticizer for hair fixative resins; liq.; sol. in water, ethanol; sp.gr. 1.05; visc. 250 cSt; HLB 9.7; flash pt. (CC) > 60.6 C; 100% conc.

Dow Corning® A Compd. [Dow] Silicone compd.; foam control agent for aq. or nonaq. systems, distillation; food grade for fermentation, beverage processing, rendering; FDA, EPA, USDA, kosher approved (food grade); med. off-wh., gray liq.; dilutable in aliphatic, aromatic or chlorinated solvs., food grade glycols (food grade); 100% act.

Dow Corning® Antifoam 1400. [Dow Corning] Silicone compd.; foam control agent for glycol scrubbing, resin mfg., oil refining, asphalts, adhesives/coatings, metalworking, pesticide/fertilizer, and detergents industries; for extreme pH conditions; med. off-wh. to wh. liq.; 100% act.

Dow Corning® Antifoam 1410. [Dow Corning] Silicone emulsion; nonionic; foam control agent for inks, textile starching/sizing, cutting oils, resin mfg., gas processing, adhesives/coatings, waste water treatment, pesticide/fertilizer industries; for extreme pH conditions; thin wh. cream; 10% act.

Dow Corning® Antifoam 1430. [Dow Corning] Silicone emulsion; nonionic; foam control agent for detergents, distillation, glycol scrubbing, latex mfg., metalworking, waste water treatment, adhesives/coatings, and textile industries; for extreme pH conditions; med. wh. cream; 30% act.

Dow Corning® Antifoam 1500. [Dow Corning] Silicone compd.; foam control agent for foods, meat and poultry prod., fermentation, pesticides/fertilizers, gas-oil separation, printing inks, asphalt; kosher and EPA approved; med. off-wh. liq.; 100% act.

Dow Corning® Antifoam 1510-US. [Dow Corning] Silicone emulsion; nonionic; foam control agent for pesticide/fertilizer, oil refining, waste water treatment, conditioners, food industries; FDA, EPA, USDA, kosher approved; med. wh. cream; 10% act.

Dow Corning® Antifoam 1520-US. [Dow Corning] Silicone emulsion; nonionic; foam control agent for food, beverage mfg., meat/poultry/seafood processing, resin mfg., waste water treatment, adhesives/coatings, metalworking, and textile industries; FDA, EPA, USDA, kosher approved; thin wh. cream; 20% act.

Dow Corning® Antifoam 2210. [Dow] Silicone emulsion; nonionic; foam control agent for distillation, glycol scrubbing, waste water and heat/cooling water treatment, adhesives/glues, latexes, cutting oils; thin wh. cream; 10% act.

Dow Corning® Antifoam A. [Dow Corning] Simethicone; foam control agent for distillation, resin sizes, textile latex backing, paper, asphalt, lubricants, detergents, pesticides, edible oils, soaps, shampoos; also avail. in food grade; med. off-wh. to gray liq.; 100% act.

Dow Corning® Antifoam AF. [Dow Corning] Simethicone; nonionic; foam control agent for food, pesticide/fertilizer, paper/printing, textile, adhesives/coatings industries; FDA, EPA, USDA, kosher approved; thick wh. cream; 30% act.

Dow Corning® Antifoam B. [Dow Corning] Silicone emulsion; nonionic; foam control agent for adhesives/glues, textile latex backing, inks, detergents, distillation, resin mfg, pesticide/fertilizer industries, cooling towers; EPA approved; thin wh. cream; 10% act.

Dow Corning® Antifoam C. [Dow Corning] Simethicone; nonionic; food grade foam control agent for food industry, paper coatings, pesticides, herbicides, fertilizers; FDA, EPA, kosher, USDA approved; med. wh. cream; 30% act.

Dow Corning® Antifoam FG-10. [Dow Corning] Silicone emulsion; nonionic; foam control agent for food industry, fermentation, high sugar-content processes, paper coatings, and waste water treatment; FDA, EPA, USDA, kosher approved; thin wh. cream; 10% act.

Dow Corning® Antifoam H-10. [Dow Corning] Silicone emulsion; nonionic; foam control agent for distillation, glycol scrubbing, detergents, waste water treatment, adhesives/coatings, resin and textile sizes, metalworking, and paper/printing industries; thin wh. cream; 10% act.

Dow Corning® Antifoam Y-30. [Dow Corning] Silicone emulsion; nonionic; foam control agent for paper/printing industry for resin sizes and inks, textiles, antifreeze, ceramic mfg., dyes; med. wh. cream; 30% act.

Dow Corning® FF-400. [Dow Corning] Silicone glycol copolymer; nonionic; lubricant; heat stabilizer, antistat, fiber finish for textiles threads; lt. amber fluid; sol. in polar solvs.; water disp.; sp.gr. 1.022; visc. 300 cs; HLB 7.0; flash pt. (OC) 177 C; 100% act.

Dow Corning® FS-1265 Fluid. [Dow Corning] Dimethyl silicone fluid; foam control agent for nonaq. systems, aromatic/chlorinated solvs., gas-oil separation, dry cleaning, metal cleaning and degreasing, oil refining; clear fluids; visc. 300 and 1000-10,000 cSt grades; 100% act.

Dow Corning® Q2-3183A Compd. [Dow] Silicone compd.; foam control agent for aq. or nonaq. systems, metal cleaning/degreasing, etching, cutting oils; med., off-wh. liq.; dilutable in water; 100% act.

Dow Corning® Q2-5160. [Dow] Silicone surfactant; for polyurethane foam industry; sp.gr. 1.03; visc. 950 cSt; flash pt. (CC) > 60.6 C.

Dow Corning® Q4-3667 Fluid. [Dow Corning] Silicone polycarbinols (ABA type); nonionic; reactive surfactant for use in org. systems to form silicone-modified polymers; clear straw liq.; sol. in hydrocarbons; visc. 320 cs; HLB 9.4; flash pt. 285 F; pour pt. 65 F; 100% conc.

Dowfax 2A0. [Dow; Dow Europe] Dodecyl diphenyloxide disulfonic acid; anionic; for use in acidic systems or for prep. of various salts; liq.; 40% conc.

Dowfax 2A1. [Dow; Dow Europe] Sodium dodecyl diphenyloxide disulfonate; anionic; detergent, emulsifier, wetting agent, solubilizer, dispersant, spreading agent, penetrant for detergent formulation, emulsion polymerization, agric., electroplating, ore flotation, drilling muds; leveling agent for acid dyeing of nylon, dyeing assistant, emulsifier for dye carriers; FDA and EPA compliance; amber clear liq.; sol. in water, aq. sol'ns. of acids, alkalis, electrolytes; m.w. 575; sp.gr. 1.16; visc. 145 cps (0.10%); HLB 16.7; surf. tens. 31 dynes/cm (0.1%); toxicology: toxic to fish; 45% act.

Dowfax 2EP. [Dow; Dow Europe] Sodium dodecyl diphenyloxide disulfonate; anionic; detergent, emulsifier, wetting agent, solubilizer, dispersant, spreading agent, penetrant for detergent formulation, emulsion polymerization, agric., electroplating, ore flotation, drilling muds; leveling agent for acid dyeing of nylon, dyeing assistant, emulsifier for dye carriers; FDA and EPA compliance; pale brn. liq.; sol. in water, 20% caustic and HCl, 35% TKPP; sl. sol. in ethanol; m.w. 575; sp.gr. 1.16; visc. 145 cps; HLB 16.7; surf. tens. 31 dynes/cm (0.1%); toxicology: toxic to fish; 45% conc.

Dowfax 3B0. [Dow] N-Decyl diphenyloxide disulfonate; anionic; for formulating cleaning prods. and agric. prods. where salts other than NaCl are required; liq.; 40% conc.

Dowfax 3B2. [Dow; Dow Europe] Sodium n-decyl diphenyloxide disulfonate; anionic; detergent, emulsifier, wetting agent, solubilizer, dispersant, spreading agent, penetrant for detergent formulation, emulsion polymerization, agric., electroplating, ore flotation, drilling muds; leveling agent for acid dyeing of nylon, dyeing assistant, emulsifier for dye carriers; biodeg.; FDA and EPA compliance; red-brn. liq.; sol. in water, 25% caustic, 15% HCl, 35% TKPP; sl. sol. in ethanol; sp.gr. 1.16; visc. 120 cps; HLB 17.8; surf. tens. 38 dynes/cm (0.1%); toxicology: toxic to fish; 45% act.

Dowfax 9N. [Dow Europe] Nonylphenol ethoxylate; nonionic; intermediate, detergent, wetting agent, dispersant for household and industrial cleaners, wool scouring, textile auxiliary, general industrial usage; liq. to waxy solid; HLB 6.1-18.0; 100% conc.

Dowfax 20A42. [Dow Europe] Alkoxylated alcohol; defoamer; liq.; 100% conc.

Dowfax 20A64. [Dow Europe] Alkoxylated alcohol; nonionic; low foaming surfactant with good wetting chars.; liq.; 100% conc.

Dowfax 30C05, 30C10, 50C15. [Dow Europe] EO/PO block copolymer; CAS 106392-12-5; nonionic; low foaming surfactant, defoamer base; liq.; 100% conc.

Dowfax 63N10. [Dow Europe] EO/PO block copolymer; nonionic; detergent in liq. dishwashing agents, glass bottle washing, defoamer; liq.; 100% conc.

Dowfax 63N30. [Dow Europe] EO/PO block copolymer; nonionic; detergent in liq. dishwashing agents, defoamer; liq.; 100% conc.

Dowfax 63N40. [Dow Europe] EO/PO block copolymer; nonionic; low foaming surfactant, defoamer base; liq.; 100% conc.

Dowfax 81N10. [Dow Europe] EO/PO block copolymer; nonionic; detergent in liq. dishwashing agent, defoamer; liq.; 100% conc.

Dowfax 92N20. [Dow Europe] EO/PO block copolymer; nonionic; detergent in liq. dishwashing agent, defoamer; liq.; 100% conc.

Dowfax 8174. [Dow] Alkylated disulfonated diphenyl oxide; detergent, emulsifier, wetting agent, solubilizer, dispersant, spreading agent, penetrant for detergent formulation, emulsion polymerization, agric., electroplating, ore flotation, drilling muds; leveling agent for acid dyeing of nylon, dyeing assistant, emulsifier for dye carriers; biodeg.; FDA and EPA compliance; amber clear liq.; sol. in 20% caustic; m.w. 575; visc. 131 cps; surf. tens. 41.1 dynes/cm (0.1%); toxicology: toxic to fish; 45% act.

Dowfax 8390. [Dow; Dow Europe] Sodium n-hexadecyl diphenyloxide disulfonate; anionic; detergent, emulsifier, wetting agent, solubilizer, dispersant, spreading agent, penetrant for detergent formulation, emulsion polymerization, agric., electroplating, ore flotation, drilling muds; leveling agent for acid dyeing of nylon, dyeing assistant, emulsifier for dye carriers; biodeg.; FDA and EPA compliance; amber clear liq.; sol. in water, 25% caustic, 20% HCl, 35% TKPP; sl. sol. in ethanol; m.w. 642; sp.gr. 1.11; visc. 10 cps; surf. tens. 42 dynes/cm (0.1%); toxicology: toxic to fish; 35% conc.

Dowfax XDS 30599. [Dow; Dow Europe] Sodium dodecyl diphenyloxide disulfonate; anionic; surfactant used in emulsion polymerization; liq.; 45% conc.

Dowfax XDS 8292.00. [Dow] Sodium hexyl diphenyloxide disulfonate; anionic; lowest foaming and highest charge density prod. in Dowfax series; high solubilizing capabilities in acids, alkalies, and electrolytes; for cleaning, latex mfg., paints, adhesives, min. and metal processing, textile applics.; amber clear liq.; sol. in water, 15% HCl, 45% TKPP, 32% NaOH; sl. sol. in ethanol; m.w. 474; sp.gr. 1.1954; visc. 50.4 cps; HLB 19.0; surf. tens. 41 dynes/cm (0.1%); 45% act.

Dowfax XDS 8390.00. [Dow; Dow Europe] Sodium n-hexadecyl diphenyloxide disulfonate; anionic; surfactant for emulsion polymerization; liq.; m.w. 632; HLB 14.4; surf. tens. 45 dynes/cm (0.1%); 35% conc.

Dowfax XU 40333.00. [Dow] Sodium dodecyl diphenyloxide disulfonate; anionic; leveling agent for acid dyeing of nylon to control barre; emulsifier for dye carriers such as butyl benzoate; for powd. cleaning compds.; solid; HLB 16.7; 92% conc.

Dowfax XU 40340.00. [Dow] Sodium n-decyl diphenyloxide disulfonate; anionic; biodeg. surfactant for cleaning prods., agric. herbicides, flowable systems, emulsion polymerization; solid; HLB 17.8; 92% conc.

Dowfax XU 40341.00. [Dow] Sodium n-hexadecyl diphenyloxide disulfonate; anionic; biodeg. surfactant for emulsion polymerization, powd. cleaning compds.; solid; 92% conc.

DRA-1500. [Toho Chem. Industry] POE disproportionated rosin ester; nonionic; emulsifier, dispersant; liq.; HLB 14.0; 100% conc.

Dresinate® 81. [Hercules] Rosin sodium soap; anionic; emulsifier ingred. and/or stabilizer in solv. cleaners and sol. oils for metalworking, disinfectants, oil-well drilling muds, drawing and grinding compds.; plasticizer; pale-colored liq.; dilutable with water and aq. sol'ns. of alcohols and glycols; dens. 8.7 lb/gal; visc. 5.7 poises (140 F); acid no. 15; 87% solids.

Dresinate® 90. [Hercules] Rosin potassium soap;

anionic; emulsifier ingred. and/or stabilizer in solv. cleaners and sol. oils for metalworking, disinfectants, oil-well drilling muds, drawing and grinding compds.; plasticizer; pale liq.; dilutable with water and aq. sol'ns. of alcohols and glycols; dens. 9.0 lb/gal; visc. 14.7 poises (140 F); acid no. 12; 90% solids.

Dresinate® 91. [Hercules; Hercules BV] Rosin potassium soap; anionic; emulsifier ingred. and/or stabilizer in solv. cleaners and sol. oils for metalworking, disinfectants, oil-well drilling muds, drawing and grinding compds.; plasticizer; pale liq.; dilutable with water and aq. sol'ns. of alcohols and glycols; dens. 8.8 lb/gal; visc. 5.3 poises (140 F); acid no. 15; 88% solids.

Dresinate® 95. [Hercules; Hercules BV] Potassium soap of dk. rosin; anionic; emulsifier ingred. and/or stabilizer in solv. cleaners and sol. oils for metalworking, disinfectants, oil-well drilling muds, drawing and grinding compds.; plasticizer; dk. liq.; dilutable with water and aq. sol'ns. of alcohols and glycols; dens. 9.1 lb/gal; visc. 4.5 poises (140 F); acid no. 15; 88% solids.

Dresinate® 214. [Hercules; Hercules BV] Potassium soap of a pale modified rosin; anionic; emulsifier, pigment wetting agent and dispersant, foaming agent, used in mfg. of adhesives, polymer emulsion syn. latices; paste; dilutable with water and aq. sol'ns. of alcohols and glycols; dens. 9.22 lb/gal (60 C); visc. 23 poises (140 F); 80% solids.

Dresinate® 515. [Hercules; Hercules BV] Potassium soap of a pale modified rosin; anionic; emulsifier; pale paste; dens. 9.24 lb/gal (60 C); visc. 64 poises (140 F); acid no. 19; 80% solids.

Dresinate® 731. [Hercules; Hercules BV] Sodium soap of a pale modified rosin; nonionic; emulsifier, pigment wetting agent and dispersant, foaming agent, used in mfg. of adhesives, polymer emulsion syn. latices; pale paste; dilutable with water and aq. sol'ns. of alcohols and glycols; dens. 9.1 lb/gal (60 C); visc. 11 poises (140 F); acid no. 11; 70% solids.

Dresinate® TX. [Hercules] Tall oil sodium soap; anionic; emulsifier for oils and asphalts; detergent aid; dispersant for pigments; stabilizer for syn. rubber latices, used for industrial and household cleaners; conditioner for sulfur dusts; dk. brn. powd.; sol. in water and aq. sol'ns. of alcohols and glycols; dens. 26 lb/ft^3; 96% min. solids.

Dresinate® TX-60W. [Hercules] Sodium soap of tall oil derivatives; used for flotations beneficiation of glass sand and phosphates and as emulsifier for asphalt; liq.; 60% solids aq. sol'n.

Dresinate® X. [Hercules] Sodium soap of a pale rosin; anionic; emulsifier for oils and asphalts; detergent aid; dispersant for pigments; stabilizer for syn. rubber latices, used for industrial and household cleaners; conditioner for sulfur dusts; pale cream, powd.; sol. in water and aq. sol'ns. of alcohols and glycols; dens. 26 lb/ft^3; 95% min. solids.

Dresinate® XX. [Hercules] Sodium soap of a dk. rosin; anionic; emulsifier for oils and asphalts; detergent aid; dispersant for pigments; stabilizer for syn. rubber latices, used for industrial and household cleaners; conditioner for sulfur dusts; lt. br. powd.; sol. in water and aq. sol'ns. of alcohols and glycols; dens. 28 lb/ft^3; 96% min. solids.

Drewclean® 26. [Drew Ind. Div.] Alkaline cleaner; cleaner for ion exchange resins and filter media fouled by oil, grease, other organics, clay or silt.

Drewfax® 0007. [Drew Ind. Div.] Sodium dioctyl sulfosuccinate; CAS 577-11-7; wetting, penetrating, surf. tens. reducing agent, dispersant for industrial coatings, adhesives, inks, pigments, textile, cosmetic, paper, metal, paint, rubber, plastics, petrol. and agric. industries; FDA compliance; colorless clear visc. liq.; sol. in water and various solvs.; sp.gr. 1.08; dens. 9.0 lb/gal; visc. 400 cps; pH 5-6 (1%); 75% solids.

Drewfax® 412. [Drew Ind. Div.] Blend of silicone derivs. and glycol ethers; nonionic; surfactant reducing surf. tens., wetting agent, flow and leveling agent, slip and mar aid for solv. and water based coatings and inks based on alkyds, epoxy, urethanes, acrylics; lt. amber clear liq.; sp.gr. 0.94-0.95; dens. 7.8-7.9 lb/gal; flash pt. (PMCC) 66 C; usage level: 0.1-0.3%.

Drewfax® 420. [Drew Ind. Div.] Blend of silicone derivs. and glycol ethers; nonionic; surfactant reducing surf. tens., antifoam, flow and leveling agent, slip and mar aid for solv. and water-based coatings and inks based on alkyds, epoxy, urethanes, acrylics; improves pigment and substrate wetting; lt. amber clear liq.; sp.gr. 0.96-0.965; dens. 8.0 lb/gal; flash pt. (PMCC) > 54 C; usage level: 0.1-0.3%.

Drewfax® 680. [Drew Ind. Div.] Silicone surfactant; nonionic; surfactant, antifoam, flow and leveling agent, air release agent for solv.-based coatings and inks based on alkyd melamine, epoxy, urethanes, acrylics; esp. suited for wood coatings; off-wh. clear liq.; disp. in water; sp.gr. 1.03; visc. 1200 cps; usage level: 0.1-0.3%.

Drewfax® 800. [Drew Ind. Div.] Surfactant, anticratering agent used in combination with silicone defoamers to maximize defoaming while minimizing film defects.

Drewfax® 818. [Drew Ind. Div.] Blend of silicone copolymers and glycol ether; nonionic; surfactant, wetting agent, flow and leveling agent, slip and mar aid, anticrater for water-based coatings and inks based on acrylics, alkyds; lt. amber clear liq.; sp.gr. 0.95; flash pt. (TCC) 60 C; usage level: 0.05-0.5%.

Drewfax® S-600. [Drew Ind. Div.] Silicone/silica derivs.; nonionic; surfactant, antifoam, flow and leveling agent for solv. and water-based coatings and inks based on alkyds, epoxy, urethanes, acrylics; esp. suited for clear lacquers, pigmented coatings, uv coatings; off-wh. opaque liq.; emulsifiable in water; sp.gr. 1.0; visc. 1500 cps; usage level: 0.1-0.3%.

Drewfax® S-700. [Drew Ind. Div.] Sodium dioctyl sulfosuccinate; anionic; surfactant, anticrater, dispersant, flow and leveling agent for high-solids and acrylic solv.-based inks and coatings; FDA approved; colorless clear visc. liq.; sol. 6.8 g/100 g water (30 C); sp.gr. 1.07-1.09; visc. 200-600 cps; pH 5.0-6.0 (1%); usage level: 0.05-0.3%; 75% solids.

Drewfax® S-800. [Drew Ind. Div.] Silicone surfactants and solvs. blend; nonionic; surfactant, antifoam, dispersant, flow and leveling agent for solv. and water-based coatings and inks based on alkyds, epoxy, urethanes, acrylics; premix vehicle and carrier; pigment wetter; clear liq.; sp.gr. 0.99; visc. 112 cps; usage level: 0.1-0.2%.

Drewmulse® DGMS. [Aquatec Quimica SA] Diethylene glycol stearate; CAS 9004-99-3; nonionic; emulsifier for w/o and o/w systems, opacifier and pearlescence agent for shampoos, additive for pigment milling; flakes; 85% act.

Drewmulse® EGDS. [Aquatec Quimica SA] Ethylene glycol distearate; nonionic; emulsifier for w/o and o/w systems, opacifier and pearlescence agent for shampoos, additive for pigment milling; flakes; 48% act.

Drewmulse® EGMS. [Aquatec Quimica SA] Ethylene glycol stearate; nonionic; emulsifier for w/o and o/w systems, opacifier and pearlescence agent for shampoos, additive for pigment milling; flakes; 48% act.

Drewmulse® GMRO. [Aquatec Quimica SA] Glyceryl ricinoleate; nonionic; emulsifier for w/o and o/w systems, opacifier and pearlescence agent for shampoos, additive for pigment milling; liq.

Drewmulse® GMS. [Aquatec Quimica SA] Glyceryl stearate; nonionic; emulsifier for w/o and o/w systems, opacifier and pearlescence agent for shampoos, additive for pigment milling; flakes; 40–45% act.

Drewmulse® PGML. [Stepan/PVO] Propylene glycol laurate; chemical specialty; liq.

Drewmulse® PGMS. [Aquatec Quimica SA] Propylene glycol stearate; nonionic; emulsifier for w/o and o/w systems, opacifier and pearlescence agent for shampoos, additive for pigment milling; flakes; 48% act.

Drewmulse® POE-SML. [Stepan/PVO] Polysorbate 20; CAS 9005-64-5; nonionic; w/o emulsifier for cosmetics and pharmaceuticals, solubilizer, dispersant, wetting agent, detergent, visc. control agent; liq.; HLB 15.1; sapon. no. 40–51; 100% conc.

Drewmulse® POE-SMO. [Stepan/PVO] Polysorbate 80; CAS 9005-65-6; nonionic; w/o emulsifier for cosmetics and pharmaceuticals, solubilizer, dispersant, wetting agent, detergent, visc. control agent; liq.; HLB 15.0; sapon. no. 45–55.

Drewmulse® POE-SMS. [Stepan/PVO] Polysorbate 60; CAS 9005-67-8; nonionic; w/o emulsifier for cosmetics and pharmaceuticals, solubilizer, dispersant, wetting agent, detergent, visc. control agent; solid; HLB 14.9; sapon. no. 45–55.

Drewmulse® POE-STS. [Stepan/PVO] Polysorbate 65; CAS 9005-71-4; nonionic; w/o emulsifier for cosmetics and pharmaceuticals, solubilizer, dispersant, wetting agent, detergent, visc. control agent; solid; HLB 10.5; sapon. no. 88–98.

Drewmulse® SML. [Stepan/PVO] Sorbitan laurate; CAS 1338-39-2; nonionic; w/o emulsifier for cosmetics and pharmaceuticals, solubilizer, dispersant, wetting agent, detergent, visc. control agent; liq.; HLB 8.6; sapon. no. 159–171; 100% conc.

Drewmulse® SMO. [Stepan/PVO] Sorbitan oleate; CAS 1338-43-8; nonionic; w/o emulsifier for cosmetics and pharmaceuticals, solubilizer, dispersant, wetting agent, detergent, visc. control agent; liq.; HLB 4.7; sapon. no. 144–156.

Drewmulse® SMS. [Stepan/PVO] Sorbitan stearate; CAS 1338-41-6; nonionic; w/o emulsifier for cosmetics and pharmaceuticals, solubilizer, dispersant, wetting agent, detergent, visc. control agent; solid; HLB 2.1; sapon. no. 147–157.

Drewmulse® STS. [Stepan/PVO] Sorbitan tristearate; CAS 26658-19-5; nonionic; w/o emulsifier for cosmetics and pharmaceuticals, solubilizer, dispersant, wetting agent, detergent, visc. control agent; solid; HLB 2.1; sapon. no. 170–190.

Drewplus® L-108. [Drew Ind. Div.] Blend of min. oils, emulsifiers, silica derivs.; defoamer for latex/rubber applics. esp. monomer stripping, acrylic, PVAc, NBR, SBR, PVC; FDA compliance; straw-colored opaque liq.; emulsifiable in water; sp.gr. 0.89; visc. 800 cps; pour pt. 9 C; flash pt. (PMCC) 93 C; usage level: 0.01-0.10%; 100% act.

Drewplus® L-121. [Drew Ind. Div.] Silica deriv.; defoamer; emulsifiable in water.

Drewplus® L-123. [Drew Ind. Div.] Nonsilicone defoamer; defoamer for latexes; FDA compliance; emulsifiable in water; sp.gr. 0.92; dens. 7.68 lb/gal.

Drewplus® L-131. [Drew Ind. Div.] Proprietary blend of min. oil, silica derivs. and surfactants; defoamer for latex/rubber applics. (SBR, PVAC, PVA, NBR, acrylic), esp. for paper coating and carpet backing; FDA compliance; yel. opaque liq.; disp. in water; sp.gr. 0.92; visc. 600 cps; pour pt. -4 C; flash pt. (PMCC) > 93 C.

Drewplus® L-139. [Drew Ind. Div.] Blend of min. oil, silica deriv., surfactants; defoamer for latexes (SBR, NBR, PVC, PVAc, PVA, SAN, polybutadiene, vinyl acrylic), monomer stripping, degassing, adhesives; FDA compliance; off-wh. translucent liq.; emulsifiable in water; sp.gr. 0.89; dens. 7.43 lb/gal; visc. 400 cps; pour pt. 8 C; flash pt. (PMCC) 94 C.

Drewplus® L-140. [Drew Ind. Div.] Blend of minerals, silica derivs., and surfactants; defoamer for latex emulsions, adhesives, inks, paint and industrial coatings; FDA compliance; gray-wh. liq.; disp. in water; sp.gr. 0.88; dens. 7.34 lb/gal; visc. 1150 cps.

Drewplus® L-156. [Drew Ind. Div.] Disp. of alcohols, fatty soaps, and surfactants; defoamer preventing air entrainment during latex monomer stripping in SBR cold emulsion polymerization, PVAC, PVA, and acrylic latexes, aq. effluent systems; FDA compliance; wh. creamy liq.; emulsifiable in water; sp.gr. 0.97; dens. 8.09 lb/gal; visc. 500 cps; pour pt. 4.4 C; flash pt. (PMCC) 93 C.

Drewplus® L-162. [Drew Ind. Div.] Silicone defoamer; defoamer for latexes; FDA compliance; emulsifiable in water; sp.gr. 1.01; dens. 8.43 lb/gal.

Drewplus® L-191. [Drew Ind. Div.] Blend of min. oils, surfactants, silica derivs.; defoamer for latexes (SBR, NBR, acrylic, vinyl acrylic); FDA compliance; straw-colored opaque liq.; disp. in water; sp.gr. 0.92; dens. 7.68 lb/gal; visc. 800 cps; pour pt. 0 C.

Drewplus® L-198. [Drew Ind. Div.] Blend of min. oils, silica derivs., and esters; defoamer for polymerization, latex stripping (SBR, PVC, EPDM, PVDC), paper coatings; FDA compliance; straw-colored opaque liq.; emulsifiable in water; sp.gr. 0.95; dens. 7.93 lb/gal; visc. 800 cps; pour pt. 10 C; flash pt. (PMCC) 148 C; usage level: 0.01-0.5%.

Drewplus® L-206. [Drew Ind. Div.] Blend of surfactants and solvs.; defoamer during mfg. of pigments such as titanium dioxide, calcium carbonate and clay, and in paper mill operation; FDA compliance; amber clear liq.; disp. in water; sp.gr. 0.915; visc. 250 cps; pH 6.5 (1% emulsion); usage level: 0.01-0.1%.

Drewplus® L-404. [Drew Ind. Div.] Modified polysiloxane copolymer; defoamer/antifoam for gloss and semigloss architectural paints and waterborne coatings without adversely affecting gloss props., and for gravure and flexographic inks; yel. translucent visc. liq., mild odor; insol. in water, disp. in surfactant systems; sp.gr. 1.00; visc. 1000 cps; flash pt. (PMCC) > 200 F.

Drewplus® L-405. [Drew Ind. Div.] Silicone/silica deriv.; nonionic; defoamer, flow and leveling agent, air release agent for clear lacquers and pigmented coatings, inks (alkyd melamine, epoxy, urethanes, acrylics, high solids, water and solv. systems); off-wh. opaque; emulsifiable in water; sp.gr. 0.98; dens.

8.18 lb/gal; pour pt. 0 C; flash pt. (PMCC) 60 C.

Drewplus® L-407. [Drew Ind. Div.] Modified polysiloxane copolymer emulsion; foam control agent for architectural paints and coatings (interior/exterior gloss and semigloss paints, waterborne industrial coatings, polymer emulsions, water-thinnable gravure and flexographic inks); milky wh. liq., mild odor; misc. with water; sp.gr. 1.0; visc. 200 cps.

Drewplus® L-418. [Drew Ind. Div.] Blend of silica derivs., hydrocarbons, and silicones; defoamer for paints/coatings, ink, adhesives; FDA compliance; off-wh. opaque liq.; disp. in water; sp.gr. 0.98; dens. 8.18 lb/gal; visc. 1250 cps; pour pt. 6 C; usage level: 1-3 lb/100 gal.

Drewplus® L-419. [Drew Ind. Div.] Blend of silica derivs., hydrocarbons, and silicones; defoamer for latex paints, water-reducible coatings, inks, adhesives, aq. systems; FDA compliance; off-wh. opaque liq.; disp. in water; sp.gr. 0.96; dens. 8.01 lb/gal.

Drewplus® L-424. [Drew Ind. Div.] Blend of glycol derivs., silica derivs., surfactants; foam control agent for aq. printing inks, overprint varnishes, lacquers, high gloss paints, industrial coatings; FDA compliance; off-wh. opaque liq.; sp.gr. 1.04-1.06; dens. 8.6-8.8 lb/gal; visc. 2000-2500 cps.

Drewplus® L-435. [Drew Ind. Div.] Blend of silica derivs., min. oils, glycols, and esters; nonionic; foam control agent for interior and exterior gloss and semigloss low VOC enamels; FDA compliance; off-wh. opaque liq.; emulsifiable in water; sp.gr. 0.921; visc. 230 cps.

Drewplus® L-464. [Drew Ind. Div.] Blend of silica and organic solids; surfactant, defoamer, air release agent effective in water-reducible coatings such as polyurethanes, interior and exterior paints; FDA compliance; off-wh. opaque liq.; disp. in surfactant systems; sp.gr. 0.88; dens. 7.34 lb/gal; visc. 1400 cps.

Drewplus® L-468. [Drew Ind. Div.] Blend of silica and organic hydrophobic solids in a min. oil carrier; foam control agent for latex paints based on vinyl acrylic, acrylic and terpolymer emulsions, aq. industrial coatings (base coats for wood, metal primers), aq. adhesives; FDA compliance; off-wh. opaque liq.; disp. in surfactant systems; sp.gr. 0.88; dens. 7.2-7.4 lb/gal.

Drewplus® L-474. [Drew Ind. Div.] Blend of alcohols, silica derivs., surfactants; defoamer for latexes (PVC, PVAc, styrene acrylic, NR), adhesives, dyestuffs, paper coatings, paints; FDA compliance; yel. opaque liq.; disp. in water; sp.gr. 0.94; dens. 7.84 lb/gal; visc. 320 cps; pour pt. 0 C.

Drewplus® L-475. [Drew Ind. Div.] Blend of min. oils and silica derivs.; nonionic; defoamer for paints/coatings, latexes (acrylic, PVAc, styrene acrylic, PVDC), adhesives, paper coatings; FDA compliance; off-wh. opaque liq.; disp. in surfactant systems; sp.gr. 0.91; dens. 7.59 lb/gal; visc. 950 cps.

Drewplus® L-477. [Drew Ind. Div.] Blend of min. oils and silica derivs.; all-purpose defoamer for interior and exterior paints incl. gloss, semigloss, and flat paints; FDA compliance; lt. amber liq.; insol. in water; disp. in surfactant systems; sp.gr. 0.90; visc. 1100 cps; pour pt. 0 C; usage level: 1-3 lb/100 gal.

Drewplus® L-483. [Drew Ind. Div.] Blend of organics and hydrocarbons; nonionic; defoamer for paints/coatings, adhesives, inks; FDA compliance; yel. opaque liq.; insol. in water; disp. in surfactant systems; sp.gr. 0.87; dens. 7.26 lb/gal; pour pt. 0 C.

Drewplus® L-493. [Drew Ind. Div.] Blend of min. oils, silica derivs., emulsifiers; defoamer for latex paints, water-reducible industrial coatings, inks, adhesives; suitable for acrylic, vinyl acrylic, and PVAc emulsions; FDA compliance; off-wh. opaque liq.; emulsifiable in water; sp.gr. 0.90; dens. 7.51 lb/gal; visc. 1400 cps; pour pt. 0 C; flash pt. 149 C.

Drewplus® L-496. [Drew Ind. Div.] Blend of min. oil, silica derivs., and surfactants; foam control agent for latex paints, emulsion and water-reducible coatings, aq. inks; suitable for acrylic, vinyl acrylic and styrene acrylic emulsions; FDA compliance; off-wh. opaque liq.; emulsifiable in water; sp.gr. 0.90; dens. 7.4-7.6 lb/gal.

Drewplus® L-523. [Drew Ind. Div.] Blend of fatty oils, surfactants, and silica derivs.; defoamer for latexes (SBR), food/fermentation applics.; FDA compliance; off-wh. opaque liq.; disp. in water; sp.gr. 0.99; dens. 8.26 lb/gal; visc. 800 cps; pour pt. 10 C.

Drewplus® L-722. [Drew Ind. Div.] Silicone defoamer; defoamer for latexes, food/fermentation applics.; FDA compliance; emulsifiable in water; sp.gr. 1.01; dens. 8.43 lb/gal.

Drewplus® L-768. [Drew Ind. Div.] Blend of silicone fluid, silica derivs. and surfactants; defoamer for industrial/chemical processes, food/fermentation and agric. applics., aq. and some nonaq. systems; FDA and EPA compliance; off-wh. opaque liq.; disp. in water; sp.gr. 1.02; dens. 8.51 lb/gal; visc. 1100 cps; pour pt. 5 C; flash pt. (COC) 93 C.

Drewplus® L-790. [Drew Ind. Div.] Blend of min. oils, silica derivs., and emulsifiers; defoamer for food/fermentation and agric. applics.; FDA and EPA compliance; off-wh. opaque liq.; disp. in water; sp.gr. 0.90; dens. 7.51 lb/gal; visc. 1400 cps; pour pt. 0 C; flash pt. > 149 C.

Drewplus® L-810. [Drew Ind. Div.] Blend of min. oils and glycols; foam control agent for fermentation and food processing; FDA compliance; lt. yel. clear to hazy liq.; disp. in water; sp.gr. 0.90; visc. 75 cps; flash pt. 77 C; usage level: 200 ppm.

Drewplus® L-813. [Drew Ind. Div.] Blend of dimethylpolysiloxane, silica, and emulsifiers; nonionic; foam control agent for industrial food processing (starch slurries, fermentation, calcium chloride brines, adhesives, glues, veg. and fruit processing, sugar, instant coffee, fruit juices, dehydrating and evaporating systems); FDA compliance; off-wh. opaque liq.; dilutable in water; sp.gr. 1.00; visc. 400 cps; 10% silicone.

Drewplus® L-833. [Drew Ind. Div.] Blend of dimethylpolysiloxane, silica, emulsifiers; nonionic; defoamer for food and industrial processing operations at dosages as low as 2 ppm (fermentation, cheese whey, starch processing, adhesives, food container coatings, fruit and veg. processing, dehydration, evaporation); effective over wide pH range and under high shear and high salt conditions; FDA compliance; wh. opaque liq.; emulsifiable in water; sp.gr. 1.01; dens. 8.43 lb/gal; visc. 1200 cps; usage level: 2-33 ppm; 30% silicone.

Drewplus® M-111. [Drew Ind. Div.] Blend of silica derivs. and surfactants; defoamer for gypsum, starch, cement, pigments, joint compd. and adhesives that require a dry defoamer; wh. free-flowing powd.; disp. in water; bulk dens. 16.2 lb/ft^3; pH 6.5 (1%); usage level: 0.1% starting pt.

Drewplus® Y-125. [Drew Ind. Div.] Blend of min. oils and fatty acid derivs.; defoamer for controlling entrained air during latex stripping, paints/coatings mfg.; suitable for SBR, PVAc, PCA emulsions, ACN resin systems; FDA compliance; tan opaque liq.; emulsifiable in water; sp.gr. 0.93; dens. 7.76 lb/gal; visc. 2200 cps; pour pt. 0 C; flash pt. (PMCC) 93 C.

Drewplus® Y-166. [Drew Ind. Div.] Blend of emulsifiable min. oils and silica derivs.; defoamer for latex/rubber applics. incl. SBR, PVA, acrylic, vinyl acrylic, and for syn. and natural adhesive formulations; FDA compliance; yel. cloudy liq.; disp. in water; sp.gr. 0.95; visc. 1500 cps; pour pt. -17.8 C; usage level: 0.05-0.10%.

Drewplus® Y-200. [Drew Ind. Div.] Nonsilicone defoamer; defoamer for paints/coatings, inks, adhesives; emulsifiable in water; sp.gr. 0.92; dens. 7.68 lb/gal.

Drewplus® Y-250. [Drew Ind. Div.] Blend of min. oils, silica derivs., and esters; defoamer for aq. industrial systems requiring quick foam knockdown, long lasting foam prevention, paints/coatings, latex (SBR, acrylic, PVAc, PVA, styrene acrylic, PVDC, vinyl acrylic), inks, adhesives, cutting oils, food/fermentation applics.; off-wh. opaque liq.; disp. in water; sp.gr. 0.91; dens. 7.59 lb/gal; visc. 1800 cps; usage level: 0.02-0.5%.

Drewplus® Y-281. [Drew Ind. Div.] Blend of min. oils, silica derivs., and surfactants; defoamer for paints, latex/rubber, industrial coatings, adhesives; FDA compliance; off-wh. liq.; disp. in water; sp.gr. 0.93; dens. 7.76 lb/gal; visc. 1000 cps; pour pt. 0 C.

Drewplus® Y-381. [Drew Ind. Div.] Blend of organics and hydrocarbons; defoamer/antifoam for latex paints (acrylic, vinyl acrylic, S/B), coatings, ink, adhesives; FDA compliance; off-wh. liq.; insol. in water; disp. in surfactant systems; sp.gr. 0.87; dens. 7.26 lb/gal; pour pt. 0 C.

Drewplus® Y-530. [Drew Ind. Div.] Silica deriv.; defoamer; nondisp. in water.

Drewplus® Y-601. [Drew Ind. Div.] Blend of min. oils, organic solids, surfactants; defoamer for latex (SBR, PVAc), paints/coatings, adhesives; off-wh. opaque liq.; disp. in water; sp.gr. 0.83; dens. 6.93 lb/gal; visc. 400 cps; pour pt. 0 C.

Drewpol® 3-1-SK. [Stepan/PVO] Polyglyceryl-3 stearate; nonionic; emulsifier; amber beads; HLB 7.0; 100% conc.

Drewpol® 10-4-OK. [Stepan/PVO] Decaglyceryl tetraoleate; emulsifier; amber liq.; HLB 6.0; sapon. no. 125–145; 100% conc.

Drewpol® HL-13788. [Stepan/PVO] Hexaglyceryl ester; nonionic; crystallization inhibitor; solid; HLB 3.2; 100% conc.

Drewsperse® S-825. [Drew Ind. Div.] High m.w. polymer; nonionic; wetting agent, dispersant, filmformer for water-based acrylic, alkyd, styrene acrylic, epoxy, alkyd-melamine, or polyurethane coatings and inks; maximizes color development; stabilizes the dispersion against flocculation and settling; clear visc. liq.; sp.gr. 1.01; usage level: 1-2%.

Droxol 200. [Henkel/Textile] PEG-200; chem. intermediate for textiles, lubricants, printing, solvs., cleaning formulations, humectants, visc. modifiers; Gardner < 1 liq.; sol. @ 5% in water, xylene; disp. in Stod.; m.w. 200; dens. 9.4 lb/gal; visc. 26 cSt (100 F); pour pt. -12 C; flash pt. 340 F.

Droxol 400. [Henkel/Textile] PEG-400; chem. intermediate for textiles, lubricants, printing, solvs., cleaning formulations, humectants, visc. modifiers; Gardner < 1 liq.; sol. @ 5% in water, xylene; disp. in Stod.; m.w. 400; dens. 9.4 lb/gal; visc. 45 cSt (100 F); pour pt. 6 C; flash pt. > 350 F.

Droxol 600. [Henkel/Textile] PEG-600; chem. intermediate for textiles, lubricants, printing, solvs., cleaning formulations, humectants, visc. modifiers; Gardner < 1 liq.; sol. @ 5% in water, xylene; disp. in Stod.; m.w. 600; dens. 9.4 lb/gal; visc. 63 cSt (100 F); pour pt. 22 C; flash pt. 475 F.

Druspin LO-VIS. [Stepan/PVO] Blended prod.; textile lubricant, antistat, emulsifier; liq.; low visc.; 100% act.

Drymet®. [Rhone-Poulenc Basic] Sodium metasilicate, anhyd.; soap builder and detergent; ingredient in industrial cleaning compds.; granules; 100% conc.

Drynol E/20, E/30, E/40. [Rhone-Poulenc Geronazzo] Calcium dodecylbenzene sulfonate with ethoxylated alkylphenols and stabilizing agent blend; nonionic/anionic; emulsifier for pesticide concs.; thick liq.; 65%, 70%, and 75% conc.

Drysperse 401. [Witco Israel] Polyoxyalkylene glycols blended with dispersing and anticaking agents; dispersant for wettable-powd. insecticides, herbicides, and fungicide formulations; powd.

Drysperse 801. [Witco Israel] Alkylaryl polyethoxyethanol; wetting agent and dispersant for wettable-powd. insecticides, herbicides, and fungicide formulations; powd.

Drysperse 902. [Witco Israel] Alkylaryl polyoxyalkylated alcohol; defoaming dispersant for wettable-powd. insecticides, herbicides, and fungicide formulations; powd.

Drysperse 908H. [Witco Israel] Polyoxyalkylene glycols blended with dispersants and anticaking resins; dispersant for wettable-powd. insecticides, herbicides, and fungicide formulations; powd.

Dulceta N25. [ICI Surf. UK] Pourable polyethylene emulsion; nonionic; used to improve the sewability and tear strength of resin-treated fabrics; liq.

Duofol T. [Hart Chem. Ltd.] Sulfated ester; anionic; wetting, rewetting agent, dispersant, penetrant, leveling agent, finishing agents for textile wet processing of cotton piece goods; stable to mild alkalies and org. acids; amber liq.; clear sol. 5% water; sp.gr. 1.068; visc. 70 cps; 65% conc.

Duomac® C. [Akzo; Akzo Chem. BV] Coco-1,3-diaminopropane diacetate; cationic; emulsifier, corrosion inhibitor, flotation agent, bactericide in drilling fluids; sol. in IPA, sparingly sol. in water; sp.gr. 0.85; HLB 11.2; m.p. 20–24 C; 89% act.

Duomac® T. [Akzo; Akzo Chem. BV] N-Tallow-1,3-propanediamine diacetate; cationic; emulsifier, corrosion inhibitor, flotation reagent; pigment flushing, flocculation; brn. paste; vinegar odor; m.w. 480; sol. in water, chloroform, ethanol, IPA; sp.gr. 0.892 (90 C); HLB 10.7; m.p. 181 F; b.p. > 300 C; flash pt. 335 F (COC); 95% conc.

Duomeen® C. [Akzo; Akzo Chem. BV] N-Coco-1,3-diaminopropane; CAS 61791-63-7; cationic; chemical intermediate; corrosion inhibitor, fuel oil additive, flotation agent; used in metals, textiles, plastics, herbicides; epoxy curing agent; dk. amber. liq.; ammonia odor; sol. in naphtha, min. oil, IPA; m.w. 276; sp.gr. 0.836; visc. 10 cP; m.p. 71 F; b.p. > 300 C; iodine no. 8; amine no. 410; flash pt. (PMCC) > 149 C; 100% conc.

Duomeen® CD. [Akzo] Coco-1,3 diaminopropane; CAS 61791-63-7; cationic; industrial surfactant; intermediate; Gardner 3 max. liq.; sp.gr. 0.84; visc.

15 cps; m.p. 20-25 C; iodine no. 8; amine no. 410; flash pt. (PMCC) > 149 C; 89% conc.

Duomeen® HT. [Akzo Chem. BV] N-Alkyl diamine; cationic; bitumen emulsifier, corrosion inhibitor, oil additive, antisettling agent, paint formulations; solid; 90% conc.

Duomeen® LT-4. [Akzo] 3-Tallowalkyl-1,3-hexahydropyrimidine; cationic; industrial surfactant; EPA listed; Gardner 8 max. liq.; sp.gr. 0.87; m.p. 12 C; iodine no. 30; amine no. 281; flash pt. (PMCC) > 121 C.

Duomeen® O. [Akzo; Akzo Chem. BV] N-Oleyl-1,3-propanediamine; cationic; bitumen emulsifier, corrosion inhibitor, oil additive, antisettling agent, paint formulations; yel. paste; ammonia odor; m.w. 350; sol. in naphtha, IPA; sp.gr. 0.841; visc. 19 cP; m.p. 59 F; b.p. > 300 C; flash pt. 370 F (COC); 89% min. act.

Duomeen® OL. [Akzo] N-Oleyl-1,3-diaminopropane; CAS 7173-62-8; cationic; industrial surfactant; Gardner 10 max. liq.; sp.gr. 0.84; iodine no. 70; amine no. 321; flash pt. (PMCC) > 149 C.

Duomeen® OTM. [Akzo] N,N,N′-Trimethyl-N′-9-octadecenyl-1,3-diaminopropane; CAS 68715-87-7; cationic; industrial surfactant; Gardner 8 max. liq.; sp.gr. 0.85; m.p. 18-25 C; iodine no. 55; amine no. 281; flash pt. (PMCC) > 149 C.

Duomeen® T. [Akzo; Akzo Chem. BV] Tallow-1,3-diaminopropane; CAS 61791-55-7; cationic; bitumen emulsifier, corrosion inhibitor, oil additive, antisettling agent; textile finishing agent, dispersant for inorg. pigments in paints, larvacidal oil additive; dk. amber paste; ammonia odor; sol. in naphtha, IPA; m.w. 330; sp.gr. 0.841; visc. 880 cP; m.p. 115 F; b.p. > 300 C; iodine no. 37; amine no. 334; flash pt. (PMCC) > 149 C; 100% conc.

Duomeen® TDO. [Akzo; Akzo Chem. BV] N-Tallow-1,3-propanediamine dioleate; CAS 61791-53-5; cationic; bitumen emulsifier, corrosion inhibitor, oil additive, antisettling agent; also in metal treatment as film and boundary lubricant, metal drawing additive; dispersant in paint industry; dk. semisolid; ammonia odor; sol. in naphtha, min. oil, IPA; insol. in water; m.w. 924; sp.gr. 0.865 (65 C); visc. 19 cps; m.p. 77 F; b.p. > 300 C; iodine no. 70; amine no. 120; flash pt. (PMCC) > 149 C; 33% min. act.

Duomeen® TTM. [Akzo] N,N,N′-Trimethyl-N′-tallow-1,3-diaminopropane; CAS 68783-25-5; cationic; industrial surfactant; Gardner 8 liq.; sp.gr. 0.83; visc. 20 cps; m.p. 10-25 C; iodine no. 25; amine no. 271; flash pt. (PMCC) > 149 C.

Duomeen® TX. [Akzo] Tallow-1,3-diaminopropane; emulsifier for bitumen emulsions, base material; antisettling agent for paints; Gardner 7 solid; ammoniacal odor; water insol.; sp.gr. 0.84; m.p. 46–55 C; flash pt. 180 C; 92% act.

Duoquad® O-50. [Akzo] N,N,N′,N′,N′-Pentamethyl-N-octadecenyl-1,3-diammonium dichloride, aq. IPA; CAS 68310-73-6; industrial surfactant; Gardner 6 max. liq.; sol. in water, alcohols, chloroform, CCl_4; flash pt. (PMCC) 15 C; pH 6–9; 50% quat. in aq. IPA.

Duoquad® T-50. [Akzo] N,N,N′,N′,N′-Pentamethyl-n-tallow-1,3-propanediammonium dichlorides, aq. IPA; CAS 68607-29-4; cationic; detergent, corrosion inhibitor, metal cleaner; emulsifier for sec. oil recovery; Gardner 13 max. liq.; sol. in water, alcohols, chloroform, CCl_4; m.w. 480; sp.gr. 0.90; HLB 14.4; f.p. –20 C; flash pt. (PMCC) 15 C; pH 6–8 (10% aq.); 48–52% act. in aq. IPA.

Duoteric MB1, MB2. [Rhone-Poulenc UK] Complex blend; anionic/nonionic; matched pair emulsifiers for agrochem. formulations; 100% conc.

Duponol® 80. [Witco/Organics] Sodium n-octyl sulfate, tech.; anionic; wetting agent, dispersant, penetrant, softening agent, emulsifier used for textile, metal, and leather processing; lt. yel. liq.; m.w. 232; sol. in presence of many multivalent ions; visc. 30 cps; surf. tens. 51.7 dynes/cm (0.1%); 33–35% act.

Duponol® D Paste. [Witco/Organics] Sodium lauryl oleyl sulfate, tech.; anionic; detergent, emulsifier, wetting agent, dispersant, dyeing assistant used for dyes, textile, metal, and leather processing; lt. yel. paste; m.w. 328; surf. tens. 30.5 dynes/cm (0.1%); 38% act.

Duponol® EP. [Witco/Organics] DEA-lauryl sulfate, tech.; anionic; foaming agent, detergent, wetting agent, emulsifier used for cosmetics, personal care prods.; very pale golden, slightly visc. liq.; bland and clean odor; sol. in water, polar solv. with some electrolyte precipitation; dens. 8.4 lb/gal; visc. 50–150 cps ; cloud pt. 5 C; 33-36% act.

Duponol® FAS. [Witco/Organics] Ethoxylated alkyl sulfate, nonionic surfactant; anionic; detergent; liq.

Duponol® G. [Witco/Organics] Fatty alcohol amine sulfate; anionic; emulsifier, wetting agent; yel. visc. translucent paste; mild fatty penetrating odor; sol. in hydrocarbons, 50% in water; dens. 8.2 lb/gal; surf. tens. 29.6 dynes/cm (0.1%); 92% act.

Duponol® LS Paste. [Witco/Organics] Sodium oleyl sulfate tech.; anionic; emulsifier, scouring agent in textile and leather industries, rewetting agent for paper; lt. tan paste; m.w. 355; surf. tens. 37.8 dynes/cm (0.1%); 26% act.

Duponol® ME Dry. [Witco/Organics] Sodium lauryl sulfate, tech.; CAS 151-21-3; anionic; detergent, wetting agent, foaming agent, emulsifier for general industrial uses; wh. to med. cream powd.; m.w. 302; 30% sol. in water; dens. 3.3 lb/gal; cloud pt. 70 F (1% aq.); surf. tens. 35 dynes/cm (0.1%); 90–96% act.

Duponol® QC. [Witco/Organics] Sodium lauryl sulfate; CAS 151-21-3; anionic; detergent, shampoo base; water wh. slightly visc. liq.; bland clean, char. odor; sol. in water, polar solv. with some electrolyte precipitation; visc. < 500 cps; cloud pt. 65 F; 29–30% act.

Duponol® RA. [Witco/Organics] Fortified sodium ether-alcohol sulfate; anionic; detergent, dispersant, emulsifier, penetrating, leveling, and wetting agent; textile processing; lt. amber clear liq.; misc. with water; sp.gr. 1.04; dens. 8.7 lb/gal; visc. 178 cps (80 F); cloud pt. 68 F; flash pt. > 212 F (TOC); surf. tens. 33 dynes/cm (0.1%).

Duponol® SP. [Witco/Organics] Sodium alcohol sulfate; anionic; wetting agent, dispersant, emulsifier for high electrolyte sol'ns.; liq.

Duponol® WA Dry. [Witco/Organics] Sodium lauryl sulfate; CAS 151-21-3; anionic; detergent, emulsifier, wetting agent, penetrant; wh. powd.; water sol., the act. ingred. is sol. in highly polar solv.; dens. 2.6 lb/gal; cloud pt. 37 F (1%); surf. tens. 33 dynes/cm (0.1%); 44% conc.

Duponol® WA Paste. [Witco/Organics] Sodium lauryl sulfate; CAS 151-21-3; anionic; detergent, shampoo base, superior cleansing and foaming, textile scouring; liq.

Duponol® WAQ. [Witco/Organics] Sodium lauryl sulfate; CAS 151-21-3; anionic; emulsifier, detergent for emulsion polymerization, general detergency; liq.

Duponol® WAQE. [Witco/Organics] Sodium lauryl sulfate, tech.; CAS 151-21-3; anionic; emulsifier for emulsion polymerization; pale yel. liq.; m.w. 302; water sol., act. ingred. is sol. in polar solv. with some electrolyte precipitation; dens. 8.6 lb/gal; visc. 500 cps; cloud pt. 65 F; 29–30.5% act.

Duponol® WN. [Witco/Organics] Sodium octyl/decyl sulfate, tech.; anionic; emulsifier, wetting and penetrating agent for acid sol'ns.; dyeing assistant; pale yel. liq; m.w. 246; visc. 30 cps; surf. tens. 52.5 dynes/cm (0.1%); 33–35% act.

Dur-Em® 117. [Van Den Bergh Foods] Glyceryl stearate; nonionic; textile lubricant and finishing agent; emulsifier for cosmetic and pharmaceutical creams and lotions, foods; lubricant for thermoplastics; dispersant for inorg. pigments; bead/flake; HLB 2.8; m.p. 62–65 C; 100% conc.; 40% min. alpha monoglyceride.

Durfax® 60. [Van Den Bergh Foods] Polysorbate 60; CAS 9005-67-8; nonionic; food emulsifier; personal care prods.; preshave beard lubricant and softener prods.; paste; sol. in water; HLB 14.0; sapon. no. 45–55; 100% conc.

Durfax® 65. [Van Den Bergh Foods] Polysorbate 65; CAS 9005-71-4; nonionic; food emulsifier; pesticide dispersant; solid; water-disp.; HLB 10.5; sapon. no. 88–98; 100% conc.

Durfax® 80. [Van Den Bergh Foods] Polysorbate 80; CAS 9005-65-6; nonionic; food emulsifier; personal care prods.; antifog agent in plastics and aerosol furniture polish; liq.; water-sol.; HLB 15.0; sapon. no. 45–55; 100% conc.

Durfax® EOM. [Van Den Bergh Foods] PEG-20 glyceryl stearate; nonionic; food emulsifier; dough conditioner; lubricant and fabric softener; plastic; HLB 13.5; sapon. no. 65–75; 100% conc.

Durlac® 100W. [Van Den Bergh Foods] Glyceryl stearate lactate; nonionic; emulsifier for food; confectionery gloss enhancer; starch gelling agent in industrial processes; flake; HLB 2.4; m.p. 46–54 C; 100% conc.

Durtan® 20. [Van Den Bergh Foods] Sorbitan laurate; CAS 1338-39-2; nonionic; used in cosmetic, toiletry, and household prods.; fabric antistat and lubricant; HLB 7.4; sapon. no. 158–170.

Durtan® 60. [Van Den Bergh Foods] Sorbitan stearate; CAS 1338-41-6; nonionic; food emulsifier, gloss enhancer for chocolate coatings; dispersant for inorganics used in thermoplastics; beads; HLB 4.7; sapon. no. 147–157; 100% conc.

Durtan® 80. [Van Den Bergh Foods] Sorbitan oleate; CAS 1338-43-8; nonionic; used in water softening compds., auto polishes, pesticide formulations; antifog in plastics; rust inhibitor additive; ink pigment dispersant; HLB 4.3; sapon. no. 148–161.

Dusoran MD. [Solvay Duphar BV] Lanolin alcohols, dist.; CAS 8027-33-6; nonionic; w/o emulsifier, stabilizer, softener, emollient for absorption bases, cosmetics, cleansing preps.; lt. yel., waxy solid; sol, in min. oil, abs. and 95% alcohols, chloroform, ether, lt. petrol., toluene; acid no. 2.0; sapon. 8.0; 100% act.

DyaFac 6-S. [Henkel] PEG-600 stearate; CAS 9004-99-3; emulsifier and lubricant for syn. fibers; water-disp.

DyaFac 1926-B. [Henkel] POE glyceride; emulsifier and lubricant for syn. fibers; water-insol.

DyaFac LA9. [Henkel] PEG-400 laurate; CAS 9004-81-3; solubilizer and emulsifier for fats and oils; water-sol.

DyaFac PEG 6DO. [Henkel/Textile] PEG-12 dioleate; CAS 9005-07-6; nonionic; surfactant for textile use; Gardner 4 liq.; sol. @ 5% in min. oil, butyl stearate, glycerol trioleate, Stod., xylene; disp. in water; dens. 8.3 lb/gal; visc. 62 cSt (100 F); HLB 10.0; pour pt. 5 C; cloud pt. < 25 C; flash pt. > 200 F.

Dyasulf 1761-A. [Henkel] Sulfated castor oil; dispersant; wetting agent, emulsifier; water-sol.

Dyasulf 2031. [Henkel] Sulfated oleic acid sodium salt; wetting agent and penetrant in textile operations; detergent and penetrant in fabrics; amber liq.; dens. 1.09 g/ml; pH 5.3 (2% aq.); 60% act.

Dyasulf 9268-A. [Henkel/Textile] Sulfonated butyl naphthalene; anionic; surfactant for textile use; tan powd.; sol. @ 5% in water; cloud pt. 22 C; flash pt. 76 F.

Dyasulf BO-65. [Henkel/Textile] Sulfated butyl oleate; anionic; wetting and rewetting agent, detergent, dyebath lubricant for textile use; Gardner 12 liq.; sol. @ 5% in water, xylene; disp. in min. oil; dens. 8.5 lb/gal; visc. 62 cSt (100 F); pour pt. -1 C; cloud pt. 1 C; flash pt. 65 F.

Dyasulf C-70. [Henkel/Textile] Sulfated castor oil; anionic; wetting and rewetting agent, detergent, dyebath lubricant for textile use; Gardner 7 liq.; sol. @ 5% in water; disp. in min. oil; dens. 8.6 lb/gal; visc. 300 cSt (100 F); pour pt. -1 C; cloud pt. 1 C; flash pt. 70 F; 70% act.

Dyasulf OA-60. [Henkel/Textile] Sulfated oleic acid; anionic; wetting and rewetting agent, detergent, dyebath lubricant for textile use; Gardner 11 liq.; sol. @ 5% in water; disp. in min. oil; dens. 9.1 lb/gal; visc. 148 cSt (100 F); pour pt. 1 C; cloud pt. 4 C; flash pt. 60 F; 60% act.

Dymsol® 31-P. [Henkel] Anionic; surfactant; biodeg. mechanical and chemical stabilizer for latex; liq.; 55% act.

Dymsol® 38-C. [Henkel/Functional Prods.] Fatty acid sulfate, sodium salt; anionic; emulsifier for polymerization; dk. amber liq.; 70% act.

Dymsol® 2031. [Henkel/Functional Prods.; Henkel-Nopco] Sulfonated fatty acid, sodium salt; anionic; primary or sec. emulsifier for emulsion polymerization; dk. amber liq.; biodeg.; 60% act.

Dymsol® L. [Henkel-Nopco] PEG fatty ester; nonionic; stabilizer, wetting agent, visc. controller for latex and resin emulsions; biodeg.; liq.; 100% act.

Dymsol® LP. [Henkel/Functional Prods.] Alkylaryl sulfonate; anionic; emulsifier for polymerization, large particle size latices; clear yel. liq.; biodeg.; 33% act.

Dymsol® PA. [Henkel/Functional Prods.] Sulfated ester; anionic; primary emulsifier for emulsion polymerization; liq.; 75% solids.

Dymsol® S. [Henkel/Functional Prods.] Sulfated syn. sperm oil; anionic; emulsion breaker, solubilizer, coupling agent; fatliquors for tanning leather; greenish-brn. liq.; 75% conc.

Dyqex®. [Georgia-Pacific] Sodium lignosulfonate; anionic; dye dispersant extender; 46% liq.; 100% powd.

E

Eccoclean C-90. [Eastern Color & Chem.] Sulfated alcohol; anionic; detergent and emulsifier for general scouring; provides foaming and foam stability; liq.

Eccoclean CR-46. [Eastern Color & Chem.] Self-emulsifying solv. cleaner; added solvency for extra cleaning power; useful on difficult cleaning problems; water-disp.

Eccoclean RPW. [Eastern Color & Chem.] Self-emulsifying solv. cleaner; provides rapid penetration and solvency of difficult cleaning problems; combustible.

Ecco Cleaner SC. [Eastern Color & Chem.] Cleaner for printing screens, printing equip., and general cleaning; does not contain toxic aromatic solvs.

Ecco Defoamer Heavy. [Eastern Color & Chem.] Silicone defoamer; nonionic; antifoam for a wide variety of systems; wh. soft paste; disp. in water.

Ecco Defoamer KD-3. [Eastern Color & Chem.] Silicone defoamer; nonionic; defoamer for use in aq. systems; wh. soft smooth paste; disp. in water at 100 F; dens. 7.9 lb/gal; pH 7.2 ± 0.2; nonirritating, nontoxic.

Ecco Defoamer KD-22. [Eastern Color & Chem.] Silicone defoamer; nonionic; defoamer for quick foam knockdown for plant processing; gives moderate visc. and good stability.

Ecco Defoamer NS-07. [Eastern Color & Chem.] Self-emulsifying hydrocarbon nonsilicone defoamer; nonionic; antifoam for jet dyeing or finishing operations; produces immediate foam knockdown; lt. straw-colored turbid liq., faint perfumed odor; forms milky emulsion in R.T. water; dens. 7.1 lb/gal; flash pt. (OC) 140 F; pH 4.2 ± 0.3 (2%).

Ecco Defoamer NSD. [Eastern Color & Chem.] Nonsilicone self-emulsifying hydrocarbon; nonionic; antifoam for dyeing processes.

Ecco Defoamer S. [Eastern Color & Chem.] Petrol./silicone; defoamer for mills waste water; clear liq.

Ecco Defoamer SD-6. [Eastern Color & Chem.] Silicone defoamer; low-cost effective defoamer for water and sewage treatment plants.

Eccoful DL Conc. [Eastern Color & Chem.] Modified amide ethoxylate; anionic; detergent, scouring and fullingagent for wool and wool blends; softener; ruby clear visc. liq.; completely sol. in water; dens. 8.2 lb/gal; pH 9.0 ± 0.3 (2%).

Eccoful FC Conc. [Eastern Color & Chem.] Amide ethoxylate, modified; anionic; detergent and scouring agent for wool and wool blends; liq.; 65% conc.

Eccoful NMR. [Eastern Color & Chem.] Nonionic; highly conc. fulling agent providing good penetration and detergency in the fulling bath.

Eccolene OW. [Eastern Color & Chem.] Sulfated fatty ester; anionic; wetting aid, dispersant, emulsifier, penetrant and scouring agent for textile processing; liq.; 65% conc.

Ecco Leveler 700. [Eastern Color & Chem.] Anionic; dye leveling agent for use with wool, synthetics, blends to give full color value and even level dyeings.

Econol 61. [Essential Industries] Foam-stabilized anionic/nonionic complex of amide sulfonate; anionic; wetting agent, detergent for liq. dishwash base, high sudsing cleaners; biodeg.; liq.; 67% conc.

Econol 66. [Essential Industries] Fatty amide condensate; nonionic; wetting agent, detergent, thickener, emulsifier for liq. detergents; biodeg.; liq.; 100% conc.

Econol 606. [Essential Industries] Org. alkylaryl sulfonate; anionic; wetting agent, detergent, foamer, surface tension reducer for liq. formulations; biodeg.; crystal clear liq.; 60% conc.

Econol 628. [Essential Industries] Coconut oil amide; nonionic; wetting agent, base for all-purpose cleaners, thickener, foam stabilizer; biodeg.; light colored liq.; 100% conc.

Econol 2818. [Essential Industries] Phosphate ester; anionic; detergent, wetting and coupling agent; liq.; 100% conc.

Econol 2833. [Essential Industries] Phosphate ester; anionic; detergent, emulsifier, wetting agent for industrial alkaline cleaners; compat. in conc. electrolyte systems; liq.; 100% conc.

Econol B. [Essential Industries] Fatty alkanolamide; nonionic; wetting agent, detergent, emulsifier for floor and wall cleaners, heavy duty formulations; coupler; biodeg.; liq.; 90% conc.

Eccoscour CB. [Eastern Color & Chem.] Sodium alkylaryl sulfonate; anionic/nonionic; detergent, wetting agent, and emulsifier for textile scouring and dyeing applics.; liq.; 30% conc.

Eccoscour D-7. [Eastern Color & Chem.] Sulfate alkyl phenol ethoxylate; anionic; surfactant and scouring agent for textile fabric and yarns; prevents redeposition of soils, dyestuffs, or pigments; clear visc. liq.; completely sol. in water; dens. 8.8 lb/gal; pH 7.0 ± 1.0 (2%); 42% conc.

Eccoscour KG. [Eastern Color & Chem.] Complex alcohol, ethoxylated; nonionic; acid type cleaning agent with nonionic surfactant; detergent, penetrant, scouring agent for removal of graphite from knit goods; clear liq.; completely sol. in water; dens. 9.0 lb/gal; pH 2.5 ± 0.5 (2%).

Eccoscour OR. [Eastern Color & Chem.] Synthetic; nonionic; detergent, emulsifier, wetting agent, scouring agent for synthetics and blends, desizing operations; exc. soil removal props.; for pre- and after-dyeing scouring; clear colorless liq.; completely sol. in water; dens. 8.3 lb/gal; pH 4.0-4.5 (2%).

Eccoscour SNP. [Eastern Color & Chem.] Anionic; detergent for general heavy-duty scouring and removal of soil and oils; phosphate-free.

Eccosolv C-14. [Eastern Color & Chem.] Self-emulsifying solvent cleaner for cleaning and degreasing of machine parts.

Eccoterge 35-S. [Eastern Color & Chem.] Sodium alkylaryl sulfonate; anionic; detergent, emulsifier for textile scouring and general detergency; liq.; 35% conc.

Eccoterge 112. [Eastern Color & Chem.] Sulfated ethoxylate; nonionic; detergent and scouring agent for textiles; liq.; 50% conc.

Eccoterge 200. [Eastern Color & Chem.] PEG 400 ester; nonionic; wetting agent, emulsifier, dispersant for solvs. in aq. systems; liq.; sol. in aliphatic and aromatic solvs., disp. in water; 100% conc.

Eccoterge ASB. [Eastern Color & Chem.] Amine alkylaryl sulfonate; anionic; detergent, wetting agent, emulsifier, scouring agent for textile applics.; lt. amber visc. liq.; 60% conc.

Eccoterge Conc. [Eastern Color & Chem.] Fatty amide condensate; anionic; emulsifier, detergent, wetting agent, scouring agent for textiles and general use; liq.; 100% conc.

Eccoterge EO. [Eastern Color & Chem.] Ethoxylated alkylphenol; nonionic; detergent, wetting and dye leveling agent for fabrics; kier boiling assistant; emulsifier for min. oils; detergent for nontextile applics.; liq.; 100% conc.

Eccoterge EO-41B. [Eastern Color & Chem.] Modified ethoxylate; nonionic; scouring agent, penetrant for textile fiber processing and dyeing; liq.

Eccoterge EO-100. [Eastern Color & Chem.] Octylphenol ethoxylate; nonionic; detergent, wetting agent, scouring agent, dye leveling agent for wool, yarns, piece goods; detergent for cotton and syn. goods; kier boiling assistant; emulsifier for min. oils and grease; general-purpose detergent for nontextile use; clear liq., odorless; dens. 8.5 lb/gal; 100% conc.

Eccoterge EOX. [Eastern Color & Chem.] Syn. detergent and scouring agent, foamer, soil-removal aid; for scouring natural fiber and processing oil from syn. fibers.

Eccoterge MV Conc. [Eastern Color & Chem.] Fatty amino condensate; anionic; detergent, emulsifier, wetting and scouring agent for textile use; ruby/amber clear visc. liq.; completely sol. in water; dens. 8.3 lb/gal; pH 9.6 ± 0.2 (2%); 92% min. solids.

Eccoterge NF-2. [Eastern Color & Chem.] Nonionic; low foaming detergent and scouring agent for textile and nontextile applics., scouring after dyeing; clear amber liq.; dens. 9.0 lb/gal; pH 7.0 ± 0.2; 62% min. act.

Eccoterge S-35. [Eastern Color & Chem.] Sodium alkylaryl sulfonate; anionic; detergent, emulsifier, wetting/rewetting agent for textiles; 35% conc.

Eccoterge SCH. [Eastern Color & Chem.] Complexed ethoxylated amide; nonionic; low foaming detergent, wetting, scouring agent, and dyeing assistant for textiles; clear, sl. visc. liq.; dens. 9.1 lb/gal; pH 10.4 ± 0.3 (5%); 44% min. act.

Eccotex P. Conc. [Eastern Color & Chem.] Alkylaryl compd., modified; anionic; dispersant, penetrant used in textile industry; wetting of dyestuffs; gel; 60% conc.

Eccowet® LF Conc. [Eastern Color & Chem.] Sodium alkyl naphthalene sulfonate; anionic; low foaming wetting agent, dispersant used in dyeing applics.; stable to acid and alkaline media; deep amber clear liq.; sol. in water; dens. 10.0 lb/gal; pH 9.0 ± 0.5; 63% act.

Eccowet® W-50. [Eastern Color & Chem.] Sodium aliphatic ester sulfonate; anionic; wetting agent, penetrant, dispersant, solubilizer, emulsifier, detergent for textiles, metal processing, disinfectants, paints, pigments, wallpaper, rubber cements, adhesives, drycleaning detergents, topical pharmaceuticals, cosmetics; colorless liq.; misc. with water; sol. in alcohol, glycols, acetones, dilute electrolyte sol'ns. (5%); visc. 55 cps; pH 8.0 ± 0.3 (1%); 50% conc.

Eccowet® W-88. [Eastern Color & Chem.] Sulfonated organic ester; wetting agent, penetrant, dispersant, solubilizer, emulsifier, detergent for textiles, metal processing, disinfectants, paints, pigments, wallpaper, rubber cements, adhesives, drycleaning detergents, topical pharmaceuticals, cosmetics; amber visc. liq.; pH 8.0 ± 0.3 (1%); 21% act.

Eccowet® Y-50. [Eastern Color & Chem.] Nonionic; wetting agent, surfactant for textile applics.; compat. with most processing sytems; good penetration.

Edaplan LA 400. [Münzing Chemie GmbH] Silicone-free leveling agent for lacquers and varnishes.

Edaplan LA 402. [Münzing Chemie GmbH] Silicone-free leveling agent for lacquers, varnishes and aq. systems; water-sol. after neutralization.

Edaplan LA 410. [Münzing Chemie GmbH] Silicone compd.; leveling agent for lacquers and varnishes; anti-crater aid; improves scratch resistance, gloss, smoothness, slip.

Edaplan LA 411. [Münzing Chemie GmbH] Silicone compd.; leveling agent for lacquers, varnishes, aq. coatings, powd. coatings.

Edaplan LA 412. [Münzing Chemie GmbH] Silicone compd.; leveling and slip agent for lacquers and varnishes; high effect on surf. tens.; improves scratch resistance and slip.

Edaplan LA 413. [Münzing Chemie GmbH] Silicone compd.; leveling and slip agent for lacquers, varnishes, aq. coatings; improves scratch resistance.

Edaplan PL 450. [Münzing Chemie GmbH] Silicone-free leveling agent for powd. coatings; improves wetting of substrate; prevents craters, fish eyes, orange peeling; anti-float; powd.

Edaplan VP LA 420. [Münzing Chemie GmbH] Silicone-free polymer; deaerator and leveling agent for coil coating systems.

Edenor GMS. [Henkel Canada] Glyceryl stearate; nonionic; consistency factor for creams and liq. emulsions; flakes; 100% conc.

Edenor ST-1. [Henkel Canada] Stearic acid; CAS 57-11-4; anionic; triple pressed NF grade; intermediate; flakes; 100% conc.

Edunine CT. [ICI Surf. UK] Fatty acid ester; nonionic; nonyel., scorch resistant softening agent for applic. to polyester-cotton blends or 100% cellulose; liq.

Edunine F. [ICI Surf. UK] Ethoxylated ester; nonionic; softening and antistatic agent used as raising agent for polyamide fabrics; paste.

Edunine PA. [ICI Surf. UK] Fatty alcohol mixed esters; anionic; antistatic softening agent for use on syn. and natural fibers; substantive on wool and nylon; liq.

Edunine RWT. [ICI Surf. UK] Ethoxylated ester/

cationic softening agent blend; nonionic/cationic; hydrophilic softening agent that does not impair the absorbency of cotton towels; aids in abrasion resistance of woolen fabrics; liq.

Edunine S82. [ICI Surf. UK] Ethoxylated ester/cationic softener/lubricant blend; nonionic/cationic; water absorbent softening agent for cellulosic fibers; liq.

Edunine SC-L. [ICI Surf. UK] Fatty amine/lubricant wax blend; cationic; softening agent giving sewing and handle props. to fabrics; liq.

Edunine SE. [ICI Surf. UK] Polysiloxane elastomer containing amino and crosslinking functionality; nonionic/cationic; softener imparting soft, silky handle to all fibers; liq.

Edunine SE 1010. [ICI Surf. UK] Amino-functional silicone elastomer fatty acid condensate; cationic; softener for soft handle and lubrication props., improved sewability and brushing performance of woven and knitted cotton and cotton blends; liq.

Edunine SE 1060. [ICI Surf. UK] Amino-functional polysiloxane elastomer modified with fatty acid condensation prods.; cationic; softener for soft handle and stretch and recovery props., for knit goods and hosiery containing elastomeric fibers; liq.

Edunine SE 2010. [ICI Surf. UK] Fatty acid ester/amino-functional polysiloxane; nonionic; softener for cotton and polyester-cotton yarns and fabrics to improve mech. props. without yellowing; liq.

Edunine SE 2060. [ICI Surf. UK] Amino-functional polysiloxane elastomer modified with fatty acid esters; nonionic; softening agent for continuous applic. to cotton and polyester-cotton fabrics; liq.

Edunine SN DF Conc. [ICI Surf. UK] PEG fatty acid esters; nonionic; softener for applic. by foam techniques; liq.

Edunine TS. [ICI Surf. UK] PEG fatty acid esters; nonionic; highly stable softening agent, lubricant, and antistatic agent; used as off-winding lubricant for yarn packages; liq.

Edunine V Fluid. [ICI Surf. UK] Fatty acid amide; cationic; substantive softening and antistatic agent for syn. and natural fibers; liq.

Eganal GES. [Hoechst AG] Nitrogen-containing ethoxylated compd.; cationic; leveling agent with affinity for dyes; used in dyeing of wool and polyamide fibers; paste.

Eganal LFI. [Hoechst AG] Fatty acid deriv.; anionic/nonionic; low-foaming lubricant and cream inhibitor for dyeing machines; liq.

Eganal PS. [Hoechst AG] Polycondensate; anionic; leveling and dispersing agent for high-temp. polyester dyeing; liq.

Eganal SME. [Hoechst AG] Ethoxylated compd. blend; nonionic; dispersant, leveling agent for disperse and Intramin dyes; permits full penetration of dyes into yarn pkgs.; liq.

Eganal SZ. [Hoechst AG] Ethoxylated nitrogen compd.; anionic; leveling agent for dyeing of polyamide and wool with acid, metal complex and reactive dyes; liq.

EG-ML. [Clark] Ethylene glycol monolaurate; coemulsifier, dispersant, defoamer, emulsifier, lubricant; yel. liq. to paste.

2-EHS Base. [Clark] Ammonium 2-ethylhexanol sulfate conc.; hydrotrope, coupler, flash foamer, detergent, wetter, anti-rust compd.; high alkali stability.

Ekaline F. [Sandoz Prods. Ltd.] Aliphatic polyglycol ether; nonionic; agent for washing off dyeings; leveling and dispersing agent; liq., flakes.

Ekaline G-80 Flakes, Liq. [Sandoz] Nonionic; detergent, wetting agent, emulsifier, for general purpose scouring; wh. flakes, colorless liq.; water sol.; 100% and 20% act.

Elec AC. [Kao Corp. SA] Amphoteric; surfactant; antistat for plastics; liq.

Elec QN. [Kao Corp. SA] Cationic; surfactant; antistat for plastics; liq.

Elec RC. [Kao Corp. SA] Anionic; surfactant; antistat for plastics; powd.

Elec TS-5, TS-6. [Kao Corp. SA] Nonionic; surfactant; antistat for plastics; needles.

Elecut S-507. [Takemoto Oil & Fat] Alkylaryl sulfonate salt; anionic; detergent; antistat for plastics; liq.; 75% conc.

Eleminol ES-70. [Sanyo Chem. Industries] POE nonylphenyl ether sulfate; anionic; primary emulsifier for emulsion polymerization; used in textile prods.; liq.; 80% conc.

Eleminol HA-100. [Sanyo Chem. Industries] POE nonylphenyl ether; nonionic; emulsifier for emulsion polymerization; protective colloidal props.; flake; HLB 18.2; 100% conc.

Eleminol HA-161. [Sanyo Chem. Industries] POE nonylphenyl ether; nonionic; primary emulsifier for emulsion polymerization; protective colloidal props.; flake; HLB 18.8; 100% conc.

Eleminol JS-2. [Sanyo Chem. Industries] Sodium alkylaryl sulfosuccinate; coemulsifier for emulsion polymerization of vinyl acetate, acrylate, vinyl chloride, and styrene/butadiene resins; liq.; 40% conc.

Eleminol MON-2. [Sanyo Chem. Industries] Sodium alkyldiphenyl ether disulfonate; emulsion polymerization of S/B and acrylate resins; liq.; 48% conc.

Eleminol MON-7. [Sanyo Chem. Industries] Sodium dodecyl diphenyl ether disulfonate; CAS 28519-02-0; anionic; emulsifier for emulsion polymerization; liq.; sol. in acidic, alkaline and electrolyte sol'ns.; 49% conc.

Eleton 1100. [Tokai Seiyu Ind.] Complex dispersion; amphoteric; antistatic agent for syn. fibers; liq.

Elfacos® C26. [Akzo Chem. BV] Hydroxyoctacosanyl hydroxystearate; nonionic; consistency regulating agent for w/o emulsions, cosmetic applics.; stabilizer; waxy substance for decorative cosmetics; pellets; 100% conc.

Elfacos® E200. [Akzo Chem. BV] Polyoxyalkylene glycol; nonionic; w/o emulsifier with high water-binding capacity; stable to pH and electrolyte; solid; 100% conc.

Elfacos® ST 9. [Akzo Chem. BV] PEG-45/dodecyl glycol copolymer; nonionic; stabilizer for w/o emulsions and emollient used in cosmetics; paste; HLB 7.0; 100% conc.

Elfacos® ST 37. [Akzo Chem. BV] PEG-22/dodecyl glycol copolymer; nonionic; stabilizer for w/o emulsions and emollient used in cosmetics; liq.; 100% conc.

Elfan® 200. [Akzo Chem. BV] Sodium lauryl sulfate; CAS 151-21-3; anionic; detergent for cosmetics, toothpaste, lt. duty detergents; wh. powd., beads; dens. 220 g/l; 90% act.

Elfan® 240. [Akzo Chem. BV] Sodium lauryl sulfate; CAS 151-21-3; anionic; personal care prods. and lt.-duty detergents; liq.; 30% conc.

Elfan® 240M. [Akzo Chem. BV] MEA lauryl sulfate; anionic; shampoos, bubble baths, lt. duty detergents; liq.; 29% conc.

Elfan® 240T. [Akzo Chem. BV] TEA lauryl sulfate;

anionic; shampoos, bubble baths, lt. duty detergents, toothpaste; liq.; 40% conc.

Elfan® 260 S. [Akzo] Sodium lauryl sulfate; CAS 151-21-3; detergent for textiles, skins, cosmetics, emulsifier for emulsion polymerization; yel. liq., paste; water sol. (30 C); sp.gr. 1.04; visc. 200 cps (30 C); f.p. 15 C; 30% act.

Elfan® 280. [Akzo Chem. BV] Coco fatty alcohol sulfate, sodium salt; anionic; general purpose heavy duty detergent, industrial, household, personal care uses; yel. paste; water sol. 700 g/l (50 C); sp.gr. 1.05; visc. 8000 cps (50 C); pH 8.5 ± 1; 42% act.

Elfan® 280 Powd. [Akzo Chem. BV] Sulfated coconut fatty alcohol, sodium salt; anionic; lt.-duty detergents and all-purpose washing agents, toothpaste; powd.; 90% conc.

Elfan® 680. [Akzo Chem. BV] Oleyl-cetyl alcohol sulfate, sodium salt; anionic; powd. detergent, cleaning agent, textile uses, shampoos, bubble baths, toothpaste; yel.-brn. paste; water sol. < 10 g/l; sp.gr. 1.05; m.p. 40 C; pH 11 ± 1; 50% act.

Elfan® KM 550. [Akzo Chem. BV] Alpha olefin sulfonate and lauryl ether sulfate; anionic; personal care prods.; dishwashing detergents, car shampoos and lt.-duty liqs.; liq.; 31% conc.

Elfan® KM 730. [Akzo Chem. BV] Alpha olefin sulfonate and lauryl ether sulfate mixture; anionic; shampoos, bubble baths, dishwashing detergents, car shampoos, lt. duty liqs.; liq.; 33% conc.

Elfan® KT 550. [Akzo Chem. BV] Sulfated coco-tallow fatty alcohol, sodium salt; anionic; heavy duty detergent, industrial, household, cosmetic uses; yel. paste; sp.gr. 1.05; visc. 10,000 cps (50 C); pH 8 ± 0.5; 42% act.

Elfan® NS. [Akzo] Lauryl ether sulfates; general purpose detergent, wetting agent; liq., paste; 30 or 70% act.

Elfan® NS 242. [Akzo Chem. BV] Sodium laureth sulfate; CAS 9004-82-4; anionic; surfactant for shampoos, bubble baths, dishwashing, lt. duty detergents, car cleaners, toiletries; liq.; 28% conc.

Elfan® NS 242 Conc. [Akzo Chem. BV] Sodium laureth sulfate (2.5 EO); CAS 9004-82-4; anionic; detergent, wetting agent used in fire extinguishers and for industrial, household, and cosmetic uses, shampoos, bubble baths, dishwashing, car cleaners, toiletries; yel. paste; water sol. 400 g/l; visc. 30,000 cps; f.p. < 0 C; pH 7–8; 70% act.

Elfan® NS 243 S. [Akzo Chem. BV] Sodium laureth sulfate (3 EO); CAS 9004-82-4; anionic; detergent, wetting agent used in fire extinguishers and for industrial, household, and cosmetic uses, shampoos, bubble baths, dishwashing, car cleaners, toiletries; clear liq.; water sol.; sp.gr. 1.04; visc. 50 cps; f.p. < 0 C; pH 6.5–7.5; 28% act.

Elfan® NS 243 S Conc. [Akzo Chem. BV] Sodium laureth sulfate (3 EO); CAS 9004-82-4; anionic; detergent, wetting agent used in fire extinguishers and for industrial, household, and cosmetic uses, shampoos, bubble baths, dishwashing, car cleaners, toiletries; yel. paste; water sol.; sp.gr. 1.12; visc. 30,000 cps; f.p. < 0 C; pH 7.5–9.5; 70% act.

Elfan® NS 252 S. [Akzo] Sodium laureth sulfate (2.5 EO); CAS 9004-82-4; anionic; detergent, wetting agent for fire extinguishers, industrial, household, and cosmetic uses; yel. liq.; water sol.; sp.gr. 1.05; visc. 150 cps; f.p. < 0 C; pH 6.5–7.5; 28% act.

Elfan® NS 252 S Conc. [Akzo] Sodium laureth sulfate (2.5 EO); CAS 9004-82-4; anionic; detergent, wetting agent for fire extinguishers, industrial, household, and cosmetic uses; yel. paste; water sol.; sp.gr. 1.11; visc. 30,000 cps; f.p. < 0 C; pH 7.5–9.5; 70% act.

Elfan® NS 423 SH. [Akzo] Sodium laureth sulfate (3 EO); CAS 9004-82-4; anionic; detergent for dishwashing, textiles, household, skins; yel. liq.; water sol.; visc. 100 cps; f.p. < 0 C; 28% act.

Elfan® NS 423 SH Conc. [Akzo] Sodium laureth sulfate (3 EO); CAS 9004-82-4; anionic; detergent for dishwashing, textiles, household, skins; yel. paste; water sol.; sp.gr. 1.10; visc. 30,000 cps; f.p. < 0 C; 70% act.

Elfan® NS 682 KS. [Akzo] Potassium tallow ether sulfate; anionic; detergent; paste; 50% conc.

Elfan® OS. [Akzo] Olefin sulfonates; detergent, wetting agent, emulsifier used for dishwashing, shampoos, textiles, laundering, household and cosmetics; liq.; 30–40% act.

Elfan® OS 46. [Akzo Chem. BV] Sodium alpha-olefin sulfonate (C14/C16); anionic; detergent for shampoos, personal care, skin, household and cosmetic cleaners; yel. liq.; water sol.; sp.gr. 1.05; visc. 250 cps; f.p. < 0 C; pH 7–9; 37% act.

Elfan® SP 325. [Akzo Chem. BV] Blend; anionic; for shampoos, bubble baths, dishwashing detergents, car shampoos, lt. duty liqs.; liq.; 33% conc.

Elfan® SP 400. [Akzo Chem. BV] Blend; anionic; for shampoos, bubble baths, dishwashing detergents, car shampoos, lt. duty liqs.; liq.; 40% conc.

Elfan® SP 500. [Akzo Chem. BV] Ether sulfate and alkylbenzene sulfonate with a nonionic surfactant; anionic; for shampoos, bubble baths, dishwashing detergents, car shampoos, lt. duty liqs.; yel. liq.; sol. 250 g/l water; sp.gr. 1.05; visc. 5000 cps; f.p. < 0 C; pH 6.5–7.5; 48% conc.

Elfan® WA. [Akzo Chem. BV] Sodium dodecylbenzene sulfonate; anionic; heavy-duty detergent for textiles, household; emulsifier for emulsion polymerization; yel. paste; sol. 250 g/l water; sp.gr. 1.08; f.p. < 0 C; m.p. 5 C; pH 8–10; 50% act.

Elfan® WA Powder. [Akzo Chem. BV] Sodium dodecylbenzene sulfonate; anionic; heavy-duty detergent for textiles, household; emulsifier for emulsion polymerization; lt. yel. beads, powd.; sol. 100 g/l water; dens. 300 g/l; pH 8–9; 80% act.

Elfan® WA Sulphonic Acid. [Akzo] Dodecylbenzene sulfonic acid; anionic; raw material; dk. brn. liq.; SO_3 odor; water sol.; sp.gr. 1.06; visc. 500 cps; flash pt. 150 C (PMCC); 98% act.

Elfan® WAT. [Akzo Chem. BV] TEA dodecylbenzene sulfonate; anionic; detergent for textile, household, personal care uses; emulsifier for insecticides; yel.-brn. liq.; water sol.; sp.gr. 1.06; visc. 1000 cps; f.p. < 0 C; pH 5.5–6.5; 50% act.

Elfanol® 510. [Akzo; Akzo Chem. BV] Sulfosuccinic acid monoester of tallow fatty acid monoethanolamide, sodium salt; anionic; lt. duty, multipurpose detergent and Syndet bars, shampoos, baby baths; lt. yel. paste; poor water sol.; sp.gr. 1.06; pH 5.5–7.5; 50% act.

Elfanol® 616. [Akzo Chem. BV] Sulfosuccinic acid monoester of an ethoxylated fatty alcohol, sodium salt (3 EO); anionic; detergent for cosmetics, shampoos, bubble and baby baths, liq. hand cleaners; clear liq.; water sol.; sp.gr. 1.05; visc. 100 cps; f.p. 0 C; pH 7; 30% act.

Elfanol® 850. [Akzo Chem. BV] Sulfosuccinic acid monoester of an ethoxylated coco fatty acid monoethanolamide, sodium salt; anionic; detergent for shampoos, bubble baths, cosmetics, liq. hand cleaners; yel.-brn. liq.; poor water sol.; sp.gr. 1.06; f.p. 14 C; pH 7; 45% act.

Elfanol® 883. [Akzo Chem. BV] Sodium dioctyl sulfosuccinate; anionic; wetting agent for technical processes, shampoos, baby baths, liq. hand cleaners; yel. liq.; water sol. 15 g/l; sp.gr. 1.08; visc. 50 cps; f.p. < 0 C; pH 7; 63% act.

Elfapur® KA 45. [Akzo; Akzo Chem. BV] Coco fatty acid monoethanolamide polyglycol ether; nonionic; detergent for textile, household, personal care use; yel. liq.; water sol 300 g/l; sp.gr. 1.05; f.p. 10–15 C; cloud pt. 74–76 C (1% aq.); 88% act.

Elfapur® LM 20. [Akzo Chem. BV] C12-14 fatty alcohol ethoxylate; nonionic; base material for prod. of anionics; liq.; 100% conc.

Elfapur® LM 25. [Akzo Chem. BV] C12-14 lauryl alcohol polyglycol ether; nonionic; see Elfapur LM 20; yel. liq.; poor water sol.; sp.gr. 1.01; f.p. 5 C; cloud pt. 56–60 C; flash pt. 180 C (10% in 25% butyl diglycol); 97% act.

Elfapur® LM 30 S. [Akzo Chem. BV] Syn. C12/C14 fatty alcohol ethoxylate; nonionic; see Elfapur LM 20; yel. liq.; poor water sol.; f.p. 5 C; cloud pt. 60–64 C (10% in 25% butyl diglycol); 97% act.

Elfapur® LM 75 S. [Akzo Chem. BV] Syn. C12-14 fatty alcohol ethoxylate; nonionic; see Elfapur LM 20; yel. liq.; water sol.; sp.gr. 1.00; f.p. 10 C; cloud pt. 60–62 C (1% aq.); 87% act.

Elfapur® LP 25 S. [Akzo Chem. BV] Syn. C12-15 fatty alcohol ethoxylate; nonionic; see Elfapur LM 20; yel. liq.; poor water sol.; sp.gr. 1.01; f.p. 5 C; cloud pt. 50–54 C (10% in 25% butyl diglycol); flash pt. 180 C; 97% act.

Elfapur® LT. [Akzo Chem. BV] Syn. fatty alcohol ethoxylate; nonionic; dishwashing detergents, all-purpose detergents, bubble baths; liq.; 90% conc.

Elfapur® ML 30 SH. [Akzo] Lauryl polyglycol ether from syn. C12/C14 fatty alcohol; base material for prod. of anionics; yel. liq.; poor water sol.; sp.gr. 1.01; f.p. 5 C; cloud pt. 62–66 C (10% in 25% butyl diglycol); PMCC flash pt. 180 C; 97% act.

Elfapur® N 50. [Akzo Chem. BV] Nonylphenol polyglycol ether; CAS 9016-45-9; nonionic; detergent, emulsifier for textiles, household and industrial uses; yel. liq.; poor water sol.; sp.gr. 1.05; f.p. –5 C; cloud pt. 60–62 C (10% in 25% butyl diglycol); PMCC flash pt. 90 C; 97% act.

Elfapur® N 70. [Akzo Chem. BV] Nonylphenol polyglycol ether; CAS 9016-45-9; nonionic; see Elfapur N 50; yel. liq., gel at 50–70% act.; water sol.; sp.gr. 1.05; f.p. 0 C; cloud pt. 22–24 C (1% aq.); PMCC flash pt. 103 C; 97% act.

Elfapur® N 90. [Akzo Chem. BV] Nonylphenol polyglycol ether; CAS 9016-45-9; nonionic; see Elfapur N 50; yel. liq., gel at 50–70% act.; water sol.; sp.gr. 1.06; f.p. 15 C; cloud pt. 60–62 C (1% aq.); PMCC flash pt. 114 C; 97% act.

Elfapur® N 120. [Akzo Chem. BV] Nonylphenol polyglycol ether; CAS 9016-45-9; nonionic; see Elfapur N 50; yel. liq., gel at 50–70% act.; water sol.; sp.gr. 1.07; cloud pt. 88–90 C (1% aq.); PMCC flash pt. 102 C; 97% act.

Elfapur® N 150. [Akzo Chem. BV] Nonylphenol polyglycol ether; CAS 9016-45-9; nonionic; see Elfapur N 50; yel. liq., gel at 40–70% act.; water sol.; sp.gr. 1.07; cloud pt. 64–66 C (1% in 10% NaCl); PMCC flash pt. 94 C; 97% act.

Elfapur® O 80. [Akzo] Oleyl-cetyl alcohol ethoxylate; nonionic; heavy-duty detergent used in textile industry; yel.-brn. liq., gel at 15–90% act.; water sol. 400 g/l; sp.gr. 1.05; m.p. 25–30 C; cloud pt. 44–46 C (1% aq.); 97% act.

Elfapur® O 160. [Akzo] Oleyl-cetyl alcohol polyglycol ether; detergent used in textile industry; water sol. 400 g/l; sp.gr. 1.06; m.p. 29–35 C; cloud pt. 64–66 C (1% in 10% NaCl); 97% act.

Elfapur® T 110. [Akzo Chem. BV] Tallow fatty alcohol polyglycol ether; nonionic; foam controlled, heavy-duty detergent powders, bubble baths, for textile use; yel. paste, gel at 30–80% act.; water sol. 300 g/l; sp.gr. 1.05; m.p. 35–40 C; cloud pt. 77–79 C (1% aq.); 97% act.

Elfapur® T 250. [Akzo] Tallow fatty alcohol polyglycol ether; nonionic; foam controlled heavy-duty detergent powds. for textile use; wh. powd., beads, gel at > 25% act.; water sol. 250 g/l; cloud pt. 77–79 C (1% in 10% NaCl); 98% act.

Elimina 254. [Marubishi Oil Chem.] POE fatty alcohol sulfate; anionic; detergent, dispersant, antistat, emulsifier used in detergent and shampoo formulations, latex, dyestuffs, pigments, oil and waxes; used for syn. fibers; paste; 30% conc.

Elimina 505. [Marubishi Oil Chem.] Fatty alcohol phosphate; anionic; textile antistat; liq.; 30% conc.

Eltesol® 4200. [Albright & Wilson UK] Xylene sulfonic acid blend; catalyst for foundry resins.

Eltesol® 4209. [Albright & Wilson UK] Aryl sulfonic acid; carpet etching agent.

Eltesol® 4402. [Albright & Wilson UK] Aromatic benzenesulfonic acid blend; anionic; highly reactive catalyst for foundry resins; liq.; 88% conc.

Eltesol® 4443M. [Albright & Wilson Am.] Benzene sulfonic acid blend; anionic; surfactant; liq.; 72% conc.

Eltesol® 5400 Series. [Albright & Wilson Am.] Phenol sulfonic acid condensates; anionic; surfactant for leather processing; liq./powd.

Eltesol® 7200 Series. [Albright & Wilson Am.] Dihydroxy diphenyl sulfonates; anionic; surfactant for leather processing; liq./powd.

Eltesol® AC60. [Albright & Wilson UK] Ammonium cumene sulfonate; anionic; surfactant, hydrotrope for agric. applics.; liq.; 60% conc.

Eltesol® ACS 60. [Albright & Wilson UK] Ammonium cumenesulfonate; hydrotrope, solubilizer, coupling agent, and visc. modifier in liq. formulations; cloud pt. depressant in detergent formulations; pale pink liq.; dens. 1.10 g/cm³; pH 7.0–8.0 (10% aq.); 60% min. act.

Eltesol® AX 40. [Albright & Wilson UK] Ammonium xylene sulfonate; anionic; hydrotrope, cloud pt. depressant used in the detergent mfg.; solubilizer, coupler; pale yel. liq.; pH 7–8.5; 41% act.

Eltesol® CA 65. [Albright & Wilson UK] Cumene sulfonic acid; CAS 28631-63-2; anionic; catalyst for foundry resins; descaling agent for metal cleaning; antistress additive and plating aid in electroplating bath; curing aid in the plastics industry; raw material in the mfg. of dyes and pigments; detergents industry; dens. 1.15 g/cm³; visc. 30 cs; 65 ± 1.0% act. in water.

Eltesol® CA 96. [Albright & Wilson UK] Cumene sulfonic acid conc.; anionic; catalyst for foundry resins; descaling agent for metal cleaning; antistress additive and plating aid in electroplating bath; curing aid in the plastics industry; raw material in the mfg. of dyes and pigments; detergents industry; brn. visc. liq.; dens. 1.25 g/cm³; visc. 1000 cs; 96.0 ± 1.0% act.

Eltesol® MGX. [Albright & Wilson UK] Magnesium

xylene sulfonate; hydrotrope for liq and spray-dried detergent formulations; powd.

Eltesol® PSA 65. [Albright & Wilson UK] Phenol sulfonic acid; CAS 1333-39-7; anionic; catalyst for foundry resins; descaling agent for metal cleaning; antistress additive and plating aid in electroplating bath; curing aid in the plastics industry; raw material in the mfg. of dyes and pigments; detergents industry; pharmaceutical chemicals and disinfectants; liq.; dens. 1.3 g/cm^3; visc. 50 cs; 65.0 ± 1.0% act. in water.

Eltesol® PT 45. [Albright & Wilson UK] Potassium toluene sulfonate; hydrotrope for liq and spray-dried detergent formulations; liq.

Eltesol® PT 93. [Albright & Wilson UK] Potassium toluene sulfonate; anionic; hydrotrope, cloud pt. depressant used in the detergent mfg.; solubilizer, coupler; off wh. powd.; dens. 0.46 g/cc; pH 8–10.5 (3%); 93% act.

Eltesol® PX 40. [Albright & Wilson UK] Potassium xylene sulfonate; anionic; hydrotrope, solubilizer, coupling agent, and visc. modifier in liq. formulations; cloud pt. depressant in detergent formulations; straw liq.; dens. 1.10 g/cm^3; pH 7.0–10.5 (10% aq.); 40.0 ± 1.0% act. in water.

Eltesol® PX 93. [Albright & Wilson UK] Potassium xylene sulfonate; anionic; hydrotrope, cloud pt. depressant used in the detergent mfg.; solubilizer, coupler; off wh. powd.; dens. 0.46 g/cc; pH 9–10.5; 93% act.

Eltesol® SC 40. [Albright & Wilson UK] Sodium cumene sulfonate; anionic; hydrotrope, solubilizer for spray-dried prods.; liq.; 40% conc.

Eltesol® SC 93. [Albright & Wilson UK] Sodium cumene sulfonate; anionic; hydrotrope for hard surf. cleaners; powd.; 93% act.

Eltesol® SC Pellets. [Albright & Wilson Am.] Sodium cumene sulfonate; anionic; surfactant; pellet; 88% conc.

Eltesol® ST 34. [Albright & Wilson UK] Sodium toluene sulfonate; CAS 657-84-1; anionic; hydrotrope, cloud pt. depressant used in the detergent mfg.; solubilizer, coupler; pale yel. liq.; pH 7–10; 34% act.

Eltesol® ST 40. [Albright & Wilson UK] Sodium toluenesulfonate; CAS 657-84-1; anionic; hydrotrope, solubilizer, coupling agent, and visc. modifier in liq. formulations; cloud pt. depressant in detergent formulations; straw liq.; dens. 1.10 g/cm^3; pH 7.0–10.5 (10% aq.); 40.0±1.0% act.

Eltesol® ST 90. [Albright & Wilson UK] Sodium toluene sulfonate; CAS 657-84-1; anionic; hydrotrope, cloud pt. depressant used in the detergent mfg.; solubilizer, coupler; off wh. powd., pellets; dens. 0.46 g/cc; pH 9–10.5 (3%); 90% act.

Eltesol® ST Pellets. [Albright & Wilson UK] Sodium toluene sulfonate; CAS 657-84-1; anionic; hydrotrope, solubilizer, coupling agent, and visc. modifier in liq. formulations; cloud pt. depressant in detergent formulations; wh. pellets; dens. 0.55 g/cm^3; pH 9.0–10.5 (3% aq.); 85.0% min. act.

Eltesol® SX 30. [Albright & Wilson UK] Sodium xylene sulfonate; CAS 1300-72-7; anionic; hydrotrope, cloud pt. depressant used in the detergent mfg.; solubilizer, coupler; pale yel. liq.; pH 7–10; 30% act.

Eltesol® SX 40. [Albright & Wilson Am.] Sodium xylene sulfonate; CAS 1300-72-7; anionic; surfactant; liq.; 40% conc.

Eltesol® SX 93. [Albright & Wilson UK] Sodium xylene sulfonate; CAS 1300-72-7; anionic; solubilizer and coupling agent in heavy-duty liq. detergents; visc. reducing agent in mfg. of powd. detergents; off wh. powd., pellets; dens. 0.46 g/cc; pH 9–10.5 (3%); 93% act.

Eltesol® SX Pellets. [Albright & Wilson UK] Sodium xylene sulfonate; CAS 1300-72-7; anionic; solubilizer and coupling agent in heavy-duty liq. detergents; visc. reducing agent in mfg. of powd. detergents; wh. pellets; dens. 0.5 g/cm^3; pH 9.0–10.5 (3% aq.); 88.0% min. act.

Eltesol® TA 65. [Albright & Wilson UK] 65% Toluene sulfonic acid and 1.4% sulfonic acid aq. sol'n.; anionic; curing agent for resins in foundry cores, plastics, coatings; intermediate; catalyst in foundry and chemical industries; hardening agent in plastics; activator for nicotine insecticides; descaling agent in metal cleaning; in electroplating baths; lt. amber clear liq.; dens. 1.2 g/cc; visc. 9–12 cs; 65% conc.

Eltesol® TA 96. [Albright & Wilson UK] Toluene sulfonic acid; anionic; curing agent for resins in foundry cores, plastics, coatings; intermediate; catalyst in foundry and chemical industries; hardening agent in plastics; activator for nicotine insecticides; descaling agent in metal cleaning; in electroplating baths; brn. thick cryst. liq.; dens. 1.3 g/cm^3; 95% act.

Eltesol® TA Series. [Albright & Wilson UK] Toluene sulfonic acid; anionic; catalyst for foundry resins; descaling agent for metal cleaning; antistress additive and plating aid in electroplating bath; curing aid in the plastics industry; raw material in the mfg. of dyes and pigments; detergents industry; hydrotrope, intermediate; used in mfg. of acrylonitrile; agric. formulations; liq.; dens. 1.2 g/cm^3; visc. 15 cs.

Eltesol® TPA. [Albright & Wilson UK] Formulated prod.; tin-plating additive; powd.; 80% conc.

Eltesol® TSX. [Albright & Wilson UK] p-Toluene sulfonic acid monohydrate BP; CAS 70788-37-3; anionic; catalyst for org. synthesis, syn. resins, mfg. of p-cresol, toluene derivs., pharmaceutical prods., dyestuffs; chemical intermediate; wh. to off-wh. cryst.; m.p. 103 C; 97% conc.

Eltesol® TSX/A. [Albright & Wilson UK] p-Toluene sulfonic acid monohydrate; anionic; catalyst for org. synthesis, syn. resins, mfg. of p-cresol, toluene derivs., pharmaceutical prods., dyestuffs; chemical intermediate; curing agent for resins and coatings; wh. crystals; m.p. 103.5 C; 95% act.

Eltesol® TSX/SF. [Albright & Wilson UK] p-Toluene sulfonic acid monohydrate; anionic; catalyst for org. synthesis, syn. resins, mfg. of p-cresol, toluene derivs., pharmaceutical prods., dyestuffs; chemical intermediate; wh. cryst.; m.p. 103.5 C; 98% act.

Eltesol® XA. [Albright & Wilson UK] Xylene sulfonic acid aq. sol'n.; anionic; catalyst in foundry and chemical industries; hydrotrope for agric. formulations; amber/brn. clear liq.; dens. 1.2 g/cm^3; visc. 15 cs; 61.0% min. act.

Eltesol® XA65. [Albright & Wilson UK] Xylene sulfonic acid; CAS 25321-41-9; anionic; intermediate, catalyst in preparation of esters, hardening agent in plastics, activator for nicotine insecticides; lt. amber, clear to slightly hazy liq.; dens. 1.2 g/cc; visc. 9–12 cs; 65% act.

Eltesol® XA90. [Albright & Wilson UK] Xylene sulfonic acid conc.; anionic; catalyst for foundry resins; descaling agent for metal cleaning; antistress additive and plating aid in electroplating bath; curing aid in the plastics industry; raw material in the mfg. of dyes and pigments; detergents industry; brn. visc. liq.; dens. 1.35 g/cm³; visc. 3000 cs; 95.0 ± 2.0% act.

EM-1. [Ferro/Keil] Lard oil; additive improving wetting and lubricity in industrial applics.; sp.gr. 0.91; dens. 7.6 lb/gal; visc. 35 cSt (40 C); pour pt. 7 C; cloud pt. 21 C; flash pt. (COC) 210 C; acid no. 25; sapon. no. 195.

EM-25-A. [Ferro/Keil] Wetting/oiliness agent for drawing compds., sol. oils, gear oils, cutting oils; highly polar; imparts wipe resistance; Gardner 8+ color; sp.gr. 1.16; dens. 9.7 lb/gal; visc. 595 cSt (40 C); pour pt. -7 C; acid no. 50.

EM-40. [Ferro/Keil] Glyceryl monotallate; nonionic; surfactant, w/o emulsifier, corrosion inhibitor for emulsions, compounded oils; Gardner 3 liq.; sol. in many alcohols, aromatic and aliphatic hydrocarbons incl. min. oils; insol. in water; sp.gr. 0.96; dens. 8.0 lb/gal; visc. 100 cSt (40 C); pour pt. < -12 C; acid no. 6; HLB 3.0; 100% conc.

EM-90. [Ferro/Keil] Fatty ester; sperm oil replacement; lubricant and wetting agent for industrial gear lubricants, cutting oils, other metalworking fluids; Gardner 1.5 liq.; oil-sol.; sp.gr. 0.90; dens. 7.5 lb/gal; visc. 45 cSt (40 F); pour pt. -1 C; cloud pt. 2 C; flash pt. (COC) 191 C; acid no. 2; sapon. no. 195.

EM-550. [Ferro/Keil] Fatty compd.; Lubricity and detergency additive for compounding water-sol. machining fluids; exc. film str.; water-sol.; sp.gr. 1.08; dens. 9.0 lb/gal; visc. 1275 cSt (40 C); pour pt. 2 C; flash pt. (COC) 188 C; acid no. 15; pH 7.6 (1% aq.).

EM-600. [Ferro/Keil] PEG 600 monotallate; CAS 61791-00-2; nonionic; surfactant, emulsifier, wetting agent, detergent for industrial applics., sol. cutting oils and drawing compds.; Gardner 2 liq.; sol. in water, aliphatic and aromatic hydrocarbons, alcohols, esters; sp.gr. 1.03; dens. 8.6 lb/gal; visc. 61 cSt (40 C); pour pt. 18 C; HLB 13.0; acid no. 7; pH 4.4 (1% aq.); 100% act.

EM-706. [Ferro/Keil] Phosphate ester; wetting, antiweld, lubricity, and antiwear additive for metalworking fluids; gives corrosion protection to freshly machined parts, seize protection at the tool/part interface; Gardner 4; water-disp.; sp.gr. 1.00; dens. 8.3 lb/gal; visc. 340 cSt (40 C); pour pt. 10 C; flash pt. (COC) 193 C; acid no. 165; pH 2.5 (1% aq.); 6% phosphorus.

EM-711. [Ferro/Keil] Phosphate ester; wetting, antiweld, lubricity, and antiwear additive for metalworking fluids (cutting oils, sol. oils); Gardner 3; oil-sol.; sp.gr. 1.01; dens. 8.3 lb/gal; visc. 95 cSt (40 C); pour pt. < -12 C; flash pt. (COC) 143 C; acid no. 330; pH 2.5 (1% aq.).

EM-900. [Ferro/Keil] Cationic; wetting agent, emulsifier for invert emulsions; surface and interfacial tension reducer for min. oils and other petrol. fractions; oil-sol.; sp.gr. 0.95; dens. 7.9 lb/gal; visc. 560 cSt (40 C); pour pt. -15 C; flash pt. (COC) 227 C; acid no. 75.

EM-965. [Ferro/Keil] Imidazoline; used for making water emulsions with asphalt; liq.; 100% conc.

EM-980. [Ferro/Keil] Mixed fatty acid diethanolamide containing excess diethanolamine; solubilizer, visc. builder, detergency and lubricity aid used in synthetic and semisynthetic metalworking fluids, floor cleaners, paint strippers, buffing compds.; contributes to corrosion resistance; sec. emulsifier for sol. oils; Gardner 4.5; water-sol.; sp.gr. 1.01; dens. 8.4 lb/gal; visc. 525 cSt (40 C); pour pt. -1 C; acid no. 50; pH 9.7 (1% aq.).

EM-985. [Ferro/Keil] Mixed fatty acid diethanolamide containing excess diethanolamine; solubilizer, visc. builder, detergency and lubricity aid used in synthetic and semisynthetic metalworking fluids, floor cleaners, paint strippers, buffing compds.; contributes to corrosion resistance; sec. emulsifier for sol. oils; Gardner 5; oil-sol.; sp.gr. 1.00; dens. 8.3 lb/gal; visc. 460 cSt (40 C); pour pt. 7 C; acid no. 5; pH 9.8 (1% aq.).

EM-9400. [Ferro/Keil] Fatty ester; blown sperm oil replacement; lubricant and wetting agent for cutting oils, drawing compds., lubricating and preservative oils; good corrosion protection; Gardner 3; sol. in min. oils and many common org. solvs.; insol. in water; sp.gr. 0.94; dens. 7.8 lb/gal; visc. 80 cSt (40 C); pour pt. 13 C; flash pt. (COC) 174 C; acid no. 25.

EM-9500. [Ferro/Keil] Fatty alcohols, fatty acids, high m.w. esters, aldehydes, and ketones; functional intermediate, performance chemical for formulating oil-sol. corrosion inhibitors; as replacement for animal and veg. based fatty acids and their derivs.; amber semisolid; sp.gr. 0.92; dens. 7.7 lb/gal; visc. 8 cSt (100 C); pour pt. 41 C; flash pt. (COC) 182 C; acid no. 90; sapon. no. 220; hyd. no. 20.

Emal 20C. [Kao Corp. SA] Sodium POE alkyl ether sulfate; anionic; detergent, emulsifier, foam stabilizer, shampoo base; scouring agent for syn. fibers; liq.; 25% conc.

Emal NC-35. [Kao Corp. SA] Sodium POE alkylaryl ether sulfate; detergent, emulsifier, foam stabilizer, shampoo base; scouring agent for syn. fibers; liq.; 35% conc.

Emalox C 102. [Yoshimura Oil Chem.] Potassium alkyl phosphate; anionic; finishing oil after dyeing for polyester and polyamide fibers; paste; 70% conc.

Emalox OEP-1. [Yoshimura Oil Chem.] Potassium alkyl phosphate; anionic; antistatic agent, after dyeing, for syn. fibers, wool, and silk; liq.; 50% conc.

Emasol L-106. [Kao Corp. SA] PEG-4 sorbitan laurate; CAS 9005-64-5; nonionic; emulsifier; solubilizer for colorants; liq.; HLB 13.3; 100% conc.

Emasol L-120. [Kao Corp. SA] PEG-20 sorbitan laurate; CAS 9005-64-5; nonionic; emulsifier for pharmaceuticals and cosmetics; stabilizer for emulsion polymerization; liq.; HLB 16.7; 100% conc.

Emasol O-15 R. [Kao Corp. SA] Sorbitan sesquioleate; CAS 8007-43-0; nonionic; emulsifier and lubricant; liq.; HLB 3.7; 100% conc.

Emasol O-105 R. [Kao Corp. SA] POE sorbitan oleate; nonionic; emulsifier and solubilizer; liq.; HLB 10.0; 100% conc.

Emasol O-106. [Kao Corp. SA] PEG-5 sorbitan oleate; CAS 9005-65-6; nonionic; emulsifier and solubilizer; liq.; HLB 10.0; 100% conc.

Emasol O-120. [Kao Corp. SA] PEG-20 sorbitan oleate; CAS 9005-65-6; nonionic; emulsifier and solubilizer; liq.; HLB 15.0; 100% conc.

Emasol O-320. [Kao Corp. SA] PEG-20 sorbitan trioleate; CAS 9005-70-3; nonionic; emulsifier and solubilizer; liq.; HLB 11.0; 100% conc.

Emasol P-120. [Kao Corp. SA] PEG-20 sorbitan palmitate; CAS 9005-66-7; nonionic; emulsifier for pharmaceuticals and cosmetics; solubilizer for colorants; stabilizer for emulsion polymerization; liq.; HLB 15.6; 100% conc.

Emasol S-20. [Kao Corp. SA] Sorbitan distearate; CAS 36521-89-8; nonionic; emulsifier and lubricant; flake; HLB 4.4; 100% conc.

Emasol S-106. [Kao Corp. SA] PEG-4 sorbitan

stearate; CAS 9005-67-8; nonionic; emulsifier and lubricant; solid; HLB 9.6; 100% conc.

Emasol S-120. [Kao Corp. SA] PEG-20 sorbitan stearate; CAS 9005-67-8; nonionic; emulsifier and lubricant; solid; HLB 14.9; 100% conc.

Emasol S-320. [Kao Corp. SA] PEG-20 sorbitan tristearate; CAS 9005-71-4; nonionic; emulsifier and lubricant; solid; HLB 10.5; 100% conc.

Emcol® 4. [Witco/Organics] Stearalkonium chloride; cationic; antistat, substantive conditioner, emollient; wh. paste; water-sol.; sp.gr. 0.99; pH 4.0; 25% solids.

Emcol® 14. [Witco/Organics; Witco SA] Polyglyceryl oleate; nonionic; w/o emulsifier; corrosion inhibitor for aerosol; spreader sticker, antifoaming agent; liq.; 100% conc.

Emcol® 150 T. [Witco/Organics; Witco SA] Surfactant blend; anionic/nonionic; specific emulsifier for highly conc. organo-phosphate insecticide emulsifiable concs.; liq.

Emcol® 226.33. [Witco/Organics; Witco SA] Propargyl alcohol ethoxylate; nonionic; used in galvano industries; liq.; 100% conc.

Emcol® 1655. [Witco/Organics] Cocamidopropyl dimethylamine propionate; cationic; cosmetics and toiletry surfactant used as antistat, conditioner, emollient, foaming and substantive agent; pale yel. liq.; water-sol.; sp.gr. 1.01; pH 6.5; 40% solids.

Emcol® 3780. [Witco/Organics] Stearamidopropyl dimethylamine lactate; cationic; cosmetics and toiletry surfactant used as antistat, conditioner, emollient, foaming and substantive agent; pale yle. liq.; water-sol.; sp.gr. 0.99; pH 4.0; 25% solids.

Emcol® 4100M. [Witco/Organics] Disodium myristamido MEA-sulfosuccinate; anionic; dispersant, wetting agent, foaming agent, detergent, emulsifier for bubble bath, shampoo, carpet and upholstery cleaners, textiles; wh. creamy semisolid; disp. in water; sp.gr. 1.01; acid no. 4.5; surf. tens. 37.8 dynes/cm (0.05%); pH 6.5 (3% aq.); 38% solids.

Emcol® 4161L. [Witco/Organics; Witco SA] Disodium oleamido MIPA sulfosuccinate; anionic; dispersant, wetting, foam booster/stabilizer, detergent, conditioner, and emulsifying agent for bubble bath, shampoos, cleansers for cosmetics and toiletries; lt. yel. clear liq.; sol. in water; sp.gr. 1.10; PMCC flash pt. 93 C; acid no. 6.0; pH 6.5 (3% aq.); surf. tens. 32.6 dynes/cm (0.5%); 38% solids.

Emcol® 4300. [Witco/Organics] Disodium C12–15 pareth sulfosuccinate; anionic; dispersant, wetting, foaming, detergent, emulsifying agent for bubble bath, shampoo, cosmetics and toiletries; emulsifier for acrylic, vinyl acetate, vinyl acrylic polymerization; lt. clear liq.; sol. in water; sp.gr. 1.08; visc. 50 cps; acid no. 5.0; pH 6.2 (3% aq.); surf. tens. 29.3 dynes/cm (1%); Ross-Miles foam 162 mm (initial, 1%, 49 C); 33% solids.

Emcol® 4500. [Witco/Organics; Witco SA] Sodium dioctyl sulfosuccinate; anionic; dispersant, detergent, wetting, foaming, emulsifying agent; for cosmetics, toiletries, textiles, industrial processing slurries; clear, lt. visc. liq.; sol. in perchloroethylene, CCl_4, kerosene, xylene, Stod.; disp. water and alcohol; sp.gr. 1.10; flash pt. > 93 C; acid no. 3.0; surf. tens. 26.3 dynes/cm (0.05%); pH 6.5 (3% aq.); 70% solids.

Emcol® 4560. [Witco/Organics] Sodium dioctyl sulfosuccinate; dispersant, wetting, foaming, detergent, emulsifying agent, solubilizer; used in drycleaning; clear, lt. visc. liq.; sol. in perchlorethylene, CCl_4, Stod.; slightly sol. in water; disp. in alcohol; sp.gr. 1.08; acid no. 3.0; pH 5.1 (3% aq. disp.); surf. tens. 27.4 dynes/cm (0.05%); 90% solids.

Emcol® 4580PG. [Witco/Organics] Sodium diester sulfosuccinate; anionic; detergent, emulsifier, and water solubilizer in dry cleaning; textile industry detergent and scouring agent for solv. systems; liq.; sol. in perchloroethylene, Stod.; disp. in water.

Emcol® 4600. [Witco/Organics; Witco SA] Sodium ditridecyl sulfosuccinate; anionic; detergent, foam modifier, wetting agent, dispersant, processing aid for pigments, in drycleaning; clear, lt. visc. liq.; sol. in perchlorethylene, CCl_4, Stod., kerosene, xylene, IPA; slightly sol. in alcohol; insol. water; sp.gr. 1.00; flash pt. > 93 C; acid no. 0.3; pH 6.6 (3% aq. disp.); 70% solids.

Emcol® 4910. [Witco/Organics] Sodium lauryl/propoxy sulfosuccinate; emulsifier for acrylic, vinyl acrylic polymerization; visc. 16,000 cps; surf. tens. 31.0 dynes/cm (1%); Ross-Miles foam 163 mm (initial, 1%, 49 C); 60% solids.

Emcol® 4930, 4940. [Witco/Organics] Asymmetrical sulfosuccinate; anionic; surfactants for emulsion polymerization, latex stabilization, wetting, emulsification; liq.

Emcol® 5430. [Witco/Organics] Cocamidopropyl betaine; amphoteric; detergent, foam booster/stabilizer, visc. modifier, wetting agent; Gardner 5 liq.; water-sol.; sp.gr. 1.04; clear pt. 0 C; 37% solids.

Emcol® 6748. [Witco/Organics] Cocamidopropyl betaine; amphoteric; detergent, foam booster/stabilizer, visc. modifier, wetting agent; Gardner 3 liq.; water-sol.; sp.gr. 1.05; clear pt. 0 C; 35% solids.

Emcol® AC 62-36. [Witco/Organics; Witco Israel] Surfactants blend; anionic/nonionic; emulsifier for arsenic-based herbicides; liq.; 100% conc.

Emcol® AC 64-6A. [Witco/Organics; Witco Israel; Witco SA] Surfactants blend; nonionic; emulsifier for Dimethoate emulsifiable concs.; liq.; 100% conc.

Emcol® AD 27-21. [Witco/Organics; Witco SA] Surfactant blend; anionic/nonionic; emulsifier for ortho dichlorobenzene; liq.

Emcol® AG 4-48N. [Witco/Organics; Witco Israel] Sulfonate/POE ether blend; anionic/nonionic; emulsifier for pesticide emulsifiable concs.; liq.; 100% conc.

Emcol® AK 16-11. [Witco/Organics; Witco SA] Surfactant blend; nonionic; solubilizer for DDVP; liq.

Emcol® AK 16-11N. [Witco/Organics; Witco SA] Surfactant blend; nonionic; emulsifier for DDVP; liq.

Emcol® AK 16-97N. [Witco/Organics; Witco Israel] Sulfonate/POE ether blend; anionic/nonionic; emulsifier for pesticide emulsifiable concs.; liq.; 100% conc.

Emcol® AK 18-72A. [Witco/Organics; Witco SA] Surfactant blend; anionic/nonionic; emulsifier for DDVP 50% w/w; liq.

Emcol® AL 26-43H. [Witco/Organics; Witco Israel] Surfactant blend; anionic/nonionic; emulsifier for Propanil emulsifiable concs.; liq.; 100% conc.

Emcol® AL 26-43L. [Witco/Organics; Witco Israel] Surfactant blend; anionic/nonionic; emulsifier for Propanil emulsifiable concs.; liq.; 100% conc.

Emcol® AL 69-49. [Witco/Organics; Witco Israel] Surfactant blend; nonionic; emulsifier for pesticide emulsifiable concs. where anionic surfactants would decompose the toxicant, e.g., for Dimethoate;

liq.; 100% conc.

Emcol® CC-9. [Witco/Organics; Witco SA] PPG-9 diethylmonium chloride; cationic; dispersant, particle suspension aid, antistat, wetting agent, o/w emulsifier, conditioner, penetrant, lubricant; for cosmetics, toiletries, textiles, industrial slurries; lt. amber clear liq.; sol. in water, IPA; sp.gr. 1.01; flash pt. > 93 C; pH 6.5 (10% aq.); 100% conc.

Emcol® CC-36. [Witco/Organics; Witco SA] PPG-25 diethylmonium chloride; cationic; dispersant, particle suspension aid, o/w emulsifier; plasticizer for hair polymers; antistat; skin cleanser for cosmetics; used in dry cleaning systems; industrial processes; lt. amber clear liq.; sol. in water, IPA; sp.gr. 1.01; flash pt. > 93 C; pH 6.7 (10% in 10:6 IPA:water); 100% conc.

Emcol® CC-37-18. [Witco/Organics] Coco-betaine; amphoteric; foaming agent and foam stabilizer for cosmetics and toiletries, specialty cleaning formulations; visc. modifier; liq.

Emcol® CC-42. [Witco/Organics] PPG-40 diethylmonium chloride; cationic; pigment dispersant, particle suspension aid, emulsifier, solv., conditioner, antistat, lubricant, corrosion inhibitor for toiletries, cosmetics, germicides, syn. fibers and plastics, textiles, industrial processes; ore flotation additive; lt. amber oily liq.; sol. in IPA, acetone, MEK, min. spirits, ethanol; partly sol. in water; sp.gr. 1.01; flash pt. > 200 C (PMCC); pH 6.5; 100% conc.

Emcol® CC-55. [Witco/Organics] Polypropoxy quat. ammonium acetate; cationic; antistat; conditioner for hair rinse preparations; emulsifier for cosmetics and textile flame retardants; solv. for phenolic-type germicides for cosmetics and toiletries; antistat for syn. fibers and plastics; fabric conditioner; lubricant for textile and industrial formulations; solv. cleaning and scouring agent; corrosion inhibitor in protective coatings; pigment dispersant in nonaq. media; o/w emulsifier; lt. amber oily liq.; sol. @ 25% in ethanol, IPA, acetone, MEK; water-disp.; sp.gr. 1.02; flash pt. > 93 C (PMCC); pH 6.5; 99% solids.

Emcol® CC-57. [Witco/Organics] Polypropoxy quat. ammonium phosphate; cationic; antistat, conditioner, emulsifier, solv., lubricant, solv. cleaning and scouring agent, corrosion inhibitor, and dispersant; used in syn. fibers and plastics,; personal care prods., germicides, flame retardants, textile and industrial applic., protective coatings, and pigments; lt. amber oily liq.; sol. in water, ethanol, IPA, acetone, MEK; sp.gr. 1.12; flash pt. > 93 C (PMCC).

Emcol® CC-59. [Witco/Organics; Witco SA] Polypropoxylated quat. ammonium phosphate; dispersant for electronics industry; liq.

Emcol® CC-422. [Witco/Organics] Polypropoxy quat. ammonium chloride; cationic; detergent and dry cleaning applics.; anticaking and cleansing agent, antistat, dispersant and o/w emulsifier for aerosol formulations; liq.; sol. in halogenated hydrocarbons, perchloroethylene, and water.

Emcol® CN-6. [Witco/Organics] Carboxylated alkyl polyether; anionic; foaming agent, foam stabilizer, visc. modifier for textile, toiletry, specialty cleaning formulations; liq.

Emcol® CS-143, -151, -165. [Witco/Organics] Org. phosphate ethers; anionic; detergent, emulsifier, dispersant, solubilizer; liq.

Emcol® CS-1361. [Witco/Organics; Witco SA] Sodium salt of complex organo-phosphate ester; wetting agent, dispersant for water-based suspension concs.; used for scale control; clear liq.; sol. in hydrocarbons, min. oil, kerosene, ethanol, water; sp.gr. 1.10; pour pt. 40 F; surf. tens. 31.9 dynes/cm (0.05%); pH 5 (3% aq.); 90% conc.

Emcol® Coco Betaine. [Witco/Organics] Cocamidopropyl betaine; amphoteric; detergent, foam booster/stabilizer, visc. modifier, wetting agent; Gardner 3 liq.; water-sol.; sp.gr. 1.04; 42% solids.

Emcol® DG. [Witco/Organics] Cocamidopropyl betaine; amphoteric; detergent, emulsifier, foaming and wetting agent, visc. builder, foam stabilizer for detergent industry; lime soap dispersant; Gardner 4 liq.; water-sol.; sp.gr. 1.03; clear pt. 0 C; 36% solids.

Emcol® DOSS. [Witco/Organics] Sodium dioctyl sulfosuccinate; emulsifier for acrylonitrile polymerization; improves surf. wetting; FDA compliance; visc. 250 cps; surf. tens. 25.7 dynes/cm (1%); Ross-Miles foam 243 mm (initial, 1%, 49 C).

Emcol® E-607L. [Witco/Organics; Witco SA] Lapyrium chloride; cationic; emollient, emulsifier, conditioner, foamer, cleanser, substantive agent, deodorant for cosmetics, toiletries, industrial applics.; hair conditioner; wh. to off-wh. cryst. powd.; sol. 37% in water and ethanol, 28% in IPA; surf. tens. 37 dynes/cm (0.1% aq.); pH 3.9 (1% aq.); 97.5% act.

Emcol® E-607S. [Witco/Organics; Witco SA] Steapyrium chloride; cationic; emollient, emulsifier, conditioner, foamer, cleanser, substantive agent, deodorant for cosmetics, toiletries, industrial applics.; hair conditioner; wh. to off-wh. powd.; sol. in oil, 2.5% in water; pH 3.4 (1% aq.); 94% act.

Emcol® H 30, H 31A, H 32. [Witco/Organics; Witco Israel] Polyglycol fatty acid esters; nonionic; emulsifier for petrol. prods.; liq.; 100% conc.

Emcol® H 400X. [Witco/Organics; Witco Israel] Sulfonate/POE ether blend; anionic/nonionic; emulsifier for pesticide emulsifiable concs.; liq.; 100% conc.

Emcol® ISML. [Witco/Organics] Isostearamidopropyl morpholine lactate; cationic; cosmetics and toiletries surfactant used as antistat, conditioner, emollient, foaming and substantive agent; nonirritating base for cream rinses and conditioning shampoos; lt. yel. liq.; water-sol.; sp.gr. 1.01; pH 4.5; 25% act.

Emcol® K8300. [Witco/Organics; Witco SA] Disodium oleamido-MIPA sulfosuccinate; anionic; dispersant, particle suspension aid, wetting agent, foam booster/stabilizer, detergent; emulsifier in emulsion polymerization of acrylic, styrene acrylic, vinyl acrylic; FDA compliance; amber liq.; sol. in water; sp.gr. 1.10; visc. 1200 cps; pH 6.5 (3% aq.); surf. tens. 30.5 dynes/cm (1%); Ross-Miles foam 158 mm (initial, 1%, 49 C); 38% solids.

Emcol® L. [Witco/Organics] Lauramine oxide; cationic; cosmetics and toiletries surfactant used as antistat, cleansing and substantive agent, emollient, lubricant, and visc. builder; detergent and foam booster/stabilizer for industrial detergents; liq.; water-sol.; 25% conc.

Emcol® LO. [Witco/Organics] Lauramine oxide; nonionic; foam booster/stabilizer, visc. modifier; liq.; water-sol.; sp.gr. 0.96.

Emcol® M. [Witco/Organics] Myristamine oxide; cationic; cosmetics and toiletries surfactant used as antistat, cleansing and substantive agent, emollient, lubricant, and visc. builder; detergent and foam booster/stabilizer for industrial detergents; liq.; water-sol.; 25% conc.

Emcol® N Series. [Witco/Organics; Witco Israel] Sulfonate/POE ether blend; anionic/nonionic;

emulsifier for pesticide emulsifiable concs.; liq.; 100% conc.

Emcol® NA-30. [Witco/Organics] Cocamidopropyl betaine; amphoteric; cosmetics/toiletries surfactant used as antistat, cleansing, foaming, spreading agent, conditioner, foam booster/stabilizer, solubilizer and visc. modifier; lime soap dispersant; Gardner 3 liq.; water-sol.; sp.gr. 1.03; clear pt. 0 C; 36% act.

Emcol® P 50-20 B. [Witco/Organics; Witco SA] Linear calcium dodecylbenzene sulfonate; anionic; lipophilic emulsifier for pesticide emulsifiable concs.; liq.; 70% conc.

Emcol® P-1020B. [Witco/Organics; Witco SA] Branched calcium dodecylbenzene sulfonate; anionic; lipophilic emulsifier for pesticide emulsifiable concs.; liq.; HLB 7.0-8.0; 70% conc.

Emcol® P-1020 BU. [Witco/Organics; Witco Israel] Calcium alkylaryl sulfonate; anionic; o/w emulsifier for industrial use; intermediate for agric. emulsifiers; amber visc. liq.; oil sol.; sp.gr. 1.01–1.03; pH 6–7.5 (3% in 20% IPA); 68–70% act.

Emcol® P-1045. [Witco/Organics] Amine salt of alkylaryl sulfonic acid; anionic; wetting agent in oil-based systems, sludge dispersant in fuel oil; liq.; oil-sol.; water-disp.; 90% act.

Emcol® P-1059B. [Witco/Organics] Amine salt of alkylaryl sulfonic acid; anionic; surfactant, water solubilizer; liq.; 90% act.

Emcol® TS-230. [Witco/Organics] Org. phosphate ester; dispersant and lubricant in oil drilling muds, o/w emulsifier, flow control agent; amber clear liq.; sol. in xylene, ethanol, CCl_4, water; sp.gr. 1.22; pour pt. < 50 F; acid no. 170; surf. tens. 49.6 dynes/cm (0.05%); pH 2 (3% aq.).

Emcon E. [Fanning] Mixt. of glycerides, phosphatides, and sterols derived from egg; nonionic; emulsifier, emollient; liq.; 100% conc.

Emerest® 1723. [Henkel/Emery] Isopropyl ester of lanolic acids; emollient, pigment dispersing and wetting agent; sol. in warm oils.

Emerest® 2301. [Henkel/Emery] Methyl oleate; nonionic; base for industrial lubricants; mold release agent, defoamer, flotation agent, plasticizer for cellulosic plastics, needle lubricants; when sulfated is useful as wetting, rewetting, and dye leveling agent in textile and leather industries; Gardner < 6 liq.; sol. 5% in min. oil, toluene, IPA, xylene; insol. in water; dens. 7.3 lb/gal; visc. 5 cSt (100 F); pour pt. –16 C; flash pt. 350 F.; 100% act.

Emerest® 2302. [Henkel/Emery; Henkel/Textile] Propyl oleate; nonionic; base for industrial lubricants; mold release agent, defoamer, flotation agent, plasticizer for cellulosic plastics, needle lubricants; when sulfated is useful as wetting, rewetting, and dye leveling agent in textile and leather industries; Gardner < 6 liq.; sol. @ 5% in min. oil, toluene, IPA, xylene; dens. 7.3 lb/gal; visc. 5 cSt (100 F); pour pt. –16 C; flash pt. 350 F; 100% act.

Emerest® 2308. [Henkel/Emery; Henkel/Textile] Tridecyl stearate; nonionic; lubricant used in sewing thread mfg. and fiber finish applics. where high heat stability is desired; Gardner 1 liq.; sol. @ 5% in min. oil, toluene, IPA, xylene; dens. 7.1 lb/gal; visc. 18 cSt (100 F); pour pt. 3 C; flash pt. 440 F; 100% act.

Emerest® 2314. [Henkel/Cospha; Henkel Canada] Isopropyl myristate; cosmetic emollient; sewing thread lubricant; Gardner 1 liq.; sol. 5% in min. oil, toluene, IPA, xylene; water-insol.; dens. 7.1 lb/gal; visc. 4 cSt (100 F); pour pt. –5 C; acid no. 1 max.; iodine no. 1 max.; flash pt. 320 F.

Emerest® 2316. [Henkel/Cospha; Henkel Canada] Isopropyl palmitate; lubricant used for syn. fibers in applics. where low friction is essential; emollient in cosmetic formulations; high purity; Gardner 1 liq.; sol. 5% in min. oil, toluene, IPA, xylene; insol. in water; dens. 7.1 lb/gal; visc. 6 cSt (100 F); acid no. 1 max.; iodine no. 1 max.; pour pt. 14 C; flash pt. 340 F.

Emerest® 2325. [Henkel/Cospha; Henkel Canada] Butyl stearate; nonionic; emollient in creams and lotions; dye solubilizer in lipsticks; lubricant; Gardner 4 liq.; sol. 5% in min. oil, toluene, IPA, xylene; water-insol.; dens. 7.1 lb/gal; visc. 7 cSt (100 F); acid no. 1 max.; iodine no. 0.5 max.; pour pt. 18 C; flash pt. 370 F.

Emerest® 2350. [Henkel/Cospha; Henkel Canada] Glycol stearate; nonionic; emulsifier, opacifying and pearlescing agent, thickener, stabilizer used in liq. cosmetic and detergent compds.; Gardner 1 beads; sol. @ 5% in IPA, toluol, min. oil, xylene; f.p. 50 C; HLB 2.2; acid no. 4 max.; iodine no. 0.5 max.; sapon. no. 185; flash pt. 390 F; 100% act.

Emerest® 2355. [Henkel/Cospha; Henkel Canada] Glycol distearate; nonionic; emulsifier, opacifier, pearlescent, thickener, stabilizer used in liq. detergent and cosmetic prods.; Gardner 4 flakes; sol. 5% in toluene, xylene; disp. in min. oil; insol. in water; f.p. 62 C; HLB 1.3; acid no. 6 max.; iodine no. 1 max.; sapon no. 195; flash pt. 455 F; 100% act.

Emerest® 2380. [Henkel/Cospha; Henkel/Textile; Henkel Canada] Propylene glycol stearate; nonionic; aux. emulsifier, opacifier, pearlescent; for lotions, makeup, textile processing; Gardner 2 solid; disp. @ 5% in min. oil, toluene, xylene; insol. in water; dens. 7.3 lb/gal (45 C); HLB 1.8; m.p. 36 C; acid no. 3 max.; iodine no. 0.5 max.; sapon. no. 171-183; flash pt. 470 F; 100% conc.

Emerest® 2400. [Henkel/Cospha; Henkel Canada] Glyceryl stearate; nonionic; emulsifier for hand creams, cosmetics, textiles, industrial lubricants, polishes, agric.; lubricant softener for textiles; opacifier and pearling agent; EPA-exempt; Gardner 1 beads; sol. 5% in IPA, hot toluol, hot min. oil; insol. in water; HLB 3.9; m.p. 52-58 C; flash pt. 415 F; acid no. 2.0; iodine no. 2.0; sapon. no. 165-175; flash pt. 415 F; 100% act.

Emerest® 2401. [Henkel/Emery; Henkel/Cospha; Henkel/Textile; Henkel Canada] Glyceryl stearate, tech. grade; nonionic; emulsifier and opacifier for hand creams, cosmetics, industrial lubricants, agric., polishes, textiles; EPA-exempt; Gardner 2 beads; sol. @ 5% in IPA, hot toluol, hot min. oil; m.p. 58 C; HLB 3.9; flash pt. 425 F; sapon. no. 153; 100% act.

Emerest® 2407. [Henkel/Emery; Henkel/Cospha; Henkel Canada] Glyceryl stearate SE; nonionic; cosmetic ester; self-emulsifying raw material for emulsions, hair conditioners, creams, lotions; industrial surfactant for agric. formulation; EPA-exempt; Gardner 3 beads; sol. in IPA; sol. hot in min. oil, toluene; disp. in water, xylene; sp.gr. 0.920; m.p. 58 C; HLB 5.1; acid no. 20 max.; iodine no. 1.0 max.; sapon. no. 148–158; cloud pt. < 25 C; flash pt. 385 F.

Emerest® 2410. [Henkel/Cospha; Henkel/Textile; Henkel Canada] Glyceryl isostearate; nonionic; emollient, lubricant, pearling agent, and w/o emulsifier for creams and lotions, textile applics.; exc. oxidation and color stability; Gardner 2 liq.; sol. @

5% in min. oil, IPA; insol. in water; dens. 7.8 lb/gal; visc. 260 cSt (100 F); HLB 2.9; acid no. 5 max.; iodine no. 3 max.; sapon. no. 162-172; pour pt. 5 C; flash pt. 400 F; 100% conc.

Emerest® 2419. [Henkel/Cospha; Henkel Canada] Glyceryl dioleate; CAS 25637-84-7; nonionic; emulsifier, lubricant, rust preventive additive, mold release agent, solv. for dyes and pigments; used in leathers, lubricants and softeners; Gardner 8 liq.; sol. in min. oil, toluene, IPA, xylene; dens. 7.7 lb/gal; visc. 55 cSt (100 F); HLB 1.6; pour pt. < 0 C; flash pt. 470 F; 100% conc.

Emerest® 2421 (See Witconol 2421). [Henkel/Emery; Henkel/Cospha; Henkel Canada] Glyceryl oleate; CAS 37220-82-9; nonionic; emulsifier for cosmetics and industrial applics., in mold release agents, anti-icing fuel additive, rust preventative; in textiles as a lubricant component in syn. fiber spin finishes; vehicle for agric. insecticides; EPA-exempt; Gardner 11 liq.; sol. 5% in toluol, min. oil, xylene; dens. 7.9 lb/gal; visc. 91 cs (38 C); f.p. 6 C; HLB 3.4; pour pt. 19 C; flash pt. 235 C; sapon no. 170; 100% act.

Emerest® 2423. [Henkel/Cospha; Henkel/Textile; Henkel Canada] Triolein; CAS 122-32-7; nonionic; lubricant, w/o emulsifier for metals, leather, textiles, cosmetics, pharmaceuticals; called syn. olive oil; sulfated form used as softener in leather and textile industries; Gardner 3 liq.; sol. 5% in min. oil, xylene; dens. 7.6 lb/gal; visc. 43 cs; HLB 0.6; pour pt. 9 C; flash pt. 293 C; sapon no. 197; 100% act.

Emerest® 2432. [Henkel/Textile] Mixed triglycerides; surfactant for textile use; Gardner 2 liq.; sol. @ 5% in min. oil, toluene, IPA, xylene; dens. 7.6 lb/gal; visc. 59 cSt (100 F); pour pt. 1 C; flash pt. 545 F.

Emerest® 2452. [Henkel/Cospha; Henkel Canada] Polyglyceryl-3 diisostearate; nonionic; emulsifier, solubilizer, dye and pigment wetter; emollient; thickener; solv.; for creams, lotions, lip prods.; Gardner 4 visc. liq.; sol. @ 5% in min. oil, triolein, IPA, IPM; disp. in glycerin; dens. 8.2 lb/gal; visc. 990 cSt (100 F); HLB 6.7; acid no. 20 max.; iodine no. 0.5 max.; sapon. no. 165-175; pour pt. 4 C; flash pt. 455 F; 100% conc.

Emerest® 2485. [Henkel/Emery] Pentaerythritol tetrapelargonate; nonionic; primary lubricant base or modifier in lubricant formulations used in metal working and syn. fiber processing; Gardner 1 liq.; sol. 5% in min. oil, toluene, IPA, xylene; insol. in water; dens. 8.0 lb/gal; visc. 35 cSt (100 F); pour pt. 10 C; flash pt. 555 F.

Emerest® 2610. [Henkel/Emery; Henkel/Textile] PEG-20 stearate; CAS 9004-99-3; nonionic; emulsifier for glyceryl stearate in nonionic textile lubricants and softeners; thickener; antigellant in starch sol'ns.; Gardner 1 solid; sol. @ 5% in water, glyceryl trioleate, xylene; HLB 15.7; m.p. 36 C; cloud pt. 86–90 C; flash pt. 430 F.

Emerest® 2617. [Henkel/Emery; Henkel/Cospha; Henkel Canada] PEG-150 oleate; CAS 9004-96-0; nonionic; strongly hydrophilic emulsifier, stabilizer, lubricant; for agric. formulations; EPA-exempt; Gardner 3 solid; sol. @ 5% in water, xylene; HLB 19.2; m.p. 58 C; cloud pt. 81 C (5% saline); flash pt. 470 F; 100% conc.

Emerest® 2618. [Henkel/Emery] PEG-75 oleate; CAS 9004-96-0; nonionic; emulsifier, plasticizer, lubricant, wetting agent, dispersant, defoamer, solubilizer, visc. modifier for household and industrial prods., agric. formulations; EPA-exempt; Gardner 3 solid; sol. in water, xylene; disp. in Stod.; m.p. 55 C; HLB 18.8; flash pt. 495 F.

Emerest® 2619. [Henkel/Emery] PEG-150 oleate; CAS 9004-96-0; nonionic; emulsifier, plasticizer, lubricant, wetting agent, dispersant, defoamer, solubilizer, visc. modifier for household and industrial prods., agric. formulations; EPA-exempt; Gardner 2 liq.; sol. in water, xylene; disp. in Stod.; dens. 8.9 lb/gal; visc. 1600 cSt (100 F); pour pt. 0 C; HLB 19.2; 50% act.

Emerest® 2620. [Henkel/Emery; Henkel/Cospha; Henkel/Textile; Henkel Canada] PEG-4 laurate; CAS 9004-81-3; nonionic; emulsifier, coupling agent, defoamer in water base coatings, visc. control additive; visc. depressant in vinyl plastisols; agric., textiles; EPA-exempt; Gardner 1 liq.; disp. in water, min. oil, butyl stearate, glyceryl trioleate, perchloroethylene, Stod.; dens. 8.2 lb/gal; visc. 40 cs; HLB 9.3; cloud pt. < 25 C; 100% act.

Emerest® 2622. [Henkel/Emery; Henkel/Cospha; Henkel Canada] PEG-4 dilaurate; CAS 9005-02-1; nonionic; coemulsifer and lubricant in SE textile and industrial oils, agric., mold release agent, visc. control agent; EPA-exempt; Gardner 2 liq.; sol. 5% in min. oil, butyl stearate, glyceryl trioleate, Stod., xylene; disp. in water; dens. 8.0 lb/gal; visc. 35 cs; HLB 7.6; cloud pt. < 25 C; flash pt. 455 F; 100% act.

Emerest® 2624. [Henkel/Emery; Henkel/Cospha; Henkel/Textile; Henkel Canada] PEG-4 oleate; CAS 9004-96-0; nonionic; lubricant component in textile processing; softener/lubricant for leather during tanning; emulsifier for min. oils, fatty oils, and solvs. for cutting oils, solvs. in metal cleaners and degreasers, w/o emulsifier for consumer pesticide aerosols; EPA-exempt; Gardner 3 liq.; sol. 5% in glyceryl trioleate, xylene; disp. in water; dens. 8.1 lb/gal; visc. 34 cSt (100 F); HLB 8.3; pour pt. < –15 C; flash pt. 415 F; cloud pt. < 25 C; 100% conc.

Emerest® 2625. [Henkel/Cospha; Henkel Canada] PEG-4 isostearate; CAS 56002-14-3; nonionic; emulsifier; component in fiber lubricants, processing aids, and conc. liq. fabric softeners; Gardner 2 liq.; sol. 5% in Stod.; water-disp.; dens. 8.2 lb/gal; visc. 50 cSt (100 F); HLB 8.3; pour pt. –8 C; flash pt. 310 F; cloud pt. < 25 C; 100% conc.

Emerest® 2630. [Henkel/Cospha; Henkel Canada] PEG-6 laurate; CAS 9004-81-3; nonionic; hydrophilic emulsifier; lubricant component and scrooping agent for textile fibers and yarns; visc. control agent for plastisols; agric. formulations; EPA-exempt; Gardner 2 liq.; sol. 5% in xylene; water-disp.; dens. 8.4 lb/gal; visc. 37 cSt (100 F); HLB 12.1; pour pt. 9 C; flash pt. 475 F; cloud pt. < 25 C; 100% conc.

Emerest® 2632. [Henkel/Emery; Henkel/Cospha; Henkel/Textile; Henkel Canada] PEG-6 oleate; CAS 9004-96-0; nonionic; emulsifier and lubricant; SE component in formulating textile softeners, agric. formulations; EPA-exempt; Gardner 3 liq.; sol. 5% in xylene; water-disp.; dens. 8.3 lb/gal; visc. 46 cSt (100 F); HLB 10.4; pour pt. –10 C; flash pt. 425 F; cloud pt. < 25 C; 100% conc.

Emerest® 2634. [Henkel/Emery; Henkel/Cospha; Henkel/Textile; Henkel Canada] PEG-6 pelargonate; nonionic; emulsifier, wetting agent, emollient, textile softener, lubricant, defoamer, stabilizer, visc. control agent, pigment wetting, mold release agent; agric. formulations, textile processing; EPA-exempt; Gardner 1 liq.; sol. in xylene; disp. in water, glycerol trioleate; dens. 8.6 lb/gal; visc. 25 cSt (100

F); HLB 12.8; pour pt. < -15 C; cloud pt. < 25 C; flash pt. 485 F; 100% conc.

Emerest® 2636. [Henkel/Emery] PEG-6 stearate; CAS 9004-99-3; nonionic; waxy emulsifier for oils and fats in industrial lubricants, agric. formulations; softener and lubricant in textiles and leather; EPA-exempt; Gardner 1 solid; sol. 5% in glyceryl trioleate, xylene; water-disp.; dens. 8.3 lb/gal; visc. 45 cSt (100 F); HLB 10.1; m.p. 29 C; flash pt. 460 F; cloud pt. < 25 C.

Emerest® 2640. [Henkel/Emery; Henkel/Textile] PEG-8 stearate; CAS 9004-99-3; nonionic; emulsifier for oils and fats in mfg. of industrial lubricants, agric., consumer prods., textile lubricants and softeners; lipophilic; thickener and stabilizer for starch coatings on paper; paper size; lubricant for channeling wire through conduit; EPA-exempt; Gardner 1 soft, waxy solid; sol. see Emerest 2632; dens. 8.5 lb/gal; visc. 57 cSt (100 F); HLB 12.0; m.p. 32 C; flash pt. 425 F; cloud pt. < 25 C; 100% act.

Emerest® 2641. [Henkel/Textile] PEG-8 stearate, anhyd.; CAS 9004-99-3; nonionic; surfactant for textile use; Gardner 1 solid; sol. @ 5% in xylene; disp. in water; dens. 8.5 lb/gal; HLB 12.0; m.p. 32 C; cloud pt. < 25 C; flash pt. 425 F.

Emerest® 2642. [Henkel/Emery; Henkel/Cospha; Henkel Canada] PEG-8 distearate; CAS 9005-08-7; nonionic; lipophilic waxy surfactant used as emulsifier and thickener in cosmetic, agric. and industrial emulsions; EPA-exempt; Gardner 2 solid; sol. (5%) in min. oil, butyl stearate, glycol trioleate, Stod., xylene; water-disp.; visc. 52 cSt (100 F); HLB 7.5; m.p. 36 C; flash pt. 470 F; cloud pt. < 25 C; 100% conc.

Emerest® 2644. [Henkel/Cospha; Henkel Canada] PEG-8 isostearate; CAS 56002-14-3; nonionic; emulsifier; component in fiber lubricants, processing aids, and conc. liq. fabric softeners; Gardner 1 liq.; disp. in water, min. oil, butyl stearate, xylene; dens. 8.5 lb/gal; visc. 70 cSt (100 F); HLB 11.3; pour pt. 10 C; flash pt. 450 F; cloud pt. < 25 C; 100% conc.

Emerest® 2646. [Henkel/Emery; Henkel/Cospha; Henkel/Textile; Henkel Canada] PEG-8 oleate; CAS 9004-96-0; nonionic; emulsifier for oil-based sol. cutting oils, specialty industrial lubricants, industrial degreasers, agric. formulations; stabilizes visc. of vinyl plastisols; component in textile softeners; drycleaning formulations; EPA-exempt; Gardner 2 liq.; sol. 5% in xylene; water-disp.; dens. 8.5 lb/gal; visc. 52 cSt (100 F); HLB 11.8; pour pt. 5 C; flash pt. 455 F; cloud pt. < 25 C; 100% conc.

Emerest® 2647. [Henkel/Emery; Henkel/Cospha; Henkel/Textile; Henkel Canada] PEG-8 sesquioleate; nonionic; emulsifier, wetting agent, emollient, textile softener, lubricant, defoamer, stabilizer, visc. control agent, pigment wetting, mold release agent, agric. formulations; EPA-exempt; Gardner 3 liq.; sol. in min. oil, butyl stearate, glycerol trioleate, xylene; disp. in water, Stod.; dens. 8.2 lb/gal; visc. 50 cSt (100 F); pour pt. -6 C; HLB 9.4; cloud pt. < 25 C; flash pt. 550 F; 100% conc.

Emerest® 2648. [Henkel/Emery; Henkel/Cospha; Henkel/Textile; Henkel Canada] PEG-8 dioleate; CAS 9005-07-6; nonionic; lipophilic emulsifier and solubilizer for min. oils, fats, and solvs.; emulsifier for kerosene in agric. and pesticide sprays; emulsification of latex paints, metalworking fluids, solvs.; speciality and industrial lubricants; textiles; EPA-exempt; Gardner 3 liq.; sol. in min. oil, glyceryl trioleate, xylene; water-disp.; dens. 8.1 lb/gal; visc. 45 cSt (100 F); HLB 8.8; pour pt. 6 C; flash pt. 515 F; cloud pt. < 25; 100% act.

Emerest® 2650. [Henkel/Emery; Henkel/Cospha; Henkel/Textile; Henkel Canada] PEG-8 laurate; CAS 9004-81-3; nonionic; hydrophilic surfactant functioning as leveling and wetting agent; defoamer in latex paints; dispersant in pigment grinding; solubilizer for oils and solvs.; antiblock agent in vinyls; textile processing; agric.; EPA-exempt; Gardner 1 liq.; sol. 5% in water, xylene; dens. 8.6 lb/gal; visc. 41 cSt (100 F); HLB 13.2; pour pt. 12 C; flash pt. 450 F; cloud pt. 33 C; 100% act.

Emerest® 2652. [Henkel/Emery; Henkel/Cospha; Henkel Canada] PEG-8 dilaurate; CAS 9005-02-1; nonionic; emulsifier, coupler, lubricant, softener, and release agent for paper, agric.; textile industry, coupler and lubricant in syn. fiber spin finishes; EPA-exempt; Gardner 2 liq.; sol. 5% in min. oil, butyl stearate, glyceryl trioleate, Stod., xylene; water-disp.; dens. 8.3 lb/gal; visc. 38 cSt (100 F); HLB 10.8; pour pt. 8 C; flash pt. 420 F; cloud pt. < 25 C; 100% act.

Emerest® 2654. [Henkel/Emery; Henkel/Cospha; Henkel/Textile; Henkel Canada] PEG-8 pelargonate; nonionic; surfactant used as base lubricant for syn. fiber spin finishes, other textile processing; coemulsifier and coupler for agric. formulations; EPA-exempt; Gardner 1 liq.; sol. 5% in water, xylene; dens. 8.7 lb/gal; visc. 34 cSt (100 F); HLB 14.3; pour pt. 5 C; flash pt. 440 F; cloud pt. 40 C (2% saline); 100% conc.

Emerest® 2658. [Henkel/Textile] PEG-8 pelargonate; nonionic; surfactant for textile use; salt-free; Gardner 1 liq.; sol. @ 5% in water, xylene; dens. 8.7 lb/gal; visc. 34 cSt (100 F); HLB 14.3; pour pt. 5 C; cloud pt. 37 C (2% saline); flash pt. 440 F; 70% act.

Emerest® 2660. [Henkel/Emery; Henkel/Cospha; Henkel/Textile; Henkel Canada] PEG-12 oleate; CAS 9004-96-0; nonionic; dye leveling agent in textiles; emulsifier in specialty lubricants, agric. formulations; detergent; acid washing of printed circuit boards; EPA-exempt; Gardner 2 liq.; sol. 5% in water, xylene; dens. 8.7 lb/gal; visc. 75 cSt (100 F); HLB 13.6; pour pt. 18 C; flash pt. 545 F; cloud pt. 47 C; 100% act.

Emerest® 2661. [Henkel/Emery; Henkel/Cospha; Henkel/Textile; Henkel Canada] PEG-12 laurate; CAS 9004-81-3; nonionic; lubricant in processing syn. fibers; dispersant, emulsifier in agric. formulations; EPA-exempt; Gardner 2 liq.; sol. 5% in water, xylene; dens. 8.6 lb/gal; visc. 60 cSt (100 F); HLB 14.8; pour pt. 14 C; flash pt. 525 F; cloud pt. 63 C; 100% conc.

Emerest® 2662. [Henkel/Emery; Henkel/Cospha; Henkel/Textile; Henkel Canada] PEG-12 stearate; CAS 9004-99-3; nonionic; emulsifier for cosmetic, agric., textile formulations; textile lubricants and softeners; visc. modifier in creams and lotions; EPA-exempt; Gardner 1 solid; sol. 5% in xylene; water-disp.; dens. 8.5 lb/gal; HLB 13.8; m.p. 40 C; flash pt. 440 F; cloud pt. 55 C; 100% conc.

Emerest® 2664. [Henkel/Emery] PEG-12 isostearate; CAS 56002-14-3; used in fiber lubricants and processing aids; emulsifier in formulations; Gardner 5 liq.; sol. 5% in glyceryl trioleate, xylene; water-disp.; dens. 8.6 lb/gal; visc. 90 cSt (100 F); HLB 13.0; pour pt. 10 C; flash pt. 490 F; cloud pt. < 25 C.

Emerest® 2665. [Henkel/Emery; Henkel/Cospha; Henkel Canada] PEG-12 dioleate; CAS 9005-07-6; nonionic; lipophilic emulsifier and solubilizer for min. oils, fats, and solvs.; emulsifier for kerosene in agric. and pesticide sprays; emulsification of latex paints, metalworking fluids, solvs.; specialty and industrial lubricants; EPA-exempt; Gardner 4 liq.; sol. in butyl stearate, glyceryl trioleate, Stod., xylene; water-disp.; dens. 8.3 lb/gal; visc. 64 cSt (100 F); HLB 10.3; pour pt. 19 C; flash pt. 530 F; cloud pt. < 25 C; 100% conc.

Emerest® 2675. [Henkel/Emery] PEG-50 stearate; CAS 9004-99-3; nonionic; hydrophilic emulsifier used for preparing solubilized oils; visc. modifier, softener or plasticizer in acrylic or vinyl resin emulsions; Gardner 1 liq.; sol. @ 5% in water; dens. 8.5 lb/gal; visc. 671 cSt (100 F); HLB 17.8; pour pt. 0 C; cloud pt. 81 C (5% saline); flash pt. 540 F; 30% act. in water.; 30% aq.

Emerest® 2704. [Henkel/Cospha; Henkel Canada] PEG-4 dilaurate; CAS 9005-02-1; nonionic; emulsifier, lubricant, dispersant for bath oils; visc. control agent for creams and lotions; for cosmetic and industrial applics.; Gardner 2 clear liq.; sol. @ 5% in min. oil, triolein, IPA; disp. in water, IPM, glycerin; dens. 8.0 lb/gal; visc. 22 cSt (100 F); HLB 7.6; acid no. 10 max.; iodine no. 10 max.; sapon. no. 175-185; pour pt. 0 C; cloud pt. < 25 C; flash pt. 455 F; 100% conc.

Emerest® 2711. [Henkel/Cospha; Henkel Canada] PEG-8 stearate; CAS 9004-99-3; nonionic; emulsifier, thickener for o/w and w/o systems, industrial and cosmetic applics.; Gardner 1 waxy solid; sol. @ 5% in glycerin; disp. in water, min. oil, triolein, IPA, IPM; dens. 8.5 lb/gal (35 C); visc. 57 cSt (100 F); HLB 11.4; m.p. 30 C; acid no. 5 max.; iodine no. 1 max.; sapon. no. 82-92; cloud pt. < 25 C; flash pt. 500 F; 100% conc.

Emerest® 2712. [Henkel/Cospha; Henkel Canada] PEG-8 distearate; CAS 9005-08-7; nonionic; emulsifier, opacifier, thickener for firm creams and high-visc. lotions; for cosmetic and industrial applics.; Gardner 2 waxy solid; sol. @ 5% in min. oil, triolein, IPA; disp. in IPM, glycerin; dens. 7.9 lb/gal (53 C); visc. 52 cSt (100 F); HLB 8.1; m.p. 36 C; acid no. 10 max.; iodine no. 1 max.; sapon. no. 115-125; cloud pt. < 25 C; flash pt. 470 F; 100% conc.

Emerest® 2715. [Henkel/Cospha; Henkel/Textile; Henkel Canada] PEG-40 stearate; CAS 9004-99-3; nonionic; emulsifier, stabilizer, antigellant, lubricant for creams, lotions, shampoos, deodorants, makeup, pharmaceuticals; textile softener; Gardner 1 waxy solid or flake; sol. @ 5% in water, IPA; dens. 8.9 lb/gal (50 C); HLB 17.3; m.p. 50 C; acid no. 1 max.; sapon. no. 25-35; cloud pt. 75–81 C (5% saline); flash pt. 515 F; 100% conc.

Emerest® 11723. [Henkel/Emery] Isopropyl ester of lanolic acids; emollient, pigment wetting, dispersing agent, conditioner; sol. in warm oils.

Emerox® 1110. [Henkel/Emery] Azelaic acid; detergent intermediate; m.p. 96-103 C; acid no. 576-591.

Emerox® 1144. [Henkel/Emery] Azelaic acid; detergent intermediate; m.p. 101-102 C; acid no. 587-594.

Emersist 7210. [Henkel/Textile] Acid dye leveling agent for nylon; optimum leveling and enhanced color yield; clear to sl. hazy liq.; sol. forms clear sol'ns. in water; pH 8.0 (2%).

Emersist 7230. [Henkel/Textile] Conc. low temp. (90-120 F) scour and wetting agent for cotton and synthetic goods; clear liq.; forms clear sol'ns.; pH 6.0 (1%).

Emersist 7233. [Henkel/Textile] Conc. high temp. (160-200 F) scour and wetting agent for cotton and synthetic goods; clear liq.; forms clear sol'ns.; pH 6.0 (1%).

Emersol® 110. [Henkel/Emery] Stearic acid; CAS 57-11-4; detergent intermediate; opacifier in cosmetics, soaps, emulsifiers, chemical specialties; acid no. 205–210; iodine no. 8–12.

Emersol® 120. [Henkel/Emery] Stearic acid; CAS 57-11-4; detergent intermediate; opacifier in cosmetics, soaps, emulsifiers, chemical specialties; acid no. 205-210; iodine no. 5-7.

Emersol® 132 NF Lily®. [Henkel/Emery] Triple pressed stearic acid; CAS 57-11-4; detergent intermediate; opacifier in cosmetics, soaps, emulsifiers, chemical specialties; acid no. 205–210; iodine no. 0.5 max.

Emersol® 143. [Henkel/Emery] Palmitic acid; CAS 57-10-3; detergent intermediate; opacifier in cosmetics, soaps, emulsifiers, chemical specialties; acid no. 215–223; iodine no. 1 max.

Emersol® 150. [Henkel/Emery] Stearic acid; CAS 57-11-4; detergent intermediate; acid no. 197-202; iodine no. 1 max.

Emersol® 152 NF, 153 NF. [Henkel/Emery] Stearic acid; CAS 57-11-4; detergent intermediate; acid no. 196-199; iodine no. 1 max.

Emersol® 210. [Henkel/Emery] Oleic acid; CAS 112-80-1; detergent intermediate for personal care, emollient, household and industrial applics.; acid no. 197-204; iodine no. 87-95.

Emersol® 213 NF. [Henkel/Emery] Low-titer oleic acid; CAS 112-80-1; detergent intermediate for personal care, emollient, household and industrial applics.; acid no. 199-204; iodine no. 88-95.

Emersol® 221 NF. [Henkel/Emery] Low-titer wh. oleic acid; CAS 112-80-1; detergent intermediate for personal care, emollient, household and industrial applics.; acid no. 199-204; iodine no. 88-95.

Emersol® 233 LL. [Henkel/Emery] Oleic acid; CAS 112-80-1; detergent intermediate for personal care, emollient, household and industrial applics.; acid no. 200-204; iodine no. 86-90.

Emersol® 315. [Henkel/Emery] Linoleic acid; detergent intermediate for personal care, emollient, household and industrial applics.; acid no. 195-202; iodine no. 145-160.

Emersol® 871. [Henkel/Emery] Isostearic acid; detergent intermediate for personal care, emollient, household and industrial applics.; acid no. 175 min.; iodine no. 12 max.

Emersol® 875. [Henkel/Emery] Isostearic acid; detergent intermediate for personal care, emollient, household and industrial applics.; acid no. 187–197; iodine no. 3 max.

Emersol® 6313 NF. [Henkel/Emery] Low-titer oleic acid USP/NF; CAS 112-80-1; food grade fatty acid; also as binder, defoamer and lubricant for pesticides; detergent intermediate in personal care, emollients, household/industrial detergents; FDA-approved; EPA-exempt; acid no. 201–204; iodine no. 88–93.

Emersol® 6320. [Henkel/Emery] Stearic acid, double pressed; CAS 57-11-4; food grade fatty acid; also as binder, defoamer, lubricant in pesticides; detergent intermediate in personal care, emollients, household/industrial detergents; FDA-approved; EPA-exempt; acid no. 205–210; iodine no. 3.5–5.0.

Emersol® 6321 NF. [Henkel/Emery] Low-titer wh. oleic acid UPS/NF; CAS 112-80-1; food grade fatty acid; also as binder, defoamer and lubricant for pesticides; detergent intermediate in personal care, emollients, household/industrial detergents; FDA-approved; EPA-exempt; acid no. 201–204; iodine no. 87–92.

Emersol® 6332 NF. [Henkel/Emery] Stearic acid, triple pressed; CAS 57-11-4; food grade fatty acid; also as binder, defoamer and lubricant in pesticides; detergent intermediate in personal care, emollients, household/industrial detergents; FDA-approved; EPA-exempt; acid no. 205–211; iodine no. 0.5 max.

Emersol® 6333 NF. [Henkel/Emery] Low-linoleic content oleic acid USP/NF; CAS 112-80-1; food grade fatty acid; also as binder, defoamer and lubricant for pesticides; detergent intermediate in personal care, emollients, household/industrial detergents; FDA-approved; EPA-exempt; acid no. 200–204; iodine no. 86–91.

Emersol® 6349. [Henkel/Emery] Stearic acid; CAS 57-11-4; food grade fatty acid; also as binder, defoamer and lubricant for pesticides; detergent intermediate in personal care, emollients, household/industrial detergents; FDA-approved; EPA-exempt; acid no. 203–206; iodine no. 0.5 max.

Emersol® 6351. [Henkel/Emery] Stearic acid; CAS 57-11-4; food grade fatty acid; also as binder, defoamer and lubricant for pesticides; detergent intermediate in personal care, emollients, household/industrial detergents; FDA-approved; EPA-exempt; acid no. 196–201; iodine no. 1.0 max.

Emersol® 7021. [Henkel/Emery] Oleic acid; CAS 112-80-1; food grade kosher fatty acid; detergent intermediate in personal care, emollients, household/industrial detergents; FDA-approved; acid no. 196-204; iodine no. 93-104.

Emery® 400. [Henkel/Emery] Stearic acid; CAS 57-11-4; detergent intermediate; acid no. 197-212; iodine no. 9.5 max.

Emery® 401. [Henkel/Emery] Fatty acid; detergent intermediate; acid no. 199-208; iodine no. 34-44.

Emery® 404. [Henkel/Emery] Stearic acid; CAS 57-11-4; detergent intermediate; acid no. 197-209; iodine no. 6-9.

Emery® 405. [Henkel/Emery] Stearic acid; CAS 57-11-4; detergent intermediate; acid no. 195-205; iodine no. 6 max.

Emery® 410. [Henkel/Emery] Stearic acid; CAS 57-11-4; detergent intermediate; acid no. 195-209; iodine no. 7 max.

Emery® 420. [Henkel/Emery] Stearic acid; CAS 57-11-4; detergent intermediate; acid no. 200-207; iodine no. 1 max.

Emery® 422. [Henkel/Emery] Stearic acid; CAS 57-11-4; detergent intermediate; acid no. 203-209; iodine no. 1 max.

Emery® 515. [Henkel/Emery] Tallow/coconut fatty acid blend; detergent intermediate; acid no. 212-218; iodine no. 44-54.

Emery® 516. [Henkel/Emery] Tallow/coconut fatty acid blend; detergent intermediate; acid no. 214-222; iodine no. 35-42.

Emery® 517. [Henkel/Emery] Tallow/coconut fatty acid blend; detergent intermediate; acid no. 216-222; iodine no. 42 max.

Emery® 531. [Henkel/Emery] Tallow fatty acid; detergent intermediate; acid no. 200-208; iodine no. 45-70.

Emery® 610. [Henkel/Emery] Soya fatty acid; detergent intermediate; Gardner 3 max. color; acid no. 195-205; iodine no. 125-138.

Emery® 618. [Henkel/Emery] Soya fatty acid; detergent intermediate; Gardner 3 max. color; acid no. 197-203; iodine no. 138-145.

Emery® 621. [Henkel/Emery] Coconut fatty acid; detergent intermediate in personal care, emollients, household/industrial detergents; Gardner 5 color; acid no. 258-268; iodine no. 5-16.

Emery® 622. [Henkel/Emery] Coconut fatty acid; detergent intermediate in personal care, emollients, household/industrial detergents; Gardner 2 color; acid no. 268-276; iodine no. 5-10.

Emery® 625. [Henkel/Emery] Partially hydrog. coconut fatty acid; detergent intermediate in personal care, emollients, household/industrial detergents; Gardner 1 color; acid no. 269-273; iodine no. 5 max.

Emery® 626. [Henkel/Emery] Low IV ultra coconut fatty acid; detergent intermediate in personal care, emollients, household/industrial detergents; Gardner 1 color; acid no. 270-276; iodine no. 1 max.

Emery® 627. [Henkel/Emery] Low IV, stripped ultra coconut fatty acid; detergent intermediate in personal care, emollients, household/industrial detergents; Gardner 1 color; acid no. 252-258; iodine no. 1 max.

Emery® 629. [Henkel/Emery] Stripped coconut fatty acid; detergent intermediate in personal care, emollients, household/industrial detergents; Gardner 1 color; acid no. 253-259; iodine no. 5-10.

Emery® 650. [Henkel/Emery] Lauric acid; CAS 143-07-7; detergent surfactant; acid no. 266-272; iodine no. 0.4 max.

Emery® 651. [Henkel/Emery] Lauric acid; CAS 143-07-7; detergent surfactant; acid no. 276-282; iodine no. 0.2 max.

Emery® 652. [Henkel/Emery] Lauric acid; CAS 143-07-7; detergent surfactant; acid no. 277-281; iodine no. 0.2 max.

Emery® 654. [Henkel/Emery] Myristic acid; CAS 544-63-8; detergent surfactant; acid no. 244-247; iodine no. 0.5 max.

Emery® 655. [Henkel/Emery] Myristic acid; CAS 544-63-8; detergent surfactant; acid no. 243-246; iodine no. 0.5 max.

Emery® 657. [Henkel/Emery] Caprylic acid; detergent surfactant; acid no. 385-390; iodine no. 0.2 max.

Emery® 658. [Henkel/Emery] Caprylic-capric acid; detergent surfactant; acid no. 356-366; iodine no. 0.3 max.

Emery® 659. [Henkel/Emery] Capric acid; detergent surfactant; acid no. 322-326; iodine no. 0.5 max.

Emery® 876. [Henkel/Emery] Fatty acid; detergent intermediate; acid no. 235-269; iodine no. 2-6.

Emery® 877. [Henkel/Emery] Fatty acid; detergent intermediate; acid no. 240-270; iodine no. 2 max.

Emery® 878. [Henkel/Emery] Fatty acid; detergent intermediate; acid no. 295-315; iodine no. 1 max.

Emery® 880. [Henkel/Emery] Fatty acid; soaps used as lubricants in metals and industrial applics.; liq.

Emery® 912. [Henkel/Emery] CP/USP glycerin; skin softener, solubilizer, visc. modifier, flavor enhancer, moisturizer, solv., humectant, thickener, and solubilizer in cosmetics, drug vehicles, food applics., glass, ceramics, agric., and adhesives; EPA-exempt; APHA 20 max. visc. liq.; odorless; sp.gr. 1.2517 min.; pour pt. 18 C; 96.0% min. glyceryl.

Emery® 916. [Henkel/Emery] CP/USP glycerin; skin softener, solubilizer, visc. modifier, flavor enhancer, moisturizer, solv., humectant, thickener and solubilizer in cosmetics, drug vehicles, food applics., glass, ceramics, agric., and adhesives; EPA-exempt; APHA 20 max. visc. liq.; odorless; sp.gr. 1.2607 min.; pour pt. 18 C; 99.5% min. glyceryl.

Emery® 918. [Henkel/Emery] CP/USP glycerin; skin softener, solubilizer, visc. modifier, flavor enhancer, moisturizer, solv., humectant, and solubilizer in cosmetics, drug vehicles, and food applics., glass, ceramics, and adhesives; APHA 8 max.; sp.gr. 1.2615; 99.8% min. glyceryl.

Emery® 1202. [Henkel/Emery] Pelargonic acid; detergent intermediate; acid no. 345-355; iodine no. 0.5 max.

Emery® 1650. [Henkel/Cospha; Henkel/Textile; Henkel Canada] Anhyd. lanolin USP; nonionic; emulsifier, emollient, conditioner, lubricant for cosmetics, sun care prods., textiles; Gardner < 9 paste; sol. @ 5% in IPM; disp. in min. oil, triolein; HLB 10.0; dens. 7.9 lb/gal; m.p. 36–42 C; flash pt. 530 F; 100% conc.

Emery® 1656. [Henkel/Cospha; Henkel/Textile; Henkel Canada] Anhyd. lanolin USP; emulsifier, emollient, conditioner, moisturizer, pigment dispersant for pharmaceutical ointments, veterinary prods., industrial hand cleansers, cosmetics, textiles; Gardner < –12 paste; sol. in IPM; disp. in min. oil, triolein; dens. 7.9 lb/gal; m.p. 36–42 C; flash pt. 530 F.

Emery® 1780. [Henkel/Cospha; Henkel Canada] Lanolin alcohol; w/o emulsifier, emollient, visc. builder, stabilizer for emulsions, personal care prods.; Gardner 10 solid; sol. @ 5% in triolein; dens. 8.2 lb/gal; m.p. 48 C; flash pt. 440 F.

Emery® 2031 Agrimul S-300. [Henkel/Emery] Blend; anionic/nonionic; for high active agric. emulsifiable concs. and phosphated toxicants; EPA-exempt; amber liq.; sol. @ 5% in water, xylene; disp. in methyl oleate; dens. 8.6 lb/gal; visc. 2240 cps. DISCONTINUED.

Emery® 2032 Agrimul A-300. [Henkel/Emery] Blend; anionic; for agric. applics.; EPA-exempt; amber liq.; sol. @ 5% in xylene, methyl oleate; disp. in water; dens. 8.8 lb/gal; visc. 1050 cps. DISCONTINUED.

Emery® 2203. [Henkel/Emery] Methyl tallowate; detergent intermediate; solv. for pesticides and herbicides; EPA-exempt; Gardner 1 color; dens. 7.2 lb/gal; visc. 6.7 cSt (100 C); m.p. 10 C; acid no. 1.0 max.; iodine no. 47-53; sapon. no. 190-200; flash pt. 355 F.

Emery® 2204. [Henkel/Emery] Hydrog. methyl tallowate; detergent intermediate; acid no. 1.0 max.; iodine no. 1 max.; sapon. no. 192-198.

Emery® 2209. [Henkel/Emery] Methyl caprylate-caprate; detergent intermediate; solv. for pesticides and herbicides; EPA-exempt; Gardner 1 color; dens. 7.1 lb/gal; visc. 1.6 cSt (100 C); m.p. -30 C; acid no. 0.5 max.; iodine no. 0.4 max.; sapon. no. 330-336; flash pt. 180 F.

Emery® 2214. [Henkel/Emery] Methyl myristate; detergent intermediate; solv. for pesticides and herbicides; EPA-exempt; Gardner 1 color; dens. 7.2 lb/gal; visc. 3.4 cSt (100 C); m.p. 17 C; acid no. 1.0 max.; iodine no. 0.6 max.; sapon. no. 230-234; flash pt. 300 F; 95% C14 ester.

Emery® 2216. [Henkel/Emery] Methyl palmitate; detergent intermediate; solv. for pesticides and herbicides; EPA-exempt; Gardner 1 color; m.p. 27 C; acid no. 0.2 max.; iodine no. 0.2 max.; sapon. no. 206-210; flash pt. 330 F; 95% C14 ester.

Emery® 2218. [Henkel/Emery] Methyl stearate; detergent intermediate; solv. for pesticides and herbicides; EPA-exempt; Gardner 1 color; m.p. 36 C; acid no. 0.5 max.; iodine no. 1.0 max.; sapon. no. 188-192; flash pt. 307 F; 95% C18 ester.

Emery® 2219. [Henkel/Emery] Methyl oleate; detergent intermediate; solv. for pesticides and herbicides; EPA-exempt; Gardner 1 color; dens. 6.8 lb/gal; m.p. 18 C; acid no. 4.0 max.; iodine no. 66-74; sapon. no. 188-192; flash pt. 345 F; 24% C18 stearate, 58% C18 oleate (unsat.), 14% C18 linoleate (unsat.).

Emery® 2253. [Henkel/Emery] Methyl coconate; detergent intermediate; solv. for pesticides and herbicides; EPA-exempt; Gardner 1 color; dens. 7.5 lb/gal; visc. 2.8 cSt (100 C); m.p. 4 C; acid no. 0.5 max.; iodine no. 7-11; sapon. no. 250-260; flash pt. 250 F.

Emery® 2254. [Henkel/Emery] Stripped methyl coconate; detergent intermediate; solv. for pesticides and herbicides; EPA-exempt; Gardner 1 color; acid no. 1.0 max.; iodine no. 5-10; sapon. no. 237-247; flash pt. 295 F.

Emery® 2255. [Henkel/Emery] Methyl palm kernelate; detergent intermediate; acid no. 1.0 max.; iodine no. 14-20; sapon. no. 230-240.

Emery® 2270. [Henkel/Emery] Methyl laurate; detergent intermediate; solv. for pesticides and herbicides; EPA-exempt; Gardner 1 color; dens. 7.2 lb/gal; visc. 2.8 cSt (100 C); m.p. -1 C; acid no. 0.5 max.; iodine no. 0.5 max.; sapon. no. 251-255; flash pt. 270 F; 70% C12, 28% C14.

Emery® 2290. [Henkel/Emery] Methyl laurate; detergent intermediate; solv. for pesticides and herbicides; EPA-exempt; Gardner 1 color; dens. 7.2 lb/gal; visc. 2.5 cSt (100 C); m.p. 2 C; acid no. 0.5 max.; iodine no. 0.5 max.; sapon. no. 258-262; flash pt. 265 F; 90% C12 ester.

Emery® 2296. [Henkel/Emery] Methyl laurate; detergent intermediate; solv. for pesticides and herbicides; EPA-exempt; Gardner 1 color; dens. 7.2 lb/gal; visc. 2.3 cSt (100 C); m.p. 5 C; acid no. 0.5 max.; iodine no. 0.5 max.; sapon. no. 260-264; flash pt. 305 F; 96% C12 ester.

Emery® 2668 Agrimul 26-B. [Henkel/Emery] Formulated prod.; nonionic; surfactant for agric. formulations; EPA-exempt; amber liq.; dens. 8.3 lb/gal; visc. 95 cps; HLB 14.0. DISCONTINUED.

Emery® 2895 Foamaster Soap L. [Henkel/Emery] Sodium tallowate; defoamer for dry agric. formulations; EPA-exempt; tan powd. flake; 93% act.

Emery® 3304. [Henkel/Emery] Behenyl alcohol (5-15% C18, 5-20% C20, ≥ 70% C22); chemical intermediate for detergent mfg.; APHA 20 fused/flakes; sp.gr. 0.800-0.810 g/cc (70 C); solid. pt. 64-67 C; b.p. 360-400 C; acid no. < 0.2; iodine no. < 1.0; sapon. no. < 0.5; hyd. no. 170-180; flash pt. 200 C.

Emery® 3310. [Henkel/Emery] Oleyl/cetyl alcohol (25-33% C16, 60-70% C18); chemical intermediate for detergent mfg.; APHA 100 solid; sp.gr. 0.820-0.830 g/cc (40 C); solid. pt. 30-37 C; b.p. 315-360 C; acid no. < 0.2; iodine no. 45-50; sapon. no. < 1.0; hyd. no. 210-220; flash pt. 180 C.

Emery® 3311. [Henkel/Emery] Oleyl/cetyl alcohol (25-35% C16, 55-75% C18); chemical intermediate for detergent mfg.; APHA 100 solid; sp.gr. 0.820-0.830 g/cc (40 C); solid. pt. 28-34 C; b.p. 315-360 C; acid no. < 0.2; iodine no. 50-55; sapon. no. < 1.0; hyd. no. 210-220; flash pt. 170 C.

Emery® 3312. [Henkel/Emery] Oleyl alcohol (2-8% C16, 87-93% C18); CAS 143-28-2; chemical intermediate for detergent mfg.; APHA 100 liq.; sp.gr. 0.830-0.840 g/cc (40 C); solid. pt. 2-6 C; b.p. 330-360 C; acid no. < 0.2; iodine no. 95-105; sapon. no. < 1.0; hyd. no. 205-220; flash pt. 190 C.

Emery® 3313. [Henkel/Emery] Oleyl/cetyl alcohol (25-35% C16, 58-68% C18); chemical intermediate for detergent mfg.; APHA 200 solid; sp.gr. 0.825-0.835 g/cc (40 C); solid. pt. 25-35 C; b.p. 315-360 C; acid no. < 0.2; iodine no. 48-55; sapon. no. < 2.0; hyd. no. 208-218; flash pt. 170 C.

Emery® 3314. [Henkel/Emery] Oleyl/cetyl alcohol (12-22% C16, 68-80% C18); chemical intermediate for detergent mfg.; APHA 100 solid/liq.; sp.gr. 0.825-0.835 g/cc (40 C); solid. pt. 18-23 C; b.p. 315-360 C; acid no. < 0.2; iodine no. 70-75; sapon. no. < 1.0; hyd. no. 208-218; flash pt. 180 C.

Emery® 3315. [Henkel/Emery] Oleyl/cetyl alcohol (8-18% C16, 70-83% C18); chemical intermediate for detergent mfg.; APHA 100 liq.; sp.gr. 0.825-0.835 g/cc (40 C); solid. pt. 3-13 C; b.p. 315-360 C; acid no. < 0.2; iodine no. 84-89; sapon. no. < 1.0; hyd. no. 205-215; flash pt. 180 C.

Emery® 3316. [Henkel/Emery] Oleyl/cetyl alcohol 20-30% C16, 55-75% C18); chemical intermediate for detergent mfg.; APHA 100 solid; sp.gr. 0.825-0.835 g/cc (40 C); solid. pt. 25-30 C; b.p. 315-360 C; acid no. < 0.2; iodine no. 60-65; sapon. no. < 1.0; hyd. no. 208-218; flash pt. 170 C.

Emery® 3317. [Henkel/Emery] Oleyl alcohol (2-10% C16, 87-95% C18); CAS 143-28-2; chemical intermediate for detergent mfg.; APHA 100 liq.; sp.gr. 0.830-0.840 g/cc (40 C); solid. pt. 2-12 C; b.p. 330-360 C; acid no. < 0.2; iodine no. 90-97; sapon. no. < 1.0; hyd. no. 205-215; flash pt. 190 C.

Emery® 3318. [Henkel/Emery] Oleyl linoleyl alcohol (5-10% C16, 90-95% C18); chemical intermediate for detergent mfg.; APHA 100 liq./solid; sp.gr. 0.830-0.840 g/cc (40 C); solid. pt. 10-21 C; b.p. 330-360 C; acid no. < 0.2; iodine no. 110-130; sapon. no. < 1.0; hyd. no. 200-220; flash pt. 190 C.

Emery® 3319. [Henkel/Emery] Sat. fatty alcohol (26-35% C16, 60-70% C18); chemical intermediate for detergent mfg.; APHA 20 fused/flakes; sp.gr. 0.805-0.815 g/cc (60 C); solid. pt. 48-52 C; b.p. 310-360 C; acid no. < 0.1; iodine no. < 1.0; sapon. no. < 0.5; hyd. no. 210-220; flash pt. 180 C.

Emery® 3320. [Henkel/Emery] Tallow alcohol (25-35% C16, 60-67% C18); chemical intermediate for detergent mfg.; APHA 20 fused/flakes; sp.gr. 0.805-0.815 g/cc (60 C); solid. pt. 48-52 C; b.p. 300-360 C; acid no. < 0.1; iodine no. < 1.0; sapon. no. < 1.2; hyd. no. 210-220; flash pt. 180 C.

Emery® 3321. [Henkel/Emery] Hexyl alcohol; chemical intermediate for detergent mfg.; APHA 10 liq.; sp.gr. 0.810-0.820 g/cc; solid. pt. 50 C; b.p. 145-160 C; acid no. < 0.1; iodine no. < 0.3; sapon. no. < 1.0; hyd. no. 530-550; flash pt. 65 C.

Emery® 3322. [Henkel/Emery] Octyl alcohol; chemical intermediate for detergent mfg.; APHA 10 liq.; sp.gr. 0.815-0.825 g/cc; solid. pt. 16 C; b.p. 185-195 C; acid no. < 0.1; iodine no. < 0.1; sapon. no.< 0.1; hyd. no. 429-431; flash pt. 90 C.

Emery® 3323. [Henkel/Emery] Decyl alcohol; chemical intermediate for detergent mfg.; APHA 10 liq.; sp.gr. 0.820-0.830 g/cc; solid. pt. 4-7 C; b.p. 220-235 C; acid no. < 0.1; iodine no. < 0.1; sapon. no.< 0.3; hyd. no. 350-355; flash pt. 110 C.

Emery® 3324. [Henkel/Emery] Octyl alcohol; chemical intermediate for detergent mfg.; APHA 10 liq.; sp.gr. 0.815-0.825 g/cc; solid. pt. 17 C; b.p. 185-200 C; acid no. < 0.1; iodine no. < 0.3; sapon. no. < 1.0; hyd. no. 420-430; flash pt. 90 C.

Emery® 3326. [Henkel/Emery] Lauryl alcohol (44-50% C12, 14-20% C14, 8-10% C16, 8-12% C18); CAS 112-53-8; chemical intermediate for detergent mfg.; APHA 20 liq./solid; sp.gr. 0.820-0.830 g/cc (30 C); solid. pt. 15-18 C; b.p. 230-250 C; acid no. < 0.1; iodine no. < 0.5; sapon. no. < 1.2; hyd. no. 280-290; flash pt. 110 C.

Emery® 3327. [Henkel/Emery] Lauryl alcohol (70-75% C12, 24-30% C14); CAS 112-53-8; chemical intermediate for detergent mfg.; APHA 20 liq./solid; sp.gr. 0.820-0.830 g/cc (30 C); solid. pt. 17-23 C; b.p. 255-295 C; acid no. < 0.1; iodine no. < 0.3; sapon. no. < 0.5; hyd. no. 285-293; flash pt. 140 C.

Emery® 3328. [Henkel/Emery] Lauryl alcohol (50-66% C8, 30-40% C10); CAS 112-53-8; chemical intermediate for detergent mfg.; APHA 10 liq.; sp.gr. 0.820-0.830 g/cc; solid. pt. 11 c; b.p. 190-260 C; acid no. < 0.1; iodine no. < 0.5; sapon. no. < 1.5; hyd. no. 385-410; flash pt. 95 C.

Emery® 3329. [Henkel/Emery] Octyl alcohol; chemical intermediate for detergent mfg.; APHA 10 liq.; sp.gr. 0.815-0.825 g/cc; solid. pt. 18 C; b.p. 185-210 C; acid no. < 0.1; iodine no. < 0.3; sapon. no. < 1.0; hyd. no. 420-423; flash pt. 90 C.

Emery® 3330. [Henkel/Emery] Decyl alcohol; chemical intermediate for detergent mfg.; APHA 10 liq.; sp.gr. 0.820-0.830 g/cc; solid. pt. 3-6 C; b.p. 220-240 C; acid no. < 0.1; iodine no. < 0.3; sapon. no.< 1.0; hyd. no. 345-356; flash pt. 110 C.

Emery® 3331. [Henkel/Emery] Lauryl alcohol; CAS 112-53-8; chemical intermediate for detergent mfg.; APHA 10 liq./solid; sp.gr. 0.815-0.825 g/cc (30 C); solid. pt. 19-23 C; b.p. 250-270 C; acid no. < 0.1; iodine no. < 0.3; sapon. no.< 0.4; hyd. no. 295-301; flash pt. 140 C.

Emery® 3332. [Henkel/Emery] Lauryl alcohol; CAS 112-53-8; chemical intermediate for detergent mfg.; APHA 10 liq./solid; sp.gr. 0.815-0.825 g/cc (30 C); solid. pt. 19-23 C; b.p. 250-270 C; acid no. < 0.1; iodine no. < 0.3; sapon. no.< 0.4; hyd. no. 297-301; flash pt. 140 C.

Emery® 3334. [Henkel/Emery] Myristyl alcohol; CAS 112-72-1; chemical intermediate for detergent mfg.; APHA 20 fused/flakes; sp.gr. 0.810-0.820 g/cc (40 C); solid. pt. 35-38 C; b.p. 280-300 C; acid no. < 0.1; iodine no. < 0.5; sapon. no.< 0.5; hyd. no. 225-262; flash pt. 155 C.

Emery® 3335 Chemically Pure. [Henkel/Emery] Lauryl alcohol; CAS 112-53-8; chemical intermediate for detergent mfg.; APHA 10 liq./solid; sp.gr. 0.815-0.825 g/cc (30 C); solid. pt. 23.7-23.9 C; b.p. 255-265 C; acid no. < 0.1; iodine no. < 0.1; sapon. no.< 0.1; hyd. no. 299-301; flash pt. 140 C.

Emery® 3336. [Henkel/Emery] Cetyl alcohol; CAS 36653-82-4; chemical intermediate for detergent mfg.; APHA 20 fused/flakes; sp.gr. 0.805-0.815 g/cc (60 C); solid. pt. 46-49 C; b.p. 305-330 C; acid no. < 0.1; iodine no. < 0.5; sapon. no.< 0.5; hyd. no. 225-235; flash pt. 180 C.

Emery® 3337. [Henkel/Emery] Cetyl alcohol; CAS 36653-82-4; chemical intermediate for detergent mfg.; APHA 20 fused/flakes; sp.gr. 0.805-0.815 g/cc (60 C); solid. pt. 46-49 C; b.p. 305-330 C; acid no. < 0.1; iodine no. < 0.5; sapon. no.< 0.5; hyd. no. 228-235; flash pt. 180 C.

Emery® 3338. [Henkel/Emery] Cetyl alcohol; CAS 36653-82-4; chemical intermediate for detergent mfg.; APHA 20 fused; sp.gr. 0.805-0.815 g/cc (60

C); solid. pt. 48.8-49.1 C; b.p. 310-325 C; acid no. < 0.1; iodine no. < 0.1; sapon. no.< 0.2; hyd. no. 231-233; flash pt. 180 C.

Emery® 3339. [Henkel/Emery] Cetyl alcohol; CAS 36653-82-4; chemical intermediate for detergent mfg.; APHA 20 fused/flakes; sp.gr. 0.805-0.815 g/cc (60 C); solid. pt. 46-49 C; b.p. 305-330 C; acid no. < 0.1; iodine no. < 0.5; sapon. no.< 0.5; hyd. no. 228-235; flash pt. 180 C.

Emery® 3343. [Henkel/Emery] Stearyl alcohol; CAS 112-92-5; chemical intermediate for detergent mfg.; APHA 20 fused/flakes; sp.gr. 0.805-0.815 g/cc (60 C); solid. pt. 55-55.7 C; b.p. 330-360 C; acid no. < 0.1; iodine no. < 1.0; sapon. no.< 0.5; hyd. no. 203-210; flash pt. 195 C.

Emery® 3344. [Henkel/Emery] Stearyl alcohol; CAS 112-92-5; chemical intermediate for detergent mfg.; APHA 20 fused/flakes; sp.gr. 0.805-0.815 g/cc (60 C); solid. pt. 55-55.7 C; b.p. 330-360 C; acid no. < 0.1; iodine no. < 1.0; sapon. no.< 0.5; hyd. no. 203-210; flash pt. 195 C.

Emery® 3345. [Henkel/Emery] Myristyl alcohol; CAS 112-72-1; chemical intermediate for detergent mfg.; APHA 20 fused/flakes; sp.gr. 0.810-0.820 g/cc (40 C); solid. pt. 36-38 C; b.p. 280-300 C; acid no. < 0.1; iodine no. < 0.3; sapon. no.< 0.5; hyd. no. 255-262; flash pt. 155 C.

Emery® 3346. [Henkel/Emery] Lauryl alcohol (48-58% C12, 19-24% C14, 9-12% C16, 11-15% C18); CAS 112-53-8; chemical intermediate for detergent mfg.; APHA 20 liq./solid; sp.gr. 0.820-0.830 g/cc (30 C); solid. pt. 18-23 C; b.p. 255-360 C; acid no. < 0.1; iodine no. < 0.5; sapon. no. < 1.2; hyd. no. 266-275; flash pt. 130 C.

Emery® 3347. [Henkel/Emery] Myristyl alcohol; CAS 112-72-1; chemical intermediate for detergent mfg.; APHA 20 fused; sp.gr. 0.810-0.820 g/cc (40 C); solid. pt. 37.4-37.7 C; b.p. 285-300 C; acid no. < 0.1; iodine no. < 0.1; sapon. no.< 0.2; hyd. no. 261-263; flash pt. 155 C.

Emery® 3348. [Henkel/Emery] Stearyl alcohol; CAS 112-92-5; chemical intermediate for detergent mfg.; APHA 20 fused; sp.gr. 0.805-0.815 g/cc (60 C); solid. pt. 57.4-57.7 C; b.p. 340-360 C; acid no. < 0.1, iodine no. < 0.2; sapon. no.< 0.2; hyd. no. 207-209; flash pt. 200 C.

Emery® 3351. [Henkel/Emery] Lauryl alcohol (81-89% C10, 6-11% C12, 4-9% C14); CAS 112-53-8; chemical intermediate for detergent mfg.; APHA 10 liq.; sp.gr. 0.820-0.830 g/cc; solid. pt. -1 to 3 C; b.p. 220-280 C; acid no. < 0.1; iodine no. < 0.3; sapon. no. < 1.0; hyd. no. 336-346; flash pt. 115 C.

Emery® 3352. [Henkel/Emery] Lauryl alcohol (60-66% C12, 21-25% C14, 10-12% C16); CAS 112-53-8; chemical intermediate for detergent mfg.; APHA 20 liq./solid; sp.gr. 0.820-0.830 g/cc (30 C); solid. pt. 18-22 C; b.p. 255-310 C; acid no. < 0.1; iodine no. < 0.5; sapon. no. < 0.5; hyd. no. 280-290; flash pt. 140 C.

Emery® 3353. [Henkel/Emery] Lauryl alcohol (36-44% C12, 16-20% C14, 8-12% C16); CAS 112-53-8; chemical intermediate for detergent mfg.; APHA 20 liq./solid; sp.gr. 0.810-0.820 g/cc (40 C); solid. pt. 25-28 C; b.p. 260-350 C; acid no. < 0.1; iodine no. < 0.5; sapon. no. < 0.8; hyd. no. 260-275; flash pt. 140 C.

Emery® 3354. [Henkel/Emery] Sat. fatty alcohol (65-72% C16, 25-33% C18); chemical intermediate for detergent mfg.; APHA 20 fused/flakes; sp.gr. 0.805-0.815 g/cc (60 C); solid. pt. 46-49 C; b.p. 305-355 C; acid no. < 0.1; iodine no. < 0.5; sapon. no. < 1.0; hyd. no. 215-225; flash pt. 160 C.

Emery® 3356. [Henkel/Emery] Sat. fatty alcohol (45-55% C16, 45-55% C18); chemical intermediate for detergent mfg.; APHA 20 fused/flakes; sp.gr. 0.805-0.815 g/cc (60 C); solid. pt. 48-52 C; b.p. 310-360 C; acid no. < 0.1; iodine no. < 0.5; sapon. no. < 1.0; hyd. no. 215-220; flash pt. 170 C.

Emery® 3357. [Henkel/Emery] Lauryl alcohol (40-48% C8, 51-59% C10); CAS 112-53-8; chemical intermediate for detergent mfg.; APHA 10 liq.; sp.gr. 0.820-0.830 g/cc; solid. pt. 7 C; b.p. 195-250 C; acid no. < 0.1; iodine no. < 0.3; sapon. no. < 1.0; hyd. no. 380-390; flash pt. 100 C.

Emery® 5314 Agriwet 1186A. [Henkel/Emery] Dioctyl sodium sulfosuccinate; anionic; wetting agent, emulsifier, solubilizer, penetrant for agric. toxicants; EPA-exempt; clear liq.; pH 7.0 (1%); 70% act. DISCONTINUED.

Emery® 5351 Lomar D. [Henkel/Emery] Highly polymerized naphthalene sulfonate; dispersant for agric. formulations; EPA-exempt; tan powd.; pH 9.5 (10%); 84% act.

Emery® 5353 Lomar PW. [Henkel/Emery] Condensed naphthalene sulfonate; dispersant for agric. formulations; EPA-exempt; tan powd.; pH 9.5 (10%); 87% act.

Emery® 5357 Lomar PL. [Henkel/Emery] Condensed naphthalene sulfonate; dispersant for agric. formulations; EPA-exempt; amber liq.; pH 9.5 (20%); 41% act. DISCONTINUED.

Emery® 5358 Lomar DL. [Henkel/Emery] Highly polymerized naphthalene sulfonate; dispersant for agric. formulations; EPA-exempt; amber liq.; pH 9.0; 30% act. DISCONTINUED.

Emery® 5366 (Lomar PWA Liq.). [Henkel/Emery] Ammonium naphthalene sulfonate; dispersant for agric. formulations; EPA-exempt; dk. brn. liq.; pH 7.3 (10%); 44% act.

Emery® 5367 Lomar PWA. [Henkel/Emery] Ammonium naphthalene sulfonate; dispersant for agric. formulations; EPA-exempt; tan powd.; pH 3.5 (10%); 92% act.

Emery® 5370 Sellogen W. [Henkel/Emery] Sodium alkyl naphthalene sulfonate; wetting agent for agric. formulations; EPA-exempt; tan powd.; pH 9.0 (5%); 65% act.

Emery® 5371 Sellogen WL. [Henkel/Emery] Sodium alkyl naphthalene sulfonate; wetting agent for agric. formulations; EPA-exempt; dk. amber liq.; flash pt. 93 C; pH 9.3 (100%); 32% act.

Emery® 5375 Sellogen DFL. [Henkel/Emery] Sodium alkyl naphthalene sulfonate; wetting agent for agric. formulations; EPA-exempt; dry tan powd.; pH 8.5 (1%); 67% act.

Emery® 5380 Sellogen HR. [Henkel/Emery] Sodium alkyl naphthalene sulfonate; wetting agent for agric. formulations; stable in high electrolyte sol'ns.; EPA-exempt; tan powd.; pH 9.0 (5%); 75% act.

Emery® 5381 Sellogen HR-90. [Henkel/Emery] Sodium alkyl naphthalene sulfonate; wetting agent for agric. formulations; EPA-exempt; dk. amber liq.; flash pt. 93 C; pH 10.2 (10%); 36% act.

Emery® 5386 Agrimul CN-4. [Henkel/Emery] Formulated prod.; emulsifier for agric. formulations; EPA-exempt; dk. liq.; dens. 8.67 lb/gal. DISCONTINUED.

Emery® 5440. [Henkel/Emery] Sodium isethionate; anionic; intermediate raw material for surfactants; liq.; 56% conc.

Emery® 5450 APG 225. [Henkel/Emery] Alkyl polyglucoside based on linear alcohol; biodeg. surfactant for agric. formulations; EPA-exempt; clear liq.; pH 6 (1%); 68% act. in water. DISCONTINUED.

Emery® 5451 APG 300. [Henkel/Emery] Alkyl polyglucoside based on linear alcohol; biodeg. surfactant for agric. formulations; EPA-exempt; clear liq.; pH 9 (1%); 63% act. in water. DISCONTINUED.

Emery® 5874. [Henkel/Cospha; Henkel Canada] Ethoxylated tridecyl alcohol; nonionic; hydrophilic emulsifier, dispersant, stabilizer, detergent; liq.; HLB 15.0; 75% conc.

Emery® 6220. [Henkel/Emery] EO/PO block polymer; nonionic; surfactant for agric. formulations; EPA-exempt; clear liq.; m.w. 2000; dens. 8.4 lb/gal; visc. 300 cps; HLB 3.0; pour pt. 0 C; cloud pt. 17 C (10%).

Emery® 6221 Monolan 2500. [Henkel/Emery] EO/PO block polymer; nonionic; surfactant for agric. formulations; EPA-exempt; clear liq.; m.w. 2500; dens. 8.6 lb/gal; visc. 400 cps; HLB 7.0; pour pt. 0 C; cloud pt. 22 C (10%).

Emery® 6222 Monolan 1030. [Henkel/Emery] EO/PO block polymer; nonionic; surfactant for agric. formulations; EPA-exempt; clear liq.; m.w. 1850; dens. 8.64 lb/gal; visc. 250 cps; HLB 12.0; pour pt. 0 C; cloud pt. 100 C (10%).

Emery® 6223 Monolan 2800. [Henkel/Emery] EO/PO block polymer; nonionic; surfactant for agric. formulations; EPA-exempt; clear liq.; m.w. 2800; dens. 8.8 lb/gal; visc. 555 cps; HLB 15.0; pour pt. 0 C; cloud pt. 60 C (10%).

Emery® 6224 Monolan 6400. [Henkel/Emery] EO/PO block polymer; nonionic; surfactant for agric. formulations; EPA-exempt; semisolid; m.w. 5800; dens. 8.7 lb/gal; HLB 13.0; pour pt. 32 C; cloud pt. 70 C (10%). DISCONTINUED.

Emery® 6225 Monolan 4500. [Henkel/Emery] EO/PO block polymer; nonionic; surfactant for agric. formulations; EPA-exempt; semisolid; m.w. 4600; dens. 8.6 lb/gal; HLB 16.0; pour pt. 40 C; cloud pt. > 100 C (10%). DISCONTINUED.

Emery® 6686. [Henkel/Emery] PEG 600; chemical intermediate for coatings, adhesives, lubricants, metalworking, paper mfg., petrol. prod., ceramics, printing, electronics, solvs., cleaners, latex paints, mold release agent, rubber; Gardner < 1 liq.; sol. in water, xylene; disp. in Stod.; dens. 9.4 lb/gal; visc. 63 cSt (100 F); pour pt. 22 C; flash pt. 475 F.

Emery® 6687. [Henkel/Emery] PEG-300; chemical intermediate for coatings, adhesives, lubricants, metalworking, paper mfg., petrol. prod., ceramics, printing, electronics, solvs., cleaners, latex paints, mold release agent, rubber, textiles; Gardner < 1 liq.; sol. in water, xylene; disp. in Stod.; dens. 9.4 lb/gal; visc. 33 cSt (100 F); pour pt. -10 C; flash pt. > 400 F.

Emery® 6744. [Henkel/Cospha; Henkel/Textile; Henkel Canada] Cocamidopropyl betaine; amphoteric; surfactant, foamer for personal care prods., shampoos, baby preps., skin cleansers; also for textile applics.; yel. clear liq.; sol. in water; disp. in glycerol trioleate; dens. 8.8 lb/gal; visc. 7 cSt (100 F); pour pt. < -2 C; 35% act.

Emery® 6750 Nopcosperse AD-6 Liq. [Henkel/Emery] Dispersant for agric. dry flowables, wettable powds., and aq. suspensions; EPA-exempt; lt. amber liq.; pH 7.8 (10%); 35% act.

Emery® 6752. [Henkel/Emery] Monocarboxylate coconut imidazolinium deriv.; amphoteric; surfactant used in cleaning compds. and shampoos, visc. improver and foam stabilizer, detergent; Gardner 9 liq.; sol. in water; dens. 8.7 lb/gal; 38% act.

Emery® 6773. [Henkel/Emery] PEG-200; chemical intermediate for coatings, adhesives, lubricants, metalworking, paper mfg., petrol. prod., ceramics, printing, electronics, solvs., cleaners, latex paints, mold release agent, rubber; Gardner < 1 clear visc. liq.; sol. @ 5% in water, xylene; m.w. 200; dens. 9.4 lb/gal; visc. 25 cSt (100 F); pour pt. < –15 C; flash pt. > 330 F.

Emery® 6779. [Henkel/Emery; Henkel/Textile] Methoxy PEG 400 oleate; lubricant, emulsifier with improved heat stability, reduced foam; for textile applics.; Gardner 4 liq.; sol. in butyl stearate, glycerol trioleate, xylene; disp. in water, min. oil, Stod.; dens. 8.4 lb/gal; visc. 29 cSt (100 F); pour pt. < 15 C; cloud pt. 42 C; flash pt. 490 F.

Emery® 6877. [Henkel/Emery] Surfactant blend; for agric. applics.; Gardner 1 liq.; dens. 8.1 lb/gal; pour pt. < -8 C; cloud pt. 32 C; flash pt. 117 F. DISCONTINUED.

Emery® 6885. [Henkel/Emery] Fatty-based; emulsifier for triglycerides, fats, oils; Gardner 3 liq.; disp. 5% in water, butyl stearate, Stod., xylene; dens. 8.3 lb/gal; visc. 149 cSt (100 F); HLB 7.8; pour pt. -14 C; flash pt. 490 F.

Emery® 6896 Foamaster FLD. [Henkel/Emery] Silicone emulsion; food-grade defoamer for aq. systems, agric. chems.; EPA-exempt; opaque wh. liq.; pH 6.0; 15% act. DISCONTINUED.

Emery® 6927 Agrimul 70-A. [Henkel/Emery] Formulated prod.; emulsifier for agric. toxicant and solv. systems; EPA-exempt; dk. amber liq.; sol. @ 5% in xylene; disp. in water, min. oil; dens. 8.8 lb/gal; visc. 8850 cps; HLB 10.5.

Emery® Methyl Lardate. [Henkel/Emery] Methyl lardate; solv.-carrier for agric. spray prods.; defoaming component in metalworking, paper deinking, pharmaceutical fermentation.

Emery® Methyl Oleate. [Henkel/Emery] Methyl oleate; Solv.-carrier for agric. spray prods.; defoaming component in metalworking, paper deinking, pharmaceutical fermentation; also for prep. of lubricants for automotive, textile, metal rolling operations.

Emgard® 2033. [Henkel/Emery] Blend; nonionic; emulsifier for methyl oleate, agric. formulations; EPA-exempt; Gardner 7 liq.; sol. @ 5% in Stod., methyl oleate; disp. in water, xylene; dens. 8.5 lb/gal; visc. 143 cSt (100 F); HLB 10.2; cloud pt. < 25 C; flash pt. 540 F.

Emgard® 2063. [Henkel/Emery] Formulated prod.; spreading and penetrating agent for agric. herbicide formulations; EPA-exempt; Gardner 1 liq.; sol. @ 5% in Stod., xylene; disp. in water, min. oil; dens. 8.0 lb/gal; visc. 21 cSt (100 F); HLB 11.3; pour pt. -2 C; cloud pt. < 25 C; flash pt. 350 F.

Emgard® 2066. [Henkel/Emery] Formulated prod.; spreader/activator and wetting agent for agric. formulations; EPA-exempt; Gardner 1 liq.; sol. @ 5% in xylene, methyl oleate; disp. in min. oil, Stod.; dens. 8.7 lb/gal; visc. 108 cSt (100 F); HLB 11.9; pour pt. -7 C; cloud pt. < 25 C; flash pt. 440 F. DISCONTINUED.

Emgard® 2067. [Henkel/Emery] Formulated prod.; spreader/activator and wetting agent for agric. formulations; EPA-exempt; Gardner 1 liq.; sol. @ 5%

in xylene, methyl oleate; disp. in min. oil, Stod.; dens. 8.2 lb/gal; visc. 24 cSt (100 F); HLB 11.9; pour pt. -10 C; cloud pt. < 25 C; flash pt. 75 F. DISCONTINUED.

Emid® 6500. [Henkel/Cospha; Henkel Canada] Cocamide MEA (1:1); CAS 68140-00-1; nonionic; thickener, foam stabilizer for shampoos, hair coloring prods., liq. detergents, and rug cleaners; Gardner 8 flaked solid; sol. (5%) in min. oil, butyl stearate, glycerol trioleate, Stod.; water-disp.; dens. 7.5 lb/gal (75 C); m.p. 72 C; acid no. 1; flash pt. 405 F; cloud pt. < 25 C; 100% act.

Emid® 6510. [Henkel/Emery] Lauramide DEA; CAS 120-40-1; nonionic; foam booster, stabilizer and thickener in shampoos, bubble baths, detergents; Gardner 2 solid; sol. in perchloroethylene, Stod., disp. in water, min. oil, butyl stearate; m.p. 42 C; 100% act.

Emid® 6511. [Henkel/Emery] Lauramide DEA; CAS 120-40-1; nonionic; emulsifier in liq. formulations; Gardner 2 solid; sol. in perchloroethylene, Stod., disp. in water, min. oil, butyl stearate; m.p. 40 C; 100% act.

Emid® 6514. [Henkel/Emery] Cocamide DEA; nonionic; thickener, foam stabilizer, and detergent component for various liq. cleaning compds.; Gardner 4 liq.; sol. (5%) in water, butyl stearate, glycerol trioleate, Stod., xylene; dens. 8.2 lb/gal; visc. 336 cSt (100 F); pour pt. 10 C; flash pt. 345 F; 100% act.

Emid® 6515. [Henkel/Cospha; Henkel/Textile; Henkel Canada] Cocamide DEA (1:1); nonionic; emulsifier, foam booster and stabilizer; inhibits redeposition of soils; thickener, superfatting agent for shampoos, bubble baths, cleansers, liq. detergents; antiredeposition agent for soils on textiles; Gardner 4 liq.; sol. (5%) in water, butyl stearate, glycerol triolate, Stod., xylene; dens. 8.3 lb/gal; visc. 390 cSt (100 F); acid no. 1; pour pt. 0 C; cloud pt. 370 C; flash pt. 370 F; 100% act.

Emid® 6518. [Henkel/Emery] Lauramide DEA; CAS 120-40-1; foam booster, stabilizer, visc. modifier for personal care prods., soaps, bath additives; Gardner < 5 liq.; sol. @ 5% in water, triolein, IPA, IPM; disp. in min. oil; dens. 8.2 lb/gal; pour pt. < –10 C; flash pt. 300 F.

Emid® 6519. [Henkel/Cospha; Henkel Canada] Lauramide DEA (1:1); CAS 120-40-1; foam booster, stabilizer, visc. modifier for personal care prods., soaps, bath additives; Gardner < 5 liq.; sol. @ 5% in water, triolein, IPA, IPM, glycerin; dens. 8.1 lb/gal (50 C); acid no. 1; pour pt. < –10 C; flash pt. 345 F.

Emid® 6521. [Henkel/Emery] Cocamide DEA (2:1); nonionic/anionic; foam booster, stabilizer, thickener in personal care prods., liq. soaps, bath additives; Gardner < 4 liq.; sol. @ 5% in triolein, IPA, IPM; disp. in water, min. oil; dens. 8.4 lb/gal; visc. 327 cSt (100 F); pour pt. –10 C; cloud pt. < 25 C; flash pt. 345 F.

Emid® 6529. [Henkel/Emery] Cocamide DEA (2:1); nonionic/anionic; thickener, foam booster, stabilizer, emulsifier; Gardner 2 liq.; sol. @ 5% in water, xylene; disp. in min. oil, butyl stearate, glyceryl trioleate; dens. 8.3 lb/gal; visc. 619 cSt (100 F); pour pt. < 25 C; flash pt. 360 F.

Emid® 6531. [Henkel/Emery] Cocamide DEA and diethanolamine; thickener, detergent, emulsifier, and foam stabilizer for cosmetics, shampoos, liq. detergents, and janitorial cleaners; deduster for powd. detergents; Gardner 3 liq.; sol. (5%) in water, Stod., xylene; water-disp.; dens. 8.4 lb/gal; visc. 360 cSt (100 F); pour pt. –10 C; flash pt. 338 F.

Emid® 6533. [Henkel/Emery; Henkel/Textile] Cocamide DEA, diethanolamine; nonionic/anionic; emulsifier, thickener, and foamer for controlled-suds detergents; fulling and scouring agent in textile industry; Gardner 5 liq.; sol. (5%) in water, Stod., xylene; water-disp.; dens. 8.7 lb/gal; visc. 345 cSt (100 F); pour pt. –15 C; flash pt. 365 F; 100% act.

Emid® 6538. [Henkel/Emery] Modified cocamide DEA; nonionic; thickener and foam stabilizer used in built, liq. detergents; Gardner 6 liq.; sol. (5%) in water, Stod., xylene; water-disp.; dens. 8.3 lb/gal; visc. 690 cSt (100 F); pour pt. –1 C; flash pt. 345 F; 100% act.

Emid® 6541. [Henkel/Emery] Lauramide DEA, diethanolamine; nonionic; emulsifier, foam stabilizer and booster in shampoos and general purpose cleaners; Gardner 3 paste; sol. (5%) in water, Stod., xylene; dens. 8.4 lb/gal; visc. 275 cSt (100 F); pour pt. –15 C; flash pt. 300 F; 100% act.

Emid® 6543. [Henkel/Emery] Stearic-oleic diethanolamide; nonionic; hydrophobic coemulsifier for min. oils, fatty esters, various solv.; Gardner 4 liq.; sol. in water, perchloroethylene, Stod., disp. in min. oil, butyl stearate; dens. 8.5 lb/gal; 100% act.

Emid® 6544. [Henkel/Emery] Capramide DEA and diethanolamine; nonionic; foam booster, wetting agent, detergency builder, used in salt and alkali systems, dedusting agent for alkaline powd.; sol. in butyl stearate, perchloroethylene, Stod., disp. in water, min. oil; m.p. 30 C; 100% act.

Emid® 6545. [Henkel/Emery; Henkel/Textile] Oleamide DEA (1:1); CAS 93-83-4; nonionic; foam suppressant in dye carrier and solv. emulsions; emulsifier for min. oils for antistatic fiber processing aids and yarn lubricants; Gardner 7 liq.; sol.in min. oil, butyl stearate, glycerol trioleate, Stod., xylene; water-disp.; dens. 7.7 lb/gal; visc. 290 cSt (100 F); pour pt. –3 C; flash pt. 475 F; cloud pt. < 25 C; 100% act.

Emigen DPR. [Hoechst AG] Acrylamide deriv.; prevents frosting effect during thermosol process on polyester/cellulosic fiber blends; liq.

Emkabase. [Emkay] Naphthenic, amino sulfonate blend; anionic; emulsifier for min. oil; liq.

Emkabase CA. [Emkay] Cationic; emulsifier for min. oil; mixed with 90 parts min. oil to yield a water-disp. substantive prod. for textile processing.

Emkabase CM-3. [Emkay] Emulsifier for min. oils; lubricant and softener for textiles; disp. in water.

Emkabase DSL. [Emkay] Cleaner for oil, grease, and wax on engines, machinery, utensils, tools, equip., trucks.

Emkabase ODC-2. [Emkay] Alkylaryl sulfonate blend; anionic; emulsifier for orthodichlorobenzene; liq.; disp. in water.

Emkacide GS-2. [Emkay] Disinfectant, cleaner, deodorizer, and fungicide with softening props.; liq.

Emka DDBSA. [Emkay] Dodecylbenzene sulfonic acid; base for compounding detergents and emulsifiers for textile processing.

Emka Defoam AA. [Emkay] Solvent type defoamer for textile processing; 100% act.

Emka Defoam BC, NC. [Emkay] Nonsilicone; nonionic; foam and froth control for industrial wastes, textile and paper processing.

Emka Defoam BCK. [Emkay] Nonsilicone; nonionic; foam and froth control agent for dyeing, printing formulations, industrial waste water.

Emka Defoam CPT. [Emkay] Nonsilicone; foam and

froth control for print pastes and for clears; wh. emulsion.

Emka Defoam DP. [Emkay] Silicone emulsion; defoamer; may be cut 50-50 with water.

Emka Defoam NSXX. [Emkay] Nonionic; foam and froth control agent for dyeing, printing, waste water, and other industrial processes; off-wh. creamy liq., very sl. pleasant odor.

Emka Defoam PKM. [Emkay] Nonsilicone; foam and froth control for high temps. and prolonged treatment; 100% act.

Emka Defoam PN. [Emkay] Nonionic; foam and froth control agent for textile processing; soft flowing paste; 100% act.

Emka Defoam PWC. [Emkay] Cationic; defoamer for textile processing; may be cut 50-50 with water before use; flowing gel.

Emka Defoam SD-100. [Emkay] Silicone; defoamer for industrial wastes; water-disp.; 100% act.

Emka Defoam SMM. [Emkay] Silicone; foam and froth control of industrial wastes; paste.

Emkafix RXC. [Emkay] Quat. ammonium deriv.; cationic; dispersant which improves fixation of direct dyes; liq.

Emkafol D. [Emkay] Highly sulfonated fatty acid ester; anionic; wetting and rewetting agent with exc. leveling props. for textile processing; also activates enzymes for starch removal; liq.

Emkafol DC. [Emkay] Wetting and rewetting agent with exc. leveling props. for textile processing; also activates enzymes for starch removal.

Emkafol OT. [Emkay] Sulfated fatty acid ester; anionic; wetting/rewetting agent for sanforizing; liq.

Emkagen 49. [Emkay] Amino condensate and alkylaryl sulfonate blend; anionic; detergent for print washing in textile applics.; liq.

Emkagen 49AM. [Emkay] Ammoniated detergent based on long chain alkylaryl sulfonate; detergent, wetting agent, emulsifier for heavy duty cleaning and scouring, industrial cleaning; resistant to hard water; readily sol. in cold or hot water.

Emkagen BT. [Emkay] Amino-condensate blend; nonionic; detergent for wool scouring; liq.

Emkagen Conc. [Emkay] Amide amino condensate; nonionic; detergent, wetting agent, emulsifier for detergent formulation; liq.; 100% conc.

Emkagen LWS. [Emkay] Blend of syn. detergents; scouring agent for wool where a low oil content and residue are required.

Emkagen RS. [Emkay] Scouring agent and dyeing assistant with chelating action to cleanse effectively in presence of metallic contaminants in hard water.

Emka Graphite Remover. [Emkay] Effective cleaner for graphited fabrics; odorless; nontoxic.

Emkal BNS. [Emkay] Butyl naphthalene sodium sulfonate; anionic; dispersant, detergent, dyeing assistant, scouring, and wetting agent, emulsifier used in textile industry; liq.; 35% conc.

Emkal BNS Acid. [Emkay] Butyl naphthalene sulfonic acid; anionic; wetting, dispersing agent, detergent, dye assistant, scouring agent, emulsifier; liq.; 90% conc.

Emkal BNX Powd. [Emkay] Butyl naphthalene sodium sulfonate; anionic; wetting agent, dispersant, detergent, dye assistant, scouring agent, emulsifier; powd.; 80% conc.

Emkal NNS. [Emkay] Nonyl naphthalene sodium sulfonate; anionic; dispersant, detergent, dyeing assistant, scouring, and wetting agent, emulsifier used in textile industry; liq.; 35% conc.

Emkal NNS Acid. [Emkay] Nonyl naphthalene sulfonic acid; anionic; dispersant, detergent, dyeing assistant, scouring, and wetting agent, emulsifier used in textile industry; liq.; 90% conc.

Emkal NOBS. [Emkay] Sodium nonyl benzene sulfonate; anionic; dispersant, detergent, dyeing assistant, scouring, and wetting agent, emulsifier used in textile industry; liq.; 40% conc.

Emkalane MF. [Emkay] Leveling agent for dyeing wool and worsted raw stock or fabrics.

Emkalane WL. [Emkay] Amino condensate; nonionic; scouring agent for wool and wool blends.

Emkalane WSDC. [Emkay] Nonionic; emulsifier for scouring and dyeing wool and wool blends in the same bath; usage level: 2-3%.

Emkalar Base C50L. [Emkay] Emulsifier; base for liq. dye carriers by mixing with ortho-phenyl phenol.

Emkalar Base E-55. [Emkay] Surfactant blend; anionic; emulsifier; base for dye carriers; resistant to acids; liq.; easily sol. in cold or warm water; pH 6.5 (1%).

Emkalar Base NC. [Emkay] Surfactant blend; anionic; used with intermediates to compound dye carriers; liq.; 100% conc.

Emkalar CHJD. [Emkay] Aromatic solvs. and nonionic emulsifiers; carrier for textile processing; gives max. color yield; thin liq., mild odor.

Emkalite BAC. [Emkay] Anionic sulfonate blend; anionic; wetting agent, partial stripping agent for cationic dyes from acrylic fibers; helps produce level redyes; liq.

Emkalite OC. [Emkay] Conc. cleaner and stripper for dried out resins, finishes and other chemicals which must be removed from equip., utensils, buckets, and pans.

Emkalite W-55. [Emkay] Mild wool detergent for home use on wool/Orlon sweaters.

Emkalon AVR. [Emkay] Quat. compound; cationic; softener, antistat used in textile applics.; paste.

Emkalon Base C-100. [Emkay] Polyethoxylated condensate; cationic; surfactant; solid; 100% act.

Emkalon KLA. [Emkay] Polyethoxylated condensate; cationic; lanolin base softener; paste.

Emkalon ML-100. [Emkay] Alkylamide blend; nonionic; surfactant for textiles; solid; 100% act.

Emkalon TN. [Emkay] Polyethoxylate; nonionic; lubricant, softener; highly compat.; paste; 25% conc.

Emkalon Wax NRF. [Emkay] Alkylamide blend; nonionic; fufll-bodied soft finish; paste.

Emkalon WN-100. [Emkay] Alkylamide blend; nonionic; surfactant; solid; 100% act.

Emkane Acid. [Emkay] Alkylaryl sulfonic acid; anionic; detergent, dyeing assistant, scouring and wetting agent; emulsifier used in the textile industry; biodeg.; liq.

Emkane HAD. [Emkay] Alkylaryl sulfonate; nonionic/anionic; detergent, dyeing assistant, scouring and wetting agent, esp. for cleansing stubborn soils; emulsifier used in the textile industry; biodeg.; paste.

Emkane HAL. [Emkay] Alkylaryl sulfonate; anionic; detergent, dyeing assistant, scouring and wetting agent; emulsifier used in the textile industry; biodeg.; liq.

Emkane HAX. [Emkay] Amine-neutralized alkylaryl sulfonate; anionic; double-strength detergent, dyeing assistant, scouring and wetting agent; emulsifier used in the textile industry; biodeg.; liq.

Emkanol NC, NCD 25, 35, 45, 55. [Emkay] Sulfonates; anionic; penetrants, wetting agents for mercerizing caustic sol'ns.

Emkanol NCDX. [Emkay] Penetrant which will accept 10% caustic soda sol'ns. at the boil; also stable in salt sol'ns.

Emkanyl 85. [Emkay] Nonionic; leveling agent for dyeing of nylon hosiery.

Emkanyl BRX. [Emkay] Nylon dyeing assistant for acid dyes; prevents barré in woven fabrics; leveling agent.

Emkapene AV, AVX. [Emkay] Low-foaming penetrant, emulsifier, scum preventatives for use in sulfur and vat dyeing; compat. with sodium sulfide.

Emkapene B. [Emkay] Low-cost pine oil soap for scouring, dyeing, wetting applics.

Emkapene RW. [Emkay] Mixt. of sulfonated oils and pine oil; leveling agent and dye assistant.

Emkapol BA, W, WS. [Emkay] Readily sol. soap jellies for scouring; good rinsability.

Emkapol PO-18. [Emkay] Potassium oleate; anionic; detergent, soap, emulsifier; stabilizer for natural latex; biodeg.; liq.; misc. in water.

Emkapon 4S, DS, SS, TS. [Emkay] Amide sulfonates; anionic; general textile detergents and emulsifiers; liq.

Emkapon 71. [Emkay] Polyglycol ether blend; nonionic; wetting and scouring agent for the textile industry; compat. with enzymes and peroxide; liq.

Emkapon AMP. [Emkay] Amphoteric; highly conc. detergent, rapid wetter, emulsifier for syn. and natural fibers, tightly woven or badly soiled fabrics; recommended for use in warm or cold water for washing printed fabrics and for soaping off vats and naphthols.

Emkapon BC. [Emkay] Wool scouring agent containing optical bleach.

Emkapon BEX. [Emkay] Bleach extender, emulsifier, detergent; accelerator and penetrant used with peroxide or chlorine bleaches.

Emkapon CN. [Emkay] Sulfated condensate; Wetting agent for cottons and cotton-polyester blends; scour for dyebaths, PVA and CMC sizing; also for peroxide bleaching; biodeg.; smooth thin paste.

Emkapon CT. [Emkay] POE deriv.; anionic; detergent, emulsifier used for textile applics.; low rewetting, good rinsing; paste.

Emkapon CTA, CTC, CTL, CTO, CTR. [Emkay] Detergent, emulsifier; low rewetting, good rinsing; for textile applics.

Emkapon DAC, DAC-50. [Emkay] Detergent/emulsified solv. blend; detergent for cleaning obstinate soils, grease, stains, wax, water repellents; recommended as Dacron scour.

Emkapon DAC-500. [Emkay] Solv. scour for removing fatty soils, oils, and waxes used in sizing; for synthetics, nylon, Dacron, Kodel, Arnel, Spandex.

Emkapon FX. [Emkay] Detergent for washing direct dyed cottons and viscose rayons; prevents bleeding and staining of colored fabrics.

Emkapon HD. [Emkay] Aromatic ethoxylate; detergent for scouring to remove fatty soils; visc. liq.; readily sol.

Emkapon Jel 500 Conc. [Emkay] Fatty methyl taurate; dispersant and surfactant for textile processing.

Emkapon Jel BS. [Emkay] Fatty amide sulfonate; anionic; detergent, dyeing assistant, scouring agent, wetting agent, emulsifier; liq.

Emkapon K. [Emkay] Sulfated fatty amides; anionic; detergent, textile agent for scouring and wetting, emulsifier, leveling agent; paste.

Emkapon KW. [Emkay] Solvs. with anionic sulfonates; anionic; emulsifier, solv. scour for removal of Dacron wax size and water repellent finishes; stable in caustic sol'n. up to 4% strength; liq.

Emkapon L. [Emkay] Amide sulfonate; anionic; detergent; paste.

Emkapon ML. [Emkay] Modified alkylaryl sulfonate; scouring and dyeing agent for synthetics.

Emkapon MP. [Emkay] Modified sulfated coco monoethanolamide; nonionic; multipurpose scouring, emulsifier, dyeing, dispersing, leveling and soaping agent; biodeg.; visc. liq.; readily sol.

Emkapon PW. [Emkay] Nonionic; detergent and print wash.

Emkapon WS. [Emkay] Alkylaryl/nonionic blend; nonionic; wool scouring agent.

Emkaron GA-1. [Emkay] Sulfated fatty acid ester; low-foaming dyeing assistant, dispersant, wetting agent for textiles; resistant to acid and alkaline media.

Emkaron N-25. [Emkay] Nonionic; dispersant, retardant for vat dyeing and printing; used with hydrosulfite as a stripping agent.

Emkasan QA-50. [Emkay] Quaternary ammonium compd.; cationic; algicide, germicide, disinfectant and deodorant for textile and industrial use.

Emkasol DE. [Emkay] Modified oleyl-lauryl alcohol sulfate; anionic; dispersant, wetting agent, detergent, dye bath stabilizer; paste; 100% conc.

Emkasol GE. [Emkay] Modified fatty alcohol sulfate; dye stripper, emulsifier, detergent.

Emkasol WAS, WAS-T. [Emkay] Lauryl alcohol sulfate; anionic; detergent and dye bath stabilizer; liq.; 30% conc.

Emkastat MLT. [Emkay] Nonionic; antistat lubricant, soil releasing agent; hard wax base which protects delicate fabrics.

Emkastat PC. [Emkay] Quat. deriv.; cationic; antistat; used in soil-resistant finishes for textiles; paste.

Emkaterge A, JC-4. [Emkay] Graphite remover, esp. useful for cleaning nylon and orlon lace; readily sol. in warm water.

Emkaterge B. [Emkay] Syn. detergents with alkyl terpenes; nonionic; wetting agent, base prod. for compounding graphite remover; liq.

Emkaterge NA-2. [Emkay] High str. liq. for mixing with three parts water to make a nylon lace graphite remover.

Emkatex 11, 21. [Emkay] POE fatty acid; nonionic; emulsifier for water kerosene formula; liq.

Emkatex 49-P. [Emkay] Synthetic detergents; color dispersant, dyebath stabilizer; amber gel.

Emkatex AA. [Emkay] Alkylaryl sulfonate; anionic; detergent, wetting agent, penetrant for textile applics.; liq.; 40% conc.

Emkatex AA-80. [Emkay] Alkylaryl sulfonate; anionic; detergent, wetting agent, penetrant for textile applics.; liq.; 80% conc.

Emkatex AES. [Emkay] Aliphatic ester sulfate; anionic; lubricant and dyeing assistant used in textile applics.; minimizes barre, cracks, and creases.

Emkatex BSC. [Emkay] Cresylic based; cleaner for removing oil residues and sizing from wool and blends.

Emkatex CF 32. [Emkay] Amino condensate with penetrants; anionic; detergent, wetting agent for continous boil-off of heavy cotton, duck, twill fabrics; prep. for subsequent padding of vat and naphthol colors; liq.

Emkatex DEM. [Emkay] Modified anionic; anionic; dispersant, dye leveling agent for textile industry; liq.

Emkatex DLW. [Emkay] Cationic; leveling and retarding agent for dyeing acrylic fibers.

Emkatex DX, DXP. [Emkay] Phosphated high m.w. alcohol; anionic; dispersant, leveling agent, penetrant, scouring agent, and textile dye bath stabilizer; for kier, bleaching processes; also increases absorbency of flame proofing finishes; liq.

Emkatex F-2. [Emkay] Blend of syn. detergents and solvs.; detergent for removal of grease and fatty soils from textiles.

Emkatex F8-B. [Emkay] Dye dispersant, wetting/rewetting agent for dyed goods where retained absorbency is important.

Emkatex GSX. [Emkay] Nonionic; dye leveler.

Emkatex LE. [Emkay] Anionic; dyeing assistant, lubricant, wetting agent, dispersant, stabilizer for vat dyeing, naphthol dyeing, sulfur dyeing of cotton fabrics.

Emkatex LS. [Emkay] POE deriv.; nonionic; detergent, dye dispersing and leveling, penetrant and retarding agent used in the textile industry; recommended for dyeing nylon carpeting with premetallized acid and acetate dyes; paste.

Emkatex N-25. [Emkay] Nonionic; dispersant, retardant for vat dyeing and printing; stripping assistant for vat and naphthol colors used with hydrosulfite and caustic soda.

Emkatex NE. [Emkay] Syn. detergents, emulsifiers, dispersants; textile auxiliary; amber clear oil; readily sol. in water.

Emkatex PX29. [Emkay] Sulfated amide; anionic; dye assistant, dispersant, emulsifier, afterwash, and anticrock used in textiles; paste; 33% conc.

Emkatex RA. [Emkay] Sulfated fatty ester; anionic; emulsifier, wetting and textile scouring agent; liq.; 41% conc.

Emkatex VO-50. [Emkay] Ethoxylated fatty alcohol; biodeg. surfactant for carpet dyeing and finishing.

Emkatex WNX. [Emkay] Sulfated esters; anionic; wetting and leveling agent for wool and nylon, lubricant for textile industry; gel.

Emkatol M. [Emkay] Graphite remover for nylon lace.

Emka Transfer Remover. [Emkay] Cleaner for redying nylon hosiery.

EM-PB. [Ferro/Keil] Lard oil; additive improving wetting and lubricity in industrial applics.; sp.gr. 0.91; dens. 7.6 lb/gal; visc. 41 cSt (40 C); pour pt. 2 C; cloud pt. 18 C; flash pt. (COC) 288 C; acid no. 2; sapon. no. 195.

Emphos AM2-10C. [Witco/Organics] Phosphate ester; anionic; emulsifier for flame retardants; liq.; 100% conc.

Emphos CS-121. [Witco/Organics] Alkylaryl ethoxylate phosphate ester; anionic; detergent, o/w emulsifier, lubricant, solubilizer, and wetting agent for metal cleaning; liq.; sol. in naphthenic and paraffinic oils; disp. in water.

Emphos CS-136. [Witco/Organics] Nonoxynol-6 phosphate; anionic; lubricant; antistat; emulsifier for cutting fluids, PVAc, and acrylic film formation; detergent for hard surfaces, metal cleaners, and dry cleaning systems; waterless hand cleaner component; FDA compliance; clear visc. liq.; sol. in water, ethanol, CCl_4, perchloroethylene, heavy aromatic naphtha, kerosene; sp.gr. 1.09; visc. 4500 cps; pour pt. 18 C; acid no. 28; pH 5.0 (3% aq.); surf. tens. 30.0 dynes/cm (1%); Ross-Miles foam 158 mm (initial, 1%, 49 C); 99% solids.

Emphos CS-141. [Witco/Organics] Nonoxynol-10 phosphate; anionic; electrolyte-compatible lubricant for water-based cutting fluids; detergent for heavy-duty, all-purpose liq. formulations; emulsifier in PVAc and acrylic film formation; textile wetting agent; water-repellent fabric finishing; foaming agent; coupling agent; FDA compliance; clear visc. liq.; sol. 25% in water, 10% sodium hydroxide, ethanol, CCl_4, perchloroethylene, xylene, heavy aromatic naphtha; sp.gr. 1.12; visc. 4500 cps; pour pt. > 4 C; acid no. 65 (to pH 5.5); 110 (to pH 9.5); pH 2.0 (3% aq.); surf. tens. 35 dynes/cm (1%); Ross-Miles foam 168 mm (initial, 1%, 49 C); 99% solids.

Emphos CS-143. [Witco/Organics] Phosphoric acid ester; emulsifier for Kelthane; clear liq.; pH 2.5–3 (3% aq.).

Emphos CS-147. [Witco/Organics] Alkylaryl ethoxylate phosphate ester; anionic; industrial detergent, o/w emulsifier, lubricant, wetting agent, solubilizer, foaming and coupling agent; liq.; water-sol.; sp.gr. 1.01; acid no. 87 (to pH 5.5), 150 (to pH 9.5); pH 2.0 (3% aq.).

Emphos CS-151. [Witco/Organics] Org. phosphate ester; anionic; electrolyte-compatible lubricant for water-based cutting fluids; detergent for heavy-duty, all-purpose liq. formulations; emulsifier in PVAc and acrylic film formation; textile wetting agent; water-repellent fabric finishing; clear visc. liq.; sol. see Emphos CS-141; sp.gr. 1.11; pour pt. > 50 F; acid no. 95 (to pH 9.5); pH 2 (3% aq.); surf. tens. 36.8 dynes/cm (0.05%).

Emphos CS-330. [Witco/Organics] Alkylaryl ethoxylate phosphate ester; anionic; detergent, o/w emulsifier, and lubricant for oils in metal processing.

Emphos CS-733. [Witco/Organics] Alkylaryl ethoxylate phosphate ester; anionic; industrial coupler, o/w emulsifier, lubricant with electrolyte tolerance; liq.; water-sol.

Emphos CS-735. [Witco/Organics] Phosphate ester; anionic; detergent, emulsifier, dispersant, and solubilizer for use with nonionic surfactants; liq.

Emphos CS-1361. [Witco/Organics] Sodium nonoxynol-9 phosphate; anionic; antistat; emulsifier for transparent gels; detergent for hard surfaces, metal cleaners, and drycleaning systems; particle dispersant for aq. systems; coupling agent; wetting agent; clear liq.; sol. @ 25% in water, ethanol, CCl_4, perchloroethylene, aromatic naphtha, kerosene; sp.gr. 1.10; pour pt. 4 C; pH 5.0 (10% aq.); surf. tens. 31 dynes/cm (0.05% aq.).

Emphos D70-30C. [Witco/Organics] Sodium glyceryl oleate phosphate; anionic; antistat, emollient, emulsifier, substantive agent, and moisture barrier for personal care prods.; aerosol formulation and food processing surfactant; food-grade mold lubricant; liq.; oil-sol.

Emphos D70-31. [Witco/Organics] Mono- and diglycerides phosphate ester; anionic; pigment dispersant for oil-base paints; w/o emulsifier, lubricant, release agent, and thickener in food processing; flake; oil-sol.; 100% conc.

Emphos F27-85. [Witco/Organics] Hydrog. veg. glycerides phosphate; anionic; emulsifier, mold lubricant, release agent for food use; pigment dispersant in oil-based systems; moisture barrier; tan soft solid;

insol. in water; sp.gr. 1.01; pour pt. 17 C; acid no. 40 (to pH 9.5); pH 6.9 (3% aq.).

Emphos PS-21A. [Witco/Organics] Alcohol ethoxylate phosphate ester; anionic; detergent base, emulsifier, foaming and wetting agent, lubricant, dispersant; used for detergent industry and industrial surfactants; clear liq.; sol. in kerosene, xylene, IPA; water-insol.; sp.gr. 1.05; acid no. 130 (to pH 5.5), 200 (to pH 9.5); flash pt. > 93 C; pH 2.0 (10% aq.).

Emphos PS-121. [Witco/Organics] Linear alcohol ethoxylate phosphate ester, acid form; anionic; emulsifier/lubricant for cutting fluids; drycleaning detergent with antistat properties; textile wetting and scouring agent; clear liq.; sol. @ 25% in ethanol, CCl_4, perchloroethylene, xylene, heavy aromatic naphtha, kerosene, min. oil; water-disp.; sp.gr. 1.06; pour pt. < 10 C; acid no. 105 (to pH 5.5); 180 (to pH 9.5); surf. tens. 28.6 dynes/cm (0.05% aq.); pH 2.0 (3% aq.).

Emphos PS-220. [Witco/Organics] Linear alcohol ethoxylate phosphate ester, acid form; anionic; lubricant; improves lubricity and load bearing of oil-based lubricants; monomer/water emulsifier; polymer particle dispersant; emulsifier for caustic sol'ns., aliphatic solvs.; textile antistat; detergent; corrosion inhibitor; clear liq.; sol. @ 25% in ethanol, CCl_4, perchloroethylene, xylene, heavy aromatic naphtha, kerosene, min. oil; insol. in water; sp.gr. 1.02; pour pt. 10 C; acid no. 105 (to pH 5.5); 185 (to pH 185); surf. tens. 32.3 dynes/cm; pH 2.0 (3% aq.).

Emphos PS-222. [Witco/Organics] Aliphatic phosphate ester; anionic; emulsifier for pesticide formulations, detergents, hydrotropes and emulsion polymerization; corrosion inhibitor; detergent; dispersant; lubricant; liq.; sol. in oil, disp. in water; sp.gr. 1.02; acid no. 105 (to pH 5.5), 155 (to pH 9.5); pH 2.4 (3% aq.).

Emphos PS-236. [Witco/Organics] Alcohol ethoxylate phosphate ester; anionic; detergent base, emulsifier, coupling, foaming and wetting agent, corrosion inhibitor, lubricant, antistat, penetrant, and solubilizer used in the detergent industry, industrial surfactants, textile systems, metal cleaning; emulsion polymerization; liq.; sol. in naphthenic oils and water; sp.gr. 1.05; acid no. 90 (to pH 5.5), 140 (to pH 9.5); pH 2.0 (3% aq.).

Emphos PS-331. [Witco/Organics] Aliphatic phosphate ester; anionic; detergent, emulsifier, dispersant, foaming and wetting agent, and solubilizer for use with nonionic surfactants; liq.; sol. in oil, water; sp.gr. 1.13; acid no. 140 (to pH 5.5), 280 (to pH 9.5); pH 1.8 (3% aq.).

Emphos PS-400. [Witco/Organics] Linear alcohol ethoxylate phosphate ester, acid form; anionic; polymer particle dispersant; monomer/water emulsifier; textile antistat; also for pesticides, detergents, emulsion polymerization; lubricant; wetting agent; clear liq.; sol. @ 25% in ethanol, CCl_4, perchloroethylene, xylene, heavy aromatic naphtha, kerosene, min. oil; water-disp.; sp.gr. 1.00; pour pt. > 10 C; acid no. 220 (to pH 5.5), 300 (to pH 9.5); surf. tens. 30.6 dynes/cm (0.05% aq.); pH 2.0 (3% aq.).

Emphos PS-410. [Witco/Organics] Aliphatic phosphate ester; anionic; emulsifier for pesticide formulations, detergents, hydrotropes and emulsion polymerization; liq.

Emphos PS-413. [Witco/Organics] Phosphate ester; anionic; detergent, emulsifier, dispersant, and solubilizer for use with nonionic surfactants; liq.

Emphos PS-415M. [Witco/Organics] Alcohol ethoxylate phosphate ester; anionic; industrial coupling and defoaming agent, lubricant, solubilizer; textile industry detergent, dispersant, scouring and wetting agent for aq. systems; liq.; water-sol.

Emphos PS-440. [Witco/Organics] Alcohol ethoxylate phosphate ester; anionic; industrial detergent, o/w emulsifier, lubricant, penetrant, solubilizer, coupling, foaming and wetting agent; metal processing surfactant; liq.; sol. in naphthenic oils and water.

Emphos PS-810. [Witco/Organics] Linear alcohol ethoxylate phosphate ester, acid form; anionic; emulsifier/lubricant for cutting fluids; lube additive; clear liq.; disp. in water; sp.gr. 1.03; pour pt. > 10 C; acid no. 85 (to pH 5.5), 150 (to pH 9.5); pH 2.0 (3% aq.); surf. tens. 37.6 dynes/cm (0.05% aq.).

Emphos PS-900. [Witco/Organics] Alcohol ethoxylate phosphate ester; anionic; detergent, o/w emulsifier, and lubricant for use in metal processing; liq.; sol. in naphthenic and paraffinic oils.

Emphos TS-230. [Witco/Organics] Phenol ethoxylate phosphate ester, acid form; anionic; lubricant; improves lubricity and load-bearing of water-based lubricants; industrial corrosion inhibitor, defoamer, metal processing surfactant for syn. oils; low-foaming hydrotrope for alkaline cleaners; amber clear liq.; sol. @ 25% in water, 10% sodium hydroxide, ethanol; sp.gr. 1.18; pour pt. < 10 C; acid no. 87 (to pH 5.5); 155 (to pH 9.5); pH 2.0 (3% aq.).

Empicol® 0185. [Albright & Wilson UK] Sodium lauryl sulfate; CAS 151-21-3; anionic; detergent, wetting and rewetting agent, emulsifier, dispersant, disintegrator; used in personal care prods., pharmaceuticals, dental care, textile printing inks, agric. and horticultural preps., electroplating baths, medical preps.; wh. powd.; dens. 0.35 g/cm³; pH 9.5–10.5 (1% aq.); 94.0% min. act.

Empicol® 0216. [Albright & Wilson UK] Fatty alcohol ethoxy phosphate ester; anionic; foam stabilizer, conditioner, and antistatic agent for toiletries; detergent used in textile processing; liq.; 95% conc.

Empicol® 0303. [Albright & Wilson UK] Sodium lauryl sulfate BP; CAS 151-21-3; anionic; surfactant in toothpaste and pharmaceutical preps.; emulsion polymerization; wh. fine powd.; dens. 0.35 g/cc; pH 8-10 (1% aq.); 95% act.

Empicol® 0303V. [Albright & Wilson UK] Sodium lauryl sulfate BP; CAS 151-21-3; anionic; surfactant in toothpaste and pharmaceutical preps.; emulsion polymerization; wh. low-dusting needles; pH 8-10 (1% aq.); 95% conc.

Empicol® 0585/A. [Albright & Wilson Am.] Sodium ethylhexyl sulfate; anionic; surfactant for toiletries; low foam degreasing agent for textiles; liq.; 40% conc.

Empicol® 0919. [Albright & Wilson UK] Sodium lauryl sulfate; CAS 151-21-3; wetting, foaming, and emulsifying agent used in personal care prods., carpet and upholstery shampoos, and specialty cleaners; sp.gr. 1.05 ± 0.005; visc. 120 cP max.; pH 8.5 ± 0.7 (10%); 30.0 ± 1.0% act.

Empicol® AL30. [Albright & Wilson/Australia] Ammonium lauryl sulfate; CAS 68081-96-9; anionic; detergent for shampoos, carpet shampoos, leather processing; amber liq./paste; m.w. 291; visc. 14,000 cs; pH 6–7 (5%); 29% act.

Empicol® CHC 30. [Albright & Wilson UK] Sodium cetyl oleyl sulfate; anionic; multipurpose textile detergent; cream soft paste; water sol.; pH 8.5 (2%);

30% act.

Empicol® DA. [Albright & Wilson Australia] DEA-lauryl sulfate; CAS 68585-44-4; anionic; ingred. in personal care prods., automobile cleaners; pale yel. visc. liq.; dilutable in water; sp.gr. 1.05; pH 7.0±0.5 (10% aq.); 34.0 ± 1.0% act.

Empicol® ESB. [Albright & Wilson Australia] Sodium laureth sulfate; CAS 9004-82-4; 68585-34-2; anionic; shampoo ingred.; mild detergent for textile and leather processing; dispersant/emulsifier for emulsion and suspension polymerization; liq.; 28% conc.

Empicol® ESB3. [Albright & Wilson Am.; Albright & Wilson UK] Sodium laureth sulfate; CAS 9004-82-4; 68585-34-2; anionic; base and foam booster/stabilizer for personal care prods.; air entraining agent for construction; colorless-pale yel. liq.; dens. 1.05 g/cm³; visc. 2600 cs (20 C); cloud pt. < 0 C; pH 6.5-7.5 (5% aq.); 27.5 ± 1.0% act.

Empicol® L Series. [Albright & Wilson UK] Alkyl sulfate; dispersant/emulsifier for emulsion and suspension polymerization, agric. formulations; foaming agent for carpet backing; air entraining agent for construction; flotation aid.

Empicol® LM45. [Albright & Wilson UK] Sodium lauryl sulfate; CAS 151-21-3; primary act. ingred., detergent, and foaming agent used in personal care and household prods.; wh. paste; dens. 0.95 g/cm³; pH 8.0–10.0 (2% aq.); 45% act.

Empicol® LM/T. [Albright & Wilson UK] Sodium lauryl sulfate; CAS 151-21-3; primary act. ingred., detergent, and foaming agent used in personal care and household prods.; wh. powd.; dens. 0.35 g/cm³; pH 9.0–10.5 (1% aq.); 94% act.

Empicol® LQ33/T. [Albright & Wilson UK] MEA-lauryl sulfate; CAS 4722-98-9; 68908-44-1; anionic; surfactant in the mfg. of personal care prods.; emulsifier in the mfg. of rubber latices and for resins; bactericidal detergents; pale amber visc. liq.; dens. 1.03 g/cm³ visc. 6000 cs; pH 7.0±0.5 (5% aq.); 33.5 ± 1.0% act.

Empicol® LS30. [Albright & Wilson UK] Sodium lauryl sulfate; CAS 151-21-3; wetting, dispersing, emulsifying, and foaming agent for industrial processes, detergent/cleaner formulations; pale yel. clear visc. liq.; visc. 750 ± 300 cps (35 C); pH 8.5 ± 1.0 (10%); 30 ± 1% act.

Empicol® LS30P. [Albright & Wilson Australia] Sodium lauryl sulfate; CAS 68585-47-7; anionic; emulsifier, wetting and foaming agent for emulsion polymerization and carpet and upholstery shampoos; base for personal care shampoos; specialty cleaners and industrial processes; pale yel. paste; sp.gr. 1.05; visc. 120 cP max.; pH 7.5 ± 0.7 (10% aq.); 30.0 ± 1.0% act.

Empicol® LX. [Albright & Wilson; Albright & Wilson UK] Sodium lauryl sulfate; CAS 68585-47-4; anionic; emulsifier in the mfg. of plastics, resins, and syn. rubbers; foaming agent for rubber foams, personal care prods., pharmaceuticals, and carpet and upholstery shampoos; lubricant in mfg. of molded rubber goods; wh. powd.; dens. 0.35 g/cm³; pH 9.5–10.5 (1% aq.); 90% act.

Empicol® LX28. [Albright & Wilson; Albright & Wilson UK] Sodium lauryl sulfate; CAS 151-21-3; anionic; emulsifier in the mfg. of plastics, resins, and syn. rubbers; foaming agent for rubber foams, personal care prods., toothpaste, and carpet and upholstery shampoos; lubricant in mfg. of molded rubber goods; pale yel. liq., paste; dens. 1.05 g/cm³; pH 8.0–9.5 (5% aq.); 28% act. aq. sol'n.

Empicol® LX32. [Albright & Wilson UK] Alkyl sulfate; wetting agent for leather processing.

Empicol® LX42. [Albright & Wilson UK] Alkyl sulfate; wetting agent for leather processing.

Empicol® LXS95. [Albright & Wilson UK] Sodium lauryl sulfate; CAS 151-21-3; anionic; detergent, foamer, emulsifier used in toothpastes, emulsion polymerization, rubber, pharmaceuticals; wh. powd.; dens 0.35 g/cm³; pH 9.6–10.5 (1% aq.); 94.0% min. act.

Empicol® LXV. [Albright & Wilson; Albright & Wilson UK] Sodium lauryl sulfate; CAS 151-21-3; anionic; emulsifier in the mfg. of plastics, resins, and syn. rubbers; foaming agent for rubber foams, personal care prods., toothpaste, and carpet and upholstery shampoos; lubricant in mfg. of molded rubber goods; wh. needles; dens. 0.5 g/cm³; pH 9.5–10.5 (1% aq.); 85% act.

Empicol® LXV/D. [Albright & Wilson UK] Sodium lauryl sulfate USP, BP; CAS 151-21-3; anionic; emulsifier in the mfg. of plastics and rubbers; foaming agent for rubber foams; lubricant for mfg. of molded rubber goods; also in mfg. of carpet and upholstery shampoos, toiletries, toothpastes; wh. needles; sp.gr. 0.5 g/cc; pH 9.5–10.5 (5% aq.); 90% act.

Empicol® LY28/S. [Albright & Wilson UK] Sodium lauryl sulfate; CAS 151-21-3; coprecipitant in mfg. of photographic film; emulsion polymerization in the plastic and rubber industries; foaming agent in carpet processes; pale yel. liq. to paste; dens. 1.05 g/cm³; visc. 28 cs; cloud pt. –1 C; pH 8.5–9.5 (5% aq.); 29.0 ± 1.0% act.

Empicol® LZ. [Albright & Wilson UK] Sodium lauryl sulfate; CAS 68955-19-1; anionic; detergent, wetting and foaming agent in personal care prods., pharmaceuticals; emulsifier in mfg. of rubbers, plastics, and resins by emulsion polymerization; foaming agent in mfg. of foam rubber goods; lubricant used in plastic goods; wh. powd.; dens. 0.35 g/cm³; pH 9.5–10.5 (1% aq.); 89.0% min. act.

Empicol® LZ/D. [Albright & Wilson UK] Sodium lauryl sulfate BP; CAS 68955-19-1; anionic; surfactant, foaming agent, emulsifier, wetting agent for dental preps., toiletries, rubber, plastics, foam rubber; lubricant in extrusion of plastic goods, e.g., PVC; wh. powd.; sp.gr. 0.35; pH 9.5–10.5 (1% aq.); 90% min. act.

Empicol® LZ/E. [Albright & Wilson UK] Sodium lauryl sulfate; CAS 151-21-3; anionic; emulsifier in mfg. of plastics, resins, and syn. rubbers by emulsion polymerization; emulsifier and dispersant in textile printing pastes; dispersant/lubricant in extrusion of PVC; wh. powd.; dens. 0.35 g/cm³; 90.0% min. act.

Empicol® LZG 30. [Albright & Wilson UK] Sodium lauryl sulfate; CAS 151-21-3; anionic; surfactant, foaming and wetting agent used in industrial processes; emulsifier in mfg. of syn. rubbers, plastics, and resins by emulsion polymerization; cream paste; dens. 1.05 g/cm³; pH 8.0–9.5 (2% aq.); 30% act.

Empicol® LZGV. [Albright & Wilson UK] Sodium lauryl sulfate; CAS 151-21-3; anionic; surfactant, foaming and wetting agent used in industrial processes; emulsifier in mfg. of syn. rubbers, plastics, and resins by emulsion polymerization; cream needles; dens. 0.5 g/cm³; pH 9.5–10.5 (1% aq.); 85% act.

Empicol® LZGV/C. [Albright & Wilson UK] Sodium lauryl sulfate; CAS 151-21-3; surfactant, foaming and wetting agent used in industrial processes; emulsifier in mfg. of syn. rubbers, plastics, and resins by emulsion polymerization; cream needles; dens. 0.5 g/cm³; pH 9.5–10.5 (1% aq.); 89% act.

Empicol® LZP. [Albright & Wilson UK] Sodium lauryl sulfate; CAS 151-21-3; detergent, wetting and foaming agent in personal care prods., pharmaceuticals; emulsifier in mfg. of rubbers, plastics, and resins by emulsion polymerization; foaming agent in mfg. of foam rubber goods; lubricant used in plastic goods; wh. powd.; dens. 0.35 g/cm³; pH 9.5–10.5 (1% aq.); 89.0% min. act.

Empicol® LZV/D. [Albright & Wilson UK] Sodium lauryl sulfate; CAS 151-21-3; anionic; surfactant, foaming agent, emulsifier, wetting agent for dental preps., toiletries, rubber, plastics, foam rubber; lubricant in extrusion of plastic goods, e.g., PVC; wh. fine needles; sp.gr. 0.5; pH 9.5–10.5 (1% aq.); 90% min. act.

Empicol® LZV/E. [Albright & Wilson UK] Sodium lauryl sulfate; CAS 151-21-3; anionic; detergent, nondusting emulsifier; wh. needles; dens. 0.5 g/cm³; 90.0% min. act.

Empicol® SDD. [Albright & Wilson UK] Disodium lauryl ethoxy sulfosuccinate; CAS 37354-45-5; anionic; mild raw material for toiletries and detergents; pale straw liq.; may separate; dens. 1.10 g/cc; visc. 100 cs (20 C); cloud pt. < 2 C; pH 5.5-7 (5%); 40% conc.

Empicol® SEE. [Albright & Wilson UK] Disodium undecylenic monoethanolamide sulfosuccinate; anionic; mild raw material for toiletries and detergents; 25% conc.

Empicol® SFF. [Albright & Wilson UK] Disodium alkyl ethoxy sulfosuccinate; anionic; mild raw material for toiletries and detergents; pale straw free-flowing liq.; dens. 1.09 g/cc; visc. 40 cs (20 C); cloud pt. < 0 C; pH 6-7 (5%); 23% active.

Empicol® SGG. [Albright & Wilson UK] Disodium alkylolamide ethoxy sulfosuccinate; anionic; mild raw material for toiletries and detergents; pale straw liq.; may separate; dens. 1.1 g/cc; visc. 150 cs (20 C); pH 6-7 (5%); 30% conc.

Empicol® SHH. [Albright & Wilson UK] Sodium monoethanolomine alkyl ethoxy sulfosuccinate; anionic; mild raw material for toiletries and detergents; colorless to pale straw free-flowing liq.; dens. 1.09 g/cc; visc. 80 cs (20 C); cloud pt. < 5 C; pH 6-7 (5%); 32% conc.

Empicol® SLL. [Albright & Wilson UK] Disodium lauryl sulfosuccinate; CAS 36409-57-1; anionic; mild raw material for toiletries and detergents; wh. paste; pH 6-7 (5%); 32% conc.

Empicol® SLL/P. [Albright & Wilson UK] Disodium lauryl sulfosuccinate; CAS 36409-57-1; anionic; mild raw material for toiletries and detergents; wh. paste; pH 6-7 (1%); 83% conc.

Empicol® STT. [Albright & Wilson UK] Disodium cetearyl sulfosuccinate; anionic; mild raw material for toiletries and detergents; wh. paste; pH 6-7 (5%); 26% conc.

Empicol® TAS30. [Albright & Wilson UK] Sodium tallow sulfate; CAS 68955-20-4; anionic; collector in the beneficiation of minerals by ore flotation; detergent active raw material; cofoaming agent for latex foam compds.; wh./straw paste; dens. 1.10 g/cm³; pH 8.0–10.0 (1%); 30% act. in water.

Empicol® TAS80V. [Albright & Wilson UK] Sodium tallow sulfate; heavy-duty detergent, wetting and emulsifying agent; cream needles; pH 10; 80% act.

Empicol® TAS90. [Albright & Wilson UK] Sodium tallow sulfate; fabric scouring agent, sizing additive; cream powd.; pH 10; 90% act.

Empicol® TL40. [Albright & Wilson UK] TEA-lauryl sulfate; detergent used in liq. and lotion shampoos; foam booster for fire fighting; pale yel., clear liq.; misc. with water; dens. 1.025 g/cc; visc. 300 cs; pH 6.8–7.1 (5%); 40% act.

Empicol® TL40/T. [Albright & Wilson UK] TEA-lauryl sulfate; CAS 139-96-8; 68908-44-1; anionic; surfactant for personal care prods.; emulsifier in the mfg. of rubber latices; pale yel. clear liq.; dens. 1.08 g/cm³; visc. 45 cs; cloud pt. < 0 C; pH 7.0 ± 0.5 (5% aq.); 41% act. in water.

Empicol® WA. [Albright & Wilson UK] Sodium lauryl sulfate; CAS 151-21-3; industrial surfactant used in rubber foams for carpet backing; pale yel. clear liq.; m.w. 292 (of act.); sp.gr. 1.04; pH 8.5-9.5 (5%); 27% act.

Empicol® WAK. [Albright & Wilson/Australia] Sodium lauryl sulfate; CAS 151-21-3; anionic; emulsifier in refining veg. oils; base for personal care prods.; wetting and foaming agent in specialty cleansers and industrial processes; pale yel. clear liq.; sp.gr. 1.050; pH 8.5 ± 0.7 (10% aq.); 35.0 ± 1.0% act.

Empicol® XM 17. [Albright & Wilson UK] Formulated prod.; anionic; conc. detergent for carpet shampoo; pale amber liq.; visc. 3000 cs; cloud pt. 5 C; pH 6–7 (5%); 33% conc.

Empicryl® 6045. [Albright & Wilson UK] Alkyl methacrylate polymer in hydrocarbon oil; visc. index improver for formulation of high visc. index hydraulic fluids; emulsifier, dispersant for emulsion polymerization; reactive diluent for adhesives and coatings; yel. visc. fluid; sp.gr. 0.90; visc. 1100 mm²/s; pour pt. –3 C; flash pt. > 100 C; 60% min. oil content.

Empicryl® 6047. [Albright & Wilson UK] Alkyl methacrylate ester; emulsifier, dispersant for emulsion polymerization; monomer for mfg. of methacrylate polymers and copolymers; internal plasticizer for methacrylate copolymers used as adhesives and waterproofing coatings; pale yel. liq.; sp.gr. 0.9; flash pt. 80 C; acid no. 0.15 max.; hyd. no. 12 max.

Empicryl® 6052. [Albright & Wilson UK] Alkyl methacrylate polymer in hydrocarbon oil; see Empicryl 6045; yel. visc. fluid; sp.gr. 0.91; visc. 1000 mm²/s (100 C); pour pt. –3 C; flash pt. > 100 C; 33% min. oil content.

Empicryl® 6054. [Albright & Wilson UK] Alkyl methacrylate polymer in hydrocarbon oil; see Empicryl 6045; yel. visc. fluid; sp.gr. 0.93; visc. 1600 mm²/s (100 C); pour pt. –12 C; flash pt. > 100 C; 20% min. oil content.

Empicryl® 6058. [Albright & Wilson UK] Alkyl methacrylate polymer in hydrocarbon oil; see Empicryl 6045; pale yel. visc. fluid; sp.gr. 0.92; visc. 1100 mm²/s (100 C); pour pt –3 C; flash pt. > 100 C; 37% min. oil content.

Empicryl® 6059. [Albright & Wilson UK] Polyalkyl methacrylate copolymer in hydrocarbon oil; see Empicryl 6045; yel. visc. fluid; sp.gr. 0.91; visc. 900 mm²/s (100 C); pour pt –3 C; flash pt. > 100 C.

Empicryl® 6070. [Albright & Wilson UK] Polyalkyl methacrylate copolymer in hydrocarbon oil; visc. index improver for formulation of hydraulic oils;

emulsifier, dispersant for emulsion polymerization; reactive diluent for adhesives and coatings; yel. visc. fluid; sp.gr. 0.94; visc. 800 mm²/s (100 C); pour pt 15 C; flash pt. > 100 C.

Empigen® 5083. [Albright & Wilson UK] Coco dimethylamine oxide; foam booster/stabilizer and visc. modifier for shampoos, foam baths, cleaners; improves conditioning in shampoos; solubilizer for liq. bleach prods.; pale straw liq.; m.w. 244; sp.gr. 0.98; visc. 35 cs; pH 7.5 ± 0.5 (5% aq.); 30% act.

Empigen® 5089. [Albright & Wilson UK] Laurtrimonium chloride; cationic; bactericide for disinfectant and sanitizer formulations for household, institutional, agric., food processing applics., antiseptic detergents in pharmaceuticals; algicide for swimming pools; wood preservatives; pale yel. liq.; sp.gr. 0.97; pH 5.5–8.5; 34 ± 2% act. in water.

Empigen® 5107. [Albright & Wilson UK] Alkyl dimethylamine betaine; amphoteric; foaming agent/stabilizer, antistat, solubilizer for shampoos, foam baths, latex foam compds., oil prod., alkaline industrial hard surf. cleaners; wetting and coupling agent for cleaners, traffic film removers; pale amber liq.; m.w. 286; sp.gr. 1.04; pH 5–9; 31 ± 2% act. in water.

Empigen® AB. [Albright & Wilson UK] Dimethyl lauramine; CAS 67700-98-5; cationic; intermediate for mfg. of high quality derivs. such as quat. ammonium compds., betaines, amine oxides; catalyst for PU foam, resin curing agent, corrosion inhibitor, and flotation aid; pale straw liq.; char. odor; m.w. 221; dens. 0.79 g/cm³; 97% conc.

Empigen® AD. [Albright & Wilson UK] Tert. alkyl dimethylamine; cationic; intermediate for amine deriv.; resin curing agents; corrosion inhibitor; ore flotation chemicals; liq.; 97% conc.

Empigen® AF. [Albright & Wilson UK] C12-16 alkyl dimethyl amine; cationic; intermediate for amine deriv.; resin curing agents; corrosion inhibitor; ore flotation chemicals; pale straw liq.; m.w. 230; sp.gr. 0.79; 97% conc.

Empigen® AG. [Albright & Wilson UK] C12-16 alkyl dimethyl amine; cationic; intermediate for amine deriv.; resin curing agent; corrosion inhibitor; ore flotation chemicals; pale straw liq.; m.w. 238; sp.gr. 0.79; 97% conc.

Empigen® AH. [Albright & Wilson UK] Dimethyl myristamine; CAS 112-75-4; cationic; intermediate for mfg. of high quality derivs. such as quat. ammonium compds., betaines, amine oxides; catalyst for PU foam, resin curing agent, corrosion inhibitor, and flotation aid; pale straw liq.; char. odor; m.w. 241; dens. 0.79 g/cm³; 97% conc.

Empigen® AM. [Albright & Wilson UK] Alkyl dimethyl amine, dist.; cationic; intermediate for mfg. of high quality derivs. such as quat. ammonium compds., betaines, amine oxides; catalyst for PU foam, resin curing agent, corrosion inhibitor, and flotation aid; pale straw waxy solid; char. odor; dens. 0.84 g/cm³; 96.0% min. tert. amine.

Empigen® AS. [Albright & Wilson UK] Cocamidopropyl dimethyl amine; cationic; industrial corrosion inhibitor for cutting oil formulations, oil drilling, and refinery operations; chemical intermediate for surfactants; amber liq., paste; char. odor; m.w. 290; dens. 0.91 g/cm³; 88% tert. amine.

Empigen® AT. [Albright & Wilson UK] Alkyl amide propyl dimethyl amine; cationic; industrial corrosion inhibitor for cutting oil formulations, oil drilling, and refinery operations; chemical intermediate for surfactants; amber liq./paste; char. odor; dens. 0.95 g/cm³; 94% conc.

Empigen® AY. [Albright & Wilson UK] Tert. alkyl ethoxy dimethyl amine; cationic; intermediate for mfg. of high quality derivs. such as quat. ammonium compds., betaines, amine oxides; catalyst for PU foam, resin curing agent, corrosion inhibitor, and flotation aid; amber liq.; dens. 0.89 g/cm³; 94.0% min. tert. amine.

Empigen® BAC50. [Albright & Wilson UK] Benzalkonium chloride NF; cationic; bactericide for disinfectant and sanitizer formulations for household, institutional, agric., food processing applics., antiseptic detergents in pharmaceuticals; algicide for swimming pools; wood preservatives; masonry biocides; permanent retarders in dyeing of acrylic fibers; phase transfer catalyst; pale straw liq.; misc. with water, alcohol, acetone; sp.gr. 0.99; pH 7–9 (5%); 50% act.

Empigen® BAC50/BP. [Albright & Wilson UK] Benzalkonium chloride BP, NF; cationic; bactericide for disinfectant and sanitizer formulations for household, institutional, agric., food processing applics., antiseptic detergents in pharmaceuticals; algicide for swimming pools; wood preservatives; pale straw liq.; sp.gr. 0.99; pH 7–9 (5%); 50% act.

Empigen® BAC80. [Albright & Wilson UK] Benzalkonium chloride; bactericide for disinfectant and sanitizer formulations for household, institutional, agric., food processing applics., antiseptic detergents in pharmaceuticals; algicide for swimming pools; wood preservatives; pale straw liq./paste; m.w. 354; sp.gr. 0.97; pH 7–9; 80% act. in org. solv./water.

Empigen® BAC90. [Albright & Wilson UK] Benzalkonium chloride; cationic; bactericide for disinfectant and sanitizer formulations for household, institutional, agric., food processing applics., antiseptic detergents in pharmaceuticals; algicide for swimming pools; wood preservatives; gel; 90% conc.

Empigen® BB. [Albright & Wilson Am.; Albright & Wilson UK] Lauryl betaine; CAS 66455-29-6; amphoteric; foam booster/stabilizer, emulsifier, dispersant, wetting agent, thickening agent, conditioner used for shampoos, detergents, latex foam compds. for carpet backing, industrial applics.; formulation of film removers; stable over wide pH range; pale straw liq.; dens. 1.03 g/cc; cloud pt. < 0 C; pH 7 ± 2 (5% aq.); toxicology: mild to skin; low irritancy to eyes; 30% act.

Empigen® BB-AU. [Albright & Wilson Australia] Cocodimethyl betaine; CAS 66455-29-6; amphoteric; foam booster/stabilizer, thickener, emulsifier, dispersant, wetting and foaming agent; used in personal care prods., industrial and institutional cleaners; pale yel. liq.; sp.gr. 1.04; cloud pt. 1 C max.; pH 8.5 ± 1.0 (5% aq.); 30% act. in water.

Empigen® BCB50. [Albright & Wilson UK] Benzalkonium chloride; CAS 68989-00-4; cationic; disinfectant for dairy, food processing, restaurant, brewing, and bottling industries; retarder in dyeing of acrylic fibers; pale yel. liq.; dens. 0.98 g/cm³; visc. 120 cs; cloud pt. < 0 C; pH 7.0–9.5 (5% aq.); 50.0 ± 1.5% act. in water.

Empigen® BCF 80. [Albright & Wilson UK] Benzalkonium chloride; cationic; bactericide, germicide; raw material; industrial applics.; liq.; 80% conc.

Empigen® BCM75, BCM75/A. [Albright & Wilson UK] Hydrog.-tallow dimethyl benzyl ammonium chloride; cationic; disinfectant for dairy/food processing, restaurant, brewing, and bottling industries, medical/pharmaceutical uses; mfg. of bentones; flotation aid; wh. waxy solid; m.p. 60 C; flash pt. 40 C (CC); pH 6.0–8.5 (5%); 75.0–80.0% act.

Empigen® BS. [Albright & Wilson UK] Cocamidopropyl dimethylamine betaine; foaming agent/stabilizer and antistat for shampoos, foam baths, latex foam compds., oil prod.; wetting and coupling agent for cleaners, traffic film removers; pale straw liq.; m.w. 350; sp.gr. 1.00; pH 4.5–8.0; 30 ± 1% act. in water.

Empigen® BS/H. [Albright & Wilson UK] Cocamidopropyl dimethylamine betaine; amphoteric; foaming agent/stabilizer and antistat for shampoos, foam baths, latex foam compds., oil prod.; wetting and coupling agent for cleaners, traffic film removers; colorless/pale sraw liq.; m.w. 350; sp.gr. 1.04; pH 4.5–8.0; 30 ± 1% act. in water.

Empigen® BS/P. [Albright & Wilson UK] Cocamidopropyl dimethylamine betaine; CAS 61789-40-0; amphoteric; foam booster and co-active agent for toiletries and detergents; pale sraw liq.; m.w. 350; sp.gr. 1.00; pH 4.0–8.0; 25.5 ± 1.5% act. in water.

Empigen® CDR10. [Albright & Wilson UK] Coconut imidazoline betaine; amphoteric; surfactant used in personal care prods., household and industrial cleaners, textile processing; amber visc. liq.; dens. 1.20 g/cm³; visc. < 500 cs; pH 8.7–9.3 (10% aq.).

Empigen® CDR30. [Albright & Wilson UK] Coconut imidazoline betaine, modified; amphoteric; surfactant used in personal care prods., household and industrial cleaners, textile processing; amber visc. liq.; dens. 1.20 g/cm³; visc. 40,000–180,000 cs; pH 8.2–8.8 (10% aq.).

Empigen® CDR40. [Albright & Wilson UK] Sodium cocoamphoacetate; CAS 68334-21-4; amphoteric; mild detergent for shampoos, skin cleansers, personal care prods., textile processing; liq.

Empigen® CDR60. [Albright & Wilson UK] Sodium cocoamphoacetate; CAS 68334-21-4; amphoteric; mild detergent for shampoos, skin cleansers, personal care prods.; textile processing; liq.; 50% conc.

Empigen® CHB. [Albright & Wilson UK] Myrtrimonium bromide; bactericide for disinfectant and sanitizer formulations for household, institutional, agric., food processing applics., antiseptic detergents in pharmaceuticals; algicide for swimming pools; wood preservatives; wh./pale cream powd.; m.w. 336; dens. 0.27 g/cc; pH 5–7; 96% act.

Empigen® CHB40. [Albright & Wilson UK] Alkyl trimethyl ammonium bromide; cationic; bactericide; liq.; 40% conc.

Empigen® CM. [Albright & Wilson UK] Tallow trimethyl ammonium methosulfate; cationic; antistat and conditioning agent for personal care prods.; emulsifier and antistat for industrial use; retarder for dyeing of acrylic fibers; pale straw cloudy liq.; dens. 0.96 g/cm³; flash pt. 40 C; cloud pt. 23 C; pH 6.5–8.5 (5% aq.); 30.0 ± 1.5% act.

Empigen® FKC75K. [Albright & Wilson UK] Alkyl diamido amine acetate; cationic; raw materical used in fabric softener formulations; paste; 75% conc.

Empigen® FKC75L. [Albright & Wilson UK] Alkyl diamido amine lactate; cationic; raw materical used in fabric softener formulations; paste; 75% conc.

Empigen® FKH75L. [Albright & Wilson UK] Dialkyl diamido amine lactate; cationic; surfactant; paste; 75% conc.

Empigen® FRB75S. [Albright & Wilson UK] Hardened tallow alkyl imidazoline methosulfate; cationic; textile conditioning agents for domestic, commercial laundry use and textile finishing; pale amber waxy solid; m.w. 723; set pt. 40 C; pH 4.0–6.5 (8% aq.); 75% act.

Empigen® FRC75S. [Albright & Wilson UK] Tallow alkyl imidazoline methosulfate; cationic; textile conditioning agent, antistat, and lubricant for fibers; raw material for fabric softener formulations; pale amber hazy liq.; disp. in water; dens. 0.95 g/cm³; visc. 300 cp; flash pt. 23 C (CC); pH 4.0–6.5 (8%); 75.0% act.

Empigen® FRC90S. [Albright & Wilson UK] Tallow alkyl imidazoline methosulfate; cationic; textile conditioning agents for domestic, commercial laundry use and textile finishing; pale amber soft paste; m.w. 723; dens. 0.95 g/cc; set pt. 20 C; pH 4.0–6.5 (8% aq.); 90% act.

Empigen® FRG75S. [Albright & Wilson UK] Partially hardened tallow alkyl imidazoline methosulfate; cationic; textile conditioning agents for domestic, commercial laundry use and textile finishing; pale amber waxy solid; m.w. 723; set pt. 30 C; pH 4.0–6.5 (8% aq.); 75% act.

Empigen® FRH75S. [Albright & Wilson UK] Oleyl alkyl imidazoline methosulfate; cationic; textile conditioning agents for domestic, commercial laundry use and textile finishing; pale amber clear or slightly hazy liq.; m.w. 739; sp.gr. 0.95; set pt. < 0 C; pH 4.0–6.5 (8% aq.); 75% act.

Empigen® OB. [Albright & Wilson; Albright & Wilson UK] Lauramine oxide; CAS 1643-20-5; 70592-80-2; nonionic; coactive, detergent, antistat, foam booster/stabilizer and visc. modifier for personal care prods., surgical scrubs, fire fighting foam concs., foamed rubbers, bleach additive; solubilizer; pale straw liq.; dens. 0.98 g/cm³; visc. 25 cs (20 C); pH 7.5 ± 0.5 (5% aq.); 30.0 ± 1.5% act.

Empigen® OC. [Albright & Wilson Am.] Alkyl dimethyl amine oxide; nonionic; surfactant; liq.; 30% conc.

Empigen® OH25. [Albright & Wilson; Albright & Wilson UK] n-Myristyl dimethyl amine oxide; CAS 3332-27-2; nonionic; foam booster/stabilizer and visc. modifier for shampoos, foam baths, detergents; improves conditioning in shampoos; solubilizer for liq. bleach prods.; pale straw liq.; m.w. 257; sp.gr. 0.99; visc. 800 cs; pH 7.5 ± 0.5 (5% aq.); 25% act.

Empigen® OS/A. [Albright & Wilson; Albright & Wilson UK] Cocamidopropyl dimethyl amine oxide; CAS 68155-09-9; nonionic; foam booster/stabilizer and visc. modifier for shampoos, foam baths, detergents; improves conditioning in shampoos; pale straw liq.; m.w. 306; sp.gr. 1.01; visc. 65 cs; pH 6.5–8.0 (5% aq.); 30.5% act.

Empigen® OY. [Albright & Wilson; Albright & Wilson UK] PEG-3 lauramine oxide; CAS 59355-61-2; nonionic; detergent, antistat, foam booster/stabilizer for foamed rubbers, fire fighting, bleach additive; pale straw liq.; dens. 1.0 g/cm³; visc. 23 cs; pH 6.5 ± 0.5 (5% aq.); 25.0 ± 1.0% act.

Empilan® 0004. [Albright & Wilson UK] Castor oil EO/PO condensate; nonionic; emulsifier and lubricant for formulation of cutting oils, grinding fluids, and textile lubricants; mfg. of brake fluids; pale amber liq.; visc. 550 cs; HLB 12.0; sapon. no. 67.5 ± 3.5.; 100% act.

Empilan® 2502. [Albright & Wilson UK] Cocamide DEA (1:1); CAS 8051-30-7; nonionic; foam booster/stabilizer; liq.; 80% conc.

Empilan® 7132. [Albright & Wilson UK] Fatty amine

alkoxylate; nonionic; surfactant.

Empilan® AM Series. [Albright & Wilson UK] Amine ethoxylates; emulsifier for agric. emulsifiable and suspension concs.; antistat for plastics.

Empilan® BQ 100. [Albright & Wilson UK] PEG-8 oleate; CAS 9004-96-0; nonionic; emulsifier for insecticides, spindle oils, industrial wetting agent, scouring agent, antifoam, PVC antistat; in laundry works, glue mfg., paper coating, hand cleaning jellies, turbine oil additive; dk. brn. liq; faint, typ. odor; m.w. 400 (of glycol base); disp. in water, yields clear sol'ns. with ethanol, xylene, oleic acid; sp.gr. 1.017; visc. 130 cps; m.p. –1 C; 100% act.

Empilan® CDE. [Albright & Wilson UK] Cocamide DEA (1:1); nonionic; foam boosting/stabilizing agent, solubilizer, detergent for use in personal care and detergent prods.; antistat in plastics; softener for leather processing; pale yel. visc. liq.; dens. 1.0 g/cm³; 100% act.

Empilan® CDX. [Albright & Wilson UK] Cocamide DEA; CAS 68603-42-9; nonionic; foam booster/stabilizer, solubilizer, and visc. modifier for toiletry and detergent formulations; antistatic agent in plastics; pale yel. visc. liq.; dens. 1.0 g/cm³.

Empilan® CIS. [Albright & Wilson UK] Cocamide MIPA; CAS 68440-05-1; nonionic; foam stabilizer, detergent, shampoo additive; cream waxy flake; dens. 0.4 g/cm³; m.p. 46 C; 100% conc.

Empilan® CM. [Albright & Wilson Australia] Cocamide MEA; CAS 68140-00-1; nonionic; foam booster/stabilizer in toiletry and detergent formulations; intermediate; cream wax-like solid; 80% conc.

Empilan® CME. [Albright & Wilson UK] Cocamide MEA; CAS 68140-00-1; nonionic; detergent, foam booster/stabilizer in detergent systems; stabilizer for hair and carpet shampoos; base for mfg. of ethoxylated alkylolamides; visc. modifier; cream waxy flake; disp. in hot water; dens. 0.4 g/cm³; m.p. 68 C; 100% act.

Empilan® CM/F. [Albright & Wilson Australia] Cocamide MEA; CAS 68140-00-1; nonionic; foam stabilizer, visc. modifier; flake; 85% conc.

Empilan® DL 40. [Albright & Wilson UK] Fatty alcohol ethoxylate; nonionic; emulsifier for fatty acids and min. oils for fiber lubricants, wetting and stabilizing agent for syn. rubber latices, textile dye leveling and dispersing agent in textile industry; antistat; pale yel. clear liq.; misc. with water; dens. 1.046 g/cc; cloud pt. > 100 C in dist. water; 40% act.

Empilan® DL 100. [Albright & Wilson UK] Fatty alcohol ethoxylate; emulsifier for fatty acids and min. oils for fiber lubricants, wetting and stabilizing agent for syn. rubber latices, textile dye leveling and dispersing agent in textile industry; antistat; stiff waxy solid; warm water sol.; m.p. 36 C; cloud pt. > 100 C (1% aq.); 100% act.

Empilan® FD. [Albright & Wilson Australia] Cocamide DEA; CAS 8051-30-7; nonionic; surfactant, wetting agent, emulsifier for mfg. of personal care prods., industrial and institutional liq. detergent systems; coactive and visc. modifier for laundry detergents and hard surface cleaners; foam stabilizer and booster; pale yel. clear visc. liq.; sp.gr. 1.00–1.01; flash pt. 95 C; 80% conc.

Empilan® FD20. [Albright & Wilson UK] Cocamide DEA; nonionic; act. ingred. in liq. detergents; foam booster and visc. modifier; pale yel. clear visc. liq.; 65% conc.

Empilan® FE. [Albright & Wilson UK] Cocamide DEA; nonionic; foam booster, visc. modifier used as coactive ingred. in toiletries, liq. detergents; pale yel. clear visc. liq.; sp.gr. 0.970; visc. 250 cP; flash pt. 28 C (PMCC); 90% conc.

Empilan® GMS LSE40. [Albright & Wilson UK] Glyceryl stearate SE; nonionic; emulsifier in the baking and food industry; lubricant in prod. of PVC sheets and expanded PS; wh. microbead powd.; dens. 0.5 g/cm³; m.p. 55–65 C; 40.0% min. monoglyceride.

Empilan® GMS LSE80. [Albright & Wilson UK] Glyceryl stearate SE; nonionic; emulsifier in the baking and food industry; lubricant in prod. of PVC sheets and expanded PS; wh. microbead powd.; dens. 0.5 g/cm³; m.p. 64 C; 80.0% min. monoglyceride.

Empilan® GMS MSE40. [Albright & Wilson UK] Glyceryl stearate SE; nonionic; emulsifier in the baking and food industry; lubricant in prod. of PVC sheets and expanded PS; wh. microbead powd.; dens. 0.5 g/cm³; m.p. 55–60 C; 36.0% min. monoglyceride.

Empilan® GMS NSE40. [Albright & Wilson UK] Glyceryl stearate SE; CAS 31566-31-1; nonionic; emulsifier in the baking and food industry; lubricant and antistat in prod. of PVC sheets and expanded PS; stabilizer; wh. microbead powd.; dens. 0.5 g/cm³; m.p. 55–60 C; pH 6–8 (10% aq.); 36-40% act.

Empilan® GMS NSE90. [Albright & Wilson UK] Glyceryl stearate SE; nonionic; emulsifier in the baking and food industry; lubricant and antistat in prod. of PVC sheets and expanded PS; stabilizer; wh. microbead powd.; dens. 0.5 g/cm³; m.p. 65 C; 90.0% min. monoglyceride.

Empilan® GMS SE32. [Albright & Wilson UK] Glyceryl stearate SE; nonionic; emulsifier; wh. wax-like powd.; m.p. 55–60 C; pH 7–9 (10% aq.); 32.5% min. monoglyceride.

Empilan® GMS SE40. [Albright & Wilson UK] Glyceryl stearate SE; CAS 31566-31-1; nonionic; emulsifier in the baking and food industry, for cosmetic and pharmaceuticals; lubricant in prod. of PVC sheets and expanded PS; wh. microbead powd.; dens. 0.5 g/cm³; m.p. 55–60 C; pH 7–9 (10% aq.); 36-40% act.

Empilan® GMS SE70. [Albright & Wilson UK] Glyceryl stearate SE; nonionic; emulsifier in the baking and food industry; lubricant in prod. of PVC sheets and expanded PS; wh. microbead powd.; dens. 0.5 g/cm³; m.p. 61 C; 70.0% min. monoglyceride.

Empilan® K Series. [Albright & Wilson UK] Alkyl ethoxylates; emulsifier, wetting agent for agric. emulsifiable and suspension concs., spinning oils in textile processing; emulsifier, dispersant for emulsion polymerization; rubber compding. aid; antistat for plastics.

Empilan® KA3. [Albright & Wilson UK] C10-12 alcohol ethoxylate; nonionic; scouring and wetting agent for textiles, emulsifier, dye leveling and dispersing agent, detergent, in metal processing, cutting oils, paper industry, paints, insecticides and pesticides; mortar plasticizer; biodeg.; pale straw clear liq.; sol. in min. oils and other org. solvs.; HLB 8.9; 100% act.

Empilan® KA5. [Albright & Wilson UK] C10-12 alcohol ethoxylate; nonionic; scouring and wetting agent for textiles, emulsifier, dye leveling and dispersing agent, detergent, in metal processing, cutting oils, paper industry, paints, insecticides and pesticides; biodeg.; pale straw clear liq.; HLB 11.5; cloud pt. 37 C; 100% act.

Empilan® KA5/90. [Albright & Wilson UK] C10-12 alcohol ethoxylate; CAS 68439-45-2; nonionic; scouring and wetting agent for textiles, emulsifier, dye leveling and dispersing agent, detergent, in metal processing, cutting oils, paper industry, paints, insecticides and pesticides; biodeg.; water wh. clear liq.; dens. 9.8 lb/gal; HLB 11.4; pour pt. 12 F; cloud pt. 95 F; 90% act.

Empilan® KA8/80. [Albright & Wilson UK] C10-12 alcohol ethoxylate; CAS 68439-45-2; nonionic; scouring and wetting agent for textiles, emulsifier, dye leveling and dispersing agent, detergent, in metal processing, cutting oils, paper industry, paints, insecticides and pesticides; biodeg.; liq.; HLB 13.6; cloud pt. 81 C; 80% act.

Empilan® KA10/80. [Albright & Wilson UK] C10-12 alcohol ethoxylate; CAS 68439-45-2; nonionic; scouring and wetting agent for textiles, emulsifier, dye leveling and dispersing agent, detergent, in metal processing, cutting oils, paper industry, paints, insecticides and pesticides; biodeg.; liq.; HLB 14.6; cloud pt. 95 C; 80% act.

Empilan® KB 2. [Albright & Wilson UK] Laureth-2; CAS 68002-97-1; nonionic; emulsifier, foam booster, superfatting agent, used in detergents and emulsifying systems; mortar plasticizer; almost colorless liq.; insol. in cold water; dens. 0.9; visc. 25 cs; 100% act.

Empilan® KB 3. [Albright & Wilson UK] Laureth-3; CAS 68002-97-1; nonionic; emulsifier, foam booster, superfatting agent, used in detergents and emulsifying systems; mortar plasticizer; liq.; HLB 8.1; 100% conc.

Empilan® KC 3. [Albright & Wilson UK] Laureth-3; nonionic; emulsifier, foam booster, superfatting agent, used in detergents and emulsifying systems; liq.; 100% conc.

Empilan® KCA Series. [Albright & Wilson UK] C12-13 alcohol ethoxylates; nonionic; surfactant; liq.; 100% conc.

Empilan® KCB Series. [Albright & Wilson UK] C11 alcohol ethoxylates; nonionic; surfactant; liq.; 100% conc.

Empilan® KCL Series. [Albright & Wilson UK] C12-15 alcohol ethoxylates; nonionic; surfactant; liq. to paste; 100% conc.

Empilan® KCMP 0703/F. [Albright & Wilson UK] Fatty alcohol alkoxylate; CAS 68551-13-3; nonionic; emulsifier, detergent, wetting agents for industrial and domestic applics.; liq.; 100% conc.

Empilan® KCMP 0705/F. [Albright & Wilson UK] Fatty alcohol mixed EO/PO condensate; CAS 68551-13-3; nonionic; emulsifier, detergent, wetting agents for industrial and domestic applics.; low foaming; pale straw clear liq.; sol. in water; dens. 0.97 g/cm^3; HLB 14.4; cloud pt. 23–27 C (1% aq.); pH 4.0–7.5 (5% aq.); 100% conc.

Empilan® KCP Series. [Albright & Wilson UK] C14-15 alcohol ethoxylates; nonionic; surfactant; liq. to paste; 100% conc.

Empilan® KCX Series. [Albright & Wilson UK] C13-15 alcohol ethoxylates; nonionic; surfactant; liq. to paste; 100% conc.

Empilan® KI Series. [Albright & Wilson UK] C13 alcohol ethoxylates; nonionic; wetting agent for leather processing; liq.; 100% conc.

Empilan® KL 6. [Albright & Wilson UK] Cetoleth-6; nonionic; emulsifier for min. oils in cosmetics, detergent and wetting agents in textile industry; cream soft paste; cold water disp.; dens. 0.9 g/cc; m.p. 35 C; pH 8–10; 100% act.

Empilan® KL 10. [Albright & Wilson UK] Cetoleth-10; nonionic; emulsifier for min. oils in cosmetics, detergent and wetting agents in textile industry; m.p. 40 C; pH 8–10; 100% act.

Empilan® KL 20. [Albright & Wilson UK] Cetoleth-20; nonionic; emulsifier for min. oils in cosmetics, detergent and wetting agents in textile industry; cream soft solid; cold water disp.; dens. 0.9 g/cc; m.p. 42 C; pH 8–10; 100% act.

Empilan® KM 11. [Albright & Wilson UK] Ceteareth-11; CAS 68439-49-6; nonionic; emulsifier, foam control agent in syn. heavy duty detergents, soap additive; dispersant for textile processing; colorless to pale straw paste; 100% act.

Empilan® KM 20. [Albright & Wilson UK] Ceteareth-20; CAS 68439-49-6; nonionic; wetting agent, detergent for industrial/domestic applics.; emulsifier, foam control agent in syn. heavy duty detergents; dispersant for textile processing; colorless to pale straw waxy flake/block; 100% act.

Empilan® KM 50. [Albright & Wilson UK] Ceteareth-50; CAS 68439-49-6; nonionic; wetting agent, detergent for industrial/domestic applics.; emulsifier, foam control agent in syn. heavy duty detergents; dispersant for textile processing, polyurethane prod.; colorless to pale straw waxy flake/block; 100% act.

Empilan® KS Series. [Albright & Wilson UK] C9-11 alcohol ethoxylates; nonionic; surfactant; liq.; 100% conc.

Empilan® LDE. [Albright & Wilson UK] Lauramide DEA (1:1); CAS 120-40-1; nonionic; foam booster/stabilizer, solubilizer, thickener, detergent used in shampoos and liq. detergent formulations; antistat for plastics; cream solid; dens. 1.0 g/cm^3; 100% act.

Empilan® LDX. [Albright & Wilson UK] Lauramide DEA and diethanolamine; nonionic; foam booster/stabilizer, solubilizer, and visc. modifier for toiletry and detergent formulations; antistatic agent in plastics; cream solid; dens. 1.0 g/cm^3; 100% conc.

Empilan® LIS. [Albright & Wilson UK] Lauramide MIPA; CAS 142-54-1; nonionic; foam booster/stabilizer for liq. and powd. detergents, shampoos; visc. modifier, base for mfg. of ethoxylated deriv.; cream waxy flake; hot water disp.; dens. 0.4 g/cc; 87.5% act.

Empilan® LME. [Albright & Wilson UK] Lauramide MEA; CAS 142-78-9; nonionic; foam booster/stabilizer in detergent systems; stabilizer for hair and carpet shampoos; base for mfg. of ethoxylated alkylolamides; visc. modifier; waxy flake; dens. 0.4 g/cm^3; m.p. 85 C; 100% conc.

Empilan® LP2. [Albright & Wilson UK] Cocamide MEA ethoxylate; nonionic; liq. detergent additive as foam stabilizer, detergent booster, solubilizer; cream yel. soft paste; faint char. odor; sapon. no. 14; 100% act.

Empilan® LP10. [Albright & Wilson UK] Ethoxylated cocamide; CAS 61791-08-0; nonionic; foam and detergent booster; paste; 100% conc.

Empilan® MAA. [Albright & Wilson UK] PEG-6 cocamide; CAS 61791-08-0; nonionic; liq. detergent additive as foam stabilizer, detergent booster, solubilizer; paste; sapon. no. 15; 100% act.

Empilan® NP9. [Albright & Wilson UK] Nonoxynol-9; CAS 9016-45-9; nonionic; wetting agent, detergent, emulsifier, solubilizer; for agric. emulsifiable and suspension concs., leather processing, emulsion and suspension polymerization; mortar plasticizer; plastics antistat; pale straw soft paste/liq.; dens. 1.0 g/cc; visc. 300 cs; cloud pt. 55 C (1% aq.); 100% act.

Empilan® OPE9.5. [Albright & Wilson UK] Octyl phenol ethoxylate; nonionic; emulsifier for agric. emulsifiable and suspension concs.; liq.; 100% conc.

Empilan® P7061. [Albright & Wilson UK] EO/PO condensate; nonionic; wetting agent for agric. emulsifiable and suspension concs.; compding. aid for pharmaceuticals; liq.; 100% conc.

Empilan® P7062. [Albright & Wilson UK] EO/PO condensate; nonionic; wetting agent for agric. emulsifiable and suspension concs.; compding. aid for pharmaceuticals; liq.; 100% conc.

Empilan® P7087. [Albright & Wilson UK] EO/PO condensate; nonionic; wetting agent for agric. emulsifiable and suspension concs.; compding. aid for pharmaceuticals; liq.; 100% conc.

Empilan® PF7158. [Albright & Wilson UK] Alcohol ethoxylate/propoxylate; nonionic; surfactant; liq.; 100% conc.

Empilan® PF7159. [Albright & Wilson UK] Alcohol ethoxylate/propoxylate; nonionic; surfactant; liq.; 100% conc.

Empilan® SM Series. [Albright & Wilson UK] Sorbitan ester ethoxylates; emulsifier for agric. emulsifiable concs.

Empimin® 3060. [Albright & Wilson UK] Surfactant blend; anionic/nonionic; high expansion fire fighting foam conc.; liq.; 25% conc.

Empimin® 3093. [Albright & Wilson UK] Anionic surfactant and solv.; defoaming agent for mfg. of wet process phosphoric acid and other aq., acidic systems; straw clear liq.; m.w. 444; sp.gr. 1.10; visc. 450 cs; flash pt. (CC) > 100 C; pH 6.0 ± 1.0 (5%); 72 ± 2% act.

Empimin® 3095. [Albright & Wilson UK] Anionic surfactant and solv.; defoaming agent for mfg. of wet process phosphoric acid and other aq., acidic systems; straw clear liq.; m.w. 444; sp.gr. 1.05; visc. 100 cs; flash pt. (CC) > 100 C; pH 6.0 ± 1.0 (5%); 44 ± 2% act.

Empimin® 3116. [Albright & Wilson UK] Sulfosuccinate blend; flotation aid for phosphate rock.

Empimin® 3631. [Albright & Wilson Australia] Foaming agent; anionic; high/med. foaming agent for extinguishing fires and for agric. marking; liq.; 20% conc.

Empimin® AQ60. [Albright & Wilson UK] Ammonium alkyl triethoxysulfate; anionic; detergent, foaming and wetting agent, dispersant in institutional and industrial cleaning specialties; pale yel. clear liq.; dilutable in water; sp.gr. 1.02; flash pt. 30 C (PMCC); pH 7.0 ± 0.6 (10% aq.); 58.0 ± 3.0% act.

Empimin® BMA. [Albright & Wilson UK] Formulated surfactant; anionic; surfactant used as air entraining agent for cementitious mixes, foaming agent for concrete, mortar plasticizer; pale yel. cloudy liq.; m.w. 433; sp.gr. 1.11; visc. 140 cs; pH 7–9 (5% aq.); 34 ± 1.5% act.

Empimin® BMB. [Albright & Wilson UK] Formulated surfactant; anionic; plasticizer/air entraining agent for construction industry.

Empimin® BMC. [Albright & Wilson UK] Formulated prod.; air entraining agent for mortar/cement; liq.; 27% conc.

Empimin® I156. [Albright & Wilson UK] Surfactant blend; anionic/nonionic; refractory foaming agent.

Empimin® KSN27. [Albright & Wilson UK] Sodium laureth sulfate (3 EO); CAS 9004-82-4; anionic; detergent raw material for high-quality liq. detergents, textile and leather processing; emulsifier, dispersant for emulsion polymerization; biodeg.; pale straw clear or slightly turbid liq.; dens. 1.05 g/cc; visc. 100 cs; cloud pt. 0 C; pH 6.5–7.5 (5% aq.); 27% aq. sol'n.

Empimin® KSN60. [Albright & Wilson UK] Sodium laureth sulfate (3 EO), water/ethanol; anionic; detergent raw material for high-quality liq. detergents, textile and leather processing; emulsifier, dispersant for emulsion polymerization; pale amber translucent liq.; dens. 1.00 g/cc; flash pt. 38 C; pH 7.0–8.0 (5% aq.); biodeg.; 60% aq. sol'n. with 10% ethanol as solubilizer.

Empimin® KSN70. [Albright & Wilson UK] Sodium laureth sulfate (3 EO); CAS 9004-82-4; anionic; detergent raw material for high-quality liq. detergents, textile and leather processing; wetting agent for agric. wettable powds.; emulsifier, dispersant for emulsion polymerization; straw opaque pourable visc. liq.; dens. 1.1 g/cc; set pt. 8 C; pH 7.0–9.0 (2% aq.); 70% aq. sol'n.

Empimin® LAM30/AU. [Albright & Wilson Australia] Ammonium alkyl ether sulfate; CAS 75422-21-0; anionic; wetting and foaming agent used in mfg. of plasterboard, geological drilling operations; liq.; 30% conc.

Empimin® LR28. [Albright & Wilson UK] Sodium lauryl sulfate; CAS 151-21-3; anionic; foaming agent in rubber latex systems; emulsifier for emulsion polymerization; yel./amber liq.; dens. 1.04 g/cm³; cloud pt. < 0 C; pH 7.0–9.0 (5% aq.); 28.0% min. act.

Empimin® LSM30. [Albright & Wilson UK] Sodium alkyl ether sulfate; anionic; foaming agent for slurries in plasterboard mfg.; foam-boosting additive in latices for carpet backing; wetting agent; ingred. in alkaline liq. cleaners; pale yel. liq.; dens. 1.06 g/cm³; visc. 40 cs; cloud pt. < 0 C; pH 8.0 ± 1.0 (5% aq.); 30.0 ± 1.0% act. in water.

Empimin® LSM30-AU. [Albright & Wilson Australia] Sodium alkyl ethoxysulfate; anionic; detergent, wetting agent; liq.; sp.gr. 1.05; visc. 40 cSt; pH 8.0 ± 1.0 (5%); cloud pt. < 0 C; 30.0 ± 1.0% act.

Empimin® MA. [Albright & Wilson UK] Sodium dihexyl sulfosuccinate; CAS 3006-15-3; anionic; emulsifier, dispersant, wetting agent for emulsion polymerization, agric. emulsifiable concs.; straw clear liq.; sol. in water; dens. 1.1 g/cm³; visc. 80 cs; flash pt. 30 C (CC); pH 6.0 ± 1.0 (5%); 63.0% min. act.

Empimin® MH. [Albright & Wilson UK] Disodium cocoyl sulfosuccinamate; anionic; foaming agent for carpet backing; liq.; 40% conc.

Empimin® MHH. [Albright & Wilson UK] Disodium N-cocoyl sulfosuccinate; foaming agent for rubber latex compds.; emulsifier in the mfg. of polymers by emulsion polymerization; dens. 1.1 g/cm³; pH 8.0 ± 0.5 (5% aq.); 32.5% min. act.

Empimin® MK/B. [Albright & Wilson UK] Disodium cetyl stearyl sulfosuccinamate; anionic; foaming agent for carpet backing; paste; 33% conc.

Empimin® MKK98. [Albright & Wilson UK] Disodium N-cetyl stearyl sulfosuccinate; anionic; foaming agent for rubber latex compds.; emulsifier in the mfg. of polymers by emulsion polymerization; pale cream powd.; dens. 0.3 g/cm³; pH 8.5 ± 0.5 (5% aq.); 98% act.

Empimin® MKK/AU. [Albright & Wilson Australia] Sodium N-tallow sulfosuccinamate; CAS 68988-69-2; anionic; foaming agent for rubber latices used in carpet backing; liq.; 30% conc.

Empimin® MKK/L. [Albright & Wilson UK] Disodium N-cetyl stearyl sulfosuccinamate; anionic; foaming agent for rubber latex compds.; emulsifier in the mfg. of polymers by emulsion polymerization; pale cream opaque liq.; dens. 1.1 g/cm³; pH 8.25 ± 0.25 (5% aq.); 28.0% min. act.

Empimin® MSS. [Albright & Wilson UK] Diammonium N-cocoyl sulfosuccinate; foaming agent for rubber latex compds.; emulsifier in the mfg. of polymers by emulsion polymerization; yel./pale cream clear/opaque liq.; dens. 1.1 g/cm³; pH 6.5 ± 0.5 (5% aq.); 31.0% min. act.

Empimin® MTT. [Albright & Wilson UK] Disodium N-oleyl sulfosuccinamate; foaming agent for rubber latices; pale cream liq.; pH 8 (5% aq.); 40% act.

Empimin® MTT/A. [Albright & Wilson UK] Disodium N-oleyl sulfosuccinate; anionic; foaming agent for rubber latex compds.; emulsifier in the mfg. of polymers by emulsion polymerization; yel./pale cream clear/opaque liq.; dens. 1.1 g/cm³; pH 8.0 ± 0.5 (5% aq.); 28.0% min. act.

Empimin® OP45. [Albright & Wilson UK] Sodium dioctyl sulfosuccinate; anionic; emulsifier, dispersant, wetting agent used for emulsion polymerization, agrochem., oil slicks, textiles; straw clear liq.; sol. in water; dens. 1.05 g/cm³; visc. 100 cs; flash pt. > 100 C (CC); pH 6.0 ± 1.0; 44% conc.

Empimin® OP70. [Albright & Wilson UK] Sodium dioctyl sulfsuccinate; CAS 1369-66-3; anionic; emulsifier, dispersant, wetting agent used for emulsion polymerization, oil slicks, textiles, agrochem.; straw clear liq.; sol. in water; dens. 1.10 g/cm³; visc. 450 cs; flash pt. > 100 C (CC); pH 6.0 ± 1.0 (5%); 72.0 ± 2.0% act.

Empimin® OT. [Albright & Wilson UK] Sodium dioctyl sulfosuccinate; CAS 577-11-7; anionic; dispersant, emulsifier, detergent, wetting agent for o/w emulsions, agrochem., emulsion polymerization, filler and extender dispersions, leather processing; straw clear liq.; dens. 1.0 g/cm³; visc. 55 cs; flash pt. 25 C (CC); pH 6.0 ± 1.0 (5%); 60.0% min. act.

Empimin® OT75. [Albright & Wilson UK] Dioctyl sodium sulfosuccinate; CAS 577-11-7; anionic; wetting agent and emulsifier for emulsion polymerization; liq.; 75% conc.

Empimin® SDS. [Albright & Wilson Australia] Sodium decyl sulfate; CAS 84501-49-5; anionic; emulsifier, dispersant, detergent, and wetting agent for industrial and institutional cleansers, mfg. of pigments, alkaline cleansers; dust suppression; pale yel. clear liq.; sp.gr. 1.06; pH 7.0 ± 0.6 (5% aq.); 30.0 ± 1.0% act.

Empimin® SQ25. [Albright & Wilson Australia] Sodium alkyl triethoxysulfate; CAS 68585-34-2; anionic; dispersant, wetting and foaming agent in institutional, household, and industrial cleaners; water-wh. to pale yel. clear liq.; dilutable in water; sp.gr. 1.043 ± 0.005; visc. 80 ± 45 cP; pH 7.0–0.7 (5% aq.); 25.0 ± 0.5% act.

Empimin® SQ70. [Albright & Wilson Australia] Sodium alkyl triethoxylsulfate; CAS 68585-34-2; anionic; see Empimin SQ25; off-wh. to pale yel. hazy visc. liq.; dens. 1.1 g/cm³; visc. 550 cP max.; cloud pt. 0 C max.; pH 8.0 ± 0.6 (2% aq.); 68.0 ± 2.0% act.

Empiphos 4KP. [Albright & Wilson UK] Tetrapotassium pyrophosphate; detergent builder for liq. detergents; pigment dispersant and stabilizer in emulsion paints; clarifying agent in liq. soaps; mfg. of syn. rubber; boiler water treatment; sol. in water; dens. 75 lb/ft³; pH 10.3.

Empiphos DF Series. [Albright & Wilson UK] Phosphate esters; emulsifier for agric. emulsifiable concs.

Empiphos OMP, OSP. [Albright & Wilson UK] Org. polyphosphates; builders and dispersants for liq. detergents; liq.

Empol® 1004. [Henkel/Emery] Dimer acid, hydrog.; surfactant for industrial applics., lubricants; Gardner 2 max. color; acid no. 190 min.

Empol® 1007. [Henkel/Emery] Dimer acid; surfactant for industrial applics., lubricants; Gardner 5 max. color; acid no. 190 min.

Empol® 1010. [Henkel/Emery] Dimer acid; polymer grade; surfactant for industrial applics., lubricants; Gardner 1 max. color; acid no. 190 min.

Empol® 1014. [Henkel/Emery] Dimer acid; surfactant for industrial applics., lubricants; Gardner 5 max. color; acid no. 194-198.

Empol® 1016. [Henkel/Emery] Dimer acid; surfactant for industrial applics., lubricants; Gardner 6 max. color; acid no. 190-198.

Empol® 1018. [Henkel/Emery] Dimer acid; surfactant for industrial applics., lubricants; Gardner 8 max. color; acid no. 190-198.

Empol® 1020. [Henkel/Emery] Dimer acid; surfactant for industrial applics., lubricants; Gardner 7 max. color; acid no. 188-196.

Empol® 1022. [Henkel/Emery] Dimer acid; surfactant for industrial applics., lubricants; Gardner 8 max. color; acid no. 189-197.

Empol® 1026. [Henkel/Emery] Dimer acid; surfactant for industrial applics., lubricants; Gardner 8 max. color; acid no. 189-197.

Empol® 1040. [Henkel/Emery] Trimer acid; surfactant for industrial applics., lubricants; acid no. 175-192.

Empol® 1041. [Henkel/Emery] Trimer acid; surfactant for industrial applics., lubricants; Gardner 11 max. color; acid no. 161-181.

Empol® 1043. [Henkel/Emery] Trimer acid; surfactant for industrial applics., lubricants; Gardner 15 max. color; acid no. 180.

Empol® 1052. [Henkel/Emery] Polybasic acid; surfactant for industrial applics., lubricants; dk. color; acid no. 250-265.

Empol® 1061. [Henkel/Emery] Dimer acid; surfactant for industrial applics., lubricants; Gardner 5 max. color; acid no. 193-201.

Emsorb® 2500. [Henkel/Emery; Henkel/Textile] Sorbitan oleate; CAS 1338-43-8; nonionic; coupler, emulsifier, lubricant, and softener for textile fibers, leather, cosmetics, agric., household prods.; for formulating petrol. oils and waxes, natural fats and waxes, and alkyl esters; EPA exempt; Gardner 8 liq.; sol. in min. oil, butyl stearate, Stod., xylene; insol. in water; dens. 8.3 lb/gal; visc. 360 cSt (100 F); HLB 4.6; pour pt. < 0 C; flash pt. 475 F; 100% conc.

Emsorb® 2502. [Henkel/Emery; Henkel/Textile] Sorbitan sesquioleate; CAS 8007-43-0; nonionic; coupler, coemulsifier for o/w systems; emulsifier for w/o systems; household aerosols, cosmetics, agric., industrial and textile oils; EPA exempt; Gardner 6 liq.; sol. in min. oil, butyl stearate, glycerol trioleate, Stod. solv., xylene; insol. in water; dens. 8.2 lb/gal; visc. 475 cSt (100 F); HLB 4.5; pour pt. < 0 C; flash pt. 470 F; 100% conc.

Emsorb® 2503. [Henkel/Emery; Henkel/Textile] Sorbitan trioleate; CAS 26266-58-0; nonionic; coupler, emulsifier, lubricant, and softener for textile fibers and leather, cosmetics, agric., household prods.; for formulating petrol. oils and waxes, natural fats and waxes, and alkyl esters; coemulsifier for min. oil; EPA exempt; Gardner 7 liq.; sol. in min. oil, butyl stearate, glycerol trioleate, Stod. solv., xylene; insol. in water; dens. 7.9 lb/gal; visc. 100 cSt (100 F); HLB 2.1; pour pt. < 0 C; flash pt. 500 F; 100% act.

Emsorb® 2505. [Henkel/Emery] Sorbitan stearate; CAS 1338-41-6; nonionic; coupler, hydrophobic emulsifier; coemulsifier for industrial oils, household prods., agric., and cosmetics; textile lubricant; paper and textile processing; EPA exempt; Gardner 4 solid; disp. in butyl stearate, perchloroethylene; insol. in water; HLB 5.2; m.p. 50 C; flash pt. 480 F; 100% conc.

Emsorb® 2507. [Henkel/Emery] Sorbitan tristearate; CAS 26658-19-5; nonionic; coupler, emulsifier, lubricant, and softener for textile fibers and leather, cosmetics, agric., household prods.; for formulating petrol. oils and waxes, natural fats and waxes, and alkyl esters; coemulsifier for min. oil; EPA-exempt; Gardner 4 waxy solid; sol. in butyl stearate, perchloroethylene, glycerol trioleate, Stod. solv., xylene; insol. in water; HLB 2.2; m.p. 54 C; flash pt. 480 F; 100% conc.

Emsorb® 2510. [Henkel/Emery] Sorbitan palmitate; CAS 26266-57-9; nonionic; coupler, emulsifier for cosmetic, agric. and household prods.; fiber-to-metal lubricant; EPA-exempt; Gardner 8 waxy solid; sol. in oil, Stod. solv., xylene; insol. in water; HLB 6.5; m.p. 47 C; flash pt. 445 F; 100% conc.

Emsorb® 2515. [Henkel/Emery; Henkel/Textile] Sorbitan laurate; CAS 1338-39-2; nonionic; coupler, emulsifier; used in household specialities, industrial oils, agric., cosmetics, and emulsion polymerization; antifoam properties; EPA-exempt; Gardner 7 liq.; sol. in min. oil, butyl stearate, glycerol trioleate, Stod. solv., xylene; water-disp.; dens. 8.8 lb/gal; visc. 1000 cSt; HLB 8.0; pour pt. 15 C; flash pt. 430 F; cloud pt. < 25 C; 100% conc.

Emsorb® 2516. [Henkel/Emery] Sorbitan isostearate; nonionic; aux. emulsifier, solubilizer, corrosion inhibitor in lubricants, metal protectants and cleaners, emulsion polymerization; Gardner 2 solid; sol. @ 5% in min. oil, Stod., xylene; disp. in water, butyl stearate, glyceryl trioleate; dens. 8.2 lb/gal; visc. 1200 cSt (100 F); HLB 4.6; m.p. 50 C; flash pt. 460 F.

Emsorb® 2518. [Henkel/Emery] Sorbitan diisostearate; nonionic; aux. emulsifier, solubilizer, corrosion inhibitor in lubricants, metal protectants and cleaners, emulsion polymerization; Gardner 5 liq.; sol. @ 5% in min. oil, butyl stearate, glyceryl trioleate, Stod., xylene; dens. 8.2 lb/gal; visc. 730 cSt (100 F); HLB 3.0; pour pt. –4 C; flash pt. 480 F; 100% conc.

Emsorb® 2722. [Henkel/Cospha; Henkel Canada] Polysorbate 80; CAS 9005-65-6; nonionic; emulsifier, coemulsifier for cosmetics; dispersant for pigments in makeup; solubilizer for oils, flavors, fragrances; Gardner 5 liq.; sol. @ 5% in water, triolein, IPA; dens. 9.0 lb/gal; visc. 200 cSt (100 F); HLB 15.0; acid no. 2 max.; sapon. no. 45-55; pour pt. –12 C; flash pt. 505 F.

Emsorb® 2726. [Henkel/Cospha; Henkel Canada] PEG-40 sorbitan diisostearate; emulsifier, solubilizer for flavors in mouthwashes, lipstick, perfume, for perfumes, germicides, and other polar substances in aq. systems; yel. liq.; sol. in water, alcohol, some cosmetic oils.

Emsorb® 2728. [Henkel/Cospha; Henkel Canada] Polysorbate 60; CAS 9005-67-8; nonionic; o/w emulsifier for cosmetics, hair straighteners, shaving prods., sun care prods.; binder in powds.; with Emsorb 2505 to stabilize wax emulsions; Gardner 3 waxy semisolid; sol. @ 5% in water, IPA; dens. 8.9 lb/gal; visc. 250 cSt (100 F); HLB 15.2; acid no. 2 max.; sapon. no. 45-55; pour pt. 25 C; flash pt. 510 F.

Emsorb® 6900. [Henkel/Emery; Henkel/Textile] PEG-20 sorbitan oleate; CAS 9005-65-6; nonionic; dispersant for pigments in coatings; solubilizer for oils and fragrances; hydrophilic emulsifier; coemulsifier for petrol. oils, fats, solvs., and waxes in cosmetics, household prods., industrial lubricants,; and textile dye carriers; emulsifier for tobacco sucker control concs.; EPA-exempt; Gardner 5 liq.; sol. in water; dens. 9.0 lb/gal; visc. 200 cSt (100 F); HLB 15.0; pour pt. –12 C; flash pt. 535 F; cloud pt. 70 C (5% saline); 100% conc.

Emsorb® 6901. [Henkel/Emery; Henkel/Textile] PEG-5 sorbitan oleate; CAS 9005-65-6; nonionic; o/w emulsifier and lubricant in industrial lubricants, textile lubricants and softeners, metal treatment, paints, emulsion polymerization, agric.; color dispersant in plastics; EPA-exempt; Gardner 6 liq.; sol. in glycerol trioleate, Stod. solv.; water-disp.; dens. 8.5 lb/gal; visc. 450 cs; HLB 10.0; cloud pt. < 25 C; 100% conc.

Emsorb® 6903. [Henkel/Emery; Henkel/Textile] PEG-20 sorbitan trioleate; CAS 9005-70-3; nonionic; o/w emulsifier for petrol. oils, fats, waxes, and alkyl esters; lubricant for metals, textiles, leather; in sol. oils for metal processing and finishing; glass fiber lubricants; automotive lubricant additives; agric. formulations; EPA-exempt; Gardner 6 liq.; sol. in butyl stearate, glycerol trioleate, Stod. solv.; water-disp.; dens. 8.6 lb/gal; visc. 300 cs; HLB 11.1; cloud pt. < 25 C; 95% act.

Emsorb® 6905. [Henkel/Emery] PEG-20 sorbitan stearate; CAS 9005-67-8; nonionic; o/w emulsifier for min. oil, fats, and waxes; fiber-to-metal lubricant and coemulsifier with Emsorb 2505 in paraffin wax emulsions for textiles, paper and wallboard coatings; useful as emulsifier in cosmetic and household prods.; Gardner 3 waxy semisolid; sol. 5% in Stod. solv. and water; dens. 8.9 lb/gal; visc. 250 cSt (100 F); HLB 15.2; pour 25 C; flash pt. 510 F; cloud pt. 70 C (5% saline); 100% conc. DISCONTINUED.

Emsorb® 6906. [Henkel/Emery] PEG-4 sorbitan stearate; CAS 9005-67-8; nonionic; w/o emulsifier used in household formulations; fiber-to-metal lubricant for syn. and cellulosic fibers and yarns; Gardner 3 solid; sol. 5% in Stod. solv.; water-disp.; HLB 9.0; m.p. 42 C; flash pt. 495 F; 100% conc.

Emsorb® 6907. [Henkel/Emery] PEG-20 sorbitan tristearate; CAS 9005-71-4; nonionic; o/w emulsifier for petrol. oils and natural fats; lubricant and softener for textile processing and finishing compds.; primary emulsifier; Gardner 5 solid; sol. 5% in min. oil, butyl stearate, glycerol trioleate, Stod. solv.; water-disp.; HLB 11.1; m.p. 34 C; flash pt. 555 F; cloud pt. < 25 C; 100% conc. DISCONTINUED.

Emsorb® 6908. [Henkel/Emery] PEG-16 sorbitan tristearate; nonionic; o/w emulsifier, lubricant, softener; for textile processing and finishing compds., agric. formulations; EPA-exempt; Gardner 4 waxy solid; sol. @ 5% in min. oil, glyceryl trioleate, Stod., xylene; disp. in water, butyl stearate; HLB 10.0; m.p. 38 C; cloud pt. < 25 C; flash pt. 540 F.

Emsorb® 6909. [Henkel/Emery] PEG-4 sorbitan stearate; CAS 9005-67-8; nonionic; emulsifier for hydraulic fluids, metal treatment, emulsion polymerization, paints; color dispersants for plastics; Gardner 6 solid; sol. in min. oil, glyceryl trioleate, Stod.; disp. in water, butyl stearate, xylene; m.p. 35 F; HLB 9.6; flash pt. 520 F.

Emsorb® 6910. [Henkel/Emery] PEG-20 sorbitan palmitate; CAS 9005-66-7; nonionic; o/w emulsifier and lubricant; Gardner 6 visc. liq.; sol. 5% in water, glycerol trioleate; dens. 9.2 lb/gal; visc. 200 cSt (100 F); HLB 15.8; pour pt. 22 C; flash pt. 485 F; cloud pt. 77 C (5%); 100% conc.

Emsorb® 6913. [Henkel/Emery] PEG-20 sorbitan trioleate; CAS 9005-70-3; nonionic; emulsifier for agric. formulations; EPA-exempt; Gardner 5 liq.; sol. @ 5% in Stod., xylene; disp. in water, methyl oleate; dens. 8.2 lb/gal; visc. 121 cSt (100 F); pour pt. -15 C; HLB 11.1; cloud pt. < 25 C; flash pt. 565 F.

Emsorb® 6915. [Henkel/Emery; Henkel/Textile] PEG-20 sorbitan laurate; CAS 9005-64-5; nonionic; o/w emulsifier and solubilizer of petrol. oils, solvs., and fats; used in cosmetic creams and lotions; visc. modifier in shampoos; emulsifier for dye carriers, antistatic scrooping agent in primary spin finishes,; and fiber processing aid in textile industry; agric. formulations; EPA-exempt; Gardner 6 liq.; sol. in water, glycerol trioleate; dens. 9.2 lb/gal; visc. 160 cSt (100 F); HLB 16.5; pour pt. –10 C; flash pt. 510 F; cloud pt. 75–85 C; 100% conc.

Emsorb® 6916. [Henkel/Emery] PEG-4 sorbitan laurate; CAS 9005-64-5; nonionic; o/w emulsifier in emulsion polymerization of PVC; Gardner 7 liq.; disp. in water, butyl stearate, perchloroethylene, Stod.; dens. 9.0 lb/gal; visc. 700 cs; HLB 12.1; cloud pt. < 25 C.

Emsorb® 6917. [Henkel/Emery] PEG-16 sorbitan trioleate; nonionic; o/w emulsifier, lubricant for metals, agric., textiles, leather, glass fiber, automotive additives; EPA-exempt; Gardner 5 liq.; sol. @ 5% in min. oil, butyl stearate, glyceryl trioleate, Stod., xylene; disp. in water; dens. 8.2 lb/gal; visc. 122 cSt (100 F); HLB 10.0; pour pt. < –10 C; cloud pt. < 25 C; flash pt. 530 F.

Emthox® 5882. [Henkel/Cospha; Henkel Canada] Laureth-4; nonionic; dispersant for bath oil; emulsifier for creams and lotions; wetting agent; for eye make-up, deodorants, hair coloring prods.; Gardner 1 liq.; sol. @ 5% in min. oil, triolein, IPA, IPM; disp. in water, glycerin; dens. 7.9 lb/gal; visc. 20 cSt (100 F); HLB 9.4; acid no. 10 max.; iodine no. 2 max.; sapon. no. 140-160; cloud pt. < 25 C; flash pt. 355 F; 100% conc.

Emthox® 5941. [Henkel/Emery] Trideceth-9; CAS 24938-91-8; nonionic; general purpose wetting agent and detergent; dispersant in bath oils; Gardner 1 liq.; sol. @ 5% in water, triolein, IPA; disp. in min. oil, glycerin; dens. 8.3 lb/gal; visc. 43 cSt (100 F); HLB 13.0; pour pt. 20 C; cloud pt. 54 C; flash pt. 390 F.

Emthox® 5942. [Henkel/Emery] Trideceth-11; CAS 24938-91-8; nonionic; general purpose wetting agent and detergent; dispersant in bath oils; Gardner 1 liq.; sol. @ 5% in water, triolein, IPA; disp. in min. oil, glycerin; dens. 8.4 lb/gal; visc. 47 cSt (100 F); HLB 13.8; pour pt. 17 C; cloud pt. 73 C; flash pt. 420 F.

Emthox® 5993. [Henkel/Emery] Trideceth-3; CAS 24938-91-8; nonionic; emulsifier, dispersant; Gardner 1 liq.; sol. @ 5% in triolein, IPA; disp. in water, min. oil; dens. 7.8 lb/gal; visc. 19 cSt (100 F); HLB 7.9; pour pt. –15 C; cloud pt. < 25 C; flash pt. 300 F.

Emthox® 6957. [Henkel/Emery] Nonoxynol-40; CAS 9016-45-9; nonionic; wetting agent, detergent, dispersant, coemulsifier; for waterless hand cleaners, hair care prods., liq. soaps, bath additives; Gardner 1 solid; sol. @ 5% in water, IPA; HLB 17.8; m.p. 40 C; cloud pt. 88–92 C (5% saline); flash pt. 560 F.

Emthox® 6960. [Henkel/Emery] Nonoxynol-1; nonionic; surfactant, coemulsifier, low foam detergent in liq. soaps; Gardner 1 liq.; sol. @ 5% in min. oil, IPA; dens. 8.3 lb/gal; visc. 150 cSt (100 F); HLB 4.6; pour pt. –10 C; cloud pt. < 25 C; flash pt. 385 F.

Emthox® 6961. [Henkel/Emery] Nonoxynol-4; CAS 9016-45-9; nonionic; surfactant, detergent, coemulsifier for bath additives, hair coloring prods., liq. soaps; Gardner 2 liq.; sol. @ 5% in IPA, IPM; disp. in min. oil, glycerin; dens. 8.5 lb/gal; HLB 8.9; pour pt. –10 C; cloud pt. < 25 C; flash pt. 430 F.

Emthox® 6962. [Henkel/Emery] Nonoxynol-6; CAS 9016-45-9; nonionic; dispersant, wetting agent, coemulsifier for hair coloring prods., soaps; Gardner 2 liq.; sol. @ 5% in IPA, IPM; disp. in water, min. oil, glycerin; dens. 8.5 lb/gal; visc. 100 cSt (100 F); HLB 10.9; pour pt. –10 C; cloud pt. < 25 C; flash pt. 515 F.

Emulamid FO-5DF. [Mayco Oil & Chem.] Fatty amide; CAS 38618-12-1; nonionic; emulsifier, lubricant, corrosion inhibitor, wetting agent for water-extendable metalworking coolants; yel. clear liq.; sp.gr. 0.97; dens. 8.08 lb/gal; acid no. 1; pH 9.9 (10%); 100% conc.

Emulamid TO-21. [Mayco Oil & Chem.] Tall oil fatty alkanolamide; CAS 68092-28-4; nonionic; emulsifier, lubricant, corrosion inhibitor, wetting agent for water extendable metalworking coolants; amber liq.; sp.gr. 0.99 (60/60 F); dens. 8.25 lb/gal (60 F); acid no. 15; pH 10 (10%); 100% conc.

Emulan® A. [BASF AG] Oleic acid oxyethylate; CAS 9004-96-0; nonionic; o/w emulsifier for min. oils and metal polishing emulsions; yel. liq.; char. odor; sol. in min. oils, fatty oils, fatty acids, polar org. solvs., emulsifiable in water; sp.gr. 0.89; visc. 120 mPa•s; HLB 11; flash pt. 210 C; sapon. no. 110; pH 6–7.5 (1% aq.); 100% act.

Emulan® AF. [BASF AG] Fatty alcohol oxyethylate; nonionic; emulsifier for paraffin wax and min. oil emulsions; wh. soft wax; sol. in min. oil and polar org. media; sp.gr. 0.91 (50 C); m.p. 42 C; HLB 11; flash pt. 190 C; sapon. no. 0; pH 6-7.5 (1% aq.); 100% act.

Emulan® AT 9. [BASF AG] Fatty alcohol ethoxylate; nonionic; emulsifier for paraffin waxes; wax; HLB 13.0; 100% conc.

Emulan® EL. [BASF AG] Castor oil oxyethylate; nonionic; emulsifier for nonaq. solvs. in water processing, lt. chemicals industry, for emulsifiable concs. for crop protection; pale yel. liq.; char. odor; sol. in water, polar org. media; sp.gr. 1.06 (30 C); visc. 200–500 mPa•s (30 C); HLB 14; flash pt. 280 C; sapon. no. 60; pH 6–7.5 (1% aq.); 100% act.

Emulan® FJ. [BASF AG] Fatty acid alkylolamide;

nonionic; w/o emulsifier for impregnation, lubrication, polishing and cleaning, o/w emulsifier in metalworking; brn. liq.; faint odor; sol. in min. oils, fatty oils and mixtures, emulsifiable in water; sp.gr. 0.96; visc. 200 mPa•s; HLB 3–5 (w/o type), 8.5 (o/w type); sapon. no. 150; pH 9 (1% aq.); 100% act.

Emulan® FM. [BASF AG] TEA monooleic ester; CAS 10277-04-0; nonionic; w/o emulsifier for impregnation, lubrication, polishing, and cleaning; o/w emulsifier for polishes; corrosion inhibitor for metals; brn. liq.; char. odor; sol. in min. oils, fatty oils and mixts.; sp.gr. 0.95; visc. 200 mPa•s; HLB 3–5 (w/o type); 9.5 (o/w type); acid no. 0–5; sapon. no. 135; pH 10 (1% aq.); flash pt. 240 C; 100% act.

Emulan® NP 2080. [BASF AG] Alkylphenol + EO; nonionic; emulsifier for o/w or w/o emulsions, emulsion polymerization; liq.; HLB 16.0; 80% conc.

Emulan® NP 3070. [BASF AG] Alkylphenol + EO; nonionic; emulsifier for emulsion polymerization; liq.; HLB 17.0; 70% conc.

Emulan® OC. [BASF AG] Fatty alcohol oxyethylate; nonionic; o/w emulsifier for dry bright emulsions, waxes; dispersant, stabilizer for emulsions; wh. or pale yel. soft wax; sol. in water, fatty acids, waxes, and polar org. solv.; sp.gr. 1.03 (50 C); HLB 17; m.p. 35–40 C; acid no. almost 0; sapon. no. 0; pH 6–7.5 (1% aq.); flash pt. 210 C; 100% act.

Emulan® OG. [BASF AG] Fatty alcohol oxyethylate; nonionic; o/w emulsifier for fatty acids, waxes, polar org. solvs.; dispersant for solid substances, grinding dyes; stabilizer for hydraulic and anticorrosion oils; waxy powd. in bead form; sol. in water and polar org. media; dens. 0.6 kg/l; HLB 17; m.p. 45–50 C; pH 6–7.5 (1% aq.); flash pt. 240 C; 100% act.

Emulan® OK 5. [BASF AG] Fatty alcohol ethoxylate; nonionic; surfactant for cleaners; liq.; HLB 11.0; 100% conc.

Emulan® OP 25. [BASF AG] Alkylphenol + EO; nonionic; emulsifier in emulsion polymerization; wax; HLB 16.5; 100% conc.

Emulan® OSN. [BASF AG] Fatty alcohol oxyethylate; nonionic; o/w emulsifier for wax emulsions; dispersant, stabilizer; wh. to pale yel. soft wax; sol. in water, fatty acids, waxes, polar org. solv.; sp.gr. 1.03 (50 C); m.p. 35–40 C; HLB 17; flash pt. 230 C; sapon no. 0; pH 6–7 (1% aq.); 100% act.

Emulan® OU. [BASF AG] Fatty alcohol oxyethylate; nonionic; o/w emulsifier for fatty acids, waxes, polar org. solvs.; dispersant for solid substances, grinding processes; stabilizer of emulsions and disps.; wh. or pale yel. soft wax; sol. in water, fatty acids, molten greases and ester waxes, polar org. solvs.; sp.gr. 1.02 (50 C); HLB 17; m.p. 40 C; solid. pt. 35 C; flash pt. 210 C; sapon. no. 0; pH 6–7.5 (1% aq.); 100% act.

Emulan® P. [BASF AG] Fatty alcohol oxyethylate; nonionic; emulsifier for min. oils; colorless clear liq.; sol. in oils; sp.gr. 0.91; HLB 11.0; flash pt. 150 C; sapon no. 0; pH 6–7.5 (1% aq.); 100% act.

Emulan® PO. [BASF AG] Alkylphenol oxyethylate; nonionic; emulsifier for cleaners, formwork release oils, drilling and cutting oils; flotation agent; colorless to pale yel. clear liq.; sol. in min. oils, fatty oils, polar org. solv., emulsifiable in water; sp.gr. 1.04; visc. 400 mPa•s; HLB 10.0; flash pt. 240 C; sapon. no. 0; pH 6–7.5 (1% aq.); 100% act.

Emulbon GB-90. [Toho Chem. Industry] Polyol borate; nonionic; sizing assistant; liq.; 100% conc.

Emulbon S-83. [Toho Chem. Industry] Diglyceryl borate sesquioleate; nonionic; emulsifier, dispersant; liq.; 100% conc.

Emulbon T-20. [Toho Chem. Industry] Diglyceryl borate POE monolaurate; nonionic; detergent, emulsifier; liq.; 100% conc.

Emulbon T-80. [Toho Chem. Industry] Diglyceryl borate POE monooleate; nonionic; emulsifier, dispersant; liq.; 100% conc.

Emulgade EO-10. [Pulcra SA] Oleth-25 and cetyl alcohol; self-emulsifying base for prep. of creams and emulsions; flakes.

Emulgane E. [Henkel-Nopco] POE veg. oil; nonionic; emulsifier for animal, veg. fats and oils; dispersant for pigments; liq.; water-sol.; 100% conc.

Emulgane FL 7. [Henkel-Nopco] Polyoxyalkylene fatty esters and esters; nonionic; emulsifier for min. oils in jute batching; liq.; 100% conc.

Emulgane O. [Henkel-Nopco] Alkylpolyglycolether; nonionic; dispersant and emulsifier; dyeing assistant in acid media; paste; 100% conc.

Emulgane OP. [Henkel-Nopco] Polyoxyalkylene fatty esters; nonionic; emulsifier for min., veg. oils and solv.; liq.; 100% conc.

Emulgateur SO. [Witco SA] Formulated emulsifier; nonionic; emulsifier for min. and veg. oils, solvs.; liq.

Emulgator 64. [Hefti Ltd.] Surfactant blend; nonionic; emulsifier for min. oils, triglycerides, textiles, cutting oils, biocides; dispersant for pigments and colorants; liq.; HLB 10.0; 100% conc.

Emulgator E-2155 SE. [Goldschmidt] Nonionic blend; nonionic; for prep. of hydrophilic oils, microemulsions; spontaneous, cold emulsifying; liq.; HLB 10.5; 100% conc.

Emulgator F-8. [Unger Fabrikker AS] Fatty alcohol ethoxylate (8 EO); nonionic; surfactant, emulsifier for liq. detergents, industrial preps.; colorless liq.; HLB 13.7; 99% act.

Emulgator U4. [Unger Fabrikker AS] Nonoxynol-4; CAS 9016-45-9; nonionic; surfactant, emulsifier, wetting agent for liq. and powd. detergents, industrial preps., textile, leather, pulp/paper, drycleaning, metal cleaners; lt. yel. liq.; sp.gr. 1.02; visc. 350 cps; HLB 8.8; cloud pt. 53-58 C (5% in 25% BDG); 99% act.

Emulgator U6. [Unger Fabrikker AS] Nonoxynol-6; CAS 9016-45-9; nonionic; surfactant, emulsifier, wetting agent for liq. and powd. detergents, industrial preps., textile, leather, pulp/paper, drycleaning, metal cleaners; lt. yel. liq.; sp.gr. 1.04; visc. 350 cps; HLB 10.0; cloud pt. 65-69 C (5% in 25% BDG); 99% act.

Emulgator U9. [Unger Fabrikker AS] Nonoxynol-9; CAS 9016-45-9; nonionic; surfactant, emulsifier, wetting agent for liq. and powd. detergents, industrial preps., textile, leather, pulp/paper, drycleaning, metal cleaners; lt. yel. liq.; sp.gr. 1.06; visc. 360 cps; HLB 12.8; cloud pt. 54 C (1% aq.); flash pt. > 100 C; surf. tens. 31-33 dynes/cm (1 g/l); Draves wetting 23-25 s (0.5 g/l); 99% act.

Emulgator U12. [Unger Fabrikker AS] Nonoxynol-12; CAS 9016-45-9; nonionic; surfactant, emulsifier, wetting agent for liq. and powd. detergents, industrial preps., textile, leather, pulp/paper, drycleaning, metal cleaners; lt. yel. liq.; sp.gr. 1.07; visc. 370 cps; HLB 13.9; cloud pt. 83 C (1% aq.); surf. tens. 34-36 dynes/cm (1 g/l); Draves wetting 26-31 s (0.5 g/l); 99% act.

Emulgeant 710. [Ceca SA] Polyethoxy ether; nonionic; emulsifier for aliphatic and aromatic solvs., paraffin; in degreasers; Gardner < 2 liq.; acid no. 40; HLB 7.7; pH 6 ± 1 (10%).

Emulgeant 900. [Ceca SA] Polyethoxy ether; nonionic; emulsifier for aliphatic solvs., wh. spirit, paraffin; liq.; HLB 10.7; pH 6 ± 1 (5%).

Emulgin PA Series. [Pulcra SA] Ethoxylated hydrog. tallow amines; CAS 61791-26-2; nonionic; detergent, emulsifier, wetting agent, dyeing assistant, scouring agent, antistat, corrosion inhibitor, dispersant, softener and latex stabilizers; liq., paste, solid; 100% conc.

Emulgit 60. [Zohar Detergent Factory] Calcium alkylbenzene sulfonate; anionic; emulsifier for pesticide concs.; liq.; 60% conc.

Emulgit 70. [Zohar Detergent Factory] Calcium alkylbenzene sulfonate; anionic; emulsifier for pesticide concs.; liq.; 70% conc.

Emulgit D. [Zohar Detergent Factory] Detergent blend; emulsifier for Dicofol emulsifiable concs.

Emulgit E. [Zohar Detergent Factory] Detergent blend; emulsifier for Endosulfan emulsifiable concs.

Emulgit M. [Zohar Detergent Factory] Detergent blend; emulsifier for Metasystox emulsifiable concs.

Emulgit OP. [Zohar Detergent Factory] Detergent blend; emulsifier for organophosphorus emulsifiable concs.

Emulgit Pr. [Zohar Detergent Factory] Detergent blend; emulsifier for Propanil emulsifiable concs.

Emulmin 40. [Sanyo Chem. Industries] POE tallow alcohol ether; CAS 61791-28-4; nonionic; emulsifier, dispersant, wetting agent, detergent; biodeg.; liq.; HLB 8.0; 100% conc.

Emulmin 50. [Sanyo Chem. Industries] POE tallow alcohol ether; CAS 61791-28-4; nonionic; emulsifier, dispersant, wetting agent, detergent; biodeg.; liq.; HLB 9.0; 100% conc.

Emulmin 70. [Sanyo Chem. Industries] POE tallow alcohol ether; CAS 61791-28-4; nonionic; emulsifier, dispersant, wetting agent, detergent; biodeg.; liq.; HLB 10.8; 100% conc.

Emulmin 140. [Sanyo Chem. Industries] Ethoxylated nonylphenol; CAS 9016-45-9; nonionic; emulsifier, dispersant, wetting agent, detergent; biodeg.; paste; HLB 14.2; 100% conc.

Emulmin 240. [Sanyo Chem. Industries] Ethoxylated nonylphenol; CAS 9016-45-9; nonionic; emulsifier, dispersant, wetting agent, detergent; biodeg.; solid; HLB 16.1; 100% conc.

Emulmin 862. [Sanyo Chem. Industries] PEG distearate; CAS 9005-08-7; nonionic; thickener for cosmetics and textile printing pastes; solid.

Emulmin L-380. [Sanyo Chem. Industries] Ethoxylated higher alcohol ether; nonionic; emulsifier, dispersant, wetting agent, detergent; biodeg.; solid; HLB 17.7; 100% conc.

Emulphogene® BC-420 (redesignated Rhodasurf® BC-420). [Rhone-Poulenc Surf.]

Emulphogene® BC-610 (redesignated Rhodasurf® BC-610). [Rhone-Poulenc Surf.]

Emulphogene® BC-720 (redesignated Rhodasurf® BC-720). [Rhone-Poulenc Surf.]

Emulphogene® BC-840 (redesignated Rhodasurf® BC-840). [Rhone-Poulenc Surf.]

Emulphogene® DA-530 (redesignated Rhodasurf® DA-530). [Rhone-Poulenc Surf.]

Emulphogene® DA-630 (redesignated Rhodasurf® DA-630). [Rhone-Poulenc Surf.]

Emulphogene® TB-970 (redesignated Rhodasurf® TB-970). [Rhone-Poulenc Surf.]

Emulphopal HC. [Stepan Canada] Proprietary; waterless hand cleaner base; 100% act.

Emulphor® EL-620 (redesignated Alkamuls® EL-620). [Rhone-Poulenc Surf.]

Emulphor® EL-620L (redesignated Alkamuls® EL-620L). [Rhone-Poulenc Surf.]

Emulphor® EL-719 (redesignated Alkamuls® EL-719). [Rhone-Poulenc Surf.]

Emulphor® EL-719L (redesignated Alkamuls® EL-719L). [Rhone-Poulenc Surf.]

Emulphor® EL-980 (redesignated Alkamuls® EL-980). [Rhone-Poulenc Surf.]

Emulphor® EL-985 (redesignated Alkamuls® EL-985). [Rhone-Poulenc Surf.]

Emulphor® LA-630. [Rhone-Poulenc Surf.] PEG-9 cocoate; nonionic; emulsifier and wetting agent, lubricant, softener for cosmetic, textile, and industrial applics.; liq.; HLB 13.2; 100% act. DISCONTINUED.

Emulphor® ON-870 (redesignated Rhodasurf® ON-870). [Rhone-Poulenc Surf.]

Emulphor® ON-877 (redesignated Rhodasurf® ON-877). [Rhone-Poulenc Surf.]

Emulphor® VN-430. [Rhone-Poulenc Surf.] PEG-5 oleate; CAS 9004-96-0; nonionic; w/o emulsifier in cosmetics, lubricant, surfactant, dyeing assistant used in textile industry; FDA compliance; dk. amber clear liq.; disp. in water, sol. in xylene, wh. min. oil, ethanol, butyl Cellosolve; sp.gr. 1.0; HLB 7.7; cloud pt. < 10 C (1%); flash pt. > 93 C (PMCC); pour pt. 13 C; surf. tens. 32 dynes/cm (0.1%); pH 7–7.5; 100% act. DISCONTINUED.

Emulphor® VT-650. [Rhone-Poulenc Surf.] PEG-9 stearate; CAS 9004-99-3; nonionic; emulsifier for silicone and nonsilicone defoamers, SE lubricant for dye carriers, textile scours, industrial applics.; VCS 2 soft wax; slight acetic acid odor; sol. in hydrocarbons, ethanol, ether; sp.gr. 0.99 (50 C); dens. 8.3 lb/gal; visc. 9 cs (100 C); HLB 11.8; flash pt. 162 F (CC); sapon. no. 83–94; 100% act. DISCONTINUED.

Emulpon A. [Witco SA] Ethoxylated fatty acid; nonionic; emulsifier for sol. min. oil; liq.; 100% conc.

Emulpon EL 18. [Witco SA] Ethoxylated castor oil; nonionic; emulsifier for min. oil; liq.; 100% conc.

Emulpon EL 20. [Witco SA] Ethoxylated castor oil; CAS 61731-12-6; nonionic; emulsifier for waxes, olein; liq.; HLB 2.7; 100% conc.

Emulpon EL 33. [Witco SA] Ethoxylated castor oil; nonionic; textile wetting agent; liq.; HLB 12.2; 100% conc.

Emulpon EL 40. [Witco SA] PEG-40 castor oil; CAS 61731-12-6; nonionic; textile wetting agent used in dyeing; liq.; 100% conc.

Emulpon HT. [Witco SA] Ethoxylated castor oil; nonionic; for dyeing at high temp.; liq.; 100% conc.

Emulsan A-67. [Reilly-Whiteman] PEG 400 mixed fatty acid ester; nonionic; emulsifier; compat. with petrol. and min. oils; liq.; 97% conc.

Emulsan BRW. [Reilly-Whiteman] Surfactant-solv. blend; nonionic; surfactant for removal of inks from printing press, blankets, and rollers; liq.; 100% conc.

Emulsan K. [Reilly-Whiteman] PEG-8 cocoate; nonionic; emulsifier for textile, leather, and other industrial applics.; liq.; 99% conc.

Emulsan O. [Reilly-Whiteman] PEG-8 oleate; CAS

9004-96-0; nonionic; emulsifier for textile, leather, and other industrial applics.; liq.; 99% conc.

Emulsifiant 33 AD. [Seppic] Ethoxylated/sulfated alkylphenol; CAS 68649-55-8; anionic; emulsifier for emulsion polymerization; liq.; 40% conc.

Emulsifier 4. [Exxon/Tomah] Dialkyl quat. ammonium chloride in water/IPA; cationic; emulsifier for nonpolar hydrophobes, e.g., min. seal oil, waxes, silicones; used for auto spraywax, carnauba spraywax, mop treatment emulsions, stainless steel cleaners, vinyl dressings; amber liq.; dens. 8.0 lb/gal; HLB 10.0; flamm. liq.; 75% act.

Emulsifier 7X. [Atsaun] Formulated blend; nonionic; emulsifier, wetting agent, antistat, lubricant; liq.; HLB 10.0; 95% conc.

Emulsifier 632/90%. [Ethox] Modified alkyl phenol ethoxylate; nonionic; low foam surfactant, dispersant for fats, oils, and waxes; liq.; HLB 12.4; cloud pt. 25 C (1% aq.); 90% conc.

Emulsifier DMR. [Hoechst AG] Fatty acid polyglycol ester; nonionic; emulsifier for mfg. of o/w emulsions which are thickeners for printing of reactive dyes; colorless to slightly yel. waxy flakes; water-sol.

Emulsifier K30 40%. [Miles/Organic Prods.] Sodium alkane sulfonates based on n-paraffin; anionic; emulsifier for emulsion polymerization; effective over wide pH range; antistat; biodeg.; FDA compliance; clear sol'n., virtually odorless; sol. in water; low sol. in alcohols; m.w. 330; visc. 5 mPa•s (25%); HLB 11-12; cloud pt. < 20 C (40%); pH alkaline; surf. tens. 34.9 dynes/cm (0.1%); toxicology: LD50 (oral, rats) 2 g/kg; 40% act.

Emulsifier K30 68%. [Miles/Organic Prods.] Sodium alkane sulfonates based on n-paraffin; anionic; emulsifier for emulsion polymerization; effective over wide pH range; antistat; biodeg.; FDA compliance; colorless to ylsh. pumpable paste, virtually odorless; sol. in water; low sol. in alcohols; m.w. 330; visc. 5 mPa•s (25%); HLB 11-12; cloud pt. < 20 C (40%); pH alkaline; surf. tens. 34.9 dynes/cm (0.1%); toxicology: LD50 (oral, rats) 2 g/kg; 68% act.

Emulsifier K30 76%. [Miles/Organic Prods.] Sodium alkane sulfonates based on n-paraffin; anionic; emulsifier for emulsion polymerization; effective over wide pH range; antistat; biodeg.; FDA compliance; wh.-ylsh. stiff paste, virtually odorless; sol. in water; low sol. in alcohols; m.w. 330; visc. 5 mPa•s (25%); HLB 11-12; cloud pt. < 20 C (40%); pH alkaline; surf. tens. 34.9 dynes/cm (0.1%); toxicology: LD50 (oral, rats) 2 g/kg; 76% act.

Emulsifier K30 95%. [Miles/Organic Prods.] Sodium alkane sulfonates based on n-paraffin; anionic; emulsifier for emulsion polymerization; effective over wide pH range; antistat; biodeg.; FDA compliance; wh.-ylsh. flakes, virtually odorless; sol. in water; low sol. in alcohols; m.w. 330; visc. 5 mPa•s (25%); HLB 11-12; cloud pt. < 20 C (40%); pH alkaline; surf. tens. 34.9 dynes/cm (0.1%); toxicology: LD50 (oral, rats) 2 g/kg; 95% act.

Emulsifier Q. [Consos] Solvent/emulsifier blend; anionic; solvent scour for heavily oiled and waxed textiles; biodeg.; liq.

Emulsifier WHC. [Stepan; Stepan Canada] Formulated emulsifier; anionic; emulsifier and detergent used in waterless hand cleaner formulations and degreasers; lt. amber liq.; disp. in water, sol. in kerosene; dens. 8.225 lb/gal; pH 8 (1% aq.).; 100% act.

Emulsit 16. [Dai-ichi Kogyo Seiyaku] POE alkylphenol ether; nonionic; emulsifier; liq.; 70% conc.

Emulsit 25. [Dai-ichi Kogyo Seiyaku] POE alkylphenol ether; nonionic; emulsifier; liq.; 60% conc.

Emulsit 49. [Dai-ichi Kogyo Seiyaku] POE alkylphenol ether; nonionic; emulsifier; liq.; 50% conc.

Emulsit 100. [Dai-ichi Kogyo Seiyaku] POE alkylphenol ether; nonionic; emulsifier; liq.; 50% conc.

Emulsogen 2144. [Hoechst Celanese/Colorants & Surf.] Ethoxylated rosin; nonionic; low foaming emulsifier, wetting agent for metalworking fluids, high acidic conditions; liq.; sol. in water; sp.gr. 1.10; pH 7.5 (1% aq.).

Emulsogen A. [Hoechst Celanese/Colorants & Surf.] Ethoxylated deriv.; nonionic; min. oil emulsifier, wetting agent; for metalworking, sol. oils, microemulsions; clear oily liq.; disp. in water; sp.gr. 0.96; pH 6 (1% aq.); 100% conc.

Emulsogen AN. [Hoechst Celanese/Colorants & Surf.] Ethoxylated deriv.; nonionic; min. oil emulsifier; clear liq.; 100% conc.

Emulsogen B2M. [Hoechst Celanese/Colorants & Surf.] Amine salt of alkyl sulfamidocarboxylic acids with partially chlorinated hydrocarbons; anionic; emulsifier, corrosion inhibitor for metal machining; lubricant; brn. clear oily liq.; dissolves in water, min. oils; sp.gr. 0.94–0.95; flash pt. 128 C (Marcusson); ref. index 1.46; pH 7.3 ± 0.2 (1% aq.).

Emulsogen BB. [Hoechst Celanese/Colorants & Surf.] Succinic acid deriv.; anionic; multifunctional additive, emulsifier, wetting agent for aq. metalworking fluids; liq.; sol. in water; sp.gr. 0.97; pH 9-10 (1% aq.).

Emulsogen CP 136. [Hoechst Celanese/Colorants & Surf.] Calcium dodecylbenzene sulfonate/castor oil ethoxylate; anionic; surfactant for agric. formulations.

Emulsogen E. [Hoechst Celanese/Colorants & Surf.] Alkylaryl polyglycol ethers with fatty amine salts; corrosion inhibitor and wetting agent for preparation of dewatering fluids for metalworking; brn. clear visc. liq.; dissolves in water.

Emulsogen EL-050. [Hoechst Celanese/Colorants & Surf.] PEG-5 castor oil; nonionic; emulsifier additive for chlorinated paraffins, triglycerides; surfactant for textile processing; yel. clear visc. liq.; visc. 700 cps; sapon. no. 145; pH 8 (1% aq.); 99% act.

Emulsogen EL-250. [Hoechst Celanese/Colorants & Surf.] PEG-25 castor oil; surfactant for textile processing; yel. visc. liq.; sapon. no. 80; pH 8 (1% aq.); 99% act.

Emulsogen EL-300. [Hoechst Celanese/Colorants & Surf.] PEG-30 castor oil; nonionic; emulsifier for fats, soaps, chlorinated paraffins, triglycerides; wetting agent; for metalworking fluids, agric. formulations; clear liq.; sol. in water; sp.gr. 1.04; pH 7-9 (1% aq.); 100% conc.

Emulsogen EL-360. [Hoechst Celanese/Colorants & Surf.] PEG-36 castor oil; surfactant for agric. formulations, textile processing; yel. visc. liq.; sapon. no. 67; pH 8 (1% aq.); 99% act.

Emulsogen EL-400. [Hoechst Celanese/Colorants & Surf.] PEG-40 castor oil; nonionic; component for selective emulsification systems; agric. formulations; textile processing; yel. visc. liq.; sapon. no. 65; pH 8 (1% aq.); 99% act.

Emulsogen EL-2000. [Hoechst Celanese/Colorants & Surf.] PEG-200 castor oil; surfactant for textile processing; sl. yel. solid; sapon. no. 16; pH 7 (1% aq.); 99% act.

Emulsogen H. [Hoechst Celanese/Colorants & Surf.] Alkyl sulfamidocarboxylic acid with chlorinated hydrocarbons; anionic; base for corrosion inhibitors; emulsifier, wetting agent; dk. brn. visc. liq.; dissolves to clear sol'n. in most min. oils, to turbid sol'n. in water; acid no. 47–52.

Emulsogen HEL-050. [Hoechst Celanese/Colorants & Surf.] PEG-5 hydrog. castor oil; CAS 61788-85-0; surfactant for textile processing; yel. visc. liq.; sapon. no. 170; pH 7 (1% aq.); 99% act.

Emulsogen HEL-160R. [Hoechst Celanese/Colorants & Surf.] PEG-16 hydrog. castor oil; CAS 61788-85-0; surfactant for textile processing; yel. visc. liq.; visc. 1100 cps; sapon. no. 100; pH 7 (1% aq.); 99% act.

Emulsogen HEL-250. [Hoechst Celanese/Colorants & Surf.] PEG-25 hydrog. castor oil; CAS 61788-85-0; surfactant for textile processing; yel. visc. liq.; visc. 2500 cps; sapon. no. 80; pH 6 (1% aq.); 99% act.

Emulsogen HFIT. [Hoechst Celanese/Colorants & Surf.] Calcium dodecylbenzene sulfonate/castor oil ethoxylate; anionic; surfactant, high flash pt. emulsifier for agric. formulations; liq.; 100% conc.

Emulsogen IC. [Hoechst Celanese/Colorants & Surf.] Calcium dodecylbenzene sulfonate/oleyl alcohol ethoxylate; anionic; surfactant for agric. formulations.

Emulsogen ICL. [Hoechst Celanese/Colorants & Surf.] Calcium dodecylbenzene sulfonate/oleyl alcohol ethoxylate; anionic; surfactant for agric. formulations.

Emulsogen IP 400. [Hoechst Celanese/Colorants & Surf.] Calcium dodecylbenzene sulfonate/castor oil ethoxylate; anionic; surfactant for agric. formulations.

Emulsogen IT. [Hoechst Celanese/Colorants & Surf.] Calcium dodecylbenzene sulfonate/castor oil ethoxylate; anionic/nonionic; surfactant, emulsifier for solv. blends, agric. formulations; liq.; 100% conc.

Emulsogen ITL. [Hoechst Celanese/Colorants & Surf.] Calcium dodecylbenzene sulfonate/castor oil ethoxylate; anionic; surfactant for agric. formulations.

Emulsogen M. [Hoechst Celanese/Colorants & Surf.] Fatty alcohol ethoxylate; nonionic; min. oil emulsifier, wetting agent; for metalworking fluids; clear liq.; disp. in water; sp.gr. 0.93; pH 5-8 (1% aq.); 100% conc.

Emulsogen N-060. [Hoechst Celanese/Colorants & Surf.; Hoechst AG] Alkylaryl polyglcyol ether; nonionic; emulsifier for mixed-base min. oils; coemulsifier for metalworking oils; wetting agent; liq.; 100% conc.

Emulsogen NP 11-300. [Hoechst Celanese/Colorants & Surf.] Nonylphenol EO/PO block copolymer; nonionic; surfactant for agric. formulations.

Emulsogen PR. [Hoechst Celanese/Colorants & Surf.] Styrenated phenol ethoxylate; surfactant for textile processing; yel. liq.; visc. < 500 cps; hyd. no. 70; pH 6 (1% aq.); 99% act.

Emulsogen S. [Hoechst Celanese/Colorants & Surf.] Inorganic/nonionic blend; emulsifier/corrosion inhibitor for metalworking oils; brn. clear visc. liq.; sol. in min. oils; dens. 0.97 g/cm³; solid. pt. -9 C; flash pt. 130 C; pH 8.5–9.5 (10% aq.).

Emulsogen STH. [Hoechst Celanese/Colorants & Surf.] Alkylsulfamido carboxylic acid deriv.; anionic; antioxidant, emulsifier for min. oil emulsions with good rust protection; wetting agent; corrosion inhibitor for aq. metalworking fluids; golden yel. clear oily liq.; sol. with turbidity in water, clear to turbid in min. oils; sp.gr. 0.96; pH 6-7 (1% aq.); 100% conc.

Emulsogen T. [Hoechst Celanese/Colorants & Surf.] Castor oil ethoxylate blend; anionic/nonionic; emulsifier for triglycerides and petrol. oils; wetting agent; for metalworking fluids; liq.; sol. in most triglycerides; disp. in water; sp.gr. 0.96; pH 5.5-6.5 (1% aq.).

Emulsol CO 45. [Atsaun] Ethoxylated castor oil; nonionic; leveling agent, dispersant, solubilizer for natural essential oils, etc.; liq.; HLB 13.5; 100% conc.

Emulson AG 2A. [Auschem SpA] Nonylphenol polyglycol ether; CAS 9016-45-9; nonionic; emulsifier, dispersant for pesticides; liq.; HLB 4.7; 100% conc.

Emulson AG 4B. [Auschem SpA] Nonylphenol polyglycol ether; CAS 9016-45-9; nonionic; emulsifier for pesticides; liq.; HLB 8.9; 100% conc.

Emulson AG 7B. [Auschem SpA] Nonylphenol polyglycol ether; CAS 9016-45-9; nonionic; wetting agent, emulsifier for pesticides; liq.; HLB 11.7; 100% conc.

Emulson AG 8A. [Auschem SpA] Polyarylphenol polyglycol ether; nonionic; wetting agent, emulsifier for pesticides; liq.; 100% conc.

Emulson AG 9B. [Auschem SpA] Nonylphenol polyglycol ether; CAS 9016-45-9; nonionic; wetting agent, emulsifier for pesticides; liq.; HLB 12.8; 100% conc.

Emulson AG 10B. [Auschem SpA] Nonylphenol polyglycol ether; CAS 9016-45-9; nonionic; wetting agent, emulsifier for pesticides; liq.; HLB 13.3; 100% conc.

Emulson AG 12B. [Auschem SpA] Nonylphenol polyglycol ether; CAS 9016-45-9; nonionic; wetting agent, emulsifier for pesticides; liq.; HLB 14.1; 100% conc.

Emulson AG 13A. [Auschem SpA] Polyarylphenol polyglycol ether; nonionic; emulsifier for emulsifiable concs., solubilizer for microemulsions, dispersant and wetting agent; liq./paste; HLB 13.1; 100% conc.

Emulson AG 15A. [Auschem SpA] Polyarylphenol polyglycol ether; nonionic; emulsifier for emulsifiable concs., solubilizer for microemulsions; paste; HLB 13.7; 100% conc.

Emulson AG 20B. [Auschem SpA] Nonylphenol polyglycol ether; CAS 9016-45-9; nonionic; emulsifier, dispersant for pesticides; solid; HLB 16.0; 100% conc.

Emulson AG 24A. [Auschem SpA] Polyarylphenol polyglycol ether; nonionic; emulsifier for pesticides; solid; HLB 15.5; 100% conc.

Emulson AG 81C. [Auschem SpA] PEG-81 castor oil; CAS 61791-12-6; nonionic; emulsifier; solid; HLB 15.6; 100% conc.

Emulson AG 255. [Auschem SpA] TEA dodecylbenzene sulfonate; CAS 27323-41-7; anionic; emulsifier for pesticides; liq.; 70% conc.

Emulson AG 3020. [Auschem SpA] Blend; anionic/nonionic; emulsifier for pesticides, microemulsions; liq.; HLB 14.5; 90% conc.

Emulson AG 3080. [Auschem SpA] Blend; anionic/nonionic; emulsifier for pesticides, microemulsions; liq.; HLB 11.5; 70% conc.

Emulson AG 3490. [Auschem SpA] Blend; anionic/nonionic; emulsifier for pesticides; liq.; HLB 10.9;

70% conc.

Emulson AG 4020. [Auschem SpA] Blend; anionic/nonionic; emulsifier for pesticides; paste; HLB 13.5; 90% conc.

Emulson AG 4080. [Auschem SpA] Blend; anionic/nonionic; emulsifier for pesticides; liq.; HLB 11.2; 70% conc.

Emulson AG 5190. [Auschem SpA] Blend; anionic/nonionic; emulsifier for pesticides; liq.; HLB 13.3; 75% conc.

Emulson AG 5590. [Auschem SpA] Blend; anionic/nonionic; emulsifier for pesticides; liq.; HLB 9.2; 70% conc.

Emulson AG 5790. [Auschem SpA] Blend; anionic/nonionic; emulsifier for pesticides; liq.; HLB 12.3; 75% conc.

Emulson AG 7000. [Auschem SpA] Nonoxynolphosphate; CAS 51811-79-1; anionic; emulsifier for pesticides, microemulsions; liq.; 100% conc.

Emulson AG 7760. [Auschem SpA] Blend; anionic/nonionic; dispersant; liq.; 100% conc.

Emulson AG/CAL. [Auschem SpA] Calcium dodecylbenzene sulfonate; CAS 26264-06-2; anionic; emulsifier for pesticides; liq.; 60% conc.

Emulson AG/CO 25. [Auschem SpA] Ethoxylated castor oil; CAS 61791-12-6; nonionic; emulsifier for pesticides; liq.; HLB 10.8; 100% conc.

Emulson AG/COH. [Auschem SpA] Ethoxylated castor oil; CAS 61791-12-6; nonionic; emulsifier for pesticides; liq.; HLB 10.8; 100% conc.

Emulson AG/EL. [Auschem SpA] Ethoxylated castor oil; CAS 61791-12-6; nonionic; emulsifier for pesticides; liq.; HLB 14.4; 100% conc.

Emulson AG/FLS. [Auschem SpA] Blend; anionic/nonionic; wetting and dispersing agent; solid; 100% conc.

Emulson AG/FN. [Auschem SpA] Blend; anionic/nonionic; emulsifier for pesticides and microemulsions; liq.; HLB 12.5; 90% conc.

Emulson AG/PE. [Auschem SpA] EO/PO alkyl ether; nonionic; emulsifier for emulsifiable concs. and microemulsions; wetting agent and dispersant; solid; 100% conc.

Emulson CO 25. [Auschem SpA] PEG-25 castor oil; CAS 61791-12-6; nonionic; emulsifier, solubilizer; liq.; HLB 11.0; 100% conc.

Emulson CO-40. [Auschem SpA] PEG-40 castor oil; CAS 61791-12-6; nonionic; emulsifier, solubilizer; liq.; HLB 13.8; 100% conc.

Emulson CO 40 N. [Auschem SpA] PEG-35 castor oil; CAS 61791-12-6; nonionic; emulsifier, solubilizer; liq.; HLB 13.0; 100% conc.

Emulson EL. [Auschem SpA] PEG-55 castor oil; CAS 61791-12-6; nonionic; emulsifier, solubilizer; paste; HLB 14.5; 100% conc.

Emulson EL 200. [Auschem SpA] PEG-200 castor oil; CAS 61791-12-6; nonionic; emulsifier, solubilizer; paste; HLB 18.0; 100% conc.

Emulson EL/H25. [Auschem SpA] PEG-25 hydrog. castor oil; CAS 61788-85-0; nonionic; emulsifier, solubilizer; liq./paste; HLB 11.0; 100% conc.

Emultex 1302. [Auschem SpA] Alkyl polyglycol ether blend; nonionic; emulsifier for paraffin, ozokerite; wax; HLB 13.0; 100% conc.

Emultex 1502, 1515. [Auschem SpA] Alkyl polyglycol ether blend; nonionic; emulsifier for paraffin, montan wax, blends; wax; HLB 15.0; 100% conc.

Emultex 1602. [Auschem SpA] Alkyl polyglycol ether blend; nonionic; emulsifier for carnauba, beeswax; flakes; HLB 16.0; 100% conc.

Emultex SMS. [Auschem SpA] Sorbitan monostearate; CAS 1338-41-6; nonionic; plasticizer, antifoamer, softener, textile lubricant, emulsifier in cosmetics; flakes; HLB 4.5; 100% conc.

Emultex WS. [Auschem SpA] Ester of a polyglycerol deriv.; nonionic; emulsifier for aq. insecticides, deodorant aerosols; liq.; HLB 4.0; 100% conc.

Emulvin S. [Miles] Polythioether; nonionic; emulsifier and latex stabilizer; ylsh. brn. oily liq.; dens. 1.13 g/cm³; pH 6.

Emulvin W. [Miles] Aromatic polyglycol ether; nonionic; emulsifier, stabilizer, wetting agent for latices; ylsh. brn. liq.; sp.gr. 1.13; pH 6.

EM-WS. [Ferro/Keil] Lard oil; additive improving wetting and lubricity in industrial applics.; sp.gr. 0.91; dens. 7.6 lb/gal; visc. 38 cSt (40 C); pour pt. 7 C; cloud pt. 13 C; flash pt. (COC) 288 C; acid no. 3; sapon. no. 195.

Endet. [Swastik] Heavy duty detergent powd. containing proteolytic enzyme; household detergent; wh. spray dried powd., colored enzyme beads; pH 9.5 (1% aq.).

Enerade 3045, 7102, 7103, 7104, 7201, 7202, 7204, 8202. [Rhone-Poulenc Oil Field Chem.] Emulsion breakers.

Ensidom O. [Henkel-Nopco] Modified fatty acids and oils; anionic; self-emulsifying wool and worsted lubricant; liq.; 100% conc.

Ensital AP 115. [Henkel-Nopco] Modified fatty acids and oils; anionic; lubricant for worsted and blends; liq.; 30% conc.

Ensital CAR. [Henkel-Nopco] Alkyl polyethoxy esters and ethers; nonionic; lubricant for wool and blends; liq.; 30% conc.

Ensital CF 2708C. [Henkel-Nopco] Polyalkoxylated ether; nonionic; lubricant for semiworsted and open end; 30% conc.

Ensital FMH. [Henkel-Nopco] Modified fatty acids and oils; nonionic; min. oil-free lubricant for worsted; liq.; 90% conc.

Ensital FPA 205. [Henkel-Nopco] Alkylpolyalkoxylated ester; nonionic; lubricant for acrylics and blends, worsted processing; 90% conc.

Epal® 6. [Ethyl] Hexanol; CAS 111-27-3; detergent and emulsifier intermediate; FDA compliance; water-wh. mobile liq., lt. fatty alcohol odor; m.w. 102; sp.gr. 0.815 g/ml; dens. 6.8 lb/gal; f.p. -49 C; b.p. 151-160 C (1 atm); flash pt. (CC) 61 C; hyd. no. 548; combustible; toxicology: temporary irritation on direct skin or eye contact; 100% conc.

Epal® 8. [Ethyl] Octanol; CAS 111-87-5; detergent and emulsifier intermediate; water-wh. mobile liq., lt. fatty alcohol odor; m.w. 130.2; sp.gr. 0.821 g/ml; dens. 6.85 lb/gal; f.p. -14 C; b.p. 184-195 C (1 atm); flash pt. (CC) 88 C; hyd. no. 431; combustible; toxicology: temporary irritation on direct skin or eye contact; 100% conc.

Epal® 10. [Ethyl] Decanol; CAS 112-30-1; detergent and emulsifier intermediate; water-wh. mobile liq., mild char. odor; m.w. 158; sp.gr. 0.827 g/ml; dens. 6.90 lb/gal; f.p. 8 C; b.p. 226-230 C (1 atm); flash pt. (PMCC) 113 C; hyd. no. 354; toxicology: temporary irritation on direct skin or eye contact; 100% conc.

Epal® 12. [Ethyl] Dodecanol; CAS 112-53-8; detergent and emulsifier intermediate; water-wh. mobile liq., mild char. odor; m.w. 186; sp.gr. 0.830 g/ml; dens. 6.93 lb/gal; f.p. 24 C; b.p. 258-264 C (1 atm); flash pt. (PMCC) 132 C; hyd. no. 301; toxicology: temporary irritation on direct skin or eye contact;

100% conc.

Epal® 12/70. [Ethyl] Dodecanol (70%), tetradecanol (29%); biodeg. detergent/emulsifier intermediate; clear liq., mild char. odor; m.w. 195; sp.gr. 0.832 g/ml; dens. 6.94 lb/gal; f.p. 18 C; b.p. 126-129 C (760 mm); flash pt. (PMCC) 135 C; hyd. no. 288; toxicology: temporary irritation on direct skin or eye contact; 100% conc.

Epal® 12/85. [Ethyl] Dodecanol (86%), tetradecanol (14%); biodeg. detergent/emulsifier intermediate; clear liq., mild char. odor; m.w. 190; sp.gr. 0.830 g/ml; dens. 6.93 lb/gal; f.p. 20 C; b.p. 126-129 C (760 mm);flash pt. (PMCC) 138 C; hyd. no. 297; toxicology: temporary irritation on direct skin or eye contact; 100% conc.

Epal® 14. [Ethyl] Tetradecanol; CAS 112-72-1; biodeg. detergent/emulsifier intermediate; wh. waxy solid, mild char. odor; m.w. 215; sp.gr. 0.822 g/ml; dens. 6.86 lb/gal; f.p. 35 C; flash pt. (PMCC) 149 C; hyd. no. 261; toxicology: temporary irritation on direct skin or eye contact; 100% conc.

Epal® 16NF. [Ethyl] Hexadecanol NF; CAS 36653-82-4; USP grade biodeg. detergent/emulsifier intermediate; wh. waxy solid, mild char. odor; m.w. 242; sp.gr. 0.818 g/ml; dens. 6.83 lb/gal; f.p. 44 C; flash pt. (PMCC) 175 C; hyd. no. 232; toxicology: temporary irritation on direct skin or eye contact; 100% conc.

Epal® 18NF. [Ethyl] Octadecanol NF; CAS 112-92-5; USP grade biodeg. detergent/emulsifier intermediate; wh. waxy solid, mild char. odor; m.w. 270; sp.gr. 0.812 g/ml; dens. 6.78 lb/gal; f.p. 54 C; flash pt. (PMCC) 191 C; hyd. no. 207; toxicology: temporary irritation on direct skin or eye contact; 100% conc.

Epal® 20+. [Ethyl] C18-C32 linear and branched alcohols (66%), C24-C40 hydrocarbons (40%); biodeg. detergent/emulsifier intermediate; off-wh. waxy solid; sp.gr. 0.83 g/ml; dens. 6.9 lb/gal; f.p. 41 C; flash pt. (PMCC) 177 C; hyd. no. 105; toxicology: temporary irritation on direct skin or eye contact; 100% conc.

Epal® 108. [Ethyl] Octanol, decanol, hexanol (55:41:4); biodeg. detergent/emulsifier intermediate; colorless clear mobile liq., mild char. odor; m.w. 139; sp.gr. 0.823 g/ml; dens. 6.87 lb/gal; f.p. -20 C; b.p. 197-237 C (1 atm); flash pt. (PMCC) 88 C; hyd. no. 404; combustible; toxicology: temporary irritation on direct skin or eye contact; 100% conc.

Epal® 610. [Ethyl] C8 alcohol (42%), decyl alcohol (54%), hexanol (4%); biodeg. detergent/emulsifier intermediate; colorless clear mobile liq., mild char. odor; m.w. 143; sp.gr. 0.824 g/ml; dens. 6.86 lb/gal; f.p. -17 C; b.p. 183-242 C (1 atm); flash pt. (PMCC) 79 C; hyd. no. 393; combustible; toxicology: temporary irritation on direct skin or eye contact; 100% conc.

Epal® 618. [Ethyl] C16-C18 linear primary alcohol; detergent and emulsifier intermediate; waxy solid; 100% conc.

Epal® 810. [Ethyl] C8 alcohol (45%), decanol (54%); biodeg. detergent/emulsifier intermediate; colorless clear mobile liq., mild char. odor; m.w. 145; sp.gr. 0.824 g/ml; dens. 6.88 lb/gal; f.p. -10 C; b.p. 178-240 C (1 atm); flash pt. (PMCC) 90 C; hyd. no. 386; combustible; toxicology: temporary irritation on direct skin or eye contact; 100% conc.

Epal® 1012. [Ethyl] Decyl alcohol (75%), dodecanol (23%); biodeg. detergent/emulsifier intermediate; water-wh. mobile liq., mild char. odor; m.w. 163; sp.gr. 0.828 g/ml; dens. 6.91 lb/gal; f.p. -5 C; b.p. 230-265 C (1 atm); flash pt. (PMCC) 113 C; hyd. no. 344; toxicology: temporary irritation on direct skin or eye contact; 100% conc.

Epal® 1214. [Ethyl] Dodecanol (66%), tetradecanol (27%), hexadecanol (6%); biodeg. detergent/emulsifier intermediate; clear sl. visc. liq., mild char. odor; m.w. 197; sp.gr. 0.830 g/ml; dens. 6.93 lb/gal; f.p. 22 C; b.p. 233-299 C (1 atm); flash pt. (PMCC) 138 C; hyd. no. 284; toxicology: temporary irritation on direct skin or eye contact; 100% conc.

Epal® 1218. [Ethyl] Dodecanol (48%), tetradecanol (20%), hexadecanol (17%), octadecanol (14%); biodeg. detergent/emulsifier intermediate; water-wh. mobile liq. at sl. above R.T., mild char. odor; m.w. 214; sp.gr. 0.825 g/ml; dens. 6.89 lb/gal; f.p. 25 C; flash pt. (PMCC) 139 C; hyd. no. 266; toxicology: temporary irritation on direct skin or eye contact; 100% conc.

Epal® 1412. [Ethyl] Tetradecanol (58%), dodecanol (40%); intermediate for surfactants, plasticizers, lubricant additives, thioesters, specialty chems.; clear sl. visc. liq., mild char. odor; m.w. 204; sp.gr. 0.820 g/ml; dens. 6.84 lb/gal; f.p. 26 C; b.p. 2 C (1 atm); flash pt. (PMCC) 137 C; hyd. no. 275; toxicology: temporary irritation on direct skin or eye contact; 100% conc.

Epal® 1416. [Ethyl] Tetradecanol (58%), dodecanol (40%); biodeg. detergent/emulsifier intermediate; wh. waxy solid, mild char. odor; m.w. 224; sp.gr. 0.825 g/ml; dens. 6.88 lb/gal; f.p. 36 C; flash pt. (CC) 143 C; hyd. no. 250; toxicology: temporary irritation on direct skin or eye contact; 100% conc.

Epal® 1416-LD. [Ethox] Tetradecanol (63.5%), hexadecanol (35.6%), low-diol blend; detergent/emulsifier intermediate; wh. waxy solid, mild char. odor; m.w. 224; sp.gr. 0.825 g/ml; dens. 6.88 lb/gal; f.p. 38 C; flash pt. (PMCC) 143 C; toxicology: temporary irritation on direct skin or eye contact; 100% conc.

Epal® 1418. [Ethyl] Tetradecanol (35%), hexadecanol (40%), octadecanol (23%); biodeg. detergent/emulsifier intermediate; wh. waxy solid, mild char. odor; m.w. 239, sp.gr. 0.825 g/ml; dens. 6.89 lb/gal; f.p. 42 C; b.p. 300-315 C (1 atm); flash pt. (PMCC) 149 C; hyd. no. 235; toxicology: temporary irritation on direct skin or eye contact; 100% conc.

Epal® 1618. [Ethyl] Hexadecanol (47%), octadecanol (50%); biodeg. detergent/emulsifier intermediate; wh. waxy solid, mild char. odor; m.w. 256; sp.gr. 0.819 g/ml; dens. 6.83 lb/gal; f.p. 46 C; flash pt. (PMCC) 202 C; hyd. no. 219; toxicology: temporary irritation on direct skin or eye contact; 100% conc.

Epal® 1618RT. [Ethyl] Hexadecanol (66%), octadecanol (32%), eicosanol (1%); biodeg. detergent/emulsifier intermediate; wh. waxy solid, mild char. odor; m.w. 250; sp.gr. 0.827 g/ml; dens. 6.90 lb/gal; f.p. 46 C; flash pt. (PMCC) > 149 C; hyd. no. 223; toxicology: temporary irritation on direct skin or eye contact; 100% conc.

Epal® 1618T. [Ethyl] Hexadecanol (32%), octadecanol (65%); intermediate for surfactants, plasticizers, lubricant additives, thioesters, specialty chems.; wh. waxy solid, mild char. odor; m.w. 260; sp.gr. 0.821 g/ml; dens. 6.86 lb/gal; f.p. 41 C; flash pt. (PMCC) 177 C; hyd. no. 216; toxicology: temporary irritation on direct skin or eye contact; 100% conc.

Epan 710. [Dai-ichi Kogyo Seiyaku] PPG PEG ether; CAS 9003-11-6; nonionic; low foaming detergent, emulsifier, dispersant for emulsion polymerization of syn. resins, latex; liq.; 100% conc.

Epan 720. [Dai-ichi Kogyo Seiyaku] PPG PEG ether; CAS 9003-11-6; nonionic; low foaming detergent, emulsifier, dispersant for emulsion polymerization of syn. resins, latex; liq.; 100% conc.

Epan 740. [Dai-ichi Kogyo Seiyaku] PPG PEG ether; CAS 9003-11-6; nonionic; low foaming detergent, emulsifier, dispersant for emulsion polymerization of syn. resins, latex; paste; 100% conc.

Epan 750. [Dai-ichi Kogyo Seiyaku] PPG PEG ether; CAS 9003-11-6; nonionic; low foaming detergent, emulsifier, dispersant for emulsion polymerization of syn. resins, latex; paste; 100% conc.

Epan 785. [Dai-ichi Kogyo Seiyaku] PPG PEG ether; CAS 106392-12-5; nonionic; low foaming detergent, emulsifier, dispersant for emulsion polymerization of syn. resins, latex; solid; 100% conc.

Epan U 102. [Dai-ichi Kogyo Seiyaku] PPG PEG ether; nonionic; emulsifier for min. oils, defoaming agent; lowers surf. tension; liq.; HLB 9.9; 99% conc.

Epan U 103. [Dai-ichi Kogyo Seiyaku] PPG PEG ether; nonionic; emulsifier, dispersant; lowers surf. tension; liq.; HLB 6.6; 99% conc.

Epan U 105. [Dai-ichi Kogyo Seiyaku] PPG PEG ether; nonionic; emulsifier, lubricant; solid; HLB 16.9; 99% conc.

Epan U 108. [Dai-ichi Kogyo Seiyaku] PPG PEG ether; nonionic; lubricant for textiles; flake; HLB 26.0; 99% conc.

Equex S. [Procter & Gamble] Sodium lauryl sulfate; CAS 151-21-3; anionic; cosmetic grade detergent, emulsifier, used in shampoos, liq. cleaners, specialty prods.; liq.; dens. 8.7 lb/gal; sp.gr. 1.04; visc. 50–200 cps; cloud pt. 54 F; pH 7.8 (10%); 29% act.

Equex SP. [Procter & Gamble] Sodium lauryl sulfate; CAS 151-21-3; anionic; polymerization grade emulsifier; pale yel. clear liq.; clean, pleasant odor; dens. 8.7 lb/gal; sp.gr. 1.04; visc. 90 cps (80 F); cloud pt. 55 F; pH 8 (10%); 29.8% act.

Equex STM. [Procter & Gamble] Sodium lauryl sulfate, cocamide MEA and TEA-lauryl sulfate; anionic; detergent, foamer for shampoos, cleaners, specialty; opaque wh. to straw thick, honey-like fluid; clean, pleasant odor; dens. 8.7 lb/gal; sp.gr. 1.04; visc. 9000 cps; cloud pt. 52–55 F; pH 8; 22.5% act.

Equex SW. [Procter & Gamble] Sodium lauryl sulfate; CAS 151-21-3; anionic; detergent, wetting agent, emulsifier for cream shampoos, paste cleaners, rug shampoos; pale yel. thick honey-like fluid; clean, pleasant odor; dens. 8.8 lb/gal; sp.gr. 1.05; visc. 40,000 cps (80 F); cloud pt. 77 F; pH 7.8 (10%); 28.5% act.

Erional® NW. [Ciba-Geigy/Dyestuffs] Aromatic sulfonic acid condensation prod.; anionic; resist agent for nylon and fixative for acid dyes on nylon; liq.

Erional® PA. [Ciba-Geigy/Dyestuffs] Aromatic sulfonic acid condensation prod.; CAS 9017-71-4; anionic; fixing agent for improving wetfastness of acid dyes on nylon; reserves nylon from direct dyes; liq.

Erional® RN. [Ciba-Geigy/Dyestuffs] Aromatic sulfonic acid condensation prod.; anionic; fixing agent for maximizing wetfastness of acid or metal complex dyes on nylon; reserves nylon from direct dyes; liq.

ES-1239. [CasChem] Dimethylaminopropyl ricinolamide benzyl chloride, propylene glycol; cationic; emulsifier with antistatic, softening, bactericidal, wetting, dispersing properties for cosmetic, textile, agric. industries; Gardner 10+ clear liq.; sol. in water, alcohol, polar solv.; sp.gr. 1.022; dens. 8.48 lb/gal; visc. 15.3 stokes; flash pt. 230 F; 70% solids in propylene glycol.

Esi-Det 21M. [Emulsion Systems] Modified alkanolamide; detergent for hard surf. cleaners; visc. builder for aq. systems; base for multipurpose cleaners, strippers, degreasers; high tolerance to alkaline builders, solvs.; amber liq.; sp.gr. 1.002; dens. 8.35 lb/gal; visc. 1300-1800 cps; pH 8.0-9.0 (1%); 22-26% free amine, 24-28% free fatty acid.

Esi-Det CDA. [Emulsion Systems] 1:1 Cocamide DEA; nonionic; detergent, foam stabilizer, thickener for personal care prods., lt. duty dishwash and household cleaners, industrial prods.; 100% biodeg.; straw color; sp.gr. 0.98; dens. 8.16 lb/gal; pH 9.0-11.0; 99% min. solids.

Esi-Det EP-20. [Emulsion Systems] Complex phosphate ester of POE ethanol; anionic; detergent for wide range of prods.; high compatibility; lt. amber visc. liq.; sp.gr. 1.13; dens. 9.7 lb/gal; acid no. 90-110; pH 2.0 ± 0.5 (1%); 99-100% act.

Esi-Terge 10. [Emulsion Systems] Cocamide DEA; nonionic; foam stabilizer, thickener for household, cosmetic, industrial prods.; liq.; 100% conc.

Esi-Terge 320. [Emulsion Systems] Phosphated nonylphenoxy polyethoxy ethanol; anionic; detergent, visc. aid, coupling and wetting agent for syn. hard surf. cleaners with high caustic conc.; high alkali stable; lt. straw liq.; mild odor; water sol.; sp.gr. 1.121; dens. 9.7 lb/gal; acid no. 100 ± 10; pH 1.5-2.5 (1%); 100% act.

Esi-Terge 330. [Emulsion Systems] Phosphated glycol ester; anionic; detergent, wetting agent, coupling and penetrating agent used in high-alkali cleaners, steam cleaners, cement cleaners, wax strippers, oven cleaners; lt. straw liq.; mild odor; water sol.; sp.gr. 1.102; dens. 9.18 lb/gal; acid no. 260-310; pH 1.5-2.5 (1%); 100% act.

Esi-Terge 40% Coconut Oil Soap. [Emulsion Systems] Cochin oil soap; anionic; detergent for liq. hand soap; mild cleaning and coupling; straw liq.; coconut odor; water sol.; dens. 8.5 lb/gal; f.p. 320 F; 40% act.

Esi-Terge B-15. [Emulsion Systems] 2:1 Cocamide DEA; nonionic; detergent for mild all-purpose cleaners, foam builder and thickener, wetting agent; for household and industrial cleaners; amber liq.; water sol.; sp.gr. 1.01; dens. 8.4 lb/gal; pH 9.8–10.8 (1%); biodeg.; 100% act.

Esi-Terge HA-20. [Emulsion Systems] Modified amine condensate; anionic/nonionic; self-coupling detergent for cleaners, detergent base; visc. builder; for low and high alkali cleaners; amber liq.; sol. in water and some solv.; sp.gr. 1.037; dens. 8.64 lb/gal; acid no. 60 ± 10; pH 7.5–8.5 (1%); 100% act.

Esi-Terge L-75. [Emulsion Systems] Blend of surfactants, chelating and coupling agents; anionic/nonionic; base for high foaming cleaners for dishwashing, carwashing, all-purpose cleaning, butyl degreasers; amber liq., mild odor; sp.gr. 1.09; dens. 9.07 lb/gal; pH 8.0-9.0 (1%); 75% conc.

Esi-Terge L-80. [Emulsion Systems] Blend; detergent for liq. dishwash compds., car washes, liq. hand soaps, household and institutional all-purpose cleaners where high foam at low concs. is important; lt. straw color; sp.gr. 1.05; pH 6.5-7.5 (1%); 60% solids.

Esi-Terge N-100. [Emulsion Systems] PEG ether surfactant; nonionic; rapid wetter for formulating mild acid cleaners as well as neutral and mildly alkaline prods.; water clear color; sp.gr. 1.05; dens. 8.75 lb/gal; cloud pt. 52-56 C (1%); pH 7.0-8.0 (1%); 99.5 ± 0.5% act.

Esi-Terge RT-61. [Emulsion Systems] Blended detergent; anionic; detergent base for emulsifying oils and fats; used in steam carpet cleaners, to replace glycol ethers; water clear color; dens. 8.00 lb/gal; visc. 30-60 cps; pH 6.5-7.5 (1%); 90% act.

Esi-Terge S-10. [Emulsion Systems] 1:1 Cocamide DEA; nonionic; detergent, emulsifier, foam stabilizer, thickener, for liq. dishwashing and car washing detergents, household, industrial, and cosmetic prods.; lt. straw liq.; sp.gr. 0.98; dens. 8.16 lb/gal; pH 9–10; biodeg.; 100% act.

Esi-Terge SXS. [Emulsion Systems] Sodium xylene sulfonate; CAS 1300-72-7; anionic; solubilizer for lt.- and heavy-duty cleaners; liq.; 40% act.

Esi-Terge T-5. [Emulsion Systems] Modified cocamide DEA; nonionic; surfactant for household and industrial cleaners; biodeg.; amber liq.; sp.gr. 0.97; dens. 8.08 lb/gal; pH 8.5-9.5 (1%); 100% conc.

Esi-Terge T-60. [Emulsion Systems] TEA dodecylbenzene sulfonate; anionic; detergent, wetting agent, foam stabilizer, for car and dishwashing detergent, syn. hand soap, household cleaners; amber liq.; dens. 9.1 lb/gal; sp.gr. 1.09; f.p. 320 F; pH 6.5–7.5 (1%); biodeg.; 60% act.

Espesilor AC Series. [Pulcra SA] Fatty acid DEAs; thickener and superfatting agents for shampoos, bubble baths, liq. detergents; perfume boosters for soaps.

Estersulf 1HH. [Climax Performance] Alkylaryl sulfonamide ester; CAS 68153-58-2; anionic; emulsifier for chlorinated materials, oils, solvs.; liq.; 100% conc.

Estrasan 1. [Reilly-Whiteman] Methyl ricinoleate; nonionic; enhances lubricity and wettability of petrol. oils for metals and other industries; liq.; 100% conc.

Estrasan 3. [Reilly-Whiteman] Methyl tallowate; nonionic; enhances lubricity and wettability of petrol. oils for metals and other industries; liq.; 100% conc.

Estrasan 4G. [Reilly-Whiteman] Methyl ester of grease; nonionic; enhances lubricity and wettability of petrol. oils for metals and other industries; liq.; 100% conc.

Estrasan 4L. [Reilly-Whiteman] Methyl lardate; nonionic; enhances lubricity and wettability of petrol. oils for metals and other industries; liq.; 100% conc.

Ethal 326. [Ethox] Laureth-3; nonionic; emulsifier for textile and industrial applics.; liq.; oil-sol.; HLB 8.0; cloud pt. < 25 C (1%).

Ethal 368. [Ethox] PEG-3 C16-18 alcohol; nonionic; emulsifier, lubricant; paste; oil-sol.; HLB 6.9; cloud pt. < 25 C (1%).

Ethal 926. [Ethox] Laureth-9; nonionic; detergent, emulsifier; paste; water-sol.; HLB 13.3; cloud pt. 69 C (1%).

Ethal 3328. [Ethox] PEG-3 C12-18 alcohol; nonionic; low HLB emulsifier and coupler; liq.; HLB 8.1; cloud pt. < 25 C (1%); 99% conc.

Ethal BPA-6. [Ethox] PEG-6 bisphenol A; monomer for polyester and urethane coatings; liq.; 100% act.

Ethal CSA-25. [Ethox] PEG-25 C16-18 alcohol; nonionic; hydrophilic emulsifier, detergent for mild acidic or alkaline sol'ns. and hot aq. systems; solid; HLB 16.3; cloud pt. > 100 C (1%).

Ethal DA-4. [Ethox] PEG-4 decyl alcohol; nonionic; wetting agent for aq. sol'ns.; intermediate for mfg. of anionic surfactants; liq.; HLB 10.5; cloud pt. < 25 C (1%); 99% conc.

Ethal DA-6. [Ethox] PEG-6 decyl alcohol; nonionic; rapid wetting agent for aq. sol'ns.; liq.; HLB 12.4; cloud pt. 42 C (1%); 99% conc.

Ethal DA-9. [Ethox] PEG-9 decyl alcohol; nonionic; rapid wetting agent for hot aq. systems; liq.; HLB 14.3; cloud pt. 80 C (1%).

Ethal DDP-7. [Ethox] Multipurpose emulsifier and detergent; exc. greaser cutter; liq.; HLB 10.7; cloud pt. 25 C (1%).

Ethal DNP-8. [Ethox] PEG-8 dinonyl phenol; CAS 9014-93-1; nonionic; low foam emulsifier for solv. systems; coupling agent for introducing water into nonaq. systems; liq.; HLB 10.1; cloud pt. < 25 C (1%).

Ethal DNP-18. [Ethox] PEG-18 dinonyl phenol; CAS 9014-93-1; nonionic; emulsifier and detergent with low foaming props.; paste; water-sol.; HLB 13.9; cloud pt. 70 C (1%).

Ethal DNP-150/50%. [Ethox] PEG-150 dinonyl phenol; CAS 9014-93-1; nonionic; hydrophilic emulsifier, dispersant for hard surf. and alkaline cleaners; aux. for acrylic dyeing systems; liq.; HLB 19.0; cloud pt. > 100 C (1%).

Ethal EH-2. [Ethox] PEG-2 2-ethylhexanol; nonionic; intermediate for anionic surfactants; component of low foam wetting systems; liq.; HLB 8.1; cloud pt. < 25 C (1%); 99% conc.

Ethal EH-5. [Ethox] PEG-5 2-ethylhexanol; nonionic; low foam wetting agent; intermediate for anionic surfactants; liq.; HLB 12.6; cloud pt. < 25 C (1%); 99% conc.

Ethal LA-4. [Ethox] Laureth-4; nonionic; general-purpose emulsifier, lubricant; liq.; oil-sol.; HLB 9.2; cloud pt. < 25 C (1%).

Ethal LA-7. [Ethox] Laureth-7; nonionic; detergent and oil dispersant for metals, textiles, and commercial cleaners; liq.; HLB 12.2; cloud pt. 49 C (1%).

Ethal NP-1.5. [Ethox] Nonoxynol-1.5; nonionic; emulsifier, intermediate; liq.; oil-sol.; HLB 4.6; cloud pt. < 25 C (1%).

Ethal NP-4. [Ethox] Nonoxynol-4; CAS 9016-45-9; nonionic; emulsifier, intermediate for anionic surfactants; liq.; oil-sol.; HLB 8.9; cloud pt. < 25 C (1%).

Ethal NP-6. [Ethox] Nonoxynol-6; CAS 9016-45-9; nonionic; coupling agent for surfactant systems, for coupling water into oil systems; liq.; oil-sol.; HLB 10.9; cloud pt. < 25 C (1%).

Ethal NP-9. [Ethox] Nonoxynol-9; CAS 9016-45-9; nonionic; general purpose wetting agent and detergent; liq.; HLB 13.0; cloud pt. 54 C (1%).

Ethal NP-20. [Ethox] Nonoxynol-20; CAS 9016-45-9; nonionic; high temp. detergent, good electrolyte stability; liq.; HLB 16.0; cloud pt. > 100 C (1%).

Ethal NP-407. [Ethox] Nonoxynol-40; CAS 9016-45-9; nonionic; stabilizer for emulsions; emulsifier for high temp. applics.; liq.; HLB 17.0; cloud pt. > 100 C (1%); 70% conc.

Ethal OA-10. [Ethox] Oleth-10; CAS 9004-98-2; nonionic; moderate HLB emulsifier for esters and oils; liq.; HLB 12.6; cloud pt. 60 C (1%).

Ethal OA-23. [Ethox] Oleth-23; CAS 9004-98-2; nonionic; hydrophilic emulsifier and detergent; useful in stabilizing dispersions and scouring textile goods; liq.; HLB 15.8; cloud pt. > 100 C (1%); 99% conc.

Ethal TDA-3. [Ethox] PEG-3 tridecyl alcohol; CAS

24938-91-8; nonionic; surfactant, coupling agent; liq.; oil-sol.; HLB 7.9; cloud pt. < 25 C (1%); 99% conc.

Ethal TDA-6. [Ethox] PEG-6 tridecyl alcohol; CAS 24938-91-8; nonionic; wetting agent, detergent, foamer, dispersant, emulsifier; liq.; HLB 11.4; cloud pt. < 25 C (1%); 99% conc.

Ethal TDA-9. [Ethox] PEG-9 tridecyl alcohol; CAS 24938-91-8; nonionic; wetting agent, detergent, foamer for high foaming cleaners; liq.; HLB 13.0; cloud pt. 54 C (1%); 99% conc.

Ethal TDA-12. [Ethox] PEG-12 tridecyl alcohol; CAS 24938-91-8; nonionic; detergent, foam builder for mildly alkaline or acidic sol'ns.; good low temp. detergency; paste; HLB 14.5; cloud pt. 70 C (1%).

Ethal TDA-18. [Ethox] PEG-18 tridecyl alcohol; CAS 24938-91-8; nonionic; high-foaming detergent for elevated temp. applics.; solid; HLB 16.0; cloud pt. > 100 C (1%).

Ethfac 102. [Ethox] Aliphatic phosphate ester; anionic; antistat, lubricant; solid; 99% act.

Ethfac 104. [Ethox] Aliphatic phosphate ester; anionic; low foaming penetrant and emulsifier; liq.; 99% act.

Ethfac 106. [Ethox] Aliphatic phosphate ester; anionic; potassium salt is an effective antistat for textiles at low levels; liq.; 99% act.

Ethfac 133. [Ethox] Aliphatic phosphate ester; anionic; lubricant; liq.; oil-sol.; 99% act.

Ethfac 136. [Ethox] Aliphatic phosphate ester; anionic; low foaming ester with hydrotropic props.; liq.; 99% act.

Ethfac 140. [Ethox] Aliphatic phosphate ester; anionic; lubricant; liq.; 99% act.

Ethfac 142W. [Ethox] Aliphatic phosphate ester; anionic; emulsifier; lubricant additive for process fluids; liq.; oil and water-disp.; 99% act.

Ethfac 161. [Ethox] Aliphatic phosphate ester; anionic; emulsifier; liq.; 99% act.

Ethfac 324. [Ethox] Aliphatic phosphate ester; anionic; detergent, wetting and scouring agent; antistat for processing textile fibers; liq.; 85% act.

Ethfac 353. [Ethox] Aliphatic phosphate ester; anionic; antistatic agent for fibers and yarns with very good emulsification props.; liq.; 90% act.

Ethfac 361. [Ethox] Aliphatic phosphate ester; anionic; detergent, emulsifier, antistat, corrosion inhibitor; liq.; 90% act.

Ethfac 363. [Ethox] Aliphatic phosphate ester; anionic; antistatic agent for fibers and yarns with very good emulsification props.; liq.; 90% act.

Ethfac 391. [Ethox] Aliphatic phosphate ester; anionic; detergent for textile scouring; effective dispersant at elevated temps.; liq.; 90% act.

Ethfac 1018. [Ethox] Aliphatic phosphate ester; anionic; antistat and softener with exc. lubricating props. when neutralized; solid; 99% act.

Ethfac NP-16. [Ethox] Aromatic phosphate ester; anionic; neutralization with various bases yields a range of emulsification props.; liq.; 99% act.

Ethfac NP-110. [Ethox] Aromatic phosphate ester; anionic; hydrophilic detergent and emulsifier, esp. for liq. detergent formulations; liq.; 99% act.

Ethfac PB-1. [Ethox] Aliphatic phosphate ester; anionic; hydrophilic surfactant with high caustic stability; for highly alkaline cleaners; liq.; 99% act.

Ethfac PB-2. [Ethox] Aliphatic phosphate ester; anionic; hydrophilic surfactant with high caustic stability; for highly alkaline cleaners; liq.; 99% act.

Ethfac PD-6. [Ethox] Aliphatic phosphate ester; anionic; higher tolerance to alkali than Ethfac 161; useful as component of alkaline cleaners; liq.; 99% act.

Ethfac PD-990. [Ethox] Aliphatic phosphate ester; anionic; dispersant for solvs. when used in a continuous aq. phase; liq.; 90% act.

Ethfac PP-16. [Ethox] Aromatic phosphate ester; anionic; low foaming hydrotrope for coupling nonionics into alkaline systems; liq.; 99% act.

Ethfac PP-36/50%. [Ethox] Aromatic phosphate ester; anionic; neutralized sol'n. of Ethfac PP-16; hydrotrope; liq.; 50% act.

Ethoduomeen® T/13. [Akzo; Akzo Chem. BV] PEG-3 tallow aminopropylamine; CAS 61790-85-0; cationic; emulsifier used in making of bitumen emulsions; dispersant for waxes; for textiles, asphalt, agric. emulsions; wetting agent, corrosion inhibitor; Gardner 18 max. liq.; sp.gr. 0.95; f.p. -20 C; m.p. 17 C; b.p. 150 C; flash pt. (PMCC) > 204 C; surf. tens. 34.5 dynes/cm (0.1%); 100% conc.

Ethoduomeen® T/20. [Akzo; Akzo Chem. BV] PEG-10 tallow aminopropylamine; CAS 61790-85-0; cationic; emulsifier, dispersant, wetting agent used in coating preparation on paperboard; corrosion inhibitor; Gardner 18 max. liq.; sp.gr. 0.99; f.p. < 25 C; b.p. 150 C; flash pt. 90 C (PMCC); surf. tens. 38.2 dynes/cm (0.1%); 95% min. act.

Ethoduomeen® T/25. [Akzo; Akzo Chem. BV] PEG-15 tallow aminopropylamine; CAS 61790-85-0; cationic; corrosion inhibitor in water treatment chemicals in sec. oil recovery; Gardner 18 min. liq.; sp.gr. 1.02; f.p. > 25 C; flash pt. 238 C (COC); surf. tens. 43.0 dynes/cm (0.1%); 95% act.

Ethoduomeen® TD/13. [Akzo] Ethoxylated diamine from tallow fatty acid; emulsifier for bitumen emulsions, dispersing waxes; Gardner 4 max. clear liq.; sp.gr. 0.95; surf. tens. 34.2 dynes/cm (0.1%).

Ethoduoquad® T/15-50. [Akzo] N,N,N′,N′,N′-Penta(2-hydroxyethyl)-N-tallowalkyl-1,3-propane diammonium diacetate, aq. IPA; industrial surfactant for agric., textiles, protective coatings, inks, pigment dispersions, acid pickling baths, metalworking, electroplating, plastics mfg.; EPA listed; Gardner 16 max. liq.; flash pt. (PMCC) 18 C; pH 6-9; 48-52% solids in aq. IPA.

Ethoduoquad® T/20. [Akzo] POE quat. ammonium salt from tallow fatty acid; cationic; antistat, emulsifier, dyeing assistant used in textile industry, as electroplating bath additive; clear liq.; pH 6–9 (10% aq.); 72% min. act.

Ethofat® 18/14. [Akzo] PEG-4 stearate; CAS 9004-99-3; nonionic; industrial surfactant; Gardner 4 max. solid; sp.gr. 0.98; acid no. 3 max.; sapon. no. 121-131; flash pt. (PMCC) 274 C; surf. tens. 39 dynes/cm (0.1%).

Ethofat® 60/15. [Akzo] PEG-5 stearate; CAS 9004-99-3; nonionic; emulsifier, detergent, wetting agent, dispersant, suspending agent, for textile, cosmetic, agric., metal and leather treating; Gardner 8 soft paste; sol. in acetone, IPA, CCl_4, benzene, disp. in water; sp.gr. 1.01; acid no. 1 max.; sapon. no. 110–120; surf. tens. 39 dynes/cm (0.1%); pH 6–7.5 (10% aq.).

Ethofat® 60/20. [Akzo] PEG-10 stearate; CAS 9004-99-3; nonionic; emulsifier, detergent, wetting agent, dispersant, suspending agent, for textile, cosmetic, agric., metal and leather treating; Gardner 8 clear liq.; sol. see Ethofat 60/15; sp.gr. 1.02; acid no. 1 max.; sapon. no. 70–80; surf. tens. 36 dynes/cm (0.1%); pH 6–7.5 (10% aq.).

Ethofat® 60/25. [Akzo] PEG-15 stearate; CAS 9004-99-3; nonionic; emulsifier, detergent, wetting agent, dispersant, suspending agent, for textile, cosmetic, agric., metal and leather treating; Gardner 11 paste; acid no. 1 max.; sapon. no. 55–65; surf. tens. 39.5 dynes/cm (0.1%); pH 6–7.5 (10% aq.).

Ethofat® 142/20. [Akzo] PEG-10 glycol tallate; emulsifier, detergent, wetting agent, dispersant, suspending agent, for textile, cosmetic, agric., metal and leather treating; Gardner 11 clear liq.; sp.gr. 1.03; acid no. 1 max.; sapon no. 72–82; surf. tens. 39.5 dynes/cm (0.1%); pH 6–7.5 (10% aq.); flash pt. (PM) > 450 F.

Ethofat® 242/25. [Akzo] PEG-15 tallate; CAS 61791-00-2; 65071-95-6; nonionic; emulsifier, detergent, dispersant; Gardner 12 clear liq.; sol. in acetone, IPA, CCl_4, dioxane, benzene, water; sp.gr. 1.08; acid no. 1 max.; sapon no. 55–65; pH 6–7.5 (10% aq.); flash pt. (PMCC) 274 C; surf. tens. 42 dynes/cm (0.1%); 100% conc.

Ethofat® 433. [Akzo] PEG-15 tallate; CAS 61791-00-2; 65071-95-6; nonionic; surfactant; Gardner 12 max. liq.; acid no. 1.2 max.; sapon. no. 46-54; 9.5-10.5% moisture.

Ethofat® C/15. [Akzo] PEG-5 cocoate; nonionic; emulsifier, detergent, dispersant; Gardner 9 clear liq.; sol. in acetone, IPA, CCl_4, benzene, disp. in water; sp.gr. 1.00; HLB 10.6; acid no. 1 max.; sapon. no. 120–130; surf. tens. 33 dynes/cm (0.1%); pH 6–7.5 (10% aq.); 100% conc.

Ethofat® C/25. [Akzo] PEG-15 cocoate; nonionic; emulsifier, detergent, wetting agent, dispersant, suspending agent, for textile, cosmetic, agric., metal and leather treating; Gardner 8 paste; sol. see Ethofat 60/15; acid no. 1 max.; sapon. no. 60–70; pH 6–7.5 (10% aq.); flash pt. (PM) > 400 F.

Ethofat® O/15. [Akzo] PEG-5 oleate; CAS 9004-96-0; nonionic; emulsifier, detergent, dispersant; Gardner 9 max. clear liq.; sol. in acetone, IPA, CCl_4, benzene, disp. in water; sp.gr. 0.99; HLB 8.6; acid no. 1 max.; sapon. no. 110–120; surf. tens. 35 dynes/cm (0.1%); pH 6–7.5 (10% aq.); flash pt. (PM) > 400 F; 100% conc.

Ethofat® O/20. [Akzo] PEG 10 oleate; CAS 9004-96-0; nonionic; emulsifier, detergent, dispersant; Gardner 9 max. clear liq.; sol. in acetone, IPA, CCl_4, benzene, disp. in water; sp.gr. 1.028; acid no. 1 max.; sapon. no. 75–85; pH 6–7.5 (10% aq.); flash pt. (PM) 485 F; surf. tens. 41 dynes/cm (0.1%); 100% conc.

Ethomeen® 18/12. [Akzo] PEG-2 stearamine; CAS 10213-78-2; cationic; emulsifier, dispersant used in textile processing; Gardner 7 solid; sol. in acetone, IPA, CCl_4, benzene; sp.gr. 0.96; flash pt. (COC) 400 F; 100% conc.

Ethomeen® 18/15. [Akzo] PEG-5 stearamine; CAS 26635-92-7; cationic; emulsifier, dispersant for textile processing; Gardner 8 solid; sol. in acetone, IPA, CCl_4, benzene; sp.gr. 0.98; surf. tens. 34 dynes/cm (0.1%); flash pt. (COC) 500 F; 100% conc.

Ethomeen® 18/20. [Akzo] PEG-10 stearamine; CAS 26635-92-7; cationic; emulsifier, dispersant used in textile processing; Gardner 8 clear liq. to paste; sp.gr. 1.02; flash pt. (COC) 540 F; surf. tens. 40 dynes/cm (0.1%).

Ethomeen® 18/25. [Akzo] PEG-15 stearamine; CAS 26635-92-7; cationic; emulsifier, dispersant used in textile processing; Gardner 8 clear liq.; sp.gr. 1.04; flash pt. (COC) 560 F; surf. tens. 43 dynes/cm (0.1%).

Ethomeen® 18/60. [Akzo] PEG-50 stearamine; CAS 26635-92-7; cationic; emulsifier, dispersant used in textile processing; prevents premature coagulation of latex rubber; wh. solid; sol. in acetone, IPA, CCl_4, benzene, water; sp.gr. 1.12; flash pt. (COC) 579 F; surf. tens. 49 dynes/cm (0.1%).

Ethomeen® C/12. [Akzo; Akzo Chem. BV] PEG-2 cocamine; CAS 61791-14-8; 61791-31-9; cationic; emulsifier, dispersant used in textile processing; Gardner 6 max. clear liq.; sol. in acetone, IPA, CCl_4, Stod., benzene; forms gel in water; sp.gr. 0.87; HLB 6.4; flash pt. (COC) 380 F; 100% conc.

Ethomeen® C/15. [Akzo; Akzo Chem. BV] PEG-5 cocamine; CAS 61791-14-8; cationic; emulsifier, dispersant for textile processing, dyeing assistant, desizing assistant, softener, antistat; Gardner 6 max. clear liq.; sol. in acetone, IPA, CCl_4, Stod., benzene, water (cloudy); sp.gr. 0.98; HLB 13.9; flash pt. (TOC) 460 F; surf. tens. 33 dynes/cm (0.1%); Draves wetting 3.5 s (0.5%); 100% conc.

Ethomeen® C/20. [Akzo] PEG-10 cocamine; CAS 61791-14-8; cationic; emulsifier, dispersant used in textile processing; Gardner 11 max. clear liq.; sol. in acetone, IPA, CCl_4, Stod., benzene, water (cloudy); sp.gr. 1.02; flash pt. (COC) 560 F; surf. tens. 39 dynes/cm (0.1%); Draves wetting 14 s (0.5%).

Ethomeen® C/25. [Akzo; Akzo Chem. BV] PEG-15 cocamine; CAS 61791-14-8; cationic; emulsifier, dispersant used in textile processing; Gardner 12 max. clear liq.; sol. in acetone, IPA, CCl_4, Stod., benzene, water; sp.gr. 1.04; HLB 19.3; flash pt. (COC) 500 F; surf. tens. 41 dynes/cm (0.1%); 100% conc.

Ethomeen® O/12. [Akzo] PEG-2 oleamine; cationic; emulsifier, dispersant used in textile processing; Gardner 8 max. clear liq.; sp.gr. 0.90; flash pt. (COC) 470 F; surf tens. 31.5 dynes/cm (0.1%); 100% conc.

Ethomeen® O/15. [Akzo] PEG-5 oleamine; cationic; emulsifier, dispersant for textile processing; Gardner 8 max. clear liq.; sp.gr. 0.96; flash pt. (COC) 540 F; surf tens. 35.3 dynes/cm (0.1%); 100% conc.

Ethomeen® O/25. [Akzo] PEG-15 oleamine; emulsifier, dispersant used in textile processing; Gardner 8 max. paste; sp.gr. 1.04; flash pt. (PM) 380 F.

Ethomeen® S/12. [Akzo; Akzo Chem. BV] PEG-2 soyamine; CAS 61791-24-0; cationic; emulsifier, dispersant used in textile processing; Gardner 14 max. clear heavy liq; sol. in acetone, IPA, CCl_4, Stod., benzene; insol. water; sp.gr. 0.91; HLB 14.6; flash pt. (TOC) 440 F; surf. tens. 31.4 dynes/cm (0.1%); 100% conc.

Ethomeen® S/15. [Akzo; Akzo Chem. BV] PEG-5 soyamine; CAS 61791-24-0; cationic; emulsifier, dispersant for textile processing; Gardner 14 max. clear heavy liq; sol. in acetone, IPA, CCl_4, Stod., benzene; forms gel or disp. in water; sp.gr. 0.95; HLB 19.0; flash pt. (TOC) 460 F; surf. tens. 33 dynes/cm (0.1%); Draves wetting 28 s (0.5%); 100% conc.

Ethomeen® S/20. [Akzo] PEG-10 soyamine; CAS 61791-24-0; cationic; emulsifier, dispersant used in textile processing; Gardner 14 max. clear heavy liq; sol. in acetone (cloudy), IPA, CCl_4, Stod., benzene, water; sp.gr. 1.02; flash pt. (COC) 540 F; surf. tens. 40 dynes/cm (0.1%); Draves wetting 29 s (0.5%).

Ethomeen® S/25. [Akzo; Akzo Chem. BV] PEG-15 soyamine; CAS 61791-24-0; cationic; emulsifier, dispersant used in textile processing; Gardner 14–18 max. clear heavy liq; sol. in acetone (cloudy), IPA, CCl_4, Stod., benzene, water; sp.gr. 1.04; HLB 9.8; flash pt. (COC) 540 F; surf. tens. 43 dynes/cm (0.1%); 100% conc.

Ethomeen® T/12. [Akzo; Akzo Chem. BV] PEG-2 tallow amine; CAS 61791-44-4; cationic; emulsifier, dispersant used in textile processing; Gardner 8 paste; sol. in acetone, IPA, CCl_4, Stod., benzene, water; sp.gr. 0.92; HLB 4.5; flash pt. (COC) 410 F; 100% conc.

Ethomeen® T/15. [Akzo; Akzo Chem. BV] PEG-5 tallow amine; CAS 61791-26-2; cationic; emulsifier, dispersant for textile processing; Gardner 8 max. paste; sol. cloudy in acetone, IPA, CCl_4, Stod., Benzene; forms gel in water; sp.gr. 0.97; flash pt. (PM) > 400 F; surf. tens. 34 dynes/cm (0.1%); Draves wetting 23 s (0.5%); 100% conc.

Ethomeen® T/25. [Akzo; Akzo Chem. BV] PEG-15 tallow amine; CAS 61791-26-2; cationic; emulsifier, dispersant used in textile processing; Gardner 8 max. clear liq.; sol. cloudy in acetone, IPA, CCl_4, benzene, water; sp.gr. 1.03; HLB 19.3; flash pt. (COC) > 500 F; surf. tens. 41 dynes/cm (0.1%); Draves wetting 53 s (0.5%); 100% conc.

Ethomeen® T/60. [Akzo] PEG-50 tallow amine; emulsifier, dispersant used in textile processing; Gardner 10 max. paste to solid; m.w. 2362–2562; sp.gr. 1.115; flash pt. > 400 F (PM).

Ethomeen® TD/15. [Akzo] Ethoxylated tert. amine from tallow fatty acid; emulsifier, dispersant used in textile processing; Gardner 4 max. clear liq.; sp.gr. 0.97; surf. tens. 35.8 dynes/cm (0.1%).

Ethomeen® TD/25. [Akzo] Ethoxylated tert. amine from tallow fatty acid; emulsifier, dispersant used in textile processing; Gardner 4 max. clear liq.; sp.gr. 1.03; surf. tens. 43.5 dynes/cm (0.1%).

Ethomid® HT/15. [Akzo Chem. BV] Ethoxylated hydrog. tallow amide; nonionic; emulsifier, dispersant, detergent; Gardner 10 max. solid; sol. in IPA; disp. in acetone, CCl_4, Stod., benzene, water; sp.gr. 1.03; HLB 13.5; surf. tens. 37 dynes/cm (0.1%); 100% conc.

Ethomid® HT/23. [Akzo] PEG-13 hydrog. tallow amide; CAS 68155-24-8; nonionic; emulsifier, dispersant, detergent, dye leveling agent; for silicone finishing agents, sizing lubricants; Gardner 8 max. solid; sp.gr. 1.028; cloud pt. 130–170 F (1%); flash pt. (PM) > 400 F; hyd. no. 105; surf. tens. 37 dynes/cm (0.1%); 100% conc.

Ethomid® HT/60. [Akzo; Akzo Chem. BV] PEG-50 hydrog. tallow amide; CAS 68155-24-8; nonionic; surfactant, emulsifier, sec. stabilizer for emulsion systems; dispersant, detergent; Gardner 11 max. hard solid; sol. in IPA, CCl_4, benzene, water; sp.gr. 1.14; HLB 19.0; hyd. no. 45; flash pt. (PMCC) 540 C; surf. tens. 47 dynes/cm (0.1%); 100% conc.

Ethomid® O/15. [Akzo Chem. BV] PEG-5 oleamide; CAS 31799-71-0; nonionic; emulsifier, dispersant, detergent; lubricant for syn. and natural fibers; Gardner 12 max. liq.; sol. in IPA, CCl_4, gels in water @ 50–80 C; sp.gr. 1.0; HLB 14.0; flash pt. (PM) 225 F; surf. tens. 35 dynes/cm (0.1%); 100% conc.

Ethomid® O/17. [Akzo] PEG-7 oleamide; CAS 26027-37-2; nonionic; emulsifier, dispersant, detergent; Gardner 8 max. liq.; sp.gr. 1.00; hyd. no. 110; flash pt. (PMCC) 107 C; surf. tens. 35 dynes/cm (0.1%); 100% conc.

Ethoquad® 18/12. [Akzo] PEG-2 stearmonium chloride and IPA; CAS 3010-24-0; 28724-32-5; cationic; antistat, emulsifier, dyeing assistant, leveling agent, antifoam used in textile industry, as electroplating bath additives; Gardner 6 max. paste; sol. in acetone, IPA, benzene, water; sp.gr. 0.919; flash pt. 71 F; surf. tens. 40.2 dynes/cm (0.1%); pH 7–8 (10% aq.); 75% act. in IPA.

Ethoquad® 18/25. [Akzo] PEG-15 stearmonium chloride; CAS 28724-32-5; cationic; antistat, emulsifier, dyeing assistant, leveling agent, antifoam used in textile industry, as electroplating bath additives; Gardner 11 max. clear liq.; sol. in acetone, IPA, benzene, water, CCl_4; sp.gr. 1.058; flash pt. (PM) 146 F; surf. tens. 50.1 dynes/cm (0.1%); pH 6–9 (10% aq.); 95% act.

Ethoquad® C/12. [Akzo] PEG-2 cocomonium chloride and IPA; CAS 70750-47-9; cationic; antistat, emulsifier, dyeing assistant, electroplating bath additive; Gardner 9 max. clear liq.; sol. in acetone, IPA, benzene, water, CCl_4; sp.gr. 0.969; flash pt. (PMCC) 20 C; surf. tens. 35.4 dynes/cm (0.1%); pH 7–8 (10% aq.); 75% act. in IPA.

Ethoquad® C/12 Nitrate. [Akzo] PEG-2 cocomethyl ammonium nitrate, IPA; CAS 71487-00-8; cationic; industrial surfactant for agric., textiles, protective coatings, inks, pigment dispersions, acid pickling baths, metalworking, electroplating, plastics mfg.; Gardner 8 max. liq.; sol. in water, alcohols, acetone, benzene, CCl_4, hexylene glycol; sp.gr. 0.975 (20 C); pH 6-7.8; flash pt. (PMCC) 20 C; 59-62% quat. in IPA.

Ethoquad® C/25. [Akzo] PEG-15 cocomonium chloride; CAS 61791-10-4; cationic; antistat, emulsifier, dyeing assistant, leveling agent, antifoam used in textile industry, as electroplating bath additives; Gardner 11 max. clear liq.; sol. in acetone, IPA, benzene, water, CCl_4; sp.gr. 1.071; flash pt. (PM) 196 F; pH 7–8 (10% aq.); surf. tens. 43.4 dynes/cm (0.1%); 95% act.

Ethoquad® CB/12. [Akzo] PEG-2 cocobenzonium chloride, IPA; CAS 61789-68-2; cationic; industrial surfactant for agric., textiles, protective coatings, inks, pigment dispersions, acid pickling baths, metalworking, electroplating, plastics mfg.; Gardner 12 max. liq.; sol. in water, alcohols, acetone, benzene, CCl_4, hexylene glycol; sp.gr. 0.970; flash pt. (PMCC) 26 C; pH 6-9; 73-77% solids in IPA.

Ethoquad® O/12. [Akzo] PEG-2 oleamonium chloride and IPA; CAS 18448-65-2; cationic; antistat, emulsifier, dyeing assistant, electroplating bath additive; Gardner 9 max. clear liq.; sol. in acetone, IPA, benzene, water, CCl_4; sp.gr. 0.932; flash pt. (PM) < 80 F; surf. tens. 40.3 dynes/cm (0.1%); pH 7–8 (10% aq.); 75% act. in IPA.

Ethoquad® O/25. [Akzo] PEG-15 oleamonium chloride; cationic; antistat, emulsifier, dyeing assistant, leveling agent, antifoam used in textile industry, as electroplating bath additives; Gardner 11 max. clear liq.; sol. in acetone, IPA, benzene, water, CCl_4; sp.gr. 1.062; flash pt. (PM) 200 F; surf. tens. 40.8 dynes/cm (0.1%); pH 7–9 (10% aq.); 95% act.

Ethoquad® T/13-50. [Akzo] PEG-3 tallow alkyl ammonium acetate, aq. IPA; industrial surfactant for agric., textiles, protective coatings, inks, pigment dispersions, acid pickling baths, metalworking, electroplating, plastics mfg.; EPA listed; Gardner 8 max. liq.; sol. in water, alcohols, acetone, benzene, CCl_4, hexylene glycol; sp.gr. 0.952 (20 C); flash pt. (PMCC) 18 C; 48-52% solids in aq. IPA.

Ethosperse® CA-2. [Lonza] Ceteth-2; nonionic; o/w emulsifier, thickener, stabilizer for hair care prods., antiperspirants; wh. solid; sol. in ethanol; disp. hot in water; HLB 6.0; m.p. 29–33 C; 100% conc.

Ethosperse® G-26. [Lonza] Glycereth-26; nonionic; emulsifier, humectant for cosmetic, pharmaceutical and industrial uses; pale straw liq.; sol. in water, methanol, ethanol, acetone, toluol; sp.gr. 1.12 (38 C); visc. 150 cps (38 C); HLB 18.0; 100% conc.

Ethosperse® LA-4. [Lonza] Laureth-4; CAS 9002-92-0; nonionic; emulsifier for cosmetic, pharmaceutical and industrial uses; pale straw liq.; sol. in methanol, ethanol, acetone, toluol, min. oil, misc. with water, veg. oils; sp.gr. 0.95; visc. 30 cps; HLB 9.5; 100% act.

Ethosperse® LA-12. [Lonza] Laureth-12; CAS 9002-92-0; nonionic; emulsifier for cosmetic, pharmaceutical and industrial uses; turbid liq. to soft paste; sol. see Ethosperse LA-4; sp.gr. 1.10; visc. 1000 cps; HLB 15.0; 100% act.

Ethosperse® LA-23. [Lonza] Laureth-23; CAS 9002-92-0; nonionic; o/w emulsifier, thickener, stabilizer for hair care prods., antiperspirants, pharmaceuticals, industrial use; wh. solid; sol. in water, hot ethanol; HLB 16.9; m.p. 30–45 C.; 100% conc.

Ethosperse® SL-20. [Lonza] Sorbeth-20; nonionic; emulsifier, humectant for in cosmetic, pharmaceutical and industrial uses; lt. yel. liq.; sol. in water, methanol, ethanol, acetone; sp.gr. 1.16; visc. 460 cps; HLB 16.6; 100% act.

Ethosperse® TDA-6. [Lonza] Trideceth-6; CAS 24938-91-8; nonionic; emulsifier for in cosmetic, pharmaceutical and industrial uses; pale straw liq.; sol. in methanol, ethanol, acetone, toluol, disp. in water; sp.gr. 0.98; visc. 80 cps; HLB 11.0; 100% act.

Ethotal CH 5. [Witco SA] Ethoxylated fatty acid; nonionic; detergent for washing machine use; wetting agent for insecticides, fungicides; liq.; 98% conc.

Ethox 25-R-8. [Ethox] Block polymer, 75% EO; dispersant, emulsifier and intermediate for esters; m.w. 9000; cloud pt. 45 C (1%).

Ethox 31-R-1. [Ethox] Block polymer, 10% EO; surfactant imparting foam control to dispersants; oil-sol.; m.w. 4000; cloud pt. 24 C (1%).

Ethox 90-R-4. [Ethox] Block polymer, 40% EO; surfactant for metalworking fluids; antimigrant for textile dyeing; m.w. 7000; cloud pt. 42 C (1%).

Ethox 1122. [Ethox] PEG-9 pelargonate; CAS 31621-91-7; high cohesion lubricant for syn. fiber prod. and processing; liq.; water-sol.; HLB 14.4.

Ethox 1212. [Ethox] Ethoxylated coconut glyceride; lubricant for textiles and metals; liq.; water-sol.; HLB 12.6.

Ethox 1345K. [Ethox] Formulated detergent; anionic/nonionic; detergent with exc. wetting props.; for textiles; liq.; cloud pt. > 100 C (1% aq.); 90% act.

Ethox 1358. [Ethox] Aliphatic phosphate ester; anionic; potassium salt as antistat and lubricant; liq.; 99% act.

Ethox 1372. [Ethox] Polyoxyalkylene fatty amine; low foam dye leveler for acid dyes; liq.; cloud pt. 61 C (1% aq.); 100% act.

Ethox 1437. [Ethox] Polyoxyalkylene glycol ether; low foam wetting agent and penetrant for textile processing; liq.; cloud pt. 20 C (1% aq.); 100% act.

Ethox 1449. [Ethox] Polyoxyalkylene glycol ether; wetting agent, detergent for low foam applics.; liq.; cloud pt. 16 C (1% aq.); 100% act.

Ethox 2156. [Ethox] Short chain triglyceride ester; lubricant for textiles and metals; liq.; water-insol.

Ethox 2191. [Ethox] Polyoxyalkylene glycol ether; component of low foam emulsifiers for solvs. and oils; good low foam wetting agent; liq.; cloud pt. 15 C (1% aq.); 100% act.

Ethox 2195. [Ethox] Aromatic phosphate ester; anionic; component of low foam solv. emulsification systems; salts are good dispersants; liq.; 99% act.

Ethox 2400. [Ethox] Polyoxyalkylene glycol ether; wetting agent effective at elevated temps.; liq.; cloud pt. 54 C (1% aq.); 100% act.

Ethox 2406. [Ethox] Polyoxyalkylene glycol ether; detergent for removal and dispersion of oils and waxes; intermediate for anionic systems; sl. higher water-sol. than Ethox LF-1226; liq.; cloud pt. 40 C (1% aq.); 100% act.

Ethox 2407. [Ethox] Polyoxyalkylene glycol ether; low foam detergent and rinse aid; for cleaning formulations, commercial dishwashing formulas; liq.; cloud pt. 30 C (1% aq.); 100% act.

Ethox 2423. [Ethox] PPG-25-laureth-25; nonionic; low foam detergent, wetting and rewetting agent, emulsifier for textiles; liq.; HLB 13.0; cloud pt. 40 C (1% aq.); 100% act.

Ethox 2440. [Ethox] Proprietary; low foam surfactant for use at lower temps.; liq.; cloud pt. 25 C (1% aq.); 100% act.

Ethox 2449. [Ethox] Modified cocamide DEA; lower foam for controlled suds detergents; liq.; cloud pt. > 100 C (1% aq.); 100% act.

Ethox 2483. [Ethox] Proprietary; low foam surfactant used for wetting, cleaning and rinsing at higher temps.; liq.; cloud pt. 40 C (1% aq.); 100% act.

Ethox 2610. [Ethox] PPG 1025 ditallate; nonionic; oil-in-oil dispersant and lubricant; for silicone and non-silicone defoamers; low foam additive for various formulations; liq.; water-insol.

Ethox 2648. [Ethox] Proprietary; formulated emulsifier for emulsification of triglycerides without coemulsifiers; liq.; 100% act.

Ethox 2650. [Ethox] Coco betaine; foaming agent and detergent; liq.; cloud pt. > 100 C (1% aq.); 35% act.

Ethox 2659. [Ethox] Ethoxylated styrenated phenol; dispersant for pigments and org. emulsions; higher HLB than Ethox 2938; liq.; cloud pt. > 100 C (1% aq.); 100% act.

Ethox 2672. [Ethox] Methoxy PEG 400 monolaurate; nonionic; low visc. lubricant; liq.; water-sol.; HLB 12.7.

Ethox 2684. [Ethox] Aliphatic phosphate ester; anionic; coupling agent for nonionic surfactants in alkali systems; liq.; 99% act.

Ethox 2928. [Ethox] Aliphatic phosphate ester; anionic; corrosion inhibitor for metalworking fluids; liq.; 75% act.

Ethox 2938. [Ethox] Ethoxylated styrenated phenol; dispersant for pigments and org. emulsions; liq.; cloud pt. 34 C (1% aq.); 100% act.

Ethox 2939. [Ethox] Proprietary; low foam detergent for metal cleaning; liq.; cloud pt. 47 C (1% aq.); 100% act.

Ethox 2966. [Ethox] PEG 400 sesquioleate; nonionic; emulsifier for oils, nonsilicone defoamers; liq.; water-disp.; HLB 9.8.

Ethox 2984. [Ethox] Stearic acid alkanolamide; dispersant, lubricant and softening agent; solid; cloud pt. < 25 C (1% aq.); 100% act.

Ethox 3092. [Ethox] Carboxylated alcohol ethoxylate, sodium salt; alkaline wetter; liq.; cloud pt. > 100 C (1% aq.); 70% act.

Ethox 3113. [Ethox] PPG-2 bisphenol A; monomer for polyester resins; liq.; 100% act.

Ethox CAM-2. [Ethox] PEG-2 cocamine; CAS 61791-

14-8; cationic; emulsifier, dispersant, textile dyeing assistant, lubricant; intermediate for amphoterics; liq.; oil-sol.; HLB 6.0; amine no. 295; 99% conc.

Ethox CAM-15. [Ethox] PEG-15 cocamine; CAS 61791-14-8; cationic; hydrophilic emulsifier, dispersant, textile dyeing assistant, lubricant, antistat; liq.; water-sol.; HLB 15.2; amine no. 867; 99% conc.

Ethox CO-5. [Ethox] PEG-5 castor oil; nonionic; emulsifier, lubricant; dispersant for pigments and clays; liq.; oil-sol.; water-insol.; HLB 4.0; 99% conc.

Ethox CO-16. [Ethox] PEG-16 castor oil; nonionic; emulsifier, lubricant with exc. softening and coating props.; liq.; water-disp.; HLB 8.6; 99% conc.

Ethox CO-25. [Ethox] PEG-25 castor oil; nonionic; emulsifier for natural and syn. oils, lubricant; liq.; water-disp.; HLB 10.8; 99% conc.

Ethox CO-30. [Ethox] PEG-30 castor oil; nonionic; emulsifier for natural fats and oils; dispersant for inorganics; stabilizer for latex emulsions; lubricant; liq.; water-sol.; HLB 11.8; 99% conc.

Ethox CO-36. [Ethox] PEG-36 castor oil; nonionic; emulsifier, lubricant, softener for textiles and leather; liq.; water-sol.; HLB 12.6; 99% conc.

Ethox CO-40. [Ethox] PEG-40 castor oil; nonionic; emulsifier, lubricant; liq.; sol. in cold water; HLB 13.0; 99% conc.

Ethox CO-81. [Ethox] PEG-80 castor oil; dyeing assistant and lubricant for textile fibers; liq.; water-sol.; HLB 15.9; 90% act.

Ethox CO-200. [Ethox] PEG-200 castor oil; nonionic; emulsifier, lubricant, antistat with good moisture regain; solid; water-sol.; HLB 18.1; 99% conc.

Ethox CO-200/50%. [Ethox] PEG-200 castor oil; nonionic; emulsifier, lubricant, antistat with good moisture regain; liq.; water-sol.; HLB 18.1; 50% conc.

Ethox COA. [Ethox] Cocamide DEA; foaming agent, detergent, and dispersant; liq.; cloud pt. > 25 C (1% aq.); 100% act.

Ethox DL-5. [Ethox] PEG 200 dilaurate; CAS 9005-02-1; lubricant and emulsifier for low foaming applics.; liq.; water-disp.; HLB 6.1.

Ethox DL-9. [Ethox] PEG 400 dilaurate; CAS 9005-02-1; lubricant and emulsifier for low foaming applics.; liq.; oil-sol.; water-disp.; HLB 10.8.

Ethox DO-9. [Ethox] PEG-8 dioleate; CAS 9005-07-6; nonionic; emulsifier for oils and solvs.; used for industrial lubricants; liq.; water-disp.; HLB 8.8; 99% conc.

Ethox DO-14. [Ethox] PEG 600 dioleate; CAS 9005-07-6; nonionic; oil and fat emulsifier and lubricant; liq.; water-disp.; HLB 10.3.

Ethox DT-15. [Ethox] PEG-15 tallow diamine; retarder for acid dyes; dispersant for use in acidic sol'ns.; liq.; water-sol.; HLB 13.1; amine no. 513.

Ethox DT-30. [Ethox] PEG-30 tallow diamine; dispersant and lubricant with anticorrosive props.; liq.; water-sol.; HLB 14.8; amine no. 830.

Ethox DTO-9A. [Ethox] PEG-8 ditallate; CAS 61791-01-3; nonionic; emulsifier for oils and solvs.; used for industrial lubricants; liq.; water-disp.; HLB 8.8; 99% conc.

Ethox HCO-16. [Ethox] PEG-16 hydrog. castor oil; CAS 61788-85-0; nonionic; heat-stable emulsifier for natural and syn. oils, lubricant; liq.; water-disp.; HLB 8.6; 99% conc.

Ethox HCO-25. [Ethox] PEG-25 hydrog. castor oil; CAS 61788-85-0; nonionic; emulsifier, lubricant for softeners; low volatility and unsaturation; liq.; water-disp.; HLB 10.8; 99% conc.

Ethox HCO-200/50%. [Ethox] PEG-200 hydrog. castor oil; CAS 61788-85-0; nonionic; emulsifier, lubricant; stabilizer and visc. control agent for emulsions; liq.; water-sol.; HLB 18.1; 99% conc.

Ethox HO-50. [Ethox] PEG-50 sorbitol hexaoleate; nonionic; emulsifier and lubricant for heat-stable systems; liq.; water-sol.; HLB 11.4.

Ethox L-61. [Ethox] Block polymer, 10% EO; low foam detergent, emulsifier, lubricant; intermediate for esters and polyesters; m.w. 2000; cloud pt. 24 C (1%).

Ethox L-62. [Ethox] Block polymer, 20% EO; low foam detergent, emulsifier, lubricant; intermediate for esters and polyesters; m.w. 2200; cloud pt. 32 C (1%).

Ethox L-121. [Ethox] Block polymer, 10% EO; emulsifier component for coatings resins; m.w. 6000; cloud pt. 14 C (1%).

Ethox L-122. [Ethox] Block polymer, 20% EO; low foam emulsifier for high m.w. reins; base for esters and functional fluids; m.w. 6000; cloud pt. 18 C (1%).

Ethox LF-1226. [Ethox] Polyoxyalkylene glycol ether; nonionic; detergent for removal and dispersion of oils and waxes; intermediate for anionic systems; liq.; cloud pt. 30 C (1% aq.); 100% act.

Ethox MA-8. [Ethox] PEG-8 monomerate; nonionic; cost-effective emulsifier, lubricant; liq.; water-disp.; HLB 11.7; 99% conc.

Ethox MA-15. [Ethox] PEG-15 monomerate; nonionic; emulsifier, lubricant; paste; water-sol.; HLB 13.7.

Ethox MI-9. [Ethox] PEG-9 isostearate; CAS 56002-14-3; emulsifier and lubricant; liq.; water-disp.; HLB 11.0.

Ethox MI-14. [Ethox] PEG-14 isostearate; CAS 56002-14-3; emulsifier and lubricant; liq.; water-sol.; HLB 13.0.

Ethox ML-5. [Ethox] PEG-5 laurate; CAS 9004-81-3; nonionic; lubricant, emulsifier, detergent, softener, coupling agent; visc. control agent for plastisol resins; liq.; water-disp.; HLB 9.3; 99% conc.

Ethox ML-9. [Ethox] PEG-9 laurate; CAS 9004-81-3; nonionic; hydrophilic surfactant, lubricant, emulsifier, detergent, softener; good leveling props. in aq. systems; liq.; water-sol.; HLB 13.1; 99% conc.

Ethox ML-14. [Ethox] PEG-14 laurate; CAS 9004-81-3; nonionic; lubricant, emulsifier, detergent, softener; good leveling in aq. systems; liq.; water-sol.; HLB 14.9; 99% conc.

Ethox MO-5. [Ethox] PEG-5 oleate; CAS 9004-96-0; emulsifier and lubricant; liq.; oil-sol.; water-disp.; HLB 8.1.

Ethox MO-9. [Ethox] PEG-9 oleate; CAS 9004-96-0; nonionic; lubricant, emulsifier for natural and syn. oils, detergent, softener; liq.; water-disp.; HLB 12.0; 99% conc.

Ethox MO-14. [Ethox] PEG-14 oleate; CAS 9004-96-0; nonionic; lubricant, emulsifier, detergent, softener; liq.; water-sol.; HLB 13.5; 99% conc.

Ethox MS-8. [Ethox] PEG-8 stearate; CAS 9004-99-3; nonionic; lubricant, wax and oil emulsifier, detergent, softener; for aq. processing; solid; water-disp.; HLB 11.4; 99% conc.

Ethox MS-14. [Ethox] PEG-14 stearate; CAS 9004-99-3; nonionic; lubricant, emulsifier, detergent, softener; solid; water-sol.; HLB 13.8; 99% conc.

Ethox MS-23. [Ethox] PEG-23 stearate; CAS 9004-99-3; nonionic; lubricant, emulsifier for glyceryl stearate, detergent, softener; thickener for aq. systems; solid; water-sol.; HLB 15.6; 99% conc.

Ethox MS-40. [Ethox] PEG-40 stearate; CAS 9004-99-3; nonionic; lubricant, emulsifier, detergent, softener; stabilizer for aq. dispersions; waxy solid or flake; sol. in water; ; HLB 17.2; 99% conc.

Ethox OAM-308. [Ethox] PEG-30 oleyl amine; cationic; emulsifier, dispersant, textile dyeing assistant and lubricant; leveling agent for dyeing nylon; antiprecipitant for cross dyeing; liq.; water-sol.; HLB 16.6; amine no. 1910; 80% conc.

Ethox PPG 1025 DTO. [Ethox] PPG-20 ditallate; nonionic; oil-in-oil dispersant and lubricant; liq.; 99% conc.

Ethox SAM-2. [Ethox] PEG-2 stearamine; cationic; dispersant, textile dyeing assistant, lubricant; emulsifier for wax systems; release agent; solid; insol. in water; HLB 4.9; amine no. 362; 99% conc.

Ethox SAM-10. [Ethox] PEG-10 stearamine; CAS 26635-92-7; cationic; moderate HLB emulsifier; component of corrosion inhibitors; liq.; water-sol.; HLB 12.3; amine no. 698.

Ethox SAM-50. [Ethox] PEG-50 stearamine; CAS 26635-92-7; cationic; emulsifier, dispersant, textile dyeing assistant, lubricant; leveling and dispersing agent for dyeing with acid, disperse, and cationic dyes; emulsifier and stabilizer for dispersions; solid; water-sol.; HLB 18.0; amine no. 2460; 99% conc.

Ethox SO-9. [Ethox] PEG-8 sesquioleate; nonionic; all-purpose oil emulsifier; liq.; HLB 9.8; 99% conc.

Ethox TAM-2. [Ethox] PEG-2 tallow amine; cationic; emulsifier, dispersant, textile dyeing assistant, lubricant; used for prep. of high pressure wax dispersions; liq.; water-insol.; HLB 4.2; amine no. 350; 99% conc.

Ethox TAM-5. [Ethox] PEG-5 tallow amine; cationic; emulsifier, dispersant, textile dyeing assistant, lubricant; coating agent; substantive to many inorganics; liq.; water-disp.; HLB 8.8; amine no. 487; 99% conc.

Ethox TAM-10. [Ethox] PEG-10 tallow amine; cationic; emulsifier; intermediate for amphoterics; liq.; water-sol.; HLB 12.5; amine no. 700.

Ethox TAM-15. [Ethox] PEG-15 tallow amine; cationic; emulsifier, dispersant, textile dyeing assistant, lubricant; intermediate for amphoteric surfactants; liq.; water-sol.; HLB 14.3; amine no. 927; 99% conc.

Ethox TAM-20. [Ethox] PEG-20 tallow amine; cationic; emulsifier, dispersant, textile dyeing assistant; lubricant, antistat for yarns; leveling agent for acid dyes on nylon; liq.; water-sol.; HLB 15.4; amine no. 1147; 99% conc.

Ethox TAM-20 DQ. [Ethox] PEG-20 tallow amine diethyl sulfate salt; lubricant, textile leveling agent, dispersant; liq.; cloud pt. > 100 C (1% aq.); 100% act.

Ethox TAM-25. [Ethox] PEG-25 tallow amine; cationic; emulsifier, dispersant, textile dyeing assistant, lubricant; leveling agent for nylon dyeing; antiprecipitant when cross dyeing; solid; water-sol.; HLB 16.0; amine no. 1367; 99% conc.

Ethox TO-8. [Ethox] PEG-8 tallate; CAS 61791-00-2; nonionic; lubricant, emulsifier, detergent, softener, dispersant; liq.; water-disp.; HLB 9.8; 99% conc.

Ethox TO-9A. [Ethox] PEG-9 tallate; CAS 61791-00-2; nonionic; cost-effective detergent, lubricant, emulsifier, softener; liq.; water-disp.; HLB 12.0; 99% conc.

Ethox TO-16. [Ethox] PEG-16 tallate; CAS 61791-00-2; nonionic; lubricant, coemulsifier for various solvs., detergent, softener; liq.; water-sol.; HLB 13.4; 99% conc.

Ethoxamine SF11. [Witco SA] Ethoxylated fatty amine; cationic; corrosion inhibitor, emulsifier, dispersant, antistat; liq.; HLB 12.8; 100% conc.

Ethoxylan® 1685. [Henkel/Cospha; Henkel/Textile; Henkel Canada] PEG-75 lanolin; CAS 61790-81-6; emollient, emulsifier, dispersant, foam stabilizer, resin plasticizer for cosmetic and pharmaceutical preparations, textile processing; Gardner < 11 solid; sol. @ 5% in water, IPA; dens. 9.6 lb/gal; m.p. 39 C; cloud pt. 85 C; flash pt. 530 F.

Ethoxylan® 1686. [Henkel/Cospha; Henkel/Textile; Henkel Canada] PEG-75 lanolin; CAS 61790-81-6; emollient, emulsifier, dispersant, foam stabilizer, resin plasticizer for cosmetic and pharmaceutical preparations, textile processing; Gardner < 10 liq.; sol. @ 5% in water, IPA; dens. 8.9 lb/gal; visc. 1767 cSt (100 F); pour pt. 1 C; cloud pt. 86 C; 50% aq. sol'n.

Ethsorbox L-20. [Ethox] PEG-20 sorbitan laurate; CAS 9005-64-5; nonionic; emulsifier for oils, solvs. and fats; lubricant for cotton and rayon; liq.; water-sol.; HLB 16.7.

Ethsorbox O-20. [Ethox] PEG-20 sorbitan oleate; CAS 9005-65-6; nonionic; emulsifier for industrial and textile lubricants; liq.; water-sol.; HLB 15.0.

Ethsorbox S-20. [Ethox] PEG-20 sorbitan stearate; CAS 9005-67-8; nonionic; waxy lubricant, softener; solid; water-sol.; HLB 15.2.

Ethsorbox TO-20. [Ethox] PEG-20 sorbitan trioleate; CAS 9005-70-3; nonionic; lubricant and emulsifier for industrial process fluids; liq.; water-disp.; HLB 11.1.

Ethsorbox TS-20. [Ethox] PEG-20 sorbitan tristearate; CAS 9005-71-4; nonionic; lubricant, softener for textile goods; emulsifier for fats and oils; solid; water-disp.; HLB 11.1.

Ethylan® 172. [Harcros UK] Cetoleth-3; nonionic; emulsifier for min. oils and hydrophobic waxes; lt. amber liq., mild fatty odor; oil-sol.; sp.gr. 0.920; visc. 57 cs; HLB 7.0; pour pt. 16 C; flash pt. (COC) > 175 C; pH 7.0 (1% aq.); 100% act.

Ethylan® 20. [Harcros UK] Nonoxynol-20; CAS 9016-45-9; nonionic; detergent, wetting agent, stabilizer, emulsifier, solubilizer in waxes, resins, hand cleaning gels, pesticides, latexes; wh. waxy solid; negligible odor; water sol.; sp.gr. 1.065 (40 C); visc. 160 cs (40 C); HLB 16.0; cloud pt. > 100 C (1% aq.); flash pt. (COC) > 200 C; pour pt. 30 C; pH 7.0 (1% aq.); surf. tens. 40 dynes/cm (0.1%); 100% act.

Ethylan® 44. [Harcros UK] Nonoxynol-4; CAS 9016-45-9; nonionic; emulsifier for paraffin hydrocarbons, silicones, min. oil, alkyd resins, etc.; formulation of degreasers, hand cleaning gels; straw clear liq.; negligible odor; water insol.; sp.gr. 1.023; visc. 340 cs; HLB 9; flash pt. (COC) > 200 C; pour pt. < 0 C; pH 7.0 (1% aq.); 100% act.

Ethylan® 55. [Harcros UK] Nonoxynol-5 (5.5 EO); CAS 9016-45-9; nonionic; emulsifier for hydrocarbons, min. oils, aliphatic solvs., silicones; formulation of degreasers and hand cleaning gels; coupling agent; straw clear liq., negligible odor; water insol.; sp.gr. 1.035; visc. 290 cs; HLB 10.5; flash pt. (COC) > 200 C; pour pt. < 0 C; pH 7.0 (1% aq.); 100% act.

Ethylan® 77. [Harcros UK] Nonoxynol-6 (6.5 EO); CAS 9016-45-9; nonionic; emulsifier for hydrocarbons, oils, waxes, silicones; formulation of degreasers, hand cleaning gels; acid wool scouring, dry cleaning; straw clear liq.; negligible odor; water disp.; sp.gr. 1.038; visc. 360 cs; HLB 10.9; pour pt. < 0 C; cloud pt. < 2 C (1% aq.); flash pt. (COC) > 200 C; pH 7.0 (1% aq.); 100% act.

Ethylan® A2. [Harcros UK] PEG-4 oleate; CAS 9004-96-0; nonionic; emulsifier for kerosene and min. oil, antifoam agent; lt. amber liq., mild fatty odor; disp. in water; sp.gr. 0.929; visc. 75 cs; HLB 7.0; flash pt. (COC) > 150 C; pH 7.0 (1% aq.); 100% act.

Ethylan® A3. [Harcros UK] PEG-6 oleate; CAS 9004-96-0; nonionic; emulsifier for kerosene, min. and veg. oils, antifoam, dispersant for industrial uses; lt. amber liq., mild fatty odor; disp. in water; sp.gr. 0.948; visc. 84 cs; HLB 8.9; flash pt. (COC) > 150 C; pH 7.0 (1% aq.); 100% act.

Ethylan® A4. [Harcros UK] PEG-8 oleate; CAS 9004-96-0; nonionic; emulsifier, antifoam agent, dispersing agent for industrial uses; plastics antistat; lt. amber liq., mild fatty odor; disp. in water; sp.gr. 0.963; visc. 120 cs; HLB 10.3; flash pt. (COC) > 150 C; pH 7.0 (1% aq.); 100% act.

Ethylan® A6. [Harcros UK] PEG-12 oleate; CAS 9004-96-0; nonionic; dispersing agent for industrial uses; emulsifier; antistat for plastics; lt. amber liq., mild fatty odor; sol. in water; sp.gr. 1.037; visc. 340 cs; HLB 12.3; cloud pt. 42 C (1% aq.); flash pt. (COC) > 175 C; pH 7.0 (1% aq.); 100% act.

Ethylan® A15. [Harcros UK] Cocamide DEA; nonionic; emulsifier used in industrial degreasing, metal cleaning; straw clear liq.; sp.gr. 1.010; visc. 880 cs; 99% act.

Ethylan® ABB10. [Harcros UK] Ethoxylated substituted aniline; nonionic; intermediate for dyestuff, pigments, general chemical mfg.; solid; 100% conc.

Ethylan® ABC20. [Harcros UK] Ethoxylated substituted aniline; nonionic; intermediate for dyestuff, pigments, general chemical mfg.; solid; 100% conc.

Ethylan® BAB20. [Harcros UK] Blocked alcohol ethoxylate; nonionic; low foaming surfactant, wetting agent for rinse aids for machine dishwashing, spray cleaning, metal cleaning; good stability on caustic powds.; biodeg.; opaque liq., faint odor; disp. in water; sp.gr. 1.032; visc. 38 cs (40 C); pour pt. 16 C; cloud pt. 16 C (1% aq.); flash pt. (PMCC) > 150 C; 100% act.

Ethylan® BBC31. [Harcros UK] Modified linear alcohol ethoxylate; nonionic; low foam iodophor intermediate; germicidal treatment of pipelines and equip. in dairy, brewery and other food processing industries; opaque liq., mild odor; sol. hazy in water; sp.gr. 0.985; visc. 80 cs; pour pt. 4 C; flash pt. (COC) > 175 C; pH 2.6 (1% aq.); low level of corrosive org. acid; toxicology: irritating to eyes and skin; 100% act.

Ethylan® BCD42. [Harcros UK] Modified phenol ethoxylate; nonionic; emulsifier for emulsion polymerization; yel. liq.; HLB 12.4; cloud pt. 55 C (1% aq.); surf. tens. 45 dynes/cm (0.1%); 100% act.

Ethylan® BCP. [Harcros UK] Nonoxynol-9; CAS 9016-45-9; nonionic; detergent, wetting agent, emulsifier for oil, antistat; wool scouring; insecticides and herbicides; colorless clear liq., negligible odor; water sol.; sp.gr. 1.058; visc. 320 cs; HLB 12.9; pour pt. 0 C; cloud pt. 50 C (1% aq.); flash pt. (COC) > 200 C; pH 7.0 (1% aq.); 100% act.

Ethylan® BKL130. [Harcros UK] Alkylaryl ethoxylate; nonionic; wetting agent, dispersant for mfg. of pigment pastes and slurries, optical brightening agents, anthraquinone; latex stabilizer and emulsifier; yel. liq.; sol. in water; sp.gr. 1.086; visc. 960 cs; HLB 13.0; pour pt. < 0 C; cloud pt. 71 C (1% aq.); flash pt. (PMCC) > 100 C; pH 7.0 (1/% aq.); surf. tens. 34 dynes/cm (0.1%); usage level: 5-20%; 85% act.

Ethylan® BNE15. [Harcros UK] Beta-naphthol ethoxylate; nonionic; intermediate for phosphation, sulfation; liq.; 100% conc.

Ethylan® BV. [Harcros UK] Nonoxynol-14; CAS 9016-45-9; nonionic; foam stabilizer and booster, solubilizer, emulsifier used in pesticides, perfumes, emulsion polymerization; EPA approved; wh. soft paste, negligible odor; water sol.; sp.gr. 1.083; visc. 380 cs; HLB 14.5; cloud pt. 90 C (1% aq.); flash pt. > (COC) > 200 C; pour pt. 20 C; pH 7.0 (1% aq.); surf. tens. 37 dynes/cm (0.1%); 100% act.

Ethylan® BZA. [Harcros UK] Modified alkylphenol ethoxylate; nonionic; low-foam detergent for mechanical dishwashing, metal cleaning; liq.; cloud pt. 18 C; 100% act.

Ethylan® C12AH. [Harcros UK] Castor oil ethoxylate; nonionic; general emulsifier for oils; mold release agent for metal forming; liq.; HLB 7.0; 100% conc.

Ethylan® C30. [Harcros UK] Castor oil ethoxylate; nonionic; emulsifier for chlorinated solvs. and veg. oils; liq.; HLB 11.6; 100% conc.

Ethylan® C35. [Harcros UK] Castor oil ethoxylate; nonionic; emulsifier for chlorinated solvs., veg. oils, agrochemicals; fiber lubricant component; liq.; HLB 12.4; 100% conc.

Ethylan® C40AH. [Harcros UK] PEG ester of unsat. fatty acid; nonionic; emulsifier for chlorinated solv., olein, veg. oils, and wax, pesticides; dye leveling agent; solubilizer for perfumes; pigment dispersant and grinding aid; EPA approved; yel. hazy liq., mild fatty odor; water-sol.; sp.gr. 0.996; visc. 357 cs (40 C); HLB 13.5; pour pt. 11 C; cloud pt. 95 C (1% aq.); flash pt. (COC) 150 C; pH 7.0 (1% aq.); 100% act.

Ethylan® C75AH. [Harcros] PEG ester of unsat. fatty acid; nonionic; emulsifier for oils; fiber lubricant; cream waxy solid, mild fatty odor; water-sol.; sp.gr. 1.022 (60 C); visc. 259 cs (60 C); HLB 15.6; pour pt. 28 C; cloud pt. > 100 C (1% aq.); flash pt. (COC) > 175 C; pH 7.0 (1% aq.); 100% act.

Ethylan® C160. [Harcros UK] PEG ester of unsat. fatty acid; nonionic; oil emulsifier for cosmetics and pharmaceutical creams; fiber lubricant; cream waxy solid, mild fatty odor; water-sol.; sp.gr. 1.080 (60 C); visc. 300 cs (60 C); HLB 17.6; pour pt. 39 C; cloud pt. > 100 C (1% aq.); flash pt. (COC) > 175 C; pH 7.0 (1% aq.); 100% act.

Ethylan® C404. [Harcros UK] Castor oil ethoxylate; nonionic; emulsifier for chlorinated solvs., veg. oils, plant protection prods.; liq.; HLB 13.5; 96% conc.

Ethylan® CD103. [Harcros UK] Isodeceth-3.2; nonionic; wetting agent, emulsifier, detergent; coemulsifier for min. oils and waxes, alkyd resins, paraffin hydrocarbons; colorless clear liq., char. fatty alcohol odor; insol. in water; sp.gr. 0.960; visc. 36 cs; HLB 9.4; pour pt. < -10 C; flash pt. (COC) 156 C; pH 7.0 (1% aq.); 100% conc.

Ethylan® CD107. [Harcros] Isodeceth-6.8; nonionic; wetting agent, emulsifier, detergent; emulsifier for paraffin waxes, hydrocarbon solvs.; base for toxicant emulsifiers; textile scouring and antistat agent; specialty cleaners; dye leveling agent; detergent sanitizers; essential oil solubilizer; colorless sl. hazy liq., char. fatty alcohol odor; sol. in water; sp.gr.

1.006; visc. 66 cs; HLB 13.1; pour pt. -7 C; cloud pt. 55 C (1% aq.); flash pt. (COC) > 210 C; pH 7.0 (1% aq.); 96% act.

Ethylan® CD109. [Harcros UK] Isodeceth-9; nonionic; wetting agent, emulsifier, detergent; sudsing agent for liq. detergents; scouring and dye leveling agent; coemulsifier for aromatic solvs., waxes, toxicants; detergent sanitizer; essential oil solubilizer; wh. opaque liq., char. fatty alcohol odor; sol. in water; sp.gr. 1.025; visc. 87 cs; HLB 14.3; pour pt. 4 C; cloud pt. 83 C (1% aq.); flash pt. (COC) > 210 C; pH 7.0 (1% aq.); 97% act.

Ethylan® CD122. [Harcros UK] C12 fatty alcohol ethoxylate; nonionic; emulsifier, wetting agent, detergent; biodeg.; liq./solid; 100% conc.

Ethylan® CD123. [Harcros UK] C12 fatty alcohol ethoxylate; nonionic; emulsifier, wetting agent, detergent; coemulsifier for min. oils and waxes, alkyd resins, paraffin hydrocarbons; biodeg.; colorless clear liq., char. odor; insol. in water; sp.gr. 0.937; visc. 145 cs; HLB 8.4; pour pt. < 0 C; cloud pt. 53 C (10%); flash pt. (PMCC) > 150 C; pH 7.0 (1% aq.); 100% conc.

Ethylan® CD124. [Harcros UK] C12 fatty alcohol ethoxylate; nonionic; emulsifier, wetting agent, detergent; coemulsifier for min. oils, waxes, alkyds, paraffin hydrocarbons; base for toxicant emulsifier; textile scouring; detergent sanitizer; biodeg.; colorless clear liq., char. odor; disp. in water; sp.gr. 0.966; visc. 55 cs; HLB 10.4; pour pt. 2 C; cloud pt. 64 C (10%); flash pt. (PMCC) > 150 C; pH 7.0 (1% aq.); 100% act.

Ethylan® CD127. [Harcros UK] C12 fatty alcohol ethoxylate; nonionic; emulsifier, wetting agent, detergent; emulsifier for paraffin waxes, hydrocarbon solvs.; base for toxicant emulsifier; textile scouring; detergent sanitizers; biodeg.; colorless clear liq., char. odor; sol. in water; sp.gr. 1.002; visc. 120 cs; HLB 12.4; pour pt. < 0 C; cloud pt. 42 C (1% aq.); flash pt. (PMCC) > 95 C; pH 7.0 (1% aq.); 100% conc.

Ethylan® CD128. [Harcros UK] C12 fatty alcohol ethoxylate; nonionic; emulsifier, wetting agent, detergent; emulsifier for paraffin waxes, hydrocarbon solvs.; base for toxicant emulsifier; textile scouring; detergent sanitizers; biodeg.; colorless clear liq., char. odor; sol. in water; sp.gr. 1.014; visc. 135 cs; HLB 13.1; pour pt. < 0 C; cloud pt. 55 C (1% aq.); flash pt. (PMCC) > 95 C; pH 7.0 (1% aq.); 95% act.

Ethylan® CD129. [Harcros UK] C12 fatty alcohol ethoxylate; nonionic; emulsifier, wetting agent, detergent; biodeg.; liq./solid; 100% conc.

Ethylan® CD175. [Harcros] Linear alkoxylate; nonionic; low foaming detergent, wetting agent for rinse aids, dishwashing, metal cleaning, spray cleaning, bottle washing; colorless clear liq., mild odor; water-sol.; sp.gr. 1.024; visc. 117 cs; pour pt. -1 C; cloud pt. 44 C (1% aq.); flash pt. (COC) > 200 C; pH 6.5 (1% aq.); 100% act.

Ethylan® CD802. [Harcros UK] Branched chain alcohol ethoxylate; nonionic; general purpose hydrophobic emulsifier; liq.; HLB 8.0; 100% conc.

Ethylan® CD913. [Harcros UK] Syn. lower fraction primary alcohol EO condensate (2.9 EO); nonionic; detergent, o/w and w/o emulsifier, wetting agent, solubilizer; emulsifier for min. oils and waxes, alkyd resins, paraffinic hydrocarbons; colorless clear liq., char. fatty odor; water insol.; sp.gr. 0.943; visc. 30 cs; HLB 8.8; flash pt. (COC) 143 C; pour pt. < 0 C; pH 7.0 (1% aq.); 97% act.

Ethylan® CD916. [Harcros UK] Syn. lower fraction primary alcohol EO condensate (6.5 EO); nonionic; coemulsifier for min. oils and waxes; emulsifier for alkyd resins, paraffinic hydrocarbons; scouring and wetting agent, textile antistat, detergent component; colorless clear liq.; char. fatty alcohol odor; water sol.; sp.gr. 0.991; visc. 53 cs; HLB 12.8; cloud pt. 58 C (1% aq.); flash pt. (COC) 179 C; pour pt. 0 C; pH 7.0 (1% aq.); 96% act.

Ethylan® CD919. [Harcros UK] Syn. lower fraction primary alcohol EO condensate (9 EO); nonionic; wetting agent, detergent, scouring agent, dye leveling; coemulsifier for aromatic solvs., waxes, toxicants; detergent sanitizers; colorless clear liq.; char. fatty alcohol odor; water sol.; sp.gr. 1.006; visc. 67 cs; HLB 14.4; pour pt. 7 C; cloud pt. 82 C (1% aq.); flash pt. (COC) 196 C; pH 7.0 (1% aq.); 97% act.

Ethylan® CD964. [Harcros UK] Linear alkoxylate; nonionic; low foaming detergent, wetting agent for rinse aids, dishwashing, metal cleaning, spray cleaning, bottle washing; wool scouring agent; colorless clear liq., mild odor; water-sol.; sp.gr. 0.995; visc. 60 cs; pour pt. < -10 C; cloud pt. 30 C (1% aq.); flash pt. (COC) > 200 C; pH 7.0 (1% aq.); 100% act.

Ethylan® CD1210. [Harcros UK] C12 fatty alcohol ethoxylate; nonionic; emulsifier, wetting agent, detergent, foaming agent; textile scouring agent; coemulsifier for aromatic solvs., waxes, toxicants; detergent sanitizers; essential oil solubilizer; emulsion polymerization; biodeg.; colorless clear liq., char. odor; sol. in water; sp.gr. 1.030; visc. 165 cs; HLB 14.1; pour pt. < 0 C; cloud pt. 82 C (1% aq.); flash pt. (PMCC) > 95 C; pH 7.0 (1% aq.); surf. tens. 32 dynes/cm (0.1%); 90% act.

Ethylan® CD1230. [Harcros UK] C12 fatty alcohol ethoxylate; nonionic; emulsifier, wetting agent, detergent; biodeg.; liq./solid; 100% conc.

Ethylan® CD1260. [Harcros UK] C12 fatty alcohol ethoxylate; nonionic; emulsifier, wetting agent, detergent; emulsion polymerization; wh. flake; HLB 18.6; cloud pt. > 100 C (1% aq.); surf. tens. 50 dynes/cm (0.1%); 100% conc.

Ethylan® CD4511. [Harcros UK] C14-15 fatty alcohol ethoxylate; nonionic; emulsifier for fatty acids, waxes, essential oils, textiles, latex coemulsifier and stabilizer; waxy solid; HLB 13.7; 100% conc.

Ethylan® CD9112. [Harcros UK] Syn. lower fraction primary alcohol EO condensate (12 EO); nonionic; wetting agent and detergent; scouring and dye leveling agent; coemulsifier for aromatic solvs., waxes, toxicants; detergent sanitizers; emulsion polymerization; essential oil solubilizer; wh. waxy solid; char. fatty alcohol odor; water sol.; sp.gr. 1.029 (40 C); visc. 53 cs (40 C); HLB 15.5; pour pt. 23 C; cloud pt. 100 C (1% aq.); flash pt. (COC) > 200 C; pH 7.0 (1% aq.); surf. tens. 35 dynes/cm (0.1%); 100% act.

Ethylan® CD9130. [Harcros UK] Syn. lower fraction primary alcohol EO condensate; nonionic; o/w emulsifier, wetting agent; wh. waxy solid; char. fatty alcohol odor; water sol.; visc. 62 cs (60 C); HLB 17.8; flash pt. > 400 F (COC); pour pt. 45 C; pH 6–8 (1% aq.); 100% act.

Ethylan® CDP2. [Harcros] Linear middle fraction fatty alcohol ethoxylate (2 EO); nonionic; surfactant, intermediate for toiletry grade sulfates and other specialty surfactants; coemulsifier for min. oils, waxes, alkyd resins, paraffinic hydrocarbons; colorless clear liq., faint odor; insol. in water; sp.gr. 0.907; HLB 6.2; pour pt. 4 C; flash pt. (COC) > 150 C; 100% act.

Ethylan® CDP3. [Harcros] Linear middle fraction fatty alcohol ethoxylate (3 EO); nonionic; surfactant, intermediate for toiletry grade sulfates and other specialty surfactants; coemulsifier for min. oils, waxes, alkyd resins, paraffinic hydrocarbons; colorless clear liq., faint odor; insol. in water; sp.gr. 0.902; HLB 8.0; pour pt. 3 C; flash pt. (COC) > 150 C; 100% act.

Ethylan® CDP16. [Harcros UK] C12-14 fatty alcohol ethoxylate; nonionic; emulsifier for waxes, essential oils; latex coemulsifier and stabilizer; waxy solid; 100% conc.

Ethylan® CF71. [Harcros UK] Coconut fatty ester; nonionic; emulsifier for cosmetics, pharmaceuticals, fiber lubricant; pale straw liq., mild fatty odor; sol. in water; sp.gr. 1.050; visc. 55 cs (40 C); HLB 14; pour pt. 14 C; cloud pt. 57 C (1% aq.); flash pt. (COC) > 175 C; pH 7.0 (1% aq.); 100% act.

Ethylan® CH. [Harcros UK] Ethoxylated coconut fatty acid alkylolamide; nonionic; foam stabilizer in liq. detergents, general purpose or hard surf. cleaners, shampoos, coemulsifier; amber clear liq.; mild odor; water sol.; sp.gr. 1.033; visc. 250 cs; pour pt. 14 C; cloud pt. 80 C (1% aq.); flash pt. (COC) > 150 C; pH 8.0 (1% aq.); 100% act.

Ethylan® CPG 630. [Harcros UK] Modified alcohol ethoxylate; nonionic; low foaming wetting agent and detergent for rinse aids, machine dishwashing, metal cleaning, iodophor formulations; colorless clear liq.; mild odor; water sol.; sp.gr. 0.984; visc. 64 cs; pour pt. -1 C; cloud pt. 38 C (1% aq.); flash pt. (COC) > 175 C; pH 6.8 (1% aq.); 99% act.

Ethylan® CPG 660. [Harcros UK] Modified alcohol ethoxylate; nonionic; low foaming wetting agent and detergent for rinse aids, machine dishwashing, spray cleaning, iodophor formulations; colorless clear liq., mild odor; water-sol.; sp.gr. 0.988; visc. 114 cs; pour pt. -10 C; cloud pt. 29 C (1% aq.); flash pt. (COC) > 175 C; pH 6.8 (1% aq.); 99% act.

Ethylan® CPG 745. [Harcros UK] Modified alcohol ethoxylate; nonionic; low foaming wetting agent, detergent for use in rinse aids, machine dishwashing, spray cleaning, iodophor formulations; water sol.; sp.gr. 0.985; visc. 84 cs; cloud pt. 37 C (1% aq.); flash pt. > 350 F (COC); pH 6–7.5 (1% aq.); 99% act.

Ethylan® CPG 945. [Harcros UK] Modified alcohol ethoxylate; nonionic; low foaming surfactant, wetting agent for rinse aids, dishwashing, metal cleaning, spray cleaning, bottle washing; colorless clear liq., mild odor; water-sol.; sp.gr. 1.005; visc. 108 cs; pour pt. 2 C; cloud pt. 47 C (1% aq.); flash pt. (COC) > 175 C; pH 6.8 (1% aq.); 99% act.

Ethylan® CPG 7545. [Harcros UK] Modified alcohol ethoxylate; nonionic; low foam wetting agent, detergent for rinse aids, machine dishwashing, spray cleaning, iodophor formulations; colorless clear liq., mild odor; water-sol.; sp.gr. 1.001; visc. 114 cs; pour pt. -6 C; cloud pt. 37 C (1% aq.); flash pt. (COC) > 175 C; pH 6.8 (1% aq.); 99% act.

Ethylan® CRS. [Harcros UK] Ethoxylated coconut fatty acid alkylolamide; nonionic; foam stabilizer in liq. detergent, general purpose, or hard surface cleaners and shampoos, coemulsifier; amber clear liq.; mild odor; water sol.; sp.gr. 1.042; visc. 270 cs; pour pt. 10 C; flash pt. (COC) > 150 C; pH 8.0 (1% aq.); 100% act.

Ethylan® CS20. [Harcros UK] Cetyl/stearyl alcohol ethoxylate; nonionic; emulsifier and stabilizer for perfumes, essential oils, waxes; latex stabilizer and coemulsifier; waxy solid; HLB 15.5; 100% conc.

Ethylan® CX138. [Harcros] Branched chain fatty alcohol ethoxylate; nonionic; detergent, emulsifier, wetting agent for hard surf. and specialty cleaners, solv. degreasers; toxicant emulsifier and agric. spray additive; EPA approved; soft paste; sp.gr. 0.9488; visc. 115 cs; pour pt. 12 C; cloud pt. 56 C (1% aq.); flash pt. (COC) > 200 C; pH 6.8 (1% aq.); 100% act.

Ethylan® CX308. [Harcros UK] Branched chain ethoxylate; nonionic; emulsifier and solubilizer for essential oils, waxes, etc.; emulsifier for plant protection chemicals; liq.; HLB 13.0; 100% conc.

Ethylan® D252. [Harcros UK] Syn. primary alcohol ethoxylate (2 EO); nonionic; o/w and w/o emulsifier for min. oils and waxes, alkyd resins, paraffin hydrocarbons; colorless clear liq., faint odor; oil sol.; sp.gr. 0.903; visc. 31 cs; HLB 5.6; pour pt. 5 C; flash pt. (COC) 155 C; pH 7.0 (1% aq.); 100% act.

Ethylan® D253. [Harcros UK] Syn. primary alcohol ethoxylate (3 EO); nonionic; o/w and w/o emulsifier for min. oils and waxes, visc. depressant for PVC plastisols; intermediate for sulfation; colorless clear liq.; faint odor; oil sol.; sp.gr. 0.920; visc. 35 cs; HLB 7.8; pour pt. 3 C; flash pt. (COC) 168 C; pH 6–8 (1% aq.); 100% act.

Ethylan® D254. [Harcros UK] Syn. primary alcohol ethoxylate (4.5 EO); nonionic; o/w emulsifier for paraffin waxes and hydrocarbon solvs.; detergent, wetting agent; base for toxicant emulsifiers; scouringagent; textile antistat; specialty cleaners; colorless clear liq.; faint odor; oil sol.; sp.gr. 0.948; visc. 45 cs; HLB 9.8; pour pt. 5 C; flash pt. (COC) 182 C; pH 7.0 (1% aq.); 97% act.

Ethylan® D256. [Harcros UK] Syn. primary alcohol ethoxylate (6 EO); nonionic; o/w emulsifier for paraffin waxes and hydrocarbon solvs.; detergent, wetting agent; base for toxicant emulsifiers; scouring agent; textile antistat; specialty cleaners; hazy liq.; faint odor; water-sol.; sp.gr. 0.964; visc. 105 cs; HLB 11.4; cloud pt. 43 C (1% aq.); pour pt. 15 C; flash pt. (COC) 190 C; pH 7.0 (1% aq.); 100% act.

Ethylan® D257. [Harcros UK] Syn. primary alcohol ethoxylate; nonionic; o/w emulsifier for waxes, detergent, wetting agent; hazy liq.; faint odor; water sol.; sp.gr. 0.974 (40 C); visc. 31 cs (40 C); HLB 12.2; cloud pt. 49 C (1% aq.); flash pt. 385 F (COC); pour pt. 21 C; pH 6–8 (1% aq.); 100% act.

Ethylan® D259. [Harcros UK] Syn. primary alcohol ethoxylate; nonionic; detergent, wetting agent, emulsifier, solubilizer for wax, oils; wh. solid; water sol.; sp.gr. 0.988; visc. 36 cs (40 C); cloud pt. 76 C; 100% act.

Ethylan® D2512. [Harcros UK] Syn. primary alcohol ethoxylate (12 EO); nonionic; o/w emulsifier, detergent, wetting agent, solubilizer; for liq. detergents; scouring and dye leveling agent; coemulsifier for aromatic solvs., waxes, toxicants; detergent sanitizer; solubilizer for essential oils; wh. waxy solid; faint odor; water sol.; sp.gr. 1.002 (40 C); visc. 82 cs (40 C); HLB 14.2; cloud pt. 92 C (1% aq.); pour pt. 29 C; flash pt. (COC) > 200 C; pH 7.0 (1% aq.); 100% act.

Ethylan® D2560 Flake. [Harcros UK] Syn. primary alcohol ethoxylate (60 EO); nonionic; emulsifier, wetting agent; latex additive; toilet block component; wh. flake; faint odor; water sol.; visc. 179 cs (60 C); HLB 18.6; cloud pt. > 100 C (1% aq.); pour pt. 47 C; flash pt. (COC) > 200 C; pH 7.0 (1% aq.); 100% act.

Ethylan® DNP16. [Harcros UK] Dinonylphenol

ethoxylate; nonionic; emulsifier, wetting agent; textile auxiliary; visc. liq.; HLB 13.4; 100% conc.

Ethylan® DP. [Harcros UK] Nonoxynol-12; CAS 9016-45-9; nonionic; foam stabilizer and booster, emulsifier, solubilizer for essential oils, perfumes; in liq. detergents, pesticides, emulsion polymerization; slight hazy visc. liq.; negligible odor; water sol.; sp.gr. 1.068; visc. 400 cs; HLB 14.0; pour pt. 13 C; cloud pt. 80 C (1% aq.); flash pt. (COC) > 200 C; pH 7.0 (1% aq.); surf. tens. 34 dynes/cm (0.1%); 100% act.

Ethylan® ENTX. [Harcros UK] Alkylphenol ethoxylate; nonionic; emulsifier for min. oil and wax, mastic plasticizer, wetting agent, dispersant for org. pigments; pale straw clear liq.; faint odor; oil sol.; sp.gr. 1.001; visc. 503 cs; HLB 9.0; flash pt. (COC) > 175 C; pour pt. 2 C; pH 7.0 (1% aq.); 100% act.

Ethylan® FO30. [Harcros UK] Fish oil ethoxylate; nonionic; component of oil spill dispersant, antifoams; liq.; HLB 6.0; 100% conc.

Ethylan® FO60. [Harcros UK] Fish oil ethoxylate; nonionic; component of oil spill dispersant, antifoams; liq.; HLB 10.7; 100% conc.

Ethylan® GD. [Harcros UK] Special fatty acid DEA; nonionic; emulsifier for oils and metal degreasing agents, antistat, lubricant; clear amber liq.; mild odor; oil sol.; sp.gr. 0.982; visc. 1200 cs; flash pt. > 300 F (COC); pour pt. 10 C; pH 8–11 (1% aq.); 75% act.

Ethylan® GEL2. [Harcros UK] Polysorbate 20; CAS 9005-64-5; nonionic; w/o emulsifier, solubilizer esp. with sorbitan esters; used in cosmetics, agric., perfumes, fiber and textile lubricants, textile antistats, polymer additives, suspension and emulsion polymerization; amber clear liq., mild odor; sol. in water, alcohols, hydrocarbons; sp.gr. 1.10; visc. 350 cs; ; HLB 16.5; pour pt. -10 C; flash pt. (PMCC) > 150 C; 97% act.

Ethylan® GEO8. [Harcros UK] Polysorbate 80; CAS 9005-65-6; nonionic; emulsifier for cosmetics, pharmaceuticals, agrochem. formulations, textile lubricants, plastic additives, emulsion and suspension polymerization; solubilizer for perfume, flavors, essential oils; amber clear liq., mild fatty odor; sol. in water, alcohols, hydrocarbons; sp.gr. 1.10; visc. 720 cs; HLB 15.0; pour pt. -20 C; flash pt. (PMCC) > 150 C; 97% act.

Ethylan® GEO81. [Harcros UK] Polysorbate 81; CAS 9005-65-6; nonionic; emulsifier for cosmetics, agric., plastic additives, textile fiber lubricants and softeners, suspension and emulsion polymerization; solubilizer for perfume, flavors, essential oils; amber clear liq., mild fatty odor; sol. in alcohols, hydrocarbons; disp. in water; sp.gr. 1.15; visc. 465 cs; HLB 10.0; pour pt. -10 C; flash pt. (PMCC) > 150 C; 100% act.

Ethylan® GEP4. [Harcros UK] Polysorbate 40; CAS 9005-66-7; nonionic; w/o emulsifier, solubilizer esp. with sorbitan esters; used in cosmetics, agric., perfumes, fiber and textile lubricants, textile antistats, polymer additives; liq.; HLB 15.5; 100% conc.

Ethylan® GES6. [Harcros UK] Polysorbate 60; CAS 9005-67-8; nonionic; general purpose, low toxicity emulsifier for cosmetics and agrochem.; textile lubricant; plastics additive; emulsion and suspension polymerization; solubilizer for perfume, flavors, essential oils; pale brn. liq./paste; sol. in water, alcohols, hydrocarbons; sp.gr. 1.07; visc. 190 cs (40 C); HLB 15.0; pour pt. 22 C; flash pt. (PMCC) > 150 C; 97% act.

Ethylan® GL20. [Harcros UK] Sorbitan laurate; CAS 1338-39-2; nonionic; emulsifier for cosmetics, pharmaceuticals, agric., plastic antifog, textile fiber lubricant/softener, suspension and emulsion polymerization; amber visc. liq., mild odor; sol. in alcohols, hydrocarbons, natural and paraffinic oils; sp.gr. 1.04; visc. 5250 cs; HLB 8.0; pour pt. 15 C; flash pt. (PMCC) > 150 C; 100% act.

Ethylan® GLE-21. [Harcros UK] Polysorbate 21; CAS 9005-64-5; nonionic; w/o emulsifier, solubilizer esp. with sorbitan esters; used in cosmetics, agric., perfumes, fiber and textile lubricants, textile antistats, polymer additives; liq.; HLB 13.3; 100% conc.

Ethylan® GMF. [Harcros UK] Alkylphenol ethoxylate; nonionic; wetting agent, emulsifier for hydrocarbon solv., emulsion polymerization; straw clear liq.; faint odor; water sol.; sp.gr. 1.050; visc. 330 cs; HLB 12.5; cloud pt. 47 C (1% aq.); flash pt. > 350 C (COC); pour pt. 7 C; pH 6–7.5 (1% aq.); 100% act.

Ethylan® GO80. [Harcros UK] Sorbitan oleate; CAS 1338-43-8; nonionic; emulsifier for cosmetics, pharmaceuticals, agric., plastic antifog, textile fiber lubricant/softener, suspension and emulsion polymerization; amber visc. liq., mild fatty odor; sol. in alcohols, hydrocarbons, natural and paraffinic oils; sp.gr. 1.00; visc. 1100 cs; HLB 4.3; pour pt. -20 C; flash pt. (PMCC) -5 C; 100% act.

Ethylan® GOE-21. [Harcros UK] Polysorbate 81; CAS 9005-65-6; nonionic; w/o emulsifier, solubilizer esp. with sorbitan esters; used in cosmetics, agric., perfumes, fiber and textile lubricants, textile antistats, polymer additives; liq.; HLB 10.0; 100% conc.

Ethylan® GP-40. [Harcros UK] Sorbitan palmitate; CAS 26266-57-9; nonionic; surfactant for use as antistat and textile fiber lubricant; solid; HLB 6.7; 100% conc.

Ethylan® GPS85. [Harcros UK] Polysorbate 85; CAS 9005-70-3; nonionic; emulsifier for cosmetics, agric., plastic additives, textile fiber lubricants and softeners, suspension and emulsion polymerization; solubilizer for perfume, flavors, essential oils; amber clear liq., mild fatty odor; sol. in alcohols, hydrocarbons; disp. in water; sp.gr. 1.15; visc. 270 cs; HLB 11.0; pour pt. -20 C; flash pt. (PMCC) > 150 C; 100% act.

Ethylan® GS60. [Harcros UK] Sorbitan stearate; CAS 1338-41-6; nonionic; emulsifier for cosmetics, pharmaceuticals, agric., plastic antifog, textile fiber lubricant/softener, suspension and emulsion polymerization; tan waxy solid, mild odor; sol. in alcohols, hydrocarbons, natural and paraffinic oils; sp.gr. 0.98; HLB 4.7; pour pt. 50 C; flash pt. (PMCC) > 150 C; 100% act.

Ethylan® GT85. [Harcros UK] Sorbitan trioleate; CAS 26266-58-0; nonionic; emulsifier for cosmetics, pharmaceuticals, agric., plastic antifog, textile fiber lubricant/softener, suspension and emulsion polymerization; amber visc. liq., mild fatty odor; sol. in alcohols, hydrocarbons, natural and paraffinic oils; sp.gr. 1.00; visc. 230 cs; HLB 1.5; pour pt. -10 C; flash pt. (PMCC) -10 C; 100% act.

Ethylan® HA Flake. [Harcros UK] Nonoxynol-35; CAS 9016-45-9; nonionic; detergent, wetting agent, latex stabilizer, emulsifier in waxes, resins, hand cleaning gels, emulsion polymerization; wh. waxy solid; negligible odor; water sol.; sp.gr. 1.064 (60

C); visc. 120 cs (60 C); HLB 17.4; cloud pt. > 100 C (1% aq.); flash pt. (COC) > 100 C; pour pt. 43 C; pH 7.0 (1% aq.); surf. tens. 42 dynes/cm (0.1%); 100% act.

Ethylan® HB Series. [Harcros UK] Aromatic alkyoxylates; nonionic; cosolv. with high flash pt. for use in iodophors, rinse aids, detergent sterilizers, glass cleaners, etc.; liq.

Ethylan® HB1. [Harcros UK] Short chain ethoxylate; nonionic; coalescing agent for surface coatings; solv. for acrylic and vinyl acetate copolymers; plasticizer; preservative effective against a variety of microorganisms; lt. amber clear liq., mild odor; sol. hazy in water; sp.gr. 1.098; visc. 27 cs; pour pt. -15 C; flash pt. (COC) > 100 C; pH 8.8 (1% aq.); toxicology: sl. toxic; LD50 (rat, oral) 2-4 ml/kg; sl. skin irritant, severe eye irritant; 100% act.

Ethylan® HB4. [Harcros UK] Aromatic ethoxylate; nonionic; penetrant, hydrotrope, cosolv., solubilizer, coupling agent; enhances detergency emulsification and wetting; for iodophors, rinse aids, detergent sterilizers, glass cleaners; water-wh. clear liq., faint odor; sol. in water; sp.gr. 1.121; visc. 64 cs; pour pt. < 0 C; cloud pt. > 100 C (1% aq.); flash pt. (COC) > 180 C; pH 6.4 (1% aq.); 100% conc.

Ethylan® HB15. [Harcros UK] Aromatic alkoxylate; nonionic; viscose processing aid; penetrant; lubricant; soft solid; HLB 17.6; 100% conc.

Ethylan® HB30. [Harcros UK] Aromatic ethoxylate; nonionic; penetrant for oil well drilling muds; solid; 100% conc.

Ethylan® HP. [Harcros UK] Nonyl phenol ethoxylate; nonionic; detergent, wetting agent, latex stabilizer, emulsifier in waxes, resins, hand cleaning gels, emulsion polymerization; wh. waxy solid; negligible odor; water sol.; sp.gr. 1.072 (40 C); visc. 78 cs (60 C); HLB 16.6; cloud pt. > 100 C (1% aq.); flash pt. > 400 F (COC); pour pt. 36 C; pH 6–8 (1% aq.); 100% act.

Ethylan® KELD. [Harcros UK] Modified fatty alkylolamide; nonionic; emulsifier for wax, kerosene and olein, solv. degreasing; clear amber liq.; mild odor; disp. in water; sp.gr. 1.005; visc. 967 cs; flash pt. (COC) > 150 C; pour pt. 5 C; pH 9.0 (1% aq.); 100% act.

Ethylan® KEO. [Harcros UK] Nonoxynol-9 (9.5 EO); CAS 9016-45-9; nonionic; detergent, wetting agent, emulsifier, foam stabilizer, solubilizer, used in pesticides, perfumes, emulsion polymerization; EPA approved; clear colorless liq.; negligible odor; water sol.; sp.gr. 1.060; visc. 331 cs; HLB 13; pour pt. 3 C; cloud pt. 54 C (1% aq.); flash pt. (COC) > 200 C; pH 6–8 (1% aq.); surf. tens. 34 dynes/cm (0.1%); 100% act.

Ethylan® LD. [Harcros UK] Cocamide DEA; nonionic; foam stabilizer, emulsifier for hand cleaning gels, hard surf. cleaners, shampoos; plastics antistat; clear straw liq.; mild odor; disp. in water; sp.gr. 0.981; visc. 1408 cs; flash pt. > 350 F (COC); pour pt. 15 C; flash pt. (COC) > 175 C; pH 9.5 (1% aq.); 90% act.

Ethylan® LDA-37. [Harcros UK] Cocamide DEA; nonionic; industrial degreasing, metal cleaning; emulsifier for solvs. and polyethylene waxes; amber clear liq., mild odor; water-sol.; sp.gr. 1.005; visc. 1469 cs; pour pt. -14 C; flash pt. (COC) > 110 C; pH 9.5 (1% aq.); 100% conc.

Ethylan® LDA-48. [Harcros] Cocamide DEA; nonionic; foam stabilizer in liq. detergents, general purpose or hard surf. cleaners, shampoos; amber clear liq., mild odor; water-sol.; sp.gr. 1.008; visc. 1342 cs; pour pt. 5 C; flash pt. (COC) > 175 C; pH 9.5 (1% aq.).

Ethylan® LDG. [Harcros UK] Cocamide DEA; nonionic; foam stabilizer for liq. detergents, general purpose and hard surf. cleaners, shampoos; amber clear liq., mild odor; disp. in water; sp.gr. 1.007; visc. 1303 cs; pour pt. 5 C; flash pt. (COC) > 175 C; 82% conc.

Ethylan® LDS. [Harcros UK] Cocamide DEA; nonionic; foam stabilizer, emulsifier for hand cleaning gels, hard surf. cleaners, shampoos; plastics antistat; straw clear liq.; mild odor; water-disp.; sp.gr. 0.990; visc. 810 cs; flash pt. > 350 F (COC); pour pt. 10 C; pH 8.5–10.5 (1% aq.); 87% act.

Ethylan® LM. [Harcros UK] Coconut monoalkanolamide; nonionic; emulsifier, antistat, foam stabilizer for liq. detergents; pale straw flake; mild odor; water insol.; sp.gr. 0.909 (80 C); visc. 40 cs (80 C); flash pt. > 350 F (COC); pour pt. 63 C; pH 8.5–10.5 (1% aq.); 94% act.

Ethylan® LM2. [Harcros UK] Ethoxylated coconut fatty acid alkylolamide; nonionic; foam stabilizer, emulsifier for detergent formulations; off-wh. soft paste; mild odor; water sol.; sp.gr. 0.963 (40 C); visc. 108 cs (40 C); flash pt. > 300 F (COC); pour pt. 2 C; pH 7–9 (1% aq.); 100% act.

Ethylan® ME. [Harcros UK] Cetoleth-5.5; nonionic; emulsifier for org. solv., veg. oils, paraffin waxes; plastics antistat; straw liq./soft paste, mild fatty odor; disp. in water; sp.gr. 0.970; visc. 81 cs; HLB 10.0; pour pt. 19 C; flash pt. (COC) > 175 C; pH 7.0 (1% aq.); 100% act.

Ethylan® MLD. [Harcros UK] Lauramide DEA; CAS 120-40-1; nonionic; foam booster and stabilizer in toiletries and detergent formulations, antistat for plastics; pale cream waxy flake; negligible odor; disp. in water; visc. 107 cs (60 C); pour pt. 40 C; flash pt. (COC) > 150 C; pH 8.5 (1% aq.); 95% act.

Ethylan® MPA. [Harcros] Dimethylaminopropyl deriv. of a coconut fatty acid amide; nonionic; intermediate for prod. of amine oxides and betaines; amber clear liq., char. odor; disp. in water; sp.gr. 0.888; visc. 116 cs; pour pt. 17 C; flash pt. (COC) > 150 C; pH 12.0 (1% aq.).

Ethylan® N30. [Harcros UK] Nonoxynol-30; CAS 9016-45-9; nonionic; detergent, wetting agent, latex stabilizer, emulsifier in waxes, resins, hand cleaning gels; wh. waxy solid; negligible odor; water sol.; sp.gr. 1.064 (60 C); visc. 90 cs (60 C); HLB 17.0; cloud pt. > 100 C (1% aq.); flash pt. (COC) > 200 C; pour pt. 39 C; pH 7.0 (1% aq.); 100% act.

Ethylan® N50. [Harcros UK] Nonyl phenol ethoxylate; nonionic; detergent, wetting agent, latex stabilizer, emulsifier in waxes, resins, hand cleaning gels; wh. waxy solid; negligible odor; water sol.; sp.gr. 1.073 (60 C); visc. 135 cs (60 C); HLB 18.2; cloud pt. > 100 C (1% aq.); flash pt. > 400 F (COC); pour pt. 43 C; pH 6–8 (1% aq.); 100% act.

Ethylan® N92. [Harcros UK] Nonyl phenol ethoxylate; nonionic; detergent, wetting agent, latex stabilizer, emulsifier in waxes, resins, hand cleaning gels; wh. waxy solid; negligible odor; water sol.; sp.gr. 1.078 (60 C); visc. 340 cs (60 C); HLB 19.0; cloud pt. > 100 C (1% aq.); flash pt. > 400 F (COC); pour pt. 50 C; pH 6–8 (1% aq.); 100% act.

Ethylan® NK4. [Harcros] Alkyl phenol ethoxylate; nonionic; emulsifier for aliphatic solvs., hand gels; clear liq., faint odor; disp. in water; sp.gr. 1.043; visc. 233 cs; HLB 11.2; pour pt. -2 C; flash pt. (COC)

> 200 C; pH 7.0 (1% aq.); 100% act.

Ethylan® NP 1. [Harcros UK] Nonoxynol-1.5; nonionic; defoamer, oil emulsifier; clear straw liq., negligible odor; oil sol.; sp.gr. 0.987; visc. 650 cs; HLB 4.5; flash pt. (COC) 149 C; pour pt. < 0 C; pH 7.0 (1% aq.); 100% act.

Ethylan® OE. [Harcros UK] Cetoleth-13; nonionic; emulsifier for fatty acids, alcohols, and waxes, oil and latex stabilizer, dye leveling agent; essential oil solubilizer; mfg. of polishes; emulsion polymerization; cream waxy solid; negligible odor; water sol.; sp.gr. 1.009 (40 C); visc. 55 cs (40 C); HLB 14; pour pt. 31 C; cloud pt. 90 C (1% aq.); flash pt. (COC) > 175 C; pH 7.0 (1% aq.); 100% act.

Ethylan® PQ. [Harcros UK] Modified alkylphenol ethoxylate; nonionic; solubilizer used in the mfg. of iodophor; water-wh. liq (clear @ 30 C); faint odor; water sol.; sp.gr. 1.045 (40 C); HLB 14.2; cloud pt. 83 C (1% aq.); flash pt. > 350 F (COC); pour pt. 15 C; pH 6–7.5 (1% aq.); 98% act.

Ethylan® R. [Harcros UK] Cetoleth-19; nonionic; biodeg. emulsifier for fatty acids, waxes; solubilizer for essential oils; latex stabilizer; polish mfg.; emulsion polymerization; dye leveling agent; cream waxy solid; negligible odor; water sol.; sp.gr. 1.023 (40 C); visc. 141 cs (40 C); HLB 17.5; pour pt. 36 C; cloud pt. > 100 C (1% aq.); flash pt. (COC) > 175 C; pH 7.0 (1% aq.); surf. tens. 40 dynes/cm (0.1%); 100% act.

Ethylan® TB345. [Harcros UK] Amine EO-PO copolymer; nonionic; low-foam surfactant for metal cleaning, dairy and brewery cleaners, rinse aids; defoamer in presence of alkaline builders; amber clear liq., faint odor; sol. in water; sp.gr. 1.034; visc. 551 cs; pour pt. -5 C; cloud pt. 40 C (10%); flash pt. (COC) > 120 C; pH 8.0 (1% aq.); surf. tens. 37.8 dynes/cm (0.1%); 100% conc.

Ethylan® TC. [Harcros UK] PEG-15 cocamine; CAS 61791-14-8; nonionic; wetting agent for metal cleaning and stripping of surface coatings, textiles, paints, agric., polishes, fiber antistat, used in cosmetics; dk. amber liq.; mild fatty amine odor; water sol.; sp.gr. 1.040; visc. 140 cs; HLB 15; flash pt. (COC) > 150 C; pour pt. < 0 C; pH 9.5 (1% aq.); 100% act.

Ethylan® TCO. [Harcros UK] Complex amine oxide; nonionic; foam and suds stabilizer in lt.- and heavy-duty detergents, cosmetic formulations, foam booster, emulsifier; straw clear to hazy liq.; mild fatty amine odor; water sol.; sp.gr. 1.008; visc. 159 cs; flash pt. > 300 F (COC); pour pt. 0 C; pH 6–8 (1% aq.); 40% act.

Ethylan® TD3. [Harcros UK] Ethoxylated alkyl diamine; nonionic; emulsifier with anticorrosive props.; demulsifier for antibiotics prod.; liq.; HLB 9.7; 100% conc.

Ethylan® TD10. [Harcros UK] Ethoxylated alkyl diamine; nonionic; emulsifier and wetter with anticorrosive and antistatic props.; oil demulsifier; liq.; HLB 11.6; 100% conc.

Ethylan® TD15. [Harcros UK] Ethoxylated alkyl diamine; nonionic; emulsifier and wetter with anticorrosive and antistatic props.; liq.; HLB 13.4; 100% conc.

Ethylan® TF. [Harcros UK] Ethoxylated coconut fatty amine; nonionic; oil emulsifier with anticorrosive properties, antistat for syn. fibers with PS; lt. amber liq.; mild fatty amine odor; disp. in water; sp.gr. 0.912; visc. 140 cs; HLB 6; flash pt. > 300 F (COC); pour pt. < 0 C; pH 8.5–10.0 (1% aq.); 100% act.

Ethylan® TH-2. [Harcros UK] Hydrog. fatty amine ethoxylate; nonionic; oil emulsifier with anticorrosive properties, antistat for syn. fibers with PS; pale brn. solid; mild fatty amine odor; disp. in water; sp.gr. 0.878 (60 C); visc. 33 cs (60 C); HLB 5; flash pt. > 300 F (COC); pour pt. 40 C; pH 8.5–10.0 (1% aq.); 100% act.

Ethylan® TH-30. [Harcros UK] Fatty amine ethoxylate (30 EO); nonionic; wetting agent in acid or alkaline formulations for metal cleaning, surf. coatings stripping, textiles, paint, agric., polish mfg.; yel. waxy solid, mild fatty amine odor; water-sol.; sp.gr. 1.035; visc. 143 cs (40 C); HLB 17.0; pour pt. 30 C; flash pt. (COC) > 150 C; pH 9.0 (1% aq.); 100% conc.

Ethylan® TLM. [Harcros UK] Ethoxylated fatty amine; nonionic; wetting agent for acid and alkaline formulations; for metal cleaning; liq.; HLB 15.6; 100% conc.

Ethylan® TN-10. [Harcros UK] PEG-10 cocamine; CAS 61791-14-8; nonionic; wetting agent for acid or alkaline metal cleaners, stripping of surf. coatings; oil emulsifier with anticorrosive properties, antistat for syn. fibers with PS; for cosmetics mfg.; dk. amber liq.; mild fatty amine odor; water sol.; sp.gr. 1.015; visc. 174 cs; HLB 14; pour pt. < 0 C; flash pt. (COC) > 150 C; pH 9.5 (1% aq.); 100% act.

Ethylan® TT-05. [Harcros UK] Ethoxylated fatty amine; nonionic; emulsifier and wetting agent with anticorrosive and antistatic props.; liq.; HLB 8.8; 100% conc.

Ethylan® TT-07. [Harcros UK] Ethoxylated fatty amine; nonionic; emulsifier and wetting agent with anticorrosive and antistatic props.; liq.; HLB 10.6; 100% conc.

Ethylan® TT-15. [Harcros UK] PEG-15 tallow amine; nonionic; wetting agent for metal cleaning and stripping of surface coatings, fiber antistat, used in cosmetics; pale brn. paste; mild fatty amine odor; water sol.; sp.gr. 1.030; visc. 252 cs; HLB 14; flash pt. (COC) > 150 C; pour pt. 0 C; pH 9.5 (1% aq.); 100% act.

Ethylan® TT-30. [Harcros UK] Ethoxylated fatty amine; nonionic; wetting agent, antistat, corrosion inhibitor for cutting oils, metal cleaning, textile auxiliaries; solid; HLB 16.6; 100% conc.

Ethylan® TT-40. [Harcros UK] Ethoxylated fatty amine; nonionic; wetting agent, antistat, corrosion inhibitor for cutting oils, metal cleaning, textile auxiliaries; solid; HLB 17.4; 100% conc.

Ethylan® TT-203. [Harcros UK] Ethoxylated fatty amine; nonionic; demulsifier in fermentation processes; oil emulsifier; fiber processing aid; internal antistat for polystyrene; liq./paste; HLB 5.0; 100% conc.

Ethylan® TU. [Harcros UK] Nonoxynol-8; CAS 9016-45-9; nonionic; textile scouring agent, dry cleaning detergent, wetting and emulsifying agent for oils and aromatic solv.; pale straw clear liq.; negligible odor; water sol.; sp.gr. 1.055; visc. 366 cs; HLB 12.2; pour pt. < 0 C; cloud pt. 32 C (1% aq.); flash pt. (COC) > 200 C; pH 7.0 (1% aq.); 100% act.

Ethylan® VPK. [Harcros UK] PEG unsat. fatty acid ester; nonionic; textile oil emulsifier, fiber lubricant; pale straw clear liq., mild fatty odor; water-sol.; sp.gr. 1.041; visc. 150 cs; HLB 12.4; pour pt. 0 C; cloud pt. 55 C (1% aq.); flash pt. (COC) 150 C; pH 7.0 (1% aq.); 100% act.

Ethyl Benzene Sulfonic Acid. [Boliden Intertrade]

Alkylaryl sulfonates; organic intermediate; catalyst, hydrotrope; liq.; 95% conc.

Ethyl® Butene-1. [Ethyl] C4 alpha olefins; intermediate for biodeg. surfactants and specialty industrial chemicals; gas; 100% conc.

Ethyl® C1416. [Ethyl] C14-C16 alpha olefins; intermediate for biodeg. surfactants and specialty industrial chemicals; liq.; 100% conc.

Ethyl® C1618. [Ethyl] C16-18 alpha olefins; intermediate for biodeg. surfactants and specialty industrial chemicals; liq.; 100% conc.

Ethyl® Decene-1. [Ethyl] C10 alpha olefins; intermediate for biodeg. surfactants and specialty industrial chemicals; liq.; 100% conc.

Ethyl® Dodecene-1. [Ethyl] C12 alpha olefins; intermediate for biodeg. surfactants and specialty industrial chemicals; liq.; 100% conc.

Ethyl Foaming Agent. [Ethyl] Alcohol ether sulfate; anionic; foaming agent for wallboard mfg., air drilling; liq.; 60% conc.

Ethyl® Hexadecene-1. [Ethyl] C16 alpha olefins; intermediate for biodeg. surfactants and specialty industrial chemicals; liq.; 100% conc.

Ethyl® Hexene-1. [Ethyl] C6 alpha olefins; intermediate for biodeg. surfactants and specialty industrial chemicals; liq.; 100% conc.

Ethyl® Octene-1. [Ethyl] C8 alpha olefins; intermediate for biodeg. surfactants and specialty industrial chemicals; liq.; 100% conc.

Ethyl® Tetradecene-1. [Ethyl] C4 alpha olefins; intermediate for biodeg. surfactants and specialty industrial chemicals; liq.; 100% conc.

Etingal® A. [BASF AG] Phosphoric acid ester; antifoam for papermaking.

Etingal® L. [BASF AG] Ethyl ether deriv. of fatty acid; foam controller for papermaking.

Etingal® S. [BASF AG] Phosphoric acid ester; foam breaker for papermaking, leather, furs.

Etocas 10. [Croda Chem. Ltd.] PEG-10 castor oil; CAS 61791-12-6; nonionic; cosmetic and essential oil solubilizer, emulsifier, lubricant, softener, leveling agent, emollient, superfatting agent, antistat, softener, detergent; used in personal care prods., fiber processing,; metalworking fluids, emulsion polymerization, insecticides; pale yel. liq.; sol. in ethanol, naphtha, MEK, trichlorethylene, disp. in water; HLB 6.3; cloud pt. < 20 C; acid no. 1.0 max.; sapon. no. 120–130; pH 6–7.5; 97% act.

Etocas 20. [Croda Chem. Ltd.] PEG-20 castor oil.; nonionic; emulsifier, wetting agent, dispersant for personal care prods., textile and metal processing; liq.; HLB 9.6; 97% conc.

Etocas 30. [Croda Chem. Ltd.] PEG-30 castor oil; humectant, o/w emulsifier, skin cleanser, conditioner, emollient, solubilizer; liq.; sol. in oil; water-disp.; HLB 11.7; 97% conc.

Etocas 35. [Croda Chem. Ltd.] PEG-35 castor oil; nonionic; cosmetic and essential oil solubilizer, emulsifier, lubricant, emollient, superfatting agent, antistat, softener, detergent; pale yel. liq.; sol. in water, alcohol, naphtha, MEK, oleic acid, trichloroethylene; HLB 12.5; cloud pt. 35–40 C; acid no. 1.0 max.; sapon. no. 62–72; surf. tens. 41.5 dynes/cm (0.1% aq.); pH 6–7.5; 97% act.

Etocas 40. [Croda Chem. Ltd.] PEG-40 castor oil; nonionic; emulsifier, wetting agent, dispersant for personal care prods., textile and metal processing; pale yel. liq.; sol. see Etocas 35; HLB 13; cloud pt. 50 C; acid no. 1.0 max.; sapon. no. 60–65; pH 6–7.5; 97% act.

Etocas 50. [Croda Chem. Ltd.] PEG-50 castor oil; nonionic; emulsifier, wetting agent, dispersant for personal care prods., textile and metal processing; liq.; 97% conc.

Etocas 60. [Croda Chem. Ltd.] PEG-60 castor oil; nonionic; cosmetic and essential oil solubilizer, emulsifier, lubricant, emollient, superfatting agent, antistat, softener, detergent; pale yel. soft paste; sol. see Etocas 35; HLB 14.7; cloud pt. 60 C; acid no. 1.0 max.; sapon. no. 45–50; surf. tens. 43.2 dynes/cm (0.1% aq.); pH 6–7.5.

Etocas 100. [Croda Chem. Ltd.] PEG-100 castor oil; nonionic; emulsifier, wetting agent, dispersant for personal care prods., textile and metal processing; humectant; pale yel. waxy solid; sol. in water, alcohol, naphtha, MEK, oleic acid, trichloroethylene; HLB 16.5; cloud pt. 66 C; acid no. 1.0 max.; sapon. no. 25–35; surf. tens. 41.6 dynes/cm (0.1% aq.); pH 6–7.5; 97% act.

Etocas 200. [Croda Chem. Ltd.] PEG-200 castor oil.; nonionic; antistat for textile processing; solid; HLB 18.0; 97% conc.

Etophen 102. [Zschimmer & Schwarz] Nonoxynol-2; CAS 9016-45-9; nonionic; detergent, dispersant, emulsifier, wetting agent for household and industrial detergents, textiles, paper, leather, and ceramic industries; liq.; 100% act.

Etophen 103. [Zschimmer & Schwarz] Nonoxynol-3; CAS 9016-45-9; nonionic; washing and cleansing agent; colorless clear liq.; insol. in water; cloud pt. 74-78 C (5 g/10 ml butyl diglycol 60%); 100% act.

Etophen 105. [Zschimmer & Schwarz] Nonoxynol-5; CAS 9016-45-9; nonionic; washing and cleansing agent; colorless clear liq.; disp. in water; cloud pt. 51-55 C (5 g/20 ml butyl diglycol 60%); 100% act.

Etophen 106. [Zschimmer & Schwarz] Nonoxynol-6; CAS 9016-45-9; nonionic; washing and cleansing agent; colorless clear liq.; disp. in water; cloud pt. 58-62 C (5 g/10 ml butyl diglycol 25%); 100% act.

Etophen 107. [Zschimmer & Schwarz] Nonoxynol-7; CAS 9016-45-9; nonionic; washing and cleansing agent; colorless clear liq.; cloud pt. 66-70 C (5 g/10 ml butyl diglycol 25%); 100% act.

Etophen 108. [Zschimmer & Schwarz] Nonoxynol-8; CAS 9016-45-9; nonionic; washing and cleansing agent; colorless clear liq.; sol. in water; cloud pt. 47-51 C (1% aq.); 100% act.

Etophen 109. [Zschimmer & Schwarz] Nonoxynol-9; CAS 9016-45-9; nonionic; washing and cleansing agent; colorless clear liq.; sol. in water; cloud pt. 56-60 C (1% aq.); 100% act.

Etophen 110. [Zschimmer & Schwarz] Nonoxynol-10; CAS 9016-45-9; nonionic; washing and cleansing agent; cloudy visc. liq.; sol. in water; cloud pt. 69-73 C (1% aq.); 100% act.

Etophen 112. [Zschimmer & Schwarz] Nonoxynol-12; CAS 9016-45-9; nonionic; washing and cleansing agent; wh. paste; sol. in water; cloud pt. 80-84 C (1% aq.); 100% act.

Etophen 114. [Zschimmer & Schwarz] Nonoxynol-14; CAS 9016-45-9; nonionic; washing and cleansing agent; wh. paste; sol. in water; cloud pt. 90-94 C (5 g/10 ml butyl diglycol 60%); 100% act.

Etophen 120. [Zschimmer & Schwarz] Nonoxynol-20; CAS 9016-45-9; nonionic; washing and cleansing agent; wh. waxy solid; sol. in water; cloud pt. 69-73 C (1% in 10% NaCl); 100% act.

Eumulgin 286. [Henkel KGaA] Nonoxynol-10; CAS 9016-45-9; nonionic; detergent, wetting agent for plant protection, pest control, dishwashing; liq.

Eumulgin 365. [Henkel KGaA] Blend; emulsifier for paraffin wax emulsions; paste; 100% conc.

Eumulgin 535. [Henkel KGaA] Fatty alcohol polyglycol ether; nonionic; emulsifier for paraffin wax emulsions; solid; 100% conc.

Eumulgin 2142. [Henkel KGaA] Mixt. of fatty acid polyglycol ester and fatty alcohol; nonionic; emulsifier for rapeseed oil for ecology-proof metalworking oils; liq.; 100% conc.

Eumulgin 2312. [Henkel KGaA] Surfactant/additive blend; nonionic; emulsifier for metalworking agents; liq.; 90% conc.

Eumulgin B1. [Henkel/Cospha; Henkel Canada; Henkel KGaA] Ceteareth-12; CAS 68439-49-6; nonionic; wetting agent and dispersant for paint systems; emulsifier for ointments, creams, low visc. emulsions, cosmetics, pharmaceuticals; waxy solid; sp.gr. 0.95; solid. pt. 50–68 F; HLB 12.0; 100% conc.

Eumulgin B2. [Henkel/Cospha; Henkel Canada; Henkel KGaA; Pulcra SA] Ceteareth-20; CAS 68439-49-6; nonionic; emulsifier for ointments, creams, low visc. emulsions, cosmetics, pharmaceuticals; waxy flakes; sp.gr. 0.95; solid. pt. 50–68 F; HLB 15.5; hyd. no. 48-54; pH 6.0-7.5 (1%); 100% conc.

Eumulgin B3. [Henkel/Cospha; Henkel Canada; Henkel KGaA; Pulcra SA] Ceteareth-30; CAS 68439-49-6; nonionic; emulsifier for transparent gels and creams, cosmetics, pharmaceuticals; waxy solid; sp.gr. 0.95; solid. pt. 50–68 F; HLB 16.7; hyd. no. 35-40; pH 6.0-7.5 (1%); 100% conc.

Eumulgin C4. [Henkel; Henkel KGaA] PEG-5 cocamide; nonionic; foaming agent for detergents; emulsifier; liq.; HLB 8.5; 100% conc.

Eumulgin C8. [Henkel KGaA] Ethoxylated fatty acid alkanolamide; nonionic; emulsifier for solvs., pesticides, fuels; paste; HLB 11.4; 100% conc.

Eumulgin EP2. [Henkel/Cospha; Henkel Canada; Henkel KGaA; Pulcra SA] Cetoleth-2; nonionic; emulsifier for min. oil, hydrocarbons, fats, metalworking oils, textile auxiliaries; liq.; HLB 5.0; hyd. no. 160-165; pH 6.5-7.5 (1%); 100% conc.

Eumulgin EP 2L. [Henkel KGaA] Ethoxylated oleyl/cetyl alcohol; nonionic; emulsifier component of pronounced cold behavior; liq.; 100% conc.

Eumulgin EP 5L. [Henkel KGaA] Ethoxylated oleyl/cetyl alcohol; nonionic; emulsifier component of pronounced cold behavior; liq.; 100% conc.

Eumulgin ET 2. [Henkel KGaA] Ethoxylated oleyl/cetyl alcohol; nonionic; emulsifier for hydrocarbons, min. oils, fats, metalworking oils, textile auxiliaries; liq.; HLB 5.6; 100% conc.

Eumulgin ET 5. [Henkel KGaA] Ethoxylated oleyl/cetyl alcohol; nonionic; emulsifier for hydrocarbons, min. oils, fats, metalworking oils, textile auxiliaries; solid; HLB 9.2; 100% conc.

Eumulgin ET 5L. [Henkel KGaA] Ethoxylated oleyl/cetyl alcohol; nonionic; emulsifier component of pronounced cold behavior; liq.; 100% conc.

Eumulgin ET 10. [Henkel KGaA] Ethoxylated oleyl/cetyl alcohol; nonionic; emulsifier for hydrocarbons, min. oils, fats, metalworking oils, textile auxiliaries; solid; HLB 12.6; 100% conc.

Eumulgin KP92. [Henkel KGaA] Mixt. of fatty acid polyglycol esters; nonionic; emulsifier for aromatic and aliphatic solvs., min. oils; liq.; HLB 12.9; 100% conc.

Eumulgin L. [Henkel/Cospha; Henkel KGaA] PPG-2-ceteareth-9; nonionic; emulsifier, solubilizer for aq. or hydroalcoholic media; for skin and hair care preps.; lt. yel. clear free-flowing liq.; acid no. < 1; 100% conc.

Eumulgin M8. [Henkel/Cospha; Henkel KGaA] Oleth-10 and oleth-5; CAS 9004-98-2; nonionic; emulsifier, solubilizer for pesticides and cosmetics; paste; HLB 11.0; 80% conc.

Eumulgin O5. [Henkel/Cospha; Henkel Canada; Henkel KGaA] Oleth-5; CAS 9004-98-2; nonionic; emulsifier, solubilizer for pesticides, perfumes, cosmetics, floor polishes, hair dressings, creams, lotions, min. oil, terpenes; dispersant improving color acceptance of emulsion paints; clear liq.; water-sol.; HLB 12.0; 100% conc.

Eumulgin O10. [Henkel/Cospha; Henkel Canada; Henkel KGaA] Oleth-10; CAS 9004-98-2; nonionic; wetting agent and dispersant for paint systems; emulsifier for creams, lotions, hair care prods.; used with min. oils and terpenes; soft waxy solid; sp.gr. 0.95; solid. pt. 50–68 F; HLB 12.0; 100% conc.

Eumulgin PA 10. [Pulcra SA] PEG-10 tallowamine; nonionic; surfactant; liq.; HLB 12.5; amine value 75-80; 100% conc.

Eumulgin PA 12. [Pulcra SA] PEG-12 cocamine; CAS 61791-14-8; nonionic; surfactant; liq.; HLB 14.3; amine value 77-85; pH 9.8-10.2 (1%); 100% conc.

Eumulgin PA 20. [Pulcra SA] PEG-20 tallowamine; nonionic; surfactant; paste; HLB 15.4; amine value 48-50; pH 9.5-10.0 (1%); 100% conc.

Eumulgin PA 30. [Pulcra SA] PEG-30 oleamine; nonionic; surfactant; paste; HLB 16.5; amine value 30-40; pH 7.7-8.2 (10%); 100% conc.

Eumulgin PAEH 4. [Pulcra SA] Ethoxylated fatty acid (4 EO); nonionic; surfactant; liq.; HLB 11.0; pH 6.0-7.5 (1%); 100% conc.

Eumulgin PC 2. [Pulcra SA] PEG-2 cocamide; CAS 61791-08-0; nonionic; detergent, foam stabilizer, solubilizer for liq. detergent systems; paste; HLB 5.8; sapon. no. < 20; pH 9.0-10.5 (1%); 100% conc.

Eumulgin PC 4. [Pulcra SA] PEG-4.5 cocamide; CAS 61791-08-0; nonionic; detergent, foam stabilizer for liq. detergent systems; liq./paste; HLB 8.8; sapon. no. < 20; pH 9.0-10.5 (1%); 100% conc.

Eumulgin PC 10. [Pulcra SA] PEG-10 cocamide; CAS 61791-08-0; nonionic; detergent, foam stabilizer, solubilizer for liq. detergent systems; liq./paste; HLB 13.5; sapon. no. < 20; pH 9.0-10.5 (5%); 100% conc.

Eumulgin PC 10/85. [Pulcra SA] PEG-10 cocamide; CAS 61791-08-0; nonionic; detergent, foam stabilizer, solubilizer for liq. detergent systems; liq.; pH 9.0-101.5 (1%); 85% conc.

Eumulgin PK 23. [Pulcra SA] PEG-23 cocoate; nonionic; surfactant; flake; HLB 16.6; hyd. no. 41-51; 100% conc.

Eumulgin PLT 4. [Pulcra SA] Ethoxylated oleic acid; CAS 9004-96-0; nonionic; emulsifier for paraffinic waxes and compds., min. oils; dyeing assistant, lubricant, antistat, emulsifier for textile industry; liq.; HLB 8.2; 100% conc.

Eumulgin PLT 5. [Pulcra SA] Ethoxylated oleic acid; CAS 9004-96-0; nonionic; emulsifier for paraffinic waxes and compds., min. oils; dyeing assistant, lubricant, antistat, emulsifier for textile industry; liq.; HLB 8.9; 100% conc.

Eumulgin PLT 6. [Pulcra SA] Ethoxylated oleic acid; CAS 9004-96-0; nonionic; emulsifier for paraffinic waxes and compds., min. oils; dyeing assistant, lubricant, antistat, emulsifier for textile industry; liq.; HLB 9.6; 100% conc.

Eumulgin PPG 40. [Pulcra SA] PPG-40; nonionic; surfactant; solid; sapon. no. 60-64; 100% conc.
Eumulgin PRT 36. [Pulcra SA] PEG-36 castor oil; nonionic; detergent, emulsifier, dispersant, solubilizer for conc. pesticides, metal, leather, cosmetics, toiletries, pharmaceuticals, textile, and polymer industries; liq.; HLB 12.7; sapon. no. 59-69; pH 6.0-7.5 (1%); 100% conc.
Eumulgin PRT 40. [Pulcra SA] PEG-40 castor oil; nonionic; detergent, emulsifier, dispersant, solubilizer for conc. pesticides, metal, leather, cosmetics, toiletries, pharmaceuticals, textile, and polymer industries; liq.; HLB 13.1; sapon. no. 55-65; pH 6.0-7.5 (1%); 100% conc.
Eumulgin PRT 56. [Pulcra SA] PEG-56 castor oil; nonionic; detergent, emulsifier, dispersant, solubilizer for conc. pesticides, metal, leather, cosmetics, toiletries, pharmaceuticals, textile, and polymer industries; paste; HLB 14.5; sapon. no. 45-55; pH 6.0-7.5 (1%); 100% conc.
Eumulgin PRT 200. [Pulcra SA] PEG-200 castor oil; nonionic; detergent, emulsifier, dispersant, solubilizer for conc. pesticides, metal, leather, cosmetics, toiletries, pharmaceuticals, textile, and polymer industries; solid; HLB 18.1; sapon. no. 16-18; pH 5.0-6.5 (10%); 100% conc.
Eumulgin PST 5. [Pulcra SA] PEG-5.2 stearate; CAS 9004-99-3; nonionic; surfactant; solid; HLB 9.1; pH 6.0-7.5 (1%); 100% conc.
Eumulgin PTL 4. [Pulcra SA] Ethoxylated fatty acid (4.2 EO); nonionic; surfactant; liq.; HLB 8.2; pH 6.0-7.5 (1%); 100% conc.
Eumulgin PTL 5. [Pulcra SA] Ethoxylated fatty acid (5 EO); nonionic; surfactant; liq.; HLB 8.9; pH 6.0-7.5 (1%); 100% conc.
Eumulgin PTL 6. [Pulcra SA] Ethoxylated fatty acid (6 EO); nonionic; surfactant; liq.; HLB 9.6; pH 6.0-7.5 (1%); 100% conc.
Eumulgin PWM2. [Pulcra SA] Oleth-2; CAS 9004-98-2; nonionic; w/o emulsifier, dispersant, lipophilic cosolv.; fragrance grade; broad pH and electrolyte tolerance; liq.; HLB 5.0; hyd. no. 155-165; pH 6.0-7.5 (1%); 100% conc.
Eumulgin PWM5. [Pulcra SA] Oleth-5; CAS 9004-98-2; nonionic; o/w emulsifier, solubilizer; stable over wide pH range; liq.; HLB 9.1; 100% conc.
Eumulgin PWM10. [Pulcra SA] Oleth-10; CAS 9004-98-2; nonionic; emulsifier for pesticides, cosmetics, floor polishes and detergents; solid; HLB 12.6; hyd. no. 77-83; pH 6.0-7.5 (1%); 100% conc.
Eumulgin PWM17. [Pulcra SA] Oleth-18; CAS 9004-98-2; nonionic; hydrophilic emulsifier, dispersant, solubilizer, detergent; dyeing assistant for wool/acrylic blends; paste; HLB 14.8; hyd. no. 54-59; pH 6.0-7.5 (1%); 100% conc.
Eumulgin PWM25. [Pulcra SA] Oleth-25; CAS 9004-98-2; nonionic; surfactant; solid; HLB 16.4; hyd. no. 41-45; pH 6.0-7.5 (1%); 100% conc.
Eumulgin RO 40. [Henkel/Cospha; Henkel Canada; Henkel KGaA] PEG-40 castor oil; nonionic; o/w emulsifier, solubilizer for perfume oils; for personal care creams and lotions; lt. yel. cloudy visc. liq.; 100% conc.
Eumulgin RT 5. [Henkel KGaA] Ethoxylated castor oil; nonionic; emulsifier component for min. oils in metalworking agents and other fields; liq.; 99% conc.
Eumulgin RT 11. [Henkel KGaA] Ethoxylated castor oil; nonionic; emulsifier component for min. oils in metalworking agents and other fields; liq.; 99% conc.
Eumulgin RT 20. [Henkel KGaA] Ethoxylated castor oil; nonionic; emulsifier for aromatic solvs., fats, oils, fatty acids, pesticides; liq.; HLB 9.6; 99% conc.
Eumulgin RT 40. [Henkel KGaA] Ethoxylated castor oil; nonionic; emulsifier for aromatic solvs., fats, oils, fatty acids, pesticides; liq.; HLB 13.0; 99% conc.
Eumulgin ST-8. [Henkel Canada] PEG-8 stearate; CAS 9004-99-3; nonionic; emulsifier for o/w liq. emulsions and creams; waxy solid.
Eumulgin TI 60. [Henkel KGaA] Fatty acid polyglycol ester; nonionic; emulsifier for aliphatic solvs.; liq.; HLB 9.9; 100% conc.
Eumulgin TL 30. [Henkel KGaA] Fatty acid polyglycol ester; nonionic; emulsifier for paraffin-based min. oils; liq.; HLB 6.3; 100% conc.
Eumulgin TL 55. [Henkel KGaA] Fatty acid polyglycol ester; nonionic; emulsifier for paraffin-based min. oils; liq.; HLB 8.9; 100% conc.
Eumulgin WM5. [Henkel KGaA; Pulcra SA] Oleth-5; CAS 9004-98-2; nonionic; raw material for foam-controlled laundry detergents; paste; HLB 9.2; hyd. no. 116-123; pH 6.0-7.5 (1%); 100% conc.
Eumulgin WM 7. [Henkel KGaA] Polyglycol ether on partially unsat. fatty alcohol; nonionic; raw material for foam-controlled laundry detergents; paste; 100% conc.
Eumulgin WM 10. [Henkel/Cospha; Henkel KGaA] Ethoxylated oleyl alcohol; nonionic; emulsifier for metalworking oils; for all-temp. laundry detergents; paste; 100% conc.
Eumulgin WO 7. [Henkel KGaA] Fatty alcohol polyglycol ether; nonionic; surfactant for foam-controlled laundry detergents; liq.; 100% conc.
Eureka 102. [Atlas Refinery] Sulfated castor oil; anionic; emulsifier, detergent; grinding aid in pigment disps.; plasticizer in finish coatings; topping oil for suede leather; lt. liq.; dens. 8.44 lb/ga; pH 7.5-8.0 (10%); 72% active.
Eureka 102-WK. [Atlas Refinery] Sulfated castor oil; anionic; used in textile industry and component for shampoo bases and cosmetic formulations; liq.; 50% conc.
Eureka 392. [Atlas Refinery] Sulfated tall oil fatty acid; anionic; emulsifier, carrier for refined oils, base for solv. fatliquor systems; for detergent formulations; penetrant, fiber lubricant for textile and leather processing; liq.; dens. 8.44 lb/gal; pH 5-6 (10%); 75% act.
Eureka 400-R. [Atlas Refinery] Sulfonated fish oil; anionic; emulsifier, lubricant, softener, fatliquor used in leather processing, fibers; lt. liq.; dens. 8.31 lb/gal; pH 5.5-6.0 (10%); 87% act.
Eureka 800-R. [Atlas Refinery] Sulfonated animal oil; anionic; fiber lubricant; fat liquor for leather tanning; liq.; 70% conc.
Eureka E-2. [Atlas Refinery] Fatty acid diamine condensate deriv.; cationic; emulsifier for raw oils, fiber lubricant, fabric softener; heavy visc. liq.; sp.gr. 0.999 (60 F); dens. 8.32 lb/gal (60 F); pH 6-6.5 (10%); 100% act.
EW-POL 8021. [Henkel/Functional Prods.] Aryl polyglycol ether; surface-active plasticizer, thickener for PVAc adhesive dispersions; liq.; 100% act.
Exameen 824 3724. [Clough] Myristalkonium chloride; germicidal quat. for hard surf. disinfection and sanitization; algicide and slimicide for swimming pool and industrial water treatment; liq.; 50% min. quat.

Exameen 2125 M 80 3709. [Clough] Myristalkonium chloride, quaternium-14; germicidal quat. for bacteriological control in disinfectant and sanitizer formulations for hospitals, nursing homes, public institutions and industry; liq.; 80% min. quat.

Exameen 2125 M 3704. [Clough] Myristalkonium chloride, quaternium-14; germicidal quat. for bacteriological control in disinfectant and sanitizer formulations for hospitals, nursing homes, public institutions and industry; liq.; 50% min. quat.

Exameen 3500 3714. [Clough] Benzalkonium chloride; germicidal quat. for hard surf. disinfection and sanitization; algicide and slimicide in swimming pool and industrial water treatment; liq.; 50% min. quat.

Exameen 3580 3719. [Clough] Benzalkonium chloride; germicidal quat. for hard surf. disinfection and sanitization; algicide and slimicide in swimming pool and industrial water treatment; liq.; 80% min. quaternary.

Exameen 8248 3729. [Clough] Myristalkonium chloride; germicidal quat. for hard surf. disinfection and sanitization; algicide and slimicide for swimming pool and industrial water treatment; liq.; 80% min. quat.

Examide-CS. [Soluol] Fatty acid deriv.; amphoteric; detergent, emulsifier and scouring agent for use in textiles; liq.; 50% conc.

Examide-DA. [Soluol] EO condensate; nonionic; detergent, wetting, emulsifier; surfactant used as scouring and leveling agent, dyeing assistant in textile industry; liq.; 50% conc.

Examide N-LS. [Soluol] Fatty acid deriv.; nonionic; detergent, emulsifier, scouring agent for Spandex fibers; liq.; 80% conc.

EZA®. [Ethyl] Zeolite A; detergent builder; solv.; anticaking agent for detergents and desiccants; powd.; 3 μ diam.; 100% act.

F

Fancol LA. [Fanning] Lanolin alcohol; nonionic; emollient, thickener, emulsifier, stabilizer, plasticizer, superfatting agent, dye dispersant, chemical intermediate, lubricant, humectant, mold release agent, conditioner for cosmetics, pharmaceuticals, soaps, industrial applics.; brn. solid wax-like; sol. in CCl_4, chloroform, IPM (@ 75 C), min. oil (@ 75 C), oleyl alcohol; insol. water; m.p. 56 C; acid no. 2 max.; sapon. no. 12 max.; 100% act.

Fancol LAO. [Fanning] Min. oil and lanolin alcohol; nonionic; conditioner, surfactant, stabilizer, moisturizer, humectant, penetrant, emollient, plasticizer, and primary emulsifier for use in cosmetics and pharmaceuticals; plasticizer in aerosol formulas; yel. clear oily liq.; odorless; sol. in oils; insol. in water; sp.gr. 0.84–0.86; sapon. no. 3.0 max.; 100% conc.

Fancol OA 95. [Fanning] Oleyl alcohol; CAS 143-28-2; nonionic; industrial plasticizer, emulsion stabilizer, antifoam and coupling agent, aerosol lubricant, petrol. additive, pigment dispersant; rust preventive; detergent, release agent, cosolvent, softener, tackifier,; spreading agent used for metalworking, petrochemicals, pulp and paper, paints and coatings, plastics and polymers, food applics., pharmaceuticals, cosmetics; chemical intermediate; liq.; sapon no. 1 max.; cloud pt. 5 C max.

Fancor LFA. [Fanning] Lanolin fatty acids; nonionic; emollient, stabilizer, emulsifier, corrosion inhibitor for personal care and pharmaceutical prods.; used in industrial leather treating, coatings, polishes, corrosion inhibitors, lubricants; wax-like solid; m.p. 57–65 C; acid no. 135–170; sapon. no. 158–175; 100% conc.

Fancoscour PO, VC. [Reilly-Whiteman] Detergent for textile scouring.

Farmin 2C. [Kao Corp. SA] Sec. fatty di-n-alkyl amine; nonionic; intermediate for textile finishing; softener; antistat; solid; 100% conc.

Farmin 20, 60, 68, 80, 86, AB, C. [Kao Corp. SA] Primary fatty amines; nonionic; emulsifier, anticaking agent, textile additive, corrosion inhibitor, flotation reagent; liq. to solid; 100% conc.

Farmin D86. [Kao Corp. SA] Sec. fatty di-n-alkyl amine; nonionic; intermediate for textile finishing; softener; antistat; solid; 100% conc.

Farmin DM20. [Kao Corp. SA] Tert. fatty n-alkyl dimethylamine; CAS 112-18-5; nonionic; intermediate for corrosion inhibitors, quaternary synthesis, benzalkonium chloride; liq.; 100% conc.

Farmin DM40, 60, 80, 86. [Kao Corp. SA] Tert. fatty n-alkyl dimethylamine; nonionic; intermediate for corrosion inhibitors, quaternary synthesis, benzalkonium chloride; liq.; 100% conc.

Farmin DMC. [Kao Corp. SA] Tert. fatty n-alkyl dimethylamine; nonionic; intermediate for corrosion inhibitors, quaternary synthesis, benzalkonium chloride; liq.; 100% conc.

Farmin HT. [Kao Corp. SA] Primary fatty amines; nonionic; emulsifier, anticaking agent, textile additive, corrosion inhibitor, flotation reagent; liq.; 100% conc.

Farmin M2C. [Kao Corp. SA] Tert. fatty di-n-alkyl methyl amine; intermediate for corrosion inhibitors, quaternary synthesis, benzalkonium chloride; paste, solid; 100% conc.

Farmin M2TH-L. [Kao Corp. SA] Tert. fatty di-n-alkyl methyl amine; intermediate for corrosion inhibitors, quaternary synthesis, benzalkonium chloride; paste/solid; 100% conc.

Farmin O. [Kao Corp. SA] Primary fatty amine; nonionic; emulsifier, anticaking agent, textile additive, corrosion inhibitor, flotation reagent; liq.; 100% conc.

Farmin R 24H, R 86H. [Kao Corp. SA] Fatty amine hydrochloride; cationic; anticaking agent for fertilizers; corrosion inhibitors; liq./solid.

Farmin R 86H. [Kao Corp. SA] Fatty amine hydrochloride; CAS 1838-08-0; cationic; anticaking agent for fertilizers; corrosion inhibitors; liq./solid.

Farmin S. [Kao Corp. SA] Primary fatty amine; nonionic; emulsifier, anticaking agent, textile additive, corrosion inhibitor, flotation reagent; liq.; 100% conc.

Farmin T. [Kao Corp. SA] Primary fatty amine; nonionic; emulsifier, anticaking agent, textile additive, corrosion inhibitor, flotation reagent; liq.; 100% conc.

Fastgene PNG-708. [Tokai Seiyu Ind.] Polyamine condensate; cationic; for fixing of direct dyes; liq.; 30% conc.

Feliderm CS. [Hoechst AG] Organic acid with additives; pickling acid with leather finishing; liq.

Feliderm M. [Hoechst AG] Alkali salt of aromatic sulfonic acids; dyeing auxiliary; hydrotrope.

Felton 3T. [Toho Chem. Industry] Blend; nonionic; detergent for felt of paper mfg. machines; liq.

Finapal E. [Finetex] Surfactant blend; anionic/nonionic; wetting and leveling agent for polyester dyeing; liq.; 50% conc.

Finazoline CA. [Finetex] Coconut hydroxyethyl imidazoline; cationic; detergent base, emollient, ore flotation agent; solid; 100% conc.

Finazoline OA. [Finetex] Oleic hydroxyethyl imidazoline; emulsifier, wetting agent, softener; intermediate for cosmetic, textile, and metal industries.

Findet A-100-UN. [Finetex] Phosphate ester, acid

form; anionic; wetting agent, dispersant, emulsifier, scouring agent in alkaline conditions for orthodichlorobenzene; liq.; 99% conc.

Findet AD-18. [Finetex] Deceth-6 phosphate; emulsifier, wetting agent, detergent for wet processing of natural and syn. fibers, heavy-duty alkaline cleaners; alkali-stable; liq.; 98.5% conc.

Findet BC. [Finetex] Complex phosphate ester, free acid form; low foaming detergent, wetting agent and emulsifier with high caustic stability and high electrolyte tolerance; liq.; 99% conc.

Findet CF-4. [Finetex] Phosphate ester; anionic; detergent, emulsifier, leveling agent for textile scouring and alkaline chemical specialty formulations; stable in caustic soda at 20% level; liq.; 99% conc.

Findet CF-440. [Finetex] Phosphate ester; anionic; detergent, emulsifier, leveling agent for textile scouring and alkaline chemical specialty formulations; stable in caustic soda at 20% level; liq.; 60% conc.

Findet DD. [Finetex] Butyl phosphate; anionic; surfactant for fiber lubricant finishes; antistat; liq.; 100% conc.

Findet NHP. [Finetex] Hexyl phosphate; anionic; surfactant for fiber lubricant finishes; antistat; liq.; 100% conc.

Findet OJP-5. [Finetex] Nonylphenol polyethoxy phosphate ester, free acid form; anionic; detergent, emulsifier, leveling agent, wetting agent, dedustant; for textile scouring and alkaline chemical specialty formulations; stable in caustic soda at 20% level; visc. liq.; 99% conc.

Findet OJP-25. [Finetex] Alkylaryl alkoxylated phosphate ester, free acid form; anionic; detergent, wetting agent, emulsifier, dispersant, dyeing assistant, antistat; waxy solid; 90% conc.

Findet SB. [Finetex] Phosphate ester; anionic; detergent, emulsifier, leveling agent for textile scouring and alkaline chemical specialty formulations, e.g., window cleaners; stable in caustic soda at 20% level; liq.; 99% conc.

Finsist C-2 Conc. [Finetex] Quat. ammonium compd.; cationic; retarder, leveling agent for basic dyes on acrylic fibers; liq.; 50% conc.

Finsist WW. [Finetex] Quat. ammonium compd.; cationic; compatibilizer; permits use of cationic and anionic dyes in same bath; liq.; 25% conc.

Finsolv® BOD. [Finetex] Octyldodecyl benzoate; emollient, solubilizer, foam modifier for syn. detergents.

Finsolv® P. [Finetex] PPG-15 stearyl ether benzoate; emollient, solubilizer, foam modifier for soap and soap/syndet systems.

Finsolv®SB. [Finetex] Isostearyl benzoate; nongreasy emollient, noncomedogenic; for cosmetics, sunscreen, antiperspirants, deodorants; lubricant; plasticizer for hair resins, in anhyd. systems; perfume solubilizer; liq.; sol. in org. solvs. and oils.

Fixogene CD Liq. [ICI Surf. UK] Dicyandiamide formaldehyde condensate; cationic; fixing agent improving wet fastness props. of direct and reactive dyes on cellulosic fibers; liq.

Fizul 201-11. [Finetex] Ethoxylated alcohol half-ester sulfosuccinate; detergent, wetting agent, solubilizer, dispersant, emulsifier, hydrotrope; surfactant for foaming wallboards, emulsion polymerization; high electrolyte tolerance; liq.; 31% conc.

Fizul 301. [Finetex] Disodium nonoxynol-10 sulfosuccinate; detergent, emulsifier, wetting agent, dispersant; imparts small particle size in acrylate emulsion polymers; emulsifier for floor polish emulsions; liq.; 35% conc.

Fizul M-440. [Finetex] Tetrasodium N-(1,2-dicarboxyethyl)-N-octadecylsulfosuccinamate; anionic; emulsifier, dispersant, solubilizer, stabilizer, suspending agent; for emulsion polymerization of vinyl chloride, styrene/acrylic, styrene/butadiene, acrylic, and vinyl acetate systems; compat. over wide pH range and in presence of high levels of electrolytes; liq.; 35% conc.

Fizul MD-318. [Finetex] Half-ester sulfosuccinamate; anionic; emulsifier for emulsion polymerization of vinyl acetate alone or in combination with other vinyl functional monomers; high tolerance to calcium carbonate and electrolytes; liq.; 35% conc.

Flexricin® 9. [CasChem] Propylene glycol ricinoleate; nonionic; wetting agent, dye solv., wax plasticizer, stabilizer for textile, household, and cosmetic applics., rewetting dried skins; FDA approval; Gardner 2+ liq.; sol. in toluene, butyl acetate, MEK, ethanol; sp.gr. 0.96; visc. 3 stokes; pour pt. -15 F; acid no. 2; iodine no. 77; sapon. no. 159; hyd. no. 288; 100% act.

Flexricin® 13. [CasChem] Glyceryl ricinoleate; nonionic; wetting agent, wax plasticizer, and mold release agent for rubber polymers, antifoam agent, household and cosmetic applics., rewetting dried skins; FDA approval; Gardner 2 min. liq.; sol. in toluene, butyl acetate, MEK, ethanol; sp.gr. 0.985; visc. 8.8 stokes; pour pt. 20 F; acid no. 5; iodine no. 77; sapon. no. 160; hyd. no. 345; 100% act.

Flexricin® 15. [CasChem] Glycol ricinoleate; nonionic; wetting agent, plasticizer, textile, household, and cosmetic applics., rewetting dried skins; chemical intermediate; FDA approval; Gardner 4 liq.; sol. in toluene, butyl acetate, MEK, ethanol; sp.gr. 0.965; visc. 3.9 stokes; 100% act.

Flexricin® 100. [CasChem] Ricinoleic acid deriv.; CAS 141-22-0; lubricant for textile, metalworking compds.; corrosion inhibitor intermediate; intermediate for water sol./disp. lubricants; Gardner 7; sol. in alcohols, ethers, ketones, aromatic hydrocarbons; sp.gr. 0.934; visc. 4 stokes; pour pt. -15 F; acid no. 120; Wijs iodine no. 90; sapon. no. 186; 100% conc.

Flocculant T-9. [Toho Chem. Industry] Complex; cationic; cohesion and sedimentation agent for TiO_2 mfg.; liq.

Flo-Mo® 1X, 2X. [Witco/Organics] Formulated prod.; anionic/nonionic; matched emulsifier pair for pesticide formulation; liq.

Flo-Mo® 8X. [Witco/Organics] Formulated prod.; anionic/nonionic; emulsifier for high conc. phosphate insecticides; used alone or in combination with Flo Mo 1X or 2X; liq.

Flo-Mo® 50H. [Witco/Organics] Calcium alkylaryl sulfonate; anionic; dispersant, emulsifier; usually used in combination with nonionic surfactants; liq.; 50% conc.

Flo-Mo® 80/20. [Witco/Organics] Modified alcohol ethoxylate; nonionic; agric. adjuvant; liq.; 80% conc.

Flo-Mo® 1002. [Witco/Organics] Formulated conc.; nonionic; emulsifier for crop oils formulated from soybean or cottonseed oils; liq.; 93% conc.

Flo-Mo® 1031. [Witco/Organics] Formulated conc.; nonionic; emulsifier for agric. spray oils for citrus fruit; liq.; 100% conc.

Flo-Mo® 1082. [Witco/Organics] Formulated conc.; emulsifier for petrol. oil-based agric. sprays; liq.;

100% conc.

Flo-Mo® 1093. [Witco/Organics] Formulated prod.; emulsifier for use with veg. oils; liq.; oil-sol.; 100% conc.

Flo-Mo® AJ-85. [Witco/Organics] Trideceth-7 (7.5 EO); CAS 24938-91-8; nonionic; detergent, wetting agent; USDA-MID approved for cleaning food handling equipment; liq.; cloud pt. 102–109 F; 85% act.

Flo-Mo® AJ-100. [Witco/Organics] Trideceth-7 (7.5 EO); CAS 24938-91-8; nonionic; detergent, wetting agent; USDA-MID approved for cleaning food handling equip.; paste; 100% act.

Flo-Mo® DEH, DEL. [Witco/Organics] Formulated prod.; anionic/nonionic; matched emulsifier pair for pesticide formulation; liq.

Flo-Mo® Low Foam. [Witco/Organics] Modified alcohol ethoxylate; nonionic; surfactant for formulation of agric. adjuvants; liq.; 100% conc.

Flo-Mo® Suspend. [Witco/Organics] Formulated prod.; anionic; compatibility agent for use in agric. formulations; liq.; 50% conc.

Flo-Mo® Suspend Plus. [Witco/Organics] Formulated prod.; anionic; compatibility agent for use in agric. formulations; liq.; 70% conc.

Flotanol Grades. [Hoechst AG] Polyglycols; flotation frothers.

Flotigam Grades. [Hoechst AG] Fatty amine derivs.; flotation collectors; antibaking agents.

Flotigol CS. [Hoechst AG] Cresylic acid (cresol/xylenol based); flotation frother; liq.

Flotinor FS-2. [Hoechst AG] Fatty acids; flotation collector for nonsulfide minerals; liq.

Flotinor S Grades. [Hoechst AG] Alkyl sulfates; flotation collector; paste.

Flotinor SM 15. [Hoechst AG] Phosphoric acid ester; flotation collector; liq.

Flotol Grades. [Hoechst AG] Pine oils; flotation frothers.

Fluilan. [Croda Inc.; Croda Chem. Ltd.] Lanolin oil; CAS 8006-54-0; nonionic; w/o emulsifier; dispersant for pigments; conditioning agent; emollient, penetrant, superfatting agent for lipsticks, baby oils, brilliantines, cleansing lotions; plasticizer for hair spray resins; moisturizer in w/o emulsions; also for soaps, shampoos, dishwashing liqs., germicidal skin cleansers; pale yel. visc. liq.; pleasant, char. odor; sol. in min. oils, IPA, fatty alcohol, hydrocarbons, and aerosol propellents; cloud pt. 18 C max.; pour pt. 8 C max.; acid no. 2 max.; iodine no. 24-40; sapon. no. 80–100; usage level: 2-10%; 100% act.

Fluilan AWS. [Croda Inc.] PPG-12-PEG-65 lanolin oil; emollient, solubilizer; plasticizer and film modifier for hair sprays; amber visc. liq., nearly odorless; sol. in water and alcohol; acid no. 3 max.; iodine no. 7–15; sapon. no. 10–25; pH 5–7 (10%); usage level: 0.1-2%.

Fluorad FC-10. [3M/Industrial Chem. Prods.] N-Ethyl-N-2-hydroxyethyl perfluorooctanesulfonamide and other primary alcohols; intermediate for prep. of fluorinated surfactants and surface treatments, e.g., ethoxylates, phosphates, esters, sulfates, urethanes; waxy solid (amber liq. @ 90 C); sol. (10 g/10 ml) in 1,1,2-trichlorotrifluoroethane, benzotrifluoride, dimethyl sulfoxide, 2-propanol, acetone, ethyl acetate; dens. 1.71 g/cc (80 C); m.p. 55-65 C; b.p. 115-120 C (1 mm); > 95% purity, 56.5% fluorine.

Fluorad FC-26. [3M/Industrial Chem. Prods.] Perfluorooctanoic acid; intermediate for prep. of monomers and surfactants; colorless to sl. yel. liq.; sol. in relatively polar org. solvs. e.g., ethers, alcohols, ketones, and in hot toluene and hot CCl_4; m.w. 414; sp.gr. 1.7 (65 C); m.p. 52-54 C; b.p. 187-189 C; flash pt. (Seta) > 200 F; 68.8% fluorine.

Fluorad FC-93. [3M/Industrial Chem. Prods.] Ammonium perfluoroalkyl sulfonate; anionic; wetting agent in etching sol'ns. in semiconductor devices; foaming agent, leveling agent, corrosion inhibitor; clear amber liq.; sol. in alcohol/water mixts.; sp.gr. 1.1; flash pt. 81 F (PMCC); pH 7 (1% aq.); 25% act. in water/IPA (73:27).

Fluorad FC-95. [3M/Industrial Chem. Prods.] Potassium perfluoroalkyl sulfonate; anionic; wetting and foaming agents for coatings, etchants, plating baths, cleaning systems; corrosion inhibitor, leveling agent; powd.; low. sol. in water and most inorg. solv.; surf. tens. 22 dynes/cm (0.1% aq.); pH 7–8 (0.1% aq.); 100% act.

Fluorad FC-98. [3M/Industrial Chem. Prods.] Potassium perfluoroalkyl sulfonate; anionic; foamer, corrosion inhibitor, and wetting and leveling agents; powd.; low. sol. in water and most inorg. solv.; surf. tens. 40 dynes/cm (0.1% aq.); pH 6–8 (0.1% aq.); 100% act.

Fluorad FC-99. [3M/Industrial Chem. Prods.] Amine perfluoroalkyl sulfonate; anionic; fluorosurfactant for electrochemical baths, chrome plating, metal and plastic etchants, acid cleaning and pickling; wetting and foaming agent in acidic systems; chemically resistant; clear yel. to amber liq.; misc. with water; sp.gr. 1.1; pH 6–7; surf. tens. 22 dynes/cm (0.4% aq.); 25% act. in water.

Fluorad FC-100. [3M/Industrial Chem. Prods.] Fluoroalkyl sulfonate, sodium salt; amphoteric; surfactant used in electrolyte systems; plating bath antimist and leak detector sol'ns.; foaming agent for metallurgical and electrochem. industries; dk. amber visc. liq.; sol. in most acidic, alkaline and neutral aq. media; sp.gr. 1.1; pour pt. 12 C; flash pt. (Seta CC) > 93 C; surf. tens. 17 dynes/cm (0.4% aq.); 25% conc. in 25% diethylene glycol butyl ether, 50% water.

Fluorad FC-109. [3M/Industrial Chem. Prods.] Potassium fluorinated alkyl carboxylates; anionic; leveling and wetting agent for coatings, floor polishes, alkaline cleaners; wetting agent where unusual chemical stability is required; yel. to amber liq.; low sol. in nonpolar org. solvs., better sol. in polar org.; sp.gr. 1.14; flash pt. (CC) > 200 F; pH 9.5 (0.8% aq.); 25% solids in 61% water, 12% dipropylene glycol methyl ether, 2% ethanol.

Fluorad FC-118. [3M/Industrial Chem. Prods.] Ammonium perfluorooctanoate; anionic; surfactant for emulsion polymerization of fluorinated monomers; lt. colored liq.; sol. in ethanol, acetone; sp.gr. 1.18; pH 5; surf. tens. 44 dynes/cm (0.5% aq.); 20% aq. sol'n.

Fluorad FC-120. [3M/Industrial Chem. Prods.] Ammonium perfluoroalkyl sulfonate; anionic; wetting, leveling agent, surfactant in aq. coating systems, clear polishes; chemical stability; lt. amber to amber clear low visc. liq.; sp.gr. 1.06; visc. 10 cps; flash pt. (Seta CC) 58 C; pH 8.5–9.5; 25% act. in 37.5% 2-butyoxyethanol, 37.5% water.

Fluorad FC-121. [3M/Industrial Chem. Prods.] Ammonium perfluoroalkyl sulfonates; anionic; wetting and leveling agent with chemical stability and activity at low-sol'n. concs.; for aq. coating systems, floor polishes; amber clear liq.; sp.gr. 1.08; flash pt. (CC) 128 F; pH 9; 25% act. in 35% propylene glycol t-butyl ether, 40% water.

Fluorad FC-126. [3M/Industrial Chem. Prods.] Ammonium perfluoroalkane carboxylate; anionic; surfactant for emulsion polymerization of fluorinated monomers; exc. surface activity in alkaline media; lt. colored free-flowing powd.; sol. > 1000 g/1000 g solv. in acetone, methanol, water; m.p. dec. at 120 C; pH 6.0; surf. tens. 44 dynes/cm (1000 ppm aq.); toxicology: moderately irritating ocularly; nonirritating dermally; LD50 (rat, ingestion) 540 mg/kg, moderately toxic; 100% act.

Fluorad FC-129. [3M/Industrial Chem. Prods.] Potassium fluorinated alkyl carboxylate; anionic; wetting and leveling agent, used in alkaline cleaners, floor polishes, coatings; for reaction systems where unusual chemical stability is required; amber liq.; limited sol. in aq. acids and bases; sp.gr. 1.3; visc. 30 cps; flash pt. (PMCC) 46 C; pH 8-11; surf. tens. 17 dynes/cm (0.1–0.2% aq.); 50% act., 32% water, 14% 2-butoxyethanol, 4% ethanol.

Fluorad FC-135. [3M/Industrial Chem. Prods.] Fluorinated alkyl quat. ammonium iodides; cationic; wetting, spreading and leveling agent, used in cleaning and polishing prods., coatings; dk. amber liq.; sol. in nonpolar org. solvs.; sp.gr. 1.2; visc. 30 cps; flash pt. (PMCC) 11 C; pH 3–5 (1% aq.); surf. tens. 17 dynes/cm (0.2% aq.); 50% act., 33% IPA, 17% water.

Fluorad FC-143. [3M/Industrial Chem. Prods.] Ammonium perfluorooctanoate; anionic; emulsifier for fluoropolymer emulsions, wetting agent in alkaline media; chemical stability; lt. colored powd.; sol. in water, methanol, ethyl acetate, acetone; bulk dens. 0.6–0.7 g/cc; decomp. @ 130 C; surf. tens. 44 dynes/cm (0.1% aq.); pH 4–6 (0.5% aq.); 100% act.

Fluorad FC-170-C. [3M/Industrial Chem. Prods.] Fluorinated alkyl POE ethanol; nonionic; wetting, spreading and leveling agent for floor polish formulations, coatings with a min. of foaming; amber visc. liq.; sol. in HCl, alcohols; sp.gr. 1.32; visc. 400 cps; cloud pt. < 0 C; flash pt. (PMCC) > 148 C; surf. tens. 20 dynes/cm (0.1% aq.); pH 7–8 (1% aq.); 95% act.

Fluorad FC-171. [3M/Industrial Chem. Prods.] Fluorinated alkyl alkoxylate; nonionic; wetting, spreading and leveling agent for floor polish formulations, coatings, cleaners; amber liq.; sol. in butyl Cellosolve, toluene, MEK, IPA, dibutyl phthalate, water; sp.gr. 1.4; visc. 150 cps; flash pt. (Seta CC) > 148.8 C; pour pt. -7 C; surf. tens. 20 dynes/cm (0.1% aq.); 100% act.

Fluorad FC-430. [3M/Industrial Chem. Prods.] Fluorinated alkyl ester; nonionic; wetting, leveling, spreading and flow control agents for water- and solv.-based coatings and polymer systems, soldering systems; clear visc. liq.; sol. > 20 g/100 ml in water, Cellosolve acetate, MEK, toluene; sp.gr. 1.15; visc. 7000 cps; flash pt. > 300 F (TOC); ref. index 1.446; surf. tens. 30.3 dynes/cm (0.1 g/100 ml aq.); 100% act.

Fluorad FC-431. [3M/Industrial Chem. Prods.] Fluorinated alkyl ester; nonionic; wetting and leveling agent for org. coatings; straw to amber thin liq.; sol. > 20 g/100 ml in butyl Cellosolve, Cellosolve acetate, MEK, toluene, dibutyl phthalate; sp.gr. 1.05; visc. 200 cp; flash pt. 40 F (TOC); surf. tens. 36.6 dynes/cm (0.1 g/100 ml aq.); ref. index 1.406; 50% act. in ethyl acetate.

Fluorad FC-740. [3M/Industrial Chem. Prods.] Fluoroaliphatic polymeric ester; nonionic; oil well stimulation surfactant; foams hydrocarbon liqs.; active in low polarity org. solvs.; lt. amber liq.; sol. in nonpolar solvs., > 80 g/100 g in kerosene, diesel fuel, xyelne, heptane; sp.gr. 1.0; visc. 150 cps; pour pt. -12 C; flash pt. (Seta CC) 55 C; 50% solids in aromatic naphtha.

Fluorad FC-742. [3M/Industrial Chem. Prods.] Fluorinated alkyl ester; nonionic; oil well stimulation surfactant; foamer for aq./alcoholic fluids; amber liq.; sp.gr. 1.05; visc. 50 cps; flash pt. (PMCC) -8 C; surf. tens. 33.6 dynes/cm (0.5% aq.); 50% act. in ethyl acetate.

Fluorad FC-750. [3M/Industrial Chem. Prods.] Fluorinated alkyl quat. ammonium iodides; cationic; oil well stimulation surfactant; offers lowest surf. tens. for aq. acids, enhances activity of org. corrosion inhibitors; brn. clear liq.; sp.gr. 1.2; flash pt. (Seta CC) 12 C; pour pt. 0-4 C; surf. tens. 17.5 dynes/cm (0.01% aq.); 50% act. in 2/1 IPA/water.

Fluorad FC-751. [3M/Industrial Chem. Prods.] Fluorochemical surfactant; amphoteric; oil well stimulation surfactant; foamer for aq. media; dk. amber liq.; sp.gr. 1.14; visc. 16 cps; pour pt. -9 C; flash pt. (PMCC) > 93 C; surf. tens. 20.5 dynes/cm (0.02% aq.); 25% act. in 25% diethylene glycol butyl ether, 50% water.

Fluorad FC-754. [3M/Industrial Chem. Prods.] Fluorinated alkyl quat. ammonium chloride; cationic; oil well stimulation surfactant; reduces surf. tens. of aq. acid and/or brine fluids or foams; brn liq.; sp.gr. 1.15; flash pt. (Seta CC) 23 C; surf. tens. 17.7 dynes/cm (0.01% aq.); 50% act. in 28% IPA, 22% water.

Fluorad FC-760. [3M/Industrial Chem. Prods.] Fluorinated alkyl alkoxylates; nonionic; oil well stimulation surfactant, useful at low levels to reduce surf. tens. of fluids used for acidizing, fracturing and water flooding; amber liq.; sp.gr. 1.4; flash pt. (Seta CC) > 149 C; pour pt. -7 CF; surf. tens. 19 dynes/cm (100 ppm aq.).

Fluorad FX-8. [3M/Industrial Chem. Prods.] Perfluorooctanesulfonyl fluoride; CAS 307-35-7; intermediate for prep. of monomers and surfactants for textile treatment, paper sizes, inert fluids; colorless clear liq.; sol. in 1,1,2-trichlorotrifluoroethane, benzotrifluoride; m.w. 502; dens. 1.81 g/cc; visc. 6 cps; b.p. 154 C; f.p. -18 to -30 C; flash pt. (Seta) > 200 F; ref. index 1.299.

Fluowet 40 M. [Hoechst Celanese/Colorants & Surf.; Hoechst AG] Fluoroaliphatic ethoxylate; anionic; surfactant for metalworking fluids; intermediate products; wetting agent for mordant baths, aq. and org. solv. systems; liq.; sol. in water; sp.gr. 0.92; pH 6-8 (1% aq.); 100% conc.

Fluowet OL. [Hoechst Celanese/Colorants & Surf.] Ammonium salt of perfluorosulfate ester; surfactant for textile processing; dk. liq.; pH 9 (1% aq.); 50% act.

Fluowet OTN. [Hoechst Celanese/Colorants & Surf.] Ethoxylated perfluorol alcohol; surfactant for textile processing; amber liq.; pH 7 (1% aq.); 99% act.

Fluowet PL. [Hoechst Celanese/Colorants & Surf.] Perfluoro alkyl phosphinate/phosphonate; anionic; surfactant for agric. formulations.

Fluowet PP. [Hoechst Celanese/Colorants & Surf.] Perfluoro alkyl phosphinate/phosphonate; anionic; surfactant for agric. formulations.

Fluowet SB. [Hoechst Celanese/Colorants & Surf.] Sodium salt of perfluorocarboxylic acid ester; surfactant for textile processing; yel. liq.; pH 6 (1% aq.); 30% act.

FMB 65-15 Quat, 65-28 Quat. [Huntington Lab]

Myristalkonium chloride; for formulation of mildewcides and swimming pool algicides; 50% and 80% act. resp.

FMB 210-8 Quat, 210-15 Quat. [Huntington Lab] Didecyldimonium chloride; for formulation of disinfectants, sanitizers, fungicides, water treatment microbicides, swimming pool algicides, mildewcides; 80 and 50% act. resp.

FMB 302-8 Quat. [Huntington Lab] Dicapryl/dicaprylyl dimonium chloride; for formulation of disinfectants, sanitizers, and fungicides; 80% act.

FMB 451-8 Quat. [Huntington Lab] Myristalkonium chloride; for formulation of disinfectants, sanitizers, fungicides, water treatment microbicides, swimming pool algicides, mildewcides; 80% act.

FMB 504-5 Quat. [Huntington Lab] Dicapryl/dicaprylyl dimonium chloride, myristalkonium chloride; for formulation of disinfectants, sanitizers, fungicides, water treatment microbicides, mildewcides; 50% act.

FMB 1210-5 Quat, 1210-8 Quat. [Huntington Lab] Dicetyldimonium chloride, myristalkonium chloride; for formulation of disinfectants, sanitizers, fungicides, water treatment microbicides, swimming pool algicides; 50 and 80% act. resp.

FMB 3328-5 Quat, 3328-8 Quat. [Huntington Lab] Myristalkonium chloride, quaternium-14; for formulation of disinfectants, sanitizers, fungicides, water treatment microbicides, swimming pool algicides; 50 and 80% act. resp.

FMB 4500-5 Quat. [Huntington Lab] Myristalkonium chloride; for formulation of disinfectants, sanitizers, fungicides, water treatment microbicides, swimming pool algicides, mildewcides; 50% act.

FMB 6075-5 Quat, 6075-8 Quat. [Huntington Lab] n-Alkyl dimethyl benzyl ammonium chloride and n-alkyl dimethyl ethylbenzyl ammonium chloride; for formulation of disinfectants, sanitizers, fungicides, water treatment microbicides, and swimming pool algicides; 50 and 80% act.

Foamaster 206-A, 267-A, 335-A. [Henkel] Nonsilicone type based on chemically modified oils and fats; nonionic; defoamer for textile industry; liq.; water disp.; 100% act.

Foamaster 340. [Henkel/Textile] Nonsilicone defoamer; for pressure dyeing applics. in textile industry; amber liq.; sol. forms stable wh. emulsions in water; pH 7.0 (10%); 100% act.

Foamaster 371-S. [Henkel/Textile] Silicone emulsion; defoamer for pressure and atmospheric processes in textile industry; wh. emulsion; sol. forms stable wh. emulsions in water; pH 8.5 (2%); 16% act.

Foamaster 1407-50. [Henkel] Nonionic; defoamer for fluidized paste; water disp.

Foamaster 1719-A. [Henkel] Nonionic; defoamer for use in gelatine; liq.; 50% act.

Foamaster 8034. [Henkel] Nonionic; defoamer for paints, adhesives, rubber, and latex; liq.; 100% act.

Foamaster A. [Henkel/Functional Prods.] Polymerized alcohol; defoamer for use in degassing and monomer stripping operations involving syn. latexes; liq.; 100% act.

Foamaster AP. [Henkel/Functional Prods.] Low silicone hydrophobic silica; defoamer for use in degassing and monomer stripping operations involving syn. latexes; liq.; 100% act.

Foamaster B. [Henkel] Nonionic; predispersible defoamer for latex systems; liq.; 100% act.

Foamaster DD-72. [Henkel] Nonionic; defoamer for emulsion paints, wallpaper coatings, inks; liq.; 100% act.

Foamaster DF-122NS. [Henkel] Fatty soaps, fatty esters in hydrocarbon, silicone-free; nonionic; defoamer for water-based coatings; water-disp.

Foamaster DF-177-F, -178. [Henkel] Nonionic; foam suppressor for coating and adhesive formulations; liq.; 100% act.

Foamaster DF-198-L. [Henkel] Defoamer for paints based on fine particle size acrylic emulsions; liq.; 100% act.

Foamaster DNH-1. [Henkel] Nonionic; defoamer for emulsion paints and polymers; liq.; 100% act.

Foamaster DR-187. [Henkel] Fatty acids, alcohols, esters in hydrocarbon; nonionic; defoamer for mfg. of paper and paperboard; water-disp.

Foamaster DRY. [Henkel] Silicone; dry defoamer for wettable powds. and dry flowable pesticide formulation; powd.

Foamaster DS. [Henkel] Silicone type; defoamer; liq.; water-disp.; 100% act.

Foamaster FLD. [Henkel/Emery] Silicone emulsion; food grade defoamer for aq. systems, agric. formulations; wh. opaque liq.; pH 6.0; 15% act. DISCONTINUED.

Foamaster G. [Henkel] Disp. latex stripping defoamer; liq.; 100% act.

Foamaster JMY. [Henkel] Nonionic; foam control agent for adhesives, rubber, and latex; liq.; 100% act.

Foamaster KF-99. [Henkel] Nonsilicone type based on chemically modified oils and fats; defoamer for textile industry; liq.; disp. in water; 100% act.

Foamaster NDW. [Henkel] Nonionic; defoamer for water-based paints, adhesive and cement screeds, rubber and latex; liq.; 100% act.

Foamaster NS-20. [Henkel] Nonionic; silicone-free defoamer for coatings or paper stock, air knife coatings; liq.; 100% act.

Foamaster NXZ. [Henkel/Functional Prods.] Low silicone, fatty acid base; defoamer for syn. latices, paints, adhesives, blade and roll coatings, emulsifiable latex stripping; liq.; 100% act.

Foamaster P. [Henkel] Nonionic; foam controller in PVAL systems in adhesive industry; liq.; 35% act.

Foamaster PD-1. [Henkel] Nonionic; defoamer for dry mixes, adhesives and cement screeds, joint cements, spackles; powd.; 67% act.

Foamaster Soap L. [Henkel/Emery] Sodium tallowate; dry defoamer for wettable powd. and dry flowable pesticide formulations; EPA-exempt; tan powd. flake; 93% act.

Foamaster TBD-1. [Henkel] Defoamer for water-sol. resins; liq.; 100% act.

Foamaster TDB. [Henkel] Nonionic; foam control agent in dry adhesives, joint cements, spackle, rubber and latex; liq.; 100% act.

Foamaster V. [Henkel] Nonionic; disp. latex stripping defoamer; liq.; 100% act.

Foamaster VC. [Henkel/Functional Prods.] Low silicone hydrophobic silica; defoamer for use in degassing and monomer stripping operations involving syn. latexes; liq.; 100% act.

Foamaster VF. [Henkel/Functional Prods.] Low silicone hydrophobic silica; defoamer for use in degassing and monomer stripping operations involving syn. latexes; liq.; 100% act.

Foamaster VL. [Henkel/Functional Prods.] Low silicone hydrophobic silica; nonionic; emulsifiable latex monomer stripping defoamer, used in latex and emulsion paints; liq.; 100% act.

Foamaster VT. [Henkel/Functional Prods.] Low silicone hydrophobic silica; defoamer for use in degassing and monomer stripping operations involving syn. latexes; liq.; 100% act.

Foamer. [Harcros] Fatty alkanolamide sulfosuccinate; anionic; foam stabilizer for lt. duty liq. detergent systems; liq.; 41% conc.

Foamer. [Hart Chem. Ltd.] Fatty alkanolamide sulfosuccinate; anionic; foam stabilizer for lt. duty liq. detergent systems; liq.; 41% conc.

Foamer AD. [Harcros UK] Proprietary blend; anionic; foamer for air mist drilling, general detergents, gypsum board prod., ether sulfates; amber clear liq.; water-sol.; sp.gr. 1.06; dens. 8.8 lb/gal (2% aq.); pour pt. < 15 F; flash pt. (PMCC) 157 F; pH 7-8 (2% aq.); 50% act.

Foamer CD. [Harcros] Proprietary blend; anionic; foamer for air mist drilling, general detergents, gypsum board prod., ether sulfates; amber clear liq.; water-sol.; sp.gr. 1.07; dens. 8.9 lb/gal (2% aq.); pour pt. 15 F; flash pt. (PMCC) 188 F; pH 7-8 (2% aq.); 55% act.

Foamex AD-50, AD-100, AD-300, J-275. [Rhone-Poulenc Surf.] Silicone defoamers.

Foamex MAB (formerly Quadefome MAB). [Rhone-Poulenc Surf.] Silicone defoamers.

Foamgard 73. [Rhone-Poulenc/Textile & Rubber] Alcohol ether; defoamer for print pastes where silicone type undesirable. DISCONTINUED.

Foamgard 161. [Rhone-Poulenc/Textile & Rubber] Nonsilicone emulsified hydrocarbon; defoamer for carpet dyeing. DISCONTINUED.

Foamgard 200. [Rhone-Poulenc/Textile & Rubber] Nonsilicone; defoamer for dye baths, scouring baths, print pastes and emulsions, paper and pulp processing, waste treatment; water-disp. DISCONTINUED.

Foamkill® 8E. [Crown Tech.] Silicone compd.; full-strength defoamer for oil or solv.-based metalworking lubricants; usage level: 0.05%.

Foamkill® 8G. [Crucible] Silicone compd.; defoamer for food applics. incl. general nonaq. systems, drug extraction and separation, drug processing, starch extractions and processing, anaerobic fermentations, vitamins, etc.

Foamkill® 8J-1. [Crucible] Silicone; defoamer for syn. metalworking lubricants; usage level: 0.5%.

Foamkill® 8R. [Crown Tech.] Silicone compd.; full-strength defoamer for oil or solv.-based metalworking lubricants, paints, coatings; higher visc. than 8E grade.

Foamkill® 30 Series. [Crucible] Org. and organo-silicone conc.; defoamer for food/pharmaceutical applics.

Foamkill® 30C. [Crucible] Organo-silicone emulsion concs.; defoamer for pulp/paper applics. incl. adhesives, casein, neoprene, and natural latices.

Foamkill® 30HP. [Crown Tech.] Organo-silicone emulsion; antifoam/defoamer for waste treatment, pulp/paper, severe foaming situations.

Foamkill® 80J Series. [Crucible] Silicone compd.; defoamer for food/pharmaceutical applics.

Foamkill® 400A. [Crucible] Organo-silicone emulsion; defoamer for pulp/paper applics. incl. adhesives backings, latex paints and coatings, water-reducible inks, floor and ceiling coatings.

Foamkill® 608. [Crucible] Nonsilicone; defoamer for paper coatings and adhesives and formulations sensitive to fish-eyeing, janitorial supply houses; readily emulsifiable; usage level: 0.5-1.0%.

Foamkill® 614. [Crown Tech.] General purpose antifoam/defoamer containing very small amt. of silicone; for inks, adhesives and coatings; suitable for SBU, PVA, PVE, polyethylene, and miscellaneous copolymers.

Foamkill® 614NS. [Crown Tech.] Organic; defoamer used for monomer stripping in ABS systems; usually diluted then fed continuously into system.

Foamkill® 618 Series. [Crucible] Org. and organo-silicone conc.; defoamer for food/pharmaceutical applics. incl. paper coatings and adhesives in contact with food.

Foamkill® 627. [Crucible] Org. conc.; defoamer for pulp/paper applics., water cooling towers, water treatment ponds; lubricant for fine papers.

Foamkill® 628A. [Crucible] Org. defoamer for pulp/paper applics.

Foamkill® 629. [Crown Tech.] Nonsilicone; higher strength defoamer than Foamkill 608; economial defoamer for adhesive and coating formulations; may be added to the grind or let-down; usage level: 0.5-1.0%.

Foamkill® 634 Series. [Crucible] Org. and organo-silicone conc.; defoamer for food/pharmaceutical applics.

Foamkill® 634B-HP. [Crucible] Org. and organo-silicone conc.; defoamer for food/pharmaceutical applics.

Foamkill® 634C. [Crucible] Nonsilicone; defoamer for food applics., canning trade, pasteurizer defoaming; FDA compliance.

Foamkill® 634D-HP. [Crucible] Org. and organo-silicone conc.; defoamer for food/pharmaceutical applics.

Foamkill® 634F-HP. [Crucible] Org. and organo-silicone conc.; defoamer for food/pharmaceutical applics.

Foamkill® 639. [Crucible] Nonsilicone; defoamer for aq. systems, pulp/paper applics., acrylic, PVC, PVA, PVPC, and other coatings, paints, inks, antifoam formulating.; usage level: 0.5-1.0%.

Foamkill® 639AA. [Crucible] 100% organic nonsilicone; defoamer for joint compd. or acrylic systems, inks, latex applics.; usage level: 0.4-0.8%.

Foamkill® 639J. [Crucible] Organo-silicone; low-priced defoamer for adhesives, coatings, paints, inks, e.g., vinyl acrylic paint; may be used in grind, let-down or post-add; highly stable and compat.; usage level: 0.3-0.75%.

Foamkill® 639J-F. [Crucible] Organo-silicone; defoamer for latex monomer stripping formulations, food applics., potato chip mfg.; FDA compliance.

Foamkill® 639JOH. [Crown Tech.] Organo-silicone; defoamer for aq. paint, ink, and coatings systems; exc. for vinyl acrylic paints and other severe foaming applics.; may be used in grind or let-down.

Foamkill® 639L. [Crucible] Org. defoamer for pulp/paper applics. incl. adhesive backings, acrylic coatings, vinyl-acrylic coatings, PVC, PVA, PVAC coatings, paper formation.

Foamkill® 639P. [Crucible] General-purpose defoamer for pulp/paper applics., acrylic/PVC/ PVA/ PVAC coatings.

Foamkill® 639Q. [Crown Tech.] Organo-silicone; defoamer for aq. systems, paints, inks, adhesives and coatings, esp. for high-foaming acrylic or Joncryl resin systems; may be used in grind or let-down; high stability; usage level: 0.2-0.5%.

Foamkill® 644 Series. [Crucible] Org. and organo-silicone conc.; defoamer for food/pharmaceutical applics. incl. paper coatings and adhesives in contact with food, cosmetics.

Foamkill® 649. [Crucible] Nonsilicone; highly compat. defoamer for aq. systems, adhesives, coatings, esp. acrylic systems, paints, inks, drawing and cutting fluids, paper applics.; very little tendency to cause fish-eyeing; usage level: 0.5-1.0%.

Foamkill® 649C. [Crown Tech.] Defoamer with very small amt. of silicone; long-lasting defoamer improving coating leveling; also for inks.

Foamkill® 649N, 649P. [Crucible] Org. defoamer for pulp/paper applics.; lubricant.

Foamkill® 652. [Crown Tech.] Nonionic; inert antifoam/defoamer for printed circuit board mfg.

Foamkill® 652B. [Crucible] 100% organic; nonionic; low emulsifier defoamer for adhesives and coatings, esp. for surfactant-loaded emulsion resin and high shear or grind phases of coatings mfg.; also for printed circuit board mfg.

Foamkill® 652H. [Crucible] Org. and organo-silicone conc.; defoamer for food/pharmaceutical applics.

Foamkill® 652-HF. [Crucible] Org. and organo-silicone conc.; defoamer for food/pharmaceutical applics. incl. general nonaq. systems.

Foamkill® 652L. [Crucible] Conc. multi-ingred. org.; defoamer for pulp/paper applics.

Foamkill® 654NS. [Crown Tech.] Nonsilicone; defoamer for drawing and cutting fluids; emulsifiable and compat. with syn. or oil-based media; usage level: 1%.

Foamkill® 660, 660F. [Crucible] Org.; defoamer for pulp/paper applics., water cooling towers, evaporators.

Foamkill® 663J. [Crucible] Organo-silicone; defoamer for food/pharmaceutical applics. incl. paper coatings and adhesives in contact with food, pulp/paper applics., water cooling towers; lubricant for fine papers; water-disp.

Foamkill® 679. [Crucible] Organo-silicone emulsion; defoamer for pulp/paper applics. incl. adhesives backings, latex paints and coatings, water-reducible inks, floor and ceiling coatings.

Foamkill® 684 Series. [Crucible] Org. and organo-silicone conc.; defoamer for food/pharmaceutical applics. incl. paper coatings and adhesives in contact with food, pulp/paper applics., water treatment ponds.

Foamkill® 684A. [Crucible] Org.; defoamer for pulp/paper applics., water cooling towers, evaporators.

Foamkill® 684P. [Crucible] Defoamer for pulp/paper applics., waste treatment, water cooling towers; lubricant.

Foamkill® 685. [Crown Tech.] Organic; defoamer for jet dyeing of textiles; exc. dye compat.; usage level: 0.1-1.0%.

Foamkill® 687D. [Crown Tech.] Organic; defoamer for waste treatment; balanced flash knockdown and staying power; for intermittent or continuous feed methods.

Foamkill® 687DS. [Crown Tech.] Organo-silicone; defoamer for waste treatment; for intermittent or continuous use.

Foamkill® 700, 700 Conc. [Crucible] All org. liq. versions of paste or brick types; defoamer for pulp/paper applics., waste treatment, water cooling towers; liq.; 60% solids (700); ≈ 100% solids (700 Conc.).

Foamkill® 810. [Crown Tech.] Silicone emulsion; defoamer for acrylic and other high foaming adhesive or coating systems; usage level: 0.2-0.5%.

Foamkill® 810F. [Crucible] Dimethicone; defoamer for food/pharmaceutical applics. incl. general aq. systems, paper coatings and adhesives in contact with food, egg washing, cleaning/sanitizing sol'ns., cosmetics, pulp/paper applics.; FDA compliance; 10% silicone emulsion.

Foamkill® 830. [Crucible] Organo-silicone emulsion; defoamer for food/pharmaceutical applics. incl. paper coatings and adhesives in contact with food, cosmetics, pulp/paper applics.; emulsion; very small particle size.; usage level: 0.1-0.4%.

Foamkill® 830F. [Crucible] Dimethicone; defoamer for food/pharmaceutical applics. incl. general aq. systems, paper coatings and adhesives in contact with food, soft drink mfg., cleaning/sanitizing sol'ns., cosmetics.; FDA compliance.

Foamkill® 836A. [Crucible] Silicone emulsions; defoamer for food/pharmaceutical applics. incl. paper coatings and adhesives in contact with food, cosmetics.

Foamkill® 836B. [Crown Tech.] Silicone emulsion; highly compat. defoamer for chemical compounding, syn. or water-based metalworking lubricants, aq. systems, inks, janitorial applics., paints, coatings, paper industry, waste treatment, severe foaming applics.; dilutable with water.

Foamkill® 852. [Crucible] Silicone emulsion; defoamer for pulp/paper applics., inks, and other applics. where foaming is a serious problem.

Foamkill® 1001 Series. [Crucible] Org. and organo-silicone conc.; defoamer for food/pharmaceutical applics. incl. paper coatings and adhesives in contact with food, pulp/paper applics.

Foamkill® 2890 Conc. [Crown Tech.] Nonsilicone; base defoamer conc. for use as is or let down with water and colloid thickeners for textile use.

Foamkill® 2947. [Crown Tech.] Organo-silicone; for formulating defoamers for wide variety of high and low temp. applics.; suitable for janitorial applics.; dilutable with water.

Foamkill® CMP. [Crown Tech.] Bis-stearamide; nonionic; efficient defoamer for paper reclaiming, caustic treatment, waste treatment, textile jet or atmospheric dyeing; usage level: 0.1-1.0%.

Foamkill® CP. [Crown Tech.] Bis-stearamide; defoamer for paper reclaiming and caustic treatment; readily emulsifiable in water with good flash knockdown and staying power.

Foamkill® CPD. [Crown Tech.] Organo-silicone emulsion; low cost defoamer for aq. systems, inks, paints, coatings, cleaners, floor polishes, carpet cleaners, textile dyeing.

Foamkill® D-1. [Crown Tech.] Org. nonsilicone conc.; defoamer for textile or pulp; dilutable with min. oil.

Foamkill® DF#4. [Crown Tech.] Nonsilicone ester-based; defoamer conc. for waste treatment, inks, and other applics. requiring defoamer active at ambient temps.; for dilution with oil; yields stable emulsions.

Foamkill® DP. [Crown Tech.] 100% organic; defoamer for textile dyeing, esp. jet dyeing of piece goods and atmospheric dyeing of carpet; exc. caustic stability.

Foamkill® EFT. [Crown Tech.] Silicone emulsion; defoamer for waste treatment; good flash knockdown and staying power; for intermittent or continuous feed.

Foamkill® FBF. [Crown Tech.] Organo-silicone emulsion; highly compat. defoamer for inks, floor polish, textile dyeing and finishing; exc. compat. with dyes; usage level: 0.1-1.0%.

Foamkill® FCD. [Crown Tech.] Organic; defoamer for jet or atmospheric dyeing of textiles; exc. dye compat.; usage level: 0.1-1.0%.

Foamkill® FPF. [Crucible] Organo-silicone emulsion; defoamer for pulp/paper applics. incl. adhesives backings, latex paints and coatings, water-reducible inks, floor and ceiling coatings.

Foamkill® GCP Series. [Crucible] Silicone fluid; defoamer for food/pharmaceutical applics., cosmetics.

Foamkill® MS. [Crown Tech.] Silicone emulsion; defoamer for waste treatment, atmospheric or pressure dyeing of textiles; for intermittent or continuous feed; usage level: 0.5-2.0%.

Foamkill® MS Conc. [Crucible] Silicone emulsion; defoamer for food/pharmaceutical applics. incl. paper coatings and adhesives in contact with food, cleaning/sanitizing sol'ns., pulp/paper applics. incl. adhesives, casein, neoprene, and natural latices, antifoam formulating, rug shampoos, cleaners.

Foamkill® MS-1. [Crucible] Organo-silicone emulsion; defoamer for pulp/paper applics. incl. adhesives backings, latex paints and coatings, water-reducible inks, floor and ceiling coatings; flow control agent.

Foamkill® MSB Conc. [Crown Tech.] High silicone conc.; used to formulate silicone defoamers for resale.

Foamkill® MSF Conc. [Crown Tech.] Organo-silicone; conc. for formulating defoamers for wide variety of high and low temp. applics., esp. food applics. (fruit and veg. washing, egg washing), water-based inks and coatings, detergents, rendering of oils and fats.

Foamkill® NSP-1, NSP-3, NSP-4, NSP-5. [Crucible] Org. defoamer for pulp/paper applics.

Foamkill® SEA. [Crown Tech.] Nonsilicone; defoamer for adhesives and coatings; highly stable and compat. with most systems; usage level: 0.5-1.0%.

Foamole M. [Van Dyk] Cocamide MEA (1:1); CAS 68140-00-1; foam booster/stabilizer, thickener, emulsifier for creams and lotions, shampoos, bubble baths, other detergents; cream colored flake; sol. in 70% ethanol, propylene glycol, oleyl alcohol; m.p. 70-74 C; acid no. 2 max.; alkali no. 12 max.; 100% conc.

Foramousse D. [Seppic] Surfactant blend; anionic/nonionic; drilling assistant; liq.; 100% conc.

Foramousse WO 107. [Seppic] Surfactant/foam stabilizer blend; amphoteric; drilling assistant; liq.; 30% conc.

Forbest 13. [Lucas Meyer] Polyester/fatty alkyls compd.; wetting and dispersing agent for water-based varnishes, highly pigmented slurries; binding agent; yel. clear thin liq.; usage level: 0.5-1.5%; 50% conc.

Forbest 18. [Lucas Meyer] Aliphatic ester; anionic; wetting and dispersing agent for solv.-based varnishes, esp. highly pigmented pumpable formulations; brn. visc. clear liq.; usage level: 1-3%; 50% conc.

Forbest 20. [Lucas Meyer] Polymeric chelating agent, neutralized; wetting and dispersing agent for water-based varnishes, highly pigmented slurries; binding agent; water clear thin liq.; usage level: 0.3-0.6%; 50% conc.

Forbest 30. [Lucas Meyer] Modified aryl-alkyl silicones; antifloating and antisilking agent for all paint and varnish systems except water-based systems; lt. blue mobile liq.; sp.gr. 0.818; flash pt. 28 C; ref. index 1.460; usage level: 0.1-0.3%; 50% act. in xylene and hydrocarbons.

Forbest 33. [Lucas Meyer] Polyalkoxy ether compd.; wetting and dispersing agent for water-thinnable systems; counteracts edge effects; brn. visc. clear liq.; usage level: 0.3-0.8%; 40% conc.

Forbest 50. [Lucas Meyer] Modified aryl-alkyl silicones; antifloating and antisilking agent for solv. systems, curtain coats; colorless mobile liq.; sp.gr. 0.818; flash pt. 28 C; usage level: 0.1-0.3%; 50% act. in xylene.

Forbest 150. [Lucas Meyer] Polyaryl ester; leveling agent for air-dry paints with high pigment loading; sp.gr. 1.085; flash pt. 88 C; 97% solids.

Forbest 560. [Lucas Meyer] Polyalkyl ester; nonionic; wetting and dispersing agent for solv.-containing paints; brn. liq.; usage level: 1-5%; 80% conc.

Forbest 610. [Lucas Meyer] Carboxylic acid diamine compd.; cationic; wetting and dispersing agent for aq. systems; recommended for air-dry and stoving water-thinnable systems based on emulsions, alkyd, and acrylic resins; brn. clear visc. liq.; sp.gr. 1.01; flash pt. 10 C; pH 8.6; 60% act. in ethanol.

Forbest 620. [Lucas Meyer] Based on polyoxy alkylic siloxanes and cyclic amide; nonionic; leveling agent for water-thinnable systems incl. varnishes, roller paints, coil coatings, emulsion paints; lt. ylsh. grn. low-visc. liq.; sp.gr. 1.0; pH 7.5; usage level: 0.3-1%.

Forbest 680. [Lucas Meyer] Alkyaryl-silicone compd.; nonionic; smoothness and leveling agent for aq. systems; recommended for water-thinnable varnishes, coil coating systems; water clear liq.; sp.gr. 1.03; visc. 1600 mPas; flash pt. 66 C; pH 7.0; usage level: 0.3-1%; 50% act. in ethanol.

Forbest 850. [Lucas Meyer] Fatty acid amino-carboxy complex; cationic; wetting agent with anticorrosion act.; for air-dry anticorrosive paints and primers, asphalt paints, marine paints, road-marking paints, caulking compds.; brn. paste; m.p. 30 C; flash pt. 150 C; 100% solids.

Forbest 1000B. [Lucas Meyer] Fatty acid/paraffin oil blend; nonionic; defoamer for solv. and aq. systems; used for solvent-based air-dry and stoving finishes, aq. systems, dip-coatings, emulsion paints, varnishes; BGA approved; ylsh. turbid liq.; sp.gr. 0.87; visc. 300–450 mPas; flash pt. 172 C; usage level: 0.1-0.5%; 100% act.

Forbest 1500W. [Lucas Meyer] Hydrocarbon-fatty acid oil-based blend; nonionic; defoamer for air-dry and stoving water-reducible systems based on emulsions, alkyds, acrylics; ylsh. turbid liq.; sp.gr. 0.84; flash pt. 170 C; usage level: 0.1-0.5%; 100% act.

Forbest 2000C. [Lucas Meyer] Min. oil/fatty acid partial ester; nonionic; defoamer for emulsion paints and varnishes, roughcasts, water-disp. alkyd emulsions, styrene-acrylates; lt. brn. cloudy, visc. liq.; sp.gr. 1.0; flash pt. > 160 C; usage level: 0.1-0.5%; 100% act.

Forbest 8209. [Lucas Meyer] High m.w. compd.; nonionic; wetting, dispersing agent, emulsifier with antistatic properties for aq. and solv. systems; used for air-dry and stoving finishes and coatings; ylsh. slightly turbid visc. liq.; sp.gr. 1.08; pH 6–8; 97% solids.

Forbest G23. [Lucas Meyer] Organo-modified polysiloxanes and leveling components in high-boiling solvs.; smoothness and leveling agent for all air-dry and stoving solv. coatings, printing inks, water-thinnable coatings; lt. yel. oily liq.; sp.gr. 1.0; visc. 21 s; flash pt. 43 C; ref. index 1.429–1.431; usage level: 0.05-0.3%; 50% solids in ethylglycol.

Forbest S 7. [Lucas Meyer] Acrylic polymerisate compd.; wetting and dispersing agent for pigment concs.; resists light and UV; brn. med.-visc. clear liq.; usage level: 2-10%; 55% conc.

Forbest VP 13. [Lucas Meyer] Polyester-fatty alcohol compd.; wetting and disp. agent , binder for all water-thinnable paints incl. leather finishes, air-dry and stoving alkyd systems, electrodeposition, emulsion paints, pigment slurries for paper coating; ylsh. clear liq.; sp.gr. 0.90; visc. 50 s; flash pt. 40 C; pH 10.5; 50% solids.

Forbest VP 18. [Lucas Meyer] Aliphatic ester; anionic; wetting and disp. agent for pigmented air-dry and stoving paints and lacquers, e.g., anticorrosive paints, alkyd enamels, zinc paints, marine paints, bituminous paints, epoxy paints, putty, printing inks, pigment concs.; brn. clear visc. liq.; sp.gr. 0.97; visc. 24 Pas; flash pt. > 250 C; 100% solids.

Forbest VP 20. [Lucas Meyer] Compd. based on neutralized chelating agents; wetting and disp. agent for water-thinnable paints, e.g., leather finishes, air-dry and stoving alkyds, electrodeposition, emulsion paints, pigment concs.; nearly colorless thin liq.; sp.gr. 1.19; pH 6; 50% solids.

Forbest VP 33. [Lucas Meyer] Modified polyalkoxyether; wetting and disp. agent for water-thinnable paints, air-dry and stoving alkyds, emulsion paints; brn. clear visc. liq.; sp.gr. 1.05; visc. 1 Pas; pH 7; 40% solids.

Forbest VP S7. [Lucas Meyer] Acrylic polymers dissolved in Shellsol A; amphoteric; wetting and disp. agent, binder, antisettling agent for pigment pastes; brn. clear liq.; sp.gr. 0.9; visc. 18–22 Pas; 55% solids in Shellsol A.

Forbest WP. [Lucas Meyer] High m.w. syn. compd.; wetting, dispersing, and antisettling agent for all air-dry and stoving solvent-based paints and systems with polar solvs.; brn. liq., mildly fragrant; sp.gr. 0.894; flash pt. 185 C; pH 9.5; 95% solids.

Forlanit P. [Henkel/Cospha; Henkel/Functional Prods.; Henkel Canada; Henkel KGaA] Sodium laureth phosphate; anionic; wetting agent, dispersant, flotation aux. for metal cleaners, galvanic baths, hot copper baths containing cyanides; liq.; 30% conc.

Forlanon. [Henkel KGaA] Sodium laureth phosphate; anionic; wetting agent, flotation aux., for hot-baths containing cyanides; liq.; 10% conc.

Fosfamide CPD-170. [Henkel/Cospha; Henkel Canada] Complex alkanolamide phosphate ester; anionic; emulsifier for prep. of metalworking fluids and greases; imparts lubricity, anticorrosion and mild EP props.; grease additive; liq.; 98% conc.

Fosfamide N. [Henkel/Cospha; Henkel Canada] Fatty amido phosphate complex; anionic; detergent, foaming and wetting agent for textile, all-purpose, institutional, and metal cleaners; liq.; 100% act.

Fosterge BA-14 Acid. [Henkel] Mono and dialkyl phenoxy POE acid phosphate; anionic; detergent intermediate for cleaners and emulsifier for min. oils and pesticides; liq.; dens. 8.50 lb/gal; 98–100% act.

Fosterge H Acid. [Henkel Canada] Mono and dialkyl phenoxy POE acid phosphate; emulsifier; liq.; dens. 8.8 lb/gal; 98–100% act.

Fosterge LF. [Henkel/Cospha; Henkel Canada] Alkylphenoxy POE acid phosphate; anionic; intermediate for foaming cleaners and solv. emulsion cleaners; emulsifier for pesticides; corrosion inhibitor; liq.; dens. 9.17 lb/gal; 98–100% act.

Fosterge LFD. [Henkel Canada] DEA salt of Fosterge LF Acid; anionic; for foaming cleaners, solv. emulsion cleaners, and emulsifiers; liq.; 100% conc.

Fosterge LFS. [Henkel Canada] Sodium salt of Fosterge LF Acid; anionic; emulsifier, dispersant, detergent; for foaming cleaners, solv. emulsion cleaners; liq.; 90% conc.

Fosterge R. [Henkel/Cospha; Henkel Canada] Mono and dialkyl phosphoric acid; anionic; intermediate for emulsifying, penetrating, and anticorrosion compds.; liq.; dens. 8.5 lb/gal; 100% act.

Fosterge RD. [Henkel Canada] Fosterge R Acid DEA salt; anionic; emulsifier, penetrant, and anticorrosion agent used in industrial cleaners; liq.; dens. 9.15 lb/gal; 86–88% act.

Fosterge W. [Henkel/Cospha; Henkel Canada] Alkyl phenoxy POE phosphoric acid; anionic; detergent, emulsifier, coupling agent and corrosion inhibitor; dedusting agent for alkaline powds.; liq.; dens. 9.29 lb/gal; 100% act.

Fuman 630. [Pulcra SA] Alkylbenzene sulfonate blend; scouring agent for textile industry, esp. for cotton, wool, polyester, and rayon fibers; liq.; 20-30% conc.

G

G-250. [ICI Am.] Polyoxyalkylene amine quat.; cationic; antistat, emulsifier; liq.; 100% conc.

G-263. [ICI Am.] N-Cetyl, N-ethyl morpholinium ethosulfate aq. sol'n.; cationic; antistat, emulsifier; liq.; HLB 30.0; 35% conc.

G-265. [ICI Am.] Fatty quat. ammonium deriv.; cationic; antistat, emulsifier; liq.; HLB 33.0; 100% conc.

G-1000-S Antifoam Compd. [Genesee Polymers] Nonsilicone; defoamer for aq. systems, industrial applics., commercial cleaners, sewage treatment, chemical processing, paints/coatings, cutting oils, metalworking fluids; rapid knockdown of foam; wh. opaque liq.; sp.gr. 0.8; dens. 6.0 lb/gal; flash pt. (TCC) 120 F; combustible; 92.5% act.

G-1045A. [ICI Am.] POE sorbitol oleate-laurate; nonionic; general purpose emulsifier for textile applics.; liq.; HLB 11.4; 100% conc.

G-1086. [ICI Am.] POE sorbitol oleate; nonionic; general purpose emulsifier for textile applics.; liq.; HLB 10.2; 100% conc.

G-1087. [ICI Am.] POE sorbitol oleate; nonionic; general purpose emulsifier for textile applics.; liq.; HLB 9.2; 100% conc.

G-1089. [ICI Am.] POE sorbitol oleate; nonionic; general purpose emulsifier for textile applics.; liq.; HLB 10.4; 100% conc.

G-1096. [ICI Am.] POE sorbitol oleate; nonionic; general purpose emulsifier for textile applics.; liq.; HLB 11.4; 100% conc.

G-1120. [ICI Am.] Acid-capped POE alcohol; nonionic; emulsifier and lubricant for textiles; liq.; HLB 12.1; 100% conc.

G-1121. [ICI Am.] Acid-capped POE alcohol; nonionic; emulsifier and lubricant for textiles; liq.; HLB 12.7; 100% conc.

G-1292. [ICI Am.] POE hydrog. castor oil; nonionic; general purpose emulsifier for textile applics.; liq.; HLB 10.8; 100% conc.

G-1293. [ICI Am.] POE castor oil; nonionic; general purpose emulsifier for textile applics.; liq.; HLB 11.5; 100% conc.

G-1300. [ICI Am.] POE castor oil; nonionic; general purpose emulsifier for textile applics.; liq.; HLB 18.1; 100% conc.

G-1556. [ICI Am.] Fatty ester blend; nonionic; fiber lubricant; liq.; HLB 11.2; 100% conc.

G-1564. [ICI Am.] Fatty ester blend; nonionic; textile softener, lubricant; solid; 100% conc.

G-2109. [ICI Am.] POE laurate; nonionic; fiber lubricant, emulsifier; liq.; HLB 13.0; 100% conc.

G-2162. [ICI Spec. Chem.] PEG-25 propylene glycol stearate; nonionic; surfactant; cream solid; sol. in water, alcohol; HLB 16.0; pour pt. 23 C.

G-2200. [ICI Am.] Potassium salt of alkyl phosphate ester; anionic; antistatic agent; liq.; 45% conc.

G-2207. [ICI Am.] Potassium salt of alkyl phosphate ester; anionic; antistatic agent; liq.; 45% conc.

G-3300. [ICI Am.] Alkylaryl sulfonate; anionic; emulsifier, dispersant; liq.; HLB 11.7; 90% conc.

G-3634A. [ICI Am.] Fatty quat. ammonium deriv.; cationic; antistatic agent, emulsifier; liq.; HLB 18.5; 100% conc.

G-3780-A. [ICI Spec. Chem.] PEG-20 tallow amine; nonionic; antistat, emulsifier; liq.; HLB 15.5.

G-3886. [ICI Am.] Alkyl-capped POE alcohol; nonionic; fiber lubricant, emulsifier; liq.; HLB 10.5; 100% conc.

G-3890. [ICI Am.] Alkyl-capped POE alcohol; nonionic; fiber lubricant, emulsifier; liq.; HLB 12.0; 100% conc.

G-4252. [ICI Spec. Chem.] PEG-80 sorbitan palmitate; nonionic; surfactant; yel. liq.; sol. in water, alcohol; disp. in propylene glycol; visc. 8500 cps; HLB 19.0.

G-4280. [ICI Spec. Chem.] PEG-80 sorbitan laurate; nonionic; surfactant; yel. liq.; sol. in water, alcohol; disp. in propylene glycol; visc. 1000 cps; HLB 19.1.

G-5000. [ICI Am.] Alkoxylated alcohol; nonionic; high performance emulsifier; solid; HLB 16.9; 100% conc.

G-7076. [ICI Am.] POE fatty ester; nonionic; fiber lubricant, emulsifier; liq.; HLB 10.6; 100% conc.

G-7205. [ICI Am.] Sulfated castor oil; anionic; wetting and dispersing agent, leveling agent for sulfur and direct dyes; liq.; HLB 11.0; 77% conc.

G-7274. [ICI Am.] Sulfated castor oil; anionic; wetting and dispersing agent, leveling agent for sulfur and direct dyes; liq.; HLB 11.0; 50% conc.

Gafac® BH-650. [Rhone-Poulenc Surf.] Free acid of complex org. phosphate ester; anionic; pesticide emulsifier, industrial and household detergent, hard surface cleaners, textile processing esp. mercerizing; clear visc. liq.; sol. in butyl Cellosolve, ethanol, water; dens. 11.4 lb/gal; sp.gr. 1.36–1.38 (50 C); pour pt. < 5 C; pH < 3 (10%); 100% act. DISCONTINUED.

Gafac® BI-729. [Rhone-Poulenc Surf.] Free acid of complex org. phosphate ester; anionic; pesticide emulsifier, industrial and household detergent, liq. drain cleaners; clear visc. liq.; sol. in butyl Cellosolve, ethanol, water; dens. 10.3 lb/gal; sp.gr. 1.23–1.25 (50 C); pour pt. < 0 C; pH 2.5 (10%); 89% act. DISCONTINUED.

Gafac® BI-750. [Rhone-Poulenc Surf.] Free acid of complex org. phosphate ester; anionic; surfactant for industrial cleaners, liq. drain cleaners; stable to caustic; clear visc. liq.; sol. in butyl Cellosolve, ethanol, water; dens. 10.6 lb/gal; sp.gr. 1.26–1.28

(50 C); pour pt. < 0 C; pH < 2.5 (10%); 100% act. DISCONTINUED.

Gafac® BP-769 (redesignated Rhodafac® BP-769). [Rhone-Poulenc Surf.]

Gafac® BX-660 (redesignated Rhodafac® BX-660). [Rhone-Poulenc Surf.]

Gafac® BX-760. [Rhone-Poulenc Surf.] Phosphate ester; anionic; hydrotrope for use in alkaline cleaning sol'ns.; emulsifier and stabilizer for emulsion polymerization; solubilizer for nonionic surfactants in highly alkaline aq. sol'ns.; liq.; 90% act. DISCONTINUED.

Gafac® GB-520 (redesignated Rhodafac® GB-520). [Rhone-Poulenc Surf.]

Gafac® LO-529 (redesignated Rhodafac® LO-529). [Rhone-Poulenc Surf.]

Gafac® MC-470 (redesignated Rhodafac® MC-470). [Rhone-Poulenc Surf.]

Gafac® PE-510 (redesignated Rhodafac® PE-510). [Rhone-Poulenc Surf.]

Gafac® RA-600 (redesignated Rhodafac® RA-600). [Rhone-Poulenc Surf.]

Gafac® RD-510 (redesignated Rhodafac® RD-510). [Rhone-Poulenc Surf.]

Gafac® RE-410 (redesignated Rhodafac® RE-410). [Rhone-Poulenc Surf.]

Gafac® RE-610 (redesignated Rhodafac® RE-610). [Rhone-Poulenc Surf.]

Gafac® RE-877. [Rhone-Poulenc Surf.] Aromatic phosphate ester; anionic; emulsifier, stabilizer, solubilizer, detergent, dispersant, wetting agent, antistat, lubricant for pesticides, textile wet processing, fiber and metal lubricants, emulsion polymerization, cosmetics; clear visc. liq.; sol. in xylene, butyl Cellosolve, ethanol, water; sol. hazy in perchloroethylene; sp.gr. 1.155; dens. 9.6 lb/gal; pour pt. 2 C; acid no. 60–74; pH < 2.5 (10%); 75% act. DISCONTINUED.

Gafac® RE-960 (redesignated Rhodafac® RE-960). [Rhone-Poulenc Surf.]

Gafac® RL-210. [Rhone-Poulenc Surf.] Aliphatic phosphate ester, free acid; anionic; mold release and defoaming agent; waxy solid; 100% act. DISCONTINUED.

Gafac® RM-510 (redesignated Rhodafac® RM-510). [Rhone-Poulenc Surf.]

Gafac® RM-710 (redesignated Rhodafac® RM-710). [Rhone-Poulenc Surf.]

Gafac® RP-710 (redesignated Rhodafac® RP-710). [Rhone-Poulenc Surf.]

Gafac® RS-410 (redesignated Rhodafac® RS-410). [Rhone-Poulenc Surf.]

Gafac® RS-610 (redesignated Rhodafac® RS-610). [Rhone-Poulenc Surf.]

Gafac® RS-710 (redesignated Rhodafac® RS-710). [Rhone-Poulenc Surf.]

Gantrez® AN. [ISP] PVM/MA copolymer; anionic; dispersant, coupling, stabilizing agent, thickener, emulsifier, solubilizer, corrosion inhibitor, film former, antistat, used in agric., paper and textile industries, chemical processing, industrial products, detergents, cosmetics; wh. fluffy powd.; sol. in water, acid, and several org. solvs.; dens. 0.32 g/cc; sp.gr. 1.37; soften. pt. 200–225 C; pH 2 (free acid, 5% aq.); 100% conc.

Gantrez® AN-119. [ISP] PVM/MA copolymer; CAS 52229-50-2; anionic; thickener, dispersant, stabilizer, gelling agent, coupler, protective colloid, suspending aid used in emulsion polymerization; adhesives, household detergents, liq. hand cleaners, acid bowl cleaners; produces clear films of high tens. and cohesive str.; powd.; sol. in water, acid, caustic, and org. solv.; 100% conc.

Gantrez® AN-139. [ISP] PVM/MA copolymer; CAS 52229-50-2; anionic; thickener, dispersant, foam stabilizer, coupling agent; for emulsion polymerization, pesticides, petrol. prod., heavy-duty liq. detergents, adhesives; powd.; sol. in water, acid, caustic, and org. solv.; 100% conc.

Gantrez® AN-149. [ISP] PVM/MA copolymer; CAS 52229-50-2; anionic; thickener, dispersant, foam stabilizer, coupling agent; for emulsion polymerization, pesticides, petrol. prod., heavy-duty liq. detergents, adhesives; powd.; sol. in water, acid, caustic, and org. solv.; 100% conc.

Gantrez® AN-169. [ISP] PVM/MA copolymer; CAS 52229-50-2; anionic; thickener, dispersant, foam stabilizer, coupling agent; for emulsion polymerization, pesticides, petrol. prod., heavy-duty liq. detergents, adhesives; powd.; sol. in water, acid, caustic, and org. solv.; 100% conc.

Gantrez® AN-179. [ISP] PVM/MA copolymer; CAS 9011-16-9; anionic; thickener, dispersant, foam stabilizer, coupling agent; for emulsion polymerization, pesticides, petrol. prod., heavy-duty liq. detergents, adhesives; powd.; sol. in water, acid, caustic, and org. solv.; 100% conc.

Gantrez® S-95. [ISP] PVM/MA copolymer; anionic; hydrolyzed low m.w. polymer; water-sol. polyelectrolyte similar to Gantrez AN series; chelating agent; powd.; rapid cold-water solubility over entire pH range; pH 2 (5% aq.); 100% conc.

Gardilene IPA/94. [Albright & Wilson Australia] Isopropylamine alkylbenzene sulfonate; CAS 68584-24-7; anionic; emulsifier, drycleaning assistant, dispersant; amber clear visc. liq.; pH 6.5; flamm.; 94% conc.

Gardilene S25L. [Albright & Wilson Australia] Sodium alkylbenzene sulfonate; CAS 68081-81-2; anionic; wetting, dispersing, and foaming agent, emulsifier in emulsion polymerization; detergent ingred.; pale yel. clear visc. liq.; visc. 3300 cps; clear pt. < 6 C; pH 7.2 (10%); 25% act. in water.

Gardinol CX. [Albright & Wilson Australia] Sodium cetyl/oleyl alcohol sulfate; anionic; detergent, wetting agent, dispersant, dyeing assistant and emulsifier used in fiber applics., hand cleaners, laundry detergents; scouring agent for sheep skins; > 80% biodeg.; off-wh. to cream soft paste, visc. liq. > 30 C; sp.gr. 1.05 (35 C); pH 7.0 ± 0.5 (10% aq.); 30.0 ± 1.0% act.

Gardinol CXM. [Albright & Wilson Australia] Sodium cetyl/oleyl sulfate, modified; anionic; textile scouring agent; for sheep skins; > 80% biodeg.; off-wh. to cream paste, visc. liq. > 30 C; dissolves in water; sp.gr. 1.02 (30 C); pH 7.5 ± 0.5 (10%); 23.0 ± 0.7% act.

Gardinol WA Paste. [Ronsheim & Moore] Sodium lauryl sulfate; CAS 151-21-3; anionic; detergent for shampoos, fabrics; emulsifier for cosmetics; wh. to cream stiff paste; 40% conc.

Gardiquat 12H. [Albright & Wilson UK] Benzalkonium chloride; bactericide in disinfectants and antiseptics; used in hospitals, food processing industries, and oil drilling muds; colorless to pale yel. clear liq.; sp.gr. 0.98; cloud pt. 0 C; pH 7.0 ± 0.5 (1% aq.); 50.0 ± 1.0% act. in water.

Gardiquat 1450. [Albright & Wilson UK] Benzalkonium chloride USP; germicide, deodorant, algicide, slimicide; almost water wh. clear liq.; cloud pt. 1 C;

pH 7.0 ± 0.5 (1%); flam.; 50% act.

Gardiquat 1480. [Albright & Wilson UK] Benzalkonium chloride USP; germicide, deodorant, algicide, slimicide; almost water wh. clear liq.; sp.gr. 0.945; cloud pt. 1 C max.; flash pt. 32 C (PMCC); pH 7 ± 0.5 (1%); 80% act.

Gardiquat SV 480. [Albright & Wilson UK] Benzalkonium chloride; germicide, deodorant, algicide, slimicide; water wh. clear liq.; sp.gr. 0.929; visc. 106 s (#4 cup); cloud pt. < 0 C; flash pt. 29 C (PMCC); pH 7.2 (1%); 80% act.

Gardisperse AC. [Albright & Wilson Australia] Sodium nonylphenol POE sulfate; CAS 9014-90-8; anionic; emulsifier, dispersant, wetting agent; amber clear liq.; visc. 120 cP max.; cloud pt. 2 C max.; pH 8.5 ± 1.0 (2%); 30.0 ± 1.0% act.

Geitol RC-100. [Tokai Seiyu Ind.] Mixt. of nonionic agent and specific builders; nonionic; detergent, reduction cleaning agent for polyester fibers; powd.; 100% conc.

Geleol. [Gattefosse; Gattefosse SA] Glyceryl stearate; CAS 31566-31-1; nonionic; emulsifier; solid; HLB 3.0; 100% conc.

Gemtex 445. [Finetex] Sodium diisobutyl sulfosuccinate; anionic; rapid wetting agent, penetrant, emulsifier, dispersant, hydrophilic wetter, coemulsifier, surfactant for emulsion polymerization of styrene/butadiene systems; liq.; 45% conc.

Gemtex 680. [Finetex] Sodium dihexyl sulfosuccinate; anionic; emulsifier, rapid wetter, penetrant, surfactant for emulsion polymerization in vinyl acedtate/acrylate, styrene/butadiene, and styrene/acrylamide systems; liq.; 80% conc.

Gemtex 691-40. [Finetex] Sodium dicyclohexyl sulfosuccinate; anionic; rapid wetting agent, penetrant, emulsifier, dispersant for emulsion polymerization of styrene/butadiene systems; post-additive for latex systems; liq.; 40% conc.

Gemtex PA-70P. [Finetex] Dioctyl sodium sulfosuccinate, propylene glycol.; anionic; wetting agent, penetrant; liq.; 70% conc.

Gemtex PA-75. [Finetex] Dioctyl sodium sulfosuccinate, isopropyl alcohol; anionic; wetting agent, penetrant for textile uses and emulsion polymerization; dye leveler; bubble baths, bath oils; liq.; 75% conc.

Gemtex PA-75E. [Finetex] Dioctyl sodium sulfosuccinate; anionic; wetting agent, penetrant.

Gemtex PA-85P. [Finetex] Dioctyl sodium sulfosuccinate, propylene glycol; anionic; wetting agent, penetrant for textiles, emulsion polymerization; dye leveler; bubble baths, bath oils; liq.; 85% conc.

Gemtex PAX-60. [Finetex] Dioctyl sodium sulfosuccinate, anhyd.; anionic; wetting agent for textile uses and emulsion polymerization; dye leveler; bubble baths, bath oils; liq.; 60% conc.

Gemtex SC-40. [Finetex] Dioctyl sodium sulfosuccinate; anionic; wetting/rewetting agent, dye leveler for textiles; liq.; 40% conc.

Gemtex SC-70. [Finetex] Dioctyl sodium sulfosuccinate; anionic; wetting and rewetting agent, dye leveler for textiles; liq.; 70% conc.

Gemtex SC-75. [Finetex] Dioctyl sodium sulfosuccinate, isopropyl alcohol; anionic; dispersant, wetting agent, pasting aid for textile dyeing applics.; antifog for glass cleaners, windshield washers; water and alcohol sol.; 75% conc.

Gemtex SC-75E, SC Powd. [Finetex] Dioctyl sodium sulfosuccinate; anionic; wetting agent for textile uses and emulsion polymerization; dye leveler; bubble baths, bath oils; liq. and powd. resp.; 75 and 83% conc. resp.

Gemtex SM-33. [Finetex] Dioctyl sodium sulfosuccinate-based; anionic; rapid wetting agent for nonwoven and textile industries; liq.; 75% conc.

Gemtex WBT. [Finetex] Sodium alkoxy carboxylate; anionic; wetting agent, improves detergency in alkaline systems; for hard surf. cleaners; pract. nonfoaming, alkali-stable; liq.; 50% conc.

Gemtex WBT-9. [Finetex] Synergistic blend of anionics; anionic; wetting agent, penetrant, mercerizing agent; boosts detergency in low to high alkaline systems; for hard surf. cleaners; alkali-stable; liq.; 50% conc.

Gemtex WNT-Conc. [Finetex] Carboxylated alcohol ethoxylate; anionic; detergent, wetting agent; liq.; 85% conc.

Genagen C-100. [Hoechst Celanese/Colorants & Surf.] PEG-10 cocoate; surfactant for textile processing; yel. liq.; visc. 90 cps; sapon. no. 90; hyd. no. 90; pH 7 (1% aq.); 99% act.

Genagen CA-050. [Hoechst Celanese/Colorants & Surf.; Hoechst AG] PEG-5 cocamide; nonionic; cleansing skin protective component for detergents and cleaning agents; 100% conc.

Genagen CD. [Hoechst Celanese/Colorants & Surf.; Hoechst AG] Fatty acid ester; nonionic; lubricant for fiber preparation; liq.

Genagen KFC-100. [Hoechst Celanese/Colorants & Surf.] Coconut fatty acid ethoxylate; surfactant for textile processing; yel. liq.; visc. 90 cps; sapon. no. 90; hyd. no. 90; pH 6 (1% aq.); 99% act.

Genagen O-090. [Hoechst Celanese/Colorants & Surf.] PEG-9 oleate; CAS 9004-96-0; surfactant for textile processing; yel. liq.; visc. 100 cps; sapon. no. 85; hyd. no. 85; pH 7 (1% aq.); 99% act.

Genagen OD-090. [Hoechst Celanese/Colorants & Surf.] C8-10 acid ethoxylate; surfactant for textile processing; yel. liq.; visc. 50 cps; sapon. no. 100; hyd. no. 100; pH 6 (1% aq.); 99% act.

Genagen P-070. [Hoechst Celanese/Colorants & Surf.] PEG-7 palmitate; surfactant for textile processing; yel. liq.; sapon. no. 100; hyd. no. 100; pH 6 (1% aq.); 99% act.

Genagen PL-090. [Hoechst Celanese/Colorants & Surf.] PEG-9 pelargonate; surfactant for textile processing; yel. liq.; visc. 60 cps; sapon. no. 100; hyd. no. 105; pH 7 (1% aq.); 98% act.

Genagen S-080. [Hoechst Celanese/Colorants & Surf.] PEG-8 stearate; CAS 9004-99-3; surfactant for textile processing; wh. solid; sapon. no. 90; hyd. no. 93; pH 7 (1% aq.); 99% act.

Genagen S-400. [Hoechst Celanese/Colorants & Surf.] PEG-40 stearate; CAS 9004-99-3; surfactant for textile processing; wh. solid; sapon. no. 30; hyd. no. 30; pH 7 (1% aq.); 99% act.

Genagen TA-080. [Hoechst Celanese/Colorants & Surf.] PEG-8 tallate; CAS 61791-00-2; surfactant for textile processing; amber liq.; visc. < 500 cps; sapon. no. 90; hyd. no. 100; pH 7 (1% aq.); 98% act.

Genagen TA-120. [Hoechst Celanese/Colorants & Surf.] PEG-12 tallate; CAS 61791-00-2; surfactant for textile processing; amber liq.; visc. < 500 cps; sapon. no. 70; hyd. no. 75; pH 7 (1% aq.); 98% act.

Genagen TA-160. [Hoechst Celanese/Colorants & Surf.] PEG-16 tallate; CAS 61791-00-2; surfactant for textile processing; amber liq.; visc. < 500 cps; sapon. no. 55; hyd. no. 60; pH 7 (1% aq.); 98% act.

Genamin C Grades. [Hoechst Celanese/Colorants & Surf.] Ethoxylated cocamine (2-25 EO); cationic; surfactant for agric. formulations.

Genamin CC Grades. [Hoechst Celanese/Colorants & Surf.] Coconut fatty acid amine; cationic; surfactant for agric. formulations.

Genamin O, S, Brands. [Hoechst Celanese/Colorants & Surf.] Fatty amine ethoxylate (2–25 EO); cationic; surface act. basic material for industrial fields; 100% conc.

Genamin T-020. [Hoechst Celanese/Colorants & Surf.] PEG-2 tallowamine; cationic; surfactant for agric. formulations, textile processing; amber paste; amine no. 163; pH 10 (50% IPA/water); 99% conc.

Genamin T-050. [Hoechst Celanese/Colorants & Surf.] PEG-5 tallowamine; cationic; surfactant for agric. formulations, textile processing; amber liq.; amine no. 112; pH 10 (1% aq.); 99% conc.

Genamin T-150. [Hoechst Celanese/Colorants & Surf.] PEG-15 tallowamine; cationic; surfactant for agric. formulations, textile processing; amber liq.; amine no. 60; pH 10 (1% aq.); 99% conc.

Genamin T-200. [Hoechst Celanese/Colorants & Surf.] PEG-20 tallowamine; cationic; surfactant for agric. formulations, textile processing; amber liq.; amine no. 48; pH 10 (1% aq.); 99% conc.

Genamin T-308. [Hoechst Celanese/Colorants & Surf.] PEG-30 tallowamine; cationic; surfactant for agric. formulations, textile processing; amber liq.; amine no. 28; pH 9 (1% aq.); 80% act.

Genamin TA Grades. [Hoechst Celanese/Colorants & Surf.] Tallow fatty acid amine; CAS 61790-33-8; cationic; surfactant for agric. formulations.

Genamin XET. [Hoechst Celanese/Colorants & Surf.] Modified ethoxylated amine; surfactant for textile processing; amber clear liq.; visc. 250 cps; pH 10 (1% aq.); 96% act.

Genamine C-020. [Hoechst Celanese] Coconut fatty amine oxethylate; cationic; raw material for min. oil additives, insecticides, pesticides, cosmetic bases, adhesives; clear liq.; sol. in min. oil, turbid in water (10 g/l); sp.gr. 0.895 (50 C); visc. 28.8 cps (50 C); flash pt. 188 C (Marcusson); surf. tens. 26 dynes/cm; pH 9–10; 100% act.

Genamine C-050. [Hoechst Celanese] Coconut fatty amine oxethylate; cationic; see Genamine C-020; clear liq.; sol. in water, min. oil; sp.gr. 0.95 (50 C); visc. 30.4 cps (50 C); flash pt. 224 C (Marcusson); surf. tens. 32 dynes/cm; pH 9–10; 100% act.

Genamine C-080. [Hoechst Celanese] Coconut fatty amine oxethylate; cationic; see Genamine C-020; clear liq.; sol. in water, turbid in min. oil; sp.gr. 0.988 (50 C); visc. 35.2 cps (50 C); flash pt. 249 C (Marcusson); surf. tens. 37 dynes/cm; pH 9–10; 100% act.

Genamine C-100. [Hoechst Celanese] Coconut fatty amine oxethylate; cationic; see Genamine C-020; clear liq.; sol. in water, turbid in min. oil; sp.gr. 1.0 (50 C); visc. 37 cps (50 C); flash pt. 273 C (Marcusson); surf. tens. 40 dynes/cm; pH 9–10; 100% act.

Genamine C-150. [Hoechst Celanese] Coconut fatty amine oxethylate; cationic; see Genamine C-020; clear liq.; sol. in water; sp.gr. 1.027 (50 C); visc. 45.5 cps (50 C); flash pt. 287 C (Marcusson); surf. tens. 44 dynes/cm; pH 9–10; 100% act.

Genamine C-200. [Hoechst Celanese] Coconut fatty amine oxethylate; cationic; see Genamine C-020; clear liq.; sol. in water, min. oil; sp.gr. 1.043 (50 C); visc. 55 cps (50 C); flash pt. 295 C (Marcusson); surf. tens. 45 dynes/cm; pH 9–10; 100% act.

Genamine C-250. [Hoechst Celanese] Coconut fatty amine oxethylate; cationic; see Genamine C-020; turbid liq.; sol. in water; sp.gr. 1.059 (50 C); visc. 65.2 cps (50 C); flash pt. 298 C (Marcusson); surf. tens. 47 dynes/cm; pH 9–10; 100% act.

Genamine O-020. [Hoechst Celanese] Oleylamine oxethylate; cationic; see Genamine C-020; turbid liq.; sol. in min. oil; sp.gr. 0.883 (50 C); visc. 37.4 cps (50 C); flash pt. 220 C (Marcusson); pH 9–10; 100% act.

Genamine O-050. [Hoechst Celanese] Oleylamine oxethylate; cationic; see Genamine C-020; clear liq.; sol. in min. oil; sp.gr. 0.956 (50 C); visc. 45.3 cps (50 C); flash pt. 269 C (Marcusson); surf. tens. 42 dynes/cm; pH 9–10; 100% act.

Genamine O-080. [Hoechst Celanese] Oleylamine oxethylate; cationic; see Genamine C-020; clear liq.; sol. in water min. oil; sp.gr. 0.983 (50 C); visc. 49.6 cps (50 C); flash pt. 282 C (Marcusson); surf. tens. 39 dynes/cm; pH 9–10; 100% act.

Genamine O-100. [Hoechst Celanese] Oleylamine oxethylate; cationic; see Genamine C-020; turbid liq.; sol. in water; sp.gr. 0.995 (50 C); visc. 52.1 cps (50 C); flash pt. 288 C (Marcusson); surf. tens. 41 dynes/cm; pH 9–10; 100% act.

Genamine O-150. [Hoechst Celanese] Oleylamine oxethylate; cationic; see Genamine C-020; turbid liq.; sol. in water; sp.gr. 1.024 (50 C); visc. 65.1 cps (50 C); flash pt. 307 C (Marcusson); surf. tens. 45 dynes/cm; pH 9–10; 100% act.

Genamine O-200. [Hoechst Celanese] Oleylamine oxethylate; cationic; see Genamine C-020; paste; sol. in water; sp.gr. 1.063 (50 C); visc. 75.9 cps (50 C); flash pt. 314 C (Marcusson); surf. tens. 48 dynes/cm; pH 9–10; 100% act.

Genamine O-250. [Hoechst Celanese] Oleylamine oxethylate; cationic; see Genamine C-020; wax; sol. in water; sp.gr. 1.08 (50 C); visc. 89.2 cps (50 C); flash pt. 317 C (Marcusson); surf. tens. 51 dynes/cm; pH 9–10; 100% act.

Genamine S-020. [Hoechst Celanese] Stearylamine oxethylate; cationic; see Genamine C-020; wax; sol. in min. oil, turbid in water; sp.gr. 0.889 (50 C); visc. 40.6 cps (50 C); flash pt. 219 C (Marcusson); pH 9–10; 100% act.

Genamine S-050. [Hoechst Celanese] Stearylamine oxethylate; cationic; see Genamine C-020; clear liq.; sol. in min. oil, turbid in water; sp.gr. 0.943 (50 C); visc. 39 cps (50 C); flash pt. 258 C (Marcusson); surf. tens. 35 dynes/cm; pH 9–10; 100% act.

Genamine S-080. [Hoechst Celanese] Stearylamine oxethylate; cationic; see Genamine C-020; turbid liq.; sol. in water, min. oil; sp.gr. 0.974 (50 C); visc. 43.3 cps (50 C); flash pt. 273 C (Marcusson); surf. tens. 39 dynes/cm; pH 9–10; 100% act.

Genamine S-100. [Hoechst Celanese] Stearylamine oxethylate; cationic; see Genamine C-020; turbid liq.; sol. in water, min. oil; sp.gr. 0.987 (50 C); visc. 46.1 cps (50 C); flash pt. 277 C (Marcusson); surf. tens. 42 dynes/cm; pH 9–10; 100% act.

Genamine S-150. [Hoechst Celanese] Stearylamine oxethylate; cationic; see Genamine C-020; turbid liq.; sol. in water; sp.gr. 1.011 (50 C); visc. 55.6 cps (50 C); flash pt. 280 C (Marcusson); surf. tens. 44 dynes/cm; pH 9–10; 100% act.

Genamine S-200. [Hoechst Celanese] Stearylamine oxethylate; cationic; see Genamine C-020; turbid liq.; sol. in water, turbid in min. oil; sp.gr. 1.033 (50 C); visc. 67.8 cps (50 C); flash pt. 281 C (Marcusson); surf. tens. 47 dynes/cm; pH 9–10; 100% act.

Genamine S-250. [Hoechst Celanese] Stearylamine

oxethylate; cationic; see Genamine C-020; paste; sol. in water; sp.gr. 1.050 (50 C); visc. 81.8 cps (50 C); flash pt. 282 C (Marcusson); surf. tens. 49 dynes/cm; pH 9–10; 100% act.

Genaminox KC. [Hoechst Celanese/Colorants & Surf.; Hoechst AG] Cocamine oxide; nonionic; foam booster/stabilizer over wide pH range, thickener for shampoos, bath prods.; surfactant for textile processing; yel. clear liq.; pH 7; 30% act.

Genapol® 24-L-3. [Hoechst Celanese/Colorants & Surf.] C12-14 pareth-2.9; nonionic; biodeg. detergent intermediate for sulfation for use in cosmetics, shampoos, lt. duty detergents; emulsifier, prewash spotter, agric. adjuvant, hydrocarbon-based cleaning systems; APHA 10 liq.; oil-sol.; m.w. 331; sp.gr. 0.93; visc. 25 cst; HLB 8.0; hyd. no. 170; pour pt. 7 C; flash pt. (FTCOC) 157 C; pH 6.5 (1% aq.); 100% conc.

Genapol® 24-L-45. [Hoechst Celanese/Colorants & Surf.] C12-14 pareth-6; nonionic; biodeg. surfactant, wetting agent, emulsifier, for general industrial and household detergents; FDA compliance; APHA 10 liq.; misc. with propylene glycol; m.w. 479; sp.gr. 0.95; visc. 45 cst; HLB 11.8; hyd. no. 117; pour pt. 15 C; cloud pt. 45 C (1% aq.); flash pt. (FTCOC) 215 C; pH 6.3 (1% aq.); surf. tens. 28 dynes/cm (0.1%); 100% conc.

Genapol® 24-L-50. [Hoechst Celanese/Colorants & Surf.] C12-14 pareth-6.9; nonionic; biodeg. surfactant, wetting agent, emulsifier for general household and industrial detergents; APHA 10 liq.; m.w. 505; sp.gr. 0.97; visc. 80 cst; HLB 12.1; hyd. no. 111; pour pt. 18 C; cloud pt. 50 C (1% aq.); flash pt. (FTCOC) 227 C; pH 6.3 (1% aq.); 100% conc.

Genapol® 24-L-60. [Hoechst Celanese/Colorants & Surf.] C12-14 pareth-7; nonionic; biodeg. surfactant, wetting agent, emulsifier for general industrial and household detergents; agric. adjuvant; FDA compliance; APHA 20 liq.; m.w. 519; sp.gr. 0.99; visc. 82 cst; HLB 12.2; hyd. no. 108; pour pt. 20 C; cloud pt. 60 C (1% aq.); flash pt. (FTCOC) 232 C; pH 6.2 (1% aq.); surf. tens. 29 dynes/cm (0.1%); 100% conc.

Genapol® 24-L-60N. [Hoechst Celanese/Colorants & Surf.] C12-14 pareth-7; nonionic; biodeg. surfactant, wetting agent, emulsifier for general industrial and household detergents; agric. adjuvant; APHA 20 liq.; misc. with propylene glycol, toluene; m.w. 510; sp.gr. 0.99; 83 cst; HLB 12.1; hyd. no. 110; pour pt. 20 C; cloud pt. 60 C (1% aq.); flash pt. (FTCOC) 232 C; pH 6.3 (1% aq.); surf. tens. 31 dynes/cm (0.1%); 100% conc.

Genapol® 24-L-75. [Hoechst Celanese/Colorants & Surf.] C12-14 pareth-8.5; nonionic; biodeg. surfactant, wetting agent, emulsifier for general industrial and household detergents; agric. adjuvant; FDA compliance; APHA 20 liq./solid; misc. with propylene glycol, ethyl acetate; m.w. 567; sp.gr. 1.00; HLB 12.9; hyd. no 99; pour pt. 22 C; cloud pt. 75 C (1% aq.); flash pt. (FTCOC) 232 C; pH 6.2 (1% aq.); surf. tens. 31 dynes/cm (0.1%); 100% conc.

Genapol® 24-L-92. [Hoechst Celanese/Colorants & Surf.] C12-14 pareth-10.5; nonionic; biodeg. surfactant, wetting agent, emulsifier for general industrial and household detergents; agric. adjuvant; FDA compliance; wh. solid; misc. with propylene glycol, toluene; m.w. 668; sp.gr. 1.00; HLB 14.0; hyd. no. 84; pour pt. 27 C; cloud pt. 92 C (1% aq.); flash pt. (FTCOC) 232 C; pH 6.3 (1% aq.); surf. tens. 34 dynes/cm (0.1%); 100% conc.

Genapol® 24-L-98N. [Hoechst Celanese/Colorants & Surf.] C12-14 pareth-11.5; nonionic; biodeg. surfactant, wetting agent, emulsifier for general industrial and household detergents; agric. adjuvant; FDA compliance; APHA 20 solid; misc. with propylene glycol, toluene; m.w. 701; sp.gr. 1.01; HLB 14.2; hyd. no. 80; pour pt. 29 C; cloud pt. 98 C (1% aq.); flash pt. (FTCOC) 232 C; pH 6.4 (1% aq.); surf. tens. 36 dynes/cm (0.1%); 100% conc.

Genapol® 26-L-1. [Hoechst Celanese/Colorants & Surf.] C12-16 pareth-1; nonionic; biodeg. detergent intermediate for sulfation for use in cosmetics, shampoos, lt. duty detergents; emulsifier, prewash spotter, agric. adjuvant, hydrocarbon-based cleaning systems; APHA 10 liq.; oil-sol.; m.w. 238; sp.gr. 0.87; HLB 3.7; hyd. no. 235; pour pt. 8 C; flash pt. (FTCOC) 149 C; pH 6.5 (1% aq.); 100% conc.

Genapol® 26-L-1.6. [Hoechst Celanese/Colorants & Surf.] C12-16 pareth-1.6; nonionic; biodeg. detergent intermediate for sulfation for use in cosmetics, shampoos, lt. duty detergents; emulsifier, prewash spotter, agric. adjuvant, hydrocarbon-based cleaning systems; FDA compliance; APHA 10 liq.; oil-sol.; m.w. 264; sp.gr. 0.9; visc. 22 cst; HLB 5.0; hyd. no. 213; pour pt. 10 C; flash pt. (FTCOC) 143 C; pH 6.6 (1% aq.); 100% conc.

Genapol® 26-L-2. [Hoechst Celanese/Colorants & Surf.] C12-16 pareth-2; nonionic; biodeg. detergent intermediate for sulfation for use in cosmetics, shampoos, lt. duty detergents; emulsifier, prewash spotter, agric. adjuvant, hydrocarbon-based cleaning systems; APHA 10 liq.; oil-sol.; m.w. 281; sp.gr. 0.91; visc. 24 cst; HLB 6.0; hyd. no. 200; pour pt. 10 C; flash pt. (FTCOC) 143 C; pH 6.4 (1% aq.); 100% conc.

Genapol® 26-L-3. [Hoechst Celanese/Colorants & Surf.] C12-16 pareth-3; nonionic; biodeg. detergent intermediate for sulfation for use in cosmetics, shampoos, lt. duty detergents; emulsifier, prewash spotter, agric. adjuvant, hydrocarbon-based cleaning systems; FDA compliance; APHA 15 liq.; oil-sol.; m.w. 328; sp.gr. 0.93; visc. 27 cst; HLB 8.0; hyd. no. 171; pour pt. 8 C; flash pt. (FTCOC) 154 C; pH 6.6 (1% aq.); 100% conc.

Genapol® 26-L-5. [Hoechst Celanese/Colorants & Surf.] C12-16 pareth-5; nonionic; biodeg. detergent intermediate for sulfation for use in cosmetics, shampoos, lt. duty detergents; emulsifier, prewash spotter, agric. adjuvant, hydrocarbon-based cleaning systems; FDA compliance; APHA 10 liq.; oil-sol.; m.w. 419; sp.gr. 0.97; visc. 29 cst; HLB 10.6; hyd. no. 134; pour pt. 8 C; flash pt. (FTCOC) 160 C; pH 6.6 (1% aq.); 100% conc.

Genapol® 26-L-45. [Hoechst Celanese/Colorants & Surf.] C12-16 pareth-6; nonionic; biodeg. detergent intermediate for sulfation for use in cosmetics, shampoos, lt. duty detergents; emulsifier, prewash spotter, agric. adjuvant, hydrocarbon-based cleaning systems; APHA 20 liq.; oil-sol.; m.w. 479; sp.gr. 0.96; visc. 49.5 cst; HLB 11.6; hyd. no. 117; pour pt. 6 C; cloud pt. 45 C (1% aq.); flash pt. (FTCOC) 216 C; pH 7.0 (1% aq.); 100% conc.

Genapol® 26-L-50. [Hoechst Celanese/Colorants & Surf.] C12-16 pareth-7; nonionic; detergent raw material; APHA 20 color; m.w. 505; sp.gr. 0.96; visc. 48.1 cst; HLB 12.2; hyd. no. 111; pour pt. 5 C; cloud pt. 50 C (1% aq.); flash pt. (FTCOC) 216 C; pH 6.1 (1% aq.); 100% conc.

Genapol® 26-L-60. [Hoechst Celanese/Colorants & Surf.] C12-16 pareth-7.3; nonionic; biodeg. surfactant, wetting agent, emulsifier for general industrial and household detergents; agric. adjuvant; APHA 20 liq.; m.w. 519; sp.gr. 0.98; visc. 53 cst; HLB 12.4; hyd. no. 108; pour pt. 18 C; cloud pt. 60 C (1% aq.); flash pt. (FTCOC) 171 C; pH 6.2 (1% aq.); 100% conc.

Genapol® 26-L-60N. [Hoechst Celanese/Colorants & Surf.] C12-16 pareth-7; nonionic; biodeg. surfactant, wetting agent, emulsifier for general industrial and household detergents; agric. adjuvant; APHA 20 liq.; m.w. 503; sp.gr. 0.98; visc. 55 cst; HLB 12.2; hyd. no. 112; pour pt. 19 C; cloud pt. 60 C (1% aq.); flash pt. (FTCOC) 182 C; pH 6.3 (1% aq.); 100% conc.

Genapol® 26-L-75. [Hoechst Celanese/Colorants & Surf.] C12-16 pareth-8.3; nonionic; surfactant; APHA 20 color; m.w. 561; sp.gr. 0.97; visc. 22.7 cst (50 C); HLB 13.0; hyd. no. 100; pour pt. 1 C; cloud pt. 75 C (1% aq.); flash pt. (FTCOC) 238 C; pH 6.1 (1% aq.); 100% conc.

Genapol® 26-L-98N. [Hoechst Celanese/Colorants & Surf.] C12-16 pareth-11.5; nonionic; biodeg. surfactant, wetting agent, emulsifier for general industrial and household detergents; agric. adjuvant; APHA 20 solid; m.w. 701; sp.gr. 1.02; HLB 14.4; hyd. no. 80; pour pt. 30 C; cloud pt. 98 C (1% aq.); flash pt. (FTCOC) 188 C; pH 6.4 (1% aq.); 100% conc.

Genapol® 42-L-3. [Hoechst Celanese/Colorants & Surf.] C12-14 pareth-3; nonionic; biodeg. detergent intermediate for sulfation for use in cosmetics, shampoos, lt. duty detergents; emulsifier, prewash spotter, agric. adjuvant, hydrocarbon-based cleaning systems; liq.; oil-sol.; HLB 7.7; 100% conc.

Genapol® 2299. [Hoechst Celanese/Colorants & Surf.] Coconut alcohol ethoxylate; surfactant for textile processing; Gardner < 1 clear liq.; visc. 40 cps; hyd. no. 150; pH 6 (1% aq.); 99% conc.

Genapol® 2317. [Hoechst Celanese/Colorants & Surf.] Alkoxylated alcohol; nonionic; multipurpose surfactant for high speed wetting in metalworking formulations; liq.; sol. in water; sp.gr. 1.00; pH 6-7 (1% aq.).

Genapol® 2822, 2908, 2909. [Hoechst AG] Oxalcohol EO/PO adduct; nonionic; low foaming surfactants.

Genapol® 3504. [Hoechst AG] Polyglycol ether; nonionic; surfactant.

Genapol® AMG. [Hoechst AG] Acylaminopolyglycol ether sulfate, magnesium salt; nonionic; surfactant.

Genapol® AMS. [Hoechst Celanese/Colorants & Surf.; Hoechst AG] TEA-PEG-3 cocamide sulfate; anionic; detergent, foaming agent used in top-grade cosmetics cleansers; lime soap dispersant; biodeg.; clear yel. low-visc. liq.; weak odor; dens. 1.0 g/cm³; visc. 200 cps max.; pH 6.5–8.0 (1%); 40% act.

Genapol® ARO. [Hoechst Celanese/Colorants & Surf.; Hoechst AG] Sodium laureth sulfate; CAS 9004-82-4; anionic; raw material for cosmetics, detergents, and cleaning agents.

Genapol® B. [Hoechst Celanese/Colorants & Surf.] Ethylenediamine alkoxylate; nonionic; wetting agent, low-foaming lubricity additive for syn. metalworking formulations; liq.; water-sol.; sp.gr. 1.06; pH 10-11 (1% aq.).

Genapol® C-050. [Hoechst Celanese/Colorants & Surf.] Coceth-5; CAS 61791-13-7; nonionic; raw material for mfg. of textile, leather, paper auxs., detergents, emulsifiers, cosmetics, agric.; turbid liq.; sol. in min. oil, benzene, turbid in water; sp. gr. 0.952 (50 C); visc. 17.6 cps (50 C); flash pt. 201 C (Marcusson); surf. tens. 32 dynes/cm; 100% act.

Genapol® C-080. [Hoechst Celanese/Colorants & Surf.] Coceth-8; CAS 61791-13-7; nonionic; raw material for mfg. of textile, leather, paper auxs., detergents, emulsifiers, cosmetics, agric.; paste; sol. in water, benzene, turbid in min. oil; sp. gr. 0.979 (50 C); visc. 25.4 cps (50 C); flash pt. 246 C (Marcusson); surf. tens. 36 dynes/cm; 100% act.

Genapol® C-100. [Hoechst Celanese/Colorants & Surf.] Coceth-10; CAS 61791-13-7; nonionic; raw material for mfg. of textile, leather, paper auxs., detergents, emulsifiers, cosmetics, agric.; paste; sol. in water, benzene, turbid in min. oil; sp. gr. 0.990 (50 C); visc. 30.4 cps (50 C); flash pt. 251 C (Marcusson); surf. tens. 38 dynes/cm; 100% act.

Genapol® C-150. [Hoechst Celanese/Colorants & Surf.] Coceth-15; CAS 61791-13-7; nonionic; raw material for mfg. of textile, leather, paper auxs., detergents, emulsifiers, cosmetics, agric.; wax; sol. in water, benzene; sp. gr. 1.027 (50 C); visc. 45.8 cps (50 C); flash pt. 260 C (Marcusson); surf. tens. 43 dynes/cm; 100% act.

Genapol® C-200. [Hoechst Celanese/Colorants & Surf.] Coceth-20; CAS 61791-13-7; nonionic; raw material for mfg. of textile, leather, paper auxs., detergents, emulsifiers, cosmetics, agric.; wax; sol. in water, benzene; sp. gr. 1.032 (50 C); visc. 61.5 cps (50 C); flash pt. 264 C (Marcusson); surf. tens. 44 dynes/cm; 100% act.

Genapol® CRT 40. [Hoechst Celanese/Colorants & Surf.; Hoechst AG] TEA alkyl sulfate; anionic; detergent raw material for cosmetics, shampoos, body cleansing agents, detergents; biodeg.; pale yel. clear liq.; slight odor; m.w. 422; sol. in water; dens. 1.0–1.05 g/cm³; visc. 100 cps max.; pH 6.5–7.5 (1%); 40% conc.

Genapol® DA-040. [Hoechst Celanese/Colorants & Surf.] Deceth-4; surfactant for textile processing; yel. liq.; visc. 40 cps; hyd. no. 170; pH 7 (1% aq.); 99% conc.

Genapol® DA-060. [Hoechst Celanese/Colorants & Surf.] Deceth-6; surfactant for textile processing; yel. liq.; hyd. no 140; pH 7 (1% aq.); 99% conc.

Genapol® GC-050. [Hoechst Celanese/Colorants & Surf.] Coceth-5; CAS 61791-13-7; surfactant for textile processing; yel. liq.; visc. 35 cps; hyd. no. 135; pH 7 (1% aq.); 99% conc.

Genapol® GEV. [Hoechst Celanese/Colorants & Surf.; Hoechst AG] Ethoxylated fatty alcohol; nonionic; lubricant for fiber preparation; liq.

Genapol® L Grades. [Hoechst Celanese/Colorants & Surf.] C12-16 alcohol polyglycol ethers (1-12 EO); nonionic; surfactant for agric. formulations.

Genapol® LRO Liq., Paste. [Hoechst Celanese/Colorants & Surf.; Hoechst AG] Sodium laureth sulfate; CAS 9004-82-4; anionic; detergent, foaming agent used in cosmetic prods., personal care prods., agric., lime soap dispersant; biodeg.; pale yel. clear liq., mobile paste resp.; slight odor; misc. with water; dens. 1.05 and 1.0 g/cm³ resp.; visc. 100 cps max. (liq.); pH 6.5–8.0 (1% aq., liq.), 7.2 ± 0.6 (1% aq., paste); 27 and 68% conc.

Genapol® O-020. [Hoechst Celanese/Colorants & Surf.] Oleth-2; CAS 9004-98-2; nonionic; raw material for mfg. of textile, leather, paper auxs., detergents, emulsifiers, cosmetics, agric.; clear liq.; sol. in min. oil, benzene, turbid in water; sp.gr. 0.894 (50 C); visc. 12.4 cps (50 C); flash pt. 186 C (Marcusson); 100% act.

Genapol® O-050. [Hoechst Celanese/Colorants & Surf.] Oleth-5; CAS 9004-98-2; nonionic; raw material for mfg. of textile, leather, paper auxs., detergents, emulsifiers, cosmetics, agric.; turbid liq.; sol. in benzene, turbid in water, min. oil; sp.gr. 0.936 (50 C); visc. 18.4 cps (50 C); flash pt. 225 C (Marcusson); surf. tens. 54 dynes/cm; 100% act.

Genapol® O-080. [Hoechst Celanese/Colorants & Surf.] Oleth-8; CAS 9004-98-2; nonionic; raw material for mfg. of textile, leather, paper auxs., detergents, emulsifiers, cosmetics, agric.; turbid liq.; sol. in water, benzene, turbid in min. oil; sp.gr. 0.960 (50 C); visc. 25.1 cps (50 C); flash pt. 246 C (Marcusson); surf. tens. 44 dynes/cm; 100% act.

Genapol® O-090. [Hoechst Celanese/Colorants & Surf.] Oleth-9; CAS 9004-98-2; nonionic; raw material for mfg. of textile, leather, paper auxs., detergents, emulsifiers, cosmetics, agric.; yel. liq.; visc. 45 cps; hyd. no. 90; pH 7 (1% aq.); 99% conc.

Genapol® O-100. [Hoechst Celanese/Colorants & Surf.] Oleth-10; CAS 9004-98-2; nonionic; raw material for mfg. of textile, leather, paper auxs., detergents, emulsifiers, cosmetics, agric.; paste; sol. in water, benzene, turbid in min. oil; sp.gr. 0.989 (50 C); visc. 33 cps (50 C); flash pt. 260 C (Marcusson); surf. tens. 41 dynes/cm; 100% act.

Genapol® O-120. [Hoechst Celanese/Colorants & Surf.] Oleth-12; CAS 9004-98-2; nonionic; raw material for mfg. of textile, leather, paper auxs., detergents, emulsifiers, cosmetics, agric.; paste; sol. in water, benzene, turbid in min. oil; sp.gr. 1.0 (50 C); visc. 42.5 cps (50 C); flash pt. 265 C (Marcusson); surf. tens. 42 dynes/cm; 100% act.

Genapol® O-150. [Hoechst Celanese/Colorants & Surf.] Oleth-15; CAS 9004-98-2; nonionic; raw material for mfg. of textile, leather, paper auxs., detergents, emulsifiers, cosmetics, agric.; wax; sol. in water, benzene, turbid in min. oil; sp.gr. 1.02 (50 C); visc. 49.1 cps (50 C); flash pt. 271 C (Marcusson); surf. tens. 45 dynes/cm; 100% act.

Genapol® O-200. [Hoechst Celanese/Colorants & Surf.] Oleth-20; CAS 9004-98-2; nonionic; raw material for mfg. of textile, leather, paper auxs., detergents, emulsifiers, cosmetics, agric.; wh. waxy solid; sol. in water, benzene, turbid in min. oil; sp.gr. 1.037 (50 C); visc. 65.9 cps (50 C); hyd. no. 50; flash pt. 278 C (Marcusson); pH 7 (1% aq.); surf. tens. 47 dynes/cm; 100% act.

Genapol® O-230. [Hoechst Celanese/Colorants & Surf.] Oleth-23; CAS 9004-98-2; nonionic; raw material for mfg. of textile, leather, paper auxs., detergents, emulsifiers, cosmetics, agric.; wh. waxy solid; sol. in water, benzene, turbid in min. oil; sp.gr. 1.042 (50 C); visc. 79.5 cps (50 C); hyd. no. 44; flash pt. 279 C (Marcusson); pH 7 (1% aq.); surf. tens. 47 dynes/cm; 100% act.

Genapol® OX Grades. [Hoechst Celanese/Colorants & Surf.] C12-15 syn. oxo alcohol polyglycol ethers (3-18 EO); nonionic; surfactant for agric. formulations.

Genapol® PAF, PL, PN Brands. [Hoechst Celanese/Colorants & Surf.] Propylene and ethylene oxide polymerization prods.; nonionic; defoamer, component for detergent, cleaning and antistat; basic material for wide variety of aux.; liq., powd.; 100% conc.

Genapol® PF 10. [Hoechst Celanese/Colorants & Surf.] EO/PO block copolymer; nonionic; surfactant for agric. formulations, textiles; Gardner < 1 clear liq.; m.w. 1800-2000; cloud pt. 20 C (1% aq.); 99% conc.

Genapol® PF 20. [Hoechst Celanese/Colorants & Surf.] EO/PO block copolymer; nonionic; surfactant for agric. formulations, textiles; Gardner < 1 clear liq.; m.w. 2500-2900; cloud pt. 59 C (in 25% aq. butyl Carbitol sol'n.); 99% conc.

Genapol® PF 40. [Hoechst Celanese/Colorants & Surf.] EO/PO block copolymer; nonionic; surfactant for agric. formulations, textiles; Gardner < 1 clear liq.; m.w. 3100-3700; cloud pt. 68 C (in 25% aq. butyl Carbitol); 99% conc.

Genapol® PF 80. [Hoechst Celanese/Colorants & Surf.] EO/PO block copolymer; nonionic; surfactant for agric. formulations, textiles; wh. flakes; m.w. 6600-9000; cloud pt. > 100 C (1% aq.); 99% conc.

Genapol® PGM Conc. [Hoechst Celanese/Colorants & Surf.; Hoechst AG] Sodium laureth sulfate, glycol distearate, and cocamide MEA; anionic; pearl-luster conc. used in shampoos, bubble baths, cosmetics, liq. soaps, detergents; biodeg.; wh. fluid paste; slight odor; misc. with water; dens. 1.0 g/cm^3; pH 7.2 ± 0.8 (1% act.); 40.0 ± 1.0% act.

Genapol® PL 120. [Hoechst AG] EO/PO fatty alcohol adduct; nonionic; low foaming surfactant.

Genapol® PN-30. [Hoechst Celanese/Colorants & Surf.; Hoechst AG] Ethylenediamine alkoxylate; nonionic; lubricant for formulating water sol. metalworking fluids; surfactant for controlled foam detergents, wetting agents; yel. liq.; cloudy sol. in water; sp.gr. 1.01 (50 C); solid. pt. < -10 C; cloud pt. 29–31 C (5 g in 25 ml of 24% butyl diglycol); ref. index 1.448 (50 C); pH 10–11 (1% aq.); 100% act.

Genapol® PS. [Hoechst Celanese/Colorants & Surf.] Random EO/PO copolymer based on pentaerythritol; nonionic; lubricant, emulsifier, wetting agent; improves lubricating properties of min. oil-free, water-sol. metalworking fluids; textile processing; ylsh. visc. liq.; sol. in water, glycols, benzene, xylene, acetone, ethyl acetate; sp.gr. 1.1; cloud pt. 80 C (1% aq.); pH 6.5 ± 0.3 (5%); 100% act.

Genapol® S-020. [Hoechst Celanese/Colorants & Surf.] Stearyl alcohol polyglycol ether; nonionic; raw material for mfg. of textile, leather, paper auxs., detergents, emulsifiers, cosmetics; wax; sol. in min. oil, benzene, turbid in water; sp.gr. 0.890 (50 C); visc. 15.2 cps (50 C); flash pt. 198 C (Marcusson); surf. tens. 50 dynes/cm; 100% act.

Genapol® S-050. [Hoechst Celanese/Colorants & Surf.] Stearyl alcohol polyglycol ether; nonionic; raw material for mfg. of textile, leather, paper auxs., detergents, emulsifiers, cosmetics; wax; sol. in min. oil, benzene, turbid in water; sp.gr. 0.934 (50 C); visc. 21.2 cps (50 C); flash pt. 224 C (Marcusson); surf. tens. 50 dynes/cm; 100% act.

Genapol® S-080. [Hoechst Celanese/Colorants & Surf.] Stearyl alcohol polyglycol ether; nonionic; raw material for mfg. of textile, leather, paper auxs., detergents, emulsifiers, cosmetics; paste; sol. in benzene, turbid in water, min. oil; sp.gr. 0.971 (50 C); visc. 28.4 cps (50 C); flash pt. 228 C (Marcusson); surf. tens. 44 dynes/cm; 100% act.

Genapol® S-100. [Hoechst Celanese/Colorants & Surf.] Stearyl alcohol polyglycol ether; nonionic; raw material for mfg. of textile, leather, paper auxs., detergents, emulsifiers, cosmetics; paste; sol. in benzene, turbid in water, min. oil; sp.gr. 0.978 (50 C); visc. 34 cps (50 C); flash pt. 239 C (Marcusson); surf. tens. 45 dynes/cm; 100% act.

Genapol® S-150. [Hoechst Celanese/Colorants & Surf.] Stearyl alcohol polyglycol ether; nonionic;

raw material for mfg. of textile, leather, paper auxs., detergents, emulsifiers, cosmetics; wax; sol. in water, benzene, turbid in min. oil; sp.gr. 1.009 (50 C); visc. 50.3 cps (50 C); flash pt. 266 C (Marcusson); surf. tens. 47 dynes/cm; 100% act.

Genapol® S-200. [Hoechst Celanese/Colorants & Surf.] Stearyl alcohol polyglycol ether; nonionic; raw material for mfg. of textile, leather, paper auxs., detergents, emulsifiers, cosmetics; wax; sol. in water, benzene, turbid in min. oil; sp.gr. 1.031 (50 C); visc. 67.4 cps (50 C); flash pt. 267 C (Marcusson); surf. tens. 48 dynes/cm; 100% act.

Genapol® S-250. [Hoechst Celanese/Colorants & Surf.] Stearyl alcohol polyglycol ether; nonionic; raw material for mfg. of textile, leather, paper auxs., detergents, emulsifiers, cosmetics; wax; sol. in water, benzene, turbid in min. oil; sp.gr. 1.049 (50 C); visc. 106 cps (50 C); flash pt. 273 C (Marcusson); surf. tens. 49 dynes/cm; 100% act.

Genapol® T Grades. [Hoechst Celanese/Colorants & Surf.; Hoechst AG] Tallow alcohol polyglycol ether (8–25 EO); nonionic; detergent base and basic material for cosmetic and specialty chemical industries, agric. formulations; biodeg.; liq., paste, wax; 100% conc.

Genapol® T-250. [Hoechst Celanese/Colorants & Surf.; Hoechst AG] Tallow alcohol ethoxylate; auxiliary for detergent bases; surfactant for textile processing; wh. powd.; hyd. no. 40; pH 7 (1% aq.); 99% conc.

Genapol® T-500. [Hoechst AG] C16-18 fatty alcohol polyalkoxylate; auxiliary for detergent bases; powd.

Genapol® UD-030. [Hoechst Celanese/Colorants & Surf.] PEG-3 fatty alcohol; surfactant for oil and gas industry; sp.gr. 0.90; HLB 8; iodine no. 1; pour pt. 0 C; cloud pt. 37-41 C; flash pt. 170 C; 100% act.

Genapol® UD-050. [Hoechst Celanese/Colorants & Surf.] PEG-5 fatty alcohol; surfactant for oil and gas industry; sp.gr. 0.93; HLB 11.0; iodine no. 1; pour pt. 10 C; cloud pt. 66-69 C; flash pt. 170 C; 100% act.

Genapol® UD-079. [Hoechst Celanese/Colorants & Surf.] PEG-7 fatty alcohol; surfactant for oil and gas industry; sp.gr. 0.97; HLB 13.0; iodine no. 1; pour pt. 0 C; cloud pt. 52-55 C; flash pt. 180 C; 90% act. in water.

Genapol® UD-080. [Hoechst Celanese/Colorants & Surf.] PEG-8 fatty alcohol; surfactant for oil and gas industry; sp.gr. 0.96; HLB 14.0; iodine no. 1; pour pt. 25 C; cloud pt. 62-65 C; flash pt. 190 C; 100% act.

Genapol® UD-088. [Hoechst Celanese/Colorants & Surf.] PEG-8 fatty alcohol; surfactant for oil and gas industry; sp.gr. 0.99; HLB 14.0; iodine no. 1; pour pt. -5 C; cloud pt. 62-65 C; flash pt. 190 C; 80% act. in water.

Genapol® UD-110. [Hoechst Celanese/Colorants & Surf.] PEG-11 fatty alcohol; surfactant for oil and gas industry; sp.gr. 0.99; HLB 15.0; iodine no. 1; pour pt. 30 C; cloud pt. 62-65 C; flash pt. 210 C; 100% act.

Genapol® V 2908. [Hoechst Celanese/Colorants & Surf.] C12-15 syn. oxo-alcohol EO/PO block copolymer; nonionic; surfactant for agric. formulations.

Genapol® V 2909. [Hoechst Celanese/Colorants & Surf.] C12-15 syn. oxo-alcohol EO/PO block copolymer; nonionic; surfactant for agric. formulations.

Genapol® X Grades. [Hoechst Celanese/Colorants & Surf.] Isotridecanol polyglycol ether (3–15 EO); nonionic; basic material for mfg. of detergents, cleaning and rinsing agents; mfg. of leather and paper aux., agric. formulations; 100% conc.

Genapol® X-040. [Hoechst Celanese/Colorants & Surf.] Trideceth-4; CAS 24938-91-8; nonionic; surfactant for textile processing; yel. liq.; visc. < 100 cps; hyd. no. 150; pH 7 (1% aq.); 99% conc.

Genapol® X-060. [Hoechst Celanese/Colorants & Surf.] Trideceth-6; CAS 24938-91-8; nonionic; component for selective emulsification systems; surfactant for textile processing; yel. liq.; visc. 50 cps; hyd. no. 120; pH 7 (1% aq.); 99% conc.

Genapol® X-080. [Hoechst Celanese/Colorants & Surf.] Trideceth-8; CAS 24938-91-8; nonionic; surfactant for textile processing; yel. liq.; hyd. no. 100; pH 7 (1% aq.); 99% conc.

Genapol® X-100. [Hoechst Celanese/Colorants & Surf.] Trideceth-10; CAS 24938-91-8; nonionic; surfactant for textile processing; wh. mass; hyd. no. 90; pH 7 (1% aq.); 99% conc.

Genapol® ZRO Liq., Paste. [Hoechst Celanese/Colorants & Surf.; Hoechst AG] Sodium laureth sulfate; CAS 9004-82-4; anionic; raw material with good foaming and cleansing for cosmetics, detergents, cleaning agents; 28% pale yel. clear liq., 70% sl. yel. mobile paste.

Generol® 122E25. [Henkel/Cospha; Henkel Canada; Henkel KGaA] PEG-25 soya sterol; nonionic; emulsifier, emollient, pigment dispersant and wetter, deflocculating agent in cosmetics; solubilizer for perfumes or preservatives; ivory hard wax; faint odor; sol. in water and ethanol; at 80 C, sol. in isopropyl esters and veg. oils; HLB 17; m.p. 45 C; pH 7 (1% aq.); 100% conc.

Genopur ASA. [Hoechst AG] Alkenyl dicarboxylic acid anhydride; solubilizer for heavy-duty liq. detergents and related prods.

Geronol ACR/4. [Rhone-Poulenc; Rhone-Poulenc Geronazzo] Disodium laurethsulfosuccinate; anionic; emulsifier for emulsion polymerization of acrylate, polyvinyl acetate; detergent base, foamer, foam stabilizer, dispersant; liq.; insol. in org. solv.; 30% conc.

Geronol ACR/9. [Rhone-Poulenc; Rhone-Poulenc Geronazzo] Disodium nonoxynol-10 sulfosuccinate; anionic; emulsifier for emulsion polymerization of acrylate, polyvinyl acetate; detergent base, foamer, foam stabilizer, dispersant; liq.; insol. in org. solv.; 30% conc.

Geronol AG-100/200 Series. [Rhone-Poulenc Surf.] Complex anionic-nonionic ethoxylate blends; anionic/nonionic; emulsifiers for toxicants, esp. organophosphate insecticides and nonsaponifiable herbicides; liqs.; HLB 9-15.

Geronol AG-821. [Rhone-Poulenc Surf.] Ethoxylate blend; nonionic; emulsifier and adjuvant for crop oil/surfactant conc.; liq.; 91% conc.

Geronol AG-900. [Rhone-Poulenc Surf.] Complex ethoxylate blends; nonionic; spreading, wetting and sticking agent for pesticide systems; liq.; HLB 10-14; 92% conc.

Geronol Aminox/3. [Rhone-Poulenc] C16-C18 alkanolamide; cationic; adjuvant for cutting oils for metals and rust protector; 100% conc.

Geronol AZ82. [Rhone-Poulenc Geronazzo] POE/POP alkylphenols and calcium dodecylbenzene sulfonate; anionic/nonionic; emulsifier for pesticide emulsifiable conc.; specific for Malathion; waxy paste; 85% conc.

Geronol FF/4. [Rhone-Poulenc Geronazzo] Calcium dodecylbenzene sulfonate, ethoxylated alkylphenol, stabilizer blend; anionic/nonionic; emulsifier for pesticide emulsifiable concs.; liq.; 65% conc.

Geronol FF/6. [Rhone-Poulenc Geronazzo] Calcium dodecylbenzene sulfonate, ethoxylated alkylphenol, stabilizer blend; anionic/nonionic; emulsifier for pesticide emulsifiable concs.; paste; 90% conc.

Geronol MOE/2/N. [Rhone-Poulenc Geronazzo] Alkylaryl polyglycolic ether and ethoxylated, modified fatty acids; nonionic; emulsifier for agric. o/w emulsions; liq.; 90% conc.

Geronol MS. [Rhone-Poulenc Geronazzo] Calcium dodecylbenzene sulfonate, ethoxylated amine, fatty acid blend; anionic/nonionic; emulsifier for pesticide emulsifiable concs.; liq.; 90% conc.

Geronol PRH/4-A, -B. [Rhone-Poulenc] POE alkylphenol phosphate; anionic; emulsifier for various compds.; liq.; 100% conc.

Geronol RE/70. [Rhone-Poulenc Geronazzo] Calcium dodecylbenzene sulfonate, ethoxylated fatty acid blend; anionic/nonionic; emulsifier for pesticide emulsifiable concs.; paste; 90% conc.

Geronol SC/120. [Rhone-Poulenc Geronazzo] Calcium dodecylbenzene sulfonate, modified POE fatty amines blend; anionic/nonionic; emulsifier for pesticide emulsifiable concs.; liq.

Geronol SC/121. [Rhone-Poulenc Geronazzo] Calcium dodecylbenzene sulfonate, alkylphenol polyglycol ester blend; anionic/nonionic; emulsifier for pesticide emulsifiable concs.; liq.; 75% conc.

Geronol SC/138. [Rhone-Poulenc] POE/POP block polymer; nonionic; emulsifier for Dimethoate E.C.; waxy paste; 100% conc.

Geronol SC/177. [Rhone-Poulenc Geronazzo] Calcium dodecylbenzene sulfonate, ethoxylated fatty acid, ethoxylated phosphated fatty acid blend; nonionic; emulsifier for min. paraffinic oil blended with phosphate ester; liq.; 99% conc.

Geronol SN. [Rhone-Poulenc Geronazzo] Calcium dodecylbenzene sulfonate, ethoxylated amine blend; anionic/nonionic; emulsifier/stabilizer for chemically unstable pesticide emulsifiable concs.; visc. liq.; 90% conc.

Geronol V/087. [Rhone-Poulenc Geronazzo] Calcium dodecylbenzene sulfonate and POE/POP alkylphenol blend; nonionic/anionic; emulsifier for pesticide conc.; specific for Alachlor; solid; 80% conc.

Geronol V/497. [Rhone-Poulenc Geronazzo] Calcium dodecylbenzene sulfonate, POE/POP alkylphenol blend; anionic/nonionic; emulsifier for emulsifiable concs.; specific for Phoxim; paste; 80% conc.

Geropon® 40/D. [Rhone-Poulenc] Alkylaryl polyglycol ether; nonionic; wetting agent for pesticides; powd.; 40% conc.

Geropon® 99 (formerly Pentex® 99). [Rhone-Poulenc Surf.] Dioctyl sodium sulfosuccinate and propylene glycol; anionic; detergent; textile scouring and dispersant for dyes; paper rewetting and felt washing surfactant; wetting agent in cosmetics; detergent additive in dry cleaning fluids; dishwashing compds.; wallpaper removers; agric. sprays; emulsion polymerization; water-based paint formulations; antifog; EPA compliance; clear liq.; water-sol.; sp.gr. 1.08–1.13 (70 F); visc. 500–1000 cps (70 F); pour pt. 40 F; 70% act.

Geropon® 111. [Rhone-Poulenc Surf.] Polymeric; anionic; surfactant, dispersant for agric. formulations; EPA compliance; liq.; 25% act.

Geropon® AB/20. [Rhone-Poulenc; Rhone-Poulenc Geronazzo] Ammonium laureth-9 sulfate; anionic; emulsifier for emulsion polymerization and min. oils; detergent, dispersant, foam stabilizer; stable to pH and temp.; liq.; 30% conc.

Geropon® AC-78 (formerly Igepon® AC-78). [Rhone-Poulenc Surf.; Rhone-Poulenc France] Sodium cocoyl isethionate; anionic; detergent, wetting agent with low salt content having good foaming, lathering, and dispersing props.; used in detergent bars, dentifrices, shampoos, bubble baths, other cosmetics;; powd.; 83% act.

Geropon® ACR/4 (formerly Geronol ACR/4). [Rhone-Poulenc France] Disodium laureth sulfosuccinate; anionic/nonionic; emulsifier for emulsion polymerization, detergent base, foamer, foam stabilizer, dispersant; liq.; 30% conc.

Geropon® ACR/9 (formerly Geronol ACR/9). [Rhone-Poulenc France] Sodium alkylaryl ethoxy half ester sulfosuccinic acid; anionic; emulsifier for emulsion polymerization, detergent base, foamer, foam stabilizer, dispersant; liq.; 30% conc.

Geropon® AS-200. [Rhone-Poulenc Surf.] Sodium cocoyl isethionate.

Geropon® AY. [Rhone-Poulenc] Sodium diamyl sulfosuccinate; anionic; wetting agent; wax; sol. in water and org. polar solv.; 70% conc.

Geropon® BIS/SODICO-2 (formerly Vanisol BIS/SODICO-2). [Rhone-Poulenc Geronazzo] Sodium bistridecyl sulfosuccinate; anionic; emulsifier and visc. depressant for emulsion polymerization of PVC; latex surf. tension stabilizer; dispersant for resins, pigments into plastics and organic media; base for rust inhibitors; liq.; oil-sol., water-disp.; 60% conc.

Geropon® CET/50/P. [Rhone-Poulenc] Fatty acids, ethoxylated; nonionic; wetting/suspending agent for pesticides; waxy powd.; 100% conc.

Geropon® CYA/45. [Rhone-Poulenc] Sodium diisobutyl sulfosuccinate; anionic; surfactant for emulsion polymerization; liq.; 45% conc.

Geropon® CYA/60. [Rhone-Poulenc Geronazzo] Sodium dioctyl sulfosuccinate; anionic; wetting agent, surf. tension reducer, visc. depressant for emulsion polymerization of PVC; liq.; 60% conc.

Geropon® CYA/DEP. [Rhone-Poulenc Geronazzo] Sodium diisooctyl sulfosuccinate; anionic; wetting agent for textile industry, pesticides, emulsion polymerization, printing inks, water paints; stable in acid media; liq.; sol. in org. solvs.; 75% conc.

Geropon® DG. [Rhone-Poulenc Surf.] Polymeric; anionic; surfactant, dispersant for agric. formulations; EPA compliance. DISCONTINUED.

Geropon® DOS (formerly Rhodiasurf DOS). [Rhone-Poulenc France] Sodium dioctyl sulfosuccinate; anionic; wetting agent, emulsifier, demulsifier for textile wet processing, specialty cleaners, dewatering agent for flotation concs., oil spill cleanup blends; limited stability in alkaline or acidic media; liq.; limited water sol.; 65% conc.

Geropon® DOS FP (formerly Rhodiasurf DOS FP). [Rhone-Poulenc France] Sodium dioctyl sulfosuccinate; anionic; wetting agent, emulsifier, demulsifier for textile wet processing, specialty cleaners, dewatering agent for flotation concs., oil spill cleanup blends; limited stability in alkaline or acidic media; liq.; limited water sol.; 60% conc.

Geropon® FA-82 (formerly Alconate FA 82). [Rhone-Poulenc Surf.] Disodium N-alkyl sulfosuccinamate; anionic; foam booster/stabilizer for carpet backing; paste; 35% conc.

Geropon® FMS. [Rhone-Poulenc] Sodium diphenylmethane sulfonate; anionic; dispersant and suspending agent for pesticide formulations; powd.; 70% conc.

Geropon® HB. [Rhone-Poulenc Surf.] Polymeric; anionic; surfactant, dispersant for agric. formulations; EPA compliance. DISCONTINUED.

Geropon® IN. [Rhone-Poulenc] Sodium diisopropyl naphthalene sulfonate; anionic; wetting agent for pesticide wettable powds.; dispersant; latex stabilizer; prevents coagulation in SBR and other elastomers; rewetting agent; humectant; powd.; 65% conc.

Geropon® K/65. [Rhone-Poulenc] Fatty acids with ethoxylated alcohols, modified; anionic/nonionic; wetting and suspending agent for Dodine wettable powds.; powd.; 85% conc.

Geropon® K/202. [Rhone-Poulenc] POE fatty alcohol; nonionic; wetting and suspending agent for Dodine wettable powds.; powd.; 50% conc.

Geropon® MLS/A. [Rhone-Poulenc Geronazzo] Sodium methallyl sulfonate; anionic; dye improver reactive comonomer for acrylic fibers polymerization; reactive emulsifier or coemulsifier in latex emulsion polymerization; powd.; 100% conc.

Geropon® NK. [Rhone-Poulenc] Sodium dibutyl naphthalene sulfonate; anionic; wetting agent for pesticides and general purpose; powd.; 65% conc.

Geropon® RM/77-D (redesignated Rhodacal® RM/77-D). [Rhone-Poulenc Surf.]

Geropon® S-1585 (formerly Pentrone S-1585). [Rhone-Poulenc France] Disodium octylphenoxy sulfosuccinate; anionic; surfactant.

Geropon® SBL-203 (formerly Alconate® SBL-203). [Rhone-Poulenc Surf.] Disodium lauramido MEA-sulfosuccinate; improves flash foam of anionic systems; produces brittle, tack-free residue for carpet shampoos; liq.; 40% conc.

Geropon® SC/211. [Rhone-Poulenc] Sodium salt of carboxylic copolymer with an anionic dispersant; anionic; dispersing/suspending/compatibility agent for pesticide wettable powds. and water-disp. gran.; powd.; 93% conc.

Geropon® SC/213. [Rhone-Poulenc] Potassium salt of carboxylic copolymer with anionic dispersant; anionic; dispersing/suspending/compatibility agent for pesticide wettable powds. and water-disp. gran.; powd.; 93% conc.

Geropon® SDS. [Rhone-Poulenc Surf.; Rhone-Poulenc France] Sodium dioctyl sulfosuccinate; anionic; dispersant and wetting agent for pigments and dyes in plastics, pesticide wettable powds.; EPA compliance; powd.; dissolves in water, partly sol. in org. solvs.; 100% conc.

Geropon® SS-L7DE (formerly Alkasurf® SS-L7DE). [Rhone-Poulenc Surf.] Sodium lauramido DEA sulfosuccinate; anionic; mild detergent, foam booster/stabilizer for liq. dish detergent and toiletry preps.; pale yel. liq.; sol. in water; dens. 1.1 g/ml; pH 5.0-7.5; 40% conc.

Geropon® SS-L9ME. [Rhone-Poulenc Surf.] Sodium lauramido MEA sulfosuccinate; anionic; high foaming and non-irritating surfactant for toiletry and carpet shampoo formulations; liq.; 40% conc. DISCONTINUED.

Geropon® SS-O-60. [Rhone-Poulenc Surf.] Sodium dioctyl sulfosuccinate; anionic; emulsifier, wetting agent for industrial applics.; freeze/thaw stability; liq.; 60% conc. DISCONTINUED.

Geropon® SS-O-70PG (formerly Alkasurf® SS-O-70PG). [Rhone-Poulenc Surf.] Sodium dioctyl sulfosuccinate; anionic; emulsifier, wetting agent for industrial applics.; high flash; liq.; 70% conc.

Geropon® SS-O-75 (formerly Alkasurf® SS-O-75). [Rhone-Poulenc Surf.] Sodium dioctyl sulfosuccinate; anionic; wetting agent, emulsifier for industrial, mining, and textile industries, resin treatments, dry cleaning systems; paint pigment dispersant; clear colorless liq.; typ. odor; 1% water sol., sol. in most org. liq.; sp.gr. 1.09; f.p. < -20 C; surf. tens. 29 dynes/cm (0.1% aq.); pH 5-6; 75% act.

Geropon® SS-TA. [Rhone-Poulenc Surf.] Sodium N-octadecyl sulfosuccinamate; anionic; emulsifier, dispersant, foamer, detergent, and solubilizer for soaps, other surfactants; paste; 35% conc. DISCONTINUED.

Geropon® T-33 (formerly Igepon® T-33). [Rhone-Poulenc Surf.; Rhone-Poulenc France] Sodium N-methyl-N-oleoyl taurate; anionic; detergent, dispersant, wetting agent for textile and general-purpose applics.; dye assistant; for kier boiling, bleaching, wetting, finishing of textiles; in industrial detergents, rug shampoos, bottle washing compds., metal cleaners, paper industry; pale yel. nearly clear liq., perfumed alcoholic odor; m.w. 425; water sol.; pH 6.5-8.0 (10%); 32% conc.

Geropon® T/36-DF. [Rhone-Poulenc] Sodium salt of polycarboxylic acid; anionic; dispersant for final solv. stripping phase in polymerization of BR-SBR; liq.; 25% conc.

Geropon® T-43 (formerly Igepon® T-43). [Rhone-Poulenc Surf.] Sodium N-methyl-N-oleoyl taurate; anionic; detergent, wetting agent, dispersant for textile and general-purpose applics.; dye assistant; for kier boiling, bleaching, wetting, finishing of textiles; in industrial detergents, rug shampoos, bottle washing compds., metal cleaners, paper industry; wh. visc. liq. slurry, fatty odor; m.w. 425; water sol.; pH 6.5-8.0 (5%); 33% act.

Geropon® T-51 (formerly Igepon® T-51). [Rhone-Poulenc Surf.] Sodium N-methyl-N-oleoyl taurate; anionic; detergent, wetting agent, dispersant for textile and general-purpose applics.; dye assistant; for kier boiling, bleaching, wetting, finishing of textiles; in industrial detergents, rug shampoos, bottle washing compds., metal cleaners, paper industry; latex stabilizer; lt. amber gel, perfumed odor; m.w. 425; water sol.; pH 6.5-8.0 (10%); 13.9% act.

Geropon® T-77 (formerly Igepon® T-77). [Rhone-Poulenc Surf.; Rhone-Poulenc France] Sodium N-methyl-N-oleoyl taurate; anionic; foamer, wetting agent, emulsifier, dispersant for textile and general-purpose applics.; dye assistant; kier boiling; in industrial detergents, rug shampoos, bottle washing compds., metal cleaners, paper industry; cream flakes, fatty odor; m.w. 425; water sol.; pH 6.5-8.0 (5%); 67% act.

Geropon® TA/72. [Rhone-Poulenc] Sodium salt of polycarboxylic acid with wetting agent; anionic; dispersing/suspending/compatibility agent for pesticide wettable powds. and water-disp. gran.; powd.; 93% conc.

Geropon® TA/72/S. [Rhone-Poulenc] Sodium salt of polycarboxylic acid; anionic; dispersant/suspending agent for pesticides and general purpose; powd.; water-sol.; 93% conc.

Geropon® TA/764. [Rhone-Poulenc] Sodium salt of polycarboxylic acid with wetting agent; anionic; dispersing/suspending/compatibility agent for pesticide wettable powds. and water-disp. gran.; powd.; 90% conc.

Geropon® TA/K. [Rhone-Poulenc] Potassium salt of polycarboxylic acid; anionic; dispersant, antisettling, general purpose; also for emulsion paints, dirt antiresettling in detergent compds.; sequestering agent for Ca and Mg salts; powd.; 90% conc.

Geropon® TBS. [Rhone-Poulenc] Alkyl polyglycolic ether, modified; anionic; detergent for fibers, cosmetics, leather, paper, and metal industry; liq.

Geropon® TC-42 (formerly Igepon® TC-42). [Rhone-Poulenc Surf.; Rhone-Poulenc France] Sodium N-methyl-N-cocoyl taurate; anionic; foamer, dispersant, detergent for detergent bars, shampoos, bubble baths, cosmetics; chemically stable; wh. soft, smooth paste; m.w. 363; pH 7.0–8.5 (10%); 24% act.

Geropon® TK-32 (formerly Igepon® TK-32). [Rhone-Poulenc Surf.] Sodium N-methyl-N-tallowyl taurate; anionic; detergent, suspending agent, dispersant; precipitation inhibitor for org. and inorg. salts of Ba, Ca, Sr; for petrol. industry; VCS 8 max. clear liq.; m.w. 439; water sol.; pH 8–10 (10%); 20% act.

Geropon® TX/99. [Rhone-Poulenc Geronazzo] Sodium-ammonium alkyl sulfosuccinate; anionic; thixotropic dispersant, emulsifier, visc. depressant, antisettling agent for solv.-based paint systems; emulsion PVC plastisols; liq.; 38% conc.

Geropon® WS-25, WS-25-I (formerly Nekal® WS-25, WS-25-I). [Rhone-Poulenc Surf.; Rhone-Poulenc France] Sodium dinonyl sulfosuccinate; anionic; rewetting agent for textile finishing, in applic. of resins, softeners, starches; wetting and dispersing agent for latex paints; liq.; 48% act.

Geropon® WT-27 (formerly Nekal® WT-27). [Rhone-Poulenc Surf.] Sodium dioctyl sulfosuccinate; anionic; high foaming wetting and rewetting agent, emulsifier, dispersant, penetrant; used in drycleaning detergents, emulsion polymerization, glass cleaners, wallpaper removers, battery separators, textiles, paper, metalworking, dyeing, fire fighting; FDA compliance; water-wh. clear liq.; water sol. up to 1.2% solids and 5.5% @ 70 C; sol. in some polar and nonpolar solvs.; flash pt. > 200 C; pH 5–6 (1%); surf. tens. 28 dynes/cm (@ CMC); 70% act.

Geropon® X2152 (formerly Rhodiasurf X2152). [Rhone-Poulenc France] Ammonium dioctyl sulfosuccinate; anionic; wetting agent, emulsifier, demulsifier, stabilizer in water treatment and petrol. processing; stable at neutral pH; liq.; insol. in water; 65% conc.

Glazamine DP2. [ICI Surf. UK] Melamine formaldehyde resin; nonionic; thermosetting resin producing durable press finishes on cellulosic fibers; stiffener for syn. fabrics; liq.

Glazamine M. [ICI Surf. UK] Stabilized hexamethylol melamine resin; nonionic; thermosetting resin producing durable press finishes on cellulosic fibers; stiffener for syn. fabrics; liq.

Glo-Break 1001. [Global United Industries] Glycol esters; nonionic; w/o demulsifier for produced crude oil and industrial emulsions; liq.; 100% conc. DISCONTINUED.

Glo-Break 1003. [Global United Industries] Glycol esters; nonionic; w/o demulsifier for produced crude oil and industrial emulsions; liq.; 100% conc. DISCONTINUED.

Glo-Break 1010. [Global United Industries] Glycol esters; nonionic; w/o demulsifier for produced crude oil and industrial emulsions; liq.; 100% conc. DISCONTINUED.

Glo-Break 1020. [Global United Industries] Glycol esters; nonionic; w/o demulsifier for produced crude oil and industrial emulsions; liq.; 100% conc. DISCONTINUED.

Glo-Break 1029. [Global United Industries] Glycol esters; nonionic; w/o demulsifier for produced crude oil and industrial emulsions; liq.; 100% conc. DISCONTINUED.

Glo-Break 1050. [Global United Industries] Modified glycol ester; nonionic; w/o demulsifier for produced crude oil and industrial emulsions; liq.; 100% conc. DISCONTINUED.

Glo-Break 1054. [Global United Industries] Modified glycol ester; nonionic; w/o demulsifier for produced crude oil and industrial emulsions; liq.; 100% conc. DISCONTINUED.

Glo-Break 1056. [Global United Industries] Modified glycol ester; nonionic; w/o demulsifier for produced crude oil and industrial emulsions; liq.; 100% conc. DISCONTINUED.

Glo-Break 1205. [Global United Industries] Phenolic resin oxyalkylates; nonionic; w/o demulsifier for produced crude oil and industrial emulsions; liq.; 100% conc. DISCONTINUED.

Glo-Break 1206. [Global United Industries] Phenolic resin oxyalkylates; nonionic; w/o demulsifier for produced crude oil and industrial emulsions; liq.; 100% conc. DISCONTINUED.

Glo-Break 1222. [Global United Industries] Phenolic resin oxyalkylates; nonionic; w/o demulsifier for produced crude oil and industrial emulsions; liq.; 100% conc. DISCONTINUED.

Glo-Break 1240. [Global United Industries] Phenolic resin oxyalkylates; nonionic; w/o demulsifier for produced crude oil and industrial emulsions; liq.; 100% conc. DISCONTINUED.

Glo-Break 1252. [Global United Industries] Phenolic resin oxyalkylates; nonionic; w/o demulsifier for produced crude oil and industrial emulsions; liq.; 100% conc. DISCONTINUED.

Glo-Break 1405. [Global United Industries] Glycol and modified glycol epoxide condensates; nonionic; w/o demulsifier for produced crude oil and industrial emulsions; liq.; 100% conc. DISCONTINUED.

Glo-Break 1420. [Global United Industries] Glycol and modified glycol epoxide condensates; nonionic; w/o demulsifier for produced crude oil and industrial emulsions; liq.; 100% conc. DISCONTINUED.

Glo-Break 1434. [Global United Industries] Glycol and modified glycol epoxide condensates; nonionic; w/o demulsifier for produced crude oil and industrial emulsions; liq.; 100% conc. DISCONTINUED.

Glo-Break 1465. [Global United Industries] Glycol and modified glycol epoxide condensates; nonionic; w/o demulsifier for produced crude oil and industrial emulsions; liq.; 100% conc. DISCONTINUED.

Glo-Break 1481. [Global United Industries] Glycol and modified glycol epoxide condensates; nonionic; w/o demulsifier for produced crude oil and industrial emulsions; liq.; 100% conc. DISCONTINUED.

Glo-Break 1602. [Global United Industries] Modified glycol ester; nonionic; w/o demulsifier for produced crude oil and industrial emulsions; liq.; 100% conc. DISCONTINUED.

Glo-Break 1611. [Global United Industries] Modified

glycol ester; nonionic; w/o demulsifier for produced crude oil and industrial emulsions; liq.; 100% conc. DISCONTINUED.

Glo-Cor 101. [Global United Industries] Amidoimidazoline; cationic; corrosion inhibitor intermediate; antistatic agent; liq.; 100% conc. DISCONTINUED.

Glo-Cor 103. [Global United Industries] Amidoimidazoline; cationic; corrosion inhibitor intermediate; liq.; 100% conc. DISCONTINUED.

Glo-Cor 105. [Global United Industries] Diimidazoline; cationic; corrosion inhibitor intermediate; antistatic agent; liq.; 100% conc. DISCONTINUED.

Glo-Cor 106. [Global United Industries] Amidoimidazoline; cationic; corrosion inhibitor intermediate; antistatic agent; high tolerance to electrolytes; liq.; 100% conc. DISCONTINUED.

Glo-Cor 107. [Global United Industries] Imidazoline; cationic; corrosion inhibitor intermediate; antistatic agent; liq.; 100% conc. DISCONTINUED.

Glo-Cor 231. [Global United Industries] Rosin oxyalkylates; nonionic/cationic; acid corrosion inhibitors, emulsifiers; liq.; 100% conc. DISCONTINUED.

Glo-Cor 232. [Global United Industries] Rosin oxyalkylates; nonionic/cationic; acid corrosion inhibitors, emulsifiers; liq.; 100% conc. DISCONTINUED.

Glo-Cor 233. [Global United Industries] Rosin oxyalkylates; nonionic/cationic; acid corrosion inhibitors, emulsifiers; liq.; 100% conc. DISCONTINUED.

Glo-Cor 240. [Global United Industries] Modified amidoimidazoline; cationic; corrosion inhibitor intermediate; oil wetting agent; antistatic agent; liq.; 100% conc. DISCONTINUED.

Glokill Series. [Rhone-Poulenc France] Quaternaries; cationic; sanitizers.

Glo-Mul 600. [Global United Industries] Metallic alkylaryl sulfonate; anionic; emulsifier for oilfield, agric., and industrial applics.; wetting agent, detergent; liq.; 100% conc. DISCONTINUED.

Glo-Mul 607. [Global United Industries] Amine alkylaryl sulfonate; anionic; emulsifier for oilfield, agric., and industrial applics.; wetting agent, detergent; liq.; 100% conc. DISCONTINUED.

Glo-Mul 609. [Global United Industries] Amine alkylaryl sulfonate; anionic; emulsifier for oilfield, agric., and industrial applics.; wetting agent, detergent; liq.; 100% conc. DISCONTINUED.

Glo-Mul 706, 706. [Global United Industries] Polyamide; nonionic; w/o emulsifier for drilling fluids, industrial emulsions; liq.; 100% conc. DISCONTINUED.

Glo-Mul 724. [Global United Industries] Modified polyamide; nonionic; w/o emulsifier for drilling fluids, industrial emulsions; liq.; 100% conc. DISCONTINUED.

Glo-Mul 740. [Global United Industries] Alkanolamide; nonionic; w/o emulsifier; oil-sol. detergent; oil solubilizer for water-sol. detergents; liq.; 100% conc. DISCONTINUED.

Glo-Mul 742. [Global United Industries] Alkanolamide; nonionic; detergent, visc. modifier; liq.; 100% conc. DISCONTINUED.

Glo-Mul 743. [Global United Industries] Alkanolamide; nonionic; lubricant, industrial cleaner, foam stabilizer, visc. modifier; liq.; 100% conc. DISCONTINUED.

Glo-Mul 771. [Global United Industries] Modified alkanolamide; nonionic; drilling fluid detergent; liq.; 100% conc. DISCONTINUED.

Glo-Mul 772. [Global United Industries] Modified alkanolamide; anionic/nonionic; detergent, wetting agent, foam stabilizer; liq.; 100% conc. DISCONTINUED.

Glo-Mul 780. [Global United Industries] Sorbitan monooleate; CAS 1338-43-8; nonionic; emulsifier for industrial, agric., and cosmetic applics.; liq.; 100% conc. DISCONTINUED.

Glo-Mul 781. [Global United Industries] PEG-400 dioleate; CAS 9005-07-6; nonionic; emulsifier for industrial, agric., and cosmetic applics.; liq.; 100% conc. DISCONTINUED.

Glo-Mul 782. [Global United Industries] PEG-400 oleate; CAS 9004-96-0; nonionic; emulsifier for industrial, agric., and cosmetic applics.; liq.; 100% conc. DISCONTINUED.

Glo-Mul 783. [Global United Industries] PEG-400 stearate; CAS 9004-99-3; nonionic; emulsifier for industrial, agric., and cosmetic applics.; solid; 100% conc. DISCONTINUED.

Glo-Mul 789. [Global United Industries] Glyceryl monostearate; nonionic; emulsifier for industrial, agric., and cosmetic applics.; solid; 100% conc. DISCONTINUED.

Glo-Mul 800. [Global United Industries] Org. phosphate ester; anionic; emulsifier for oilfield, agric. and industrial applics.; lubricant, gellant, wetting agent, dispersant, solubilizer, antistat; liq.; 100% conc. DISCONTINUED.

Glo-Mul 802. [Global United Industries] Org. phosphate ester; anionic; emulsifier for oilfield, agric. and industrial applics.; lubricant, gellant, wetting agent, dispersant, solubilizer, antistat; liq.; 100% conc. DISCONTINUED.

Glo-Mul 811. [Global United Industries] Org. phosphate ester; anionic; emulsifier for oilfield, agric. and industrial applics.; lubricant, gellant, wetting agent, dispersant, solubilizer, antistat; liq.; 100% conc. DISCONTINUED.

Glo-Mul 814. [Global United Industries] Org. phosphate ester; anionic; emulsifier for oilfield, agric. and industrial applics.; lubricant, gellant, wetting agent, dispersant, solubilizer, antistat; liq.; 100% conc. DISCONTINUED.

Glo-Mul 822. [Global United Industries] Org. phosphate ester; anionic; emulsifier for oilfield, agric. and industrial applics.; lubricant, gellant, wetting agent, dispersant, solubilizer, antistat; liq.; 100% conc. DISCONTINUED.

Glo-Mul 4001. [Global United Industries] Sodium dialkylsulfosuccinate; anionic; wetting agent, dispersant, foaming agent, emulsifier; liq.; 100% conc. DISCONTINUED.

Glo-Mul 4007. [Global United Industries] Disodium alkylsulfosuccinate; anionic; wetting agent, dispersant, foaming agent, emulsifier; liq.; 100% conc. DISCONTINUED.

Glo-Sperse 60. [Global United Industries] Quat. amine oxyalkylate; cationic; dispersant, wetting agent, antistat, emulsifier, corrosion inhibitor; liq.; 100% conc. DISCONTINUED.

Glo-Sperse 67. [Global United Industries] Quat. amine oxyalkylate; cationic; dispersant, wetting agent, antistat, emulsifier, corrosion inhibitor; liq.; 100% conc. DISCONTINUED.

Glo-Sperse 68. [Global United Industries] Quat. amine oxyalkylate; cationic; dispersant, wetting agent, antistat, emulsifier, corrosion inhibitor; liq.; 100% conc. DISCONTINUED.

Glo-Verse 2010. [Global United Industries] Modified polymeric amine; cationic; o/w demulsifier for crude oil, industrial emulsions; liq.; 100% conc. DISCONTINUED.

Glo-Verse 2011. [Global United Industries] Modified polymeric amine; cationic; o/w demulsifier for crude oil, industrial emulsions; liq.; 100% conc. DISCONTINUED.

Glo-Verse 2040. [Global United Industries] Modified polymeric amine; cationic; o/w demulsifier for crude oil, industrial emulsions; liq.; 100% conc. DISCONTINUED.

Glucam® E-10. [Amerchol] Methyl gluceth-10; nonionic; humectant for personal care prods.; freezing pt. depressant; emollient in aq. and hydroalcoholic prods.; moisturizer; foam modifier in detergent and shampoo systems; solv. and solubilizer for topical pharmaceuticals; used in emulsions, toilet articles; adds gloss, conditioning; pale yel. med. visc. syrup; odorless; sol. in water, alcohol, hydroalcoholic systems; acid no. 1.5 max.; sapon. no. 1.5 max; 100% conc.

Glucam® E-20. [Amerchol] Methyl gluceth-20; humectant for personal care prods.; freezing pt. depressant; emollient in aq. and hydroalcoholic prods.; moisturizer; foam modifier in detergent and shampoo systems; solv. and solubilizer for topical pharmaceuticals; pale yel. thin syrup; odorless; sol. see Glucam E-10; acid no. 1.0 max.; sapon. no. 1.0 max.; 100% conc.

Glucam® P-10. [Amerchol] PPG-10 methyl glucose ether; nonionic; humectant for personal care prods.; freezing pt. depressant; emollient in aq. and hydroalcoholic prods.; moisturizer; foam modifier in detergent and shampoo systems; solv. and solubilizer for topical pharmaceuticals; pale yel. heavy visc. syrups; odorless; sol. in water, alcohol, and hydroalcoholic systems, castor oil, IPM, IPP; visc. 8500 cps; acid no. 1.0 max.; sapon. no. 1.0 max.; 100% conc.

Glucam® P-20. [Amerchol] PPG-20 methyl glucose ether; nonionic; humectant for personal care prods.; freezing pt. depressant; emollient in aq. and hydroalcoholic prods.; moisturizer; foam modifier in detergent and shampoo systems; solv. and solubilizer for topical pharmaceuticals; pale yel. med. visc. syrup; odorless; sol. see Glucam P-10; visc. 1700 cps; acid no. 1.0 max.; sapon. no. 1.0 max.; 100% conc.

Glucamate® SSE-20. [Amerchol; Amerchol Europe] PEG-20 methyl glucose sesquistearate; CAS 68389-70-8; nonionic; o/w emulsifier, solubilizer used with Glucate SS; effective at low concs.; pale yel. soft solid; sol. in water, IPA, ethanol, castor oil, corn oil; HLB 15.0; cloud pt. 74 C (1% in 5% NaCl); flash pt. 570 F (OC); sapon. no. 47; pH 6.5 (10% aq.); 100% conc.

Glucate® DO. [Amerchol; Amerchol Europe] Methyl glucoside dioleate; CAS 83933-91-3; nonionic; w/o emulsifier, aux. emulsifier for o/w systems; conditioner, emollient, lubricant, plasticizer, and pigment dispersant; liq.; HLB 5.0; 100% conc.

Glucate® SS. [Amerchol; Amerchol Europe] Methyl glucose sesquistearate; CAS 68936-95-8; nonionic; w/o emulsifier used with Glucamate SSE-20 to provide visc. stability, mildness; off wh. flakes; sol. in IPA, misc. with common oil phase ingred., water insol.; m.p. 51 C; HLB 6.0; flash pt. 530 F (OC); sapon. no. 136; pH 5.5 (10% in 1:1 IPA:water); 100% conc.

Glucopon 225. [Henkel/Emery] C8,10 alkyl polyglycoside; CAS 68515-73-1; nonionic; surfactant, detergent, wetting agent, surf. tens. reducer, hydrotrope, dispersant for laundry detergents, liq. cleaners, hard surf. cleaners, institutional and industrial cleaners; biodeg.; lt. yel. hazy liq.; dens. 9.8 lb/gal; visc. 4800 cps; HLB 13.6; pour pt. -10 C; cloud pt. > 100 C (1%); flash pt. (CC) > 93 C; surf. tens. 27.8 dynes/cm (0.1%); toxicology: LD50 (oral) > 5 g/kg, (dermal) > 2 g/kg; 70% act.

Glucopon 425. [Henkel/Emery] C8-16 alkyl polyglycoside; CAS 68515-73-1, 110615-47-9; nonionic; surfactant, detergent, wetting agent, surf. tens. reducer, hydrotrope, dispersant for laundry detergents, liq. cleaners, hard surf. cleaners, institutional and industrial cleaners; biodeg.; lt. yel. hazy liq.; dens. 9.2 lb/gal; visc. 420 cps; HLB 13.1; pour pt. -4 C; cloud pt. > 100 C (1%); flash pt. (CC) > 93 C; surf. tens. 28.9 dynes/cm (0.1%); toxicology: LD50 (oral) > 5 g/kg, (dermal) > 2 g/kg; 50% act.

Glucopon 600. [Henkel/Emery] C12-16 alkyl polyglycoside; CAS 110615-47-9; nonionic; surfactant, detergent, wetting agent, surf. tens. reducer, hydrotrope, dispersant for laundry detergents, liq. cleaners, hard surf. cleaners, institutional and industrial cleaners; biodeg.; lt. yel. hazy liq.; dens. 8.9 lb/gal; visc. 17,000 cps; HLB 11.6; pour pt. 10 C; cloud pt. > 100 C (1%); flash pt. (CC) > 93 C; surf. tens. 29.4 dynes/cm (0.1%); toxicology: LD50 (oral) > 5 g/kg, (dermal) > 2 g/kg; 50% act.

Glucopon 625. [Henkel/Emery] C12-16 alkyl polyglycoside; CAS 110615-47-9; nonionic; surfactant, detergent, wetting agent, surf. tens. reducer, hydrotrope, dispersant for laundry detergents, liq. cleaners, hard surf. cleaners, institutional and industrial cleaners; biodeg.; lt. yel. hazy liq.; dens. 9.0 lb/gal; visc. 21,500 cps; HLB 12.1; pour pt. 12 C; cloud pt. > 100 C (1%); flash pt. (CC) > 93 C; surf. tens. 29.4 dynes/cm (0.1%); toxicology: LD50 (oral) > 5 g/kg, (dermal) > 2 g/kg; 50% act.

Glutrin. [LignoTech] Calcium lignosulfonate; anionic; binder for foundry, ceramic, and refractory prods.; powd.; 100% conc.

Glycidol Surfactant 10G. [Olin] para-Nonylphenoxy-polyglycidol; nonionic; surfactant used in agric. chemical emulsions, leather processing, paint, emulsion polymerization, waste paper deinking, alkaline cleaners, photographic film emulsion and coating formulations; amber liq.; mild odor; sol. in alcohols, polyethers, acids; sp.gr. 1.032–1.114; dens. 8.6–9.3 lb/gal; visc. 110–195 cps; pour pt. –9 to –14 C; b.p. 93–103 C; cloud pt. > 100 C (0.5% aq.); pH 7.1–7.9 (1% aq.); surf. tens. 27–30 dynes/cm (0.1% act.); 50% act.

Glycomul® L. [Lonza] Sorbitan laurate; CAS 1338-39-2; nonionic; emulsifier for edible, cosmetic, industrial, pharmaceutical uses; amber liq.; sol. in methanol, ethanol, naphtha; sp.gr. 1.0; visc. 4500 cps; HLB 8.6; acid no. 5; sapon. no. 157–171; 100% conc.

Glycomul® MA. [Lonza] Sorbitan ester mixed fatty acids; nonionic; emulsifier for edible, cosmetic, industrial, pharmaceutical uses; dk. amber liq.; sol. in methanol, ethanol, naphtha, acetone, toluol, min. and veg. oil; sp.gr. 1.0; visc. 1000 cps; sapon. no. 139–156; 100% conc.

Glycomul® O. [Lonza] Sorbitan oleate; CAS 1338-43-8; nonionic; emulsifier for cosmetic, pharmaceutical, and industrial applics.; amber liq.; sol. in ethyl acetate, min. and veg. oils, disp. in water; sp.gr. 1.0; visc. 1000 cps.; HLB 4.3; sapon. no. 148–161; 100% conc.

Glycomul® P. [Lonza] Sorbitan palmitate; CAS 26266-57-9; nonionic; emulsifier for cosmetic, pharmaceutical, and industrial applics.; cream beads; sol. in veg. and min. oil, ethyl acetate, ethanol, acetone, toluol; HLB 6.7; sapon. no. 139–150; 100% conc.

Glycomul® S. [Lonza] Sorbitan stearate (also avail. in veg. and kosher grade); CAS 1338-41-6; nonionic; emulsifier for cosmetic, pharmaceutical, and industrial applics.; cream beads; sol. in veg. oil; HLB 5.0; sapon. no. 146–158; 100% conc.

Glycomul® S FG. [Lonza] Sorbitan stearate; CAS 1338-41-6; nonionic; emulsifier for food, cosmetic, household and industrial use; Gardner 5 beads; m.p. 53 C; HLB 5; acid no. 5.

Glycomul® S KFG. [Lonza] Sorbitan stearate; CAS 1338-41-6; nonionic; emulsifier for food, cosmetic, household and industrial prods.; Gardner 5 beads; m.p. 53 C; HLB 5; acid no. 5.

Glycomul® SOC. [Lonza] Sorbitan sesquioleate; CAS 8007-43-0; nonionic; emulsifier for cosmetic, pharmaceutical, and industrial applics.; cream beads; sol. in methanol, ethanol, ethyl acetate; sp.gr. 1.0; visc. 1000 cps; HLB 4.0; sapon. no. 149–166; 100% conc.

Glycomul® TAO. [Lonza] Sorbitan ester mixed resin and fatty acids; nonionic; emulsifier for edible, cosmetic, industrial, pharmaceutical uses; amber liq.; sol. in methanol, ethanol, toluol, naphtha, min. and veg. oils, disp. in water; sp.gr. 1.02; visc. 1300 cps.; HLB 4.3; sapon. no. 135–148; 100% conc.

Glycomul® TO. [Lonza] Sorbitan trioleate; CAS 26266-58-0; nonionic; emulsifier for cosmetic, pharmaceutical, and industrial applics.; amber, oily liq.; sol. in ethyl acetate, toluol, naphtha, min. and veg. oils, disp. in water; sp.gr. 0.95; visc. 200 cps; HLB 1.8; sapon. no. 171–185; 100% conc.

Glycomul® TS. [Lonza] Sorbitan tristearate; CAS 26658-19-5; nonionic; emulsifier for cosmetic, pharmaceutical, and industrial applics.; lt. tan beads; poorly sol. in ethyl acetate, toluol, disp. in acetone, naphtha, min. and veg. oils; HLB 2.1; sapon. no. 175–190; 100% conc.

Glycomul® TS KFG. [Lonza] Sorbitan tristearate; CAS 26658-19-5; nonionic; emulsifier for food, cosmetic, household and industrial applics.; Gardner 2 beads; m.p. 55 C; HLB 2; acid no. 14.

Glycosperse® HTO-40. [Lonza] PEG-40 sorbitan hexatallate; emulsifier for food, cosmetic, household or industrial applics.; Gardner 7 liq.; HLB 10; acid no. 10.

Glycosperse® L-10. [Lonza] PEG-10 sorbitan laurate; emulsifier for food, cosmetic, household or industrial applics.; Gardner 5 liq.; HLB 8; acid no. 2.

Glycosperse® L-20. [Lonza] Polysorbate 20; CAS 9005-64-5; 9062-73-1; nonionic; emulsifier for food, cosmetic, pharmaceutical, and industrial uses; flavor solubilizer and dispersant; yel. liq.; sol. in water, alcohol, acetone; sp.gr. 1.1; visc. 400 cps; HLB 16.7; sapon. no. 39–52; 100% conc.

Glycosperse® O-5. [Lonza] Polysorbate 81; CAS 9005-65-6; nonionic; flavor solubilizer and dispersant; emulsifier for cosmetic, pharmaceutical, and industrial use; amber liq.; sol. in alcohol, ethyl acetate, min. oil; disp. in water; sp.gr. 1.0; visc. 450 cps; HLB 10.0; sapon. no. 95–105; 100% conc.

Glycosperse® O-20. [Lonza] Polysorbate 80; CAS 9005-65-6; nonionic; emulsifier for food, cosmetic, pharmaceutical, and industrial uses; flavor solubilizer and dispersant; also antifogging agent; yel. liq.; sol. in water, alcohol, ethyl acetate, toluol, veg. oil; sp.gr. 1.0; visc. 400 cps; HLB 15; sapon. no. 44–56; 100% conc.

Glycosperse® O-20 FG, O-20 KFG. [Lonza] Polysorbate 80; CAS 9005-65-6; nonionic; emulsifier for ice cream, frozen desserts; solubilizer and dispersant for shortenings; adjuvant for herbicides and plant growth regulators; Gardner 5 liq.; HLB 15.0; acid no. 2; 100% conc.

Glycosperse® O-20 Veg. [Lonza] Polysorbate 80, veg. grade; CAS 9005-65-6; nonionic; emulsifier for food, cosmetic, pharmaceutical, and industrial uses; flavor solubilizer and dispersant; yel. liq.; sol. in water, alcohol, ethyl acetate, toluol, veg. oil; sp.gr. 1.0; visc. 400 cps; sapon. no. 44–56; 100% conc.

Glycosperse® O-20X. [Lonza] Polysorbate 80, anhyd.; CAS 9005-65-6; nonionic; emulsifier for food, cosmetic, pharmaceutical, and industrial uses; flavor solubilizer and dispersant; yel. liq.; sol. in water, alcohol, ethyl acetate, toluol, veg. oil; sp.gr. 1.0; visc. 600 cps; sapon. no. 44–56; 100% conc.

Glycosperse® P-20. [Lonza] Polysorbate 40; CAS 9005-66-7; nonionic; emulsifier for food, cosmetic, pharmaceutical, and industrial uses; flavor solubilizer and dispersant; yel. liq.; sol. in water, methanol, ethanol, acetone, ethyl acetate; sp.gr. 1.0; visc. 550 cps; HLB 15.6; sapon. no. 40–53; 100% conc.

Glycosperse® S-20. [Lonza] Polysorbate 60; CAS 9005-67-8; nonionic; emulsifier for food, cosmetic, pharmaceutical, and industrial uses; flavor solubilizer and dispersant; yel. liq.; sol. in water, ethyl acetate, toluol; sp.gr. 1.1; HLB 15.0; sapon. no. 44–56; 100% conc.

Glycosperse® TO-20. [Lonza] Polysorbate 85; CAS 9005-70-3; nonionic; emulsifier for food, cosmetic, pharmaceutical, and industrial uses; flavor solubilizer and dispersant; yel. liq., gels on standing; sol. in ethanol, methanol, ethyl acetate; water disp.; sp.gr. 1.0; visc. 300 cps; HLB 11.0; sapon. no. 82–95; 100% conc.

Glycosperse® TS-20. [Lonza] Polysorbate 65; CAS 9005-71-4; nonionic; emulsifier for food, cosmetic, pharmaceutical, and industrial uses; flavor solubilizer and dispersant; tan waxy solid; sol. in ethanol, methanol, acetone, ethyl acetate, naphtha, min. and veg. oils; disp. water, toluol; sp.gr. 1.05; HLB 11.0; sapon. no. 88–98; 100% conc.

Glycox® PETC. [Lonza] Proprietary; nonionic; textile spin finish lubricant/emulsifier, antistat; liq./solid; 100% conc.

Glytex® 203. [Lonza] Polyol ester; heat-stable nylon and polyester lubricant, emulsifier, antistat.

Glytex® 213. [Lonza] Polyol ester; nylon filament yarn lubricant, emulsifier, antistat.

Glytex® 273. [Lonza] Polyol ester; nylon filament yarn lubricant, emulsifier, antistat.

Glytex® 513. [Lonza] Ester ethoxylate; nylon and polyester tire cord finish emulsifier, lubricant, antistat.

Glytex® 558. [Lonza] Ester ethoxylate; nylon and polyester tire cord and yarn finish emulsifier, lubricant, antistat.

Glytex® 663. [Lonza] Hydantoin ester; nylon and polyester industrial yarn lubricant, emulsifier, antistat.

Glytex® 1085. [Lonza] Alcohol ester; textile spin finish lubricant, emulsifier, antistat.

Glytex® EL 176. [Lonza] Used in textile spinning process as lubricants, emulsifiers, and antistats for polyester and nylon fibers; Gardner 1; smoke pt. 140 C; acid no. 1; sapon. no. 93; hyd. no. 86.

Glytex® EL 882. [Lonza] Used in textile spinning process as lubricants, emulsifiers, and antistats for polyester and nylon fibers; Gardner 3+; smoke pt. 134 C; acid no. 9; sapon. no. 105; hyd. no. 32.

Glytex® EL 905. [Lonza] Used in textile spinning process as lubricants, emulsifiers, and antistats for polyester and nylon fibers; APHA 20; smoke pt. 146 C; acid no. 1; sapon. no. 46; hyd. no. 83.

Glytex® L 154. [Lonza] Used in textile spinning process as lubricants and antistats for polyester and nylon fibers; Gardner 2; smoke pt. 127 C; acid no. 0.5; sapon. no. 197; hyd. no. 5.

Glytex® L 203. [Lonza] Used in textile spinning process as lubricants and antistats for polyester and nylon fibers; APHA 50; smoke pt. 138 C; acid no. 0.5; sapon. no. 327; hyd. no. 5.

GMS Base. [Clark] Glyceryl stearate; base for nonionic textile softeners.

GMS/SE Base. [Clark] Glyceryl stearate SE; base for nonionic textile softeners; disp. readily in hot water.

Good-rite® K-702. [BFGoodrich] Polyacrylic acid aq. sol'n.; detergent assistant, soap builder, particulate soil dispersant, sequesterant for calcium, magnesium, iron; scale inhibitor; for laundry, dishwash, consumer/institutional cleaning prods.; amber clear to hazy sol'n., odorless; water-sol.; m.w. 240,000; sp.gr. 1.09; visc. 400-1200 cP; pH 3.0 (1% aq.); 25% solids.

Good-rite® K-732. [BFGoodrich/Spec. Polymers] Polyacrylic acid aq. sol'n.; anionic; dispersant resin used to disperse pigments, fillers, clay, silt, other suspended matter in water; detergent assistant, soap builder, particulate soil dispersant, sequesterant for calcium, magnesium, iron; scale inhibitor; for laundry, dishwash, consumer/institutional cleaning prods.; amber clear to hazy sol'n., odorless; water-sol.; m.w. 5100; sp.gr. 1.18 g/cc; visc. 250-500 cP; pH 3.0 (1% aq.); 50% solids.

Good-rite® K-739. [BFGoodrich] Sodium polyacrylate; detergent assistant, soap builder, particulate soil dispersant, sequesterant for calcium, magnesium, iron; scale inhibitor; for laundry, dishwash, consumer/institutional cleaning prods.; wh. free-flowing powd.; water-sol.; m.w. 5100; bulk dens. 37.6 lb/ft^3; pH 6-9 (1% aq.).

Good-rite® K-752. [BFGoodrich/Spec. Polymers] Polyacrylic acid aq. sol'n.; detergent assistant, soap builder, particulate soil dispersant, sequesterant for calcium, magnesium, iron; scale inhibitor; for laundry, dishwash, consumer/institutional cleaning prods.; amber clear to hazy sol'n., odorless; water-sol.; sp.gr. 1.23 g/cc; visc. 400-1400 cP; pH 3.0 (1% aq.); 63% solids.

Good-rite® K-759. [BFGoodrich] Sodium polyacrylate; detergent assistant, soap builder, particulate soil dispersant, sequesterant for calcium, magnesium, iron; scale inhibitor; for laundry, dishwash, consumer/institutional cleaning prods.; wh. free-flowing powd., odorless; water-sol.; m.w. 2100; bulk dens. 38.2 lb/ft^3; pH 6-9 (1% aq.).

Good-rite® K-7058. [BFGoodrich/Spec. Polymers] Low to med. m.w. linear polyacrylic acids and their sodium salts; anionic; co-builder in laundry, auto-dish, and misc. cleaners; provides detergency boosting, antisoil redeposition, antiscaling, antifilming/spotting; process and granulating aid in spray-dried detergents; chelating agent; water sol'n. and dry powd.; wh./amber sl. hazy sol'n., odorless; water-sol.; m.w. 5800; sp.gr. 1.23 g/cc; visc. 300-700 cP; pH 2.2-3.0; 50% solids.

Good-rite® K-7058D. [BFGoodrich] Sodium polyacrylate; detergent assistant, soap builder, particulate soil dispersant, sequesterant for calcium, magnesium, iron; scale inhibitor; for laundry, dishwash, consumer/institutional cleaning prods.; off-wh. free-flowing powd., odorless; water-sol.; m.w. 5800; bulk dens. 32.8 lb/ft^3; pH 6-9 (1% aq.).

Good-rite® K-7058N. [BFGoodrich] Polyacrylic acid aq. sol'n.; detergent assistant, soap builder, particulate soil dispersant, sequesterant for calcium, magnesium, iron; scale inhibitor; for laundry, dishwash, consumer/institutional cleaning prods.; amber clear sol'n., odorless; water-sol.; m.w. 5800; sp.gr. 1.30 g/cc; visc. 500-3500 cP; pH 9.0 (1% aq.); 45% solids.

Good-rite® K-7200N. [BFGoodrich/Spec. Polymers] Polyacrylic acid aq. sol'n.; anionic; co-builder in laundry, auto-dish, and misc. cleaners; provides detergency boosting, antisoil redeposition, antiscaling, antifilming/spotting; process and granulating aid in spray-dried detergents; water sol'n. or dry powd.; wh./amber sl. hazy sol'n., odorless; water-sol.; m.w. 20,000; sp.gr. 1.30 g/cc; visc. 600-1600 cP; pH 9.1 (1% aq.); 40% solids.

Good-rite® K-7600N. [BFGoodrich/Spec. Polymers] Polyacrylic acid aq. sol'n.; anionic; co-builder in laundry, auto-dish and misc. cleaners; provides detergency boosting, antisoil redeposition, antiscaling, antifilming/spotting; anticaking agent; process and granulating aid in spray-dried detergents; water sol'n. or dry powd. forms; wh./amber sl. hazy sol'n., odorless; water-sol.; m.w. 60,000; sp.gr. 1.24 g/cc; visc. 1400-3400 cP; pH 10.0 (1% aq.); 35% solids.

Good-rite® K-7658. [BFGoodrich] Polyacrylic acid aq. sol'n.; detergent assistant, soap builder, particulate soil dispersant, sequesterant for calcium, magnesium, iron; scale inhibitor; for laundry, dishwash, consumer/institutional cleaning prods.; wh./amber sl. hazy sol'n., odorless; water-sol.; m.w. 46,000; sp.gr. 1.24 g/cc; visc. 800 cP; pH 4.0-5.0; 39% solids.

Goulac. [LignoTech] Calcium lignosulfonate; anionic; binder for foundry, ceramic, and refractory prods.; liq.

GP-209. [Genesee Polymers] Dimethyl silicone EO/PO block copolymer; nonionic; emulsifier, wetting agent, pigment dispersant, leveling agent, profoaming additive for PU foams, hard surf. cleaners, polishes, cosmetic formulations; inverse sol. suggests use as defoamer for hot aq. surfactant sol'ns.; lt. straw clear liq.; sol. in water; m.w. 7800; sp.gr. 1.03; dens. 8.5 lb/gal; visc. 2600 cst; f.p. -50 F; flash pt. (PMCC) > 300 F; 100% act., 15% silicone.

GP-210 Silicone Antifoam Emulsion. [Genesee Polymers] Silicone emulsion; nonionic; defoamer for hot and cold foaming systems, industrial applics., commercial cleaning compds., latex stripping, adhesives, cutting oils, leather treating, sewage treatment, chemical processing, paints/coatings; wh. liq.; sp.gr. 0.98; dens. 8.0 lb/gal; 10% silicone.

GP-214. [Genesee Polymers] Dimethyl silicone EO/PO block copolymer; nonionic; profoaming additive in plastisol and organosol formulations, polyether PU foams; pigment dispersant, wetting agent, internal lubricant for plastics, leveling and flow control agent for solv. coatings; lt. straw clear liq.; insol. in water; m.w. 11,500; sp.gr. 1.03; dens. 8.5 lb/gal; visc. 6500 cst; f.p. -50 F; flash pt. (PMCC) > 300 F; 100% act., 31% silicone.

GP-215. [Genesee Polymers] Dimethyl silicone EO/PO block copolymer; emulsifier, wetting agent, pigment dispersant, leveling agent, profoaming additive for PU foams, hard surf. cleaners, polishes, cosmetics; inverse sol. suggests use as defoamer for hot aq. surfactant sol'ns.; lt. straw clear liq.; sol. in water; m.w. 9800; sp.gr. 1.03; dens. 8.5 lb/gal; visc. 2000 cst; f.p. -50 F; flash pt. (PMCC) > 300 F; 100% act., 18% silicone.

GP-217. [Genesee Polymers] Dimethylpolysiloxane EO block copolymer; wetting agent, emulsifier for water-based coatings, inks, polishes, hard surf. cleaners; dispersant for clays, pigments; thread lubricant; leveling and flow control agent; profoaming additive in aq. systems; lt. straw clear liq.; sol. in water; m.w. 3800; sp.gr. 1.05; dens. 8.75 lb/gal; visc. 240 cst; f.p. 65 F; flash pt. (PMCC) > 300 F; 100% act., 33% silicone.

GP-218. [Genesee Polymers] Dimethylpolysiloxane PO block copolymer; wetting, leveling, flow control agent, lubricant for solv.-based coatings, industrial finishes; profoamer additive in PU foams; textile and thread lubricants; internal lubricant for plastics; base for aq. defoamers; pigment dispersant; release agent; colorless clear liq.; sol. in aliphatic, aromatic, and chlorinated hydrocarbons, alcohols; insol. in water; m.w. 11,000; sp.gr. 0.98; dens. 8.0 lb/gal; visc. 1500 cst; f.p. 65 F; flash pt. (CC) > 300 F; 100% act., 32% silicone.

GP-219. [Genesee Polymers] Dimethylpolysiloxane EO block copolymer; emulsifier for w/o emulsions; pigment dispersant for thermoplastics or solv.-based coatings; wetting agent; thread lubricant; leveling and flow control agent; lt. straw clear liq.; insol. in water; m.w. 6500; sp.gr. 1.1; dens. 8.4 lb/gal; visc. 440 cst; HLB 8.4; f.p. 65 F; flash pt. (PMCC) > 300 F; 100% act., 58% silicone.

GP-226. [Genesee Polymers] Dimethylpolysiloxane EO block copolymer; wetting agent, emulsifier for water-based coatings, inks, polishes, hard surf. cleaners; pigment/clay dispersant; thread lubricant; leveling and flow control agent; profoaming additive for aq. systems; lt. straw clear liq.; disp. in water; m.w. 4340; sp.gr. 1.03; dens. 8.5 lb/gal; visc. 150 cst; f.p. 32 F; flash pt. (PMCC) > 300 F; 100% act., 42% silicone.

GP-227. [Genesee Polymers] Organo-modified dimethylsilicone polymer; surfactant, w/o emulsifier for emulsions, auto and furniture polishes, vinyl conditioners; gloss aid; adds detergent resistance; straw clear liq.; sp.gr. 0.8; dens. 6.7 lb/gal; visc. 6 cst; f.p. < 0 F; flash pt. 172 F; combustible liq.; 10% NV.

GP-262 Defoamer. [Genesee Polymers] Nonsilicone; defoamer for aq. systems, metalworking fluids, cleaner formulations, paints, coatings, paper/paperboard, waste water treatment; alkali resistant; tan liq.; sp.gr. 0.85; dens. 7.0 lb/gal; flash pt. > 200 F; 100% act.

GP-295 Defoamer. [Genesee Polymers] Nonsilicone; defoamer for aq. systems, metalworking fluids, cleaner formulations, paints, coatings, paper/paperboard, waste water treatment; alkali resistant; FDA approved; tan liq.; sp.gr. 0.85; dens. 7.0 lb/gal; flash pt. > 200 F; 100% act.

GP-300-I Antifoam Compd. [Genesee Polymers] Silicone; defoamer for hot and cold foaming systems, industrial applics., commercial cleaning, latex stripping, cutting oils, paper processing, paints/coatings, solv.-based coatings, metalworking fluids; wh. visc. liq.; sp.gr. 1.0; dens. 8.3 lb/gal; flash pt. (COC) > 600 F; 100% silicone.

GP-310-I. [Genesee Polymers] Silicone aq. emulsion; nonionic; dilutable antifoam for industrial processing in hot, cold, alkaline or aq. systems; for chemical processing, cleaning prods., paints, paper and latex processing; wh. thixotropic liq.; sp.gr. 1.0; dens. 8.3 lb/gal; 10% silicone.

GP-330-I Antifoam Emulsion. [Genesee Polymers] Silicone aq. emulsion; nonionic; antifoam for industrial processing in hot, cold, alkaline or high shear systems; for sewage treatment, insecticides, adhesives, cutting oils, latex stripping, brine sol'ns., cleaning compds.; wh. thixotropic liq.; sp.gr. 1.0; dens. 8.3 lb/gal; 30% silicone.

GP-7000. [Genesee Polymers] Methyl alkyl dimethyl silicone fluid; surf. tens. reducer for nonaq. solv. and oil systems; wetting and leveling agent in inks, coatings, plastisols; antifoam, mold release agent; internal release agent for plastics and rubber; textile lubricant; straw clear liq., odorless; sol. in aliphatic, aromatic, and chlorinated solvs.; sp.gr. 0.94; dens. 7.8 lb/gal; visc. 100 cst; flash pt. (PMCC) > 300 F; 100% act.

Gradonic 400-ML. [Graden] PEG 400 laurate; CAS 9004-81-3; nonionic; emulsifier for oils and solvs.; plasticizer, visc. reducer for starches; stabilizer, wetting agent for latex paints; liq.; 100% conc.

Gradonic FA-20. [Graden] Coconut oil alkylolamide; nonionic; liq. detergent; base for syn. and soap syn. floor, wall, car cleaners; liq.; 100% conc.

Gradonic LFA Series. [Graden] Linear alcohol alkoxylates; nonionic; low foaming surfactants for mech. dishwashing, metal cleaning, pulp and paper, textiles, etc.; liq.; 100% conc.

Gradonic N-95. [Graden] Nonoxynol-9.5; CAS 9016-45-9; nonionic; detergent, wetting agent, dispersant, emulsifier; for household and industrial cleaners, metal cleaning, wool scouring, insecticides and herbicides; almost colorless liq.; mild odor; sol. in water, dilute inorg. salt and caustic sol'ns., min. acid sol'ns.; cloud pt. 130 F (5%); pH 7 (2%); 100% act.

Gran UP A-600. [Sanyo Chem. Industries] Surfactant blend; anionic/nonionic; scouring agent for cotton or cotton/polyester blended fabrics; alkali resistant; liq.

Gran UP AX-15. [Sanyo Chem. Industries] Surfactant blend; anionic/nonionic; low foaming soaping agent for polyester or cotton/polyester blends; liq.

Gran UP CS-500. [Sanyo Chem. Industries] Surfactant blend; anionic/nonionic; scouring agent for cotton or cotton/polyester blends; liq.

Gran UP CS-700F. [Sanyo Chem. Industries] Surfactant blend; anionic/nonionic; scouring agent for cotton or cotton/polyester blends; low foaming, penetrating; liq.

Gran UP US-800. [Sanyo Chem. Industries] Surfactant blend; anionic/nonionic; low foaming penetrating and scouring agent for continuous desizing/scouring operation; liq.

Grilloten LSE87. [RITA] Sucrose laurate; nonionic; o/w emulsifier, solubilizer, foam booster, counterirritant; stable over wide pH range; liq.; HLB 12.5;

100% conc.

Grilloten LSE87K. [RITA] Sucrose cocoate; nonionic; o/w emulsifier, solubilizer, foam booster, counter-irritant; stable over wide pH range; liq.; HLB 12.5; 100% conc.

Grilloten PSE141G. [RITA] Sucrose stearate; nonionic; o/w emulsifier, solubilizer; stable over wide pH range; liq.; HLB 15.0; 100% conc.

Grilloten ZT12, ZT40, ZT80. [RITA] Sucrose ricinoleate; o/w emulsifier, solubilizer, foam booster, counter-irritant; stable over wide pH range; liq.; HLB 9.0, 8.5, and 9.5 resp.; 100% conc.

Grindtek CA-P. [Grindsted Prods.] Hydrog. tallow glyceride citrate; o/w emulsifier; cream powd.; sol. warm in toluene, wh. spirit, propylene glycol; HLB 11.

Grindtek DAT-L. [Grindsted Prods.] Diacetylated tartaric acid esters of glycerol (oleic/stearic) fatty acid esters; o/w emulsifier; yel.-brn. visc. liq.; sol. in ethanol, toluene, peanut oil, disp. in water; HLB 7.

Grindtek DAT-S. [Grindsted Prods.] Diacetylated tartaric acid ester of glycerol (stearic/palmitic) fatty acid esters; o/w emulsifier; cream flakes; sol. in toluene, wh. spirit, peanut oil; HLB 8.

Grindtek DAT-VO. [Grindsted Prods.] Diacetylated tartaric acid esters of glycerol (oleic/stearic) fatty acid esters; w/o emulsifier; yel.-br. liq.; sol. in peanut oil, toluene, wh. spirit, disp. in water, propylene glycol; HLB 6.

Grindtek FAL 1. [Grindsted Prods.] Sodium stearoyl lactylate; o/w emulsifier; cream powd.; sol. warm in ethanol, propylene glycol, toluene, wh. spirit, paraffin oil; HLB 10.

Grindtek FAL 2. [Grindsted Prods.] Calcium stearoyl lactylate; w/o emulsifier; cream flake; sol. warm in toluene, wh. spirit; HLB 2.

Grindtek FAL 3. [Grindsted Prods.] Stearoyl lactylic acid; o/w emulsifier; tan block; sol. in toluene; sol. warm in ethanol, propylene glycol, peanut oil; HLB 2.

Grindtek FAL 4. [Grindsted Prods.] Caprylic fatty acid ester of lactylic acid and its salts; o/w emulsifier with fungistatic properties; yel./brn. plastic; sol. in ethanol, propylene glycol, toluene, warm water, wh. spirit; HLB > 13.

Grindtek PGE 25. [Grindsted Prods.] Polyglyceryl-3 oleate; o/w emulsifier, antifogging agent for plastics; yel.-brn. liq.; sol. in ethanol, warm in peanut oil, toluene, paraffin oil, disp. in water, propylene glycol; HLB 5.5.

Grindtek PGE 55. [Grindsted Prods.] Polyglyceryl-3 stearate; o/w emulsifier, antifogging agent for plastics; wh. powd., tan block; sol. warm in toluene; HLB 6.8.

Grindtek PGE 55-6. [Grindsted Prods.] Polyglyceryl-3 stearate SE; o/w emulsifier, antifogging agent for plastics; wh. powd., tan block; sol. warm in ethanol; HLB 7.4.

Grindtek PGE-DSO. [Grindsted Prods.] Oleic/linoleic fatty acid ester of polyglycerol; o/w emulsifier, antifogging agent for plastics, effective for w/o emulsifications with veg. oils; yel.-brn. visc. liq.; sol. in toluene, warm in peanut oil, wh. spirit, paraffin oil; HLB 2.8.

Grindtek PGMS 90. [Grindsted Prods.] Propylene glycol stearate; w/o emulsifier; whitish block; sol. in ethanol, toluene; sol. warm in propylene glycol, peanut oil, wh. spirit, paraffin oil; HLB 3.5.

Grindtek SMS. [Grindsted Prods.] Sorbitan stearate; CAS 1338-41-6; nonionic; w/o and o/w emulsifier; tan powd.; sol. warm in toluene; HLB 5.2.

Grindtek STS. [Grindsted Prods.] Sorbitan tristearate; CAS 26658-19-5; nonionic; w/o and o/w emulsifier; tan powd.; sol. warm in toluene, wh. spirit; HLB 2.3.

Griton Series. [Toho Chem. Industry] Min. oil/nonionic surfactant; nonionic; water-sol. cutting oil; liq.

Gulftene 4. [Chevron] C4 alpha olefins; CAS 106-98-9; intermediate for biodeg. surfactants and specialty industrial chemicals; gas; 100% conc.

Gulftene 6. [Chevron] C6 alpha olefins; CAS 592-41-6; intermediate for biodeg. surfactants and specialty industrial chemicals; liq.

Gultfene 8. [Chevron] C8 alpha olefins; CAS 111-66-0; intermediate for biodeg. surfactants and specialty industrial chemicals; liq.

Gulftene 10. [Chevron] C10 alpha olefins; CAS 872-05-9; intermediate for biodeg. surfactants and specialty industrial chemicals; liq.

Gulftene 12. [Chevron] C12 alpha olefins; CAS 112-41-4; intermediate for biodeg. surfactants and specialty industrial chemicals; liq.

Gulftene 14. [Chevron] C14 alpha olefins; CAS 1120-36-1; intermediate for biodeg. surfactants and specialty industrial chemicals; liq.

Gulftene 16. [Chevron] C16 alpha olefins; CAS 629-73-2; intermediate for biodeg. surfactants and specialty industrial chemicals; liq.

Gulftene 18. [Chevron] C18 alpha olefins; CAS 112-88-9; intermediate for biodeg. surfactants and specialty industrial chemicals; liq.

Gulftene 20-24. [Chevron] C20-24 alpha olefins; intermediate for biodeg. surfactants and specialty industrial chemicals; solid.

Gulftene 24-28. [Chevron] C24-28 alpha olefins; intermediate for biodeg. surfactants and specialty industrial chemicals; solid.

Gulftene 30+. [Chevron] C30 alpha olefin; intermediate for biodeg. surfactants and specialty industrial chemicals; solid.

H

Hallcomid® M-18-OL. [C.P. Hall] N,N-dimethyl oleamide; nonionic; solubilizer, solv., dispersant, wetting agent; slightly yel. liq.; m.w. 310; sol. see Hallcomid® M-18; sp.gr. 0.875; visc. 18.5 cps; f.p. –8 C; b.p. 196–211 C (3 mm Hg); HLB 7.0; flash pt. 410 F; 95% conc.

Hampfoam 35. [W.R. Grace/Organics] Sodium cocoyl sarcosinate; anionic; surfactant used in alkaline industrial formulations, textile applics.; forms stable small bubbles; pale yel. clear liq.; visc. 100 cps; HLB 27.0; pH 8.0–9.5; 35.0% min. solids.

Hamposyl® AC-30. [W.R. Grace] Ammonium cocoyl sarcosinate; anionic; surfactant.

Hamposyl® AL-30. [W.R. Grace/Organics] Ammonium lauroyl sarcosinate; anionic; surfactant for shampoos, skin cleansers, bath gels; sec. emulsifier for emulsion polymerization; liq.; HLB 29.0; 30% conc.

Hamposyl® C. [W.R. Grace/Organics] Cocoyl sarcosine; anionic; detergent, wetting and foaming agent, foam stabilizer, emulsifier, anticorrosive agent, conditioner for hair and rug shampoos, cosmetics, skin cleansers; biodeg.; pale yel. liq.; sol. in most org. solv.; m.w. 280; sp.gr. 0.97–0.99; HLB 10.0; m.p. 23–26 C; 100% conc.

Hamposyl® C-30. [W.R. Grace/Organics] Sodium cocoyl sarcosinate; anionic; detergent, wetting and foaming agent, foam stabilizer, emulsifier, anticorrosive agent, conditioner for hair and rug shampoos, cosmetics, skin cleansers; biodeg.; colorless to very pale yel. liq.; misc. in water; m.w. 302; sp.gr. 1.02–1.03; visc. 30 cps; HLB 27.0; f.p. –1 C; pH 7.5-8.5 (10%); surf. tens. 30 dynes/cm; 30% act.

Hamposyl® L. [W.R. Grace/Organics] Lauroyl sarcosine; anionic; detergent, wetting and foaming agent, foam stabilizer, emulsifier, anticorrosive agent, conditioner for hair and rug shampoos, cosmetics, skin cleansers; biodeg.; wh. waxy solid; sol. in most org. solvs.; m.w. 270; sp.gr. 0.97–0.99; HLB 13.0; m.p. 34–37 C; 94% act.

Hamposyl® L-30. [W.R. Grace/Organics] Sodium lauroyl sarcosinate; anionic; detergent, wetting and foaming agent, foam stabilizer, emulsifier, anticorrosive agent, conditioner for hair and rug shampoos, cosmetics, skin cleansers; biodeg.; colorless liq.; misc. in water; m.w. 292; sp.gr. 1.02–1.03; visc. 30 cps; HLB 30.0; f.p. –1 C; pH 7.5-8.5 (10%); surf. tens. 30 dynes/cm; 30% act.

Hamposyl® L-95. [W.R. Grace/Organics] Sodium lauroyl sarcosinate; anionic; detergent, wetting and foaming agent, foam stabilizer, emulsifier, anticorrosive agent, conditioner for hair and rug shampoos, cosmetics, skin cleansers; biodeg.; wh. powd.; sol. in water; m.w. 292; dens. 25 lb/ft^3; HLB 30.0; pH 7.5-8.5 (10%); 94% act.

Hamposyl® M. [W.R. Grace/Organics] Myristoyl sarcosine; anionic; detergent, wetting and foaming agent, foam stabilizer, emulsifier, anticorrosive agent, conditioner for hair and rug shampoos, cosmetics, skin cleansers; biodeg.; wh. waxy solid; sol. in most org. solv.; m.w. 298; sp.gr. 0.97–0.99; m.p. 48–53 C; 94% act.

Hamposyl® M-30. [W.R. Grace/Organics] Sodium myristoyl sarcosinate; anionic; detergent, wetting and foaming agent, foam stabilizer, emulsifier, anticorrosive agent, conditioner for hair and rug shampoos, cosmetics, skin cleansers; biodeg.; colorless liq.; misc. in water; m.w. 320; sp.gr. 1.02–1.03; visc. 30 cps; f.p. –1 C; pH 7.5-8.5 (10%); surf. tens. 30 dynes/cm; 30% act.

Hamposyl® O. [W.R. Grace/Organics] Oleoyl sarcosine; anionic; detergent, wetting and foaming agent, foam stabilizer, emulsifier, corrosion inhibitor, mold release agent, conditioner for hair and rug shampoos, cosmetics, skin cleansers; ceramic dispersant; biodeg.; yel. liq.; sol. in most org. solv.; m.w. 349; sp.gr. 0.95–0.97; visc. 250 cps; HLB 10.0; 94% act.

Hamposyl® S. [W.R. Grace/Organics] Stearoyl sarcosine; anionic; detergent, wetting and foaming agent, foam stabilizer, emulsifier, anticorrosive agent, conditioner for hair and rug shampoos, cosmetics, skin cleansers; biodeg.; wh. waxy solid; sol. in most org. solv.; m.w. 338; sp.gr. 0.96–0.98; m.p. 53–58 C; 94% act.

Haroil SCO-65, -7525. [Graden] Sulfated castor oil; anionic; superfatting agent for cosmetic creams, emulsifier for cosmetic formulations, plasticizer for adhesives, antisagging and antisettling agent for paints and stains; amber liq.; pH 7.0 ± 0.5 (SCO-65, 2% aq.); dens. 1.025 (SCO-7525); 50%, 75% act.

Haroil SCO-75. [Graden] Sulfated castor oil; anionic; plasticizer for animal glue, starch, and casein adhesives; antisagging and antisettling agent for paints and pigments; liq.; 75% conc.

Harol RG-71L. [Graden] Sodium polynaphthalene sulfonate; anionic; dye retardant, leveler; pitch dispersant for pulp and paper; liq.; 47.5% conc.

Hartaine CB-40. [Hart Prods. Corp.] Betaine; amphoteric; detergent, emulsifier; liq.; 40% conc.

Hartamide 9137. [Hart Chem. Ltd.] Oleamide DEA; anionic; coupling agent, emulsion stabilizer, lubricant and antistat; liq.; 80% amide.

Hartamide AD. [Hart Prods. Corp.] Coconut alkanolamide; nonionic; foam booster, solubilizer, emulsifier; liq.; 100% conc.

Hartamide KL. [Hart Prods. Corp.] Modified coconut DEA; anionic; foam booster, detergent; liq.; 100%

conc.

Hartamide LDA. [Hart Chem. Ltd.] Lauramide DEA; CAS 120-40-1; nonionic; detergent, foam stabilizer and thickener for liq. and powd. detergent systems, shampoos, bubble baths; wh. solid; 100% act.

Hartamide LMEA. [Hart Chem. Ltd.] Lauramide MEA; CAS 142-78-9; anionic; foam stabilizer for spray-dried powd. detergents and bubble bath preparations; visc. modifier for detergents, shampoos, bubble baths; wh. solid; 95% amide.

Hartamide OD. [Hart Chem. Ltd.] Cocamide DEA; nonionic; detergent, foam stabilizer and visc. regulator for liq. and powd. detergent systems, shampoos, bubble baths; yel. liq.; dens. 1.02 lb/gal; visc. 900 cps; 100% act.

Hartasist 16. [Hart Chem. Ltd.] Silicone-based; defoamer for oil and gas industry; liq.; 11% act.

Hartasist 20, 37. [Hart Chem. Ltd.] Proprietary blend; defoaming agents for drilling mud; liq.; 100% act.

Hartasist 46. [Hart Chem. Ltd.] Specialty surfactant; cold flow improver for oil deposit; heavy-duty cleaner; liq.; 95% act.

Hartasist DF-28. [Hart Chem. Ltd.] Anionic; defoamer for cementing and other applics. in oil and gas industry; liq.; 100% act.

Hartbreak Series. [Hart Chem. Ltd.] Phenolic resins, polyol ester, alkylaryl sulfonate; demulsifiers for w/o emulsions in oil and gas industry; liq.; 70-100% act.

Hartenol LAS-30. [Hart Prods. Corp.] Sodium lauryl sulfate; CAS 151-21-3; anionic; detergent, wetting agent, emulsifier, foaming agent for lt. duty household detergents; liq.; 30% conc.

Hartenol LES 60. [Hart Prods. Corp.] Sodium laureth sulfate; CAS 9004-82-4; anionic; detergent, wetting agent, emulsifier, foaming agent for lt. duty household detergents; liq.; 60% conc.

Hartofix 2X. [Hart Chem. Ltd.] Cationic resin; cationic; fixative for direct and reactive dyes; improves fastness of most substantive colors to water, soaps, and detergents; pale yel. liq.; 40% conc.

Hartofol 40. [Hart Prods. Corp.] Sodium dodecylbenzene sulfonate; anionic; surfactant; liq.; 40% conc.

Hartofol 60T. [Hart Prods. Corp.] TEA dodecylbenzene sulfonate; anionic; surfactant; liq.; 60% conc.

Hartolon 1328. [Hart Chem. Ltd.] Fatty acid ester blend; cationic; nonyel. softener for ring spun yarn fabric; liq.

Hartolon 5683. [Hart Chem. Ltd.] Polyethylene and selected esters; nonionic; nonyel. softener for 100% cotton toweling; provides static control and absorbency; liq.

Hartolon AL. [Hart Chem. Ltd.] Fatty acid condensate; nonionic; softener, antistat; minimizes crocking on polyester; paste.

Hartolon HVH Base. [Hart Chem. Ltd.] Fatty acid esters; nonionic; softener base for softener formulations for all fibers; liq.; 100% conc.

Hartolon NA. [Hart Chem. Ltd.] Sat. esters and alkyl condensate; nonionic; softener, lubricant, antistat used in textiles; wh. paste; 25% act.

Hartolon PC. [Hart Chem. Ltd.] Ester blend; nonionic; softener for increased lubricity on sheeting, corduroy, and cotton fabrics; liq.

Hartomer 4900. [Hart Chem. Ltd.] Sodium vinyl sulfonate; anionic; monomer for latex emulsion polymerization systems; liq.; 25% act.

Hartomer GP 2164. [Hart Chem. Ltd.] Phosphate ester; surfactant for water-based adhesives and latex emulsion polymerization; liq.; 100% act.

Hartomer GP 4935. [Hart Chem. Ltd.] Complex phosphate ester; low foaming surfactant for latex emulsion polymerization; effective in acidic and alkaline media; liq.; 100% act.

Hartomer JV 4091. [Hart Chem. Ltd.] Blend; anionic/nonionic; surfactant for latex emulsion polymerization; liq.; 20% act.

Hartomul PE-30. [Hart Chem. Ltd.] Blend of polyethylene and polyether derivs.; nonionic; softener for use with resin finishes on lightweight fabrics; mold release agent for brake linings, urethane foam; liq.; 30% act.

Hartonyl L531. [Hart Chem. Ltd.] Ethoxylated fatty amine; cationic; leveling aid for dyeing polyamides; migrating leveling agent and dispersant for acid and disperse dye stuffs; increases contrast and clarity between nylon fibers of different affinities; effective in continuous and winch dyeing of carpets; yel. liq.; ref. 1.407; pH alkaline; 45% act.

Hartonyl L535. [Hart Chem. Ltd.] Ethoxylated fatty amine; cationic; dyeing assistant, antiprecipitant, dispersant, leveling agent; dyeing assistant for leveling and dispersing of acid and disperse dye stuffs; promotes contrast; yel. liq.; sp.gr. 1.025; pH acid; 33% act.

Hartonyl L537. [Hart Chem. Ltd.] Ethoxylated fatty amine; cationic; leveling and retarding agent for acid dyes; used for dyeing of dyeable nylon carpet; effective in continuous and winch dyeing; yel. liq.; sp.gr. 0.999; ref. index 1.412; pH alkaline; 60% act.

Hartopol 25R2. [Hart Chem. Ltd.] Polyoxyalkylene glycol; nonionic; low foaming surfactant for rinse aids and windshield washer fluids; liq.; cloud pt. 24–32 C (1% aq.); 100% act.

Hartopol 31R1. [Hart Chem. Ltd.] Polyoxyalkylene glycol; nonionic; low foaming surfactant for rinse aids and windshield washer fluids; liq.; cloud pt. 23-26 C (1% aq.); 100% act.

Hartopol L42. [Hart Chem. Ltd.] EO/PO block copolymer; nonionic; low foaming rinse aid, defoamer, emulsifier, detergent, dispersant for many industries; liq.; HLB 8.0; cloud pt. 36–40 C (1% aq.); 100% act.

Hartopol L44. [Hart Chem. Ltd.] Polyoxyalkylene glycol; nonionic; dispersant, demulsifier, defoamer, emulsifier, detergent for many industries; liq.; HLB 16.0; cloud pt. 69–73 C (1% aq.); 100% act.

Hartopol L62. [Hart Chem. Ltd.] Polyoxyalkylene glycol; nonionic; defoamer, emulsifier, detergent, dispersant used in rinse aids and industrial cleaners; liq.; HLB 7.0; cloud pt. 31–33 C (1% aq.); 100% act.

Hartopol L62LF. [Hart Chem. Ltd.] EO/PO block copolymer; nonionic; defoamer, emulsifier, low-foaming detergent, dispersant used in rinse aids and industrial formulations; liq.; HLB 6.6; cloud pt. 27-29 C (1% aq.); 100% act.

Hartopol L64. [Hart Chem. Ltd.] Polyoxyalkylene glycol; nonionic; dispersant and emulsifier in emulsion polymerization; wetting agent for pulp/paper industry; defoamer, detergent for many industries; liq.; HLB 15.0; cloud pt. 58–62 C (1% aq.); 100% act.

Hartopol L81. [Hart Chem. Ltd.] Polyoxyalkylene glycol; nonionic; industrial defoamer, emulsifier, detergent, dispersant; liq.; HLB 2.0; cloud pt. 18–22 C (1% aq.); 100% act.

Hartopol LF-1. [Hart Chem. Ltd.] Polyoxalkylene glycols; nonionic; detergent, emulsifier, defoamer for low-foam dishwash, rinse aids, general household and industrial use; clear liq.; sp.gr. 1.01–1.035;

visc. 320–400 cps; HLB 3.0; cloud pt. 22-26 C (1% aq.); 100% act.

Hartopol LF-2. [Hart Chem. Ltd.] Polyoxyalkylene glycol; nonionic; low foaming surfactant for mech. dishwashing, rinse aids, general household and industrial use; liq.; HLB 7.0; cloud pt. 33-37 C (1% aq.); 100% act.

Hartopol LF-5. [Hart Chem. Ltd.] Polyoxyalkylene glycol; nonionic; mechanical dishwashing rinse aid conc. blend; liq.; HLB 6.8; cloud pt. 22–26 C (1% aq.); 87% act.

Hartopol P65. [Hart Chem. Ltd.] Polyoxyalkylene glycol; nonionic; dispersant, demulsifier, defoamer, emulsifier, detergent for many industries; paste; HLB 17.0; cloud pt. 80–84 C (1% aq.); 100% act.

Hartopol P85. [Hart Chem. Ltd.] EO/PO block copolymer; nonionic; defoamer, emulsifier, detergent, dispersant used in many industries; paste; HLB 16.0; cloud pt. 83-87 C (1% aq.); 100% conc.

Hartosoft 171. [Hart Chem. Ltd.] Esters and quat. amines; cationic; nonyel. softener with antistatic and lubricating props. for velours and toweling; wh. liq.; visc. 1400 cps; pH acid; 15% act.

Hartosoft CN. [Hart Chem. Ltd.] Ester and amine blend; cationic; nonyel. softener for napping of cotton and cotton/polyester fabrics; paste; 25% conc.

Hartosoft GF. [Hart Chem. Ltd.] Proprietary; cationic; softener, lubricant, antistat for needle-punched acrylic, nylon, polyester, and wool substrates; liq.

Hartosoft S5793. [Hart Chem. Ltd.] Amino functional silicone emulsion; cationic; softener providing lubricity to cotton and cotton blended knits and woven fabrics; liq.

Hartosolve OL. [Hart Prods. Corp.] Emulsified solvents; anionic; solvent scour and cleaner; liq.; 100% conc.

Hartotrope AXS. [Hart Chem. Ltd.] Ammonium xylene sulfonate; anionic; detergent, solubilizer and cloud pt. depressant for lt. duty and built liq. detergent systems; clear liq.; sp.gr. 1.125; 40% act.

Hartotrope KTS 50. [Hart Chem. Ltd.] Potassium toluene sulfonate; anionic; detergent, solubilizer and cloud pt. depressant for lt. duty and built liq. detergent systems; clear liq.; sp.gr. 1.24; 50% act.

Hartotrope STS-40, Powd. [Hart Chem. Ltd.] Sodium toluene sulfonate; anionic; detergent, solubilizer and cloud pt. depressant for lt. duty and built liq. detergent systems; clear liq., wh. powd. resp.; dens. 1.19 lb/gal (liq.); 40%, 93% act. resp.

Hartotrope SXS 40, Powd. [Hart Chem. Ltd.] Sodium xylene sulfonate; CAS 1300-72-7; anionic; detergent, solubilizer and cloud pt. depressant for lt. duty and built liq. detergent systems; clear liq., powd. resp.; sp.gr. 1.19 (liq.); 40 and 90% act.

Hartowet 5917. [Hart Chem. Ltd.] Blend; anionic/nonionic; wetting aid for all fabrics; contains no phosphates; milky wh. liq.; 55% act.

Hartowet CW. [Hart Chem. Ltd.] Nonionic; wetting agent for prep. of recycled box board; liq.; 100% act.

Hartowet MSW. [Hart Chem. Ltd.] Blend; anionic; low foaming wetting aid with wax emulsification and antiredeposition props.; for continuous scouring and bleaching operations; caustic-stable; milky wh. liq. emulsion; 55% act.

Hartowet SLN. [Hart Chem. Ltd.] Alcohol condensate; nonionic; wetting agent for finishing formulations in textiles; clear liq.; 60% conc.

Hartox DMCD. [Hart Prods. Corp.] Amine oxide; nonionic/cationic; wetting and foaming agent, foam stabilizer; liq.; 40% conc.

HD-Ocenol 45/50. [Henkel/Emery] Oleyl/cetyl alcohol (25-33% C16, 60-70% C18); intermediate for surfactant mfg.; APHA 100 solid; sp.gr. 0.820-0.830 g/cc (40 C); solid. pt. 30-37 C; b.p. 315-360 C; acid no. < 0.2; iodine no. 45-50; sapon. no. < 1.0; hyd. no. 210-220; flash pt. 180 C; 100% conc.

HD-Ocenol 50/55. [Henkel/Emery] Oleyl/cetyl alcohol (25-35% C16, 55-75% C18); intermediate for surfactant mfg.; APHA 100 solid; sp.gr. 0.820-0.830 g/cc (40 C); solid. pt. 28-34 C; b.p. 315-360 C; acid no. < 0.2; iodine no. 50-55; sapon. no. < 1.0; hyd. no. 210-220; flash pt. 170 C; 100% conc.

HD-Ocenol 50/55III. [Henkel/Emery] Oleyl/cetyl alcohol (25-35% C16, 58-68% C18); chemical intermediate for detergent mfg.; APHA 200 solid; sp.gr. 0.825-0.835 g/cc (40 C); solid. pt. 25-35 C; b.p. 315-360 C; acid no. < 0.2; iodine no. 48-55; sapon. no. < 2.0; hyd. no. 208-218; flash pt. 170 C.

HD-Ocenol 60/65. [Henkel/Emery] Oleyl cetyl alcohol (20-30% C16, 55-75% C18); intermediate for surfactant mfg.; APHA 100 solid; sp.gr. 0.825-0.835 g/cc (40 C); solid. pt. 25-30 C; b.p. 315-360 C; acid no. < 0.2; iodine no. 60-65; sapon. no. < 1.0; hyd. no. 208-218; flash pt. 170 C; 100% conc.

HD-Ocenol 70/75. [Henkel/Emery] Oleyl cetyl alcohol (12-22% C16, 68-80% C18); intermediate for surfactant mfg.; APHA 100 solid/liq.; sp.gr. 0.825-0.835 g/cc (40 C); solid. pt. 18-23 C; b.p. 315-360 C; acid no. < 0.2; iodine no. 70-75; sapon. no. < 1.0; hyd. no. 208-218; flash pt. 180 C; 100% conc.

HD-Ocenol 80/85. [Henkel/Emery] Oleyl cetyl alcohol (8-18% C16, 70-83% C18); intermediate for surfactant mfg.; APHA 100 liq.; sp.gr. 0.825-0.835 g/cc (40 C); solid. pt. 3-13 C; b.p. 315-360 C; acid no. < 0.2; iodine no. 84-89; sapon. no. < 1.0; hyd. no. 205-215; flash pt. 180 C; 100% conc.

HD-Ocenol 90/95. [Henkel/Emery] Oleyl alcohol (2-10% C16, 87-95% C18); CAS 143-28-2; emollient, superfatting agent, carrier for cosmetics; intermediate for surfactant mfg.; APHA 100 liq.; oil-sol.; sp.gr. 0.830-0.840 g/cc (40 C); solid. pt. 2-12 C; b.p. 330-360 C; acid no. < 0.2; iodine no. 90-97; sapon. no. < 1.0; hyd. no. 205-215; flash pt. 190 C; 100% conc.

HD-Ocenol 92/96. [Henkel/Emery] Oleyl alcohol (2-8% C16, 87-93% C18); CAS 143-28-2; emollient, superfatting agent for alcohol preparations and emulsions; intermediate for surfactant mfg.; clear oily liq.; sp.gr. 0.830-0.840 g/cc (40 C); solid. pt. 2-6 C; b.p. 330-360 C; acid no. < 0.2; iodine no. 95-105; sapon. no. < 1.0; hyd. no. 205-220; flash pt. 190 C; 100% conc.

HD-Ocenol 110/130. [Henkel/Emery] Oleyl linoleyl alcohol (5-10% C16, 90-95% C18); intermediate for surfactant mfg.; APHA 100 liq./solid; sp.gr. 0.830-0.840 g/cc (40 C); solid. pt. 10-21 C; b.p. 330-360 C; acid no. < 0.2; iodine no. 110-130; sapon. no. < 1.0; hyd. no. 200-220; flash pt. 190 C; 100% conc.

Heavy Duty Cleaner HDC. [Emkay] All-purpose degreaser for removing persistent stains and deposits of oil and grease and for cleaning tanks.

Hercules® 4 Defoamer. [Hercules] High-effiency defoamer for acid and alkaline papermaking systems, deinking systems; FDA compliance; brick, soapy odor; sp.gr. 0.96; m.p. 54 C; 100% conc.

Hercules® 137 Defoamer. [Hercules] Silica org.; defoamer used in paper and food pkg. applic.; pulp washing; lt. tan liq.; dens. 0.88 kg/l; visc. 1500–2500 cps.

Hercules® 187 Defoamer. [Hercules] Hydrocarbon oil-based; defoamer for kraft pulpmill brown stock washing systems; used in food pkg.; lt. brn. liq.; dens. 0.92 kg/l; visc. 1500–3000 cps.

Hercules® 356 Defoamer. [Hercules] Defoamer for wet-end use in acid or alkaline papermaking systems; FDA compliance; straw liq.; disp. in hard water; dens. 7.3 lb/gal.

Hercules® 388 Defoamer. [Hercules] Hydrocarbon oil-based; defoamer in aq. systems; used in mill effluent and waste treatment systems; kraft pulpmill screening and bleaching operations, papermaking systems; food pkg.; tan oily liq.; water-disp.; dens. 0.864 kg/l; 100% conc.

Hercules® 491 Defoamer. [Hercules] High-efficiency defoamer for acid and alkaline papermaking systems, size press sol'ns., and other aq. foaming systems; FDA compliance; wh. fluid paste, hydrocarbon odor; dens. 7.65 lb/gal; visc. 1800 cps; pH 7.4.

Hercules® 492 Defoamer. [Hercules] High-efficiency water-based defoamer for acid and alkaline papermaking systems, size press sol'ns., and other aq. foaming systems; FDA compliance; wh. liq., hydrocarbon odor; dens. 7.65 lb/gal; visc. 250 cps; pH 7.4.

Hercules® 831 Defoamer. [Hercules] Hydrocarbon oil based; defoamer used in size press applic. and paper making and coatings; food pkg.; grnsh. brn. oily liq.; dens. 0.91 kg/l; visc. 800–1000 cps; 100% conc.

Hercules® 845 Defoamer. [Hercules] Hydrocarbon oil-based; quick-acting defoamer for coatings in paper processing; FDA compliance; lt. gray liq.; dens. 7.7 lb/gal; visc. 750 cps.

Hercules® 1512 Defoamer. [Hercules] Hydrocarbon oil based; see Hercules 187 Defoamer; gray liq.; dens. 0.92 kg/l; visc. 600 cps.

Hercules® 2051GS Defoamer. [Hercules] Hydrocarbon oil-based; defoamer for use as drainage aid and foam killer in kraft pulpmill brownstock washing systems; FDA compliance; gray liq.; dens. 7.7 lb/gal; visc. 4000 cps.

Hercules® 2470 Defoamer. [Hercules] Silica-org.; defoamer for pulp washing; food grade paper, paperboards; lt. tan liq.; dens. 0.88 kg/l; visc. 2500 cps.

Hercules® AR150. [Hercules] PEG-15 rosinate; nonionic; low foaming surfactant, detergent, emulsifier, wetting agent, suspending agent, dispersant for industrial cleaners; FDA compliance; Gardner 13 liq. to soft wax; sol. in water, IPA, benzene, toluene, ester solvs.; visc. 300 cps (38 C); cloud pt. 60 C (2%); pH 9 (1%); surf. tens. 39 dynes/cm (0.1% aq.).

Hercules® AR160. [Hercules] PEG-16 rosinate; nonionic; low foaming surfactant, detergent, emulsifier, wetting agent, suspending agent, dispersant for industrial cleaners; FDA compliance; Gardner 14 soft wax; sol. in water, IPA, benzene, toluene, ester solvs.; visc. 250 cps (38 C); cloud pt. 68 C (2%); pH 8 (1%); surf. tens. 37.6 dynes/cm (0.1% aq.).

Hercules® Eff-101 Defoamer. [Hercules] Hydrocarbon oil based; defoamer for plant effluent and waste treatment systems; kraft pulpmill screening and pulpmill bleaching operations; mfg. of food paper applic.; grnsh. brn. oily liq.; water-disp.; dens. 0.91 kg/l; visc. 800–1000 cps; 100% conc.

Hercules® Surfactant AR 150. [Hercules] PEG ester; nonionic; low-foaming detergent; Gardner 10 liq./soft wax; sol. in water, IPA, benzene, toluene, ester solv.; visc. 300 cps (38 C); cloud pt. 60 C (1%); surf. tens. 39 dynes/cm (0.01% aq.); pH 9 (1%).

Hercules® Surfactant AR 160. [Hercules] Rosin oxyethylene glycol ester; nonionic; detergent, emulsifier for detergents and food related areas; liq., wax; 100% conc.

Hetamide 1069. [Heterene] Capramide DEA; pale yel. liq.; acid no. 0.5 max.

Hetamide CMA. [Heterene] Modified cocamide MEA; CAS 68140-00-1; thickener, foam booster, foam stabilizer, emulsifier; Gardner 8 max. flake; m.p. 68-74 C; acid no. 5.0 max.; 99% conc.

Hetamide CME. [Heterene] Cocamide MEA; CAS 68140-00-1; thickener, foam booster, foam stabilizer, emulsifier; tan flake; pH 7.5-9.0 (5%); 99% conc.

Hetamide CME-CO. [Heterene] Cocamide MEA; CAS 68140-00-1; thickener, foam booster/stabilizer, emulsifier; Gardner 5 max. flake; acid no. 1.0 max.; pH 9.5-10.5 (5%).

Hetamide DO. [Heterene] Oleamide DEA, diethanolamine; Gardner 7 liq.; acid no. 12-16; pH 9.0-10.5 (5%).

Hetamide DS. [Heterene] Stearamide DEA; thickener, foam booster, foam stabilizer, emulsifier; tan flake; acid no. 5 max.; pH 9.0-10.5 (5%); 99% conc.

Hetamide DSUC. [Heterene] Modified cocamide DEA; yel. to lt. amber clear liq.; acid no. 18-22.

Hetamide LA. [Heterene] Lauric acid; CAS 143-07-7; nonionic; foam stabilizer, emulsifier, visc. builder, detergent for car shampoos, cleaning, cosmetics, dispersants, lubricants; liq.; sp.gr. 1.00; acid no. 10–14; pH 9–10 (1% aq.); 100% act.

Hetamide LL. [Heterene] Lauramide DEA; CAS 120-40-1; thickener, foam booster, foam stabilizer, emulsifier; Gardner 2 max. solid; acid no. 2.0 max.; pH 9.5-10.5 (5%); 99% conc.

Hetamide LML. [Heterene] Lauramide/linoleamide DEA; thickener, foam booster, foam stabilizer, emulsifier; Gardner 8 max. liq.; 99% conc.

Hetamide LN. [Heterene] Linoleamide DEA; thickener, foam booster, foam stabilizer, emulsifier; Gardner 8 max. liq.; acid no. 2 max.; pH 9.5-11.0 (5%); 99% conc.

Hetamide LNO. [Heterene] Linoleamide DEA; Gardner 9 max. liq.; acid no. 1.0 max.

Hetamide M. [Heterene] Myristamide DEA; thickener, foam booster, foam stabilizer, emulsifier; Gardner 2 max. solid; acid no. 2 max.; pH 9.5-10.5 (5%); 99% conc.

Hetamide MA. [Heterene] Acetamide MEA; thickener, foam booster, foam stabilizer, emulsifier; Gardner 1 max. liq.; acid no. 15 max.

Hetamide MC. [Heterene] Cocamide DEA; nonionic; thickener, foam booster, foam stabilizer, emulsifier; Gardner 5 max. liq.; sp.gr. 0.98; pH 9.5–11.0 (5%); 100% act.

Hetamide MCS. [Heterene] Cocamide DEA; thickener, foam booster, foam stabilizer, emulsifier; liq.

Hetamide ML. [Heterene] Lauramide/myristamide DEA; CAS 120-40-1; nonionic; thickener, foam booster, foam stabilizer, emulsifier; Gardner 5 max. solid; sp.gr. 0.96; m.p. 35 C; acid no. 1 max.; pH 9.5-11.0 (5%); 100% act.

Hetamide MMC, OC. [Heterene] Mixed fatty acids; nonionic; foam stabilizer, emulsifier, visc. builder, detergent, for car shampoos, cleaning, cosmetics, dispersants, lubricants; liq.; sp.gr. 1.01; pH 9–10.5 (1% aq.); 100% act.

Hetamide MOC. [Heterene] Lauramide DEA; CAS

120-40-1; thickener, foam booster, foam stabilizer, emulsifier; Gardner 5 max. liq.; acid no. 2 max.; pH 9.5-10.5 (5%); 99% conc.

Hetamide OC. [Heterene] Oleamide DEA; thickener, foam booster, foam stabilizer, emulsifier; liq.

Hetamide RC. [Heterene] Cocamide DEA; nonionic; thickener, foam booster, foam stabilizer, emulsifier; Gardner 8 max. liq.; sp.gr. 1.01; pH 9.5-11.0 (5%); 100% act.

Hetan SL. [Heterene] Sorbitan laurate; CAS 1338-39-2; nonionic; lipophilic emulsifier; flake.

Hetan SO. [Heterene] Sorbitan oleate; CAS 1338-43-8; nonionic; lipophilic emulsifier; flake.

Hetan SS. [Heterene] Sorbitan stearate; CAS 1338-41-6; nonionic; lipophilic emulsifier; ivory flake; acid no. 11 max.; sapon. no. 140-154; hyd. no. 230-260; 99% conc.

Hetester 412. [Heterene] Stearyl stearate; wh. to off-wh. flake; m.p. 52-56 C; acid no. 2 max.; iodine no. 1 max.; sapon. no. 100-115.

Hetester PCA. [Heterene; Bernel] Propylene glycol ceteth-3 acetate; nonionic; emulsifier, pigment wetter, antichalking agent, emollient used in personal care prods., antiperspirants; colorless clear liq.; sol. @ 5% in water, 95% ethanol, most cosmetic oils; cloud pt. 15 C; acid no. 2 max.; sapon. no. 110–130; hyd. no. 10 max.; pH 6.0–7.0; 100% conc.

Hetester PHA. [Heterene; Bernel] Propylene glycol isoceteth-3 acetate; nonionic; emulsifier, pigment wetter, antichalking agent, emollient used in personal care prods., antiperspirants; wh. to pale yel. clear liq.; self-emulsifying in water; sol. in 95% ethanol, most cosmetic oils; acid no. 2 max.; sapon. no. 110–130; hyd. no. 10 max.; pH 6.0–7.0; 100% conc.

Hetoxamate 200 DL. [Heterene] PEG-4 dilaurate; CAS 9005-02-1; nonionic; thickener, foam booster, foam stabilizer, emulsifier; Gardner 4 max. liq.; sol. in IPA, min. oil; acid no. 10 max.; sapon. no. 165-180.

Hetoxamate 400 DS. [Heterene] PEG-8 distearate; CAS 9005-08-7; thickener, foam booster, foam stabilizer, emulsifier; Gardner 3 max. solid; disp. in water, min. oil; sol. in IPA; acid no. 10 max.; sapon. no. 120-130.

Hetoxamate FA-5. [Heterene] PEG-5 tallate; CAS 61791-00-2; nonionic; detergent, emulsifier, lubricant, softener, coupling agent for cosmetics, textiles, leather, metal cleaning; Gardner 7 max. liq.; sol. in IPA, min. oil, disp. in water; HLB 8.8; acid no. 2.0 max.; sapon. no. 100–120; 99% conc.

Hetoxamate FA-20. [Heterene] PEG-20 tallate; CAS 61791-00-2; nonionic; detergent, emulsifier, lubricant, softener, coupling agent for cosmetics, textiles, leather, metal cleaning; semisolid; sol. in water, IPA; disp. min. oil; HLB 14.9; acid no. 2.0; sapon. no. 50–60.

Hetoxamate LA-5. [Heterene] PEG-5 laurate; CAS 9004-81-3; nonionic; detergent, emulsifier, lubricant, softener, coupling agent for cosmetics, textiles, leather, metal cleaning; Gardner 1 max. liq.; sol. in IPA; disp. in water; HLB 10.5; acid no. 2.0 max. ; sapon. no. 125–145; 99% conc.

Hetoxamate LA-9. [Heterene] PEG-9 laurate; CAS 9004-81-3; nonionic; detergent, emulsifier, lubricant, softener, coupling agent for cosmetics, textiles, leather, metal cleaning; Gardner 1 max. liq.; sol. in water, IPA; HLB 13.3; acid no. 2.0 max.; sapon. no. 90-100; 99% conc.

Hetoxamate MO-2. [Heterene] PEG-2 oleate; CAS 9004-96-0; detergent, emulsifier for personal care prods.; softener for leather; Gardner 2 max. liq.; sol. in IPA, min. oil; insol. in water; HLB 5.3; acid no. 1 max.; sapon. no. 145–160; 99% conc.

Hetoxamate MO-4. [Heterene] PEG-4 oleate; CAS 9004-96-0; softener, emulsifier, coupling agent, lubricant.

Hetoxamate MO-5. [Heterene] PEG-5 oleate; CAS 9004-96-0; nonionic; detergent, emulsifier, lubricant, softener, coupling agent for cosmetics, textiles, leather, metal cleaning; Gardner 3 max. liq.; sol. in IPA; disp. water; HLB 8.8; acid no. 2.0 max. ; sapon. no. 115–125.

Hetoxamate MO-9. [Heterene] PEG-9 oleate; CAS 9004-96-0; nonionic; detergent, emulsifier, lubricant, softener, coupling agent for cosmetics, textiles, leather, metal cleaning; Gardner 4 max. liq.; sol. in IPA; disp. water; HLB 11.7; acid no. 2.0 max.; sapon. no. 80–88; 99% conc.

Hetoxamate MO-15. [Heterene] PEG-15 oleate; CAS 9004-96-0; nonionic; detergent, emulsifier, lubricant, softener, coupling agent for cosmetics, textiles, leather, metal cleaning; liq.; sol. in water, IPA; HLB 13.5; acid no. 2.0; sapon. no. 60–70.

Hetoxamate SA-5. [Heterene] PEG-5 stearate; CAS 9004-99-3; nonionic; detergent, emulsifier, lubricant, softener, coupling agent for cosmetics, textiles, leather, metal cleaning; Gardner 2 max. solid; sol. in IPA; disp. hot water, min. oil; HLB 8.8; acid no. 2.0 max. ; sapon. no. 105-120.

Hetoxamate SA-7. [Heterene] PEG-7 stearate; CAS 9004-99-3; nonionic; detergent, emulsifier for personal care prods.; softener for leather; Gardner 1 max. solid; sol. in IPA; disp. in water; HLB 10.5; acid no. 2 max.; sapon. no. 90–100.

Hetoxamate SA-9. [Heterene] PEG-9 stearate; CAS 9004-99-3; nonionic; see Hetoxamate FA-5; Gardner 1 max. solid; sol. in IPA; disp. hot water; HLB 11.5; acid no. 2.0 max.; sapon. no. 80-90.

Hetoxamate SA-13. [Heterene] PEG-12 stearate; CAS 9004-99-3; detergent, emulsifier for personal care prods.; softener for leather; Gardner 2 max. liq.; sol. in water, IPA; HLB 13.4; acid no. 2 max.; sapon. no. 60–70.

Hetoxamate SA-23. [Heterene] PEG-20 stearate; CAS 9004-99-3; detergent, emulsifier for personal care prods.; softener for leather; Gardner 2 max. solid; sol. in water, IPA; HLB 15.6; acid no. 2 max.; sapon. no. 39–49.

Hetoxamate SA-35. [Heterene] PEG-35 stearate, dioxane-free; CAS 9004-99-3; nonionic; see Hetoxamate FA-5; Gardner 2 max. flake; sol. in water, IPA; HLB 16.9; acid no. 2.0 max.; sapon. no. 24–34; hyd. no. 33-43.

Hetoxamate SA-40. [Heterene] PEG-40 stearate; CAS 9004-99-3; nonionic; detergent, emulsifier for personal care prods.; softener for leather; Gardner 2 max. flake; sol. in water, IPA; HLB 17.2; sapon. no. 24–34.

Hetoxamate SA-90. [Heterene] PEG-90 stearate; CAS 9004-99-3; nonionic; see Hetoxamate FA-5; Gardner 3 max. solid; sol. in water IPA; HLB 18.6; sapon. no. 9-20; hyd. no. 11-18.

Hetoxamate SA-90F. [Heterene] PEG-90 stearate; CAS 9004-99-3; nonionic; see Hetoxamate FA-5; off-wh. to tan flakes; sol. in water, IPA; acid no. 2 max.; sapon. no. 9-20; hyd. no. 11-18.

Hetoxamine C-2. [Heterene] PEG-2 cocamine; CAS 61791-14-8; cationic; emulsifier, softener, antistat, water repellent, desizing agent in agriculture, waxes, oils, textile/leather, metal cleaning; Gardner 8 max. liq.; sol. in IPA, min. oil; gels in water; m.w. 285; 95% tert. amine.

Hetoxamine C-5. [Heterene] PEG-5 cocamine; CAS 61791-14-8; cationic; emulsifier, softener, antistat, water repellent, desizing agent in agriculture, waxes, oils, textile/leather, metal cleaning; Gardner 8 max. liq.; sol. in water, IPA, min. oil; m.w. 425; 95% tert. amine.

Hetoxamine C-15. [Heterene] PEG-15 cocamine; CAS 61791-14-8; cationic; emulsifier, softener, antistat, water repellent, desizing agent in agriculture, waxes, oils, textile/leather, metal cleaning; Gardner 12 max. liq.; sol. in water, IPA; m.w. 860; 95% tert. amine.

Hetoxamine O-2. [Heterene] PEG-2 oleamine; nonionic; emulsifier, softener, antistat, water repellent, desizing agent in agriculture, waxes, oils, textile/leather, metal cleaning; liq.; m.w. 350; sol. in IPA, min. oil; 99% conc.

Hetoxamine O-5. [Heterene] PEG-5 oleamine; nonionic; emulsifier, softener, antistat, water repellent, desizing agent in agriculture, waxes, oils, textile/leather, metal cleaning; liq.; m.w. 492; sol. in IPA, min. oil; 99% conc.

Hetoxamine O-15. [Heterene] PEG-15 oleamine; nonionic; emulsifier, softener, antistat, water repellent, desizing agent in agriculture, waxes, oils, textile/leather, metal cleaning; liq.; m.w. 930; sol. in water, IPA.

Hetoxamine S-2. [Heterene] PEG-2 soyamine; CAS 61791-24-0; cationic; emulsifier, softener, antistat, water repellent, desizing agent in agriculture, waxes, oils, textile/leather, metal cleaning; liq.; m.w. 350; sol. in IPA, min. oil.

Hetoxamine S-5. [Heterene] PEG-5 soyamine; CAS 61791-24-0; cationic; emulsifier, softener, antistat, water repellent, desizing agent in agriculture, waxes, oils, textile/leather, metal cleaning; liq.; m.w. 480; sol. in IPA; partly sol. in min. oil; gel in water.

Hetoxamine S-15. [Heterene] PEG-15 soyamine; CAS 61791-24-0; cationic; desizing agent, antistat; emulsifier in agriculture, waxes and oils, leather processing, and metal cleaning industries; water repellent and wet spinning assistant in textile industries; Gardner 12 max. liq., solid; sol. see Hetoxamate SA-13; 95.0% min. tert. amine.

Hetoxamine ST-2. [Heterene] PEG-2 stearamine; nonionic; emulsifier, softener, antistat, water repellent, desizing agent in agriculture, waxes, oils, textile/leather, metal cleaning; Gardner 7 max. solid; sol. in IPA, min. oil; m.w. 388; 95% tert. amine.

Hetoxamine ST-5. [Heterene] PEG-5 stearamine; CAS 26635-92-7; nonionic; emulsifier, softener, antistat, water repellent, desizing agent in agriculture, waxes, oils, textile/leather, metal cleaning; Gardner 8 max. solid; sol. in IPA, min. oil; insol. in water; m.w. 520; 95% tert. amine.

Hetoxamine ST-15. [Heterene] PEG-15 stearamine; CAS 26635-92-7; nonionic; desizing agent, antistat; emulsifier in agriculture, waxes and oils, leather processing, and metal cleaning industries; water repellent and wet spinning assistant in textile industries; Gardner 8 max. solid; sol. in water, IPA; 95% min. tert. amine.

Hetoxamine ST-50. [Heterene] PEG-50 stearamine; CAS 26635-92-7; nonionic; desizing agent, antistat; emulsifier in agriculture, waxes and oils, leather processing, and metal cleaning industries; water repellent and wet spinning assistant in textile industries; Gardner 8 max. semisolid; sol. in water, IPA.

Hetoxamine T-2. [Heterene] PEG-2 tallow amine; cationic; emulsifier, softener, antistat, water repellent, desizing agent in agriculture, waxes, oils, textile/leather, metal cleaning; Gardner 10 max. liq.; sol. in IPA, min. oil; insol. in water; m.w. 350; 95% tert. amine.

Hetoxamine T-5. [Heterene] PEG-5 tallow amine; cationic; emulsifier, softener, antistat, water repellent, desizing agent in agriculture, waxes, oils, textile/leather, metal cleaning; Gardner 10 max. semisolid; sol. in water, IPA, min. oil; m.w. 490; 95% tert. amine.

Hetoxamine T-15. [Heterene] PEG-15 tallow amine; nonionic; emulsifier, softener, antistat, water repellent, desizing agent in agriculture, waxes, oils, textile/leather, metal cleaning; Gardner 10 max. semisolid; sol. in water, IPA; m.w. 925; 95% tert. amine.

Hetoxamine T-20. [Heterene] PEG-20 tallow amine; nonionic; emulsifier, softener, antistat, water repellent, desizing agent in agriculture, waxes, oils, textile/leather, metal cleaning; Gardner 10 max. semisolid; sol. in water, IPA; m.w. 1150; 95% tert. amine.

Hetoxamine T-30. [Heterene] PEG-30 tallow amine; nonionic; emulsifier, softener, antistat, water repellent, desizing agent in agriculture, waxes, oils, textile/leather, metal cleaning; Gardner 10 max. semisolid; sol. in water, IPA; 90% tert. amine.

Hetoxamine T-50. [Heterene] PEG-50 tallowamine; emulsifier, antistat, dyeing assistant, softener, detergent; Gardner 10 max. solid; sol. in water, IPA; 90% tert. amine.

Hetoxamine T-50-70%. [Heterene] PEG-50 tallowamine; emulsifier, softener, antistat, water repellent, desizing agent in agriculture, waxes, oils, textile/leather, metal cleaning; Gardner 10 max. liq.; sol. in water, IPA; 70% act.

Hetoxide BN-13. [Heterene] PEG-13 beta-naphthol ether; CAS 35545-57-4; nonionic; emollient, emulsifier, visc. control agent, lubricant, pigment dispersant, perfume solubilizer, used in cosmetics, household, textile industry, metal treating and plating; intermediate; Gardner 8 max. paste; sol. in water, IPA, min. oil; HLB 16.0; acid no. 1.0; hyd. no. 73-83; 99% conc.

Hetoxide BP-3. [Heterene] Modified butanol ethoxylate; nonionic; emollient, emulsifier, used in cosmetics, household, textile industry; liq.; sol. in water, IPA.

Hetoxide BY-1.8. [Heterene] PEG-1.8 butynediol; metal plating emulsifier; liq.; sol. in water, IPA; insol. in min. oil.

Hetoxide BY-3. [Heterene] PEG-3 butynediol; nonionic; emollient, emulsifier used in cosmetics, household, textile industry; liq.; sol. in water, IPA.

Hetoxide C-2. [Heterene] PEG-2 castor oil; nonionic; emollient, emulsifier, solubilizer, pigment dispersant, detergent used in cosmetics, household, textile industry; Gardner 6 max. liq.; sol. in IPA, min. oil; insol. in water; HLB 1.7; acid no. 2 max.; sapon. no. 155–170; 99% conc.

Hetoxide C-9. [Heterene] PEG-9 castor oil; nonionic; emollient, emulsifier, solubilizer, pigment dispersant, detergent used in cosmetics, household, textile industry; Gardner 5 max. liq.; sol. in IPA, min. oil; disp. in water; HLB 6.0; acid no. 2 max.; sapon. no. 120–136.

Hetoxide C-15. [Heterene] PEG-15 castor oil; nonionic; emollient, emulsifier, solubilizer, pigment dispersant, detergent, lubricant used in cosmetics, household, textile industry; Gardner 5 max. liq.; sol. in IPA, min. oil; disp. in water; HLB 8.3; acid no. 2 max.; sapon. no. 95–105.

Hetoxide C-25. [Heterene] PEG-25 castor oil; nonionic; emollient, emulsifier, lubricant, solubilizer, pigment dispersant, detergent used in cosmetics, household, textile industry; Gardner 5 max. liq.; sol. in IPA; disp. in water, min. oil; HLB 10.8; acid no. 2 max.; sapon. no. 74–82.

Hetoxide C-30. [Heterene] PEG-30 castor oil; nonionic; perfume solubilizer, emollient, emulsifier, visc. control and scouring agent, lubricant, dispersant for cosmetic formulations; dyeing assistant, dye carrier for textiles; used in household cleaning comps., metal treatment,; metal plating, chemical intermediate; Gardner 5 max. liq.; sol. in IPA; disp. in water, min. oil; HLB 11.7; acid no. 2 max.; sapon. no. 65–75.

Hetoxide C-40. [Heterene] PEG-40 castor oil; nonionic; emollient, emulsifier, lubricant, solubilizer, pigment dispersant, detergent used in cosmetics, household, textile industry; Gardner 4 max. paste; sol. in water, IPA; HLB 13.1; acid no. 1 max.; sapon. no. 55–65; 99% conc.

Hetoxide C-60. [Heterene] PEG-60 castor oil; nonionic; emollient, emulsifier, solubilizer, pigment dispersant, detergent used in cosmetics, household, textile industry; Gardner 4 max. soft solid; sol. in water, IPA; HLB 14.8; acid no. 2 max.; sapon. no. 41-51; pH 6-8 (3%).

Hetoxide C-200. [Heterene] PEG-200 castor oil; nonionic; emollient, emulsifier, lubricant, wetting agent, solubilizer, pigment dispersant, detergent used in cosmetics, household, textile industry; Gardner 4 max. solid; sol. in water, IPA; acid no. 1 max.; sapon. no. 16–18.

Hetoxide C-200-50%. [Heterene] PEG-200 castor oil; lubricant, emulsifier, solubilizer; Gardner 4 max. liq.; sol. in water, IPA; acid no. 1 max.; sapon no. 7–10; 50% conc.

Hetoxide DNP-4. [Heterene] Nonyl nonoxynol-4; CAS 9014-93-1; nonionic; intermediate, emulsifier; Gardner 4 max. liq.; sol. in IPA, min. oil; disp. in water; HLB 6.7; hyd. no. 120-140; 99% conc.

Hetoxide DNP-5. [Heterene] Nonyl nonoxynol-5; CAS 9014-93-1; nonionic; emollient, emulsifier, solubilizer, pigment dispersant, detergent used in cosmetics, household, textile industry; liq.; sol. in IPA, min. oil, disp. in water.

Hetoxide DNP-9.6. [Heterene] Nonyl nonoxynol-9.6; CAS 9014-93-1; nonionic; intermediate, emulsifier, surfactant; Gardner 5 max. liq.; sol. in IPA, min. oil; disp. in water; HLB 11.0; hyd. no. 65-75; 99% conc.

Hetoxide DNP-10. [Heterene] Nonyl nonoxynol-10; CAS 9014-93-1; nonionic; emollient, emulsifier, solubilizer, pigment dispersant, detergent used in cosmetics, household, textile industry; liq.; sol. in IPA, min. oil, disp. in water.

Hetoxide G-7. [Heterene] Glycereth-7; nonionic; emulsifier; Gardner 2 max. liq.; sol. in IPA, water; acid no. 1.0 max.; hyd. no. 412-428; 99% conc.

Hetoxide G-26. [Heterene] Glycereth-26; nonionic; humectant for pressure-sensitive adhesives; emulsifier; Gardner 2 max. liq.; sol. in water, IPA; acid no. 2.0 max.; hyd. no. 127-137; 99% conc.

Hetoxide HC-16. [Heterene] PEG-16 hydrog. castor oil; CAS 61788-85-0; nonionic; emollient, emulsifier, solubilizer, pigment dispersant, detergent used in cosmetics, household, textile industry; Gardner 4 max. liq.; sol. in IPA, min. oil; HLB 8.6; acid no. 1.5 max.; sapon. no. 85–95; 99% conc.

Hetoxide HC-25. [Heterene] PEG-25 hydrog. castor oil; CAS 61788-85-0; nonionic; Gardner 4 max. semisolid; sol. in water, IPA; HLB 10.8; acid no. 1.5 max.; sapon. no. 67-77; 99% conc.

Hetoxide HC-40. [Heterene] PEG-40 hydrog. castor oil; CAS 61788-85-0; nonionic; emulsifier, lubricant; Gardner 4 max. semisolid; sol. in water, IPA; HLB 13.1; acid no. 1.5 max.; sapon. no. 50–60; 99% conc.

Hetoxide HC-60. [Heterene] PEG-60 hydrog. castor oil; CAS 61788-85-0; emulsifier, lubricant; Gardner 4 max. solid; sol. in water, IPA; HLB 14.8; acid no. 1.5 max.; sapon. no. 41–51; hyd. no. 39-49; 99% conc.

Hetoxide NP-4. [Heterene] Nonoxynol-4; CAS 9016-45-9; nonionic; detergent and emulsifier; Gardner 1 max. liq.; sol. in IPA, min. oil; insol. in water; HLB 8.7; hyd. no. 135-139; 99% conc.

Hetoxide NP-6. [Heterene] Nonoxynol-6; CAS 9016-45-9; nonionic; Gardner 1 max. liq.; sol. in IPA; disp. in water, min. oil; HLB 10.7; hyd. no. 113-119.

Hetoxide NP-9. [Heterene] Nonoxynol-9; CAS 9016-45-9; nonionic; Gardner 1 max. liq.; sol. in water, IPA; HLB 12.7; hyd. no. 88-93; 99% conc.

Hetoxide NP-10. [Heterene] Nonoxynol-10; CAS 9016-45-9; nonionic; Gardner 1 max. liq.; sol. in water, IPA; acid no. 1 max.; hyd. no. 81-95.

Hetoxide NP-12. [Heterene] Nonoxynol-12; CAS 9016-45-9; nonionic; Gardner 1 max. semisolid; sol. in water, IPA; HLB 15.7; hyd. no. 72-78; 99% conc.

Hetoxide NP-15-85%. [Heterene] Nonoxynol-15; CAS 9016-45-9; nonionic; Gardner 1 max. liq.; sol. in water, IPA; HLB 16.4; 85% conc.

Hetoxide NP-30. [Heterene] Nonoxynol-30; CAS 9016-45-9; nonionic; emulsifier for emulsion polymerization; Gardner 4 max. solid; sol. in water, IPA; HLB 17.0; hyd. no. 35-45; 99% conc.

Hetoxide NP-40. [Heterene] Nonoxynol-40; CAS 9016-45-9; nonionic; Gardner 2 max. solid; sol. in water, IPA; HLB 17.7; 99% conc.

Hetoxide NP-50. [Heterene] Nonoxynol-50; CAS 9016-45-9; nonionic; Gardner 2 max. solid; sol. in water, IPA; 99% conc.

Hetoxide P-3. [Heterene] PEG-3 phenyl ether; Gardner 1 max. liq.; sol. in water, IPA; hyd. no. 240-260.

Hetoxide P-6. [Heterene] PEG-6 phenyl ether; APHA 120 max. liq.; sol. in water, IPA; HLB 14.7; acid no. 2 max.; hyd. no. 152-164.

Hetoxide P12 PG21. [Heterene] PEG-12 PPG-21; Gardner 2 max. liq.; sol. in water, IPA; hyd. no. 70-85; cloud pt. 48-52 C (1% aq.); 99% conc.

Hetoxol 15 CSA. [Heterene] Ceteareth-15; CAS 68439-49-6; detergent, emulsifier, leveling agent, intermediate for cosmetics, textiles, scouring agents, dyes, household prods., silicone emulsification, surfactants; Gardner 1 max. solid; sol. in IPA, water; acid no. 1 max.; hyd. no. 59-69; 99% conc.

Hetoxol 916P. [Heterene] PEG-6 PPG-2.5 C9–C11 alcohols ether; surfactant for industrial applics.; Gardner 2 max. liq.; sol. in water, IPA; acid no. 1 max.; hyd. no. 105-120; cloud pt. 42–45 C (1% aq.).; 99% conc.

Hetoxol C-24. [Heterene] Oleth-24 and ceteth-24; CAS 9004-98-2; nonionic; intermediate, emulsifier, wetting agent, solubilizer, coupling agent; flake; 100% conc.

Hetoxol CA-2. [Heterene] Ceteth-2; nonionic; detergent, emulsifier, leveling agent, intermediate, used for cosmetics, household formulations, silicone emulsification, textile processing; Gardner 1 max. solid; sol. in IPA, min. oil; insol. in water; HLB 5.1; acid no. 1 max.; hyd. no. 160–180; 99% conc.

Hetoxol CA-10. [Heterene] Ceteth-10; nonionic; detergent, emulsifier, leveling agent, intermediate, used for cosmetics, household formulations, silicone emulsification, textile processing; Gardner 1 max. solid; sol. in water, IPA; HLB 12.7; acid no. 1 max.; hyd. no. 75-90; 99% conc.

Hetoxol CA-20. [Heterene] Ceteth-20; nonionic; detergent, emulsifier, leveling agent, intermediate, used for cosmetics, household formulations, silicone emulsification, textile processing; Gardner 1 max. solid; sol. in water; HLB 15.5; acid no. 1 max.; hyd. no. 45–60; 97% conc.

Hetoxol CAWS. [Heterene] PPG-5, ceteth-20; nonionic; intermediate, emulsifier, wetting agent, solubilizer, coupling agent; Gardner 1 max. liq.; sol. in water, IPA; disp. in min. oil; HLB 14.4; acid no. 1 max.; hyd. no. 40-60.

Hetoxol CD-3. [Heterene] PEG-3 2-ethylhexyl ether; intermediate, emulsifier, wetting agent, solubilizer, coupling agent; liq.; HLB 11.5; 99% conc.

Hetoxol CD-4. [Heterene] PEG-4 2-ethyl hexyl ether; nonionic; intermediate, emulsifier, wetting agent, solubilizer, coupling agent; APHA 100 max. liq.; sol. in water, IPA; disp. in min. oil; HLB 11.5; hyd. no. 160-172; 99% conc.

Hetoxol CS. [Heterene] Cetearyl alcohol; Gardner 1 max. flake; acid no. 1 max.; iodine no. 2 max.; sapon. no. 2 max.; hyd. no. 204-218; 99% conc.

Hetoxol CS-4. [Heterene] Ceteareth-4; CAS 68439-49-6; detergent, emulsifier, leveling agent, intermediate for personal care prods., wax, oils, textiles, scouring agents, dyes, household prods., silicone emulsification, surfactdants; Gardner 1 max. solid; sol. in IPA; disp. in water, min. oil; HLB 8.2; acid no. 1 max.; hyd. no. 130-150; 99% conc.

Hetoxol CS-5. [Heterene] Ceteareth-5; CAS 68439-49-6; detergent, emulsifier, leveling agent, intermediate for personal care prods., wax, oils, textiles, scouring agents, dyes, household prods., silicone emulsification, surfactants; Gardner 1 max. solid; sol. in IPA; disp. in water, min. oil; HLB 9.2; acid no. 1 max.; hyd. no. 115-130; 99% conc.

Hetoxol CS-9. [Heterene] Ceteareth-9; CAS 68439-49-6; nonionic; detergent, emulsifier, leveling agent, intermediate, used for cosmetics, household formulations, silicone emulsification, textile processing; Gardner 1 max. solid; sol. in water, IPA; HLB 12.2; acid no. 1 max.; hyd. no. 85–90; 99% conc.

Hetoxol CS-15. [Heterene] Ceteareth-15; CAS 68439-49-6; nonionic; detergent, emulsifier, leveling agent, intermediate used for cosmetics, household formulations, silicone emulsification, textile processing; Gardner 1 max. solid; sol. water, IPA; HLB 14.2; acid no. 2 max.; hyd. no. 65-73; 98.5% conc.

Hetoxol CS-20. [Heterene] Ceteareth-20; CAS 68439-49-6; nonionic; detergent, emulsifier, leveling agent, intermediate, used for cosmetics, household formulations, silicone emulsification, textile processing; Gardner 2 max. solid; sol. in water, IPA; HLB 15.4; acid no. 2 max.; hyd. no. 50–70; 99% conc.

Hetoxol CS-20D. [Heterene] Cetearyl alcohol and ceteareth-20; intermediate, emulsifier, wetting agent, solubilizer, coupling agent; flake; 99% conc.

Hetoxol CS-25. [Heterene] Ceteareth-25; CAS 68439-49-6; intermediate, emulsifier, wetting agent, solubilizer, coupling agent.

Hetoxol CS-30. [Heterene] Ceteareth-30; CAS 68439-49-6; nonionic; detergent, emulsifier, leveling agent, intermediate, used for cosmetics, household formulations, silicone emulsification, textile processing; Gardner 2 max. solid; sol. in water, IPA; HLB 16.7; acid no. 2 max.; hyd. no. 40–52; 99% conc.

Hetoxol CS-40W. [Heterene] Ceteareth-40; CAS 68439-49-6; Gardner 1 max. solid; sol. in IPA; gels in water; acid no. 1 max.; hyd. no. 25-30; 99% conc.

Hetoxol CS-50. [Heterene] Ceteareth-50; CAS 68439-49-6; nonionic; detergent, emulsifier, leveling agent, intermediate, used for cosmetics, household formulations, silicone emulsification, textile processing; Gardner 2 max. flake; sol. in water, IPA; acid no. 2 max.; hyd. no. 20-40; 99% conc.

Hetoxol CS-50 Special. [Heterene] Ceteareth-50; CAS 68439-49-6; detergent, emulsifier, leveling agent, intermediate, used for cosmetics, household formulations, silicone emulsification, textile processing; Gardner 2 max. flake; sol. in IPA; gels in water; set pt. 44-48 C; hyd. no. 18-24; 99% conc.

Hetoxol CSA-15. [Heterene] Ceteareth-15; CAS 68439-49-6; nonionic; detergent, emulsifier, leveling agent, intermediate; used in personal care prods., wax, oil and textiles, scouring agents, dyes, household formulations, silicone emulsification, surfactants; Gardner 1 max. solid; HLB 14.2.

Hetoxol D. [Heterene] Cetearyl alcohol and ceteareth-20; nonionic; intermediate, emulsifier, wetting agent, solubilizer, coupling agent; for cosmetics, paper, textile industries; flake; m.p. 47-55 C; acid no. 1 max.; iodine no. 2 max.; sapon. no. 2 max.; 99% conc.

Hetoxol G. [Heterene] Stearyl alcohol and ceteareth-20; nonionic; intermediate, emulsifier, wetting agent, solubilizer, coupling agent; flake; acid no. 1 max.; sapon. no. 2 max.; 99% conc.

Hetoxol IS-2. [Heterene] Isosteareth-2; CAS 52292-17-8; Gardner 1 max. solid (clear on remelt); acid no. 1 max.; hyd. no. 140-150; pH 5.5-7.5 (3% aq.).

Hetoxol J. [Heterene] Cetearyl alcohol and ceteareth-20; nonionic; intermediate, emulsifier, wetting agent, solubilizer, coupling agent; flake; m.p. 47-55 C; acid no. 1 max.; iodine no. 2 max.; sapon. no. 2 max.; hyd. no. 166-178; 99% conc.

Hetoxol L. [Heterene] Cetearyl alcohol and ceteareth-30; nonionic; intermediate, emulsifier, wetting agent, solubilizer, coupling agent; flake; acid no. 1 max.; iodine no. 2 max.; sapon. no. 2 max.; hyd. no. 178-192; 99% conc.

Hetoxol L-1. [Heterene] Laureth-1; Gardner 1 max. liq.; sol. in IPA, min. oil; insol. in water; HLB 3.6; hyd. no. 231-238; 99% conc.

Hetoxol L-2. [Heterene] Laureth-2; Gardner 1 max. liq.; sol. in IPA, min. oil; insol. in water; HLB 6.1; hyd. no. 191-201; 99% conc.

Hetoxol L-3N. [Heterene] Laureth-3; nonionic; detergent, emulsifier, leveling agent, intermediate, used for cosmetics, household formulations, silicone emulsification, textile processing; liq.; sol. in IPA, min. oil; HLB 7.9; hyd. no. 170–176.

Hetoxol L-4. [Heterene] Laureth-4; CAS 5274-68-0; intermediate, emulsifier, wetting agent, solubilizer, coupling agent; Gardner 1 max. liq.; sol. in IPA, min. oil; disp. in water; HLB 9.4; acid no. 2 max.;

hyd. no. 145-160; 99% conc.

Hetoxol L-4N. [Heterene] Laureth-4; nonionic; detergent, emulsifier, leveling agent, intermediate, used for cosmetics, household formulations, silicone emulsification, textile processing; liq.; sol. in IPA, min. oil; HLB 9.7; hyd. no. 145–165.

Hetoxol L-9. [Heterene] Laureth-9; intermediate, emulsifier, wetting agent, solubilizer, coupling agent; Gardner 1 max. liq.; sol. in water, IPA; HLB 13.3; acid no. 1 max.; hyd. no. 95-105; 99% conc.

Hetoxol L-9N. [Heterene] Laureth-9; nonionic; detergent, emulsifier, leveling agent, intermediate, used for cosmetics, household formulations, silicone emulsification, textile processing; liq.; sol. in water, IPA; HLB 11.8; hyd. no. 90–110.

Hetoxol L-23. [Heterene] Laureth-23; nonionic; detergent, emulsifier, leveling agent, intermediate, used for cosmetics, household formulations, silicone emulsification, textile processing; Gardner 1 max. solid; sol. in water, IPA; HLB 16.7; acid no. 0.5 max.; hyd. no. 42-52.

Hetoxol LS-9. [Heterene] Laureth-9, steareth-9; nonionic; detergent, emulsifier, leveling agent, intermediate, used for cosmetics, household formulations, silicone emulsification, textile processing; APHA 75 max. semisolid; sol. in water, IPA; HLB 12.6; hyd. no. 94-104; cloud pt. 64-69 C (1% aq.); pH 4.5-7.5 (1% aq.); 99% conc.

Hetoxol M-3. [Heterene] Myreth-3; CAS 27306-79-2; emulsifier and pigment dispersant in makeup; intermediate, wetting agent, solubilizer, coupling agent; Gardner 1 max. clear thin liq.; sol. in IPA; disp. in water, min. oil; HLB 7.6; acid no. 0.2 max.; hyd. no. 150-162; pH 5.5–7.0 (5% in IPA/water 1:1); 97% conc.

Hetoxol MP-3. [Heterene] PPG-3 myristyl ether; Gardner 1 max. liq.; sol. in IPA, min. oil; insol. in water; acid no. 1.5 max.; sapon. no. 2 max.; hyd. no. 135-155; 99% conc.

Hetoxol OA-3 Special. [Heterene] Oleth-3; CAS 9004-98-2; nonionic; emulsifier and pigment dispersant for cosmetic applics.; pale yel. liq., low odor; sol. in IPA, min. oil; insol. in water; HLB 6.4; acid no. 2 max.; iodine no. 57-62; hyd. no. 135-150.

Hetoxol OA-5 Special. [Heterene] Oleth-5; CAS 9004-98-2; nonionic; intermediate, emulsifier, wetting agent, solubilizer, coupling agent; pale yel. liq., low odor; sol. in IPA, min. oil; disp. in water; HLB 9.0; acid no. 2 max.; iodine no. 40-52; hyd. no. 120-135; 99% conc.

Hetoxol OA-10 Special. [Heterene] Oleth-10; CAS 9004-98-2; nonionic; emulsifier and pigment dispersant for cosmetic applics.; wh. semisolid, low odor; sol. in water, IPA; HLB 12.4; acid no. 2 max.; iodine no. 31-37; hyd. no. 79-91.

Hetoxol OA-20 Special. [Heterene] Oleth-20; CAS 9004-98-2; nonionic; emulsifier and pigment dispersant for cosmetic applics.; wh. solid; sol. in water, IPA; HLB 15.3; acid no. 2 max.; iodine no. 18-25; hyd. no. 50-58.

Hetoxol OL-2. [Heterene] Oleth-2; CAS 9004-92-2; nonionic; intermediate, emulsifier, wetting agent, solubilizer, coupling agent; Gardner 2 max. liq.; sol. in IPA, min. oil; insol. in water; HLB 4.9; acid no. 1 max.; hyd. no. 160-180; 99% conc.

Hetoxol OL-4. [Heterene] Oleth-4; CAS 9004-98-2; nonionic; intermediate, emulsifier, wetting agent, solubilizer, coupling agent; Gardner 4 max. liq.; sol. in IPA, min. oil; insol. in water; HLB 7.9; hyd. no. 123-133; 99% conc.

Hetoxol OL-5. [Heterene] Oleth-5; CAS 9004-98-2; nonionic; intermediate, emulsifier, wetting agent, solubilizer, coupling agent; Gardner 4 max. liq.; sol. in IPA; disp. in water, min. oil; HLB 9.0; hyd. no. 110–121; 99% conc.

Hetoxol OL-10. [Heterene] Oleth-10; CAS 9004-98-2; nonionic; surfactant for cosmetic formulations; intermediate, emulsifier, wetting agent, solubilizer, coupling agent; Gardner 3 semisolid; sol. in water, IPA; HLB 12.4; hyd. no. 77-83; 99% conc.

Hetoxol OL-10H. [Heterene] Oleth-10; CAS 9004-98-2; nonionic; surfactant for cosmetic formulations; intermediate, emulsifier, wetting agent, solubilizer, coupling agent; Gardner 1 max. semisolid; sol. in water, IPA; HLB 12.8; acid no. 0.5 max.; iodine no. 25-37; hyd. no. 75-88; cloud pt. 71–77 C (1% aq.).; 99% conc.

Hetoxol OL-20. [Heterene] Oleth-20; CAS 9004-98-2; nonionic; surfactant for cosmetic formulations; intermediate, emulsifier, wetting agent, solubilizer, coupling agent; Gardner 2 max. solid; sol. in water, IPA; HLB 15.3; acid no. 2 max.; hyd. no. 50-58; 99% conc.

Hetoxol OL-23. [Heterene] Oleth-23; CAS 9004-98-2; nonionic; intermediate, emulsifier, wetting agent, solubilizer, coupling agent, lubricant; Gardner 2 max. semisolid; sol. in water, IPA; HLB 15.8; acid no. 2 max.; hyd. no. 47–62; 99% conc.

Hetoxol OL-40. [Heterene] Oleth-40; CAS 9004-98-2; nonionic; intermediate, emulsifier, wetting agent, solubilizer, coupling agent; Gardner 2 max. solid; sol. in water, IPA; HLB 17.4; hyd. no. 28-38; 99% conc.

Hetoxol P. [Heterene] Emulsifying wax NF; emulsion base; flake; acid no. 2 max.; sapon. no. 14.0 max.; hyd. no. 178-192; 99% conc.

Hetoxol PA-1. [Heterene] Ethylene glycol propargyl ether; metal plating emulsifier, wetting agent; liq.; sol. in water, IPA.

Hetoxol PLA. [Heterene] PPG-30 lanolin ether; nonionic; oily emollient, intermediate, emulsifier, wetting agent, solubilizer, coupling agent; liq.; sol. in min. oil, IPA; insol. in water; acid no. 1.0 max.

Hetoxol SP-15. [Heterene] PPG-15 stearyl ether; nonionic; oily emollient material in cosmetics; ASTM 100 max. liq.; sol. in IPA, min. oil; insol. in water; acid no. 1.5 max.; sapon. no. 2.0 max.; hyd. no. 62–70; 99% conc.

Hetoxol STA-2. [Heterene] Steareth-2; CAS 9005-00-9; nonionic; intermediate, emulsifier, wetting agent, solubilizer, coupling agent; Gardner 1 max. solid; sol. in IPA, min. oil; disp. in water; HLB 4.9; acid no. 2 max.; hyd. no. 155–165; 99% conc.

Hetoxol STA-10. [Heterene] Steareth-10; CAS 9005-00-9; nonionic; intermediate, emulsifier, wetting agent, solubilizer, coupling agent; Gardner 1 max. solid; sol. in water, IPA; HLB 12.4; acid no. 1 max.; hyd. no. 75-90; 99% conc.

Hetoxol STA-20. [Heterene] Steareth-20; CAS 9005-00-9; nonionic; intermediate, emulsifier, wetting agent, solubilizer, coupling agent; Gardner 1 max. solid; sol. in water, IPA; insol. in min. oil; HLB 15.3; acid no. 2 max.; hyd. no. 45-60 C; 99% conc.

Hetoxol STA-30. [Heterene] Steareth-30; CAS 9005-00-9; nonionic; intermediate, emulsifier, wetting agent, solubilizer, coupling agent; Gardner 2 max. flake; sol. in water, IPA; HLB 16.6; m.p. 46–50 C; acid no. 2 max.; hyd. no. 35-45; 99% conc.

Hetoxol TD-3. [Heterene] Trideceth-3; CAS 24938-91-8; nonionic; detergent, emulsifier, leveling agent, intermediate, used for cosmetics, household formulations, silicone emulsification, textile processing; Gardner 1 max. liq.; sol. in IPA, min. oil; insol. in water; HLB 7.9; acid no. 2 max.; hyd. no. 165–175; 99% conc.

Hetoxol TD-6. [Heterene] Trideceth-6; CAS 24938-91-8; nonionic; detergent, emulsifier, leveling agent, intermediate, used for cosmetics, household formulations, silicone emulsification, textile processing; Gardner 1 max. liq.; sol. in IPA, water; HLB 11.3; acid no. 2 max.; hyd. no. 115-1205; 99% conc.

Hetoxol TD-9. [Heterene] Trideceth-9; CAS 24938-91-8; nonionic; wetting agent, industrial surfactant, solubilizer; Gardner 1 max. liq.; sol. in water, IPA; HLB 13.2; acid no. 2 max.; hyd. no. 90-100; 99% conc.

Hetoxol TD-12. [Heterene] Trideceth-12; CAS 24938-91-8; nonionic; detergent, emulsifier, leveling agent, intermediate, used for cosmetics, household formulations, silicone emulsification, textile processing; Gardner 1 max. liq.; sol. in IPA, water; HLB 14.5; acid no. 2 max.; hyd. no. 72–87; 99% conc.

Hetoxol TD-18. [Heterene] Trideceth-18; CAS 24938-91-8; nonionic; leveling agent, solubilizer; Gardner 1 max. solid; sol. in water, IPA; HLB 15.7; hyd. no. 54-62; 99% conc.

Hetoxol TD-25. [Heterene] Trideceth-25; CAS 24938-91-8; nonionic; wetting agent, industrial surfactant, solubilizer; Gardner 2 max. semisolid; sol. in water, IPA; HLB 16.9; acid no. 2 max.; hyd. no. 40-60; 99% conc.

Hetoxol TDEP-15. [Heterene] PEG-10 PPG-15 tridecyl ether; nonionic; surfactant for industrial use, automatic dishwashing formulations; Gardner 2 max. liq.; sol. in water, IPA; acid no. 2 max.; hyd. no. 40-60; cloud pt. 21–23 C (1% aq.).; 99% conc.

Hetoxol TDEP-63. [Heterene] PEG-6 PPG-3 tridecyl ether; nonionic; surfactant for industrial use, automatic dishwashing formulations; Gardner 2 max. liq.; sol. in water, IPA; hyd. no. 80-115; cloud pt. 27–33 C (1% aq.).

Hetsorb L-4. [Heterene] Polysorbate 21; CAS 9005-64-5; nonionic; emulsifier, lubricant, thickener, corrosion inhibitor; Gardner 8 max. liq.; sol. in IPA; disp. in water, min. oil; HLB 13.3; acid no. 3 max.; sapon. no. 100–115; hyd. no. 225-255; 99% conc.

Hetsorb L-10. [Heterene] PEG-10 sorbitan laurate; nonionic; detergent, emulsifier, lubricant for cosmetics; corrosion inhibitor; Gardner 7 max. liq.; sol. in water, IPA; HLB 8.4; acid no. 2 max.; sapon. no. 66–76; hyd. no. 150-170; 97% conc.

Hetsorb L-20. [Heterene] Polysorbate 20; CAS 9005-64-5; nonionic; detergent, emulsifier, lubricant for cosmetics; corrosion inhibitor; Gardner 6 max. liq.; sol. in water, IPA; HLB 16.7; acid no. 2 max.; sapon. no. 41-50; hyd. no. 97-108; 97% conc.

Hetsorb O-20. [Heterene] Polysorbate 80; CAS 9005-65-6; nonionic; emulsifier, lubricant for cosmetics; corrosion inhibitor; Gardner 7 max. liq.; sol. in water, IPA; HLB 15.0; acid no. 2 max.; sapon. no. 45–55; hyd. no. 65-80; 97% conc.

Hetsorb P-20. [Heterene] Polysorbate 40; CAS 9005-66-7; nonionic; general purpose emulsifier; Gardner 6 max. liq. to semisolid; sol. in water, IPA; disp. in min. oil; HLB 15.6; acid no. 2 max.; sapon. no. 40–53; hyd. no. 91-107; 97% conc.

Hetsorb S-4. [Heterene] PEG-4 sorbitan stearate; CAS 9005-67-8; nonionic; emulsifier; amber semisolid; sol. in IPA; disp. in water, min. oil; acid no. 3 max.; sapon. no. 95–115; hyd. no. 165-200; 97% conc.

Hetsorb TO-20. [Heterene] Polysorbate 85; CAS 9005-70-3; nonionic; general purpose emulsifier, thickener, lubricant, corrosion inhibitor; Gardner 8 max. liq.; sol. in IPA, min. oil; disp. in water; HLB 11.0; acid no. 2 max.; sapon. no. 82–95; hyd. no. 39-52; 95% conc.

Hetsorb TS-20. [Heterene] Polysorbate 65; CAS 9005-71-4; nonionic; general purpose emulsifier, thickener, lubricant, corrosion inhibitor; Gardner 7 max. solid; sol. in IPA, hot min. oil; disp. in water; HLB 10.5; acid no. 2 max.; sapon. no. 88–98; hyd. no. 44-60; 97% conc.

Hetsulf 40, 40X. [Heterene] Sodium dodecylbenzene sulfonate; anionic; wetting agent, emulsifier, dispersant, intermediate, detergent, liq. formulation syndet; paste, liq.; pH 7–8.5 (5% aq.); biodeg.; 38% and 35% act.

Hetsulf 50A. [Heterene] Ammonium dodecylbenzene sulfonate; anionic; wetting agent, emulsifier, dispersant, for lt. duty detergent formulations; slurry; pH 6.0–7.0 (5% aq.); biodeg.; 48% act.

Hetsulf 60S. [Heterene] Sodium dodecylbenzene sulfonate; anionic; wetting agent, emulsifier, dispersant, base for formulated prods.; slurry; pH 7.0–8.0 (5% aq.); biodeg.; 57% act.

Hetsulf Acid. [Heterene] Sodium dodecylbenzene sulfonate; anionic; wetting agent, emulsifier, dispersant, intermediate, base for neutralized surfactant; liq.; dens. 1.05; biodeg.; 97% act.

Hetsulf IPA. [Heterene] MIPA-dodecylbenzenesulfonate; anionic; wetting agent, emulsifier, dispersant; liq.; pH 5.0–6.5 (5% aq.); biodeg.

Hexaryl D 60 L. [Witco SA] TEA-dodecylbenzenesulfonate; CAS 27323-41-7; anionic; multipurpose detergent; liq.

Hi-Fluid. [Takemoto Oil & Fat] Alkyaryl sulfonate/formaldehyde condensate; anionic; water reducing agent for superplasticized concrete; liq.

Hipochem ADN. [High Point] Quat. ammonium complex; nonionic; dyeing assistant, leveling agent of acid dyes on nylon; amber liq.; mild alcohol odor; sol. in water; dens. 8.73 lb/gal; sp.gr. 1.048; 50% act.

Hipochem AM-99. [High Point] Super amide condensate; nonionic; multipurpose detergent, wetting agent, emulsifier, foam builder; amber liq.; mild amine odor; water sol.; dens. 8.89 lb/gal; sp.gr. 1.06; 99% act.

Hipochem AMC. [High Point] Amine condensate; nonionic; multipurpose detergent, wetting agent, emulsifier, foam builder; dk. br. liq.; mild amine odor; water sol.; dens. 8.16 lb/gal; sp.gr. .98; 99–100% act.

Hipochem AR-100. [High Point] Alkylphenol polyethoxide ether; detergent and textile scouring agent; liq.; 100% conc.

Hipochem B-3-M. [High Point] Emulsified butyl benzoate; anionic/nonionic; self-emulsifying carrier for polyester and triacetate fibers; liq.; 100% conc.

Hipochem Base MC. [High Point] Surfactant blend; nonionic; prohibits blocking of dyesites on cationic dyeable fibers; liq.; 100% conc.

Hipochem BSM. [High Point] Quat. fatty deriv.; cationic; softener for fibers; liq.; 20% conc.

Hipochem C-95. [High Point] Quat. ammonium compd.; cationic; dyeing assistant, retarder for cationic dyes on acrylics; amber liq.; mild alcohol odor; water sol.; dens. 8.29 lb/gal; sp.gr. 0.995; 50% act.

Hipochem CAD. [High Point] Aromatic ester, emulsified; amphoteric; dyeing assistant for textiles; liq.;

100% conc.

Hipochem CDL. [High Point] Fatty alcohol ethoxylate; nonionic; dyeing assistant and scouring agent for textiles; liq.; 30% conc.

Hipochem CRD. [High Point] Emulsified oils; defoamer; liq.

Hipochem Carrier 761. [High Point] Emulsified aromatic hydrocarbon; nonionic; dye carrier for poly/wool blends; liq.; 100% conc.

Hipochem Carrier TA-3. [High Point] Emulsified hydrocarbon solvs.; dye carrier for poly/wool blends; liq.; 100% conc.

Hipochem Compatibilizer WMC. [High Point] Surfactant blend; nonionic; compatibilizer for cationic/disperse dyes; thick liq.; 80% conc.

Hipochem D-6-H. [High Point] Emulsified blend of aromatic esters; nonionic; carrier for basic dyeable polyesters; liq.; 100% conc.

Hipochem Dispersol GTO. [High Point] Sulfated glyceryl trioleate; anionic; detergent, wetting agent, emulsifier, dyeing assistant; dk. amber liq.; mild sulfate odor; water sol.; dens. 8.66 lb/gal; sp.gr. 1.04; 55% act.

Hipochem Dispersol SB. [High Point] Sulfated butyl oleate; anionic; detergent, wetting agent, emulsifier, dyeing assistant; dk. amber liq.; mild sulfate odor; water disp.; dens. 8.41 lb/gal; sp.gr. 1.01; 55% act.

Hipochem Dispersol SCO. [High Point] Sulfated castor oil; anionic; detergent, wetting agent, emulsifier, dyeing assistant; dk. amber liq.; mild sulfate odor; water disp.; dens. 8.33 lb/gal; sp.gr. 1.0; 70% act.

Hipochem Dispersol SP. [High Point] Sulfated propyl oleate; anionic; dyeing assistant, detergent, emulsifier used in textiles; liq.; 60% conc.

Hipochem DZ. [High Point] Quat. ammonium compd.; cationic; dyeing assistant, retarder for cationic dyes on acrylics; amber liq.; mild alcohol odor; water sol.; dens. 8.08 lb/gal; sp.gr. 0.97; 40% act.

Hipochem EFK. [High Point] Aliphatic sulfate; detergent, textile lubricant, substantive to cellulosic fibers; wh. paste; water sol.; 40% act.

Hipochem EK-6. [High Point] Long chain phosphated alcohol; anionic; wetting agent for desizing, dyeing, bleaching; low foaming; lt. amber liq.; mild alcohol odor; water sol.; dens. 8.70 lb/gal; sp.gr. 1.044; 60% act.

Hipochem EK-18. [High Point] Sodium dioctyl sulfosuccinate; anionic; wetting, rewetting, desizing, bleaching, scouring for textiles; lt. amber liq.; slight alcohol odor; water sol.; dens. 8.91 lb/gal; sp.gr. 1.07; 75% act.

Hipochem FNL. [High Point] Modified alkylaryl sulfonate; anionic; one-piece dyeing assistant for acid dyes on nylon; liq.; 40% conc.

Hipochem Finish 178. [High Point] Complex quat. compd.; cationic; softener and lubricant for acrylic/blend fibers; liq.; 14% conc.

Hipochem GM. [High Point] Emulsified 1,2,4-trichlorobenzene; anionic; dyeing assistant; carrier for polyester in removal of trimers; liq.; 100% conc.

Hipochem HPEL. [High Point] Modified ethoxylated fatty acid; nonionic; multipurpose dyeing assistant and lubricant; liq.; 40% conc.

Hipochem Jet Dye T. [High Point] Emulsified trichlorobenzene; nonionic; dyeing assistant; carrier for jet dyeing machine; aids in removal of trimers; liq.; 100% conc.

Hipochem Jet Scour. [High Point] Modified ethoxylated alcohol; nonionic; low foaming scour for all fibers; good wetting; exc. removal of min. oil, butyl stearate, greases; water-wh. liq., odorless; sol. in cold water; dens. 8.66 lb/gal; cloud pt. 85 F (1%); pH 6-8 (1%).

Hipochem JN-6. [High Point] Emulsified aromatic and aliphatic solvs. and esters; nonionic; carrier for polyester dyed in pressure machines; liq.; 100% conc.

Hipochem LCA. [High Point] Sulfonated oleyl alcohol; anionic; detergent, lubricant, textile dye leveling agent; wh. paste; mild sulfate odor; disp. in water (10%); 40% act.

Hipochem LH-Soap. [High Point] Coconut oil soap; anionic; detergent, wetting agent, emulsifier, developer and brightener; water sol.; dens. 7.77 lb/gal; sp.gr. 0.93; 40% act.

Hipochem M-51. [High Point] Nonsurface active quat. ammonium compd.; cationic; dyeing assistant; migrating agent for dyeing basic dyeable fibers; liq.; 60% conc.

Hipochem MS-BW. [High Point] Emulsified aromatic and aliphatic solvs.; nonionic; low foaming detergent, solvent scour for removal of oils, grease, waxes, and fatty materials in textile operations; colorless clear liq., mild aromatic odor; pH 6-8 (1%); 100% conc.

Hipochem MS-LF. [High Point] Emulsified hydrocarbons; nonionic; nonfoaming scouring agent esp. for jet machine operations; liq.; 100% conc.

Hipochem MT-1. [High Point] Phosphate ester; anionic; bleach bath wetting agent; continuous dyeing assistant; liq.; 35% conc.

Hipochem MTD. [High Point] Complex phosphate ester; wetting agent, leveling agent in caustic boil-off, bleach baths, other wet processing of piece goods, hosiery, knit goods; dispersant, antiredeposition aid; straw clear liq.; sol. in cold or hot water, caustic concs. to 7%; pH 9-10; 70 ± 1% act.

Hipochem Migrator J. [High Point] Benzyl trimethyl ammonium chloride; cationic; migrating agent for basic dyes; liq.; 60% conc.

Hipochem NAC. [High Point] Natural and syn. sulfonate blend; anionic; low foam dyeing assistant for tricot and hosiery; liq.; 40% conc.

Hipochem No. 3. [High Point] Sulfated oil plus soap; anionic; detergent and dye leveling agent used in textiles; liq.; 45% conc.

Hipochem No. 40-L. [High Point] Modified alkylbenzenesulfonate; anionic; detergent, wetting agent, emulsifier, dyeing penetrant; dk. amber liq.; water sol.; dens. 8.33 lb/gal; sp.gr. 1.0; 40% act.

Hipochem No. 641. [High Point] Alkyldiaryl sulfonate; anionic; dyeing assistant for polyamide fibers; lt. brn. liq.; water sol.; dens. 9.7 lb/gal; sp.gr. 1.16; 45% act.

Hipochem NOC. [High Point] Proprietary aromatic; nonionic; no-odor carrier for polyester; liq.; 100% conc.

Hipochem PDO. [High Point] Optically brightened detergent powd. for improving wash effectiveness on denim; wh. powd.; pH 12 ± 0.25 (1%).

Hipochem PND-11. [High Point] Phosphate ester deriv.; anionic; wool dyeing assistant and reserving agent in multifiber cross dyeings.

Hipochem Retarder CJ. [High Point] Quat. ammonium compd.; cationic; dyeing assistant; retarder for cationic dyes on acrylic fibers; dye retarder for jet machines; amber liq.; mild amine odor; water-sol.; sp.gr. 0.985; dens. 8.21 lb/gal; pH acid; 40% act.

Hipochem RPS. [High Point] Emulsified blend of

aromatic solvs.; anionic/nonionic; carrier for polyester and blends; liq.; 100% conc.

Hipochem SB-40. [High Point] Fatty ester base; nonionic; softener, fiber lubricant; paste; 20% conc.

Hipochem SO. [High Point] Detergent combined with mild bleaching agents; anionic; powd.; 100% conc.

Hipochem SRC. [High Point] Self-emulsifying solv. systems; anionic; detergent; removes dyes, oil, grease, pigment, binders from pads and machinery; liq.; 100% conc.

Hipochem TXF-1. [High Point] Modified ester; nonionic; dyeing assistant for polyester and cotton fibers; liq.; 100% conc.

Hipochem WSS. [High Point] Surfactant/inorg. salt blend; nonionic; stabilizer for peroxide bleach baths; liq.; 55% conc.

Hipofix 491. [High Point] Methyl methylol resin; cationic; fixing agent for direct dyes on cellulosics; liq.

Hipolon New. [High Point] Modified cationic surfactant; cationic; detergent, dye retarder, leveler; liq.; 25% conc.

Hiposcour® 1. [High Point] Modified ethoxylate; nonionic; detergent for scouring syn. fibers; liq.; 100% conc.

Hiposcour® 3-80. [High Point] Modified ethoxylate; nonionic; scouring agent for synthetics and cellulosics; detergent with nonredeposition props. for min. oils, motor oils, grease, butyl stearate stains; biodeg.; colorless clear liq.; dens. 8.8 lb/gal; 80% act.

Hiposcour® 6. [High Point] Ethoxylated alcohol; nonionic; low foaming scouring agent for textiles; liq.; 100% conc.

Hiposcour® ARG-2. [High Point] Modified ethoxylated alcohol; anionic/nonionic; low foaming scour for all fibers; good wetting; exc. removal of oils, waxes, and dirt; wh. clear liq., sl. alcoholic odor; sol. in cold water; dens. 8.5 lb/gal; pH 5.5-6.5 (1%).

Hiposcour® BFS. [High Point] Complex Varsol/surfactant blend; anionic; detergent scour for removal of oil and grease from cotton and blends; wetting agent; stable over broad pH range, to caustic; biodeg.; clear liq.; easily disp. in cold water, sp.gr. 0.8; pH 10-11 (1% aq.).

Hiposcour® LC-JET. [High Point] Modified ethoxylated alcohol; nonionic; low foaming scour for all fibers; good wetting; exc. removal of oils, waxes, and dirt; wh. clear liq., odorless; sol. in cold water; dens. 8.59 lb/gal; cloud pt. 85 F (1%); pH 4-5 (1%).

Hiposcour® NFMS-2. [High Point] Complex solv./surfactant blend; anionic; solvent and detergent scour for removal of oil and grease; stable over broad pH range; can be used in sodium peroxide and sodium hypochlorite bleach baths; amber clear liq., mild alcoholic odor; sp.gr. 0.93; dens. 7.75 lb/gal; pH 6.0-6.5.

Hipowet IBS. [High Point] Complex modified ethoxylate; anionic; wetting and scouring agent, detergent, suspending agent; caustic stable; stable with enzymes; straw clear liq., mild alcohol odor; sp.gr. 1.06; pH 5.5 -7.5 (1%); 95 ± 2% act.

HK-1618. [Procter & Gamble] Palmitic/oleic acid; CAS 67701-08-0; intermediate; m.w. 276; acid no. 200-209; iodine no. 55-65; sapon. no. 200-210.

Hodag 20-L. [Calgene] PEG-4 laurate; CAS 9004-81-3; nonionic; emulsifier, wetting agent, plasticizer for cosmetic, pharmaceutical and other uses; liq.; HLB 10.0; 100% conc.

Hodag 22-L. [Calgene] PEG-4 dilaurate; CAS 9005-02-1; nonionic; emulsifier, wetting agent, plasticizer for cosmetic, pharmaceutical and other uses; liq.; HLB 6.6; 100% conc.

Hodag 40-L. [Calgene] PEG-8 laurate; CAS 9004-81-3; nonionic; emulsifier, wetting agent, plasticizer for cosmetic, pharmaceutical and other uses; liq.; HLB 12.8; 100% conc.

Hodag 40-O. [Calgene] PEG-8 oleate; CAS 9004-96-0; nonionic; emulsifier, wetting agent, plasticizer for cosmetics, pharmaceuticals, other uses; liq.; HLB 11.4; 100% conc.

Hodag 40-R. [Calgene] PEG-8 ricinoleate; CAS 9004-97-1; nonionic; emulsifier, wetting agent, plasticizer for gneral cosmetic, pharmaceutical and other uses; liq.; HLB 11.6; 100% conc.

Hodag 40-S. [Calgene] PEG-8 stearate; CAS 9004-99-3; nonionic; emulsifier, wetting agent, plasticizer for cosmetic, pharmaceutical and other uses; paste; HLB 11.1; 100% conc.

Hodag 42-L. [Calgene] PEG-8 dilaurate; CAS 9005-02-1; nonionic; emulsifier, wetting agent, plasticizer for cosmetic, pharmaceutical and other uses; liq.; HLB 10.0; 100% conc.

Hodag 42-O. [Calgene] PEG-8 dioleate; CAS 9005-07-6; nonionic; emulsifier, wetting agent, plasticizer for cosmetic, pharmaceutical and other uses; liq.; HLB 8.4; 100% conc.

Hodag 42-S. [Calgene] PEG-8 distearate; CAS 9005-08-7; nonionic; emulsifier, wetting agent, plasticizer for cosmetic, pharmaceutical and other uses; paste; HLB 8.2; 100% conc.

Hodag 60-L. [Calgene] PEG-12 laurate; CAS 9004-81-3; nonionic; emulsifier, wetting agent, plasticizer for cosmetic, pharmaceutical and other uses; liq.; HLB 14.8; 100% conc.

Hodag 60-S. [Calgene] PEG-12 stearate; CAS 9004-99-3; nonionic; emulsifier, wetting agent, plasticizer for cosmetic, pharmaceutical and other uses; solid; HLB 13.6; 100% conc.

Hodag 62-O. [Calgene] PEG-12 dioleate; CAS 9005-07-6; nonionic; emulsifier, wetting agent, plasticizer for cosmetic, pharmaceutical and other uses; liq.; HLB 10.0; 100% conc.

Hodag 100-S. [Calgene] PEG-20 stearate; CAS 9004-99-3; nonionic; emulsifier, wetting agent, plasticizer for cosmetic, pharmaceutical and other uses; solid; HLB 15.6; 100% conc.

Hodag 150-S. [Calgene] PEG-6-32 stearate; CAS 9004-99-3; nonionic; emulsifier, wetting agent, plasticizer for cosmetic, pharmaceutical and other uses; solid; HLB 16.8; 100% conc.

Hodag Amine C-100-L. [Calgene] Lauric deriv.; cationic; dispersant, emulsifier, corrosion inhibitor; for textile, paper, metal applics.; paste; 100% conc.

Hodag Amine C-100-O. [Calgene] Cationic; dispersant, emulsifier, corrosion inhibitor; for textile, paper, metal applics.; paste; 100% conc.

Hodag Amine C-100-S. [Calgene] Stearic deriv.; cationic; dispersant, emulsifier, corrosion inhibitor; for textile, paper, metal applics.; paste; 100% conc.

Hodag DGL. [Calgene] PEG-2 laurate; CAS 9004-81-3; nonionic; emulsifier; liq.; HLB 6.5; 100% conc.

Hodag DGO. [Calgene] PEG-2 oleate; CAS 9004-96-0; nonionic; lubricant; liq.; HLB 4.7; 100% conc.

Hodag DGS. [Calgene] PEG-2 stearate; CAS 9004-99-3; nonionic; emulsifier, lubricant, and plasticizer for cosmetic and industrial use; solid; HLB 4.7; 100% conc.

Hodag DOSS-70. [Calgene] Dioctyl sodium sulfosuccinate; anionic; wetting agent, surface tension reducer; liq.; sol. in polar and nonpolar org. solvs.; water-disp.; 70% conc.

Hodag DOSS-75. [Calgene] Dioctyl sodium sulfosuccinate; anionic; wetting agent, surface tension reducer; liq.; 75% conc.

Hodag DTSS-70. [Calgene] Sodium sulfosuccinate type; anionic; wetting agent, surface tension reducer; liq.; sol. in polar and nonpolar org. solvs.; water-disp.; 70% conc.

Hodag EGMS. [Calgene] Glycol stearate; nonionic; wetting agent, surface tension reducer; solid; sol. in polar and nonpolar org. solvs.; water-disp.; HLB 3.5; 100% conc.

Hodag FD Series. [Calgene] Silicones; food and industrial defoamers; liq.

Hodag Nonionic 1017-R. [Calgene] Meroxapol 171; nonionic; detergent, antifoam, dispersant, demulsifier, dishwashing rinse visc. control agent; liq.; 100% conc.

Hodag Nonionic 1025-R. [Calgene] EO/PO block polymer; nonionic; detergent, antifoam, rinse aids (dishwashing), visc. control, dispersant, demulsifier; liq.; 100% conc.

Hodag Nonionic 1035-L. [Calgene] EO/PO block polymer; nonionic; detergent, antifoam, wetting agent, emulsifier, antistat, demulsifier, visc. modifier, deduster, gelation aid, metalworking lubricants, dispersants; liq.; 100% conc.

Hodag Nonionic 1044-L. [Calgene] EO/PO block polymer; nonionic; detergent, antifoam, wetting agent, emulsifier, antistat, demulsifier, visc. modifier, deduster, gelation aid, metalworking lubricants, dispersants; liq.; 100% conc.

Hodag Nonionic 1061-L. [Calgene] EO/PO block polymer; nonionic; detergent, antifoam, wetting agent, emulsifier, antistat, demulsifier, visc. modifier, deduster, gelation aid, metalworking lubricants, dispersants; liq.; 100% conc.

Hodag Nonionic 1062-L. [Calgene] EO/PO block polymer; nonionic; detergent, antifoam, wetting agent, emulsifier, antistat, demulsifier, visc. modifier, deduster, gelation aid, metalworking lubricants, dispersants; liq.; 100% conc.

Hodag Nonionic 1064-L. [Calgene] EO/PO block polymer; nonionic; detergent, antifoam, wetting agent, emulsifier, antistat, demulsifier, visc. modifier, deduster, gelation aid, metalworking lubricants, dispersants; liq.; 100% conc.

Hodag Nonionic 1068-F. [Calgene] EO/PO block polymer; nonionic; detergent, antifoam, wetting agent, emulsifier, antistat, demulsifier, visc. modifier, deduster, gelation aid, metalworking lubricants, dispersants; flake; 100% conc.

Hodag Nonionic 1088-F. [Calgene] EO/PO block polymer; nonionic; detergent, antifoam, wetting agent, emulsifier, antistat, demulsifier, visc. modifier, deduster, gelation aid, metalworking lubricants, dispersants; flake; 100% conc.

Hodag Nonionic 2017-R. [Calgene] Meroxapol 172; nonionic; detergent, antifoam, dispersant, demulsifier, dishwashing rinse visc. control agent; liq.; 100% conc.

Hodag Nonionic 2025-R. [Calgene] EO/PO block polymer; nonionic; detergent, antifoam, dispersant, demulsifier, dishwashing rinse visc. control agent; liq.; 100% conc.

Hodag Nonionic 4017-R. [Calgene] EO/PO block polymer; nonionic; detergent, antifoam, dispersant, demulsifier, dishwashing rinse visc. control agent; liq.; 100% conc.

Hodag Nonionic 4025-R. [Calgene] EO/PO block polymer; nonionic; detergent, antifoam, dispersant, demulsifier, dishwashing rinse visc. control agent; liq.; 100% conc.

Hodag Nonionic 5025-R. [Calgene] EO/PO block polymer; nonionic; detergent, antifoam, dispersant, demulsifier, dishwashing rinse visc. control agent; liq.; 100% conc.

Hodag Nonionic E-5. [Calgene] POE alkylaryl ether; nonionic; detergent and wetting agent for cosmetics, insecticides and other formulations; liq.; 100% conc.

Hodag Nonionic E-6. [Calgene] POE alkylaryl ether; nonionic; detergent and wetting agent for cosmetics, insecticides and other formulations; liq.; HLB 10.6; 100% conc.

Hodag Nonionic E-7. [Calgene] POE alkylaryl ether; nonionic; detergent and wetting agent for cosmetics, insecticides and other formulations; liq.; 100% conc.

Hodag Nonionic E-10. [Calgene] POE alkylaryl ether; nonionic; detergent and wetting agent for cosmetics, insecticides and other formulations; liq.; HLB 12.6; 100% conc.

Hodag Nonionic E-12. [Calgene] POE alkylaryl ether; nonionic; detergent and wetting agent for cosmetics, insecticides and other formulations; liq.; HLB 14.0; 100% conc.

Hodag Nonionic E-20. [Calgene] POE alkylaryl ether; nonionic; detergent and wetting agent for cosmetics, insecticides and other formulations; solid; HLB 17.0; 100% conc.

Hodag Nonionic E-30. [Calgene] POE alkylaryl ether; nonionic; detergent and wetting agent for cosmetics, insecticides and other formulations; solid; HLB 17.5; 100% conc.

Hodag PE-005. [Calgene] Alkylaryl phosphate ester; anionic; emulsifier, EP lube additive, antistat, corrosion inhibitor, surfactant; liq.; 100% conc.

Hodag PE-104. [Calgene] Alkylaryl phosphate ester; anionic; emulsifier, EP lube additive, antistat, corrosion inhibitor, surfactant; liq.; 100% conc.

Hodag PE-106. [Calgene] Alkylaryl phosphate ester; anionic; emulsifier, EP lube additive, antistat, corrosion inhibitor, surfactant; liq.; 100% conc.

Hodag PE-109. [Calgene] Alkylaryl phosphate ester; anionic; emulsifier, EP lube additive, antistat, corrosion inhibitor, surfactant; liq.; 100% conc.

Hodag PE-206. [Calgene] Alkylaryl phosphate ester; anionic; emulsifier, EP lube additive, antistat, corrosion inhibitor, surfactant; liq.; 100% conc.

Hodag PE-209. [Calgene] Alkylaryl phosphate ester; anionic; emulsifier, EP lube additive, antistat, corrosion inhibitor, surfactant; liq.; 100% conc.

Hodag PGO. [Calgene] Fatty glyceride ester; nonionic; antifoamer for carbohydrate systems; liq.; 100% conc.

Hodag PGS. [Calgene] Propylene glycol stearate; CAS 1323-39-3; nonionic; emulsifier for food processing; solid; 100% conc.

Hodag PSML-20. [Calgene] Polysorbate 20; CAS 9005-64-5; nonionic; emulsifier; liq.; HLB 16.7; 100% conc.

Hodag PSMO-20. [Calgene] Polysorbate 80; CAS 9005-65-6; nonionic; emulsifier for food processing, industrial applics.; HLB 15.0; 100% conc.

Hodag PSMP-20. [Calgene] Polysorbate 40; CAS 9005-66-7; nonionic; emulsifier for food processing, industrial applics.; liq.; HLB 15.6; 100% conc.

Hodag PSMS-20. [Calgene] Polysorbate 60; CAS

9005-67-8; nonionic; emulsifier for food processing, industrial applics.; solid; HLB 14.8; 100% conc.

Hodag PSTS-20. [Calgene] Polysorbate 65; CAS 9005-71-4; nonionic; emulsifier for food processing, industrial applics.; solid; HLB 10.5; 100% conc.

Hodag SML. [Calgene] Sorbitan laurate; CAS 1338-39-2; nonionic; emulsifier, oil additive, corrosion inhibitor; liq.; oil-sol.; HLB 8.6; 100% conc.

Hodag SMO. [Calgene] Sorbitan oleate; CAS 1338-43-8; nonionic; emulsifier, oil additive, corrosion inhibitor; liq.; oil-sol.; HLB 4.3; 100% conc.

Hodag SMP. [Calgene] Sorbitan palmitate; CAS 26266-57-9; nonionic; emulsifier, oil additive, corrosion inhibitor; solid; oil-sol.; HLB 6.7; 100% conc.

Hodag STO. [Calgene] Sorbitan trioleate; CAS 26266-58-0; nonionic; emulsifier, oil additive, corrosion inhibitor; liq.; oil-sol.; HLB 1.8; 100% conc.

Hodag STS. [Calgene] Sorbitan tristearate; CAS 26658-19-5; nonionic; emulsifier, oil additive, corrosion inhibitor; solid; oil-sol.; HLB 2.1; 100% conc.

Hodag Sole-Mulse B. [Calgene] Modified ethoxylate; anionic; emulsifier for emulsion degreasers, industrial cleaners; liq.; 100% conc.

Hoe S 1816. [Hoechst Celanese/Colorants & Surf.] EO/PO block copolymer; nonionic; surfactant for agric. formulations; m.w. 4000-4500.

Hoe S 1816-1. [Hoechst Celanese/Colorants & Surf.] EO/PO block copolymer; nonionic; surfactant for agric. formulations; m.w. 4000-4500.

Hoe S 1816-2. [Hoechst Celanese/Colorants & Surf.] EO/PO block copolymer; nonionic; surfactant for agric. formulations; m.w. 7000-8000.

Hoe S 1984 (TP 2279). [Hoechst Celanese/Colorants & Surf.] Calcium dodecylbenzene sulfonate/nonylphenol ethoxylate; anionic; surfactant for agric. formulations.

Hoe S 1984 (TP 2283). [Hoechst Celanese/Colorants & Surf.] Calcium dodecylbenzene sulfonate/alcohol ethoxylate; anionic; surfactant for agric. formulations.

Hoe S 2713. [Hoechst Celanese/Colorants & Surf.] Sodium dodecylbenzene sulfonate; anionic; surfactant for agric. formulations.

Hoe S 2713 HF. [Hoechst Celanese/Colorants & Surf.] Sodium dodecylbenzene sulfonate; anionic; surfactant for agric. formulations; flash pt. > 100 C.

Hoe S 2749. [Hoechst Celanese/Colorants & Surf.] TEA dodecylbenzene sulfonate; anionic; surfactant for agric. formulations.

Hoe S 2817. [Hoechst Celanese/Colorants & Surf.] Alkyl substituted dicarboxylic anhydride; nonionic; hydrotrope/solubilizer for formulations with high electrolyte concs.; liq.; sol. in min. oil, gasolines; insol. in water; 100% act.

Hoe S 2895. [Hoechst Celanese/Colorants & Surf.] Calcium dodecylbenzene sulfonate, nonylphenol ethoxylate, EO/PO polymer; anionic; surfactant for agric. formulations.

Hoe S 2896. [Hoechst Celanese/Colorants & Surf.] Calcium dodecylbenzene sulfonate, nonylphenol ethoxylate; anionic; surfactant for agric. formulations.

Hoe S 3435. [Hoechst Celanese/Colorants & Surf.] C12-15 syn. oxo alcohol polyglycol ether; nonionic; surfactant for agric. formulations.

Hoe S 3510. [Hoechst Celanese/Colorants & Surf.] EO/PO block copolymer based on butyl alcohol; nonionic; surfactant for agric. formulations.

Hoe S 3618. [Hoechst Celanese/Colorants & Surf.] EO/PO block copolymer bis-mono-phosphate ester; anionic; surfactant for agric. formulations.

Hoe S 3680. [Hoechst Celanese/Colorants & Surf.] Epoxidized soybean oil; nonionic; surfactant for agric. formulations.

Honol 405. [Takemoto Oil & Fat] Sulfated oil POE alkyl ether; anionic/nonionic; finishing oil for rayon filament; liq.

Honol GA. [Takemoto Oil & Fat] POE alkylether; nonionic; finishing oil for rayon filament; liq.; 100% conc.

Honol MGR. [Takemoto Oil & Fat] Sulfated oil; anionic/nonionic; finishing oil for rayon staple; paste; 90% conc.

Honoralin PL. [Takemoto Oil & Fat] Sulfated oil; anionic; wetting and penetrating agent; dyeing aux. for cotton and wool; stable to acid, alkali, and hard water; liq.

Hostacerin DGI. [Hoechst AG] Fatty acid polyglyceryl ester; emulsifier for w/o emulsions; liq.

Hostacerin O-20. [Hoechst Celanese/Colorants & Surf.] Oleth-20; CAS 9004-98-2; nonionic; surfactant; wax; 100% conc.

Hostacor 2098. [Hoechst Celanese/Colorants & Surf.] Complex carboxylic acid; anionic; wetting agent, corrosion inhibitor in aq. systems, syn. metalworking and cleaning formulations; visc. liq.; sol. in water; sp.gr. 1.00; pH 6.5-7.5 (1% aq.).

Hostacor 2125. [Hoechst Celanese/Colorants & Surf.] Carboxylic acid complex; anionic; corrosion inhibitor for synthetics on ferrous and nonferrous metals; visc. liq.; water-sol.; sp.gr. 1.00; pH 7.2-7.3 (1% aq.).

Hostacor 2270. [Hoechst Celanese/Colorants & Surf.] Carboxylic acid complex; anionic; corrosion inhibitor for syn. metalworking fluids for use with ferrous metals; amber liq.; water-sol.; sp.gr. 1.12; pH 5.0-6.0 (1% aq.).

Hostacor 2272. [Hoechst Celanese/Colorants & Surf.] Boron alkanolamide; anionic; emulsifier, wetting agent; surfactant offering bio-resistance, hard water stability, lubricity and low foam to metalworking fluids; amber liq.; water-sol.; sp.gr. 1.06; pH 9.4 (1% aq.).

Hostacor 2291. [Hoechst Celanese/Colorants & Surf.] Monocarboxylic acid blend; anionic; surfactant for amine-free metalworking coolants; low foam corrosion inhibitor for microemulsions; liq.; water-sol. as salt; sp.gr. 0.90; pH 3-4 (1% aq.).

Hostacor 2292. [Hoechst Celanese/Colorants & Surf.] Carboxylic acid blend; anionic; corrosion inhibitor for ferrous, aluminum, and magnesium; yel. liq.; water-sol. as salt; sp.gr. 0.8-1.0; pH 3-5 (1% aq.).

Hostacor 2732. [Hoechst Celanese/Colorants & Surf.] Alkylamidocarboxylic acid; anionic; corrosion inhibitor for steel, galvanized, and aluminum; visc. liq.; water-sol. as salt; sp.gr. 0.98; pH 5-6 (1% aq.).

Hostacor BBM. [Hoechst Celanese/Colorants & Surf.] Boric acid/carboxylic acid amine condensate; anionic; emulsifier, wetting agent, and corrosion inhibitor for semi-synthetic metalworking fluids; visc. liq.; sol. in water; sp.gr. 1.18; pH 9 (1% aq.).

Hostacor BF. [Hoechst Celanese/Colorants & Surf.] Alkanolamine borate condensate; anionic; rust protection and buffering capacity in a microbial-resistant package for metalworking formulation; visc. liq.; sol. in water; sp.gr. 1.10; pH 9.8 (1% aq.).

Hostacor BK. [Hoechst Celanese/Colorants & Surf.] Complex carboxylic acid condensate; anionic; corrosion inhibitor, emulsifier, wetting agent, and lubricant for formulation of drilling aids and aq. cooling systems; brn. clear visc. liq.; sol. in water, min. oil, glycols; sp.gr. 1.05; pH 9.2-9.5 (1% aq.); 85% act.

Hostacor BM. [Hoechst Celanese/Colorants & Surf.] Boron complex; anionic; emulsifier, wetting agent, and corrosion inhibitor in aq. systems, metalworking fluids, coolants; yel., brn. clear liq.; water sol; sp.gr. 1.1; pour pt. 14 F; pH 9.0-9.4 (1%).

Hostacor BS. [Hoechst Celanese/Colorants & Surf.] Boron complex; anionic; emulsifier and corrosion inhibitor in coolants, syn. cutting fluids, aq. systems; yel., brn. clear liq.; sol. in water, min. oils; sp.gr. 1.1; pour pt. 14 F; flash pt. 302 F; pH 9.0-10.0 (1% aq.); 70% conc.

Hostacor DT. [Hoechst Celanese/Colorants & Surf.] Fatty acid alkanolamide; nonionic; emulsifier, wetting agent, corrosion inhibitor, lubricant used in aq. metalworking fluids; brn. oily visc. liq.; sol. in water, min. oils; sp.gr. 1.01; flash pt. 210 C (Marcusson); pH 9.5-10.5 (1% aq.).; 100% conc.

Hostacor E. [Hoechst Celanese/Colorants & Surf.] Alkylsulfonamidocarboxylic acid salt; anionic; corrosion inhibitor for min. oils, sol. oils, semi-synthetic metalworking formulations; visc. oily liq.; insol. in water; sp.gr. 0.95.

Hostacor H Liq. N. [Hoechst Celanese/Colorants & Surf.] Arylsulfonamidocarboxylic acid; anionic; chemical intermediate for mfg. of corrosion inhibitors for aq. systems, metalworking fluids; yel. visc. liq.; sol. in ethanol, glycol, water insol.; sp.gr. 1.17 ± 0.05 (122 F); pH 5-6 (1% aq.).

Hostacor TP 2445. [Hoechst Celanese/Colorants & Surf.] Boric acid/carboxylic acid amine condensate; anionic; high boron content corrosion inhibitor with surface active props.; wetting agent; liq.; water-sol.; sp.gr. 1.15; pH 9.3 (1% aq.).

Hostadrill Brands. [Hoechst AG] Vinylamide/vinylsulfonic acid copolymers; filtration agents.

Hostaflot L Grades. [Hoechst AG] Aliphatic dithiophosphates; flotation collector for sulfide minerals; liq.

Hostaflot X Grades. [Hoechst AG] Thionocarbamate; flotation collectors; liq.

Hostamer Brands. [Hoechst AG] Vinylamide/vinylsulfonic acid copolymers; polymers for enhanced oil recovery, esp. at high temps. and salinity.

Hostapal 2345. [Hoechst Celanese/Colorants & Surf.] Nonylphenol polyglycol ether carboxylate; anionic; surfactant for agric. formulations.

Hostapal 3634 Highly Conc. [Hoechst AG] Alkylaryl polyglycol ether; nonionic; wetting agent and detergent used in fibers, leather; liq.

Hostapal BV Conc. [Hoechst Celanese/Colorants & Surf.; Hoechst AG] Alkylaryl polyglycol ether sulfate, sodium salt; anionic; dispersant for pigments, wetting agent, detergent for fibers, emulsifier for emulsion polymerization; yel. brn. gelatinous paste; misc. with water; dens. 9.10 lb/gal; 49–51% act.

Hostapal CV Brands. [Hoechst AG] Alkylaryl polyglycol ether; nonionic; wetting agent and detergent for textile industry; stiff-flowing liq.

Hostapal CVH. [Hoechst Celanese/Colorants & Surf.] Nonylphenol ethoxylate; surfactant for textile processing; sl. yel. liq.; visc. < 500 cps; hyd. no. 90; pH 6 (1% aq.); 99% act.

Hostapal CVS. [Hoechst Celanese/Colorants & Surf.] Alkylphenol polyglycol ether; anionic; detergent for household and industrial cleaners; wetting agent, dispersant for org. and inorganic pigment preparations, for use with anionic dispersing agents; slighty yel. visc. liq.; dens. 8.85 lb/gal; cloud pt. 52–56 C (1% aq.); 100% act.

Hostapal N-040. [Hoechst Celanese/Colorants & Surf.] Nonoxynol-4; CAS 9016-45-9; nonionic; detergent, wetting agent, emulsifier for general industrial applics., agric., textiles; sl. yel. liq.; visc. < 500 cps; hyd. no 150; pH 7 (1% aq.); 99% act.

Hostapal N-060. [Hoechst Celanese/Colorants & Surf.] Nonoxynol-6; CAS 9016-45-9; nonionic; detergent, wetting agent, emulsifier for general industrial applics., agric.; liq.; 100% conc.

Hostapal N-060R. [Hoechst Celanese/Colorants & Surf.] Nonoxynol-6; CAS 9016-45-9; nonionic; surfactant for textile processing; sl. yel. liq.; visc. < 500 cps; hyd. no. 115; pH 6 (1% aq.); 99% act.

Hostapal N-080. [Hoechst Celanese/Colorants & Surf.] Nonoxynol-8.; CAS 9016-45-9; nonionic.

Hostapal N-090. [Hoechst Celanese/Colorants & Surf.] Nonoxynol-9; CAS 9016-45-9; nonionic; detergent, wetting agent, emulsifier for general industrial applics., agric.; liq.; 100% conc.

Hostapal N-100. [Hoechst Celanese/Colorants & Surf.] Nonoxynol-10; CAS 9016-45-9; nonionic; detergent, wetting agent, emulsifier for general industrial applics., agric.; liq.; 100% conc.

Hostapal N-110. [Hoechst Celanese/Colorants & Surf.] Nonoxynol-11.; CAS 9016-45-9; nonionic.

Hostapal N-130. [Hoechst Celanese/Colorants & Surf.] Nonoxynol-13.; CAS 9016-45-9; nonionic.

Hostapal N-230. [Hoechst Celanese/Colorants & Surf.] Nonoxynol-23.; CAS 9016-45-9; nonionic.

Hostapal N-300. [Hoechst Celanese/Colorants & Surf.] Nonoxynol-30; CAS 9016-45-9; nonionic; surfactant for textile processing; wh. solid; hyd. no. 36; pH 7 (1% aq.); 99% act.

Hostapal SF. [Hoechst AG] Detergent blend; nonionic; detergent and stain removing agent for textiles; liq.

Hostapal TP 2347. [Hoechst Celanese/Colorants & Surf.] Nonyl phenol ether sulfate, amine salt; anionic; surfactant for agric. formulations.

Hostaphat 2122. [Hoechst Celanese/Colorants & Surf.] 2-Ethylhexyl phosphate; surfactant for textile processing; yel. liq.; visc. 400 cps; pH 2.2 (5% aq.); 99% act.

Hostaphat 2188. [Hoechst Celanese/Colorants & Surf.] Ethoxylated C12-14 phosphate; surfactant for textile processing; yel. liq.; visc. 800 cps; pH 2.5 (5% aq.); 99% act.

Hostaphat 2204. [Hoechst Celanese/Colorants & Surf.] Ethoxylated tridecyl phosphate; surfactant for textile processing; yel. liq.; visc. 600 cps; pH 2.0 (5% aq.); 99% act.

Hostaphat AR K. [Hoechst Celanese/Colorants & Surf.] Nonylphenol ethoxylate phosphate ester, potassium salt; anionic; surfactant for agric. formulations.

Hostaphat AW. [Hoechst Celanese/Colorants & Surf.] Aromatic phosphate ester; low foaming wetting agent, corrosion inhibitor and lubricant for aq. metalworking systems; visc. liq.; sol. in water, acetone, ethanol, benzene; sp.gr. 1.22-1.23; pH 2.5-3.0 (1% aq.).

Hostaphat F Brands. [Hoechst Celanese/Colorants & Surf.; Hoechst AG] Phosphoric acid complex org.

ester; anionic; antistat finish for syn. fibers; liq.; 95% conc.

Hostaphat FL-340 N. [Hoechst Celanese/Colorants & Surf.] Ethoxylated lauryl phosphate; surfactant for textile processing; yel. liq.; visc. 250 cps; pH 6.5 (5% aq.); 97% act.

Hostaphat FO-380. [Hoechst Celanese/Colorants & Surf.] Ethoxylated oleyl phosphate; surfactant for textile processing; yel. liq.; visc. 100 cps; pH 5.0 (5% aq.); 98% act.

Hostaphat HI. [Hoechst Celanese/Colorants & Surf.] Hexyl phosphate; surfactant for textile processing; Gardner 1 clear liq.; visc. 100 cps; pH 2 (5% aq.); 96% act.

Hostaphat K Grades. [Hoechst AG] Organic phosphoric acid esters; emulsifiers; liq. to paste.

Hostaphat LPKN158. [Hoechst Celanese/Colorants & Surf.] Phosphate ester; nonionic; surfactant; wh. wax; 100% conc.

Hostaphat MDAR Grades. [Hoechst Celanese/Colorants & Surf.] Nonylphenol ethoxylate phosphate esters; anionic; surfactant for agric. formulations.

Hostaphat MDIT. [Hoechst Celanese/Colorants & Surf.] Tridecyl phosphate; surfactant for textile processing; yel. liq.; visc. 500 cps; pH 2.5 (5% aq.); 99% act.

Hostaphat MDL. [Hoechst Celanese/Colorants & Surf.] Dodecyl phosphate; surfactant for textile processing; wh. solid; pH 2 (5% aq.); 99% act.

Hostaphat MDLZ. [Hoechst Celanese/Colorants & Surf.] C12 alcohol ethoxylate phosphate ester; anionic; surfactant for agric. formulations.

Hostaphat OD. [Hoechst Celanese/Colorants & Surf.] C8-10 phosphate; surfactant for textile processing; yel. liq.; visc. 150 cps; pH 2 (5% aq.); 99% act.

Hostapon IDC. [Hoechst Celanese/Colorants & Surf.] Sodium N-palmityl N-cyclohexyl taurine; surfactant for textile processing; beige wax; pH 9 (1% aq.); 45% act.

Hostapon KA Powd. [Hoechst Celanese/Colorants & Surf.; Hoechst AG] Sodium cocoyl isethionate; anionic; detergent base for cosmetic industry, detergent bars; foamer, dispersant; wh. powd.; 80% conc.

Hostapon T Powd. [Hoechst Celanese/Colorants & Surf.; Hoechst AG] Sodium methyl oleoyl taurate; CAS 137-20-2; anionic; cleaner for textile, leather, household prods., agric., cosmetics, etc.; powd.; 63% conc.

Hostapur CX Highly Conc. [Hoechst Celanese/Colorants & Surf.] Alkyl polyglycol ether; CAS 24938-91-8; wetting agent and detergent for textiles; liq.

Hostapur DOS Hi Conc. [Hoechst Celanese/Colorants & Surf.] Dialkyl sodium sulfosuccinate; surfactant for textile processing; yel. liq.; visc. 150 cps; pH 7 (1% aq.); 80% act.

Hostapur DTC. [Hoechst Celanese/Colorants & Surf.] Isotridecyl alcohol polyglycol ether carboxylate; anionic; surfactant for agric. formulations.

Hostapur DTC FA. [Hoechst Celanese/Colorants & Surf.] Isotridecyl alcohol polyglycol ether carboxylate, free acid; anionic; surfactant for agric. formulations.

Hostapur OS Brands. [Hoechst Celanese/Colorants & Surf.; Hoechst AG] Sodium alpha-olefin sulfonate; anionic; basic materials for detergents and cleaning agents; carpet shampoos; liq., powd.

Hostapur SAS 60. [Hoechst Celanese/Colorants & Surf.; Hoechst AG] Sodium C14–17 alkyl sec sulfonate; detergent base for mfg. of washing-up liq., detergent, shampoo, wetting agent; surfactant for textile processing; biodeg.; colorless to weakly yel. clear liq., yel. soft paste, flake; virtually odorless; m.w. 328; sp.gr. 1.048–1.087; visc. 26–6500 cps; 60% act.

Hostarex Grades. [Hoechst AG] Sec. and tert. amines; anion/cation exchangers; liq.

Hostawet TDC. [Hoechst Celanese/Colorants & Surf.] Ether carboxylate; anionic; emulsifier, wetting agent for metal cleaning formulations; visc. liq.; water-sol.; sp.gr. 1.03; pH 3 (1% aq.).

12-HSA. [CasChem] Hydroxystearic acid; chemical intermediate; solid; 90% conc.

Humectol C, C Highly Conc. [Hoechst AG] Sulfonated fatty acid amines; anionic; low foaming wetting and leveling agent, dispersant for dyeing and printing; liq.

Hybase® C-300. [Witco/Sonneborn] Overbased calcium sulfonate, highly basic oil-sol.; anionic; detergent, rust inhibitor, lube oil additive used in marine or stationary diesels; dk. brn. liq.; sol. in org. solvs.; dens. 9.4 lb/gal; sp.gr. 1.13 (60 F); visc. 800 SUS (210 F); flash pt. 380 F; 29% act.

Hybase® C-400. [Witco/Sonneborn] Overbased calcium sulfonate; anionic; lube oil additive component for marine and diesel use; rust preventive, industrial detergent; liq.; 20% conc.

Hybase® C-500. [Witco/Sonneborn] Overbased calcium sulfonate; anionic; lube oil additive component for marine and diesel use; rust preventive, industrial detergent; liq.; 20% conc.

Hybase® M-12. [Witco/Sonneborn] Overbased magnesium sulfonate in No. 2 fuel oil; anionic; fuel oil additive (vanadium scavenger, sulfuric acid neutralizer, detergent); liq.; 15% conc.

Hybase® M-300. [Witco/Sonneborn] Overbased magnesium sulfonate; anionic; highly basic oil-sol.; lubricant, corrosive inhibitor, used in marine or stationary diesels, automotive use; 30% conc.

Hybase® M-400. [Witco/Sonneborn] Overbased magnesium sulfonate; anionic; lube oil additive component for diesel and automotive use; rust preventive, fuel oil additive; liq.; 27% conc.

Hydrenol D. [Henkel/Cospha; Henkel KGaA] Tallow alcohol (25-35% C16, 60-67% C18); wetting agent, emulsifier, emollient, consistency giving agent for skin creams and lotions; intermediate for surfactant mfg.; APHA 20 fused/flakes; sp.gr. 0.805-0.815 g/cc (60 C); solid. pt. 48-52 C; b.p. 300-360 C; acid no. < 0.1; iodine no. < 1.0; sapon. no. < 1.2; hyd. no. 210-220; flash pt. 180 C; 100% conc.

Hydrenol DD. [Henkel/Emery; Henkel KGaA] Linear sat. fatty alcohols (26-35% C16, 60-70% C18); intermediate for surfactant mfg.; APHA 20 fused/flakes; sp.gr. 0.805-0.815 g/cc (60 C); solid. pt. 48-52 C; b.p. 310-360 C; acid no. < 0.1; iodine no. < 1.0; sapon. no. < 0.5; hyd. no. 210-220; flash pt. 180 C; 100% conc.

Hydrine. [Gattefosse; Gattefosse SA] PEG-2 stearate; CAS 9004-99-3; 106-11-6; consistency stabilizer for ointments, cream lotions; opacifier in shampoos, liq. soaps; solid; HLB 5.0; 100% conc.

Hydroace Series. [Toho Chem. Industry] Anionic/nonionic; emulsifier for emulsion type fuel oils; liq.

Hydrolene. [Reilly-Whiteman] Sulfated veg. oils; surfactant used in metalworking, textile, leather, printing inks, paints, detergents; liq.; 100% conc.

Hydrotriticum QL. [Croda Inc.] Laurdimonium hydroxypropyl hydrolyzed wheat protein; cationic; film-forming conditioner for hair and skin care prods., e.g., waving systems, activated conditioner treatments, shampoos, styling mousses, hair coloring, wrinkle remover creams and lotions, liq. soap, facial scrubs, skin cleansers, bath prods.; lt. amber clear liq., mild char. odor; sol. @ 10% in water, water/ethanol, glycerin, propylene glycol, surfactants; sp.gr. 1.05; b.p. > 300 F; pH 4.0-5.0; toxicology: may be a skin and eye irritant; 30% act.

Hydrotriticum QM. [Croda Inc.] Cocodimonium hydroxypropyl hydrolyzed wheat protein; cationic; film-forming conditioner for hair and skin care prods., e.g., waving systems, activated conditioner treatments, shampoos, styling mousses, hair coloring, wrinkle remover creams and lotions, liq. soap, facial scrubs, skin cleansers, bath prods.; lt. amber clear liq., mild char. odor; sol. in water, water/ethanol, glycerin, propylene glycol, surfactants; sp.gr. 1.05; b.p. > 300 F; pH 4.0-5.0; toxicology: may be a skin and eye irritant; 28-32% solids.

Hydrotriticum QS. [Croda Inc.] Steardimonium hydroxypropyl hydrolyzed wheat protein; cationic; film-forming conditioner for hair and skin care prods., e.g., waving systems, activated conditioner treatments, shampoos, styling mousses, hair coloring, wrinkle remover creams and lotions, liq. soap, facial scrubs, skin cleansers, bath prods.; lt. amber opaque paste, mild char. odor; sol. in water, water/ethanol, glycerin, propylene glycol; sp.gr. 1.05; b.p. > 300 F; pH 4.0-5.0; toxicology: may be a skin and eye irritant; 28-32% solids.

Hydroxylan. [Fanning] Hydroxylated lanolin; nonionic; emulsifier, conditioner, emollient; solid; 100% conc.

Hymolon CWC. [Hart Chem. Ltd.] Cocamide DEA; anionic; fulling agent for wool and household cleaners; liq.

Hymolon K90. [Hart Chem. Ltd.] Cocamide DEA; nonionic; detergent and emulsifier for household, alkaline and heavy-duty cleaners, scouring agent for wool, cotton, polyester; yel. liq.; dens. 1.04 lb/gal; 91% act.

Hyonic 407. [Henkel/Functional Prods.] Alkyl phenoxy POE ethanol; nonionic; surfactant, emulsifier, wetting agent, dispersant, detergent; liq.; 70% conc.

Hyonic CPG 745. [Henkel] Linear aliphatic alcohol block polymer; nonionic; surfactant and wetting agent used in rinse aids, machine dishwashing, spray cleaning, bottle washing, metal cleaning; clear liq.; sp.gr. 110; dens. 9.2 lb/gal; visc. 59 cP; pour pt. 0 C; cloud pt. 2 C (10%); 99% act.

Hyonic GL 400. [Henkel/Functional Prods.] Polyethoxy alkyl phenol; nonionic; surfactant, coemulsifier for specialty polymerizations; liq.; HLB 18.0; 70% act.

Hyonic NP-40. [Henkel/Functional Prods.; Henkel/Organic Prods.] Nonoxynol-4; CAS 9016-45-9; nonionic; emulsifier for solv. cleaning compds., waterless hand cleaners, agric. emulsifiable concs., silicone prods., detergent for petrol. oils; visc. reducer for plastisols; intermediate for prod. of anionic surfactants; metal cleaners; food contact applics.; clear liq.; insol. in water; dens. 1.02 g/ml; HLB 8.9; pour pt. 26 C; pH 7.0 (1% aq.); > 99% act.

Hyonic NP-60. [Henkel/Functional Prods.; Henkel/Organic Prods.; Henkel Canada] Nonoxynol-6; CAS 9016-45-9; nonionic; emulsifier, wetting agent, detergent for textile lubricants, agric. emulsifiable concs.; intermediate for textile antistats; color enhancer in latex paint; post-polymerization stabilizer; food applics.; clear liq.; disp. in water; sol. in oil; dens. 1.04 g/ml; HLB 10.9; pour pt. –32 C; cloud pt. < 0 C (1% aq.); pH 7.0 (1% aq.); surf. tens. 30 dynes/cm (0.01%); > 99% act.

Hyonic NP-90. [Henkel/Functional Prods.; Henkel/Organic Prods.; Henkel Canada] Nonoxynol-9; CAS 9016-45-9; nonionic; detergent and penetrant for household and industrial cleaning compds., textile processing; degreasing compds.; wetting agent for paper toweling or tissue; emulsifier in agric. applics.; clear liq.; sol. in water; dens. 1.06 g/ml; HLB 13; pour pt. 2 C; cloud pt. 54 C (1% aq.); pH 7.0 (1% aq.); surf. tens. 31 dynes/cm (0.01%); > 99% act.

Hyonic NP-100. [Henkel/Functional Prods.; Henkel/Organic Prods.; Henkel Canada] Nonoxynol-10; CAS 9016-45-9; nonionic; surfactant, wetting agent, emulsifier; base for liq. dishwashing detergents, laundry preparations, household and industrial cleaners; agric. applics.; penetrant for corrosion inhibitors; stabilizer for latex; rewetting agent for paper toweling and tissue; food applics.; clear liq.; sol. in water; dens. 1.07 g/ml; HLB 13.2; pour pt. 4 C; cloud pt. 68 C (1% aq.); pH 7.0 (1% aq.); surf. tens. 32 dynes/cm (0.01%); > 99% act.

Hyonic NP-110. [Henkel/Functional Prods.; Henkel Canada] Nonoxynol-11; CAS 9016-45-9; nonionic; surfactant, wetting agent, emulsifier; base for liq. dishwashing detergents, laundry preparations, household and industrial cleaners; agric. applics.; penetrant for corrosion inhibitors; stabilizer for latex; rewetting agent for paper toweling and tissue; food applics.; clear liq.; sol. in water; dens. 1.07 g/ml; HLB 13.8; pour pt. 13 C; cloud pt. 72 C (1% aq.); pH 7.0 (1% aq.); surf. tens. 34 dynes/cm (0.01%); > 99% act.

Hyonic NP-120. [Henkel/Functional Prods.; Henkel/Organic Prods.] Nonoxynol-12; CAS 9016-45-9; nonionic; detergent, wetting agent, penetrant, coemulsifier; food and agric. applics.; liq., semisolid; sol. in water; dens. 1.07 g/ml; HLB 14.1; pour pt. 15 C; cloud pt. 91 C (1% aq.); pH 7.0 (1% aq.); surf. tens. 36 dynes/cm (0.01%); > 99% act.

Hyonic NP-407. [Henkel Canada] Nonoxynol-40; CAS 9016-45-9; nonionic; stabilizer for water dispersions; primary emulsifier in mfg. of vinyl acetate or acrylic latex; wetting agent for min. acid and alkali sol'ns.; coemulsifier for agric. formulations; food applics.; floor polishes, high temp. detergents; clear liq.; sol. in water; dens. 1.10 g/ml; HLB 17.6; pour pt. –6 C; cloud pt. 100 C (1% aq.); pH 7.0 (1% aq.); 70% act.

Hyonic NP-500. [Henkel] Nonoxynol-50; CAS 9016-45-9; nonionic; stabilizer for aq. disps.; emulsifier for vinyl acetate or acrylic latex; wetting agent for min. acid and alkali sol'ns.; coemulsifier for agric. formulations; food applics.; clear solid; sol. in water; dens. 1.08 g/ml; HLB 18; pour pt. 24 C; cloud pt. 100 C (1% aq.); pH 7.0 (1% aq.); > 99% act.

Hyonic OP-7. [Henkel/Textile] Octoxynol-7; CAS 9002-93-1; nonionic; surfactant for textile use; Gardner 1 liq.; sol. @ 5% in water, xylene; dens. 8.3 lb/gal; visc. 115 cSt (100 F); HLB 12.0; pour pt. -2 C; cloud pt. < 25 C; flash pt. > 500 F.

Hyonic OP-10. [Henkel/Textile] Octoxynol-10; CAS 9002-93-1; nonionic; surfactant for textile use; Gardner 1 liq.; sol. @ 5% in water, xylene; dens. 8.6 lb/gal; visc. 117 cSt (100 F); HLB 13.6; pour pt. 4 C; cloud pt. 64 C; flash pt. 425 F.

Hyonic OP-40. [Henkel] Octoxynol-4; CAS 9002-93-1; nonionic; emulsifier in solv. cleaning compds., waterless hand cleaners, agric. emulsifiable concs., silicone prods.; metal cleaners; detergent for petrol. oils; visc. reducer for plastisols; intermediate for prod. of anionic surfactants; agric. use; clear liq.; disp. in water; dens. 1.02 g/ml; HLB 8; pour pt. -26 C; cloud pt. 53 C (1% aq.); pH 7.0 (1% aq.); surf. tens. 28 dynes/cm (0.01%); > 99% act.

Hyonic OP-55. [Henkel/Textile] Nonionic; scouring for textile desizing; emulsifier for waxes and oils preventing redeposition; clear liq.; forms translucent sol'ns.; pH 6.5 (5%).

Hyonic OP-70. [Henkel/Functional Prods.; Henkel/Organic Prods.] Octoxynol-7; CAS 9002-93-1; nonionic; surfactant, emulsifier, wetting agent, dispersant, detergent, intermediate for textile lubricants, agric. emulsifiable concs.; color development enhancer in latex paint; post-polymerization stabilizer; clear liq.; disp. in water; dens. 1.06 g/ml; HLB 12; pour pt. 11 C; cloud pt. 23 C (1% aq.); pH 7.0 (1% aq.); surf. tens. 27 dynes/cm (0.01%); > 99% act.

Hyonic OP-100. [Henkel/Functional Prods.; Henkel/Organic Prods.] Octoxynol-9; CAS 9002-93-1; nonionic; wetting agent, coemulsifier; detergent for household and industrial cleaning compds. and textile processing; penetrant; degreasing compds.; wetting agent for paper toweling/tissue; agric. applics.; clear liq.; sol. in water; dens. 1.07 g/ml; HLB 13; pour pt. 10 C; cloud pt. 65 C (1% aq.); pH 7.0 (1% aq.); surf. tens. 30 dynes/cm (0.01%); > 99% act.

Hyonic OP-407. [Henkel] Octoxynol-40; CAS 9002-93-1; nonionic; stabilizer for water dispersions; mfg. and stabilize latexes; wetting agent in min. acid and alkali formulations; coemulsifier and stabilizer in agric. formulations; clear liq.; sol. in water; dens. 0.91 g/ml; HLB 18; pour pt. -2 C; cloud pt. 73 C (1% aq.); pH 7.0 (1% aq.); surf. tens. 48 dynes/cm (0.01%); 70% act.

Hyonic OP-705. [Henkel] Octoxynol-70; CAS 9002-93-1; nonionic; stabilizer for water dispersions; mfg. and stabilize latexes; wetting agent in min. acid and alkali formulations; coemulsifier and stabilizer in agric. formulations; clear liq.; sol. in water; dens. 1.10 g/ml; HLB 18.5; pour pt. -12 C; cloud pt. 91 C (1% aq.); pH 7.0 (1% aq.); surf. tens. 49 dynes/cm (0.01%); 50% act.

Hyonic PE-40. [Henkel] Nonyl phenol ethoxylate; nonionic; emulsifier for solv. cleaning compds., waterless hand cleaners, metal cleaners, surfactant intermediate; clear liq.; dens. 8.5 lb/gal; cloud pt. 0 C (1% aq.); pour pt. -29 C; surf. tens. 28 dynes/cm (0.01%); pH 7.0 (1% aq.); > 99% act.

Hyonic PE-90. [Henkel] Nonyl phenol ethoxylate; nonionic; detergent, wetter, emulsifier, stabilizer for household and industrial cleaning, textile, paper processing, latex adhesives; clear liq.; sp.gr. 1.06; dens. 8.8 lb/gal; cloud pt. 54 C (1% aq.); pour pt. 2 C; surf. tens. 31 dynes/cm (0.01%); pH 7.0 (1% aq.); > 99% act.

Hyonic PE-100. [Henkel] Nonoxynol-10; CAS 9016-45-9; nonionic; detergent, wetter, emulsifier, base, penetrant for household and industrial cleaners, paper toweling; latex stabilizer; clear liq.; sp.gr. 1.068; dens. 8.9 lb/gal; cloud pt. 68 C (1% aq.); pour pt. 4 C; surf. tens. 32 dynes/cm (0.01%); pH 7.0 (1% aq.); > 99% act.

Hyonic PE-120. [Henkel] Nonyl phenol ethoxylate; CAS 9016-45-9; nonionic; detergent, penetrant, wetting agent, coemulsifier; liq./semisolid; dens. 8.9 lb/gal; cloud pt. 91 C (1% aq.); pour pt. 15 C; surf. tens. 36 dynes/cm (0.01%); pH 7.0 (1% aq.); > 99% act.

Hyonic PE-240, -260. [Henkel/Process] Alkyl phenol POE ethanol; CAS 9036-19-5; nonionic; wetting and foaming agent; clear liq.; dens. 8.8 and 8.9 lb/gal resp.; cloud pt. 25 and 65 C resp.; pour pt. -9 and 7 C resp.; 100% act.

Hyonic PE-250. [Henkel/Process] Octoxynol-9; CAS 9002-93-1; nonionic; wetting and foaming agent; clear liq.; dens. 8.9 lb/gal; cloud pt. 52 C; pour pt. 5 C; 100% act.

Hyonic PE-360. [Henkel/Process] Tridecyl alcohol; nonionic; detergent, emulsifier, wetting agent; semisolid; dens. 8.1 lb/gal; cloud pt. 0 C; pour pt. 18 C.

Hyonic TD 60. [Henkel/Functional Prods.] Alcohol ethoxylate; nonionic; surfactant; liq.; sol. in water and oil; 100% conc.

Hypan®. [Lipo] Polymer; amphoteric; gel-forming polymer; as primary emulsifier, emollient; forms stable complex emulsions; powd.

Hypan® SA100H. [Lipo] Polymer; amphoteric; gel-forming polymer; as primary emulsifier, emollient; forms stable complex emulsions; powd.

Hystrene® 1835. [Witco/Humko] Mixt. tallow/coconut acid (CTFA); CAS 67701-05-7; chemical intermediate, emulsifier; used for personal care prods., soaps, waxes, textile aux., pharmaceuticals; paste; solid. pt. 40 C max.; acid no. 214-222; iodine no. 36-42; sapon. no. 211-220.

Hystrene® 3022. [Witco/Humko] Hydrog. menhaden acid; chemical intermediate, emulsifier; used for personal care prods., waxes, greases, textile aux., pharmaceuticals; solid; solid. pt. 50-54 C; acid no. 193-202; iodine no. 5; sapon. no. 193-202; 100% conc.

Hystrene® 3675. [Witco/Humko] 75% Dimer acid; corrosion inhibitor, intermediate; derivs. used as syn. lube components, corrosion inhibitors for petrol. processing, as extenders and crosslinking agents for high polymeric systems; mildness additive in detergents; Gardner 9 max. liq.; visc. 9000 cSt; acid no. 189–197; sapon. no. 189–199; 100% conc.

Hystrene® 3675C. [Witco/Humko] 75% Dimer acid, 3% monomer [dimer acid (CTFA)]; corrosion inhibitor, intermediate; derivs. used as syn. lube components, corrosion inhibitors for petrol. processing, as extenders and crosslinking agents for high polymeric systems; mildness additive in detergents; Gardner 9 max. liq.; visc. 7500 cSt; acid no. 189–197; sapon. no. 189–199; 100% conc.

Hystrene® 3680. [Witco/Humko] 80% Dimer acid; corrosion inhibitor, intermediate; derivs. used as syn. lube components, corrosion inhibitors for petrol. processing, as extenders and crosslinking agents for high polymeric systems; mildness additive for detergents; Gardner 8 max. liq.; visc. 8500 cSt; acid no. 190–197; sapon. no. 190–199; 100% conc.

Hystrene® 3687. [Witco/Humko] 87% Dimer acid; corrosion inhibitor, intermediate; derivs. used as syn. lube components, corrosion inhibitors for petrol. processing, as extenders and crosslinking agents for high polymeric systems; mildness additive for detergents; liq.; 100% conc.

Hystrene® 3695. [Witco/Humko] 95% Dimer acid; corrosion inhibitor, intermediate; derivs. used as syn. lube components, corrosion inhibitors for petrol. processing, as extenders and crosslinking agents for high polymeric systems; mildness additive in detergents; Gardner 6 max. liq.; visc. 7500 cSt; acid no. 190–196; sapon. no. 190–202.

Hystrene® 4516. [Witco/Humko] Stearic acid; CAS 57-11-4; lubricant, textile aux., emulsifier, plasticizer, intermediate, used in cosmetics, shampoos, pharmaceuticals; solid; acid no. 203-209; iodine no. 1 max.; sapon. no. 204-210; 100% conc.

Hystrene® 5012. [Witco/Humko] Hydrog. stripped coconut acid; chemical intermediate, emulsifier; used for personal care prods., soaps, lubricants, waxes, textile aux., pharmaceuticals; liq.; solid. pt. 24–33 C; acid no. 250-266; iodine no. 2 max.; sapon. no. 250-266; 100% conc.

Hystrene® 5016. [Witco/Humko] Stearic acid, triple pressed; CAS 57-11-4; stabilizer, lubricant, textile aux., emulsifier, plasticizer, intermediate, used in cosmetics, shampoos, pharmaceuticals; solid; acid no. 206-210; iodine no. 0.5 max.; sapon. no. 206-211; 100% conc.

Hystrene® 5016 NF. [Witco/Humko] Stearic acid, triple pressed; CAS 57-11-4; food grade acids used as lubricants, release agents, binders, and defoamers, and in components for producing other food grade additives; Lovibond 1.0Y-0.1R solid; solid pt. 54.5–56.5 C; acid no. 206–210; iodine no. 0.5 max.; sapon. no. 206–211; 100% conc.

Hystrene® 5460. [Witco/Humko] Trilinoleic acid; CAS 68937-90-6; corrosion inhibitor, lubricant; intermediate for mfg. of soaps, emulsions, creams, lotions, ethoxylates, buffing compds., lubricants; Gardner > 18 liq.; visc. 30,000 cs; acid no. 182–190; sapon. no. 190–198; 100% conc.

Hystrene® 5522. [Witco/Humko] Behenic acid; CAS 112-85-6; chemical intermediate, emulsifier; used for personal care prods., soaps, waxes, textile aux., pharmaceuticals; solid. pt. 60–63 C; acid no. 178-185; iodine no. 5; sapon. no. 179-186; 100% conc.

Hystrene® 7018. [Witco/Humko] Stearic acid (70%); CAS 57-11-4; lubricant, textile aux., emulsifier, plasticizer, intermediate, used in cosmetics, shampoos, pharmaceuticals; solid; acid no. 200-206; iodine no. 0.8 max.; sapon. no. 200-207; 100% conc.

Hystrene® 7018 FG. [Witco/Humko] Stearic acid; CAS 57-11-4; food grade acids used as lubricants, release agents, binders, and defoamers, and in components for producing other food grade additives; Lovibond 1.0Y-0.1R solid; solid. pt. 58.0-62.5 C; acid no. 200–205; iodine no. 0.8 max.; sapon. no. 200–206; 100% conc.

Hystrene® 7022. [Witco/Humko] Behenic fatty acid; CAS 112-85-6; lubricant, textile aux., emulsifier, plasticizer, intermediate, used in cosmetics, shampoos, pharmaceuticals; solid; acid no. 170-180; iodine no. 3.5; sapon. no. 170-181; 100% conc.

Hystrene® 8016. [Witco/Humko] 80% Palmitic acid; CAS 57-10-3; intermediate for mfg. of soaps, emulsions, creams, lotions, ethoxylates, buffing compds., lubricants; solid; 100% conc.

Hystrene® 8018. [Witco/Humko] 80% Stearic acid; CAS 57-11-4; intermediate for mfg. of soaps, emulsions, creams, lotions, ethoxylates, buffing compds., lubricants; solid; 100% conc.

Hystrene® 8718 FG. [Witco/Humko] Stearic acid; CAS 57-11-4; food grade acids used as lubricants, release agents, binders, and defoamers; intermediate for producing other food grade emulsifiers; solid; 100% conc.

Hystrene® 9014. [Witco/Humko] Myristic acid (90%); CAS 544-63-8; lubricant, textile aux., emulsifier, plasticizer, intermediate, used in cosmetics, shampoos, pharmaceuticals; solid; acid no. 238-243; iodine no. 0.5 max.; sapon. no. 238-246; 100% conc.

Hystrene® 9016. [Witco/Humko] Palmitic acid (90%); CAS 57-10-3; lubricant, textile aux., emulsifier, plasticizer, intermediate, used in cosmetics, shampoos, pharmaceuticals; solid; acid no. 216-220; iodine no. 0.5 max.; sapon. no. 216-221; 100% conc.

Hystrene® 9022. [Witco/Humko] Behenic acid (90%); lubricant, textile aux., emulsifier, plasticizer, intermediate, used in cosmetics, shampoos, pharmaceuticals; solid; acid no. 165-175; iodine no. 3; sapon. no. 165-176; 100% conc.

Hystrene® 9512. [Witco/Humko] Lauric acid (95%); CAS 143-07-7; lubricant, textile aux., emulsifier, plasticizer, intermediate, used in cosmetics, shampoos, pharmaceuticals; solid; acid no. 276-281; iodine no. 0.5 max.; sapon. no. 276-282; 100% conc.

Hystrene® 9514. [Witco/Humko] Myristic acid (95%); CAS 544-63-8; chemical intermediate, emulsifier; used for personal care prods., waxes, textile aux., pharmaceuticals; solid. pt. 51.9–53.5 C; acid no. 243-246; iodine no. 0.5 max.; sapon. no. 243-247.

Hystrene® 9718. [Witco/Humko] Stearic acid (92%); CAS 57-11-4; lubricant, textile aux., emulsifier, plasticizer, intermediate, used in cosmetics, shampoos, pharmaceuticals; solid; acid no. 196-201; iodineno. 0.8 max.; sapon. no. 196-202; 100% conc.

Hystrene® 9718 NF FG. [Witco/Humko] Stearic acid NF (92%); CAS 57-11-4; food grade acids used as lubricants, release agents, binders, and defoamers; intermediate for producing food grade emulsifiers; solid; solid. pt. 66.5-68.5 C; acid no. 196–201; iodine no. 0.8 max.; sapon no. 196–201; 100% conc.

Hystrene® 9912. [Witco/Humko] 99% Lauric acid; CAS 143-07-7; chemical intermediate, emulsifier; used for personal care prods., waxes, textile aux., pharmaceuticals; solid. pt. 43–44 C; acid no. 277-281; iodine no. 0.2 max.; sapon. no. 278-281.

I

Iberpal B.I.G. [A. Harrison] Ethoxylated alkyl phenol; nonionic; detergent and scouring agent; liq.; 100% conc.

Iberpenetrant-114. [A. Harrison] Alkylphenol polyethylene sulfonate; anionic; textile wetting agent, dispersant, foaming agent; liq.

Iberpol Gel. [A. Harrison] Aminoethyl sulfonate, sodium salt; anionic; detergent and textile scouring acid; liq.; 32% conc.

Iberpon W. [A. Harrison] Alkylaryl sodium sulfonate; anionic; detergent; textile scouring agent; liq.; 25% conc.

Iberscour AC. [A. Harrison] Amine condensate; anionic; surface act. agent; wetting agent and detergent; gives soft hand to most fibers; alkali stable; liq.

Iberscour P. [A. Harrison] Ethoxylated alkyl phenol; nonionic; wetting agent, detergent, emulsifier used in polyester scouring; liq.

Iberscour W Conc. [A. Harrison] Alkyphenol EO ether; nonionic; detergent; textile wetting agent; liq.; 100% conc.

Iberterge 65. [A. Harrison] Alcohol straight chain ethoxylate, modified; detergent, emulsifier; textile scouring agent; liq.; 35% conc.

Iberterge CO-40. [A. Harrison] Sodium alkylaryl sulfonate; anionic; detergent; textile scouring agent; flake, powd.; 40% conc.

Iberwet B.I.G. [A. Harrison] Alkylaryl sulfonate, modified; anionic; wetting agent; liq.

Iberwet BO. [A. Harrison] Butyl oleate and builders; nonionic; rewetting agent for textile use; sanforizing assistant; liq.

Iberwet E. [A. Harrison] Sodium alkylaryl sulfonate; detergent; textile scouring aid; liq.

Iberwet W-100. [A. Harrison] EO nonylphenol; nonionic; detergent; textile wetting agent; liq.; 100% conc.

Ice #2. [Van Den Bergh Foods] Glyceryl stearate and polysorbate 80; nonionic; stabilizer and emulsifier for the food industry; lubricant for textiles and plastics; fabric softener; bead; HLB 5.2; m.p. 59–63 C; 32–38% alpha monoglyceride.

Icinol H260. [ICI Australia] Polyoxyalkylene glycol ether; syn. lubricant, defoamer; high visc. index solv.; liq.; water-sol.; visc. 260 SUS (100 F).

Icinol H280X. [ICI Australia] Polyoxyalkylene glycol ether; inhibited version of Icinol H260; syn. lubricant, defoamer; high visc. index solv.

Icinol H660. [ICI Australia] Polyoxyalkylene glycol ether; syn. lubricant, defoamer; high visc. index solv.; liq.; water-sol.; visc. 660 SUS.

Icinol H660YA. [ICI Australia] Polyoxyalkylene glycol ether; inhibited version of Icinol H660; syn. lubricant, defoamer; high visc. index solv.

Icinol H5100. [ICI Australia] Polyoxyalkylene glycol ether; syn. lubricant, defoamer; high visc. index solv.; liq.; water-sol.; visc. 5100 SUS.

Icinol L65. [ICI Australia] Polyoxyalkylene glycol ether; syn. lubricant, defoamer; high visc. index solv.; liq.; water-insol.; visc. 65 SUS (100 F).

Icinol L285. [ICI Australia] Polyoxyalkylene glycol ether; syn. lubricant, defoamer; high visc. index solv.; liq.; water-insol.; visc. 285 SUS.

Icinol L300X. [ICI Australia] Polyoxyalkylene glycol ether; inhibited version of Icinol L285.

Icinol L385. [ICI Australia] Polyoxyalkylene glycol ether; syn. lubricant, defoamer; high visc. index solv.; liq.; water-insol.; visc. 385 SUS.

Icinol L625. [ICI Australia] Polyoxyalkylene glycol ether; syn. lubricant, defoamer; high visc. index solv.; liq.; water-insol.; visc. 625 SUS.

Icinol L1715. [ICI Australia] Polyoxyalkylene glycol ether; syn. lubricant, defoamer; high visc. index solv.; liq.; water-insol.; visc. 1715 SUS.

Icomeen® 18-5. [BASF] PEG-5 stearamine; CAS 26635-92-7; cationic/nonionic; wetting agent, penetrant, emulsifier, stabilizer, dispersant, antistat, lubricant, solubilizer; Gardner 8 max. cast solid; disp. in water; sp.gr. 0.98; HLB 8.9; pH 9 (5% aq.); 100% act.

Icomeen® 18-50. [BASF] PEG-50 stearamine; CAS 26635-92-7; cationic; surfactant; Gardner 9 max. cast solid; water-sol.; sp.gr. 1.07; HLB 17.7; pH 9 (5%); 100% act.

Icomeen® O-30. [BASF] PEG-30 oleamine; cationic/nonionic; wetting agent, penetrant, emulsifier, stabilizer, dispersant, antistat, lubricant, solubilizer; Gardner 10 max. paste; m.w. 1590; water-sol.; sp.gr. 1.05; HLB 16.2; pH 8.0–9.5 (5% aq.); 100% act.

Icomeen® O-30-80%. [BASF] PEG-30 oleamine; cationic/nonionic; wetting agent, penetrant, emulsifier, stabilizer, dispersant, antistat, lubricant, solubilizer; Gardner 10 max. liq.; water-sol.; sp.gr. 1.05; HLB 16.2; pH 7.5–9.0 (5% aq.); 80% act.

Icomeen® S-5. [BASF] PEG-5 soyamine; CAS 61791-24-0; cationic/nonionic; wetting agent, penetrant, emulsifier, stabilizer, dispersant, antistat, lubricant, solubilizer; disp. in water; sp.gr. 0.98; HLB 9.0; pH 9 (5% aq.); 100% act.

Icomeen® T-2. [BASF] PEG-2 tallow amine; cationic/nonionic; wetting agent, penetrant, emulsifier, stabilizer, dispersant, antistat, lubricant, solubilizer; Gardner 7 max. paste; insol. in water; m.w. 350; sp.gr. 0.92; HLB 5.0; pH 8.5-10.5 (5% aq.); 100% act.

Icomeen® T-5. [BASF] PEG-5 tallow amine; cationic/nonionic; wetting agent, penetrant, emulsifier, stabilizer, dispersant, antistat, lubricant, solubilizer;

Gardner 8 max. liq., paste; water-sol.; sp.gr. 0.96; HLB 9.2; cloud pt. 45 C (1% aq.); pH 9 (5% aq.); surf. tens. 33 dynes/cm (0.1% aq.); 100% act.

Icomeen® T-7. [BASF] PEG-7 tallow amine; cationic/nonionic; wetting agent, penetrant, emulsifier, stabilizer, dispersant, antistat, lubricant, solubilizer; Gardner 8 max. liq.; water-sol.; m.w. 570; sp.gr. 0.98; HLB 10.8; cloud pt. 85 C (1% aq.); pH 9 (5% aq.); surf. tens. 36 dynes cm (0.1% aq.); 100% act.

Icomeen® T-15. [BASF] PEG-15 tallow amine; cationic; surfactant; Gardner 7 max. liq.; water-sol.; m.w. 925; sp.gr. 1.03; HLB 14.1; pH 9 (5% aq.); surf. tens. 41 dynes/cm (0.1% aq.); 100% act.

Icomeen® T-20. [BASF] PEG-20 tallow amine; cationic/nonionic; wetting agent, penetrant, emulsifier, stabilizer, dispersant, antistat, lubricant, solubilizer; Gardner 8 max. liq.; water-sol.; m.w. 1145; sp.gr. 1.04; HLB 15.4; pH 9 (5% aq.); 100% act.

Icomeen® T-25. [BASF] PEG-25 tallow amine; cationic/nonionic; wetting agent, penetrant, emulsifier, stabilizer, dispersant, antistat, lubricant, solubilizer; Gardner 8 max. liq.; water-sol.; m.w. 1365; sp.gr. 1.05; HLB 16.0; pH 9 (5% aq.); 100% act.

Icomeen® T-25 CWS. [BASF] Tallow amine ethoxylate, modified; cationic/nonionic; surfactant for gel formation; wetting agent, penetrant, emulsifier, stabilizer, dispersant, antistat, lubricant, solubilizer; Gardner 8 max. liq.; water-sol.; sp.gr. 1.05; HLB 16.0; pH 7.5–9.0 (5% aq.); 100% act.

Icomeen® T-40. [BASF] PEG-40 tallow amine; cationic/nonionic; wetting agent, penetrant, emulsifier, stabilizer, dispersant, antistat, lubricant, solubilizer; Gardner 8 max. cast solid; water-sol.; m.w. 2030; sp.gr. 1.09; HLB 17.2; m.p. 33 C; pH 8.0–9.5 (5% aq.); 100% act.

Icomeen® T-40-80%. [BASF] PEG-40 tallow amine; cationic/nonionic; wetting agent, penetrant, emulsifier, stabilizer, dispersant, antistat, lubricant, solubilizer; Gardner 8 max. liq.; water-sol.; m.w. 2030; sp.gr. 1.08; HLB 17.2; pH 8.0–9.5 (5% aq.); 80% act.

Iconol 28. [BASF] Cocamide DEA; nonionic; emulsifier, foam stabilizer, finishing agent, lubricant; Gardner 12 max. liq.; water-sol.; sp.gr. 1.01; visc. 2300 cps; cloud pt. > 100 C; pH 8.0–10.0 (5% aq.); 100% act.

Iconol CNP-1, CNP-10. [BASF] Nonylphenol fatty esters, ethoxylated; surfactant; Gardner 8 and 5 max. resp. liqs.; disp. in water; sp.gr. 1.03; visc. 200 and 700 cps resp.; cloud pt. < 25 C; 100% act.

Iconol COA. [BASF] Cocamide DEA; nonionic; emulsifier, foam stabilizer, finishing agent, lubricant; Gardner 10 max. liq.; water-sol.; sp.gr. 1.02; visc. 1300 cps; cloud pt. > 100 C; pH 8.5–11.0 (5% aq.); 100% act.

Iconol DA-4. [BASF] Deceth-4; nonionic; wetting agent, detergent, emulsifier, dispersant, solubilizer for textile scouring and dyeing, industrial and institutional cleaners, household cleaning prods. and specialties; APHA 70 max. liq.; disp. in water; m.w. 330; sp.gr. 0.958; visc. 70 cps; HLB 11.0; pour pt. -24 C; cloud pt. < 0 C (1% aq.); pH 6.0–7.5 (5% aq.); surf. tens. 27 dynes/cm (0.1% aq.); Draves wetting < 1 s (0.1% aq.); Ross-Miles foam 30 mm (0.1% aq., 50 C); 100% act.

Iconol DA-6. [BASF] Deceth-6; nonionic; wetting agent, detergent, emulsifier, dispersant, solubilizer for textile scouring and dyeing, industrial and institutional cleaners, household cleaning prods. and specialties; APHA 70 max. liq.; water-sol.; m.w. 390; sp.gr. 0.98; visc. 20 cps; HLB 13.0; pour pt. 8 C; cloud pt. 26–34 C (1% aq.); pH 6.0–7.5 (5% aq.); 100% act.

Iconol DA-6-90%. [BASF] Deceth-6; nonionic; wetting agent, detergent, emulsifier, dispersant, solubilizer for textile scouring and dyeing, industrial and institutional cleaners, household cleaning prods. and specialties; APHA 50 max. liq.; water-sol.; m.w. 410; sp.gr. 0.99; HLB 12.5; pour pt. < 0 C; cloud pt. 34–40 C (1% aq.); pH 5.5–7.5 (5% aq.); 90% act.

Iconol DA-9. [BASF] Deceth-9; nonionic; wetting agent, detergent, emulsifier, dispersant, solubilizer for textile scouring and dyeing, industrial and institutional cleaners, household cleaning prods. and specialties; APHA 70 max. paste; water-sol.; m.w. 550; sp.gr. 1.02; HLB 14.0; pour pt. 10 C; cloud pt. 79–84 C (1% aq.); pH 6.0–7.5 (5% aq.); 100% act.

Iconol DDP-10. [BASF] Dodoxynol-10; nonionic; emulsifier, wetting and cleaning agent, penetrant, detergent, stabilizer, dispersant, coemulsifier; Gardner 1 max. liq.; water-sol.; m.w. 725; sp.gr. 1.04; visc. 400 cps; HLB 12.6; pour pt. 10 C; cloud pt. 39–44 C (1% aq.); pH 6.0–7.5 (5% aq.); 100% act.

Iconol DNP-8. [BASF] Nonyl nonoxynol-8; CAS 9014-93-1; nonionic; emulsifier, wetting and cleaning agent, penetrant, detergent, stabilizer, dispersant, coemulsifier; Gardner 4 max. clear liq.; disp. in water; m.w. 700; sp.gr. 1.0; visc. 500 cps; HLB 10.0; pour pt. 9 C; cloud pt. < 25 C (1% aq.); pH 6.0–7.5 (5% aq.); 100% act.

Iconol DNP-24. [BASF] Nonyl nonoxynol-24; CAS 9014-93-1; nonionic; emulsifier, wetting and cleaning agent, penetrant, detergent, stabilizer, dispersant, coemulsifier; Gardner 1 max. paste; water-sol.; m.w. 1400; HLB 15.0; m.p. 33 C; cloud pt. > 95 C (1% aq.); pH 6.0–7.5 (5% aq.); 100% act.

Iconol DNP-150. [BASF] Nonyl nonoxynol-150; CAS 9014-93-1; nonionic; emulsifier, wetting and cleaning agent, penetrant, detergent, stabilizer, dispersant, coemulsifier; Gardner 1 max. flakes, cast solid; water-sol.; m.w. 6900; sp.gr. 1.05; HLB 19.0; m.p. 55 C; cloud pt. > 100 C (1% aq.); pH 6.0–7.5 (5% aq.); 100% act.

Iconol NP-1.5. [BASF] Nonoxynol-1.5; nonionic; surfactant; Gardner 2 max. clear liq.; oil-sol.; insol. in water; m.w. 281; sp.gr. 0.99; visc. 350 cps; HLB 4.6; pour pt. –11 C; pH 6.0–7.5 (5% aq.); 1.0% max. water.

Iconol NP-4. [BASF] Nonoxynol-4; CAS 9016-45-9; nonionic; surfactant; APHA 100 max. liq.; oil-sol.; insol. in water; m.w. 391; sp.gr. 1.02; visc. 350 cps; HLB 8.9; pour pt. –27 C; pH 6.0–7.5 (5% aq.); 1.0% max. water.

Iconol NP-5. [BASF] Nonoxynol-5; CAS 9016-45-9; nonionic; surfactant; APHA 100 max. liq.; disp. in water; m.w. 435; sp.gr. 1.03; visc. 300 cps; HLB 10.0; pour pt. –27 C; pH 6.0–7.5 (5% aq.); surf. tens. 29 dynes/cm (0.1% aq.); 1.0% max. water.

Iconol NP-6. [BASF] Nonoxynol-6; CAS 9016-45-9; nonionic; surfactant; APHA 100 max. liq.; disp. in water; m.w. 479; sp.gr. 1.04; visc. 300 cps; HLB 10.8; pour pt. –28 C; pH 6.0–7.5 (5% aq.); surf. tens. 29 dynes/cm (0.1% aq.); 1.0% max. water.

Iconol NP-7. [BASF] Nonoxynol-7; CAS 9016-45-9; nonionic; wetting and cleaning agent, penetrant, detergent, emulsifier, stabilizer, dispersant, coemulsifier; used in food; APHA 100 max. liq.; water-sol.; m.w. 523; sp.gr. 1.05; visc. 300 cps; HLB

11.9; pour pt. 5 C; cloud pt. 22–27 C (1.0% aq.); surf. tens. 30 dynes/cm (0.1% aq.); 100% conc.

Iconol NP-9. [BASF] Nonoxynol-9; CAS 9016-45-9; nonionic; surfactant; APHA 70 max. liq.; water-sol.; m.w. 611; sp.gr. 1.06; visc. 300 cps; HLB 13.0; pour pt. 4 C; cloud pt. 52–56 C (1.0% aq.); pH 6.0–7.5 (5% aq.); surf. tens. 32 dynes/cm (0.1% aq.); 1.0% max. water.

Iconol NP-10. [BASF] Nonoxynol-10; CAS 9016-45-9; nonionic; surfactant; APHA 70 max. liq.; water-sol.; m.w. 655; sp.gr. 1.06; visc. 300 cps; HLB 13.5; pour pt. 9 C; cloud pt. 60–65 C (1.0 aq.); pH 6.0–7.5 (5% aq.); surf. tens. 34 dynes/cm (0.1% aq.); 1.0% max. water.

Iconol NP-12. [BASF] Nonoxynol-12; CAS 9016-45-9; nonionic; see Iconol NP-7; liq.; HLB 14.0; 100% conc.

Iconol NP-15. [BASF] Nonoxynol-15; CAS 9016-45-9; nonionic; surfactant; APHA opaque visc. liq.; water-sol.; m.w. 875; sp.gr. 1.07; visc. 700 cps; HLB 15.0; pour pt. 26 C; cloud pt. 95–100 C (1.0% aq.); pH 6.0–7.5 (5% aq.); surf. tens. 39 dynes/cm (0.1% aq.); 1.0% max. water.

Iconol NP-20. [BASF] Nonoxynol-20; CAS 9016-45-9; nonionic; surfactant; water-sol.; m.w. 1095; sp.gr. 1.08 (50/25 C); HLB 16.0; m.p. 30 C; cloud pt. > 100 C (1% aq.); pH 6.0–7.5 (5% aq.); surf. tens. 40 dynes/cm (0.1% aq.); 1.0% max. water.

Iconol NP-30. [BASF] Nonoxynol-30; CAS 9016-45-9; nonionic; wetting agent, penetrant, detergent, cleaning agent, emulsifier, latex stabilizer, dispersant for industrial, institutional and household cleaning prods.; emulsifier for emulsion polymerization, asphalt emulsions; wh. wax; water-sol.; m.w. 1535; sp.gr. 1.08 (50/25 C); visc. 30 cps (100 C); HLB 17.0; m.p. 41 C; cloud pt. > 100 C (1% aq.); pH 6.0–7.5 (5% aq.); surf. tens. 42 dynes/cm (0.1% aq.); 100% conc.

Iconol NP-30-70%. [BASF] Nonoxynol-30; CAS 9016-45-9; nonionic; wetting agent, penetrant, detergent, cleaning agent, emulsifier, latex stabilizer, dispersant for industrial, institutional and household cleaning prods.; emulsifier for emulsion polymerization, asphalt emulsions; APHA 100 max. liq.; water-sol.; m.w. 1535 (act.); sp.gr. 1.09; visc. 1100 cps; HLB 17.0; pour pt. 0 C; cloud pt. > 100 C (1% aq.); pH 6.0–7.5 (5% aq.); surf. tens. 42 dynes/cm (0.1% aq.); 28.5–31.5% water.

Iconol NP-40. [BASF] Nonoxynol-40; CAS 9016-45-9; nonionic; wetting agent, penetrant, detergent, cleaning agent, emulsifier, latex stabilizer, dispersant for industrial, institutional and household cleaning prods.; emulsifier for emulsion polymerization, asphalt emulsions; wh. wax; water-sol.; m.w. 1975; sp.gr. 1.09 (50/25 C); visc. 40 cps (100 C); HLB 18.0; m.p. 48 C; cloud pt. > 100 (1% aq.); pH 6.0–7.5 (5% aq.); 100% conc.

Iconol NP-40-70%. [BASF] Nonoxynol-40; CAS 9016-45-9; nonionic; wetting agent, penetrant, detergent, cleaning agent, emulsifier, latex stabilizer, dispersant for industrial, institutional and household cleaning prods.; emulsifier for emulsion polymerization, asphalt emulsions; water-sol.; m.w. 1975 (act.); sp.gr. 1.1; visc. 1400 cps; HLB 18.0; pour pt. 7 C; cloud pt. > 100 C (1% aq.); pH 6.0–7.5 (5% aq.); 28.5–31.5% water.

Iconol NP-50. [BASF] Nonoxynol-50; CAS 9016-45-9; nonionic; wetting agent, penetrant, detergent, cleaning agent, emulsifier, stabilizer, dispersant, coemulsifier; wh. wax; water-sol.; m.w. 2415; sp.gr. 1.1 (50/25 C); visc. 60 cps (100 C); HLB 19.0; m.p. 49 C; cloud pt. > 100 C (1% aq.); pH 6.0–7.5 (5% aq.); 100% conc.

Iconol NP-50-70%. [BASF] Nonoxynol-50; CAS 9016-45-9; nonionic; wetting agent, penetrant, detergent, cleaning agent, emulsifier, latex stabilizer, dispersant for industrial, institutional and household cleaning prods.; emulsifier for emulsion polymerization, asphalt emulsions; liq.; HLB 19.0; 70% conc.

Iconol NP-70. [BASF] Nonoxynol-70; CAS 9016-45-9; nonionic; wetting agent, penetrant, detergent, cleaning agent, emulsifier, latex stabilizer, dispersant for industrial, institutional and household cleaning prods.; emulsifier for emulsion polymerization, asphalt emulsions; wh. wax; water-sol.; m.w. 3300; sp.gr. 1.1 (50/25 C); visc. 80 cps (100 C); HLB 18.6; pour pt. 51 C; cloud pt. > 100 C (1% aq.); pH 6.0–7.5 (5% aq.); 100% conc.

Iconol NP-70-70%. [BASF] Nonoxynol-70; CAS 9016-45-9; nonionic; wetting agent, penetrant, detergent, cleaning agent, emulsifier, latex stabilizer, dispersant for industrial, institutional and household cleaning prods.; emulsifier for emulsion polymerization, asphalt emulsions; APHA 100 max. liq.; water-sol.; m.w. 3300; sp.gr. 1.1; visc. 3600 cps; HLB 18.6; pour pt. 17 C; cloud pt. > 100 C (1% aq.); pH 6.0–7.5 (5% aq.); 70% conc.

Iconol NP-100. [BASF] Nonoxynol-100; CAS 9016-45-9; 26027-38-3; nonionic; wetting agent, penetrant, detergent, cleaning agent, emulsifier, latex stabilizer, dispersant for industrial, institutional and household cleaning prods.; emulsifier for emulsion polymerization, asphalt emulsions; Gardner 1 max. flake, cast solid; water-sol.; m.w. 4315; sp.gr. 1.12 (50/25 C); visc. 150 cps (100 C); HLB 19.0; m.p. 52 C; cloud pt. > 100 C (1% aq.); pH 6.0–7.5 (5% aq.); 100% conc.

Iconol NP-100-70%. [BASF] Nonoxynol-100; CAS 9016-45-9; nonionic; wetting agent, penetrant, detergent, cleaning agent, emulsifier, latex stabilizer, dispersant for industrial, institutional and household cleaning prods.; emulsifier for emulsion polymerization, asphalt emulsions; liq.; HLB 19.0; cloud pt. > 100 C (1% aq.); 70% conc.

Iconol NP-915. [BASF] Alkylphenoxypoly (ethylenoxy) ethanol; nonionic; surfactant; Gardner 1 liq., paste; water-sol.; sp.gr. 1.06; visc. 400 cps; pour pt. 16 C; cloud pt. 70–78 C (1% aq.); pH 6.0–7.5 (5% aq.); 100% conc.

Iconol OP-5. [BASF] Octoxynol-5; CAS 9002-93-1; nonionic; surfactant; APHA 100 max. liq.; disp. in water; m.w. 425; sp.gr. 1.04; visc. 300 cps; HLB 10.0; pour pt. < 0 C; cloud pt. < 25 C (1% aq.); pH 6.0–7.5 (5% aq.); surf. tens. 28 dynes/cm (0.1% aq.); 1.0% max. water.

Iconol OP-7. [BASF] Octoxynol-7; CAS 9002-93-1; nonionic; surfactant; APHA 100 max. liq.; water-sol.; m.w. 515; sp.gr. 1.05; visc. 300 cps; HLB 12.4; pour pt. –9 C; cloud pt. 21–25 C (1% aq.); pH 6.0–7.5 (5% aq.); surf. tens. 29 dynes/cm (0.1% aq.); 1.0% max. water.

Iconol OP-10. [BASF] Octoxynol-10; CAS 9002-93-1; nonionic; wetting agent, penetrant, detergent, cleaning agent, emulsifier, latex stabilizer, dispersant for industrial, institutional and household cleaning prods.; emulsifier for emulsion polymerization, asphalt emulsions; APHA 100 max. liq.; water-sol.; m.w. 650; sp.gr. 1.06; visc. 250 cps; HLB 14.0; pour pt. 7 C; cloud pt. 63–67 C (1% aq.); surf. tens. 30

dynes/cm (0.1% aq.); 100% conc.

Iconol OP-30. [BASF] Octoxynol-30; CAS 9002-93-1; nonionic; emulsifier, wetting agent, dispersant, syn. latex stabilizer, detergent in formulating industrial, institutional, and household cleaning prods.; primary emulsifier for acrylic and vinyl emulsion polymerization and for asphalt emulsion systems; solid; HLB 17; cloud pt. > 100 C (1% aq.); surf. tens. 46.4 dynes/cm (0.1%); 100% conc.

Iconol OP-30-70%. [BASF] Octoxynol-30; CAS 9002-93-1; nonionic; emulsifier, wetting agent, dispersant, syn. latex stabilizer, detergent in formulating industrial, institutional, and household cleaning prods.; primary emulsifier for acrylic and vinyl emulsion polymerization and for asphalt emulsion systems; liq.; HLB 17; cloud pt. > 100 C (1% aq.); 70% conc.

Iconol OP-40. [BASF] Octoxynol-40; CAS 9002-93-1; nonionic; wetting agent, penetrant, detergent, cleaning agent, emulsifier, latex stabilizer, dispersant for industrial, institutional and household cleaning prods.; emulsifier for emulsion polymerization, asphalt emulsions; Gardner 1 max. cast solid; water-sol.; m.w. 1970; sp.gr. 1.09 (50/25 C); visc. 40 cps (100 C); HLB 18.0; m.p. 50 C; cloud pt. > 100 C (1% aq.); pH 6.0–7.5 (5% aq.); surf. tens. 42 dynes/cm (0.1% aq.); 100% conc.

Iconol OP-40-70%. [BASF] Octoxynol-40; CAS 9002-93-1; nonionic; wetting agent, penetrant, detergent, cleaning agent, emulsifier, latex stabilizer, dispersant for industrial, institutional and household cleaning prods.; emulsifier for emulsion polymerization, asphalt emulsions; APHA 100 max. liq.; water-sol.; m.w. 1970; sp.gr. 1.1; visc. 500 cps; HLB 18.0; pour pt. –4 C; cloud pt. > 100 C (1% aq.); pH 6.0–7.5 (5% aq.); 70% conc.

Iconol P-5. [BASF] Ethoxylated phenol (5 EO); surfactant; Gardner 1 max. liq.; water-sol.; m.w. 314; sp.gr. 1.11; visc. 100 cps; cloud pt. > 100 C (1% aq.); pH 6.0–7.5 (5% aq.); 100% act.

Iconol P-6. [BASF] Ethoxylated phenol (6 EO); surfactant; Gardner 1 max. liq.; water-sol.; m.w. 358; sp.gr. 1.11; visc. 100 cps; cloud pt. > 100 C (1% aq.); pH 6.0–7.5 (5% aq.); 100% act.

Iconol PD-8-90%. [BASF] Decyl alcohol alkoxylate; nonionic; surfactant; APHA 70 max. liq.; water-sol.; sp.gr. 1.0; HLB 11.0; pour pt. < –10 C; cloud pt. 45–50 C (1% aq.); pH 6.0–7.5 (5% aq.); 9–11% water.

Iconol TDA-3. [BASF] Trideceth-3; CAS 24938-91-8; nonionic; wetting agent, detergent, emulsifier, dispersant, solubilizer for textile scouring and dyeing, industrial and institutional cleaners, household cleaning prods. and specialties; APHA 70 max. liq.; disp. in water; m.w. 325; sp.gr. 0.95; visc. 30 cps; HLB 8.0; pour pt. –15 C; cloud pt. < 25 C (1% aq.); pH 6.0–7.5 (5% aq.); 100% act.

Iconol TDA-6. [BASF] Trideceth-6; CAS 24938-91-8; nonionic; wetting agent, detergent, emulsifier, dispersant, solubilizer for textile scouring and dyeing, industrial and institutional cleaners, household cleaning prods. and specialties; APHA 70 max. liq.; disp. in water; m.w. 505; sp.gr. 0.98; visc. 50 cps; HLB 11.4; pour pt. 10 C; cloud pt. < 25 C (1% aq.); pH 6.0–7.5 (5% aq.); 100% act.

Iconol TDA-8. [BASF] Trideceth-8; CAS 24938-91-8; nonionic; wetting agent, detergent, emulsifier, dispersant, solubilizer for textile scouring and dyeing, industrial and institutional cleaners, household cleaning prods. and specialties; APHA 70 max. liq.; water-sol.; m.w. 550; sp.gr. 1.0; HLB 12.7; pour pt. 10 C; cloud pt. 40–45 C (1% aq.); pH 6.0–7.5 (5% aq.); 100% act.

Iconol TDA-8-90%. [BASF] Trideceth-8; CAS 24938-91-8; nonionic; wetting agent, detergent, emulsifier, dispersant, solubilizer for textile scouring and dyeing, industrial and institutional cleaners, household cleaning prods. and specialties; APHA 70 max. liq.; water-sol.; m.w. 550 (act.); sp.gr. 1.0; visc. 60 cps; HLB 12.7; pour pt. < 0 C; cloud pt. 40–45 C (1% aq.); pH 6.0–7.5 (5% aq.); 90% conc.

Iconol TDA-9. [BASF] Trideceth-9; CAS 24938-91-8; nonionic; wetting agent, degreaser, emulsifier, detergent; APHA 70 max. liq.; water-sol.; m.w. 590; sp.gr. 1.0; visc. 75 cps; HLB 13.2; pour pt. 20 C; cloud pt. 55–62 C (1% aq.); pH 6.0–7.5 (5% aq.); 100% act.

Iconol TDA-10. [BASF] Trideceth-10; CAS 24938-91-8; nonionic; wetting agent, detergent, emulsifier, dispersant, solubilizer for textile scouring and dyeing, industrial and institutional cleaners, household cleaning prods. and specialties; APHA 70 max. liq., paste; water-sol.; m.w. 640; sp.gr. 1.03; HLB 13.7; pour pt. 20 C; cloud pt. 77–88 C (1% aq.); pH 6.0–7.5 (5% aq.); 100% act.

Iconol TDA-18-80%. [BASF] Trideceth-18; CAS 24938-91-8; nonionic; surfactant; APHA 70 max. liq.; water-sol.; m.w. 1000 (act.); sp.gr. 1.07; HLB 16.0; pour pt. 5 C; cloud pt. > 100 C (1% aq.); pH 6.0–7.5 (5% aq.); 19–21% water.

Iconol TDA-29-80%. [BASF] Trideceth-29; CAS 24938-91-8; nonionic; surfactant; APHA 70 max. liq.; water-sol.; m.w. 1475 (act.); sp.gr. 1.08; HLB 17.4; pour pt. 7 C; cloud pt. > 100 C (1% aq.); pH 6.0–7.5 (5% aq.); 19–21% water.

Iconol WA-1. [BASF] Alkoxylated phenolic; nonionic; surfactant; pigment dispersant for aq. systems; Gardner 6 clear liq.; sol. in water, alcohols, toluene; sp.gr. 1.02; visc. 300 cps; HLB 13; cloud pt. 53–58 C (1% aq.); pH 6.0–7.5 (5% aq.); 90% act.

Iconol WA-4. [BASF] Alkoxylated phenolic; nonionic; surfactant; dispersant; Gardner 6 clear liq.; sol. in water, alcohols, toluene; sp.gr. 1.02; visc. 300 cps; HLB 13; cloud pt. 53–58 C (1% aq.); pH 6.0–7.5 (5% aq.); 90% act.

Idet 5L. [Swastik] Sodium salt of alkylaryl sulfonate; anionic; wetting agent, detergent; clear yel. liq.; pH 6.8 (1%); 17% conc.

Idet 5LP. [Swastik] Alkylaryl sulfonate; anionic; dispersant and wetting agent used in paints and emulsion latices; liq.; 25% conc.

Idet 5L SP NF. [Swastik] Specially formulated alkylaryl sulfonate; anionic; detergent and wetting agent for wet processes in textile industry, dispersing and penetrating agent; yel. clear visc. liq.; pH 6.8 (1%); 25% conc.

Idet 10, 20–P. [Swastik] Alkylaryl sulfonate sodium salt; anionic; wetting agent in textile wet processing, in formulations of water disp. powd., detergent; spray–dried powd.; pH 7.0–9.0 (1% aq.); 37%, 76% act.

Igepal® 131 (formerly Rhodiasurf 131). [Rhone-Poulenc France] beta-Naphthol + 8-9 EO; nonionic; surfactant.

Igepal® 132. [Rhone-Poulenc France] beta-Naphthol + 6-7 EO; nonionic; surfactant.

Igepal® CA-210. [Rhone-Poulenc Surf.; Rhone-Poulenc France] Octoxynol-1.5; CAS 9002-93-1; nonionic; emulsifier for solv. cleaners, drycleaning, pesticides, floor polish; defoamer; FDA, EPA compliance; yel. liq.; aromatic odor; sol. in wh. min. oil, kerosene, Stod., xylene, butyl Cellosolve, perchloroethylene, ethanol; sp.gr. 0.99; visc. 500–800 cps; HLB 4.8; flash pt. > 200 F (PMCC); pour pt. 23 ± 2 F; 100% act.

Igepal® CA-420. [Rhone-Poulenc Surf.] Octoxynol-3; CAS 9002-93-1; nonionic; detergent, emulsifier, dispersant for solvent emulsion cleaners, drycleaning detergents; FDA, EPA compliance; pale yel. liq.; aromatic odor; sol. in kerosene, Stod., xylene, butyl Cellosolve, perchloroethylene, ethanol; sp.gr. 1.02; visc. 310–400 cps; HLB 8.0; flash pt. > 200 F (PMCC); pour pt. –10 ± 2 F; 100% act.

Igepal® CA-520. [Rhone-Poulenc Surf.] Octoxynol-5; CAS 9002-93-1; nonionic; emulsifier, dispersant, detergent, wetting agent; for wax polishes, cleaners, agric., emulsion cleaner formulations; FDA, EPA compliance; pale yel. liq.; aromatic odor; sol. in Stod., xylene, butyl Cellosolve, perchloroethylene, ethanol; disp. in water; sp.gr. 1.04; visc. 250–280 cps; HLB 10.0; flash pt. > 200 F (PMCC); pour pt. –15 ± 2 F; surf. tens. 30 dynes/cm (0.01%); 100% act.

Igepal® CA-620. [Rhone-Poulenc Surf.] Octoxynol-7; CAS 9002-93-1; nonionic; detergent, wetter, emulsifier for household and industrial detergents, textile, paper, metal and acid cleaning compds., fine fabric detergents, agric. formulations; FDA, EPA compliance; pale yel. liq.; aromatic odor; sol. in xylene, butyl Cellosolve, perchloroethylene, ethanol, water; sp.gr. 1.05; visc. 240–260 cps; HLB 12.0; cloud pt. 70–75 F (1%); flash pt. > 200 F (PMCC); pour pt. 15 ± 2 F; surf. tens. 30 dynes/cm (0.01%); 100% act.

Igepal® CA-630. [Rhone-Poulenc Surf.; Rhone-Poulenc France] Octoxynol-9; CAS 9002-93-1; nonionic; detergent, wetting agent, emulsifier for metal processing, emulsion cleaners, agric. formulations; as wetting agent with min. acids and corrosion inhibitors; FDA, EPA compliance; pale yel. liq.; aromatic odor; sol. in xylene, butyl Cellosolve, perchloroethylene, ethanol, water; sp.gr. 1.06; visc. 230–260 cps; HLB 13.0; cloud pt. 63-67 C (1% aq.); flash pt. > 200 F (PMCC); pour pt. 45 ± 2 F; surf. tens. 31 dynes/cm (0.01%); 100% act.

Igepal® CA-720. [Rhone-Poulenc Surf.; Rhone-Poulenc France] Octoxynol-12.5; CAS 9002-93-1; nonionic; detergent, wetter, emulsifier for industrial cleaning, metal pickling, electrolytic cleaning, agric. formulations; stable to strong acids and many alkalies; FDA, EPA compliance; opaque liq.; aromatic odor; sol. in xylene, butyl Cellosolve, perchloroethylene, ethanol, water; sp.gr. 1.07; visc. 70–76 cps (50 C); HLB 14.6; cloud pt. 86-90 C (1% aq.); flash pt. > 200 F (PMCC); pour pt. 69 ± 2 F; surf. tens. 32 dynes/cm (0.01%); 100% act.

Igepal® CA-730. [Rhone-Poulenc Surf.] Octoxynol-15; CAS 9002-93-1; nonionic; emulsifier for agric. emulsifiable concs. DISCONTINUED.

Igepal® CA-877. [Rhone-Poulenc Surf.] Nonoxynol-25; CAS 9016-45-9; nonionic; surfactant; 70% act.

Igepal® CA-880. [Rhone-Poulenc Surf.] Octoxynol-30; CAS 9002-93-1; nonionic; emulsifier for fats and oils, vinyl acetate and acrylate polymerization; post-stabilizer for syn. latices; dyeing assistant; FDA compliance; solid; HLB 17.4; surf. tens. 38 dynes/cm (at CMC); 100% conc.

Igepal® CA-887. [Rhone-Poulenc Surf.] Octoxynol-30 aq. sol'n.; CAS 9002-93-1; nonionic; emulsifier for emulsion polymerization, stabilizer for plastics; dyeing assistant; FDA compliance; pale yel. liq.; aromatic odor; sp.gr. 1.10; HLB 17.4; cloud pt. > 212 F (1% aq.); flash pt. > 200 F (PMCC); pour pt. 36 ± 2 F; surf. tens. 39 dynes/cm (0.01%); 70% act.

Igepal® CA-890. [Rhone-Poulenc Surf.; Rhone-Poulenc France] Octoxynol-40; CAS 9002-93-1; nonionic; emulsifier for emulsion polymerization, stabilizer for plastics; dyeing assistant; FDA compliance; off-wh. wax; aromatic odor; sol. in butyl Cellosolve, ethanol, water; sp.gr. 1.08 (50 C); HLB 18.0; cloud pt. > 212 F (1% aq.); flash pt. > 200 F (PMCC); pour pt. 115 ± 2 F; surf. tens. 42 dynes/cm (0.01%); 100% act.

Igepal® CA-897. [Rhone-Poulenc Surf.; Rhone-Poulenc France] Octoxynol-40; CAS 9002-93-1; nonionic; emulsifier, stabilizer for vinyl acetate and acrylate polymerization; dyeing assistant; emulsifier for fats and waxes; FDA compliance; pale yel. liq.; aromatic odor; sp.gr. 1.10; HLB 18.0; cloud pt. > 212 F (1% aq.); flash pt. > 200 F (PMCC); pour pt. 25 ± 2 F; surf. tens. 48 dynes/cm (0.01%); 70% act.

Igepal® Cephene Distilled (formerly Cephene Distilled). [Rhone-Poulenc France] Phenoxyethanol; nonionic; surfactant.

Igepal® CO-210. [Rhone-Poulenc Surf.; Rhone-Poulenc France] Nonoxynol-2 (1.5 EO); CAS 9016-45-9; nonionic; foam and emulsion stabilizer, detergent, coemulsifier, intermediate, defoamer, dispersant; for metalworking; biodeg.; FDA compliance; yel. liq., aromatic odor; sol. in wh. min. oil, kerosene, Stod., naphtha, xylene, butyl Cellosolve, perchloroethylene, ethanol; sp.gr. 0.99; visc. 440-540 cps; solid. pt. 8 F; HLB 4.6; flash pt. > 200 F (PMCC); pour pt. 13 ± 2 F; toxicology: minimal eye irritation; LD50 (oral, rat) 3.55 g/kg; 100% act.

Igepal® CO-430. [Rhone-Poulenc Surf.; Rhone-Poulenc France] Nonoxynol-4; CAS 9016-45-9; nonionic; coemulsifier, plasticizer, stabilizer, antistat, detergent, dispersant; for plastics, latex emulsions, petrol. oils, agric.; intermediate for anionic surfactants; biodeg.; FDA, EPA compliance; pale yel. liq., aromatic odor; sol. in kerosene, Stod., naphtha, xylene, butyl Cellosolve, perchloroethylene, ethanol; sp.gr. 1.02; visc. 160-260 cps; solid. pt. -21 F; HLB 8.8; flash pt. > 200 F (PMCC); pour pt. –16 ± 2 F; toxicology: severe eye irritant; LD50 (oral, rat) 7.4 g/kg; 100% act.

Igepal® CO-520. [Rhone-Poulenc Surf.; Rhone-Poulenc France] Nonoxynol-5; CAS 9016-45-9; nonionic; intermediate, emulsifier, coupling agent, dispersant; rust inhibitor in jet aircraft and automotive fuels, metalworking fluids, agric. formulations; biodeg.; FDA, EPA compliance; pale yel. liq., aromatic odor; sol. in Stod., naphtha, xylene, butyl Cellosolve, perchloroethylene, ethanol; sp.gr. 1.03; visc. 170-270 cps; solid. pt. -22 F; HLB 10.0; flash pt. > 200 F (PMCC); pour pt. –24 ± 2 F; surf. tens. 30 dynes/cm (0.01%); Ross-Miles foam 8 mm (0.1% aq., initial); 100% act.

Igepal® CO-530. [Rhone-Poulenc Surf.; Rhone-Poulenc France] Nonoxynol-6; CAS 9016-45-9; nonionic; emulsifier for silicones, detergent, dispersant for agric., petrol., paper, plastics, metalworking industries; de-icing fluid for jet aircraft fuels and automotive gasoline; biodeg.; FDA, EPA compliance; pale yel. liq., aromatic odor; sol. in Stod., naphtha, xylene, butyl Cellosolve, perchloroethylene, ethanol; sp.gr. 1.04; visc. 230–300 cps; HLB 10.8; flash pt. > 200 F (PMCC); pour pt. –26 ± 2 F; surf. tens. 28 dynes/cm (0.01%); Ross-Miles foam 15 mm (0.1% aq., initial); toxicology: severe eye irritant; LD50 (oral, rat) 1.98 g/kg; 100% act.

Igepal® CO-610. [Rhone-Poulenc Surf.; Rhone-Poulenc France] Nonoxynol-8 (7–8 EO); CAS 9016-45-9; nonionic; low foaming detergent, wetting agent, emulsifier, lubricant; for metalworking; biodeg.; FDA compliance; pale yel. liq., aromatic odor; sol. in naphtha, xylene, butyl Cellosolve, perchloroethylene, ethanol, water; sp.gr. 1.05; visc. 230–290 cps; HLB 12.2; cloud pt. 72–82 F (1%); flash pt. > 200 F (PMCC); pour pt. 37 ± 2 F; surf. tens. 30 dynes/cm (0.01%); Ross-Miles foam 55 mm (0.1% aq., initial); 100% act.

Igepal® CO-620. [Rhone-Poulenc Surf.] Nonoxynol-8 (8.5 EO); CAS 9016-45-9; nonionic; detergent, wetting agent, emulsifier; surfactant used in heavy-duty liq. detergent compds.; biodeg.; FDA compliance; pale yel. liq., aromatic odor; sol. in water, xylene, dibutyl phthalate, perchloroethylene, ethanol, ethylene glycol; sp.gr. 1.05; visc. 180-280 cps; solid. pt. 31 F; HLB 12.6; pour pt. 36 F; cloud pt. 104-115 F (1% aq.); flash pt. (PMCC) > 200 F; surf. tens. 32 dynes/cm (0.01%); Ross-Miles foam 69 mm (0.1% aq., initial); 100% conc. DISCONTINUED.

Igepal® CO-630. [Rhone-Poulenc Surf.; Rhone-Poulenc France] Nonoxynol-9; CAS 9016-45-9; nonionic; detergent, wetting and rewetting agent, corrosion inhibitor, penetrant, emulsifier, dispersant for textile, paper, leather, household/industrial cleaners, agric., paints, metal processing, emulsion cleaning; biodeg.; FDA, EPA compliance; almost colorless liq., aromatic odor; sol. in naphtha, xylene, butyl Cellosolve, perchloroethylene, ethanol, water; sp.gr. 1.06; visc. 225–300 cps; HLB 13.0; cloud pt. 126–133 F (1%); flash pt. > 200 F (PMCC); pour pt. 31 ± 2 F; surf. tens. 31 dynes/cm (0.01%); Ross-Miles foam 80 mm (0.1% aq., initial); toxicology: severe eye irritant; LD50 (oral, rat) 3 g/kg; 100% act.

Igepal® CO-660. [Rhone-Poulenc Surf.; Rhone-Poulenc France] Nonoxynol-10; CAS 9016-45-9; nonionic; detergent, wetting and rewetting agent, corrosion inhibitor, penetrant, emulsifier for textile, paper, leather, household/industrial cleaners, agric., paints, metal processing, emulsion cleaning; biodeg.; FDA, EPA compliance; pale yel. liq., aromatic odor; sol. in naphtha, xylene, butyl Cellosolve, perchloroethylene, ethanol, water; sp.gr. 1.06; visc. 225–275 cps; HLB 13.2; cloud pt. 140–149 F (1%); flash pt. > 200 F (PMCC); pour pt. 46 ± 2 F; surf. tens. 31 dynes/cm (0.01%); Ross-Miles foam 82 mm (0.1% aq., initial); 100% act.

Igepal® CO-710. [Rhone-Poulenc Surf.; Rhone-Poulenc France] Nonoxynol-10.5; CAS 9016-45-9; nonionic; detergent, wetting and rewetting agent, corrosion inhibitor, penetrant, emulsifier for textile, paper, leather, household/industrial cleaners, agric., paints, metal processing, emulsion cleaning; biodeg.; FDA, EPA compliance; pale yel. liq., aromatic odor; sol. in naphtha, xylene, butyl Cellosolve, perchloroethylene, ethanol, water; sp.gr. 1.06; visc. 240–300 cps; HLB 13.6; cloud pt. 158–165 F (1%); flash pt. > 200 F (PMCC); pour pt. 49 ± 2 F; surf. tens. 32 dynes/cm (0.01%); Ross-Miles foam 110 mm (0.1% aq., initial); 100% act.

Igepal® CO-720. [Rhone-Poulenc Surf.; Rhone-Poulenc France] Nonoxynol-12; CAS 9016-45-9; nonionic; detergent, wetting and rewetting agent, corrosion inhibitor, penetrant, emulsifier, dispersant for textile, paper, leather, household/industrial cleaners, agric., paints, metal processing, emulsion cleaning; biodeg.; FDA, EPA compliance; opaque liq., aromatic odor; sol. in naphtha, xylene, butyl Cellosolve, perchloroethylene, ethanol, water; sp.gr. 1.06; visc. 260–340 cps; HLB 14.2; flash pt. > 200 F (PMCC); pour pt. 62 ± 2 F; surf. tens. 34 dynes/cm (0.01%); Ross-Miles foam 110 mm (0.1% aq., initial); toxicology: severe eye irritant; LD50 (oral, rat) > 1.5 g/kg; 100% act.

Igepal® CO-730. [Rhone-Poulenc Surf.; Rhone-Poulenc France] Nonoxynol-15; CAS 9016-45-9; nonionic; detergent, dispersant, wetting agent, penetrant, emulsifier for industrial cleaners, metalworking fluids, agric. formulations; effective at high temps. and in high electrolyte media; biodeg.; FDA, EPA compliance; yel. liq., aromatic odor; sol. in naphtha, xylene, butyl Cellosolve, perchloroethylene, ethanol, water; sp.gr. 1.07; visc. 450–550 cps; HLB 15.0; cloud pt. 203–212 F (1%); flash pt. > 200 F (PMCC); pour pt. 71 ± 2 F; surf. tens. 36 dynes/cm (0.01%); Ross-Miles foam 130 mm (0.1% aq., initial); toxicology: moderate eye irritant; LD50 (oral, rat) 2.5 g/kg; 100% act.

Igepal® CO-738. [Rhone-Poulenc Surf.] Nonoxynol-15; CAS 9016-45-9; nonionic; detergent, dispersant, wetting agent, penetrant, emulsifier for industrial cleaners; liq.; HLB 15.0; 80% act. DISCONTINUED.

Igepal® CO-850. [Rhone-Poulenc Surf.; Rhone-Poulenc France] Nonoxynol-20; CAS 9016-45-9; nonionic; detergent, wetting agent, dispersant, emulsifier, for industrial cleaners, metalworking fluids, polyester resins; latex stabilizer; demulsifier for crude petrol. oil emulsions; glass mold release agent; FDA compliance; pale yel. wax; aromatic odor; sol. in naphtha, xylene, butyl Cellosolve, perchloroethylene, ethanol, water; sp.gr. 1.08 (50 C); HLB 16.0; cloud pt. > 212 F (1% aq.); flash pt. > 200 F (PMCC); pour pt. 91 ± 2 F; surf. tens. 39 dynes/cm (0.01%); Ross-Miles foam 120 mm (0.1% aq., initial); 100% act.

Igepal® CO-880. [Rhone-Poulenc Surf.; Rhone-Poulenc France] Nonoxynol-30; CAS 9016-45-9; nonionic; detergent, wetting agent, dispersant, emulsifier, for industrial cleaners, polyester resins, agric.; latex stabilizer; demulsifier for crude petrol. oil emulsions; also textile scouring, solubilizer; FDA, EPA compliance; pale yel. wax; aromatic odor; sol. in naphtha, xylene, butyl Cellosolve, perchloroethylene, ethanol, water; sp.gr. 1.08 (50 C); HLB 17.2; cloud pt. > 212 F (1% aq.); flash pt. > 200 F (PMCC); pour pt. 109 ± 2 F; surf. tens. 43 dynes/cm (0.01%); Ross-Miles foam 120 mm (0.1% aq., initial); toxicology: minimal eye irritant; LD50 (oral, rat) > 16 g/kg; 100% act.

Igepal® CO-887. [Rhone-Poulenc Surf.; Rhone-Poulenc France] Nonoxynol-30; CAS 9016-45-9; nonionic; detergent, wetting agent, dispersant, emulsifier, for industrial cleaners, polyester resins; latex stabilizer; demulsifier for crude petrol. oil emulsions; also textile scouring, solubilizer; FDA, EPA compliance; pale yel. liq.; aromatic odor; sp.gr. 1.09; solid. pt. 28 F; HLB 17.2; cloud pt. clear at 212 F (1% aq.); flash pt. > 200 F (PMCC); pour pt. 34 ± 2 F; 70% act.

Igepal® CO-890. [Rhone-Poulenc Surf.] Nonoxynol-40; CAS 9016-45-9; nonionic; emulsifier, stabilizer, wetting agent, dyeing assistant for plastics, agric., latexes, floor polishes, etc.; FDA, EPA compliance; off-wh. wax; aromatic odor; sol. in ethanol, ethylene dichloride, water; sp.gr. 1.09 (50 C); HLB 17.8; cloud pt. > 212 F (1% aq.); flash pt. > 200 F (PMCC); pour pt. 112 ± 2 F; Ross-Miles foam 115 mm (0.1% aq., initial); 100% act.

Igepal® CO-897. [Rhone-Poulenc Surf.; Rhone-Poulenc France] Nonoxynol-40; CAS 9016-45-9; nonionic; emulsifier, stabilizer, wetting agent, dyeing assistant for plastics, latexes, floor polishes, etc.; FDA, EPA compliance; pale yel. liq.; aromatic odor; sp.gr. 1.10; solid. pt. 40 F; HLB 17.8; flash pt. > 200 F (PMCC); pour pt. 46 ± 2 F; 70% act.

Igepal® CO-970. [Rhone-Poulenc Surf.; Rhone-Poulenc France] Nonoxynol-50; CAS 9016-45-9; nonionic; emulsifier, stabilizer, wetting agent, dyeing assistant for plastics, agric., latexes, floor polishes, etc.; FDA, EPA compliance; off-wh. wax; aromatic odor; sol. in ethanol, water, ethylene dichloride; sp.gr. 1.10 (50 C); HLB 18.2; flash pt. > 200 F (PMCC); pour pt. 114 ± 2 F; Ross-Miles foam 100 mm (0.1% aq., initial); 100% conc.

Igepal® CO-977. [Rhone-Poulenc Surf.; Rhone-Poulenc France] Nonoxynol-50; CAS 9016-45-9; nonionic; emulsifier, stabilizer, wetting agent, dyeing assistant for plastics, latexes, floor polishes, etc.; FDA, EPA compliance; pale yel. liq.; aromatic odor; sol. in ethanol, water; sp.gr. 1.0 (50 C); HLB 18.2; flash pt. > 200 F (PMCC); pour pt. 52 ± 2 F; surf. tens. 44 dynes/cm (@ CMC); 70% conc.

Igepal® CO-980. [Rhone-Poulenc Surf.] Nonoxynol-70; CAS 9016-45-9; nonionic; stabilizer and dyeing assistant; polymerization emulsifier for vinyl acetate and acrylic emulsions; solid; water-sol.; HLB 18.7; cloud pt. > 212 F (1% aq.); 100% conc.

Igepal® CO-987. [Rhone-Poulenc Surf.] Nonoxynol-70; CAS 9016-45-9; nonionic; stabilizer and dyeing assistant; polymerization emulsifier for vinyl acetate and acrylic emulsions; off-wh. paste, aromatic odor; water-sol.; sp.gr. 1.10 (50 C); solid. pt. 58 F; HLB 18.6; pour pt. 63 F; cloud pt. > 212 F (1% aq.); flash pt. (PMCC) > 200 F; 70% conc.

Igepal® CO-990. [Rhone-Poulenc Surf.; Rhone-Poulenc France] Nonoxynol-100; CAS 9016-45-9; nonionic, emulsifier, stabilizer, wetting agent, dyeing assistant for plastics, agric., latexes, floor polishes, etc.; off-wh. wax, flake; aromatic odor; sol. in ethanol, water; sp.gr. 1.12 (50 C); HLB 19.0; cloud pt. > 212 F (1% aq.); flash pt. > 200 F (PMCC); pour pt. 122 ± 2 F; surf. tens. 50 dynes/cm (@ CMC); 100% act.

Igepal® CO-997. [Rhone-Poulenc Surf.; Rhone-Poulenc France] Nonoxynol-100; CAS 9016-45-9; nonionic; emulsifier, stabilizer, wetting agent, dyeing assistant for plastics, latexes, floor polishes, etc.; EPA compliance; pale yel. paste; aromatic odor; sp.gr. 1.11; solid. pt. 65 F; HLB 19.0; cloud pt. > 212 F (1% aq.); flash pt. > 200 F (PMCC); pour pt. 68 ± 2 F; surf. tens. 50 dynes/cm (@ CMC); 70% act. in water.

Igepal® CTA-639W. [Rhone-Poulenc Surf.] Alkylphenol ethoxylated (9 EO); nonionic; detergent, wetting; emulsifier in rubber latex emulsion paints; surfactant; liq.; water-sol; surf. tens. 34 dynes/cm (@ CMC); 100% conc.

Igepal® DM-430. [Rhone-Poulenc Surf.; Rhone-Poulenc France] Nonyl nonoxynol-7; CAS 9014-93-1; nonionic; emulsifier for agric., emulsion polymerization, leather industries; EPA compliance; yel. slightly hazy liq.; sol. in naphtha, xylene, perchloroethylene, ethanol; dens. 8.3 lb/gal; sp.gr. 0.995; visc. 13 cps (100 C); HLB 9.4; cloud pt. 51-55 C (1% in propanol/NaCl); flash pt. 500–520 F (COC); pour pt. –7 C; 100% act.

Igepal® DM-530. [Rhone-Poulenc Surf.; Rhone-Poulenc France] Nonyl nonoxynol-9; CAS 9014-93-1; nonionic; emulsifier for cutting oils, agric., textile finishing, acid bowl cleaners; EPA compliance; yel. slightly hazy liq.; sol. in naphtha, xylene, perchloroethylene, ethanol; partly sol. water; dens. 8.43 lb/gal; sp.gr. 1.010; visc. 15 cps (100 C); HLB 10.6; cloud pt. 72-78 C (1% in propanol/NaCl); flash pt. 500–520 F (COC); pour pt. 0 C; surf. tens. 29 dynes/cm (0.01%); 100% act.

Igepal® DM-710. [Rhone-Poulenc Surf.; Rhone-Poulenc France] Nonyl noxoxynol-15; CAS 9014-93-1; nonionic; detergent, emulsifier, dispersant, wetting agent, antistat for industrial cleaners, textiles, leather, metal cleaners, paper, latex, pesticides; EPA compliance; pale yel. opaque liq.; aromatic odor; sol. in naphtha, xylene, perchloroethylene, ethanol, water; dens. 8.72 lb/gal; sp.gr. 1.045; visc. 19 cps (100 C); HLB 13.0; cloud pt. 48–52 C (1%); flash pt. 500–520 F (COC); pour pt. 18 C; surf. tens. 29 dynes/cm (0.01%); 100% act.

Igepal® DM-730. [Rhone-Poulenc Surf.] Nonyl nonoxynol-24; CAS 9014-93-1; nonionic; detergent, emulsifier for textiles, leather, metal cleaners, paper, latex, pesticides, emulsion polymerization, latex stabilization; pale yel. paste; aromatic odor; sol. in xylene, ethanol, water; dens. 8.75 lb/gal; sp.gr. 1.049; visc. 25 cps (100 C); HLB 15.1; cloud pt. > 100 C (1%); flash pt. 500–520 F (COC); pour pt. 25 C; surf. tens. 33 dynes/cm (0.01%); 100% act.

Igepal® DM-880. [Rhone-Poulenc Surf.] Nonyl nonoxynol-49; CAS 9014-93-1; nonionic; emulsifier, solubilizer for emulsion polymerization, latex stabilization, pesticides, essential oils, cleaners; pale yel. wax; aromatic odor; sol. in xylene, ethanol, water; sp.gr. 1.050 (50 C); visc. 55 cps (100 C); HLB 17.2; cloud pt. > 100 C (1%); flash pt. 500–520 F (COC); pour pt. 47 C; surf. tens. 38 dynes/cm (0.01%); 100% act.

Igepal® DM-970 FLK. [Rhone-Poulenc Surf.; Rhone-Poulenc France] Nonyl nonoxynol-150; CAS 9014-93-1; nonionic; detergent, dispersant, wetter, stabilizer for laundry, household, textile, hard-surface detergents, cosmetics, insecticides, paper, petrol., paints; EPA compliance; wh. flakes; sol. in water, aromatic solv., methanol, ethanol; sp.gr. 1.05 (80 C); HLB 19; cloud pt. > 100 C (1%); 100% act.

Igepal® DX-430. [Rhone-Poulenc France] Dialkylphenol ethoxylate; nonionic; emulsifier for herbicide or insecticide emulsifiable concs., emulsion polymerization, fatliquoring of leather; liq.; oil-sol.; HLB 9.4; 99% conc.

Igepal® ES (formerly Rhodiasurf ES). [Rhone-Poulenc France] Nonyl phenol ethoxylate; nonionic; surfactant.

Igepal® F-85 (formerly Texafor F-85). [Rhone-Poulenc France] Octoxynol-9; CAS 9002-93-1; nonionic; surfactant.

Igepal® F 707. [Rhone-Poulenc France] Octoxynol-70; CAS 9002-93-1; nonionic; pigment dispersant for coatings; coemulsifier for emulsion polymerization and other industrial applics.; liq.; HLB 18.7; 70% conc. DISCONTINUED.

Igepal® F-920 (formerly Texafor F-920). [Rhone-Poulenc France] Octoxynol-92; CAS 9002-93-1; nonionic; surfactant.

Igepal® FN-9.5 (formerly Texafor FN-9.5). [Rhone-Poulenc France] Nonoxynol-9; CAS 9016-45-9;

nonionic; surfactant.

Igepal® FN-11 (formerly Texafor FN-11). [Rhone-Poulenc France] Nonoxynol-11; CAS 9016-45-9; nonionic; surfactant.

Igepal® FN-30 (formerly Texafor FN-30). [Rhone-Poulenc France] Nonoxynol-30; CAS 9016-45-9; nonionic; surfactant.

Igepal® KA (formerly Rhodiasurf KA). [Rhone-Poulenc France] Octoxynol-9; CAS 9002-93-1; nonionic; detergent, wetting agent, emulsifier, dispersant; for textiles, cosmetics, institutional cleaners, herbicide/insecticide emulsifiable concs., metal degreasers; liq.; HLB 13.1; 99% conc.

Igepal® LAVE. [Rhone-Poulenc France] Phenol ethoxylate; nonionic; solv. for vinyl, phenolic, polyester, nitrocellulose, cellulose acetate, and other resins; ingred. for metal cleaners, paint strippers, cleaning compds.; vehicle for inks; liq.; 99% conc.

Igepal® NP-2 (formerly Rhodiasurf NP-2). [Rhone-Poulenc France] Nonylphenol ethoxylate; nonionic; antifoaming co-surfactant in industrial cleaners, metalworking compds., textile processing; liq.; HLB 5.2; 99% conc.

Igepal® NP-4 (formerly Rhodiasurf NP-4). [Rhone-Poulenc France] Nonylphenol ethoxylate; nonionic; lipophilic coemulsifier for min. oil and solv. emulsions; intermediate for ethoxy sulfates; liq.; HLB 8.8; 99% conc.

Igepal® NP-5 (formerly Rhodiasurf NP-5). [Rhone-Poulenc France] Nonylphenol ethoxylate; nonionic; emulsifier/coemulsifier for institutional cleaners, textile processing compds.; liq.; HLB 10.0; 99% conc.

Igepal® NP-6 (formerly Rhodiasurf NP-6). [Rhone-Poulenc France] Nonylphenol ethoxylate; nonionic; emulsifier, solubilizer for sol. oils used in metalworking and solv. cleaners; liq.; HLB 10.9; 99% conc.

Igepal® NP-8 (formerly Rhodiasurf NP-8). [Rhone-Poulenc France] Nonylphenol ethoxylate; nonionic; emulsifier, wetting agent, detergent for household and industrial cleaners, textile processing compds., insecticides; liq.; HLB 12.2; 99% conc.

Igepal® NP-9 (formerly Rhodiasurf NP-9). [Rhone-Poulenc France] Nonylphenol ethoxylate; nonionic; emulsifier, wetting agent, dispersant for textile scouring, dyeing and finishing, household/industrial cleaners, alkaline degreasers, acid pickling in metalworking; emulsifier/stabilizer for emulsion polymerization; pitch dispersant in pulp prod.; liq.; HLB 13.0; 99% conc.

Igepal® NP-10 (formerly Rhodiasurf NP-10). [Rhone-Poulenc France] Nonylphenol ethoxylate; nonionic; detergent, emulsifier, wetting agent, dispersant for textile scouring, dyeing and finishing, household/industrial cleaners, alkaline degreasers, acid pickling in metalworking; emulsifier/stabilizer for emulsion polymerization; pitch dispersant in pulp prod.; liq.; HLB 13.0; 99% conc.

Igepal® NP-12 (formerly Rhodiasurf NP-12). [Rhone-Poulenc France] Nonylphenol ethoxylate; nonionic; detergent, emulsifier, wetting agent for institutional cleaners, metalworking compds., textile processing; paste; HLB 16.0; 99% conc.

Igepal® NP-14 (formerly Rhodiasurf NP-14). [Rhone-Poulenc France] Nonylphenol ethoxylate; nonionic; wetting agent, detergent for industrial cleaners to be used at higher temps.; paste; HLB 14.9; 99% conc.

Igepal® NP-17 (formerly Rhodiasurf NP-17). [Rhone-Poulenc France] Nonylphenol ethoxylate; nonionic; emulsifier, dispersant, detergent for industrial cleaners for high temp. applic., pigment dispersions; paste; HLB 15.5; 99% conc.

Igepal® NP-20 (formerly Rhodiasurf NP-20). [Rhone-Poulenc France] Nonylphenol ethoxylate; nonionic; dispersant, solubilizer, emulsifier, detergent for industrial alkaline built cleaners, textile dyeing and printing, metal cleaners; stabilizer for latex and pigment dispersions, emulsions of resins, polyesters, waxes, nitrocellulose; solid; HLB 16.0; 99% conc.

Igepal® NP-30 (formerly Rhodiasurf NP-30). [Rhone-Poulenc France] Nonylphenol ethoxylate; nonionic; coemulsifier, dispersant for alkaline industrial cleaners, resins, waxes; vinyl/acrylic latex stabilizer; solid; HLB 17.1; 99% conc.

Igepal® O (formerly Rhodiasurf O). [Rhone-Poulenc France] Octoxynol-10; CAS 9002-93-1; nonionic; detergent, emulsifier, wetting agent, dispersant for textile processing, industrial cleaners, metalworking compds., rubber and plastics mfg.; liq.; HLB 13.6; 99% conc.

Igepal® OD-410. [Rhone-Poulenc Surf.; Rhone-Poulenc France] Ethoxylated alkylphenol (1.0 EO); nonionic; solv. for vinyl, phenolic, polyester, alkyd, NC, and cellulose acetate resins; ingred. of metal cleaners, paint strippers, and cleaning compds. where solvs. and solv. boosters are required; used as ink vehicle; liq.; 100% act.

Igepal® RC-520. [Rhone-Poulenc Surf.; Rhone-Poulenc France] Dodoxynol-6; nonionic; rewetting agent, emulsifier for industrial cleaners, paper, textiles, agric., aerosols, drycleaning soaps, etc.; FDA, EPA compliance; VCS 2 max., slightly visc. liq.; sol. in most hydrocarbon and polar solv.; sp.gr. 1.03; visc. 360 cps (Ostwald-Fenske); HLB 10.0; pour pt. –10 C; 100% act.

Igepal® RC-620. [Rhone-Poulenc Surf.; Rhone-Poulenc France] Dodoxynol-10; nonionic; detergent, wetting agent, emulsifier, dispersant for agric., household/industrial cleaners; dedusting agent; FDA, EPA compliance; APHA 200 max. liq.; HLB 12.6; cloud pt. 38–42 C (1% aq.); 100% act.

Igepal® RC-630. [Rhone-Poulenc Surf.; Rhone-Poulenc France] Dodoxynol-10; nonionic; detergent, wetting agent, emulsifier, dispersant for household/industrial cleaners, agric. formulations; FDA, EPA compliance; APHA 200 max. liq.; HLB 12.7; cloud pt. 62–66 C (1% aq.); 100% act.

Igepal® SS-837 (formerly Diazopon® SS-837). [Rhone-Poulenc Surf.; Rhone-Poulenc France] Ethoxylated alkylphenol; nonionic; surfactant for elevated temp. applics.; anticrock and soaping agent for naphthol dyes; dyeing assistant; solubilizer and stabilizer for fast color salts; accelerates diazotization of fast color bases; liq.; 70% conc.

Igepon® AC-78 (redesignated Geropon® AC-78). [Rhone-Poulenc Surf.; Rhone-Poulenc France]

Igepon® T-22A. [Rhone-Poulenc Surf.] Sodium oleoyl taurate; anionic; surfactant, dispersant for agric. formulations; EPA compliance; liq. DISCONTINUED.

Igepon® T-33 (redesignated Geropon® T-33). [Rhone-Poulenc Surf.]

Igepon® T-43 (redesignated Geropon® T-43). [Rhone-Poulenc Surf.]

Igepon® T-51 (redesignated Geropon® T-51). [Rhone-Poulenc Surf.]

Igepon® T-77 (redesignated Geropon® T-77). [Rhone-Poulenc Surf.]

Igepon® TK-32 (redesignated Geropon® TK-32). [Rhone-Poulenc Surf.]

Imacol JN. [Sandoz Prods. Ltd.] Fatty acid ester; nonionic; foam reducing lubricant for preventing creases and abrasions in wet finishing treatments of cellulosic fibers and blended fabrics; liq.

Imacol S Liq. [Sandoz Prods. Ltd.] Polyglycol ether deriv.; cationic; low foaming lubricant for preventing creases and abrasions in wet finishing treatments of syn. fibers and blended fabrics; liq.

Imbirol OT/Na 70. [Auschem SpA] Sodium di-2-ethylhexyl sulfosuccinate; CAS 577-11-7; anionic; detergent, wetting agent; liq.; 70% conc.

Imerol NCP, NCP Liq. [Sandoz Prods. Ltd.] Sulfonate containing solvs.; wetting agent, emulsifier, dispersant used for print paste dyes; sp.gr. 1.09; 57% act.

Imerol SS Liq. [Sandoz Prods. Ltd.] Detergent, used for solv. scouring in textile industry; yel. liq.; misc. with water; sp.gr. 0.874; 88% act.

Imerol VLF Liq. [Sandoz Prods. Ltd.] Detergent, used for solv. scouring in textile industry; yel. liq.; disp. in water @ 70–100 F; sp.gr. 0.94; 100% act.

Imerol XN. [Sandoz Prods. Ltd.] Modified polyglycol ether; anionic/nonionic; odorless scouring and stain removing agent; liq.; sol. in solv.

Imperon Binder Brands. [Hoechst AG] Self-cross-linking acrylate-based copolymer disp.; for pigment dyeing method with fastness to dry cleaning; dispersion.

Imperon Emulsifier 774. [Hoechst AG] Fatty acid condensate; nonionic; visc. moderator for o/w emulsions in printing inks; liq.

Imperon Softener D. [Hoechst AG] Dioctyl phthalate; softener for pigment printing.

Imperon Thickener A. [Hoechst AG] Acrylate polymer; nonionic; thickener for pigment printing with little or no white spirit; emulsion.

Imwitor® 191. [Hüls Am.; Hüls AG] Glyceryl stearate; CAS 31566-31-1; 68308-54-3; nonionic; coemulsifier, dispersant for personal care prods.; emulsifier in o/w and w/o emulsions; lubricants and binders used in the pharmaceutical industry; suspending agent, stabilizer, thickener; food emulsifier; Gardner 4 max. powd.; sol. in oils, molten fats; HLB 4.0; m.p. 56–61 C; solid. pt. 63–68 C; sapon. no. 155–170; 90% 1-monoglycerides.

Imwitor® 312. [Hüls Am.] Glyceryl laurate; nonionic; coemulsifier for o/w emulsions; solubilizer, carrier for lipophilic drugs; wh. cryst. mass; HLB 4.0; 90% conc.

Imwitor® 370. [Hüls Am.; Hüls AG] Glyceryl stearate citrate; CAS 91744-38-6; nonionic; food emulsifier; o/w emulsifier for very polar oils, fats and liq. wax esters; solid; HLB 13.0; 100% conc.

Imwitor® 375. [Hüls Am.; Hüls AG] Glyceryl citrate/lactate/linoleate/oleate.; nonionic; o/w emulsifier for very polar oils and fats; for oil baths, cream baths; solid; HLB 11.0; 100% conc.

Imwitor® 595. [Hüls Am.] Monoglyceride, molecular dist.; nonionic; stabilizer and plasticizer for emulsions; dispersant; food emulsifier; powd.; HLB 4.4; 100% conc.

Imwitor® 742. [Hüls Am.; Hüls AG] Caprylic/capric glycerides; CAS 26402-26-6; nonionic; plasticizer for hard fats, solv. for lipophilic ingreds.; emollient; coemulsifier, solubilizer, carrier for lipophilic drugs; paste; HLB 4.0; 100% conc.

Imwitor® 780. [Hüls Am.] Glyceryl isostearate; nonionic; w/o emulsifier; liq.; HLB 3.7; 100% conc.

Imwitor® 780 K. [Hüls Am.; Hüls AG] Isostearyl diglyceryl succinate; CAS 66085-00-5; nonionic; w/o emulsifier for polar oils and fats, cosmetic and pharmaceutical preparations; yel., med. visc. liq.; slight char. odor; sol. in chloroform, benzene, 96% alcohol, oils; sp.gr. 0.96–0.98; visc. 700–900 cps; HLB 3.7; sapon. no. 240–260; 100% conc.

Imwitor® 900. [Hüls Am.; Hüls AG] Glyceryl stearate; nonionic; coemulsifier, dispersant for personal care prods.; emulsifier in o/w and w/o emulsions; lubricants and binders used in the pharmaceutical industry; suspending agent, stabilizer, thickener; Gardner 4 max. powd.; sol. in fats, oils, waxes; HLB 3.0; m.p. 56–61 C; solid. pt. 56–61 C; sapon. no. 162–173; 100% conc., 40–50% 1-monoglyceride.

Imwitor® 908. [Hüls Am.] Glyceryl caprylate; nonionic; coemulsifier for o/w emulsions, lipophilic materials; ylsh. oily liq.; HLB 3.0; 45% conc.

Imwitor® 910. [Hüls Am.] Glyceryl caprate; nonionic; food emulsifier; coemulsifier for o/w emulsions, lipophilic materials; powd.; HLB 3.0; 100% conc.

Imwitor® 914. [Hüls Am.] Glyceryl myristate; nonionic; coemulsifier for o/w emulsions; solubilizer, carrier for lipophilic drugs; ylsh. cryst. mass; 45% conc.

Imwitor® 940. [Hüls Am.] Palm oil glycerides; nonionic; stabilizer for cosmetics; dispersant for clay systems; sec. emulsifier and stabilizer for asphalt emulsions; flakes, powd.; sol. in fats, oils, waxes; HLB 3.0; 100% conc.

Imwitor® 940 K. [Hüls Am.] Glyceryl stearate/palmitate Ph. Eur.; nonionic; coemulsifier, dispersant for personal care prods.; emulsifier in o/w and w/o emulsions; lubricants and binders used in the pharmaceutical industry; suspending agent, stabilizer, thickener; food emulsifier; Gardner 4 max. powd.; m.p. 53–57 C; solid. pt. 54–60; sapon. no. 165–178; 42–48% 1-monoglyceride.

Incrocas 30. [Croda Inc.] PEG-30 castor oil; nonionic; emulsifier, solubilizer, emollient, superfatting agent, lubricant for personal care prods., detergents, metalworking fluids, insecticides, herbicides, household prods.; lubricant, antistat, softener, dye leveling agent for textiles; also lime soap dispersant in alkaline scouring systems; pale yel. liq.; sol. in water, ethanol, oleyl alcohol, naphtha, MEK, oleic acid, trichloroethylene; HLB 11.7; cloud pt. 35–40 C (1% in 10% brine); sapon. no. 72–82; pH 6.0–7.5 (3% aq.); surf. tens. 41.5 dynes/cm (0.1% DW); usage level: 0.5-5%; 100% act.

Incrocas 40. [Croda Inc.] PEG-40 castor oil; nonionic; emulsifier, solubilizer, emollient, superfatting agent, lubricant for personal care prods., detergents, metalworking fluids, insecticides, herbicides, household prods.; lubricant, antistat, softener, dye leveling agent for textiles; pale yel. liq.; sol. @ 10% see Incrocas 30; HLB 13.0; cloud pt. 50 C (1% in 10% brine); sapon. no. 60–65; pH 6.0–7.5 (3% aq.); surf. tens. 40.90 dynes/cm (0.1% DW); usage level: 0.5-5%; 100% act.

Incrodet TD7-C. [Croda Inc.] Trideceth-7 carboxylic acid; anionic; mild surfactant for shampoos, bath gels, cleansers; neutralizer for trace caustic or thioglycolate residues present after hair permanents or relaxers; in neutralizing shampoos which follow ethnic hair straighteners; lime soap dispersant; emulsifier; stable at low and high pH; pale yel. clear liq.; sol. in water/alcohol, ethanol, propylene glycol;

disp. in water, glycerin; acid no. 45-52; cloud pt. 67-80 C; usage level: 1-12%; 90% act.

Incromate CDL. [Croda Inc.] Cocamidopropyl dimethylamine lactate; cationic; surfactant, foamer, conditioner for personal care prods., conditioners; base for cationic emulsions; yel. visc. liq.; sol. in water; pH 6–7 (5% sol'n.); usage level: 0.5-5%; 95–97% solids.

Incromate SDL. [Croda Inc.] Stearamidopropyl dimethylamine lactate; cationic; visc. builder, opacifier, softener for hair shampoos, conditioners, fabrics; emulsifier for hand creams, cleansers, lotions; raw material; Gardner 3 max. slurry; pH 4–5 (10%); usage level: 1-10%; 24–26% conc.

Incromide BAD. [Croda Inc.] Babassuamide DEA (1:1); nonionic; visc. builder, foam stabilizer, emulsifier; for shampoos, bubble baths, soaps, bath prods.; clear liq.; usage level: 1-10%; 100% conc.

Incromide CA. [Croda Inc.] Cocamide DEA; nonionic; surfactant, foam stabilizer, emulsifier and thickener used in household, cosmetic, and industrial formulations; clear liq.; usage level: 1-10%; 100% act.

Incromide CM. [Croda Inc.] Cocamide MEA; CAS 68140-00-1; nonionic; foam stabilizer, thickener, opacifier for cosmetic, industrial, household formulations; flake; 100% conc.

Incromide L-90. [Croda Inc.] Lauramide DEA; CAS 120-40-1; nonionic; foam stabilizer, thickener, detergent, and foaming agent in household, industrial and institutional cleaning comps., car washes, rug and upholstery cleaners, and personal care prods.; Gardner 2 max. solid; mild fatty odor; sol. in alcohol, chlorinated and aromatic hydrocarbons; disp. in water; usage level: 1-10%; 100% act.

Incromide LA. [Croda Inc.] Linoleamide DEA; nonionic; superfatting agent and thickener for personal care and household prods.; useful in increasing visc. of various sulfate and ether sulfate dilutions; conditioner and lubricant; Gardner 5 max. clear liq.; bland odor; usage level: 1-10%; 100% act.

Incromide LCL. [Croda Inc.] Lauramide MEA; CAS 142-78-9; surfactant; yel. liq.

Incromide LM-70. [Croda Inc.] Lauramide DEA; CAS 120-40-1; nonionic; foam stabilizer, thickener, detergent, and foaming agent in household, industrial and institutional cleaning comps., car washes, rug and upholstery cleaners, and personal care prods.; Gardner 2 max. solid; usage level: 1-10%; 100% act.

Incromide LR. [Croda Inc.] Lauramide DEA; CAS 120-40-1; nonionic; thickener, foam stabilizer used in personal care prods., dishwash detergents and industrial cleaners; Gardner 5 max. liq.; bland odor; usage level: 1-10%; 100% act.

Incromide Mink D. [Croda Inc.] Minkamide DEA; nonionic; visc. builder, foam stabilizer for shampoos, bubble baths, liq. soaps, bath prods.; clear liq.; usage level: 1-10%; 100% conc.

Incromide OD. [Croda Inc.] Oleamide DEA (1:1); nonionic; low color visc. builder with good sol. in anionic surfactants; for shampoos, bubble baths, soaps, bath prods.; clear liq.; usage level: 0.5-5%; 100% conc.

Incromide OLD. [Croda Inc.] Olivamide DEA (1:1); nonionic; visc. builder, foam stabilizer; for shampoos, bubble baths, soaps, bath prods.; clear liq.; usage level: 0.5-10%; 100% conc.

Incromide SED. [Croda Inc.] Sesamide DEA (1:1); nonionic; visc. builder and foam stabilizer; for shampoos, bubble baths, soaps, bath prods.; clear liq.; usage level: 0.5-5%; 100% conc.

Incromide WGD. [Croda Inc.] Wheat germamide DEA (1:1); nonionic; visc. builder and foam stabilizer; for shampoos, bubble baths, soaps, bath prods.; clear liq.; usage level: 1-10%; 100% conc.

Incromine Oxide I. [Croda Inc.; Croda Surf. Ltd.] Isostearamidopropylamine oxide; nonionic; foam stabilizer, thickener, lubricant, and visc. builder used in cosmetic, household, and janitorial prods.; wetting agent for conc. electrolyte sol'ns.; Gardner 1 max. gel; pH 6.5–7.5; usage level: 0.5-10%; 24.5–26.5% act.

Incromine Oxide ISMO. [Croda Inc.] Isostearamidopropyl morpholine oxide; nonionic; foam stabilizer, thickener, lubricant, and visc. builder used in cosmetic, household, and janitorial prods.; wetting agent for conc. electrolyte sol'ns.; Gardner 2 max. liq.; pH 6.5–7.5; usage level: 0.5-10%; 24.5–26.5% amine oxide.

Incromine Oxide L. [Croda Inc.; Croda Surf. Ltd.] Lauramine oxide; nonionic; foam booster/stabilizer, degreaser for dishwashing comps., household prods., hair conditioner used in personal care prods.; essentially colorless clear liq.; pH 7.0–8.0 (5% aq.); usage level: 0.5-20%; 29.0–31.0% amine oxide.

Incromine Oxide L-40. [Croda Inc.] Lauramine oxide; nonionic; foaming agent, foam stabilizer, degreaser; for shampoos and lt.-duty liqs.; essentially colorless liq.; usage level: 0.5-20%; 40% act.

Incromine Oxide M. [Croda Inc.; Croda Surf. Ltd.] Myristamine oxide; CAS 3332-27-2; nonionic; surfactant, emulsifier, emollient, conditioner, visc. builder, foam booster used in personal care prods.; essentially colorless liq.; pH 7.0–8.0 (5%); usage level: 0.5-20%; 29.5–30.5% amine oxide.

Incromine Oxide MC. [Croda Inc.] Myristamine oxide and cetamine oxide; nonionic; foam stabilizer and visc. builder used in personal, household and industrial applics.; wetting agent in electrolyte sol'ns.; colorless liq.; usage level: 0.5-10%; 30% act.

Incromine Oxide Mink. [Croda Inc.] Minkamidopropylamine oxide; nonionic; foaming agent, visc. builder; pale yel. gel; usage level: 0.5-20%; 25% act.

Incromine Oxide O. [Croda Inc.; Croda Surf. Ltd.] Oleamidopropylamine oxide; CAS 25159-40-4; nonionic; foam booster/stabilizer, thickener, lubricant, and visc. builder used in cosmetic, household, and janitorial prods.; wetting agent for conc. electrolyte sol'ns.; colorless gel; pH 6.5–7.5; usage level: 0.5-10%; 24.5–26.5% amine oxide.

Incromine Oxide OD-50. [Croda Surf. Ltd.] Oleyl dimethylamine oxide; CAS 14351-50-9; nonionic; foam stabilizer, visc. builder for cosmetic and household prods.; essentially colorless liq.; usage level: 0.5-10%; 50% conc.

Incromine Oxide OL. [Croda Inc.] Olivamidopropylamine oxide; nonionic; thickener for clear systems; yel. gel; usage level: 0.5-10%; 25% act.

Incromine Oxide S. [Croda Inc.; Croda Surf. Ltd.] Stearamine oxide; CAS 2571-88-2; nonionic; conditioner, emulsifier, visc. builder for personal care prods.; wh. paste; usage level: 0.5-10%; 24.5–26.5% amine oxide.

Incromine Oxide SE. [Croda Inc.] Sesamidopropylamine oxide; nonionic; visc. builder, conditioner for hair care prods.; pale yel. gel; usage level: 0.5-10%; 25% act.

Incromine Oxide WG. [Croda Inc.] Wheat

germamidopropylamine oxide; nonionic; visc. builder, conditioner for hair care prods.; yel. gel; usage level: 0.5-10%; 25% act.

Incronam 30. [Croda Inc.; Croda Surf. Ltd.] Cocamidopropyl betaine; CAS 61789-40-0; amphoteric; surfactant, emulsifier, coupling agent, visc. builder, foam detergent for personal care prods., chemical specialities, rug and upholstery shampoos, dishwashing compds.; Gardner 5 max. clear liq.; sol. in water, alcohol; pH 5.5–7.5 (10% aq.); usage level: 2-20%; 30% act.

Incronam CD-30. [Croda Surf. Ltd.] Coco betaine; CAS 68424-94-2; amphoteric; mild, high foaming surfactant for detergent systems; liq.; 30% conc.

Incropol CS-20. [Croda Inc.] Ceteareth-20; CAS 68439-49-6; nonionic; surfactant, emulsifier, lubricant, detergent for industrial and household prods.; coupling agent, antistat, fiber lubricant and solubilizer for personal care prods.; Gardner 1 max. solid; water-sol.; pH 5.5–7.5 (3%); HLB 15.5; usage level: 0.5-10%; 100% conc.

Incropol CS-50. [Croda Inc.] Ceteareth-50; CAS 68439-49-6; nonionic; surfactant, emulsifier, lubricant, detergent, foamer, wetting agent for industrial and household prods., toilet bowl cleaners, cistern blocks; coupling agent, antistat, fiber lubricant and solubilizer for personal care prods.; Gardner 1 max. flake; water-sol.; pH 6–8 (3%); HLB 17.9; usage level: 0.5-10%; 100% act.

Incropol L-7. [Croda Inc.] Laureth-7; nonionic; wetting agent for hard surf. cleaners; emulsifier for oils, emulsion polymerization; liq.; water-sol.; HLB 11.8; usage level: 0.5-10%.

Incroquat Behenyl BDQ/P. [Croda Inc.; Croda Universal Ltd.] Propylene glycol and behenalkonium chloride; cationic; o/w emulsifier and conditioner; paste; 25% conc.

Incroquat Behenyl TMC/P. [Croda Inc.; Croda Universal Ltd.] Propylene glycol and behenalkonium chloride; cationic; o/w emulsifier and conditioner; paste; 25% conc.

Incroquat SDQ-25. [Croda Inc.; Croda Universal Ltd.] Stearalkonium chloride; CAS 122-19-0; cationic; surfactant used as ingred. in personal care prods., textile and paper; dispersant for pigments and dyestuffs; antistat for fibers and synthetics; hair conditioner; wh. paste; m.p. 140 F; pH 3.0–4.0 (1%); usage level: 2-10%; 25% conc.

Incrosoft 100. [Croda Inc.] Methyl bis (hydrog. tallow amido ethyl) 2-hydroxyethyl ammonium chloride; fabric softener with good hand and antistatic properties; for home, commercial or industrial laundry applics., textile and finishing operations; pale yel. flake, typ. bland odor; pH 6.0-8.0 (1%); usage level: 5-30%; 100% act.

Incrosoft 100P. [Croda Inc.] Methyl bis (hydrogenated tallowamidoethyl) 2-hydroxyethyl ammonium chloride; cationic; fabric softener for home, commerical or industrial laundry applics., textile and finishing operations; Gardner 5 max. powd., typ. bland odor; disp. in water; nonirritating; 97% min. act.

Incrosoft 248. [Croda Inc.] Quaternium-72; softener for dryer sheets; yel. solid; usage level: 5-10%; 90% act.

Incrosoft CFI-75. [Croda Inc.] Quaternium-72; cationic; used in detergent and fabric softener systems; Gardner 7 max. clear liq.; pH 5–7.5 (5% in 50% IPA); usage level: 5-10%; 74% min. act.

Incrosoft S-75. [Croda Inc.] Quaternium-27; CAS 86088-85-9; softener base, lubricant, antistat and rewetting agent for fabrics and syns.; Gardner 5 max. liq.; water-sol.; pH 4–7 (5%); usage level: 5-10%; 74–76% act.

Incrosoft S-90. [Croda Inc.] Quaternium-27; CAS 86088-85-9; softener base, lubricant, antistat and rewetting agent for fabrics and syns.; also used in textile maintenance and finishing; yel. semisolid; disp. in water; pH 5.0–7.0 (5% in DW); usage level: 5-10%; 85–88% act.

Incrosoft S-90M. [Croda Inc.] Quaternium-27; CAS 86088-85-9; fabric softener with improved sol. in detergent softener systems; superior hard water tolerance; yel. visc. paste; usage level: 5-10%; 90% act.

Incrosoft T-75. [Croda Inc.] Quaternium-53; CAS 130124-24-2; fabric softener, lubricant and antistat for home and commercial laundry prods.; yel. liq.; disp. in water; usage level: 5-10%; 75% act.

Incrosoft T-90. [Croda Inc.] Quaternium-53; CAS 130124-24-2; fabric softener, lubricant and antistat for home and commercial laundry prods.; good rewetting; Gardner 7 max. visc. liq., alcoholic odor; disp. in water; pH 4–7 (3%); usage level: 5-10%; 88–90% act.

Incrosoft T-90HV. [Croda Inc.] Quaternium-53; CAS 130124-24-2; softening agent with exc. rewetting; general purpose fabric softener; high visc. version of Incrosoft T-90; yel. paste; usage level: 5-10%; 90% act.

Incrosul LMS. [Croda Inc.] Disodium lauramido MEA-sulfosuccinate; anionic; mild foaming agent and cleanser used in personal care prods. and carpet shampoos; lime soap dispersant; Gardner 4 max. liq./slurry; pH 6–7 (3%); usage level: 5-50%; 39–41% act.

Incrosul LSA. [Croda Inc.] Diammonium lauryl sulfosuccinate; anionic; mild, high foaming surfactant used in personal care prods.; lime soap dispersant; Gardner 1 max. liq.; mild char. odor; sol. in water; pH 6–7 (3%); usage level: 5-50%; 39–41% act.

Indulin® 201. [Westvaco] Tall oil fatty acid; emulsifier for soap sol'ns. for high float emulsions; sp.gr. 0.980; dens. 8.17 lb/gal; visc. 940 cps; pour pt. 16 C; flash pt. 232 C.

Indulin® 202. [Westvaco] Tall oil fatty acid; emulsifier for soap sol'ns. for high float emulsions; sp.gr. 0.943; dens. 7.87 lb/gal; visc. 90 cps; pour pt. 4 C; flash pt. 191 C.

Indulin® 206. [Westvaco] Fatty ester amine; asphalt additive producing high float emulsion when used with Indulin 201 or 202 as emulsifier; brn. liq., ammoniacal odor; insol. in water; sp.gr. 0.950; dens. 7.92 lb/gal; visc. 145 cps; b.p. > 500 F; pour pt. -21 C; flash pt. > 150 C; pH 8.5 (10% aq. susp.); toxicology: not established; may be skin or eye irritant.

Indulin® AQS. [Westvaco] Tall oil amino polycarboxylic acid; amphoteric; emulsifier for asphalt emulsions; brn. visc. liq., IPA odor; sp.gr. 1.05; dens. 8.76 lb/gal; visc. 2000 cps; b.p. > 212 F; pour pt. 10 C; flash pt. (PM) 29 C; pH 7.9 (10%); toxicology: LD50 (oral, rat) > 10 g/kg; severe eye irritant.

Indulin® AQS-IM. [Westvaco] Amine derivs.; anionic; asphalt emulsifier for quick set slurry seal applics.; liq.; sp.gr. 1.09; dens. 9.09 lb/gal; visc. 2600 cps; pour pt. 25 C; flash pt. (PM) > 150 C; 100% conc.

Indulin® AS-1. [Westvaco] Amine derivs.; cationic;

antistripping additive for highway paving applics.; asphalt adhesion agent; dk. brn. visc. liq.; sp.gr. 0.970; dens. 8.10 lb/gal; visc. 2450 cps; pour pt. 55 F; flash pt. (COC) 320 F; 100% act.

Indulin® AS-Special. [Westvaco] Amine deriv.; cationic; antistripping additive for highway paving applics.; asphalt adhesion agent; dk. brn. liq.; sp.gr. 0.985; dens. 8.22 lb/gal; visc. 370 cps; pour pt. 54 F; flash pt. (COC) 280 F; 100% act.

Indulin® AT. [Westvaco] Kraft pine lignin; anionic; surfactant; powd.; sol. in aq. alkaline sol'ns. above pH 9.0; insol. in hydrocarbon solvs.; 100% conc.

Indulin® MQK. [Westvaco] C21 dicarboxylic amido alkyl amine; cationic; asphalt emulsifier for quick set slurry seal applics.; amber moderately visc. liq., ammoniacal amine odor; moderate water sol.; sp.gr. 0.984; dens. 8.20 lb/gal; visc. 4000 cSt; b.p. 180 F; pour pt. 4.4 C; flash pt. (PM) 20 C; pH 10.5 (10%); toxicology: LD50 (oral, rat) 0.37 g/kg; toxic by oral ingestion; corrosive to eyes; 100% conc.

Indulin® MQK-IM. [Westvaco] Amine derivs.; cationic; asphalt emulsifier for quick set slurry seal applics.; liq.; 100% conc.

Indulin® SA-L. [Westvaco] Sodium lignate; anionic; emulsifier, stabilizer, retarder for asphalt emulsions; brn. liq., bland vanilla odor; sp.gr. 1.148; dens. 9.6 lb/gal; visc. 800 cps; b.p. > 212 F; pH 11 (2%%); toxicology: LD50 (oral, rat) > 19 g/kg; 40% solids.

Indulin® W-1. [Westvaco] Amine deriv. of pine lignin; CAS 110152-58-4; cationic; emulsifier, stabilizer in asphalt emulsions, retarding agent in cement; also for oilfield chems.; brn. powd., fishy amine odor; water sol. ≥ pH 8.3 and ≤ pH 4.2; sp.gr. 1.2-1.3; dens. 22 lb/ft^3 (loose); visc. 100 cps @ 20% solids; pH 10.8 (2% aq.); surf. tens. 47 dynes (1% aq.); toxicology: LD50 (oral, rat) > 12 g/kg (pract. nontoxic); sl. irritating to eyes, abbraded skin; 95% conc.

Indulin® W-3. [Westvaco] Lignin amine; cationic; primary emulsifier, stabilizer, and retarder in asphalt emulsions, slurry seal applics.; brn. liq.; amine odor; sol. in water and acid sol'ns.; sp.gr. 1.090; dens. 9.1 lb/gal; visc. 450 cps; b.p. > 212 F; pH 11 (2.8%); toxicology: LD50 (oral, rat) > 19 g/kg; 35% solids.

Indulin® XD-70. [Westvaco] Nonylphenol ethoxylate; CAS 9016-45-9; nonionic; coemulsifier for anionic or cationic slow-setting asphalt emulsions; clear liq., mild odor; sol. in water; sp.gr. 1.09; dens. 9.1 lb/gal; visc. 890 cps; b.p. > 212 F; pour pt. -6.7 C; flash pt. (PM) 200+ C; pH 8.0; toxicology: LD50 (oral, rat) 4-5 g/kg (sl. toxic); may cause eye irritation on prolonged contact; mild skin irritant; 70% act.

Industrene® 104. [Witco/Humko] Oleic acid, low titer; CAS 112-80-1; chemical intermediate; solid. pt. 4 C max.; acid no. 198-204; iodine no. 96 max.; sapon. no. 190-205.

Industrene® 105. [Witco/Humko] Oleic acid; CAS 112-80-1; intermediate used in alkyd resins, rubber compding., water repellents, polishes, soaps, abrasives, cutting oils, candles, crayons, emulsifiers, personal care prods.; Gardner 6 liq.; solid. pt. 145 C max.; acid no. 195-204; iodine no. 85-95; sapon. no. 195-205; 100% conc.

Industrene® 106. [Witco/Humko] Oleic acid NF, low titer; CAS 112-80-1; chemical intermediate; personal care prods.; solid. pt. 6 C max.; acid no. 198-204; iodine no. 95 max.; sapon. no. 199-205.

Industrene® 120. [Witco/Humko] Linolenic acid; chemical intermediate; solid. pt. 15-22 C; acid no. 197–202; iodine no. 180-195; sapon. no. 197–203.

Industrene® 126. [Witco/Humko] Soya acid; chemical intermediate; solid. pt. 24–29 C; acid no. 196–205; iodine no. 125-135; sapon. no. 196–206.

Industrene® 130. [Witco/Humko] Linolenic acid; chemical intermediate; solid. pt. 30–38 C; acid no. 199–206; iodine no. 90-112; sapon. no. 199–207.

Industrene® 143. [Witco/Humko] Tallow acid; CAS 61790-37-2; intermediate used in alkyd resins, rubber compding., water repellents, polishes, soaps, abrasives, cutting oils, candles, crayons, emulsifiers; FG grades as lubricant, release agent, binder, defoamer in foods, intermediate for food emulsifiers; Gardner 5 paste; solid. pt. 39-43 C; acid no. 202-206; iodine no. 50-65; sapon. no. 202-207; 100% conc.

Industrene® 145. [Witco/Humko] Tallow acid; chemical intermediate; solid. pt. 44–50 C; acid no. 195 min.; iodine no. 38-45; sapon. no. 195 min.

Industrene® 205. [Witco/Humko] Oleic acid; CAS 112-80-1; intermediate used in alkyd resins, rubber compding., water repellents, polishes, soaps, abrasives, cutting oils, candles, crayons, emulsifiers; FG grades as lubricant, release agent, binder, defoamer in foods, intermediate for food emulsifiers; wh. liq.; acid no. 195-204; iodine no. 85-95; sapon. no. 195-205; 100% conc.

Industrene® 206. [Witco/Humko] Oleic acid NF; CAS 112-80-1; intermediate used in alkyd resins, rubber compding., water repellents, polishes, soaps, abrasives, cutting oils, candles, crayons, emulsifiers, personal care prods.; liq.; solid. pt. 6 C max.; acid no. 199-204; iodine no. 95 max.; sapon. no. 200-205; 100% conc.

Industrene® 206LP. [Witco/Humko] Oleic acid; CAS 112-80-1; chemical intermediate; solid. pt. 6 C max.; acid no. 199-205; iodine no. 92 max.; sapon. no. 200-206.

Industrene® 223. [Witco/Humko] Hydrog. coconut acid; chemical intermediate, emulsifier; used for personal care prods., waxes, textile aux., pharmaceuticals; solid. pt. 23–26 C; acid no. 266-274; iodine no. 3 max.; sapon. no. 267-276.

Industrene® 224. [Witco/Humko] Oleic-linoleic acid; chemical intermediate; solid. pt. 17–22 C; acid no. 195–204; iodine no. 125-140; sapon. no. 195–205.

Industrene® 225. [Witco/Humko] Soya acid, dist.; CAS 67701-08-0; intermediate used in alkyd resins, rubber compding., water repellents, polishes, soaps, abrasives, cutting oils, candles, crayons, emulsifiers; FG grades as lubricant, release agent, binder, defoamer in foods, intermediate for food emulsifiers; Gardner 3–4 liq.; solid. pt. 25 C max.; acid no. 195–201; iodine no. 135-145; sapon. no. 197–204; 100% conc.

Industrene® 226. [Witco/Humko] Soya acid, dist.; CAS 67701-08-0; intermediate used in alkyd resins, rubber compding., water repellents, polishes, soaps, abrasives, cutting oils, candles, crayons, emulsifiers; FG grades as lubricant, release agent, binder, defoamer in foods, intermediate for food emulsifiers; Gardner 3–4 liq.; solid. pt. 26 C max.; acid no. 195–203; iodine no. 125-135; sapon. no. 197–204; 100% conc.

Industrene® 226 FG. [Witco/Humko] Soya acid, dist.; lubricant, release agent, binder, defoaming agent and intermediate for food additives; liq.; Lovibond 25.0Y/2.5R; solid. pt. 26 C max.; acid no.

195–203; iodine no. 125-135; sapon. no. 195–204; 100% conc.

Industrene® 325. [Witco/Humko] Dist. coconut acid; CAS 61788-47-4; intermediate used in alkyd resins, rubber compding., water repellents, polishes, soaps, abrasives, cutting oils, candles, crayons, emulsifiers, personal care prods.; paste; acid no. 265-277; iodine no. 6-15; sapon. no. 265-278; 100% conc.

Industrene® 328. [Witco/Humko] Stripped coconut acid; CAS 61788-47-4; intermediate used in alkyd resins, rubber compding., water repellents, polishes, soaps, abrasives, cutting oils, candles, crayons, emulsifiers, personal care prods.; paste; acid no. 253-260; iodine no. 5-14; sapon. no. 253-260; 100% conc.

Industrene® 365. [Witco/Humko] Mixt. caprylic/capric acid; intermediate used in alkyd resins, rubber compding., water repellents, polishes, soaps, abrasives, cutting oils, candles, crayons, emulsifiers; FG grades as lubricant, release agent, binder, defoamer in foods, intermediate for food emulsifiers; acid no. 355-369; iodine no. 1 max.; sapon. no. 355-374; 100% conc.

Industrene® 4516. [Witco/Humko] 45% Palmitic acid; CAS 57-10-3; intermediate for mfg. of soaps, emulsions, creams, lotions, ethoxylates, buffing compds., lubricants; solid; acid no. 205-211; iodine no. 3 max.; sapon. no. 205-212; 100% conc.

Industrene® 4518. [Witco/Humko] Single pressed stearic acid; CAS 57-11-4; intermediate used in alkyd resins, rubber compding., water repellents, polishes, soaps, abrasives, cutting oils, candles, crayons, emulsifiers; FG grades as lubricant, release agent, binder, defoamer in foods, intermediate for food emulsifiers; Gardner 3 solid; acid no. 204-211; iodine no. 8-11; sapon. no. 204-212; 100% conc.

Industrene® 5016. [Witco/Humko] Double pressed stearic acid; CAS 57-11-4; intermediate used in alkyd resins, rubber compding., water repellents, polishes, soaps, abrasives, cutting oils, candles, crayons, emulsifiers; FG grades as lubricant, release agent, binder, defoamer in foods, intermediate for food emulsifiers; solid; acid no. 207-210; iodine no. 4-7; sapon. no. 207-211; 100% conc.

Industrene® 5016 FG. [Witco/Humko] Stearic acid; CAS 57-11-4; lubricant, release agent, binder, defoamer in foods; intermediate for food-grade emulsifiers.

Industrene® 6018. [Witco/Humko] Hydrog. fatty acid, buffing grade; intermediate for mfg. of soaps, emulsions, creams, lotions, ethoxylates, buffing compds., lubricants; solid; 100% conc.

Industrene® 7018. [Witco/Humko] 70% Stearic acid; CAS 57-11-4; intermediate used in alkyd resins, rubber compding., water repellents, polishes, soaps, abrasives, cutting oils, candles, crayons, emulsifiers; FG grades as lubricant, release agent, binder, defoamer in foods, intermediate for food emulsifiers; solid; acid no. 200-207; iodine no. 1 max.; sapon. no. 200-208; 100% conc.

Industrene® 9018. [Witco/Humko] 90% Stearic acid; CAS 57-11-4; intermediate used in alkyd resins, rubber compding., water repellents, polishes, soaps, abrasives, cutting oils, candles, crayons, emulsifiers, personal care prods.; solid; acid no. 196-201; iodine no. 2 max.; sapon. no. 196-202; 100% conc.

Industrene® B. [Witco/Humko] Hydrog. stearic acid; CAS 57-11-4; intermediate used in alkyd resins, rubber compding., water repellents, polishes, soaps, abrasives, cutting oils, candles, crayons, emulsifiers; FG grades as lubricant, release agent, binder, defoamer in foods, intermediate for food emulsifiers; solid; acid no. 199-207; iodine no. 3 max.; sapon. no. 199-208; 100% conc.

Industrene® D. [Witco/Humko] 40% Dimer acid; intermediate for lubricants, corrosion inhibitors for petrol. industry, extenders and cross-linking agents for high polymeric systems, in hot-melt adhesives, epoxy curing agents; Gardner 8 color; visc. 600 c?St; acid no. 184-188; sapon. no. 184-197.

Industrene® M. [Witco/Humko] Oleic-stearic acid; chemical intermediate; solid. pt. 35 C max.; acid no. 175-190; iodine no. 80-105; sapon. no. 175-200.

Industrene® R. [Witco/Humko] Hydrog. rubber grade stearic acid; CAS 57-11-4; intermediate used in alkyd resins, rubber compding., water repellents, polishes, soaps, abrasives, cutting oils, candles, crayons, emulsifiers; FG grades as lubricant, release agent, binder, defoamer in foods, intermediate for food emulsifiers; Gardner 10 solid; acid no. 193-213; iodine no. 10 max.; sapon. no. 193-214; 100% conc.

Industrol® 68. [BASF] Sorbitol stearate ethoxylate; emulsifier; wax; sp.gr. 1.1; m.p. 35 C; sapon. no. 60–75; 100% act.

Industrol® 400-MOT. [BASF] PEG-8 tallate; CAS 61791-00-2; nonionic; surfactant; Gardner 10 max. liq.; disp. in water; sp.gr. 1.04; visc. 400 cps; HLB 11.0; pour pt. –5 C; cloud pt. < 25 C (1% aq.); pH 6.0–7.5 (5% aq.); 100% act.

Industrol® 1025-DT. [BASF] POP 1000 ditallate; nonionic; surfactant; Gardner 6 max. liq.; insol. in water; sp.gr. 0.98; pour pt. < –10 C; cloud pt. < 25 C (1% aq.); 100% act.

Industrol® 1186. [BASF] Sorbitol oleate ethoxylate; emulsifier; Gardner 7 max. liq.; disp. in water; sp.gr. 1.0; visc. 450 cps; cloud pt. < 25 C (1% aq.); sapon. no. 75–95; pH 5.5–7.0 (5% aq.); 100% act.

Industrol® CO-25. [BASF] PEG-25 castor oil; nonionic; surfactant; Gardner 5 max. liq.; water-sol.; m.w. 2100 (theoret.); sp.gr. 1.04; HLB 10.8; pour pt. –5 C; pH 6.0–7.5 (5% aq.); 100% act.

Industrol® CO-30. [BASF] PEG-30 castor oil; nonionic; surfactant; Gardner 5 max. liq.; water-sol.; m.w. 2350 (theoret.); sp.gr. 1.04; visc. 800 cps; HLB 11.7; pour pt. 9 C; pH 6.0–7.5 (5% aq.); 100% act.

Industrol® CO-36. [BASF] PEG-36 castor oil; nonionic; surfactant; Gardner 5 max. liq.; water-sol.; m.w. 2600 (theoret.); sp.gr. 1.05; visc. 700 cps; HLB 12.5; pour pt. 18 C; pH 5.5–7.0 (5% aq.); 100% act.

Industrol® CO-40. [BASF] PEG-40 castor oil; nonionic; surfactant; Gardner 5 max. liq.; water-sol.; m.w. 2780 (theoret.); sp.gr. 1.05; visc. 600 cps; HLB 12.9; pour pt. 18 C; pH 6.0–7.5 (5% aq.); 100% act.

Industrol® CO-80-80%. [BASF] PEG-80 castor oil; nonionic; surfactant; water-sol.; m.w. 4500 (theoret.); sp.gr. 1.08; HLB 15.5; pH 6.0–7.5 (5% aq.); 80% act.

Industrol® CO-200. [BASF] PEG-200 castor oil; nonionic; surfactant; wh. wax; water-sol.; m.w. 9800 (theoret.); sp.gr. 1.08; HLB 18.1; m.p. 40 C; pH 6.0–7.5 (5% aq.); 100% act.

Industrol® CO-200-50%. [BASF] PEG-200 castor oil; nonionic; surfactant; APHA 200 max. liq.; water-sol.; sp.gr. 1.04; visc. 2000 cps; HLB 18.1; pour pt. 7 C; pH 5.0–7.0 (2% aq.); 50% act.

Industrol® COH-25. [BASF] PEG-25 hydrog. castor oil; CAS 61788-85-0; nonionic; surfactant; Gardner 3 max. liq.; water-sol.; m.w. 2140 (theoret.); sp.gr. 1.03; HLB 11.0; pour pt. 5 C; pH 6.0–7.5 (5% aq.);

100% act.

Industrol® COH-200. [BASF] PEG-200 hydrog. castor oil; CAS 61788-85-0; nonionic; surfactant; Gardner 2 max. wax; water-sol.; m.w. 9800 (theoret.); HLB 18.1; m.p. 45 C; pH 6.0–7.5 (5% aq.); 100% act.

Industrol® CSS-25. [BASF] Ceteareth-25; CAS 68439-49-6; nonionic; surfactant; Gardner 1 max. flake/wax; water-sol.; m.w. 1340; HLB 16.1; m.p. 41 C; cloud pt. > 100 C (1% aq.); pH 5.0–6.5 (5% aq.); 100% act.

Industrol® DL-9. [BASF] PEG-8 dilaurate; CAS 9005-02-1; nonionic; surfactant; Gardner 3 max. liq.; insol. in water; m.w. 815; sp.gr. 0.99; HLB 10.8; pour pt. 8 C; cloud pt. < 25 C (1% aq.); pH 6.0–7.5 (5% aq.); 100% act.

Industrol® DO-9. [BASF] PEG 400 dioleate; CAS 9005-07-6; nonionic; surfactant; Gardner 4 max. liq.; disp. in water; m.w. 960; sp.gr. 0.97; HLB 8.5; pour pt. 6 C; cloud pt. < 25 C; pH 5.0–7.0 (5% aq.); 100% act.

Industrol® DO-13. [BASF] PEG 600 dioleate; CAS 9005-07-6; nonionic; surfactant; Gardner 4 max. liq.; disp. in water; m.w. 1160; sp.gr. 1.0; HLB 10.6; pour pt. 20 C; cloud pt. < 25 C (1% aq.); pH 5.0–7.0 (5% aq.); 100% act.

Industrol® DT-13. [BASF] PEG 600 ditallate; CAS 61791-01-3; nonionic; surfactant; Gardner 4 max. liq.; disp. in water; sp.gr. 1.00; HLB 10.5; pour pt. 10 C; cloud pt. < 25 C (1% aq.); pH 6.0–7.5 (5% aq.); 100% act.

Industrol® DW-5. [BASF] Modified oxyethylated alcohol; nonionic; biodeg., chlorine-stable, low-foaming detergent, dispersant, wetting agent, emulsifier for home/commercial machine dishwash, spray cleaners; defoamer for protein soils; APHA 100 max. sl. hazy liq.; m.w. 1400; sp.gr. 1.01; visc. 180 cps; HLB 6.0; pour pt. 2 C; cloud pt. 20 C (1% aq.); pH 6.9 (1% aq.); surf. tens. 29.8 dynes/cm (0.1% aq.); Draves wetting 13 s (0.1% aq.); Ross-Miles foam 0 mm (0.1% aq., initial, 50 C); 100% act.

Industrol® L-20-S. [BASF] Polysorbate 20; CAS 9005-64-5; nonionic; surfactant; liq.; water-sol.; sp.gr. 1.13; visc. 400 cps; HLB 16.7; sapon. no. 40–50; 3.0% max. water.

Industrol® LG-70. [BASF] Oleth-20, modified; CAS 9004-98-2; nonionic; surfactant to prevent gelling in aq. sol'ns.; Gardner 1 max. liq.; water-sol.; sp.gr. 1.07; HLB 16.0; pour pt. 18 C; cloud pt. > 100 C (1% aq.); pH 5.0–7.5 (5% aq.); 70% act.

Industrol® LG-100. [BASF] Oleth-20, modified; CAS 9004-98-2; nonionic; surfactant to prevent gelling in aq. sol'ns.; Gardner 1 max. wax; water-sol.; HLB 16.0; m.p. 32 C; cloud pt. > 100 C; pH 6.0–7.5 (5% aq.); 100% act.

Industrol® MIS-9. [BASF] PEG-9 isostearate; CAS 56002-14-3; nonionic; surfactant; Gardner 7 max. liq.; disp. in water; sp.gr. 1.03; visc. 100 cps; HLB 11.8; pour pt. 10 C; pH 6.0–7.5 (5% aq.); 100% act.

Industrol® ML-5. [BASF] PEG-5 laurate; CAS 9004-81-3; nonionic; surfactant; Gardner 4 max. liq.; disp. in water; m.w. 425; sp.gr. 1.02; visc. 100 cps; HLB 9.8; pour pt. 9 C; cloud pt. < 25 C (1% aq.); pH 6.0–7.5 (5% aq.); 100% act.

Industrol® ML-9. [BASF] PEG-9 laurate; CAS 9004-81-3; nonionic; surfactant; Gardner 4 max. liq.; water-sol.; m.w. 600; sp.gr. 1.04; HLB 13.1; pour pt. 12 C; cloud pt. 40 C (1% aq.); pH 6.0–7.5 (5% aq.); 100% act.

Industrol® ML-14. [BASF] PEG-14 laurate; CAS 9004-81-3; nonionic; surfactant; Gardner 4 max. liq.; water-sol.; sp.gr. 1.05; HLB 14.9; pour pt. 14 C; cloud pt. 64 C; pH 6.0–7.5 (5% aq.); 100% act.

Industrol® MO-5. [BASF] PEG-5 oleate; CAS 9004-96-0; nonionic; surfactant; Gardner 6 max. liq.; disp. in water; m.w. 485; sp.gr. 0.99; HLB 8.3; pour pt. –10 C; pH 6.0–7.5 (5% aq.); 100% act.

Industrol® MO-6. [BASF] PEG-6 oleate; CAS 9004-96-0; nonionic; surfactant; Gardner 6 max. liq.; disp. in water; m.w. 544; sp.gr. 0.99; HLB 9.7; pour pt. –10 C; pH 6.0–7.5 (5% aq.); 100% act.

Industrol® MO-7. [BASF] PEG-7 oleate; CAS 9004-96-0; surfactant; Gardner 6 max. liq.; disp. in water; m.w. 585; sp.gr. 1.0; HLB 10.4; pour pt. –10 C; pH 6.0–7.5 (5% aq.); 100% act.

Industrol® MO-9. [BASF] PEG-9 oleate; CAS 9004-96-0; surfactant; Gardner 6 max. liq.; disp. in water; m.w. 675; sp.gr. 1.01; HLB 11.7; pour pt. 5 C; pH 5.5–7.0 (5% aq.); 100% act.

Industrol® MO-13. [BASF] PEG-12 oleate; CAS 9004-96-0; nonionic; surfactant; Gardner 7 max. liq.; water-sol.; m.w. 850; sp.gr. 1.04; HLB 13.6; pour pt. 18 C; pH 6.0–7.5 (5% aq.); 100% act.

Industrol® MS-5. [BASF] PEG-5 stearate; CAS 9004-99-3; nonionic; surfactant; Gardner 3 max. wax; insol. in water; m.w. 500; HLB 9.1; m.p. 28 C; pH 5.5–7.0 (5% aq.); 100% act.

Industrol® MS-7. [BASF] PEG-7 stearate; CAS 9004-99-3; nonionic; surfactant; Gardner 2 max. wax; disp. in water; m.w. 575; HLB 10.1; m.p. 29 C; pH 5.5–7.0 (5% aq.); 100% act.

Industrol® MS-8. [BASF] PEG-8 stearate; CAS 9004-99-3; nonionic; surfactant; Gardner 2 max. wax; disp. in water; m.w. 620; HLB 11.4; m.p. 30 C; pH 5.5–7.0 (5% aq.); 100% act.

Industrol® MS-9. [BASF] PEG-9 stearate; CAS 9004-99-3; nonionic; surfactant; Gardner 2 max. wax; disp. in water; m.w. 665; HLB 12.0; m.p. 30 C; pH 5.5–7.0 (5% aq.); 100% act.

Industrol® MS-23. [BASF] PEG-23 stearate; CAS 9004-99-3; nonionic; surfactant; Gardner 1 max. wax; water-sol.; m.w. 1280; HLB 15.7; m.p. 42 C; pH 5.5–7.0 (5% aq.); 100% act.

Industrol® MS-40. [BASF] PEG-40 stearate; CAS 9004-99-3; nonionic; surfactant; Gardner 1 max. wax.; water-sol.; m.w. 2030; HLB 16.9; m.p. 48 C; pH 5.5–7.0 (5% aq.); 100% act.

Industrol® MS-100. [BASF] PEG-100 stearate; CAS 9004-99-3; nonionic; surfactant; Gardner 1 max. flake/wax; water-sol.; m.w. 4670; HLB 18.8; m.p. 57 C; pH 5.5–7.0 (5% aq.); 100% act.

Industrol® N2. [BASF] EO/PO block copolymer; detergent, dispersant, wetting agent for low to high temp. rinse aids where low foam and sheeting are important; liq.

Industrol® N3. [BASF] EO/PO block copolymer; CAS 9003-11-6; detergent, dispersant, wetting agent for low to high temp. rinse aids where low foam and sheeting are important; liq.

Industrol® O-20-S. [BASF] Polysorbate 80; CAS 9005-65-6; nonionic; surfactant; liq.; water-sol.; sp.gr. 1.08; visc. 400 cps; HLB 15; sapon. no. 45–55; 3.0% max. water.

Industrol® OAL-20. [BASF] Oleth-20; CAS 9004-98-2; nonionic; surfactant; Gardner 1 max. wax; m.w. 1150; sp.gr. 1.03; HLB 15.8; m.p. 32 C; cloud pt. > 100 C (1% aq.); pH 6.0–7.5 (5% aq.); 100% act.

Industrol® S-20-S. [BASF] Polysorbate 60; CAS 9005-67-8; nonionic; surfactant; water-sol.; sp.gr. 1.13; visc. 600 cps; HLB 14.9; sapon. no. 45–55;

3.0% max. water.
Industrol® SO-13. [BASF] PEG 600 sesquioleate; nonionic; surfactant; Gardner 4 max. liq.; disp. in water; m.w. 1020; sp.gr. 1.0; HLB 12.1; pour pt. 19 C; cloud pt. < 25 C (1% aq.); 100% act.
Industrol® STS-20-S. [BASF] Polysorbate 65; CAS 9005-71-4; nonionic; surfactant; tan solid; disp. in water; sp.gr. 1.05; HLB 10.5; sapon. no. 88–98; 3.0% max. water.
Industrol® TFA-8. [BASF] Fatty acid ethoxylate (8 EO); nonionic; surfactant; Gardner 8 max. liq.; disp. in water; m.w. 600; sp.gr. 1.02; HLB 10.0; pour pt. 0 C; pH 7.0–8.5 (5% aq.); 100% act.
Industrol® TFA-15, TFA-15-80%. [BASF] Fatty acid ethoxylate (15 EO); nonionic; surfactant; Gardner 7 max. solid and liq. resp.; water-sol.; sp.gr. 1.08 and 1.06 resp.; HLB 14.0; pH 6.0–7.5 (5% aq.); 100 and 80% act. resp.
Industrol® TO-10. [BASF] PEG-10 tallate; CAS 61791-00-2; nonionic; surfactant; Gardner 10 max. liq.; disp. in water; m.w. 660; sp.gr. 1.03; HLB 11.2; pour pt. 0 C; pH 6.0–7.5 (5% aq.); 100% act.
Industrol® TO-16. [BASF] PEG-16 tallate; CAS 61791-00-2; nonionic; surfactant; Gardner 10 max. liq.; water-sol.; m.w. 900; sp.gr. 1.05; HLB 13.5; pour pt. 16 C; pH 6.0–7.5 (5% aq.); 100% act.
Infrasan C. [Reilly-Whiteman] Coconut fatty acid alkanolamide (2:1); nonionic; detergent, wetting agent; liq.; 100% conc.
Inhibitor 60Q. [Exxon/Tomah] Surfactant blend; cationic; inhibitor for acid cleaners; liq.; 90% conc.
Inhibitor 212. [Zschimmer & Schwarz] Fatty alcohol alkanolamide; corrosion inhibitor additive; liq.; 100% act.
Inipol OO2. [Ceca SA] N-Oleyl propylene diamine dioleate; CAS 34140-91-5; wetting agent, dispersant, rust inhibitor; paint and fuel additive; liq.
Inipol OT2. [Ceca SA] N-Oleyl propylene diamine ditallate; wetting agent, dispersant, rust inhibitor, paint additive; liq.
Inipol S 43. [Ceca SA] Salt of fatty acid, polyalkylated N-alkyl propylene diamine; wetting agent, dispersant, lubricant additive, paint additive, rust inhibitor, fuel additive; liq.
Inset XL-300. [Hart Prods. Corp.] Resin condensate; cationic; fixative for direct dyes; liq.; 50% conc.
Intermediate 300. [Witco/Humko] Coco/tallow MEA; nonionic; thickener, foam booster, superfatting agent; solid; 100% conc.
Intermediate 325. [Witco/Humko] Coco/tallow MEA; nonionic; thickener, foam booster, superfatting agent; solid; 100% conc.
Intermediate 512. [Witco/Humko] Lauramide DEA; CAS 120-40-1; nonionic; thickener, foam booster, superfatting agent; solid; 100% conc.
Interwet® 33. [Akzo] Nonionic; emulsifier, penetrant for coatings, adhesives, insecticides, wax dispersions, bitumen emulsions; yel. liq., mild fatty odor; sol. in alcohols, acetones, toluol, disp. in water; sp.gr. 1.01; visc. 85 cps; f.p. 2 C; b.p. 290 C; HLB 11.5; 100% act.
Intex Scour 707. [Intex] Phosphate; anionic; detergent, emulsifier, fabric preparations, scour after dyeing; liq.; 90% conc.
Intex Scour Base 706. [Intex] Alkanolamide; nonionic; foam additive, scour, pulling soap, shampoo base; liq.; 100% conc.
Intracarrier® ATM. [Crompton & Knowles] Blend of esters and aromatic hydrocarbon derivs. with nonionic emulsifier; biodeg. dyeing accelerant for disperse dyeing of syn. fibers and their blends; carrier for polyester and triacetate; moderate foaming; high flash pt.; off-wh. to yel. liq., mild odor; sp.gr. 1.085 g/ml.
Intrafomil® AK. [Crompton & Knowles] Nonsilicone; defoamer for textile operations, esp. jet equip.; ylsh. to amber opaque liq.; emulsifiable in water; dens. 8.7 lb/gal; visc. 150 cps; neutral pH; usage level: 2-5 oz/100 gal.
Intralan® Salt HA. [Crompton & Knowles] Condensed sodium alkylaryl sulfonate; anionic; low foaming dispersing and leveling agent for disperse dyes and dyeing of syn. fibers; esp. for beam, package and jet dyeing; tan powd.: sol. in water; pH 7-8 (10%); brn. liq.: misc. with water; pH 7-9.
Intralan® Salt N. [Crompton & Knowles] EO condensate; nonionic; dye assistant, leveling agent, penetrant for textiles; yel. liq.; sol. in boiling water.
Intraphasol COP. [Crompton & Knowles] Hydrocarbons, solubilized, aliphatic and sulfonic acid salts; anionic; emulsifier, wetting agent assistant for dyeing of fabrics and carpets; liq.
Intraphasol PC. [Crompton & Knowles] Sulfuric acid ester salt of polyglycol ether compd.; anionic; wetting agent, emulsifier, dispersant used in printing of polyamide and polyester carpets; amber liq.; alcoholic odor; water sol.; dens. 8.7 lb/gal.
Intrapol 1014. [Henkel-Nopco] Solv. and alkylpolyglycol ether blend; nonionic; low foaming detergent and dispersant; kier boiling assistant; liq.; 80% conc.
Intraquest® TA Sol'n. [Crompton & Knowles] Tetrasodium EDTA; sequestering agent for scouring, bleaching, dyeing, and other wet finishing textile operations; softens water by complexing calcium, magnesium, and divalent heavy metal ions over broad pH range; detergent builder for scouring; straw clear liq.; misc. with water; dens. 10.6 lb/gal; pH 11-12 (1%); chel. value 100-102 mg $CaCO_3$/g.
Intrassist® LA-LF. [Crompton & Knowles] Quat. compd.; cationic; nonretarding, low-foaming leveling agent for Sevron® cationic dyes on acrylic fibers; clear liq.; misc. with water; pH 5.2 ± 0.5.
Intratex® A. [Crompton & Knowles] Complex amino condensate; amphoteric; nonfoaming leveling agent for dyeing wool; tan liq.; water sol.; dens. 9 lb/gal.
Intratex® AN. [Crompton & Knowles] EO condensate; nonionic; leveling agent for wool dyeing, dye stripping; liq.; water sol.
Intratex® B. [Crompton & Knowles] Polyglycol ether deriv.; amphoteric; leveling agent, detergent, for dye leveling of wool; tan liq.; water sol.; dens. 8.7 lb/gal.
Intratex® BD. [Crompton & Knowles] Sodium salt of an aromatic nitro sulfonic acid; anionic; dyeing and printing assistant for cellulose and viscose; antireduction agent for use with direct, reactive, disperse, and vat dyes; sol'ns. are stable to acids, alkalis, and hard water; straw gran.; sol. in hot water; pH 9-10 (1%).
Intratex® C. [Crompton & Knowles] Complex amino condensate; amphoteric; leveling agent for dyeing wool; clear liq.; dens. 8 lb/gal.
Intratex® CA-2. [Crompton & Knowles] Aliphatic ethoxylates; cationic; compatibilizer; liq.; 90% conc.
Intratex® DD. [Crompton & Knowles] Surfactant blend; amphoteric; leveling agent for disperse, direct and acid dyes with penetrating, emulsifying, and compatibilizing props.; for cotton and blends

with polyester and nylon; ylsh. clear to sl. hazy pourable liq.; misc. with water; pH 6.5 ± 1.0.

Intratex® DD-LF. [Crompton & Knowles] Surfactant blend; amphoteric/anionic; low-foaming leveling agent for direct and disperse dyes on jet equip.; suitable for cotton and blends with polyester and nylon; creamy pourable paste; disp. in water; pH 6.5 ± 1.0.

Intratex® JD. [Crompton & Knowles] Surfactant blend; anionic; leveling agent for high temp. jet dyeing with disperse and direct dyes; for polyester and polyester/cotton blends; lubricant during dye cycle; low foaming, biodeg.; ylsh. clear to opalescent liq.; readily sol. in water, yields opalescent sol'ns. or emulsions at higher concs.; pH 5-7.

Intratex® JD-E. [Crompton & Knowles] Surfactant blend; anionic; leveling agent for high temp. jet dyeing with disperse and direct dyes; for polyester and polyester/cotton blends; lubricant during dye cycle; low foaming, biodeg.; wh. fluid paste; readily emulsifiable in water; pH 5-7.

Intratex® N. [Crompton & Knowles] Phenolic condensate; anionic; dyebath auxiliary, leveling and aftertreating agent for acid dyes on nylon carpet and apparel; dk. brn. liq.; misc. with water; pH 10.5 ± 0.5.

Intratex® N-1. [Crompton & Knowles] Phenolic condensate; anionic; dyebath auxiliary, leveling and aftertreating agent for acid dyes on nylon carpet and apparel; dk. brn. liq.; misc. with water; pH 8.0 ± 0.2.

Intratex® OR. [Crompton & Knowles] Low molecular polyamide; nonionic; leveling and penetrating agent for vat and direct dyes; liq.

Intratex® POK. [Crompton & Knowles] Sulfonated benzimidazole deriv.; anionic; dispersant, leveling agent for vat and disperse dyes; yel. brn. liq.; water sol.

Intratex® SCS. [Crompton & Knowles] Metallic salt; nonionic; for tannic acid after-treatment of acid dyed nylon.

Intratex® W New. [Crompton & Knowles] EO condensate; nonionic/cationic; leveling agent, penetrant, for wool dyeing; antiprecipitant for dyeing acrylic blends with anionic-cationic dyes by one-bath method; yel. liq.; water sol.; dens. 8.5 lb/gal.

Intravon® AN. [Crompton & Knowles] Fatty acid deriv.; scouring and emulsifying agent for wool, cotton, synthetics and blends, detergent in laundering, dyeing, household and cosmetic applics.; amber liq.; water sol.; dens. 8.4 lb/gal; sp.gr. 1.01.

Intravon® JET. [Crompton & Knowles] Surfactant blend; anionic/nonionic; detergent, emulsifier, low foaming scouring agent for textiles; for jet dyeing machines; yel. brn. liq.; water sol.

Intravon® JF. [Crompton & Knowles] EO condensate; nonionic; detergent, emulsifier, dispersant, wetting agent, penetrant for textile, household and cosmetic applics.; yel. liq.; water sol.; dens. 8.4–8.5 lb/gal.

Intravon® JU. [Crompton & Knowles] EO condensate; nonionic; detergent, emulsifier, dispersant, wetting agent, penetrant for textile, household and cosmetic applics.; antiprecipitant for dyeing acrylic blends with anionic/cationic dyes in a one-bath method; colorless to pale yel. visc. liq.; water-sol.; sp.gr. 1.02; dens. 8.5 lb/gal; pH 7.0 ± 0.5 (1%).

Intravon® NI. [Crompton & Knowles] Straight chain aliphatic EO condensate; nonionic; scouring, dyeing assistant for dyeing all fibers; low foaming, good detergency with alkaline builders; antiprecipitant for dyeing acrylic blends with anionic/cationic dyes in a one-bath method; off-wh. to yel. clear visc. liq.; misc. with water; pH 6.0-8.0.

Intravon® SO. [Crompton & Knowles] Surfactant blend; anionic/nonionic; detergent for laundering and household uses, wetting agent, emulsifier for scouring and dyeing textiles; for removal of oil, grease, and graphite; yel. clear visc. liq.; misc. with water; dens. 8.6 lb/gal; nonflamm.

Intravon® SOL. [Crompton & Knowles] Aromatic solv./surfactant blend; anionic/nonionic; solvent scouring detergent for syn. and natural fibers and their blends, one-bath dyeing/scouring processes; aftersoaping agent, wetting agent, penetrant; alkali stable; biodeg.; off-wh. to yel. gel; self-emulsifiable in water; flash pt. (PMCC) 52 C; pH 6.0 ± 0.5; toxicology: defatting on prolonged skin contact.

Intravon® SOL-N. [Crompton & Knowles] Aliphatic solv./surfactant blend; anionic/nonionic; low odor solv.-based scouring detergent for all fibers, in one-bath dyeing/scouring processes; oil solubilizer; good wetting and penetration, moderate foamer; stable to alkali; off-wh. to yel. flowable gel, low odor; water sol./emulsifiable at all temps.; f.p. < -3 C; flash pt. (PMCC) > 95 C; pH 6.0 ± 0.5; toxicology: defatting to skin on prolonged contact; eye irritant.

Intravon® SOL-W. [Crompton & Knowles] Aromatic solv./surfactant blend; anionic/nonionic; fast wetting and penetrating solv.-based scouring detergent for syn. and natural fibers and their blends, in one-bath dyeing/scouring processes; oil solubilizer; alkali stable; biodeg.; off-wh. to yel. gel; self-emulsifiable in water; flash pt. (PMCC) 52 C; pH 6.0 ± 0.5; toxicology: defatting to skin on prolonged contact.

Invermul. [NL Baroid] Modified calcium salts of higher org. acids; emulsifier for w/o emulsion drilling fluids, reduces filtration; powd.; 100% act.

Inversol 140. [Ferro/Keil] Complex polyglycol ester; nonionic; inversely sol. lubricity agent for metalworking liqs.; reduces surf. tens. and coeff. of friction; Gardner 4 liq.; sp.gr. 1.06; dens. 8.8 lb/gal; visc. 3200 cSt (40 C); pour pt. 16 C; flash pt. (COC) 191 C; acid no. 15; pH 7.0 (1% aq.); 100% conc.

Inversol 170. [Ferro/Keil] Complex polyglycol ester; nonionic; inversely sol. lubricity agent for metalworking liqs.; reduces surf. tens. and coeff. of friction; Gardner 5 liq.; sp.gr. 1.06; dens. 8.8 lb/gal; visc. 2750 cSt (40 C); pour pt. 13 C; flash pt. (COC) 191 C; acid no. 15; pH 7.1 (1% aq.); 100% conc.

Ionet 300. [Sanyo Chem. Industries] POE alkyl ether type surfactant; anionic; scouring agent for fibers; leveling agent; penetrant; dispersant; liq.

Ionet AT-140. [Sanyo Chem. Industries] POE amine; nonionic; emulsifier for chlorinated solvs.; corrosion inhibitor; hard water resistance; paste; 100% conc.

Ionet DL-200. [Sanyo Chem. Industries] POE dilaurate; CAS 9005-02-1; nonionic; emulsifier for emulsion polymerization, metal processing lubricant and personal care prods.; liq.; HLB 6.6; 100% conc.

Ionet DO-200. [Sanyo Chem. Industries] POE dioleate; CAS 9005-07-6; nonionic; emulsifier for emulsion polymerization of vinyl resins, for metal processing lubricants, cosmetics; liq.; HLB 5.3; 100% conc.

Ionet DO-400. [Sanyo Chem. Industries] POE dioleate; CAS 9005-07-6; nonionic; emulsifier for emulsion polymerization of vinyl resins, for metal processing lubricants, cosmetics; liq.; HLB 8.4; 100% conc.

Ionet DO-600. [Sanyo Chem. Industries] POE dioleate; CAS 9005-07-6; nonionic; emulsifier for emulsion polymerization of vinyl resins, for metal processing lubricants, cosmetics; liq.; HLB 10.4; 100% conc.

Ionet DO-1000. [Sanyo Chem. Industries] POE dioleate; CAS 9005-07-6; nonionic; emulsifier for emulsion polymerization of vinyl resins, for metal processing lubricants, cosmetics; solid; HLB 12.9; 100% conc.

Ionet DS-300. [Sanyo Chem. Industries] POE distearate; CAS 9005-08-7; nonionic; emulsifier for emulsion polymerization of vinyl resins, for metal processing lubricants, cosmetics; solid; HLB 7.3; 100% conc.

Ionet DS-400. [Sanyo Chem. Industries] POE distearate; CAS 9005-08-7; nonionic; emulsifier for emulsion polymerization of vinyl resins, for metal processing lubricants, cosmetics; solid; HLB 8.5; 100% conc.

Ionet LD-7-200. [Sanyo Chem. Industries] Surfactant; anionic; leveling agent for high temp. dyeing of polyester fibers; improves dispersibility of disperse dyes; liq.

Ionet MO-200. [Sanyo Chem. Industries] POE oleate; CAS 9004-96-0; nonionic; emulsifier for min. oil, pigment dispersions; base material for detergents; improves lubricity; liq.; HLB 8.4; 100% conc.

Ionet MO-400. [Sanyo Chem. Industries] POE oleate; nonionic; emulsifier for min. oil, pigment dispersants; base material for detergents; improves lubricity; liq.; HLB 11.8; 100% conc.

Ionet MS-400. [Sanyo Chem. Industries] POE stearate; CAS 9004-99-3; nonionic; emulsifier for min. oil, pigment dispersants; base material for detergents; improves lubricity; solid; HLB 11.9; 100% conc.

Ionet MS-1000. [Sanyo Chem. Industries] POE stearate; nonionic; emulsifier for min. oil, pigment dispersants; base material for detergents; improves lubricity; solid; HLB 15.7; 100% conc.

Ionet RAP-50. [Sanyo Chem. Industries] Nonionic/anionic surfactant blend; anionic; low foaming leveling agent for high temp. dyeing of polyester fibers; liq.

Ionet RAP-80. [Sanyo Chem. Industries] Nonionic/anionic surfactant blend; anionic; leveling agent for high temp. dyeing of polyester fibers; liq.

Ionet RAP-250. [Sanyo Chem. Industries] Nonionic/anionic surfactant blend; anionic; leveling agent, foam depressant, dispersant for high temp. dyeing of polyester fibers; liq.

Ionet S-20. [Sanyo Chem. Industries] Sorbitan laurate; CAS 1338-39-2; nonionic; emulsifier for personal care prods.; lubricant, rust inhibitor; pigment dispersant; spreading agent for agric. pesticides; base for textile lubricants; liq.; HLB 8.6; 100% conc.

Ionet S-60 C. [Sanyo Chem. Industries] Sorbitan stearate; CAS 1338-41-6; nonionic; see Ionet S-20; solid; HLB 4.7; 100% conc.

Ionet S-80. [Sanyo Chem. Industries] Sorbitan oleate; CAS 1338-43-8; nonionic; see Ionet S-20; liq.; HLB 4.3; 100% conc.

Ionet S-85. [Sanyo Chem. Industries] Sorbitan trioleate; CAS 26266-58-0; nonionic; see Ionet S-20; liq.; HLB 1.8; 100% conc.

Ionet T-20 C. [Sanyo Chem. Industries] Polysorbate 20; CAS 9005-64-5; nonionic; base and emulsifier for personal care prods., metal processing, lubricant and rust inhibitor; pigment dispersant; spreader sticker for agric. pesticides; base for textile lubricants; liq.; HLB 16.7; 100% conc.

Ionet T-60 C. [Sanyo Chem. Industries] POE sorbitan monostearate; CAS 9005-67-8; nonionic; see Ionet T-20 C; liq.; HLB 14.9; 100% conc.

Ionet T-80 C. [Sanyo Chem. Industries] POE sorbitan monooleate; CAS 9005-65-6; nonionic; see Ionet T-20 C; liq.; HLB 15.0; 100% conc.

Irgalev® PBF. [Ciba-Geigy/Dyestuffs] Diphenyl deriv.; anionic; dyeing assistant for nylon to minimize barre effects and promote level dyeing; liq.

Irgalube® 53. [Ciba-Geigy] POE alkyl ester and fatty ester; nonionic; dyeing assistant for fabrics; dye bath lubricant; paste.

Irgasol® DA Liq., Powd. [Ciba-Geigy/Dyestuffs] High m.w. aromatic condensate; CAS 26545-58-4; anionic; dyestuff dispersing agent; liq., powd.; 40% conc.

J

JAQ Powdered Quat. [Huntington Lab] Myristalkonium chloride; for formulation of disinfectants, sanitizers, and swimming pool algicides; 95% act.

Jet Amine DC. [Jetco] Coco diamine; CAS 61791-63-7; cationic; emulsifier, fuel oil and gasoline additives, corrosion inhibitors, mineral flotation; paste; 100% conc.

Jet Amine DE-13. [Jetco] Tridecyl ether diamine; CAS 22023-23-0; emulsifier, corrosion inhibitor; Gardner 5 max. liq.; pour pt. -33 F; 90% act.

Jet Amine DE 810. [Jetco] Octadecyl ether diamine; cationic; emulsifier, corrosion inhibitor; Gardner 5 max. liq.; pour pt. -30 F; 100% conc.

Jet Amine DMCD. [Jetco] Dimethyl coco amine; cationic; intermediate for quats., surfactants, agric., and detergent formulations; liq.; 100% conc.

Jet Amine DMOD. [Jetco] Dimethyl oleyl amine; CAS 14727-68-5; cationic; intermediate for quats., surfactants, agric., and detergent formulations; liq.; 100% conc.

Jet Amine DMSD. [Jetco] Dimethyl soya amine; cationic; intermediate for quats., surfactants, agric., and detergent formulations; liq.; 100% conc.

Jet Amine DMTD. [Jetco] Dimethyl tallow amine; cationic; intermediate for quats., surfactants, agric., and detergent formulations; liq.; 100% conc.

Jet Amine DO. [Jetco] Oleyl diamine; CAS 7173-62-8; cationic; emulsifier, fuel oil and gasoline additives, corrosion inhibitors, mineral flotation; liq.; 100% conc.

Jet Amine DT. [Jetco] Tallow diamine; CAS 61791-55-7; cationic; emulsifier, fuel oil and gasoline additives, corrosion inhibitors, mineral flotation; paste; 100% conc.

Jet Amine M2C. [Jetco] Methyl dicocamine; CAS 61788-62-3; cationic; intermediate for quats., surfactants, agric., and detergent formulations; liq.; 100% conc.

Jet Amine PC. [Jetco] Cocamine; CAS 61788-46-3; cationic; corrosion inhibitor, flotation agent, emulsifier, mold release agent, lube oil additive, fertilizer anticaking agent, fabric finishing; liq.; 100% conc.

Jet Amine PCD. [Jetco] Distilled cocamine; CAS 61788-46-3; cationic; corrosion inhibitor, flotation agent, emulsifier, mold release agent, lube oil additive, fertilizer anticaking agent; liq.; 100% conc.

Jet Amine PE 08/10. [Jetco] Ether amine; cationic; ore flotation, emulsifier, corrosion inhibitor, fuel oil additive; liq.; 100% conc.

Jet Amine PE 1214. [Jetco] Dodecyl/tetradecyl ether amine; CAS 68511-41-1; emulsifier, corrosion inhibitor; Gardner 2 max. liq.; pour pt. 4 F; 90% act.

Jet Amine PHT. [Jetco] Hydrog. tallow amine; CAS 61788-45-2; cationic; corrosion inhibitor, flotation agent, emulsifier, mold release agent, lube oil additive, fertilizer anticaking agent; solid; 100% conc.

Jet Amine PHTD. [Jetco] Distilled hydrog. tallow amine; CAS 61788-45-2; cationic; corrosion inhibitor, flotation agent, emulsifier, mold release agent, lube oil additive, fertilizer anticaking agent; solid; 100% conc.

Jet Amine PO. [Jetco] Oleamine; CAS 112-90-3; cationic; corrosion inhibitor, flotation agent, emulsifier, mold release agent, lube oil additive, fertilizer anticaking agent; liq.; 100% conc.

Jet Amine POD. [Jetco] Distilled oleamine; CAS 112-90-3; cationic; corrosion inhibitor, flotation agent, emulsifier, mold release agent, lube oil additive, fertilizer anticaking agent; liq.; 100% conc.

Jet Amine PS. [Jetco] Soyamine; CAS 61790-18-9; cationic; corrosion inhibitor, flotation agent, emulsifier, mold release agent, lube oil additive, fertilizer anticaking agent; paste; 100% conc.

Jet Amine PSD. [Jetco] Distilled soyamine; CAS 61790-18-9; cationic; corrosion inhibitor, flotation agent, emulsifier, mold release agent, lube oil additive, fertilizer anticaking agent; paste; 100% conc.

Jet Amine PT. [Jetco] Tallowamine; CAS 61790-33-8; cationic; corrosion inhibitor, flotation agent, emulsifier, mold release agent, lube oil additive, fertilizer anticaking agent; solid; 100% conc.

Jet Amine PTD. [Jetco] Distilled tallowamine; CAS 61790-33-8; cationic; corrosion inhibitor, flotation agent, emulsifier, mold release agent, lube oil additive, fertilizer anticaking agent; solid; 100% conc.

Jet Amine TET. [Jetco] Tallow tetramine; CAS 68911-79-5; gasoline detergent, corrosion inhibitor, in petrol. prods., dispersion agents for min. pigments in org. vehicles, asphalt emulsifier; Gardner 7 max. paste.

Jet Amine TRT. [Jetco] Tallow triamine; CAS 61791-57-9; gasoline detergent, corrosion inhibitor, in petrol. prods., dispersion agents for min. pigments in org. vehicles, asphalt emulsifier; Gardner 5 paste.

Jet Amine TT. [Jetco] Tallow pentamine; gasoline detergent, corrosion inhibitor, in petrol. prods., dispersion agents for min. pigments in org. vehicles, asphalt emulsifier; Gardner 8 max. paste.

Jet Quat 2C-75. [Jetco] Dicoco dimethyl ammonium chloride; CAS 61789-77-3; cationic; bactericide, textile softener, asphalt emulsifier, petrol. processing; Gardner 5 max. liq.; pH 6-9; 75% conc. in IPA/water.

Jet Quat 2HT-75. [Jetco] Dihydrog. tallow dimethyl ammonium chloride; CAS 61789-80-8; cationic; bactericide, textile softener, asphalt emulsifier, petrol. processing; liq.; 75% conc.

Jet Quat C-50. [Jetco] Coco trimethyl ammonium

chloride; CAS 61789-18-2; cationic; bactericide, textile softener, asphalt emulsifier, petrol. processing; home and personal care prods.; Gardner 5 max. liq.; pH 6-9; 50% conc. in IPA/water.

Jet Quat DT-50. [Jetco] Methyl quat. of tallow diamine; CAS 68607-29-4; cationic; bactericide, textile softener, asphalt emulsifier, petrol. processing; Gardner 8 max. liq.; pH 6-9; 50% conc. in IPA/water.

Jet Quat S-2C-50. [Jetco] Soya and dicocoammonium chlorides blend; cationic; textile softener, corrosion inhibitor, petrol. processing emulsifier, antistat, hair conditoner, home laundry softener; liq.; 50% conc.

Jet Quat S-50. [Jetco] Soya trimethyl ammonium chloride; CAS 61790-41-8; cationic; bactericide, textile softener, asphalt emulsifier, petrol. processing; Gardner 6 max. liq.; pH 6-9; 50% conc. in IPA/water.

Jet Quat T-2C-50. [Jetco] Tallow dicoco quat. ammonium chloride; cationic; bactericide, textile softener, asphalt emulsifier, petrol. processing; liq.; 50% conc.

Jet Quat T-27W. [Jetco] Tallow trimethyl ammonium chloride; CAS 8030-78-2; cationic; bactericide, textile softener, asphalt emulsifier, petrol. processing; Gardner 4 max. liq.; pH 6-9; 27% conc. in IPA/water.

Jet Quat T-50. [Jetco] Tallow trimethyl ammonium chloride; CAS 8030-78-2; cationic; bactericide, textile softener, asphalt emulsifier, petrol. processing; Gardner 6 max. liq.; pH 6-9; 50% conc. in IPA/water.

Jordapon® CI-60 Flake. [PPG/Specialty Chem.] Sodium cocoyl isethionate; anionic; mild, high foaming surfactant for syn. detergent soap bars; lime soap dispersant; flake; 50% act.

Jordapon® CI-Powd. [PPG/Specialty Chem.] Sodium cocoyl isethionate; anionic; mild high foaming detergent; lime soap dispersant; for syn. detergent soap bars; powd.; 82% act.

Juniorlan 1664. [Henkel/Organic Prods.] Lanolin wax and triolein; nonionic; emulsifying properties; solid; 100% conc.

J Wet 19A. [Sybron] anionic; nonrewetting wetting agent for flame retardant textile finishes.

K

Kadif 50 Flakes. [Witco UK] Sodium dodecylbenzene sulfonate; general purpose wetting and penetrating agent, detergent for industrial, institutional and household uses; lt. flakes; practically odorless; water sol.; pH 8.0 (1% aq.); 50% act.

Kalcohl 5-24, 6-24, 7-24. [Kao Corp. SA] Lauryl myristyl alcohol; raw material for sodium lauryl sulfate; metal rolling agent; liq.; 100% conc.

Kalcohl 10H. [Kao Corp. SA] n-Decyl alcohol; tobacco-offshoots controller; raw material for antioxidant, alkyl phosphate, ethoxylate, chloride; liq.; 100% conc.

Kalcohl 20. [Kao Corp. SA] Lauryl alcohol; CAS 112-53-8; tobacco-offshoots controller; raw material for antioxidant, alkyl phosphate, ethoxylate, chloride; liq.; 100% conc.

Kalcohl 40. [Kao Corp. SA] Myristyl alcohol; CAS 112-72-1; tobacco-offshoots controller; raw material for antioxidant, alkyl phosphate, ethoxylate, chloride; solid; 100% conc.

Kalcohl 60. [Kao Corp. SA] Cetyl alcohol; CAS 36653-82-4; tobacco-offshoots controller; raw material for antioxidant, alkyl phosphate, ethoxylate, chloride; beads or solid; 100% conc.

Kamar BL. [Finetex] Sulfonate; anionic; emulsifier for trichlorobenzene and other solvents; for textile scours and industrial cleaners; liq.; 85% conc.

Kamar L-15. [Finetex] Surfactant blend; nonionic; emulsifier for xylene, Varsol; liq.; 97% conc.

Kamar LF-7. [Finetex] Surfactant blend; nonionic; low foam emulsifier for trichlorobenzene, other solvs.; for jet carriers, low foam scours; paste; 97% conc.

Karafac 78. [Clark] Conc. surfactant; anionic; neutralized (potassium salt) version of Karaphos XFA.

Karafac 78 (LF). [Clark] Conc. surfactant; anionic; lower foam version of Karafac 78.

Karafac IBO. [Clark] Alkanolamine salt of 2-ethyl hexanol phosphate (Base 104); surfactant conc.

Kara Lube AL. [Rhone-Poulenc/Textile & Rubber] Surfactant blend; anionic; lubricant and dispersant for dyestuffs; liq.

Kara Lube AL-Conc. [Rhone-Poulenc/Textile & Rubber] Formulated prod.; anionic; fiber lubricant and dyestuff dispersant; 70% conc.

Karamide 121. [Clark] Cocamide DEA (1:1); heavy-duty detergent, thickener, foam stabilizer in cleaning compds., shampoos, textile scours; as lubricant, antistat.

Karamide 221. [Clark] Cocamide DEA (2:1); used in heavy-duty cleaners, lt.-duty high-foam detergents.

Karamide 363. [Clark] Cocamide DEA (1:1); heavy-duty detergent, thickener, foam stabilizer in cleaning compds., shampoos, textile scours; as lubricant, antistat.

Karamide 442-M. [Clark] Modified cocamide DEA (2:1); used in heavy-duty cleaners, lt.-duty high-foam detergents, lubricants, antistats.

Karamide CO2A. [Clark] Cocamide DEA (2:1); nonionic; detergent, base for floor cleaners, all-purpose cleaners; straw liq.

Karamide CO9A. [Clark] Cocamide DEA (2:1); nonionic; detergent, base for floor cleaners, all-purpose cleaners; amber liq.

Karamide CO22. [Clark] Cocamide DEA (2:1); nonionic; detergent, base to solubilize high alkaline salt content floor cleaners; amber liq.

Karamide HTDA. [Clark] Fatty alkanolamide; co-emulsifier, thickener for emulsions of oils and waxes; thickener for cleaners, shampoos; as textile softener base.

Karamide SDA. [Clark] Veg. oil alkanolamide; co-emulsifier for o/w systems; conditioner for natural fibers; oil-sol. lubricant.

Karamide ST-DEA. [Clark] Stearamide DEA; nonionic; thickener, emulsifier for veg. oil, min. oil, microcrystalline wax; cream solid.

Karapeg 200-DL. [Clark] PEG-4 dilaurate; CAS 9005-02-1; w/o emulsifier, dispersant, defoamer, coemulsifier, lubricant; yel. liq. to paste.

Karapeg 200-DO. [Clark] PEG-4 dioleate; CAS 9005-07-6; w/o emulsifier, dispersant, defoamer, coemulsifier, lubricant; amber liq.; oil-sol.

Karapeg 200-ML. [Clark] PEG-4 laurate; CAS 9004-81-3; emulsifier, coupling agent, solubilizer; yel. liq.

Karapeg 200-MO. [Clark] PEG-4 oleate; CAS 9004-96-0; w/o emulsifier for min. oil, fatty acids, solvents applics.; for solv. cleaners, degreasers, etc.; yel. liq.; oil-sol.

Karapeg 200-MS. [Clark] PEG-4 stearate; CAS 9004-99-3; emulsifier for fats and oils; provides softening and lubricating props.; wh. solid.

Karapeg 400-DL. [Clark] PEG-8 dilaurate; CAS 9005-02-1; lipophilic emulsifier used to solubilize min. oils, fats, and solvs.; yel. liq. to paste.

Karapeg 400-DO. [Clark] PEG-8 dioleate; CAS 9005-07-6; lipophilic emulsifier used to solubilize min. oils, fats, and solvs.; yel. liq. to paste.

Karapeg 400-DS. [Clark] PEG-8 distearate; CAS 9005-08-7; emulsifier, thickener; cream solid; oil-sol.

Karapeg 400-ML. [Clark] PEG-8 laurate; CAS 9004-81-3; emulsifier, coemulsifier for toiletry prep.; defoamer and leveling agent for latex paints; yel. liq.

Karapeg 400-MO. [Clark] PEG-8 oleate; CAS 9004-96-0; moderately water-sol. surfactant; yel. liq.

Karapeg 600-DL. [Clark] PEG-12 dilaurate; CAS

9005-02-1; lipophilic emulsifier used to solubilize min. oils, fats, and solvs.; yel. liq.

Karapeg 600-DO. [Clark] PEG-12 dioleate; CAS 9005-07-6; lipophilic emulsifier used to solubilize min. oils, fats, and solvs.; amber liq.

Karapeg 600-ML. [Clark] PEG-12 laurate; CAS 9004-81-3; emulsifier and coemulsifier for toiletry preps.; defoamer and leveling agent for latex paints; yel. liq.

Karapeg 600-MO. [Clark] PEG-12 oleate; CAS 9004-96-0; surfactant; yel. liq.; water-sol.

Karapeg 6000-DS. [Clark] PEG-150 distearate; CAS 9005-08-7; hydrophilic emulsifier; cream solid.

Karapeg DEG-DO. [Clark] PEG-2 dioleate; CAS 9005-07-6; w/o emulsifier, dispersant, defoamer, coemulsifier, lubricant; amber liq.; oil-sol.

Karapeg DEG-MO. [Clark] PEG-2 oleate; CAS 9004-96-0; coemulsifier for w/o or o/w emulsions; lubricant component in textile processing; yel. liq.

Karapeg DEG-MS. [Clark] PEG-2 stearate; CAS 9004-99-3; opacifier, pearlescent for liq. cosmetic and detergent compds.; wh. solid.

Karaphos HSPE. [Clark] Blended phosphate ester, free acid; hydrotrope and wetter in heavy-duty detergents; high caustic-stable wetter and penetrant in textile formulations; as antistat; as rust preventative in metal cleaners; very low foaming; exc. alkali stability.

Karaphos SWPE. [Clark] Blended phosphate ester, free acid; fast wetting agent for detergents, cleaners, textile scours, penetrants, metal prep. compds., antistats.

Karaphos XFA. [Clark] Phosphate co-ester, free acid; moderately low foaming base for detergents, metalworking and cleaning compds., antistats, textile scours, lubricants, etc.

Karasoft YB-11. [Clark] Conc. self-emulsifying softener base; produces soft, dry, fluffy hand when applied to cotton, blends, and synthetics.

Karasurf AS-26. [Clark] Ammonium 2-ethyl hexanol sulfate; wetter, penetrant, hydrotrope, solubilizer for industrial cleaners, fire fighting foams, etc.

Karawet CS. [Rhone-Poulenc/Textile & Rubber] Org. phosphate ester, complex salt; anionic; emulsifier; detergent, and wetting agent; liq.; 40% conc.

Karawet DOSS. [Rhone-Poulenc/Textile & Rubber] Sodium dioctyl sulfosuccinate; anionic; wetting and rewetting agent, dispersant, spreading agent, detergent for textile applications, emulsion polymerization; liq.; water-sol.; 70% conc.

Karawet SB. [Rhone-Poulenc/Textile & Rubber] Org. phosphate ester, complex salt; alkaline stable emulsifier, detergent and wetter; liq.; 50% conc.

Karox AO-30. [Clark] Cocamine oxide; high foaming, mild surfactant, wetter, emulsifier, and coupling agent with exc. alkali tolerance.

Karox LO. [Clark] Dimethyl lauramine oxide; high foaming, mild detergent, coupling agent, wetter, emulsifier with good alkali and bleach stability.

Katapol® OA-860. [Rhone-Poulenc Surf.] PEG-30 oleamine; hydrophilic emulsifier, leveler; textile dyeing assistant; antiprecipitant for dyeing processes; solid; water-sol.; 99% act. DISCONTINUED.

Katapol® OA-910 (redesignated Rhodameen® OA-910). [Rhone-Poulenc Surf.]

Katapol® PN-430 (redesignated Rhodameen® PN-430). [Rhone-Poulenc Surf.]

Katapol® PN-730. [Rhone-Poulenc Surf.] PEG-15 tallow amine; cationic; emulsifier, dye assistant, softener, antistat, lubricant for textiles and leathers; amber clear liq.; water sol.; HLB 14.0; 100% act. DISCONTINUED.

Katapol® PN-810 (redesignated Rhodameen® PN-810). [Rhone-Poulenc Surf.] PEG-20 tallow amine; cationic; emulsifier, dye assistant, softener, antistat, lubricant for textiles, leathers, agric. formulations; amber clear liq.; water sol.; 99% min. act. DISCONTINUED.

Katapol® SPB. [Rhone-Poulenc Surf.] PEG alkyl amine; cationic; dispersant for fiberglass strands in mfg. of fiberglass mats; retardant in applic. of cationic dyes to acrylic fibers; antiprecipitant for acid and cationic dyes; liq. DISCONTINUED.

Kelacid®. [Kelco] Alginic acid; used as gelling agent, emulsifier and stabilizer in food, pharmaceutical, and industrial applics.; stabilizer in paper and textile industry; wh. fibrous particles; sol. in alkaline sol'n.; swells in water; pH 2.9 (1% aq.); surf. tens. 53 dynes/cm; 7% moisture.

Kelco® HV. [Kelco] Low-calcium sodium alginates; used as gelling agent, emulsifier and stabilizer in food, pharmaceutical, and industrial applics.; stabilizer in paper and textile industry; cream fibrous particles; sp.gr. 1.64; dens. 43.38 lb/ft^3; visc. 400 cps; ref. index 1.3342; pH 7.2; 9% moisture.

Kelco® LV. [Kelco] Low-calcium sodium alginates; used as gelling agent, emulsifier and stabilizer in food, pharmaceutical, and industrial applics.; stabilizer in paper and textile industry; cream fibrous particles; sp.gr. 1.64; dens. 43.38 lb/ft^3; visc. 50 cps; ref. index 1.3342; pH 7.2; 9% moisture.

Kelcoloid® D. [Kelco] Propylene glycol alginate; used as gelling agent, emulsifier and stabilizer in food, pharmaceutical, and industrial applics.; stabilizer in paper and textile industry; cream fibrous particles; sp.gr. 1.46; dens. 33.71 lb/ft^3; visc. 170 cps; ref. index 1.3343; pH 4.4; surf. tens. 58 dynes/cm; 13% max. moisture.

Kelcoloid® DH, DO, DSF. [Kelco] Propylene glycol alginate; used as gelling agent, emulsifier and stabilizer in food, pharmaceutical, and industrial applics.; stabilizer in paper and textile industry; cream agglomerated; sp.gr. 1.46; dens. 33.71 lb/ft^3; visc. 400, 25, 20 cps resp.; ref. index 1.3343; pH 4.0, 4.3, 4.0 resp.; surf. tens. 58 dynes/cm.

Kelcoloid® HVF, LVF, O, S. [Kelco] Propylene glycol alginate; used as gelling agent, emulsifier and stabilizer in food, pharmaceutical, and industrial applics.; stabilizer in paper and textile industry; cream fibrous particles; sp.gr. 1.46; dens. 33.71 lb/ft^3; visc. 400, 120, 25, 20 cps resp.; ref. index 1.3343; pH 4.0, 4.0, 4.3, 4.0 resp.; surf. tens. 58 dynes/cm.

Kelcosol®. [Kelco] Algin; used as gelling agent, emulsifier and stabilizer in food, pharmaceutical, and industrial applics.; stabilizer in paper and textile industry; cream fibrous particles; water-sol.; sp.gr. 1.64; dens. 43.38 lb/ft^3; visc. 1300 cps; pH 7.2; surf. tens. 70 dynes/cm; 9% moisture.

Kelgin® F. [Kelco] Algin, refined; used as gelling agent, emulsifier and stabilizer in food, pharmaceutical, and industrial applics.; stabilizer in paper and textile industry; ivory gran. particles; sp.gr. 1.59; dens. 54.62 lb/ft^3; visc. 300 cps; ref. index 1.3343; surf. tens. 62 dynes/cm; 13% moisture.

Kelgin® HV, LV, MV. [Kelco] Algin; used as gelling agent, emulsifier and stabilizer in food, pharmaceutical, and industrial applics.; stabilizer in paper and textile industry; ivory gran. particles; sp.gr. 1.59; dens. 54.62 lb/ft^3; visc. 800, 60, 400 cps resp.; ref.

index 1.3343; pH 7.5; surf. tens. 62 dynes/cm; 13% moisture.

Kelgin® QL. [Kelco] Treated sodium alginate; used as gelling agent, emulsifier and stabilizer in food, pharmaceutical, and industrial applics.; stabilizer in paper and textile industry; improved disp.; ivory gran. particles; visc. 30 cps; pH neutral.

Kelgin® XL. [Kelco] Refined sodium alginate; used as gelling agent, emulsifier and stabilizer in food, pharmaceutical, and industrial applics.; stabilizer in paper and textile industry; ivory gran. particles; sp.gr. 1.59; dens. 54.62 lb/ft^3; visc. 30 cps; ref. index 1.3343; pH 7.5; surf. tens. 62 dynes/cm; 13% moisture.

Kelig 100. [LignoTech] Sodium lignosulfonate; anionic; metal cleaning component for industrial cleaning applics.; corrosion inhibitor; powd., liq.; 55% liq., 95% powd.

Kelmar®. [Kelco] Potassium alginate; gellant, emulsifier, and stabilizer in food and indust. applic.; gum, bodying agent for creams and lotions, dental impression materials; used for water holding in foods and industry; cream gran. and fibrous particles resp.; water-sol.; visc. 270 and 400 cps resp.; pH neutral.

Kelmar® Improved. [Kelco] Potassium alginate; gellant, emulsifier, and stabilizer in food and indust. applic.; used for water holding in foods and industry; cream granular particles; visc. 400 cps; pH neutral.

Kelset®. [Kelco] Alginate; used as gelling agent, emulsifier and stabilizer in food, pharmaceutical, and industrial applics.; stabilizer in paper and textile industry; self-gelling gum; lt. ivory fibrous particles; water-sol.; pH neutral.

Keltose®. [Kelco] Calcium alginate and ammonium alginate; gellant, binder, emulsifier, and stabilizer in food and indust. applic.; ivory gran. particles; water-sol.; visc. 250 cps; pH neutral.

Kelvis®. [Kelco] Sodium alginate, refined; gellant, emulsifier, and stabilizer in food and indust. applic.; ivory gran. particles; water-sol.; sp.gr. 1.59; dens. 54.62 lb/ft^3; visc. 760 cps; ref. index 1.3343; pH 7.5; surf. tens. 62 dynes/cm; 13% moisture.

Kemamide® B. [Witco/Humko] Behenamide; lubricant, slip, antiblock, and mold release agent for plastics, crayons, petrol. prods., asphalts, inks, metals, textiles; mold release agent for thermoplastic resins in inj. molding; defoamer and water repellent in industrial/household applic.; corrosion inhibitor; pigment grinding aid and dyestuff dispersant in paints, enamels, varnishes, and lacquers; intermediate for textile emulsifiers and softeners; foam stabilizer in household detergents; FDA compliance; Gardner 4 max. waxy solid, powd., and pellet; m.w. 312; sol. (g/100 g solv. @ 60 C); > 10 g in IPA, MEK, 10 g in methanol and toluene; water-insol.; dens. 0.807 g/ml (130 C); visc. 6.5 cP (130 C); m.p. 98–106 C; acid no. 4 max.; iodine no. 4 max.; flash pt. 257 C (COC); 95% conc.

Kemamide® E. [Witco/Humko] Erucamide; see Kemamide B; FDA compliance; Gardner 5 max solid; m.w. 335; sol. (g/100 g solv.) 25 g in chloroform, 8 g in IPA, 4 g in MEK, 3 g in methyl alcohol and toluene; water insol.; m.p. 76–86 C; acid no. 4 max.; iodine no. 70-80; 100% conc.

Kemamide® O. [Witco/Humko] Oleamide, tech.; see Kemamide B; FDA compliance; Gardner 7 max. solid; m.p. 68-78 C; acid no. 5 max.; iodine no. 72-90; 100% conc.

Kemamide® S. [Witco/Humko] Stearamide; see Kemamide B; FDA compliance; Gardner 4 max. waxy solid, powd., and pellet; m.w. 278; sol. (g/100 g solv. @ 50 C) > 10 g in chloroform, 10 g in IPA, 4 g in MEK, 3 g in methyl alcohol, 2 g in toluene; dens. 0.809 g/ml (130 C); visc. 5.8 cP (130 C); m.p. 98–108 C; acid no. 4 max.; iodine no. 3 max.; flash pt. 246 C (COC); fire pt. 268 C (COC); 95% min. amide.

Kemamide® S-221. [Witco/Humko] Erucyl stearamide; lubricant additive, friction modifier for high-temp. plastics applics.; wh. powd.; sol. (g/100 g solv. @ 50 C) < 50 g in chloroform and toluene, 20 g in VM&P naphtha, 14 g in dichloroethane and IPA, 7 g in MEK; dens. 0.7877 g/ml (110 C); m.p. 72–75 C; acid no. 8 max.; iodine no. 46; flash pt. 268 C (CC).

Kemamide® U. [Witco/Humko] Oleamide; see Kemamide B; FDA compliance; Gardner 5 max. waxy solid, powd., and pellet; m.w. 275; sol. (g/100 g solv. @ 30 C) > 30 g in IPA, 28 g in methyl alcohol, 25 g in toluene, > 20 g in MEK; dens. 0.823 g/ml (130 C); visc. 5.5 cP (130 C); m.p. 68–78 C; acid no. 4 max.; iodine no. 72-90; flash pt. 245 C (COC); 95% min. amide.

Kemamide® W-20. [Witco/Humko] Ethylene dioleamide; see Kemamide B; also as internal and external lubricants in ABS, PS, polyethylene, PP, PVC, nylon, cellulose acetate, PVAc, and phenolic resins; defoamer in paper industry blk. liquoring, fabric dyeing, latex systems; metal processing, asphalts; FDA compliance; Gardner 6 max. powd. and flake; sol. (g/100 g solv. @ 35 C) > 20 g in toluene, 12 g in IPA, 4 g in dichloroethane, 3 g in MEK; m.p. 120 C; acid no. 10 max.; flash pt. 296 C (COC); fire pt. 315 C (COC).

Kemamide® W-39. [Witco/Humko] Ethylene distearamide; see Kemamide B and W-20; FDA compliance; Gardner 18 max. flakes; sol. (g/100 g solv. @ 70 C) 2.0 g toluene, 1.6 g dichloroethane; 1.4 g IPA, 0.9 g MEK; m.p. 140 C; acid no. 10 max.; flash pt. 299 C (COC); fire pt. 315 C (COC).

Kemamide® W-40. [Witco/Humko] Ethylene distearamide; see Kemamide B and W-20; FDA compliance; Gardner 3 max. powd. and flake; sol. (g/100 g solv. @ 70 C) 2.0 g toluene, 1.6 g dichloroethane; 1.4 g IPA, 0.9 g MEK; m.p. 140 C; acid no. 10 max.; flash pt. 299 C (COC); fire pt. 315 C (COC).

Kemamide® W-40/300. [Witco/Humko] Ethylene distearamide; see Kemamide B; Gardner 3 max. atomized very fine powd.; m.p. 140 C; acid no. 10 max.

Kemamide® W-45. [Witco/Humko] Ethylene distearamide; see Kemamide B and W-20; FDA compliance; Gardner 3 max. powd. and flake; sol. (g/100 g solv. @ 70 C) 2.0 g toluene, 1.6 g dichloroethane; 1.4 g IPA, 0.9 g MEK; m.p. 145 C; acid no. 10 max.; flash pt. 304 C (COC); fire pt. 322 C.

Kemamine® BQ-2802C. [Witco/Humko] Behenalkonium chloride; cationic; antistat, textile softener, dyeing aid, corrosion inhibitor, emulsifier; used in personal care prods., e.g., creams, lotions, shampoos, hair conditioners; Gardner 4 max.; water-disp. or misc.; m.w. 475; pH 9 max. (5%); 75% min. act.

Kemamine® BQ-9702C. [Witco/Humko] Dimethyl hydrog. tallow benzyl ammonium chloride; cationic; germicide, sanitizer, slimicide, antistat, textile softener, dyeing aid, corrosion inhibitor, emulsifier; also for personal care prods.; Gardner 4 max. liq.; m.w. 420; disp. in water; pH 9 max. (5%); 75% act.

Kemamine® BQ-9742C. [Witco/Humko] Tallowalkonium chloride; cationic; antistat, textile softening agent, dyeing aid, corrosion inhibitor, emulsifier, leveling agent, shampoo and cream rinse conditioner; Gardner 6 max.; water-misc.; m.w. 420; pH 9 (5%); 75% min. act.

Kemamine® D-190. [Witco/Humko] Arachidyl-behenyl 1,3-propylenediamine; cationic; gasoline detergent, bactericide, corrosion inhibitor in petrol. prod., epoxy hardener, dispersant, asphalt emulsifier; Gardner 9 max. solid; iodine no. 10 max.; 88% conc.

Kemamine® D-650. [Witco/Humko] N-coconut 1,3-propylenediamine; cationic; gasoline detergent, bactericide, corrosion inhibitor in petrol. prod., epoxy hardener, dispersant, asphalt emulsifier; Gardner 6 max. paste; 88% conc.

Kemamine® D-970. [Witco/Humko] N-hydrog. tallow 1,3-propylenediamine; cationic; gasoline detergent, bactericide, corrosion inhibitor in petrol. prod., epoxy hardener, dispersant, asphalt emulsifier; Gardner 6 max. solid; 88% conc.

Kemamine® D-974. [Witco/Humko] Tallowaminopropylamine; cationic; gasoline detergent, bactericide, corrosion inhibitor in petrol. prod., epoxy hardener, dispersant, asphalt emulsifier; Gardner 12 max. solid; 88% conc.

Kemamine® D-989. [Witco/Humko] N-Oleyl 1,3-propylenediamine; cationic; gasoline detergent, bactericide, corrosion inhibitor in petrol. prod., epoxy hardener, dispersant, asphalt emulsifier; Gardner 7 max. liq.; 88% conc.

Kemamine® D-999. [Witco/Humko] Soyaminopropylamine; gasoline detergent, bactericide, corrosion inhibitor in petrol. prod., epoxy hardener, dispersant, asphalt emulsifier; Gardner 8 max. paste; iodine no. 70 min.; 85% min. act.

Kemamine® P-150, P-150D. [Witco/Humko] 50% Arachidyl-behenyl primary amine (P-150D—dist.); cationic; emulsifier, flotation agent, dispersing and flushing agent, intermediate, used in metalworking oils, as fuel oil additive; mold release for rubber and plastics; lubricant and spinning aid in metalworking oils; Gardner 3 and 1 resp., solid; sol. in common org. solv.; 93 and 97% conc.

Kemamine® P-190, P-190D. [Witco/Humko] 90% Arachidyl-behenyl primary amine (P-190D—dist.); cationic; emulsifier, flotation agent, dispersing and flushing agent, intermediate, used in metalworking oils, as fuel oil additive; mold release for rubber and plastics; lubricant and spinning aid in metalworking oils; Gardner 3 and 1 resp., solid; sol. in common org. solv.; 93, 97% conc.

Kemamine® P-650D. [Witco/Humko] Cocamine, dist.; cationic; emulsifier, flotation agent, dispersing and flushing agent, intermediate, used in metalworking oils, as fuel oil additive; mold release for rubber and plastics; lubricant and spinning aid in metalworking oils; Gardner 1 liq.; sol. in common org. solv.; iodine no. 12 max.; 97% conc.

Kemamine® P-880, P-880D. [Witco/Humko] Palmityl primary amine (tech. and dist. resp.); cationic; emulsifier, flotation agent, dispersing and flushing agent, intermediate, used in metalworking oils, as fuel oil additive; mold release for rubber and plastics; lubricant and spinning aid in metalworking oils; Gardner 3 and 1 resp., solid; sol. in common org. solv.; 93, 97% conc.

Kemamine® P-970. [Witco/Humko] Hydrog. tallow amine (tech.); CAS 61788-45-2; cationic; emulsifier, flotation agent, dispersing and flushing agent, intermediate, used in metalworking oils, as fuel oil additive; mold release for rubber and plastics; lubricant and spinning aid in metalworking oils; Garder 3 solid; sol. in common org. solvs.; iodine no. 3 max.; 93% conc.

Kemamine® P-970D. [Witco/Humko] Hydrog. tallow amine (dist.); CAS 61788-45-2; cationic; emulsifier, flotation agent, dispersing and flushing agent, intermediate, used in metalworking oils, as fuel oil additive; mold release for rubber and plastics; lubricant and spinning aid in metalworking oils; Gardner 1 solid; sol. in common org. solv.; iodine no. 3 max.; 93, 97% conc.

Kemamine® P-974D. [Witco/Humko] Tallow amine (dist.); CAS 61790-33-8; cationic; emulsifier, flotation agent, dispersing and flushing agent, intermediate, used in metalworking oils, as fuel oil additive; mold release for rubber and plastics; lubricant and spinning aid in metalworking oils; Gardner 1 paste; sol. in common org. solv.; iodine no. 38 min.; 97% conc.

Kemamine® P-989D. [Witco/Humko] Oleamine, dist.; CAS 112-90-3; cationic; emulsifier, flotation agent, dispersing and flushing agent, intermediate, used in metalworking oils, as fuel oil additive; mold release for rubber and plastics; lubricant and spinning aid in metalworking oils; Gardner 1 liq.; sol. in common org. solv.; iodine no. 70 min.; 97% conc.

Kemamine® P-990, P-990D. [Witco/Humko] Stearamine (tech. and dist. resp.); cationic; emulsifier, flotation agent, dispersing and flushing agent, intermediate, used in metalworking oils, as fuel oil additive; mold release for rubber and plastics; lubricant and spinning aid in metalworking oils; Gardner 3 and 1 resp., solid; sol. in common org. solv.; 93, 97% conc.

Kemamine® P-999. [Witco/Humko] Tech. oleic-linoleic amine; emulsifier, flotation agent, dispersing and flushing agent, intermediate, used in metalworking oils, as fuel oil additive; mold release for rubber and plastics; lubricant and spinning aid in metalworking oils; Gardner 5 max.; iodine no. 85 min.; 93% conc.

Kemamine® Q-1902C. [Witco/Humko] Dibehenyl/diarachidyl dimonium chloride; cationic; germicide, sanitizer, slimicide, antistat, textile softening agent, dyeing aid, corrosion inhibitor, emulsifier; Gardner 4 max. solid; m.w. 680; disp. in water; pH 9 max (5%); 75% act.

Kemamine® Q-2802C. [Witco/Humko] Dibehenyl-dimonium chloride; antistat, textile softener, dyeing aid, corrosion inhibitor, emulsifier; used in personal care prods., e.g., creams, lotions, shampoos, hair conditioners; Gardner 4 max.; m.w. 690; pH 9 max. (5%); 75% min. act.

Kemamine® Q-6502C. [Witco/Humko] Dimethyl dicoconut ammonium chloride; cationic; germicide, sanitizer, slimicide, antistat, textile softener, dyeing aid, corrosion inhibitor, emulsifier; also for personal care prods.; Gardner 4 max. liq.; m.w. 465; disp. in water; pH 9 max. (5%); 75% act.

Kemamine® Q-9702C. [Witco/Humko] Quaternium-18; CAS 61789-80-8; cationic; antistat, textile softener, dyeing aid, corrosion inhibitor, emulsifier for personal care prods., e.g., creams, lotions, shampoos, hair conditioners; Gardner 2 max. solid; disp. in water; m.w. 575; pH 9 max. (5%); 75% min. act.

Kemamine® Q-9743C. [Witco/Humko] Tallowtrimonium chloride; cationic; antistat, textile softening agent, dyeing aid, corrosion inhibitor, emulsifier; Gardner 4 max.; water-misc.; m.w. 335–355; pH 9 (5%); 65% min. act.

Kemamine® Q-9743CHGW. [Witco/Humko] Tallowtrimonium chloride; cationic; antistat, textile softening agent, dyeing aid, corrosion inhibitor, emulsifier; Gardner 4 max.; water-misc.; m.w. 335–355; pH 9 (5%); 65% min. act.

Kemamine® T-1902D. [Witco/Humko] Dist. dimethyl-90% arachidyl-behenyl tert. amine; cationic; chemical intermediate for quat. ammonium derivs. used for cosmetics and textiles; acid scavenger in petrol. prods.; epoxy hardener, catalyst in mfg. of flexible PU foams; Gardner 1 max. liq.; 95% conc.

Kemamine® T-2801. [Witco/Humko] Dibehenyl methylamine.

Kemamine® T-2802D. [Witco/Humko] Dimethyl behenamine.

Kemamine® T-6501. [Witco/Humko] Methyl dicoconut tert. amine; cationic; chemical intermediate for quat. ammonium derivs. used for cosmetics and textiles; acid scavenger in petrol. prods.; epoxy hardener, catalyst in mfg. of flexible PU foams; Gardner 3 max. liq.; 95% conc.

Kemamine® T-6502. [Witco/Humko] Dimethyl cocamine.

Kemamine® T-6502D. [Witco/Humko] Dist. dimethyl cocamine; cationic; chemical intermediate for quat. ammonium derivs. used for cosmetics and textiles; acid scavenger in petrol. prods.; epoxy hardener, catalyst in mfg. of flexible PU foams; Gardner 1 max.; 95% conc.

Kemamine® T-8902. [Witco/Humko] Dimethyl palmitamine.

Kemamine® T-9701. [Witco/Humko] Dihydrog. tallow methylamine; CAS 61788-63-4; cationic; chemical intermediate for quat. ammonium derivs. used for cosmetics and textiles; acid scavenger in petrol. prods.; epoxy hardener, catalyst in mfg. of flexible PU foams; corrosion inhibitor; gasoline additive; Gardner 3 max. liq.; 95% conc.

Kemamine® T-9702D. [Witco/Humko] Dist. dimethyl hydrog. tallow amine; cationic; chemical intermediate for quat. ammonium derivs. used for cosmetics and textiles; acid scavenger in petrol. prods.; epoxy hardener, catalyst in mfg. of flexible PU foams; Gardner 1 max. liq.; 95% conc.

Kemamine® T-9742D. [Witco/Humko] Dimethyl hydrog. tallow amine; CAS 68391-07-1; cationic; intermediate for mfg. of quat. ammonium chlorides; corrosion inhibitor; gasoline additive; liq.; 95% conc.

Kemamine® T-9892D. [Witco/Humko] Dist. dimethyl oleamine; intermediate for quats. used in cosmetics and textiles; acid scavenger in petrol. prods.; Gardner 1 max. color; 95% tert. amine.

Kemamine® T-9902. [Witco/Humko] Dimethyl stearamine; intermediate for quats. used in cosmetics and textiles; acid scavenger in petrol. prods.; Gardner 1 max. color; 95% min. tert. amine.

Kemamine® T-9902D. [Witco/Humko] Dist. dimethyl stearamine; CAS 124-28-7; cationic; chemical intermediate for quat. ammonium derivs.; acid scavenger in petrol. prods.; epoxy hardener, catalyst in mfg. of flexible PU foams; corrosion inhibitor; gasoline additive; Gardner 1 max. liq.; 95% conc.

Kemamine® T-9972. [Witco/Humko] Dimethyl soyamine.

Kemamine® T-9972D. [Witco/Humko] Dist. dimethyl soyamine; cationic; chemical intermediate for quat. ammonium derivs. used for cosmetics and textiles; acid scavenger in petrol. prods.; epoxy hardener, catalyst in mfg. of flexible PU foams; Gardner 1 max. liq.; 95% conc.

Kemamine® T-9992D. [Witco/Humko] Dist. dimethyl oleic-linolenic amine; CAS 68037-96-7; intermediate for quats. used in cosmetics and textiles; acid scavenger in petrol. prods.; corrosion inhibitor; gasoline additive; Gardner 2 max. color; 95% tert. amine.

Kemester® 104. [Witco/Humko] Methyl oleate; nonionic; emulsifier, emollient for cosmetics; lubricant for leather; carrier for agric. spray prods.; Gardner 2 max. liq.; acid no. 2 max.; iodine no. 84-92; sapon. no. 190–200; 100% conc.

Kemester® 115. [Witco/Humko] Methyl oleate; intermediate in prod. of superamides, in metalworking lubricants, as solv.; Gardner 4 max.; acid no. 4.0 max.; iodine no. 90 max.; sapon. no. 185–205.

Kemester® 143. [Witco/Humko] Methyl tallowate; intermediate in prod. of superamides, in metalworking lubricants, as solv.; lubricant, plasticizer for cosmetics, leather, rubber prods.; Gardner 8 max.; acid no. 4.0 max.; iodine no. 60 max.; sapon. no. 195 min.

Kemester® 205. [Witco/Humko] Methyl oleate; intermediate in prod. of superamides, in metalworking lubricants, as specialized solv.; opacifier, visc. control agent; Gardner 1 liq.; acid no. 4.0 max.; iodine no. 80-90; sapon. no. 194–203; 100% conc.

Kemester® 213. [Witco/Humko] Methyl oleate/linoleate; intermediate in prod. of superamides, in metalworking lubricants, as solv.; Gardner 3 max.; acid no. 10.0 max.; iodine no. 130-145; sapon. no. 194–204.

Kemester® 226. [Witco/Humko] Methyl soyate; intermediate in prod. of superamides, in metalworking lubricants, as specialized solv.; opacifier, visc. control agent; Gardner 4 max. liq.; acid no. 7.0 max.; iodine no. 125-135; sapon. no. 195–200; 100% conc.

Kemester® 1000. [Witco/Humko] Triolein; nonionic; emollient used in cosmetics, textiles, leather, metalworking lubricants; base for sulfation; Gardner 6 max. liq.; m.p. –8 C; acid no. 5 max.; iodine no. 85-90; sapon. no. 190–198; 100% conc.

Kemester® 2000. [Witco/Humko] Glycerol oleate; CAS 111-03-5; nonionic; emollient, emulsifier, stabilizer, wetting agent for textiles, personal care prods.; Gardner 6 max. liq.; m.p. 25 C; acid no. 3 max.; iodine no. 73-83; sapon. no. 160–175; 100% conc.

Kemester® 2050. [Witco/Humko] Methyl eicosenate; intermediate in prod. of alkanolamides, in metalworking lubricants, as specialized solvs.; foam depressant and nutrient in fermentation; Gardner 14 max. color; acid no. 25 max.; iodine no. 90-110; sapon. no. 180–190.

Kemester® 4000. [Witco/Humko] Butyl oleate; nonionic; emollient, wetting agent; plasticizer for textiles, leathers, elastomers, personal care prods.; Gardner 2 max. liq.; m.p. 2 C; acid no. 2 max.; iodine no. 72-81; sapon. no. 164–172; 100% conc.

Kemester® 4516. [Witco/Humko] Methyl stearate; intermediate in prod. of alkanolamides, in metalworking lubricants, as specialized solvs.; foam depressant and nutrient in fermentation; Gardner 1 max. color; acid no. 3 max.; iodine no. 2.5 max.; sapon. no. 192–202.

Kemester® 5221SE. [Witco/Humko] PEG-2 stearate

SE; CAS 9004-99-3; anionic; emollient, emulsifier, plasticizer, lubricant for cosmetics, rubber, textiles; Gardner 3 max. flake; m.p. 46–56 C; acid no. 95–105; iodine no. 3 max.; sapon. no. 163–178; 100% conc.

Kemester® 5500. [Witco/Humko] Glyceryl stearate; nonionic; emollient, emulsifier; stabilizer, plasticizer, lubricant for cosmetic, paper, textile, and industrial uses; Gardner 3 max. bead, flake; m.p. 56–60 C; acid no. 3 max.; iodine no. 0.5 max.; sapon. no. 164–177; 100% conc.

Kemester® 6000. [Witco/Humko] Glyceryl stearate; nonionic; cosmetic and industrial emulsifier, emollient; plasticizer for elastomers; Gardner 3 max. bead; m.p. 57–60 C; acid no. 3 max.; iodine no. 2 max.; sapon. no. 162–176; 100% conc.

Kemester® 6000SE. [Witco/Humko] Glyceryl stearate SE; anionic; emulsifier, emollient, lubricant, plasticizer for cosmetic, paper and textiles; Gardner 3 max. beads; m.p. 56–61 C; acid no. 10 max.; iodine no. 3 max.; sapon. no. 140–156; 100% conc.

Kemester® 9022. [Witco/Humko] Methyl behenate; intermediate in prod. of superamides, in metalworking lubricants, as solv.; Gardner 2 max.; acid no. 20 max.; iodine no. 1 max.; sapon. no. 150–160.

Kemester® EGDS. [Witco/Humko] Glycol distearate; intermediate in prod. of superamides, in metalworking lubricants, specialized solv.; industrial lubricant; opacifier, pearling additive, thickener for cosmetics and pharmaceuticals; Gardner 2 max. solid; m.p. 60–63 C; acid no. 6 max.; iodine no. 1 max.; sapon. no. 190–200; 100% conc.

Kemester® EGMS. [Witco/Humko] Glycol stearate; intermediate in prod. of superamides, in metalworking lubricants, specialized solv.; industrial lubricant; opacifier, pearling additive, thickener for cosmetics and pharmaceuticals; Gardner 1 solid; m.p. 56–60 C; acid no. 1; iodine no. 1 max.; sapon. no. 179–195.

Ke-Mul® 181. [Georgia-Pacific] Sodium lignosulfonate; anionic; asphalt emulsifier and stabilizer; liq.; 45% conc.

Ke-Mul® A97. [Georgia-Pacific] Sodium lignosulfonate; anionic; emulsifier for o/w emulsions; liq.; 45% conc.

Kerapol 791. [Henkel-Nopco] Coconut alkylolamide; nonionic; detergent; foam booster; liq.; 100% conc.

Kerasol 1014. [Henkel-Nopco] Solv. and alkylolamide; nonionic; detergent; used in textile processing; kier boiling assistant; liq.; 100% conc.

Kerasol 1398. [Henkel-Nopco] Solvs., organic sulfates, and nonionic blend; anionic/nonionic; cleaning agent for flame suppressors of diesel locomotives used in coal and other mines; liq.; 100% conc.

Kerinol 2012 F. [Henkel-Nopco] Fatty acid alkanolamide; nonionic; detergent, foam stabilizer and thickener; liq.; 100% conc.

Kerinol C 109. [Henkel-Nopco] Alkanolamide, modified; nonionic; synergist and scouring agent for wool; liq.; 100% conc.

Kessco® 653. [Stepan; Stepan Canada] Cetyl palmitate; nonionic; syn. spermaceti wax, thickener, visc. booster for pharmaceutical and cosmetic prods.; w/o emulsifier, lubricant for metalworking fluids; wh. flakes; sol. in boiling alcohol, ether, chloroform, other waxes, oils, hydrocarbons; insol. in water; m.p. 51–55 C; acid no. 2.0 max.; sapon. no. 109–117.

Kessco® 874. [Stepan; Stepan Canada] Pentaerythritol tetracaprylate/caprate; high temp. stable lubricant base for textile and metalworking applics.; clear liq.; f.p. 21 F; acid no. 1.0; sapon. no. 350; flash pt. 500 F; 100% act.

Kessco® 887. [Stepan; Stepan Canada] Trimethylolpropane tricaprylate/caprate; high temp. stable lubricant base for textile and metalworking applics.; clear liq.; f.p. -60 F; acid no. 1.0; sapon. no. 310; flash pt. 500 F; 100% act.

Kessco® 891. [Stepan; Stepan Canada] EO/PO monooleate; lubricant, emulsifier, coupling agent for textile finishing applics., as replacement for min. oil in metalworking fluids; yel. liq.; water-sol.; HLB 12.8; acid no. 1.0; sapon. no. 65; 100% act.

Kessco® 894. [Stepan; Stepan Canada] EO/PO dioleate; lubricant, emulsifier, coupling agent, replacement for min. oil in metalworking fluids; yel. liq.; water-disp.; HLB 8.8; acid no. 1.0; sapon. no. 100; 100% act.

Kessco® BS. [Stepan; Stepan Canada] Butyl stearate; biodeg. replacement for min. oil; used as lubricants in textile spin finish, coning oils, carding, dye bath; liq.; 100% act.

Kessco® EGDS. [Stepan; Stepan Canada] Glycol distearate; pearlescent, emollient, emulsifier; wh. flakes; 100% conc.

Kessco® EGMS. [Stepan; Stepan Canada] Glycol stearate; pearlescent, bodying agent, emulsion stabilizer; wh. flakes; 100% conc.

Kessco® GDL. [Stepan; Stepan Canada] Glyceryl dilaurate; surfactant for free-flowing lotions; wh. solid; 100% conc.

Kessco® GDS 386F. [Stepan; Stepan Canada] Glyceryl distearate; CAS 1323-83-7; emulsifier, opacifier, bodying agent; waxy flake; 100% conc.

Kessco® GML. [Stepan; Stepan Canada] Glyceryl laurate; primary emulsifier for w/o emulsions; wh. solid; 100% conc.

Kessco® GMO. [Stepan; Stepan Canada] Glyceryl oleate; CAS 111-03-5; w/o emulsifier, emollient, spreading agent, pigment dispersant; lubricant for textiles, metalworking compds.; sperm oil replacement; liq.; HLB 3.8; acid no. 3.0; 100% act.

Kessco GMS. [Akzo Chem. BV] Glyceryl stearate; nonionic; bodying agent, emulsifier, and thickener in o/w emulsions; flakes.

Kessco® GMS. [Stepan; Stepan Canada] Glyceryl stearate; emulsifier, opacifier, bodying agent; lubricant for textiles, metalworking compds.; flakes; 100% act.

Kessco® GMS SE. [Stepan; Stepan Canada] Glyceryl stearate SE; for use in o/w emulsions at pH 5-9; wh. flakes; 100% conc.

Kessco® GMS SE/AS. [Stepan; Stepan Canada] Glyceryl stearate, PEG-100 stearate; emulsifier, self-emulsifying cream base, hair and skin conditioner; used at pH 3-5; wh. flakes; 100% conc.

Kessco® IPM. [Stepan; Stepan Canada] Isopropyl myristate; biodeg. replacement for min. oil; used as lubricants in textile spin finish, coning oils, carding, dye bath; liq.; 100% act.

Kessco® IPP. [Stepan; Stepan Canada] Isopropyl palmitate; biodeg. replacement for min. oil; used as lubricants in textile spin finish, coning oils, carding, dye bath; liq.; 100% act.

Kessco® OP. [Stepan; Stepan Canada] Octyl palmitate; biodeg. replacement for min. oil; used as lubricants in textile spin finish, coning oils, carding, dye bath; liq.; 100% act.

Kessco® PEG 200 DL. [Stepan; Stepan Canada] PEG-4 dilaurate; CAS 9005-02-1; lubricant, emulsifier, softener for textile and metalworking applics.; lt. yel. liq.; sol. in IPA, acetone, CCl_4, ethyl acetate, toluol, IPM, wh. oil; water-disp.; sp.gr. 0.951; dens. 7.9 lb/gal; f.p. < 9C; HLB 5.9; acid no. 10.0 max.; iodine no. 9.0; sapon. no. 176–186; flash pt. (COC) 460 F; fire pt. 510 F; 100% act.

Kessco® PEG 200 DO. [Stepan; Stepan Canada] PEG-4 dioleate; CAS 9005-07-6; surfactants for cosmetics, pharmaceuticals, food, agric., plastic, and other industries; thickener, solubilizer; lt. amber liq.; f.p. < –15 C; sol. in naptha, kerosene, IPA, acetone, CCl_4, ethyl acetate, toluol, IPM, peanut oil, wh. oil; water-disp.; sp.gr. 0.942; dens. 7.9 lb/gal; HLB 6.0; acid no. 10.0 max; sapon. no. 148–158; pH 5.0 (3%); flash pt. 545 F (COC).

Kessco® PEG 200 DS. [Stepan; Stepan Canada] PEG-4 distearate; CAS 9005-08-7; surfactants for cosmetics, pharmaceuticals, food, agric., plastic, and other industries; thickener, solubilizer; wh. to cream soft solid; sol. in naptha, kerosene, IPA, acetone, CCl_4, ethyl acetate, toluol, IPM, peanut oil, wh. oil; water-disp.; sp.gr. 0.9060 (65 C); HLB 5.0; m.p. 34 C; acid no. 10.0 max.; sapon. no. 153–162; pH 5.0 (3%); flash pt. 475 F (COC).

Kessco® PEG 200 ML. [Stepan; Stepan Canada] PEG-4 laurate; CAS 9004-81-3; surfactants for cosmetics, pharmaceuticals, food, agric., plastic, and other industries; thickener, solubilizer; lt. yel. liq.; f.p. < 5 C; sol. in IPA, acetone, CCl_4, ethyl acetate, toluol, IPM, water-disp.; sp.gr. 0.985; dens. 8.2 lb/gal; HLB 9.8; acid no. 5.0 max.; sapon. no. 132–142; pH 4.5; flash pt. 385 F (COC).

Kessco® PEG 200 MO. [Stepan; Stepan Canada] PEG-4 oleate; CAS 9004-96-0; surfactants for cosmetics, pharmaceuticals, food, agric., plastic, and other industries; thickener, solubilizer; lt. amber liq.; f.p. < –15 C; sol. in IPA, acetone, CCl_4, ethyl acetate, toluol, water-disp.; sp.gr. 0.973; dens. 8.1 lb/gal; HLB 8.0; acid no. 5.0 max.; sapon. no. 115–124; pH 5.0; flash pt. 395 F (COC).

Kessco® PEG 200 MS. [Stepan; Stepan Canada] PEG-4 stearate; CAS 9004-99-3; surfactants for cosmetics, pharmaceuticals, food, agric., plastic, and other industries; thickener, solubilizer; wh. to cream soft solid; sol. in IPA, acetone, CCl_4, ethyl acetate, toluol, IPM, peanut oil; water-disp.; sp.gr. 0.9360 (65 C); HLB 7.9; m.p. 31 C; acid no. 5.0 max.; sapon. no. 120–129; flash pt. 410 F (COC).

Kessco® PEG 300 DL. [Stepan; Stepan Canada] PEG-6 dilaurate; CAS 9005-02-1; surfactants for cosmetics, pharmaceuticals, food, agric., plastic, and other industries; thickener, solubilizer; lt. yel. liq.; f.p. < 13 C; sol. in naptha, IPA, acetone, toluol, IPM; water-disp.; sp.gr. 0.975; dens. 8.1 lb/gal; HLB 9.8; acid no. 10.0 max.; sapon. no. 148–158; flash pt. 475 F (COC).

Kessco® PEG 300 DO. [Stepan; Stepan Canada] PEG-6 dioleate; CAS 9005-07-6; surfactants for cosmetics, pharmaceuticals, food, agric., plastic, and other industries; thickener, solubilizer; lt. amber liq.; f.p. < –5 C; sol. in IPA, acetone, CCl_4, ethyl acetate, toluol, IPM, peanut oil; water-disp.; sp.gr. 0.962; dens. 8.0 lb/gal; HLB 7.2; acid no. 10.0 max.; sapon. no. 128–137; pH 5.0; flash pt. 510 F (COC).

Kessco® PEG 300 DS. [Stepan; Stepan Canada] PEG-6 distearate; CAS 9005-08-7; surfactants for cosmetics, pharmaceuticals, food, agric., plastic, and other industries; thickener, solubilizer; wh. to cream soft solid; sol. in naptha, kerosene, IPA, acetone, CCl_4, ethyl acetate, toluol, IPM, peanut oil, wh. oil; water-disp.; HLB 6.5; m.p. 32 C; acid no. 10.0 max.; sapon. no. 130–139; pH 5.0.

Kessco® PEG 300 ML. [Stepan; Stepan Canada] PEG-6 laurate; CAS 9004-81-3; surfactants for cosmetics, pharmaceuticals, food, agric., plastic, and other industries; thickener, solubilizer; lt. yel. liq.; f.p. < 8C; sol. in IPA, acetone, CCl_4, ethyl acetate, toluol; water-disp.; sp.gr. 1.011; dens. 8.4 lb/gal; HLB 11.4; acid no. 5.0 max.; sapon. no. 104–114; pH 4.5; flash pt. 445 F (COC).

Kessco® PEG 300 MO. [Stepan; Stepan Canada] PEG-6 oleate; CAS 9004-96-0; surfactants for cosmetics, pharmaceuticals, food, agric., plastic, and other industries; thickener, solubilizer; lt. amber liq.; f.p. < –5 C; sol. in IPA, acetone, CCl_4, ethyl acetate, toluol; water-disp.; sp.gr. 0.998; dens. 8.3 lb/gal; HLB 9.6; acid no. 5.0 max.; sapon. no. 94–102; pH 5.0 (3% disp.); flash pt. 450 F (COC).

Kessco® PEG 300 MS. [Stepan; Stepan Canada] PEG-6 stearate; CAS 9004-99-3; surfactants for cosmetics, pharmaceuticals, food, agric., plastic, and other industries; thickener, solubilizer; wh. to cream soft solid; sol. in IPA, acetone, CCl_4, ethyl acetate, toluol, IPM, peanut oil; water-disp.; sp.gr. 0.9660 (65 C); HLB 9.7; m.p. 28 C; acid no. 5.0 max.; sapon. no. 97–105; pH 5.0 (3% disp.); flash pt. 475 F (COC).

Kessco® PEG 400 DL. [Stepan; Stepan Canada] PEG-8 dilaurate; CAS 9005-02-1; surfactants for cosmetics, pharmaceuticals, food, agric., plastic, and other industries; thickener, solubilizer; lt. yel. liq.; f.p. 18 C; sol. in naptha, IPA, acetone, CCl_4, ethyl acetate, toluol, IPM, peanut oil; water-disp.; sp.gr. 0.990; dens. 8.3 lb/gal; HLB 9.8; acid no. 10.0 max.; sapon. no. 127–137; flash pt. 480 F (COC).

Kessco® PEG 400 DO. [Stepan; Stepan Canada] PEG-8 dioleate; CAS 9005-07-6; surfactants for cosmetics, pharmaceuticals, food, agric., plastic, and other industries; thickener, solubilizer; lt. amber liq.; f.p. < 7 C; sol. in naptha, IPA, acetone, CCl_4, ethyl acetate, toluol, IPM, peanut oil; water-disp.; sp.gr. 0.977; dens. 8.1 lb/gal; HLB 8.5; acid no. 10.0 max.; sapon. no. 113–122; pH 5.0; flash pt. 520 F (COC).

Kessco® PEG 400 DS. [Stepan; Stepan Canada] PEG-8 distearate; CAS 9005-08-7; lubricant, emulsifier, softener for textile and metalworking applics.; wh. to cream soft solid; sol. in naptha, IPA, acetone, CCl_4, ethyl acetate, toluol, IPM, peanut oil, wh. oil; water-disp; sp.gr. 0.9390 (65 C); HLB 8.5; m.p. 36 C; acid no. 10.0 max.; sapon. no. 115–124; pH 5.0 (3% disp.); flash pt. 500 F (COC); 100% act.

Kessco® PEG 400 ML. [Stepan; Stepan Canada] PEG-8 laurate; CAS 9004-81-3; surfactants for cosmetics, pharmaceuticals, food, agric., plastic, and other industries; thickener, solubilizer; lt. yel. liq.; f.p. 12 C; sol. in water, IPA, acetone, CCl_4, ethyl acetate, toluol; sp.gr. 1.028; dens. 8.6 lb/gal; HLB 13.1; acid no. 5.0 max.; sapon. no. 86–96; flash pt. 475 F (COC).

Kessco® PEG 400 MO. [Stepan; Stepan Canada] PEG-8 oleate; CAS 9004-96-0; surfactants for cosmetics, pharmaceuticals, food, agric., plastic, and other industries; thickener, solubilizer; lt. amber liq.; f.p. < 10 C; sol. in IPA, acetone, CCl_4, ethyl acetate, toluol; water-disp.; sp.gr. 1.013; dens. 8.4 lb/gal; HLB 11.4; acid no. 5.0 max.; sapon. no. 80–89; pH 5.0 (3% disp.); flash pt. 510 F (COC).

Kessco® PEG 400 MS. [Stepan; Stepan Canada] PEG-8 stearate; CAS 9004-99-3; surfactants for cosmetics, pharmaceuticals, food, agric., plastic, and other industries; thickener, solubilizer; wh. to cream soft solid; sol. in IPA, acetone, CCl_4, ethyl acetate, toluol; water-disp.; sp.gr. 0.9780 (65 C); HLB 11.6; m.p. 32 C; acid no. 5.0 max.; sapon. no. 83–92; pH 5.0 (3% disp.); flash pt. 480 F (COC).

Kessco® PEG 600 DL. [Stepan; Stepan Canada] PEG-12 dilaurate; CAS 9005-02-1; lubricant, emulsifier, softener for textile and metalworking applics.; liq.; sol. in IPA, acetone, CCl_4, ethyl acetate, toluol, IPM; water-disp.; sp.gr. 0.9820 (65 C); f.p. 24 C; HLB 11.7; acid no. 10.0 max.; sapon. no. 102–112; flash pt. 465 F (COC); 100% act.

Kessco® PEG 600 DO. [Stepan; Stepan Canada] PEG-12 dioleate; CAS 9005-07-6; surfactants for cosmetics, pharmaceuticals, food, agric., plastic, and other industries; thickener, solubilizer; lt. amber liq.; f.p. 19 C; sol. in IPA, acetone, CCl_4, ethyl acetate, toluol, IPM, peanut oil; water-disp.; sp.gr. 1.001; dens. 8.3 lb/gal; HLB 10.5; acid no. 10.0 max.; sapon. no. 92–102; pH 5.0 (3% disp.); flash pt. 495 F (COC).

Kessco® PEG 600 DS. [Stepan; Stepan Canada] PEG-12 distearate; CAS 9005-08-7; lubricant, emulsifier, softener for textile and metalworking applics.; wh. to cream soft solid; sol. in IPA, acetone, CCl_4, ethyl acetate, toluol, IPM, peanut oil; water-disp.; sp.gr. 0.9670 (65 C); HLB 10.7; m.p. 39 C; acid no. 10.0 max.; sapon. no. 93–102; pH 5.0 (3% disp.); flash pt. 490 F (COC); 100% act.

Kessco® PEG 600 ML. [Stepan; Stepan Canada] PEG-12 laurate; CAS 9004-81-3; lubricant, emulsifier, softener for textile and metalworking applics.; lt. yel. liq.; sol. in water, Na_2SO_4, IPA, acetone, CCl_4, ethyl acetate, toluol; sp.gr. 1.050; dens. 8.8 lb/gal; f.p. 23 C; HLB 14.6; acid no. 5.0 max.; sapon. no. 64–74; flash pt. 475 F (COC); 100% act.

Kessco® PEG 600 MO. [Stepan; Stepan Canada] PEG-12 oleate; CAS 9004-96-0; surfactants for cosmetics, pharmaceuticals, food, agric., plastic, and other industries; thickener, solubilizer; lt. amber liq.; f.p. 23 C; sol. in water, IPA, acetone, CCl_4, ethyl acetate, toluol; sp.gr. 1.037; dens. 8.7 lb/gal; HLB 13.5; acid no. 5.0 max.; sapon. no. 60–69; pH 5.0 (3% disp.); flash pt. 525 F (COC).

Kessco® PEG 600 MS. [Stepan; Stepan Canada] PEG-12 stearate; CAS 9004-99-3; surfactants for cosmetics, pharmaceuticals, food, agric., plastic, and other industries; thickener, solubilizer; wh. to cream soft solid; sol. in water, IPA, acetone, CCl_4, ethyl acetate, toluol; sp.gr. 1.000 (65 C); HLB 13.6; m.p. 37 C; acid no. 5.0 max.; sapon. no. 61–70; pH 5.0 (3% disp.); flash pt. 480 F (COC).

Kessco® PEG 1000 DL. [Stepan; Stepan Canada] PEG-20 dilaurate; CAS 9005-02-1; surfactants for cosmetics, pharmaceuticals, food, agric., plastic, and other industries; thickener, solubilizer; cream soft solid; f.p. 38 C; sol. in water, IPA, acetone, CCl_4, ethyl acetate; toluol, IPM; sp.gr. 1.015 (65 C); HLB 14.5; acid no. 10.0 max.; sapon. no. 68–78; flash pt. 475 (COC).

Kessco® PEG 1000 DO. [Stepan; Stepan Canada] PEG-20 dioleate; CAS 9005-07-6; surfactants for cosmetics, pharmaceuticals, food, agric., plastic, and other industries; thickener, solubilizer; cream soft solid; f.p. 37 C; sol. in water, IPA, acetone, CCl_4, ethyl acetate, toluol; sp.gr. 1.005 (65 C); HLB 13.1; acid no. 10.0 max.; sapon. no. 64–74; pH 5.0 (3% disp.); flash pt. 505 F (COC).

Kessco® PEG 1000 DS. [Stepan; Stepan Canada] PEG-20 distearate; CAS 9005-08-7; surfactants for cosmetics, pharmaceuticals, food, agric., plastic, and other industries; thickener, solubilizer; cream wax; sol. in water, IPA, acetone, CCl_4, ethyl acetate, toluol; sp.gr. 1.005 (65 C); HLB 12.3; m.p. 40 C; acid no. 10.0 max.; sapon. no. 65–74; pH 5.0 (3% disp.); flash pt. 485 F (COC).

Kessco® PEG 1000 ML. [Stepan; Stepan Canada] PEG-20 laurate; CAS 9004-81-3; surfactants for cosmetics, pharmaceuticals, food, agric., plastic, and other industries; thickener, solubilizer; cream soft solid; f.p. 40 C; sol. in water, propylene glycol (hot), Na_2SO_4, IPA, acetone, CCl_4, ethyl acetate, toluol; sp.gr. 1.035 (65 C); HLB 16.5; acid no. 5.0 max.; sapon. no. 41–51; flash pt. 490 (COC).

Kessco® PEG 1000 MO. [Stepan; Stepan Canada] PEG-20 oleate; CAS 9004-96-0; surfactants for cosmetics, pharmaceuticals, food, agric., plastic, and other industries; thickener, solubilizer; cream soft solid; f.p. 39 C; sol. in water, Na_2SO_4 (5%), IPA, acetone, CCl_4, ethyl acetate, toluol; sp.gr. 1.035 (65 C); HLB 15.4; acid no. 5.0 max.; sapon. no. 40–49; pH 5.0; flash pt. 515 F (COC).

Kessco® PEG 1000 MS. [Stepan; Stepan Canada] PEG-20 stearate; CAS 9004-99-3; surfactants for cosmetics, pharmaceuticals, food, agric., plastic, and other industries; thickener, solubilizer; cream wax; sol. in water, Na_2SO_4 (5%), IPA, acetone, CCl_4, ethyl acetate, toluol; sp.gr. 1.030 (65 C); HLB 15.6; m.p. 41 C; acid no. 5.0 max.; sapon. no. 40–48; pH 5.0 (3% disp.); flash pt. 475 F (COC).

Kessco® PEG 1540 DL. [Stepan; Stepan Canada] PEG-32 dilaurate; CAS 9005-02-1; surfactants for cosmetics, pharmaceuticals, food, agric., plastic, and other industries; thickener, solubilizer; cream wax; f.p. 42 C; sol. in water, Na_2SO_4 (5%); hot in propylene glycol, IPA, acetone, CCl_4, ethyl acetate, toluol; sp.gr. 1.04 (65 C); HLB 15.7; acid no. 10.0 max.; sapon. no. 48–56; pH 4.5 (3% disp.); flash pt. 450 F (COC).

Kessco® PEG 1540 DO. [Stepan; Stepan Canada] PEG-32 dioleate; CAS 9005-07-6; surfactants for cosmetics, pharmaceuticals, food, agric., plastic, and other industries; thickener, solubilizer; cream wax; f.p. 44 C; sol. in water, propylene glycol, Na_2SO_4; hot in IPA, acetone, CCl_4, ethyl acetate, toluol; sp.gr. 1.025 (65 C); HLB 15.0; acid no. 10.0 max.; sapon. no. 45–55; pH 5.0 (3% disp.); flash pt. 480 F (COC).

Kessco® PEG 1540 DS. [Stepan; Stepan Canada] PEG-32 distearate; CAS 9005-08-7; surfactants for cosmetics, pharmaceuticals, food, agric., plastic, and other industries; thickener, solubilizer; cream wax; sol. in water, IPA, acetone, CCl_4, ethyl acetate, toluol; sp.gr. 1.015 (65 C); HLB 14,8; m.p. 45 C; acid no. 10.0 max.; sapon. no. 49–58; pH 5.0; flash pt. 490 F (COC).

Kessco® PEG 1540 ML. [Stepan; Stepan Canada] PEG-32 laurate; CAS 9004-81-3; surfactants for cosmetics, pharmaceuticals, food, agric., plastic, and other industries; thickener, solubilizer; cream wax; f.p. 46 C; sol. in water, Na_2SO_4 (5%); sol. hot in propylene glycol, IPA, acetone, CCl_4, ethyl acetate, toluol; sp.gr. 1.06 (65 C); HLB 17.6; acid no. 5.0 max.; sapon. no. 26–36; pH 4.5 (3% disp.); flash pt. 445 F (COC).

Kessco® PEG 1540 MO. [Stepan; Stepan Canada] PEG-32 oleate; CAS 9004-96-0; surfactants for cosmetics, pharmaceuticals, food, agric., plastic, and other industries; thickener, solubilizer; cream wax; f.p. 45 C; sol. in water, propylene glycol, Na_2SO_4; sol. hot in IPA, acetone, CCl_4, ethyl acetate, toluol; sp.gr. 1.050 (65 C); HLB 17.0; f.p. 47 C; acid no. 5.0 max.; sapon. no. 28–37; pH 5.0 (3% disp.); flash pt. 520 F (COC).

Kessco® PEG 1540 MS. [Stepan; Stepan Canada] PEG-32 stearate; CAS 9004-99-3; surfactant for cosmetics, pharmaceuticals, food, agric., plastic, and other industries; thickener, solubilizer; cream wax; sol. in water, Na_2SO_4 (5%), IPA, acetone, CCl_4, ethyl acetate, toluol; sp.gr. 1.050 (65 C); HLB 17.3; m.p. 47 C; acid no. 5.0 max.; sapon. no. 27–36; pH 5.0; flash pt. 495 F (COC).

Kessco® PEG 4000 DL. [Stepan; Stepan Canada] PEG-75 dilaurate; CAS 9005-02-1; surfactant for cosmetics, pharmaceuticals, food, agric., plastic, and other industries; thickener, solubilizer; cream wax; f.p. 52 C; sol. in water, Na_2SO_4 (5%); sol. hot in propylene glycol, IPA, acetone, CCl_4, ethyl acetate, toluol; sp.gr. 1.065 (65 C); HLB 17.6; acid no. 5.0 max.; sapon. no. 20–30; pH 4.5 (3% disp.); flash pt. 495 F (COC).

Kessco® PEG 4000 DO. [Stepan; Stepan Canada] PEG-75 dioleate; CAS 9005-07-6; surfactant for cosmetics, pharmaceuticals, food, agric., plastic, and other industries; thickener, solubilizer; cream wax; f.p. 49 C; sol. in water, propylene glycol, Na_2SO_4; sol. hot in IPA, acetone, CCl_4, ethyl acetate, toluol; sp.gr. 1.060 (65 C); HLB 17.8; acid no. 5.0 max.; sapon. no. 19–27; pH 5.0; flash pt. 500 F (COC).

Kessco® PEG 4000 DS. [Stepan; Stepan Canada] PEG-75 distearate; CAS 9005-08-7; surfactant for cosmetics, pharmaceuticals, food, agric., plastic, and other industries; thickener, solubilizer; cream wax; sol. in water, Na_2SO_4 (5%), IPA, acetone, CCl_4, ethyl acetate, toluol; sp.gr. 1.060 (65 C); HLB 17.3; m.p. 51 C; acid no. 5.0 max.; sapon. no. 19–27; pH 5.0 (3% disp.); flash pt. 515 F (COC).

Kessco® PEG 4000 ML. [Stepan; Stepan Canada] PEG-75 laurate; CAS 9004-81-3; surfactant for cosmetics, pharmaceuticals, food, agric., plastic, and other industries; thickener, solubilizer; cream wax; f.p. 55 C; sol. in water, Na_2SO_4 (5%); sol. hot in propylene glycol, IPA, acetone, CCl_4, ethyl acetate, toluol; sp.gr. 1.075 (65 C); HLB 18.8; acid no. 5.0 max.; sapon. no. 9–18; pH 4.5; flash pt. 515 F (COC).

Kessco® PEG 4000 MO. [Stepan; Stepan Canada] PEG-75 oleate; CAS 9004-96-0; surfactant for cosmetics, pharmaceuticals, food, agric., plastic, and other industries; thickener, solubilizer; cream wax; f.p. 55 C; sol. in water, propylene glycol, Na_2SO_4; sol. hot in IPA, acetone, CCl_4, ethyl acetate, toluol; sp.gr. 1.075 (65 C); HLB 18.3; acid no. 5.0 max.; sapon. no. 10–18; pH 5.0; flash pt. 495 F (COC).

Kessco® PEG 4000 MS. [Stepan; Stepan Canada] PEG-75 stearate; CAS 9004-99-3; surfactants for cosmetics, pharmaceuticals, food, agric., plastic, and other industries; thickener, solubilizer; cream wax; sol. in water, Na_2SO_4 (5%), IPA, acetone, CCl_4, ethyl acetate, toluol; sp.gr. 1.075 (64 C); HLB 18.6; m.p. 56 C; acid no. 5.0 max.; sapon. no. 10–18; pH 5.0; flash pt. 465 F (COC).

Kessco® PEG 6000 DL. [Stepan; Stepan Canada] PEG-150 dilaurate; CAS 9005-02-1; surfactants for cosmetics, pharmaceuticals, food, agric., plastic, and other industries; thickener, solubilizer; cream wax; f.p. 57 C; sol. in water, Na_2SO_4 (5%); sol. hot in propylene glycol, IPA, acetone, CCl_4, ethyl acetate, toluol; sp.gr. 1.077 (65 C); HLB 18.7; m.p. 56 C; acid no. 9.0 max.; sapon. no. 12–20; pH 4.5 (3% disp.); flash pt. 435 F (COC).

Kessco® PEG 6000 DO. [Stepan; Stepan Canada] PEG-150 dioleate; CAS 9005-07-6; surfactant for cosmetics, pharmaceuticals, food, agric., plastic, and other industries; thickener, solubilizer; cream wax; f.p. 56 C; sol. in water, propylene glycol, Na_2SO_4 (5%); sol. hot in IPA, acetone, CCl_4, ethyl acetate, toluol; sp.gr. 1.070 (65 C); HLB 18.3; acid no. 9.0 max.; sapon. no. 13–21; pH 5.0 (3% disp.); flash pt. 500 F.

Kessco® PEG 6000 DS. [Stepan; Stepan Canada] PEG-150 distearate; CAS 9005-08-7; lubricant, emulsifier, softener for textile and metalworking applics.; cream wax; sol. in propylene glycol, Na_2SO_4 (5%), IPA, acetone, CCl_4, ethyl acetate, toluol; sp.gr. 1.075 (65 C); HLB 18.4; m.p. 55 C; acid no. 9.0 max.; sapon. no. 14–20; pH 5.0 (3% disp.); flash pt. 475 F (COC); 100% act.

Kessco® PEG 6000 ML. [Stepan; Stepan Canada] PEG-150 laurate; CAS 9004-81-3; surfactants for cosmetics, pharmaceuticals, food, agric., plastic, and other industries; thickener, solubilizer; cream wax; f.p. 61 C; sol. in water, Na_2SO_4 (5%); sol. hot in propylene glycol, IPA, acetone, CCl_4, ethyl acetate, toluol; sp.gr. 1.085 (65 C); HLB 19.2; acid no. 5.0 max.; sapon. no. 7–13; pH 4.5.

Kessco® PEG 6000 MO. [Stepan; Stepan Canada] PEG-150 oleate; CAS 9004-96-0; surfactants for cosmetics, pharmaceuticals, food, agric., plastic, and other industries; thickener, solubilizer; cream wax; f.p. 59 C; sol. in water, propylene glycol, Na_2SO_4 (5%); sol. hot in IPA, acetone, CCl_4, ethyl acetate, toluol; sp.gr. 1.085 (65 C); HLB 19.0; acid no. 5.0 max.; sapon. no. 7–13; pH 5.0; flash pt. 470 F.

Kessco® PEG 6000 MS. [Stepan; Stepan Canada] PEG-150 stearate; CAS 9004-99-3; surfactants for cosmetics, pharmaceuticals, food, agric., plastic, and other industries; thickener, solubilizer; cream wax; sol. in water, propylene glycol, Na_2SO_4 (5%), IPA, acetone, CCl_4, ethyl acetate, toluol; sp.gr. 1.080 (65 C); HLB 18.8; m.p. 61 C; acid no. 5.0 max.; sapon. no. 7–13; pH 5.0 (3% disp.); flash pt. 480 F (COC).

Kessco® PGML. [Stepan; Stepan Canada] Propylene glycol laurate; emollient, emulsifier; clear liq.; 100% conc.

Kessco PGMS. [Akzo Chem. BV] Propylene glycol stearate; nonionic; solubilizer for Eosin dyes in lipsticks; bodying agent in emulsions; flakes.

Kessco® PGMS. [Stepan; Stepan Canada] Propylene glycol stearate; emulsifier with m.p. near body temp.; wh. flakes; 100% conc.

Kieralon® B, B Highly Conc. [BASF AG] Nonionic/anionic; surfactants; detergent, wetting agent, dispersant for textile processing; brnish. liq.; water-sol.; dens. 1.03 and 1.05 g/ml resp.; set pt. –3 and –16 C.

Kieralon® C. [BASF/Fibers] Detergent blend; anionic; detergent for scouring; used in neutral and high strength caustic; liq.; 60% conc.

Kieralon® D. [BASF AG] Nonionic; detergent for textile and dye operations; colorless to lt. yel. clear visc. liq.; water-sol.; dens. 1.01 g/ml; pour pt. 6–7 C.

Kieralon® ED. [BASF] Blend; anionic/nonionic; low-foaming detergent, emulsifier, dispersant for scouring cotton, syn. fibers and blends; clear to sl. opaque liq.; sol. in cold water; dens. 1.03 g/cc; pour pt. -5 C; pH 3.5 (1%).

Kieralon® JET. [BASF AG] Nonionic; wetting agent

and detergent with minimum foam for textile finishing; wh. visc. liq.; dens. 1 g/ml; pour pt. 3 C.

Kieralon® KB. [BASF AG] Anionic/nonionic; detergent, wetting agent for textiles; ylsh. clear liq.; water-sol.; dens. 1.08 g/ml; pour pt. –5 C.

Kieralon® NB-150. [BASF/Fibers] Low-foaming nonionic emulsion; nonionic; nonfoaming scouring agent for desizing, bleaching processes in high turbulence equip.; milky wh. liq.; water-sol.; pH 5-6 (10% aq.).

Kieralon® NB-ED. [BASF/Fibers] Nonionic/anionic formulation; anionic/nonionic; low foaming detergent with emulsifying and dispersing props.; for desizing, scouring, caustic boil-off, bleaching processes; liq.; 30% conc.

Kieralon® NB-OL. [BASF/Fibers] Synergistic nonionic/anionic prod.; anionic; scouring agent for removing oily, fatty, waxy impurities, for washing out spun finishes, water-sol. sizes; esp. for polyester and blends; low to moderate foaming; stable to hard water, acids, alkalies, salts, reduction or oxidation bleaching agents; lt. amber clear liq.; readily sol. in water; sp.gr. 1.017; dens. 8.48 lb/gal; toxicology: skin and mucous membrane irritant on repeated contact; 56% active.

Kieralon® OL. [BASF AG] Nonionic/anionic; detergent for washing out spin finishes and sizes; stain remover; lt. yel. pourable paste; water-sol.

Kieralon® TX-199. [BASF/Fibers] Emulsion; nonionic; low foaming wetting/scouring additive for pretreatment, bleaching of all fabrics; for use in high turbulence equip.; milky wh. liq.; water-sol.; pH 5-6 (10% aq.); 38% conc.

Kieralon® TX-410 Conc. [BASF/Fibers] Polyoxyethylated blend; cationic; afterscouring agent, wetting agent, detergent for dyed and printed fabrics; stable to hard water, moderate concs. of acids and alkalies; lt. yel. clear visc. liq.; sol. in water; pH 8.5 (10% aq.); toxicology: skin and mucous membrane irritant; 70% conc.

Kilfoam. [Arol Chem. Prods.] Silicone, emulsifiers, and stabilizer emulsion; nonionic/anionic; defoamer for textile and aq. industrial processing, dye and finish baths; emulsifier for paints; milky wh. emulsion, mild odor; disp. in water; sp.gr. 0.97; dens. 8.1 lb/gal; pH 6.8-7.3 (1%).

Kingoil S-10. [Sanyo Chem. Industries] Highly sulfated fatty acid; anionic; wetting, penetrating, dispersing, and leveling agent for bleaching, scouring, and dyeing applics.; low foaming; liq.

Kito 703. [Van Waters & Rogers] Linear alcohol, ethoxylated polyhydric; nonionic; wetting agent, emulsifier, detergent, defoamer; rewetting in papermaking; colorless liq.; mild odor; f.p. –32 C; sol. in oils and org. media; water-insol.; sp.gr. 0.924; visc. 37 cs; HLB 9.0; pH neutral; flash pt. 370 F; biodeg.; 100% act.

Klearfac® 870. [BASF] Block copolymer phosphate ester; surfactant; solid.

Klearfac® AA040. [BASF] Phosphate ester; anionic; solubilizer for nonionics in high electrolyte sol'ns.; for nonionic surfactants and dedusters for powd. alkalies; emulsifier; antistat used with syn. fiber yarns; used in textile processing to improve effectiveness of kier boiling, scouring,; bleaching, soaping, and print-washing operations; used in heavy-duty cleaning formulations, pesticides, herbicides, insecticides, liq. fertilizers, and plastics; liq.; sp.gr. 1.112; visc. 240 cps; pour pt. 12.8 C; cloud pt. 2 C (1% aq.); biodeg.; 60% min. act.

Klearfac® AA270. [BASF] Alcohol alkoxylate phosphate ester, free acid; anionic; emulsifier, solubilizer, dedusting agent, hydrotrope, metal cleaner; liq.; sp.gr. 1.165; visc. 5900 cps; pour pt. –3.9 C; cloud pt. 95 C (1% aq.); 90% min. act.

Klearfac® AA420. [BASF] Phosphate ester; anionic; solubilizer for nonionics in high electrolyte sol'ns.; for nonionic surfactants and dedusters for powd. alkalies; emulsifier; antistat used with syn. fiber yarns; used in textile processing to improve effectiveness of kier boiling, scouring,; liq.; sp.gr. 1.118; visc. 2950 cps; pour pt. –6.7 C; cloud pt. 67 C (1% aq.); 90% min. act.

Klearfac® AB270. [BASF] Phosphate ester; anionic; solubilizer for nonionics in high electrolyte sol'ns.; for nonionic surfactants and dedusters for powd. alkalies; emulsifier; antistat used with syn. fiber yarns; used in textile processing to improve effectiveness of kier boiling, scouring,; liq.; sp.gr. 1.088; visc. 800 cps; pour pt. –15 C; cloud pt. 88 C (1% aq.); 95% min. act.

Kleen-Paste. [Witco Israel] Sodium alkylaryl sulfonate; anionic; raw material for detergent industry; paste.

Klucel® E, G, H, J, L, M. [Aqualon] Hydroxypropylcellulose, standard grades; CAS 9004-64-2; nonionic; surface active thickener, stabilizer, film-former, suspending agent, protective colloid for coatings, adhesives, extrusions, moldings, paper, paint removers, encapsulations, inks; off-wh. powd., tasteless; 99% thru 20 mesh; sol. in water and many polar org. solvs.; m.w. 80,000; visc. 150–700 cps (10% aq.); bulk dens. 0.5 g/ml; soften. pt. 100-150 C.

Korantin® BH Liq. [BASF AG] But-2-yne-1,4-diol; emulsifier, corrosion inhibitor in acid pickles and cleaners; water-misc.; 33% conc.

Korantin® BH Solid. [BASF AG] But-2-yne-1,4-diol; see Korantin BH Liq.; water-sol.

Korantin® CD. [BASF AG] Fatty acid DEA condensate; corrosion inhibitor, emulsifier, solubilizer for aq. alkaline systems, e.g., metal treating fluids; liq.; 98% conc.

Korantin® LUB. [BASF AG] Polyether acid phosphate; corrosion inhibitor in weakly alkaline systems and cleaners; lubricant for aq. cutting fluids; liq.; water-disp.; 100% conc.

Korantin® MAT. [BASF AG] Alkanolamine salt of an org. acid containing nitrogen; corrosion inhibitor in neutral and weakly alkaline systems and cleaners; liq.; water-misc.; 100% conc.

Korantin® PA. [BASF AG] Org. acid containing nitrogen; corrosion inhibitor as salts for alkaline aq. systems, e.g., cleaners, cutting fluids; liq.; 80% conc.

Korantin® PAT. [BASF AG] TEA salt of a nitrogen-containing org. acid; corrosion inhibitor for alkaline aq. systems, e.g., cleaners, cutting fluids; liq.; 80% conc.

Korantin® SH. [BASF AG] Fatty acid condensate; anionic; anticorrosive agent for metal treating as the alkanolamine salt; liq.; 100% conc.

Korantin® SMK. [BASF AG] Phosphoric acid monoester; corrosion inhibitor in aq. alkaline systems, cleaners, detergents; liq.; 100% conc.

Korantin® TD. [BASF AG] Fatty acid DEA condensate; corrosion inhibitor and emulsifier for aq. alkaline systems, e.g., cleaners, cutting fluids; liq.; 99% conc.

L

Labrafil ISO. [Gattefosse SA] Isostearic ethoxylated glycerides; nonionic; hydrophilic oil; liq.; HLB 4.0; 100% conc.

Labrafil WL 1958 CS. [Gattefosse SA] Ricinoleic ethoxylated glycerides; nonionic; hydrophilic wax; emulsion stabilizer; liq.; 100% conc.

Labrasol. [Gattefosse; Gattefosse SA] PEG-8 caprylic/capric glycerides; CAS 57307-99-0; nonionic; hydrophilic oil; excipient, solubilizer for pharmaceutical and cosmetic formulations; surfactant for microemulsions; wetting agent; penetration enhancer; Gardner < 5 oily liq., faint odor; sol. in water; very sol. in ethanol, chloroform, methylene chloride; sp.gr. 1.060-1.070; visc. 80-110 mPa•s; HLB 14.0; acid no. < 1; iodine no. < 2; sapon. no. 85-105; ref. index 1.450-1.470; toxicology: sl. ocular irritant at 0.1 ml; 100% conc.

LABS-100. [Zohar Detergent Factory] Linear alkylbenzene sulfonic acid; anionic; detergent intermediate; liq.; 100% conc.

LABS 100/H.V. [Zohar Detergent Factory] Tridecylbenzene sulfonic acid; anionic; higher visc. detergent intermediate; liq.; 100% conc.

LABS-100 SP. [Zohar Detergent Factory] Alkylbenzene and toluene cosulfonic acid; anionic; detergent intermediate for spray drying; liq.; 100% conc.

Lactimon®. [Byk-Chemie USA] Sol'n. of a partial amide and alkylammonium salt of a higher m.w. unsat. polycarboxylic acid and a polysiloxane copolymer; anionic; wetting and dispersing additive to prevent settling of pigments; for solv. or solv.-free coating systems; sp.gr. 0.90-1.92 g/cc; dens. 7.51-7.68 lb/gal; acid no. 50-70; flash pt. (Seta) 24 C; ref. index 1.481-1.491; 48-52% NV in xylene/isobutanol (5/1).

Lactimon®-WS. [Byk-Chemie USA] Sol'n. of partially neutralized alkylammonium salt of higher m.w. polycarboxylic acid and a polysiloxane copolymer; anionic; wetting and dispersing additive to prevent settling of pigments; for water-based coating systems; sp.gr. 0.94-0.96 g/cc; dens. 7.84-8.01 lb/gal; acid no. 38-48; flash pt. 42 C; ref. index 1.439-1.449; 48-52% NV in 2-butoxyethanol/isobutanol/water (5/4/1).

Lactomul 461. [Grünau] Monodistearate POE glycol; nonionic; emulsifier for calf-milk replacer; paste; 100% conc.

Lactomul 463. [Grünau] Glycerol polyethylene tallow fatty acid ester; nonionic; emulsifier for calf-milk replacer; liq.; 100% conc.

Lactomul 466. [Henkel/Functional Prods.] Blended glyceride; nonionic; for animal milk replacer prods.; emulsion stability; liq.; 100% conc.

Lactomul 468. [Henkel/Functional Prods.] Blended glyceride; nonionic; for animal milk replacer prods.; suitable for wet process concs.; soft solid; 100% conc.

Lactomul 843. [Grünau] POE glycol monodioleate; nonionic; emulsifier for calf-milk replacer; liq.; 100% conc.

Lactomul 925. [Grünau] Hydrog. lard monoglyceride, distilled; nonionic; emulsifier for calf-milk replacer; powd.; 100% conc.

Lactomul CN-28. [Henkel/Functional Prods.] Blended glyceride; nonionic; for milk replacer prods. with spray drying processes; waxy solid; 100% conc.

Lamacit AP 6. [Grünau] Fatty acid and POE derivs.; anionic; emulsifier for paraffins used in particle and fiber boards; paper mfg.; solid; 100% conc.

Lamacit GML 20. [Henkel Canada] PEG-20 glyceryl laurate; CAS 51248-32-9; nonionic; o/w emulsifier for creams and lotions; solubilizer for essential oils in aq./alcoholic systems; liq.; HLB 17.0; 100% conc.

Lamepon 287 SF. [Grünau] Sulfonic acid; anionic; dispersant, wetting agent, protective colloid for dyestuffs; liq.; 40% conc.

Lamepon A. [Grünau] Protein fatty acid; anionic; dispersant, protective colloid for dyes; stabilizer for peroxide bleaching; liq.; 45% conc.

Lamigen ES 30. [Dai-ichi Kogyo Seiyaku] PEG lanolin fatty acid ester; nonionic; softener and antistat for textile; oiling agent for leather; emulsifier for emulsion polymerization; additive to lubricating oil and water sol. paint; paste; HLB 11.0; 100% conc.

Lamigen ES 60. [Dai-ichi Kogyo Seiyaku] PEG lanolin fatty acid ester; nonionic; softener and antistat for textile; oiling agent for leather; emulsifier for emulsion polymerization; additive to lubricating oil and water sol. paint; solid; HLB 13.0; 100% conc.

Lamigen ES 100. [Dai-ichi Kogyo Seiyaku] PEG lanolin fatty acid ester; nonionic; softener and antistat for textile; oiling agent for leather; emulsifier for emulsion polymerization; additive to lubricating oil and water sol. paint; solid; HLB 14.5; 100% conc.

Lamigen ET 20. [Dai-ichi Kogyo Seiyaku] PEG lanolin fatty alcohol ether; nonionic; softener and antistat for textile; oiling agent for leather; emulsifier for emulsion polymerization; additive to lubricating oil and water sol. paint; paste; HLB 12.0; 100% conc.

Lamigen ET 70. [Dai-ichi Kogyo Seiyaku] PEG lanolin fatty acid ester; nonionic; softener and antistat for textile; oiling agent for leather; emulsifier for emulsion polymerization; additive to lubricating oil and water sol. paint; solid; HLB 14.0; 100% conc.

Lamigen ET 90. [Dai-ichi Kogyo Seiyaku] PEG

lanolin fatty acid ester; nonionic; softener and antistat for textile; oiling agent for leather; emulsifier for emulsion polymerization; additive to lubricating oil and water sol. paint; solid; HLB 15.0; 100% conc.

Lamigen ET 180. [Dai-ichi Kogyo Seiyaku] PEG lanolin fatty acid ester; nonionic; softener and antistat for textile; oiling agent for leather; emulsifier for emulsion polymerization; additive to lubricating oil and water sol. paint; solid; HLB 16.0; 100% conc.

LAN-401. [Nikko Chem. Co. Ltd.] Min. oil and surfactant; nonionic; coning oil for syn. fibers; liq.; 100% conc.

Lanapex HTS. [ICI Surf. UK] Organic phosphoric acid ester; anionic; sequesterant for calcium, magnesium, and heavy metal ions used at high temps. and in presence of high electrolyte and alkali; liq.

Lanapex JS Conc. [ICI Surf. UK] Wetting agent/detergent/solv. blend; anionic/nonionic; low foaming wetting agent and detergent for scouring syn. fibers in jet dyeing machines or other circulating liquor machines; liq.

Lanapex R. [ICI Surf. UK] Modified phosphate ester, polyphosphate blend; anionic; after-clearing/soaping agent for removal of hydrolyzed reactive dye from cellulosic fibers; liq.

Lanapex TM. [ICI Surf. UK] Ethoxylate/non-chlorinated biodeg. solv. blend; nonionic; detergent for raw wool and yarn scouring; liq.

Lan-Aqua-Sol 50. [Fanning] PEG-75 lanolin; CAS 61790-81-6; anionic; emulsifier for cosmetic and pharmaceutical emulsions; emollient, superfatting agent, conditioner for skin and hair care prods., household detergents; solubilizer, wetting agent, dispersing aid.

Lan-Aqua-Sol 100. [Fanning] PEG-75 lanolin; CAS 61790-81-6; anionic; emulsifier for cosmetic and pharmaceutical emulsions; emollient, superfatting agent, conditioner for skin and hair care prods., household detergents; solubilizer, wetting agent, dispersing aid.

Lancare. [Henkel-Nopco] Anionic/nonionic; surfactant; used for a range of formulated shampoos, bubble baths, mild abrasive cleaners; liq.

Landemul DCC Series. [Harcros UK] EO/PO copolymers; nonionic; demulsifier; liq.

Lanesta G. [Westbrook Lanolin] Glyceryl lanolate; CAS 97404-50-7; nonionic; emollient, emulsifier; forms stable w/o emulsions; soft wax; HLB 4.5; 100% conc.

Laneto 40. [RITA] PEG-40 lanolin; CAS 61790-81-6; nonionic; moisturizer, lubricant, emulsifier, and solubilizer surfactant for soap and detergent systems; emollient, resin modifier and solubilizer for personal care prods.; glossing agent for hair; plasticizer; Gardner 12 max.; char. odor; sol. in alcohol; slightly sol. in water; 50 ± 1% water.

Laneto 50. [RITA] PEG-75 lanolin; CAS 61790-81-6; nonionic; moisturizer, lubricant, emulsifier, emollient, plasticizer, solubilizer for soap and detergent systems, personal care prods., fragrances; glossing agent for hair; Gardner 12 max. liq.; char. odor; water-sol.; sp.gr. 1.00–1.10; sapon. no. 1.00–1.10; 50 ± 1% water.

Laneto 60. [RITA] PEG-60 lanolin; CAS 61790-81-6; moisturizer, lubricant, emulsifier, and solubilizer surfactant for soap and detergent systems; emollient, resin modifier and solubilizer for personal care prods.; glossing agent for hair; plasticizer; m.p. 48–52 C; sapon. no. 10–16.

Laneto AWS. [RITA] PPG-12-PEG-50 lanolin; nonionic; aux. emulsifier, moisturizer, emollient; plasticizer for hair spray, resins; liq.; sol. in water and alcohol; HLB 16.0; 100% conc.

Lanette E. [Henkel/Cospha; Henkel/Functional Prods.; Henkel Canada; Henkel KGaA] Sodium cetearyl sulfate; anionic; emulsifier and wetting agent for o/w emulsions, personal care prods.; wh. powd.; 88% act.

Lanexol AWS. [Croda Inc.] PPG-12-PEG-50 lanolin; nonionic; emollient, conditioner, superfatting agent, foam stabilizer, and lubricant for alcoholic and aq. compositions, plasticizer for hair sprays, o/w emulsifier, solubilizer; amber visc. liq.; sol. in oil, water, ethanol and mixts.; cloud pt. 65–80 C (1% aq.); pour pt. 13 C max.; acid no. 2 max.; iodine no. 10 max.; sapon. no. 10–20; pH 6.0–7.0 (1% aq.); usage level: 1-5%; 97% conc.

Lanfrax®. [Henkel/Emery] Lanolin wax; nonionic; cosmetic emulsion stabilizer, w/o emulsifier and waxing agent, o/w aux. emulsifier; slip reducing agent; for floor finishing prods., polishes; wax; 100% conc.

Lanfrax® 1776. [Henkel/Cospha; Henkel Canada] USP lanolin wax fraction; nonionic; emulsifier, emollient; waxing agent, o/w aux. emulsifier; slip reducing agent in floor finishing compds.; w/o emulsion stabilizer and thickener; wax; 100% conc.

Lankrocell® D15L. [Harcros] Blend of org. surfactants; mechanical foam promoter for PVC for carpet and floor backing applics.; amber liq.; misc. with DOP, BBP, chlorinated paraffin, phosphate plasticizers; sp.gr. 1.07; usage level: 4 phr.

Lankrocell® KLOP. [Harcros] Surfactant and Ba/Cd/Zn metal soap stabilizer; stabilizer and mechanical foam promoter for foamed PVC plastisols used for carpet and floor backing applics.; heat stabilizer for plastisol during gelation step; brn. liq.; sp.gr. 1.5; usage level: 4-6 phr.

Lankrocell® KLOP/CV. [Harcros] Surfactant and Ba/Zn metal soap stabilizer; stabilizer and mechanical foam promoter for foamed PVC plastisols used for carpet and floor backing applics.; heat stabilizer for plastisol during gelation step; brn. liq.; sp.gr. 1.0.

Lankro Mud-Aids. [Harcros UK] Series of general workability aids for drilling muds incl. mud surfactants and defoamers.

Lankro Mud Detergents. [Harcros UK] Anionic or nonionic detergent for well drilling mud applics..

Lankro Mud-Emuls. [Harcros UK] Series of w/o and o/w emulsifiers for water-based and invert muds.

Lankromul OSD. [Harcros UK] Emulsifier in aliphatic hydrocarbon solv.; emulsifier and dispersant for oil spills; straw clear liq.; mild, fatty odor; disp. in water; sp.gr. 0.808 (10 C); visc. 3.5 cs; flash pt. 88 C (COC); pour pt. < –20 C.

Lankropol® ADF. [Harcros UK] Modified sulfosuccinamate; anionic; foaming agent used in foamed aq. polymer systems, textile, leather, upholstery industries; lt. br. clear liq.; char. odor; sol. in water; sp.gr. 1.078; visc. 11 cs; flash pt. > 200 F (COC); pour pt. –5 C; pH 6–8 (1% aq.); 35% act.

Lankropol® ATE. [Harcros UK] Tetrasodium N-(1,2-dicarboxyethyl)-N-octadecyl sulfosuccinamate; anionic; primary emulsifier and mechanical stabilizer for emulsion polymers, aux. foaming agent, solubilizing agent; amber slightly hazy liq.; ethanolic odor; sp.gr. 1.119; visc. 36 cs; flash pt. 73 F (Abel CC); pour pt. 3 C; pH 7.0–8.5 (1% aq.); 35% act.

Lankropol® KMA. [Harcros UK] Sodium dihexyl

sulfosuccinate, ethanol; CAS 6001-97-4; anionic; emulsifier, wetting agent esp. in sol'ns. of electrolytes; solubilizer for soaps, emulsion polymerization aid; pale straw hazy liq.; ethanolic odor; sol. in water; sp.gr. 1.082; visc. 31 cs; flash pt. 91 F (Abel CC); pour pt. < 0 C; pH 6.0–7.5 (1% aq.); surf. tens. 46 dynes/cm (0.1%); 60% act. in ethanol.

Lankropol® KN51. [Harcros UK] Sodium dicyclohexylalkoxide sulfosuccinate aq. sol'n.; anionic; low foam primary emulsifier for emulsion polymerization affording poor film rewettability; amber liq., char. odor; sol. in water; sp.gr. 1.063; visc. 200 cps; pour pt. -8 C; flash pt. (Abel CC) > 100 C; pH 7.0 (1% aq.); surf. tens. 42 dynes/cm (0.1%); 50% act.

Lankropol® KNB22. [Harcros UK] Monoalkyl sulfosuccinate; anionic; foaming agent for personal care prods.; primary emulsifier for latex prod.; cement foaming agent; liq.; 29% conc.

Lankropol® KO. [Harcros UK] Sodium dioctyl sulfosuccinate, min. oil; anionic; solv. emulsifier, water carrier in dry cleaning formulations, dewatering aid, emulsifier for min. oil with nonionics; pale straw clear/sl. hazy liq.; mild odor; disp. in water; sp.gr. 1.012; visc. 1250 cs; flash pt. (COC) > 100 C; pour pt. < 0 C; pH 6.5 (1% aq.); 60% act. in min. oil.

Lankropol® KO2. [Harcros UK] Sodium dioctyl sulfosuccinate, ethanol; CAS 577-11-7; anionic; wetting agent, emulsifier for emulsion polymerization; pale straw clear liq.; ethanolic odor; sol. up to 0.5% in water; sp.gr. 0.996; visc. 43 cs; flash pt. (Abel CC) 27 C; pour pt. < 0 C; pH 6.5 (1% aq.); surf. tens. 32 dynes/cm (0.1%); 60% act. in ethanol.

Lankropol® KSG72. [Harcros UK] Monoester sulfosuccinate; anionic; mild foaming agent for toiletries; primary emulsifier for latex prod.; cement foaming agents; liq.; 35% conc.

Lankropol® ODS/LS. [Harcros UK] Disodium octadecyl sulfosuccinamate; CAS 14481-60-8; anionic; foaming agent for aq. polymer dispersions; lt. amber hazy liq., paste; mild odor; sol. in water; sp.gr. 1.066–1.082; visc. 12 cs; flash pt. > 200 F (COC); pour pt. –5 to 4 C; pH 7–9 (1% aq.); 35% act.

Lankropol® ODS/PT. [Harcros UK] Disodium N-octadecyl sulfosuccinamate; anionic; foaming agent for aq. polymer dispersions; lt. amber hazy liq., paste; mild odor; sol. in water; sp.gr. 1.066–1.082; visc. 5400 cs; flash pt. > 200 F (COC); pour pt. –5 to 4 C; pH 7–9 (1% aq.); 35% act.

Lankropol® OPA. [Harcros UK] Potassium salt of a fatty acid sulfonate; anionic; surfactant used in metal industry and for household use; wetting agent, detergent, dispersant; emulsifier for emulsion polymerization; electrolyte stable; biodeg.; dk. amber clear liq.; mild fatty odor; sol. in water; sp.gr. 1.110; visc. 223 cs; flash pt. (COC) > 95 C; pour pt. < 0 C; pH 6.0 (1% aq.); surf. tens. 37 dynes/cm (0.1%); 50% act.

Lankropol® WA. [Harcros UK] Ammonium salt of a sulfated monoester of fatty acid; anionic; low foaming wetting agent in textile industry, mercerizing, dispersion aid in paint industry; solv. emulsifier in degreasing formulations and herbicides, detergent; biodeg.; dk. amber clear liq.; char. odor; sol. in water; sp.gr. 1.020; visc. 55 cs; flash pt. (COC) > 95 C; pour pt. < 0 C; pH 6.5 (1% aq.); 50% act.

Lankropol® WN. [Harcros UK] Sodium salt of a sulfated monoester of fatty acid; anionic; low foaming wetting agent in textile industry, dispersion aid in paint industry; solv. emulsifier in degreasing formulations and herbicides, detergent; biodeg.; dk. amber clear liq.; char. odor; sol. in water; sp.gr. 1.024; visc. 63 cs; flash pt. (COC) > 95 C; pour pt. < 0 C; pH 6.5 (1% aq.); 50% act.

Lankrosol HS101. [Harcros UK] Potassium salt of phosphate ester condensate; anionic; low foam hydrotrope for use in highly built liqs.; stable to acids, alkalis, electrolytes; clear liq., char. odor; sol. in water; sp.gr. 1.222; visc. 39 cs; pour pt. < 0 C; flash pt. (COC) > 95 C; pH 7.0 (1% aq.); 50% act.

Lankrosol HS112. [Harcros UK] Potassium salt of carboxylic acid compd.; anionic; hydrotrope for use in highly built liqs.; clear liq., mild odor; sol. in water; sp.gr. 1.114; visc. 70 cs; pour pt. < 0 C; flash pt. (COC) > 95 C; pH 7.5 (1% aq.); 36% act.

Lankrosol SXS-30. [Harcros UK] Sodium xylene sulfonate; CAS 1300-72-7; anionic; hydrotrope and visc. modifier for high act. liq. detergents, solubilizer for anionic surfactants; pale straw clear liq.; negligible odor; sol. in water; sp.gr. 1.128; visc. 2.7 cs; flash pt. > 200 F (COC); pour pt. < 0 C; pH 7 (1% aq.); 30% act.

Lanogel® 21. [Amerchol] PEG-27 lanolin; CAS 61790-81-6; nonionic; emollient, emulsifier, dispersant, wetting agent, solubilizer, foam stabilizer, used in cosmetics, personal care prods., pharmaceuticals, facial tissues, antiperspirants, germicidal hand soaps; ASTM 3 max. gel; HLB 15.0; 50% act.

Lanogel® 31. [Amerchol] PEG-40 lanolin; CAS 61790-81-6; nonionic; emollient, emulsifier, dispersant, wetting agent, solubilizer, foam stabilizer, used in cosmetics, personal care prods., pharmaceuticals, facial tissues, antiperspirants, germicidal hand soaps; ASTM 3 max. gel; 50% act.

Lanogel® 41. [Amerchol] PEG-75 lanolin; CAS 61790-81-6; nonionic; emollient, emulsifier, dispersant, wetting agent, solubilizer, foam stabilizer, used in cosmetics, personal care prods., pharmaceuticals, facial tissues, antiperspirants, germicidal hand soaps; ASTM 3 max. gel; HLB 15.0; 50% act.

Lanogel® 61. [Amerchol] PEG-85 lanolin; CAS 61790-81-6; nonionic; emollient, emulsifier, dispersant, wetting agent, solubilizer, foam stabilizer, used in cosmetics, personal care prods., pharmaceuticals, facial tissues, antiperspirants, germicidal hand soaps; ASTM 3 max. gel; 50% act.

Lanolin Alcohols LO. [Solvay Duphar BV] Lanolin; CAS 8006-54-0; nonionic; w/o emulsifier, stabilizer, emollient for textile, wood, and paper industries; brn. soft solid; sol. in hydrocarbons, chlorinated hydrocarbons, veg. and min. oils, insol. in water; sapon. no. 20–40; 100% conc.

Lanolin Fatty Acids O. [Solvay Duphar BV] Lanolin fatty acids; CAS 8020-84-6; anionic; emulsifier, used in lubricating greases, polishing compositions, anticorrosion compds.; wax, brown and bleached resp.; sol. in min. oils and many org. solv.; sapon. no. 140–165; 100% conc.

Lanoxyl 30. [Witco SA] Lanolin, ethoxylated; nonionic; emulsifier; paste; 100% conc.

Lanpolamide 5. [Croda Inc.] PEG-5 lanolinamide, PEG-5 lanolate; w/o emulsifier forming stable emulsions; emulsion stabilizer, corrosion inhibitor for aerosols; amber soft solid; sol. with sl. haze at 1% in min. oil; water-insol.; usage level: 0.1-0.5%.

Lanquell 206, 217. [Harcros UK] Polyglycol-based; low-toxicity antifoam for food industry, used in mfg. of paper for food pkg.; pale yel. liq.; faint odor; disp. in water; sp.gr. 0.984 and 1.005 resp.; visc. 180 cs and 355 cs resp.; flash pt. > 150 and > 200 F resp.

(Abel CC); pour pt. –28 and –13 C resp.; pH 7 (1% aq.); 92 and 98% act.

Lantrol® PLN. [Pulcra SA] Ethoxylated lanolin; CAS 8039-09-6; nonionic; cleaning and wetting agent; solubilizer for perfumes and germicides; conditioner for shampoos; superfatting agent for soaps; o/w emulsifier for nonfatty preps.; plasticizer for aerosols and hair sprays; solid; HLB 17.0; sapon. no. 10-15; pH 6.0-7.5 (1%); 100% conc.

Lantrol® PLN/50. [Pulcra SA] Ethoxylated lanolin; nonionic; cleaning and wetting agent; solubilizer for perfumes and germicides; conditioner for shampoos; superfatting agent for soaps; o/w emulsifier for nonfatty preps.; plasticizer for aerosols and hair sprays; paste; sapon. no. 4-10; pH 6.0-7.5 (1%); 50% conc.

Larosol ALM-1. [PPG/Specialty Chem.] Proprietary blend; cationic; leveling agent for acrylics; liq.; 40% conc.

Larosol DBL. [PPG/Specialty Chem.] Proprietary blend; anionic; dyebath lubricant for syn. fibers; liq.; 23% conc.

Larosol DBL-3. [PPG/Specialty Chem.] Proprietary blend; anionic; dyebath lubricant for syn. fibers; emulsion; 23% conc.

Larosol DBL-3 Conc. [PPG/Specialty Chem.] Proprietary blend; anionic; dyebath lubricant for syn. fibers; emulsion; 65% conc.

Larosol NLA-25. [PPG/Specialty Chem.] Proprietary blend; nonionic; leveling agent, scouring agent, antiprecipitant; liq.; 25% conc.

Larosol NRL-40. [PPG/Specialty Chem.] Sodium alkyl diphenyl oxide disulfonate; anionic; leveling, transfer agent with streak coverage; nylon carpet dyeing; emulsifier for dye carriers; liq.; 45% conc.

Larosol NRL Conc. [PPG/Specialty Chem.] Proprietary blend; anionic; nylon leveling agent; liq.; 64% conc.

Larosol PDQ-2. [PPG/Specialty Chem.] Alkyl nitrogen deriv.; cationic; leveler and retarder for acrylics; liq.; 40% conc.

Larosol PNC. [PPG/Specialty Chem.] Proprietary blend; nonionic; leveling and dispersing agent for disperse dyes on polyester and nylon; liq.; 25% conc.

Larostat® 88. [PPG/Specialty Chem.] Modified soyadimethylethyl ammonium ethosulfate; cationic; noncorrosive mold release agent; antistat; surface active; liq.; 10% act.

Larostat® 92. [PPG/Specialty Chem.] Surface active antistat; liq.; 100% act.

Larostat® 143. [PPG/Specialty Chem.] Oleyldimethylethyl ammonium ethosulfate; cationic; surface active antistat; liq.; 100% act.

Larostat® 264 A. [PPG/Specialty Chem.] Modified soyadimethylethyl ammonium ethosulfate; cationic; antistat for syn. fibers, fiberglass, plastic, polyethylene; surface active; yel. clear liq.; 34.0–36.0% act.

Larostat® 264 A Anhyd. [PPG/Specialty Chem.] Modified soyadimethylethyl ammonium ethosulfate; antistat for syn. fibers, fiberglass, plastic, polyethylene; surface active; gel-wax; 100% act.

Larostat® 264 A Conc. [PPG/Specialty Chem.] Modified soyadimethylethyl ammonium ethosulfate; cationic; noncorrosive mold release agent; antistat; surface active; liq.; 90% act.

Larostat® 300 A. [PPG/Specialty Chem.] Potassium alkyl phosphate ester; surface active antistat for syn. fibers; high thermal stability; nonyel.; liq.; 100% act.

Larostat® 300 I. [PPG/Specialty Chem.] Surface active antistat for syn. fibers; high thermal stability; built-in corrosion inhibitor; liq.; 45% act.

Larostat® 300 I-325. [PPG/Specialty Chem.] Surface active antistat for syn. fibers; high thermal stability; liq.; 25% act.

Larostat® 377 DPG. [PPG/Specialty Chem.] Lauric myristic dimethylethyl ammonium ethosulfate, dipropylene glycol; noncorrosive mold release agent; antistat; useful in polyurethanes; surface active; anhyd. liq.; 80% act.

Larostat® 377 FR. [PPG/Specialty Chem.] Flame-retardant surface active antistat for polyurethane foam; liq.; 50% act.

Larostat® 451. [PPG/Specialty Chem.] Stearyldimethylethyl ammonium ethosulfate; cationic; noncorrosive release agent forming a hard film; imparts gloss and antistatic properties; post treatment in polystyrenes and fiberglass; surface active; liq.; 50% act.

Larostat® 477. [PPG/Specialty Chem.] Surface active antistat with improved stability; liq.; water-sol.; 100% act.

Larostat® 1084. [PPG/Specialty Chem.] Surface active antistat with improved stability; liq.; water-sol.; 100% act.

Larostat® JMR. [PPG/Specialty Chem.] Surface active antistat, lubricant; liq.; water-sol.; 50% act.

Larostat® JMR I Conc. [PPG/Specialty Chem.] Surface active antistat, lubricant; liq.; water-sol.; 80% act.

L.A.S. [Gattefosse SA] PEG-8 caprylic/capric glycerides; nonionic; nontoxic excipient for creams, lotions; surfactant for microemulsions; Gardner < 5 oily liq., faint odor; sol. in water; very sol. in ethanol, chloroform, methylene chloride; sp.gr. 1.060-1.070; visc. 80-110 mPa•s; HLB 14.0; acid no. < 1; iodine no. < 2; sapon. no. 85-105; ref. index 1.450 1.470; toxicology: sl. ocular irritant at 0.1 ml.

Lathanol® LAL. [Stepan; Stepan Canada; Stepan Europe] Sodium lauryl sulfoacetate; anionic; emulsifier, wetting agent, detergent, foaming agent, thickener used in cosmetics and personal care prods.; wh. flake, powd.; faint lauryl alcohol odor; slightly acrid taste; water sol. 3.5 g/100 ml; surf. tens. 30.1 dynes/cm (0.1%) pH 5.0–7.5 (5% conc.); 70% act.

Latol MOD. [Climax Performance] Blown tall oil; primary emulsifier for oil muds.

Latol MTO. [Climax Performance] Blown tall oil; primary emulsifier for oil muds.

Laural D. [Ceca SA] TEA lauryl sulfate; anionic; foaming shampoo; detergent for wool and syn. fibers; liq.; 50% conc.

Laural EC. [Ceca SA] Ammonium laureth sulfate; anionic; surfactant; liq.

Laural ED. [Ceca SA] Sodium/TEA laureth sulfate; anionic; surfactant; liq.

Laural LS. [Ceca SA] Sodium laureth sulfate; CAS 9004-82-4; anionic; foaming shampoo; detergent for household prods.; emulsifier for emulsion polymerization; gel; 27% conc.

Laural P. [Ceca SA] Sodium lauryl sulfate; CAS 151-21-3; anionic; surfactant; paste.

Lauramide 11. [Zohar Detergent Factory] Cocamide DEA; nonionic; foam booster, thickener, superfatting agent; liq.; 100% conc.

Lauramide D. [Zohar Detergent Factory] Cocamide DEA; nonionic; foam booster, thickener, and superfatting agent; liq.

Lauramide EG. [Zohar Detergent Factory] Ethylene glycol stearate; nonionic; opacifying agent; paste; 50% conc.

Lauramide ME. [Zohar Detergent Factory] Cocamide DEA; nonionic; foam booster, thickener, superfatting agent; liq. to paste; 100% conc.

Lauramide R. [Zohar Detergent Factory] Fatty acid DEA; nonionic; foam booster, thickener, superfatting agent; liq.; 100% conc.

Lauramide S. [Zohar Detergent Factory] Unsat. fatty polyalkanolamide; nonionic; additive in metal cleaning compds.; liq.

Lauramide Special. [Zohar Detergent Factory] Fatty acid diethanolamide; nonionic; foam booster, thickener, superfatting agent; liq.; 100% conc.

Laurel M-10-257. [Reilly-Whiteman] Sulfated tall oil fatty acid; anionic; replacement for sulfated castor oil in textile and metalworking compds.; liq.; 75% conc.

Laurel PDW. [Reilly-Whiteman] Surfactant blend; anionic/nonionic; base for liq. dishwash formulations; liq.; 48% conc.

Laurel PEG 400 DT. [Reilly-Whiteman] PEG-8 ditallate; CAS 61791-01-3; nonionic; emulsifier and solubilizer for min. oils, fats, solvs.; for latex paints, metalworking fluids, industrial lubricants, textile specialties; liq.; HLB 8.6; 100% conc.

Laurel PEG 400 MO. [Reilly-Whiteman] PEG-8 oleate; CAS 9004-96-0; emulsifier for metalworking, paints, solvs., textile chem. specialties.

Laurel PEG 400 MT. [Reilly-Whiteman] PEG-8 tallate; CAS 61791-00-2; nonionic; emulsifier for sol. oils, industrial lubricants; softener component for textile industry; liq.; HLB 12.0; 100% conc.

Laurel PEG 600 DT. [Reilly-Whiteman] PEG-12 ditallate; CAS 61791-01-3; nonionic; emulsifier and solubilizer for min. oils, fats, solvs.; for latex paints, metalworking fluids, industrial lubricants, textile specialties; liq.; HLB 10.6; 100% conc.

Laurel PEG 600 MT. [Reilly-Whiteman] PEG-12 tallate; CAS 61791-00-2; nonionic; emulsifier for metalworking lubricants, specialty textile formulations; liq.; water-sol.; HLB 13.6; 100% conc.

Laurel R-50. [Reilly-Whiteman] Sulfated castor oil; anionic; penetrant, lubricant, emulsifier used in detergent cleaners, metalworking lubricants, paint, textile lubricants, low-irritation and ethnic hair preps. and dyes, skin cleaners and lotions; liq.; 50% conc.

Laurel R-75. [Reilly-Whiteman] Sulfated castor oil; anionic; penetrant, lubricant, emulsifier for textile wet processing, metalworking, paints, coatings; liq.; 75% conc.

Laurel SBT. [Reilly-Whiteman] Sulfated butyl tallate; anionic; industrial lubricant, emulsifier in textile formulations, rewetting agent for corrugated medium; liq.; 60% conc.

Laurel SD-101. [Reilly-Whiteman] Cocamide DEA (2:1); nonionic; detergent, visc. builder, foamer for hard surf. cleaners, laundry prods., textile scouring, metal cleaners; lt. amber liq.; acid no. 5 max.; 100% act.

Laurel SD-120. [Reilly-Whiteman] Coconut oil/oleic alkanolamide (2:1); nonionic; visc. builder, detergent for household, hard surf. cleaners, waterless hand cleaners; liq.; 100% conc.

Laurel SD-140N. [Reilly-Whiteman] Coconut/mixed fatty acids alkanolamide (2:1); nonionic; ingred. in wax strippers, degreasers, hard surf. and alkaline cleaners; liq.; 100% conc.

Laurel SD-150. [Reilly-Whiteman] Coconut oil/ DDBSA alkanolamide; anionic; detergent, foam booster for lt. duty liqs.; stable foaming props. in hard water; liq.; 100% conc.

Laurel SD-180. [Reilly-Whiteman] Coconut/tall oil fatty acid alkanolamide (2:1); nonionic; ingred. in wax strippers, degreasers, high alkaline cleaners; liq.; 100% conc.

Laurel SD-300. [Reilly-Whiteman] Cocamide DEA (2:1); nonionic; detergent, emulsifier, visc. builder, anticorrosive for metalworking formulations; dk. amber liq.; acid no. 16-24; 100% act.

Laurel SD-350. [Reilly-Whiteman] Modified alkanolamide (2:1); surfactant for hard surf. cleaners, wax strippers, degreasers, alkaline cleaners; visc. builder; raises cloud pt. of conc. detergent systems; dk. amber liq.; acid no. 34-40; 100% act.

Laurel SD-400. [Reilly-Whiteman] Oleamide DEA (2:1); nonionic; emulsifier, visc. builder in metalworking fluids; liq.; sol. in industrial oils and solvs.; 100% conc.

Laurel SD-520T. [Reilly-Whiteman] ErucamideTEA (2:1); nonionic; surfactant, EP lubricant for metalworking formulations; amber clear liq.; acid no. 2-6; pH 9-10 (5%); 100% act.

Laurel SD-570. [Reilly-Whiteman] Coconut/canola oil TEA alkanolamide (2:1); surfactant, EP lubricant with improved corrosion protection for metalworking formulations; dk. amber liq.; acid no. 5 max.; pH 9-10 (5%).

Laurel SD-580. [Reilly-Whiteman] Tall oil TEA alkanolamide (2:1); surfactant, emulsifier, EP lubricant for metalworking formulations; dk. amber liq.; acid no. 25-35; pH 9.5-10.5 (5%).

Laurel SD-590. [Reilly-Whiteman] Cocamide TEA (2:1); surfactant for metalworking formulations; dk. amber liq.; acid no. 8-14; pH 9-10 (5%).

Laurel SD-750. [Reilly-Whiteman] Tall oil DEA (2:1); nonionic; surfactant, emulsifier, detergent, lubricant improving corrosion protection in metalworking fluids; dk. amber liq.; acid no. 10-14; 100% act.

Laurel SD-800. [Reilly-Whiteman] Adipic acid alkanolamide (2:1); nonionic; low foaming amide for metal cleaning and lubricant formulations; visc. liq.; 100% conc.

Laurel SD-900M. [Reilly-Whiteman] Cocamide DEA (1:1); nonionic; surfactant, visc. builder, foam stabilizer for lt. duty liqs., dishwash, shampoos; lt. amber liq.; 100% act.

Laurel SD-950. [Reilly-Whiteman] Coconut fatty acid DEA (1:1); nonionic; visc. builder; liq.; 100% conc.

Laurel SD-1000. [Reilly-Whiteman] Coconut oil MEA; CAS 68606-27-3; nonionic; foam booster for liq. and powd. formulations; visc. builder; flake; 100% conc.

Laurel SD-1031. [Reilly-Whiteman] Mixed fatty alkanolamide; nonionic; surfactant for hard surf. cleaners, disinfectant cleaners; paste; 100% conc.

Laurel SD-1050. [Reilly-Whiteman] Cocamide MEA; CAS 68140-00-1; nonionic; visc. builder; flake; 100% conc.

Laurel SD-1500. [Reilly-Whiteman] Coconut superamide (1:1); thickener, w/o emulsifier, visc. bulder for shampoos, lt. duty liqs.; corrosion inhibitor for sol. oils; liq.; 100% conc.

Laurel SD-LOA. [Reilly-Whiteman] Lard oil alkanolamide (2:1); nonionic; lower foam amide for metalworking compds.; anticorrosive emulsifier; liq.;

100% conc.

Laurel SDW. [Reilly-Whiteman] Surfactant blend; anionic/nonionic; base for lt. duty liq. detergents; high foaming; maintains visc. when diluted; liq.; 50% conc.

Laurel SMR. [Reilly-Whiteman] Sulfated methyl rapeseed ester; fiber-to-metal lubricant for textile fibers; improved wetting, high smoke pt.

Laurel SRO. [Reilly-Whiteman] Sulfated rapeseed oil; anionic; used in metalworking lubricants and extreme pressure lubricants; liq.; 75% conc.

Laurelox 12. [Reilly-Whiteman] Lauryl dimethylamine oxide; amphoteric; foam booster/stabilizer for lt. duty liqs.; nonwetting agent in textile applics.; liq.; 30% conc.

Laurelphos 39. [Reilly-Whiteman] Phosphate ester, free acid; surfactant for oil and water-based metalworking lubricants and syn. cutting fluids; EP lubricant; neutralized with caustic as wetting agent; dk. amber liq.; sp.gr. 1.04; acid no. 315-335; pH < 2.0 (10%).

Laurelphos 60G. [Reilly-Whiteman] Org. phosphate ester, free acid; anionic; detergent, hydrotrope for strong alkali formulations; improves flowability of aq. systems at lower temps.; liq.; 100% conc.

Laurelphos 400. [Reilly-Whiteman] Aromatic phosphate ester, free acid; CAS 39464-69-2; anionic; lubricant additive, emulsifier, rust inhibitor for metalworking lubricants; dk. amber visc. liq.; sol. in oil and water; sp.gr. 1.02; acid no. 140-150; pH < 2.5 (10%); 100% conc.

Laurelphos A-600. [Reilly-Whiteman] Aliphatic phosphate ester, free acid; anionic; detergent, hydrotrope for liq. high alkaline cleaners; for self coupling surfactants in carpet cleaning compds.; lt. straw liq.; sp.gr. 1.07; acid no. 210-230; pH < 2.0 (10%); 100% act.

Laurelphos C-30LF. [Reilly-Whiteman] Aromatic phosphate ester, free acid; anionic; low foam hydrotrope for coupling nonionics into conc. caustic sol'ns. (to 25%); yel. liq.; sol. in high concs. of alkali; sp.gr. 1.2; acid no. 400-420; pH < 2.0 (10%); 100% act.

Laurelphos D 44 B. [Reilly-Whiteman] Org. phosphate ester, free acid; CAS 51811-79-1; anionic; detergent/hydrotrope for liq. alkaline cleaners; coupling limited to conc. of alkalies; liq.; 100% conc.

Laurelphos E-61. [Reilly-Whiteman] Org. phosphate ester, free acid (CAS #51811-79-1); anionic; detergent for liq. formulations; deduster for alkaline powds.; for dry cleaning, textile scouring; liq.; 100% conc.

Laurelphos L-50. [Reilly-Whiteman] Phosphate ester, free acid; surfactant additive in oil and water-sol. metalworking lubricants; emulsifier for min. oils; rust inhibitor.

Laurelphos LFH. [Reilly-Whiteman] Phosphate ester, neutralized; low foam hydrotrope for coupling nonionics into very high alkaline formulations.

Laurelphos OL-529. [Reilly-Whiteman] Org. phosphate ester, partial sodium salt; anionic; surfactant for floor cleaners, hard surf. detergents; visc. liq.; sol. in water, many solvs.; 88% conc.

Laurelphos P-71. [Reilly-Whiteman] Aromatic phosphate ester, free acid; anionic; low foaming hydrotrope, detergent; solubilizes nonionics into high alkaline cleaners; hard water tolerance and lubricating props. in aq. metalworking formulations; yel. clear liq.; sp.gr. 1.23; acid no. 150-180; pH < 2.0 (10%); 100% act.

Laurelphos R-6. [Reilly-Whiteman] Aliphatic phosphate ester, free acid; surfactant with improved coupling for heavy caustic sol'ns.; amber clear liq.; sp.gr. 1.06-1.08; acid no. 150-175; pH < 2.5 (10%); 100% act.

Laurelphos RH-44. [Reilly-Whiteman] Aliphatic phosphate ester, free acid; CAS 52623-95-7; anionic; detergent, hydrotrope, visc. builder for strong alkali formulations; deduster for powd. alkalis; amber clear visc. liq.; sp.gr. 1.18; acid no. 260-280; pH 1.3-2.0 (10%); 80% act.

Laurelterge 837, 1390. [Reilly-Whiteman] Detergent for textile scouring.

Laureltex 308, 308 LF. [Reilly-Whiteman] Detergent for textile scouring.

Laureltex 6030, 6030S. [Reilly-Whiteman] Detergent for textile scouring.

Laureltex CW. [Reilly-Whiteman] Detergent for textile scouring.

Laureltex FMC. [Reilly-Whiteman] Additive for cleaning of pressure equip.

Laurex® 16/18, 16/18D. [Albright & Wilson UK] Primary fatty alcohol derived from naturally occurring oils and fats; (16/18D—dist. grade); raw material for mfg. of sulfated derivs. and additives for lubricating oils; wh. solid; m.p. 50–54 C; sapon. no. 1.5 and 0.8 resp.

Laurex® 810. [Albright & Wilson UK] Octyl-decyl alcohol fraction; raw material in the mfg. of alkyl phthalates, phosphoro diethionates, alkyl methacrylate monomers; defoaming agent in drilling muds and fermentation broths; colorless clear liq.; sp.gr. 0.83; flash pt. 92 C; acid no. 0.15; sapon. no. 1.0 max.

Laurex® 4526. [Albright & Wilson UK] Primary fatty alcohol blend; lubricant for rigid PVC for inj. molding processes; feedstock for ethoxylation; wh. waxy flake; dens. 0.45 g/cc; m.p. 48–53 C; flash pt. 202 C.

Laurex® 4550 [Albright & Wilson Am.] Cetearyl alcohol; for mfg. of surfactants; wh. waxy flakes; dens. 0.4 g/cc (20 C); m.p. 48-53 C; acid no. 0.5 max.; sapon. no. 2.0 max; 100% act.

Laurex® CH. [Albright & Wilson UK] Coconut alcohol; superfatting agent in shampoos, raw material for sulfation, ethoxylation; wh. soft solid; m.p. 18–23 C; sapon. no. 1.5.

Laurex® CS [Albright & Wilson Am.; Albright & Wilson UK] Cetearyl alcohol BP; nonionic; mfg. of surfactants; raw material for ethoxylation, sulfation, etc.; stabilizer in emulsion polymerization; lubricant in rigid PVC, also for pharmaceutical creams, hand lotions, bath oils, shaving creams; wh. waxy flake; dens. 0.4 g/cc; m.p. 48–53 C; acid no. 0.5 max.; sapon. no. 2.0 max.; flash pt. 150 C; 100% act.

Laurex® CS/D. [Albright & Wilson UK] Cetearyl alcohol BP; nonionic; raw material for ethoxylation, sulfation, etc.; also for pharmaceutical creams, hand lotions, bath oils, shaving creams; wh. waxy flake; dens. 0.4 g/cc; m.p. 48–53 C; acid no. 0.28 max.; sapon. no. 2.0 max.; flash pt. 150 C; 100% conc.

Laurex® CS/W. [Albright & Wilson UK] Cetearyl alcohol; nonionic; surfactant; liq.; 100% conc.

Laurex® L1. [Albright & Wilson UK] Lauryl alcohol; CAS 112-53-8; raw material for ethoxylation, sulfation, etc.; superfatting agent for shampoos; wh. soft solid; dens. 0.84 g/cc; m.p. 20–25 C; acid no. 0.2 max.; sapon. no. 1.0 max.; flash pt. 130 C.

Laurex® NC. [Albright & Wilson UK] Lauryl alcohol; CAS 112-53-8; raw material for ethoxylation, sulfation, etc.; stabilizer in emulsion polymerization; foam stabilizer for fire-fighting foams; superfatting agent for shampoos; wh. soft solid; dens. 0.84 g/cc; m.p. 20–25 C; acid no. 0.2 max.; sapon. no. 1.0 max.; flash pt. 132 C.

Laurex® PKH. [Albright & Wilson UK] Palm kernel alcohol; superfatting agent in shampoos, raw material for sulfation, ethoxylation; wh. soft solid; m.p. 18–23 C; sapon. no. 1.5.

Lauridit® KD, KDG. [Akzo Chem. BV] Cocamide DEA; nonionic; detergent, emulsifier for cosmetic and household applics., visc. increasing additive; yel. liq.; water sol. 40 and 800 g/l resp.; sp.gr. 1.02; flash pt. (PMCC) 104 and 114 C resp.; pH 9.0–9.5; 90 and 85% act.

Lauridit® KM. [Akzo Chem. BV] Cocamide MEA; CAS 68140-00-1; nonionic; detergent for textile, household and cosmetic applics., foam stabilizer for detergents, shampoos, bubble baths; yel. flakes; poor water sol.; sp.gr. 0.97; m.p. 65–70 C; sapon. no. 14 max.; 93% act.

Lauridit® LM. [Akzo Chem. BV] Lauramide MEA; CAS 142-78-9; nonionic; detergent for textile, household and cosmetic applics., foam stabilizer for detergents, shampoos, bubble baths; ylsh. flakes; poor water sol.; sp.gr. 1.01 (80 C); m.p. 80–84 C; sapon. no. 20 max; pH 9; 93% act.

Lauridit® LMI. [Akzo Chem. BV] LauramideMIPA; CAS 142-54-1; nonionic; detergent for soaps, toiletries; flakes; 95% conc.

Lauridit® OD. [Akzo Chem. BV] Oleamide DEA; nonionic; detergent, emulsifier for cosmetic and household applics., visc. increasing additive; yel. liq.; water sol. 100 g/l; sp.gr. 0.97; f.p. 0 C; flash pt. 100 C (PMCC); pH 9; 85% act.

Lauridit® PD. [Akzo Chem. BV] Fatty acid poly DEA/fatty acid DEA 1:2; household detergent, corrosion inhibitor for cutting oils, additive for acid cleaners, emulsifier; brn. liq.; water sol.; sp.gr. 1.02; flash pt. 105 C (PMCC); 65% act.

Lauridit® PPD. [Akzo Chem. BV] Palm kernelamide DEA; nonionic; detergent for personal care prods., dishwashing agent, dissolving intermediary for perfumes and liq. hand cleaners; liq.; 100% conc.

Lauridit® SDG. [Akzo Chem. BV] Soyamide DEA; nonionic; surfactant for detergent shampoos, bubble baths, dishwashing agents; dissolving intermediary for perufmes, liq. hand cleaners; liq.; 85% conc.

Lauropal 2. [Witco SA] Ethoxylated fatty alcohol; CAS 3055-33-4; nonionic; base for liq. detergents and shampoos; liq.; HLB 6.2; 100% conc.

Lauropal 9. [Witco SA] Ethoxylated fatty alcohol; CAS 9002-92-0; nonionic; wetting agent, emulsifier for textiles; wax; HLB 13.2; 100% conc.

Lauropal 11. [Witco SA] Ethoxylated fatty alcohol; CAS 9002-92-0; nonionic; wetting agent and emulsifier for textiles, metal cleaning; wax; HLB 14.4; 100% conc.

Lauropal 0205. [Witco SA] Ethoxylated fatty alcohol; nonionic; biodeg. detergent for textiles; liq.; HLB 11.4; 90% conc.

Lauropal 0207L. [Witco SA] Ethoxylated fatty alcohol; nonionic; biodeg. detergent; liq.; HLB 13.0; 80% conc.

Lauropal 0227. [Witco SA] Ethoxylated fatty alcohol; nonionic; wetting agent and detergent for industrial use; stable in acid media; liq.

Lauropal 950. [Witco SA] Fatty alcohol, ethoxylated; nonionic; wetting agent, emulsifier for textiles; gel.

Lauropal 1150. [Witco SA] Ethoxylated fatty alcohol; CAS 9002-32-0; nonionic; wetting agent, emulsifier for textiles, metal cleaning; liq.; HLB 14.4; 50% conc.

Lauropal X 1003. [Witco SA] Ethoxylated alcohol; nonionic; wetting agent, detergent, emulsifier; liq.; HLB 9.1; 100% conc.

Lauropal X 1005. [Witco SA] Ethoxylated alcohol; nonionic; wetting agent, detergent, emulsifier; liq.; HLB 11.6; 100% conc.

Lauropal X 1007. [Witco SA] Ethoxylated alcohol; nonionic; wetting agent, detergent, emulsifier; liq.; HLB 13.2; 100% conc.

Lauropal X 1103. [Witco SA] Ethoxylated alcohol; nonionic; wetting agent, detergent, emulsifier; liq.; HLB 8.7; 100% conc.

Lauropal X 1105. [Witco SA] Ethoxylated branched fatty alcohol; nonionic; wetting agent, detergent; liq.; HLB 11.0; 100% conc.

Lauropal X 1107. [Witco SA] Ethoxylated branched fatty alcohol; nonionic; wetting agent, detergent; liq.; HLB 12.8; 100% conc.

Lauropal X 1203. [Witco SA] Ethoxylated alcohol; nonionic; wetting agent, detergent, emulsifier; liq.; HLB 8.3; 100% conc.

Lauropal X 1207. [Witco SA] Ethoxylated branched fatty alcohol; nonionic; wetting agent, detergent; liq.; HLB 12.5; 100% conc.

Lauroxal 3. [Witco SA] Ethoxylated fatty alcohol; nonionic; wetting agent and detergent for industrial use; liq.; HLB 8.0; 100% conc.

Lauroxal 6. [Witco SA] Ethoxylated fatty alcohol; nonionic; wetting agent and detergent for textiles; liq.; HLB 11.5; 100% conc.

Lauroxal 8. [Witco SA] Ethoxylated fatty alcohol; nonionic; wetting agent and detergent for metal cleaning and leather; liq.; HLB 13.0; 100% conc.

Laventin® CW. [BASF AG] Alkanol polyglycol ether; nonionic; water-free wetting agent and detergent used in textile industry; colorless to slightly yel. clear, low visc. liq.; water sol.; dens. 0.99 g/ml; pH 6.5–7.5 (1%); biodeg.

Laventin® W. [BASF AG] Oxyethylated fatty acid deriv.; nonionic; water-free detergent and wetting agent used in textile industry; yel.-brn., visc. liq. or semisolid paste; solidifies at low temps.; water sol.

Lebon 15. [Sanyo Chem. Industries] Sodium alkyl diaminoethyl glycine; germicide with detergency; for dairy farming, wide range of applics.; liq.

Lebon 101H, 105. [Sanyo Chem. Industries] Alkylimidazoline type surfactant; amphoteric; nontoxic and nonirritating shampoo base; lt.-duty detergents; good foam and foam stability; stable to acids, alkalies, and hard water; liq.; 36 and 35% conc. resp.

Lebon 2000. [Sanyo Chem. Industries] Cocamidopropyl betaine; amphoteric; low irritation surfactant for shampoos and lt. duty detergents; liq.; 30% conc.

Lebon A-5000. [Sanyo Chem. Industries] Disodium lauroyl ethanolamine POE sulfosuccinate; anionic; raw material for mild shampoos, bubble baths, cleansing agents; liq.; 30% conc.

Lebon GM. [Sanyo Chem. Industries] Benzalkonium chloride; germicide; liq.

Lecithin W.D. [Troy] Lecithin prod.; CAS 8002-43-5; wetting agent, dispersant for pigments in water-based paints; 90% conc.

Lenetol 416. [ICI Surf. UK] Detergent/ethoxylated alcohol blend; anionic/nonionic; biodeg. detergent, emulsifier, soil-suspending agent; liq.

Lenetol 527. [ICI Surf. UK] Detergent/ethoxylated alcohol blend; anionic/nonionic; biodeg. detergent, emulsifier, soil-suspending agent; liq.

Lenetol B Conc. [ICI Surf. UK] Ethoxylated fatty alcohol blend; nonionic; biodeg. detergent for wool scouring; liq.

Lenetol HP-LFN. [ICI Surf. UK] Ethoxylated fatty alcohol; nonionic; high performance, low foaming detergent for scouring and bleaching of cotton and syn. fibers in jet and pkg. dyeing machines; liq.

Lenetol KWB. [ICI Surf. UK] Detergent/buffering agent blend; anionic/nonionic; scouring agent for use on woolen garments, wool hosiery, and hand knitting yarn; liq.

Lenetol PS. [ICI Surf. UK] Complexing agent/phosphate ester; anionic; wetting agent, detergent, stabilizer for peroxide bleaching, esp. in winch machines; alkali-stable; liq.

Lenetol WLF 125. [ICI Surf. UK] Modified phosphate esters; anionic; low foaming wetting agent and detergent with penetrating props.; for prep. of cotton and cotton blends; alkali-stable; mobile gel.

Leocon 1020B. [Lion] Polyoxyalkylene glycol; nonionic; defoamer for industrial effluents; liq.

Leocon 1070B. [Lion] Polyoxyalkylene glycol; nonionic; defoamer for polymerization processes; liq.

Leocon PL-71L. [Lion] Polyoxyalkylene glycol; nonionic; defoamer for fermentation processes; liq.

Leomin AN. [Hoechst Celanese/Colorants & Surf.; Hoechst AG] Alkyl phosphonate; anionic; surfactant for textile processing; antistat for fiber mfg. and processing, plastics processing; sl. yel. clear liq.; visc. 200 cps; pH 8 (1% aq.); 87% act.

Leomin CN. [Hoechst Celanese/Colorants & Surf.] Fatty imidazoline condensate; surfactant for textile processing; beige wax; pH 5 (1% aq.); 99% act.

Leomin FA. [Hoechst AG] Quaternary nitrogen compd.; cationic; preparation, antistatic agent, and softener for fiber mfg., processing, textile finishing; liq.; 40% conc.

Leomin FANF. [Hoechst Celanese/Colorants & Surf.] Fatty alkyl quaternary ammonium salt; cationic; surfactant for textile processing; amber liq.; visc. 5 cps; pH 6.0-7.5 (1% aq.); 20% act.

Leomin HSG. [Hoechst AG] Fatty acid polyglycol ester; nonionic; preparation agent and softener for fiber mfg. and processing for a silk-like scroopy handle; dispersant for removing polyester oligomers during reduction cleaning of polyester dyeings; paste; 90% conc.

Leomin KP. [Hoechst AG] Quaternary fatty acid deriv.; cationic; preparation and antistatic finishing agent and softener for fiber mfg. and processing; for electrostatic flocking; paste; 50% conc.

Leomin LS. [Hoechst Celanese/Colorants & Surf.] Fatty acid ethoxylate; surfactant for textile processing; amber liq.; visc. 100 cps; pH 7 (1% aq.); 96% act.

Leomin OR. [Hoechst AG] Fatty acid polyglycol ester; nonionic; dispersant and preparation agent for textile applics.; liq.; 90% conc.

Leomin TR. [Hoechst AG] Fatty amine ethoxylate; cationic; cleaning agent for HT apparatus; liq.

Leonil DB Powd. [Hoechst Celanese/Colorants & Surf.; Hoechst AG] Sodium diisobutyl naphthalene sulfonate; anionic; wetting agent and dyeing auxiliaries for textiles, agric.; powd.

Leonil EBL. [Hoechst AG] Oxidizing agents in detergents; anionic; fast desizing and wetting agent for cellulosic fibers and cellulosic/synthetic blends; paste.

Leonil KS. [Hoechst AG] Surfactant blend; anionic; low foaming wetting agent, padding auxiliary; liq.

Leonil L. [Hoechst AG] Surfactant blend; anionic/nonionic; detergent, wetting agent, dispersant for textile industry; liq.

Leonil OS. [Hoechst Celanese/Colorants & Surf.] Sodium dioctyl sulfosuccinate; anionic; surfactant for agric. formulations.

Leonil UN. [Hoechst AG] Surfactant blend; anionic/nonionic; wetting agent and detergent for pretreatment and bleaching processes; paste.

Leophen® BN. [BASF AG] Alcohol sulfonate; anionic; foam-free wetting agent in mercerizing textiles; yel.-brn. clear low-visc. liq.; misc. with water and dilute caustic soda; dens. 1.09 g/ml; biodeg.

Leophen® LG. [BASF AG] Mixt. of sodium salts of aliphatic sulfonic acids; anionic; wetting agent used in caustic baths; yel. almost clear slightly visc. liq.

Leophen® M. [BASF AG] Neutral phosphoric acid ester with nonionic emulsifiers; anionic; wetting agent and detergent in textile industry, antifoam; SE; colorless to weak yel. liq.; misc. in water; dens. 1.0 g/ml.

Leophen® ML. [BASF AG] Alkylsulfate; anionic; wetting agent for yarns, fabrics, knitgoods with min. foam formation; yel.-brn. liq.; dens. 1 g/ml; pour pt. -3 C.

Leophen® RA. [BASF AG] Sodium salt of a sulfosuccinate; anionic; wetting/rewetting agent for fibers and textiles; yel. liq.; water sol.; pH 5.5–7.0 (1% aq.).

Leophen® RBD. [BASF AG] Dialkyl sulfimide; anionic; wetting and rewetting agent used in bleaching and dyeing fibers; golden yel. clear visc. liq.; water sol.; dens. 1.1 g/ml.

Leophen® U. [BASF AG] Surfactant blend; nonionic/anionic; wetting agent, detergent and dispersant used in desizing, treating with alkali and bleaching; yel. paste; water sol.; pH 7.5–8.0 (10 g/l).

Leukonöl LBA-2. [Münzing Chemie GmbH] Anionic; emulsifier for emulsion polymers; wetting agent for alkaline systems to pH 13; BGA compliance.

Levelan® A0192. [Harcros UK] Fatty alcohol ethoxylate; nonionic; leveling agent for direct dyestuffs; clear liq.; faint odor; sol. in water; sp.gr. 1.018; visc. 13 cs; pour pt. < 0 C; pH 6.0–8.0 (1% aq.); flash pt. > 200 F (COC); cloud pt. > 100 (1% aq.); 20% act.

Levelan NKD. [Marubishi Oil Chem.] Amino carboxylic acid; amphoteric; leveling agent for polyamide fibers; eliminates barre marks; liq.

Levelan® NKS. [Harcros UK] Amino carboxylic acid; amphoteric; leveling agent for textiles; paste.

Levelan® P148. [Harcros UK] Nonyl phenol EO condensate; nonionic; primary emulsifier in emulsion polymer industry; latex stabilizer; clear liq.; faint odor; sol. in water; sp.gr. 1.069; visc. 490 cs; cloud pt. > 100 C (1% aq.); flash pt. (COC) > 200 F; pour pt. < 0 C; 80% act.

Levelan® P208. [Harcros UK] Nonyl phenol ethoxylate aq. dilution; nonionic; wetting agent, scouring agent, antistat, dye leveling agent at high temp. and high electrolyte conc. in textile industry; latex stabilizer and emulsifier in emulsion polymer industry; clear liq.; faint odor; sol. in water; sp. gr. 1.086; visc. 460 cs; HLB 16.0; cloud pt. > 100 C (1% aq.); flash pt. (COC) > 95 C; pour pt. < 0 C; pH 7.0 (1% aq.); surf. tens. 40 dynes/cm (0.1%); 80% act.

Levelan® P307. [Harcros UK] Alkylphenol ethoxylate; wetting agent for high temp. applics.; liq.; 70% act.

Levelan® P357. [Harcros UK] Nonyl phenol EO

condensate; nonionic; see Levelan P208; clear liq.; faint odor; sol. in water; sp.gr. 1.090; visc. 960 cs; cloud pt. > 100 C (1% aq.); flash pt. (COC) > 200 F; pour pt. 2 C; pH 6.0–8.0 (1% aq.); 70% act.

Levelan® PE 304. [Harcros UK] Ethoxylated aromatic aq. dilution; lubricant, dye carrier, antistat for textiles and fibers; liq.; 35% conc.

Levelan® PG 434. [Harcros UK] PEG 4000 aq. dilution; nonionic; textile fiber lubricant, antistat, mold release; surfactant intermediate and solubilizer for essential oils; liq.; 35% conc.

Levelan R-15, -200. [Marubishi Oil Chem.] Quat. ammonium compd.; cationic; leveling agent for dyestuffs, dyeing of polyacrylonitrile fiber; liq.

Levelan WS. [Marubishi Oil Chem.] Modified fatty amine condensate; amphoteric; leveling agent for acid dyestuffs and metal complex dyes, dyeing of polyamide fibers and animal fibers; liq.; 50% conc.

Levelene. [Ciba-Geigy/Dyestuffs] POE alkyl ether, modified; nonionic; rewetting and leveling agent, penetrant and stripper used in texile applics.; liq.

Levelox TYF-10. [Yoshimura Oil Chem.] Surfactant blend; anionic/nonionic; leveling agent; liq.

Levenol A Conc. [Kao Corp. SA] POE alkyl amine; nonionic; retarding and stripping agent for textiles; paste; 100% conc.

Levenol DS-1. [Kao Corp. SA] POE alkyl amine; nonionic; stripping agent for textiles; liq.; 100% conc.

Levenol PW. [Kao Corp. SA] POE alkyl ether; nonionic; leveling agent; flake; 100% conc.

Levenol RK. [Kao Corp. SA] Quat. ammonium halide; cationic; retarding agent for dyeing of acrylics; liq.; 41% conc.

Levenol TD-326. [Kao Corp. SA] Quat. ammonium halide; nonionic; leveling and dispersant for polyester dyeing; liq.

Levenol WX, WZ. [Kao Corp. SA] POE alkyl or alkylaryl ether sulfate; nonionic; leveling agent and dye coagulation preventing agent for dyeing applics.; liq.; 25% conc.

Lexaine® C. [Inolex] Cocamidopropyl betaine; amphoteric; visc. builder, foam booster, thickener in conditioners, specialty shampoos, personal care prods., dishwash, sanitizers; Gardner 4 max. clear liq.; sp.gr. 1.044; f.p. -4 C; pH 4.5–5.5; 30% act.

Lexaine® CG-30. [Inolex] Cocamidopropyl betaine; amphoteric; surfactant, foam and visc. modifier, lime soap dispersant, thickener used in light duty liq. detergent systems, personal care prods.; straw to lt. amber clear nonvisc. liq.; bland odor; sp.gr. 1.06; f.p. -6 C; pH 6.5–8.0; 29% act.

Lexaine® CS. [Inolex] Cocamidopropyl betaine; amphoteric; mild surfactant for shampoo, bubble bath, dishwash, conditioners, sanitizers, creams and lotions; stable in acid and alkaline systems; clear liq.; sp.gr. 1.044; f.p. -4 C; pH 4.5-5.5; 30% conc.

Lexaine® CSB-50. [Inolex] Cocamidopropylhydroxysultaine; amphoteric; surfactant, foaming and wetting agent used in personal care prods. (shampoos, conditioners, bath prods.), industrial (heavy-duty alkaline cleaners, paint strippers, metal cleaners); stable in systems containing acids, alkali, and electrolytes; lime soap dispersant; clear liq.; sp.gr. 1.10; f.p. -11 C; pH 7-9; 48% conc.

Lexaine® LM. [Inolex] Lauramidopropyl betaine; CAS 4292-10-8; amphoteric; mild surfactant, foam booster for bath prods., shampoos, liq. soaps, dishwash liqs.; pale yel. clear liq.; sp.gr. 1.045; pH 4.5-8.0; 30% conc.

Lexate BPQ. [Inolex] Lauramidopropyl betaine, TEA-coco-hydrolyzed collagen, oleamidopropyl dihydroxypropyl dimonium chloride; blended detergent, conditioner, and protein used as economical base for shampoo, bath gel, liq. soaps, dishwash, bubble baths, cleansers; lt. amber liq., char. odor; pH 5.5-6.5; 35% solids; 4% protein.

Lexein® S620TA. [Inolex] TEA-coco-hydrolyzed collagen; anionic; visc. builder, foam modifier; aq. sol'n.

Lexemul® 55SE. [Inolex] Glyceryl stearate SE; anionic; primary emulsifier for o/w emulsion systems, cosmetic creams, lotions, hair dressing, shave creams; flakes, mild fatty char. odor; sol. @ 60 C in CCl_4, ethyl acetate, IPA, min. oil; disp. in water @ 60 C; m.p. 56 C; HLB 5.4; acid no. 16-20; iodine no. 1 max.; sapon. no. 148-156; pH 8.5.

Lexemul® 503. [Inolex] Glyceryl stearate; nonionic; emulsifier, stabilizer, thickener, opacifier in emulsions or surfactant systems, cosmetics, toiletries, pharmaceuticals; flakes, mild fatty char. odor; m.p. 57-60 C; HLB 3.9; acid no. 2 max.; iodine no. 3 max.; sapon. no. 158-168; pH 6.5 (3% aq. disp.); 100% conc.

Lexemul® 515. [Inolex] Glyceryl stearate; nonionic; emulsifier, stabilizer, thickener, opacifier in emulsions or surfactant systems, cosmetics, topical pharmaceuticals; emollient in nonaq. oil and wax-based systems, e.g., lipsticks; flakes, mild fatty char. odor; m.p. 60 C; HLB 3.2; acid no. 16-20; iodine no. 3 max.; sapon. no. 166-176; 100% conc.

Lexemul® 530. [Inolex] Glyceryl stearate SE; anionic; primary emulsifier in o/w systems, cosmetic creams and lotions, hair dressings, shave creams; wh. to cream flakes; low odor; sol. @ 60 C in CCl_4, ethyl acetate, IPA, min. oil; disp. in water; HLB 5.2; m.p. 56 C; acid no. 16-20; iodine no. 1 max.; sapon. no. 146–154; pH 8.5 (3% aq. disp.); 100% conc.

Lexemul® EGDS. [Inolex] Glycol distearate; nonionic; lubricant, opacifier and pearling agent for cosmetic surfactant systems, liq. hand soaps, lt. duty liqs.; flakes, mild fatty char. odor; m.p. 60 C; HLB 1.3; acid no. 3-6; iodine no. 1 max.; sapon. no. 188-198; hyd. no. 33-43; pH 6.0 (3% aq. susp.); 100% conc.

Lexemul® EGMS. [Inolex] Glycol stearate; nonionic; opacifier and pearling agent for personal care prods., lt. duty liqs.; emulsifier for creams, lotions, topicals; sec. suspending agent in o/w systems; flakes, mild fatty char. odor; sol. in hot min. and veg. oils; water-insol.; m.p. 57 C; HLB 2.2; acid no. 2 max.; iodine no. 1 max.; sapon. no. 180-190; 100% conc.

Lexemul® PEG-200 DL. [Inolex] PEG-4 dilaurate; CAS 9005-02-1; nonionic; emulsifier, emollient, lubricant for cosmetics, pharmaceuticals, metalworking fluids, paints, polishes and misc. industrial formulations; straw to yel. liq., typ. mild fatty odor; water-disp.; sp.gr. 0.954; m.p. 2-3 C; HLB 5.9; acid no. 5 max.; iodine no. 8 max.; sapon. no. 170-180.

Lexemul® PEG-400 DL. [Inolex] PEG-8 dilaurate; CAS 9005-02-1; nonionic; emulsifier, emollient, lubricant for bath oils, suppositories, creams, lotions for cosmetic, pharmaceutical and industrial formulations; straw to yel. liq., typ. mild fatty odor; water-disp.; sp.gr. 0.988; m.p. 10-11 C; HLB 9.8; acid no. 10 max.; iodine no. 10 max.; sapon. no. 127-137.

Lexemul® PEG-400ML. [Inolex] PEG-8 laurate; CAS 9004-81-3; nonionic; emulsifier, emollient, lubricant dispersant for creams, lotions, spreading bath oils, cosmetic, pharmaceutical and industrial applics.; pale yel. clear liq., typ. mild fatty odor; water-disp.; sp.gr. 1.024; m.p. 5-6 C; HLB 13.1; acid no. 4 max.; iodine no. 5 max.; sapon. no. 86-96; 100% conc.

Lexemul® T. [Inolex] Glyceryl stearate SE; anionic; for use as emulsifier, opacifier, stabilizer, and emollient in alkaline anionic systems, cosmetics; flakes, mild fatty char. odor; m.p. 60 C; HLB 5.3; acid no. 16-20; iodine no. 3 max.; sapon. no. 146-154; pH 8.5 (3% aq. disp.); 100% conc.

Lexin K. [Am. Lecithin] Lecithin; CAS 8002-43-5; nonionic; emulsifier; semisolid, low odor and taste.

LHS 40% Coconut Oil Soap. [Emulsion Systems] Highly refined coconut oil, caustic potash; soap; sl. amber color, coconut odor; dens. 8.33 lb/gal; pH 9.5-10.0; 37.5% min. solids.

Licowet F 1. [Hoechst AG] Fluoro surfactants; wetting and flow agents for lustering emulsions.

Lignosite®. [Georgia-Pacific] Calcium lignosulfonate; anionic; dispersant, emulsifier and emulsion stabilizer; 50% liq., 100% powd.

Lignosite® 17. [Georgia-Pacific] Ammonium lignosulfonate; CAS 8061-53-8; anionic; dispersant, emulsifier, emulsion stabilizer; 48% liq., 100% powd.

Lignosite® 231. [Georgia-Pacific] Purified sodium lignosulfonate; CAS 8061-51-6; anionic; dispersant, emulsifier, emulsion stabilizer; 40% liq., 100% powd.

Lignosite® 260. [Georgia-Pacific] Modified lignosulfonate; anionic; emulsifier, dispersant for pigments, insecticides; o/w emulsifier; tan fine powd.; 100% conc.

Lignosite® 401. [Georgia-Pacific] Calcium lignosulfonate; anionic; dispersant, emulsifier, emulsion stabilizer; 40% liq., 100% powd.

Lignosite® 431. [Georgia-Pacific] Sodium lignosulfonate; CAS 8061-51-6; anionic; dispersant for pigments, wettable powds., wax emulsions, industrial cleaners; brn. fine powd. or liq.; sol. in water; dens. 10.5 lb/gal; pH 4 (10%); 80% conc. (powd.), 50% conc. (liq.).

Lignosite® 458. [Georgia-Pacific] Sodium lignosulfonate; CAS 8061-51-6; anionic; emulsifier, emulsion stabilizer, dispersant for pigments, mfg. of concrete admixtures, wax emulsions, wettable powds., industrial cleaners; brn. fine powd. or liq.; sol. in hot or cold water; dens. 23 lb/ft³ (powd.), 10.3 lb/gal (liq.); 80% act. (powd.), 46% solids (liq.).

Lignosite® 823. [Georgia-Pacific] Sodium lignosulfonate; CAS 8061-51-6; anionic; emulsifier, emulsion stabilizer, dispersant for dyestuffs and pigments, mfg. of wax emulsions, wettable powds.; sol. in cold or hot water; dens. 23 lb/ft³ ; pH 7-8 (104% aq.); 100% powd., 47% liq.

Lignosite® AC. [Georgia-Pacific] Lignosulfonate, modified; anionic; dispersant, emulsifier, emulsion stabilizer; 44% liq., 100% powd.

Lignosite® L. [Georgia-Pacific] Lignosulfonate formulation; anionic; dispersant, emulsifier, emulsion stabilizer; liq.; 50% conc.

Lignosol AXD. [LignoTech] Sodium lignosulfonate; CAS 8061-51-6; anionic; wetting agent for insecticides, herbicides and fungicides, emulsifier, dispersant; boiler feedwater treatment, industrial cleaners, gypsum walboard additives; brn. powd.; water sol.; dens. 28–32 lb/ft³; pH 4.9 (27%); 95% act.

Lignosol B. [LignoTech] Calcium lignosulfonate; anionic; wetting agent, emulsifier, dispersant, binder; used in refractories, bricks, construction, insecticides, herbicides, fungicides; soil stabilizer; blk. liq.; sp.gr. 1.25; visc. 200 cps (80 F); f.p. –10 C; b.p. 105 C; 50% act.

Lignosol BD. [LignoTech] Calcium lignosulfonate; anionic; wetting agent, emulsifier, dispersant, binder; used in refractories, bricks, construction, insecticides, herbicides, fungicides; soil stabilizer; yel. powd.; water sol.; dens. 28–32 lb/ft³; pH 4.5 (27%); 95% act.

Lignosol D-10, D-30. [LignoTech] Sodium lignosulfonate; CAS 8061-51-6; anionic; wetting agent, dispersant for disperse dyes; brn. powd.; water sol.; 95% act.

Lignosol DXD. [LignoTech] Sodium lignosulfonate; CAS 8061-51-6; anionic; wetting agent and dispersant in industrial cleaners, insecticides, herbicides and fungicides, emulsifier, emulsion stabilizer for asphalt emulsions; blk. powd. or liq.; water sol.; dens. 30–32 lb/ft³; pH 9.0 (27%); 95% act. (powd.) or 42% act. (liq.).

Lignosol FTA. [LignoTech] Sodium lignosulfonate; CAS 8061-51-6; anionic; wetting agent, primary dispersant for disperse dye; brn. powd.; water sol.; dens. 30–32 lb/ft³ ; pH 9.5 (27%); 95% act.

Lignosol HCX. [LignoTech] Sodium lignosulfonate; CAS 8061-51-6; anionic; wetting agent, dispersant, emulsifier; used in insecticides, herbicides, fungicides; chelating agent for 24D and 45T amines; brn. powd.; water sol.; 95% act.

Lignosol NSX 110. [LignoTech] Sodium lignosulfonate; CAS 8061-51-6; anionic; primary dyestuff dispersant; powd.; 100% act.

Lignosol NSX 120. [LignoTech] Sodium lignosulfonate, modified; CAS 8061-51-6; primary dyestuff dispersant; powd.

Lignosol SFX. [LignoTech] Calcium lignosulfonate; CAS 8061-52-7; anionic; dispersant for concrete admixtures, dyestuff dispersant and extender, sludge conditioner; brn. powd.; dens. 28–32 lb/ft³ ; pH 6.8 (27%); 95% act.

Lignosol TS. [LignoTech] Ammonium lignosulfonate; CAS 8061-51-6; anionic; wetting agent, emulsifier, dispersant, tanning extract, slurry water reducer and grinding aid in cement mfg.; blk. liq.; water sol.; sp.gr. 1.23; f.p. –10C; 47% act.

Lignosol TSD. [LignoTech] Ammonium lignosulfonate; anionic; wetting agent, leather retanning agent, emulsifier, dispersant, slurry water reducer and grinding aid in cement mfg.; yel. powd.; water sol.; dens. 28–32 lb/ft³; pH 4.5 (27%); 95% act.

Lignosol WT. [LignoTech] Sodium lignosulfonate; CAS 8061-51-6; anionic; wetting agent, dispersant, sludge conditioner; yel. powd.; water sol.; 95% act.

Lignosol X. [LignoTech] Sodium lignosulfonate; CAS 8061-51-6; anionic; dispersant, wetting agent, emulsifier, tanning and retanning agent, used for dye leveling, water reducer, grinding aid; blk. liq.; water sol.; sp.gr. 1.24; f.p. –5 C; b.p. 105 C; 47% act.

Lignosol XD. [LignoTech] Sodium lignosulfonate; CAS 8061-51-6; anionic; dispersant, wetting agent, emulsifier for wax emulsions, used for retanning leather, water treatment; yel. powd.; water sol.; dens. 28–32 lb/ft³; pH 6.5 (27%); 95% act.

Lilamac S1. [Berol Nobel AB] Amine acetate; cationic; flotation agent; flakes.

Lilamin VP75. [Berol Nobel AB] Polyamine blend; cationic; bitumen adhesion agent; liq.

Lilaminox M4. [Berol Nobel AB] Tetradecyl dimethylamine oxide; CAS 3332-27-2; amphoteric; detergency booster, thickener for household bleaches based on sodium hypochlorite; foaming agent for hair shampoos; clear liq.; sol. in water and polar solvs.; dens. 973 kg/m³; visc. 1500 mPa•s; pour pt. 3 C; pH 7±1 (20%); surf. tens. 32 mN/m (0.1%); Ross-Miles foam 85 ml (initial, 20 ml of 1% aq.); 24-26% act.

Lilaminox M24. [Berol Nobel AB] C12-14 alkyl dimethylamine oxide; CAS 85408-49-7; amphoteric; foam booster/stabilizer, antistat, softener for hair shampoos; thickener for household bleaches based on sodium hypochlorite, hard surf. cleaners; clear liq.; sol. in water and polar solvs.; dens. 969 kg/m³; visc. 3 mPa•s; pour pt. 1 C; pH 7±1 (20%); surf. tens. 32 mN/m (0.1%); Ross-Miles foam 100 ml (initial, 20 ml of 1% aq.); 30-32% act.

Lilamuls BG. [Berol Nobel AB] N-Tallow propylene diamine; CAS 61791-55-7; cationic; emulsifier and adhesion agent; paste.

Lilamuls EM 24, EM 26, EM 33. [Berol Nobel AB] Amine-based; cationic; emulsifier for bitumen; liq.

Lipal DGMS. [Aquatec Quimica SA] PEG-2 stearate; CAS 9004-99-3; 106-11-6; nonionic; emulsifier for lotions and creams; flakes; 100% conc.

Lipo DGLS. [Lipo] PEG-2 laurate SE; CAS 9004-81-3; nonionic; spreading agent, emulsifier, dispersant, lubricant, opacifier, emulsion stabilizer, emollient, visc. builder used in bath oils, creams, lotions; defoamer for process applics.; yel. liq.; water-disp.; HLB 8.3 ± 1; acid no. 4 max.; sapon. no. 160–170; 100% act.

Lipo DGS-SE. [Lipo] Diethylene glycol stearate SE; CAS 9004-99-3; spreading agent, emulsifier, dispersant, lubricant, opacifier, emulsion stabilizer, emollient, visc. builder used in bath oils, creams, lotions; defoamer for process applics.; wh./off wh. beads, flakes; HLB 4.0±1; acid no. 90–110; sapon. no. 160–180.

Lipo Diglycol Laurate. [Lipo] Diglycol laurate; CAS 9004-81-3; emulsifier, defoamer; liq.; 100% conc.

Lipo EGDS. [Lipo] Glycol distearate; spreading agent, emulsifier, dispersant, lubricant, opacifier, emulsion stabilizer, emollient, visc. builder used in bath oils, creams, lotions; defoamer for process applics.; wh./off wh. beads, flakes; HLB 1.0 ± 1; acid no. 7 max.; sapon. no. 190–205.

Lipo EGMS. [Lipo] Glycol stearate; nonionic; opacifier, pearlizer for shampoos, detergents; w/o emulsifier; stabilizer for o/w systems; wh./off wh. beads, flakes; HLB 2.0 ± 1; acid no. 6 max.; sapon. no. 175–190.; 100% conc.

Lipocol C-2. [Lipo] Ceteth-2; nonionic; emulsifier, defoamer, wetting agent, solubilizer, conditioning agent for personal care prods. and pigment disp.; wh., solid wax; HLB 5.3; acid no. 1; 100% act.

Lipocol C-10. [Lipo] Ceteth-10; nonionic; emulsifier, defoamer, wetting agent, solubilizer, conditioning agent for personal care prods. and pigment disp.; wh., solid wax; HLB 12.9; acid no. 1; 100% act.

Lipocol C-20. [Lipo] Ceteth-20; CAS 9004-95-9; nonionic; emulsifier, defoamer, wetting agent, solubilizer, conditioning agent for personal care prods. and pigment disp.; wh., solid wax; HLB 15.7; acid no. 2 max; 100% act.

Lipocol L-4. [Lipo] Laureth-4; nonionic; surfactant for pigment dispersions, antiperspirants, depilatories, creams, lotions; antistat, emulsifier; liq.; HLB 9.7; acid no. 2 max; 100% act.

Lipocol L-12. [Lipo] Laureth-12; nonionic; surfactant for pigment dispersions, antiperspirants, depilatories, creams, lotions; antistat, emulsifier; solid; HLB 14.5; acid no. 1 max; 100% act.

Lipocol L-23. [Lipo] Laureth-23; nonionic; surfactant for pigment dispersions, antiperspirants, depilatories, creams, lotions; antistat, emulsifier; solid; HLB 16.9; acid no. 2 max; 100% act.

Lipocol O-2. [Lipo] Oleth-2; CAS 9004-98-2; nonionic; surfactant for pigment dispersions, antiperspirants, depilatories, creams, lotions; antistat, emulsifier; yel. liq.; HLB 4.9; acid no. 1 max; 100% act.

Lipocol O-5. [Lipo] Oleth-5; CAS 9004-98-2; nonionic; w/o emulsifier for cold waves, bleaches, dyes, depilatories; liq./solid; HLB 8.8.

Lipocol O-10. [Lipo] Oleth-10; CAS 9004-98-2; nonionic; surfactant for pigment dispersions, antiperspirants, depilatories, creams, lotions; antistat, emulsifier; yel. liq.; HLB 12.4; acid no. 2 max; 100% act.

Lipocol O-20. [Lipo] Oleth-20; CAS 9004-98-2; nonionic; surfactant for pigment dispersions, antiperspirants, depilatories, creams, lotions; antistat, emulsifier; yel. liq.; HLB 15.3; acid no. 2 max; 100% act.

Lipocol S-2. [Lipo] Steareth-2; CAS 9005-00-9; nonionic; surfactant for pigment dispersions, antiperspirants, depilatories, creams, lotions; antistat, emulsifier; wh. solid wax; HLB 4.9; acid no. 1 max; 100% act.

Lipocol S-10. [Lipo] Steareth-10; CAS 9005-00-9; nonionic; surfactant for pigment dispersions, antiperspirants, depilatories, creams, lotions; antistat, emulsifier; wh. solid wax; HLB 12.4; acid no. 1 max; 100% act.

Lipocol S-20. [Lipo] Steareth-20; CAS 9005-00-9; nonionic; surfactant for pigment dispersions, antiperspirants, depilatories, creams, lotions; antistat, emulsifier; wh. solid wax; HLB 15.3; acid no. 1 max; 100% act.

Lipocol SC-4. [Lipo] Ceteareth-4; CAS 68439-49-6; nonionic; surfactant for pigment dispersions, antiperspirants, depilatories, creams, lotions; antistat, emulsifier; solid; HLB 8.0; 100% conc.

Lipocol SC-15. [Lipo] Ceteareth-15; CAS 68439-49-6; nonionic; surfactant for pigment dispersions, antiperspirants, depilatories, creams, lotions; antistat, emulsifier; solid; HLB 14.3; 100% conc.

Lipocol SC-20. [Lipo] Ceteareth-20; CAS 68439-49-6; nonionic; surfactant for pigment dispersions, antiperspirants, depilatories, creams, lotions; antistat, emulsifier; solid; HLB 15.4; 100% conc.

Lipocol TD-6. [Lipo] Trideceth-6; CAS 24938-91-8; nonionic; emulsifier, wetting and scouring agent, dispersant for essential oils; raw material for sulfation and phosphation; liq.; 100% act.

Lipocol TD-12. [Lipo] Trideceth-12; CAS 24938-91-8; nonionic; emulsifier, wetting and scouring agent, dispersant for essential oils; raw material for sulfation and phosphation; solubilizer; paste; HLB 14.6 ± 1; 100% act.

Lipolan. [Lipo] Hydrog. lanolin; nonionic; aux. w/o emulsifier; emollient; conditioner; lubricant; wh. to off-wh. paste; mild char. odor; water-insol.; m.p. 37–45; acid no. 1 max.; sapon. no. 5 max.; 100% conc.

Lipolan 327 F. [Lion] alpha-Olefin sulfonate; anionic; detergent base; paste; 37% conc.

Lipolan 1400. [Lion] alpha-Olefin sulfonate; anionic; emulsifier for cosmetics; emulsifier, dispersant for emulsion polymerization; powd.; 100% conc.

Lipolan AO. [Lion] alpha-Olefin sulfonate; anionic; milling and scouring agent for wool; liq.; 30% conc.

Lipolan AOL. [Lion] alpha-Olefin sulfonate; anionic; scouring agent for wool and feathers; liq.; 28% conc.

Lipolan G. [Lion] alpha-Olefin sulfonate; anionic; deinking agent for waste paper; liq.; 30% conc.

Lipolan LB-440. [Lion] alpha-Olefin sulfonate; anionic; detergent base; emulsifier for cosmetics and emulsion polymerization; liq.; 37% conc.

Lipolan LB-840. [Lion] alpha-Olefin sulfonate; anionic; detergent base; paste; 37% conc.

Lipolan PB-800. [Lion] alpha-Olefin sulfonate; anionic; detergent, emulsifier, dispersant; powd.; 95% conc.

Lipolan PJ-400. [Lion] alpha-Olefin sulfonate; anionic; emulsifier for cosmetics; dispersant for emulsion polymerization; powd.; 100% conc.

Lipolan TE, TE(P). [Lion] Alkyl methyl tauride; anionic; scouring agent for textiles; dispersant for pigment; paste, powd. resp.; 25 and 28% conc. resp.

Lipomin CH. [Lion] Imidazoline; amphoteric; milling and scouring agent for wool knit goods; liq.; 20% conc.

Lipomix G. [Lion] alpha-Olefin sulfonate; anionic; deinking agent for waste paper; liq.; 30% conc.

Lipomulse 165. [Lipo] Glyceryl stearate and PEG-100 stearate; nonionic; general purpose emulsifier, emollient, opacifier and visc. builder in creams and lotions; wh. bead or flake; HLB 11.0 ± 1; acid no. 2 max.; sapon. no. 90–100; 100% act.

Lipon PS-206. [Lion] Alkyl benzene sulfonate; anionic; detergent; powd.; 60% conc.

Liponic EG-1. [Lipo] Glycereth-26; CAS 31694-55-0; nonionic; humectant in creams and lotions, lubricant, plasticizer for hair resins, foam stabilizer, pigment dispersant, hair conditioner, foam modifier, antistat; used in personal care prods.; colorless, clear to slightly hazy visc. liq.; sol. in water, alcohol, acetone and ethyl acetate; acid no. 0.5 max.; 100% act.

Liponox LCF. [Lion] POE alkyl ether; nonionic; wetting agent; paste; 100% conc.

Liponox NC 2Y. [Lion] Alkylphenol ether, ethoxylated; nonionic; emulsifier for emulsion polymerization; flakes; 100% conc.

Liponox NC 6E, NCG, NCI, NCT. [Lion] POE alkylphenol ether; nonionic; detergent, penetrant, emulsifier, dispersant used in dyeing applic. for textiles, min. oil; wax, liq.; 100% conc.

Liponox NC-70. [Lion] POE alkylphenol ether; nonionic; detergent, penetrant, emulsifier, dispersant for bleaching, dyeing operations; emulsifier for min. oil; liq.; HLB 11.9; 100% conc.

Liponox NC-95. [Lion] POE alkylphenol ether; nonionic; detergent, penetrant, emulsifier, dispersant for bleaching, dyeing operations; emulsifier for min. oil; liq.; HLB 13.0; 100% conc.

Liponox NC-200. [Lion] POE alkylphenol ether; nonionic; detergent, penetrant, emulsifier, dispersant for bleaching, dyeing operations; emulsifier for min. oil; wax; HLB 15.8; 100% conc.

Liponox NC-300. [Lion] POE alkylphenol ether; nonionic; detergent, penetrant, emulsifier, dispersant for bleaching, dyeing operations; emulsifier for min. oil; wax; HLB 17.1; 100% conc.

Liponox NC-500. [Lion] POE alkylphenol ether; nonionic; emulsifier for emulsion polymerization; flakes; HLB 18.2; 100% conc.

Liponox OCS. [Lion] POE alkyl ether; nonionic; dye leveling agent; wax; 100% conc.

Lipophos PE9. [Lipo] Nonylphenol ether phosphate ester acid form; anionic; coupling agent for textile scouring, emulsion polymerization, emulsifier and wetting agent; liq.; water-sol.; 100% conc.

Lipophos PL6. [Lipo] Linear alcohol ether phosphate ester, acid form; anionic; coupling agent for textile scouring, emulsion polymerization, emulsifier and wetting agent; liq.; water-sol.; 100% conc.

Lipoproteol LCO. [Seppic] Sodium/TEA-lauroyl hydrolyzed collagen amino acid; nonionic; mild additive with good foaming props. for rich lather shampoos, facial cleaners, infant shampoos, detergents; liq.; 22% conc.

Lipoproteol LCOK. [Seppic] Lipoaminoacid salt; anionic; shampoo base and detergent; liq.; 22–25% conc.

Lipoproteol LK. [Seppic] Sodium/TEA-lauroyl hydrolyzed keratin amino acids; anionic; shampoo base; liq.; 25% conc.

Lipoproteol UCO. [Seppic] Sodium/TEA-undecylenoyl animal collagen amino acids; anionic; shampoo base for oily hair; antiseborrheic; liq.; 23% conc.

Liposorb L. [Lipo] Sorbitan laurate; CAS 1338-39-2; nonionic; emulsifier, thickener, lubricant, antistat, all-purpose lipophilic surfactant used with POE Liposorb series; also used in defoamers, aerosol w/o emulsions, corrosion inhibition; amber liq.; HLB 8.6 ± 1; sapon. no. 158–170; 100% act.

Liposorb L-10. [Lipo] PEG-10 sorbitan laurate; nonionic; o/w emulsifier, lubricant, antistat, all-purpose hydrophilic surfactant used for solubilizing oils and in conjunction with Liposorb esters; yel. liq.; HLB 14.9 ± 1; sapon. no. 66–76; 100% act.

Liposorb L-20. [Lipo] Polysorbate 20; CAS 9005-64-5; nonionic; o/w emulsifier, lubricant, antistat, all-purpose hydrophilic surfactant used for solubilizing oils and in conjunction with Liposorb esters; yel. liq.; HLB 16.7 ± 1; sapon. no. 40–50; 100% act.

Liposorb O. [Lipo] Sorbitan oleate; CAS 1338-43-8; nonionic; emulsifier, thickener, lubricant, antistat, all-purpose lipophilic surfactant used with POE Liposorb series; also used in defoamers, aerosol w/o emulsions, corrosion inhibition; yel./amber liq.; HLB 4.3 ± 1; sapon. no. 145–160; 100% act.

Liposorb O-5. [Lipo] Polysorbate 81; CAS 9005-65-6; nonionic; hydrophilic surfactant used for solubilizing oils; emulsifier, lubricant, antistat; liq.; HLB 10.0; 100% conc.

Liposorb P. [Lipo] Sorbitan palmitate; CAS 26266-57-9; nonionic; lipophilic surfactant, emulsifier, thickener, lubricant, antistat; also in defoamers, aerosol w/o emulsions, corrosion inhibition; tan beads or flakes; HLB 6.7 ± 1; sapon. no. 139–151; 100% act.

Liposorb P-20. [Lipo] Polysorbate 40; CAS 9005-66-7; nonionic; hydrophilic surfactant, solubilizer for oils, emulsifier, lubricant, antistat; yel. liq.; HLB 15.6 ± 1; sapon. no. 40–53; 100% act.

Liposorb S. [Lipo] Sorbitan stearate; CAS 1338-41-6; nonionic; emulsifier, thickener, lubricant, antistat, all-purpose lipophilic surfactant used with POE Liposorb series; also for defoamers, aerosol w/o emulsions, corrosion inhibition; cream beads or flakes; HLB 4.7 ± 1; sapon. no. 147–157; 100% act.

Liposorb SQO. [Lipo] Sorbitan sesquioleate; CAS 8007-43-0; nonionic; emulsifier, thickener, lubricant, antistat, all-purpose lipophilic surfactant used with POE Liposorb series; amber liq.; HLB 3.7 ± 1; sapon. no. 149–160; 100% act.

Liposorb TO. [Lipo] Sorbitan trioleate; CAS 26266-58-0; nonionic; emulsifier, thickener, lubricant, antistat, all-purpose lipophilic surfactant used with POE Liposorb series; amber liq.; HLB 1.8 ± 1; sapon. no. 171–185; 100% act.

Liposorb TO-20. [Lipo] Polysorbate 85; CAS 9005-70-3; nonionic; o/w emulsifier, lubricant, antistat, all-purpose hydrophilic surfactant used for solubilizing oils and in conjunction with Liposorb esters; yel. liq.; HLB 11.0 ± 1; sapon. no. 82–95; 100% act.

Liposorb TS. [Lipo] Sorbitan tristearate; CAS 26658-19-5; nonionic; emulsifier, thickener, lubricant, antistat, all-purpose lipophilic surfactant used with POE Liposorb series; cream flakes or beads; HLB 2.1 ± 1; sapon. no. 175–190; 100% act.

Lipotac TE. [Lion] Alkyl methyl tauride; anionic; scouring agent for cotton, wool; pigment dispersant; paste; 25% conc.

Lipotac TE-P. [Lion] Alkyl methyl tauride; anionic; scouring agent for cotton, wool; pigment dispersant; powd.; 28% conc.

Lipotin 100, 100J, SB. [Lucas Meyer] Refined soy lecithin, filtrated, deodorized; CAS 8002-43-5; wetting and dispersing agents preventing sedimentation in paints; raw material for textile and leather industry compds.; visc. 10, 20, and 12 Pas resp.; acid no. 30, 34, and 30 max. resp.; usage level: 0.7-3%.

Lipotin A. [Lucas Meyer] Phosphoamino compd.; amphoteric; emulsifier, wetting and dispersing agent for aq. systems, stabilizer for latex and emulsion paints, leather finishes, water-disp. and -reducible air-dry and stoving alkyds, offset printing inks, textile auxs.; lt. brn. visc. liq.; visc. 15 Pas; usage level: 1-5%; 99% solids.

Lipotin H. [Lucas Meyer] Hydroxylated soy lecithin; amphoteric; wetting and dispersing agent for air-drying and stoving paints; raw material for textile and leather compds.; emulsifier; almost clear high visc. liq.; usage level: 0.7-3%.

Liptol 40C. [Lion] Alkyl ethoxylate; nonionic; deinking agent for flotation process; liq.; 93% conc.

Liptol R-4000. [Lion] Polyoxyalkylene glycol; nonionic; deinking agent for flotation process; liq.; 80% conc.

Liptol S-2800. [Lion] Polyoxyalkylene glycol; nonionic; deinking agent for flotation process; liq.; 80% conc.

Liqui-Nox®. [Alconox] Blend; anionic/nonionic; phosphate-free detergent for critical manual and ultrasonic cleaning applics.; yel. liq., pract. odorless; sol. in water; sp.gr. 1.075; b.p. 214 F; flash pt. none; pH 8.5; toxicology: skin irritant; ingestion may cause discomfort or diarrhea.

Lomar® D. [Henkel/Textile] Sodium naphthalene sulfonate; anionic; dispersant for disperse and vat dyes; tan powd.; sol. @ 5% in water; disp. in glycerol trioleate; cloud pt. 11 C; flash pt. 84 F; pH 9.5 (10%).

Lomar® D SOL. [Henkel/Textile] Sodium naphthalene sulfonate; anionic; dispersant for disperse dyes; Gardner 18 liq.; sol. @ 5% in water; disp. in glycerol trioleate; dens. 9.8 lb/gal; visc. 9 cSt (100 F); pour pt. -1 C; cloud pt. 4 C; flash pt. 30 F.

Lomar® HP. [Henkel/Functional Prods.] Condensed potassium naphthalene sulfonate; anionic; dispersant; sec. emulsifier for emulsion polymerization; powd.; 94% conc.

Lomar® LS. [Henkel/Functional Prods.; Henkel/Textile] Condensed sodium naphthalene sulfonate; CAS 9084-06-4; anionic; dispersant, emulsifier for emulsion polymerization, dyestuff mfg., agric. formulations; leveling agent for dyeing fibers; low salt; tan powd.; sol. @ 5% in water; cloud pt. 2 C; flash pt. 92 F; pH 9.5 (10% aq.); 95% act.

Lomar® LS Liq. [Henkel/Functional Prods.] Condensed sodium naphthalene sulfonate; anionic; dispersant; primary emulsifier for emulsion polymerization; liq.; 46% solids.

Lomar® PL. [Henkel/Emery; Henkel/Functional Prods.; Henkel/Textile] Condensed sodium naphthalene sulfonate; anionic; dispersant for pigments, extenders, and fillers in aq. media; used in dyeing syn. and natural fibers; emulsifier in emulsion polymerization; ceramics; gypsum board, for pigments, printing, rubber and wet milling; food pkg. applics.; agric. prods.; EPA-exempt; dk. amber liq.; water-sol. @ 5%; sp.gr. 1.25; dens 10 lb/gal; visc. 11 cSt (100 F); pour pt. -1 C; cloud pt. 6 C; flash pt. 41 F; pH 9.5 (20%); 46% solids. DISCONTINUED.

Lomar® PW. [Henkel/Emery; Henkel/Functional Prods.; Henkel/Textile] Condensed sodium naphthalene sulfonate; CAS 9084-06-4; anionic; dispersant for pigments, extenders, and fillers in aq. media; used in dyeing syn. and natural fibers; emulsifier in emulsion polymerization; ceramics; gypsum board, for pigments, printing, rubber and wet milling; food pkg. applics.; agric. prods.; suspending agent, stabilizer for paint and paper industries; EPA-exempt; tan powd.; water-sol. @ 5%; dens. 0.66 g/cc; cloud pt. 5 C; flash pt. 87 F; pH 9.5 (10%); 87% act.

Lomar® PWA. [Henkel/Emery; Henkel/Functional Prods.] Condensed ammonium naphthalene sulfonate; anionic; visc. depressant; for molding and extruding operations in ceramics; dispersant for emulsion paints, agric. formulations; emulsifier for emulsion polymerization of syn. elastomers; visc. reducer for pigment slurries; stabilizer; EPA-exempt; lt. tan powd.; sol. in water; pH 3.5 (10%); 92% act.

Lomar® PWA Liq. [Henkel/Emery; Henkel/Functional Prods.] Condensed ammonium naphthalene sulfonate; anionic; dispersant; sec. emulsifier for emulsion polymerization, agric. formulations; EPA-exempt; dk. brn. liq.; pH 7.3 (10%); 44% act.

Lonzaine® 12C. [Lonza] Coco betaine; amphoteric; conditioner, foaming agent, visc. modifier, irritation mitigant used in personal care prods. and industrial applics.; biodeg.; liq.; visc. 14 cps; pour pt. 3 C; pH 7.5 (3%); surf. tens. 34.4 dynes/cm (0.1%); Draves wetting 11 s; 35% conc.

Lonzaine® 16SP. [Lonza] Cetyl betaine; surfactant; solid; visc. 720 cps; surf. tens. 32 dynes/cm (0.1%); Draves wetting 11 s; 35% solids.

Lonzaine® 18S. [Lonza] Stearyl betaine; CAS 820-66-6; surfactant.

Lonzaine® C. [Lonza] Cocoamidopropyl betaine; amphoteric; foaming agent, conditioner, visc. booster, wetting agent, used in cosmetics, toiletries, detergents, metal finishing, textile finishing, etc.; biodeg.; Gardner 2 nonvisc. liq.; char. odor; visc. 29 cps; pour pt. 3 C; pH 4.5–5.5 (10%); surf. tens. 33.6 dynes/cm (0.1%); Draves wetting 28 s; 30% act.

Lonzaine® CO. [Lonza] Cocamidopropyl betaine; amphoteric; foaming agent, conditioner, visc. booster, wetting agent, used in cosmetics, toiletries, detergents, metal finishing, textile finishing, etc.; biodeg.; Gardner 2 nonvisc. liq.; char. odor; visc. 28 cps; pour pt. 3 C; pH 6–8 (10%); surf. tens. 33.9 dynes/cm (0.1%); Draves wetting 16 s; 35% solids.

Lonzaine® CS. [Lonza] Cocamidopropylhydroxysultaine; amphoteric; conditioner used in personal care prods. and industrial applics.; liq.; water-sol.; visc. 189 cps; pour pt. 9 C; pH 8.0 (10%); surf. tens. 35.2 dynes/cm (0.1%); Draves wetting > 2 min; 50% conc.

Lonzaine® JS. [Lonza] Cocamidopropylhydroxysultaine; amphoteric; foaming agent, visc. modifier, irritation mitigant for personal care and industrial applics.; liq.; 50% solids.

Lonzest® PEG 4-DO. [Lonza] PEG-8 dioleate; CAS 9005-07-6; nonionic; detergent, lubricant, dispersant, anticorrosive, emulsifier, defoamer, used in textile, petrol., insecticide industries; yel. liq.; sol. in min. oil, IPA, toluol, disp. in water; sp.gr. 0.98; m.p. 0 C; HLB 9.9; acid no. 5–10; sapon. no. 120–130; 100% conc.

Lonzest® PEG 4-L. [Lonza] PEG-8 laurate; CAS 9004-81-3; nonionic; dispersant, rewetting agent, stabilizer, used in cosmetics, pharmaceuticals, textile, paper, agric. industries; yel. liq.; sol. in water, IPA, toluol; sp.gr. 1.03; m.p. 10 C; HLB 13.1; acid no. 5–10; sapon. no. 90–100; 100% conc.

Lonzest® PEG 4-O. [Lonza] PEG-8 oleate; CAS 9004-96-0; nonionic; detergent, emulsifier, polish remover in pharmaceuticals and cosmetics, rust preventative oils and degreasers; yel. liq.; sol. in IPA, toluol, min. oil, water disp.; sp.gr. 1.01; m.p. 0 C; HLB 11.4; sapon. no. 86–96; 100% conc.

Lonzest® SML. [Lonza] Sorbitan laurate; CAS 1338-39-2; nonionic; emulsifier for o/w systems; fiber lubricant and corrosion inhibitor used in food industry; liq.; 100% conc.

Lonzest® SML-20. [Lonza] Polysorbate 20; CAS 9005-64-5; nonionic; emulsifier used where o/w and w/o emulsions are required; solubilizer; liq.; HLB 17.0; 100% conc.

Lonzest® SMO. [Lonza] Sorbitan oleate; CAS 1338-43-8; nonionic; emulsifier for o/w systems; fiber lubricant and corrosion inhibitor used in food industry; liq.; 100% conc.

Lonzest® SMO-20. [Lonza] Polysorbate 80; CAS 9005-65-6; nonionic; emulsifier, solubilizer and stabilizer used in foods, personal care prods. and industrial applics.; liq.; HLB 15.0; 100% conc.

Lonzest® SMP. [Lonza] Sorbitan palmitate; CAS 26266-57-9; nonionic; emulsifier for o/w systems; fiber lubricant and corrosion inhibitor used in food industry; solid; 100% conc.

Lonzest® SMP-20. [Lonza] Polysorbate 40; CAS 9005-66-7; nonionic; emulsifier, solubilizer and stabilizer used in foods, personal care prods. and industrial applics.; liq.; HLB 16.0; 100% conc.

Lonzest® SMS. [Lonza] Sorbitan stearate; CAS 1338-41-6; nonionic; emulsifier for o/w systems; fiber lubricant and corrosion inhibitor used in food industry; flake; 100% conc.

Lonzest® SMS-20. [Lonza] Polysorbate 60; CAS 9005-67-8; nonionic; emulsifier, solubilizer and stabilizer used in foods, personal care prods. and industrial applics.; liq.; HLB 15.0; 100% conc.

Lonzest® STO. [Lonza] Sorbitan trioleate; CAS 26266-58-0; nonionic; emulsifier for o/w systems; fiber lubricant and corrosion inhibitor used in food industry; liq.; 100% conc.

Lonzest® STO-20. [Lonza] Polysorbate 85; CAS 9005-70-3; nonionic; emulsifier, solubilizer and stabilizer used in foods, personal care prods. and industrial applics.; liq.; HLB 11.0; 100% conc.

Lonzest® STS. [Lonza] Sorbitan tristearate; CAS 26658-19-5; nonionic; emulsifier for o/w systems; fiber lubricant and corrosion inhibitor used in food industry; solid; 100% conc.

Lonzest® STS-20. [Lonza] Polysorbate 65; CAS 9005-71-4; nonionic; emulsifier, solubilizer and stabilizer used in foods, personal care prods. and industrial applics.; solid; HLB 11.0; 100% conc.

Lorol C6. [Henkel/Emery] Hexyl alcohol; chemical intermediate for detergent mfg.; APHA 10 liq.; sp.gr. 0.810-0.820 g/cc; solid. pt. 50 C; b.p. 145-160 C; acid no. < 0.1; iodine no. < 0.3; sapon. no. < 1.0; hyd. no. 530-550; flash pt. 65 C.

Lorol C8. [Henkel/Emery] Octyl alcohol; intermediate for surfactant mfg.; in rolling oils and other lubricants; APHA 10 liq.; sp.gr. 0.815-0.825 g/cc; solid. pt. 18 C; b.p. 185-210 C; acid no. < 0.1; iodine no. < 0.3; sapon. no. < 1.0; hyd. no. 420-423; flash pt. 90 C; 100% conc.

Lorol C8-98. [Henkel/Emery] Octyl alcohol; intermediate for surfactant mfg.; component in rolling oils, lubricants; APHA 10 liq.; sp.gr. 0.815-0.825 g/cc; solid. pt. 17 C; b.p. 185-200 C; acid no. < 0.1; iodine no. < 0.3; sapon. no. < 1.0; hyd. no. 420-430; flash pt. 90 C; 100% conc.

Lorol C8-C10. [Henkel/Emery] Lauryl alcohol (50-66% C8, 30-40% C10); CAS 112-53-8; intermediate for detergent mfg.; component in rolling oils, lubricants; APHA 10 liq.; sp.gr. 0.820-0.830 g/cc; solid. pt. 11 c; b.p. 190-260 C; acid no. < 0.1; iodine no. < 0.5; sapon. no. < 1.5; hyd. no. 385-410; flash pt. 95 C; 100% conc.

Lorol C8-C10 Special. [Henkel/Emery] Lauryl alcohol (40-48% C8, 51-59% C10); CAS 112-53-8; chemical intermediate for detergent mfg.; APHA 10 liq.; sp.gr. 0.820-0.830 g/cc; solid. pt. 7 C; b.p. 195-250 C; acid no. < 0.1; iodine no. < 0.3; sapon. no. < 1.0; hyd. no. 380-390; flash pt. 100 C.

Lorol C8-C18. [Henkel/Emery] Lauryl alcohol (44-50% C12, 14-20% C14, 8-10% C16, 8-12% C18); CAS 112-53-8; intermediate for detergent mfg.; component in lubricants; APHA 20 liq./solid; sp.gr. 0.820-0.830 g/cc (30 C); solid. pt. 15-18 C; b.p. 230-250 C; acid no. < 0.1; iodine no. < 0.5; sapon. no. < 1.2; hyd. no. 280-290; flash pt. 110 C; 100% conc.

Lorol C8 Chemically Pure. [Henkel/Emery] Octyl alcohol; chemical intermediate for detergent mfg.; APHA 10 liq.; sp.gr. 0.815-0.825 g/cc; solid. pt. 16 C; b.p. 185-195 C; acid no. < 0.1; iodine no. < 0.1; sapon. no.< 0.1; hyd. no. 429-431; flash pt. 90 C.

Lorol C10. [Henkel/Emery] Decyl alcohol; intermediate for surfactant mfg.; component in rolling oils, lubricants; APHA 10 liq.; sp.gr. 0.820-0.830 g/cc; solid. pt. 3-6 C; b.p. 220-240 C; acid no. < 0.1; iodine no. < 0.3; sapon. no.< 1.0; hyd. no. 345-356; flash pt. 110 C; 100% conc.

Lorol C10-98. [Henkel/Emery] Decyl alcohol; intermediate for surfactant mfg.; component in rolling oils, lubricants; APHA 10 liq.; sp.gr. 0.820-0.830 g/cc; solid. pt. 4-7 C; b.p. 220-235 C; acid no. < 0.1; iodine no. < 0.1; sapon. no.< 0.3; hyd. no. 350-355; flash pt. 110 C; 100% conc.

Lorol C10-C12. [Henkel/Emery] Lauryl alcohol (81-89% C10, 6-11% C12, 4-9% C14); CAS 112-53-8; intermediate for surfactant mfg.; component in rolling oils, lubricants; APHA 10 liq.; sp.gr. 0.820-0.830 g/cc; solid. pt. -1 to 3 C; b.p. 220-280 C; acid no. < 0.1; iodine no. < 0.3; sapon. no. < 1.0; hyd. no. 336-346; flash pt. 115 C; 100% conc.

Lorol C12. [Henkel/Emery] Lauryl alcohol; CAS 112-53-8; intermediate for surfactant mfg.; component in rolling oils, lubricants; APHA 10 liq./solid; sp.gr. 0.815-0.825 g/cc (30 C); solid. pt. 19-23 C; b.p. 250-

270 C; acid no. < 0.1; iodine no. < 0.3; sapon. no.< 0.4; hyd. no. 295-301; flash pt. 140 C; 100% conc.

Lorol C12-99. [Henkel/Emery] Lauryl alcohol; CAS 112-53-8; intermediate for surfactant mfg.; component in rolling oils, lubricants; APHA 10 liq./solid; sp.gr. 0.815-0.825 g/cc (30 C); solid. pt. 19-23 C; b.p. 250-270 C; acid no. < 0.1; iodine no. < 0.3; sapon. no.< 0.4; hyd. no. 297-301; flash pt. 140 C; 100% conc.

Lorol C12-C14. [Henkel/Emery] Lauryl alcohol (65-69% C12, 24-28% C14, 4-8% C16); CAS 112-53-8; intermediate for surfactant mfg.; component in rolling oils, lubricants; APHA 20 liq./solid; sp.gr. 0.820-0.830 g/cc (30 C); solid. pt. 18-21 C; b.p. 255-305 C; acid no. < 0.1; iodine no. < 0.3; sapon. no. < 0.5; hyd. no. 280-290; flash pt. 140 C; 100% conc.

Lorol C12-C16. [Henkel/Emery] Lauryl alcohol (60-66% C12, 21-25% C14, 10-12% C16); CAS 112-53-8; intermediate for surfactant mfg.; component in lubricants; APHA 20 liq./solid; sp.gr. 0.820-0.830 g/cc (30 C); solid. pt. 18-22 C; b.p. 255-310 C; acid no. < 0.1; iodine no. < 0.5; sapon. no. < 0.5; hyd. no. 280-290; flash pt. 140 C; 100% conc.

Lorol C12-C18. [Henkel/Emery] Lauryl alcohol (36-44% C12, 16-20% C14, 8-12% C16); CAS 112-53-8; intermediate for surfactant mfg.; component in lubricants; APHA 20 liq./solid; sp.gr. 0.810-0.820 g/cc (40 C); solid. pt. 25-28 C; b.p. 260-350 C; acid no. < 0.1; iodine no. < 0.5; sapon. no. < 0.8; hyd. no. 260-275; flash pt. 140 C; 100% conc.

Lorol C12 Chemically Pure. [Henkel/Emery] Lauryl alcohol; CAS 112-53-8; chemical intermediate for detergent mfg.; APHA 10 liq./solid; sp.gr. 0.815-0.825 g/cc (30 C); solid. pt. 23.7-23.9 C; b.p. 255-265 C; acid no. < 0.1; iodine no. < 0.1; sapon. no.< 0.1; hyd. no. 299-301; flash pt. 140 C.

Lorol C14. [Henkel/Emery] Myristyl alcohol; CAS 112-72-1; intermediate for surfactant mfg.; component in rolling oils, lubricants; APHA 20 fused/flakes; sp.gr. 0.810-0.820 g/cc (40 C); solid. pt. 35-38 C; b.p. 280-300 C; acid no. < 0.1; iodine no. < 0.5; sapon. no.< 0.5; hyd. no. 225-262; flash pt. 155 C; 100% conc.

Lorol C14-98. [Henkel/Emery] Myristyl alcohol; CAS 112-72-1; intermediate for surfactant mfg.; component in rolling oils, lubricants; APHA 20 fused/flakes; sp.gr. 0.810-0.820 g/cc (40 C); solid. pt. 36-38 C; b.p. 280-300 C; acid no. < 0.1; iodine no. < 0.3; sapon. no.< 0.5; hyd. no. 255-262; flash pt. 155 C; 100% conc.

Lorol C14 Chemically Pure. [Henkel/Emery] Myristyl alcohol; CAS 112-72-1; chemical intermediate for detergent mfg.; APHA 20 fused; sp.gr. 0.810-0.820 g/cc (40 C); solid. pt. 37.4-37.7 C; b.p. 285-300 C; acid no. < 0.1; iodine no. < 0.1; sapon. no.< 0.2; hyd. no. 261-263; flash pt. 155 C.

Lorol C16. [Henkel/Emery] Cetyl alcohol; CAS 36653-82-4; intermediate for surfactant mfg.; component in rolling oils, lubricants; APHA 20 fused/flakes; sp.gr. 0.805-0.815 g/cc (60 C); solid. pt. 46-49 C; b.p. 305-330 C; acid no. < 0.1; iodine no. < 0.5; sapon. no.< 0.5; hyd. no. 225-235; flash pt. 180 C; 100% conc.

Lorol C16-95. [Henkel/Emery] Cetyl alcohol; CAS 36653-82-4; intermediate for surfactant mfg.; component in rolling oils, lubricants; APHA 20 fused/flakes; sp.gr. 0.805-0.815 g/cc (60 C); solid. pt. 46-49 C; b.p. 305-330 C; acid no. < 0.1; iodine no. < 0.5; sapon. no.< 0.5; hyd. no. 228-235; flash pt. 180 C; 100% conc.

Lorol C16-98. [Henkel/Emery] Cetyl alcohol; CAS 36653-82-4; intermediate for surfactant mfg.; component in rolling oils, lubricants; APHA 20 fused/flakes; sp.gr. 0.805-0.815 g/cc (60 C); solid. pt. 46-49 C; b.p. 305-330 C; acid no. < 0.1; iodine no. < 0.5; sapon. no.< 0.5; hyd. no. 228-235; flash pt. 180 C; 100% conc.

Lorol C16 Chemically Pure. [Henkel/Emery] Cetyl alcohol; CAS 36653-82-4; chemical intermediate for detergent mfg.; APHA 20 fused; sp.gr. 0.805-0.815 g/cc (60 C); solid. pt. 48.8-49.1 C; b.p. 310-325 C; acid no. < 0.1; iodine no. < 0.1; sapon. no.< 0.2; hyd. no. 231-233; flash pt. 180 C.

Lorol C18. [Henkel/Emery] Stearyl alcohol; CAS 112-92-5; intermediate for surfactant mfg.; component in rolling oils, lubricants; APHA 20 fused/flakes; sp.gr. 0.805-0.815 g/cc (60 C); solid. pt. 55-55.7 C; b.p. 330-360 C; acid no. < 0.1; iodine no. < 1.0; sapon. no.< 0.5; hyd. no. 203-210; flash pt. 195 C; 100% conc.

Lorol C18-98. [Henkel/Emery] Stearyl alcohol; CAS 112-92-5; chemical intermediate for detergent mfg.; APHA 20 fused/flakes; sp.gr. 0.805-0.815 g/cc (60 C); solid. pt. 55-55.7 C; b.p. 330-360 C; acid no. < 0.1; iodine no. < 1.0; sapon. no.< 0.5; hyd. no. 203-210; flash pt. 195 C.

Lorol C18 Chemically Pure. [Henkel/Emery] Stearyl alcohol; CAS 112-92-5; chemical intermediate for detergent mfg.; APHA 20 fused; sp.gr. 0.805-0.815 g/cc (60 C); solid. pt. 57.4-57.7 C; b.p. 340-360 C; acid no. < 0.1; iodine no. < 0.2; sapon. no.< 0.2; hyd. no. 207-209; flash pt. 200 C.

Lorol Lauryl Alcohol Tech. [Henkel/Emery] Lauryl alcohol (48-58% C12, 19-24% C14, 9-12% C16, 11-15% C18); CAS 112-53-8; chemical intermediate for detergent mfg.; APHA 20 liq./solid; sp.gr. 0.820-0.830 g/cc (30 C); solid. pt. 18-23 C; b.p. 255-360 C; acid no. < 0.1; iodine no. < 0.5; sapon. no. < 1.2; hyd. no. 266-275; flash pt. 130 C.

Lorol Special. [Henkel/Emery] Lauryl alcohol (70-75% C12, 24-30% C14); CAS 112-53-8; chemical intermediate for detergent mfg.; APHA 20 liq./solid; sp.gr. 0.820-0.830 g/cc (30 C); solid. pt. 17-23 C; b.p. 255-295 C; acid no. < 0.1; iodine no. < 0.3; sapon. no. < 0.5; hyd. no. 285-293; flash pt. 140 C.

Loropan CME. [Triantaphyllou] Cocamide MEA; CAS 68140-00-1; nonionic; foam stabilizer and thickening agent for shampoos and detergents; bead; 92–96% conc.

Loropan KD. [Triantaphyllou] Cocamide DEA; nonionic; emulsifier, stabilizer, thickener for shampoos and bubble baths; liq., paste; 85–90% conc.

Loropan LD. [Triantaphyllou] Lauramide DEA; CAS 120-40-1; nonionic; foam stabilizer for shampoos, detergents, fortifier for perfumes in soap; solid; 90% conc.

Loropan LM. [Triantaphyllou] Lauramide MEA; CAS 142-78-9; nonionic; foam stabilizer and thickening agent for shampoos and detergents; bead; 92–96% conc.

Loropan LMD. [Triantaphyllou] Lauric/myristic DEA; nonionic; foam stabilizer; superfatting and thickening agent for shampoos; fortifier for perfume in soaps; paste; 90% conc.

Loxiol G 52. [Henkel/Functional Prods.] C16-18 fatty alcohol; surfactant for polymerization; solid/flakes; m.p. 48-52 C.

Loxiol G 53. [Henkel/Functional Prods.] C16-18 fatty alcohol; surfactant for polymerizations; internal lubricant for rigid PVC; inj. molding; solid; sp.gr.

0.790–0.802 (80 C); visc. 4–6 mPa.s (80 C); m.p. 48-51 C; acid no. < 0.2; ref. index 1.427–1.430 (80 C); flash pt. > 160 C; 100% conc.

Loxiol P 1420. [Henkel/Functional Prods.] C16-18 fatty alcohol; surfactant for polymerization; solid/flakes; m.p. 48-52 C.

Loxiol VPG 1354. [Henkel/Functional Prods.] Stearyl alcohol; CAS 112-92-5; nonionic; surfactant for polymerization; flakes; m.p. 55-57 C; 100% conc.

Loxiol VPG 1451. [Henkel/Functional Prods.] Behenyl alcohol; surfactant for polymerization; solid/flakes; m.p. 63-65 C.

Loxiol VPG 1496. [Henkel/Functional Prods.] C12-18 fatty alcohol; surfactant for polymerization; liq./solid; m.p. 18-23 C; 100% act.

Loxiol VPG 1743. [Henkel/Functional Prods.] Cetyl alcohol; CAS 36653-82-4; nonionic; surfactant for polymerization; flakes; m.p. 46-49 C; 100% conc.

LSF-54. [Witco UK] Anionic; detergent, wetting agent, high flame resistance; foam; sp.gr. 1.032; visc. 5.0 cs; f.p. 5 C; pH 6.5–8.0.

LSP 33. [Berol Nobel AB] N,N-Bis(3-aminopropyl tallowamine; CAS 61791-57-9; cationic; intermediate, wetting agent, pigment grinding aid, dispersant; flushing agent; corrosion inhibitor; polymer industry; liq.

Lubran 145. [Toho Chem. Industry] Alkyl methacrylate polymer; pour pt. depressant, visc. index improver for lubricating oils; liq.

Lubran 170. [Toho Chem. Industry] Alkyl methacrylate polymer; pour pt. depressant, visc. index improver for lubricating oils; liq.

Lubran AD. [Toho Chem. Industry] Hydrocarbon wax/naphthalene condensate; pour pt. depressant, dewaxing aid for lubricating oils; liq.

Lubran BS. [Toho Chem. Industry] Hydrocarbon wax/naphthalene condensate; dewaxing aid for lubricating oils; paste.

Lubran BSP. [Toho Chem. Industry] Hydrocarbon wax/naphthalene condensate; dewaxing aid for lubricating oils; paste.

Lubran FPD Series. [Toho Chem. Industry] Surfactant/polymer blends; flow improver for fuel oils; pour pt. depressant; liq./paste.

Lubrhophos® HR-719 (formerly Antara® HR-719). [Rhone-Poulenc Surf.] Aliphatic phosphate ester, acid form; anionic; biodeg. low foaming emulsifier, EP agent, corrosion inhibitor, cleaning and lubricity aids for metalworking fluids; liq.; water-sol.; dens. 10.912 lb/gal; flash pt. > 200 C; 90% act. DISCONTINUED.

Lubrhophos® LB-400 (formerly Antara® LB-400) [Rhone-Poulenc Surf.; Rhone-Poulenc France] Org. phosphate ester, free acid; anionic; lubricity and EP additive, rust inhibitor, wetting agent, emulsifier, detergent, dispersant; for lubricating and rolling oils, hydraulic and water-based cutting fluids, glass cutting and polishing lubricants; amber opaque visc. liq.; sol. in oil, water, and aliphatic and aromatic solvs.; sp.gr. 1.04; dens. 8.663 lb/gal; 98% act.

Lubrhophos® LE-500 (formerly Antara LE-500). [Rhone-Poulenc Surf.; Rhone-Poulenc France] Aromatic phosphate ester; anionic; EP agent, emulsifier, corrosion inhibitor, detergent, lubricant for metalworking fluids; yel. clear visc. liq.; sol. in oil and water, aliphatic and aromatic solvs.; sp.gr. 1.11; dens. 9.246 lb/gal; flash pt. > 200 C; 100% act.

Lubrhophos® LE-600 (formerly Antara® LE-600). [Rhone-Poulenc Surf.; Rhone-Poulenc France] Phosphate acid ester; aromatic base; anionic; lubricant and emulsifier for cutting fluids, hydraulic fluids, rolling oils; FDA compliance; liq.; sol. in oil and water; 100% act.

Lubrhophos® LE-700 (formerly Antara® LE-700). [Rhone-Poulenc Surf.] Aromatic phosphate ester, free acid; anionic; high foaming lubricant, emulsifier, EP agent, corrosion inhibitor for water-based cutting fluids; FDA compliance; VCS 8 sl. cloudy visc. liq.; water-sol.; dens. 9.330 lb/gal; flash pt. > 200 C; 100% act.

Lubrhophos® LF-200 (formerly Antara® LF-200). [Rhone-Poulenc Surf.; Rhone-Poulenc France] Org. phosphate ester, free acid; anionic; lubricant, emulsifier for oil-based cutting fluids, hydraulic fluids, rolling oils, slushing compds.; amber visc. liq.; sol. in oil and water; 100% act.

Lubrhophos® LF-700 (formerly Antara® LF-700). [Rhone-Poulenc Surf.] Org. phosphate ester, free acid; anionic; low foaming lubricant, pressure additive for syn. aq. cutting and grinding fluids; liq.; water-sol.; 100% conc. DISCONTINUED.

Lubrhophos® LK-500 (formerly Antara® LK-500). [Rhone-Poulenc Surf.] Org. phosphate ester, free acid; anionic; low-foaming lubricant, emulsifier for metalworking fluids, with extreme pressure, antiwear, and rust inhibition properties; pale yel., slightly visc. clear liq.; sol. in xylene, ethanol, ethylene glycol, carbon tetrachloride, butyl Cellosolve, water; sp.gr. 1.11; visc. 21 cSt (100 C); biodeg.; 100% act.

Lubrhophos® LL-550. [Rhone-Poulenc Surf.] Anionic; emulsifier, EP agent, corrosion inhibitor, lubricant for metalworking fluids; water-sol.; dens. 8.630 lb/gal; flash pt. > 200 C; 98% act.

Lubrhophos® LM-400 (formerly Antara® LM-400). [Rhone-Poulenc Surf.] Aromatic phosphate ester, free acid; anionic; lubricant, emulsifier, EP agent, corrosion inhibitor, detergent for cutting fluids, hydraulic fluids, rolling oils; yel. clear liq.; sol. in oil, aliphatic and aromatic solvs.; disp. in water; sp.gr. 1.05; dens. 8.746 lb/gal; flash pt. > 200 C; 100% act.

Lubrhophos® LM-600 (formerly Antara® LM-600). [Rhone-Poulenc Surf.] Aromatic phosphate ester; anionic; EP agent, corrosion inhibitor, detergent, lubricant for oil- and water-based metalworking fluids; yel. clear visc. liq.; sol. in oil, water, aliphatic and aromatic solvs.; sp.gr. 1.06; dens. 8.830 lb/gal; flash pt. > 200 C; 100% act.

Lubrhophos® LP-700 (formerly Antara® LP-700). [Rhone-Poulenc Surf.; Rhone-Poulenc France] Complex org. phosphate ester, free acid; anionic; low foaming emulsifier, EP agent, corrosion inhibitor, detergent, lubricant for water-based metalworking fluids; yel. clear visc. liq.; sol. in water, aliphatic and aromatic solvs.; sp.gr. 1.20; dens. 9.996 lb/gal; flash pt. > 200 C; 100% act.

Lubrhophos® LS-500 (formerly Antara® LS-500). [Rhone-Poulenc Surf.; Rhone-Poulenc France] Phosphate acid ester; aliphatic hydrophobic base; anionic; lubricant, emulsifier for oil and water-based cutting fluids, hydraulic fluids, rolling oils; liq.; sol. in oil, water, aliphatic and aromatic solv.; 100% act.

Lubricant EHS. [Hoechst Celanese/Colorants & Surf.] Fatty acid ester; anionic; heat-stable boundary lubricant for microemulsions for heavy-duty machining of aluminum; liq.; water-insol.; sp.gr. 0.85.

Lubrizol® 2152. [Lubrizol] Calcium sulfonate; anionic; pigment dispersant and wetting agent for color concs., paints, coatings, inks; pigment flushing aid for org. and inorg. pigments; visc. stabilizer/reducer in plastisols; amber liq.; sp.gr. 0.969-0.999; dens. 8.08-8.33 lb/gal; visc. 2000-6000 cps; flash pt. (PMCC) 155 C; 100% act.

Lubrizol® 2153. [Lubrizol] Succinimide; nonionic; pigment dispersant and wetting agent for color concs., inks, plastisols and organosols; amber liq.; sp.gr. 0.910-0.940; dens. 7.59-7.84 lb/gal; visc. 8000-24,000 cps; flash pt. (PMCC) 160 C; 100% act.

Lubrizol® 2155. [Lubrizol] Succinimide; nonionic; pigment dispersant for color concs., inks, plastisols and organosols; wetting agent for carbon fibers; amber liq.; sp.gr. 0.909-0.939; dens. 7.58-7.83 lb/gal; visc. 13,000-23,000 cps; flash pt. (PMCC) 160 C; 100% act.

Lubrizol® 5369, 5372. [Lubrizol] Sodium sulfonate/carboxylate mixture; anionic; general-purpose sol. oil emulsifier pkg. for wide range of paraffinic base stocks; liq.; 100% conc.

Lubrizol® 5375. [Lubrizol] Sodium sulfonate/carboxylate mixture; anionic; sol. oil emulsifier/corrosion inhibitor pkg. for wide range of paraffinic stocks; liq.; 100% conc.

Lubrol N5. [ICI Surf. UK] Alkyl polyglycol ether; nonionic; emulsifier for removing oil and grease stains caused by loom or knitting machine oil; liq.

Lufibrol® E. [BASF] extraction agent for textile desizing and scouring.

Lufibrol® FW. [BASF] Bleaching agent, wool fiber protection agent for prevention of yellowing during dyeing.

Lufibrol® KB Liq. [BASF] Anionic; textile aux. for extraction scouring and bleaching of cotton and cotton blends.

Lufibrol® KE. [BASF] Anionic; surfactant-free textile aux. for alkaline boil-off treatment of cotton and PES/cotton.

Lufibrol® NB-7. [BASF] Extracting and dispersing agent for detaching and washing impurities in desizing, scouring, and caustic boil-off operations.

Lufibrol® NB-T. [BASF] Textile aux. for alkaline boil-off of cotton and polyester/cotton blends.

Lufibrol® O. [BASF] Extraction aux. for scouring and bleaching processes.

Lumiten® E. [BASF AG] Foam suppressants for aq. emulsion paints, cement systems, dispersion adhesives.

Lumiten® I. [BASF AG] Anionic; wetting agent for emulsion paints, for the modification of cement systems, and for adhesives.

Lumiten® N. [BASF AG] Nonionic; wetting agent for emulsion paints, for the modification of cement systems, and for adhesives.

Lumorol 4153. [Zschimmer & Schwarz] Nonionic/anionic surfactants with phosphate and dissolving agents; anionic/nonionic; cleansing agents; liq.; 57% act.

Lumorol 4154. [Zschimmer & Schwarz] Nonionic/anionic surfactants; anionic/nonionic; cleansing agents; liq.; 50% act.

Lumorol 4192. [Zschimmer & Schwarz] Nonionic/anionic surfactants with phosphate and dissolving agents; anionic/nonionic; cleansing agents; liq.; 50% act.

Lumorol 4290. [Zschimmer & Schwarz] Surfactants; anionic; cleansing agents; liq.; 30% act.

Lumorol BZ. [Zschimmer & Schwarz] Surfactants; liq.; 18% act.

Lumorol GG 65. [Zschimmer & Schwarz] Surfactant blend; anionic/nonionic; detergent for household and industrial applics., all-purpose cleaners; paste; 65% conc.

Lumorol RK. [Zschimmer & Schwarz] Surfactant blend; anionic/amphoteric; detergent for cosmetics, shampoos, bath preps., household cleaners; liq.; 28% conc.

Lumorol W 5058. [Zschimmer & Schwarz] Surfactants; anionic/nonionic; cleansing agents; liq.; 93% act.

Lumorol W 5157. [Zschimmer & Schwarz] Surfactants; anionic/nonionic; cleansing agents; liq.; 83% act.

Lumo Stabil S 80. [Zschimmer & Schwarz] Sodium dodecylbenzene sulfonate; anionic; washing and cleansing agents, industrial cleaners; powd.; 80% act.

Lumo WW 75. [Zschimmer & Schwarz] Sodium dodecylbenzene sulfonate; anionic; heavy and lt. duty detergent; paste; 75% conc.

Luprintol® PE. [BASF AG] Aryl polyglycol ether; nonionic; emulsifier for prep. of o/w emulsions, solv.-containing or solv.-free thickenings for textile pigment printing; liq.; misc. with water; dens. 1.11 g/cc; b.p. 100 C; pH 7-8 (50 g/l water); toxicology: avoid breathing vapors from hot liquors; use gloves and goggles handling conc. prod.

Lutensit® A-BO. [BASF AG] Sodium dioctylsulfosuccinate in water-neopentyl glycol; anionic; surfactant for chemical industry; liq.; 60% conc.

Lutensit® A-EP. [BASF AG] Fatty alcohol alkoxylate acid phosphate ester; anionic; emulsifier, wetting agent, dispersant, hydrotrope, solubilizer, and cleaner used in industrial and household cleaning agents; biodeg.; ylsh. liq.; sol. in water, 5% caustic soda, 5% HCl, IPA, wh. spirit, trichlorethylene, xylene; dens. 1.07 g/ml; surf. tens. 39 dynes/cm; pH 2 (0.1% aq.); 100% act.

Lutensit® A-ES. [BASF AG] Alkyl phenol ether sulfate, sodium salt; anionic; wetting and dispersing agents for engineering, industrial, and household cleaners; biodeg.; ylsh. liq.; sol. in water, 5% caustic soda, 5% HCl, IPA; dens. 1.09 g/ml; surf. tens. 31 dynes/cm (1 g/l water); pH 7–9 (0.1% aq.); 40% act.

Lutensit® A-FK. [BASF AG] Fatty acid condensation prod., sodium salt; anionic; wetting and dispersing agents for engineering, industrial, and household cleaners; biodeg.; ylsh. liq.; sol. in water, trichlorethylene, xylene; dens. 1.00 g/ml; pour pt. 5 C; surf. tens. 36 dynes/cm (1 g/l water); pH 8–9.5 (0.1% water); 55% act.

Lutensit® A-LBA. [BASF AG] Alkylbenzene sulfonate, alkanolamine salt; anionic; surfactant for chemical industry, prod. of detergents; biodeg.; ylsh. liq.; sol. in water, 5% caustic soda, 5% HCl, IPA, trichlorethylene, xylene; dens. 1.08 g/ml; pour pt. –5 C; surf. tens. 33 dynes/cm (1 g/l water); pH 7–9 (0.1% aq.); 55% act.

Lutensit® A-PS. [BASF AG] Sodium alkylsulfonate; anionic; wetting, dispersing and cleansing agents for industrial and household uses; biodeg.; yel. liq.; sol. in water, 5% caustic soda, 5% HCl; dens. 1.2 g/ml; surf. tens. 34 dynes/cm (1 g/l water); pH 7–8 (0.1% aq.); 65% act.

Lutensit® AN 10. [BASF AG] Mixts. of various nonionic alkoxylates and an anionic, acidic, readily neutralizable surfactant; anionic/nonionic; degreasing agent, cold cleaners; ylsh. liq.; dens. 0.96–0.97 g/cm³; flash pt. 120 C; pour pt. < −15 C; surf. tens. 31 mN/m (1 g/l sodium salt); pH 4–5 (0.1% aq.); 100% act.

Lutensit® AN 30. [BASF AG] Surfactant; anionic/nonionic; for solv.-based cleaners; liq.; 100% conc.

Lutensit® AN 40. [BASF AG] Nonionic surfactant/alkylcarboxylic acid blend; anionic; for alkaline cleaners; liq.; 80% conc.

Lutensit® AS 2230, 2270. [BASF AG] Sulfated natural alcohol polyglycol ether, sodium salt; anionic; foaming surfactant for chemical and cosmetic industry; liq., paste resp.; 28 and 70% conc. resp.

Lutensit® AS 3330. [BASF AG] Sodium salt of sulfated oxoalcohol polyglycol ether; anionic; foaming agent, surfactant for the lt. chemical and cosmetics industry; liq.; 28% act.

Lutensit® AS 3334. [BASF AG] Sodium salt of sulfated fatty alcohol polyglycol ether; anionic; foaming agent, surfactant for the lt. chemical and cosmetics industry; liq.; 28% act.

Lutensit® K-HP. [BASF AG] Fatty acid condensation prod.; cationic; detergent, hydrophobic rinse aid for car washing; liq.; 100% act.

Lutensit® K-LC, K-LC 80. [BASF AG] Benzalkonium chloride; cationic; biocidal, wetting and dispersing agents for prod. of disinfectant cleaners for beverages and foodstuffs, trading concerns and household; clear to ylsh. liq.; sol. in water, 5% sodium hydroxide, 5% HCl, 5% sodium chloride, alcohol, chlorinated hydrocarbons, oil, wh. spirit; dens. 0.98 g/cm³; visc. 80 and 500 mPa•s resp.; cloud pt. −4 and −10 C; surf. tens. 37 mN/m (1 g/l); pH 6 (1% aq.); 80% biodeg.; 50 and 80% act.

Lutensit® K-OC. [BASF AG] Benzalkonium chloride; cationic; biocidal, wetting and dispersing agents for prod. of disinfectant cleaners for beverages and foodstuffs, trading concerns and household; ylsh. liq.; sol. in water, 5% sodium hydroxide, alcohol; dens. 0.98 g/cm³; visc. 100 mPa•s; cloud pt. −3 C; surf. tens. 31 mN/m; pH 6 (1% aq.); 80% biodeg.; 50% act.

Lutensit® K-TI. [BASF AG] Iodine surfactant complex; disinfectant, biocidal for cleaners; liq.

Lutensol® A 7. [BASF AG] PEG-7 C16C18 fatty alcohol; water-sol. detergent for prod. of detergents and in chemical processing industry; liq.; 100% act.

Lutensol® A 8. [BASF AG] PEG-8 C12-14 fatty alcohol; nonionic; detergent, wetting, dispersing, and emulsifying agent; for chemical industry, detergents; clear liq.; sol. in water, 5% caustic soda, 5% HCl, min. oil, gasoline, chlorinated hydrocarbon; cloud pt. 55 C (0.1% aq.); pH 6.0–7.5 (1% aq.); 90% act.

Lutensol® A 80. [BASF AG] Tallow alcohol ethoxylate; nonionic; detergent, wetting, dispersing, and emulsifying agent; wh. or lt. ylsh., waxy powd.; sol. in water, 5% caustic soda, 5% HCl, alcohol, ketones, benzene, xylene, chlorinated hydrocarbon; cloud pt. > 100 C (1% aq.); pH 6.0–7.5 (1% aq.); 100% act.

Lutensol® AO 3. [BASF AG] PEG-3 straight chain syn. C13-15 fatty alcohol; nonionic; detergent, emulsifying and dispersing agent used in household and industrial detergents, chemical industry; > 80% biodeg.; clear, dull liq.; sol. in petrol. fractions, alcohols, aromatic hydrocarbons; slightly sol. water; dens. 0.93 g/cm³; visc. 40 mPa•s; flash pt. 130 C; pH 7 (1% aq.); 100% act.

Lutensol® AO 4. [BASF AG] PEG-4 straight chain C13-15 fatty alcohol; nonionic; surfactant for chemical industry, prod. of detergents; liq.; 100% conc.

Lutensol® AO 5. [BASF AG] PEG-5 straight chain syn. C13-15 fatty alcohol; nonionic; detergent, emulsifying and dispersing agent used in household and industrial detergents, chemical industry; > 80% biodeg.; clear, dull liq.; sol. in petrol. fractions, alcohols, aromatic hydrocarbons; slightly sol. water; dens. 0.96 g/cm³; visc. 40 mPa•s; flash pt. 150 C; pH 7 (1% aq.); 100% act.

Lutensol® AO 7. [BASF AG] PEG-7 straight chain syn. C13-15 fatty alcohol; nonionic; detergent, emulsifying and dispersing agent used in household and industrial detergents, chemical industry; > 80% biodeg.; clear, dull liq.; sol. water, 5% HCl, petrol. fractions, alcohols, aromatic and chlorinated hydrocarbons; dens. 0.98 g/cm³; visc. 120 mPa•s (60 C); cloud pt. 43 C (1%); flash pt. 190 C; pH 7 (1% aq.); 100% act.

Lutensol® AO 8. [BASF AG] PEG-8 straight chain syn. C13-15 fatty alcohol; nonionic; detergent, emulsifying and dispersing agent used in household and industrial detergents, chemical industry; > 80% biodeg.; clear, dull liq.; sol. water, 5% HCl, petrol. fractions, alcohols, aromatic and chlorinated hydrocarbons; dens. 0.96 g/cm³; visc. 30 mPa•s (60 C); cloud pt. 52 C (1%); flash pt. 200 C; pH 7 (1% aq.); 100% act.

Lutensol® AO 10. [BASF AG] PEG-10 straight chain syn. C13-15 fatty alcohol; nonionic; detergent, emulsifying and dispersing agent used in household and industrial detergents, chemical industry; > 80% biodeg.; wh. soft paste; sol. water, 5% HCl, alcohols, aromatic and chlorinated hydrocarbons; dens. 0.98 g/cm³ (60 C); visc. 40 cps (60 C); cloud pt. 80 C (1%); flash pt. 200 C; pH 7 (1% aq.); 100% act.

Lutensol® AO 11. [BASF AG] PEG-11 straight chain syn. C13-15 fatty alcohol; nonionic; detergent, emulsifying and dispersing agent used in household and industrial detergents, chemical industry; > 80% biodeg.; wh. soft paste; sol. water, 5% HCl, alcohols, aromatic and chlorinated hydrocarbons; dens. 0.99 g/cm³ (60 C); visc. 40 cps (60 C); cloud pt. 86 C (1%); flash pt. 200 C; pH 7 (1% aq.); 100% act.

Lutensol® AO 12. [BASF AG] PEG-12 straight chain syn. C13-15 fatty alcohol; nonionic; detergent, emulsifying and dispersing agent used in household and industrial detergents, chemical industry; > 80% biodeg.; wh. solid paste; sol. water, 5% HCl, alcohols, aromatic and chlorinated hydrocarbons; dens. 1.00 g/cm³ (60 C); visc. 40 cps (60 C); cloud pt. 91 C (1%); flash pt. 200 C; pH 7 (1% aq.); 100% act.

Lutensol® AO 30. [BASF AG] PEG-30 straight chain syn. C13-15 fatty alcohol; nonionic; detergent, emulsifying and dispersing agent used in household and industrial detergents, chemical industry; > 80% biodeg.; wax; 100% conc.

Lutensol® AO 109. [BASF AG] PEG-10 straight chain syn. C13-15 fatty alcohol; nonionic; detergent, emulsifying and dispersing agent used in household and industrial detergents, chemical industry; > 80% biodeg.; clear, dull liq.; sol. water, 5% HCl, petrol. fractions, alcohols, aromatic and chlorinated hydrocarbons; dens. 1.03 g/cm³; visc. 200 cps; cloud pt. 80 C (1%); flash pt. 190 C; pH 7 (1% aq.); 90% act.

Lutensol® AO 3109. [BASF AG] Fatty alcohol ethoxylate; nonionic; low foaming detergent for household laundry prods.; liq.; 90% conc.

Lutensol® AP 6. [BASF AG] PEG-6 alkylphenol; nonionic; detergent, wetting, emulsifying and dispersing agent, used in cleaners, detergents, leather, fur, paper, paint and dye industries; ylsh. liq.; sol. in min. oil, alcohols, ketones, aromatic and chlorinated hydrocarbons; sp.gr. 1.05; visc. 400 cps; surf. tens. 31 dynes/cm; 100% act.

Lutensol® AP 7. [BASF AG] PEG-7 alkylphenol; nonionic; detergent, wetting, emulsifying and dispersing agent, used in cleaners, detergents, leather, fur, paper, paint and dye industries; clear liq.; sol. in min. oil, alcohols, ketones, aromatic and chlorinated hydrocarbons; sp.gr. 1.04; visc. 350 cps; surf. tens. 31 dynes/cm; 100% act.

Lutensol® AP 8. [BASF AG] PEG-8 alkylphenol; nonionic; detergent, wetting, emulsifying and dispersing agent, used in cleaners, detergents, leather, fur, paper, paint and dye industries; clear liq.; sol. in water, 5% HCl, alcohols, ketones, aromatic and chlorinated hydrocarbons; sp.gr. 1.05; visc. 350 cps; surf. tens. 31 dynes/cm; 100% act.

Lutensol® AP 9. [BASF AG] PEG-9 alkylphenol; nonionic; detergent, wetting, emulsifying and dispersing agent, used in cleaners, detergents, leather, fur, paper, paint and dye industries; clear liq.; sol. in water, 5% HCl, alcohols, ketones, aromatic and chlorinated hydrocarbons; sp.gr. 1.05; visc. 350 cps; surf. tens. 31 dynes/cm; 100% act.

Lutensol® AP 10. [BASF AG] Nonoxynol-10; CAS 9016-45-9; nonionic; detergent, wetting, emulsifying and dispersing agent, used in cleaners, detergents, leather, fur, paper, paint and dye industries; clear liq.; sol. in water, 5% HCl, alcohols, ketones, aromatic and chlorinated hydrocarbons; sp.gr. 1.06; visc. 350 cps; surf. tens. 32 dynes/cm; 100% act.

Lutensol® AP 14. [BASF AG] PEG-14 alkylphenol; nonionic; detergent, wetting, emulsifying and dispersing agent, used in cleaners, detergents, leather, fur, paper, paint and dye industries; clear liq.; sol. in water, 5% HCl, alcohols, ketones, aromatic and chlorinated hydrocarbons; sp.gr. 1.05 (50 C); visc. 250 cps (30 C); surf. tens. 37 dynes/cm; 100% act.

Lutensol® AP 20. [BASF AG] Nonoxynol-20; CAS 9016-45-9; nonionic; detergent, wetting, emulsifying and dispersing agent, used in cleaners, detergents, leather, fur, paper, paint and dye industries; wh. soft wax; sol. in water, 5% HCl, alcohols, ketones, aromatic and chlorinated hydrocarbons; sp.gr. 1.06 (50 C); surf. tens. 42 dynes/cm; 100% act.

Lutensol® AP 30. [BASF AG] PEG-30 alkylphenol; nonionic; detergent, wetting, emulsifying and dispersing agent, used in cleaners, detergents, leather, fur, paper, paint and dye industries; wax; 100% act.

Lutensol® AT 11. [BASF AG] PEG-11 sat. C16-C18 alcohol; nonionic; detergent, wetting, dispersing and emulsifying agent used in chemical industry, for household and industrial detergents; wh. to pale ylsh. paste; sol. in dist. water; visc. 30 cps (60 C); cloud pt. 86 C (0.5% aq.); pH practically neutral (1%); 100% act.

Lutensol® AT 18. [BASF AG] PEG-18 sat. C16-C18 alcohol; nonionic; detergent, wetting, dispersing and emulsifying agent used in chemical industry, for household and industrial detergents; paste; 100% act.

Lutensol® AT 25. [BASF AG] PEG-25 sat. C16-C18 alcohol; nonionic; detergent, wetting, dispersing and emulsifying agent used in chemical industry, for household and industrial detergents; wh. powd.; sol. in dist. water, benzene, chlorinated hydrocarbons, ethanol, IPA; visc. 70 cps (60 C); cloud pt. > 100 C (0.5% aq.); pH practically neutral (1%); 100% act.

Lutensol® AT 50. [BASF AG] PEG-50 sat. C16-C18 alcohol; nonionic; detergent, wetting, dispersing and emulsifying agent used in chemical industry, for household and industrial detergents; wh. powd.; sol. in dist. water, benzene, chlorinated hydrocarbons, ethanol; visc. 125 cps (60 C); cloud pt. > 100 C (0.5% aq.); pH practically neutral (1%); 100% act.

Lutensol® AT 80. [BASF AG] PEG-80 sat. C16-C18 alcohol; nonionic; wetting and dispersing agents for engineering, industrial, and household cleaners; wax; 100% conc.

Lutensol® ED 140. [BASF AG] EO-PO ethylene diamine compd.; nonionic; antistat, detergent, dispersant, defoamer, wetting agent, gelling agent, solubilizer, emulsifier, demulsifier, lubricant, foam suppressor for household and industrial uses; clear ylsh. liq.; sol. @ 10% in water, 5% HCl, benzene, ethanol, IPA, trichlorethylene; dens. 1.05 g/cm^3; visc. 750 cps; cloud pt. 80 C (1% aq.); pour pt. –20 C; surf. tens. 44 mN/m (1 g/l); pH 9 (5% aq.); 100% act.

Lutensol® ED 310. [BASF AG] EO-PO ethylene diamine compd.; CAS 52503-47-6; nonionic; antistat, detergent, dispersant, defoamer, wetting agent, gelling agent, solubilizer, emulsifier, demulsifier, lubricant, foam suppressor for household and industrial uses; clear to dull ylsh. liq.; sol. @ 10% in benzene, ethanol, IPA; dens. 1.00 g/cm^3; visc. 800 cps; cloud pt. 20 C (1% aq.); pour pt. –25 C; surf. tens. 33 mN/m (1 g/l); pH 9 (5% aq.); 100% act.

Lutensol® ED 370. [BASF AG] EO-PO ethylene diamine compd.; CAS 52503-47-6; nonionic; antistat, detergent, dispersant, defoamer, wetting agent, gelling agent, solubilizer, emulsifier, demulsifier, lubricant, foam suppressor for household and industrial uses; wh. to ylsh. fine powd.; sol. @ 10% in water, 5% HCl, ethanol, IPA, trichlorethylene; dens. 1.05 g/cm^3; visc. 600 cps; m.p. 50 C; cloud pt. > 100 C (1% aq.); surf. tens. 44 mN/m (1 g/l); pH 8 (5% aq.); 100% act.

Lutensol® ED 610. [BASF AG] EO-PO ethylene diamine compd.; nonionic; antistat, detergent, dispersant, defoamer, wetting agent, gelling agent, solubilizer, emulsifier, demulsifier, lubricant, foam suppressor for household and industrial uses; colorless to ylsh. dull liq.; sol. @ 10% in benzene, ethanol, IPA; dens. 1.02 g/cm^3; visc. 1300 cps; cloud pt. 15 C (1% aq.); pour pt. –25 C; surf. tens. 33 mN/m (1 g/l); pH 8 (5% aq.); 100% act.

Lutensol® FA 12. [BASF AG] Oleyl amine ethoxylate; nonionic; emulsifier, dispersant, wetting and degreasing agent used in heavy-duty and other detergents, shampoos; > 80% biodeg.; ylsh. liq.; sol. @ 10% in dist. water, 5% caustic soda, 5% HCl, alcohol; dens. 1.02 g/cm^3 (20 C); visc. 200 cps; surf. tens. 35 dynes/cm (1 g/l aq.); pH 8 (1% aq.); 100% act.

Lutensol® FSA 10. [BASF AG] Oleic acid amide ethoxylate; CAS 31799-71-0; nonionic; emulsifier, dispersant, detergent, wetting agent, used in fine detergents, hand cleaners, fur dressing; > 80% biodeg.; rdsh. liq. with some sediment; sol. @ 10% in dist. water, 5% caustic soda, 5% HCl, alcohol; dens. 1.01 (60 C); visc. 70 cps (60 C); cloud pt. 80 C (1% aq.); surf. tens. 35 dynes/cm (1 g/l aq.); pH 8 (1% aq.); 100% act.

Lutensol® GD 50, GD 70. [BASF AG] Alkyl glucoside; nonionic; foaming surfactant for formulation of cleaners and degreasing agents; liq.; 50 and 70% conc. resp.

Lutensol® LF 220, 221, 223, 224. [BASF AG] Alkoxylated straight chain alcohol; nonionic; surfactant for dishwashing powds. and rinse aids; liq.; 95, 95, 98, and 100% conc. resp.

Lutensol® LF 400. [BASF AG] Alkoxylated straight chain alcohol; nonionic; low-foaming detergent, wetting and dispersing agent, acidic rinse aid, dishwashing, glass cleaners, metal cleaning; colorless clear liq.; sol. @ 10% in water, 5% HCl, wh. spirit, benzene, ethanol, IPA, trichlorethylene; dens. 0.97 g/cm³; visc. 60 cps; cloud pt. 31 C (1% aq.); surf. tens. 30 dynes/cm (0.1% aq.); pH 7 (5% aq.); 100% act.

Lutensol® LF 401. [BASF AG] Alkoxylated straight chain alcohol; nonionic; low-foaming detergent, wetting and dispersing agent for heavy-duty detergents, acid pickling, metal cleaning; colorless clear liq.; sol. @ 10% in water, 5% HCl, wh. spirit, benzene, ethanol, IPA, trichlorethylene; dens. 1.05 g/cm³; visc. 330 cps; cloud pt. 77 C (1% aq.); surf. tens. 40 dynes/cm (0.1% aq.); pH 7 (5% aq.); 100% act.

Lutensol® LF 403. [BASF AG] Alkoxylated straight chain alcohol; nonionic; surfactant for dishwashing powds.; liq.; 100% conc.

Lutensol® LF 404, 405. [BASF AG] Alkoxylated straight chain alcohol; nonionic; surfactant for dishwashing powds. and rinse aids; liq.; 100 and 95% conc.

Lutensol® LF 431. [BASF AG] Alkoxylated straight chain alcohol; nonionic; surfactant for dishwashing powds, rinse aids and bottlewashing; liq.; 100% conc.

Lutensol® LF 600. [BASF AG] Alkoxylated straight chain alcohol; nonionic; low-foaming detergent, wetting and dispersing agent for car washing, hard surface cleaners; colorless clear liq.; sol. @ 10% in water, 5% HCl, wh. spirit, benzene, ethanol, IPA, trichlorethylene; dens. 0.97 g/cm³; visc. 110 cps; cloud pt. 52 C (1% aq.); surf. tens. 40 dynes/cm (0.1% aq.); pH 7 (5% aq.); 100% act.

Lutensol® LF 700. [BASF AG] Alkoxylated straight chain alcohol; nonionic; low-foaming detergent, wetting and dispersing agent for dishwashing, rinse aids, dairy cleaners, household and industrial cleaners; slightly ylsh. clear liq.; sol. @ 10% in 5% HCl, wh. spirit, benzene, ethanol, IPA, trichlorethylene; dens. 0.96 g/cm³; visc. 110 cps; cloud pt. 77 C (1% aq.); surf. tens. 34 dynes/cm (0.1% aq.); pH 7 (5% aq.); 100% act.

Lutensol® LF 711. [BASF AG] Alkoxylated straight chain alcohol; nonionic; low-foaming detergent, wetting and dispersing agent for dishwashing machines, rinse aids, metal cleaning; colorless clear liq.; sol. @ 10% in water, 5% HCl, wh. spirit, benzene, ethanol, IPA, trichlorethylene; dens. 1.0 g/cm³; visc. 110 cps; cloud pt. 36 C (1% aq.); surf. tens. 28 dynes/cm (0.1% aq.); pH 7 (5% aq.); 100% act.

Lutensol® LF 1300. [BASF AG] Alkoxylated straight chain alcohol; nonionic; low-foaming detergent, wetting and dispersing agent for dishwashing, rinse aids, dairy cleaners, household and industrial cleaners; slightly ylsh. clear liq.; sol. @ 10% in wh. spirit, benzene, ethanol, IPA, trichlorethylene; dens. 0.97 g/cm³; visc. 140 cps; surf. tens. 35 dynes/cm (0.1% aq.); pH 7 (5% aq.); 100% act.

Lutensol® LSV. [BASF AG] Alkoxylated straight chain alcohol; nonionic; low-foaming detergent, wetting and dispersing agent for bottle washing, lime and scale dispersant; slightly ylsh. clear liq.; sol. @ 10% in benzene, 5% HCl, ethanol, IPA, trichlorethylene; sol. opaque in water; dens. 1.04 g/cm³; visc. 500 cps; cloud pt. 30 C (1% aq.); surf. tens. 42 dynes/cm (0.1% aq.); pH 8–9 (5% aq.); 100% act.

Lutensol® LT 30. [BASF AG] Specially alkoxylated fatty alcohol; nonionic; surfactant for acid and alkaline low-foaming cleaners; liq.; 100% act.

Lutensol® ON 30. [BASF AG] PEG-3 C10 oxo-alcohol; nonionic; surfactant for chemical industry and prod. of detergents; liq.; 100% conc.

Lutensol® ON 50. [BASF AG] PEG-5 C10 oxo-alcohol; nonionic; surfactant for chemical industry and prod. of detergents; liq.; 100% conc.

Lutensol® ON 60. [BASF AG] PEG-6 C10 oxo-alcohol; nonionic; surfactant for chemical industry and prod. of detergents; liq.; 100% conc.

Lutensol® ON 70. [BASF AG] PEG-7 C10 oxo-alcohol; nonionic; surfactant for chemical industry and prod. of detergents; liq.; 100% conc.

Lutensol® ON 80. [BASF AG] PEG-8 C10 oxo-alcohol; nonionic; surfactant for chemical industry and prod. of detergents; liq.; 100% conc.

Lutensol® ON 110. [BASF AG] PEG-11 C10 oxo-alcohol; nonionic; surfactant for chemical industry and prod. of detergents; liq.; 100% conc.

Lutensol® TO 3. [BASF AG] PEG-3 C13 oxo-alcohol; nonionic; surfactant for chemical industry and prod. of detergents; liq.; 100% conc.

Lutensol® TO 5. [BASF AG] PEG-5 C13 oxo-alcohol; nonionic; surfactant for chemical industry and prod. of detergents; liq.; 100% conc.

Lutensol® TO 7. [BASF AG] PEG-7 C13 oxo-alcohol; nonionic; surfactant for chemical industry and prod. of detergents; liq.; 100% conc.

Lutensol® TO 8. [BASF AG] PEG-8 C13 oxo-alcohol; nonionic; surfactant for chemical industry and prod. of detergents; liq.; 100% conc.

Lutensol® TO 10. [BASF AG] PEG-10 C13 oxo-alcohol; nonionic; surfactant for chemical industry and prod. of detergents; paste; 100% conc.

Lutensol® TO 12. [BASF AG] PEG-12 C13 oxo-alcohol; nonionic; surfactant for chemical industry and prod. of detergents; paste; 100% conc.

Lutensol® TO 15. [BASF AG] PEG-15 C13 oxo-alcohol; nonionic; surfactant for chemical industry and prod. of detergents; paste; 100% conc.

Lutensol® TO 20. [BASF AG] PEG-20 C13 oxo-alcohol; nonionic; surfactant for chemical industry and prod. of detergents; paste; 100% conc.

Lutensol® TO 89. [BASF AG] PEG-8 C13 oxo-alcohol; nonionic; surfactant for chemical industry and prod. of detergents; liq.; 90% conc.

Lutensol® TO 109. [BASF AG] PEG-10 C13 oxo-alcohol; nonionic; surfactant for chemical industry and prod. of detergents; liq.; 90% conc.

Lutensol® TO 129. [BASF AG] PEG-12 C13 oxo-alcohol; nonionic; surfactant for chemical industry and prod. of detergents; liq.; 88% conc.

Lutensol® TO 389. [BASF AG] Fatty alcohol ethoxylate; nonionic; surfactant for chemical industry and prod. of detergents; liq.; 90% conc.

Lutopon SN. [Henkel-Nopco] Linear alkylaryl sulfonate and solv.; anionic; detergent used in textile processing; degreaser; liq.; 100% conc.

Lutostat 171. [Henkel-Nopco] Betaine; amphoteric; textile antistat; liq.; 30% conc.

Lutrol® E 300. [BASF AG] PEG-6; emulsifier, emollient, lubricant, and solv. for liq. preps.; liq.;

100% conc.

Lutrol® E 400. [BASF AG] PEG-8; CAS 25322-68-3; emulsifier, emollient, lubricant, and solv. for liq. preps.; liq.; 100% conc.

Lutrol® E 1500. [BASF AG] PEG-32; nonionic; emulsifier, binder, solubilizer, resorption promoter for substances insol. or sparingly sol. in water; microbeads; 100% conc.

Lutrol® E 4000. [BASF AG] PEG-75; CAS 25322-68-3; nonionic; emulsifier, binder, solubilizer, resorption promoter for substances insol. or sparingly sol. in water; microbeads; 100% conc.

Lutrol® E 6000. [BASF AG] PEG-150; nonionic; emulsifier, binder, solubilizer, resorption promoter for substances insol. or sparingly sol. in water; microbeads; 100% conc.

Lutrol® E 8000. [BASF AG] PEG-150.

Lutrol® F 127. [BASF AG] EO/PO block copolymer; CAS 106392-12-5; nonionic; thickening and gelling agent; microbeads; 100% conc.

Lutrol® OP-2000. [BASF] PPG-26 oleate.

Lutrol® W-3520. [BASF] PPG-28-buteth-35.

Lyogen AFS Liq. [Sandoz] Detergent used for scouring, leveling and finishing of nylon hosiery; brn. liq.; turbid sol. in water; sp.gr. 1.01; 85–86% act.

Lyogen BE Liq. [Sandoz] Dyeing assistant used on nylon; yel. liq.; 55% sol. in water @ 70 F; sp.gr. 1.16; 45% act.

Lyogen DFT Liq. [Sandoz] Detergent, wetting agent, emulsifier, dispersant, dyeing assistant for polyester; yel. liq.; misc. with water; sp.gr. 1.05; 30% act.

Lyogen F Liq. [Sandoz] Leveling agent, dyeing assistant used on wool; brn. liq.; misc. with water; sp.gr. 1.06; 85% act.

Lyogen KF Liq. [Sandoz] Detergent, wetting agent, emulsifier, dispersant, used with vat dyes; yel. liq.; sp.gr. 1.05; 17% act.

Lyogen MS Liq. [Sandoz] Dyeing and leveling agent for wool and nylon; yel. liq.; misc. with water; sp.gr. 1.08; 33% act.

Lyogen NL Liq., P Liq. [Sandoz] Dyeing assistant used on nylon with acid dyes; amber and yel. resp. liq.; misc. with water (NL Liq.); sp.gr. 1.05 and 1.09 resp.; 34 and 35% act. resp.

Lyogen PAA Liq. [Sandoz] Antiprecipitant used in dyeing acrylic fibers and blends with wool and nylon where cationic and anionic dyes are used simultaneously; amber liq.; misc. with water; 43% act.

Lyogen SF Liq. [Sandoz] Dyeing assistant used in wool, silk, nylon; yel. liq.; sp.gr. 1.03; 25% act.

Lyogen SMK-40 Liq., SMK Paste. [Sandoz] Dye leveling agent, antiprecipitating agent for dyes; amber liq., brn. gel, paste; misc. with water; sp.gr. 1.03; 20–49.5% act.

Lyogen V (U) Liq. [Sandoz] Detergent, wetting agent, surfactant and leveling agent for scouring and printing of polyester and nylon; amber liq.; water disp.; sp.gr. 1.04.

Lyogen WD Liq. [Sandoz] Dyeing assistant, leveling agent; amber liq.; misc. with water; sp.gr. 1.05; 40% act.

LZ 5362 A. [Lubrizol] Proprietary; anionic; w/o emulsifier for industrial applics.; liq.; sol. in naphthenic base oils; 100% conc.

LZ 5364. [Lubrizol] Proprietary; anionic; w/o emulsifier for industrial applics.; liq.; sol. in naphthenic and paraffinic oils; 100% conc.

M

Mackadet 40K. [McIntyre] Potassium coconate; anionic; liq. hand soap; liq.; pH 9.0; 38% conc.
Mackadet RS. [McIntyre] Blend; rug shampoo conc. that leaves dry residue; liq.; pH 7.0; 35% conc.
Mackadet WHC. [McIntyre] Blend; waterless hand cleaner conc.; liq.; pH 8.0; 100% conc.
Mackam 1C. [McIntyre] Sodium cocoamphoacetate; amphoteric; surfactant for high alkaline cleansers, shampoos, baby shampoos; liq.; pH 11; 45% conc.
Mackam 1C-SF. [McIntyre] Coco imidazoline carboxylate; amphoteric; salt-free surfactant for shampoos, cleaner, dishwash formulation; liq.; 38% conc.
Mackam 1L. [McIntyre] Sodium lauramphoacetate; amphoteric; surfactant for nonirritant shampoos, cleaners; liq.; pH 10.0; 38% conc.
Mackam 1L-30. [McIntyre] Sodium lauroamphoacetate; amphoteric; surfactant for high alkaline cleansers, shampoos, baby shampoos; liq.; 36% conc.
Mackam 2C. [McIntyre] Disodium cocoamphodiacetate; amphoteric; surfactant for baby shampoo, high alkaline cleaners; liq.; pH 8.5; 50% conc.
Mackam 2C-75. [McIntyre] Disodium cocoamphodiacetate; amphoteric; surfactant for baby shampoo, high alkaline cleansers; liq.; pH 8.0; 37% conc.
Mackam 2C-SF. [McIntyre] Disodium cocoamphodipropionate; amphoteric; salt-free surfactant for shampoos, cleaners, dishwash formulations; emulsifier for aerosol and emulsion systems; liq.; pH 10.0; 39% conc.
Mackam 2CT. [McIntyre] Disodium caproamphodiacetate, sodium trideceth sulfate, and hexylene glycol; amphoteric; nonirritating shampoo; liq.; 50% conc.
Mackam 2CY. [McIntyre] Disodium capryloamphodiacetate; amphoteric; low-foaming alkaline cleaner; liq.; pH 11; 50% conc.
Mackam 2CYSF. [McIntyre] Disodium capryloamphodipropionate; amphoteric; surfactant for heavy duty, all-purpose and metal cleaners; emulsifier for aerosol and emulsion systems; liq.; pH 10.0; 50% conc.
Mackam 2L. [McIntyre] Disodium lauroamphodiacetate; amphoteric; surfactant for nonirritating shampoos, cleaners; liq.; pH 9.0; 38% conc.
Mackam 2LSF. [McIntyre] Disodium lauroamphodipropionate; amphoteric; surfactant for heavy duty cleaners, metal and all-purpose cleaners, nonirritating shampoos; emulsifier for aerosol and emulsion systems; liq.; pH 10.0; 39% conc.
Mackam 2W. [McIntyre] Disodium wheat germamphodiacetate; amphoteric; mild surfactant for baby shampoos; wetting agent in caustic cleaners; liq.; pH 9.5; 35% conc.
Mackam 35. [McIntyre] Cocoamidopropyl betaine; amphoteric; foamer for shampoos, bubble baths, dishwash; liq.; pH 6.0; 35% conc.
Mackam 35 HP. [McIntyre] Cocamidopropyl betaine; amphoteric; foamer, visc. builder for shampoos, bubble baths, dishwash; liq.; pH 6.0; 35% conc.
Mackam 160C. [McIntyre] Sodium lauriminodipropionate; amphoteric; surfactants for personal care and industrial applics.; liq.; pH 7.0; 38% conc.
Mackam CAP. [McIntyre] Cocamidopropyl dimethylamine propionate; anionic; surfactant for industrial cleaners and personal care prods.; liq.; 30% conc.
Mackam CB-35. [McIntyre] Coco-betaine; amphoteric; surfactant, conditioner, visc. builder, foam booster for shampoos, skin cleansers, bath prods., heavy duty cleaners; liq.; pH 8.0; 35% conc.
Mackam CB-LS. [McIntyre] Coco betaine, low salt; amphoteric; surfactant, conditioner, visc. builder, foam booster for shampoos, skin cleansers, bath prods., heavy duty cleaners; liq.; pH 7.5; 33% conc.
Mackam CET. [McIntyre] Cetyl betaine; surfactant, conditioner, visc. builder, foam booster for shampoos, skin cleansers, bath prods., heavy duty cleaners; gel; pH 7.0; 33% conc.
Mackam CSF. [McIntyre] Sodium cocoamphopropionate; amphoteric; surfactant for heavy duty cleanser, metal and all-purpose cleaner, nonirritating shampoos; liq.; pH 10.0; 39% conc.
Mackam HV. [McIntyre] Oleamidopropyl betaine; amphoteric; hair conditioner, foamer, emollient, visc. builder; liq.; pH 6.5; 35% conc.
Mackam ISA. [McIntyre] Isostearamidopropyl betaine; amphoteric; surfactant, conditioner, visc. builder, foam booster for shampoos, skin cleansers, bath prods., heavy duty cleaners; liq.; water-sol.; pH 7.5; 33% conc.
Mackam ISP. [McIntyre] Isostearamido dimethylaminopropionate; anionic; salt-free betaine; liq.; 30% conc.
Mackam J. [McIntyre] Cocamidopropyl betaine; surfactant, conditioner, visc. builder, foam booster for shampoos, skin cleansers, bath prods., heavy duty cleaners; liq.; pH 6.0; 35% conc.
Mackam LAP. [McIntyre] Lauramidopropyl dimethylamine propionate; anionic; surfactant for industrial cleaners and personal care prods.; liq.; 30% conc.
Mackam LMB. [McIntyre] Lauramidopropyl betaine; amphoteric; surfactant for shampoos, bubble baths, dishwash formulations; liq.; pH 6.0; 35% conc.
Mackam LT. [McIntyre] Disodium lauroamphodiacetate and sodium trideceth sulfonate; amphoteric;

shampoos, cleaners, detergents; liq.; 38 and 35% conc. resp.

Mackam MEJ. [McIntyre] Mixed alkylamphocarboxylate; amphoteric; surfactants for personal care and industrial applics.; liq.; pH 10.0; 34% conc.

Mackam MLT. [McIntyre] Sodium lauroamphoacetate and sodium trideceth sulfate; amphoteric; surfactant for shampoos, cleaners, detergents; liq.; pH 10.0; 35% conc.

Mackam OB-30. [McIntyre] Oleyl betaine; amphoteric; visc. builder for alkaline cleanser; amber liq.; pH 7.0; 30% conc.

Mackam RA. [McIntyre] Ricinoleamidopropyl betaine; surfactant, conditioner, visc. builder, foam booster for shampoos, skin cleansers, bath prods., heavy duty cleaners; amber liq.; pH 6.5; 35% conc.

Mackam TM. [McIntyre] Dihydroxyethyl tallow glycinate; amphoteric; surfactant, conditioner for shampoos; thickener for alkaline oven cleaners, acid bowl cleaners; liq.; water-sol.; pH 5.0; 40% conc.

Mackam WGB. [McIntyre] Wheat germamidopropyl betaine; surfactant, conditioner, visc. builder, foam booster for shampoos, skin cleansers, bath prods., heavy duty cleaners; liq.; pH 6.5; 34% conc.

Mackamide 100-A. [McIntyre] Cocamide DEA; nonionic; foam stabilizer and thickener; liq.; 100% conc.

Mackamide AME-75, AME-100. [McIntyre] Acetamide MEA; nonionic; humectant; surfactant, thickener, foam booster/stabilizer for personal care and industrial applics.; liq.; pH 7.0; 75 and 100% conc.

Mackamide AN55. [McIntyre] Alkylolamide DEA; nonionic; foam enhancer, visc. builder; liq.; 100% conc.

Mackamide C. [McIntyre] Cocamide DEA (1:1); nonionic; foam stabilizer and thickener for shampoos, industrial applics.; liq.; pH 10.0; 100% conc.

Mackamide CD. [McIntyre] Cocamide DEA (2:1); nonionic; surfactant, thickener, foam booster/stabilizer for personal care and industrial applics.; liq.; pH 10.0; 100% conc.

Mackamide CD-6. [McIntyre] Cocamide DEA, DEA caprate; nonionic; degreaser, rust inhibitor, visc. builder; liq.; pH 9.0; 100% conc.

Mackamide CD-8. [McIntyre] Cocamide DEA and mixed soaps; nonionic; degreaser, rust inhibitor, visc. builder; liq.; pH 9.0; 100% conc.

Mackamide CD-10. [McIntyre] Capramide DEA; nonionic; surfactant, thickener, foam booster/stabilizer for personal care and industrial applics.; liq.; pH 10.0; 100% conc.

Mackamide CD-25. [McIntyre] Cocamide DEA, tall oil soap; nonionic; degreaser, rust inhibitor, visc. builder; liq.; pH 9.0; 100% conc.

Mackamide CDC. [McIntyre] Cocamide DEA, DEA coconate; nonionic; surfactant, thickener, foam booster/stabilizer for personal care and industrial applics.; liq.; pH 9.0; 100% conc.

Mackamide CDM. [McIntyre] Cocamide DEA, DEA oleate; nonionic; degreaser, rust inhibitor, visc. builder; liq.; pH 9.0; 100% conc.

Mackamide CDS-80. [McIntyre] Cocamide DEA, DEA dodecylbenzene sulfonate; nonionic; surfactant, thickener, foam booster/stabilizer for personal care and industrial applics.; liq.; pH 9.0; 80% conc.

Mackamide CDT. [McIntyre] Cocamide DEA, tall oil soap; nonionic; degreaser, rust inhibitor, visc. builder; liq.; pH 9.0; 100% conc.

Mackamide CDX. [McIntyre] Amide, modified; nonionic; degreaser, rust inhibitor, visc. builder; liq.; 100% conc.

Mackamide CMA. [McIntyre] Cocamide MEA; CAS 68140-00-1; nonionic; surfactant, thickener, foam booster/stabilizer for personal care and industrial applics.; flake; pH 10.0; 100% conc.

Mackamide CS. [McIntyre] Cocamide DEA (1:1); nonionic; surfactant, thickener, foam booster/stabilizer for personal care and industrial applics.; liq.; pH 10.0; 100% conc.

Mackamide CSA. [McIntyre] Cocamide, modified; nonionic; degreaser, rust inhibitor, visc. builder; liq.; 100% conc.

Mackamide EC. [McIntyre] Cocamide DEA; nonionic; surfactant, thickener, foam booster/stabilizer for personal care and industrial applics.; liq.; 100% conc.

Mackamide ISA. [McIntyre] Isotearamide DEA; nonionic; lubricant, surfactant, thickener, foam booster/stabilizer for personal care and industrial applics.; liq.; pH 10.0; 100% conc.

Mackamide L10. [McIntyre] Lauramide DEA; CAS 120-40-1; nonionic; surfactant, thickener, foam booster/stabilizer for personal care and industrial applics.; liq.; pH 10.0; 100% conc.

Mackamide L95. [McIntyre] Lauramide DEA (95% lauric); CAS 120-40-1; nonionic; surfactant, thickener, foam booster/stabilizer for personal care and industrial applics.; solid; pH 10.0; 100% conc.

Mackamide LLM. [McIntyre] Lauramide DEA; CAS 120-40-1; nonionic; surfactant, thickener, foam booster/stabilizer for personal care and industrial applics.; liq.; pH 10.0; 100% conc.

Mackamide LMD. [McIntyre] Lauramide DEA (70% lauric); CAS 120-40-1; nonionic; surfactant, thickener, foam booster/stabilizer for personal care and industrial applics.; solid; pH 10.0; 100% conc.

Mackamide LME. [McIntyre] Lactamide MEA; nonionic; conditioner for shampoos; surfactant, thickener, foam booster/stabilizer for personal care and industrial applics.; liq.; water-sol.; pH 5.0; 100% conc.

Mackamide LMM. [McIntyre] Lauramide MEA; CAS 142-78-9; nonionic; surfactant, thickener, foam booster/stabilizer for personal care and industrial applics.; flake; pH 10.0; 100% conc.

Mackamide LOL. [McIntyre] Linoleamide DEA; nonionic; surfactant, thickener, foam booster/stabilizer for personal care and industrial applics.; liq.; pH 10.0; 100% conc.

Mackamide MC. [McIntyre] Cocamide DEA (1:1); nonionic; surfactant, thickener, foam booster/stabilizer for personal care and industrial applics.; liq.; pH 10.0; 100% conc.

Mackamide MO. [McIntyre] Oleamide DEA (1:1); nonionic; surfactant, thickener, foam booster/stabilizer for personal care and industrial applics.; liq.; pH 10.0; 100% conc.

Mackamide NOA. [McIntyre] Oleamide DEA (1:1); nonionic; surfactant, thickener, foam booster/stabilizer for personal care and industrial applics.; liq.; pH 10.0; 100% conc.

Mackamide O. [McIntyre] Oleamide DEA (2:1); nonionic; surfactant, thickener, foam booster/stabilizer for personal care and industrial applics.; solv. degreaser; liq.; pH 10.0; 100% conc.

Mackamide ODM. [McIntyre] Oleamide DEA, DEA oleate; nonionic; surfactant, thickener, foam booster/stabilizer for personal care and industrial applics.; gel; pH 9.0; 100% conc.

Mackamide OP. [McIntyre] Oleamide MIPA; nonionic; surfactant, thickener, foam booster/stabilizer for personal care and industrial applics.; paste; pH 10.0; 100% conc.

Mackamide PK. [McIntyre] Palm kernelamide DEA; nonionic; surfactant, thickener, foam booster/stabilizer for personal care and industrial applics.; liq.; pH 10.0; 100% conc.

Mackamide PKM. [McIntyre] Palm kernelamide MEA; nonionic; surfactant, thickener, foam booster/stabilizer for personal care and industrial applics.; flake; pH 10.0; 100% conc.

Mackamide R. [McIntyre] Ricinoleamide DEA; CAS 40716-42-5; nonionic; emulsifier; softener; surfactant, thickener, foam booster/stabilizer for personal care and industrial applics.; liq.; pH 10.0; 100% conc.

Mackamide S. [McIntyre] Soyamide DEA (1:1); nonionic; surfactant, thickener, foam booster/stabilizer for personal care and industrial applics.; hair conditioner; liq.; pH 10.0; 100% conc.

Mackamide SD. [McIntyre] Soyamide DEA (2:1); nonionic; surfactant, thickener, foam booster/stabilizer for personal care and industrial applics.; liq.; pH 10.0; 100% conc.

Mackamide SMA. [McIntyre] Stearamide MEA; nonionic; surfactant, thickener, foam booster/stabilizer for personal care and industrial applics.; flake; pH 10.0; 100% conc.

Mackamine CAO. [McIntyre] Cocamidopropylamine oxide; cationic; detergent, wetting agent for heavy-duty cleaners; hair conditioner, visc. builder, foam booster for personal care prods.; liq.; pH 7.0; 30% conc.

Mackamine CO. [McIntyre] Cocamine oxide; amphoteric; detergent, wetting agent for heavy-duty cleaners; hair conditioner, visc. builder, foam booster/stabilizer for personal care prods.; liq.; pH 7.0; 30% conc.

Mackamine IAO. [McIntyre] Isostearamidopropylamine oxide; nonionic; detergent, wetting agent for heavy-duty cleaners; hair conditioner, visc. builder, foam booster for personal care prods.; gel; pH 7.0; 30% conc.

Mackamine ISMO. [McIntyre] Isostearamidopropyl morpholine oxide; detergent, wetting agent for heavy-duty cleaners; hair conditioner, visc. builder, foam booster for personal care prods.; liq.; pH 7.0; 30% conc.

Mackamine LAO. [McIntyre] Lauramidopropylamine oxide; nonionic; detergent, wetting agent for heavy-duty cleaners; hair conditioner, visc. builder, foam booster for personal care prods.; water-wh. liq.; pH 7.0; 30% conc.

Mackamine LO. [McIntyre] Lauramine oxide; nonionic; detergent, wetting agent for heavy-duty cleaners; hair conditioner, visc. builder, foam booster for personal care prods.; water-wh. liq.; pH 7.0; 30% conc.

Mackamine O2. [McIntyre] Oleamine oxide; nonionic; detergent, wetting agent for heavy-duty cleaners; hair conditioner, visc. builder, foam booster for personal care prods.; amber liq.; pH 7.5; 35% conc.

Mackamine OAO. [McIntyre] Oleamidopropylamine oxide; nonionic; detergent, wetting agent for heavy-duty cleaners; hair conditioner, visc. builder, foam booster for personal care prods.; amber gel; pH 7.0; 50% conc.

Mackamine SAO. [McIntyre] Stearamidopropylamine oxide; detergent, wetting agent for heavy-duty cleaners; hair conditioner, visc. builder, foam booster for personal care prods.; paste; pH 7.0; 25% conc.

Mackamine SO. [McIntyre] Stearamine oxide; detergent, wetting agent for heavy-duty cleaners; hair conditioner, visc. builder, foam booster for personal care prods.; paste; pH 7.0; 25% conc.

Mackamine WGO. [McIntyre] Wheat germamidopropylamine oxide; detergent, wetting agent for heavy-duty cleaners; hair conditioner, visc. builder, foam booster for personal care prods.; amber gel; pH 7.0; 30% conc.

Mackanate AY-65TD. [McIntyre] Diamyl sodium sulfosuccinate, tridecyl alcohol; surfactant, wetting agent for industrial applics.; liq.; pH 6.0; 65% conc.

Mackanate CM. [McIntyre] Disodium cocamido MEA-sulfosuccinate; anionic; base for rug cleaners, shampoos, bubble baths; liq.; pH 6.0; 40% conc.

Mackanate CM-100. [McIntyre] Disodium cocamido MEA-sulfosuccinate; anionic; base for rug cleaners, shampoos, bubble baths; powd.; pH 6.0; 100% conc.

Mackanate DOS-40. [McIntyre] Dioctyl sodium sulfosuccinate; surfactant, wetting agent for industrial applics.; liq.; pH 6.0; 40% conc.

Mackanate DOS-70. [McIntyre] Dioctyl sodium sulfosuccinate; surfactant, wetting agent for industrial applics.; liq.; pH 6.0; 70% conc.

Mackanate DOS-70BC. [McIntyre] Dioctyl sodium sulfosuccinate, butyl Carbitol; surfactant, wetting agent for industrial applics.; liq.; pH 6.0; 70% conc.

Mackanate DOS-70DEG. [McIntyre] Dioctyl sodium sulfosuccinate, diethylene glycol; surfactant, wetting agent for industrial applics.; liq.; pH 6.0; 70% conc.

Mackanate DOS-70MS. [McIntyre] Dioctyl sodium sulfosuccinate, mineral spirits; anionic; emulsifier for drycleaning and agric. use; liq.; pH 6.0; 70% conc.

Mackanate DOS-70N. [McIntyre] Dioctyl sodium sulfosuccinate, nonoxynol-9; surfactant, wetting agent for industrial applics.; liq.; pH 6.0; 70% conc.

Mackanate DOS-70PG. [McIntyre] Dioctyl sodium sulfosuccinate, propylene glycol; surfactant, wetting agent for industrial applics.; liq.; pH 6.0; 70% conc.

Mackanate DOS-75. [McIntyre] Dioctyl sodium sulfosuccinate; anionic; wetting agent, dispersant, and penetrant; for drycleaning, pesticide formulations; liq.; pH 6.0; 75% conc.

Mackanate L-101, -102. [McIntyre] Disodium laureth sulfosuccinate; anionic; surfactant for powds.; 100% conc.

Mackazoline C. [McIntyre] Cocoyl hydroxyethyl imidazoline; nonionic; emulsifier; corrosion inhibitor for acid bowl cleaners, pickling systems; salts as antistats, water displacer; amber liq.; pH 11.5; 100% conc.

Mackazoline CY. [McIntyre] Capryl hydroxyethyl imidazoline; emulsifier; corrosion inhibitor for acid bowl cleaners, pickling systems; salts as antistats, water displacer; liq.; pH 11.5; 100% conc.

Mackazoline L. [McIntyre] Lauryl hydroxyethyl imidazoline; nonionic; emulsifier; corrosion inhibitor for acid bowl cleaners, pickling systems; salts as antistats, water displacer; amber liq.; 100% conc.

Mackazoline O. [McIntyre] Oleyl hydroxyethyl imidazoline; nonionic; emulsifier; corrosion inhibitor for acid bowl cleaners, pickling systems; salts as antistats, water displacer; amber liq.; 100% conc.

Mackester EGDS. [McIntyre] Glycol distearate;

emulsifier, lubricant, antistat, defoamer for metalworking, textile lubricants, plastics, paper; emulsifier, pearlescent, emollient for cosmetics; flake; 100% conc.

Mackester EGMS. [McIntyre] Glycol stearate; emulsifier, lubricant, antistat, defoamer for metalworking, textile lubricants, plastics, paper; emulsifier, pearlescent, emollient for cosmetics; flake; 100% conc.

Mackester IDO. [McIntyre] Isodecyl oleate; emulsifier, lubricant, antistat, defoamer for metalworking, textile lubricants, plastics, paper; emulsifier, pearlescent, emollient for cosmetics; liq.; 100% conc.

Mackester IP. [McIntyre] Glycol stearate, other ingreds.; emulsifier, lubricant, antistat, defoamer for metalworking, textile lubricants, plastics, paper; emulsifier, pearlescent, emollient for cosmetics; flake; 100% conc.

Mackester SP. [McIntyre] Glycol stearate; emulsifier, lubricant, antistat, defoamer for metalworking, textile lubricants, plastics, paper; emulsifier, pearlescent, emollient for cosmetics; flake; 100% conc.

Mackester TD-88. [McIntyre] Triethylene glycol dioctoate; emulsifier, lubricant, antistat, defoamer for metalworking, textile lubricants, plastics, paper; emulsifier, pearlescent, emollient for cosmetics; liq.; 100% conc.

Mackine 101. [McIntyre] Cocamidopropyl dimethylamine; cationic; intermediate for cationic surfactants, chemical specialties, hair conditioners, mild cleansers; corrosion inhibitor; salts as emulsifier for acid systems; liq.; 100% act.

Mackine 201. [McIntyre] Ricinoleamidopropyl dimethylamine; cationic; intermediate for cationic surfactants, chemical specialties, hair conditioners, mild cleansers; corrosion inhibitor; salts as emulsifier for acid systems; liq.; 100% act.

Mackine 301. [McIntyre] Stearamidopropyl dimethylamine; cationic; intermediate for cationic surfactants, chemical specialties, hair conditioners, mild cleansers; corrosion inhibitor; salts as emulsifier for acid systems; flake; 100% act.

Mackine 321. [McIntyre] Stearamidopropyl morpholine; cationic; intermediate for cationic surfactants, chemical specialties, hair conditioners, mild cleansers; corrosion inhibitor; salts as emulsifier for acid systems; flake; 100% act.

Mackine 401. [McIntyre] Isostearamidopropyl dimethylamine; cationic; intermediate for cationic surfactants, chemical specialties, hair conditioners, mild cleansers; corrosion inhibitor; salts as emulsifier for acid systems; liq.; 100% act.

Mackine 421. [McIntyre] Isostearamidopropyl morpholine; cationic; intermediate for cationic surfactants, chemical specialties, hair conditioners, mild cleansers; corrosion inhibitor; salts as emulsifier for acid systems; liq.; 100% act.

Mackine 501. [McIntyre] Oleamidopropyl dimethylamine; cationic; intermediate for cationic surfactants, chemical specialties, hair conditioners, mild cleansers; corrosion inhibitor; salts as emulsifier for acid systems; amber liq.; 100% conc.

Mackine 601. [McIntyre] Behenamidopropyl dimethylamine; nonionic; intermediate for cationic surfactants, chemical specialties, hair conditioners, mild cleansers; corrosion inhibitor; salts as emulsifier for acid systems; flake; 100% act.

Mackine 701. [McIntyre] Wheat germamidopropyl dimethylamine; intermediate for cationic surfactants, chemical specialties, hair conditioners, mild cleansers; corrosion inhibitor; salts as emulsifier for acid systems; amber paste; 100% conc.

Mackine 801. [McIntyre] Lauramidopropyl dimethylamine; intermediate for cationic surfactants, chemical specialties, hair conditioners, mild cleansers; corrosion inhibitor; salts as emulsifier for acid systems; amber solid; 100% conc.

Mackine 901. [McIntyre] Soyamidopropyl dimethylamine; intermediate for cationic surfactants, chemical specialties, hair conditioners, mild cleansers; corrosion inhibitor; salts as emulsifier for acid systems; amber paste; 100% conc.

Macol® 1. [PPG/Specialty Chem.] Poloxamer 181; nonionic; defoamer, deduster, emulsifier, detergent, dispersant, dye leveler, gellant, antistat, solubilizer, dispersant, wetting agent, lubricant base for metalworking and textile lubricants, cosmetics, medical, paper, pharmaceutical, chemical intermediates; APHA < 100 liq.; water-sol.; m.w. 2000; sp.gr. 1.015; dens. 8.5 lb/gal; visc. 285 cps; HLB 3.0; pour pt. –29 C; flash pt. (PMCC) 455 F; cloud pt. 24 C (1% aq.); ref. index 1.4520; Ross-Miles foam 10 mm (0.1%); 100% conc.

Macol® 2. [PPG/Specialty Chem.] Poloxamer 182; nonionic; detergent, emulsifier, wetting agent, dispersant, antistat, defoamer, gellant, solubilizer, lubricant base for cosmetic, medical, paper, metalworking, pharmaceutical and textile industries; liq.; sol. in aromatic solvs.; m.w. 2500; sp.gr. 1.03; visc. 415 cps; HLB 7.0; pour pt. -4 C; cloud pt. 32 C (1% aq.); surf. tens. 42.8 dynes/cm (0.1%); Ross-Miles foam 35 mm (0.1%); 100% conc.

Macol® 2D. [PPG/Specialty Chem.] EO/PO block copolymer; nonionic; wetting agent, dispersant, antistat, defoamer, gellant, solubilizer, lubricant base for cosmetic, medical, paper, metalworking, pharmaceutical and textile industries; liq.; sol. in aromatic solvs.; m.w. 2360; sp.gr. 1.03; visc. 400 cps; HLB 7.6; pour pt. -1 C; cloud pt. 35 C (1% aq.); surf. tens. 43.0 dynes/cm (0.1%); Ross-Miles foam 15 mm (0.1%); 100% conc.

Macol® 2LF. [PPG/Specialty Chem.] Block polymer; nonionic; detergent, wetting agent, dispersant, antistat, defoamer, gellant, solubilizer, lubricant base for cosmetic, medical, paper, metalworking, pharmaceutical and textile industries; liq.; sol. in aromatic solvs.; m.w. 2300; sp.gr. 1.02; visc. 400 cps; HLB 6.6; pour pt. -7 C; cloud pt. 28 C (1% aq.); surf. tens. 41.2 dynes/cm (0.1%); Ross-Miles foam 26 mm (0.1%); 100% conc.

Macol® 4. [PPG/Specialty Chem.] Poloxamer 184; nonionic; detegent, foaming agent, emulsifier, wetting agent, dispersant, antistat, defoamer, gellant, solubilizer, lubricant base for cosmetic, medical, paper, pharmaceutical and textile industries; liq.; sol. in aromatic solvs.; m.w. 2900; sp.gr. 1.05; visc. 800 cps; HLB 15.0; pour pt. 16 C; cloud pt. 60 C (1% aq.); surf. tens. 43.2 dynes/cm (0.1%); Ross-Miles foam > 600 mm (0.1%); 100% conc.

Macol® 8. [PPG/Specialty Chem.] Poloxamer 188; nonionic; detergent, foaming agent, wetting agent, dispersant, antistat, gellant, solubilizer, lubricant base for cosmetic, medical, paper, pharmaceutical and textile industries; flake; sol. in aromatic solvs.; m.w. 8500; sp.gr. 1.06; visc. 1100 cps; HLB 29.0; pour pt. 52 C; cloud pt. > 100 C (1% aq.); surf. tens. 50.3 dynes/cm (0.1%); Ross-Miles foam > 600 mm (0.1%); 100% conc.

Macol® 10. [PPG/Specialty Chem.] PO-EO block copolymers; nonionic; wetting agent, dispersant, antistat, defoamer, gellant, solubilizer, lubricant base for cosmetic, medical, paper, pharmaceutical and textile industries; liq.; sol. in aromatic solvs.; m.w. 3200; sp.gr. 1.04; visc. 660 cps; HLB 4.5; pour pt. -5 C; cloud pt. 32 C (1% aq.); surf. tens. 40.6 dynes/cm (0.1%); Ross-Miles foam 90 mm (0.1%); 100% conc.

Macol® 15. [PPG/Specialty Chem.] Meroxapol 105; nonionic; defoamer, detergent, emulsifier, pulp and paper additive, dispersant, lubricant, leveling aid, wetting agent; liq.; sol. in aromatic solvs.; m.w. 2000; sp.gr. 1.06; visc. 420 cps; HLB 15.0; pour pt. 15 C; cloud pt. 69 C (1% aq.); surf. tens. 50.9 dynes/cm (0.1%); Ross-Miles foam 45 mm (0.1%); 100% conc.

Macol® 15-20. [PPG/Specialty Chem.] EO/PO block copolymer; nonionic; defoamer, dispersant, lubricant, leveling aid, wetting agent; pulp and paper additive; liq.; HLB 15.0; 100% conc.

Macol® 16. [PPG/Specialty Chem.] Meroxapol 108; defoamer, deduster, detergent, emulsifier, dispersant, dye leveler, gellant, antistat; flake; sol. in aromatic solvs.; m.w. 4600; sp.gr. 1.06; visc. 400 cps; pour pt. 46 C; cloud pt. 95 C (1% aq.); surf. tens. 54.1 dynes/cm (0.1%); Ross-Miles foam 120 mm (0.1%); 100% conc.

Macol® 18. [PPG/Specialty Chem.] Meroxapol 171; defoamer, wetting agent, deduster, demulsifier, detergent, dispersant, dye leveler, gellant, antistat; syn. lubricant base fluid for metalworking and textile lubricants, chemical intermediates; pulp and paper additive; APHA < 100 liq.; sol. in water, aromatic solvs.; m.w. 1900; sp.gr. 1.018; dens. 8.5 lb/gal; visc. 285 cps; HLB 6.0; pour pt. –27 C; cloud pt. 32 C (1% aq.); flash pt. (PMCC) > 450 F; ref. index 1.4516; surf. tens. 33.1 dynes/cm (0.1%); Ross-Miles foam 5 mm (0.1%); 100% conc.

Macol® 19. [PPG/Specialty Chem.] Meroxapol 172; nonionic; defoamer, wetting agent, deduster, demulsifier, detergent, dispersant, dye leveler, gellant, antistat; syn. lubricant base fluid for metalworking and textile lubricants, chemical intermediates; pulp and paper additive; APHA < 100 liq.; sol. in water, aromatic solvs.; m.w. 2200; sp.gr. 1.030; dens. 8.6 lb/gal; visc. 450 cps; HLB 8.0; pour pt. –25 C; cloud pt. 36 C (1% aq.); flash pt. (PMCC) > 450 F; ref. index 1.4535; surf. tens. 42.0 dynes/cm (0.1%); Ross-Miles foam 5 mm (0.1%); 100% conc.

Macol® 20. [PPG/Specialty Chem.] EO/PO block copolymer; nonionic; defoamer, dispersant, lubricant, leveling aid, wetting agent; pulp and paper additive; liq.; HLB 3.0; 100% conc.

Macol® 21. [PPG/Specialty Chem.] Modified oxethylated straight chain alcohol; nonionic; biodeg. defoamer, detergent, wetting agent, solubilizer, rinse additive for low foam applics., e.g., mech. dishwash, cold phosphate cleaners, car wash, hard surf. cleaners; liq.; m.w. 820; sp.gr. 0.97; visc. 70 cps; HLB 7.0; pour pt. -27 C; cloud pt. 24 C (1% aq.); surf. tens. 30.3 dynes/cm (0.1%); Ross-Miles foam 5 mm (0.1%); 100% conc.

Macol® 22. [PPG/Specialty Chem.] Block polymer; nonionic; wetting agent, dispersant, antistat, defoamer, gellant, solubilizer, lubricant base for cosmetic, medical, paper, metalworking, pharmaceutical and textile industries; liq.; sol. in aromatic solvs.; m.w. 2000; sp.gr. 1.01; visc. 520 cps; pour pt. -10 C; cloud pt. 17 C (1% aq.); 100% conc.

Macol® 23. [PPG/Specialty Chem.] Poloxamer 403; nonionic; foaming agent, emulsifier, wetting agent, dispersant, antistat, gellant, solubilizer, lubricant base for cosmetic, medical, paper, pharmaceutical and textile industries; paste; sol. in aromatic solvs.; m.w. 5600; sp.gr. 1.01 (60 C); visc. 350 cps (60 C); pour pt. 31 C; cloud pt. 90 C (1% aq.); surf. tens. 34.1 dynes/cm (0.1%); Ross-Miles foam 360 mm (0.1%).

Macol® 24. [PPG/Specialty Chem.] Surfactant; biodeg. defoamer, detergent, wetting agent, solubilizer, rinse additive for low foam applics., e.g., mech. dishwash, cold phosphate cleaners, car wash, hard surf. cleaners; liq.; water-sol.; m.w. 800; sp.gr. 0.99; visc. 80 cps; pour pt. 8 C; cloud pt. 45 C; surf. tens. 33.0 dynes/cm (0.1%); Ross-Miles foam 35 mm (0.1%).

Macol® 25. [PPG/Specialty Chem.] Surfactant; biodeg. defoamer, detergent, wetting agent, solubilizer, rinse additive for low foam applics., e.g., mech. dishwash, cold phosphate cleaners, car wash, hard surf. cleaners; liq.; water-sol.; m.w. 1000; sp.gr. 1.00; visc. 100 cps; pour pt. -20 C; cloud pt. 59 C; surf. tens. 34.3 dynes/cm (0.1%); Ross-Miles foam 600 mm (0.1%).

Macol® 26. [PPG/Specialty Chem.] Surfactant; biodeg. defoamer, detergent, wetting agent, solubilizer, rinse additive for low foam applics., e.g., mech. dishwash, cold phosphate cleaners, car wash, hard surf. cleaners; liq.; water-sol.; m.w. 850; sp.gr. 0.98; visc. 200 cps; pour pt. -6 C; cloud pt. 24 C; surf. tens. 30.5 dynes/cm (0.1%); Ross-Miles foam 5 mm (0.1%).

Macol® 27. [PPG/Specialty Chem.] Poloxamer 407; nonionic; emulsifier, wetting agent, dispersant, antistat, defoamer, gellant, solubilizer, lubricant base for cosmetic, medical, paper, pharmaceutical and textile industries; flake; sol. in aromatic solvs.; m.w. 12,500; sp.gr. 1.05 (77 C); visc. 3100 cps (77 C); HLB 2.2; pour pt. 56 C; cloud pt. > 100 C (1% aq.); surf. tens. 40.7 dynes/cm (0.1%); Ross-Miles foam > 600 mm (0.1%); 100% conc.

Macol® 30. [PPG/Specialty Chem.] Modified oxethylated straight chain alcohol; nonionic; biodeg. defoamer, detergent, wetting agent, solubilizer, rinse additive for low foam applics., e.g., mech. dishwash, cold phosphate cleaners, car wash, hard surf. cleaners; liq.; water-sol.; m.w. 600; sp.gr. 0.97; visc. 65 cps; HLB 9.0; pour pt. 10 C; cloud pt. 30 C (1% aq.); surf. tens. 30.7 dynes/cm (0.1%); Ross-Miles foam 20 mm (0.1%); 100% conc.

Macol® 31. [PPG/Specialty Chem.] EO-PO block copolymer; nonionic; wetting agent, dispersant, antistat, defoamer, gellant, solubilizer, lubricant base for cosmetic, medical, paper, pharmaceutical and textile industries; APHA < 100 liq.; sol. in water, aromatic solvs.; m.w. 3300; sp.gr. 1.05; dens. 8.5 lb/gal; visc. 950 cps; HLB 6.3; pour pt. 24 C; cloud pt. 40 C (1% aq.); flash pt. (PMCC) 439 F; ref. index 1.4515.

Macol® 32. [PPG/Specialty Chem.] Meroxapol 251; nonionic; wetting agent, dispersant, antistat, defoamer, gellant, solubilizer, lubricant base for cosmetic, medical, paper, pharmaceutical and textile industries; APHA < 100 liq.; sol. in water, aromatic solvs.; m.w. 2700; sp.gr. 1.017; dens. 8.5 lb/gal; visc. 460 cps; HLB 4.0; pour pt. –27 C; cloud pt. 26 C (1% aq.); flash pt. (PMCC) < 450 F; ref. index 1.4521; surf. tens. 36.3 dynes/cm (0.1%); Ross-Miles foam < 5 mm (0.1%).

Macol® 33. [PPG/Specialty Chem.] Meroxapol 311; nonionic; wetting agent, dispersant, antistat, defoamer, gellant, solubilizer, lubricant base for cosmetic, medical, paper, metalworking, pharmaceutical and textile industries; APHA < 100 liq.; sol. in water, aromatic solvs.; m.w. 3200; sp.gr. 1.018; dens. 8.5 lb/gal; visc. 578 cps; HLB 4.0; pour pt. – 25 C; cloud pt. 25 C (1% aq.); flash pt. (PMCC) < 450 F; ref. index 1.4522; surf. tens. 34.1 dynes/cm (0.1%); Ross-Miles foam < 5 mm (0.1%); 100% conc.

Macol® 34. [PPG/Specialty Chem.] Meroxapol 254; nonionic; wetting agent, dispersant, antistat, defoamer, gellant, solubilizer, lubricant base for cosmetic, medical, paper, pharmaceutical and textile industries; paste; sol. in aromatic solvs.; m.w. 3600; sp.gr. 1.05 (60 C); visc. 1100 cps (60 C); HLB 10.0; pour pt. 25 C; cloud pt. 40 C (1% aq.); surf. tens. 41.0 dynes/cm (0.1%); Ross-Miles foam 70 mm (0.1%); 100% conc.

Macol® 35. [PPG/Specialty Chem.] Poloxamer 105; nonionic; wetting agent, dispersant, antistat, defoamer, gellant, solubilizer, lubricant base for cosmetic, medical, paper, pharmaceutical and textile industries; liq.; sol. in aromatic solvs.; m.w. 1900; sp.gr. 1.06; visc. 375 cps; HLB 8.0; pour pt. 7 C; cloud pt. 78 C (1% aq.); surf. tens. 48.8 dynes/cm (0.1%); Ross-Miles foam 145 mm (0.1%); 100% conc.

Macol® 40. [PPG/Specialty Chem.] Meroxapol 252; nonionic; wetting agent, dispersant, antistat, defoamer, gellant, solubilizer, lubricant base for cosmetic, medical, paper, metalworking, pharmaceutical and textile industries; liq.; sol. in aromatic solvs.; m.w. 3100; sp.gr. 1.03; visc. 700 cps; pour pt. -5 C; cloud pt. 29 C (1% aq.); surf. tens. 37.5 dynes/cm (0.1%); Ross-Miles foam < 5 mm (0.1%); 100% conc.

Macol® 42. [PPG/Specialty Chem.] Poloxamer 122; nonionic; wetting agent, dispersant, antistat, defoamer, gellant, solubilizer, lubricant base for cosmetic, medical, paper, pharmaceutical and textile industries; APHA < 100 liq.; sol. in water, aromatic solvs.; m.w. 1600; sp.gr. 1.03; dens. 8.5 lb/gal; visc. 250 cps; HLB 12.0; pour pt. –26 C; cloud pt. 37 C (1% aq.); flash pt. (PMCC) 450 F; ref. index 1.4541; surf. tens. 46.5 dynes/cm (0.1%); Ross-Miles foam 10 mm (0.1%); 100% conc.

Macol® 44. [PPG/Specialty Chem.] Poloxamer 124; nonionic; wetting agent, dispersant, antistat, defoamer, gellant, solubilizer, lubricant base for cosmetic, medical, paper, pharmaceutical and textile industries; liq.; sol. in aromatic solvs.; m.w. 2200; sp.gr. 1.05; visc. 450 cps; HLB 3.0; pour pt. 15 C; cloud pt. 68 C (1% aq.); surf. tens. 45.4 dynes/cm (0.1%); Ross-Miles foam 360 mm (0.1%); 100% conc.

Macol® 45. [PPG/Specialty Chem.] Surfactant; biodeg. defoamer, detergent, wetting agent, solubilizer, rinse additive for low foam applics., e.g., mech. dishwash, cold phosphate cleaners, car wash, hard surf. cleaners; liq.; water-sol.; m.w. 1100; sp.gr. 0.98; visc. 180 cps; pour pt. -18 C; cloud pt. 20 C; surf. tens. 31.0 dynes/cm (0.1%); Ross-Miles foam 5 mm (0.1%).

Macol® 46. [PPG/Specialty Chem.] Poloxamer 101; nonionic; wetting agent, dispersant, antistat, defoamer, gellant, solubilizer, lubricant base for cosmetic, medical, paper, pharmaceutical and textile industries; liq.; sol. in aromatic solvs.; m.w. 1100; sp.gr. 1.02; visc. 180 cps; HLB 18.5; pour pt. -32 C; cloud pt. 36 C (1% aq.); surf. tens. 47.1 dynes/cm (0.1%); Ross-Miles foam 18 mm (0.1%); 100% conc.

Macol® 65. [PPG/Specialty Chem.] Functional fluid with lubricating, antiwear props. for hydraulic systems, cosmetics; chemical intermediate for prep. of resins, plasticizers, modifiers, and surfactants; in ink and dye solvs.; liq.; sol. @ 5% in alcohols, ketones, kerosene, min. oil; disp. in water; sp.gr. 0.960; visc. 12 cst (100 F); pour pt. -58 C; flash pt. (PMCC) 245 F; ref. index 1.440.

Macol® 72. [PPG/Specialty Chem.] Poloxamer 212; nonionic; wetting agent, dispersant, antistat, defoamer, gellant, solubilizer, lubricant base for cosmetic, medical, paper, pharmaceutical and textile industries; liq.; sol. in aromatic solvs.; m.w. 2750; sp.gr. 1.03; visc. 500 cps; HLB 2.0; pour pt. -7 C; cloud pt. 25 C (1% aq.); surf. tens. 39 dynes/cm (0.1%); Ross-Miles foam 20 mm (0.1%); 100% conc.

Macol® 77. [PPG/Specialty Chem.] Poloxamer 217; nonionic; wetting agent, dispersant, antistat, defoamer, gellant, solubilizer, lubricant base for cosmetic, medical, paper, pharmaceutical and textile industries; flake; sol. in aromatic solvs.; m.w. 6600; sp.gr. 1.04; visc. 475 cps; HLB 24.5; pour pt. 48 C; cloud pt. > 100 C (1% aq.); surf. tens. 47.0 dynes/cm (0.1%); Ross-Miles foam > 600 mm (0.1%); 100% conc.

Macol® 85. [PPG/Specialty Chem.] Poloxamer 235; nonionic; wetting agent, dispersant, antistat, defoamer, gellant, solubilizer, lubricant base for cosmetic, medical, paper, pharmaceutical and textile industries; paste; sol. in aromatic solvs.; m.w. 4600; sp.gr. 1.04 (60 C); visc. 320 cps (60 C); HLB 16.0; pour pt. 29 C; cloud pt. 85 C (1% aq.); surf. tens. 42.5 dynes/cm (0.1%); Ross-Miles foam > 600 mm (0.1%); 100% conc.

Macol® 88. [PPG/Specialty Chem.] Block polymer; nonionic; wetting agent, dispersant, antistat, defoamer, gellant, solubilizer, lubricant base for cosmetic, medical, paper, pharmaceutical and textile industries; flake; sol. in aromatic solvs.; m.w. 11,500; sp.gr. 1.06 (77 C); visc. 2300 cps (77 C); pour pt. 54 C; cloud pt. > 100 C (1% aq.); surf. tens. 48.5 dynes/cm (0.1%); Ross-Miles foam > 600 mm (0.1%).

Macol® 90. [PPG/Specialty Chem.] Functional fluid with lubricating, antiwear props. for hydraulic systems, cosmetics; chemical intermediate for prep. of resins, plasticizers, modifiers, and surfactants; in ink and dye solvs.; liq.; sol. @ 5% in water, alcohols, ketones; disp. in min. oil; sp.gr. 1.097; visc. 19,500 cst (100 F); pour pt. 5 C; flash pt. (PMCC) 350 F; ref. index 1.465; 100% act.

Macol® 90(70). [PPG/Specialty Chem.] Functional fluid with lubricating, antiwear props. for hydraulic systems, cosmetics; chemical intermediate for prep. of resins, plasticizers, modifiers, and surfactants; in ink and dye solvs.; liq.; sol. @ 5% in water, alcohols, ketones; disp. in min. oil; sp.gr. 1.049; visc. 1700 cst (100 F); pour pt. 1 C; flash pt. none; ref. index 1.418; 70% act.

Macol® 97. [PPG/Specialty Chem.] Surfactant; nonionic; emulsifier for detergents and cleaning; liq.; 100% conc.

Macol® 99A. [PPG/Specialty Chem.] Surfactant; nonionic; emulsifier for printing ink and other pigmented systems; liq.; 100% conc.

Macol® 101. [PPG/Specialty Chem.] Poloxamer 331; nonionic; wetting agent, dispersant, antistat, defoamer, gellant, solubilizer, lubricant base for cosmetic, medical, paper, pharmaceutical and textile industries; APHA < 100 liq.; sol. in water, aromatic solvs.; sp.gr. 1.020; dens. 8.5 lb/gal; visc. 800 cps; HLB 1.0; pour pt. –23 C; cloud pt. 15 C (1% aq.); flash pt. (PMCC) < 450 F; ref. index 1.4524; Ross-Miles foam 10 mm (0.1%); 100% conc.

Macol® 108. [PPG/Specialty Chem.] Poloxamer 338; nonionic; wetting agent, dispersant, antistat, defoamer, gellant, solubilizer, lubricant base for cosmetic, medical, paper, pharmaceutical and textile industries; flake; sol. in aromatic solvs.; m.w. 14,600; sp.gr. 1.06 (77 C); visc. 3000 cps (77 C); HLB 17.5; pour pt. 57 C; cloud pt. > 100 C (1% aq.); surf. tens. 41.2 dynes/cm (0.1%); Ross-Miles foam > 600 mm (0.1%); 100% conc.

Macol® 300. [PPG/Specialty Chem.] PPG-7 buteth-10; nonionic; detergent for toilet cleaners, laundry detergents, emulsion polymerization, defoamers, metalworking fluids, hydraulic fluids; liq.; sol. @ 5% in water, alcohols, ketones; disp. in min. oil; sp.gr. 1.033; visc. 60 cst (100 F); pour pt. -40 C; flash pt. (PMCC) 340 F; ref. index 1.456; 100% conc.

Macol® 625. [PPG/Specialty Chem.] Functional fluid with lubricating, antiwear props. for hydraulic systems, cosmetics; chemical intermediate for prep. of resins, plasticizers, modifiers, and surfactants; in ink and dye solvs.; liq.; sol. @ 5% in alcohols, ketones, kerosene, min. oil; disp. in water; sp.gr. 1.000; visc. 135 cst (100 F); pour pt. -31 C; flash pt. (PMCC) 360 F; ref. index 1.450.

Macol® 626. [PPG/Specialty Chem.] Functional fluid with lubricating, antiwear props. for hydraulic systems, cosmetics; chemical intermediate for prep. of resins, plasticizers, modifiers, and surfactants; in ink and dye solvs.; liq.; sol. @ 5% in alcohols, ketones, kerosene, min. oil; disp. in water; sp.gr. 1.002; visc. 300 cst (100 F); pour pt. -27 C; flash pt. (PMCC) 360 F; ref. index 1.451.

Macol® 627. [PPG/Specialty Chem.] Functional fluid with lubricating, antiwear props. for hydraulic systems, cosmetics; chemical intermediate for prep. of resins, plasticizers, modifiers, and surfactants; in ink and dye solvs.; liq.; sol. @ 5% in alcohols, ketones, kerosene, min. oil; disp. in water; sp.gr. 1.002; visc. 370 cst (100 F); pour pt. -23 C; flash pt. (PMCC) 360 F; ref. index 1.452.

Macol® 660. [PPG/Specialty Chem.] PPG-12 buteth-16; nonionic; detergent for toilet bowl cleaners, laundry; defoamer, rubber lubricant, intermediate; hydraulic, heat transfer, and metal working fluids; mold release agent; emulsion polymerization; APHA 100 max. clear visc. liq.; sol. in acetone, propylene glycol, oleic acid, castor oil, water, toluene, IPA; sp.gr. 1.047; dens. 8.72 lb/gal; visc. 140 cst (100 F); acid no. 0–1; pour pt. -37 C; cloud pt. 61 C (1% aq.); flash pt. 430 F (COC); ref. index 1.459; pH 5–7; 100% conc.

Macol® 3520. [PPG/Specialty Chem.] PPG-28 buteth-35; nonionic; detergent for toilet bowl cleaners, laundry; defoamer, rubber lubricant, intermediate; hydraulic, heat transfer, and metal working fluids; mold release agent; emulsion polymerization; APHA 100 max. clear visc. liq.; sol. in acetone, propylene glycol, oleic acid, castor oil, water, toluene, IPA; sp.gr. 1.050; dens. 8.75 lb/gal; visc. 760 cst (100 F); pour pt. -28 C; acid no. O–1; ref. index 1.461; pH 5–7; flash pt. 430 F (COC); 100% conc.

Macol® 5100. [PPG/Specialty Chem.] PPG-33 buteth-45; nonionic; detergent for toilet bowl cleaners, laundry; defoamer, rubber lubricant, intermediate; hydraulic, heat transfer, and metal working fluids; mold release agent; emulsion polymerization; food processing; FDA compliance; APHA 100 max. clear visc. liq.; sol. in acetone, propylene glycol, oleic acid, castor oil, water, toluene, IPA; sp.gr. 1.050; dens. 8.75 lb/gal; visc. 1100 cst (100 F); acid no. 0–1; pour pt. -28 C; cloud pt. 55 C (1% aq.); flash pt. 430 F (COC); ref. index 1.462; pH 5–7; 100% conc.

Macol® CA-2. [PPG/Specialty Chem.] Ceteth-2; nonionic; emulsifier, detergent, wetting agent, dispersant, solubilizer, coupling agent for cosmetics, textile, metalworking, household, industrial and other applics.; solid; sol. @ 5% in IPA; insol. in water; m.p. 38 C; HLB 4.9; iodine no. 0.5; hyd. no. 170; flash pt. (PMCC) > 350 F.

Macol® CA-10. [PPG/Specialty Chem.] Ceteth-10; nonionic; emulsifier, detergent, wetting agent, dispersant, solubilizer, coupling agent for cosmetics, textile, metalworking, household, industrial and other applics.; solid; sol. @ 5% in IPA; disp. in water, min. spirits; m.p. 41 C; HLB 12.3; iodine no. 0.5; hyd. no. 95; flash pt. (PMCC) > 350 F.

Macol® CSA-2. [PPG/Specialty Chem.] Ceteareth-2; CAS 68439-49-6; nonionic; detergent, wetting agent, emulsifier, dispersant, solubilizer, and coupling agent for cosmetics, textiles, metalworking lubricants, household prods., and industrial applics.; wh. solid; sol. @ 5% in IPA; insol. in water; m.p. 39 C; HLB 5.1; iodine no. 0.5; hyd. no. 160; flash pt. (PMCC) > 350 F.

Macol® CSA-4. [PPG/Specialty Chem.] Ceteareth-4; CAS 68439-49-6; nonionic; see Macol CSA-2; wh. solid; sol. @ 5% in IPA; insol. water and min. oil; m.p. 38 C; HLB 7.9; iodine no. 0.5; hyd. no. 128; flash pt. (PMCC) > 350 F.

Macol® CSA-10. [PPG/Specialty Chem.] Ceteareth-10; CAS 68439-49-6; nonionic; see Macol CSA-2; wh. solid; sol. @ 5% in IPA; disp. in water, min. spirits; m.p. 38 C; HLB 12.3; iodine no. 0.5; hyd. no. 80; flash pt. (PMCC) > 350 F.

Macol® CSA-15. [PPG/Specialty Chem.] Ceteareth-15; CAS 68439-49-6; nonionic; see Macol CSA-2; wh. solid; sol. @ 5% in water, IPA; disp. in min. spirits; insol. min. oil; m.p. 38 C; HLB 14.2; iodine no. 0.5; hyd. no. 65; flash pt. (PMCC) > 350 F.

Macol® CSA-20. [PPG/Specialty Chem.] Ceteareth-20; CAS 68439-49-6; nonionic; see Macol CSA-2; wh. solid; sol. @ 5% in water, IPA; insol. min. oil; m.p. 40 C; HLB 15.2; iodine no. 0.5; hyd. no. 52; flash pt. (PMCC) > 350 F; 100% conc.

Macol® CSA-40. [PPG/Specialty Chem.] Ceteareth-40; CAS 68439-49-6; nonionic; see Macol CSA-2; solid; sol. @ 5% in water, IPA; m.p. 40 C; HLB 16.8; iodine no. 0.5; hyd. no. 30; flash pt. (PMCC) > 350 F.

Macol® CSA-50. [PPG/Specialty Chem.] Ceteareth-50; CAS 68439-49-6; nonionic; see Macol CSA-2.

Macol® DNP-5. [PPG/Specialty Chem.] Nonyl nonoxynol-5; CAS 9014-93-1; nonionic; emulsifier, detergent, wetting agent, dispersant, solubilizer, coupling agent for cosmetics, textile, metalworking, household, industrial and other applics.; liq.; sol. @ 5% in toluene, perchloroethylene; disp. in water, min. oil, min. spirits; sp.gr. 0.97; visc. 385 cps; HLB 8.2; pour pt. -10 C; flash pt. (PMCC) > 350 F.

Macol® DNP-10. [PPG/Specialty Chem.] Nonyl nonoxynol-10; CAS 9014-93-1; nonionic; see Macol DNP-5; liq.; sol. @ 5% in toluene, perchloroethylene; disp. in water, min. oil, min. spirits; sp.gr. 1.00; visc. 390 cps; HLB 11.3; pour pt. 0 C; flash pt. (PMCC) > 350 F; 100% conc.

Macol® DNP-15. [PPG/Specialty Chem.] Nonyl nonoxynol-15; CAS 9014-93-1; nonionic; see Macol DNP-5; paste; sol. @ 5% in toluene, perchloroethylene; disp. in water, min. oil, min. spirits; sp.gr. 1.02; HLB 13.0; pour pt. 30 C; flash pt. (PMCC) > 350 F.

Macol® DNP-21. [PPG/Specialty Chem.] Nonyl nonoxynol-21; CAS 9014-93-1; nonionic; see Macol DNP-5; solid; HLB 14.8; cloud pt. 91 C (1% aq.).

Macol® DNP-150. [PPG/Specialty Chem.] Nonyl nonoxynol-150; CAS 9014-93-1; nonionic; dispersant and wetting agent for pesticides, heavy-duty detergents, alkaline cleaners, dairy detergents; flake; sol. @ 5% in water, toluene; sp.gr. 1.06; HLB 19.0; pour pt. 55 C; cloud pt. > 100 C (1% aq.); flash pt. (PMCC) > 350 F; 100% conc.

Macol® LA-4. [PPG/Specialty Chem.] Laureth-4; nonionic; detergent, wetting agent, emulsifier, dispersant, solubilizer, stabilizer, coupling agent for cosmetics, textiles, metalworking lubricants, household prods., industrial uses; colorless liq.; sol. @ 5% in IPA, min. oil; disp. in min. spirits; m.p. 12 C; HLB 9.5; iodine no. 0.1; hyd. no. 155; flash pt. (PMCC) > 350 F; 100% conc.

Macol® LA-9. [PPG/Specialty Chem.] Laureth-9; nonionic; see Macol LA-4; colorless paste; sol. @ 5% in water; disp. in min. spirits; HLB 13.3; pour pt. 26 C; iodine no. 0.1; hyd. no. 95; flash pt. (PMCC) > 350 F; 100% conc.

Macol® LA-12. [PPG/Specialty Chem.] Laureth-12; nonionic; see Macol LA-4; wh. solid; sol. @ 5% in water; disp. in min. spirits; m.p. 30 C; HLB 14.6; iodine no. 0.1; hyd. no. 75; flash pt. (PMCC) > 350 F; 100% conc.

Macol® LA-23. [PPG/Specialty Chem.] Laureth-23; nonionic; see Macol LA-4; wh. solid; sol. @ 5% in water; disp. in min. spirits; m.p. 40 C; HLB 16.4; iodine no. 0.1; hyd. no. 47; flash pt. (PMCC) > 350 F; 100% conc.

Macol® LA-790. [PPG/Specialty Chem.] Laureth-7; nonionic; detergent, wetting agent, emulsifier, dispersant, solubilizer, and coupling agent for cosmetics, textiles, metalworking lubricants, household prods., and industrial applics.; colorless liq.; sol. @ 5% in water, IPA; disp. in min. spirits; insol. min. oil; HLB 10.8; pour pt. 5 C; iodine no. 0.1; flash pt. (PMCC) > 350 F; 90% act. in water.

Macol® LF-110. [PPG/Specialty Chem.] Polyalkoxylated aliphatic ether; nonionic; biodeg. low foam wetting aid and low surface tension surfactant; syn. lubricant base fluid for metalworking, hard-surface cleaning, and metal cleaning and degreasing; used in cleaners and rinse aids; APHA < 100 liq.; water-sol.; m.w. 1100; sp.gr. 1.04; dens. 8.4 lb/gal; visc. 140 cps; pour pt. -9 C; flash pt. (PMCC) 375 F; cloud pt. 12 C (1% aq.); surf. tens. 32.8 dynes/cm (0.1%); Ross-Miles foam 5 mm (0.1%); 100% act.

Macol® LF-111. [PPG/Specialty Chem.] Polyalkoxylated aliphatic ether; nonionic; biodeg. low foam wetting aid and low surface tension surfactant; syn. lubricant base fluid for metalworking, hard-surface cleaning, and metal cleaning and degreasing; used in cleaners and rinse aids; APHA < 100 liq.; water-sol.; m.w. 1000; sp.gr. 1.040; dens. 8.4 lb/gal; visc. 100 cps; pour pt. -12 C; flash pt. (PMCC) 380 F; cloud pt. 12 C (1% aq.); surf. tens. 32.8 dynes/cm (0.1%); Ross-Miles foam 5 mm (0.1%); 100% act.

Macol® LF-115. [PPG/Specialty Chem.] Surfactant; biodeg. defoamer, detergent, wetting agent, solubilizer, rinse additive for low foam applics., e.g., mech. dishwash, cold phosphate cleaners, car wash, hard surf. cleaners; liq.; water-sol.; m.w. 1000; sp.gr. 1.04; visc. 100 cps; pour pt. -12 C; cloud pt. 15 C; surf. tens. 32.8 dynes/cm (0.1%); Ross-Miles foam 5 mm (0.1%).

Macol® LF-120. [PPG/Specialty Chem.] Polyalkoxylated aliphatic ether; nonionic; biodeg. wetting aid and low surface tension surfactant; syn. lubricant base fluid for metalworking, hard-surface cleaning, and metal cleaning and degreasing; used in cleaners and rinse aids; metalworking base fluid; APHA 150 liq.; water-sol.; m.w. 1100; sp.gr. 1.04; dens. 8.4 lb/gal; visc. 140 cps; pour pt. -9 C; flash pt. (PMCC) 310 F; cloud pt. 18 C (1% aq.); surf. tens. 32.8 dynes/cm (0.1%); Ross-Miles foam 5 mm (0.1%); 100% act.

Macol® LF-125. [PPG/Specialty Chem.] Surfactant; biodeg. defoamer, detergent, wetting agent, solubilizer, rinse additive for low foam applics., e.g., mech. dishwash, cold phosphate cleaners, car wash, hard surf. cleaners; liq.; water-sol.; m.w. 540; sp.gr. 1.01; visc. 120 cps; pour pt. -16 C; cloud pt. 40 C; surf. tens. 31.5 dynes/cm (0.1%); Ross-Miles foam 5 mm (0.1%).

Macol® NP-4. [PPG/Specialty Chem.] Nonoxynol-4; CAS 9016-45-9; nonionic; emulsifier, detergent, wetting agent, dispersant, solubilizer, coupling agent for cosmetics, textile, metalworking, household, industrial and other applics.; liq.; sol. @ 5% in min. oil, min. spirits, toluene, perchloroethylene; sp.gr. 1.02; visc. 350 cps; HLB 8.9; pour pt. -27 C; flash pt. (PMCC) > 350 F; 100% conc.

Macol® NP-5. [PPG/Specialty Chem.] Nonoxynol-5; CAS 9016-45-9; nonionic; see Macol NP-4; liq.; sol. @ 5% in min. oil, min. spirits, toluene, perchloroethylene; sp.gr. 1.03; visc. 320 cps; HLB 10.0; pour pt. -27 C; flash pt. (PMCC) > 350 F.

Macol® NP-6. [PPG/Specialty Chem.] Nonoxynol-6; CAS 9016-45-9; nonionic; see Macol NP-4; liq.; sol. @ 5% in min. spirits, toluene, perchloroethylene; disp. in water, min. oil; sp.gr. 1.04; visc. 300 cps; HLB 10.9; pour pt. -28 C; flash pt. (PMCC) > 350 F; 100% conc.

Macol® NP-8. [PPG/Specialty Chem.] Nonoxynol-8; CAS 9016-45-9; nonionic; see Macol NP-4; liq.; sol. @ 5% in water, min. spirits, toluene, perchloroethylene; sp.gr. 1.05; visc. 260 cps; HLB 12.3; pour pt. 5 C; cloud pt. 25 C (1% aq.); flash pt. (PMCC) > 350 F.

Macol® NP-9.5. [PPG/Specialty Chem.] Nonoxynol-9.5; CAS 9016-45-9; nonionic; detergent, emulsifier, solubilizer, wetting agent, stabilizer; surfactant used in paints, textile and pesticide formulation; liq.; sol. @ 5% in water, toluene, perchloroethylene; sp.gr. 1.06; visc. 275 cps; HLB 12.9; pour pt. 5 C; cloud pt. 55 C (1% aq.); flash pt. (PMCC) > 350 F; 100% conc.

Macol® NP-11. [PPG/Specialty Chem.] Nonoxynol–11; CAS 9016-45-9; nonionic; see Macol NP-4; liq.; sol. @ 5% in water, toluene, perchloroethylene; sp.gr. 1.06; visc. 275 cps; HLB 13.7; pour pt. 14 C; cloud pt. 74 C (1% aq.); flash pt. (PMCC) > 350 F; 100% conc.

Macol® NP-12. [PPG/Specialty Chem.] Nonoxynol-12; CAS 9016-45-9; nonionic; see Macol NP-4; liq.; sol. @ 5% in water, toluene, perchloroethylene; sp.gr. 1.06; visc. 325 cps; HLB 14.0; pour pt. 17 C;

cloud pt. 81 C (1% aq.); flash pt. (PMCC) > 350 F; 100% conc.

Macol® NP-15. [PPG/Specialty Chem.] Nonoxynol-15; CAS 9016-45-9; nonionic; see Macol NP-4; paste; sol. @ 5% in water; disp. in toluene, perchloroethylene; sp.gr. 1.07; HLB 15.0; pour pt. 26 C; cloud pt. 65 C (1% in 10% NaCl); flash pt. (PMCC) > 350 F.

Macol® NP-20. [PPG/Specialty Chem.] Nonoxynol–20; CAS 9016-45-9; nonionic; see Macol NP-4; solid; sol. @ 5% in water; sp.gr. 1.08; HLB 16.0; pour pt. 30 C; cloud pt. 70 C (1% in 10% NaCl); flash pt. (PMCC) > 350 F; 100% conc.

Macol® NP-20(70). [PPG/Specialty Chem.] Nonoxynol-20; CAS 9016-45-9; nonionic; see Macol NP-4; liq.; sol. @ 5% in water; sp.gr. 1.06; visc. 900 cps; HLB 16.0; pour pt. 0 C; cloud pt. 70 C (1% in 10% NaCl); flash pt. (PMCC) > 350 F; 70% act. in water.

Macol® NP-30(70). [PPG/Specialty Chem.] Nonoxynol-30; CAS 9016-45-9; nonionic; see Macol NP-4; liq.; sol. @ 5% in water; sp.gr. 1.06; visc. 1100 cps; HLB 17.2; pour pt. 4 C; cloud pt. 75 C (1% in 10% NaCl); flash pt. (PMCC) > 350 F; 70% act. in water.

Macol® NP-100. [PPG/Specialty Chem.] Nonoxynol–100; CAS 9016-45-9; nonionic; see Macol NP-4; flake; sol. @ 5% in water; sp.gr. 1.11; HLB 19.0; pour pt. 54 C; cloud pt. > 100 C (1% aq.); flash pt. (PMCC) > 350 F; 100% conc.

Macol® OA-2. [PPG/Specialty Chem.] Oleth-2; CAS 9004-98-2; nonionic; detergent, wetting agent, emulsifier, dispersant, solubilizer, stabilizer, coupling agent for cosmetics, textiles, metalworking lubricants, household prods., industrial uses; colorless liq.; sol. @ 5% in IPA, min. oil; HLB 3.8; pour pt. < 0 C; iodine no. 70; hyd. no. 170; flash pt. (PMCC) > 350 F; 100% conc.

Macol® OA-4. [PPG/Specialty Chem.] Oleth-4; CAS 9004-98-2; nonionic; see Macol LA-4; colorless liq.; sol. @ 5% in IPA, min. oil; disp. in min. spirits; HLB 8.0; pour pt. < 0 C; iodine no. 53; hyd. no. 128; flash pt. (PMCC) > 350 F; 100% conc.

Macol® OA-5. [PPG/Specialty Chem.] Oleth-5; CAS 9004-98-2; nonionic; see Macol OA-2; colorless liq.; sol. @ 5% in IPA, min. oil; disp. in min. spirits; insol. in water; HLB 8.2; pour pt. 5 C; iodine no. 53; hyd. no. 125; flash pt. (PMCC) > 350 F; 100% conc.

Macol® OA-10. [PPG/Specialty Chem.] Oleth-10; CAS 9004-98-2; nonionic; see Macol OA-2; liq.; sol. @ 5% in water, IPA; disp. min. oil, min. spirits; HLB 12.5; pour pt. 16 C; iodine no. 33; hyd. no. 80; flash pt. (PMCC) > 350 F; 100% conc.

Macol® OA-20. [PPG/Specialty Chem.] Oleth-20; CAS 9004-98-2; nonionic; see Macol OA-2; cream solid; sol. @ 5% in water, IPA; m.p. 30 C; HLB 14.7; iodine no. 23; hyd. no. 58; flash pt. (PMCC) > 350 F; 100% conc.

Macol® OP-3. [PPG/Specialty Chem.] Octoxynol-3; CAS 9002-93-1; nonionic; emulsifier, detergent, wetting agent, dispersant, solubilizer, coupling agent for cosmetics, textile, metalworking, household, industrial and other applics.; liq.; sol. @ 5% in min. oil, min. spirits, toluene, perchloroethylene; sp.gr. 1.02; visc. 350 cps; HLB 7.8; pour pt. -23 C; flash pt. (PMCC) > 350 F; 100% conc.

Macol® OP-5. [PPG/Specialty Chem.] Octoxynol-5; CAS 9002-93-1; nonionic; see Macol OP-3; liq.; sol. @ 5% in min. spirits, toluene, perchloroethylene; disp. in min. oil; sp.gr. 1.04; visc. 300 cps; HLB 10.4; pour pt. -26 C; flash pt. (PMCC) > 350 F; 100% conc.

Macol® OP-8. [PPG/Specialty Chem.] Octoxynol-8; CAS 9002-93-1; nonionic; see Macol OP-3; liq.; sol. @ 5% in water, toluene, perchloroethylene; sp.gr. 1.05; visc. 275 cps; HLB 12.3; pour pt. 5 C; cloud pt. 23 C (1% aq.); flash pt. (PMCC) > 350 F.

Macol® OP-10. [PPG/Specialty Chem.] Octoxynol-10; CAS 9002-93-1; nonionic; see Macol OP-3; liq.; sol. @ 5% in toluene, perchloroethylene; disp. in water; sp.gr. 1.06; visc. 250 cps; HLB 13.4; pour pt. 8 C; cloud pt. 65 C (1% aq.); flash pt. (PMCC) > 350 F; 100% conc.

Macol® OP-10 SP. [PPG/Specialty Chem.] Modified octoxynol-10; CAS 9002-93-1; nonionic; see Macol OP-3; visc. 250 cps; HLB 13.4; cloud pt. 65 C (1% aq.).

Macol® OP-12. [PPG/Specialty Chem.] Octoxynol-12; CAS 9002-93-1; nonionic; see Macol OP-3; liq.; sol. @ 5% in water, toluene, perchloroethylene; sp.gr. 1.07; visc. 335 cps; HLB 14.6; pour pt. 16 C; cloud pt. 88 C (1% aq.); flash pt. (PMCC) > 350 F; 100% conc.

Macol® OP-16(75). [PPG/Specialty Chem.] Octoxynol-16; CAS 9002-93-1; nonionic; see Macol OP-3; liq.; sol. @ 5% in water; sp.gr. 1.08; visc. 540 cps; HLB 15.8; pour pt. 13 C; cloud pt. > 100 C (1% aq.); flash pt. (PMCC) > 350 F; 75% act. in water.

Macol® OP-30(70). [PPG/Specialty Chem.] Octoxynol-30; CAS 9002-93-1; nonionic; see Macol OP-3; liq.; sol. @ 5% in water; sp.gr. 1.10; visc. 470 cps; HLB 17.3; pour pt. 2 C; cloud pt. > 100 C (1% aq.); flash pt. (PMCC) > 350 F; 70% act. in water.

Macol® OP-40(70). [PPG/Specialty Chem.] Octoxynol-40; CAS 9002-93-1; nonionic; see Macol OP-3; emulsifier for vinyl acetate and acrylate polymerization; liq.; sol. @ 5% in water; sp.gr. 1.10; visc. 490 cps; HLB 17.9; pour pt. 4 C; cloud pt. > 100 C (1% aq.); flash pt. (PMCC) > 350 F; 70% act. in water.

Macol® P-500. [PPG/Specialty Chem.] PPG-9; nonionic; defoamer for aq. systems, in mold release applics., lubricant bases for textile, paper, metalworking formulations, chemical intermediates for fatty acid esters, components for urethane resins; liq.; sol. @ 5% in water, min. oils, perchloroethylene; m.w. 500; sp.gr. 1.005; visc. 50 cps; pour pt. -49 C; flash pt. (PMCC) 330 F; ref. index 1.448.

Macol® P-1200. [PPG/Specialty Chem.] PPG-20; nonionic; see Macol P-500; liq.; sol. @ 5% in min. oil, perchloroethylene; m.w. 1200; sp.gr. 1.007; visc. 160 cps; pour pt. -40 C; flash pt. (PMCC) > 350 F; ref. index 1.449.

Macol® P-1750. [PPG/Specialty Chem.] PPG; see Macol P-500; liq.; sol. @ 5% in min. oils, min. spirits, aromatic solvs., perchloroethylene; m.w. 1750; sp.gr. 1.005; visc. 195 cps; pour pt. -35 C; flash pt. (PMCC) > 350 F; ref. index 1.450.

Macol® P-2000. [PPG/Specialty Chem.] PPG-26; nonionic; see Macol P-500; liq.; sol. @ 5% in min. oils, min. spirits, aromatic solvs., perchloroethylene; m.w. 2000; sp.gr. 1.002; visc. 230 cps; pour pt. -31 C; flash pt. (PMCC) > 350 F; ref. index 1.450.

Macol® P-3000. [PPG/Specialty Chem.] PPG; see Macol P-500; liq.; sol. @ 5% in min. oils, min. spirits, perchloroethylene; m.w. 3000; sp.gr. 1.002; visc. 600 cps; pour pt. -26 C; flash pt. (PMCC) > 350 F; ref. index 1.450.

Macol® P-4000. [PPG/Specialty Chem.] PPG-30; nonionic; see Macol P-500; liq.; sol. @ 5% in min. oils, min. spirits, perchloroethylene; m.w. 4000; sp.gr. 1.002; visc. 1150 cps; pour pt. -20 C; flash pt. (PMCC) > 350 F; ref. index 1.450.

Macol® SA-2. [PPG/Specialty Chem.] Steareth-2; CAS 9005-00-9; nonionic; detergent, wetting agent, emulsifier, dispersant, solubilizer, stabilizer, coupling agent for cosmetics, textiles, metalworking lubricants, household prods., industrial uses; wh. solid; sol. @ 5% in IPA; m.p. 43 C; HLB 4.7; iodine no. 0.1; hyd. no. 158; flash pt. (PMCC) > 350 F; 100% conc.

Macol® SA-5. [PPG/Specialty Chem.] Steareth-5; CAS 9005-00-9; nonionic; see Macol SA-2; wh. solid; sol. @ 5% in IPA; m.p. 41 C; HLB 9.0; iodine no. 0.1; hyd. no. 116; flash pt. (PMCC) > 350 F; 100% conc.

Macol® SA-10. [PPG/Specialty Chem.] Steareth-10; CAS 9005-00-9; nonionic; see Macol SA-2; wh. solid; sol. @ 5% in IPA; disp. in water, min. spirits; m.p. 40 C; HLB 12.3; iodine no. 0.1; hyd. no. 80; flash pt. (PMCC) > 350 F; 100% conc.

Macol® SA-15. [PPG/Specialty Chem.] Steareth-15; CAS 9005-00-9; nonionic; see Macol SA-2; wh. solid; sol. @ 5% in water, IPA; disp. in min. spirits; m.p. 38 C; HLB 14.3; iodine no. 0.1; hyd. no. 64; flash pt. (PMCC) > 350 F; 100% conc.

Macol® SA-20. [PPG/Specialty Chem.] Steareth-20; CAS 9005-00-9; nonionic; see Macol SA-2; wh. solid; sol. @ 5% in water, IPA; m.p. 39 C; HLB 15.4; iodine no. 0.1; hyd. no. 52; flash pt. (PMCC) > 350 F; 100% conc.

Macol® SA-40. [PPG/Specialty Chem.] Steareth-40; CAS 9005-00-9; nonionic; see Macol SA-2; wh. solid; sol. @ 5% in water, IPA; m.p. 40 C; HLB 17.4; iodine no. 0.1; hyd. no. 32; flash pt. (PMCC) > 350 F; 100% conc.

Macol® TD-3. [PPG/Specialty Chem.] Trideceth-3; CAS 24938-91-8; nonionic; detergent, wetting agent, emulsifier, dispersant, solubilizer, stabilizer, coupling agent for cosmetics, textiles, metalworking lubricants, household prods., industrial uses; liq.; sol. @ 5% in min. spirits, toluene, perchloroethylene; disp. in water, min. oil; sp.gr. 0.96; visc. 17 cps; HLB 8.0; pour pt. -32 C; flash pt. (PMCC) > 250 F; 100% conc.

Macol® TD-8. [PPG/Specialty Chem.] Trideceth-8; CAS 24938-91-8; nonionic; see Macol TD-3; liq.; sol. @ 5% in water, min. oil, min. spirits, toluene, perchloroethylene; sp.gr. 1.02; visc. 50 cps; HLB 12.4; pour pt. 8 C; cloud pt. 55 C (1% aq.); flash pt. (PMCC) > 275 F; 100% conc.

Macol® TD-10. [PPG/Specialty Chem.] Trideceth-10; CAS 24938-91-8; nonionic; see Macol TD-3; liq.; sol. @ 5% in water, min. spirits, toluene, perchloroethylene; disp. in min. oil; sp.gr. 1.02; visc. 60 cps; HLB 13.6; pour pt. 10 C; cloud pt. 76 C (1% aq.); flash pt. (PMCC) > 300 F; 100% conc.

Macol® TD-12. [PPG/Specialty Chem.] Trideceth-12; CAS 24938-91-8; nonionic; see Macol TD-3; liq.; sol. @ 5% in water, toluene, perchloroethylene; disp. in min. spirits; sp.gr. 1.03; visc. 540 cps; HLB 14.1; pour pt. 14 C; cloud pt. 91 C (1% aq.); flash pt. (PMCC) > 300 F; 100% conc.

Macol® TD-15. [PPG/Specialty Chem.] Trideceth-15; CAS 24938-91-8; nonionic; see Macol TD-3.

Macol® TD-100. [PPG/Specialty Chem.] Trideceth-100; CAS 24938-91-8; nonionic; see Macol TD-3; solid; sol. @ 5% in water; sp.gr. 1.06; HLB 18.9; pour pt. 55 C; cloud pt. > 100 C (1% aq.); flash pt. (PMCC) > 300 F; 100% conc.

Macol® TD-610. [PPG/Specialty Chem.] Trideceth-6; CAS 24938-91-8; nonionic; see Macol TD-3; liq.; sol. @ 5% in min. spirits, toluene, perchloroethylene; disp. in water, min. oil; sp.gr. 0.98; visc. 115 cps; HLB 11.2; pour pt. 6 C; cloud pt. 41 C (1% aq.); flash pt. (PMCC) > 275 F; 86% conc.

Madeol AG 1989 N. [Auschem SpA] Blend; anionic; dispersing and wetting agent for pesticides; fine powd.; 100% conc.

Madeol AG BX. [Auschem SpA] Alkylnaphthalene sulfonate; anionic; wetting agent for pesticides; fine powd.; 60% conc.

Madeol AG/TR 8, AG/TR 12. [Auschem SpA] Ethoxylated alcohol, inert absorbed; nonionic; wetting agent for pesticides; fine powd.; 50% conc.

Mafo® 13. [PPG/Specialty Chem.] Potassium salt of complex n-stearyl amino acid; amphoteric; biodeg. emulsifier, wetting agent, corrosion inhibitor, suspending agent, solubilizer for difficult materials, dairy chain and metal-to-metal lubricant, emollient; for burnishing and polishing compds.; clear amber liq.; sol. in water and most popular solvs.; sp.gr. 1.015; visc. 140–200 cps; b.p. 230 F; flash pt. 230 F; pour pt. 10 C; pH 10–11 (100%); biodeg.; 70% act. in water.

Mafo® 13 MOD 1. [PPG/Specialty Chem.] Potassium salt of complex fatty amine carboxylate; amphoteric; biodeg. detergent, chelating agent, wetting agent, lubricant, emulsifier; for metal-to-metal lubricants, metal burnishing and polishing compds.; compat. with strong acid and alkali systems; liq.; water-sol.; sp.gr. 1.005; 90% solids in water.

Mafo® CAB. [PPG/Specialty Chem.] Cocamidopropyl betaine; amphoteric; detergent, dispersant, surfactant, conditioner, foam and visc. stabilizer, chelating agent, wetting agent, solubilizer, lubricant, emulsifier; used in personal care, dishwashing, rug and carpet cleaning, metalworking applics.; liq.; water-sol.; sp.gr. 1.010; 35% solids in water.

Mafo® CAB 425. [PPG/Specialty Chem.] Cocamidopropyl betaine; amphoteric; biodeg. solubilizer, wetting agent, emollient, coupling agent, emulsifier, foam booster for shampoos; liq.; sp.gr. 1.040; 42.5% solids in water.

Mafo® CAB SP. [PPG/Specialty Chem.] Cocamidopropyl betaine; amphoteric; biodeg. solubilizer, emollient, coupling agent, emulsifier, foam booster for shampoos, metalworking formulations; sp.gr. 1.050; 43% solids in water.

Mafo® CB 40. [PPG/Specialty Chem.] Coco betaine; amphoteric; biodeg. foam booster, visc. builder, solubilizer, emollient, coupling agent, emulsifier for shampoos; liq.; sp.gr. 1.040; 40% solids in water.

Mafo® CFA 35. [PPG/Specialty Chem.] Cocamidopropyl betaine; amphoteric; biodeg. solubilizer, emollient, coupling agent, emulsifier, foam booster for shampoos; sp.gr. 1.040; 35% solids in water.

Mafo® CSB 50. [PPG/Specialty Chem.] Cocamidopropyl hydroxysultaine; amphoteric; biodeg. solubilizer, emollient, coupling agent, emulsifier, foam booster, visc. builder for shampoos; liq.; sp.gr. 1.100; 50% solids in water.

Mafo® SBAO 110. [PPG/Specialty Chem.] Amphoteric; biodeg. solubilizer, emollient, coupling agent, emulsifier, foam booster for shampoos; lime soap dispersant; sp.gr. 1.049; 42% solids in water.

Makon® 4. [Stepan; Stepan Canada; Stepan Europe] Nonoxynol-4; CAS 9016-45-9; nonionic; detergent, emulsifier used in chemical specialties, cosmetic, agric., industrial and metal cleaners, textile, paper and petrol. industries; lt. straw liq.; oil-sol.; dens. 8.5 lb/gal; visc. 260 cps; pour pt. –20 C; pH 7.7 (1%); 100% act.

Makon® 6. [Stepan; Stepan Canada; Stepan Europe] Nonoxynol-6; CAS 9016-45-9; nonionic; detergent, emulsifier used in chemical specialties, cosmetic, agric., industrial and metal cleaners, textile, paper and petrol. industries; lt. straw liq.; disp. in water; dens. 8.67 lb/gal; visc. 255 cps; pour pt. –29 C; pH 7.9 (1%); 100% act.

Makon® 7. [Stepan Europe] Nonoxynol-7; CAS 9016-45-9; nonionic; detergent, wetting agent, emulsifier for all-purpose cleaners, agric., oilfield, textile formulations; water-wh. to yel. liq.; 100% act.

Makon® 8. [Stepan; Stepan Canada; Stepan Europe] Nonoxynol-8; CAS 9016-45-9; nonionic; detergent, emulsifier used in chemical specialties, cosmetic, agric., industrial and metal cleaners, textile, paper and petrol. industries; lt. straw liq.; sol. in water; dens. 8.76 lb/gal; visc. 205 cps; cloud pt. 24 C (1%); pour pt. –5 C; pH 7.0 (1%); 100% act.

Makon® 10. [Stepan; Stepan Canada; Stepan Europe] Nonoxynol-10; CAS 9016-45-9; nonionic; detergent, emulsifier used in chemical specialties, cosmetic, agric., industrial and metal cleaners, textile, paper and petrol. industries; lt. straw liq.; sol. in water; dens. 8.85 lb/gal; visc. 235 cps; cloud pt. 54 C (1%); pour pt. 2.8 C; pH 8.2 (1%); 100% act.

Makon® 11. [Stepan Europe] Nonoxynol-11; CAS 9016-45-9; nonionic; detergent, wetting agent, emulsifier for all-purpose cleaners, agric., oilfield, textile formulations; water-wh. to yel. liq.; 100% act.

Makon® 12. [Stepan; Stepan Canada; Stepan Europe] Nonoxynol-12; CAS 9016-45-9; nonionic; detergent, emulsifier used in chemical specialties, cosmetic, agric., industrial and metal cleaners, textile, paper and petrol. industries; lt. straw liq.; sol. in water; dens. 8.9 lb/gal; visc. 300 cps; cloud pt. 81 C (1%); pour pt. 12.2 C; pH 7.2 (1%); 100% act.

Makon® 14. [Stepan; Stepan Canada; Stepan Europe] Nonoxynol-14; CAS 9016-45-9; nonionic; detergent, emulsifier used in chemical specialties, cosmetic, agric., industrial and metal cleaners, textile, paper and petrol. industries; lt. straw liq.; sol. in water; dens. 8.9 lb/gal; visc. 520 cps; cloud pt. 94 C (1%); pour pt. 18.9 C; pH 7.2 (1%); 100% act.

Makon® 30. [Stepan; Stepan Canada; Stepan Europe] Nonoxynol-30; CAS 9016-45-9; nonionic; detergent, emulsifier used in chemical specialties, cosmetic, agric., industrial and metal cleaners, textile, paper and petrol. industries; off-wh. waxy solid; sol. in water; dens. 9.0 lb/gal; pour pt. 40 C; pH 8.5 (1%); 100% act.

Makon® 40. [Stepan Europe] Alkyl phenol ethoxylate (40 EO); nonionic; additive for formulation of polyvinyl emulsions for textiles; wh. flakes; 100% act.

Makon® 50. [Stepan Europe] Nonoxynol-50; CAS 9016-45-9; nonionic; detergent, dispersant, emulsifier for all-purpose cleaners and toilet deodorizing blocks, agric. formulations; water-wh. to yel. solid; 100% act.

Makon® 8240. [Stepan; Stepan Canada] PEG-36 castor oil; wetting agent, lubricant, coupling agent, defoamer for metalworking fluids, corrosion inhibitors; liq.; HLB 13.0; pH 5-7 (1%); 100% act.

Makon® NF-5. [Stepan; Stepan Canada; Stepan Europe] Polyalkoxylated aliphatic base; nonionic; low foaming detergent, wetting agent, penetrant for dishwashing, metal cleaning, metalworking fluids, corrosion inhibitors, bottle and spray washing, household/industrial cleaners, oilfield prod., textile scouring and dye baths, etc.; colorless to pale straw liq.; bland odor; sp.gr. 1.021; pour pt. < 40 F; pH 9; 97+% act.

Makon® NF-12. [Stepan; Stepan Canada; Stepan Europe] Polyalkoxylated aliphatic base; nonionic; low foaming wetting agent, detergent, lubricant, coupling agent, defoamer, and penetrant for textile applics., metalworking fluids, corrosion inhibitors, household/industrial cleaners, oilfield prod.; acid and alkali stable; water-wh. to turbid liq., bland odor; sp.gr. 0.995; pour pt. < 15 F; cloud pt. 15 C; pH 9; 100% act.

Makon® NI 10, NI 20, NI 30. [Stepan Europe] Alkylphenol alkoxylate; nonionic; emulsifier, dispersant for agric. microemulsions and emulsifiable concs.; water-wh. to yel. solid; 100% act.

Makon® OP-6. [Stepan Europe] Octoxynol-6; CAS 9002-93-1; nonionic; detergent, wetting agent, emulsifier for household/industrial cleaners; water-wh. to yel. liq.; 100% act.

Makon® OP-9. [Stepan; Stepan Canada; Stepan Europe] Octoxynol-9; CAS 9002-93-1; nonionic; surfactant for industrial cleaners, household prods., metal cleaners, janitorial prods., dairy cleaners, acid and alkaline cleaners, floor cleaners, sanitizers, laundry, textile, automotive specialties, waterless hand cleaners; water-wh. to yel. liq.; HLB 13.5; 100% act.

Manoxol IB. [Manchem] Sodium diisobutyl sulfosuccinate; anionic; wetting agent used in dispersion, agriculture, battery separators, dry cleaning, laundry; m.w. 332; sol. 76.0 g/100 ml in water; sp.gr. 1.06 (60%); surf. tens. 49.4 dynes/cm (0.1%); 100% solid.

Manoxol MA. [Manchem] Sodium dimethylamyl sulfosuccinate; anionic; wetting agent used in dispersion, agriculture, battery separators, dry cleaning, laundry; sol. 34.3 g/100 ml in water; sp.gr. 1.06 (60%); surf. tens. 49.1 dynes/cm (0.1%); 60% act.

Manoxol N. [Manchem] Sodium dinonyl sulfosuccinate; anionic; wetting agent used in dispersion, agriculture, battery separators, dry cleaning, laundry; m.w. 472; sol. 11.5 g/100 ml in water; surf. tens. 27.3 dynes/cm (0.1%); 100% solid and 60% water/alcohol sol'n.

Manoxol OT. [Manchem] Dioctyl sodium sulfosuccinate; anionic; wetting agent used in dispersion, agriculture, battery separators, dry cleaning, laundry; m.w. 444; sol. in most solvs., oils, org. media, 1.5% sol. in water; sp.gr. 1.1; surf. tens. 30 dynes/cm (0.1%); 100% act.

Manoxol OT 60%. [Manchem] Dioctyl sodium sulfosuccinate (Manoxol OT) sol'n., methylated spirits; anionic; wetting agent used in dispersion, agriculture, battery separators, dry cleaning, laundry; water-wh. liq.; water disp.; sp.gr. 1.05; flash pt. 80 F; 60% act. in 15% industrial methylated spirits and 25% water.

Manoxol OT/B. [Manchem] Dioctyl sodium sulfosuccinate (85%) and sodium benzoate (15%); anionic; wetting agent used in dispersion, agriculture, battery separators, dry cleaning, laundry; wettable powds.; powd.; 85% act.

Manro ADS 35. [Manro Prods. Ltd.] Ammonium alcohol sulfate; surfactant for household and industrial cleaning; liq.; 35% act.

Manro ALEC 27. [Manro Prods. Ltd.] Ammonium laureth sulfate (3 mole), natural based; surfactant for hair care prods., bath prods., household detergents; liq.; 27% act.

Manro ALES 60. [Manro Prods. Ltd.] Ammonium

laureth sulfate (3 mole); anionic; surfactant used in liq. detergents, cleaning prods., fire fighting foams, hair care prods.; biodeg.; amber, slightly hazy, mobile gel; slight alcoholic odor; sp.gr. 1.033; pH 7–8 (10% aq.); 60% act.

Manro ALS 30. [Manro Prods. Ltd.] Ammonium lauryl sulfate; anionic; used in cosmetics and toiletries, esp. shampoos, and household detergents; biodeg.; pale yel. clear to slightly hazy, visc. liq.; mild odor; sp.gr. 1.02; visc. 5000 cps; pH 6–7 (10% aq.); 28% min. act.

Manro AO 3OC. [Manro Prods. Ltd.] N-Alkyl dimethyl amine oxide; amphoteric; foam booster, detergent for toiletries, household and industrial cleaners; thickener for bleaches; liq.; 30% act.

Manro AT 1200. [Manro Prods. Ltd.] Blend; amphoteric; acid thickener for industrial and household cleaners, metalworking; visc. liq.; 40% act.

Manro BA Acid. [Manro Prods. Ltd.] Linear dodecylbenzene sulfonic acid; anionic; raw material used in emulsifiers, heavy and lt. duty detergents, hand cleaning gels, machine degreasers, tank cleaners; emulsion polymerization; catalyst; metalworking; dk. brn. visc. liq.; char. odor; sp.gr. 1.04; visc. 1500 cps; biodeg.; 96% conc.

Manro BES 27. [Manro Prods. Ltd.] Sodium laureth sulfate (3 mole); CAS 9004-82-4; anionic; foaming agent, emulsifier used in cosmetics, toiletries, household detergent and industrial prods.; biodeg.; pale yel., mobile liq.; mild odor; sp.gr. 1.04; visc. 200 cps; pH 6.5–7.5 (10% aq.); 26.5% min. act.

Manro BES 60. [Manro Prods. Ltd.] Sodium laureth sulfate (3 mole); CAS 9004-82-4; anionic; high foaming surfactant used in liq. detergents, cleaning prods., fire fighting foams, personal care prods.; biodeg.; amber, slightly hazy, mobile gel; slight alcoholic odor; sp.gr. 1.090; pH 7–8 (10% aq.); 60% act.

Manro BES 70. [Manro Prods. Ltd.] Sodium laureth sulfate (3 mole); CAS 9004-82-4; anionic; foaming agent, emulsifier for chlorinated solvs., lime soap dispersant, intermediate used in cosmetics, household cleaning prods. and fire fighting foams; biodeg.; pale yel., mobile gel; mild odor; sp.gr. 1.05; pH 7–9 (10% aq.); 67.5% min. act.

Manro CD. [Manro Prods. Ltd.] Cocamide DEA; anionic; solubilizer, foam stabilizer, detergent for shampoos, liq. detergents, hand cleaners; liq.; 90% conc.

Manro CD/G. [Manro Prods. Ltd.] Cocamide DEA, up to 10% glycerin; anionic; foam stabilizer and solubilizer for liq. detergents, shampoos, bubble baths; liq.; 78% conc.

Manro CDS. [Manro Prods. Ltd.] Cocamide DEA; anionic; biodeg. foam stabilizer and solubilizer for liq. detergents, shampoos, bubble baths; liq.; 85% conc.

Manro CDX. [Manro Prods. Ltd.] Cocamide DEA, 25% free amine; anionic; foam stabilizer and solubilizer; liq.; 70% conc.

Manro CMEA. [Manro Prods. Ltd.] Cocamide MEA; CAS 68140-00-1; anionic; foam booster/stabilizer for powd. and liq. detergents, hair shampoos; flake; 94% conc.

Manro DB 30. [Manro Prods. Ltd.] Blend; anionic/nonionic; detergent, foaming agent used in liq. detergents, hard surface cleaners for household and industrial use; clear golden yel., visc. liq.; negligible odor; sp.gr. 1.048; visc. 500 cps; cloud pt. 0 C; pH 6.2–6.8; 30% act.

Manro DB 56. [Manro Prods. Ltd.] Blend; anionic; for liq. detergent preps., hard surf. cleaners for household and industrial use; soft gel; 56% conc.

Manro DB 98. [Manro Prods. Ltd.] Detergent conc.; anionic/nonionic; blend for liq. detergents, hard surf. cleaners for household and industrial use; visc. liq.; 98% conc.

Manro DES 32. [Manro Prods. Ltd.] Sodium alcohol ether sulfate (2.5 mole); anionic; foamer for wallboard, plasterboard; household and industrial cleaning; liq.; 32% conc.

Manro DL 28. [Manro Prods. Ltd.] Sodium lauryl sulfate, modified; CAS 151-21-3; anionic; foaming and wetting agent used in foamable precoats and no-gel foam compds. in latex industry, emulsion polymerization; biodeg.; pale yel. clear liq.; mild char. odor; sp.gr. 1.05; visc. 50 cps; cloud pt. 0 C; pH 7–8 (10% aq.); 28% min. act.

Manro DL 32. [Manro Prods. Ltd.] Sodium dodecylbenzene sulfonate; anionic; emulsion polymerization surfactant; liq.; 32% conc.

Manro DNNS/B. [Manro Prods. Ltd.] Barium dinonylnaphthalene sulfonate; surfactant for metalworking; visc. liq.; 50% act.

Manro DNNS/C. [Manro Prods. Ltd.] Calcium dinonylnaphthalene sulfonate; surfactant for metalworking; visc. liq.; 50% act.

Manro DPM 2169. [Manro Prods. Ltd.] Blend; amphoteric; thickener for phosphoric and citirc acids; 55% conc.

Manro DS 35. [Manro Prods. Ltd.] Sodium alcohol sulfate; anionic; emulsifier, solubilizer, wetting agent used in household and industrial cleaners, metal cleaners; clear pale yel. liq.; char. odor; sp.gr. 1.07; visc. 50 cps; pH 7–8 (10% aq.); 35% act.

Manro HA Acid. [Manro Prods. Ltd.] Branched dodecylbenzene sulfonic acid; anionic; raw material for prod. of emulsifiers, heavy and lt. duty detergents, hand cleaning gels, machine degreasers, tank cleaners; catalyst; metalworking; dk. visc. liq.; char. odor; sp.gr. 1.06; visc. 18,000 cps; 95% min. act.

Manro HCS. [Manro Prods. Ltd.] Isopropylamine dodecylbenzene sulfonate; anionic; emulsifier in the mfg. of kerosene-based hand cleaning preparations, solvent degreasing, engine cleaners; amber clear visc. liq.; mild odor; sp.gr. 1.03; visc. 3000 cps; pH 5–6 (2% aq.); 89.5% min. act.

Manro MA 35. [Manro Prods. Ltd.] Sodium tallow sulfosuccinamate; anionic; foaming agent used in no-gel foam systems based on high solids, noncarboxylated SBR latex or natural rubber latex; lt. cream fluid; mild, char. odor; sp.gr. 1.05; pH 7.5–9.5 (10% aq.); 35% solids.

Manro MO 70S. [Manro Prods. Ltd.] Dioctyl sulfosuccinate in solv.; dewatering aid in oil-based systems; 70% conc.

Manro NA Acid. [Manro Prods. Ltd.] Linear dodecylbenzene sulfonic acid; anionic; raw material used as base for liq. heavy and lt. duty detergents, hand cleaning gels, machine degreasers, tank cleaners; catalyst; metalworking; dk. br. visc. liq.; char. odor; sp.gr. 1.06; visc. 2000 cps; 96% min. act.

Manro PTSA/C. [Manro Prods. Ltd.] Para toluene sulfonic acid monohydrate; catalyst in prod. of esters; curing agent for coating resins; adhesive systems; mfg. of dyes and pigments; wh. to pink crystals; 97% min. act.

Manro PTSA/E. [Manro Prods. Ltd.] Toluene sulfonic acid; anionic; catalyst for prod. of resin bound sand castings, mfg. of esters, resin prod.; intermediate for hydrotrope prod.; amber clear liq.; slight odor; sp.gr. 1.24; visc. 10 cps; 65% act.

Manro PTSA/H. [Manro Prods. Ltd.] Toluene sulfonic acid; anionic; catalyst for prod. of resin bound sand castings, mfg. of esters, resin prod.; intermediate for hydrotrope prod.; liq.; 62.5% conc.

Manro PTSA/LS. [Manro Prods. Ltd.] Toluene sulfonic acid; anionic; catalyst for prod. of resin bound sand castings, mfg. of esters, resin prod.; intermediate for hydrotrope prod.; amber clear liq., sl. odor; sp.gr. 1.24; visc. 10 cps; 65% conc.

Manro SDBS 25/30. [Manro Prods. Ltd.] Sodium dodecylbenzene sulfonate; anionic; detergent, emulsifier, foaming agent, wetting agent used in liq. detergents and cleaning prods., emulsion polymerization, rubber, plastics, textiles, insecticides; biodeg.; amber clear visc. liq.; mild odor; sp.gr. 1.05 (30 C); visc. 1000 cps (30 C); pH 7–9 (10% aq.); 29% min. act.

Manro SDBS 60. [Manro Prods. Ltd.] Sodium dodecylbenzene sulfonate, narrow cut; anionic; surfactant for liq. detergents, cleaning compds., household and industrial cleaning; paste; 60% act.

Manro SLS 28. [Manro Prods. Ltd.] Sodium lauryl sulfate; CAS 151-21-3; anionic; foaming and wetting agent used in cosmetics, toiletries, emulsion polymerization, plastics, rubber, foam rubber, carpet and upholstery shampoos; biodeg.; very pale yel. clear liq.; mild odor; sp.gr. 1.04; visc. 200–7000 cps; pH 7–8 (10% aq.); 28% min. act.

Manro TDBS 60. [Manro Prods. Ltd.] TEA dodecylbenzene sulfonate; anionic; mild detergent used in car shampoos, bubble baths, emulsion polymerization, household and industrial cleaners; amber clear visc. liq.; slight, alcoholic odor; sp.gr. 1.06; visc. 400 cps; pH 6.5–7.5 (10% aq.); 60% act.

Manro TL 40. [Manro Prods. Ltd.] TEA lauryl sulfate; anionic; foaming and wetting agent used in shampoos, bubble baths, household detergents; biodeg.; pale yel. clear liq.; mild odor; sp.gr. 1.025; visc. 100 cps; pH 6.5–7.0 (10% aq.); 40% act.

Manromid 150-ADY. [Manro Prods. Ltd.] Soyamide DEA (1:1); nonionic; thickener, emulsifier, corrosion inhibitor; for personal care prods., household and industrial cleaning, metalworking; liq.; 78% act.

Manromid 853. [Manro Prods. Ltd.] Cocamide DEA (2:1); surfactant for household and industrial cleaning, metalworking; liq.; 70% act.

Manromid 1224. [Manro Prods. Ltd.] Lauramide DEA; CAS 120-40-1; nonionic; foam booster/stabilizer for cosmetics and toiletries, household and industrial cleaning; liq.; 82% act.

Manromid CD. [Manro Prods. Ltd.] Cocamide DEA (1:1); surfactant for personal care prods., household and industrial cleaning, metalworking; liq.; 92% act.

Manromid CDG. [Manro Prods. Ltd.] Cocamide DEA (1:1); surfactant for personal care prods., household and industrial cleaning; liq.; 78% act.

Manromid CDS. [Manro Prods. Ltd.] Cocamide DEA (1:1); surfactant for personal care prods., household and industrial cleaning; liq.; 85% act.

Manromid CDX. [Manro Prods. Ltd.] Cocamide DEA (2:1); surfactant for household and industrial cleaning, metalworking; liq.; 70% act.

Manromid CMEA. [Manro Prods. Ltd.] Cocamide MEA (1:1); CAS 68140-00-1; surfactant for hair care and bath prods., household and industrial cleaning; flake; 95% act.

Manromid LMA. [Manro Prods. Ltd.] Lauramide MEA (1:1); CAS 142-78-9; nonionic; foam booster/stabilizer for solid or powd. detergents for household and industrial cleaning, hair care and bath prods.; flake; 95% act.

Manromine 853. [Manro Prods. Ltd.] Modified alkanolamide; nonionic; surfactant for waterless hand cleaners; liq.; 100% conc.

Manrosol ACS60. [Manro Prods. Ltd.] Ammonium cumene sulfonate; hydrotrope for household and industrial cleaning; liq.; 60% act.

Manrosol SCS40. [Manro Prods. Ltd.] Sodium cumene sulfonate; hydrotrope for household and industrial cleaning; liq.; 40% act.

Manrosol SCS93. [Manro Prods. Ltd.] Sodium cumene sulfonate; hydrotrope for household and industrial cleaning; powd.; 93% act.

Manrosol STS40. [Manro Prods. Ltd.] Sodium toluene sulfonate; hydrotrope for household and industrial cleaning; liq.; 40% act.

Manrosol STS90. [Manro Prods. Ltd.] Sodium toluene sulfonate; hydrotrope for household and industrial cleaning; powd.; 90% act.

Manrosol SXS30. [Manro Prods. Ltd.] Sodium xylene sulfonate; CAS 1300-72-7; anionic; hydrotrope for personal care prods., household and industrial cleaning; liq.; 30% act.

Manrosol SXS40. [Manro Prods. Ltd.] Sodium xylene sulfonate; CAS 1300-72-7; anionic; hydrotrope for personal care prods., household and industrial cleaning; liq.; 40% act.

Manrosol SXS93. [Manro Prods. Ltd.] Sodium xylene sulfonate; CAS 1300-72-7; anionic; hydrotrope for household and industrial cleaning prods.; powd.; 93% act.

Manroteric CAB. [Manro Prods. Ltd.] Cocamidopropylbetaine; amphoteric; mild high foaming base surfactant for shampoos, bubble baths, household and industrial cleaning; liq.; 30% act.

Manroteric CDX38. [Manro Prods. Ltd.] Cocoamphocarboxy glycinate; amphoteric; mild detergent for personal care prods., household detergents; visc. liq.; 39% act.

Manroteric CEM38. [Manro Prods. Ltd.] Cocoamphopropionate; amphoteric; surfactant for household and industrial detergents, metalworking; liq.; 38% act.

Manroteric CyNa50. [Manro Prods. Ltd.] Capryloamphopropionate; amphoteric; surfactant for industrial cleaning, metalworking; liq.; 50% act.

Manroteric NAB. [Manro Prods. Ltd.] N-Alkyl dimethyl betaine; amphoteric; mild surfactant, foam booster for shampoos, household and industrial cleaners; liq.; 30% act.

Manroteric SAB. [Manro Prods. Ltd.] N-Alkyl dimethyl betaine; amphoteric; surfactant for personal care prods., household and industrial cleaners; liq.; 30% act.

Manrowet MO70S. [Manro Prods. Ltd.] Dioctylsulfosuccinate, anhyd.; surfactant for industrial cleaning; liq.; 70% act.

Mapeg® 200 DC. [PPG/Specialty Chem.] PEG-4 dicocoate; nonionic; emulsifier, wetting agent, lubricant, defoamer additive; liq.; HLB 7.0; 100% conc.

Mapeg® 200 DL. [PPG/Specialty Chem.] PEG-4 dilaurate; CAS 9005-02-1; nonionic; emulsifier, dispersant used in cosmetics, pharmaceuticals, metalworking and fiber lubricants, etc.; yel. clear liq.; sol. in IPA, toluol, soybean and min. oil, water disp.;

sp.gr. 0.95; m.p. 10 C; HLB 7.6; acid no. 10 max.; sapon. no. 176–192; flash pt. (PMCC) > 350 F; 100% conc.

Mapeg® 200 DO. [PPG/Specialty Chem.] PEG-4 dioleate; CAS 9005-07-6; nonionic; emulsifier, dispersant used in cosmetics, pharmaceuticals, metalworking and fiber lubricants, etc.; yel. clear liq.; sol. in IPA, toluol, soybean and min. oil, water disp.; sp.gr. 0.95; m.p. < –10 C; HLB 6.0; acid no. 10 max.; sapon. no. 148–158; flash pt. (PMCC) > 350 F; 100% conc.

Mapeg® 200 DOT. [PPG/Specialty Chem.] PEG-4 ditallate; CAS 61791-01-3; nonionic; surfactant, emulsifier for metalworking lubricants; emollient for hair preps., creams and lotions; solubilizer for bath oils and fragrances; liq.; sol. in IPA, min. spirits, toluene, min. oil; disp. in water; sp.gr. 0.95; HLB 6.0; pour pt. –18 C; acid no. 10 max.; sapon. no. 150; flash pt. (PMCC) > 350 F.

Mapeg® 200 DS. [PPG/Specialty Chem.] PEG-4 distearate; CAS 9005-08-7; nonionic; emulsifier, dispersant used in cosmetics, pharmaceuticals, metalworking and fiber lubricants, etc.; wh. solid; sol. in IPA, toluol, soybean and min. oil; disp. in hot water; m.p. 34 C; HLB 4.7; acid no. 10 max.; sapon. no. 155–165; flash pt. (PMCC) > 350 F; 100% conc.

Mapeg® 200 ML. [PPG/Specialty Chem.] PEG-4 laurate; CAS 9004-81-3; nonionic; emulsifier, dispersant used in cosmetics, pharmaceuticals, metalworking and fiber lubricants, etc.; yel. clear liq.; sol. in IPA, toluol, soybean and min. oil, water disp.; sp.gr. 0.991; m.p. 5 C; HLB 9.3; acid no. 5 max.; sapon. no. 139–159; flash pt. (PMCC) > 350 F; 100% conc.

Mapeg® 200 MO. [PPG/Specialty Chem.] PEG-4 oleate; CAS 9004-96-0; nonionic; emulsifier, dispersant used in cosmetics, pharmaceuticals, metalworking and fiber lubricants, etc.; yel. clear liq.; sol. in IPA, toluol, soybean oil; disp. hot in water; sp.gr. 0.97; m.p. < –10 C; HLB 8.3; acid no. 5 max.; sapon. no. 115–125; flash pt. (PMCC) > 350 F; 100% conc.

Mapeg® 200 MOT. [PPG/Specialty Chem.] PEG-4 tallate; CAS 61791-00-2; nonionic; emulsifier, dispersant used in cosmetics, pharmaceuticals, metalworking and fiber lubricants, etc.; liq.; sol. @ 5% in IPA, toluene; disp. in water, min. oil, min. spirits; sp.gr. 0.98; HLB 8.3; acid no. 5 max.; sapon. no. 120; pour pt. -22 C; flash pt. (PMCC) > 350 F; 100% conc.

Mapeg® 200 MS. [PPG/Specialty Chem.] PEG-4 stearate; CAS 9004-99-3; nonionic; emulsifier, dispersant used in cosmetics, pharmaceuticals, metalworking and fiber lubricants, etc.; wh. solid; sol. in IPA, toluol, soybean oil; disp. hot in water; m.p. 33 C; HLB 8.0; acid no. 5 max.; sapon. no. 120–130; flash pt. (PMCC) > 350 F; 100% conc.

Mapeg® 400 DL. [PPG/Specialty Chem.] PEG-8 dilaurate; CAS 9005-02-1; nonionic; emulsifier, dispersant used in cosmetics, pharmaceuticals, metalworking and fiber lubricants, etc.; lt. yel. liq.; sol. in IPA, toluol, soybean oil, water disp.; sp.gr. 0.98; m.p. 18 C; HLB 10.8; acid no. 10 max.; sapon. no. 130–140; flash pt. (PMCC) > 350 F; 100% conc.

Mapeg® 400 DO. [PPG/Specialty Chem.] PEG-8 dioleate; CAS 9005-07-6; nonionic; emulsifier, dispersant used in cosmetics, pharmaceuticals, metalworking and fiber lubricants, etc.; yel. liq.; sol. in IPA, toluol, soybean and min. oil, water disp.; sp.gr. 0.98; m.p. < 7 C; HLB 8.8; acid no. 10 max.; sapon. no. 114–122; flash pt. (PMCC) > 350 F; 100% conc.

Mapeg® 400 DOT. [PPG/Specialty Chem.] PEG-8 ditallate; CAS 61791-01-3; nonionic; emulsifier, dispersant used in cosmetics, pharmaceuticals, metalworking and fiber lubricants, etc.; liq.; sol. @ 5% in IPA, min. spirits, toluene, min. oil; disp. in water; sp.gr. 0.98; HLB 8.8; acid no. 10 max.; sapon. no. 118; pour pt. 6 C; flash pt. (PMCC) > 350 F; 100% conc.

Mapeg® 400 DS. [PPG/Specialty Chem.] PEG-8 distearate; CAS 9005-08-7; nonionic; emulsifier, dispersant used in cosmetics, pharmaceuticals, metalworking and fiber lubricants, etc.; wh. solid; sol. in IPA, toluol, soybean and min. oil, hot water disp.; m.p. 36 C; HLB 8.1; acid no. 10 max.; sapon. no. 116–125; flash pt. (PMCC) > 350 F; 100% conc.

Mapeg® 400 ML. [PPG/Specialty Chem.] PEG-8 laurate; CAS 9004-81-3; nonionic; emulsifier, dispersant used in cosmetics, pharmaceuticals, metalworking and fiber lubricants, etc.; lt. yel., liq.; sol. in IPA, toluol, water; disp. in min. spirits; sp.gr. 1.01; m.p. 12 C; HLB 13.2; acid no. 5 max.; sapon. no. 89–96; flash pt. (PMCC) > 350 F; 100% conc.

Mapeg® 400 MO. [PPG/Specialty Chem.] PEG-8 oleate; CAS 9004-96-0; nonionic; emulsifier, dispersant used in cosmetics, pharmaceuticals, metalworking and fiber lubricants, etc.; yel. liq.; sol. in IPA, toluol, soybean oil, water disp.; sp.gr. 1.01; m.p. < 10 C; HLB 11.8; acid no. 5 max.; sapon. no. 80–88; flash pt. (PMCC) > 350 F; 100% conc.

Mapeg® 400 MOT. [PPG/Specialty Chem.] PEG-8 tallate; CAS 61791-00-2; nonionic; emulsifier, dispersant used in cosmetics, pharmaceuticals, metalworking and fiber lubricants, etc.; liq.; sol. @ 5% in IPA, toluene; disp. in water, min. spirits, min. oil; sp.gr. 1.01; HLB 11.8; acid no. 5 max.; sapon. no. 84; pour pt. 5 C; flash pt. (PMCC) > 350 F; 100% conc.

Mapeg® 400 MS. [PPG/Specialty Chem.] PEG-8 stearate; CAS 9004-99-3; nonionic; emulsifier, dispersant used in cosmetics, pharmaceuticals, metalworking and fiber lubricants, etc.; wh. solid; sol. in IPA, toluol, soybean oil, hot water disp.; m.p. 33 C; HLB 11.5; acid no. 5 max.; sapon. no. 84–93; flash pt. (PMCC) > 350 F; 100% conc.

Mapeg® 600 DL. [PPG/Specialty Chem.] PEG-12 dilaurate; CAS 9005-02-1; nonionic; dispersant and emulsifier for metalworking lubricants, fiber lubricants and softeners, solubilizers, defoamers, antistats, cosmetics, pharmaceuticals, and chemical intermediates; lt. yel. semisolid; sol. in IPA, toluol, soybean oil; partly sol. min. oil; disp. water; sp.gr. 0.99; HLB 12.2; m.p. 24 C; acid no. 10 max.; sapon. no. 102–112; flash pt. (PMCC) > 350 F.

Mapeg® 600 DO. [PPG/Specialty Chem.] PEG-12 dioleate; CAS 9005-07-6; nonionic; emulsifier, dispersant used in cosmetics, pharmaceuticals, metalworking and fiber lubricants, etc.; yel. liq.; sol. in IPA, toluol, soybean oil, water disp.; sp.gr. 1.00; m.p. 20 C; HLB 10.3; acid no. 10 max.; sapon. no. 92–102; flash pt. (PMCC) > 350 F; 100% conc.

Mapeg® 600 DOT. [PPG/Specialty Chem.] PEG-12 ditallate; CAS 61791-01-3; nonionic; dispersant and emulsifier for metalworking lubricants, fiber lubricants and softeners, solubilizers, defoamers, antistats, cosmetics, pharmaceuticals, and chemical intermediates; amber liq.; sol. IPA, toluol, soybean oil; disp. water; sp.gr. 1.00; HLB 10.3; m.p. 15 C; acid no. 10 max.; sapon. no. 85–95; flash pt. (PMCC) > 350 F; 100% conc.

Mapeg® 600 DS. [PPG/Specialty Chem.] PEG-12

distearate; CAS 9005-08-7; nonionic; emulsifier, dispersant used in cosmetics, pharmaceuticals, metalworking and fiber lubricants, etc.; wh. solid or flake; sol. in IPA, toluol, soybean oil, hot water disp.; m.p. 41 C; HLB 10.6; acid no. 10 max.; sapon. no. 94–104; flash pt. (PMCC) > 350 F; 100% conc.

Mapeg® 600 ML. [PPG/Specialty Chem.] PEG-12 laurate; CAS 9004-81-3; nonionic; dispersant and emulsifier for metalworking lubricants, fiber lubricants and softeners, solubilizers, defoamers, antistats, cosmetics, pharmaceuticals, and chemical intermediates; lt. yel. liq.; sol. in water, IPA, toluol; sp.gr. 1.02; HLB 14.8; m.p. 23 C; acid no. 5 max.; sapon. no. 64–74; flash pt. (PMCC) > 350 F.

Mapeg® 600 MO. [PPG/Specialty Chem.] PEG-12 oleate; CAS 9004-96-0; nonionic; emulsifier, dispersant used in cosmetics, pharmaceuticals, metalworking and fiber lubricants, etc.; yel. liq.; sol. in IPA, toluol, soybean oil, water; sp.gr. 1.03; m.p. 25 C; HLB 13.6; acid no. 5 max.; sapon. no. 60–70; flash pt. (PMCC) > 350 F; 100% conc.

Mapeg® 600 MOT. [PPG/Specialty Chem.] PEG-12 tallate; CAS 61791-00-2; nonionic; emulsifier, dispersant used in cosmetics, pharmaceuticals, metalworking and fiber lubricants, etc.; visc. 175 cps; HLB 13.6.

Mapeg® 600 MS. [PPG/Specialty Chem.] PEG-12 stearate; CAS 9004-99-3; nonionic; emulsifier, dispersant used in cosmetics, pharmaceuticals, metalworking and fiber lubricants, etc.; wh. solid; sol. in IPA, toluol, soybean oil, water, propylene glycol; disp. hot in min. oil; m.p. 36 C; HLB 13.6; acid no. 5 max.; sapon. no. 62–70; flash pt. (PMCC) > 350 F; 100% conc.

Mapeg® 1000 MS. [PPG/Specialty Chem.] PEG-20 stearate; CAS 9004-99-3; nonionic; emulsifier, dispersant used in cosmetics, pharmaceuticals, metalworking and fiber lubricants, etc.; wh. solid or flake; sol. in IPA, toluol, propylene glycol, water; m.p. 42 C; HLB 15.7; acid no. 5; sapon. no. 41–49; flash pt. (PMCC) > 350 F; 100% conc.

Mapeg® 1500 MS. [PPG/Specialty Chem.] PEG-6-32 stearate; CAS 9004-99-3; nonionic; emulsifier, dispersant used in cosmetics, pharmaceuticals, metalworking and fiber lubricants, etc.; solid; sol. @ 5% in water, IPA, toluene, propylene glycol; disp. in min. spirits; HLB 16.1; pour pt. 37 C; acid no. 5 max.; sapon. no. 62; flash pt. (PMCC) > 350 F; 100% conc.

Mapeg® 1540 DS. [PPG/Specialty Chem.] PEG-32 distearate; CAS 9005-08-7; nonionic; emulsifier, dispersant used in cosmetics, pharmaceuticals, metalworking and fiber lubricants, etc.; wh. solid or flake; sol. in IPA, toluol, soybean oil, propylene glycol, water; m.p. 45 C; HLB 14.8; acid no. 10 max.; sapon. no. 49–58; flash pt. (PMCC) > 350 F; 100% conc.

Mapeg® 6000 DS. [PPG/Specialty Chem.] PEG-150 distearate; CAS 9005-08-7; nonionic; emulsifier, dispersant used in cosmetics, pharmaceuticals, metalworking and fiber lubricants, etc.; wh. solid or flake; sol. in IPA, toluol, propylene glycol, water; m.p. 55 C; HLB 18.4; acid no. 9 max.; sapon. no. 14–20; flash pt. (PMCC) > 350 F; 100% conc.

Mapeg® CO-16. [PPG/Specialty Chem.] PEG-16 castor oil; nonionic; surfactant, emulsifier, dispersant, wetting agent, emollient for hair preps., creams and lotions; solubilizer for bath oils and fragrances.

Mapeg® CO-16H. [PPG/Specialty Chem.] PEG-16 hydrog. castor oil; CAS 61788-85-0; nonionic; surfactant, emulsifier, dispersant, wetting agent, emollient for hair preps., creams and lotions; solubilizer for bath oils and fragrances; liq.; sol. in toluene, min. oil; disp. in water, IPA, min. spirits; sp.gr. 1.010; HLB 8.6; pour pt. 7 C; acid no. 2 max.; sapon. no. 105; flash pt. (PMCC) > 350 F; 100% conc.

Mapeg® CO-25. [PPG/Specialty Chem.] PEG-25 castor oil; nonionic; surfactant; solubilizer, coupling agent for oils, solvs., waxes; for cosmetic, paper, metalworking fluid, and emulsion polymerization; liq.; sol. @ 5% in IPA; disp. in water, min. spirits, toluene; sp.gr. 1.040; HLB 10.8; acid no. 2 max.; sapon. no. 83; pour pt. 5 C; flash pt. (PMCC) > 350 F.

Mapeg® CO-25H. [PPG/Specialty Chem.] PEG-25 hydrog. castor oil; CAS 61788-85-0; nonionic; surfactant for formulation of gels; solubilizer, coupling agent; for cosmetic, paper, metalworking fluids, and emulsion polymerization; liq.; sol. @ 5% in IPA, toluene, min. oil; disp. in water, min. spirits; sp.gr. 1.040; HLB 10.8; acid no. 2 max.; sapon. no. 82; pour pt. 5 C; flash pt. (PMCC) > 350 F; 100% conc.

Mapeg® CO-30. [PPG/Specialty Chem.] PEG-30 castor oil; nonionic; surfactant, emulsifier, emollient for hair preps., creams and lotions; solubilizer for bath oils and fragrances; liq.; sol. in water, IPA, toluene; sp.gr. 1.046; HLB 11.8; pour pt. 9 C; acid no. 2 max.; sapon. no. 75; flash pt. (PMCC) > 350 F.

Mapeg® CO-36. [PPG/Specialty Chem.] PEG-36 castor oil; nonionic; surfactant, emulsifier, emollient for hair preps., creams and lotions; solubilizer for bath oils and fragrances; liq.; sol. in water, IPA, toluene; sp.gr. 1.055; HLB 12.6; pour pt. 12 C; acid no. 2 max.; sapon. no. 73; flash pt. (PMCC) > 350 F.

Mapeg® CO-200. [PPG/Specialty Chem.] PEG-200 castor oil; nonionic; surfactant, emulsifier, emollient for hair preps., creams and lotions; solubilizer for bath oils and fragrances; solid; sol. in water, IPA; HLB 18.1; pour pt. 50 C; acid no. 2 max.; sapon. no. 17.5; flash pt. (PMCC) > 350 F.

Mapeg® DGLD. [PPG/Specialty Chem.] Diethylene glycol laurate; CAS 9004-81-3; nonionic; surfactant for formation of gels; liq.; HLB 8.3; 100% conc.

Mapeg® EGDS. [PPG/Specialty Chem.] Glycol distearate; nonionic; emulsifier, dispersant used in cosmetics, pharmaceuticals, metalworking and fiber lubricants, etc.; thickener, opacifier, pearling additive; wh. solid or flake; sol. in IPA, toluol, soybean and min. oil; disp. in min. spirits; m.p. 63 C; HLB 1.4; acid no. 6 max.; sapon. no. 190–199; flash pt. (PMCC) > 350 F; 100% conc.

Mapeg® EGMS. [PPG/Specialty Chem.] Glycol stearate; nonionic; emulsifier, dispersant used in cosmetics, pharmaceuticals, metalworking and fiber lubricants, etc.; thickener, opacifier, pearling additive; wh. to cream solid or flake; sol. @ 5% in IPA, toluol, soybean and min. oil; disp. in min. spirits; m.p. 56 C; HLB 2.9; acid no. 4 max.; sapon. no. 180–188; flash pt. (PMCC) > 350 F; 100% conc.

Mapeg® EGMS-K. [PPG/Specialty Chem.] Ethylene glycol monostearate; emulsifier, dispersant; flake; 100% conc.

Mapeg® S-40. [PPG/Specialty Chem.] PEG-40 stearate; CAS 9004-99-3; nonionic; emulsifier, dispersant used in cosmetics, pharmaceuticals, metalworking and fiber lubricants, etc.; wh. solid or flake; sol. in IPA, toluol, propylene glycol, water; m.p. 48 C; HLB 17.2; acid no. 1 max.; sapon. no. 25–35; flash pt. (PMCC) > 350 F; 100% conc.

Mapeg® S-40K. [PPG/Specialty Chem.] PEG-40

stearate, kosher; CAS 9004-99-3; nonionic; emulsifier for defoamer formulations; solid/flake; HLB 16.9; 100% conc.

Mapeg® TAO-15. [PPG/Specialty Chem.] PEG-660 tallate; CAS 61791-00-2; nonionic; emulsifier, dispersant used in cosmetics, pharmaceuticals, metalworking and fiber lubricants, etc.; clear amber liq.; sol. in IPA, toluol, water; sp.gr. 1.03; m.p. 20 C; HLB 13.8; acid no. 5 max.; sapon. no. 55–65; flash pt. (PMCC) > 350 F; 100% conc.

Maphos® 15. [PPG/Specialty Chem.] Aromatic phosphate ester; nonionic/anionic; surfactant, hydrotrope, detergency aid, antistat for drycleaning and lubricant systems, emulsion polymerization; sp.gr. 1.08; acid no. 55 (to pH 5.2); flash pt. (PMCC) > 300 F; pH 2 (1% aq.); toxicology: skin irritant; 70% act., 2.8% phosphorus.

Maphos® 17. [PPG/Specialty Chem.] Aromatic phosphate ester; anionic; emulsifier for emulsion polymerization; solubilizer; yel. clear visc. liq.; sol. @ 5% in water @ pH 2.0 and 9.5; sp.gr. 1.11; acid no. 70 (to pH 5.2); pour pt. < 0 C; flash pt. (PMCC) > 300 F; pH 1.5–2.5 (1% aq.); toxicology: skin irritant; 70% act., 3.8% phosphorus.

Maphos® 18. [PPG/Specialty Chem.] Aliphatic phosphate ester; nonionic/anionic; surfactant, hydrotrope, detergency aid, antistat for drycleaning and lubricant systems, emulsion polymerization; sp.gr. 1.16; acid no. 120 (to pH 5.2); flash pt. (PMCC) > 300 F; pH 2 (1% aq.); toxicology: skin irritant; 99.5% act., 6.7% phosphorus.

Maphos® 30. [PPG/Specialty Chem.] Aliphatic phosphate ester; nonionic/anionic; surfactant, hydrotrope, detergency aid, antistat for drycleaning and lubricant systems, emulsion polymerization, metalworking; liq.; water-sol.; sp.gr. 1.10; acid no. 190 (to pH 5.2); flash pt. (PMCC) > 200 F; pH 2 (1% aq.); toxicology: skin irritant; 99.5% act., 11.1% phosphorus.

Maphos® 33. [PPG/Specialty Chem.] Aliphatic phosphate ester; nonionic/anionic; emulsifier, lubricant with anticorrosive/antifrictional props. for oil and water-sol. lubricant systems, e.g., greases, syn. cutting oils, drawing compds., chain-belt lubricants, gear oils, rust preventatives; pale yel. paste; sp.gr. 1.12; acid no. 215 (to pH 5.2); flash pt. (PMCC) > 300 F; pH 2 (1% aq.); toxicology: skin irritant; 99.5% act., 12.6% phosphorus.

Maphos® 41A. [PPG/Specialty Chem.] Aromatic phosphate ester; nonionic/anionic; emulsifier, lubricant with anticorrosive/antifrictional props. for oil and water-sol. lubricant systems, e.g., greases, syn. cutting oils, drawing compds., chain-belt lubricants, gear oils, rust preventatives; particle size modifier for emulsion polymerization; soft, waxy paste; sp.gr. 1.12; acid no. 73 (to pH 5.2); flash pt. (PMCC) > 300 F; pH 2 (1% aq.); toxicology: skin irritant; 82% act., 4.6% phosphorus.

Maphos® 54. [PPG/Specialty Chem.] Aromatic phosphate ester; nonionic/anionic; surfactant, hydrotrope, detergency aid, antistat for drycleaning and lubricant systems, emulsion polymerization; liq.; oil-sol.; sp.gr. 1.08; acid no. 70 (to pH 5.2); flash pt. (PMCC) > 300 F; pH 2 (1% aq.); toxicology: skin irritant; 99.5% act., 4.1% phosphorus.

Maphos® 55. [PPG/Specialty Chem.] Aromatic phosphate ester; nonionic/anionic; surfactant, hydrotrope, detergent, antistat for drycleaning, metalworking lubricants, emulsion polymerization; liq.; oil-sol.; sp.gr. 1.14; acid no. 110 (to pH 5.2); flash pt. (PMCC) > 300 F; pH 2 (1% aq.); toxicology: skin irritant; 80% act., 8.2% phosphorus.

Maphos® 56. [PPG/Specialty Chem.] Aliphatic phosphate ester; nonionic/anionic; surfactant, hydrotrope, detergent, dispersant, wetting agent, antistat for drycleaning, metalworking lubricants, emulsion polymerization; liq.; oil-sol.; sp.gr. 1.16; acid no. 130 (to pH 5.2); flash pt. (PMCC) > 300 F; pH 2 (1% aq.); 90% act., 7.4% phosphorus.

Maphos® 58. [PPG/Specialty Chem.] Aliphatic phosphate ester; nonionic/anionic; emulsifier, hydrotrope, dispersant, lubricant with anticorrosive/antifrictional props. for oil and water-sol. lubricant systems, e.g., greases, syn. cutting oils, drawing compds., chain-belt lubricants, gear oils, rust preventatives; lt. yel. liq.; sp.gr. 1.265; acid no. 155 (to pH 5.2); flash pt. (PMCC) > 300 F; pH 2 (1% aq.); toxicology: skin irritant; 90% act., 10% phosphorus.

Maphos® 60A. [PPG/Specialty Chem.] Complex org. phosphate acid ester; nonionic/anionic; textile wetting, hard surface detergent; lubricant, anticorrosive, dispersant, hydrotrope, solubilizer, emulsifier; metalworking; yel. clear liq.; water-sol.; sp.gr. 1.160; acid no. 175 (to pH 5.2); flash pt. (PMCC) > 300 F; pH 2 (1% aq.); toxicology: skin irritant; 99.5% act., 10.4% phosphorus.

Maphos® 66H. [PPG/Specialty Chem.] Phosphate acid ester, neutralized; nonionic/anionic; hard surface cleaning hydrotrope; antistat for solubilization of low foam and conventional surfactants in alkaline liqs.; metalworking; liq.; sp.gr. 1.15; flash pt. (PMCC) > 300 F; pH 5 (1% aq.); 50% act., 4.8% phosphorus.

Maphos® 76. [PPG/Specialty Chem.] Aromatic phosphate ester; nonionic/anionic; detergent, antistat, anticorrosive agent, emulsifer, solubilizer, hydrotrope used in drycleaning formulations, hard surface cleaners, emulsion polymerization for PVAc, acrylate polymers; yel. clear visc. liq.; sol. in dist. water @ pH 9.5 (@ 5%), in perchloroethylene (@ 10%); sp.gr. 1.09; acid no. 55 (to pH 5.2); pour pt. < 0 C; flash pt. (PMCC) > 300 F; pH 1.5–2.5 (1% aq.); toxicology: skin irritant; 99.5% act., 3.1% phosphorus.

Maphos® 76 NA. [PPG/Specialty Chem.] Partial sodium salt of aromatic phosphate ester; anionic; detergent, antistat, anticorrosive agent, solubilizer used in drycleaning formulations; emulsifier for PVAc and acrylate polymerization; lt. yel. clear liq.; sol. in dist. water @ pH 2.0 and 9.5 (@ 5%), in perchloroethylene (@ 10%); sp.gr. 1.08; pour pt. 8 C; flash pt. (PMCC) > 300 F; pH 5.0–6.0 (1% aq.); 90% act., 2.6% phosphorus.

Maphos® 77. [PPG/Specialty Chem.] Aliphatic phosphate ester; nonionic/anionic; surfactant, hydrotrope, detergency aid, antistat for drycleaning and lubricant systems, emulsion polymerization; lt. yel. clear liq.; sp.gr. 1.010; acid no. 34 (to pH 5.2); flash pt. (PMCC) > 300 F; pH 2 (1% aq.); toxicology: skin irritant; 99.5% act., 1.9% phosphorus.

Maphos® 78. [PPG/Specialty Chem.] Complex org. phosphate acid ester; anionic; hydrotrope, detergent; liq.; 100% conc.

Maphos® 79. [PPG/Specialty Chem.] Aliphatic phosphate ester; nonionic/anionic; emulsifier, lubricant with anticorrosive/antifrictional props. for oil and water-sol. lubricant systems, e.g., greases, syn. cutting oils, drawing compds., chain-belt lubricants, gear oils, rust preventatives; water-wh. clear liq.; oil-sol.; sp.gr. 1.130; acid no. 0 (to pH 5.2); flash pt.

(PMCC) > 300 F; pH 4 (1% aq.); toxicology: skin irritant; 99.5% act., 7% phosphorus.

Maphos® 91. [PPG/Specialty Chem.] Aromatic phosphate ester; nonionic/anionic; emulsifier, detergent, wetting agent, lubricant, anticorrosive, and antifrictional additive for greases, syn. cutting oils (water- and oil-based), drawing compds., chain belt lubricant, gear oils, rust preventatives; solubilizer; yel. clear visc. liq.; sol. in dist. water @ pH 2.0 and 9.5 (@ 5%), perchloroethylene (@ 10%); xylene (@ 10%); clear to boiling in 10% STPP and 5% TSP (@ 1%); clear to 30 C in 16% NaOH (@ 1%); sp.gr. 1.10; acid no. 65 (to pH 5.2); pour pt. 2 C; flash pt. (PMCC) > 300 F; pH 1.5–2.5 (1% aq.); toxicology: skin irritant; 99.5% act., 2.9% phosphorus.

Maphos® 96. [PPG/Specialty Chem.] Phosphate ester; emulsifier, detergent, lubricant with anticorrosive and antifrictional props., good high temp. stability; moderate foamer; for metalworking formulations, hard surf. cleaners, built detergents; acid no. 125 (to pH 5.2); 6.8% phosphorus.

Maphos® 151. [PPG/Specialty Chem.] Aromatic phosphate ester; nonionic/anionic; surfactant, hydrotrope, detergency aid, antistat for drycleaning and lubricant systems, emulsion polymerization; sp.gr. 1.09; acid no. 50 (to pH 5.2); flash pt. (PMCC) > 300 F; pH 2 (1% aq.); toxicology: skin irritant; 99.5% act., 2.6% phosphorus.

Maphos® 236. [PPG/Specialty Chem.] Aliphatic phosphate ester; nonionic/anionic; surfactant, hydrotrope, detergency aid, antistat for drycleaning and lubricant systems, emulsion polymerization; sp.gr. 1.04; acid no. 95 (to pH 5.2); flash pt. (PMCC) > 300 F; pH 2 (1% aq.); toxicology: skin irritant; 99.5% act., 5.4% phosphorus.

Maphos® 8135. [PPG/Specialty Chem.] Aromatic phosphate ester; nonionic/anionic; dispersant, hydrotrope, emulsifier, EP lubricant additive for greases, syn. cutting oils, drawing compds., and hard surf. cleaners; lt. amber visc. liq.; sol. in dist. water @ pH 2.0 and 9.5 (@ 5%); sp.gr. 1.07; acid no. 95 (to pH 5.2); pour pt. 2 C; flash pt. (PMCC) > 350 F; pH 1.5–2.5 (1% aq.); toxicology: skin irritant; 99.5% act., 5.1% phosphorus.

Maphos® DT. [PPG/Specialty Chem.] Aliphatic phosphate ester; nonionic/anionic; surfactant for hard surf. cleaning, textile scouring; liq.; sp.gr. 1.05; acid no. 85 (to pH 5.2); flash pt. (PMCC) > 300 F; pH 2 (1% aq.); toxicology: skin irritant; 99.5% act., 4.7% phosphorus.

Maphos® FDEO. [PPG/Specialty Chem.] Phosphate ester; nonionic/anionic; surfactant, hydrotrope, detergency aid, antistat for drycleaning and lubricant systems, emulsion polymerization; sp.gr. 1.12; acid no. 105 (to pH 5.2); flash pt. (PMCC) > 200 F; pH 2 (1% aq.); toxicology: skin irritant; 90% act., 5.7% phosphorus.

Maphos® JA 60. [PPG/Specialty Chem.] Aliphatic phosphate ester; nonionic/anionic; detergent, coupling agent; compat. with builders; for hard surf. cleaners, built detergents, metalworking fluids; liq.; sp.gr. 1.00; acid no. 110 (to pH 5.2); flash pt. (PMCC) > 300 F; pH 2 (1% aq.); toxicology: skin irritant; 99.5% act., 6.2% phosphorus.

Maphos® JM 51. [PPG/Specialty Chem.] Aliphatic phosphate ester; nonionic/anionic; emulsifier for aliphatic solvs., chlorinated solvs., aerosol propellants used in cutting oils and lubricants; liq.; sp.gr. 1.06; acid no. 50 (to pH 5.2); flash pt. (PMCC) > 300 F; pH 2 (1% aq.); toxicology: skin irritant; 99.5% act., 2.6% phosphorus.

Maphos® JM 71. [PPG/Specialty Chem.] Aromatic phosphate ester; nonionic/anionic; emulsifier for aliphatic solvs., chlorinated solvs., aerosol propellants used in cutting oils and lubricants; liq.; water-sol.; sp.gr. 1.06; acid no. 37 (to pH 5.2); flash pt. (PMCC) > 300 F; pH 2 (1% aq.); toxicology: skin irritant; 99.5% act., 2.0% phosphorus.

Maphos® JP 70. [PPG/Specialty Chem.] Aromatic phosphate ester; anionic; nonfoaming lubricant, hydrotrope, extreme pressure additive for alkaline cleaning systems, metalworking fluids; liq.; water-sol.; acid no. 97 (to pH 5.2); 100% conc.; 5.1% phosphorus.

Maphos® L 4. [PPG/Specialty Chem.] Aromatic phosphate ester; nonionic/anionic; surfactant, lubricant, anticorrosive, and antifrictional additive for greases, syn. cutting oils (water- and oil-based), drawing compds., chain belt lubricant, gear oils, rust preventatives; solubilizer; yel. clear visc. liq.; sol. in perchloroethylene, min. oil, and xylene (@ 10%); sp.gr. 1.09; acid no. 77 (to pH 5.2); pour pt. < 0 C; flash pt. (PMCC) > 300 F; pH 1.5–2.5 (1% aq.); toxicology: skin irritant; 99.5% act., 4.0% phosphorus.

Maphos® L-6. [PPG/Specialty Chem.] Complex org. phosphate acid ester; anionic; o/w dispersant, emulsion polymerization, syn. cutting fluids; yel. liq.; 100% conc.

Maphos® L 13. [PPG/Specialty Chem.] Aliphatic phosphate ester; nonionic/anionic; surfactant, lubricant, anticorrosive, coupling agent for metalworking, dry cleaning, hard surf. cleaning, dedusting, lubrication, emulsion polymerization; lt. yel. amber clear liq.; sol. in dist. water @ pH 9.5 (@ 5%), in perchloroethylene, min. oil, and xylene (@ 10%), clear to boiling in 5% TSP (@ 1%); sp.gr. 1.015; acid no. 90 (to pH 5.2); pour pt. 3 C; flash pt. (PMCC) > 300 F; pH 1.5–2.5 (1% aq.); toxicology: skin irritant; 99.5% act., 5.4% phosphorus.

Maprosyl® 30. [Stepan; Stepan Canada] Sodium n-lauroyl sarcosinate; anionic; detergent, wetting and foaming agent used in personal care and household detergent prods.; anticorrosive props.; liq.; sp.gr. 1.03; flash pt. > 200 F; 29.5–30.5% solids.

Maquat 4450-E. [Mason] Didecyl dimonium chloride; disinfectants, algicides, sanitizers, deodorant; 50% quat.

Maquat DLC-1214. [Mason] n-Alkyl dimethyl dichlorobenzyl ammonium chloride; germicide with good wetting and penetration; APHA 100 max. liq.; m.w. 425; sol. in water and most polar solvs.; dens. 8.0–8.4 lb/gal; pH 7–8 (10% sol'n.); 50 and 80% act. in water.

Maquat LC-12S. [Mason] Benzalkonium chloride; antimicrobial; sol. in water, polar solvs.; 50 and 80% act.

Maquat MC-1412. [Mason] Benzalkonium chloride; germicide with good wetting and penetration; APHA 100 max. liq.; m.w. 358; sol. in water and most polar solvs.; dens. 8.0–8.4 lb/gal; pH 7–8 (10%); 50% act. in water, 80% act. in IPA or ethanol.

Maquat MC-1416. [Mason] Benzalkonium chloride; germicide with good wetting and penetration; APHA 100 max. liq.; m.w. 380; sol. in water and most polar solvs.; dens. 7.8–8.2 lb/gal; pH 7–8 (10% sol'n.); 50% act. in water; 80% act. in IPA or ethanol.

Maquat MC-6025-50%. [Mason] Benzalkonium chloride; germicide with good wetting and penetration; liq.; 50% act.

Maquat MQ-2525. [Mason] Benzalkonium chloride (25%), alkyl dimethyl ethylbenzyl ammonium chloride (25%); antimicrobial with hard water tolerance; sol. in water and polar solvs.; 50 and 80% act.

Maquat MQ-2525M. [Mason] Benzalkonium chloride (A) and n-alkyl dimethyl ethylbenzyl ammonium chloride (B); germicide with good wetting and penetration; APHA 100 max. liq.; m.w. 384; sol. in water, polar solvs.; dens. 7.8–8.2 lb/gal; pH 7–8 (10% sol'n.); 50% act. (25% A, 25% B) in water or 80% act. (40% A, 40% B) in 20% IPA.

Maquat SC-18. [Mason] Stearalkonium chloride; germicide with good wetting and penetration; paste, flake, or powd.; 25%, 85% and 94% act. resp.

Maquat SC-1632. [Mason] Cetearyl alcohol, PEG-40 castor oil, stearalkonium chloride; germicide with good wetting and penetration; sol. in water, polar solvs.

Maquat TC-76. [Mason] Benzalkonium chloride (42%) and n-dialkyl methyl benzyl ammonium chloride (8%); germicide with good wetting and penetration; liq.; 50% act.

Maramul SS. [LignoTech] Lignosulfonate; anionic; emulsifier for asphalt emulsions; powd., liq.; 45-100% conc.

Maranil A. [Henkel Canada] Linear dodecylbenzene sulfonate; anionic; detergent base for household and industrial cleansers; dishwashing liq.; powd.; 90% conc.

Maranil ABS. [Henkel Canada] Dodecylbenzene sulfonic acid; anionic; dispersant, emulsifier, emulsion breaker; in household and industrial cleaners; biodeg.; liq.; 96% conc.

Maranil CB-22. [Pulcra SA] TEA dodecylbenzene sulfonate; anionic; surfactant for foam baths, liq. detergents, carwashes, heavy duty detergents; emulsifier for carnauba wax and pine oil, emulsion polymerization; pigment dispersant; liq.; 50% conc.

Maranil DBS. [Henkel/Functional Prods.; Pulcra SA] Dodecylbenzene sulfonic acid; anionic; intermediate for mfg. of liq., powd., or paste detergents, emulsifiers, textile auxiliaries; acid catalyst; liq.; > 98% solids.

Maranil Paste A 55, A 75. [Henkel KGaA] Sodium dodecylbenzene sulfonate; anionic; base for mfg. detergents, dishwashing, cleaning agents; wetting agent; paste, powd.; 55, 75% conc. resp.

Maranil Powd. A. [Henkel/Functional Prods.] Sodium dodecylbenzene sulfonate; anionic; base for mfg. detergents, dishwashes, cleaning agents; wetting agent; emulsifier for PVC copolymers, carboxylated S/B latexes; powd.; 86-90% act.

Marasperse 52 CP. [LignoTech] Lignosulfonate; anionic; dispersant for dyestuffs; low stain, high heat stability; powd.; 100% conc.

Marasperse N-22. [LignoTech] Sodium lignosulfonate; anionic; dispersant, o/w emulsion stabilizer, emulsifier; mfg. of disperse dyes for dyeing acetate and polyesters; dispersant and sequestering agent in cooling water treatments; agric. chemical formulations; gypsum board additive; industrial cleaners; brn. powd.; water-sol.; dens. 35–40 lb/ft^3; surf. tens. 52.8 dynes/cm (1%); pH 7.5–8.5 (3%); 100% conc.

Marchon® DC 1102. [Albright & Wilson UK] Formulated prod.; nonionic; shale and cuttings wash cleaner for oilfield industry; liq.; 81% conc.

Markwet 851. [Ivax Industries] Amphoteric; thermodegradable surfactant, emulsifier, wetting agent, stabilizer for water repellent and fluorochemical formulations, foam finish applics.; clear to lt. golden liq.; sol. in water; pH 7.5 ± 1; usage level: 1-3%.

Markwet 3003. [Ivax Industries] Surfactant; anionic; surfactant blend giving fast wetting with low foaming in alkaline dyebaths; dispersant for dyestuffs; esp. for continuous dyeing of vat, sulfur or indigo dyes; clear liq.; disp. in water; dens. 9.1 lb/gal; pH 10 ± 0.5; usage level: 0.75-1.25%.

Markwet FA-43. [Ivax Industries] Foaming aid for foam finishing and water repellent applics.; reduces color transfer on polyester/cotton dyed yarn; clear to sl. hazy golden liq.; sol. in warm water; pH 7 ± 1; usage level: 1-3%.

Markwet NR-25. [Ivax Industries] Ethoxylated alcohol blend; nonionic; textile penetrant and detergent imparting nonrewetting chars.; wetting agent and detergent in alkaline, acid, and neutral media; does not promote color transfer of dispersed dyes; water clear liq.; sol. in hot or cold water; pH 4 ± 0.5; usage level: 0.25-0.5%.

Markwet WL-12. [Ivax Industries] Surfactant; anionic; wetting agent, dispersant for dyestuffs, pigments and dirt; solubilizer; stable in alkaline and mild acid media; water clear to lt. straw liq.; sol. in water; dens. 8.8 lb/gal; pH 9-10; usage level: 0.1-2.5%.

Marlamid® A 18. [Hüls Am.; Hüls AG] Stearamidoethyl ethanolamine; CAS 141-21-9; cationic; base for textile aux. agents and softeners; waxy flakes; 100% act.

Marlamid® A 18 E. [Hüls Am.; Hüls AG] Stearic acid alkanolamide; cationic; base for textile aux. agents and softeners; waxy flakes; 100% act.

Marlamid® AS 18. [Hüls Am.; Hüls AG] Stearic acid alkanolamide/fatty alcohol polyglycol ether; cationic; base for textile aux. agents and softeners; waxy flakes; 100% act.

Marlamid® D 1218. [Hüls Am.; Hüls AG] Cocamide DEA; nonionic; foam stabilizer, thickener, superfatting agent for liq. detergents, shampoos; increases soil suspending power in felting auxiliaries; liq.; 100% conc.

Marlamid® D 1885. [Hüls Am.; Hüls AG] Oleamide DEA; CAS 93-83-4; nonionic; foam stabilizer, thickener, superfatting agent for liq. detergents, shampoos; liq.; water-sol.; 100% act.

Marlamid® DF 1218. [Hüls Am.; Hüls AG] Cocamide DEA; CAS 61790-63-4; nonionic; foam stabilizer, thickener, superfatting agent for liq. detergents, shampoos; liq.; water-sol; 100% act.

Marlamid® DF 1818. [Hüls Am.; Hüls AG] Soyamide DEA; CAS 68425-47-8; nonionic; foam stabilizer, thickener, superfatting agent for liq. detergents, shampoos; liq.; 100% act.

Marlamid® KL. [Hüls Am.; Hüls AG] Cocamidopropyl lauryl ether; nonionic; pearlescent surfactant, foam stabilizer, thickener, opacifier for liq. and paste detergents, shampoos; flakes; insol. in water; 100% conc.

Marlamid® KLP. [Hüls Am.; Hüls AG] Cocamidopropyl lauryl ether, sodium laureth sulfate; anionic/nonionic; pearlescent base for shampoos, bubble baths, liq. soaps; fluid paste; 30% act.

Marlamid® M 1218. [Hüls Am.; Hüls AG] Cocamide MEA; CAS 68140-00-1; nonionic; foam stabilizer, thickener in household, personal, industrial detergents, superfatting agent; flake; water-insol.; 100% act.

Marlamid® M 1618. [Hüls Am.; Hüls AG] Tallowamide MEA; CAS 68153-63-9; nonionic;

foam stabilizer, thickener for household, personal, and industrial detergents; flake; 100% act.

Marlamid® O 18. [Hüls AG] Fatty acid alkanolamide; intermediate for textile preparing agents and softeners.

Marlamid® OS 18. [Hüls AG] Fatty acid alkanolamide; starting material for mfg. of fabric softeners.

Marlamid® PG 20. [Hüls Am.; Hüls AG] Cocamide MEA, glycol ditallowate; nonionic; pumpable pearlescent base for shampoos, bubble baths, liq. soaps; fluid paste; 21% act.

Marlamid® PGT. [Hüls AG] Fatty acid alkanolamide; pearlescent.

Marlazin® 7102. [Hüls Am.; Hüls AG] Fatty amine/fatty alcohol polyglycol ether blend; cationic; thickener, wetting agent for mfg. of visc. acid cleaners, sanitary cleaners; liq.; 100% act.

Marlazin® 7265. [Hüls Am.; Hüls AG] Fatty alkylimidazline/fatty alkyl ammonium salt blend; cationic; surfactant used in the mfg. of prods. having water-repellent properties; hydrophobic agent for painted surfaces and car rinses; liq.; 81% act.

Marlazin® 8567. [Hüls Am.] Aromatic amine; cationic; surfactant used in floor cleaners and polymer strippers; liq.; 100% act.

Marlazin® KC 21/50. [Hüls Am.; Hüls AG] Cocoalkonium chloride; cationic; bactericide; prod. of disinfectant cleaning agents; liq.; 48% act.

Marlazin® L 2. [Hüls Am.; Hüls AG] Lauryl amine polyglycol ether; nonionic; industrial cleaner, dyeing assistant; alkali and acid resistant; liq.; 100% conc.

Marlazin® L 10. [Hüls Am.; Hüls AG] PEG-10 lauramine; cationic; surfactant for prod. of low-foaming acidic cleaners, textile auxiliaries; very high acid resist.; liq.; cloud pt. 65 C (0.1% in 6% NaOH); 100% act.

Marlazin® L 410. [Hüls Am.] Fatty amine polyglycol ether; nonionic; surfactant for industrial cleaners, in dye and fiber auxiliaries; liq.; 50% act.

Marlazin® OK 1. [Hüls Am.] Alkanolamine fatty condensate; cationic; surfactant for the mfg. of prods. having water-repellent properties; solid; 100% act.

Marlazin® OL 2. [Hüls Am.; Hüls AG] PEG-2 oleamine; CAS 26635-93-8; cationic; detergent for industrial cleaners, acid cleaners, textile auxiliaries; resistant to acids and alkalies; liq.; 100% act.

Marlazin® OL 20. [Hüls Am.; Hüls AG] PEG-20 oleamine; cationic; surfactant for prod. of acid cleaners, preparation agents; liq.; cloud pt. 70 C (2% in 10% NaCl); 100% act.

Marlazin® S 10. [Hüls Am.; Hüls AG] PEG-10 stearamine; CAS 26635-92-7; cationic; detergent for industrial cleaners, acid cleaners, textile auxiliaries; resistant to acids and alkalies; liq.; cloud pt. 72 C (2% in 10% NaCl); 100% act.

Marlazin® S 40. [Hüls Am.; Hüls AG] PEG-40 stearamine; CAS 26635-92-7; cationic; detergent for industrial cleaners, textile auxiliaries; resistant to acids and alkalies; liq.; cloud pt. 84 C (2% in 10% NaCl); 100% act.

Marlazin® T 10. [Hüls Am.; Hüls AG] PEG-10 tallowamine; cationic; detergent for industrial cleaners, textile and dyeing auxiliaries; resistant to acids and alkalies; liq.; cloud pt. 71 C (2% in 10% NaCl); 100% act.

Marlazin® T 15/2. [Hüls AG] PEG-15 tallowamine; cationic; surfactant for prod. of textile and dyeing auxiliaries; liq./paste; 100% act.

Marlazin® T 16/1. [Hüls AG] PEG-16 tallowamine; cationic; surfactant for prod. of textile and dyeing auxiliaries; solid; 100% act.

Marlazin® T 50. [Hüls Am.; Hüls AG] PEG-50 tallowamine; nonionic; detergent, dispersant for industrial cleaners, textile auxiliaries, fabric softeners; resistant to acids and alkalies; liq.; cloud pt. 85 C (2% in 10% NaCl); 70% act.

Marlazin® T 410. [Hüls Am.] Fatty amine polyglycol ether; nonionic; detergent for industrial cleaners, textile auxiliaries; resistant to acids and alkalies; liq.; 50% act.

Marlican®. [Hüls AG] Straight-chain dodecylbenzene; CAS 67774-74-7; anionic; detergent intermediate, solubilizer; sec. plasticizer; reduces visc. of PVC pastes; solv. for carbonless copy papers; biodeg.; liq.; 100% act.

Marlinat® 24/28. [Hüls AG] Anionic; surfactant for detergents, cleaners, cosmetics.

Marlinat® 24/70. [Hüls AG] Anionic; surfactant for detergents, cleaners, cosmetics.

Marlinat® 242/28. [Hüls Am.; Hüls AG] Sodium laureth (2) sulfate; CAS 9004-82-4; anionic; strongly foaming base surfactant for detergents, shampoos, liq. soaps; liq.; 28% act.

Marlinat® 242/70. [Hüls Am.; Hüls AG] Sodium laureth (2 EO) sulfate; CAS 9004-82-4; anionic; strongly foaming base surfactant for detergents, shampoos, liq. soaps; liq./paste; 70% act.

Marlinat® 242/70 S. [Hüls AG] Sodium laureth (2 EO) sulfate; CAS 9004-82-4; anionic; strongly foaming base surfactant for detergents, shampoos, liq. soaps; dioxane content < 10 ppm; liq./paste; 70% act.

Marlinat® 243/28. [Hüls Am.; Hüls AG] Sodium laureth (3) sulfate; CAS 9004-82-4; anionic; strong foaming surfactant for detergents, shampoos, cleaners; liq.; 28% act.

Marlinat® 243/70. [Hüls Am.; Hüls AG] Sodium laureth (3) sulfate; CAS 9004-82-4; anionic; strongly foaming surfactant for detergents, shampoos, cleaners; liq./paste; 70% act.

Marlinat® 5303. [Hüls AG] Wetting agent for high electrolyte conc. sol'ns.

Marlinat® CM 20. [Hüls Am.; Hüls AG] C12-14 fatty alcohol ether carboxylic acid (2 EO); anionic; surfactant for mild detergents, shampoos, cleaners; liq.; 90% act.

Marlinat® CM 40. [Hüls Am.; Hüls AG] Laureth-5 carboxylic acid; anionic; surfactant for mild detergents, shampoos, cleaners; liq.; 90% act.

Marlinat® CM 45. [Hüls Am.; Hüls AG] Sodium laureth-5 carboxylate; anionic; surfactant for prod. of very mild detergents; insensitive to water hardness; liq.; 23% act.

Marlinat® CM 100. [Hüls Am.; Hüls AG] Laureth-11 carboxylic acid; anionic; surfactant for mild detergents, shampoos, cleaners; 90% act.

Marlinat® CM 105. [Hüls Am.; Hüls AG] Sodium laureth-11 carboxylate; anionic; surfactant for prod. of very mild detergents; insensitive to water hardness; liq.; 23% act.

Marlinat® CM 105/80. [Hüls AG] Sodium laureth-11 carboxylate; anionic; surfactant for prod. of very mild detergents; insensitive to water hardness; liq.; 80% act.

Marlinat® DF 8. [Hüls Am.; Hüls AG] Dioctyl sodium sulfosuccinate; CAS 577-11-7; anionic; highly act. wetting agent for textile, paint and paper industries used in cleaners, cosmetic preparations;

liq.; 65% act.

Marlinat® DFK 30. [Hüls Am.; Hüls AG] Sodium lauryl sulfate; CAS 151-21-3; anionic; base surfactant for hair shampoos, foam baths, shower foams, liq. soaps; liq.; 30% act.

Marlinat® DFL 40. [Hüls AG] TEA lauryl sulfate; anionic; finely porous foaming surfactant; liq.; 40% act.

Marlinat® DFN 30. [Hüls Am.; Hüls AG] Ammonium lauryl sulfate; anionic; base surfactant for hair shampoos, foam baths, shower foams, liq. soaps; liq.; 30% act.

Marlinat® HA 12. [Hüls AG] Sodium salt of lauric acid monoethanolamide sulfosuccinate; anionic; detergent raw material used in cleaners of sensitive textiles, cosmetic detergents; paste; 45% act.

Marlinat® SL 3/40. [Hüls Am.; Hüls AG] Disodium laureth sulfosuccinate; anionic; base surfactant for hair shampoos, foam baths, shower foams, liq. soaps; liq.; 30% act.

Marlinat® SRN 30. [Hüls Am.; Hüls AG] Disodium lauramido MEA-sulfosuccinate, sodium C12-14 olefin sulfonate; anionic; base surfactant for carpet and upholstery cleaners; liq.; 30% act.

Marlipal® 1/12. [Hüls AG] PEG methyl ether; nonionic; surfactant for prod. of methyl-terminated fatty acid esters; liq.; 100% act.

Marlipal® 011/30. [Hüls AG] C11-oxoalcohol polyglycol ether; nonionic; wetting agent, detergent, dispersant; homogenizing capacity; liq.; HLB 8.6; 100% conc.

Marlipal® 011/50. [Hüls AG] C11-oxoalcohol polyglycol ether; nonionic; wetting agent, detergent, dispersant; homogenizing capacity; liq.; HLB 11.2; 100% conc.

Marlipal® 011/79. [Hüls AG] C11-oxoalcohol polyglycol ether; nonionic; wetting agent, detergent, dispersant; homogenizing capacity; liq.; HLB 12.8; 90% conc.

Marlipal® 011/88. [Hüls AG] C11-oxoalcohol polyglycol ether; nonionic; wetting agent, detergent, dispersant; homogenizing capacity; liq.; HLB 13.4; 80% conc.

Marlipal® 011/110. [Hüls AG] C11-oxoalcohol polyglycol ether; nonionic; wetting agent, detergent, dispersant; homogenizing capacity; paste; HLB 14.8; 100% conc.

Marlipal® 013/20. [Hüls Am.; Hüls AG] C13-oxoalcohol polyglycol ether; CAS 63011-36-5; nonionic; surfactant for detergents; emulsifier and aux. agent in textile processing; liq.; HLB 6.1; 100% conc.

Marlipal® 013/30. [Hüls Am.; Hüls AG] C13-oxoalcohol polyglycol ether; nonionic; surfactant for detergents; emulsifier and aux. agent in textile processing; liq.; HLB 8.0; 100% conc.

Marlipal® 013/50. [Hüls Am.; Hüls AG] C13-oxoalcohol polyglycol ether; nonionic; surfactant for detergents; emulsifier and aux. agent in textile processing; liq.; HLB 10.5; 100% conc.

Marlipal® 013/60. [Hüls Am.; Hüls AG] C13-oxoalcohol polyglycol ether; nonionic; surfactant for detergents; emulsifier and aux. agent in textile processing; liq.; HLB 11.4; 100% conc.

Marlipal® 013/70. [Hüls Am.; Hüls AG] C13-oxoalcohol polyglycol ether; nonionic; surfactant for detergents; emulsifier and aux. agent in textile processing; liq.; HLB 12.1; 100% conc.

Marlipal® 013/80. [Hüls Am.; Hüls AG] C13-oxoalcohol polyglycol ether; nonionic; surfactant for detergents; emulsifier and aux. agent in textile processing; liq.; HLB 12.8; 100% conc.

Marlipal® 013/90. [Hüls Am.; Hüls AG] C13-oxoalcohol polyglycol ether; nonionic; surfactant for detergents; emulsifier and aux. agent in textile processing; liq.; HLB 13.3; 100% conc.

Marlipal® 013/100. [Hüls Am.; Hüls AG] C13-oxoalcohol polyglycol ether; CAS 9043-30-5; nonionic; surfactant for detergents; emulsifier and aux. agent in textile processing; liq.; HLB 13.7; 100% conc.

Marlipal® 013/120. [Hüls Am.; Hüls AG] C13-oxoalcohol polyglycol ether; nonionic; surfactant for detergents; emulsifier and aux. agent in textile processing; liq.; HLB 14.5; 100% conc.

Marlipal® 013/150. [Hüls Am.; Hüls AG] C13-oxoalcohol polyglycol ether; nonionic; surfactant for detergents; emulsifier and aux. agent in textile processing; paste; HLB 15.3; 100% conc.

Marlipal® 013/170. [Hüls Am.; Hüls AG] C13-oxoalcohol polyglycol ether; nonionic; surfactant for detergents; emulsifier and aux. agent in textile processing; liq.; HLB 15.8; 100% conc.

Marlipal® 013/200. [Hüls Am.; Hüls AG] C13-oxoalcohol polyglycol ether; nonionic; surfactant for detergents; emulsifier and aux. agent in textile processing; wax; HLB 16.3; 100% conc.

Marlipal® 013/400. [Hüls Am.; Hüls AG] C13-oxoalcohol polyglycol ethers; nonionic; surfactant in detergents; emulsifier and aux. agent in textile processing; finishing and wetting agent; wax; HLB 18.0; 100% conc.

Marlipal® 013/939. [Hüls AG] C13-oxoalcohol polyglycol ether blend; nonionic; low foaming surfactant for washing powds.; liq.; HLB 18.0; 90% conc.

Marlipal® 24/20. [Hüls Am.; Hüls AG] Laureth-2; CAS 68439-50-9; nonionic; dispersant, wetting agent, emulsifier, detergent for washing, cleaning, soil suspension, textile pretreating and dyeing; liq.; HLB 6.2; cloud pt. 50 C (10% in 25% BDG); 100% act.

Marlipal® 24/30. [Hüls Am.; Hüls AG] Laureth-3; nonionic; dispersant, wetting agent, emulsifier, detergent for washing, cleaning, soil suspension, textiles; liq.; HLB 8.1; cloud pt. 60 C (10% in 25% BDG); 100% act.

Marlipal® 24/40. [Hüls Am.; Hüls AG] Laureth-4; nonionic; dispersant, wetting agent, emulsifier, detergent for washing, cleaning, soil suspension, textiles; liq.; HLB 9.5; cloud pt. 67 C (10% in 25% BDG); 100% act.

Marlipal® 24/50. [Hüls Am.; Hüls AG] Laureth-5; nonionic; dispersant, wetting agent, emulsifier, detergent for washing, cleaning, soil suspending, and homogenizing applics., textile processing; liq.; HLB 10.6; cloud pt. 73 C (10% in 25% BDG); 100% act.

Marlipal® 24/60. [Hüls Am.; Hüls AG] Laureth-6; nonionic; dispersant, wetting agent, emulsifier, detergent for washing, cleaning, soil suspending, and homogenizing applics., textile processing; liq.; HLB 11.5; cloud pt. 77 C (10% in 25% BDG); 100% act.

Marlipal® 24/70. [Hüls Am.; Hüls AG] Laureth-7; nonionic; dispersant, wetting agent, emulsifier, detergent for washing, cleaning, soil suspension, textiles; liq.; HLB 12.3; cloud pt. 54 C (2% aq.); 100% act.

Marlipal® 24/80. [Hüls Am.; Hüls AG] Laureth-8;

nonionic; dispersant, wetting agent, emulsifier, detergent for washing, cleaning, soil suspending, and homogenizing applics., textiles; liq.; HLB 12.9; cloud pt. 67 C (2% aq.); 100% act.

Marlipal® 24/90. [Hüls Am.; Hüls AG] Laureth-9; nonionic; dispersant, wetting agent, emulsifier, detergent for washing, cleaning, soil suspending, and homogenizing applics., textiles; liq./paste; HLB 13.4; cloud pt. 82 C (2% aq.); 100% act.

Marlipal® 24/100. [Hüls Am; Hüls AG] Laureth-10; nonionic; dispersant, wetting agent, emulsifier, detergent for washing, cleaning, soil suspension, textiles; paste; HLB 13.9; cloud pt. 56 C (2% in 10% NaCl); 100% act.

Marlipal® 24/110. [Hüls Am.; Hüls AG] Laureth-11; nonionic; dispersant, wetting agent, for washing, cleaning, soil suspending, and homogenizing applics.; paste; HLB 14.3; 100% conc.

Marlipal® 24/120. [Hüls Am.; Hüls AG] Laureth-12; nonionic; dispersant, wetting agent, for washing, cleaning, soil suspending, and homogenizing applics.; paste; HLB 14.6; 100% conc.

Marlipal® 24/140. [Hüls Am.; Hüls AG] Laureth-14; nonionic; dispersant, wetting agent, for washing, cleaning, soil suspending, and homogenizing applics.; paste; HLB 15.2; 100% conc.

Marlipal® 24/150. [Hüls AG] Laureth-15; nonionic; dispersant, wetting agent, emulsifier, detergent for washing, cleaning, soil suspension, textiles; paste; cloud pt. 73 C (2% in 10% NaCl); 100% act.

Marlipal® 24/200. [Hüls Am.; Hüls AG] Laureth-20; nonionic; dispersant, wetting agent, emulsifier, detergent for washing, cleaning, soil suspension, textiles; waxy solid; HLB 16.4; cloud pt. 76 C (2% in 10% NaCl); 100% act.

Marlipal® 24/300. [Hüls Am.; Hüls AG] Laureth-30; nonionic; dispersant, wetting agent, emulsifier, detergent for washing, cleaning, soil suspension, textiles; waxy solid; HLB 17.4; cloud pt. 76 C (2% in 10% NaCl); 100% act.

Marlipal® 24/939. [Hüls AG] C12-14 alcohol ethoxylate blend; nonionic; dispersant, wetting agent, emulsifier, detergent for washing, cleaning, soil suspension, textiles; liq.; cloud pt. 75 C (10% in 25% BDG); 90% act.

Marlipal® 34/30, /50, /60, /70, /79, /99, /100, /109, / 110, /119, /120, /140, /200, /300, /1039. [Hüls Am.] C13/C14-oxoalcohol polyglycol ethers; nonionic; surfactants in powd. and liq. detergents for household and industry, emulsifiers and wetting agents in textile industry; liq. to solid; 100% act.

Marlipal® 104. [Hüls Am.; Hüls AG] Linear C10 alcohol polyglycol ether; CAS 26183-52-8; nonionic; wetting surfactant; esp. for hard surf. cleaning; liq.; HLB 10.5; cloud pt. 71 C (5% in 25% BDG); 100% act.

Marlipal® 124. [Hüls Am.; Hüls AG] Laureth-4; CAS 9002-92-0; nonionic; surfactant for cosmetics, textile aux. agents; solubilizer for oils and perfumes; liq.; HLB 9.7; cloud pt. 68 C (10% in 25% BDG); 100% act.

Marlipal® 129. [Hüls Am.; Hüls AG] Laureth-9; nonionic; surfactant; solubilizer for oils and perfumes; paste; sol. in oil and water; cloud pt. 71 C (2% aq.); 100% act.

Marlipal® 1012/4. [Hüls Am.] Deceth-4; nonionic; wetting surfactant, esp. for hard surf. cleaning; liq.; cloud pt. 66 C (2% in 10% NaCl); 100% act.

Marlipal® 1012/6. [Hüls Am.; Hüls AG] Linear C10-12 alcohol polyglycol ether (6 EO); nonionic; wetting surfactant, esp. for hard surf. cleaning; liq.; HLB 12.1; cloud pt. 53 C (2% aq.); 100% act.

Marlipal® 1218/5. [Hüls Am.] Fatty alcohol polyglycol ethers; nonionic; surfactants; liq. to wax; sol. in oil and water; 100% act.

Marlipal® 1618/6. [Hüls AG] Ceteareth-6; CAS 68439-49-6; nonionic; dispersant; prod. of powd. detergents; binding agent and base material for solid cleaning agents (toilet sticks); coating material for foam suppressant, enzymes, etc.; dyeing auxiliaries; solid; cloud pt. 75 C (10% in 25% BDG); 100% act.

Marlipal® 1618/8. [Hüls Am.; Hüls AG] Ceteareth-8; CAS 68439-49-6; nonionic; dispersant; prod. of powd. detergents; binding agent and base material for solid cleaning agents (toilet sticks); coating material for foam suppressant, enzymes, etc.; dyeing auxiliaries; waxy solid; HLB 11.6; cloud pt. 83 C (10% in 25% BDG); 100% act.

Marlipal® 1618/10. [Hüls Am.; Hüls AG] Ceteareth-10; CAS 68439-49-6; nonionic; dispersant; prod. of powd. detergents; binding agent and base material for solid cleaning agents (toilet sticks); coating material for foam suppressant, enzymes, etc.; dyeing auxiliaries; waxy solid; HLB 12.6; cloud pt. 70 C (2% aq.); 100% act.

Marlipal® 1618/11. [Hüls Am.; Hüls AG] Ceteareth-11; CAS 68439-49-6; nonionic; dispersant; prod. of powd. detergents; binding agent and base material for solid cleaning agents (toilet sticks); coating material for foam suppressant, enzymes, etc.; dyeing auxiliaries; waxy solid; HLB 13.1; cloud pt. 87 C (2% aq.); 100% act.

Marlipal® 1618/18. [Hüls AG] Ceteareth-18; CAS 68439-49-6; nonionic; dispersant; prod. of powd. detergents; binding agent and base material for solid cleaning agents (toilet sticks); coating material for foam suppressant, enzymes, etc.; dyeing auxiliaries; solid; cloud pt. 74 C (2% in 10% NaCl); 100% act.

Marlipal® 1618/25. [Hüls Am.; Hüls AG] Ceteareth-25; CAS 68439-49-6; nonionic; surfactant for mfg. of cleaners in block form, dyeing assistants; wax flakes or powd.; HLB 16.2; cloud pt. 77 C (2% in 10% NaCl); 100% act.

Marlipal® 1618/25 P 6000. [Hüls Am.; Hüls AG] Ceteareth-25/polyethylene glycol blend; nonionic; surfactant for mfg. of textile processing agents; spray-dried powd.; sol. in cold water; cloud pt. 77 C (2% in 10% NaCl); 100% act.

Marlipal® 1618/40. [Hüls Am.; Hüls AG] Ceteareth-40; CAS 68439-49-6; nonionic; surfactant for mfg. of cleaners in block form, dyeing assistants; wax flakes and powd.; HLB 17.5; cloud pt. 77 C (2% in 10% NaCl); 100% act.

Marlipal® 1618/80. [Hüls AG] Ceteareth-80; CAS 68439-49-6; nonionic; dispersant; prod. of powd. detergents; binding agent and base material for solid cleaning agents (toilet sticks); coating material for foam suppressant, enzymes, etc.; dyeing auxiliaries; flakes; cloud pt. 77 C (2% in 10% NaCl); 100% act.

Marlipal® 1850/5. [Hüls Am.; Hüls AG] Oleth-5; CAS 9004-98-2; nonionic; surfactant for mfg. of washing powds., textile auxiliaries; liq.; cloud pt. 74 C (5% in 25% BDG); 100% act.

Marlipal® 1850/10. [Hüls Am.; Hüls AG] Oleth-10; CAS 9004-98-2; nonionic; surfactant for mfg. of washing powds., textile auxiliaries; paste; cloud pt. 70 C (2% aq.); 100% act.

Marlipal® 1850/30. [Hüls Am.; Hüls AG] Oleth-30; CAS 9004-98-2; nonionic; surfactant for mfg. of washing powds., textile auxiliaries; waxy flakes;

cloud pt. 75 C (2% in 10% NaCl); 100% act.

Marlipal® 1850/40. [Hüls AG] Oleth-40; CAS 9004-98-2; nonionic; surfactant for mfg. of washing powds., textile auxiliaries; flakes; cloud pt. 74 C (2% in 10% NaCl); 100% act.

Marlipal® 1850/80. [Hüls Am.; Hüls AG] Oleth-80; CAS 9004-98-2; nonionic; surfactant for mfg. of washing powds., textile auxiliaries; waxy flakes; cloud pt. 71 C (2% aq.); 100% act.

Marlipal® BS. [Hüls Am.] Fatty acid polyglycol ester; nonionic; surfactant, superfatting agent, thickener in detergents, dishwashing and cosmetic preparations; wax; 100% act.

Marlipal® FS. [Hüls Am.] PEG-12 dioleate; CAS 9005-07-6; nonionic; surfactant used for liq. and paste detergents, esp. hair shampoos, cosmetic preparations; liq.; 100% act.

Marlipal® KE. [Hüls Am.] Fatty alcohol polyglycol ethers; nonionic; surfactant, detergent raw material, and additive used in dishwashing agents, in mfg. of tablet soaps; liq. to solid; 100% act.

Marlipal® KF. [Hüls Am.; Hüls AG] Deceth-6; CAS 66455-15-0; nonionic; surfactant, detergent raw material, wetting agent, and additive used in dishwashing agents, glass and hard surf. cleaners, in mfg. of tablet soaps; degreaser for textiles and leather; liq.; cloud pt. 53 C (2% aq.); 100% act.

Marlipal® ML. [Hüls Am.] Fatty alcohol polyglycol ethers; nonionic; surfactant, detergent raw material, and additive used in dishwashing agents, in mfg. of tablet soaps; liq. to solid; 100% act.

Marlipal® NE. [Hüls Am.; Hüls AG] Oxo alcohol polyglycol ether (8.3 EO); nonionic; wetting agent for textile processing; detergent for dishwash, bar soaps; liq.; cloud pt. 53 C (2% aq.); 100% act.

Marlipal® O11/30. [Hüls AG] C11-oxo alcohol ethoxylate (3 EO); nonionic; wetting agent for hard surf. cleaners, textile pretreatment; liq.; cloud pt. 52 C (10% in 25% BDG); 100% act.

Marlipal® O11/50. [Hüls AG] C11-oxo alcohol ethoxylate (5 EO); nonionic; wetting agent for hard surf. cleaners, textile pretreatment; liq.; cloud pt. 72 C (10% in 25% BDG); 100% act.

Marlipal® O11/79. [Hüls AG] C11-oxo alcohol ethoxylate (7 EO); nonionic; wetting agent for hard surf. cleaners, textile pretreatment; liq.; cloud pt. 53 C (2% aq.); 90% act.

Marlipal® O11/88. [Hüls AG] C11-oxo alcohol ethoxylate (8 EO); nonionic; wetting agent for hard surf. cleaners, textile pretreatment; liq.; cloud pt. 61 C (2% aq.); 80% act.

Marlipal® O11/110. [Hüls AG] C11-oxo alcohol ethoxylate (11 EO); nonionic; wetting agent for hard surf. cleaners, textile pretreatment; solid; cloud pt. 63 C (2% in 10% NaCl); 100% act.

Marlipal® O13/20. [Hüls AG] C13-oxo alcohol ethoxylate (2 EO); nonionic; dispersant, wetting agent, detergent, cleaning, soil suspending and homogenizing agent, emulsifier for textiles; liq.; cloud pt. 38 C (10% in 25% BDG); 100% act.

Marlipal® O13/30. [Hüls AG] C13-oxo alcohol ethoxylate (3 EO); nonionic; dispersant, wetting agent, detergent, cleaning, soil suspending and homogenizing agent, emulsifier for textiles; liq.; cloud pt. 50 C (10% in 25% BDG); 100% act.

Marlipal® O13/40. [Hüls AG] C13-oxo alcohol ethoxylate (4 EO); nonionic; dispersant, wetting agent, detergent, cleaning, soil suspending and homogenizing agent, emulsifier for textiles; liq.; cloud pt. 60 C (10% in 25% BDG); 100% act.

Marlipal® O13/50. [Hüls AG] C13-oxo alcohol ethoxylate (5 EO); nonionic; dispersant, wetting agent, detergent, cleaning, soil suspending and homogenizing agent, emulsifier for textiles; liq.; cloud pt. 65 C (10% in 25% BDG); 100% act.

Marlipal® O13/60. [Hüls AG] C13-oxo alcohol ethoxylate (6 EO); nonionic; dispersant, wetting agent, detergent, cleaning, soil suspending and homogenizing agent, emulsifier for textiles; liq.; cloud pt. 70 C (10% in 25% BDG); 100% act.

Marlipal® O13/70. [Hüls AG] C13-oxo alcohol ethoxylate (7 EO); nonionic; dispersant, wetting agent, detergent, cleaning, soil suspending and homogenizing agent, emulsifier for textiles; liq.; cloud pt. 73 C (10% in 25% BDG); 100% act.

Marlipal® O13/80. [Hüls AG] C13-oxo alcohol ethoxylate (8 EO); nonionic; dispersant, wetting agent, detergent, cleaning, soil suspending and homogenizing agent, emulsifier for textiles; liq.; cloud pt. 47 C (2% aq.); 100% act.

Marlipal® O13/90. [Hüls AG] C13-oxo alcohol ethoxylate (9 EO); nonionic; dispersant, wetting agent, detergent, cleaning, soil suspending and homogenizing agent, emulsifier for textiles; liq.; cloud pt. 57 C (2% aq.); 100% act.

Marlipal® O13/100. [Hüls AG] C13-oxo alcohol ethoxylate (10 EO); nonionic; dispersant, wetting agent, detergent, cleaning, soil suspending and homogenizing agent, emulsifier for textiles; liq./paste; cloud pt. 75 C (2% aq.); 100% act.

Marlipal® O13/120. [Hüls AG] C13-oxo alcohol ethoxylate (10 EO); nonionic; dispersant, wetting agent, detergent, cleaning, soil suspending and homogenizing agent, emulsifier for textiles; liq./paste; cloud pt. 55 C (2% in 10% NaCl); 100% act.

Marlipal® O13/150. [Hüls AG] C13-oxo alcohol ethoxylate (15 EO); nonionic; dispersant, wetting agent, detergent, cleaning, soil suspending and homogenizing agent, emulsifier for textiles; paste; cloud pt. 67 C (2% in 10% NaCl); 100% act.

Marlipal® O13/170. [Hüls AG] C13-oxo alcohol ethoxylate (17 EO); nonionic; dispersant, wetting agent, detergent, cleaning, soil suspending and homogenizing agent, emulsifier for textiles; solid; cloud pt. 72 C (2% in 10% NaCl); 100% act.

Marlipal® O13/200. [Hüls AG] C13-oxo alcohol ethoxylate (20 EO); nonionic; dispersant, wetting agent, detergent, cleaning, soil suspending and homogenizing agent, emulsifier for textiles; solid; cloud pt. 73 C (2% in 10% NaCl); 100% act.

Marlipal® O13/400. [Hüls AG] C13-oxo alcohol ethoxylate (40 EO); nonionic; dispersant, wetting agent, detergent, cleaning, soil suspending and homogenizing agent, emulsifier for textiles; flakes; cloud pt. 74 C (2% in 10% NaCl); 100% act.

Marlipal® O13/500. [Hüls AG] C13-oxo alcohol ethoxylate (50 EO); nonionic; dispersant, wetting agent, detergent, cleaning, soil suspending and homogenizing agent, emulsifier for textiles; flakes; cloud pt. 74 C (2% in 10% NaCl); 100% act.

Marlipal® O13/939. [Hüls AG] C13-oxo alcohol ethoxylate blend; nonionic; dispersant, wetting agent, detergent, cleaning, soil suspending and homogenizing agent, emulsifier for textiles; liq.; cloud pt. 71 C (10% in 25% BDG); 90% act.

Marlipal® SU. [Hüls Am.; Hüls AG] Cetoleth-25; CAS 54045-08-8; nonionic; surfactant for mfg. of washing powds., dyeing assistants; plasticizer for soap prod.; yields finely porous foam; waxy solid; cloud pt. 76 C (2% in 10% NaCl); 100% act.

Marlon® A 350. [Hüls Am.; Hüls AG] Sodium dodecylbenzene sulfonate; anionic; detergent raw materials for household, industrial, and institutional detergents, textile auxiliaries; biodeg.; liq./paste; 50% act.

Marlon® A 360. [Hüls Am.; Hüls AG] Sodium dodecylbenzene sulfonate; anionic; surfactant for prod. of detergents, cleaning agents, textile auxiliaries; biodeg.; liq./paste; 60% act.

Marlon® A 365. [Hüls Am.; Hüls AG] Sodium dodecylbenzene sulfonate; CAS 38411-30-3; anionic; surfactant for prod. of detergents, cleaning agents, textile auxiliaries; biodeg.; liq./paste; 65% act.

Marlon® A 375. [Hüls Am.; Hüls AG] Sodium dodecylbenzene sulfonate; anionic; surfactant for prod. of detergents, cleaning agents, textile auxiliaries; biodeg.; paste; 75% act.

Marlon® A 390. [Hüls Am.; Hüls AG] Sodium dodecylbenzene sulfonate; anionic; surfactant for prod. of powd. detergents; biodeg.; powd.; 90% act.

Marlon® AFM 40, 40 N, 43, 50N. [Hüls Am.] Sodium dodecylbenzene sulfonate, alkyl ether sulfate, and nonionic; anionic/nonionic; base for dishwashing and cleaning agents and industrial cleaners; visc. liq.; 40–50% act.

Marlon® AFO 40. [Hüls Am.; Hüls AG] Sodium dodecylbenzene sulfonate, alkyl ether sulfate, nonionic surfactant blend; anionic/nonionic; mild, high foaming detergent for domestic and industrial applics.; liq.; 40% act.

Marlon® AFO 50. [Hüls Am.; Hüls AG] Sodium dodecylbenzene sulfonate, alkyl ether sulfate, nonionic surfactant blend; anionic/nonionic; mild, high foaming detergent for domestic and industrial applics.; liq.; 50% act.

Marlon® AFR. [Hüls Am.] Sodium dodecylbenzene sulfonate, modified with urea; anionic; base material for dishwashing agents, cosmetic detergents, car shampoos; liq.; 30% act.

Marlon® AM 80. [Hüls Am.; Hüls AG] Sodium dodecylbenzene sulfonate/nonionic blend; anionic/nonionic; detergent for domestic and industrial use, liq. preps., degreasers, washing-up liq. base; liq.; 80% act.

Marlon® AMX. [Hüls Am.; Hüls AG] Amine dodecylbenzene sulfonate; anionic; base material for prod. of drycleaning detergents, degreasing agents for metal industry, floor cleaners; biodeg.; liq.; 90% act.

Marlon® ARL. [Hüls Am.; Hüls AG] Sodium dodecylbenzene sulfonate and sodium toluene sulfonate; anionic; detergent component, foaming, wetting agent; for powd. detergents and scouring powds.; wh.-yel. powd.; dens. 470 g/l; pH 8; 80% act.

Marlon® AS3. [Hüls Am.; Hüls AG] Dodecylbenzene sulfonic acid; CAS 85536-14-7; anionic; intermediate for mfg. of anionic surfactants, detergents, sulfonates, textile auxiliaries; biodeg.; liq.; 98% act.

Marlon® AS3-R. [Hüls AG] Alkylbenzene sulfonic acid; anionic; surfactant for prod. of detergents, cleaning agents, textile auxiliaries; liq.; 98% act.

Marlon® PF 40. [Hüls Am.; Hüls AG] Sodium C13-17 alkane sulfonate, sodium laureth sulfate; anionic/nonionic; mild, high foaming detergent for domestic and industrial applics.; biodeg.; liq.; 40% act.

Marlon® PS 30. [Hüls Am.; Hüls AG] Sodium C13-17 alkane sulfonate; anionic; surfactant for prod. of liq., conc., mild cleaning agents, textile auxiliaries; liq.; 30% act.

Marlon® PS 60. [Hüls Am.; Hüls AG] Sodium C13-17 alkane sulfonate; anionic; base for detergents, cleaners, textile auxiliaries; liq./paste; 60% act.

Marlon® PS 60 W. [Hüls Am.; Hüls AG] Sodium paraffin sulfonate; anionic; mfg. of liq. conc. cleaners, textile auxiliaries; liq./paste; 60% act.

Marlon® PS 65. [Hüls Am.; Hüls AG] Sodium C13-17 alkane sulfonate; anionic; mfg. of liq. conc. cleaners, textile auxiliaries; liq./paste; 65% act.

Marlophen® 41, 47. [Hüls AG] Alkylphenol polyglcyol ether; solubilizers.

Marlophen® 81. [Hüls Am.; Hüls AG] Alkylphenol polyglycol ether; nonionic; wetting agent, dispersant with washing and homogenizing capacity; liq.; HLB 3.3; 100% conc.

Marlophen® 81N. [Hüls AG] Nonoxynol-1; CAS 85005-55-6; nonionic; wetting agent, detergent, dispersant; homogenizing capacity; polymer remover in floor care; liq.; water-insol.; HLB 3.3; cloud pt. 66 C (2% in 25% BDG); 100% act.

Marlophen® 82. [Hüls Am.; Hüls AG] Alkylphenol polyglycol ether; nonionic; wetting agent, dispersant with washing and homogenizing capacity; liq.; HLB 5.7; 100% conc.

Marlophen® 82N. [Hüls AG] Nonoxynol-2; CAS 9016-45-9; 37205-87-1; nonionic; wetting agent, detergent, dispersant; homogenizing capacity; polymer remover in floor care; liq.; water-insol.; HLB 5.7; cloud pt. 29 C (10% in 25% BDG); 100% act.

Marlophen® 83. [Hüls Am.; Hüls AG] Nonylphenol polyglycol ether; nonionic; detergent, wetting agent, dispersant with washing and homogenizing capacity; liq.; HLB 7.5; 100% conc.

Marlophen® 83N. [Hüls AG] Nonoxynol-3; CAS 9016-45-9; 37205-87-1; nonionic; low foam wetting agent, detergent, dispersant; homogenizing capacity; for solv. cleaners; liq.; water-insol.; HLB 7.5; cloud pt. 44 C (10% in 25% BDG); 100% act.

Marlophen® 84. [Hüls Am.; Hüls AG] Nonylphenol polyglycol ether; nonionic; detergent, wetting agent, dispersant with washing and homogenizing capacity; liq.; HLB 8.9; 100% conc.

Marlophen® 84N. [Hüls AG] Nonoxynol-4; CAS 9016-45-9; 37205-87-1; nonionic; low foam wetting agent, detergent, dispersant; homogenizing capacity; for solv. cleaners; liq.; water-insol.; HLB 8.9; cloud pt. 55 C (10% in 25% BDG); 100% act.

Marlophen® 85. [Hüls Am.; Hüls AG] Octoxynol-5; CAS 9002-93-1; nonionic; detergent, wetting agent, dispersant with washing and homogenizing capacity; for solv. cleaners; liq.; insol. in water; HLB 10.0; cloud pt. 61 C (10% in 25% BDG); 100% act.

Marlophen® 85N. [Hüls AG] Nonoxynol-5; CAS 9016-45-9; 37205-87-1; nonionic; low foam wetting agent, detergent, dispersant; homogenizing capacity; for solv. cleaners; liq.; water-insol.; HLB 10.0; cloud pt. 61 C (10% in 25% BDG); 100% act.

Marlophen® 86. [Hüls Am.; Hüls AG] Octoxynol-6.5; CAS 9002-93-1; nonionic; detergent, wetting agent, dispersant with washing and homogenizing capacity; for solv. cleaners; liq.; insol. in water; HLB 10.9; cloud pt. 72 C (10% in 25% BDG); 100% act.

Marlophen® 86N. [Hüls AG] Nonoxynol-6.5; CAS 9016-45-9; 37205-87-1; nonionic; low foam wetting agent, detergent, dispersant; homogenizing capacity; for solv. cleaners; liq.; water-insol.; HLB 10.9; cloud pt. 72 C (10% in 25% BDG); 100% act.

Marlophen® 86N/S. [Hüls AG] Nonoxynol-6; CAS 9016-45-9; nonionic; low foam wetting agent, detergent, dispersant; homogenizing capacity; for solv. cleaners; liq.; cloud pt. 68 C (10% in 25% BDG); 100% act.

Marlophen® 87. [Hüls Am.; Hüls AG] Octoxynol-7; CAS 9002-93-1; nonionic; wetting agent for acidic, neutral, and alkaline cleaning agents; prod. of textile auxiliaries; liq.; water-sol.; HLB 11.7; cloud pt. 23 C (2% aq.); 100% act.

Marlophen® 87N. [Hüls AG] Nonoxynol-7; CAS 9016-45-9; 37205-87-1; nonionic; wetting agent for acidic, neutral and alkaline cleaners, textile auxiliaries; liq.; water-sol.; HLB 11.7; cloud pt. 23 C (2% aq.); 100% act.

Marlophen® 88. [Hüls Am.; Hüls AG] Octoxynol-8; CAS 9002-93-1; nonionic; wetting agent for acidic, neutral, and alkaline cleaning agents; prod. of textile auxiliaries; liq.; water-sol.; HLB 12.3; cloud pt. 42 C (2% aq.); 100% act.

Marlophen® 88N. [Hüls AG] Nonoxynol-8; CAS 9016-45-9; 37205-87-1; nonionic; wetting agent, detergent, dispersant; homogenizing capacity; for acidic, neutral and alkaline cleaners, textile auxiliaries; liq.; water-sol.; HLB 12.3; cloud pt. 42 C (2% aq.); 100% act.

Marlophen® 89. [Hüls Am.; Hüls AG] Octoxynol-9.5; CAS 9002-93-1; nonionic; wetting agent for acidic, neutral, and alkaline cleaning agents; prod. of textile auxiliaries; liq.; water-sol.; HLB 12.8; cloud pt. 61 C (2% aq.); 100% act.

Marlophen® 89N. [Hüls AG] Nonoxynol-9.5; CAS 9016-45-9; 37205-87-1; nonionic; wetting agent, detergent, dispersant; homogenizing capacity; for acidic, neutral and alkaline cleaners, textile auxiliaries; liq.; water-sol.; HLB 12.8; cloud pt. 61 C (2% aq.); 100% act.

Marlophen® 89.5N. [Hüls AG] Nonylphenol polyglycol ether; CAS 37205-87-1; nonionic; wetting agent, detergent, dispersant; homogenizing capacity; liq.; HLB 13.0; 100% conc.

Marlophen® 810. [Hüls Am.; Hüls AG] Octoxynol-10; CAS 9002-93-1; nonionic; wetting agent for acidic, neutral, and alkaline cleaning agents; prod. of textile auxiliaries; liq./paste; water-sol.; HLB 13.3; cloud pt. 72 C (2% aq.); 100% act.

Marlophen® 810N. [Hüls AG] Nonoxynol-10; CAS 9016-45-9; 37205-87-1; nonionic; wetting agent, detergent, dispersant; homogenizing capacity; for acidic, neutral and alkaline cleaners, textile auxiliaries; liq./paste; water-sol.; HLB 13.3; cloud pt. 72 C (2% aq.); 100% act.

Marlophen® 812. [Hüls Am.; Hüls AG] Octoxynol-12; CAS 9002-93-1; nonionic; wetting agent for acidic, neutral, and alkaline cleaning agents; prod. of textile auxiliaries; liq./paste; water-sol.; HLB 14.1; cloud pt. 85 C (2% aq.); 100% act.

Marlophen® 812N. [Hüls AG] Nonoxynol-12; CAS 9016-45-9; 37205-87-1; nonionic; wetting agent, detergent, dispersant; homogenizing capacity; for acidic, neutral and alkaline cleaners, textile auxiliaries; liq./paste; water-sol.; HLB 14.1; cloud pt. 85 C (2% aq.); 100% act.

Marlophen® 814. [Hüls Am.; Hüls AG] Octoxynol-14; CAS 9002-93-1; nonionic; binder for pasty cleaning agents; emulsifier for waxes; paste; HLB 14.7; cloud pt. 61 C (2% in 10% NaCl); 100% act.

Marlophen® 814N. [Hüls AG] Nonoxynol-14; CAS 9016-45-9; 37205-87-1; nonionic; wetting agent, detergent, dispersant; homogenizing capacity; binding agent for pasty cleaners; emulsifier for waxes; paste; HLB 14.7; cloud pt. 61 C (2% in 10% NaCl); 100% act.

Marlophen® 820. [Hüls Am.; Hüls AG] Octoxynol-20; CAS 9002-93-1; nonionic; binder for pasty cleaning agents; emulsifier for waxes; waxy paste; HLB 16.0; cloud pt. 73 C (2% in 10% NaCl); 100% act.

Marlophen® 820N. [Hüls AG] Nonoxynol-20; CAS 9016-45-9; 37205-87-1; nonionic; wetting agent, detergent, dispersant; homogenizing capacity; binding agent for pasty cleaners; emulsifier for waxes; waxy paste; HLB 16.0; cloud pt. 73 C (2% in 10% NaCl); 100% act.

Marlophen® 825. [Hüls Am.; Hüls AG] Alkylphenol polyglycol ether; nonionic; detergent, wetting agent, dispersant with washing and homogenizing capacity; wax; HLB 16.7; 100% conc.

Marlophen® 830. [Hüls Am.; Hüls AG] Alkylphenol polyglycol ether; nonionic; detergent, wetting agent, dispersant with washing and homogenizing capacity; wax; HLB 17.1; 100% conc.

Marlophen® 830N. [Hüls AG] Nonoxynol-30; CAS 9016-45-9; 37205-87-1; nonionic; wetting agent, detergent, dispersant; homogenizing capacity; binding agent and base material for solid cleaning agents such as toilet sticks; waxy solid; HLB 17.1; cloud pt. 73 C (2% in 10% NaCl); 100% act.

Marlophen® 840N. [Hüls AG] Nonoxynol-40; CAS 9016-45-9; 37205-87-1; nonionic; wetting agent, detergent, dispersant; homogenizing capacity; binding agent and base material for solid cleaning agents, e.g., toilet sticks; waxy flakes; HLB 17.8; cloud pt. 73 C (2% in 10% NaCl); 100% act.

Marlophen® 850. [Hüls Am.; Hüls AG] Alkylphenol polyglycol ether; nonionic; detergent, wetting agent, dispersant with washing and homogenizing capacity; wax; HLB 18.2; 100% conc.

Marlophen® 850N. [Hüls AG] Nonoxynol-50; CAS 9016-45-9; 37205-87-1; nonionic; wetting agent, detergent, dispersant; homogenizing capacity; binding agent and base material for solid cleaning agents, e.g., toilet sticks; waxy flakes; HLB 18.2; cloud pt. 73 C (2% in 10% NaCl); 100% act.

Marlophen® 890. [Hüls Am.; Hüls AG] Alkylphenol polyglycol ethers; nonionic; detergent, wetting and dispersing agent; wax; HLB 18.9; 100% act.

Marlophen® 890N. [Hüls AG] Nonylphenol polyglycol ether; CAS 37205-87-1; nonionic; wetting agent, detergent, dispersant; homogenizing capacity; wax; HLB 18.9; 100% conc.

Marlophen® 1028. [Hüls Am.; Hüls AG] Octoxynol-9; CAS 9002-93-1; nonionic; wetting agent for acidic, neutral, and alkaline cleaning agents; prod. of textile auxiliaries; liq.; water-sol.; HLB 12.6; cloud pt. 54 C (2% aq.); 100% act.

Marlophen® 1028N. [Hüls AG] Nonoxynol-9; CAS 9016-45-9; 37205-87-1; nonionic; wetting agent, detergent, dispersant; homogenizing capacity; for acidic, neutral and alkaline cleaners, textile auxiliaries; liq.; water-sol.; HLB 12.6; cloud pt. 54 C (2% aq.); 100% act.

Marlophen® DNP 16. [Hüls Am.; Hüls AG] Nonyl nonoxynol-16; CAS 9014-93-1; nonionic; raw material for textile and paper auxiliaries, dispersant; waxy solid; HLB 13.4; cloud pt. 58 C (2% aq.); 100% act.

Marlophen® DNP 18. [Hüls Am.; Hüls AG] Nonyl nonoxynol-18; CAS 9014-93-1; nonionic; raw material for textile and paper auxiliaries, dispersant; waxy solid; HLB 13.9; cloud pt. 75 C (2% aq.); 100% act.

Marlophen® DNP 30. [Hüls Am.; Hüls AG] Nonyl nonoxynol-30; CAS 9014-93-1; nonionic; raw material for textile and paper auxiliaries, dispersant; waxy solid; HLB 15.8; cloud pt. 70 C (2% in 10% NaCl); 100% act.

Marlophen® DNP 100. [Hüls AG] Textile and paper auxiliaries with dispersing props.

Marlophen® DNP 150. [Hüls AG] Nonyl nonoxynol-150; CAS 9014-93-1; nonionic; surfactant, dispersant for textile and paper auxiliaries; flakes; cloud pt. 71 C (2% in 10% NaCl); 100% act.

Marlophen® DP 9, DP 11. [Hüls AG] Dispersants.

Marlophen® P 1. [Hüls Am.; Hüls AG] Phenol ethoxylate (1 EO); nonionic; solubilizer, solvent; liq.; cloud pt. 76 C (5% in 25% BDG); 100% act.

Marlophen® P 4. [Hüls AG] Phenol ethoxylate (4 EO); nonionic; solubilizer; liq.; cloud pt. 33 C (2% in 10% NaCl); 100% act.

Marlophen® P 7. [Hüls Am.; Hüls AG] Phenol ethoxylate (7 EO); CAS 9004-78-8; nonionic; solubilizer, solvent; liq.; cloud pt. 65 C (2% in 10% NaCl); 100% act.

Marlophen® X. [Hüls AG] Alkylphenol polyglycol ether; nonionic; wetting agent for use in binders for coal dust, pigments, and in concrete mfg.; liq.; 100% act.

Marlophor® AS-Acid. [Hüls Am.; Hüls AG] Phosphate ester; CAS 76483-21-1; anionic; base compd. for formulating acid cleaners for glass, metal, and ceramics; anticorrosive and rust-removing props.; liq.; 100% conc.

Marlophor® CS-Acid. [Hüls Am.; Hüls AG] Isopropyl phosphate ester; CAS 76483-21-1; anionic; base compd. for formulating acid cleaners for glass, metal, and ceramics, flamer retardants, mercerizing wetting agents and antitstats; anticorrosive and rust-removing props.; liq.; 100% act.

Marlophor® DG-Acid. [Hüls Am.] Phosphate ester; anionic; wetting/antistatic agent for drycleaning detergents; liq.; 100% conc.

Marlophor® DS-Acid. [Hüls Am.; Hüls AG] n-Butyl phosphate ester; CAS 68439-39-4; anionic; surfactant used as base material for acid cleaners for glass, metal, and ceramics, flame retardants, mercerizing wetting agents and antistats; anticorrosive and rust-removing props.; liq.; 100% act.

Marlophor® F1-Acid. [Hüls Am.; Hüls AG] n-Hexyl PEG ether phosphate ester; anionic; low foam wetting agent for textile and paper industries; antistat, drycleaning detergent; liq.; water-sol.; 100% act.

Marlophor® FC-Acid. [Hüls Am.; Hüls AG] 2-Ethylhexyl PEG ether phosphate ester, acid form; CAS 68439-39-4; anionic; surfactant; base for acid cleaners; low-foam wetting agent for alkaline sol'ns., high pressure cleaners, metal degreasing, causticizing and kier boiling agents; anticorrosive props.; liq.; 100% act.

Marlophor® FC-Sodium Salt. [Hüls Am.; Hüls AG] Sodium 2-ethylhexyl polyglycol ether phosphate; CAS 111798-26-6; anionic; detergent, wetting agent for alkaline sol'ns.; for machine dishwashing, industrial cleaners, textile and cellulose industry applics.; liq.; 85% act.

Marlophor® HS-Acid. [Hüls Am.; Hüls AG] n-Octyl phosphate ester; CAS 39407-03-9; anionic; emulsifier for silicone oils; liq.; 100% act.

Marlophor® ID-Acid. [Hüls Am.] Phosphate ester; anionic; for mfg. of highly alkaline resistant wetting agents and special purpose acid cleaners; liq.; 100% conc.

Marlophor® IH-Acid. [Hüls Am.; Hüls AG] 2-Ethylhexyl phosphate ester; CAS 12645-31-7; anionic; low foam wetting agent for weakly to med. strong alkaline range in textile pretreatment and finishing; liq.; water-sol.; 100% act.

Marlophor® LN-Acid. [Hüls Am.] Trilauryl phosphate; anionic; wetting agent for textile, paper, and personal care industries; antistat, drycleaning detergent; liq.; 100% conc.

Marlophor® MN-60. [Hüls Am.] Surfactant blend; anionic; for mfg. of highly alkaline resistant wetting agents and special purpose acid cleaners; liq.; 60% conc.

Marlophor® MO 3-Acid. [Hüls Am.; Hüls AG] Laureth-3 phosphate; anionic; surfactant for prod. of textile and dyeing auxiliaries, antistats, drycleaning detergents; liq.; 100% act.

Marlophor® N5-Acid. [Hüls Am.; Hüls AG] Isononyl PEG ether phosphate ester; anionic; low foam wetting agent for textile and paper industry; antistat, drycleaning detergent; liq.; water-sol.; 100% act.

Marlophor® ND. [Hüls Am.] Partial phosphate ester; anionic; wetting agent for textile and paper industries, antistat for natural and syn. fibers used in drycleaning detergents; liq.; water- and oil-sol.; 100% act.

Marlophor® ND-Acid. [Hüls Am.; Hüls AG] Isoalkylphosphate ester/alkylphenol PEG ether blend; anionic; detergent; wetting agent; antistat; component for textile aux. agents, drycleaning formulations, paper industry; liq.; sol. in water, solvs., min. oil; 100% act.

Marlophor® ND DEA Salt. [Hüls Am.; Hüls AG] Isoalkyl phosphate ester/alkylphenol PEG ether blend; anionic/nonionic; detergent, wetting agent, antistat; component for textile aux. agents, drycleaning formulations, paper industry; liq.; 100% act.

Marlophor® ND NA-Salt. [Hüls Am.] Org. phosphate ester/alkylphenol polyglycol ether; anionic/nonionic; detergent, wetting agent, antistat; component for textile aux. agents and drycleaning formulations; liq.; 92% conc.

Marlophor® NP5-Acid, NP6-Acid, NP7-Acid. [Hüls Am.; Hüls AG] Nonylphenol PEG ether phosphate esters; anionic; low foam wetting agent, base for mfg. of alkaline or weakly acid cleaning agents, drycleaning formulations with antistatic properties, flame retardants, flotation aids; dispersing capacity; liq.; water-sol.; 100% conc.

Marlophor® OC5-Acid. [Hüls AG] Olein-PEG ether phosphate ester; anionic; surfactant for prod. of textile and dyeing auxiliaries, antistats, drycleaning detergents; liq.; 100% act.

Marlophor® ON3-Acid, ON5-Acid, ON7-Acid. [Hüls AG] Isotridecyl PEG ether phosphate ester; anionic; low foam wetting agents for neutral to weakly alkaline range; in salt form used as emulsifers and antistats for textiles; liq.; water-sol.; 100% act.

Marlophor® T6-Acid. [Hüls AG] Tallow PEG ether phosphate ester; anionic; surfactant for prod. of textile and dyeing auxiliaries, antistats, drycleaning detergents; liq.; 100% act.

Marlophor® T10-Acid. [Hüls Am.; Hüls AG] Diceteareth-10 phospate; anionic; special liq. formulation for prod. of drycleaning detergents with antistatic props., textile and dyeing auxiliaries; liq.; 100% act.

Marlophor® T10-DEA Salt. [Hüls Am.] Diethanolamine alkyl polyglycol ether phosphate; anionic; surfactant base for drycleaning formulations with antistatic props., acid cleaners; liq.; 100% conc.

Marlophor® T10-Sodium Salt. [Hüls Am.] Sodium diceteareth-10 phosphate; anionic; special liq. formulation for prod. of drycleaning detergents; liq.; 100% act.

Marlophor® UW12-Acid. [Hüls Am.; Hüls AG] Alkyl PEG ether phosphate ester; anionic; wetting agent, antistat for drycleaning detergents, textile and dyeing auxiliaries; wax; 100% act.

Marlopon® ADS 50. [Hüls Am.] DEA dodecylbenzene sulfonate; anionic; detergent raw material for liq. phosphate-containing detergents and cleaners; biodeg.; fluid paste; 50% act.

Marlopon® ADS 65. [Hüls Am.; Hüls AG] DEA dodecylbenzene sulfonate; anionic; detergent raw material for liq. and paste detergent formulations with corrosion protection; biodeg.; liq.; 65% act.

Marlopon® AMS 60. [Hüls Am.; Hüls AG] Amine dodecylbenzene sulfonate/nonionic blend; anionic/nonionic; surfactant used for detergents, dishwashing agents, industrial cleaners, car shampoos, hand cleaners; fluid paste; 60% act.

Marlopon® AT. [Hüls Am.] TEA-dodecylbenzene sulfonate.; CAS 29381-93-9; surfactant for mfg. of cosmetic detergents, dishwashing agents.

Marlopon® AT 50. [Hüls Am.; Hüls AG] TEA-dodecylbenzene sulfonate; CAS 68411-31-4; anionic; surfactant used as neutral detergent base; for hair and bath shampoos, dishwashes, car shampoos, toilet preps.; biodeg.; liq.; 50% act.

Marlopon® CA. [Hüls Am.; Hüls AG] TEA-dodecylbenzene sulfonate, modified; anionic; detergent; superfatting and dishwashing agent; personal care prods.; liq.; 60% act.

Marlosoft® A 18 M. [Hüls AG] Blend of fatty acid alkanolamide acetate and fatty acid polyglycol ester; for prod. of cold water-sol. textile softeners; flakes; 97% act.

Marlosoft® B 18 M. [Hüls AG] Blend of fatty acid alkanolamide acetate and fatty acid polyglycol ester; for prod. of cold water sol. textile softeners; flakes; 100% act.

Marlosoft® IQ 75. [Hüls Am.] Imidazolinium methosulfate; cationic; base for fabric softeners; liq.; 75% act.

Marlosoft® IQ 90. [Hüls AG] Quaternized tallow fatty imidazolinium methosulfate; cationic; surfactant for prod. of fabric softeners; liq./paste; 90% act.

Marlosol® 183. [Hüls Am.; Hüls AG] PEG-3 stearate; CAS 9004-99-3; nonionic; raw material for finishing agents in the syn. fiber industry; emulsifier; waxy solid; water-insol.; cloud pt. 65 C (2% in 60% BDG); 100% act.

Marlosol® 186. [Hüls Am.; Hüls AG] PEG-6 stearate; CAS 9004-99-3; nonionic; raw material for processing agents for syn. fibers industry; emulsifier; paste; water-insol.; cloud pt. 87 C (5 g/20 g BDG + 25 g water); 100% act.

Marlosol® 188. [Hüls Am.; Hüls AG] PEG-8 stearate; CAS 9004-99-3; nonionic; raw material for processing agents for syn. fiber industry; emulsifier; paste; water-insol.; cloud pt. 65 C (10% in 25% BDG); 100% act.

Marlosol® 189. [Hüls Am.; Hüls AG] PEG-9 stearate; CAS 9004-99-3; nonionic; raw material for finishing agents in the syn. fiber industry; emulsifier; paste; 100% act.

Marlosol® 1820. [Hüls Am.; Hüls AG] PEG-20 stearate; CAS 9004-99-3; nonionic; raw material for finishing agents in the syn. fiber industry, fabric softeners; emulsifier; waxy solid; water-sol.; cloud pt. 80 C (2% aq.); 100% act.

Marlosol® 1825. [Hüls Am.] Stearic acid polyglycol esters; nonionic; raw material for finishing agents in the syn. fiber industry; emulsifier; wax; 100% act.

Marlosol® BS. [Hüls Am.; Hüls AG] PEG-12 distearate; CAS 9005-08-7; nonionic; superfatting agent, visc. enhancer for hair shampoos, cosmetic preps., fabric softeners, syn. fiber finishing; waxy solid; 100% act.

Marlosol® F08. [Hüls AG] PEG-12 dioleate/olein polyglycol ester blend; nonionic; refatting and thickening agent for shampoos; raw materials for prep. agents for syn. fibers, fabric softeners; liq.; 100% act.

Marlosol® FS. [Hüls Am.; Hüls AG] PEG-12 dioleate; CAS 9005-07-6; 52668-97-0; nonionic; superfatting agent, visc. enhancer for hair shampoos, cosmetic preps., fabric softeners, syn. fiber finishing; liq.; 100% act.

Marlosol® OL2. [Hüls AG] PEG-2 oleate; CAS 9004-96-0; nonionic; raw material for prep. agents for syn. fiber industry; liq.; 100% act.

Marlosol® OL7. [Hüls Am.; Hüls AG] PEG-7 oleate; CAS 9004-96-0; nonionic; raw material for finishing agents in the syn. fiber industry; liq.; cloud pt. 62 C (10% in 25% BDG); 100% act.

Marlosol® OL8. [Hüls Am.] PEG-8 oleate; CAS 9004-96-0; nonionic; raw material for finishing agents in the syn. fiber industry; liq.; 100% act.

Marlosol® OL10. [Hüls Am.; Hüls AG] PEG-10 oleate; CAS 9004-96-0; nonionic; raw material for finishing agents in the syn. fiber industry; liq./paste; cloud pt. 70 C (10% in 25% BDG); 100% act.

Marlosol® OL15. [Hüls Am.; Hüls AG] PEG-15 oleate; CAS 9004-96-0; nonionic; raw material for finishing agents in the syn. fiber industry; paste; cloud pt. 46 C (2% in 5% NaCl); 100% act.

Marlosol® OL20. [Hüls Am.; Hüls AG] PEG-20 oleate; CAS 9004-96-0; nonionic; raw material for finishing agents in the syn. fiber industry; paste; cloud pt. 62 C (2% in 5% NaCl); 100% act.

Marlosol® R70. [Hüls Am.; Hüls AG] PEG-70 castor oil; CAS 61791-12-6; nonionic; thickener, conditioner for toilet sticks; flakes; 100% act.

Marlosol® RF3. [Hüls AG] PEG-3 rapeseed fatty acid ester; nonionic; preparation agent; liq.; 100% act.

Marlosol® TF3. [Hüls AG] PEG-3 tallate; CAS 61791-00-2; nonionic; raw material for prep. agent for syn. fiber industry; liq.; 100% act.

Marlosol® TF4. [Hüls AG] PEG-4 tallate; CAS 61791-00-2; nonionic; raw material for prep. agent for syn. fiber industry; liq.; 100% act.

Marlowet® 1072. [Hüls Am.; Hüls AG] C12-14 alcohol polyglycol ether carboxylic acid; anionic; emulsifier for metalworking, water-miscible cooling lubricants, drilling oils, textile auxiliaries; liq.; 90% act.

Marlowet® 4508. [Hüls Am.] Org. ammonium salt; anionic; emulsifier for wh. spirit, kerosene, for environmentally compatible cold cleaners; liq.; 100% act.

Marlowet® 4530. [Hüls Am.; Hüls AG] Dinonylphenol polyglycol ether carboxylic acid; anionic; emulsifier for min. oils; used in metalworking, water-miscible cooling lubricants, drilling oils; liq.; 90% act.

Marlowet® 4530 LF. [Hüls AG] Dinonylphenol

polyglycol ether carboxylic acid; low-foaming emulsifier for water-miscible cooling lubricants and drilling oils; liq.; 90% act.

Marlowet® 4534. [Hüls Am.] Alkyl PEG acetic acid; anionic; surfactant for metalworking; liq.; 90% conc.

Marlowet® 4536. [Hüls AG] Nonylphenol polyglycol ether carboxylic acid; emulsifier for water-miscible cooling lubricants and drilling oils, prod. of textile auxiliaries; liq.; 90% act.

Marlowet® 4538. [Hüls Am.; Hüls AG] C13 oxo-alcohol polyglycol ether carboxylic acid; anionic; emulsifier for metalworking, water-miscible cooling lubricants, drilling oils, textile auxiliaries; liq.; 90% act.

Marlowet® 4539. [Hüls Am.; Hüls AG] C9 oxo-alcohol polyglycol ether carboxylic acid; anionic; surfactant for metalworking, water-miscible cooling lubricants, drilling oils; liq.; 90% act.

Marlowet® 4539 LF. [Hüls AG] C9 oxo-alcohol polyglycol ether carboxylic acid; low-foaming emulsifier for water-miscible cooling lubricants and drilling oils; liq.; 90% act.

Marlowet® 4540. [Hüls AG] Sodium nonylphenol polyglycol ether sulfate; emulsifier for polymer dispersions; liq.; 35% act.

Marlowet® 4541. [Hüls AG] C12-14 alcohol polyglycol ether carboxylic acid; emulsifier for water-miscible cooling lubricants, drilling oils, textile auxiliaries; liq.; 92% act.

Marlowet® 4603. [Hüls Am.] Carboxylic acid polyglycol ester; nonionic; emulsifer for min. oils, textile finishes; liq.; 100% act.

Marlowet® 4700. [Hüls Am.; Hüls AG] Carboxylic acid polyglycol ester; nonionic; emulsifier for min. oils; used in metalworking; wax; HLB 17.1; 100% conc.

Marlowet® 4702. [Hüls Am.; Hüls AG] C18 fatty acid polyglycol ester; CAS 52668-97-0; nonionic; emulsifer for min. oils, spindle oils, metalworking, textile lubricants; liq.; cloud pt. 26 C (10% in 25% BDG); 100% act.

Marlowet® 4703. [Hüls Am.; Hüls AG] C18 fatty acid polyglycol ester; CAS 52668-97-0; nonionic; emulsifer for solvs., min. oils, syn. and natural waxes, metal cleaning, metalworking, textile lubricants; liq.; cloud pt. 39 C (10% in 25% BDG); 100% act.

Marlowet® 4800. [Hüls Am.; Hüls AG] C16-18 alcohol polyglycol ether; CAS 68439-49-6; nonionic; emulsifier for waxes, car and furniture polishes, textile lubricants, leather care; waxy flakes; cloud pt. 77 C (2% in 10% NaCl); 100% act.

Marlowet® 4857. [Hüls Am.; Hüls AG] C16-18 alcohol polyglycol ether; CAS 68439-49-6; nonionic; emulsifier for paraffins, lanolin, waxes, silicone oils for textile, wood, paper, mold release, polishes; paste; cloud pt. 76 C (10% in 25% BDG); 100% act.

Marlowet® 4862. [Hüls Am.] Alkyl polyglycol ether; nonionic; emulsifier for waxes, paraffins, polishes; flakes; 100% act.

Marlowet® 4900. [Hüls Am.; Hüls AG] Nonylphenol PEG ether; nonionic; emulsifier for min. oils used in metalworking and in cold cleaners; liq.; HLB 10.9; cloud pt. 72 C (10% in 25% BDG); 100% act.

Marlowet® 4901. [Hüls Am.; Hüls AG] Nonylphenol PEG ether; nonionic; emulsifier of solvs., pesticides, cold cleaners; liq.; HLB 11.7; cloud pt. 23 C (2% aq.); 100% act.

Marlowet® 4902. [Hüls Am.; Hüls AG] Nonylphenol PEG ether; nonionic; emulsifier for solvs., pesticides, cold cleaners; liq.; HLB 12.8; cloud pt. 51 C (2% aq.); 100% act.

Marlowet® 4930. [Hüls Am.; Hüls AG] Nonylphenol PEG ether; nonionic; dispersant; liq.; HLB 17.1; cloud pt. 73 C (2% in 10% NaCl); 100% conc.

Marlowet® 4938. [Hüls Am.; Hüls AG] Nonylphenol PEG ether; nonionic; emulsifier for min. oils, metalworking; liq.; HLB 8.9; cloud pt. 55 C (10% in 25% BDG); 100% act.

Marlowet® 4939. [Hüls Am.; Hüls AG] Nonylphenol PEG ether; nonionic; emulsifier for min. oils; used in w/o emulsion, as defoaming agent; liq.; HLB 7.5; 100% conc.

Marlowet® 4940. [Hüls Am.; Hüls AG] Nonylphenol PEG ether; nonionic; emulsifier for min. oils; metalworking; wetting agent; liq.; HLB 13.3; cloud pt. 72 C (2% aq.); 100% act.

Marlowet® 4941. [Hüls Am.; Hüls AG] Nonylphenol PEG ether; nonionic; dispersant; liq.; HLB 16.0; cloud pt. 73 C (2% in 10% NaCl); 100% conc.

Marlowet® 5001. [Hüls Am.; Hüls AG] C18 alcohol polyalkylene glycol ether; nonionic; emulsifier for min. oils, wh. oils, textile aux., cosmetics, textile lubricants and finishes; liq.; cloud pt. 61 C (10% in 25% BDG); 100% act.

Marlowet® 5165. [Hüls Am.; Hüls AG] Alkyldiamine alkylene oxide addition prod.; cationic; lubrication aid for metalworking and foam control; liq.; cloud pt. 39 C (10% in 25% BDG); 100% act.

Marlowet® 5301. [Hüls Am.; Hüls AG] Coconut PEG ether phosphate ester; anionic; emulsifier for min. oils, metalworking, metal cleaning; liq.; 100% act.

Marlowet® 5311. [Hüls Am.; Hüls AG] Isononanol phosphate ester; CAS 97999-44-5; anionic; emulsifier used in paint removers; wetting agent for chlorinated hydrocarbons; prod. of textile auxiliaries; liq.; 100% act.

Marlowet® 5320. [Hüls Am.; Hüls AG] Nonylphenol polyglycol ether phosphate ester; anionic; emulsifier for chlorinated paraffins, min. oils, metal processing; liq.; 100% act.

Marlowet® 5324. [Hüls Am.; Hüls AG] Phenol polyglycol ether phosphate ester; CAS 39464-70-5; anionic; emulsifier for min. oils; lubricant auxiliaries; metal processing; liq.; 100% act.

Marlowet® 5361. [Hüls AG] Alkyl polyglycol ether phosphate ester; emulsifier; lubricant auxiliaries; metal processing; liq.; 100% act.

Marlowet® 5400. [Hüls Am.; Hüls AG] Alkylamine polyglycol ether; CAS 26635-93-8; nonionic; emulsifier for min. oils, paraffin oils, car wash rinses, furniture polishes, textile auxiliaries; liq.; 100% act.

Marlowet® 5401. [Hüls Am.; Hüls AG] Ethoxylated alkylamine, acetic acid salt; cationic; emulsifier for mixts. of waxes and wh. spirits; paper and textile impregnation; liq.; 100% act.

Marlowet® 5440. [Hüls Am.; Hüls AG] Substituted imidazoline; CAS 95-38-5; cationic; emulsifier for min. oils, corrosion protection, car wash rinses; liq.; 100% act.

Marlowet® 5459. [Hüls Am.; Hüls AG] Fatty acid DEA; nonionic; corrosion inhibitor for metalworking; liq.; 100% conc.

Marlowet® 5480. [Hüls Am.] Carboxylic acid alkanolamide; nonionic; emulsifier for min. oils; anticorrosion agent; liq.; 100% conc.

Marlowet® 5600. [Hüls Am.] Amine salt of carboxylic acid; anionic; emulsifier for waxes, leather care agents, wood and paper impregnation agents; solid; 100% act.

Marlowet® 5606. [Hüls Am.] Amine salt of carboxylic acid; anionic; emulsifier for mfg. of water-repellent chipboard; wax; 100% act.

Marlowet® 5609. [Hüls Am.; Hüls AG] Amine salt of carboxylic acid; anionic; emulsifier for wh. spirit, trichlorethylene; dewatering agents; cold cleaners; liq.; 100% act.

Marlowet® 5622. [Hüls Am.; Hüls AG] TEA salt of oleic acid; anionic; emulsifier for solvs., min. oils, metalworking, corrosion protection, floor care; liq.; 100% act.

Marlowet® 5626. [Hüls Am.; Hüls AG] Carboxylic acid amine salt; anionic; emulsifier for use in floor care prods.; liq.; 100% conc.

Marlowet® 5635. [Hüls Am.; Hüls AG] TEA salt of oleic acid and solubilizer; anionic; emulsifier for use in floor care prods.; liq.; 90% conc.

Marlowet® 5641. [Hüls Am.; Hüls AG] Modified dinonylphenol polyglycol ether; nonionic; dispersant for use in coal/water slurries; liq.; 45% act.

Marlowet® BIK. [Hüls Am.; Hüls AG] Mixt. of carboxylic acid amine salt and alkyl polyglycol ether; anionic/nonionic; emulsifier for min. oils, hydrocarbons, bitumen, for metalworking, road construction; liq.; 100% act.

Marlowet® BL. [Hüls Am.; Hüls AG] C12 alcohol polyglycol ether; CAS 9002-92-0; nonionic; emulsifier for min. oils, spindle oils, textile lubricants, bitumen; liq.; cloud pt. 68 C (10% in 25% BDG); 100% act.

Marlowet® CA 5. [Hüls Am.] Coceth-5.; CAS 61791-13-7.

Marlowet® CA 10. [Hüls Am.] Coceth-10.; CAS 61791-13-7.

Marlowet® CA 12. [Hüls Am.] Laureth-12.

Marlowet® EF. [Hüls Am.] Mixt. of carboxylic acid polyglycol esters; nonionic; emulsifier for textile lubricants, pest control, cosmetic preparations; paste; 100% act.

Marlowet® FOX. [Hüls Am.; Hüls AG] Ceteareth-28; CAS 68439-49-6; 68920-66-1; nonionic; emulsifier for oleic acid and waxes, textile lubricants, car and furniture polishes, leather care; waxy solid; cloud pt. 76 C (2% in 10% NaCl); 100% act.

Marlowet® G 12 DO. [Hüls Am.] PEG-12 glyceryl dioleate.

Marlowet® GDO 4. [Hüls Am.] PEG-5 glyceryl sesquioleate.

Marlowet® GFN. [Hüls Am.] C18 alcohol polyglycol ether; nonionic; emulsifier for waxes, ester waxes, furniture and floor care, emulsion polishes; paste; cloud pt. 70 C (2% in 10% NaCl); 100% act.

Marlowet® GFW. [Hüls Am.; Hüls AG] C16-18 alcohol polyglycol ether; nonionic; emulsifier for syn. and natural waxes, oleic acid, car and furniture polishes, emulsion polishes, textile lubricants; flake; cloud pt. 74 C (2% in 10% NaCl); 100% act.

Marlowet® IHF. [Hüls Am.; Hüls AG] Mixt. of n-alkylbenzene sulfonate, alkyl polyglycol ether, and IPA; nonionic; emulsifier for insecticides, herbicides, pesticides, cleaning agents; liq.; 80% act.

Marlowet® ISM. [Hüls Am.; Hüls AG] Nonylphenol polyglycol ether; CAS 37205-87-1; nonionic; emulsifier for solvs., for pesticides, cold cleaners; liq.; cloud pt. 42 C (2% aq.); 100% act.

Marlowet® LA 4. [Hüls Am.] Laureth-4.

Marlowet® LA 7. [Hüls Am.] Laureth-7.

Marlowet® LMA 2. [Hüls Am.] Laureth-2.

Marlowet® LMA 3. [Hüls Am.] Laureth-3.

Marlowet® LMA 4. [Hüls Am.] Laureth-4.

Marlowet® LMA 7. [Hüls Am.] Laureth-7.

Marlowet® LMA 10. [Hüls Am.] Laureth-10.

Marlowet® LMA 20. [Hüls Am.] Laureth-20.

Marlowet® LVS. [Hüls Am.; Hüls AG] C18 fatty acid ester of ethoxylated castor oil; nonionic; emulsifier for veg. oils; used in metalworking, leather auxiliaries, release agents; liq.; 100% act.

Marlowet® LVS/K. [Hüls Am.] PEG-18 castor oil dioleate.

Marlowet® LVX. [Hüls Am.; Hüls AG] Mixt. of C18 fatty acid esters; nonionic; emulsifier for animal and vegetable oils, for leather auxiliaries, textile lubricants; biodeg.; liq.; 100% act.

Marlowet® LVX/K. [Hüls Am.] PEG-18 castor oil dioleate, PEG-12 dioleate.

Marlowet® MA. [Hüls Am.; Hüls AG] Mixt. of carboxylic acid polyglycol esters and alkylolamides; nonionic; emulsifier for min. oils, spindle oils, textile lubricants, solvs., metal processing; liq.; cloud pt. 90 C (5% in 25% BDG); 100% act.

Marlowet® NF. [Hüls Am.] Mixt. of carboxylic acid polyglycol esters; nonionic; emulsifier for leather, auxiliaries, textile lubricants; liq.; 100% act.

Marlowet® OA 4/1. [Hüls Am.] Propylene glycol oleth-5.

Marlowet® OA 5. [Hüls Am.] Oleth-5.; CAS 9004-98-2; nonionic.

Marlowet® OA 10. [Hüls Am.] Oleth-10.; CAS 9004-98-2; nonionic.

Marlowet® OA 30. [Hüls Am.] Oleth-30.; CAS 9004-98-2; nonionic.

Marlowet® OAM, OAM Spec. [Hüls AG] Alkylamine polyglycol ether; anticorrosive agent for water-misc. cooling lubricants; liq.; 100% act.

Marlowet® OCM. [Hüls AG] Fatty acid alkanolamide polyglycol ether; anticorrosive agent for water-misc. cooling lubricants; liq.; 100% act.

Marlowet® OFA. [Hüls Am.; Hüls AG] Mixt. of n-alkyl benzene sulfonate, carboxylic acid polyglycol esters, and alkyl polyglycol ethers; anionic/nonionic; emulsifier for oleic acid, solvs., textile lubricants, pesticides, metal cleaning; biodeg.; liq.; 100% act.

Marlowet® OFW. [Hüls Am.] Mixt. of carboxylic acid polyglycol esters and alkyl polyglycol ethers; nonionic; emulsifier for textile lubricants, metalworking; liq.; 100% act.

Marlowet® OTS. [Hüls Am.; Hüls AG] Carboxylic acid polyglycol esters; nonionic; emulsifier for min. oils, solvs., metalworking and cleaning, textile lubricants; liq.; cloud pt. 78 C (10% in 25% BDG); 100% act.

Marlowet® PW. [Hüls Am.; Hüls AG] C16-18 alcohol polyglycol ether; CAS 68439-49-6; nonionic; surfactant, emulsifier for paraffin, lanolin waxes, textile and paper impregnation, mold release agents, furniture polishes; paste; cloud pt. 78 C (10% in 25% BDG); 100% act.

Marlowet® R 11. [Hüls Am.; Hüls AG] Ethoxylated castor oil; CAS 61791-12-6; nonionic; emulsifier for animal and veg. oils and neutral fats, leather auxiliaries; liq.; cloud pt. 52 C (10% in 25% BDG); 100% act.

Marlowet® R 11/K. [Hüls Am.] PEG-11 castor oil.

Marlowet® R 20. [Hüls AG] Ethoxylated castor oil; emulsifier for fatty acids, solvs., cosmetic oils; textile lubricants, dyeing auxiliaries, pesticides, creams; liq.; 100% act.

Marlowet® R 22. [Hüls AG] Ethoxylated castor oil; emulsifier for fatty acids, solvs., cosmetic oils; textile lubricants, dyeing auxiliaries, pesticides, creams; liq.; 100% act.
Marlowet® R 25. [Hüls AG] Ethoxylated castor oil; emulsifier for fatty acids, solvs., cosmetic oils; textile lubricants, dyeing auxiliaries, pesticides, creams; liq.; 100% act.
Marlowet® R 32. [Hüls AG] Ethoxylated castor oil; emulsifier for fatty acids, solvs., cosmetic oils; textile lubricants, dyeing auxiliaries, pesticides, creams; liq.; 100% act.
Marlowet® R 36. [Hüls AG] Ethoxylated castor oil; emulsifier for fatty acids, solvs., cosmetic oils; textile lubricants, dyeing auxiliaries, pesticides, creams; liq.; 100% act.
Marlowet® R 40. [Hüls Am.; Hüls AG] Ethoxylated castor oil; nonionic; emulsifier for fatty acids, solvs., cosmetic oils; textile lubricants, dyeing auxiliaries, pesticides, creams; biodeg.; liq.; cloud pt. 50 C (2% in 10% NaCl); 100% act.
Marlowet® R 40/K. [Hüls Am.] PEG-40 castor oil.
Marlowet® R 54. [Hüls AG] Ethoxylated castor oil; emulsifier for fatty acids, solvs., cosmetic oils; textile lubricants, dyeing auxiliaries, pesticides, creams; liq.; 100% act.
Marlowet® RA. [Hüls Am.; Hüls AG] Mixt. of n-alkylbenzene sulfonate and carboxylic acid polyglycol esters; anionic/nonionic; emulsifier for essential oils, cosmetic preparations; biodeg.; liq.; 70% act.
Marlowet® RNP. [Hüls Am.; Hüls AG] Mixt. of carboxylic acid polyglycol ether and a glycol deriv.; nonionic; emulsifier for essential oils; liq.; cloud pt. 68 C (2% aq.); 100% act.
Marlowet® RNP/K. [Hüls Am.] Octoxynol-9, PEG-40 castor oil, ethoxydiglycol.
Marlowet® SAF. [Hüls Am.; Hüls AG] Mixt. of alkyl polyglycol ethers and glycol deriv.; nonionic; emulsifier for min. oils, esp. wh. oils, textile lubricants, metal processing; liq.; cloud pt. 90 C (10% in 25% BDG); 90% act.
Marlowet® SAF/K. [Hüls Am.] Oleth-5, laureth-4.; nonionic.
Marlowet® SDT. [Hüls AG] Fatty acid diethanolamide; corrosion-inhibiting additive for metal processing fluids; liq.; 100% act.
Marlowet® SLM. [Hüls Am.] Mixt. of carboxylic acid polyglycol esters and alkyloamides; nonionic; emulsifier for textile lubricants; liq.; 100% act.
Marlowet® SLS. [Hüls Am.; Hüls AG] C18 fatty acid ester of ethoxylated castor oil; nonionic; emulsifier for min. oils, solvs., wh. spirit, bitumen, textile lubricants, cold cleaners, road construction; liq.; cloud pt. 65 C (10% in 25% BDG); 100% act.
Marlowet® SW. [Hüls Am.; Hüls AG] Mixt. of carboxylic acid polyglycol ester and alkyl polyglycol ether; nonionic; emulsifier for min. oils, textile lubricants, metalworking; liq.; cloud pt. 90 C (10% in 25% BDG); 100% act.
Marlowet® SWN. [Hüls Am.; Hüls AG] Mixt. of carboxylic acid polyglycol ester and alkyl polyglycol ether; nonionic; emulsifier for spindle and wh. oils, solvs., for easily washed-out textile lubricants, metal cleaning; liq.; cloud pt. 70 C (10% in 25% BDG); 100% act.
Marlowet® T. [Hüls Am.; Hüls AG] TEA salt of oleic acid; anionic; emulsifier for min. oils, wh. spirit, bitumen, for metalworking, road construction, floor care prods.; paste; 100% act.
Marlowet® TA 6. [Hüls Am.] Ceteareth-6.; CAS 68439-49-6.
Marlowet® TA 8. [Hüls Am.] Ceteareth-8.; CAS 68439-49-6.
Marlowet® TA 10. [Hüls Am.] Ceteareth-10.; CAS 68439-49-6.
Marlowet® TA 18. [Hüls Am.] Ceteareth-18.; CAS 68439-49-6.
Marlowet® TA 25. [Hüls Am.] Ceteareth-25.; CAS 68439-49-6.
Marlowet® TM. [Hüls Am.; Hüls AG] Nonylphenol polyglycol ether; nonionic; emulsifier for min. oils, for metalworking, cold cleaners; liq.; HLB 10.0; cloud pt. 61 C (10% in 25% BDG); 100% act.
Marlowet® WOE. [Hüls Am.; Hüls AG] Oleth-5; CAS 9004-98-2; nonionic; emulsifier for paraffinic min. oils, textile lubricants; liq.; cloud pt. 69 C (10% in 25% BDG); 100% act.
Marlowet® WSD. [Hüls Am.] Mixt. of alkyl and alkylaryl polyglycol ethers; nonionic; emulsifier for min. oils, textile lubricants; liq.; 100% act.
Marlox® 3000. [Hüls Am.; Hüls AG] Butyl glycol/alkylene oxide addition prod.; nonionic; surfactant for mfg. of textile auxiliaries and finishing agents; liq.; cloud pt. 64 C (2% in 10% NaCl); 100% act.
Marlox® B 24/50. [Hüls AG] C12-14 alcohol ethoxylate (5 EO), tert-butyl blocked; nonionic; low foaming surfactant, wetting agent for prod. of textile auxiliaries, bottle cleaners; strongly alkali-resistant; liq.; cloud pt. 36 C (10% in 25% BDG); 100% act.
Marlox® B 24/60. [Hüls Am.] C12-14 alcohol ethoxylate (6 EO), tert-butyl blocked; nonionic; low foaming surfactant, wetting agent for prod. of textile auxiliaries, bottle cleaners; strongly alkali-resistant; liq.; cloud pt. 40 C (10% in 25% BDG); 100% act.
Marlox® B 24/80. [Hüls Am.] C12-14 alcohol ethoxylate (8 EO), tert-butyl blocked; nonionic; low foaming surfactant, wetting agent for prod. of textile auxiliaries, bottle cleaners; strongly alkali-resistant; liq.; cloud pt. 36 C (2% aq.); 100% act.
Marlox® FK 14. [Hüls Am.; Hüls AG] Propylene glycol capreth-4; nonionic; detergent, antistat, foam controller; for low-foaming detergents and cleaners, dishwash, industrial cleaners, textile auxiliaries; liq.; cloud pt. 45 C (2% aq.); 100% act.
Marlox® FK 64. [Hüls Am.; Hüls AG] PPG-6 deceth-4; CAS 68154-97-2; nonionic; detergent, antistat, foam controller; for low-foaming detergents and cleaners, dishwash, industrial cleaners, textile auxiliaries; liq.; cloud pt. 55 C (10% in 25% BDG); 100% act.
Marlox® FK 69. [Hüls Am.; Hüls AG] PPG-6 deceth-9; CAS 68154-97-2; nonionic; detergent, antistat, foam controller; for low-foaming detergents and cleaners, dishwash, industrial cleaners, textile auxiliaries; liq.; cloud pt. 43 C (2% aq.); 100% act.
Marlox® FK 86. [Hüls Am.; Hüls AG] PPG-8 deceth-6; CAS 68154-97-2; nonionic; detergent, antistat, foam controller; for low-foaming detergents and cleaners, dishwash, industrial cleaners, textile auxiliaries; liq.; cloud pt. 22 C (2% aq.); 100% act.
Marlox® FK 1614. [Hüls AG] C10-12 fatty alcohol/alkylene oxide condensate; nonionic; detergent, antistat, foam controller; for low-foaming detergents and cleaners, dishwash, industrial cleaners, textile auxiliaries; liq.; cloud pt. 26 C; 100% act.
Marlox® L 6. [Hüls Am.; Hüls AG] PPG-7 lauryl ether; CAS 9064-14-6; nonionic; detergent, antistat, foam controller, textile auxiliary agent; liq.; cloud pt. 27 C (10% in 25%); 100% act.

Marlox® LM 25/30. [Hüls Am.] Alkylene oxide addition prod.; nonionic; detergent, antistat for controlled foam detergents and cleaners; liq.; 100% conc.

Marlox® LM 55/18. [Hüls Am.; Hüls AG] Alkylene oxide addition prod.; nonionic; component for low foaming industrial cleaners; liq.; 100% conc.

Marlox® LM 75/30. [Hüls Am.; Hüls AG] Alkylene oxide addition prod.; nonionic; detergent, antistat, foam controller for controlled foam detergents and cleaners; liq.; cloud pt. 42 C (2% aq.); 100% act.

Marlox® LP 90/20. [Hüls Am.; Hüls AG] Alkylene oxide addition prod.; nonionic; detergent, antistat, foam controller for controlled foam detergents and cleaners; liq.; cloud pt. 19 C (2% aq.); 100% act.

Marlox® M 606/1. [Hüls Am.; Hüls AG] TEA soap blend; anionic/nonionic; ingred. for liq. detergents and low foam cleaners; liq.; 95% act.

Marlox® M 606/2. [Hüls Am.; Hüls AG] TEA soap blend; anionic/nonionic; ingred. for liq. detergents, low foam cleaners, high pressure cleaners, lubricants; liq.; 95% act.

Marlox® MO 124. [Hüls Am.; Hüls AG] PPG-4 laureth-2; CAS 68439-51-0; nonionic; component for low foaming detergents and cleaners, automatic dishwasher formulations, industrial cleaners, textile auxiliaries; liq.; cloud pt. 39 C (10% in 25% BDG); 100% act.

Marlox® MO 145. [Hüls AG] C12-14 fatty alcohol/alkylene oxide condensate; nonionic; component for low-foaming detergents and cleaners, dishwash, industrial cleaners, textile auxiliaries; liq.; cloud pt. 43 C (10% in 25% BDG); 100% act.

Marlox® MO 154. [Hüls Am.; Hüls AG] PPG-4 laureth-5; CAS 68439-51-0; nonionic; component for low foaming detergents and cleaners, automatic dishwasher formulations, industrial cleaners; liq.; cloud pt. 42 C; 100% act.

Marlox® MO 174. [Hüls Am.; Hüls AG] PPG-4 laureth-7; CAS 68439-51-0; nonionic; component for low foaming detergents and cleaners, automatic dishwasher formulations, industrial cleaners, textile auxiliaries; liq.; cloud pt. 41 C (2% aq.); 100% act.

Marlox® MO 244. [Hüls AG] C12-14 fatty alcohol/alkylene oxide condensate; nonionic; component for low-foaming detergents and cleaners, dishwash, industrial cleaners, textile auxiliaries; liq.; cloud pt. 55 C (10% in 25% BDG); 100% act.

Marlox® MS 48. [Hüls Am.; Hüls AG] C12-18 fatty alcohol alkylene oxide addition prod.; CAS 69227-21-0; nonionic; surfactant for mfg. of textile auxiliaries and finishing agents; liq.; cloud pt. 72 C (10% in 25% BDG); 90% act.

Marlox® ND 121. [Hüls Am.; Hüls AG] Alkylene oxide addition prod.; nonionic; surfactant for mfg. of textile auxiliaries and finishing agents; liq.; 100% conc.

Marlox® NP 109. [Hüls Am.; Hüls AG] Nonylphenol/alkylene oxide addition prod.; nonionic; surfactant, detergent, antistat for low-foam cleaning formulations; liq.; cloud pt. 29 C (2% aq.); 100% act.

Marlox® OD 105. [Hüls Am.; Hüls AG] C8 fatty alcohol/alkylene oxide addition prod.; nonionic; surfactant, foam regulator for mfg. of textile auxiliaries and finishing agents; liq.; cloud pt. 20 C (2% aq.); 100% act.

Marlox® Q 286. [Hüls AG] C16-18 fatty alcohol/alkylene oxide condensate; nonionic; surfactant for prod. of textile auxiliaries; liq.; cloud pt. 62 C (10% in 25% BDG); 100% act.

Marlox® S 58. [Hüls Am.; Hüls AG] C13-15 fatty alcohol/alkylene oxide addition prods.; nonionic; detergent, foam regulator in low-foam formulations, automatic dishwashing, industrial cleaners; liq.; cloud pt. 39 C (10% in 25% BDG); 100% act.

Marlox® T 50/5. [Hüls Am.; Hüls AG] C16-18 fatty alcohol alkylene oxide addition prod.; nonionic; surfactant for mfg. of textile auxiliaries and finishing agents; liq.; cloud pt. 54 C (10% in 25% BDG); 90% act.

Marvanfix® ATA. [Marlowe-Van Loan] Nonformaldehyde fixing agent for improved wetfastness of direct dyes on cellulosic fibers; clear liq., mild odor; misc. with water; pH 4.5 (1%).

Marvanfix® C. [Marlowe-Van Loan] Quat. ammonium resin complex; textile fixative; improves wetfastness on direct and reactive dyes.

Marvanfix® FNC. [Marlowe-Van Loan] Polyquat. amine; nonformaldehyde textile fixative; minimal shade change.

Marvanfix® NDF. [Marlowe-Van Loan] Sodium salt of sulfonated phenolic deriv.; nylon fixative; reserving agent.

Marvanlube 1031. [Marlowe-Van Loan] Complex polymeric; nonionic; lubricant, scouring agent, dyeing assistant for cellulosics and syn. blends.

Marvanol® Aftertreat 2AF. [Marlowe-Van Loan] Emulsion polymer; dyeing assistant; exhaust pigment aftertreat for garments, hosiery.

Marvanol® BAN. [Marlowe-Van Loan] Sulfated veg. oil; dye dispersant; minimizes barré.

Marvanol® BVD. [Marlowe-Van Loan] Sulfated/phosphated esters; wetting agent; leveling agent for direct dyes on cellulosics.

Marvanol® CO. [Marlowe-Van Loan] Ethoxylated phenol; textiles detergent, emulsifier, wetting agent in finishing operations; colorless liq.; 99+% act.

Marvanol® DC. [Marlowe-Van Loan] Ethoxylated alcohol; nonionic; textile fiber wetting and dyeing agent; for 100% cotton, polyester/cotton blend and package, beck, jigs, and paddle operations; colorless liq.; 35% act.

Marvanol® Defoamer AM-2. [Marlowe-Van Loan] Conc. silicone emulsion; nonionic; foam inhibitor and depressant for textile applics.

Marvanol® Defoamer MOB. [Marlowe-Van Loan] Nonsilicone; nonionic; biodeg. textile defoamer for high temps., jet, package and pressure becks.

Marvanol® Defoamer MR-30A. [Marlowe-Van Loan] Silicone emulsion; textile defoamer; salt, acid, and alkali stable.

Marvanol® Defoamer S-22. [Marlowe-Van Loan] Silicone emulsion; textile defoamer, antifoam.

Marvanol® GAW. [Marlowe-Van Loan] Ethoxylated org. deriv.; textile leveling agent; compatibilizer for basic and acid dyes.

Marvanol® GC. [Marlowe-Van Loan] Sulfosuccinamide; anionic; textile fiber wetting and dye leveling agent; amber gel; 65% act.

Marvanol® KMA. [Marlowe-Van Loan] Cryptocationic blended org. derivs.; dyeing assistant; minimizes strike differential of basic dyes.

Marvanol® KXL. [Marlowe-Van Loan] Org. esters; nonionic; textile leveling agent, lubricant for disperse dyes on polyester and nylon.

Marvanol® LSL. [Marlowe-Van Loan] Ethoxylated surfactant blend; nonionic; compatibilizer for basic and acid dyes on synthetics.

Marvanol® Leveler DL. [Marlowe-Van Loan] Cationic surfactant; cationic; leveler, nonretarding migrator for basic dyes on acrylic and basic dyeable polyester.

Marvanol® Leveltone 7.5. [Marlowe-Van Loan] Sulfated fatty ester; anionic; mild scouring agent, penetrant, leveler for disperse dyes on nylon.

Marvanol® MRB. [Marlowe-Van Loan] Neutralized phosphated alcohol ethoxylate; nonionic; textile scouring agent, wetting agent, dyeing assistant for direct, vat or reactive dyes on cellulosic blends with nylon; clear liq.; sp.gr. 1.02; pH 6.5 (1%); usage level: 1-2%.

Marvanol® NHM. [Marlowe-Van Loan] Ethoxylated org. deriv.; nylon dye leveler; migrator.

Marvanol® Penetrant 35. [Marlowe-Van Loan] Amine neutralized sulfonic acid; anionic; detergent, wetting agent, textile dyeing, scouring agent, finishing, leveling and retarding agents for acid dyes; amber liq.; 40% act.

Marvanol® POL-41. [Marlowe-Van Loan] Surfactants; nonionic; high temp. leveling agent for polyester and its blends.

Marvanol® Pretreat GD-P. [Marlowe-Van Loan] Polymeric resin; dyeing assistant; exhaust pigment pretreat for garments, hosiery.

Marvanol® Pretreat HPC. [Marlowe-Van Loan] Polymeric amine deriv.; dyeing assistant; exhaust pigment pretreat for garments, hosiery; for heavy shades.

Marvanol® RD2-1852. [Marlowe-Van Loan] Sulfated ester blend; dyeing assistant; dyebath lubricant for direct dyes on cotton.

Marvanol® RD2-2284. [Marlowe-Van Loan] Copolymers sol'n.; dyeing assistant for cellulosics; exc. in low liquor ratios.

Marvanol® RD2-2581. [Marlowe-Van Loan] Org. esters/ethers; nonionic; biodeg. leveling agent; scour for nylon hosiery and piece goods.

Marvanol® RDF. [Marlowe-Van Loan] Sol'n. of org. and inorg. bases; liq. alkali dyeing assistant for use with fiber reactive dyes; clear liq., odorless; sol. in cold or warm water; dens. 12 lb/gal; pH > 12.0 (5%); corrosive.

Marvanol® RE-1274. [Marlowe-Van Loan] 1:1 coconut fatty amide; cationic; emulsifer, visc. builder, dye leveler; amber liq.; sp.gr. 0.950; 97% act.

Marvanol® RE-1281. [Marlowe-Van Loan] 2:1 coconut fatty amide; cationic; detergent, emulsifer, visc. builder, dye leveler, fulling and scouring agent for most textiles; amber liq.; sp.gr. 1.02; 98% act.

Marvanol® RE-1824. [Marlowe-Van Loan] POE mono-ester of fatty acids; nonionic; dyeing assistant for disperse dyes on nylon and polyester.

Marvanol® REAC A-213. [Marlowe-Van Loan] Polymer salt; anionic; low foaming dispersing agent, textile scouring agent for fiber reactives; improves dye fixation; sl. yel. clear liq.; water-sol.; visc. 200-600 mPa•s; pH 8-10 (1%).

Marvanol® REACT 1051. [Marlowe-Van Loan] Org. blend; dispersing agent, textile scouring agent for fiber reactives; improves solubility and cleanup.

Marvanol® SBO (60%). [Marlowe-Van Loan] Sulfated butyl oleate, sodium salt; anionic; detergent, wetting and leveling agent, emulsifier, dyeing assistant, lubricant; amber liq.; 60% act.

Marvanol® SCO (50%). [Marlowe-Van Loan] Sulfated castor oil; anionic; detergent, wetting agent, dyeing assistant and lubricant used in finishing operations; leveling agent for cotton; amber liq.; 50% act.

Marvanol® Scour 2 Base. [Marlowe-Van Loan] Alcohol ether condensate; nonionic; multifunctional biodeg. textile scour.

Marvanol® Scour 05. [Marlowe-Van Loan] Ethoxylate; nonionic; cold water scouring agent for all natural and syn. fibers.

Marvanol® Scour 2582. [Marlowe-Van Loan] Ethoxylated alcohol; nonionic; biodeg. textile scouring agent for natural and syn. fibers; acid and alkali stable.

Marvanol® Scour FRM. [Marlowe-Van Loan] Blend; anionic; afterscour for soaping off fiber reactive, vat, direct and naphthol dyes on 100% cotton and rayon blends; for package and piece dyeing operations; stable to alkali; amber clear liq.; pH 8.5 (1%); usage level: 1-2%.

Marvanol® Scour LF. [Marlowe-Van Loan] Alkoxylated alcohol; nonionic; biodeg. low foaming scouring agent for cotton and blends; for atmospheric and pressure applics.

Marvanol® Scour PCO. [Marlowe-Van Loan] Modified polyglycol ether/sulfated ester; scouring agent, wetting agent for all textiles, continuous and batch operations.

Marvanol® Solvent Scour 34. [Marlowe-Van Loan] Solvs. and detergents; anionic; solv. scour; amber liq.; 100% act.

Marvanol® SOR. [Marlowe-Van Loan] Hydrophilic copolymer; multipurpose soil release for PES/ dyebath or pad; yields soft hand, rewetting and improved sewability.

Marvanol® SPO (60%). [Marlowe-Van Loan] Sulfated propyl oleate, sodium salt; anionic; detergent, wetting and leveling agent, dyeing assistant, lubricant for textiles; amber liq.; 60% act.

Marvanscour® KW. [Marlowe-Van Loan] Solvs. and detergents; nonionic; detergent for prescour and afterscour; straw liq.; 83% act.

Marvanscour® LF. [Marlowe-Van Loan] Phosphate ester; anionic/nonionic; low-foaming wetting agent and dyeing assistant for jet dyeing cotton and blends; clear liq.; completely sol. in water; sp.gr. 1.03; dens. 8.59 lb/gal; pH 4.2-4.8; corrosive.

Marvantex BS. [Marlowe-Van Loan] Alkaline built blend; anionic; detergent, alkaline scour for dye stripping; yel. to wh. powd.; 100% act.

Marvantex GS. [Marlowe-Van Loan] Alkaline scour; anionic; detergent used for oil, grease and graphite removal; powd.; 100% act.

Marvantex RBDS. [Marlowe-Van Loan] Inorg. salts; one-bath scour, bleaching agent, and dye.

Marvantex T-100. [Marlowe-Van Loan] Anionic surfactant/alkaline builders; for prescouring of syn. fibers; powd.

Marvelin W-50. [Matsumoto Yushi-Seiyaku] Alkylene oxide addition prods.; nonionic; dye leveling agent of wool; flake; 100% conc.

Masil® 173. [PPG/Specialty Chem.] Silicone fluid dispersed in halogenated org. solv.; defoamer for chem. processing, refiners, solv. cleaning; esp. for defoaming light hydrocarbon solvs.; liq.; sp.gr. 1.32; visc. 60 cSt; flash pt. (PMCC) > 300 F; 5% act.

Masil® 1066C. [PPG/Specialty Chem.] Dimethicone copolyol; lubricant and antistat for plastics, textiles, metal processing; wetting and leveling char.; antifog for glass cleaners; liq.; water-sol.; sp.gr. 1.02; visc. 1800 cSt; cloud pt. 42 C (1% aq.); flash pt. (PMCC) > 300 F; pour pt. –50 C; surf. tens. 25.9 dynes/cm (1% aq.).

Masil® 1066D. [PPG/Specialty Chem.] Dimethicone copolyol; lubricant and antistat for plastics, textiles, and metal processing; wetting props.; liq.; water-sol.; sp.gr. 1.03; visc. 1050 cSt; cloud pt. 37 C (1% aq.); flash pt. (PMCC) > 300 F; pour pt. –33 C; surf. tens. 25.2 dynes/cm (1% aq.).

Masil® 2132. [PPG/Specialty Chem.] Silicone glycol; antistat and wetting agent for personal care prods., textile, plastics, lubricants and formulations; solv. based coating, dispersant, and antifoam; liq.; sp.gr. 1.03; visc. 2000 cSt; cloud pt. 38 C (1% aq.); flash pt. (PMCC) > 300 F; pour pt. –40 C; surf. tens. 26.5 dynes/cm (1% aq.); 100% conc.

Masil® 2133. [PPG/Specialty Chem.] Silicone glycol; antistat and wetting agent for personal care prods., textile, plastics, lubricants and formulations; solv. based coating, dispersant, and antifoam; liq.; sp.gr. 1.03; visc. 1050 cSt; cloud pt. 37 C (1% aq.); flash pt. (PMCC) > 300 F; pour pt. -33 C; surf. tens. 25.8 dynes/cm (1% aq.).

Masil® 2134. [PPG/Specialty Chem.] Silicone glycol; antistat and wetting agent for personal care prods., textile, plastics, lubricants and formulations; solv. based coating, dispersant, and antifoam; liq.; sp.gr. 1.02; visc. 1800 cSt; cloud pt. 39 C (1% aq.); flash pt. (PMCC) > 300 F; pour pt. -50 C; surf. tens. 32.0 dynes/cm (1% aq.).

Masil® SF 5. [PPG/Specialty Chem.] Dimethicone; release aid, defoamer for nonaq. processes, esp. in the petrol., foods, and printing inks industries; internal lubricant for plastics, rubber, and metal; also in furniture and auto-wax polishes, household and personal care prods.; textile lubricant; lower visc. fluids recommended for cosmetic applications; higher visc. fluids (> 10,000 cSt) for formulating lubricants and mold releases in mfg. of plastics, rubber parts; water-wh. oily, clear liq.; odorless, tasteless; sp.gr. 0.916; visc. 5 cSt; pour pt. -65 C; flash pt. (PMCC) 280 F; ref. index 1.3970.

Masil® SF 10. [PPG/Specialty Chem.] Dimethicone; see Masil SF 5; sp.gr. 0.940; pour pt. -65 C; flash pt. (PMCC) 320 F; ref. index 1.3990.

Masil® SF 20. [PPG/Specialty Chem.] Dimethicone; see Masil SF 5; water-wh. oily, clear liq.; odorless, tasteless; sp.gr. 0.953; visc. 20 cSt; pour pt. –65 C; flash pt. (PMCC) 395 F; ref. index 1.4010.

Masil® SF 50. [PPG/Specialty Chem.] Dimethicone; see Masil SF 5; water-wh. oily, clear liq.; odorless, tasteless; sp.gr. 0.963; visc. 50 cps; pour pt. –55 C; flash pt. (PMCC) 460 F; ref. index 1.4020.

Masil® SF 100. [PPG/Specialty Chem.] Dimethicone; see Masil SF 5; water-wh. oily, clear liq.; odorless, tasteless; sp.gr. 0.968; visc. 100 cps; pour pt. –55 C; flash pt. (PMCC) 461 C; ref. index 1.4030.

Masil® SF 200. [PPG/Specialty Chem.] Dimethicone; see Masil SF 5; water-wh. oily, clear liq.; odorless, tasteless; sp.gr. 0.972; visc. 200 cSt; pour pt. –50 C; flash pt. (PMCC) 460 F; ref. index 1.4031.

Masil® SF 350. [PPG/Specialty Chem.] Dimethicone; see Masil SF 5; water-wh. oily, clear liq.; odorless, tasteless; sp.gr. 0.973; visc. 350 cSt; pour pt. –50 C; flash pt. (PMCC) 500 F; ref. index 1.4032.

Masil® SF 350 FG. [PPG/Specialty Chem.] Dimethicone; see Masil SF 5; sp.gr. 0.973; pour pt. -50 C; flash pt. (PMCC) 500 F; ref. index 1.4032.

Masil® SF 500. [PPG/Specialty Chem.] Dimethicone; see Masil SF 5; water-wh. oily, clear liq.; odorless, tasteless; sp.gr. 0.973; visc. 500 cSt; pour pt. –50 C; flash pt. (PMCC) 500 F; ref. index 1.4033.

Masil® SF 1000. [PPG/Specialty Chem.] Dimethicone; see Masil SF 5; water-wh. oily, clear liq.; odorless, tasteless; sp.gr. 0.974; visc. 1000 cSt; pour pt. –50 C; flash pt. (PMCC) 500 F; ref. index 1.4035; CC flash pt. 260 C.

Masil® SF 5000. [PPG/Specialty Chem.] Dimethicone; see Masil SF 5; water-wh. oily, clear liq.; odorless, tasteless; sp.gr. 0.975; visc. 5000 cSt; pour pt. –49 C; flash pt. (PMCC) 500 F; ref. index 1.4035.

Masil® SF 10,000. [PPG/Specialty Chem.] Dimethicone; see Masil SF 5; water-wh. oily, clear liq.; odorless, tasteless; sp.gr. 0.975; visc. 10,000 cSt; pour pt. –47 C; flash pt. (PMCC) 500 F; ref. index 1.4035.

Masil® SF 12,500. [PPG/Specialty Chem.] Dimethicone; see Masil SF 5; water-wh. oily, clear liq.; odorless, tasteless; sp.gr. 0.975; visc. 12,500 cSt; pour pt. –47 C; flash pt. (PMCC) 500 F; ref. index 1.4035.

Masil® SF 30,000. [PPG/Specialty Chem.] Dimethicone; see Masil SF 5; water-wh. oily, clear liq.; odorless, tasteless; sp.gr. 0.976; visc. 30,000 cSt; pour pt. –46 C; flash pt. (PMCC) 500 F; ref. index 1.4035.

Masil® SF 60,000. [PPG/Specialty Chem.] Dimethicone; see Masil SF 5; water-wh. oily, clear liq.; odorless, tasteless; sp.gr. 0.977; visc. 60,000 cSt; pour pt. –44 C; flash pt. (PMCC) 500 F; ref. index 1.4035.

Masil® SF 100,000. [PPG/Specialty Chem.] Dimethicone; see Masil SF 5; water-wh. oily, clear liq.; odorless, tasteless; sp.gr. 0.978; visc. 100,000 cSt; pour pt. –40 C; flash pt. (PMCC) 500 F; ref. index 1.4035.

Masil® SF 300,000. [PPG/Specialty Chem.] Dimethicone; see Masil SF 5; water-wh. oily, clear liq.; odorless, tasteless; sp.gr. 0.978; visc. 300,000 cSt; pour pt. –40 C; flash pt. (PMCC) 500 F; ref. index 1.4035.

Masil® SF 500,000. [PPG/Specialty Chem.] Dimethicone; see Masil SF 5; sp.gr. 0.978; pour pt. -40 C; flash pt. (PMCC) 500 F; ref. index 1.4035.

Masil® SF 600,000. [PPG/Specialty Chem.] Dimethicone; see Masil SF 5; water-wh. oily, clear liq.; odorless, tasteless; sp.gr. 0.979; visc. 600,000 cSt; pour pt. –34 C; flash pt. (PMCC) 500 F; ref. index 1.4035.

Masil® SF 1,000,000. [PPG/Specialty Chem.] Dimethicone; see Masil SF 5; sp.gr. 0.979; pour pt. -25 C; flash pt. (PMCC) 500 F; ref. index 1.4035.

Matexil AA-NS. [ICI Surf. UK] Polyglycol and fatty acid surfactants in min. hydrocarbon; anionic; deaerating and antifoaming agent for textile processes; liq.

Matexil BA-PK. [ICI Surf. UK] Alkaline sol'n. of inorganic sodium salts; anionic; neutralizer for residual peroxide from hydrogen peroxide bleaching baths before dyeing cellulosic fibers with reactive or direct dyes; liq.

Matexil Binder AS. [ICI Surf. UK] Reactive acrylic copolymer aq. disp.; anionic/nonionic; film-forming resin for pigment printing of fabrics; liq.

Matexil Binder BD. [ICI Surf. UK] Carboxylated butadiene/styrene/acrylonitrile latex with nonstaining antioxidant; anionic; binder for pigment printing systems, esp. on fashion prints where handle is important; liq.

Matexil BN PA. [ICI Surf. UK] Substituted biguanamide; nonionic; protecting agent preventing degradation of polyamide fibers during peroxide bleaching of polyamide-cotton blends; powd.

Matexil CA DPL. [ICI Surf. UK] Diphenyl; anionic; carrier, leveling agent for use in high-temp. dyieng

of polyester/cellulose with disperse dyes; liq.
Matexil CA MN. [ICI Surf. UK] Methyl naphthalene; anionic; carrier for dyeing of polyester and blends with disperse dyes; liq.
Matexil DA N. [ICI Surf. UK] Naphthalene sulfonic acid salt condensate; anionic; dispersant for dyeing of polyester and blends with disperse dyes; liq.
Matexil DN VL 200. [ICI Surf. UK] Fatty alcohol ethoxylate; nonionic; dispersant applied in the alkaline reduction clear after treatment of dyeings with disperse dyes on polyester fibers; antiprecipitant; antistat on polyamide; liq.
Matexil FA MIV. [ICI Surf. UK] Polymeric acrylic aq. sol'n.; anionic; migration inhibitor for use in continuous dyeing of cellulose and polyester/cellulose blends with vat and disperse dyes; liq.
Matexil FA N. [ICI Surf. UK] Heterocyclic carboxylic acid; anionic; acid catalyst permitting fixation of dyes on polyester-cellulose blends from a neutral print paste; also for cotton, wool, silk and blends; powd.
Matexil FA SN Liq. [ICI Surf. UK] Sulfonated aromatic condensate; anionic; resist agent used in prod. of multi-color effects in printing of nylon carpets; syntan and restraining agent for dyeing wool/nylon blends; liq.
Matexil FC ER. [ICI Surf. UK] Low m.w. resin aq. sol'n.; cationic; fixing agent improving wetfastness of reactive dyed cellulose; prevents migration of residual reactive dyes during processing; liq.
Matexil FC PN. [ICI Surf. UK] Dicyandiamide-formaldehyde resin condensate; CAS 26591-12-8; cationic; fixing agent improving wetfastness of direct and reactive dyes on cellulosic fibers; liq.
Matexil Fixer SF. [ICI Surf. UK] Heterocyclic aminoplast resin; nonionic; crosslinking agent for binders used in pigment printing systems; liq.
Matexil LA NS. [ICI Surf. UK] Sodium salt of highly sulfonated oil/ethylene oxide condensate; anionic; dyebath leveling agent minimizing barre in dyeing polyamide fibers with acid dyes; liq.
Matexil LC CWL. [ICI Surf. UK] Ethylene oxide condensate; cationic; leveling agent for dyeing polyamide with anionic dyestuffs; liq.
Matexil LC RA. [ICI Surf. UK] Quaternary ammonium compd. with small amt. of isopropanol; cationic; nonhydrolyzable retarding agent for ensuring level dyeing of modified basic dyes on acrylic fibers; liq.
Matexil LN RD. [ICI Surf. UK] Fatty acid/ethylene oxide condensate; nonionic; leveling and stripping agent for disperse dyes on polyester; liq.
Matexil PA Liq. [ICI Surf. UK] Sodium meta-nitrobenzene sulfonate; anionic; mild oxidizing agent for protecting dyestuffs against reduction and for use in oxidation of vat dyes; liq.
Matexil PA SNX Liq. [ICI Surf. UK] Phenol sulfonate; anionic; resist agent for prod. of multi-color effects in printing of nylon carpets; liq.
Matexil PN. [ICI Surf. UK] Ethoxylate/polyol blend; nonionic; for alkaline discharge of Dispersol PC dyes on polyester fabrics; liq.
Matexil PN DG. [ICI Surf. UK] Ethoxylate blend; nonionic; discharge printing assistant to permit alkali discharge printing of polyester ground shades dyed by conventional exhaust methods with Dispersol PC dyes; liq.
Matexil PN MFC. [ICI Surf. UK] Ethoxylate blend; nonionic; multi-fiber fixation assistant for use with disperse dyes in textile printing; liq.
Matexil PN PR. [ICI Surf. UK] Fatty alcohol ethoxylate; nonionic; emulsifier for prep. of o/w emulsion printing thickeners for printing reactive dyes on cellulose, wool, silk, and disperse dyes on syn. fabrics; flakes.
Matexil Softener GK. [ICI Surf. UK] Polymeric plasticizer/silicone aq. emulsion; nonionic; softener for incorporation into pigment printing systems to give improved handle and rub fastness; liq.
Matexil Thickener CP. [ICI Surf. UK] Acrylic copolymer blend in org. solv.; nonionic; thickener for pigment printing; liq.
Matexil WA HS. [ICI Surf. UK] Sodium salt of highly sulfonated oil in aq. sol'n.; anionic; biodeg. wetting agent for use with vat, reactive, sulfur, or direct dyes; promotes wetting and penetration for textile processing; liq.
Matexil WA KBN. [ICI Surf. UK] Sulfonated syn. fish oil/pine oil aq. emulsion; anionic; wetting agent, lubricant, mild antifoam for textile print pastes containing reactive, disperse, acid, or premetallized dyes; liq.
Matexil WN PB. [ICI Surf. UK] Modified ethoxylate; nonionic; wetting agent, penetrant for pad-batch applic. of reactive dyes; penetrant for cellulosic substrates of poor absorbency; liq.
Maxitol No. 10. [Dexter] Sulfated ricinoleic ester; anionic; wetting and protective agent for the treatment of animal fibers; high resistance to sol'ns. of strong acids, alkalies, lime, and magnesium; liq.; 45% conc.
Mayco Base BFO. [Mayco Oil & Chem.] Fatty ester; nonionic; lubricant additive; liq.; 100% conc.
Mayphos 45. [Mayco Oil & Chem.] Org. phosphate ester, free acid; anionic; EP lubricant additive, metal wetting agent, surfactant; noncorrosive to ferrous and nonferrous metals; prevents water-induced plugging of diatomaceous earth filter; amber visc. liq.; sol. in min. oils; sp.gr. 1.06 (60/60 F); dens. 8.8 lb/gal; visc. 400 cSt (100 C); acid no. 200; flash pt. (COC) 193 C; pH 2.1 (2% in water/IPA); usage level: 0.5-1.%; 100% conc.
Maypon 4C. [Inolex] Potassium coco-hydrolyzed collagen; anionic; detergent used in personal care prods., general purpose cleansers; clear to slightly hazy amber liq.; misc. with water; visc. 500 cps max; pH 6.7–7.3; 34-36% solids.
Maypon 4CT. [Inolex] TEA-coco-hydrolyzed collagen; visc. builder, foam modifier; clear to slightly hazy amber liq.; misc. with water; visc. 150 cps max; pH 6.7–7.3; 38–40% solids.
May-Tein C. [Maybrook] Potassium cocoyl hydrolyzed collagen; anionic; mild surfactant, conditioner, softener, moisturizer, lubricant, antistat, anti-irritant for hair relaxers, shampoos, conditioners, liq. soaps, depilatories; biodeg.; liq.; 36% conc.
May-Tein CT. [Maybrook] TEA-cocoyl hydrolyzed collagen; anionic; mild surfactant, conditioner, anti-irritant for shampoos, conditioners, baby prods., mousses, makeup removers, shave creams, liq. soaps; biodeg.; liq.; 38% conc.
May-Tein SK. [Maybrook] Sodium cocoyl hydrolyzed collagen; anionic; mild surfactant, conditioner, antistat, lubricant, dye leveler, textile softener for shampoos, conditioners, hair coloring prods., fine woolen, silk, and hand washables; biodeg.; visc. liq.; sol. in cold water; 44% conc.
Mazamide® 65. [PPG/Specialty Chem.] 2:1 mixed fatty acid DEA; nonionic/anionic; detergent, thickener for liq. detergent systems and shampoos, emul-

sifier, foam booster, rust inhibitor; used in hard surf. cleaners, metalworking fluids/syn. coolants, waterless hand cleaners, automotive specialties; also solubilizer; biodeg.; liq.; sp.gr. 0.99–1.02; flash pt. (PMCC) > 300 F; pH 9.0–10.0; 100% conc.

Mazamide® 66. [PPG/Specialty Chem.] Mixed fatty acid alkanolamide; nonionic; detergent, thickener, emulsifier for cleaning compds., floor cleaners with stability at high pH; liq.; 100% conc.

Mazamide® 68. [PPG/Specialty Chem.] Cocamide DEA (2:1); nonionic; biodeg. thickener, emulsifier, foam builder; for hard surf. cleaners, dishwash, shampoos, metalworking fluids, waterless hand cleaners, automotive specialties; liq.; sp.gr. 0.99; flash pt. (PMCC) > 300 F; 100% conc.

Mazamide® 70. [PPG/Specialty Chem.] 2:1 Cocamide DEA and diethanolamine; nonionic; emulsifier, detergent, solubilizer, thickener used in hard surface cleaners, dishwash liq., metalworking fluids; corrosion inhibitor for sol. oils; biodeg.; liq.; sp.gr. 1.01; pH 9–10; flash pt. (PMCC) > 300 F; 100% conc.

Mazamide® 80. [PPG/Specialty Chem.] 1:1 Cocamide DEA; nonionic; thickener, foam stabilizer, solubilizer, emulsifier used in cosmetic and toiletry formulations, hard surface wetters and cleaners, household and industrial detergents; biodeg.; liq.; sp.gr. 0.98–1.00; pH 9–10.5; flash pt. (PMCC) > 300 F; 100% conc.

Mazamide® 524. [PPG/Specialty Chem.] Cocamide DEA (2:1); biodeg. thickener, emulsifier, foam builder; for hard surf. cleaners, dishwash, shampoos, metalworking fluids, waterless hand cleaners, automotive specialties; liq.; sp.gr. 1.01; flash pt. (PMCC) > 300 F; 100% conc.

Mazamide® 1214. [PPG/Specialty Chem.] Lauramide DEA (2:1); CAS 120-40-1; biodeg. thickener, emulsifier, foam builder; for hard surf. cleaners, dishwash, shampoos, metalworking fluids, waterless hand cleaners, automotive specialties; liq.; sp.gr. 1.02; flash pt. (PMCC) > 300 F.

Mazamide® 1281. [PPG/Specialty Chem.] Cocamide DEA (2:1); nonionic; biodeg. foam booster/stabilizer, visc. builder, detergent, emulsifier; for hard surf. cleaners, dishwash, shampoos, metalworking fluids, waterless hand cleaners, automotive specialties; liq.; sp.gr. 0.99; flash pt. (PMCC) > 300 F; 100% conc.

Mazamide® C-5. [PPG/Specialty Chem.] PEG-6 cocamide MEA; nonionic; emulsifier, lubricant, rust inhibitor, buffing compd., thickener, foam booster, detergent, solubilizer, cosmetics; biodeg.; liq.; sp.gr. 1.05; flash pt. (PMCC) > 300 F; pH 9.5–10.5; 100% conc.

Mazamide® CCO. [PPG/Specialty Chem.] Cocamide DEA (2:1); nonionic; biodeg. emulsifier, wetting agent for heavy and lt. duty cleansing applics., hard surf. cleaners, dishwash, shampoos, metalworking fluids, waterless hand cleaners, automotive specialties; good visc. at low concs.; liq.; sp.gr. 1.02; flash pt. (PMCC) > 300 F; 100% conc.

Mazamide® CFAM. [PPG/Specialty Chem.] Cocamide MEA (1:1); CAS 68140-00-1; nonionic; biodeg. foam stabilizer, visc. builder for cosmetic and pharmaceutical shampoos; flake; sp.gr. 0.93; solid. pt. 70 C; flash pt. (PMCC) > 300 F; 100% conc.

Mazamide® CMEA Extra. [PPG/Specialty Chem.] Cocamide MEA (1:1); CAS 68140-00-1; nonionic; biodeg. foam stabilizer, thickener, emulsifier for shampoos, bubble bath, dishwash, household and cosmetic preps.; flake; sp.gr. 0.93; solid. pt. 61 C; flash pt. (PMCC) > 300 F; 100% conc., 85% amide.

Mazamide® CS 148. [PPG/Specialty Chem.] 1:1 Cocamide DEA; nonionic; thickener, foam stabilizer, emulsifier, solubilizer used in cosmetic and toiletry formulations, industrial and household detergents; biodeg.; liq.; sp.gr. 0.98–1.00; flash pt. (PMCC) > 300 F; pH 9.0–10.5; 100% conc.

Mazamide® J 10. [PPG/Specialty Chem.] Mixed alkanolamide (2:1); biodeg. thickener, emulsifier, foam builder; for hard surf. cleaners, dishwash, shampoos, metalworking fluids, waterless hand cleaners, automotive specialties; liq.; sp.gr. 1.01; flash pt. (PMCC) > 300 F.

Mazamide® JR 100. [PPG/Specialty Chem.] Modified coconut alkanolamide (2:1); nonionic; biodeg. emulsifier providing high visc. and low foam for hard surf. cleaners; liq.; sp.gr. 0.93; flash pt. (PMCC) > 300 F; 100% conc.

Mazamide® JR 300. [PPG/Specialty Chem.] Mixed alkanolamide (2:1); biodeg. thickener, emulsifier, foam builder; for hard surf. cleaners, dishwash, shampoos, metalworking fluids, waterless hand cleaners, automotive specialties; liq.; sp.gr. 1.01; flash pt. (PMCC) > 300 F.

Mazamide® JR 400. [PPG/Specialty Chem.] Mixed alkanolamide (2:1); biodeg. thickener, emulsifier, foam builder; for hard surf. cleaners, dishwash, shampoos, metalworking fluids, waterless hand cleaners, automotive specialties; liq.; sp.gr. 1.00; flash pt. (PMCC) > 300 F.

Mazamide® JT 128. [PPG/Specialty Chem.] Cocamide DEA (2:1); nonionic; biodeg. foam stabilizer, thickener, emulsifier for shampoos, bubble bath, dishwash, household and cosmetic preps.; liq.; sp.gr. 0.99; flash pt. (PMCC) > 300 F; 100% conc.

Mazamide® L-5. [PPG/Specialty Chem.] PEG-6 lauramide DEA; nonionic; emulsifier, lubricant, rust inhibitor, buffing compd., detergent, foam builder, stabilizer for cosmetic and pharmaceutical creams, lotions, bath oils, shampoos; biodeg.; solid; sp.gr. 1.05; flash pt. (PMCC) > 300 F; pH 9.5–10.5; 100% conc.

Mazamide® L-298. [PPG/Specialty Chem.] 2:1 Lauramide DEA; CAS 120-40-1; nonionic/anionic; foam builder and stabilizer, emulsifier, dispersant, visc. builder, solubilizer for hard surface cleaners, dishwashing, shampoos, metalworking fluids, automotive specialties, fiber and hair conditioners, dry cleaning, agric. sprays,; leather/fur preparations, emulsifiable waxes, rust inhibitors, polishes, paint removers, rug shampoos, fuel oil additives, textile detergents; biodeg.; solid; water sol.; sp.gr. 1.00; flash pt. (PMCC) > 300 F; pH 9.4 (5% aq.).

Mazamide® LLD. [PPG/Specialty Chem.] Linoleamide DEA (2:1); biodeg. thickener, emulsifier, foam builder; for hard surf. cleaners, dishwash, shampoos, metalworking fluids, waterless hand cleaners, automotive specialties; liq.; sp.gr. 0.97; flash pt. (PMCC) > 300 F.

Mazamide® LM. [PPG/Specialty Chem.] Lauramide DEA (2:1); CAS 120-40-1; biodeg. thickener, emulsifier, foam builder; for hard surf. cleaners, dishwash, shampoos, metalworking fluids, waterless hand cleaners, automotive specialties; liq.; sp.gr. 1.00; flash pt. (PMCC) > 300 F.

Mazamide® LM 20. [PPG/Specialty Chem.] 2:1 Lauramide DEA; CAS 120-40-1; nonionic/anionic; biodeg. thickener, emulsifier, foam builder; for hard surf. cleaners, dishwash, shampoos, metalworking fluids, waterless hand cleaners, automotive specialties; liq.; sp.gr. 0.99–1.01; flash pt. (PMCC) > 300 F; pH 9.2–10.0; 100% conc.

Mazamide® O 20. [PPG/Specialty Chem.] 2:1 Oleamide DEA; nonionic; biodeg. thickener, emulsifier, foam builder, corrosion inhibitor, dispersant; for hard surf. cleaners, dishwash, shampoos, metalworking fluids, waterless hand cleaners, automotive specialties; liq.; sp.gr. 0.99–1.01; flash pt. (PMCC) > 300 F; pH 9–10; 100% conc.

Mazamide® PCS. [PPG/Specialty Chem.] Mixed alkanolamide (2:1); biodeg. thickener, emulsifier, foam builder; for hard surf. cleaners, dishwash, shampoos, metalworking fluids, waterless hand cleaners, automotive specialties; liq.; sp.gr. 0.99; flash pt. (PMCC) > 300 F.

Mazamide® RO. [PPG/Specialty Chem.] Mixed alkanolamide (2:1); biodeg. thickener, emulsifier, foam builder; for hard surf. cleaners, dishwash, shampoos, metalworking fluids, waterless hand cleaners, automotive specialties; liq.; sp.gr. 0.99; flash pt. (PMCC) > 300 F.

Mazamide® SCD. [PPG/Specialty Chem.] Mixed alkanolamide (1:1); biodeg. foam booster/stabilizer, thickener, emulsifier for cosmetic and pharmaceutical shampoos; liq.; sp.gr. 1.00; flash pt. (PMCC) > 300 F.

Mazamide® SMEA. [PPG/Specialty Chem.] Stearamide MEA (1:1); biodeg. emulsifier, pearlescent for syndet soap bars; flake; sp.gr. 0.89; flash pt. (PMCC) > 300 F.

Mazamide® SS 10. [PPG/Specialty Chem.] 1:1 Linoleamide DEA; nonionic; emulsifier, lubricant, thickener, solubilizer, corrosion inhibitor, buffing compd., metalworking; biodeg.; liq.; sp.gr. 0.98–1.00; flash pt. (PMCC) > 300 F; pH 9.0–10.5; 100% conc.

Mazamide® SS 20. [PPG/Specialty Chem.] 2:1 Linoleamide DEA; nonionic/anionic; biodeg. thickener, emulsifier, foam builder, corrosion inhibitor; for hard surf. cleaners, dishwash, shampoos, metalworking fluids, waterless hand cleaners, automotive specialties; liq.; water sol.; sp.gr. 1.01; flash pt. (PMCC) > 300 F; pH 9.8 (5% aq.); 100% conc.

Mazamide® T 20. [PPG/Specialty Chem.] 2:1 tall oil alkanolamide; nonionic/anionic; thickener for aq. systems; w/o emulsifier, corrosion inhibitor for sol. oils; metalworking; biodeg.; liq.; water sol.; sp.gr. 1.00; flash pt. (PMCC) > 300 F; pH 9.5 (5% aq.); 100% conc.

Mazamide® TC. [PPG/Specialty Chem.] Lauric alkanolamide (2:1); biodeg. thickener, emulsifier, foam builder; for hard surf. cleaners, dishwash, shampoos, metalworking fluids, waterless hand cleaners, automotive specialties; liq.; sp.gr. 1.01; flash pt. (PMCC) > 300 F.

Mazamide® WC Conc. [PPG/Specialty Chem.] Cocamide DEA (1:1); nonionic; biodeg. foam stabilizer, thickener, emulsifier for shampoos, bubble baths, dishwash, household and cosmetic preps.; liq.; sp.gr. 0.99; flash pt. (PMCC) > 300 F; 100% conc.

Mazawet® 36. [PPG/Specialty Chem.] Surfactant; nonionic; low foaming wetting agent, detergent, and dispersant; for machine dishwashing rinse additives, penetrant for textile finishing and scouring operations; liq.; sp.gr. 1.04; cloud pt. 34 C (1% aq.); surf. tens. 32 dynes/cm (0.1% aq.); Draves wetting 10 s (0.1% aq.); 92% act.

Mazawet® 77. [PPG/Specialty Chem.] Alkyl polyoxyalkylene ether; nonionic; wetting agent, surfactant used in metalworking, textile processing, spray-dried detergents, emulsion polymerization, drycleaning systems, paints, inks, hard surface cleaners, deresination of sulfite pulp; Gardner 1 liq.; m.w. 426 avg.; sol. @ 5% in toluene, perchloroethylene, Stod., propylene glycol, water; sp.gr. 0.965; pour pt. 11 C; cloud pt. 45 C (1% aq.); flash pt. 340 F (PMCC); pH 6.5 (3% aq.); surf. tens. 28.1 dynes/cm (0.1%); Draves wetting 4 s (0.1% aq.); 100% act.

Mazawet® DF. [PPG/Specialty Chem.] Alkyl polyoxyalkylene ether; nonionic; low foam wetting surfactant for metalworking fluids, paint, printing inks, polishes, floor waxes, hard surface cleaning, and emulsion polymerization; liq.; sp.gr. 1.06; cloud pt. 60 C (1% aq.); surf. tens. 27.3 dynes/cm (0.1% aq.); Draves wetting 6 s (0.1% aq.); 100% act.

Mazawet® DOSS 70. [PPG/Specialty Chem.] Dioctyl sodium sulfosuccinate; anionic; fast wetting surfactant, emulsifier, dispersant for drycleaning, paper and textile processing, control agent in coal dedusting, antifog for glass cleaners; liq.; sp.gr. 1.08; cloud pt. -5 C (1% aq.); surf. tens. 29.0 dynes/cm (0.1% aq.); Draves wetting 1 s (0.1% aq.); 70% act. in water/alcohol.

Mazclean EP. [PPG/Specialty Chem.] Proprietary emulsifier; microemulsifier package for use in preparing terpene-based cleaning formulations.

Mazclean W. [PPG/Specialty Chem.] Terpene-based surfactant blend; nonionic; moderate foaming detergent base; hard surface cleaner; used as replacement for chlorinate solvs. in some metalworking applics.; liq.; 75% conc.

Mazclean W-10. [PPG/Specialty Chem.] Proprietary; water-diluted version of Mazclean W; moderate foaming detergent base; hard surface cleaner; used as replacement for chlorinate solvs. in some metalworking applics.

Mazclean WRI. [PPG/Specialty Chem.] Proprietary; nonionic; detergent base with rust inhibitor for multi-metal protection against flash rusting.

Mazeen® 173. [PPG/Specialty Chem.] Tetrahydroxypropyl ethylenediamine; insecticide and herbicide emulsifier, antistat and rewetting agent, grease additive, textile lubricant; emulsifier for lubricants, inks, and cosmetics; chelant in electroless deposition formulations for electronics industry; crosslinker for rigid polyurethane; Gardner 2 liq.; sol. in water, benzene, acetone, IPA, min. spirits, toluene; m.w. 292; sp.gr. 1.01; flash pt. (PMCC) > 350 F; surf. tens. 51.4 dynes/cm (0.1%).

Mazeen® 174. [PPG/Specialty Chem.] Tetrahydroxypropyl ethylenediamine; cationic; insecticide and herbicide emulsifier, antistat and rewetting agent, grease additive, textile lubricant; emulsifier for lubricants, inks, and cosmetics; chelant in electroless deposition formulations for electronics industry; crosslinker for rigid polyurethane; Gardner < 1 liq.; sol. in water, acetone, IPA, min. spirits, toluene; m.w. 292; sp.gr. 1.00; flash pt. (PMCC) > 350 F; surf. tens. 54.2 dynes/cm (0.1%).

Mazeen® 174-75. [PPG/Specialty Chem.] Tetrahydroxypropyl ethylenediamine; cationic; insecticide and herbicide emulsifier, antistat and rewetting agent, grease additive, textile lubricant; emulsifier for lubricants, inks, and cosmetics; in electroless deposition formulations for electronics industry; Gardner < 1 liq.; sol. in water, acetone, IPA, min. spirits, toluene; m.w. 292; sp.gr. 1.00; flash pt.

(PMCC) > 350 F; surf. tens. 54.2 dynes/cm (0.1%).

Mazeen® 175. [PPG/Specialty Chem.] Alkoxylated diamine; cationic; low foam surfactant, emulsion stabilizer, dispersant, wetting agent, antistat for rigid PVC, visc. control agent; liq.; sol. in water, min. spirits, toluene, IPA; m.w. 8900; visc. 1500 cps; pour pt. 18 C; cloud pt. 65 C (1% aq.); flash pt. (PMCC) > 350 F; surf. tens. 34.1 dynes/cm (0.1%); Ross-Miles foam 35 mm (0.1%).

Mazeen® 176. [PPG/Specialty Chem.] Alkoxylated diamine; cationic; low foam surfactant, emulsion stabilizer, dispersant, wetting agent, antistat for rigid PVC, visc. control agent; liq.; sol. in water, min. spirits, toluene, IPA; m.w. 5500; visc. 1000 cps; pour pt. 18 C; cloud pt. 63 C (1% aq.); flash pt. (PMCC) > 350 F; surf. tens. 40.3 dynes/cm (0.1%); Ross-Miles foam 600 mm (0.1%).

Mazeen® 241-3. [PPG/Specialty Chem.] High m.w. alkoxylated diamine; cationic; lubricant, emulsifier; chelant for electroplating processes; dispersant for textile/water-based ink systems; liq.; sol. in water.

Mazeen® C-2. [PPG/Specialty Chem.] PEG-2 cocamine; CAS 61791-14-8; cationic; emulsifier, rewetting agent, lubricant, coupler used in insecticides and herbicides, grease additives, textile lubricants, water-based inks, cosmetics; plastics antistat; visc. modifier and rust inhibitor in acid media for metalworking; Gardner 10 liq.; sol. in benzene, acetone, IPA, min. oil, toluene, min. spirits, forms gel in water; m.w. 285; sp.gr. 0.874; flash pt. (PMCC) > 350 F; surf. tens. 28 dynes/cm (0.1%); 100% conc.

Mazeen® C-5. [PPG/Specialty Chem.] PEG-5 cocamine; CAS 61791-14-8; cationic; emulsifier, rewetting agent, lubricant, coupler used in insecticides and herbicides, grease additives, textile lubricants, water-based inks, cosmetics; plastics antistat; rust inhibitor in acid media for metalworking; Gardner 10 liq.; sol. in water, benzene, acetone, IPA, min. oil, toluene; disp. in min. spirits; m.w. 425; sp.gr. 0.976; flash pt. (PMCC) > 350 F; surf. tens. 33 dynes/cm (0.1%); 100% conc.

Mazeen® C-10. [PPG/Specialty Chem.] PEG-10 cocamine; CAS 61791-14-8; cationic; emulsifier, rewetting agent, lubricant, coupler used in insecticides and herbicides, grease additives, textile lubricants, water-based inks, cosmetics; plastics antistat; Gardner 11 liq.; sol. in water, benzene, acetone, IPA, toluene; m.w. 645; sp.gr. 1.017; flash pt. (PMCC) > 350 F; surf. tens. 39 dynes/cm (0.1%); 100% conc.

Mazeen® C-15. [PPG/Specialty Chem.] PEG-15 cocamine; CAS 61791-14-8; cationic; emulsifier, rewetting agent, lubricant, coupler used in metalworking, insecticides and herbicides, grease additives, textile lubricants, water-based inks, cosmetics; plastics antistat; Gardner 9 liq.; sol. in water, benzene, acetone, IPA, toluene; m.w. 860; sp.gr. 1.042; flash pt. (PMCC) > 350 F; surf. tens. 41 dynes/cm (0.1%); 100% conc.

Mazeen® DBA-1. [PPG/Specialty Chem.] Polyalkoxylated amine; cationic; surfactant and wetting agent for paints and inks; Gardner 2 liq.; sol. in min. oil, min. spirits, toluene, IPA; m.w. 172; sp.gr. 0.860; flash pt. (PMCC) > 350 F; surf. tens. 41 dynes/cm (0.1%); 100% conc.

Mazeen® S-2. [PPG/Specialty Chem.] PEG-2 soyamine; CAS 61791-24-0; cationic; emulsifier, rewetting agent, lubricant, coupler used in insecticides and herbicides, grease additives, textile lubricants, water-based inks, cosmetics; plastics antistat; Gardner 14 liq.; sol. in benzene, acetone, IPA, min. oil; disp. in min. spirits, toluene; m.w. 350; sp.gr. 0.911; flash pt. (PMCC) > 350 F; surf. tens. 26 dynes/cm (0.1%); 100% conc.

Mazeen® S-5. [PPG/Specialty Chem.] PEG-5 soyamine; CAS 61791-24-0; cationic; emulsifier, rewetting agent, lubricant, coupler used in insecticides and herbicides, grease additives, textile lubricants, water-based inks, cosmetics; plastics antistat; Gardner 14 liq.;sol. in benzene, IPA; partly sol. acetone, min. oil; m.w. 480; sp.gr. 0.951; flash pt. (PMCC) > 350 F; surf. tens. 33 dynes/cm (0.1%); 100% conc.

Mazeen® S-10. [PPG/Specialty Chem.] PEG-10 soyamine; CAS 61791-24-0; cationic; emulsifier, rewetting agent, lubricant, coupler used in insecticides and herbicides, grease additives, textile lubricants, water-based inks, cosmetics; plastics antistat; Gardner 14 liq.; sol. in water, benzene, acetone, IPA, toluene; m.w. 710; sp.gr. 1.020; flash pt. (PMCC) > 350 F; surf. tens. 40 dynes/cm (0.1%); 100% conc.

Mazeen® S-15. [PPG/Specialty Chem.] PEG-15 soyamine; CAS 61791-24-0; cationic; emulsifier, rewetting agent, lubricant, coupler used in insecticides and herbicides, grease additives, textile lubricants, water-based inks, cosmetics; plastics antistat; Gardner 18 liq.; sol. in water, benzene, acetone, IPA, toluene; m.w. 930; sp.gr. 1.040; flash pt. (PMCC) > 350 F; surf. tens. 43 dynes/cm (0.1%); 100% conc.

Mazeen® T-2. [PPG/Specialty Chem.] PEG-2 tallow amine; cationic; emulsifier, rewetting agent, lubricant, coupler used in insecticides and herbicides, grease additives, textile lubricants, water-based inks, cosmetics, metalworking; plastics antistat; visc. modifier, rust inhibitor in acid media; Gardner 11 liq.; sol. in benzene, acetone, IPA, min. oil; m.w. 350; sp.gr. 0.916; flash pt. (PMCC) > 350 F; surf. tens. 29 dynes/cm (0.1%); 100% conc.

Mazeen® T-3.5. [PPG/Specialty Chem.] PEG-3.5 tallowamine; cationic; visc. modifier and rust inhibitor in acid media for metalworking applics.; liq.; gels in water.

Mazeen® T-5. [PPG/Specialty Chem.] PEG-5 tallow amine; cationic; emulsifier, rewetting agent, lubricant, coupler used in insecticides and herbicides, grease additives, textile lubricants, water-based inks, cosmetics, metalworking; plastics antistat; Gardner 12 liq.; sol. in benzene, acetone, IPA, min. oil; gels in water; m.w. 480; sp.gr. 0.966; flash pt. (PMCC) > 350 F; surf. tens. 34 dynes/cm (0.1%); 100% conc.

Mazeen® T-10. [PPG/Specialty Chem.] PEG-10 tallow amine; cationic; emulsifier, rewetting agent, lubricant, coupler used in insecticides and herbicides, grease additives, textile lubricants, water-based inks, cosmetics; plastics antistat.

Mazeen® T-15. [PPG/Specialty Chem.] PEG-15 tallow amine; cationic; emulsifier, rewetting agent, lubricant, coupler used in insecticides and herbicides, grease additives, textile lubricants, water-based inks, cosmetics; plastics antistat; Gardner 18 liq.; sol. in water, benzene, acetone, IPA; m.w. 925; sp.gr. 1.028; flash pt. (PMCC) > 350 F; surf. tens. 41 dynes/cm (0.1%); 100% conc.

Mazol® 159. [PPG/Specialty Chem.] PEG-7 glyceryl cocoate; emulsifier for food prods.; emollient; used in cosmetics, toiletries, pharmaceuticals, lubricants, mold release compds.; plasticizer in syn. fabrics and plastics; amber liq.; sol. in water, min. oil; disp. in min. spirits, toluene, IPA; HLB 13.0; acid no. 5

max.; sapon. no. 82-98; flash pt. (PMCC) > 350 F.

Mazol® 300 K. [PPG/Specialty Chem.] Glyceryl oleate, kosher; CAS 111-03-5; nonionic; GRAS dispersant for oil or solv. systems; antifoam for sugar and protein processing; coemulsifier with T-Maz 60 or 80; amber liq.; sol. in min. oil, toluene, IPA; disp. in min. spirits; insol. in water; HLB 3.8; acid no. 2 max.; flash pt. (PMCC) > 350 F; 100% conc.

Mazol® GMO. [PPG/Specialty Chem.] Glyceryl oleate; CAS 111-03-5; nonionic; GRAS dispersant for oil or solv. systems; antifoam for food processing; base for cosmetic creams, lotions, ointments; w/o emulsifier with emolliency, thickening properties; plasticizer; lubricant; antifog for PVC; emulsifier, coupling agent for metalworking applics.; yel. liq.; sol. in soybean oil, min. oil, toluene, IPA; disp. in ethanol, propylene glycol; HLB 3.8; acid no. 2 max.; sapon. no. 150-170; flash pt. (PMCC) > 350 F; 100% conc.

Mazol® GMO #1. [PPG/Specialty Chem.] Glyceryl oleate; CAS 111-03-5; nonionic; industrial antifoam, coemulsifier for metalworking applics.; dk. liq.; sol. in min. oil, toluene, IPA; disp. in min. spirits; HLB 3.8; acid no. 3 max.; sapon. no. 145-175; flash pt. (PMCC) > 350 F; 100% conc.

Mazol® GMO Ind. [PPG/Specialty Chem.] Glyceryl oleate; industrial version of Mazol GMO #1; emulsifier, coupling agent for metalworking applics.; acid no. 5 max.

Mazol® GMO K. [PPG/Specialty Chem.] Glyceryl oleate, kosher; CAS 111-03-5; GRAS-type antistat for plastic fibers in contact with food; dispersant for oil or solv. systems; antifoam for sugar and protein processing; coemulsifier with T-Maz 60 or 80; Gardner 2 paste; sol. in min. oil, toluene, IPA; disp. in min. spirits; HLB 3.8; acid no. 2 max.; flash pt. (PMCC) > 350 F; 100% act.

Mazol® GMR. [PPG/Specialty Chem.] Glyceryl ricinoleate; base for cosmetic creams, lotions, ointments; w/o emulsifier with emolliency, thickening properties; plasticizer; dk. liq.; sol. in toluene, IPA; disp. in min. spirits; HLB 6.0; acid no. 7 max.; sapon. no. 138–145; flash pt. (PMCC) > 350 F.

Mazol® GMS-90. [PPG/Specialty Chem.] Glyceryl stearate; emulsifier for food prods., cosmetics, toiletries, pharmaceuticals, lubricants, mold release compds., in plasticizers for syn. fabrics and plastics; tan flake; sol. in IPA; sol. hot in min. oil, toluene; HLB 3.9; acid no. 2 max.

Mazol® GMS-D. [PPG/Specialty Chem.] Glyceryl stearate SE; emulsifier for food prods.; emollient; used in cosmetics, toiletries, pharmaceuticals, lubricants, mold release compds., metalworking applics.; plasticizer in syn. fabrics and plastics; tan flake; sol. in IPA; sol. hot in min. oil, toluene; disp. in water; HLB 6.0; acid no. 3.5 max.; sapon. no. 145-160; flash pt. (PMCC) > 350 F.

Mazol® PETO. [PPG/Specialty Chem.] Pentaerythritol tetraoleate; emulsifier, coupling agent for metalworking applics.

Mazol® PETO Mod 1. [PPG/Specialty Chem.] Pentaerythritol tetraoleate; emulsifier, coupling agent for metalworking applics.; additive improving high temp. performance of oil-based rolling lubricants.

Mazol® PGO-31 K. [PPG/Specialty Chem.] Triglyceryl oleate; nonionic; emulsifier for food prods., in cosmetics, toiletries, pharmaceuticals, lubricants, mold release compds., plasticizers; Gardner 8 max. liq.; HLB 6.2; acid no. 4 max.; sapon. no. 125–150; 100% conc.

Mazol® PGO-104. [PPG/Specialty Chem.] Decaglyceryl tetraoleate; nonionic; emulsifier for food prods., cosmetics, lubricants, etc.; also food-grade solubilizer and carrier for essential oils and flavors; Gardner 8 max. liq.; water-disp.; HLB 6.2; acid no. 8 max.; sapon. no. 125–150; flash pt. (PMCC) > 350 F.

Mazon® 18A. [PPG/Specialty Chem.] Proprietary surfactant; nonionic; visc. booster for aq. systems; liq.; sp.gr. 0.998; flash pt. (PMCC) > 350 F; 100% act.

Mazon® 21. [PPG/Specialty Chem.] Alkyl polyoxyalkylene ether; anionic; emulsifier for degreasers and industrial cleaner formulations; liq.; sp.gr. 1.04; flash pt. (PMCC) > 350 F; 100% act.

Mazon® 23. [PPG/Specialty Chem.] Proprietary surfactant; detergent base for lt. duty hard surf. cleaning; liq.; sp.gr. 1.08; flash pt. (PMCC) > 350 F; 70% act.

Mazon® 27. [PPG/Specialty Chem.] Proprietary surfactant; wetting agent for lt. duty cleaning applics.; liq.; sp.gr. 1.00; flash pt. (PMCC) > 350 F; 60% act.

Mazon® 29. [PPG/Specialty Chem.] Proprietary surfactant; cleaner conc. for soils and oily residues; liq.; sp.gr. 1.07; flash pt. (PMCC) > 350 F.

Mazon® 40. [PPG/Specialty Chem.] Alkyl glucoside; nonionic; caustic coupling surfactant, wetting agent for alkaline cleaning compds., bottle washes; biodeg.; liq.; sol. in 50% sodium hydroxide sol'ns.; sp.gr. 1.15; flash pt. (PMCC) > 350 F; 70% act.

Mazon® 40A. [PPG/Specialty Chem.] Modified alkyl glycoside; nonionic; low-foam caustic coupling surfactant, wetting agent for alkaline cleaning compds., caustic bottle wash compositions; liq.; sp.gr. 1.07; flash pt. (PMCC) > 350 F; 76% act.

Mazon® 41. [PPG/Specialty Chem.] Ammonium salt of alkylphenol ethoxylate; surfactant for cleaning formulations; liq.; sp.gr. 1.065; flash pt. (PMCC) none; 60% act.

Mazon® 43, 43LF. [PPG/Specialty Chem.] Proprietary surfactant; low foaming deinking aid for recycled paper and board; highly stable to salts, pH changes, temp.; liq.; sp.gr. 1.110; flash pt. (PMCC) > 300 F; 97% act.

Mazon® 60T. [PPG/Specialty Chem.] TEA dodecylbenzene sulfonate; anionic; high foaming formulated detergent, emulsifier for shampoos, dishwashing, car wash, textile, hard surf. cleaners, metal cleaners and other formulations; liq.; sp.gr. 1.08; flash pt. (PMCC) > 350 F; 60% act.

Mazon® 61. [PPG/Specialty Chem.] Proprietary surfactant; deinking aid; liq.; sp.gr. 1.058; flash pt. (PMCC) > 300 F; 96% act.

Mazon® 70. [PPG/Specialty Chem.] Amine/amide surfactant; anionic; reduced foam detergent conc. for use in industrial and institutional cleaners; liq.; water-dilutable; sp.gr. 1.05; flash pt. (PMCC) > 350 F; 100% act.

Mazon® 85. [PPG/Specialty Chem.] Modified dodecylbenzene sulfonic acid; anionic; self-emulsifying degreaser, detergent, emulsifier for cleaning applics.; used with hydrocarbon solvs.; liq.; sol. in oil and aq. media; 98% conc.

Mazon® 86. [PPG/Specialty Chem.] Polyoxyalkylene ether ester; nonionic; extreme pressure surfactant for oil and water sol. metalworking lubricants; liq.; 100% conc.

Mazon® 114. [PPG/Specialty Chem.] Proprietary surfactant; amphoteric; surfactant with corrosion inhibiting props.; protects ferrous metals; liq.; sp.gr. 0.99; visc. 9500 cps; flash pt. (PMCC) > 350 F; 100% act.

Mazon® 114A. [PPG/Specialty Chem.] Surface active corrosion inhibitor protecting ferrous metals; reduced water sol. compared to Mazon 114; paste.

Mazon® 1045A. [PPG/Specialty Chem.] POE sorbitol fatty acid ester; nonionic; emulsifier for pesticide, herbicide, metalworking, die-cast lubricant formulations, and emulsion polymerization; humectant, emollient; liq.; water-sol.; sp.gr. 1.02; visc. 260 cps; HLB 13.0; sapon. no. 90; flash pt. (PMCC) > 350 F; 100% conc.

Mazon® 1086. [PPG/Specialty Chem.] POE sorbitol ester; nonionic; emulsifier, coupling agent for agric. pesticide and herbicide formulations, emulsion polymerization, metalworking lubricants, die-cast lubricants; liq.; water-sol.; sp.gr. 1.02; visc. 200 cps; HLB 10.4; sapon. no. 97; flash pt. (PMCC) > 350 F; 100% conc.

Mazon® 1096. [PPG/Specialty Chem.] POE sorbitol ester; nonionic; emulsifier, coupling agent for agric. pesticide and herbicide formulations, emulsion polymerization, metalworking lubricants, die-cast lubricants; liq.; water-sol.; sp.gr. 1.02; visc. 240 cps; HLB 11.2; sapon. no. 85; flash pt. (PMCC) > 350 F; 100% conc.

Mazon® DWD-100. [PPG/Specialty Chem.] Proprietary; anionic/nonionic; base for household and industrial hard surface cleaners; liq.; 100% conc.

Mazon® RI 6. [PPG/Specialty Chem.] Surface active corrosion inhibitor protecting ferrous, aluminum, brass, and copper surfs.; visc. 2500 cps; 6.8% phosphorus.

Mazon® RI 37. [PPG/Specialty Chem.] Surface active corrosion inhibitor protecting ferrous, aluminum, brass and copper metals; visc. 100 cps.

Mazox® CAPA. [PPG/Specialty Chem.] Cocamidopropylamine oxide; nonionic/cationic; surfactant, conditioner, emollient, emulsifier, foam booster, visc. builder, lime soap dispersant for cosmetics, toiletries, household and industrial uses; liq.; sp.gr. 1.02; flash pt. > 200 F; 30% amine oxide.

Mazox® CAPA-37. [PPG/Specialty Chem.] Cocamidopropylamine oxide; nonionic/cationic; surfactant, conditioner, emollient, emulsifier, foam booster, visc. builder for cosmetics, toiletries, household and industrial uses.

Mazox® CDA. [PPG/Specialty Chem.] Palmitamine oxide; nonionic/cationic; surfactant, conditioner, emollient, emulsifier, foam booster, visc. builder for cosmetics, toiletries, household and industrial uses; textile softener; liq.; sp.gr. 0.96; flash pt. > 200 F; 30% amine oxide.

Mazox® KCAO. [PPG/Specialty Chem.] Potassium dihydroxyethyl cocamine oxide phosphate; nonionic/cationic; foam booster/stabilizer, detergent, emollient, lime soap dispersant for dishwash, heavy-duty detergents, caustic sol'ns.; stable in caustic soda; compat. with most surfactants, builders, many solvs.; liq.; sp.gr. 1.003; flash pt. > 200 F; 33% amine oxide.

Mazox® LDA. [PPG/Specialty Chem.] Lauramine oxide; nonionic/cationic; surfactant, conditioner, emollient, emulsifier, foam booster, visc. builder for cosmetics, toiletries, household and industrial uses; liq.; sp.gr. 0.96; flash pt. > 200 F; 30% amine oxide.

Mazox® MDA. [PPG/Specialty Chem.] Myristamine oxide; nonionic/cationic; surfactant, conditioner, emollient, emulsifier, foam booster, visc. builder for cosmetics, toiletries, household and industrial uses; liq.; sp.gr. 0.96; flash pt. > 200 F; 30% amine oxide.

Mazox® ODA. [PPG/Specialty Chem.] Oleamine oxide; nonionic/cationic; surfactant, conditioner, emollient, emulsifier, foam booster, visc. builder for cosmetics, toiletries, household and industrial uses; textile softener; liq.; sp.gr. 0.90; flash pt. 120 F; 50% amine oxide.

Mazox® SDA. [PPG/Specialty Chem.] Stearamine oxide; nonionic/cationic; surfactant, conditioner, emollient, emulsifier, foam booster, visc. builder for cosmetics, toiletries, household and industrial uses; textile softener; paste; sp.gr. 0.99; flash pt. > 200 F; 25% amine oxide.

Maztreat 246. [PPG/Specialty Chem.] Organic defoamer; defoamer for metalworking formulations; water-disp.

Mazu® 43 C. [PPG/Specialty Chem.] Organic defoamer; defoamer for enzyme, effluent wastewater applics.; liq.; sp.gr. 0.878; visc. 45 cps; flash pt. (PMCC) 300 F.

Mazu® 68 C. [PPG/Specialty Chem.] Organic defoamer; defoamer for effluent, latex coatings; liq.; sp.gr. 0.862; visc. 40 cps; flash pt. (PMCC) 300 F.

Mazu® 108 L. [PPG/Specialty Chem.] Organic defoamer; defoamer for brownstock, drainage aid, latex paints; liq.; sp.gr. 0.910; visc. 2000 cps; flash pt. (PMCC) > 300 F.

Mazu® 112. [PPG/Specialty Chem.] Organic defoamer; defoamer for effluent wastewater; liq.; sp.gr. 0.876; visc. 50 cps; flash pt. (PMCC) 300 F.

Mazu® 140. [PPG/Specialty Chem.] Organic defoamer; defoamer for whitewater, screen room, bleach plant; emulsion; sp.gr. 0.822; visc. 500 cps; flash pt. (PMCC) none.

Mazu® 141. [PPG/Specialty Chem.] Organic defoamer; defoamer for whitewater, screen room, bleach plant; emulsion; sp.gr. 0.96; visc. 500 cps; flash pt. (PMCC) none.

Mazu® 142. [PPG/Specialty Chem.] Organic defoamer; defoamer for whitewater, screen room, bleach plant; emulsion; sp.gr. 0.973; visc. 400 cps; flash pt. (PMCC) none.

Mazu® 151 PY. [PPG/Specialty Chem.] Organic defoamer; defoamer for adhesive emulsions, latex coatings; liq.; sp.gr. 0.892; visc. 100 cps; flash pt. (PMCC) 300 F.

Mazu® 160 CA. [PPG/Specialty Chem.] Organic defoamer; defoamer for paint pigment flotation, flocculant aid; liq.; sp.gr. 0.844; visc. 25 cps; flash pt. (PMCC) 72 F.

Mazu® 161. [PPG/Specialty Chem.] Organic defoamer; defoamer for latex paint; liq.; sp.gr. 0.860; visc. 1250 cps; flash pt. (PMCC) 300 F.

Mazu® 197. [PPG/Specialty Chem.] Organic defoamer; defoamer for metal oxide slurries, metalworking; liq.; sp.gr. 1.050; visc. 1000 cps; flash pt. (PMCC) 300 F.

Mazu® 198. [PPG/Specialty Chem.] Organic defoamer; defoamer for jet dyeing, hot scouring; liq.; sp.gr. 1.030; visc. 2000 cps; flash pt. (PMCC) 300 F.

Mazu® 201. [PPG/Specialty Chem.] Organic defoamer; defoamer for protein and starch; liq.; sp.gr. 0.915; visc. 130 cps; flash pt. (PMCC) > 300 F.

Mazu® 201 A. [PPG/Specialty Chem.] Organic defoamer; defoamer for protein and starch; liq.; sp.gr. 0.950; visc. 1000 cps; flash pt. (PMCC) > 300 F.

Mazu® 201 B. [PPG/Specialty Chem.] Organic defoamer; defoamer for protein and starch, distillation; liq.; sp.gr. 0.956; visc. 700 cps; flash pt.

(PMCC) > 300 F.

Mazu® 201 PM. [PPG/Specialty Chem.] Organic defoamer; defoamer for protein and starch; liq.; sp.gr. 0.966; visc. 400 cps; flash pt. (PMCC) > 300 F.

Mazu® 208. [PPG/Specialty Chem.] Organic defoamer; defoamer for protein and starch; emulsion; sp.gr. 0.971; visc. 2000 cps; flash pt. (PMCC) none.

Mazu® 208 L. [PPG/Specialty Chem.] Organic defoamer; defoamer for brownstock, latex paints; liq.; sp.gr. 0.950; visc. 3500 cps; flash pt. (PMCC) > 300 F.

Mazu® 208 LPD. [PPG/Specialty Chem.] Organic defoamer; defoamer for starch and agric. powds.; powd.; dens. 25 lb/ft^3; flash pt. (PMCC) > 300 F.

Mazu® 251. [PPG/Specialty Chem.] Organic defoamer; defoamer for brownstock, adhesives, latex paints; liq.; sp.gr. 0.920; visc. 500 cps; flash pt. (PMCC) > 300 F.

Mazu® 252 A. [PPG/Specialty Chem.] Organic defoamer; defoamer for brownstock, adhesives, latex paints; liq.; sp.gr. 0.920; visc. 500 cps; flash pt. (PMCC) > 300 F.

Mazu® 252. [PPG/Specialty Chem.] Organic defoamer; defoamer for brownstock, adhesives, latex paints; liq.; sp.gr. 0.920; visc. 500 cps; flash pt. (PMCC) > 300 F.

Mazu® 255. [PPG/Specialty Chem.] Organic defoamer; defoamer for brownstock, latex, adhesives; liq.; sp.gr. 0.91; visc. 500 cps; flash pt. (PMCC) 300 F.

Mazu® 289. [PPG/Specialty Chem.] Organic defoamer; defoamer for distillation, fermentation; liq.; sp.gr. 1.010; visc. 500 cps; flash pt. (PMCC) > 350 F.

Mazu® 290. [PPG/Specialty Chem.] Organic defoamer; defoamer for latex paints, adhesives; liq.; sp.gr. 0.993; visc. 850 cps; flash pt. (PMCC) 350 F.

Mazu® 306. [PPG/Specialty Chem.] Organic defoamer; general purpose water-dilutable defoamer; emulsion; sp.gr. 1.010; visc. 8500 cps; flash pt. (PMCC) none.

Mazu® 307. [PPG/Specialty Chem.] Organic defoamer; defoamer for hot processing, food applics.; emulsion; sp.gr. 0.99; visc. 1300 cps; flash pt. (PMCC) none.

Mazu® 309. [PPG/Specialty Chem.] Organic defoamer; defoamer for food processing, latex paints; emulsion; sp.gr. 1.001; visc. 6200 cps; flash pt. (PMCC) none.

Mazu® 319. [PPG/Specialty Chem.] Organic defoamer; defoamer for commercial cleaning compds.; emulsion; sp.gr. 1.050; visc. 1000 cps; flash pt. (PMCC) none.

Mazu® 320. [PPG/Specialty Chem.] Organic defoamer; defoamer for latex and acrylic systems; emulsion; sp.gr. 1.050; visc. 3000 cps; flash pt. (PMCC) none.

Mazu® 321. [PPG/Specialty Chem.] Organic defoamer; defoamer for latex and adhesives; liq.; sp.gr. 0.971; visc. 400 cps; flash pt. (PMCC) 300 F.

Mazu® 650. [PPG/Specialty Chem.] Organic defoamer; defoamer for hot alkaline systems; liq.; sp.gr. 0.950; visc. 100 cps; flash pt. (PMCC) 300 F.

Mazu® 2501. [PPG/Specialty Chem.] Organic defoamer; defoamer for brownstock, adhesives, latex paints; liq.; sp.gr. 0.920; visc. 750 cps; flash pt. (PMCC) > 300 F.

Mazu® 2502. [PPG/Specialty Chem.] Organic defoamer; defoamer for brownstock, adhesives, latex paints; liq.; sp.gr. 0.920; visc. 1200 cps; flash pt. (PMCC) > 300 F.

Mazu® 5116. [PPG/Specialty Chem.] Organic defoamer; defoamer for delayed coking; liq.; sp.gr. 0.92; visc. 1200 cps; flash pt. (PMCC) 175 F; 50% act.

Mazu® 5117. [PPG/Specialty Chem.] Organic defoamer; defoamer for delayed coking; liq.; sp.gr. 0.894; visc. 645 cps; flash pt. (PMCC) 175 F; 40% act.

Mazu® 5118. [PPG/Specialty Chem.] Organic defoamer; defoamer for delayed coking; liq.; sp.gr. 0.870; visc. 200 cps; flash pt. (PMCC) 175 F; 25% act.

Mazu® 5119. [PPG/Specialty Chem.] Organic defoamer; defoamer for delayed coking; liq.; sp.gr. 0.885; visc. 800 cps; flash pt. (PMCC) 175 F; 50% act.

Mazu® 5120. [PPG/Specialty Chem.] Organic defoamer; defoamer for delayed coking; liq.; sp.gr. 1.00; visc. 1300 cps; flash pt. (PMCC) 250 F; 100% act.

Mazu® DF 197. [PPG/Specialty Chem.] Defoamer for metalworking concs.; opaque liq.; water-disp.

Mazu® DF 200SX. [PPG/Specialty Chem.] Silicone compd.; defoamer for industrial use; used in adhesives, solv.-based inks and paints, insecticides, resin polymerization, and petrol. industry; liq.; sp.gr. 0.99; dens. 8.3 lb/gal; visc. 2000 cSt; flash pt. (PMCC) > 350 F; 100% silicone.

Mazu® DF 200SXSP. [PPG/Specialty Chem.] Silicone compd.; defoamer for industrial use; used in adhesives, solv.-based inks and paints, insecticides, resin polymerization, and petrol. industry; sp.gr. 0.99; dens. 8.3 lb/gal; visc. 2500 cSt; flash pt. (PMCC) > 350 F; 100% silicone.

Mazu® DF 204. [PPG/Specialty Chem.] Defoamer for metalworking concs.; modest wetting chars.; clear liq.; water-disp.

Mazu® DF 205SX. [PPG/Specialty Chem.] Silicone emulsion; industrial defoamer formulated with low visc. for ease of handling and pumping; for leather finishing, metalworking, carpet cleaning, waste treatment; sp.gr. 1.00; flash pt. (PMCC) none; pH 7.0; 5% silicone.

Mazu® DF 210SX Mod 1. [PPG/Specialty Chem.] Silicone emulsion; nonionic; emulsifier; defoamer for adhesives mfg., antifreeze, hot aq. systems, water-based inks and paints, insecticides, vinyl latex binders and emulsions, petrol., textile and paper applic., carpet cleaning applic.; liq.; sp.gr. 1.00; dens. 8.3 lb/gal; visc. 1000 cSt; flash pt. (PMCC) none; pH 7.0; 3.5% silicone.

Mazu® DF 210SX. [PPG/Specialty Chem.] Silicone emulsion; nonionic; emulsifier; defoamer for metalworking, adhesives mfg., antifreeze, hot aq. systems, water-based inks and paints, insecticides, vinyl latex binders and emulsions, petrol., textile and paper applic.; liq.; water-disp.; sp.gr. 1.00; dens. 8.3 lb/gal; visc. 1900 cSt; flash pt. (PMCC) none; pH 7.0; 10% silicone.

Mazu® DF 210SXSP. [PPG/Specialty Chem.] Silicone emulsion; industrial defoamer for highly acidic and alkaline systems, adhesives, antifreeze, brines, hot aq. systems, inks, insecticides, paints, resin polymerization, starch processing, petrol., latex binders and emulsions, textile, paper; sp.gr. 1.00; flash pt. (PMCC) none; pH 7.0; 10% silicone.

Mazu® DF 215SX. [PPG/Specialty Chem.] Silicone emulsion; industrial grade defoamer for dilution and repkg.; for leather finishing, metalworking, carpet cleaning, waste treatment; liq.; sp.gr. 1.00; flash pt. (PMCC) none; pH 7.0; 15% silicone.

Mazu® DF 230S. [PPG/Specialty Chem.] Silicone emulsion; nonionic; emulsifier; defoamer used as direct food additive; latex and food processing, boiler water defoaming, and waste treatment; FDA compliance; liq.; sp.gr. 1.00; dens. 8.3 lb/gal; visc. 1500 cSt; flash pt. (PMCC) none; pH 7.0; 30% silicone.

Mazu® DF 230SP. [PPG/Specialty Chem.] Simethicone emulsion; defoamer for food processing, fermentation, wine making, yeast processing, latex processing, waste treatment; FDA compliance; sp.gr. 1.00; flash pt. (PMCC) none; pH 4.5; 30% silicone.

Mazu® DF 230SX. [PPG/Specialty Chem.] Silicone emulsion; nonionic; emulsifier; defoamer for adhesive, water-based paints; soap mfg., antifreeze, hot aq. systems, insecticides, textile, paper, petrol., vinyl latex binders and emulsions,; boiler water defoaming, leather finishing, metal working, and waste treatment; liq.; sp.gr. 1.00; dens. 8.3 lb/gal; visc. 3000 cSt; flash pt. (PMCC) none; pH 7.0; 30% silicone.

Mazu® DF 230SXSP. [PPG/Specialty Chem.] Silicone emulsion; defoamer for highly acidic and alkaline systems, adhesives, antifreeze, brines, hot aq. systems, inks, insecticides, paints, soap mfg., starch processing, petrol., textile, paper, leather, metalworking, waste treatment; sp.gr. 1.00; flash pt. (PMCC) none; pH 7.0; 30% silicone.

Mazu® DF 243. [PPG/Specialty Chem.] Silicone emulsion; nonionic; emulsifier; defoamer for adhesive, water-based paints; soap mfg., antifreeze, hot aq. systems, insecticides, textile, paper, petrol., vinyl latex binders and emulsions, waste treatment; formulated for rigid dilution requirements; liq.; sp.gr. 1.00; dens. 8.3 lb/gal; visc. 3000 cSt; flash pt. (PMCC) none; pH 7.0; 30% silicone.

Mazu® DF 255. [PPG/Specialty Chem.] Organic defoamer; defoamer for institutional cleaners, metalworking formulations.

M-C-Thin. [Lucas Meyer] Lecithin, standardized; CAS 8002-43-5; amphoteric; used for food and tech. applics.; liq., paste; 100% conc.

M-C-Thin 45. [Lucas Meyer] Lecithin; CAS 8002-43-5; nonionic; release aid; liq.; 30-40% conc.

Medialan Brands. [Hoechst Celanese/Colorants & Surf.; Hoechst AG] Fatty acid sarcosides, sodium salt; anionic; basic material for cosmetic industry; detergent for textile industry; paste; 50% conc.

Melatex AS-80. [Tokai Seiyu Ind.] Polyacrylic resin; anionic; hand modifying agent for syn. knitting clothes; liq.; 30% conc.

Mellorian 118. [Ceca SA] Sodium cetearyl sulfate; anionic; degreaser, ore flotation for barite and fluorspar; paste; 42% conc.

Merbron R. [Sybron] Cationic/nonionic; machine cleaner for aftercleaning of dyed polyester, removal of trimers from fibers or equip.

Merce Assist ADB. [Sybron] Anionic; low foaming mercerization penetrant.

Mercerol AW-LF. [Sandoz Prods. Ltd.] Cresol-free sulfonate; nonionic; wetting agent for mercerizing liquors; liq.

Mercerol GVC-65 Liq. [Sandoz Prods. Ltd.] Wetting agent, mercerizing penetrant for efficient wetout in caustic soda saturator; yel. liq.; disp. in water; sp.gr. 1.03; 65% act.

Mercerol NL. [Sandoz Prods. Ltd.] Sulfonate, cresol-free, containing solvs.; anionic; wetting agent for causticizing baths; liq.

Mercerol QW. [Sandoz Prods. Ltd.] Cresol-free sulfonate and solvs.; anionic; wetting agent for mercerizing liquors; liq.

Mercerol SM. [Sandoz Prods. Ltd.] Sulfonate in aq./org. sol'n.; anionic; wetting agent for mercerizing; liq.

Mergital OC 30E. [Pulcra SA] Ethoxylated oleo-cetyl alcohol (30 EO); nonionic; surfactant; flakes; HLB 16.7; hyd. no. 35-42; pH 6.0-7.5 (5%); 100% conc.

Mergital ST 30/E. [Pulcra SA] PEG-30 stearate; CAS 9004-99-3; nonionic; surfactant; flake; HLB 16.8; pH 6-8 (5%); 100% conc.

Merinol LH-200. [Takemoto Oil & Fat] Surfactant blend; anionic/nonionic; surfactant for worsted oil; liq.; 90% conc.

Merpol® 100. [DuPont] Octylphenol ethoxylate; nonionic; surfactant, emulsifier, wetting agent, detergent, solubilizer, emulsion stabilizer for metal cleaning, industrial and household detergents, agric.; stable to acids, bases, heat, freezing; clear liq., mild odor; sol. in water, polar and nonpolar solvs.; sp.gr. 1.06 g/mL; dens. 8.8 lb/gal; visc. 222 cP; HLB 13.5; cloud pt. 63.69 C (upper, 1%); flash pt. (PMCC) 165 C; pH 5-7; surf. tens. 33 dynes/cm (0.1%); 100% act.

Merpol® A. [DuPont] Ethoxylated phosphate; nonionic; low foaming wetting agent, surf. tens. reducer for chemical mfg., cosmetics, metal processing, paper, petrol., inks, plastics, soaps, syn. fibers, textiles; stable to acids, bases, heat to 100 C, freezing; colorless to pale yel. liq., mild odor; sol. in polar solvs., 0.1-1% in nonpolar solvs.; disp. in water to 1%; sp.gr. 1.07 g/mL; dens. 8.9 lb/gal; visc. 104 cP; HLB 6.7; cloud pt. < 25 C (upper, 1%); flash pt. (TCC) 138 C; pH 5-7 (1% emulsion); surf. tens. 26 dynes/cm (0.1%); 100% act.

Merpol® C. [DuPont] Long-chain alcohol sulfate; anionic; wetting agent, detergent, penetrant, emulsifier used in textile and paper industries; lt. amber liq.; misc. with water; dens. 7.5 lb/gal; pH 8 (1%).

Merpol® CH-196. [DuPont] Modified EO condensate; nonionic; detergent, wetting agent used in textile industry; clear, colorless liq.; disp. in water; dens. 8.1 lb/gal; sp.gr. 0.973; visc. 24 cps; HLB 13.0; surf. tens. 27.2 dynes/cm (0.1 g/100 ml); pH 6–8.

Merpol® DA. [DuPont] Amine ethoxylate; nonionic; low-foaming leveling agent and dyeing assistant for textile dyes; leveling agent for disperse, acid, premetallized and cationic dyes; stable to acids, bases, heat, freezing; lt. yel. clear, sl. visc. liq.; alcoholic odor; sol. in water; > 50% sol. in ethanol, ethylene glycol, cellosolve, acetone, MEK, oleic acid; sp.gr. 1.03 g/mL; dens. 8.6 lb/gal; visc. 117 cps; HLB 17.9; cloud pt. 10 C (lower), > 100 C (upper, 1%); clear pt. 12 C; flash pt. (PMCC) 46 C; pH 8-9 (1%); surf. tens. 50 dynes/cm (0.1%); combustible; 60% act.

Merpol® HCS. [DuPont] Alcohol ethoxylate; nonionic; wetting agent, detergent, emulsifier, penetrant, antistat, leveling agent, dyeing assistant, stabilizer used in textiles, leather, paper, metal processing, rubber, emulsion polymerization, paints, inks, medicinal ointments, antiperspirants,; cutting oils, polishes, cosmetics; pigment dispersant; lt. yel. clear, visc. liq.; mild fatty odor; 40% sol. in water, sol. in org. solvs. that are misc. with water; sp.gr.

1.03 g/mL; dens. 8.63 lb/gal; HLB 15.3; cloud pt. > 100 C (upper, 1%); flash pt. > 235 F; pH 6–8 (10%); surf. tens. 42.9 dynes/cm (0.1%); 60% act.

Merpol® HCW. [DuPont] Higher fatty alcohol EO condensate; nonionic; emulsifier, leveling agent, stabilizer, wetting agent, dispersant, detergent, dyebath additive for all fibers; clear, visc. liq.; mild, alcoholic odor; water sol.; dens. 8.32 lb/gal; f.p. < 3 F; cloud pt. > 212 F; flash pt. > 80 F; pH 6–7 (10%); 30% act.

Merpol® LF-H. [DuPont] Alcohol ethoxylate/ propoxylate; nonionic; detergent, dispersant, lubricant, leveling agent, emulsifier, nonfoam wetting, scouring and dyeing assistant for textiles, emulsifier, lubricant; biodeg.; stable to acids, bases, heat; pale yel. clear liq.; mild odor; 40% max. sol. in water below 70 F; sol. in polar solvs.; sp.gr. 0.96 g/mL; dens. 7.96 lb/gal; visc. 28 cps; HLB 10.0; cloud pt. < 25 C (1%); flash pt. > 200 F (TCC); pH 6–8; surf. tens. 32 dynes/cm (0.1%).

Merpol® OJ. [DuPont] Alcohol ethoxylate; CAS 9004-98-2; nonionic; wetting agent, dispersant, emulsifier; stable to acids, bases, heat, freezing; wh. to pale yel. paste, mild fatty odor; sol. to 30% in water; sol. in polar solvs.; sp.gr. 1.00 g/mL; dens. 8.3 lb/gal; HLB 12.5; cloud pt. 65 C (upper, 1%); flash pt. (PMCC) > 110 C; pH 6-8 (1%); surf. tens. 26 dynes/cm (0.1%); 100% conc.

Merpol® OJS. [DuPont] EO condensate; nonionic; wetting and dispersing agent for textiles, emulsifier, antistat, detergent; sl. yel. clear liq.; mild fatty odor; sol. in water, sol. or disp. in alcohol, ketones, aliphatic and aromatic hydrocarbons; dens. 8.04 lb/gal; sp.gr. 0.9648; visc. 46 cps; HLB 12.7; cloud pt. 30 F; flash pt. 167 F (TOC); surf. tens. 33.5 dynes/cm (0.1%); pH 6–8; 60% act.

Merpol® SE. [DuPont] Long-chain fatty alcohol ethoxylate; nonionic; low foaming wetting and rewetting agent, emulsifier, dispersant for textiles, paper, o/w emulsions, asbestos; stable to acids, bases, heat, freezing; colorless to pale yel. clear liq., mild odor; 0.1% sol. in water, cloudy disp. at higher conc.; sol. in polar solvs.; sp.gr. 0.97 g/mL; dens. 8.1 lb/gal; visc. 60 cP; HLB 10.5; cloud pt. < 25 C (upper, 1%); flash pt. (PMCC) 93 C; pH 6-8 (1% emulsion); surf. tens. 27 dynes/cm (0.1%); 95% act.

Merpol® SH. [DuPont] Long-chain alcohol ethoxylate; CAS 24938-91-8; nonionic; detergent, wetting agent for textiles, hard surf. cleaning, paper, metal processing; biodeg.; stable to acids, bases, heat, freezing to 5 C; colorless clear liq.; alcoholic odor; misc. with water, sol. or disp. in alcohols, acetone, MEK, ethylene glycol, glycerin, oleic acid; sp.gr. 0.96 g/mL; dens. 8.0 lb/gal; visc. 24 cps; HLB 12.8; cloud pt. 50 C (1%); flash pt. (TCC) 41 C; surf. tens. 27.4 dynes/cm (0.1 g/100 ml); pH 6-8 (10% aq.); 50% act.

Mersitol 2434 AP. [Seppic] Alkyl sulfate; anionic; wetting agent for mercerizing.

Mersolat H 30. [Miles/Organic Prods.] Sodium alkane sulfonates based on n-paraffin; anionic; biodeg. detergent base, wetting agent for mfg. electrolyte-resistant textile and leather auxiliaries, alkaline and acid detergents, floor cleaners, bubble baths, disinfectants, car shampoos; colorless clear sol'n.; m.w. 310-320; visc. 15 mPa•s; cloud pt. < 10 C; pH alkaline; surf. tens. 34.9 dynes/cm (0.1%); 30% act.

Mersolat H 40. [Miles/Organic Prods.] Sodium alkane sulfonates based on n-paraffin; anionic; biodeg. detergent base, wetting agent for mfg. electrolyte-resistant textile and leather auxiliaries, alkaline and acid detergents, floor cleaners, bubble baths, disinfectants, car shampoos; colorless clear sol'n.; m.w. 310-320; visc. 80-160 mPa•s; cloud pt. < 20 C; pH alkaline; surf. tens. 34.9 dynes/cm (0.1%); 40% act.

Mersolat H 68. [Miles/Organic Prods.] Sodium alkane sulfonates based on n-paraffin; anionic; biodeg. detergent base, wetting agent for mfg. electrolyte-resistant textile and leather auxiliaries, alkaline and acid detergents, floor cleaners, bubble baths, disinfectants, car shampoos; colorless-ylsh. pumpable paste; m.w. 310-320; pH alkaline; surf. tens. 34.9 dynes/cm (0.1%); 68% act.

Mersolat H 76. [Miles/Organic Prods.] Sodium alkane sulfonates based on n-paraffin; anionic; biodeg. detergent base, wetting agent for mfg. electrolyte-resistant textile and leather auxiliaries, alkaline and acid detergents, floor cleaners, bubble baths, disinfectants, car shampoos; wh.-ylsh. stiff paste, virtually odorless; m.w. 310-320; pH alkaline; surf. tens. 34.9 dynes/cm (0.1%); 76% act.

Mersolat H 95. [Miles/Organic Prods.] Sodium alkane sulfonates based on n-paraffin; anionic; biodeg. detergent base, wetting agent for mfg. electrolyte-resistant textile and leather auxiliaries, alkaline and acid detergents, floor cleaners, bubble baths, disinfectants, car shampoos; wh.-ylsh. flakes; m.w. 310-320; pH alkaline; surf. tens. 34.9 dynes/cm (0.1%); toxicology: LD50 (oral, rats) 2100 mg/kg; 95% act.

Mersolat W 40. [Miles/Organic Prods.] Sodium alkane sulfonates based on n-paraffin; anionic; biodeg. detergent base, wetting agent for mfg. electrolyte-resistant textile and leather auxiliaries, alkaline and acid detergents, floor cleaners, bubble baths, disinfectants, car shampoos; colorless clear sol'n.; m.w. 290; visc. 50 mPa•s; cloud pt. < 10 C; surf. tens. 35.0 dynes/cm (0.1%); 40%% act.

Mersolat W 68. [Miles/Organic Prods.] Sodium alkane sulfonates based on n-paraffin; anionic; biodeg. detergent base, wetting agent for mfg. electrolyte-resistant textile and leather auxiliaries, alkaline and acid detergents, floor cleaners, bubble baths, disinfectants, car shampoos; colorless-ylsh. pumpable paste; m.w. 290; surf. tens. 35.0 dynes/cm (0.1%); 68% act.

Mersolat W 76. [Miles/Organic Prods.] Sodium alkane sulfonates based on n-paraffin; anionic; biodeg. detergent base, wetting agent for mfg. electrolyte-resistant textile and leather auxiliaries, alkaline and acid detergents, floor cleaners, bubble baths, disinfectants, car shampoos; wh.-ylsh. stiff paste, virtually odorless; m.w. 290; pH alkaline; surf. tens. 35.0 dynes/cm (0.1%); 76% act.

Mersolat W 93. [Miles/Organic Prods.] Sodium alkane sulfonates based on n-paraffin; anionic; biodeg. detergent base, wetting agent for mfg. electrolyte-resistant textile and leather auxiliaries, alkaline and acid detergents, floor cleaners, bubble baths, disinfectants, car shampoos; wh.-ylsh. flakes; m.w. 290; surf. tens. 35.0 dynes/cm (0.1%); toxicology: LD50 (oral, rats) 2000 mg/kg; 93% act.

Metachloron® A4 Liq. [Crompton & Knowles] Organo metallic compd.; color bleed assistant for direct dyes on wet cotton and rayon goods; off-wh. clear to hazy liq.; misc. with water; pH 6.5 ± 0.5.

Metasol HP-500. [Marubishi Oil Chem.] Compounded prod.; nonionic; heavy duty and low foaming detergent for metal cleaning; powd.

Methyl Ester B. [Mayco Oil & Chem.] Fatty esters; CAS 68990-52-3; nonionic; intermediate, lubricant, wetting agent; liq.; 100% conc.

Metolat FC 355. [Münzing Chemie GmbH] Nonionic; low foaming wetting agent for color paints, tinting pastes, org./inorg. pigments; improves color acceptance in emulsion paints.

Metolat FC 388. [Münzing Chemie GmbH] Nonionic; low foaming wetting agent for aq. and nonaq. systems, biological and natural paints; biodeg.; BGA compliance.

Metolat FC 514. [Münzing Chemie GmbH] Anionic; dispersing for extenders and pigments, emulsion paints; BGA compliance.

Metolat FC 515. [Münzing Chemie GmbH] Anionic; dispersing for extenders and pigments, emulsion paints; BGA compliance.

Metolat FC 530. [Münzing Chemie GmbH] Anionic; dispersing and grinding agent for pigments and fillers in aq. systems; BGA compliance; powd.

Metolat LA 524. [Münzing Chemie GmbH] Grinding and wetting agent for lacquers and varnishes; wetting agent for organically modified bentonite.

Metolat LA 571. [Münzing Chemie GmbH] Anionic; surface-act. antisettling agent for lacquers and varnishes, esp. for applic. of heavy pigments.

Metolat LA 573. [Münzing Chemie GmbH] Anionic; grinding aid for pigment pastes in org. media.

Metolat P 853. [Münzing Chemie GmbH] Nonionic; wetting agent for dry paints, syn. renderings, cements; powd.

Metolat TH 75. [Münzing Chemie GmbH] Anionic; emulsifier for min. oil, esp. for cutting oils.

Metrosol AZ. [Reilly-Whiteman] Detergent for textile scouring.

Michelene 10. [M. Michel] Diamine fatty acid condensate; cationic; softener for textile finishing; paste.

Michelene 15. [M. Michel] Diamine fatty acid condensate; cationic; softener for textile finishing; paste.

Micromulse WIO. [Norman, Fox] Designed to form stable w/o emulsions or o/w microemulsions.

Micro-Step® H-301. [Stepan; Stepan Canada; Stepan Europe] Sulfonate/nonionic blend; nonionic; microemulsifier for broad range of pesticides; amber liq.; sol. in xylene, water; HLB 12.0.

Micro-Step® H-302. [Stepan; Stepan Canada; Stepan Europe] Sulfonate/nonionic blend; nonionic; microemulsifier for broad range of pesticides; amber liq.; sol. in xylene, water; HLB 12.0.

Micro-Step® H-303. [Stepan; Stepan Europe] Surfactant blend; nonionic; microemulsifier for broad range of pesticides; amber hazy liq.; sol. in xylene, water; HLB 13.0.

Micro-Step® H-304. [Stepan; Stepan Europe] Sulfonate/nonionic blend; nonionic; microemulsifier for broad range of pesticides; amber liq.; sol. in xylene, water; HLB 12.0.

Micro-Step® H-305. [Stepan; Stepan Europe] Sulfonate/nonionic blend; nonionic; microemulsifier for broad range of pesticides; amber liq.; sol. in xylene, water; HLB 12.0.

Micro-Step® H-306. [Stepan; Stepan Europe] Sulfonate/nonionic blend; nonionic; microemulsifier for various active chloropyrifos formulations; lt. amber liq.; sol. in xylene, water; HLB 12.0.

Micro-Step® H-307. [Stepan; Stepan Europe] Sulfonate/nonionic blend; nonionic; emulsifier for agric. microemulsions; amber liq.; sol. in xylene, water; HLB 12.0.

Miglyol® 812. [Hüls Am.; Hüls AG] Caprylic/capric triglyceride; CAS 65381-09-1; nonionic; dispersant, lubricant, anticaking agent, carrier, solv., solubilizer, suspending agent for cosmetics, dietetic prods.; liq.; sol. in alcohol, min. oil, acetone; 100% conc.

Migrassist® D. [Sybron] Benzyl trimethyl ammonium chloride; cationic; migrator/dye leveling agent for cationic dyes; nonretarding.

Migrassist® NYL. [Sybron] Mildly cationic; leveling agent for acid dyes.

Migregal 2N. [Nippon Senka] Alkyl sulfate; anionic; level dyeing of nylon piece goods with acid dyes; liq.

Migregal NC-2. [Nippon Senka] Alkyl sulfate and nonionic surfactant; anionic/nonionic; leveling agent for fabrics; liq.

Millifoam ODE-60A. [Stepan; Stepan Canada] Ammonium ether sulfate; gypsum board foamer; liq.; 60% conc.

Mindust Series. [Hart Chem. Ltd.] Blend; nonionic/anionic; dedusting and wetting agent for use in coal prep.; liq.

Minemal 320, 325, 330. [Nippon Nyukazai] POE alkyl phenyl ether; nonionic; degreaser for leather; liq.; 100% conc.

Minemal 350. [Nippon Nyukazai] Nonionic/anionic; emulsifier for machine oil; liq.

Miracare® 2MHT (formerly Miranol® 2MHT). [Rhone-Poulenc Surf.; Rhone-Poulenc France] Disodium lauroamphodiacetate, sodium trideceth sulfate and hexylene glycol; amphoteric; detergent base; surfactant for nonirritating and non-eye-sting shampoos; liq.; 42% conc.

Miracare® BC-10. [Rhone-Poulenc Surf.; Rhone-Poulenc France] PEG-80 sorbitan laurate, cocamidopropyl betaine, sodium trideceth sulfate, sodium lauroamphoacetate, PEG-150 distearate, sodium laureth-13 carboxylate; conc. for prep. of baby shampoo, bubble bath, bath gel and liq. hand soap prods. requiring mildness.

Miracare® MS-1 (formerly Compound MS-1). [Rhone-Poulenc Surf.] PEG-80 sorbitan laurate, sodium trideceth sulfate, PEG-150 distearate, disodium lauraminopropionate, cocamidopropyl hydroxysultaine, sodium laureth-13 carboxylate; anionic/nonionic; conc. for prep. of baby shampoo, bubble bath, bath gel and liq. hand soap prods. requiring mildness; liq.

Miracare® MS-2 (formerly Compound MS-2). [Rhone-Poulenc Surf.] PEG-80 sorbitan laurate, sodium trideceth sulfate, PEG-150 distearate, disodium lauraminopropionate, cocamidopropyl hydroxysultaine, sodium laureth-13 carboxylate; anionic/nonionic; conc. for prep. of baby shampoo, bubble bath, bath gel and liq. hand soap prods. requiring mildness; liq.

Miracare® MS-4. [Rhone-Poulenc Surf.] PEG-80 sorbitan laurate, sodium trideceth sulfate, PEG-150 distearate, disodium lauraminopropionate, cocamidopropyl hydroxysultaine, sodium laureth-13 carboxylate; anionic/nonionic; conc. for prep. of baby shampoo, bubble bath, bath gel and liq. hand soap prods. requiring mildness; liq.

Miracare® SMC. [Rhone-Poulenc Surf.] Amidononionic-anionic complex; nonionic/anionic; high conc. base for laundry detergents and all-purpose hard surf. cleaners; liq.; 98% conc. DISCONTINUED.

Miramine® C (formerly Alkazine® C). [Rhone-Poulenc Surf.] Coco hydroxyethyl imidazoline; cationic; emulsifier, corrosion inhibitor, softener, antistat, for textiles, asphalt, plastics, petrol. industry, cutting oils; water repellent treatment of cement, concrete, and plaster; antifungal agent for wood; tar emulsion breaker; slime control additive in paperboard; brn. paste; amine odor; sol. in aq. acid; disp. in water; dens. 0.96 g/ml; f.p. 20 C; 85% act.

Miramine® CC. [Rhone-Poulenc Surf.] Cocoyl hydroxyethyl imidazoline; cationic; biodeg. emulsifier, corrosion inhibitor in cutting oils, lubricant for metal surfaces, used in acid shampoos, with min. spirits to emulsify grease and oil; amber liq.; bland to faintly ammoniacal odor; sol. in many org. solvs., disp. in water; sp.gr. 0.940; dens. 7.747 lb/gal; flash pt. > 200 C; 99% act. DISCONTINUED.

Miramine® GS. [Rhone-Poulenc France] Alkyl carboxyl deriv.; CAS 67990-17-4; anionic; wetting agent for foam wax stripper, degreaser, cleaning formulations; stable to acid and alkali; liq.; 49% conc. DISCONTINUED.

Miramine® O (formerly Alkazine® O). [Rhone-Poulenc Surf.] Oleyl hydroxyethyl imidazoline; cationic; emulsifier, corrosion inhibitor, softener, antistat, wetting and flocculating agent, lubricant for textiles, asphalt, car wax emulsions, cleaners, paint mfg., agric. applics., syn. coolants; dispersant for clay and pigments in solv. systems; tar emulsion breaker; brn. liq.; amine odor; sol. in min. oil and spirits, aromatic solv., perchloroethylene; water-disp.; dens. 0.94 g/ml; f.p. 5 C; 85% act.

Miramine® OC. [Rhone-Poulenc Surf.] Oleyl hydroxyethyl imidazoline; cationic; detergent, emulsifier, wetting agent for industrial cleaners, thickener of HCl, corrosion inhibitor in metal treating and cutting fluids; emulsifier; lubricant; fabric softener; biodeg.; amber liq. to pasty solid; ammoniacal odor; oil sol. and water disp.; sp.gr. 0.930; dens. 7.747 lb/gal; flash pt. > 200 C; 99% min. act.

Miramine® SC. [Rhone-Poulenc Surf.] Soya hydroxyethyl imidazoline; cationic; emulsifier and coemulsifier in solv./water systems, corrosion inhibitor in cutting oils, lubricant for metal surfaces; amber liq.; ammoniacal odor; sol. in many org. solvs., disp. in water; sp.gr. 0.930; biodeg.; 99% act. DISCONTINUED.

Miramine® SODI (formerly Alkamide® SODI). [Rhone-Poulenc Surf.; Rhone-Poulenc France] Stearamidopropyl dimethylamine; emulsifier, conditioner; produces cationic emulsions; off-wh. solid; 100% act., 98% amide.

Miramine® TO (formerly Alkazine® TO). [Rhone-Poulenc Surf.] Tall oil hydroxyethyl imidazoline; cationic; emulsifier, corrosion inhibitor in oil burning systems, pickling bath operations, asphalt emulsions; dispersant for clay and pigments in solv. systems; brn. liq.; amine odor; sol. in acid systems; pH alkaline; 80% act.

Miramine® TO-A. [Rhone-Poulenc Surf.] Tall oil aminoethyl imidazoline; cationic; aggregate wetting agent, detergent, lubricant, emulsifier for cationic asphalt emulsions, metalworking fluids; liq.; sol. in oil; dens. 7.747 lb/gal; flash pt. > 200 C; 99% act. DISCONTINUED.

Miramine® TOC. [Rhone-Poulenc Surf.] Tall oil hydroxyethyl imidazoline; cationic; emulsifier, corrosion inhibitor in cutting oils, lubricant for metal surfaces, used in acid shampoos, with min. spirits to emulsify grease and oil; solid; water-disp.; 100% conc. DISCONTINUED.

Miranate® B (formerly Mirawet® B). [Rhone-Poulenc Surf.; Rhone-Poulenc France] Sodium butoxyethoxy acetate; CAS 67990-17-4; anionic; low foaming wetting agent, emulsifier, lubricant for wax strippers, degreasers, metalworking fluids, other cleaner formulations; compat. with high concs. of electrolytes; stable to acid and alkali media; liq.; water-sol.; dens. 9.163 lb/gal; flash pt. > 200 C; 49% act.

Miranate® LEC. [Rhone-Poulenc Surf.] Sodium laureth-13 carboxylate; anionic; aux. detergent for shampoo systems, lime soap dispersant; emulsifier, lubricant for metalworking fluids; hazy gel; water-sol.; dens. 8.996 lb/gal; flash pt. > 200 C; pH 8.0 (10%); 70% act.

Miranol® 2CIB (formerly Alkateric® 2CIB). [Rhone-Poulenc Surf.; Rhone-Poulenc France] Disodium cocoamphodiacetate; amphoteric; detergent for high foaming, nonirritating shampoos, skin cleansers, cosmetics, industrial detergents; emulsifier, solubilizer, coupling agent for heavy-duty liq. cleaners; amber clear liq.; sol. in water; dens. 1.18 g/ml; pH 8-9.5; 49.5-50.5% solids.

Miranol® 2MCA (redesignated Miracare® 2 MCA). [Rhone-Poulenc Surf.]

Miranol® C2M Anhyd. Acid. [Rhone-Poulenc Surf.; Rhone-Poulenc France] Cocoamphodipropionic acid; CAS 68919-40-4; amphoteric; detergent, wetting agent, high foaming surfactant for disp. on caustic soda and on powd. mixes, etc., leveling agent in tin plating from acid baths; metal cleaning; industrial cleaning; amber stiff gel; sol. in water and alcohol, slightly sol. in nonpolar org. solvs.; pH 4.6 (50%); 100% act.

Miranol® C2M Conc. NP. [Rhone-Poulenc Surf.; Rhone-Poulenc France] Disodium cocoamphodiacetate; CAS 68650-39-5; amphoteric; detergent used in nonirritating shampoos, skin cleansers, make-up removers, emulsifier, solubilizer and stabilizer in pharmaceuticals, household and industrial cleaners; clear, visc. liq.; sol. in water; pH 8.0–8.5 (20% aq.); 38% act.

Miranol® C2M Conc. NP LV. [Rhone-Poulenc Surf.; Rhone-Poulenc France] Sodium coco dicarboxylate; amphoteric; high foaming detergent, wetting agent for dispersion on caustic soda and powd. mixes; in alkaline built metal cleaners; liq.; 50% conc.

Miranol® C2M Conc. NP-PG. [Rhone-Poulenc Surf.] Disodium cocoamphodiacetate, propylene glycol; surfactant.

Miranol® C2M Conc. OP. [Rhone-Poulenc Surf.; Rhone-Poulenc France] Disodium cocoamphodiacetate; amphoteric; surfactant for nonirritating shampoos, skin cleansers, medicated cosmetics, medium-duty liq. cleaners; liq.; 50% conc.

Miranol® C2M NP LA. [Rhone-Poulenc Surf.; Rhone-Poulenc France] Disodium cocoamphodiacetate; CAS 68650-39-5; amphoteric; high foaming detergent, wetting agent for dispersion on caustic soda and powd. mixes; in alkaline built metal cleaners; liq.; 50% conc.

Miranol® C2M-SF 70%. [Rhone-Poulenc Surf.; Rhone-Poulenc France] Disodium cocoamphodipropionate; CAS 68604-71-7; amphoteric; detergent used for heavy-duty liq. cleaning compds., steam cleaners, nonirritating shampoos, medicated cosmetics; paste at R.T., pumps and pours at 50 C; pH 9–10 (10% aq.); 70% act.

Miranol® C2M-SF Conc. [Rhone-Poulenc Surf.; Rhone-Poulenc France] Disodium cocoamphodipropionate; CAS 68604-71-7; amphoteric; detergent for extra heavy duty cleaners, steam, pressure, metal, all-purpose cleaners, shampoos, medicated cosmetics; also coupling agent, solubilizer, wetting agent; biodeg.; amber clear liq., fruity odor; sol. in water and alcohol; sp.gr. 1.07; f.p. < 0 C; b.p. 98 C; flash pt. 144 F; pH 9.4–9.8; toxicology: minimally irritating to skin and eyes; LD50 (oral, mouse) > 5 ml/kg; 39% act.

Miranol® CM Conc. NP. [Rhone-Poulenc Surf.; Rhone-Poulenc France] Sodium cocoamphoacetate; CAS 68608-65-1; amphoteric; detergent, wetting and foaming agent, sequestrant, emulsifier, dispersant, germicidal, visc. builder; for extra heavy duty cleaners, steam, pressure, metal, all-purpose cleaners; biodeg.; lt. amber visc. liq.; pH 9.0–13.0; 37% conc.

Miranol® CM Conc. OP. [Rhone-Poulenc Surf.; Rhone-Poulenc France] Sodium cocoamphoacetate; CAS 68608-65-1; amphoteric; surfactant for extra heavy duty liq. cleaners, steam, pressure, metal, and all-purpose cleaners; liq.; 42% conc.

Miranol® CM-SF Conc. [Rhone-Poulenc Surf.; Rhone-Poulenc France] Sodium cocoamphopropionate; CAS 68919-41-5; amphoteric; coemulsifier for emulsion polymerization; emulsifier, wetting agent for industrial, institutional and household cleaners; biodeg.; lt. amber clear liq.; sol. in water and alcohol; pH 9.5–10.5; 36–38% solids.

Miranol® CS Conc. [Rhone-Poulenc Surf.; Rhone-Poulenc France] Sodium cocoamphohydroxypropylsulfonate; CAS 68604-73-9; amphoteric; detergent, wetting agent, corrosion inhibitor, emulsifier, sequestrant, foaming agent for shampoos, cold water fabrics, household and industrial cleaners; biodeg.; yel. visc. liq., fruity odor; water-sol.; sp.gr. 1.16; f.p. < 0 C; b.p. 102 C; pH 8.0; toxicology: minimally irritating to skin and eyes; LD50 (oral, mouse) > 5 ml/kg; 38% conc.

Miranol® DM. [Rhone-Poulenc Surf.; Rhone-Poulenc France] Sodium stearoamphoacetate; amphoteric; antistat, household softener, lubricant, dispersant used in textile industry, hair conditioners; biodeg.; wh. creamy paste, readily pourable @ 60 C; pH 5.4–6.0 (65 C); 25–27% solids.

Miranol® DM Conc. 45%. [Rhone-Poulenc Surf.; Rhone-Poulenc France] Sodium stearoamphoacetate; amphoteric; viscosifier, lubricant, softener, conditioner for cosmetics, textiles, industrial, institutional and household cleaners; paste; water-disp.; 45% conc.

Miranol® FA-NP (formerly Mirapon FA-NP). [Rhone-Poulenc Surf.; Rhone-Poulenc France] Sodium cocoamphoacetate; CAS 68608-65-1; amphoteric; surfactant for extra-heavy duty liq. cleaning compds., steam, pressure, metal, and all-purpose cleaners; liq.; 44% conc.

Miranol® FAS (formerly Mirapon FAS). [Rhone-Poulenc Surf.] Sodium cocoamphopropionate; amphoteric; surfactant for industrial, institutional and household cleaners; liq.

Miranol® FB-NP. [Rhone-Poulenc Surf.; Rhone-Poulenc France] Disodium cocoamphodiacetate; CAS 68650-39-5; amphoteric; surfactant for nonirritating shampoos, skin cleansers, medicated cosmetics, med.-duty liq. cleaners; liq.; 50% conc.

Miranol® FBS. [Rhone-Poulenc Surf.; Rhone-Poulenc France] Disodium cocoamphodipropionate; CAS 68604-71-7; amphoteric; surfactant for extra heavy duty liq. cleaning compds., e.g., steam, pressure, metal, and all-purpose cleaners; liq.; 39% conc.

Miranol® H2C-HA (formerly Alkateric® A2P-LPS). [Rhone-Poulenc Surf.] Disodium N-lauryl iminodipropionate; amphoteric; foamer and wetting agent used in alkaline cleaners, fire fighting compds.; pale yel. liq.; 29-31% solids.

Miranol® HMA. [Rhone-Poulenc Surf.] Imidazoline deriv.; amphoteric; surfactant.

Miranol® HM Conc. [Rhone-Poulenc Surf.; Rhone-Poulenc France] Sodium lauroamphoacetate; CAS 68608-66-2; amphoteric; detergent, wetting and foaming agent, sequestrant, emulsifier, dispersant, germicidal for shampoos, dishwashing, paints; biodeg.; lt. amber, visc. liq.; pH 9.0–9.5; 43–45% solids.

Miranol® HM-SF Conc. [Rhone-Poulenc Surf.] Sodium lauroamphopropionate; amphoteric; emulsion polymerization; liq.; 37% conc. DISCONTINUED.

Miranol® J2M Conc. [Rhone-Poulenc Surf.; Rhone-Poulenc France] Disodium capryloamphodiacetate; CAS 68608-64-0; amphoteric; emulsifier, caustic soda wetting agent, food washing and peeling, industrial, institutional and household cleaners, wax stripper, emulsion polymerization of syn. rubbers and resins; biodeg.; clear liq.; pH 8.2–8.6; 48–50% solids.

Miranol® J2M-SF Conc. [Rhone-Poulenc Surf.] Disodium capryloamphodipropionate; amphoteric; salt free version of Miranol J2M Conc.; emulsifier, wetting agent, coupling agent, solubilizer; for dispersion on caustic soda and on powd. mixes, metal cleaning, industrial cleaning; higher tolerance for alkalies and/or electrolytes; biodeg.; liq.; pH 8.8–9.3; 38–40% solids.

Miranol® JA (formerly Mirapon JA). [Rhone-Poulenc Surf.; Rhone-Poulenc France] Dicarboxylic octoic deriv., sodium salt; amphoteric; wetting agent in bottle washing compds., alkaline metal cleaners; for dispersion on caustic soda and powd. mixes; liq.; 34% conc.

Miranol® JAS-50 (formerly Mirapon JAS-50). [Rhone-Poulenc Surf.] Capryloamphopropionate; amphoteric; emulsifier, wetting agent for industrial, institutional and household cleaners; liq.; 50% conc.

Miranol® JB (formerly Mirapon JB). [Rhone-Poulenc Surf.; Rhone-Poulenc France] Disodium capryloamphodiacetate; CAS 68608-64-0; amphoteric; wetting agent used in caustic soda based cleaners used for food washing and peeling; also in wax stripper formulations; liq.; 49% conc.

Miranol® JBS (formerly Mirapon JBS). [Rhone-Poulenc Surf.; Rhone-Poulenc France] Disodium capryloamphodipropionate; amphoteric; low foaming surfactant for medicated shampoos and skin cleansers; emulsifier, wetting agent for industrial cleaners; liq.; 38% conc.

Miranol® JEM Conc. (formerly Mirapon JEM Conc.). [Rhone-Poulenc Surf.; Rhone-Poulenc France] Sodium octoic dicarboxylate; amphoteric; emulsifier, detergent, lubricant, wetting agent for bottle washing compds., alkaline metal cleaning; for dispersion on caustic soda on powd. mixes; liq.; sol. in water; dens. 9.413 lb/gal; flash pt. > 200 C; 34% act.

Miranol® JEM-SF. [Rhone-Poulenc Surf.] Imidazoline deriv., salt-free; amphoteric; surfactant.

Miranol® JS Conc. (formerly Mirapon JS Conc.) [Rhone-Poulenc Surf.; Rhone-Poulenc France] Sodium capryloamphohydroxypropylsulfonate; CAS 68610-39-9; amphoteric; emulsifier, wetting agent, corrosion inhibitor used in pickling acids, low foam, acid and alkali stable industrial cleaners; biodeg.; liq.; water-sol.; 40% conc.

Miranol® L2M-SF Conc. (formerly Mirapon L2M-SF Conc.) [Rhone-Poulenc Surf.] Disodium tallamphodipropionate; amphoteric; high foaming detergent, emulsifier for oil, wash and wax prods., heavy-duty detergents, metalworking fluids; liq.; water-sol.; dens. 8.830 lb/gal; flash pt. > 200 C; 39% act.

Miranol® LB (formerly Mirapon LB). [Rhone-Poulenc Surf.] Disodium lauroamphodiacetate; amphoteric; high foaming emulsifier, wetting agent for industrial, institutional and household cleaners; liq.

Miranol® LM-SF Conc. [Rhone-Poulenc Surf.] Sodium tallamphopropionate; foamer with good oil emulsifying props., suggested as lubricant for ferrous and nonferrous metals; visc. liq. which sets to a pasty gel on aging at R.T., pourable @ 60 C; pH 9.5–10.5; biodeg.; 36–38% solids. DISCONTINUED.

Miranol® OM-SF Conc. [Rhone-Poulenc Surf.] Sodium oleoamphopropionate; amphoteric; wetting and foaming agent, sequestrant, emulsifier, dispersant, germicidal; sol. in water, alcohol; pH 9.8–10.2; biodeg.; 36–38% solids. DISCONTINUED.

Miranol® S2M Conc. [Rhone-Poulenc Surf.] Disodium caproamphodiacetate; amphoteric; wetting and foaming agent, sequestrant, emulsifier, dispersant, germicidal; clear liq.; pH 8.1–8.3; biodeg.; 50–52% solids. DISCONTINUED.

Miranol® S2M-SF Conc. [Rhone-Poulenc Surf.; Rhone-Poulenc France] Disodium caproamphodipropionate; CAS 68815-45-2; amphoteric; low wetting surfactant for medicated shampoos, cleansers; biodeg.; liq.; pH 8.8–9.2; 38–40% solids. DISCONTINUED.

Miranol® SM Conc. [Rhone-Poulenc Surf.; Rhone-Poulenc France] Sodium caproamphoacetate; CAS 68608-61-7; amphoteric; wetting agent, foaming agent, detergent used in medicated and germicidal shampoos and hand soaps, rug and upholstery shampoos, in emulsion polymerization; biodeg.; clear liq.; pH 9.0–9.5; 40–42% solids.

Miranol® TM. [Rhone-Poulenc Surf.] Dihydroxyethyl tallow glycinate; visc. building agent for industrial applics.; conditioner for premium shampoos; 40% act.

Mirapol® A-15. [Rhone-Poulenc Surf.; Rhone-Poulenc France] Polyquaternium-2; cationic; softening, conditioning, lubricant, antistat, surface modifying agent used in cream rinses, conditioning-type shampoos, textile processing; amber visc. liq.; m.w. 2260; dissolves readily in water; pH 8.5; 64% act. in water.

Mirapol® WT. [Rhone-Poulenc Surf.] Polyquaternary ammonium chloride; cationic; antiscaling compd. in high salinity aq. systems, corrosion inhibitor, textile softener; liq.; 63% conc.

Mirasheen® 202 (formerly Cyclosheen 202). [Rhone-Poulenc Surf.; Rhone-Poulenc France] Glycol stearate, lauramide DEA, cocamidopropyl betaine, glycerin; pearl conc. for cold blend cosmetic formulations, liq. soaps; contains visc. building, foam boosting, and mild conditioning agents; lotion/paste; 42% conc.

Mirataine® A2P-LPS. [Rhone-Poulenc Surf.] Disodium N-lauryl aminodipropionate; amphoteric; foamer, wetting agent with tolerance to alkali; for use in cleaners; liq.; 30% conc. DISCONTINUED.

Mirataine® A2P-TS-30 (formerly Alkateric® A2P-TS). [Rhone-Poulenc Surf.] Disodium N-tallow aminodipropionate; amphoteric; foamer, wetting agent with tolerance to alkali; for use in alkaline cleaners; yel. liq.; 39-31% solids.

Mirataine® AP-C. [Rhone-Poulenc Surf.] N-Coco beta aminopropionic acid; amphoteric; substantivity agent, detergent for conditioning shampoos, creams and lotions, heavy-duty and specialty industrial cleaners; liq.; 40% conc. DISCONTINUED.

Mirataine® ASC (formerly Mirawet ASC). [Rhone-Poulenc Surf.; Rhone-Poulenc France] Alkylether hydroxyl sultaine; amphoteric; nonfoaming wetting agent for industrial, institutional and household cleaners; stable in 50% NaOH and strong min. acids; liq.; 50% conc.

Mirataine® BB. [Rhone-Poulenc Surf.; Rhone-Poulenc France] Lauramidopropyl betaine; amphoteric; mild substantive surfactant, visc. builder and foam booster, wetting agent for shampoos and dishwashing liqs.; conditioner, antistat, emollient; as solubilizer, visc. builder, foam booster with lauryl sulfates; stable to acid and alkali media; clear liq.; 30% conc.

Mirataine® BD-R. [Rhone-Poulenc Surf.] Cocamidopropyl betaine.

Mirataine® BET-C-30 (formerly Cycloteric BET-C-30). [Rhone-Poulenc Surf.] Cocamidopropyl betaine; amphoteric; mild foaming agent, conditioner, detergent, emulsifier, foam booster/stabilizer, visc. builder for shampoos, liq. soaps, facial cleansers, bath gels, bubble baths; yel. liq.; pH 4.5-5.5; 29-31% act.

Mirataine® BET-O-30 (formerly Cycloteric BET-O-30). [Rhone-Poulenc Surf.; Rhone-Poulenc France] Oleamidopropyl betaine; amphoteric; conditioner, detergent, emulsifier, foam booster/stabilizer, visc. builder for shampoos, liq. soaps, facial cleansers, bath gels, bubble baths; yel. liq.; pH 5.5-6.5; 29-31% act.

Mirataine® BET-P-30. [Rhone-Poulenc Surf.] Cetyl betaine; foaming agent, conditioner, detergent, emulsifier, foam booster/stabilizer, visc. builder for shampoos, liq. soaps, facial cleansers, bath gels, bubble baths; 35% conc. DISCONTINUED.

Mirataine® BET-W (formerly Cycloteric BET-W). [Rhone-Poulenc Surf.] Cocamidopropyl betaine, tech.; foam booster, foaming agent, thickener, conditioner for shampoos, cosmetic, and industrial applics.; irritation mollifying agent for baby shampoos; 30% act.

Mirataine® CBC. [Rhone-Poulenc Surf.; Rhone-Poulenc France] Cocamidopropyl betaine; amphoteric; high foaming visc. builder, foam booster, emulsifier, wetting agent, dispersant for shampoos and dishwashing liqs.; liq.; 30% conc.

Mirataine® CB/M. [Rhone-Poulenc Surf.; Rhone-Poulenc France] Cocamidopropyl betaine; amphoteric; mild, high foaming, substantive surfactant, emulsifier, wetting agent, dispersant, visc. builder, foam booster for shampoo formulations, dishwashing liqs.; conditioner, antistat, emollient, solubilizer; stable to acid and alkaline media; pale-colored clear liq.; dens. 8.71 lb/gal; sp.gr. 1.045; pH 8.5; 35% act. in water.

Mirataine® CBR. [Rhone-Poulenc Surf.] Cocamidopropyl betaine; amphoteric; visc. builder and foam booster for shampoos and dishwashing liqs.; liq.; 30% conc.

Mirataine® CBS, CBS Mod. [Rhone-Poulenc Surf.;

Rhone-Poulenc France] Cocamidopropyl hydroxysultaine; CAS 70851-08-0; amphoteric; mild high foaming surfactant, visc. builder, foam booster, emulsifier, wetting agent for shampoo formulations, liq. bubble baths, industrial cleaners; liq.; water-sol.; dens. 9.1 lb/gal; cloud pt. ≤ -12 C; pH 8.2; 50% act. in water.

Mirataine® CCB. [Rhone-Poulenc Surf.] Cocamidopropyl betaine.

Mirataine® D-40 (formerly Ambiteric D-40). [Rhone-Poulenc France] Coconut dimethyl betaine; amphoteric; surfactant; 41% conc.

Mirataine® FM (formerly Mirapon FM). [Rhone-Poulenc Surf.] Potassium salt of alkylamino acid; CAS 68909-63-7; amphoteric; detergent, emulsifier, wetting agent, lubricant for industrial, institutional and household cleaners; acid and alkali stable; liq.; 70% conc.

Mirataine® H2C. [Rhone-Poulenc Surf.] Disodium lauriminodipropionate; amphoteric; detergent, lubricant, antistat, corrosion inhibitor, and solubilizer used in detergent and industrial formulations, personal care prods., leather, and textile fibers; yel. thin liq.; pH 10.5; flash pt. (PM) 132 C; biodeg.; 30.0% act. DISCONTINUED.

Mirataine® H2C-HA. [Rhone-Poulenc Surf.; Rhone-Poulenc France] Sodium lauriminodipropionate; CAS 3655-00-3; amphoteric; high foaming surfactant, foam booster, wetting agent, dispersant for shampoos and skin cleansers, and hard surface cleaners; liq.; 30% conc.

Mirataine® JC-HA. [Rhone-Poulenc Surf.] Amphoteric; low foam hydrotrope.

Mirataine® T2C-30. [Rhone-Poulenc Surf.] Disodium tallowiminodipropionate; amphoteric; detergent, solubilizer, moderate foaming surfactant used in textile, leather, metalworking, industrial and personal care prods.; liq.; sol. in water; dens. 8.746 lb/gal; surf. tens. 31.6 dynes/cm (1% aq.); pH 11.5; biodeg.; 30% act. in water.

Mirataine® TM. [Rhone-Poulenc Surf.; Rhone-Poulenc France] Dihydroxyethyl tallow glycinate; amphoteric; wetting agent, viscosifier for industrial and household cleaners; conditioner for shampoos; HCl thickener; lt. amber slightly hazy visc. liq.; pH 5.1; 35% conc.

Miravon B12DF. [Rhone-Poulenc France; Rhone-Poulenc UK] Ethylene/propylene oxide adducts; nonionic; biodeg. surfactant, defoamer for dishwash rinse aids, dairy cleaning, metal degreasing, hard surf. cleaners, biocides; colorless to pale yel. liq., may become turbid at low temps.; sol. in cold water below the cloud pt., lower alcohols, chloroform, xylene; sp.gr. 0.99; cloud pt. 18-22 C; flash pt. (PMCC) > 100 C; pH 5-8 (5% aq.); 100% act.

Miravon B79R. [Rhone-Poulenc France; Rhone-Poulenc UK] Ethylene/propylene oxide adducts; nonionic; biodeg. surfactant, wetting agent for dishwash rinse aids, dairy cleaning, hard surf. cleaners, metal degreasing; colorless to pale yel. liq., may become turbid at low temps.; sol. in cold water below the cloud pt., lower alcohols, chloroform, xylene; sp.gr. 1.00; cloud pt. 31-35 C; flash pt. (PMCC) > 100 C; pH 5-8 (5% aq.); surf. tens. 32.1 dynes/cm (0.1% aq.); Draves wetting 12 s (0.1% aq.); Ross-Miles foam 28 mm (0.1% aq.); 100% act.

Mirawet® B (redesignated Miranate® B). [Rhone-Poulenc Surf.]

Mirawet® FL (redesignated Rhodaterge® FL). [Rhone-Poulenc Surf.]

Mitin® FF High Conc. [Ciba-Geigy/Dyestuffs] Phenylsulfonic acid deriv.; CAS 3567-25-7; anionic; patented durable mothproofing agent for wool; powd.; 80% conc.

ML-33-F. [Hefti Ltd.] Sorbitan laurate; CAS 1338-39-2; nonionic; emulsifier for creams, sun lotions, ointments; coemulsifier for latex stabilization; lubricant; liq.; HLB 8.0; 100% conc.

ML-55-F. [Hefti Ltd.] Polysorbate 20; CAS 9005-64-5; nonionic; solubilizer for essential oils, fragrances, vitamins; emulsifier for cosmetics, pharmaceuticals; stabilizer in syrups; liq.; HLB 16.5; 100% conc.

ML-55-F-4. [Hefti Ltd.] Polysorbate 21; CAS 9005-64-5; nonionic; coemulsifier with ML-55-F for creams and ointments, pigment dispersions, color pastes, fiber preps., polishes, hydraulic fluids; liq.; HLB 12.0; 100% conc.

MO-33-F. [Hefti Ltd.] Sorbitan oleate; CAS 1338-43-8; nonionic; emulsifier for w/o creams, lotions, shampoos, ointments, biocides, paints, varnishes, anticorrosion oils, polishes, cutting oils; liq.; HLB 4.4; 100% conc.

Modicol L. [Henkel/Functional Prods.] PEG fatty ester; nonionic; chemical and mech. stabilizer, coagulant, wetting, visc. control, and dispersing agent, used in latex and resin emulsions; clear yel. liq.; sol. in ethanol, acetone, ethyl acetate, toluol, veg. oil, water, min. oil; dens. 8.6 lb/gal; acid no. 1.0; pH 7.5 (2%); 99% act.

Modicol N. [Henkel/Functional Prods.] Alkanolamide; nonionic; stabilizer for natural and syn. latex; liq.; 98% act.

Mona AT-1200. [Mona Industries] Surfactant conc.; amphoteric; surfactant, thickener, corrosion inhibitor, surf. tens. reducer for acids; used in acid bowl cleaners, metal cleaners, pickling baths, petrol. acidizing sol'ns., etc.; biodeg.; Gardner 9 clear visc. liq.; pH 5.5; 40% act.

Mona NF-10. [Mona Industries] Anionic; low-foaming detergent, wetting agent, solubilizer for spray, soak tank, in-place pipeline cleaners, floor scrubbing formulations; Gardner < 1 clear liq.; sol. in conc. alkaline builder sol'ns.; sp.gr. 1.06; visc. 1000 cP; cloud pt. 32 C (0.5%); pH 9.2; surf. tens. 33 dynes/cm (0.5%); Draves wetting 14 s (0.5%); 50% solids.

Mona NF-15. [Mona Industries] Anionic; low-foaming detergent, wetting agent, solubilizer for spray, soak tank, in-place pipeline cleaners, floor scrubbing formulations; Gardner < 1 clear liq.; sol. in conc. alkaline builder sol'ns.; sp.gr. 1.16; visc. 900 cP; cloud pt. 32 C (0.5%); pH 7.5; surf. tens. 33.6 dynes/cm (0.5%); Draves wetting 25 s (0.5%); 50% solids.

Mona NF-25. [Mona Industries] Anionic; low-foaming detergent, wetting agent, solubilizer for spray, soak tank, in-place pipeline cleaners, floor scrubbing formulations; Gardner < 1 clear liq.; sol. in conc. alkaline builder sol'ns.; sp.gr. 1.22; visc. 900 cP; cloud pt. 31 C (0.5%); pH 7.9; surf. tens. 33.7 dynes/cm (0.5%); Draves wetting 25 s (0.5%); 50% solids.

Monafax 057. [Mona Industries] Aromatic-based phosphate ester; anionic; low foaming EP lubricant in metalworking fluids and high performance syn. coolants; corrosion inhibitor; low coeff. of friction, good antiwear and antiweld properties; hydrotrope; used in cutting fluids, chain lubricants, hydraulic fluids, rust preventives, metal and maintenance cleaners; lt. yel. clear visc. liq.; sol. (@ 10%) in water, ethanol, butyl Carbitol, aromatic hydrocarbon, chlorinated paraffin, and naphthenic oil (100 SSU); sp.gr. 1.2; dens. 10 lb/gal; acid no. 165 ± 15 (pH 9.5-9.8); pH < 2.5 (10% aq.); 100% act.

Monafax 060. [Mona Industries] Aliphatic-based phosphate ester; anionic; moderate foaming detergent, hydrotrope and solubilizer for nonionics in alkaline systems; used in soak tank, alkaline, and steam cleaners; amber clear visc. liq.; sol. @ 10% in water, ethanol, ethylene glycol monobutyl ether; sp.gr. 1.335; dens. 11.3 lb/gal; acid no. 335 ± 5 (pH 5.0-5.5); pH 2.5 (10%); 100% act.

Monafax 785. [Mona Industries] Nonoxynol-9 phosphate; anionic; emulsifier, lubricant, antistat, detergent, corrosion inhibitor for emulsion polymerization, agric., metalworking lubricants, alkaline cleaners, industrial use; antisoil redeposition for dry cleaning; clear to slightly hazy visc. liq.; sol. in water, ethanol, perchlorethylene, xylene; sp.gr. 1.115; dens. 9.2 lb/gal; acid no. 70 ± 3; pH < 2.5 (10%); 100% act.

Monafax 786. [Mona Industries] Nonoxynol-6 phosphate; anionic; emulsifier, lubricant, antistat, detergent, corrosion inhibitor for agric., metalworking lubricants, alkaline cleaners, industrial use; antisoil redeposition for dry cleaning; clear to slightly hazy visc. liq.; sol. @ 10% in ethanol, perchloroethylene, Stod., xylene; disp. water; sp.gr. 1.085; dens. 9.0 lb/gal; acid no. 57 ± 3 (pH 5.0-5.5); pH < 2.5 (10%); 100% act.

Monafax 794. [Mona Industries] Mixt. of mono and diphosphate esters; anionic; emulsifier for polymerization, corrosion inhibitor, lubricant, antisoil redeposition aid, antistat; for agric., alkaline cleaners, metalworking lubricants, drycleaning systems; soft waxy paste; sol. @ 10% in water, ethanol, perchloroethylene, xylene; dens. 9.8 lb/gal; sp.gr. 1.175; acid no. 98 ± 5; pH < 2.5 (10%); 100% act.

Monafax 831. [Mona Industries] Deceth-4 phosphate; anionic; emulsifier, lubricant, antistat, detergent, corrosion inhibitor, coupler for cleaning and industrial use; clear visc. liq.; sol. @ 10% in water, ethanol, perchloroethylene, min. oil, Stod., xylene; sp.gr. 1.08; dens. 9.0 lb/gal; acid no. 110±3 (pH 5.0-5.5); pH < 2.5 (10%); 100% act.

Monafax 872. [Mona Industries] Org. phosphate ester, potassium salt; anionic; emulsifier, lubricant, antistat, detergent, corrosion inhibitor for agric., metalworking lubricants, alkaline cleaners, industrial use; antisoil redeposition for dry cleaning; hydrotrope and solubilizer for detergent systems; lt. straw clear liq.; sol. @ 10% in water, disp. in ethanol, ethylene glycol monobutyl ether, cottonseed oil; sp.gr. 1.35; dens. 11.24 lb/gal; acid no. 12-22 (pH 5.0-5.5); alkali no. 90-100; pH 7–9 (10%); 50% act.

Monafax 939. [Mona Industries] Alkyl phosphate ester in acid form; anionic; emulsifier, lubricant, corrosion inhibitor, EP additive used in metalworking lubricants and syn. cutting fluids; low coeff. of friction, reduced surf. tension, antiwear, and antiweld properties; used in water and oil-based formulations; amber clear liq.; m.w. 297; sol. at 1.0% and 10% in ethanol, chlorinated and aromatic hydrocarbons, min. spirits, kerosene, and min. oil; insol. in water; sp.gr. 1.04; dens. 8.6 lb/gal; acid no. 220 ± 3 (pH 5.0-5.5); pH 2.5 max.; toxicology: irritating to skin and eyes as free acid; DEA salt (10%) is very mild skin, mild eye irritant, oral > 5 ml/kg; 100% act.

Monafax 1214. [Mona Industries] Deceth-4 phosphate; anionic; hypochlorite-stable surfactant, surf. tens. reducer, wetting agent for mildew removers, tile cleaners, bowl cleaner, tire cleaners, bleaching of paper pulp and textiles, dairy cleaners, hard surf. cleaners; Gardner 6 clear visc. liq.; sol. @ 10% in water, ethanol, perchlorethylene, ethylene glycol butyl ether; disp. in min. oil, xylene; sp.gr. 1.22; dens. 10.1 lb/gal; acid no. 238 (pH 5); pH 2.0 (10%); 100% act.

Monafax 1293. [Mona Industries] Org. phosphate ester; anionic; biodeg. hydrotrope, coupling agent for nonionics and other detergents which are only slightly sol. in high electrolyte concs.; surfactant, wetting agent, antistat, dispersant; for household and industrial hard surf. detergents, metal cleaners, agric., textile, paper/pulp, drycleaning, emulsion polymerization; Gardner 3 clear liq.; sol. in water, ethanol, trichlorethylene, kerosene, toluene, hexane; disp. in min. oil, min. spirits; sp.gr. 1.07; dens. 8.9 lb/gal; acid no. 111 (@ pH 4.5), 185 (@ pH 9.4); pH < 2 (10%); 98% act.

Monalube 780. [Mona Industries] Modified alkanolamide; nonionic/anionic; lubricant for copper drawing; corrosion inhibitor; amber liq.; disp. in water; dens. 8.2 lb/gal; acid no. 7.5; 100% act.

Monamate C-1142. [Mona Industries] Disodium cocamido MIPA sulfosuccinate; foaming/cleaning surfactants for personal care and household prods.; liq.; toxicology: low skin and eye irritation; 40% act.

Monamate CPA-40. [Mona Industries] Disodium cocamido MIPA-sulfosuccinate; anionic; high foaming, nonirritating surfactant for personal care and household prods.; biodeg.; Gardner 6 clear to hazy liq.; pH 6; Ross-Miles foam: 210 mm (1%, 120 F); toxicology: low eye and skin irritation; LD50 (oral) > 5 g/kg; 40% act.

Monamate CPA-100. [Mona Industries] Disodium cocamido MIPA sulfosuccinate; anionic; foaming/cleaning surfactants for personal care and household prods.; powd.; toxicology: low skin and eye irritation; 100% act.

Monamate LA-100. [Mona Industries] Disodium lauryl sulfosuccinate; anionic; high foaming, low irritation surfactant for personal care and household prods.; wh. fine powd.; pH 6.0–7.0 (10%); toxicology: low eye irritation, nonirritating to skin; LD50 (oral) > 5 g/kg; 85% act.

Monamate OPA-30. [Mona Industries] Disodium oleamido PEG-2 sulfosuccinate; anionic; high foaming, nonirritating surfactant for shampoos, bubble baths, soap bars; biodeg.; lt. yel. liq.; sp.gr. 1.06; dens. 8.85 lb/gal; pH 5.6; toxicology: nonirritating to eyes and skin @ 15% act.; 30% solids.

Monamate OPA-100. [Mona Industries] Disodium oleamido PEG-2 sulfosuccinate; anionic; high foaming, nonirritating surfactant for shampoos, bubble baths, soap bars; powd.; toxicology: low skin and eye irritation; 100% conc.

Monamid® 7-100. [Mona Industries] Cocamide DEA (1:1); nonionic; foam booster/stabilizer for industrial and household detergents; biodeg.; Gardner 6 liq.; sol. @ 10% in ethanol, chlorinated and aromatic hydrocarbons, min. spirits, kerosene, wh. min. oil, natural oils and fats; disp. in water; sp.gr. 0.96; dens. 8.00 lb/gal; acid no. 0-2; alkali no. 5-20; pH 8-9 (10%); 100% act. DISCONTINUED.

Monamid® 7-153 CS. [Mona Industries] Modified cocamide DEA (1:1); nonionic/anionic; emulsifier, detergent for drycleaning detergents and other systems; biodeg.; Gvcs-33 3 max. liq.; sol. @ 10% in ethanol, chlorinated and aromatic hydrocarbons, min. spirits, kerosene, wh. min. oil, disp. in water and natural oils and fats; dens. 8.75 lb/gal; sp.gr. 1.05; acid no. 0–2; alkali no. 4-10; pH 8.5–9.5 (10%); 100% act.

Monamid® 15-70W. [Mona Industries] 1:1 Linoleamide DEA; nonionic; thickener, visc. builder, hair and fiber conditioner; biodeg.; Gvcs-33 11 max. liq.; sol. @ 10% in ethanol, chlorinated and aromatic hydrocarbons, min. spirits, kerosene, natural oils and fats, gels in water; dens. 8.00 lb/gal; sp.gr. 0.96; acid no. 0–1; alkali no. 25-40; pH 10–11 (10%); 100% act.

Monamid® 150-AD. [Mona Industries] 1:1 Cocamide DEA; nonionic; foam booster/stabilizer for industrial and household detergents; biodeg.; Gvcs-33 11 liq.; sol. @ 10% in water, ethanol, chlorinated and aromatic hydrocarbons, natural oils and fats; dens. 8.25 lb/gal; sp.gr. 0.99; acid no. 0–3; alkali no. 55-70; pH 9.8–10.8 (10%); 100% act.

Monamid® 150-ADD. [Mona Industries] Cocamide DEA (1:1); nonionic; foam booster/stabilizer for industrial and household detergents; biodeg.; Gardner 4 liq.; sol. @ 10% in water, ethanol, chlorinated and aromatic hydrocarbons, natural oils and fats; sp.gr. 1.00; dens. 8.3 lb/gal; acid no. 0-3; alkali no. 58-68; pH 10-11 (10%); 100% act.

Monamid® 150-ADY. [Mona Industries] 1:1 Linoleamide DEA; nonionic; thickener for aq. systems; w/o emulsifier, corrosion inhibitor for sol. oils; biodeg.; Gvcs-33 10 max. liq.; sol. @ 10% in ethanol, chlorinated and aromatic hydrocarbons, min. spirits, kerosene, gels in water; dens. 8.20 lb/gal; sp.gr. 0.97; acid no. 0–1; alkali no. 30-45; pH 10–11 (10%); 100% act.

Monamid® 150-DR. [Mona Industries] 1:1 Cocamide DEA; nonionic; foam booster/stabilizer for industrial and household detergents; biodeg.; Gvcs-33 5 max. liq.; sol. @ 10% in ethanol, min. spirits, kerosene, wh. min. oil, natural oils and fats, disp. in water, chlorinated and aromatic hydrocarbons; dens. 8.20 lb/gal; sp.gr. 0.98; acid no. 0–5; alkali no. 10-25; pH 8.5–9.5 (10%); 100% act.

Monamid® 150-IS. [Mona Industries] 1:1 Isostearamide DEA; nonionic; lubricant, slip agent for cosmetic and industrial prods.; emulsifier, corrosion inhibitor and lubricant in syn. coolants; biodeg.; lt. amber liq.; sol. 10% in ethanol, chlorinated and aromatic hydrocarbons, min. spirits, kerosene, wh. min. oil, natural oils and fats, water disp.; dens. 8.00 lb/gal; sp.gr. 0.96; acid no. 5–10; alkali no. 30-60; cloud pt. < -5 C; pH 8.8–9.8 (10%); 100% act.

Monamid® 150-LMWC. [Mona Industries] Lauramide DEA; CAS 120-40-1; nonionic; foam booster/stabilizer for industrial and household detergents; biodeg.; Gvcs-33 3 max. solid; sol. @ 10% in ethanol, chlorinated and aromatic hydrocarbons, min. spirits, gels in water; dens. 8.20 lb/gal; sp.gr. 0.98 (40 C); acid no. 0–1; alkali no. 30-45; pH 10.2–11.2 (10%); 100% act.

Monamid® 150-LW. [Mona Industries] 1:1 Lauramide DEA; CAS 120-40-1; nonionic; foam booster/stabilizer for industrial and household detergents; biodeg.; Gvcs-33 3 max. solid; sol. @ 10% in ethanol, chlorinated and aromatic hydrocarbons, disp. in water; dens. 8.20 lb/gal; sp.gr. 0.98 (40 C); acid no. 0–1; alkali no. 30-45; pH 10–11 (10%); 100% act.

Monamid® 150-LWA. [Mona Industries] Cocamide DEA (1:1); nonionic; foam booster/stabilizer for industrial and household detergents; biodeg.; Gardner 4 liq.; sol. @ 10% in water, ethanol, chlorinated and aromatic hydrocarbons, natural oils and fats; sp.gr. 1.00; dens. 8.3 lb/gal; acid no. 0-3; alkali no. 58-68; pH 10-11 (10%); 100% act.

Monamid® 150-MW. [Mona Industries] 1:1 Myristamide DEA; nonionic; nonirritating foam stabilizer, emulsifier for aq. or nonaq. cosmetics and toiletries; thickener for systems containing sodium ions; biodeg.; wh. solid wax; sol. @ 10% in ethanol, chlorinated and aromatic hydrocarbons, disp. in water; dens. 8.20 lb/gal; sp.gr. 0.98 (45 C); m.p. 50 C; acid no. 0-3; alkali no. 35-50; pH 9.5-10.5 (10%); 100% act.

Monamid® 664-MC. [Mona Industries] Coconut alkanolamide; nonionic; foam booster, stabilizer and thickener for industrial and household detergents; liq.; 100% conc.

Monamid® 716. [Mona Industries] Lauramide DEA; CAS 120-40-1; nonionic; detergent, foam booster and stabilizer, visc. builder, solubilizer, coupler, wetting agent, used in shampoos, bubble baths, and household liq. detergent formulations; biodeg.; lt. amber clear liq.; sol. @ 10% in water, ethanol, chlorinated and aromatic hydrocarbons, natural oils and fats; sp.gr. 0.98; dens. 8.2 lb/gal; cloud pt. < 1 C; acid no. 0-3; alkali no. 45-60; pH 10.8 (10%); 100% act.

Monamid® 718. [Mona Industries] Stearamide DEA; nonionic; emulsifier, thickener, opacifier for creams and lotions; biodeg.; Gvcs-33 4 max. solid; sol. @ 10% in ethanol, aromatic hydrocarbons, gels in water; dens. 8.00 lb/gal; sp.gr. 0.96 (45 C); acid no. 21 ± 3; alkali no. 45-65; pH 9.3–10.3 (10%); 100% act.

Monamid® 770. [Mona Industries] Modified cocamide DEA (1:1); nonionic; foam booster/stabilizer for industrial and household detergents; biodeg.; Gardner 4 liq.; sol. @ 10% in water, ethanol, chlorinated and aromatic hydrocarbons, natural oils and fats; sp.gr. 0.99; dens. 8.25 lb/gal; acid no. 0-1; alkali no. 35-45; pH 9.2-10.2 (10%); 85% act. DISCONTINUED.

Monamid® 853. [Mona Industries] Alkanolamide, modified; nonionic; detergent, wetting emulsifier, foam stabilizer used in shampoos, bubble baths, skin cleansers, liq. household and industrial detergents, dairy cleaners, wool scouring, waterless hand cleaners; liq.; 100% conc.

Monamid® 1007. [Mona Industries] 1:1 Lauramide DEA and linoleamide DEA; foam booster and stabilizer, visc. builder, used in shampoos, bubble baths, skin cleansers, household and industrial liq. detergents; amber clear liq.; dens. 8.16 lb/gal; acid no. 1; alkali no. 35; iodine no. 40; pH 10.5 (10% aq.); usage level: 4-6%; 100% act.

Monamid® CMA. [Mona Industries] 1:1 Cocamide MEA; CAS 68140-00-1; nonionic; foamer, thickener for liq. and powd. detergents; biodeg.; tan granular; sol. @ 10% in ethanol, disp. in water, chlorinated and aromatic hydrocarbons, min. spirits, kerosene, wh. min. oil, natural oils and fats; solid. pt. 63 ± 2 C; acid no. 0–1; alkali no. 6-12; pH 9.4–10.8 (10%); 100% act.

Monamid® LIPA. [Mona Industries] 1:1 Lauramide MIPA; CAS 142-54-1; nonionic; foamer, thickener for liq. and powd. detergents; biodeg.; tan granular; sol. @ 10% in ethanol, disp. in water, kerosene, wh. min. oil, natural fats and oils; solid. pt. 55 ± 3 C; acid no. 0–1; alkali no. 12-22; pH 10.3–11.3 (10%);

100% act.

Monamid® LMA. [Mona Industries] 1:1 Lauramide MEA; CAS 142-78-9; nonionic; foamer, thickener for liq. and powd. detergents; biodeg.; tan granular; sol. @10 in ethanol, disp. in water, aromatic hydrocarbons, kerosene, wh. min. oil, natural fats and oils; solid. pt. 80 ± 2 C; acid no. 0–1; alkali no. 5-12; pH 10–11 (10%); 100% act.

Monamid® LMIPA. [Mona Industries] Lauramide MIPA; CAS 142-54-1; nonionic; foam booster, thickener used in personal care prods., household, and industrial detergent systems; off-wh. solid; m.p. 61 ± 2 C; acid no. 1 max.; alkali no. 10–25; pH 10.5 ± 0.5 (10% aq.); 100% act.

Monamid® LMMA. [Mona Industries] 1:1 Lauramide MEA; CAS 142-78-9; nonionic; foam booster, thickener for cosmetic, household and industrial detergents; biodeg.; tan granular; sol. @ 10% in ethanol, disp. in water, aromatic hydrocarbons, kerosene, wh. min. oil, natural oils and fats; solid. pt. 80 ± 2 C; acid no. 0–1; alkali no. 5-12; pH 9.7–10.7 (10%); 100% act.

Monamid® S. [Mona Industries] 1:1 Stearamide MEA; nonionic; emulsifier, thickener for kerosene, min. oils; biodeg.; tan granular; disp. @ 10% in water, chlorinated and aromatic hydrocarbons, min. spirits, kerosene, wh. min. oil, natural oils and fats; solid. pt. 87 ± 2 C; acid no. 0–1; alkali no. 5-18; pH 9.5–11.0 (10%); 100% act.

Monamide. [Zohar Detergent Factory] Cocamide MEA; CAS 68140-00-1; nonionic; foam booster, thickener, superfatting agent; flakes; 100% conc.

Monamine 779. [Mona Industries] Cocamide DEA and DEA-laureth sulfate; nonionic/anionic; foaming agent, visc. builder, detergent, soil suspending agent, solubilizer, wetting and penetrating agent used in shampoos, bubble baths, household and industrial cleaners, germicides, uv absorbers; biodeg.; lt. yel. clear to hazy visc. liq.; sol. @ 10% in water, polar solvs., alcohol, glycols, fatty acid esters, chlorinated and aromatic hydrocarbons, dens. 8.75 lb/gal; sp.gr. 1.05 ± 0.1; cloud pt. 2 C; pH 9.25 ± 0.5 (10%); surf. tens. 29 dynes/cm (0.1%); 100% act.

Monamine 853. [Mona Industries] Modified alkanolamide; detergent for waterless hand cleaners; amber clear liq.; dens. 8.4 lb/gal; acid no. 8 max.; alkali no. 105 ± 10; 100% act. DISCONTINUED.

Monamine 1255. [Mona Industries] Specialty alkanolamide; emulsifier for lotion or gel waterless hand cleaners; improves cleaning and rinsability; emulsifies wide range of solvs., e.g., odorless min. spirits, deodorized kerosene, d-limonene; Gardner 11 clear liq.; sp.gr. 1.00; acid no. 55; alkali no. 85; pH 8.7; 100% act.

Monamine AA-100. [Mona Industries] 1:2 Dist. cocamide DEA and diethanolamine; nonionic/anionic; detergent, emulsifier, thickener; biodeg.; Gvcs-33 7 max. liq.; sol. @ 10% in water, ethanol, chlorinated and aromatic hydrocarbons; dens. 8.30 lb/gal; sp.gr. 1.00; acid no. 28–32; alkali no. 165-185; pH 9.5–10.5 (10%); 100% act.

Monamine AC-100. [Mona Industries] Cocamide DEA and linoleamide DEA; nonionic/anionic; detergent, emulsifier, thickener; biodeg.; Gvcs-33 13 max. liq.; sol. @ 10% in ethanol, chlorinated and aromatic hydrocarbons, gels in water; dens. 8.30 lb/gal; sp.gr. 1.00; acid no. 22–32; alkali no. 170-190; pH 9.5–10.5 (10%); 100% act.

Monamine ACO-100. [Mona Industries] Lauramide DEA and diethanolamine; nonionic/anionic; detergent, emulsifier, thickener; biodeg.; Gvcs-33 5 max. paste; sol. @ 10% in water, ethanol, chlorinated and aromatic hydrocarbons; dens. 8.40 lb/gal; sp.gr. 1.01; acid no. 10–14; alkali no. 180-200; pH 9.5–10.5 (10%); 100% act.

Monamine AD-100. [Mona Industries] Cocamide DEA and diethanolamine; nonionic; detergent, emulsifier, thickener; biodeg.; Gvcs-33 11 liq.; sol. @ 10% in water, ethanol, chlorinated hydrocarbons, disp. in aromatic hydrocarbons; dens. 8.50 lb/gal; sp.gr. 1.02; acid no. 2-8; alkali no. 105-125; pH 9.5–10.5 (10%); 100% act.

Monamine ADD-100. [Mona Industries] Cocamide DEA (1:2); nonionic; detergent, emulsifier, thickener; biodeg.; Gardner 3 liq.; sol. @ 10% in water, ethanol, chlorinated hydrocarbins; disp. in aromatic hydrocarbons; sp.gr. 1.02; dens. 8.5 lb/gal; acid no. 2-6; alkali no. 105-125; pH 9.5-10.5; 100% act.

Monamine ADS-100. [Mona Industries] Cocamide DEA; nonionic/anionic; detergent, emulsifier, thickener; biodeg.; Gvcs-33 11 max. liq.; sol. @ 10% in water, ethanol, chlorinated and aromatic hydrocarbons; dens. 8.30 lb/gal; sp.gr. 1.00; acid no. 48–52; alkali no. 110-125; pH 9.0–10.0 (10%); 100% act.

Monamine ADY-100. [Mona Industries] Linoleamide DEA and diethanolamine; nonionic; detergent, emulsifier, thickener; biodeg.; Gvcs-33 9 max. liq.; sol. @ 10% in ethanol, min. spirits, kerosene; disp. in water, chlorinated and aromatic hydrocarbons; dens. 8.20 lb/gal; sp.gr. 0.98; acid no. 0–2; alkali no. 110-130; pH 10.5–11.5 (10%); 100% act.

Monamine ALX-80SS. [Mona Industries] Modified coconut alkanolamide; nonionic/anionic; detergent for industrial and household cleaners, dishwashing compds.; Gardner 6 liq.; sol. @ 10% in water, ethanol; sp.gr. 1.05; dens. 8.75 lb/gal; acid no. 52-60; alkali no. 65-75; pH 8.5-9.5 (10%); 80% conc.

Monamine ALX-100 S. [Mona Industries] Cocamide DEA, DEA-dodecylbenzene sulfonate, and diethanolamine; nonionic/anionic; detergent for general industrial and household cleaners, dishwash compds., textile detergents; biodeg.; Gvcs-33 8 max. liq.; sol. @ 10% in water, ethanol; dens. 8.85; sp.gr. 1.06; acid no. 62–70; alkali no. 70-90; pH 8.5–9.5 (10%); 100% act.

Monamine CD-100. [Mona Industries] Linoleamide DEA and diethanolamine; nonionic; lubricant and emulsifier for syn. cutting and drawing fluids; metal corrosion inhibitor; biodeg.; dk. br. visc. liq.; water-disp.; acid no. 10–15; alkali no. 118; pH 9.5–10.5 (10% disp.); 100% act.

Monamine CF-100 M. [Mona Industries] Cocamide DEA and diethanolamine; nonionic/anionic; detergent, emulsifier, thickener; biodeg.; Gvcs-33 11 max. liq.; sol. @ 10% in water, ethanol, chlorinated and aromatic hydrocarbons; dens. 8.40 lb/gal; sp.gr. 1.01; acid no. 56–64; alkali no. 110-120; pH 8.5–9.5 (10%); 100% act.

Monamine I-76. [Mona Industries] 1:2 Cocamide DEA; nonionic/anionic; foam booster/stabilizer, emulsifier, detergent, wetting agent for consumer and industrial prods.; biodeg.; Gvcs-33 12 max. liq.; sol. @ 10% in water, ethanol, chlorinated and aromatic hydrocarbons; dens. 8.30 lb/gal; sp.gr. 1.00; acid no. 45–55; alkali no. 80-100; pH 8.5–9.5 (10%); 100% act.

Monamine LM-100. [Mona Industries] Lauramide DEA and diethanolamine; nonionic/anionic; detergent, wetting agent, emulsifier, thickener, corrosion inhibitor; biodeg.; Gvcs-33 4 max. liq.; sol. @ 10% in water, ethanol, chlorinated and aromatic hydrocarbons; dens. 8.50 lb/gal; sp.gr. 1.02; acid no. 18–23; alkali no. 160-175; pH 9.5–10.5 (10%); 100% act.

Monamine R8-26. [Mona Industries] Linoleamide DEA and diethanolamine; nonionic/anionic; detergent, wetting agent, foam booster, thickener for household detergents; biodeg.; amber liq.; sol. @ 10% in water, ethanol, chlorinated and aromatic hydrocarbons; dens. 8.25 lb./gal; sp.gr. 0.99; acid no. 75–85; alkali no. 180; pH 9.0–10.0 (10%); 100% act.

Monamine T-100. [Mona Industries] Tallamide DEA and diethanolamine; nonionic/anionic; detergent, wetting agent, foam booster, thickener for household detergents; biodeg.; Gvcs-33 10 max. liq.; sol. @ 10% in ethanol, chlorinated and aromatic hydrocarbons, min. spirits, kerosene, disp. in water; dens. 8.10 lb/gal; sp.gr. 0.97; acid no. 10–16; alkali no. 100-120; pH 10.0–11.0 (10%); 100% act.

Monamulse 653-C. [Mona Industries] Alkanolamide, modified; anionic/nonionic; emulsifier, solubilizer, degreaser for solvs. such as min. spirits, Stod., kerosene, pine oil; used for cleaners for engine blocks, garage floor, truck bodies, silk screens, mechanics hand cleaners; amber clear visc. liq.; sp.gr. 1.04; dens. 8.66 lb/gal; acid no. 72; alkali no. 17; pH 8.8 (10%); 100% act.

Monamulse 748. [Mona Industries] Alkanolamide, modified; nonionic; emulsifier and dispersant for crude oil; liq.; toxicology: nontoxic to fish; 100% conc.

Monamulse 947. [Mona Industries] Alkylaryl sulfonate; anionic; emulsifier, solubilizer for aromatic and chlorinated solvs.; for cleaning engine blocks, truck bodies, garage floors, silk screens, paint booths; clear to hazy visc. liq.; sp.gr. 1.11; dens. 9.25 lb/gal; acid no. 86; alkali no. 100; pH 9.5 (10%); 100% act.

Monamulse CI. [Mona Industries] Imidazoline, modified; corrosion inhibitor improving water resistance in greases, oil-based lubricant systems, and on cast iron; improves emulsion stability; penetrant; amber clear liq.; sol. in alcohol, min. spirits, kerosene, wh. min. oil, natural fats and oils, and chlorinated and aromatic hydrocarbons; sp.gr. 0.959 ± 0.005; acid no. 0–2; alkali no. 118.

Monamulse dL-1273. [Mona Industries] Surfactant blend; anionic/nonionic; emulsifier for polar and nonpolar solvs., improves detergency, wetting, and rinsability for degreasers, all-purpose cleaners, waterless hand cleaners, deodorizers; biodeg.; Gardner 3-4 clear liq.; acid no. 2; alkali no. 8; pH 8.9 (10%).

Monaquat AT-1074. [Mona Industries] Quat. compd.; cationic; thickener with foaming and cleaning properties for HCl sol'ns.; used in acid bowl and pipe line cleaners, rust removers, and oil well acidizing applics.; Gvcs-33 3 clear to slightly hazy liq.; sol. in HCl; sp.gr. 0.9864; dens. 8.22 lb/gal; solid pt. < 0 C; pH 5.5 (10% aq.); 30% act.

Monaquat ISIES. [Mona Industries] Isostearyl ethylimidonium ethosulfate; cationic; antistat, lubricant, softener, corrosion inhibitor used in cosmetic industry, in industrial and textile applics.; biodeg.; amber liq.; sol. @ 10% in water, ethanol, butyl Cellosolve, aromatic, chlorinated and fluorinated hydrocarbons; sp.gr. 1.03; pH 6.9 (10%); 100% act.

Monaquat TG. [Mona Industries] Bishydroxyethyl dihydroxypropyl stearaminium chloride; cationic; surfactant for personal care prods.; conditioner for hair rinses; antistat and foaming used in fabric laundering/softening prods.; thickener for acid bowl cleaners, naval gels; Gardner 3 clear to slightly hazy liq.; sp.gr. 1.011; dens. 8.39 lb/gal; solid. pt. –10 to –12 C; cloud pt. 3 C; pH 5.5 (10%); biodeg.; toxicology: moderate skin irritant, nonirritating to eyes @ 3%; LD50 (oral) > 5 g/kg (10%); 30% act.

Monastat 1195. [Mona Industries] Formulated conc.; cationic; antistat/cleaner for glass and plastic, e.g., TV screens, computers, medical diagnostic equip., safety goggles; clear amber liq.; sol. in water, IPA, ethanol, methanol, propylene glycol, ethylene glycol, perchloroethylene; dens. 8.2 lb/gal; pH 6.6 (10%); usage level: 1-4 oz/gal; 80% act.

Monaterge 85. [Mona Industries] Fatty acid amido complex; nonionic/anionic; detergent, wetting agent, emulsifier, solubilizer for chlorinated solvs., used in alkaline industrial and laundry cleaners; high tolerance to electrolyte systems; biodeg.; amber clear liq.; sol. in water, ethanol, aromatic hydrocarbons, natural oils and fats; sp.gr. 1.02; dens. 8.5 lb/gal; acid no. 45; alkali no. 84; pH 8.9 (10%); surf. tens. 27.9 dynes/cm (0.05%); Draves wetting 8 s (0.25%); 85% act.

Monaterge 85 HF. [Mona Industries] Cocamide DEA, DEA-acrylinoleate, and DEA-dodecylbenzene sulfonate; anionic/nonionic; high foaming detergent and wetting agent, hydrotrope; stable in high electrolyte systems; liq.; 85% conc.

Monaterge LF-945. [Mona Industries] Modified imidazoline; amphoteric; low foaming detergent for heavy duty alkaline and acid cleaners at low temps., e.g., spray tank cleaners, steam cleaners, soak tank cleaners; lt. yel. clear to hazy liq.; sp.gr. 1.00; dens. 8.3 lb/gal; pH 10–11 (10%); 100% act.

Monateric 810-A-50. [Mona Industries] Caprylic/capric carboxylic propionate, imidazoline-derived, salt-free; amphoteric; surfactant for industrial detergents, cleaners, cosmetics; hydrotrope, coupling agent, and/or solubilizer; corrosion inhibitor in metalworking systems, oil well flooding, and aerosol pkgs.; brn. clear liq.; sp.gr. 1.07; dens. 8.9 lb/gal; pH 4.4 (10%); Draves wetting 5 s (1%); 50% act. DISCONTINUED.

Monateric 811. [Mona Industries] Disodium capryloamphodipropionate; amphoteric; corrosion inhibitor, detergent, wetting agent in noncorrosive cleaners and industrial detergents; biodeg.; amber clear to hazy liq.; water sol.; sp.gr. 1.04; dens. 8.7 lb/gal; pH 11.4 (10%); Draves wetting instantaneous (1%); 50% act.

Monateric 1000. [Mona Industries] Disodium capryloamphodipropionate; amphoteric; corrosion inhibitor, detergent, wetting agent for metal cleaning, cutting fluids, syn. lubricants; biodeg.; lt. amber clear to hazy liq.; water sol.; sp.gr. 1.05; dens. 8.8 lb/gal; pH 11.2 (10%); Draves wetting 20 s (0.1%); 50% act. in water.

Monateric 1188M. [Mona Industries] Disodium lauryl beta-iminodipropionate; amphoteric; high foaming surfactant, hydrotrope for household and industrial hard surf. cleaners, shampoos, bubble bath, mild skin cleansers, down hole foamers, air drilling; textile wetter; biodeg.; stable to acid and alkali; Gardner 3 clear liq.; pH 10; Draves wetting 43 s (0.1% aq.); Ross-Miles foam 165 mm (0.1%, initial); 30% act. in water.

Monateric ADA. [Mona Industries] Cocamidopropyl betaine; amphoteric; high foaming surfactant used in air drilling, foam drilling, foam blanketing, air entraining agent for cement, gypsum, wallboard; for use in presence of brine and oil; biodeg.; clear amber liq.; sp.gr. 1.05; dens. 8.8 lb/gal; pH 7.7 (10%); Draves wetting 13 s (1%); 33% act., 38% total solids.

Monateric ADFA. [Mona Industries] Cocamidopropyl betaine; amphoteric; surfactant for air and foam drilling; amber clear liq.; sp.gr. 1.03; dens. 8.6 lb/gal; pH 7.8 (10%); Draves wetting 10 s (1%); 31% act., 34% total solids. DISCONTINUED.

Monateric CA-35. [Mona Industries] Sodium cocoamphopropionate; amphoteric; detergent, wetting agent, emulsifier, dispersant, foaming agent used in cosmetic, household, and industrial prods.; coupling agent, solubilizer; biodeg.; stable to acid, alkali, electrolytes; amber clear liq.; sol. in high concs. of phosphates, silicates, carbonates; m.w. 360; sp.gr. 1.02; dens. 8.5 lb/gal; pH 5.7 (10%); surf. tens. 29.5 dynes/cm (0.1% conc.); Draves wetting 11 s (1%); Ross-Miles foam 160 mm (0.1% aq., initial); toxicology: low skin and eye irritation; 35% act. in water.

Monateric CAM-40. [Mona Industries] Sodium cocoamphopropionate; amphoteric; surfactant; amber clear liq.; sp.gr. 1.05; dens. 8.8 lb/gal; pH 9.3 (10%); Draves wetting 40 s (1%); 40% act. DISCONTINUED.

Monateric CDS. [Mona Industries] Disodium cocoamphodiacetate, sodium lauryl sulfate; amphoteric; surfactant; yel. clear liq.; sp.gr. 1.09; dens. 9.1 lb/gal; pH 8.5 (10%); Draves wetting 3 s (1%); 31% act., 37% total solids. DISCONTINUED.

Monateric CDTD. [Mona Industries] Disodium cocoamphodiacetate, sodium trideceth sulfate; amphoteric; surfactant for industrial detergents, cleaners, cosmetics; hydrotrope, coupling agent, and/or solubilizer; corrosion inhibitor in metalworking systems, oil well flooding, and aerosol pkgs.; yel. clear liq.; sp.gr. 1.11; dens. 9.2 lb/gal; pH 8.3 (10%); Draves wetting instantaneous (1%); 44% act., 50% total solids. DISCONTINUED.

Monateric CDX-38. [Mona Industries] Disodium cocoamphodiacetate; amphoteric; detergent, foaming agent, mild base surfactant for shampoos and skin cleansers; biodeg.; lt. amber visc. liq.; sp.gr. 1.18; dens. 9.8 lb/gal; pH 8.4 (10%); Draves wetting 18 s (1%); 39% act., 50% total solids.

Monateric CDX-38 Mod. [Mona Industries] Disodium cocoamphodiacetate; amphoteric; high foaming detergent for nonirritating shampoos, skin cleansers, cosmetics; biodeg.; yel. clear liq.; sp.gr. 1.18; dens. 9.8 lb/gal; pH 8.8 (10%); Draves wetting 20 s (1%); 39% act., 50% total solids.

Monateric CEM-38. [Mona Industries] Disodium cocoamphodipropionate; amphoteric; detergent, wetting agent, emulsifier, dispersant, solubilizer, hydrotrope used in liq. detergent systems, heavy-duty detergents, acid bowl cleaners, cosmetics; biodeg.; amber clear to hazy liq.; sp.gr. 1.05; dens. 8.75 lb/gal; pH 8.5 ± 0.5 (10%); surf. tens. 35 dynes/cm (0.1%); Draves wetting 4.5 min (1%); 39% act.

Monateric CEM-38CG. [Mona Industries] Disodium cocoamphodipropionate; amphoteric; surfactant for industrial detergents, cleaners, cosmetics; hydrotrope, coupling agent, and/or solubilizer; corrosion inhibitor in metalworking systems, oil well flooding, and aerosol pkgs.; amber clear to hazy liq.; sp.gr. 1.07; dens. 8.9 lb/gal; pH 9.8 (10%); Draves wetting 10 min (1%); 38% act. DISCONTINUED.

Monateric CM-36S. [Mona Industries] Sodium cocoamphoacetate; amphoteric; foaming agent, emulsifier, high foaming detergent, wetting agent, solubilizer, conditioner, coupling agent, fulling agent used in cosmetic and textile industries; biodeg.; clear yel. liq.; water sol.; sp.gr. 1.10; dens. 9.2 lb/gal; pH 11.9 (10%); surf. tens. 32 dynes/cm (0.015%); Draves wetting 5 s (1%); 36% act., 42% total solids.

Monateric CNa-40. [Mona Industries] Coconut monocarboxylic propionate, imidazoline-derived, salt-free; amphoteric; surfactant for nonirritating cosmetics; wetting agent and detergent in high electrolyte systems; amber clear liq.; sp.gr. 1.09; dens. 9.1 lb/gal; pH 10.9 (10%); Draves wetting 6 s (1%); 40% act.

Monateric COAB. [Mona Industries] Cocamidopropyl betaine; amphoteric; surfactant for industrial detergents, cleaners, cosmetics; hydrotrope, coupling agent, and/or solubilizer; corrosion inhibitor in metalworking systems, oil well flooding, and aerosol pkgs.; yel. clear liq.; sp.gr. 1.04; dens. 8.7 lb/gal; pH 7.9 (10%); Draves wetting 10 s (1%); 32% act., 37% total solids.

Monateric CyA-50. [Mona Industries] Caprylic dicarboxylic propionate, imadazoline-derived, salt-free; amphoteric; surfactant for industrial detergents, cleaners, cosmetics; hydrotrope, coupling agent, and/or solubilizer; corrosion inhibitor in metalworking systems, oil well flooding, and aerosol pkgs.; dk. br. clear liq.; sp.gr. 1.07; dens. 8.9 lb/gal; pH 5.6 (10%); Draves wetting 9 s (1%); 50% act.

Monateric CyMM-40. [Mona Industries] Caprylic dicarboxylic propionate, imidazoline-derived, salt-free; amphoteric; surfactant for industrial detergents, cleaners, cosmetics; hydrotrope, coupling agent, and/or solubilizer; corrosion inhibitor in metalworking systems, oil well flooding, and aerosol pkgs.; amber clear liq.; sp.gr. 1.10; dens. 9.2 lb/gal; pH 9.8 (10%); Draves wetting 4.5 min (1%); 40% act. DISCONTINUED.

Monateric CyNa-50. [Mona Industries] Sodium capryloamphopropionate; amphoteric; detergent, emulsifier, coupling agent, solubilizer, wetting agent, low to moderate foaming surfactant used in conc. electrolyte systems, bottle washing, wax strippers, steam cleaners, industrial cleaning, textile processing, acid pickling baths; biodeg.; stable to acid, alkali, electrolytes; dk. amber liq.; sp.gr. 1.10; dens. 9.2 lb/gal; alkali no. 177; pH 10.5 ± 0.5 (10%); surf. tens. 26.9 dynes/cm (0.1% aq.); Draves wetting 29 s (1%); Ross-Miles foam 35 mm (0.1% aq.); 50% act. in water.

Monateric ISA-35. [Mona Industries] Sodium isostearoamphopropionate; amphoteric; surfactant used in cosmetic and industrial prods.; conditioner, lubricant, thickener; biodeg.; amber clear to hazy flowable gel; sol. @ 10% in water, alcohol, glycol ethers and alkanolamines, disp. in min. oil and natural oils and fats; sp.gr. 1.01; dens. 8.4 lb/gal; pH 5–6 (10%); surf. tens. 30 dynes/cm (0.1%); Draves wetting 10 min (1%); 35% act.

Monateric LF-100. [Mona Industries] Imidazoline-derived propionate; amphoteric; low foaming, biodeg. surfactact, detergent for alkaline and acid cleaners, industrial cleaning; brn. clear liq.; sol. in aq. acid systems; m.w. 290; sp.gr. 1.04; dens. 8.7 lb/gal; alkali no. 194; pH 11.7 (10%); Draves wetting 25 s (1%); 100% act. DISCONTINUED.

Monateric LFNa-50. [Mona Industries] Sodium salt of 2-alkyl imidazoline; amphoteric; very low foaming detergent for use in low or high temp. alkaline or acid cleaners, spray tank cleaners, industrial carpet and hard surf. detergents; biodeg.; brn. clear liq.; sp.gr. 1.09; dens. 9.1 lb/gal; alkali no. 189; pH 11.0 ± 0.5 (10%); Draves wetting 10 min (1%); 50% act.

Monateric TA-35. [Mona Industries] Sodium tallamphopropionate; amphoteric; surfactant for industrial detergents, cleaners, cosmetics; hydrotrope, coupling agent, and/or solubilizer; corrosion inhibitor in metalworking systems, oil well flooding, and aerosol pkgs.; dk. brn. gel; sp.gr. 1.02; dens. 8.5 lb/gal; pH 5.2 (10%); Draves wetting 7 min (1%); 35% act. DISCONTINUED.

Monateric TDB-35. [Mona Industries] Disodium tallow beta-iminodipropionate; CAS 61791-56-8; amphoteric; hydrotrope and detergent for high electrolyte systems such as heavy-duty liq. cleaners and wax strippers; yel. clear-hazy thin liq.; sp.gr. 1.03; dens. 8.6 lb/gal; pH 11 (1%); 35% act.

Monatrope 1250. [Mona Industries] Sodium alkanoate; anionic; surfactant hydrotrope for formulating alkaline built liq. detergent concs.; coupling agent for nonionic and other surfactants in high concs. of electrolytes; for household and industrial detergents, spray washes,; textiles, hypochlorite detergents/sanitizers, dishwash liqs.; Gardner 1 clear liq.; sol. in 10% sodium hydroxide, 20% potassium hydroxide; pH 10; Ross-Miles foam 5 mm (0.1% aq., initial); 45% solids.

Monatrope 1296. [Mona Industries] Org. phosphate ester; anionic; low foaming hydrotrope for formulating highly built liq. detergents for household or industrial use; solubilizer for nonionics in high electrolyte systems; used for spray and soak tank cleaners, liq. laundry detergents, mech. dishwashing; chlorine-stable; Gardner 2 clear liq.; sol. in 27% sodium hydroxide, 35% potassium hydroxide; sp.gr. 1.23; pH 5; 50% solids in water.

Monawet 1240. [Mona Industries] Disodium nonoxynol-10 sulfosuccinate; anionic; wetting agent for emulsion polymerization, adhesives, paints; FDA compliance; APHA 25 clear liq.; sp.gr. 1.11; dens. 9.25 lb/gal; pH 7.0 (10%); Ross-Miles foam 275 mm (0.1%); 35% total solids.

Monawet DL-30. [Mona Industries] Disodium alkoxy sulfosuccinate; anionic; wetting agent; APHA 150 clear liq.; sp.gr. 1.08; dens. 9.0 lb/gal; pH 5.5 (10%); Ross-Miles foam 140 mm (0.1%); 30% total solids. DISCONTINUED.

Monawet MB-45. [Mona Industries] Diisobutyl sodium sulfosuccinate; anionic; wetting, dispersing, emulsifying, penetrating and solubilizing agent used in emulsion polymerization of S/B for rug backing, paper coating, water treatment; EPA and FDA compliance; APHA 25 clear colorless liq.; m.w. 332; sol. in water, fairly sol. in polar and nonpolar solvs.; sp.gr. 1.12; dens. 9.3 lb/gal; cloud pt. 13 C; flash pt. 215 F (PMCC); surf. tens. 54 dynes/cm (0.1%); pH 6.0 (10%); Ross-Miles foam 30 mm (0.1%); 45% act. in water.

Monawet MB-100. [Mona Industries] Sodium diisobutyl sulfosuccinate; anionic; wetting agent for agric., water treatment applics.; EPA and FDA compliance; wh. powd.; pH 5.5 (10%); Ross-Miles foam 30 mm (0.1%); 100% act.

Monawet MM-80. [Mona Industries] Dihexyl sodium sulfosuccinate, 15% water, 5% IPA; anionic; wetting agent, detergent for emulsion and suspension polymerization, rug backing, paper coating, textiles, paint, agric., cosmetic, detergent, mining, water treatment, electroplating baths, and food industries; electrolyte tolerant; clear colorless liq.; m.w. 388; sol. 33 g/100 g water, in polar and nonpolar solvs.; sp.gr. 1.10; dens. 9.2 lb/gal; cloud pt. < 0 C; flash pt. (PMCC) 110 F; pH 6 ± 1 (10%); surf. tens. 46 dynes/cm (0.1%); Ross-Miles foam 60 mm (0.1%); 80% act.

Monawet MO-65-150. [Mona Industries] Dioctyl sodium sulfosuccinate, anhyd.; anionic; wetting, penetrating and spreading agent, emulsifier used in oil well cleaning, drycleaning detergents, solv. cleaners and strippers, lubricants, agric., paints, mining, water treatment, cosmetic applics.; APHA 50 clear colorless liq.; sp.gr. 1.05; dens. 8.75 lb/gal; acid no. 0.5; flash pt. (COC) 42–43 C; pH 5.5 (10%); 65% act.

Monawet MO-65 PEG. [Mona Industries] Dioctyl sodium sulfosuccinate; wetting agent for agric., paints, mining, water treatment, cosmetic applics.

Monawet MO-70. [Mona Industries] Dioctyl sodium sulfosuccinate, 20% water, 10% diethylene glycol butyl ether; anionic; wetting, dispersing, emulsifying, penetrating and solubilizing agent used in emulsion and suspension polymerization, adhesives, paints, textile, fertilizer, mining, water treatment, fire fighting, cosmetic, food industries; EPA and FDA compliance; clear colorless liq.; m.w. 444; sol. in polar and nonpolar solvs.; sp.gr. 1.08; dens. 9.0 lb/gal; cloud pt. < –5 C; flash pt. (PMCC) 325 F; pH 6.0 (10%); surf. tens. 29 dynes/cm (0.1%); Ross-Miles foam 225 mm (0.1%); 70% act.

Monawet MO-70-150. [Mona Industries] Dioctyl sodium sulfosuccinate; anionic; emulsifier, wetting agent, spreading agent, penetrant for agric., paints, mining, water treatment, cosmetic applics.; liq.; sol. in oils, solvs.; 70% conc.

Monawet MO-70E. [Mona Industries] Dioctyl sodium sulfosuccinate, 19% water, 11% ethanol; anionic; wetting agent for industrial applics., emulsion polymerization, adhesives, paints, textiles, agric., cosmetics, glass cleaners, mining, water treatment, wall paper removal, food pkg. plants; EPA and FDA compliance; clear colorless liq.; m.w. 444; sol. in polar and nonpolar solvs.; sp.gr. 1.08; dens. 9.0 lb/gal; cloud pt. < –5 C; flash pt. (PMCC) 82 F; pH 6.0 (10%); surf. tens. 29 dynes/cm (0.1%); Ross-Miles foam 220 mm (0.1%); 70% act.

Monawet MO-70 PEG. [Mona Industries] Dioctyl sodium sulfosuccinate; anionic; wetting agent for agric., paints, mining, water treatment, cosmetic applics.

Monawet MO-70R. [Mona Industries] Dioctyl sodium sulfosuccinate, 15% water, 15% propylene glycol; anionic; wetting, dispersing, emulsifying, penetrating and solubilizing agent for textile, printing, agric., paints, mining, water treatment, fire fighting, cosmetic, food industries; EPA and FDA compliance; clear colorless liq.; m.w. 444; sol. in polar and nonpolar solvs.; sp.gr. 1.06; dens. 8.8 lb/gal; cloud pt. < –5 C; flash pt. (PMCC) 280 F; pH 6.0 (10%); surf. tens. 29 dynes/cm (0.1%); Ross-Miles foam 210 mm (0.1%); 70% act.

Monawet MO-70S. [Mona Industries] Dioctyl sodium sulfosuccinate, 30% odorless min. spirits; anionic; wetting agent for dry cleaning soaps, spotting compds; wetting agent and emulsifier; agric., paints, mining, water treatment, cosmetics applics.; EPA and FDA compliance; APHA 25 clear liq.; oil and solv. sol.; sp.gr. 1.08; dens. 9.0 lb/gal; pH 5.5 (10%); Ross-Miles foam 190 mm (0.1%); 70% conc.

Monawet MO-75E. [Mona Industries] Dioctyl sodium sulfosuccinate, 18% water, 7% ethanol; anionic; wetting agent for industrial, agric., paints, mining, water treatment, cosmetics applics.; EPA and FDA compliance; clear colorless liq.; m.w. 444; sol. in polar and nonpolar solvs.; sp.gr. 1.08; dens. 9.0 lb/gal; cloud pt. < -5 C; flash pt. (PMCC) 80 F; pH 6.0 (10%); surf. tens. 29 dynes/cm (0.1%); Ross-Miles foam 220 mm (0.1%); 75% act.

Monawet MO-84R2W. [Mona Industries] Dioctyl sodium sulfosuccinate, 16% propylene glycol; anionic; wetting agent for general use, agric., paints, mining, water treatment, cosmetics applics.; EPA and FDA compliance; lt. yel. clear visc. liq.; m.w. 444; sol. in polar and nonpolar solvs.; sp.gr. 1.10; dens. 9.16 lb/gal; b.p. 298 F; cloud pt. < -10 C; flash pt. (PMCC) 223 F; pH 5.5 (10%); surf. tens. 29 dynes/cm (0.1%); Ross-Miles foam 190 mm (0.1%); toxicology: severe eye and skin irritant on overexposure; 84% act.

Monawet MO-85P. [Mona Industries] Dioctyl sodium sulfosuccinate, 15% sodium benzoate; anionic; wetting for wettable powds., pigments, paints, mining, water treatment, cosmetics; EPA and FDA compliance; wh. fine powd.; 0.85% sol. in water; pH 5.5 (10%); surf. tens. 26 dynes/cm (0.1%); Ross-Miles foam 200 mm (0.1%); 85% act.

Monawet MT-70. [Mona Industries] Ditridecyl sodium sulfosuccinate, 12% water, 18% hexylene glycol; anionic; wetting, dispersing, emulsifying, penetrating and solubilizing agent used in emulsion and suspension polymerization, paints, coatings, indirect food additives; EPA and FDA compliance; lt. straw clear liq.; m.w. 584; sol. in polar and nonpolar solvs.; sp.gr. 1.02; dens. 8.5 lb/gal; cloud pt. -2 C; flash pt. (PMCC) 230 F; pH 6.0 (10%); surf. tens. 29 dynes/cm (0.1%); 70% act.

Monawet MT-70E. [Mona Industries] Ditridecyl sodium sulfosuccinate; anionic; wetting, dispersing, emulsifying, penetrating and solubilizing agent used in emulsion and suspension polymerization, paints, coatings, indirect food additives; lt. straw clear liq.; m.w. 584; sol. in polar and nonpolar solvs.; sp.gr. 1.01; dens. 8.4 lb/gal; cloud pt. -15 C; flash pt. 86 F (PMCC); surf. tens. 29 dynes/cm (0.1%); pH 6 ± 1; 70% act.

Monawet MT-80H2W. [Mona Industries] Ditridecyl sodium sulfosuccinate, 20% hexylene glycol; anionic; wetting, dispersing, emulsifying, penetrating and solubilizing agent used in printing inks, indirect food additives; EPA and FDA compliance; lt. yel. clear liq.; m.w. 584; sol. in polar and nonpolar solvs.; sp.gr. 1.02; dens. 8.5 lb/gal; cloud pt. < 0 C; flash pt. (PMCC) 225 F; pH 5.5 (10%); surf. tens. 29 dynes/cm (0.1%); 80% act.

Monawet SNO-35. [Mona Industries] Tetrasodium dicarboxyethyl stearyl sulfosuccinamate; anionic; wetting agent, solubilizer, emulsifier, dispersant, visc. depressant, mild detergent used in polymerization, paints, coatings, textile, cosmetic, agric. prods.; biodeg.; FDA and EPA compliance; lt. amber clear liq.; m.w. 653; sol. in water, high electrolyte salt sol'ns.; sp.gr. 1.14; dens. 9.5 lb/gal; visc. 16–18 s (#2 Zahn cup); f.p. 45 ± 5 F; acid no. 2.0; iodine no. 0.5; pH 7.5; surf. tens. 43 dynes/cm (0.1%); Draves wetting 232 s (0.25%, 30 C); Ross-Miles foam 185 mm (0.1%); toxicology: very low acute oral toxicity; mild eye irritant; 35% solids.

Monawet TD-30. [Mona Industries] Disodium deceth-6 sulfosuccinate; anionic; surfactant, emulsifier, foaming agent used in emulsion polymerization, cosmetic and textile industries; lt. yel. liq.; sp.gr. 1.08; dens. 9.0 lb/gal; visc. 45 cps; cloud pt. -3 C; pH 5–6; surf. tens. 32.9 dynes/cm (0.1%); Ross-Miles foam 60 mm (0.1% DW, initial); 30% act.

Monazoline C. [Mona Industries] Cocoyl hydroxyethyl imidazoline; cationic; wetting agent, emulsifier for nonpolar liq., detergent, thickener, corrosion inhibitor, antistat, softener, bactericide used in paint and textile industries; also dispersant for clay and pigments, in acid dairy cleaners, in oil well acidifying and sec. recovery operations; biodeg.; amber liq., may crystallize on aging; m.w. 282; sol. @ 10% in ethanol, chlorinated hydrocarbons, min. oil, toluene, kerosene, min. spirits, veg. oil, disp. in water; sp.gr. 0.93; dens. 7.75 lb/gal; acid no. 1 max.; alkali no. 205; pH 10.5–12.0 (10% disp.); 90% min. imidazoline.

Monazoline CY. [Mona Industries] Capryl hydroxyethyl imidazoline; cationic; wetting agent, emulsifier for nonpolar liq., detergent, thickener, corrosion inhibitor, antistat, softener, bactericide used in paint and textile industries; biodeg.; amber liq., may crystallize on aging; m.w. 212; sol. @ 10% in ethanol, chlorinated hydrocarbons, toluene, veg. oil; sp.gr. 0.99; dens. 8.25 lb/gal; acid no. 1 max.; alkali no. 270; pH 10.5–12.0 (10% disp.); 90% min. imidazoline.

Monazoline IS. [Mona Industries] Isostearyl hydroxyethyl imidazoline; cationic; corrosion inhibitor and lubricant; liq.; 100% conc.

Monazoline O. [Mona Industries] Oleyl hydroxyethyl imidazoline; cationic; wetting agent, emulsifier for nonpolar liq., detergent, thickener, corrosion inhibitor, antistat, softener, bactericide used in paint and textile industries; also dispersant for clays and pigments, in agric. preparations, acid dairy cleaners; biodeg.; amber liq., may crystallize on aging; m.w. 345; sol. @ 10% in ethanol, chlorinated hydrocarbons, min. oil, toluene, kerosene, min. spirits, veg. oil; sp.gr. 0.92; dens. 7.66 lb/gal; acid no. 1 max.; alkali no. 168; pH 10.0–11.5 (10% disp.); 90% min. imidazoline.

Monazoline T. [Mona Industries] Tall oil hydroxyethyl imidazoline; cationic; wetting agent, emulsifier for nonpolar liq., detergent, thickener, corrosion inhibitor, antistat, softener, bactericide used in paint and textile industries; also dispersant for clays and pigments, aids gravel-to-asphalt bonding, rinse aid for automatic car washes, printing ink additive, protective metal coatings; biodeg.; amber liq., may crystallize on aging; m.w. 350; sol. @ 10% in ethanol, chlorinated hydrocarbons, min. oil, toluene, kerosene, min. spirits, veg. oil, disp. in water; sp.gr. 0.93; dens. 7.75 lb/gal; acid no. 1 max.; alkali no. 165; pH 10.0–11.5 (10% disp.); 90% min. imidazoline.

Monogen. [Dai-ichi Kogyo Seiyaku] Sulfated fatty alcohol, sodium salt; anionic; textile scouring and washing agent; dye dispersant and leveling agent; paste; 60% conc.

Monolan® 1030. [Henkel/Emery; Henkel Canada] EO/PO block copolymer; nonionic; low foam wetting agent in liq. rinse prods., agric. formulations; defoamer for processing industries; EPA-exempt; clear liq.; m.w. 1850; dens. 8.64 lb/gal; visc. 250 cps; HLB 12; pour pt. 0 C; cloud pt. 100 C (10%); 100% conc.

Monolan® 1206/2. [Harcros UK] EO-PO block polymer; nonionic; antifoam for detergents, wetting agents, syn. lubricants; colorless clear liq., faint odor; water-insol.; sp.gr. 1.003; visc. 445 cs; flash pt. > 400 F (COC); pour pt. < 0 C; pH 6–8 (1% aq.); 100% act.

Monolan® 2000. [Henkel/Emery; Henkel/Functional Prods.] EO/PO block polymer; nonionic; defoamer, wetting agent, deduster for processing industries, industrial and household cleaning prods., agric. formulations; EPA-exempt; clear liq.; m.w. 2000; dens. 8.4 lb/gal; visc. 300 cps; HLB 3; pour pt. 0 C; cloud pt. 17 C (10%); 100% conc.

Monolan® 2000 E/12. [Harcros UK] EO-PO block polymer; nonionic; low-foam detergent, wetting agent, defoaming agent, dispersant, rinse aid; colorless clear liq., faint odor; water-insol.; sp.gr. 1.020; visc. 390 cs; flash pt. (COC) > 200 C; pour pt. < 0 C; pH 7.0 (1% aq.); 100% act.

Monolan® 2500. [Henkel/Emery; Henkel/Functional Prods.; Henkel Canada] EO/PO block polymer; nonionic; low foam defoamer, wetting agent, deduster for processing industries, paint strippers, dishwash, spray cleaners, industrial and household cleaning prods., agric. formulations; EPA-exempt; clear liq.; m.w. 2500; dens. 8.6 lb/gal; visc. 400 cps; HLB 7.0; pour pt. 0 C; cloud pt. 22 C (10%); 100% conc.

Monolan® 2500 E/30. [Harcros UK] EO-PO block polymer; nonionic; detergent, wetting agent for emulsion polymerization, metal cleaning, resin plasticizers, latex stabilization, textile processing; coemulsifier for phosphate toxicants; colorless clear liq., faint odor; water-sol.; sp.gr. 1.043; visc. 650 cs; HLB 6.0; flash pt. (COC) > 200 C; pour pt. < 0 C; pH 7.0 (1% aq.); surf. tens. 43 dynes/cm (0.1%); 100% act.

Monolan® 2800. [Henkel/Emery; Henkel/Functional Prods.; Henkel Canada] EO/PO block polymer; CAS 106392-12-5; nonionic; low foam defoamer, wetting agent, deduster for processing industries, industrial and household cleaning prods.; dispersant for pigments and pesticides; EPA-exempt; clear liq.; m.w. 2800; dens. 8.8 lb/gal; visc. 555 cps; HLB 15.0; pour pt. 0 C; cloud pt. 60 C (10%); 100% conc.

Monolan® 3000 E/50. [Harcros UK] EO-PO copolymer; nonionic; detergent and rinse aid; solid; 100% conc.

Monolan® 3000 E/60, 8000 E/80. [Harcros UK] EO-PO block polymer; nonionic; detergent, wetting agent, antifoam for industrial and domestic detergents, emulsion polymerization, metal cleaning, resin plasticizers, latex stabilization, textile processing; wh. solid and wh. flake resp., faint odor; water-sol.; sp.gr. 1.052 and 1.070; visc. 355 cs and 1100 cs (60 C) resp.; flash pt. > 400 F (COC); pour pt. 31 and 49 C; pH 6–8 and 7–8 (1% aq.); 100% act.

Monolan® 4500. [Henkel/Functional Prods.] POE POP copolymer; nonionic; defoamer, wetting agent, deduster for processing industries, industrial and household cleaning prods., agric. formulations; EPA-exempt; semisolid; m.w. 4600; dens. 8.6 lb/gal; HLB 16.0; pour pt. 40 C; cloud pt. > 100 C (10%); 100% conc. DISCONTINUED.

Monolan® 6400. [Henkel/Functional Prods.] EO/PO block polymer; nonionic; defoamer, wetting agent, deduster for processing industries, industrial and household cleaning prods., agric. formulations; EPA-exempt; semisolid; m.w. 5800; dens. 8.7 lb/gal; HLB 13.0; pour pt. 32 C; cloud pt. 70 C (10%); 100% conc. DISCONTINUED.

Monolan® 8000 E/80. [Harcros UK] EO/PO block polymer; CAS 9003-11-6; nonionic; low foam wetting agent and detergent, dispersant; primary emulsifier in emulsion polymerization; wh. flake, faint odor; water-sol.; sp.gr. 1.070; visc. 1100 cs (60 C); HLB 16.0; pour pt. 49 C; cloud pt. 100 C (1% aq.); flash pt. (COC) > 200 C; pH 7.0 (1% aq.); surf. tens. 52 dynes/cm (0.1%); 100% act.

Monolan® 12,000 E/80. [Harcros UK] EO-PO copolymer; CAS 9003-11-6; nonionic; latex stabilizer and sec. emulsifier; toilet block component; high foaming; solid; HLB 16.0; 100% conc.

Monolan® M. [Harcros UK] EO-PO block polymer; nonionic; low-foam emulsifier for herbicides and pesticides; colorless clear liq., faint odor; water-sol.; sp.gr. 1.020; visc. 470 cs; cloud pt. 24 C (1% aq.); flash pt. > 400 F (COC); pour pt. < 0 C; pH 6–8 (1% aq.); 100% act.

Monolan® O Range. [Harcros UK] EO/PO copolymer; nonionic; low foam surfactants and defoamers for agric., textiles, emulsion polymerization, syn. lubricants.

Monolan® OM 48. [Harcros UK] Complex alkoxylates; nonionic; components for defoaming and antifoaming agents; liq.; 100% conc.

Monolan® OM 59. [Harcros UK] High m.w. alkoxylate; nonionic; lubricant additive, defoamer component; liq.; 100% conc.

Monolan® OM 81. [Harcros UK] High m.w. alkoxylate; nonionic; lubricant additive, defoamer component; liq.; 100% conc.

Monolan® P222. [Harcros UK] EO-PO block polymer; nonionic; low foaming detergent, antifoam for detergents, wetting agents, syn. lubricants; colorless clear liq., faint odor; water-insol.; sp.gr. 1.004; visc. 440 cs; flash pt. (COC) > 200 C; pour pt. < 0 C; pH 7.0 (1% aq.); 100% act.

Monolan® PB. [Harcros UK] Complex EO–PO copolymer; nonionic; wetting agent, low foam detergent and rinse aid, pigment dispersant; antifoam; dye leveling agent; for car shampoos, window cleaners, yarn lubricants; colorless liq.; faint odor; disp. water; sp.gr. 1.020; visc. 680 cs; cloud pt. 19 C (1% aq.); flash pt. (COC) > 200 C; pour pt. < 0 C; pH 6–8 (1% aq.); 100% act.

Monolan® PC. [Harcros UK] EO/PO block polymer; CAS 9003-11-6; nonionic; low foaming wetting agent, detergent, pigment dispersant, rinse aid; spreader/sticker for agric. sprays; colorless liq., faint odor; water-sol.; sp.gr. 1.040; visc. 800 cs; pour pt. 8 C; cloud pt. 44 C (1% aq.); flash pt. (COC) > 200 C; 100% act.

Monolan® PEG 300. [Harcros UK] PEG; nonionic; surfactant intermediate; textile fiber lubricant, dye carrier; antistatic and mold release agent; liq./solid; 100% conc.

Monolan® PEG 400. [Harcros UK] PEG; nonionic; surfactant intermediate; textile fiber lubricant, dye carrier; antistatic and mold release agent; liq./solid; 100% conc.

Monolan® PEG 600. [Harcros UK] PEG; nonionic; surfactant intermediate; textile fiber lubricant, dye carrier; antistatic and mold release agent; liq./solid; 100% conc.

Monolan® PEG 1000. [Harcros UK] PEG; nonionic; surfactant intermediate; textile fiber lubricant, dye carrier; antistatic and mold release agent; liq./solid; 100% conc.

Monolan® PEG 1500. [Harcros UK] PEG; nonionic;

surfactant intermediate; textile fiber lubricant, dye carrier; antistatic and mold release agent; liq./solid; 100% conc.

Monolan® PEG 4000. [Harcros UK] PEG; nonionic; surfactant intermediate; textile fiber lubricant, dye carrier; antistatic and mold release agent; liq./solid; 100% conc.

Monolan® PEG 6000. [Harcros UK] PEG; nonionic; surfactant intermediate; textile fiber lubricant, dye carrier; antistatic and mold release agent; liq./solid; 100% conc.

Monolan® PK. [Harcros UK] EO/PO block polymer; nonionic; lubricant and hydraulic fluid additive, defoamer component; pale yel. clear liq., mild odor; sp.gr. 1.031; visc. 500 cs; flash pt. (PMCC) > 150 C; 100% act.

Monolan® PL. [Harcros] High m.w. EO/PO copolymer; nonionic; defoamer component; intermediate for syn. lubricants and hydraulic fluids; pale yel. clear liq., mild odor; sp.gr. 0.973; visc. 930 cs; flash pt. (PMCC) > 150 C; 100% act.

Monolan® PM7. [Harcros UK] EO-PO copolymer on an org. substrate; nonionic; low foam detergent and wetting agent for use in bottle washing formulations, rinse aid; clear, colorless liq.; faint odor; sol. in water; sp.gr. 1.050; visc. 710 cs; cloud pt. 41 C (1% aq.); flash pt. > 400 F (COC); pour pt. < 0 C; pH 6–8 (1% aq.); 100% act.

Monolan® PPG440, PPG1100, PPG2200. [Harcros UK] PPG; nonionic; lubricant, antistat, plasticizer, cosolvs. in dyestuff, ink, resin, rubber industries, cosmetic preparations, intermediates in surfactants and plastic prod.; colorless liq.; faint odor; sol. in water (PPG440), insol. in water (PPG1100, 2200); m.w. 400, 1000, 2000 resp.; sp.gr. 1.010, 1.005, and 1.004; visc. 80, 180, and 450 cs; flash pt. > 450 F (COC); pour pt. < 0 C; 100% act.

Monolan® PT. [Harcros UK] Complex EO–PO copolymer; nonionic; low foam detergent and rinse aid, pigment dispersant; antifoam; dye leveling agent; car shampoos, window cleaners, yarn lubricants; colorless liq.; faint odor; water sol.; sp.gr. 1.037; visc. 750 cs; cloud pt. 28 C (1% aq.); flash pt. (COC) > 200 C; pour pt. < 0 C; pH 6–8 (1% aq.); 100% act.

Monomuls 90-25. [Henkel; Henkel KGaA] Dist. hydrog. tallow glyceride; nonionic; consistency factor, emulsifier for o/w emulsions and stick preps.; powd.; HLB 3.8; 100% conc.

Monopole Oil. [Henkel/Organic Prods.] Sulfated castor oil; anionic; wetting agent, penetrant, leveling agent, softener/lubricant or softener additive for cellulosic fibers, used in dyeing; lt. yel., clear liq.; sp.gr. 1.05; 75% act.

Monosteol. [Gattefosse SA] Propylene glycol palmitostearate; nonionic; emulsifier; solid; HLB 4.0; 100% conc.

Monosulf. [Henkel/Organic Prods.] Sulfated castor oil; anionic; penetrant, emulsifier; textile dyeing assistant; fat liquor for suede leather; paper coating evener; plasticizer for starch, glues; emulsifier for latex; liq.; 68% conc.

Montaline 1054. [Seppic] Surfactant blend; anionic; textile wetting agent; paste; 55% conc.

Montaline 9575 M. [Seppic] Anionic/nonionic surfactant blend; anionic; textile wetting agent; paste; 50% conc.

Montaline RH. [Seppic] Castor oil sulfonated ester; anionic; wetting agent; liq.; 60% conc.

Montaline SPCV. [Seppic] POE fatty alcohol; nonionic; aux. for prod. of phosphoric acid and superphosphate; liq.; 50% conc.

Montane 20. [Seppic] Sorbitan laurate; CAS 1338-39-2; nonionic; emulsifier; liq.; HLB 8.6; 100% conc.

Montane 40. [Seppic] Sorbitan palmitate; CAS 26266-57-9; nonionic; emulsifier; solid; HLB 6.7; 100% conc.

Montane 60. [Seppic] Sorbitan stearate; CAS 1338-41-6; nonionic; emulsifier; solid; HLB 4.7; 100% conc.

Montane 65. [Seppic] Sorbitan tristearate; CAS 26658-19-5; nonionic; emulsifier; solid; HLB 2.7; 100% conc.

Montane 70. [Seppic] Sorbitan isostearate; CAS 71902-01-7; nonionic; emulsifier; liq.; 100% conc.

Montane 73. [Seppic] Sorbitan sesquiisostearate.

Montane 80. [Seppic] Sorbitan oleate; CAS 1338-43-8; nonionic; emulsifier; liq.; HLB 4.3; 100% conc.

Montane 83. [Seppic] Sorbitan sesquioleate; CAS 8007-43-0; nonionic; emulsifier; liq.; HLB 3.7; 100% conc.

Montane 85. [Seppic] Sorbitan trioleate; CAS 26266-58-0; nonionic; emulsifier; liq.; HLB 1.8; 100% conc.

Montanox 20 DF. [Seppic] Polysorbate 20; CAS 9005-64-5; nonionic; emulsifier; liq.; HLB 16.7; 100% conc.

Montanox 21. [Seppic] PEG-5 sorbitan laurate; nonionic; emulsifier; liq.; HLB 13.3; 100% conc.

Montanox 40 DF. [Seppic] Polysorbate 40; CAS 9005-66-7; nonionic; emulsifier; gel; HLB 15.6; 100% conc.

Montanox 60 DF. [Seppic] Polysorbate 60; CAS 9005-67-8; nonionic; emulsifier; gel; HLB 14.9; 100% conc.

Montanox 61. [Seppic] PEG-4 sorbitan stearate; CAS 9005-67-8; nonionic; emulsifier; solid; HLB 9.6; 100% conc.

Montanox 65. [Seppic] Polysorbate 65; CAS 9005-71-4; nonionic; emulsifier; solid; HLB 10.5; 100% conc.

Montanox 70. [Seppic] PEG-20 sorbitan isostearate; nonionic; emulsifier; liq.; 100% conc.

Montanox 71. [Seppic] PEG-5 sorbitan isostearate.

Montanox 80 DF. [Seppic] Polysorbate 80; CAS 9005-65-6; nonionic; emulsifier; liq.; HLB 15.0; 100% conc.

Montanox 81. [Seppic] Polysorbate 81; CAS 9005-65-6; nonionic; emulsifier; liq.; HLB 10.0; 100% conc.

Montanox 85. [Seppic] Polysorbate 85; CAS 9005-70-3; nonionic; emulsifier; liq.; HLB 11.0; 100% conc.

Montapol CST. [Seppic] Oleo/cetyl sulfate; anionic; detergent; paste; 35% conc.

Montegal 150 RG. [Seppic] Alkylpolyethoxy ether; nonionic; textile leveling agent; liq.; 50% conc.

Montegal AP 80. [Seppic] Alkyl polyethoxyether; nonionic; textile leveling agent; flakes; 100% conc.

Montegal OL 50. [Seppic] Ethoxylated castor oil; nonionic; textile leveling agent; liq.; 100% conc.

Montegal SH 25. [Seppic] Ethoxylated fatty amine; nonionic; textile leveling agent; liq.; 45% conc.

Monthyle. [Gattefosse; Gattefosse SA] Glycol stearate; CAS 111-60-4; nonionic; emulsifier, stabilizer for ointments, cream lotions; solid; HLB 3.0; 100% conc.

Montopol CST. [Seppic] Oleo/cetyl sulfate; anionic; detergent; paste; 35% conc.

Montosol IL-13. [Pulcra SA] IPA laureth-2.7 sulfate; anionic; used in personal care prods., dishwashing agents, and fire-fighting foam concs.; liq.; 30 ± 1%

conc.

Montosol IQ-15. [Pulcra SA] Ammonium alcohol ether (4) sulfate; anionic; foaming agent; liq.; 35 ± 1% conc.

Montosol PB-25. [Pulcra SA] Sodium nonoxynol-25 sulfate; anionic; emulsifier for emulsion polymerization; liq.; 35% conc.

Montosol PF-16, -18. [Pulcra SA] Ammonium laureth-1.5 sulfate; anionic; wetting agent used in personal care prods., detergents and window cleaners; liq.; 27 ± 1% conc.

Montosol PG-12. [Pulcra SA] Ammonium laureth sulfate; anionic; wetting agent used in personal care prods., detergents and window cleaners; liq.; 27 ± 1% conc.

Montosol PL-14. [Pulcra SA] TEA laureth-2 sulfate; anionic; used in personal care prods., dishwashing agents, and fire-fighting foam concs.; liq.; 35 ± 1% conc.

Montosol PQ-11, -15. [Pulcra SA] Ammonium laureth-3 sulfate; anionic; wetting agent used in personal care prods., detergents and window cleaners; liq.; 60 ± 2 and 27 ± 1% conc. resp.

Montosol TQ-11. [Pulcra SA] Ammonium laureth-2.4 sulfate; anionic; wetting agent used in personal care prods., detergents and window cleaners; liq.; 25 ± 1% conc.

Montovol GJ-12. [Pulcra SA] Sodium tallow alcohol sulfate; anionic; detergent, dispersant; paste; 34 ± 1% conc.

Montovol GL-13. [Pulcra SA] Magnesium lauryl sulfate; anionic; detergent, rug shampoos, wetting agent; liq.; 27 ± 1% conc.

Montovol RF-10. [Pulcra SA] Sodium lauryl sulfate; CAS 151-21-3; anionic; emulsifier; additive for emulsion polymerization; used in shampoos and specialty cleaners; liq., paste; 29 ± 1% conc.

Montovol RF-11. [Pulcra SA] TEA lauryl sulfate; anionic; hair and rug shampoos; liq.; 40 ± 1% conc.

Montovol RJ-13. [Pulcra SA] Sodium fatty alcohol sulfate; anionic; emulsifier used in cleaners and detergents; paste; 40 ± 1% conc.

Montovol RL-10. [Pulcra SA] TEA lauryl sulfate; anionic; hair and rug shampoos and specialty cleaners; liq.; 40 ± 1% conc.

Montoxyl NM Conc. [Seppic] Ethoxylated propoxylated fatty alcohol; nonionic; low foam wetting agent; liq.; 100% conc.

Morwet® B. [Witco/Organics] Sodium n-butyl naphthalene sulfonate; anionic; wetting agent for pesticides and dyestuffs; powd.; 75% conc.

Morwet® DB. [Witco/Organics] Sodium di-n-butyl naphthalene sulfonate; anionic; wetting agent, dispersant for pesticides, etc.; powd.; 100% conc.

Morwet® EFW. [Witco/Organics] Proprietary; anionic; wetting agent; powd.; 75% conc.

Morwet® IP. [Witco/Organics] Sodium diisopropyl naphthalene sulfonate; anionic; wetting agent for pesticides and dyestuffs; powd.; 75% conc.

Morwet® M. [Witco/Organics] Sodium dimethyl naphthalene sulfonate; anionic; wetting agent for pesticide formulations; powd.; 95% conc.

MP-33-F. [Hefti Ltd.] Sorbitan palmitate; CAS 26266-57-9; nonionic; surfactant for w/o emulsions based on min. oils and triglycerides, cosmetic creams, sun lotions, deodorant sticks, gels, ointments; softener and lubricant for polymers; flakes; HLB 6.5; 100% conc.

M-Quat® 32. [PPG/Specialty Chem.] Octadecyl diethanol methyl ammonium chloride; cationic; emulsifier, defoamer, coagulant; liq.; 50% conc.

M-Quat® 257. [PPG/Specialty Chem.] Quaternium-18, isopropyl alcohol; CAS 61789-80-8; quat. for use as textile softener; liq.; 75% act.

M-Quat® 620. [PPG/Specialty Chem.] Quat. ammonium compd.; textile softener applics.; liq.; 90% act.

M-Quat® 2475. [PPG/Specialty Chem.] Dicocodimonium chloride and isopropanol; cationic; emulsifier, defoamer, coagulant; for auto spray wax; liq.; 75% act.

MS-33-F. [Hefti Ltd.] Sorbitan stearate; CAS 1338-41-6; nonionic; surfactant for w/o emulsions based on min. oils and saponifiable fats and oils, cosmetic and pharmaceutical creams, emulsifier for milk replacers; coemulsifier in latex stabilization; flakes; HLB 5.0; 100% conc.

MS-55-F. [Hefti Ltd.] Polysorbate 60; CAS 9005-67-8; nonionic; o/w emulsifier for cosmetics, pharmaceuticals, tech. applics.; latex stabilizer for emulsion polymerization; antistatic agent; liq./paste; HLB 15.0; 100% conc.

M-Soft-1. [Zohar Detergent Factory] Silicone-enriched cationic; cationic; softener for all fibers, yarns, fabrics; esp. for brushed or raised fabrics; wh. o/w emulsion; pH 5.

M-Soft-10. [Zohar Detergent Factory] Amino functional silicone plus cationic; cationic; softener producing an elastomeric silicone finish with exc. soft handle on all kinds of fibers, yarns, and fabrics; wh. o/w emulsion; pH 5.

M-Soft-J. [Zohar Detergent Factory] Quat.; cationic; softener for jeans, for domestic and laundry use and in textile denim finishing; wh. o/w emulsion; pH 4-5.

M-Soft-JHV. [Zohar Detergent Factory] Quat.; cationic; softener for jeans, for domestic and laundry use and in textile denim finishing; wh. visc. liq. to paste (o/w emulsion); pH 4-5.

M-Soft-J-Super. [Zohar Detergent Factory] Quat.; cationic; softener for bed linen; wh. visc. liq. to paste (o/w emulsion); pH 4-5.

Mulsifan ABN. [Zschimmer & Schwarz] Surfactant blend; anionic/nonionic; emulsifier for solvs. for formulation of cold cleaning agents and steam jet cleaning agents for cleaning of transportation and equip.; brn. clear liq.; sol. in kerosene, diesel oil, petrol.; dens. 1.01 g/cc; HLB 9-10; pH 5.0-6.5 (1%); 94% act.

Mulsifan HF. [Zschimmer & Schwarz] Fatty alcohol polyglycol ether, fatty acid and fatty alcohol blend; emulsifier for paraffin oils; wax; 100% act.

Mulsifan K 326 Spezial. [Zschimmer & Schwarz] Surfactant blend; anionic/nonionic; emulsifier for solvs. for mfg. of cold cleaners for engines, chassis and machinery equip., transportation cleaners, steam jet cleaners; straw-colored liq.; sol. in petrol., benzene, kerosene, chlorinated hydrocarbons, turpentine, toluene, xylene; disp. in water; dens. 1.06 g/cc; HLB 10.0; pH 6-7 (10%); 83% act.

Mulsifan RT 1. [Zschimmer & Schwarz] Fatty acid polyglycol ester; nonionic; emulsifier for min. oils, textile processing, metal processing, drilling and cutting oils, chemo-tech. prods.; brn. liq.; sol. in min. oil; disp. in water; HLB 10.0; pH 5-7 (10% aq.); 100% act.

Mulsifan RT 2. [Zschimmer & Schwarz] Fatty acid polyglycol ester; nonionic; emulsifier for min. oils, textiles, metal processing, drilling and cutting oils, chemo-tech. prods.; yel. visc. liq.; sol. in min. oils; disp. in water; dens. 0.99 g/cc; HLB 10.0; pH 5-7

(10%); 100% act.

Mulsifan RT 7. [Zschimmer & Schwarz] Ethoxylated triglyceride; nonionic; emulsifier for triglycerides of veg. and animal origin, chem-tech. prods. for treatment of metals, textiles, leather; yel. visc. liq.; sol. in ethanol, IPA, benzene, toluene, xylene, chlorinated hydrocarbons; disp. in water; dens. 1.03 g/cc; HLB 10.0; pH 7-9 (10%); 100% act.

Mulsifan RT 11. [Zschimmer & Schwarz] Fatty alcohol polyglycol ether; nonionic; emulsifier for fatty acids and alcohols, triglycerides, waxes; wh. to ylsh. wax; sol. in water, ethanol, IPA, acetone, benzene, toluene, xylene, chlorinated hydrocarbons; dens. 1.03 g/cc; HLB 16.0; pH 5-7 (10%); 100% act.

Mulsifan RT 18. [Zschimmer & Schwarz] Alkylaryl polyglycol ether; nonionic; emulsifier for solvs.; solubilizer for perfumes and essential oils; for cleaning of metal, engines, machinery, leather degreasing, drycleaning booster; colorless visc. liq.; sol. in water, ethanol, petrol., benzene, toluene, xylene, acetone, chlorinated hydrocarbons; dens. 1.06 g/cc; HLB 13; pH 4-6 (10%); 100% act.

Mulsifan RT 19. [Zschimmer & Schwarz] Fatty alcohol polyglycol ether; nonionic; emulsifier for waxes, chem-tech. prods.; paste; HLB 13.0; 100% act.

Mulsifan RT 23. [Zschimmer & Schwarz] Laureth-5; CAS 3055-95-6; nonionic; emulsifier for paraffin oils, wh. oils for formulation of lubricating agents, spin finishes, coning oils, emulsions for cosmetics; colorless liq.; sol. in wh. oils, min. oils, petrol., benzene, toluene, CCl_4, kerosene, perchloroethylene, turpentine; disp. in water; dens. 0.96 g/cc; HLB 11.0; pH 5-7 (10%); 100% act.

Mulsifan RT 24. [Zschimmer & Schwarz] Fatty alcohol polyglycol ether; nonionic; emulsifier for wh. and paraffin oils, lubricating agents, spin finishes, coning oils; colorless liq.; sol. in wh. oils, min. oils, petrol., benzene, toluene, CCl_4, kerosene, perchloroethylene, turpentine; disp. in water; dens. 0.94 g/cc; HLB 9.0; pH 5-7 (10%); 100% act.

Mulsifan RT 27. [Zschimmer & Schwarz] Oleth-25; CAS 9004-98-2; nonionic; emulsifier for fatty acids, fatty alcohols, triglycerides; for formulation of polishing, latex and wax emulsions; wh. to ylsh. wax; sol. in water, ethanol, acetone, petrol., toluene, xylene, chlorinated hydrocarbons; dens. 1.04 g/cc; HLB 16.0; pH 5-7 (10%); 100% act.

Mulsifan RT 37. [Zschimmer & Schwarz] Alkylphenol polyglycol ether; nonionic; emulsifier for min. oils and solvs., lubricating agents, drilling and cutting oils, textiles, metal degreasing, chemo-tech. prods.; colorless to ylsh. visc. liq.; sol. in benzene, toluene, xylene, petrol., chlorinated hydrocarbons, turpentine, min. oil; disp. in water; dens. 1.04 g/cc; HLB 11.0; pH 4-6 (10% aq.); 100% act.

Mulsifan RT 63. [Zschimmer & Schwarz] Fatty amine ethoxylate; nonionic; detergent for cleaning agents; liq.; 100% conc.

Mulsifan RT 69. [Zschimmer & Schwarz] PEG-40 castor oil; CAS 61791-12-6; nonionic; emulsifier for olein, fats, and oils; solubilizer for perfumes and volatile oils; yel. paste; sol. in water, ethanol, IPA, toluene, xylene, chlorinated hydrocarbons; dens. 1.09 g/cc; HLB 13.0; pH 6-8 (10%); 100% act.

Mulsifan RT 72. [Zschimmer & Schwarz] Coco fatty acid MEA ethoxylate; washing agent; liq.; 100% act.

Mulsifan RT 110. [Zschimmer & Schwarz] Fatty alcohol polyglycol ether blend; nonionic; emulsifier for paraffins, microparaffin waxes for textile, paper, ceramics, cleaning and polishing agents; wh. wax; sol. in acetone, ethanol, IPA, benzene, toluene, xylene, chlorinated hydrocarbons; disp. in water; dens. 0.94 g/cc; HLB 10.0; pH 5-7 (10%); 100% act.

Mulsifan RT 113. [Zschimmer & Schwarz] Fatty acid polyglycol ester; nonionic; emulsifier for min. oils, textiles, metal processing, drilling and cutting oils, chemo-tech. prods.; yel. visc. liq.; sol. in ethanol, IPA, acetone, benzene, toluene, xylene, petrol., chlorinated hydrocarbons, min. oils; dens. 1.00 g/cc; HLB 9.0; pH 5-7 (10%); 100% act.

Mulsifan RT 125. [Zschimmer & Schwarz] Fatty alcohol polyglycol ether; emulsifier for waxes, fatty acids, triglycerides; wax; HLB 16.0; 100% act.

Mulsifan RT 157. [Zschimmer & Schwarz] Fatty alcohol polyglycol ether; emulsifier for waxes and their blends; for polishing agents; ylsh. wax; sol. in hot water; dens. 1.1 g/cc; HLB 15.0; pH 5-7 (10%); 100% act.

Mulsifan RT 163. [Zschimmer & Schwarz] Ethoxylated triglyceride; nonionic; emulsifier for triglycerides, chem-tech. prods.; yel. visc. liq.; sol. in ethanol, IPA, acetone; disp. in water; dens. 1.0 g/cc; HLB 6.0; pH 7-9 (10%); 100% act.

Mulsifan RT 231. [Zschimmer & Schwarz] Alkylaryl and alkyl polyglycol ether blend; nonionic; emulsifier for self-brilliant floor waxes; wh. wax; sol. in ethanol, IPA, acetone, toluene, chlorinated hydrocarbons; disp. in water; dens. 1.01 g/cc; HLB 12.0; pH 5-7 (10%); 100% act.

Mulsifan RT 237. [Zschimmer & Schwarz] Alkylaryl polyalkylene glycol ether; industrial surfactant; liq.; 100% act.

Mulsifan RT 245. [Zschimmer & Schwarz] Alkylaryl polyglycol ether; industrial surfactant; liq.; 100% act.

Mulsifan RT 248. [Zschimmer & Schwarz] Fatty acid polyglycol ester; nonionic; emulsifier for solvs., min. oils for cleaners; ylsh. brn. liq.; sol. in ethanol, IPA, trichloroethylene; disp. in water; dens. 1.03 g/cc; HLB 12.0; pH 5-7 (10%); 100% act.

Mulsifan RT 258. [Zschimmer & Schwarz] Fatty alcohol polyglycol ether blend; nonionic; emulsifier for paraffins for textile, paper, and ceramic industries and for polishing agents; ylsh. wax; disp. in water, cloudy sol. in molten paraffin; dens. 0.94 g/cc; HLB 10.0; pH 5-7 (10%); 100% act.

Mulsifan RT 269. [Zschimmer & Schwarz] Fatty alcohol polyglycol ether; coemulsifier; liq.; 100% act.

Mulsifan RT 282. [Zschimmer & Schwarz] Alkyl and alkylaryl polyglycol combination; nonionic; emulsifier for wh. oils; liq.; HLB 8.0; 100% conc.

Mulsifan RT 324. [Zschimmer & Schwarz] Alkyl and alkylaryl polyglycol blend; nonionic; emulsifier for wh. oils; liq.; HLB 8.0; 100% act.

Mulsifan RT 359. [Zschimmer & Schwarz] Alkyl polyglycol ether blend; emulsifier for self-glossing waxes; wax; HLB 13.0; 100% act.

Mulsifan STK. [Zschimmer & Schwarz] Surfactant blend; anionic/nonionic; emulsifier for solvs. in mfg. fast separating cold cleaners for degreasing motors, chassis, machines, etc.; brn. clear liq.; sol. in wh. spirit, petrol., diesel oil, pine oil, chlorinated hydrocarbons, glycols, alcohols; disp. in water; dens. 0.985 g/cc; visc. 750 mPa•s; HLB 5.0; cloud pt. < 0 C; ref. index 1.500 (20 C); pH 6-7 (10%); 100% act.

Multinol C. [Nippon Senka] Special sulfonic acid deriv.; anionic; one-bath scouring, bleaching and dyeing agent; paste.

Multisperse CP. [BASF] Highly effective silicate dispersant for prevention of silicate scale build-up in continuous bleaching equip.

MY Silicone A-08. [Yoshimura Oil Chem.] Silicone emulsion; cationic; softening agent; liq.; 18% conc.

MY Silicone A-13. [Yoshimura Oil Chem.] Silicone surfactant; softening agent; liq.; 35% conc.

MY Silicone A-16. [Yoshimura Oil Chem.] Silicone surfactant; softening agent; liq.; 35% conc.

MY Silicone FT-80. [Yoshimura Oil Chem.] Silicone emulsion; finishing oil for polyester fiber fills; liq.; 30% conc.

MY Silicone R-10, R-80. [Yoshimura Oil Chem.] Silicone; for spinneret spray.

N

Nacconol® 35SL. [Stepan; Stepan Canada] Alkyaryl sulfonate; anionic; detergent, wetting agent; pale clear, low visc. liq.; very low haze pt.

Nacconol® 40G. [Stepan; Stepan Canada] Sodium dodecylbenzene sulfonate; anionic; foamer, dispersant, wetting agent, detergent for agric., cement, dyeing, emulsion polymerization, textile, metal cleaning, metalworking, mining, paper industries; biodeg.; cream gran. powd.; sol. 2 g/100 ml in water; dens. 35 lb/ft^3; pH 6.4–7.6 (1% aq.); 38–42% act.

Nacconol® 90G. [Stepan; Stepan Canada; Stepan Europe] Sodium dodecylbenzene sulfonate; anionic; foamer, dispersant, wetting agent, detergent for agric., cement, dyeing, emulsion and latex polymerization, textile, metal cleaning, laundry prods., metalworking, mining, paper industries; biodeg.; wh. to beige gran.; sol. 15 g/100 ml in water; dens. 30 lb/ft^3; pH 6.0–7.5 (1% aq.); 90% act.

Na Cumene Sulfonate 40, Sulfonate Powd. [Hüls AG] Sodium cumene sulfonate; anionic; hydrotrope, solubilizer, coupling agent, sol'n. aid for liq. detergents and slurries; 40% act. liq., 96% act. powd. resp.

Nalco® 2300. [Nalco] Silica/silicone; defoamer for coatings, latex and high solids systems; wh. liq.; dens. 7.8 lb/gal; flash pt. (PMCC) 160 F.

Nalco® 2301. [Nalco] Silicone-organic; antifoam for solv.-based coatings and adhesives; water-wh. liq.; dens. 8.0 lb/gal; flash pt. (PMCC) > 200 F.

Nalco® 2305. [Nalco] Silica/silicone-organic; antifoam for water-based coatings and adhesives; esp. for formulations sensitive to silicone-caused surf. imperfections; milky wh. liq.; water-disp.; dens. 8.2 lb/gal; flash pt. (PMCC) > 200 F.

Nalco® 2311. [Nalco] Silica-organic; nonsilicone antifoam for aq. coatings and adhesives; for short term persistency applics., low visc. systems; cream; dens. 7.8 lb/gal; flash pt. (PMCC) > 200 F.

Nalco® 2314. [Nalco] Silica-silicone-organic; antifoam for aq./nonaq. coatings and adhesives, water-reducible systems; hazy wh.; dens. 7.6 lb/gal; flash pt. (PMCC) 151 F.

Nalco® 2340. [Nalco] Silica-organic; nonsilicone antifoam for aq. resin and polymer processing; short term persistency; cream appearance; dens. 7.8 lb/gal; flash pt. (PMCC) > 200 F.

Nalco® 2341. [Nalco] Silica-organic; nonsilicone antifoam for aq. stripping or caustic processing; effective over wide pH range; tan appearance; dens. 7.4 lb/gal; flash pt. (PMCC) > 200 F.

Nalco® 2342. [Nalco] Organic; nonsilicone antifoam for aq. glycol processing applics.; effective over wide temp. range; wh. appearance; dens. 8.6 lb/gal.

Nalco® 2343. [Nalco] Polyglycol and fatty type surfactants; nonionic; nonsilicone antifoam for aq. industrial processing, esp. low visc. applics.; quick foam kill with short term persistency; pale straw liq.; dens. 7.2 lb/gal; flash pt. (PMCC) > 200 F.

Nalco® 8638. [Nalco] Organic; nonsilicone antifoam for aq. gypsum board edge hardening, low visc. processing applics.; quick foam kill; pale straw appearance; dens. 7.4 lb/gal; flash pt. (PMCC) > 200 F.

Nalco® 8639. [Nalco] Organic; nonsilicone antifoam for aq. fiberglass mat whitewater processing, high solids systems; wh. appearance; dens. 7.9 lb/gal; flash pt. (PMCC) > 200 F.

Nalco® 8669. [Nalco] Silica-organic; nonsilicone antifoam for aq. fiberglass mat resin processing; effective over wide temp. range; straw appearance; dens. 7.6 lb/gal; flash pt. (PMCC) > 200 F.

Nalkylene® 500. [Vista] C10-12 linear undecylbenzene; surfactant intermediate for prod. of biodegrad. detergent prods., foaming agents, industrial and specialty chemicals; Saybolt 30+ liq.; odorless; m.w. 237; sp.grt. 0.8654; dens. 7.209 lb/gal (60 F); visc. 4.5 cSt (100 F); b.p. 536 F; f.p. < -70 F; flash pt. (PM) 280 F; 100% act.

Nalkylene® 515, 550. [Vista] Linear alkylbenzene; surfactant intermediate for prod. of biodegrad. detergent prods., foaming agents, industrial and specialty chemicals; water-wh. oily liq. and Saybolt 30+ liq. resp., odorless; sp.gr. 0.861 and 0.862; m.p. < -90 F; 100% act. (550).

Nalkylene® 550L. [Vista] C10-14 linear dodecylbenzene; surfactant intermediate for prod. of biodegrad. detergent prods., foaming agents, industrial and specialty chemicals; Saybolt 30+ liq.; m.w. 241; sp.gr. 0.860 (60 F); dens. 7.2 lb/gal (60 F); visc. 4.3 cSt (100 F); b.p. 521 F; flash pt. (PM) 298 F; 100% act.

Nalkylene® 575L. [Vista] C12-14 tridecylbenzene; sulfonation feedstock for producing biodeg. surfactants, powd. detergents, industrial and specialty chems.; liq.; 100% conc.

Nalkylene® 600. [Vista] Linear alkylbenzene; surfactant intermediate for prod. of biodegrad. detergent prods., foaming agents, industrial and specialty chemicals; Saybolt 30+ liq.; odorless; sp.gr. 0.865; visc. 46.9 SSU (100 F); m.p. < -90 F; 100% act.

Nalkylene® 600L. [Vista] C12-14 tridecylbenzene; intermediate to produce surfactants, powd. detergents, industrial and specialty chemicals; liq.; 100% act.

Nansa® 1042. [Albright & Wilson UK] Dodecylbenzene sulfonic acid; CAS 68584-22-5; anionic; intermediate used in mfg. of detergents, laundry prods., emulsifiers for emulsion polymerization; dk. br.

visc. liq.; sp.gr. 1.05 g/cc; visc. 1900 cs (20 C); 95% act.

Nansa® 1042/P. [Albright & Wilson UK] Dodecylbenzene sulfonic acid; CAS 27176-87-0; anionic; intermediate for neutralization; sodium salts as emulsifiers in emulsion polymerization of plastics and syn. rubbers; dk. br. visc. liq.; sp.gr. 1.05 g/cc; visc. 1900 cs (20 C); 95% act.

Nansa® 1106/P. [Albright & Wilson UK] Sodium dodecylbenzene sulfonate; surfactant for emulsion polymerization; golden yel. clear liq.; sp.gr. 1.0 g/cc; visc. 2500 cs (20 C); cloud pt. 11 C; pH 8.0 ± 1 (5% aq.); 30 ± 1% act.

Nansa® 1169/P. [Albright & Wilson UK] Sodium dodecylbenzene sulfonate; surfactant for emulsion polymerization; golden yel. cloudy liq.; dens. 1.0 g/cc; visc. 3000 cs (20 C); cloud pt. 25 C; pH 8 ± 1 (5% aq.); 30% act.

Nansa® 1192. [Albright & Wilson UK] Foaming agent for gypsum slurries in plaster board prod.; golden yel. clear liq.; dens. 1.03 g/cc; visc. 290 cs; cloud pt. < 0 C; pH 8 (5%).

Nansa® 1339. [Albright & Wilson UK] Formulated prod.; nonbiological low-foam detergent powd.

Nansa® 1340. [Albright & Wilson UK] Formulated prod.; biological low-foam detergent powd. with softener.

Nansa® 1347. [Albright & Wilson UK] Formulated prod.; biological low-foam detergent powd.

Nansa® 1385. [Albright & Wilson UK] Formulated prod.; phosphate-free biological low-foam detergent powd.

Nansa® 1389. [Albright & Wilson UK] Formulated prod.; economy nonbiological low-foam detergent powd.

Nansa® 1390/E. [Albright & Wilson UK] Formulated prod.; economy biological low-foam detergent powd.

Nansa® 1400 Series. [Albright & Wilson UK] Formulated prod.; heavy-duty laundry liqs.

Nansa® 1909. [Albright & Wilson UK] Dodecylbenzene sulfonic acid; surfactant for high-visc. specialty industrial, institutional cleaners; reddish-br. visc. liq.; sp.gr. 1.04; visc. 1450 cP; > 80% biodeg.; 94% act.

Nansa® 7052. [Albright & Wilson UK] Formulated prod.; conc. nonbiological detergent powd.

Nansa® 7053/E. [Albright & Wilson UK] Formulated prod.; conc. biological detergent powd.

Nansa® 7069. [Albright & Wilson UK] Formulated prod.; high foam detergent powd.

Nansa® AS 40. [Albright & Wilson UK] Ammonium dodecylbenzene sulfonate; anionic; formulation of domestic and industrial liq. detergents; golden yel. opaque gel; pH 6.0–6.5 (2% aq.); 40% act.

Nansa® BMC. [Albright & Wilson UK] Blend; anionic; foaming agent, air entraining agent for mortar and concrete; clear yel. liq.; m.w. 336; sp.gr. 1.045; visc. 800 cs; cloud pt. 15 C; set pt. –6 C; pH 7.0 (5% aq.); 27 ± 1% act.

Nansa® BXS. [Albright & Wilson UK] Sodium alkyl naphthalene sulfonate; anionic; wetting and dispersing agent for colloidal systems without detergent properties, used in agric., leather, paper and pulp, textile industries; cream-colored powd.; pH 7.5–9.5 (1%); 38% act.

Nansa® EVM50. [Albright & Wilson UK] Calcium dodecylbenzene sulfonate in aromatic solv.; CAS 68953-96-8; anionic; emulsifier, dispersant for agrochemicals, textiles, surf. coatings, polymerization, leather industries; brn. visc. liq.; flash pt. (CC) > 48 C; pH 5.5–7.5 (3%); 50 ± 1.5% act.

Nansa® EVM62/H. [Albright & Wilson UK] Calcium dodecylbenzene sulfonate in isobutanol; anionic; emulsifier for agric. chemicals; wetting agent for oil-based drilling fluids; brn. visc. liq.; flash pt. (CC) > 28 C; pH 6.5–8.0 (3%).; 62 ± 2% act.

Nansa® EVM70. [Albright & Wilson UK] Calcium dodecylbenzene sulfonate in isobutanol; anionic; emulsifier for agric. chemicals; wetting agent for oil-based drilling fluids; brn. visc. liq.; flash pt. (CC) > 28 C; pH 6.5–8.0 (3%); 68.5 ± 1.5% act.

Nansa® EVM70/B. [Albright & Wilson UK] Calcium dodecylbenzene sulfonate in hexanol; anionic; emulsifier, dispersant for agrochemicals, textiles, surf. coatings, polymerization, leather industries; brn. visc. liq.; flash pt. (CC) > 28 C; pH 5.5–7.5 (3%); 68.5 ± 1.5% act.

Nansa® EVM70/E. [Albright & Wilson UK] Calcium dodecylbenzene sulfonate in isobutanol; anionic; emulsifier, dispersant for agrochemicals, textiles, surf. coatings, polymerization, leather industries; brn. visc. liq.; flash pt. (CC) > 57 C; pH 5.5–7.5 (3%); 67 ± 2% act.

Nansa® HAD. [Albright & Wilson UK] Blended detergent conc.; anionic/nonionic; liq. detergent conc. for dishwashing; pale amber hazy liq.; dens. 1.08; visc. 2000 cps (20 C); pH 6.5 ± 0.5 (2% aq.); 56% conc.

Nansa® HS40-AU. [Albright & Wilson UK] Sodium dodecylbenzene sulfonate; wetting agent, dispersant in scouring powds., industrial applics., automotive, floor, wall, and hard surface cleaners; cream coarse powd.; dens. 0.55 g/cm^3; pH 8.5 ± 1.0 (1%); > 80% biodeg.; 38 + 3% act.

Nansa® HS40/S. [Albright & Wilson UK] Sodium dodecylbenzene sulfonate; anionic; wetting and dispersing agent, emulsifier used in detergent compd. for domestic and industrial use; cream-colored, free-flowing coarse powd.; dens. 0.55 g/cm^3; pH 8.5 ± 1 (1%); > 80% biodeg.; 38 ± 3% act.

Nansa® HS80-AU. [Albright & Wilson Australia] Sodium alkylbenzene sulfonate; CAS 68081-81-2; anionic; wetting agent, dispersant in laundry, detergent compds., insecticides, metal pickling sol'ns.; > 80% biodeg.; cream coarse powd.; dens. 0.55 g/cm^3; pH 8.5 ± 1.0 (1%); 78 + 3% act.

Nansa® HS80P. [Albright & Wilson Australia] Sodium dodecylbenzene sulfonate; CAS 68081-81-2; anionic; ingred. in detergent compds.; > 80% biodeg.; cream-colored powd.; dens. 0.52 g/cm^3; pH 8.5 ± 1 (1%); 80 ± 3% act.

Nansa® HS80S. [Albright & Wilson UK] Sodium dodecylbenzene sulfonate; CAS 25155-30-0; anionic; formulation of detergents, hard surface and bottle cleaners; metal treatment and paper processing; scouring and wetting agent for textile industry; foamer and mortar plasticizer in building industry; cream powd.; sp.gr. 0.65; pH 9–11 (1%); 80% act.

Nansa® HS80/SF. [Albright & Wilson UK] Linear alkylbenzene sulfonate; grinding agent for construction industry.

Nansa® HS80SK. [Albright & Wilson UK] Sodium dodecylbenzene sulfonate; built detergent, wetting, dispersing agent for dairy, laundry, heavy-duty, hard surface detergents; off-wh. to cream coarse powd.; dens. 0.55 g/cm^3; pH 9.0 ± 0.8 (1%); > 80% biodeg.; 75 + 3% act.

Nansa® HS85. [Albright & Wilson UK] Sodium dodecylbenzene sulfonate; CAS 85117-50-6; anionic; general purpose detergent base; flake; 85% conc.

Nansa® HS85S. [Albright & Wilson UK] Sodium dodecylbenzene sulfonate; CAS 25155-30-0; anionic; formulation of detergents, hard surface and bottle cleaners; metal treatment and paper processing; scouring and wetting agent for textile industry; foamer and mortar plasticizer in building industry; rubber/plastics emulsifier; biodeg.; cream flake; dens. 0.5 g/cm³; pH 9–11 (1%); 85 ± 3% act.

Nansa® HSA/L. [Albright & Wilson UK] Alkylbenzene sulfonic acid; anionic; basic ingredient in low act. lt. duty liq. detergents; dk. brn. visc. liq.; > 80% biodeg.; 86% min. act.

Nansa® LES 42. [Albright & Wilson UK] Blended detergent conc.; used in dishwashing and lt. duty clothes washing formulations; golden clear liq.; visc. 250 cs; cloud pt. < 0 C; surf. tens. 30 dynes/cm (1 g/l); pH 6.2–6.8 (5%); 37% act.

Nansa® LSS38/A. [Albright & Wilson UK] Sodium C14-16 olefin sulfonate; anionic; general purpose detergent base; wetting agent for agric. wettable powds., textile processing; emulsifier for enhanced oil recovery; liq.; 38% conc.

Nansa® MA30. [Albright & Wilson UK] Sodium dodecylbenzene sulfonate and ethoxylated nonionic blend; anionic/nonionic; detergent base for dishwashing and hard surf. cleansers; scouring agent for textiles; mortar plasticizer in mfg. of masonry cement; yel liq.; sp.gr. 1.05; visc. 400 cs; cloud pt. 0 C; pH 6.2–7.2 (5% aq.); 20% act., 10% nonionic in water.

Nansa® MS45. [Albright & Wilson UK] Magnesium dodecylbenzene sulfonate; anionic; detergent ingred.; liq.; 45% conc.

Nansa® S40/S. [Albright & Wilson UK] Sodium alkylbenzene sulfonate; anionic; detergent, wetting agent in insecticides, metal pickling, printing inks, paper processing; wh./pale cream powd.; pH 8–9 (1% aq.); 40% act.

Nansa® SB 30. [Albright & Wilson UK] Branched sodium alkylbenzene sulfonate; anionic; surfactant; paste; 30% conc.

Nansa® SB 62. [Albright & Wilson UK] Sodium dodecylbenzene sulfonate; anionic; detergent ingred.; paste; 62% conc.

Nansa® SBA. [Albright & Wilson UK] Branched dodecylbenzene sulfonic acid; CAS 68608-88-8; anionic; detergent intermediate; dk. br. liq.; visc. 20,000 cs; 96% act.

Nansa® SL 30. [Albright & Wilson UK] Sodium dodecylbenzene sulfonate; anionic; surfactant used in detergent formulations; pale yel. liq.; dens. 1.05 g/cc; visc. 2400 cs (20 C); cloud pt. 15 C; pH 6.8-7.8 (5%); 30% aq. sol'n.

Nansa® SS 30. [Albright & Wilson UK] Sodium dodecylbenzene sulfonate; anionic; surfactant for detergent formulations; emulsifier for agric. emulsifiable concs.; curing/foaming agent for thermosets; golden liq.; visc. 250 cs; cloud pt. 0 C; pH 6.0–6.8 (2%); 30% conc.

Nansa® SS 50. [Albright & Wilson UK] Sodium alkylbenzene sulfonate; anionic; emulsifier for agric. emulsifiable concs.; curing/foaming agent for thermosets; paste; 50% conc.

Nansa® SS 55. [Albright & Wilson UK] Sodium alkylbenzene sulfonate; anionic; emulsifier for agric. emulsifiable concs.; curing/foaming agent for thermosets; paste; 55% conc.

Nansa® SS 60. [Albright & Wilson UK] Sodium dodecylbenzene sulfonate; CAS 68081-81-2; anionic; detergent intermediate; emulsifier for agric. emulsifiable concs.; curing/foaming agent for thermosets; cream paste; pH 7.0–9.0 (2%); 60% act.

Nansa® SSA. [Albright & Wilson UK] Dodecylbenzene sulfonic acid; CAS 68584-22-5; anionic; detergent intermediate; in prep. of emulsifiers for emulsion polymerization; dk. br. visc. liq.; dens. 1.05 g/cc; visc. 1900 cs (20 C); 96% act.

Nansa® SSAL. [Albright & Wilson UK] Alkylbenzene sulfonic acid; anionic; surfactant, basic ingredient in lt.-duty liq. detergents and cleaners used in mfg. of wetting, emulsifying, and foaming agents; dk. br. visc. liq.; > 80% biodeg.; 94% act.

Nansa® SSA/P. [Albright & Wilson UK] Dodecylbenzene sulfonic acid; intermediate for neutralization; sodium salts as emulsifiers in emulsion polymerization of plastics and syn. rubbers; dk. br. visc. liq.; sp.gr. 1.05 g/cc; visc. 1900 cs (20 C); 95% act.

Nansa® SS A/S. [Albright & Wilson Australia] Alkylbenzene sulfonic acid; CAS 68584-22-4; anionic; detergent intermediate; emulsifier for agric. emulsifiable concs.; liq.; 96% conc.

Nansa® TDB. [Albright & Wilson UK] Tridecylbenzene sulfonic acid; anionic; detergent ingred.; liq.; 90% conc.

Nansa® TS 50. [Albright & Wilson UK] TEA alkylbenzene sulfonate; CAS 68584-25-8; anionic; detergent raw material; liq.; 50% conc.

Nansa® TS 60. [Albright & Wilson UK] TEA-dodecylbenzene sulfonate; anionic; detergent, emulsifier, pigment dispersant used in cleaners, desizing agent for syn. fibers; med. br. visc. liq.; sp.gr. 1.09; visc. 9000 cs; cloud pt. < 0 C (50%); pH 6.6–7.0 (2% aq.); 60% act.

Nansa® UCA/S, UCP/S. [Albright & Wilson UK] Built powders based on sodium alkylbenzene sulfonate; anionic; domestic and industrial detergent powd.; blue powd.; sol. in hot or cold water; biodeg.; 17 and 15% act. resp.

Nansa® YS94. [Albright & Wilson UK] Isopropylamine dodecylbenzene sulfonate; CAS 68584-24-7; anionic; emulsifier for solv.-based hand cleaners, agric. emulsifiable concs.; coupling agent for water in charge detergent systems; biodeg.; amber visc. liq.; sp.gr. 1.02; visc. 6000 cs; cloud pt. < 0 C; pH 5–8 (2% aq.); 94% act. in water.

Na Toluene Sulfonate 30, 40. [Hüls AG] Sodium toluene sulfonate; hydrotrope, solubilizer, and coupling agent for lt.-duty detergents and heavy-duty detergent slurries; sol. in water.

Natrex D 3. [Henkel-Nopco] Modified phosphated ester; anionic/nonionic; detergent, wetting agent, emulsifier for processing cotton and syns.; alkali-stable; liq.; 60% conc.

Natrex DSP 213. [Henkel-Nopco] Anionic/nonionic surfactant blend; anionic/nonionic; detergent, wetting agent, hydrophiling agent for cotton; liq.; 50% conc.

Natrex EFB 171. [Henkel-Nopco] Betaine; amphoteric; foamer; liq.; 30% conc.

Natrex GA 251. [Henkel-Nopco] Surfactant blend; nonionic; after-dyeing scouring agent; liq.; 40% conc.

Natrex J 3. [Henkel-Nopco] Modified phosphated ester; anionic; textile wetting agent; detergent; liq.; 65% conc.

Natrex SO. [Henkel-Nopco] Anionic/nonionic surfactant blend; anionic; detergent and wetting agent for kier boiling and continuous alkaline processing;

liq.; 35% conc.

Naturechem® EGHS. [CasChem] Glycol hydroxystearate; nonionic; aux. emulsifier, emollient, thickener, opacifier for cosmetics, household prods.; wh. flakes; m.p. 66 C; HLB 2.0; acid no. 3; iodine no. < 5; hyd. no. 266; 100% conc.

Naturechem® GMHS. [CasChem] Glyceryl hydroxystearate; nonionic; aux. emulsifier, emollient, opacifier, bodying and thickening agent for cosmetics, household prods.; wh. flakes; m.p. 69 C; HLB 3.4; acid no. 6; iodine no. < 5; hyd. no. 320; 100% conc.

Naturechem® GTR. [CasChem] Glyceryl triacetyl ricinoleate; mild emollient, pigment wetter, cosolv.; softener for waxes and resins; Gardner 2+ color; m.p. -40 C; acid no. < 1; iodine no. 76; hyd. no. 5.

Naturechem® MHS. [CasChem] Methyl hydroxystearate; opacifiier, pearlescent, emulsifier, visc. builder for surfactant systems; Gardner 1 color; m.p. 52 C; acid no. 2; iodine no. < 5; hyd. no. 164.

Naturechem® OHS. [CasChem] Octyl hydroxystearate; emollient, softener for cosmetics; refatting additive for soaps, cleansers; lt. yel. liq.; m.p. 5 C; acid no. 1; iodine no. < 5; hyd. no. 75; 100% act.

Naturechem® PGHS [CasChem] Propylene glycol hydroxystearate; nonionic; aux. emulsifier, dispersant, opacifier, thickener, emollient, stabilizer for cosmetics, household prods.; wh. flakes; m.p. 53 C; HLB 2.6; acid no. 2; iodine no. < 5; hyd. no. 289; 100% conc.

Naturechem® PGR. [CasChem] Propylene glycol ricinoleate; wetting agent, stabilizer, pigment/dye dispersant providing emolliency, gloss, plasticization to cosmetics, household prods.; pale yel. liq.; m.p. -26 C; acid no. 2.5; iodine no. 76; hyd. no. 296; 100% act.

Naturechem® THS-200. [CasChem] PEG-200 trihydroxystearin; nonionic; emulsifier, emollient, thickener, stabilizer for cosmetics, household prods.; stable over broad pH range; wh. wax-like solid; sol. in water and alcohol; m.p. 52 C; HLB 18.0; acid no. 1.2; iodine no. < 5; 100% conc.

Naxchem CD-6M. [Ruetgers-Nease] Surfactant blend; detergent, foam booster/stabilizer for cosmetics, carpet shampoos, dishwash, laundry detergents, textile processing, pigment dispersions; lt. amber clear liq.; 100% act.

Naxchem Detergent CNB. [Ruetgers-Nease] Surfactant blend; detergent, foam booster/stabilizer for lotions and creams, carept and upholstery shampoos, dishwash, laundry, textile processing, pigment disps.; visc. liq.; 65-67% solids.

Naxchem Dispersant K. [Ruetgers-Nease] Blend of alkanolamides and syndets; biodeg. dispersant, emulsifier for oil slicks, naphthas, kerosene, other solvs.; general cleaning of crude oil, marine and land transport, and mfg. plants; liq.; sp.gr. 1.0173; visc. 706.3 SSU (37.8 C); pour pt. -9.4 C; flash pt. 162.8 C; pH 10.0; 99% act.

Naxchem Emulsifier 700. [Ruetgers-Nease] Blend of esters and surfactants; nonionic; antistatic base for emulsifying low visc. pale min. oils; for metal cutting, textile fiber lubricants; liq.; sp.gr. 1.02; dens. 8.5 lb/gal; 100% act.

Naxchem N-Foam 802. [Ruetgers-Nease] Blend of crosslinking and emulsifying agents; curing and foaming agent in mixts. with urea-formaldehyde resins to mask odors from landfills and for foam insulation; as cleaning compd. when neutralized; yel.-gold clear liq.; 48-50% solids.

Naxel AAS-40S. [Ruetgers-Nease] Sodium dodecylbenzene sulfonate; anionic; biodeg. surfactant, foamer, wetting agent, detergent for household and industrial detergents, rug shampoos, textile wet processing, metal cleaners, dairy cleaners, cosmetics; 40% act.

Naxel AAS-45S. [Ruetgers-Nease] Sodium dodecylbenzene sulfonate; anionic; surfactant for household and industrial detergents, dairy cleaners, textile scouring compds., car wash prods., leather prods., cosmetics, rug shampoos, aircraft cleaners; gold liq.; 40-43% act.

Naxel AAS-60S. [Ruetgers-Nease] TEA dodecylbenzene sulfonate; anionic; foaming agent for bubble baths, shampoos, household and industrial detergents, textile dyeing compds., etc.; gold liq.; sol. in water, methanol; dens. 9.3 lb/gal; 60% act.

Naxel AAS-98S. [Ruetgers-Nease] Dodecylbenzenesulfonic acid; anionic; biodeg. detergent intermediate, wetting agent, emulsifier; Klett 70 max. liq.; dens. 8.82 lb/gal; visc. 478 cSt (100 F); 96% act.

Naxel AAS-Special 3. [Ruetgers-Nease] Isopropylamine dodecylbenzene sulfonate; anionic; emulsifier for drycleaning, metal cleaning, emulsifiable solv. cleaners, fuel oil additives, mop treatments; amber liq.; sol. in kerosene, alcohol, pine oil, min. spirits, chlorinated and aromatic solvs.; dens. 8.4 lb/gal; 95% act.

Naxel DDB 500. [Ruetgers-Nease] Dodecylbenzene; biodeg. detergent intermediate for prod. of dishwash, laundry, all-purpose and industrial cleaners, in specialty coatings and other industrial applics.; m.w. 231-241; sp.gr. 0.865; dens. 7.209 lb/gal; visc. 4.5 cSt (100 F); ibp 536 F; flash pt. (PM) 280 F.

Naxide 1230. [Ruetgers-Nease] Cocamine oxide; CAS 61788-90-7; detergent, foam booster/stabilizer, visc. builder, conditioner for shampoos, hand cleaners, dishwash, lt. duty detergents, textiles, lubricants, paper coatings; APHA 125 max. color; 30% min. act.

Naxolate WA-97. [Ruetgers-Nease] Sodium lauryl sulfate USP, BP; CAS 151-21-3; anionic; biodeg. detergent, wetting agent, foamer, emulsifier for cosmetic and household prods. incl. shampoos, bubble baths, rug shampoos, toothpaste, dishwash, laundry detergents; wh. powd.; dens. 200 ± 50 g/l; 96-98% act.

Naxolate WAG. [Ruetgers-Nease] Sodium lauryl sulfate USP, BP; CAS 151-21-3; anionic; biodeg. detergent, wetting agent, foamer, emulsifier for cosmetic and household prods. incl. shampoos, bubble baths, rug shampoos, toothpaste, dishwash, laundry detergents, textile scouring; ivory wh. needles; dens. 500 ± 50 g/l; pH 7-10 (1% aq.); 89% act.

Naxolate WA Special. [Ruetgers-Nease] Sodium lauryl sulfate; CAS 151-21-3; anionic; detergent, wetting agent, foamer for cosmetic and household prods. incl. shampoos, bubble baths, rug shampoos, toothpaste, acid soap processing, pigment dispersion, compded. detergents; lt. yel. liq., pleasant odor; visc. 50-300 cps; cloud pt. 22 C max.; 30% act.

Naxonac 510. [Ruetgers-Nease] Nonyl phenol ether phosphate; anionic; detergent, wetting agent, emulsifier, lubricant, hydrotrope for heavy-duty and household detergents, waterless hand cleaners, solv. degreasers, emulsion polymerization, paint and wax strippers, electrolytic cleaners; clear liq.; sol. and stable in aq. sol'ns. with high alkaline concs.; pH 2; 99% conc.

Naxonac 600. [Ruetgers-Nease] Alcohol ether phosphate; anionic; emulsifier for aromatic and chlorinated hydrocarbons; hydrotrope, wetting agent for acid cleaners; solubilizer for nonionics; detergent; for industrial and household cleaners, electrolytic cleaners, dairy cleaners, acid cleaners, solv. cleaners; clear visc. liq., mild odor; sol. and stable in aq. sol'ns. with high alkaline concs.; pH 1.5-2.5; 99% solids.

Naxonac 610. [Ruetgers-Nease] Nonyl phenol ether phosphate; anionic; detergent, emulsifier, wetting agent, stabilizer; solubilizer for nonionics; for household and industrial cleaners, electrolytic cleaning, dairy cleaning, acid cleaners, solv. cleaners; clear visc. liq., mild odor; sol. and stable in aq. sol'ns. with high concs. of alkaline builders; pH 1.5-2.5; 99% solids.

Naxonac 690-70. [Ruetgers-Nease] Nonyl phenol ether phosphate; anionic; emulsifier imparting corrosion resistance; for emulsion polymerization of vinyl acetate and highly carboxylated acrylics; does not discolor when used to dedust alkaline powds.; colorless to pale yel. visc. liq.; sp.gr. 1.07; pH 1.7 (10%); 70% act.

Naxonic NI-40. [Ruetgers-Nease] Nonoxynol-4; CAS 9016-45-9; nonionic; wetting agent, dispersant, penetrant, emulsifier, detergent, solubilizer for textile wet processing, agric., cosmetic, industrial and household detergents, latex and polymers, wax/polishes; demulsifier for petrol.; clear liq.; oil-sol.; sp.gr. 1.02-1.04.

Naxonic NI-60. [Ruetgers-Nease] Nonoxynol-6; CAS 9016-45-9; nonionic; wetting agent, dispersant, penetrant, emulsifier, detergent, solubilizer for textile wet processing, agric., cosmetic, industrial and household detergents, latex and polymers, wax/polishes; demulsifier for petrol.; clear liq.; oil-sol.; sp.gr. 1.03-1.05; cloud pt. 32 F.

Naxonic NI-100. [Ruetgers-Nease] Nonoxynol-10; CAS 9016-45-9; nonionic; wetting agent, dispersant, penetrant, emulsifier, detergent, solubilizer for textile wet processing, agric., cosmetic, industrial and household detergents, latex and polymers, wax/polishes; demulsifier for petrol.; clear liq.; water-sol.; sp.gr. 1.05-1.07; cloud pt. 52-56 C.

Naxonol CO. [Ruetgers-Nease] Cocamide DEA (2:1); nonionic; detergent, wetting agent, thickener for household and industrial detergents, shampoos, textile wet processing, leather and fur processing, metal cleaning, solv. emulsification; amber liq.; sol. in water; sp.gr. 2.47 g/ml; 97% conc.

Naxonol PN 66. [Ruetgers-Nease] Cocamide DEA; nonionic; wetting agent, foam booster/stabilizer, emulsifier, thickener, detergent for household and heavy-duty cleaners, shampoos, leather and fur processing, car wash, textile wet processing, metal cleaners; amber liq.; 98% act.

Naxonol PO. [Ruetgers-Nease] Cocamide DEA (1:1); nonionic; deterent, emulsifier, wetting agent, foam booster, thickener, lubricant, dispersant for household and industrial cleaners, cosmetics, buffing and polishing compds., textiles; lt. yel. visc. liq.; sp.gr. 2.46 g/ml; 85% conc.

Neatsan D. [Reilly-Whiteman] Sulfated oil; anionic; emulsifiable oil for processing leather; liq.; 75% conc.

Nekal® BA-77 (redesignated Rhodacal® BA-77). [Rhone-Poulenc Surf.]

Nekal® BX-78 (redesignated Rhodacal® BX-78). [Rhone-Poulenc Surf.]

Nekal® BX Conc. Paste. [BASF AG] Sodium alkyl naphthalene sulfonate; anionic; surfactant for chemical industry; paste; 60% conc.

Nekal® BX Dry. [BASF AG] Sodium alkyl naphthalene sulfonate; anionic; surfactant for chemical industry; powd.; 65% conc.

Nekal® WS-25, WS-25-I (redesignated Geropon® WS-25, WS-25-I). [Rhone-Poulenc Surf.]

Nekal® WT-27 (redesignated Geropon® WT-27). [Rhone-Poulenc Surf.]

Nekanil® 907. [BASF AG] Low-ethoxylated alkyl phenol; nonionic; detergent, wetting agent used in textile industry; 100% act.

Nekanil® 910. [BASF AG] Alkyl phenol adduct (9–10 moles EO); nonionic; detergent, wetting and dispersing agent used in textile industry; 100% act.

Nekanil® LN. [BASF AG] Low-ethoxylated alkyl phenol; CAS 9036-19-5; nonionic; detergent, wetting agent used in textile industry; 100% act.

Neocation G. [Nikko Chem. Co. Ltd.] Quat. ammonium salt; cationic; surfactant, leveling agent for cationic dyes; liq.; 50% conc.

Neodene® 6. [Shell] Linear C6 alpha olefin; CAS 592-41-6; intermediate for biodeg. surfactants and specialty industrial chemicals; liq.; 100% conc.

Neodene® 8. [Shell] Linear C8 alpha olefin; CAS 111-66-0; intermediate for biodeg. surfactants and specialty industrial chemicals; liq.; 100% conc.

Neodene® 10. [Shell] Linear C10 alpha olefin; CAS 872-05-9; intermediate for biodeg. surfactants and specialty industrial chemicals; liq.; 100% conc.

Neodene® 12. [Shell] Linear C12 alpha olefin; CAS 112-41-4; intermediate for biodeg. surfactants and specialty industrial chemicals; liq.; 100% conc.

Neodene® 14. [Shell] Linear C14 alpha olefin; CAS 1120-36-1; intermediate for biodeg. surfactants and specialty industrial chemicals; liq.; 100% conc.

Neodene® 14/16/18. [Shell] C14-18 alpha olefins; CAS 64743-02-8; intermediate for biodeg. surfactants and specialty industrial chemicals; liq.; 100% conc.

Neodene® 16. [Shell] Linear C16 alpha olefin; CAS 629-73-2; intermediate for biodeg. surfactants and specialty industrial chemicals; liq.; 100% conc.

Neodene® 18. [Shell] Linear C18 alpha olefin; CAS 112-88-9; intermediate for biodeg. surfactants and specialty industrial chemicals; liq.; 100% conc.

Neodene® 20. [Shell] Linear C20 alpha olefin; CAS 3452-07-1; intermediate for biodeg. surfactants and specialty industrial chemicals; wax; 100% conc.

Neodene® 1112 IO. [Shell] C11-12 internal olefins; CAS 68411-00-7; intermediate for biodeg. surfactants and specialty industrial chemicals; liq.; 100% conc.

Neodene® 1214. [Shell] C12-C14 alpha olefin; CAS 64743-02-8; intermediate for biodeg. surfactants and specialty industrial chemicals; liq.; 100% conc.

Neodene® 1314 IO. [Shell] C13-14 internal olefins; CAS 68411-00-7; intermediate for biodeg. surfactants and specialty industrial chemicals; liq.; 100% conc.

Neodene® 1416. [Shell] C14-C16 alpha olefin; CAS 64743-02-8; intermediate for biodeg. surfactants and specialty industrial chemicals; liq.; 100% conc.

Neodene® 1418. [Shell] C14-C18 alpha olefin; intermediate for biodeg. surfactants and specialty industrial chemicals; liq.; 100% conc.

Neodene® 1518 IO. [Shell] C15-18 internal olefins; CAS 68411-00-7; intermediate for biodeg. surfactants and specialty industrial chemicals; liq.; 100% conc.

Neodene® 1618. [Shell] C16-C18 alpha olefin; CAS 64743-02-8; intermediate for biodeg. surfactants and specialty industrial chemicals; liq.; 100% conc.

Neodol® 1. [Shell] Undecyl alcohol; CAS 112-42-5; nonionic; detergent intermediate; clear liq.; m.w. 173; sp.gr. 0.831; visc. 11 cSt (100 F); m.p. 42–57 F; acid no. < 0.001; sapon. no. < 0.001; hyd. no. 323-327; pour pt. 52 F; flash pt. (PMCC) 250 F; ref. index 1.4379; toxicology: low acute oral toxicity; irritating to skin and eyes in undiluted form; 100% act.

Neodol® 1-3. [Shell] Undeceth-3; nonionic; detergent intermediate, emulsifier for general industrial usage, hard surf. cleaning; biodeg.; colorless clear to sl. hazy liq.; oil-sol.; m.w. 305; sp.gr. 0.936; dens. 7.7 lb/gal; visc. 10 cSt; m.p. -31 to 4 C; HLB 8.7; acid no. < 0.001; hyd. no. 184; pour pt. -7 C; flash pt. (PMCC) 142 C; Draves wetting 12 s (0.1% aq.); toxicology: low acute oral toxicity; severely irritating to skin and eyes in undiluted form; 100% conc.

Neodol® 1-5. [Shell] Undeceth-5; nonionic; detergent intermediate, surfactant for general industrial usage incl. textile, carwash, laundry, and hard surf. cleaning; colorless liq.; misc. with many hydrocarbon-based formulations; m.w. 392; sp.gr. 0.9663; dens. 8.1 lb/gal; visc. 21 cSt (38 C); HLB 11.2; m.p. –4 to +12 C; pour pt. 6 C; cloud pt. 18 C (1% aq.); flash pt. (PMCC) 143 C; acid no. < 0.001; hyd. no. 143; pH 6.0 (1% aq.); Draves wetting 5 s (0.1% aq.); toxicology: low acute oral toxicity; severely irritating to skin and eyes in undiluted form; 100% conc.

Neodol® 1-7. [Shell] Undeceth-7; nonionic; detergent, wetting agent for general industrial use, hard surf. cleaners, laundry detergents, liq. detergents; biodeg.; hazy liq.; water-sol.; m.w. 479; sp.gr. 0.996; dens 8.2 lb/gal; visc. 28 cSt (38 C); m.p. -9 to 17 C; HLB 12.9; acid no. < 0.001; hyd. no. 117; pour pt. 7 C; cloud pt. 58 C (1% aq.); flash pt. (PMCC) 165 C; Draves wettomg 5 s (0.1% aq.); toxicology: low acute oral toxicity; severely irritating to skin and eyes in undiluted form; 100% conc.

Neodol® 1-9. [Shell] Undeceth-9; nonionic; detergent, wetting agent for general industrial use, hard surf. cleaners, laundry detergents, liq. detergents; biodeg.; hazy liq. to wh. pasty solid; water-sol.; m.w. 569; sp.gr. 1.011; dens. 8.3 lb/gal; visc. 31 cSt (38 C); HLB 13.9; acid no. < 0.001; hyd. no. 99; pour pt. 18 C; cloud pt. 74 C (1% aq.); flash pt. (PMCC) 176 C; Draves wetting 9 s (0.1% aq.); toxicology: low acute oral toxicity; severely irritating to skin and eyes in undiluted form; 100% conc.

Neodol® 3. [Shell] Tridecyl alcohol; CAS 112-70-9; detergent intermediate; paste; 100% conc.

Neodol® 5. [Shell] Pentadecyl alcohol; CAS 629-76-5; detergent intermediate; paste; 100% conc.

Neodol® 23. [Shell] C12-13 alcohols; CAS 75782-86-4; detergent intermediate; liq.; m.w. 194; sp.gr. 0.833; visc. 18.9 cSt (100 F); m.p. 18-22 C; pour pt. 63 F; flash pt. (PMCC) 132 C; acid no. < 0.001; hyd. no. 289; toxicology: LD50 (oral, rat) 28.2 g/kg; practically nonirritating to eyes, mild skin irritant; 100% act.

Neodol® 23-1. [Shell] C12-13 pareth-1; nonionic; detergent intermediate used in preparation of sulfates for high-foaming liq. detergents; emulsifier; for cosmetic, specialty industrial, dishwashing applics.; biodeg.; APHA 5–10 color.; m.w. 238; sp.gr. 0.873; visc. 13 cSt (100 F); m.p. 27–48 F; HLB 3.7; cloud pt. 13.6 F (1% aq.); flash pt. 289 F (PMCC); pour pt. 41 F; acid no. < 0.001; pH 10.1 (1%); 100% act.

Neodol® 23-3. [Shell] C12–13 pareth-3; nonionic; detergent intermediate used in preparation of sulfates for high-foaming liq. detergents; emulsifier; for cosmetic, specialty industrial, dishwashing applics.; biodeg.; APHA 50 max., clear to slightly hazy liq.; m.w. 310–342; dens. 7.7 lb/gal; sp.gr. 0.925; visc. 19 cs (100 F); m.p. 5–6 C; HLB 7.9; flash pt. 300 F (PMCC); pour pt. 4 C; 100% act.

Neodol® 23-5. [Shell] C12-13 pareth-5; nonionic; detergent intermediate used in preparation of sulfates for high-foaming liq. detergents, household and industrial use; biodeg.; APHA 5–10 color.; m.w. 413; sp.gr. 0.965; visc. 23 cSt (100 F); m.p. 27–61 F; HLB 10.7; flash pt. 315 F (PMCC); pour pt. 45 F; acid no. < 0.001; pH 6.0 (1%); 100% act.

Neodol® 23-6.5. [Shell] C12-13 pareth-7; nonionic; detergent, wetting agent, dispersant for general industrial usage, household detergent prods.; biodeg.; practically colorless liq., mild odor; m.w. 484; dens. 8.08 lb/gal (100 F); sp.gr. 0.963; visc. 29.5 cs (100 F); m.p. 11–15 C; HLB 12.0; flash pt. 330 F (PMCC); pour pt. 16 C; toxicology: LD50 (oral, rat) 4.6 g/kg; severe eye irritant, mild skin irritant; 100% act.

Neodol® 23-6.5T. [Shell] C12-13 pareth-7; nonionic; detergent intermediate used in preparation of sulfates for high-foaming liq. detergents; wetting agent, dispersant; for general industrial and household detergent prods.; biodeg.; APHA 10–15 color.; m.w. 529; sp.gr. 0.993; visc. 33 cSt (100 F); m.p. 36–66 F; HLB 12.6; cloud pt. 147 F (1% aq.); flash pt. 289 F (PMCC); pour pt. 61 F; acid no. < 0.001; pH 6.5 (1%); 100% act.

Neodol® 23-12. [Shell] C12-13 pareth-12; nonionic; detergent intermediate used in preparation of sulfates for high-foaming liq. detergents; wetting agent, dispersant; for general industrial and household detergent prods.; biodeg.; APHA 10–20 color.; m.w. 719; sp.gr. 1.006 (122/77 F); visc. 53 cSt (100 F); m.p. 63–90 F; HLB 14.6; cloud pt. 177 F (5% aq. NaCl); flash pt. 399 F (PMCC); pour pt. 79 F; acid no. < 0.001; pH 10.1; 100% act.

Neodol® 25. [Shell] C12-15 alcohols; CAS 63393-82-8; detergent, emulsifier intermediate; liq.; m.w. 203; sp.gr. 0.834; visc. 18.3 cSt (100 F); m.p. 18-22 C; pour pt. 66 F; flash pt. (PMCC) 138 C; acid no. < 0.001; HYD. NO. 277; toxicology: LD50 (oral, rat) > 23.1 g/kg; minimal eye irritant; mild skin irritant; 100% act.

Neodol® 25-3. [Shell] C12–15 pareth-3; nonionic; detergent intermediate used in preparation of sulfates for high-foaming liq. detergents; emulsifier; for cosmetic, industrial, dishwashing and liq. detergents; biodeg.; colorless, clear to slightly hazy liq.; mild odor; m.w. 336; sp.gr. 0.925; dens. 7.70 lb/gal; visc. 19 cs (100 F); m.p. 5–6 C; HLB 7.9; hyd. no. 167; flash pt. 315 F (PMCC); pour pt. 4 C; toxicology: LD50 (oral, rat) 2.5 g/kg; severe eye irritant, extreme skin irritant; 100% act.

Neodol® 25-3A. [Shell] Ammonium C12–15 pareth sulfate; anionic; detergent used in high-foaming liq. detergents; lt. clear visc. liq.; mild ethanol odor; m.w. 432; dens. 8.5 lb/gal (60 F); sp.gr. 1.02; visc. 45 cs (100 F); flash pt. 74 F (PMCC); pH 7.3; biodeg.; toxicology: LD50 (oral, rat) 10.2 g/kg; severe eye irritant, mild skin irritant; 59% act.

Neodol® 25-3S. [Shell] Sodium C12–15 pareth sulfate; anionic; detergent used in high-foaming liq. detergents; lt., visc. liq.; mild ethanol odor; m.w. 437; dens. 8.76 lb/gal (60 F); sp.gr. 1.02; visc. 45 cs

(100 F); flash pt. 73 F (PMCC); pH 7.5–9.0; toxicology: LD50 (oral, rat) 10.2 g/kg; severe eye irritant, mild skin irritant; 59% act.

Neodol® 25-7. [Shell] C12-15 pareth-7; nonionic; detergent intermediate used in preparation of sulfates for high-foaming liq. detergents; wetting agent, dispersant; for general industrial and household detergents; biodeg.; APHA 5–10 paste-like, mild odor.; m.w. 524; sp.gr. 0.965 (122/77 F); visc. 34 cSt (100 F); m.p. 36–70 F; HLB 12.2; cloud pt. 121 F (1% aq.); flash pt. 367 F (PMCC); pour pt. 66 F; acid no. < 0.001; pH 6.0 (1%); toxicology: LD50 (oral, rat) 2.7 g/kg; moderate eye irritant, mild skin irritant; 100% act.

Neodol® 25-9. [Shell] C12-15 pareth-9; nonionic; detergent intermediate used in preparation of sulfates for high-foaming liq. detergents; wetting agent, dispersant; for general industrial and household detergents; biodeg.; APHA 5–10 paste-like.; m.w. 597; sp.gr. 0.982 (122/77 F); visc. 41 cSt (100 F); m.p. 57–77 F; HLB 13.3; cloud pt. 163 F (1% aq.); flash pt. 370 F (PMCC); pour pt. 70 F; acid no. < 0.001; pH 6.0 (1%); toxicology: LD50 (oral, rat) 1.6 g/kg; extreme eye irritant, severe skin irritant; 100% act.

Neodol® 25-12. [Shell] C12-15 pareth-12; nonionic; detergent intermediate used in preparation of sulfates for high-foaming liq. detergents; wetting agent, dispersant, emulsifier; for general industrial and household detergent prods.; biodeg.; APHA 5–10 color.; m.w. 729; sp.gr. 0.999 (122/77 F); visc. 53 cSt (100 F); m.p. 68–86 F; HLB 14.4; cloud pt. 173 F (5% aq. NaCl); flash pt. 433 F (PMCC); pour pt. 81 F; acid no. < 0.001; pH 6.0 (1%); toxicology: LD50 (oral, rat) 1.8 g/kg; severe eye irritant, minimal skin irritant; 100% act.

Neodol® 45. [Shell] C14-15 alcohol; CAS 75782-87-5; detergent intermediate; liq.; m.w. 218; sp.gr. 0.834; visc. 29.3 cSt (100 F); m.p. 29-32 C; pour pt. 84 F; flash pt. (PMCC) 152 C; acid no. < 0.001; hyd. no. 257; toxicology: LD50 (oral, rat) 26.4 g/kg; minimal eye irritant; mild skin irritant; 100% act.

Neodol® 45-2.25. [Shell] C14-15 pareth-2.25; non ionic; detergent intermediate used in preparation of sulfates for high-foaming liq. detergents; emulsifier; for cosmetic, specialty industrial, dishwashing, liq. detergent applics.; biodeg.; APHA 5–10 color.; m.w. 319; sp.gr. 0.903; visc. 19 cSt (100 F); m.p. 48–68 F; HLB 6.3; cloud pt. 21 F (1% aq.); flash pt. 336 F (PMCC); pour pt. 59 F; acid no. < 0.001; pH 6.5 (1%); 100% act.

Neodol® 45-7. [Shell] C14-15 pareth-7; nonionic; detergent, wetting agent, emulsifier for general industrial and household detergent prods.; biodeg.; APHA 10 color; m.w. 539; sol. in alcohols, esters, ketones, chlorinated solvs., aromatic and aliphatic hydrocarbons; sp.gr. 0.967; dens. 8.20 lb/gal; visc. 35 cSt (38 C); HLB 11.8; m.p. 21–25 C; flash pt. 204 C (PMCC); cloud pt. 46 C (1% aq.); surf. tens. 29 dynes/cm (0.1%); 100% act.

Neodol® 45-7T. [Shell] C14-15 pareth-7; nonionic; detergent intermediate used in preparation of sulfates for high-foaming liq. detergents; wetting agent, dispersant for general industrial and household detergent prods.; biodeg.; APHA 10–15 color.; m.w. 567; sp.gr. 0.966 (122/77 F); visc. 39 cSt (100 F); m.p. 46–73 F; HLB 12.3; cloud pt. 131 F (1% aq.); flash pt. 441 F (PMCC); pour pt. 66 F; acid no. < 0.001; pH 6.8 (1%); 100% act.

Neodol® 45-11. [Shell] C14-15 pareth-11; nonionic; detergent; m.w. 702; sp.gr. 0.993 (50/25 C); visc. 51 cst (38 C); HLB 13.8; hyd. no. 80; pour pt. 29 C; cloud pt. 88 C (1% aq.); flash pt. (PMCC) 232 C; pH 6.0 (1% aq.); 100% act.

Neodol® 45-12T. [Shell] C14-15 pareth-12; nonionic; detergent, wetting agent, dispersant, emulsifier for general industrial and household detergent prods.; biodeg.; paste; HLB 14.3; 100% conc.

Neodol® 45-13. [Shell] C14-15 pareth-13; nonionic; surfactant, detergent, wetting agent, emulsifier for general industrial and household detergent prods.; biodeg.; APHA 10 color; m.w. 790; sol. in alcohols, esters, ketones, chlorinated solvs., aromatic and aliphatic hydrocarbons; sp.gr. 1.008 (50/25 C); dens. 8.39 lb/gal (49 C); visc. 58 cSt (38 C); HLB 14.4; m.p. 31–36 C; flash pt. 241 C (PMCC); cloud pt. 78 C (5% aq. NaCl); surf. tens. 34 dynes/cm (0.1%); 100% act.

Neodol® 91. [Shell] C9-11 alcohols; CAS 66455-17-2; nonionic; detergent intermediate; clear liq.; m.w. 160; sp.gr. 0.835; dens. 6.95 lb/gal; visc. 9 cSt (100 F); m.p. -9 C; pour pt. 10 F; flash pt. (PMCC) 107 C; acid no. < 0.001; hyd. no. 350; toxicology: LD50 (oral, rat) > 10 g/kg; moderate eye irritant, mild skin irritant; 100% act.

Neodol® 91-2.5 [Shell] C9–11 pareth-3 (2.5 EO); nonionic; oil-sol. emulsifier and wetting agent, detergent intermediate, dispersant, surfactant; for general industrial usage, hard surf. cleaning; biodeg.; colorless liq.; mild odor; m.w. 270; sol. in alcohols, esters, ketones, chlorinated solvs., aromatic hydrocarbons; partly sol. in water; dens. 7.73 lb/gal; sp.gr. 0.934; visc. 12 cs (100 F); HLB 8.1; hyd. no. 208; pour pt. -15 C; flash pt. (PMCC) 138 C; pH 6.0 (1% aq. disp.); toxicology: LD50 (oral, rat) 2.7 g/kg; mild eye irritant, severe skin irritant; 100% act.

Neodol® 91-6. [Shell] C9–11 pareth-6; nonionic; oil-sol. emulsifier and wetting agent, intermediate, dispersant, surfactant; for general industrial usage, hard surf. cleaners, liq. detergents; biodeg.; colorless liq.; mild odor; m.w. 424; sol. in wide range of oxygenated and hydrocarbon solvs.; dens. 8.24 lb/gal; sp.gr. 0.991; visc. 23 cs (100 F); HLB 12.5; flash pt. 334 F (PMCC); pour pt. 7 C; pH 6.0 (1% aq. disp.); toxicology: LD50 (oral, rat) 1.2 g/kg; severe eye irritant, strong skin irritant; 100% act.

Neodol® 91-8. [Shell] C9–11 pareth-8 (8.4 EO); nonionic; oil-sol. emulsifier and wetting agent, intermediate, dispersant, surfactant, solubilizer; for general industrial, hard surf. cleaners, liq. detergents; biodeg.; colorless liq.; mild odor; m.w. 529; sol. in wide range of oxygenated and hydrocarbon solvs., readily disp. in water; dens. 8.42 lb/gal; sp.gr. 1.002; visc. 30 cs (100 F); HLB 14.0; m.p. 45–68 F; flash pt. 349 F (PMCC); pour pt. 16 C; acid no. < 0.001; pH 6.0 (1% aq. disp.); toxicology: LD50 (oral, rat) 1.0 g/kg; severe eye irritant, severe skin irritant; 100% act.

Neolisal HCN. [Seppic] Ethoxylated fatty amine; nonionic; stripping agent; liq.; 100% conc.

Neopelex FS. [Kao Corp. SA] Alkylbenzene sulfonic acid; anionic; detergent base; liq.; 96% conc.

Neopelex No. 6, No. 25, No. 6F Powder, F-25, F-65. [Kao Corp. SA] Sodium alkylaryl sulfonate; anionic; general purpose detergent; paste, liq., beads, liq., paste resp.; 60, 25, 60, 25, and 65% conc. resp.

Neoscoa 203C. [Toho Chem. Industry] POE alkyl ether; nonionic; detergent, scouring agent for cotton, wool, and syn. fibers; deinking agent for paper;

liq.; 52% act.

Neoscoa 363. [Toho Chem. Industry] Blend; nonionic; detergent, scouring agent for raw wool and textiles, metal cleaning; deinking agent for paper; liq.; 100% act.

Neoscoa 500C. [Toho Chem. Industry] Blend; anionic/nonionic; detergent, scouring agent for cotton, wool, and syn. fibers; deinking agent for paper; liq.; 50% act.

Neoscoa 2326. [Toho Chem. Industry] Surfactant; nonionic; low foaming scouring agent for syn. fibers; deinking agent for paper; liq.; 100% conc.

Neoscoa CM-40. [Toho Chem. Industry] Blend; anionic; wetting and penetrating agent for high alkali sol'n. (15-25%) for cotton and polyester fibers; deinking agent for paper; liq.; 40% conc.

Neoscoa CM-57. [Toho Chem. Industry] Blend; anionic; wetting and penetrating agent for high alkali sol'n. (1-10%) for cotton fibers; deinking agent for paper; liq.; 55% conc.

Neoscoa ED-201C. [Toho Chem. Industry] Blend; anionic; wetting and penetrating agent for low alkali sol'n. (1-3%) for cotton fibers; deinking agent for paper; liq.; 60% conc.

Neoscoa FS-100. [Toho Chem. Industry] Blend; nonionic; deinking agent for paper prod. by flotation method; liq.

Neoscoa GF 3C. [Toho Chem. Industry] POE alkylaryl ether, POE alkylaryl ether phosphate; nonionic/anionic; detergent, scouring and wetting agent for cotton and syn. fibers; deinking agent for paper; liq.; 80% act.

Neoscoa GF-2000. [Toho Chem. Industry] Blend; nonionic; aftersoaping agent for polyester fibers; deinking agent for paper; liq.

Neoscoa MSC-80. [Toho Chem. Industry] Blend; nonionic; cleaner for dyeing machines; deinking agent for paper; liq.; 80% conc.

Neoscoa OT-80E. [Toho Chem. Industry] Blend; anionic; scouring agent for cotton, wool, and polyester fibers; deinking agent for paper; liq.; 80% conc.

Neoscoa PRA-8C. [Toho Chem. Industry] Blend; anionic/nonionic; aftersoaping agent for polyester fibers; deinking agent for paper; liq.; 54% act.

Neoscoa SS-10. [Toho Chem. Industry] Blend; nonionic; deinking agent for paper prod. by washing method; liq.

Neoscoa TH-102. [Toho Chem. Industry] Nonionic/anionic; detergent, dyeing machine cleaner for disperse dyeing; deinking agent for paper; liq.; 50% act.

Neosolve® AD-1. [M.S. Paisner] Wetting agent for asbestos.

Neospinol 264. [Toho Chem. Industry] Blend of special anionics, nonionics, and min. oil; anionic; dispersant for sulfur in rayon prod.; spinning oil for nylon; liq.; 100% conc.

Neospinol 358. [Toho Chem. Industry] Special nonionic complex (nitrogen compd.); nonionic; dispersant for sulfur in rayon prod.; liq.; 100% conc.

Neustrene® 045. [Witco/Humko] Hydrog. menhaden oil; CAS 68002-72-2; textile lubricant, pharmaceutical intermediate, emulsifier, mold release agent, buffing compd.; solid; acid no. 5–6; iodine no. 18-30; sapon. no. 188–201; 100% conc.

Neustrene® 053. [Witco/Humko] Hydrog. menhaden oil; CAS 68002-72-2; used in mfg. of alkali metal soaps, monoglycerides, textile auxiliaries, greases, personal care prods.; solid; acid no. 5 max.; iodine no. 5 max.; sapon. no. 186-201; 100% conc.

Neustrene® 059. [Witco/Humko] Hydrog. tallow glycerides; CAS 67701-27-3; used in mfg. of alkali metal soaps, monoglycerides, textile auxiliaries, greases; solid; acid no. 10 max.; iodine no. 5 max.; sapon. no. 193-205; 100% conc.

Neustrene® 060. [Witco/Humko] Refined hydrog. tallow glycerides; CAS 67701-27-3; textile lubricant, pharmaceutical intermediate, emulsifier, mold release agent, buffing compd.; Gardner 5 max. (059) solids; acid no. 2.5 max.; iodine no. 1 max.; sapon. no. 193–205; 100% conc.

Neustrene® 064. [Witco/Humko] Hydrog. soybean oil; CAS 68002-71-1; textile lubricant, pharmaceutical intermediate, emulsifier, mold release agent, buffing compd.; solid; acid no. 4 max.; iodine no. 2 max.; sapon. no. 188–200; 100% conc.

Neutronyx® 656. [Stepan; Stepan Canada] Nonoxynol-11; CAS 9016-45-9; nonionic; detergent, dispersant, wetting agent, emulsifier for household detergents, dishwashing, fine fabrics, metal cleaning and degreasing, industrial cleaning, sanitizers, insecticides, herbicides; silicone emulsifier; lime dispersant; stable in acid, alkali and hard water; liq.; sp.gr. 1.05; cloud pt. 54 and 71 C; flash pt. > 200 F; 100% act.

Newcol 3-80, 3-85. [Nippon Nyukazai] Sorbitan oleate; CAS 1338-43-8; nonionic; antistat, lubricant, emulsifier, corrosion inhibitor, emulsion solubilizer; liq.; 100% conc.

Newcol 20. [Nippon Nyukazai] Sorbitan laurate; CAS 1338-39-2; nonionic; antistat, emulsifier, corrosion inhibitor, EP lubricant, emulsion solubilizer; solid; 100% conc.

Newcol 25. [Nippon Nyukazai] POE sorbitan laurate; nonionic; antistat, emulsifier, corrosion inhibitor, EP lubricant, emulsion solubilizer; liq.; 100% conc.

Newcol 40. [Nippon Nyukazai] Sorbitan palmitate; CAS 26266-57-9; nonionic; antistat, emulsifier, corrosion inhibitor, EP lubricant, emulsion solubilizer; solid; 100% conc.

Newcol 45. [Nippon Nyukazai] POE sorbitan palmitate; nonionic; see Newcol 20; solid; 100% conc.

Newcol 60. [Nippon Nyukazai] Sorbitan stearate; CAS 1338-41-6; nonionic; antistat, emulsifier, corrosion inhibitor, EP lubricant, emulsion solubilizer; solid; 100% conc.

Newcol 65. [Nippon Nyukazai] POE sorbitan stearate; nonionic; antistat, emulsifier, corrosion inhibitor, EP lubricant, emulsion solubilizer; solid; 100% conc.

Newcol 80. [Nippon Nyukazai] Sorbitan oleate; CAS 1338-43-8; nonionic; antistat, emulsifier, corrosion inhibitor, EP lubricant, emulsion solubilizer; liq.; 100% conc.

Newcol 85. [Nippon Nyukazai] POE sorbitan oleate; nonionic; antistat, emulsifier, corrosion inhibitor, EP lubricant, emulsion solubilizer; liq.; 100% conc.

Newcol 150. [Nippon Nyukazai] POE laurate; nonionic; emulsifier, dispersant, lubricant; liq.; 100% conc.

Newcol 170. [Nippon Nyukazai] POE oleate; nonionic; emulsifier, dispersant, lubricant; liq.; 100% conc.

Newcol 180. [Nippon Nyukazai] POE stearate; nonionic; emulsifier, dispersant, lubricant; solid; 100% conc.

Newcol 180T. [Nippon Nyukazai] POE stearate; nonionic; surfactant used in emulsion polymerization; solid; 100% conc.

Newcol 210. [Nippon Nyukazai] Ammonium dodecylbenzene sulfonate; anionic; wetting agent, emulsifier, detergent; liq.; 50% conc.

Newcol 261A, 271A. [Nippon Nyukazai] Sodium alkyl diphenyl ether disulfonate; anionic; detergent, emulsifier used in emulsion polymerization; liq.; 45% conc.

Newcol 290K, 290M, 291PG. [Nippon Nyukazai] Sodium alkyl sulfosuccinate; anionic; wetting and dispersing agent, solubilizer, penetrant; liq.; 75, 75, and 70% conc.

Newcol 405, 410, 420. [Nippon Nyukazai] POE lauryl amine; nonionic; corrosion inhibitor, intermediate for textile industry; liq., solid; 100% conc.

Newcol 506. [Nippon Nyukazai] POE nonylphenyl ether; nonionic; antifoam, mold lubricant, wetting agent, penetrant, spreading agent, dispersant, emulsifier for detergents, agric. chemicals, machine oils, emulsion polymerization; solid; 100% conc.

Newcol 508. [Nippon Nyukazai] POE nonylphenyl ether; nonionic; antifoam, mold lubricant, wetting agent, penetrant, spreading agent, dispersant, emulsifier for detergents, agric. chemicals, machine oils, emulsion polymerization; solid; 100% conc.

Newcol 560. [Nippon Nyukazai] POE nonylphenyl ether; nonionic; antifoam, mold lubricant, wetting agent, penetrant, spreading agent, dispersant, emulsifier for detergents, agric. chemicals, machine oils; liq.; 100% conc.

Newcol 560SF. [Nippon Nyukazai] Ammonium POE nonylphenyl ether sulfate; anionic; emulsifier, wetting agent, penetrant, detergent used in emulsion polymerization; liq.; 50% conc.

Newcol 560SN. [Nippon Nyukazai] Sodium POE nonylphenyl ether sulfate; anionic; emulsifier, wetting agent, penetrant, detergent used in emulsion polymerization; liq.; 30% conc.

Newcol 561H, 562, 564, 565. [Nippon Nyukazai] POE nonylphenyl ether; nonionic; antifoam, mold lubricant, wetting agent, penetrant, spreading agent, dispersant, emulsifier for detergents, agric. chemicals, machine oils; liq.; 100% conc.

Newcol 565FH. [Nippon Nyukazai] Polyoxyalkylene nonylphenyl ether; nonionic; low-foaming detergent; liq.; 100% conc.

Newcol 566. [Nippon Nyukazai] POE nonylphenyl ether; nonionic; wetting agent, penetrant, spreading and dispersing agent, emulsifier, base material for detergents, agric. chemicals, machine oils; liq.; 100% conc.

Newcol 568. [Nippon Nyukazai] POE nonylphenyl ether; nonionic; low-foaming detergent; surfactant for emulsion polymerization; solid; 100% conc.

Newcol 569E. [Nippon Nyukazai] POE nonylphenyl ether deriv.; nonionic; low-foaming detergent; paste; 100% conc.

Newcol 607, 610, 614, 623. [Nippon Nyukazai] POE alkylaryl ether; nonionic; emulsifier, solubilizer, detergent used in emulsion polymerization; liqs. and solids; 100% conc.

Newcol 704, 707. [Nippon Nyukazai] POE alkylaryl ether; nonionic; emulsifier, solubilizer, detergent for emulsion polymerization; liq.; 100% conc.

Newcol 707SF. [Nippon Nyukazai] POE alkylaryl ether; CAS 104042-16-2; anionic; emulsifier, detergent for emulsion polymerization; liq.; 30% conc.

Newcol 710, 714, 723. [Nippon Nyukazai] POE alkylaryl ether; CAS 104042-16-2; anionic; emulsifier, solubilizer, detergent for emulsion polymerization; liq.; 30% conc.

Newcol 804, 808, 860. [Nippon Nyukazai] POE octylphenyl ether; nonionic; antifoam, mold lubricant, wetting agent, penetrant, spreading agent, dispersant, emulsifier for detergents, agric. chemicals, machine oils; liq., solid; 100% conc.

Newcol 861S. [Nippon Nyukazai] Sodium octylphenoxyethoxyethyl sulfonate; anionic; emulsifier, wetting agent, penetrant, detergent used in emulsion polymerization; paste; 30% conc.

Newcol 862. [Nippon Nyukazai] POE octylphenyl ether; nonionic; antifoam, mold lubricant, wetting agent, penetrant, spreading agent, dispersant, emulsifier for detergents, agric. chemicals, machine oils; liq.; 100% conc.

Newcol 864. [Nippon Nyukazai] POE octylphenyl ether; nonionic; antifoam, mold lubricant, wetting agent, penetrant, spreading agent, dispersant, emulsifier for detergents, agric. chemicals, machine oils; solid; 100% conc.

Newcol 865. [Nippon Nyukazai] POE octylphenyl ether; nonionic; wetting, penetrating, spreading, and dispersing agent, emulsifier, base for detergents, agric. chemicals, machine oils; solid; 100% conc.

Newcol 1010, 1020. [Nippon Nyukazai] POE octyl ether; nonionic; latex stabilizer; solid; 100% conc.

Newcol 1100. [Nippon Nyukazai] POE lauryl ether; nonionic; emulsifier, lubricant, detergent, dyeing additive; emulsifier for min. oil; solid; 100% conc.

Newcol 1105. [Nippon Nyukazai] POE lauryl ether; nonionic; emulsifier of min. oil, lubricant, detergent, dyeing additive agent; liq.; 100% conc.

Newcol 1110. [Nippon Nyukazai] POE lauryl ether; nonionic; emulsifier, lubricant, detergent, dyeing additive; emulsifier for min. oil; solid; 100% conc.

Newcol 1120. [Nippon Nyukazai] POE lauryl ether; nonionic; emulsifier, lubricant, detergent, dyeing additive; emulsifier for min. oil; solid; 100% conc.

Newcol 1200. [Nippon Nyukazai] POE oleyl ether; nonionic; emulsifier of min. oil, dyeing additive agent; solid; 100% conc.

Newcol 1203. [Nippon Nyukazai] POE oleyl ether; nonionic; emulsifier of min. oil, dyeing additive agent; liq.; 100% conc.

Newcol 1204, 1208, 1210. [Nippon Nyukazai] POE oleyl ether; nonionic; emulsifier of min. oil, dyeing additive agent; liq. (1204), solid (1208, 1210); 100% conc.

Newcol 1305. [Nippon Nyukazai] POE tridecyl ether; nonionic; emulsifier for silicone, dyeing additive agent, lubricant; solid; 100% conc.

Newcol 1305SN, 1310SN. [Nippon Nyukazai] Sodium POE tridecyl ether sulfate; anionic; emulsifier, wetting agent, penetrant, detergent used in emulsion polymerization; liq.; 30% conc.

Newcol 1310. [Nippon Nyukazai] POE tridecyl ether; nonionic; emulsifier of silicone, dyeing assistant; solid; 100% conc.

Newcol 1515, 1525, 1545. [Nippon Nyukazai] POE alkyl ether; nonionic; emulsifier, lubricant; liq., liq., and solid resp.; 100% conc.

Newcol 1610, 1620. [Nippon Nyukazai] POE cetyl ether; nonionic; emulsifier of paraffin wax, latex stabilizer; solid; 100% conc.

Newcol 1807, 1820. [Nippon Nyukazai] POE stearyl ether; nonionic; emulsifier of paraffin wax, latex stabilizer; solid; 100% conc.

Newcol B4, B10, B18. [Nippon Nyukazai] POE naphthyl ether; nonionic; dispersant, suspending and diffusion agent; liqs. (B4, B10), solid (B18); 100% conc.

Newkalgen 135R. [Takemoto Oil & Fat] POE alkylaryl ether, POE alkyl ester and org. sulfonate; anionic/nonionic; emulsifier for spray oil; liq.; 100% conc.

Newkalgen 2360X1. [Takemoto Oil & Fat] POE alkylaryl ether, POE alkylaryl polymer and org. sulfonate; anionic/nonionic; emulsifier for agric. emulsion concs.; liq.; 80% conc.

Newkalgen 2720X75. [Takemoto Oil & Fat] POE alkylaryl ether, POE alkyl ether and org. sulfonate; anionic/nonionic; emulsifier for Diazinon emulsion concs.; liq., 75% conc.

Newkalgen 3000 A & B. [Takemoto Oil & Fat] POE alkylaryl ether, POE alkyl polymer, POE alkyl ether and org. sulfonate; anionic/nonionic; emulsifier for agric. emulsion concs.; liq.; 80% conc.

Newlon K-1. [Takemoto Oil & Fat] Polyglycol ether; nonionic; raw wool scouring agent; liq.; 100% conc.

Newpol PE-61. [Sanyo Chem. Industries] EO/PO block copolymer; CAS 9003-11-6; nonionic; base material for household and industrial detergents; plasticizer, antistat for phenol resins; emulsifier for agric. pesticides and emulsion polymerization; pigment and pitch dispersant; liq.; HLB 5.8; 100% conc.

Newpol PE-62. [Sanyo Chem. Industries] EO/PO block copolymer; CAS 9003-11-6; nonionic; base material for household and industrial detergents; plasticizer, antistat for phenol resins; emulsifier for agric. pesticides and emulsion polymerization; pigment and pitch dispersant; liq.; HLB 6.3; 100% conc.

Newpol PE-64. [Sanyo Chem. Industries] EO/PO block copolymer; CAS 9003-11-6; nonionic; base material for household and industrial detergents; plasticizer, antistat for phenol resins; emulsifier for agric. pesticides and emulsion polymerization; pigment and pitch dispersant; liq.; HLB 10.1; 100% conc.

Newpol PE-68. [Sanyo Chem. Industries] EO/PO block copolymer; CAS 9003-11-6; nonionic; base material for household and industrial detergents; plasticizer, antistat for phenol resins; emulsifier for agric. pesticides and emulsion polymerization; pigment and pitch dispersant; solid; HLB 14.9; 100% conc.

Newpol PE-74. [Sanyo Chem. Industries] EO/PO block copolymer; CAS 9003-11-6; nonionic; base material for household and industrial detergents; plasticizer, antistat for phenol resins; emulsifier for agric. pesticides and emulsion polymerization; pigment and pitch dispersant; liq.; HLB 10.1; 100% conc.

Newpol PE-75. [Sanyo Chem. Industries] EO/PO block copolymer; CAS 9003-11-6; nonionic; base material for household and industrial detergents; plasticizer, antistat for phenol resins; emulsifier for agric. pesticides and emulsion polymerization; pigment and pitch dispersant; liq.; HLB 10.7; 100% conc.

Newpol PE-78. [Sanyo Chem. Industries] EO/PO block copolymer; CAS 9003-11-6; nonionic; base material for household and industrial detergents; plasticizer, antistat for phenol resins; emulsifier for agric. pesticides and emulsion polymerization; pigment and pitch dispersant; solid; HLB 14.8; 100% conc.

Newpol PE-88. [Sanyo Chem. Industries] EO/PO block copolymer; CAS 9003-11-6; nonionic; base material for household and industrial detergents; plasticizer, antistat for phenol resins; emulsifier for agric. pesticides and emulsion polymerization; pigment and pitch dispersant; solid; HLB 14.6; 100% conc.

Niaproof® Anionic Surfactant 4. [Niacet] Sodium tetradecyl sulfate; CAS 139-88-8; anionic; detergent, wetting agent, penetrant, emulsifier used in adhesives and sealants, coatings, photo chemicals, emulsion polymerization, metal processing, electrolytic cleaning, pickling baths, plating, pharmaceuticals, leather, textiles; FDA compliance; colorless liq., mild char. odor; misc. with water; sp.gr. 1.031; dens. 8.58 lb/gal; b.p. 92 C; COC flash pt. none; pH 8.5 (0.1% aq.); surf. tens. 47 dynes/cm (0.1% aq.); Draves wetting 20 s (0.26%); Ross-Miles foam 10 mm; corrosive, slippery; toxicology: moderate oral and skin toxicity; eye irritant; LD50 (oral, rats) 4.95 ml/kg; 27% act. in water.

Niaproof® Anionic Surfactant 7. [Niacet] Sodium heptadecyl sulfate; anionic; detergent, wetting agent, emulsifier used in adhesives, metal finishing, leather treating; colorless liq.; misc. with water; dens. 8.82 lb/gal; sp.gr. 1.060; surf. tens. 34 dynes/cm (0.1% aq.); pH 5.3 (0.1% aq.); 26% act.

Niaproof® Anionic Surfactant 08. [Niacet] Sodium 2-ethylhexyl sulfate; CAS 126-92-1; anionic; detergent, wetting agent, penetrant, emulsifier used in textile mercerizing, metal cleaning, electroplating, photo chemicals, adhesives, emulsion polymerization, household and industrial cleaners, agric., pharmaceuticals; stable to high concs. of electrolytes; FDA compliance; colorless liq., mild char. odor; misc. with water; sp.gr. 1.109; dens. 9.23 lb/gal; b.p. 95 C; COC flash pt. none; pH 7.3 (0.1% aq.); surf. tens. 63 dynes/cm (0.1% aq.); Ross-Miles foam 10 mm (initial); toxicology: moderate oral and skin toxicity; eye irritant; LD50 (oral, rats) 7.27 ml/kg; 39% act.

Nikkol BC-1SY thru BC-8SY [Nikko Chem. Co. Ltd.] Ethoxylated n-hexadecyl ether; nonionic; used in industrial applics.; solid; 100% conc.

Nikkol BC-15TX, -15TX(FF). [Nikko Chem. Co. Ltd.] Ceteth-15; nonionic; emulsifier for cosmetics and pharmaceuticals; used in refining techniques; solid; HLB 15.5; 100% conc.

Nikkol BD-1SY thru BD-8SY. [Nikko Chem. Co. Ltd.] Ethoxylated n-decyl ether; nonionic; used in industrial applics.; liq., solid; 100% conc.

Nikkol BL-1SY. [Nikko Chem. Co. Ltd.] Laureth-1; nonionic; compd. for use in research on ether-type nonionic surfactants; m.w. 230.4; sp.gr. 0.8681 m.p. 19.5; b.p. 135 C (1 mm Hg); 98% pure.

Nikkol BL-2. [Nikko Chem. Co. Ltd.] Laureth-2; nonionic; for emulsifying prods. requiring abundant liquidity; liq.; HLB 9.5; 100% conc.

Nikkol BL-2SY. [Nikko Chem. Co. Ltd.] Laureth-2; nonionic; compd. for use in research on ether-type nonionic surfactants; m.w. 274.4; sp.gr. 0.8996; m.p. 17.6; b.p. 164 C (1 mm Hg); 98% pure.

Nikkol BL-3SY. [Nikko Chem. Co. Ltd.] Laureth-3; nonionic; compd. for use in research on ether-type nonionic surfactants; m.w. 318.5; sp.gr. 0.9216; m.p. 16.0; b.p. 194–197 C (1.5 mm Hg); 98% pure.

Nikkol BL-4SY. [Nikko Chem. Co. Ltd.] Laureth-4; nonionic; compd. for use in research on ether-type nonionic surfactants; m.w. 362.6; sp.gr. 0.9442; m.p. 18.7; b.p. 214–215 C (2 mm Hg); 98% pure.

Nikkol BL-4.2. [Nikko Chem. Co. Ltd.] Laureth-4 (4.2 EO); nonionic; for emulsifying prods. requiring abundant liquidity; liq.; HLB 11.5; 100% conc.

Nikkol BL-5SY. [Nikko Chem. Co. Ltd.] Laureth-5; nonionic; compd. for use in research on ether-type nonionic surfactants; m.w. 406.6; sp.gr. 0.9586; m.p. 22.8 C; b.p. 225–229 C (2 mm Hg); cloud pt. 30.5 C (0.1%); surf. tens. 29.6 dynes/cm; 98% pure.

Nikkol BL-6SY. [Nikko Chem. Co. Ltd.] Laureth-6; nonionic; compd. for use in research on ether-type nonionic surfactants; m.w. 450.6; sp.gr. 0.9756; m.p. 24.8; cloud pt. 55 C (0.1%); surf. tens. 31.5 dynes/cm; 98% pure.

Nikkol BL-7SY. [Nikko Chem. Co. Ltd.] Laureth-7; nonionic; compd. for use in research on ether-type nonionic surfactants; m.w. 494.6; sp.gr. 0.9837; m.p. 26.1; cloud pt. 70 C (0.1%); surf. tens. 33.4 dynes/cm; 98% pure.

Nikkol BL-8SY. [Nikko Chem. Co. Ltd.] Laureth-8; nonionic; compd. for use in research on ether-type nonionic surfactants; m.w. 538.7; sp.gr. 0.9951; m.p. 31.4; cloud pt. 79 C (0.1%); surf. tens. 34.6 dynes/cm; 98% pure.

Nikkol BL-9EX, -9EX(FF). [Nikko Chem. Co. Ltd.] Laureth-9; nonionic; for emulsifying prods. requiring abundant liquidity; liq.; HLB 14.5; 100% conc.

Nikkol BL-21. [Nikko Chem. Co. Ltd.] Laureth-21; nonionic; for emulsifying prods. requiring abundant liquidity; solid; HLB 19.0; 100% conc.

Nikkol BL-25. [Nikko Chem. Co. Ltd.] Laureth-25; nonionic; for emulsifying prods. requiring abundant liquidity; solid; HLB 19.5; 100% conc.

Nikkol BM-1SY thru BM-8SY. [Nikko Chem. Co. Ltd.] Ethoxylated n-tetradecyl ether; nonionic; compd. for use in research on ether-type nonionic surfactants; solid; 100% conc.

Nikkol BO-2. [Nikko Chem. Co. Ltd.] Oleth-2; CAS 9004-98-2; nonionic; emulsifier for preparations requiring abundant liquidity; liq.; HLB 7.5; 100% conc.

Nikkol BO-7. [Nikko Chem. Co. Ltd.] Oleth-7; CAS 9004-98-2; nonionic; emulsifier for preparations requiring abundant liquidity; liq.; HLB 10.5; 100% conc.

Nikkol BO-10TX. [Nikko Chem. Co. Ltd.] Oleth-10; CAS 9004-98-2; nonionic; emulsifier for preparations requiring abundant liquidity; liq.; HLB 14.5; 100% conc.

Nikkol BO-15TX. [Nikko Chem. Co. Ltd.] Oleth-15; CAS 9004-98-2; nonionic; emulsifier for preparations requiring abundant liquidity; paste; HLB 16.0; 100% conc.

Nikkol BO-20TX. [Nikko Chem. Co. Ltd.] Oleth-20; CAS 9004-98-2; nonionic; emulsifier for preparations requiring abundant liquidity; solid; HLB 17.0; 100% conc.

Nikkol BO-50. [Nikko Chem. Co. Ltd.] Oleth-50; CAS 9004-98-2; nonionic; emulsifier for preparations requiring abundant liquidity; flake; HLB 18.0; 100% conc.

Nikkol BPS-5. [Nikko Chem. Co. Ltd.] PEG-5 phytosterol; nonionic; emulsifier for o/w and w/o compds., solubilizer, dispersant, emollient, foam stabilizer, visc. modifier, conditioner used in cosmetics, pharmaceuticals; wh. to pale yel. paste; sol. in propylene glycol, ethanol; partly sol. in water; HLB 9.5; acid no. 0.25 max.; pH 4.8 (5%); 100% conc.

Nikkol BPS-10. [Nikko Chem. Co. Ltd.] PEG-10 phytosterol; nonionic; see Nikkol BPS-5; wh. to pale yel. paste or solid; sol. in water, propylene glycol, ethanol; HLB 12.5; acid no. 0.18 max.; pH 5.2 (5%); 100% conc.

Nikkol BPS-15. [Nikko Chem. Co. Ltd.] PEG-15 phytosterol; nonionic; see Nikkol BPS-5; wh. to pale yel. wax-like solid; sol. in water, propylene glycol, ethanol; HLB 15.0; acid no. 0.06 max; pH 5.0 (5%).

Nikkol BPS-20. [Nikko Chem. Co. Ltd.] PEG-20 phytosterol; nonionic; see Nikkol BPS-5; wh. to pale yel. wax-like solid; sol. in water, propylene glycol, ethanol; HLB 15.5; acid no. 0.07 max.; pH 5.5 (5%); 100% conc.

Nikkol BPS-25. [Nikko Chem. Co. Ltd.] PEG-25 phytosterol; nonionic; see Nikkol BPS-5; wh. to pale yel. wax-like solid; sol. in water, propylene glycol, ethanol; HLB 18.0; acid no. 0.11 max.; pH 5.7 (5%).

Nikkol BPS-30. [Nikko Chem. Co. Ltd.] PEG-30 phytosterol; nonionic; see Nikkol BPS-5; wh. to pale yel. wax-like solid; sol. in water, propylene glycol, ethanol; HLB 18.0; acid no. 0.09 max.; pH 5.7 (5%); 100% conc.

Nikkol CCK-40. [Nikko Chem. Co. Ltd.] Potassium coco-hydrolyzed animal protein; anionic; detergent used in shampoos, cleansing cream, kitchen cleaning agents; liq.; 30% conc.

Nikkol CCN-40. [Nikko Chem. Co. Ltd.] Sodium coco-hydrolyzed animal protein; anionic; detergent for shampoos, cleansing creams, kitchen cleaning agents; liq.; 30% conc.

Nikkol CMT-30. [Nikko Chem. Co. Ltd.] Sodium methyl cocoyl taurate; anionic; shampoo base, foamer and detergent; liq.; 30% conc.

Nikkol DDP-2. [Nikko Chem. Co. Ltd.] Di-PEG-2 alkyl ether phosphate; anionic; emulsifier, stabilizer, dispersant, anticorrosive agent and detergent used in cosmetics, drugs, agric. chemicals and general industrial use; liq.; HLB 6.5; 100% conc.

Nikkol DDP-4. [Nikko Chem. Co. Ltd.] Di-PEG-4 alkyl ether phosphate; anionic; see Nikkol DDP-2; liq.; HLB 9.0; 100% conc.

Nikkol DDP-6. [Nikko Chem. Co. Ltd.] Di-PEG-6 alkyl ether phosphate; anionic; see Nikkol DDP-2; liq.; HLB 9.0; 100% conc.

Nikkol DDP-8. [Nikko Chem. Co. Ltd.] Di-PEG-8 alkyl ether phosphate; anionic; see Nikkol DDP-2; semisolid; HLB 11.5; 100% conc.

Nikkol DDP-10. [Nikko Chem. Co. Ltd.] Di-PEG-10 alkyl ether phosphate; anionic; see Nikkol DDP-2; semisolid; HLB 13.5; 100% conc.

Nikkol Decaglyn 2-IS. [Nikko Chem. Co. Ltd.] Polyglyceryl-10 diisostearate; nonionic; emulsifier, lubricant, coating agent and anticrystallization agent.

Nikkol Decaglyn 3-O. [Nikko Chem. Co. Ltd.] Polyglyceryl-10 trioleate; nonionic; o/w and w/o emulsifier; gelling agent for hydrocarbon; liq.; HLB 6.5; 100% conc.

Nikkol Decaglyn 3-S. [Nikko Chem. Co. Ltd.] Polyglyceryl-10 tristearate; nonionic; o/w and w/o emulsifier; gelling agent for hydrocarbons; flake; HLB 6.5; 100% conc.

Nikkol Decaglyn 5-IS. [Nikko Chem. Co. Ltd.] Polyglyceryl-10 pentaisostearate; nonionic; emulsifier, lubricant, coating agent and anticrystallization agent; liq.; HLB 3.5; 100% conc.

Nikkol Decaglyn 5-O. [Nikko Chem. Co. Ltd.] Polyglyceryl-10 pentaoleate; nonionic; emulsifier, lubricant, coating agent and anticrystallization agent; liq.; HLB 4.0; 100% conc.

Nikkol Decaglyn 5-S. [Nikko Chem. Co. Ltd.] Polyglyceryl-10 pentastearate; nonionic; emulsifier, lubricant, coating agent and anticrystallization agent; solid; HLB 3.5; 100% conc.

Nikkol Decaglyn 7-IS. [Nikko Chem. Co. Ltd.] Polyglyceryl-10 heptaisostearate; nonionic; emulsifier, lubricant, coating agent and anticrystallization agent; liq.; 100% conc.

Nikkol Decaglyn 7-O. [Nikko Chem. Co. Ltd.] Polyglyceryl-10 heptaoleate; nonionic; emulsifier, lubricant, coating agent and anticrystallization agent; liq.; 100% conc.

Nikkol Decaglyn 7-S. [Nikko Chem. Co. Ltd.] Polyglyceryl-10 heptastearate; nonionic; emulsifier, lubricant, coating agent and anticrystallization agent; solid; 100% conc.

Nikkol Decaglyn 10-IS. [Nikko Chem. Co. Ltd.] Polyglyceryl-10 decaisostearate; nonionic; emulsifier, lubricant, coating agent and anticrystallization agent; liq.; 100% conc.

Nikkol Decaglyn 10-O. [Nikko Chem. Co. Ltd.] Polyglyceryl-10 decaoleate; nonionic; emulsifier, lubricant, coating agent and anticrystallization agent; liq.; 100% conc.

Nikkol Decaglyn 10-S. [Nikko Chem. Co. Ltd.] Polyglyceryl-10 decastearate; nonionic; emulsifier, lubricant, coating agent and anticrystallization agent; solid; 100% conc.

Nikkol DGO-80. [Nikko Chem. Co. Ltd.] Glyceryl dioleate; CAS 25637-84-7; nonionic; emollient, emulsifier; liq.; 100% conc.

Nikkol DGS-80. [Nikko Chem. Co. Ltd.] Glyceryl distearate; CAS 1323-83-7; nonionic; emollient, emulsifier; powd.; 100% conc.

Nikkol DLP-10. [Nikko Chem. Co. Ltd.] Dilaureth-10 phosphate; anionic; emulsifier, dispersant, hydrotrope; surfactant, solubilizer; paste; 100% conc.

Nikkol ECT-3NEX, ECTD-3NEX. [Nikko Chem. Co. Ltd.] Sodium trideceth-3 carboxylate; anionic; shampoo base, foamer, detergent, and emulsifier; liq.; 85% conc.

Nikkol ECTD-6NEX. [Nikko Chem. Co. Ltd.] Sodium trideceth-6 carboxylate; anionic; shampoo base, foamer, detergent and emulsifier; liq.; 85% conc.

Nikkol GO-430. [Nikko Chem. Co. Ltd.] PEG-30 sorbitan tetraoleate; nonionic; emulsifier, solubilizer, superfatting agent used in drugs and cosmetics, for emulsion polymerization, agric. chemicals, printing inks; pale yel. liq.; sol. in ethanol, ethyl acetate, xylene; partly sol. in water; sp.gr. 1.048; HLB 11.5; ref. index 1.4727; 100% conc.

Nikkol GO-440. [Nikko Chem. Co. Ltd.] PEG-40 sorbitan tetraoleate; nonionic; see Nikkol GO-430; pale yel. liq.; sol. in ethanol, ethyl acetate, xylene; partly sol. in water, propylene glycol; sp.gr. 1.054; HLB 12.5; 100% conc.

Nikkol GO-460. [Nikko Chem. Co. Ltd.] PEG-60 sorbitan tetraoleate; nonionic; see Nikkol GO-430; pale yel. liq.; sol. in water, ethanol, ethyl acetate, xylene; partly sol. in propylene glycol; sp.gr. 1.060; HLB 14.0; 100% conc.

Nikkol Hexaglyn 1-L. [Nikko Chem. Co. Ltd.] Polyglyceryl-6 laurate; nonionic; o/w emulsifier, anticrystallizing agent; liq; HLB 13.0; 100% conc.

Nikkol Hexaglyn 1-O. [Nikko Chem. Co. Ltd.] Polyglyceryl-6 oleate; nonionic; o/w emulsifier, anticrystallizing agent; liq.; HLB 9.5; 100% conc.

Nikkol Hexaglyn 1-S. [Nikko Chem. Co. Ltd.] Polyglyceryl-6 stearate; nonionic; emulsifier, dispersant, lubricant; solid; HLB 9.5; 100% conc.

Nikkol Hexaglyn 3-S. [Nikko Chem. Co. Ltd.] Polyglyceryl-6 tristearate; nonionic; o/w emulsifier, anticrystallizing agent; flake; HLB 2.5; 100% conc.

Nikkol Hexaglyn 5-O. [Nikko Chem. Co. Ltd.] Polyglyceryl-6 pentaoleate; nonionic; o/w emulsifier, anticrystallizing agent; liq.; HLB 6.0; 100% conc.

Nikkol Hexaglyn 5-S. [Nikko Chem. Co. Ltd.] Polyglyceryl-6 pentastearate; nonionic; o/w emulsifier, anticrystallizing agent; flake; 100% conc.

Nikkol Hexaglyn PR-15. [Nikko Chem. Co. Ltd.] Polyglyceryl-6 polyricinoleate; nonionic; o/w emulsifier, anticrystallizing agent; liq.; 100% conc.

Nikkol LSA. [Nikko Chem. Co. Ltd.] Sodium lauryl sulfoacetate; anionic; acid detergent base, foamer; solid; 92.5% conc.

Nikkol NP-5. [Nikko Chem. Co. Ltd.] Nonoxynol-5; CAS 9016-45-9; nonionic; emulsifier, detergent, solubilizer, and wetting agent used in cosmetics, insecticide, and other industrial applics.

Nikkol NP-7.5. [Nikko Chem. Co. Ltd.] Nonoxynol-8 (7.5 EO); CAS 9016-45-9; nonionic; emulsifier, detergent, solubilizer, and wetting agent used in cosmetics, insecticide, and other industrial applics.; liq.; HLB 14.0; 100% conc.

Nikkol NP-10. [Nikko Chem. Co. Ltd.] Nonoxynol-10; CAS 9016-45-9; nonionic; emulsifier, detergent, solubilizer, and wetting agent used in cosmetics, insecticide, and other industrial applics.; liq.; HLB 16.5; 100% conc.

Nikkol NP-15. [Nikko Chem. Co. Ltd.] Nonoxynol-15; CAS 9016-45-9; nonionic; emulsifier, detergent, solubilizer, and wetting agent used in cosmetics, insecticide, and other industrial applics.; liq.; HLB 18.0; 100% conc.

Nikkol NP-18TX. [Nikko Chem. Co. Ltd.] Nonoxynol-18; CAS 9016-45-9; nonionic; emulsifier, detergent, solubilizer, and wetting agent used in cosmetics, insecticide, and other industrial applics.; paste; HLB 19.0; 100% conc.

Nikkol OP-3. [Nikko Chem. Co. Ltd.] Octoxynol-3; CAS 9002-93-1; nonionic; emulsifier, detergent, solubilizer, and wetting agent used in cosmetics, insecticide and other industrial applics.; liq.; HLB 6.0; 100% conc.

Nikkol OP-10. [Nikko Chem. Co. Ltd.] Octoxynol-10; CAS 9002-93-1; nonionic; emulsifier, detergent, solubilizer, and wetting agent used in cosmetics, insecticide and other industrial applics.; liq.; HLB 11.5; 100% conc.

Nikkol OP-30. [Nikko Chem. Co. Ltd.] Octoxynol-30; CAS 9002-93-1; nonionic; emulsifier, detergent, solubilizer, and wetting agent used in cosmetics, insecticide and other industrial applics.; paste; HLB 17.0; 100% conc.

Nikkol OTP-100S. [Nikko Chem. Co. Ltd.] Sodium dioctyl sulfosuccinate USP; anionic; dispersant, wetting agent; sponge-like solid; 98% conc.

Nikkol PBC-31. [Nikko Chem. Co. Ltd.] PPG-4-ceteth-1; nonionic; emulsifier, solubilizer, dispersant used in cosmetic, pharmaceuticals and other industrial applics.; liq.; HLB 9.4; 100% conc.

Nikkol PBC-33. [Nikko Chem. Co. Ltd.] PPG-4-ceteth-10; nonionic; see Nikkol PBC-31; paste; HLB 10.5; 100% conc.

Nikkol PBC-34. [Nikko Chem. Co. Ltd.] PPG-4-ceteth-20; nonionic; see Nikkol PBC-31; solid; HLB 16.5; 100% conc.

Nikkol PBC-41. [Nikko Chem. Co. Ltd.] PPG-8-ceteth-1; nonionic; see Nikkol PBC-31; liq.; HLB 9.5; 100% conc.

Nikkol PBC-44. [Nikko Chem. Co. Ltd.] PPG-8-

ceteth-20; nonionic; see Nikkol PBC-31; HLB 12.5; 100% conc.

Nikkol PBC-44(FF). [Nikko Chem. Co. Ltd.] PPG-8-ceteth-20; nonionic; see Nikkol PBC-31; solid; 100% conc.

Nikkol SCS. [Nikko Chem. Co. Ltd.] Sodium cetyl sulfate; anionic; detergent; sodium chloride-free; powd.; 100% conc.

Nikkol SGC-80N. [Nikko Chem. Co. Ltd.] Sodium cocomonoglyceride sulfate; anionic; foamer and detergent for shampoos, facial cleansers, soap; non-irritating to skin; powd.; 70% conc.

Nikkol SLS. [Nikko Chem. Co. Ltd.] Sodium lauryl sulfate; CAS 151-21-3; anionic; detergent; sodium chloride-free; powd.; 97% conc.

Nikkol SMT. [Nikko Chem. Co. Ltd.] Sodium N-stearoyl methyl taurate; anionic; shampoo base, foamer and detergent; solid; 92% conc.

Nikkol SO-10. [Nikko Chem. Co. Ltd.] Sorbitan oleate; CAS 1338-43-8; nonionic; lipophilic emulsifier; food emulsifier; liq.; HLB 5.0; 100% conc.

Nikkol SO-15. [Nikko Chem. Co. Ltd.] Sorbitan sesquioleate; CAS 8007-43-0; nonionic; lipophilic emulsifier; food emulsifier; liq.; HLB 4.5; 100% conc.

Nikkol SO-30. [Nikko Chem. Co. Ltd.] Sorbitan trioleate; CAS 26266-58-0; nonionic; lipophilic emulsifier; liq.; HLB 4.0; 100% conc.

Nikkol SS-10. [Nikko Chem. Co. Ltd.] Sorbitan stearate; CAS 1338-41-6; nonionic; lipophilic emulsifier; flake; HLB 4.7; 100% conc.

Nikkol SS-30. [Nikko Chem. Co. Ltd.] Sorbitan tristearate; CAS 26658-19-5; nonionic; lipophilic emulsifier; flake; HLB 2.1; 100% conc.

Nikkol TDP-2. [Nikko Chem. Co. Ltd.] C12-15 pareth-2 phosphate; anionic; emulsifier and solubilizer for cosmetics, drugs, agric. chemicals, dispersant, anticorrosive agent and detergent for general industrial use; liq.; HLB 7.0; 100% conc.

Nikkol TDP-4. [Nikko Chem. Co. Ltd.] Tri-PEG-4 alkyl ether phosphate; anionic; see Nikkol TDP-2; liq.; HLB 7.0; 100% conc.

Nikkol TDP-6. [Nikko Chem. Co. Ltd.] Tri-PEG-6 alkyl ether phosphate; anionic; see Nikkol TDP-2; liq.; HLB 8.0; 100% conc.

Nikkol TDP-8. [Nikko Chem. Co. Ltd.] Tri-PEG-8 alkyl ether phosphate; anionic; see Nikkol TDP-2; liq.; HLB 11.5; 100% conc.

Nikkol TDP-10. [Nikko Chem. Co. Ltd.] Tri-PEG-10 alkyl ether phosphate; anionic; see Nikkol TDP-2; semisolid; HLB 14.0; 100% conc.

Nikkol Tetraglyn 1-O. [Nikko Chem. Co. Ltd.] Polyglyceryl-4 oleate; nonionic; o/w emulsifier, anticrystallizing agent; food emulsifier; liq.; HLB 6.0; 100% conc.

Nikkol Tetraglyn 1-S. [Nikko Chem. Co. Ltd.] Polyglyceryl-4 stearate; nonionic; see Nikkol Tetraglyn 1-O; flake; HLB 6.0; 100% conc.

Nikkol Tetraglyn 3-S. [Nikko Chem. Co. Ltd.] Polyglyceryl-4 tristearate; nonionic; see Nikkol Tetraglyn 1-O; flake; 100% conc.

Nikkol Tetraglyn 5-O. [Nikko Chem. Co. Ltd.] Polyglyceryl-4 pentaoleate; nonionic; see Nikkol Tetraglyn 1-O; liq.; 100% conc.

Nikkol Tetraglyn 5-S. [Nikko Chem. Co. Ltd.] Polyglyceryl-4 pentastearate; nonionic; see Nikkol Tetraglyn 1-O; flake; 100% conc.

Nilfom 2X. [Am. Emulsions] Nonsilicone defoamer; nonionic; defoamer for atmospheric and pressure dyeing; compat. with all systems.

Nilfom DF-155. [Am. Emulsions] Nonsilicone defoamer; defoamer which does not contribute to flamm.; water-disp.

Nilfom DF-230. [Am. Emulsions] Org. waxes; nonsilicone defoamer designed to sink carpet.

Nilo VON. [Sandoz Prods. Ltd.] EO sulfonate; nonionic; emulsifier for min. oils, benzene, and petroleum; liq; 69% conc.

Nimco® 1780. [Henkel/Emery] Lanolin alcohols; nonionic; emulsifier, stabilizer, emollient for o/w systems, creams, lipsticks; solid; sol. in ethanol, IPM, castor oil; HLB 8.9; 100% conc.

Nimcolan® 1740. [Henkel/Emery] Absorp. base of lanolin esters, alcohols, and sterols; nonionic; emollient, w/o emulsifier, aux. o/w emulsifier; soft solid; 100% act.

Nimcolan® 1747. [Henkel/Emery] Petrolatum, lanolin and lanolin alcohol; nonionic; emollient w/o emulsifier and aux. o/w emulsifier; soft solid; 100% conc.

Ninate® 401. [Stepan; Stepan Canada; Stepan Europe] Calcium alkylbenzene sulfonate; anionic; emulsifier, dispersant used in pesticide formulations and in self-dispersing liq.; foaming agent; dk. amber visc. liq.; sol. in kerosene, xylene and aromatic naphtha, insol. in water and min. oil; sp.gr. 1.00; visc. 2500 cps; flash pt. 108 F (Seta CC); pH 4.5–5.5 (5% in 50% aq. IPA); 65% act.

Ninate® 401-A. [Stepan; Stepan Canada; Stepan Europe] Calcium alkylbenzene sulfonate; emulsifier for agric. formulations; liq.; sol. in xylene.

Ninate® 411. [Stepan; Stepan Canada] Amine dodecylbenzene sulfonate; anionic; emulsifier, solv. degreaser, drycleaning detergent, surf. tens. reducer, defoamer; emulsifier used in emulsifiable kerosene formulations, agric. formulations, textiles, metalworking; yel. visc. liq.; m.w. 385; sol. in kerosene, Stod., chlorinated and aromatic solvs., alcohol, pine oil, insol. in min. oil, forms clear sol'n. in water at higher concs.; dens. 8.5 lb/gal; surf. tens. 30 dynes/cm (0.1% aq.); pH 3.6 (20% in 1:1 IPA/water); 95% act.

Ninate® DS 70. [Stepan Europe] Sodium dioctyl sulfosuccinate; anionic; dispersant, wetting agent, emulsifier for agric. flowables, suspension and emulsifiable concs., gran. and powd. formulations; water-wh. to pale yel. liq.; 70% act.

Ninate® PA. [Stepan Europe] Sodium polycarboxylate; anionic; dispersant, fluidifying, stabilizing agent for agric. emulsions and suspension concs.; orange visc. liq.; 44% act.

Ninol® 11-CM. [Stepan; Stepan Canada] Cocamide DEA, modified; nonionic; detergent base for syn. cleaners; emulsifier, lubricant, antistat for textile applics.; emulsifier, corrosion inhibitor in cutting fluids, drawing compds., metal cleaning; lt. color liq.; HLB 14.5; 100% act.

Ninol® 30-LL. [Stepan; Stepan Canada] Lauramide DEA; CAS 120-40-1; nonionic; foam booster/stabilizer, visc. builder/modifier for liq. detergents, shampoos, hand soaps, bath prods.; amber clear liq.; 100% act.

Ninol® 40-CO. [Stepan; Stepan Canada] Cocamide DEA; nonionic; foam booster/stabilizer, visc. booster for liq. detergents, textile applics.; lt. color liq.; 100% act.

Ninol® 49-CE (formerly Onyxol SD). [Stepan; Stepan Canada] Cocamide DEA; nonionic; detergent, foam booster and stabilizer, thickener in detergent formulations; lt. color liq.; 100% act.

Ninol® 55-LL. [Stepan; Stepan Canada] Lauramide DEA (1:1); CAS 120-40-1; nonionic; surfactant for improved visc. in AOS systems; Gardner 3 liq.; 100% conc.

Ninol® 70-SL. [Stepan; Stepan Canada] Lauramide DEA; CAS 120-40-1; nonionic; foam stabilizer, thickener; gel; 100% conc.

Ninol® 96-SL. [Stepan; Stepan Canada] Lauramide DEA; CAS 120-40-1; nonionic; thickener, foam stabilizer and booster for liq. detergents; lt. color wax; 100% act.

Ninol® 201. [Stepan; Stepan Canada] Oleamide DEA; nonionic; emulsifier, corrosion inhibitor in industrial lubricant systems, cutting fluids, drawing compds., metal cleaners; thickener for personal care and liq. detergent prods.; emulsifier, lubricant, antistat for textiles; amber liq.; sol. in oils, disp. in water; dens. 8.23 lb/gal; pH 9.5–10.5; 100% act.

Ninol® 1281. [Stepan; Stepan Canada; Stepan Europe] Fatty acid alkylolamide; nonionic; detergent used in floor cleaners and all purpose detergent formulations; emulsifier, corrosion inhibitor in cutting fluids, drawing compds., metal cleaning; lubricant for surf. treatments; amber visc. liq.; dens. 8.5 lb/gal; pH 9.0 (1%); 100% act.

Ninol® 1285. [Stepan; Stepan Canada] Modified fatty alkanolamide; detergent base for syn. cleaners, high alkali strippers and degreasers; lt. straw visc liq.; dens. 8.35 lb/gal; pH 9.0 (1%); 100% act.

Ninol® 1301. [Stepan; Stepan Canada] Fatty acid alkylolamide derived from coconut oil; nonionic; detergent for germicidal floor cleaners, wax strippers, alkali cleaners, sanitizers, textile applics.; emulsifier, corrosion inhibitor in cutting fluids, drawing compds., metal cleaning; lt. tan soft wax; sol. in hard water; congeal pt. 110–120 F; pH 9.5–10.5; 100% act.

Ninol® 4821 F. [Stepan Europe] Cocamide DEA; nonionic; thickener, foam booster/stabilizer, emollient for household/industrial cleaners; yel. visc. liq.; 98% act.

Ninol® 5024. [Stepan; Stepan Canada] Mixed fatty acid DEA; nonionic; detergent, visc. builder for general purpose and hard surf. cleaners; highly salt tolerant; dk. amber liq.; 100% act.

Ninol® A-10MM. [Stepan; Stepan Canada] Modified alkanolamide; nonionic; detergent for hard surf. cleaners; self-coupling; develops high visc.; liq.; 100% conc.

Ninol® AX. [Stepan Canada] Lauramide DEA; CAS 120-40-1; surfactant; 100% act.

Ninol® B. [Stepan Canada] Amide blend; nonionic; detergent; base for mfg. of floor and general purpose cleaners; liq.; 100% act.

Ninol® CMP. [Stepan; Stepan Canada] Cocamide MEA; CAS 68140-00-1; foam booster, visc. builder for liq. detergents, detergent blocks or bars; wh. beads; 100% act.

Ninol® CX. [Stepan Canada] Alkanolamide; nonionic; foam and visc. booster; liq.; 100% conc.

Ninol® LDL 2. [Stepan Europe] Modified alkylolamide; nonionic; thickener, foam booster/stabilizer, detergent for liq. soaps, toilet deodorizing blocks, surf. treatments; orange-yel. visc. liq.; 96% act.

Ninol® LMP. [Stepan; Stepan Canada; Stepan Europe] Lauramide MEA; CAS 142-78-9; nonionic; foam booster/stabilizer, thickener, emollient, detergent for dishwash, liq. detergents, detergent blocks or bars; wh. beads; 100% act.

Ninol® SR-100. [Stepan; Stepan Canada] Coconut/oleic DEA; nonionic; detergent, emulsifier, low-foaming visc. builder for hard-surf. cleaners; emulsifier, corrosion inhibitor in cutting fluids, drawing compds., metal cleaning; liq.; 100% act.

Ninox® FCA. [Stepan Europe] Cocamidopropylamine oxide; nonionic; thickener, foam booster/stabilizer, detergent, antistat for scale-removing liqs., liq. soaps, cleaning foams, personal care prods.; water-wh. to pale yel. liq.; 33% act.

Ninox® L. [Stepan Europe] Lauramine oxide; nonionic; thickener, foam booster/stabilizer, detergent, antistat for scale-removing liqs., liq. soaps, cleaning foams; water-wh. to pale yel. liq.; 31% act.

Ninox® M. [Stepan Europe] Myristamine oxide; nonionic; thickener, foam booster/stabilizer, detergent, antistat for scale-removing liqs., liq. soaps, cleaning foams; water-wh. to pale yel. liq.; 30% act.

Ninox® SO. [Stepan Europe] Stearamine oxide; nonionic; thickener, emollient, mild detergent for scouring pastes or creams; additive for hypochlorite; wh. paste; 24% act.

Niox AK-40. [Pulcra SA] PEG-200 oleate; CAS 9004-96-0; spreading agent, antifoamer, kerosene and min. oil emulsifier, dispersant; water-disp.

Niox AK-44. [Pulcra SA] PEG-400 oleate; CAS 9004-96-0; lubricant, min. and veg. oil emulsifier, dispersant; water-disp.; sol. in min. oil.

Niox EO-12. [Pulcra SA] Linear fatty alcohol, ethoxylated sat.; nonionic; surface act. agent for detergents; dispersant; soap additive; flakes; HLB 16.0; 100% conc.

Niox EO-13. [Pulcra SA] Linear fatty alcohol, ethoxylated sat.; nonionic; powd. detergent mixtures; flakes; HLB 18.6; 100% conc.

Niox EO-14, 23. [Pulcra SA] Linear fatty alcohol ethoxylated sat.; nonionic; foam controller in detergents; dispersant; flakes; HLB 17.8 and 17.4 resp.; 100% conc.

Niox EO-32, -35. [Pulcra SA] Linear fatty alcohol ethoxylated sat.; nonionic; surface act. agent for detergents; dispersant; soap additive; flakes; HLB 16.2 and 15.5 resp.; 100% conc.

Niox KF-12. [Pulcra SA] Linear fatty alcohol ethoxylated sat.; nonionic; detergent, wetting agent, dispersant; emulsifier for waxes, oils, and fats; stabilizer for syn. resins; solid; HLB 14.5; 100% conc.

Niox KF-13. [Pulcra SA] Linear fatty alcohol ethoxylate; nonionic; emulsifier for min. oils, aliphatic solvs.; visc. depressant; intermediate for sulfonation; liq.; HLB 7.6; 100% conc.

Niox KF-17. [Pulcra SA] Linear fatty alcohol ethoxylate; nonionic; emulsifier for min. oils; additive for detergents; intermediate for sulfonation; liq.; oil-sol.; HLB 6.1; 100% conc.

Niox KF-18. [Pulcra SA] Linear fatty alcohol ethoxylate; nonionic; wetting agent and detergent for liq. formulations; wax emulsifier; paste; oil-sol.; HLB 13.3; 100% conc.

Niox KF-22. [Pulcra SA] Linear fatty alcohol ethoxylate; nonionic; emulsifier for silicones and solvs.; assistant in dry cleaning formulations; liq.; HLB 9.4; 100% conc.

Niox KF-25. [Pulcra SA] Linear fatty alcohol ethoxylate; nonionic; detergent and emulsifier for oils, waxes and solvs.; wool scouring and wetting agent; liq., paste; HLB 11.4; 100% conc.

Niox KF-26. [Pulcra SA] Linear fatty alcohol ethoxylate; nonionic; emulsifier for min. oils; additive for detergents; intermediate for sulfonation;

liq.; HLB 6.8; 100% conc.

Niox KG-11. [Pulcra SA] Linear fatty alcohol ethoxylate; nonionic; basic prod. for sulfates; visc. depressant for plastisols; emulsifier for cosmetics; liq.; HLB 5.0; 100% conc.

Niox KG-14. [Pulcra SA] Linear fatty alcohol ethoxylate; nonionic; wetting agent for detergents, textile liq. formulations.

Niox KH Series. [Pulcra SA] Fatty acid, ethoxylated; nonionic; raw material used as emulsifier and coemulsifier for paraffinic waxes and compds., dyeing assistant, lubricant, antistat; emulsifier for the textile industry, min. oils, defoamer, softener, detergent and solubilizer for all industries; 100% conc.

Niox KI Series. [Pulcra SA] Ethoxylated castor oil; nonionic; surface act. agent used as detergent, emulsifier, dispersant, and hydrosolubilizer for conc. pesticides in the metal, leather, cosmetics, toiletries, pharmaceutical and polymer industries; textile assistant; 100% conc.

Niox KJ-55. [Pulcra SA] Linear fatty alcohol, ethoxylated sat.; nonionic; surface act. agent for detergents; soap additive; dispersant; solid; HLB 13.2; 100% conc.

Niox KJ-56. [Pulcra SA] Linear fatty alcohol ethoxylated sat.; nonionic; emulsifier intermediate raw material for mfg. of ethoxysulfates used in detergent and industrial specialties; solid; HLB 8.1; 100% conc.

Niox KJ-61. [Pulcra SA] Linear fatty alcohol ethoxylated sat.; nonionic; surface act. agent for detergents; soap additive; dispersant; solid; HLB 14.1; 100% conc.

Niox KJ-66. [Pulcra SA] Linear fatty alcohol ethoxylated sat.; nonionic; emulsifier intermediate raw material for mfg. of ethoxysulfates used in detergent and industrial specialties; solid; HLB 10.2; 100% conc.

Niox KL-16. [Pulcra SA] Linear fatty alcohol ethoxylated sat.; nonionic; scouring agent for the textile industry; soaking agent for the leather industry; household and industrial detergent; paste; HLB 12.8; 100% conc.

Niox KL-19. [Pulcra SA] Linear fatty alcohol ethoxylated sat.; nonionic; emulsifier for min. oils; additive for detergents; intermediate for sulfonation; liq.; HLB 8.0; 100% conc.

Niox KP-62, -63, -67. [Pulcra SA] Isotridecanol, ethoxylated; nonionic; emulsifier for cosmetic and pharmaceutical preparations; wetting agent for industrial applics.; liq.; HLB 11.7, 10.4, and 8.0 resp.; 100% conc.

Niox KP-68. [Pulcra SA] Isotridecanol, ethoxylated; nonionic; detergent, foamer, solubilizer; scouring and leveling agent for the textile industry; paste; HLB 14.1; 100% conc.

Niox KP-69. [Pulcra SA] Isotridecanol, ethoxylated; nonionic; detergent, foamer, solubilizer used in foaming detergents; emulsifier and wetting agent for textile fiber finishing; liq.; HLB 12.7; 100% conc.

Niox KQ-20. [Pulcra SA] Fatty oxo-alcohol, ethoxylated; nonionic; aux. for wool scouring; emulsifier for hydrocarbon solv.; wetting and dispersant for detergents; paste; HLB 13.0; 100% conc.

Niox KQ-30. [Pulcra SA] Fatty oxo-alcohol, ethoxylated; nonionic; emulsifier for min. oils; additive for detergents; intermediate for sulfonation; liq.; HLB 7.8; 100% conc.

Niox KQ-32. [Pulcra SA] Fatty oxo-alcohol, ethoxylated; nonionic; aux. for wool scouring; emulsifier for hydrocarbon solv.; wetting and dispersant for detergents; paste; HLB 12.0; 100% conc.

Niox KQ-33. [Pulcra SA] Fatty oxo-alcohol, ethoxylated; nonionic; emulsifier for silicones and hydrocarbon solvs.; dry cleaning assistant; intermediate raw material for sulfonation; liq.; HLB 9.2; 100% conc.

Niox KQ-34. [Pulcra SA] Oxo-alcohol, ethoxylated; nonionic; wetting agent; emulsifier for fats and oleins; aux. prod. for emulsion polymerization; solid; HLB 15.4; 100% conc.

Niox KQ-36. [Pulcra SA] Fatty oxo-alcohol, ethoxylated; nonionic; aux. for wool scouring; emulsifier for hydrocarbon solv.; wetting and dispersant for detergents; paste; HLB 11.0; 100% conc.

Niox KQ-51. [Pulcra SA] Fatty oxo-alcohol, ethoxylated; nonionic; w/o emulsifier; coemulsifier of min. oils, waxes and paraffinic hydrocarbons; liq.; HLB 11.6; 100% conc.

Niox KQ-55, -56. [Pulcra SA] Fatty oxo-alcohol, ethoxylated; nonionic; o/w emulsifier for hydrocarbon solvs. and min. oils; detergent; wetting agent and antistat for textile industry; assistant for wool scouring and specialty cleaners; liq., paste resp.; water-sol.; HLB 12.8 and 13.7 resp.; 100% conc.

Niox KQ-70, LQ-13. [Pulcra SA] Fatty oxo-alcohol, ethoxylated; nonionic; aux. for wool scouring; emulsifier for hydrocarbon solv.; wetting and dispersant for detergents; paste, liq. resp.; 100 and 88% conc. resp.

Nipol®2782. [Stepan; Stepan Canada] Block polymer; emulsifier for agric. formulations; tan solid; sol. in xylene, water; HLB 14.0.

Nipol®4472. [Stepan; Stepan Canada] Block polymer; emulsifier for agric. formulations; tan solid; sol. in xylene, water; HLB 12.5.

Nipol®5595. [Stepan; Stepan Canada] Block polymer; emulsifier for agric. formulations; tan solid; sol. in xylene, water; HLB 15.0.

Nissan Amine 2-OLR. [Nippon Oils & Fats] Cationic; intermediate for cationic surfactants; anticorrosion agent for metals; solid; 100% conc.

Nissan Amine AB. [Nippon Oils & Fats] Stearyl amine; intermediate for cationic surfactants; anticorrosive agent, germicide, wetting agent, mold release agent, softener, emulsifier, dispersant, intermediate used in textiles, water treatment, concrete, asphalt, agriculture, ceramics; wh. waxy solid or flake; solid. pt. 47–53 C; 98% min. primary amine.

Nissan Amine ABT. [Nippon Oils & Fats] Primary hydrog. tallow amine; intermediate for cationic surfactants; anticorrosive agent, germicide, wetting agent, mold release agent, softener, emulsifier, dispersant, intermediate used in textiles, water treatment, concrete, asphalt, agriculture, ceramics; wh. waxy solid or flake; solid. pt. 40–46 C; 98% min. primary amine.

Nissan Amine BB. [Nippon Oils & Fats] Primary dodecylamine; intermediate for cationic surfactants; anticorrosive agent, germicide, wetting agent, mold release agent, softener, emulsifier, dispersant, intermediate used in textiles, water treatment, concrete, asphalt, agriculture, ceramics; APHA 120 max. liq. in summer, solid in winter; solid. pt. 24–28 C; 98% min. primary amine.

Nissan Amine DT. [Nippon Oils & Fats] Tallow-alkyl propylenediamine; asphalt emulsifier, corrosion inhibitor, wetting agent, dispersant used in water treatment, pigment flushing and dispersing, ore flotation, gasoline, grease, fuel oil additive; yel. br. solid; solid. pt. 25–34 C.

Nissan Amine DTH. [Nippon Oils & Fats] Tallow-hydrog.-alkyl propylenediamine; asphalt emulsifier, corrosion inhibitor, wetting agent, dispersant used in water treatment, pigment flushing and dispersing, ore flotation, gasoline, grease, fuel oil additive; ylsh. br. flake.

Nissan Amine FB. [Nippon Oils & Fats] Primary cocoalkylamine; intermediate for cationic surfactants; anticorrosive agent, germicide, wetting agent, mold release agent, softener, emulsifier, dispersant, intermediate used in textiles, water treatment, concrete, asphalt, agriculture, ceramics; APHA 120 max., liq. in summer, solid in winter; solid. pt. 16–22 C; 98% min. primary amine.

Nissan Amine MB. [Nippon Oils & Fats] Primary tetradecylamine; intermediate for cationic surfactants; anticorrosive agent, germicide, wetting agent, mold release agent, softener, emulsifier, dispersant, intermediate used in textiles, water treatment, concrete, asphalt, agriculture, ceramics; wh. waxy solid; solid. pt. 31–35 C; 98% min. primary amine.

Nissan Amine OB. [Nippon Oils & Fats] Primary oleylamine; CAS 112-90-3; intermediate for cationic surfactants; anticorrosive agent, germicide, wetting agent, mold release agent, softener, emulsifier, dispersant, intermediate used in textiles, water treatment, concrete, asphalt, agriculture, ceramics; APHA 140 max liq. in summer, solid in winter; solid. pt. 30 C max.; 98% min. primary amine.

Nissan Amine PB. [Nippon Oils & Fats] Primary hexadecylamine; intermediate for cationic surfactants; anticorrosive agent, germicide, wetting agent, mold release agent, softener, emulsifier, dispersant, intermediate used in textiles, water treatment, concrete, asphalt, agriculture, ceramics; wh. waxy solid; solid. pt. 38–46 C; 98% min. primary amine.

Nissan Amine SB. [Nippon Oils & Fats] Primary soybean alkylamine; CAS 61790-18-9; intermediate for cationic surfactants; anticorrosive agent, germicide, wetting agent, mold release agent, softener, emulsifier, dispersant, intermediate used in textiles, water treatment, concrete, asphalt, agriculture, ceramics; liq., solid; 100% conc.

Nissan Amine VB. [Nippon Oils & Fats] Behenylamine; intermediate for cationic surfactants; anticorrosive agent, germicide, wetting agent, mold release agent, softener, emulsifier, dispersant, intermediate used in textiles, water treatment, concrete, asphalt, agriculture, ceramics; solid; 100% conc.

Nissan Anon BF. [Nippon Oils & Fats] Dimethyl cocoalkyl betaine; amphoteric; plastic antistatic agent, softener, germicide in foods and cleaning industries, asphalt antistripping agent, textile treatment; lt. yel. liq.; 25% min act.

Nissan Asphasol 10. [Nippon Oils & Fats] Alkyl diamine; cationic; emulsifier for asphalt; flake; 100% conc.

Nissan Asphasol 20. [Nippon Oils & Fats] Alkyl diamine; cationic; asphalt emulsifier for penetrating and mixing processes; yel.-br. semisolid, reddish-br. liq.; sol. in aq. sol'n. of org. acid, insol. in water; 100% conc.

Nissan Cation AB. [Nippon Oils & Fats] Octadecyl trimethyl ammonium chloride; cationic; germicide in water treatment, petrol., paper, foods and textile industries, antistat in plastics, pulp and paper industries, rinse agent, pigment coating agent, dispersant, coagulant, softener; lt. yel. liq.; 23% min. act.

Nissan Cation ABT-350, 500. [Nippon Oils & Fats] Tallow-hydrog. alkyl trimethyl ammonium chloride; cationic; intermediate for cationic surfactants; anticorrosive agent, germicide, wetting agent, mold release agent, softener, emulsifier, dispersant, intermediate used in textiles, water treatment, concrete, asphalt, agriculture, ceramics; lt. yel. liq.; 33 and 50% act.

Nissan Cation AR-4. [Nippon Oils & Fats] Alkyl imidazoline chloride; cationic; softener, rinse agent; yel.-br. liq.; 35% min. act.

Nissan Cation BB. [Nippon Oils & Fats] Dodecyl trimethyl ammonium chloride; cationic; softener for textiles; detergent sanitizer; lt. yel. liq.; 30% min. act.

Nissan Cation F2-10R, -20R, -40E, -50. [Nippon Oils & Fats] Coco-alkyl dimethyl benzyl ammonium chloride; cationic; intermediate for cationic surfactants; anticorrosive agent, germicide, wetting agent, mold release agent, softener, emulsifier, dispersant, intermediate used in textiles, water treatment, concrete, asphalt, agriculture, ceramics; colorless to lt. yel. liq.; 9.5, 19–21, 40–44, and 50% act. resp.

Nissan Cation FB, FB-500. [Nippon Oils & Fats] Coco-alkyl trimethyl ammonium chloride; cationic; intermediate for cationic surfactants; anticorrosive agent, germicide, wetting agent, mold release agent, softener, emulsifier, dispersant, intermediate used in textiles, water treatment, concrete, asphalt, agriculture, ceramics; lt. yel. liq.; 30 and 50% min act.

Nissan Cation L-207. [Nippon Oils & Fats] POE dodecyl monomethyl-ammonium chloride; cationic; intermediate for cationic surfactants; anticorrosive agent, germicide, wetting agent, mold release agent, softener, emulsifier, dispersant, intermediate used in textiles, water treatment, concrete, asphalt, agriculture, ceramics; ylsh.-br., visc. liq.; 90% min. act.

Nissan Cation M2-100. [Nippon Oils & Fats] Tetradecyl dimethyl benzyl ammonium chloride; cationic; intermediate for cationic surfactants; anticorrosive agent, germicide, wetting agent, mold release agent, softener, emulsifier, dispersant, intermediate used in textiles, water treatment, concrete, asphalt, agriculture, ceramics; wh. or lt. yel. cryst. powd.; 85% min. act.

Nissan Cation PB-40, -300. [Nippon Oils & Fats] Hexadecyl trimethyl ammonium chloride; cationic; intermediate for cationic surfactants; anticorrosive agent, germicide, wetting agent, mold release agent, softener, emulsifier, dispersant, intermediate used in textiles, water treatment, concrete, asphalt, agriculture, ceramics; lt. yel. liq.; 40 and 27% act. resp.

Nissan Cation S2-100. [Nippon Oils & Fats] Octadecyl dimethyl benzyl ammonium chloride; cationic; intermediate for cationic surfactants; anticorrosive agent, germicide, wetting agent, mold release agent, softener, emulsifier, dispersant, intermediate used in textiles, water treatment, concrete, asphalt, agriculture, ceramics; wh. or lt. yel. cryst. powd.; 85% min. act.

Nissan Chloropearl. [Nippon Oils & Fats] Nonionic/cationic blend; nonionic/cationic; drycleaning soap for 1,1,1-trichloroethane type solv.; liq.

Nissan Diapon K. [Nippon Oils & Fats] Sodium methyl coco taurate; anionic; detergent for shampoo, household, and industrial applics.; paste; 28% conc.

Nissan Diapon T. [Nippon Oils & Fats] Sodium N-methyl tallow taurate; anionic; dyeing aux. for hair dye detergent, emulsifier; powd.; 30% conc.

Nissan Diapon TO. [Nippon Oils & Fats] Sodium N-methyl oleoyl taurate; anionic; detergent, emulsifier; scouring agent for wool; dyeing aux.; powd.; 30% conc.

Nissan Disfoam C Series. [Nippon Oils & Fats] Polyalkylene glycol derivs.; defoamers for fermentation field, pulp, synthetics, etc.; liq.

Nissan Dispanol 16A. [Nippon Oils & Fats] POE alkyl ether; nonionic; emulsifier for paraffin; APHA 120 max. liq.; sol. in methanol, xylene, kerosene, ethyl ether, disp. in water; HLB 10.7; cloud pt. 0 C max. (1% aq.); 100% conc.

Nissan Dispanol LS-100. [Nippon Oils & Fats] Polyoxyalkylene alkylaryl ether; nonionic; low-foam detergent for syn. fibers; APHA 50 max. liq.; sol. in methanol, ethyl ether, water, disp. in xylene; cloud pt. 23–29 C (1% aq.); 100% conc.

Nissan Dispanol N-100. [Nippon Oils & Fats] POE alkylaryl ether; nonionic; detergent for animal, cellulosic, and syn. fibers; APHA 120 max. liq.; sol. in water, methanol, ethyl ether, xylene; HLB 12.5; cloud pt. 33–41 C (1% aq.); 100% conc.

Nissan Dispanol TOC. [Nippon Oils & Fats] POE alkyl ether; nonionic; detergent, wetting agent, emulsifier for wool fiber; APHA 120 max. liq.; sol. in water, methanol, ethyl ether, disp. in xylene; HLB 12.0; cloud pt. 45–53 C (1% aq.).; 100% conc.

Nissan New Royal P. [Nippon Oils & Fats] Blend; cationic/nonionic; drycleaning soap for perchlorethylene type solv.; liq.

Nissan Newrex. [Nippon Oils & Fats] Alkylbenzene sulfonate; anionic; general detergent; paste, powd.; 30-60% conc.

Nissan Nonion DS-60HN. [Nippon Oils & Fats] PEG distearate; nonionic; emulsifier, thickener used in cosmetics, textile, industrial uses; solid; sol. in water, methanol, warm in diethylene glycol; m.p. 54–62 C; HLB 19; acid no. 2 max.; 100% conc.

Nissan Nonion E-205. [Nippon Oils & Fats] POE oleyl ether; nonionic; emulsifier, dispersant, detergent, wetting agent used in textile processing, cosmetics, metalworking, agric. preparations, industrial cleaners; APHA 140 max. liq.; sol. in xylene, kerosene, methanol, disp. in water; HLB 9.0; cloud pt. 0 C (1% aq.); 100% conc.

Nissan Nonion E-206. [Nippon Oils & Fats] POE oleyl ether; nonionic; emulsifier, dispersant, detergent, wetting agent used in textile processing, cosmetics, metalworking, agric. preparations, industrial cleaners; liq.; HLB 9.9; 100% conc.

Nissan Nonion E-208. [Nippon Oils & Fats] POE oleyl ether; nonionic; emulsifier, dispersant, detergent, wetting agent used in textile processing, cosmetics, metalworking, agric. preparations, industrial cleaners; liq.; 100% conc.

Nissan Nonion E-215. [Nippon Oils & Fats] POE oleyl ether; nonionic; emulsifier, dispersant, detergent, wetting agent used in textile processing, cosmetics, metalworking, agric. preparations, industrial cleaners; APHA 140 max. semisolid; sol. in water, methanol, cloudy in xylene; HLB 14.2; cloud pt. 95 C (1% aq.); 100% conc.

Nissan Nonion E-220. [Nippon Oils & Fats] POE oleyl ether; nonionic; emulsifier, dispersant, detergent, wetting agent used in textile processing, cosmetics, metalworking, agric. preparations, industrial cleaners; solid; HLB 15.3; 100% conc.

Nissan Nonion E-230. [Nippon Oils & Fats] POE oleyl ether; nonionic; emulsifier, dispersant, detergent, wetting agent used in textile processing, cosmetics, metalworking, agric. preparations, industrial cleaners; APHA 140 max. solid; sol. in water, methanol, cloudy in xylene; HLB 16.6; cloud pt. 100 C resp. (1% aq.); 100% conc.

Nissan Nonion HS-204.5. [Nippon Oils & Fats] POE octylphenyl ether; nonionic; detergent, emulsifier for household and industrial cleaners and detergents, textile processing, paints, printing inks, agric., latex polymerization; APHA 120 liq.; sol. in xylene, kerosene, ether, methanol; disp. in water; HLB 9.8; cloud pt. 0 C (1% aq.); 100% conc.

Nissan Nonion HS-206. [Nippon Oils & Fats] POE octylphenyl ether; nonionic; detergent, emulsifier for household and industrial cleaners and detergents, textile processing, paints, printing inks, agric., latex polymerization; APHA 120 liq.; sol. in xylene, kerosene, ether, methanol; disp. in water; HLB 11.2; 100% conc.

Nissan Nonion HS-208. [Nippon Oils & Fats] POE octylphenyl ether; nonionic; detergent, emulsifier for household and industrial cleaners and detergents, textile processing, paints, printing inks, agric., latex polymerization; APHA 120 liq.; sol. in xylene, ether, methanol; disp. in water; HLB 12.6; 100% conc.

Nissan Nonion HS-210. [Nippon Oils & Fats] POE octylphenyl ether; nonionic; detergent, emulsifier for household and industrial cleaners and detergents, textile processing, paints, printing inks, agric., latex polymerization; APHA 120 liq.; sol. in water, xylene, ether, methanol; HLB 13.3; cloud pt. 70–78 C (1% aq.); 100% conc.

Nissan Nonion HS-215. [Nippon Oils & Fats] POE octylphenyl ether; nonionic; detergent, emulsifier for household and industrial cleaners and detergents, textile processing, paints, printing inks, agric., latex polymerization; APHA 120 liq.; sol. in water, xylene, ether, methanol; HLB 15.0; cloud pt. 91–99 C (1% aq.); 100% conc.

Nissan Nonion HS-220. [Nippon Oils & Fats] POE octylphenyl ether; nonionic; detergent, emulsifier for household and industrial cleaners and detergents, textile processing, paints, printing inks, agric., latex polymerization; APHA 120 liq.; sol. in water, methanol, cloudy in xylene, ether; HLB 16.2; cloud pt. 100 C min. (1% aq.); 100% conc.

Nissan Nonion HS-240. [Nippon Oils & Fats] PEG-40 octylphenyl ether; CAS 9002-93-1; nonionic; detergent, emulsifier for household and industrial cleaners and detergents, textile processing, paints, printing inks, agric., latex polymerization; APHA 120 liq.; sol. in water, methanol; HLB 17.9; cloud pt. 100 C min. (1% aq.); 100% conc.

Nissan Nonion HS-270. [Nippon Oils & Fats] PEG-70 octylphenol ether; CAS 9002-93-1; nonionic; detergent, emulsifier for household and industrial cleaners and detergents, textile processing, paints, printing inks, agric., latex polymerization; semisolid; HLB 18.7; 100% conc.

Nissan Nonion K-202. [Nippon Oils & Fats] POE lauryl ether; nonionic; emulsifier, dispersant, detergent, wetting agent used in textile processing, cosmetics, metalworking, agric. preparations, industrial cleaners; APHA 200 max. liq.; sol. in xylene, ether, methanol; disp. in kerosene; HLB 6.0; cloud pt. 0 C max. (1% aq.); 100% conc.

Nissan Nonion K-203. [Nippon Oils & Fats] POE lauryl ether; nonionic; emulsifier, dispersant, detergent, wetting agent used in textile processing, cosmetics, metalworking, agric. preparations, industrial cleaners; APHA 200 max. liq.; sol. in xylene, ether, methanol; disp. in kerosene; HLB 8.0; cloud pt. 0 C max. (1% aq.); 100% conc.

Nissan Nonion K-204. [Nippon Oils & Fats] POE lauryl ether; nonionic; emulsifier, dispersant, detergent, wetting agent used in textile processing, cosmetics, metalworking, agric. preparations, industrial cleaners; APHA 200 max. semisolid; sol. in xylene, ether, methanol; sol. warm in kerosene; disp. in water; HLB 9.2; cloud pt. 0 C max. (1% aq.); 100% conc.

Nissan Nonion K-207. [Nippon Oils & Fats] POE lauryl ether; nonionic; emulsifier, dispersant, detergent, wetting agent used in textile processing, cosmetics, metalworking, agric. preparations, industrial cleaners; APHA 50 max. semisolid; sol. in water, xylene, methanol; sol. warm in ether; HLB 12.1; cloud pt. 55–63 C max. (1% aq.); 100% conc.

Nissan Nonion K-211. [Nippon Oils & Fats] POE lauryl ether; nonionic; emulsifier, dispersant, detergent, wetting agent used in textile processing, cosmetics, metalworking, agric. preparations, industrial cleaners; APHA 50 max. semisolid; sol. in water, xylene, methanol; HLB 14.1; cloud pt. 90–98 C max. (1% aq.); 100% conc.

Nissan Nonion K-215. [Nippon Oils & Fats] POE lauryl ether; nonionic; emulsifier, dispersant, detergent, wetting agent used in textile processing, cosmetics, metalworking, agric. preparations, industrial cleaners; APHA 140 max. semisolid; sol. in water, xylene, methanol; HLB 15.2; cloud pt. 100 C min. (1% aq.); 100% conc.

Nissan Nonion K-220. [Nippon Oils & Fats] POE lauryl ether; nonionic; emulsifier, dispersant, detergent, wetting agent used in textile processing, cosmetics, metalworking, agric. preparations, industrial cleaners; APHA 120 max. solid; sol. in water, xylene, methanol; HLB 16.2; cloud pt. 100 C min. (1% aq.); 100% conc.

Nissan Nonion K-230. [Nippon Oils & Fats] POE lauryl ether; nonionic; emulsifier, dispersant, detergent, wetting agent used in textile processing, cosmetics, metalworking, agric. preparations, industrial cleaners; APHA 120 max. semisolid; sol. in water, xylene, methanol; HLB 17.3; cloud pt. 100 C min. (1% aq.); 100% conc.

Nissan Nonion L-2. [Nippon Oils & Fats] PEG laurate; nonionic; general emulsifier for textile, leather, industrial uses; Gardner 4 max. liq.; sol. in methanol, sol. warm in kerosene, disp. in water; HLB 9.5; 100% conc.

Nissan Nonion L-4. [Nippon Oils & Fats] PEG laurate; nonionic; general emulsifier for textile, leather, industrial uses; Gardner 4 max. liq.; sol. in water, ether, methanol; HLB 13.1; 100% conc.

Nissan Nonion LP-20R, LP-20RS. [Nippon Oils & Fats] Sorbitan laurate; CAS 1338-39-2; nonionic; emulsifier for cosmetic, pharmaceutical and food applics., o/w emulsion stabilizer and thickener, fiber lubricant and softener; Gardner 5 max. oily liq.; sol. in methanol, ethanol, acetone, xylene, ethyl ether, kerosene, disp. in water; HLB 8.6; 100% conc.

Nissan Nonion LT-221. [Nippon Oils & Fats] Polysorbate 20; CAS 9005-64-5; nonionic; emulsifier for cosmetic, pharmaceutical and food applics.; Gardner 6 max. oily liq.; sol. in water, methanol, ethanol, acetone, xylene, ethyl ether, ethylene glycol, HLB 16.7; 100% conc.

Nissan Nonion NS-202. [Nippon Oils & Fats] POE nonylphenyl ether; nonionic; detergent, emulsifier for household and industrial cleaners, textile processing, paints, printing inks, agric. preparations, and latex polymerization; APHA 120 max. liq.; sol. in xylene, kerosene, ether, methanol; disp. in water; HLB 5.7; cloud pt. 0 C (1% aq.); 100% conc.

Nissan Nonion NS-204.5. [Nippon Oils & Fats] POE nonylphenyl ether; nonionic; see Nissan Nonion NS-202; APHA 120 max. liq.; sol. in xylene, kerosene, ether, methanol; disp. in water; HLB 9.5; cloud pt. 0 C max. (1% aq.); 100% conc.

Nissan Nonion NS-206. [Nippon Oils & Fats] PEG-6 nonylphenyl ether; CAS 9016-45-9; nonionic; see Nissan Nonion NS-202; APHA 120 max. liq.; sol. in xylene, ether, methanol; disp. in water; HLB 10.9; cloud pt. 0 C max. (1% aq.); 100% conc.

Nissan Nonion NS-208.5. [Nippon Oils & Fats] POE nonylphenyl ether; nonionic; see Nissan Nonion NS-202; APHA 120 max. liq.; sol. in xylene, ether, methanol; disp. in water; HLB 12.6; cloud pt. 33–41 C (1% aq.); 100% conc.

Nissan Nonion NS-209. [Nippon Oils & Fats] POE nonylphenyl ether; nonionic; see Nissan Nonion NS-202; liq.; HLB 12.9; 100% conc.

Nissan Nonion NS-210. [Nippon Oils & Fats] POE nonylphenyl ether; nonionic; see Nissan Nonion NS-202; APHA 120 max. liq.; sol. in water, xylene, ether, methanol; HLB 13.3; cloud pt. 46–54 C (1% aq.); 100% conc.

Nissan Nonion NS-212. [Nippon Oils & Fats] PEG-12 nonylphenyl ether; CAS 9016-45-9; nonionic; see Nissan Nonion NS-202; APHA 120 max. liq.; sol. in water, xylene, ether, methanol; HLB 14.1; cloud pt. 73–84 C (1% aq.); 100% conc.

Nissan Nonion NS-215. [Nippon Oils & Fats] POE nonylphenyl ether; nonionic; see Nissan Nonion NS-202; APHA 120 max. liq.; sol. in water, xylene, ether, methanol; HLB 15; cloud pt. 89–97 C (1% aq.); 100% conc.

Nissan Nonion NS-220. [Nippon Oils & Fats] POE nonylphenyl ether; nonionic; see Nissan Nonion NS-202; APHA 120 max. semisolid; sol. in water, xylene, ether, methanol; HLB 16; cloud pt. 100 C min. (1% aq.); 100% conc.

Nissan Nonion NS-230. [Nippon Oils & Fats] POE nonylphenyl ether; nonionic; see Nissan Nonion NS-202; APHA 120 max. solid; sol. in water, xylene, methanol, cloudy in ether; HLB 17.1; cloud pt. 100 C min. (1% aq.); 100% conc.

Nissan Nonion NS-240. [Nippon Oils & Fats] POE nonylphenyl ether; nonionic; see Nissan Nonion NS-202; APHA 120 max. solid; sol. in water, methanol; HLB 17.8; cloud pt. 100 C min. (1% aq.); 100% conc.

Nissan Nonion NS-250. [Nippon Oils & Fats] POE nonylphenyl ether; nonionic; see Nissan Nonion NS-202; liq.; HLB 18.2; 100% conc.

Nissan Nonion NS-270. [Nippon Oils & Fats] POE nonylphenyl ether; nonionic; see Nissan Nonion NS-202; liq.; HLB 18.7; 100% conc.

Nissan Nonion O-2. [Nippon Oils & Fats] PEG oleate; nonionic; general emulsifier for textile, leather, industrial uses; Gardner 10 max. liq.; sol. in kerosene, methanol; disp. in water, xylene, ether; HLB 7.9; 100% conc.

Nissan Nonion O-3. [Nippon Oils & Fats] PEG oleate; nonionic; emulsifier for textile, leather, industrial uses; Gardner 10 max. liq.; sol. in kerosene, methanol; disp. in water, xylene, ether; HLB 10.3; 100% conc.

Nissan Nonion O-4. [Nippon Oils & Fats] PEG oleate; nonionic; general emulsifier for textile, leather, industrial uses; Gardner 9 max. liq.; sol. in xylene, kerosene, methanol; sol. warm in ether; disp. in water; HLB 11.5; 100% conc.

Nissan Nonion O-6. [Nippon Oils & Fats] PEG oleate; nonionic; general emulsifier for textile, leather, industrial uses; Gardner 9 max. liq.; sol. in water, xylene, kerosene, ether, methanol; HLB 13.5; 100% conc.

Nissan Nonion P-6. [Nippon Oils & Fats] PEG monopalmitate; nonionic; emulsifier and thickener for cosmetic applics., textile and other industrial uses; semisolid; HLB 13.8; 100% conc.

Nissan Nonion P-208. [Nippon Oils & Fats] POE cetyl ether; nonionic; emulsifier, dispersant, detergent, wetting agent used in textile processing, cosmetics, metalworking, agric. preparations, industrial cleaners; APHA 120 liq.; sol. in water, xylene, methanol; HLB 11.9; cloud pt. 43–53 C (1% aq.); 100% conc.

Nissan Nonion P-210. [Nippon Oils & Fats] POE cetyl ether; nonionic; emulsifier, dispersant, detergent, wetting agent used in textile processing, cosmetics, metalworking, agric. preparations, industrial cleaners; APHA 120 max. semisolid; sol. in water, xylene, methanol; HLB 12.9; cloud pt. 64–74 C (1% aq.); 100% conc.

Nissan Nonion P-213. [Nippon Oils & Fats] POE cetyl ether; nonionic; emulsifier, dispersant, detergent, wetting agent used in textile processing, cosmetics, metalworking, agric. preparations, industrial cleaners; APHA 120 max. semisolid; sol. in water, kerosene, ether, methanol; HLB 14.1; cloud pt. 87–97 C (1% aq.); 100% conc.

Nissan Nonion S-2. [Nippon Oils & Fats] PEG monostearate; nonionic; emulsifier, thickener used in cosmetics, textile, industrial uses; Gardner 4 max. semisolid; sol. in methanol, warm in xylene, disp. in water, kerosene, ether; m.p. 33–41 C; HLB 8.0; 100% conc.

Nissan Nonion S-4. [Nippon Oils & Fats] PEG monostearate; nonionic; emulsifier, thickener used in cosmetics, textile, industrial uses; Gardner 4 max. semisolid; sol. in kerosene, ether, methanol, warm in xylene, disp. in water; m.p. 30–40 C; HLB 11.6; 100% conc.

Nissan Nonion S-6. [Nippon Oils & Fats] PEG monostearate; nonionic; emulsifier, thickener used in cosmetics, textile, industrial uses; Gardner 4 max. semisolid; sol. in water, methanol, warm in xylene, ether, cloudy in kerosene; m.p. 35–41 C; HLB 13.6; 100% conc.

Nissan Nonion S-10. [Nippon Oils & Fats] PEG monostearate; nonionic; emulsifier, thickener used in cosmetics, textile, industrial uses; Gardner 4 max. semisolid; sol. in water, xylene, ether, methanol; m.p. 38–44 C; HLB 15.2; 100% conc.

Nissan Nonion S-15. [Nippon Oils & Fats] PEG monostearate; nonionic; emulsifier, thickener used in cosmetics, textile, industrial uses; Gardner 4 max. semisolid; sol. in kerosene, ether, methanol, warm in xylene; disp. in water; m.p. 39–45 C; HLB 12.8; 100% conc.

Nissan Nonion S-15.4. [Nippon Oils & Fats] PEG monostearate; nonionic; emulsifier, thickener used in cosmetics, textile, industrial uses; Gardner 4 max. semisolid; sol. in water, xylene, ether, methanol; m.p. 42–48 C; HLB 16.7; 100% conc.

Nissan Nonion S-40. [Nippon Oils & Fats] PEG monostearate; nonionic; emulsifier, thickener used in cosmetics, textile, industrial uses; Gardner 6 max. semisolid; sol. in water, xylene, ether, methanol; m.p. 49–55 C; HLB 18.2; 100% conc.

Nissan Nonion S-206. [Nippon Oils & Fats] POE stearyl ether; nonionic; emulsifier, dispersant, detergent, wetting agent for textile processing, cosmetics, metalworking, agric., industrial cleaners; solid; HLB 9.9; 100% conc.

Nissan Nonion S-207. [Nippon Oils & Fats] POE stearyl ether; nonionic; emulsifier, dispersant, detergent, wetting agent for textile processing, cosmetics, metalworking, agric., industrial cleaners; APHA 120 max. semisolid; sol. in xylene, kerosene, liq. paraffin, soybean oil, tetrachloromethan, methanol, diethylene glycol; water-disp.; HLB 10.7; cloud pt. 0 C max. (1% aq.); 100% conc.

Nissan Nonion S-215. [Nippon Oils & Fats] POE stearyl ether; nonionic; emulsifier, dispersant, detergent, wetting agent for textile processing, cosmetics, metalworking, agric., industrial cleaners; APHA 120 max. semisolid; sol. in water, xylene, kerosene, soybean oil, ether, tetrachloromethan, methanol, diethylene glycol; HLB 14.2; cloud pt. 100 C min. (1% aq.); 100% conc.

Nissan Nonion S-220. [Nippon Oils & Fats] POE stearyl ether; nonionic; emulsifier, dispersant, detergent, wetting agent for textile processing, cosmetics, metalworking, agric., industrial cleaners; APHA 120 max. semisolid; sol. in water, xylene, kerosene, soybean oil, ether, tetrachloromethan, methanol, diethylene glycol; HLB 15.3; cloud pt. 100 C min. (1% aq.); 100% conc.

Nissan Nonion T-15. [Nippon Oils & Fats] PEG monotallow acid ester; nonionic; emulsifier and thickener for cosmetics, textiles, industrial uses; liq.; HLB 12.8; 100% conc.

Nissan Nymeen DT-203, -208. [Nippon Oils & Fats] POE tallow-alkylpropylene diamine; nonionic; pigment wetting and disp. agent, emulsifier and acid cleaner additive for metals, textile antistatic agent and auxiliary; lt. br. liq.; cloud pt. 0 and 100 C resp.

Nissan Nymeen L-201. [Nippon Oils & Fats] Oxyethylene dodecylamine; nonionic; see Nissan Nymeen DT-203; colorless to lt. yel. liq.; cloud pt. 0 C max.

Nissan Nymeen L-202, 207. [Nippon Oils & Fats] POE dodecylamine; nonionic; see Nissan Nymeen DT-203; lt. yel. and lt. br. liq. resp.; cloud pt. 0 and 80 C.

Nissan Nymeen S-202, S-204. [Nippon Oils & Fats] POE octadecylamine; nonionic; see Nissan Nymeen DT-203; ylsh.-br. solid; cloud pt. 0 C max.

Nissan Nymeen S-210. [Nippon Oils & Fats] POE octadecylamine; nonionic; see Nissan Nymeen DT-203; lt. br. solid in winter, liq. in summer; cloud pt. 95 C min.

Nissan Nymeen S-215, -220. [Nippon Oils & Fats] POE octadecylamine; nonionic; see Nissan Nymeen DT-203; ylsh.-br. solid; cloud pt. 100 C min.

Nissan Nymeen T2-206, -210. [Nippon Oils & Fats] POE tallow-alkylamine; nonionic; see Nissan Nymeen DT-203; lt. br. liq.; cloud pt. 95 C.

Nissan Nymeen T2-230, -260. [Nippon Oils & Fats] POE tallow-alkylamine; nonionic; see Nissan Nymeen DT-203; lt. br. solid; cloud pt. 100 C.

Nissan Nymide MT-215. [Nippon Oils & Fats] POE alkylamide; nonionic; modifier for soaps; solid; 100% conc.

Nissan Ohsen A. [Nippon Oils & Fats] Alkylbenzene sulfonate; anionic; general purpose detergent and emulsifier; paste; 60% conc.

Nissan Panacete 810. [Nippon Oils & Fats] Med. chain triglyceride; nonionic; diluent for perfumes; raw material for pharmaceuticals and special foods; liq.; 100% conc.

Nissan Persoft EK. [Nippon Oils & Fats] Sulfated fatty alcohol ethoxylate, sodium salt; anionic; emulsifier for vinyl polymerization; textile detergent; dyeing assistant; degreaser; liq.; 30% conc.

Nissan Persoft NK-60, -100. [Nippon Oils & Fats] Ethoxylated fatty alcohols; nonionic; detergent, emulsifier, wetting agent; textile detergent; dyeing assistant; base for liq. detergent; dispersant for pulp pitch; lubricant for bottle cleaning processing; biodeg.; liq.; 90% conc.

Nissan Persoft SK. [Nippon Oils & Fats] Sulfated fatty alcohol, sodium salt; anionic; detergent, emulsifier, dispersant, wetting agent; base for shampoo and liq. detergent; biodeg.; liq.; 30% conc.

Nissan Plonon 102. [Nippon Oils & Fats] POE-POP ether; nonionic; emulsifier, solubilizer, dispersant, detergent, antifoaming agent, wetting agent used in soaps, syn. resins, metal cleaning, emulsion polymerization, fermentation, paper industries; colorless liq.; odorless and tasteless; m.w. 1250; sol. in water, ethanol, acetone, benzol, dioxane; visc. 105 cs (37.8 C); HLB 7; cloud pt. 22 C (10% aq.); surf. tens. 41.1 dynes/cm (0.1%); 100% conc.

Nissan Plonon 104. [Nippon Oils & Fats] POE-POP ether; nonionic; emulsifier, solubilizer, dispersant, detergent, antifoaming agent, wetting agent used in soaps, syn. resins, metal cleaning, emulsion polymerization, fermentation, paper industries; colorless liq.; odorless and tasteless; m.w. 1650; sol. in water, ethanol, acetone, benzol, dioxane; visc. 105 cs (37.8 C); HLB 15.0; cloud pt. 64 C (10% aq.); surf. tens. 43.5 dynes/cm (0.1%); 100% conc.

Nissan Plonon 108. [Nippon Oils & Fats] POE-POP ether; nonionic; emulsifier, solubilizer, dispersant, detergent, antifoaming agent, wetting agent used in soaps, syn. resins, metal cleaning, emulsion polymerization, fermentation, paper industries; m.w. 4000; HLB 30.5; cloud pt. 100 C (10% aq.); 100% conc.

Nissan Plonon 171. [Nippon Oils & Fats] POE-POP ether; nonionic; emulsifier, solubilizer, dispersant, detergent, antifoaming agent, wetting agent used in soaps, syn. resins, metal cleaning, emulsion polymerization, fermentation, paper industries; m.w. 2000; sol. in ethanol, acetone, benzol, dioxane; visc. 160 cs (37.8 C); HLB 3; cloud pt. 24 C (10% aq.); surf. tens. 39.5 dynes/cm (0.1%); 100% conc.

Nissan Plonon 172. [Nippon Oils & Fats] POE-POP ether; nonionic; emulsifier, solubilizer, dispersant, detergent, antifoaming agent, wetting agent used in soaps, syn. resins, metal cleaning, emulsion polymerization, fermentation, paper industries; m.w. 2400; sol. in ethanol, acetone, benzol, dioxane; visc. 210 cs (37.8 C); HLB 7; cloud pt. 28 C (10% aq.); 100% conc.

Nissan Plonon 201. [Nippon Oils & Fats] POE-POP ether; nonionic; emulsifier, solubilizer, dispersant, detergent, antifoaming agent, wetting agent used in soaps, syn. resins, metal cleaning, emulsion polymerization, fermentation, paper industries; colorless liq.; odorless and tasteless; m.w. 2200; sol. in ethanol, acetone, benzol, dioxane; visc. 190 cs (37.8 C); HLB 3; cloud pt. 21 C (10% aq.); surf. tens. 38 dynes/cm (0.1%); 100% conc.

Nissan Plonon 204. [Nippon Oils & Fats] POE-POP ether; nonionic; emulsifier, solubilizer, dispersant, detergent, antifoaming agent, wetting agent used in soaps, syn. resins, metal cleaning, emulsion polymerization, fermentation, paper industries; colorless paste; odorless and tasteless; m.w. 3400; sol. in water, ethanol; visc. 370 cs (37.8 C); HLB 13.5; cloud pt. 64 C (10% aq.); surf. tens. 39.8 dynes/cm (0.1%); 100% conc.

Nissan Plonon 208. [Nippon Oils & Fats] POE-POP ether; nonionic; emulsifier, solubilizer, dispersant, detergent, antifoaming agent, wetting agent used in soaps, syn. resins, metal cleaning, emulsion polymerization, fermentation, paper industries; colorless paste; odorless and tasteless; m.w. 8000; sol. in water; HLB 28; cloud pt. 100 C (10% aq.); surf. tens. 44.1 dynes/cm (0.1%); 100% conc.

Nissan Rapisol B-30, B-80, C-70. [Nippon Oils & Fats] Sodium dioctyl sulfosuccinate; anionic; wetting agent, polymerization agent for PVC, dyeing aux.; liq.; 30, 80, and 70% conc. resp.

Nissan Softer 1000. [Nippon Oils & Fats] Amide amine; cationic; softener for textiles; flake; 100% conc.

Nissan Softer 706. [Nippon Oils & Fats] Amide amine; cationic; softening and finishing agent for syn. fibers; paste; 30% conc.

Nissan Soft Osen 550A. [Nippon Oils & Fats] Linear alkylbenzene sulfonate; anionic; general purpose detergent, emulsifier; paste; 60% conc.

Nissan Stafoam DF. [Nippon Oils & Fats] Cocamide DEA; nonionic; detergent and scouring agent for wool and syn. fibers; liq.; 100% conc.

Nissan Stafoam DL. [Nippon Oils & Fats] Lauramide DEA; CAS 120-40-1; nonionic; foam stabilizer, thickener for liq. detergents; liq.; 100% conc.

Nissan Stafoam DO, DOS. [Nippon Oils & Fats] Oleamide DEA; nonionic; thickener, foam stabilizer for shampoo and liq. detergents; liq.; 100% conc.

Nissan Stafoam L. [Nippon Oils & Fats] Lauramide DEA (1:2); CAS 120-40-1; nonionic; foam stabilizer for liq. detergents; liq.; 100% conc.

Nissan Sunalpha T. [Nippon Oils & Fats] Alpha-sulfonated fatty acid ester blend, sodium salt; anionic; detergent for household and industrial cleaning; biodeg.; good detergency in cold and hard water; gran.; sol. in water; 100% conc.

Nissan Sunamide C-3, CF-3, CF-10. [Nippon Oils & Fats] Fatty alkylolamide ether sulfate; anionic; base for preparation of low irritation and biodeg. shampoos and dishwashing detergents; liq.; 30, 30 and 35% conc. resp.

Nissan Sunbase, Powder. [Nippon Oils & Fats] Alpha-sulfonated fatty acid ester, sodium salt; anionic; lime soap dispersant; nonphosphate detergent base; builder for household detergent; emulsifier and dispersant; biodeg.; paste, gran.; 30 and 40% conc. resp.

Nissan Sun Flora. [Nippon Oils & Fats] Blend; nonionic/cationic; drycleaning soap for petroleum type solv.; liq.

Nissan Tert. Amine AB. [Nippon Oils & Fats] Octadecyl-dimethylamine; intermediate for various surfactants, visc. index improver for lubricating oil, curing catalyst for epoxy resin, corrosion inhibitor, germicide; lt. yel. liq. or half-solid; 95% min. tert. amine.

Nissan Tert. Amine ABT. [Nippon Oils & Fats] Tallow-hydrog. alkyl dimethylamine; see Nissan Tert. Amine AB; lt. yel. liq. or half-solid; 95% min.

tert. amine.

Nissan Tert. Amine BB. [Nippon Oils & Fats] Dodecyl-dimethylamine; see Nissan Tert. Amine AB; lt. yel. liq.; 95% min. tert. amine.

Nissan Tert. Amine FB. [Nippon Oils & Fats] Cocoalkyl dimethylamine; see Nissan Tert. Amine AB; lt. yel. liq.; 95% min. tert. amine.

Nissan Tert. Amine MB. [Nippon Oils & Fats] Tetradecyl dimethylamine; see Nissan Tert. Amine AB; lt. yel. liq. or half-solid; 95% min. tert. amine.

Nissan Tert. Amine PB. [Nippon Oils & Fats] Hexadecyl-dimethylamine; see Nissan Tert. Amine AB; lt. yel. liq. or half-solid; 95% min. tert. amine.

Nissan Trax H-45. [Nippon Oils & Fats] POE alkylaryl ether sulfate; anionic; penetrant, detergent, wetting agent; scouring agent for wool; dyeing aux. and emulsifier for syn. resins; liq.; 30% conc.

Nissan Trax K-40. [Nippon Oils & Fats] POE alkyl ether sulfate; anionic; penetrant, detergent, wetting agent; scouring agent for wool; dyeing auxiliary and emulsifier for syn. resins; liq.; 30% conc.

Nissan Trax K-300. [Nippon Oils & Fats] POE alkyl ether sulfate; anionic; emulsifier for emulsion polymerization; liq.; 34% conc.

Nissan Trax N-300. [Nippon Oils & Fats] POE alkylaryl ether sulfate; anionic; emulsifier for emulsion polymerization; liq.; 34% conc.

Nissan Unisafe A-LE. [Nippon Oils & Fats] Alkyl dihydroxy ethyl amine oxide; nonionic; foam stabilizer, detergent; stable over wide pH range; liq.; 40% conc.

Nissan Unisafe A-LM. [Nippon Oils & Fats] Alkyl dimethyl amine oxide; nonionic; foam stabilizer, detergent; stable over wide pH range; liq.; 35% conc.

Nissan Unister. [Nippon Oils & Fats] Neopentylpolyol fatty acid ester; nonionic; lubricating oil, cutting oil, cosmetics base; plasticizer; thermal stability; liq.; 100% conc.

Nitrene 100 SD. [Henkel] Linoleamide DEA; nonionic; thickener, foamer, corrosion inhibitor; emulsifier for oils/greases used in industrial and household cleaners; liq.; 100% conc.

Nitrene 11120. [Henkel] Modified cocamide DEA (2:1); detergent for truck washes, household, alkaline industrial cleaners; liq.; dens. 8.3 lb/gal; 100% conc.

Nitrene 11230. [Henkel/Cospha; Henkel Canada] Modified cocamide DEA (2:1); anionic/nonionic; detergent, wetting agent, emulsifier, thickener, foam stabilizer for industrial and specialty cleaners, floor strippers and degreasers, household hard surface cleaners; liq.; sol. in alcohols, glycols, esters; dens. 8.3 lb/gal; 100% conc.

Nitrene 13026. [Henkel] Modified cocamide DEA; detergent, wetting agent, emulsifier for industrial, household, and specialty cleaners; liq.; sol. in alcohols, glycols, esters; 100% conc.

Nitrene A-309. [Henkel] Modified cocamide DEA (2:1); detergent, wetting agent, emulsifier for industrial, household, and specialty cleaners; chain lubricant; liq.; sol. in alcohols, glycols, esters; dens. 8.4 lb/gal; 100% conc.

Nitrene A-567. [Henkel] Modified cocamide DEA (2:1); detergent, emulsifier for alkaline industrial and household cleaners; liq.; dens. 8.3 lb/gal; 100% conc.

Nitrene C. [Henkel/Cospha; Henkel Canada] Cocamide DEA (2:1); nonionic; detergent, emulsifier, dispersant, wetting agent, foam booster, thickener, foam stabilizer for liq. detergents, car shampoos, solv. cleaners, drycleaning, industrial and household cleaners; lt. amber liq.; sol. in water, lower alcohols, PEG, acetone, toluene, xylene; dens. 8.3 lb/gal; pH 9–11 (10%); 100% conc.

Nitrene C Extra. [Henkel/Cospha; Henkel Canada] Cocamide DEA (1:1); nonionic; emulsifier, dispersant, wetting agent, foam booster/stabilizer, detergent, and visc. builder for industrial/household detergents, metalworking; intermediate for liq. detergents; clarifier for liq. soaps; solv. cleaners, drycleaning; lt. amber liq.; sol. in water, alcohol, PEG, acetone, toluene, xylene; dens. 8.35 lb/gal; pH 9-11 (10%); 100% conc.

Nitrene L-76. [Henkel] Lauric-myristic (70/30) DEA (1:1); nonionic; wetting, foaming, dispersing agent, foam stabilizer, thickener for household, institutional, and industrial cleaners; pigment dispersant; soft solid to lt. amber liq.; sol. in lower alcohols, propylene glycol, PEG; disp. water; dens. 8.1 lb/gal; pH 9–11 (10%); biodeg.; 100% conc.

Nitrene L-90. [Henkel] Lauramide DEA (1:1); CAS 120-40-1; nonionic; detergent, thickener, foam stabilizer for liq. detergents; solid; sol. in alcohols, glycols, esters; dens. 8.3 lb/gal; 100% conc.

Nitrene N. [Henkel] Cocamide DEA; nonionic; emulsifier, dispersant, wetting agent, foam booster, visc. builder for liq. detergents, car shampoos, solv. cleaners, drycleaning; liq.; 97% active.

Nitrene OE. [Henkel/Cospha; Henkel Canada] Polyoxyalkylene DEA amide ester; nonionic; emulsifier for sol. oils and lubricants, metalworking compds.; liq.; dens. 8.2 lb/gal; 100% conc.

Nofome 2510. [Sybron] Nonsilicone water-based; textile defoamer for atmospheric and pressure applics.

Noigen 140L. [Dai-ichi Kogyo Seiyaku] POE octylphenol ether; nonionic; emulsifier; liq.; HLB 14.0; 100% conc.

Noigen EA 33. [Dai-ichi Kogyo Seiyaku] POE dodecyl phenol ether; CAS 9014-92-0; nonionic; emulsifier; liq.; HLB 4.1; 100% conc.

Noigen EA 50. [Dai-ichi Kogyo Seiyaku] POE nonylphenol ether; CAS 9016-45-9; nonionic; emulsifier; liq.; HLB 6.0; 100% conc.

Noigen EA 70. [Dai-ichi Kogyo Seiyaku] PEG nonylphenol ether; CAS 9016-45-9; nonionic; emulsifier; liq.; HLB 8.0; 100% conc.

Noigen EA 73. [Dai-ichi Kogyo Seiyaku] PEG dodecylphenol ether; CAS 9014-92-0; nonionic; emulsifier; liq.; HLB 9.0; 100% conc.

Noigen EA 80. [Dai-ichi Kogyo Seiyaku] PEG nonylphenol ether; CAS 9014-92-0; nonionic; emulsifier; liq.; HLB 10.0; 100% conc.

Noigen EA 80E. [Dai-ichi Kogyo Seiyaku] PEG nonylphenol ether; nonionic; emulsifier; liq.; HLB 9.0; 100% conc.

Noigen EA 83. [Dai-ichi Kogyo Seiyaku] PEG dodecylphenol ether; nonionic; emulsifier; liq.; HLB 10.0; 100% conc.

Noigen EA 92. [Dai-ichi Kogyo Seiyaku] PEG octyl phenol ether; CAS 9036-19-5; nonionic; emulsifier; liq.; HLB 9.0; 100% conc.

Noigen EA 102. [Dai-ichi Kogyo Seiyaku] POE octylphenol ether; CAS 9036-19-5; nonionic; emulsifier; liq.; HLB 11.0; 100% conc.

Noigen EA 110. [Dai-ichi Kogyo Seiyaku] POE octylphenol ether; CAS 9036-19-5; nonionic; emulsifier; liq.; HLB 11.0; 100% conc.

Noigen EA 112. [Dai-ichi Kogyo Seiyaku] POE

octylphenol ether; CAS 9036-19-5; nonionic; emulsifier; liq.; HLB 12.0; 100% conc.

Noigen EA 120. [Dai-ichi Kogyo Seiyaku] POE octylphenol ether; CAS 9036-19-5; nonionic; emulsifier; liq.; HLB 12.0; 100% conc.

Noigen EA 120B. [Dai-ichi Kogyo Seiyaku] POE octylphenol ether; CAS 9036-19-5; nonionic; emulsifier; liq.; HLB 12.0; 100% conc.

Noigen EA 130T. [Dai-ichi Kogyo Seiyaku] POE octylphenol ether; CAS 9016-45-9; nonionic; emulsifier; liq.; HLB 13.0; 100% conc.

Noigen EA 140. [Dai-ichi Kogyo Seiyaku] POE octylphenol ether; CAS 9036-19-5; nonionic; emulsifier; liq.; HLB 14.0; 100% conc.

Noigen EA 142. [Dai-ichi Kogyo Seiyaku] POE octylphenol ether; CAS 9036-19-5; nonionic; emulsifier; liq.; HLB 14.0; 100% conc.

Noigen EA 143. [Dai-ichi Kogyo Seiyaku] POE dodecylphenol ether; CAS 9014-92-0; nonionic; emulsifier; liq.; HLB 14.0; 100% conc.

Noigen EA 150. [Dai-ichi Kogyo Seiyaku] POE octylphenol ether; CAS 9016-45-9; nonionic; emulsifier; paste; HLB 15.0; 100% conc.

Noigen EA 152. [Dai-ichi Kogyo Seiyaku] POE octylphenol ether; nonionic; emulsifier; paste; HLB 15.0; 100% conc.

Noigen EA 160. [Dai-ichi Kogyo Seiyaku] POE octylphenol ether; CAS 9036-19-5; nonionic; emulsifier; paste; HLB 16.0; 100% conc.

Noigen EA 160P. [Dai-ichi Kogyo Seiyaku] POE octylphenol ether; nonionic; emulsifier; solid; HLB 16.0; 100% conc.

Noigen EA 170. [Dai-ichi Kogyo Seiyaku] POE octylphenol ether; CAS 9036-19-5; nonionic; emulsifier; solid; HLB 17.0; 100% conc.

Noigen EA 190D. [Dai-ichi Kogyo Seiyaku] POE octylphenol ether; CAS 9036-19-5; nonionic; emulsifier; flake; HLB 19.0; 100% conc.

Noigen ES 90. [Dai-ichi Kogyo Seiyaku] PEG oleic acid ester; CAS 9004-96-0; nonionic; emulsifier for animal, veg. and min. oil; dispersant for paint and pigments; liq.; HLB 9.0; 100% conc.

Noigen ES 120. [Dai-ichi Kogyo Seiyaku] PEG oleic acid ester; CAS 9004-96-0; nonionic; emulsifier for animal, veg. and min. oil; dispersant for paint and pigments; liq.; HLB 12.0; 100% conc.

Noigen ES 140. [Dai-ichi Kogyo Seiyaku] PEG oleic acid ester; CAS 9004-96-0; nonionic; emulsifier for animal, veg. and min. oil; dispersant for paint and pigments; liq.; HLB 14.0; 100% conc.

Noigen ES 160. [Dai-ichi Kogyo Seiyaku] PEG oleic acid ester; CAS 9004-96-0; nonionic; emulsifier for animal, veg. and min. oil; dispersant for paint and pigments; liq.; HLB 16.0; 100% conc.

Noigen ET60. [Dai-ichi Kogyo Seiyaku] PEG oleyl ether; CAS 9004-98-2; nonionic; emulsifier; liq.; oil-sol.; HLB 6.0; 100% conc.

Noigen ET65. [Dai-ichi Kogyo Seiyaku] PEG alkyl ether; nonionic; emulsifier; liq.; oil-sol.; HLB 6.0; 100% conc.

Noigen ET77. [Dai-ichi Kogyo Seiyaku] PEG alkyl ether; CAS 9004-98-2; nonionic; emulsifier; liq.; HLB 7.0; 100% conc.

Noigen ET80. [Dai-ichi Kogyo Seiyaku] PEG oleyl ether; CAS 9004-98-2; nonionic; emulsifier; liq.; oil-sol.; HLB 8.0; 100% conc.

Noigen ET83. [Dai-ichi Kogyo Seiyaku] PEG lauryl ether; CAS 9002-92-0; nonionic; emulsifier; liq.; HLB 8.0; 100% conc.

Noigen ET95. [Dai-ichi Kogyo Seiyaku] PEG alkyl ether; nonionic; emulsifier; liq.; HLB 9.0; 100% conc.

Noigen ET97. [Dai-ichi Kogyo Seiyaku] PEG alkyl ether; nonionic; emulsifier; liq.; HLB 9.0; 100% conc.

Noigen ET100. [Dai-ichi Kogyo Seiyaku] PEG oleyl ether; CAS 9004-98-2; nonionic; emulsifier; liq.; HLB 10.0; 100% conc.

Noigen ET102. [Dai-ichi Kogyo Seiyaku] PEG lauryl ether; CAS 9002-92-0; nonionic; emulsifier; liq.; HLB 10.0; 100% conc.

Noigen ET107. [Dai-ichi Kogyo Seiyaku] PEG alkyl ether; nonionic; emulsifier; liq.; HLB 10.0; 100% conc.

Noigen ET115. [Dai-ichi Kogyo Seiyaku] PEG alkyl ether; nonionic; emulsifier; liq.; HLB 11.0; 100% conc.

Noigen ET120. [Dai-ichi Kogyo Seiyaku] PEG oleyl ether; CAS 9004-98-2; nonionic; emulsifier; paste; HLB 12.0; 100% conc.

Noigen ET127. [Dai-ichi Kogyo Seiyaku] PEG alkyl ether; nonionic; emulsifier; liq.; HLB 12.0; 100% conc.

Noigen ET135. [Dai-ichi Kogyo Seiyaku] PEG alkyl ether; nonionic; emulsifier; liq.; HLB 13.0; 100% conc.

Noigen ET140. [Dai-ichi Kogyo Seiyaku] PEG oleyl ether; CAS 9004-98-2; nonionic; emulsifier; solid; HLB 14.0; 100% conc.

Noigen ET143. [Dai-ichi Kogyo Seiyaku] PEG lauryl ether; CAS 9002-92-0; nonionic; emulsifier; liq.; HLB 14.0; 100% conc.

Noigen ET147. [Dai-ichi Kogyo Seiyaku] PEG alkyl ether; nonionic; emulsifier; liq.; HLB 14.0; 100% conc.

Noigen ET150. [Dai-ichi Kogyo Seiyaku] PEG oleyl ether; nonionic; emulsifier; liq.; HLB 15.0; 100% conc.

Noigen ET157. [Dai-ichi Kogyo Seiyaku] PEG alkyl ether; nonionic; emulsifier; liq.; HLB 15.0; 100% conc.

Noigen ET160. [Dai-ichi Kogyo Seiyaku] PEG lauryl ether; CAS 9002-92-0; nonionic; emulsifier; liq.; HLB 16.0; 100% conc.

Noigen ET165. [Dai-ichi Kogyo Seiyaku] PEG alkyl ether; nonionic; emulsifier; liq.; HLB 16.0; 100% conc.

Noigen ET167. [Dai-ichi Kogyo Seiyaku] PEG lauryl ether; nonionic; emulsifier; solid; HLB 16.0; 100% conc.

Noigen ET170. [Dai-ichi Kogyo Seiyaku] PEG lauryl ether; CAS 9002-92-0; nonionic; emulsifier; solid; HLB 17.0; 100% conc.

Noigen ET180. [Dai-ichi Kogyo Seiyaku] PEG oleyl ether; CAS 9004-98-2; nonionic; emulsifier; solid; HLB 18.0; 100% conc.

Noigen ET187. [Dai-ichi Kogyo Seiyaku] PEG alkyl ether; nonionic; emulsifier; solid; HLB 18.0; 100% conc.

Noigen ET190. [Dai-ichi Kogyo Seiyaku] PEG lauryl ether; CAS 9002-92-0; nonionic; emulsifier; solid; HLB 19.0; 100% conc.

Noigen ET190S. [Dai-ichi Kogyo Seiyaku] PEG oleyl ether; CAS 9004-98-2; nonionic; emulsifier; solid; HLB 19.0; 100% conc.

Noigen ET207. [Dai-ichi Kogyo Seiyaku] PEG alkyl ether; nonionic; emulsifier; solid; HLB 20.0; 100% conc.

Noigen O100. [Dai-ichi Kogyo Seiyaku] PEG oleyl ether; nonionic; emulsifier; solid; water-sol.; HLB

20.0; 100% conc.

Noigen YX400, YX500. [Dai-ichi Kogyo Seiyaku] PEG lauryl ether; CAS 9002-92-0; nonionic; emulsifier; solid; water-sol.; HLB 20.0; 100% conc.

Noiox AK-40. [Pulcra SA] PEG oleate; nonionic; spreading agent, defoamer, kerosene and min. oil emulsifier; liq.; 100% conc.

Noiox AK-41. [Pulcra SA] PEG oleate; nonionic; lubricant, solubilizer and emulsifier; wetting agent and pigment dispersant; antistat for plastics and aux. dispersant; liq.; 100% conc.

Noiox AK-44. [Pulcra SA] PEG 400 oleate; CAS 9004-96-0; nonionic; lubricant, emulsifier, dispersant for min. and veg. oils; antifoamer for dispersing assistant; liq.; sol. in min. oil; disp. in water; 100% conc.

Non Ionic Emulsifier T-9. [Werner G. Smith] Proprietary; nonionic; low foam emulsifier for oils and emulsion systems; 100% biodeg.; Gardner 6-12 color; sol. in paraffinic oils, disp. in water; sp.gr. 0.967-0.976 (60 F); visc. 195-250 SSU (100 F); HLB 12.0; acid no. 9-15; iodine no. 35-45; sapon. no. 60-70; pH 6.5-7.0 (1%); 100% act.

Nonal 206. [Toho Chem. Industry] Nonylphenol ethoxylate; nonionic; detergent, penetrant, emulsifier, scouring agent, wetting agent, dispersant; for textiles; liq.; HLB 10.9; 100% conc.

Nonal 208. [Toho Chem. Industry] Nonylphenol ethoxylate; nonionic; detergent, penetrant, emulsifier, scouring agent, wetting agent, dispersant; for textiles; liq.; HLB 12.3; 100% conc.

Nonal 210. [Toho Chem. Industry] Nonylphenol ethoxylate; nonionic; detergent, penetrant, emulsifier, scouring agent, wetting agent, dispersant; for textiles; liq.; HLB 13.3; 100% conc.

Nonal 310. [Toho Chem. Industry] Octylphenol ethoxylate; nonionic; detergent, penetrant, emulsifier, scouring agent, wetting agent, dispersant; for textiles; liq.; HLB 13.7; 100% conc.

Nonarox 575, 730. [Seppic] Alkylphenyl polyethoxy ether; nonionic; wetting agent; liq.; 100% conc.

Nonarox 1030, 1230. [Seppic] Alkylphenyl polyethoxy ether; nonionic; detergent; liq.; 100% conc.

Nonasol 3922. [Hart Chem. Ltd.] Blend of sulfates, sulfonates, and amide; anionic; detergent base for car wash prods., etc.; yel. liq.; sp.gr. 1.03; 60% act.

Nonasol LD-50. [Hart Chem. Ltd.] Blend; anionic; liq. detergent conc.; liq.; 50% conc.

Nonasol LD-51. [Hart Chem. Ltd.] Blend; anionic; liq. detergent conc.; liq.; 50% conc.

Nonasol MBS. [Hart Chem. Ltd.] Proprietary; anionic; post-scour agent for most dyestuffs; yel. liq.

Nonasol N4AS. [Hart Chem. Ltd.] Ammonium laureth sulfate; anionic; detergent base for dishwash, hard surf. cleaners, shampoos; wetting agent; biodeg.; yel. liq.; sp.gr. 1.02; 58% act.

Nonasol N4SS. [Hart Chem. Ltd.] Sodium laureth sulfate; CAS 9004-82-4; anionic; detergent base for high foaming liq. detergents, dishwash, hard surf. cleaners, shampoos; wetting agent; biodeg.; yel. liq.; sp.gr. 1.04; 60% act.

Nonasol SNS-30. [Hart Chem. Ltd.] Sodium alkyl benzene sulfonate; anionic; raw material for detergent formulations; liq.; 30% act.

Nonatell 1002. [Shell] Proprietary; deinking surfactant for use on xerographic finish; liq.; 100% conc.

Nonatell 1003. [Shell] Proprietary; surfactant for use in pulp deresination; paste; 100% conc.

Nonatell 1038. [Shell] Proprietary; deinking surfactant; paste; 100% conc.

Nonatell 1052. [Shell] Proprietary; deinking surfactant for use on groundwood finish; liq.; 100% conc.

Nonatell 1056. [Shell] Proprietary; digester process aids used to lower resin content and increase chip yield in kraft and sulfite processes; paste; 100% conc.

Nonatell 1058. [Shell] Proprietary; digester process aids used to lower resin content and increase chip yield in kraft and sulfite processes; paste; 100% conc.

Nonatell 1061, 1075, 1088, 1089, 1092. [Shell] Proprietary; deinking surfactants; liqs. (exc. 1075, paste); 100% conc.

Nonex C5E. [Hart Chem. Ltd.] PEG 500 cocoate; lubricant for textile applics.; liq.; water-sol.; HLB 13.0; 100% act.

Nonex DL-2. [Hart Chem. Ltd.] PEG 200 dilaurate; CAS 9005-02-1; nonionic; coemulsifier and lubricant in industrial and textile oils; liq.; HLB 7.4; 100% act.

Nonex DO-4. [Hart Chem. Ltd.] PEG 400 dioleate; CAS 9005-07-6; nonionic; emulsifier, solubilizer for oils, fats, solvs.; liq.; HLB 7.2; 100% act.

Nonex O4E. [Hart Chem. Ltd.] PEG 400 oleate; CAS 9004-96-0; nonionic; emulsifier for min. and animal oils; liq.; HLB 11.0; 100% act.

Nonex S3E. [Hart Chem. Ltd.] PEG 300 stearate; CAS 9004-99-3; nonionic; lubricant and softener for textiles; solid; HLB 9.6; 100% act.

Nonfome IDC. [Sybron] Nonsilicone oil-based; textile defoamer for atmospheric use.

Nonicol 100. [Atsaun] Ethoxylated alkylphenol; nonionic; wetting agent, emulsifier; liq.; HLB 1.3; 98+% conc.

Nonicol 190. [Atsaun] Ethoxylated alkylphenol; nonionic; emulsifier; solid; HLB 16.0; 100% conc.

Nonionic E-4. [Calgene] POE alkylaryl ethers; nonionic; detergent and wetting agent used in cosmetics, insecticide and other formulations; liq.; HLB 8.6; 100% conc.

Nonipol 20. [Sanyo Chem. Industries] POE nonyl phenyl ether; CAS 9016-45-9; nonionic; penetrant, wetting agent, spreader-sticker; base material for detergents; emulsifier for agric. pesticides and emulsion polymerization, org. solv., machine oils, liq. paraffins; liq.; HLB 5.7; 100% conc.

Nonipol 40. [Sanyo Chem. Industries] POE nonylphenyl ether; nonionic; see Nonipol 20; liq.; HLB 8.9; 100% conc.

Nonipol 55. [Sanyo Chem. Industries] POE nonylphenyl ether; nonionic; see Nonipol 20; liq.; HLB 10.5; 100% conc.

Nonipol 60. [Sanyo Chem. Industries] POE nonylphenyl ether; nonionic; see Nonipol 20; liq.; HLB 10.9; 100% conc.

Nonipol 70. [Sanyo Chem. Industries] POE nonylphenyl ether; nonionic; see Nonipol 20; liq.; HLB 11.7; 100% conc.

Nonipol 85. [Sanyo Chem. Industries] POE nonylphenyl ether; nonionic; see Nonipol 20; liq.; HLB 12.6; 100% conc.

Nonipol 95. [Sanyo Chem. Industries] POE nonyl phenyl ether; see Nonipol 20; liq.; HLB 13.1; 100% conc.

Nonipol 100. [Sanyo Chem. Industries] POE nonylphenyl ether; nonionic; see Nonipol 20; liq.; HLB 13.1; 100% conc.

Nonipol 110. [Sanyo Chem. Industries] POE nonylphenyl ether; nonionic; see Nonipol 20; liq.; HLB

13.8; 100% conc.

Nonipol 120. [Sanyo Chem. Industries] POE nonylphenyl ether; nonionic; see Nonipol 20; liq.; HLB 14.1; 100% conc.

Nonipol 130. [Sanyo Chem. Industries] POE nonylphenyl ether; nonionic; see Nonipol 20; liq.; HLB 14.5; 100% conc.

Nonipol 160. [Sanyo Chem. Industries] POE nonylphenyl ether; nonionic; see Nonipol 20; solid; HLB 15.2; 100% conc.

Nonipol 200. [Sanyo Chem. Industries] POE nonylphenyl ether; nonionic; see Nonipol 20; solid; HLB 16.0; 100% conc.

Nonipol 400. [Sanyo Chem. Industries] POE nonylphenyl ether; nonionic; see Nonipol 20; solid; HLB 17.8; 100% conc.

Nonipol BX. [Sanyo Chem. Industries] POE alkylphenyl ether; nonionic; textile scouring agent and penetrant; effective at low dose over wide temp. range; liq.; 100% conc.

Nonipol D-160. [Sanyo Chem. Industries] POE dinonyl phenyl ether; CAS 9014-93-1; nonionic; textile scouring and penetrating agent; solid; HLB 13.3.

Nonipol Soft SS-50. [Sanyo Chem. Industries] POE C11-15 alcohol ether; CAS 68131-40-8; nonionic; penetrant, detergent, emulsifier, dispersant, wetting agent, or base material for household liq. detergent; biodeg.; liq.; HLB 10.5; 100% conc.

Nonipol Soft SS-70. [Sanyo Chem. Industries] POE C11-15 alcohol ether; CAS 68131-40-8; nonionic; penetrant, detergent, emulsifier, dispersant, wetting agent, or base material for household liq. detergent; biodeg.; liq.; HLB 12.8; 100% conc.

Nonipol Soft SS-90. [Sanyo Chem. Industries] POE C11-15 alcohol ether; CAS 68131-40-8; nonionic; penetrant, detergent, emulsifier, dispersant, wetting agent, or base material for household liq. detergent; biodeg.; liq.; HLB 13.2; 100% conc.

Nonipol T-20. [Toho Chem. Industry] Nonionic/anionic/min. oil blend; anionic/nonionic; spinning oil for wool; liq.; 100% conc.

Nonipol T-28. [Toho Chem. Industry] Nonionics and min. oil; nonionic; surfactant used in spinning oil for wool; liq.; 100% act.

Nonipol T-100. [Toho Chem. Industry] Nonionic/anionic/min. oil blend; anionic/nonionic; spinning oil for wool; liq.; 100% conc.

Nonipol TH. [Toho Chem. Industry] Nonionic/anionic/min. oil blend; anionic/nonionic; spinning oil for rough wool; liq.; 100% conc.

Nonisol 100. [Ciba-Geigy] PEG-8 laurate; CAS 9004-81-3; nonionic; emulsifier, solubilizing and wetting agent, thickener, used in textiles, cosmetics, hand cleaners, spreading agents; yel. liq.; water-sol.; 100% conc.

Nonisol 210. [Ciba-Geigy] PEG-8 dioleate; CAS 9005-07-6; nonionic; toxicant emulsifier, dispersant of sludge in fuel oil, textile lubricant; lt. amber liq.; sol. in kerosene; 100% conc.

Nonisol 300. [Ciba-Geigy] PEG-8 stearate; CAS 9004-99-3; nonionic; emulsifier, wetting agent, starch stabilizer, base for ointments, creams, suppositories, hair dressing, liq. makeup; wh. paste; sol. in xylene, disp. in water; m.p. 30 C; 100% conc.

Nopalcol 1-L. [Henkel/Functional Prods.] PEG-2 laurate; CAS 9004-81-3; nonionic; general purpose emulsifier, plasticizer, lubricant, wetting agent, dispersant, binding and thickening agent for emulsion polymerization, dry cleaning, leather, min. oil emulsions, paper industry, wall-tile mastics, solv. emulsions; liq.; HLB 6.0; 98% act.

Nopalcol 1-S. [Henkel/Functional Prods.] Diethylene glycol stearate; CAS 9004-99-3; nonionic; emulsifier, plasticizer, lubricant, wetting agent, defoamer, binding and thickening agent, used in cosmetics, dry cleaning, leather, textile industries; solid; HLB 3.8; 98% conc.

Nopalcol 1-TW. [Henkel/Functional Prods.] PEG-2 tallowate; nonionic; emulsifier, plasticizer, lubricant, wetting agent, defoamer, binding and thickening agent, used in cosmetics, dry cleaning, leather, textile industries; solid; HLB 4.1; 99% act.

Nopalcol 2-DL. [Henkel/Functional Prods.] PEG-4 dilaurate; CAS 9005-02-1; nonionic; general purpose emulsifier, plasticizer, lubricant, wetting agent, dispersant, binding and thickening agent for emulsion polymerization, dry cleaning, leather, min. oil emulsions, paper industry, wall-tile mastics, solv. emulsions; liq.; HLB 5.9; 99% act.

Nopalcol 2-L. [Henkel/Functional Prods.] PEG-4 laurate; CAS 9004-81-3; nonionic; emulsifier; liq.; HLB 10.5; 100% conc.

Nopalcol 4-C. [Henkel/Functional Prods.] PEG-8 cocoate; nonionic; general purpose emulsifier, plasticizer, lubricant, wetting agent, dispersant, binding and thickening agent for emulsion polymerization, dry cleaning, leather, min. oil emulsions, paper industry, wall-tile mastics, solv. emulsions; liq.; HLB 13.9; 99% act.

Nopalcol 4-CH. [Henkel/Functional Prods.] PEG-8 cocoate (hyd.); nonionic; general purpose emulsifier, plasticizer, lubricant, wetting agent, dispersant, binding and thickening agent for emulsion polymerization, dry cleaning, leather, min. oil emulsions, paper industry, wall-tile mastics, solv. emulsions; liq.; HLB 13.9; 99% act.

Nopalcol 4-DTW. [Henkel/Functional Prods.] PEG-8 ditallowate; nonionic; emulsifier; solid; HLB 9.1; 99% conc.

Nopalcol 4-L. [Henkel/Functional Prods.] PEG-8 laurate; CAS 9004-81-3; nonionic; general purpose emulsifier, plasticizer, lubricant, wetting agent, dispersant, binding and thickening agent for emulsion polymerization, dry cleaning, leather, min. oil emulsions, paper industry, wall-tile mastics, solv. emulsions; liq.; HLB 13.1; 99% act.

Nopalcol 4-O. [Henkel/Functional Prods.] PEG-8 oleate; CAS 9004-96-0; nonionic; emulsifier; dispersant for leather pigments; paper coating defoamer, plasticizer and leveling agent; liq.; HLB 11.7; 95% act.

Nopalcol 4-S. [Henkel/Functional Prods.] PEG-8 stearate; CAS 9004-99-3; nonionic; general purpose emulsifier, plasticizer, lubricant, wetting agent, dispersant, binding and thickening agent for emulsion polymerization, dry cleaning, leather, min. oil emulsions, paper industry, wall-tile mastics, solv. emulsions; paste; HLB 11.3; 99% act.

Nopalcol 6-DO. [Henkel/Functional Prods.] PEG-12 dioleate; CAS 9005-07-6; nonionic; general purpose emulsifier, plasticizer, lubricant, wetting agent, dispersant, binding and thickening agent for emulsion polymerization, dry cleaning, leather, min. oil emulsions, paper industry, wall-tile mastics, solv. emulsions; liq.; HLB 12.7; 98% act.

Nopalcol 6-DTW. [Henkel/Functional Prods.] PEG-12 ditallowate; nonionic; general purpose emulsifier, plasticizer, lubricant, wetting agent, dispersant, binding and thickening agent for emulsion polymerization, dry cleaning, leather, min. oil emulsions, paper industry, wall-tile mastics, solv. emulsions; paste; HLB 11.2; 99% act.

Nopalcol 6-L. [Henkel/Functional Prods.] PEG-12 laurate; CAS 9004-81-3; nonionic; general purpose emulsifier, plasticizer, lubricant, wetting agent, dispersant, binding and thickening agent for emulsion polymerization, dry cleaning, leather, min. oil emulsions, paper industry, wall-tile mastics, solv. emulsions; liq.; HLB 15.0; 99% act.

Nopalcol 6-O. [Henkel/Functional Prods.] PEG-12 oleate; CAS 9004-96-0; nonionic; emulsifier, wetting agent for oil phase systems; liq.; HLB 13.6; 99% act.

Nopalcol 6-R. [Henkel/Functional Prods.] PEG-12 ricinoleate; CAS 9004-97-1; nonionic; general purpose emulsifier, plasticizer, lubricant, wetting agent, dispersant, binding and thickening agent for emulsion polymerization, dry cleaning, leather, min. oil emulsions, paper industry, wall-tile mastics, solv. emulsions; liq.; HLB 13.9; 99% act.

Nopalcol 6-S. [Henkel/Functional Prods.] PEG-12 stearate; CAS 9004-99-3; nonionic; general purpose emulsifier, plasticizer, lubricant, wetting agent, dispersant, binding and thickening agent for emulsion polymerization, dry cleaning, leather, min. oil emulsions, paper industry, wall-tile mastics, solv. emulsions; paste; HLB 13.2; 99% act.

Nopalcol 10-CO. [Henkel/Functional Prods.] PEG-20 castor oil; nonionic; emulsifier; liq.; HLB 10.3; 99% conc.

Nopalcol 10-COH. [Henkel/Functional Prods.] PEG-20 hydrog. castor oil; CAS 61788-85-0; nonionic; general purpose emulsifier, plasticizer, lubricant, wetting agent, dispersant, binding and thickening agent for emulsion polymerization, dry cleaning, leather, min. oil emulsions, paper industry, wall-tile mastics, solv. emulsions; liq.; HLB 10.3; 99% act.

Nopalcol 12-CO. [Henkel/Functional Prods.] PEG 1200 castor oil; nonionic; general purpose emulsifier, plasticizer, lubricant, wetting agent, dispersant, binding and thickening agent for emulsion polymerization, dry cleaning, leather, min. oil emulsions, paper industry, wall-tile mastics, solv. emulsions; liq.; HLB 11.3; 99% act.

Nopalcol 12-COH. [Henkel/Functional Prods.] PEG 1200 hydrog. castor oil; CAS 61788-85-0; nonionic; general purpose emulsifier, plasticizer, lubricant, wetting agent, dispersant, binding and thickening agent for emulsion polymerization, dry cleaning, leather, min. oil emulsions, paper industry, wall-tile mastics, solv. emulsions; liq.; HLB 11.3; 99% act.

Nopalcol 12-R. [Henkel/Functional Prods.] PEG 1200 castor oil; nonionic; emulsifier; liq.; HLB 16.0; 100% conc.

Nopalcol 19-CO. [Henkel/Functional Prods.] PEG 1900 castor oil; nonionic; general purpose emulsifier, plasticizer, lubricant, wetting agent, dispersant, binding and thickening agent for emulsion polymerization, dry cleaning, leather, min. oil emulsions, paper industry, wall-tile mastics, solv. emulsions; liq.; HLB 14.9; 99% conc.

Nopalcol 30-S. [Henkel/Functional Prods.] PEG 3000 stearate; CAS 9004-99-3; nonionic; emulsifier; solid; HLB 18.5; 100% conc.

Nopalcol 30-TWH. [Henkel/Functional Prods.] PEG-60 hydrog. tallowate; nonionic; general purpose emulsifier, plasticizer, lubricant, wetting agent, dispersant, binding and thickening agent for emulsion polymerization, dry cleaning, leather, min. oil emulsions, paper industry, wall-tile mastics, solv. emulsions; solid; HLB 18.0; 99% act.

Nopalcol 200. [Henkel/Functional Prods.] PEG 200; nonionic; general purpose emulsifier, plasticizer, lubricant, wetting agent, dispersant, binding and thickening agent for emulsion polymerization, dry cleaning, leather, min. oil emulsions, paper industry, wall-tile mastics, solv. emulsions; liq.; 99% act.

Nopalcol 400. [Henkel/Functional Prods.] PEG 400; nonionic; general purpose emulsifier, plasticizer, lubricant, wetting agent, dispersant, binding and thickening agent for emulsion polymerization, dry cleaning, leather, min. oil emulsions, paper industry, wall-tile mastics, solv. emulsions; liq.; 99% act.

Nopalcol 600. [Henkel/Functional Prods.] PEG 600; nonionic; general purpose emulsifier, plasticizer, lubricant, wetting agent, dispersant, binding and thickening agent for emulsion polymerization, dry cleaning, leather, min. oil emulsions, paper industry, wall-tile mastics, solv. emulsions; paste; 99% act.

Nopalcol Series. [Henkel/Functional Prods.] PEG fatty esters; dispersant, wetting agent, emulsifier, plasticizer, lubricant, binder, thickener for cosmetics, distillation defoamers, dry cleaning agents, leather and paper industries; solv., textiles, wall tile mastic; liq. to solid; 100% conc.

Nopco® 1179. [Henkel/Functional Prods.] Fatty amido condensate; nonionic; wool and worsted fulling and scouring; general purpose detergent; liq.; HLB 14.0; 100% conc.

Nopco® 1186A. [Henkel/Functional Prods.] Sulfonated alkyl ester; anionic; wetting agent for fibrous materials; emulsifier in latex mfg.; liq.; 65% conc.

Nopco® 1419-A. [Henkel] Defoamer for SBR, PVAc, acrylic paint and adhesive systems; amber liq.; forms milky emulsion in water; sp.gr. 0.85; dens. 7.1 lb/gal; pH 6.5 (2%); flash pt. 47 C; 99% act.

Nopco® 1471. [Henkel/Functional Prods.; Henkel-Nopco] High sulfated veg. oil; anionic; emulsifier for min. and veg. oils; liq.; 68% conc.

Nopco® 2031. [Henkel/Functional Prods.; Henkel-Nopco] Sulfonated fatty prod., sodium neutralized; anionic; emulsifier for emulsion polymerization; liq.; 60% conc.

Nopco® 2272-R. [Henkel/Functional Prods.; Henkel-Nopco] Highly sulfated fatty ester; anionic; wetting agent in paper towels, latex impregnation and dipping, in metal treating and textile dyeing; liq.; 65% conc.

Nopco® JMY. [Henkel] Defoamer for PVAc, acrylic paint and adhesive systems; amber liq.; forms milky emulsion in water; sp.gr. 0.88; dens. 7.3 lb/gal; pH 6.5 (2%); flash pt. 116 C; 99% act.

Nopco® NDW. [Henkel] Defoamer for SBR, PVAc, acrylic, water-sol. resins paint and adhesive systems; amber slightly hazy liq.; disp. in all latex and adhesive systems; sp.gr. 0.90; dens. 7.5 lb/gal; pH 6.5 (2%); flash pt. 79 C; 99.9% act.

Nopco® NXZ. [Henkel] Defoamer for syn. latex emulsions, paint and adhesives from SBR, PVAc, acrylic, water-sol. resins; amber hazy liq.; forms milky emulsion in water; sp.gr. 0.91; dens. 7.6 lb/gal; pH 7 (2%); flash pt. 171 C; 99% act.

Nopco® PD#1-D. [Henkel] Defoamer for adhesives, paints, joint compds., plaster; off-wh. powd.; water-wettable; dens. 22–26 lb/ft3; pH 8.7 (1%); 65% act.

Nopco® RDY. [Henkel] Nonoxynol-50; CAS 9016-45-9; nonionic; agric. surfactant; solid; cloud pt. 100

C; pour pt. 46 C.

Nopcocastor. [Henkel/Functional Prods.] Sulfated castor oil; anionic; emulsifier; superfatting agent for cosmetics; liq.; 75% conc.

Nopcocaster L. [Henkel/Functional Prods.] Sulfated castor oil; anionic; emulsifier, ironing aid; liq.; 50% conc.

Nopco® Colorsperse 188-A. [Henkel-Nopco] Ethoxylated fatty acid; nonionic; pigment wetting and dispersing; liq.; 100% conc.

Nopcogen 14-L. [Henkel/Functional Prods.; Henkel-Nopco] Lauric alkylolamine condensate; detergent, wetting, emulsifier; liq.; 100% conc.

Nopcogen 14-LT. [Henkel-Nopco] Coconut fatty acid alkylolamide; nonionic; detergent; cleaning and sterilizer of surgical instruments; liq.; 99+% conc.

Nopcogen 14-S. [Henkel-Nopco] Stearamide DEA; textile softener; solid; m.p. 48 C; 100% conc.

Nopcogen 16-L. [Henkel-Nopco] Lauric polyamine condensate; CAS 2016-56-0; emulsifier and textile finishing agent; solid; m.p. 60 C; 100% conc.

Nopcogen 16-O. [Henkel-Nopco] Oleic polyamine condensate; emulsifier; w/o emulsions with min. and veg. oils; solid; m.p. 44 C; 100% conc.

Nopcogen 22-O. [Henkel/Functional Prods.] Oleyl imidazoline; cationic; emulsifier for min. oil, kerosene, wetting agent, corrosion inhibitor, used in textile, asphalt, paper, agric. industries, car wax formulations, acid detergents; amber visc. liq.; sol. in water, acetic acid, HCl; misc. with pine oil and min. oil; cloud pt. 0 C; 90% act.

Nopcosperse 28-B. [Henkel/Textile] Nonionic; textile dispersant for pigments providing high grinding efficiency and superior color yield; tan powd.; sol. forms clear sol'ns. water; pH 7.0 (5%).

Nopcosperse AD-6. [Henkel/Emery] Polyacrylate; anionic; dispersant for dry flowables, wettable powds., aq. suspension pesticide formulations; lt. amber liq.; pH 7.8 (10%); 35% act.

Nopcosperse WEZ. [Henkel/Textile] Nonionic; textile dispersant for pigments providing high grinding efficiency and superior color yield at higher temps. than Nopcosperse 28-B; lt. yel. clear liq.; sol. forms clear sol'ns. water; pH 7.0 (5%).

Nopcosulf CA-60, -70. [Henkel-Nopco] Sulfated castor oil; anionic; softener in finishing starches, gums; plasticizer for starches; furniture base polish; liq.; 68 and 63% conc. resp.

Nopcosulf TA-30. [Henkel/Functional Prods.] Sulfated tallow; anionic; softener for cotton goods; paste; 44% conc.

Nopcosulf TA-45V. [Henkel/Functional Prods.; Henkel-Nopco] Sulfated tallow; anionic; softener for cotton goods, in finishing starches and gums; plasticizer for starches; furniture polish base; paste; 78% conc.

Nopcosurf CA. [Henkel/Functional Prods.] Sulfated castor oil; anionic; softener in finishing starches, gums; plasticizer for starches; furniture polish base; liq.; 68% conc.

Noram 2C. [Ceca SA] N-Dicoco amine; CAS 61789-75-2; cationic; industrial detergent; synthesis intermediate, anticaking agent, flotation, antistripping for road making, soil stabilization; auxs. for fuel additives, rust inhibition, paint; chemical intermediate for quats., betaines, amine oxides; solid.

Noram 2SH. [Ceca SA] N-Dihydrog. tallow amine; CAS 61789-79-5; cationic; industrial detergent; synthesis intermediate; oil industry; anti-caking agent; solid.

Noram C. [Ceca SA] N-Coco amine; CAS 61788-66-3; cationic; chemical intermediate for quaternaries, betaines; industrial detergent; liq.

Noram DMC. [Ceca SA] N-Coco dimethylamine; cationic; industrial detergent; synthesis intermediate, anticaking agent, flotation, antistripping for road making, soil stabilization; auxs. for fuel additives, rust inhibition, paint, cosmetics; chemical intermediate for quats., betaines, amine oxides; liq.

Noram DMCD. [Ceca SA] N-Coco dimethylamine; CAS 61788-93-0; cationic; industrial detergent; chemical intermediate for quaternaries, betaines, amine oxides; liq.

Noram DMS. [Ceca SA] N-Tallow dimethylamine; see Noram 2 C; liq.

Noram DMSD. [Ceca SA] N-Tallow dimethylamine; CAS 68814-69-7; cationic; industrial detergent; chemical intermediate for quaternaries, betaines, amine oxides; liq.

Noram DMSH D. [Ceca SA] N-hydrog. tallow dimethylamine; CAS 61788-95-2; cationic; industrial detergent; chemical intermediate for quaternaries, betaines, amine oxides; cosmetics; pasty.

Noram M2C. [Ceca SA] N-Dicoco methylamine; CAS 61788-62-3; cationic; industrial detergent; synthesis and chemical intermediate for quaternaries; pasty.

Noram M2SH. [Ceca SA] N-Dihydrog. tallow methylamine; CAS 61788-63-6; cationic; industrial detergent; synthesis intermediate, anticaking agent, flotation, antistripping for road making, soil stabilization; auxs. for fuel additives, rust inhibition, paint; chemical intermediate for quats., betaines, amine oxides; solid.

Noram O. [Ceca SA] Oleamine; CAS 112-90-3; cationic; industrial detergent; synthesis intermediate, lubricant and textile industries; liq., pasty.

Noram S. [Ceca SA] N-tallow amine; CAS 61790-33-8; cationic; industrial detergent; synthesis intermediate; lubricant, textile, and oil industries; pasty.

Noram SH. [Ceca SA] N-Hydrog. tallow amine; CAS 61788-45-2; cationic; industrial detergent; anti-caking agent, corrosion inhibitor; flakes.

Noramac C. [Ceca SA] Cocamine acetate; cationic; dispersant, bactericide, emulsifier, anticaking, soil stabilization, flotation collector, flocculation, corrosion inhibitor; industrial detergent.

Noramac C 26. [Ceca SA] N-coco amine acetate; CAS 61790-57-6; cationic; dispersant, bactericide, emulsifier, anticaking, soil stabilization, flocculation, corrosion inhibitor; industrial detergent; liq.; 26% conc.

Noramac O. [Ceca SA] Oleamine acetate; cationic; flotation collector; industrial detergent.

Noramac S. [Ceca SA] Tallowamine acetate; CAS 2190-04-7; cationic; flotation collector; industrial detergent.

Noramac SH. [Ceca SA] N-hydrog. tallow amine acetate; CAS 61790-59-8; cationic; flotation agent, bactericide, emulsifier, anticaking agent, soil stabilizer, flocculant, corrosion inhibitor; industrial detergent; solid flakes.

Noramium DA.50. [Ceca SA] Coco dimethyl benzyl ammonium chloride; CAS 139-07-1; cationic; bactericide, fungicide, demulsification of hydrocarbons, cosmetics, latex coagulation, flotation, electrostatic paints; 50% conc. in water.

Noramium MC 50. [Ceca SA] Coco trimethyl ammonium chloride; CAS 61789-18-2; cationic; additive for antibiotics mfg.; liq.; 50% conc.

Noramium MO 50. [Ceca SA] Oleyl trimethyl ammonium chloride; cationic; additive for antibiotics mfg.; liq.

Noramium MS 50. [Ceca SA] Tallow trimethyl ammonium chloride; CAS 8030-78-2; cationic; emulsifier; pharmaceuticals; liq.; 50% conc.

Noramium MSH 50. [Ceca SA] Hydrog. tallow trimethyl ammonium chloride; cationic; additive for antibiotics mfg.; liq.; 75% conc.

Noramium S 75. [Ceca SA] Tallow dimethyl benzyl ammonium chloride; cationic; additive for antibiotics mfg.

Noramox C2. [Ceca SA] PEG-2 cocamine; CAS 61791-14-8; cationic; emulsifier, drying assistant, viscose additive, metal treatment, rust inhibitor; industrial detergent; antistat for ABS, PS; liq.

Noramox C5. [Ceca SA] PEG-5 cocamine; CAS 61791-14-8; cationic; industrial detergent.

Noramox C11. [Ceca SA] PEG-11 cocamine; CAS 61791-14-8; cationic; industrial detergent.

Noramox C15. [Ceca SA] PEG-15 cocamine; CAS 61791-14-8; cationic; industrial detergent.

Noramox O2. [Ceca SA] PEG-2 oleamine; cationic; industrial detergent.

Noramox O5. [Ceca SA] PEG-5 oleamine; cationic; industrial detergent.

Noramox O11. [Ceca SA] PEG-11 oleamine; cationic; industrial detergent.

Noramox O15. [Ceca SA] PEG-15 oleamine; cationic; industrial detergent.

Noramox O20. [Ceca SA] PEG-20 oleamine; cationic; industrial detergent.

Noramox S1. [Ceca SA] PEG-1 tallowamine; cationic; see Noramox C2; pasty.

Noramox S2. [Ceca SA] PEG-2 tallowamine; cationic; industrial detergent.

Noramox S5. [Ceca SA] PEG-5 tallowamine; cationic; industrial detergent.

Noramox S7. [Ceca SA] PEG-7 tallowamine; cationic; industrial detergent.

Noramox S11. [Ceca SA] PEG-11 tallowamine; cationic; industrial detergent.

Noramox S15. [Ceca SA] PEG-15 tallowamine; cationic; industrial detergent.

Noramox S20. [Ceca SA] PEG-20 tallowamine; cationic; industrial detergent.

Norfox® 1 Polyol. [Norman, Fox] Hydrophobic EO/PO block polymer; nonionic; defoamer base for surfactant mixtures; liq.; HLB 3.0; 100% conc.

Norfox® 2LF. [Norman, Fox] Low foam EO/PO block polymer; nonionic; detergent base for low foam; liq.; HLB 7.0; 100% conc.

Norfox® 2 Polyol. [Norman, Fox] EO/PO block polymer; nonionic; detergent base for controlled foam; liq.; HLB 7.0; 100% conc.

Norfox® 4 Polyol. [Norman, Fox] Hydrophilic EO/PO block polymer; nonionic; controlled foam solubilizer; liq.; HLB 11.0; 100% conc.

Norfox® 40. [Norman, Fox] Sodium dodecylbenzene sulfonate; anionic; base for industrial and agric. wetting agents and detergents; dry blending applics.; beads; 40% conc.

Norfox® 85. [Norman, Fox] Sodium dodecylbenzene sulfonate; anionic; base for industrial and agric. wetting agents and detergents; flakes, gran.; 85% conc.

Norfox® 90. [Norman, Fox] Sodium dodecylbenzene sulfonate; anionic; base for industrial and agric. wetting agents and detergents; also used in paper industry; flakes, gran.; HLB 20.0; 90% conc.

Norfox® 243. [Norman, Fox] Silicone emulsion; defoamer for industrial applics.; liq.; 30% act.

Norfox® 916. [Norman, Fox] Ethoxylated linear primary alcohol; nonionic; wetting agent; wet processing aid for paper, textiles and leather; liq.; HLB 12.5; 100% conc.

Norfox® 1101. [Norman, Fox] Potassium cocoate; CAS 61789-30-8; anionic; flash foamer, emulsifier for shampoo bases, liq. hand soaps; lubricant for conveyors; coupling agent for heavily built liq. alkali systems such as steam cleaners and whitewall tire cleaners; liq.; HLB 20.0; sapon. no. 250 min.; pH 10 (1%); 40% solids.

Norfox® 1115. [Norman, Fox] Potassium cocoate; CAS 61789-30-8; anionic; soap base; emulsifier, flash foamer, lubricant for liq. soap for dispensers, hand soap conc.; liq.; HLB 20.0; sapon. no. 225 min.; pH 10 (1%); 35-37% solids.

Norfox® Agent 2A-2S. [Norman, Fox] Air entraining agent, wetting agent for prod. of mortar, stucco and plastic cement; straw-colored visc. liq.; dens. 8.7 lb/gal; pour pt. 40 F; usage level: 0.7-1.0 lb/ton.

Norfox® ALES-60. [Norman, Fox] Ammonium pareth sulfate; anionic; window cleaner surfactant; liq.; 60% conc.

Norfox® ALKA. [Harcros UK] Fatty amidoalkyl betaine; amphoteric; substantive surfactant, flash foamer for industrial, household and personal care prods. incl. heavy-duty caustic steam cleaners, baby shampoos, bubble baths; exc. electrolyte tolerance; yel. clear liq.; sp.gr. 1.04; cloud pt. 0 C; pH 6.0-7.5 (1% aq.); 29-32% act., 34-37% total solids.

Norfox® ALPHA XL. [Norman, Fox] Sodium C14-16 alpha olefin sulfonate; anionic; base for shampoos, hand soaps, bath prods., home and janitorial cleaners, dishwash, and lt. duty liqs.; Klett 100 max. liq.; pH 7-9 (5%); 38-41% act.

Norfox® ALS. [Norman, Fox] Ammonium lauryl sulfate; anionic; base for woolen carpet shampoo, hair shampoo; liq.; 30% conc.

Norfox® Anionic 27. [Norman, Fox] Sodium 2-ethylhexyl sulfate; anionic; wetting agent and peeling aid; liq.; 27% conc.

Norfox® CMA. [Norman, Fox] Cocamide MEA; CAS 68140-00-1; nonionic; surfactant for dry fabric cleaners, carpet shampoos; flake; 100% conc.

Norfox® Coco Powder. [Norman, Fox] Sodium cocoate; anionic; soap base; stabilizer; gelling agent; for powd. hand soaps; gran.; 92% conc.

Norfox® DC. [Norman, Fox] Cocamide DEA and diethanolamine; CAS 61791-31-9; nonionic; intermediate for detergent mfg., liq. dishwash, cosmetics; suds stabilizer and dedusting agent for dry prods.; straw-colored visc. liq.; sp.gr. 1.02; pour pt. -8 C; flash pt. 166 C; pH 9.5 (1%); 100% conc.

Norfox® DC-38. [Norman, Fox] Modified cocamide DEA; primary surfactant, thickener, stabilizer for home care and janitorial formulations, wax strippers, hard surf. cleaners; straw-colored visc. liq.; dens. 8.3 lb/gal; flash pt. (COC) 345 F; pH 9.2 (1%).

Norfox® DCO. [Norman, Fox] Oleic acid modified coconut oil DEA condensate; anionic/nonionic; biodeg. cleaning agent, lubricant, visc. builder; metal weaving, drawing, stamping, burnishing or cold rolling lubricant; heavy duty cleaner for soft water areas; amber visc. liq.; sp.gr. 1.002; dens. 8.34 lb/gal; visc. 19.8 poises; pour pt. 10 F; ref. index 1.4730; pH 9.15 (5%); 100% conc.

Norfox® DCS. [Norman, Fox] Alkylaryl sulfonate modified coconut oil DEA condensate; anionic/

nonionic; biodeg. wetting agent and detergent base for textiles, paper, auto and home care prods.; shampoos, bubble bath, liq. dishwash, hand soap, car wash; dye dispersant; demulsifier; straw-colored visc. liq., bland odor; sp.gr. 1.05; pour pt. 40 F; pH 9.0 (1-5%); 100% act.

Norfox® DCSA. [Norman, Fox] Cocamide DEA; CAS 61791-31-9; nonionic; foam booster/stabilizer, detergent, wetting agent, visc. builder for household/industrial cleaners, drycleaning, cosmetics, lt. duty detergents, dishwash, shampoos, bubble bath; Gardner < 4 liq.; sol. in water, alcohol, chlorinated hydrocarbons, aromatic hydrocarbons, natural fats and oils; disp. in min. oil; sp.gr. 1.0016; dens. 8.3 lb/gal; visc. 11.5 poises; m.p. 0 C; acid no. < 2.0; amine no. < 38; flash pt. (OC) 405 F; fire pt. 415 F; pH 10 (1%); 100% conc.

Norfox® DESA. [Norman, Fox] Cocamide DEA; nonionic; visc. and foam enhancer; liq.; 100% conc.

Norfox® DF210SX. [Norman, Fox] Silicone emulsion; defoamer for spas; liq.; 10% act.

Norfox® DOSA. [Norman, Fox] Cocamide DEA; nonionic; foam stabilizer, emollient, conditioner, emulsifier, corrosion inhibitor; visc. builder for use at reduced concs.; for shampoos, bubble baths, industrial applics.; lubricant for auto-wash brushes; degreaser for heavy-duty cleaners; lt. yel. oily liq.; amine no. < 2.0; pH 10 (1% aq.); 100% conc.

Norfox® EGMS. [Norman, Fox] Glycol stearate; nonionic; pearlescent ingred. in liq. shampoos and detergents; emulsification aid and bodying agent; beads; 100% org.

Norfox® F-221. [Norman, Fox] Oleamide DEA; nonionic; invert emulsifier with hydraulic props.; wetting agent, penetrant; salt tolerant when dissolved in oil phase over a wide temp. range; for herbicides, drilling fluids for petrol. industry, penetrating oils, specialty lubricants; liq.; HLB 5.0; usage level: 0.5-5.0%; 100% conc.

Norfox® F-342. [Norman, Fox] Amine condensate blend; nonionic; o/w emulsifier, wetting agent, penetrant for oil-based systems, drycleaning, spot removers; lt. amber liq., char. odor; sol. in kerosene, diesel fuel, paraffin oil, ethanol, IPA; disp. in water; dens. 8 lb/gal; pour pt. -6 F; flash pt. (CC) 210 F; toxicology: LD50 (rat) 9.74 cc/kg; 100% conc.

Norfox® GMS. [Norman, Fox] Glyceryl stearate; nonionic; lotion and cream base in cosmetics; opacifier in liq. shampoos and detergents; emulsion stabilizer; beads; 40–80% alpha mono, 100% org.

Norfox® Hercules Conc. [Norman, Fox] Built detergent blend; anionic/nonionic; heavy-duty cleaning, truck, auto washing; degreaser; liq.; 25% conc.

Norfox® IM-38. [Norman, Fox] N-Oleyl imidazolinium hydrochloride; cationic; high-foaming wetting agent and emulsifier; base for automotive rinse-wax formulations, corrosion inhibition compds.; amber liq., mild char. odor; sol. in water, HCl, min. oil, kerosene, benzene; sp.gr. 0.95 (30/30 C); pour pt. -20 F; pH 6.1 (5%); 52% act.

Norfox® KD. [Norman, Fox] Cocamide DEA and diethanolamine; CAS 61791-31-9; nonionic; visc. enhancer and foam booster/stabilizer for shampoos or cleaners; straw-colored liq.; dens. 8.2 lb/gal; pH 9.4 (1%); 100% conc.

Norfox® KO. [Norman, Fox] Potassium oleate; anionic; liq. soap for hand cleaners, tire mounting lubricant; emulsifier and corrosion control in paint strippers; surfactant in insecticides; liq.; HLB 20.0; 80% conc.

Norfox® LAS-99. [Norman, Fox] Sodium dodecylbenzene sulfonate; anionic; used as is in acidic media and/or neutralized in neutral/alkaline applics.; liq.; HLB 20.0; 97% conc.

Norfox® MLS. [Norman, Fox] Magnesium lauryl sulfate; anionic; carpet shampoo base; liq.; 30% conc.

Norfox® NP-1. [Norman, Fox] Nonoxynol-1; nonionic; emulsion stabilizer and defoamer; liq.; oil-sol.; HLB 4.5; 100% conc.

Norfox® NP-4. [Norman, Fox] Nonoxynol-4; CAS 9016-45-9; nonionic; emulsifier, detergent and dispersant for petrol. based lubricants; intermediate for sulfonation to produce foaming agent; liq.; oil-sol.; HLB 9.0; 100% conc.

Norfox® NP-6. [Norman, Fox] Nonoxynol-6; CAS 9016-45-9; nonionic; emulsifier; wetting agent for oil-based systems; liq.; HLB 11.0; 100% conc.

Norfox® NP-7. [Norman, Fox] Nonoxynol-7 (7.5 EO); CAS 9016-45-9; nonionic; detergent for phosphoric, citric or lactic acid based cleaners, emulsion cleaners; liq.; HLB 12-13; 100% conc.

Norfox® NP-9. [Norman, Fox] Nonoxynol-9; CAS 9016-45-9; nonionic; formulation of iodophor sanitizers; replacement for dedusting oil in heavy-duty detergent; liq.; HLB 13.0; 100% conc.

Norfox® NP-11. [Norman, Fox] Nonoxynol-11; CAS 9016-45-9; nonionic; formulation of iodophor sanitizers; replacement for dedusting oil in heavy-duty cleaners; liq.; HLB 13.0-14.0; 100% conc.

Norfox® Oleic Flakes. [Norman, Fox] Sodium oleate; anionic; textile scouring, fulling, dye leveling agent; flake; HLB 20.0; 90% conc.

Norfox® OP-45. [Norman, Fox] Ethoxylated octylphenol; nonionic; emulsifier for oil systems, polishes, drycleaning; liq.; HLB 10.4; 100% conc.

Norfox® OP-100. [Norman, Fox] Ethoxylated octylphenol; nonionic; high foaming surfactant for household and industrial cleaners; liq.; HLB 13.5; cloud pt. 60-70 C; 100% conc.

Norfox® OP-102. [Norman, Fox] Ethoxylated octylphenol; nonionic; surfactant for hot cleaning systems; liq.; HLB 14.6; cloud pt. 86-90 C; 100% conc.

Norfox® OP-114. [Norman, Fox] Ethoxylated octylphenol; nonionic; surfactant for soak tank, household and industrial cleaners; liq.; HLB 12.4; cloud pt. 22-26 C; 100% conc.

Norfox® PE-600. [Norman, Fox] Acid phosphate ester of ethoxylated alcohol (CAS #68909-65-9); anionic; coupling agent and surfactant for conc. builder sol'ns.; liq.; 100% conc.

Norfox® PEA-N. [Norman, Fox] Cocamide DEA, phosphate-modified; anionic; emulsifier, thickener, degreaser, detergent, wetting agent for textiles, metal cleaners, institutional cleaners; emulsifier for water/solv. systems; amber visc. liq.; sol. in water, alcohols, aromatic and chlorinated solvs.; sp.gr. 1.015; dens. 8.46 lb/gal; acid no. 38-43; amine no. 82-92; 100% act.

Norfox® PE-LF. [Norman, Fox] Ethoxylated alcohol phosphate (CAS #51811-79-1); anionic; controlled foam coupling agent for conc. builder sol'n.; liq.; HLB 20.0; 100% conc.

Norfox® PE-W. [Norman, Fox] Mono- and dialkyl acid phosphate (CAS #51811-79-1); anionic; detergent, emulsifier, coupling agent, corrosion inhibitor; liq.; 100% conc.

Norfox® SEHS. [Norman, Fox] Sodium 2-ethylhexyl sulfate; anionic; low foamer with wetting and detergency; coupling agent; stable in high electrolyte media; liq.; 40% conc.

Norfox® SLES-02. [Norman, Fox] Sodium laureth sulfate; CAS 9004-82-4; anionic; foaming and wetting agent for household and cosmetic specialties; liq.; 28% conc.

Norfox® SLES-03. [Norman, Fox] Sodium laureth sulfate; CAS 9004-82-4; anionic; foaming agent; liq.; 28% conc.

Norfox® SLES-30. [Norman, Fox] Sodium pareth sulfate; anionic; foaming detergent base; ingred. in window cleaners, car wash base; liq.; 30% conc.

Norfox® SLES-60. [Norman, Fox] Sodium laureth sulfate, 14% denatured alcohol; CAS 9004-82-4; anionic; base for shampoos and lt. duty liq. formulators; flash foam enhancer for automotive, household, personal care and industrial cleaners; in mfg. of gypsum wallboard, gas drilling of deep wells; APHA 100 liq., mild alcoholic odor; sp.gr. 1.05; pH 7.7 (10%); 59% act.

Norfox® SLS. [Norman, Fox] Sodium lauryl sulfate; CAS 151-21-3; anionic; biodeg. foamer and wetting agent for household, industrial and personal care prods., shampoos, hand and body soaps, fabric care prods.; Gardner 1 max. paste; sp.gr. 1.02; dens. 8.5 lb/gal; HLB 20.0; 28-30% act.

Norfox® Sorbo S-80. [Norman, Fox] Sorbitan oleate; CAS 1338-43-8; nonionic; hydrophobic emulsifier; liq.; HLB 4.3; 100% conc.

Norfox® Sorbo T-20. [Norman, Fox] Polysorbate 20; CAS 9005-64-5; nonionic; flavor and fragrance solubilizer; liq.; HLB 16.7; 100% act.

Norfox® SXS40, SXS96. [Norman, Fox] Sodium xylene sulfonate; CAS 1300-72-7; anionic; coupling agent for detergent applics.; liq., flaked solid resp.

Norfox® T-60. [Norman, Fox] TEA dodecylbenzene sulfonate; anionic; detergent, wetting agent and foamer for agric. and industrial/household use, lt. duty detergents, hard surf. cleaners, shampoos, wool and fine fabric washing; yel. clear visc. liq., mild odor; dens. 9.1 lb/gal; HLB 20.0; pH 5.7; 60% act.

Norfox® TB Granules. [Norman, Fox] Sodium tallow/coco soap; anionic; soap base for mfg. specialty bar soap; ingred. in glycerine bar soap; gran.; 92% conc.

Norfox® TLS. [Norman, Fox] TEA lauryl sulfate; CAS 139-96-8; biodeg. mild ingred. for shampoo, lt. duty liqs., fine fabric detergents; Gardner 2 max. liq.; sp.gr. 1.05; dens. 8.8 lb/gal; visc. 100 cps (30 C); HLB 20.0; pH 7.2 (10% aq.); 39-42% act.

Norfox® Unimulse OW. [Norman, Fox] Polymer/surfactant; nonionic/anionic; emulsifier for oily liqs. into oil and water emulsions; soft paste; usage level: 0.3-1.0%; 50% act.

Norfox® Vertex Flakes. [Norman, Fox] Sodium oleate; anionic; textile scouring, fulling, dye leveler; flake; HLB 20.0; 90% conc.

Norfox® X. [Norman, Fox] Cocamide DEA; modified; cationic/nonionic; used with acidic builders; wash-wax formulations, glass and appliance cleaners; liq.; sp.gr. 1.05; pH 5.5-7.0 (10% aq.); 93% conc.

Novel® 1412-70. [Vista] Ethoxylated alcohol (11.0 EO); surfactant intermediate; wh. solid; HLB 14.0.

Novel® 1412-70 Ethoxylate. [Vista] Linear alcohol, ethoxylated (70% EO); nonionic; surfactant, wetting agent, emulsifier, detergent formulation; wh. solid; HLB 11.0; 100% conc.

Noxamine C2-30. [Ceca SA] Cocamine oxide; detergent additive.

Noxamine CA 30. [Ceca SA] Cocamine oxide; CAS 61788-90-7; nonionic; foaming and degreasing agent; detergent additive; liq.; 30% conc.

Noxamine O2-30. [Ceca SA] Oleamine oxide; CAS 61791-66-6; nonionic; foaming and rheolabic agent; detergent additive; liq.; 30% conc.

Noxamine S2-30. [Ceca SA] Tallowamine oxide; detergent additive.

Noxamium C2-15. [Ceca SA] Ethoxylated cocamine, ammonium quat. deriv.; cationic; antistat, bactericidal emulsifier; liq.; 50% conc.

Noxamium S2-11. [Ceca SA] Tallow amines, ethoxylated, ammonium quat. deriv.; cationic; antistat, bactericidal emulsifier; liq.; 50% conc.

Noxamium S2-50. [Ceca SA] Ammonium quat. deriv. of ethoxylated tallow amines; cationic; antistat, bactericide, emulsifier; liq.; 50% conc.

Noxamium Series. [Ceca SA] Quat., ethoxylated; cationic; emulsifier, corrosion; liq.

NP-55-60. [Hefti Ltd.] Nonoxynol-6; CAS 9016-45-9; 27177-05-5; nonionic; wetting agent for agric. chemicals, color pigments; detergent component for prods. containing solvs.; emulsifier for veg. oils, castor oil; liq.; HLB 11.0; 100% conc.

NP-55-80. [Hefti Ltd.] Nonoxynol-8; CAS 9016-45-9; 27177-05-5; nonionic; low foaming cold wetting agent for textile industry, coloring pigments; detergent base material for industrial and household detergents; emulsifier; liq.; HLB 12.0; 100% conc.

NP-55-85. [Hefti Ltd.] Nonoxynol-8.5; CAS 9016-45-9; 27177-05-5; nonionic; low foaming wetting agent for textile industry; detergent base material; emulsifier; liq.; HLB 12.5; 100% conc.

NP-55-90. [Hefti Ltd.] Nonoxynol-9; CAS 9016-45-9; 27177-05-5; nonionic; wetting agent for textile, paper, leather, and other industries; detergent base material; emulsifier; liq.; HLB 13.0; 100% conc.

NSA-17. [Sanyo Chem. Industries] Proprietary; nonionic; additive for heavy-duty powd. detergent; increases detergency; powd.; 100% conc.

Nutrapon AL 2. [Clough] Ammonium lauryl ether (2) sulfate; anionic; surfactant for shampoos, bubble baths, dishwashing detergents; liq.; 25% conc.

Nutrapon AL 30. [Clough] Ammonium lauryl ether (3) sulfate; anionic; surfactant for shampoos, bubble baths, dishwashing, and lt. duty detergents; liq.; 26-28% conc.

Nutrapon AL 60. [Clough] Ammonium laureth-3 sulfate; anionic; surfactant for shampoo concs., bubble baths, dishwashing, and lt. duty detergents; liq.; 57-60% conc.

Nutrapon B 1365. [Clough] Sodium lauryl sulfate and ethylene glycol stearate; anionic/nonionic; shampoo and detergent blend with pearlizing agent for pearlescent formulations; for shampoos, bubble baths, liq. hand soaps; liq.; 26.5-28.5% act.

Nutrapon DL 3891. [Clough] Sodium lauryl sulfate; CAS 151-21-3; high foaming surfactant for personal care and industrial formulations; liq.; 28-30% act.

Nutrapon DW 0266. [Clough] Proprietary blend; liq. dishwashing conc.; liq.; 43-45% act.

Nutrapon ES-60 3568. [Clough] Sodium laureth (3) sulfate; CAS 9004-82-4; anionic; mild surfactant for personal care and industrial prods.; flash foam in hard water; liq.; 56-59% conc.

Nutrapon FA-50 0066. [Clough] Ammonium deceth sulfate; high foaming surfactant for specialty applics., e.g., sec. oil recovery; liq.; 49-51% act.

Nutrapon HA 3841. [Clough] Ammonium lauryl sulfate; anionic; surfactant for personal care prods.

and detergent cleaners; liq.; 27.0-28.5% conc.

Nutrapon KPC 0156. [Clough] Sodium laureth-3 sulfate; CAS 9004-82-4; anionic; surfactant for shampoos, bubble bath, dishwashing detergents; liq.; 27-28% act.

Nutrapon LD 0206. [Clough] Proprietary blend; liq. laundry detergent conc.; liq.; 40-42% act.

Nutrapon PP 3563. [Clough] Ammonium lauryl sulfate; used in nonalkaline shampoos, bubble baths, mild detergents and cleaners below pH 7.0; visc. liq.; 27-30% act.

Nutrapon RS 1147. [Clough] Sodium lauryl sulfate, sodium cocoyl sarcosinate; anionic; surfactant for rug and upholstery shampoo; high foaming, min. wetting, low cloud pt.; leaves dry residue for easy removal; liq.; 28-31% conc.

Nutrapon TD 3792. [Clough] Mixture based on tridecyl sulfate; anionic; wetting agent, emulsifier; high tolerance to electrolytes; liq.; 30% min. act.

Nutrapon TLS-500. [Clough] TEA-lauryl sulfate; anionic; mild ingred. in cosmetic and industrial formulations; liq.; 39% min. act.

Nutrapon W 1367. [Clough] Sodium lauryl sulfate; CAS 151-21-3; anionic; high foaming surfactant for personal care and industrial formulations; liq.; 28-30% conc.

Nutrapon WAC 3005. [Clough] Sodium lauryl sulfate; CAS 151-21-3; anionic; surfactant for personal care and industrial formulations; visc. liq.; 28-30% conc.

Nutrapon WAQ. [Clough] Sodium lauryl sulfate; CAS 151-21-3; anionic; surfactant for personal care prods. and industrial formulations; very visc. liq.; 28-30% conc.

Nutrapon WAQE 2364. [Clough] Sodium lauryl sulfate; CAS 151-21-3; high foaming surfactant for personal care and industrial formulations; liq.; 28.5-29.5% act.

Nutrol 100. [Clough] Octoxynol-9; CAS 9002-93-1; nonionic; wetting agent, dispersant for metal and acid cleaners, pesticides; liq.; HLB 13.0; 99% min. conc.

Nutrol 600. [Clough] Nonoxynol-9; CAS 9016-45-9; nonionic; detergent, emulsifier, wetting agent, dispersant; liq.; HLB 13.0; 99% min. conc.

Nutrol 611. [Clough] Nonoxynol-8; CAS 9016-45-9; nonionic; detergent, emulsifier, wetting agent, dispersant; liq.; HLB 12.2; 99% min. conc.

Nutrol 622. [Clough] Nonoxynol-4; CAS 9016-45-9; nonionic; emulsifier, lt. duty detergent, moderate foaming agent; liq.; HLB 8.8; 99% min. conc.

Nutrol 640. [Clough] Nonoxynol-15; CAS 9016-45-9; nonionic; emulsifier, lt. duty detergent, moderate foaming agent; liq.; HLB 15.0; 99% min. conc.

Nutrol 656. [Clough] Nonoxynol-11; CAS 9016-45-9; nonionic; emulsifier, lt. duty detergent, moderate foaming agent; liq.; HLB 13.6; 99% min. conc.

Nutrol Betaine OL 3798. [Clough] Cocamidopropyl betaine; amphoteric; foamer, foam stabilizer, wetting agent for industrial and household cleaners, dishwashing, liq. hand soaps; liq.; 40-44% conc.

Nutrol S-60 5350. [Clough] Ammonium nonyl phenoxy polyethoxy sulfate; anionic; emulsifier for use in polymers; high foaming detergent, wetting agent for textiles; liq.; 58 ± 1% conc.

Nutrol SXS 5418. [Clough] Sodium xylene sulfonate; CAS 1300-72-7; anionic; hydrotrope, coupling agent, solubilizer; liq.; pH 7-10; 39–42% conc.

Nuva F, FH. [Hoechst AG] Fluorocopolymer disp.; anionic; water and oil repellent finishing of textiles from syn. and natural fibers; ylsh. opalescent liq.

Nuva L. [Hoechst AG] Fluorine derivs.; oleophobic fur and leather finishing agents.

Nylomine Assistant DN. [ICI Colors] Blend of surfactants; anionic; dyeing assistant, leveling agent for dyeing fibers; yel. liq.; sol. in water; sp.gr. 1.0; f.p. 0 C; b.p. 100 C; pH 6.0–7.5 (1%); 55% act.

Nysist. [Eastern Color & Chem.] Anionic; surfactant improving dyeing of nylon with neutral metallic and acid dyestuffs.

Nysist LSO. [Eastern Color & Chem.] Ethoxylate; anionic; leveling agent and retarder producing uniformity in dyeing and good color value.

O

Oakite Defoamant. [Oakite Prods.] O/w emulsion containing org. esters, alcohols, silicone, hydrocarbons, and stabilizers; foam control agent for industrial applics., e.g., paper mill stock systems, gas dehydration units, amine scrubbing units, propane deasphalting units; milky wh. visc. liq., bland odor; sp.gr. 1.002; dens. 8.4 lb/gal; visc. 1000 cps; flash pt. none; pH 9.6; usage level: 25-200 ppm.

Oakite Defoamant RC. [Oakite Prods.] Silicone/nonionic emulsifiers; nonionic; foam control agent for plastic recycling systems, food processing, textile, pulp and paper, other industrial applics.; milky wh. emulsion, mild odor; disp. in cold or hot water; sp.gr. 1.010; dens. 8.4 lb/gal; flash pt. none; usage level: 15-100 ppm.

Oakite Ladd. [Oakite Prods.] Blend of org. defoamer, detergent, solvs.; low-temp. cleaning/foam controlling additive for acidic or alkaline detergents, spray washing; pale straw-yel. liq., solv.-like odor; sp.gr. 1.060; dens. 8.8 lb/gal; visc. 15 cps; flash pt. none; toxicology: eye and skin irritant; harmful if swallowed.

Obanol 516. [Toho Chem. Industry] Polyoctyl polyamino ethyl glycine and POE alkylphenol ether; germicide, disinfectant, deodorant, fungicidal cleaning aid; liq.

Obazoline 662Y. [Toho Chem. Industry] Imidazoline deriv.; amphoteric; antistat and softener for syn. fibers; base material for shampoos and hair rinse; liq.; 35% conc.

Ocenol 50/55+2EO. [Henkel/Cospha; Henkel Canada] Ethoxylated oleyl alcohol; nonionic; emulsifier and lubricant aid for metalworking fluids; liq.; 100% conc.

Ocetox 5525. [Witco SA] Ethoxylated fatty alcohol; nonionic; emulsifier for oleine, veg. oils; detergent; paste.

Octapol 60. [Sanyo Chem. Industries] POE octylphenyl ether; CAS 9063-89-2; nonionic; base material for detergents; biodeg.; liq.; HLB 11.3; 100% conc.

Octapol 100. [Sanyo Chem. Industries] POE octylphenyl ether; CAS 9063-89-2; nonionic; base material for detergents; biodeg.; liq.; HLB 13.6; 100% conc.

Octapol 300. [Sanyo Chem. Industries] POE octylphenyl ether; CAS 9063-89-2; nonionic; base material for detergents; biodeg.; solid; HLB 17.3; 100% conc.

Octapol 400. [Sanyo Chem. Industries] POE octylphenyl ether; CAS 9063-89-2; nonionic; base material for detergents; biodeg.; solid; HLB 17.9; 100% conc.

Octaron PS 80. [Seppic] POE nonylphenol ether sulfate, ammonium salt; anionic; detergent; paste; 80% conc.

Octasol IB-45. [Tiarco] Sodium diisobutyl sulfosuccinate; anionic; surfactant for emulsion polymerization; liq.; 45% conc.

Octenyl Succinic Anhydride, n-. [Humphrey] Octenyl succinic anhydride; CAS 26680-54-6; starch modifier used as thickener, emulsifier and opacifier for food mixes; alkali metal salts as detergents in industrial cleaning formulations; liq.; 100% conc.

Octosol 400. [Tiarco] Alkyl trimethyl ammonium chloride; cationic; gel sensitizer for latex foam rubber.

Octosol 449. [Tiarco] Potassium oleate; anionic; foaming agent, stabilizer, emulsifier, dispersant; primary frothing aid in gelled latex foam compds.; clear liq.; dens. 8.75 lb/gal; pH 9.8-10.2; 16.5-17.5% solids.

Octosol 474. [Tiarco] Octadecyl trimethyl ammonium chloride; cationic; emulsifier for cationic or cationic/anionic emulsion systems.

Octosol 496. [Tiarco] Proprietary surfactant; anionic; emulsifier, dispersant, stabilizer, foaming agent for frothed latex adhesive compds. for carpet backing and laminating; amber clear liq.; sol. in water; pH 7-10; 29-31% solids in water.

Octosol 562. [Tiarco] Lauryl trimethyl ammonium chloride; cationic; gel sensitizer for latex foam rubber; 33% solids.

Octosol 571. [Tiarco] Lauryl trimethyl ammonium chloride; cationic; emulsifier for cationic or blended ionic emulsion systems; frothing agent, gel sensitizer in latex foam rubber; Gardner 1 max. liq.; sol. in water; pH 7-9; 49-51% solids in aq. IPA.

Octosol 1006. [Tiarco] Alkanolamine salt of dodecylbenzenesulfonic acid; emulsifier, dispersant.

Octosol A-1. [Tiarco] Disodium N-[3-(dodecyloxy) propyl] sulfosuccinamate; CAS 58353-68-7; anionic; emulsifier, dispersant, wetting agent, foaming agent for frothed latex compds. and adhesives; suspending agent in emulsion polymerization; textile softener; straw yel. to lt. amber liq.; sol. in warm water; pH 9-10; 33-35% solids.

Octosol A-18. [Tiarco] Disodium N-octadecyl sulfosuccinamate; anionic; emulsifier, dispersant, foaming agent for latex compds., cleaners; textile softener; suspending agent in emulsion polymerization; stable in acid and alkaline aq. systems; wh. to cream-colored paste; sol. in warm water; pH 7-9; 34-36% solids in aq. disp.

Octosol A-18-A. [Tiarco] Diammonium N-octadecyl sulfosuccinamate; anionic; emulsifier, stabilizer, foaming agent for acrylic latex frothed compds., formulations where reduced sodium ion content is desirable; suspending agent in emulsion polymerization; clear to creamy wh. liq.; sol. in warm water; pH 8-10; 34-36% solids.

Octosol ALS-28. [Tiarco] Ammonium lauryl sulfate; anionic; stabilizer, emulsifier, dispersant, foaming agent for industrial aq. systems; generates stable, high foam; pale clear liq.; sol. in water; sp.gr. 1.04; pH 6-8; 27-29% act. in water.

Octosol HA-80. [Tiarco] Sodium dihexyl sulfosuccinate; anionic; emulsifier for latex emulsion polymerization; liq.; 80% conc.

Octosol IB-45. [Tiarco] Sodium diisobutyl sulfosuccinate; emulsifier for latex emulsion polymerization.

Octosol SLS. [Tiarco] Sodium lauryl sulfate; CAS 151-21-3; anionic; stabilizer, frothing aid, emulsifier, dispersant for latex compds.; low cloud pt., low salt content; lt. color clear liq., low odor; sol. in water; sp.gr. 1.03-1.08; visc. < 500 cps; cloud pt. < 18 C; pH 7-9; 28-30% solids in water.

Octosol SLS-1. [Tiarco] Sodium lauryl sulfate; CAS 151-21-3; anionic; stabilizer, frothing aid for latex compds.; emulsifier, dispersant; low cloud pt., low salt content; clear liq.; sol. in water; sp.gr. 1.03-1.08; dens. 8.67 lb/gal; visc. < 800 cps; cloud pt. 40 F; pH 7.5-9.0; 28-31% solids in water.

Octosol TH-40. [Tiarco] Sodium dicyclohexyl sulfosuccinate; anionic; surfactant for emulsion polymerization, mfg. of carboxylated latexes; paste; 40% conc.

Octosperse TS-10. [Tiarco] Silicone emulsion; defoamer; 10% act.

Octosperse TS-20. [Tiarco] Silicone emulsion; defoamer; 20% act.

Octosperse TS-30. [Tiarco] Silicone emulsion; defoamer; 30% act.

Octosperse TS-50. [Tiarco] Silicone emulsion; defoamer; 50% act.

Octowet 40. [Tiarco] Sodium dioctyl sulfosuccinate; anionic; wetting agent, emulsifier, penetrant for textile washing and dyeing operations, agric., mining, paper, printing; colorless liq.; m.w. 444.63; dens. 8.6 lb/gal; visc. 200-400 cps; acid no. < 2; iodine no. < 0.15; flash pt. > 120 F; pH 5-7; Draves wetting < 5 s (0.075%); 39-41% act. in water, alcohol, glycol.

Octowet 55. [Tiarco] Sodium dioctyl sulfosuccinate; anionic; wetting agent, emulsifier, penetrant for textile washing and dyeing operations, agric., mining, paper, printing; colorless liq.; m.w. 444.63; dens. 8.7 lb/gal; visc. 200-400 cps; acid no. < 2; iodine no. < 0.15; flash pt. > 120 F; pH 5-7; Draves wetting < 5 s (0.075%); 54-56% act. in water, alcohol, glycol.

Octowet 60-I. [Tiarco] Sodium dioctyl sulfosuccinate; anionic; high speed wetting agent effective at low concs.; penetrant, emulsifier for coatings, textile finishing, agric., mining, paper, printing; colorless liq.; m.w. 444.63; dens. 8.6 lb/gal; visc. 200-400 cps; acid no. < 2; iodine no. < 0.15; flash pt. < 100 F; pH 5-7; Draves wetting < 5 s (0.075%); 59-61% act. in water, 20% IPA.

Octowet 60. [Tiarco] Sodium dioctyl sulfosuccinate; anionic; wetting agent, emulsifier, penetrant for textile dyeing and washing operations, agric., mining, paper, printing; colorless liq.; m.w. 444.63; dens. 8.8 lb/gal; visc. 200-400 cps; acid no. < 2; iodine no. < 0.15; flash pt. > 120 F; pH 5-7; Draves wetting < 5 s (0.075%); 59-61% act. in water, alcohol, glycol.

Octowet 65. [Tiarco] Sodium dioctyl sulfosuccinate; anionic; wetting agent, emulsifier, penetrant for textile dyeing and washing operations, agric., mining, paper, printing; colorless liq.; m.w. 444.63; dens. 9.0 lb/gal; visc. 200-400 cps; acid no. < 2; iodine no. < 0.15; flash pt. > 120 F; pH 5-7; Draves wetting < 5 s (0.075%); 64-66% act. in water, alcohol, glycol.

Octowet 70. [Tiarco] Sodium dioctyl sulfosuccinate; anionic; wetting agent, penetrant, emulsifier for textile dyeing and washing operations, agric., mining, paper, printing, industrial applics.; colorless liq.; m.w. 444.63; dens. 9.06 lb/gal; visc. 200-400 cps; acid no. < 2; iodine no. < 0.15; flash pt. > 120 f; pH 4-7; Draves wetting < 5 s (0.075%); 69-71% act. in water, alcohol, glycol.

Octowet 70A. [Tiarco] Ammonium dioctyl sulfosuccinate; anionic; high speed wetting agent, solubilizer, penetrant for textile processing, agric., mining, paper, printing; colorless liq.; pH 7-9; 69-71% act. in water, alcohol, glycol.

Octowet 70BC. [Tiarco] Sodium dioctyl sulfosuccinate; anionic; high speed wetting agent effective at low concs.; penetrant, emulsifier for paints/coatings, polymerization, textiles, agric., mining, paper, printing, and industrial applics. where high flash is important; colorless liq.; m.w. 444.63; dens. 9.08 lb/gal; visc. 200-400 cps; acid no. < 2; iodine no. < 0.15; flash pt. > 212 F; pH 5-7; Draves wetting < 5 s (0.075%); 69-71% act. in water, butyl Carbitol.

Octowet 70PG. [Tiarco] Sodium dioctyl sulfosuccinate; anionic; high speed wetting agent, emulsifier, penetrant for textile processing, paints, coatings, polymerization, agric., mining, paper, printing, industrial applics. where high flash is important; colorless liq.; m.w. 444.63; dens. 9.08 lb/gal; visc. 200-400 cps; acid no. < 2; iodine no. < 0.15; flash pt. > 212 F; pH 5-7; Draves wetting < 5 s (0.075%); 69-71% act. in water, propylene glycol.

Octowet 75. [Tiarco] Sodium dioctyl sulfosuccinate; anionic; wetting agent, penetrant, emulsifier for textile dyeing and washing operations, agric., mining, paper, printing, industrial applics.; colorless liq.; m.w. 444.63; dens. 9.18 lb/gal; visc. 200-400 cps; acid no. < 2; iodine no. < 0.15; pH 5-7; flash pt. > 105 F; Draves wetting < 5 s (0.075%); 74-76% act. in water, alcohol, glycol.

Octowet 75E. [Tiarco] Sodium dioctyl sulfosuccinate; anionic; wetting agent, emulsifier, penetrant for textile dyeing and washing operations, agric., mining, paper, printing, industrial applics.; colorless liq.; m.w. 444.63; dens. 9.18 lb/gal; visc. 200-400 cps; acid no. < 2; iodine no. < 0.15; flash pt. > 105 F; pH 5-7; Draves wetting < 5 s (0.075%); 74-76% act. in water, ethanol, glycol.

Ofax® Series. [Stepan; Stepan Canada] Special blends; high foaming surfactants for oilfield applics.; liq.

Ogtac-85. [Chem-Y GmbH] Glycidyl trimethyl ammonium chloride; cationic; starch modifier, intermediate for paper, textile and cosmetic industry; liq.; 70% conc.

Ogtac 85 V. [Chem-Y GmbH] Glycidyl trimethyl ammonium chloride; CAS 3033-77-0; cationic; intermediate for cationic surfactants; modifier for syn. polymers such as starch, cellulose, gelatins, polyacrylic acids, epoxy resins; used for prod. of emulsion layers on photographic plates; liq.; 70% act.

OHlan®. [Amerchol; Amerchol Europe] Hydroxylated lanolin; nonionic; primary w/o emulsifier, aux.

emulsifier and stabilizer, pigment wetting and dispersing agent, emollient and conditioner in personal care prods., absorp. bases, pharmaceuticals; yel.-amber to lt. tan waxy solid; misc. with common oil phase ingredients, sol. at low levels in castor oil; oil-misc.; m.p. 39–46 C; HLB 4; acid no. 10 max.; sapon. no. 95–110; 100% conc.

#1 Oil. [CasChem] Castor oil, tech.; tech. grade for industrial use; Gardner 2+ color; sp.gr. 0.959; visc. 7.5 stokes; pour pt. -10 F; acid no. 2; iodine no. 86; sapon. no. 180; hyd. no. 158.

#15 Oil. [CasChem] Polymerized castor oil; pigment wetting/dispersing agent; plasticizer for resins, gums, polymers; lubricant, penetrant; coupling solv.; adhesion promoter; for cellulose lacquers, inks, adhesives, industrial lubricants, polishes, caulks, leather dressing, hydraulic fluids,; rubber compding.; FDA approval; Gardner 11 color; sp.gr. 1.013; visc. 250 stokes; pour pt. 35 F; acid no. 14; iodine no. 64; sapon. no. 220; hyd. no. 137.

#30 Oil. [CasChem] Polymerized castor oil; see #15 Oil; FDA approval; Gardner 13 color; sp.gr. 1.019; visc. 500 stokes; pour pt. 45 F; acid no. 13; iodine no. 63; sapon. no. 220; hyd. no. 136.

#40 Oil. [CasChem] Polymerized castor oil; see #15 Oil; FDA approval; Gardner 14 color; sp.gr. 1.020; visc. 800 stokes; pour pt. 50 F; acid no. 13; iodine no. 60; sapon. no. 225; hyd. no. 135.

Olapon ND-9, SW. [Reilly-Whiteman] Detergent for textile scouring.

Oleine D. [Ceca SA] Oleic acid; CAS 112-80-1; anionic; surfactant.

Olitex 75. [Reilly-Whiteman] Sulfated olive oil; anionic; similar to sulfated castor oil with improved lubrication; liq.; 75% conc.

Ombrelub FC 533. [Münzing Chemie GmbH] Nonionic; waterproofing agent for aq. systems, e.g., printing inks, cement and concrete mixts.; BGA compliance; aq. suspension.

Onyxide® 200. [Stepan; Stepan Canada] Hexahydro-1.3.5-tris (2-hydroxethyl)-s-triazine; preservative for sol. cutting fluids and coolants; bactericide for oilfield drilling fluids, enhanced oil recovery operations; liq.; sp.gr. 1.15; flash pt. > 200 F; 79% act.

Onyxol® SD (redesignated Ninol 49CE). [Stepan] Lauramide DEA; CAS 120-40-1; nonionic; foam stabilizer and thickener for personal care prods. and lt.-duty detergents; liq.; 100% conc. DISCONTINUED.

Orapol HC. [Seppic] Fatty DEA; nonionic; detergent; liq.; 100% conc.

Orapret WTNB 25. [Ceca SA] Alkylaryl sulfate/min. oil complex; anionic; surfactant; paste.

Orgozon CC 1118. [Clough] Fatty phosphate ester, potassium salt; anionic; detergent and emulsifier for use in mild to strong alkaline conditions; liq.; pH 4-6; 35-36% conc.

Orgozon Conc. 0680. [Clough] Phosphate ester, free acid form; anionic; detergent, emulsifier for use in mild to strong alkaline conditions; may be neutralized with sodium hydroxide, potassium hydroxide, etc.; hard paste; pH 1.5-2.0; 97% min. conc.

Orzan® AE. [Rayonier] Lignosulfonate base; SS & CSS asphalt emulsifier, dispersant, suspending agent; liq.

Orzan® CD. [Rayonier] Sodium lignosulfonate; dyestuff extender with min. foaming tendency; reduces dusting in dye powd.; powd.; 100% conc.

Orzan® CG. [Rayonier] Lignosulfonate; dispersant for gypsum board mfg.; liq. and powd.; 40 and 95% conc.

Orzan® LS. [Rayonier] Sodium lignosulfonate; dispersant, suspending agent in wettable pesticides; sequestrant in water treatment; emulsifier, stabilizer in oil and wax emulsions; industrial cleaners; ore flotation; biodeg.; yel. powd.; rapidly sol. in water; bulk dens. 30 lb/ft^3; pH 6.5 (25%); 59% lignosulfonate.

Orzan® LS-50. [Rayonier] Sodium lignosulfonate aq. sol'n.; for use with anionic surfactants in formulating cleaning compds.; biodeg.; liq.; sp.gr. 1.255–1.265; visc. 100 cps (20 C); pH 7.0 (25%); 47% sol'n., 59% lignosulfonate.

OS-2. [Werner G. Smith] Emulsifiable ester; forms stable emulsions for buffing, lapping, drawing, grinding compds.; rust preventative; dk. brn. solid fat; visc. 75-90 SSU (210 F); soften. pt. 86-95 F; acid no. nil; iodine no. 40-50; sapon. no. 145-165; flash pt. (COC) 445-465 F; pH 9.0-10.5 (10% aq.); 100% act.

Osimol Grunau 109. [Grünau] Alkylaryl sulfonate; anionic; leveling agent for dyeing polyamide; liq.; 60% conc.

Osimol Grunau 110. [Grünau] Alkylaryl sulfonate and ethylene oxide deriv.; anionic; leveling agent for dyeing polyamide; liq.; 50% conc.

Osimol Grunau DP. [Grünau] Sulfonic acid; anionic; dispersant for dyeing polyester; liq.; 40% conc.

Osimol Grunau EFA. [Grünau] Amide amine deriv.; cationic; leveling agent for dyeing acrylic; liq.; 50% conc.

Osimol Grunau MA. [Grünau] Amide amine deriv.; cationic; migrating agent for dyeing acrylic; liq.; 55% conc.

Osimol Grunau PHT. [Grünau] EO deriv.; cationic; dispersant and leveling agent for polyester dyeing; liq.; 100% conc.

Osimol Grunau RAC. [Grünau] Amide amine deriv.; cationic; leveling agent for dyeing acrylic; liq.; 50% conc.

Osimol Grunau SF. [Grünau] Sulfonic acid; anionic; dispersant and protective colloid for dyeing applics.; liq.; 45% conc.

Ospin L-800. [Tokai Seiyu Ind.] Nonionic; leveling agent for acid and chrome colors; liq.; 70% conc.

Ospin Salt ON. [Tokai Seiyu Ind.] Quat. ammonium salt; cationic; migrator for dyeing acrylic fibers with cationic dyes; liq.; 50% conc.

Ospin TAN. [Tokai Seiyu Ind.] Quat. ammonium salt; cationic; retarder in dyeing of acrylic fibers with cationic dyes; liq.; 50% conc.

Ospol 790. [Tokai Seiyu Ind.] Nonionic; scouring and washing agent; liq.; HLB 11.8; 90% conc.

Oxamin LO. [ICI Australia] Lauryl dimethyl amine oxide; nonionic; detergent, foamer and foam stabilizer for personal care prods. and detergent formulations; biodeg.; liq.; 30% conc.

Oxetal 500/85. [Zschimmer & Schwarz] Fatty alcohol polyglycol ether (5 EO); nonionic; detergent, dispersant, emulsifier, wetting agent for household and industrial use, textile, paper, and leather industries; liq.; cloud pt. 65-68 C (5 g/20 ml butyl diglycol 25%); 85% conc.

Oxetal 800/85. [Zschimmer & Schwarz] Fatty alcohol polyglycol ether (8 EO); nonionic; washing and cleansing agent; liq.; cloud pt. 66-70 C (1% aq.); 85% act.

Oxetal C 110. [Zschimmer & Schwarz] Coceth-10; CAS 61791-13-7; washing and cleansing agent; wh. paste; cloud pt. 75-78 C (1% aq.); 100% act.

Oxetal D 104. [Zschimmer & Schwarz] Deceth-4; nonionic; detergent, dispersant, emulsifier, wetting agent for household and industrial use, textile, paper, and leather industries; liq.; cloud pt. 58-61 C (5 g/10 ml butyl diglycol 25%); 100% act.

Oxetal ID 104. [Zschimmer & Schwarz] Isodeceth-4; nonionic; detergent, dispersant, emulsifier, wetting agent for household and industrial use, textile, paper, and leather industries; colorless liq.; cloud pt. 50-53 C (5 g/20 ml butyl diglycol 25%); 100% act.

Oxetal O 108. [Zschimmer & Schwarz] Cetoleth-8; nonionic; detergent, dispersant, emulsifier, wetting agent for household and industrial use, textile, paper, and leather industries; wh. paste; cloud pt. 45-50 C (1% aq.); 100% act.

Oxetal O 112. [Zschimmer & Schwarz] Cetoleth-12; nonionic; detergent, dispersant, emulsifier, wetting agent for household and industrial use, textile, paper, and leather industries; wh. paste; cloud pt. 83-87 C (1% aq.); 100% act.

Oxetal T 106. [Zschimmer & Schwarz] Isotrideceth-6; washing and cleansing agent; colorless liq.; cloud pt. 40-44 C (1% aq.); 100% act.

Oxetal T 110. [Zschimmer & Schwarz] Isotrideceth-10; washing and cleansing agent; paste; cloud pt. 69-74 C (1% aq.); 100% act.

Oxetal TG 111. [Zschimmer & Schwarz] Talloweth-11; CAS 61791-28-4; nonionic; detergent, dispersant, emulsifier, wetting agent for household and industrial use, textile, paper, and leather industries; wh. wax; cloud pt. 70-75 C (1% aq.); 100% act.

Oxetal TG 118. [Zschimmer & Schwarz] Talloweth-18; CAS 61791-28-4; nonionic; detergent, dispersant, emulsifier, wetting agent for household and industrial use, textile, paper, and leather industries; wax; cloud pt. 72-76 C (1% in 10% NaCl); 100% act.

Oxetal VD 20. [Zschimmer & Schwarz] Laureth-2; nonionic; washing and cleansing agent; liq.; 100% act.

Oxetal VD 28. [Zschimmer & Schwarz] Laureth-3; nonionic; washing and cleansing agent; liq.; 100% act.

P

P®-10 Acid. [CasChem] Ricinoleic acid; CAS 141-22-0; chemical intermediate; imparts lubricity and rust-proofing to sol. cutting oils; basis for grease, soaps, resin plasticizers, and ethoxylated derivs.; FDA approval; Gardner 5 liq.; sp.gr. 0.940; visc. 4 stokes; pour pt. 10 F; acid no. 180; Wijs iodine no. 89; sapon. no. 186.

Pale 4. [CasChem] Polymerized castor oil; pigment wetting/dispersing agent; plasticizer for resins, gums, polymers; lubricant, penetrant; coupling solv.; adhesion promoter; for cellulose lacquers, inks, adhesives, industrial lubricants, polishes, caulks, leather dressing, hydraulic fluids; rubber compding., gasket cement; FDA approval; Gardner 4 color; sp.gr. 0.998; visc. 32 stokes; pour pt. 5 F; acid no. 16; iodine no. 70; sapon. no. 212; hyd. no. 158.

Pale 16. [CasChem] Polymerized castor oil; pigment wetting/dispersing agent; plasticizer for resins, gums, polymers; lubricant, penetrant; coupling solv.; adhesion promoter; for cellulose lacquers, inks, adhesives, industrial lubricants, polishes, caulks, leather dressing, hydraulic fluids,; rubber compding., gasket cement; FDA approval; Gardner 9 color; sp.gr. 1.025; visc. 250 stokes; pour pt. 25 F; acid no. 24; iodine no. 56; sapon. no. 237; hyd.no. 136.

Pale 170. [CasChem] Polymerized castor oil; pigment wetting/dispersing agent; plasticizer for resins, gums, polymers; lubricant, penetrant; coupling solv.; adhesion promoter; for cellulose lacquers, inks, adhesives, industrial lubricants, polishes, caulks, leather dressing, hydraulic fluids,; rubber compding., gasket cement; FDA approval; Gardner 2 color; sp.gr. 0.970; visc. 11 stokes; pour pt. -5 F; acid no. 4; iodine no. 80; sapon. no. 184; hyd. no. 160.

Pale 1000. [CasChem] Polymerized castor oil; pigment wetting/dispersing agent; plasticizer for resins, gums, polymers; lubricant, penetrant; coupling solv.; adhesion promoter; for cellulose lacquers, inks, adhesives, industrial lubricants, polishes, caulks, leather dressing, hydraulic fluids,; rubber compding., gasket cement; FDA approval; Gardner 9 color; sp.gr. 1.018; visc. 120 stokes; pour pt. 25 F; acid no. 20; iodine no. 59; sapon. no. 230; hyd. no. 139.

Pamolyn® 100. [Hercules] Oleic acid; CAS 112-80-1; detergent intermediate, emulsifier, fiber lubricant, textile processing aid, defoamer, emulsion breaker; alkali-sol.

Pantex. [Texo] Alkylaryl sulfonate; for institutional hand washing; biodeg.; powd.; 10% conc.

Paracol AL, AH. [Nippon Nyukazai] Blend; anionic/nonionic; emulsifier for organic phosphate insecticides; liq.; 100% conc.

Paracol OP. [Nippon Nyukazai] Alkyl phosphate; anionic; anticorrosion agent, emulsion breaker; liq.; 100% conc.

Paracol SV. [Nippon Nyukazai] Alkyl phosphate; anionic/nonionic; emulsifier for org. phosphate insecticides (DDVP); liq.; 100% conc.

Paraffin Wax Emulsifier CB0674. [Croda Chem. Ltd.] POE fatty alcohols; nonionic; paraffin wax emulsifier for high conc. liq. emulsions; solid; HLB 10.6; 100% conc.

Paraffin Wax Emulsifier CB0680. [Croda Chem. Ltd.] POE fatty alcohol; nonionic; paraffin wax emulsifier for high conc. liq. emulsions; solid; HLB 10.5; 100% conc.

Paralan. [Croda Chem. Ltd.] Tech. lanolin; nonionic; emulsifier for liq. emulsions; solid.

Paramul® J. [Bernel] Cetearyl alcohol, ceteareth-20; nonionic; emulsifier; approved for use in Japan, EEC, US; colorless solid; sol. in oil; disp. in water; usage level: 0.5-4.0%; 100% conc.

Paramul® SAS. [Bernel] Stearamide DIBA-stearate; nonionic; emulsifier, pearlizing agent, opacifier; yel. solid; sol. in oil; usage level: 1-3%; 100% conc.

Partsprep Degreaser. [ISP] N-Methyl-2-pyrrolidone and surfactants; degreaser formulations; clear liq.; sp.gr. 1.02-1.04; ref. index 1.46-1.48.

Patcote® 305. [Am. Ingredients/Patco] Silicone emulsion, 10% filled; defoamer for general food and industrial applics.; FDA compliance; milky wh. emulsion; sp.gr. 1.002; dens. 8.35 lb/gal; pour pt. 30 F; usage level: 50-150 ppm.

Patcote® 306. [Am. Ingredients/Patco] 20% Filled silicone emulsion; defoamer for general food and industrial applics.; FDA compliance; milky wh. emulsion; sp.gr. 1.004; dens. 8.36 lb/gal; pour pt. 1-3 C; usage level: 10-50 ppm.

Patcote® 307. [Am. Ingredients/Patco] 30% Filled silicone emulsion; defoamer for food and industrial applics.; FDA compliance; milky wh. emulsion; sp.gr. 1.008; dens. 8.40 lb/gal; pour pt. 1-3 C; usage level: 10-33 ppm.

Patcote® 309. [Am. Ingredients/Patco] Nonsilicone aq. emulsion; defoamer for food and industrial use; FDA compliance; milky wh. emulsion; sp.gr. 0.9076; dens. 7.56 lb/gal; pour pt. 30 F; usage level: 50-150 ppm.

Patcote® 310. [Am. Ingredients/Patco] Nonsilicone aq. emulsion; defoamer for general food and industrial applics.; esp. for starchy applics., e.g., potatoes; FDA compliance; milky wh. emulsion; sp.gr. 0.9478; dens. 8.12 lb/gal; pour pt. 30 F; usage level: 75-200 ppm.

Patcote® 311. [Am. Ingredients/Patco] 100% Filled silicone; defoamer for food and nonfood processing; FDA compliance; wh. semitransparent emulsion; sp.gr. 0.994; dens. 8.28 lb/gal; pour pt. 25 F; flash pt. > 300 F; usage level: 5-10 ppm; toxicology: eye or skin irritant on prolonged contact; do not take internally.

Patcote® 315. [Am. Ingredients/Patco] 10% Silicone emulsion; defoamer for food and industrial applics.; FDA compliance; milky blue emulsion; sp.gr. 0.960; dens. 8.00 lb/gal; pour pt. 30 F; usage level: 50-100 ppm.

Patcote® 460. [Am. Ingredients/Patco] Nonsilicone; defoamer for solv.-based high-solids coatings; alleviates air entrapment and foam problems; clear yel. liq.; sp.gr. 0.874; dens. 7.29 lb/gal; pour pt. < 30 F; flash pt. (PMCC) 105 F; usage level: 0.2-1.0%; toxicology: eye or skin irritant on prolonged contact; do not take internally; 100% active.

Patcote® 512. [Am. Ingredients/Patco] Silicone-containing; defoamer for use in urethane-modified resins; lt. amber liq.; sp.gr. 0.892 ± 0.012; dens. 7.43 ± 0.1 lb/gal; pour pt. < –30 F; PMCC flash pt. 143 F; usage level: 0.25-0.5%; toxicology: eye or skin irritant on prolonged contact; do not take internally; 100% act.

Patcote® 513. [Am. Ingredients/Patco] Silicone-containing; defoamer for use in water-reducible acrylic coatings; wh. cloudy liq.; sp.gr. 0.830 ± 0.012; dens. 6.91 ± 0.1 lb/gal; pour pt. < –30 F; PMCC flash pt. 145 F; usage level: 0.5%; toxicology: eye or skin irritant on prolonged contact; do not take internally; 100% act.

Patcote® 519. [Am. Ingredients/Patco] Silicone-containing; defoamer for use in trade sales and industrial acrylic lacquers; wh. cloudy liq.; sp.gr. 0.840 ± 0.012; dens. 7.00 ± 0.1 lb/gal; pour pt. < –30 F; PMCC flash pt. 148 F; usage level: 0.2-0.5%; toxicology: eye or skin irritant on prolonged contact; do not take internally; 100% act.

Patcote® 520. [Am. Ingredients/Patco] Silicone-containing; defoamer for use in water-reducible alkyd, industrial acrylic, and urethane resins; wh. cloudy liq.; sp.gr. 0.839 ± 0.012; dens. 6.99 ± 0.1 lb/gal; pour pt. < –30 F; PMCC flash pt. 160 F; cloud pt. 77–85 F; usage level: 0.25-0.5%; toxicology: eye or skin irritant on prolonged contact; do not take internally; 100% act.

Patcote® 525. [Am. Ingredients/Patco] Silicone-containing; defoamer for use in water-reducible alkyd and industrial acrylic systems; clear, slightly opalescent liq.; sp.gr. 0.822 ± 0.012; dens. 6.85 ± 0.1 lb/gal; pour pt. < –30 F; PMCC flash pt. 158 F; usage level: 0.2-0.5%; toxicology: eye or skin irritant on prolonged contact; do not take internally; 100% act.

Patcote® 531. [Am. Ingredients/Patco] Silicone-containing; defoamer for water-reducible acrylic systems; wh. cloudy liq.; dens. 6.98 lb/gal; pour pt. < 30 F; flash pt. (PMCC) 158 F; usage level: 0.5-0.75%; toxicology: eye or skin irritant on prolonged contact; do not take internally; 100% active.

Patcote® 550. [Am. Ingredients/Patco] Silicone-containing; defoamer for use in water-reducible alkyds and acrylics as well as acrylic latices for both industrial and trade sales formulations; wh. slightly cloudy liq.; sp.gr. 0.874 ± 0.012; dens. 7.28 ± 0.1 lb/gal; pour pt. < –30 F; PMCC flash pt. 150 F; usage level: 0.1-0.5%; toxicology: eye or skin irritant on prolonged contact; do not take internally; 100% act.

Patcote® 555. [Am. Ingredients/Patco] Silicone-containing; defoamer for acrylic latex systems for trade sales; milky wh. liq.; sp.gr. 0.994; dens. 8.28 lb/gal; pour pt. 30 F; flash pt. (PMCC) < 300 F; usage level: 0.5-2 lb/100 gal; toxicology: eye or skin irritant on prolonged contact; do not take internally; 100% active.

Patcote® 555K. [Am. Ingredients/Patco] 100% Filled silicone; kosher grade; defoamer for food and nonfood processing; FDA compliance; wh. semitransparent; sp.gr. 0.994; dens. 8.28 lb/gal; pour pt. < -30 F; flash pt. > 300 F; usage level: 5-10 ppm; toxicology: eye or skin irritant on prolonged contact; do not take internally.

Patcote® 577. [Am. Ingredients/Patco] Silicone-containing; defoamer for use in PVA resins for trade sales and water-reducible alkyds for industrial coatings; lt. amber cloudy liq.; sp.gr. 0.826±0.012; dens. 6.88 ± 0.1 lb/gal; pour pt. < –30 F; PMCC flash pt. 140 F; usage level: 3-10 lb/100 gal; toxicology: eye or skin irritant on prolonged contact; do not take internally; 100% act.

Patcote® 597. [Am. Ingredients/Patco] Silicone-containing; defoamer for water-reducible alkyds and acrylic emulsions; sl. cloudy liq.; sp.gr. 0.814; dens. 6.78 lb/gal; pour pt. < 30 F; flash pt. (PMCC) 155 F; usage level: 0.5-1.0%; toxicology: eye or skin irritant on prolonged contact; do not take internally; 100% active.

Patcote® 801. [Am. Ingredients/Patco] Nonsilicone; defoamer for use in PVA-acrylic copolymers and terpolymer emulsions for trade sales; FDA clearance; wh. opaque liq.; sp.gr. 0.911 ± 0.012; dens. 7.59 ± 0.1 lb/gal; pour pt. –5 F; flash pt. (PMCC) 300 F; usage level: 5-10 lb/100 gal; toxicology: eye or skin irritant on prolonged contact; do not take internally; 100% act.

Patcote® 803. [Am. Ingredients/Patco] Nonsilicone; defoamer for acrylic and terpolymer emulsions for trade sales; FDA compliance; wh. opaque liq.; sp.gr. 0.896; dens. 7.47 lb/gal; pour pt. 5 F; flash pt. (PMCC) > 300 F; usage level: 5-10 lb/100 gal; toxicology: eye or skin irritant on prolonged contact; do not take internally; 100% active.

Patcote® 806. [Am. Ingredients/Patco] Nonsilicone; defoamer for acrylic and terpolymer emulsions in trade sales and some industrial applics.; FDA clearance; lt. amber liq.; sp.gr. 0.8787; dens. 7.32 lb/gal; pour pt. 5 F; flash pt. (PMCC) > 300 F; usage level: 5-10 lb/100 gal; toxicology: eye or skin irritant on prolonged contact; do not take internally; 100% active.

Patcote® 811. [Am. Ingredients/Patco] Nonsilicone; defoamer for graphic arts water-based acrylic systems; creamy opaque; mild oily odor; sp.gr. 0.924; dens. 7.70 lb/gal; pour pt. 34-37 F; flash pt. (PMCC) > 300 F; usage level: 2-10 lb/100 gal; toxicology: eye or skin irritant on prolonged contact; do not take internally; 100% active.

Patcote® 812. [Am. Ingredients/Patco] Nonsilicone; defoamer for nonionic and anionic acrylic emulsion systems; creamy opaque; mild oily odor; sp.gr. 0.924; dens. 7.70 lb/gal; pour pt. 34-37 F; flash pt. (PMCC) > 300 F; usage level: 2-10 lb/100 gal; toxicology: eye or skin irritant on prolonged contact; do not take internally; 100% active.

Patcote® 841M. [Am. Ingredients/Patco] Nonsilicone; defoamer for acrylic, terpolymer, and PVC systems for trade sales paints, screen printing inks; amber opaque; sp.gr. 0.879; dens. 7.32 lb/gal; pour pt. -5 F; flash pt. (PMCC) > 300 F; usage level: 3-6 lb/100 gal; toxicology: eye or skin irritant on prolonged contact; do not take internally.

Patcote® 845. [Am. Ingredients/Patco] Nonsilicone; defoamer for solvent, high solids and water-based acrylic systems; clear liq.; sp.gr. 0.8643; dens. 7.20 lb/gal; pour pt. < -30 F; flash pt. (SCC) 97 F; flamm.; usage level: 0.05-0.5%; toxicology: eye or skin irritant on prolonged contact; do not take internally; 100% active.

Patcote® 847. [Am. Ingredients/Patco] Nonsilicone; defoamer for use in both solvent- and water-based alkyds and high solids systems; clear liq.; sp.gr. 0.864 ± 0.012; dens. 7.20 ± 0.1 lb/gal; pour pt. < -30 F; flash pt. (PMCC) 107 F; flamm.; usage level: 0.1-0.6%; toxicology: eye or skin irritant on prolonged contact; do not take internally; 100% act.

Patcote® 883. [Am. Ingredients/Patco] Nonsilicone; defoamer for use in trade sales acrylic emulsions; amber cloudy liq.; sp.gr. 0.871 ± 0.012; dens. 7.26 ± 0.1 lb/gal; pour pt. 20 F; flash pt. (PMCC) 395 F; usage level: 4 lb/100 gal; toxicology: eye or skin irritant on prolonged contact; do not take internally; 100% act.

Pationic® 901. [Am. Ingredients/Patco] Glyceryl stearate, distilled; antistat, mold release for polyolefins; internal lubricant for PVC; melt flow enhancer, pigment wetting/dispersion for polymer systems; FDA compliance; GRAS; ivory wh. fine beads; m.p. 72 C; 96% monoglyceride.

Pationic® 902. [Am. Ingredients/Patco] Glyceryl stearate, distilled; antistat and mold release agent for polyolefins; internal lubricant for rigid PVC; melt flow enhancer, pigment dispersion/wetting for polymer systems; FDA compliance; GRAS; ivory wh. fine beads; m.p. 68 C; 96% monoglyceride.

Pationic® 907. [Am. Ingredients/Patco] Unsat. dist. glycerol ester derived from veg. oil; internal lubricant for rigid PVC; antifog for PVC and polyolefins; antistat for polyolefins; FDA compliance; GRAS; ivory wh. liq. @ 40-45 C, soft paste at ambient temps.; 96% monoglyceride.

Pationic® 909. [Am. Ingredients/Patco] Glyceryl stearate, distilled; colorant dispersant, lubricant, binder, anticaking aid, processing aid for expandable polystyrene and other polymer systems where it is dry blended or surface coated onto substrate; FDA compliance; GRAS; ivory wh. fine beads; 98% thru 100 mesh; m.p. 72 C; 96% monoglyceride.

Pationic® 910. [Am. Ingredients/Patco] Dist. polyol-based ester derived from veg. oils; vehicle, dispersant, wetting agent, binder, process aid for pigments and colorant systems; antistat, antifog, melt flow enhancer, filler for polymers; FDA compliance; GRAS; ivory wh. liq. above 40 C, soft paste at ambient temps.

Pationic® 919. [Am. Ingredients/Patco] Glyceryl tristearate; lubricant, anticaking agent, processing aid, dispersant for colorants, expandable polystyrene; for dry blending or surface coating onto substrates; ivory wh. fine beads; 98% thru 100 mesh; m.p. 65 C.

Pationic® 920. [Am. Ingredients/Patco] Sodium stearoyl lactylate; acid and catalyst scavenger/neutralizer, internal/external lubricant for flame retardant compds., filled and unfilled polymers, color concs.; FDA compliance; off-wh. powd.; 99% thru 20 mesh; m.p. 47-53 C; sapon. no. 210-235; 3.5-5.0% sodium.

Pationic® 925. [Am. Ingredients/Patco] Calcium/sodium stearoyl lactylate; acid neutralizer/acceptor/scavenger for polyolefins, flame retardants, pigments; lubricant for PVC, polyolefins, filled compds.; off-wh. powd.; 99% thru 20 mesh; m.p. 45-55 C; sapon. no. 195-230.

Pationic® 930. [Am. Ingredients/Patco] Calcium stearoyl-2-lactylate; acid neutralizer/acceptor/scavenger for polyolefins, flame retardants, pigments; lubricant for PVC, polyolefins, filled compds.; FDA compliance; off-wh. powd.; 99% thru 20 mesh; m.p. 45-55 C; sapon. no. 195-230; 4.2-5.2% calcium.

Pationic® 940. [Am. Ingredients/Patco] Calcium stearoyl-2-lactylate; acid/catalyst neutralizer, lubricant for polyolefins and other thermoplastic systems; neutralizes corrosive acids; melt flow enhancer and stabilizer; FDA compliance; ivory free-flowing powd.; 14.5% calcium.

Pationic® 1042. [Am. Ingredients/Patco] Glyceryl stearate; antistat, mold release agent, flow modifier, internal lubricant for polyolefins, styrenics, PVC, TPEs, and TPOs; FDA compliance; GRAS; ivory wh. very fine flakes; m.p. 60-63 C; 42% monoglyceride.

Pationic® 1052. [Am. Ingredients/Patco] Glyceryl stearate; antistat, mold release agent, flow modifier, internal lubricant for polyolefins, styrenics, PVC, TPEs, and TPOs; FDA compliance; GRAS; ivory wh. very fine flakes; m.p. 61-63 C; 52% monoglyceride.

Pationic® 1064. [Am. Ingredients/Patco] Glyceryl oleate; CAS 111-03-5; internal lubricant, antifog aid, antistat, flow modifier, additive dispersant for PVC, polyolefins, and styrenics; FDA compliance; GRAS; ivory wh. liq. above 39-41 C; 42% monoglyceride.

Pationic® 1074. [Am. Ingredients/Patco] Glyceryl oleate; CAS 111-03-5; internal lubricant, antifog aid, antistat, flow modifier, additive dispersant for PVC, polyolefins, and styrenics; FDA compliance; GRAS; ivory wh. liq. above 48-50 C; 42% monoglyceride.

Pationic® 1230. [Am. Ingredients/Patco] Calcium lactylate; acid/catalyst neutralizer for polyolefins and other thermoplastic systems; neutralizes corrosive acids in polymers; melt flow stabilizer; minimizes polymer color development; FDA compliance; GRAS; wh. free-flowing powd.; pH 7.1 (1%); 13.6% calcium.

Pationic® 1240. [Am. Ingredients/Patco] Calcium lactylate; acid/catalyst neutralizer for polyolefins and other thermoplastic systems; neutralizes corrosive acids in polymers; melt flow stabilizer; minimizes polymer color development; FDA compliance; GRAS; wh. free-flowing powd.; 22% calcium.

Pationic® 1264. [Am. Ingredients/Patco] Zinc lactylate; additive for polymer and plastics industry; wh. free-flowing powd.; 100% thru 80 mesh; pH 5.25 (10%); 22.9% zinc.

Pationic ISL. [RITA] Sodium isostearoyl-2-lactylate; anionic; surfactant, emulsifier for cosmetics; perfume solubilizer; substantive conditioner for hair and skin; straw, honey clear visc. liq.; sol. in min. oil, propylene glycol, IPM, IPA; disp. in water; HLB 5.9; sapon. no. 205–225; surf. tens. 26.28 dynes/cm; pH 6.30 (2% aq.); 100% act.

Patlac® LA. [Am. Ingredients/Patco; RITA] Lactic acid; moisture binder, humectant; clear liq.

Patlac® NAL. [Am. Ingredients/Patco; RITA] Sodium lactate; pH buffer, humectant, stabilizer, component of stratum corneium; for food, pharmaceutical, and cosmetic industries; clear liq.

Patogen 311. [Yorkshire Pat-Chem] Nonionic; low foam wetting agent/scour for bleach baths.

Patogen 345. [Yorkshire Pat-Chem] Nonionic; low foam wetting agent/scour for bleach baths.

Patogen 353. [Yorkshire Pat-Chem] Wetting and scouring agent for bleach baths, dye baths.

Patogen 378. [Yorkshire Pat-Chem] Scouring agent with exc. caustic stability and scouring props. equal to solv. scours.

Patogen 393. [Yorkshire Pat-Chem] Anionic; low foaming scour with exc. caustic stability; esp. for continuous prep.

Patogen AO-30. [Yorkshire Pat-Chem] Cocamidopropyl dimethylamine oxide; amphoteric; fugitive wetter, foam booster, detergent, emulsifier, foaming agent for textile applics.; liq.; 30% conc.

Patogen CAC. [Yorkshire Pat-Chem] Coconut amine condensate; surfactant for textile scouring.

Patogen P-10 Acid. [Yorkshire Pat-Chem] Alkyl phosphate; anionic; scouring agent for textiles; liq.; 90% conc.

Patogen PD-3. [Yorkshire Pat-Chem] High flash pt., non "red label" solv. scour for removal of stubborn sizes; high caustic stability.

Patogen SME. [Yorkshire Pat-Chem] Anionic; surfactant, emulsifier for textile scouring; suited for chemical compounders.

Pat-Wet LF-55. [Yorkshire Pat-Chem] Nonionic; low foaming wetting agent with non-rewetting props. for textile processing.

Pat-Wet Q-4. [Yorkshire Pat-Chem] Low foaming wetting agent for vat and other high alkali dyes.

Pat-Wet SP. [Yorkshire Pat-Chem] Alkyl phosphate; anionic; low foaming wetting agent, bath stabilizer for textile dyeing; exc. for sulfur and indigo dye baths; liq.; 30-35% conc.

Pat-Wet SW. [Yorkshire Pat-Chem] Alkyl phosphate; anionic; wetting and scouring agent for dyeing and textile prep.; liq.; 80-85% conc.

Peceol Isostearique. [Gattefosse SA] Glyceryl isostearate; CAS 66085-00-5; nonionic; w/o coemulsifier; pigment dispersant; additive for lipsticks; superfatting agent for emulsified preps.; liq.; HLB 3.0; 100% conc.

Pegnol C-14. [Toho Chem. Industry] POE cetyl ether; nonionic; emulsifier, dispersant; solid; HLB 14.4; 100% conc.

Pegnol C-18. [Toho Chem. Industry] POE cetyl ether; nonionic; emulsifier, dispersant; solid; HLB 15.3; 100% conc.

Pegnol C-20. [Toho Chem. Industry] POE cetyl ether; nonionic; emulsifier, dispersant; solid; HLB 15.7; 100% conc.

Pegnol HA-120. [Toho Chem. Industry] POE alkyl amine; nonionic; emulsifier; liq.; 100% act.

Pegnol L-6. [Toho Chem. Industry] POE lauryl ether; nonionic; emulsifier, dispersant; paste; HLB 11.7; 100% conc.

Pegnol L-8. [Toho Chem. Industry] POE lauryl ether; nonionic; emulsifier, dispersant; solid; HLB 13.1; 100% conc.

Pegnol L-10. [Toho Chem. Industry] POE lauryl ether; nonionic; emulsifier, dispersant; solid; HLB 14.1; 100% conc.

Pegnol L-12. [Toho Chem. Industry] POE lauryl ether; nonionic; emulsifier, dispersant; solid; HLB 14.8; 100% conc.

Pegnol L-15. [Toho Chem. Industry] POE lauryl ether; nonionic; emulsifier, dispersant; solid; HLB 15.6; 100% conc.

Pegnol L-20. [Toho Chem. Industry] POE lauryl ether; nonionic; emulsifier, dispersant; solid; HLB 16.5; 100% conc.

Pegnol O-6. [Toho Chem. Industry] POE oleyl ether; nonionic; emulsifier, dispersant; solid; HLB 9.6; 100% conc.

Pegnol O-16. [Toho Chem. Industry] POE lauryl ether; nonionic; emulsifier, dispersant; solid; HLB 14.5; 100% conc.

Pegnol OA-400. [Toho Chem. Industry] POE alkyl amine; nonionic; antistat for syn. fibers, emulsifier, dispersant, leveling agent for polyester and polyamide fabrics; solid; HLB 17.4; 100% conc.

Pegnol PDS-60. [Toho Chem. Industry] Fatty acid ester; nonionic; thickener in printing of fabrics; emulsifier, dispersant; solid (block); sol. in water; 100% conc.

Pegnol PDS-60A. [Toho Chem. Industry] Fatty acid ester; nonionic; emulsifier, dispersant; solid; 100% conc.

Pegol® 10R8. [Rhone-Poulenc Surf.] EO/PO block copolymer; nonionic; surfactant for fermentation, paper processing, rinse aids, automatic dishwashing, metal cleaning; HLB 18-23; pour pt. 46 C; cloud pt. 98 C (1% aq.). DISCONTINUED.

Pegol® 17R1. [Rhone-Poulenc Surf.] EO/PO block copolymer; nonionic; surfactant for fermentation, paper processing, rinse aids, automatic dishwashing, metal cleaning; liq.; HLB 6.0; pour pt. -27 C; cloud pt. 32 C (1% aq.); 100% conc. DISCONTINUED.

Pegol® 17R2 (redesignated Antarox® 17-R-2). [Rhone-Poulenc Surf.]

Pegol® 17R4. [Rhone-Poulenc Surf.] EO/PO block copolymer; nonionic; surfactant for laundry, hard surf. cleaning; HLB 7-12; pour pt. -18 C; cloud pt. 47 C (1% aq.). DISCONTINUED.

Pegol® 17R8. [Rhone-Poulenc Surf.] EO/PO block copolymer; nonionic; surfactant for laundry, hard surf. cleaning; HLB 12-18; pour pt. 53 C; cloud pt. 81 C (1% aq.). DISCONTINUED.

Pegol® 25R1. [Rhone-Poulenc Surf.] EO/PO block copolymer; nonionic; surfactant for rinse aids, automatic dishwashing, paper processing, metal cleaning, fermentation; HLB 4.0; pour pt. -27 C; cloud pt. 28 C (1% aq.). DISCONTINUED.

Pegol® 25R2 (redesignated Antarox® 25-R-2. [Rhone-Poulenc Surf.]

Pegol® 25R8. [Rhone-Poulenc Surf.] EO/PO block copolymer; nonionic; surfactant for rinse aids, automatic dishwashing, paper processing, metal cleaning, fermentation; HLB 12-18; pour pt. 54 C; cloud pt. 80 C (1% aq.). DISCONTINUED.

Pegol® 31R1 (redesignated Antarox® 31-R-1). [Rhone-Poulenc Surf.]

Pegol® E-200. [Rhone-Poulenc Surf.] PEG; nonionic; surfactant intermediate; binder and lubricant in compressed tablets; softener for paper, plasticizer for starch pastes and polyethylene films; liq.; 100% conc. DISCONTINUED.

Pegol® E-300. [Rhone-Poulenc Surf.] PEG; nonionic; surfactant intermediate; binder and lubricant in compressed tablets; softener for paper, plasticizer for starch pastes and polyethylene films; liq.; pour pt. -10 C; 100% conc. DISCONTINUED.

Pegol® E-400 (redesignated Rhodasurf® E 400). [Rhone-Poulenc Surf.]

Pegol® E-600 (redesignated Rhodasurf® E 600). [Rhone-Poulenc Surf.]

Pegol® E-1000. [Rhone-Poulenc Surf.] PEG; nonionic; surfactant intermediate; binder and lubricant in tablets; liq.; pour pt. 37 C; 100% conc. DISCONTINUED.

Pegol® E-1500. [Rhone-Poulenc Surf.] PEG; nonionic; surfactant intermediate; binder and lubricant in tablets; liq.; pour pt. 45 C; 100% conc. DISCONTINUED.

Pegol® E-4000. [Rhone-Poulenc Surf.] PEG; nonionic; surfactant intermediate; mold release agent for rubber prods.; liq.; pour pt. 56 C; 100% conc. DISCONTINUED.

Pegol® F-68. [Rhone-Poulenc Surf.] EO/PO block copolymer; nonionic; defoamer, dispersant, wetting agent, emulsifier, demulsifier, leveling agent and detergent; for agric. formulations, hard surf. cleaning, laundry, shampoos, toilet tank blocks, lt. duty liqs., syndet bars; EPA compliance; solid; HLB 29.0; m.p. 52 C; cloud pt. > 100 C (1% aq.); 100% conc. DISCONTINUED.

Pegol® F-68LF. [Rhone-Poulenc Surf.] EO/PO block copolymer; nonionic; surfactant for hard surf. cleaning, laundry, shampoos, toilet tank blocks, lt. duty liqs., syndet bars; HLB 26.0; m.p. 50 C; cloud pt. 32 C (1% aq.). DISCONTINUED.

Pegol® F-87. [Rhone-Poulenc Surf.] EO/PO block copolymer; nonionic; surfactant for hard surf. cleaning, laundry, shampoos, electrolytic cleaning; HLB 24.0; pour pt. 87 C; cloud pt. > 100 C (1% aq.). DISCONTINUED.

Pegol® F-88 (redesignated Antarox® F88). [Rhone-Poulenc Surf.]

Pegol® F-108. [Rhone-Poulenc Surf.] EO/PO block copolymer; nonionic; surfactant, dispersant for agric. formulations; EPA compliance; HLB 27.0; pour pt. 57 C; cloud pt. > 100 C (1% aq.). DISCONTINUED.

Pegol® F-127. [Rhone-Poulenc Surf.] EO/PO block copolymer; nonionic; surfactant for toilet tank blocks, syndet bars; HLB 22.0; pour pt. 56 C; cloud pt. 100 C (1% aq.). DISCONTINUED.

Pegol® L-10. [Rhone-Poulenc Surf.] EO/PO block copolymer; nonionic; emulsifier for agric. formulations, cutting and grinding fluids, asphalt emulsions; EPA compliance; HLB 14.0; pour pt. -5 C; cloud pt. 49 C (1% aq.). DISCONTINUED.

Pegol® L-31. [Rhone-Poulenc Surf.] EO/PO block copolymer; nonionic; surfactant for cutting and grinding fluids, emulsifiable concs., asphalt emulsions; HLB 4.5; pour pt. -32 F; cloud pt. 37 C (1% aq.); 100% conc. DISCONTINUED.

Pegol® L-35. [Rhone-Poulenc Surf.] EO/PO block copolymer; nonionic; surfactant for rinse aids, automatic dishwashing, metal treatment; HLB 18.5; pour pt. 7 C; cloud pt. 77 C (1% aq.); 100% conc. DISCONTINUED.

Pegol® L-42. [Rhone-Poulenc Surf.] EO/PO block copolymer; nonionic; surfactant for rinse aids, automatic dishwashing, metal treatment; HLB 8.0; pour pt. -26 C; cloud pt. 37 C (1% aq.); 100% conc. DISCONTINUED.

Pegol® L-43. [Rhone-Poulenc Surf.] EO/PO block copolymer; nonionic; surfactant for rinse aids, automatic dishwashing, metal treatment, skin care prods., emulsifiable concs.; HLB 12.0; pour pt. -1 C; cloud pt. 42 C (1% aq.); 100% conc. DISCONTINUED.

Pegol® L-44. [Rhone-Poulenc Surf.] EO/PO block copolymer; nonionic; surfactant for hard surf. cleaning, laundry, skin care prods., emulsifiable concs.; HLB 16.0; pour pt. 16 C; cloud pt. 65 C (1% aq.); 100% conc. DISCONTINUED.

Pegol® L-61 (redesignated Antarox® L-61). [Rhone-Poulenc Surf.]

Pegol® L-62 (redesignated Antarox® L-62). [Rhone-Poulenc Surf.]

Pegol® L-62D (see Antarox® PGP 18-2D). [Rhone-Poulenc Surf.]

Pegol® L-62LF (redesignated Antarox® L-62 LF). [Rhone-Poulenc Surf.]

Pegol® L-64 (redesignated Antarox® L-64). [Rhone-Poulenc Surf.]

Pegol® L-72. [Rhone-Poulenc Surf.] EO/PO block copolymer; nonionic; surfactant for skin care, emulsion polymerization, cutting fluids, pharmaceutical prods.; HLB 6.5; pour pt. -7 C; cloud pt. 25 C (1% aq.). DISCONTINUED.

Pegol® L-81. [Rhone-Poulenc Surf.] EO/PO block copolymer; nonionic; surfactant for rinse aids, automatic dishwashing, metal treatment, water treatment, asphalt systems, w/o emulsions; liq.; HLB 2.0; pour pt. -37 C; cloud pt. 20 C (1% aq.); 100% conc. DISCONTINUED.

Pegol® L-92. [Rhone-Poulenc Surf.] EO/PO block copolymer; nonionic; surfactant for w/o emulsions; HLB 5.5; pour pt. 7 C; cloud pt. 26 C (1% aq.). DISCONTINUED.

Pegol® L-101. [Rhone-Poulenc Surf.] EO/PO block copolymer; nonionic; surfactant for rinse aids, automatic dishwashing, metal treatment, water treatment, w/o emulsions; liq.; HLB 1.0; pour pt. -23 C; cloud pt. 15 C (1% aq.); 100% conc. DISCONTINUED.

Pegol® L-121. [Rhone-Poulenc Surf.] EO/PO block copolymer; nonionic; surfactant for rinse aids, automatic dishwashing, metal treatment, water treatment, asphalt w/o emulsions; liq.; HLB 0.5; pour pt. 5 C; cloud pt. 14 C (1% aq.); 100% conc. DISCONTINUED.

Pegol® P-65. [Rhone-Poulenc Surf.] EO/PO block copolymer; nonionic; surfactant for hard surf. cleaning, laundry, shampoos, lt. duty liqs., electrolytic cleaners; HLB 17.0; pour pt. 27 C; cloud pt. 82 C (1% aq.). DISCONTINUED.

Pegol® P-75. [Rhone-Poulenc Surf.] EO/PO block copolymer; nonionic; surfactant for hard surf. cleaning, laundry, shampoos, lt. duty liqs., electrolytic cleaners; HLB 16.8; pour pt. 27 C; cloud pt. 82 C (1% aq.). DISCONTINUED.

Pegol® P-84 (redesignated Antarox® P-84). [Rhone-Poulenc Surf.]

Pegol® P-85. [Rhone-Poulenc Surf.] EO/PO block copolymer; nonionic; surfactant for hard surf. cleaning, laundry, shampoos, skin care prods.; HLB 16.0; pour pt. 34 C; cloud pt. 85 C (1% aq.); 100% conc. DISCONTINUED.

Pegol® P-104 (redesignated Antarox® P-104). [Rhone-Poulenc Surf.]

Pegol® P-105. [Rhone-Poulenc Surf.] EO/PO block copolymer; nonionic; emulsifier, dipsersant for agric. formulations; EPA compliance. DISCONTINUED.

Pegol® P-1000. [Rhone-Poulenc Surf.] PPG; nonionic; antifoam for industrial applics., additive for tire lubricants; carrier and cosolv. for personal care prods.; liq.; pour pt. < -25 C; 100% conc. DISCONTINUED.

Pegol® P-2000. [Rhone-Poulenc Surf.] PPG; nonionic; antifoam for industrial applics., additive for tire lubricants; carrier and cosolv. for personal care prods.; liq.; pour pt. < -25 C; 100% conc. DISCONTINUED.

Pegosperse® 50 DS. [Lonza] Glycol distearate; nonionic; emulsifier, opacifier, stabilizer for suspensions and dispersions; emollient, lubricant and pigment dispersant in pharmaceuticals and cosmetics; thickener, wetting agent and plasticizer in hair prods.; wh. flakes; sol. in ethanol, min. and veg. oil; insol. in water; HLB 1; m.p. 58–63 C; sapon. no. 185–200; 100% conc.

Pegosperse® 50 MS. [Lonza] PEG stearate; CAS 9004-99-3; nonionic; dispersant, emulsifier for o/w emulsions for industrial use; wh. flakes; sol. in methanol, ethanol, acetone, ethyl acetate, toluol, naphtha, min. and veg. oils; sp.gr. 0.96; m.p. 55–60 C; HLB 2; acid no. < 5; sapon. no. 180–187; pH 4.0–6.0 (3% aq.); 100% conc.

Pegosperse® 100 L. [Lonza] PEG-2 laurate; CAS 9004-81-3; nonionic; emulsifier for o/w emulsions; dispersant; for industrial use; straw liq.; sol. in ethanol, toluol, naphtha, min. oil; disp. in water; sp.gr. 0.97; HLB 7.4 ± 0.5; solid. pt. < 13 C; acid no. < 4; sapon. no. 160–170; pH 8–10 (5% aq.); 100% conc.

Pegosperse® 100 ML. [Lonza] PEG-2 laurate; CAS 9004-81-3; nonionic; dispersant, emulsifier for o/w emulsions for industrial use; straw liq.; sol. in methanol, ethanol, acetone, ethyl acetate; sp.gr. 0.96; HLB 6.0 ± 0.5; solid. pt. 16–19 C; acid no. < 4; sapon. no. 180–190; pH 3.0–6.0 (5% aq.); 100% conc.

Pegosperse® 100 MR. [Lonza] PEG-2 ricinoleate; CAS 9004-97-1; nonionic; dispersant, emulsifier for o/w emulsions for industrial use; amber liq.; sol. in methanol, ethanol, acetone, ethyl acetate; sp.gr. 0.98; HLB 4.8 ± 0.5; solid. pt. < 0 C; acid no. < 10; sapon. no. 140–150; pH 3–5 (5% aq.); 100% conc.

Pegosperse® 100 O. [Lonza] PEG-2 oleate; CAS 9004-96-0; nonionic; emulsifier for o/w emulsions; dispersant; for industrial use; amber liq.; sol. in methanol, ethanol, toluol, naphtha; disp. in water; sp.gr. 0.93; HLB 3.5 ± 0.5; solid. pt. < 0 C; acid no. 80–95; sapon. no. 160–175; pH 8–9 (5% aq.); 100% conc.

Pegosperse® 100 S. [Lonza] PEG-2 stearate; CAS 9004-99-3; nonionic; emulsifier for o/w emulsions; dispersant; for industrial use; wh. beads; sol. in methanol, ethanol, toluol, naphtha, min. and veg. oils; disp. in water; sp.gr. 0.96; m.p. 46–54 C; HLB 3.8 ± 0.5; acid no. 95–105; sapon. no. 165–175; pH 6.8–7.5 (3% aq.); 100% conc.

Pegosperse® 200 DL. [Lonza] PEG-4 dilaurate; CAS 9005-02-1; nonionic; emulsifier, dispersant, opacifier, visc. control agent, defoamer for cosmetics, household prods., textiles, plastics, water treatment; lt. yel. liq.; sol. in ethanol, min. and veg. oil; disp. in water; sp.gr. 0.96; HLB 7 ± 1; solid. pt. 3 C; sapon. no. 170–185.

Pegosperse® 200 ML. [Lonza] PEG-4 laurate; CAS 9004-81-3; nonionic; emulsifier for o/w emulsions; dispersant; for industrial use; yel. liq.; sol. in methanol, ethanol, acetone, ethyl acetate, toluol; misc. with water, sp.gr. 0.99; HLB 8.6 ± 0.5; solid pt. < 5; acid no. > 5; sapon. no. 149–159; pH 4.0–6.5 (5% aq.); 100% conc.

Pegosperse® 400 DL. [Lonza] PEG-8 dilaurate; CAS 9005-02-1; nonionic; emulsifier for o/w emulsions; dispersant; for industrial use; yel. liq.; sol. in methanol, ethanol, acetone, ethyl acetate, toluol, naphtha, min. and veg. oils; sp.gr. 0.99; HLB 10.0 ± 0.5; solid. pt. 5–12 C; acid no. < 5; sapon. no. 127–140; pH 3–5 (5% aq.); 100% conc.

Pegosperse® 400 DO. [Lonza] PEG-8 dioleate; CAS 9005-07-6; nonionic; emulsifier for o/w emulsions; dispersant; for industrial use; amber liq.; sol. in methanol, ethanol, ethyl acetate, toluol, naphtha, min. and veg. oils; disp. in water; sp.gr. 0.97; HLB 8.0 ± 0.5; solid. pt. < 0 C; acid no. < 10; sapon. no. 115–125; pH 4.5–6.5 (5% aq.); 100% conc.

Pegosperse® 400 DOT. [Lonza] PEG-8 ditallate; CAS 61791-01-3; nonionic; emulsifier, dispersant, opacifier, visc. control agent, defoamer for cosmetic, household prods., textiles, paper, water treatment; Gardner 10 liq.; HLB 8; acid no. 15; sapon. no. 121.

Pegosperse® 400 DS. [Lonza] PEG-8 distearate; CAS 9005-08-7; nonionic; emulsifier for o/w emulsions; dispersant; for industrial use; cream soft solid; sol. in ethanol, ethyl acetate, toluol, naphtha, min. and veg. oil, methanol; disp. in water; sp.gr. 0.98; m.p. 29–37 C; HLB 7.8 ± 0.5; acid no. < 10; sapon. no. 115–125; pH 4.0–6.5 (5% aq.); 100% conc.

Pegosperse® 400 DTR. [Lonza] PEG-8 ditriricinoleate; nonionic; dispersant, emulsifier for o/w emulsions for industrial use; amber liq.; sol. in acetone, ethyl acetate, toluol, naphtha, min. and veg. oils; disp. in water; sp.gr. 0.95–0.97; HLB 1.8 ± 0.5; solid. pt. < 10 C; acid no. < 12; sapon. no. 156–166 pH 3.5–5.0 (5% aq.); 100% conc.

Pegosperse® 400 MC. [Lonza] PEG-8 cocoate; nonionic; dispersant, emulsifier for o/w emulsions for industrial use; yel. liq.; sol. in methanol, ethanol, acetone, ethyl acetate, toluol, veg. and min. oil; misc. with water; sp.gr. 1.03; HLB 13.9 ± 0.5; solid. pt. < 6 C; acid no. < 4; sapon. no. 88–100; pH 4–6 (5% aq.); 100% conc.

Pegosperse® 400 ML. [Lonza] PEG-8 laurate; CAS 9004-81-3; nonionic; emulsifier for o/w emulsions; dispersant; for industrial use; straw liq.;sol. in methanol, ethanol, acetone, ethyl acetate, toluol, veg. and min. oil; misc. with water; sp.gr. 1.03; HLB 13.9 ± 0.5; solid. pt. < 7 C; acid no. < 3; sapon. no. 90–100; pH 4.0–6.0 (5% aq.); 100% conc.

Pegosperse® 400 MO. [Lonza] PEG-8 oleate; CAS 9004-96-0; nonionic; emulsifier for o/w emulsions; dispersant; for industrial use; amber liq.; sol. in methanol, ethanol, acetone, ethyl acetate, toluol, naphtha; disp. in water; sp.gr. 1.01; HLB 11.0 ± 0.5; solid. pt. < 0 C; acid no. > 5; sapon. no. 80–88; pH 4–6 (5% aq.); 100% conc.

Pegosperse® 400 MOT. [Lonza] PEG-8 tallate; CAS 61791-00-2; nonionic; emulsifier for o/w emulsions; dispersant; for industrial use; yel. liq.; sol. in methanol, ethanol, acetone, ethyl acetate, toluol, naphtha, min. and veg. oils; disp. in water; sp.gr. 1.02; HLB 11.0 ± 0.5; solid. pt. < 0 C; acid no. < 5–8; sapon. no. 83–89; pH 3.5–5.0 (5% aq.); 100% conc.

Pegosperse® 400 MS. [Lonza] PEG-8 stearate; CAS 9004-99-3; nonionic; emulsifier for o/w emulsions; dispersant; for industrial use; wh. soft solid; sol. in methanol, ethanol, acetone, ethyl acetate, toluol, naphtha, min. and veg. oils; disp. in water; ; sp.gr. 1.0; m.p. 30 C min.; HLB 11.2 ± 0.5; acid no. > 3; sapon. no. 83–94; pH 3.5–6.0 (5% aq.); 100% conc.

Pegosperse® 600 DOT. [Lonza] PEG-12 ditallate; CAS 61791-01-3; nonionic; emulsifier, dispersant, opacifier, visc. control agent, and defoamer for cosmetic, household prods., textiles, paper, and water treatment; Gardner 10 liq.; HLB 9; acid no. 15;

sapon. no. 101.

Pegosperse® 600 ML. [Lonza] PEG-12 laurate; CAS 9004-81-3; nonionic; emulsifier for o/w emulsions; dispersant; for industrial use; yel. liq.; sol. in water, methanol, ethanol, acetone, ethyl acetate, naphtha; sp.gr. 1.01; HLB 14.6 ± 0.5; solid. pt. 16–21 C; acid no. < 1; sapon. no. 65–75; pH 6–8 (5% aq.); 100% conc.

Pegosperse® 600 MS. [Lonza] PEG-12 stearate; CAS 9004-99-3; nonionic; emulsifier for o/w emulsions; dispersant; for industrial use; wh. soft solid; sol. in methanol, ethanol, acetone, ethyl acetate, toluol, veg. oil; disp. in water; sp.gr. 1.01; m.p. 27–32 C; HLB 13.2 ± 0.5; acid no. < 8.5; sapon. no. 72–78; pH 3–5 (5% aq.); 100% conc.

Pegosperse® 700 TO. [Lonza] PEG-14 oleate; CAS 9004-96-0; nonionic; dispersant, emulsifier for o/w emulsions for industrial use; amber liq.; sol. in water, methanol, ethanol, acetone, ethyl acetate, toluol, veg. oil; sp.gr. 1.10; HLB 13.5 ± 0.5; solid. pt. < 15 C; acid no. < 2; sapon. value 35–45; pH 6–8 (5% aq.); 100% conc.

Pegosperse® 1000 MS. [Lonza] PEG-20 stearate; CAS 9004-99-3; nonionic; dispersant, emulsifier for o/w emulsions for industrial use; cream solid; sol. in methanol, ethanol, acetone, ethyl acetate, toluol, naphtha, veg. oil; disp. in water; sp.gr. 1.02; m.p. 37–43 C; HLB 15.2 ± 0.5; acid no. < 3; sapon. no. 45–55; pH 3–5 (5% aq.); 100% conc.

Pegosperse® 1500 DO. [Lonza] PEG-6-32 dioleate; CAS 9005-07-6; nonionic; dispersant, emulsifier for o/w emulsions for industrial use; amber soft solid; sol. in methanol, ethanol, acetone, ethyl acetate, toluol, veg. oil; disp. in water; sp.gr. 1.05; m.p. 30–38 C; HLB 7.8 ± 0.5; acid no. < 11; sapon. no. 104–112; pH 3.5–5.0 (5% aq.); 100% conc.

Pegosperse® 1500 MS. [Lonza] PEG-6-32 stearate; CAS 9004-99-3; nonionic; emulsifier for o/w emulsions; dispersant; for industrial use; cream waxy solid; sol. in ethanol, ethyl acetate, toluol, veg. oil, methanol, naphtha; misc. with water; sp.gr. 1.05; m.p. 27–31 C; HLB 13.8 ± 0.5; acid no. < 3.5; sapon. no. 57–67; pH 3–5 (5% aq.); 100% conc.

Pegosperse® 1750 MS. [Lonza] PEG-40 stearate; CAS 9004-99-3; nonionic; emulsifier for o/w emulsions; dispersant; for industrial use; wh. flakes; sol. in water, ethanol; HLB 18 ± 1; m.p. 46 C; sapon. no. 25–35; 100% conc.

Pegosperse® 4000 MS. [Lonza] PEG-75 stearate; CAS 9004-99-3; nonionic; dispersant, emulsifier for o/w emulsions for industrial use; cream waxy solid; sol. in water, methanol, ethanol, acetone, ethyl acetate, toluol, veg. oil; sp.gr. 1.10; m.p. 54–61; HLB 18.0 ± 0.5; acid no. < 6; sapon. no. 17–22; pH 3–5 (5% aq.); 100% conc.

Pegosperse® 6000 DS. [Lonza] PEG-150 distearate; CAS 9005-08-7; nonionic; emulsifier, opacifier, stabilizer for suspensions and dispersions; emollient, lubricant and pigment dispersant in pharmaceuticals and cosmetics; thickener, wetting agent and plasticizer in hair prods.; sol. in water, ethanol; HLB 18 ± 1; m.p. 55 C; sapon. no. 17.

Pegosperse® 9000 CO. [Lonza] PEG 9000 castor oil; nonionic; dispersant, emulsifier for o/w emulsions for industrial use; tan solid; sol. in water, methanol, ethanol, acetone; sp.gr. 1.08; m.p. 41–44 C; HLB 18.1 ± 0.5; acid no. < 1; sapon. no. 14–20; pH 6.0–7.0 (3% aq.); 100% conc.

Pegosperse® EGMS-70. [Lonza] Glycol stearate; emulsifier, opacifier, stabilizer for suspensions and dispersions; emollient, lubricant and pigment dispersant in pharmaceuticals and cosmetics; thickener, wetting agent and plasticizer in hair prods.; wh. flakes; sol. in ethanol, min. and veg. oil; insol. in water; HLB 2 ± 1; m.p. 52–56 C; sapon. no. 180–188.

Pegosperse® MFE. [Lonza] Mixed PEG esters; nonionic; dispersant, emulsifier for o/w emulsions for industrial use; brn. soft solid; sol. in methanol, ethanol, veg. and min. oil, water; sp.gr. 1.05; HLB 13.0 ± 0.5; m.p. 46–54 C; acid no. < 10; sapon. no. 60–70; pH 3–5 (5% aq.); 100% conc.

Pegosperse® PMS CG. [Lonza] Propylene glycol stearate; nonionic; emulsifier, dispersant, opacifier, visc. control agent, defoamer for cosmetics, household prods., textiles, plastics, water treatment; Gardner 2 flakes; m.p. 51 C; HLB 3; acid no. 3; sapon. no. 180.

Pelex NBL, NB Paste. [Kao Corp. SA] Sodium alkyl naphthalene sulfonate; anionic; wetting agent and penetrant; liq. and paste resp.; 35 and 50% conc. resp.

Pelex OT-P. [Kao Corp. SA] Sodium dialkyl sulfosuccinate; anionic; wetting agent for textile processing; emulsifier for emulsion polymerization; liq.; 70% conc.

Pelex SS-H. [Kao Corp. SA] Sodium dodecyl diphenyl ether disulfonate; anionic; emulsifier for syn. latex and emulsion polymerization; paste; 50% conc.

Pelex SS-L. [Kao Corp. SA] Sodium dodecyl diphenyloxide disulfonate; anionic; emulsifier for syn. latex and emulsion polymerization; paste; 50% conc.

Pelex TA. [Kao Corp. SA] Sodium dialkyl sulfosuccinate; anionic; foam controller for latex; paste; 35% conc.

Peltex. [LignoTech] Ferro-chromium lignosulfate, modified; anionic; wetting agent, emulsifier, dispersant; oil well drilling mud thinner; blk. powd.; sol. in water; dens. 0.5; 95% act.

Penestrol N-160. [Tokai Seiyu Ind.] Sec. alcohol ethoxylate; nonionic; penetrant and wetting agent; bleaching assistant; liq.; 60% conc.

Penetral NA 20. [Ceca SA] Anionic; low-foam wetting agent for textile mercerizing; stable in strongly alkaline sol'n.

Penetrol 2-EHS. [Clark] Ammonium 2-ethylhexyl sulfate; solubilizer and wetting agent esp. at high pH and temp.; lt. yel. liq.

Penetron OT-30. [Hart Prods. Corp.] 2-Ethylhexyl sulfosuccinate; anionic; penetrant, wetting agent; liq.; 25-30% conc.

Pentex® 40. [Rhone-Poulenc Surf.] Alkylaryl sulfonate; anionic; wetting agent, penetrant, emulsifier; textile and industrial processing; improves dye uniformity; aids rewetting of leather, paper, and paper-mill felts; red-br. liq.; sp.gr. 1.15; pour pt. 0 C; pH 9.5 (5%). DISCONTINUED.

Pentine 1185 5432. [Clough] Isopropylamine dodecylbenzene sulfonate; anionic; emulsifier for emulsion degreasers, dry cleaning soaps, agric. emulsifier; visc. liq.; pH 4-6; 95–97% conc.

Pentine Acid 5431. [Clough] Dodecylbenzene sulfonic acid; CAS 27176-87-0; anionic; base for dishwashing and laundry detergents; visc. liq.; 95-97% conc.

Pentine Acid 5431. [Clough] Dodecylbenzene sulfonic acid; for use in mfg. household, industrial, and institutional cleaners; liq.; pH < 1.0; 100% solids.

Pentrone S127. [Rhone-Poulenc UK] Disodium

alkoxy sulfosuccinate deriv.; anionic; surfactant for aq. systems; liq.; 22% conc.

Pepol A-0638. [Toho Chem. Industry] Oxyethylated fatty alcohol, modified; nonionic; detergent, wetting agent; liq.; 100% conc.

Pepol A-0858. [Toho Chem. Industry] Oxyethylated fatty alcohol, modified; nonionic; detergent, penetrant; paste; 100% conc.

Pepol AX-1. [Toho Chem. Industry] Oxyethylated fatty alcohol, modified; nonionic; low foaming detergent, wetting agent; liq.; 100% conc.

Pepol B. [Toho Chem. Industry] EO condensate; nonionic; antistat, deduster, defoamer, demulsifier, detergent, emulsifier, wetting agent, dye leveler, gellant, foam controller; liq.; HLB 3.0; 100% conc.

Pepol B-182. [Toho Chem. Industry] EO condensate; nonionic; antistat, deduster, defoamer, demulsifier, detergent, emulsifier, wetting agent, dye leveler, gellant, foam controller; liq.; HLB 7.0; 100% conc.

Pepol B-184. [Toho Chem. Industry] EO condensate; nonionic; antistat, deduster, defoamer, demulsifier, detergent, emulsifier, wetting agent, dye leveler, gellant, foam controller; liq.; HLB 15.0; 100% conc.

Pepol BS-184. [Toho Chem. Industry] PO condensate; nonionic; defoamer, foam controller, dispersant, dye leveler, wetting agent for pulp, textiles, antifreeze, oil-sol. compositions; liq.; 100% conc.

Pepol BS-201. [Toho Chem. Industry] PO condensate; nonionic; defoamer, foam controller, dispersant, dye leveler, wetting agent for pulp, textiles, antifreeze, oil-sol. compositions; liq.; 100% conc.

Pepol BS-403. [Toho Chem. Industry] PO condensate; nonionic; defoamer, foam controller, dispersant, dye leveler, wetting agent for pulp, textiles, antifreeze, oil-sol. compositions; paste; 100% conc.

Pepol D-301. [Toho Chem. Industry] EO condensate; nonionic; emulsion stabilizer, dispersant, wetting agent, defoamer, demulsifier, visc. controller; liq.; HLB 7.0; 100% conc.

Pepol D-304. [Toho Chem. Industry] EO condensate; nonionic; emulsion stabilizer, dispersant, wetting agent, defoamer, demulsifier, visc. controller; liq.; HLB 15.0; 100% conc.

Peramit MLN. [Toho Chem. Industry] Ammonium lauryl sulfate; anionic; surfactant for lt. duty detergents, hand cleaning pastes; paste; 30% conc.

Peregal®ST. [ISP] PVP polymer; cationic; dispersant, stripping assistant, detergent; used for dyes; used in long liquors, pkg. machines, or jigs; fulling soap additive; rag-stripping assistant in paper industry; detergent and soil-suspending agent for washing yarns; VCS 5 max. liq.; water-sol.; pH 5–7 (5% aq.); 30–32% act.

Perenol EI. [Henkel] Polyvinyl isobutyl ether; defoamer for petrol. solv.-based paints.

Perlankrol® ACM2. [Harcros UK] Sodium fatty alkylolamide ether sulfate; anionic; base; preparation of personal care prods.; 30% act.

Perlankrol® ATL40. [Harcros UK] TEA lauryl sulfate; CAS 139-96-8; anionic; biodeg. surfactant for prep. of high foaming shampoos and toiletries; emulsifier for emulsion polymerization; pale straw clear liq.; mild odor; water-sol.; sp.gr. 1.048; visc. 60 cs.; flash pt. (COC) > 95 C; pour pt. < 0 C; pH 7.5 (1% aq.); surf. tens. 32 dynes/cm (0.1%); 40% act.

Perlankrol® DAF25. [Harcros UK] Ammonium lauryl sulfate; anionic; foaming agent, base; formulation of toiletries and carpet shampoos; pale yel. mobile gel, mild nonammoniacal odor; water-sol.; sp.gr. 1.007; visc. 1600 cs; pour pt. 5 C; flash pt. (COC) > 95 C; pH 6.4 (1% aq.); 25% act.

Perlankrol® DGS. [Harcros UK] Sodium primary alcohol sulfate; anionic; solubilizer, textile aux.; detergent base in household and industrial cleaners; foaming agent in syn. latex industry; primary emulsifier in emulsion polymerization; wh. slurry; faint fatty alcohol odor; water-sol.; sp.gr. 1.041; visc. 220 cs (40 C); flash pt. > 200 F (COC); pour pt. 12 C; pH 7.0–8.5 (1% aq.); 28% act.

Perlankrol® DSA. [Harcros UK] Sodium lauryl sulfate; CAS 151-21-3; anionic; foaming agent for syn. latexes, emulsion polymerization aid; base for prep. of high foaming shampoos and toiletries; wetting agent; industrial detergent additive; wh. slurry; faint fatty alcohol odor; water-sol.; sp.gr. 1.041; visc. 220 cs (40 C); flash pt. (COC) > 95 C; pour pt. 12 C; pH 8.0 (1% aq.); surf. tens. 30 dynes/cm (0.1%); 28% act.

Perlankrol® EAD60. [Harcros UK] Ammonium alcohol ether sulfate, ethanol; anionic; stabilizer, detergent for industrial and household cleaners; foaming agent for fire fighting foam compds.; dispersant and emulsifier for chlorinated solvs.; emulsifier for emulsion polymerization; clear mobile straw gel; faint alcoholic odor; water-sol.; sp.gr. 1.033; visc. 240 cs; flash pt. (Abel CC) 85 F; pour pt. < 0 C; pH 7–9 (1% aq.); surf. tens. 34 dynes/cm (0.1%); 60% act. in ethanol.

Perlankrol® EP12. [Harcros UK] Fatty alcohol ether sulfate; anionic; emulsifier for emulsion polymerization; biodeg.; pale yel. clear to hazy liq.; surf. tens. 39 dynes/cm (0.1%); 20% act.

Perlankrol® EP24. [Harcros UK] Fatty alcohol ether sulfate; anionic; emulsifier for emulsion polymerization; biodeg.; straw clear liq.; surf. tens. 42 dynes/cm (0.1%); 28% act.

Perlankrol® EP36. [Harcros UK] Fatty alcohol ether sulfate; anionic; emulsifier for emulsion polymerization; biodeg.; straw clear liq.; surf. tens. 40 dynes/cm (0.1%); 28% act.

Perlankrol® ESD. [Harcros UK] Syn. primary alcohol ether sulfate, sodium salt; anionic; biodeg. high foam additive for liq. household and industrial detergents; emulsifier for emulsion polymerization; water-wh. clear liq.; faint odor; water-sol.; sp.gr. 1.043; visc. 26 cs; flash pt. (COC) > 95 C; pour pt. < 0 C; pH 8.0 (1% aq.); surf. tens. 40 dynes/cm (0.1%); 27% act. in water.

Perlankrol® ESD60. [Harcros] Sodium fatty alcohol ether sulfate; anionic; biodeg. foam booster for liq. detergent formulations and shampoos; emulsifier for emulsion polymerization; water-wh. clear to hazy mobile gel; faint alcoholic odor; water-sol.; sp.gr. 1.046; visc. 210 cs; flash pt. (Abel CC) 90 F; pour pt. < 0 C; pH 8.0 (1% aq.); surf. tens. 39 dynes/cm (0.1%); 60% act. in aq. alcohol.

Perlankrol® ESK29. [Harcros UK] Syn. primary alcohol ether sulfate, sodium salt; anionic; biodeg. solubilizer for disinfectants; foam additive for industrial detergents; concrete foaming agent; yel. clear liq., faint odor; water-sol.; sp.gr. 1.052; visc. 19 cs; pour pt. -2 C; flash pt. (COC) > 95 C; pH 7.5 (1% aq.); 30% act. in water.

Perlankrol® ESS25. [Harcros UK] Syn. primary alcohol ether sulfate, sodium salt; anionic; stabilizer, detergent for industrial and household cleaners; foaming agent for fire fighting foam compds.; dispersant and emulsifier for chlorinated solvs.; water-wh. clear liq.; faint odor; water-sol.; sp.gr. 1.040; visc. 65 cs; flash pt. > 200 F (COC); pour pt.

< 0 C; pH 7–9 (1% aq.); 25% act.

Perlankrol® FB25. [Harcros UK] Alkyl phenol ether sulfate, modified, sodium salt; anionic; primary emulsifier for emulsion polymerization; clear straw liq.; faint alcoholic odor; water-sol.; sp.gr. 1.038; visc. 48 cs; flash pt. 86 F (Abel CC); pour pt. –3 C; pH 6.0–8.5 (1% aq.); 25% act.

Perlankrol® FD63. [Harcros UK] Alkyl phenol ether sulfate, ammonium salt; anionic; primary emulsifier for emulsion polymerization; clear straw liq.; faint alcoholic odor; water-sol.; sp.gr. 1.068; visc. 132 cs; flash pt. (Abel CC) 29 C; pour pt. < -5 C; pH 7.5 (1% aq.); surf. tens. 29 dynes/cm (0.1%); 63% act. in aq. alcohol.

Perlankrol® FF. [Harcros UK] Alkyl phenol ether sulfate, ammonium salt; anionic; foam booster/stabilizer for industrial detergents; emulsifier for emulsion polymerization; amber hazy visc. liq.; faint ammoniacal odor; water-sol.; sp.gr. 1.123; visc. 4200 cs; flash pt. (Abel CC) 30 C; pour pt. 5 C; pH 7.5 (1% aq.); surf. tens. 29 dynes/cm (0.1%); 90% act. in aq. alcohol.

Perlankrol® FN65. [Harcros UK] Alkyl phenol ether sulfate, sodium salt; anionic; detergent base, foam stabilizer used in specialty detergent formulations; primary emulsifier in emulsion polymerization; clear straw liq.; faint alcoholic odor; water-sol.; sp.gr. 1.123; visc. 398 cs; flash pt. (Abel CC) 30 C; pour pt. < -5 C; pH 7.5 (1% aq.); surf. tens. 27 dynes/cm (0.1%); 65% act. in aq. ethanol.

Perlankrol® FT58. [Harcros UK] Ammonium alkyl phenol ether sulfate; anionic; primary emulsifier for emulsion polymerization; detergent base, foam stabilizer/booster; emulsifier for cresylic acid; amber clear/sl. hazy liq., faint alcoholic odor; water-sol.; sp.gr. 1.058; visc. 119 cs; pour pt. < -5 C; flash pt. (Abel CC) 32 C; pH 7.8 (1% aq.); 59% act. in aq. alcohol.

Perlankrol® FV70. [Harcros] Ammonium alkyl phenol ether sulfate; anionic; primary emulsifier in emulsion polymerization, latex compding., textile auxiliaries, oilfield applics.; straw clear/sl. hazy liq., faint alcoholic odor; water-sol.; sp.gr. 1.097; visc. 848 cs; pour pt. -15 C; flash pt. (Abel CC) 29 C; pH 7.5 (1% aq.); 70% act. in aq. alcohol.

Perlankrol® FX35. [Harcros UK] Alkylphenol ether sulfate, sodium salt; anionic; emulsifier for emulsion polymerization; pale yel. clear liq.; surf. tens. 33 dynes/cm (0.1%); 35% act.

Perlankrol® O. [Harcros UK] Sodium alkyl sulfate; anionic; clarifying and wetting agent, stabilizer, visc. modifier used in detergent and cleaning formulations; clear lt. amber liq.; char. odor; water-sol.; sp.gr. 1.120; visc. 44 cs; flash pt. > 200 F (COC); pour pt. < 0 C; pH 7.0–8.5 (1% aq.); 40% act.

Perlankrol® PA Conc. [Harcros UK] Alkyl phenol ether sulfate, ammonium salt; CAS 30416-77-4; anionic; foam booster/stabilizer, detergent base, frothing agent used in liq. detergent formulations; emulsifier for cresylic acid, emulsion polymerization; amber hazy visc. liq.; faint ammoniacal odor; water-sol.; sp.gr. 1.110; visc. 6000 cs; flash pt. (Abel CC) 30 C; pour pt. 7 C; pH 7.5 (1% aq.); surf. tens. 37 dynes/cm (0.1%); 90% act. in aq. alcohol.

Perlankrol® RN75. [Harcros UK] Alkyl phenol ether sulfate, sodium salt; anionic; speciality foaming and wetting agent in high electrolyte concs.; primary emulsifier in emulsion polymerization; straw clear liq.; faint alcoholic odor; water-sol. sp.gr. 1.118; visc. 347 cs; flash pt. (Abel CC) 31 C; pour pt. < -5 C; pH 7.5 (1% aq.); surf. tens. 32 dynes/cm (0.1%); 75% act. in aq. alcohol.

Perlankrol® SN. [Harcros UK] Alkyl phenol ether sulfate, sodium salt; anionic; foam booster and stabilizer, wetting agent, emulsifier in emulsion polymerization; industrial detergent formulations; straw clear liq.; faint odor; water-sol.; sp.gr. 1.060; visc. 116 cs; flash pt. (COC) > 95 C; pour pt. < -5 C; pH 7.5 (1% aq.); surf. tens. 30 dynes/cm (0.1%); 30% act. in water.

Permalene A-100. [Rhone-Poulenc/Textile & Rubber] Amine-coconut oil condensation prod.; nonionic; detergent, dispersant, dye leveler and lubricant for rayon, acetate dyeing; liq.; 100% conc.

Permalose TM. [ICI Surf. UK] Hydrophilic copolymer aq. disp.; nonionic; soil release/antistatic agent for treatment of polyester fabrics and blends; liq.

Peronal MTB. [Seppic] Blend; anionic; textile detergent; liq.; 60% conc.

Peroxal 36. [Seppic] Blend; anionic/nonionic; textile detergent; liq.; 50% conc.

Pestilizer® B Series. [Stepan Europe] Amphiphilic alkyl sulfosuccinate; emulsifiers for liq. fertilizers; high stability in electrolytic systems; visc. liq.

Petro® 11. [Witco/Organics] Sodium alkyl naphthalene sulfonate; anionic; hydrotrope, surfactant, germicidal agent, liquid hand soaps, wetting agent in plating and electrolytic baths; 50% act. liq., 95% act. powd.; water-sol.

Petro® 22. [Witco/Organics] Sodium alkyl naphthalene sulfonate; anionic; surfactant for carpet steam cleaners, tank soak compds., high alkaline and acid cleaners; 50% act. liq., 95% act. powd.

Petro® AA. [Witco/Organics] Sodium alkyl naphthalene sulfonate; hydrotrope, detergent; stable in strong acids, alkalies, salts, detergent builders; 50% act. liq., 95% act. powd.

Petro® BA. [Witco/Organics] Sodium alkyl naphthalene sulfonate; hydrotrope, surfactant for high-solids liquid laundry detergents, electrolytic baths, metal and dairy cleaners; 50% act. liq., 95% act. powd.; water-sol.

Petro® BAF. [Witco/Organics] Sodium alkyl naphthalene sulfonate; surfactant for rug and upholstery shampoos, car wash, window cleaners; 50% act. liq., 95% act. powd.

Petro® Dispersant 98. [Witco/Organics] Sodium alkyl naphthalene sulfonate; dispersant for dyestuffs.

Petro® Dispersant 425. [Witco/Organics] Sodium naphthalene-formaldehyde sulfonate; dispersant for dyestuff pigments.

Petro® LBA. [Witco/Organics] Sodium alkyl naphthalene sulfonate; lighter color form of Petro BA; 50% act. liq., 95% act. powd.

Petro® LBAF. [Witco/Organics] Alkyl naphthalene sulfonate; lighter color form of Petro BAF; liq.; 50% act.

Petro® P. [Witco/Organics] Alkyl naphthalene sodium sulfonate; anionic; high-foaming detergents, high-alkaline and acid cleaners; 50% act. liq., 95% act. powd.

Petro® S. [Witco/Organics] Alkyl naphthalene sodium sulfonate; anionic; soil conditioner; 50% act. liq.; 95% act. powd.

Petro® ULF. [Witco/Organics] Alkyl naphthalene sodium sulfonate; anionic; wetting agent for jet-beck dyeing, bottle washing; liq.; 50% act.

Petro® WP. [Witco/Organics] Alkyl naphthalene sodium sulfonate, modified; anionic; wetting agent used in pesticides and industrial cleaners; powd.; 95% conc.

Petromix®. [Witco/Sonneborn] Petrol. sulfonate, modified; anionic; emulsifier; base for preparation of cutting oils, solv. degreasers; liq.

Petronate® 25C, 25H. [Witco/Sonneborn] Calcium petrol. sulfonate; anionic; detergent and rust inhibitor component in lube oil additives; emulsifier for w/o systems; ASTM 3.5 liq.; dens. 8.2 lb/gal; visc. 170 SSU (100 C); flash pt. 360 F; 43–46% calcium sulfonate.

Petronate® Basic. [Witco/Sonneborn] Barium petrol. sulfonate; anionic; oil additive for lube, industrial oils, fuels, greases; ASTM 7.0 liq.; visc. 250 SSU (100 C); 40–45% barium sulfonate.

Petronate® CR. [Witco/Sonneborn] Sodium petrol. sulfonate; anionic; wetting agent, emulsifier, dispersant, rust preventative for cutting oils, leather oils, lube oil additives, textile oils, enhanced oil recovery, oil flotation; dk. brn. visc. liq.; petrol. odor; sol. in org. solvs.; dens. 8.5 lb/gal; sp.gr. 1.02 (60 F); visc. 175 s (Furol, 100 C); flash pt. 380 F; 61–63% act.

Petronate® HL. [Witco/Sonneborn] Sodium petrol. sulfonate; anionic; wetting agent, emulsifier, dispersant, rust preventative for cutting oils, leather oils, lube oil additives, textile oils, enhanced oil recovery, oil flotation; dk. brn. visc. liq.; petrol. odor; sol. in oil., org. solvs.; dens. 8.5 lb/gal; sp.gr. 1.1 (60 F); visc. 135 s (Furol, 100 C); flash pt. 380 F; 61–63% act.

Petronate® HMW. [Witco/Sonneborn] Calcium petrol. sulfonate; detergent, rust inhibitor; lubricating oil additives; dens. 8.2 lb/gal; visc. 200 SSU (100 C); flash pt. 350 F; 45–46% calcium sulfonate.

Petronate® K. [Witco/Sonneborn] Sodium petrol. sulfonate; anionic; dispersant, emulsifier and wetting agent in dry cleaning soaps, textile processing; lt. br. visc. liq.; oil-sol.; dens. 8.3 lb/gal; sp.gr. 1.1 (60 F); visc. 85 s (Furol, 100 C); flash pt. 380 F; 61–63% act.

Petronate® L. [Witco/Sonneborn] Sodium petrol. sulfonate; anionic; wetting agent, emulsifier, dispersant, rust preventative for cutting oils, leather oils, lube oil additives, textile oils, enhanced oil recovery, oil flotation; dk. brn. visc. liq.; petrol. odor; sol. in oil and org. solvs.; dens. 8.5 lb/gal; sp.gr. 1.1 (60 F); visc. 125 s (Furol, 100 C); flash pt. 380 F; 61–63% act.

Petronate® Neutral (50-S). [Witco/Sonneborn] Barium petrol. sulfonate; anionic; oil additive for lube, industrial and fuel oils, undercoating, and greases; ASTM 3.5 liq.; visc. 100 SSU (100 C); 50–52% barium sulfonate.

Petronate® S. [Witco/Sonneborn] Sodium didodecylbenzene sulfonate; anionic; wetting agent, emulsifier, dispersant, rust preventative for cutting oils, leather oils, lube oil additives, textile oils, drycleaning soaps, oil flotation; liq.; 62% conc.

Petrosan 102. [Reilly-Whiteman] Methyl palmitate/oleate; nonionic; lubricant, wetting agent for petrol. oils for metals; liq.; 100% conc.

Petrostep 420. [Stepan; Stepan Canada] Petroleum sulfonate; anionic; for enhanced oil recovery operations; liq.; 50% conc.

Petrostep 465. [Stepan; Stepan Canada] Petroleum sulfonate; anionic; for enhanced oil recovery operations; liq.; 50% conc.

Petrostep HMW. [Stepan; Stepan Canada] Petroleum sulfonate; anionic; for enhanced oil recovery operations; liq.; 50% conc.

Petrostep LMC. [Stepan; Stepan Canada] Petroleum sulfonate; anionic; for enhanced oil recovery operations; liq.; 28% conc.

Petrostep MMW. [Stepan; Stepan Canada] Petroleum sulfonate; anionic; for enhanced oil recovery operations; liq.; 50% conc.

Petrosul® H-50. [Penreco] Sodium petrol. sulfonate; anionic; surfactant and corrosion inhibitor; used as motor and fuel oil additives, rustproofing formulations; also in dry cleaning solvs., leather processing, printing inks, oil well drilling fluids; m.w. 510–550; dens. 8.6 lb/gal; 50% conc.

Petrosul® H-60. [Penreco] Sodium petroleum sulfonate; anionic; surfactant and corrosion inhibitor; used in ore flotation applics., dry cleaning solvs., leather processing, printing inks, oil well drilling fluids, textile oils, demulsifier; liq.; m.w. 510-550; dens. 8.5 lb/gal; 60% conc.

Petrosul® H-70. [Penreco] Sodium petroleum sulfonate; anionic; surfactant and corrosion inhibitor; used in ore flotation applics., dry cleaning solvs., leather processing, printing inks, oil well drilling fluids, textile oils, demulsifier; liq.; m.w. 510-540; dens. 8.6 lb/gal; 70% conc.

Petrosul® HM-62, HM-70. [Penreco] Sodium petrol. sulfonate; anionic; surfactant and corrosion inhibitor; used in ore flotation applics., dry cleaning solvs., leather processing, printing inks, oil well drilling fluids, textile oils, demulsifier; m.w. 485–495; dens. 8.5 and 8.6 lb/gal resp.; 61 and 70% min. sulfonate.

Petrosul® L-60. [Penreco] Sodium petroleum sulfonate; anionic; surfactant and corrosion inhibitor; used in drycleaning detergents, cutting oils, textile oils, leather oils, demulsifier, fuel and motor oil additives; liq.; 60% conc.

Petrosul® M-50, M-60, M-70. [Penreco] Sodium petrol. sulfonate; anionic; surfactant, emulsifier, and corrosion inhibitor for metalworking fluids; also for dry cleaning solvs., leather processing, textile oils, printing inks, oil well drilling fluids; m.w. 440–475; dens. 8.2 lb/gal; 50% conc.

Petrosul® M-60. [Penreco] Sodium petroleum sulfonate; anionic; surfactant, emulsifier, and corrosion inhibitor for metalworking fluids; also for dry cleaning solvs., leather processing, textile oils, printing inks, oil well drilling fluids; liq.; m.w. 440-475; dens. 8.4 lb/gal; 60% conc.

Petrosul® M-70. [Penreco] Sodium petroleum sulfonate; anionic; surfactant, emulsifier, and corrosion inhibitor for metalworking fluids; also for dry cleaning solvs., leather processing, textile oils, printing inks, oil well drilling fluids; liq.; m.w. 450-470; dens. 8.6 lb/gal; 70% conc.

Petrowet® R. [DuPont] Sodium alkyl sulfonate; anionic; wetting agent, detergent, dispersant, penetrant, foamer for petrol., metalworking, textile, and paper industries, chem. mfg., oil well servicing; industrial formulations and cleaning; yel. to amber clear liq.; alcoholic odor; water-sol.; sp.gr. 1.08 g/mL; dens. 9 lb/gal; visc. 15 cps (27 C); cloud pt. < 0 C; flash pt. (PMCC) 26 C; surf. tens. 43 dynes/cm (0.1%); pH 4.0–6.0 (1% aq.); flamm.; 22% act.; contains IPA.

P & G Amide No. 27. [Procter & Gamble] Cocamide MEA; CAS 68140-00-1; nonionic; foam stabilizer, thickener; foam booster and visc. builder for detergents and shampoos; flakes; 100% conc.

PGE-400-DS. [Hefti Ltd.] PEG-8 distearate; CAS

9005-08-7; nonionic; surfactant for pharmaceuticals, cosmetics, to incorporate fats and ester waxes in creams; homogenizer for suppository bases; plasticizer for plastics; solid; HLB 8.0; 100% conc.

PGE-400-MS. [Hefti Ltd.] PEG-8 stearate; CAS 9004-99-3; nonionic; o/w emulsifier for cosmetic creams, pharmaceutical ointments, active ingreds.; paper industry deinking; solid; HLB 11.5; 100% conc.

PGE-600-DS. [Hefti Ltd.] PEG-12 distearate; CAS 9005-08-7; nonionic; o/w emulsifier for cosmetics, pharmaceuticals; plasticizer for plastics, various tech. applics.; flakes; HLB 10.5; 100% conc.

PGE-600-ML. [Hefti Ltd.] PEG-12 laurate; CAS 9004-81-3; nonionic; emulsifier, solubilizer for pharmaceuticals, cosmetics, textile oils, insecticides, pesticides; antistat for plastics; liq.; HLB 15.0; 100% conc.

PGE-600-MS. [Hefti Ltd.] PEG-12 stearate; CAS 9004-99-3; nonionic; emulsifier, solubilizer for pharmaceuticals, cosmetics, textile oils, insecticides, pesticides; solid; HLB 13.5; 100% conc.

PG No. 4. [Hart Prods. Corp.] PEG-8 stearate; CAS 9004-99-3; nonionic; emulsifier, base for textile softeners; solid; 100% conc.

Phenyl Sulphonate CA, CAL. [Hoechst Celanese/Colorants & Surf.; Hoechst AG] Calcium dodecylbenzene sulfonate; anionic; basic material for mfg. of biocide emulsifiers; liq.; 70% conc.

Phenyl Sulphonate HFCA. [Hoechst Celanese/Colorants & Surf.] Calcium dodecylbenzene sulfonate; anionic; surfactant for agric. formulations; flash pt. > 100 C.

Phenyl Sulphonate HSR. [Hoechst Celanese/Colorants & Surf.] Sodium dodecylbenzene sulfonate; CAS 25155-30-0; anionic; surfactant for agric. formulations.

Philacid 0810. [United Coconut Chem.] Caprylic/capric acid (C8-10); intermediate for mfg. of syn. specialty detergents, esters for use as lubricants and emollients; liq.; acid no. 355-365; iodine no. 0.8 max.; sapon. no. 356-366.

Philacid 0818. [United Coconut Chem.] Whole distilled coconut fatty acid (C8-18); intermediate for mfg. of soap, syn. detergents, fatty amines; liq.; acid no. 268-274; iodine no. 6-10; sapon. no. 269-275.

Philacid 1200. [United Coconut Chem.] Lauric acid (C12); CAS 143-07-7; intermediate for mfg. of toilet soaps, syn. detergents, cosmetics and pharmaceuticals; solid, flakes; acid no. 279-282; iodine no. 0.3 max.; sapon. no. 279-282.

Philacid 1214. [United Coconut Chem.] Lauric/myristic acid (C12-14); intermediate for mfg. of soaps, detergents; acid no. 268-273; iodine no. 0.5 max.; sapon. no. 268-273.

Philacid 1218. [United Coconut Chem.] Topped coconut fatty acid (C12-C18); intermediate for mfg. of toilet soaps, syn. detergents; liq.; acid no. 254-260; iodine no. 8-12; sapon. no. 254-260.

Philacid 1400. [United Coconut Chem.] Myristic acid (C14); CAS 544-63-8; intermediate for mfg. of toilet soaps, cosmetics, esters for use as flavors and perfumes; solid, flakes; acid no. 244-247; iodine no. 0.3 max.; sapon. no. 244-247.

Philacid 1618. [United Coconut Chem.] Palmitic stearic-oleic linoleic acid (C16-18); intermediate for mfg. of soaps, detergents; acid no. 205-215; iodine no. 36-46; sapon. no. 206-216.

Philcohol 1200. [United Coconut Chem.] Lauryl alcohol (C12); CAS 112-53-8; intermediate for mfg. of syn. detergents (e.g., lauryl sulfates, ethoxylated lauryl alcohol, lauryl ether sulfates, lauric alkanolamides, etc.); APHA 10 max. liq.; acid no. 0.1 max.; iodine no. 0.3 max.; sapon. no. 0.5 max.; hyd. no. 299-302.

Philcohol 1214. [United Coconut Chem.] C12-14 coconut fatty alcohol; intermediate for surfactant mfg.; APHA 10 max. liq.; acid no. 0.1 max.; iodine no. 0.3 max.; sapon. no. 1 max.; hyd. no. 282-291.

Philcohol 1218. [United Coconut Chem.] C12-18 coconut fatty alcohol; intermediate for surfactant mfg.; APHA 10 max. liq.; acid no. 0.1 max.; iodine no. 0.3 max.; sapon. no. 1.5 max.; hyd. no. 267-279.

Philcohol 1400. [United Coconut Chem.] C14 alcohols; CAS 112-72-1; intermediate for surfactant mfg.; APHA 10 max. color; acid no. 0.1 max.; iodine no. 0.3 max.; sapon. no. 1 max.; hyd. no. 258-264.

Philcohol 1600. [United Coconut Chem.] C16 alcohols; CAS 36653-82-4; intermediate for surfactant mfg.; APHA 10 max. flakes; acid no. 0.2 max.; iodine no. 0.3 max.; sapon. no. 0.5 max.; hyd. no. 230-233.

Philcohol 1618. [United Coconut Chem.] C18-18 alcohols; intermediate for surfactant mfg.; APHA 10 max. flakes; acid no. 0.2 max.; iodine no. 0.3 max.; sapon. no. 2 max.; hyd. no. 211-220.

Philcohol 1800. [United Coconut Chem.] C18 alcohols; CAS 112-92-5; intermediate for surfactant mfg.; APHA 10 max. flakes; acid no. 0.2 max.; iodine no. 0.3 max.; sapon. no. 1.5 max.; hyd. no. 206-209.

Phos-Ad 100. [Chemron] Tributyl phosphate; defoamer conc.; for drilling fluids; lt. yel. liq.; 100% conc.

Phosfac 1004, 1006, 1044, 1044FA, 1066, 1066FA, 1068FA. [Rhone-Poulenc/Textile & Rubber] Phosphate ester; anionic; emulsifier, penetrant, antistat, wetting agent, detergent, lubricant, corrosion inhibitor and dispersant used in liq. detergent formulations; liq.; 99, 99, 99, 90, 90, 99, and 99% conc. resp.

Phosfac 1068 C-FA. [Rhone-Poulenc/Textile & Rubber] Org. phosphate ester; controlled foam lubricant for textiles, metalworking fluids, aq. systems; biodeg.; liq.; 100% act.

Phosfac 5500. [Rhone-Poulenc/Textile & Rubber] Phosphate ester; anionic; wetting agent, dispersant for dye leveling, scouring, bleaching, etc.; liq.; 100% conc.

Phosfac 5508. [Rhone-Poulenc/Textile & Rubber] Phosphate ester; anionic; wetting agent, dispersant for dye leveling, scouring, bleaching, etc.; liq.; 60% conc.

Phosfac 5513. [Rhone-Poulenc/Textile & Rubber] Org. phosphate ester, potassium salt; anionic; wetting agent and dispersant for metallic pigments in aq. latex systems, carbon blk.; paste; 70% act.

Phosfac 5520. [Rhone-Poulenc/Textile & Rubber] Phosphate ester; anionic; controlled foam detergent, wetting agent, antistat for lubricant formulations; liq.; 100% conc.

Phosfac 8000. [Rhone-Poulenc/Textile & Rubber] Phosphate ester; anionic; emulsifier, penetrant, antistat, wetting agent, detergent, lubricant, corrosion inhibitor, dispersant; liq.; 99% conc.

Phosfac 8608. [Rhone-Poulenc/Textile & Rubber] Phosphate ester; anionic; emulsifier, penetrant, antistat, wetting agent, detergent, lubricant, corrosion inhibitor, dispersant; liq.; 99% conc.

Phosfac 9604. [Rhone-Poulenc/Textile & Rubber] Phosphate ester; anionic; emulsifier, penetrant, antistat, wetting agent, detergent, lubricant, corrosion inhibitor, dispersant; liq.; 99% conc.

Phosfac 9609. [Rhone-Poulenc/Textile & Rubber] Phosphate ester; anionic; emulsifier, penetrant, antistat, wetting agent, detergent, lubricant, corrosion inhibitor, dispersant; liq.; 99% conc.

Phosfetal 201. [Zschimmer & Schwarz] Alkyl polyglycol ether phosphate, acid; cleansing agent; liq.; 100% act.

Phosfetal 204. [Zschimmer & Schwarz] Isoalkyl polyglycol ether phosphate, acid; cleansing agent; liq.; 100% act.

Phosfetal 205. [Zschimmer & Schwarz] Alkyl polyglycol ether phosphate, acid; cleansing agent; wax; 100% act.

Phosfetal 600. [Zschimmer & Schwarz] Alkylaryl polyglycol ether phosphate, acid; cleansing agent; liq.; 100% act.

Phosfetal 601. [Zschimmer & Schwarz] Alkylaryl polyglycol ether phosphate, acid; anionic; basic material for formulation of cleansing agents, acidic metal cleaners, alkaline degreasers, metallic mordants; colorless to ylsh. clear liq.; sol. in ethanol, IPA, perchloroethylene, min. oil; disp. in water; dens. 1.07 g/cc; visc. 6200 mPa•s; acid no. 80-90; pH 1.9-2.2 (10%); 100% act.

Phosfetal 602. [Zschimmer & Schwarz] Alkylaryl polyglycol ether phosphate, acid; cleansing agent; liq.; 100% act.

Phosfetal 603. [Zschimmer & Schwarz] Alkylaryl polyglycol ether phosphate, acid; cleansing agent; liq.; 100% act.

Phosokresol Grades. [Hoechst AG] Aromatic dithiophosphates; flotation collectors for sulfide minerals.

Phosphanol Series. [Toho Chem. Industry] Phosphate ester; emulsifier for textile spinning oils, polymerization; antistat; anticorrosive agent; liq., paste, solid.

Phospholan® AD-1. [Henkel/Process] Complex phosphate ester, free acid; anionic; surfactant for liq. detergents; amber clear liq.; dens. 1.08 g/ml; pour pt. < 0 C; pH 2.9 (10%).

Phospholan® ALF5. [Harcros UK] Tallow fatty alcohol phosphate ester on sodium carbonate carrier; anionic; foam limiter for use with anionic surfactants; off-wh. solid; mild odor; insol. in water; visc. 120 cs (80 C); flash pt. (COC) > 150 C; pour pt. 65 C; pH 2.5 (1% aq.); corrosive org. acid; toxicology: extremely irritating to eyes and skin; 100% act.

Phospholan® ALF16. [Harcros] Phosphate ester on sodium carbonate carrier; anionic; foam limiter for use with anionic surfactants; wh. fine powd., mild odor; partly sol. in water; bulk dens. 704 g/l; pH 10.3 (1% aq.); corrosive org. acid; toxicology: extremely irritating to eyes and skin; 50% act. on sodium carbonate.

Phospholan® BH14. [Harcros UK] Complex phosphate ester, free acid; anionic; surfactant for cleaners; coupling agent; lt. yel. clear visc. liq.; dens. 1.27 g/ml; pour pt. < 0 C; pH 2.6 (10%).

Phospholan® KPE4 [Henkel/Cospha; Harcros UK] Arylethoxy phosphate, potassium salt; CAS 36863-52-2; anionic; hydrotrope for solubilization of low foaming surfactants into highly built liqs.; industrial and domestic detergents and cleaners; fertilizer and toxicant formulations; pale straw liq.; mild odor; water-sol.; sp.gr. 1.294; visc. 180 cs; flash pt. (COC) > 95 C; pour pt. -18 C; pH 7.0 (1% aq.); 65% act.

Phospholan® PDB3. [Harcros UK] Fatty alcohol ethoxylate phosphate ester; anionic; wetting agent, detergent component, textiles, drycleaning; metal cleaning; emulsifier for phenols and in emulsion polymerization; agric.; waterless hand cleaners; metalworking lubricants with anticorrosive properties; iodophors; antistat; pale straw liq.; mild odor; disp. in water; sp.gr. 1.000; visc. 2400 cs; flash pt. (COC) > 150 C; pour pt. 10 C; pH 2.5 (1% aq.); surf. tens. 30 dynes/cm (0.1%); corrosive org. acid; toxicology: extremely irritating to eyes and skin; 100% act.

Phospholan® PHB14. [Harcros UK] Phosphate ester, free acid; anionic; hydrotrope and compatibility agent for agrochemical formulations; solubilizers anionic and nonionic surfactants into high electrolyte concs.; amber liq., mild odor; water-sol.; sp.gr. 1.285; visc. 4200 cs (40 C); pour pt. 2 C; flash pt. (COC) > 150 C; pH 2.0 (1% aq.); toxicology: corrosive org. acid; extremely irritating to skin and eyes; 100% act.

Phospholan® PNP9. [Harcros UK] Nonylphenol ethoxylate phosphate ester, acid form; CAS 51811-79-1; anionic; electrolyte sol. wetting agent and emulsifier; for alkaline and industrial cleaners, textiles, dyestuff carrier, metal cleaning, emulsion polymerization, agric., waterless hand cleaners, metalworking lubes, iodophors; dk. amber visc. liq.; mild odor; water-sol.; sp.gr. 1.200; visc. 11,500 cs (40 C); flash pt. (COC) > 150 C; pour pt. 15 C; pH 2.5 (1% aq.); surf. tens. 36 dynes/cm; corrosive org. acid; toxicology: extremely irritating to eyes and skin; 100% act.

Phospholan® PRP5. [Harcros UK] Complex phosphate ester; anionic; dispersant for org. pigments, agrochemical toxicants, latex stabilization; lubricant and cutting oil additive; liq.; 100% conc.

Phospholan® PSP6. [Harcros UK] Phosphate ester; anionic; med./low foam wetting agent for industrial cleaning, textile applics.; pigment dispersant for gypsum and other minerals; liq.; 82% conc.

Phospholipid PTD. [Mona Industries] Lauramidopropyl PEG-dimonium chloride phosphate; cationic; bactericidal, conditioner, antistat, detergent, foamer, emulsifier, solubilizer, dispersant, thickener and wetting agent for personal care, household, pharmaceutical, veterinary prods., fire fighting foams, petrol. prod., photographic processes,; agric., mining, textiles; amber clear liq.; sp.gr. 1.05; dens. 8.7 lb/gal; pH 7.5 (10%); surf. tens. 43.1 dynes/cm (1%); 34% act.

Phospholipid PTL. [Mona Industries] Lauramphο PEG-glycinate phosphate; cationic; bactericidal, conditioner, antistat, detergent, foamer, emulsifier, solubilizer, dispersant, thickener and wetting agent for personal care, household, pharmaceutical, veterinary prods., fire fighting foams, petrol. prod., photographic processes,; agric., mining, textiles; clear to opaque visc. liq.; sp.gr. 1.10; dens. 9.1 lb/gal; pH 7.5 (10%); surf. tens. 33.7 dynes/cm (1%); 30% act.

Phospholipid PTS. [Mona Industries] Stearamidopropyl PG-dimonium chloride phosphate; cationic; patented mild substantivity agent, skin conditioner, thickener for personal care prods.; forms stable, low pH, smooth and elegant cosmetic emulsions; lt. yel. paste; sp.gr. 1.01; dens. 8.4 lb/gal; HLB 14; m.p. 40 C; pH 7.6 (10% in 50/50 IPA/water); toxicology: extremely mild to skin and eyes; 35% solids.

Phospholipid PTZ. [Mona Industries] Cocohydroxyethyl PEG-imidazolinium chloride phosphate; cationic; bactericidal, conditioner, antistat, detergent, foamer, emulsifier, solubilizer, dispersant, thickener and wetting agent for personal care, household, pharmaceutical, veterinary prods., fire fighting foams, petrol. prod., photographic processes,; agric., mining, textiles; amber clear liq.; sp.gr. 1.07; dens. 8.9 lb/gal; pH 7.0 (10%); surf. tens. 37.6 dynes/cm (1%); 30% act.

Phospholipon 50G. [Am. Lecithin] Lecithin; CAS 8002-43-5; nonionic; gran.; 100% conc.

Phosphoteric® T-C6. [Mona Industries] Sodium dicarboxyethylcoco phosphoethyl imidazoline; amphoteric; patented surfactant, hydrotrope; synergizes detergency with ethoxylated nonionics; improves wetting, penetrating, and detergency; for high electrolyte industrial cleaners; clear thin amber liq.; sp.gr. 1.09; dens. 9.1 lb/gal; pH 7.0 (10%); Draves wetting 40 s (1% in 100 ppm hard water); toxicology: very low eye irritation, low to moderate skin irritation; acute oral > 5.0 ml/kg; 35% act.

Pinamine A. [Consos] Pine oil, solv. and surfactant; anionic; scouring compd. for cotton and syn. fibers; liq.; biodeg.

Pinamine K. [Consos] Pine oil and surfactant; nonionic; scour and detergent for soaping off dyed cotton yarn and knit goods; gel; biodeg.

Pineotrene K. [Reilly-Whiteman] Detergent for textile scouring.

Plantaren 600 CS UP. [Henkel KGaA] C12-14 fatty alcohol glycoside; nonionic; surfactant for dishwashing agents, laundry detergents, cleaners; liq.; 50% conc.

Plantaren APG 225. [Henkel KGaA] C8-10 fatty alcohol glycoside; nonionic; solubilizer, wetting agent for alkali, neutral, and acidic cleaners; liq.; 70% conc.

Plantaren CG 60. [Henkel KGaA] C8-10 fatty alcohol glycoside; nonionic; solubilizer, wetting agent for alkali bottle washing, neutral and acidic cleaners; liq.; 60% conc.

Plasfalt HM, SS. [Kao Corp. SA] Amine-based; cationic; surfactant for slurry seal additives; liq./paste; 100% conc.

Pluracol® E200. [BASF] PEG-4; intermediate for preparation of nonionic surfactants; binder, base, coating, stabilizer, solv., vehicle, extender, and coupling agent for pharmaceutical, cosmetic, and toiletries; lubricant for metal applics., rubber industry; wood treatment; textile conditioning, antistat, and sizing agent, softener; colorless clear liq.; m.w. 200; sol. in water and org. solvs. except aliphatic hydrocarbons; dens. 9.4 lb/gal; sp.gr. 1.12; visc. 4.36 cs (210 F); flash pt. 360 F; surf. tens. 57.2 dynes/cm (1%); pH 6.5 (5% aq.).

Pluracol® E300. [BASF] PEG-6; see Pluracol E200; also dispersant in food tablets and preparations; plasticizer; colorless clear liq.; m.w. 300; sol. in water and org. solvs. except aliphatic hydrocarbons; dens. 9.4 lb/gal; sp.gr. 1.12; visc. 5.9 cs (99 C); flash pt. (COC) 210 C; pour pt. -13 C; surf. tens. 62.9 dynes/cm (1%); pH 5.7 (5% aq.).

Pluracol® E400. [BASF] PEG-8; CAS 25322-68-3; see Pluracol E300; colorless clear liq.; sol. in water and org. solvs. except aliphatic hydrocarbons; dens. 9.4 lb/gal; sp.gr. 1.12; visc. 7.39 cs (210 F); flash pt. 460 F; surf. tens. 66.6 dynes/cm (1%); pH 6.2 (5% aq.).

Pluracol® E400 NF. [BASF] PEG-8; chemical intermediate, base, coupler, thickener, lubricant, mold release agent, defoamer, softener, conditioner, antistat, sizing agent, dispersant for pharmaceutical, cosmetic, and oral care preparations,; in metal polishing and cleaning formulations, rubber prods., paper and wood prods., textile processing, ink formulations; liq.; m.w. 400; visc. 7.4 cs (99 C); pour pt. 5 C; flash pt. (COC) 182 C.

Pluracol® E600. [BASF] PEG-12; see Pluracol E300; colorless clear liq.; sol. in water and org. solvs. except aliphatic hydrocarbons; dens. 9.4 lb/gal; sp.gr. 1.12; visc. 10.83 cs (210 F); flash pt. 480 F; surf. tens. 65.2 dynes/cm (1%); pH 5.3 (5% aq.).

Pluracol® E600 NF. [BASF] PEG-12; see Pluracol E400 NF; liq.; m.w. 600; visc. 10.8 cs (99 C); pour pt. 20 C; flash pt. 249 C (COC).

Pluracol® E1000. [BASF] PEG-20; see Pluracol E400 NF; solid; m.w. 1000; visc. 17.5 cs (99 C); m.p. 38 C; flash pt. 255 C (COC).

Pluracol® E1450. [BASF] PEG; see Pluracol E400 NF; solid; m.w. 1450; visc. 28.5 cs (99 C); m.p. 45 C; flash pt. 255 C (COC).

Pluracol® E1450 NF. [BASF] PEG, NF grade; see Pluracol E400 NF; solid; m.w. 600; visc. 28.5 cs (99 C); m.p. 45 C; flash pt. 255 C (COC).

Pluracol® E1500. [BASF] PEG-6-32; see Pluracol E300; wh. waxy solid; sol. in water and org. solvs. except aliphatic hydrocarbons; dens. 10.0 lb/gal; sp.gr. 1.20; m.p. 46.0–47.5 C; flash pt. > 490 F; surf. tens. 62.8 dynes/cm (1%); pH 6.7 (5% aq.).

Pluracol® E2000. [BASF] PEG-40; see Pluracol E400 NF; solid; m.w. 2000; visc. 43.5 cs (99 C); m.p. 52 C; flash pt. > 260 C (COC).

Pluracol® E4000. [BASF] PEG-75; see Pluracol E300; wh. waxy solid; sol. in water and org. solvs. except aliphatic hydrocarbons; dens. 10.0 lb/gal; sp.gr. 1.20; m.p. 59.5 C; flash pt. > 490 F; surf. tens. 61.9 dynes/cm (1%); pH 6.7 (5% aq.).

Pluracol® E4000 NF. [BASF] PEG-75; see Pluracol E400 NF; solid; m.w. 4000; visc. 134 cs (99 C); m.p. 59 C; flash pt. > 260 C (COC).

Pluracol® E4500. [BASF] PEG; see Pluracol E400 NF; solid; m.w. 4500; visc. 170 cs (99 C); m.p. 60 C; flash pt. > 260 C (COC).

Pluracol® E6000. [BASF] PEG-150; see Pluracol E200; wh. waxy solid; sol. in water and org. solvs. except aliphatic hydrocarbons; dens. 10.0 lb/gal; sp.gr. 1.21; m.p. 61.0 C; flash pt. > 490 F; surf. tens. 62.1 dynes/cm (1%); pH 6.7 (5% aq.).

Pluracol® E8000. [BASF] PEG; CAS 25322-68-3; see Pluracol E400 NF; solid; m.w. 8000; visc. 750 cs (99 C); m.p. 61 C; flash pt. > 260 C (COC).

Pluracol® E8000 NF. [BASF] PEG, NF grade; see Pluracol E400 NF; solid; m.w. 8000; visc. 750 cs (99 C); m.p. 61 C; flash pt > 260 C (COC).

Pluracol® P-410. [BASF] PPG-9; chemical intermediate, antifoam agent in fermentation and in paint formulations, antiblooming agent for pentachlorophenol-treated wood; binder and lubricant for ceramics; plasticizer of resin-treated papers; preparation of PU foams; hydraulic and grinding fluids; ore flotation; water-wh. liq.; slight ether-like odor; water-sol.; m.w. 425; dens. 8.36 lb/gal; sp.gr. 1.005; visc. 35 cps (100 F); flash pt. > 400 F (PMCC); pour pt. –35 F; pH 6–7.

Pluracol® P-710. [BASF] PPG-12; see Pluracol P-410; water-wh. liq.; slight ether-like odor; water-sol.; m.w. 775; dens. 8.35 lb/gal; sp.gr. 1.004; visc. 65 cps (100 F); flash pt. > 400 F (PMCC); pour pt. 35 F; pH 6–7.

Pluracol® P-1010. [BASF] PPG-17; see Pluracol P-410; water-wh. liq.; slight ether-like odor; insol. in water; m.w. 1050; dens. 8.38 lb/gal; sp.gr. 1.007; visc. 80 cps (100 F); flash pt. > 400 F (PMCC); pour pt. –35 F; pH 6–7.

Pluracol® P-2010. [BASF] PPG-26; see Pluracol P-410; water-wh. liq.; slight ether-like odor; insol. in water; m.w. 2000; dens. 8.34 lb/gal; sp.gr. 1.002; visc. 175 cps (100 F); flash pt. > 400 F (PMCC); pour pt. –35 F; pH 6–7.

Pluracol® P-3010. [BASF] PPG; see Pluracol P-410; water-wh. liq.; slight ether-like odor; insol. in water; m.w. 3000; dens. 8.32 lb/gal; sp.gr. 1.00; visc. 295 cps (100 F); flash pt. > 400 (PMCC); pour pt. –20 F; pH 6–7.

Pluracol® P-4010. [BASF] PPG-30; see Pluracol P-410; water-wh. liq.; slight ether-like odor; insol. in water; m.w. 4000; dens. 8.33 lb/gal; sp.gr. 1.00; visc. 550 cps (100 F); flash pt. > 400 (PMCC); pour pt. –20 F; pH 6–7.

Pluracol® W170. [BASF] PPG-5-buteth-7; CAS 74623-31-7; component in demulsifying and wetting formulations; brake and metalworking fluids; rubber and fiber lubricant; textile applics.; defoamer for hot and cold applics., food and chemical processing; cosmetic formulations; APHA 50 max. visc. liq.; sol. in water, alcohols, ketones, esters, benzene, toluene, glycol ethers, chlorinated solvs.; dens. 8.58 lb/gal; sp.gr. 1.03; visc. 160–180 SUS (100 F); cloud pt. 73 C (1%); flash pt. 360 F (OC); pour pt. –45 F; pH 5.5–7.0 (10% aq.).

Pluracol® W260. [BASF] Polyalkoxylated polyether; see Pluracol W170; liq.; m.w. 1000; water-sol.; visc. 260 SUS (37.8 C); pour pt. –40 C; flash pt. (COC) 222 C.

Pluracol® W660. [BASF] PPG-12-buteth-16; CAS 74623-31-7; see Pluracol W170; APHA 50 max. visc. liq.; sol. in water, alcohols, ketones, esters, benzene, toluene, glycol ethers, chlorinated solvs.; dens. 8.79 lb/gal; sp.gr. 1.055; visc. 660 SUS (100 F); cloud pt. 60.5 C (1%); flash pt. 440 F (OC); pour pt. –34 F; pH 5.5–7.5 (10% aq.).

Pluracol® W2000. [BASF] PPG-20-buteth-30; CAS 74623-31-7; see Pluracol W170; APHA 50 max. visc. liq.; sol. in water, alcohols, ketones, esters, benzene, toluene, glycol ethers, chlorinated solvs.; dens. 8.83 lb/gal; sp.gr. 1.06; visc. 2000 SUS (100 F); cloud pt. 57.0 C (1%); flash pt. 440 F (OC); pour pt. –25 F; pH 5.5–7.5 (10% aq.).

Pluracol® W3520N. [BASF] PPG; CAS 9038-95-3; see Pluracol W170; APHA 40 max. visc. liq.; sol. in water, alcohols, ketones, esters, benzene, toluene, glycol ethers, chlorinated solvs.; dens. 8.83 lb/gal; sp.gr. 1.06; visc. 3520 SUS (100 F); cloud pt. 57 C (1%); flash pt. 437 F (OC); pour pt. –20 F; pH 6.0–7.5 (10% aq.).

Pluracol® W3520N-RL. [BASF] PPG; see Pluracol W170; APHA 200 max. visc. liq.; sol. in water, alcohols, ketones, esters, benzene, toluene, glycol ethers, chlorinated solvs.; dens. 8.87 lb/gal; sp.gr. 1.065; visc. 1752–2500 SUS (100 F); cloud pt. 55.5 C (1%); flash pt. 440 F (OC); pour pt. 28.4 F; pH 6.0–7.5 (10% aq.).

Pluracol® W5100N. [BASF] PPG-33-buteth-45; CAS 74623-31-7; see Pluracol W170; APHA 40 max. visc. liq.; sol. in water, alcohols, ketones, esters, benzene, toluene, glycol ethers, chlorinated solvs.; dens. 8.83 lb/gal; sp.gr. 1.06; visc. 5100 SUS (100 F); cloud pt. 55 C (1%); flash pt. 437 F (OC); pour pt. –20 F; pH 5.5–7.0 (10% aq.).

Pluracol® WD90K. [BASF] Polyalkoxylated polyether; chemical intermediate for halides, ethers, esters, urethane derivs.; defoamer for boiler water treatment; in compressor lubricants, hydraulic, brake, metalworking, and heat transfer fluids, rubber lubricants, demulsifiers, textile lubricants; liq.; m.w. 13,000; visc. 110,000 SUS (37.8 C); pour pt. 5 C; flash pt. (COC) 280 C.

Pluracol® WD1400. [BASF] Polyalkoxylated polyether; see Pluracol W170; liq.; m.w. 2500; water-sol.; visc. 1400 SUS (37.8 C); pour pt. –20 C; flash pt. (COC) 255 C.

Pluradot HA-410. [BASF] Trifunctional polyoxyalkylene glycol; nonionic; emulsifier, solubilizer, deduster, wetting agent, demulsifier, detergent, dispersant; surfactant for hard surface cleaning, machine dishwashing; liq.; m.w. 3200; sp.gr. 1.03; visc. 450 cps; cloud pt. 25 C (1%); pour pt. –15 F; surf. tens. 36 dynes/cm (0.1%); 100% conc.

Pluradot HA-420. [BASF] Trifunctional polyoxyalkylene glycol; see Pluradot HA-410; liq.; m.w. 3600; sp.gr. 1.04; visc. 565 cps; cloud pt. 28 C (1%); pour pt. –10 F; surf. tens. 37 dynes/cm (0.1%); 100% conc.

Pluradot HA-430. [BASF] Trifunctional polyoxyalkylene glycol; CAS 107498-00-0; see Pluradot HA-410; liq.; m.w. 3900; sp.gr. 1.05; visc. 700 cps; cloud pt. 43 C (1%); pour pt. 20 C; surf. tens. 38 dynes/cm (0.1%); 100% conc.

Pluradot HA-433. [BASF] Trifunctional polyoxyalkylene glycol; CAS 52624-57-4; see Pluradot HA-410; liq.; m.w. 3900; sp.gr. 1.05; visc. 700 cps; pour pt. 20 F; 100% conc.

Pluradot HA-440. [BASF] Trifunctional polyoxyalkylene glycol; see Pluradot HA-410; liq.; m.w. 4450; sp.gr. 1.05; visc. 750 cps; cloud pt. 54 C (1%); pour pt. 25 C; surf. tens. 39 dynes/cm (0.1%); 100% conc.

Pluradot HA-450. [BASF] Trifunctional polyoxyalkylene glycol; see Pluradot HA-410; paste; m.w. 5700; sp.gr. 1.06; cloud pt. 74 C (1%); pour pt. 60 F; surf. tens. 41 dynes/cm (0.1%).

Pluradot HA-510. [BASF] Trifunctional polyoxyalkylene glycol; CAS 52624-57-4; see Pluradot HA-410; liq.; m.w. 4600; sp.gr. 1.03; visc. 710 cps; cloud pt. 24 C (1%); pour pt. –15 F; surf. tens. 35 dynes/cm; 100% conc.

Pluradot HA-520. [BASF] Trifunctional polyoxyalkylene glycol; see Pluradot HA-410; liq.; m.w. 5000; sp.gr. 1.03; visc. 760 cps; cloud pt. 27 C (1%); pour pt. 5 F; surf. tens. 36 dynes/cm (0.1%); 100% conc.

Pluradot HA-530. [BASF] Trifunctional polyoxyalkylene glycol; CAS 52624-57-4; see Pluradot HA-410; liq.; m.w. 5300; sp.gr. 1.04; visc. 850 cps; cloud pt. 42 C (1%); pour pt. 30 F; surf. tens. 37 dynes/cm (0.1%); 100% conc.

Pluradot HA-540. [BASF] Trifunctional polyoxyalkylene glycol; see Pluradot HA-410; liq.; m.w. 6000; sp.gr. 1.05; visc. 1240 cps; cloud pt. 54 C (1%); pour pt. 40 F; surf. tens. 38 dynes/cm (0.1%); 100% conc.

Pluradot HA-550. [BASF] Trifunctional polyoxyalkylene glycol; see Pluradot HA-410; paste; m.w. 7500; sp.gr. 1.06; cloud pt. 77 C (1%); pour pt. 75 F; surf. tens. 39 dynes/cm (0.1%); 100% conc.

Plurafac® A-24. [BASF] Straight chain, primary aliphatic oxyalkylated alcohol; nonionic; detergent for commercial and home cleaning formulations; colorless liq.; ref. index 1.4501; sol. in alcohols, acetone, MEK, butyl Cellosolve, propylene glycol, chloroform, perchloroethylene, toluene, xylene, kerosene, water; dens. 7.7 lb/gal; sp.gr. 0.927; visc. 27 cps; HLB 6; cloud pt. 0 C (1%); flash pt. 400 F; pour pt. 56 F; surf. tens. 29.4 dynes/cm; pH 6–7 (1%); 100% conc.

Plurafac® A-27. [BASF] Straight chain, primary aliphatic oxyalkylated alcohol; nonionic; see Plurafac A-24; wh. paste; sol. in water, methanol, ethanol, acetone, MEK, butyl Cellosolve, propylene glycol, chloroform, toluene, xylene; dens. 8.4 lb/gal; HLB 14.5; cloud pt. 100 C (1%); flash pt. 460 F; pour pt. 75 F; surf. tens. 31.0 dynes/cm (0.1%); pH 6–7 (1%); 100% conc.

Plurafac® A-38. [BASF] Ceteareth-27; CAS 68439-49-6; nonionic; detergent, dispersant, wetting agent, emulsifier for heavy-duty detergents, metal cleaners, detergent tablets, electrolytic cleaning; biodeg.; wh. solid; sol. in water, methyl, ethyl, IPA, acetone, MEK, butyl Cellosolve, chloroform, perchloroethylene, toluene, xylene; m.p. 114 F; HLB 19; cloud pt. > 100 C (1%); flash pt. 480 F; surf. tens. 43.6 dynes/cm (0.1%); pH 6–7; 100% conc.

Plurafac® A-39. [BASF] Ceteareth-55; CAS 68439-49-6; nonionic; detergent, dispersant, wetting agent, emulsifier for heavy-duty detergents, metal cleaners, detergent tablets, electrolytic cleaning; biodeg.; wh. solid; sol. in water, chloroform; m.p. 132 F; HLB 24.0; cloud pt. > 100 C (1%); flash pt. 480 F; surf. tens. 53.2 dynes/cm (0.1%); pH 6–7; 100% conc.

Plurafac® A-46. [BASF] Straight chain primary aliphatic oxyalkylated alcohol; see Plurafac A-24; wh. paste; sol. in water, methyl, ethyl, IPA, acetone, MEK, butyl Cellosolve, chloroform, perchloroethylene, toluene, xylene; dens. 8.2 lb/gal; HLB 13.0; cloud pt. 72 C (1%); flash pt. 475 F; pour pt. 75 F; surf. tens. 32.5 dynes/cm (0.1%); pH 6–7 (1%); 100% conc.

Plurafac® B-25-5. [BASF] Straight chain primary aliphatic oxyalkylated alcohol; nonionic; detergent, dispersant, wetting agent, emulsifier for heavy-duty detergents, all-purpose liqs., detergent tablets, sanitizers, metal cleaners; biodeg.; colorless clear liq.; ref. index 1.4570; sol. in water, methyl, ethyl, IPA, acetone, MEK, butyl Cellosolve, chloroform, perchloroethylene, toluene, xylene; dens. 8.2 lb/gal; sp.gr. 1.002; visc. 100 cps; HLB 12; cloud pt. 65 C (1%); pour pt. 9 F; surf. tens. 36.4 dynes/cm (0.1%); pH 6–7 (1%); 100% conc.

Plurafac® B-26. [BASF] Straight chain primary aliphatic oxyalkylated alcohol; nonionic; detergent, dispersant, wetting agent, emulsifier for heavy-duty detergents, all-purpose liqs., detergent tablets, sanitizers, metal cleaners; biodeg.; colorless clear liq.; ref. index 1.4587; sol. in water, methyl, ethyl, IPA, acetone, MEK, butyl Cellosolve, chloroform, perchloroethylene, toluene, xylene; dens. 8.5 lb/gal; sp.gr. 1.020; visc. 114 cps; HLB 14; cloud pt. 73 C (1%); flash pt. 475 F; pour pt. 35 F; surf. tens. 36.3 dynes/cm (0.1%); pH 6–7 (1%); 100% conc.

Plurafac® C-17. [BASF] Straight chain primary aliphatic oxyalkylated alcohol; nonionic; see Plurafac A-24; wh. opaque liq.; ref. index 1.4526; sol. in water, methyl, ethyl, IPA, acetone, MEK, butyl Cellosolve, chloroform, perchloroethylene, toluene, xylene; dens. 8.6 lb/gal; sp.gr. 1.030; visc. 100 cps; HLB 16; cloud pt. 81 C (1%); flash pt. 520 F; pour pt. 56 F; surf. tens. 34.4 dynes/cm (0.1%); pH 6–7 (1%); 100% conc.

Plurafac® D-25. [BASF] PPG-6 C12-18 pareth-11; nonionic; detergent, dispersant, wetting agent, emulsifier, deduster for heavy-duty detergents, all-purpose liqs., detergent tablets, sanitizers, metal cleaners, hard surf. cleaners; biodeg.; slightly cloudy liq.; ref. index 1.4560; sol. in water, methyl, ethyl, IPA, acetone, MEK, butyl Cellosolve, chloroform, perchloroethylene, toluene, xylene; dens. 8.4 lb/gal; sp.gr. 1.007; visc. 95 cps; HLB 10.0; cloud pt. 59 C (1%); flash pt. 465 F; pour pt. 0 F; surf. tens. 34.3 dynes/cm (0.1%); pH 6–7 (1%); 100% conc.

Plurafac® LF 120. [BASF AG] Alkoxylated fatty alcohol; nonionic; surfactant for acid and alkaline low foaming cleaners; liq.; 100% conc.

Plurafac® LF 131. [BASF AG] Alkoxylated fatty alcohol; nonionic; low foaming surfactant for washing powds., rinse aids, bottle washing; alkali-stable; liq.; 100% conc.

Plurafac® LF 132. [BASF AG] Alkoxylated fatty alcohol; nonionic; low foaming surfactant for washing powds., rinse aids, bottle washing; alkali-stable; liq.; 100% conc.

Plurafac® LF 220. [BASF AG] Alkoxylated straight chain alcohol; nonionic; low foaming surfactant for dishwashing powds., rinse aids; liq.; 95% conc.

Plurafac® LF 221. [BASF AG] Alkoxylated straight chain alcohol; nonionic; low foaming surfactant for dishwashing powds., rinse aids; liq.; 95% conc.

Plurafac® LF 223. [BASF AG] Alkoxylated straight chain alcohol; nonionic; low foaming surfactant for dishwashing powds., rinse aids; liq.; 98% conc.

Plurafac® LF 224. [BASF AG] Alkoxylated straight chain alcohol; nonionic; low foaming surfactant for dishwashing powds., rinse aids; liq.; 100% conc.

Plurafac® LF 231. [BASF AG] Alkoxylated straight chain alcohol; nonionic; low foaming surfactant for washing powds., rinse aids, bottle washing; alkali-stable; liq.; 100% conc.

Plurafac® LF 400. [BASF AG] Alkoxylated straight chain alcohol; nonionic; low foaming surfactant; liq.; 100% conc.

Plurafac® LF 401, LF 403. [BASF AG] Alkoxylated straight chain alcohol; nonionic; low foaming surfactant for washing powds.; liq.; 100% conc.

Plurafac® LF 404, LF 405. [BASF AG] Alkoxylated straight chain alcohol; nonionic; low foaming surfactant for washing powds., rinse aids; liq.; 100 and 95% conc. resp.

Plurafac® LF 431. [BASF AG] Alkoxylated straight chain alcohol; nonionic; low foaming surfactant for washing powds., rinse aids, bottle washing; alkali-stable; liq.; 100% conc.

Plurafac® LF 600. [BASF AG] Alkoxylated straight chain alcohol; nonionic; low foaming surfactant for high-pressure cleaners; liq.; 100% conc.

Plurafac® LF 700, LF 711. [BASF AG] Alkoxylated straight chain alcohol; nonionic; low foaming surfactant for strong acid cleaners; liq.; 100% conc.

Plurafac® LF 1300. [BASF AG] Alkoxylated straight chain alcohol; nonionic; low foaming surfactant; defoamer for sugar industry; liq.; 100% conc.

Plurafac® LF 1430. [BASF AG] Alkoxylate; nonionic; low foaming surfactant for bottlewashing; liq.; 100% conc.

Plurafac® RA-20. [BASF] Straight chain primary aliphatic oxyalkylated alcohol; nonionic; detergent, dispersant, wetting agent, emulsifier, defoamer, deduster used in rinse aids and dishwashing prods.; colorless clear liq.; ref. index 1.4530; sol. in alcohols, acetone, MEK, butyl Cellosolve, propylene glycol, chloroform, perchloroethylene, toluene, xylene, kerosene, water; dens. 8.3 lb/gal; sp.gr. 0.992;

visc. 83 cps; HLB 10; cloud pt. 45 C (1%); flash pt. 475 F; pour pt. 45 F; surf. tens. 32.8 dynes/cm (0.1%); pH 6–7 (1%); 100% conc.

Plurafac® RA-30. [BASF] Straight chain primary aliphatic oxyalkylated alcohol; nonionic; detergent, dispersant, wetting agent, emulsifier, defoamer, deduster used in rinse aids and dishwashing prods.; colorless clear liq.; ref. index 1.4534; sol. in alcohols, acetone, MEK, butyl Cellosolve, propylene glycol, chloroform, perchloroethylene, toluene, xylene, kerosene, water; dens. 8.1 lb/gal; sp.gr. 0.973; visc. 56 cps; HLB 9; cloud pt. 35 C (1%); flash pt. 455 F; pour pt. 50 F; surf. tens. 30.6 dynes/cm (0.1%); pH 6–7 (1%); 100% conc.

Plurafac® RA-40. [BASF] Straight chain primary aliphatic oxyalkylated alcohol; nonionic; detergent, dispersant, wetting agent, emulsifier, defoamer, deduster used in rinse aids and dishwashing prods.; colorless clear liq.; ref. index 1.4510; sol. in alcohols, acetone, MEK, butyl Cellosolve, propylene glycol, chloroform, perchloroethylene, toluene, xylene, kerosene, water; dens. 8.1 lb/gal; sp.gr. 0.970; visc. 91 cps; HLB 7; cloud pt. 25 C (1%); flash pt. 437 F; pour pt. –15 F; surf. tens. 30.5 dynes/cm (0.1%); pH 6–7 (1%); 100% conc.

Plurafac® RA-43. [BASF] Straight chain primary aliphatic oxyalkylated alcohol; nonionic; detergent, dispersant, wetting agent, emulsifier, defoamer, deduster used in rinse aids and dishwashing prods.; wh. opaque liq.; dens. 7.9 lb/gal; sp.gr. 0.950; visc. 120 cps; HLB 7; cloud pt. 24 C (1%); flash pt. 437 F; pour pt. 21 F; surf. tens. 30.5 dynes/cm (0.1%); pH 3–4 (1%); 100% conc.

Plurafac® RA-50. [BASF] Straight chain, primary oxyalkylated alcohol; nonionic; surfactant, dispersant, wetting agent, emulsifier, demulsifying agent; for industrial and household detergents; colorless liq.; sol. in benzene, xylene, acetone, CCl_4, perchlorethylene, alcohols, kerosene, butyl Cellosolve, chloroform, MEK, propylene glycol; sol. < 10% in water; visc. 90 cps; sp.gr. 0.992; HLB 9; cloud pt. 40 C (1% aq.); flash pt 451 F; pour pt 13 C; surf. tens. 31.5 dynes/cm (0.1%); pH 6–7 (1% aq.); biodeg.; 100% act.

Plurafac® T-55. [BASF] Oxyalkylated primary aliphatic alcohol; nonionic; agric. spray adjuvant, emulsifier; laundry and car wash detergent, general purpose cleaner, degreaser; pale yel. liq.; water-sol.; sp.gr. 1.012; visc. 164 cps; cloud pt. 57 C (1%); pour pt. 34 F; surf. tens. 36.2 dynes/cm (0.1%).

Pluraflo® E4A. [BASF] Formulated prod.; nonionic; dispersant, wetting agent for pesticides; EPA clearance; liq.; visc. 820 cps; pour pt. 7 C; flash pt. 69 F (CC); surf. tens. 42.8 dynes/cm (0.1%).

Pluraflo® E4B. [BASF] Formulated prod.; nonionic; dispersant, wetting agent for pesticides; EPA clearance; liq.; visc. 740 cps; pour pt. 11 C; flash pt. 116 F (CC); surf. tens. 42.3 dynes/cm (0.1%).

Pluraflo® E5A. [BASF] Formulated prod.; nonionic; dispersant, wetting agent for pesticides; EPA clearance; liq.; visc. 430 cps; pour pt. 5 C; flash pt. 61 F (CC); surf. tens. 42.5 dynes/cm (0.1%).

Pluraflo® E5B. [BASF] Formulated prod.; nonionic; dispersant, wetting agent for pesticides; EPA clearance; liq.; visc. 520 cps; pour pt. 3 C; flash pt. 110 F (CC); surf. tens. 42.7 dynes/cm (0.1%).

Pluraflo® E5BG. [BASF] Formulated prod.; nonionic; dispersant, wetting agent for pesticides; EPA clearance; liq.; visc. 250 cps; pour pt. –19 C; flash pt. 138 F (CC); surf. tens. 43.8 dynes/cm (0.1%).

Pluraflo® E5G. [BASF] Formulated prod.; nonionic; wetting agent for pesticides; dispersant; EPA clearance; liq.; visc. 260 cps; pour pt. –22 C; flash pt. 126 F (CC); surf. tens. 43.9 dynes/cm (0.1%).

Pluraflo® N5G. [BASF] Formulated prod.; nonionic; dispersant, wetting agent for pesticides; EPA clearance; liq.; visc. 410 cps; pour pt. –21 C; flash pt. 126 F (CC); surf. tens. 46.0 dynes/cm (0.1%).

Pluriol® E 200. [BASF AG] PEG; nonionic; solubilizer, impregnating agent, humectant, mold release agent; flow improver, thermal and hydraulic fluid, org. intermediate; detergent and cleaner; dye and pigment dispersant; inks; textile and coatings industry; coloring ceramics; softener in paper industry; plasticizer in adhesives industry and in prod. of cellulose film; ceramics and metalworking lubricant; clear colorless liq.; m.w. 200; water-sol.; dens. 1.12 g/cc; visc. 55–65 cs.; m.p. < –40 C; flash pt. > 150 C; pH 6.0–7.5 (1% aq.); 100% conc.

Pluriol® E 300. [BASF AG] PEG; nonionic; solubilizer, impregnating agent, humectant, mold release agent; flow improver, thermal and hydraulic fluid, org. intermediate; detergent and cleaner; dye and pigment dispersant; inks; textile and coatings industry; coloring ceramics; colorless clear liq.; m.w. 300; water-sol.; dens. 1.13 g/cc; visc. 75–95 cs; m.p. –20 to –10 C; flash pt. > 150 C; pH 6.0–7.5 (1% aq.); 100% conc.

Pluriol® E 400. [BASF AG] PEG; nonionic; solubilizer, impregnating agent, humectant, mold release agent; flow improver, thermal and hydraulic fluid, org. intermediate; detergent and cleaner; dye and pigment dispersant; inks; textile and coatings industry; coloring ceramics; colorless clear liq.; m.w. 400; water-sol.; dens. 1.13 g/cc; visc. 105–120 cs; m.p. –10 to 10 C; flash pt. > 150 C; pH 6.0–7.5 (1% aq.); 100% conc.

Pluriol® E 600. [BASF AG] PEG; nonionic; solubilizer, impregnating agent, humectant, mold release agent; flow improver, thermal and hydraulic fluid, org. intermediate; detergent and cleaner; dye and pigment dispersant; inks; textile and coatings industry; coloring ceramics; semisolid; m.w. 600; water-sol.; dens. 1.15 g/cc; visc. 8–11 cs (99 C); m.p. 10–20 C; flash pt. > 220 C; pH 6.0–7.5 (1% aq.); 100% conc.

Pluriol® E 1500. [BASF AG] PEG; nonionic; solubilizer, humectant; binder and hardener in personal care prods.; dispersing dyes and pigments; inks; textile and coating industry; coloring ceramics; paper industry softener; plasticizer in adhesives industry and prod. of cellulose film; ceramics and metalworking lubricants; aux. for copper and nickel electroplating baths; electrolytic polishing of steel; wh. fine powd.; m.w. 1500; water-sol.; bulk dens. 0.6 kg/l; visc. 25–35 cs (99 C); m.p. 45 C; flash pt. > 230 C; pH 6.0–7.5 (1% aq.); 100% conc.

Pluriol® E 4000. [BASF AG] PEG; nonionic; solubilizer, humectant; binder and hardener in personal care prods.; dispersing dyes and pigments; inks; textile and coating industry; coloring ceramics; paper industry softener; plasticizer in adhesives industry and prod. of cellulose film; wh. fine powd.; m.w. 4000; water-sol.; bulk dens. 0.6 kg/l; visc. 80–130 cs (99 C); m.p. 55 C; flash pt. > 240 C; pH 6.0–7.5 (1% aq.); 100% conc.

Pluriol® E 6000. [BASF AG] PEG; nonionic; solubilizer, humectant; binder and hardener in personal care prods.; dispersing dyes and pigments; inks; textile and coating industry; coloring ceramics; paper industry softener; plasticizer in adhesives industry and prod. of cellulose film; wh. fine powd.; m.w. 6000; water-sol.; bulk dens. 0.6 kg/l; visc. 300 cs (99 C); m.p. 60 C; flash pt. > 240 C; pH 6.0–7.5 (1% aq.); 100% conc.

Pluriol® E 9000. [BASF AG] PEG; nonionic; solubilizer, humectant; binder/hardener in personal care prods.; dispersing dyes/pigments; inks; textiles; coatings; ceramics; paper industry softener; plasticizer in adhesives industry and prod. of cellulose film; mold release for rubbers; wh. fine powd.; m.w. 9000; water-sol.; bulk dens. 0.6 kg/l; visc. 1900 cs (99 C); m.p. 65 C; flash pt. > 240 C; pH 6.0–7.5 (1% aq.); 100% conc.

Pluriol® P 600. [BASF AG] PPG; nonionic; mold release agent, additive for oils and fluids; lubricant and antifoam for rubber; consistency improver and solubilizer; intermediate in industrial applics.; colorless clear liq.; misc. with water and oil; m.w. 600; dens. 1.0 g/cc; visc. 130 cs; flash pt. 216 C; pour pt. –43 C; pH 6.5–7.5 (1% aq.); 100% conc.

Pluriol® P 900. [BASF AG] PPG; nonionic; mold release agent, additive for oils and fluids; lubricant and antifoam for rubber; consistency improver and solubilizer; intermediate in industrial applics.; solv.-type cleaners; colorless clear liq.; m.w. 900; dens. 1.0 g/cc; visc. 180 cs; cloud pt. 33 C (1% aq.); flash pt. 220 C; pour pt. –38 C; pH 6.5–7.5 (1% aq.); 100% conc.

Pluriol® P 2000. [BASF AG] PPG; nonionic; mold release agent, additive for oils and fluids; lubricant and antifoam for rubber; consistency improver and solubilizer; intermediate in industrial applics.; colorless clear liq.; m.w. 2000; dens. 1.0 g/cc; visc. 440 cs; flash pt. 222 C; pour pt. –35 C; pH 6.5–7.5 (1% aq.); 100% conc.

Pluriol® P 4000. [BASF AG] PPG; defoamer for tech. applics.; liq.

Pluriol® PE 3100. [BASF AG] PO/EO block polymer; nonionic; surfactant; dispersant for colorants; defoamer in various applics.; clear colorless liq.; m.w. 1100; ref. index 1.45; sol. in ethanol, 2-propanol, toluene, tetrachloroethylene, min. spirit; dens. 1.02 g/cc; visc. 175 cps: cloud pt. 40 C (1% aq.); pour pt. –20 C; surf. tens. 45 dynes/cm; pH 7 (5% aq.); 100% conc.

Pluriol® PE 4300. [BASF AG] PO/EO block polymer; nonionic; surfactant; dishwashing detergent and rinse aid; colorless clear liq.; m.w. 1700; ref. index 1.45; sol. in water, 10% HCl, ethanol, 2-propanol, tetrachloroethylene; dens. 1.04 g/cc; visc. 300 cps; cloud pt. 42 C (1% aq.); pour pt. –20 C; surf. tens. 43 dynes/cm; pH 7 (5% aq.); 100% conc.

Pluriol® PE 6100. [BASF AG] PO/EO block polymer; nonionic; defoamer in household dishwashing machines, bottlewashing plants, metal cleaners, boiler feedwater, acid dyebaths; antifreeze; colorless clear liq.; m.w. 2000; ref. index 1.45; sol. in ethanol, 2-propanol, toluene, tetrachloroethylene, min. spirit; dens. 1.02 g/cc; visc. 350 cps; cloud pt. 24 C (1% aq.); pour pt. –20 C; surf. tens. 44 dynes/cm; pH 7 (5% aq.); 100% conc.

Pluriol® PE 6101. [BASF AG] PO/EO block polymer; nonionic; defoamer; rinse aid; clear, slightly dull liq.; m.w. 2000; ref. index 1.45; sol. in 10% HCl, ethanol, 2-propanol, toluene, tetrachloroethylene, min. spirit; dens. 1.02 g/cc; visc. 350 cps; cloud pt. 28 C (1% aq.); pour pt. –10 C; surf. tens. 44 dynes/cm; pH 7 (5% aq.); 100% conc.

Pluriol® PE 6200. [BASF AG] PO/EO block polymer; nonionic; wetting agent, emulsifier, demulsifier used in mechanical cleaning processes and splitting crude oil emulsions; household rinse aids; phosphatizing baths; emulsifier in polymerization processes; colorless clear liq.; m.w. 2500; ref. index 1.45; sol. in water, 10% HCl, ethanol, 2-propanol, toluene and tetrachloroethylene; dens. 1.04 g/cc; visc. 500 cps; cloud pt. 32 C (1% aq.); pour pt. –14 C; surf. tens. 44 dynes/cm; pH 7 (5% aq.); 100% conc.

Pluriol® PE 6400. [BASF AG] PO/EO block polymer; nonionic; detergent for dishwashing machines, dairy cleaners; coolant and lubricant for grinding, drilling and cutting; dispersant, emulsifier; emulsion polymerization; slightly dull liq.; m.w. 3000; ref. index 1.45; sol. in Pluriol PE 6200; dens. 1.05 g/cc; visc. 850 cps; cloud pt. 59 C (1% aq.); pour pt. 16 C; surf. tens. 44 dynes/cm; pH 7 (5% aq.); 100% conc.

Pluriol® PE 6800. [BASF AG] PO/EO block polymer; nonionic; powd. detergent for household and industry; dust binder, dispersant, solubilizer for cosmetics, emulsifier, prod. of emulsion polymers; wh. fine powd.; m.w. 8500; ref. index 1.45 (70 C); sol. in water, 10% HCl, ethanol, toluene; dens. 1.06 g/cc (70 C); cloud pt. > 100 (1% aq.); surf. tens. 48 dynes/cm; pH 7 (5% aq.); 100% conc.

Pluriol® PE 8100. [BASF AG] PO/EO polymer; nonionic; defoamer; surfactant for cleaning processes; breweries and dairies; colorless to clear-slightly dull liq.; m.w. 2600; ref. index 1.45 (70 C); sol. in 2-propanol, toluene, tetrachloroethylene; dens. 1.04 g/cc; visc. 600 cps; cloud pt. 18 C (1% aq.); pour pt. –30 C; surf. tens. 35 dynes/cm; pH 7 (5% aq.); 100% conc.

Pluriol® PE 9200. [BASF AG] PO/EO polymer; nonionic; wetting agent for cleaning processes; surfactant in dishwashing machine detergents for household and canteens; rinse aid; clear to slightly dull liq.; m.w. 3650; ref. index 1.45 (70 C); sol. see Pluriol PE 6200; dens. 1.03 g/cc; visc. 700 cps; cloud pt. 22 C (1% aq.); pour pt. –3 C; surf. tens. 35 dynes/cm; pH 7 (5% aq.); 100% conc.

Pluriol® PE 9400. [BASF AG] PO/EO polymer; nonionic; emulsifier in oil industry and prod. of aq. disps., dispersant; dishwashing machines, bottle washing lines and dairy cleaners; wh. waxy; m.w. 4600; ref. index 1.45 (70 C); sol. see Pluriol PE 6200; surf. tens. 40 dynes/cm; pH 7 (5% aq.); 100% conc.

Pluriol® PE 10100. [BASF AG] PO/EO block polymer; nonionic; wetting and antifoam agent; cleaners and rinse aids for dishwashing machines; clear to slightly dull liq.; m.w. 3550; ref. index 1.45; sol. in 10% HCl, ethanol, 2-propanol, toluene, tetrachloroethylene, min. spirit; dens. 1.02 g/cc; visc. 1000 cps; cloud pt. 15 C (1% aq.); pour pt. –30 C; surf. tens. 35 dynes/cm; pH 7 (5% aq.); 100% conc.

Pluriol® PE 10500. [BASF AG] PO/EO block polymer; nonionic; emulsifier for agrochemicals; household and industrial cleaners; emulsion polymerization; wh. waxy; m.w. 6500; ref. index 1.45 (70 C); sol. in water, 10% HCl, ethanol, 2-propanol; dens. 1.03 g/cc; cloud pt. > 100 C (1% aq.); surf. tens. 36 dynes/cm; pH 7 (5% aq.); 100% conc.

Pluriol® RPE 2540. [BASF AG] PEG-PPG; nonionic; surfactant; defoaming and foam controlled detergents; liq.; dens. 1.0 g/cc; visc. 1500 cps; surf. tens. 37 dynes/cm; pH 9 (5% aq.); 100% conc.

Pluriol® RPE 3110. [BASF AG] PEG-PPG; nonionic;

defoaming and foam controlled detergents; liq.; 100% conc.

Plurol Isostearique. [Gattefosse SA] Polyglyceryl-6 isostearate; nonionic; emulsifier; cosurfactant for microemulsions; liq.; HLB 10.0; 100% conc.

Plurol Oleique WL 1173. [Gattefosse SA] Polyglyceryl-6 dioleate; nonionic; emulsifier; cosurfactant for microemulsions; liq.; 100% conc.

Plurol Stearique WL 1009. [Gattefosse SA] Polyglyceryl-6 distearate; nonionic; consistency agent and stabilizer for heated o/w emulsions; food emulsifier; FCC listed; Color < 10 waxy solid, faint odor; sol. in chloroform, methylene chloride; partly sol. in ethanol; disp. in water; HLB 9.0; drop ot. 48-53 C; acid no. < 5; iodine no. < 3; sapon. no. 120-140; pH 7.0-9.5 (10% aq.); 100% conc.

Pluronic® 10R5. [BASF] Meroxapol 105; nonionic; emulsifier, wetting agent, binder, stabilizer, plasticizer, lubricant, solubilizer, dispersant, visc. control agent, defoamer, intermediate for hard surface detergents, rinse aids, automatic dishwashing, textile processing; cosmetics; pharmaceuticals, pulp, paper, and petrol. industries, agric. prods., in iodophors, water treating systems, fermentation, cutting and grinding fluids; liq.; m.w. 1970; ref. index. 1.4587; sol. in water, propylene glycol, xylene, IPA, ethyl acetate, perchloroethylene, IPM; sp.gr. 1.058; visc. 400 cps; HLB 21.0; cloud pt. 69 C (1% aq.); flash pt. > 450 F (COC); pour pt. 15 C; surf. tens. 50.9 dynes/cm; toxicology: minimal to mild eye and skin irritation; 100% act.

Pluronic® 10R8. [BASF] Meroxapol 108; nonionic; see Pluronic 10R5; flakable solid; m.w. 5000; sol. in water, xylene, ethyl acetate; sp.gr. 1.062 (77 C); m.p. 46 C; HLB 33.0; cloud pt. 99 C (1% aq.); flash pt. > 450 F (COC); surf. tens. 54.1 dynes/cm (0.1%); toxicology: minimal to mild eye and skin irritation; 100% act.

Pluronic® 12R3. [BASF] PO/EO block copolymer; nonionic; defoamer, emulsifier, demulsifier, solubilizer, detergent, dispersant, binder, stabilizer, gelling agent, wetting agent, rinse aid, chemical intermediate in cosmetic, drug, textile, paper, petrol., paint,; detergent, and metal cleaning industries; liq.; m.w. 1800; visc. 340 cps; HLB 5.0; pour pt. –20 C; cloud pt. 53 C (1% aq.); surf. tens. 43 dynes/cm (0.1%); toxicology: minimal to mild eye and skin irritation; 100% act.

Pluronic® 17R1. [BASF] Meroxapol 171; nonionic; see Pluronic 10R5; also for foam control in paper sizing operations; liq.; m.w. 1950; ref. index 1.4516; sol. in xylene, IPA, ethyl acetate, min. oil, perchloroethylene, IPM, trichlorotrifluoroethane; sp.gr. 1.018; visc. 300 cps; HLB 4.0; cloud pt. 32 C (1% aq.); flash pt. > 450 F (COC); pour pt. –27 C; surf. tens. 33.0 dynes/cm (0.1%); toxicology: minimal to mild eye and skin irritation; 100% act.

Pluronic® 17R2. [BASF] Meroxapol 172; nonionic; see Pluronic 10R5; liq.; m.w. 2100; ref. index 1.4535; sol. in water, xylene, IPA; ethyl acetate, perchloroethylene, IPM, trichlorotrifluoroethane; sp.gr. 1.030; visc. 350 cps; HLB 8.0; cloud pt. 39 C (1% aq.); flash pt. > 450 F (COC); pour pt. –25 C; surf. tens. 41.9 dynes/cm (0.1%); toxicology: minimal to mild eye and skin irritation; 100% act.

Pluronic® 17R4. [BASF] Meroxapol 174; nonionic; see Pluronic 10R5; liq.; m.w. 2700; ref. index. 1.4572; sol. in water, propylene glycol, xylene, IPA, ethyl acetate, perchloroethylene, IPM, trichlorotrifluoroethane; sp.gr. 1.048; visc. 560 cps; HLB 16.0; cloud pt. 47 C (1% aq.); flash pt. > 450 F (COC); pour pt. 18 C; surf. tens. 44.1 dynes/cm (0.1%); toxicology: minimal to mild eye and skin irritation; 100% act.

Pluronic® 17R8. [BASF] Meroxapol 178; see Pluronic 10R5; also dry toilet bowl cleaners, dye levelers, solubilizer of drugs, stick type cosmetics; soap bars, dispersant in deinking operations; flakable solid; m.w. 7500; sol. in water, ethyl acetate, IPM; sp.gr. 1.064 (77 C); m.p. 53 C; HLB 32.0; cloud pt. 81 C (1% aq.); flash pt. > 450 F (COC); surf. tens. 47.3 dynes/cm (0.1%); toxicology: minimal to mild eye and skin irritation; 100% act.

Pluronic® 22R4. [BASF] Block copolymer; see Pluronic 10R5; liq.; m.w. 3350; visc. 950 cps; HLB 1-7; m.p. 24 C; cloud pt. 40 C (1% aq.); surf. tens. 43 dynes/cm (0.1%); toxicology: minimal to mild eye and skin irritation.

Pluronic® 25R1. [BASF] Meroxapol 251; nonionic; see Pluronic 10R5; also foam control in paper sizing operations and antifreeze; liq.; m.w. 2800; ref. index 1.4521; sol. in xylene, IPA, ethyl acetate, perchloroethylene, IPM, trichlorotrifluoroethane; sp.gr. 1.017; visc. 460 cps; HLB 2.3; cloud pt. 28 C (1% aq.); flash pt. > 450 F (COC); pour pt. –27 C; surf. tens. 36.3 dynes/cm (0.1%); toxicology: minimal to mild eye and skin irritation; 100% act.

Pluronic® 25R2. [BASF] Meroxapol 252; nonionic; see Pluronic 10R5; also wetting and rinse aid; lubricant and leveling agent for paper coating; liq.; m.w. 3120; ref. index 1.4541; sol. see Pluronic 25R1; sp.gr. 1.039; visc. 680 cps; HLB 6.3; cloud pt. 33 C (1% aq.); flash pt. > 450 F (COC); pour pt. –5 C; surf. tens. 37.5 dynes/cm (0.1%); toxicology: minimal to mild eye and skin irritation; 100% act.

Pluronic® 25R4. [BASF] Meroxapol 254; nonionic; see Pluronic 10R5; liq.; m.w. 3800; ref. index 1.4574; sol. in water, propylene glycol, ethyl acetate, perchloroethylene, IPM, trichlorotrifluoroethane; sp.gr. 1.046; visc. 1110 cps; HLB 14.3; cloud pt. 40 C (1% aq.); flash pt. > 450 F (COC); pour pt. 25 C; surf. tens. 40.9 dynes/cm (0.1%); toxicology: minimal to mild eye and skin irritation; 100% act.

Pluronic® 25R5. [BASF] Meroxapol 255; nonionic; see Pluronic 10R5; also thickener for cosmetic pastes and creams; paste; m.w. 4500; sol. see Pluronic 25R4; sp.gr. 1.036 (60 C); m.p. 33 C; HLB 18.3; cloud pt. 44 C (1% aq.); flash pt. > 450 F (COC); surf. tens. 43.5 dynes/cm (0.1%); toxicology: minimal to mild eye and skin irritation; 100% act.

Pluronic® 25R8. [BASF] Meroxapol 258; nonionic; see Pluronic 10R5; also dry toilet bowl cleaners, dye levelers for fabrics; solubilizer for drugs; thickener for cosmetics; deinking operations; felt washing operations; flakable solid; m.w. 9000; sol. in water; sp.gr. 1.062; m.p. 56 C; HLB 30.3; cloud pt. 80 C (1% aq.); flash pt. > 450 F (COC); surf. tens. 46.1 dynes/cm (0.1%); toxicology: minimal to mild eye and skin irritation; 100% act.

Pluronic® 31R1. [BASF] Meroxapol 311; nonionic; see Pluronic 10R5; also floating bath oils; foam control in antifreeze; liq.; m.w. 3200; ref. index 1.4522; sol. Pluronic® 25R1; sp.gr. 1.018; visc. 578 cps; HLB 1.7; cloud pt. 25 C (1% aq.); flash pt. > 450 F (COC); pour pt. –25 C; surf. tens. 34.1 dynes/cm (0.1%); toxicology: minimal to mild eye and skin irritation; 100% act.

Pluronic® 31R2. [BASF] Meroxapol 312; nonionic; see Pluronic 10R5; paper coating color additive;

deinking and felt washing operations; liq.; m.w. 3400; ref. index 1.4542; sol. see Pluronic 25R4; sp.gr. 1.030; visc. 818 cps; HLB 5.7; cloud pt. 30 C (1% aq.); flash pt. > 450 F (COC); pour pt. 9 C; surf. tens. 38.9 dynes/cm (0.1%); toxicology: minimal to mild eye and skin irritation; 100% act.

Pluronic® 31R4. [BASF] Meroxapol 314; nonionic; see Pluronic 10R5; paste; m.w. 4300; sol. see Pluronic 25R1; sp.gr. 1.028 (60 C); m.p. 26 C; HLB 13.7; cloud pt. 31 C (1% aq.); flash pt. > 450 F (COC); pour pt. 26 C; surf. tens. 41.2 dynes/cm (0.1%); toxicology: minimal to mild eye and skin irritation; 100% act.

Pluronic® F38. [BASF] Poloxamer 108; nonionic; wetting agent, emulsifier, demulsifier, foam and visc. control agent, dispersant, antistat, gelling agent, dyeing assistant, leveler, lubricant for agric., cosmetics, pharmaceuticals, metal cleaning, pulp/paper, textile scouring, water treatment; prilled; m.w. 5000; sol. in ethanol, propylene glycol, water, toluene, dens. 8.9 lb/gal (77 C); sp.gr. 1.07 (77 C); visc. 260 cps (77 C); m.p. 48 C; HLB 30.5; cloud pt. > 100 C (1% aq.); flash pt. 505 F (COC); surf. tens. 52.2 dynes/cm (0.1% conc.); toxicology: none to sl. eye and skin irritation; 100% act.

Pluronic® F68. [BASF] Poloxamer 188; nonionic; wetting agent, emulsifier, demulsifier, foam and visc. control agent, dispersant, antistat, gelling agent, dyeing assistant, leveler, lubricant for agric., cosmetics, pharmaceuticals, metal cleaning, pulp/paper, textile scouring, water treatment; prilled; m.w. 8350; sol. in ethanol, water; dens. 8.8 lb/gal (77 C); sp.gr. 1.06 (77 C); visc. 1000 cps (77 C); m.p. 52 C; HLB 29.0; cloud pt. > 100 C (1% aq.); flash pt. 500 F (COC); surf. tens. 50.3 dynes/cm (0.1%); toxicology: none to sl. eye and skin irritation; 100% act.

Pluronic® F68LF. [BASF] Poloxamer 108; nonionic; wetting agent, emulsifier, demulsifier, foam and visc. control agent, dispersant, antistat, gelling agent, dyeing assistant, leveler, lubricant for agric., cosmetics, pharmaceuticals, metal cleaning, pulp/paper, textile scouring, water treatment; flake; m.w. 7700; sol. in toluene; sp.gr. 1.06 (77 C); dens. 8.7 lb/gal (77 C); visc. 850 cps (77 C); m.p. 50 C; cloud pt. 32 C (1% aq.); toxicology: none to sl. eye and skin irritation; 100% conc.

Pluronic® F77. [BASF] Poloxamer 217; nonionic; wetting agent, emulsifier, demulsifier, foam and visc. control agent, dispersant, antistat, gelling agent, dyeing assistant, leveler, lubricant for agric., cosmetics, pharmaceuticals, metal cleaning, pulp/paper, textile scouring, water treatment; prilled; m.w. 6600; sol. in ethanol, water, toluene; dens. 8.7 lb/gal (77 C); sp.gr. 1.04 (77 C); m.p. 48 C; HLB 24.5; cloud pt. > 100 C (1% aq.); flash pt. 485 F (COC); surf. tens. 47.0 dynes/cm (0.1%); toxicology: none to sl. eye and skin irritation; 100% act.

Pluronic® F87. [BASF] Poloxamer 237; nonionic; wetting agent, emulsifier, demulsifier, foam and visc. control agent, dispersant, antistat, gelling agent, dyeing assistant, leveler, lubricant for agric., cosmetics, pharmaceuticals, metal cleaning, pulp/paper, textile scouring, water treatment; prilled; m.w. 7700; sol. see Pluronic F77; dens. 8.7 lb/gal (77 C); sp.gr. 1.04 (77 C); visc. 700 cps. (77 C); m.p. 49 C; HLB 24.0; cloud pt. > 100 C (1% aq.); flash pt. 472 F (COC); surf. tens. 44.0 dynes/cm (0.1%); toxicology: none to sl. eye and skin irritation; 100% act.

Pluronic® F88. [BASF] Poloxamer 238; nonionic; wetting agent, emulsifier, demulsifier, foam and visc. control agent, dispersant, antistat, gelling agent, dyeing assistant, leveler, lubricant for agric., cosmetics, pharmaceuticals, metal cleaning, pulp/paper, textile scouring, water treatment; prilled; m.w. 10,800; sol. in ethanol and water; dens. 8.8 lb/gal (77 C); sp.gr. 1.06 (77 C); visc. 2300 cps (77 C); m.p. 54 C; HLB 28.0; cloud pt. > 100 C (1% aq.); surf. tens. 48.5 dynes/cm (0.1%); toxicology: none to sl. eye and skin irritation; 100% act.

Pluronic® F98. [BASF] Poloxamer 288; nonionic; wetting agent, emulsifier, demulsifier, foam and visc. control agent, dispersant, antistat, gelling agent, dyeing assistant, leveler, lubricant for agric., cosmetics, pharmaceuticals, metal cleaning, pulp/paper, textile scouring, water treatment; prilled; m.w. 13,000; sol. in ethanol, water, perchloroethylene; dens. 8.8 lb/gal (77 C); sp.gr. 1.06 (77 C); visc. 2700 cps (77 C); m.p. 55 C; HLB 27.5; cloud pt. > 100 C (1% aq.); flash pt. 491 F (COC); surf. tens. 43.0 dynes/cm (0.1%); toxicology: none to sl. eye and skin irritation; 100% act.

Pluronic® F108. [BASF] Poloxamer 338; nonionic; see Pluronic F38; prilled; m.w. 14,000; sol. in ethanol, water; dens. 8.8 lb/gal (77 C); sp.gr. 1.06 (77 C); visc. 8000 cps (77 C); m.p. 57 C; HLB 27.0; cloud pt. > 100 C (1% aq.); flash pt. 495 F (COC); surf. tens. 41.2 dynes/cm (0.1%); toxicology: none to sl. eye and skin irritation; 100% act.

Pluronic® F127. [BASF] Poloxamer 407; nonionic; see Pluronic F38; prilled; m.w. 12,500; sol. in ethanol, water, toluene, perchloroethylene; dens. 8.8 lb/gal (77 C); sp.gr. 1.05 (77 C); visc. 3100 cps (77 C); m.p. 56 C; HLB 22.0; cloud pt. > 100 C (1% aq.); surf. tens. 40.6 dynes/cm (0.1%); toxicology: none to sl. eye and skin irritation; 100% act.

Pluronic® L10. [BASF] PO/EO block copolymer; nonionic; defoamer, emulsifier, demulsifier, solubilizer, detergent, dispersant, binder, stabilizer, gelling agent, wetting agent, rinse aid, chemical intermediate in cosmetic, drug, textile, paper, petrol., paint, metal cleaning; liq.; m.w. 3200; visc. 660 cps; HLB 14.0; pour pt. -5 C, cloud pt. 32 C (1% aq.); surf. tens. 41 dynes/cm (0.1%); toxicology: none to sl. eye and skin irritation; 100% act.

Pluronic® L31. [BASF] Poloxamer 101; nonionic; wetting agent, emulsifier, demulsifier, foam and visc. control agent, dispersant, antistat, gelling agent, dyeing assistant, leveler, lubricant for agric., cosmetics, pharmaceuticals, metal cleaning, pulp/paper, textile scouring, water treatment; liq.; m.w. 1100; ref. index 1.4515; sol. in ethanol, propylene glycol, water, toluene, xylene, perchloroethylene; dens. 8.5 lb/gal; sp.gr. 1.02; visc. 165 cps; HLB 4.5; cloud pt. 37 C (1% aq.); flash pt. 439 F (COC); pour pt. -32 C; surf. tens. 46.9 dynes/cm (0.1%); toxicology: none to sl. eye and skin irritation; 100% act.

Pluronic® L35. [BASF] Poloxamer 105; nonionic; wetting agent, emulsifier, demulsifier, foam and visc. control agent, dispersant, antistat, gelling agent, dyeing assistant, leveler, lubricant for agric., cosmetics, pharmaceuticals, metal cleaning, pulp/paper, textile scouring, water treatment; liq.; m.w. 1900; sol. in ethanol, propylene glycol, water, toluene, xylene, perchloroethylene; dens. 8.8 lb/gal; sp.gr. 1.06; visc. 340 cps; HLB 18.5; cloud pt. 77 C (1% aq.); pour pt. 7 C; surf. tens. 48.8 dynes/cm (0.1%); toxicology: none to sl. eye and skin irritation; 100% act.

Pluronic® L42. [BASF] Poloxamer 122; nonionic; wetting agent, emulsifier, demulsifier, foam and visc. control agent, dispersant, antistat, gelling agent, dyeing assistant, leveler, lubricant for agric., cosmetics, pharmaceuticals, metal cleaning, pulp/ paper, textile scouring, water treatment; liq.; m.w. 1630; ref. index 1.4541; sol. in ethanol, propylene glycol, water, toluene, xylene, perchloroethylene; dens. 8.6 lb/gal; sp.gr. 1.03; visc. 250 cps; HLB 8.0; cloud pt. 37 C (1% aq.); flash pt. 450 F (COC); pour pt. –26 C; surf. tens. 46.5 dynes/cm (0.1%); toxicology: none to sl. eye and skin irritation; 100% act.

Pluronic® L43. [BASF] Poloxamer 123; nonionic; wetting agent, emulsifier, demulsifier, foam and visc. control agent, dispersant, antistat, gelling agent, dyeing assistant, leveler, lubricant for agric., cosmetics, pharmaceuticals, metal cleaning, pulp/ paper, textile scouring, water treatment; liq.; m.w. 1850; ref. index 1.4563; sol. in ethanol, propylene glycol, water, toluene, xylene, perchloroethylene; dens. 8.7 lb/gal; sp.gr. 1.04; visc. 310 cps; HLB 12.0; cloud pt. 42 C (1% aq.); pour pt. –1 C; surf. tens. 47.3 dynes/cm (0.1%); toxicology: none to sl. eye and skin irritation; 100% act.

Pluronic® L44. [BASF] Poloxamer 124; nonionic; wetting agent, emulsifier, demulsifier, foam and visc. control agent, dispersant, antistat, gelling agent, dyeing assistant, leveler, lubricant for agric., cosmetics, pharmaceuticals, metal cleaning, pulp/ paper, textile scouring, water treatment; liq.; m.w. 2200; ref. index 1.4580; dens. 8.8 lb/gal; sp.gr. 1.05; visc. 440 cps; HLB 16.0; cloud pt. 65 C (1% aq.); flash pt. 464 F (COC); pour pt. 16 C; surf. tens. 45.3 dynes/cm (0.1%); toxicology: none to sl. eye and skin irritation; 100% act.

Pluronic® L61. [BASF] Poloxamer 181; nonionic; wetting agent, emulsifier, demulsifier, foam and visc. control agent, dispersant, antistat, gelling agent, dyeing assistant, leveler, lubricant for agric., cosmetics, pharmaceuticals, metal cleaning, pulp/ paper, textile scouring, water treatment; liq.; m.w. 2000; ref. index 1.4520; sol. in ethanol, toluene, xylene, perchloroethylene; dens. 8.4 lb/gal; sp.gr. 1.01; visc. 285 cps; HLB 3.0; cloud pt. 24 C (1% aq.); flash pt. 455 F (COC); pour pt. –29 C; toxicology: none to sl. eye and skin irritation; 100% act.

Pluronic® L62. [BASF] Poloxamer 182; nonionic; wetting agent, emulsifier, demulsifier, foam and visc. control agent, dispersant, antistat, gelling agent, dyeing assistant, leveler, lubricant for agric., cosmetics, pharmaceuticals, metal cleaning, pulp/ paper, textile scouring, water treatment; liq.; m.w. 2500; sol. in ethanol, propylene glycol, water, toluene, xylene, perchloroethylene; dens. 8.6 lb/gal; sp.gr. 1.03; visc. 400 cps; HLB 7.0; cloud pt. 32 C (1% aq.); flash pt. 466 F (COC); pour pt. –4 C; surf. tens. 42.8 dynes/cm (0.1%); toxicology: none to sl. eye and skin irritation; 100% act.

Pluronic® L62D. [BASF] Poloxamer 108; nonionic; wetting agent, emulsifier, demulsifier, foam and visc. control agent, dispersant, antistat, gelling agent, dyeing assistant, leveler, lubricant for agric., cosmetics, pharmaceuticals, metal cleaning, pulp/ paper, textile scouring, water treatment; liq.; m.w. 2350; sol. in ethanol, propylene glycol, water, toluene, xylene, perchloroethylene; sp.gr. 1.04; dens. 8.7 lb/gal; visc. 385 cps; pour pt. –1 C; ref. index 1.4557; cloud pt. 35 C (1% aq.); toxicology: none to sl. eye and skin irritation; 100% conc.

Pluronic® L62LF. [BASF] Poloxamer 108; nonionic; wetting agent, emulsifier, demulsifier, foam and visc. control agent, dispersant, antistat, gelling agent, dyeing assistant, leveler, lubricant for agric., cosmetics, pharmaceuticals, metal cleaning, pulp/ paper, textile scouring, water treatment; liq.; m.w. 2450; sol. in ethanol, propylene glycol, water, toluene, xylene, perchloroethylene; sp.gr. 1.03; dens. 8.6 lb/gal; visc. 400 cps; pour pt. –10 C; ref. index 1.4546; cloud pt. 28 C (1% aq.); toxicology: none to sl. eye and skin irritation; 100% act.

Pluronic® L63. [BASF] Poloxamer 183; nonionic; wetting agent, emulsifier, demulsifier, foam and visc. control agent, dispersant, antistat, gelling agent, dyeing assistant, leveler, lubricant for agric., cosmetics, pharmaceuticals, metal cleaning, pulp/ paper, textile scouring, water treatment; liq.; m.w. 2650; ref. index 1.4562; sol. see Pluronic L31; dens. 8.7 lb/gal; sp.gr. 1.04; visc. 490 cps; HLB 11.0; cloud pt. 34 C (1% aq.); pour pt. 10 C; surf. tens. 43.3 dynes/cm (0.1%); toxicology: none to sl. eye and skin irritation; 100% act.

Pluronic® L64. [BASF] Poloxamer 184; nonionic; wetting agent, emulsifier, demulsifier, foam and visc. control agent, dispersant, antistat, gelling agent, dyeing assistant, leveler, lubricant for agric., cosmetics, pharmaceuticals, metal cleaning, pulp/ paper, textile scouring, water treatment; liq.; m.w. 2900; ref. index 1.4575; sol. in ethanol, propylene glycol, water, toluene, xylene, perchloroethylene; dens. 8.8 lb/gal; sp.gr. 1.05; visc. 550 cps; HLB 15.0; cloud pt. 61 C (1% aq.); flash pt. 485 F (COC); pour pt. 16 C; surf. tens. 43.2 dynes/cm (0.1%); toxicology: none to sl. eye and skin irritation; 100% act.

Pluronic® L72. [BASF] Poloxamer 212; nonionic; wetting agent, emulsifier, demulsifier, foam and visc. control agent, dispersant, antistat, gelling agent, dyeing assistant, leveler, lubricant for agric., cosmetics, pharmaceuticals, metal cleaning, pulp/ paper, textile scouring, water treatment; liq.; m.w. 2750; ref. index 1.4542; sol. in ethanol, water, toluene, xylene, perchloroethylene; dens. 8.6 lb/gal; sp.gr. 1.03; visc. 510 cps; HLB 6.5; cloud pt. 25 C (1% aq.); flash pt. 442 F (COC); surf. tens. 39.0 dynes/cm (0.1%); toxicology: none to sl. eye and skin irritation; 100% act.

Pluronic® L81. [BASF] Poloxamer 231; nonionic; wetting agent, emulsifier, demulsifier, foam and visc. control agent, dispersant, antistat, gelling agent, dyeing assistant, leveler, lubricant for agric., cosmetics, pharmaceuticals, metal cleaning, pulp/ paper, textile scouring, water treatment; liq.; m.w. 2750; ref. index 1.4526; sol. in ethanol, toluene, xylene, perchloroethylene; dens. 8.5 lb/gal; sp.gr. 1.02; visc. 475 cps; HLB 2.0; cloud pt. 20 C (1% aq.); pour pt. –37 C; toxicology: none to sl. eye and skin irritation; 100% act.

Pluronic® L92. [BASF] Poloxamer 282; nonionic; wetting agent, emulsifier, demulsifier, foam and visc. control agent, dispersant, antistat, gelling agent, dyeing assistant, leveler, lubricant for agric., cosmetics, pharmaceuticals, metal cleaning, pulp/ paper, textile scouring, water treatment; liq.; m.w. 3650; ref. index 1.4547; sol. in ethanol, toluene, xylene, perchloroethylene; dens. 8.6 lb/gal; sp.gr. 1.03; visc. 700 cps; HLB 5.5; cloud pt. 26 C (1% aq.); flash pt. 445 F (COC); pour pt. 7 C; surf. tens. 35.9 dynes/cm (0.1%); toxicology: none to sl. eye and skin irritation; 100% act.

Pluronic® L101. [BASF] Poloxamer 331; nonionic;

wetting agent, emulsifier, demulsifier, foam and visc. control agent, dispersant, antistat, gelling agent, dyeing assistant, leveler, lubricant for agric., cosmetics, pharmaceuticals, metal cleaning, pulp/paper, textile scouring, water treatment; liq.; m.w. 3800; ref. index 1.4524; sol. in ethanol, toluene, xylene, perchloroethylene; dens. 8.5 lb/gal; sp.gr. 1.02; visc. 800 cps; HLB 1.0; cloud pt. 15 C (1% aq.); pour pt. –23 C; toxicology: none to sl. eye and skin irritation; 100% act.

Pluronic® L121. [BASF] Poloxamer 401; nonionic; wetting agent, emulsifier, demulsifier, foam and visc. control agent, dispersant, antistat, gelling agent, dyeing assistant, leveler, lubricant for agric., cosmetics, pharmaceuticals, metal cleaning, pulp/paper, textile scouring, water treatment; liq.; m.w. 4400; ref. index 1.4527; sol. in ethanol, toluene, xylene, perchloroethylene; dens. 8.4 lb/gal; sp.gr. 1.01; visc. 1200 cps; HLB 5.0; cloud pt. 14 C (1% aq.); pour pt. 5 C; surf. tens. 33.0 dynes/cm (0.1%); toxicology: none to sl. eye and skin irritation; 100% act.

Pluronic® L122. [BASF] Poloxamer 402; nonionic; wetting agent, emulsifier, demulsifier, foam and visc. control agent, dispersant, antistat, gelling agent, dyeing assistant, leveler, lubricant for agric., cosmetics, pharmaceuticals, metal cleaning, pulp/paper, textile scouring, water treatment; liq.; m.w. 5000; ref. index 1.4558; sol. in ethanol, toluene, xylene, perchloroethylene; dens. 8.6 lb/gal; sp.gr. 1.03; visc. 1750 cps; HLB 4.0; cloud pt. 19 C (1% aq.); flash pt. 490 F (COC); pour pt. 20 C; surf. tens. 33.0 dynes/cm (0.1%); toxicology: none to sl. eye and skin irritation; 100% act.

Pluronic® P65. [BASF] Poloxamer 185; nonionic; wetting agent, emulsifier, demulsifier, foam and visc. control agent, dispersant, antistat, gelling agent, dyeing assistant, leveler, lubricant for agric., cosmetics, pharmaceuticals, metal cleaning, pulp/paper, textile scouring, water treatment; paste; sol. > 10 g/100 ml in 95% ethanol, propylene glycol, water, toluene, xylene, perchloroethylene; m.w. 3400; dens. 8.8 lb/gal (60 C); sp.gr. 1.06 (60 C); visc. 180 cps (60 C); m.p. 30 C; HLB 17.0; cloud pt. 82 C (1% aq.); pour pt. 27 C; surf. tens. 46.3 dynes/cm (0.1%); toxicology: none to sl. eye and skin irritation; 100% act.

Pluronic® P75. [BASF] Poloxamer 215; nonionic; wetting agent, emulsifier, demulsifier, foam and visc. control agent, dispersant, antistat, gelling agent, dyeing assistant, leveler, lubricant for agric., cosmetics, pharmaceuticals, metal cleaning, pulp/paper, textile scouring, water treatment; paste; m.w. 4150; sol. see Pluronic F31; dens. 8.8 lb/gal (60 C); sp.gr. 1.06 (60 C); visc. 250 cps (60 C); m.p. 34 C; HLB 16.5; cloud pt. 82 C (1% aq.); pour pt. 27 C; surf. tens. 42.8 dynes/cm (0.1%); toxicology: none to sl. eye and skin irritation; 100% act.

Pluronic® P84. [BASF] Poloxamer 234; nonionic; wetting agent, emulsifier, demulsifier, foam and visc. control agent, dispersant, antistat, gelling agent, dyeing assistant, leveler, lubricant for agric., cosmetics, pharmaceuticals, metal cleaning, pulp/paper, textile scouring, water treatment; paste; m.w. 4200; dens. 8.6 lb/gal (60 C); sp.gr. 1.03 (60 C); visc. 265 cps (60 C); m.p. 34 C; HLB 14.0; cloud pt. 74 C (1% aq.); flash pt. 442 F (COC); pour pt. 18 C; surf. tens. 42.0 dynes/cm (0.1%); toxicology: none to sl. eye and skin irritation; 100% act.

Pluronic® P85. [BASF] Poloxamer 235; nonionic; wetting agent, emulsifier, demulsifier, foam and visc. control agent, dispersant, antistat, gelling agent, dyeing assistant, leveler, lubricant for agric., cosmetics, pharmaceuticals, metal cleaning, pulp/paper, textile scouring, water treatment; paste; m.w. 4600; sol. in ethanol, propylene glycol, water, toluene, xylene, perchloroethylene; dens. 8.7 lb/gal (60 C); sp.gr. 1.04 (60 C); visc. 310 cps (60 C); m.p. 40 C; HLB 16.0; cloud pt. 85 C (1% aq.); pour pt. 29 C; surf. tens. 42.5 dynes/cm (0.1%); toxicology: none to sl. eye and skin irritation; 100% act.

Pluronic® P94. [BASF] Poloxamer 284; nonionic; wetting agent, emulsifier, demulsifier, foam and visc. control agent, dispersant, antistat, gelling agent, dyeing assistant, leveler, lubricant for agric., cosmetics, pharmaceuticals, metal cleaning, pulp/paper, textile scouring, water treatment; paste; m.w. 5100; sol. in ethanol, propylene glycol, water, toluene, xylene, perchloroethylene; dens. 8.7 lb/gal (60 C); sp.gr. 1.04 (60 C); visc. 110 cps (60 C); m.p. 37 C; HLB 13.5; cloud pt. 76 C (1% aq.); pour pt. 33 C; surf. tens. 39.6 dynes/cm (0.1%); toxicology: none to sl. eye and skin irritation; 100% act.

Pluronic® P103. [BASF] Poloxamer 333; nonionic; wetting agent, emulsifier, demulsifier, foam and visc. control agent, dispersant, antistat, gelling agent, dyeing assistant, leveler, lubricant for agric., cosmetics, pharmaceuticals, metal cleaning, pulp/paper, textile scouring, water treatment; paste; m.w. 4950; sol. in ethanol, toluene, xylene, perchloroethylene; dens. 8.7 lb/gal (60 C); sp.gr. 1.04 (60 C); visc. 285 cps (60 C); m.p. 30 C; HLB 9.0; cloud pt. 86 C (1% aq.); pour pt. 21 C; surf. tens. 34.4 dynes/cm (0.1%); toxicology: none to sl. eye and skin irritation; 100% act.

Pluronic® P104. [BASF] Poloxamer 334; nonionic; wetting agent, emulsifier, demulsifier, foam and visc. control agent, dispersant, antistat, gelling agent, dyeing assistant, leveler, lubricant for agric., cosmetics, pharmaceuticals, metal cleaning, pulp/paper, textile scouring, water treatment; paste; m.w. 5850; sol. in ethanol, toluene, xylene, perchloroethylene; dens. 8.7 lb/gal (60 C); sp.gr. 1.04 (60 C); visc. 550 cps (60 C); m.p. 37.5; HLB 13.0; cloud pt. 81 C (1% aq.); flash pt. 448 F (COC); pour pt. 32 C; surf. tens. 33.1 dynes/cm (0.1%); toxicology: none to sl. eye and skin irritation; 100% act.

Pluronic® P105. [BASF] Poloxamer 335; nonionic; wetting agent, emulsifier, demulsifier, foam and visc. control agent, dispersant, antistat, gelling agent, dyeing assistant, leveler, lubricant for agric., cosmetics, pharmaceuticals, metal cleaning, pulp/paper, textile scouring, water treatment; paste; m.w. 6500; sol. in ethanol, toluene, xylene, perchloroethylene; dens. 8.8 lb/gal (60 C); sp.gr. 1.05 (60 C); visc. 800 cps (60 C); m.p. 42 C; HLB 15.0; cloud pt. 91 C (1% aq.); pour pt. 35 C; surf. tens. 39.1 dynes/cm (0.1%); toxicology: none to sl. eye and skin irritation; 100% act.

Pluronic® P123. [BASF] Poloxamer 403; nonionic; wetting agent, emulsifier, demulsifier, foam and visc. control agent, dispersant, antistat, gelling agent, dyeing assistant, leveler, lubricant for agric., cosmetics, pharmaceuticals, metal cleaning, pulp/paper, textile scouring, water treatment; paste; m.w. 5750; sol. in ethanol, toluene, xylene, perchloroethylene; dens. 8.5 lb/gal (60 C); sp.gr. 1.02 (60 C); visc. 350 cps (60 C); m.p. 31 C; HLB 8.0; cloud pt. 90 C (1% aq.); pour pt. 31 C; surf. tens. 34.1 dynes/cm (0.1%); toxicology: none to sl. eye and skin irritation; 100% act.

Pluronic® PE 3100. [BASF AG] PO/EO block polymer; nonionic; for defoaming and controlled foam detergents; liq.; 100% conc.

Pluronic® PE 4300. [BASF AG] PO/EO block polymer; nonionic; for defoaming and controlled foam detergents; liq.; 100% conc.

Pluronic® PE 6100. [BASF AG] PO/EO block polymer; nonionic; for defoaming and controlled foam detergents; defoamer for sugar industry; liq.; 100% conc.

Pluronic® PE 6200. [BASF AG] PO/EO block polymer; nonionic; for defoaming and controlled foam detergents; defoamer for sugar industry; liq.; 100% conc.

Pluronic® PE 6400. [BASF AG] PO/EO block polymer; nonionic; for defoaming and controlled foam detergents; defoamer for sugar industry; liq.; 100% conc.

Pluronic® PE 6800. [BASF AG] PO/EO block polymer; nonionic; for defoaming and controlled foam detergents; solid; 100% conc.

Pluronic® PE 8100. [BASF AG] PO/EO block polymer; nonionic; for defoaming and controlled foam detergents; defoamer for sugar industry; liq.; 100% conc.

Pluronic® PE 9200. [BASF AG] PO/EO block polymer; nonionic; for defoaming and controlled foam detergents; liq.; 100% conc.

Pluronic® PE 9400. [BASF AG] PO/EO block polymer; nonionic; for defoaming and controlled foam detergents; solid; 100% conc.

Pluronic® PE 10100. [BASF AG] PO/EO block polymer; nonionic; for defoaming and controlled foam detergents; liq.; 100% conc.

Pluronic® PE 10500. [BASF AG] PO/EO block polymer; nonionic; for defoaming and controlled foam detergents; dispersant for crop protection suspension concs.; solid; 100% conc.

Pluronic® RPE 2520. [BASF AG] EO/PO block polymer; nonionic; for defoaming and controlled foam detergents; defoamer for sugar industry; liq.; 100% conc.

Pluronic® RPE 3110. [BASF AG] EO/PO block polymer; nonionic; for defoaming and controlled foam detergents; liq.; 100% conc.

Plysurf A207H. [Dai-ichi Kogyo Seiyaku] Complex org. phosphate ester, free acid; CAS 39464-64-7; anionic; antistat, emulsifier for agric., metal cleaning, emulsion polymerization; pigment dispersant; general purpose detergent; liq.; HLB 7.1; 99% conc.

Plysurf A208B. [Dai-ichi Kogyo Seiyaku] Complex org. phosphate ester, free acid; anionic; antistat, emulsifier for agric., metal cleaning, emulsion polymerization; pigment dispersant; general purpose detergent; liq.; HLB 6.6; 99% conc.

Plysurf A208S. [Dai-ichi Kogyo Seiyaku] Complex org. phosphate ester, free acid; anionic; antistat, emulsifier for agric., metal cleaning, emulsion polymerization; pigment dispersant; general purpose detergent; liq.; HLB 7.0; 99% conc.

Plysurf A210G. [Dai-ichi Kogyo Seiyaku] Complex org. phosphate ester, free acid; anionic; antistat, emulsifier for agric., metal cleaning, emulsion polymerization; pigment dispersant; general purpose detergent; liq.; HLB 9.6; 99% conc.

Plysurf A212C. [Dai-ichi Kogyo Seiyaku] Complex org. phosphate ester, free acid; CAS 9046-01-9; anionic; antistat, emulsifier for agric., metal cleaning, emulsion polymerization; pigment dispersant; general purpose detergent; paste; HLB 9.4; 99% conc.

Plysurf A215C. [Dai-ichi Kogyo Seiyaku] Complex org. phosphate ester, free acid; anionic; antistat, emulsifier for agric., metal cleaning, emulsion polymerization; pigment dispersant; general purpose detergent; liq.; HLB 11.5; 99% conc.

Plysurf A216B. [Dai-ichi Kogyo Seiyaku] Complex org. phosphate ester, free acid; anionic; antistat, emulsifier for agric., metal cleaning, emulsion polymerization; pigment dispersant; general purpose detergent; paste; HLB 14.4; 99% conc.

Plysurf A217E. [Dai-ichi Kogyo Seiyaku] Complex org. phosphate ester, free acid; anionic; antistat, emulsifier for agric., metal cleaning, emulsion polymerization; pigment dispersant; general purpose detergent; solid; HLB 14.9; 99% conc.

Plysurf A219B. [Dai-ichi Kogyo Seiyaku] Complex org. phosphate ester, free acid; anionic; antistat, emulsifier for agric., metal cleaning, emulsion polymerization; pigment dispersant; general purpose detergent; solid; HLB 16.2; 99% conc.

Plysurf AL. [Dai-ichi Kogyo Seiyaku] Complex org. phosphate ester, free acid; anionic; antistat, emulsifier for agric., metal cleaning, emulsion polymerization; pigment dispersant; general purpose detergent; liq.; HLB 5.6; 98.5% conc.

Pogol 200. [Hart Chem. Ltd.] PEG; nonionic; intermediate for fatty acid esters; solubilizer, antistat, softener, humectant, fiber and metal lubricant, plasticizer, tablet binder; for pharmaceuticals, cosmetics; clear liq.; m.w. 190-210; sp.gr. 1.14; 100% act.

Pogol 300. [Hart Chem. Ltd.] PEG; nonionic; solubilizer, dispersant, emulsifier; herbicides and pesticides; plasticizer for starch, paste and polyethylene sheet; clear liq.; m.w. 285-315; sp.gr. 1.13; 100% act.

Pogol 400 NF. [Hart Chem. Ltd.] PEG, NF grade; nonionic; solubilizer, solv. for pharmaceuticals; mold release agent and lubricant for both natural and syn. prods.; dye carrier; clear liq.; m.w. 380-420; sp.gr. 1.12; visc; 92 cps; 100% act.

Pogol 600. [Hart Chem. Ltd.] PEG; nonionic; intermediate for fatty acid esters; solubilizer, lubricant; used in paper coating mixes; antisticking agents and oil resistors; clear paste; m.w. 570-630; sp.gr. 1.13; visc. 130 cps; ref. index 1.4699; 100% act.

Pogol 1000. [Hart Chem. Ltd.] PEG; nonionic; lubricant; anticracking agent for green wood; ester intermediate; solid; m.w. 950-1050; 100% act.

Pogol 1500. [Hart Chem. Ltd.] PEG; nonionic; lubricant for rubber industry; ester intermediate; solid; m.w. 1350-1600; 100% act.

Polefine 510N. [Takemoto Oil & Fat] Alkyl naphthalene sulfonate/formaldehyde condensates; anionic; high water reducing agent for concrete; liq.

Polirol 1BS. [Auschem SpA] POE/POP alkyl ether; nonionic; emulsifier for emulsion polymerization; solid; 100% conc.

Polirol 4, 6. [Auschem SpA] Ammonium alkylphenol ether sulfate; anionic; emulsifier for emulsion polymerization; liq.; 25% conc.

Polirol 10. [Auschem SpA] Alkylaryl phosphate, acid form; CAS 51811-79-1; anionic; emulsifier for emulsion polymerization; liq.; 100% conc.

Polirol 23. [Auschem SpA] Sodium alkyl ether sulfate; nonionic; emulsifier for emulsion polymerization; liq.; 25% conc.

Polirol 215. [Auschem SpA] POE/POP alkyl ether; CAS 68002-96-0; nonionic; emulsifier for emulsion polymerization; solid; 100% conc.

Polirol C5. [Auschem SpA] PEG-200 cocoate; CAS 67762-35-0; nonionic; emulsifier for emulsion polymerization; liq.; HLB 11.0; 100% conc.

Polirol DS. [Auschem SpA] Alkylaryl sulfonate; CAS 27176-87-0; anionic; emulsifier for emulsion polymerization; liq.; 100% conc.

Polirol L400. [Auschem SpA] Ethoxylated lauric acid; CAS 9004-81-3; nonionic; emulsifier for emulsion polymerization; liq.; HLB 14.0; 100% conc.

Polirol LS. [Auschem SpA] Sodium lauryl sulfate; CAS 151-21-3; anionic; emulsifier for emulsion polymerization; liq./paste; 28% conc.

Polirol NF80. [Auschem SpA] Nonylphenol polyglycol ether; CAS 9016-45-9; nonionic; emulsifier for emulsion polymerization; flakes; 100% conc.

Polirol O55. [Auschem SpA] Cetoleth-55; CAS 68920-66-1; nonionic; emulsifier for emulsion polymerization; flakes; 100% conc.

Polirol SE 301. [Auschem SpA] Disodium sulfosuccinate monoester; anionic; emulsifier for emulsion polymerization; liq.; 30% conc.

Polirol TR/LNA. [Auschem SpA] Ditridecyl sodium sulfosuccinate; CAS 2673-22-5; anionic; emulsifier for emulsion polymerization; liq.; 60% conc.

Polyaldo® DGHO. [Lonza] Polyglyceryl-10 hexaoleate; nonionic; emulsifier, emollient, lubricant for cosmetics, toiletries, pharmaceuticals, household specialty prods.; amber clear liq.; sol. in ethanol, min. and veg. oils; insol. in water; HLB 5 ± 1; sapon. no. 130–160.

Polyanthrene KS Liq. New. [Crompton & Knowles] Polymeric aliphatic compd.; cationic; formaldehyde-free fixing agent for improving washfastness of direct dyes on cellulose; lt. amber hazy to clear liq.; misc. with water; pH 7.0 ± 0.5.

Polycarboxylate AMC 60. [Hüls AG] Acrylic acid-maleic acid copolymer; high calcium dispersion power; co-builder in phosphate-free washing powds.; liq.; molar mass 80,000 g/mole; 40% act.

Polychol 5. [Croda Inc.; Croda Chem. Ltd.] Laneth-5; CAS 61791-20-6; nonionic; o/w emulsifier, dispersant, gellant, emollient, solubilizer for personal care prods., hair straighteners, pharmaceuticals; bromo dye solv.; Gardner 13 max. wax; disp. in water; HLB 7.3;acid no. 5 max.; pH 4.5–6.0 (10% aq.); usage level: 1-10%; 100% conc.

Polychol 10. [Croda Inc.; Croda Chem. Ltd.] Laneth-10; CAS 61791-20-6; nonionic; emollient, o/w emulsifier, dispersant, gellant, solubilizer for cosmetics and pharmaceuticals; soft golden yel. wax; water-disp.; HLB 10.7; usage level: 1-10%; 97% conc.

Polychol 20. [Croda Inc.; Croda Chem. Ltd.] Laneth-20; CAS 61791-20-6; nonionic; emollient, emulsifier, dispersant, gellant, solubilizer for cosmetics and pharmaceuticals; yel. semi-hard wax; almost totally water-sol.; HLB 14.0; usage level: 1-10%; 97% conc.

Polychol 40. [Croda Inc.; Croda Chem. Ltd.] Laneth-40; CAS 61791-20-6; nonionic; emollient, emulsifier, dispersant, gellant, solubilizer for cosmetics and pharmaceuticals; yel. hard wax; water-sol.; HLB 16.4; usage level: 1-10%; 97% conc.

Polyester 1606. [Chemron] Complex polyester; nonionic; emulsion preventer/breaker base for oil treatment; base for oil-emulsion prevention; amber visc. liq.; oil-sol.; 100% act.

Polyester N-95. [Chemron] Surfactant; nonionic; emulsion preventer/breaker base for oil treatment; well stimulation additive; clear yel. liq.; sol. in water or HCl; 60% act.

Poly-G® 200. [Olin] PEG 200; chemical intermediate for prod. of surfactants for cleaners, textiles, paper, cosmetics; carrier for pharmaceuticals; also in cosmetics and personal care prods., textiles, rubber mold releases, printing inks and dyes, metalworking fluids,; foods, paints, paper, wood prods., adhesives, agric. prods., ceramics, elec. equipment, petrol. prods., photographic prods., resins; APHA 25 max. liq.; m.w. 200; sol. in water, acetone, ethanol, ethyl acetate, toluene; sp.gr. 1.125; dens. 9.38 lb/gal; visc. 4.3 cs (99 C); flash pt. 171 C (COC).

Poly-G® 300. [Olin] PEG 300; see Poly-G 200; APHA 25 max. liq.; m.w. 300; sol. in water, acetone, ethanol, ethyl acetate, toluene; sp.gr. 1.125; dens. 9.38 lb/gal; visc. 5.8 cs (99 C); f.p. –15 to –8 C; flash pt. 196 C (COC).

Poly-G® 400. [Olin] PEG 400; see Poly-G 200; APHA 25 max. liq.; m.w. 400; sol. in water, acetone, ethanol, ethyl acetate, toluene; sp.gr. 1.127; dens. 9.4 lb/gal; visc. 7.3 cs (99 C); f.p. 4–10 C; flash pt. 224 C (COC); pour pt. 3–10 C.

Poly-G® 600. [Olin] PEG 600; see Poly-G 200; APHA 25 max. liq.; m.w. 600; sol. in water, acetone, ethanol, ethyl acetate, toluene; sp.gr. 1.127; dens. 9.4 lb/gal; visc. 10.5 cs (99 C); f.p. 20–25 C; flash pt. 246 C (COC); pour pt. 19–24 C.

Poly-G® 1000. [Olin] PEG 1000; see Poly-G 200; wh. waxy solid; m.w. 1000; somewhat less sol. in water than liq. glycols; sp.gr. 1.104 (50/20 C); dens. 9.20 lb/gal (50/20 C); visc. 17.4 cs (99 C); f.p. 38–41 C; flash pt. 260 C (COC); pour pt. 40 C.

Poly-G® 1500. [Olin] PEG 1500; see Poly-G 200; wh. waxy solid; m.w. 1500; somewhat less sol. in water than liq. glycols; sp.gr. 1.104 (50/20 C); dens. 9.20 lb/gal (50/20 C); visc. 28 cs (99 C); f.p. 43–46 C; flash pt. 266 C (COC); pour pt. 45 C.

Poly-G® 2000. [Olin] PEG 2000; see Poly-G 200; APHA 25 max. liq.; m.w. 2000; somewhat less sol. in water than lower m.w. glycols; sp.gr. 1.113; dens. 9.26 lb/gal; visc. 11.7 cs (99 C); f.p. –20 C; 60% aq. sol'n.

Poly-G® B1530. [Olin] PEG 500-600; see Poly-G 200; wh. waxy solid; m.w. 900; somewhat less sol. in water than liq. glycols; sp.gr. 1.104 (50/20 C); dens. 9.20 lb/gal (50/20 C); visc. 15 cs (99 C); f.p. 38–41 C; flash pt. 254 C (COC); pour pt. 38 C.

Polyglycol B-11-50. [Hoechst Celanese/Colorants & Surf.] EO/PO random copolymer based on butanol; surfactant for textile processing; Gardner < 1 clear liq.; cloud pt. 58 C (1% aq.); 99% conc.

Polyglycol B-11-100. [Hoechst Celanese/Colorants & Surf.] EO/PO random copolymer based on butanol; surfactant for textile processing; Gardner < 1 clear liq.; cloud pt. 70 C (1% aq.); 99% conc.

Polyglycol B-11-150. [Hoechst Celanese/Colorants & Surf.] EO/PO random copolymer based on butanol; surfactant for textile processing; Gardner < 1 clear liq.; cloud pt. 58 C (1% aq.); 99% conc.

Polyglycol B-11-260. [Hoechst Celanese/Colorants & Surf.] EO/PO random copolymer based on butanol; surfactant for textile processing; Gardner < 1 clear liq.; cloud pt. 65 C (1% aq.); 99% conc.

Polyglycol B-11-660. [Hoechst Celanese/Colorants & Surf.] EO/PO random copolymer based on butanol; surfactant for textile processing; Gardner < 1 clear liq.; cloud pt. 54 C (1% aq.); 99% conc.

Polyglycol P-41-300. [Hoechst Celanese/Colorants &

Surf.] EO/PO random copolymer based on pentaerythritol; surfactant for textile processing; Gardner < 1 clear liq.; cloud pt. 90 C (1% aq.); 99% conc.

Polylev #745. [Polymer Research Corp. of Am.] Hydroxyethylated compds.; dispersant, penetrant, and leveling agent for dyestuffs; misc. with hot water.

Polylube #745. [Polymer Research Corp. of Am.] Hydroxyethylated compds.; dispersant, penetrant, and leveling agent for dyestuffs; misc. with hot water.

Polylube 4507. [Hart Chem. Ltd.] Fatty acid ester and polyoxyalkylene derivs.; nonionic; nonyel. lubricant for polyproylene fiber and slit film; liq.; 90% conc.

Polylube 5713. [Hart Chem. Ltd.] Blend of fatty acid esters; nonionic; heat-stable lubricant for high-speed applics. in spinning and twisting; liq.; 85% act.

Polylube DDL. [Hart Chem. Ltd.] Blend of polyoxyalkylene alcohol derivs. and polyalkylene glycol ethers; nonionic; heat-stable lubricant for glass extrusion applics.; liq.; 35% conc.

Polylube GK. [Hart Chem. Ltd.] Polyoxyalkylene condensate; nonionic; syn. fiber lubricant, antistat for wool and syn. fibers; liq.; water-sol.; ref. index 1.467; sp.gr. 1.09; visc. 330 cps; 100% act.

Polylube RE. [Hart Chem. Ltd.] Blend of polyoxyalkylene derivs.; nonionic; lubricant, antistat for acetate, nylon; liq.; 85% act.

Polylube SC. [Hart Chem. Ltd.] Fatty acid condensate; nonionic; fiber lubricant, antistat for polyester, acrylic; liq.; water-sol.; ref. index 1.467; sp.gr. 1.02; visc. 110 cps; 87% act.

Polylube WS. [Hart Chem. Ltd.] Polyalkylene glycol ether; nonionic; nonyel. heat-stable lubricant for spinning nylon and wool yarns; clear liq.; water-sol.; ref. index 1.4642; sp.gr. 1.086; visc. 500 cps; 99% act.

Polylube Wax. [Hart Chem. Ltd.] Blend of polyoxyalkylene derivs.; wax for sizing operations; cohesive agent for short staple fiber in woolen systems; water-sol.; 50% act.

Polypeg E-400. [Olin] PEG fatty ester; visc. control agent; dispersing and surface active agent; liq.

Polyram C. [Ceca SA] Coco polypropylene polyamine; cationic.

Polyram O. [Ceca SA] Oleic polypropylene polyamine; cationic.

Polyram S. [Ceca SA] N-tallow polypropylene polyamine; CAS 68911-79-5; cationic; emulsifier for bitumen; antistripping agent for road making; paste.

Polysol J. [ICI Surf. UK] Low visc. polyvinyl alcohol sol'n.; nonionic; used for producing firm finishes on cellulose fabrics; liq.

Polyspin MP-7-29. [Hart Chem. Ltd.] Polyoxyalkylene deriv.; nonionic; heat-stable lubricant for polypropylene filament; liq.; 75% act.

Polyspin PA. [Hart Chem. Ltd.] Blend of syn. oils and polyoxyethylene derivs.; nonionic; spin finish for nylon 6 and polypropylene; liq.; 85% conc.

Polyspin PF. [Hart Chem. Ltd.] Blend of polyoxyalkylene derivs.; nonionic; heat-stable lubricant, antistat for polypropylene filament; liq.; 75% act.

Polystep® A-4. [Stepan; Stepan Canada] Sodium linear alkyl (C12) benzene sulfonate; anionic; emulsifier for emulsion polymerization, S/B, vinyl chloride, and vinylidene chloride latexes; FDA compliance; yel. turbid liq.; 50% act.

Polystep® A-7. [Stepan; Stepan Canada] Sodium dodecylbenzene sulfonate, linear; anionic; emulsifier for emulsion polymerization, S/B, vinyl chloride, vinylidene chloride latexes; FDA compliance; yel. slurry; 39% conc.

Polystep® A-11 [Stepan; Stepan Canada; Stepan Europe] Isopropylamine dodecylbenzene sulfonate, branched; anionic; emulsifier, pigment dispersant; emulsion polymerization (S/B, vinyl chloride, vinylidene chloride latexes); FDA compliance; pale clear visc. liq.; oil-sol.; 88% act.

Polystep® A-13. [Stepan; Stepan Canada] Linear dodecylbenzene sulfonic acid; emulsion polymerization surfactant; catalyst in acid catalyzed reactions; FDA compliance; dk. visc. liq.; 97% act.

Polystep® A-15. [Stepan; Stepan Canada] Sodium linear dodecylbenzene sulfonate; anionic; emulsifier for emulsion and styrene polymerization; FDA compliance; pale clear to hazy liq.; 22% conc.

Polystep® A-15-30K. [Stepan; Stepan Canada] Potassium dodecylbenzene sulfonate, linear; anionic; surfactant for styrene-butadiene, vinyl chloride, and vinylidene chloride latexes; thermal and hydrolytic stability; FDA compliance; hazy slurry; 30% conc.

Polystep® A-16. [Stepan; Stepan Canada] Sodium dodecylbenzene sulfonate, branched; anionic; emulsifier for styrene polymerization; FDA compliance; pale turbid liq.; 30% conc.

Polystep® A-16-22. [Stepan; Stepan Canada] Sodium dodecylbenzene sulfonate, branched; anionic; emulsifier for styrene, vinyl chloride polymerization; FDA compliance; pale turbid liq.; 22% conc.

Polystep® A-17. [Stepan; Stepan Canada] Dodecylbenzene sulfonic acid, branched; anionic; intermediate for the prod. of sodium salts; neutralized acid as emulsifier; salts as emulsifiers for SBR latex; emulsion polymerization; catalyst for acid catalyzed reactions; FDA compliance; dk. visc. liq.; water-sol.; dens. 9.1 lb/gal; sp.gr. 1.09; visc. 9000 cps (30 C); 97.4% act.

Polystep® A-18. [Stepan; Stepan Canada; Stepan Europe] Sodium alpha olefin (C14, C16) sulfonate; anionic; surfactant for vinyl and vinylidene chloride, acrylic, styrene-acrylaic, SBR polymerization; FDA compliance; pale yel. clear liq.; 40% act.

Polystep® B-1. [Stepan; Stepan Canada] Ammonium nonoxynol-4 sulfate; anionic; emulsifier for emulsion polymerization; FDA compliance; yel. visc. liq.; 60% act.

Polystep® B-3. [Stepan; Stepan Canada; Stepan Europe] Sodium lauryl sulfate; CAS 151-21-3; anionic; emulsifier for emulsion polymerization, vinyl chloride, sol. acrylics; FDA compliance; wh. powd.; 97.5% act.

Polystep® B-5. [Stepan; Stepan Canada; Stepan Europe] Sodium lauryl sulfate; CAS 151-21-3; anionic; emulsifier for emulsion polymers, incl. vinyl chloride, S/B, acrylic; FDA compliance; pale yel. clear liq.; 30% act.

Polystep® B-7. [Stepan; Stepan Canada] Ammonium lauryl sulfate; anionic; emulsion polymerization surfactant; latex foaming agent; water resistance in coatings; FDA compliance; pale clear liq.; 30% act.

Polystep® B-11. [Stepan; Stepan Canada; Stepan Europe] Ammonium laureth sulfate (4 EO); anionic; emulsifier for emulsion polymerization (acrylics, styrene-acrylic, vinyl acrylics); FDA compliance; pale clear liq.; 59% act.

Polystep® B-12. [Stepan; Stepan Canada; Stepan Europe] Sodium laureth sulfate (4 EO); CAS 9004-82-

4; anionic; emulsifier for polymerization (acrylics, styrene-acrylics, vinyl acrylics); FDA compliance; pale clear liq.; 60% act.

Polystep® B-19. [Stepan; Stepan Canada] Sodium laureth sulfate (30 EO); anionic; emulsifier for emulsion polymerization (acrylics, styrene-acrylics, vinyl acrylics), floor polish, finish polymer and latexes; good water resistance; FDA compliance; pale clear liq.; 26% act.

Polystep® B-20. [Stepan; Stepan Canada] Ammonium laureth sulfate (30 EO); anionic; emulsifier for emulsion polymerization (acrylics, styrene-acrylics, vinyl acrylics), floor finish latexes; good water resistance; FDA compliance; pale clear liq.; 30% act.

Polystep® B-22. [Stepan; Stepan Canada] Ammonium laureth sulfate (12 EO); anionic; emulsifier for acrylic copolymers; FDA compliance; pale clear liq.; 30% act.

Polystep® B-23. [Stepan; Stepan Canada; Stepan Europe] Sodium laureth-12 sulfate; anionic; high active emulsifier for emulsion polymerization (acrylics, styrene-acrylics, vinyl acrylics); somewhat monomer sol.; FDA compliance; amber hazy liq.; oil-sol.; 60% act.

Polystep® B-24. [Stepan; Stepan Canada; Stepan Europe] Sodium lauryl sulfate; CAS 151-21-3; anionic; emulsifier with low cloud pt.; for emulsion polymerization (S/B, vinyl chloride, acrylic, styrene acrylic); FDA compliance; pale clear liq.; 30% act.

Polystep® B-25. [Stepan; Stepan Canada; Stepan Europe] Sodium decyl sulfate; anionic; emulsifier for emulsion polymerization (S/B, vinyl chloride, acrylic), high surf. tens. latex; hydrophilic; FDA compliance; pale clear liq.; 38% act.

Polystep® B-27. [Stepan; Stepan Canada; Stepan Europe] Sodium nonoxynol-4 sulfate; anionic; emulsifier for acrylics, SBR, vinyl chloride, butyl rubber; FDA compliance; pale yel. clear liq.; 30% act.

Polystep® B-29. [Stepan; Stepan Canada; Stepan Europe] Sodium octyl sulfate; anionic; low-foaming emulsifier for vinyl chloride systems; FDA compliance; water-wh. to pale yel. clear liq.; 33% act.

Polystep® B-LCP. [Stepan; Stepan Canada] Sodium alkyl sulfate; anionic; surfactant for vinyl chloride systems; FDA compliance; pale clear liq.; 30% conc.

Polystep® CM 4S. [Stepan Europe] Sodium alkylphenol ether sulfate; anionic; emulsifier for various systems; gives latex with large particles; visc. disp.; 30% conc.

Polystep® C-OP3S. [Stepan; Stepan Canada; Stepan Europe] Sodium octoxynol-3 sulfate; anionic; emulsifier for controlling particle size in vinyl acetate specialty copolymers; FDA compliance; wh. visc. disp.; 20% act.

Polystep® F-1. [Stepan; Stepan Canada; Stepan Europe] Nonoxynol-4; CAS 9016-45-9; nonionic; nonfoaming pigment dispersant; emulsifier for emulsion polymerization (styrene acrylic, acrylic, vinyl acrylic, S/B); FDA compliance; water wh. to pale yel. clear liq.; oil-sol.; 100% act.

Polystep® F-2. [Stepan; Stepan Canada; Stepan Europe] Nonoxynol-6; CAS 9016-45-9; nonionic; pigment dispersant; contributes mech. and freeze/thaw stability to latexes; allows control of particle size; FDA compliance; pale clear liq.; oil-sol.; 100% act.

Polystep® F-3. [Stepan; Stepan Canada; Stepan Europe] Nonoxynol-8; CAS 9016-45-9; nonionic; pigment dispersant, emulsifier for polymerization (styrene acrylic, acrylic, vinyl acrylic, S/B); contributes mech. and freeze/thaw stability to latexes; allows control of particle size; FDA compliance; water-wh. to pale yel. clear liq.; water-sol.; 100% act.

Polystep® F-4. [Stepan; Stepan Canada; Stepan Europe] Nonoxynol-10; CAS 9016-45-9; nonionic; emulsifier for emulsion polymerization (styrene acrylic, acrylic, vinyl acrylic, S/B); FDA compliance; water-wh. to pale yel. clear liq.; 100% act.

Polystep® F-5. [Stepan; Stepan Canada; Stepan Europe] Nonoxynol-12; CAS 9016-45-9; nonionic; emulsifier and stabilizer for emulsion polymerization (styrene acrylic, acrylic, vinyl acrylic, S/B); allows control of particle size; FDA compliance; water-wh. to pale yel. clear to turbid liq.; 100% act.

Polystep® F-6. [Stepan; Stepan Canada; Stepan Europe] Nonoxynol-14; CAS 9016-45-9; nonionic; emulsifier for emulsion polymerization (styrene acrylic, acrylic, vinyl acrylic, S/B); FDA compliance; water-wh. to pale yel. clear to turbid liq.; 100% act.

Polystep® F-9. [Stepan; Stepan Canada; Stepan Europe] Nonoxynol-30; CAS 9016-45-9; nonionic; emulsifier for acrylic systems; FDA compliance; water-wh. to pale yel. clear to hazy liq. or gel; 70% act.

Polystep® F-10, F-10 Ec. [Stepan; Stepan Canada; Stepan Europe] Nonoxynol-40; CAS 9016-45-9; nonionic; emulsifier for acrylics and vinyl acetate; FDA compliance; clear to hazy liq. or gel; 70 and 100% conc. resp.

Polystep® F-95B. [Stepan; Stepan Canada; Stepan Europe] Nonoxynol-34; CAS 9016-45-9; nonionic; emulsifier for acrylics and vinyl acetate; blended with other surfactants to increase the latex particle size; FDA compliance; clear liq.; 70% act.

Polystep® PN 209. [Stepan; Stepan Canada; Stepan Europe] Complexed phosphate ester, acid form; anionic; emulsifier, dispersant, anticorrosive agent, compatibilizing agent, deresinator for emulsion polymerization, hydrotrope for nonionics in alkaline cleaners, derusters, etc.; water-wh. to yel. liq.; 99% act.

Polysurf A212E. [Dai-ichi Kogyo Seiyaku] Complex org. phosphate ester, free acid; anionic; antistat, emulsifier for agric., metal cleaning, emulsion polymerization; pigment dispersant; general purpose detergent; liq.; HLB 10.3; 99% conc.

Polyterge PAT. [CNC Int'l.] Surfactant for removal and suspension of residual disperse dyes on polyester.

Poly-Tergent® 2A1 Acid. [Olin] Dodecyl diphenyl ether disulfonic acid; anionic; industrial cleaners for textile industry; dk. brn. liq.; sol. in water, alcohols; sp.gr. 1.1; dens. 9.15 lb/gal; visc. 500 cp; surf. tens. 31.7 dynes/cm (0.1%); Draves wetting 152 s (0.5%); Ross-Miles foam 140 mm (1%, initial); toxicology: LD50 (oral, rats) 2 g/kg; severe eye irritant; possible skin burns; 40% act.

Poly-Tergent® 2A1-L. [Olin] Sodium dodecyl diphenyl ether disulfonate; anionic; industrial cleaners for textile industry; yel. to brn. liq.; sol. in water, alcohols; sp.gr. 1.16; dens. 9.65 lb/gal; visc. 150 cp; surf. tens. 45.1 dynes/cm (0.1%); Ross-Miles foam 120 mm (1%, initial); toxicology: LD50 (oral, rats) 2 g/kg; severe eye irritant; possible skin burns; 45% act.

Poly-Tergent® 2EP. [Olin] Sodium dodecyl diphenyl

ether disulfonate; anionic; surfactant for emulsion polymerization; yel. to brn. liq.; sol. in water, alcohols; sp.gr. 1.16; dens. 9.65 lb/gal; visc. 150 cp; surf. tens. 31.7 dynes/cm (0.1%); Draves wetting 152 s (0.5%); Ross-Miles foam 140 mm (1%, initial); toxicology: LD50 (oral, rats) 2 g/kg; severe eye irritant; possible skin burns; 45% act.

Poly-Tergent® 3B2. [Olin] Sodium decyl diphenyl ether disulfonate; anionic; surfactant for industrial and institutional detergents and textiles; yel. to brn. liq.; sol. in water, alcohols; sp.gr. 1.16; dens. 9.65 lb/gal; visc. 120 cp; surf. tens. 43.0 dynes/cm (0.1%); Draves wetting 278 s (0.5%); Ross-Miles foam 135 mm (1%, initial); toxicology: LD50 (oral, rats) 3 g/kg; severe eye irritant; possible skin burns; 45% act.

Poly-Tergent® 3B2 Acid. [Olin] Decyl diphenyl ether disulfonic acid; anionic; surfactant for industrial and institutional detergents and textiles; yel. to brn. liq.; sol. in water, alcohols; sp.gr. 1.16; dens. 9.65 lb/gal; visc. 150 cp; surf. tens. 43.0 dynes/cm (0.1%); Draves wetting 278 s (0.5%); Ross-Miles foam 135 mm (1%, initial); toxicology: LD50 (oral, rats) 3 g/kg; severe eye irritant; possible skin burns; 45% act.

Poly-Tergent® 4C3. [Olin] Sodium alkyl diphenyl ether disulfonate; anionic; surfactant for paper, textiles, metalworking, industrial and household cleaners; lt. yel. liq.; sol. in water, alcohols; sp.gr. 1.12; dens. 9.32 lb/gal; visc. 150 cp; surf. tens. 47.6 dynes/cm (0.1%); Draves wetting 7300 s (0.5%); Ross-Miles foam 140 mm (1%, initial); toxicology: LD50 (oral, rats) > 5 g/kg; severe eye irritant; possible skin burns; 35% act.

Poly-Tergent® B-150. [Olin] Nonoxynol-4.5; CAS 9016-45-9; nonionic; surfactant for a wide variety of applics.; pale yel. liq.; mild odor; sol. in CCl_4, ethanol, kerosene, min. spirits, xylene; disp. in water; sp.gr. 1.02; dens. 8.51 lb/gal; HLB 9.5; f.p. –20 C; flash pt. (COC) 243 C; cloud pt. < 0 C (1% aq.); surf. tens. 27 dynes/cm (0.1%); toxicology: LD50 (oral, rats) ≈ 3 g/kg; nonirritating to skin; irritating to eyes; 100% act.

Poly-Tergent® B-200. [Olin] Nonoxynol-6; CAS 9016-45-9; nonionic; surfactant for a wide variety of applics.; pale yel. liq.; mild odor; sol. in CCl_4, ethanol, kerosene, min. spirits, xylene; disp. in water; sp.gr. 1.035–1.045; dens. 8.64 lb/gal; HLB 10.9; pour pt. –30 C; flash pt. 260 C (COC); cloud pt. < O (1% aq.); surf. tens. 28 dynes/cm (0.1%); toxicology: LD50 (oral, rats) ≈ 3 g/kg; nonirritating to skin; irritating to eyes; 100% act.

Poly-Tergent® B-300. [Olin] Nonoxynol-9; CAS 9016-45-9; nonionic; surfactant for a wide variety of applics.; pale yel. liq.; mild odor; sol. in CCl_4, ethanol, ethylene glycol, water, xylene; sp.gr. 1.050–1.065; dens. 8.75 lb/gal; HLB 12.9; pour pt. 3 C; flash pt. 288 C (COC); cloud pt. 58 C (1% aq.); surf. tens. 29 dynes/cm (0.1%); Draves wetting 11 s (0.1%); Ross-Miles foam 105 mm (0.1%, initial); toxicology: LD50 (oral, rats) 3.8 g/kg; nonirritating to skin; irritating to eyes; 100% act.

Poly-Tergent® B-315. [Olin] Nonoxynol-9.5; CAS 9016-45-9; nonionic; surfactant for a wide variety of applics.; APHA 50 max. clear liq.; mild odor; sol. in water, alcohols, glycol ethers, aromatic hydrocarbons, chlorinated solvs.; sp.gr. 1.06; dens. 8.8 lb/gal; HLB 13.1; pour pt. 5 C; flash pt. 282 C (COC); cloud pt. 59 C (1% aq.); pH 5.0–7.0 (1% aq.); surf. tens. 31 dynes/cm (0.1%); Draves wetting 11 s (0.1%); Ross-Miles foam 105 mm (0.1%, initial); toxicology: LD50 (oral, rats) ≈ 3 g/kg; nonirritating to skin; irritating to eyes; 100% act.

Poly-Tergent® B-350. [Olin] Nonoxynol-10.5; CAS 9016-45-9; nonionic; surfactant for a wide variety of applics.; pale yel. liq.; mild odor; sol. in CCl_4, ethanol, ethylene glycol, water, xylene; sp.gr. 1.060–1.070; dens. 8.83 lb/gal; HLB 13.5; pour pt. 10 C; flash pt. 288 C (COC); cloud pt. 73 C (1% aq.); surf. tens. 30 dynes/cm (0.1%); Draves wetting 14 s (0.1%); Ross-Miles foam 130 mm (0.1%, initial); toxicology: LD50 (oral, rats) ≈ 3 g/kg; nonirritating to skin; irritating to eyes; 100% act.

Poly-Tergent® B-500. [Olin] Nonoxynol-15; CAS 9016-45-9; nonionic; surfactant for a wide variety of applics.; APHA 100 clear slush, mild odor; sol. in water, alcohols, glycol ethers, aromatic hydrocarbons, chlorinated solvs.; sp.gr. 1.07; dens. 9.0 lb/gal; f.p. 22 C; HLB 15.0; cloud pt. 97 C (1% aq.); flash pt. (COC) 288 C; pH 5-7 (1% aq.); surf. tens. 36 dynes/cm (0.1%); Draves wetting 54 s (0.1%); Ross-Miles foam 140 mm (0.1%, initial); toxicology: LD50 (oral, rats) ≈ 3 g/kg; nonirritating to skin; irritating to eyes; 100% act.

Poly-Tergent® CS-1. [Olin] Carboxylated linear alcohol alkoxylate, sodium salt; anionic; builder surfactant, emulsifier, sequestrant for laundry detergents, hard surf. cleaners, bottle washing, dairy and food service, alkaline, metal, transportation cleaners; high temp. electrolyte and alkaline stability; yel. to brn. liq.; sol. in water, alcohols; sp.gr. 1.18; dens. 9.82 lb/gal; visc. 275 cp; surf. tens. 30.6 dynes/cm (0.1%); Draves wetting 180 s (0.5%); Ross-Miles foam 35 mm (1%, initial); toxicology: LD50 (oral, rats) > 5 g/kg; irritating to skin and eyes; 50% act.

Poly-Tergent® E-17A. [Olin] EO/PO block polymer; nonionic; foam control agent, solubilizer, dispersant, wetting agent, spreading agent for automatic dishwashing, rinse aids, industrial laundry, metal cleaning, water treatment, textiles, agric., paper processing; APHA 50 clear liq., mild odor; sol. in water, alcohols, glycol ethers, aromatic and aliphatic hydrocarbons, chlorinated solvs.; sp.gr. 1.02; dens. 8.5 lb/gal; f.p. -30 C; cloud pt. 32 C (1% aq.); flash pt. (COC) 232 C; pH 4.5-7.5 (1% aq.); surf. tens. 42 dynes/cm (0.1%); Draves wetting 43 s (0.1%); Ross-Miles foam 18 mm (0.1%, initial); toxicology: LD50 (oral, rats) 2.5 g/kg; nonirritating to skin and eyes; 100% act.

Poly-Tergent® E-17B. [Olin] EO/PO block polymer; nonionic; foam control agent, solubilizer, dispersant, wetting agent, spreading agent for automatic dishwashing, rinse aids, industrial laundry, metal cleaning, water treatment, textiles, agric., paper processing; APHA 50 clear liq., mild odor; sol. in water, alcohols, glycol ethers, aromatic and aliphatic hydrocarbons, chlorinated solvs.; sp.gr. 1.02; dens. 8.5 lb/gal; f.p. -35 C; cloud pt. 35 C (1% aq.); flash pt. (COC) 232 C; pH 4.5-7.5 (1% aq.); surf. tens. 44 dynes/cm (0.1%); Draves wetting > 300 s (0.1%); Ross-Miles foam 25 mm (0.1%, initial); toxicology: LD50 (oral, rats) ≈ 2.5 g/kg; nonirritating to skin and eyes; 100% act.

Poly-Tergent® E-25B. [Olin] EO/PO block polymer; nonionic; foam control agent, solubilizer, dispersant, wetting agent, spreading agent for automatic dishwashing, rinse aids, industrial laundry, metal cleaning, water treatment, textiles, agric., paper processing; APHA 50 clear liq., mild odor; sol. in water, alcohols, glycol ethers, aromatic and aliphatic hydrocarbons, chlorinated solvs.; sp.gr. 1.03; dens. 8.6 lb/gal; f.p. -10 C; cloud pt. 29 C (1% aq.); flash pt.

(COC) 232 C; pH 4.5-7.5 (1% aq.); surf. tens. 42 dynes/cm (0.1%); Draves wetting > 300 s (0.1%); Ross-Miles foam 20 mm (0.1%, initial); toxicology: LD50 (oral, rats) 2.5 g/kg; nonirritating to skin and eyes; 100% act.

Poly-Tergent® J-200. [Olin] Branched chain alcohols, ethoxylated; nonionic; wetting agent for textile and paper industries; APHA 100 max. clear liq.; mild odor; sol. in alcohols, glycol ethers, aromatic hydrocarbons, chlorinated solvs.; disp. in water; sp.gr. 1.01; dens. 8.4 lb/gal; HLB 11.9; flash pt. 166 C (COC); cloud pt. < 0 C (1% aq.); pH 5.0–7.0 (1% aq.); surf. tens. 28 (0.1%); 100% act.

Poly-Tergent® J-300. [Olin] Branched chain alcohols, ethoxylated; nonionic; see Poly-Tergent J-200; APHA 100 max. clear liq.; mild odor; sol. in water, alcohols, glycol ethers, aromatic hydrocarbons, chlorinated solvs.; sp.gr. 1.03; dens. 8.6 lb/gal; HLB 13.3; flash pt. 185 C (COC); cloud pt. 56 C (1% aq.); pH 6.0–7.0 (1% aq.); surf. tens. 27 dynes/cm (0.1%); 100% act.

Poly-Tergent® P-9E. [Olin] EO/PO block polymer; nonionic; defoamer, deduster, emulsifier for rinse aids, machine dishwashing, laundry detergents; APHA 50 clear liq., mild odor; sol. in water, alcohols, glycol ethers, aromatic and chlorinated hydrocarbons; sp.gr. 1.06; dens. 8.8 lb/gal; f.p. 12 C; cloud pt. 75 C (1% aq.); flash pt. (COC) 243 C; pH 4.5-7.5 (1% aq.); surf. tens. 42 dynes/cm (0.1%); Draves wetting > 300 s (0.1%); Ross-Miles foam 45 mm (0.1%, initial); toxicology: LD50 (oral, rats) 2.5 g/kg; nonirritating to skin and eyes; 100% act.

Poly-Tergent® P-17A. [Olin] EO/PO block polymer; nonionic; emulsifier, detergent, wetting agent, foam control agent, deduster, binder for rinse aids, automatic dishwashing, coatings, sizes, water treatment, egg washing, textiles, foods; APHA 50 max. clear liq.; mild odor; sol. in water, alcohols, glycol ethers, aromatic and aliphatic hydrocarbons, chlorinated solvs.; sp.gr. 1.01; dens. 8.4 lb/gal; flash pt. 241 C (COC); cloud pt. 28 C (1% aq.); pH 5.0–7.5 (1% aq.); surf. tens. 37 dynes/cm (0.1%); Draves wetting 35 s (0.25%); Ross-Miles foam 0 mm (0.1%, initial); toxicology: LD50 (oral, rats) ≈ 2.5 g/kg; nonirritating to skin and eyes; 100% act.

Poly-Tergent® P-17B. [Olin] EO/PO block polymer; nonionic; emulsifier, detergent, wetting agent, foam control agent, deduster, binder for rinse aids, automatic dishwashing, coatings, sizes, water treatment, egg washing, textiles, foods; APHA 50 max. clear liq.; mild odor; sol. in water, alcohols, glycol ethers, aromatic hydrocarbons, chlorinated solvs.; sp.gr. 1.04; dens. 8.6 lb/gal; flash pt. 246 C (COC); cloud pt. 31 (1% aq.); pH 5.0–7.5 (1% aq.); surf. tens. 40 dynes/cm (0.1%); Draves wetting > 180 s (0.1%); Ross-Miles foam 30 mm (0.1%, initial); toxicology: LD50 (oral, rats) ≈ 2.5 g/kg; nonirritating to skin and eyes; 100% act.

Poly-Tergent® P-17BLF. [Olin] EO/PO block polymer; nonionic; emulsifier, detergent, wetting agent, foam control agent, deduster, binder for rinse aids, automatic dishwashing, coatings, sizes, water treatment, egg washing, textiles, foods; APHA 50 clear liq., mild odor; sol. in water, alcohols, glycol ethers, aromatic and chlorinated hydrocarbons; sp.gr. 1.02; dens. 8.5 lb/gal; f.p. -10 C; cloud pt. 31 C (1% aq.); flash pt. (COC) 238 C; pH 4.5-7.5 (1% aq.); surf. tens. 40 dynes/cm (0.1%); Draves wetting > 300 s (0.1%); Ross-Miles foam 15 mm (0.1%, initial); toxicology: LD50 (oral, rats) ≈ 2.5 g/kg; nonirritating to skin and eyes; 100% act.

Poly-Tergent® P-17BX. [Olin] EO/PO block polymer; nonionic; emulsifier, detergent, wetting agent, foam control agent, deduster, binder for rinse aids, automatic dishwashing, coatings, sizes, water treatment, egg washing, textiles, foods; APHA 50 clear liq., mild odor; sol. in water, alcohols, glycol ethers, aromatic and chlorinated solvs.; sp.gr. 1.02; dens. 8.5 lb/gal; f.p. -12 C; cloud pt. 32 C (1% aq.); flash pt. (COC) 232 C; pH 4.5-7.5 (1% aq.); surf. tens. 40 dynes/cm (0.1%); Draves wetting > 300 s (0.1%); Ross-Miles foam 35 mm (0.1%, initial); toxicology: LD50 (oral, rats) ≈ 3 g/kg; nonirritating to skin; eye irritant; 100% act.

Poly-Tergent® P-17D. [Olin] EO/PO block polymer; nonionic; emulsifier, detergent, wetting agent, foam control agent, deduster, binder for rinse aids, automatic dishwashing, coatings, sizes, water treatment, egg washing, textiles, foods, metal cleaning, papermaking; APHA 50 clear liq., mild odor; sol. in water, alcohols, glycol ethers, aromatic and chlorinated solvs.; sp.gr. 1.05; dens. 8.7 lb/gal; f.p. 9 C; cloud pt. 59 C (1% aq.); flash pt. (COC) 238 C; pH 4.5-7.5 (1% aq.); surf. tens. 41 dynes/cm (0.1%); Draves wetting 265 s (0.1%); Ross-Miles foam 50 mm (0.1%, initial); toxicology: LD50 (oral, rats) 1.85 g/kg; nonirritating to skin; irritating to eyes; 100% act.

Poly-Tergent® P-22A. [Olin] EO/PO block polymer; nonionic; demulsifier, dedusting agent, dispersant for rinse aids, dishwashing, metal cleaning, papermaking; APHA 50 clear liq., mild odor; sol. in alcohols, glycol ethers, aromatic, chlorinated, and aliphatic hydrocarbons; insol. in water; sp.gr. 1.02; dens. 8.5 lb/gal; f.p. -25 C; cloud pt. 20 C (1% aq.); flash pt. (COC) 232 C; pH 4.5-7.5 (1% aq.); surf. tens. 41 dynes/cm (0.1%); Draves wetting 102 s (0.1%); Ross-Miles foam 10 mm (0.1%, initial); toxicology: LD50 (oral, rats) 2.5 g/kg; nonirritating to skin and eyes; 100% act.

Poly-Tergent® P-32A. [Olin] EO/PO block polymer; nonionic; demulsifier, dedusting agent, dispersant for rinse aids, dishwashing, metal cleaning, papermaking; APHA 50 max. clear liq.; mild odor; sol. in alcohols, glycol ethers, aromatic hydrocarbons, chlorinated solvs.; insol. in water; sp.gr. 1.02; dens. 8.5 lb/gal; flash pt. 229 C (COC); cloud pt. 15 C (1% aq.); pH 5.0–7.5 (1% aq.); surf. tens. 43 dynes/cm (0.1%); toxicology: LD50 (oral, rats) 1.85 g/kg; nonirritating to skin and eyes; 100% act.

Poly-Tergent® P-32D. [Olin] EO/PO block polymer; nonionic; emulsifier, detergent, wetting agent, foam control agent, deduster, binder for rinse aids, automatic dishwashing, coatings, sizes, water treatment, egg washing, textiles, foods; APHA 50 clear semisolid, mild odor; sol. in water, alcohols, glycol ethers, aromatic hydrocarbons, chlorinated solvs.; sp.gr. 1.04; dens. 8.6 lb/gal; f.p. 15 C; cloud pt. 81 C (1% aq.); flash pt. (COC) 232 C; pH 4.5-7.5 (1% aq.); surf. tens. 34 dynes/cm (0.1%); Draves wetting 60 s (0.1%); Ross-Miles foam 48 mm (0.1%, initial); toxicology: LD50 (oral, rats) ≈ 2.5 g/kg; nonirritating to skin and eyes; 100% act.

Poly-Tergent® RCS-43. [Olin] Proprietary; nonionic; formulated rinse aid conc. for industrial and institutional dishwashing, low foam cleaners at temps. less than 40 C; liq.; 80% conc.

Poly-Tergent® RCS-48. [Olin] Linear alcohol alkoxylate; nonionic; rinse aid conc.; liq.; 80% conc.

Poly-Tergent® S-205LF. [Olin] Alkoxylated linear alcohol; nonionic; surfactant for dishwashing applics., metal cleaning baths; foam depressant in hard surface cleaners; cleaners for dairy industry; rewetting agents for paper toweling; APHA 100 max. clear liq.; mild odor; sol. in water, alcohols, glycol ethers, aromatic hydrocarbons, chlorinated solvs.; sp.gr. 1.00; visc. 192 cs; flash pt. 229 C (COC); cloud pt. 10 C (1% aq.); pH 5.5–7.0 (1% aq.); surf. tens. 32 dynes/cm (0.1%); 100% act.

Poly-Tergent® S-305LF. [Olin] Alkoxylated linear alcohol; nonionic; defoamer, wetting agent, dispersant for rinse aids, dairy cleaners, textiles, hard surf. cleaners, automatic dishwashing, metal cleaning; biodeg.; APHA max. clear liq.; mild odor; sol. in alcohols, glycol ethers, aromatic hydrocarbons, chlorinated solvs.; disp. in water; sp.gr. 1.00; dens. 8.3 lb/gal; visc. 171 cs; flash pt. 224 C (COC); cloud pt. 19 C (1% aq.); pH 5.5–7.0 (1% aq.); surf. tens. 31 dynes/cm (0.1%); Draves wetting 17 s (0.1%); Ross-Miles foam 15 mm (0.1%, initial); toxicology: LD50 (oral, rats) 3 g/kg; nonirritating to skin and eyes; 100% act.

Poly-Tergent® S-405LF. [Olin] Alkoxylated linear alcohol; nonionic; defoamer, wetting agent, dispersant for rinse aids, dairy cleaners, textiles, hard surf. cleaners, automatic dishwashing, metal cleaning; biodeg.; APHA 100 max. clear liq.; mild odor; sol. in water, alcohols, glycol ethers, aromatic hydrocarbons, chlorinated solvs.; sp.gr. 1.01; dens. 8.4 lb/gal; visc. 189 cs; flash pt. 227 C (COC); cloud pt. 28 C (1% aq.); pH 5.5–7.0 (1% aq.); surf. tens. 32 dynes/cm (0.1%); Draves wetting 12 s (0.1%); Ross-Miles foam 20 mm (0.1%, initial); toxicology: LD50 (oral, rats) 3.4 g/kg; nonirritating to skin and eyes; 100% act.

Poly-Tergent® S-505LF. [Olin] Alkoxylated linear alcohol; nonionic; defoamer, wetting agent, dispersant for rinse aids, dairy cleaners, textiles, hard surf. cleaners, automatic dishwashing, metal cleaning; biodeg.; sol. in water, alcohols, glycol ethers, aromatic hydrocarbons, chlorinated solvs.; APHA 100 max. clear liq.; mild odor; sol. in water, alcohols, glycol ethers, aromatic hydrocarbons, chlorinated solvs.; sp.gr. 1.02; dens. 8.5 lb/gal; visc. 125 cs; flash pt. 232 C (COC); cloud pt. 47 C (1% aq.); pH 5.5–7.0 (1% aq.); surf. tens. 33 dynes/cm (0.1%); Draves wetting 18 s (0.1%); Ross-Miles foam 75 mm (0.1%, initial); toxicology: LD50 (oral, rats) 2.6 g/kg; nonirritating to skin; severe eye irritant; 100% act.

Poly-Tergent® SL-42. [Olin] Alkoxylated linear alcohol; nonionic; wetting agent, emulsifier, for prespots, laundry detergents, transportation cleaners, toilet bowl cleaners, glass cleaners, textiles, metal cleaners, degreasers, abrasive cleaners; biodeg.; APHA 100 max. clear liq.; mild odor; sol. in water, alcohols, glycol ethers, aromatic hydrocarbons, chlorinated solvs.; sp.gr. 1.01; dens. 8.4 lb/gal; HLB 13.0; flash pt. 182 C (COC); cloud pt. 42 C (1% aq.); pH 6.0–8.0 (1% aq.); surf. tens. 28 dynes/cm (0.1%); Draves wetting 12 s (0.1%); Ross-Miles foam 125 mm (0.1%, initial); toxicology: LD50 (oral, rat) 3 g/kg; irritating to eyes; possible skin irritant; 100% act.

Poly-Tergent® SL-62. [Olin] Alkoxylated linear alcohol; nonionic; moderate foaming wetting agent, emulsifier, for prespots, laundry detergents, transportation cleaners, toilet bowl cleaners, glass cleaners, textiles, metal cleaners, degreasers, abrasive cleaners; biodeg.; APHA 100 max. clear visc. liq.; mild odor; sol. in water, alcohols, glycol ethers, aromatic hydrocarbons, chlorinated solvs.; sp.gr. 1.02; dens. 8.5 lb/gal; HLB 14.0; flash pt. 204 C (COC); cloud pt. 62 C (1% aq.); pH 6.0–8.0 (1% aq.); surf. tens. 29 dynes/cm (0.1%); Draves wetting 16 s (0.1%); Ross-Miles foam 140 mm (0.1%, initial); toxicology: LD50 (oral, rats) 2.48 g/kg; sl. skin irritant, severe eye irritant; 100% act.

Poly-Tergent® SL-92. [Olin] Alkoxylated linear alcohol; nonionic; wetting agent, emulsifier, for prespots, laundry detergents, transportation cleaners, toilet bowl cleaners, glass cleaners, textiles, metal cleaners, degreasers, abrasive cleaners; biodeg.; APHA 100 max. slush to solid; mild odor; sol. in water, alcohols, glycol ethers, aromatic hydrocarbons, chlorinated solvs.; sp.gr. 1.05; dens. 8.7 lb/gal; HLB 15.0; flash pt. 274 C (COC); cloud pt. 92 C (1% aq.); pH 6.0–8.0 (1% aq.); surf. tens. 31 dynes/cm (0.1%); Draves wetting 42 s (0.1%); Ross-Miles foam 140 mm (0.1%, initial); toxicology: LD50 (oral, rats) 3.7 g/kg; nonirritating to skin; irritating to eyes; 100% act.

Poly-Tergent® SLF-18. [Olin] Alkoxylated linear alcohol; nonionic; low-foaming biodeg. detergent, dispersant, wetting agent, emulsifier, deduster; for commercial detergents, rinsing aid in machine dishwashing; used in removing protein-type soils; APHA 100 max. clear liq.; mild odor; sol. in water, chlorinated, aromatic and polar solvs.; sp.gr. 1.02; dens. 8.8 lb/gal; flash pt. 229 C (COC); cloud pt. 17 C (1% aq.); pH 6.0–8.0 (1% aq.); surf. tens. 33 dynes/cm (0.1%); Draves wetting 27 s (0.1%); Ross-Miles foam 15 mm (0.1%, initial); toxicology: LD50 (oral, rat) 2.74 g/kg; eye irritant; nonirritating to skin; 100% act.

Polywet AX-7. [Uniroyal] Functionalized oligomer, ammonium salt; anionic; emulsifier for emulsion polymerization of vinyl acetate/acrylate polymers and other acidic systems; liq.; 45% conc.

Polywet KX-3. [Uniroyal] Functionalized oligomer, potassium salt; anionic; emulsifier; mfg. of S/B latices; amber liq.; mild thiol odor; water-sol.; sp.gr. 1.19; visc. 380 cps; HLB 16; pour pt. 5 F; surf. tens. 47 dynes/cm (0.1%) pH 7.0–8.0; 46% act.

Polywet KX-4. [Uniroyal] Functionalized oligomer, potassium salt; mfg. of acrylate latices; amber liq.; mild thiol odor; water-sol.; sp.gr. 1.177; visc. 5000 cps (15 C); HLB 17; pour pt. 15 F; surf. tens. 47 dynes/cm (0.1%); pH 7–8; 40% act.

Polywet ND-1. [Uniroyal] Polyfunctional oligomer, sodium salt; anionic; dispersant for mins., pigments and fillers in paints, paper, latex compds., and water treatment applics.; slight yel. liq.; sp.gr. 1.13; visc. 35 cps; pH 6.8–7.8; 25% act.

Polywet ND-2. [Uniroyal] Functionalized oligomer, sodium salt; anionic; dispersant for pigments, mins., extenders, fillers in aq. systems; latex paints and enamels; coatings, adhesives, paper and paperboard; boiler water; colorless to slight yel. liq.; sp.gr. 1.16; visc. 32 cps; pH 6.8–7.8; 25% act.

Polywet Z1766. [Uniroyal] Sodium salt of a polyfunctional oligomer; anionic; dispersant for titanium dioxide, other pigments; liq.; 35% act.

Prechem 90. [Ivax Industries] Phosphate detergent; anionic; detergent, wetting agent, dispersant, soil suspending agent for textile prep. and dyeing; leveling agent; penetrant; straw clear liq.; sol. in water; dens. 8.5 lb/gal; pH 8-9; usage level: 0.1-3%.

Prechem 120. [Ivax Industries] Anionic; detergent for textile dyeing, bleaching, printing and finishing operations; scouring agent for cotton, wool and synthetics; stable to alkalis, hard water; amber clear liq.; sol. in water; dens. 8.5 lb/gal; pH 9.0 ± 1; usage level: 0.25-1.25 oz/gal.

Prechem 2000. [Ivax Industries] Blend; anionic/nonionic; wetting and scouring agent for processing cellulose and cellulose/synthetic blends; dispersant, penetrant; stable to caustic, hydrogen peroxide; lt. golden clear liq.; sol. in water; f.p. < 32 F; pH 7.5 ± 1.0.

Prechem NFE. [Ivax Industries] Nonionic; nonfoaming detergent for use as prescour and alkali extender, emulsifier and dispersant for waxes, pectins, and oils; in caustic and peroxide saturating in jet, beck, and package dyeing; wh. visc. liq.; dens. 8.5 lb/gal; pH 7 ± 1.

Prechem NPX. [Ivax Industries] Nonylphenoxy polyethoxy alcohol; nonionic; detergent conc., wetting agent, emulsifier for textile wet processing of cotton, wool and synthetics; aids dyebath stability; water clear appearance; sol. in water; dens. 8.35 lb/gal; pH 6.0-7.5 (1% aq.); 100% act.

Prestogen® K. [BASF] Stabilizer for low silicate peroxide bleaching of cellulosic fibers and blends.

Prifac 5902, 5904, 5905. [Unichema] Partially hydrog. tallow and palm oil fatty acids; emulsifier for emulsion polymerization of SBR.

Prifac 7920. [Unichema] Tallow acid; raw material for surfactants, soaps, nitrogen derivs., buffing formulations; low melting solid; acid no. 200-206; iodine no. 50-62; sapon. no. 201-207.

Prifac 7935. [Unichema] Dist. tallow fatty acid; raw material for surfactants, soaps, nitrogen derivs., buffing formulations; low melting solid; acid no. 198-204; iodine no. 52-62; sapon. no. 200-208.

Prifac 9428. [Unichema] Hydrog. tallow acid; chemical intermediate for surfactants, stabilizers, detergents, fabric softeners; solid; acid no. 200-210; iodine no. 5 max.; sapon. no. 200-210.

Prifac 9429. [Unichema] Hydrog. tallow acid; chemical intermediate for surfactants, stabilizers, detergents, fabric softeners; solid; acid no. 192-210; iodine no. 11 max.; sapon. no. 196-210.

Prifrac 2901. [Unichema] Caprylic acid; intermediate for ester prod., syn. lubes, latex stabilizers, substituted glycerides used as skin protectors; liq.; acid no. 385-390; iodine no. 0.5 max.; 98% conc.

Prifrac 2910. [Unichema] Caprylic acid; intermediate for ester prod., syn. lubes, and substituted glycerides used as skin protectors; liq.; 98% conc.

Prifrac 2912. [Unichema] Caprylic/capric acid; intermediate for ester prod., syn. lubes; in surf. coatings; liq.; acid no. 355-372; iodine no. 1 max.

Prifrac 2920. [Unichema] Lauric acid; CAS 143-07-7; emulsifier for hot emulsion polymerization of NBR, NR latex stabilization; acid no. 277-282; iodine no. 0.5 max.; 92-94% conc.

Prifrac 2922. [Unichema] Lauric acid; CAS 143-07-7; emulsifier for mfg. of nitrile rubbers, stabilization of NR latex; acid no. 278-282; iodine no. 0.2 max.; 98-100% conc.

Prifrac 2940. [Unichema] Myristic acid; CAS 544-63-8; intermediate for ester, detergent, and surfactant prods.; solid; acid no. 243-248; iodine no. 0.5 max.; 92-94% conc.

Prifrac 2942. [Unichema] Myristic acid; CAS 544-63-8; intermediate for ester, detergent and surfactant prods.; solid; acid no. 244-248; iodine no. 0.5 max.; 98-100% conc.

Prifrac 2960. [Unichema] Palmitic acid; CAS 57-10-3; intermediate for surfactants, soap and cosmetic formulations; cryst. solid; acid no. 216-220; iodine no. 1 max.; 92% min. conc.

Prifrac 2980. [Unichema] Stearic acid; CAS 57-11-4; intermediate for surfactants, soap and cosmetic formulations; cryst. solid; acid no. 195-199; iodine no. 2 max.; 93% min. conc.

Prifrac 2981. [Unichema] Stearic acid; CAS 57-11-4; intermediate for surfactants, soap and cosmetic formulations; cryst. solid; acid no. 194-198; iodine no. 2 max.; 98-100% conc.

Prifrac 2989. [Unichema] Behenic acid; intermediate for surfactants, soap and cosmetic formulations; solid; acid no. 160-166; iodine no. 2 max.; 85-90% conc.

Primasol® FP. [BASF/Fibers; BASF AG] Aliphatic sulfonic acids, organic phosphate aq. sol'n.; anionic; detergent, wetting agent, textile aux., pigment padding, mercerizing and bleaching processes; migration inhibitor; low foaming; stable to electrolytes, alkalies; low visc. yel.-brn. liq.; sol. in water, alkali; dens. 1.06 g/cc; visc. 75 s (Ford cup); set pt. -15 C; b.p. 100 C; flash pt. > 100 C; 75% conc.

Primasol® KW. [BASF AG] Mixt. of anionic and nonionic compds., polyfunctional, synergistic; textile aux., wetting agent, detergent, stabilizer; dyeing aux. for continuous dyeing of wool, polyamide, and polyester fibers; used for textile floor coverings, space dyeing; ylsh.-brn.; dissolves readily in cold or warm water; pH 6–6.5.

Primasol® NB-NF. [BASF/Fibers] Phosphate ester; anionic; low foaming wetting and deaerating agent for dyeing, desizing, bleaching, pad bath applic.; modified surf. tension to minimize foam formation; colorless visc. liq.; sol. in water; pH 7.5-8.0 (1% aq.); toxicology: skin irritant on prolonged contact with conc. form; 50% conc.

Primasol® NF. [BASF] Phosphated alcohol; anionic; wetting and deaerating agent, pad bath applic.; liq.; 50% conc.

Primasol® SD. [BASF AG] Mixt. of oxyethylation prods.; nonionic; textile aux., stabilizer, wetting agent; aux. for space dyeing of polyamide and acrylic carpet yarns with acid, metal complex, and basic dyes; colorless or lt. ylsh. liq.; 6.5–7.5 (10% sol'n.).

Printac CN-230. [Toho Chem. Industry] Complex; anionic/nonionic; leveling and penetrating agent for acrylic fabric printing; liq.; 50% conc.

Priolene 6900. [Unichema] Oleic acid; CAS 112-80-1; intermediate for ethoxylates, esters, nitrogen derivs., surfactants; used in soaps, personal care prods., lubricant and metalworking fluids, for NR latex stabilization; liq.; acid no. 196-204; iodine no. 89-97; sapon. no. 197-205.

Priolene 6901. [Unichema] Oleic acid; CAS 112-80-1; emulsifier for cold emulsion polymerization of SBR latex; foaming agent for NR, SBR latex, NBR, CR rubbers.

Priolene 6905. [Unichema] Oleic acid; CAS 112-80-1; intermediate for ethoxylates, esters, nitrogen derivs., surfactants; used in soaps, personal care prods., lubricant and metalworking fluids; liq.; acid no. 199-204; iodine no. 95 max.; sapon. no. 201-206.

Priolene 6906. [Unichema] Oleic acid; CAS 112-80-1; intermediate for ethoxylates, esters, nitrogen derivs., surfactants; used in soaps, personal care prods., lubricant and metalworking fluids; liq.; acid no. 199-204; iodine no. 95 max.; sapon. no. 201-206.

Priolene 6907. [Unichema] Oleic acid; CAS 112-80-1; emulsifier for cold emulsion polymerization of SBR latex.
Priolene 6910. [Unichema] Oleic acid; CAS 112-80-1; intermediate for ethoxylates, esters, nitrogen derivs., surfactants; used in soaps, personal care prods., lubricant and metalworking fluids; liq.; acid no. 196-204; iodine no. 89-97; sapon. no. 197-205.
Priolene 6911. [Unichema] primary foaming agent for NR, SBR latex, NBR, CR rubbers.
Priolene 6922. [Unichema] Oleic acid, tech.; CAS 112-80-1; intermediate; liq.
Priolene 6930. [Unichema] Oleic acid; CAS 112-80-1; emulsifier for cold emulsion polymerization of SBR latex.
Priolene 6933. [Unichema] Oleic acid; CAS 112-80-1; intermediate for ethoxylates, esters, nitrogen derivs., surfactants; used in soaps, personal care prods., lubricant and metalworking fluids; liq.; acid no. 197-203; iodine no. 89-95; sapon. no. 199-205.
Priosorine 3501. [Unichema] Isostearic acid; intermediate; polyol ester in syn. lubricants; emulsifier for esters, soaps; textile softener and antistat; anticorrosion additive; liq.
Priosorine 3505. [Unichema] Isostearic acid; emulsifier for o/w emulsions; provides surfactant and structuring props.; liq.
Pripol 1004. [Unichema] C44 dimer acid; modifier for nylon; polyester fibers; polyamide for hot melts; urethane elastomer; APHA 100 max. liq., faint odor; insol. in water; sp.gr. 0.92; visc. 4100 mPa•s; b.p. > 200 C; acid no. 159-164; iodine no. 7; sapon. no. 160-166; pour pt. -10 C; flash pt. (COC) 325 C; ref. index 1.475.
Pripol 1017, 1022. [Unichema] Dimer acids; aux. emulsifier for chloroprene rubbers.
Pripol 1025. [Unichema] Hydrog. dimer acid; for hot melt adhesives; polyamide for thermographic inks; Gardner 3 max. visc. liq.; dens. 0.95; acid no. 192-197; sapon. no. 195-202; flash pt. (COC) 275 C.
Prisavon 1981. [Unichema] Sodium soap of tallow and coconut fatty acids; soap base for prep. of specialty soaps; translucent solid noodle.
Prisavon 9220, 9230, 9240, 9242. [Unichema] Sodium soap of palm/palm kernel/coconut fatty acids; veg. dervied soap base for prep. of specialty soaps; solid noodle.
Pristerene 4904. [Unichema] Stearic acid; CAS 57-11-4; intermediate for ethoxylates, esters, nitrogen derivs., personal prod. formulations; heat-stable; wh. cryst. solid; acid no. 206-212; iodine no. 0.5 max.; sapon.no. 208-214.
Pristerene 4905. [Unichema] Stearic acid, triple pressed; CAS 57-11-4; intermediate for ethoxylates, esters, nitrogen derivs., personal prod. formulations; wh. cryst. solid; acid no. 206-212; iodine no. 0.5 max.; sapon. no. 208-214.
Pristerene 4910. [Unichema] Stearic acid; CAS 57-11-4; intermediate for ethoxylates, esters, nitrogen derivs., personal prod. formulations; wh. solid; acid no. 200-208; iodine no. 2 max.; sapon. no. 201-209.
Pristerene 4911. [Unichema] Stearic acid, triple pressed; CAS 57-11-4; intermediate for ethoxylates, esters, nitrogen derivs., personal prod. formulations; wh. solid; acid no. 206-210; iodine no. 0.5 max.; sapon. no. 207-211.
Pristerene 4915. [Unichema] Stearic acid, triple pressed; CAS 57-11-4; intermediate for ethoxylates, esters, nitrogen derivs., personal prod. formulations; heat-stable; wh. solid; acid no. 206-210; iodine no. 0.5 max.; sapon. no. 207-211.
Pristerene 4921. [Unichema] Stearic acid, double pressed; CAS 57-11-4; intermediate for ethoxylates, esters, nitrogen derivs., personal prod. formulations; heat-stable; wh. solid; acid no. 203-208; iodine no. 7 max.; sapon. no. 205-210.
Procetyl 10. [Croda Inc.] PPG-10 cetyl ether; CAS 9035-85-2; nonionic; emollient, coupler, cosolvent, plasticizer, superfatting, wetting and spreading agent, penetrant; lubricant in cosmetics and personal care prods.; alcoholic and aq. alcoholic compositions; APHA 150 max. clear liq.; faint, char. sweet odor; sol. in min. oil, acetone, IPM, lanolin oil, alcohol; acid no. 1 max.; pH 6.0–7.5 (3% disp.); usage level: 5-30%; 100% act.
Procetyl 30. [Croda Inc.] PPG-30 cetyl ether; CAS 9035-85-2; emollient, coupler, cosolvent, plasticizer, superfatting, wetting and spreading agent, penetrant; lubricant in cosmetics and personal care prods.; alcoholic and aq. alcoholic compositions; APHA 150 max. clear liq.; faint, char. sweet odor; pH 6.0–7.5 (3% disp.); 100% act.
Procetyl 50. [Croda Inc.] PPG-50 cetyl ether; CAS 9035-85-2; emollient, coupler, cosolvent, plasticizer, superfatting, wetting and spreading agent, penetrant; lubricant in cosmetics and personal care prods.; alcoholic and aq. alcoholic compositions; APHA 150 max. clear liq.; faint, char. sweet odor; sol. in PEG-200, lanolin oil, alcohol; acid no. 1 max.; pH 6–7.5 (3% disp); usage level: 5-30%; 100% act.
Procetyl AWS. [Croda Inc.; Croda Chem. Ltd.] PPG-5 ceteth-20; CAS 9087-53-0; nonionic; emulsifier, plasticizer, coupler, humectant, dispersant, emollient, and fragrance solubilizer in aq. and aq. alcoholic systems, personal care prods., antiperspirants, bath oils; water-wh. turbid oily liq.; char. sweet odor; sol. in water, alcohol, IPM, PEG-200, glycerin, propylene glycol, lanolin oil; acid no. 1 max.; HLB 16.0; pH 6.0–7.5 (30% aq.); usage level: 0.5-5%; 100% act.
Procoal 20. [Tokai Seiyu Ind.] Surfactant; nonionic; deairing, wetting and penetrating agent; low foaming; liq.
Product BCO. [DuPont] Cetyl betaine; amphoteric; wetting agent, detergent, emulsifier, dispersant, surfactant; dyeing applics.; softener for textiles; leveling and rewetting agent in the paper industry; dyeing assistant and degreaser in the leather industry; antistat on plastic films; brn. clear liq.; mild odor; misc. in water; dens. 8.4 lb/gal; visc. free-flowing; surf. tens. 27.8 dynes/cm (32 C, 0.1%); 33% solids.
Product VN-11. [Henkel/Cospha; Henkel Canada] Ethoxylated oleyl alcohol; nonionic; emulsifier for metalworking fluids; liq.; 100% conc.
Produkt 2058. [Zschimmer & Schwarz] Fatty alcohol/fatty acid ester; industrial surfactant; liq.; 100% act.
Produkt GM 4111. [Zschimmer & Schwarz] Blend; nonionic; industrial surfactant; liq.; 45% act.
Produkt GM 4210. [Zschimmer & Schwarz] MIPA-ammonium fatty alcohol ether sulfate, propylene glycol; industrial surfactant; liq.; 90% act.
Produkt GM 4300. [Zschimmer & Schwarz] Surfactant blend; anionic/amphoteric; industrial surfactant; liq.; 29% act.
Produkt GM 6115. [Zschimmer & Schwarz] Surfactant blend; nonionic/cationic; industrial surfactant; liq. to paste; 96% act.
Produkt GS 5001. [Zschimmer & Schwarz] Fatty acid DEA, modified; for hair shampoos, bath additives,

dishwash, cleansing agents; liq.; 100% act.

Produkt RT 63. [Zschimmer & Schwarz] Alkyl polyglycolamine; nonionic; detergent for cleaning agents; liq.; 100% conc.

Profan ME-20. [Sanyo Chem. Industries] POE coconut fatty acid amide; nonionic; used in pH balanced shampoos, dishwashing detergents and acid cleansers; stable in acid or alkali media; liq.; 97% conc.

Progalan X-13. [Rhone-Poulenc/Textile & Rubber] Ammonium alkylaryl ether sulfate; anionic; foaming and scouring agent for textile applics. and detergent formulations; liq.; 60% conc.

Progasol 40. [Rhone-Poulenc/Textile & Rubber] Alkylaryl sulfonate; anionic; wetting, leveling, and scouring agent; emulsifier used in textiles; liq.; 40% conc.

Progasol 230. [Rhone-Poulenc/Textile & Rubber] Fatty alkanolamide; nonionic; textile scouring agent for fibers and formulation of cleaners and detergents; liq.; 100% conc.

Progasol 443, 457. [Rhone-Poulenc/Textile & Rubber] Fatty alkanolamide; nonionic; textile scouring agent for fibers and formulation of cleaners and detergents; liq.; 100% conc.

Progasol COG. [Rhone-Poulenc/Textile & Rubber] Nonionic; nonfoaming detergent; liq.; 40% conc.

Progasol FSD (formerly Dianol FSD). [Rhone-Poulenc Surf.] Textile dye auxiliary.

Progasol SN. [Rhone-Poulenc/Textile & Rubber] Solv./surfactant blend; solv. scouring agent, detergent; liq.; 84% conc.

Promyristyl PM-3. [Croda Inc.] PPG-3 myristyl ether; low-visc. emollient for clear analgesic, deodorant, and fragrance sticks; Gardner 1 max. clear liq.; sol. in oil, alcohol, lanolin oil; acid no. 1 max.; usage level: 5-15%.

Pronal 502, 502A. [Toho Chem. Industry] Polyalkylene glycol ester; nonionic; defoaming agent for paper, latex, and water paint mfg.; liq.

Pronal 2200. [Toho Chem. Industry] Blend; nonionic; antifoaming agent for kraft paper mfg.; liq.

Pronal 3300. [Toho Chem. Industry] Blend; nonionic; antifoaming agent for kraft paper mfg.; liq.

Pronal EX-100. [Toho Chem. Industry] Silicone emulsion; antifoaming agent for petrochem. industry; liq.

Pronal EX-300. [Toho Chem. Industry] Silicone emulsion; antifoaming agent for natural gas and fertilizer mfg.; liq.

Pronal P-805. [Toho Chem. Industry] Blend; nonionic; antifoaming agent for coating color formulation in paper industry; liq.

Propetal 99. [Zschimmer & Schwarz] Fatty alcohol EO/PO adduct; nonionic; detergent, wetting agent, emulsifier, antifoaming additive; base material for low-foaming cleansing agents; liq.; cloud pt. 18-22 C (1% aq.); 100% act.

Propetal 103. [Zschimmer & Schwarz] Fatty alcohol EO/PO adduct; nonionic; base material for low-foaming industrial detergents; wetting agent; colorless to ylsh. liq.; sol. in most org. solvs., e.g., benzene, toluene, perchloroethylene; disp. in water; cloud pt. 44-48 C (5 g/20 ml butyl diglycol 25%); pH 5-7 (10%); 100% act.

Propetal 241. [Zschimmer & Schwarz] Fatty alcohol EO/PO adduct; nonionic; biodeg. detergent, wetting agent, emulsifier, antifoam; intermediate for prep. of low foaming detergents for household and industry, dishwash, sanitary and floor tile cleaners, industrial spray cleaners, metal pickling, paper, textile, leather; colorless to ylsh. liq.; sol. in water, alcohols, most org. solvs.; dens. 0.99 g/cc; cloud pt. 37-41 C (1% aq.); pH 5-7 (10%); 100% act.

Propetal 254. [Zschimmer & Schwarz] Fatty alcohol EO/PO adduct; nonionic; emulsifier, antifoaming additive; liq.; cloud pt. 5-8 C (1% aq.); 100% act.

Propetal 281. [Zschimmer & Schwarz] Fatty alcohol EO/PO adduct; nonionic; detergent and wetting agent; base material for low-foaming cleansing agents; liq.; cloud pt. 30-33 C (1% aq.); 100% act.

Propetal 281. [Zschimmer & Schwarz] Fatty alcohol polyalkylene glycol ether; nonionic; detergent, wetting agent, emulsifier, antifoaming additive; intermediate for prep. of low foaming detergents; liq.; 100% conc.

Propetal 340. [Zschimmer & Schwarz] Fatty alcohol EO/PO adduct; nonionic; detergent, wetting agent, emulsifier, antifoaming additive; intermediate for prep. of low foaming detergents; liq.; cloud pt. 48-52 C (1% aq.); 100% act.

Propetal 341. [Zschimmer & Schwarz] Fatty alcohol EO/PO adduct; nonionic; detergent, wetting agent, emulsifier, antifoaming additive; intermediate for prep. of low foaming detergents; liq.; cloud pt. 58-62 C (1% aq.); 100% act.

Propomeen 2HT-11. [Akzo] Bis-hydrog. tallowalkyl-2-hydroxypropyl amines; CAS 71060-61-2; industrial surfactant; Gardner 7 max. color; amine no. 99; flash pt. (PMCC) > 204 C; 95% tert. amine.

Propomeen C/12. [Akzo] Dipropylene glycol cocamine; CAS 68516-06-3; cationic; industrial surfactant; Gardner 5 max. hazy liq.; amine no. 179; 95% min. tert. amine.

Propomeen HT/12. [Akzo] Dipropylene glycol hydrog. tallowamine.

Propomeen T/12. [Akzo] Dipropylene glycol tallowamine; CAS 68951-72-4; cationic; industrial surfactant; Gardner 5 clear liq.; amine no. 148; 95% min. tert. amine.

Propoquad® 2HT/11. [Akzo] Di(hydrogenated tallowalkyl) (2-hydroxy-2-methylethyl) quaternary ammonium chloride, aq. IPA; CAS 68554-09-6; cationic; industrial surfactant; Gardner 8 max.; sol. in water, alcohols, acetone, benzene, CCl_4, hexylene glycol; flash pt. (PMCC) 46 C; pH 5.5-7.5; 82-85% quat. in aq. IPA.

Propoquad® T/12 [Akzo] Tallowalkylmethyl-bis(2-hydroxy-2-methylethyl)quaternary ammonium methylsulfates, diethylene glycol; CAS 79770-97-1; industrial surfactant; Gardner 6 max.; sol. in water, alcohols, acetone, benzene, CCl_4, hexylene glycol; flash pt. (PMCC) > 100 C; pH 4.5-6.5; 70-76% quat. in diethylene glycol.

Prosol 518. [Hart Chem. Ltd] Refined min. oil and fatty acid condensates; nonionic; min. oil lubricant for processing of wool and wool syn. blends; amber liq.; sp.gr. 0.895; visc. 65 cps; ref. index 1.471; pH acid; 100% act.

Prosol 525. [Hart Chem. Ltd.] Emulsified mineral oil with selected antistatic agents; antistatic oil for woolen and worsted processes; liq.; 100% conc.

Prosol CT30. [Hart Chem. Ltd.] Emulsified min. oil; nonyel. texturizing coning and knitting oil for nylon, polyester; liq.; 100% act.

Prosol RS. [Hart Chem. Ltd.] Fatty esters and polyalkylene derivs. in min. oil; high performance oil/syn. lubricant for woolen and worsted systems; liq.; 100% act.

Prostearyl 15. [Croda Inc.] PPG-15 stearyl ether; emollient, lubricant for cosmetics, bath oils, sunscreens, hair prods., aerosol antiperspirants, hand and body lotions; coupler for fragrances; clear liq.; sol. in oil, alcohol; usage level: 5-15%.

Protamine-45. [Nat'l. Starch] Amide-amine fatty derivative; cationic; emulsifier, softener and lubricant base; solid; 100% conc.

Protectol® KLC 50, 80. [BASF AG] Benzalkonium chloride; cationic; biocidal surfactant for chemical industry, detergent mfg.; liq.; 50 and 80% conc. resp.

Prote-pon P 2 EHA-02-K30. [Protex] Alkyl ether potassium phosphate; wetting agent, detergent, hydrotrope, emulsifier, rust inhibitor, EP lubricant for alkaline detergents, metal cleaners/lubricants, hard surf. cleaners, textile scours/lubricants, emulsion polymerization, agric., and drycleaning formulations; liq.; pH 7.2 ± 0.3.

Prote-pon P 2 EHA-02-Z. [Protex] Alkyl ether phosphoric acid; wetting agent, detergent, hydrotrope, emulsifier, rust inhibitor, EP lubricant for alkaline detergents, metal cleaners/lubricants, hard surf. cleaners, textile scours/lubricants, emulsion polymerization, agric., and drycleaning formulations; liq.

Prote-pon P 2 EHA-Z. [Protex] Alkyl ether phosphoric acid; wetting agent, detergent, hydrotrope, emulsifier, rust inhibitor, EP lubricant for alkaline detergents, metal cleaners/lubricants, hard surf. cleaners, textile scours/lubricants, emulsion polymerization, agric., and drycleaning formulations; liq.

Prote-pon P-0101-02-Z. [Protex] Alkyl ether phosphoric acid; wetting agent, detergent, hydrotrope, emulsifier, rust inhibitor, EP lubricant for alkaline detergents, metal cleaners/lubricants, hard surf. cleaners, textile scours/lubricants, emulsion polymerization, agric., and drycleaning formulations; liq.

Prote-pon P-L 201-02-K30. [Protex] Alkyl ether potassium phosphate; wetting agent, detergent, hydrotrope, emulsifier, rust inhibitor, EP lubricant for alkaline detergents, metal cleaners/lubricants, hard surf. cleaners, textile scours/lubricants, emulsion polymerization, agric., and drycleaning formulations; emulsion; pH 7.0 ± 0.5.

Prote-pon P-L 201-02-Z. [Protex] Alkyl ether phosphoric acid; wetting agent, detergent, hydrotrope, emulsifier, rust inhibitor, EP lubricant for alkaline detergents, metal cleaners/lubricants, hard surf. cleaners, textile scours/lubricants, emulsion polymerization, agric., and drycleaning formulations; wax.

Prote-pon P-NP-06-K30. [Protex] Alkyl ether potassium phosphate; wetting agent, detergent, hydrotrope, emulsifier, rust inhibitor, EP lubricant for alkaline detergents, metal cleaners/lubricants, hard surf. cleaners, textile scours/lubricants, emulsion polymerization, agric., and drycleaning formulations; liq.; pH 7.2 ± 0.3.

Prote-pon P-NP-06-Z. [Protex] Alkyl ether phosphoric acid; wetting agent, detergent, hydrotrope, emulsifier, rust inhibitor, EP lubricant for alkaline detergents, metal cleaners/lubricants, hard surf. cleaners, textile scours/lubricants, emulsion polymerization, agric., and drycleaning formulations; liq.

Prote-pon P-NP-10-K30. [Protex] Alkyl ether potassium phosphate; wetting agent, detergent, hydrotrope, emulsifier, rust inhibitor, EP lubricant for alkaline detergents, metal cleaners/lubricants, hard surf. cleaners, textile scours/lubricants, emulsion polymerization, agric., and drycleaning formulations; liq.; pH 7.2 ± 0.3.

Prote-pon P-NP-10-MZ. [Protex] Alkyl ether phosphoric acid; wetting agent, detergent, hydrotrope, emulsifier, rust inhibitor, EP lubricant for alkaline detergents, metal cleaners/lubricants, hard surf. cleaners, textile scours/lubricants, emulsion polymerization, agric., and drycleaning formulations; liq.

Prote-pon P-NP-10-Z. [Protex] Alkyl ether phosphoric acid; wetting agent, detergent, hydrotrope, emulsifier, rust inhibitor, EP lubricant for alkaline detergents, metal cleaners/lubricants, hard surf. cleaners, textile scours/lubricants, emulsion polymerization, agric., and drycleaning formulations; liq.

Prote-pon P-OX 101-02-K75. [Protex] Alkyl ether potassium phosphate; wetting agent, detergent, hydrotrope, emulsifier, rust inhibitor, EP lubricant for alkaline detergents, metal cleaners/lubricants, hard surf. cleaners, textile scours/lubricants, emulsion polymerization, agric., and drycleaning formulations; liq.; pH 6.5 ± 0.3.

Prote-pon P-TD-06-K13. [Protex] Alkyl ether potassium phosphate; wetting agent, detergent, hydrotrope, emulsifier, rust inhibitor, EP lubricant for alkaline detergents, metal cleaners/lubricants, hard surf. cleaners, textile scours/lubricants, emulsion polymerization, agric., and drycleaning formulations; liq.; pH 7.5 ± 0.5.

Prote-pon P-TD-06-K30. [Protex] Alkyl ether potassium phosphate; wetting agent, detergent, hydrotrope, emulsifier, rust inhibitor, EP lubricant for alkaline detergents, metal cleaners/lubricants, hard surf. cleaners, textile scours/lubricants, emulsion polymerization, agric., and drycleaning formulations; liq.; pH 7.2 ± 0.3.

Prote-pon P-TD-06-K60. [Protex] Alkyl ether potassium phosphate; wetting agent, detergent, hydrotrope, emulsifier, rust inhibitor, EP lubricant for alkaline detergents, metal cleaners/lubricants, hard surf. cleaners, textile scours/lubricants, emulsion polymerization, agric., and drycleaning formulations; gel; pH 7.5 ± 0.5.

Prote-pon P-TD-06-Z. [Protex] Alkyl ether phosphoric acid; wetting agent, detergent, hydrotrope, emulsifier, rust inhibitor, EP lubricant for alkaline detergents, metal cleaners/lubricants, hard surf. cleaners, textile scours/lubricants, emulsion polymerization, agric., and drycleaning formulations; liq.

Prote-pon P-TD-09-Z. [Protex] Alkyl ether phosphoric acid; wetting agent, detergent, hydrotrope, emulsifier, rust inhibitor, EP lubricant for alkaline detergents, metal cleaners/lubricants, hard surf. cleaners, textile scours/lubricants, emulsion polymerization, agric., and drycleaning formulations; paste.

Prote-pon P-TD-12-Z. [Protex] Alkyl ether phosphoric acid; wetting agent, detergent, hydrotrope, emulsifier, rust inhibitor, EP lubricant for alkaline detergents, metal cleaners/lubricants, hard surf. cleaners, textile scours/lubricants, emulsion polymerization, agric., and drycleaning formulations; paste.

Prote-pon TD-09-K30. [Protex] Alkyl ether potassium phosphate; wetting agent, detergent, hydrotrope, emulsifier, rust inhibitor, EP lubricant for alkaline detergents, metal cleaners/lubricants, hard surf. cleaners, textile scours/lubricants, emulsion polymerization, agric., and drycleaning formulations; liq.; pH 7.0 ± 0.3.

Prote-sorb SML. [Protex] Sorbitan laurate; CAS 1338-39-2; nonionic; emulsifier for food, cosmetic, household and industrial, agric., leather, metalworking, and textile industries; liq.; HLB 8.6; sapon. no. 162.

Prote-sorb SMO. [Protex] Sorbitan oleate; CAS 1338-43-8; nonionic; emulsifier for food, cosmetic, household and industrial, agric., leather, metalworking, and textile industries; liq.; HLB 4.3; sapon. no. 153.

Prote-sorb SMP. [Protex] Sorbitan palmitate; CAS 26266-57-9; nonionic; emulsifier for food, cosmetic, household and industrial, agric., leather, metalworking, and textile industries; solid; HLB 6.7; sapon. no. 155.

Prote-sorb SMS. [Protex] Sorbitan stearate; CAS 1338-41-6; nonionic; emulsifier for food, cosmetic, household and industrial, agric., leather, metalworking, and textile industries; solid; HLB 4.7; sapon. no. 152.

Prote-sorb STO. [Protex] Sorbitan trioleate; CAS 26266-58-0; nonionic; emulsifier for food, cosmetic, household and industrial, agric., leather, metalworking, and textile industries; liq.; HLB 1.8; sapon. no. 180.

Prote-sorb STS. [Protex] Sorbitan tristearate; CAS 26658-19-5; emulsifier for food, cosmetic, household and industrial, agric., leather, metalworking, and textile industries; solid; HLB 2.1; sapon. no. 182.

Protowet 100. [Nat'l. Starch] EO condensate; nonionic; wetting, rewetting and scouring agent used in textile industry; liq.; 100% conc.

Protowet 5171. [Nat'l. Starch] Sodium di(2-ethylhexyl) sulfosuccinate; anionic; textile wetting and rewetting and dyeing applics.; emulsifier for insol. fatty prods.; metal and glass cleaning, pigment disp., dry cleaning comps.; lt. visc. liq.

Protowet 5218. [Nat'l. Starch] Surfactant blend; anionic/nonionic; scouring agent; stable in up to 14% caustic soda; liq.; 45% conc.

Protowet 5799. [Nat'l. Starch] Surfactant blend; detergent, scouring agent with antiredeposition properties; liq.; 80% conc.

Protowet C. [Nat'l. Starch] Fatty acid ester, sulfonated; anionic; emulsifier, textile wet processing, dyeing and other applics.; preparation of wettable insecticide powds.; clear amber oil; dissolves in water; sol. in org. solvs.; 60% conc.

Protowet D-75. [Nat'l. Starch] Sodium di(2-ethylhexyl) sulfosuccinate; anionic; rewetting agent for textiles; 75% conc.

Protowet E-4. [Nat'l. Starch] EO condensate; nonionic; low foaming penetrant, dyeing assistant, stabilizer, wetting, nonrewetting, emulsifying, scouring agent used in textiles; liq.; 25% conc.

Protowet MB. [Nat'l. Starch] Ethoxylated alcohol sulfates; anionic; wetting, leveling and dyeing assistant; biodeg.; liq.; 68% conc.

Protowet NBF. [Nat'l. Starch] Surfactant blend; anionic; emulsifier, scouring agent, redeposition props.; liq.; 50% conc.

Protowet XL. [Nat'l. Starch] Sulfonated fatty acid ester; anionic; wetting and rewetting agent for dyeing and finishing; emulsifier for solvs.; amber clear oil; dissolves in water; sol. in org. solvs.; 62% act.

Provol 50. [Croda Inc.] PPG-50 oleyl ether; emollient, superfatting agent, lubricant, pigment dispersant, coupler for personal care prods., ethnic hair prods.; aids spreading and pigment dispersion in make-up systems; Gardner 10 max. liq.; sol. in acetone, IPM, lanolin oil; insol. in water; pH 5.5–7.5 (3% aq.).; usage level: 5-20%.

Prox-onic 2EHA-1/02. [Protex] PEG-2 2-ethylhexyl ether; detergent, wetting agent, emulsifier, dispersant, solubilizer, defoamer for textiles, metal cleaners, industrial, institutional and household cleaners, hand cleaners; liq.; HLB 8.0.

Prox-onic 2EHA-1/05. [Protex] PEG-5 2-ethylhexyl ether; detergent, wetting agent, emulsifier, dispersant, solubilizer, defoamer for textiles, metal cleaners, industrial, institutional and household cleaners, hand cleaners; liq.; HLB 12.6; cloud pt. < 25 C (1% aq.).

Prox-onic CC-05. [Protex] PEG-5 cocoate; emulsifier, lubricant additive for metalworking, textiles, cosmetics, defoamers; visc. control agent in plastisols; liq.; sapon. no. 135.

Prox-onic CC-09. [Protex] PEG-9 cocoate; emulsifier, lubricant additive for metalworking, textiles, cosmetics, defoamers; visc. control agent in plastisols; liq.; sapon. no. 90.

Prox-onic CC-014. [Protex] PEG-14 cocoate; emulsifier, lubricant additive for metalworking, textiles, cosmetics, defoamers; visc. control agent in plastisols; liq.; sapon. no. 68.

Prox-onic CSA-1/04. [Protex] Ceteareth-4; CAS 68439-49-6; emulsifier, emulsion stabilizer, detergent, wetting agent, dispersant, solubilizer, defoamer, dye assistant, leveling agent for cosmetics, textiles, metal cleaners, industrial, institutional and household cleaners, emulsion polymerization; solid; HLB 8.0.

Prox-onic CSA-1/06. [Protex] Ceteareth-6; CAS 68439-49-6; see Prox-onic CSA-1/04; solid; HLB 10.1.

Prox-onic CSA-1/010. [Protex] Ceteareth-10; CAS 68439-49-6; see Prox-onic CSA-1/04; solid; HLB 12.4; cloud pt. 73 C (1% aq.).

Prox-onic CSA-1/015. [Protex] Ceteareth-15; CAS 68439-49-6; see Prox-onic CSA-1/04; paste; HLB 14.3; cloud pt. 95-100 C (1% aq.).

Prox-onic CSA-1/020. [Protex] Ceteareth-20; CAS 68439-49-6; see Prox-onic CSA-1/04; solid; cloud pt. > 100 C (1% aq.).

Prox-onic CSA-1/030. [Protex] Ceteareth-30; CAS 68439-49-6; see Prox-onic CSA 1/04; solid; cloud pt. > 100 C (1% aq.).

Prox-onic CSA-1/050. [Protex] Ceteareth-50; CAS 68439-49-6; see Prox-onic CSA-1/04; solid; HLB 16.9; cloud pt. < 100 C (1% aq.).

Prox-onic DA-1/04. [Protex] Deceth-4; wetting agent, emulsifier, detergent, dispersant, solubilizer, defoamer for textiles, metal cleaners, industrial, institutional and household cleaners; liq.; HLB 10.5; cloud pt. < 25 C (1% aq.).

Prox-onic DA-1/06. [Protex] Deceth-6; see Prox-onic DA-1/04; liq.; HLB 12.4; cloud pt. 41 C (1% aq.).

Prox-onic DA-1/09. [Protex] Deceth-9; see Prox-onic DA-1/04; liq.; HLB 14.3; cloud pt. 80 C (1% aq.).

Prox-onic DDP-09. [Protex] Dodoxynol-9; detergent, wetting agent for industrial and heavy-duty detergents; liq.; cloud pt. 18 C (1% aq.).

Prox-onic DDP-012. [Protex] Dodoxynol-12; detergent, wetting agent for industrial and heavy-duty detergents; liq.; cloud pt. 76 C (1% aq.).

Prox-onic DNP-08. [Protex] Nonyl nonoxynol-8; CAS 9014-93-1; nonionic; emulsifier, detergent, wetting agent, dispersant, solubilizer, coupling agent for textiles, metalworking, household, industrial, agric., paper, paint and other industries; liq.; HLB 10.4; cloud pt. < 25 C (1% aq.).

Prox-onic DNP-0150. [Protex] Nonyl nonoxynol-150;

CAS 9014-93-1; nonionic; emulsifier, detergent, wetting agent, dispersant, solubilizer, coupling agent for textiles, metalworking, household, industrial, agric., paper, paint and other industries; solid; HLB 19.0; cloud pt. > 100 C (1% aq.).

Prox-onic DNP-0150/50. [Protex] Nonyl nonoxynol-150; CAS 9014-93-1; nonionic; emulsifier, detergent, wetting agent, dispersant, solubilizer, coupling agent for textiles, metalworking, household, industrial, agric., paper, paint and other industries; liq.; HLB 19.0; cloud pt. > 100 C (1% aq.).

Prox-onic DT-03. [Protex] PEG-3 tallow diamine; cationic; surfactant, emulsifier, lubricant additive, antistat, detergent for textile, metal, plastics, dyeing assistants, degreasers, corrosion inhibitor, agric.; intermediate for quats.; liq.; m.w. 475; HLB 5.5.

Prox-onic DT-015. [Protex] PEG-15 tallow diamine; cationic; surfactant, emulsifier, lubricant additive, antistat, detergent for textile, metal, plastics, dyeing assistants, degreasers, corrosion inhibitor, agric.; intermediate for quats.; liq.; m.w. 1020; HLB 13.0.

Prox-onic DT-030. [Protex] PEG-30 tallow diamine; cationic; surfactant, emulsifier, lubricant additive, antistat, detergent for textile, metal, plastics, dyeing assistants, degreasers, corrosion inhibitor, agric.; intermediate for quats.; liq.; m.w. 1665; HLB 15.9.

Prox-onic EP 1090-1. [Protex] Difunctional block polymer ending in primary hydroxyl groups; nonionic; defoamer for metalworking, cosmetic, paper, textiles; base for low foaming surfactants, antifoams, dishwash, dispersing and wetting agents for paints, drilling muds, emulsifiers, petrol. demulsifiers, emulsion polymerization; liq.; m.w. 2000; HLB 3.0; cloud pt. 24 C (1% aq.); 100% act.

Prox-onic EP 1090-2. [Protex] Difunctional block polymer ending in primary hydroxyl groups; nonionic; defoamer for metalworking, cosmetic, paper, textiles; base for low foaming surfactants, antifoams, dishwash, dispersing and wetting agents for paints, drilling muds, emulsifiers, petrol. demulsifiers, emulsion polymerization; liq.; m.w. 2600; HLB 6.5; cloud pt. 28 C (1% aq.); 100% act.

Prox-onic EP 2080-1. [Protex] Difunctional block polymer ending in primary hydroxyl groups; nonionic; defoamer for metalworking, cosmetic, paper, textiles; base for low foaming surfactants, antifoams, dishwash, dispersing and wetting agents for paints, drilling muds, emulsifiers, petrol. demulsifiers, emulsion polymerization; liq.; m.w. 2500; HLB 7.0; cloud pt. 30 C (1% aq.); 100% act.

Prox-onic EP 4060-1. [Protex] Difunctional block polymer ending in primary hydroxyl groups; nonionic; defoamer for metalworking, cosmetic, paper, textiles; base for low foaming surfactants, antifoams, dishwash, dispersing and wetting agents for paints, drilling muds, emulsifiers, petrol. demulsifiers, emulsion polymerization; liq.; m.w. 3000; HLB 1.0; cloud pt. 16 C (1% aq.); 100% act.

Prox-onic HR-05. [Protex] PEG-5 castor oil; nonionic; emulsifier, pigment dispersant, leveling agent, softener, rewetting agent, degreaser, antistat, emulsion stabilizer, lubricant for leather, paint, paper, plastics, textile, and cosmetics industries; solubilizer for perfumes; liq.; HLB 3.8; sapon. no. 145.

Prox-onic HR-016. [Protex] PEG-16 castor oil; nonionic; see Prox-onic HR-05; liq.; HLB 8.6; sapon. no. 100.

Prox-onic HR-025. [Protex] PEG-25 castor oil; nonionic; see Prox-onic HR-05; liq.; HLB 10.8; sapon. no. 80.

Prox-onic HR-030. [Protex] PEG-30 castor oil; nonionic; see Prox-onic HR-05; liq.; HLB 11.7; sapon. no. 73.

Prox-onic HR-036. [Protex] PEG-36 castor oil; nonionic; see Prox-onic HR-05; liq.; HLB 12.6; sapon. no. 68.

Prox-onic HR-040. [Protex] PEG-40 castor oil; nonionic; see Prox-onic HR-05; liq.; HLB 12.9; sapon. no. 61.

Prox-onic HR-080. [Protex] PEG-80 castor oil; nonionic; see Prox-onic HR-05; solid; HLB 15.8; sapon. no. 34.

Prox-onic HR-0200. [Protex] PEG-200 castor oil; nonionic; see Prox-onic HR-05; liq.; HLB 18.1; sapon. no. 16.

Prox-onic HR-0200/50. [Protex] PEG-200 castor oil; nonionic; see Prox-onic HR-05; liq.; HLB 18.1; sapon. no. 16; 50% act.

Prox-onic HRH-05. [Protex] PEG-5 hydrogenated castor oil; CAS 61788-85-0; nonionic; emulsifier, pigment dispersant, leveling agent, softener, rewetting agent, degreaser, antistat, emulsion stabilizer, lubricant for leather, paint, paper, plastics, textile, and cosmetics industries; solubilizer for perfumes; liq.; HLB 3.8; sapon. no. 142.

Prox-onic HRH-016. [Protex] PEG-16 hydrogenated castor oil; CAS 61788-85-0; nonionic; see Prox-onic HRH-05; liq.; HLB 8.6; sapon. no. 100.

Prox-onic HRH-025. [Protex] PEG-25 hydrogenated castor oil; CAS 61788-85-0; nonionic; see Prox-onic HRH-05; liq.; HLB 10.8; sapon. no. 80.

Prox-onic HRH-0200. [Protex] PEG-200 hydrogenated castor oil; CAS 61788-85-0; nonionic; see Prox-onic HRH-05; liq.; HLB 18.1; sapon. no. 17.

Prox-onic HRH-0200/50. [Protex] PEG-200 hydrogenated castor oil; CAS 61788-85-0; nonionic; see Prox-onic HRH-05; liq.; HLB 18.1; sapon. no. 17; 50% act.

Prox-onic L 081-05. [Protex] POE (5) linear alcohol ether; biodeg. low foam detergent, wetting agent, emulsifier for household, agric. and industrial cleaners; coupling agent and solubilizer for perfumes and org. additives.

Prox-onic L 101-05. [Protex] POE (5) linear alcohol ether; biodeg. low foam detergent, wetting agent, emulsifier for household, agric. and industrial cleaners; coupling agent and solubilizer for perfumes and org. additives.

Prox-onic L 102-02. [Protex] POE (2) linear alcohol ether; biodeg. low foam detergent, wetting agent, emulsifier for household, agric. and industrial cleaners; coupling agent and solubilizer for perfumes and org. additives.

Prox-onic L 121-09. [Protex] POE (9) linear alcohol ether; biodeg. low foam detergent, wetting agent, emulsifier for household, agric. and industrial cleaners; coupling agent and solubilizer for perfumes and org. additives.

Prox-onic L 161-05. [Protex] POE (5) linear alcohol ether; biodeg. low foam detergent, wetting agent, emulsifier for household, agric. and industrial cleaners; coupling agent and solubilizer for perfumes and org. additives.

Prox-onic L 181-05. [Protex] POE (5) linear alcohol ether; biodeg. low foam detergent, wetting agent, emulsifier for household, agric. and industrial cleaners; coupling agent and solubilizer for perfumes and org. additives.

Prox-onic L 201-02. [Protex] POE (2.5) linear alcohol ether; biodeg. low foam detergent, wetting agent, emulsifier for household, agric. and industrial cleaners; coupling agent and solubilizer for perfumes and org. additives.

Prox-onic LA-1/02. [Protex] Laureth-2; coupling agent, solubilizer, emulsion stabilizer for cosmetic and hair care prods.; with anionic surfactants for emulsion polymerization; in coning and textile spin finishes; liq.; HLB 6.4.

Prox-onic LA-1/04. [Protex] Laureth-4; see Prox-onic LA-1/02; liq.; HLB 9.2; cloud pt. 52 C (1% aq.).

Prox-onic LA-1/09. [Protex] Laureth-9; see Prox-onic LA-1/02; liq.; HLB 13.3; cloud pt. 73-76 C (1% aq.).

Prox-onic LA-1/012. [Protex] Laureth-12; see Prox-onic LA-1/02; solid; HLB 14.5; cloud pt. < 100 C (1% aq.).

Prox-onic LA-1/023. [Protex] Laureth-23; see Prox-onic LA-1/02; solid; HLB 16.7.

Prox-onic MC-02. [Protex] PEG-2 cocamine; CAS 61791-14-8; cationic; surfactant, emulsifier, lubricant additive, antistat, detergent for textile, metal, plastics, dyeing assistants, degreasers, corrosion inhibitor, agric.; intermediate for quats.; liq.; m.w. 290; HLB 6.1.

Prox-onic MC-05. [Protex] PEG-5 cocamine; CAS 61791-14-8; cationic; surfactant, emulsifier, lubricant additive, antistat, detergent for textile, metal, plastics, dyeing assistants, degreasers, corrosion inhibitor, agric.; intermediate for quats.; liq.; m.w. 425; HLB 10.4.

Prox-onic MC-015. [Protex] PEG-15 cocamine; CAS 61791-14-8; cationic; surfactant, emulsifier, lubricant additive, antistat, detergent for textile, metal, plastics, dyeing assistants, degreasers, corrosion inhibitor, agric.; intermediate for quats.; liq.; m.w. 890; HLB 15.0.

Prox-onic MHT-05. [Protex] PEG-5 hydrog. tallow amine; cationic; surfactant, emulsifier, lubricant additive, antistat, detergent for textile, metal, plastics, dyeing assistants, degreasers, corrosion inhibitor, agric.; intermediate for quats.; paste; m.w. 495; HLB 9.0.

Prox-onic MHT-015. [Protex] PEG-15 hydrog. tallow amine; cationic; surfactant, emulsifier, lubricant additive, antistat, detergent for textile, metal, plastics, dyeing assistants, degreasers, corrosion inhibitor, agric.; intermediate for quats.; liq.; m.w. 925; HLB 14.3.

Prox-onic MO-02. [Protex] PEG-2 oleamine; cationic; surfactant, emulsifier, lubricant additive, antistat, detergent for textile, metal, plastics, dyeing assistants, degreasers, corrosion inhibitor, agric.; intermediate for quats.; liq.; m.w. 358; HLB 4.1.

Prox-onic MO-015. [Protex] PEG-15 oleamine; cationic; surfactant, emulsifier, lubricant additive, antistat, detergent for textile, metal, plastics, dyeing assistants, degreasers, corrosion inhibitor, agric.; intermediate for quats.; liq.; m.w. 930; HLB 14.2.

Prox-onic MO-030. [Protex] PEG-30 oleamine; cationic; surfactant, emulsifier, lubricant additive, antistat, detergent for textile, metal, plastics, dyeing assistants, degreasers, corrosion inhibitor, agric.; intermediate for quats.; liq.; m.w. 1600; HLB 16.5.

Prox-onic MO-030-80. [Protex] PEG-30 oleamine; cationic; surfactant, emulsifier, lubricant additive, antistat, detergent for textile, metal, plastics, dyeing assistants, degreasers, corrosion inhibitor, agric.; intermediate for quats.; 80% act.

Prox-onic MS-02. [Protex] PEG-2 stearamine; cationic; surfactant, emulsifier, lubricant additive, antistat, detergent for textile, metal, plastics, dyeing assistants, degreasers, corrosion inhibitor, agric.; intermediate for quats.; solid; m.w. 353; HLB 6.9.

Prox-onic MS-05. [Protex] PEG-5 stearamine; CAS 26635-92-7; cationic; surfactant, emulsifier, lubricant additive, antistat, detergent for textile, metal, plastics, dyeing assistants, degreasers, corrosion inhibitor, agric.; intermediate for quats.; solid; m.w. 485; HLB 9.0.

Prox-onic MS-011. [Protex] PEG-11 stearamine; CAS 26635-92-7; cationic; surfactant, emulsifier, lubricant additive, antistat, detergent for textile, metal, plastics, dyeing assistants, degreasers, corrosion inhibitor, agric.; intermediate for quats.; liq.; m.w. 750; HLB 12.9.

Prox-onic MS-050. [Protex] PEG-50 stearamine; CAS 26635-92-7; cationic; surfactant, emulsifier, lubricant additive, antistat, detergent for textile, metal, plastics, dyeing assistants, degreasers, corrosion inhibitor, agric.; intermediate for quats.; solid; m.w. 2465; HLB 17.8.

Prox-onic MT-02. [Protex] PEG-2 tallow amine; cationic; surfactant, emulsifier, lubricant additive, antistat, detergent for textile, metal, plastics, dyeing assistants, degreasers, corrosion inhibitor, agric.; intermediate for quats.; paste; m.w. 350; HLB 5.0.

Prox-onic MT-05. [Protex] PEG-5 tallow amine; cationic; surfactant, emulsifier, lubricant additive, antistat, detergent for textile, metal, plastics, dyeing assistants, degreasers, corrosion inhibitor, agric.; intermediate for quats.; liq.; m.w. 490; HLB 9.0.

Prox-onic MT-015. [Protex] PEG-15 tallow amine; cationic; surfactant, emulsifier, lubricant additive, antistat, detergent for textile, metal, plastics, dyeing assistants, degreasers, corrosion inhibitor, agric.; intermediate for quats.; liq.; m.w. 930; HLB 14.3.

Prox-onic MT-020. [Protex] PEG-20 tallow amine; cationic; surfactant, emulsifier, lubricant additive, antistat, detergent for textile, metal, plastics, dyeing assistants, degreasers, corrosion inhibitor, agric.; intermediate for quats.; liq.; m.w. 1150; HLB 15.7.

Prox-onic NP-1.5. [Protex] Nonoxynol-1; nonionic; defoamer, detergent, emulsifier; liq.; oil-sol.; HLB 4.6.

Prox-onic NP-04. [Protex] Nonoxynol-4; CAS 9016-45-9; nonionic; surfactant, detergent, defoamer; liq.; HLB 8.9.

Prox-onic NP-06. [Protex] Nonoxynol-6; CAS 9016-45-9; nonionic; emulsifier, detergent; liq.; HLB 10.9; cloud pt. < 25 C (1% aq.).

Prox-onic NP-09. [Protex] Nonoxynol-9; CAS 9016-45-9; nonionic; detergent, wetting agent, emulsifier for textile, paper, metal cleaning industries; liq.; HLB 13.0; cloud pt. 54 C (1% aq.).

Prox-onic NP-010. [Protex] Nonoxynol-10; CAS 9016-45-9; nonionic; detergent, wetting agent, emulsifier for textile, paper, metal cleaning industries; liq.; HLB 13.5; cloud pt. 72 C (1% aq.).

Prox-onic NP-015. [Protex] Nonoxynol-15; CAS 9016-45-9; nonionic; detergent, wetting agent at elevated temps. and electrolyte; for metal cleaning; paste; HLB 15.0; cloud pt. 96 C (1% aq.).

Prox-onic NP-020. [Protex] Nonoxynol-20; CAS 9016-45-9; nonionic; detergent, wetting agent at elevated temps. and electrolyte; solid; HLB 16.0; cloud pt. > 100 C (1% aq.).

Prox-onic NP-030. [Protex] Nonoxynol-30; CAS 9016-45-9; nonionic; emulsifier for fats, oils, and waxes; solid; HLB 17.1; cloud pt. > 100 C (1% aq.).

Prox-onic NP-030/70. [Protex] Nonoxynol-30; CAS 9016-45-9; nonionic; emulsifier for fats, oils, and waxes; liq.; HLB 17.1; cloud pt. > 100 C (1% aq.); 70% act.

Prox-onic NP-040. [Protex] Nonoxynol-40; CAS 9016-45-9; nonionic; surfactant, emulsifier for high temps. and electrolyte use; solid; HLB 17.8; cloud pt. > 100 C (1% aq.).

Prox-onic NP-040/70. [Protex] Nonoxynol-40; CAS 9016-45-9; nonionic; surfactant, emulsifier for high temps. and electrolytes; liq.; HLB 17.8; cloud pt. > 100 C (1% aq.); 70% act.

Prox-onic NP-050. [Protex] Nonoxynol-50; CAS 9016-45-9; nonionic; surfactant for high temps. and electrolyte; emulsifier for waxes and polishes; solid; HLB 18.2; cloud pt. > 100 C (1% aq.).

Prox-onic NP-050/70. [Protex] Nonoxynol-50; CAS 9016-45-9; nonionic; surfactant for high temps. and electrolyte; emulsifier for waxes and polishes; liq.; HLB 18.2; cloud pt. > 100 C (1% aq.); 70% act.

Prox-onic NP-0100. [Protex] Nonoxynol-100; CAS 9016-45-9; nonionic; surfactant for high temps. and electrolyte; emulsifier for waxes and polishes; solid; HLB 19.0; cloud pt. > 100 C (1% aq.).

Prox-onic NP-0100/70. [Protex] Nonoxynol-100; CAS 9016-45-9; nonionic; surfactant for high temps. and electrolyte; emulsifier for waxes and polishes; liq.; HLB 19.0; cloud pt. > 100 C (1% aq.); 70% act.

Prox-onic OA-1/04. [Protex] Oleth-4; CAS 9004-98-2; nonionic; coupling agent, solubilizer, emulsion stabilizer for cosmetic and hair care prods.; with anionic surfactants for emulsion polymerization; in coning and textile spin finishes; liq.; HLB 7.9.

Prox-onic OA-1/09. [Protex] Oleth-9; CAS 9004-98-2; nonionic; see Prox-onic OA-1/04; liq.; HLB 11.9; cloud pt. 52 C (1% aq.).

Prox-onic OA-1/020. [Protex] Oleth-20; CAS 9004-98-2; nonionic; emulsifier for min. oils, fatty acids, waxes; for polishes, cosmetics, polyethylene aq. disps.; stabilizer for rubber latex; solubilizer/emulsifier for essential oils, pharmaceuticals; wetting agent in metal cleaners; dyeing assistant; dyestuff dispersant for leather; solid; HLB 15.3; cloud pt. > 100 C (1% aq.).

Prox-onic OA-2/020. [Protex] Oleth-20; CAS 9004-98-2; nonionic; emulsifier for min. oils, fatty acids, waxes; for polishes, cosmetics, polyethylene aq. disps.; stabilizer for rubber latex; solubilizer/emulsifier for essential oils, pharmaceuticals; wetting agent in metal cleaners; dyeing assistant; dyestuff dispersant for leather; liq.; HLB 15.3; cloud pt. > 100 C (1% aq.).

Prox-onic OCA-1/06. [Protex] Cetoleth-6; detergent, wetting agent, emulsifier, dispersant, solubilizer, defoamer for textiles, metal cleaners, industrial, institutional and household cleaners, hand cleaners; paste; HLB 10.7.

Prox-onic OL-1/05. [Protex] PEG-5 oleate; CAS 9004-96-0; surfactant for cutting oils, degreasing solvs., metal cleaners, textiles, leather, cosmetics; dyeing assistant; emulsifier for min. oils, fatty oils; liq.; sapon. no. 118.

Prox-onic OL-1/09. [Protex] PEG-9 oleate; CAS 9004-96-0; see Prox-onic OL-1/05; liq.; sapon. no. 85.

Prox-onic OL-1/014. [Protex] PEG-14 oleate; CAS 9004-96-0; see Prox-onic OL-1/05; liq.; sapon. no. 65.

Prox-onic OP-09. [Protex] Octoxynol-9; CAS 9002-93-1; nonionic; emulsifier for metal, textile processing, household and industrial cleaners, vinyl and acrylic polymerization; liq.; HLB 13.5; cloud pt. 65 C (1% aq.).

Prox-onic OP-016. [Protex] Octoxynol-16; CAS 9002-93-1; nonionic; emulsifier, detergent, wetting agent; for hot cleaning and pickling systems, hard surf. cleaners, household and industrial cleaners; paste; HLB 15.8; cloud pt. > 100 C (1% aq.).

Prox-onic OP-030/70. [Protex] Octoxynol-30; CAS 9002-93-1; nonionic; coemulsifier for vinyl and acrylic polymerization; dye assistant; liq.; HLB 17.3; cloud pt. > 100 C (1% aq.); 70% act.

Prox-onic OP-040/70. [Protex] Octoxynol-40; CAS 9002-93-1; nonionic; coemulsifier for vinyl and acrylic polymerization; dye assistant; liq.; HLB 17.9; cloud pt. > 100 C (1% aq.); 70% act.

Prox-onic PEG-2000. [Protex] PEG-2M; low foam wetting in paper pulping; emulsifier for metal degreasing; bottle cleaner defoamer; binder for tobacco; polyurethane mfg.; mold release agent; agric.; m.w. 2000; solid. pt. 48-52 C; hyd. no. 51-62.

Prox-onic PEG-4000. [Protex] PEG-4M; low foam wetting in paper pulping; emulsifier for metal degreasing; bottle cleaner defoamer; binder for ceramics, tobacco; in depilatories; agric.; m.w. 4000; solid. pt. 53-58 C; hyd. no. 25-30.

Prox-onic PEG-6000. [Protex] PEG-6M; low foam wetting in paper pulping; emulsifier for metal degreasing; bottle cleaner defoamer; dispersant, binder for ceramics; plasticizer for cement; anticaking agent for fertilizer; soap molding; in depilatories; agric.; m.w. 6000; solid. pt. 55-60 C; hyd. no. 16-20.

Prox-onic PEG-10,000. [Protex] PEG-10M; low foam wetting in paper pulping; emulsifier for metal degreasing; plasticizer for cement; ceramics binder; bottle cleaning defoamer; agric.; m.w. 10,000; solid. pt. 55-60 C; hyd. no. 9-12.

Prox-onic PEG-20,000. [Protex] PEG-20M; low foam wetting in paper pulping; emulsifier for metal degreasing; bottle cleaner defoamer; plasticizer for cement; ceramics binder; mold release for rubber; agric.; m.w. 20,000; solid. pt. 60 C; hyd. no. 7-11.

Prox-onic PEG-35,000. [Protex] PEG-35M; low foam wetting in paper pulping; emulsifier for metal degreasing; bottle cleaner defoamer; plasticizer for cement; ceramics binder; agric.; m.w. 35,000; solid. pt. 60 C; hyd. no. < 7.

Prox-onic PPG-900. [Protex] PPG (900); agric. emulsifiable concs.; m.w. 900; hyd. no. 112-134.

Prox-onic PPG-1800. [Protex] PPG (1800); agric. emulsifiable concs.; m.w. 1800; hyd. no. 55-65.

Prox-onic PPG-4000. [Protex] PPG (4000); agric. emulsifiable concs.; m.w. 4000; hyd. no. 25-30.

Prox-onic SA-1/02. [Protex] Steareth-2; CAS 9005-00-9; nonionic; detergent, wetting agent, emulsifier, dispersant, solubilizer, defoamer for textiles, metal cleaners, industrial, institutional and household cleaners, hand cleaners; solid; HLB 4.9; cloud pt. 57-61 C (1% aq.).

Prox-onic SA-1/010. [Protex] Steareth-10; CAS 9005-00-9; nonionic; see Prox-onic SA-1/02; solid; HLB 12.4; cloud pt. 60-64 C (1% aq.).

Prox-onic SA1-015/P. [Protex] POP (15) stearyl alcohol; surfactant; liq.

Prox-onic SA-1/020. [Protex] Steareth-20; CAS 9005-00-9; nonionic; see Prox-onic SA-1/020; solid; HLB 15.3; cloud pt. 73-77 C (1% aq.).

Prox-onic SML-020. [Protex] Polysorbate 20; CAS 9005-64-5; nonionic; emulsifier, solubilizer for petrol. oils, solvs., veg. oils, waxes, silicones, etc.;

for agric., cosmetic, leather, metalworking and textile industries; liq.; HLB 16.7; sapon. no. 45.

Prox-onic SMO-05. [Protex] Polysorbate 81; CAS 9005-65-6; emulsifier, solubilizer for petrol. oils, solvs., veg. oils, waxes, silicones, etc.; for agric., cosmetic, leather, metalworking and textile industries; liq.; HLB 10.0; sapon. no. 100.

Prox-onic SMO-020. [Protex] Polysorbate 80; CAS 9005-65-6; emulsifier, solubilizer for petrol. oils, solvs., veg. oils, waxes, silicones, etc.; for agric., cosmetic, leather, metalworking and textile industries; liq.; HLB 15.0; sapon. no. 50.

Prox-onic SMP-020. [Protex] Polysorbate 40; CAS 9005-66-7; nonionic; emulsifier, solubilizer for petrol. oils, solvs., veg. oils, waxes, silicones, etc.; for agric., cosmetic, leather, metalworking and textile industries; liq.; HLB 15.6; sapon. no. 46.

Prox-onic SMS-020. [Protex] Polysorbate 60; CAS 9005-67-8; emulsifier, solubilizer for petrol. oils, solvs., veg. oils, waxes, silicones, etc.; for agric., cosmetic, leather, metalworking and textile industries; soft paste; HLB 14.9; sapon. no. 50.

Prox-onic ST-05. [Protex] PEG-5 stearate; CAS 9004-99-3; emulsifier, lubricant additive for metalworking, textiles, cosmetics, defoamers; visc. control agent in plastisols; paste; sapon. no. 112.

Prox-onic ST-09. [Protex] PEG-9 stearate; CAS 9004-99-3; emulsifier, lubricant additive for metalworking, textiles, cosmetics, defoamers; visc. control agent in plastisols; paste; sapon. no. 87.

Prox-onic ST-014. [Protex] PEG-14 stearate; CAS 9004-99-3; emulsifier, lubricant additive for metalworking, textiles, cosmetics, defoamers; visc. control agent in plastisols; paste; sapon. no. 62.

Prox-onic ST-023. [Protex] PEG-23 stearate; CAS 9004-99-3; emulsifier, lubricant additive for metalworking, textiles, cosmetics, defoamers; visc. control agent in plastisols; solid; sapon. no. 43.

Prox-onic STO-020. [Protex] Polysorbate 85; CAS 9005-70-3; nonionic; emulsifier, solubilizer for petrol. oils, solvs., veg. oils, waxes, silicones, etc.; for agric., cosmetic, leather, metalworking and textile industries; soft paste; HLB 11.0; sapon. no. 88.

Prox-onic STS-020. [Protex] Polysorbate 65; CAS 9005-71-4; nonionic; emulsifier, solubilizer for petrol. oils, solvs., veg. oils, waxes, silicones, etc.; for agric., cosmetic, leather, metalworking and textile industries; solid; HLB 10.5; sapon. no. 93.

Prox-onic TA-1/08. [Protex] PEG-8 tallate; CAS 61791-00-2; emulsifier, detergent for degreasers and neutral or mildly alkaline detergents; liq.; HLB 10.5; cloud pt. < 25 C (1% aq.).

Prox-onic TA-1/010. [Protex] PEG-10 tallate; CAS 61791-00-2; emulsifier, detergent for degreasers and neutral or mildly alkaline detergents; liq.; HLB 11.5; cloud pt. 25 C (1% aq.).

Prox-onic TA-1/016. [Protex] PEG-16 tallate; CAS 61791-00-2; emulsifier, detergent for degreasers and neutral or mildly alkaline detergents; liq.; HLB 13.4; cloud pt. 70 C (1% aq.).

Prox-onic TBP-08. [Protex] POE (8) tert. butyl phenol; emulsifier, detergent, wetting agent, dispersant, solubilizer, coupling agent for textiles, metalworking, household, industrial, agric., paper, paint and other industries.

Prox-onic TBP-030. [Protex] POE (30) tert. butyl phenol; emulsifier, detergent, wetting agent, dispersant, solubilizer, coupling agent for textiles, metalworking, household, industrial, agric., paper, paint and other industries.

Prox-onic TD-1/03. [Protex] Trideceth-3; CAS 24938-91-8; nonionic; intermediate for surfactants; emulsifier, detergent, foam builder, dispersant, wetting agent, solubilizer for mech. dishwash, alkaline cleaners, pulp and paper, textiles, corrosion inhibition; liq.; HLB 7.9.

Prox-onic TD-1/06. [Protex] Trideceth-6; CAS 24938-91-8; nonionic; see Prox-onic TD-1/03; liq.; HLB 11.4.

Prox-onic TD-1/09. [Protex] Trideceth-9; CAS 24938-91-8; nonionic; see Prox-onic TD-1/03; liq.; HLB 13.0; cloud pt. 54 C (1% aq.).

Prox-onic TD-1/012. [Protex] Trideceth-12; CAS 24938-91-8; nonionic; see Prox-onic TD-1/03; paste; HLB 14.5; cloud pt. 70 C (1% aq.).

Prox-onic TM-06. [Protex] POE (6) isolauryl mercaptan; wetting agent, surfactant, detergent for metal cleaning, household and industrial cleaners/scours, degreasers; emulsifier for herbicides/insecticides, greases, soils; liq.; HLB 11.0.

Prox-onic TM-08. [Protex] POE (8) isolauryl mercaptan; wetting agent, surfactant, detergent for metal cleaning, household and industrial cleaners/scours, degreasers; emulsifier for herbicides/insecticides, greases, soils; liq.; HLB 12.7; cloud pt. 28 C (1% aq.).

Prox-onic TM-010. [Protex] POE (10) isolauryl mercaptan; wetting agent, surfactant, detergent for metal cleaning, household and industrial cleaners/scours, degreasers; emulsifier for herbicides/insecticides, greases, soils; liq.; HLB 13.9; cloud pt. 52 C (1% aq.).

Prox-onic UA-03. [Protex] Isoundeceth-3; detergent, wetting agent, emulsifier, dispersant, solubilizer, defoamer for textiles, metal cleaners, industrial, institutional and household cleaners, hand cleaners; liq.; cloud pt. < 100 C (1% aq.).

Prox-onic UA-06. [Protex] Isoundeceth-6; see Prox-onic UA-03; liq.

Prox-onic UA-09. [Protex] Isoundeceth-9; see Prox-onic UA-03; liq.

Prox-onic UA-012. [Protex] Isoundeceth-12; see Prox-onic UA-03; liq.

Purton CFD. [Zschimmer & Schwarz] Cocamide DEA; nonionic; foam stabilizer, thickener and superfatting agent for cosmetics, cleaners; liq.; 100% conc.

Purton SFD. [Zschimmer & Schwarz] Linoleamide DEA; nonionic; foam stabilizer, thickener and superfatting agent for cosmetics, cleaners; liq.; 100% conc.

Puxol CB-22. [Pulcra SA] TEA dodecylbenzene sulfonate; anionic; used for formulation of liq. detergents, personal care prods., car washing shampoos; emulsifier for emulsion polymerization reactions; pigment dispersant; liq.; 50 ± 1% conc.

Puxol FB-11. [Pulcra SA] Sodium dodecylbenzene sulfonate; anionic; scouring and wetting agent in the textile industry and emulsifier for emulsion polymerization in the rubber and plastic industry; paste; 65 ± 2% conc.

Puxol XB-10. [Pulcra SA] Dodecylbenzene sulfonic acid; anionic; intermediate for mfg. of detergents; liq.; 95% conc.

Pyronate® 40. [Witco/Sonneborn] Sodium petrol. sulfonate; anionic; wetting agent, dispersant; oil and froth flotation; dk. liq.; water-sol.; dens. 9.6 lb/gal; sp.gr. 1.18 (60 F); 42% act.

Pyroter CPI-40. [Ajinomoto; Nihon Emulsion] PEG-40 hydrog. castor oil PCA isostearate; nonionic;

emulsifier, solubilizer and thickener used in personal care prods. and detergents; low irritation, nontoxic; liq.; water-sol.; HLB 9.4; 100% conc.

Pyroter GPI-25. [Ajinomoto; Nihon Emulsion] Glycereth-25 PCA isostearate; nonionic; moisturizer, emulsifier, solubilizer, dispersant and emollient; liq.; water-sol.; HLB 13.4.

Q

Quadrilan® AT. [Harcros UK] Quat. fatty amine ethoxylate; cationic; surfactant, antistat used in PVC and other polymers, textile auxiliary, additive for detergents and rinse aids; lt. brn. clear visc. liq.; char. odor; water-sol.; sp.gr. 1.081; visc. 400 cs; flash pt. (COC) > 150 C; pour pt. < O C; pH 7.5 (1% aq.); biodeg.; 100% act.

Quadrilan® BC. [Harcros UK] Benzalkonium chloride BP grade; cationic; bactericide, fungicide, germicide in disinfectants and detergent sanitizers, emulsifier in the dyeing industry; biodeg.; pale yel. liq.; mild aromatic odor; sol. in water; sp.gr. 0.972; visc. 120 cs; flash pt. > 200 F (COC); pour pt. < O C; pH 7–8 (1% aq.); 50% act.

Quadrilan® MY211. [Harcros UK] Blend; cationic; surfactant for high pressure spraying soil removal, alkaline spray cleaning concs., vehicle cleaning conc.; antistat; improved rinsing; amber hazy liq., char. odor; sol. in water; sp.gr. 1.100; visc. 71 cst; pour pt. < 0 C; flash pt. (COC) > 100 C; pH 7.0 (1% aq.); 50% act.

Quadrilan® SK. [Harcros UK] Quat. fatty amine ethoxylate; cationic; antistat used in polymers; liq.; 100% conc.

Quartamin 86P. [Kao Corp. SA] N-Alkyl trimethyl ammonium chlorides; cationic; emulsifier, sanitizing agent, antistat, textile softener; liq.; biodeg.; 50% conc.

Quartamin CPR. [Kao Corp. SA] Coco trimethyl ammonium chloride; cationic; emulsifier; biodeg.; liq.; 50%.

Quartamin D86P, D86PL, D86PI. [Kao Corp. SA] Dihydrog. tallow dimethyl ammonium chloride; CAS 61789-80-8; cationic; emulsifier; paste/liq.; 75% conc.

Quartamin DCP. [Kao Corp. SA] Dicoco dimethyl ammonium chloride; cationic; emulsifier and corrosion inhibitor; liq.; 75% conc.

Quartamin HTPR. [Kao Corp. SA] Hydrog. tallow trimethyl ammonium chloride; cationic; emulsifier; biodeg.; liq.; 50% conc.

Quartamin TPR. [Kao Corp. SA] Tallow trimethyl ammonium chloride; cationic; emulsifier; biodeg.; liq.; 50% conc.

Quaternary O. [Ciba-Geigy] Quat. oleyl imidazoline; cationic; detergent, wetting agent, penetrant, antistat used in acids, solvs., germicides, fungistats, polishes, acid cleaners and corrosion inhibitor formulations, ore flotation, asphalt wetting; amber liq.; 100% conc.

Quatrene 7670. [Henkel] Fatty amidoamine quat. compd.; cationic; wetting agent, demulsifier, and corrosion inhibitor for petrol. prod.; liq.; 60% act.

Quatrene CE. [Henkel] Coco dialkyl benzyl ammonium chloride; cationic; wetting agent, demulsifier, and corrosion inhibitor for petrol. prod.; liq., paste; 75% act.

Quatrex 152. [Chemron] Quaternized imidazoline; emulsifier, stabilizer for oilfield applics.; amber liq.; 60% act.

Quatrex 162. [Chemron] Surfactant; cationic; wetting agent, additive for reverse emulsion breakers, oil treating; emulsion preventer in acids; clay stabilizer; amber liq. or solid; water-sol.; 75% act.

Quatrex 172. [Chemron] Surfactant; cationic; emulsion preventer for acidizing; emulsion breaker for oil treatment; intermediate for reverse emulsion breakers; dk. amber liq.; 60% act.

Quatrex 182. [Chemron] Quaternized imidazoline; cationic; corrosion inhibitor conc.; surfactant, solubilizer for other corrosion inhibitor components; oilfield applics.; emulsion preventer; clay stabilizer; dark visc. liq.; water-sol.; 60% act.

Querton 14Br-40. [Berol Nobel AB] Myrtrimonium bromide; CAS 1119-97-7; cationic; germicidal disinfectant, detergent for hospital and industrial cleaning and disinfection; water-wh. to yel. clear visc. liq., mild char. odor; misc. with water, ethanol, acetone; m.w. 336.4; dens. 1.00 kg/m^3; visc. 125 mPa•s; pH 5-7 (2% aq.); 39-41% act., 5-7% ethanol.

Querton 16Cl-29. [Berol Nobel AB] Cetrimonium chloride; CAS 112-02-7; cationic; emulsifier, dispersant, antistat for hair conditiners, shampoos, detergent sanitizers; clear liq.; sol. in water, alcohols, other polar solvs.; dens. 964 kg/m^3; visc. 20 mPa•s; pour pt. 2 C; pH 6.5 ± 1 (20%); surf. tens. 41 mN/m (0.1%); 29% act., 3% ethanol.

Querton 16Cl-50. [Berol Nobel AB] Cetrimonium chloride; CAS 112-02-7; cationic; emulsifier, dispersant, antistat for hair conditiners, shampoos, detergent sanitizers; clear liq.; sol. in water, alcohols, other polar solvs.; dens. 899 kg/m^3; visc. 30 mPa•s; pour pt. 6 C; pH 6.5 ± 1 (20%); surf. tens. 42 mN/m (0.1%); 50% act., 30-35% ethanol.

Querton 280. [Berol Nobel AB] Trimethyl alkyl ammonium chloride; CAS 68391-03-7; cationic; general purpose quat. surfactant; colorless to pale yel. clear liq.; misc. with water, lower alcohols (ethanol, IPA); m.w. 280; dens. 970 kg/m^3; visc. 30 mPa•s; solid. pt. -5 C; pH 5.5-8.5; 33-37% act. in water.

Querton 441-BC. [Berol Nobel AB] Hydrog. tallow dimethylbenzyl ammonium chloride; cationic; agent for organophilic bentonites prod.; liq.; 76% conc.

Querton KKBCl-50. [Berol Nobel AB] Cocoalkyl dimethylbenzyl ammonium chloride; CAS 61789-71-7; cationic; corrosion inhibitor for oil drilling;

emulsifier and dispersant for sludge in oil drilling and waste water treatment; wh. to lt. yel. clear visc. liq., mild char. odor; misc. with water, alcohols, acetone; m.w. 360; dens. 980 kg/m³; visc. 100 mPa•s; solid. pt. -4 C; pH 6.5-8.0 (1% aq.); 50-51% act.

Quimipol 9106B. [Quimigal-Quimica] Syn. primary alcohol ethoxylate; nonionic; coemulsifier for min. oil and waxes; emulsifier for alkyl resins; liq.; HLB 12.6; 100% conc.

Quimipol 9108. [Quimigal-Quimica] Syn. primary alcohol ethoxylate; nonionic; component for dishwashing liqs. and surfactants; liq.; HLB 13.8; 100% conc.

Quimipol DEA OC. [Quimigal-Quimica] Cocamide DEA; nonionic; foam booster/stabilizer, emulsifier, thickener; liq.; 85% conc.

Quimipol EA 2503. [Quimigal-Quimica] Syn. primary alcohol ethoxylate; nonionic; emulsifier for min. oils and aliphatic solvs.; visc. depressant for PVC plastisols; liq.; HLB 7.8; 100% conc.

Quimipol EA 2504. [Quimigal-Quimica] Syn. primary alcohol ethoxylate; nonionic; emulsifier for silicones and hydrocarbon solvs.; drycleaning assistant; liq.; HLB 9.8; 97% conc.

Quimipol EA 2506. [Quimigal-Quimica] Syn. primary alcohol ethoxylate; nonionic; wool scouring and hydrocarbon solv. emulsifier; wetting agent; liq.; HLB 11.4; 100% conc.

Quimipol EA 2507. [Quimigal-Quimica] Syn. primary alcohol ethoxylate; nonionic; wool scouring, antistat, detergent, wetting agent, emulsifier; liq.; HLB 12.2; 100% conc.

Quimipol EA 2508. [Quimigal-Quimica] Syn. primary alcohol ethoxylate; nonionic; emulsifier for hydrocarbon solvs.; paste; HLB 12.6; 100% conc.x.

Quimipol EA 2509. [Quimigal-Quimica] Syn. primary alcohol ethoxylate; nonionic; wetting agent and detergent in liq. formulations; wax emulsifier; solid; HLB 13.2; 100% conc.

Quimipol EA 2512. [Quimigal-Quimica] Syn. primary alcohol ethoxylate; nonionic; wetting agent at elevated temps.; cosmetic emulsifier; solid; HLB 14.2; 100% conc.

Quimipol EA 4505. [Quimigal-Quimica] Syn. primary alcohol ethoxylate; nonionic; clothes washing powds.; liq.; HLB 10.0; 100% conc.

Quimipol EA 4508. [Quimigal-Quimica] Syn. primary alcohol ethoxylate; nonionic; clothes washing powds.; solid; HLB 12.3; 100% conc.

Quimipol EA 6801. [Quimigal-Quimica] Syn. primary alcohol ethoxylate; nonionic; clothes washing powds.; solid; HLB 2.8; 100% conc.

Quimipol EA 6802. [Quimigal-Quimica] Syn. primary alcohol ethoxylate; nonionic; clothes washing powds.; solid; HLB 5.0; 100% conc.

Quimipol EA 6803. [Quimigal-Quimica] Syn. primary alcohol ethoxylate; nonionic; emulsifier and textile lubricant; solid; HLB 6.7; 100% conc.

Quimipol EA 6804. [Quimigal-Quimica] Syn. primary alcohol ethoxylate; nonionic; emulsifier and textile lubricant; solid; HLB 8.0; 100% conc.

Quimipol EA 6806. [Quimigal-Quimica] Syn. primary alcohol ethoxylate; nonionic; emulsifier and textile lubricant; solid; HLB 10.0; 100% conc.

Quimipol EA 6807. [Quimigal-Quimica] Syn. primary alcohol ethoxylate; nonionic; emulsifier and textile lubricant; solid; HLB 11.0; 100% conc.

Quimipol EA 6808. [Quimigal-Quimica] Syn. primary alcohol ethoxylate; nonionic; emulsifier and textile lubricant; solid; HLB 11.5; 100% conc.

Quimipol EA 6810. [Quimigal-Quimica] Syn. primary alcohol ethoxylate; nonionic; solubilizer, emulsifier, pigment dispersant, and detergent; solid; HLB 12.5; 100% conc.

Quimipol EA 6812. [Quimigal-Quimica] Syn. primary alcohol ethoxylate; nonionic; solubilizer, emulsifier, pigment dispersant, detergent; solid; HLB 13.5; 100% conc.

Quimipol EA 6814. [Quimigal-Quimica] Syn. primary alcohol ethoxylate; nonionic; solubilizer, emulsifier, pigment dispersant, detergent; solid; HLB 14.0; 100% conc.

Quimipol EA 6818. [Quimigal-Quimica] Syn. primary alcohol ethoxylate; nonionic; solubilizer, emulsifier, pigment dispersant, detergent; solid; HLB 15.0; 100% conc.

Quimipol EA 6820. [Quimigal-Quimica] Syn. primary alcohol ethoxylate; nonionic; solubilizer, emulsifier, pigment dispersant, detergent; solid; HLB 15.5; 100% conc.

Quimipol EA 6823. [Quimigal-Quimica] Syn. primary alcohol ethoxylate; nonionic; solubilizer, emulsifier, pigment dispersant, detergent; solid; HLB 15.8; 100% conc.

Quimipol EA 6825. [Quimigal-Quimica] Syn. primary alcohol ethoxylate; nonionic; solubilizer, emulsifier, pigment dispersant, detergent; solid; HLB 16.2; 100% conc.

Quimipol EA 6850. [Quimigal-Quimica] Syn. primary alcohol ethoxylate; nonionic; solubilizer, emulsifier, pigment dispersant, detergent; solid; HLB 16.7; 100% conc.

Quimipol EA 9105. [Quimigal-Quimica] Syn. primary alcohol ethoxylate; nonionic; wetting agent and emulsifier; liq.; HLB 11.6; 100% conc.

Quimipol EA 9106. [Quimigal-Quimica] Syn. primary alcohol ethoxylate; nonionic; detergent, wetting agent in textile processing; degreaser in metal cleaning; liq.; HLB 12.5; 100% conc.

Quimipol EA 9106B. [Quimigal-Quimica] Syn. primary alcohol ethoxylate; nonionic; coemulsifier for min. oil and waxes; emulsifier for alkyl resins; liq.; HLB 12.6; 100% conc.

Quimipol EA 9108. [Quimigal-Quimica] Syn. primary alcohol ethoxylate; nonionic; component of dishwashing liqs., etc.; liq.; HLB 13.8; 100% conc.

Quimipol ED 2021. [Quimigal-Quimica] Polypropylene glycol ethoxylate; nonionic; antistat, deduster, defoamer, demulsifier, detergent, dispersant; liq.; HLB 3.0; 100% conc.

Quimipol ED 2022. [Quimigal-Quimica] Propylene glycol ethoxylate; nonionic; antistat, deduster, defoamer, demulsifier, detergent, dispersant; liq.; HLB 7.0; 100% conc.

Quimipol ENF 15. [Quimigal-Quimica] Nonylphenol ethoxylate; nonionic; emulsifier and degreaser; liq.; HLB 4.5; 100% conc.

Quimipol ENF 20. [Quimigal-Quimica] Nonylphenol ethoxylate; nonionic; emulsifier, degreaser; liq.; HLB 5.8; 100% conc.

Quimipol ENF 30. [Quimigal-Quimica] Nonylphenol ethoxylate; nonionic; emulsifier, degreaser; liq.; HLB 7.6; 100% conc.

Quimipol ENF 40. [Quimigal-Quimica] Nonylphenol ethoxylate; nonionic; emulsifier for min. oils, alkyl resins and paraffin hydrocarbons; liq.; HLB 9.0; 100% conc.

Quimipol ENF 55. [Quimigal-Quimica] Nonylphenol ethoxylate; nonionic; emulsifier for min. oil and aliphatic solvs.; coupling agent; liq.; HLB 10.5; 100% conc.

Quimipol ENF 65. [Quimigal-Quimica] Nonylphenol ethoxylate; nonionc; acid wool scouring, drycleaning and emulsifier for oils, hydrocarbon solvs. and waxes; liq.; HLB 10.9; 100% conc.

Quimipol ENF 80. [Quimigal-Quimica] Nonoxynol-8; CAS 9016-45-9; nonionic; wool scouring, oil emulsification and drycleaning detergents; liq.; HLB 12.2; 100% conc.

Quimipol ENF 90. [Quimigal-Quimica] Nonylphenol ethoxylate; nonionic; wool scouring, oil emulsification, antistat, detergent, and wetting agent; liq.; HLB 12.9; 100% conc.

Quimipol ENF 95. [Quimigal-Quimica] Nonylphenol ethoxylate; nonionic; textile scouring agent, emulsifier, detergent and wetting agent; liq.; HLB 13.0; 100% conc.

Quimipol ENF 100. [Quimigal-Quimica] Nonylphenol ethoxylate; nonionic; textile scouring and emulsifying agent, detergent, wetting agent; liq.; HLB 13.4; 100% conc.

Quimipol ENF 110. [Quimigal-Quimica] Nonylphenol ethoxylate; nonionic; textile scouring and emulsifying agent, detergent, wetting agent; liq.; HLB 13.8; 100% conc.

Quimipol ENF 120. [Quimigal-Quimica] Nonylphenol ethoxylate; nonionic; foam stabilizer and suspending agent for liq. detergents; liq.; HLB 14.0; 100% conc.

Quimipol ENF 140. [Quimigal-Quimica] Nonylphenol ethoxylate; nonionic; solubilizer for alkylaryl sulfonates and essential oils, emulsifier for pesticides and herbicides; paste; HLB 14.5; 100% conc.

Quimipol ENF 170. [Quimigal-Quimica] Nonylphenol ethoxylate; nonionic; solubilizer for alkylaryl sulfonates and essential oils; emulsifier for pesticides and herbicides; solid; HLB 15.4; 100% conc.

Quimipol ENF 200. [Quimigal-Quimica] Nonylphenol ethoxylate; nonionic; emulsifier for emulsion polymerization; wetting agent and detergent at high temps. and high electrolyte concs.; solid; HLB 16.0; 100% conc.

Quimipol ENF 230. [Quimigal-Quimica] Nonylphenol ethoxylate; nonionic; emulsifier for emulsion polymerization; wetting agent and detergent at high temps. and high electrolyte concs.; solid; HLB 16.4; 100% conc.

Quimipol ENF 300. [Quimigal-Quimica] Nonylphenol ethoxylate; nonionic; emulsifier for emulsion polymerization; wetting agent and detergent at high temps. and high electrolyte concs.; solid; HLB 17.1; 100% conc.

R

R-60 Z-5. [Werner G. Smith] Fish oil deriv. containing 40% C-20 and C-22 acids; wetter for hot metal surfs.; high water absorptive capacity; easy emulsification; for marine lubricants, Degras replacement in metalworking oils and rust preventatives; scavenger in chlorinated paraffin mixts.; Gardner 18 color; sol. in solvs. and naphthenic base stocks; sp.gr. 0.97 (60/60 F); visc. 700-750 SSU (210 F); acid no. 6-15; sapon. no. 190-205.

R3124 Ester. [Reilly-Whiteman] PEG-12 ditallate; CAS 61791-01-3; nonionic; multipurpose emulsifier; liq.; 99% conc.

RAC-100. [Tokai Seiyu Ind.] Polyamine deriv.; cationic; lubricating and softening agent; liq.

Radiamac 6149. [Fina Chemicals] Hydrog. tallow amine acetate; cationic; flotation reagent, anticaking aid for fertilizer, corrosion inhibitor, emulsifier; solid; HLB 11; m.p. 55-65 C; acid no. 160; amine no. 170; 100% conc.

Radiamac 6159. [Fina Chemicals] Proprietary amine acetate; cationic; flotation reagent, anticaking aid for fertilizer, corrosion inhibitor, emulsifier; solid; m.p. 55-65 C; acid no. 155; amine no. 165.

Radiamac 6169. [Fina Chemicals] Coconut oil amine acetate; cationic; flotation reagent, anticaking aid for fertilizer, corrosion inhibitor, emulsifier; paste; HLB 11; acid no. 200; amine no. 210; 100% conc.

Radiamac 6179. [Fina Chemicals] Tallow amine acetate; cationic; flotation reagent, anticaking aid for fertilizer, corrosion inhibitor, emulsifier; solid; HLB 11; 100% conc.

Radiamine 6140. [Fina Chemicals] Hydrog. tallow amine; CAS 61788-45-2; cationic; min. flotation, corrosion inhibitor, pigment dispersant; cosmetics; lubricant and mold release for hard rubber, textile chemical, chemical synthesis; antistat and antifog additive for plastic foils; Gardner 3 max. solid; m.p. 46 C; amine no. 210; iodine no. < 3; 100% conc.

Radiamine 6141. [Fina Chemicals] Hydrog. tallow amine, distilled; CAS 61788-45-2; cationic; min. flotation, corrosion inhibitor, pigment dispersant; cosmetics; lubricant and mold release for hard rubber, textile chemical, chemical synthesis; antistat and antifog additive for plastic foils; Gardner 1 max. solid; m.p. 48 C; amine no. 212; iodine no. < 3; 98% amine.

Radiamine 6160. [Fina Chemicals] Coconut oil amine; CAS 61788-46-3; cationic; min. flotation, corrosion inhibitor, pigment dispersant; cosmetics; lubricant and mold release for hard rubber, textile chemical, chemical synthesis; antistat and antifog additive for plastic foils; Gardner 3 max. liq.; m.p. 16 C; amine no. 275; iodine no. < 13; 100% conc.

Radiamine 6161. [Fina Chemicals] Coconut oil amine, distilled; CAS 61788-46-3; cationic; min. flotation, corrosion inhibitor, pigment dispersant; cosmetics; lubricant and mold release for hard rubber, textile chemical, chemical synthesis; antistat and antifog additive for plastic foils; Gardner 1 max. liq.; m.p. 15 C; amine no. 282; iodine no. < 12; 98% amine.

Radiamine 6163, 6164. [Fina Chemicals] Laurylamine (6164—dist.); cationic; min. flotation, corrosion inhibitor, pigment dispersant; cosmetics; lubricant and mold release for hard rubber, textile chemical, chemical synthesis; antistat and antifog additive for plastic foils; liq., solid; 100% conc.

Radiamine 6170. [Fina Chemicals] Tallow amine; CAS 61790-33-8; cationic; min. flotation, corrosion inhibitor, pigment dispersant; cosmetics; lubricant and mold release for hard rubber, textile chemical, chemical synthesis; antistat and antifog additive for plastic foils; Gardner 3 max. paste; m.p. 32 C; amine no. 210; iodine no. 40; 100% conc.

Radiamine 6171. [Fina Chemicals] Tallow amine, distilled; CAS 61790-33-8; cationic; min. flotation, corrosion inhibitor, pigment dispersant; cosmetics; lubricant and mold release for hard rubber, textile chemical, chemical synthesis; antistat and antifog additive for plastic foils; Gardner 1 max. liq.; m.p. 30 C; amine no. 212; iodine no. 40; 98% amine.

Radiamine 6172. [Fina Chemicals] Oleyl amine; CAS 112-90-3; cationic; min. flotation, corrosion inhibitor, pigment dispersant; cosmetics; lubricant and mold release for hard rubber, textile chemical, chemical synthesis; antistat and antifog additive for plastic foils; Gardner 2 max. liq.; cloud pt. < 20 C; amine no. 207; iodine no. 82; 95% amine.

Radiamine 6173. [Fina Chemicals] Oleamine, distilled; CAS 112-90-3; cationic; min. flotation, corrosion inhibitor, pigment dispersant; cosmetics; lubricant and mold release for hard rubber, textile chemical, chemical synthesis; antistat and antifog additive for plastic foils; Gardner 2 max. liq.; cloud pt. < 20 C; amine no. 209; iodine no. 80; 97% amine.

Radiamine 6540. [Fina Chemicals] Hydrog. tallow propanediamine; cationic; corrosion inhibitor, dispersant, emulsifier, intermediate for chemical synthesis; Gardner 3 solid; m.p. 45 C; amine no. 330; iodine no. < 3; 100% conc.

Radiamine 6560. [Fina Chemicals] Coconut oil propane diamine; CAS 61791-63-7; cationic; corrosion inhibitor, dispersant, emulsifier, intermediate for chemical synthesis; Gardner 6 max. paste; amine no. 400; iodine no. < 10; 98% amine.

Radiamine 6570. [Fina Chemicals] Tallow propane diamine; CAS 68439-73-6; cationic; corrosion inhibitor, dispersant, emulsifier, intermediate for chemical synthesis; Gardner 5 paste; amine no. 330;

iodine no. > 35; 98% amine.

Radiamine 6572. [Fina Chemicals] Oleyl propane diamine; CAS 68037-97-8; cationic; corrosion inhibitor, dispersant, emulsifier, intermediate for chemical synthesis; Gardner 6 max. liq.; amine no. 325; iodine no. > 60; 100% conc.

Radiamine AA 23, 27, 57. [Fina Chemicals] n-Alkyl propylene diamine; cationic; bitumen emulsifier in road construction; paste; 100% conc.

Radiamine AA 60. [Fina Chemicals] n-Alkyl propylene diamine; cationic; bitumen emulsifier in road construction; liq.; 100% conc.

Radiaquat 6410. [Fina Chemicals] Didecyl dimethyl ammonium chloride; cationic; bactericidal quat. for hard surf. cleaners, mildew preventers for commercial and industrial laundries; Gardner 2 max. liq.; 50% min. conc.

Radiaquat 6412. [Fina Chemicals] Didecyl dimethyl ammonium chloride; cationic; bactericidal quat. for hard surf. cleaners, mildew preventers for commercial and industrial laundries; Gardner 2 max. liq.; 70% min. conc.

Radiaquat 6442. [Fina Chemicals] Dihydrog. tallow dimethyl ammonium chloride; CAS 61789-80-8; cationic; surfactant, softener for laundry applics.; Gardner 3 max. paste; 74% conc.

Radiaquat 6443. [Fina Chemicals] Dialkyl dimethyl ammonium chloride; cationic; softener; liq.; 75% conc.

Radiaquat 6444. [Fina Chemicals] Palmityl trimethyl ammonium chloride; CAS 112-02-7; cationic; softener, antistat, detergent for textile, leather, detergent, and cosmetic industries; clay modifier; Gardner 3 max. liq.; 50% conc.

Radiaquat 6462. [Fina Chemicals] Dicoco dimethyl ammonium chloride; CAS 61789-77-3; cationic; detergent, antistat, softener, bactericide for detergents, textiles, fabric softeners, cosmetics; Gardner 4 max. paste; 75% conc.

Radiaquat 6470. [Fina Chemicals] Ditallow dimethyl ammonium chloride; CAS 68153-32-2; cationic; surfactant, softener for laundry applics.; Gardner 4 max. liq. (30 C); 75% conc.

Radiaquat 6471. [Fina Chemicals] Tallow trimethyl ammonium chloride; CAS 68002-61-9; cationic; detergent, antistat, softener for textile, leather, detergent and cosmetic prods.; Gardner 5 max. liq.; 50% conc.

Radiaquat 6475. [Fina Chemicals] Dihydrog. tallow dimethyl ammonium chloride; CAS 61789-80-8; cationic; surfactant, softener for laundry prods.; Gardner 3 max. liq.; 76% min. conc.

Radiaquat 6480. [Fina Chemicals] Dihydrog. tallow dimethyl ammonium chloride; CAS 61789-80-8; cationic; surfactant, softener for laundry prods.; Gardner 3 max. liq.; 80% min. conc.

Radiasurf® 7000. [Fina Chemicals] PEG-20 glyceryl stearate; nonionic; chemical intermediate, emulsifier, detergent, lubricant, wetting agent; chemical synthesis; o/w or w/o cosmetic and pharmaceutical emulsions; pearlescent shampoos; cleaning prods.; formulation of cutting, lamination, and textile oils; rust inhibitors; pigment wetting, faster grinding, and improved gloss in paints, lacquers, and printing inks; lubricant and mold release in plastics; m.w. 1395; sol. (@ 10%) in trichlorethylene, hexane, benzene, IPA, min. and veg. oils; water-disp.; sp.gr. 1.032 (37.8 C); visc. 113.30 cps (37.8 C); HLB 13.0; ref. index 1.4672; flash pt. 244 C (COC); cloud pt. 27.5 C.

Radiasurf® 7125. [Fina Chemicals] Sorbitan laurate; CAS 1338-39-2; nonionic; emulsifier, descouring aid, antistat; anticorrosive agent for pipelines; cleaner for metallic surfaces; superfatting, bodying and antifog aid; pigment dispersant; detergent; emulsion of solv.; cutting oils; textile lubricant additive; concrete and paper additives; leather aux.; cosmetics and pharmaceuticals; plastics; pesticides; dry cleaning formulations; amber liq.; m.w. 450; sol. in hexane, trichlorethylene, min. and veg. oil, benzene, IPA; disp. in water; sp.gr. 1.025; visc. 1193.0 cps (37.8 C); HLB 7.6; acid no. 7 max.; iodine no. < 10; sapon. no. 155–175; ref. index 1.4718; flash pt. 198 C (COC); cloud pt. 23 C.

Radiasurf® 7135. [Fina Chemicals] Sorbitan palmitate; CAS 26266-57-9; nonionic; emulsifier, descouring aid, antistat; anticorrosive agent for pipelines; cleaner for metallic surfaces; superfatting, bodying and antifog aid; pigment dispersant; detergent; emulsion of solv.; cutting oils; textile lubricant additive,; wh. solid; m.w. 500; sol. cloudy in IPA, trichlorethylene, min. oil; disp. in water; sp.gr. 0.943 (98.9 C); visc. 51.70 cps (98.9 C); HLB 6.3; m.p. 50 C; acid no. 7 max.; iodine no. < 1; sapon. no. 140–160; flash pt. 215 C (COC).

Radiasurf® 7136. [Fina Chemicals] POE sorbitan palmitate; nonionic; emulsifier, wetting agent, defoamer, rust inhibitor, pigment grinding, antistat; liq.; HLB 15.5.

Radiasurf® 7137. [Fina Chemicals] Polysorbate 20; CAS 9005-64-5; nonionic; emulsifier, descouring aid, antistat; anticorrosive agent for pipelines; cleaner for metallic surfaces; superfatting, bodying and antifog aid; pigment dispersant; detergent; emulsion of solv.; cutting oils; textile lubricant additive,; amber liq.; m.w. 1340; sol. in water, benzene, IPA, trichlorethylene, min. and veg. oil; sp.gr. 1.095 (37.8 C); visc. 172.3 cps (37.8 C); HLB 16.5; pour pt. -12 C; cloud pt. -10 C; flash pt. 250 C (COC); acid no. 2 max.; iodine no. < 3; sapon. no. 40–55; ref. index 1.4712.

Radiasurf® 7140. [Fina Chemicals] Glyceryl stearate; nonionic; emulsifier, wetting agent, defoamer, rust inhibitor, pigment grinding, antistat; internal lubricant for PVC; Lovibond 5Y-1R flakes, powd.; sol. (@ 75 C) in benzene, IPA, trichlorethylene, min. and veg. oils, cloudy in hexane; insol. in water; sp.gr. 885 kg/m^3 (100 C); HLB 2.9; m.p. 58 C; flash pt. (COC) 235 C; acid no. 3 max.; iodine no. < 1.

Radiasurf® 7141. [Fina Chemicals] Glyceryl stearate SE; nonionic; emulsifier, wetting agent, defoamer, rust inhibitor, pigment grinding, antistat; powd., flakes.

Radiasurf® 7144. [Fina Chemicals] Glyceryl monococoate; nonionic; emulsifier, wetting agent, defoamer, rust inhibitor, pigment grinding, antistat; paste; HLB 3.0.

Radiasurf® 7145. [Fina Chemicals] Sorbitan stearate; CAS 1338-41-6; nonionic; emulsifier, descouring aid, antistat; anticorrosive agent for pipelines; cleaner for metallic surfaces; superfatting, bodying and antifog aid; pigment dispersant; detergent; emulsion of solv.; cutting oils; textile lubricant additive,; wh. solid; m.w. 510; sol. in trichlorethylene; disp. in water; sp.gr. 0.943 (98.9 C); visc. 48.20 cps (98.9 C); HLB 5; m.p. 55 C; acid no. 7 max.; iodine no. 1 max.; sapon. no. 146–158; flash pt. 232 C (COC).

Radiasurf® 7146. [Fina Chemicals] Glyceryl mono-12-hydroxystearate; nonionic; chemical intermediate, emulsifier, detergent, wetting agent, lubricant; chemical synthesis; cosmetics and pharmaceuticals; pearlescent shampoo formulations; detergency and cleaning prods.; cutting, lamination, and textile oils; rust inhibitor; pigment wetting, grinding, and improved gloss in paints, lacquers, printing inks; paper industry; lubricant and mold release in plastics; textile and leather processing; m.w. 550; sol. (@ 10%); @ 75 C in benzene, IPA, trichlorethylene; sp.gr. 0.925 (98.9 C); visc. 25.80 cps (98.9 C); HLB 1.9; m.p. 70 C; flash pt. 260 C (COC).

Radiasurf® 7147. [Fina Chemicals] Polysorbate 60; CAS 9005-67-8; nonionic; emulsifier, descouring aid, antistat; anticorrosive agent for pipelines; cleaner for metallic surfaces; superfatting, bodying and antifog aid; pigment dispersant; detergent; emulsion of solv.; cutting oils; textile lubricant additive,; amber paste; m.w. 1400; sol. in trichlorethylene, min. and veg. oil, water, benzene, IPA; sp.gr. 1.068 (37.8 C); visc. 202.6 cps (37.8 C); HLB 15.1; m.p. 21 C; acid no. 2 max.; sapon. no. 40–56; ref. index 1.4694; flash pt. 260 C (COC).

Radiasurf® 7150. [Fina Chemicals] Glyceryl oleate; CAS 111-03-5; internal lubricant for PVC; biodeg. surfactant, wetting agent, emulsifier for cosmetics, pharmaceuticals, agric., chem. synthesis, explosives, polymers, glass fibers, surface coatings, textiles and leather; Lovibond 15Y-3R liq.; sol. in hexane, benzene, IPA, min. oil; sol. cloudy in trichlorethylene, veg. oil; sp.gr. 934 kg/m^3; HLB 2.7; pour pt. 3 C; cloud pt. < 10 C; flash pt. (COC) 230 C; acid no. 3 max.; iodine no. 75-85; 40% alpha monoglycerides.

Radiasurf® 7151. [Fina Chemicals] Glyceryl oleate, SE; nonionic; chemical intermediate, emulsifier, detergent, wetting agent, lubricant; chemical synthesis; cosmetic and pharmaceuticals; detergency and cleaning prods.; cutting, lamination, and textile oils; rust inhibitors; pigment wetting,; grinding, and improved gloss in paints, lacquers, and printing inks; paper industry; lubricant and mold release in plastics; textile and leather processing; liq.; m.w. 515; sol. in IPA, min. and veg. oil, hexane, benzene, trichlorethylene; sp.gr. 0.947 (37.8 C); visc. 137.70 cps (37.8 C); HLB 3.3; ref. index 1.4696; flash pt. 208 C (COC); cloud pt. 12.0 C.

Radiasurf® 7152. [Fina Chemicals] Glyceryl oleate; CAS 111-03-5; nonionic; chemical intermediate, emulsifier, detergent, wetting agent, lubricant; chemical synthesis; cosmetic and pharmaceuticals; detergency and cleaning prods.; cutting, lamination, and textile oils; rust inhibitors; pigment wetting,; liq.; m.w. 521; sol. in IPA, min. and veg. oil, hexane, benzene, trichlorethylene; sp.gr. 0.934 (37.8 C); visc. 77.10 cps (37.8 C); HLB 2.8; ref. index 1.4697; flash pt. 232 C (COC); cloud pt. 7.5 C.

Radiasurf® 7153. [Fina Chemicals] Glyceryl ricinoleate; nonionic; chemical intermediate, emulsifier, detergent, wetting agent, lubricant; cosmetic and pharmaceuticals; detergency and cleaning prods.; cutting, lamination, and textile oils; rust inhibitors; pigment wetting; antistat/antifog in plastics; m.w. 554; sol. (@ 10%); in benzene, IPA, trichlorethylene, min. and veg. oils; sp.gr. 0.982 (37.8 C); visc. 356.20 cps (37.8 C); HLB 2.5; ref. index 1.4774; flash pt. 229 C (COC); cloud pt. –16.0 C; 40% conc.

Radiasurf® 7155. [Fina Chemicals] Sorbitan oleate; CAS 1338-43-8; nonionic; emulsifier, descouring aid, antistat; anticorrosive agent for pipelines; cleaner for metallic surfaces; superfatting, bodying and antifog aid; pigment dispersant; detergent; emulsion of solv.; cutting oils; textile lubricant additive,; amber liq.; m.w. 510; sol. in hexane, IPA, trichlorethylene, min. and veg. oils; sp.gr. 0.987 (37.8 C); visc. 395.2 cps (37.8 C); HLB 4.7; pour pt. -10 C; cloud pt. 7 C; acid no. 7 max.; sapon. no. 145–161; ref. index 1.4778; flash pt. 250 C (COC).

Radiasurf® 7156. [Fina Chemicals] Pentaerythritol oleate; nonionic; corrosion inhibitor for lubricating oils and greases; biodeg. surfactant, wetting agent, emulsifier for cosmetics, pharmaceuticals, agric., chem. synthesis, explosives, polymers, glass fibers, surface coatings, textiles and leather; Lovibond 15Y-3R liq.; sol. in bezene, IPA, trichlorethylene, min. and veg. oils; sol. cloudy in hexane; sp.gr. 938 kg/m^3; HLB 2.1; cloud pt. < 0 C; flash pt. (COC) 280 C; acid no. 3 max.; iodine no. 80-90.

Radiasurf® 7157. [Fina Chemicals] Polysorbate 80; CAS 9005-65-6; nonionic; emulsifier, descouring aid, antistat; anticorrosive agent for pipelines; cleaner for metallic surfaces; superfatting, bodying and antifog aid; pigment dispersant; detergent; emulsion of solv.; cutting oils; textile lubricant additive,; amber liq.; m.w. 400; sol. in water, benzene, trichlorethylene, IPA, min. and veg. oils; sp.gr. 1.071 (37.8 C); visc. 240.4 cps (37.8 C); HLB 14.9; cloud pt. 5.5 C; flash pt. 287 C (COC); acid no. 2 max.; sapon. no. 44–56; ref. index 1.4726.

Radiasurf® 7175. [Fina Chemicals] Pentaerythritol stearate; nonionic; chemical intermediate, emulsifier, detergent, wetting agent, lubricant; chemical synthesis; cosmetics and pharmaceuticals; pearlescent shampoo formulations; detergency and cleaning prods.; cutting, lamination, and textile oils; rust inhibitor; pigment wetting, grinding, and improved gloss in paints, lacquers, printing inks; paper industry; lubricant and mold release in plastics; textile and leather processing; flakes; m.w. 675; sol. in trichlorethylene; sp.gr. 0.886 (98.9 C); visc. 17.60 cps (98.9 C); HLB 2.3; m.p. 52 C; flash pt. 260 C (COC); acid no. 2 max.; iodine no. < 1.

Radiasurf® 7196. [Fina Chemicals] Propylene glycol myristate; nonionic; wetting aid, lubricant, opacifier, antistat, dispersant, w/o emulgent, scouring and detergent aid, defoamer, plasticizer, rust inhibitor; cosmetics and pharmaceuticals, lubricating and cutting oils, textile and leather aids, pigment grinding; in paints and printing inks, latex and emulsion paints, plastics, waxes, and maintenance prods., insecticides; wh. to amber liq.; m.w. 378; sol. in trichlorethylene, hexane, benzene, IPA, min. and veg. oils; disp. in water; sp.gr. 0.902 (37.8 C); visc. 17.30 cps (37.8 C); HLB 3.9; acid no. 5 max.; sapon. no. 190–205; ref. index 1.4379; flash pt. 145 C (COC).

Radiasurf® 7201. [Fina Chemicals] Propylene glycol stearate; nonionic; wetting aid, lubricant, opacifier, antistat, dispersant, w/o emulgent, scouring and detergent aid, defoamer, plasticizer, rust inhibitor; cosmetics and pharmaceuticals, lubricating and cutting oils, textile and leather aids, pigment grinding; wh. to cream solid; m.w. 457; sol. (@ 10%) in hexane, trichlorethylene, min. oil; sp.gr. 0.849 (98.9 C); visc. 4.50 cps (98.9 C); HLB 1.9; acid no. 2 max.; sapon. no. 177; flash pt. 206 C (COC).

Radiasurf® 7206. [Fina Chemicals] Propylene glycol oleate; nonionic; wetting aid, lubricant, opacifier, antistat, dispersant, w/o emulgent, scouring and detergent aid, defoamer, plasticizer, rust inhibitor;

cosmetics and pharmaceuticals, lubricating and cutting oils, textile and leather aids, pigment grinding; amber liq.; m.w. 460; sol. in benzene, IPA, trichlorethylene, veg. and min. oil, hexane; sp.gr. 0.900 (37.8 C); visc. 18.80 cps (37.8 C); HLB 2.3; acid no. 2 max.; sapon. no. 172–180; ref. index. 1.4606; flash pt. 157 C (COC); cloud pt. –9 C.

Radiasurf® 7269. [Fina Chemicals] Ethylene glycol distearate; nonionic; wetting aid, lubricant, opacifier, antistat, dispersant, w/o emulgent, scouring and detergent aid, defoamer, plasticizer, rust inhibitor; cosmetics and pharmaceuticals, lubricating and cutting oils, textile and leather aids, pigment grinding; wh. solid; sol. in trichlorethylene, hexane, benzene, IPA, min. and veg. oils; HLB 1.5; m.p. 65 C; acid no. 3 max.; sapon. no. 190–200.

Radiasurf® 7270. [Fina Chemicals] Ethylene glycol stearate; CAS 97281-23-7; nonionic; wetting aid, lubricant, opacifier, antistat, dispersant, w/o emulgent, scouring and detergent aid, defoamer, plasticizer, rust inhibitor; cosmetics and pharmaceuticals, lubricating and cutting oils, textile and leather aids, pigment grinding; wh. flakes; m.w. 445; sol. (@ 10% and 75 C) in benzene, IPA, trichlorethylene, veg. oil; sp.gr. 0.851 (98.9 C); visc. 4.80 cps (98.9 C); HLB 2.1; m.p. 57 C; acid no. 2 max.; sapon. no. 180–190; flash pt. 214 C (COC).

Radiasurf® 7345. [Fina Chemicals] Sorbitan tristearate; CAS 26658-19-5; nonionic; emulsifier, descouring aid, antistat; anticorrosive agent for pipelines; cleaner for metallic surfaces; superfatting, bodying and antifog aid; pigment dispersant; detergent; emulsion of solv.; cutting oils; textile lubricant additive,; solid; m.w. 900; sol. in trichlorethylene, min. and veg. oils; sp.gr. 0.893 (98.9 C); visc. 19.50 cps (98.9 C); HLB 2.7; m.p. 60 C; flash pt. 253 C (COC).

Radiasurf® 7372. [Fina Chemicals] Trimethylolpropane oleate; nonionic; wetting agent, defoamer, rust inhibitor, pigment grinder, emulsifier, antistatic agent; liq.; m.w. 514; sol. in hexane, benzene, IPA, trichlorethylene, min. and veg. oil; sp.gr. 0.935 (37.8 C); visc. 90 cps (37.8 C); HLB 4.4; ref. index 1.4705; flash pt. 224 C (COC); cloud pt. –18.5 C.

Radiasurf® 7400. [Fina Chemicals] Diethylene glycol oleate; CAS 9004-96-0; 93455-78-8; nonionic; emulsifier, wetting agent, defoamer, rust inhibitor, pigment grinder, antistat; amber liq.; m.w. 478; sol. in trichlorethylene, hexane, benzene, IPA, min. and veg. oil; disp. in water; sp.gr. 0.929 (37.8 C); visc. 58.50 cps (37.8 C); HLB 3.8; pour pt. -12 C; cloud pt. < -2 C;acid no. 4 max.; iodine no. 75-85; sapon. no. 155–168; ref. index 1.4642; flash pt. 180 C (COC).

Radiasurf® 7402. [Fina Chemicals] PEG-4 oleate; CAS 9004-96-0; nonionic; wetting aid, lubricant, opacifier, antistat, dispersant, o/w emulgent, scouring and detergent aid, defoamer, plasticizer, rust inhibitor, visc. modifier, antifog aid; cosmetics and pharmaceuticals; lubricating and cutting oils; textile and leather aids; pigment grinding aids in paints and printing inks; latex and emulsion paints; plastics; waxes and maintenance prods.; glass fiber; insecticides; silicones; amber liq.; sol. (@ 10%) in benzene, isopropyl, trichlorethylene, hexane, min. and veg. oils; water-disp.; sp.gr. 0.962 (37.8 C); visc. 34.00 cps (37.8 C); HLB 7.4; acid no. < 3; sapon. no. 105–125; iodine no. 55-65; ref. index 1.4645; flash pt. 218 C (COC); cloud pt. –10 C.

Radiasurf® 7403. [Fina Chemicals] PEG-8 oleate; CAS 9004-96-0; nonionic; wetting aid, lubricant, opacifier, antistat, dispersant, o/w emulgent, scouring and detergent aid, defoamer, plasticizer, rust inhibitor, visc. modifier, antifog aid; cosmetics and pharmaceuticals; lubricating and cutting oils; amber liq.; m.w. 728; sol. in trichlorethylene, hexane, benzene, IPA, min. and veg. oil; disp. in water; sp.gr. 1.007 (37.8 C); visc. 49.50 cps (37.8 C); HLB 11.5; acid no. < 3; sapon. no. 75–90; ref. index 1.4672; flash pt. 261 C (COC); cloud pt. –1 C.

Radiasurf® 7404. [Fina Chemicals] PEG-12 oleate; CAS 9004-96-0; nonionic; wetting aid, lubricant, opacifier, antistat, dispersant, o/w emulgent, scouring and detergent aid, defoamer, plasticizer, rust inhibitor, visc. modifier, antifog aid; cosmetics and pharmaceuticals; lubricating and cutting oils; amber liq.; m.w. 920; sol. (@ 10%) in water, benzene, IPA, trichlorethylene, hexane, min. and veg. oils; sp.gr. 1.030 (37.8 C); visc. 66.30 cps (37.8 C); HLB 13.2; acid no. < 3; sapon. no. 60–75; ref. index 1.4669; flash pt. 254 C (COC); cloud pt. 19.

Radiasurf® 7410. [Fina Chemicals] Diethylene glycol stearate; CAS 9004-99-3; 85116-97-8; nonionic; wetter, lubricant, opacifier, antistat, dispersant, detergent, defoamer, plasticizer, rust inhibitor; for cosmetics, pharmaceuticals, textiles, leather, paints, inks, plastic, waxes, maintenance prods., insecticides; wh. paste; m.w. 475; sol. in trichlorethylene, hexane, benzene, IPA, min. and veg. oils; disp. in water; sp.gr. 0.873 (100 C); visc. 5.95 cps (98.9 C); HLB 3.5; m.p. 49 C; acid no. 3 max.; sapon. no. 160–175; flash pt. 191 C (COC).

Radiasurf® 7411. [Fina Chemicals] Diethylene glycol stearate, SE; CAS 9004-99-3; nonionic; wetter, lubricant, opacifier, antistat, dispersant, detergent, defoamer, plasticizer, rust inhibitor; for cosmetics, pharmaceuticals, textiles, leather, paints, inks, plastic, waxes, maintenance prods., insecticides; wh. solid; m.w. 470; sol. in trichlorethylene, hexane, benzene, IPA, min. and veg. oils; disp. in water; sp.gr. 0.878 (98.9 C); visc. 5.90 cps (98.9 C); HLB 4.1; m.p. 52 C; acid no. 3 max.; sapon. no. 160–170; flash pt. 176 C (COC).

Radiasurf® 7412. [Fina Chemicals] PEG-4 stearate; CAS 9004-99-3; nonionic; wetting agent, defoamer, rust inhibitor, pigment grinder, emulsifier, antistat; wh. paste; m.w. 522; sol. (@ 10%) in benzene, IPA, trichlorethylene, hexane, min. oil; sp.gr. 0.913 (98.9 C); visc. 6.70 cps (98.9 C); HLB 7.5; m.p. 36 C; acid no. < 3; sapon. no. 120–135; flash pt. 186 C (COC).

Radiasurf® 7413. [Fina Chemicals] PEG-8 stearate; CAS 9004-99-3; nonionic; wetting aid, lubricant, opacifier, antistat, dispersant, o/w emulgent, scouring and detergent aid, defoamer, plasticizer, rust inhibitor, visc. modifier, antifog aid; cosmetics and pharmaceuticals; lubricating and cutting oils; wh. paste; m.w. 722; sol. (@ 10%) in benzene, trichlorethylene, min. oil, hexane; sp.gr. 0.951 (98.9 C); visc. 9.45 cps (98.9 C); HLB 11.9; m.p. 34 C; acid no. < 3; sapon. no. 75–90; flash pt. 248 C (COC).

Radiasurf® 7414. [Fina Chemicals] PEG-12 stearate; CAS 9004-99-3; nonionic; wetting aid, lubricant, opacifier, antistat, dispersant, o/w emulgent, scouring and detergent aid, defoamer, plasticizer, rust inhibitor, visc. modifier, antifog aid; cosmetics and pharmaceuticals; lubricating and cutting oils; wh. paste; m.w. 906; sol. in benzene, IPA, hexane, min. and veg. oil, trichlorethylene; sp.gr. 0.981 (98.9 C); visc. 12.60 cps (98.9 C); HLB 13.5; m.p. 38 C; acid no. < 3; sapon. no. 60–75; flash pt. 241 C (COC).

Radiasurf® 7417. [Fina Chemicals] PEG-1500 stearate; CAS 9004-99-3; nonionic; wetting aid, lubricant, opacifier, antistat, dispersant, o/w emulgent, scouring and detergent aid, defoamer, plasticizer, rust inhibitor, visc. modifier, antifog aid; cosmetics and pharmaceuticals; lubricating and cutting oils; wh. flakes; m.w. 1812; sol. in benzene, trichlorethylene; sp.gr. 1.023 (98.9 C); visc. 28.40 cps (98.9 C); HLB 15.5; m.p. 47 C; acid no. < 3; sapon. no. 30–40; flash pt. 248 C (COC).

Radiasurf® 7420. [Fina Chemicals] Diethylene glycol laurate; CAS 9004-81-3; nonionic; wetting aid, lubricant, opacifier, antistat, dispersant, w/o emulgent, scouring and detergent aid, defoamer, plasticizer, rust inhibitor; cosmetics and pharmaceuticals, lubricating and cutting oils, textile and leather aids, pigment grinding; wh. paste; m.w. 352; sol. in IPA, trichlorethylene, hexane, min. and veg. oil; sp.gr. 0.942 (37.8 C); visc. 15.20 cps (37.8 C); HLB 5.7; acid no. 3 max.; sapon. no. 195–205; flash pt. 164 C (COC); cloud pt. 28.5 C.

Radiasurf® 7421. [Fina Chemicals] Diethylene glycol laurate; CAS 9004-81-3; nonionic; wetter, defoamer, rust inhibitor, pigment grinder, emulsifier, antistat; wh. liq.; m.w. 346; sol. in trichlorethylene, hexane, benzene, IPA, min. and veg. oil; sp.gr. 0.901 (98.9 C); visc. 3.60 cps (98.9 C); HLB 6.3; acid no. 3 max.; sapon. no. 183–193; flash pt. 173 C (COC); cloud pt. 29.5 C.

Radiasurf® 7422. [Fina Chemicals] PEG-4 laurate; CAS 9004-81-3; nonionic; wetting aid, lubricant, opacifier, antistat, dispersant, o/w emulgent, scouring and detergent aid, defoamer, plasticizer, rust inhibitor, visc. modifier, antifog aid; cosmetics and pharmaceuticals; lubricating and cutting oils; wh. liq.; m.w. 379; sol. in trichlorethylene, hexane, benzene, IPA, min. and veg. oil; disp. in water; sp.gr. 0.983 (37.8 C); visc. 23.40 cps (37.8 C); HLB 9.6; acid no. < 3; sapon. no. 140–150; ref. index 1.4542; flash pt. 188 C (COC); cloud pt. –3 C.

Radiasurf® 7423. [Fina Chemicals] PEG-8 laurate; CAS 9004-81-3; nonionic; wetting aid, lubricant, opacifier, antistat, dispersant, o/w emulgent, scouring and detergent aid, defoamer, plasticizer, rust inhibitor, visc. modifier, antifog aid; cosmetics and pharmaceuticals; lubricating and cutting oils; wh. liq.; m.w. 610; sol. in benzene, IPA, trichlorethylene, water, hexane, min. and veg. oil; sp.gr. 1.023 (37.8 C); visc. 41.70 cps (37.8 C); HLB 12.8; acid no. < 3; sapon. no. 90–100; ref. index 1.4596; flash pt. 238 C (COC); cloud pt. 4 C.

Radiasurf® 7431. [Fina Chemicals] PEG-6 oleate; CAS 9004-96-0; nonionic; wetting aid, lubricant, opacifier, antistat, dispersant, o/w emulgent, scouring and detergent aid, defoamer, plasticizer, rust inhibitor, visc. modifier, antifog aid; cosmetics and pharmaceuticals; lubricating and cutting oils; amber liq.; sol. in trichlorethylene, hexane, benzene, IPA, min. and veg. oil; disp. in water; HLB 9.5; acid no. < 3; sapon. no. 95–115; cloud pt. –8 C.

Radiasurf® 7432. [Fina Chemicals] PEG-6 stearate; CAS 9004-99-3; nonionic; wetting aid, lubricant, opacifier, antistat, dispersant, o/w emulgent, scouring and detergent aid, defoamer, plasticizer, rust inhibitor, visc. modifier, antifog aid; cosmetics and pharmaceuticals; lubricating and cutting oils; wh. paste; sol. in benzene, IPA, trichlorethylene, min. oil, hexane; HLB 9.7; m.p. 28–35 C; acid no. < 3; sapon. no. 95–115.

Radiasurf® 7443. [Fina Chemicals] PEG-8 dioleate; CAS 9005-07-6; nonionic; wetting aid, lubricant, opacifier, antistat, dispersant, o/w emulgent, scouring and detergent aid, defoamer, plasticizer, rust inhibitor, visc. modifier, antifog aid; cosmetics and pharmaceuticals; lubricating and cutting oils; amber liq.; m.w. 911; sol. in trichlorethylene, hexane, benzene, IPA, min. and veg. oil; disp. in water; sp.gr. 0.962 (37.8 C); visc. 47.00 cps (37.8 C); HLB 7.4; acid no. < 5; sapon. no. 120–130; ref. index 1.4655; flash pt. 248 C (COC); cloud pt. –3 C.

Radiasurf® 7444. [Fina Chemicals] PEG-12 dioleate; CAS 9005-07-6; nonionic; biodeg. surfactant, wetting agent, emulsifier for cosmetics, pharmaceuticals, agric., chem. synthesis, explosives, polymers, glass fibers, surface coatings, textiles and leather; Lovibond 10Y-1.5R liq.; sol. cloudy in water, min. oil; sol. clear in hexane, benzene, IPA, trichlorethylene, veg. oil; sp.gr. 992 kg/m^3; HLB 9.9; pour pt. 2 C; cloud pt. < 15 C; flash pt. (COC) 275 C; acid no. 5 max.; iodine no. 45-55.

Radiasurf® 7453. [Fina Chemicals] PEG-8 distearate; CAS 9005-08-7; nonionic; wetting aid, lubricant, opacifier, antistat, dispersant, o/w emulgent, scouring and detergent aid, defoamer, plasticizer, rust inhibitor, visc. modifier, antifog aid; cosmetics and pharmaceuticals; lubricating and cutting oils; wh. paste; m.w. 901; sol. (@ 10%) in benzene, IPA, trichlorethylene; sp.gr. 0.920 (98.9 C); visc. 9.85 cps (98.9 C); HLB 7.8; m.p. 39 C; acid no. < 5; sapon. no. 120–130; flash pt. 242 C (COC).

Radiasurf® 7454. [Fina Chemicals] PEG-12 distearate; CAS 9005-08-7; nonionic; wetting aid, lubricant, opacifier, antistat, dispersant, o/w emulgent, scouring and detergent aid, defoamer, plasticizer, rust inhibitor, visc. modifier, antifog aid; cosmetics and pharmaceuticals; lubricating and cutting oils; wh. paste; m.w. 1101; sol. in benzene, trichlorethylene, hexane, IPA, min. oil; sp.gr. 0.940 (98.9 C); visc. 12.20 cps (98.9 C); HLB 10.6; m.p. 40 C; acid no. < 5; sapon. no. 100–110; flash pt. 249 C.

Radiasurf® 7600. [Fina Chemicals] Glyceryl stearate; nonionic; wetter, defoamer, rust inhibitor, pigment grinder, emulsifier, antistat; flakes, powd.; sol. (@ 75 C) in hexane, benzene, IPA, trichlorethylene, min. and veg. oils; sp.gr. 893 kg/m^3 (100 C); HLB 3.3; m.p. 58 C; flash pt. (COC) 225 C; acid no. 3 max.; iodine no. < 1; 60% conc.

Radiasurf® 7900. [Fina Chemicals] Glyceryl stearate; nonionic; biodeg. surfactant, wetting agent, emulsifier for cosmetics, pharmaceuticals, agric., chem. synthesis, explosives, polymers, glass fibers, surface coatings, textiles and leather; Lovibond 5Y-1R powd.; sol. (@ 75 C) in hexane, benzene, IPA, trichlorethylene, min. and veg. oils; sp.gr. 902 kg/m^3; HLB 4.8; m.p. 68 C; flash pt. (COC) 220 C; acid no. 3 max.; iodine no. < 2; 90% alpha monoglycerides.

Ralufon® 414. [Raschig] Cocamidopropyl betaine; amphoteric; surfactant.

Ralufon® CA. [Raschig] N,N-Dimethyl-N-lauric acid-amidopropyl-N-(3-sulfopropyl)-ammonium betaine; amphoteric; surfactant.

Ralufon® CM 15-13. [Raschig] Mixt. of sulfopropylated alkoxylate and an alkanol polyethylene oxide; anionic/nonionic; surfactant.

Ralufon® DCH. [Raschig] Fatty acids of coconut-N,N-dimethyl-N-(3-sulfopropyl)-ammonium betaine; amphoteric; surfactant.

Ralufon® DL. [Raschig] N,N-Dimethyl-N-lauryl-N-(3-sulfopropyl)-ammonium betaine; amphoteric;

surfactant.

Ralufon® DM. [Raschig] N,N-Dimethyl-N-myristyl-N-(3-sulfopropyl)-ammonium betaine; amphoteric; surfactant.

Ralufon® DP. [Raschig] N,N-Dimethyl-N-palmityl-N-(3-sulfopropyl)-ammonium betaine; amphoteric; surfactant.

Ralufon® DS. [Raschig] N,N-Dimethyl-N-stearyl-N-(3-sulfopropyl)-ammonium betaine; amphoteric; surfactant.

Ralufon® DT. [Raschig] N,N-Dimethyl-N-tallow-N-(3-sulfopropyl)-ammonium betaine; amphoteric; surfactant.

Ralufon® EA 15-90. [Raschig] Sulfopropylated alkyl alkoxylate; anionic; surfactant.

Ralufon® EN 16-80. [Raschig] Alkyl alkoxylate; nonionic; surfactant.

Ralufon® F. [Raschig] Alkanol polyethylene oxide sulfopropyl ether, potassium salt; anionic; surfactant.

Ralufon® MDS. [Raschig] N,N-Distearyl-N-methyl-N-(3-sulfopropyl)-ammonium betaine; amphoteric; surfactant.

Ralufon® N. [Raschig] Nonylphenol polyethylene oxide sulfopropyl ether, potassium salt; anionic; surfactant.

Ralufon® NAPE 14-90. [Raschig] Sulfoalkylated polyalkoxylated naphthol, alkali salt; anionic; surfactant.

Ralufon® TA. [Raschig] N,N-Dimethyl-N-stearic acid-amidopropyl-N-(3-sulfopropyl)-ammonium betaine; surfactant.

Ram Polymer 110. [Atramax] Ammonium salt of acrylic terpolymer; CAS 30643-08-4; anionic; crosslinkable dispersant for pigments used in textiles; vehicle for flexo inks; avail. as metal or amine salt; cures at R.T. or with heat; aq. sol'n.; 35% conc.

Reax® 45A. [Westvaco] Sodium lignosulfonate with anionic wetting agent; anionic; wetting agent and dispersant for pesticides; recommended for use with chlorinated hydrocarbons and organophosphates; EPA exempt; brn. powd., scorched vanilla odor; water-sol.; sp.gr. 1.2-1.3; bulk dens. 27.1 lb/ft^3 (packed); pH 10 (2% aq.).; usage level: 1-6%; toxicology: LD50 (oral, rat) 10.1 g/kg (pract. nontoxic); may be irritating to eyes, skin on prolonged exposure.

Reax® 45DA. [Westvaco] Sodium lignosulfonate with anionic wetting agent; anionic; dispersant/wetting agent for pesticides; effective with pH-sensitive organophosphate, carbamates, difficult-to-wet components; EPA exempt; brn. powd., scorched vanilla odor; water-sol.; sp.gr. 1.2-1.3; bulk dens. 27.1 lb/ft^3 (packed); pH 6.5-7.0 (2% aq.); usage level: 0.5-8%; toxicology: not established; may be skin or eye irritant.

Reax® 45DTC. [Westvaco] Sodium lignosulfonate with anionic wetting agent; anionic; dispersant/wetting agent for pesticides; effective for organometallic pesticides such as maneb, zineb, mancozeb, and triphenyl tin hydroxide; EPA exempt; brn. powd., scorched vanilla odor; water-sol.; sp.gr. 1.2-1.3; bulk dens. 27.1 lb/ft^3 (packed); pH 7.5 (2% aq.); toxicology: not established; may be skin or eye irritant.

Reax® 45L. [Westvaco] Sodium lignosulfonate with anionic wetting agent; anionic; dispersant providing controlled wetting; suspending agent; for pesticides; recommended for use with triazines, urea derivs., fluorinated hydrocarbons, phenoxy compds.; EPA exempt; brn. powd., scorced vanilla odor; sol. in water and alkaline sol'ns.; some sol. in acidic sol'ns.; sp.gr. 1.2-1.3; bulk dens. 27 lb/ft^3 (packed); pH 11 (2% aq.); usage level: 2-6%; toxicology: not extablished; probably irritating to eyes.

Reax® 45T. [Westvaco] Sodium lignosulfonate with two anionic wetting agents; anionic; dispersant/wetting agent for pesticides; effective with triazine, urea derivs., carbamates, organophosphates, chlorinated hydrocarbons; EPA exempt; brn. powd., scorched vanilla odor; sp.gr. 1.2; bulk dens. 27 lb/ft^3 (packed); pH 10 (2% aq.); usage level: 2-6%; toxicology: not established; probably skin or eye irritant.

Redicote AD-130. [Akzo/Asphalt] Amine-based additive; cationic; antistripping agent, asphalt modifier; liq./paste; 50-100% conc.

Redicote AD-141. [Akzo/Asphalt] Amine-based additive; cationic; antistripping agent, asphalt modifier; liq.; 50-100% conc.

Redicote AD-142. [Akzo/Asphalt] Amine-based additive; cationic; antistripping agent, asphalt modifier; liq.; 50-100% conc.

Redicote AD-150. [Akzo/Asphalt] Amine-based additive; cationic; antistripping agent, asphalt modifier; liq./paste; 50-100% conc.

Redicote AD-164. [Akzo/Asphalt] Amine-based additive; cationic; emulsion antistripping additive; liq./paste; 50-100% conc.

Redicote AD-170. [Akzo/Asphalt] Amine-based additive; cationic; antistripping agent, asphalt modifier; liq./paste; 50-100% conc.

Redicote AE 1A. [Akzo/Asphalt] Amine-based emulsifier; cationic; asphalt emulsifier; liq./solid; 50-100% conc.

Redicote AE 6. [Akzo/Asphalt] Amine-based additive; cationic; asphalt emulsifier; liq./solid; 50-100% conc.

Redicote AE 6-A. [Akzo/Asphalt] Amine-based additive; cationic; asphalt emulsifier; liq./solid; 50-100% conc.

Redicote AE 7. [Akzo/Asphalt] Amine-based additive; cationic; asphalt emulsifier; liq./solid; water-sol.; 50-100% conc.

Redicote AE 9. [Akzo/Asphalt] Amine-based additive; cationic; asphalt emulsifier; promotes emulsion stability, antistripping, coating and mixing props.; liq./solid; 100% conc.

Redicote AE 22. [Akzo/Asphalt] Amine-based additive; cationic; asphalt emulsifier with low pour pt.; liq./solid; 50-100% conc.

Redicote AE 26. [Akzo/Asphalt] Amine-based additive; cationic; asphalt emulsifier; liq./solid; 50-100% conc.

Redicote AE 45. [Akzo/Asphalt] Amine-based additive; cationic; asphalt emulsifier; liq./solid; 50-100% conc.

Redicote AE 55. [Akzo/Asphalt] Amine-based additive; cationic; asphalt emulsifier; promotes emulsion stability, antistripping, coating, and mixing props.; liq./solid; 100% conc.

Redicote AE 80. [Akzo/Asphalt] Tall oil deriv.; anionic; asphalt emulsifier; liq.; 35-40% conc.

Redicote AE 91. [Akzo/Asphalt] Amine-based additive; cationic; asphalt emulsifier; liq./paste; 50-100% conc.

Redicote AE 93. [Akzo/Asphalt] Amine-based additive; cationic; asphalt emulsifier; liq./paste; 50-100% conc.

Redicote Series. [Akzo Chem. BV] Amine-based;

cationic; bitumen emulsifier for rapid break and slurry seal bitumen emulsions; heat-stable adhesion agent in hot mixes; liq./solid; 100% conc.

Redicote TR-1114X. [Akzo/Asphalt] Amine-based surfactant; cationic; coal tar modifier, enhances adhesion, storage stability and watering of coal tar emulsions; liq./paste; 100% conc.

Redicote TR-1130. [Akzo/Asphalt] Amine-based surfactant; cationic; coal tar modifier, promotes coating, storage stability and adhesion of coal tar emulsions; liq./solid; 50-100% conc.

Redicote TR-1134. [Akzo/Asphalt] Surfactant; cationic; coal tar modifier, promotes coating, storage stability and adhesion of coal tar emulsions; liq./solid; 50-100% conc.

Reginol 2701. [Hart Prods. Corp.] Surfactant blend; nonionic; scouring agent for dyeing; liq.; 40% conc.

Regitant S-78. [Tokai Seiyu Ind.] Phenol deriv.; anionic; resisting agent for dyeing nylon and nylon/wool blends with acid and 1:2 metal complex dyes; liq.; 40% conc.

Remanol 1300/35. [Kempen] Collagen protein hydrolysate; used in personal care products and building industry; liq.; m.w. 1300; 35% conc.

Rematard Grades. [Hoechst AG] Quat. ammonium compd.; cationic; retarder for dyeing of acrylic fiber with carbonic dyes; liq.

Remazol Salt FD. [Hoechst AG] Sodium chlorinated carboxylate; printing auxiliary for Remazol dyes; gran.

Remcopal 4. [Ceca SA] Laureth-4; nonionic; emulsifier for beeswax; intermediate for polyethoxy ether sulfate; liq.; HLB 9.7; 100% conc.

Remcopal 6. [Ceca SA] PEG-6 oleate; CAS 9004-96-0; nonionic; emulsifier for min. oil and solvs.; stabilizer for polyurethane foams; liq.; HLB 10.7; 100% conc.

Remcopal 10. [Ceca SA] Cetoleth-10; nonionic; base for emulsifiers and visc. spindle oils; paste; HLB 12.2; 100% conc.

Remcopal 18. [Ceca SA] Cetoleth-18; nonionic; retarder for colors; emulsifier for fatty alcohol and olein; paste; HLB 14.7; 100% conc.

Remcopal 20. [Ceca SA] Laureth-20; nonionic; emulsifier for fatty alcohol and olein; solid; HLB 16.2; 100% conc.

Remcopal 25. [Ceca SA] Cetoleth-25; nonionic; raw material for degreaser compds.; emulsifier for fatty alcohol, olein, and waxes; paste; HLB 16.0; 100% conc.

Remcopal 29. [Ceca SA] Nonoxynol-8.5; CAS 9016-45-9; nonionic; emulsifier; liq.; HLB 12.6; 100% conc.

Remcopal 40. [Ceca SA] PEG-31 castor oil; nonionic; emulsifier for olein; liq.; HLB 11.9; 95% conc.

Remcopal 40 S3. [Ceca SA] PEG-40 castor oil; nonionic; emulsifier for vitamins and essential oils; paste; HLB 13.0; 97% conc.

Remcopal 40 S3 LE. [Ceca SA] PEG-40 castor oil; nonionic; surfactant; liq.; HLB 13.0; 90% conc.

Remcopal 121. [Ceca SA] Laureth-3; nonionic; intermediate for sulfation; liq.; sol. in alcohol; insol. in water, min. oil; HLB 8.3; pH 6.5; 100% conc.

Remcopal 207. [Ceca SA] PEG-4.5 oleate; CAS 9004-96-0; nonionic; emulsifier for vitamins; liq.; HLB 8.0; 100% conc.

Remcopal 220. [Ceca SA] Cetoleth-25; nonionic; emulsifier; paste; HLB 16.0; 80% conc.

Remcopal 229. [Ceca SA] Ceteareth-25; CAS 68439-49-6; nonionic; emulsifier for emulsion polymerization, washing detergents; solid; HLB 15.2; 100% conc.

Remcopal 234. [Ceca SA] Cetoleth-4; nonionic; surfactant; paste; HLB 8.0; 100% conc.

Remcopal 238. [Ceca SA] Ceteareth-20; CAS 68439-49-6; nonionic; surfactant for degreaser compds.; paste; HLB 15.1; 100% conc.

Remcopal 258. [Ceca SA] Laureth-9; nonionic; wetting agent; paste; HLB 13.0; 100% conc.

Remcopal 273. [Ceca SA] Isodeceth-3; nonionic; wetting agent for pigments and fillers; liq.; HLB 8.7; 100% conc.

Remcopal 306. [Ceca SA] Octoxynol-5.5; CAS 9002-93-1; nonionic; wetting agent for carbon blk., emulsifier for aromatic hydrocarbons, turpentine oil, and tallow; liq.; HLB 10.5; 100% conc.

Remcopal 334. [Ceca SA] Nonoxynol-4; CAS 9016-45-9; nonionic; emulsifier, dispersant for pigments; liq.; HLB 8.5; 100% conc.

Remcopal 349. [Ceca SA] Nonoxynol-8; CAS 9016-45-9; nonionic; emulsifier, dispersant for pigments; liq.; HLB 12.3; 100% conc.

Remcopal 666. [Ceca SA] Nonoxynol-6; CAS 9016-45-9; nonionic; emulsifier for turpentine oil, heavy aromatic solvs.; wetting agent and dispersant for pigments; Gardner < 2 liq.; sol. in aromatic and chlorinated solvs., min. and veg. oils, water; HLB 10.3; pH 6 ± 1; 100% conc.

Remcopal 3112. [Ceca SA] Nonoxynol-2; CAS 9016-45-9; nonionic; antifoam, emulsifier; liq.; HLB 5.3; 100% conc.

Remcopal 3515. [Ceca SA] Nonoxynol-12.5; CAS 9016-45-9; nonionic; emulsifier; visc. liq.; HLB 14.3; 100% conc.

Remcopal 3712. [Ceca SA] Nonoxynol-12; CAS 9016-45-9; nonionic; emulsifier, wetting agent; visc. liq.; HLB 13.9; 97% conc.

Remcopal 3820. [Ceca SA] Nonoxynol-20; CAS 9016-45-9; nonionic; surfactant; paste; HLB 16; 100% conc.

Remcopal 4000. [Ceca SA] PEG-31 castor oil; nonionic; surfactant; liq.; HLB 12.9; 100% conc.

Remcopal 4018. [Ceca SA] PEG-23 castor oil; nonionic; surfactant; liq.; HLB 10.4; 100% conc.

Remcopal 6110. [Ceca SA] Nonoxynol-9; CAS 9016-45-9; nonionic; emulsifier, wetting agent; liq.; HLB 12.8; 100% conc.

Remcopal 21411. [Ceca SA] Laureth-11; nonionic; emulsifier for lt. hydrocarbons, raw material shampoos; paste; HLB 13.5; 100% conc.

Remcopal 21912 AL. [Ceca SA] Laureth-12; nonionic; surfactant; paste; HLB 14.8; 100% conc.

Remcopal 31250. [Ceca SA] Nonoxynol-50; CAS 9016-45-9; nonionic; emulsifier for epoxy resin; solid; HLB 18.2; 100% conc.

Remcopal 33820. [Ceca SA] Nonoxynol-20; CAS 9016-45-9; nonionic; emulsifier for emulsion polymerization; paste; HLB 15.5; 100% conc.

Remcopal D. [Ceca SA] Cetoleth-23; nonionic; surfactant; gel; HLB 16.0; 33% conc.

Remcopal HC 7. [Ceca SA] PEG-7 hydrog. castor oil; CAS 61788-85-0; nonionic; surfactant; visc. liq.; HLB 1.8; 100% conc.

Remcopal HC 20. [Ceca SA] PEG-20 hydrog. castor oil; CAS 61788-85-0; nonionic; surfactant; paste; HLB 9.5; 100% conc.

Remcopal HC 33. [Ceca SA] PEG-33 hydrog. castor oil; CAS 61788-85-0; nonionic; surfactant; HLB 12.0; 100% conc.

Remcopal HC 40. [Ceca SA] PEG-40 hydrog. castor

oil; CAS 61788-85-0; nonionic; surfactant; solubilizer for essential oils and vitamins; Gardner < 4 paste; sol. in water, ethyl acetate, aromatic and chlorinated solvs., alcohols, veg. oil; disp. in min. oil, aliphatic solvs.; HLB 12.9; acid no. 1.5; iodine no. 0.2; sapon. no. 64; hyd. no. 65; ref. index 1.467; pH 6.5 ± 0.5 (10%); 100% conc.

Remcopal HC 60. [Ceca SA] PEG-60 hydrog. castor oil; CAS 61788-85-0; nonionic; surfactant; HLB 14.6; 100% conc.

Remcopal L9. [Ceca SA] Laureth-9; nonionic; emulsifier for heavy hydrocarbons, degreaser, antistat for syn. fibers; wh. fluid paste; sol. in water, aromatic and chlorinated solvs.; HLB 13.6; pH 4-7 (10% aq.); 100% conc.

Remcopal L12. [Ceca SA] Laureth-10.5; nonionic; surfactant; solid; HLB 14.0; 100% conc.

Remcopal L30. [Ceca SA] Nonoxynol-27; CAS 9016-45-9; nonionic; surfactant; liq.; HLB 16.8; 70% conc.

Remcopal LC. [Ceca SA] Laureth-9; nonionic; washing agent for wool-polyester; gel; HLB 12.8; 50% conc.

Remcopal LO 2B. [Ceca SA] Isodeceth-3; nonionic; surfactant; liq.; HLB 8.0; 65% conc.

Remcopal LP. [Ceca SA] Laureth-9; nonionic; surfactant; liq.; HLB 12.8; 32% conc.

Remcopal NP 30. [Ceca SA] Nonoxynol-27; CAS 9016-45-9; nonionic; emulsifier; solid; HLB 16.8; 100% conc.

Remcopal O9. [Ceca SA] Octoxynol-9; CAS 9002-93-1; nonionic; solubilizer for essential oils; emulsifier; liq.; HLB 13.0; 100% conc.

Remcopal O11. [Ceca SA] Octoxynol-11; CAS 9002-93-1; nonionic; solubilizer for essential oils; emulsifier; liq.; HLB 14.0; 100% conc.

Remcopal O12. [Ceca SA] Octoxynol-12; CAS 9002-93-1; nonionic; solubilizer for essential oils; emulsifier; liq.; HLB 14.5; 100% conc.

Remcopal PONF. [Ceca SA] Nonoxynol-11; CAS 9016-45-9; nonionic; emulsifier; liq.; HLB 13.5; 100% conc.

Remol Brands. [Hoechst AG] Aromatic hydrocarbon; carrier for dyeing of polyester fibers; liq.

Remolgan CX, PM. [Hoechst AG] Oxethylate prods.; degreasing agent for leather; liq.

Renex® 20. [ICI Am.; ICI Surf. Belgium] PEG-16 tallate; CAS 61791-00-2; nonionic; detergent for home laundry, textile scouring, spotting agent, rug shampoos; clear amber liq.; sol. in water, lower alcohols, toluene, acetone, diethyl ether, dioxane, Cellosolve, ethyl acetate, aniline; sp.gr. 1.1; visc. 350 cps; HLB 13.8; cloud pt. 126 F (1% aq.); flash pt. > 300 F; pour pt. 55 F; surf. tens. 41 dynes/cm (0.01%); 100% act.

Renex® 22. [ICI Am.] Mixed fatty and resin acids POE ester (12 EO); detergent for home laundry, textile scouring, rug shampoos, mechanical dishwashing, metal cleaning; amber liq.; sol. in acetone, ethyl acetate, dioxane, Cellosolve, aniline, ethyl alcohol; disp. in water; sp.gr. 1.1; visc. 300 cps; HLB 12.2; cloud pt. 32 F (1% aq.); flash pt. > 300 F; pour pt. 35 F; surf. tens. 39 dynes/cm (0.01%).

Renex® 25. [ICI Am.] PEG-16 tallate and urea; nonionic; detergent; formulates cleaning products; used in home laundry, rug shampoos; tan powd.; sol. in water, lower alcohols, Cellosolve; HLB 13.5; cloud pt. 124 F (1%); flash pt. > 300 F; surf. tens. 39 dynes/cm (0.01%); pH 7 (1% aq.); 50% act.

Renex® 26. [ICI Am.] Fatty acid ester, ethoxylated; nonionic; detergent; amber liq.; sol. in water, petroleum solv. and cleaners' naphtha; visc. 600 cps; flash pt. > 300 F.

Renex® 30. [ICI Am.; ICI Surf. Belgium] Trideceth-12; CAS 24938-91-8; nonionic; detergent and wetting agent; used for raw wool scouring and carbonizing, metal cleaning, bottle washing, home laundry; clear-hazy colorless liq.; sol. in water, lower alcohols, xylene, butyl Cellosolve, carbon tetrachloride, ethylene glycol, propylene glycol; sp.gr. 1.0; visc. 60 cps; HLB 14.5; cloud pt. 183 F (1% aq.); flash pt. > 300 F; pour pt. 55 F; surf. tens. 28 dynes/cm (0.01%); pH 6 (1%); 100% conc.

Renex® 31. [ICI Am.] Trideceth-15; CAS 24938-91-8; nonionic; detergent used for liq. hand dishwashing formulations and metal cleaning; wh. soft paste; sol. in water, propylene glycol, lower alcohols, butyl Cellosolve, carbon tetrachloride, xylene; sp.gr. 1.0; visc. 130 cps; HLB 15.4; cloud pt. 210 F (1% aq.); flash pt. > 300 F; pour pt. 61 F; surf. tens. 34 dynes/cm (0.01%); 100% conc.

Renex® 35. [ICI Am.] Trideceth-12 and urea complex; detergent, wetting agent cleaning prods., raw wool scourng, metal cleaning, bottle washing, home laundry; ivory wh. powd.; sol. in water, ethylene glycol, lower alcohols; m.p. 257 F; HLB 14.5; cloud pt. 180 F (1% aq.); flash pt. > 300 F; surf. tens. 33 dynes/cm (0.01%); pH 7 (1% aq.); 50% conc.

Renex® 36. [ICI Am.; ICI Surf. Belgium] Trideceth-6; CAS 24938-91-8; nonionic; detergent, wetting agent, dye leveling agent used in alkaline cleaners, dishwashing, textiles, dairy cleaners, sanitizers; dispersant for solids and paint pigments; cloudy-clear colorless thin oily liq.; sol. in acetone, CCl_4, ethyl acetate, Cellosolve, ethanol, toluene, aniline; disp. in water; sp.gr. 1.0; visc. 80 cps.; HLB 11.4; cloud pt. < 32 F (1% aq.); flash pt. > 300 F; surf. tens. 27 dynes/cm (0.01%); pH 6 (1%); 100% conc.

Renex® 647. [ICI Am.; ICI Surf. Belgium] Nonoxynol-4; CAS 9016-45-9; nonionic; emulsifier, detergent; liq.; HLB 8.9; 100% conc.

Renex® 648. [ICI Am.; ICI Surf. Belgium] Nonoxynol-5; CAS 9016-45-9; nonionic; detergent, solv. and emulsion cleaning applications; metal cleaning, clear liq.; sol. in propylene glycol, perchlorethylene, IPA, cottonseed oil; sp.gr. 1.03; visc. 223 cps; HLB 10; flash pt. > 200 F; pour pt. –20 F; 100% act.

Renex® 649. [ICI Am.; ICI Surf. Belgium] Nonoxynol-20; CAS 9016-45-9; nonionic; emulsifier, detergent; solid; HLB 16.0; 100% conc.

Renex® 650. [ICI Am.; ICI Surf. Belgium] Nonoxynol-30; CAS 9016-45-9; nonionic; detergent, emulsifier, stabilizer; surfactant for detergent use; emulsification, and latex stabilizer; wh. solid; sol. in water, propylene glycol, IPA glycol, IPA, xylene; sp.gr. 1.15; HLB 17.1; cloud pt. 212 F; flash pt. > 300 F; pour pt. 91 F; surf. tens. 42 dynes/cm (0.01%); 100% act.

Renex® 678. [ICI Am.; ICI Surf. Belgium] Nonoxynol-15; CAS 9016-45-9; nonionic; detergent, wetting agent used in dishwashing; clear liq.; sol. in water, propylene glycol, perchloroethylene, IPA, xylene; sp.gr 1.08; visc. 319 cps; HLB 15; cloud pt. 211 F (1%); flash pt. > 300 F; pour pt. 69 F; surf. tens. 33 dynes/cm (0.01%); 100% act.

Renex® 679. [ICI Am.; ICI Surf. Belgium] Nonoxynol-13; CAS 9016-45-9; nonionic; emulsifier, detergent; liq.; HLB 14.4; 100% conc.

Renex® 682. [ICI Am.; ICI Surf. Belgium] Nonoxynol-12; CAS 9016-45-9; nonionic; emulsifier and detergent; liq.; HLB 13.9; 100% conc.

Renex® 688. [ICI Am.; ICI Surf. Belgium] Nonoxynol-8; CAS 9016-45-9; nonionic; detergent, wetting agent; surfactant for dishwashing, textile scouring; liq.; sol. in propylene glycol, perchloroethylene, IPA, xylene, water and cottonseed oil; sp.gr. 1.05; visc. 240 cps; HLB 12.3; cloud pt. 87 F (1%); flash pt. > 300 F; pour pt. 12 F; surf. tens. 30 dynes/cm (0.01%); 100% act.

Renex® 690. [ICI Am.; ICI Surf. Belgium] Nonoxynol-10; CAS 9016-45-9; nonionic; wetting agent, detergent for home and commercial laundry products, textile scouring, dyeing and desizing, acid metal cleaner, sanitizer, and hard surface cleaning; liq.; sol. see Renex 688; sp.gr. 1.06; visc. 260 cps; HLB 13.3; cloud pt. 150 F (1%); flash pt. > 300 F; pour pt. 33 F; surf. tens. 31 dynes/cm (0.01%); 100% act.

Renex® 697. [ICI Am.; ICI Surf. Belgium] Nonoxynol-6; CAS 9016-45-9; nonionic; emulsifier, detergent; clear liq.; sol. in propylene glycol, perchloroethylene, IPA, xylene, cottonseed oil; sp.gr. 1.04; visc. 250 cps; HLB 10.9; cloud pt. < 32 F (1%); flash pt. > 300 F; pour pt. -21 F; surf. tens. 30 dynes/cm (0.01%); 100% act.

Renex® 698. [ICI Am.; ICI Surf. Belgium] Nonoxynol-9 (9–9.5 EO); CAS 9016-45-9; nonionic; emulsifier, detergent; clear liq.; sol. in propylene glycol, perchloroethylene, IPA, xylene, water, cottonseed oil; sp.gr. 1.06; visc. 245 cps; HLB 13; cloud pt. 129 F (1%); flash pt. > 300 F; pour pt. 28 F; surf. tens. 30 dynes/cm (0.01%); 100% act.

Renex® 702. [ICI Am.; ICI Surf. Belgium] PEG-2 syn. primary C13–15 alcohol; nonionic; general purpose detergent, emulsifier; biodeg.; liq.; HLB 5.9; 100% conc.

Renex® 703. [ICI Am.; ICI Surf. Belgium] PEG-3 syn. primary C13–15 alcohol; nonionic; general purpose detergent, emulsifier; biodeg.; liq.; HLB 7.8; 100% conc.

Renex® 704. [ICI Am.] POE alkyl ether; nonionic; general purpose detergent, emulsifier; liq.; biodeg.; 100% conc.

Renex® 707. [ICI Am.; ICI Surf. Belgium] PEG-7 syn. primary C13–15 alcohol; nonionic; general purpose detergent, emulsifier; biodeg.; liq.; HLB 12.2; 100% conc.

Renex® 711. [ICI Am.; ICI Surf. Belgium] PEG-11 syn. primary C13–15 alcohol; nonionic; general detergent, emulsifier; biodeg.; liq.; HLB 13.9; 100% conc.

Renex® 714. [ICI Am.] POE alkyl ether; nonionic; general purpose detergent, emulsifier; liq.; biodeg.; 100% conc.

Renex® 720. [ICI Am.; ICI Surf. Belgium] PEG-20 syn. primary C13–15 alcohol; nonionic; general detergent, emulsifier; biodeg.; liq.; HLB 16.2; 100% conc.

Renex® 750. [ICI Am.; ICI Surf. Belgium] Octoxynol-10; CAS 9002-93-1; nonionic; emulsifier, detergent; biodeg.; liq.; HLB 13.6; 100% conc.

Renex® 751. [ICI Am.] POE octyl phenol ether; nonionic; lubricant, emulsifier, textile processing; liq.; 100% conc.

Reserve Salt Flake. [Ciba-Geigy/Dyestuffs] Sodium m-nitrobenzene sulfonate; anionic; stabilizer for fiber reactive dyeing systems; assistant in discharge printing; flake.

Resogen® 35 Conc. [Crompton & Knowles] Aminoaldehyde condensate; cationic; aftertreatment fixing agent for improving wetfastness of direct dyes on cellulose; good penetration, exc. lightfastness; low formaldehyde (< 1.5%); yel.-orange clear to sl. hazy liq.; sol. in water @ 3%, disp. @ 5-10%; pH 3.2-3.5.

Resogen® DM. [Crompton & Knowles] Aminoaldehyde condensate; cationic; aftertreatment fixing agent to improve wetfastness of direct dyes on cellulosics; good penetration; off-wh. clear to sl. hazy liq.; readily sol. in water; pH 2.5 ± 0.5.

Resolvyl 610. [ICI Surf. UK] Polyvinyl acetate resin aq. disp.; nonionic; gives filled, stiff finishes without yellowing; paste.

Resolvyl BC. [ICI Surf. UK] Polyvinyl acetate resin aq. disp.; nonionic; gives filled, stiff finishes without yellowing; paste.

Retarder N. [Hart Prods. Corp.] Dimethyl lauryl benzyl ammonium chloride; amphoteric; retarder in dyeing; antistatic agent; liq.; 40% conc.

Retarder N-85. [Hart Prods. Corp.] Dimethyl lauryl benzyl ammonium chloride; amphoteric; retarder in dyeing; antistatic agent; liq.; 85% conc.

Retardine. [Henkel/Textile] Retarder/leveling agent for dyeing cellulosics with vat, direct and sulfur dyes; dk. brn. liq.; sol. forms clear sol'ns. in water; pH 3.5 (2%).

Retentol RM. [Zschimmer & Schwarz] Quat. ammonium; cationic; disinfectant; liq.; 50% act.

Rewocid® U 185. [Rewo GmbH] Undecylenamide MEA; nonionic; fungicide, anti-mycotic agent; detergent; improves foam quality, stability, superfatting, increases visc.; yel. flakes; strong odor; sol. in alcohol, glycol, disp. in water; m.p. 50 C; 100% act.

Rewocoros RAB 90. [Rewo GmbH] Modified boric DEA; anionic; low foaming corrosion inhibitor for sol. aq. metalworking oils, syn. cold lubricants, water glycol hydraulic fluids, grinding lubricants; liq.; 100% conc.

Rewoderm® S 1333. [Rewo GmbH] Disodium ricinoleamido MEA-sulfosuccinate; anionic; detergent; used for skin protection, washing up liqs., personal care products; decreases irritancy of alkylbenzene sulfonate and other surfactants; emulsifier for emulsion polymerization; amber liq.; pH 6.5–7.5 (5%); surf. tens. 28 mN/m; 40% act.

Rewoderm® S 1333 P. [Rewo GmbH] Disodium ricinoleamido MEA-sulfosuccinate; anionic; surfactant for syndet soaps, detergent powds.; powd.; 95% conc.

Rewolub KSM 14. [Rewo GmbH] Dicarboxylic acid ethoxylate; nonionic; surfactant for syn. cooling oils, lubricant; visc. liq.; 100% conc.

Rewolub KSM 80. [Rewo GmbH] Dicarboxylic acid diamide modified; nonionic; surfactant for syn. cooling oils, metalworking lubricants; visc. liq.; 83% conc.

Rewomat B 2003. [Rewo GmbH] Tetrasodium (1,2-dicarboxyethyl)-N-alkyl sulfosuccinamide; anionic; detergent, foaming agent and stabilizer; emulsion polymerization additive; industrial, metal, household and all-purpose cleaner; solubilizer for detergent raw materials; soldering aid; pigment dispersant; emulsifier for wax and oil; cosmetics; amber clear liq.; sol. in water, alkali and electrolytes; pH 7–9 (10% solids); 35% solids.

Rewomat TMS. [Rewo GmbH] Disodium alkyl sulfosuccinamide; anionic; foaming agent, stabilizer, emulsifier, foam additive for natural and syn. latexes; emulsion polymerization; wetting agent for latex impregnation; soft creamy paste or liq.; pH 9–

10 (10% solids); 35% solids.

Rewomid® AC 28. [Rewo GmbH] Groundnut fatty acid polydialkanolamide; nonionic; detergent, emulsifer used as an anticorrosion additive for metal-working lubricants and oils; brn. liq.; dens. 0.98 g/cm^3; sp.gr. 1.0; visc. med.; 100% act.

Rewomid® C 212. [Rewo GmbH] Cocamide MEA; CAS 68140-00-1; nonionic; detergent, thickener, foam booster/stabilizer, superfatting agent used in detergent products; stabilizer of emulsions; off-wh. flakes; amidic slight odor; m.p. 73 C; 100% act.

Rewomid® C 220SE. [Rewo GmbH] Cocamide DEA; nonionic; detergent, foam stabilizer, detergent builder; liq.; 100% conc.

Rewomid® CD. [Rewo GmbH] Cocamide MEA; CAS 68140-00-1; nonionic; detergent, foam stabilizer, detergent builder; liq.; 100% conc.

Rewomid® DC 212 S. [Rewo GmbH] Cocamide DEA; nonionic; detergent, foam stabilizer and visc. modifier used for shampoos and all-purpose detergents; lt. yel. visc. liq.; slight odor; dens. 0.99 g/cm^3; 100% act.

Rewomid® DC 220 SE. [Rewo GmbH] Cocamide DEA; nonionic; detergent, thickener, foam stabilizer and superfatting agent; soft paste; 100% act.

Rewomid® DL 203. [Rewo GmbH] Lauramide DEA and diethanolamine; nonionic; detergent; increases visc.; good superfatting; stabilizer of emulsions; fixation of perfumes; wh. wax; dens. 0.97 g/cm^3; sp.gr. 1.0; 100% act.

Rewomid® DL 203 S. [Rewo GmbH] Lauramide DEA and diethanolamine; nonionic; foam booster/stabilizer, superfatting agent, and thickener for personal care prods., floor cleaners, general purpose cleaners, textile lubricants; wax; 100% conc.

Rewomid® DL 240. [Rewo GmbH] Cocamide DEA (2:1) and diethanolamine; nonionic; detergent added to cosmetic preparations and cleaning and washing agents; stabilizer of emulsions; superfatting agent; yel. liq.; faint odor; visc. med.; 100% act.

Rewomid® DLMS. [Rewo GmbH] Lauramide DEA; CAS 120-40-1; nonionic; foam stabilizer and thickener in detergent systems; wax; 100% conc.

Rewomid® DO 280. [Rewo GmbH] Oleamide DEA and diethanolamine; nonionic; detergent; products in the cosmetic, cleaning, and detergent industries; amber liq.; slight odor; sp.gr. 1.0; 100% act.

Rewomid® DO 280 S. [Rewo GmbH] Mixed fatty acid DEA; nonionic; emulsifier; superfatter; aerosol anticorrosive agent; hair cosmetic specialty; cream bath additive; dye and perfume solubilizer; paste; 100% conc.

Rewomid® DO 280 SE. [Rewo GmbH] Oleamide DEA (1:1); nonionic; detergent, thickener, foam booster/stabilizer, superfatting agent for cosmetic products; conditioner for shampoos, bath oils; honey liq.; dens. 0.96 g/cm^3; 100% act.

Rewomid® IPL 203. [Rewo GmbH] Lauramide MIPA; CAS 142-54-1; nonionic; foam stabilizer, detergent for shampoos, shaving foams, and dishwashing liqs.; additive for solid and paste end prods.; improved washing power; stabilizer of emulsions; wh. flakes; slight odor; m.p. 55 C; 100% act.

Rewomid® IPP 240. [Rewo GmbH] Cocamide MIPA; nonionic; detergent, foam stabilizer, thickener, additive for solid and paste end prods.; improved washing power; emulsion stabilizer; wh. flakes; slight odor; m.p. 50 C; 100% act.

Rewomid® L 203. [Rewo GmbH] Lauramide MEA; CAS 142-78-9; nonionic; detergent, thickener, foam booster/stabilizer, superfatting agent for detergent preparations; fixation of perfumes; stabilizer of emulsions; wh. flakes; slight odor; m.p. 84 C; 100% act.

Rewomid® OM 101/G. [Rewo GmbH] Peanutamide MEA; nonionic; emulsifier, detergent intermediate; solv.; specialty cosmetic prod.; wax; 100% conc.

Rewomid® OM 101/IG. [Rewo GmbH] Peanutamide MIPA; nonionic; emulsifier, detergent intermediate; solv.; specialty cosmetic prod.; wax; 100% conc.

Rewomid® R 280. [Rewo GmbH] Ricinoleamide MEA; nonionic; surfactant for syn. soap bars; foam stabilizer, thickener; wax; 100% conc.

Rewomid® RE. [Rewo GmbH] Blend of anionics and alkylolamides; emulsifier and corrosion inhibitor for metalworking lubricants; liq.; 100% conc.

Rewomid® S 280. [Rewo GmbH] Stearamide MEA; nonionic; detergent, thickener, foam booster, superfatting agent; stabilizer of emulsions; syn. soap bars; anti-inflammatory agent; wh. flakes; slight odor; m.p. 92 C; 100% act.

Rewomid® U 185. [Rewo GmbH] Undecylenamide MEA; nonionic; fungicide; antimycotic; flake; 100% conc.

Rewomine IM-BT. [Rewo GmbH] Imidazoline deriv.; cationic; adhesive additive/agent and emulsifier in acidic bitumen emulsions; rust inhibitor in acidic media; Gardner 13 max. gr.-brn. liq.; amine odor; dens. 0.94 g/cm^3; 100% act.

Rewomine IM-CA. [Rewo GmbH] 1-Hydroxyethyl-2-heptadecenyl imidazoline; cationic; corrosion inhibitor, emulsifier, penetrant, wetting agent; used in leather and metalworking industry, paint and dyes, for carbonization baths; yel. wax; m.w. 285; sol. in nonpolar solvs.; disp. in water; 100% act.

Rewomine IM-OA. [Rewo GmbH] 1-Hydroxyethyl-2-heptadecenyl imidazoline; cationic; coupling agent for rust prevention; emulsifier for the emulsification of oils and bitumens; antistat; raw material for quat. reactions; paint industry; amber liq.; m.w. 345; sol. in polar and nonpolar solvs.; disp. in water; dens. 0.94 g/cm^3; visc. 800 cps; pH 10–11 (1% aq.), 100% act.

Rewopal® BN 13. [Rewo GmbH] PEG-13 naphthole; nonionic; wetting agent for electroplating baths; liq.; 100% conc.

Rewopal® C 6. [Rewo GmbH] PEG-6 cocamide; nonionic; dispersant, emulsifier, foam booster, wetting agent for calcium soap, personal care prods.; solubilizer for perfume oils; amber liq.; cloud pt. 80–90 C (2%); pH 8–10 (1% solids); 100% act.

Rewopal® CSF 20. [Rewo GmbH] Ceteareth-20; CAS 68439-49-6; nonionic; flake; 100% conc.

Rewopal® EO 70. [Rewo GmbH] Oleic acid polyglycol ester; nonionic; emulsifier for min. oils (textile oils, spinning oils, cutting and grinding fluids, drawing and rolling oils); med. visc. liq.; 100% conc.

Rewopal® HV 4. [Rewo GmbH] Nonoxynol-4; CAS 9016-45-9; nonionic; emulsifier for min. oils, petroleum, aliphatic hydrocarbons; med. visc. liq.; 100% conc.

Rewopal® HV 5. [Rewo GmbH] Nonoxynol-5; CAS 9016-45-9; nonionic; emulsifier, solubilizer used in emulsion polymerization and insecticides, raw material for rinsing, washing, and cleaning agents; wetting agent in the paper and cellulose industry, degreaser used in pickling and alkaline immersion-bath cleaners; BGA and FDA compliance; Gardner 2 max. liq.; sol. in min. oil, alcohol, ketone, toluene, xylene, chlorinated hydrocarbons; cloud pt. 59–62 C; pH 5–7 (1%); surf. tens. 30 dynes/cm; 100% act.

Rewopal® HV 6. [Rewo GmbH] Nonoxynol-6; CAS 9016-45-9; nonionic; emulsifier for min. oils, petroleum, aliphatic hydrocarbons in degreasers; med. visc. liq.; 100% conc.

Rewopal® HV 8. [Rewo GmbH] Nonoxynol-8; CAS 9016-45-9; nonionic; emulsifier for min. oils, petrol., aliphatic hydrocarbons in degreasers (e.g., wool scouring); BGA and FDA compliance; Gardner 2 max. liq.; sol. in min. oil, alcohol, ketone, toluene, xylene, chlorinated hydrocarbons; cloud pt. 41–45 C; pH 5–7 (1% solids); surf. tens. 32 dynes/cm; 100% act.

Rewopal® HV 9. [Rewo GmbH] Nonoxynol-9; CAS 9016-45-9; nonionic; emulsifier, solubilizer used in emulsion polymerization and insecticides, raw material for rinsing, washing, and cleaning agents; wetting agent in the paper and cellulose industry, degreaser used in pickling and alkaline immersion-bath cleaners; Gardner 2 max. liq.; sol. in min. oil, alcohol, ketone, toluene, xylene, chlorinated hydrocarbons; cloud pt. 59–62 C (2%); pH 5–7 (1% solids).

Rewopal® HV 10. [Rewo GmbH] Nonoxynol-10; CAS 9016-45-9; nonionic; see Rewopal HV 5; BGA and FDA compliance; Gardner 2 max. liq.; sol. in min. oil, alcohol, ketone, toluene, xylene, chlorinated hydrocarbons; cloud pt. 70–73 C (2%); pH 5–7 (1% solids); surf. tens. 34 dynes/cm; 100% act.

Rewopal® HV 14. [Rewo GmbH] Nonoxynol-14; CAS 9016-45-9; nonionic; surfactant for detergents and cleaners with high electrolyte conc. (household and industrial detergents); dye leveling agent; wh. soft wax; sol. in min. oil, alcohol, ketone, toluene, xylene, chlorinated hydrocarbons; cloud pt. 60–63 C (2%); pH 5–7 (1% solids); 100% act.

Rewopal® HV 25. [Rewo GmbH] Nonoxynol-25; CAS 9016-45-9; nonionic; see Rewopal HV 5; BGA and FDA compliance; wh. wax; sol. in min. oil, alcohol, ketone, toluene, xylene, chlorinated hydrocarbons; cloud pt. 73–76 C (2%); pH 5–7 (1% solids); surf. tens. 42 dynes/cm; 100% act.

Rewopal® HV 50. [Rewo GmbH] Nonoxynol-50; CAS 9016-45-9; nonionic; emulsifier for emulsion polymerization; wax; 100% conc.

Rewopal® LA 3. [Rewo GmbH] Laureth-3; nonionic; emulsifier for solvs. (cold water detergents), cosmetic oils (bath oils), min. oils (textile aux., metalworking), coupler, raw material for the prod. of ether sulfates; shampoos; clear liq.; sol. in min. and org. solvs.; pH 5–7 (1% solids); biodeg.; 100% conc.

Rewopal® LA 6. [Rewo GmbH] Laureth-6; nonionic; detergent; soft wax; 100% conc.

Rewopal® LA 6-90. [Rewo GmbH] Laureth-6; nonionic; detergent; turbid liq.; 90% conc.

Rewopal® LA 10. [Rewo GmbH] Laureth-10; nonionic; detergent; soft wax; 100% conc.

Rewopal® LA 10-80. [Rewo GmbH] Laureth-10; nonionic; detergent; turbid liq.; 80% conc.

Rewopal® M 365. [Rewo GmbH] Ricinoleic acid polyglycol ester; nonionic; additive for metal working fluids and oils; med. visc. liq.; 100% conc.

Rewopal® MT 65. [Rewo GmbH] Fatty alcohol/PEG methyl ether; nonionic; low foaming detergent for strong acidic cleaners, textile auxiliaries; acid-stable; visc. liq.; HLB 12.0; 100% conc.

Rewopal® MT 2455. [Rewo GmbH] Fatty alcohol EO/PO methyl ether; nonionic; fiber lubricant, detergent; liq.; 100% conc.

Rewopal® MT 2540. [Rewo GmbH] Fatty alcohol EO/PO methyl ether; nonionic; low foaming detergent, syn. fiber lubricant; textile applics.; visc. liq; 100% conc.

Rewopal® MT 5722. [Rewo GmbH] Fatty alcohol EO/PO methyl ether; nonionic; syn. fiber lubricant, low foaming detergent; liq.; 100% conc.

Rewopal® O 8. [Rewo GmbH] PEG-9 oleamide; nonionic; detergent; wetting agent, o/w emulsifier, and dispersant for calcium soap; suitable for machine washing formulations; brn. liq.; sol. in water, alcohol, ketone, ester, chlorinated hydrocarbons, benzene, fatty oils; cloud pt. 75–85 C; pH 8–10 (1%); 100% act.

Rewopal® O 15. [Rewo GmbH] PEG-15 oleamide; nonionic; detergent; waxy; 100% conc.

Rewopal® PO. [Rewo GmbH] EO/PO block polymers; nonionic; wetting agent for nonfoaming cleaning agents; med. visc. liq.; 100% conc.

Rewopal® RO 40. [Rewo GmbH] Castor oil ethoxylate; nonionic; emulsifier for metalworking fluids, textile auxiliaries, agric. chemicals; liq.; 100% conc.

Rewopal® TA 11. [Rewo GmbH] Talloweth-11; CAS 61791-28-4; nonionic; detergent, wash-act. base, wetting agent, dispersant, emulsifier for waxes; biodeg.; wh. soft wax; sol. in water and org. solvs.; cloud pt. 69–73 C; pH 5–7 (1% solids).

Rewopal® TA 25. [Rewo GmbH] Talloweth-25; CAS 61791-28-4; nonionic; detergent, dispersant, emulsifier, solubilizer for cleansing agents; wh. powd.; sol. in water and org. solvs.; cloud pt. 75–90 C (2%); pH 5–7 (1% solids); 100% conc.

Rewopal® TA 25/S. [Rewo GmbH] Talloweth-25; CAS 61791-28-4; nonionic; detergents surfactant, dispersant; flakes; 100% conc.

Rewopal® TA 50. [Rewo GmbH] Talloweth-50; CAS 61791-28-4; nonionic; detergent, dispersant; wh. powd.; cloud pt. 72–76 C (2%); pH 5–7 (1% solids); 100% conc.

Rewopal® TPD 30. [Rewo GmbH] Fatty amine ethoxylate; nonionic; emulsifier for min. oils; visc. liq.; 100% conc.

Rewophat E 1027. [Rewo GmbH] Alkylphenol polyglycol ether phosphate; anionic; corrosion inhibitor, emulsifier for min. oils, dispersant, wetting agent; for emulsion polymerization; metalworking fluids; antistat; high pressure additive; textile antistat; yel. liq.; sol. in org. solvs.; sol. < 5 g/1000 ml water; pH 2-4 (1%); surf. tens. 32 dynes/cm; 100% act.

Rewophat EAK 8190. [Rewo GmbH] Laureth-3 phosphate; anionic; corrosion inhibitor, emulsifier, dispersant, wetting agent; for metalworking fluids, textile auxiliaries and cleaners; antistat; high pressure additive; biodeg.; yel. liq.; disp. in water; 100% act.

Rewophat NP 90. [Rewo GmbH] Nonylphenol polyglycol ether phosphate; anionic; emulsifier for emulsion polymerization; textile auxiliaries; antistat, raw material for industrial cleaners; visc. liq.; sol. in water, org. solvs.; pH 2-4 (1%); surf. tens. 34 dynes/cm; 100% conc.

Rewophat OP 80. [Rewo GmbH] Fatty alcohol polyglycol ether phosphate; anionic; additive for syn. fiber; min. oil emulsifier, antistat; paste; 100% conc.

Rewophat TD 40. [Rewo GmbH] Fatty alcohol

polyglycol ether phosphate; anionic; hydrotrope; solubilizer for acid cleaners; liq.; 100% conc.

Rewophat TD 70. [Rewo GmbH] Fatty alcohol polyglycol ether phosphate; anionic; raw material for chemical and acid cleaners; antisoiling finish, flame retarder; liq.; 100% conc.

Rewopol® 15/L. [Rewo GmbH] Sodium lauryl sulfate; CAS 151-21-3; anionic; surfactant for emulsion polymerization; liq.; 15% conc.

Rewopol® AL 3. [Rewo GmbH] Ammonium laureth sulfate; anionic; raw material for dishwashing and cleansing agents; visc. liq.; 60% conc.

Rewopol® B 1003. [Rewo GmbH] Disodium tallow sulfosuccinamate; anionic; foaming and antigelling agent for latex foam backings and coatings; emulsifier for emulsion polymerization; flotation agent; FDA compliance; paste; sol. 35 g/1000 ml water (40 C); pH 8-10 (1%); surf. tens. 42 dynes/cm; 35% conc.

Rewopol® B 2003. [Rewo GmbH] Tetrasodium dicarboxyethyl stearyl sulfosuccinamate; anionic; flotation reagent; emulsifier for emulsion polymerization; foaming agent for latex emulsion (carpet backing); antigelling agent, cleaning agent for paper mill felts; FDA compliance; low visc. liq.; sol. in water; partly sol. in org. solvs.; pH 7-8 (1%); surf. tens. 40 dynes/cm; 35% conc.

Rewopol® BW. [Rewo GmbH] Blend of nonionics; nonionic; special emulsifier for cleaning agents containing solvs.; for hand cleaning, engine cleaning; clear yel. visc. liq.; oil sol.; 100% act.

Rewopol® BWA. [Rewo GmbH] Blend of nonionics; nonionic; emulsifier for cleaning agents containing solvs., e.g., hand cleaning, metal cleaning, solv. degreaser, paraffin wax and cold cleaners to dewax cars; clear yel. visc. liq.; 100% act.

Rewopol® CHT 12. [Rewo GmbH] Coco-EDTA-amide; anionic; sequestering agent, complexing surfactant for detergents; visc. liq.; 40% conc.

Rewopol® CL 30. [Rewo GmbH] Laureth-3 carboxylic acid; anionic; surfactant for household cleaners and toiletries; liq.; 90% conc.

Rewopol® CT 65. [Rewo GmbH] Trideceth-7 carboxylic acid; anionic; acid-stable cleaner for household and industrial use, textile auxiliaries, min. oil emulsions, tert. oil recovery; emulsifier, wetting agent for personal care products; opaque liq.; 65% conc.

Rewopol® CTN. [Rewo GmbH] Sodium trideceth-7 carboxylate; anionic; surfactant for textile auxs.; liq.; 60% conc.

Rewopol® DBC. [Rewo GmbH] Dodecyl benzyl chloride; nonionic; alkylating agent; visc. liq.; 100% conc.

Rewopol® FBR. [Rewo GmbH] Blend of nonionic surfactants and alkylolamides; nonionic; surfactant for floor cleaners, aluminum cleaners, all-purpose cleaners; liq.; 98% conc.

Rewopol® HD 50 L. [Rewo GmbH] Surfactant blend; anionic; preconc. for liq. heavy-duty detergents; liq.; 50% conc.

Rewopol® MLS 30. [Rewo GmbH] MEA-lauryl sulfate; anionic; surfactant for personal care products, foam baths, shampoos, liq. detergents; detergent raw material; visc. liq.; 30% conc.

Rewopol® MLS 35. [Rewo GmbH] MEA-lauryl sulfate; anionic; raw material for liq. detergents, foam baths and shampoos; liq.; 35% conc.

Rewopol® NEHS 40. [Rewo GmbH] Sodium octyl sulfate; anionic; low foaming wetting agent for alkaline cleaners, mercerizing, electroplating; hydrotrope; liq.; 40% conc.

Rewopol® NI 56. [Rewo GmbH] Sodium 2-hydroxyethyl sulfate; anionic; surfactant for syndet soaps, textile aux.; low visc. liq.; 56% conc.

Rewopol® NL 2. [Rewo GmbH] Sodium laureth sulfate; CAS 9004-82-4; surfactant for shampoos, washing liqs., detergents; liq.; 28% conc.

Rewopol® NL 2-28. [Rewo GmbH] Sodium laureth sulfate; CAS 9004-82-4; anionic; surfactant for shampoos, shower gels, foam baths, liq. soaps, dishwashing liqs., emulsion polymerization, air entrainment agent, textile auxiliaries; liq.; 28% conc.

Rewopol® NL 3. [Rewo GmbH] Sodium laureth sulfate; CAS 9004-82-4; anionic; detergent, foamer; cleaning formulations; personal care products; emulsifier for emulsion polymerization; BGA compliance; colorless clear liq.; low odor; visc. 50–150 cps; pH 6.5-7.5 (1%); surf. tens. 38 mN/m; 28% act.

Rewopol® NL 3-28. [Rewo GmbH] Sodium laureth sulfate; CAS 9004-82-4; anionic; surfactant for shampoos, shower gels, foam baths, liq. soaps, dishwashing liqs., emulsion polymerization, air entrainment agent, textile auxiliaries; liq.; 28% conc.

Rewopol® NL 3-70. [Rewo GmbH] Sodium laureth sulfate; CAS 9004-82-4; anionic; surfactant for shampoos, shower gels, foam baths, liq. soaps, dishwashing liqs., emulsion polymerization, air entrainment agent, textile auxiliaries; paste; HLB 18.0; 70% conc.

Rewopol® NLS 15 L. [Rewo GmbH] Sodium lauryl sulfate; CAS 151-21-3; anionic; emulsifier for emulsion polymerization; BGA and FDA compliance; liq.; water-sol.; pH 9-9.5 (1%); surf. tens. 30 mN/m; 15% conc.

Rewopol® NLS 28. [Rewo GmbH] Sodium lauryl sulfate; CAS 151-21-3; anionic; detergent, emulsifier in emulsion polymerization; raw material for mild detergents, shampoos, detergent pastes, cosmetics; BGA and FDA compliance; lt. yel. low visc. liq.; water-sol.; m.w. 300; pH 7–8 (1%); surf. tens. 30 mN/m; 28% act.

Rewopol® NLS 30 L. [Rewo GmbH] Sodium lauryl sulfate; CAS 151-21-3; anionic; emulsifier for emulsion polymerization; BGA and FDA compliance; liq.; water-sol.; pH 9-9.5 (1%); surf. tens. 30 mN/m; 30% conc.

Rewopol® NLS 90. [Rewo GmbH] Sodium lauryl sulfate; CAS 151-21-3; anionic; raw material for mild detergents, syndet soaps, shampoos in powd. form; wh.-lt. yel. powd.; m.w. 300; pH 7–9 (1% solids); 95% conc.

Rewopol® NOS 5. [Rewo GmbH] Nonyl phenol polyglycol ether sulfate; anionic; emulsion polymerization surfactant esp. for styrene; BGA and FDA compliance; liq.; sol. in water; pH 6-8 (1%); surf. tens. 29 mN/m; 30% conc.

Rewopol® NOS 8. [Rewo GmbH] Nonyl phenol polyglycol ether sulfate; anionic; emulsifier for styrene polymerization and all other monomers; BGA and FDA compliance; liq.; water-sol.; pH 7-8 (1%); surf. tens. 35 mN/m; 33% conc.

Rewopol® NOS 10. [Rewo GmbH] Nonyl phenol polyglycol ether sulfate; anionic; emulsifier for styrene polymerization and all other monomers; BGA and FDA compliance; liq.; water-sol.; pH 7-8 (1%); surf. tens. 36 mN/m; 35% conc.

Rewopol® NOS 25. [Rewo GmbH] Nonyl phenol polyglycol ether sulfate; anionic; emulsifier for styrene polymerization and all other monomers; BGA and FDA compliance; liq.; water-sol.; pH 7-8 (1%); surf. tens. 40 mN/m; 35% conc.

Rewopol® PGK 2000. [Rewo GmbH] Surfactant blend with pearlizing agent; anionic; pearlizing agent for shampoos, foam baths, household cleaners; liq.; 35% conc.

Rewopol® S 1954. [Rewo GmbH] Surfactant blend; anionic; dishwashing liq. conc., all-purpose cleaner; liq.; 45% conc.

Rewopol® S 2311. [Rewo GmbH] Surfactant blend; anionic; dishwashing liq. conc., all-purpose cleaner; liq.; 40% conc.

Rewopol® SBC 212. [Rewo GmbH] Disodium cocamido MEA-sulfosuccinate; anionic; surfactant for foam cleaners, lt. duty detergents, and personal care products; low visc. liq.; 40% conc.

Rewopol® SBC 212 G. [Rewo GmbH] Disodium cocamido MEA-sulfosuccinate; anionic; surfactant for foam cleaners, lt. duty detergents, and personal care products; granules; 95% conc.

Rewopol® SBDB 45. [Rewo GmbH] Diisobutyl sodium sulfosuccinate; anionic; emulsifier for emulsion polymerization; stabilizer for disps.; pigment dispersant; BGA and FDA compliance; liq.; sol. 240 g/1000 ml water; pH 5-7 (1%); surf. tens. 49 dynes/cm; 45% conc.

Rewopol® SBDC 40. [Rewo GmbH] Dicyclohexyl sodium sulfosuccinate; anionic; emulsifier for emulsion polymerization; stabilizer for disps.; pigment dispersant; BGA and FDA compliance; paste; sol. 10 g/1000 ml water; sol. hot in org. solvs.; pH 5-7 (1%); surf. tens. 40 dynes/cm; 40% conc.

Rewopol® SBDD 65. [Rewo GmbH] Diisodecyl sodium sulfosuccinate; anionic; emulsifier for emulsion polymierzation; stabilizer for disps.; pigment dispersant; liq.; sol. 3 g/1000 ml water; very sol. in org. solvs.; pH 5-6 (1%); surf. tens. 31.5 dynes/cm; 65% conc.

Rewopol® SBDO 70. [Rewo GmbH] Dioctyl sodium sulfosuccinate; anionic; wetting agent, solubilizer; cosmetic and personal care preparations, household and metal cleaners; used in paper, textile, paint, and dye industries; dry cleaning; colorless liq.; surf. tens. < 30 dynes/cm (0.1%); pH 6.5–7.5 (5% solids); biodeg.; 64% min. act.

Rewopol® SBDO 75. [Rewo GmbH] Dioctyl sodium sulfosuccinate; anionic; wetting agent, solubilizer; emulsion polymerization; BGA and FDA compliance; visc. liq.; sol. 10 g/1000 ml water; very sol. in org. solvs.; pH 5-7 (1%); surf. tens. 27 dynes/cm; 75% conc.

Rewopol® SBF 12. [Rewo GmbH] Disodium lauryl sulfosuccinate; anionic; detergent raw material for personal care products; carpet and upholstery shampoos; toilet and syndet soaps; wh. paste; pH 6.5–7.5 (5% solids); 39% solids.

Rewopol® SBF 12 P. [Rewo GmbH] Disodium lauryl sulfosuccinate; anionic; surfactant for mild detergents, soaps, spray drying additive; powd.; 95% conc.

Rewopol® SBF 18. [Rewo GmbH] Disodium stearyl sulfosuccinamate; anionic; raw material for detergents; additive to syndet soap; flakes; 95% conc.

Rewopol® SBFA 30. [Rewo GmbH] Disodium laureth sulfosuccinate; anionic; detergent raw material for personal care products, cleansing agents; colorless clear liq.; visc. 100–299 cps; pH 6.5–7.5 (5% solids); 39% min. solids.

Rewopol® SBFA 50. [Rewo GmbH] Fatty alcohol polyglycol ether sulfosuccinate; anionic; emulsion polymerization emulsifier, post-stabilizer, dispersant for pigments; best for acrylates and vinyl acetates; FDA compliance; low visc. liq.; HLB 14.0; pH 5-7 (1%); surf. tens. 30 mN/m; 30% conc.

Rewopol® SBL 203. [Rewo GmbH] Disodium lauramido MEA-sulfosuccinate; anionic; detergent for cosmetics; aids spray-drying; carpet and upholstery shampoos; soaps; yel. turbid liq.; pH 6.5–7.5 (5% solids); 39% min. solids.

Rewopol® SBL 203 G, 203 P. [Rewo GmbH] Disodium lauramido MEA-sulfosuccinate; anionic; raw material for carpet shampoos, soaps, and household cleaners; granules and powd. resp.; 95% conc.

Rewopol® SBLC, SBLC G. [Rewo GmbH] Fatty acid alkylolamide sulfosuccinate; anionic; surfactant for foam cleaners, lt. duty detergents, shampoos; paste, granules resp.; 40 and 95% conc.

Rewopol® SBMB 80. [Rewo GmbH] Diisohexyl sulfosuccinate; anionic; emulsion polymerization surfactant; stabilizer for disps.; pigment dispersant; BGA and FDA compliance; visc. liq.; sol. 400 g/1000 ml water; sol. in org. solvs.; pH 5-7 (1%); surf. tens. 30 dynes/cm; 80% conc.

Rewopol® SBR 12-Powder. [Rewo GmbH] Disodium lauryl sulfosuccinate; anionic; detergent for soaps, shampoos, moisture sensitive tech. products; wh., lt. yel. powd.; water-sol.; dens. 250 g/l; pH 6.5–7.5 (5% solids); 95% solids.

Rewopol® SBV. [Rewo GmbH] Disodium laureth sulfosuccinate, sodium laureth sulfate, cocamide DEA; anionic; mild raw material for bubble baths, shampoos, dishwashing and liq. detergents; lt. yel. clear liq.; visc. 2500 cps; pH 6.5–7.5 (1% solids); 28% min. solids.

Rewopol® SBZ. [Rewo GmbH] Disodium PEG-4 cocamido MIPA-sulfosuccinate; anionic; surfactant for mild foam baths, shampoos, lt. duty detergents; low visc. liq.; 50% conc.

Rewopol® SCK 2040. [Rewo GmbH] Detergent blend; anionic; base for liq. dishwashing and cleansing agents; visc. liq.; 40% conc.

Rewopol® SK 275. [Rewo GmbH] Oleyl sarcosinic acid; anionic; emulsifier and corrosion inhibitor for min. oils; liq.; 100% conc.

Rewopol® SMS 35. [Rewo GmbH] Alkyl disodium sulfosuccinamate; anionic; emulsifier for emulsion polymerization; foaming agent for latex emulsion; antigelling and cleaning agent for paper mill felts; paste; 35% conc.

Rewopol® TLS 40. [Rewo GmbH] TEA-lauryl sulfate; anionic; surfactant raw material for shampoos, foam baths, liq. detergents; liq.; 40% conc.

Rewopol® TLS 45/B. [Rewo GmbH] TIPA-lauryl sulfate; anionic; surfactant for fire fighting foams; textile aux.; liq.; 45% conc.

Rewopol® TMSF. [Rewo GmbH] Alkyl sulfosuccinamate; anionic; emulsifier for emulsion polymerization; foaming agent for latex emulsions; antigelling agent; liq.; 31% conc.

Rewopol® TS 25. [Rewo GmbH] Surfactant blend; anionic; for carpet, upholstery and fabric shampoos, foam cleaners; liq.; 25% conc.

Rewopol® TS 35. [Rewo GmbH] Surfactant blend; anionic; for carpet shampoos, upholstery shampoos, foam cleaners; low visc. liq.; 35% conc.

Rewopol® TS 40 P. [Rewo GmbH] Surfactant blend and polymers; anionic; anti-resoiling carpet shampoos; emulsion; 30% conc.

Rewopol® TS 100. [Rewo GmbH] Surfactant blend;

anionic; for carpet shampoos, upholstery shampoos, foam cleaners; powd.; 100% conc.

Rewopol® TSK 30. [Rewo GmbH] Surfactant blend and polymers; anionic; carpet shampoos for aerosol systems; liq.; 30% conc.

Rewopol® TSSP 25. [Rewo GmbH] Surfactant blend and polymers; anionic; carpet shampoos for spray extraction systems; emulsion; 25% conc.

Rewopol® WS 11. [Rewo GmbH] Surfactant blend; anionic; wool washing liq.; liq.; 30% conc.

Rewopol® WS 12. [Rewo GmbH] Surfactant blend; anionic; wool washing liq.; liq.; 30% conc.

Rewopon® AM-V. [Rewo GmbH] Fatty acid-based ampholyte, short-chain; amphoteric; detergent; preparation of cleaning agents; wetting agent for acidic and alkaline cleaners; dk. amber liq.; pH 10–12 (10% solids); 28% act.

Rewopon® IM-CA. [Rewo GmbH] 1-Hydroxyethyl-2-alkyl-imidazoline; cationic; corrosion inhibitor, antistat, lubricant adhesive, raw material for quarternization; firm paste; 100% conc.

Rewopon® IM OA. [Rewo GmbH] 1-Hydroxyethyl-2-alkyl-imidazoline; cationic; corrosion inhibitor, emulsifier, antistat; med. visc. liq.; 100% conc.

Rewopon® JMBT. [Rewo GmbH] Long-chain imidazoline deriv.; cationic; adhesion agent and emulsifier for acid asphalt; liq.; 100% conc.

Rewopon® JMCA. [Rewo GmbH] Coconut imidazoline deriv.; cationic; emulsifier, anticorrosive agent; firm paste; 100% conc.

Rewopon® JMOA. [Rewo GmbH] Oleic imidazoline deriv.; cationic; rust inhibitor, emulsifier for antistatic preparations; liq.; 100% conc.

Rewoquat CPEM. [Rewo GmbH] PEG-5 cocomonium methosulfate; cationic; hair conditioner for shampoos, emulsifier in emulsion polymerization, antistat; visc. liq.; 100% conc.

Rewoquat CR 3099. [Rewo GmbH] Difatty acid isopropyl ester dimethyl ammonium methylsulfate; cationic; fabric softener, dry cleaning agent; visc. liq.; 95% conc.

Rewoquat DQ 35. [Rewo GmbH] PEG-3 tallow propylenedimonium dimethosulfate; cationic; antistat and wetting agent; liq.; 35% conc.

Rewoquat UTM 185. [Rewo GmbH] Undecylenic quat. ammonium methosulfate; bacteriostat and fungicide for deodorants, shampoos, liq. soaps, antimycobic foot preps.; liq.; 50% conc.

Reworyl® ACS 60. [Rewo GmbH] Ammonium cumene sulfonate; anionic; hydrotrope for detergent and cleaning systems; liq.; 60% conc.

Reworyl® K. [Rewo GmbH] Dodecylbenzene sulfonic acid; anionic; biodeg. raw material for anionic detergent systems; liq.; 97% conc.

Reworyl® NCS 40. [Rewo GmbH] Sodium cumene sulfonate; anionic; hydrotrope for detergent and cleaning systems; liq.; 40% conc.

Reworyl® NKS 50. [Rewo GmbH] Sodium dodecylbenzene sulfonate; anionic; biodeg. raw material for detergents and cleaners; textile industry; emulsifier for styrene polymerization; BGA and FDA compliance; paste; sol. 250 g/1000 ml water; pH 7-8 (1%); surf. tens. 38 mN/m; 50% conc.

Reworyl® NKS 100. [Rewo GmbH] Sodium dodecylbenzene sulfonate; anionic; detergent and cleaner, textile auxiliaries; powd.; 98% conc.

Reworyl® NTS 40. [Rewo GmbH] Sodium toluene sulfonate; anionic; hydrotrope for detergent and cleaning systems; liq.; 40% conc.

Reworyl® NXS 40. [Rewo GmbH] Sodium xylene sulfonate; CAS 1300-72-7; anionic; hydrotrope for detergent systems; liq.; 40% conc.

Reworyl® Sulfonic Acid K. [Rewo GmbH] Dodecylbenzene sulfonic acid; anionic; raw material for anionic detergent systems; med. visc. liq.; 97% conc.

Reworyl® TKS 90 F. [Rewo GmbH] Trialkanolammonium dodecylbenzene sulfonate; anionic; detergent raw material for cleansing and defatting agents, car shampoos, dishwash, ball bearing polishes, chain lubricants; dk. amber visc. liq.; pH 7.0–7.5 (1% solids); 90% conc.

Reworyl® TKS 90/L. [Rewo GmbH] TEA dodecylbenzene sulfonate; anionic; biodeg. raw material for liq. detergents; visc. liq.; 90% conc.

Rewoteric AM B13. [Rewo GmbH] Cocamidopropyl betaine; CAS 61789-40-0; amphoteric; mild raw material, foam booster for personal care products, liq. soaps, all-purpose cleaners; low visc. liq.; pH 5-6 (10%); wetting 1 min 34 s; Ross-Miles foam 200/185 mm; 35% conc.

Rewoteric AM B14. [Rewo GmbH] Cocamidopropyl betaine; CAS 61789-40-0; amphoteric; mild raw material for shampoos, bubble baths, liq. soaps, all-purpose cleaners; liq.; 35% conc.

Rewoteric AM CAS. [Rewo GmbH] Cocamidopropyl hydroxysultaine; CAS 68139-30-0; amphoteric; surfactant used in personal care products and acid and alkaline cleaners; low visc. liq.; pH 7-8 (10%); wetting 1 min 34 s; Ross-Miles foam 170/165 mm; 50% conc.

Rewoteric AM DML. [Rewo GmbH] Lauryl betaine; CAS 11140-78-6; amphoteric; surfactant for baby shampoos, hard surf. cleaners, steam jet cleaners, and pickling baths; Gardner 1 max. liq.; pH 7-8 (10%); 40% conc.

Rewoteric AM HC. [Rewo GmbH] Lauryl dimethyl hydroxypropyl sulfobetaine; CAS 13197-76-7; amphoteric; mild detergent; liq.; 45% conc.

Rewoteric AM KSF 40. [Rewo GmbH] Sodium cocoamphopropionate, salt-free; CAS 93820-52-1; amphoteric; raw material, wetting agent used in personal care products and industrial cleaners; visc. liq.; pH 9-10 (10%), wetting 1 min 12 s; Ross-Miles foam 175/165 mm; 40% conc.

Rewoteric AM TEG. [Rewo GmbH] Dihydroxyethyl tallow glycinate; CAS 61791-25-1; cationic; surfactant for acidic and alkaline cleaning products, conditioning shampoos; thickener; Gardner 6 max. liq.; pH 4.5-5.5 (10%); wetting 2 min 30 s; Ross-Miles foam 145/135 mm; 40% conc.

Rewoteric AM V. [Rewo GmbH] Sodium capryloamphoacetate; amphoteric; raw material, wetting agent, rust inhibitor for pickling baths, acid and alkaline cleaners; low visc. liq.; 35% conc.

Rewoteric AM VSF. [Rewo GmbH] Sodium capryloamphopropionate, salt-free; amphoteric; wetting agent for alkaline and acidic cleaners; liq.; 50% conc.

Rewoteric QAM 50. [Rewo GmbH] Cocobetainamido amphopropionate; cationic/amphoteric; surfactant for hard surf. disinfectants, deodorants, cleaning and bacteriocidal prods. for use in industrial catering, hotels, hospitals, and households; biodeg.; straw clear liq.; pH 8-9 (10% aq.); Draves wetting 1 min 4 s; Ross-Miles foam 180/0; toxicology: LD50 (acute oral, rat) 5100 mg/kg; 50% conc.

Rexfoam 150-A. [Graden] Hydrophobic silica; defoamer for gloss and semigloss latex paints; poorly disp. in water.

Rexfoam B, C. [Graden] Silicone emulsion; defoamer for paint and ink, chemical processing, pulp and paper, adhesive formulations, water treatment; disp. in water; dens. 8.3 lb/gal; 10 and 30% act. resp.

Rexfoam D. [Graden] Silicone emulsion; nonionic; defoamer for paint, ink, pulp and paper mfg., petrol., chemical, and textile processing, resin polymerization, adhesives, metal and water treatment; opaque liq.; sol. in aliphatic, aromatic, and chlorinated solv.; dens. 8.4 lb/gal; 100% conc.

Rexobase BAT. [Emkay] Alkylaryl detergent; economical detergent and emulsifier which can be cut 50-50 with water for use as detergent and 50-50 with varsol to make a solv. scour.

Rexobase EN. [Emkay] Emulsifier for naphtha or toluol.

Rexobase GB-XX. [Emkay] Emulsifier for min. spirits used in 1:9 ratio with the solv.; gives a stable emulsion in water.

Rexobase HD. [Emkay] Nonionic; degreaser for hide tanning; stable in brine sol'n.; misc. with kerosene or Stod.

Rexobase LN. [Emkay] Nonionic; emulsifier for varsol and kerosene; exc. detergent.

Rexobase MX. [Emkay] Emulsifier for xylene, high flash naphtha, and other solvs.; for cleaners, blanket washes, solv. scours; disp. in water.

Rexobase NC. [Emkay] Emulsifier for kerosene and varsol; produces sol. oils used as scouring agent for obstinate oil and grease on rayon, nylon, and acetate fabrics; also emulsifies pine oil.

Rexobase OAC. [Emkay] Emulsifier for chlorinated solvs. which gives a mixt. stable in oxalic acid.

Rexobase PBX. [Emkay] Nonionic; emulsifier for perchloroethylene, xylene, butyl benzoate; low foaming; for mfg. of solv. scours, butyl benzoate carriers.

Rexobase PCL. [Emkay] Anionic; emulsifier for perchlorethylene; liq.; 100% conc.

Rexobase PW. [Emkay] Naphthenic deriv.; cationic; emulsifier for min. oil and paraffin; paste.

Rexobase PWX. [Emkay] Emulsifier for paraffin wax.

Rexobase TR. [Emkay] Emulsifier for pine oil and petrol. solvs.; used for making an oil and tar remover for dyers of woolen blankets and piece goods.

Rexobase XX. [Emkay] Amide condensate; anionic; emulsifier for varsol or Stod.; forms bright sol. oil used in water sol'ns. for the removal of stubborn soil and grease; liq.; 100% act.

Rexoclean. [Emkay] Syn. detergents, terpenes, and solvs.; complete scouring agent for stubborn oil and grease stains and for removal of nylon warp size; 85% act.

Rexoclean 25X. [Emkay] Blend of amino condensate and nonionic detergent; scouring agent for wool and blends; exc. rinsability.

Rexoclean 200N. [Emkay] Emulsified blend of solvs. and detergents; multipurpose cleaner for heavy duty cleaning incl. rubber marks and stains, in soap dispensers.

Rexoclean APC. [Emkay] Alkylaryl sulfonate; all-purpose cleaner.

Rexoclean APXX. [Emkay] all-purpose industrial cleaner.

Rexoclean DBJ. [Emkay] Cleaner for removing residual dyes and dried films form dyeing equip., in dyebaths; water disp.

Rexoclean GA. [Emkay] Solv. scour, detergent for low or high pH applics, bleaching, scouring, and cleansing obstinate fatty soils.

Rexoclean HAC. [Emkay] Highly alkaline cleaner for cleaning machines, equip., walls, etc.; scouring agent for fabrics not sensitive to high alkalinity; removes oil and grease, dried films.

Rexoclean JA. [Emkay] Emulsified solvs./detergent blend; anionic; scouring agent for oil and grease soils and fatty sizing; liq.; 35% conc.

Rexoclean NB Conc. [Emkay] Mixt. of solvs., emulsifiers, and detergents; can be cut 50-50 with water to make an efficient blanket cleaner.

Rexoclean NFC. [Emkay] Nonflamm. cleaner for rubber blankets, blanket wash, for cleaning rollers and equip.

Rexoclean PCL. [Emkay] Emulsified perchloroethylene; anionic; cleaner and dye carrier for polyester fabrics; liq.; 100% conc.

Rexoclean RC Base. [Emkay] Emulsifier for Solvesso 100.

Rexoclean RC, RCA, RCN. [Emkay] Roller and blanket cleaners for removal of films and pigments; emulsifiable in water.

Rexoclean SA. [Emkay] Blend of long chain highly sulfonated detergents, surfactants, and alkalies; detergent, emulsifier, penetrant for heavy-duty scouring and cleansing.

Rexoclean SRT. [Emkay] Heavy-duty cleaner for removing oils and waxes to prescour and to give a good foundation for dyeing.

Rexol 25/1. [Hart Chem. Ltd.] Nonoxynol-1; nonionic; defoamer for detergent systems; coemulsifier for surfactant and solv. blends; liq.; HLB 4.6; 100% act.

Rexol 25/4. [Hart Chem. Ltd.] Nonoxynol-4; CAS 9016-45-9; nonionic; detergent, dispersant, stabilizer; low foaming emulsifier for oils and petrol. solvs., emulsion polymerization; intermediate for anionic sulfonates; pigment dispersant; clear liq.; oil-sol.; sp.gr. 1.02; visc. 200 cps; HLB 8.6; ref. index 1.5000; 100% act.

Rexol 25/6. [Hart Chem. Ltd.] Nonoxynol-6; CAS 9016-45-9; 26027-38-3; nonionic; dispersant in petrol. systems, for pigments; intermediate for anionic sulfates; emulsifier for oils and waxes, emulsion polymerization; antistat plasticizer for plastics; clear liq.; oil-sol.; sp.gr. 1.04; visc. 250 cps; HLB 10.8; cloud pt. 68-72 C (10% in 25% diethylene glycol butyl ether); ref. index 1.4909; 100% act.

Rexol 25/7. [Hart Chem. Ltd.] Nonoxynol-7; CAS 9016-45-9; nonionic; emulsifier for herbicides and pesticides; intermediate for anionic sulfates; antistatic plasticizer for plastics; clear liq.; HLB 11.6; cloud pt. 76-78 C (10% in 25% in diethylene glycol butyl ether); 100% act.

Rexol 25/8. [Hart Chem. Ltd.] Nonoxynol-8; CAS 9016-45-9; nonionic; detergent, wetting agent, emulsifier, dispersant; base surfactant for household and industrial detergents; used for leather, paint, textile, pesticides, pulp and paper industries, emulsion polymerization; clear liq.; water-sol.; HLB 12.4; cloud pt. 23-27 C (1% aq.); 100% act.

Rexol 25/9. [Hart Chem. Ltd.] Nonoxynol-9; CAS 9016-45-9; nonionic; detergent, dispersant, emulsifier and wetting agent; used in leather, paint, textile, pulp and paper industries, emulsion polymerization; base for industrial and household detergents; liq.; water-sol.; HLB 13.0; cloud pt. 52-56 C (1% aq.); 99% act.

Rexol 25/10. [Hart Chem. Ltd.] Nonoxynol-10; CAS 9016-45-9; nonionic; detergent, dispersant, emulsifier, wetting agent; for paint, textiles, pulp/paper industries, emulsion polymerization; degreaser for leather; scouring agent for raw wool; liq.; water-sol.; HLB 13.4; cloud pt. 63-66 C (1% aq.); 99% act.

Rexol 25/11. [Hart Chem. Ltd.] Nonoxynol-11; CAS 9016-45-9; nonionic; detergent, wetting agent; scouring agent for natural and syn. textile fibers; coemulsifier with oil-sol. surfactants; clear liq.; water-sol.; sp.gr. 1.06; visc. 314 cps; HLB 13.8; cloud pt. 71-74 C (1% aq.); ref. index 1.4893; 99% act.

Rexol 25/12. [Hart Chem. Ltd.] Nonionic; emulsifier, wetting agent, stabilizer for emulsion polymerization; liq.; 100% act.

Rexol 25/14. [Hart Chem. Ltd.] Nonoxynol-14; CAS 9016-45-9; nonionic; detergent and wetting agent used for metal cleaning, alkaline systems; paste; sp.gr. 1.055; visc. 250 cps; HLB 14.6; cloud pt. 87-89 C (1% aq.); ref. index 1.4848; 100% act.

Rexol 25/15. [Hart Chem. Ltd.] Nonoxynol-15; CAS 9016-45-9; nonionic; wetting agent and detergent for hard surfaces, metal cleaning, bottle washing; effective at elevated temps. and in electrolyte systems; clear paste; sp.gr. 1.06; visc. 400 cps (35 C); HLB 14.8; cloud pt. 95-99 C (1% aq.); ref. index 1.4870; 100% act.

Rexol 25/20. [Hart Chem. Ltd.] Nonoxynol-20; CAS 9016-45-9; nonionic; particle dispersant in aq. systems; effective at high temps. and electrolyte concs.; solid; HLB 16.0; cloud pt. 72–74 C (1% in 10% NaCl); 100% act.

Rexol 25/30. [Hart Chem. Ltd.] Nonoxynol-30; CAS 9016-45-9; nonionic; solubilizer for essential oils and pesticides; solid; HLB 17.2; cloud pt. 74–76 C (1% in 10% NaCl); 100% act.

Rexol 25/40. [Hart Chem. Ltd.] Nonoxynol-40; CAS 9016-45-9; nonionic; emulsifier used in polymer emulsification for paints and coatings; solid; HLB 17.8; cloud pt. 74–76 C (1% in 10% NaCl); 100% act.

Rexol 25/50. [Hart Chem. Ltd.] Nonoxynol-50; CAS 9016-45-9; nonionic; emulsifier and stabilizer used in floor finishes; solid; HLB 18.2; cloud pt. 74–76 C (1% in 10% NaCl); 100% act.

Rexol 25/100-70%. [Hart Chem. Ltd.] Nonoxynol-100; CAS 9016-45-9; nonionic; demulsifier for petrol. oils; pressured textile scouring; emulsifier/stabilizer for latexes, asphalt, and floor finishes; liq.; HLB 19.0; cloud pt. 74–76 C (1% in 10% NaCl); 70% act.

Rexol 25/307. [Hart Chem. Ltd.] Nonoxynol-30; CAS 9016-45-9; nonionic; solubilizer for essential oils and pesticides; surfactant for latex emulsion polymerization; liq.; HLB 17.2; cloud pt. 74–76 C (1% in 10% NaCl); 70% act. in water.

Rexol 25/407. [Hart Chem. Ltd.] Nonoxynol-40; CAS 9016-45-9; nonionic; detergent, wetting agent; emulsifier for vinyl acetate and acrylate polymerization; demulsifier of petrol. oils; textile scouring; stabilizer for syn. latexes; yel. liq.; water-sol.; sp.gr. 1.10; HLB 17.8; cloud pt. 74-76 C (1% in 10% NaCl); 70% act. in water.

Rexol 25/507. [Hart Chem. Ltd.] Nonoxynol-50; CAS 9016-45-9; nonionic; emulsifier, stabilizer for prep. of floor finishes; liq.; HLB 18.2; cloud pt. 74–76 C (1% in 10% NaCl); 70% act. in water.

Rexol 25J. [Hart Chem. Ltd.] Nonyl phenol ethoxylate; CAS 9016-45-9; nonionic; detergent and wetting agent, pitch dispersant for textile, pulp and paper, leather, paint, and pesticide industries; emulsifier; base surfactant for household and industrial detergents; clear liq.; sp.gr. 1.06; visc. 250 cps; HLB 13.0; cloud pt. 52-56 C (1% aq.); ref. index 1.4823; 99% act.

Rexol 25JM1. [Hart Chem. Ltd.] Alkyl phenol ethoxylate; nonionic; detergent, wetting agent, emulsifier used in Kraft and sulfite digesters, screen rooms, bleach plants and paper machines; lower foam; yel. liq.; sp.gr. 1.05; visc. 290 cps; 100% act.

Rexol 25JWC. [Hart Chem. Ltd.] Nonyl phenol ethoxylate; nonionic; degreasing agent for leather; scouring agent for raw wool; liq.; HLB 13.4; cloud pt. 63-66 C (1% aq.); 99% act.

Rexol 35/3. [Hart Chem. Ltd.] Alcohol ethoxylate; nonionic; emulsifier for o/w; intermediate for sulfation and phosphation; liq.; HLB 8.0; cloud pt. 49–51 C (1% in 25% diethylene glycol butyl ether); 100% act.

Rexol 35/5. [Hart Chem. Ltd.] Alcohol ethoxylate; nonionic; cold water detergent for textile and leather applications; liq.; HLB 10.6; cloud pt. 64–65 C (1% in 25% diethylene glycol butyl ether); 100% act.

Rexol 35/8. [Hart Chem. Ltd.] Alcohol ethoxylate; nonionic; detergent intermediate; wetting agent used in textile applics.; liq.; HLB 12.8; cloud pt. 79–83 C (1% in 25% diethylene glycol butyl ether); 85% act.

Rexol 35/11. [Hart Chem. Ltd.] Alcohol ethoxylate; nonionic; hard surface cleaner used in bottle washing and metal cleaning; paste; HLB 14.2; cloud pt. 72–76 C (1% aq.); 100% act.

Rexol 35/100. [Hart Chem. Ltd.] Fatty alcohol ethoxylate; nonionic; detergent, wetting agent used in scouring of paper machine felts; hard surf. and industrial detergents; emulsifier, dispersant; clear paste; sp.gr. 1.003; visc. 100 cps; HLB 13.0; cloud pt. 52-56 C (1% aq.); 100% act.

Rexol 45/1. [Hart Chem. Ltd.] Octoxynol-1; CAS 9002-93-1; nonionic; emulsifier, detergent, dispersant; coemulsifier for surfactant and solv. preparation blends; liq.; oil-sol.; HLB 3.6; 100% act.

Rexol 45/3. [Hart Chem. Ltd.] Octoxynol-3; CAS 9002-93-1; nonionic; emulsifier, detergent, dispersant, coemulsifier for petrol. oils and solvs.; liq.; HLB 7.8; 100% act.

Rexol 45/5. [Hart Chem. Ltd.] Octoxynol 5; CAS 9002-93-1; nonionic; emulsifier for fats and oils, detergent, dispersant; liq.; HLB 10.4; cloud pt. 63–66 C (10% in 25% diethylene glycol butyl ether); 100% act.

Rexol 45/7. [Hart Chem. Ltd.] Octoxynol-7; CAS 9002-93-1; nonionic; low foaming hard surface detergent; emulsifier, dispersant; liq.; HLB 12.4; cloud pt. 21–23 C (1% aq.); 100% act.

Rexol 45/10. [Hart Chem. Ltd.] Octoxynol-10; CAS 9002-93-1; nonionic; detergent and wetting agent used in textile, paint, plastic, pulp and paper industries; base surfactant for household and industrial detergents; agricultural formulations; liq.; HLB 13.5; cloud pt. 64–67 C (1% aq.); 100% act.

Rexol 45/12. [Hart Chem. Ltd.] Octoxynol-12; CAS 9002-93-1; nonionic; emulsifier, detergent, dispersant used in metal cleaning, industrial and household liq. detergents and cleaners; liq.; HLB 14.5; cloud pt. 86–90 C (1% aq.); 100% act.

Rexol 45/16. [Hart Chem. Ltd.] Octoxynol-16; CAS 9002-93-1; nonionic; detergent and wetting agent used in metal cleaning and bottle washing; emulsifier; dispersant; effective at high temps. and electrolyte concs.; paste; HLB 15.8; cloud pt. 65–69 C (1% in 10% NaCl); 100% act.

Rexol 45/307. [Hart Chem. Ltd.] Octoxynol-30; CAS

9002-93-1; nonionic; emulsifier for vinyl acetate and acrylate emulsion polymerization; detergent, dispersant; liq.; HLB 17.4; cloud pt. 73–77 C (1% in 10% NaCl); 70% act.

Rexol 45/407. [Hart Chem. Ltd.] Octoxynol-40; CAS 9002-93-1; nonionic; emulsifier for vinyl acetate and acrylate emulsion polymerization; liq.; HLB 18.0; cloud pt. 73–77 C (1% in 10% NaCl); 70% act.

Rexol 65/4. [Hart Chem. Ltd.] Dodoxynol-4; nonionic; ingred. for solv. cleaner and drycleaning formulations; emulsifier for agric. oils; degreaser; liq.; HLB 8.0; cloud pt. 52-56 C (10% in 25% diethylene glycol butyl ether); 100% act.

Rexol 65/6. [Hart Chem. Ltd.] Dodocynol-6; nonionic; ingred. for solv. cleaner and drycleaning formulations; emulsifier for agric. oils; degreaser; liq.; HLB 10.0; cloud pt. 67 C (10% in 25% diethylene glycol butyl ether); 100% act.

Rexol 65/9. [Hart Chem. Ltd.] Dodoxynol-9; nonionic; general purpose cleaner, degreaser, emulsifier, low foamer; liq.; HLB 12.0; cloud pt. 78 C (10% in 25% diethylene glycol butyl ether); 100% act.

Rexol 65/10. [Hart Chem. Ltd.] Dodoxynol-10; nonionic; general purpose cleaner, degreaser, emulsifier, low foamer; liq.; HLB 13.1; cloud pt. 39-45 C (1% aq.); 100% act.

Rexol 65/11. [Hart Chem. Ltd.] Dodoxynol-11; nonionic; general purpose cleaner, degreaser, emulsifier, low foamer; liq.; HLB 13.5; cloud pt. 56 C (1% aq.); 100% act.

Rexol 65/14. [Hart Chem. Ltd.] Dodoxynol-14; nonionic; general purpose cleaner, degreaser, emulsifier; liq.; HLB 14.5; cloud pt. 87 C (1% aq.); 100% act.

Rexol 130. [Hart Chem. Ltd.] Alcohol ethoxylate; nonionic; emulsifier, dyeing assistant, leveling agent for dyeing vats, naphthols, reactive and direct dyes; clear liq.; ref. index 1.3850; sp.gr. 1.02; visc. 93 cps; 35% act.

Rexol 2000 HWM. [Hart Chem. Ltd.] Amide ethoxylate, modified; nonionic; antistatic emulsifier for min. oils, mfg. of textile lubricants; anticorrosive and cohesive props.; amber liq.; ref. index 1.4618; sp.gr. 0.982 visc. 300 cps; 85% act.

Rexol 4736. [Hart Chem. Ltd.] Alcohol ethoxylate, branch chained; nonionic; surfactant with solv. and oil emulsification props. for prescouring operations; low foaming wetting agent for continuous carpet dyeing; clear liq.; water-sol.; 85% act.

Rexol AE-1. [Hart Chem. Ltd.] Linear alcohol ethoxylate; nonionic; intermediate for shampoo and detergent mfg.; liq.; HLB 4.2; cloud pt. 36–39 C (1% in 25% diethylene glycol butyl ether); 100% act.

Rexol AE-2. [Hart Chem. Ltd.] Linear alcohol ethoxylate; nonionic; intermediate for shampoo and detergent mfg.; liq.; HLB 6.1; cloud pt. 50-52 C (1% in 25% diethylene glycol butyl ether); 100% act.

Rexol AE-3. [Hart Chem. Ltd.] Linear alcohol ethoxylate; nonionic; detergent; intermediate for mfg. of ether sulfates; emulsifier for hydrocarbons; clear liq.; sp.gr. 0.901; visc. 24 cps; HLB 8.2; cloud pt. 59-61 C (1% in 25% diethylene glycol butyl ether); ref. index 1.4494; 100% act.

Rexol CCN. [Hart Chem. Ltd.] Castor oil ethoxylate; nonionic; emulsifier, antistat, demulsifier, leveling agent for dyeing of syn. fibers; liq.; 90% conc.

Rexonic 1006. [Hart Chem. Ltd.] Linear alcohol ethoxylate; nonionic; wetting and scouring agent for natural and syn. fibers; leveling and dispersing aid for acid and disperse dyes; foaming agent for Kuster dyeing of carpets; compat. with dyestuffs over wide pH range; biodeg.; clear liq.; 75% act.

Rexonic 1012-6. [Hart Chem. Ltd.] Linear alcohol ethoxylate; nonionic; emulsifier, detergent, dispersant, and wetting agent for scouring of natural and syn. textile fibers, leather, paint, textile, and pulp and paper industries; base surfactant for household and industrial detergents; clear liq.; sp.gr. 0.986; visc. 34 cps; HLB 12; cloud pt. 48-52 C (1% aq.); ref. index 1.4532; 100% act.

Rexonic 1218-6. [Hart Chem. Ltd.] Linear alcohol ethoxylate; nonionic; detergent and wetting agent for textile, leather, paint, pulp/paper, household and industrial detergents; clear liq.; sp.gr. 1.002; visc. 140 cps; HLB 12.6; cloud pt. 64-68 C (1% aq.); ref. index 1.4492; 90% act.

Rexonic L125-9. [Hart Chem. Ltd.] Alcohol ethoxylate; detergent and wetting agent for textile, pulp and paper, paint, and pesticide industries; solid; HLB 13.1; cloud pt. 60-65 C (1% aq.); 100% act.

Rexonic N23-3. [Hart Chem. Ltd.] Linear alcohol ethoxylate; nonionic; detergent, emulsifier, dispersant, and chemical intermediate used in solv. emulsion cleaners and dry cleaning detergents; biodeg.; clear liq.; oil-sol.; sp.gr. 0.93; visc. 19 cps; HLB 8.1; cloud pt. 59-61 C (10% in 25% diethylene glycol butyl ether); 100% act.

Rexonic N23-6.5. [Hart Chem. Ltd.] Linear alcohol ethoxylate; nonionic; detergent, wetting agent, emulsifier; household cleaners, industrial maintenance, metal cleaning, paper, paint, and leather processing; paste; sp.gr. 0.96; visc. 30 cps; HLB 12.0; cloud pt. 43-47 C (1% aq.); 100% act.

Rexonic N25-3. [Hart Chem. Ltd.] Alcohol ethoxylate; nonionic; emulsifier for oil; wetting agent, dispersant, desizing agent for recycled waste paper; absorbency aid for tissue; intermediate for ether sulfate detergents; liq.; HLB 7.8; cloud pt. 59–61 C (10% in 25% diethylene glycol butyl ether); 100% act.

Rexonic N25-7. [Hart Chem. Ltd.] Linear alcohol ethoxylate; nonionic; detergent, dispersant, emulsifier, wetting agent for formulation of foaming detergents, leather, paint, textile, and pulp and paper specialties; biodeg.; clear paste; sp.gr. 0.958; HLB 12.1; cloud pt. 51-53 C (1% aq.); 100% act.

Rexonic N25-9. [Hart Chem. Ltd.] Alcohol ethoxylate; nonionic; detergent and wetting agent for textile, pulp and paper, paint and pesticide industries, hard surf. detergents; solid; HLB 13.1; cloud pt. 69–78 C (1% aq.); 100% act.

Rexonic N25-9(85%). [Hart Chem. Ltd.] Alcohol ethoxylate; nonionic; detergent, wetting agent for textile, pulp and paper, paint and pesticide industries, hard surf. detergents; liq.; HLB 13.1; cloud pt. 69-78 C (1% aq.); 85% act.

Rexonic N25-12. [Hart Chem. Ltd.] Alcohol ethoxylate; nonionic; detergent used in strong electrolytic sol'ns.; solid; HLB 14.4; cloud pt. 90–95 C (1% aq.); 100% act.

Rexonic N25-14. [Hart Chem. Ltd.] Alcohol ethoxylate; nonionic; detergent for metal cleaning and bottle washing; solid; HLB 15.2; cloud pt. 83–86 C (1% in 5% NaCl); 100% act.

Rexonic N25-14(85%). [Hart Chem. Ltd.] Linear alcohol ethoxylate; nonionic; detergent for scouring textile fibers; base for mfg. of ether sulfates; dispersant, emulsifier, and wetting agent; used in leather, paint, textile, pulp and paper industries; base surfactant for household and industrial cleaners; liq.; water-sol.; HLB 15.2; cloud pt. 83-86 C (1% in 5%

NaCl); 85% act.

Rexonic N91-1.6. [Hart Chem. Ltd.] Alcohol ethoxylate; nonionic; intermediate for mfg. of ethoxysulfates; emulsifier, detergent, dispersant; liq.; HLB 6.1; cloud pt. 39–41 C (10% in 25% diethylene glycol butyl ether); 100% act.

Rexonic N91-2.5. [Hart Chem. Ltd.] Alcohol ethoxylate; nonionic; emulsifier for oils and solvs.; intermediate for ethoxysulfates; detergent, dispersant; biodeg.; liq.; HLB 8.1; cloud pt. 50–54 C (10% in 25% diethylene glycol butyl ether); 100% act.

Rexonic N91-6. [Hart Chem. Ltd.] Alcohol ethoxylate; nonionic; detergent for industrial use; emulsifier; shampoo intermediate, dispersant; biodeg.; liq.; water-sol.; HLB 12.5; cloud pt. 51–53 C (1% aq.); 100% act.

Rexonic N91-8. [Hart Chem. Ltd.] Alcohol ethoxylate; nonionic; detergent, emulsifier, dispersant for general industrial usage, hard surf. cleaners, liq. detergents; shampoo intermediate; biodeg.; liq.; water-sol.; HLB 14.1; cloud pt. 80-82 C (1% aq.); 100% act.

Rexonic P-1. [Hart Chem. Ltd.] Propoxylated ethoxylated linear alcohol; nonionic; low-foaming detergent, wetter, emulsifier, dispersant for natural and syn. fibers, industrial and commercial low foam detergents; base for rinse aids and machine dishwashing formulations; clear liq.; water-sol.; sp.gr. 0.97; visc. 90 cps; HLB 7.0; cloud pt. 25 C (1% aq.); ref. index 1.455; 100% act.

Rexonic P-3. [Hart Chem. Ltd.] Polyoxyalkylene alcohol; nonionic; detergent, emulsifier with controlled foam; liq.; HLB 9.0; cloud pt. 36 C (1% aq.); 100% act.

Rexonic P-4. [Hart Chem. Ltd.] Polyoxyalkylene alcohol; nonionic; heavy-duty detergent used in detergent tablets, sanitizers, and metal cleaners; liq.; HLB 10.0; cloud pt. 57 C (1% aq.); 100% act.

Rexonic P-5. [Hart Chem. Ltd.] Polyoxyalkylene alcohol; nonionic; detergent used in machine dishwashing and rinse aids; liq.; HLB 10.0; cloud pt. 45 C (1% aq.); 100% act.

Rexonic P-6. [Hart Chem. Ltd.] Polyoxyalkylene alcohol; nonionic; low foaming detergent for mechanical dishwashing; liq.; cloud pt. 35 C (10% in 25% diethylene glycol butyl ether); 100% act.

Rexonic P-9. [Hart Chem. Ltd.] Polyoxyalkylene alcohol; nonionic; industrial cleaner, metal cleaner, hard surf. cleaner; liq.; HLB 13.1; cloud pt. 56-60 C; 100% act.

Rexonic RL. [Hart Chem. Ltd.] Alcohol ethoxylate; nonionic; detergent, post-scouring agent; clear liq.; 25% act.

Rexopal 3928. [Hart Chem. Ltd.] Modified linear alcohol ethoxylate; nonionic; low foaming wetting and scouring agent for natural and syn. fibers; biodeg.; liq.; 85% conc.

Rexopal EP-1. [Hart Chem. Ltd.] Linear alcohol polyoxyalkylene; nonionic; low foaming emulsifier and scouring agent for syn. fibers; liq.

Rexopal SM-5. [Hart Chem. Ltd.] Nonionic; nonfoaming surfactant, wetting agent, detergent for prescouring; clear yel. liq.; 30% act.

Rexopene. [Emkay] Sodium alkylaryl sulfonate; anionic; wetting, dye leveling agent, penetrant, scouring assistant; liq.

Rexophos 25/67. [Hart Chem. Ltd.] Alkylaryl phosphate ester; anionic; emulsifier for solvs. used over wide temp. range; surfactant for hard surf. cleaners, alkaline metal cleaners; liq.; 100% act.

Rexophos 25/97. [Hart Chem. Ltd.] Alkylaryl phosphate ester; anionic; detergent for liq. industrial alkaline cleaners; prescouring aid for cotton and syn. fabrics; emulsifier for min. oils and aromatic solvs.; pale yel. liq.; 100% act.

Rexophos 35/98. [Hart Chem. Ltd.] Phosphated nonionic, free acid form; anionic; emulsifier for min. oils and aromatic solvs.; scouring agent for cellulosic and syn. fibers; liq.; 100% conc.

Rexophos 4668. [Hart Chem. Ltd.] Linear alcohol phosphate ester; anionic; detergent; stable to highly alkaline cleaning sol'ns.; liq.; 80% act.

Rexophos BP-2. [Hart Chem. Ltd.] Linear alcohol phosphate ester; anionic; scouring agent and cleaner for textile industry; high alkaline stable; liq.; 100% act.

Rexophos JV 5015. [Hart Chem. Ltd.] Linear alcohol phosphate ester; anionic; detergent for liq. industrial alkaline cleaners; liq.; 100% act.

Rexophos N25-38. [Hart Chem. Ltd.] Linear alcohol phosphate ethoxylate; anionic; emulsifier for solvs. and oils; liq.; 100% act.

Rexopon E. [Emkay] Amide amine condensate, modified; nonionic; syn. detergent; afterwash for prints; liq.

Rexopon RS. [Emkay] Rug shampoo diluted with 4-6 parts water; high foaming, good cleaner.

Rexopon RSN. [Emkay] Nonionic; rapid scour, wetting agent for heavy-duty scouring, in dye liquor, as desizing agent; stable to acid and alkaline conditions.

Rexopon SK. [Emkay] Silk degumming and scouring agent; used with soap or detergent to accelerate degumming and boil-off of silk goods.

Rexopon V. [Emkay] Alkylaryl sulfonate sulfonated amide condensate; anionic; syn. detergent, foamer; scouring agent for syn. fabrics; used for boil-off, afterwash for vats and naphthols, kierboil assistant; paste.

Rexoscour. [Emkay] Soap/solv./fatty acid blend; anionic; detergent, kier boiling assistant, fulling agent, desizing agent for continuous machines; paste.

Rexoscour SF. [Emkay] Scour and fulling agent for wool.

Rexosolve. [Emkay] Blend of petrol. solvs. and terpenes; cleaner for oil or grease soils, dirty knots, etc. in textile industry; sol. in water.

Rexosolve 150. [Emkay] Sulfonated oils/emulsifier blend; anionic; detergent, scouring assistant; oil and grease remover; liq.; easily disp. in water; 100% act.

Rexosolve BCT. [Emkay] Dewaxing agent and solv. scour; used with caustic for cleaning rubber print blankets and to remove metallic pigment dispersions; leveling agent for overdyed Dacron; stripper on many colors.

Rexosolve CR. [Emkay] Emulsified blend of solvs. and cresylic acid; wool scouring agent; for removing min. oil.

Rexosolve EP, EPN. [Emkay] Blend of aromatic solvs.; cleaners for oily and greasy soils; recommended for Burlington machines; misc. with hot or cold water to produce highly stable emulsions.

Rexosolve GA, GAX. [Emkay] Solvent scours for bleach, scour and dye baths to remove fatty soils; stable in acid and alkaline baths.

Rexosolve HP. [Emkay] Conc. cleaner containing high flash solvs.; cleaner for removing stubborn tars and greases.

Rexosolve MS. [Emkay] Solv. scour used as general cleaner and washing compd. for back grays; softens and removes dried pigment; usually used with 2.5% caustic at 140 F; 100% act.

Rexosolve OA, OAC. [Emkay] Solv. scours for cleaning stains, rust and grease in one bath; compat. with oxalic acid.

Rexosolve RF-2. [Emkay] Blend of org. solvs. with syn. emulsifiers and detergents; cleaner for removal of stubborn tars, greases and stains; lt. free-flowing liq.; disp. readily in water to give stable emulsion.

Rexosolve SL. [Emkay] Stain and soil remover for use in neutral or alkaline baths; may be used in the dye liquor.

Rexowet 77. [Emkay] Heptadecyl sodium sulfate; anionic; fast wetting and penetrating agent for textiles; resistant to acid and alkali media; lt. amber oil; easily dissolved in water; 30% conc.

Rexowet 500. [Emkay] Sulfonated dioctyl succinate; surf. tens. depressant; stable in acid, alkaline and neutral baths.

Rexowet A, A-25, A Conc. [Emkay] Dioctyl succinate sulfonates; surf. tens. depressants for textile processing; stable in acid, alkaline, and neutral baths.

Rexowet ASG-81. [Emkay] Sodium dioctyl sulfosuccinate; wetting agent for textile use; 80% act.

Rexowet CR. [Emkay] Sulfonated isopropyl oleate and cresylic acid; anionic; wetting agent and penetrant for cotton fabrics; scouring agent for removing min. oil stains; liq.

Rexowet GA-1. [Emkay] Sulfated fatty acid ester; low foaming wetting agent, dyeing assistant, dispersant for textile use; resistant to acid and alkaline conditions; clear liq.

Rexowet GR. [Emkay] Sulfated diester; anionic; fast penetrant, wetting agent for piece, raw stock, package and skein dyeing; liq.

Rexowet GR-LF. [Emkay] Low-foaming penetrant for textile processing.

Rexowet GRS. [Emkay] fast wetting agent for piece, raw stock, package, and skein dyeing.

Rexowet LS. [Emkay] Sanforizing penetrant, wetting/rewetting agent.

Rexowet MS. [Emkay] Sodium alkyl naphthalene sulfonate; anionic; low foaming penetrant for warp sizing formulations; liq.

Rexowet NF, NFX. [Emkay] Alcohol sulfate salt; anionic; penetrant, wetting agent; textile dyeing assistant; low foaming; liq.

Rexowet RW. [Emkay] Aliphatic mono and diester, sulfonated; anionic; wetting, rewetting and leveling agent, penetrant; dyeing assistant; enzyme activator; stable to salts, acids, alkalis; gel.

Rexowet RWF, RW Conc. [Emkay] Wetting/rewetting agent for use as general dyeing assistants; stable to salts, acids, alkalis.

Rexowet VL. [Emkay] Penetrant for textile applics.

Rheodol SP-O10. [Kao Corp. SA] Sorbitan oleate; CAS 1338-43-8; nonionic; emulsifier and dispersant for printing inks, pastes, and paints; liq.; HLB 4.3; 100% conc.

Rheodol SP-O30. [Kao Corp. SA] Sorbitan trioleate; CAS 26266-58-0; nonionic; emulsifier and dispersant for printing inks, pastes, and paints; liq.; HLB 1.8; 100% conc.

Rheodol SP-S10. [Kao Corp. SA] Sorbitan stearate; CAS 1338-41-6; nonionic; emulsifier and dispersant for printing inks, pastes, and paints; beads; HLB 4.7; 100% conc.

Rheodol SP-S30. [Kao Corp. SA] Sorbitan tristearate; CAS 26658-19-5; nonionic; emulsifier and dispersant for printing inks, pastes, and paints; beads; HLB 2.1; 100% conc.

Rheodol TW-P120. [Kao Corp. SA] PEG-20 sorbitan palmitate; CAS 9005-66-7; nonionic; emulsifier for pharmaceuticals, cosmetics; solubilizer for colorants; stabilizer for emulsion polymerization; liq.; HLB 15.6; 100% conc.

Rheotol. [R.T. Vanderbilt] Polymerized alkyl phosphate; dispersant and wetting agent for pigmented coatings; pale liq.; dens. 0.97 mg/m^3.

Rhodacal® 70/B (formerly Soprophor® 70/B). [Rhone-Poulenc Surf.; Rhone-Poulenc Geronazzo] Calcium dodecylbenzene sulfonate; anionic; emulsifier, dispersant for herbicides and pesticides; EPA compliance; liq.; 70% conc.

Rhodacal® 301-10 (formerly Siponate® 301-10). [Rhone-Poulenc Surf.] Sodium C14-16 olefin sulfonate; anionic; emulsion polymerization surfactant; liq.; surf. tens. 29 dynes/cm (@ CMC); 40% solids.

Rhodacal® 301-10F (formerly Siponate® 301-10F). [Rhone-Poulenc Surf.; Rhone-Poulenc France] Sodium C14-16 olefin sulfonate; anionic; detergent, foam booster, emulsifier, wetting agent, dispersant for dry bubble baths, powd. hand soaps, lt. duty and fine fabric detergents, metal soak cleaners, aerosol rug and upholstery cleaners; flake; 90% conc. DISCONTINUED.

Rhodacal® 330 (formerly Siponate® 330). [Rhone-Poulenc Surf.] Isopropylamine dodecylbenzene sulfonate; anionic; emulsifier, wetting agent, grease/pigment dispersant, lubricant, solubilizer, solv., penetrant, high foaming base for shampoos and cleaners, metalworking fluids, agric. pesticides, emulsion polymerization, drycleaning, latex paints, metal cleaning; amber liq.; oil-sol.; sp.gr. 1.03; dens. 8.497 lb/gal; visc. 6500 cps; HLB 11.7; flash pt. > 200 C; pH 3-6 (5%); 90% act.

Rhodacal® 2283 (formerly Soprophor® 2283). [Rhone-Poulenc Surf.; Rhone-Poulenc Geronazzo] Amine dodecylbenzene sulfonate; anionic; emulsifier for agric. formulations; EPA compliance.

Rhodacal® A-246L (formerly Siponate® A-246L). [Rhone-Poulenc Surf.] Sodium C14-16 olefin sulfonate; anionic; detergent, foaming agent, emulsifier, wetting agent, dispersant for hair shampoos, liq. soaps, skin cleansers, industrial cleaners, emulsion polymerization; liq.; surf. tens. 29 dynes/cm (@ CMC); 40% act.

Rhodacal® BA-77 (formerly Nekal® BA-77). [Rhone-Poulenc Surf.; Rhone-Poulenc France] Sodium alkylnaphthalene sulfonate; anionic; emulsifier, wetting agent, dispersant without detergent props.; for industrial cleaners, leather dyeing, textile processing, paints, inks, pesticides, syn. latex emulsions, plastics; latex stabilizer; prevents coagulation of syn. rubbers; FDA compliance; tan powd.; m.w 314; pH 8-10 (5%); 75% act.

Rhodacal® BX-78 (formerly Nekal® BX-78). [Rhone-Poulenc Surf.; Rhone-Poulenc France] Sodium alkylnaphthalene sulfonate; anionic; emulsifier, wetting agent, penetrant, dispersant for industrial cleaning, textiles, dyeing, leather, insecticides, herbicides, paper, dyes/pigments, wallpaper pastes, rubber, latex polymerization; cream to tan powd.; m.w. 326; water-sol.; pH 6–8 (5%); surf. tens. 36 dynes/cm (@ CMC); 75% solids.

Rhodacal® CA, 70% (formerly Alkasurf® CA). [Rhone-Poulenc Surf.; Rhone-Poulenc Surf. Canada] Calcium dodecylbenzene sulfonate; anionic; biodeg. o/w emulsifier for agric., industrial applics.; dispersant for polyester yarn dyeing; amber hazy liq.; sol. in min. oil, min. spirits, aliphatic and aromatic solvs., perchlorethylene; disp. in water; dens. 1.05 g/ml; HLB 10.5; flash pt. 33 C; pH 5-7; 69-71% act. in isobutanol.

Rhodacal® DDB-40 (formerly Siponate® DDB-40). [Rhone-Poulenc Surf.] Sodium dodecylbenzene sulfonate; anionic; high foaming emulsifier, dispersant, wetting agent for industrial, institutional and household cleaners, agric. formulations; EPA compliance; liq.

Rhodacal® DOV (formerly Siponate® DOV). [Rhone-Poulenc Surf.] TEA dodecylbenzene sulfonate; anionic; emulsion polymerization surfactant; liq.; 40% solids.

Rhodacal® DS-4 (formerly Siponate® DS-4). [Rhone-Poulenc Surf.; Rhone-Poulenc France] Sodium dodecylbenzene sulfonate; anionic; emulsifier for emulsion polymerization; food pkg. applics.; surfactant for washing fruits and vegs.; FDA compliance; dk. brn. liq.; surf. tens. 32 dynes/cm (@ CMC); 23% act.

Rhodacal® DS-10 (formerly Siponate® DS-10). [Rhone-Poulenc Surf.; Rhone-Poulenc France] Sodium dodecylbenzene sulfonate; anionic; emulsifier for emulsion polymerization; emulsifier, dispersant for agric. formulations; FDA, EPA compliance; flake; surf. tens. 32 dynes/cm (@ CMC); 98% conc.

Rhodacal® DSB (formerly Siponate® DSB). [Rhone-Poulenc Surf.; Rhone-Poulenc France] Sodium dodecyl diphenyloxide disulfonate; anionic; detergent, foamer, emulsifier, textile dye leveling agent, coupling agent, solubilizer; dispersant in metal and other industrial cleaners; coemulsifier in emulsion polymerization; wetting agent, dispersant for agric. formulations; alkaline and chlorine stable; FDA, EPA compliance; liq.; surf. tens. 32 dynes/cm (@ CMC); 45% solids.

Rhodacal® IN (formerly Geropon IN). [Rhone-Poulenc France] Sodium isopropyl naphthalene sulfonate; anionic; dispersant and suspending agent for pesticide formulations; powd.; 65% conc.

Rhodacal® IPAM (formerly Alkasurf® IPAM). [Rhone-Poulenc Surf.] Isopropylamine dodecylbenzene sulfonate; anionic; emulsifier for drycleaning charge soaps, solv. degreasers; solubilizer in fuel oil; forms clear blends of water and kerosene; pale yel. clear visc. liq.; water insol., sol. in most commonly used solv.; sp.gr. 1.06; 95% act.

Rhodacal® LA Acid (formerly Alkasurf® LA Acid). [Rhone-Poulenc Surf.] Dodecylbenzene sulfonic acid; anionic; biodeg. detergent, wetter, emulsifier, penetrant, foamer, intermediate used in liq. dishwashing detergents, cleaners, personal care prods., and degreasers; brn. visc. liq.; sol. in water, min. oil, perchloroethylene; dens. 1.06 g/ml.

Rhodacal® LDS-22 (formerly Siponate® LDS-22). [Rhone-Poulenc Surf.] Linear sodium dodecylbenzene sulfonate; anionic; emulsion polymerization surfactant; FDA compliance; liq.; surf. tens. 32 dynes/cm (@ CMC); 23% solids.

Rhodacal® N (formerly Blancol® N). [Rhone-Poulenc Surf.] Sodium polynaphthalene sulfonate; anionic; process aid in paper, leather; dispersant for pulp/paper, metal cleaning; emulsifier, wetting agent for industrial cleaners, pesticides; moisture reducer in concrete; textile dye leveling agent; tan to brn. gran. powd.; odorless; sol. in water; dens. 0.65–0.75 g/ml; 88-90% act.

Rhodacal® NK (formerly Geropon NK). [Rhone-Poulenc France] Sodium dibutyl naphthalene sulfonate; anionic; wetting agent for pesticides and general purpose applics.; powd.; 65% conc.

Rhodacal® RM/77-D (formerly Geropon® RM/77-D). [Rhone-Poulenc Geronazzo] Dinaphthalene-methane sulfonate sodium salt; anionic; dispersant, protective colloid, stabilizer of nat. and syn. elastomers; pigment grinding aid for paints; powd.; 95% conc.

Rhodacal® RM/210 (formerly Geropon® RM/210). [Rhone-Poulenc Geronazzo] Polynaphthalene-methane sulfonate, sodium salt; anionic; fluidizing and plasticizer agent for concrete and mortar; powd.; 92% conc.

Rhodacal® T. [Rhone-Poulenc Surf.] TEA dodecylbenzene sulfonate; anionic; base for clear liq. detergents and bubble baths; liq.; 60% conc. DISCONTINUED.

Rhodafac® B6-56A. [Rhone-Poulenc Surf.] Phosphate ester, free acid; anionic; detergent, emulsifier, wetting agent, antistat, lubricant for textiles, drycleaning, yarns and fibers; liq.; 100% conc. DISCONTINUED.

Rhodafac® BG-510 (formerly Gafac® BG-510). [Rhone-Poulenc Surf.; Rhone-Poulenc France] Aliphatic phosphate ester, free acid; anionic; detergent, emulsifier, wetting agent, dispersant for liq. industrial alkaline cleaners, hard surf. detergents, soak-tank metal cleaning, steam cleaning, household cleaning; compat. in conc. electrolyte systems; liq.; 100% conc.

Rhodafac® BG-520. [Rhone-Poulenc France] Partial sodium salt of a complex org. phosphate ester; anionic; emulsifier, lubricant, softener, finishing agent for wool and syn. fibers; liq.; 98% conc. DISCONTINUED.

Rhodafac® BI-750. [Rhone-Poulenc Surf.] Complex org. phosphate ester, free acid; anionic; hydrophilic surfactant for liq. drain cleaners; stable and sol. in caustic; liq.; 100% conc. DISCONTINUED.

Rhodafac® BP-769 (formerly Gafac® BP-769). [Rhone-Poulenc Surf.] Aromatic phosphate ester; anionic; low foaming surfactant, hydrotrope for industrial, institutional and household cleaners with electrolyte content; good alkali stability; liq.; 90% act.

Rhodafac® BX-660 (formerly Gafac® BX-660). [Rhone-Poulenc Surf.] Aromatic phosphate ester; anionic; emulsifier, stabilizer for emulsion polymerization; emulsifier, wetting agent, hydrotrope for use in alkaline cleaning sol'ns.; liq.; 80% act.

Rhodafac® CB (formerly Rhodiasurf CB). [Rhone-Poulenc France] Complex phosphate ester; anionic; surfactant.

Rhodafac® GB-520 (formerly Gafac® GB-520). [Rhone-Poulenc Surf.; Rhone-Poulenc France] Phosphate ester, partial sodium salt; anionic; emulsifier, lubricant, softener, textile finishing aid for wool and syn. fibers, metalworking fluids; hazy visc. liq.; sol. in xylene, butyl Cellosolve, perchloroethylene, ethanol; disp. in water; sp.gr. 1.03-1.04; dens. 8.621 lb/gal; pour pt. 9 C; flash pt. > 200 C; pH 2.5-4.5 (10%); 98% act.

Rhodafac® HA-70 (formerly Intrex HA-70). [Rhone-Poulenc France] Phosphate ester; anionic; low foam surfactant.

Rhodafac® L3-15A (formerly Alkaphos L3-15A). [Rhone-Poulenc Surf.] Phosphate ester, free acid; anionic; detergent, wetting agent, coupling agent, solubilizer for alkaline cleaners, electrolyte sol'ns.; visc. liq.; 100% conc.

Rhodafac® L3-64A (formerly Alkaphos L3-64A). [Rhone-Poulenc Surf.] Phosphate ester, free acid; anionic; emulsifier, antistat, detergent for drycleaning charge soaps; textile yarn lubricant; liq.; oil-sol.; 100% conc.

Rhodafac® L4-27A (formerly Alkaphos L4-27A). [Rhone-Poulenc Surf.] Phosphate ester, free acid; anionic; detergent, emulsifier, wetting agent, solubilizer in highly alkaline liq. cleaning compds. and electrolyte sol'ns.; liq.; 100% conc.

Rhodafac® L6-36A (formerly Alkaphos L6-36A). [Rhone-Poulenc Surf.] Phosphate ester, free acid; anionic; detergent, hydrotrope for hard surf. detergents and metal cleaners; sol. and stable in strong alkali sol'ns.; liq.; 80% conc.

Rhodafac® LO-11A (formerly Alkaphos LO-11A). [Rhone-Poulenc Surf.] Phosphate ester of mixed linear alcohols; surfactant.

Rhodafac® LO-529 (formerly Gafac® LO-529). [Rhone-Poulenc Surf.; Rhone-Poulenc France] Phosphate ester, sodium salt; anionic; detergent, foamer, corrosion inhibitor, emulsifier for detergent concs., floor cleaners, agric. concs., metalworking fluids; FDA, EPA compliance; visc. liq.; sol. in water, kerosene, Stod., xylene, butyl Cellosolve, perchloroethylene; sp.gr. 1.05-1.15; dens. 9.163 lb/gal; pour pt. 6 C; flash pt. > 200 C; pH 5-6 (10%); 88% act.

Rhodafac® MB (formerly Soprophor MB). [Rhone-Poulenc France] Complex phosphate ester; anionic; surfactant.

Rhodafac® MC-470 (formerly Gafac® MC-470). [Rhone-Poulenc Surf.] Sodium laureth-4 phosphate; anionic; detergent, emulsifier, visc. builder for creams and lotions, polymerization and stabilization of latexes; fatliquoring of leathers; metalworking fluids; textile antistat, lubricant, softener; emulsifier for min. oils; clear visc. liq.; sol. in min. oil, kerosene, xylene, perchloroethylene, disp. in water; sp.gr. 1.02-1.04; dens. 8.580 lb/gal; pour pt. 0 C; flash pt. > 200 C; pH 5.0-6.5 (10%); 95% act.

Rhodafac® MD-12-116 (formerly Alkaphos MD-12-116). [Rhone-Poulenc Surf.] Aliphatic phosphate ester, free acid; anionic; low foaming emulsifier, dispersant, wetting agent with good rinsability; for industrial cleaners, pesticides, oil well cleanout; hydrotrope for metal cleaners, rinse aids; pale brn. solid; sol. in water and most aromatic solvs.; insol. in min. oil, most aliphatic solvs.; m.p. 45 C; acid no. 70-90 (to pH 9-9.5); pH 2.5 max. (1% aq.); 99% min. act.

Rhodafac® PA-15 (formerly Rhodiasurf PA-15). [Rhone-Poulenc France] Complex org. phosphate ester, free acid; anionic; hydrophilic detergent for liq. industrial cleaners; low foaming; stable to alkalies; liq.; 99% conc.

Rhodafac® PA-17 (formerly Rhodiasurf PA-17). [Rhone-Poulenc France] Complex org. phosphate ester, free acid; anionic; emulsifier, dispersant for household and industrial cleaners, textile wet processing, metalworking compds.; compat. with alkali, anionic and nonionic surfactants; liq.; 99% conc.

Rhodafac® PA-19 (formerly Rhodiasurf PA-19). [Rhone-Poulenc France] Complex org. phosphate ester, free acid; anionic; detergent, emulsifier, dispersant for household and industrial cleaners, textile wet processing, metalworking compds.; acid and alkali stability; liq.; 99% conc.

Rhodafac® PA-23 (formerly Rhodiasurf PA-23). [Rhone-Poulenc France] Complex org. phosphate ester, free acid; anionic; detergent, emulsifier, dispersant for alkaline built household or industrial cleaners; alkali stable; liq.; sol. in water, ethanol, most aromatic and chlorinated solvs.; 99% conc.

Rhodafac® PA-35 (formerly Rhodiasurf PA-35). [Rhone-Poulenc France] Complex phosphate ester; anionic; surfactant.

Rhodafac® PE-9 (see Rhodafac® RE-610). [Rhone-Poulenc Surf.] Phosphate ester; industrial surfactant.

Rhodafac® PE-510 (formerly Gafac® PE-510). [Rhone-Poulenc Surf.; Rhone-Poulenc France] Phosphate ester, free acid; anionic; detergent for drycleaning, waterless hand cleaners; intermediate for textile lubricants; emulsifier for emulsion polymerization (PVAc and acrylic films); emulsifier, dispersant for agric. formulations; corrosion inhibitor; good electrolyte tolerance; FDA, EPA compliance; lt. colored clear to sl. hazy visc. liq.; sol. in kerosene, Stod., xylene, butyl Cellosolve, perchloroethylene, ethanol; disp. in water; sp.gr. 1.08-1.09; dens. 9.0 lb/gal; pour pt. 20 C; pH < 2.5 (10%); 100% act.

Rhodafac® PEH (formerly Pegafac PEH). [Rhone-Poulenc Surf.] Complex org. phosphate ester; nonionic; detergent, dispersant, and wetting agent in textile wet processing; liq.; 100% conc.

Rhodafac® PL-6 (see Rhodafac® RA-600). [Rhone-Poulenc Surf.] Phosphate ester; industrial surfactant.

Rhodafac® PL-620 (formerly Gafac® PL-620). [Rhone-Poulenc Surf.] Complex phosphate ester of a linear alcohol alkoxylate; surfactant.

Rhodafac® PS-17 (formerly Rhodiasurf PS-17, Soprophor PS-17). [Rhone-Poulenc France] Complex org. phosphate ester, potassium salt; anionic; detergent, wetting agent, dispersant, foamer, foam stabilizer for heavy-duty liq. detergents, paraffin emulsions, emulsion polymerization, textile wet processing; stable to electrolytes; limited acid compat.; visc. liq.; sol. in water, ethanol, some aromatic and aliphatic solvs. and trichlorethylene; 80% conc.

Rhodafac® PS-19 (formerly Rhodiasurf PS-19, Soprhophor PS-19). [Rhone-Poulenc France] Complex org. phosphate ester, potassium salt; anionic; detergent, wetting agent, dispersant, foamer, foam stabilizer for emulsion polymerization, textile wet processing; alkaline compat.; limited acid compat.; visc. liq.; 80% conc.

Rhodafac® PS-23 (formerly Rhodiasurf PS-23). [Rhone-Poulenc France] Complex org. phosphate ester, potassium salt; anionic; detergent, wetting agent, dispersant for heavy-duty liq. detergents; compat. with alkali and electrolyte sol'ns.; visc. liq.; 70% conc.

Rhodafac® PV-27 (formerly Solumin PV-27). [Rhone-Poulenc France] Complex phosphate ester; anionic; surfactant.

Rhodafac® R5-09/S (formerly Alkaphos R5-09S). [Rhone-Poulenc Surf.] Aromatic phosphate ester, potassium salt; anionic; low foam detergent, wetting agent, dispersant, solubilizer, coupling agent for high alkaline content industrial cleaners, metal cleaners, floor cleaners, rinse aid concs.; pale yel. clear liq.; sol. in water, strong alkalis and electrolytes, alcohols, glycols, other polar solvs.; insol. in min. oil; dens. 1.27 g/ml; visc. 120 cps; pH 7.0-8.5 (5% aq.); 50% min. act.

Rhodafac® R9-47A (formerly Alkaphos R9-47A). [Rhone-Poulenc Surf.] Phosphate ester, free acid; anionic; detergent, wetting agent, emulsifier for heavy-duty all-purpose detergents, drycleaning formulations, pesticides; dedusting agent for alkaline powds.; lt. colored liq.; water-sol.; 100% conc.

Rhodafac® RA-600 (formerly Gafac® RA-600). [Rhone-Poulenc Surf.; Rhone-Poulenc France] Deceth-4 phosphate; anionic; detergent, emulsifier, wetting agent, foamer, dispersant for hard surf. cleaners, industrial alkaline detergents, textile wet processing; coupler used in liq. alkali detergents; clear to slightly hazy visc. liq.; sol. in kerosene, xylene, butyl Cellosolve, perchloroethylene, ethanol, water; dens. 8.9 lb/gal; sp.gr. 1.06–1.08; pour pt. < 0 C; pH < 2.5 (10%); 98.50% act.

Rhodafac® RB-400 (formerly Gafac® RB-400). [Rhone-Poulenc Surf.] Free acid of complex org. phosphate ester; anionic; lubricant, antistat, emulsifier for fibers and metal; opaque visc. liq.; sol. in min. oil, kerosene, Stod., xylene, butyl Cellosolve, perchloroethylene, ethanol; disp. in water; dens. 8.6 lb/gal; sp.gr. 1.03–1.04; pour pt. 18 C; pH < 2.5 (10%); 98% act.

Rhodafac® RD-510 (formerly Gafac® RD-510). [Rhone-Poulenc Surf.] Laureth-4 phosphate; anionic; emulsifier, antistat, lubricant, solubilizer for fibers, metals, cosmetics, agric. formulations; FDA, EPA compliance; clear visc. liq.; sol. in min. oil, kerosene, xylene, butyl Cellosolve, perchloroethylene, ethanol; disp. in water; sp.gr. 1.05–1.06; dens. 8.8 lb/gal; pour pt. 13 C; pH < 2.5 (10%); 98% act.

Rhodafac® RE-410 (formerly Gafac® RE-410). [Rhone-Poulenc Surf.; Rhone-Poulenc France] Phosphate ester, free acid; anionic; detergent, emulsifier, wetting agent, dispersant for industrial, institutional and household cleaners, agric. formulations; EPA compliance; clear visc. liq.; sol. in kerosene, Stod., xylene, butyl Cellosolve, perchloroethylene, ethanol; disp. in water; dens. 9.1 lb/gal; sp.gr. 1.00–1.20; pour pt. 18 C; pH < 2.5 (10%); 100% act.

Rhodafac® RE-610 (formerly Gafac® RE-610). [Rhone-Poulenc Surf.; Rhone-Poulenc France] Nonoxynol-9 phosphate; anionic; detergent, emulsifier, wetting agent, dispersant, antistat, lubricant, dedusting agent for drycleaning, pesticides, emulsion polymerization, textile wet processing, metals, household and industrial detergents; FDA, EPA compliance; slightly hazy visc. liq.; sol. in xylene, butyl Cellosolve, perchloroethylene, ethanol, water; dens. 9.2 lb/gal; sp.gr. 1.1–1.12; pour pt. < 0 C; surf. tens. 34 dynes/cm (@ CMC); 100% act.

Rhodafac® RE-870 (formerly Gafac® RE-870). [Rhone-Poulenc Surf.] Free acid of complex org. phosphate ester; anionic; detergent, emulsifier, wetting agent, dispersant, antistat, lubricant for pesticides, drycleaning, textile wet processing, metals, cosmetics; EPA compliance; soft waxy paste; sol. in xylene, butyl Cellosolve, perchloroethylene, ethanol, water; sp.gr. 1.1-1.2 (50 C); dens. 9.6 lb/gal; pour pt. 29 C; pH < 2.5 (10%); 99% act. DISCONTINUED.

Rhodafac® RE-960 (formerly Gafac® RE-960). [Rhone-Poulenc Surf.; Rhone-Poulenc France] Aromatic phosphate ester, free acid; anionic; emulsifier, wetting agent, dispersant for industrial cleaners, emulsion polymerization, pesticides, drycleaning, textile wet processing, metals; promotes freeze/thaw and mech. stability in PVAc and acrylic latexes; FDA compliance; soft waxy paste; sol. in butyl Cellosolve, ethanol, water; dens. 9.8 lb/gal; sp.gr. 1.17–1.18 (50 C); pour pt. 20 C; pH < 2.5 (10%); surf. tens. 43 dynes/cm (@ CMC); 90% act.

Rhodafac® RK-500 (formerly Gafac® RK-500). [Rhone-Poulenc Surf.] Aliphatic phosphate ester, free acid; anionic; low foaming hydrotrope, lubricant, antistat, emulsifier for metal treatment; clear, visc. liq.; sol. in xylene, butyl Cellosolve, perchloroethylene, ethanol, water; dens. 9.1 lb/gal; sp.gr. 1.05–1.15; pour pt. < 0 C; pH < 2.5 (10%); 100% act.

Rhodafac® RM-410 (formerly Gafac® RM-410). [Rhone-Poulenc Surf.] Nonyl nonoxynol-7 phosphate; anionic; detergent, emulsifier, antistat for drycleaning, pesticides; rust inhibitor; textile wetting agent; slightly hazy, visc. liq.; sol. in min. oil, kerosene, Stod., xylene, butyl Cellosolve, perchloroethylene, ethanol; disp. in water; dens. 8.8 lb/gal; sp.gr. 1.05–1.07; pour pt. 19 C; pH < 2.5 (10%); 100% act.

Rhodafac® RM-510 (formerly Gafac® RM-510). [Rhone-Poulenc Surf.; Rhone-Poulenc France] Nonyl nonoxynol-10 phosphate; anionic; emulsifier for chlorinated and phosphated pesticide concs., polyethylene; wetting agent, emulsifier for industrial cleaners, drycleaning, cosmetics; EPA compliance; hazy, visc. liq.; sol. in kerosene, Stod., xylene, butyl Cellosolve, perchloroethylene, ethanol; disp. in water; sp.gr. 1.05–1.07; dens. 8.8 lb/gal; pour pt. 5 C; pH < 2.5 (10%); 100% act.

Rhodafac® RM-710 (formerly Gafac® RM-710). [Rhone-Poulenc Surf.; Rhone-Poulenc France] Nonyl nonoxynol-15 phosphate; anionic; emulsifier for chlorinated and phosphated pesticide concs. and polyethylene; wetting agent, emulsifier for industrial cleaners, drycleaning, cosmetics; EPA compliance; hazy, visc. liq.; sol. in xylene, butyl Cellosolve, perchloroethylene, ethanol, water; sp.gr. 1.06–1.08; dens. 8.9 lb/gal; pour pt. 15 C; acid no. 33–41; pH < 3 (10%); 100% conc.

Rhodafac® RP-710 (formerly Gafac® RP-710). [Rhone-Poulenc Surf.; Rhone-Poulenc France] Aromatic phosphate ester, free acid; anionic; detergent, emulsifier, wetting agent, low foaming hydrotrope for industrial and household cleaners, hard surf. cleaning; for solubilizing low foaming nonionic surfactants; slightly hazy visc. liq.; sol. in xylene, butyl Cellosolve, perchloroethylene, ethanol, water; dens. 10.2; sp.gr. 1.22–1.23; pour pt. –7 C; pH < 2.5 (10%); 100% act.

Rhodafac® RS-410 (formerly Gafac® RS-410). [Rhone-Poulenc Surf.; Rhone-Poulenc France] Phosphate ester, free acid; anionic; emulsifier, detergent, wetting agent, dispersant for industrial, institutional and household cleaners, pesticides, drycleaning, textiles, metal treatment; EPA compliance; hazy visc. liq.; sol. in min. oil, kerosene, Stod., xylene, butyl Cellosolve, perchloroethylene, ethanol, disp. in water; sp.gr. 1.03–1.04; dens. 8.6 lb/gal; pour pt. –15 C; pH < 2.5 (10%); 100% act.

Rhodafac® RS-610 (formerly Gafac® RS-610). [Rhone-Poulenc Surf.; Rhone-Poulenc France] Trideceth-6 phosphate; anionic; emulsifier for emulsion polymerization, waterless hand cleaners, pesticides; detergent for drycleaning formulations, textile wetting agent, lubricant for fiber and metal treatment, paper-mill felt washing; antistat for aerosols; FDA, EPA compliance; hazy, visc. liq.; sol. in kerosene, Stod., xylene, butyl Cellosolve, perchloroethylene, ethanol; disp. in water; sp.gr. 1.04–1.06; dens. 8.7 lb/gal; pour pt. < 0 C; pH < 2.5 (10%); 100% act.

Rhodafac® RS-710 (formerly Gafac® RS-710). [Rhone-Poulenc Surf.; Rhone-Poulenc France] Phosphate ester, free acid; anionic; detergent, wetting agent, emulsifier, dispersant for textile wet processing, industrial cleaners, pesticides, drycleaning, metal treatment; FDA, EPA compliance; opaque visc. liq.; sol. in Stod., xylene, butyl Cellosolve, perchloroethylene, ethanol, water; sp.gr. 1.04–1.06; dens. 8.7 lb/gal; pour pt. 18 C; pH < 2.5 (10%); 100% act.

Rhodameen® 220 (formerly Rhodiasurf 220). [Rhone-Poulenc France] Fatty amine ethoxylate; cationic; surfactant.

Rhodameen® HT-50. [Rhone-Poulenc Surf.] Ethoxylated fatty amine; nonionic/cationic; wetting agent, penetrant, emulsifier, stabilizer, dispersant, antistat, lubricant; solid; HLB 17.9; 100% conc. DISCONTINUED.

Rhodameen® O-2 (formerly Catafor O-2). [Rhone-Poulenc France] PEG-2 oleamine; cationic; surfactant.

Rhodameen® O-6. [Rhone-Poulenc France] PEG-6 oleamine; cationic; surfactant.

Rhodameen® O-12 (formerly Rhodiasurf O-12). [Rhone-Poulenc France] PEG-2 oleamine; cationic; coemulsifier in emulsifiable concs. (herbicides), w/o emulsions of solvs., waxes, and oils, catiaonic asphalt emulsions; liq.; 99% conc.

Rhodameen® OA-910 (formerly Katapol® OA-910). [Rhone-Poulenc Surf.] PEG-30 oleamine; cationic; hydrophilic emulsifier and textile dyeing assistant; antiprecipitant to prevent cross-staining; stripping agent and dye leveler for acid dyes; liq.; HLB 16.4; 100% act.

Rhodameen® OS-12 (formerly Rhodiasurf OS-12). [Rhone-Poulenc France] PEG-2 oleyl/stearyl amine; cationic; coemulsifier and antistat for cosmetics, textile and plastic processing; flowable paste; 99% conc.

Rhodameen® PN-430 (formerly Katapol® PN-430). [Rhone-Poulenc Surf.; Rhone-Poulenc France] PEG-5 tallowamine; cationic; emulsifier, corrosion inhibitor, lubricant for metalworking fluids, agric. formulations; acid corrosion inhibitor for ferrous alloys; brn. liq.; sol. in oil; dens. 7.830 lb/gal; HLB 10.0; flash pt. > 200 C; pH 8-10 (10% aq.); 100% act.

Rhodameen® PN-810 (formerly Katapol® PN-810). [Rhone-Poulenc Surf.; Rhone-Poulenc France] PEG-20 tallowamine; cationic; emulsifier, dye assistant, softener, antistat, lubricant for textiles, leather, agric. formulations; amber clear liq.; water-sol.; HLB 15.4; 100% act. DISCONTINUED.

Rhodameen® RAM/7 Base (formerly Soprophor RAM/7). [Rhone-Poulenc France] PEG-15 oleamine; cationic; surfactant.

Rhodameen® RAM/8 Base (formerly Soprophor RAM/8). [Rhone-Poulenc France] PEG-25 oleamine; cationic; surfactant.

Rhodameen® S-20 (formerly Rhodiasurf S-20). [Rhone-Poulenc France] PEG-10 tallowamine; cationic; emulsifier, dispersant for emulsifiable conc. herbicides, asphalt emulsions, emulsions of solvs., oils, and waxes; liq.; 99% conc.

Rhodameen® S-25 (formerly Rhodiasurf S-25). [Rhone-Poulenc France] PEG-15 tallow amine; cationic; surfactant.

Rhodameen® T-5 (formerly Alkaminox® T-5). [Rhone-Poulenc Surf.] PEG-5 tallow amine; cationic; emulsifier, dispersant; textile scouring and desizing assistant; softener and antistatic agent; liq./paste; HLB 9.0; 100% conc.

Rhodameen® T-12 (formerly Alkaminox® T-12). [Rhone-Poulenc Surf.] PEG-12 tallowamine; cationic; emulsifier, dispersant; textile scouring, leveling, and desizing assistant; softener and antistatic agent; slow release agent for agric. formulations; amber liq.; sol. in water; sp.gr. 1.03 g/ml; HLB 13.5; 90% conc.

Rhodameen® T-15 (formerly Alkaminox® T-15). [Rhone-Poulenc Surf.] PEG-15 tallow amine; cationic; emulsifier, dispersant; textile scouring and desizing assistant; softener and antistatic agent; corrosion inhibitor for steam generating and circulating systems; emulsifier for wax emulsions; Gardner 12 paste; water-sol.; sp.gr. 1.03; HLB 14.2; 100% conc.

Rhodameen® T-30-45% (formerly Alkaminox® T-30). [Rhone-Poulenc Surf.] PEG-30 tallowamine; cationic; coemulsifier, leveling agent for the differential dyeing of nylon; activating agent for agric. formulations; EPA compliance; solid; 45% conc.

Rhodameen® T-30-90% (formerly Alkaminox® T-30). [Rhone-Poulenc Surf.] PEG-30 tallowamine; cationic; coemulsifier, leveling agent for differential dyeing of nylon; activating agent for agric. formulations; EPA compliance; amber liq.; HLB 16.6; 90% conc.

Rhodameen® T-50 (formerly Alkaminox® T-50). [Rhone-Poulenc Surf.] PEG-50 tallowamine; cationic; coemulsifier, leveling agent for differential dyeing of nylon; yel. solid; water-sol.; HLB 17.5; 100% conc.

Rhodameen® VP-532/SPB (formerly Katapol® VP-532). [Rhone-Poulenc Surf.] POE alkylamine; nonionic/cationic; wetting agent, penetrant, emulsifier, stabilizer, antistat, and lubricant; retarder, antiprecipitant, leveling agent in dyeing; dispersant for fiberglass mat mfg.; liq.; HLB 15.0; 20% act.

Rhodamox® CAPO (formerly Alkamox® CAPO). [Rhone-Poulenc Surf.] Cocamidopropyl dimethylamine oxide; nonionic/cationic; foaming agent/stabilizer, thickener, emollient for shampoos, bath prods., dishwash, rug shampoos, fine fabric detergents, shaving creams, lotions, foam rubber, electroplating, paper coatings; used in toiletries for mildness; colorless clear liq.; faint odor; water sol.; disp. in min. oil; dens. 1.0 g/ml; f.p. < 0 C; 30% conc.

Rhodamox® LO (formerly Alkamox® LO, Cyclomox® L). [Rhone-Poulenc Surf.; Rhone-Poulenc France] Lauryl dimethylamine oxide; cationic; foaming agent/stabilizer, thickener, emollient for shampoos, bath prods., dishwash, fine fabric detergents, shaving creams, lotions, textile softeners, foam rubber, in electroplating, paper coatings; used in toiletries for mildness; colorless clear liq.; faint odor; water sol.; disp. in min. oil; dens. 1.0 g/ml; f.p. < 0 C; 30% conc.

Rhodapex® AB-20 (formerly Geropon AB/20). [Rhone-Poulenc France] Ammonium laureth sulfate; anionic; emulsifier for emulsion polymerization and min. oils; detergent; liq.; 30% conc.

Rhodapex® CD-128 (formerly Alipal® CD-128). [Rhone-Poulenc Surf.; Rhone-Poulenc France] Ammonium alcohol ethoxylate sulfate; anionic; high foaming emulsifier, wetting agent, dispersant for aq. high electrolyte systems, built detergents; air-entraining agent, foamer, frothing agent for concrete, cement, gypsum wallboard; solubilizer; straw liq.; good sol. in electrolyte systems; sp.gr. 1.056; pour pt. -5 C; 56% act.

Rhodapex® CO-433 (formerly Alipal® CO-433). [Rhone-Poulenc Surf.; Rhone-Poulenc France] Sodium nonoxynol-4 sulfate; anionic; high foaming detergent, wetting agent, dispersant for dishwashing, scrub soaps, car washes, rug and hair shampoos; emulsifier for emulsion polymerization, petrol. waxes; antistat for plastics and syn. fibers; lime soap dispersant; FDA compliance; Varnish 4 max. clear liq.; mild aromatic odor; water-sol.; sp.gr. 1.065; dens. 8.9 lb/gal; visc. 2500 cps; surf. tens. 32 dynes/cm (1%); 28% act.

Rhodapex® CO-436 (formerly Alipal® CO-436). [Rhone-Poulenc Surf.; Rhone-Poulenc France] Ammonium nonoxynol-4 sulfate; anionic; high foaming detergent, wetting agent, dispersant for dishwashing, scrub soaps, car washes, rug and hair shampoos; emulsifier for emulsion polymerization, petrol. waxes; antistat for plastics and syn. fibers; lime soap dispersant; FDA, EPA compliance; Varnish 5 max clear liq.; alcoholic odor; water-sol.; sp.gr. 1.065; dens. 8.9 lb/gal; visc. 100 cps; surf. tens. 34 dynes/cm (1%); 58% act.

Rhodapex® EA. [Rhone-Poulenc Surf.] Ammonium laureth (3) sulfate; anionic; high foaming emulsifier for industrial, institutional and household cleaners; liq.. DISCONTINUED.

Rhodapex® ES (formerly Sipon® ES). [Rhone-Poulenc Surf.] Sodium laureth (3) sulfate; CAS 9004-82-4; anionic; emulsifier, high foaming base for shampoos, lt. duty detergents, bubble baths; polymerization surfactant; liq.; 27% conc.

Rhodapex® EST-30 (formerly Sipex® EST-30). [Rhone-Poulenc Surf.; Rhone-Poulenc France] Sodium trideth sulfate; anionic; emulsifier, wetting agent, dispersant for baby shampoo, other personal care prods., household, industrial, institutional and industrial formulations, emulsion polymerization of styrene systems, textile scouring, dishwash; FDA compliance; Gardner 2 max. liq.; cloud pt. 14 C max.; pH 7.5–8.5 (10%); surf. tens. 33 dynes/cm (@ CMC); 29-30% act.

Rhodapex® ESY (formerly Sipon® ESY). [Rhone-Poulenc Surf.; Rhone-Poulenc France] Sodium laureth (1) sulfate; CAS 9004-82-4; anionic; base, cosurfactant, emulsifier for shampoos, lt. duty detergents; liq.; 30% conc.

Rhodapex® F-85/SD (formerly Solumin F-85/SD). [Rhone-Poulenc France] Sodium octylphenol ethoxy sulfate; anionic; surfactant; 25% conc.

Rhodapex® F-775/SD (formerly Solumin F-775/SD). [Rhone-Poulenc France] Sodium octylphenol ethoxy sulfate; anionic; surfactant; 25% conc.

Rhodapex® MA360 (formerly Sipon® MA360). [Rhone-Poulenc Surf.] Ammonium laureth (3) sulfate; anionic; high active base for lt. duty dish and fine fabric cleaners at neutral pH; emulsifier; liq.; 60% conc.

Rhodapex® NA 61. [Rhone-Poulenc Surf.] Sodium laureth (3) sulfate; CAS 9004-82-4; anionic; emulsifier, high foaming base for dishwasher, textile scours, carwashing compds.; liq.; 60% conc. DISCONTINUED.

Rhodapex® NA 61 CG. [Rhone-Poulenc Surf.] Sodium laureth sulfate; CAS 9004-82-4; anionic; high activity colorless detergent conc.; colorless liq.; 60% conc. DISCONTINUED.

Rhodapon® BOS (formerly Sipex® BOS). [Rhone-Poulenc Surf.; Rhone-Poulenc France] Sodium 2-ethylhexylsulfate; anionic; biodeg. wetting agent, emulsifier, detergent, foamer with high electrolyte tolerance for industrial, institutional and household cleaners; latex stabilizer, nickel brightener; metal treatment; textile and plywood mfg.; fruit and veg. washing; FDA compliance; clear liq.; visc. 50 cps; cloud pt. < 10 C; pH 9.5–10.5 (10%); surf. tens. 33 dynes/cm (@ CMC); 39-40% act.

Rhodapon® CAV (formerly Sipex® CAV). [Rhone-Poulenc Surf.] Sodium isodecyl sulfate; anionic; wetting agent, emulsifier, detergent, foamer, rinse aid, visc. control agent; post-stabilizer in latex paints; metal treatment; textile and plywood mfg.; fruit/veg. washing; hard surface cleaners; emulsion polymerization of vinyl, acrylic, SBR, PVC; clear liq.; dens. 8.95 lb/gal; visc. 150 cps; cloud pt. < 10 C: pH 9.5–10.5 (10%); surf. tens. 34 dynes/cm (@ CMC); 39-40% act.

Rhodapon® EC111 (formerly Sipex® EC111). [Rhone-Poulenc Surf.; Rhone-Poulenc France] Sodium cetyl/stearyl sulfate; anionic; emulsifier, detergent, flotation agent; collector in ore flotation; softener/lubricant for textiles; cosmetics and toiletries; wh. thick paste; pH 8 (10%); 25% act.

Rhodapon® L-22, L-22/C (formerly Sipon® L-22). [Rhone-Poulenc Surf.; Rhone-Poulenc France] Ammonium lauryl sulfate; anionic; high foaming detergent, emulsifier for shampoo, bubble bath, pet shampoos, industrial and institutional cleaners, wool scouring, fire fighting foams, assistant for pigment dispersion; emulsion polymerization aid; clear visc. liq.; visc. 1000 cps; HLB 31.0; cloud pt. 14 C; pH 6.8 (10%); 28% act.

Rhodapon® L-22HNC (formerly Sipon® L-22HNC). [Rhone-Poulenc Surf.] Ammonium lauryl sulfate; surfactant for personal care prods. and SBR latex froth applics.; liq.; 30% act.

Rhodapon® LCP (formerly Sipon® LCP). [Rhone-Poulenc Surf.] Sodium lauryl sulfate; CAS 151-21-3; anionic; low cloud pt. emulsion polymerization surfactant; FDA compliance; liq.; surf. tens. 30 dynes/cm (@ CMC); 30% solids.

Rhodapon® LM (formerly Sipon® LM). [Rhone-Poulenc Surf.] Magnesium lauryl sulfate; anionic; high-foaming emulsifier, wetting agent, dispersant, detergent for rug and upholstery shampoos, bubble baths, shampoos; food pkg. applics.; clear liq.; visc. 50 cps; cloud pt. 2 C; pH 6.5 (10%); 27% act.

Rhodapon® LSB, LSB/CT (formerly Sipon® LSB). [Rhone-Poulenc Surf.] Sodium lauryl sulfate; CAS 151-21-3; anionic; high foaming detergent, emulsifier for hair shampoo, lotion, pastes, industrial, institutional and household cleaners; pale clear liq.; visc. 150 cps; HLB 40; cloud pt. 20 C; pH 8 (10%); 29% act.

Rhodapon® LT-6 (formerly Sipon® LT-6). [Rhone-Poulenc Surf.] TEA lauryl sulfate; anionic; emulsifier, high foaming base for industrial and household detergents, shampoos, bubble baths; obtains creamy, mild lather; pale yel. clear liq.; visc. 50 cps; HLB 34.0; cloud pt. 2 C; pH 7 (10%); 40% act.

Rhodapon® OLS (formerly Sipex® OLS). [Rhone-Poulenc Surf.] Sodium octyl sulfate; anionic; wetting agent; rinse aid for industrial, institutional and household cleaners; mercerizing agent for cotton goods; surfactant in electrolyte baths for metal cleaning; hard surface cleaning; neoprene dispersant; emulsifier for emulsion polymerization; clear liq.; visc. 100 cps; pH 8 (10%); surf. tens. 33 dynes/cm (@ CMC); 33% act.

Rhodapon® OS (formerly Sipex® OS). [Rhone-Poulenc Surf.] Sodium oleyl sulfate; anionic; specialty emulsifier for emulsion polymerization; FDA compliance; liq.; surf. tens. 34 dynes/cm (@ CMC); 26% solids.

Rhodapon® SB (formerly Sipon® SB). [Rhone-Poulenc Surf.] Sodium lauryl sulfate; CAS 151-21-3; anionic; emulsifier for emulsion polymerization; liq.; HLB 40.0; surf. tens. 31 dynes/cm (@ CMC); 29% conc.

Rhodapon® SB-8208/S (formerly Sipon® 21LS). [Rhone-Poulenc Surf.] Sodium lauryl sulfate; CAS 151-21-3; detergent/emulsifier and foaming agent for shampoos, liq. soaps, bath gels, bubble baths; 30% conc.

Rhodapon® SM Special (formerly Sipon® SM Spec.). [Rhone-Poulenc Surf.; Rhone-Poulenc France] Sodium lauryl sulfate; CAS 151-21-3; anionic; cosurfactant for household, institutional and industrial cleaner formulations, personal care prods.; liq.; 35% conc.

Rhodaquat® A-50 (formerly Cequartyl A-50). [Rhone-Poulenc France] C13-15 benzalkonium chloride; cationic; surfactant; 50% conc.

Rhodaquat® DAET-90 (formerly Alkaquat® DAET-90). [Rhone-Poulenc Surf.] Ditallow quat. compd.; cationic; substantive antistat, softener for home and commercial laundering; yel. paste; dens. 0.99 g/ml; pH 4.0-7.0; 90% act.

Rhodaquat® T (formerly Alkaquat® T). [Rhone-Poulenc Surf.] Ditallow imidazolinium quat.; cationic; textile surfactant, softening agent, antistat in home and commercial laundries; pale yel. slightly visc. liq.; typ. odor; sol. in min. oil, perchlorethylene; disp. in water, min. spirits, aromatic solv.; sp.gr. 0.95; pH 6.0–6.8; 74-76% act.

Rhodaquat® VV-328 (formerly Katapone VV-328). [Rhone-Poulenc Surf.] Quat. ammonium chloride; cationic; surfactant; deep amber clear visc. liq.; sol. in water, xylene, butyl Cellosolve; sp.gr. 0.985-0.995; visc. 2700 cps; flash pt. (COC) 125 F; fire pt. 205 F; pH 6.5-8.5 (10%); surf. tens. 32.8 dynes/cm (0.1% aq.).

Rhodasurf® 25-7 (formerly Siponic® 25-7). [Rhone-Poulenc Surf.] C12-15 pareth-7; biodeg. detergent, wetting agent, emulsifier for household and industrial cleaners; coupler and solubilizer for perfumes and org. additives; HLB 12.2; 100% act.

Rhodasurf® 91-6 (formerly Siponic® 91-6). [Rhone-Poulenc Surf.] C9-11 pareth-6; biodeg. detergent, wetting agent, emulsifier for household and industrial cleaners; coupler and solubilizer for perfumes and org. additives; HLB 12.5; 100% act.

Rhodasurf® 110 (formerly Rhodiasurf 110). [Rhone-Poulenc France] Mixed alcohol ethoxylate; nonionic; surfactant.

Rhodasurf® 840 (formerly Rhodiasurf 840). [Rhone-Poulenc France] Branched syn. alcohol ethoxylate; nonionic; coemulsifier for oil-sol. emulsifier blends; intermediate for ether sulfate prod.; liq.; HLB 8.3; 99% conc.

Rhodasurf® 860/P (formerly Soprophor® 860/P). [Rhone-Poulenc France] Isodeceth-6; nonionic; wetting agent, foamer, emulsifier, detergent for textile processing, metal degreasing, herbicides, waxes, resins; stable to strong acidic and alkaline media; EPA compliance; liq.; HLB 11.4; 90% conc.

Rhodasurf® 870 (formerly Rhodiasurf 870). [Rhone-Poulenc France] Branched chain syn. alcohol ethoxylate; nonionic; wetting agent, foaming agent, emulsifier, detergent for textile processing, household detergent formulations; stable to alkaline and acidic media; flowable paste; HLB 13.7; 99% conc.

Rhodasurf® 1012-6. [Rhone-Poulenc Surf.] Linear C10-12 alcohol ethoxylate; nonionic; surfactant for household and industrial cleaners; biodeg.; liq.; HLB 12.0; 100% conc.

Rhodasurf® A-1P (formerly Texafor A-1P). [Rhone-Poulenc France] Ceteareth-23; CAS 68439-49-6; nonionic; surfactant.

Rhodasurf® A-2 (formerly Texafor A-2). [Rhone-Poulenc France] Ceteareth-2; CAS 68439-49-6; nonionic; surfactant.

Rhodasurf® A-6 (formerly Texafor A-6). [Rhone-Poulenc France] Ceteareth-6; CAS 68439-49-6; nonionic; surfactant.

Rhodasurf® A 24 (formerly Emulphogene® A 24). [Rhone-Poulenc Surf.] Ethoxylated alcohol; nonionic; wetting agent, emulsifier, dispersant, detergent; solid; HLB 7.3; 100% conc.

Rhodasurf® A-60 (formerly Texafor A-60). [Rhone-Poulenc France] Ceteareth-60; CAS 68439-49-6; nonionic; surfactant.

Rhodasurf® B-1 (formerly Texafor B-1). [Rhone-Poulenc France] Laureth-16; nonionic; surfactant.

Rhodasurf® B-7 (formerly Texafor B-7). [Rhone-Poulenc France] Laureth-7; nonionic; surfactant.

Rhodasurf® BC-420 (formerly Emulphogene® BC-420). [Rhone-Poulenc Surf.] Trideceth-3; CAS 24938-91-8; nonionic; emulsifier, detergent, dispersant for petrol. oils, agric. formulations; intermediate for mfg. of high-foaming anionic surfactants; EPA compliance; VSC 2 (50 C) cloudy, sl. visc. pourable liq.; mild, pleasant odor; sol. in toluene; insol. in water; sp.gr. 0.92-0.94; pour pt. -7 to -1 C; flash pt. (PMCC) > 200 F; pH 6.5-8.5 (10%); 100% act.

Rhodasurf® BC-610 (formerly Emulphogene® BC-610). [Rhone-Poulenc Surf.; Rhone-Poulenc France] Trideceth-6; CAS 24938-91-8; nonionic; detergent, foamer, wetting agent, dispersant, solubilizer for alkylaryl sulfonate; for lt. and heavy duty high foaming detergent formulations (dishwash, alkaline cleaners); emulsifier for silicones, petrol. oils, agric. formulations; EPA compliance; VCS 2 (50 C) cloudy sl. visc. pourable liq.; sol. in alcohols, acetone, toluene; disp. in water; sp.gr. 0.97-0.99; HLB 11.4; pour pt. 4.5-7.5 C; cloud pt. 38 C (1% in NaCl); flash pt. (PMCC) > 200 F; pH 6-8 (10%); 100% conc.

Rhodasurf® BC-630 (formerly Emulphogene® BC-630). [Rhone-Poulenc Surf.] Trideceth-7; CAS 24938-91-8; nonionic; detergent, wetting agent for dishwash, alkaline cleaners; emulsifier for silicones and petrol. oils; cloudy, sl. visc. pourable liq., mild pleasant odor; sol. in alcohols, acetone, toluene, tetrachloroethylene; sp.gr. 0.99; pour pt. 15-20 C; cloud pt. 50-60 C (1%); flash pt. (PMCC) > 200 F; pH 6-8 (10%); 100% act.

Rhodasurf® BC-720 (formerly Emulphogene® BC-720). [Rhone-Poulenc Surf.; Rhone-Poulenc France] Trideceth-10; CAS 24938-91-8; nonionic; emulsifier, stabilizer, foam builder, detergent, wetting agent, dispersant; solubilizer for alkylaryl sulfonates; for lt. and heavy duty high foaming detergent, agric. formulations; EPA compliance; VCS 2 (50 C) opaque visc. pourable liq.; mild, pleasant odor; sol. in water, alcohol, acetone, toluene, tetrachloroethylene; sp.gr. 1.00-1.03 (40 C); HLB 13.8;

pour pt. 20 C; cloud pt. 60 C (1% aq.); flash pt. (PMCC) > 200 F; pH 6-8 (10%); 100% conc.

Rhodasurf® BC-737 (formerly Emulphogene® BC-737). [Rhone-Poulenc Surf.] Trideceth-14; CAS 24938-91-8; nonionic; surfactant.

Rhodasurf® BC-840 (formerly Emulphogene® BC-840). [Rhone-Poulenc Surf.; Rhone-Poulenc France] Trideceth-15; CAS 24938-91-8; nonionic; foam builder, detergent; solubilizer for alkylaryl sulfonates; for lt. and heavy duty high foaming detergent, agric. formulations; EPA compliance; paste; sp.gr. 1.01-1.04 (40 C); pour pt. 33-40 C; cloud pt. > 95 C (1%); flash pt. (PMCC) > 200 F; pH 6-8 (10%); 100% act.

Rhodasurf® DA-4 (formerly Alkasurf® DA-4). [Rhone-Poulenc Surf. Canada] Decyl alcohol ethoxylate; nonionic; wetting agent, penetrant for pressure dyeing of fabrics, textile and industrial applics.; hazy liq.; water-disp.; HLB 10.5; 100% conc.

Rhodasurf® DA-6 (formerly Alkasurf® DA-6). [Rhone-Poulenc Surf. Canada] Decyl alcohol ethoxylate; nonionic; wetting agent, penetrant for pressure dyeing of fabrics, textile and industrial applics.; hazy liq.; water-disp.; HLB 12.5; 100% conc.

Rhodasurf® DA-530 (formerly Emulphogene® DA-530) [Rhone-Poulenc Surf.] Isodeceth-4; nonionic; low foaming rapid wetting agent for industrial, institutional and household cleaners, textile compding.; scouring agent, emulsifier for defoamers; liq.; HLB 10.5; 100% conc.

Rhodasurf® DA-630 (formerly Emulphogene® DA-630). [Rhone-Poulenc Surf.] Isodeceth-6; nonionic; low foaming wetting agent, dispersant for industrial, institutional and household cleaners, textile scouring; intermediate for mfg. of esters for textile and industrial applics.; emulsifier for defoamers; acid and alkali stable; liq.; HLB 12.5; cloud pt. 42 C (1% aq.); 100% conc.

Rhodasurf® DA-639 (formerly Emulphogene® DA-639). [Rhone-Poulenc Surf.] Deceth-6; nonionic; rapid, low foaming wetting/rewetting agent for textile compding.; scouring agent, emulsifier for defoamers; liq.; 90% act.

Rhodasurf® DB 311 (formerly Rhodiasurf DB 311). [Rhone-Poulenc France] Linear syn. alcohol ethoxylate; nonionic; biodeg. surfactant for household and industrial cleaner formulations; liq.; 99% conc.

Rhodasurf® E 300. [Rhone-Poulenc Surf.] PEG; nonionic; surfactant intermediate; liq.; 100% conc. DISCONTINUED.

Rhodasurf® E 400 (formerly Pegol® E-400). [Rhone-Poulenc Surf.] PEG; nonionic; surfactant intermediate; binder and lubricant in compressed tablets; softener for paper, plasticizer for starch pastes and polyethylene films; coupling agent for skin care lotions; liq.; pour pt. 6 C; 100% conc.

Rhodasurf® E 600 (formerly Pegol® E-600). [Rhone-Poulenc Surf.] PEG; nonionic; surfactant intermediate; binder and lubricant in compressed tablets; color stabilizer for fuel oils; liq.; pour pt. 22 C; 100% conc.

Rhodasurf® L-4 (formerly Siponic® L-4). [Rhone-Poulenc Surf.] Laureth-4; nonionic; emulsifier, thickener, wetting agent, pigment dispersant, lubricant, solubilizer for cosmetic and industrial emulsions; textile scouring agent, emulsion polymerization, metal cleaning, monomer systems, floor waxes, paper finishes, rubber; emollient for pharmaceuticals; liq.; HLB 9.7; pH 6.5 (1%); 100% conc.

Rhodasurf® L-25 (formerly Siponic® L-25). [Rhone-Poulenc Surf.] Laureth-23; nonionic; surfactant for cosmetic and industrial emulsions; wetting agent, pigment dispersant, lubricant, solubilizer, textile scouring agent, emulsion polymerization, metal cleaning, rubber and monomer systems; stabilizer for emulsion polymers; floor waxes, paper finishes, emollient for pharmaceuticals; wax; HLB 16.9; cloud pt. > 95 C (1%); pH 6.5 (1%); 100% act.

Rhodasurf® L-790 (formerly Siponic® L-7-90). [Rhone-Poulenc Surf.; Rhone-Poulenc France] Laureth-7; nonionic; surfactant for cosmetic and industrial emulsions; wetting agent, pigment dispersant, lubricant, solubilizer, textile scouring agent, emulsion polymerization, metal cleaning; liq.; HLB 12.1; 90% act.

Rhodasurf® LA-3 (formerly Alkasurf® LA-3). [Rhone-Poulenc Surf.] C12-15 pareth-3; nonionic; detergent base and emulsifier for dishwash, personal care prods., industrial applics.; textile lubricant; translucent to opaque liq.; sol. in min. oil, aromatic solv, perchloroethylene; sp.gr. 0.93; visc. 19 cps (100 F); HLB 7.8; cloud pt. 58-62 C; 100% act.

Rhodasurf® LA-7 (formerly Alkasurf® LA-7). [Rhone-Poulenc Surf.] C12-15 pareth-7; nonionic; biodeg. detergent, dispersant, wetting agent, emulsifier for heavy-duty detergents, all-purpose liqs., detergent tablets, sanitizers, metal cleaners, hard surf. cleaners, transportation cleaners; lt. liq.; water sol.; sp.gr. 0.96; visc. 32 cps (100 F); m.p. 11-15 C; HLB 12; cloud pt. 44-48 C; flash pt. 440 F (COC); pour pt. 70 F; 100% act.

Rhodasurf® LA-9 (formerly Alkasurf® LA-9). [Rhone-Poulenc Surf.] Straight chain fatty alcohol ethoxylate; nonionic; detergent, dispersant, wetting agent, emulsifier for heavy-duty detergents, all-purpose liqs., detergent tabelts, sanitizers, metal cleaners, hard surf. cleaners, transportation cleaners; paste; HLB 13.1.

Rhodasurf® LA-12 (formerly Alkasurf® LA-12). [Rhone-Poulenc Surf.] C12-15 pareth-12; nonionic; detergent, dispersant, wetting agent, emulsifier for heavy-duty detergents, all-purpose liqs., detergent tablets, sanitizers, metal cleaners, hard surf. cleaners, transportation cleaners; wh. paste, low odor; water-sol.; sp.gr. 1.0; HLB 14.4; 100% act.

Rhodasurf® LA-15 (formerly Alkasurf® LA-15). [Rhone-Poulenc Surf.] Straight chain fatty alcohol ethoxylate; nonionic; detergent, dispersant, wetting agent, emulsifier for heavy-duty detergents, all-purpose liqs., detergent tabelts, sanitizers, metal cleaners, hard surf. cleaners, transportation cleaners; solid; HLB 15.3.

Rhodasurf® LA-23-6.5. [Rhone-Poulenc Surf.] Straight chain fatty alcohol ethoxylate; nonionic; detergent for general industrial and household usage; liq.; HLB 12.0; 100% conc. DISCONTINUED.

Rhodasurf® LA-30 (formerly Rhodiasurf LA-30). [Rhone-Poulenc France] Linear syn. alcohol ethoxylate; nonionic; coemulsifier for pesticides, cosmetics, min. oil emulsions for metalworking, textile processing, degreasers; intermediate for biodeg. ether sulfate prod.; liq.; HLB 8.0; 99% conc.

Rhodasurf® LA-42 (formerly Rhodiasurf LA-42). [Rhone-Poulenc France] Fatty alcohol ethoxylate; nonionic; coemulsifier for fats and oils; degreasing agent in textile processing; liq.; HLB 9.6; 99% conc.

Rhodasurf® LA-90 (formerly Rhodiasurf LA-90). [Rhone-Poulenc France] Linear syn. alcohol ethoxylate; nonionic; wetting agent for textile wet processing, cosmetic preps.; solid; HLB 13.5; 99% conc.

Rhodasurf® LAN-03. [Rhone-Poulenc Surf.] Natural C12-14 fatty alcohol ethoxylate; nonionic; detergent intermediate; liq.; HLB 8.0; 100% conc. DISCONTINUED.

Rhodasurf® LAN-23-75% (formerly Alkasurf® LAN-23). [Rhone-Poulenc Surf.] Laureth-23; nonionic; biodeg. emulsifier, stabilizer, solubilizer for cosmetics; post add stabilizer for syn. latexes; almost colorless solid; char. odor; sol. in water, aromatic solv., perchloroethylene; HLB 17; cloud pt. 90.5–93.5 C; 75% act.

Rhodasurf® N-12 (formerly Texafor N-12). [Rhone-Poulenc France] PEG-12 octyl alcohol; nonionic; surfactant.

Rhodasurf® ON-870 (formerly Emulphor® ON-870). [Rhone-Poulenc Surf.; Rhone-Poulenc France] Oleth-20; CAS 9004-98-2; nonionic; high foaming emulsifier, stabilizer, dispersant, wetting agent, solubilizer for min. oils, fatty acids, waxes; for industrial cleaners, metal cleaners, agric., paints, adhesives, textile, leather, cosmetic, pharmaceutical industries; FDA, EPA compliance; wh. solid wax; sol. in water, xylene, ethanol, ethylene glycol, butyl Cellosolve; sp.gr. 1.04; HLB 15.4; pour pt. 46 C; cloud pt. > 100 C; flash pt. (PMCC) 93 C; surf. tens. 37 dynes/cm (0.1%); 100% act.

Rhodasurf® ON-877 (formerly Emulphor® ON-877). [Rhone-Poulenc Surf.] Oleth-20; CAS 9004-98-2; nonionic; stabilizer for natural and syn. rubber latex emulsions; acid degreaser, dyeing assistant for textiles; dyestuff dispersant, tanning assistant, fat liquor, degreaser for leathers; solubilizer and emulsifier for essential oils and pharmaceuticals; emulsifier for aq. dispersions of polyethylene; liq.; 70% conc.

Rhodasurf® PEG 600 (formerly Alkapol PEG 600). [Rhone-Poulenc Surf. Canada] PEG-14; plasticizer, solv.; lubricant, binder for pharmaceuticals; color stabilizer for fuel oils; intermediate for processing surfactants; wh. liq.; m.w. 570–630; water sol.; sp.gr. 1.13 (30/15.5 C); f.p. 20-25 C; pH 5–8; 100% conc.

Rhodasurf® PEG 3350. [Rhone-Poulenc Surf.] PEG 3350; mold release and antistat for rubber prods.; binder; waxy flakes; 100% conc. DISCONTINUED.

Rhodasurf® PEG 4000. [Rhone-Poulenc Surf.] PEG 4000; solv. or cosolv. for industrial applics. where lubricity, water sol., low toxicity, and chemical stability are important; solid; 100% conc. DISCONTINUED.

Rhodasurf® PEG 8000. [Rhone-Poulenc Surf.] PEG 8000; mold release and antistat for rubber; binder; waxy flakes; 100% conc.

Rhodasurf® RHS (formerly Soprofor RHS). [Rhone-Poulenc Geronazzo] Alkyl polyglycolic ether; nonionic; emulsifier for wh. oils; liq.; 100% conc.

Rhodasurf® ROX (formerly Soprophor ROX). [Rhone-Poulenc France] Trideceth-8-9; CAS 24938-91-8; nonionic; surfactant; 85% act.

Rhodasurf® T (formerly Rhodiasurf T). [Rhone-Poulenc France] Unsat. fatty alcohol ethoxylate; nonionic; detergent, emulsifier, dispersant, wetting agent for laundry powds., textile processing, color concs., metalworking compds.; wax; 99% conc.

Rhodasurf® T50 (formerly Rhodiasurf T50). [Rhone-Poulenc France] Unsat. fatty alcohol ethoxylate; nonionic; detergent, emulsifier, dispersant, wetting agent for laundry powds., textile processing, color concs., metalworking compds.; liq.; 50% conc.

Rhodasurf® T-95 (formerly Texafor T-95). [Rhone-Poulenc France] Trideceth-10; CAS 24938-91-8; nonionic; surfactant; 85% act.

Rhodasurf® TB-970 (formerly Emulphogene® TB-970). [Rhone-Poulenc Surf.; Rhone-Poulenc France] Steareth-200; CAS 9005-00-9; nonionic; detergent, viscosifier, foam booster, dispersant, surfactant for most dry blending operations in detergent formulations, sanitizers; controls dissolution rate of solid or block type hard surf. cleaners; textile wet processing aid; biodeg.; off-wh. waxy solid and flakes; sol. in water, ethanol, CCl_4, Cellosolve, xylene; HLB 18.0; cloud pt. > 100 C (1% aq.); surf. tens. 50.8 dynes/cm (0.1%); 100% act.

Rhodasurf® TDA-5 (formerly Alkasurf® TDA-5). [Rhone-Poulenc Surf. Canada] Ethoxylated tridecyl alcohol; nonionic; coemulsifier, wetting agent, dispersant for cutting oils, low temp. wool scouring; liq.; sol. in min. oil, min. spirits, aromatic solvs., perchloroethylene; disp. in water; dens. 0.96 g/ml; HLB 10.6; cloud pt. 65-68 C; 100% conc.

Rhodasurf® TDA-6 (formerly Alkasurf® TDA-6). [Rhone-Poulenc Surf. Canada] Ethoxylated tridecyl alcohol; nonionic; coemulsifier, wetting agent, dispersant for cutting oils, low temp. wool scouring; liq.; HLB 10.6; 100% conc.

Rhodasurf® TDA-8.5 (formerly Alkasurf® TDA-8.5). [Rhone-Poulenc Surf. Canada] Ethoxylated tridecyl alcohol; nonionic; emulsifier, strong rewetting agent for textile and paper prods.; liq.; sol. in water, min. oil, min. spirits, aromatic solvs., perchloroethylene; dens. 0.98 g/ml; HLB 12.5; cloud pt. 52-56 C; 100% conc.

Rhodasurf® TR Series (formerly Soprofor TR/Series). [Rhone-Poulenc Geronazzo] Tridecyl alcohol, ethoxylated; nonionic; emulsifier for Vaseline and min. oil; liq.; 100% conc.

Rhodaterge® 206C (formerly Cycloryl 206C). [Rhone-Poulenc Surf.] Proprietary; nonionic; high foaming upholstery cleaner conc.; liq.

Rhodaterge® CAN (formerly Cycloryl CAN). [Rhone-Poulenc Surf.] Proprietary; nonionic; high foaming heavy-duty cleaner conc.; liq.

Rhodaterge® DCA (formerly Cycloryl DCA). [Rhone-Poulenc Surf.] Modified coconut polydiethanolamide; nonionic/anionic; high foaming industrial/household cleaner conc.; for human and pet shampoos; visc. builder; solubilizer for difficult ingreds.; thickener for agric. biocidal agents; amber liq.; 98-100% act.

Rhodaterge® DCB-100 (formerly Alkadet DCB-100). [Rhone-Poulenc Surf.] Formulated prod.; anionic/nonionic; base for drycleaning charge soap yielding good soil suspension, water tolerance, cleaning efficiency, antistatic props.; liq.; 100% conc.

Rhodaterge® EPS. [Rhone-Poulenc France] Modified alkylphenol ethoxylate; nonionic; emulsifier for prep. of stable emulsions of solvs. and min. oils; liq.; 99% conc.

Rhodaterge® FL (formerly Mirawet® FL). [Rhone-Poulenc Surf.] Blend; amphoteric; controlled foaming wetting agent, emulsifier, detergent, dispersant, lubricant for heavy-duty cleaners, metalworking fluids; stable in presence of caustic and electrolytes; liq.; sol. in water; dens. 8.413 lb/gal; 85% act.

Rhodaterge® FSC. [Rhone-Poulenc Surf.] Proprietary; nonionic; low foaming fabric softener conc.; liq. DISCONTINUED.

Rhodaterge® LD-50Q (formerly Alkadet LD-50Q). [Rhone-Poulenc Surf. Canada] Blend; anionic/nonionic; detergent conc. for dishwashing and general purpose cleaning; relatively mild to skin; amber clear liq.; dens. 1.075 g/ml; visc. 450 cps; cloud pt. < 30 F; pH 7.0-7.5 (10% aq.); 49.5-50.5% act.

Rhodaterge® LD-60. [Rhone-Poulenc Surf. Canada] Blend; anionic/nonionic; detergent conc. for dishwashing and general purpose cleaning; relatively mild to skin; straw-colored clear liq.; dens. 1.05 g/ml; visc. 100-200 cps; haze pt. 0 C max.; pH 7-8 (10% aq.); 59-61% act.

Rhodaterge® RS-25 (formerly Cycloryl RS-25). [Rhone-Poulenc Surf.] Proprietary; nonionic; high foaming rug shampoo conc.; liq.

Rhodaterge® SMC (formerly Cycloryl SMC). [Rhone-Poulenc Surf.] Proprietary; nonionic; high foaming laundry and dishwashing conc.; liq.

Rhodaterge® SSB (formerly Miranate® SSB). [Rhone-Poulenc Surf.] Sodium lauryl sulfate, disodium lauryl sulfosuccinate, propylene glycol; surfactant.

Rhodaterge® WHC-347 (formerly Alkasurf® WHC). [Rhone-Poulenc Surf.] Formulated blend; anionic/nonionic; high foaming surfactant for formulating waterless hand cleaners; wetting agent; amber clear visc. liq., typ. odor; 100% act.

Rhodiasurf B (redesignated Alkamuls® B). [Rhone-Poulenc] PEG-33 castor oil; nonionic; emulsifier, dispersant for textiles, metallurgy, metal degreasing, personal care prods.; dye leveler, fabric softener; liq.; HLB 11.5; 96% conc. DISCONTINUED.

Rhodopol® 23. [Rhone-Poulenc Surf.; Rhone-Poulenc France] Xanthan gum; thickener for agric. formulations; EPA compliance.

Rhodorsil® AF 422. [Rhone-Poulenc Silicones] Dimethylpolysiloxane emulsion; nonionic; antifoam for aq. systems, agric., textile, chemical, rubber, metallurgy industries; EPA compliance; wh. to wh.-gray milky liq.; faint odor; sp.gr. 1.0; pH 5-6; usage level: 10-1000 mg/kg; 10% act. in water.

Rhodorsil® AF 426R. [Rhone-Poulenc Silicones] Silicone emulsion; antifoam for agric. formulation; EPA compliance; emulsion; 30% act.

Rhodorsil® AF 454. [Rhone-Poulenc Silicones] Polydimethylsiloxane oil compd. with min. filler; antifoam for aq. and anhydrous media, crop protection prods., natural or refinery gases, detergents, chemicals; EPA compliance; grayish-wh. opalescent liq., odorless; sp.gr. 1.0; usage level: 1-100 mg/kg; toxicology: eye irritant; 100% act.

Rhodorsil® AF 461 LV. [Rhone-Poulenc Silicones] Polydimethylsiloxane oil compd.; antifoam for detergents; effective in alkaline medium and at small doses; grayish-wh. opalescent liq.; none to very sl. odor; sp.gr. 1.0; visc. 3200 mPa•s; 100% act.

Rhodorsil® AF 20432. [Rhone-Poulenc Silicones] Polydimethylsiloxane emulsion; nonionic; antifoam for aq. media, textiles, detergents, gas and oil, crop protection applics.; wh. to grayish wh. milky liq., sl. odor; sp.gr. 1.0; pH 6; usage level: 10-1000 mg/kg; 30% act. in water.

Rhodorsil® AF 20441. [Rhone-Poulenc Silicones] Silicone; antifoam for agric. formulation; EPA compliance; powd.; 11% act.

Rhodorsil® EP 6703. [Rhone-Poulenc Silicones] Silicone; antifoam for agric. formulation; EPA compliance; powd.; 25% act.

Rhodorsil® SC 5020. [Rhone-Poulenc Silicones] Silicone emulsion; antifoam for agric. formulation; EPA compliance; emulsion; 20% act.

Rhodorsil® SC 5021. [Rhone-Poulenc Silicones] Silicone emulsion; antifoam for agric. formulation; EPA compliance; emulsion; 10% act.

Ridafoam NS-221. [PPG/Specialty Chem.] Surfactant blend; nonionic; dyeing and finishing agent for textiles; wetting agent; nonsilicone defoamer; liq.; 100% conc.

Rilanit G 16. [Henkel KGaA] Isocetyl alcohol; nonionic; lubricant, mold release for copper slabbing; liq.; 100% conc.

Rioklen NF 2. [Auschem SpA] Nonylphenol polyglycol ether; CAS 9016-45-9; nonionic; emulsifier, detergent, wetting agent; liq.; HLB 5.5; 100% conc.

Rioklen NF 4. [Auschem SpA] Nonylphenol polyglycol ether; CAS 9016-45-9; nonionic; emulsifier, detergent, wetting agent; liq.; HLB 9.1; 100% conc.

Rioklen NF 6. [Auschem SpA] Nonylphenol polyglycol ether; CAS 9016-45-9; nonionic; emulsifier, detergent, wetting agent; liq.; HLB 10.2; 100% conc.

Rioklen NF 7. [Auschem SpA] Nonylphenol polyglycol ether; CAS 9016-45-9; nonionic; emulsifier, detergent, wetting agent; liq.; HLB 11.8; 100% conc.

Rioklen NF 8. [Auschem SpA] Nonylphenol polyglycol ether; CAS 9016-45-9; nonionic; emulsifier, detergent, wetting agent; liq.; HLB 12.1; 100% conc.

Rioklen NF 9. [Auschem SpA] Nonylphenol polyglycol ether; CAS 9016-45-9; nonionic; emulsifier, detergent, wetting agent; liq.; HLB 13.0; 100% conc.

Rioklen NF 10. [Auschem SpA] Nonylphenol polyglycol ether; CAS 9016-45-9; nonionic; emulsifier, detergent, wetting agent; liq.; HLB 13.4; 100% conc.

Rioklen NF 12. [Auschem SpA] Nonylphenol polyglycol ether; CAS 9016-45-9; nonionic; emulsifier, detergent, wetting agent; paste; HLB 14.0; 100% conc.

Rioklen NF 15. [Auschem SpA] Nonylphenol polyglycol ether; CAS 9016-45-9; nonionic; emulsifier, detergent, wetting agent; solid; HLB 15.3; 100% conc.

Rioklen NF 20. [Auschem SpA] Nonylphenol polyglycol ether; CAS 9016-45-9; nonionic; emulsifier, detergent, wetting agent; solid; HLB 16.2; 100% conc.

Rioklen NF 30. [Auschem SpA] Nonylphenol polyglycol ether; CAS 9016-45-9; nonionic; emulsifier, detergent, wetting agent; flakes; HLB 17.3; 100% conc.

Rioklen NF 40. [Auschem SpA] Nonylphenol polyglycol ether; CAS 9016-45-9; nonionic; emulsifier, detergent, wetting agent; flakes; 100% conc.

Ritacetyl®. [RITA] Acetylated lanolin; nonionic; superfatting agent for soaps, shampoos; film-former for creams and lotions, water resistant films; solid; oil-sol.; 100% conc.

Ritachol® 1000. [RITA] Cetearyl alcohol, polysorbate

60, PEG-150 stearate, steareth-20; nonionic; emulsifer for personal care prods., pharmaceuticals, and household specialties; wh. waxy, flakes; low odor; m.p. 48–52 C; sapon. no. 9–14; 100% conc.

Ritachol®3000. [RITA] Cetearyl alcohol, polysorbate 60, PEG-150 stearate, steareth-20, cetyl alcohol, laneth-16, PEG-60 lanolin.; nonionic; emulsifying wax; flake; 100% conc.

Ritachol® 4000. [RITA] Cetyl alcohol, steareth-20, PEG-60 lanolin.; nonionic; emulsifying wax; flake; 100% conc.

Ritapeg 100 DS. [RITA] Ethoxylated distearic acid ester; nonionic; thickener, emulsifier; flake; 100% conc.

Ritapro 100. [RITA] Cetearyl alcohol, steareth-20, and steareth-10; nonionic; o/w emulsifier; solid; 100% conc.

Ritasynt IP. [RITA] Glycol stearate; nonionic; foam booster, thickener, opacifier; flake; 100% conc.

Ritoleth 2. [RITA] Oleth-2; CAS 9004-98-2; nonionic; o/w emulsifier, solubilizer; stable over wide pH range; liq.; HLB 4.9; 100% conc.

Ritoleth 5. [RITA] Oleth-5; CAS 9004-98-2; nonionic; o/w emulsifier, solubilizer; stable over wide pH range; liq.; HLB 9.6; 100% conc.

Ritoleth 10. [RITA] Oleth-10; CAS 9004-98-2; nonionic; o/w emulsifier, solubilizer; stable over wide pH range; semisolid; HLB 12.4; 100% conc.

Ritoleth 20. [RITA] Oleth-20; CAS 9004-98-2; nonionic; o/w emulsifier, solubilizer; stable over wide pH range; semisolid; HLB 15.1; 100% conc.

Rolamet C 11. [Auschem SpA] Cocamine polyglycol ether; CAS 61791-14-8; nonionic; emulsifier for cationic emulsions; codetergent for industrial use; liq.; 100% conc.

Rolfat 5347. [Auschem SpA] Ethoxylated monoglyceride; CAS 51158-08-8; nonionic; emulsifier; liq.; 100% conc.

Rolfat C 9. [Auschem SpA] PEG-9 cocoate; CAS 67762-35-0; nonionic; emulsifier; liq.; HLB 12.9; 100% conc.

Rolfat OL 6. [Auschem SpA] PEG-6 oleate; CAS 9004-96-0; nonionic; emulsfier, codetergent for industrial use; liq.; HLB 9.2; 100% conc.

Rolfat OL 7. [Auschem SpA] PEG-7 oleate; CAS 9004-96-0; nonionic; emulsfier, codetergent for industrial use; liq.; HLB 10.5; 100% conc.

Rolfat OL 9. [Auschem SpA] PEG-9 oleate; CAS 9004-96-0; nonionic; emulsfier, codetergent for industrial use; liq.; HLB 11.3; 100% conc.

Rolfat R 106. [Auschem SpA] PEG-6 tallate; CAS 61791-00-2; nonionic; emulsfier, codetergent for industrial use; liq.; HLB 11.3; 100% conc.

Rolfat SG. [Auschem SpA] PEG-9 stearate; CAS 9004-99-3; nonionic; emulsifier; solid; HLB 11.6; 100% conc.

Rolfat SO 6. [Auschem SpA] PEG-6 soyate; CAS 61791-07-9; nonionic; emulsifier; solid; HLB 11.6; 100% conc.

Rolfat ST 40. [Auschem SpA] PEG-40 stearate; CAS 9004-99-3; nonionic; emulsifier; flakes; HLB 17.3; 100% conc.

Rolfat ST 100. [Auschem SpA] PEG-100 stearate; CAS 9004-99-3; nonionic; emulsifier, thickener; flakes; HLB 18.8; 100% conc.

Rolfor 162. [Auschem SpA] Ethoxylated cetyl alcohol; CAS 9004-95-9; nonionic; w/o emulsifier; solid; HLB 5.3; 100% conc.

Rolfor C. [Auschem SpA] Ethoxylated monoglyceride; nonionic; emulsifier; liq.; 100% conc.

Rolfor CO 11. [Auschem SpA] Cetoleth-11; CAS 68920-66-1; nonionic; emulsifier, detergent; solid; HLB 12.9; 100% conc.

Rolfor COH 40. [Auschem SpA] PEG-40 hydrog. castor oil; CAS 61788-85-0; nonionic; emulsifier, solubilizer; paste; HLB 14.0; 100% conc.

Rolfor E 270, E 527. [Auschem SpA] Ethoxylated oxo-alcohol; CAS 68439-54-3; nonionic; emulsifier, detergent for industrial use; liq.; 100% conc.

Rolfor EP 237, EP 526. [Auschem SpA] POE/POP alkyl ether; CAS 68551-13-3; nonionic; low foam detergent; liq.; 100% conc.

Rolfor HT 11. [Auschem SpA] Ceteareth-11; CAS 68439-49-6; nonionic; emulsifier, detergent, heavy-duty powds.; solid; HLB 13.1; 100% conc.

Rolfor HT 25. [Auschem SpA] Ceteareth-25; CAS 68439-49-6; nonionic; emulsifier, detergent, heavy-duty powds.; flakes; HLB 15.5; 100% conc.

Rolfor N 24/2. [Auschem SpA] Laureth-2; CAS 68439-50-9; nonionic; emulsifier; liq.; HLB 6.4; 100% conc.

Rolfor N 24/3. [Auschem SpA] Laureth-3; CAS 68439-50-9; nonionic; emulsifier; liq.; HLB 8.3; 100% conc.

Rolfor O 8. [Auschem SpA] Oleth-8; CAS 9004-98-2; nonionic; emulsifier; solid; HLB 11.6; 20% conc.

Rolfor O 22. [Auschem SpA] Oleth-22; CAS 9004-98-2; nonionic; emulsifier; flakes; HLB 15.5; 100% conc.

Rolfor TR 6. [Auschem SpA] Trideceth-6; CAS 24938-91-8; nonionic; emulsifier, detergent; liq.; HLB 11.4; 20% conc.

Rolfor TR 9. [Auschem SpA] Trideceth-9; CAS 24938-91-8; nonionic; emulsifier, detergent; liq.; HLB 13.8; 20% conc.

Rolfor TR 12. [Auschem SpA] Trideceth-12; CAS 24938-91-8; nonionic; emulsifier, detergent; liq./paste; HLB 14.6; 20% conc.

Rolfor Z 24/9. [Auschem SpA] Laureth-9; CAS 68439-50-9; nonionic; emulsifier, detergent; paste; HLB 13.4; 20% conc.

Rolfor Z 24/23. [Auschem SpA] Laureth-23; CAS 68439-50-9; nonionic; emulsifier, detergent; solid; HLB 16.8; 20% conc.

Rolpon 24/230. [Auschem SpA] Sodium laureth-2 sulfate; CAS 9004-82-4; 15826-16-1; anionic; raw material for use in detergent toiletry preps.; liq.; 27% conc.

Rolpon 24/270. [Auschem SpA] Sodium laureth-2 sulfate; CAS 9004-82-4; 15826-16-1; anionic; raw material for use in detergent toiletry preps.; paste; 70% conc.

Rolpon 24/330. [Auschem SpA] Sodium alkyl ether sulfate; CAS 91648-56-5; anionic; detergent; liq.; 27% conc.

Rolpon 24/330 N. [Auschem SpA] Sodium laureth-3 sulfate; CAS 9004-82-4; 15826-16-1; anionic; raw material for detergent toiletry preps.; liq.; 27% conc.

Rolpon 24/370. [Auschem SpA] Sodium alkyl ether sulfate; CAS 91648-56-5; anionic; detergent; paste; 70% conc.

Rolpon C 200. [Auschem SpA] Sodium trideceth-7 carboxylate; anionic; raw material for use in detergent toiletry preps.; liq.; 30% conc.

Rolpon LSX. [Auschem SpA] Sodium lauryl sulfate; CAS 151-21-3; anionic; raw material for use in detergent toiletry preps.; liq./paste; 28% conc.

Rolpon SE 138. [Auschem SpA] Disodium laureth sulfosuccinate; CAS 39354-45-5; anionic; raw material for use in detergent toiletry preps.; liq.; 40%

conc.

Romie 802. [Tokai Seiyu Ind.] Polyquaternary ammonium salt; cationic; softener for acrylic and other syn. fibers; paste; 15% conc.

Ross Chem 743. [Ross Chem.] Copolymer; nonionic; low foaming surfactant; used to increase starch penetration in boxboard mfg.; liq.; 100% conc.

Ross Chem PEG 600 DT. [Ross Chem.] Polyol diester; nonionic; general purpose emulsifier, dispersant; liq.; HLB 9.2; 100% conc.

Ross Chem PPG 1025DT. [Ross Chem.] PPG 1000 ditallate; nonionic; oil-in-oil dispersant, lubricant; syn. high temp. oils; liq.; 100% conc.

Ross Emulsifier DF. [Ross Chem.] Ethoxylate blend; nonionic; low foaming surfactant used in prep. of organic defoamers; liq.; HLB 9.5; 100% conc.

Ross Emulsifier LA-4. [Ross Chem.] Ethoxylated alcohol; nonionic; low foaming emulsifier; liq.; HLB 9.7; 100% conc.

Rotolan®. [RITA] Proprietary lanolin deriv.; aids in pigment dispersion for use in inks; paste; 100% conc.

RS-55-40. [Hefti Ltd.] PEG-40 stearate; CAS 9004-99-3; 31791-00-2; nonionic; o/w emulsifier for cosmetic creams, lotions, pharmaceutical ointments; glass surface finishing agent; for silicone oil, cooling emulsions; solid; HLB 17.0; 100% conc.

Rueterg 60-T. [Finetex] TEA-dodecylbenzene sulfonate; anionic; household and industrial detergent base.

Rueterg 97-G. [Finetex] MEA-dodecylbenzene sulfonate; anionic; detergent base for laundry, dishwashing, car washes, household detergents; emulsifier for oils, dyeing aid, wetting agent; liq.; 95% conc.

Rueterg 97-S. [Finetex] Alkylaryl sulfonate, amine salt; anionic; detergent base for laundry, dishwashing and car washes; liq.; 95% conc.

Rueterg 97-T. [Finetex] TEA-dodecylbenzene sulfonate; anionic; household and industrial detergent base; liq.; 95% conc.

Rueterg IPA-HP. [Finetex] Alkylaryl sulfonate, amine salt; anionic; detergent and car wash, dry cleaning; liq.; 96% conc.

Rueterg SA. [Finetex] Dodecylbenzene sulfonic acid; anionic; intermediate for detergents and emulsifiers; liq.; 97% conc.

Runox 1000C. [Toho Chem. Industry] Salt of condensed naphthalene sulfonic acid; anionic; dispersant for dyestuff, pigment, agric. wettable powds.; powd.

Runox Series. [Toho Chem. Industry] Surfactants; anionic/nonionic; detergent for metal surf. treatment, rust preventive, rust and paint remover; liq. and powd.

Rychem® 17. [Reilly-Whiteman] Polyglycol ester; nonionic; emulsifier for fatty and min. oils; lubricant, softener, rewetting agent; liq.; 90-95% conc.

Rychem® 21. [Reilly-Whiteman] Polyglycol ester; nonionic; oil emulsifier for textile, leather and other industrial uses; liq.; 95% conc.

Rychem® 33. [Reilly-Whiteman] Polyglycol ester; nonionic; oil emulsifier for textile, leather and other industrial uses; more hydrophilic than Rychem 21; liq.; 95% conc.

Rycofax® 618. [Reilly-Whiteman] Fatty amido-amine salt; cationic; softener, debonder, release agent for tissue mfg.; liq.; 98% conc.

Rycofax® O. [Reilly-Whiteman] PEG ester; nonionic; improves wetting and rewetting in pulp/paper and other wet processes for textile, metal industries; emulsifier for oils, hydrocarbons; liq.; 100% conc.

Rycomid 2120. [Reilly-Whiteman] Fatty ester; nonionic; foam stabilizer for detergent formulations; solid; 100% conc.

S

S-210. [Procter & Gamble] Soya fatty acid; CAS 67701-08-0; intermediate for mfg. of soaps, amides, esters, alcoholamides; raw material for non-surfactant applics.; Gardner 2 max. liq.; m.w. 279; acid no. 197-203; iodine no. 133 min.; sapon. no. 199-205; 100% conc.

Sactol 2 OS 2. [Lever Industriel] Sodium laureth sulfate; CAS 9004-82-4; anionic; base for detergents and shampoos; liq.; 50% conc.

Sactol 2 OS 28. [Lever Industriel] Sodium laureth sulfate; CAS 9004-82-4; anionic; base for detergents and shampoos; liq.; 28% conc.

Sactol 2 OT. [Lever Industriel] TEA laureth sulfate; anionic; base for detergents and shampoos; liq.; 35% conc.

Sactol 2 S 3. [Lever Industriel] Sodium lauryl sulfate; CAS 151-21-3; anionic; base for detergents and shampoos; paste; 49% conc.

Sactol 2 T. [Lever Industriel] TEA lauryl sulfate; anionic; base for detergents and shampoos; liq.; 35% conc.

Salt 100, 200, 300M. [Tokai Seiyu Ind.] Mixture of high polymeric active agent and solv.; nonionic; dyeing assistant for dyeing of wool at low temp.; liq.; 100% conc.

Sanac C. [Karlshamns] Complex carboxylated coco quaternary; CAS 62705-16-2; amphoteric; detergent, wetting and foaming agent, solubilizer of organics and inorganics; biodeg.; liq.; 60% conc.

Sanac S. [Karlshamns] Complex carboxylated soya quaternary; amphoteric; detergent, wetting and foaming agent, solubilizer of organics and inorganics; biodeg.; liq.; 60% conc.

Sanac T. [Karlshamns] Complex carboxylated tallow quaternary; amphoteric; detergent, wetting and foaming agent, solubilizer of organics and inorganics; biodeg.; liq.; 60% conc.

Sandet 60. [Sanyo Chem. Industries] Sodium dodecylbenzene sulfonate; CAS 25155-30-0; anionic; detergent base and emulsifier for emulsion polymerization of vinyl resin; liq.

Sandet ALH. [Sanyo Chem. Industries] Alkyl diphenyl ether sodium disulfate; anionic; wetting agent, penetrant and base material for making liq. bleaching detergents in combination with sodium hypochlorite and caustic soda; liq.; 30% conc.

Sandogen C-CM Liq. [Sandoz] Dye leveling agent; antiprecipitant for dyeing nylon carpet; yel. liq.; misc. with water; sp.gr. 1.04; 31% act.

Sandolube NV. [Sandoz Prods. Ltd.] Aliphatic hydrocarbon aq. disp.; amphoteric; enhances sewing props. of knitwear and textiles; liq.

Sandolube NVJ. [Sandoz Prods. Ltd.] Aliphatic hydrocarbon aq. disp.; cationic; enhances sewing props. of knitwear and textiles; esp. suited for jet dyeing machines; liq.; 100% conc.

Sandolube NVN. [Sandoz Prods. Ltd.] Aliphatic hydrocarbon aq. disp.; nonionic; enhances sewing props. of knitwear and textiles; esp. suited for jet dyeing machines; liq.; 100% conc.

Sandolube NVS. [Sandoz Prods. Ltd.] Aliphatic hydrocarbon aq. disp.; nonionic; enhances sewing props. of knitwear and textiles; esp. suited for jet dyeing machines; liq.

Sandolube TRA. [Sandoz Prods. Ltd.] Aliphatic hydrocarbon aq. disp.; amphoteric; raising agent; liq.

Sandopan® 2N. [Sandoz Prods. Ltd.] Polyglycol ether, modified; anionic/nonionic; scouring agent and emulsifier for min. oils; liq., solid.

Sandopan® B. [Sandoz] Carboxylated C4 paraffinic ethoxylate; anionic; nonfoaming industrial surfactant, hydrotrope; caustic stable; lt. amber clear liq.; HLB 6.0; pH 8.0 (10%); Ross-Miles foam 0 mm (0.1%, 40 C); 40 ± 5% solids.

Sandopan® BFN. [Sandoz Prods. Ltd.] Phosphoric ester; anionic/nonionic; low foaming prod. for continuous peroxide bleaching of cotton; liq.

Sandopan® CBH Paste. [Sandoz] Sulfonate EO adduct, modified; anionic; detergent, emulsifier, wetting agent used for applics. in the fabric industry; yel. paste; mild odor; misc. with water; 51% act.

Sandopan® CBN. [Sandoz Prods. Ltd.] Phosphate ester; anionic; low foaming wetting, scouring and dispersing agent for pretreatment of cellulose and blends with syn. fibers; liq.

Sandopan® DTC. [Sandoz; Sandoz Prods. Ltd.] Sodium trideceth-7 carboxylate; anionic; detergent, emulsifier, wetting agent, solubilizer; cosmetics, household, specialty, industrial and institutional cleaner; dishwashing detergent; lime soap dispersant; clear gel; water-sol.; sp.gr. 1.06; visc. 5000 cps; b.p. 220 F; HLB 16.0 (@ pH 8); pH 6.5 (10%); Ross-Miles foam 150 mm (0.1%, 40 C, initial); 68% act.

Sandopan® DTC-Acid. [Sandoz] Trideceth-7 carboxylic acid; anionic; detergent, emulsifier, wetting agent for industrial and personal care, conditioning shampoos, liq. soaps, household and industrial cleaners; clear liq.; sol. in oils, solvs.; HLB 13.0 (@ pH 2.5); pH 2.5 (10%); Ross-Miles foam 145 mm (0.1%, 40 C, initial); 90% conc.

Sandopan® DTC-100. [Sandoz] Sodium trideceth-7 carboxylate; anionic; detergent, emulsifier, wetting agent for liq. detergents, solv. cleaners, all-purpose cleaners, germicidal cleaners, shampoos, bubble baths; oil solubilizer for aq. systems, acid bowl cleaners; yel. clear liq.; HLB 15.0 (@ pH 8); pH 7.0 (10%); Ross-Miles foam 150 mm (0.1%, 40 C, initial); 70 ± 2% solids.

Sandopan® DTC Linear P. [Sandoz] Sodium C12-15 pareth-6 carboxylate; anionic; detergent, emulsifier, wetting agent, solubilizer, visc. booster for industrial, personal care, and household prods.; stable in alkali high temps.; wh. opaque paste; HLB 15.0 (pH 8); pH 8.5 (10%); Ross-Miles foam 130 mm (0.1%, 40 C, initial); 70% conc.

Sandopan® DTC Linear P Acid. [Sandoz] C12–15 pareth-7 carboxylic acid; anionic; detergent, emulsifier, wetting agent for industrial, personal care, and household use; oil solubilizer; clear liq.; HLB 10 (pH 2.5); pH 3.5 (10%); Ross-Miles foam 120 mm (0.1%, 40 C, initial); 90% conc.

Sandopan® JA-36. [Sandoz] Trideceth-19 carboxylic acid; anionic; moderate foaming mild surfactant, oil solubilizer, wetting agent cosmetics/toiletries, laundry prods., cleaners, industrial specialties;; clear to sl. hazy liq.; HLB 16 (pH 2.5); pH 2.5 (10% aq.); Ross-Miles foam 130 mm (0.1%, 40 C, initial); 90 ± 2% solids.

Sandopan® KD. [Sandoz Prods. Ltd.] Sodium alkyl sulfate, modified; anionic; scouring, milling, wetting agent and dispersant used in textiles; liq.

Sandopan® KD (U) Liquid. [Sandoz] Sodium alkyl sulfate, modified; anionic; detergent; scouring of fibers; fulling of wool; clear liq.; mild odor; misc. with water; sp.gr. 1.02; 21% act.

Sandopan® KST. [Sandoz] Sodium ceteth-13 carboxylate; anionic; mild emulsifier, detergent, lime soap dispersant for use in sticks, bar soaps, antiperspirants, other personal care prods., industrial specialties, laundry prods.; inhibits sodium stearate crystal formation; solid; HLB 11 (pH 8); pH 6.5 (10% aq.); Ross-Miles 70 mm (0.1%, 40 C, initial); 97 ± 2% solids.

Sandopan® LA-8. [Sandoz] Laureth-5 carboxylic acid; anionic; surfactant for cosmetics/toiletries, laundry prods., cleaners, industrial specialties.

Sandopan® LF Liquid. [Sandoz] EO adduct; nonionic; detergent, emulsifier for scouring/dyeing and prescouring of polyester knits; clear liq.; misc. with water; mild odor; sp.gr. 1.02; pH 6.0; 25% act.

Sandopan® LFW. [Sandoz Prods. Ltd.] EO adduct; nonionic; detergent and wetting agent for polyester and cellulosic fibers; liq.; 100% conc.

Sandopan® MA-18. [Sandoz] Nonoxynol-10 carboxylic acid; anionic; detergent, wetting agent, solubilizer for cosmetics/toiletries, laundry prods., cleaners, industrial specialties; clear liq.; oil-sol.; pH 2.5 (10% aq.); Ross-Miles foam 105 mm (0.1%, 40 C, initial); 90 ± 3% solids.

Sandopan® RS-8. [Sandoz] Sodium C16-20 ethoxylate carboxylate; anionic; emulsifier, solubilizer; off-wh. firm paste; disp. in water; HLB 13 (pH 8); pH 8 (10% aq.); Ross-Miles foam 63 mm (0.1%, 40 C, initial); 70 ± 2% solids.

Sandopan® TFL Conc. [Sandoz] Sodium oleoamphohydroxypropylsulfonate; amphoteric; detergent, cosmetic and toiletries prods.; leather industry; amber gel; mild fatty odor; water-sol.; sp.gr. 1.14; b.p. 220 F; 48% conc.

Sandopan® TFL Liquid. [Sandoz] Fatty acid deriv.; amphoteric; detergent, wetting agent, emulsifier, lubricant, dispersant; dyeing nylon carpets with disperse dyes; lubricant and dispersant in boil-off and dyeing of syn. fabrics; yel. liq.; mild odor; water-misc.; sp.gr. 1.03; pH 9.0.

Sandopur DK. [Sandoz Prods. Ltd.] Fatty alkyl polyglycol ether, modified; amphoteric; washing-off agent for textiles; liq., paste.

Sandoteric CFL. [Sandoz] Cocamphohydroxypropyl sulfonate; extremely mild surfactant producing synergistic visc. increase with alkyl sulfates; weak ampholyte.

Sandoteric TFL Conc. [Sandoz] Sodium oleoamphohydroxypropyl sulfonate; extremely mild surfactant producing synergistic visc. increase with alkyl sulfates; weak ampholyte; toxicology: zero skin and eye irritation in rabbits.

Sandoxylate® 206. [Sandoz] Alkoxylated alcohol; nonionic; wetting agent, dispersant used in cleaning and treatment prods. and processes; liq.; HLB 12–14; 100% conc.

Sandoxylate® 224. [Sandoz] Alkoxylated alcohol; nonionic; wetting agent, dispersant used in cleaning and treatment prods. and processes; semisolid; HLB 16–17; 100% conc.

Sandoxylate® 408. [Sandoz] Alkoxylated alcohol; nonionic; wetting agent, dispersant used in cleaning and treatment prods. and processes; liq.; HLB 10–12; 100% conc.

Sandoxylate® 412. [Sandoz] Alkoxylated alcohol; nonionic; wetting agent, dispersant used in cleaning and treatment prods. and processes; liq.; HLB 12–14; 100% conc.

Sandoxylate® 418. [Sandoz] Alkoxylated alcohol; nonionic; fragrance solubilizer, wetting agent, dispersant used in cleaning and treatment prods. and processes, household and industrial prods.; liq.; HLB 16.0; 100% conc.

Sandoxylate® 424. [Sandoz] Alkoxylated alcohol; nonionic; wetting agent, dispersant used in cleaning and treatment prods. and processes; semisolid; HLB 16.0; 100% conc.

Sandoxylate® AC-9. [Sandoz] Alcohol ethoxylate; nonionic; wetting agent, dispersant used in cleaning and treatment prods. and processes; solid; HLB 12.0; 100% conc.

Sandoxylate® AC-24. [Sandoz] Alcohol ethoxylate; nonionic; wetting agent, dispersant used in cleaning and treatment prods. and processes; solid, flakes; HLB 15.8; 100% conc.

Sandoxylate® AC-46. [Sandoz] Alcohol ethoxylate; nonionic; nonfoaming wetting agent, dispersant for cleaning prods. and processes; stable in alkaline and acid media; solid.

Sandoxylate® AD-4. [Sandoz] Alcohol ethoxylate; nonionic; nonfoaming wetting agent, dispersant for cleaning prods. and processes; stable in alkaline and acid media; liq.; 100% conc.

Sandoxylate® AD-6. [Sandoz] Alcohol ethoxylate; nonionic; nonfoaming wetting agent, dispersant for cleaning prods. and processes; stable in alkaline and acid media; liq.; 100% conc.

Sandoxylate® AD-9. [Sandoz] Alcohol ethoxylate; nonionic; wetting agent, dispersant for cleaning prods. and processes; stable in alkaline and acid media; liq.; 100% conc.

Sandoxylate® AL-4. [Sandoz] Alcohol ethoxylate; nonionic; wetting agent, dispersant for cleaning prods. and processes; stable in alkaline and acid media; liq.; 100% conc.

Sandoxylate® AO-12. [Sandoz] Alcohol ethoxylate; nonionic; wetting agent, dispersant used in cleaning and treatment prods. and processes; liq.; HLB 12.7; 100% conc.

Sandoxylate® AO-20. [Sandoz] Alcohol ethoxylate; nonionic; wetting agent, dispersant used in cleaning and treatment prods. and processes; semisolid; HLB

15.4; 100% conc.

Sandoxylate® AO-60. [Sandoz] Alcohol ethoxylate; nonionic; wetting agent, dispersant used in cleaning and treatment prods. and processes; solid; HLB 18.9; 100% conc.

Sandoxylate® AT-6.5. [Sandoz] Alcohol ethoxylate; nonionic; wetting agent, dispersant for cleaning prods. and processes; stable in alkaline and acid media; liq.; HLB 11.5; 100% conc.

Sandoxylate® AT-12. [Sandoz] Alcohol ethoxylate; nonionic; wetting agent, dispersant used in cleaning and treatment prods. and processes; liq.; HLB 14.5; 100% conc.

Sandoxylate® C-10. [Sandoz] Castor oil ethoxylate; nonionic; wetting agent, dispersant used in cleaning and treatment prods. and processes; liq.; 100% conc.

Sandoxylate® C-15. [Sandoz] Castor oil ethoxylate; nonionic; wetting agent, dispersant used in cleaning and treatment prods. and processes; liq.; HLB 13.5; 100% conc.

Sandoxylate® C-32. [Sandoz] Castor oil ethoxylate; nonionic; nonfoaming wetting agent, dispersant for cleaning prods. and processes; stable in alkaline and acid media; liq.; HLB 15.0; 100% conc.

Sandoxylate® FO-9. [Sandoz] Fatty acid ethoxylate; nonionic; surfactant, wetting agent, dispersant used in cleaning and treatment prods. and processes; liq.; HLB 11.9; 70% conc.

Sandoxylate® FO-30/70. [Sandoz] Fatty acid ethoxylate; nonionic; surfactant, wetting agent, dispersant used in cleaning and treatment prods. and processes; liq.; HLB 13.9; 100% conc.

Sandoxylate® FS-9. [Sandoz] Fatty acid ethoxylate; nonionic; surfactant, wetting agent, dispersant used in cleaning and treatment prods. and processes; solid; HLB 11.5; 100% conc.

Sandoxylate® FS-35. [Sandoz] Fatty acid ethoxylate; nonionic; surfactant, wetting agent, dispersant used in cleaning and treatment prods. and processes; solid; HLB 16.9; 100% conc.

Sandoxylate® NC-5. [Sandoz] Amine ethoxylate; nonionic; surfactant, wetting agent, dispersant used in cleaning and treatment prods. and processes; liq.; 100% conc.

Sandoxylate® NC-15. [Sandoz] Amine ethoxylate; nonionic; surfactant, wetting agent, dispersant used in cleaning and treatment prods. and processes; liq.; 100% conc.

Sandoxylate® NSO-30. [Sandoz] Amine ethoxylate; nonionic; surfactant, wetting agent, dispersant used in cleaning and treatment prods. and processes; liq.; 100% conc.

Sandoxylate® NT-5. [Sandoz] Amine ethoxylate; nonionic; surfactant, wetting agent, dispersant used in cleaning and treatment prods. and processes; liq., paste; 100% conc.

Sandoxylate® NT-15. [Sandoz] Amine ethoxylate; nonionic; nonfoaming wetting agent, dispersant for cleaning prods. and processes; stable in alkaline and acid media; liq., paste; 100% conc.

Sandoxylate® PDN-7. [Sandoz] Amine ethoxylate; nonionic; wetting agent, dispersant for cleaning prods. and processes; stable in alkaline and acid media; liq.; HLB 9.9; 100% conc.

Sandoxylate® PN-6. [Sandoz] Amine ethoxylate; nonionic; wetting agent, dispersant for cleaning prods. and processes; stable in alkaline and acid media; liq.; 100% conc.

Sandoxylate® PN-9. [Sandoz] Alkyl phenol ethoxylate; nonionic; nonfoaming wetting agent, dispersant for cleaning prods. and processes; stable in alkaline and acid media; liq.; HLB 12.9; 100% conc.

Sandoxylate® PN-10.9. [Sandoz] Alkyl phenol ethoxylate; nonionic; nonfoaming wetting agent, dispersant for cleaning prods. and processes; stable in alkaline and acid media; liq.; 100% conc.

Sandoxylate® PO-5. [Sandoz] Alkyl phenol ethoxylate; nonionic; nonfoaming wetting agent, dispersant for cleaning prods. and processes; stable in alkaline and acid media; liq.; HLB 10.4; 100% conc.

Sandoxylate® SX-208. [Sandoz] PO/EO alkoxylate; nonionic; wetting agent for household and industrial applics., textile wet processing; intermediate for anionic surfactants; Gardner < 1 liq.; water-sol.; sp.gr. 0.9625; HLB 11; cloud pt. 22 ± 2 C (1% aq.); pH 6.5 ± 1 (1%); surf. tens. 28.0 dynes/cm; 98 ± 2% act.

Sandoxylate® SX-224. [Sandoz] PO/EO alkoxylate; nonionic; wetting agent for household and industrial applics., textile wet processing; intermediate for anionic surfactants; Gardner < 1 liq.; water-sol.; sp.gr. 1.0262; HLB 15; cloud pt. 93 ± 2 C (1% aq.); pH 6.5 ± 1 (1%); surf. tens. 30.2 dynes/cm; 98 ± 2% act.

Sandoxylate® SX-408. [Sandoz] PPG-2-isodeceth-4; nonionic; wetting agent for household and industrial applics., textile wet processing; intermediate for anionic surfactants; Gardner < 1 liq.; water-sol.; sp.gr. 0.9650; HLB 11; cloud pt. 22 ± 2 C (1% aq.); pH 6.5 ± 1 (1%); surf. tens. 28.8 dynes/cm; 98 ± 2% act.

Sandoxylate® SX-412. [Sandoz] PPG-2-isodeceth-12; nonionic; wetting agent for household and industrial applics., textile wet processing; intermediate for anionic surfactants; Gardner < 1 liq.; water-sol.; sp.gr. 0.9822; HLB 13; cloud pt. 38 ± 2 C (1% aq.); pH 6.5 ± 1 (1%); surf. tens. 28.6 dynes/cm; 98 ± 2% act.

Sandoxylate® SX-418. [Sandoz] PPG-2-isodeceth-9; nonionic; wetting agent for household and industrial applics., textile wet processing; intermediate for anionic surfactants; Gardner < 1 liq.; water-sol.; sp.gr. 1.007; HLB 16.0; cloud pt. 65 ± 2 C (1% aq.); pH 6.5 ± 1 (1%); surf. tens. 29.4 dynes/cm; 98 ± 2% act.

Sandoxylate® SX-424. [Sandoz] PPG-2-isodeceth-12; nonionic; wetting agent for household and industrial applics., textile wet processing; intermediate for anionic surfactants; Gardner < 1 solid; water-sol.; sp.gr. 1.0267; HLB 16.0; cloud pt. 89 ± 2 C (1% aq.); pH 6.5 ± 1 (1%); surf. tens. 30.3 dynes/cm; 98 ± 2% act.

Sandoxylate® SX-602. [Sandoz] PPG-3-isodeceth-1; nonionic; fragrance solubilizer, wetting agent, low/nonfoaming surfactant for household and industrial prods., liq. laundry, hard surf. cleaners; liq.; HLB 16.0; 100% conc.

Sandoz Amide CO. [Sandoz] Coconut alkanolamide (2:1); detergent, foamer, wetting and rewetting agent, visc. modifier used in metal cleaning, heavy duty detergents, hard surface cleaning, textile wet processing; amber liq.; 97% act.

Sandoz Amide NP. [Sandoz] Coconut alkanolamide (2:1); detergent, foamer, wetting and rewetting agent, visc. modifier used in metal cleaning, heavy duty detergents, hard surface cleaning, textile wet processing; amber liq.; 97% act.

Sandoz Amide NT. [Sandoz] Alkanolamide; nonionic; surfactant used as a detergent, emulsifier and visc. aid for household care and cosmetic applics.; liq.; 100% conc.

Sandoz Amide PE. [Sandoz] Lauryl alkanolamide (1:1); detergent, foamer, wetting and rewetting agent, visc. modifier used for liq. dishwashing prods. and personal care prods.; wh. solid; 90–95% act.

Sandoz Amide PL. [Sandoz] Lauryl alkanolamide (2:1); detergent, foamer, wetting and rewetting agent, visc. modifier used for liq. dishwashing prods. and personal care prods.; amber liq.; 100% act.

Sandoz Amide PN. [Sandoz] Coconut alkanolamide (2:1); detergent, foamer, wetting and rewetting agent, visc. modifier used for heavy duty cleaning formulations, oil emulsification; amber liq.; 97% act.

Sandoz Amide PO. [Sandoz] Coconut alkanolamide (1:1); detergent, foamer, wetting and rewetting agent, visc. modifier used for personal care prods. and cleaning formulations; lt. yel. visc. liq.; 85% act.

Sandoz Amide PS. [Sandoz] Coconut alkanolamide (2:1); detergent, foamer, wetting and rewetting agent, visc. modifier used in hard surface cleaners, detergents, dishwashing compds.; textile scouring; amber liq.; 97% act.

Sandoz Amine Oxide XA-C. [Sandoz] Cetyl amine oxide; nonionic; surfactant, foam booster, visc. improver for personal care and liq. dishwashing formulations; antistat; textile wet processing; softener and emollient; amber liq.; sol. in water, lower alcohols, glycols; biodeg.; 20% act.

Sandoz Amine Oxide XA-L. [Sandoz] Lauryl amine oxide; nonionic; surfactant, foam booster, visc. improver for personal care and liq. dishwashing formulations; antistat, textile wet processing; amber liq.; sol. in water, lower alcohols, glycols; biodeg.; 30% act.

Sandoz Amine Oxide XA-M. [Sandoz] Myristyl amine oxide; nonionic; surfactant, foam booster, visc. improver for personal care and liq. dishwashing formulations; antistat, textile wet processing; amber liq.; sol. in water, lower alcohols, glycols; biodeg.; 30% act.

Sandoz Amine XA-Q. [Sandoz] Cetylamine oxide; amphtoeric; high foaming surfactant, foam stabilizer; biodeg.; liq.; 30% conc.

Sandozin® AM. [Sandoz Prods. Ltd.] Alkylaryl sulfonate; anionic; wetting agent, textile prods.; paste.

Sandozin® AMP. [Sandoz Prods. Ltd.] Alkyl sulfonate; anionic; low foaming wetting agent; liq.

Sandozin® NA Liq. [Sandoz Prods. Ltd.] Phosphoric acid ester; nonionic; wetting agent for pretreatment of textiles; liq.

Sandozin® NE. [Sandoz Prods. Ltd.] Sulfonated fatty acid deriv.; anionic; wetting agent for hypochlorite bleaching in textile applics.; liq.

Sandozin® NIT. [Sandoz Prods. Ltd.] EO adduct; nonionic; wetting agent and detergent for textiles; liq.

Sandozol KB. [Sandoz Prods. Ltd.] Sulfonated oil; anionic; finishing and sanforizing oil; wetting and dispersing agent; liq.

Sandoz Phosphorester 510. [Sandoz] Complex phosphate ester; anionic; detergent, emulsifier in waterless hand cleaners, dry cleaning detergents; antistat, corrosion inhibitor in textile fiber lubricants; primary emulsifier in emulsion polymerization of PVAc, acrylic; clear visc. liq.; water-disp.; sol. in conc. electrolyte systems; pH 1.5–2.5 (10%); 99% solids.

Sandoz Phosphorester 600. [Sandoz] Complex phosphate ester; hydrotrope; solubilizer for nonionic surfactants; wetting agent used in acid and solv. cleaners; emulsifier for solvs.; clear visc. liq.; sol. in conc. electrolyte systems; pH 1.5–2.5 (10%); 99% solids.

Sandoz Phosphorester 610. [Sandoz] Complex phosphate ester; emulsifier for aromatic and chlorinated solvs.; solubilizer for nonionic surfactants; emulsion polymerization surfactant used in vinyl acetate and acrylic systems; clear visc. liq.; sol. in conc. electrolyte systems; pH 1.5–2.5 (10%); 99% solids.

Sandoz Phosphorester 690. [Sandoz] Complex phosphate ester; emulsifier for emulsion polymerization; household and industrial cleaners; soft waxy paste; pH 1.5–2.5; 95% solids.

Sandoz Sulfate 216. [Sandoz] Ammonium laureth sulfate (3 EO); anionic; foamer and visc. modifier for personal care prods.; basic ingred. in shampoos and liq. dishwashing prods.; textile scouring; emulsifier and suspending agent in fabrics; pale yel. liq.; visc. 1000 cps max.; cloud pt. 15 C; pH 6.5–7.5 (10%); 58.5–60% act.

Sandoz Sulfate 219. [Sandoz] Sodium laureth sulfate (3 EO); CAS 9004-82-4; anionic; foamer and visc. modifier in personal care prods.; textile scouring; pale yel. liq.; visc. 1000 cps max.; cloud pt. 15 C; pH 7.5–8.5 (15%); 58.5–60% act.

Sandoz Sulfate 830. [Sandoz] Sodium N-octyl sulfate; anionic; surfactant for household and industrial cleaning formulations; liq.; 30% conc.

Sandoz Sulfate 1030. [Sandoz] Sodium decyl sulfate; anionic; surfactant for household and industrial cleaning formulations; liq.; 30% conc.

Sandoz Sulfate A. [Sandoz] Ammonium lauryl sulfate; foamer and visc. modifier in personal care prods.; basic ingred. in shampoos and liq. dishwashing prods.; lt. yel. liq.; visc. 6500 cps; cloud pt. 5 C; pH 6.3–7.0 (10%); 28–30% act.

Sandoz Sulfate ES-3. [Sandoz] Sodium laureth sulfate (3 EO); CAS 9004-82-4; foamer and visc. modifier for personal care prods.; textile scouring; wetting agent; emulsifier and suspending agent in fabrics; pale yel. liq.; visc. 1000 cps max.; cloud pt. 15 C; pH 7.5–8.5 (10%); 28–30% act.

Sandoz Sulfate K. [Sandoz] Ammonium mixed alkyl sulfate; foamer and visc. modifier in personal care prods.; basic ingred. in shampoos and liq. dishwashing prods.; brn. liq.; visc. 1000 cps max.; cloud pt. 15 C; pH 6.5–7.0 (10%); 45% act.

Sandoz Sulfate W2-30. [Sandoz] Sodium 2-ethylhexyl sulfate; anionic; surfactant for household and industrial cleaning formulations; liq.; 30% conc.

Sandoz Sulfate WA Dry. [Sandoz] Sodium lauryl sulfate; CAS 151-21-3; foamer and visc. modifier for personal care, dental, and household prods.; wh. powd.; pH 6.5–7.5 (10%); 90–96% act.

Sandoz Sulfate WAG. [Sandoz] Sodium lauryl sulfate; CAS 151-21-3; foamer and visc. modifier for personal care, dental, and household prods.; wh. needles; pH 6.5–7.5 (10%); 90–96% act.

Sandoz Sulfate WAS. [Sandoz] Sodium lauryl sulfate; CAS 151-21-3; foamer, visc. modifier for personal care, household prods.; lt. yel. liq.; visc. 1000–3000 cps; cloud pt. 17–23 C; pH 7.4–8.6 (10%); 27.5–30.5% act.

Sandoz Sulfate WA Special. [Sandoz] Sodium lauryl sulfate; CAS 151-21-3; foamer, visc. modifier for personal care, household prods.; wh. paste; visc. 1000–5000 cps; cloud pt. 17–21 C; pH 7.4–8.6 (10%); 27.5–30.5% act.

Sandoz Sulfate WE. [Sandoz] Sodium laureth sulfate (3.5 EO); CAS 9004-82-4; foamer and visc. modifier for personal care prods.; textile scouring; colorless liq.; visc. 750 cps; cloud pt. 13 C; pH 7.8–8.5 (10%); 27–29% act.

Sandoz Sulfonate 2A1. [Sandoz] Sodium dodecyl diphenyl ether sulfonate; anionic; solubilizer; household detergents, industrial, disinfectant, and metal cleaners, emulsifier in emulsion polymerization; textile aux.; also avail in acid form; yel./brn. liq.; sol. in electrolyte sol'ns., caustic, HCl, TKPP; sp.gr. 1.16; visc. 145 cps; 45% act.

Sandoz Sulfonate 3B2. [Sandoz] Decyl diphenyl ether sulfonate, sodium salt; anionic; wetting agent; agric. formulations; spreader and penetrant for farm usage; also avail. in acid form; amber liq.; sol. see Sandoz Sulfonate 2A1; sp.gr. 1.16; visc. 120 cps; 45% act.

Sandoz Sulfonate AAS 35S. [Sandoz] Linear alkyl (C12) benzene sulfonate, sodium salt; anionic; wetting agent and foamer for household and industrial detergent systems; hard surface cleaners; liq.; biodeg.; 31.5% act.

Sandoz Sulfonate AAS 40FS. [Sandoz] Linear alkyl (C12) benzene sulfonate, sodium salt; anionic; wetting agent and foamer for household and industrial detergent systems; hard surface cleaners; powd.; 40% act.

Sandoz Sulfonate AAS 45S. [Sandoz] Linear alkyl (C12) benzene sulfonate, sodium salt; anionic; wetting agent and foamer for household and industrial detergent systems; hard surface cleaners; textile wet processing; liq.; 41–43% act.

Sandoz Sulfonate AAS 50MS. [Sandoz] Linear alkyl (C12) benzene sulfonate, ammonium salt; anionic; wetting agent and foamer for household and industrial detergent systems; hard surface cleaners; dishwashing liq.; metal cleaning; liq.; 40% act.

Sandoz Sulfonate AAS 60S. [Sandoz] Linear alkyl (C12) benzene sulfonate, TEA salt; anionic; wetting agent and foamer for household and industrial detergent systems; hard surface cleaners; liq. detergents, personal and animal care prods.; textile wet processing; liq.; 60% act.

Sandoz Sulfonate AAS 70S. [Sandoz] Linear alkyl (C12) benzene sulfonate, sodium salt; anionic; wetting agent and foamer for household and industrial detergent systems; liq. detergents, metal and hard surface cleaners; liq.; 60% act.

Sandoz Sulfonate AAS 75S. [Sandoz] Linear alkyl (C12) benzene sulfonate, calcium salt; anionic; wetting agent and foamer for household and industrial detergent systems; hard surface cleaners; agric. prods.; liq.; oil-sol.; 60% act.

Sandoz Sulfonate AAS 90. [Sandoz] Linear alkyl (C12) benzene sulfonate, sodium salt; anionic; wetting agent and foamer for household and industrial detergent systems; hard surface cleaners; powd.; 85% act.

Sandoz Sulfonate AAS 98S. [Sandoz] DDBSA; anionic; conc. which is neutralized to any of a variety of salts as required; liq.; 96–98% act.

Sandoz Sulfonate AAS-Special 3. [Sandoz] Linear alkyl (C12) benzene sulfonate, isopropylamine salt; anionic; surfactant with good detergency; metal cleaning, dry cleaning systems; liq.; oil-sol.; 97% min. act.

Sanfix 555. [Sanyo Chem. Industries] Cationic resin; cationic; fixing agent for direct and reactive dyes; improves colorfastness of dyed fibers; powd.

Sanfix 555-200. [Sanyo Chem. Industries] Cationic resin; cationic; fixing agent for direct and reactive dyes; powd.

Sanfix 555C. [Sanyo Chem. Industries] Cationic resin containing copper salt; cationic; fixing agent for direct dyes requiring after-treatment with copper compds.; improves colorfastness of dyed fibers; powd.

Sanfix 555C-200. [Sanyo Chem. Industries] Cationic resin containing copper salt; cationic; fixing agent for direct dyes; powd.

Sanimal 55HX, 55LX, 250C, K51. [Nippon Nyukazai] Anionic/nonionic; emulsifier for sanitary pesticides; powd.

Sanimal L, M, H. [Nippon Nyukazai] Mixt. of POE alkylaryl ether, POE alkyl ether and anionic sulfonate; anionic/nonionic; emulsifier for organophosphorus insecticides, carbonate insecticides; liq.; 100% conc.

Sanleaf CL-533M. [Sanyo Chem. Industries] Surfactant blend; cationic; yarn finishing agent for cheese-oiling applic.; for all types of fibers; liq.

Sanmorin 11. [Sanyo Chem. Industries] POE alkyl ether; nonionic; penetrant and wetting agent for textile processing; for applic. on thick, dense fabrics and hard twisted yarns; liq.

Sanmorin AM. [Sanyo Chem. Industries] Surfactant; anionic; low foaming penetrating agent for mercerizing process; liq.

Sanmorin OT 70. [Sanyo Chem. Industries] Sodium dioctyl sulfosuccinate; anionic; wetting agent and penetrant for industrial uses; liq.; 70% conc.

Sansilic 11. [Ceca SA] Silicone-free; anionic; antifoaming agent for textile processing, all fibers.

Sanstat 230. [Sanyo Chem. Industries] Phosphate surfactant; anionic; antistat for polyester dyed fabrics; paste.

Sanstat 1200. [Sanyo Chem. Industries] Quat. ammonium salt; cationic; antistat for polyester fibers; liq.

Santone® 3-1-SH. [Van Den Bergh Foods] Polyglyceryl-3 oleate; nonionic; emulsifier and aerating agent used in food industry, textile and plastic lubricant; color dispersant; plastic; HLB 7.2; sapon no. 115–135.; 100% conc.

Santone® 8-1-O. [Van Den Bergh Foods] Polyglyceryl-8 oleate; nonionic; emulsifier for personal care prods.; visc. reducer; emulsion stabilizer; beverage clouding agent; liq.; HLB 13.0; sapon. no. 75–85.; 100% conc.

Sapogenat T Brands. [Hoechst AG] Tributyl phenol polyglycol ether (4–50 EO); nonionic; surface act. basic material; 100% conc.

Sarkosyl® L. [Ciba-Geigy AG] Lauroyl sarcosine; anionic; detergent, corrosion inhibitor, foam booster and stabilizer, wetting agent, lubricant, emulsifier used dentifrices, personal care, and household cleaning prods., pharmaceuticals, metal processing and finishing, metalworking and cutting oils; powd.; m.w. 264–285; sol. in org. solvs.; insol. in water; sp.gr. 0.969; m.p. 35–37 C; 94% min. purity.

Sarkosyl® LC. [Ciba-Geigy AG] Cocoyl sarcosine; see Sarkosyl L; powd.; m.w. 285–300; sol. in org. solvs.; insol. in water; sp.gr. 0.969; m.p. 23–28 C; 94% min. purity.

Sarkosyl® NL-30. [Ciba-Geigy AG] Sodium lauroyl sarcosinate; see Sarkosyl L; dentifrices, shampoos, rug and window cleaners and fine fabric detergents; colorless liq.; pH 7.5–8.5 (1%); 30% act.

Sarkosyl® O. [Ciba-Geigy AG] Oleoyl sarcosine; see Sarkosyl L; powd.; m.w. 340–360; sol. in org. solvs.; insol. in water; sp.gr. 0.948; 94% min. purity.

Satexlan 20. [Croda Inc.] PEG-20 hydrog. lanolin; nonionic; o/w emulsifier, emollient, thickener; perfume solubilizer; imparts superfatting properties; Gardner 4 max.; sol. in water and aq./alcoholic sol'ns.; m.p. 42–50 C; sapon no. 7 max.

Savosellig REAC 4. [Ceca SA] Anionic; soaping agent for fiber reactive dyes on cotton.

Sawaclean AO. [Lion] Alpha-olefin sulfonate; anionic; milling and scouring agent for wool and raw wool; liq.; 30% conc.

Sawaclean AOL. [Lion] Alpha-olefin sulfonate; anionic; scouring agent for feathers, wool, raw wool; liq.; 28% conc.

Schercamox C-AA. [Scher] Cocamidopropylamine oxide; nonionic; conditioner, detergent, wetting agent, antistat used in personal care prods. and lt. dishwashing detergents; biodeg.; Gardner 4.0 max. clear to slightly hazy liq.; m.w. 320; sol. in water and hydrophilic solvs.; sp.gr. 0.986±0.01; pH 7.0±0.5; 30% min. act.

Schercamox DMA. [Scher] Myristamine oxide; CAS 3332-27-2; nonionic; wetting and foaming agent; foam booster for shampoos, bubble baths, dishwashing compds.; visc. liq.; 30% conc.

Schercamox DMC. [Scher] Cocamine oxide; CAS 61788-90-7; wetting agent, foam stabilizer, visc. enhancer; yel. clear liq.; 29% amine oxide.

Schercamox DML. [Scher] Lauramine oxide; nonionic; antistat; emulsifier, emulsion stabilizer for used in cosmetics industry; foam booster and visc. builder for shampoos; Gardner 1.0 max. clear liq.; mild odor; m.w. 235; sol. in water and hydrophilic solvs.; sp.gr. 0.99 ± 0.01; pH 7.0 ± 1.0 (1%); 30% min. conc.

Schercamox DMM. [Scher] Myristamine oxide; CAS 3332-27-2; nonionic; wetting and foaming agent, surfactant for lt. duty dishwashing compds., shampoos; emulsifier and emulsion stabilizer for min. oils; Gardner 1.0 max. clear liq.; mild odor; m.w. 263; sol. in water and hydrophilic solvs.; sp.gr. 0.98; pH 7.0 ± 1.0 (1%); 29% min. conc.

Schercodine C. [Scher] Cocamidopropyl dimethylamine; CAS 68140-01-2; cationic; good foaming surfactant for hair and bath preps.; emulsifier, intermediate for betaine amphoterics; tan soft solid; m.w. 304; alkali no. 177-187; 100% conc.

Schercodine L. [Scher] Lauramidopropyl dimethylamine; CAS 3179-80-4; cationic; emulsifier; surfactant, intermediate for betaine amphoterics; lt. tan solid; m.w. 284; m.p. 35–40 C; alkali value 196–206; 98% min. amide.

Schercodine O. [Scher] Oleamidopropyl dimethylamine; CAS 109-28-4; cationic; o/w emulsifier; emollient conditioner for hair and skin preparations; amber liq.; m.w. 366; alkali value 150–160; 98% min. amide.

Schercomid 1-102. [Scher] Cocamide DEA; nonionic; thickener, foam booster, and emulsifier for cosmetic and detergent applics.; exhibits detergency when incorporated into bubble baths and detergent formulations; offers low cloud pt.; clear yel. visc. liq.; mild odor; sol. in water, alcohols, glycols, polyols, glycol ethers, aliphatic (lower members), aromatic, and chlorinated hydrocarbons; disp. in min. oil; sp.gr. 0.99; flash pt. (OC) > 180 C; acid no. 4.0 max.

Schercomid 304. [Scher] Coco amide, modified; nonionic; dry-cleaning detergent; liq.; 98% act.

Schercomid 1214. [Scher] Lauramide DEA and diethanolamine; nonionic; foam booster/stabilizer for detergent compositions; good detergency by itself and works synergistically with other surfactants; thickening agent and visc. builder; used in personal care items and cleaners for hard surfaces; clear yel. liq.; sl. typ. odor; sol. in water and in most org. solvs.; disp. in aliphatic hydrocarbons, min. oil, and natural fats; sp.gr. 1.01; dens. 8.4 lb/gal; flash pt. (OC) > 170 C; acid no. 12–16; 100% act.; 60% min. amide content.

Schercomid AME. [Scher] Acetamide MEA; nonionic; solubilizer, humectant, skin and hair conditioner, intermediate, coupling agent, pigment dispersant, solubilizer; Gardner 2.0 max. clear liq.; mild organoleptic odor; sol. in most alcohols, glycols, diols, polyols, glycol ethers, ketones, and water; sp.gr. 1.120; dens. 9.3 lb/gal; acid no. 10.0 max.; alkali no. 15.0 max.; ref. index 1.4700; pH 6.0–8.5 (50% aq.); flash pt. > 180 C (OC); 100% conc.

Schercomid CCD. [Scher] Cocamide DEA and diethanolamine; nonionic; emulsifier, thickener, detergent, soil dispersing and suspending agent; for floor cleaners, liq. hand dishwashing, car waxes, liq. hand soaps, all-purpose industrial and household cleaners, lubricants, emulsifiable oils, agric. sprays; dk. amber visc. liq.; sol. in water, alcohols, diols, triols, glycol ethers, polyols, aromatic and chlorinated hydrocarbons; sp.gr. 0.99 ± 0.01; dens. 8.3 lb/gal; acid no. 15–20; alkali no. 140-160; flash pt. (OC) > 170 C; pH 9.9 ± 0.5 (10%); 100% act., 60% min. amide.

Schercomid CDA. [Scher] Cocamide DEA and diethanolamine; nonionic/anionic; foam stabilizer, soil suspender, lime soap dispersant, and detergency booster for industrial and household cleaners; lt. amber clear liq.; sol. in water, alcohols, diols, triols, glycol ethers, polyols, aromatic and chlorinated hydrocarbons; sp.gr. 1.01 ± 0.01; dens. 8.33 lb/gal; visc. 1000 cps min.; acid no. 40–50; alkali value 150–170; flash pt. (OC) > 170 C; pH 9.6±0.5 (10%); 100% act., 60% min. amide.

Schercomid CDO-Extra. [Scher] Cocoamide DEA and diethanolamine; nonionic/anionic; detergent, wetting agent, foam stabilizer, visc. builder; shampoos, household cleaners, industrial cleaners with builders, liq. dishwashing compds., floor cleaners, personal care formulations; visc. yel. liq.; slt. odor; sol. in water, alcohols, diols, triols, glycol ethers, polyols, aromatic and chlorinated hydrocarbons; sp.gr. 1.00±0.01; visc. 950 cps min.; flash pt. (OC) > 170 C; acid no. 4–5; alkali no. 110-140; pH 10 ± 0.5 (10% sol'n.); 100% act., 65% min. amide.

Schercomid CME. [Scher] Cocamide MEA; CAS 68140-00-1; nonionic; foam booster and visc. builder for shampoos, bubble baths, and powded. detergent compositions; emulsifier for creams and lotions, esp. in cream hair dye formulations; tan wax; ammoniacal odor; sol. in alcohols, glycols, glycol ethers, aliphatic, aromatic, and chlorinated hydrocarbons; disp. in water; m.p. 63 ± 3 C; flash pt. (OC) > 180 C; acid no. 1.0 max.; 100% active; 85% min. amide content.

Schercomid CMI. [Scher] Cocamide MIPA; nonionic; foam booster and stabilizer, antidefatting agent for detergents; lt. tan wax; acid no. 2 max.; alkali value 20 max.; 90% min. amide.

Schercomid EAC. [Scher] Cocamide DEA; nonionic;

wetting agent, emulsifier; detergent for scouring textile fibers; wool scouring and fulling agent effective at low temps.; amber liq.; mild ammoniacal odor; water-sol.; dens. 8.16 lb/gal (50 C); sp.gr. 0.98; pH 9.8 ± 1.0 (1%); 100% act.

Schercomid EAC-S. [Scher] Cocamide DEA; nonionic; cold water detergent for general textile scouring and wool fulling; liq.; 30% act.

Schercomid HT-60. [Scher] PEG-50 hydrog. tallow amide; CAS 68783-22-2; nonionic; thickener, detergent, emulsifier, disperant with foam char.; Gardner 7 max. (molten) hard, waxy solid; ammoniacal odor; sol. in alcohols, glycols, triols, polyols, glycol ethers, water, in some aromatic and chlorinated hydrocarbons; sp.gr. 1.064 (60 C); dens. 9.6 lb/gal; acid no. 2 max.; alkali no. 10; m.p. 50–55 C; flash pt. (OC) > 180 C; pH 9.0–10.0 (10% aq.); 100% act.

Schercomid LD. [Scher] Lauramide DEA and diethanolamine; foam booster/stabilizer for detergent compositions; thickening agent and visc. builder; used in personal care items and hard surf. cleaners; yel. visc. liq.; slt., typ. odor; sol.: see Schercomid 1214; sp.gr. 1.01 ± 0.01; dens. 8.4 lb/gal; visc. 1000 min.; flash pt. (OC) > 170 C; acid no. 20–26; pH 9.7 ± 0.5 (10% sol'n.); 100% act.; 50% min. amide content.

Schercomid LME. [Scher] Lactamide MEA; nonionic; humectant, skin and hair conditioner, coupling agent, emollient; liq.; 75% conc.

Schercomid MME. [Scher] Myristamide MEA; nonionic; emulsifier, pearlescent, thickener, opacifier in cosmetic and pharmaceutical formulations; visc. builder and foam booster in shampoos, liq. dishwashing compds., and powd. detergents; pale wax; slight ammoniacal odor; m.w 271; sol. in most org. solvs.; water-disp.; m.p. 88 ± 4.0 C; acid no. 2.0 max.; flash pt. > 170 C (OC); 100% act.

Schercomid ODA. [Scher] Oleamide DEA and diethanolamine; nonionic/anionic; w/o emulsifier, pigment dispersant, conditioner, corrosion inhibitor, and visc. builder; emulsifier for aromatic and aliphatic hydrocarbon solv.; used in gel-type pine cleaners, shampoo formulations, hair conditioning agent; amber liq.; mild, char. odor; sol. in alcohols, glycols, glycol ethers, aliphatic and chlorinated hydrocarbons; water-disp.; sp.gr. 0.950 ± 0.01; dens. 7.9 lb/gal; visc. 1200 cps min.; flash pt. (OC) > 170 C; acid no. 12–16; pH 9.9 ± 0.5 (10% disp.); 100% active; 60% min. amide content.

Schercomid OMI. [Scher] Oleamide MIPA; CAS 111-05-7; nonionic; thickener, foam stabilizer for shampoos; hair conditioning agent; emulsifier for min. oil, IPP, IPM, butyl stearate, creams and lotions; imparts slip, lubrication, some emolliency, and softening effects upon the skin; amber liq. to soft solid; mild ammoniacal odor; sol. in alcohols, esters, glycol ethers, min. and veg. oils, aliphatic, aromatic, and chlorinated hydrocarbons; sp.gr. 0.90 ± 0.01; dens. 7.5 lb/gal; flash pt. (OC) 180 C; acid no. 10 max.; alkali no. 7-17; 100% active; 85% min. amide.

Schercomid SCE. [Scher] Cocamide DEA (1:1); CASs 68603-42-9; nonionic; detergent, visc. builder and foam stabilizer for cosmetic formulations, bubble baths, liq. dish wash detergents, and rug shampoos; lt. amber clean, visc. liq.; mild odor; sol. in alcohols, diols, triols, polyols, glycol ethers, aliphatic, aromatic, and chlorinated hydrocarbons; water-disp.; sp.gr. 0.97; dens. 8.1 lb/gal; flash pt. (OC) > 180 C; acid no. 2.0 max.; alkali no. 20-40; 100% active; 88% min. amide content.

Schercomid SCO-Extra. [Scher] Cocamide DEA (1:); CAS 68603-42-9; nonionic; thickener, foam builder, foam stabilizer, and detergency booster for alkylaryl sulfonates and lauryl sulfates; used in shampoos, bubble baths, floor cleaners and liq. and powd. dishwashing detergents; for making self-emulsifiable solv. from aliphatic, aromatic, or chlorinated hydrocarbons, used in waterless hand cleaners, engine shampoos, concrete floor cleaners, wax strippers, and tar removers; clear lt. amber liq.; mild odor; sol. in water, alcohols, glycols, polyols, glycol ethers, aliphatic (lower members), aromatic, chlorinated hydrocarbons, and natural fats; disp. in min. oil; sp.gr. 0.99; dens. 8.25 lb/gal; flash pt. (OC) > 180 C; acid no. 3.0 max.; alkali no. 20-40; 100% active; 77% min. amide content.

Schercomid SLA. [Scher] Lauric/myristic/palmitic DEA; nonionic; visc. builder for shampoos; thick, copious, stable foam useful in shampoos, facial scrubs, bubble baths, hand soaps, and detergent compositions; yel. liq. when fresh (cryst. on aging); mild, fruity odor; sol. in alcohols, glycols, glycol ethers, polyols, aliphatic (lower members), aromatic, and chlorinated hydrocarbons, and natural fats; disp. in water and min. oil; sp.gr. 0.96 ± 0.01; dens. 8.0 lb/gal; visc. 2500 cps min. (10% disp.); flash pt. (OC) > 170 C; acid no. 1.0 max.; pH 10.2 ± 0.5 (10% disp.); 100% active; 88% min. amide content.

Schercomid SLE. [Scher] Linoleamide DEA (1:1); CAS 56863-02-6; nonionic; solubilizer, thickener, w/o emulsifier, conditioner, and emollient for personal care prods.; emulsion stabilizer for o/w emulsions; dk. amber clear liq.; mild, typical odor; sol. in most org. solvs., min. and veg. oil; disp. in water; sp.gr. 0.965; dens. 8.0 lb/gal; acid no. 1.0 max.; alkali no. 20-40; flash pt. > 180 C (OC); 100% act., 87% min. amide.

Schercomid SL-Extra. [Scher] Lauramide DEA (1:1); CAS 120-40-1; nonionic; thickener, foam booster/stabilizer for hair shampoos, soaps, syn. detergent formulations; bubble bath applics.; industrial applics. incl. manual dishwashing formulations, liq. heavy-duty laundry detergents, all-purpose cleaning prods.; emulsifier for aliphatic, aromatic hydrocarbons and oils for o/w emulsions; off-wh. cryst. solid; mild odor; sol. in alcohols, glycols, glycol ethers, polyols, aliphatic (lower members); aromatic, and chlorinated hydrocarbons, and natural fats and oils; disp. in water and min. oils; sp.gr. 0.97 (45 C); dens. 8.1 lb/gal (45 C); m.p. 42 C; flash pt. (OC) > 170 C; acid no. 1.0 max.; alkali no. 20-40; 100% active; 88% min. amide content.

Schercomid SLM-C. [Scher] Lauramide DEA (1:1); CAS 120-40-1; nonionic; thickener and foam stabilizer with wetting properties; amber clear liq.; acid no. 1 max.; alkali value 30–50; 85% min. amide.

Schercomid SLMC-75. [Scher] Lauramide DEA; CAS 120-40-1; visc. builder which generates a thick, copious foam; mild shampoos, bubble baths, aerosol shave creams, and other detergent compositions; liq. at ambient temps. to permit ease of handling; clear yel. liq.; mild odor; sol. in alcohols, glycols, glycol ethers, polyols, aliphatic, aromatic, and chlorinated hydrocarbons, natural fats and oils; disp. in min. oil and water; sp.gr. 0.99; dens. 8.0 lb/gal; flash pt. (OC) > 180 C; acid no. 2.0 max.; 100% active; 80% min. amide content.

Schercomid SL-ML, SL-ML-LC. [Scher] Lauramide DEA (1:1); CAS 120-40-1; nonionic; visc. builder which generates thick, copious foam; mild shampoos, bubble baths, aerosol shave creams, and other detergent compositions; o/w and w/o emulsifier in creams and lotions; clear amber liq., clear yel. liq., clear amber liq., resp.; mild odor; sol. in alcohols, glycols, glycol ethers, polyols, aliphatic, aromatic, and chlorinated hydrocarbons, natural fats and oils; disp. in min. oil and water; sp.gr. 0.97, 0.98 resp.; dens. 8.0 lb/gal; flash pt. (OC) > 180 C; acid no. 1.5, 1.5 max. resp.; 100% act.; 87% min. amide content.

Schercomid SLM-LC. [Scher] Lauramide DEA; CAS 120-40-1; nonionic; thick foamer, visc. builder; liq.; sol. in alcohols, glycols, glycol ethers, polyols, aliphatic, aromatic, and chlorinated hydrocarbons, natural fats and oils; disp. in min. oil and water; sp.gr. 0.99; dens. 8.0 lb/gal; flash pt. (OC) > 180 C; acid no. 2.0 max.; 100% conc.

Schercomid SLM-S. [Scher] Lauramide DEA (1:1); CAS 120-40-1; nonionic; visc. builder for the cosmetic industry; thick, copious foam in shampoos, facial scrubs, bubble baths, hand soaps, etc.; lt. yel. liq. when fresh (cryst. on aging); mild odor; sol. in alcohols, glycols, glycol ethers, polyols, aliphatic, aromatic, and chlorinated hydrocarbons, natural fats and oils; disp. in min. oil and water; sp.gr. 0.97 ± 0.01 (45 C); dens. 8.1 lb/gal; flash pt. (OC) > 170 C; acid no. 1.0 max.; 100% act.; 88% min. amide content.

Schercomid SO-A. [Scher] Oleamide DEA (1:1); CAS 93-83-4; nonionic/anionic; w/o emulsifier, lubricant, conditioner; lt. amber clear liq.; slight typ. oleic odor; sol. in most org. solvs.; water-disp.; sp.gr. 0.95 ± 0.01; dens. 7.9 lb/gal; visc. 450 cps min.; acid no. 8.0 max.; alkali no. 40-60; flash pt. > 180 C (OC); 100% act.

Schercomid SO-T. [Scher] Tallamide DEA (1:1); CAS 68155-20-4; anionic/nonionic; w/o emulsifier; amber clear liq.; acid no. 15 max.; alkali no. 40-50; 100% conc., 85% min. amide.

Schercomid TO-2. [Scher] Tallamide DEA and diethanolamine; nonionic/anionic; w/o emulsifier, visc. builder, pigment and min. clay dispersant, corrosion inhibitor; emulsifier for aromatic and aliphatic hydrocarbon solv.; used in shampoos where it generates a creamy, luxurious foam; stabilizes foam when used with surfactants and detergents; hair conditioning agent; clear dk. amber liq.; mild, char. odor; sol. in alcohols, glycols, glycol ethers, aliphatic and chlorinated hydrocarbons; disp. in water; sp.gr. 0.990; dens. 8.25 lb/gal; acid no. 13–17; alkali no. 130-150; flash pt. (OC) > 170 C; 100% act.; 65% min. amide.

Schercomul G. [Scher] Emulsifier blend; nonionic; w/o emulsifier for min. spirits; liq.; 100% conc.

Schercomul H. [Scher] Fatty acid DEA; nonionic; w/o emulsifier for kerosene; liq.; 100% conc.

Schercomul K. [Scher] Fatty acid DEA; anionic; emulsifier for perchloroethylene; liq.; 100% conc.

Schercomul QW. [Scher] Surfactant blend; nonionic/cationic; w/o and o/w emulsifier, detergent; imparts antistatic properties; liq.; 95% conc.

Schercophos L. [Scher] Complex phosphate; anionic; detergent builder and water conditioner for the textile industry; liq.; 60% act.

Schercophos NP-6. [Scher] Nonoxynol-6 phosphate; anionic; emulsifier for polar and nonpolar solvs., emulsion polymerization; corrosion inhibitor; liq.; 100% conc.

Schercophos NP-9. [Scher] Nonoxynol-9 phosphate; anionic; low to moderate foam liq. detergent and emulsifier for heavy-duty all-purpose liq. detergents; solv. cleaner for drycleaning; liq.; 100% conc.

Schercopol CMS-Na. [Scher] Disodium cocamido MEA-sulfosuccinate; CAS 68784-08-7; anionic; nonirritating surfactant, foam-stabilizer, solubilizer, softener for personal care prods.; home and industrial detergent cleaning formulations; anti-irritant for other surfactants; biodeg.; m.w. 477; sol. in water; partly sol. in most org. solvs.; sp.gr. 1.12; dens. 9.3 lb/gal; visc. 100 cps max.; pH 5–7; cloud pt. 5.0 C; 30% solids.

Schercopol DOS-70. [Scher] Dioctyl sodium sulfosuccinate; CAS 577-11-7; anionic; wetting agent, surf. tens. depressant; visc. liq.; pH 6-8; 70% solids.

Schercopol DOS-PG-70. [Scher] Dioctyl sodium sulfosuccinate; wetting agent, surf. tens. depressant; visc. liq.; pH 6–8; 70% min. dry solids.

Schercopol DOS-PG-85. [Scher] Dioctyl sodium sulfosuccinate; CAS 577-11-7; anionic; wetting agent, surf. tens. depressant; visc. liq.; 85% min. dry solids.

Schercopol DS-120. [Scher] Modified ethoxylated alkylamine; nonionic/cationic; dispersant for glass fibers in aq. media; for textile industry; liq.; 20% act.

Schercopol DS-140. [Scher] Modified ethoxylated alkylamine; nonionic/cationic; dispersant for glass fibers in aq. media; for textile industry; liq.; 40% act.

Schercopol LPS. [Scher] Disodium laureth sulfosuccinate; CAS 39354-45-5; mild high foaming surfactant; visc. enhancer; yel. clear liq.; pH 5–7; 39% min. dry solids.

Schercopol OMS-Na. [Scher] Disodium oleamido MEA-sulfosuccinate; CAS 68479-64-1; anionic; solubilizer; nonirritating surfactant imparting soft, emollient feel on skin and conditioning effect on hair; foamer in toiletries, hand dishwashing detergents, and personal care prods.; yel. clear liq.; mild char. odor; sol. in water; partly sol. in most org. solvs.; sp.gr. 1.10 ± 0.05; dens. 9.16 lb/gal; visc. 1000 cps max.; pH 6.0 ± 1.0; cloud pt. 5.0 C max.; 34% min. solids conc.

Schercopol RMS-Na. [Scher] Disodium ricinoleamido MEA-sulfosuccinate; anionic; surfactant; yel. clear liq.; mild char. odor; m.w. 565; sp.gr. 1.120; dens. 9.3 lb/gal; visc. 500 cps. max.; cloud pt. 5.0 C max.; pH 5–7; 40% min. solids.

Schercopol UMS-Na. [Scher] Disodium undecylenamido MEA-sulfosuccinate; anionic; surfactant; amber pearlescent liq.; mild char. odor; m.w. 453; sp.gr. 1.16; dens. 9.7 lb/gal; cloud pt. 40 C max.; pH 5–7; 43% min. solids.

Schercopon 2WD. [Scher] Ethoxylated sulfosuccinate; anionic/nonionic; detergent and wetting agent for high electrolyte cleansers, drycleaning detergent; water-wh. liq.; 90% act.

Schercoquat 2IAE. [Scher] Hydroxypropyl bisisostearamidopropyldimonium chloride; cationic; w/o and o/w emulsifier; conditioner for hair and skin care prods.; antistat; liq.; 85% conc.

Schercoquat 2IAP. [Scher] Hydroxypropyl bisisostearamidopropyldimonium chloride; cationic; o/w and o/w emulsifier; conditioner for hair and skin care prods.; liq.; 85% conc.

Schercoquat ALA. [Scher] Dilauryl acetyl dimonium chloride; cationic; w/o and o/w emulsifier; conditioner for hair and skin prods.; liq.; 80% conc.

Schercotaine APAB. [Scher] Apricotamidopropyl

betaine; CAS 133934-08-4; amphoteric; mild detergent, conditioner, emollient, visc. enhancer; amber clear liq.; pH 5-7; 35% solids.

Schercotaine CAB-A. [Scher] Cocamidopropyl betaine, ammonium chloride; amphoteric; mild surfactant with higher foam than the sodium counterpart, decreased defatting properties; lt. yel. clear liq.; pH 5-7; 45% solids.

Schercotaine CAB-K. [Scher] Cocamidopropyl betaine, potassium chloride; amphoteric; mild detergent, surfactant, foamer with decreased defatting props.; visc. stabilizer in natural soap systems; lt. yel. clear liq.; pH 5-7; 45% solids.

Schercotaine CAB-KG. [Scher] N-[3-(cocoamido)-propyl]-N,N-dimethyl betaine, potassium salt; amphoteric; surfactant with increased solubility and lower cloud pt.; visc. stabilizer for natural soap systems; yel. clear liq.; typ. odor; m.w. 401; sp.gr. 1.06; dens. 8.8 lb/gal; pH 5.0–7.0; 40% min. act.

Schercotaine IAB. [Scher] Isostearamidopropyl betaine; CAS 6179-44-8; amphoteric; conditioner and detergent for shampoos and emollient body treatments; visc. control agent; textile softener; amber visc. liq., soft opaque gel; slight char. odor; m.w. 477; sol. in aq. alcohol, glycols; sp.gr. 1.05; dens. 8.75 lb/gal; pH 5.5 ± 1.0; 34.0% min. solid.; 30% min. act.

Schercotaine OAB. [Scher] Oleamidopropyl betaine; amphoteric; visc. control agent, textile softener and conditioner; liq.; 35% conc.

Schercotaine SCAB-A. [Scher] Cocamidopropylhydroxysultaine, ammonium chloride; amphoteric; low cloud pt. surfactant with higher foam and decreased defatting properties; lt. yel. clear liq.; pH 5-7; 50% conc.

Schercotaine SCAB-K. [Scher] Cocamidopropyl hydroxysultaine, potassium salt; amphoteric; low cloud pt. surfactant, visc. stabilizer in natural soap systems; lt. yel. clear liq.; pH 5-7; 50% conc.

Schercotaine SCAB-KG. [Scher] Cocamidopropylhydroxysultaine, potassium chloride; visc. stabilizer in natural soap systems; low cloud pt. surfactant; lt. yel. clear liq.; typ. odor; m.w. 480; sp.gr. 1.10; dens. 9.1 lb/gal; pH 5–7; 50% min. dry solids.

Schercotarder. [Scher] Fatty amine; cationic; retarding agent for cationic dyes in the textile industry; liq.; 40% act.

Schercoterge 140. [Scher] Ethoxylated amide; nonionic; detergent, wetting and textile scouring agent, emulsifier, wool fulling; dyeing assistant; post scouring agent; dye bath stabilizer; lemon liq.; mild ammoniacal odor; water-sol.; dens. 8.3 lb/gal; sp.gr. 1.0; pH 9.5 ± 0.5 (1%); 98% act.

Schercoteric CY-2. [Scher] Disodium capryloamphodiacetate; CAS 7702-01-4; amphoteric; low foaming surfactant for household and industrial cleaning prods.; amber clear liq.; 50% min. dry solids.

Schercoteric CY-SF-2. [Scher] Disodium capryloamphodipropionate; amphoteric; low foaming surfactant for household and industrial cleaning prods.; amber soft paste; 75% min. dry solids.

Schercoteric I-AA. [Scher] Sodium isostearoamphopropionate; CAS 68630-96-6; amphoteric; surfactant for cosmetic and industrial cleaners; amber visc. liq.; 34% min. dry solids.

Schercoteric LS. [Scher] Lauroamphoacetate; amphoteric; detergent for shampoos, dishwashing compds., textiles; yel. clear visc. liq.; 45% min. dry solids.

Schercoteric LS-2. [Scher] Lauroamphodiacetate; amphoteric; detergent for shampoos, dishwashing compds., textiles; yel. clear visc. liq.; 50% min. dry solids.

Schercoteric LS-EP. [Scher] Lauroamphohydroxypropylsulfonate; amphoteric; detergent for shampoos, dishwashing compds., textiles; yel. clear liq.; 45% min. dry solids.

Schercoteric MS. [Scher] Sodium cocoamphoacetate; CAS 68334-21-4; amphoteric; foamer, mild detergent, conditioner used in personal care prods. and industrial cleaners; amber clear visc. liq.; 45% min. dry solids.

Schercoteric MS-2. [Scher] Disodium cocoamphodiacetate; CAS 68650-39-5; amphoteric; mild detergent used in personal care prods. and industrial cleaners; amber clear visc. liq.; 50% min. dry solids.

Schercoteric MS-2ES Modified. [Scher] Disodium cocoamphodiacetate, sodium laureth sulfate; amphoteric; base for personal care prods. and cleaning preparations; foamer; yel. clear liq.; 40% min. dry solids.

Schercoteric MS-2 Modified. [Scher] Disodium cocoamphodiacetate, sodium lauryl sulfate, hexylene glycol; amphoteric; base for personal care prods. and cleaning preparations; foamer; yel. clear liq.; 37% min. dry solids.

Schercoteric MS-2TE Modified. [Scher] Disodium cocoamphodiacetate, sodium trideceth sulfate; amphoteric; base for personal care prods. and cleaning preparations; foamer; yel. clear liq.; 32% min. dry solids.

Schercoteric MS-SF. [Scher] Sodium cocoamphopropionate; amphoteric; mild surfactant for shampoos, industrial cleaners, dishwashing compds.; liq.; 38% conc.

Schercoteric MS-SF (38%). [Scher] Coco imidazolinium deriv. monocarboxylate; amphoteric; surfactant for shampoos, industrial cleaners, dishwashing compds.; yel. clear liq.; mild, typ. odor; m.w. 406; dens. 8.7 lb/gal; sp.gr. 1.05; pH 8–10; 38% act.

Schercoteric MS-SF (70%). [Scher] Coco imidazolinium deriv. monocarboxylate; amphoteric; surfactant for shampoos, industrial cleaners, dishwashing compds.; lt. amber soft gel; mild, typ. odor; sol. in water; m.w. 406; dens. 9.1 lb/gal; sp.gr. 1.1; pH 8–10; 70% act.

Schercoteric O-AA. [Scher] Sodium oleoamphopropionate; CAS 67892-37-9; amphoteric; surfactant for drycleaning industry, other industrial cleaners; amber clear liq.; 80% min. dry solids.

Schercoteric OS-SF. [Scher] Oleoamphopropionate; amphoteric; surfactant for cosmetic and industrial cleaners; amber soft paste; 38% min. dry solids.

Schercowet DOS-70. [Scher] Sodium dioctyl sulfosuccinate; anionic; emulsifier for emulsion polymerization; wetting and rewetting agent for textile wet processing; liq.; 70% act.

Schercozoline B. [Scher] Behenyl hydroxyethyl imidazoline; cationic; antistat, dispersant, wetting agent, emulsifier, microbicide used in acid and emulsion cleaning, cleaners, polishes, surf. treatment, textile and leather processing, agriculture and cosmetic intermediate; tan solid; m.w. 383; m.p. 45–49 C; 90% min. imidazoline.

Schercozoline C. [Scher] Cocoyl hydroxyethyl imidazoline; CAS 61791-38-6; cationic; antistat, dispersant, wetting agent, emulsifier, microbicide, detergent, intermediate for quat. ammonium compds., primer paints, emulsion cleaning, cleaners, polishes, surf. treatment, textile and leather processing, agriculture and cosmetic; tan semisolid; m.w. 278; alkali no. 200-214; 90% min. imidazoline.

Schercozoline I. [Scher] Isostearyl hydroxyethyl imidazoline; CAS 68966-38-1; cationic; surfactant, softener, antistat, dye assistant for textiles, paper, cutting oils, metal lubricants, polishes, cosmetics, agric., corrosion inhibitors, building materials; amber clear liq.; m.w. 378; alkali no. 150-160; 90% min. imidazoline.

Schercozoline L. [Scher] Lauryl hydroxyethyl imidazoline; CAS 136-99-2; cationic; surfactant, softener, dye assistant, antistat for textiles, paper, cutting oils, metal lubricants, polishes, cosmetics, corrosion inhibitors, building materials; intermediate for quats.; cream solid; m.w. 268; m.p. 38–42 C; alkali no. 204-214; 90% min. imidazoline.

Schercozoline O. [Scher] Oleyl hydroxyethyl imidazoline; CAS 95-38-5; cationic; surfactant, softener, dye assistant, antistat, w/o emulsifier, corrosion inhibitor, intermediate for quat. ammonium compds., textiles, paper, cutting oils, metal lubricants, polishes, cosmetics, agric., building materials; dk. amber liq.; m.w. 350; alkali no. 160-170; 90% min. imidazoline.

Schercozoline S. [Scher] Stearyl imidazoline; cationic; surfactant, softener, antistat, dye assistant for textiles, paper, cutting oils, metal lubricants, polishes, cosmetics, agric., corrosion inhibitors, building materials; cream solid; m.w. 336; m.p. 44–48 C; 90% min. imidazoline.

Scour 1161. [Catawba-Charlab] Nonionic; low foaming multipurpose surfactant.

Scour ALF. [Catawba-Charlab] Low foaming wetting and scouring agent.

Scour KSV, KSV Special. [Catawba-Charlab] Emulsified solv. containing stain remover and wetting and scouring agent.

Scour MEC. [Catawba-Charlab] Alkali-stable low-foaming wetting and scouring agent.

Scour MS. [Catawba-Charlab] Low foaming non-phenolic wetting and scouring agent.

Scour NFP. [Sybron] High flash pt. solv. scour for textile roller cleaner, screen clear applics.

Scourol 700. [Kao Corp. SA] POE alkyl ether; nonionic; textile processing assistant; scouring and desizing agent; liq.

Scripset 500. [Monsanto] Styrene/maleic anhydride, sodium salt; anionic; emulsifier, dispersant, thickener; powd.; 100% conc.

Scripset 520. [Monsanto] Styrene/maleic anhydride copolymer; nonionic; emulsifier, binder, sizing agent, visc. modifier, stabilizer; starch modifier; pigment dispersant, protective colloid, sizing, coating, water-paint calsomines, adhesives, printing, preparation of emulsifier paints; off-wh. powd.; faint aromatic odor; m.w. 50,000; sol. in alkaline aq. systems; 100% conc.

Scripset 540. [Monsanto] Styrene/maleic anhydride copolymer; emulsifier, dispersant, protective colloid, stabilizer; off-wh. powd.; faint aromatic odor; m.w. 20,000; sol. in alkaline aq. systems, org. solvs.; 100% conc.

Scripset 550. [Monsanto] Styrene/maleic anhydride copolymer; emulsifier, dispersant, protective colloid, stabilizer; off-wh. powd.; faint aromatic odor; m.w. 10,000; sol. in org. solvs., alkaline aq. sol'ns.; 100% conc.

Scripset 700. [Monsanto] Styrene/maleic anhydride copolymer; anionic; emulsifier, dispersant, thickener; liq.; faint aromatic odor; m.w. 50,000; water-sol.; dens. 9.5 lb/gal; 30% act.

Scripset 720. [Monsanto] Styrene/maleic anhydride copolymer; anionic; emulsifier, dispersant, thickener for high solids systems, latex mfg.; liq.; faint aromatic odor; m.w. 50,000; water-sol.; 25% act.

Scripset 808. [Monsanto] Styrene maleic anhydride amide/NH_4OH acid salt; emulsifier, dispersant, thickener; compat. in systems with ≥ pH 6.0; powd.; water-sol.; 100% conc.

Seapol-06. [Sanyo Chem. Industries] Surfactants and solv.; low toxic oil spill remover; liq.

Sebase. [Westbrook Lanolin] Ethoxylated lanolin plus fatty alcohols and hydrocarbons; nonionic; base, emollient, lubricant for o/w emulsions; visc. stabilizer for cosmetics; paste; HLB 8.0; 100% conc.

Secolat. [Stepan Europe] Alkyl disodium sulfosuccinamate; anionic; foamer for latex emulsions; emulsifier for emulsion polymerization; liq.; 35% conc.

Secomine TA 02. [Stepan Europe] Tallow amine ethoxylate; nonionic; lubricating, wetting, antistatic agent, emulsifier, anticorrosive agent for detergents, surf. treatments; beige liq. to paste; 100% act.

Secomix® E40. [Stepan Europe] Surfactant blend; nonionic; detergent for industrial/household cleaners; yel. visc. liq.; 23% act.

Secosol® AL/959. [Stepan Europe] Sodium lauryl sulfosuccinate; anionic; mild foamer for shampoos, bubble baths, liq. soaps, shower gels, bath salts; paste; 25% conc.

Secosol® ALL40. [Stepan Europe] Sodium laureth sulfosuccinate; anionic; mild foamer for shampoos, bubble baths, liq. soaps, shower gels, bath salts; emulsifier for emulsion polymerization; water-wh. to pale yel. liq.; 31% act.

Secosol® AS. [Stepan Europe] Sodium cetyl-stearyl sulfosuccinate; anionic; additive to toilet and syndet soaps; cleaning agent; paste; 45% conc.

Secosol® DOS 70. [Stepan Europe] Sodium dioctyl sulfosuccinate; anionic; emulsifier, dispersant and wetting/rewetting agent for household/industrial cleaners, emulsion polymerization, paints, inks, oilfield prod.; textile additive; water-wh. to pale yel. liq.; 70% act.

Secosol® EA/40. [Stepan Europe] Sodium mono alkylethanolamide sulfosuccinate; anionic; rug and hair shampoos, foam baths; paste; 40% conc.

Secoster® 874. [Stepan Europe] Pentaerythritol ester; syn. lubricating agent with high thermal stability for metalworking, marine and jet engine lubricants; water-wh. liq.; 100% act.

Secoster® 887. [Stepan Europe] Trimethylolpropane ester; syn. lubricating agent with high thermal stability for metalworking, marine and jet engine lubricants; water-wh. liq.; 100% act.

Secoster® A. [Stepan Europe] Fatty acid ethoxylate; nonionic; dispersant, solv. and oil emulsifier for household/industrial cleaners, oilfield prod.; yel. to brn. liq.; 100% act.

Secoster® CL 10. [Stepan Europe] PEG-20 sorbitan laurate; CAS 9005-64-5; nonionic; emulsifier, dispersant, solubilizer; liq.; 100% conc.

Secoster® CP 10. [Stepan Europe] PEG-20 sorbitan palmitate; CAS 9005-66-7; nonionic; emulsifier, dispersant, solubilizer; liq.; 100% conc.

Secoster® CS 10. [Stepan Europe] PEG-20 sorbitan stearate; CAS 9005-67-8; nonionic; emulsifier, dispersant, solubilizer; liq.; 100% conc.

Secoster® DO 600. [Stepan Europe] PEG 600 dioleate; CAS 9005-07-6; nonionic; additive for cutting oils; solv., emulsifier for solvs. and oils, creams, cleansing milks, pesticides, textile lubricants, oilfield prod.; dispersant; for household/industrial cleaners; pale yel. to amber liq.; 100% act.

Secoster® KL 10. [Stepan Europe] Sorbitan laurate; CAS 1338-39-2; nonionic; emulsifier, dispersant, solubilizer; liq.; 100% conc.

Secoster® KP 10. [Stepan Europe] Sorbitan palmitate; CAS 26266-57-9; nonionic; emulsifier, dispersant, solubilizer; solid; 100% conc.

Secoster® KS 10. [Stepan Europe] Sorbitan stearate; CAS 1338-41-6; nonionic; emulsifier, dispersant, solubilizer; solid; 100% conc.

Secoster® MA 300. [Stepan Europe] PEG 300 abietate; nonionic; additive for cutting oils; solv., emulsifier for solvs. and oils, creams, cleansing milks and pesticides; dispersant; for household/industrial cleaners; brn. liq.; 100% act.

Secoster® ML 300. [Stepan Europe] PEG 300 laurate; CAS 9004-81-3; nonionic; emulsifier, dispersant; liq.; 100% conc.

Secoster® ML 4000. [Stepan Europe] PEG 4000 laurate; CAS 9004-81-3; nonionic; emulsifier, dispersant; solid; 100% conc.

Secoster® MO 100. [Stepan Europe] Diethylene glycol oleate; CAS 9004-96-0; nonionic; emulsifier, dispersant; liq.; 100% conc.

Secoster® MO 400. [Stepan Europe] PEG 400 oleate; CAS 9004-96-0; nonionic; lubricant, antistat, solvent and oil emulsifier; for household/industrial cleaners, oilfield prod.; pale yel. to amber liq.; 100% act.

Secoster® SDG. [Stepan Europe] Glyceryl stearate; nonionic; antistat and lubricant for polyolefins; wh. solid; 100% act.

Secosyl. [Stepan Europe] Sodium N-lauroyl sarcosinate; anionic; detergent, foaming agent, base, anticorrosion additive for rug shampoos, mild dishwash, household cleaners, personal care prods.; stable in hard water; water-wh. to pale yel. liq.; 30% act.

Sedipol®. [BASF AG] Higher aliphatic alcohols; antifoams for industrial and communal sewage works.

Sedoran FF-180, FF-200, FF-210, FF-220. [Sanyo Chem. Industries] Nonionic; low foaming base materials for detergents; liq.; 100% conc.

Sellig 10 Mode. [Ceca SA] Anionic; low-foam detergent for deoiling, pre-dyeing for all fibers.

Sellig 13. [Ceca SA] Anionic; low-foam detergent for deoiling, pre-dyeing for all fibers.

Sellig AO 6 100. [Ceca SA] PEG-6 oleate; CAS 9004-96-0; nonionic; surfactant; liq.; HLB 9.6; 100% conc.

Sellig AO 9 100. [Ceca SA] PEG-8.5 oleate; CAS 9004-96-0; nonionic; surfactant; liq.; HLB 11.4; 100% conc.

Sellig AO 15 100. [Ceca SA] PEG-15 oleate; CAS 9004-96-0; nonionic; surfactant for emulsions, detergents; visc. liq.; sol. in water, chlorinated and aromatic solvs., veg. oil; HLB 14.0; sp.gr. 1.02; iodine no. 23-27; pH 7 ± 0.5 (10% aq.); 100% conc.

Sellig AO 25 100. [Ceca SA] PEG-25 oleate; CAS 9004-96-0; nonionic; surfactant; Gardner 6.5 solid; HLB 16.0; pH 6.45 (10% aq.); 99% active.

Sellig Antimousse S. [Ceca SA] Silicone-based; nonionic; antifoaming agent for textiles.

Sellig Dispersant FPZ. [Ceca SA] Anionic; for washing out of dispersed dyes on polyester.

Sellig DN 10 100. [Ceca SA] Nonyl nonoxynol-10; CAS 9014-93-1; nonionic; surfactant; HLB 10.0; 100% conc.

Sellig DN 22 100. [Ceca SA] Nonyl nonoxynol-22; CAS 9014-93-1; nonionic; surfactant; HLB 13.8; 100% conc.

Sellig HR 18 100. [Ceca SA] PEG-21 castor oil; nonionic; surfactant; liq.; HLB 10.4; 100% conc.

Sellig Hydrophilisant. [Ceca SA] Nonionic; low-foam antiredeposition soaping agent for textile dyeing and printing.

Sellig JN 25. [Ceca SA] Anionic; wetting agent for textile scouring of cellulose fibers.

Sellig LA 9 100. [Ceca SA] Laureth-9.5; nonionic; surfactant; paste; HLB 13.5; 100% conc.

Sellig LA 1150. [Ceca SA] Laureth-20; nonionic; surfactant for shampoos, degreaser, textile applics.; liq.; sp.gr. 1.0 ± 0.2; HLB 14.5; pH 7 ± 0.5 (10% aq.); 50% conc.

Sellig LA 11 100. [Ceca SA] Laureth-20; nonionic; surfactant; liq. at 37 C; HLB 16.0; 100% conc.

Sellig Mouillant 9083 NI/AL. [Ceca SA] Nonionic; low-foaming wetting agent for use in cold or hot textile processing at low rates of addition.

Sellig Mouillant EMPT 80. [Ceca SA] Nonionic; wetting agent for use in cold or hot mercerizing treatments.

Sellig MRC. [Ceca SA] Anionic; low-foaming wetting agent and detergent.

Sellig N 1050. [Ceca SA] Nonoxynol-40; CAS 9016-45-9; nonionic; surfactant; liq.; HLB 17.8; 50% conc.

Sellig N 1780. [Ceca SA] Nonoxynol-17; CAS 9016-45-9; nonionic; surfactant; liq.; HLB 15.3; 80% conc.

Sellig N 20 80. [Ceca SA] Nonoxynol-19; CAS 9016-45-9; nonionic; surfactant; liq.; HLB 15.8; 80% conc.

Sellig N 30 70. [Ceca SA] Nonoxynol-25; CAS 9016-45-9; nonionic; surfactant; liq.; HLB 16.6; 70% conc.

Sellig N 4 100. [Ceca SA] Nonoxynol-4; CAS 9016-45-9; nonionic; surfactant; liq.; HLB 8.5; 100% conc.

Sellig N 5 100. [Ceca SA] Nonoxynol-5; CAS 9016-45-9; nonionic; surfactant; liq.; HLB 10.2; 100% conc.

Sellig N 6 100. [Ceca SA] Nonoxynol-6; CAS 9016-45-9; nonionic; surfactant for emulsions, petrol. prods., chlorinated and aromatic solvs., greases, silicones; pale yel. liq.; sol. in hydrocarbons, chlorinated and aromatic solvs.; insol. in water; sp.gr. 1.03; HLB 11.2; pH 7.5 ± 0.5 (10% aq. disp.); 100% conc.

Sellig N 8 100. [Ceca SA] Nonoxynol-8; CAS 9016-45-9; nonionic; surfactant for emulsions, petrol. prods., min. oils, greases, silicones; base for low foam household detergents; pale yel. liq.; sol. in water, chlorinated and aromatic solvs.; insol. in hydrocarbons; sp.gr. 1.03; HLB 12.3; pH 7.5 ± 0.5 (10% aq.); 100% conc.

Sellig N 9 100. [Ceca SA] Nonoxynol-9; CAS 9016-45-9; nonionic; surfactant for emulsions, min. oils, greases, silicones; base for household, industrial and textile detergents; pale yel. liq.; sol. in water, chlorinated and aromatic solvs.; sp.gr. 1.03; HLB 12.7; pH 7.5 ± 0.5 (10% aq.); 100% conc.

Sellig N 10 100. [Ceca SA] Nonoxynol-10; CAS 9016-45-9; nonionic; surfactant; liq.; HLB 13.1; 100% conc.

Sellig N 11 100. [Ceca SA] Nonoxynol-11; CAS 9016-45-9; nonionic; surfactant; liq.; HLB 13.5; 100% conc.

Sellig N 12 100. [Ceca SA] Nonoxynol-12; CAS 9016-

45-9; nonionic; surfactant; liq.; HLB 14.2; 100% conc.

Sellig N 15 100. [Ceca SA] Nonoxynol-16; CAS 9016-45-9; nonionic; surfactant for emulsions, oils, greases; cream-colored paste; sp.gr. 1.07; HLB 15.1; pH 7.5 ± 0.5 (10% aq.); 100% conc.

Sellig N 17 100. [Ceca SA] Nonoxynol-17; CAS 9016-45-9; nonionic; surfactant; solid; HLB 15.3; 100% conc.

Sellig N 20 100. [Ceca SA] Nonoxynol-19; CAS 9016-45-9; nonionic; surfactant for emulsions, oils, solvs., greases; cream-colored paste; sol. in chlorinated and aromatic solvs.; insol. in hydrocarbons; sp.gr. 1.07; ; HLB 15.8; pH 7.5 ± 0.5 (10% aq.); 100% conc.

Sellig N 30 100. [Ceca SA] Nonoxynol-25; CAS 9016-45-9; nonionic; surfactant; liq.; HLB 16.6; 100% conc.

Sellig N 50 100. [Ceca SA] Nonoxynol-52; CAS 9016-45-9; nonionic; surfactant; solid; HLB 18.2; 100% conc.

Sellig NK 59. [Ceca SA] Nonionic; wetting agent for textile carbonizing; for use in acidic sol'n. for keratinous fibers.

Sellig NK 83. [Ceca SA] Nonionic; low-foam detergent for degreasing, pre- and post-dyeing for all fibers.

Sellig NK 729064. [Ceca SA] Nonionic; wetting agent and detergent for textile scouring; stable in alkaline sol'n.

Sellig O 4 100. [Ceca SA] Octoxynol-4; CAS 9002-93-1; nonionic; surfactant; liq.; HLB 9.3; 100% conc.

Sellig O 5 100. [Ceca SA] Octoxynol-5; CAS 9002-93-1; nonionic; surfactant; liq.; HLB 10.4; 100% conc.

Sellig O 6 100. [Ceca SA] Octoxynol-6; CAS 9002-93-1; nonionic; surfactant; liq.; HLB 10.8; 100% conc.

Sellig O 8 100. [Ceca SA] Octoxynol-8; CAS 9002-93-1; nonionic; surfactant; liq.; HLB 12.6; 100% conc.

Sellig O 9 100. [Ceca SA] Octoxynol-9; CAS 9002-93-1; nonionic; surfactant; liq.; HLB 13.1; 100% conc.

Sellig O 11 100. [Ceca SA] Octoxynol-11; CAS 9002-93-1; nonionic; surfactant; liq.; HLB 14.0; 100% conc.

Sellig O 12 100. [Ceca SA] Octoxynol-12; CAS 9002-93-1; nonionic; surfactant; liq.; HLB 14.6; 100% conc.

Sellig O 20 100. [Ceca SA] Octoxynol-20; CAS 9002-93-1; nonionic; surfactant; liq.; HLB 16.2; 100% conc.

Sellig R 3395. [Ceca SA] PEG-33 castor oil; nonionic; surfactant; liq.; HLB 12.0; 95% conc.

Sellig R 3395-C435. [Ceca SA] PEG-32 castor oil; nonionic; surfactant; liq.; HLB 11.2; 95% conc.

Sellig R 3395 SP. [Ceca SA] PEG-30 castor oil; nonionic; surfactant; liq.; HLB 11.6; 95% conc.

Sellig R 4095. [Ceca SA] PEG-40 castor oil; nonionic; surfactant; liq.; HLB 12.9; 95% conc.

Sellig R 4495. [Ceca SA] PEG-44 castor oil; nonionic; surfactant; liq.; HLB 13.3; 95% conc.

Sellig R 20 100. [Ceca SA] PEG-20 castor oil; nonionic; surfactant; liq.; HLB 9.5; 100% conc.

Sellig S 30 100. [Ceca SA] PEG-30 stearate; CAS 9004-99-3; nonionic; surfactant; solid; HLB 11.9; 100% conc.

Sellig SP 25 50. [Ceca SA] Cetoleth-27; nonionic; surfactant; liq.; HLB 16.4; 50% conc.

Sellig SP 3020. [Ceca SA] Cetoleth-30; nonionic; surfactant; liq.; HLB 16.7; 20% conc.

Sellig SP 8 100. [Ceca SA] Cetoleth-8; nonionic; surfactant for o/w emulsions, oils, org. solvs., min. oil, paraffin; stable to electrolytes; cream-colored paste; sol. in chlorinated and aromatic solvs.; disp. in water, hydrocarbons, min. and veg. oils; sp.gr. 1.02; HLB 12.0; pH 7 ± 0.5 (10% aq.); 100% conc.

Sellig SP 16 100. [Ceca SA] Cetoleth-16; nonionic; surfactant for emulsions; cream-colored liq.; sp.gr. 1.0; HLB 14.2; pH 7.5 ± 0.5 (10% aq.); 100% conc.

Sellig SP 20 100. [Ceca SA] Cetoleth-18; nonionic; surfactant; solid; HLB 15.2; 100% conc.

Sellig SP 25 100. [Ceca SA] Cetoleth-27; nonionic; surfactant; solid; HLB 16.4; 100% conc.

Sellig SP 30 100. [Ceca SA] Cetoleth-30; nonionic; surfactant; solid; HLB 16.7; 100% conc.

Sellig Stearo 6. [Ceca SA] PEG-6 stearate; CAS 9004-99-3; nonionic; surfactant; solid; HLB 11.9; 100% conc.

Sellig SU 4 100. [Ceca SA] Ceteareth-4; CAS 68439-49-6; nonionic; surfactant for emulsions, oils; pale yel.; sp.gr. 0.95; HLB 7.8; pH 7 ± 0.5 (1% aq. disp.); 99% active.

Sellig SU 18 100. [Ceca SA] Ceteareth-18; CAS 68439-49-6; nonionic; surfactant; paste; HLB 14.9; 100% conc.

Sellig SU 25 100. [Ceca SA] Ceteareth-20; CAS 68439-49-6; nonionic; surfactant; paste; 100% conc.

Sellig SU 30 100. [Ceca SA] Ceteareth-32; CAS 68439-49-6; nonionic; surfactant, dispersant, household and industrial detergent base, emulsions; cream-colored solid; sp.gr. 1.02; HLB 16.6; pH 7 ± 0.5 (10% aq.); 100% conc.

Sellig SU 50 100. [Ceca SA] Ceteareth-46; CAS 68439-49-6; nonionic; surfactant; solid; HLB 17.6; 100% conc.

Sellig T 1790. [Ceca SA] PEG-17 tallate; CAS 61791-00-2; nonionic; surfactant; visc. liq.; HLB 14.2; 90% conc.

Sellig T 3 100. [Ceca SA] PEG-3 tallate; CAS 61791-00-2; nonionic; surfactant; yel. amber liq.; sp.gr. 1.05; HLB 6.0; pH 7.5 ± 0.5 (10% aq. disp.); 100% conc.

Sellig T 14 100. [Ceca SA] PEG-14 tallate; CAS 61791-00-2; nonionic; detergent base with controlled foam, solubilizer for essential oils; Gardner < 6 liq., char. odor; sp.gr. 1.05; HLB 13.4; pH 7-9 (10%); 100% active.

Selligon SP. [Ceca SA] Anionic; softening detergent for wool and cotton.

Selligor 860 SP. [Ceca SA] Nonionic; for washing out of acidic dyes.

Selligor SAN 1. [Ceca SA] Nonionic; wetting agent for sanforizing; improves hydrophilicity of the fabric.

Sellogen 641. [Henkel/Functional Prods.] Alkyloxyalkyl sodium sulfate; anionic; textile detergent, dispersant and emulsifier; pickling; yel. clear liq.; cloud pt. 5 C; flash pt. 43 C (PMCC); pH 8.5 (2%); biodeg.; 30% act.

Sellogen DFL. [Henkel/Functional Prods.] Sodium alkyl naphthalene sulfonate; anionic; wetting and dispersing agent for pesticide formulations; EPA-exempt; dry tan powd.; pH 8.5 (1%); 67% act.

Sellogen HR. [Henkel/Organic Prods.; Henkel/Textile; Henkel Canada; Henkel-Nopco] Sodium dialkyl naphthalene sulfonate; anionic; wetting agent, dispersant; heavy duty household and industrial cleaners; fire control, paint strippers, agric. chemical formulations, emulsion and suspension polymerization aids, latex paints, textile dyeing on cellulosics; stable in high concs. of electrolytes; EPA-exempt; lt. tan powd.; water-sol.; dens. 0.40 g/cc; surf. tens.

44.1 dynes/cm (0.25%); pH 9.0 (5%); 75% act.

Sellogen HR-90. [Henkel/Functional Prods.] Sodium alkyl naphthalene sulfonate; anionic; wetting and dispersing agent for insecticides, acid and alkaline media; EPA-exempt; dk. amber liq.; flash pt. 93 C; pH 10.2 (10%); 35.5% conc.

Sellogen NS-50. [Henkel/Functional Prods.] Sodium alkyl aromatic sulfonate; anionic; anticaking agent in fertilizer and clay; used in phosphate rock acidulation; slurry; 50% conc.

Sellogen W. [Henkel/Functional Prods.; Henkel Canada; Henkel-Nopco] Sodium dialkyl naphthalene sulfonate; anionic; wetting and dispersing agent for detergents, pesticides; tolerant to electrolytes, extremes of pH and temp.; EPA-exempt; tan powd.; water-sol.; dens. 0.38 g/cc; surf. tens. 42.1 dynes/cm (0.25%); pH 9.0 (5%); 65% act.

Sellogen WL Acid. [Henkel/Functional Prods.; Henkel/Organic Prods.] Alkyl naphthalene sulfonic acid; anionic; dye assistant; liq.; 100% conc.

Sellogen WL Liq. [Henkel/Functional Prods.; Henkel/Organic Prods.] Sodium alkyl naphthalene sulfonate; anionic; wetting agent for insecticides, acid and alkaline media; EPA-exempt; dk. amber liq.; water-sol.; sp.gr. 1.140; dens. 9.5 lb/gal; flash pt. 93 C; pH 9.3 (100%); surf. tens. 48.9 dynes/cm (0.25%); 32% act.

SEM-35. [Harcros] Dimethyl silicone fluid-water emulsion; nonionic; emulsifier; lubricant, mold release for rubber and plastics, letterpress and lithographic printing; also for textile softeners, cosmetics; milky wh. liq. emulsion; water-disp.; sp.gr. 0.99; dens. 8.2 lb/gal; f.p. 30 F; flash pt. (PMCC) > 212 F; pH 6.0–8.0; 35% act.

Senka Antifoam 800. [Nippon Senka] Silicone modified oil emulsion; antifoaming agent; milky wh. liq.

Sepabase A Grades. [BASF AG] High m.w. EO/PO adduct; nonionic; for dehydration of crude oil emulsions and removal of residual salts; liq.; 100% conc.

Sepabase B Grades. [BASF AG] High m.w. EO/PO adduct; nonionic; for dehydration of crude oil emulsions and removal of residual salts; liq.; 100% conc.

Sepabase R Grades. [BASF AG] High m.w. EO/PO adduct; nonionic; for dehydration of crude oil emulsions and removal of residual salts; liq.; 80% conc.

Separol AF 27. [BASF AG] High m.w. EO/PO adduct; nonionic; demulsifier used in crude oil emulsions and dehydrating equipment; yel. clear liq.; sol. in alcohols; dens. 0.96 g/cc; visc. 30 cs; flash pt. > 23 C; pour pt. –30 C.

Separol WF 22. [BASF AG] High m.w. EO/PO adduct; nonionic; demulsifier used in crude oil emulsions and dehydrating equipment; yel. clear liq.; sol. in water and alcohols; dens. 1.02 g/cc; visc. 200 cs; flash pt. > 23 C; pour pt. –50 C.

Separol WF 34. [BASF AG] High m.w. EO/PO adduct; CAS 9003-11-6; nonionic; demulsifier used in crude oil emulsions and dehydrating equipment; yel. clear liq.; sol. in alcohols; dens. approx 0.95 g/cc; visc. 200 cs; flash pt. 16 C; pour pt. –50 C.

Separol WF 41. [BASF AG] High m.w. EO/PO adduct; CAS 9003-11-6; nonionic; demulsifier used in crude oil emulsions and dehydrating equipment; yel. clear liq.; sol. in alcohols; dens. 0.95 g/cc; visc. 50 cs; flash pt. 17 C; pour pt. < –50 C.

Separol WF 221. [BASF AG] High m.w. EO/PO adduct; nonionic; demulsifier used in crude oil emulsions and dehydrating equipment; yel. clear liq.; sol. in alcohols and water; dens. 1.02 g/cc; visc. 200 cs; flash pt. > 23 C; pour pt. –50 C.

Separol WK 25. [BASF AG] High m.w. EO/PO adduct; nonionic; demulsifier used in crude oil emulsions and dehydrating equipment; yel. clear liq.; sol. in alcohols and water; dens. 1 g/cc; visc. 300 cs; flash pt. 24 C; pour pt. –50 C.

Sepawet®. [BASF AG] Surfactants for petrol. prod. and pipeline transport.

Ser-Ad FA 153. [Servo Delden BV] Surfactant; anionic; wetting agent and emulsifier for use as paint additives; liq.

Ser-Ad FN Series. [Servo Delden BV] Nonionic; wetting agent, emulsifier for the paint industry; liq.

Serdas GBS, GBU. [Servo Delden BV] Surfactants in min. oil; defoamer for paint industry; water-disp.

Serdas GE 4010. [Servo Delden BV] Oil and silicone-free defoamer for aq. paint systems; emulsion.

Serdas GE 4050. [Servo Delden BV] Oil-free silicone-containing composition; antifoaming agent in aq. paint systems; emulsion.

Serdas GLN. [Servo Delden BV] Surfactants in min. oil; defoamer for paper, sugar, textile, leather, paint industries; disp. in water.

Serdet DCK 3/70. [Servo Delden BV] Sodium laureth sulfate (2 EO); CAS 9004-82-4; anionic; used in personal care prods. and dishwashing formulations; liq.; biodeg.; 70% conc.

Serdet DCK 30. [Servo Delden BV] Sodium laureth sulfate (3 EO); CAS 9004-82-4; anionic; used in personal care prods. and dishwashing formulations; liq.; biodeg.; 27% conc.

Serdet DCN 30. [Servo Delden BV] Ammonium C12–C14 alkyl ether sulfate (3 EO); anionic; used in personal care prods. and dishwashing formulations; liq.; biodeg.; 30% conc.

Serdet DFK 40. [Servo Delden BV] Sodium lauryl sulfate; CAS 151-21-3; anionic; personal care and dishwashing formulations; emulsifier in emulsion polymerization; paste; biodeg.; 41% conc.

Serdet DJK 30. [Servo Delden BV] Sodium decyl sulfate; anionic; emulsifier for emulsion polymerization; liq.; 28% conc.

Serdet DM. [Servo Delden BV] Dodecylbenzene sulfonic acid; anionic; used in scouring powds. and liq. detergents; emulsifier for emulsion polymerization; liq.; 97% conc.

Serdet DNK 30. [Servo Delden BV] Sodium nonoxynol-4 sulfate; anionic; detergent base; emulsifier for emulsion polymerization; liq.; 31% conc.

Serdet DPK 3/70. [Servo Delden BV] Sodium C12–15 alkyl ether sulfate (3 EO); anionic; personal care prods. and dishwashing formulations; paste; 70% conc.

Serdet DPK 30. [Servo Delden BV] Sodium C12–15 pareth sulfate; anionic; used in personal care prods. and dishwashing formulations; liq.; 27% conc.

Serdet DSK 40. [Servo Delden BV] Sodium 2-ethylhexyl sulfate; anionic; wetting agent; latex stabilizer; spreading, dispersing agent, detergent; emulsion polymerization; liq.; water-sol.; 40% conc.

Serdet Perle Conc. [Servo Delden BV] Lauryl ether sulfate and alkylolamide; anionic; wetting agent; latex stabilizer; spreading, dispersing agent, detergent; emulsion polymerization; liq.; 34% conc.

Serdolamide POF 61. [Servo Delden BV] Oleic acid polydiethanolamide; nonionic; foam stabilizer, refatting agent, visc. modifier in detergents and personal care prods.; liq.; 100% conc.

Serdolamide POF 61 C. [Servo Delden BV] Oleic acid polydiethanolamide; nonionic; component in emulsifiers for cutting oils; corrosion inhibitor; liq.; 100% conc.

Serdox NBS 4. [Servo Delden BV] C9–11 alkyl polyglycol ether (4 EO); nonionic; scouring agent for textiles; soaking assistant in leather industry; liq.; HLB 10.5; 100% conc.

Serdox NBS 5.5. [Servo Delden BV] C9–11 alkyl polyglycol ether (5.5 EO); nonionic; scouring agent for textiles; soaking assistant in leather industry; liq.; HLB 12.0; 100% conc.

Serdox NBS 6. [Servo Delden BV] C9–11 alkyl polyglycol ether (6 EO); nonionic; scouring agent for textiles; soaking assistant in leather industry; liq.; HLB 12.5; 100% conc.

Serdox NBS 6.6. [Servo Delden BV] C9–11 alkyl polyglycol ether (6.6 EO); nonionic; scouring agent for textiles; soaking assistant in leather industry; liq.; HLB 13.0; 100% conc.

Serdox NBS 8.5. [Servo Delden BV] C9–11 alkyl polyglycol ether (8.5 EO); nonionic; detergent and wetting agent used in electrolyte concentrations; paste; HLB 14.0; 100% conc.

Serdox NBSQ 5/5. [Servo Delden BV] Alkyl polyglycol ether; nonionic; foamer and wetting agent; liq.; biodeg.; 100% conc.

Serdox NDI 100. [Servo Delden BV] Dinonylphenol polyglycol ether (100 EO); CAS 9014-93-1; nonionic; intermediate for textile auxs.; HLB 18.5; 100% conc.

Serdox NEL 12/80. [Servo Delden BV] C12–14 alkyl polyglycol ether (12 EO); nonionic; emulsifier for fats, oils, and waxes; stabilizer for syn. latices; liq.; HLB 14.5; biodeg.; 100% conc.

Serdox NES 8/85. [Servo Delden BV] C12–15 alkyl polyglycol ether (8 EO); nonionic; scouring agent for textiles; soaking agent in leather industry; detergent for household and industrial purposes; solid; HLB 12.5; biodeg.; 85% conc.

Serdox NJAD 15. [Servo Delden BV] Tallow amine polyglycol ether, dist. (15 EO); nonionic; emulsifier and dispersant in acid and neutral formulations, oil and wax emulsifier; wetting agent in metal cleaning compds.; antistat; solid; 100% conc.

Serdox NJAD 20. [Servo Delden BV] Tallow amine polyglycol ether, dist. (20 EO); nonionic; emulsifier and dispersant in acid and neutral formulations, oil and wax emulsifier; wetting agent in metal cleaning compds.; antistat; solid; 100% conc.

Serdox NJAD 30. [Servo Delden BV] Tallow amine polyglycol ether, dist. (30 EO); nonionic; emulsifier and dispersant in acid and neutral formulations, oil and wax emulsifier; wetting agent in metal cleaning compds.; antistat; solid; 100% conc.

Serdox NNP 4. [Servo Delden BV] Nonoxynol-4; CAS 9016-45-9; nonionic; detergent; dispersant; emulsifier for insecticides and herbicides; liq.; oil-sol.; HLB 9.0; biodeg.; 100% conc.

Serdox NNP 5. [Servo Delden BV] Nonoxynol-5; CAS 9016-45-9; nonionic; detergent; emulsifier for insecticides and herbicides; liq.; HLB 10.0; biodeg.; 100% conc.

Serdox NNP 6. [Servo Delden BV] Nonoxynol-6; CAS 9016-45-9; nonionic; detergent; emulsifier for insecticides and herbicides; liq.; oil-sol.; HLB 11.0; biodeg.; 100% conc.

Serdox NNP 7. [Servo Delden BV] Nonoxynol-7; CAS 9016-45-9; nonionic; detergent; emulsifier for insecticides and herbicides; liq.; oil-sol.; HLB 12.0; biodeg.; 100% conc.

Serdox NNP 8.5. [Servo Delden BV] Nonoxynol-8 (8.5 EO); CAS 9016-45-9; nonionic; scouring agent for textiles; soaking assistant in the leather industry; detergent for household and industrial purposes; emulsifier for insecticides and herbicides; plasticizer for mortar and concrete; liq.; HLB 12.5; 100% conc.

Serdox NNP 10. [Servo Delden BV] Nonoxynol-10; CAS 9016-45-9; nonionic; used in textiles; rinsing agent for washing paper machine felts; rewetting agent for paper towels and tissues; liq.; HLB 13.5; biodeg.; 100% conc.

Serdox NNP 11. [Servo Delden BV] Nonoxynol-11; CAS 9016-45-9; nonionic; used in textiles; rinsing agent for washing paper machine felts; rewetting agent for paper towels and tissues; liq.; HLB 14.0; biodeg.; 100% conc.

Serdox NNP 12. [Servo Delden BV] Nonoxynol-12; CAS 9016-45-9; nonionic; scouring agent for textiles; soaking assistant in leather industry; detergent for household and industrial purposes; emulsifier for insecticides and herbicides; liq.; HLB 14.0; 100% conc.

Serdox NNP 13. [Servo Delden BV] Nonoxynol-13; CAS 9016-45-9; nonionic; scouring agent for textiles; soaking assistant in leather industry; detergent for household and industrial purposes; emulsifier for insecticides and herbicides; liq.; HLB 14.5; 100% conc.

Serdox NNP 14. [Servo Delden BV] Nonoxynol-14; CAS 9016-45-9; nonionic; detergent and wetting agent for electrolyte concentrations; emulsifier for fatty acids and waxes; liq.; HLB 14.5; biodeg.; 100% conc.

Serdox NNP 15. [Servo Delden BV] Nonoxynol-15; CAS 9016-45-9; nonionic; detergent and wetting agent for electrolyte concentrations; emulsifier for fatty acids and waxes; liq.; HLB 15.0; biodeg.; 100% conc.

Serdox NNP 25. [Servo Delden BV] Nonoxynol-25; CAS 9016-45-9; nonionic; emulsifier for emulsion polymerization; solid; HLB 16.5; 100% conc.

Serdox NNP 30/70. [Servo Delden BV] Nonoxynol-30; CAS 9016-45-9; nonionic; dyeing assistant; lime soap dispersant; emulsifier and stabilizer for emulsion polymerization; liq.; HLB 17.0; 70% conc.

Serdox NNPQ 7/11. [Servo Delden BV] Alkylphenol polyglycol ether; nonionic; wetting agent; liq.; 100% conc.

Serdox NOG 200, 400. [Servo Delden BV] Fatty acid, ethoxylated; nonionic; emulsifier for fats and oils; water-sol.; 100% conc.

Serdox NOG 440. [Servo Delden BV] PEG-10 oleate; CAS 9004-96-0; nonionic; emulsifier for crude and veg. oils; liq.; biodeg.; 100% conc.

Serdox NOL 2. [Servo Delden BV] Oleth-2; CAS 9004-98-2; nonionic; co-emulsifier for min. oils and non-polar solvs.; liq.; HLB 5.0; biodeg.; 100% conc.

Serdox NOL 8. [Servo Delden BV] Oleth-8; CAS 9004-98-2; nonionic; co-emulsifier for min. oils and non-polar solvs.; liq.; HLB 11.5; biodeg.; 100% conc.

Serdox NOL 15. [Servo Delden BV] Oleth-15; CAS 9004-98-2; nonionic; co-emulsifier for min. oils and non-polar solvs.; paste; HLB 14.5; biodeg.; 100% conc.

Serdox NOP 9. [Servo Delden BV] Octoxynol-9; CAS 9002-93-1; nonionic; used in textiles; rinsing agent for washing paper machine felts; rewetting agent for paper towels and tissues; liq.; HLB 13.0; 100%

conc.

Serdox NOP 30/70. [Servo Delden BV] Octoxynol-30; CAS 9002-93-1; nonionic; emulsifier and stabilizer for emulsion polymerization; liq.; HLB 17.5; 70% conc.

Serdox NOP 40/70. [Servo Delden BV] Octoxynol-40; CAS 9002-93-1; nonionic; emulsifier and stabilizer for emulsion polymerization; liq.; HLB 18.0; 70% conc.

Serdox NSG 400. [Servo Delden BV] PEG-9 stearate; CAS 9004-99-3; nonionic; biodeg. surfactant for textile processing and cosmetic emulsions; antistat for plastics; solid; water-disp.; 100% conc.

Serica 300. [Tokai Seiyu Ind.] Mixt. of refined veg., min. oil and modified fatty alcohol; nonionic; soaking agent for raw silk; liq.; 100% conc.

Serica 830. [Tokai Seiyu Ind.] Mixt. of refined veg., min. oil and modified fatty alcohol; nonionic; soaking agent for raw silk; paste; 100% conc.

Sermul EA 30. [Servo Delden BV] Sodium laureth-3 sulfate; CAS 9004-82-4; 68891-38-3; anionic; emulsifier for emulsion polymerization; liq.; 27% conc.

Sermul EA 54. [Servo Delden BV] Sodium nonoxynol-4 sulfate; CAS 9014-90-8; anionic; emulsifier for emulsion polymerization; liq.; 31% conc.

Sermul EA 88. [Servo Delden BV] Calcium dodecylbenzene sulfonate; CAS 26264-06-2; anionic; emulsifier for pesticide formulations; biodeg.; liq.; 65% conc.

Sermul EA 129. [Servo Delden BV] Ammonium lauryl sulfate; CAS 2235-54-3; anionic; emulsifier for emulsion polymerization; biodeg.; liq.; 27% conc.

Sermul EA 136. [Servo Delden BV] Nonoxynol-15 phosphate; anionic; emulsifier for emulsion polymerization; liq.; 100% conc.

Sermul EA 139. [Servo Delden BV] 2-Ethylhexyl polyglycol ether (3 EO) phosphate, acid form; anionic; emulsifier for emulsion polymerization; biodeg.; liq.; 100% conc.

Sermul EA 146. [Servo Delden BV] Sodium nonylphenol polyglycol ether (15 EO) sulfate; CAS 9014-90-8; anionic; emulsifier for emulsion polymerization; liq.; 35% conc.

Sermul EA 150. [Servo Delden BV] Sodium lauryl sulfate; CAS 151-21-3; anionic; emulsifier for emulsion polymerization; biodeg.; paste; 41% conc.

Sermul EA 151. [Servo Delden BV] Sodium nonoxynol-10 sulfate; CAS 9014-90-8; anionic; emulsifier for emulsion polymerization; liq.; 35% conc.

Sermul EA 152. [Servo Delden BV] Ammonium nonoxynol-10 sulfate; anionic; emulsifier for emulsion polymerization; liq.; 35% conc.

Sermul EA 176. [Servo Delden BV] Sodium nonoxynol-10 sulfosuccinate; anionic; emulsifier for emulsion polymerization; biodeg.; liq.; 35% conc.

Sermul EA 188. [Servo Delden BV] Nonoxynol-10 phosphate, acid form; anionic; emulsifier for emulsion polymerization; liq.; 100% conc.

Sermul EA 205. [Servo Delden BV] Nonoxynol-50 phosphate, acid form; anionic; emulsifier for emulsion polymerization; liq.; 100% conc.

Sermul EA 211. [Servo Delden BV] Nonoxynol-6 phosphate; anionic; emulsifier for emulsion polymerization; liq.; 100% conc.

Sermul EA 214. [Servo Delden BV] Sodium alpha olefin sulfonate; CAS 68439-57-6; anionic; emulsifier for emulsion polymerization; liq.; 37% conc.

Sermul EA 221. [Servo Delden BV] Sodium C12–15 pareth-3 sulfosuccinate; anionic; emulsifier for emulsion polymerization; liq.; 40% conc.

Sermul EA 224. [Servo Delden BV] Oleyl polyglycol ether phosphate; anionic; emulsifier for emulsion polymerization; liq.; 100% conc.

Sermul EA 242. [Servo Delden BV] Dodecylbenzene sulfonate; anionic; emulsifier for emulsion polymerization; liq.; 100% conc.

Sermul EA 370. [Servo Delden BV] Sodium laureth sulfate; CAS 9004-82-4; anionic; emulsifier for emulsion polymerization; liq.; 28% conc.

Sermul EK 330. [Servo Delden BV] N-Talloil aminopropyl N,N-dimethylamine; cationic; emulsifier for natural resins, min. and natural oils; intermediate; liq.; 100% conc.

Sermul EN 3. [Servo Delden BV] Nonoxynol-3; CAS 9016-45-9; nonionic; emulsifier for min. oils and solvs.; liq.; 100% conc.

Sermul EN 7. [Servo Delden BV] Surfactant; anionic; emulsifier for wh. spirit, turpentine; biodeg.; liq.; 100% conc.

Sermul EN 15. [Servo Delden BV] Nonoxynol-15; CAS 9016-45-9; nonionic; emulsifier for natural resins; liq.; 100% conc.

Sermul EN 20/70. [Servo Delden BV] Nonoxynol-20; CAS 9016-45-9; nonionic; post stabilizer for emulsion polymerization; solid; 70% conc.

Sermul EN 25. [Servo Delden BV] Nonionic; emulsifier for veg. and animal oils, in pesticide formulations; biodeg.; liq.; 100% conc.

Sermul EN 26. [Servo Delden BV] Nonionic; emulsifier for veg. and animal oils, in pesticide formulations; biodeg.; liq.; 100% conc.

Sermul EN 30/70. [Servo Delden BV] Nonoxynol-30; CAS 9016-45-9; nonionic; post stabilizer for emulsion polymerization; solid; 70% conc.

Sermul EN 134. [Servo Delden BV] Fatty amine polyglycol ether; cationic; emulsifier for natural and min. oils and solvs.; liq.; 100% conc.

Sermul EN 145. [Servo Delden BV] Nonoxynol-30; CAS 9016-45-9; nonionic; post-stabilizer for emulsion polymerization; liq.; 70% conc.

Sermul EN 155. [Servo Delden BV] Alkyl polyglycol ether; nonionic; emulsifier for emulsion polymerization; liq.; 100% conc.

Sermul EN 229. [Servo Delden BV] Nonoxynol-12; CAS 9016-45-9; nonionic; emulsifier for natural resins; liq.; 100% conc.

Sermul EN 237. [Servo Delden BV] Alkyl polyglycol ether; nonionic; emulsifier for emulsion polymerization; liq.; 100% conc.

Sermul EN 259. [Servo Delden BV] Surfactant; nonionic; emulsifier for solvs. and min. oils; liq.; 95% conc.

Sermul EN 312. [Servo Delden BV] Alkyl polyglycol ether; nonionic; emulsifier for emulsion polymerization; liq.; 100% conc.

Sermulen 56. [Servo Delden BV] Oleyl polyglycol ether; nonionic; coemulsifier for min. and veg. oils; liq.; 100% conc.

Servamine KAC 412. [Servo Delden BV] N-Coco-N,N,N-trimethyl ammonium chloride; antistat and lubricant for syn. fibers; bactericide, fungicide, disinfectant, sanitizer; liq.; 50% conc.

Servamine KAC 422. [Servo Delden BV] N-Coco-N,N-dimethyl-N-benzyl ammonium chloride; germicide, disinfectant, sanitizer; liq.; 50% conc.

Servamine KEP 4527. [Servo Delden BV] N (Palmityl amido propyl) N,N,N trimethyl ammonium chloride; cationic; emulsifier with bactericide properties; liq.; biodeg.; 50% conc.

Servamine KET 4542. [Servo Delden BV] Tall oil amidopropyl dimethylethyl ammonium ethosulfate; cationic; emulsifier; liq.; sol. in water, IPA, wh. spirits; biodeg.; 50% conc.

Servamine KOO 330. [Servo Delden BV] Aminoethyl imidazoline; corrosion inhibitor, intermediate, emulsifier; sol. in oil, xylene, IPA, kerosene; disp. in water.

Servamine KOO 330 B. [Servo Delden BV] Oleyl amido ethyl oleyl imidazoline; CAS 68310-76-9; cationic; basic material in mfg. of quat. imidazoline; biodeg.; liq.; 100% conc.

Servirox OEG 45. [Servo Delden BV] PEG-17 castor oil; nonionic; used in textile, phytopharmaceutical, leather, paper and cosmetic industries; liq.; 100% conc.

Servirox OEG 55. [Servo Delden BV] PEG-26 castor oil; nonionic; used in textile, phytopharmaceutical, leather, paper and cosmetic industries; liq.; 100% conc.

Servirox OEG 65. [Servo Delden BV] PEG-32 castor oil; nonionic; used in textile, phytopharmaceutical, leather, paper and cosmetic industries; liq.; 100% conc.

Servirox OEG 90/50. [Servo Delden BV] PEG-180 castor oil; nonionic; used in textile, phytopharmaceutical, leather, paper and cosmetic industries; liq.; 50% conc.

Servo CK 492, CK 494. [Servo Delden BV] Org. quat. compd.; algicide, fungicide, bactericide for oil industry; sol. in water, IPA, xylene.

Servo CK 601. [Servo Delden BV] Org. compds.; algicide, fungicide, bactericide for oil industry; sol. in water, IPA, xylene.

Servoxyl VLA 2170. [Servo Delden BV] Sodium di-2-ethylhexyl sulfosuccinate; anionic; wetting and rewetting agent; liq.; biodeg.; 70% conc.

Servoxyl VLB 1123. [Servo Delden BV] Sodium monoalkyl polyglycol ether sulfosuccinate; anionic; raw material for personal care prods.; cleaning agent; liq.; biodeg.; 40% conc.

Servoxyl VPGZ 7/100. [Servo Delden BV] Cetyl oleyl ether phosphate, acid form (7 EO); anionic; detergent in drycleaning, emulsifier in formulation of metal cleaners in emulsion polymerization; pesticide and cosmetic preparations; liq.; 100% conc.

Servoxyl VPIZ 100. [Servo Delden BV] Acid butyl phosphate; detergent in drycleaning, emulsifier in formulation of metal cleaners in emulsion polymerization; pesticide and cosmetic preparations; liq.; 100% conc.

Servoxyl VPNZ 10/100. [Servo Delden BV] Nonylphenol ether phosphate, acid form (10 EO); anionic; detergent in drycleaning, emulsifier in formulation of metal cleaners in emulsion polymerization; pesticide and cosmetic preparations; liq.; 100% conc.

Servoxyl VPQZ 9/100. [Servo Delden BV] Dinonylphenol polyglycol ether phosphate, acid form (9 EO); anionic; detergent in drycleaning, emulsifier in formulation of metal cleaners in emulsion polymerization; pesticide and cosmetic preparations; liq.; 100% conc.

Servoxyl VPTZ 3/100. [Servo Delden BV] 2-Ethylhexyl polyglycol ether phosphate, acid form; anionic; detergent in drycleaning, emulsifier in formulation of metal cleaners in emulsion polymerization; pesticide and cosmetic preparations; liq.; 100% conc.

Servoxyl VPTZ 100. [Servo Delden BV] 2-Ethylhexyl phosphate, acid form; anionic; detergent in drycleaning, emulsifier in formulation of metal cleaners in emulsion polymerization; pesticide and cosmetic preparations; liq.; 100% conc.

Servoxyl VPUZ. [Servo Delden BV] Acid methyl phosphate; detergent in drycleaning, emulsifier in formulation of metal cleaners in emulsion polymerization; pesticide and cosmetic preparations; liq.; 100% conc.

Servoxyl VPYZ 500. [Servo Delden BV] Acid phosphate ester, modified; anionic; detergent in drycleaning, emulsifier in formulation of metal cleaners in emulsion polymerization; pesticide and cosmetic preparations; liq.; 100% conc.

Serwet WH 170. [Servo Delden BV] Sodium di-2-ethylhexyl sulfosuccinate; anionic; wetting agent; dispersant for pigments; sol. in water and alcohol; 65% conc.

Setacin 103 Spezial. [Zschimmer & Schwarz] Disodium laurethsulfosuccinate; anionic; detergent for personal care prods.; cleaning agent; liq.; 40% conc.

Setacin F Spezial Paste. [Zschimmer & Schwarz] Disodium lauryl sulfosuccinate; anionic; detergent, cosmetics, washing and cleaning agent; paste; 40% conc.

Setacin M. [Zschimmer & Schwarz] Semisulfosuccinate of ethoxylated fatty alcohols/fatty acid derivs., neutralized; anionic; detergent, personal care prods. and cleaning agents; yel. liq.; sol. in water, alcohol; pH 6-7; 42% act. in water.

Sevefilm 20. [Stepan Europe] Surfactant blend; cationic; film-forming, anticorrosive, antiredeposition agent for water cooling circuit treatment prods.; milky liq.; 20% act.

Sevestat ML 300. [Stepan Europe] Surfactant blend; nonionic; lubricant, antistat, emulsifier for textile lubricants for min. and syn. fibers; yel. amber liq.; 42% act.

Sevestat NDE. [Stepan Europe] Phosphoric ester/nonionic blend; nonionic; antistat for syn. fibers; visc. liq.; 100% act.

SF18. [GE Silicones] Dimethicone fluid; lubricant, antifoam, mold release; rubber and plastic lubricant; base fluid for grease; mold release for rubber, plastic, and food applic.; antifoam in food applic. and aq. defoaming formulations; FDA compliance; water-wh. clear oily liq., tasteless, odorless; disp. in org. solvs.; sp.gr. 0.973; dens. 8.0 lb/gal; visc. 350 cs; pour pt. –58 F; flash pt. (PMCC) 204 C; ref. index 1.4030.

SF81-50. [GE Silicones] Dimethyl silicone; defoamers, release agents, in cosmetics, polishes, paint additives, and mechanical devices; lubricant in rubber or plastic-to-metal applics.; for damping and heat transfer in mechanical/elec. applics.; water-wh. clear, oily fluid; sp.gr. 0.972; visc. 50 cstk; pour pt. -120 F; ref. index 1.4030; flash pt. 460 F; surf. tens. 21.0 dynes/cm; conduct. 0.087 Btu/h-°F ft²/ft; sp. heat 0.36 Btu/lb/°F; dissip. factor 0.0001; dielec. str. 35.0 kV; dielec. const. 2.74; vol. resist. 1×10^{14} ohm-cm.

SF96®. [GE Silicones] Polydimethylsiloxane; nonionic; emollient, lubricant for polishes, antifoams, textiles, chemical specialties; plastic and rubber lubrication; dampening or heat transfer fluids; oil defoamer, paint additives; mold release for tires, rubber, plastics; textile softener/modifier; water-wh. liq.; sol. in aliphatic, aromatic, and chlorinated hydrocarbons, alcohols, and ketones, higher hydrocarbons; sp.gr. 0.916–0.974; dens. 8.0 lb/gal; visc.

50–1000 cs; pour pt. –120 to –58 F; flash pt. (PMCC) 204 C; ref. index 1.3970–1.4035; 100% act.

Silicone AF-10 FG. [Harcros] Polydimethylsiloxane; nonionic; emulsifier; antifoamer for aq. systems; food applic.; chemical processing (adhesive, ink, and soap mfg., latex and starch processing); textiles and paper, leather finishing, metal working; wh. liq. lt. cream; water-disp.; sp.gr. 1.0; pH 4–5; 10% act.

Silicone AF-10 IND. [Harcros] Polydimethylsiloxane; nonionic; emulsifier; antifoamer for aq. systems; used in water-based industrial processes, coolants, detergents, insecticides, abrasive slurries, cutting oils, adhesive and ink mfg., latex and starch processing, pulp slurries; wh. liq. lt. cream; water disp.; sp.gr. 1.0; pH 4–5.

Silicone AF-30 FG. [Harcros] Polydimethylsiloxane; nonionic; emulsifier; antifoamer for food and latex processing, adhesive, water-base ink, paint and soap mfg., dyeing and sizing, and boiler feed water; wh. creamy liq.; water-disp.; sp.gr. 1.01; pH 4–5; 30% act.

Silicone AF-30 IND. [Harcros] Polydimethylsiloxane; nonionic; emulsifier; antifoamer for antifreeze, cutting oils, industrial soaps, adhesive, and ink mfg.; latex and starch processing, sizes, insecticides, and metal working; wh. creamy liq.; water-disp.; sp.gr. 1.01; pH 4–5; 30% act.

Silicone AF-100 FG. [Harcros] Compd. silicone; antifoamer for aq. and nonaq. systems; used in mfg. of food pkg. materials and direct food additive; adhesive, ink, paint, and corn oil mfg., resin polymerization; translucent syrupy fluid; sol. in aliphatic, aromatic, and chlorinated solvs.; sp.gr. 1.01; flash pt. 316 C min. (OC); 100% act.

Silicone AF-100 IND. [Harcros] Compd. silicone; antifoamer for aq. and nonaq. systems; used in adhesive, paint, and ink mfg., sizes, industrial cooking processes, vacuum distillations, insecticides, deasphalting, and resin polymerization; translucent syrupy fluid; sol. see Silicone AF-100 FG; sp.gr. 1.01; flash pt. 316 C min. (OC); 100% act.

Silicone Defoamer #5037. [Polymer Research Corp. of Am.] Emulsion based on polymeric silicone fluids; defoamer for aq. foaming; used in fermentation applic.; wh. emulsion; water-disp.; sp.gr. 0.980; dens. 8.2 lb/gal; 15% act.

Silicopearl SR. [Tokai Seiyu Ind.] Mixt. of silicone, neutral oil, and nonionic active agents; nonionic; lubricant for syn. fibers; liq.

Silgen® MA. [BASF] Wetting agent for min. applic. techniques.

Silvatol® AS Conc. [Ciba-Geigy/Dyestuffs] Organic phosphate ester blend; anionic; high detergent scouring agent; afterscouring agent for cellulosics dyed with fiber reactive dyes; alkaline stable; liq.

Silvatol® PBS. [Ciba-Geigy/Dyestuffs] Surfactant/stabilizer blend; anionic; wetting agent, detergent, stabilizer for peroxide bleaching; liq.

Silvatol® SO. [Ciba-Geigy/Dyestuffs] Proprietary blend; anionic; scouring agent for syn. fibers for removal of oils, grease, graphite; liq.

Silwet® L-77. [Union Carbide] Polyalkylene oxide-modified polymethylsiloxane; nonionic; surfactant, flow/leveling agent, antistat, antifog, dispersant, wetting agent, flotation agent, spreading agent for coatings, printing inks, adhesives, agric., automotive, cleaners, antifogging agent, mining, paper, pharmaceutical applics.; pale amber clear liq.; m.w. 600; sol. in methanol, IPA, acetone, xylene, methylene chloride; disp. in water; sp.gr. 1.007; dens. 8.37 lb/gal; visc. 20 cSt; HLB 5–8; cloud pt. < 10 C (0.1%); flash pt. (PMCC) 240 F; pour pt. 35 F; surf. tens. 20.5 dyne/cm (0.1% aq.); Draves wetting 8 s (0.1%); Ross-Miles foam 33 mm (0.1%, initial); 100% act.

Silwet® L-720. [Union Carbide] Dimethicone copolyol; nonionic; surfactant; anticaking agent; slip additive for paper; also for pharmaceutical use, printing inks; colorless clear liq.; m.w. 12,000; sol. in water, methanol, IPA, acetone, xylene, methylene chloride; sp.gr. 1.039; dens. 8.64 lb/gal; visc. 1100 cSt; HLB 9–12; cloud pt. 42 C (1%); flash pt. (PMCC) 205 F; pour pt. –30 F; surf. tens. 29.3 dyne/cm (0.1% aq.); Draves wetting > 300 s (0.1%); Ross-Miles foam 43 mm (0.1%, initial); 50% act.

Silwet® L-722. [Union Carbide] Dimethicone copolyol; nonionic; low surf. tension and high wetting surfactant, lubricant, antistat, dispersant, emulsifier; liq.; HLB 5-8.

Silwet® L-7001. [Union Carbide] Dimethicone copolyol; nonionic; surfactant, dispersant, emulsifier, leveling/flow control agent, antifog, lubricant, antiblock, slip additive for adhesives, agric., automotive, coatings, printing inks, textiles, household specialties, cutting fluids,; petrol. extraction, paper, personal care prods., plastics and rubber; pale yel. clear liq.; m.w. 20,000; sol. in water, methanol, IPA, acetone, xylene, methylene chloride; sp.gr. 1.023; dens. 8.50 lb/gal; visc. 1700 cSt; HLB 9–12; cloud pt. 39 C (1%); flash pt. (PMCC) 206 F; pour pt. –55 F; surf. tens. 30.5 dyne/cm (0.1% aq.); Draves wetting > 300 s (0.25%); Ross-Miles foam 46 mm (0.1%, initial); 75% act.

Silwet® L-7002. [Union Carbide] Dimethicone copolyol; nonionic; low surf. tension, high wetting surfactant, lubricant, antistat, dispersant, emulsifier; liq.; HLB 5-8; 100% conc.

Silwet® L-7004. [Union Carbide] Dimethicone copolyol; nonionic.

Silwet® L-7200. [Union Carbide] Silicone glycol copolymer; nonionic; flow, wetting, slip, dispersion, gloss agent, emulsifier for industrial coatings, household and institutional prods., textiles, inks; liq.

Silwet® L-7210. [Union Carbide] Silicone glycol copolymer; nonionic; flow, wetting, slip, dispersion, gloss agent, emulsifier for industrial coatings, household and institutional prods., textiles, inks; liq.

Silwet® L-7230. [Union Carbide] Silicone glycol copolymer; nonionic; flow, wetting, slip, dispersion, gloss agent, emulsifier for industrial coatings, household and institutional prods., textiles, inks; liq.

Silwet® L-7500. [Union Carbide] Dimethicone copolyol; nonionic; surfactant, antifoam, dispersant, emulsifier, leveling and flow control agent, lubricant, slip additive for adhesives, automotive, chemical processing, coatings, petrol. extraction, paper, personal care prods.,; plastics and rubber, pharmaceutical, textile applics.; lt. yel. clear liq.; m.w. 3000; sol. in methanol, IPA, acetone, xylene, hexanes, methylene chloride; sp.gr.0.982; dens. 8.16 lb/gal; visc.140 cSt; HLB 5–8; flash pt. (PMCC) 250 F; pour pt. –45 F; 100% act.

Silwet® L-7600. [Union Carbide] Dimethicone copolyol; nonionic; surfactant, wetting agent for adhesives, window cleaners, textiles, personal care prods.; internal lubricant for plastics and rubber; lt. amber clear liq.; m.w. 4000; sol. in water, methanol, IPA, acetone, xylene, methylene chloride; sp.gr. 1.066; dens. 8.86 lb/gal; visc. 110 cSt; HLB 13–17; cloud pt. 64 C(1%); flash pt. (PMCC) 165 F; pour pt.35 F; surf. tens. 25.1 dyne/cm; Draves wetting 131 s (0.1%); Ross-Miles foam 94 mm (0.1%, initial); 100% act.

Silwet® L-7602. [Union Carbide] Dimethicone copolyol; nonionic; surfactant, defoamer, dispersant, emulsifier, leveling and flow control agent, gloss agent, lubricant, release agent, antiblock and slip additive, wetting agent for adhesives, agric., automotive specialties, chemical processing,; coatings, petrol. extraction, skin care prods., urethane bubble release, pharmaceutical, printing inks, textiles; pale yel. clear liq.; m.w. 3000; sol. in methanol, IPA, acetone, xylene, methylene chloride; disp. in water; sp.gr. 1.027; dens. 8.54 lb/gal; visc. 100 cSt; HLB 5–8; flash pt. (PMCC) 260 F; pour pt. 5 F; surf. tens. 26.6 dyne/cm; 100% act.

Silwet® L-7604. [Union Carbide] Dimethicone copolyol; nonionic; surfactant, dispersant, emulsifier, wetting agent, flotation agent, spreading agent for automotive specialties, coatings, window cleaners, mining, personal care prods., textiles; pale yel. clear liq.; m.w. 4000; sol. in water, methanol, IPA, acetone, xylene, methylene chloride; sp.gr. 1.063; dens. 8.84 lb/gal; visc.420 cSt; HLB 13–17; cloud pt. 50 C (1%); flash pt. (PMCC) 175 F; pour pt. 30 F; surf. tens. 25.4 dyne/cm; Draves wetting 210 s (0.1%); Ross-Miles foam 71 mm (0.1%, initial); 100% act.

Silwet® L-7605. [Union Carbide] Dimethicone copolyol; nonionic; defoamer, slip additive for chemical processing, coatings; lt. tan waxy solid, melts to clear amber liq. at 32 C; m.w. 6000; sol. in water, methanol, IPA, acetone, xylene, methylene chloride; sp.gr. 1.068 (35/25 C); dens. 8.88 lb/gal; visc. 210 cSt; HLB 13–17; cloud pt. 93 F (1%); flash pt. (PMCC) 280 F; pour pt.90 F; surf. tens. 30.2 dyne/cm; Draves wetting 240 s (1%); Ross-Miles foam 80 mm (0.1%, initial); 100% act.

Silwet® L-7607. [Union Carbide] Polyalkylene oxide-modified polymethylsiloxane; nonionic; surfactant, wetting agent, leveling and flow control agent, grease cleaner, flotation and spreading agent for adhesives, agric., automotive specialties, chemical processing, carpet antistat, mining, metal processing,; petrol. extraction, printing inks, textiles; lt. amber clear liq.; m.w. 1000; sol. in water, methanol, IPA, acetone, xylene, methylene chloride; sp.gr. 1.050; dens. 8.73 lb/gal; visc. 50 cSt; HLB 13–17; cloud pt. 58 C (1%); flash pt. (PMCC) 200 F; pour pt. 10 F; surf. tens. 23.4 dyne/cm; Draves wetting 7 s (0.1%); Ross-Miles foam 109 mm (0.1%, initial); 100% act.

Silwet® L-7614. [Union Carbide] Dimethicone copolyol; nonionic; low surf. tension and high wetting surfactant, lubricant, antistat, dispersant, emulsifier; for personal care prods.; pale yel. clear liq.; sol. in water, methanol, IPA, acetone, xylene, methylene chloride; m.w. 5000; sp.gr. 1.084; dens. 9.03 lb/ga; visc. 480 cSt; HLB 5-8; pour pt. 55 F; cloud pt. 88 C (1% aq.); flash pt. (PMCC) 174 F; surf. tens. 27.6 dynes/cm (0.1% aq.); Draves wetting > 300 s (0.1%); Ross-Miles foam 84 mm (0.1%, initial); 100% act.

Silwet® L-7622. [Union Carbide] Silicone glycol copolymer; nonionic; flow, wetting, slip, dispersion, gloss agent, emulsifier for industrial coatings, household and institutional compds., textiles, inks; liq.

Simulsol 52. [Seppic] Ceteth-2; nonionic; emulsifier; solid; HLB 5.3; 100% conc.

Simulsol 56. [Seppic] Ceteth-10; nonionic; emulsifier; solid; HLB 12.9; 100% conc.

Simulsol 72. [Seppic] Steareth-2; CAS 9005-00-9; nonionic; emulsifier; solid; HLB 4.9; 100% conc.

Simulsol 76. [Seppic] Steareth-10; CAS 9005-00-9; nonionic; emulsifier; solid; HLB 12.4; 100% conc.

Simulsol 78. [Seppic] Steareth-20; CAS 9005-00-9; nonionic; emulsifier; solid; HLB 15.3; 100% conc.

Simulsol 92. [Seppic] Oleth-2; CAS 9004-98-2; nonionic; emulsifier; liq.; HLB 4.9; 100% conc.

Simulsol 96. [Seppic] Oleth-10; CAS 9004-98-2; nonionic; emulsifier; liq.; HLB 12.4; 100% conc.

Simulsol 98. [Seppic] Oleth-20; CAS 9004-98-2; nonionic; cosmetic emulsifier; solid; HLB 15.3; 100% conc.

Simulsol 220 TM. [Seppic] PEG-200 glyceryl stearate.

Simulsol 1030 NP. [Seppic] Nonoxynol-10.; CAS 9016-45-9; nonionic.

Simulsol 1285. [Seppic] PEG-60 castor oil; solubilizer; liq.

Simulsol A. [Seppic] Alkyl polyethoxy ester; nonionic; emulsifier; liq.; 100% conc.

Simulsol A 686. [Seppic] Alkyl polyethoxy ester; nonionic; emulsifier; powd.; 100% conc.

Simulsol CG. [Seppic] PEG-78 glyceryl cocoate.

Simulsol G 101. [Seppic] Surfactant blend; anionic; foaming agent for plasterboard; liq.; 80% conc.

Simulsol M 45. [Seppic] PEG-8 stearate; CAS 9004-99-3; nonionic; emulsifier; solid; HLB 11.1; 100% conc.

Simulsol M 49. [Seppic] PEG-20 stearate; CAS 9004-99-3; nonionic; emulsifier; solid; HLB 15.0; 100% conc.

Simulsol M 51. [Seppic] PEG-30 stearate; CAS 9004-99-3; nonionic; emulsifier; solid; HLB 16.0; 100% conc.

Simulsol M 52. [Seppic] PEG-40 stearate; CAS 9004-99-3; nonionic; emulsifier; solid; HLB 16.9; 100% conc.

Simulsol M 53. [Seppic] PEG-50 stearate; CAS 9004-99-3; nonionic; emulsifier; solid; HLB 17.9; 100% conc.

Simulsol M 59. [Seppic] PEG-100 stearate; CAS 9004-99-3; nonionic; emulsifier; solid; HLB 18.8; 100% conc.

Simulsol NP 575. [Seppic] Alkylphenyl polyethoxy ether; nonionic; emulsifier; liq.; 100% conc.

Simulsol O. [Seppic] Alkyl polyethoxy ether; nonionic; emulsifier; paste; 100% conc.

Simulsol OL 50. [Seppic] PEG-40 castor oil.

Simulsol P4. [Seppic] Laureth-4; nonionic; emulsifier; liq.; HLB 9.7; 100% conc.

Simulsol PS20. [Seppic] PEG-25 propylene glycol stearate.

Sinochem GMS. [Sino-Japan] Glyceryl monostearate; nonionic; emulsifier; flake; HLB 4.2; 99.5% conc.

Sinocol L. [Sino-Japan] Nonionic/anionic; emulsifier for parathion, methyl parathion, sumicidin, dimethoate, phosphate compds., etc.; liq.; HLB 13.7.

Sinocol MBX. [Sino-Japan] Nonionic/anionic; emulsifier for parathion, methyl parathion, sumicidin, dimethoate, phosphate compds., etc.; liq.; HLB 14.0.

Sinocol PQ2. [Sino-Japan] Cationic/nonionic; penetrant, sticking/spreading agent for paraquat; liq.; HLB 14.1.

Sinocol RP-16. [Sino-Japan] Ethoxylated amine; penetrant, sticking/spreading agent for glyphosate; liq.

Sinol LDA-6. [Sino-Japan] Nonionic/anionic; degreaser, emulsifier, moisture regaining agent for leather treatment; HLB 12.6.

Sinol NPX. [Sino-Japan] Nonionic/anionic; degreaser, emulsifier, moisture regaining agent for leather treatment; HLB 12.1; 95.5% conc.

Sinol T4-1. [Sino-Japan] Nonionic; emulsifier for paraffin; HLB 10.7; 99% conc.

Sinol TWS. [Sino-Japan] Nonionic/anionic; emulsifier and leveling agent.

Sinonate 263B. [Sino-Japan] Calcium alkylbenzene sulfonate in butanol solv.; anionic; emulsifier, anticorrosive agent; 70% conc.

Sinonate 263M. [Sino-Japan] Calcium alkylbenzene sulfonate in methanol solv.; anionic; emulsifier, anticorrosive agent; 70% conc.

Sinonate 290M. [Sino-Japan] Sodium dialkyl sulfosuccinate; anionic; wetting agent, penetrant, dispersant, solubilizer; liq.; 75% conc.

Sinonate 290MH. [Sino-Japan] Sodium dialkyl sulfosuccinate; anionic; wetting agent, penetrant, dispersant, solubilizer; liq.; 70% conc.

Sinonate 960SF. [Sino-Japan] Ethoxylated alkylphenol sulfate; anionic; detergent, foaming agent, wetting agent, penetrant; for emulsion polymerization; paste; 50% conc.

Sinonate 960SN. [Sino-Japan] Ethoxylated alkylphenol sulfate; anionic; detergent, foaming agent, wetting agent, penetrant; for emulsion polymerization; liq.; 30% conc.

Sinopol 20. [Sino-Japan] Sorbitan deriv.; nonionic; emulsifier, solubilizer, dispersant, foam inhibitor, corrosion inhibitor; extreme pressure additive; paste; HLB 8.1; 100% conc.

Sinopol 25. [Sino-Japan] Ethoxylated sorbitan ester; nonionic; emulsifier, solubilizer, dispersant; liq.; 100% conc.

Sinopol 60. [Sino-Japan] Sorbitan deriv.; nonionic; emulsifier, solubilizer, dispersant, foam inhibitor, corrosion inhibitor, EP additive; solid; HLB 6.4; 100% conc.

Sinopol 65. [Sino-Japan] Ethoxylated sorbitan ester; nonionic; emulsifier, solubilizer, dispersant; solid; 100% conc.

Sinopol 80. [Sino-Japan] Sorbitan deriv.; nonionic; emulsifier, solubilizer, dispersant, foam inhibitor, corrosion inhibitor, EP additive; liq.; HLB 6.4; 100% conc.

Sinopol 85. [Sino-Japan] Ethoxylated sorbitan ester; nonionic; emulsifier, solubilizer, dispersant; liq.; 100% conc.

Sinopol 150. [Sino-Japan] POE laurate; nonionic; emulsifier, dispersant, lubricant; liq.; HLB 9.4; 100% conc.

Sinopol 150T. [Sino-Japan] POE laurate; nonionic; emulsifier, dispersant, lubricant; liq.; HLB 13.7; 100% conc.

Sinopol 170. [Sino-Japan] POE oleate; nonionic; mold release, emulsifier, dispersant; liq.; HLB 7.7; 100% conc.

Sinopol 170F. [Sino-Japan] POE oleate; nonionic; mold release, emulsifier, dispersant; liq.; HLB 9.7; 100% conc.

Sinopol 170I. [Sino-Japan] POE oleate; nonionic; dyeing auxiliary; HLB 11.7; 100% conc.

Sinopol 170I2. [Sino-Japan] POE dioleate; nonionic; emulsifier; liq.

Sinopol 170N. [Sino-Japan] POE oleate; nonionic; dyeing auxiliary; HLB 13.7; 100% conc.

Sinopol 170N2. [Sino-Japan] POE dioleate; nonionic; emulsifier; liq.

Sinopol 180I. [Sino-Japan] POE stearate; nonionic; dyeing auxiliary; solid; HLB 11.6.

Sinopol 180I2. [Sino-Japan] POE distearate; nonionic; emulsifier; solid; 100% conc.

Sinopol 180N. [Sino-Japan] POE stearate; nonionic; dyeing auxiliary; solid; HLB 11.7.

Sinopol 180N2. [Sino-Japan] POE distearate; nonionic; emulsifier; solid; 100% conc.

Sinopol 254A. [Sino-Japan] Nonionic/anionic; scouring agent for polyester; liq.

Sinopol 405, 410. [Sino-Japan] POE laurylamine; nonionic; dyeing auxiliary, corrosion inhibitor; liq.; 100% conc.

Sinopol 410ST, 415ST. [Sino-Japan] POE alkylamine; nonionic; dyeing auxiliary; liq.; 100% conc.

Sinopol 420. [Sino-Japan] POE laurylamine; nonionic; dyeing auxiliary, corrosion inhibitor; solid; 100% conc.

Sinopol 430ST. [Sino-Japan] POE alkylamine; nonionic; dyeing auxiliary; paste; 100% conc.

Sinopol 610. [Sino-Japan] Surfactant; nonionic; low foaming emulsifier, solubilizer, detergent; for emulsion polymerization; liq.; HLB 13.6; 100% conc.

Sinopol 623. [Sino-Japan] Surfactant; nonionic; low foaming emulsifier, solubilizer, detergent; for emulsion polymerization; liq.; HLB 16.6; 100% conc.

Sinopol 707. [Sino-Japan] Surfactant; nonionic; low foaming emulsifier for emulsion polymerization; detergent, cleaning agent; liq.; HLB 12.0; 100% conc.

Sinopol 714. [Sino-Japan] Surfactant; nonionic; low foaming emulsifier for emulsion polymerization; detergent, cleaning agent; paste; HLB 15.0; 100% conc.

Sinopol 806. [Sino-Japan] POE octylphenyl ether; nonionic; stabilizer, dispersant, suspending agent; for emulsion polymerization; solid; HLB 17.3; 100% conc.

Sinopol 808. [Sino-Japan] POE octylphenyl ether; nonionic; stabilizer, dispersant, suspending agent; for emulsion polymerization; solid; HLB 17.9; 100% conc.

Sinopol 860. [Sino-Japan] POE octylphenyl ether; nonionic; emulsifier, solubilizer, wetting agent, spreading agent, dispersant, penetrant, foaming agent, antidusting and anticaking props.; textile auxiliaries; liq.; HLB 11.2; 100% conc.

Sinopol 864. [Sino-Japan] POE octylphenyl ether; nonionic; emulsifier, solubilizer, wetting agent, spreading agent, dispersant, penetrant, foaming agent, antidusting and anticaking props.; textile auxiliaries; liq.; HLB 12.6; 100% conc.

Sinopol 864H. [Sino-Japan] POE octylphenyl ether; nonionic; emulsifier, solubilizer, wetting agent, spreading agent, dispersant, penetrant, foaming agent, antidusting and anticaking props.; textile auxiliaries; liq.; HLB 13.2; 100% conc.

Sinopol 865. [Sino-Japan] POE octylphenyl ether; nonionic; emulsifier, solubilizer, wetting agent, spreading agent, dispersant, penetrant, foaming agent, antidusting and anticaking props.; textile auxiliaries; liq.; HLB 13.6; 100% conc.

Sinopol 908. [Sino-Japan] POE nonylphenyl ether; nonionic; stabilizer, dispersant, suspending agent; for emulsion polymerization; flake; HLB 17.8; 100% conc.

Sinopol 910. [Sino-Japan] POE nonylphenyl ether; nonionic; stabilizer, dispersant, suspending agent; for emulsion polymerization; flake; HLB 18.2; 100% conc.

Sinopol 960. [Sino-Japan] POE nonylphenyl ether; nonionic; emulsifier, wetting and spreading agent, dispersant; liq.; HLB 10.9; 100% conc.

Sinopol 964. [Sino-Japan] POE nonylphenyl ether; nonionic; emulsifier, solubilizer, wetting and spreading agent, dispersant, penetrant, foaming agent, detergent, antidusting and anticaking props.; textile auxiliaries; liq.; HLB 12.5; 100% conc.

Sinopol 964H. [Sino-Japan] POE nonylphenyl ether; nonionic; emulsifier, solubilizer, wetting and spreading agent, dispersant, penetrant, foaming agent, detergent, antidusting and anticaking props.; textile auxiliaries; liq.; HLB 12.9; 100% conc.

Sinopol 965. [Sino-Japan] POE nonylphenyl ether; nonionic; emulsifier, solubilizer, wetting and spreading agent, dispersant, penetrant, foaming agent, detergent, antidusting and anticaking props.; textile auxiliaries; liq.; HLB 13.3; 100% conc.

Sinopol 965FH. [Sino-Japan] Propoxylated/ethoxylated alkylphenol; nonionic; low foaming emulsifier, wetting agent, detergent; HLB 12.5.

Sinopol 966FH. [Sino-Japan] Propoxylated/ethoxylated alkylphenol; nonionic; low foaming emulsifier, wetting agent, detergent; HLB 13.0.

Sinopol 1100H. [Sino-Japan] POE lauryl ether; nonionic; emulsifier for silicone, min. oil; liq.; HLB 12.9; 100% conc.

Sinopol 1102. [Sino-Japan] POE lauryl ether; nonionic; emulsifier for silicone, min. oil; liq.; HLB 6.2; 100% conc.

Sinopol 1103. [Sino-Japan] POE lauryl ether; nonionic; emulsifier for silicone, min. oil; liq.; HLB 8.3; 100% conc.

Sinopol 1109. [Sino-Japan] POE lauryl ether; nonionic; emulsifier, wetting agent, detergent, dyeing auxiliary; paste; HLB 13.6; 100% conc.

Sinopol 1112. [Sino-Japan] POE lauryl ether; nonionic; emulsifier, wetting agent, detergent, dyeing auxiliary; solid; HLB 14.6; 100% conc.

Sinopol 1120. [Sino-Japan] POE lauryl ether; nonionic; emulsion stabilizer for latex, suspending agent; solid; HLB 16.5; 100% conc.

Sinopol 1203. [Sino-Japan] POE oleyl/cetyl ether; nonionic; emulsifier for min. oil; dyeing auxiliary; liq.; HLB 6.6; 100% conc.

Sinopol 1207. [Sino-Japan] POE oleyl/cetyl ether; nonionic; emulsifier for min. oil; dyeing auxiliary; liq.; HLB 10.7; 100% conc.

Sinopol 1210. [Sino-Japan] POE oleyl/cetyl ether; nonionic; emulsifier, wetting agent, detergent, dyeing auxiliary; solid; HLB 12.4; 100% conc.

Sinopol 1225. [Sino-Japan] POE oleyl/cetyl ether; nonionic; emulsifier, wetting agent, detergent, dyeing auxiliary; solid; HLB 14.6; 100% conc.

Sinopol 1305. [Sino-Japan] POE tridecyl ether; nonionic; emulsifier for silicone; wetting agent, detergent; dyeing auxiliary; liq.; HLB 10.5; 100% conc.

Sinopol 1310. [Sino-Japan] POE tridecyl ether; nonionic; emulsifier for silicone; wetting agent, detergent; dyeing auxiliary; solid; HLB 13.8; 100% conc.

Sinopol 1510. [Sino-Japan] POE alkyl ether; nonionic; emulsifier, solubilizer, dispersant, suspending agent, lubricant; liq.; HLB 6.6; 100% conc.

Sinopol 1525. [Sino-Japan] POE alkyl ether; nonionic; emulsifier, solubilizer, dispersant, suspending agent, lubricant; liq.; HLB 11.0; 100% conc.

Sinopol 1545. [Sino-Japan] POE alkyl ether; nonionic; emulsifier, solubilizer, dispersant, suspending agent, lubricant; liq.; HLB 13.8; 100% conc.

Sinopol 1610. [Sino-Japan] POE cetyl ether; nonionic; emulsifier for paraffin; dyeing auxiliary; solid; HLB 12.9; 100% conc.

Sinopol 1620. [Sino-Japan] POE cetyl ether; nonionic; emulsifier for wax; latex stabilizer; dyeing auxiliary; solid; HLB 15.7; 100% conc.

Sinopol 1807. [Sino-Japan] POE stearyl ether; nonionic; emulsifier for paraffin; dyeing auxiliary; solid; HLB 10.7; 100% conc.

Sinopol 1830. [Sino-Japan] POE stearyl ether; nonionic; emulsifier for wax; latex stabilizer; dyeing auxiliary; solid; HLB 16.6; 100% conc.

Sinopol NN-15. [Sino-Japan] Surfactant; nonionic; emulsifier, dispersant, suspending agent; paste; HLB 16.0.

Sinoponic PE 61. [Sino-Japan] POP/POE ether; nonionic; low foaming detergent, wetting agent for phosphatizing bath; emulsifier for acrylic polymers; liq.; HLB 3.0; 100% conc.

Sinoponic PE 62. [Sino-Japan] POP/POE ether; nonionic; low foaming detergent, wetting agent for phosphatizing bath; emulsifier for acrylic polymers; liq.; HLB 7.0; 100% conc.

Sinoponic PE 64. [Sino-Japan] POP/POE ether; nonionic; low foaming detergent, wetting agent for phosphatizing bath; emulsifier for acrylic polymers; liq.; HLB 15.0; 100% conc.

Sinoponic PE 68. [Sino-Japan] POP/POE ether; nonionic; low foaming detergent, wetting agent for phosphatizing bath; emulsifier for acrylic polymers; liq.; HLB 29.0; 100% conc.

Sipex® 280. [Rhone-Poulenc Surf.] Ammonium nonoxynol-4 sulfate; high foaming surfactant, wetting agent, dispersant, emulsifier for shampoos, skin cleansers, lt. duty cleaners, emulsion polymerization; liq.; 58% act. DISCONTINUED.

Sipex® BOS (redesignated Rhodapon® BOS). [Rhone-Poulenc Surf.]

Sipex® CAV (redesignated Rhodapon® CAV). [Rhone-Poulenc Surf.]

Sipex® EC-111 (redesignated Rhodapon® EC111). [Rhone-Poulenc Surf.]

Sipex® EST-30 (redesignated Rhodapex® EST-30). [Rhone-Poulenc Surf.]

Sipex® EST-75. [Rhone-Poulenc Surf.] Sodium trideceth sulfate; anionic; emulsifier, wetting agent; personal care and cleaning prods.; FDA compliance; liq.; pH 9.0–10.5 (10%); surf. tens. 33 dynes/cm (@ CMC); 75% solids. DISCONTINUED.

Sipex® NB60. [Rhone-Poulenc Surf.] Sodium alkyl ether sulfate; industrial grade foaming agent for dedusting treatments, air drilling, wallboard foaming, and brine water baths; liq.; 60% act. DISCONTINUED.

Sipex® OLS (redesignated Rhodapon® OLS). [Rhone-Poulenc Surf.]

Sipex® OS (redesignated Rhodapon® OS). [Rhone-Poulenc Surf.]

Sipex® TDS (redesignated Rhodapon® TDS). [Rhone-Poulenc Surf.]

Sipon® 101-10. [Rhone-Poulenc Surf.] Sodium lauryl sulfate; CAS 151-21-3; detergent/emulsifier and foaming agent for shampoos, liq. soaps, bath gels, bubble baths; 30% conc. DISCONTINUED.

Sipon® EA. [Rhone-Poulenc Surf.] Ammonium laureth sulfate; anionic; detergent, wetting agent, foamer, solubilizer, penetrant, personal care prods. and pet shampoos; liq. detergent for fabrics; liq.; mild, pleasant odor; visc. 300 cps; cloud pt. –5 C; pH 6.7 (10%); 27% act. DISCONTINUED.

Sipon® EAY. [Rhone-Poulenc Surf.] Ammonium

laureth sulfate; anionic; detergent, wetting agent, foamer, solubilizer, penetrant, personal care prods. and pet shampoos; liq. detergent for fabrics; thixotropic product; cloud pt. 0 C; pH 6.7 (10%); 26% act. DISCONTINUED.

Sipon® ES (redesignated Rhodapex® ES). [Rhone-Poulenc Surf.]

Sipon® L-22 (redesignated Rhodapon® L-22). [Rhone-Poulenc Surf.]

Sipon® LCP (redesignated Rhodapon® LCP). [Rhone-Poulenc Surf.]

Sipon® LM (redesignated Rhodapon® LM). [Rhone-Poulenc Surf.]

Sipon® SB (redesignated Rhodapon® SB). [Rhone-Poulenc Surf.]

Sipon® UB (redesignated Rhodapon® UB). [Rhone-Poulenc Surf.]

Siponate® 301-10 (redesignated Rhodacal® 301-10). [Rhone-Poulenc Surf.]

Siponate® 301-10F. [Rhone-Poulenc Surf.] Sodium C14–16 olefin sulfonate; anionic; detergent for personal care prods., soaps, and fabrics; emulsion polymerization surfactant; flake; 90% act. DISCONTINUED.

Siponate® 301-10P. [Rhone-Poulenc Surf.] Sodium C14-16 olefin sulfonate; anionic; emulsion polymerization surfactant. DISCONTINUED.

Siponate® 330 (redesignated Rhodacal® 330). [Rhone-Poulenc Surf.]

Siponate® A-246 L (redesignated Rhodacal® A-246 L). [Rhone-Poulenc Surf.]

Siponate® A-246 LX. [Rhone-Poulenc Surf.] Sodium C14-16 olefin sulfonate; detergent, visc. builder; liq.; 40% act. DISCONTINUED.

Siponate® DDB-40 (redesignated Rhodacal® DDB-40). [Rhone-Poulenc Surf.]

Siponate® DS-4 (redesignated Rhodacal® DS-4). [Rhone-Poulenc Surf.]

Siponate® DS-10 (redesignated Rhodacal® DS-10). [Rhone-Poulenc Surf.]

Siponate® DSB (redesignated Rhodacal® DSB). [Rhone-Poulenc Surf.]

Siponate® DSB-85. [Rhone-Poulenc Surf.] Polysulfonate; anionic; wetting agent, dispersant for agric. formulations; EPA compliance; spray-dried; 85% act. DISCONTINUED.

Siponate® LDS-10. [Rhone-Poulenc Surf.] Sodium dodecylbenzene sulfonate; anionic; emulsifier and surfactant for emulsion polymerization; heavy duty household and institutional cleaning prods.; oil recovery operations; washing fruits and vegs.; food pkg. materials; flake; pH 7.5 (10%); surf. tens. 32 dynes/cm (@ CMC); 98% act. DISCONTINUED.

Siponic® 25-3. [Rhone-Poulenc Surf.] POE (3) C12–C15; biodeg. detergent, wetting agent, emulsifier for household and industrial cleaners; coupler and solubilizer for perfumes and org. additives; HLB 7.9; 100% act. DISCONTINUED.

Siponic® 25-9. [Rhone-Poulenc Surf.] POE (9) C12–C15; biodeg. detergent, wetting agent, emulsifier for household and industrial cleaners; coupler and solubilizer for perfumes and org. additives; HLB 13.3; 100% act. DISCONTINUED.

Siponic® 260 (redesignated Alcodet® 260). [Rhone-Poulenc Surf.]

Siponic® E-10. [Rhone-Poulenc Surf.] Ceteareth-20; CAS 68439-49-6; nonionic; emulsifier, lubricant, emollient, and conditioner for personal care prods.; dyeing assistant, leveling agent, emulsion polymerization; HLB 15.3; 100% act. DISCONTINUED.

Siponic® E-2. [Rhone-Poulenc Surf.] Ceteareth-4; CAS 68439-49-6; nonionic; emulsifier, lubricant, emollient, and conditioner for personal care prods.; dyeing assistant, leveling agent, emulsion polymerization; HLB 8.0; 100% act. DISCONTINUED.

Siponic® E-3. [Rhone-Poulenc Surf.] Ceteareth-6; CAS 68439-49-6; nonionic; emulsifier, lubricant, emollient, and conditioner for personal care prods.; dyeing assistant, leveling agent, emulsion polymerization; HLB 10.1; 100% act. DISCONTINUED.

Siponic® E-5. [Rhone-Poulenc Surf.] Ceteareth-10; CAS 68439-49-6; nonionic; rubber emulsifier, lubricant, dye and fulling assistant; conditioner and emollient for personal care prods.; emulsion polymerization; floor waxes, paper finishes, pharmaceuticals; wax; HLB 12.4; cloud pt. 73 C (1%); pH 6.5 (1%); 100% act. DISCONTINUED.

Siponic® E-7. [Rhone-Poulenc Surf.] Ceteareth-14; CAS 68439-49-6; nonionic; rubber emulsifier, lubricant, textile assistant; conditioner and emollient for personal care prods.; emulsion polymerization; floor waxes, paper finishes, pharmaceuticals; HLB 14.1; cloud pt. 93 C (1%); pH 6.5 (1%); 100% act. DISCONTINUED.

Siponic® E-15. [Rhone-Poulenc Surf.] Ceteareth-30; CAS 68439-49-6; nonionic; rubber emulsifier, lubricant, dye and fulling assistant; conditioner and emollient for personal care prods.; emulsion polymerization; floor waxes, paper finishes, pharmaceuticals; wax; HLB 16.7; cloud pt. > 95 C (1%); pH 6.5 (1%); 100% act. DISCONTINUED.

Siponic® F-160. [Rhone-Poulenc Surf.] Octoxynol-16; CAS 9002-93-1; nonionic; primary surfactant in emulsion polymerization of latices; textile scouring and dye coupling applics.; cosmetic prods., household and industrial cleaners; HLB 15.8; 100% act. DISCONTINUED.

Siponic® F-300 (see Igepal® CA-887). [Rhone-Poulenc Surf.] Octoxynol-30; CAS 9002-93-1; nonionic; primary surfactant in emulsion polymerization of latices; textile scouring and dye coupling applics.; cosmetic prods., household and industrial cleaners; FDA compliance; liq.; HLB 17.3; surf. tens. 38 dynes/cm (@ CMC); 70% act.

Siponic® F-400 (see Igepal® CA-897). [Rhone-Poulenc Surf.] Octoxynol-40; CAS 9002-93-1; nonionic; primary surfactant in emulsion polymerization of latices; textile scouring and dye coupling applics.; cosmetic prods., household and industrial cleaners; FDA compliance; liq.; HLB 17.9; surf. tens. 34 dynes/cm (@ CMC); 70% act.

Siponic® F-707. [Rhone-Poulenc Surf.] Octoxynol-70; CAS 9002-93-1; nonionic; primary surfactant in emulsion polymerization of latices; textile scouring and dye coupling applics.; cosmetic prods., household and industrial cleaners; liq.; HLB 18.7; surf. tens. 30 dynes/cm (@ CMC); 70% act. DISCONTINUED.

Siponic® L-1. [Rhone-Poulenc Surf.] Laureth-1; nonionic; thickener in shampoos and bubble baths; stabilizer for emulsion polymers; emulsifier for monomer systems, floor waxes, paper finishes, rubber; emollient for pharmaceuticals; liq.; HLB 3.7 (1%); pH 6.5 (1%); 100% act. DISCONTINUED.

Siponic® L-4 (redesignated Rhodasurf® L-4). [Rhone-Poulenc Surf.]

Siponic® L-7-90 (redesignated Rhodasurf® L-790). [Rhone-Poulenc Surf.]

Siponic® L-12 (see Rhodasurf® LA-12). [Rhone-Poulenc Surf.] Laureth-12; nonionic; emulsifier for personal care prods., rubber and monomer systems; stabilizer for emulsion polymers; floor waxes, paper finishes, emollient for pharmaceuticals; wax; HLB 14.6; cloud pt. 89 C (1%); pH 6.5 (1%); 100% act.

Siponic® L-25 (redesignated Rhodasurf® L-25). [Rhone-Poulenc Surf.]

Siponic® NP-4 (see Igepal® CO-430). [Rhone-Poulenc Surf.] Nonoxynol-4; CAS 9016-45-9; nonionic; emulsifier; detergent, wetting agent, coupler, dispersant; liq.; oil-sol.; HLB 8.1; 100% conc.

Siponic® NP-6 (see Igepal® CO-530). [Rhone-Poulenc Surf.] Nonoxynol-6; CAS 9016-45-9; nonionic; coemulsifier; intermediate for the synthesis of anionic surfactants; liq.; HLB 10.7; 100% conc.

Siponic® NP-7. [Rhone-Poulenc Surf.] Nonoxynol-7; CAS 9016-45-9; nonionic; emulsifier, detergent, wetting agent, dispersant, solubilizer, coupler for cosmetic and industrial applics. and in emulsion polymerization; HLB 11.7; 100% act. DISCONTINUED.

Siponic® NP-8 (see Alkasurf® NP-8). [Rhone-Poulenc Surf.] Nonoxynol-8; CAS 9016-45-9; nonionic; emulsifier, detergent, wetting agent, dispersant, solubilizer, coupler for cosmetic and industrial applics. and in emulsion polymerization; HLB 12.3; 100% act.

Siponic® NP-9 (see Igepal® CO-630). [Rhone-Poulenc Surf.] Nonoxynol-9; CAS 9016-45-9; nonionic; emulsifier, detergent, wetting agent; used in polymerization, paints, textiles, leather, paper, cleaning compds.; oil-sol.; HLB 12.7; 100% conc.

Siponic® NP-9.5 (see Igepal® CO-630). [Rhone-Poulenc Surf.] Nonoxynol-9.5; CAS 9016-45-9; nonionic; emulsifier, detergent, wetting agent, dispersant, solubilizer, coupler for cosmetic and industrial applics. and in emulsion polymerization; HLB 13.2; 100% act.

Siponic® NP-10 (see Igepal® CO-660). [Rhone-Poulenc Surf.] Nonoxynol-10; CAS 9016-45-9; nonionic; emulsifier, detergent, wetting agent, dispersant, solubilizer, coupler for cosmetic and industrial applics. and in emulsion polymerization; HLB 13.6; 100% act.

Siponic® NP-13 (see Igepal® CO-720). [Rhone-Poulenc Surf.] Nonoxynol-13; CAS 9016-45-9; nonionic; emulsifier, detergent, wetting agent, dispersant, solubilizer, coupler for cosmetic and industrial applics. and in emulsion polymerization; HLB 14.4; 100% act.

Siponic® NP-15 (see Igepal® CO-730). [Rhone-Poulenc Surf.] Nonoxynol-15; CAS 9016-45-9; nonionic; emulsifier, detergent, wetting agent; used in polymerization, paints, textiles, leather, paper, cleaning compds.; liq.; HLB 14.9; 100% conc.

Siponic® NP-40 (see Igepal® CO-890). [Rhone-Poulenc Surf.] Nonoxynol-40; CAS 9016-45-9; nonionic; detergent, wetting agent; coemulsifier and vinyl/acrylic latex stabilizer; emulsifier for floor waxes and polishes; liq.; water-sol.; HLB 17.7; 70% conc.

Siponic® NP-407 (see Igepal® CO-897). [Rhone-Poulenc Surf.] Nonoxynol-40; CAS 9016-45-9; nonionic; emulsifier, detergent, wetting agent, dispersant, solubilizer, coupler for cosmetic and industrial applics. and in emulsion polymerization; HLB 17.8; 70% act.

Siponic® SK (redesignated Alcodet® SK). [Rhone-Poulenc Surf.] PEG-8 isolauryl thioether; nonionic; wetting agent, metal cleaning; heavy duty detergent; inhibitor in steel processing; scouring of textiles; agric. formulations; antiskinning agent for paints; hair prods.; wood and paper industry; emulsifier for petrol. oils, chlorinated solvs., silicones, metallic soaps, and emulsion polymers; Gardner 6 max. liq.; sol. in water; dens. 8.5 lb/gal; sp.gr. 1.03; cloud pt. 28 C (1%); surf. tens. 31 dynes/cm (0.05%); 98% act. DISCONTINUED.

Siponic® TD-3 (see Rhodasurf® BC-420). [Rhone-Poulenc Surf.] Trideceth-3; CAS 24938-91-8; nonionic; emulsifier used in textile scouring, household and metal cleaners and industrial applics.; surfactant in cosmetic emulsions; solubilizer for fragrance and oils; HLB 7.9; 100% act.

Siponic® TD-6 (see Rhodasurf® BC-610). [Rhone-Poulenc Surf.] Trideceth-6; CAS 24938-91-8; nonionic; emulsifier, solubilizer used in dry cleaning solvs., agric. sprays; degreaser; metal cleaning; clear oily liq.; oil-sol.; dens. 8.2 lb/gal; sp.gr. 0.98; visc. 80 cps; HLB 11.4; flash pt. 355 F; pH 6 (1%); 99% act.

Siponic® TD-12. [Rhone-Poulenc Surf.] Trideceth-12; CAS 24938-91-8; nonionic; emulsifier used in textile scouring, household and metal cleaners and industrial applics.; surfactant in cosmetic emulsions; solubilizer for fragrance and oils; HLB 14.5; 100% act. DISCONTINUED.

Siponic® Y500-70. [Rhone-Poulenc Surf.] Oleth-25; CAS 9004-98-2; nonionic; wetting agent, emulsifier for polymerization; stabilizer for latices; dispersant in PU foam; textile industry; foamer for latex; floor waxes; pharmaceutical emulsions; metal cleaning; leather industry; food pkg. applics.; liq.; mild odor; visc. 1650 cps; HLB 16.2; cloud pt. 6 C (1%); pH 7.5 (1%); 70% act. DISCONTINUED.

Sipothix® 1941. [Rhone-Poulenc Surf.] Acrylates/steareth-20 methylacrylate copolymer; pH sensitive thickening agent, emulsifier for liq. detergents, shampoos, and cosmetics; optimum thickening at slightly alkaline pH; opaque liq.; 30% solids.

Sipothix® H-65. [Rhone-Poulenc Surf.] Ethylacrylate/methylacrylic acid copolymer; pH sensitive thickening agent, emulsifier for liq. detergents, shampoos, and cosmetics; optimum thickening at slightly alkaline pH; opaque liq.; 30% solids. DISCONTINUED.

Size CB. [BASF] Polyacrylate; CAS 9003-03-6; sizing agent for cellulosic fibers and their blends; water-sol.

Sizing Wax PA, PT, SM. [Hüls AG] PEGs; for prod. of water-sol. sizing auxiliaries; antistats, dyeing auxiliaries; flakes; 100% act.

SK Flot 1. [Kao Corp. SA] Amine base, acetate salt; cationic; min. flotation reagent; solid, paste; 100% conc.

SK Flot 2. [Kao Corp. SA] Amine base, acetate salt; cationic; min. flotation reagent; paste; 100% conc.

SK Flot 3. [Kao Corp. SA] Amine base, acetate salt; cationic; min. flotation reagent; solid; 100% conc.

SK Flot 4. [Kao Corp. SA] Amine base, acetate salt; cationic; min. flotation reagent; solid, flakes; 100% conc.

SK Flot FA 1. [Kao Corp. SA] Amine-based; cationic; min. flotation reagent; liq./solid; 100% conc.

SK Flot FA 2. [Kao Corp. SA] Amine-based; cationic; min. flotation reagent; liq./solid; 100% conc.

SK Flot FA 3. [Kao Corp. SA] Amine-based; cationic; min. flotation reagent; solid/paste; 100% conc.

SK Flot FA 4. [Kao Corp. SA] Amine-based; cationic; min. flotation reagent; solid/flakes; 100% conc.

Skliro Distilled. [Croda Inc.] Lanolin acid; emollient,

superfatting agent for aerosol shave foams, hand soaps; water repellent films; w/o emulsifier; stable emulsions in preparation of waterproof makeup; yel. waxy solid; char. odor; sol. in oils, esters; m.w. 45–60 C; acid no. 140–165; sapon. no. 155–185; usage level: 1-10%.

S-Maz® 20. [PPG/Specialty Chem.] Sorbitan laurate; CAS 1338-39-2; nonionic; lubricant, antistat, textile softener, process defoamer, opacifier, coemulsifier, solubilizer, dispersant, suspending agent, coupler; prepares exc. w/o emulsions; with T-Maz Series used as o/w emulsifiers in cosmetics, food formulations, industrial oils, and household prods.; lipophilic; amber liq.; sol. in ethanol, naphtha, min. oils, toluene, veg. oil; water-disp.; sp.gr. 1.0; visc. 4500 cps; HLB 8.0; acid no. 7 max.; sapon. no. 158–170; hyd. no. 330-358; flash pt. (PMCC) > 350 F; 100% conc.

S-Maz® 40. [PPG/Specialty Chem.] Sorbitan palmitate; CAS 26266-57-9; nonionic; see S-Maz 20; tan flakes; partly sol. in hot ethanol, acetone, toluol, min. and veg. oils; m.p. 45–46 C; HLB 6.5; acid no. 7.5 max.; sapon. no. 140–150; hyd. no. 275-305; flash pt. (PMCC) > 350 F; 100% conc.

S-Maz® 60. [PPG/Specialty Chem.] Sorbitan stearate; CAS 1338-41-6; nonionic; see S-Maz 20; cream flakes; insol. in water, min. oils; m.p. 50–53 C; HLB 4.7; acid no. 10 max.; sapon. no. 147–157; hyd. no. 235-260; flash pt. (PMCC) > 350 F; 100% conc.

S-Maz® 60KHM. [PPG/Specialty Chem.] Sorbitan stearate; CAS 1338-41-6; nonionic; see S-Maz 20; kosher high melt grade; Gardner 3 flake; insol. in water, min. oils; HLB 4.7; acid no. 10 max.; sapon. no. 147-157; hyd. no. 235-260; flash pt. (PMCC) > 350 F.

S-Maz® 65K. [PPG/Specialty Chem.] Sorbitan tristearate; CAS 26658-19-5; nonionic; see S-Maz 20; Gardner 3 flake; sol. in min. spirits, toluene; disp. in min. and veg. oils; HLB 2.2; acid no. 15 max.; sapon. no. 176-188; hyd. no. 66-80; flash pt. (PMCC) > 350 F; 100% conc.

S-Maz® 80. [PPG/Specialty Chem.] Sorbitan oleate; CAS 1338-43-8; nonionic; see S-Maz 20; also as lubricant, rust inhibitor, penetrant in metalworking formulations; amber liq.; sol. in ethanol, naphtha, min. oils, min. spirits, toluene, veg. oil; disp. in water; sp.gr. 1.0; visc. 1000 cps; HLB 4.6; acid no. 7.5 max.; sapon. no. 149–160; hyd. no. 193-209; flash pt. (PMCC) > 350 F; 100% conc.

S-Maz® 80K. [PPG/Specialty Chem.] Sorbitan oleate; CAS 1338-43-8; nonionic; see S-Maz 20; kosher grade; Gardner 6 liq.; sol. in min. oils, min. spirits, toluene, veg. oil; insol. in water; HLB 4.6; acid no. 8 max.; sapon. no. 149-160; hyd. no. 193-209; flash pt. (PMCC) > 350 F.

S-Maz® 83R. [PPG/Specialty Chem.] Sorbitan sesquioleate; CAS 8007-43-0; nonionic; solubilizer, emulsifier and dispersant; Gardner 5 liq.; sol. in oils and solvs.; HLB 4.6; acid no. 12 max.; sapon. no. 145-160; hyd. no. 185-215; flash pt. (PMCC) > 350 F; 100% conc.

S-Maz® 85. [PPG/Specialty Chem.] Sorbitan trioleate; CAS 26266-58-0; nonionic; see S-Maz 20; also as lubricant, rust inhibitor, penetrant in metalworking formulations; amber liq.; sol. in toluol, naphtha, min. and veg. oils; water-disp.; sp.gr. 1.0; visc. 200 cps; HLB 2.1; acid no. 14 max.; sapon. no. 172–186; hyd. no. 56-68; flash pt. (PMCC) > 350 F; 100% conc.

S-Maz® 85K. [PPG/Specialty Chem.] Sorbitan trioleate; CAS 26266-58-0; nonionic; see S-Maz 20; kosher grade; Gardner 6 liq.; sol. in min. oils, min. spirits, toluene, veg. oil; insol. in water; HLB 2.1; acid no. 14 max.; sapon. no. 172-186; hyd. no. 56-68; flash pt. (PMCC) > 350 F.

S-Maz® 90. [PPG/Specialty Chem.] Sorbitan tallate; nonionic; see S-Maz 20; also as lubricant, rust inhibitor, penetrant in metalworking formulations; amber liq.; sol. in min. and veg. oils; water-disp.; sp.gr. 1.0; visc. 800 cps; HLB 4.3; acid no. 10 max.; sapon. no. 145–160; hyd. no. 180-210; flash pt. (PMCC) > 350 F; 100% conc.

S-Maz® 95. [PPG/Specialty Chem.] Sorbitan tritallate; nonionic; antistat, textile softener, lubricant, process defoamer, opacifier, coemulsifier; prepares w/o emulsions; together with T-Maz series, as o/w emulsifier in metalworking fluids and coolants, semi-syn. and oil-based metalworking lubricants; and coolants, cosmetics, food formulations, industrial oils, and household prods.; amber liq.; sol. in toluol, naphtha, min. oil, veg. oil; misc. with ethanol, acetone; disp. in water; sp.gr. 0.9; visc. 200 cps; HLB 1.9; acid no. 15 max.; sapon. no. 168–186; hyd. no. 55-85; flash pt. (PMCC) 350 F.

Smithol 22LD. [Werner G. Smith] Ester; metal and fiber wetting agent; demulsifier; forms continuous monomolecular film with rust prevention props.; stabilizer for chlorinated systems; for drawing, stamping, rolling lubricants; waste water stripping oils; leather treatment; textile spin finishes; Gardner 10 max. color; visc. 100-140 SUS (100 F); acid no. 6-10; iodine no. 100-130; sapon. no. 170-180; hyd. no. 21-35; pour pt. -6 to 6 F; cloud pt. 40 F max.; flash pt. (COC) 450 F min.

Smithol 50. [Werner G. Smith] Syn. diester; lubricant base with metal wetting chars.; Gardner 8 color; sp.gr. 0.892 (60 F); visc. 100-140 SUS (100 F); acid no. 8 max.; iodine no. 6 max.; pour pt. -14 to -18 F; cloud pt. -8 to 10 F; flash pt. (COC) 320 F.

Smoothar DT-130. [Yoshimura Oil Chem.] Mixt. of oil and nonionic surfactant; nonionic; coning oil, lubricant for the double twister; liq.; 100% conc.

Smoothar F828. [Yoshimura Oil Chem.] Mixt. of syn. lubricant, anionic and nonionic surfactants; nonionic; lubricant for dyed cheese or hank before knitting; liq.; 100% conc.

Sochamine A 8955. [Witco SA] Disodium capryloamphodiacetate; CAS 7702-01-4; amphoteric; low foaming and wetting agent in alkaline formulations; liq.; 38% conc.

Sochamine OX 30. [Witco SA] Cocamine oxide; CAS 61788-90-7; cationic; for lt. duty detergent; liq.; 30% conc.

Sodium Octyl Sulfate Powd. [Henkel/Cospha; Henkel Canada] Sodium octyl sulfate; anionic; wetting agent for cleaning prods., reprographic coating baths and electroplating; powd.; 90% conc.

Sofbon C-1-S. [Takemoto Oil & Fat] Quat. ammonium compd.; cationic; softener for polyester jersey; paste.

Sofnon 105G. [Toho Chem. Industry] Imidazoline deriv.; cationic; softener for cotton, wool, and syn. fibers; wax; 80% conc.

Sofnon B-3. [Toho Chem. Industry] Anionic/nonionic; softener for cotton and cellulose fibers; paste.

Sofnon GF-2, GF-5. [Toho Chem. Industry] Anionic; softener for cotton, polyester, sheep wool, silk fibers; liq.

Sofnon HG-180. [Toho Chem. Industry] Alkyl amide imidazoline acetate; cationic; softening agent for glass fibers; liq.

Sofnon LA-75. [Toho Chem. Industry] Imidazoline derivs. and nonionics; nonionic/amphoteric; softening and lubricating agent for cotton and syn. fibers; wax; 75% conc.

Sofnon SP-1000. [Toho Chem. Industry] Anionics/lubricating oil blend; anionic; softening and lubricating agent for cotton and syn. fibers; paste.

Sofnon SP-852R. [Toho Chem. Industry] Silicone/polyethylene coemulsion; nonionic; softener for cotton and cotton/polyester fibers; liq.

Sofnon SP-9400. [Toho Chem. Industry] Silicone emulsion; nonionic; softening and lubricating agent for cotton and syn. fibers; liq.; 40% conc.

Sofnon TK-11. [Toho Chem. Industry] Blend of nonionics, anionics, wool fat, and min. oil; anionic/nonionic; spinning oil for rough wool; liq.; 100% conc.

Softal 300. [Tokai Seiyu Ind.] Polyamine; cationic; softener and lubricant for syn. fibers; liq.; 15% conc.

Softal A-3. [Tokai Seiyu Ind.] Aliphatic amine deriv.; cationic; softener for acrylic fibers; paste; 15% conc.

Softal IT-7. [Toho Chem. Industry] Fatty acid deriv.; nonionic; softening agent for various fibers; liq.; 18% conc.

Softal MR-30. [Tokai Seiyu Ind.] Special polyamide deriv.; cationic/nonionic; water absorptive softener for syn. fibers and blends; liq.

Soft Detergent 60. [Lion] Alkyl benzene sulfonate; anionic; general use detergent; powd.; 60% conc.

Soft Detergent 95. [Lion] alpha-Olefin sulfonate; anionic; emulsifier and dispersant for emulsion polymerization; powd.; 95% conc.

Softenol® 3900. [Hüls Am.] Glyceryl stearate; nonionic; emulsifier, lubricant, antistat, demolding and antiblocking agent, greaser, plasticizer, transparency and pigment dispersant used in mfg. and processing of plastics, textiles, leather, oils and polishing materials; powd.; neutral odor and taste; dens. 0.893 g/cm^3 (80 C); visc. 22.9 mm^2/s (80 C); m.p. 58–65 C; sapon. no. 155–175; biodeg.

Softenol® 3991. [Hüls Am.] Glyceryl stearate; nonionic; lubricant, emulsifier, antistat, pigment dispersant, plasticizer, antiblocking agent for plastics, textile and leather processing aids, drilling and cutting oils, polishes, rubber; biodeg.; powd.; neutral odor and taste; dens. 0.914 g/cm^3 (80 C); visc. 34.5 mm^2/s (80 C); m.p. 65 C; sapon. no. 155–165; biodeg.; 90% 1-monoester.

Softigen® 767. [Hüls Am.; Hüls AG] PEG-6 caprylic/capric glycerides; CAS 52504-24-2; nonionic; refatting and wetting agent, solubilizer, emollient used in cosmetics and pharmaceuticals, liq. soaps; yel. visc. oily liq.; char. odor; sol. in water, acetone, ethyl and butyl acetate; dens. 1.068 g/ml; visc. 110–130 cps; HLB 19; acid no. 1 max.; sapon. no. 90–100; surf. tens. 29.7 dynes/cm (1%); 100% act.

Softlon AC-4. [Yoshimura Oil Chem.] Polyamide type surfactant; cationic; softener producing durable effect on dyed hank of syn. fibers; liq.; 12% conc.

Softlon FC-28. [Yoshimura Oil Chem.] Polyamide type surfactant and syn. lubricant; cationic; softener and lubricant for dyed cheese or hank of syn. fibers before knitting; liq.; 15% conc.

Softlon FC-136. [Yoshimura Oil Chem.] Polyamide type surfactant and wax; cationic; softener and lubricant for dyed cheese or hank of syn. fibers before knitting; liq.; 25% conc.

Softyne CLS. [Hart Prods. Corp.] Amido-fatty quaternary; cationic; textile softening agent; liq.; 14% conc.

Softyne CSN. [Hart Prods. Corp.] Amido-fatty quaternary; cationic; textile softening agent; liq.; 20% conc.

Softyne H. [Hart Prods. Corp.] Amido-fatty quaternary; cationic; textile softening agent; liq.; 20% conc.

Sol Mul 130. [Societa Ital. Emulsionanti] Sodium petrol. sulfonates, rust inhibitors, solvents blend; anionic/nonionic; emulsifier and rust inhibitor for naphthenic cutting oils; liq.

Sol Mul 235. [Societa Ital. Emulsionanti] Sodium petrol. sulfonates, rust inhibitors, solvents blend; anionic/nonionic; emulsifier and rust inhibitor for paraffinic cutting oils; liq.

Soltem 5 C/70. [Societa Ital. Emulsionanti] Sodium dodecylbenzene sulfonate; anionic; surfactant for pesticides; liq.; HLB 4.0; 70% conc.

Soltem 8 FL/N. [Societa Ital. Emulsionanti] Phosphate derivate, amminic salt; anionic; wetting agent and dispersant for flowable pesticide formulations; thick liq.; 100% conc.

Soltem 13 FL/N. [Societa Ital. Emulsionanti] Phosphate derivate, amminic salt; anionic; wetting agent and dispersant for flowable pesticide formulations; wax; 100% conc.

Soltem 70. [Societa Ital. Emulsionanti] Sodium polycarboxylate; dispersing suspending agent for W.P. pesticides; powd.

Soltem 101. [Societa Ital. Emulsionanti] Ethoxylated phenolic compd. and org. sulfonate; anionic/nonionic; emulsifier for emulsifiable conc. pesticides; thick liq.; HLB 11.5; 90% conc.

Soltem 207. [Societa Ital. Emulsionanti] Sodium methane naphthalene sulfonate; dispersing suspending agent for W.P. pesticides; powd.

Soltem 251. [Societa Ital. Emulsionanti] Ethoxylated fatty acid and org. sulfonate; anionic/nonionic; emulsifier for emulsifiable conc. pesticides (org. chlorinated, phosphates); thick liq.; HLB 11.0; 87% conc.

Soltem 258. [Societa Ital. Emulsionanti] Ethoxylated fatty acid and org. sulfonate; anionic/nonionic; emulsifier forDicofol and Tetradifon formulations; liq.; 80% conc.

Soltem 480. [Societa Ital. Emulsionanti] Ethoxylated, modified fatty acid and org. sulfate; anionic/nonionic; stabilizer and emulsifier for emulsifiable conc. pesticides; thick liq.; 95% conc.

Soltem 520. [Societa Ital. Emulsionanti] Modified ethoxylates and org. sulfonates blend; anionic/nonionic; emulsifier for emulsifiable conc. pesticides; liq.; HLB 5.0; 63% conc.

Soltem B. [Societa Ital. Emulsionanti] Sodium alkylnaphthalene sulfonate, modified; anionic; nonfoaming wetting agent for pesticides; powd.; 50% conc.

Soltem SC/70. [Societa Ital. Emulsionanti] Sodium dodecylbenzene sulfonate; anionic; surfactant for emulsifiable conc. pesticides; liq.; 70% conc.

Soltem SFL/1. [Societa Ital. Emulsionanti] Polyglycol ethers; nonionic; wetting agent and dispersant for pesticide formulations; liq.; 100% conc.

Sokalan® HP 22. [BASF] Acetic acid ethenyl ester, polymers, polymer with oxirane, graft; CAS 109464-53-1; nonionic; antiredeposition agent and soil shield polymer for laundry prods.; liq.; m.w. 24,000; visc. 53 cps; pH 6.0; 20% conc.

Sokalan® HP 50. [BASF; BASF AG] Polyvinylpyrrolidone; CAS 9003-39-8; antiredeposition inhibitor for detergents in laundry applics.; dispersant; solid; 95% conc.

Sokalan® HP 53. [BASF; BASF AG] Polyvinylpyrrolidone; antiredeposition inhibitor for detergents; carbon dispersant; liq.; m.w. 40,000; visc. 100 cps; pH 8.0; 30% conc.

Solan 50. [Croda Inc.] PEG-60 lanolin; CAS 61790-81-6; nonionic; hydrophilic surfactant, emollient, conditioner, thickener, superfatting agent, foam stabilizer, plasticizer, humectant for personal care prods., soaps, pharmaceuticals, chemical specialties; fragrance solubilizer; Gardner 11 max. visc. liq.; water-sol.; acid no. 2 max.; iodine no. 6 max.; sapon. no. 8 max.; pH 5.5–7.0 (1% aq.); usage level: 1-10%; 50% act.

Solan. [Croda Inc.] PEG-75 lanolin; CAS 61790-81-6; nonionic; surfactant, emollient, conditioner, superfatting agent, emulsifier, solubilizer, foam stabilizer, plasticizer, humectant for soaps, detergent bars, shampoos, skin cleansers, hair sprays, deodorants, chemical specialties; Gardner 11 max. color; m.p. 46-54 C; acid no. 2 max.; iodine no. 10 max.; sapon. no. 8-16; pH 5.5-7.0 (1% aq.); 100% act.

Solan E. [Croda Inc.; Croda Chem. Ltd.] PEG-55 lanolin; CAS 61790-81-6; nonionic; emollient and emulsifier; solid; water-sol.; HLB 16.2; 100% conc.

Solangel 401. [Croda Inc.] PEG-75 lanolin; CAS 61790-81-6; nonionic; emulsifier, humectant for soap; liq.; HLB 16.0; 50% conc.

Solar Soap Powd. [Guelph Soap] Soap; used for blending hand soaps or laundry prods.; solid, powd., flakes; sol. in water.

Solegal W Conc. [Hoechst AG] Alkyl phenyl polyglycol ester; nonionic; emulsifier for textile printing; liq.

Sole-Onic CDS. [Calgene] Diethylene glycol laurate; CAS 9004-81-3; anionic; defoamer, emulsifier for emulsion paints, industrial oil emulsions, protein systems; self-emulsifying; liq.; 100% conc.

Sole-Terge 8. [Calgene] Oleic acid isopropanolamide sulfosuccinate; nonionic; high foaming detergent, emulsifier for personal care prods.; ore flotation and emulsion polymerization; liq.; 100% conc.

Sole-Terge TS-2-S. [Calgene] Sodium 2-ethylhexyl sulfate; anionic; wetting agent, industrial penetrant for textiles; liq.; 35% conc.

Solidegal SR. [Hoechst AG] Higher amine-based prep.; cationic; leveling agent for vat dyes; liq.

Solidokoll Brands. [Hoechst AG] Polyacrylamide; nonionic; auxiliaries to inhibit migration of dyes; liq.

Solocod G. [Reilly-Whiteman] Sulfated fish oil; anionic; emulsifiable oil for leather processing; liq.; 75% conc.

Solricin® 135. [CasChem] Potassium ricinoleate; anionic; detergent, emulsifier, mild germicide, glycerized rubber lubricant, foam stabilizer in foamed rubber; making of cutting and sol. oils, household and cosmetic prods.; FDA approval; Gardner 2 liq.; sp.gr. 1.034; dens. 8.6 lb/gal; visc. 0.9 stokes; 32% act.

Solricin® 235. [CasChem] Potassium castor soap sol'n. in water; emulsifier, dispersant, mild germicide; glycerized rubber lubricant; emulsifier, foam stabilizer for foamed rubber; Gardner 4 liq.; sp.gr. 1.034; visc. 0.9 stokes; 35% solids.

Solricin® 285. [CasChem] Ammonium ricinoleate sol'n. in water; emulsifier, dispersant, rustproofing agent (leaves corrosion resistant film on exposure to air); lubricant; for both oil and water systems; amber paste; sp.gr. 0.967; 84% solids.

Solricin® 435. [CasChem] Sodium ricinoleate aq. sol'n.; emulsifier, stabilizer, defoamer for emulsion polymerization of resins (PVC, PVAc); FDA approval; Gardner 4 liq.; sp.gr. 1.022; visc. 2 stokes; 35% solids.

Solricin® 535. [CasChem] Sodium ricinoleate sol'n. in water; mild germicide; glycerized rubber lubricant; emulsifier and foam stabilizer for foamed rubber; FDA approval; Gardner 4 liq.; sp.gr. 1.025; visc. 1 stokes; 35% solids.

Solton Series. [Toho Chem. Industry] Surfactant/EP agents blend; water-sol. cutting oil; liq.

Solulan® 5. [Amerchol; Amerchol Europe] Laneth-5, ceteth-5, oleth-5 and steareth-5; nonionic; emulsifier, wetting agent, dispersant, lubricant, solv., conditioner, plasticizer, emollient used in personal care and dermatology prods.; foam stabilizer for detergent systems; amber semisolid; faint pleasant odor; sol. @ 5% in anhyd. ethanol; disp. in water; HLB 8.0; acid no. 3 max.; sapon. no. 14 max.; pH 4.5–7.0 (10% aq.); 100% conc.

Solulan® 16. [Amerchol; Amerchol Europe] Laneth-16, ceteth-16, oleth-16 and steareth-16; nonionic; emulsifier, wetting agent, dispersant, lubricant, solv., conditioner, plasticizer, emollient used in personal care and dermatology prods.; foam stabilizer for detergent systems; lt. tan waxy solid; sol. @ 5% in water, 30% ethanol; HLB 15; cloud pt. 64–70 C; acid no. 3 max.; sapon. no. 8 max.; pH 4.5–7.5 (10% aq.); 100% conc.

Solulan® 75. [Amerchol; Amerchol Europe] PEG-75 lanolin; CAS 61790-81-6; nonionic; emulsifier, wetting agent, dispersant, lubricant, solv., conditioner, plasticizer, emollient used in personal care and dermatology prods.; foam stabilizer for detergent systems; lt. yel.-amber waxy solid; faint pleasant odor; sol. see Solulan 16; HLB 15; cloud pt. 80–87 C; acid no. 3 max.; sapon. no. 10–20; pH 4.5–7.5 (10% aq.); 100% conc.

Solulan® C-24. [Amerchol; Amerchol Europe] Choleth-24 and ceteth-24; nonionic; emulsifier, wetting agent, dispersant, lubricant, solv., conditioner, plasticizer, emollient used in personal care and dermatology prods.; foam stabilizer for detergent systems; off-wh. to pale yel. waxy solid; faint pleasant odor; sol. @ 5% in water; HLB 14; cloud pt. 88–95 C; acid no. 1; sapon. no. 3 max.; 100% conc.

Solulan® L-575. [Amerchol; Amerchol Europe] PEG-75 lanolin; CAS 61790-81-6; nonionic; emulsifier, wetting agent, dispersant, lubricant, solv., conditioner, plasticizer, emollient used in personal care and dermatology prods.; foam stabilizer for detergent systems; lt. yel.-amber visc. liq.; faint pleasant odor; sol. @ 5% in water, 30% ethanol; HLB 15; cloud pt. 80–87 C; acid no. 1 max.; sapon. no. 10 max.; 50% conc.

Solulan® PB-2. [Amerchol; Amerchol Europe] PPG-2 lanolin alcohol ether; CAS 68439-53-2; nonionic; spreading agent, dispersant, plasticizer; emollient and conditioner for personal care prods., detergents, pharmaceuticals, waxes, polishes, leather treatment; amber semisolid, liq.; sol. in castor oil, IPM, IPP, anhyd. ethanol; HLB 8.0; acid no. 3 max.; sapon. no. 7 max.; 100% conc.

Solulan® PB-5. [Amerchol; Amerchol Europe] PPG-5 lanolin alcohol ether; nonionic; spreading agent, dispersant, plasticizer; emollient and conditioner for personal care prods., detergents, pharmaceuticals, waxes, polishes, leather treatment; lt. amber clear visc. liq.; sol. see Solulan PB-2; HLB 10.0; acid no. 2 max.; sapon. no. 6 max.; 100% conc.

Solulan® PB-10. [Amerchol; Amerchol Europe] PPG-10 lanolin alcohol ether; nonionic; spreading agent, dispersant, plasticizer; emollient and conditioner for personal care prods., detergents, pharmaceuticals, waxes, polishes, leather treatment; straw clear heavy-visc. liq.; sol. see Solulan PB-2; HLB 12.0; acid no. 1 max.; sapon. no. 4 max.; 100% conc.

Solulan® PB-20. [Amerchol; Amerchol Europe] PPG-20 lanolin alcohol ether; nonionic; spreading agent, dispersant, plasticizer; emollient and conditioner for personal care prods., detergents, pharmaceuticals, waxes, polishes, leather treatment; lt. straw clear med-visc. liq.; sol. see Solulan PB-2; HLB 14.0; acid no. 0.75 max.; sapon. no. 3 max.; 100% conc.

Solumin F Range. [Rhone-Poulenc UK] Ether sulfate; anionic; surfactant for aq. systems; liq.; 25% conc.

Solupret Brands. [Hoechst AG] Polysiloxane in solvents; water repellent finishing agent, coatings for woven or knit goods.

Solusoft NK, WA, WL. [Hoechst AG] Silicone elastomer; finishing agent for textiles; imparts soft and springy fabric feel; liq.

Solutene TER. [ICI Surf. UK] Self-emulsifiable dichlorobenzene; anionic; carrier for dyeing polyester fibers and blends with disperse dyes; liq.

Solvent Scour 25/27. [Hart Chem. Ltd.] Alkanolamides and alkyl phenol ethoxylate; nonionic; scouring agent and prescour for greasy and oil soiled fabrics; yel. liq.; dens. 0.899; 100% act.

Solvent-Scour 263. [ICI Surf. UK] Chlorinated solv. and emulsifier; anionic/nonionic; general purpose scouring agent for all fibers; liq.

Solvent-Scour 880. [ICI Surf. UK] Chlorinated solv., detergent, and emulsifier; nonionic; general purpose scouring agent for all fibers; liq.

Sopralub ACR 265. [Henkel-Nopco] Polyoxyalkylene fatty glycerides; nonionic; textile softener, plasticizer; liq.; 30% conc.

Sopralub ACR 275. [Henkel-Nopco] Polyoxyalkylene fatty glycerides; nonionic; softener, plasticizer for textile finishes and coatings; fluid cream; 30% conc.

Soprofor 3D33. [Rhone-Poulenc Geronazzo] POE arylphenol phosphate; anionic; dispersing/suspending agent for pesticide flowable formulations; thick liq.; 99% conc.

Soprofor BC/Series. [Rhone-Poulenc Geronazzo] Octylphenol, ethoxylated; nonionic; emulsifier, detergent, wetting agent; 100% conc.

Soprofor DO/64. [Rhone-Poulenc Geronazzo] Block polymer dioleate; nonionic; antifoamer; intermediate for industrial applics.; liq.; 100% conc.

Soprofor FL. [Rhone-Poulenc Geronazzo] POE arylphenol phosphate, amine salt; anionic; dispersing/suspending agent for pesticide flowable formulations; thick liq.; 99% conc.

Soprofor M/52. [Rhone-Poulenc Geronazzo] PEG 220 monooleate; CAS 9004-96-0; nonionic; o/w emulsifier for min. oils, greases, aromatics, silicone oil; oil spill dispersant; solid; 100% conc.

Soprofor NP/20. [Rhone-Poulenc Geronazzo] Nonoxynol-20; CAS 9016-45-9; nonionic; finishing auxs. used as antifoaming agents, detergents, dispersants, emulsifiers, softeners, and dyeing assistants; solid; 100% conc.

Soprofor NP/30. [Rhone-Poulenc Geronazzo] Nonoxynol-30; CAS 9016-45-9; nonionic; finishing auxs. used as antifoaming agents, detergents, dispersants, emulsifiers, softeners, and dyeing assistants; solid; 100% conc.

Soprofor NP/100. [Rhone-Poulenc Geronazzo] Nonoxynol-100; CAS 9016-45-9; nonionic; finishing auxs. used as antifoaming agents, detergents, dispersants, emulsifiers, softeners, and dyeing assistants; powd.; 100% conc.

Soprofor NPF/10. [Rhone-Poulenc Geronazzo] Acid nonylphenyl ether phosphate; anionic; detergent in industrial cleaners; household detergents; paint stripping agent; dedusting; emulsifier for acrylic, PVAc emulsion polymerization; liq.; 100% conc.

Soprofor PA/17, PA/19. [Rhone-Poulenc Geronazzo] Alkylpolyethoxyether phosphate; anionic; dispersing/suspending agent for pesticide flowable formulations; liq.; 100% conc.

Soprofor PS/17, PS/19, PS/21 (see Rhodafac® PS-17, PS-19, PS-21). [Rhone-Poulenc Geronazzo] Alkylpolyethoxyether phosphate salt form; anionic; dispersing/suspending agent for pesticide flowable formulations; liq.; 80% conc.

Soprophor® 4D384. [Rhone-Poulenc Surf.; Rhone-Poulenc Geronazzo] Ethoxylated tristyrylphenol sulfate; anionic; emulsifier, dispersant for agric. formulations; EPA compliance pending.

Soprophor® 37. [Rhone-Poulenc Geronazzo] Ethoxylated polystyrylphenol; nonionic; emulsifier for styrene copolymer; latex stabilizer; dispersant and antioxidant for phenolic slurries; paste; 100% conc.

Soprophor® 40/D. [Rhone-Poulenc Geronazzo] Alkylaryl polyglycol ether; nonionic; wetting agent for wettable powds. for pesticides; powd.; 40% conc.

Soprophor® 497/P. [Rhone-Poulenc Surf.; Rhone-Poulenc Geronazzo] EO/PO alkylphenol block polymer; nonionic; emulsifier, dispersant used in agric. industry for prep. of emulsifiable concs. and toxicant flowable systems; EPA compliance; solid; 99% conc.

Soprophor® 724/P. [Rhone-Poulenc Surf.; Rhone-Poulenc Geronazzo] EO/PO alkylphenol block polymer; nonionic; emulsifier, dispersant used in agric. industry for prep. of emulsifiable concs. and toxicant flowable systems; EPA compliance; solid; 99% conc.

Soprophor® 796/P. [Rhone-Poulenc Surf.; Rhone-Poulenc Geronazzo] Ethoxylated propoxylated tristyrylphenol; nonionic; emulsifier for agric. formulations; EPA compliance pending.

Soprophor® BO/318. [Rhone-Poulenc Geronazzo] Ethoxy propoxylated alcohol; nonionic; biodeg. surfactant for controlled foam alkaline degreasing compds., acid pickling compds.; liq.; 100% conc.

Soprophor® BSU. [Rhone-Poulenc Surf.; Rhone-Poulenc Geronazzo] PEG-16 tristyrylphenol; nonionic; emulsifier, dispersant for agric. formulations; EPA compliance pending.

Soprophor® CY/8. [Rhone-Poulenc Surf.; Rhone-Poulenc Geronazzo] PEG-20 tristyrylphenol; nonionic; emulsifier, dispersant for agric. formulations; EPA compliance pending.

Soprophor® FLK. [Rhone-Poulenc Surf.; Rhone-Poulenc Geronazzo] Ethoxylated tristyrylphenol phosphate, potassium salt; anionic; surfactant, dispersant for agric. formulations; EPA compliance pending; 40% act.

Soprophor® K/202. [Rhone-Poulenc Geronazzo] POE fatty alcohol; nonionic; wetting/suspending agent for wettable powd. pesticides; specific for Dodine; powd.; 50% conc.

Soprophor® S/25. [Rhone-Poulenc Surf.; Rhone-Poulenc Geronazzo] PEG-25 tristyrylphenol; nonionic; emulsifier, dispersant for agric. formulations; EPA compliance pending.

Soprophor® S40-P. [Rhone-Poulenc Surf.; Rhone-Poulenc Geronazzo] PEG-40 tristyrylphenol; nonionic; surfactant, dispersant for agric. formulations; EPA compliance pending.

Soprophor® SC/167. [Rhone-Poulenc Geronazzo] EO/PO alkylphenol block polymer; nonionic; emulsifier for dimethoate emulsifiable concs.; solid paste; 99% conc.

Sorban AL. [Witco SA] Sorbitan ester; nonionic; antifoaming; wax; 100% conc.

Sorban AO. [Witco SA] Sorbitan oleate; CAS 1338-43-8; nonionic; emulsifier; liq.; HLB 4.5; 100% conc.

Sorban AO 1. [Witco SA] Sorbitan ester; nonionic; oil additive, corrosion inhibitor; liq.; 100% conc.

Sorban AST. [Witco SA] Sorbitan stearate; CAS 1338-41-6; nonionic; emulsifier for oils and greases; antifoamer, corrosion inhibitor; cosmetic applics.; flakes; 100% conc.

Sorban CO. [Witco SA] Sorbitan trioleate; CAS 26266-58-0; nonionic; antifoamer; corrosion inhibitor; liq.; HLB 2.0; 100% conc.

Sorbanox AL. [Witco SA] Sorbitan ester, ethoxylated; nonionic; detergent used in veg. cleaning, antistat; liq.; 100% conc.

Sorbanox AOM. [Witco SA] Ethoxylated sorbitan ester; CAS 9005-65-6; nonionic; emulsifier; solubilizer of vitamins; liq.; HLB 15.0; 100% conc.

Sorbanox AP. [Witco SA] Ethoxylated sorbitan ester; nonionic; emulsifier; liq.; 100% conc.

Sorbanox AST. [Witco SA] Ethoxylated sorbitan ester; CAS 9005-67-8; nonionic; emulsifier for paraffin and min. oil; insecticides, cosmetic applics.; liq.; 100% conc.

Sorbanox CO. [Witco SA] Polysorbate 85; CAS 9005-70-3; nonionic; emulsifier for min. oil; liq.; HLB 11.0; 100% conc.

Sorbax HO-40. [Chemax] POE sorbitol ester; emulsifier for agric. pesticide/herbicide, emulsion polymerization, metalworking lubricants, die-cast lubricants; liq.; sapon. no. 97; HLB 10.4.

Sorbax HO-50. [Chemax] POE sorbitol ester; o/w emulsifier for solvs., veg. and petrol. oils used in the textile, agric., emulsion polymerization, and metal lubricant industries; liq.; sapon. no. 85; HLB 11.4.

Sorbax MO-40. [Chemax] POE sorbitol ester; o/w emulsifier for solvs., veg. and petrol. oils used in the textile, agric., emulsion polymerization, and metal lubricant industries; liq.; HLB 10.2.

Sorbax PML-20. [Chemax] Polysorbate 20; CAS 9005-64-5; nonionic; o/w emulsifier, solubilizer for perfumes, flavors; for agric., cosmetic, leather, metalworking, and textile industries; liq.; water-sol.; sapon. no. 45; HLB 16.7; 100% conc.

Sorbax PMO-5. [Chemax] Polysorbate 81; CAS 9005-65-6; nonionic; o/w emulsifier, solubilizer for perfumes and flavors; emulsifier for petrol. oils, solvs., veg. oils, waxes, silicones in agric., cosmetic, leather, metalworking, and textile industries; liq.; water-sol.; sapon. no. 100; HLB 10.0; 100% conc.

Sorbax PMO-20. [Chemax] Polysorbate 80; CAS 9005-65-6; nonionic; o/w emulsifier, solubilizer for perfumes and flavors; emulsifier for petrol. oils, solvs., veg. oils, waxes, silicones in agric., cosmetic, leather, metalworking, and textile industries; liq.; water-sol.; sapon. no. 50; HLB 15.0; 100% conc.

Sorbax PMP-20. [Chemax] Polysorbate 40; CAS 9005-66-7; nonionic; o/w emulsifier, solubilizer for perfumes and flavors; emulsifier for petrol. oils, solvs., veg. oils, waxes, silicones in agric., cosmetic, leather, metalworking, and textile industries; liq.; water-sol.; sapon. no. 46; HLB 15.6; 100% conc.

Sorbax PMS-20. [Chemax] Polysorbate 60; CAS 9005-67-8; nonionic; o/w emulsifier, solubilizer for perfumes and flavors; emulsifier for petrol. oils, solvs., veg. oils, waxes, silicones in agric., cosmetic, leather, metalworking, and textile industries; soft paste; water-sol.; sapon. no. 50; HLB 14.9; 100% conc.

Sorbax PTO-20. [Chemax] Polysorbate 85; CAS 9005-70-3; nonionic; o/w emulsifier, solubilizer for perfumes and flavors; emulsifier for petrol. oils, solvs., veg. oils, waxes, silicones in agric., cosmetic, leather, metalworking, and textile industries; liq., gel; water-sol.; sapon. no. 88; HLB 11.0; 100% conc.

Sorbax PTS-20. [Chemax] Polysorbate 65; CAS 9005-71-4; nonionic; o/w emulsifier, solubilizer for perfumes and flavors; emulsifier for petrol. oils, solvs., veg. oils, waxes, silicones in agric., cosmetic, leather, metalworking, and textile industries; solid; water-sol.; sapon. no. 93; HLB 10.5; 100% conc.

Sorbax SML. [Chemax] Sorbitan laurate; CAS 1338-39-2; nonionic; emulsifier for petrol. oils, solvs., veg. oils, waxes, silicones in agric., cosmetic, leather, metalworking, and textile industries; liq.; oil-sol.; sapon. no. 162; HLB 8.6.

Sorbax SMO. [Chemax] Sorbitan oleate; CAS 1338-43-8; nonionic; emulsifier, surfactant for o/w emulsion stabilizers and thickeners used in cosmetic, agric., metalworking, leather, and textile industries; liq.; oil-sol.; sapon. no. 153; HLB 4.3; 100% conc.

Sorbax SMP. [Chemax] Sorbitan palmitate; CAS 26266-57-9; nonionic; emulsifier, surfactant for o/w emulsion stabilizers and thickeners used in cosmetic, agric., leather, metalworking, and textile industries; solid; oil-sol.; sapon. no. 155; HLB 6.7; 100% conc.

Sorbax SMS. [Chemax] Sorbitan stearate; CAS 1338-41-6; nonionic; emulsifier, surfactant for o/w emulsion stabilizers and thickeners used in cosmetic, agric., leather, metalworking, and textile industries; solid; oil-sol.; sapon. no. 152; HLB 4.7; 100% conc.

Sorbax STO. [Chemax] Sorbitan trioleate; CAS 26266-58-0; nonionic; emulsifier, surfactant for o/w emulsion stabilizers and thickeners used in cosmetic, agric., metalworking, leather, and textile industries; liq.; oil-sol.; sapon. no. 180; HLB 1.8; 100% conc.

Sorbax STS. [Chemax] Sorbitan tristearate; CAS 26658-19-5; nonionic; emulsifier, surfactant for o/w emulsion stabilizers and thickeners used in cosmetic, agric., leather, metalworking, and textile industries; solid; oil-sol.; sapon. no. 182; HLB 2.1; 100% conc.

Sorbeth 40. [Croda Surf. Ltd.] PEG-40 sorbitol hexaoleate; nonionic; o/w emulsifier; liq.; 100% conc.

Sorbeth 40HO. [Croda Surf. Ltd.] PEG-40 sorbitol hexaoleate; nonionic; emulsifier/dispersant for agric. chemicals and the oil industry; liq.; HLB 10.5; 100% conc.

Sorbeth 55. [Croda Surf. Ltd.] PEG-55 sorbitol; nonionic; o/w emulsifier; liq.; 100% conc.

Sorbeth 55HO. [Croda Surf. Ltd.] PEG-55 sorbitol hexaoleate; nonionic; emulsifier/dispersant for agric. chemicals and the oil industry; liq.; HLB 11.5; 100% conc.

Sorbilene O. [Auschem SpA] Polysorbate 80; CAS 9005-65-6; nonionic; emulsifier and solubilizer for cosmetics and pharmaceuticals, emulsifiable concs.; liq.; HLB 15.0; 100% conc.

Sorbirol O. [Auschem SpA] Sorbitan oleate; CAS 1338-43-8; nonionic; emulsifier for cosmetics and pharmaceuticals, emulsifiable concs. in agric. preps.; liq.; HLB 4.3; 100% conc.

Sorbit P. [Ciba-Geigy AG] Alkyl naphthalene sulfonate; anionic; wetting agent and foamer; emulsifier, detergent; germicides, textile leveling; powd.; 65% conc.

Sorbon S-20. [Toho Chem. Industry] Sorbitan laurate; CAS 1338-39-2; nonionic; emulsifier, dispersant for w/o emulsion; liq.; 100% conc.

Sorbon S-40. [Toho Chem. Industry] Sorbitan palmitate; CAS 26266-57-9; nonionic; used with Sorbon T series; solid; 100% conc.

Sorbon S-60. [Toho Chem. Industry] Sorbitan stearate; CAS 1338-41-6; nonionic; used with Sorbon T series; solid; 100% conc.

Sorbon S-80. [Toho Chem. Industry] Sorbitan oleate; CAS 1338-43-8; nonionic; used with Sorbon T series; liq.; 100% conc.

Sorbon T-20. [Toho Chem. Industry] PEG-20 sorbitan laurate; CAS 9005-64-5; nonionic; emulsifier, dispersant, solubilizer for o/w emulsion; liq.; HLB 16.7; 100% conc.

Sorbon T-40. [Toho Chem. Industry] POE sorbitan palmitate; nonionic; used with Sorbon S series; liq.; HLB 15.7; 100% conc.

Sorbon T-60. [Toho Chem. Industry] POE sorbitan stearate; nonionic; used with Sorbon S series; liq.; HLB 15.3; 100% conc.

Sorbon T-80. [Toho Chem. Industry] POE sorbitan oleate; nonionic; used with Sorbon S series; liq.; 100% conc.

Sorbon TR 814. [Toho Chem. Industry] POE sorbitol oleate; nonionic; detergent, emulsifier for cosmetics; liq.; HLB 17.5; 100% conc.

Sorbon TR 843. [Toho Chem. Industry] POE sorbitol oleate; nonionic; detergent, emulsifier for cosmetics; liq.; HLB 11.2; 100% conc.

Sorpol 230. [Toho Chem. Industry] Nonionic/org. sulfonate; nonionic; emulsifier for agric. emulsifiable concs.; liq.

Sorpol 320. [Toho Chem. Industry] POE alkylaryl ether, POE sorbitan alkylate, and sulfonate; anionic/nonionic; emulsifier for Malathion emulsifiable concs. and other pesticides; liq.; 100% conc.

Sorpol 355. [Toho Chem. Industry] POE alkylaryl ether, POE sorbitan alkylate, and sulfonate; anionic/nonionic; emulsifier for highly conc. Sumithion emulsifiable conc. (agric.); liq.; 100% conc.

Sorpol 560. [Toho Chem. Industry] POE alkylaryl ether, POE sorbitan alkylate, and sulfonate; anionic/nonionic; emulsifier for Malathion emulsifiable concs.; liq.; 100% conc.

Sorpol 900A, 1200, 2020K, 3005X. [Toho Chem. Industry] POE alkylaryl ether, POE sorbitan alkylate, and sulfonate; anionic/nonionic; emulsifier for pesticides; liq.; 100% conc.

Sorpol 900H. [Toho Chem. Industry] Nonionic/org. sulfonate; nonionic; matched pair emulsifier with Sorpol 900L for agric. pesticide emulsifiable concs.; liq.

Sorpol 900L. [Toho Chem. Industry] Nonionic/org. sulfonate; nonionic; matched pair emulsifier with Sorpol 900H for agric. pesticide emulsifiable concs.; liq.

Sorpol 1200K. [Toho Chem. Industry] Nonionic/org. sulfonate; nonionic; emulsifier for agric. emulsifiable concs.; liq.

Sorpol 2495 G. [Toho Chem. Industry] Sulfonate; anionic; dispersant and wetting agent for Sumithion wettable powd.; liq.; 100% conc.

Sorpol 2676S. [Toho Chem. Industry] Nonionic/org. sulfonate; nonionic; emulsifier for sanitizers using kerosene as solv., agric. pesticides; liq.

Sorpol 2678S. [Toho Chem. Industry] Nonionic/org. sulfonate; nonionic; emulsifier for sanitizers using kerosene as solv., agric. pesticides; liq.

Sorpol 3005X. [Toho Chem. Industry] Nonionic/org. sulfonate; nonionic; emulsifier for org. phosphated pesticides; liq.

Sorpol 3044. [Toho Chem. Industry] POE alkylaryl ether, POE fatty acid ether, and sulfonate; anionic/nonionic; emulsifier for parathion, methyl parathion, Sumithion, Malathion, EPN and others; liq.; 100% conc.

Sorpol 3370. [Toho Chem. Industry] Nonionic/org. sulfonate; nonionic; emulsifier for agric. emulsifiable concs. containing 70-90% organo-phosphorous pesticides; liq.

Sorpol 5037. [Toho Chem. Industry] Modified anionics; anionic; dispersant for agric. wettable powds. used in hard water areas; powd.

Sorpol 5039. [Toho Chem. Industry] Sulfonate; anionic; dispersant and wetting agent for agric. wettable powd. for Milbex, MTMC, Sumithion, Dimethoate, etc.; powd.; 100% conc.

Sorpol 5060. [Toho Chem. Industry] Anionic complex; anionic; collapsing and spreading agent for granular agric. pesticides; powd.

Sorpol 8070. [Toho Chem. Industry] Anionic; dispersant for agric. wettable powds. and gran.; powd.

Sorpol 9939. [Toho Chem. Industry] Esters; nonionic; defoaming spreaders for agric. formulation; liq.

Sorpol H-770. [Toho Chem. Industry] POE alkylaryl ether, POE sorbitan alkylate, and sulfonate; anionic/nonionic; emulsifier for emulsifiable concs. of organic phosphated and chlorinated pesticides; liq.; 100% conc.

Sorpol L-550. [Toho Chem. Industry] POE alkylaryl ether, POE sorbitan alkylate, and sulfonate; anionic/nonionic; emulsifier for emulsifiable concs. of organic phosphated and chlorinated pesticides; liq.

Sorpol W-150. [Toho Chem. Industry] Nonionic complex; nonionic; spreading agent for agric. wettable powds.; liq.

So/San 30M. [Stepan; Stepan Canada] Specialty prod. containing BTC 2125M as active biocide; fabric softener-sanitizer; EPA registered; Gardner 3 liq.; dens. 8.20 lb/gal; visc. 700 cps (45 F); pour pt. < 40 F; pH 4 (10% water/IPA); 50% solids.

Sovatex C Series. [Standard Chem. UK] Dodecylbenzene sulfonate blends; anionic; detergent, solvent scour, wetting agent, emulsifier for textiles, laundry, drycleaning, leather, and chemical industries; liq., paste, solid; water-disp.

Sovatex DS/C5. [Standard Chem. UK] Sodium dodecylbenzene sulfonate and fatty alcohol ethoxylate; anionic/nonionic; conc. detergent, wetting and dye control agent for textile processes; pale amber clear liq.; water-sol.; sp.gr. 1.04; pH 7–8.5; 50% solids.

Sovatex IM Series. [Standard Chem. UK] Fatty acid

imidazolines; intermediate for prep. of cationic or amphoteric surfactants; corrosion inhibitor, lubricant; liq./solid; 100% conc.

Sovatex MP/1. [Standard Chem. UK] Imidazoline; amphoteric; detergent, wetting agent, emulsifier, lubricant; liq.; 30% conc.

Span® 20. [ICI Spec. Chem.; ICI Surf. Belgium] Sorbitan laurate NF; CAS 1338-39-2; nonionic; emulsifier, stabilizer, thickener, lubricant, softener, antistatic agent; foods, pharmaceuticals, cosmetics, cleaning compds., textiles; amber liq.; sol. (@ 1%) in IPA, perchloroethylene, xylene, cottonseed oil, min. oil; sol. (hazy) in propylene glycol; visc. 4250 cps; HLB 8.6; 100% act.

Span® 40. [ICI Spec. Chem.; ICI Surf. Belgium] Sorbitan palmitate NF; CAS 26266-57-9; nonionic; see Span 20; tan solid; sol. (@ 1%) in IPA, xylene; sol. (hazy) in perchloroethylene; HLB 6.7; pour pt. 48 C; 100% act.

Span® 60, 60K. [ICI Spec. Chem.; ICI Surf. Belgium] Sorbitan stearate NF; CAS 1338-41-6; nonionic; see Span 20; also dispersant for inorg. pigments in thermoplastics; tan solid; sol. (@1%): sol. in IPA; sol. (hazy) in perchloroethylene, xylene; HLB 4.7; pour pt. 53 C; 100% act.

Span® 65. [ICI Spec. Chem.; ICI Surf. Belgium] Sorbitan tristearate; CAS 26658-19-5; nonionic; see Span 20; cream solid; sol. (@ 1%) in IPA, perchloroethylene, xylene; HLB 2.1; pour pt. 53 C; 100% act.

Span® 80. [ICI Spec. Chem.; ICI Surf. Belgium] Sorbitan oleate NF; CAS 1338-43-8; nonionic; w/o emulsifier, oil additive for corrosion inhibition; fiber lubricant and softener; amber liq.; sol. (@ 1%) in IPA, perchloroethylene, xylene, cottonseed and min. oils; visc. 1000 cps; HLB 4.3.; 100% act.

Span® 85. [ICI Spec. Chem.; ICI Surf. Belgium] Sorbitan trioleate; CAS 26266-58-0; nonionic; see Span 20; amber liq.; sol. (@ 1%) in IPA, perchloroethylene, xylene, cottonseed and min. oils; visc. 210 cps; HLB 1.8; 100% act.

Spanscour EFS. [CNC Int'l.] Surfactant; anionic; detergent for Spandex scouring.

Spanscour GR. [CNC Int'l.] Scour and inhibitor combination for nylon/Spandex fabrics with a tendency to yellow due to atmospheric conditions.

Spanscour N20. [Reilly-Whiteman] Detergent for textile scouring.

Sperse Polymer IV. [ICI Spec. Chem.] Ammonium polycarboxylate; anionic; dispersant for pigments used in textiles; vehicle for flexo inks; heat curable; avail. as metal or amine salt; aq. sol'n.; 25% conc.

Sponto® 101. [Witco/Organics] Sulfonate/POE ether blend; anionic/nonionic; emulsifier for phosphate toxicant mixtures; EPA clearance; liq.

Sponto® 102. [Witco/Organics] Metal sulfonate and POE ether blend; anionic/nonionic; emulsifier for phosphated/chlorinated insecticides; liq.; oil-sol.

Sponto® 140T. [Witco/Organics] Sulfonate/POE ether blend; anionic/nonionic; agric. emulsifier for highly conc. organophosphate pesticides; EPA clearance; liq.

Sponto® 150T. [Witco/Organics] Sulfonate/POE ether blend; anionic/nonionic; agric. emulsifier for highly conc. organophosphate pesticides; EPA clearance; liq.

Sponto® 150 TH. [Witco/Organics; Witco Israel] Sulfonate and POE ether; anionic/nonionic; agric. emulsifier for organophosphate pesticides; liq.; oil-sol.; 100% conc.

Sponto® 150 TL. [Witco/Organics; Witco Israel] Sulfonate and POE ether; anionic/nonionic; agric. emulsifier for organophosphate pesticides; liq.; 100% conc.

Sponto® 168-D. [Witco/Organics] Phosphoric acid mono and diesters/alkylphenoxy polyethoxyethanol blend; anionic; emulsifier and compatibility agent for pesticides and fertilizers, high electrolyte systems; EPA clearance; liq.; 100% conc.

Sponto® 169-T. [Witco/Organics; Witco SA] Surfactant blend; anionic; hydrophilic emulsifier for insecticide emulsifiable concs., esp. for Dicofol; EPA clearance; liq.

Sponto® 200. [Witco/Organics] POE ether; nonionic; emulsifier for citrus and crop oils; liq.

Sponto® 203. [Witco/Organics] Sulfonate/POE ether blend; anionic/nonionic; broad range agric. emulsifier for organic phosphate toxicants when used with Sponto 101; EPA clearance; liq.

Sponto® 207. [Witco/Organics] Sulfonate blend; anionic; emulsifier for pentachlorophenol; liq.

Sponto® 217. [Witco/Organics; Witco Israel] Sulfonate and POE ether; nonionic; emulsifier for pesticide emulsifiable concs., esp. for thiocarbamates; EPA clearance; liq.; 100% conc.

Sponto® 221. [Witco/Organics; Witco Israel] Sulfonate and POE ether; anionic/nonionic; emulsifier for pesticide emulsifiable concs., esp. for thiocarbamates; liq.; 100% conc.

Sponto® 232, 234. [Witco/Organics; Witco SA] Surfactant blend; anionic/nonionic; matched pair emulsifiers for emulsifiable concs.; liq.

Sponto® 232T. [Witco/Organics; Witco Israel] Sulfonate and POE ether blend; anionic/nonionic; emulsifier for pesticide emulsifiable concs., esp. for thiocarbamates; EPA clearance; liq.; oil-sol.; 100% conc.

Sponto® 234T. [Witco/Organics] Sulfonate and POE ether blend; anionic/nonionic; emulsifier for pesticides; EPA clearance; liq.; oil-sol.

Sponto® 300T. [Witco/Organics] Sulfonate/POE ether blend; anionic/nonionic; emulsifier for pesticides; EPA clearance; liq.

Sponto® 305. [Witco/Organics] Sulfonate/POE ether blend; anionic/nonionic; agric. emulsifier for highly conc. organic phosphate insecticides; liq.

Sponto® 500T. [Witco/Organics] Sulfonate/POE ether blend; anionic/nonionic; emulsifier for agric. pesticides; EPA clearance; liq.

Sponto® 710T. [Witco/Organics] Sulfonate and POE ether blend; anionic/nonionic; emulsifier for phenoxy-type herbicides; EPA clearance; liq.; oil-sol.

Sponto® 712T. [Witco/Organics] Sulfonate and POE ether blend; anionic/nonionic; emulsifier for phenoxy-type herbicides; EPA clearance; liq.; oil-sol.

Sponto® 714T. [Witco/Organics] Sulfonate and POE ether blend; anionic/nonionic; emulsifier for phenoxy-type herbicides; EPA clearance; liq.; oil-sol.

Sponto® 723T. [Witco/Organics] Surfactant blend; anionic/nonionic; surfactant for herbicides; EPA clearance; liq.

Sponto® 2174 H. [Witco/Organics; Witco Israel] Sulfonate and POE ether; anionic/nonionic; emulsifier for flowable pesticide formulations; liq.; 100% conc.

Sponto® 2224T. [Witco/Organics] Sulfonate and POE ether blend; anionic/nonionic; emulsifier for chlorinated insecticides; liq.

Sponto® 4648-23A, 4648-23B. [Witco/Organics; Witco Israel] Sulfonate and POE ether; anionic/nonionic; emulsifier for agric. formulations; specific for Toxaphene; liq.; 100% conc.

Sponto® AC60-02D. [Witco/Organics] POE ether blend; nonionic; emulsifier for crop and citrus oils; liq.

Sponto® AD4-10N. [Witco/Organics] Sulfonate and POE ether blend; anionic/nonionic; emulsifier for aromatic solvs. in weed control; liq.; oil-sol.

Sponto® AD6-39A. [Witco/Organics] Sulfonate and POE ether blend; anionic/nonionic; aux. for Sponto matched-pair emulsifiers; hydrophilic; liq.; oil-sol.

Sponto® AG3-55T. [Witco/Organics] Sulfonate and POE ether blend; nonionic; specialty emulsifier for phenoxy-ester herbicides; liq.; oil-sol.

Sponto® AG-540. [Witco/Organics] Surfactant blend; nonionic; veg. oil emulsifier (agric.); EPA clearance; liq.

Sponto® AG-1040. [Witco/Organics] Surfactant blend; nonionic; surfactant for orchard and crop oils; EPA clearance; liq.

Sponto® AG-1265. [Witco/Organics] Surfactant blend; nonionic; surfactant for herbicides; EPA clearance; liq.

Sponto® AK30-02BT. [Witco/Organics] Surfactant blend; nonionic; surfactant for insecticides; EPA clearance; liq.

Sponto® AK30-23. [Witco/Organics] Phosphoric acid mono- and diester and alkylphenoxy; anionic; emulsifier for Vapona; liq.

Sponto® AK31-53. [Witco/Organics] Surfactant blend; nonionic; veg. oil emulsifier (agric.); EPA clearance; liq.

Sponto® AK31-56. [Witco/Organics] Surfactant blend; nonionic; surfactant for insecticides; EPA clearance; liq.

Sponto® AK31-64. [Witco/Organics] Surfactant blend; nonionic; agric. foaming agent; EPA clearance; liq.

Sponto® AK31-66. [Witco/Organics] Surfactant blend; nonionic; surfactant for insecticides; EPA clearance; liq.

Sponto® AK31-69. [Witco/Organics] Surfactant blend; nonionic; surfactant for orchard and crop oils; EPA clearance; liq.

Sponto® AL69-49. [Witco/Organics] POE ether blend; nonionic; emulsifier for Dimethoate insecticides; EPA clearance; liq.

Sponto® AM2-07. [Witco/Organics] Sulfonate, phosphate ester and POE ether; anionic/nonionic; agric. emulsifier for propanil concs.; liq.; oil-sol.

Sponto® CA-861. [Witco/Organics] Surfactant blend; nonionic; agric. surfactant, compatibility agent; EPA clearance; liq.

Sponto® H-44-C. [Witco/Organics] Polyglycol fatty acid ester; nonionic; agric. emulsifier for w/o systems, natural pyrethrum systems; compatibility agent; EPA clearance; liq.

Sponto® N-140B. [Witco/Organics] Sulfonate and POE ether; nonionic; emulsifier for organic phosphate insecticides; liq.; oil-sol.

Spraywax 660-A Conc. [Finetex] Proprietary imidazolinium; industrial car wax sprays.

Spreading Agent ET0672. [Croda Chem. Ltd.] Complex alkoxylate; nonionic; surface spreading agent for min. oils used for mosquito control, in bath oils; liq.; HLB 7.5; 100% conc.

Stabilisal S Liq. [Hoechst AG] Polysulfide sol'n.; anionic; stabilizer for sulfur dyeing; orange-red liq.

Stabilizer AWN. [Sandoz Prods. Ltd.] Inorganic salt/wetting agent aq. sol'n.; anionic/nonionic; stabilizer for hydrogen peroxide bleaching baths; liq.

Stabilizer CB. [ICI Surf. UK] Sodium polycarboxylate; anionic; stabilizer controlling decay of hydrogen peroxide during bleaching process; liq.

Stabilizer CS. [Sandoz Prods. Ltd.] Fatty acid deriv.; anionic; stabilizer for hydrogen peroxide bleaching baths; liq.

Stabilizer SIFA. [Sandoz Prods. Ltd.] Carbonic acid deriv.; anionic; stabilizer for peroxide bleaching; liq.; 100% conc.

Stafoam. [Nippon Oils & Fats] Fatty acid DEA; detergent, rust inhibitor; liq. detergent for laundry or drycleaning use; textile softener aux.; dyeing aux.; metal cleaning agent; liq. soaps; dirt dispersant; lt. yel. liq.; water-sol.; sp.gr. 0.996 (30 C); visc. 810 cps (30 C); pH 9.5–10.5 (1% aq.); surf. tens. 8 dynes/cm (0.1%).

Stafoam DF-1. [Nippon Oils & Fats] Cocamide DEA (1:1); detergent, foam booster, softener, antistat; used in shampoo, dishwashing, laundry detergent, drycleaning, textile softener aux., and metal cleaning; lt. yel. liq.; sol. in ethanol, IPA, propylene glycol, diethylene glycol, petrol. ether, tetrachloroethylene, benzene, xylene; sp.gr. 0.981 (30 C); visc. 625 cps (30 C); pH 9.6–10.6 (1% aq.); surf. tens. 14 dynes/cm (0.1%).

Stafoam DF-4. [Nippon Oils & Fats] Cocamide DEA (1:1); detergent, foam booster, softener, antistat; used in shampoo, dishwashing, laundry detergent, drycleaning, textile softener aux., and metal cleaning; lt. yel. liq.; sol. see Stafoam DF-1; sp.gr. 0.974; visc. 675 cps (30 C); m.p. 15 C max.; surf. tens. 16 dynes/cm (0.1%); pH 9.6–10.6 (1% aq.).

Stafoam DL. [Nippon Oils & Fats] Lauramide DEA (1:1); CAS 120-40-1; detergent, foam stabilizer and booster, visc. modifier, softener; shampoos, drycleaning, cosmetics components, textile softening and dyeing auxs., anticlouding agents; wh. solid; sol. > 2% in ethanol, IPA, propylene glycol, diethylene glycol, tetrachloroethylene, benzene, xylene; sp.gr. 0.969 (50 C), visc. 179 cps (50 C); m.p. 42–47 C; surf. tens. 5 dynes/cm (0.1%).

Stafoam F. [Nippon Oils & Fats] Cocamide DEA (1:2); detergent, rust inhibitor, foamer, penetrant; liq. detergents, metal cleaning, dirty-dispersal aux., textile processing; dye aux.; lt. yel. liq.; sol. see Stafoam; sp.gr. 0.995 (30 C); visc. 755 cps (30 C); surf. tens. 7 dynes/cm (0.1%); pH 9.5–10.5 (1% aq.).

Stamid HT 3901. [Clough] Cocamide DEA; nonionic; foam stabilizer, emulsifier and thickener used in a variety of household, industrial and cosmetic formulations; liq.; pH 9.5-10.5; 85% conc.

Standamid® 100. [Henkel] Cocamide MEA; CAS 68140-00-1; foam stabilizer; superfatting and thickening agent for medicated and cosmetic shampoos, bubble baths, and other detergents; consistency-giving factor for cosmetic preps. in stick form; skin protecting component for industrial ointments; additive to hair treatment preps.; solid; 90% amide content.

Standamid® CD. [Henkel/Cospha; Henkel Canada] Capramide DEA (2:1) and diethanolamine; CAS 136-26-5; nonionic; detergent, foam enhancer for anionic systems; solubilizes fragrances into hydroalcoholic systems; sec. emulsifier in o/w systems, perfume stabilizer; personal care prods., industrial cleaners; amber liq.; sol. in water, propylene glycol, PEG-8, SD-40 alcohol; dens. 8.19 lb/gal; visc. 675 cps; acid no. 15-20; cloud pt. 0 C; gel pt. 0 C; 98% conc.

Standamid® CM. [Henkel] Cocamide MEA; CAS 68140-00-1; nonionic; foaming agent and stabilizer; powded. or beaded bubble baths; gran.; m.p. 140–148 F; acid no. 8.0–12.0; pH 9.5–10.5 (10% sol'n.); 100% conc.

Standamid® CMG. [Henkel] Cocamide MEA; CAS 68140-00-1; super amide (1:1) derived from whole coconut oil; foam boosters, stabilizer, thickener; shampoos, bubble baths, dishwashing detergents, contains no amide ester; solid; 100% conc.

Standamid® ID. [Henkel/Emery] Isostearamide DEA (1:1); nonionic; visc. builder esp. in sodium lauryl sulfate systems; foam enhancer; hair conditioning agent; lubricant, mold release agent, emulsifier; amber liq.; sol. in SD-40 alcohol; sp.gr. 0.964; dens. 8.04 lb/gal; visc. 1600 cps; cloud pt. < 0 C; gel pt. < 0 C; pH 8–10 (1.0%); 100% act.

Standamid® KD. [Henkel/Cospha; Henkel Canada] Cocamide DEA (1:1); nonionic; foaming and thickening agent for liq. or gel shampoos and bubble baths, industrial cleaners, etc.; superfatting agent, foam stabilizer, emulsifier, intermediate; detergent and solubilizer for oily components; hair conditioner; visc. amber liq.; sol. in propylene glycol., PEG-8, SD-40 alcohol; insol. in water; dens. 8.13 lb/gal; visc. 1150 cps; m.p. 100 F; acid no. 1.0 max.; cloud pt. 11 C; gel pt. 5 C; pH 9.0–9.5 (10%); 85–90% amide content.

Standamid® KDM. [Henkel] Cocamide DEA (1:2); nonionic; foam builder and stabilizer, thickener, wetting agent, dispersant; shampoos, bubble baths, household detergents, liq. soaps; good lime-soap dispersing and antiredeposition properties; amber visc. liq.; readily sol. in water, alcohol, acetone, and glycols; acid no. 0–5; pH 9.5–10.0 (1%); 100% conc.

Standamid® KDO. [Henkel/Cospha; Henkel Canada] Cocamide DEA (1:1); nonionic; visc. builder for anionic systems, good foam enhancement, low odor, good color, broad compatibility, and good handling qualities; used in shampoos, bath prods., liq. hand soaps, industrial cleaners; amber liq.; sol. in lower alcohols, propylene glycol, and polyethylene glycols; sp.gr. 0.982; dens. 8.20 lb/gal; visc. 1050 cps; acid no. 2.0 max.; cloud pt. 12 C; gel pt. 4 C; pH 9–11 (1.0%); 100% conc.

Standamid® KDS. [Henkel/Cospha; Henkel Canada] Lauramide DEA (1:1); CAS 120-40-1; nonionic; detergent, solubilizer for oils; enhances foam dens., lubricity, stability for shampoos, skin cleansers, bath and shower prods.; lt. amber liq.; dens. 8.20 lb/gal; 80% conc.

Standamid® KM. [Henkel] Cocamide MEA (1:1); CAS 68140-00-1; visc. builder for anionic systems; produces lubricious foam for shampoo and bath prods., industrial cleaners; flakes, beads; sol. in lower alcohols; sp.gr. 0.884; m.p. 80 C; acid no. 2.0 max.; cloud pt. 64 C; gel pt. 59 C; pH 9–11 (1.0%).

Standamid® LD. [Henkel/Cospha; Henkel Canada] Lauramide DEA (1:1); CAS 120-40-1; nonionic; foam booster/stabilizer, wetting agent, superfatting agent, thickening agent, emulsifier for oils and grease, shampoos, bath prods., industrial cleaners, fortifier for perfumes in soaps; waxy solid; sol. in propylene glycol, PEG-8, SD-40 alcohol; insol. in water; dens. 8.20 lb/gal; m.p. 105 F; acid no. 0–1.0; cloud pt. 1 C; gel pt. < 0 C; pH 9.5–10.5 (10% sol'n.); > 90% amide content.

Standamid® LD 80/20. [Henkel/Cospha; Henkel Canada] Modified lauric/myristic acid DEA; nonionic; foam booster/stabilizer, detergency builder, emulsifier, thickener, conditioner; shampoos, bubble baths, dishwashing detergents, creams, and lotions; clear amber liq.; dens. 8.27 lb/gal; cloud pt. 0 C; acid no. 0.5 max.; pH 9.5–10.5 (10% sol'n.); 80% conc.

Standamid® LDO. [Henkel/Cospha; Henkel Canada] Lauramide DEA (1:1); CAS 120-40-1; nonionic; detergent, foam and visc. builder for shampoos, bath prods., cleansers, industrial cleaners; amber liq.; sol. in propylene glycol, PEG-8, SD-40 alcohol; insol. in water; sp.gr. 0.985; dens. 8.22 lb/gal; visc. 850 cps; acid no. 2.0 max.; cloud pt. < 0 C; gel pt. < 0 C; pH 9–11 (1.0%); 100% conc.

Standamid® LDS. [Henkel/Cospha; Henkel Canada] Lauramide DEA (1:1); CAS 120-40-1; nonionic; foam booster/stabilizer, thickener; solubilizer for liqs., oils, perfumes; used in personal care prods., dishwashing detergents; lt. amber liq.; sol. in lower alcohols, propylene glycol, polyethylene glycol; disp. in water; sp.gr. 0.98; dens. 8.2 lb/gal; visc. 1000 cps; cloud pt. < 0 C; gel pt. < 0 C; pH 9–11 (1.0%); 100% conc.

Standamid® LM. [Henkel] Lauramide MEA (1:1); CAS 142-78-9; visc. builder for anionic systems; produces lubricious foam for shampoo and bath prods., industrial cleaners; flakes, beads; sol. in lower alcohols; insol. in water; sp.gr. 0.915; m.p. 80 C; acid no. 2.0 max.; cloud pt. 83 C; gel pt. 73 C; pH 9–11 (1.0 %).

Standamid® LP. [Henkel] Lauramide MIPA; CAS 142-54-1; nonionic; foam stabilizer, superfatting and thickening agent; shampoos and other detergents; fortifier for perfumes in soaps; solid; acid no. < 2%; > 90% amide content.

Standamid® PD. [Henkel/Cospha; Henkel Canada] Cocamide DEA (2:1) and diethanolamine; nonionic; more efficient foam builder, less efficient visc. builder than 1:1 superamides; stabilizer; used in shampoos, bubble baths with anionics, nonionics, and amphoterics; industrial cleaning; solubilizer for oily additives; amber visc. liq.; sol. in water, propylene glycol, PEG-8, SD-40 alcohol; dens. 8.17 lb/gal; visc. 1050 cps; cloud pt. 0 C; gel pt. 0 C; 100% conc.

Standamid® PK-KD. [Henkel/Cospha; Henkel Canada] Cocamide DEA superamide; nonionic; visc. builder, foamer; lt. amber liq.

Standamid® PK-KDO. [Henkel/Cospha; Henkel Canada] Cocamide DEA superamide; nonionic; enhances foam for shampoos, liq. hand soaps, etc.; lt. amber liq.

Standamid® PK-KDS. [Henkel/Cospha; Henkel Canada] Cocamide DEA superamide; nonionic; solubilizer, foam booster; lt. amber liq.

Standamid® PK-SD. [Henkel/Cospha; Henkel Canada] Cocamide DEA superamide; nonionic; for liq. conditioners, foaming prods.; lt. amber liq.

Standamid® SD. [Henkel/Cospha; Henkel Canada] Cocamide DEA (1:1); nonionic; high foaming detergent and foam stabilizer with pronounced visc. build-up; used in bath gels, liq. bubble baths, shampoos, dishwashing and laundry detergents; derived from whole coconut oil; contains no amide ester; low cost alkanolamide; amber visc. liq.; sol. in propylene glycol, PEG-8, SD-40 alcohol; insol. in water; visc. 1225 cps; m.p. 25 F; acid no. 0–1.0; cloud pt. < 0 C; gel pt. < 0 C; pH 9.5–10.5 (10%

sol'n.); 100% conc.

Standamid® SDO. [Henkel/Cospha; Henkel Canada] Cocamide DEA (1:1); nonionic; visc. builder and foam enhancer for aq. systems, shampoo, bubble baths, liq. hand soaps, industrial cleaners, offers excellent handling chars.; residual glycerin to enhance humectancy and hair conditioning effects; amber liq.; sol. in lower alcohols, propylene glycol, polyethylene glycols; sp.gr. 0.991; dens. 8.27 lb/gal; visc. 950 cps; acid no. 2.0 max.; cloud pt. < 0 C; gel pt. < 0 C; pH 9–11 (1.0%); 100% conc.

Standamid® SM. [Henkel/Cospha; Henkel Canada] Cocamide MEA (1:1); CAS 68140-00-1; visc. builder for anionic systems; wetting agent; produces lubricious foam for shampoo and bath prods., industrial cleaners; flakes, beads; sol. in lower alcohols; insol. in water; sp.gr. 0.934; m.p. 62 C; acid no. 1.0 max.; cloud pt. 63 C; gel pt. 56 C; pH 9–11 (1.0%); 100% conc.

Standamid® SOD. [Henkel] Linoleamide DEA (1:1); nonionic; visc. enhancer, thickener in industrial, institutional and household cleaners; corrosion inhibitor and emulsifier for oils and lubricants; dk. amber liq.; sol. in lower alcohols, glycols, acetone, toluene, xylene; disp. in water; dens. 8.20 lb/gal; pH 9–11 (1%); 100% conc.

Standamid® SOMD. [Henkel/Cospha] Linoleamide DEA (1:1); nonionic; visc. builder with foam enhancement chars.; produces esp. high visc. when used with ethoxylated anionics; used in gel shampoos, bubble baths, liq. hand soaps, industrial cleaners, and formulation; where the amt. of electrolyte must be kept to a min.; low-cost shampoo concentrates; waxy solid; sol. in lower alcohols; sp.gr. 0.935; acid no. 2 max.; cloud pt. 38 C; gel pt. 34 C; pH 9–11 (1.0%); 100% conc.

Standamox C 30. [Henkel] Coco amido amine oxide; wetting agent, foam stabilizer, visc. builder; shampoos, bubble baths, bath oils, dishwash concentrates, and other detergents; liq.; pH 6–7 (as is); 28–30% act.

Standamox CAW. [Henkel/Cospha; Henkel Canada] Cocamidopropylamine oxide; nonionic; wetting agent, foam builder, stabilizer, thickener, lubricant, emollient for low irritation baby shampoos, bubble baths, skin care preps.; clear liq.; 30% conc.

Standamox LAO-30. [Henkel/Cospha; Henkel Canada] Lauryl/myristyl dimethylamine oxide; cationic; foaming and conditioning agent for household and industrial cleaners; liq.; 30% conc.

Standamox LMW-30. [Henkel] Lauryl/myristyl dimethylamine oxide; cationic; foamer and conditioner in household and industrial cleaners; liq.; 30% conc.

Standamox O1. [Henkel/Cospha; Henkel Canada] Oleamine oxide; nonionic; biodeg. thickener, bacteriostat, dye assistant, lubricant, softener used in industrial applics., hair prods., plating compds., lube oils; amber clear liq.; 55% conc.

Standamox PCAW. [Pulcra SA] Cocamidopropyl dimethylamine oxide; CAS 68155-09-9; nonionic; detergent, emulsifier, wetting agent, foam stabilizer; stable over wide pH range; liq.; m.w. 307; pH 5-7 (5%); 29-31% conc.

Standamox PL. [Pulcra SA] Lauramine oxide and myristamine oxide; nonionic; detergent, emulsifier, wetting agent, foam stabilizer; stable over wide pH range; liq.; m.w. 246; pH 7.0-7.3 (5%); 29-31% act.

Standamox PS. [Pulcra SA] Stearyl dimethylamine oxide; CAS 2571-88-2; nonionic; conditioner, softener for cream rinses, lotions, liq. bath prods.; paste; 25% conc.

Standamul® B-1. [Henkel] Ceteareth-12; CAS 68439-49-6; nonionic; emulsifier and solubilizer for personal care and household prods., essential oils; waxy solid; sp.gr. 0.97 (70 C); HLB 12; solid. pt. 35 C; 99.5–100% solids.

Standamul® B-2. [Henkel] Ceteareth-20; CAS 68439-49-6; nonionic; emulsifier and solubilizer for personal care prods.; additive in the paper and textile industry; household cleaning and leather prods.; waxy solid; sp.gr. 1.0 (70 C); HLB 14; solid. pt. 40 C; 99.7–100% solids.

Standamul® B-3. [Henkel] Ceteareth-30; CAS 68439-49-6; nonionic; emulsifier and solubilizer for personal care prods.; additive in the paper and textile industry; household cleaning and leather prods.; waxy solid; sp.gr. 1.023 (70 C); HLB 15; solid. pt. 43–46 C; 99.5–100% solids.

Standamul® STC-25. [Henkel] Stearalkonium chloride; cationic; antistat, surfactant for personal care prods. and fabrics; wh. slurry; 25% conc.

Standapol® 1610. [Henkel/Textile] Sulfated peanut oil; anionic; emulsifier, wetting agent, dispersant, antistat for nylon finishes; amber clear oil; disp. in water; dens. 8.5 lb/gal; visc. 800 cps; 77% act.

Standapol® A. [Henkel/Cospha; Henkel Canada] Ammonium lauryl sulfate; anionic; detergent, foamer, suspending agent, base for shampoos, cleaning compds. with near neutral pH; water-wh. visc. liq.; visc. 1500–3000 cps; pH 6.5–7.0 (10% aq.); 27.5–28.5% act.

Standapol® AK-43. [Pulcra SA] POE stearate; nonionic; lubricant, antistat for textile industry, for aftertreatment of dyed polyester fibers; preparation agent for syn. fibers, esp. viscose rayon staple and wool-type polyester; paste; 90% conc.

Standapol® AL-60. [Henkel] Ammonium laureth sulfate; surfactant for personal care, household and industrial cleaning prods.; yel. clear liq.; water-sol.; dens. 1.025 g/ml; visc. 100–200 cP; cloud pt. 10 C max.; pH 6.0–7.0; 58–60% act.

Standapol® AP-60. [Henkel] Ammonium C12-15 pareth sulfate; surfactant for personal care, household and industrial cleaning prods.; yel. clear visc. liq.; water-sol.; dens. 1.025 g/ml; visc. 100–200 cps; cloud pt. 10 C max.; pH 6.0–7.0 (10%); 58–60% act.

Standapol® AP Blend. [Henkel/Cospha; Henkel Canada] Sodium laureth sulfate, cocamide DEA, cocamidopropyl betaine; conc. for shampoo, bath and cleansing prods., liq. soaps; exc. foaming and visc. response; lt. amber liq.; 35-40% conc.

Standapol® BW. [Henkel/Cospha; Henkel Canada] Sodium lauryl sulfate, cocamide DEA, cocamidopropyl; anionic/nonionic; formulated conc. for personal care cleaning and bath prods.; base for liq. dishwash; liq.; 37% conc.

Standapol® EA-K. [Henkel/Cospha; Henkel Canada] Ammonium myreth sulfate, cocamide DEA; base for manual dishwash, pet shampoos, bath prods., lt. duty cleaners, body cleansers; exc. visc. and foam performance; liq.; 61% conc.

Standapol® ES-50. [Henkel/Cospha; Henkel Canada] Sodium myreth sulfate; anionic; surfactant for liq. shampoos, skin cleansers, cleaning prods., foams; cloudy visc. liq.; 50% conc.

Standapol® ES-250. [Henkel/Cospha; Henkel Canada] Sodium laureth sulfate; CAS 9004-82-4; anionic; surfactant for shampoos, cleaning prods.; pale yel. clear liq.; 53% act.

Standapol® ES-350. [Henkel/Cospha; Henkel Canada] Sodium laureth sulfate; CAS 9004-82-4; anionic; surfactant for shampoos, bath prods., detergent cleaning prods.; pale yel. clear liq.; 53% act.

Standapol® LF. [Henkel/Cospha; Henkel Canada] Sodium octyl sulfate; anionic; wetting agent for metal degreasers, hard surface cleaners, food equipment cleaners, dust control; solubilizer, hydrotrope; resistant to hard water; liq.; 33% act.

Standapol® S. [Henkel/Cospha; Henkel Canada] Sodium lauryl sulfate, sodium laureth sulfate, lauramide MIPA, cocamide MEA, glycol stearate, and coceth-8; anionic/nonionic; pearlescent shampoo and liq. soap base; pearly-wh. visc. liq.; visc. 1200–1700 cps; pH 6.6–7.4 (10% aq.); 36.8–37.3% act.

Standapol® SL-60. [Henkel] Sodium laureth sulfate; CAS 9004-82-4; surfactant for personal care, household, and industrial cleansing prods.; lt. yel. clear liq.; water-sol.; dens. 1.050 g/ml; visc. 100–200 cps; cloud pt. 10 C max.; pH 7.0–9.0 (10%); 58–60% act.

Standapol® SP-60. [Henkel] Sodium C12–15 pareth sulfate; surfactant for personal care, household, and industrial cleansing prods.; yel. clear liq.; water-sol.; dens. 1.050 g/ml; visc. 100–200 cps; cloud pt. 10 C max.; pH 7.0–9.0 (10%); 58–60% act.

Standapol® WAQ-LC. [Henkel/Cospha; Henkel Canada] Sodium lauryl sulfate; CAS 151-21-3; foaming agent, detergent, suspending agent for personal care prods., liq. cleaners; low salt content for improved corrosion resistance; water-wh. liq.; visc. 100 cps max.; cloud pt. 8 C max.; pH 7–9 (10% aq.); 28.5–31.5% conc.

Standapol® WAQ-LCX. [Henkel/Cospha; Henkel Canada] Sodium lauryl sulfate; CAS 151-21-3; anionic; low cloud pt. surfactant used in rug shampoos; liq.; 28–30% conc.

Standapon 4149 Conc. [Henkel/Textile] Anionic; conc. scour for soaping-off dyed or printed goods; clear liq.; forms clear sol'ns.; pH 7.5 (5%).

Stanlev R-276. [Henkel/Textile] Leveling agent for use with disperse or neutral acid dyes on nylon; clear liq.; sol. forms clear sol'ns. in water; pH 7.5 (2%).

Stanol 212F. [Henkel-Nopco] Organic sulfates with nonionics; anionic; scouring and fulling agent for wool; shampoo base; bubble baths; liq.; 40% conc.

Stansperse 506. [Henkel/Textile] Nonionic; dyebath stabilizer improving stability of softeners in textile dyebaths; clear liq.; sol. forms clear sol'ns. water; pH 6.0 (1%).

Stantex 322. [Henkel Canada] Ester sulfate; anionic; wetting and rewetting agent for dyeing and finishing; liq.; 64% conc.

Stantex Antistat F. [Henkel/Textile] Nonionic; low m.p. antistat and emulsifier for nylon finishes; off-wh. paste; sol. in water; dens. 8.8 lb/gal; 99% act.

Stantex MOR. [Henkel Canada] Solvent/detergent blend; anionic; detergent for removal of grease, tar, and other impurities from cotton piece goods in prep. for bleaching and dyeing; liq.; 95% conc.

Stantex MOR Special. [Henkel Canada] Solvent/detergent blend; anionic; detergent for removal of grease, tar, and other impurities from cotton piece goods in prep. for bleaching and dyeing; liq.; 100% conc.

Stantex PENE 20. [Henkel/Textile] Anionic; wetting agent for latex carpet backing; clear liq.; disp. in 120 F water; pH 7.0 (1%).

Stantex PENE 40-DF. [Henkel/Textile] Anionic; wetting and rewetting agent for textile processing under acid and alkaline conditions; clear liq.; disp. in 120 F water; pH 7.0 (1%).

Stantex T-14 DF. [Henkel/Textile] Anionic; low foaming wetting and rewetting agent for textile processing; clear visc. liq.; disp. in 120 F water; pH 7.0 (1%).

Statik-Blok® FDA-3. [Amstat Industries] Polyether type; nonionic; surface-active antistatic sol'n. for food contact surfaces; clear, colorless liq., pract. odorless; sol. in water, IPA; pH 7.

Stearate PEG 1000. [Ceca SA] PEG-21 stearate; CAS 9004-99-3; nonionic; surfactant; solid; HLB 15.3; 100% conc.

Stenol 16-65. [Henkel/Emery] Sat. fatty alcohol (65-72% C16, 25-33% C18); chemical intermediate for detergent mfg.; APHA 20 fused/flakes; sp.gr. 0.805-0.815 g/cc (60 C); solid. pt. 46-49 C; b.p. 305-355 C; acid no. < 0.1; iodine no. < 0.5; sapon. no. < 1.0; hyd. no. 215-225; flash pt. 160 C.

Stenol 1222. [Henkel/Emery] Sat. fatty alcohol (25-35% C16, 17-27% C18, 10-20% C20, 10-20% C22); chemical intermediate for detergent mfg.; APHA 20 fused/flakes; sp.gr. 0.810-0.820 g/cc (60 C); solid. pt. 45-50 C; b.p. 310-390 C; acid no. < 0.1; iodine no. < 2.0; sapon. no. < 1.2; hyd. no. 205-215; flash pt. 200 C.

Stenol 1618. [Henkel/Emery] Sat. fatty alcohols (45-55% C16, 45-55% C18); intermediate for surfactant mfg.; APHA 20 fused/flakes; sp.gr. 0.805-0.815 g/cc (60 C); solid. pt. 48-52 C; b.p. 310-360 C; acid no. < 0.1; iodine no. < 0.5; sapon. no. < 1.0; hyd. no. 215-220; flash pt. 170 C; 100% conc.

Stenol 1822 SR. [Henkel/Emery] Sat. fatty alcohol (40-50% C18, 5-20% C20, 42-50% C22); chemical intermediate for detergent mfg.; APHA 20 fused/flakes; sp.gr. 0.800-0.810 g/cc (70 C); solid. pt. 55-60 C; b.p. 340-390 C; acid no. < 0.2; iodine no. < 1.0; sapon. no. < 0.5; hyd. no. 180-190; flash pt. 200 C.

Steol® 4N. [Stepan; Stepan Canada] Sodium laureth sulfate; CAS 9004-82-4; anionic; detergent, emulsifier, foamer, wetting agent used in personal care prods.; car wash, dishwash; textile mill applics.; emulsion polymerization; water-wh. liq.; water-sol.; sp.gr. 1.045; visc. 31 cps; cloud pt. –4 C; pH 8.0; 28% act.

Steol® CA-460. [Stepan; Stepan Canada] Ammonium laureth sulfate; anionic; detergent, emulsifier, foamer, dispersant, and wetting agent used in shampoos, dishwashers, car washers, textile mill applics., emulsion polymerization; pale yel. liq.; water-sol.; sp.gr. 1.016; visc. 67 cps; cloud pt. 19 C; pH 7.0; 60% conc.

Steol® CO 436. [Stepan Canada] Ammonium nonoxynol sulfate; surfactant; 60% act.

Steol® COS 433. [Stepan Canada] Sodium nonoxynol-4 sulfate; anionic; emulsifier for acrylics, SBR, vinyl chloride, butyl rubber; pale yel. clear liq.; 30% act.

Steol® CS-260. [Stepan Canada] Sodium laureth sulfate; CAS 9004-82-4; surfactant; 60% act.

Steol® CS-460. [Stepan; Stepan Canada] Sodium laureth sulfate; CAS 9004-82-4; anionic; surfactant, foamer, emulsifier for liq. detergents, carwash, dishwash liq., laundry, alkaline detergents, textiles; pale yel. liq.; water-sol.; sp.gr. 1.037; visc. 69 cps; cloud pt. 17 C; pH 8.0 (1% aq.); 59.5% act.

Steol® EP-110. [Stepan Canada] Ammonium nonoxynol sulfate; surfactant; 30% act.

Steol® EP-115. [Stepan Canada] Ammonium nonoxynol sulfate; surfactant; 30% act.

Steol® EP-120. [Stepan Canada] Ammonium nonoxynol sulfate; surfactant; 30% act.

Steol® KA-460. [Stepan; Stepan Canada] Ammonium ether sulfate; anionic; detergent, emulsifier, foamer and wetting agent used in liq. detergents, shampoos, carwash, textile detergents; suitable for soft and hard waters; biodeg.; straw liq.; water-sol.; dens. 8.65 lb/gal; cloud pt. 0 C; pH 7.0 (10% aq.); 60% act.

Steol® KS-460. [Stepan; Stepan Canada] Linear fatty alcohol ether sulfate, sodium salt; anionic; surfactant, foamer, emulsifier for liq. detergents, shampoos, carwash, textile detergent; biodeg.; straw liq.; water-sol.; dens. 8.94 lb/gal; cloud pt. 0 C; pH 8.0 (10% aq.); 60% act.

Steol® OS 28. [Stepan Europe] Sodium laureth sulfate; CAS 9004-82-4; anionic; detergent, foaming agent for all-purpose and specialty household/industrial cleaners; water-wh. to yel. visc. liq.; 28% act.

Stepan C-40. [Stepan; Stepan Canada] Methyl laurate; intermediate for mfg. of detergents, emulsifiers, wetting agents, stabilizers, lubricants, plasticizers, resins, and textile specialties; lubricant for metalworking formulations; water-wh.; m.p. 5 C; acid no. 0.5 max.; iodine no. 0.1; sapon. no. 260–264.

Stepan C-65. [Stepan; Stepan Canada] Methyl palmitate-oleate; intermediate for mfg. of detergents, emulsifiers, wetting agents, stabilizers, lubricants, plasticizers, resins, and textile specialties; lubricant for metalworking formulations; Gardner 1 max. liq.; m.p. 14 C; acid no. 1.0 max.; iodine no. 50; sapon. no. 197–203.

Stepan C-68. [Stepan; Stepan Canada] Methyl oleate/stearate; intermediate for mfg. of detergents, emulsifiers, wetting agents, stabilizers, lubricants, plasticizers, resins, and textile specialties; lubricant for metalworking compds.; Gardner 2 max. liq.; m.p. 16 C; acid no. 2.0 max.; iodine no. 75; sapon. no. 191–197; 100% act.

Stepanflo 20. [Stepan; Stepan Canada] Proprietary; anionic; surfactant for enhanced oil recovery operations where high temp. and pressure stability are necessary; liq.; 40% conc.

Stepanflo 30. [Stepan; Stepan Canada] Proprietary; anionic; surfactant for enhanced oil recovery operations where high temp. and pressure stability are necessary; liq.; 30% conc.

Stepanflo 40. [Stepan; Stepan Canada] Proprietary; anionic; surfactant for enhanced oil recovery operations where high temp. and pressure stability are necessary; liq.; 60% conc.

Stepanflo 50. [Stepan; Stepan Canada] Proprietary; anionic; surfactant for enhanced oil recovery operations where high temp. and pressure stability are necessary; liq.; 60% conc.

Stepanflote® 24. [Stepan; Stepan Canada] Sodium alkyl sulfate; anionic; flotation reagent for nonmetallic ores; tan paste; 25% conc.

Stepanflote® 85L. [Stepan; Stepan Canada] Sodium alkyl ether sulfate; anionic; flotation reagent for molybdenum ore; clear liq.; 30% act.

Stepanflote® 97A. [Stepan; Stepan Canada] Sodium alkyl ether sulfate; anionic; flotation reagent for nonmetallic ore; clear liq.; 42% act.

Stepanform® 1050. [Stepan; Stepan Canada] Blend; anionic; drilling foamer; clear liq.; 47% act.

Stepanform® 1440. [Stepan; Stepan Canada] Blend; anionic; foaming agent, air entrainer for cellular concrete; clear liq.; 37% act.

Stepanform® 1750. [Stepan; Stepan Canada] Blend; anionic; foamer for drilling applics.; clear liq.; 50% act.

Stepanform® 1850. [Stepan; Stepan Canada] Blend; anionic; foamer for drilling applics.; clear liq.; 50% act.

Stepanform® 2160. [Stepan; Stepan Canada] Blend; anionic; foamer for drilling applics.; clear liq.; 55% act.

Stepanform® 3040. [Stepan; Stepan Canada] Blend; anionic; high expansion foamer for dust control applics.; clear liq.; 35% act.

Stepanform® 5012. [Stepan; Stepan Canada] Blend; anionic; very stable high expansion foamer; low cloud pt.; amber clear liq.; 30% act.

Stepanform® 5040. [Stepan; Stepan Canada] Blend; anionic; very stable high expansion foamer; clear liq.; 30% act.

Stepanform® HP-95. [Stepan; Stepan Canada] Blend; anionic/nonionic; emulsifier for vegetable oils in nonedible applics.; clear visc. liq.; 98% conc.

Stepanform® HP-116. [Stepan; Stepan Canada] Blend; anionic/nonionic; emulsifier for hydrocarbon liqs.; clear visc. liq.; 98% conc.

Stepan-Mild® LSB. [Stepan; Stepan Canada] Sodium lauryl sulfoacetate, disodium laurethsulfosuccinate; surfactant for shampoos, hand soaps, bubble baths, facial cleansers, baby prods., sensitive skin prods.; clear liq.

Stepan-Mild® SL3. [Stepan; Stepan Canada] Disodium laurethsulfosuccinate; surfactant for low-irritation shampoos, bubble baths, dishwashing detergents; liq.; 30-34% conc.

Stepanol® AEG. [Stepan; Stepan Canada] Ammonium lauryl sulfate, ammonium laureth sulfate, cocamidopropyl betaine, cocamide DEA; anionic; base for liq. hand and body soaps, shampoos; liq.; 41% conc.

Stepanol® AM. [Stepan; Stepan Canada] Ammonium lauryl sulfate; anionic; detergent, foamer used in personal care prods.; rug and upholstery shampoos; household, metal, and industrial cleaners; fruit washing; insecticides; textile and leather processing; pharmaceuticals; pale yel. visc. liq.; water-sol.; pH 6–7 (10%); 28–30% act.

Stepanol® AM-V. [Stepan; Stepan Canada] Ammonium lauryl sulfate; anionic; detergent, foamer used in personal care prods.; rug and upholstery shampoos; household, metal, and industrial cleaners; fruit washing; insecticides; textile and leather processing; pharmaceuticals; visc. liq., lt. gel; water-sol.; pH 6–7 (10%); 27–29% act.

Stepanol® DEA. [Stepan; Stepan Canada] DEA-lauryl sulfate; anionic; detergent, foamer used in personal care prods.; rug and upholstery shampoos; household, metal, and industrial cleaners; fruit washing; insecticides; textile and leather processing; pharmaceuticals; pale yel. clear liq.; water-sol.; pH 7.3–7.7 (10%); 33–35% act.

Stepanol® DFS. [Stepan; Stepan Canada] Formulated prod.; anionic; rug shampoos, hard surface cleaners.

Stepanol® GP-3 Conc. [Stepan Canada] Surfactant blend; general purpose cleaner conc.

Stepanol® HDL-50. [Stepan Canada] Surfactant blend; heavy-duty liq. blend.

Stepanol® LDL-3. [Stepan Canada] Surfactant blend; lt.-duty liq. blend.

Stepanol® ME Dry. [Stepan; Stepan Canada] Sodium lauryl sulfate; CAS 151-21-3; anionic; detergent, foamer used in personal care prods.; rug and upholstery shampoos; household, metal, and industrial cleaners; fruit washing; insecticides; textile and leather processing; pharmaceuticals; wh. powd.; water-sol.; pH 7.5–1.0 (10%); 93% min. act.

Stepanol® MG. [Stepan; Stepan Canada] Magnesium lauryl sulfate; anionic; detergent, foamer used in personal care prods.; rug and upholstery shampoos; household, metal, and industrial cleaners; fruit washing; insecticides; textile and leather processing; pharmaceuticals; water-sol.; pH 6.5–7.5 (10%); 28–30% act.

Stepanol® RS. [Stepan; Stepan Canada] Blend; anionic; foaming and wetting agent for rug and upholstery shampoos; low cloud pt.

Stepanol® SPT. [Stepan Europe] TEA lauryl sulfate; anionic; mild detergent, foaming agent for household cleaners, liq. soaps; yel. visc. liq.; 40% act.

Stepanol® WA-100. [Stepan; Stepan Canada; Stepan Europe] Sodium lauryl sulfate; CAS 151-21-3; anionic; detergent, foamer, wetting and suspending agent; dentrifrice formulations; pharmaceuticals; emulsion polymerization; clay dispersions; wh. powd.; water-sol.; pH 7.5–10 (10%); 98.5% min. act.

Stepanol® WAC. [Stepan; Stepan Canada; Stepan Europe] Sodium lauryl sulfate; CAS 151-21-3; anionic; detergent, foamer used in personal care prods.; rug and upholstery shampoos; household, metal, and industrial cleaners; fruit washing; insecticides; textile and leather processing; pharmaceuticals; pale yel. clear liq.; water-sol.; pH 7.5–8.5 (10%); 28–30% act.

Stepanol® WAC-P. [Stepan Canada] Sodium lauryl sulfate; CAS 151-21-3; surfactant; 30% act.

Stepanol® WA Extra. [Stepan; Stepan Canada] Sodium lauryl sulfate; CAS 151-21-3; anionic; detergent, foamer used in personal care prods.; rug and upholstery shampoos; household, metal, and industrial cleaners; fruit washing; insecticides; textile and leather processing; pharmaceuticals; clear liq.; water-sol.; pH 7.5–8.5 (10%); 28–30% act.

Stepanol® WA Paste. [Stepan; Stepan Canada] Sodium lauryl sulfate; CAS 151-21-3; anionic; detergent, foamer used in personal care prods.; rug and upholstery shampoos; household, metal, and industrial cleaners; fruit washing; insecticides; textile and leather processing; pharmaceuticals; water-wh. clear paste; water-sol.; pH 7.5–8.5 (10%); 28–30% act.

Stepanol® WAQ. [Stepan; Stepan Canada] Sodium lauryl sulfate; CAS 151-21-3; anionic; detergent, foamer used in personal care prods.; rug and upholstery shampoos; household, metal, and industrial cleaners; fruit washing; insecticides; textile and leather processing; pharmaceuticals; water-wh. clear visc. liq.; water-sol.; pH 7.5–8.5 (10%); 28–30% act.

Stepanol® WA Special. [Stepan; Stepan Canada] Sodium lauryl sulfate; CAS 151-21-3; anionic; detergent, foamer used in personal care prods.; rug and upholstery shampoos; household, metal, and industrial cleaners; fruit washing; insecticides; textile and leather processing; pharmaceuticals; water-wh. clear liq.; water-sol.; pH 7.5–8.5 (10%); 28–30% act.

Stepanol® WAT. [Stepan; Stepan Canada] TEA-lauryl sulfate; anionic; detergent, foamer used in personal care prods.; rug and upholstery shampoos; household, metal, and industrial cleaners; fruit washing; insecticides; textile and leather processing; pharmaceuticals; water-wh. clear liq.; water-sol.; pH 7.0–8.5 (10%); 39–41.5% act.

Stepanon CG. [Stepan; Stepan Canada] Surfactant; amphoteric; brine tolerant foamer for acid or alkaline media; clear liq.; 30% act.

Stepanquat® F/T. [Stepan Europe] Alkyl ammonium methoxysulfate; cationic; fludifying, dispersing, antistatic agent; additive for stabilizing and lower visc. in emulsions and cationic disps.; for household/industrial cleaners; water-wh. to yel. visc. liq.; 85-100% act.

Stepantan® AS-12. [Stepan; Stepan Canada] Sodium alpha olefin sulfonate; foamer for soft and hard waters, fresh water, and moderate brine conditions; amber clear liq.; 40% act.

Stepantan® AS-12 Flake. [Stepan; Stepan Canada] Sodium alpha olefin sulfonate; foaming surfactant for soapstick applics.; amber flakes; 97% act.

Stepantan® AS-40. [Stepan; Stepan Canada] Sodium alpha olefin sulfonate; anionic; foaming agent yielding expanded foam volume with stability; for firefighting, cellular concrete, air and mist drilling; yel. clear liq.; 40% act.

Stepantan® AS-90 Beads. [Stepan; Stepan Canada] Sodium alpha olefin sulfonate; surfactant for foaming in fresh water; solid beads; 90% act.

Stepantan® CG. [Stepan; Stepan Canada] Anionic; brine-tolerant foamer for acid or alkaline media; clear liq.; 30% conc.

Stepantan® DS-40. [Stepan; Stepan Canada] Sodium dodecylbenzene sulfonate; anionic; detergent, wetting agent, foaming agent; used specifically in dust control applics.; clear liq.; 40% conc.

Stepantan® DT-60. [Stepan; Stepan Canada] TEAdodecylbenzenesulfonate; anionic; detergent, wetting agent, foaming agent; used specifically as air entraining agent in concrete applics.; yel. clear liq.; 60% conc.

Stepantan® H-100. [Stepan; Stepan Canada] Dodecylbenzene sulfonic acid; anionic; detergent intermediate; emulsifier for oils, solvs., waxes and oil field applics.; air entraining agent and foamer for cellular concrete; wetting agent; dk. visc. liq.; water-sol.; sp.gr. 1.09; dens. 9.1 lb/gal; visc. 9000 cps; 97% act.

Stepantan® HP-90. [Stepan; Stepan Canada] Alkylamine alkylaryl sulfonate; anionic; emulsifier, detergent; visc. amber liq.; oil-sol.; 90% conc.

Stepantan® NP 80. [Stepan Europe] Condensed naphthalene sulfonate; dispersant, wetting agent for wettable powds.; powd.

Stepantex® 130. [Stepan; Stepan Canada] Blend; anionic/nonionic; high caustic textile scouring agent; liq.; 40% act.

Stepantex® B-29. [Stepan; Stepan Canada] Sodium octyl sulfate; wetting and mercerizing agent for textiles; clear liq.; 32% act.

Stepantex® CO-30. [Stepan; Stepan Canada] PEG-30 castor oil; nonionic; lubricant, emulsifier for fiber finish and wet processing; liq.; 100% act.

Stepantex® CO-36. [Stepan; Stepan Canada] PEG-36 castor oil; nonionic; lubricant, emulsifier for fiber finish and wet processing; liq.; 100% act.

Stepantex® CO-40. [Stepan; Stepan Canada] PEG-40 castor oil; nonionic; lubricant, emulsifier for fiber finish and wet processing; liq.; 100% act.

Stepantex® DA-6. [Stepan; Stepan Canada] Alcohol ethoxylate; low foaming scouring, wetting agent, leveling agent, penetrant for textiles; liq.; 100% act.

Stepantex® DO 90. [Stepan Europe] Dialkyl ammonium methoxysulfate; cationic; antistat, hydrophobing agent for metal surf. treatments; softening agent; pale yel. to beige liq.; 90% act.

Stepantex® GS 90. [Stepan; Stepan Canada] Veg. derived diester quat. ammonium methyl sulfate; cationic; good hand, exc. rewet and antistatic props. for household and commercial rinse-added fabric softeners; textile processing aid for natural and syn. fibers; Gardner 3 paste; dens. 8.07 lb/gal (50 C); visc. 1200 cps (105 F); pour pt. 97 F; pH 3 (10% aq. IPA); 90% solids.

Stepantex® TD14. [Stepan; Stepan Canada] Tall oil ester; biodeg. textile aux.; liq.; 100% act.

Stepantex® TM10. [Stepan; Stepan Canada] Tall oil ester; biodeg. textile aux.; liq.; 100% act.

Stepantex® TM15. [Stepan; Stepan Canada] Tall oil ester; biodeg. textile aux.; liq.; 100% act.

Stepantex® VS 90. [Stepan; Stepan Canada] Diester quat. ammonium methyl sulfate; cationic; good hand, exc. rewet and antistatic props. for household and commercial rinse-added fabric softeners; textile processing aid for natural and syn. fibers; Gardner 3 paste; dens. 8.08 lb/gal (50 C); visc. 1000 cps (105 F); pour pt. 95 F; pH 3 (10% aq. IPA); 90% solids.

Stepfac® 8170. [Stepan; Stepan Canada] Ethoxylated nonylphenol phosphate; anionic; hydrotrope for nonionics; emulsifier for agric., emulsion polymerization, oils, metalworking lubricants, corrosion inhibitors, pigment dispersants; heavy-duty industrial/household alkali cleaners; compatibility agent for liq. fertilizers; amber clear liq.; sol. in xylene, water; pH 2.0 (1%); 100% act.

Stepfac® 8171. [Stepan; Stepan Canada] Acid form; anionic; compatibility agent for agric. formulations; pale yel. clear liq.; sol. in xylene.

Stepfac® 8172. [Stepan; Stepan Canada] Acid form; anionic; compatibility agent for agric. formulations; pale yel. clear liq.; sol. in xylene.

Stepfac® 8173. [Stepan; Stepan Canada] Acid form; anionic; compatibility agent for agric. formulations; pale yel. clear liq.; sol. in xylene.

Stepfac® PN 10. [Stepan Europe] Phosphate ester; anionic; emulsifier, wetting and dispersing agent for agric. wettable powds. and flowables; water-wh. to yel. visc. liq.; 100% conc.

Step-Flow 21. [Stepan; Stepan Europe] Nonionic dispersant; nonionic; surfactant for aq. agric. flowables; lt. amber liq.; sol. in xylene, water; HLB 15.0.

Step-Flow 22. [Stepan; Stepan Europe] Nonionic dispersant; nonionic; surfactant for aq. agric. flowables; off-wh. paste; sol. in xylene, water; HLB 17.0.

Step-Flow 23. [Stepan; Stepan Europe] Nonionic dispersant; nonionic; surfactant for aq. agric. flowables; off-wh. paste; sol. in xylene, water; HLB 12.0.

Step-Flow 24. [Stepan; Stepan Europe] Nonionic dispersant; nonionic; surfactant for aq. agric. flowables; clear liq.; sol. in xylene, water; HLB 13.0.

Step-Flow 25. [Stepan; Stepan Europe] Nonionic dispersant; nonionic; surfactant for aq. agric. flowables; off-wh. paste; sol. in xylene, water; HLB 14.0.

Step-Flow 26. [Stepan; Stepan Europe] Nonionic dispersant; nonionic; surfactant for aq. agric. flowables; off-wh. paste; sol. in xylene, water; HLB 13.0.

Step-Flow 41. [Stepan; Stepan Europe] Anionic dispersant; anionic; surfactant for aq. agric. flowables; dk. liq.; sol. in water.

Step-Flow 42. [Stepan; Stepan Europe] Anionic dispersant; anionic; surfactant for aq. agric. flowables; yel. clear liq.; sol. in xylene, water.

Step-Flow 61. [Stepan; Stepan Europe] Anionic stabilizing surfactant; anionic; surfactant for aq. agric. flowables; pale yel. clear liq.; sol. in xylene, water.

Step-Flow 63. [Stepan; Stepan Europe] Nonionic stabilizing surfactant; nonionic; surfactant for aq. agric. flowables; pale yel. clear liq.; sol. in water; HLB 11.0.

Steposol® CA-60H. [Stepan; Stepan Canada] Ammonium ether sulfate; high flash pt. foamer for heavy brine conditions; amber clear liq.; 60% act.

Steposol® CA-207. [Stepan; Stepan Canada; Stepan Europe] Ammonium ether sulfate; anionic; foaming agent used for oilfield applics.; also for gypsum board, cellular concrete, air drilling, foam cleaners; exc. for heavy brine conditions; stable in soft or hard water; pale yel. liq.; f.p. < 0 F; 60% act.

Steposol® CA-319. [Stepan; Stepan Canada] Ammonium ether sulfate; foamer for heavy brine and hydrocarbon conditions; for oilfield applics.; yel. clear liq.; 60% act.

Stepsperse® DF-100. [Stepan; Stepan Europe] Surfactant blend; anionic/nonionic; dispersant for agric. flowables and dry flowables; dk. brn. powd.; sol. in water.

Stepsperse® DF-200. [Stepan; Stepan Europe] Surfactant blend; anionic; dispersant for agric. flowables and dry flowables; brn. powd.; sol. in water.

Stepsperse® DF-300. [Stepan; Stepan Europe] Surfactant blend; anionic; dispersant for agric. flowables and dry flowables; brn. powd.; sol. in water.

Stepsperse® DF-400. [Stepan; Stepan Europe] Surfactant blend; anionic/nonionic; dispersant for agric. flowables and dry flowables; dk. brn. powd.; sol. in water.

Stepwet® DF-60. [Stepan; Stepan Europe] Surfactant blend; anionic; wetting agent for agric. flowables and dry flowables; brn. powd.; sol. in water.

Stepwet® DF-90. [Stepan; Stepan Europe] Surfactant blend; anionic; wetting agent for agric. flowables and dry flowables; pale yel. ground flake; sol. in water.

Steramine 49. [Henkel-Nopco] Fatty amidoamine/nonionics blend; nonionic; antistatic softener for cotton and syn. fibers; fluid paste; 25% conc.

Steramine CD. [Henkel-Nopco] Polyoxyalkylene fatty deriv.; nonionic; softener assistant in textile prods., starch and thermosetting resin finishes; fluid paste; 26% conc.

Steramine CGL. [Henkel-Nopco] Nonionic and quat. amine derivs. blend; cationic; biodeg. textile softener; liq. cream; 23% conc.

Steramine CR 25. [Henkel-Nopco] Nonionic and quat. amine derivs. blend; cationic; antistatic softener for syn. fibers, esp. polyamide and acrylic; paste; 25% conc.

Steramine FPA 197. [Henkel-Nopco] Alkyl amido amine; cationic; antistat, lubricant, softener for acrylics and blends; fluid paste; 21% conc.

Steramine GS. [Henkel-Nopco] Fatty amine and amide deriv.; cationic; antistatic lubricant for acrylics; fluid paste; 20% conc.

Steramine PNA 75. [Henkel-Nopco] Alkyl amido amine; antistat, lubricant for acrylics and blends; fluid paste; 20% conc.

Steramine S2. [Henkel-Nopco] Fatty alcohol and acid EO adduct; nonionic; finishing agent for textile prods.; wax; 100% conc.

Steramine TV. [Henkel-Nopco] Polyoxyalkylene fatty deriv.; nonionic; softener assistant for textile prods., starch and thermosetting resin finishes; fluid paste; 32% conc.

Sterol CC 595. [Auschem SpA] PEG-6 caprylic/capric glycerides; nonionic; superfatting agent; liq.; 100% conc.

Sterol GMS. [Auschem SpA] Glyceryl stearate; CAS 31566-31-1; nonionic; w/o emulsifier, thickener for o/w emulsions; flakes; HLB 3.2; 100% conc.

Sterol LG 491. [Auschem SpA] PEG-8 glyceryl laurate; nonionic; superfatting agent; liq.; water-sol.; HLB 11.0; 100% conc.

Sterol ST 1. [Auschem SpA] PEG-1 stearate; CAS 111-60-4; nonionic; pearling agent, w/o emulsifier; flakes; HLB 2.7; 100% conc.

Sterol ST 2. [Auschem SpA] Diethylene glycol stearate; CAS 9004-99-3; 106-11-6; nonionic; pearling agent, w/o emulsifier; solid; HLB 4.7; 100% conc.

Straight Alkaline Cleaner SAC. [Emkay] Liq. alkaline cleaner for removing residues from tanks after carrying veg., animal and fish oils.

Strodex® MO-100. [Dexter] 2-Ethylhexyl polyphosphoric ester acid anhydride; anionic; emulsifier; pigment dispersant used in oil-based paints and leather coating specialties; Gardner 14 max. liq.; sol. in polar and nonpolar solvs.; insol. in water; sp.gr. 1.08; visc. 1300 ± 300 cps; flash pt. (TCC) 160 F; pH 1.9; 98.5 ± 0.5% act.

Strodex® MOK-70. [Dexter] Phosphated alcohol; anionic; dispersant/stabilizer for carbon blk. pigments, zinc pigments, lead silicate in exterior latex paints; wetting and rewetting agent; rust inhibitor; Gardner 3 max. gel; sol. in water, ethylene glycol; sp.gr. 1.14; flash pt. (TCC) 120 F; pH 8.6 ± 0.2; surf. tens. 43.2 dynes/cm; 70% act.

Strodex® MR-100. [Dexter] Phosphated aromatic ethoxylate acid anhydride; anionic; emulsifier, dispersant for org. pigments in polar and nonpolar solvs.; Gardner 4 max. visc. liq.; sol. in min. spirits, diethyl phthalate, IPA, butyl Cellosolve; sp.gr. 1.06; visc. 2500 ± 500 cps; flash pt. (TCC) 300+ F; pH 2.0 ± 0.2; 99 ± 0.5% act.

Strodex® MRK-98. [Dexter] Phosphated aromatic ethoxylate; anionic; dispersant, emulsifier for grinding and dispersing org. pigments; for heavy-duty maintenance finishes; aids stability and color development; Gardner 4 max. hazy visc. liq.; sol. in ethylene glycol, diethyl phthalate; sp.gr. 1.08; visc. 4500 ± 1000 cps; b.p. 286 F; pH 8.2 ± 0.2; surf. tens. 33.3 dynes/cm; 96 ± 1% act.

Strodex® P-100. [Dexter] Phosphated coester of alcohol and aliphatic ethoxylate acid anhydride; anionic; emulsifier; dispersant for pigments in polar and nonpolar solvs.; coupler for emulsifiers in aq. and nonaq. systems; used in metal and tile cleaners; pigment grinding aid; Gardner 3 max. clear liq.; sol. in min. spirits, IPA, butyl Cellosolve; sp.gr. 1.04; visc. 600 ± 100 cps; flash pt. (TCC) 235 F; pH 2 ± 0.2; 98 ± 1% act.

Strodex® PK-80A. [Dexter] Phosphated coester of alcohol and aliphatic ethoxylate; dispersant, wetting agent for pigments, in gloss or semigloss paints; anti-rusting props.; Gardner 4 max. clear liq.; sol. in water, ethylene glycol; sp.gr. 1.07; visc. 110 ± 40 cps; flash pt. (TCC) 120 F; pH 12 ± 1; surf. tens. 31.7 dynes/cm; 67 ± 2% act.

Strodex® PK-90. [Dexter] Phosphated coester of alcohol and aliphatic ethoxylate; anionic; emulsifier, dispersant, wetting agent for extender pigments in latex paints, barium sulfate, and iron oxides; oxidation-corrosion inhibitor; used in heavy-duty alkaline cleaners; Gardner 2 max. visc. liq.; sol. in high electrolyte concs., water, min. spirits, ethylene glycol; sp.gr. 1.15; visc. 7000 ± 2000 cps; b.p. 262 F; pH 10.5 ± 0.5; surf. tens. 32.3 dynes/cm; 90 ± 2% act.

Strodex® PK-95G. [Dexter] Phosphated coester of alcohol and aliphatic ethoxylate; dispersant, wetting agent for pigments, in gloss or semigloss paints; anti-rusting props.; Gardner 2 max. clear liq.; sol. in water, ethylene glycol; sp.gr. 1.15; visc. 1700 ± 300 cps; b.p. 262 F; pH 10.5 ± 0.5; surf. tens. 32.3 dynes/cm; 81 ± 2% act.

Strodex® PSK-28. [Dexter] Phosphated coester of alcohol and aliphatic ethoxylate; anionic; emulsifier, wetting agent, pigment dispersant; Gardner 2 max. liq.; sol. in water, ethylene glycol; sp.gr. 1.03; visc. 90 ± 20 cps; flash pt. (TCC) 125 F; pH 8.5 ± 0.5; surf. tens. 30.2 dynes/cm; 57 ± 2% act.

Strodex® SE-100. [Dexter] Phosphated aliphatic ethoxylate acid anhydride; anionic; dispersant, emulsifier and coemulsifier in aq. and nonaq. systems; Gardner 2 max. hazy liq.; sol. in diethyl phthalate, ethylene glycol; sp.gr. 1.01; visc. 200 ± 100 cps; flash pt. (TCC) 300+ F; pH 1.4 ± 0.2; 96 ± 1% act.

Strodex® SEK-50. [Dexter] Phosphated aliphatic ethoxylate; anionic; dispersant, emulsifier, stabilizer for latex emulsions, paints, titanium dioxide and extender pigments; Gardner 2 max. clear liq.; sol. in water, butyl Cellosolve, ethylene glycol; sp.gr. 1.00; visc. 80 ± 10 cps; flash pt. (TCC) 120 F; pH 7.0 ± 0.2; surf. tens. 29.7 dynes/cm; 50 ± 1% act.

Strodex® Super V-8. [Dexter] Phosphate ester, potassium salt; anionic; patented emulsifier, detergent, dispersant, wetting agent used in textile processing applics.; clear liq.; sol. in water, aliphatic, aromatic, and chlorinated solvs.; sp.gr. 1.09–1.11; pH 7–8 (1%); 75–80% act.

Sufatol LS/3. [Standard Chem. UK] Sulfated fatty alcohol and fatty alcohol ethoxylate; anionic/nonionic; scouring, wetting and washing-off agent for textiles; liq.; 25% conc.

Sufatol LX/B. [Standard Chem. UK] Sulfated fatty alcohol; anionic; scouring agent, softener, and dyebath aux. for textiles; paste; 25% conc.

Sufatol LX/C. [Standard Chem. UK] Sulfated fatty alcohol; detergent; textile scouring; paste.

Sulfetal 4069. [Zschimmer & Schwarz] Sodium alkylsulfate (C8-10); anionic; wetting agent for strongly acid and alkaline media; liq.; 45% conc.

Sulfetal 4105. [Zschimmer & Schwarz] Sodium isooctyl sulfate; anionic; low-foaming detergent, wetting agent in acid and alkaline media; for industrial and metal cleaners; liq.; 40% conc.

Sulfetal 4187. [Zschimmer & Schwarz] Sodium isooctyl sulfate; anionic; detergent, wetting agent for industrial and metal cleaners; resistant to alkali; liq.; 40% conc.

Sulfetal AF. [Zschimmer & Schwarz] Sodium fatty alcohol sulfate (C16–C24); flotation agent, cleansing pastes; paste; 50% act.

Sulfetal C 38. [Zschimmer & Schwarz] Sodium lauryl sulfate; CAS 151-21-3; anionic; lt. duty detergent, dishwashing agent, cold washagent; paste; 38% conc.

Sulfetal C 90. [Zschimmer & Schwarz] Sodium lauryl sulfate; CAS 151-21-3; anionic; detergent, washing and cleansing agent; wetting agent in galvanic baths; wh. to lt. straw-colored spray-dried powd.; bulk dens. 250 g/l; pH 6-9 (1%); 90% act.

Sulfetal CCO 50. [Zschimmer & Schwarz] Sodium coco/oleyl alcohol sulfate; base material for detergent mfg., high temp. laundry wash; wh. to ylsh. paste; pH 8 (1%); 50% act.

Sulfetal CJOT 38. [Zschimmer & Schwarz] MIPA-lauryl sulfate; CAS 21142-28-9; anionic; basic material for hair shampoos, bath additives, cosmetics, liq. detergents; straw-colored clear visc. liq.; dens. 1.02 g/cc; visc. 5000-10,000 mPa•s; cloud pt. 8 C; pH 6.5-7.0 (10%); 38% act. in water.

Sulfetal FA 40. [Zschimmer & Schwarz] Sodium isooctyl sulfate, modified; anionic; low-foaming detergent and wetting agent; used in industrial cleaners, metal and tank cleaning; highly resistant to alkalies; liq.; 40% conc.

Sulfetal KT 400. [Zschimmer & Schwarz] TEA-lauryl sulfate; anionic; for detergents, cosmetics, hair shampoos; liq.; 40% conc.

Sulfetal MG 30. [Zschimmer & Schwarz] Magnesium lauryl sulfate; anionic; for detergents, cosmetics; liq.; 30% conc.

Sulfetal TC 50. [Zschimmer & Schwarz] Sodium tallow sulfate and sodium lauryl sulfate; anionic; detergent; washing and cleaning agents; hand cleaning pastes; paste; 50% conc.

Sulfetal TC 50 W. [Zschimmer & Schwarz] Sodium tallow/coco fatty alcohol sulfate; hand cleaning pastes, detergents, cleansing agents; paste; 55% act.

Sulfochem 25-3A. [Chemron] Ammonium alkyl ether (3) sulfate; flash foamer, detergent, wetting agent and emulsifier for industrial and household cleaning compds., e.g., liq. dishwash; esp. suitable for formulations at pH 6-7; pale yel. liq.; 58-60% act.

Sulfochem 25-3S. [Chemron] Sodium alkyl ether (3) sulfate; flash foamer, detergent, wetting agent and emulsifier for light-duty dishwashing, laundry, heavy-duty industrial cleaning applics.; pale yel. liq.; 58-60% act.

Sulfochem 436. [Chemron] Ammonium nonoxynol-4 sulfate; anionic; high-foaming detergent for dishwashing, carwash, and carpet shampoo formulations; emulsion polymerization surfactant for SBR, acrylic, vinyl acrylic systems; pale yel. liq.; 58-60% act.

Sulfochem 437. [Chemron] Ammonium nonylphenol ether sulfate; anionic; emulsion polymerization surfactant for SBR, acrylic, vinyl acrylic, PVAc, and copolymer systems; pale yel. liq.; 30% act.

Sulfochem 438. [Chemron] Ammonium nonylphenol ether sulfate; anionic; emulsion polymerization surfactant for SBR, acrylic, vinyl acrylic, PVAc, and copolymer systems; pale yel. liq.; 30% act.

Sulfochem ALS. [Chemron] Ammonium lauryl sulfate; anionic; surfactant, mild detergent for use in low pH systems; foaming and suspending agent; for personal care and industrial applics.; water-wh. liq.; 29% conc.

Sulfochem B-221. [Chemron] Blend; anionic; base with dense, long-lasting foam; for shampoo, body soap, liq. hand cleaner formulations; liq.; 34% conc.

Sulfochem B-221OP. [Chemron] Blend; anionic; base with dense, long-lasting foam; for shampoo, body soap, liq. hand cleaner formulations; pearlized version of Sulfochem B-221 permitting cold blend procedures; liq.; 37% conc.

Sulfochem ES-2. [Chemron] Sodium laureth-2 sulfate; CAS 9004-82-4; anionic; flash foamer and detergent for personal cleansing prods., specialty cleaning prods.; water-wh. liq.; 27% conc.

Sulfochem ES-70. [Chemron] Sodium laureth-2 sulfate; CAS 9004-82-4; anionic; surfactant for toiletries, cosmetics, specialty industrial compds.; pale yel. flowing gel; 69% conc.

Sulfochem K. [Chemron] Potassium lauryl sulfate; anionic; detergent, foamer; liq.; 30% conc.

Sulfochem MG. [Chemron] Magnesium lauryl sulfate; anionic; mild detergent, foamer for bubblebath, shampoos, cleansing preps., carpet shampoos; liq.; 30% conc.

Sulfochem PA. [Chemron] Alkyl ether sulfate; drilling foamer; nontoxic, biodeg. and acceptable for water well drilling; wetting agent, anionic emulsion preventer; compat. with 2% KCl brine; lt. yel. liq.; 50% act.

Sulfochem RF. [Chemron] Alkyl ether sulfate; biodeg. foaming agent for brines; drilling foamer; aid for unloading gas wells; compat. with polymers for added foam stability; lt. yel. liq.; 50% act.

Sulfochem SLN. [Chemron] Sodium lauryl sulfate; CAS 151-21-3; foamer, dispersant, wetting agent, detergent for dry blends used in cleaning compds., carpet shampoos, shampoo concs., bubble baths, cosmetic cleansers; wh. needles; 88% act.

Sulfochem SLP. [Chemron] Sodium lauryl sulfate; CAS 151-21-3; foamer, dispersant, wetting agent, detergent for high-act., cleaning concs., dentifrices, bath prods., cleansers; wh. powd.; 90% act.

Sulfochem SLP-95. [Chemron] Sodium lauryl sulfate; CAS 151-21-3; foamer, dispersant, wetting agent, detergent for high-act., cleaning concs., dentifrices, high purity cleansers; wh. powd.; 95% act.

Sulfochem SLS. [Chemron] Sodium lauryl sulfate; CAS 151-21-3; anionic; detergent, foamer, suspending agent for hard surf. cleaners, carpet shampoos, upholstery cleaners, spot removers, personal care prods.; water-wh. liq.; 29% conc.

Sulfochem SLX. [Chemron] Sodium lauryl sulfate; CAS 151-21-3; detergent, foaming and suspending agent for rug and upholstery shampoos, spot removers; water-wh. liq.; 28-30% act.

Sulfochem TLS. [Chemron] TEA-lauryl sulfate; anionic; surfactant foamer, wetting agent with soap-like chars.; for liq. soaps and shampoos, industrial applics.; good tolerance to hard water; water-wh. liq.; 40% conc.

Sul-fon-ate AA-9. [Boliden Intertrade] Sodium dodecylbenzene sulfonate; anionic; wetting agent; air pollution control; cement, food, commercial laundry and industrial industries; cosmetics; fertilizers, insecticides; leather, paper, petrol., and rubber processing; metal cleaning, electroplating, etching, pickling; mining; wh. crisp flake; odorless; surf. tens. 30.6 dynes/cm (86 F); pH 7–8 (1%); 90% act.

Sul-fon-ate AA-10. [Boliden Intertrade] Sodium dodecylbenzene sulfonate; anionic; wetting agent; air pollution control; cement, food, commercial laundry and industrial industries; cosmetics; fertilizers, insecticides; leather, paper, petrol., and rubber processing; metal cleaning, electroplating, etching, pickling; mining; wh. crisp flake; sol. in water; pH 6–8 (1%); surf. tens. 33.2 dynes/cm (0.05%, 86 F); toxicology: toxic orally; LD50 (male rat, oral) 1-5 g/kg; eye and skin irritant but nontoxic dermally; 96% act.

Sul-fon-ate LA-10. [Boliden Intertrade] Sodium dodecylbenzene sulfonate; surfactant; wh. crisp flake; water-sol.; pH 6.

Sul-fon-ate OA-5. [Boliden Intertrade] Sulfonated oleic acid, sodium salt; anionic; wetting agent; antifoam/defoamer, corrosion inhibitor, acid-stable surfactant, solubilizer; stable; biodeg.; lt. brn. liq., sl. odor; sol. in water in all proportions, acid sol'ns.; sp.gr. 1.11; b.p. 207-214 F; m.p. 25 F; surf. tens. 32.7 dynes/cm (0.05%); toxicology: irritating to eyes; nontoxic orally, dermally, by inhalation; LD50 (rat, oral) > 5 g/kg; 41% act.

Sul-fon-ate OA-5R. [Boliden Intertrade] Sulfonated oleic acid, sodium salt; surfactant; lt. brn. liq.; visc. 200 cps; cloud pt. 50–56 F; pH 5.5–6.5; 42% act.

Sul-fon-ate OE-500. [Boliden Intertrade] Sulfonated oleic acid, sodium salt, amyl-ester; anionic; surfactant base, coupler, solubilizer, emulsifier for dyes and additives used in polyesters; biodeg.; liq.

Sulfonated Castor Oil 50%. [Nat'l. Starch] Fatty glyceride sulfate; anionic; emulsifier, lubricant, dyeing assistant, dye dispersant; biodeg.; liq.; 50% conc.

Sulfonated Castor Oil 75%. [Nat'l. Starch] Fatty glyceride sulfate; anionic; emulsifier, lubricant, dyeing assistant, dye dispersant; liq.; 70% conc.

Sulfonated Castor Oil GTO. [Nat'l. Starch] Fatty glycerine sulfate; anionic; emulsifier, lubricant, dyeing assistant, dye dispersant and leveler; liq.; biodeg.; 77% conc.

Sulfonated Red Oil. [Nat'l. Starch] Sulfated fatty prod.; anionic; dyeing assistant for cellulosics or acid colors; biodeg.; liq.; 70% conc.

Sulfonic 800. [Boliden Intertrade] Sulfonated oleic acid; anionic; intermediate used in textile and industrial cleaning; wetting agent; biodeg.; dk. brn. visc. liq.; water-sol.; sp.gr. 1.09; m.p. < 16 F; b.p. 248–257 F; flash pt.> 230 F (PMCC); pH 0.6; 50–55% act.

Sulfonic 864. [Boliden Intertrade] Sulfonic acids; anionic; detergent and penetrant; liq.; biodeg.

Sulfonic Acid LS. [Hart Chem. Ltd.] Linear alkylbenzene sulfonic acid; anionic; intermediate for liq. and powd. detergent formulations; liq.; 96% conc.

Sulfopon 101, 101 Special. [Henkel KGaA; Pulcra SA] Sodium lauryl sulfate; CAS 151-21-3; anionic; surfactant for personal care prods. and lt.-duty detergents; paste; 30% conc.

Sulfopon 101/POL. [Pulcra SA] Sodium lauryl sulfate; CAS 151-21-3; anionic; surfactant for emulsion polymerization; liq.; m.w. 302; visc. < 500 mPa•s; pH 7.5-8.5 (10%); 28-30% act.

Sulfopon 101 Spez. [Pulcra SA] Sodium lauryl sulfate; CAS 151-21-3; anionic; emulsifier for emulsion polymerization; surfactant for shampoos, specialty cleaners; low freezing pt.; liq.; m.w. 302; pH 7.5-8.5 (10%); 29-31% act.

Sulfopon 102. [Henkel KGaA] Sodium lauryl sulfate C10-C16; CAS 151-21-3; anionic; detergent, emulsifier for emulsion polymerization of syn. rubber, PVC, and other polymers; liq.; 30% conc.

Sulfopon K35. [Henkel KGaA] Sodium lauryl sulfate; CAS 151-21-3; anionic; base for lt. duty detergents, washing and cleansing pastes; paste; 35% conc.

Sulfopon KT 115. [Henkel KGaA] Coconut/tallow fatty alcohol sulfate; anionic; base for heavy-duty detergents and hand cleaning pastes; paste; 40–42% conc.

Sulfopon KT 115-50. [Henkel KGaA] Coconut/C16-18 fatty alcohol sulfate; anionic; for heavy-duty detergents, hand washing pastes; paste; 50% conc.

Sulfopon O. [Henkel KGaA] Oleyl/cetyl sulfate; anionic; detergent; paste; 25% conc.

Sulfopon O 680. [Henkel/Functional Prods.] Sodium oleyl sulfate; anionic; low foaming surfactant; paste; 50% solids.

Sulfopon P-40. [Pulcra SA] Sodium lauryl sulfate; CAS 151-21-3; anionic; dispersant and emulsifier for acrylates, styrene acrylic, vinyl chloride, vinyl acetate copolymers; also for cream shampoos, specialty cleaners, rug shampoos; paste; m.w. 302; pH 6.0-7.0 (10%); 38-42% act.

Sulfopon T 55. [Henkel KGaA] Tallow fatty alcohol sulfate; anionic; detergent; for heavy-duty detergents, hand washing pastes; paste; 55% conc.

Sulfopon T Powd. [Henkel KGaA] C16-18 fatty alcohol sulfate; anionic; for handwashing pastes, laundry detergents; powd.; 90% conc.

Sulfopon WA 1. [Henkel Canada] Sodium alkyl sulfate; anionic; detergent, wetting, emulsifier, dispersant; solubilizer for emulsion polymers; base for rug and upholstery shampoos; liq.; 30% conc.

Sulfopon WAQ LCX. [Henkel Canada] Sodium lauryl sulfate; CAS 151-21-3; anionic; detergent, foamer; base for shampoos and bubble baths; rug shampoos; liq.; 30% conc.

Sulfopon WAQ Special. [Henkel Canada] Sodium lauryl sulfate; CAS 151-21-3; anionic; surfactant for personal care prods., rug shampoos, lt. duty detergents; water-wh. liq.; 28–30% act.

Sulfosil P-491. [Witco] Sodium silicate/sodium sulfate-based builder; detergent; booster for household and industrial cleaners; powd.; 85% conc.

Sulfotex 110. [Henkel/Cospha; Henkel Canada] Sodium n-decyl sulfate; anionic; wetting agent and emulsifier used in cleaning formulations, rug shampoos; used in sealing food containers; effective in cold and hard water; biodeg.; pale amber liq.; dens. 8.8 lb/gal; pH 9–10 (10%); 30–32% act.

Sulfotex 130. [Henkel Canada] Sulfated castor oil; anionic; wetting agent; liq.; 75% conc.

Sulfotex 6040. [Henkel/Cospha; Henkel Canada] Sodium laureth sulfate; CAS 9004-82-4; anionic; surfactant for soap and detergent applics., industrial use; liq.; 60% conc.

Sulfotex A. [Henkel/Cospha; Henkel Canada] Blend; anionic; detergent base for liq. manual dishwash formulations; liq.; 53% conc.

Sulfotex DOS. [Henkel Canada] Disodium oleamido PEG sulfosuccinate; anionic; detergent, emulsifier, lathering agent for skin cleansers; foamer for carpet backing and fire fighting; emulsifier for polymerization; liq.; 28% conc.

Sulfotex LAS-90. [Henkel/Cospha; Henkel Canada] Sodium dodecylbenzene sulfonate; surfactant for all-purpose cleaners, laundry prods., hard surf. cleaners; penetrates and removes grease; flakes; 90% conc.

Sulfotex LCX. [Henkel/Cospha; Henkel Canada] Sodium lauryl sulfate; CAS 151-21-3; anionic; detergent, foaming agent, wetting agent and emulsifier; primary surfactant for rug and upholstery shampoos, hard surface cleaners; biodeg.; water-wh. liq.; dens. 8.73 lb/gal; cloud pt. 2 C; pH 7.0–9.0 (10%); 28.5–30.5% act.

Sulfotex LMS-E. [Henkel/Cospha; Henkel Canada] Sodium lauryl ether sulfate; CAS 9004-82-4; anionic; detergent, wetting agent, foamer for industrial applics., liq. dishwashing detergents; biodeg.; pale yel. clear liq.; dens. 8.75 lb/gal; cloud pt. 10 C; pH 7.0–9.0 (10%); 58–60% act.

Sulfotex OA. [Henkel/Cospha; Henkel Canada] Sodium octyl sulfate; anionic; foaming agent, wetting agent for animal glue on paper and paperboard, high electrolyte concs., and food processing; detergent and dispersant in industrial cleaners; mercerizing agent in dyeing of fibers; biodeg.; lt. amber; dens. 9.2 lb/gal; pH 8.0–10.0 (10%); 43-46% act.

Sulfotex OT. [Henkel/Cospha; Henkel Canada] Ammonium lauryl ether sulfate; anionic; detergent, emulsifier, foaming agent used in household and industrial detergents, car shampoos; biodeg.; pale yel. visc. liq.; dens. 8.54 lb/gal; cloud pt. 10 C; pH 6.0–7.0 (10%); 58–60% act.

Sulfotex PAI. [Henkel/Cospha; Henkel Canada] Ammonium alkyl ether sulfate; anionic; emulsifier, detergent, foamer, and dispersant used in liq. detergents, car shampoos, household, industrial cleaners, petrol industry applics.; effective foaming in high electrolyte concs.; biodeg.; pale yel. liq.; dens. 8.31 lb/gal; pH 7.0–7.5; 46–50% act.

Sulfotex PAI-S. [Henkel/Cospha; Henkel Canada] Sodium alkyl ether sulfate; anionic; emulsifier, detergent, foamer, and dispersant used in liq. detergents, car shampoos, household, industrial cleaners, petrol industry applics., cleaners requiring good salt tolerance; biodeg.; pale yel. liq.; dens. 8.69 lb/gal; pH 7.5–8.5; 45–49% act.

Sulfotex PAW. [Henkel Canada] Modified sulfated alcohol ethoxylate; anionic; foaming agent for foam drilling, fresh water brines, calcium contaminated brines, oil contaminated systems; liq.

Sulfotex RAW. [Henkel Canada] Modified sulfated alcohol ethoxylate; anionic; foaming agent for foam drilling, fresh water brines, calcium contaminated brines, oil contaminated systems; liq.

Sulfotex RIF. [Henkel Canada] Ammonium alkyl ether sulfate; anionic; foamer for brines, foam drilling, petrol. industry applics., household and industrial cleaners; biodeg.; liq.; 47% act.

Sulfotex SAL, SAT. [Henkel] Sulfated linear alcohol ethoxylate, ammonium salt; anionic; emulsifier, wetting agent, and foamer for liq. detergents, car shampoos, household and industrial cleaners, petrol. industry applics.; liq.; biodeg.; 58% act.

Sulfotex SXS-40. [Henkel/Cospha; Henkel Canada] Sodium xylene sulfonate; CAS 1300-72-7; anionic; hydrotrope, solubilizer for liq. detergents, pine oil in water, inks to prevent gumming; liq.; biodeg.; 40% act.

Sulfotex UBL-100. [Henkel/Cospha; Henkel Canada] Linear dodecylbenzene sulfonic acid; anionic; detergent intermediate when neutralized; for laundry, hard surfaces, stripping, wetting, foaming applics.; thick liq.; 98% act.

Sulfotex WA. [Henkel Canada] Sodium lauryl sulfate; CAS 151-21-3; anionic; detergent base for personal care prods. and liq. cleaners; liq.; 28% conc.

Sulfotex WAQ-LCX. [Henkel] Sodium lauryl sulfate, sodium decyl sulfate; anionic; surfactant for rug and upholstery shampoos, hard surface cleaners, spot removers; water-wh. to slightly yel. liq.; min. odor; water-sol.; dens. 1.046 g/ml; visc. 100 cps; cloud pt. 5 C; pH 7.0–9.0 (10%); biodeg.; 29.5 ± 1.0% act.

Sulfotex WAT. [Henkel] TEA lauryl sulfate; anionic; detergent base, foamer and wetting agent for soaps, shampoos, and detergents; liq.; biodeg.; 40% act.

Sulframin 40. [Witco] Linear alkylaryl sodium sulfonate; anionic; detergent base and wetting agent for household and industrial specialty compds., personal care prods., metal cleaning; lubricant, emulsifier, dye dispersant, penetrant, scouring agent, antistat in textile industry; defoamer for petrol. industry; flakes, gran.; 40% conc.

Sulframin 40DA. [Witco] Linear alkylaryl sodium sulfonate; anionic; detergent for car washing, household and industrial cleaning; wetting agent, lubricant, emulsifier, antistat; metal cleaning; textile surfactant; beads; 40% conc.

Sulframin 40RA. [Witco] Linear alkylaryl sodium sulfonate; anionic; detergent for car washing, household and industrial cleaning; wetting agent, lubricant, emulsifier, antistat; metal cleaning; textile surfactant; base in personal care prods.; beads; 40% conc.

Sulframin 40T. [Witco] Linear alkylaryl TEA sulfonate; anionic; detergent for car washing, household and industrial cleaning; wetting agent, lubricant, emulsifier, antistat; metal cleaning; textile surfactant; liq.; 40% conc.

Sulframin 45. [Witco] Linear alkylaryl sodium sulfonate; anionic; detergent, wetting agent, base for liq. detergent, household, and industrial specialty compds.; metal cleaning; textile surfactant; liq.; cloud pt. 14 C; 45% conc.

Sulframin 45LX. [Witco] Linear alkylaryl sodium sulfonate; anionic; detergent, wetting agent, base for liq. detergent, household, and industrial specialty compds.; metal cleaning; textile surfactant; liq.; 44% conc.

Sulframin 60T. [Witco] Linear alkylaryl TEA sulfonate; detergent for car washing, household and industrial cleaning; wetting agent, lubricant, emulsifier, antistat; metal cleaning; textile surfactant; personal care prods.; liq.; 60% conc.

Sulframin 85. [Witco] Linear alkylaryl sodium sulfonate; anionic; detergent, lubricant, emulsifier, antistat, wetting agent for household and industrial specialty compds., metal cleaning, personal care prods., textiles; flakes, powd.; 85% conc.

Sulframin 90. [Witco] Linear alkylaryl sodium sulfonate; anionic; detergent, lubricant, emulsifier, antistat, wetting agent for household and industrial specialty compds., metal cleaning, personal care prods., textiles; emulsion polymerization and latex stabilization; flakes; 91% conc.

Sulframin 1230. [Witco SA] Sodium dodecylbenzenesulfonate; anionic; base for liq. detergents; emulsion polymerization; liq.; 30% conc.

Sulframin 1240, 1245. [Witco] Linear alkylaryl sodium sulfonate; anionic; detergent, lubricant, emulsifier, antistat, wetting agent for household and industrial specialty compds., metal cleaning, personal care prods., textiles; slurry; 40 and 45% conc. resp.

Sulframin 1250, 1260. [Witco] Linear alkylaryl sodium sulfonate; anionic; detergent for car washing, household and industrial cleaning; wetting agent, lubricant, emulsifier, antistat; metal cleaning; textile surfactant; personal care prods.; slurry; 51 and 60% conc. resp.

Sulframin 1255. [Witco SA] Sodium dodecylbenzenesulfonate; CAS 25155-30-0; anionic; base for liq. detergents; emulsion polymerization; paste; 55% conc.

Sulframin 1288. [Witco] Linear alkylaryl sulfonic acid; anionic; penetrant, lubricant, dispersant, detergent, antistat, intermediate for wetting agents, emulsifiers, and detergents for household and industrial specialties, personal care prods.; textile surfactant; liq.; 88% conc.

Sulframin 1298. [Witco] Linear alkylaryl sulfonic acid; anionic; penetrant, lubricant, dispersant, detergent, antistat, intermediate for wetting agents, emulsifiers, and detergents for household and industrial specialties, personal care prods.; textile surfactant; emulsion polymerization; liq.; 98% conc.
Sulframin 1388. [Witco] Linear alkylaryl sulfonic acid; anionic; penetrant, lubricant, dispersant, detergent, antistat, intermediate for wetting agents, emulsifiers, and detergents for household and industrial specialties, personal care prods.; textile surfactant; liq.; 88% conc.
Sulframin 4010D, 4010R. [Witco] Linear alkylaryl sodium sulfonate; detergent for car washing, household and industrial cleaning; beads.
Sulframin AOS. [Witco SA] Sodium alpha olefin sulfonate; CAS 68439-57-6; anionic; detergent base for personal care prods.; foaming agent; surfactant for emulsion polymerization and latex stabilization; liq.; 38% conc.
Sulframin Acide B. [Witco SA] Straight chain dodecylbenzene sulfonic acid; CAS 27176-87-0; anionic; intermediate for prod. of biodeg. sulfonates; liq.; 97% conc.
Sulframin Acide TPB. [Witco SA] Branched chain dodecylbenzene sulfonic acid.; CAS 68411-32-5; anionic; intermediate for prod. of branched sulfonates for polymerization and pesticides; liq.; 97% conc.
Sulframin CSA. [Witco] Cumene sulfonic acid; hydrotrope, coupler, solubilizer, catalyst; used in liq. detergents.
Sulframin HD, Low-Foam HD. [Witco] Linear alkylaryl sodium sulfonate; laundry and all-purpose detergent; beads.
Sulframin LX. [Witco] Linear alkylaryl sodium sulfonate, modified; anionic; detergent base and wetting agent for household and industrial specialty prods.; metal cleaning; textile surfactant; flakes; 85% conc.
Sulframin Phos-Free HD. [Witco] Linear alkylaryl sodium sulfonate; laundry and all-purpose detergent; beads.
Sulframin TX. [Witco] Toluene sulfonic acid, modified; hydrotrope; coupler, catalyst; solubilizer for liq. detergent.
Sulphonic Acid LS. [Hart Chem. Ltd.] Alkylbenzene sulfonic acid; anionic; base for wide variety of detergents; can be neutralized with metallic bases or amines; liq.; 96% act.
Sunaptol DL Conc. [ICI Surf. UK] Alkyl polyglycol ester; nonionic; detergent, emulsifier for desizing, scouring and bleaching operations; emulsifier for min. oils and waxes; liq.
Sunaptol LT. [ICI Surf. UK] Ethoxylated castor oil; nonionic; dispersant, leveling agent for use in dyeing polyester fibers at high temp.; liq.
Sunaptol P Extra Liq. [ICI Surf. UK] Ethoxylated fatty alcohol; nonionic; leveling agent for dyeing polyamides with acid, disperse and disperse premetallized dyes, or cotton with direct and vat dyes; liq.
Sunnol 710 H. [Lion] POE alkyl ether sulfate; anionic; detergent and personal care prods.; paste; 70% conc.
Sunnol CM-1470. [Lion] POE laureth sulfate; anionic; shampoos, detergents; paste; 70% conc.
Sunnol DL-1430. [Lion] POE alkyl ether sulfate; anionic; shampoos, detergents, bubble baths; liq.; 27% conc.
Sunnol DOS. [Lion] POE alkylphenyl ether sulfate; anionic; emulsion polymerization agent; liq.; 30% conc.
Sunnol DP-2630. [Lion] POE alkylphenyl ether sulfate; anionic; emulsion polymerization agent; liq.; 30% conc.
Sunnol LM-1130. [Lion] Lauryl sulfate; anionic; foaming agent for flotation; liq.; 30% conc.
Sunnol LST. [Lion] TEA lauryl sulfate; anionic; detergents and personal care prods.; liq.; 39–41% conc.
Sunnol NES. [Lion] POE alkylphenyl ether sulfate; anionic; dyeing assistant in textiles; liq.; 30% conc.
Sunnol NP-2030. [Lion] POE alkylphenol ether sulfate; anionic; dyeing assistant; stable in alkali; liq.; 30% conc.
Sunsoflon CK. [Nikko Chem. Co. Ltd.] Alkyl polyamide; cationic; surfactant, softener for textiles; liq.
Sunsoflon K-2. [Nikko Chem. Co. Ltd.] Alkyl polyamide; cationic; see Sunsoflon CK; liq.; 18% conc.
Sunsoflon MT-100. [Nikko Chem. Co. Ltd.] Alkyl polyamide; cationic; see Sunsoflon CK; liq.
Sunsolt RZ-2, -6. [Nikko Chem. Co. Ltd.] Alkylaryl ester; anionic; dispersant and leveling agent for textiles; liq.; 26 and 40% conc. resp.
Superkleen C. [CNC Int'l.] Polymer disp.; imparts durable hydrophilic and soil release props. to 100% polyester.
Superloid®. [Kelco] Refined ammonium alginate; gum used as gelling agent, thickener, emulsifier, film-forming agent, suspending agent, and stabilizer in food, pharmaceutical, and industrial applics.; stabilizer in paper and textile industry; tan gran. particles; water-sol.; sp.gr. 1.73; dens. 56.62 lb/ft^3; visc. 1500 cps; ref. index 1.3347; pH 5.5; 13% moisture.
Supermontaline SLT65. [Seppic] Dioctyl sodium sulfosuccinate; anionic; wetting agent and emulsifier; liq.; 65% conc.
Superol. [Procter & Gamble] Glycerin USP; CAS 56-81-5; humectant; in pharmaceuticals, toiletries, tobacco, alkyds, food prods., explosives, cellophane, urethane foam, other industries; APHA 10 max. color; sp.gr. 1.2612.
Super-Sat AWS-4. [RITA] PEG-20 hydrog. lanolin; nonionic; emollient, emulsifier, plasticizer for emulsion systems; amber solid; slightly water-sol.; 100% conc.
Super-Sat AWS-24. [RITA] PEG-24 hydrog. lanolin; nonionic; emollient, plasticizer, emulsifier; flake; 100% conc.
Supersol ICS. [Sybron] Anionic/nonionic; textile solv. scour to remove wax and oil; caustic stable.
Super Solan Flaked. [Croda Inc.] PEG-75 lanolin; CAS 61790-81-6; emollient, conditioner, superfatting agent, solubilizer; yel. flake; water-sol.; usage level: 1-10%.
Supersurf AFX. [Am. Emulsions] Blend; nonionic; wetting agent for fourth generation nylon.
Supersurf FL/4. [Am. Emulsions] Penetrant, wetting/rewetting agent, emulsifier for bleaching, dyeing, finishing.
Supersurf JM. [Am. Emulsions] Blend; nonionic/anionic; wetting and leveling agent for stack dyeing of nylon with acid dyes.
Supersurf WAF. [Am. Emulsions] Complex phosphate; nonionic; wetting and leveling agent for fifth generation nylon.
SuperWash DZ. [DeeZee] Detergent blend; textile scouring agent and laundry detergent; powd.; 100% conc.

Super Wet. [W.A. Cleary] Nonionic; wetting agent and emulsifier for pesticides; soil penetrant; enhances pesticide and fertilizer efficiency; liq.; 100% conc.

Super Wet Granular 15G. [W.A. Cleary] Nonionic; wetting agent formulated with high dens. granular clay to allow ease of applic., promote exc. soil wetting and drainage in agric. use; usage level: 3.5 lb/1000 ft².

Supragil® GN. [Rhone-Poulenc Geronazzo] Sodium phenyl sulfonate; anionic; dispersant, suspending agent for pesticide wettable powds. and water-disp. gran.; nonfoaming; powd.; 70% conc.

Supragil® MNS/90. [Rhone-Poulenc Surf.; Rhone-Poulenc France; Rhone-Poulenc Geronazzo] Sodium methyl naphthalene sulfonate condensate; anionic; surfactant, dispersing/suspending agent for pesticide wettable powds. and water-disp. gran.; powd.; 90% conc.

Supragil® NK. [Rhone-Poulenc Surf.; Rhone-Poulenc Geronazzo] Sodium alkyl naphthalene sulfonate; anionic; emulsifier, wetting and dispersing agent, foamer for dyes, pigments, emulsion polymerization, industrial use; EPA compliance; cream to tan powd.; m.w. 326; water-sol.; pH 6-8 (5%); surf. tens. 36 dynes/cm (@ CMC); 75% solids.

Supragil® NS/90. [Rhone-Poulenc Surf.; Rhone-Poulenc Geronazzo] Alkylnaphthalene sulfonate salt; anionic; surfactant, dispersant for agric. formulations; EPA compliance.

Supragil® WP. [Rhone-Poulenc Surf.; Rhone-Poulenc Geronazzo] Sodium diisopropyl naphthalene sulfonate; anionic; wetting agent, dispersant, stabilizer without detergent properties, foaming agent used in textiles, paints, pesticides, latex emulsions; EPA compliance; tan powd.; m.w. 314; pH 8-10 (5%); 75% act.

Surfac® P14B. [Sherex Polymers] Primary ether amine acetate; cationic; surfactant, ore flotation agent, corrosion inhibitor; m.w. 268; disp. in water; dens. 7.60 lb/gal; 100% conc.

Surfac® P24M. [Sherex Polymers] Primary ether amine acetate; cationic; surfactant, ore flotation agent, corrosion inhibitor; m.w. 261; water-disp.; dens. 7.61 lb/gal; 100% conc.

Surfactol® 13. [CasChem] Castor oil, modified; CAS 1323-38-2; nonionic; wetting agent, emulsifier, solubilizer, dispersant, wax plasticizer, mold release agent, antifoamer for textiles, leather, paints, household, cosmetics, dyeing, tanning, finishing, sizing, making of cutting and sol. oils, oilfield chems.; biodeg.; FDA approval; Gardner 4 liq.; faint odor; water-sol.; sp.gr. 1.005; dens. 8.36 lb/gal; visc. 17 stokes; iodine no. 70; sapon. no. 130; hyd. no. 425; flash pt. 385 F; toxicology: sl. skin or eye irritant; low oral toxicity; 100% conc.

Surfactol® 318. [CasChem] PEG-5 castor oil; CAS 61791-12-6; nonionic; emulsifier for oils, waxes; solubilizer for fragrances; emollient; used in textiles, paints, household, cosmetics, dyeing, tanning, finishing, sizing, insecticides, herbicides, fungicides, kier boiling, making of cutting and sol. oils; dispersing waxes, pigments, resins, rewetting dried skins; FDA approval; Gardner 3 liq.; faint odor; sol. in toluene, butyl acetate, MEK; sp.gr. 0.984; dens. 8.2 lb/gal; visc. 690 cps; pour pt. -25 C; HLB 3.8; acid no. 0.2; iodine no. 70; sapon. no. 148; hyd. no. 141; flash pt. 525 F; toxicology: sl. eye or skin irritant; low oral toxicity; 100% conc.

Surfactol® 340. [CasChem] Ethoxylated castor oil; CAS 61791-12-6; wetting agent, wax plasticizer, mold release agent, antifoam for textiles, leather, paints, household, cosmetics, dyeing, tanning, finishing, sizing, cutting and sol. oils, dispersing waxes, pigments, resins; Gardner 4 color; sp.gr. 1.015; dens. 8.4 lb/gal; visc. 5 stokes; pour pt. -30 F; acid no. 0.3; iodine no. 51; sapon. no. 108; hyd. no. 112; flash pt. 520 F; toxicology: low oral toxicity.

Surfactol® 365. [CasChem] PEG-40 castor oil; CAS 61791-12-6; nonionic; see Surfactol 318; FDA approval; clear amber liq., char. odor; sol. in toluene, butyl acetate, MEK, ethanol, and water; sp.gr. 1.054; dens. 8.8 lb/gal; visc. 500 cps; m.p. 10 C; HLB 13.3; acid no. 0.2; iodine no. 36; sapon. no. 68; hyd. no. 80; flash pt. (PMCC) 520 F; toxicology: sl. eye or skin irritation; low oral toxicity; 100% conc.

Surfactol® 380. [CasChem] Ethoxylated castor oil; CAS 61791-12-6; wetting agent, wax plasticizer, mold release agent, antifoam for textiles, leather, paints, household, cosmetics, dyeing, tanning, finishing, sizing, cutting and sol. oils, dispersing waxes, pigments, resins; Gardner 3 paste; water-sol.; sp.gr. 1.076; dens. 9.0 lb/gal; acid no. 0.2; iodine no. 23; sapon. no. 36; hyd. no. 50; toxicology: sl. eye or skin irritant; low oral toxicity.

Surfactol® 575. [CasChem] PEG-66 trihydroxystearin; CAS 61788-85-0; see Surfactol 318; wh. solid; water-sol.; alcohol; sp.gr. 1.08; dens. 9.0 lb/gal; m.p. 39 C; HLB 15.; toxicology: skin and eye irritant; inhalation may cause severe choking reaction.

Surfactol® 590. [CasChem] PEG-200 trihydroxystearin; CAS 61788-85-0; nonionic; see Surfactol 318; wh. solid; sol. in water, alcohol; sp.gr. 1.18; dens. 9.8 lb/gal; m.p. 53 C; HLB 18; acid no. 1.2; iodine no. 50; sapon. no. 18; hyd. no. 24; 100% conc.

Surfagene FAD 105. [Chem-Y GmbH] Nonyl phenol deriv.; anionic; emulsifier for emulsion polymerization; Gardner 3 clear liq.; sp.gr. 1.01; visc. 10,000 mPa•s; pH 2 (10%); 99% act.

Surfagene FAD 106. [Chem-Y GmbH] Sodium nonoxynol-6 phosphate; wetting agent for weakly acidic and alkaline cleaners, metalworking cooling lubricants; liq.; 99% act.

Surfagene FAZ 109. [Chem-Y GmbH] Nonoxynol-9 phosphate; anionic; emulsifier for emulsion polymerization; Gardner 3 clear liq.; sp.gr. 1.1; visc. 10,000 mPa•s; pH 2 (10%); 99% act.

Surfagene FAZ 109 NV. [Chem-Y GmbH] Nonoxynol-9 phosphate; anionic; emulsifier for emulsion polymerization; liq.; 20% act.

Surfagene FDD 402. [Chem-Y GmbH] Laureth-2 phosphate; anionic; emulsifier for cosmetic formulations, oil baths; additive for mineral collectors; liq.; 99% act.

Surfagene FPT. [Chem-Y GmbH] TEA phosphate ester; scale inhibitor for cooling water systems; liq.; 75% act.

Surfam P5 Dist. [Sherex Polymers] Methoxypropylamine; intermediate; emulsifier in wax emulsions for floor care prods.; textile finishing, paints, and metal working oils; insecticide emulsions; Gardner 2 max. clear liq.; m.w. 90; sol. in water and org. solvs.; dens. 7.29 lb/gal; b.p. 115–123 C; flash pt. 102 F; 98% min. act.

Surfam P5 Tech. [Sherex Polymers] Methoxypropylamine; intermediate; emulsifier in wax emulsions for floor care prods.; textile finishing, paints, and metal working oils; insecticide emulsions; Gardner 7 max. clear liq.; sol. see Surfam P5 Dist.;

m.w. 91; dens. 7.30 lb/gal; 90% min. act.

Surfam P10. [Sherex Polymers] Primary ether amine; cationic; corrosion inhibitor; emulsifier and lubricant in fiber and fabric processing; raw material in industrial applics.; intermediate in formulation of anticaking agents, specialty surfactant; gasoline and fuel additive; Gardner 4 max. clear liq.; m.w. 169; dens. 7.09 lb/gal; 100% conc.

Surfam P12B. [Sherex Polymers] Primary ether amine; cationic; see Surfam P10; Gardner 4 max. clear liq.; m.w. 200; dens. 7.11 lb/gal; 100% conc.

Surfam P14B. [Sherex Polymers] Primary ether amine; cationic; see Surfam P10; Gardner 4 max. clear liq.; m.w. 229; dens. 7.11 lb/gal; 100% conc.

Surfam P17B. [Sherex Polymers] Primary ether amine; cationic; see Surfam P10; Gardner 4 max. clear liq.; m.w. 279; dens. 7.30 lb/gal; 100% conc.

Surfam P24M. [Sherex Polymers] Primary ether amine; cationic; see Surfam P10; Gardner 4 max. clear liq.; m.w. 212; dens. 7.10 lb/gal; 100% conc.

Surfam P86M. [Sherex Polymers] Primary ether amine; cationic; see Surfam P10; Gardner 4 max. clear liq.; m.w. 280; dens. 7.05 lb/gal; 100% conc.

Surfam P89MB. [Sherex Polymers] Primary ether amine; cationic; see Surfam P10; Gardner 4 max. clear liq.; m.w. 305; dens. 7.10 lb/gal; 100% conc.

Surfam P-MEPA. [Sherex Polymers] Methoxyethoxypropylamine; specialty chemical; used in textile dye chemicals, emulsifier intermediate in floor waxes and agric. prods.; Gardner 1 liq.; dens. 7.82 lb/gal; sp.gr. 0.93–0.94; b.p. 210–230 C; flash pt. < 212 F (COC); 97% min. act.

Surfax 150. [Aquatec Quimica SA] Phosphate ester, free acid; anionic; detergent, emulsifier, wetting agent for industrial alkaline cleaners; liq.; 100% conc.

Surfax 215. [Aquatec Quimica SA] Ammonium lauryl sulfate; CAS 2235-54-3; anionic; emulsion and suspension polymerization of polystyrene; liq.; 29% conc.

Surfax 218. [Aquatec Quimica SA] Sodium laureth sulfate; CAS 9004-82-4; 3088-31-1; anionic; foaming agent for rubber systems; liq.; 27% conc.

Surfax 220. [Aquatec Quimica SA] Sodium alkyl sulfate; anionic; emulsifier for emulsion polymerization of styrene butadiene, vinyl acetate, vinyl chloride, acrylic copolymers; liq.; 28% conc.

Surfax 250. [Aquatec Quimica SA] Sodium lauryl sulfate; CAS 151-21-3; anionic; wetting agent for wettable powds. used in agric. formulations; powd.; 89% min. conc.

Surfax 345. [Aquatec Quimica SA] Surfactant blend; anionic; wetting and dispersing agent for wettable powds. used in agric. formulations; powd.; 91% conc.

Surfax 495. [Aquatec Quimica SA] Sodium alkylaryl sulfonate; anionic; emulsifier for emulsion polymerization of vinyl acetate; liq.; 95% conc.

Surfax 500. [Aquatec Quimica SA] Disodium N-octadecyl sulfosuccinamate; anionic; emulsifier, dispersant, foaming agent for industrial processes; collecting agent for flotation of phosphate ores; liq.; 95% conc.

Surfax 502. [Aquatec Quimica SA] Disodium ethoxylated alcohol half ester of sulfosuccinic acid; anionic; emulsifier for emulsion polymerization of vinyl acetate and acrylates; 34% conc.

Surfax 536. [Aquatec Quimica SA] Disodium ethoxylated alcohol half ester of sulfosuccinic acid; anionic; emulsifier for emulsion polymerization of vinyl acetate and acrylates; 33% conc.

Surfax 539. [Aquatec Quimica SA] Disodium ethoxylated alcohol half ester of sulfosuccinic acid; anionic; emulsifier for emulsion polymerization of vinyl acetate and acrylates; 34% conc.

Surfax 550. [Aquatec Quimica SA] Sodium dioctyl sulfosuccinate; anionic; for emulsion polymerization of vinyl acetate; wetting agent for pigments; liq.; 68% conc.

Surfax 560. [Aquatec Quimica SA] Sodium dialkyl sulfosuccinate; anionic; for emulsion polymerization of modified styrene-butadiene; powd.; 83% conc.

Surfax 585. [Aquatec Quimica SA] Dioctyl sodium sulfosuccinate; anionic; wetting agent for agric. wettable powd.; powd.; 75% conc.

Surfax 8916/A. [E.F. Houghton] PEG-8 dioleate; CAS 9005-07-6; nonionic; surfactant for use in ferrous and nonferrous metalworking fluids, for low-foaming detergents in textile spinning, dyeing, and oil removal operations; liq.; HLB 7.1; 100% conc.

Surfax ACB. [Aquatec Quimica SA] Coco betaine; CAS 68424-94-2; amphoteric; mild detergent for shampoos, foam baths, other foaming toiletries; liq.; 30% conc.

Surfax CN. [Aquatec Quimica SA] Sodium lauryl sulfate; CAS 151-21-3; anionic; detergent, dispersant, foaming and wetting agent for shampoos, shaving creams; base for rug and household cleaners; liq.; 30% conc.

Surfax MG. [Aquatec Quimica SA] Magnesium lauryl sulfate; anionic; detergent and foaming agent for shampoos; base for rug and household cleaners; liq.; 29% conc.

Surfax WO. [E.F. Houghton] Sulfated ester; anionic; penetrant, softener, wetting and rewetting agent; liq.; 60% conc.

Surfine AZI-A. [Finetex] Nonoxynol-10 carboxylic acid; CAS 53610-02-9; nonionic; detergent, wetter, lime soap dispersant, emulsifier, solubilizer, coupler; for liq. detergents, household, personal care, industrial and institutional prods.; liq.; 90% conc.

Surfine T-A. [Finetex] Trideceth-7 carboxylic acid; nonionic; detergent, wetting agent, lime soap dispersant for cosmetic and household formulations; liq.; 90% conc.

Surfine WLG-A. [Finetex] Ethoxylated alcohol carboxylic acid; coupler, solubilizer, lime soap dispersant for hard surf. cleaners, personal care prods.

Surfine WLL. [Finetex] Sodium laureth-13 carboxylate; anionic; solubilizer and coupler, lime soap dispersant for mild hair and body shampoos; improves formulation stability; gel; 70% conc.

Surfine WNG-A. [Finetex] C14-15 pareth-8 carboxylic acid; nonionic; detergent, wetting agent, emulsifier, solubilizer and coupler used in detergents; liq.; 90% conc.

Surfine WNT-A. [Finetex] C12-15 pareth-7 carboxylic acid; nonionic; detergent, wetting agent, lime soap dispersant used in cosmetic and household formulations; liq.; 90% conc.

Surfine WNT Conc. [Finetex] Carboxylated alcohol ethoxylate; anionic; detergent, wetting agent; dispersant for household and cosmetic formulations; liq.; 85% conc.

Surfine WNT Gel. [Finetex] Sodium C12–15 pareth-7 carboxylic acid; anionic; detergent, wetting agent, lime soap dispersant for cosmetic and household formulations; gel; 60% conc.

Surfine WNT LC. [Finetex] Sodium C12–15 pareth-

7 carboxylate; anionic; detergent, wetting agent, dispersant for cosmetic and household formulations; liq.; 50% conc.

Surfine WNT-LS. [Finetex] Sodium C12-15 pareth-7 carboxylate; anionic; detergent, wetting agent, lime soap dispersant for cosmetic and household formulations; liq.; 95% conc.

Surflo® 390. [Exxon] Alcohol ethoxylate sulfated, ammonium salt; anionic; foamer used in air drilling operations; liq.; dens. 8.46 lb/gal; sp.gr. 1.016; visc. 195 SUS; flash pt. 82 F (PMCC); pour pt. –10 F; pH 6.4 (1%); 38% conc.

Surflo® CW3. [Exxon] Alcohol ethoxylate, sulfated, ammonium salt; anionic; surfactant used in drilling operations; liq.; 40% conc.

Surflo® OW-1. [Exxon] Fatty acid glycerol ester, oxyethylated; nonionic; o/w emulsifier for elimination of oil slicks; liq.; sol. in IPA and water; dens. 8.56 lb/gal; sp.gr. 1.028 (65 F); visc. 160 SUS; flash pt. > 175 F (TOC); pour pt. 5 F; pH 7.5; 58% conc.

Surflo® S24. [Exxon] Alkylphenol and alcohol, oxyethylated; nonionic; dispersant, detergent, wetting and surf. and interfacial tens. reducing agent; used in producing wells and flotation units; liq.; water-sol.; dens. 8.30 lb/gal; sp.gr. 0.996 (68 F); visc. 125 SUS; flash pt. 82 F (PMCC); pour pt. < –10 F; 30% conc.

Surflo® S30. [Exxon] Sulfated alcohol ethoxylate, ammonium salt; anionic; detergent, foamer, surfactant and foamer in wells, prod. equipment, and air drilling applics.; liq.; water-sol.; dens. 8.16 lb/gal; sp.gr. 0.979 (68 F); visc. 52 SUS; flash pt. 81 F (TOC); pour pt. –3 F; biodeg.; 40% conc.

Surflo® S32. [Exxon] Sodium alkylaryl sulfonate; anionic; detergent, dispersant, emulsifier; surfactant in oil industry; liq.; water-sol.; dens. 8.16 lb/gal; sp.gr. 0.980 (68 F); visc. 44 SUS; flash pt. 80 F (TOC); pour pt. < –10 F; 20% conc.

Surflo® S40. [Exxon] Polyglycol and alcohol ethoxylate sulfated, ammonium salt; anionic; demulsifier, antifoulant, surfactant in oil industry; liq.; water-sol.; dens. 8.22 lb/gal; sp.gr. 0.9873 (68 F); visc. 43 SUS; flash pt. 90 F (TCC); pour pt. –10 F; pH 7.3 (1%); 27% conc.

Surflo® S41. [Exxon] Polyglycol; demulsifier, antifoulant, surfactant in oil industry; liq.; water-sol.; dens. 8.04 lb/gal; sp.gr. 0.9662 (68 F); visc. 47 SUS; flash pt. 120 F (TOC); pour pt. –5 F; pH 5.5 (1%); 20% conc.

Surflo® S242. [Exxon] Oxyethylated alkylphenol and oxyethylated alcohol; nonionic; surfactant in oil prod. operations; liq.; 40% conc.

Surflo® S302. [Exxon] Sulfated alcohol ethoxylate, ammonium salt; anionic; surfactant in oil prod. operations; liq.; 94% conc.

Surflo® S362. [Exxon] Sulfated alcohol ethoxylate, ammonium salt; anionic; foamer used in air drilling operations; liq.; water-sol.; dens. 8.65 lb/gal; sp.gr. 1.038 (68 F); visc. 95 SUS; flash pt. 79 F (TCC); pour pt. –10 F; pH 8.8 (5%); 100% conc.

Surflo® S375. [Exxon] Sulfated alcohol ethoxylate, ammonium salt; anionic; foamer used in air drilling operations; liq.; dens. 8.50 lb/gal; sp.gr. 1.020 (68 F); visc. 192 SUS; flash pt. 74 F (PMCC); pour pt. –10 F; pH 6.4 (1%); 53% conc.

Surflo® S398. [Exxon] alpha-Olefin sulfonate; anionic; foamer in air drilling operations; liq.; 40% conc.

Surflo® S412. [Exxon] Polyglycol; anionic; surfactant in oil prod. operations; liq.; 40% conc.

Surflo® S568. [Exxon] Sodium alkylaryl sulfonate; anionic; surfactant for oil prod. operations; liq.; 10% conc.

Surflo® S586. [Exxon] Ammonium alcohol ethoxylate sulfate; anionic; surfactant for oil prod. operations; biodeg.; liq.; 20% conc.

Surflo® S596. [Exxon] Ethoxylated alkyl phenol and ethoxylated alcohol; nonionic; surfactant for oil prod. operations; liq.; 30% conc.

Surflo® S970. [Exxon] alpha-Olefin sulfonate and ethoxylated alcohol sulfate; anionic; foaming agent for stimulating gas wells; liq.; 12% conc.

Surflo® S1143. [Exxon] Ammonium alcohol ethoxylate sulfate; anionic; surfactant for oil prod. operations; biodeg.; liq.; 60% conc.

Surflo® S1259. [Exxon] Polyglycol; nonionic; surfactant for oil prod. operations; liq.; 18% conc.

Surflo® S1720. [Exxon] Sulfated alcohol ethoxylate, ammonium salt; anionic; foamer in air drilling operations; liq.; 41% conc.

Surflo® S1748. [Exxon] Ammonium alcohol ethoxylate sulfate; anionic; foaming agent for air drilling operations; liq.; 53% conc.

Surflo® S1815. [Exxon] alpha-Olefin sulfonate; anionic; foaming agent for air drilling operations; liq.; 40% conc.

Surflo® S2023. [Exxon] Ethoxylated alcohol sulfate; anionic; foaming agent for stimulating gas wells; liq.; 9% conc.

Surflo® S2024. [Exxon] Alkyl quaternary ammonium chloride and amine oxide; cationic; foaming agent for stimulating gas wells; liq.; 15% conc.

Surflo® S2181. [Exxon] Polyglycol; nonionic; surfactant for oil prod. operations; liq.; 20% conc.

Surflo® S2221. [Exxon] Ethoxylated alkylphenol; nonionic; surfactant for oil prod. operations; biodeg.; liq.; 37% conc.

Surflo® S3902. [Exxon] Sulfated alcohol ethoxylate, ammonium salt; anionic; foamer in air drilling operations; liq.; 51% conc.

Surflo® S7650. [Exxon] Ethoxylated alcohol; nonionic; surfactant for oil prod. operations; liq.; 40% conc.

Surflo® S7655. [Exxon] Alkyl quaternary ammonium chloride and ethoxylated alkylamine; cationic; surfactant for oil prod. operations; liq.; 52% conc.

Surflo® S8536. [Exxon] Ethoxylated alcohol; nonionic; surfactant for oil prod. operations; liq.; 100% conc.

Surfonic® HDL. [Texaco] Nonoxynol-8 (86%) and triethanolamine (14%); CAS 9016-45-9; nonionic; biodeg. emulsifier, wetting agent, dry cleaning detergent, penetrant, solubilizer, lime soap dispersant, antifoamer used in agric., cosmetics, industrial cleaners, ceramics, concrete, dust control, wallpaper removal, photographic film developing; fire fighting, emulsion polymerization, indirect food additives, cutting oil emulsifiers; stabilizer for rubber latex and drilling-mud additives; degreaser for leather industry; sol. in water; sp.gr. 1.065; dens. 8.8 lb/gal; visc. 68 SUS (210 F); HLB 17.1; f.p. –7 C; cloud pt. 43 C (1% aq.); flash pt. (PMCC) 425 F; surf. tens. 31.5 dynes/cm (0.1%); ref. index 1.4888; toxicology: LD50 (oral, rat) 1938 mg/kg; mod. toxic by ingestion; severely irritating to eyes; 100% act.

Surfonic® HDL-1. [Texaco] Nonoxynol-10 (86%) and 14% triethanolamine; CAS 9016-45-9; nonionic; biodeg. emulsifier, wetting agent, detergent, penetrant, solubilizer, dispersant for household cleaners, textile, agric., metal cleaning, petrol, cosmetic, latex paint, cutting oil, janitorial supply industries; liq.; dens. 8.8 lb/gal; f.p. -6 C; flash pt. (PMCC) 485 F; 100% act.

Surfonic® JL-80X. [Texaco] Alkoxypolyalkoxyethanol; nonionic; biodeg. surfactant used as emulsifier, wetting agent, detergent, penetrant, solubilizer, dispersant for household detergents, industrial prods., agric. sprays, dry cleaning, metal cleaners, ceramics, concrete, textile processing, paper mfg.; clear liq.; sp.gr. 1.0072; dens. 8.4 lb/gal; visc. 159.4 SUS (100 F); HLB 13.1; f.p. -5 C; hyd. no. 92; cloud pt. 56–63 C (1% aq.); flash pt. (PMCC) 320 F; pH 6.0–7.5 (1% aq.); surf. tens. 29.24 dynes/cm (0.1%); Draves wetting < 1.0 s (0.25%); Ross-Miles foam 75 mm (initial, 0.1%, 120 F); toxicology: LD50 (oral, rat) 2.6 ml/kg; sl. toxic by ingestion; irritating to eyes; 100% act.

Surfonic® L10-3. [Texaco] Ethoxylated alcohol; nonionic; surfactant for hard surf. cleaners, gypsum board mfg., laundry prespotters; liq.; HLB 9.1; 100% conc.

Surfonic® L12-3. [Texaco] C10-12 pareth-3; nonionic; biodeg. surfactant for detergent, laundry prespotters, hard surf. cleaners, emulsifiers, personal care prods., agric. pesticides; clear to sl. turbid liq.; water-insol.; m.w. 295; dens. 7.8 lb/gal; HLB 9.0; hyd. no. 190; pour pt. -15 C; flash pt. (PMCC) 280 F; pH 5.5-6.5 (1% in aq. IPA); 100% conc.

Surfonic® L12-6. [Texaco] C10-12 pareth-6; nonionic; biodeg. surfactant for detergent, laundry prespotters, hard surf. cleaners, emulsifiers, personal care prods., agric. pesticides; clear to sl. turbid liq.; water-sol.; m.w. 428; dens. 8.2 lb/gal; HLB 12.4; hyd. no. 133; pour pt. 10 C; cloud pt. 48-52 C (1% aq.); flash pt. (PMCC) 325 F; pH 5.5-6.5 (1% in aq. IPA); 100% conc.

Surfonic® L12-8. [Texaco] C10-12 pareth-8; nonionic; surfactant for detergent, laundry prespotters, hard surf. cleaners, emulsifiers, personal care prods., agric. pesticides; liq.; HLB 13.7; 100% conc.

Surfonic® L24-1. [Texaco] Ethoxylated alcohol; nonionic; surfactant for detergents, personal care prods.; liq.; HLB 4.5; 100% conc.

Surfonic® L24-2. [Texaco] C12-14 pareth-2; nonionic; biodeg. surfactant for detergents, laundry prespotters, hard surf. cleaners, personal care prods., agric. pesticides; clear to sl. turbid liq.; water-insol.; m.w. 267; dens. 7.5 lb/gal; HLB 5.3; hyd. no. 210; pour pt. 10 C; flash pt. (PMCC) 300 F; pH 5.5-6.5 (1% in aq. IPA); 100% conc.

Surfonic® L24-3. [Texaco] C12-14 pareth-3; nonionic; biodeg. surfactant for detergents, laundry prespotters, hard surf. cleaners, personal care prods., agric. pesticides; clear to sl. turbid liq.; water-insol.; m.w. 330; dens. 7.8 lb/gal; HLB 8.0; hyd. no. 170; pour pt. 5 C; flash pt. (PMCC) 305 F; pH 5.5-6.5 (1% in aq. IPA); 100% conc.

Surfonic® L24-4. [Texaco] C12-14 pareth-4; nonionic; surfactant for detergent, emulsifier, laundry prespotters; liq.; HLB 9.4; 100% conc.

Surfonic® L24-7. [Texaco] C12-14 pareth-7; nonionic; biodeg. surfactant for detergents, laundry prespotters, hard surf. cleaners, personal care prods., agric. pesticides; clear to sl. turbid liq.; water-sol.; m.w. 487; dens. 8.1 lb/gal; HLB 11.9; hyd. no. 117; pour pt. 15 C; cloud pt. 48-52 C (1% aq.); flash pt. (PMCC) 375 F; pH 5.5-6.5 (1% in aq. IPA); 100% conc.

Surfonic® L24-9. [Texaco] C12-14 pareth-8.2; nonionic; biodeg. surfactant for detergents, laundry prespotters, hard surf. cleaners, personal care prods., agric. pesticides; clear to sl. turbid semiliq.; water-sol.; m.w. 561; dens. 8.3 lb/gal; HLB 13.0; hyd. no. 100; pour pt. 20 C; cloud pt. 73-77 C (1% aq.); flash pt. (PMCC) 325 F; pH 5.5-6.5 (1% in aq. IPA); 100% conc.

Surfonic® L24-12. [Texaco] C12-14 pareth-10.8; nonionic; biodeg. surfactant for detergents, laundry prespotters, hard surf. cleaners, personal care prods., agric. pesticides; waxy solid; water-sol.; m.w. 703; dens. 8.4 lb/gal; HLB 14.4; hyd. no. 83; pour pt. 28 C; cloud pt. 65-71 C (1% in 10% NaCl); flash pt. (PMCC) 400 F; pH 5.5-6.5 (1% in aq. IPA); 100% conc.

Surfonic® L42-3. [Texaco] Ethoxylated alcohol; nonionic; surfactant for detergents, personal care prods.; liq.; HLB 7.6; 100% conc.

Surfonic® L46-7. [Texaco] C14-16 pareth-7; nonionic; biodeg. surfactant, detergent, emulsifier, foamer, penetrant, intermediate for detergent, laundry prespotters, hard surf. cleaners, personal care prods., agric. pesticides; clear to sl. turbid liq.; water-sol.; m.w. 534; dens. 8.0 lb/gal; HLB 11.6; hyd. no. 107; pour pt. 22 C; cloud pt. 48-52 C (1% aq.); flash pt. (PMCC) 400 F; pH 5.5-6.5 (1% in aq. IPA); 100% conc.

Surfonic® L46-8X. [Texaco] Alkyl polyoxyalkylene ether; nonionic; low pour pt. surfactant; liq.; 100% conc.

Surfonic® L610-3. [Texaco] Ethoxylated alcohol; nonionic; surfactant for hard surf. cleaners, gypsum board mfg., laundry prespotters; liq.; HLB 9.4; 100% conc.

Surfonic® LF-17. [Texaco] Primary alcohol-EO adduct, modified; CAS 69013-18-9; nonionic; biodeg. low foaming wetting agent, detergent for aq. systems, metal cleaners, latex paints, textiles, paper, rinse aids, industrial and home mech. dishwashing compds.; defoamer in some systems; pale yel. clear liq.; sol. in water; sp.gr. 1.003; visc. 54.5 cSt (100 F); f.p. -2 C; HLB 12.2; cloud pt. 32-37 C (1% aq.); flash pt. (PMCC) 295 F; surf. tens. 33.0 dynes/cm (0.1%); Draves wetting 7.2 s (0.25%); Ross-Miles foam 6 mm (initial, 0.1%, 120 F); toxicology: LD50 (oral, rat) 3.53 g/kg; sl. toxic by ingestion, skin absorption; mod. irritating to eyes and skin; 100% act.

Surfonic® LF-27. [Texaco] Alkyl polyoxyalkylene ether; nonionic; low foaming surfactant; liq.; 100% conc.

Surfonic® LF-37. [Texaco] Alkyl polyoxyalkylene ether; nonionic; low foaming surfactant, wetting agent, detergent; liq.; 100% conc.

Surfonic® N-10. [Texaco] Nonoxynol-1; nonionic; biodeg. emulsifier, wetting agent, detergent, penetrant, solubilizer, dispersant for household cleaners, textile, agric., metal cleaning, petrol, cosmetic, latex paint, cutting oil, janitorial supply industries; FDA compliance; clear sl. visc. liq.; water-insol.; oil-sol.; m.w. 264; dens. 8.1 lb/gal; HLB 3.4; flash pt. (PMCC) 340 F; Ross-Miles foam 5 mm (initial, 0.1%, 120 F); 100% act.

Surfonic® N-31.5. [Texaco] Nonoxynol-3; CAS 9016-45-9; nonionic; biodeg. emulsifier, wetting agent, detergent, penetrant, solubilizer, dispersant for household cleaners, textile, agric., metal cleaning, petrol, cosmetic, latex paint, cutting oil, janitorial supply industries; FDA compliance; clear sl. visc. liq.; water-insol., oil-sol.; m.w. 358; sp.gr. 1.01; dens. 8.4 lb/gal; visc. 480 SUS (100 F); HLB 7.7; flash pt. (PMCC) 405 F; ref. index 1.4950;

Ross-Miles foam 8 mm (initial, 0.1%, 120 F); 100% act.

Surfonic® N-40. [Texaco] Nonoxynol-4; CAS 9016-45-9; nonionic; biodeg. emulsifier, wetting agent, detergent, penetrant, solubilizer, dispersant for household cleaners, textile, agric., metal cleaning, petrol, cosmetic, latex paint, cutting oil, janitorial supply industries; FDA compliance; clear sl. visc. liq.; sol. in acetone, methanol, xylene, CCl_4, Stoddard; m.w. 396; sp.gr. 1.026; dens. 8.5 lb/gal; visc. 445 SUS (100 F); HLB 8.9; flash pt.(PMCC) 360 F; ref. index 1.4979; surf. tens. 27.5 dynes/cm (0.1%); Ross-Miles foam 10 mm (initial, 0.1%, 120 F); 100% act.

Surfonic® N-60. [Texaco] Nonoxynol-6; CAS 9016-45-9; nonionic; biodeg. emulsifier, wetting agent, detergent, penetrant, solubilizer, dispersant for household cleaners, textile, agric., metal cleaning, petrol, cosmetic, latex paint, cutting oil, janitorial supply industries; FDA compliance; clear sl. visc. liq.; sol. in acetone, methanol, xylene, CCl_4, Stoddard solv.; disp. in water; m.w. 484; sp.gr. 1.041; dens. 8.7 lb/gal; visc. 440 SUS (100 F); f.p. < 0 C; HLB 10.9; flash pt. (PMCC) 430 F; ref. index 1.4938; surf. tens. 28.7 dynes/cm (0.1%); Draves wetting 6 s (0.25%); Ross-Miles foam 12 mm (initial, 0.1%, 120 F); 100% act.

Surfonic® N-85. [Texaco] Nonoxynol-8.5; CAS 9016-45-9; nonionic; biodeg. emulsifier, wetting agent, detergent, penetrant, solubilizer, dispersant for household cleaners, textile, agric., metal cleaning, petrol, cosmetic, latex paint, cutting oil, janitorial supply industries; FDA compliance; liq.; sol. in acetone, methanol, xylene, CCl_4, water; m.w. 596; sp.gr. 1.056; dens. 8.8 lb/gal; visc. 485 SUS (100 F); f.p. -1 C; HLB 12.6; cloud pt. 44 C (1% aq.); flash pt. (PMCC) 470 F; ref. index 1.4923; surf. tens. 30.5 dynes/cm (0.1%); Draves wetting 3.5 s (0.25%); Ross-Miles foam 40 mm (initial, 0.1%, 120 F); 100% act.

Surfonic® N-95. [Texaco] Nonoxynol-10; CAS 9016-45-9; nonionic; biodeg. emulsifier, wetting agent, detergent, penetrant, solubilizer, dispersant for household cleaners, textile, agric., metal cleaning, petrol, cosmetic, latex paint, cutting oil, janitorial supply industries; FDA compliance; clear sl. visc. liq.; sol. in acetone, methanol, xylene, CCl_4, water; m.w. 632; sp.gr. 1.061; dens. 8.8 lb/gal; visc. 510 SUS (100 F); f.p. 5 C; HLB 12.9; cloud pt. 54.2 (1% aq.); flash pt. (PMCC) 460 F; ref. index 1.4893; surf. tens. 30.8 dynes/cm (0.1%); Draves wetting 3 s (0.25%); Ross-Miles foam 80 mm (initial, 0.1%, 120 F); 100% act.

Surfonic® N-100. [Texaco] Nonoxynol-10; CAS 9016-45-9; nonionic; biodeg. emulsifier, wetting agent, detergent, penetrant, solubilizer, dispersant for household cleaners, textile, agric., metal cleaning, petrol, cosmetic, latex paint, cutting oil, janitorial supply industries; FDA compliance; clear sl. visc. liq.; sol. in acetone, methanol, xylene, CCl_4, water; m.w. 660; sp.gr. 1.064; dens. 8.8 lb/gal; visc. 69 SUS (210 F); f.p. 8 C; HLB 13.2; cloud pt. 65 C (1% aq.); flash pt. (PMCC) 415 F; ref. index 1.4888; surf. tens. 31.0 dynes/cm (0.1%); Draves wetting 4 s (0.25%); Ross-Miles foam 85 mm (initial, 0.1%, 120 F); 100% act.

Surfonic® N-102. [Texaco] Nonoxynol-10; CAS 9016-45-9; nonionic; biodeg. emulsifier, wetting agent, detergent, penetrant, solubilizer, dispersant for household cleaners, textile, agric., metal cleaning, petrol, cosmetic, latex paint, cutting oil, janitorial supply industries; clear sl. visc. liq.; sol. in acetone, methanol, xylene, CCl_4, water; m.w. 668; sp.gr. 1.065; dens. 8.8 lb/gal; f.p. 9 C; HLB 13.4; cloud pt. 81 C (1% aq.); flash pt. (PMCC) 500 F; ref. index 1.4884; surf. tens. 31.2 dynes/cm (0.1%); Draves wetting 4.5 s (0.25%); Ross-Miles foam 85 mm (initial, 0.1%, 120 F); 100% act.

Surfonic® N-120. [Texaco] Nonoxynol-12; CAS 9016-45-9; nonionic; biodeg. emulsifier, wetting agent, detergent, penetrant, solubilizer, dispersant for household cleaners, textile, agric., metal cleaning, petrol, cosmetic, latex paint, cutting oil, janitorial supply industries; FDA compliance; clear sl. visc. liq.; sol. in acetone, methanol, xylene, CCl_4, water; m.w. 748; sp.gr. 1.070; dens. 8.9 lb/gal; visc. 560 SUS (100 F); f.p. 14 C; HLB 14.1; cloud pt. 81 C (1% aq.); flash pt. (PMCC) 400 F; ref. index 1.4869; surf. tens. 32.3 dynes/cm (0.1%); Draves wetting 7 s (0.25%); Ross-Miles foam 110 mm (initial, 0.1%, 120 F); 100% act.

Surfonic® N-130. [Texaco] Nonoxynol-13; CAS 9016-45-9; nonionic; biodeg. emulsifier, wetting agent, detergent, penetrant, solubilizer, dispersant for household cleaners, textile, agric., metal cleaning, petrol, cosmetic, latex paint, cutting oil, janitorial supply industries; sl. turbid, sl. visc. liq.; water-sol.; m.w. 792; dens. 8.9 lb/gal; f.p. 18 C; HLB 14.4; cloud pt. 56-60 C (1% in 10% brine); flash pt. (PMCC) 750 F; 100% act.

Surfonic® N-150. [Texaco] Nonoxynol-15; CAS 9016-45-9; nonionic; biodeg. emulsifier, wetting agent, detergent, penetrant, solubilizer, dispersant for household cleaners, textile, agric., metal cleaning, petrol, cosmetic, latex paint, cutting oil, janitorial supply industries; FDA compliance; wh. waxy solid; sol. in acetone, methanol, xylene, CCl_4, water; m.w. 880; sp.gr. 1.065 (30/4 C); dens. 8.95 lb/gal; visc. 89 SUS (210 F); f.p. 23 C; HLB 15.0; cloud pt. 94 C (1% aq.); flash pt. (PMCC) 355 F; ref. index 1.4815 (30 C); surf. tens. 34.2 dynes/cm (0.1%); Draves wetting 17 s (0.25%); Ross-Miles foam 120 mm (initial, 0.1%, 120 F); 100% act.

Surfonic® N-200. [Texaco] Nonoxynol-20; CAS 9016-45-9; nonionic; biodeg. emulsifier, wetting agent, detergent, penetrant, solubilizer, dispersant for household cleaners, textile, agric., metal cleaning, petrol, cosmetic, latex paint, cutting oil, janitorial supply industries; FDA compliance; waxy solid; water-sol.; m.w. 1100; dens. 9.0 lb/gal; f.p. 34 C; HLB 15.8; cloud pt. 72-73 C (1% in 10% brine); flash pt. (PMCC) 450 F; 100% act.

Surfonic® N-300. [Texaco] Nonoxynol-30; CAS 9016-45-9; nonionic; biodeg. emulsifier, wetting agent, detergent, penetrant, solubilizer, dispersant for household cleaners, textile, agric., metal cleaning, petrol, cosmetic, latex paint, cutting oil, janitorial supply industries; FDA compliance; sol. in water; m.w. 1540; dens. 9.1 lb/gal; visc. 150-170 SUS (210 F); f.p. 44 C; HLB 17.1; cloud pt. > 100 C (1% aq.); flash pt. (PMCC) 385 F; ref. index 1.4690 (50 C); 100% act.

Surfonic® N-557. [Texaco] Nonoxynol-55; CAS 9016-45-9; nonionic; emulsifier for emulsion polymerization; emulsifier/stabilizer for asphalt emulsions; liq.; HLB 18.3; 70% conc.

Surfonic® NB-5. [Texaco] Nonoxynol-30; CAS 9016-45-9; nonionic; biodeg. emulsifier, wetting agent, detergent, penetrant, solubilizer, dispersant for household cleaners, textile, agric., metal cleaning, petrol, cosmetic, latex paint, cutting oil, janitorial supply industries; clear liq.; 70% act. in water.

Surfynol® 61. [Air Prods.] 3,5-Dimethyl 1-hexyn-3-ol; CAS 107-54-0; nonionic; surfactant, wetting agent used for paper coatings, inks, floor polishes, and glass cleaning formulations; cleaner in silicon wafer industry; colorless clear liq., camphoric odor; misc. with acetone, benzene, CCl_4, diethylene glycol, ethanol, ethyl acetate, kerosene, MEK, min. oil, min. spirits, petrol. ether, Stod., naphtha; sp.gr. 0.8545; dens. 7.2 lb/gal; f.p. -68 C; b.p. 151 C; HLB 6-7; flash pt. (TOC) 135 F; ref. index 1.4353; surf. tens. 56.4 (0.1% aq.); Draves wetting < 300 s (0.1%); 100% act.

Surfynol® 82. [Air Prods.] 3,6-Dimethyl-4-octyne-3,6-diol; CAS 78-66-0; nonionic; surfactant, defoamer, wetting agent, visc. reducer used in aq. systems, pesticide concs., shampoos, vinyl plastisols, agric., aq. starch sol'ns., flexographic inks, electroplating baths; detergent for radiator cleaners, descaling compds.; EPA compliance; wh. cryst. powd.; very sol. in acetone, benzene, CCl_4, Cellosolve, ethylene glycol, ethyl acetate, ethanol, MEK;sp.gr. 0.933; m.p. 49-51 C; b.p. 222 C; surf. tens. 55.3 (0.1% aq.); toxicology: LD50 (oral, rat) 1400 mg/k, (dermal, rabbit) > 1000 mg/k; not considered toxic; 100% act.

Surfynol® 82S. [Air Prods.] 3,6-Dimethyl-4-octyne-3,6-diol on amorphous silica carrier; nonionic; defoamer/wetting agent in pesticide wettable powds., electroplating baths, cement, plastics, coatings; solubilizer and clarifier in shampoos; EPA compliance; wh. free-flowing powd.; disp. in water; sp.gr. 0.47 g/ml; dens. 3.9 lb/gal; pH 7.6 (5% aq.); Draves wetting 4 s (4.35%); 46% conc.

Surfynol® 104. [Air Prods.] Tetramethyl decynediol; CAS 126-86-3; nonionic; defoamer and dye dispersant in paints, inks, dyestuffs, pesticides; surfactant in rinse aids; substrate pigment wetting agent for industrial coatings and adhesives; wetting agent for industrial cleaners; visc. reducer for vinyl dispersions; FDA, EPA compliance; wh. waxy solid; sp.gr. 0.893; HLB 4.0; m.p. 37 C; dens. 7.4 lb/gal; m.p. 37 C; b.p. 260 C; surf. tens. 31.6 (0.1% aq.); Draves wetting 15 s; Ross-Miles foam 0 mm (initial); toxicology: LD50 (rat, oral) 1.0 g/kg; LD50 (inhalation) > 20 mg/l; 100% conc.

Surfynol® 104A. [Air Prods.] Tetramethyl decynediol and 2-ethyl hexanol; CAS 126-86-3; nonionic; defoamer, wetting agent for pesticides, coatings, dyestuffs, aq. systems; lubricity additive for metalworking formulations; EPA approved; lt. yel. liq.; sp.gr. 0.869; dens. 7.2 lb/gal; HLB 4.0; m.p. < 0 C; surf. tens. 33.0 (0.1% aq.); 50% conc.

Surfynol® 104BC. [Air Prods.] Tetramethyl decynediol in 2-butoxy ethanol; CAS 126-86-3; nonionic; wetting and defoaming agent in water-based systems, e.g., coatings, adhesives, inks, cements, metalworking fluids, latex dipping and paper coatings; lt. yel. liq.; sp.gr. 0.898 (25 C); dens. 7.6 lb/gal; HLB 4.0; m.p. < -40 C; 50% conc.

Surfynol® 104E. [Air Prods.] Tetramethyl decynediol and ethylene glycol; CAS 126-86-3; nonionic; wetting agent, defoamer, dispersant, visc. stabilizer; EPA approved; lt. yel. clear liq.; sp.gr. 1.001; dens. 8.3 lb/gal; HLB 4.0; m.p. < 0 C; surf. tens. 36.2 (0.1% aq.); 50% conc.

Surfynol® 104H. [Air Prods.] Tetramethyl decynediol in ethylene glycol; CAS 126-86-3; nonionic; wetting and defoaming agent in water-based systems, e.g., coatings, adhesives, inks, cements, metalworking fluids, latex dipping and paper coatings; lt. yel. clear liq.; sp.gr. 0.951; dens. 7.9 lb/gal; HLB 4.0; m.p. 10 C; surf. tens. 33.8 (0.1% aq.); 75% conc.

Surfynol® 104PA. [Air Prods.] Tetramethyl decynediol in IPA; CAS 126-86-3; nonionic; wetting and defoaming agent in water-based systems, e.g., coatings, adhesives, inks, cements, metalworking fluids, latex dipping and paper coatings; lt. yel. liq.; sp.gr. 0.839 (25 C); dens. 8.1 lb/gal; HLB 4.0; m.p. -16 C; 50% conc.

Surfynol® 104PG. [Air Prods.] Tetramethyl decynediol in propylene glycol; CAS 126-86-3; nonionic; wetting and defoaming agent in water-based systems, e.g., coatings, adhesives, inks, cements, metalworking fluids, latex dipping and paper coatings; lt. yel. semisolid; sp.gr. 0.899 (50 C); dens. 7.7 lb/gal; HLB 4.0; m.p. 35 C; 85% conc.

Surfynol® 104S. [Air Prods.] Tetramethyl decynediol on amorphous silica; CAS 126-86-3; nonionic; wetting agent, defoamer, dispersant, visc. stabilizer for agric. formulations; EPA approved; free-flowing powd.; sp.gr. 0.457; dens. 3.9 lb/gal; HLB 4.0; surf. tens. 31.4 dynes/cm (0.1% aq.); 46% conc.

Surfynol® 420. [Air Prods.] Ethoxylated tetramethyl decynediol (1.3 EO); nonionic; wetting agent, defoamer, dispersant for aq. coatings, inks, adhesives, agric., electroplating, oilfield chems., paper coatings; FDA compliance; straw-colored liq.; sol. in CCl_4, xylene, ethylene glycol, min. oil; sl. sol. in dist. water; sp.gr. 0.943; visc. < 250 cps; HLB 4.0; pour pt. -13 F; pH 6-8 (1% aq.); surf. tens. 32 dynes/cm (0.1%).

Surfynol® 440. [Air Prods.] PEG-3.5 tetramethyl decynediol; CAS 9014-85-1; nonionic; water-based industrial finishes; defoamer, rewetting, and leveling agent for paperboard coatings, agric. formulations; metal cleaning and plating bath additive; FDA, EPA compliance; straw clear liq.; sol. in CCl_4, xylene, ethylene glycol, min. oil; sl. sol. in dist. water; sp.gr. 0.982; HLB 8.0; pour pt. -55 F; pH 6-8 (1% aq.); surf. tens. 33.2 dynes/cm (0.1% aq.); 100% conc.

Surfynol® 465. [Air Prods.] PEG-10 tetramethyl decynediol; CAS 9014-85-1; nonionic; wetting agent, defoamer for aq. coatings, inks, adhesives; surfactant for emulsion polymerization; electroplating additive; FDA, EPA compliance; straw clear liq.; sol. in water, CCl_4, xylene, ethylene glycol, min. oil; sp.gr. 1.038; visc. < 200 cps; HLB 13.0; pour pt. 44 F; cloud pt. 63 C (5%); pH 6-8 (1% aq.); surf. tens. 33.2 (0.1% aq.); 100% conc.

Surfynol® 485. [Air Prods.] PEG-30 tetramethyl decynediol; CAS 9014-85-1; nonionic; wetting agent, defoamer for aq. coatings, inks, adhesives, agric., electroplating, oilfield chems., paper coatings; FDA, EPA compliance; straw clear liq.; sol. in water, CCl_4, xylene, ethylene glycol, min. oil; sp.gr. 1.080; visc. < 350 cps; HLB 17.0; pour pt. 85 F; cloud pt. > 100 C (5%); pH 6-8 (1% aq.); surf. tens. 40.1 (0.1% aq.); 100% conc.

Surfynol® CT-136. [Air Prods.] Proprietary blend; nonionic/anionic; wetting agent, defoamer, grind aid and dispersant for water and glycol-based inks and pigments; lt. yel. liq.; dens. 8.8 lb/gal; visc. 465 cps; f.p. -10 C; b.p. 68 F; HLB 13.0; cloud pt. 167 F (15% aq.); flash pt. 200 F; pH 8.0 (5%); 60% act.

Surfynol® D-101. [Air Prods.; Air Prods. Nederland BV] Nonsilicone; defoamer/antifoamer for aq. systems, coatings, adhesives; effective for PVAc, ethylene-vinyl acetate systems; FDA compliance; yel. clear liq.; emulsifiable in water; sp.gr. 0.92; b.p. 171 C (109 mm); cloud pt. < 25 F; flash pt. (PMCC) 165 F; pH 5 (5% aq.); 100% act.

Surfynol® D-201. [Air Prods.; Air Prods. Nederland BV] Nonsilicone; defoamer/antifoamer for aq. systems, coatings, adhesives; effective for PVAc, ethylene-vinyl acetate systems; FDA compliance; yel. clear liq.; emulsifiable in water; sp.gr. 0.93; b.p. 204 C (17 mm); flash pt. (PMCC) > 25 F; pH 5 (5% aq.); 100% act.

Surfynol® DF-08. [Air Prods.; Air Prods. Nederland BV] Surfactant; wetting agent, defoamer for waterborne coatings, inks, adhesives, and latex dipping.

Surfynol® DF-34. [Air Prods.] Proprietary blend; nonionic; wetting agent, defoamer for water-based coatings, inks, and adhesives; dewebbing agent and defoamer for latex gloves and other dipped goods; liq.

Surfynol® DF-37. [Air Prods.; Air Prods. Nederland BV] Silicone-free org. defoamer; defoamer minimizing web formation during latex glove, waterborne coating dipping processes, and other aq. systems (inks, adhesives, coatings, agric., cement, metalworking fluids); wetting agent for low energy substrates; FDA compliance; translucent pale yel. liq.; emulsifiable in water; sp.gr. 0.941; b.p. > 380 F; flash pt. > 280 F; 100% act.

Surfynol® DF-58. [Air Prods.] Organo-modified silicone-based defoamer; self-emulsifying defoamer used in water-based applics, e.g., inks, coatings, adhesives, cements, latex-containing formulations; pale tan hazy liq.; sp.gr. 0.995; visc. 100 cps; flash pt. (Seta) 80 F; pH 4.4; 100% act.

Surfynol® DF-60. [Air Prods.] Silicone-based defoamer; defoamer for aq. coating and ink systems, esp. for grinding and dispersion stage; pale yel. liq.; dens. 8.4 lb/gal; b.p. 523 F; flash pt. 140 F; ref. index 1.443.

Surfynol® DF-70. [Air Prods.] Organic defoamer; defoamer for aq. systems, printing inks, coatings, adhesives, esp. acrylic and styrene-acrylic formulations; FDA compliance; pale tan opaque liq.; disp. in water; sp.gr. 0.90; visc. 800 cps; b.p. 106 C; flash pt. (Seta) > 229 F; pH 6.9 (5% disp.); 100% act.

Surfynol® DF-75. [Air Prods.; Air Prods. Nederland BV] Silicone-free; defoamer for water-based systems, e.g., inks, adhesives, coatings, overprint varnishes, paper coatings; esp. effective in acrylic systems; FDA compliance; amber opaque liq.; disp. in water; sp.gr. 0.99; b.p. 240 F; flash pt. > 400 F; pH 8.0 (5% disp.); 100% act.

Surfynol® DF-110, DF-110S. [Air Prods.; Air Prods. Nederland BV] Silicone-free acetylenic derivs.; defoamer, de-air entrainment agent for inks, metalworking fluids, coatings, cement, ceramics, adhesives; waxy solid and free-flowing powd. resp.

Surfynol® DF-110D. [Air Prods.] Higher m.w. acetylenic glycol in dipropylene glycol; nonionic; defoamer, de-air entrainment aid for water-based coatings, inks, adhesives, and highly pigmented systems (concrete, paper coatings, grouts, ceramics); liq.; HLB 3.0.

Surfynol® DF-110L. [Air Prods.; Air Prods. Nederland BV] Higher m.w. acetylenic glycol in mixed glycols; nonionic; defoamer, de-air entrainment aid for water-based coatings, inks, adhesives, and highly pigmented systems (concrete, paper coatings, grouts, ceramics); liq.; HLB 3.0.

Surfynol® DF-210. [Air Prods.] Silicone-free organic defoamer; defoamer for aq. formulations, printing inks, adhesives, coatings; tan liq.; self-emulsifiable in water; sp.gr. 0.912; visc. < 600 cps; m.p. -7 C; pour pt. < 2 C; flash pt. (PMCC) 110 C; 100% act.

Surfynol® DF-574. [Air Prods.] Organo and organo-modified silicone compd.; rapid knockdown defoamer for aq. coatings and inks; sl. yel. liq.; sp.gr. 0.94; visc. 35 cps; flash pt. (Seta) 230 F; pH 6.0.

Surfynol® DF-695. [Air Prods.] Silicone emulsion; defoamer for aq. ink systems; FDA compliance; milky emulsion; sp.gr. 1.04; visc. 9000 cps; pH 7.3.

Surfynol® GA. [Air Prods.] Acetylenic diol; nonionic; pigment wetting agent, grinding aid in coatings and other pigmented systems; straw yel. clear liq., mild fatty acid odor; misc. in water; sp.gr. 1.051; dens. 8.76 lb/gal; visc. 140 cps; f.p. -5 C; HLB 13.0; cloud pt. 57 C (5%); flash pt. > 200 F; pH 6.5 (5%); surf. tens. 33 dynes/cm (0.1%); Draves wetting 19 s (0.1%).

Surfynol® PC. [Air Prods.] Acetylenic glycol; nonionic; defoamer used in paper coating and adhesive latexes; antishock agent for paper coatings; lt. yel. clear mobile liq.; sp.gr. 0.985; dens. 8.0 lb/gal; pour pt. 15 F; cloud pt. 47 F; 100% conc.

Surfynol® PG-50. [Air Prods.] Tetramethyl decynediol in propylene glycol; nonionic; wetting and defoaming agent in water-based systems, e.g., coatings, adhesives, inks, cements, metalworking fluids, latex dipping and paper coatings; 50% conc.

Surfynol® PSA-204. [Air Prods.; Air Prods. Nederland BV] Surfactant; low or nonfoaming wetting agent for pressure-sensitive adhesives.

Surfynol® PSA-216. [Air Prods.; Air Prods. Nederland BV] Surfactant; defoaming wetting agent for pressure-sensitive adhesives.

Surfynol® SE. [Air Prods.] Acetylenic diol; nonionic; wetting agent, foam control agent in pressure-sensitive adhesives, aq. lubricants, water based paints, inks, dye processing, agric. formulations, and paper coatings; EPA approved; lt. yel. liq.; misc. @ 50% in acetone, xylene, kerosene, perchloroethylene, IPA, butyl Cellosolve, propylene glycol; sp.gr. 0.94; dens. 7.8 lb/gal; visc. 160 cps; HLB 4-5; surf. tens. 32 dynes/cm (0.1%); Draves wetting 17 s (0.1%).

Surfynol® SE-F. [Air Prods.] Surfactant blend; nonionic; wetting agent, foam control agent for sensitive adhesives, aq. lubricants, paints, paper coatings, dye processing, and aq. inks; liq.; HLB 4-5.

Surfynol® TG. [Air Prods.] Tetramethyl decynediol and ethylene glycol; CAS 126-86-3; nonionic; pigment and substrate wetting agent used in latex and water-reducible paints, adhesives, paper coatings, and pigmented aq. systems; lt. yel. clear liq.; sp.gr. 1.000; HLB 9-10; surf. tens. 27.6 (0.1% aq.); 83% conc.

Surfynol® TG-E. [Air Prods.] Acetylenic glycol and propylene glycol; nonionic; low foaming and wetting agent used in pesticide formulations; EPA compliant; APHA 300 max. clear liq.; misc. with xylene, heavy aromatic naphtha, IPA, MIBK; sol. 0.4% in water; sp.gr. 0.99; visc. 145-160 cps; HLB 9-10; flash pt. (TOC) > 200 F; surf. tens. 28 dynes/cm (0.1%); Draves wetting 8 s (0.1%); toxicology: LD50 (oral) 5.4 g/kg, (dermal) > 8 g/kg; not considered toxic; 83% act.

Surmax® CS-504. [Chemax] Surfactant; modified anionic; alkaline-stable surfactant for formulating detergent concs.; moderate foamer, wetting agent, detergent; for metalworking formulas, paint strippers, tire cleaners, transportation cleaners, dairy and food plant cleaners, paper felt washing; sanitizers, wax strippers; biodeg.; clear amber to brn. liq.; mild odor; sol. in min. acids; sp.gr. 1.19; visc. 1390 cps; pour pt. < 15 F; cloud pt. > 100 C (1% aq.); flash pt. (COC) > 100 C; pH 7 (5% aq.); surf. tens. 38 dynes/cm (0.05% aq.); toxicology: nontoxic; moderate to severe eye and skin irritant; LD50 (oral) > 5 g/kg; LD50 (dermal) > 2 g/kg; 80% active in water.

Surmax® CS-515. [Chemax] Surfactant; modified anionic; alkaline-stable surfactant for formulating detergent concs.; moderate foamer, wetting agent, detergent; for metalworking formulas, paint strippers, tire cleaners, transportation cleaners, dairy and food plant cleaners, paper felt washing; sanitizers, wax strippers; biodeg.; clear amber to brn. liq.; mild odor; sol. in min. acids; sp.gr. 1.20; visc. 1490 cps; pour pt. < 15 F; cloud pt. > 100 C (1% aq.); flash pt. (COC) > 100 C; pH 7 (5% aq.); surf. tens. 40 dynes/cm (0.05% aq.); toxicology: nontoxic; moderate to severe eye and skin irritant; LD50 (oral) > 5 g/kg; LD50 (dermal) > 2 g/kg; 80% active in water.

Surmax® CS-521. [Chemax] Surfactant; modified anionic; alkaline-stable surfactant for formulating detergent concs.; low foaming wetting agent, detergent; for metalworking formulas, paint strippers, tire cleaners, transportation cleaners, dairy and food plant cleaners, paper felt washing; sanitizers, wax strippers; biodeg.; clear amber to brn. liq.; mild odor; sol. in min. acids; sp.gr. 1.17; visc. 1280 cps; pour pt. < 15 F; cloud pt. > 100 C (1% aq.); flash pt. (COC) > 100 C; pH 7 (5% aq.); surf. tens. 57 dynes/cm (0.05% aq.); toxicology: nontoxic; moderate to severe eye and skin irritant; LD50 (oral) > 5 g/kg; LD50 (dermal) > 2 g/kg; 80% active in water.

Surmax® CS-522. [Chemax] Surfactant; modified anionic; alkaline-stable surfactant for formulating detergent concs.; low foaming wetting agent, detergent; for metalworking formulas, paint strippers, tire cleaners, transportation cleaners, dairy and food plant cleaners, paper felt washing; sanitizers, wax strippers; biodeg.; clear amber to brn. liq.; mild odor; sol. in min. acids; sp.gr. 1.18; visc. 1550 cps; pour pt. < 15 F; cloud pt. > 100 C (1% aq.); flash pt. (COC) > 100 C; pH 7 (5% aq.); surf. tens. 51 dynes/cm (0.05% aq.); toxicology: nontoxic; moderate to severe eye and skin irritant; LD50 (oral) > 5 g/kg; LD50 (dermal) > 2 g/kg; 80% active in water.

Surmax® CS-555. [Chemax] Surfactant; modified anionic; alkaline-stable surfactant for formulating detergent concs.; low foaming wetting agent, detergent; for metalworking formulas, paint strippers, tire cleaners, transportation cleaners, dairy and food plant cleaners, paper felt washing; sanitizers, wax strippers; biodeg.; liq.; 60% act.

Surmax® CS-586. [Chemax] Surfactant; modified anionic; alkaline-stable surfactant for formulating detergent concs.; high foaming wetting agent, detergent; for metalworking formulas, paint strippers, tire cleaners, transportation cleaners, dairy and food plant cleaners, paper felt washing; sanitizers, wax strippers; biodeg.; liq.; 80% act.

Surpasol 53. [Climax Performance] Sec. emulsifier for oil muds.

Surpasol E-436. [Climax Performance] Esters in a hydrocarbon base; paper mill defoamer; water emulsifiable.

Surpasol NT-57. [Climax Performance] Sec. emulsifier for oil muds; low toxicity.

Swanic 6L, 7L. [Swastik] Nonylphenol ethoxylate; nonionic; wetting agent, emulsifier and detergent used in wool scouring, paint formulation and textile processing; clear liq.; pH 6–7 (1% in dist. water); 100 and 25% act. resp.

Swanic 51. [Swastik] Coconut MEA (oil-based); CAS 68140-00-1; foam booster/stabilizer in detergent powds.; wh. solid paste.

Swanic 52L. [Swastik] Coconut alkylolamide; anionic; foam booster, visc. controller used in shampoos and liq. detergent formulations; yel. liq.; 100% conc.

Swanic 110. [Swastik] Alcohol ethoxylate; nonionic; retarder and dye leveling agent used in textiles; liq.; 30% conc.

Swanic C-30. [Swastik] Alcohol ethoxylate; nonionic; detergent; liq.; 100% conc.

Swanic CD-48. [Swastik] Alkyl phenol ethoxylate; nonionic; emulsifier; liq.; 100% conc.

Swanic CD-72. [Swastik] Alkyl phenol ethoxylate; nonionic; emulsifier; paste; 100% conc.

Swanic CWT. [Swastik] Sulfonated ester; anionic; wetting and rewetting agent used in textiles; liq.; 30% conc.

Swanic D-16, -24, -52, -60, -80, -120. [Swastik] Alkyl phenol POE ether; nonionic; wetting agent, detergent and emulsifier ingred. in formulation of pesticides; liq., paste, solid.

Swanic SD-36. [Swastik] Alkyl phenol POE ether; nonionic; surfactant for pigment paste preparation; liq.

Swanic T-51. [Swastik] Alcohol sulfate; anionic; wetting agent for textiles; liq.; 30% conc.

Swanic X-100. [Swastik] Alkyl phenol POE ether; nonionic; wetting agent; liq.; 100% conc.

Swanol AM-301. [Nikko Chem. Co. Ltd.] Lauryl betaine; amphoteric; detergent, wetting, dispersant; liq.; 35% conc.

Swascol 1P. [Swastik] Sodium lauryl sulfate; CAS 151-21-3; anionic; detergent for toothpaste mfg.; emulsifier in emulsion polymerization; powd.; 95% conc.

Swascol 1PC. [Swastik] Sodium lauryl sulfate; CAS 151-21-3; anionic; wetting agent and foamer used in personal care prods., liq. detergents; powd.; 96% act.

Swascol 3L. [Swastik] Sodium lauryl sulfate; CAS 151-21-3; anionic; wetting agent, foamer and detergent used in personal care prods. and liq. detergents; liq., paste; 30% act.

Swascol 4L. [Swastik] TEA lauryl sulfate; anionic; foamer, detergent, base material for shampoos and liq. detergents; liq.; 40% act.

Swastik Detergent Powder. [Swastik] Sodium alkylaryl sulfonate with alkaline builders, STPP, etc.; detergent powd. for heavily soiled clothes; blue powd.

Sway. [Swastik] Sodium alkylaryl sulfonate with STPP, alkaline builders, etc.; heavy duty detergent for cottons; blue powd.

Sykanol DKM 45, 80. [Henkel KGaA] Sulfated castor oil; anionic; wetting agent for mfg of bath preparations; liq.; 45 and 80% conc. resp.

Sylfam 2042. [Arizona] Corrosion inhibitor for oilfield servicing, refinery operations and oil transport pipes, surfactant in metalworking formulations, pigment dispersant, chem. intermediate in the synthesis of amines; Gardner 6 color; sp.gr. 0.99; visc. 930 cps; acid no. 7.5; amine no. 100; flash pt. (OC) > 232 C; 100% solids.

Sylfam 2082. [Arizona] Imidazoline; corrosion inhibitor for oilfield servicing, refinery operations and oil transport pipes, surfactant in metalworking formulations, pigment dispersant, chem. intermediate in the synthesis of amines; sp.gr. 0.96; visc. 12,680 cps; acid no. 15; amine no. 108; flash pt. (CC) 27 C; 85% solids.

Sylfan 20. [Arizona] Corrosion inhibitor for oilfield servicing, refinery operations and oil transport pipes, surfactant in metalworking formulations, pigment dispersant, chem. intermediate in the synthesis of amines; Gardner 4+ color; sp.gr. 0.85; visc. 20 cps; acid no. 0.4; flash pt. (CC) 149 C; 100% solids.

Sylfan 96. [Arizona] Corrosion inhibitor for oilfield servicing, refinery operations and oil transport pipes, surfactant in metalworking formulations, pigment dispersant, chem. intermediate in the synthesis of amines; Gardner 3 color; sp.gr. 0.85; visc. 20 cps; acid no. 0.3; flash pt. (CC) 149 C; 100% solids.

Sylfat® D-1. [Arizona] Monomeric fatty acid distillate from dimerization of tall oil fatty acid; dispersant for drilling muds, component of cutting oils, grinding compds., textile drawing lubricants, defoamers, plasticizers, thickeners in greases and lubricant oil formulations, oilfield applics.; Gardner 4 color; sp.gr. 0.91; acid no. 172; iodine no. 70; sapon. no. 185; flash pt. (OC) 184 C.

Sylfat® DX. [Arizona] Tall oil fatty acid ester; for plasticizers, extenders, surface-act. agents in grinding and cutting oils, specialty lubricant additives, corrosion inhibitors, specialty solvs. for printing inks, metalworking, and oil well servicing; Gardner 5+ color; sp.gr. 0.94; visc. 700 cps; acid no. 12; flash pt. (OC) 190 C.

Sylfat® MM. [Arizona] Tall oil fatty acid ester; for plasticizers, extenders, surface-act. agents in grinding and cutting oils, specialty lubricant additives, corrosion inhibitors, specialty solvs. for printing inks, metalworking, and oil well servicing; Gardner 9 color; sp.gr. 0.91; visc. 20 cps; acid no. 5; iodine no. 105; flash pt. (OC) 190 C.

Sylfat® RD-1. [Arizona] Monomeric fatty acid distillate from dimerization of tall oil fatty acid; dispersant for drilling muds, component of cutting oils, grinding compds., textile drawing lubricants, defoamers, plasticizers, thickeners in greases and lubricant oil formulations, oilfield applics.; Gardner 3 color; sp.gr. 0.91; acid no. 174; iodine no. 70; sapon. no. 180; flash pt. (OC) 184 C.

Sylvacote® K. [Arizona] Corrosion inhibitor for oilfield servicing, refinery operations and oil transport pipes, surfactant in metalworking formulations, pigment dispersant, chem. intermediate in the synthesis of amines; Gardner 5 color; sp.gr. 1.04; visc. 370-500 cps; acid no. 177-187; flash pt. (OC) 218 C; 100% solids.

Sylvaros® 20. [Arizona] Tall oil rosin; printing ink binder as resin or salt, paper sizing agent, emulsifier for SBR polymerization as soap, tackifier resin in adhesives, imidazoline modifier in corrosion inhibitors, elastomer modifier in emulsion polymerization, dust control additive; film former/plasticizer in lacquers and varnishes; Gardner 6- color; sp.gr. 1.01; soften. pt. (R&B) 75 C; acid no. 176; flash pt. (OC) 226 C.

Sylvaros® 80. [Arizona] Tall oil rosin; printing ink binder as resin or salt, paper sizing agent, emulsifier for SBR polymerization as soap, tackifier resin in adhesives, imidazoline modifier in corrosion inhibitors, elastomer modifier in emulsion polymerization, dust control additive; film former/plasticizer in lacquers and varnishes; Gardner 6 color; sp.gr. 1.01; soften. pt. (R&B) 75 C; acid no. 171; flash pt. (OC) 226 C.

Syncrowax AW1-C. [Croda Inc.] C18-36 acid; anionic; emulsifier, emollient, opacifier; pale cream waxy flakes, mild waxy odor; oil-sol.; HLB 2.0; m.p. 69–74 C; acid no. 155–170; iodine no. 3 max.; sapon. no. 165–175; usage level: 1-15%; 100% conc.

Syncrowax BB4. [Croda Inc.] Syn. beeswax; see Syncrowax AW1-C; also suspending agent for anhyd. systems, aux. w/o emulsifier, thickener for oils and waxes; used in creams and sticks; pale cream waxy flakes; oil-sol.; m.p. 60–65 C; usage level: 1-5%.

Syncrowax ERL-C. [Croda Inc.] C18-36 acid glycol ester; see Syncrowax AW1-C; also lubricant, stabilizer, suspending agent for anhyd. systems, thickener, reducer of bleeding and sweating; gloss improver; sticks, creams; lt. tan waxy flakes; mild waxy odor; m.p. 70–75 C; acid no. 10–15; iodine no. 3 max.; sapon. no. 155–160; usage level: 1-4%.

Syncrowax HGL-C. [Croda Inc.] C18-36 acid triglyceride; see Syncrowax AW1-C; also lubricant, suspending agent, strength improver, stabilizer, gloss improver; used in cosmetic makeup; lt. tan waxy flakes; mild waxy odor; m.p. 70–75 C; acid no. 6–12; iodine no. 3 max.; sapon. no. 160–175; usage level: 4-12%.

Syncrowax HR-C. [Croda Inc.] Glyceryl tribehenate; CAS 18641-57-1; see Syncrowax AW1-C; also suspending agent, thickener, gloss improver used in personal care prods.; off-wh. waxy pastilles/powd.; mild waxy odor; m.p. 60–65 C; acid no. 10 max.; iodine no. 3 max.; sapon. no. 170–175; usage level: 4-12%.

Syncrowax HRS-C. [Croda Inc.] Glyceryl tribehenate, calcium behenate; see Syncrowax AW1-C; also suspending agent for anhyd. systems, aux. w/o emulsifier, gellant, thickener for oils and waxes; pale cream flakes; mild waxy odor; oil-sol.; m.p. 105–115 C; acid no. 8–13; iodine no. 3 max.; sapon. no. 115–125; usage level: 2-10%.

Syn Fac® 334. [Milliken] Aryl POE ether; nonionic; emulsifier, dispersant for pigments, insecticides, solvs., cleaning compds. and latex paint; liq. and semiliq. resp.; HLB 11.6; 87% conc.

Syn Fac® 905. [Milliken] Nonoxynol-9 (9.5 EO); CAS 9016-45-9; nonionic; wetting agent, dispersant, emulsifier; liq.; 100% conc.

Syn Fac® 8210. [Milliken] Polyoxyaryl ether; nonionic; low foaming emulsifier, dispersant for pigments; liq.; 100% act.

Syn Fac® 8216. [Milliken] Aryl POE ether; nonionic; pigment dispersant for paints, coatings, and textile print systems; solid; HLB 15.0; 100% conc.

Syn Fac® TDA-92. [Milliken] PEG-8 triethanolamine; nonionic; emulsifier and wetting agent used in fiber and fabric applics.; liq.; HLB 12.5; 100% conc.

Syn Fac® TEA-97. [Milliken] Ethoxylated amine; nonionic; solv. and dispersant in the dye industry; liq.; 100% conc.

Syn Lube 106 (60%). [Milliken] PEG-200 glyceride ester; nonionic; emulsifier, lubricant, softener; spin finish and viscose additive, tufting aid; liq.; 60% conc.

Syn Lube 107. [Milliken] PEG-25 glyceride ester; nonionic; emulsifier, lubricant, softener; spin finish and viscose additive, tufting aid; liq.; HLB 15.5;

100% conc.

Syn Lube 728. [Milliken] Modified ethoxylated castor oil; nonionic; emulsifier, lubricant for use in finishes, dye carriers, and solv. emulsion formulations; liq.; HLB 9.0; 100% conc.

Syn Lube 1632H. [Milliken] Modified ethoxylated castor oil; nonionic; emulsifier, coemulsifier for difficult systems containing halogenated flame retardants and hydrog. natural esters; liq.; HLB 6.0; 100% conc.

Syn Lube 6277-A. [Milliken] Modified ethoxylated glyceride; nonionic; finish additive for syn. fibers and tufing aid for carpet mfg.; liq.; HLB 12.0; 100% conc.

Synotol 119 N. [Aquatec Quimica SA] Cocamide DEA; nonionic; detergent, wetting agent, thickener, foam booster and stabilizer, lime soap dispersant; used in personal care prods.; liq.; 50% amide.

Synotol CN 20. [Aquatec Quimica SA] Cocamide DEA; nonionic; foam stabilizer, thickener, superfatting agent for cosmetic and household prods.; liq.; 90% conc.

Synotol CN 60. [Aquatec Quimica SA] Cocamide DEA; CAS 68603-42-9; nonionic; foam stabilizer, thickener, superfatting agent for cosmetic and household prods.; liq.; 60% conc.

Synotol CN 80. [Aquatec Quimica SA] Cocamide DEA; nonionic; foam stabilizer, thickener, superfatting agent for cosmetic and household prods.; liq.; 80% conc.

Synotol CN 90. [Aquatec Quimica SA] Cocamide DEA; nonionic; foam stabilizer, thickener, superfatting agent for cosmetic and household prods.; liq.; 90% conc.

Synotol Detergent E. [Aquatec Quimica SA] Cocamide DEA; detergent, wetting agent, thickener, foam booster and stabilizer, lime soap dispersant; used in personal care prods.; liq.; 50% amide.

Synotol L 60. [Aquatec Quimica SA] Lauramide DEA; CAS 120-40-1; detergent, wetting agent, thickener, foam booster and stabilizer, lime soap dispersant; used in personal care prods.; liq.; 50% amide.

Synotol L 90. [Aquatec Quimica SA] Lauramide DEA; CAS 120-40-1; detergent, wetting agent, thickener, foam booster and stabilizer, lime soap dispersant; used in personal care prods.; solid; 88% conc.

Synotol LM 60. [Aquatec Quimica SA] Lauramide DEA; CAS 120-40-1; detergent, wetting agent, thickener, foam booster and stabilizer, lime soap dispersant; used in personal care prods.; liq.; 50% amide.

Synotol LM 90. [Aquatec Quimica SA] Lauric/myristic acid DEA; detergent, wetting agent, thickener, foam booster and stabilizer, lime soap dispersant; used in personal care prods.; paste; 88% conc.

Synotol ME 90. [Aquatec Quimica SA] Cocamide MEA; CAS 68140-00-1; nonionic; foam booster, wetting agent, thickener, foam stabilizer, and superfatting agent for cosmetic and household preparations; flakes; 90% conc.

Synperonic 3S60A. [ICI PLC] Syn. primary alcohol ethoxylate sulfate, ammonium salt; anionic; detergent, foamer; used in household formulations; yel. clear to hazy mobile liq.; dens. 1.020 g/ml; surf. tens. 39.4 dynes/cm (0.1%); pH 6.5–8.0 (5% aq.); 59–61% act.

Synperonic 3S60S. [ICI PLC] Syn. primary alcohol ethoxylate sulfate, sodium salt; detergent, foamer; used in household formulations; colorless to pale yel. mobile liq.; dens. 1.046 g/ml; surf. tens. 40.5 dynes/cm (0.1%); pH 7.0–8.5 (5% aq.); 59–61% act.

Synperonic 3S70. [ICI PLC] Syn. primary alcohol ethoxylate; anionic; dishwashing liqs. and personal care prods.; visc. liq.; 70% conc.

Synperonic 10/3-90%. [ICI PLC] Syn. primary alcohol ethoxylate; CAS 61827-42-7; nonionic; emulsifier; liq.; oil-sol.; 90% conc.

Synperonic 10/3-100%. [ICI PLC] Syn. primary alcohol ethoxylate; CAS 61827-42-7; nonionic; emulsifier; liq.; oil-sol.; HLB 10.0; 100% conc.

Synperonic 10/5-90%. [ICI PLC] Syn. primary alcohol ethoxylate; CAS 61827-42-7; nonionic; wetting agent, wool scouring, dyeing agent; liq.; water-sol.; 90% conc.

Synperonic 10/5-100%. [ICI PLC] Syn. primary alcohol ethoxylate; CAS 61827-42-7; nonionic; wetting agent, wool scouring, dyeing agent; liq.; water-sol.; HLB 12.0; 100% conc.

Synperonic 10/6-90%. [ICI PLC] Syn. primary alcohol ethoxylate; CAS 61827-42-7; nonionic; wetting agent, wool scouring, dyeing agent; liq.; water-sol.; 90% conc.

Synperonic 10/6-100%. [ICI PLC] Syn. primary alcohol ethoxylate; CAS 61827-42-7; nonionic; wetting agent, wool scouring, dyeing agent; liq.; water-sol.; HLB 12.5; 100% conc.

Synperonic 10/7-90%. [ICI PLC] Syn. primary alcohol ethoxylate; CAS 61827-42-7; nonionic; wetting agent, detergent; liq.; water-sol.; 90% conc.

Synperonic 10/7-100%. [ICI PLC] Syn. primary alcohol ethoxylate; CAS 61827-42-7; nonionic; wetting agent, detergent; liq.; water-sol.; HLB 13.5; 100% conc.

Synperonic 10/8-90%. [ICI PLC] Syn. primary alcohol ethoxylate; CAS 61827-42-7; nonionic; wetting agent, detergent; liq.; water-sol.; 90% conc.

Synperonic 10/8-100%. [ICI PLC] Syn. primary alcohol ethoxylate; CAS 61827-42-7; nonionic; wetting agent, detergent; liq.; water-sol.; HLB 14.2; 100% conc.

Synperonic 10/11-90%. [ICI PLC] Syn. primary alcohol ethoxylate; CAS 61827-42-7; nonionic; wetting agent, detergent; liq.; water-sol.; 90% conc.

Synperonic 10/11-100%. [ICI PLC] Syn. primary alcohol ethoxylate; CAS 61827-42-7; nonionic; wetting agent, detergent; liq.; water-sol.; HLB 15.5; 100% conc.

Synperonic 13/3. [ICI PLC] Syn. primary alcohol ethoxylate; CAS 24938-91-8; nonionic; detergent, emulsifier for oils, waxes and solvs.; intermediate for sulfation; liq.; HLB 8.6; 100% conc.

Synperonic 13/5. [ICI PLC] Syn. primary alcohol ethoxylate; CAS 24938-91-8; nonionic; detergent, emulsifier for oils, waxes and solvs.; intermediate for sulfation; liq.; HLB 11.2; 100% conc.

Synperonic 13/6.5. [ICI PLC] Syn. primary alcohol ethoxylate; CAS 24938-91-8; nonionic; industrial emulsifier, wetting agent, degreaser; intermediate for sulfation; liq.; HLB 12.5; 100% conc.

Synperonic 13/8. [ICI PLC] Syn. primary alcohol ethoxylate; CAS 24938-91-8; nonionic; wetting agent, general purpose detergent, solubilizer, emulsifier; liq.; water-sol.; HLB 13.3; 100% conc.

Synperonic 13/9. [ICI PLC] Syn. primary alcohol ethoxylate; CAS 24938-91-8; nonionic; wetting agent, general purpose detergent; scouring and dye wetting agent effective at low temps.; solubilizer and emulsifier for essential oils, aromatic solvs., fats, and waxes; liq.; water-sol.; HLB 13.6; 100% conc.

Synperonic 13/10. [ICI PLC] Syn. primary alcohol ethoxylate; CAS 24938-91-8; nonionic; wetting agent, general purpose detergent, solubilizer, emulsifier; liq.; water-sol.; HLB 14.1; 100% conc.

Synperonic 13/12. [ICI PLC] Syn. primary alcohol ethoxylate; CAS 24938-91-8; nonionic; wetting agent, general purpose detergent, solubilizer, emulsifier; liq.; water-sol.; HLB 14.8; 100% conc.

Synperonic 13/15. [ICI PLC] Syn. primary alcohol ethoxylate; CAS 24938-91-8; nonionic; foam builder, detergent, solubilizer, emulsifier, stabilizer; solid; water-sol.; HLB 15.5; 100% conc.

Synperonic 13/18. [ICI PLC] Syn. primary alcohol ethoxylate; CAS 24938-91-8; nonionic; foam builder, detergent, solubilizer, emulsifier, stabilizer; solid; water-sol.; HLB 16.2; 100% conc.

Synperonic 13/20. [ICI PLC] Syn. primary alcohol ethoxylate; CAS 24938-91-8; nonionic; foam builder, detergent, solubilizer, emulsifier, stabilizer; solid; water-sol.; HLB 16.4; 100% conc.

Synperonic 87K. [ICI PLC] Syn. primary alcohol ethoxylate; CAS 69227-21-0; nonionic; low foaming detergent for textile, dairy, and other cleaning applics.; liq.; water-sol.; HLB 12.0; 100% conc.

Synperonic 91/2.5. [ICI PLC] Syn. primary C9-11 alcohol ethoxylate; CAS 68439-46-3; nonionic; detergent, emulsifier; intermediate for sulfation giving high foaming anionic suitable for lt. duty liqs.; liq.; oil-sol.; HLB 8.2; 100% conc.

Synperonic 91/4. [ICI PLC] Syn. primary C9-11 alcohol ethoxylate; CAS 68439-46-3; nonionic; detergent, emulsifier; intermediate for sulfation giving high foaming anionic suitable for lt. duty liqs.; liq.; oil-sol.; HLB 10.8; 100% conc.

Synperonic 91/5. [ICI PLC] Syn. primary C9-11 alcohol ethoxylate; CAS 68439-46-3; nonionic; wetting and scouring agent for wool textile processing, degreasing hard surface cleaners, alkaline detergents, fabric washing liqs.; coupler; o/w emulsifier; liq.; HLB 11.8; 100% conc.

Synperonic 91/6. [ICI PLC] Syn. primary C9-11 alcohol ethoxylate; nonionic; wetting and scouring agent for wool textile processing, degreasing hard surface cleaners, alkaline detergents, fabric washing liqs.; coupler; o/w emulsifier; liq.; HLB 12.5; 100% conc.

Synperonic 91/7. [ICI PLC] Syn. primary C9-11 alcohol ethoxylate; CAS 68439-46-3; nonionic; wetting and scouring agent for wool textile processing, degreasing hard surface cleaners, alkaline detergents, fabric washing liqs.; coupler; o/w emulsifier; liq.; HLB 13.4; 100% conc.

Synperonic 91/8. [ICI PLC] Syn. primary C9-11 alcohol ethoxylate; nonionic; wetting and scouring agent for wool textile processing, degreasing hard surface cleaners, alkaline detergents, fabric washing liqs.; coupler; o/w emulsifier; liq.; HLB 13.9; 100% conc.

Synperonic 91/10. [ICI PLC] Syn. primary C9-11 alcohol ethoxylate; nonionic; wetting and scouring agent for wool textile processing, degreasing hard surface cleaners, alkaline detergents, fabric washing liqs.; coupler; o/w emulsifier; liq.; HLB 14.7; 100% conc.

Synperonic 91/12. [ICI PLC] Syn. primary C9-11 alcohol ethoxylate; CAS 68439-46-3; nonionic; wetting and scouring agent for wool textile processing, degreasing hard surface cleaners, alkaline detergents, fabric washing liqs.; coupler; o/w emulsifier; liq.; 100% conc.

Synperonic 91/20. [ICI PLC] Syn. primary C9-11 alcohol ethoxylate; CAS 68439-46-3; nonionic; detergent used as detergent solubilizer, dispersant, and stabilizer; liq.; HLB 16.9; 100% conc.

Synperonic A2. [ICI Am.; ICI PLC] Trideceth-2; CAS 678213-23-0; nonionic; detergent, emulsifier, coemulsifier; base for emulsifiers, textile antistats, scouring and wetting agents, and specialty cleaners; intermediate for sulfation used in personal care prods.; Hazen 50 max. liq.; sol. in alcohol, glycol ethers, kerosene, and min. oil; dens. 0.897 g/ml; visc. 25 cps; HLB 5.9; pour pt. 2 C; pH 6–8 (1% aq.); 99% min. act.

Synperonic A3. [ICI Am.; ICI PLC] Trideceth-3; CAS 24938-91-8; 68213-23-0; nonionic; see Synperonic A2; Hazen 50 max. liq.; sol. in alcohol, glycol ethers, kerosene, and min. oil; dens. 0.918 g/ml; visc. 29 cps; HLB 7.8; pour pt. 5 C; pH 6–8 (1% aq.); 99% min. act.

Synperonic A4. [ICI Am.; ICI PLC] Syn. primary alcohol ethoxylate; CAS 68213-23-0; nonionic; see Synperonic A2; Hazen 50 max. liq.; sol. in alcohol, glycol ethers, kerosene; sol./disp. in min. oil; dens. 0.937 g/ml; visc. 34 cps; HLB 9.1; pour pt. 9 C; pH 6–8 (1% aq.); 99% min. act.

Synperonic A6. [ICI Am.] Syn. primary alcohol ethoxylate; detergent, emulsifier; base for emulsifiers, textile antistats, scouring and wettingagents, in specialty cleaners; intermediate for sulfation for personal care prods.; Hazen 50 max. liq.; sol. in alcohol, glycol ethers; dens. 0.966 g/ml; visc. 46 cps; HLB 11.2; pour pt. 17 C; surf. tens. 29.8 dynes/cm (0.1%); pH 6–8 (1% aq.); 99% min. act.

Synperonic A7. [ICI Am.; ICI PLC] Trideceth-7; CAS 24938-91-8; 68213-23-0; nonionic; see Synperonic A2; Hazen 50 max. visc. liq.; sol. in water, glycol ethers and alcohol; dens. 0.977 g/ml; visc. 60 cps; HLB 12.2; cloud pt. 44–48 C (1% aq.); pour pt. 21 C; surf. tens. 30.2 dynes/cm (0.1%); pH 6–8 (1% aq.); 99% min. act.

Synperonic A9. [ICI Am.; ICI PLC] Trideceth-9; CAS 24938-91-8; 68213-23-0; nonionic; see Synperonic A2; wh. paste; sol. in water, glycol ethers and alcohol; dens. 0.984 g/ml (40 C); visc. 35 cps (40 C); HLB 12.5; cloud pt. 62–68 C (1% aq.); pour pt. 24 C; surf. tens. 31.4 dynes/cm (0.1%); pH 6–8 (1% aq.); 99% min. act.

Synperonic A11. [ICI Am. ; ICI PLC] Trideceth-11; CAS 24938-91-8; 68213-23-0; nonionic; see Synperonic A2; wh. paste; sol. in water, glycol ethers and alcohol; dens. 0.998 g/ml (40 C); visc. 42 cps (40 C); HLB 13.9; cloud pt. 85–89 C (1% aq.); pour pt. 27 C; surf. tens. 33.2 dynes/cm (0.1%); pH 6–8 (1% aq.); 99% min. act.

Synperonic A14. [ICI Am.] Syn. primary alcohol ethoxylate; nonionic; detergent, solubilizer, dispersant, stabilizer; wh. wax; sol. see Synperonic A7; dens. 1.014 g/ml (40 C); visc. 55 cps (40 C); HLB 14.9; cloud pt. > 100 C (1% aq.); pour pt. 30 C; surf. tens. 36.5 dynes/cm (0.1%); pH 6–8 (1% aq.); 99% min. act.

Synperonic A20. [ICI Am.; ICI PLC] Trideceth-20; CAS 24938-91-8; nonionic; detergent, solubilizer, dispersant, stabilizer; used for high temp. and alkaline applics.; wh. hard wax; sol. in water, glycol ethers and alcohol; dens. 1.028 g/ml (50 C); visc. 58 cps (50 C); HLB 16.2; cloud pt. > 100 C (1% aq.); pour pt. 37 C; surf. tens. 40.1 dynes/cm (0.1%); pH 6–8 (1% aq.); 99% min. act.

Synperonic A50. [ICI Am.; ICI PLC] Trideceth-50;

CAS 24938-91-8; nonionic; detergent, solubilizer, dispersant, stabilizer; used for high temp. and alkaline applics.; solid; water-sol.; HLB 18.3; 100% conc.

Synperonic BD100. [ICI PLC] Syn. primary alcohol ethoxylate; CAS 69227-21-0; nonionic; low foaming detergent, emulsifier; for textile and raw wool scouring; Hazen 150 max. liq.; sol. in water, alcohol, glycol ethers, kerosene; dens. 0.985 g/ml; visc. 92 cps; HLB 12.0; pour pt. 5 C; surf. tens. 29.7 dynes/cm (0.1%); pH 6–8 (1% aq.); 95% min. act.

Synperonic FS. [ICI PLC] Syn. quat. ammonium compd.; cationic; textile prods.; liq.; 80% conc.

Synperonic K87. [ICI PLC] Syn. primary alcohol ethoxylate with PO; nonionic; detergent for textile and raw wool scouring; Hazen 150 max. liq.; sol. see Synperonic A7; dens. 0.980 g/ml; visc. 72 cps; HLB 12.0; cloud pt. 44–49 C; pour pt. 11 C; surf. tens. 29.5 dynes/cm (0.1%); pH 6–8 (1% aq.); 99% min. act.

Synperonic KB. [ICI PLC] Alcohol ethoxylate; CAS 61827-42-7; nonionic; wetting agent; wool scouring; liq.; water-sol.; HLB 12.3; 100% conc.

Synperonic LF/B26. [ICI PLC] Fatty alcohol alkoxylate; CAS 69227-21-0; nonionic; biodeg. low foaming detergent for laundry, sanitizers, carpet cleaners, acid pickling of metals; wetter for cotton desizing, general finishing, dyeing aid; liq.; 100% conc.

Synperonic LF/D25. [ICI PLC] Fatty alcohol alkoxylate; CAS 69227-21-0; nonionic; biodeg. low foaming detergent for laundry, sanitizers, carpet cleaners, acid pickling of metals; wetter for cotton desizing, general finishing, dyeing aid; liq.; 100% conc.

Synperonic LF/RA30. [ICI PLC] Fatty alcohol alkoxylate; CAS 69227-21-0; nonionic; biodeg. low foaming wetter for domestic automatic dishwash rinse aids, heavy-duty laundry liqs., alkaline cleaning, metal degreasing; liq.; 100% conc.

Synperonic LF/RA40. [ICI PLC] Fatty alcohol alkoxylate; CAS 69227-21-0; nonionic; low foam detergent, defoamer, wetter for domestic and institutional auto dishwash powds., rinse aids, alkaline cleaning; liq.; 100% conc.

Synperonic LF/RA43. [ICI PLC] Fatty alcohol alkoxylate; CAS 69227-21-0; nonionic; defoamer for domestic and institutional auto dishwash powds., washing of fruits and vegs.; liq.; 100% conc.

Synperonic LF/RA50. [ICI PLC] Fatty alcohol alkoxylate; CAS 69227-21-0; nonionic; defoamer for institutional auto dishwash rinse aids; liq.; 100% conc.

Synperonic LF/RA260. [ICI PLC] Fatty alcohol alkoxylate; CAS 69227-21-0; nonionic; biodeg. low foam wetter for domestic automatic dishwasher rinse aids, heavy-duty laundry liqs., alkaline cleaners, metal degreasing; liq.; 100% conc.

Synperonic LF/RA290. [ICI PLC] Fatty alcohol alkoxylate; CAS 69227-21-0; nonionic; biodeg. low foam wetter for domestic automatic dishwasher rinse aids, heavy-duty laundry liqs., alkaline cleaners, metal degreasing; liq.; 100% conc.

Synperonic LF/RA343. [ICI PLC] Fatty alcohol alkoxylate; CAS 69227-21-0; nonionic; defoamer for domestic automatic dishwash powds.; liq.; 100% conc.

Synperonic M2. [ICI PLC] Syn. primary alcohol EO condensate; CAS 68213-23-0; nonionic; intermediate for mfg. of ether sulfates; liq.; HLB 6.1; 100% conc.

Synperonic M3. [ICI PLC] Syn. primary alcohol EO condensate; CAS 68213-23-0; nonionic; intermediate for mfg. of ether sulfates; liq.; HLB 7.9; 100% conc.

Synperonic N. [ICI PLC] Nonylphenol ethoxylate; CAS 68412-54-4; nonionic; wetting agent and detergent for textile, metal treatment, dust suppression, and general cleaning applics.; liq.; 27% conc.

Synperonic NP4. [ICI Am.; ICI PLC] Nonoxynol-4; CAS 9016-45-9; 68412-54-4; nonionic; detergent, emulsifier, coemulsifier; intermediate for sulfation in mfg. lubricants and antistats; Hazen 150 max. liq.; sol. see Synperonic A3; dens. 1.022 g/ml; visc. 400 cps; HLB 8.9; pour pt. < 0 C; pH 6–8 (1% aq.); 99% min. act.

Synperonic NP5. [ICI Am.; ICI PLC] Nonoxynol-5; CAS 9016-45-9; nonionic; detergent, emulsifier, coemulsifier; intermediate for sulfation in mfg. lubricants and antistats; Hazen 150 max. liq.; sol. in alcohol, glycol ethers, kerosene, min. oil; insol./disp. in water; dens. 1.035 g/ml; visc. 350 cps; HLB 10.5; pour pt. < 0 C; pH 6–8 (1% aq.); 99% min. act.

Synperonic NP6. [ICI Am.; ICI PLC] Nonoxynol-6; CAS 9016-45-9; nonionic; detergent, emulsifier, coemulsifier; intermediate for sulfation in mfg. lubricants and antistats; Hazen 150 max. liq.; sol. in alcohol, glycol ethers, min. oil; disp. in water; dens. 1.041 g/ml; visc. 355 cps; HLB 10.9; pour pt. < 0 C; pH 6–8 (1% aq.); 99% min. act.

Synperonic NP7. [ICI Am.; ICI PLC] Nonoxynol-7; CAS 9016-45-9; nonionic; detergent, emulsifier for oils, waxes, solvs.; liq.; oil-sol.; HLB 11.7; 100% conc.

Synperonic NP8. [ICI Am.; ICI PLC] Nonoxynol-8; CAS 9016-45-9; nonionic; detergent for textile scouring; wetting agent; emulsifier for med. polarity oils and solvs.; Hazen 150 max. liq.; sol. in water, alcohol, glycol ethers; dens. 1.053 g/ml; visc. 355 cps; HLB 12.3; cloud pt. 30–34 C (1% aq.); pour pt. < 0 C; surf. tens. 29.4 dynes/cm (0.1%); pH 6–8 (1% aq.); 99% min. act.

Synperonic NP8.75. [ICI PLC] Nonylphenol ethoxylate; nonionic; detergent, wetting agent, emulsifier; liq.; water-sol.; HLB 12.7; 100% conc.

Synperonic NP9. [ICI Am.; ICI PLC] Nonoxynol-9; CAS 9016-45-9; nonionic; detergent for textile scouring; wetting agent; emulsifier for med. polarity oils and solvs.; Hazen 150 max. liq.; sol. in water, alcohol, glycol ethers; dens. 1.058 g/ml; visc. 340 cps; HLB 12.8; cloud pt. 51–56 C (1% aq.); pour pt. 0 C; surf. tens. 30.6 dynes/cm; pH 6–8 (1% aq.); 99% min. act.

Synperonic NP10. [ICI Am.; ICI PLC] Nonoxynol-10; CAS 9016-45-9; nonionic; detergent, solubilizer, dispersant, stabilizer; Hazen 150 max. liq.; sol. in water, alcohol, glycol ethers; dens. 1.061 g/ml; visc. 360 cps; HLB 13.3; cloud pt. 62–67 C (1% aq.); pour pt. 5 C; surf. tens. 30.6 dynes/cm; pH 6–8 (1% aq.); 99% min. act.

Synperonic NP12. [ICI Am.; ICI PLC] Nonoxynol-12; CAS 9016-45-9; nonionic; detergent, solubilizer, dispersant, stabilizer; Hazen 150 max. liq.; sol. in water, alcohol, glycol ethers; dens. 1.062 g/ml; visc. 265 cps; HLB 13.9; cloud pt. 79–84 C (1% aq.); pour pt. 14 C; surf. tens. 35.2 dynes/cm; pH 6–8 (1% aq.); 99% min. act.

Synperonic NP13. [ICI Am.; ICI PLC] Nonoxynol-13; CAS 9016-45-9; nonionic; emulsifier and wetting agent for solvs. and agrochemical pesticides and herbicides; detergent additive, solubilizer, and dispersant; Hazen 150 max. paste; sol. in water, alcohol, glycol ethers; dens. 1.068 g/ml; visc. 280 cps; HLB 14.4; cloud pt. 87–92 C (1% aq.); pour pt. 17 C; surf. tens. 34.8 dynes/cm; pH 6–8 (1% aq.); 99% min. act.

Synperonic NP15. [ICI PLC] Nonoxynol-15; CAS 9016-45-9; nonionic; emulsifier and wetting agent for agrochem. powds. and solvs.; detergent additive, solubilizer, dispersant; Hazen 150 max. paste; sol. see Synperonic NP9; dens. 1.058 g/ml (40 C); visc. 125 cps (40 C); HLB 15.0; cloud pt. 95–99 C (1% aq.); pour pt. 21 C; surf. tens. 33.4 dynes/cm; pH 6–8 (1% aq.); 99% min. act.

Synperonic NP20. [ICI PLC] Nonoxynol-20; CAS 9016-45-9; nonionic; solubilizer and emulsifier for polar substrates; Hazen 150 max. solid; sol. in water, alcohol, glycol ethers; dens. 1.073 g/ml (40 C); visc. 168 cps (40 C); HLB 16.0; cloud pt. > 100 C (1% aq.); pour pt. 30 C; surf. tens. 41.7 dynes/cm; pH 6–8 (1% aq.); 99% min. act.

Synperonic NP30. [ICI PLC] Nonoxynol-30; CAS 9016-45-9; nonionic; solubilizer and emulsifier for high polarity materials; Hazen 150 max. solid; sol. in water, alcohol, glycol ethers; dens. 1.074 g/ml (50 C); visc. 150 cps (50 C); HLB 7.1; cloud pt. > 100 C (1% aq.); pour pt. 40 C; surf. tens. 42.8 dynes/cm; pH 6–8 (1% aq.); 99% min. act.

Synperonic NP30/70. [ICI Am.] Nonoxynol-30; CAS 9016-45-9; nonionic; solubilizer and emulsifier for high polarity materials.

Synperonic NP35. [ICI PLC] Nonoxynol-35; CAS 9016-45-9; nonionic; solubilizer and emulsifier for high polarity materials; solid; HLB 17.4; 100% conc.

Synperonic NP40. [ICI Am.; ICI PLC] Nonoxynol-40; CAS 9016-45-9; nonionic; solubilizer and emulsifier for high polarity materials; solid; HLB 17.8; 100% conc.

Synperonic NP50. [ICI Am.; ICI PLC] Nonoxynol-50; CAS 9016-45-9; nonionic; solubilizer and emulsifier for high polarity materials; solid; HLB 18.2; 100% conc.

Synperonic NPE1800. [ICI PLC] Nonylphenol ethoxylate; CAS 68891-11-2; nonionic; used in agric. industry for prep. of pesticide emulsifiable concs.; solid; 100% conc.

Synperonic NX. [ICI PLC] Nonylphenol ethoxylate; CAS 68412-54-4; nonionic; detergent, wetting agent, emulsifier for textile processing; emulsification of med. polarity oils and solvs.; liq.; water-sol.; HLB 12.3; 100% conc.

Synperonic NXP. [ICI PLC] Nonylphenol ethoxylate; CAS 68412-54-4; nonionic; general purpose cleaning applics.; liq.; HLB 13.3; 100% conc.

Synperonic OP3. [ICI Am.; ICI PLC] Octoxynol-3; CAS 9002-93-1; 68987-90-6; nonionic; detergent and emulsifier for oils, waxes, and solvs.; liq.; oil-sol.; HLB 7.1; 100% conc.

Synperonic OP4.5. [ICI Am.; ICI PLC] Octoxynol-5; CAS 9002-93-1; nonionic; detergent and emulsifier for oils, waxes, and solvs.; liq.; oil-sol.; HLB 9.4; 100% conc.

Synperonic OP7.5. [ICI Am.] Octoxynol-8.; CAS 9002-93-1; nonionic; detergent for textile and paper processing; liq.; HLB 11.7; 100% conc.

Synperonic OP8. [ICI PLC] Octylphenol ethoxylate; nonionic; detergent for textile and paper processing; liq.; HLB 12.6; 100% conc.

Synperonic OP10. [ICI Am.; ICI PLC] Octoxynol-10; CAS 9002-93-1; nonionic; detergent, wetting agent, emulsifier; for textile and paper processing; Hazen 150 max. liq.; sol. in water, alcohol, glycol ethers; dens. 1.062 g/ml; visc. 393 cps; HLB 13.6; cloud pt. 63–67 C (1% aq.); pour pt. 7 C; surf. tens. 31.8 dynes/cm; pH 6–8 (1% aq.); 99% min. act.

Synperonic OP10.5. [ICI PLC] Octylphenol ethoxylate; nonionic; detergent for textile and paper processing; liq.; HLB 13.5; 100% conc.

Synperonic OP11. [ICI Am.; ICI PLC] Octoxynol-11; CAS 9002-93-1; nonionic; detergent for textile and paper processing; Hazen 150 max. liq.; sol. in water, alcohol, glycol ethers; dens. 1.067 g/ml; visc. 371 cps; HLB 14.0; cloud pt. 79–82 C (1% aq.); pour pt. 8 C; surf. tens. 31.0 dynes/cm; pH 6–8 (1% aq.); 99% min. act.

Synperonic OP12.5. [ICI Am.; ICI PLC] Octoxynol-13; CAS 9002-93-1; nonionic; detergent for hard surf. cleaners; liq.; water-sol.; HLB 14.3; 100% conc.

Synperonic OP16. [ICI Am.] Octoxynol-16; CAS 9002-93-1; nonionic; emulsion polymerization; liq.; HLB 15.6; 70% conc.

Synperonic OP16.5. [ICI PLC] Octylphenol ethoxylate; nonionic; detergent for hard surf. cleaners; liq.; HLB 15.3; 100% conc.

Synperonic OP20. [ICI Am.] Octoxynol-20.; CAS 9002-93-1; nonionic.

Synperonic OP25. [ICI Am.] Octoxynol-25.; CAS 9002-93-1; nonionic.

Synperonic OP30. [ICI Am.; ICI PLC] Octoxynol-30; CAS 9002-93-1; nonionic; for emulsion polymerization; solid; HLB 17.2; 100% conc.

Synperonic OP40. [ICI Am.; ICI PLC] Octoxynol-40; CAS 9002-93-1; nonionic; solubilizer, dispersant, emulsifier for highly polar substrates; liq.; HLB 17.4; 100% conc.

Synperonic OP40/70. [ICI PLC] Octoxynol-40; CAS 9002-93-1; nonionic; solubilizer, dispersant, emulsifier for highly polar substrates; liq.; HLB 17.4; 70% conc.

Synperonic P105. [ICI PLC] POP/POE block copolymer; CAS 9003-11-6; nonionic; emulsifier, demulsifier; paste; 100% conc.

Synperonic PE30/10. [ICI PLC] PO/EO block copolymer, nonionic, defoamer, demulsifier, wetting agent, dispersant, rinse aid for machine dishwashing; liq.; m.w. 1900; sp.gr. 1.017; visc. 325 cs; m.p. < 0 C; HLB 3; cloud pt. 24 C (1% aq.); pH 6–8 (2.5%); 99% min. act.

Synperonic PE30/20. [ICI PLC] PO/EO block copolymer; nonionic; defoamer, demulsifier, wetting agent, dispersant, rinse aid for machine dishwashing; liq.; m.w. 2500; sp.gr. 1.032; visc. 420 cs; m.p. < 0 C; HLB 7; cloud pt. 32 C (1% aq.); surf. tens. 42 dynes/cm; pH 6–8 (2.5%); 99% min. act.

Synperonic PE30/40. [ICI PLC] PO/EO block copolymer; nonionic; defoamer, demulsifier, wetting agent, dispersant, rinse aid for machine dishwashing; visc. liq.; m.w. 2900; sp.gr. 1.05; visc. 600 cs; m.p. < 0 C; HLB 15; cloud pt. 58 C (1% aq.); surf. tens. 41.1 dynes/cm; pH 6–8 (2.5%); 99% act.

Synperonic PE30/80. [ICI PLC] PO/EO block copolymer; nonionic; emulsifier, dispersant, stabilizer; emulsion polymerization; used in pigments in emulsion paints, inks; flake; m.w. 8500; m.p. 50 C; HLB 29; cloud pt. > 100 C (1% aq.); surf. tens. 48 dynes/cm; pH 6–8 (2.5%); 99% min. act.

Synperonic PE39/70. [ICI PLC] PO/EO block copolymer; nonionic; emulsifier, dispersant; wh. flake; m.w. 7500; m.p. 49 C; HLB 24; cloud pt. > 100 C (1% aq.); surf. tens. 45.2 dynes/cm; pH 6–8 (2.5%);

99% min. act.

Synperonic PE/F38. [ICI Am.; ICI PLC] Poloxamer 108; CAS 9003-11-6; nonionic; emulsifier, dispersant; flake; 100% conc.

Synperonic PE/F68. [ICI Am.; ICI PLC] Poloxamer 188; nonionic; emulsifier, dispersant; flake; 100% conc.

Synperonic PE/F87. [ICI Am.; ICI PLC] Poloxamer 237; nonionic; emulsifier, dispersant; flake; 100% conc.

Synperonic PE/F88. [ICI Am.; ICI PLC] Poloxamer 238; nonionic; emulsifier, dispersant; flake; 100% conc.

Synperonic PE/F98. [ICI Am.] Poloxamer 288.

Synperonic PE/F108. [ICI Am.; ICI PLC] Poloxamer 338; nonionic; emulsifier, dispersant; flake; 100% conc.

Synperonic PE/F127. [ICI Am.; ICI PLC] Poloxamer 407; nonionic; emulsifier, dispersant; flake; 100% conc.

Synperonic PE/L31. [ICI PLC] POP/POE block copolymer; CAS 9003-11-6; nonionic; low foaming rinse aid additive, wetting agent; liq.; 100% conc.

Synperonic PE/L35. [ICI Am.; ICI PLC] Poloxamer 105; nonionic; emulsifier, dispersant, detergent; liq.; 100% conc.

Synperonic PE/L42. [ICI Am.; ICI PLC] Poloxamer 122; nonionic; low foaming rinse aid additive, wetting agent; liq.; 100% conc.

Synperonic PE/L43. [ICI Am.; ICI PLC] Poloxamer 123; nonionic; emulsifier, dispersant, detergent; liq.; 100% conc.

Synperonic PE/L44. [ICI Am.; ICI PLC] Poloxamer 124; nonionic; emulsifier, dispersant, detergent; liq.; 100% conc.

Synperonic PE/L61. [ICI Am.; ICI PLC] Poloxamer 181; nonionic; multiuse antifoam, demulsifier, liq. rinse aid; liq.; 100% conc.

Synperonic PE/L62. [ICI Am.; ICI PLC] Poloxamer 182; nonionic; low foaming rinse aid additives, wetting agents; liq.; 100% conc.

Synperonic PE/L62LF. [ICI PLC] POP/POE block copolymer; CAS 9003-11-6; nonionic; low foaming rinse aid additive, wetting agent; liq.; 100% conc.

Synperonic PE/L64. [ICI Am.; ICI PLC] Poloxamer 184; nonionic; emulsifier, dispersant, detergent; liq.; 100% conc.

Synperonic PE/L81. [ICI Am.; ICI PLC] Poloxamer 231; nonionic; multiuse antifoam, demulsifier, liq. rinse aid; liq.; 100% conc.

Synperonic PE/L92. [ICI Am.; ICI PLC] Poloxamer 282; nonionic; low foaming rinse aid additive, wetting agent; liq.; 100% conc.

Synperonic PE/L101. [ICI Am.; ICI PLC] Poloxamer 331; nonionic; multiuse antifoam, demulsifier, liq. rinse aid; liq.; 100% conc.

Synperonic PE/L121. [ICI Am.; ICI PLC] Poloxamer 401; nonionic; multiuse antifoam, demulsifier, liq. rinse aid; liq.; 100% conc.

Synperonic PE/P75. [ICI Am.; ICI PLC] Poloxamer 215; CAS 9003-11-6; nonionic; emulsifier, demulsifier; paste; 100% conc.

Synperonic PE/P84. [ICI Am.; ICI PLC] Poloxamer 234; nonionic; emulsifier, demulsifier; paste; 100% conc.

Synperonic PE/P85. [ICI Am.; ICI PLC] Poloxamer 235; nonionic; emulsifier, demulsifier; paste; 100% conc.

Synperonic PE/P94. [ICI Am.; ICI PLC] Poloxamer 284; nonionic; emulsifier, demulsifier; paste; 100% conc.

Synperonic PE/P103. [ICI Am.; ICI PLC] Poloxamer 333; nonionic; emulsifier, demulsifier; paste; 100% conc.

Synperonic PE/P104. [ICI Am.] Poloxamer 334.

Synperonic T/304. [ICI Am.; ICI PLC] Poloxamine 304; CAS 11111-34-5; nonionic; low foaming agent, wetting agent, dispersant, demulsifier; liq.; 100% conc.

Synperonic T/701. [ICI Am.; ICI PLC] Poloxamine 701; nonionic; low foaming agent, wetting agent, dispersant, demulsifier; liq.; 100% conc.

Synperonic T/707. [ICI Am.; ICI PLC] Poloxamine 707; nonionic; detergent, emulsifier, demulsifier, wetting agent; flaked solid; 100% conc.

Synperonic T/904. [ICI Am.; ICI PLC] Poloxamine 904; nonionic; detergent, emulsifier, demulsifier, wetting agent; paste; 100% conc.

Synperonic T/908. [ICI Am.; ICI PLC] Poloxamine 908; nonionic; detergent, emulsifier, demulsifier, wetting agent; flaked solids; 100% conc.

Synperonic T/1301. [ICI Am.; ICI PLC] Poloxamine 1301; nonionic; low foaming agent, wetting agent, dispersant, demulsifier; liq.; 100% conc.

Synperonic T/1302. [ICI Am.; ICI PLC] Poloxamine 1302; nonionic; low foaming agent, wetting agent, dispersant, demulsifier; liq.; 100% conc.

Synperonic TO5. [ICI PLC] Acid ethoxylate; nonionic; general purpose emulsification; liq.; 100% conc.

Synprol Alcohol. [ICI Am.] Tridecyl alcohol; intermediate for prod. of ethoxylates, sulfates and ether sulfates; ref. index. 1.4457; sp.gr. 0.837; visc. 29.1 cps; flash pt. 290 F (PMCC); 99% act.

Synprolam 35. [ICI PLC] C13–15 alkyl primary amine; CAS 68155-27-1; cationic; anticaking agent for fertilizer; flotation agent, corrosion inhibitor; metal cleaning formulations, pigment dispersions, rubber processing auxs., bitumen emulsifier; intermediate for prod. of amine salts, ethoxylates and sulfosuccinates; Hazen 50 clear liq.; sp.gr. 0.8; visc. 4.5 cps; flash pt. 127 C (PMCC); 99% primary amine.

Synprolam 35A. [ICI PLC] Syn. C13-15 alkyl primary amine; cationic; emulsifier, fertilizer anticaking agent; min. flotation; solid; 100% conc.

Synprolam 35BQC (50). [ICI PLC] Syn. C13-15 dimethyl tert. amine benzyl ammonium chloride; cationic; emulsifier, sanitizer, biocide, corrosion inhibitor, textile drying aux., timber preservative; liq.; 50% conc.

Synprolam 35BQC (80). [ICI PLC] Syn. C13-15 dimethyl tert. amine benzyl ammonium chloride; cationic; emulsifier, sanitizer, biocide, corrosion inhibitor, textile drying aux., timber preservative; liq.; 80% conc.

Synprolam 35DM. [ICI PLC] Syn. C13-15 dimethyl tert. fatty amine; CAS 68391-04-8; cationic; chemical intermediate for prod. of quat. salts and amine oxides used in cosmetics; catalyst for PU foam; corrosion inhibitor; Hazen 50 clear liq.; sp.gr. 0.79; visc. 3.9 cps; flash pt. 143 C (PMCC); 97% tert. amine.

Synprolam 35DMA. [ICI PLC] Syn. C13–15 dimethyl tert. amine, acetate salt; cationic; emulsifier, biocide, timber preservative; liq.; 100% conc.

Synprolam 35DMBQC. [ICI PLC] C13–15 alkyl dimethyl benzyl ammonium chloride; CAS 68391-01-5; cationic; emulsifier, disinfectant, algicide, fungicide, germicide; corrosion inhibitor; general antistat for syn. and nat. fibers; fiber conditioner; textile dyeing auxiliary; timber preservative; liq.; 50 or 80% conc.

Synprolam 35M. [ICI PLC] Syn. (C13-15) alkyl methyl sec. amine; cationic; surfactant intermediate, corrosion inhibitor; liq.; 100% conc.

Synprolam 35MX1. [ICI PLC] PEG-1 C13-15 alkyl methyl amine; CAS 92112-62-4; cationic; emulsifier, textile processing aid, wetting agent; antistat for polyolefin and PVC; liq.; 100% conc.

Synprolam 35MX1/O. [ICI PLC] C13–15 alkyl methyl hydroxyethyl amine oxide; nonionic; foam stabilizer, visc. builder for shampoos, bubble baths, dishwashing liqs., detergents; liq.; 30% act.

Synprolam 35MX1QC. [ICI PLC] C13–15 alkyl dimethyl hydroxy ethyl ammonium chloride; CAS 85736-63-6; antistat and conditioner for syn. and nat. fibers; biocide, emulsifier; liq.; 30% act.

Synprolam 35MX3. [ICI PLC] PEG-3 C13–15 alkyl methyl amine; cationic; antistat for polyolefin and PVC; liq.; 100% conc.

Synprolam 35MX5. [ICI PLC] PEG-5 C13–15 alkyl methyl amines; cationic; antistat for polyolefin and PVC; liq.; 100% conc.

Synprolam 35N3. [ICI PLC] N-(C13-15) alkyl-1,3-propane diamine; cationic; corrosion inhibitor, bitumen adhesion agent, emulsifier, intermediate; liq.; 100% conc.

Synprolam 35N3X3, 35N3X5, 35N3X10, 35N3X15, 35N3X25. [ICI PLC] N-(C13–15) alkyl-1,3-propane diamine ethoxylate; cationic; emulsifier, corrosion inhibitor, textile processing aids, wetting agent and antistat; liq.; 100% conc.

Synprolam 35TMQC. [ICI PLC] (C13–15) alkyl trimethyl ammonium chloride; cationic; biocide, emulsifier, wetting agent and antistat; liq.; 100% conc.

Synprolam 35X2. [ICI PLC] PEG-2 C13-15 alkyl amine; cationic; emulsifier, PU catalyst, textile and plastic processing aid, textile dyeing aux., antistat; liq.; 100% conc.

Synprolam 35X2/O. [ICI PLC] C13-15 alkyl bis (2-hydroxyethyl) amine oxide; nonionic; foam stabilizer for personal care prods. and dishwashing detergents; liq.; 30% conc.

Synprolam 35X2QS, 35X5QS, 35X10QS. [ICI PLC] PEG-2, -5, and -10 C13–15 alkyl methyl ammonium methosulfate; cationic; antistat for PVC and PU; hydrotrope, wetting agent, conditioner, textile processing aid; alkali-stable; liq.; 90–100% conc.

Synprolam 35X5. [ICI PLC] PEG-5 C13-15 alkyl amine; cationic; emulsifier, PU catalyst, textile and plastic processing aid, textile dyeing aux., antistat; 100% conc.

Synprolam 35X10. [ICI PLC] PEG-10 C13-15 alkyl amine; cationic; emulsifier, PU catalyst, textile and plastic processing aid, textile dyeing aux., antistat; 100% conc.

Synprolam 35X15. [ICI PLC] PEG-15 C13-15 alkyl amine; cationic; emulsifier, PU catalyst, textile and plastic processing aid, textile dyeing aux., antistat; 100% conc.

Synprolam 35X20. [ICI PLC] PEG-20 C13-15 alkyl amine; cationic; emulsifier, PU catalyst, textile and plastic processing aid, textile dyeing aux., antistat; 100% conc.

Synprolam 35X25. [ICI PLC] PEG-25 C13-15 alkyl amine; cationic; emulsifier, PU catalyst, textile and plastic processing aid, textile dyeing aux., antistat; 100% conc.

Synprolam 35X35. [ICI PLC] PEG-35 C13-15 alkyl amine; cationic; emulsifier, PU catalyst, textile and plastic processing aid, textile dyeing aux., antistat; 100% conc.

Synprolam 35X50. [ICI PLC] PEG-50 C13-15 alkyl amine; cationic; emulsifier, PU catalyst, textile and plastic processing aid, textile dyeing auxs., antistat; 100% conc.

Synprolam FS. [ICI PLC] Syn. quat. ammonium compd.; cationic; fabric softener and antistat with rewet properties for textile treatment; liq.; 80% conc.

Synprolam VC. [ICI PLC] Syn. quat. ammonium compd.; cationic; vehicle cleaning formulation component; liq.; 100% conc.

Syntaryl Series. [Witco SA] Blend; anionic; multipurpose detergent, base for liq. detergents; liq.

Syntens KMA 55. [Hefti Ltd.] PEG-5.5 alkyl ether alcohol; CAS 68439-45-2; nonionic; cold wetting agent for textiles, paper, and leather industries; detergent base for industrial and household detergents; emulsifier; liq.; HLB 12.5; 100% conc.

Syntens KMA 70-W-02. [Hefti Ltd.] PO/EO polymer of alcohol; CAS 68937-66-6; nonionic; low foaming, biodeg. surfactant for household and industrial cleaners, textiles; liq.; 100% conc.

Syntens KMA 70-W-07. [Hefti Ltd.] PO/EO polymer of alcohol; CAS 68937-66-6; nonionic; low foaming, biodeg. surfactant for household and industrial cleaners, textiles; liq.; 100% conc.

Syntens KMA 70-W-12. [Hefti Ltd.] PO/EO polymer of alcohol; CAS 68937-66-6; nonionic; low foaming, biodeg. surfactant for household and industrial cleaners, textiles; liq.; 100% conc.

Syntens KMA 70-W-17. [Hefti Ltd.] PO/EO polymer of alcohol; CAS 68937-66-6; nonionic; low foaming, biodeg. surfactant for household and industrial cleaners, textiles; liq.; 100% conc.

Syntens KMA 70-W-22. [Hefti Ltd.] PO/EO polymer of alcohol; CAS 68937-66-6; nonionic; low foaming, biodeg. surfactant for household and industrial cleaners, textiles; liq.; 100% conc.

Syntens KMA 70-W-32. [Hefti Ltd.] PO/EO polymer of alcohol; CAS 68937-66-6; nonionic; low foaming, biodeg. surfactant for household and industrial cleaners, textiles; liq.; 100% conc.

Syntergent 55-A. [Henkel/Textile] fulling and scouring agent for wool and wool/syn. blends; provides exc. alkaline stability, effective felting action and detergency; brn. clear liq.; forms sl. hazy sol'ns.; pH 7.5 (5%).

Syntergent 647-M. [Henkel/Textile] Nylon scouring agent providing detergency for all syn. fabrics; clear liq.; forms clear sol'ns.; pH 7.0 (2%).

Syntergent DMC. [Henkel/Textile] Machine cleaner for pressure dye machines; removes dye tar and oil buildup while removing polyester trimer; pale clear liq.; forms clear sol'ns.; pH 8.0 (2%).

Syntergent K. [Henkel/Organic Prods.] Fatty amido condensate; nonionic; detergent for textile, leather, paper, and metalworking; liq.; 30% conc.

Syntergent SF. [Henkel/Textile] Scouring agent for nylon; clear liq.; forms clear sol'ns.; pH 7.0 (1%).

Syntergent TER-1. [Henkel/Textile; Henkel-Nopco] Phosphate ester; anionic; low foam wetting agent, textile scouring agent and penetrant for cotton and blends; calcium sulfate dispersant; yel. sl. hazy liq.; disp. in warm or cold water; pH 6.5 (5%); 82% conc.

Synthionic Series. [Witco SA] Block polymer; nonionic; low foaming wetting agent, detergent, rinse aid, defoamer; liq.

Synthrapol KB. [ICI Am.] Aliphatic alcohol ethoxylate; nonionic; wetting and rewetting agent, detergent; used in textile processing; colorless clear liq.; alcohol odor; water-sol.; sp.gr. 1.0; cloud pt. 40 C; pH 5.5–7.0 (1%); 96 ± 2% act.

Synthrapol LA-DN. [ICI Am.] Surfactant blend; anionic; wetting agent, level dyeing assistant for acid dyes on nylon; liq.; 55% conc.

Synthrapol LN-VLA. [ICI Am.] Formulated prod.; nonionic; reduction clearing assistant, leveling agent for disperse dyes on nylon; liq.

Synthrapol N. [ICI Am.] Nonyl phenol ethoxylate; CAS 9016-45-9; nonionic; detergent, wetting agent, emulsifier, penetrant/dyeing assistant used in textile processing; colorless liq.; mild pleasant odor; sp.gr. 1.0; b.p. 100 C; pH 5.5–7.0; 26% conc.

Synthrapol SP. [ICI Am.] Surfactant blend; anionic; detergent and wetting agent for scouring and soaping of textile after dyeing; liq.

Synthrapol WN-K. [ICI Am.] Formulated prod.; nonionic; wetting agent for carpet dyeing; pre- or post-scouring agent for carpets; liq.

Syntofor A03, A04, AB03, AB04, AL3, AL4, B03, B04. [Witco SA] Polyglycol fatty ester; nonionic; emulsifier for textiles and cosmetics; auxs. for textile, leather, metal industries, insecticides; liq.

Syntophos O6N. [Witco SA] Phosphate ester; anionic; wetting agent and detergent for textiles; liq.

Syntopon 8 A. [Witco SA] Octyl phenol ethoxylate; anionic; wetting agent for textile and leather; detergent; emulsifier; liq.; HLB 11.2; 100% conc.

Syntopon 8 B. [Witco SA] Octyl phenol ethoxylate; anionic; detergent for metal cleaning, low temp. applics.; liq.; HLB 12.3; 100% conc.

Syntopon 8 C. [Witco SA] Octoxynol-9; CAS 9002-93-1; 9010-43-9; anionic; detergent for wool; liq.; HLB 13.0; 100% conc.

Syntopon 8 D. [Witco SA] Octyl phenol ethoxylate; anionic; detergent for degreasing; wetting agent; emulsifier for insecticides; liq.; 100% conc.

Syntopon A. [Witco SA] Nonoxynol-6; CAS 9016-45-9; 977057-32-1; nonionic; detergent; emulsifier for solvs. and min. oils; wetting agent; liq.; HLB 10.9; 100% conc.

Syntopon A 100. [Witco SA] Nonoxynol-4; CAS 9016-45-9; nonionic; detergent, emulsifier for latex; wetting agent; liq.; HLB 9.0; 100% conc.

Syntopon B. [Witco SA] Nonoxynol-8; CAS 9016-45-9; 977057-34-3; nonionic; wetting agent, detergent for textiles; emulsifier; liq.; HLB 12.6; 100% conc.

Syntopon B 300. [Witco SA] Nonylphenol ethoxylate; nonionic; detergent for metal cleaning, low temp. applics.; liq.; 100% conc.

Syntopon C. [Witco SA] Nonoxynol-9; CAS 9016-45-9; nonionic; wetting agent and detergent for textiles; liq.; HLB 12.7; 100% conc.

Syntopon D. [Witco SA] Nonoxynol-10; CAS 9016-45-9; 977057-35-4; nonionic; detergent, emulsifier for perfume; liq.; HLB 13.2; 100% conc.

Syntopon E. [Witco SA] Nonoxynol-13; CAS 9016-45-9; nonionic; detergent, emulsifier for metal cleaning, high temp. applics.; dispersant; wax; HLB 14.0; 100% conc.

Syntopon F. [Witco SA] Nonoxynol-15; CAS 9016-45-9; nonionic; detergent, emulsifier, cleaning agent; wax; HLB 15.0; 100% conc.

Syntopon G. [Witco SA] Nonoxynol-20; CAS 9016-45-9; nonionic; detergent, emulsifier, cleaning agent; wax; HLB 16.0; 100% conc.

Syntopon H. [Witco SA] Nonoxynol-30; CAS 9016-45-9; nonionic; crude oil demulsifier; wax; HLB 17.0; 100% conc.

Syntopon S 493, 630, 1030. [Witco SA] Alkylaryl ether sulfate; anionic; household detergent; paste, liq., liq. resp.

T

T-9. [Werner G. Smith] Ethoxylated linear alcohol tallate; CAS 84501-87-1; nonionic; low foam, low freezing pt. surfactant; liq.; disp. in water; HLB 12.0; 100% conc.

T-11. [Procter & Gamble] Tallow fatty acid; CAS 67701-06-8; intermediate for mfg. of soaps, amides, esters, alcoholamides; raw material for non-surfactant applics.; paste; m.w. 276; acid no. 200-206; iodine no. 38-45; sapon. no. 200-208; 100% conc.

T-18. [Procter & Gamble] Tallow acid; CAS 67701-06-8; intermediate for mfg. of soaps, amides, esters, alcoholamides; raw material for non-surfactant applics.; Gardner 5 max. paste; m.w. 275; acid no. 200-208; iodine no. 40-55; 100% conc.

T-20. [Procter & Gamble] Tallow fatty acid; CAS 67701-06-8; intermediate for mfg. of soaps, amides, esters, alcoholamides; raw material for non-surfactant applics.; Gardner 7 max. paste; m.w. 274; acid no. 200-210; iodine no. 40-55; sapon. no. 206; 100% conc.

T-22. [Procter & Gamble] Tallow fatty acid; CAS 67701-06-8; intermediate for mfg. of soaps, amides, esters, alcoholamides; raw material for non-surfactant applics.; Gardner 7 max. paste; m.w. 274; acid no. 200-210; iodine no. 45-70; 100% conc.

TA-1618. [Procter & Gamble] Cetearyl alcohol; CAS 67762-30-5; intermediate; wh. waxy solid; sp.gr. 0.810 (65 C); m.p. 50 C; acid no. 1 max.; iodine no. 1 max.; sapon. no. 3 max.; hyd. no. 204-216.

Tagat® I. [Goldschmidt; Goldschmidt AG] PEG-30 glyceryl isostearate; CAS 69468-44-6; nonionic; preparation of o/w emulsions; solubilizer for flavors, perfumes, vitamin oils; dispersant and antistat; liq.; HLB 15.6; 100% conc.

Tagat® I2. [Goldschmidt; Goldschmidt AG] PEG-20 glyceryl isostearate; CAS 69468-44-6; nonionic; preparation of o/w emulsions; solubilizer for flavors, perfumes, vitamin oils; dispersant and antistat; liq.; HLB 14.2; 100% conc.

Tagat® L. [Goldschmidt; Goldschmidt AG] PEG-30 glyceryl laurate; CAS 51248-32-9; nonionic; preparation of o/w emulsions; solubilizer for flavors, perfumes, vitamin oils; dispersant and antistat; liq.; HLB 17.0; 100% conc.

Tagat® L2. [Goldschmidt; Goldschmidt AG] PEG-20 glyceryl laurate; CAS 51248-32-9; nonionic; preparation of o/w emulsions; solubilizer for flavors, perfumes, vitamin oils; dispersant and antistat; liq.; HLB 15.7; 100% conc.

Tagat® O. [Goldschmidt; Goldschmidt AG] PEG-30 glyceryl oleate; CAS 51192-09-7; nonionic; preparation of o/w emulsions; solubilizer for flavors, perfumes, vitamin oils; dispersant and antistat; liq.; HLB 16.4; 100% conc.

Tagat® O2. [Goldschmidt; Goldschmidt AG] PEG-20 glyceryl oleate; CAS 51192-09-7; nonionic; preparation of o/w emulsions; solubilizer for flavors, perfumes, vitamin oils; dispersant and antistat; liq.; HLB 15.0; 100% conc.

Tagat® R40. [Goldschmidt; Goldschmidt AG] PEG-40 hydrog. castor oil; CAS 61788-85-0; nonionic; solubilizer for water-insol. substances, e.g., essential oils, perfumes, vitamins, cosmetic/pharmaceutical active ingreds.; coemulsifier for o/w emulsions; solid; HLB 13.0; 100% conc.

Tagat® R60. [Goldschmidt; Goldschmidt AG] PEG-60 hydrog. castor oil; CAS 61788-85-0; nonionic; solubilizer for water-insol. substances, e.g., essential oils, perfumes, vitamins, cosmetic/pharmaceutical active ingreds.; coemulsifier for o/w emulsions; solid; HLB 15.0; 100% conc.

Tagat® R63. [Goldschmidt; Goldschmidt AG] PEG-60 hydrog. castor oil and propylene glycol; CAS 61788-85-0; nonionic; solubilizer for water-insol. substances, e.g., essential oils, perfumes, vitamins, cosmetic/pharmaceutical active ingreds.; coemulsifier for o/w emulsions; liq.

Tagat® RI. [Goldschmidt] PEG-15 glyceryl ricinoleate; CAS 39310-72-0; nonionic; preparation of o/w emulsions; solubilizer for flavors, perfumes, vitamin oils; dispersant and antistat; liq.; HLB 14.0; 100% conc.

Tagat® S. [Goldschmidt; Goldschmidt AG] PEG-30 glyceryl stearate; CAS 51158-08-8; nonionic; preparation of o/w emulsions; solubilizer for flavors, perfumes, vitamin oils; dispersant and antistat; liq.; HLB 16.4; 100% conc.

Tagat® S2. [Goldschmidt; Goldschmidt AG] PEG-20 glyceryl stearate; CAS 51158-08-8; nonionic; preparation of o/w emulsions; solubilizer for flavors, perfumes, vitamin oils; dispersant and antistat; liq.; HLB 15.0; 100% conc.

Tagat® TO. [Goldschmidt; Goldschmidt AG] PEG-25 glyceryl trioleate; CAS 68958-64-5; nonionic; preparation of o/w emulsions; solubilizer for flavors, perfumes, vitamin oils; dispersant and antistat; refatting agent for hair/bath preps.; liq.; HLB 11.3; 100% conc.

Tallates, K. [Murphy-Phoenix] Potassium salt of tall oil fatty acid; anionic; detergent, emulsifier for petrol. and agric. prods.

Tallow Amine. [Exxon/Tomah] Tallow primary amine, tech.; CAS 61790-33-8; cationic; surfactant, wetting agent, ore flotation agent, asphalt emulsifier, corrosion inhibitor, anticaking agent, pigment modifier; oil prod.; petrol. prod. additives; Gardner 5 max. solid.; iodine no. 35 min.; amine no. 205-218; alkaline corrosive liq.; 100% conc.

Tallow Diamine. [Exxon/Tomah] Tallow diamine; cationic; surfactant, wetting agent, ore flotation agent, asphalt emulsifier, corrosion inhibitor, anticaking agent, pigment modifier; oil prod.; petrol. prod. additives; Gardner 8 max. solid; iodine no. 30 min.; amine no. 320-350; alkaline corrosive liq.; 100% conc.

Tallow Tetramine. [Exxon/Tomah] Tallow tetramine; surfactant, wetting agent, ore flotation agent, asphalt emulsifier, corrosion inhibitor, anticaking agent, pigment modifier; oil prod.; petrol. prod. additives; Gardner 7 max. color; iodine no. 15 min.; amine no. 440 min.; alkaline corrosive liq.

Tallow Triamine. [Exxon/Tomah] Tallow triamine; cationic; surfactant, wetting agent, ore flotation agent, asphalt emulsifier, corrosion inhibitor, anticaking agent, pigment modifier; oil prod.; petrol. prod. additives; Gardner 7 max. solid; iodine no. 20 min.; amine no. 380 min.; alkaline corrosive liq.; 100% conc.

Tally® 100 Plus. [Van Den Bergh Foods] Glyceryl stearate, PEG-20 glyceryl stearate, hydrog. soybean oil; nonionic; emulsifier, dough strengthener and crumb softener used in breads; also as textile lubricant and fabric softener; bead; HLB 6.1; m.p. 55–57 C; 100% conc.

Tanabron W. [Sybron] Anionic/nonionic; wetter, detergent with good oil and wax emulsification; textile aux.; caustic stable.

Tanalev® 221. [Sybron] Anionic polymer; anionic; direct dye leveler with retarding props. over 1% owg use level.

Tanapal® LD-3, LD-3T. [Sybron] Nonionic; leveling agent for disperse dyes.

Tanapal® ME Acid. [Sybron] Phosphate ester, free acid; anionic; leveling agent for disperse dyes when neutralized; liq.; 100% conc.

Tanapal® NC. [Sybron] Nonionic; low foam disperse dye leveling agent; recommended for carrierless polyester; 100% act.

Tanapal® TTD. [Sybron] Nonionic; extremely low foam, high temp. textile leveling agent with good lubricating props.

Tanapon NF-200. [Sybron] Phosphate ester; anionic; low foaming wetting agent for alkaline cleaning compds., pigments, adhesive; liq.; 60% conc.

Tanapure® AC. [Sybron] Anionic; leveling agent for acid dyes on nylon fibers.

Tanaquad T-85. [Sybron] Quat. ammonium compd.; cationic; wetting agent, bactericide, sanitizer, dye retarder, antistat; liq.; 85% conc.

Tanassist® JCR. [Sybron] Mildly cationic; low foaming dye migrator for pressure equip.

Tanaterge FTD. [Sybron] Phosphated deriv.; anionic; afterwashing surfactant for dyed or printed goods; liq.; 67% conc.

Tanaterge PE-67. [Sybron] Phosphated deriv.; anionic; detergent for industrial cleaners; liq.; 67% conc.

Tanawet® AR. [Sybron] Nonionic; nonrewetting wetting agent for water repellent and fluorocarbon textile finishes.

Tanawet® FBW. [Sybron] Nonionic; nonrewetting wetting agent for textile finishes.

Tanawet® NRWG. [Sybron] Nonionic; low foaming nonrewetting penetrant for textile finishes.

Tanawet® RCN. [Sybron] Anionic; wetter and scouring agent for textile processing, peroxide bleaching; removes oils and waxes.

Taski TR101. [Lever Industrial] Highly conc. dry foam carpet and upholstery shampoo; toxicology: mild eye irritant.

Tauranol I-78. [Finetex] Sodium cocoyl isethionate; anionic; mild detergent, foamer, dispersant for soap-syndet, toilet bars, shampoos, bubble baths, mild foaming facial cleansers; imparts conditioning to hair and skin; emulsifier in creams and lotions; powd.; 83% conc.

Tauranol I-78-3. [Finetex] Sodium cocoyl isethionate; anionic; mild detergent, foamer, dispersant for soap-syndet, toilet bars, shampoos, bubble baths, mild foaming facial cleansers; imparts conditioning to hair and skin; emulsifier in creams and lotions; paste; 27% conc.

Tauranol I-78-6. [Finetex] Sodium cocoyl isethionate; anionic; mild detergent, foamer, dispersant for soap-syndet, toilet bars, shampoos, bubble baths, mild foaming facial cleansers; imparts conditioning to hair and skin; emulsifier in creams and lotions; paste; 50% conc.

Tauranol I-78/80. [Finetex] Sodium cocoyl isethionate, stearic acid; anionic; mild detergent, foamer, dispersant for soap-syndet, toilet bars, shampoos, bubble baths, mild foaming facial cleansers; imparts conditioning to hair and skin; emulsifier in creams and lotions.

Tauranol I-78E, I-78E Flakes. [Finetex] Sodium cocoyl isethionate; anionic; mild detergent, foamer, dispersant for soap-syndet, toilet bars, shampoos, bubble baths, mild foaming facial cleansers; imparts conditioning to hair and skin; emulsifier in creams and lotions.

Tauranol I-78 Flakes. [Finetex] Sodium cocoyl isethionate; anionic; mild detergent, foamer, dispersant for soap-syndet, toilet bars, shampoos, bubble baths, mild foaming facial cleansers; imparts conditioning to hair and skin; emulsifier in creams and lotions; flakes; 83% conc.

Tauranol M-35. [Finetex] Sodium N-methyl-N oleoyl taurate; anionic; detergent, wetting agent, all-purpose surfactant for textile processes, soaping prints, naphthols and vats, kier boil-off, bleaching, dyeing, shampoos, bubble baths, household cleaners; liq.; 23% conc.

Tauranol ML. [Finetex] Sodium methyl oleoyl taurate and isopropyl alcohol; anionic; detergent, dispersant for hard surf. cleaners, personal care prods.; lime soap dispersant; liq.; 33% conc.

Tauranol MS. [Finetex] Sodium methyl oleoyl taurate; anionic; wetting agent and detergent scour; for personal care prods., household cleaners; paste; 33% conc.

Tauranol T-Gel. [Finetex] Sodium N-methyl-N oleoyl taurate; anionic; wetting and detergent scour for textiles, personal care prods.; soaping off agent; gel; 14% conc.

Tauranol WS, WS Conc. [Finetex] Sodium methyl cocoyl taurate; anionic; detergent, foamer used in cosmetic preparations, household cleaners; slurry; 24 and 31% conc. resp.

Tauranol WSP. [Finetex] Sodium methyl cocoyl taurate; anionic; detergent, foaming agent for cosmetic preps., household cleaners; powd.; 70% conc.

TBEP. [Rhone-Poulenc Surf.; Rhone-Poulenc France] Tributoxyethylphosphate; nonfoaming emulsifier, wetting agent, dispersant, leveling agent; acid and electrolyte stable; liq.

TC-1005. [Procter & Gamble] Tallow/coconut fatty acids; intermediate for mfg. of soaps, amides, esters, alcoholamides; raw material for non-surfactant applics.; paste; m.w. 259; acid no. 214-218; iodine no. 35-42; 100% conc.

TC-1010. [Procter & Gamble] Tallow/coconut fatty acids; intermediate for mfg. of soaps, amides, esters, alcoholamides; raw material for non-surfactant applics.; paste; m.w. 256; acid no. 216-222; iodine no. 42 max.; sapon. no. 216-223; 100% conc.

TC-1010T. [Procter & Gamble] Tallow/coconut fatty acids; intermediate for mfg. of soaps, amides, esters, alcoholamides; raw material for non-surfactant applics.; paste; m.w. 259; acid no. 214-218; iodine no. 36-41; 100% conc.

T-Det® 25-3A. [Harcros] Ammonium C12-15 alkyl ether sulfate (3 EO); anionic; high foaming detergent, general industrial cleaner, dishwashes, laundry prods., car wash; pale yel. liq.; water-sol.; sp.gr. 1.02; dens. 8.5 lb/gal; pour pt. 28 F; flash pt. (PMCC) 74 F; pH 7-8 (1% aq.); 59% act.

T-Det® 25-3S. [Harcros] Sodium C12-15 alkyl ether sulfate (3 EO); anionic; high foaming detergent, general industrial cleaner, dishwashes, laundry prods., car wash; pale yel. liq.; water-sol.; sp.gr. 1.05; dens. 8.7 lb/gal; pour pt. 30 F; flash pt. (PMCC) 73 F; pH 7-8 (1% aq.); 59% act.

T-Det® A-026. [Harcros] EO and straight-chain primary alcohols (C10-C12); nonionic; wetting agent, detergent; textile processing; surfactant for formulated detergents for laundry use and cleaners; APHA 100 max. clear liq.; sol. in water, ethanol, glycol ether, HAN, isopropanol, xylene; dens. 8.2 lb/gal (60/60 F); sp.gr. 0.98 (60/60 F); visc. 21 cps (100 F); HLB 13.5; cloud pt. 36 C (1%); flash pt. 250 F (OC); pH 6–8 (1%); biodeg.; 100% conc.

T-Det® BP-1. [Harcros] PO/EO block copolymer; nonionic; emulsifier, detergent, dispersant and polymerization agent; pesticide formulations; FDA compliance; wh. solid; water-sol.; sp.gr. 1.04 (105 F); dens. 8.7 lb/gal (105 F); HLB 15.7; pour pt. 85 F; cloud pt. 154 F (1%); flash pt. (PMCC) > 400 F; pH 4-7 (1% aq.); 99.5% act.

T-Det® C-18. [Harcros] Ethoxylated castor oil; nonionic; emulsifier for industrial and agric. formulations; liq.; water-sol.; HLB 13.0; 100% conc.

T-Det® C-40. [Harcros] PEG-40 castor oil; nonionic; emulsifier, solubilizer, degreaser, lubricant, dispersant, penetrant used in leather, paper, textile and metal processing, rubber, paint; dispersant for pigment slurries; leveling agent, defoamer, stabilizer; wax and polish preparations; FDA compliance; yel. liq.; mild, oily odor; sol. in acetone, butyl Cellosolve, CCl_4, ethanol, ether, ethylene glycol, methanol, veg. oil, water, xylene; sp.gr. 1.05; dens. 8.7 lb/gal; HLB 14.2; pour pt. 60 F; flash pt. (PMCC) > 200 F; pH 6.0 (1%); 99.5% min. act.

T-Det® D-70. [Harcros] PEG-70 dinonyl phenyl ether; CAS 9014-93-1; nonionic; emulsifier, detergent, surfactant; tolerant to electrolyte conc.; solid; HLB 18.0; 100% conc.

T-Det® D-150. [Harcros] PEG-150 dinonyl phenyl ether; CAS 9014-93-1; nonionic; surfactant used in hard surface cleaners; solid; HLB 19.0; 100% conc.

T-Det® DD-5. [Harcros] Dodoxynol-5; CAS 9014-92-0; nonionic; emulsifier, surfactant for aromatic and aliphatic hydrocarbon solvs., solvent and emulsion cleaners, agric., degreasers; dry cleaning soap additive; FDA compliance; pale yel. liq.; mild aromatic odor; oil-sol.; sp.gr. 1.01; dens. 8.42 lb/gal; HLB 9.6; flash pt. (PMCC) > 300 F; pour pt. < 0 F; pH 5-7 (1% aq.); 99.5% min. act.

T-Det® DD-7. [Harcros] Dodoxynol-7; CAS 9014-92-0; nonionic; emulsifier; surfactant for solvs.; solv. and emulsion cleaning, agric., degreasers; FDA compliance; pale yel. liq.; mild aromatic odor; disp. in water; sp.gr. 1.04; dens. 8.66 lb/gal; HLB 11.1; flash pt. (PMCC) > 300 F; pour pt. < 0 F; pH 5-7 (1%); 99.5% min. act.

T-Det® DD-9. [Harcros] Dodoxynol-9; CAS 9014-92-0; nonionic; low foaming detergent, emulsifier, wetting agent; general purpose cleaners, degreasing; FDA compliance; pale yel. liq.; mild aromatic odor; disp. in water; sol. in aromatic hydrocarbons; sp.gr. 1.05; dens. 8.7 lb/gal; HLB 12.0; flash pt. (PMCC) > 300 F; pour pt. 42 F; pH 5-7 (1%); 99.5% min. act.

T-Det® DD-10. [Harcros] Dodoxynol-10; CAS 9014-92-0; nonionic; emulsifier; surfactant for solvs.; low foaming detergent for paper, textile and industrial cleaner formulations; paint pigment dispersant; FDA compliance; pale yel. liq.; mild aromatic odor; disp. in water; sp.gr. 1.05; dens. 8.75 lb/gal; HLB 13.1; cloud pt. 112 F (1%); flash pt. (PMCC) > 300 F; pour pt. 48 F; pH 7 (1%); 99.5% min. act.

T-Det® DD-11. [Harcros] Dodoxynol-11; CAS 9014-92-0; nonionic; surfactant, detergent, emulsifier; pale yel. liq.; water-sol.; sp.gr. 1.06 (68 F); dens. 8.8 lb/gal (68 F); HLB 13.5; pour pt. 53 F; flash pt. (PMCC) > 300 F; cloud pt. 134 F (1%); pH 5–7 (1% aq.); 99.5% min. act.

T-Det® DD-14. [Harcros] Dodoxynol-14; CAS 9014-92-0; nonionic; surfactant, detergent, emulsifier; pale yel. semisolid; water-sol.; sp.gr. 1.05 (68 F); dens. 8.7 lb/gal (68 F); HLB 14.5; pour pt. 75 F; flash pt. (PMCC) > 400 F; cloud pt. 190 F (1%); pH 5–7; 99.5% min. act.

T-Det® EPO-35. [Harcros] PO/EO block copolymer; nonionic; detergent, wetting agent, dispersant, rinse additive for low foam applics.; liq.; 100% conc.

T-Det® EPO-61. [Harcros] EO/PO block copolymer; nonionic; wetting agent, defoamer for paper mfg.; water treating compds., metal cleaning, rinse aids, mech. dishwash, textile; crude oil demulsifier; emulsifier and wetting agent in agric. toxicant formulations; FDA compliance; clear liq.; sol. in ethanol, toluene, xylene, perchlorethylene; disp. in water; sp. gr. 1.01; dens. 8.34 lb/gal; visc. 271 cps; HLB 3.0; pour pt. -28 C; cloud pt. 23 C (1%); flash pt. > 200 C; pH 5-7 (1% aq.); 99.5% min. act.

T-Det® EPO-62L. [Harcros] PO/EO block copolymer; nonionic; low foam surfactant, emulsifier, defoamer for metal cleaners, degreasers, rinse aids, mech. dishwash, hard surf. cleaners, textile, paper; FDA compliance; clear liq.; water-sol.; sp.gr. 1.03; dens. 8.6 lb/gal; HLB 7.0; pour pt. 27 F; cloud pt. 89 F (1%); flash pt. (PMCC) > 250 F; pH 4-8 (1% aq.); 100% conc.

T-Det® EPO-64L. [Harcros] PO/EO block copolymer; nonionic; wetting agent, low foaming surfactant, defoamer for metal cleaning compds., rinse aids, mech. dishwash, hard surf. cleaners, paper, textiles; emulsifier for agric. toxicants; demulsifier; FDA compliance; clear liq.; water-sol.; sp.gr. 1.04; dens. 8.7 lb/gal; HLB 15.0; pour pt. 48 F; cloud pt. 138 F (1%); flash pt. (PMCC) > 250 F; pH 4-8 (1% aq.); 100% conc.

T-Det® LF-416. [Harcros] Alkoxylated alcohol; nonionic; low foam detergent, wetting agent for machine dishwashing, rinse aids, metal and hard surf. cleaners; clear liq.; water-sol.; sp.gr. 1.06; dens. 8.8 lb/gal; pour pt. 40 F; cloud pt. 63 F (1%); flash pt. (PMCC) > 350 F; pH 5-7 (1% aq.); 100% conc.

T-Det® N 1.5. [Harcros] Nonoxynol-1 (1.5 EO); nonionic; surfactant, stabilizer, coemulsifier, defoamer, intermediate for petrol. oils applics., latex paints, solv. cleaners; FDA compliance; pale yel. liq.; oil-sol.; sp.gr. 0.99 (68 F); dens. 8.2 lb/gal (68 F); HLB 4.6; pour pt. < 0 F; flash pt. (PMCC) 340 F; pH 5–7 (1% aq.); 99.5% min. act.

T-Det® N-4. [Harcros] Nonoxynol-4; CAS 9016-45-9; nonionic; detergent, wetting agent; agric. toxicant formulations; intermediate for household detergents; dry cleaning soap additive, sludge dispersant additive in fuel oils; gasoline additive; solv. cleaner for metal processing; color development aid, stabilizer and emulsifier for latex paints; FDA compliance; pale yel. oily liq.; mild aromatic odor; sol. in butyl Cellosolve, CCl_4, min. and corn oil, ethanol, kerosene, min. spirits, perchloroethylene, Stod., toluene; dens. 8.5 lb/gal; sp.gr. 1.02; visc. 52 SUS (210 F); HLB 8.9; flash pt. > 400 F (OC); pour pt. < O F; pH 7; 99.5% min. act.

T-Det® N-6. [Harcros] Nonoxynol-6; CAS 9016-45-9; nonionic; emulsifier for silicones, household waxes and polishes; wetting agent; detergent and dispersant for petrol. and fuel oils; agric. toxicant formulations; dry cleaning soap additive; leather and metal processing; solv. cleaner; deinking agent for paper; FDA compliance; pale yel. liq.; mild aromatic odor; sol. in butyl Cellosolve, CCl_4, corn oil, diesel fuel, ethanol, kerosene, Stod., xylene; dens. 8.7 lb/gal; sp.gr. 1.04; visc. 58 SUS (210 F); HLB 10.9; flash pt. > 400 F (OC); pour pt. < 0 F; pH 7; 99.5% min. act.

T-Det® N-8. [Harcros] Nonoxynol-8; CAS 9016-45-9; nonionic; detergent, wetting agent, emulsifier, solubilizer, degreaser; textile, leather and paint processing; agric. toxicant, household and industrial cleaning formulations; pigment dispersant; rewetting, deinking; felt washing, leveling in paper mfg.; oil recovery operations; FDA compliance; clear visc. liq.; mild aromatic odor; dens. 8.75 lb/gal; sp.gr. 1.05; visc. 63 SUS (210 F); HLB 12.4; cloud pt. 78 F (1%); flash pt. 500 F (OC); pH 6 (1%); 99.5% min. act.

T-Det® N-9.5. [Harcros] Nonoxynol-10; CAS 9016-45-9; nonionic; detergent, wetting agent, emulsifier, solubilizer, degreaser, foamer; textile, leather and paint processing; agric. toxicant, household and industrial cleaning formulations; pigment dispersant; rewetting, deinking; felt washing, leveling in paper mfg.; oil recovery operations; FDA compliance; clear visc. liq.; mild aromatic odor; sol. in butyl Cellosolve, CCl_4, ethanol, ethylene glycol, HAN, toluene, water, xylene; dens. 8.8 lb/gal; sp.gr. 1.06; visc. 68 SUS (210 F); HLB 13.1; cloud pt. 135 F (1%); flash pt. 500 F (OC); pH 7 (1%); 99.5% min. act.

T-Det® N-10.5. [Harcros] Nonoxynol-11; CAS 9016-45-9; nonionic; detergent, wetting agent, dispersant, emulsifier, solubilizer; paints; textile processing; household and industrial cleaning prods.; FDA compliance; clear visc. liq.; mild aromatic odor; sol. in butyl Cellosolve, ethanol, ethylene glycol, HAN, toluene, water, xylene; dens. 8.8 lb/gal; sp.gr. 1.06; HLB 13.4; cloud pt. 155 F; flash pt. 500 F (OC); pH 7 (1%); 99.5% min. act.

T-Det® N-12. [Harcros] Nonoxynol-12; CAS 9016-45-9; nonionic; detergent; base for detergent and sanitizer formulations; wetting agent used in metal cleaning; FDA compliance; opaque liq.; mild aromatic odor; water-sol.; sp.gr. 1.07; dens. 8.9 lb/gal; HLB 14.1; cloud pt. 180 F (1%); flash pt. > 400 F (OC); pour pt. 57 F; pH 5-7 (1% aq.); 99.5% min. act.

T-Det® N-14. [Harcros] Nonoxynol-14; CAS 9016-45-9; nonionic; detergent; emulsifier for oils, fats and waxes; wetting agent for metal cleaners; penetrant; corrosion inhibitor; industrial and household cleaners, agric., textiles, petrol. processing; FDA compliance; off-wh. semisolid; mild aromatic odor; sol. in butyl Cellosolve, CCl_4, corn oil, ethanol, ethylene glycol, water, xylene; dens. 8.9 lb/gal (68 F); sp.gr. 1.07 (68 F); visc. 89 SUS (210 F); HLB 14.7; cloud pt. 203 F (1%); flash pt. > 400 F (OC); pour pt. 67 F; pH 5-7 (1% aq.); 99.5% min. act.

T-Det® N-20. [Harcros] Nonoxynol-20; CAS 9016-45-9; nonionic; surfactant for high temp. detergents, emulsion polymerization, petrol. processing; emulsifier; FDA compliance; opaque solid; water-sol.; sp.gr. 1.08 (105 F); dens. 9.0 lb/gal (105 F); HLB 16.0; pour pt. 90 F; flash pt. (PMCC) > 500 F; cloud pt. > 212 F (1%); pH 5–7 (1% aq.); 99.5% min. act.

T-Det® N-30. [Harcros] Nonoxynol-30; CAS 9016-45-9; nonionic; detergent, stabilizer, pressure scouring textile operations; emulsifier for oils, fats and waxes; wetting agent in electrolytes; FDA compliance; pale yel. waxy solid; mild aromatic odor; sol. in butyl Cellosolve, ethanol, ethylene glycol, water, xylene; sp.gr. 1.07 (132 F); dens. 8.9 lb/gal (132 F); HLB 17.0; cloud pt. > 212 F; flash pt. > 500 F; pour pt. 110 F; pH 6.5–8.0 (1%); 99.5% min. act.

T-Det® N-40. [Harcros] Nonoxynol-40; CAS 9016-45-9; nonionic; emulsifier for emulsion polymerization, asphalt, elevated temp. applics.; FDA compliance; pale yel. waxy solid; mild aromatic odor; sol. in water; sp.gr. 1.07 (132 F); HLB 17.7; cloud pt. 212 F (1%); flash pt. > 500 F; pour pt. 120 F; pH 6.5–8.0 (1%); 99.5% min. act.

T-Det® N-50. [Harcros] Nonoxynol-50; CAS 9016-45-9; nonionic; detergent, emulsifier for emulsion polymerization, asphalt, textiles, agric., elevated temp. applics.; FDA compliance; wh. solid; water-sol.; sp.gr. 1.07 (132 F); dens. 8.9 lb/gal (132 F); HLB 18.0; pour pt. 120 F; cloud pt. > 212 F (1%); flash pt. (PMCC) > 500 F; pH 5-7 (1% aq.); 99.5% min. act.

T-Det® N-70. [Harcros] Nonoxynol-70; CAS 9016-45-9; nonionic; high temp. detergent, emulsifier for oils, fats and waxes; stabilizer in latex and alkyd emulsions; emulsion polymerization; asphalt; scouring textile operations; FDA compliance; pale yel. waxy solid; mild aromatic odor; dens. 9.0 lb/gal (135 F); sp.gr. 1.08 (135 F); HLB 18.7; cloud pt. 212 F (1%); flash pt. > 500 F; pour pt. 125 F; pH 6.0–7.0 (1%); 99.5% min. act.

T-Det® N-100. [Harcros] Nonoxynol-100; CAS 9016-45-9; nonionic; emulsifier for asphalt, emulsion polymerization, textiles, toilet block prods.; high temp. and high electrolyte applics.; FDA compliance; wh. waxy solid; mild aromatic solid; sol. in ethylene dichloride, ethanol and water; sp.gr. 1.08 (135 F); dens. 9.0 lb/gal (132 F); HLB 19.0; cloud pt. 212 F (1%); flash pt. (PMCC) > 500 F; pour pt. 127 F; pH 6–7 (1%); 99.5% min. act.

T-Det® N-307. [Harcros] Nonoxynol-30; CAS 9016-45-9; nonionic; high temp. detergent, emulsifier for emulsion polymerization, asphalt, textiles; stabilizer; FDA compliance; clear liq.; water-sol.; sp.gr. 1.10; dens. 9.2 lb/gal; visc. 806 cps; HLB 17.0; cloud pt. > 100 C (1%); flash pt. > 212 F (TOC); pour pt. 30 F; 70 ± 0.5% act.

T-Det® N-407. [Harcros] Nonoxynol-40; CAS 9016-

45-9; nonionic; emulsifier for emulsion polymerization, asphalt, paint systems, elevated temp. applics.; FDA compliance; clear liq.; dens. 9.0 lb/gal; sp.gr. 1.08; visc. 900 cps; HLB 17.7; cloud pt. > 212 F (1%); flash pt. > 212 F (TOC); pour pt. 30 F; 70 ± 0.5% act.

T-Det® N-507. [Harcros] Nonoxynol-50; CAS 9016-45-9; nonionic; detergent, emulsifier for emulsion polymerization, asphalt, paint systems, textiles, agric., elevated temp. applics.; FDA compliance; APHA 100 opaque liq.; sp.gr. 1.09; dens. 9.1 lb/gal; visc. 760 cps; HLB 18.0; cloud pt. > 100 C (1%); flash pt. > 212 F (TOC); pour pt. 40 F; 70 ± 0.5% act.

T-Det® N-705. [Harcros] Nonoxynol-70; CAS 9016-45-9; nonionic; high temp. detergent, emulsifier for oils, fats and waxes; stabilizer in latex and alkyd emulsions; emulsion polymerization; asphalt; scouring textile operations; clear to hazy liq.; mild aromatic odor; dens. 9.0 lb/gal; sp.gr. 1.08; cloud pt. 212 F (1%); flash pt. > 500 F; pour pt. 40 F; pH 5.0–6.5 (1%); 50% min. act.

T-Det® N-707. [Harcros] Nonoxynol-70; CAS 9016-45-9; nonionic; detergent, emulsifier for oils, fats and waxes; stabilizer in latex and alkyd emulsions; emulsion polymerization; scouring textile operations; FDA compliance; clear to hazy liq.; mild aromatic odor; dens. 9.0 lb/gal; sp.gr. 1.08; HLB 18.7; cloud pt. 212 F (1%); flash pt. > 500 F; pour pt. 40 F; pH 5.0–6.5 (1%); 50% min. act.

T-Det® N-1007. [Harcros] Nonoxynol-100; CAS 9016-45-9; nonionic; emulsifier for asphalt, emulsion polymerization, textiles, high temp. and high electrolyte applics.; FDA compliance; clear-opaque visc. liq.; mild aromatic odor; water-sol.; sp.gr. 1.1; dens. 9.2 lb/gal; HLB 19.0; cloud pt. 212 F (1%); flash pt. (PMCC) > 500 F; pour pt. 62 F; pH 6–7 (1%); 70 ± 0.5% act.

T-Det® O-4. [Harcros] Octoxynol-4; CAS 9002-93-1; nonionic; surfactant for agric., leather and metal processing, paint, drycleaning detergents, degreasing applics.; FDA compliance; clear liq.; water-disp.; sp.gr. 1.03 (68 F); dens. 8.6 lb/gal (68 F); HLB 8.5; pour pt. < 0 F; flash pt. (PMCC) > 500 F; pH 6–8 (1%); 99.5% min. act.

T-Det® O-6. [Harcros] Octoxynol-6; CAS 9002-93-1; nonionic; detergent, dispersant, wetting agent, emulsifier for detergents, textile and leather processing, agric. formulations; FDA compliance; clear liq.; water-disp.; sp.gr. 1.05 (68 F); dens. 8.7 lb/gal (68 F); HLB 11.6; pour pt. < 0 F; flash pt. (PMCC) > 500 F; pH 6–8 (1%); 99.5% min. act.

T-Det® O-8. [Harcros] Octoxynol-8; CAS 9002-93-1; nonionic; surfactant for detergents, agric. formulations, oil well drilling, leather processing, paints; FDA compliance; clear liq.; water-sol.; sp.gr. 1.04 (68 F); dens. 8.7 lb/gal (68 F); HLB 12.4; pour pt. 15 F; flash pt. (PMCC) > 500 F; cloud pt. 70 F (1%); pH 5–7 (1%); 99.5% min. act.

T-Det® O-9. [Harcros] Octoxynol-9; CAS 9002-93-1; nonionic; controlled foam detergent, dispersant, emulsifier, wetting agent for general purpose applics., household and industrial cleaners; FDA compliance; clear liq.; water-sol.; sp.gr. 1.06 (68 F); dens. 8.8 lb/gal (68 F); HLB 13.5; pour pt. 45 F; flash pt. (PMCC) > 500 F; cloud pt. 150 F (1%); pH 5–7 (1% aq.); 99.5% min. act.

T-Det® O-12. [Harcros] Octoxynol-12; CAS 9002-93-1; nonionic; detergent, emulsifier for household and industrial cleaners; liq.; 100% conc.

T-Det® O-40. [Harcros] Octoxynol-40; CAS 9002-93-1; nonionic; emulsifier, stabilizer for agric. formulations, detergents, textiles, emulsion polymerization; FDA compliance; wh. solid; water-sol.; sp.gr. 1.06 (150 F); dens. 8.8 lb/gal (150 F); HLB 17.9; pour pt. 120 F; flash pt. (PMCC) > 500 F; cloud pt. > 212 F (1%); 99.5% min. act.

T-Det® O-307. [Harcros] Octoxynol-30; CAS 9002-93-1; nonionic; emulsifier, stabilizer for emulsion polymerization, industrial intermediate and agric. applics.; liq.; HLB 17.3; 70% conc.

T-Det® O-407. [Harcros] Octoxynol-40; CAS 9002-93-1; nonionic; detergent, emulsifier, stabilizer for various applics.; FDA compliance; APHA 100 max. liq.; water-sol.; sp.gr. 1.10; dens. 9.1 lb/gal; visc. 600 cps; HLB 17.9; cloud pt. > 100 C (1%); flash pt. > 100 C (TOC); pour pt. 30 F; pH 7 (1%); 70 ± 0.5% act.

T-Det® RQ1. [Harcros UK] Block copolymer; nonionic; low foam surfactant for mech. dishwash, metal cleaners, rinse aids, hard surf. cleaners, textile, paper; defoamer; FDA compliance; clear liq.; water-sol.; sp.gr. 1.02; dens. 8.5 lb/gal; HLB 3.0; pour pt. -22 F; cloud pt. 77 F (1%); flash pt. (PMCC) > 250 F; pH 4-8 (1% aq.); 99.5% min. act.

T-Det® RY2. [Harcros UK] Block copolymer; nonionic; low foam surfactant for mech. dishwash, metal cleaners, rinse aids, hard surf. cleaners, textile, paper; defoamer; FDA compliance; clear liq.; water-sol.; sp.gr. 1.03; dens. 8.6 lb/gal; HLB 4.0; pour pt. 14 F; cloud pt. 84 F (1%); flash pt. (PMCC) > 250 F; pH 4-8 (1% aq.); 99.5% min. act.

T-Det® TDA-60. [Harcros] Tridecyl alcohol/60% EO adduct; nonionic; detergent, dispersant, wetting agent; liq.; water-sol.; HLB 12.0; 100% conc.

T-Det® TDA-65. [Harcros] Tridecyl alcohol/65% EO adduct; nonionic; emulsifier, solubilizer; textile processing; hard surface detergent; disperses and suspends solids; wetting and rewetting agent for dust settling, corrosion inhibitor, metal cleaning and pickling; hazy liq.; mild aliphatic odor; water-sol.; dens. 8.20 lb/gal; HLB 13.0; cloud pt. 130 F (1%); flash pt. > 290 F (PMCC); pour pt. 62 F; pH 5–7; 99.5% min. act.

T-Det® TDA-70. [Harcros] Tridecyl alcohol/70% EO adduct; nonionic; detergent, dispersant, wetting agent; paste; water-sol.; HLB 14.0; 100% conc.

T-Det® TDA-150. [Harcros] Trideceth-150; CAS 24938-91-8; nonionic; controls dissolution rate of solid or block type hard surface cleaners; solid; HLB 19.4; 100% conc.

Teepol CM 44. [Shell] Alkylbenzene sulfonate, sodium salt; anionic; detergent, emulsifier; used in dishwashing and general cleaning liqs.; pale amber liq.; misc. with water; dens. 1.05 kg/l; visc. 100–500 cs; clear pt. > 16 C; pH 8.0–9.5 (5% aq.); biodeg.; > 25% act.

Teepol GC 56P. [Shell] Alkylbenzene sulfonate and primary alcohol ethoxysulfate; detergent; used in dishwashing and general cleaning applics.; clear liq.; misc. with water; dens. 1.06 kg/l; visc. 1900 cs; cloud pt. 1 C; pH 6.5–8.0 (5%); biodeg.; > 33% act.

Teepol GD53. [Shell] Alkylbenzene sulfonate and primary alcohol ethoxysulfate; anionic/nonionic; detergent; used in dishwashing and general cleaning applics.; clear liq.; misc. with water; dens. 1.07 kg/l; visc. 24–300 cs; cloud pt. 1 C; pH 6.5–8.0 (5%); > 30% act.

Teepol GD7P. [Shell] Alkylbenzene sulfonate and primary alcohol ethoxysulfate; anionic/nonionic; detergent; used in dishwashing and general cleaning applics.; clear liq.; misc. with water; dens. 1.07 kg/ 1; visc. 620 cs; cloud pt. 2 C; pH 6.5–8.0 (5%); > 40% act.

Teepol HB 6. [Shell] Primary alcohol, sulfated, sodium salt; solubilizer, emulsifier; component of hard surface and germicidal cleaners; pale amber liq.; water-sol.; dens. 1.06 kg/1; visc. 20 cs; clear pt. 10 C; pH 8.0 (5% aq.); biodeg.; 34.2% act.

Teepol HB 7. [Shell] C9-C13 primary alcohol sulfate, sodium salt; anionic; solubilizer and emulsifier component of hard surface and germicidal cleaners; liq.; 40% conc.

Teepol PB. [Shell] Primary alcohol ethoxysulfate (sodium cation); foamer; used in plasterboard mfg.; clear liq.; misc. with water; dens. 1.06 kg/1; visc. 50 cs; clear pt. 0 C; pH 6.5–8.0 (5%); biodeg.; > 32% act.

Tegacid® Regular VA. [Goldschmidt] Glyceryl stearate, stearamidoethyldiethylamine; cationic; self-emulsifying; forms stable emulsions for creams and lotions; acid stable; flake; HLB 4.2; 100% conc.

Tegacid® Special VA. [Goldschmidt] Glyceryl stearate SE; anionic; self-emulsifying, acid stable surfactant; flake; HLB 4.9; 100% conc.

Tegamineoxide WS-35. [Goldschmidt] Fatty acid amido alkyl dimethylamine oxide; nonionic; foam builder/stabilizer for shampoos, bath preps., liq. detergents; liq.; 35% conc.

Tegiloxan®. [Goldschmidt] Methylsilicone oils; antifoams for min. oils, in rubber and plastics industry; lubricant for tire prod.; additive for polishes.

Tegin® C-63. [Goldschmidt] Monodiglyceride citric acid ester; anionic; hydrophilic emulsifier; powd.; HLB 10.0; 100% conc.

Tegin® C-611. [Goldschmidt] Glyceryl stearate SE and hydrog. tallow glyceride citrate; anionic; o/w for acidic emulsions; powd.; HLB 11.0; 100% conc.

Tegin® D 6100. [Goldschmidt; Goldschmidt AG] PEG-2 stearate; CAS 9004-99-3; nonionic; lipophilic coemulsifier for o/w emulsions; waxy; HLB 2.8; 100% conc.

Tegin® DGS. [Goldschmidt] Diglycol stearate; CAS 9004-99-3; anionic; coemulsifier for o/w emulsions; solid; HLB 2.8; 100% conc.

Tegin® E-41. [Goldschmidt] Acetylated hydrog. tallow glyceride; nonionic; emulsifier for cosmetic and technical prods.; emollient, plasticizer, food additive; solid; HLB 2.0–3.0; 100% conc.

Tegin® E-41 NSE. [Goldschmidt] Acetylated hydrog. tallow glycerides; nonionic; emulsifier for cosmetic and technical prods.; emollient/plasticizer food additives; semisolid; HLB 2.0–3.0; 100% conc.

Tegin® E-61. [Goldschmidt] Acetylated hydrog. lard glyceride; nonionic; emulsifier for cosmetic and technical prods.; emollient, plasticizer, food additive; solid; HLB 2.0–3.0; 100% conc.

Tegin® E-61 NSE. [Goldschmidt] Acetylated hydrog. lard glyceride; nonionic; emulsifier for cosmetic and technical prods.; emollient/plasticizer, food additives; semisolid; HLB 2.0–3.0; 100% conc.

Tegin® E-66. [Goldschmidt] Acetylated lard glyceride; nonionic; emulsifier for cosmetic and technical prods.; emollient, plasticizer, food additive; liq.; HLB 2.0–3.0; 100% conc.

Tegin® E-66 NSE. [Goldschmidt] Acetylated lard glyceride; nonionic; emulsifier for cosmetic and technical prods.; emollient/plasticizer, food additives; liq.; HLB 2.0–3.0; 100% conc.

Tegin® G 6100. [Goldschmidt; Goldschmidt AG] Glycol stearate; CAS 111-60-4; nonionic; lipophilic coemulsifier for o/w emulsions; opacifier for shampoos; powd.; HLB 3.2; 100% conc.

Tegin® GO. [Goldschmidt] Ethylene glycol oleate; anionic; coemulsifier and stabilizer; liq.; HLB 2.8; 100% conc.

Tegin® GRB. [Goldschmidt] Glyceryl stearate; nonionic; emulsifier for o/w systems; powd.; 100% conc.

Tegin® ISO. [Goldschmidt; Goldschmidt AG] Glyceryl mono/diisostearate; CAS 68958-48-5; anionic; w/o emulsifier for creams; paste; HLB 3.4; 100% conc.

Tegin® M. [Goldschmidt; Goldschmidt AG] Glyceryl stearate; anionic; emulsifier for o/w systems; solid; HLB 3.8; 100% conc.

Tegin® O. [Goldschmidt; Goldschmidt AG] Glyceryl mono/dioleate; CAS 25496-72-4; nonionic; emulsifier for w/o emulsions; paste; HLB 3.3; 100% conc.

Tegin® O NSE. [Goldschmidt] Glyceryl oleate; CAS 111-03-5; anionic; surfactant for w/o emulsions with paraffin hydrocarbons, fat, oil, and wax; semisolid; HLB 3.3; 100% conc.

Tegin® O Special. [Goldschmidt] Glyceryl oleate; CAS 111-03-5; anionic; emulsifier for o/w systems; paste; HLB 3.3; 100% conc.

Tegin® O Special NSE. [Goldschmidt] Glyceryl oleate; CAS 111-03-5; anionic; w/o emulsions; semisolid; HLB 3.3; 100% conc.

Tegin® P. [Goldschmidt; Goldschmidt AG] Propylene glycol stearate SE; anionic; emulsifier for o/w lotions and creams; solubilizer for dyestuffs; solid; HLB 4.4; 100% conc.

Tegin® P-411. [Goldschmidt] Propylene glycol stearate; nonionic; lipophilic coemulsifier for o/w emulsions; waxy; HLB 2.8; 100% conc.

Tegin® RZ. [Goldschmidt] Glyceryl ricinoleate; nonionic; emulsifier, solubilizer, emollient, plasticizer; paste; HLB 3.0; 100% conc.

Tegin® RZ NSE. [Goldschmidt] Glyceryl ricinoleate; nonionic; o/w emulsifier, solubilizer, emollient, plasticizer; semisolid; HLB 3.0; 100% conc.

Tegin® T 4753. [Goldschmidt] Polyglyceryl-4 isostearate; nonionic; emulsifier and coemulsifier for cosmetic w/o emulsions; liq.; 100% conc.

Teginacid®. [Goldschmidt; Goldschmidt AG] Glyceryl stearate and ceteareth-20; nonionic; emulsifier for o/w emulsions; powd.; HLB 12.0; 100% conc.

Teginacid® H. [Goldschmidt; Goldschmidt AG] Glyceryl stearate and ceteth-20; nonionic; emulsifier for o/w emulsions, acid and salt resisting emulsions; self-emulsifying; powd.; HLB 11.2; 100% conc.

Teginacid® H-SE. [Goldschmidt] Glyceryl monodistearate and other nonionics; nonionic; self-emulsifying; for acid and salt resistant o/w emulsions with liq. or creamy consistency; powd.; HLB 11.2; 100% conc.

Teginacid® ML. [Goldschmidt] Glyceryl stearate and PEG-40 stearate; emulsifier for finely dispersed emulsions.

Teginacid® ML-SE. [Goldschmidt] Glyceryl monodistearate and PEG-40 stearate; cationic; o/w emulsions; powd.; HLB 10.6; 100% conc.

Teginacid® SE. [Goldschmidt] Glyceryl monodistearate; nonionic; self-emulsifying; for acid and salt resistant o/w emulsions with liq. or creamy consistency; powd.; HLB 12.0; 100% conc.

Teginacid® Special SE. [Goldschmidt] Glyceryl stearate; anionic; o/w emulsions; powd.; HLB 12.0; 100% conc.

Teginacid® X. [Goldschmidt; Goldschmidt AG] Glyceryl stearate and ceteareth-20; nonionic; emulsifier for o/w emulsions; solid; HLB 12.0; 100% conc.

Teginacid® X-SE. [Goldschmidt] Glyceryl monodistearate; nonionic; self-emulsifying; for acid and salt resistant o/w emulsions with liq. or creamy consistency; powd.; HLB 12.0; 100% conc.

Tego® Antifoam. [Goldschmidt] Org. antifoam for waste water treatment.

Tego®-Betaine C. [Goldschmidt] Cocamidopropyl betaine; amphoteric; surfactant used as foam stabilizer and visc. builder in personal care prods., dishwash, liq. soap; liq.; 30% conc.

Tego®-Betaine F. [Goldschmidt; Goldschmidt AG] Cocamidopropyl betaine; CAS 61789-40-0; amphoteric; surfactant for nonirritating shampoos, bubble baths, hygiene and baby care prods.; liq.; 30% conc.

Tego®-Betaine L-90. [Goldschmidt] Lauramidopropyl betaine; amphoteric; surfactant, foam stabilizer, visc. builder for shampoos, bath, dishwash, liq. soap, and cream and lotion prods.; liq.; 30% conc.

Tego®-Betaine L-5351. [Goldschmidt; Goldschmidt AG] Cocamidopropyl betaine; CAS 61789-40-0; amphoteric; low-salt surfactant for aerosols and electrolyte-sensitive formulations, e.g., hair dyes, fixatives, and reactive prods.; liq.; 30% conc.

Tego®-Betaine N-192. [Goldschmidt; Goldschmidt AG] Dihydroxyethyl tallow glycinate; CAS 61791-25-1; amphoteric; surfactant for acid and alkali-stable formulations; highly visc. liq.; 35% conc.

Tego®-Betaine S. [Goldschmidt] Cocamidopropyl betaine; amphoteric; surfactant, foam stabilizer, visc. builder for shampoos, bath, dishwash, liq. soap, cream and lotion prods.; liq.; 30% conc.

Tego®-Betaine T. [Goldschmidt] Cocamidopropyl betaine; amphoteric; household cleansers, floor cleaner, car wash; liq.; 25% conc.

Tego® Dispers 610. [Tego] Higher m.w. unsaturated polycarboxylic acid; wetting and dispersion additive to counter sedimentation and flooding of pigments; produces selective flocculation of pigments and extenders; stabilizer for pigment dispersions; used in binder systems such as alkyd, acrylate, polyester/melamine; acrylate, polyisocyanate, nitrocellulose paints, chlorinated polymers; lt. brn. clear liq.; sp.gr. 0.93 g/cc; flash pt. 28 C; usage level: 0.1-1.0%; 50 ± 2% act. in xylene/diisobutyl ketone (9:1).

Tego® Dispers 610 S. [Tego] Higher m.w. unsaturated polycarboxylic acid and organically modified siloxane copolymer; wetting and dispersion additive to prevent flooding; selective flocculates pigments and extenders; antifoam; improves gloss and leveling; for binder systems such as alkyd, acrylate, polyester/melamine, polyisocyanate, nitrocellulose-based paints; chlorinated polymers; lt. brn. clear liq.; sp.gr. 0.93 g/cc; flash pt. 28 C; usage level: 0.1-1.0%; 50 ± 2% act. in xylene/diisobutyl ketone (9:1).

Tego® Dispers 620. [Tego] Higher m.w. unsaturated polycarboxylic acid; wetting and dispersion additive to prevent flooding; produces selective flocculation of pigments and extenders, stabilizes pigment dispersion; for binder systems such as alkyds, acrylate polymers, chlorinated polymers, polyester/melamine; polyisocyanate systems; lt. brn. clear liq.; sp.gr. 0.93 g/cc; flash pt. 28 C; usage level: 0.1-1.0%; 50±2% act. in xylene/diisobutyl ketone (9:1).

Tego® Dispers 620 S. [Tego] Higher m.w. unsaturated polycarboxylic acid and organically modified siloxane copolymer; wetting and dispersion additive to prevent flooding; produces selective flocculation of pigments and extenders; antifloat agent; for binder systems such as alkyds, acrylate polymers, chlorinated polymers, polyester/melamine; polyisocyanate systems; lt. brn. clear liq.; sp.gr. 0.93 g/cc; flash pt. 28 C; usage level: 0.1-1.0%; 50 ± 2% act. in xylene/diisobutylketone (9:1).

Tego® Dispers 630. [Tego] Salt of a higher m.w. polycarboxylic acid and amine deriv.; wetting and dispersion additive against sediment, sagging, and flooding; produces selective flocculation of pigments and extenders, stabilizes pigment dispersion; for binder systems such as alkyd, acrylate, polyester/melamine, epoxy; chlorinated polymers; lt. brn. clear liq.; sp.gr. 0.88 g/cc; flash pt. 42 C; usage level: 0.1-0.8%; 50 ± 3% act. in higher boiling aromatics.

Tego® Dispers 700. [Tego] Salt of a high m.w. fatty acid deriv.; wetting and dispersing additive for solv.-based paint systems; deflocculant, antiflooding agent; for alkyd, epoxide, chlorinated rubber, bitumen, polyurethane, and polyester systems; sp.gr. 0.93 g/cc; flash pt. 25 C; usage level: 0.4-4.0% on pigment; 50% act. in xylene.

Tego® Dispers 705. [Tego] Salt of a polycarboxylic acid with an amine deriv.; wetting and dispersion additive for solv.-based paints; deflocculant; sp.gr. 0.93 g/cc; flash pt. 25 C; usage level: 0.1-1.0%; 50 ± 1% act. in xylene/diisobutyl ketone (4:1).

Tego® Flow 425. [Tego] Polysiloxane-polyether copolymer; flow and leveling additive for clear water and solv.-based paint systems (alkyd, acrylate, PU, alkyd-melamine, PU-acrylic), building protection paints, wood/furniture varnishes; sol. in polar and nonpolar solvs.; sp.gr. 1.0 g/cc; visc. 90 ± 30 mm^2 s^{-1}; f.p. 17 C; flash pt. 71 C; usage level: 0.05-0.15%; 100% act.

Tego® Foamex 800. [Tego] Polysiloxane-polyether copolymer o/w emulsion; nonionic; defoamer for water-based emulsion paints and water-thinnable systems, PU paints, PU-acrylics, furniture paint, wood varnish, adhesives, dispersion paints, polymer dispersions; sp.gr. 1.0 g/cc; usage level: 0.5-1.5%; 20% act. in water.

Tego® Foamex 805. [Tego] Polysiloxane-polyether copolymer o/w emulsion; nonionic; defoamer for water-based emulsion paints and water-thinnable systems, PU paints, PU-acrylics, furniture paint, wood varnish, adhesives, dispersion paints, polymer dispersions; sp.gr. 1.0 g/cc; usage level: 0.5-1.5%; 20% act. in water.

Tego® Foamex 1435. [Tego] Polysiloxane w/o emulsion; nonionic; defoamer for water thinnable printing inks, high visc. emulsion paints and renders; sp.gr. 1.0 g/cc; usage level: 0.25-0.75%; 20% act. in water.

Tego® Foamex 1488. [Tego] Dimethicone copolyol emulsion; nonionic/anionic; defoamer for water-based emulsion paints and water-thinnable systems, building protection coatings (acrylic, styrene-acrylic, PVA), wood/furniture varnishes, anticorrosive coatings/industrial paints (alkyd, polyester, acrylic), polymer emulsions; sp.gr. 1.0 g/cc; usage level: 0.2-0.6%; 20% act. in water.

Tego® Foamex 3062. [Tego] Dimethicone copolyol; CAS 67762-96-3; defoamer for water-based emulsion paints and water-thinnable systems, building protection coatings (acrylic, styrene-acrylic, PVA), wood/furniture varnishes, printing inks; Gardner 9 max. color; sp.gr. 1.0 g/cc; usage level: 0.05-0.15%; 100% act.

Tego® Foamex 7447. [Tego] Dimethicone copolyol emulsion; nonionic; defoamer for water-based emulsion paints and water-thinnable systems, building protection coatings (acrylic, styrene-acrylic, PVAc), wood/furniture varnishes, anticorrosion/industrial paints (alkyd, polyester, acrylic), printing inks, polymer disps.; sp.gr. 1.0 g/cc; usage level: 0.05-0.5%; 20% act. in water.

Tego® Foamex KS 6. [Tego] Paraffin-based mineral oil, emulsifiers, trace silicone; defoamer for emulsion paints, building protection coatings (styrene-acrylic, PVAc, vinyl propionate, S/B), gravure and flexographic inks; sp.gr. 1.0 g/cc; flash pt. > 130 C; usage level: 0.1-0.5%; 100% act.

Tego® Foamex KS 10. [Tego] Paraffin-based mineral oil, polysiloxane polyether copolymer; defoamer for emulsion paints, building protection coatings (styrene-acrylate, PVAc, vinyl propionate, S/B), wood/furniture varnishes; sp.gr. 0.94 g/cc; flash pt. > 155 C; usage level: 0.1-0.5%; 100% act.

Tego® Foamex L 808. [Tego] Polysiloxane polyether copolymer o/w emulsion; nonionic; defoamer for aq. systems incl. acrylic emulsions, wax emulsions, pigment pastes; misc. with water; sp.gr. 1.0 g/cc; usage level: 0.2-0.65%; 20% act. in water.

Tego® Foamex L 822. [Tego] Polysiloxane polyether copolymer; defoamer for water-based finishes, acrylic or wax emulsions, pigment pastes; Gardner 9 max. color; sp.gr. 1.0 g/cc; usage level: 0.05-0.15%; 100% act.

Tego® Foamex N. [Tego] Simethicone; defoamer conc. for med., high solid and solv.-free paint systems, building protection coatings, anticorrosive coatings, high visc. paints; sp.gr. 1.0 g/cc; flash pt. > 200 C; usage level: 0.05-2.0% of 1% sol; 100% act.

Tego® Glide 100. [Tego] Dimethicone copolyol; surfactant, mar resistant and flow additive for water and solv.-based paints (alkyd, sat. polyester, polyacrylate), building protection coatings, anticorrosive paints, car finishes, printing inks; BGA approved; Gardner < 5 color; sp.gr. 1.0 g/cc; visc. 800 ± 150 $mm^2 s^{-1}$; flash pt. > 41 C; surf. tens. 21 $mN \cdot m^{-1}$; usage level: 0.05-0.15%; 100% act.

Tego® Glide 405. [Tego] Dimethicone copolyol; surfactant, mar resistant additive for water- and solv.-based paints (alkyd, alkyd-melamine, PU, vinyl), building protection coatings, anticorrosive paints, wood/furniture varnishes, printing inks; hydrolytically stable; Gardner < 3 color; sp.gr. 0.96 g/cc; flash pt. 65 C; usage level: 0.1-0.4%; 50% act. in butyl glycol.

Tego® Glide 406. [Tego] Dimethicone copolyol; surfactant, mar resistant additive for water- and solv.-based paints (alkyd, alkyd-melamine, PU, vinyl), building protection coatings, anticorrosive paints, wood/furniture varnishes, printing inks; Gardner < 3 color; sp.gr. 0.99 g/cc; flash pt. 65 C; usage level: 0.1-0.4%; 50% act. in dipropylene glycol methyl ether.

Tego® Glide 410. [Tego] Dimethicone copolyol; mar resistant additive for solv.-based paints, primers, extenders, automotive paints, building protection coatings, anticorrosion paints, wood/furniture varnishes, industrial paints, lacquers, aq. and solv.-based printing inks; reduces surf. tens., improves leveling; BGA approved; sp.gr. 1.01 g/cc; flash pt. 81 C; usage level: 0.005-0.5; 100% act.

Tego® Glide 470. [Tego] Dimethicone copolyol; mar resistant additive for water- and solv.-based paints (alkyd-melamine, polyester-melamine, acrylates, NC-PU), automobile paints, wood/furniture varnishes, printing inks, industrial paints, pkg. lacquers; BGA approved; sp.gr. 1.0 g/cc; flash pt. > 80 C; usage level: usage level: 0.005-0.1; 100% act.

Tego® Glide A 115. [Tego] Organically modified polysiloxane; mar resistant, flow and leveling additive for solv.-based paints (PU, vinyl, alkyd-melamine, acrylate-isocyanate), one-coat systems, anticorrosive paints, wood/furniture varnishes, printing inks; reduces surf. tens., aids wetting; sol. in aliphatic, aromatic, and chlorinated hydrocarbons, alcohols, ethers, and esters; sp.gr. 1.0 g/cc; visc. 100 ±30 $mm^2 s^{-1}$; flash pt. 60 C; usage level: 0.005-0.5%; 100% act.

Tego® Glide B 1484. [Tego] Dimethicone copolyol; additive to improve scratch resistance, flow, and leveling of solv.-based systems; defoamer; deaerator for epoxy resins; used for epoxide, PU, acrylate/isocyanate, alkyd/melamine paint systems, floor coatings, car finishes, wood/furniture varnishes; BGA approved; Gardner < 6 color; sol. in most org. solvs.; insol. in water; sp.gr. 1.00 g/cc; visc. 550 ± 200 $mm^2 s^{-1}$; flash pt. 65 C; usage level: 0.05-0.15%; 100% act.

Tego® Glide ZG 400. [Tego] Dimethicone copolyol; surfactant, mar resistant additive for water- and solv.-based paints (alkyd, alkyd-melamine, PU, vinyl), building protection coatings, anticorrosive paints, wood/furniture varnishes, printing inks; hydrolytically stable; BGA approved; Gardner < 3 color; sp.gr. 1.035 g/cc; visc. 850 ± 200 $mm^2 s^{-1}$; flash pt. > 65 C; usage level: 0.05-1.0%; 100% act.

Tego® Hammer 300000. [Tego] Methyl silicone oil; hammer finish additive for solv.-based paints; low surf. tens.; BGA and FDA approved; sol. in aliphatic and aromatic hydrocarbons; sp.gr. 0.98 g/cc; flash pt. 350 C; surf. tens. 20.5 $mN \cdot m^{-1}$; usage level: 0.01-0.3%; 100% act.

Tegopren®. [Goldschmidt] Organo modified siloxanes; surfactants used as antistats, wetting and leveling agents, emulsifiers, dispersants, and for the improvement of lubricity; additive for polishes.

Tegotain A 4080. [Goldschmidt AG] Alkyl dimethylamine oxide; nonionic; foam booster, antistat, alkaline and acid cleaner; liq.; 30% conc.

Tegotain D. [Goldschmidt AG] Cocamidopropyl betaine; amphoteric; mild surfactant; spray-dried powd.; 85% conc.

Tegotain L 5351. [Goldschmidt AG] Cocamidopropyl betaine; amphoteric; low salt, mild surfactant; esp. for electrolyte-sensitive formulations; liq.; 30% conc.

Tegotain L 7. [Goldschmidt AG] Cocamidopropyl betaine; amphoteric; surfactant; improves skin compatibility of anionic surfactants; good biodeg.; liq.; 30% conc.

Tegotain N 192. [Goldschmidt AG] Dihydroxyethyl alkyl betaine; amphoteric; mild surfactant, thickener; stable in acid media; liq.; 35% conc.

Tegotain S. [Goldschmidt AG] Cocamidopropyl betaine; amphoteric; surfactant improving skin compatibility of anionic surfactants; good biodeg.; liq.; 30% conc.

Tegotain WS 35. [Goldschmidt AG] Alkylamido amine oxide; amphoteric; foam booster, antistat; liq.; 35% conc.

Tegotens 4100. [Goldschmidt AG] Palmitic/stearic

acid mono/diglycerides; nonionic; surface coating for expanded polystyrene beads containing a propellant; antistat for PE/PP; fabric softener; coemulsifier; powd.

Tegotens BL 130. [Goldschmidt AG] Palmitic/stearic acid mono/diglycerides; nonionic; surface coating for expanded polystyrene beads containing a propellant; powd.

Tegotens BL 150. [Goldschmidt AG] Glycerol triester of fatty acids; nonionic; surface coating for expanded polystyrene beads containing a propellant; powd.

Tegotens I. [Goldschmidt AG] POE glyceryl isostearate; nonionic; solubilizer, coemulsifier, wetting agent; liq.; 100% conc.

Tegotens I 2. [Goldschmidt AG] POE glyceryl isostearate; nonionic; solubilizer, coemulsifier, wetting agent; liq.; HLB 14.2; 100% conc.

Tegotens L. [Goldschmidt AG] POE glyceryl laurate; nonionic; solubilizer, coemulsifier, wetting agent; liq.; HLB 17.0; 100% conc.

Tegotens L 2. [Goldschmidt AG] POE glyceryl laurate; nonionic; solubilizer, coemulsifier, wetting agent; liq.; HLB 15.7; 100% conc.

Tegotens O. [Goldschmidt AG] POE glyceryl oleate; nonionic; solubilizer, coemulsifier, wetting agent; liq.; HLB 16.4; 100% conc.

Tegotens O 2. [Goldschmidt AG] POE glyceryl oleate; nonionic; solubilizer, coemulsifier, wetting agent; liq.; 100% conc.

Tegotens R 1. [Goldschmidt AG] POE glyceryl ricinoleate; nonionic; solubilizer, coemulsifier, wetting agent; liq.; HLB 14.0; 100% conc.

Tegotens R 40. [Goldschmidt AG] Hydrog. castor oil EO deriv.; nonionic; solubilizer, coemulsifier, wetting agent; for prep. of emulsions of fatty acids, oils, org. solvs., polymers, resins, colors, etc.; highly visc. liq.; HLB 13.0; 100% conc.

Tegotens R 60. [Goldschmidt AG] Hydrog. castor oil EO deriv.; nonionic; solubilizer, coemulsifier, wetting agent; for prep. of emulsions of fatty acids, oils, org. solvs., polymers, resins, colors, etc.; solid; HLB 15.0; 100% conc.

Tegotens S. [Goldschmidt AG] POE glyceryl stearate; nonionic; solubilizer, coemulsifier, wetting agent; semisolid; HLB 16.4; 100% conc.

Tegotens S 2. [Goldschmidt AG] POE glyceryl stearate; nonionic; solubilizer, coemulsifier, wetting agent; semisolid; HLB 15.0; 100% conc.

Tegotens TO. [Goldschmidt AG] POE glyceryl trioleate; nonionic; solubilizer, emulsifier, wetting agent; emulsifier for prep. of self-emulsifying min. oil concs.; liq.; HLB 11.3; 100% conc.

Tego® Wet KL 245. [Tego] Dimethicone copolyol; substrate wetting and spreading additive for solv. and water-based systems, wood/furniture varnishes, industrial/household appliance paints, heat-resist. coatings; sp.gr. 1.04 g/cc; flash pt. > 65 C; surf. tens. 25 mN•m^{-1}; usage level: 0.01-1.0%; 100% act.

Tego® Wet ZFS 453. [Tego] Surfactant; nonionic; substrate wetting additive for solv.-based systems, building protection coatings, anticorrosive coatings, industrial paints, printing inks; sp.gr. 0.92 g/cc; flash pt. 26 C; usage level: 0.2-0.6%; 23 ± 3% act. in xylene.

Tego® Wet ZFS 454. [Tego] Low polymer methylpolysiloxane and nonionic surfactant; nonionic; substrate wetting and flow additive for solv.-based systems, building protection coatings, anticorrosive coatings, industrial paints, printing inks; sp.gr. 0.90 g/cc; flash pt. 26 C; usage level: 0.2-0.6%; 20.5 ± 2.5% act. in xylene.

Tekstim 8741. [Exxon/Tomah] Nonionic; surfactant, self-demulsifying detergent for truckwash applics.; hazy liq.; sp.gr. 0.99; 100% act.

Tek-Wet 951. [Van Waters & Rogers] PEG; nonionic; leather wetting and emulsifying agent; dispersant; sol'n.; biodeg.

Tek-Wet 955. [Van Waters & Rogers] PEG nonylphenol; nonionic; leather wetting and emulsifying; dispersant; sol'n.

Tembind A 002. [Temfibre] Ammonium lignosulfonate; anionic; emulsifier and stabilizer for asphalt; liq.; 50% conc.

Temsperse S 001. [Temfibre] Sodium lignosulfonate; anionic; dyestuff dispersant, water reducer, slurry thinner; powd.; 95% conc.

Temsperse S 003. [Temfibre] Modified lignosulfonate; anionic; dyestuff dispersant; powd.; 95% conc.

Tenax 2010. [Westvaco] Tall oil fatty acid, maleated; CAS 68139-89-9; anionic; hydrotrope for nonionics in strongly alkaline systems; intermediate for alkali metal salts, amine salts (used as corrosion inhibitors); biodeg.; liq.; nontoxic; 100% conc.

Tensagex DLM 627. [ICI PLC] Sodium trideceth sulfate; anionic; for liq. detergent blends, shampoos, bubble baths, liq. soaps; liq.; 26-28% conc.

Tensagex DLM 670. [ICI Am.; ICI PLC] Sodium trideceth sulfate; anionic; for liq. detergent blends, shampoos, bubble baths, liq. soaps; liq.; 68-72% conc.

Tensagex DLM 927. [ICI PLC] Sodium trideceth sulfate; anionic; for liq. detergent blends, shampoos, bubble baths, liq. soaps; liq.; 26-28% conc.

Tensagex DLM 970. [ICI Am.; ICI PLC] Sodium trideceth sulfate; anionic; for liq. detergent blends, shampoos, bubble baths, liq. soaps; liq.; 68-72% conc.

Tensagex DLS 670. [ICI Am.; ICI PLC] Sodium trideceth sulfate; anionic; for liq. detergent blends, shampoos, bubble baths, liq. soaps; liq.; 68-72% conc.

Tensagex DLS 970. [ICI Am.; ICI PLC] Sodium trideceth sulfate; anionic; for liq. detergent blends, shampoos, bubble baths, liq. soaps; liq.; 68-72% conc.

Tensagex EOC 628. [ICI PLC] Sodium coconut fatty alcohol ethoxylate sulfate; CAS 68585-34-2; anionic; for liq. detergent blends, shampoos, bubble baths, liq. soaps; liq.; 26-28% conc.

Tensagex EOC 670. [ICI PLC] Sodium coconut fatty alcohol ethoxylate sulfate; CAS 68585-34-2; anionic; for liq. detergent blends, shampoos, bubble baths, liq. soaps; liq.; 68-72% conc.

Tensianol 399 ISL. [ICI PLC] Blend; anionic; raw material for mfg. of alkali-free soap bars; gran.; 36-38% conc.

Tensianol 399 KS1. [ICI PLC] Blend; anionic; raw material for mfg. of alkali-free soap bars; gran.; 46-48% conc.

Tensianol 399 N1. [ICI PLC] Blend; anionic; raw material for mfg. of alkali-free soap bars; gran.; 39-41% conc.

Tensianol 399 SCI-L. [ICI PLC] Blend; anionic; raw material for mfg. of alkali-free soap bars; gran.; 32-34% conc.

Tensiofix 20200. [OmniChem NV] Surfactant; nonionic; special emulsifier for Dimethoate emulsifiable conc.; wax; 100% conc.

Tensiofix 35300. [OmniChem NV] Surfactant; nonionic; surfactant for agrochem. formulations, mainly for pesticide suspension concs.; wax; 100% conc.

Tensiofix 35600. [OmniChem NV] Surfactant; nonionic; surfactant for agrochem. formulations, mainly for pesticide suspension concs.; wax; 100% conc.

Tensiofix AS. [OmniChem NV] Surfactant blend; anionic/nonionic; general purpose emulsifier for pesticide emulsifiable concs.; liq.; 75% conc.

Tensiofix B7416. [OmniChem NV] Surfactant blend; anionic/nonionic; general purpose emulsifier for pesticide emulsifiable concs.; liq.; 91% conc.

Tensiofix B7438. [OmniChem NV] Surfactant blend; anionic/nonionic; general purpose emulsifier for pesticide emulsifiable concs.; liq.; 75% conc.

Tensiofix B7453. [OmniChem NV] Surfactant blend; anionic/nonionic; general purpose emulsifier for pesticide emulsifiable concs.; liq.; 70% conc.

Tensiofix B7504. [OmniChem NV] Surfactant blend; anionic/nonionic; special emulsifier for Dicofol emulsifiable concs.; wax; 90% conc.

Tensiofix B8426. [OmniChem NV] Surfactant blend; anionic/nonionic; emulsifier for pesticide microemulsions and emulsifiable concs.; liq.; 76% conc.

Tensiofix B8427. [OmniChem NV] Surfactant blend; anionic/nonionic; emulsifier for pesticide microemulsions and emulsifiable concs.; liq.; 76% conc.

Tensiofix BC222. [OmniChem NV] Surfactant blend; anionic/nonionic; wetting/dispersing agent for pesticide wettable powds.; powd.; 60% conc.

Tensiofix BCZ. [OmniChem NV] Alcohol sulfate; wetting/dispersing agent for pesticide wettable powds.; powd.; 90% conc.

Tensiofix BS. [OmniChem NV] Surfactant blend; anionic/nonionic; general purpose emulsifier for pesticide emulsifiable concs.; liq.; 75% conc.

Tensiofix CG11. [OmniChem NV] Surfactant blend; anionic/nonionic; wetting/dispersing agent for pesticide suspension concs.; liq.; 75% conc.

Tensiofix CG21. [OmniChem NV] Surfactant blend; anionic/nonionic; wetting/dispersing agent for pesticide suspension concs.; liq.; 100% conc.

Tensiofix CS. [OmniChem NV] Surfactant blend; anionic/nonionic; general purpose emulsifier for pesticide emulsifiable concs.; wax; 75% conc.

Tensiofix D40 Bio-Act. [OmniChem NV] Surfactant blend; cationic/nonionic; adjuvant for agrochemical formulations; liq.; 78% conc.

Tensiofix EDS. [OmniChem NV] Surfactant blend; anionic/nonionic; emulsifier for agric. oil formulations; liq.; 77% conc.

Tensiofix EDS3. [OmniChem NV] Surfactant blend; anionic/nonionic; emulsifier for agric. oil formulations; liq.; 80% conc.

Tensiofix LX Special. [OmniChem NV] Lignosulfonate; anionic; wetting/dispersing agent for pesticide wettable powds.; powd.; 90% conc.

Tensiofix PO120, PO132. [OmniChem NV] Surfactant; nonionic; emulsifier for agric. oil formulations; liq.; 100% conc.

Tensiofix XN6, XN10. [OmniChem NV] Phosphate ester of nonionic surfactant; wetting/dispersing agent for pesticide suspension concs.; emulsifier for pesticide emulsifiable concs.; liq.; 100% conc.

Tensopol A 79. [ICI PLC] Sodium fatty alcohol sulfate; CAS 73296-89-6; anionic; surfactant for toiletries, emulsion polymerization, pigment dispersion; needles; 89-91% conc.

Tensopol A 795. [ICI PLC] Sodium fatty alcohol sulfate; CAS 73296-89-6; anionic; surfactant for toiletries, emulsion polymerization, pigment dispersion; needles; 93-95% conc.

Tensopol ACL 79. [ICI PLC] Sodium fatty alcohol sulfate; CAS 68955-19-1; anionic; surfactant for toiletries, emulsion polymerization, pigment dispersion; needles; 89-91% conc.

Tensopol PCL 94. [ICI PLC] Sodium fatty alcohol sulfate; CAS 73296-89-6; anionic; used in detergents; powd.; 93-95% conc.

Tensopol S 30 LS. [ICI PLC] Sodium fatty alcohol sulfate; CAS 73296-89-6; anionic; used in latex foam, emulsion polymerization; paste; 29-31% conc.

Tephal Grunau AB Conc. [Chem-Y BV] Protein fatty acid condensate; anionic; washing, milling and wetting agent for textile prods.; paste; 65% conc.

Tephal Grunau FL Conc. [Chem-Y BV] Protein fatty acid condensate; anionic; washing, milling and wetting agent; liq.; 65% conc.

Tequat RO. [Tessilchimica] Acyl amido-amine; cationic; car washing; paste; 100% conc.

Terg-A-Zyme®. [Alconox] Alkylaryl sulfonate, lauryl alcohol sulfate, phosphate, carbonate, and protease enzyme; anionic; biodeg. detergent, wetting agent, sequestering and synergistic agents; used in hospitals, laboratories, dairies; cleaning agent in dairy and pollution processing; wh. powd. with cream and brown specks, odorless; sol. in water; flash pt. none; pH 9.0–9.5 (1%); surf. tens. 32 dyne/cm (1%); toxicology: nontoxic orally; mild to moderate eye irritant if not rinsed; powd. potential irritant by inhalation; 100% active.

Tergenol 1122. [Hart Chem. Ltd.] Blend; anionic; heavy-duty detergent for heavily soiled woven and knitted fabrics; stable to acids, alkalies, and oxidizing agents; yel. liq.; 70% act.

Tergenol 3964. [Hart Chem. Ltd.] Blend; nonionic; detergent, emulsifier, dispersant for laundry powd. compds.; liq.; 90% conc.

Tergenol G. [Hart Prods. Corp.] Sodium oleyl methyl taurate; detergent, textile scouring and dye leveling agent; gel.

Tergenol S Liq. [Hart Prods. Corp.] Sodium oleyl methyl taurate; anionic; detergent, textile scouring and dye leveling agent; liq.

Tergenol Slurry. [Hart Prods. Corp.] Sodium oleyl methyl taurate; anionic; detergent, textile scouring and dye leveling agent; paste; 40% conc.

Tergitol® 15-S-3. [Union Carbide] C11-15 pareth-3; CAS 68131-40-8; nonionic; biodeg. detergent, emulsifier, wetter, defoamer for aq. systems, intermediate used in textiles, solv. cleaners, drycleaning, metalworking fluids, water treatment, oilfield chems., pulp/paper deinking, latex emulsions, plastics antistat, agric.; FDA, EPA compliance; clear liq.; sol. in oil, chlorinated solvs., most org. solvs.; m.w. 336; sp.gr. 0.930; dens. 7.74 lb/gal; visc. 26 cs; HLB 8.3; hyd. no. 167; pour pt. -46 C; cloud pt. < 0 (1% aq.); flash pt. (PMCC) 174 C; pH 4.0 (1% in aq. IPA); 100% act.

Tergitol® 15-S-5. [Union Carbide] C11-15 pareth-5; CAS 68131-40-8; nonionic; biodeg. detergent, emulsifier, wetting agent, intermediate for household and industrial detergents, textiles, drycleaning, water treatment, metalworking/cleaning, oilfield chems., pulp/paper deinking, agric.; latex emulsion stabilizer; plastics antistat; defoamer for aq. systems; fuel de-icing; FDA, EPA compliance; clear liq.; sol. in oils, chlorinated solvs., most org. solvs.;

m.w. 415; sp.gr. 0.965; dens. 8.03 lb/gal; visc. 35 cP; HLB 10.6; hyd. no. 135; cloud pt. < 0 C (1% aq.); flash pt. (PMCC) 178 C; pour pt. –24 C; pH 6–8 (1%); 100% act.

Tergitol® 15-S-7. [Union Carbide] C11-15 pareth-7; CAS 68131-40-8; nonionic; biodeg. detergent, emulsifier, wetting agent, dye leveling agent, coupler, dispersant, stabilizer for household and industrial detergents, paper, textiles, leather, paints, agric., metal cleaners, oilfield chems., water treatment, electronics; intermediate; FDA, EPA compliance; clear liq.; sol. in water, chlorinated solvs., most org. solvs.; m.w. 515; sp.gr. 0.992; dens. 8.26 lb/gal; visc. 51 cP; HLB 12.4; hyd. no. 109; pour pt. 2 C; cloud pt. 37 C (1% aq.); flash pt. (PMCC) 187 C; pH 6.8 (1% aq.); surf. tens. 28 dynes/cm (0.1% aq.); Draves wetting 7 s (0.1%); Ross Miles foam 125 mm (1% aq.); 100% act.

Tergitol® 15-S-9. [Union Carbide] C11-15 pareth-9; CAS 68131-40-8; nonionic; biodeg. detergent, emulsifier, wetting agent, dye leveling agent, coupling agent for household and industrial detergents, paper, textiles, leather, paints, agric., metal cleaning, oilfield chems., water treatment, electomics; intermediate; FDA, EPA compliance; clear liq.; sol. in water, chlorinated solvs., most polar org. solvs.; m.w. 584; sp.gr. 1.006; dens. 8.37 lb/gal; visc. 60 cP; HLB 13.3; hyd. no. 96; pour pt. 9 C; cloud pt. 60 C (1% aq.); flash pt. (PMCC) 193 C; pH 7.1 (1%); surf. tens. 30 dynes/cm (0.1% aq.); Draves wetting 8 s (0.1%); Ross-Miles foam 95 mm (0.1% aq.); 100% act.

Tergitol® 15-S-12. [Union Carbide] C11-15 pareth-12; CAS 68131-40-8; nonionic; biodeg. detergent, wetting agent, emulsifier, coupler, dispersant, dye leveling agent, stabilizer for household/industrial detergents, paper, textiles, fiber lubricants, paints, agric., metal cleaners, oilfield chems., water treatment; electronics, leather, elevated temp. and high electrolyte applics.; FDA, EPA compliance; clear liq.; sol. in water, chlorinated solvs., most polar org. solvs.; m.w. 738; sp.gr. 1.020; dens. 8.49 lb/gal; visc. 85 cP; HLB 14.7; hyd. no. 76; pour pt. 20 C; cloud pt. 88 C (1% aq.); flash pt. (PMCC) > 227 C; pH 6.2 (1%); surf. tens. 31 dynes/cm (0.1% aq.); Draves wetting 24 s (0.1%); Ross-Miles foam 125 mm (1%, initial); 100% act.

Tergitol® 15-S-15. [Union Carbide] C11-15 pareth-15; CAS 68131-40-8; nonionic; biodeg. emulsifier, detergent, wetting agent for use at elevated temps., in presence of electrolytes; for alkaline industrial cleaners/degreasers, textile scouring, dye carriers, demulsification of petrol. oil emulsions, oilfield chems.; pulp/paper processing aid; stabilizer for syn. latexes; FDA, EPA compliance; wh. waxy semisolid; sol. in water, chlorinates solvs, polar org. solvs.; m.w. 877; sp.gr. 1.009 (55/20 C); dens. 8.40 (55 C); visc. 43 cP (50 C); HLB 15.6; hyd. no. 64; pour pt. 28 C; cloud pt. > 100 C (1% aq.); flash pt. (PMCC) 246 C; pH 6.4 (1% aq.); surf. tens. 34 dynes/cm (0.1% aq.); Ross-Miles foam 130 mm (0.1% aq., initial); 100% act.

Tergitol® 15-S-20. [Union Carbide] C11-15 pareth-20; CAS 68131-40-8; nonionic; biodeg. emulsifier, detergent, wetting agent for use at elevated temps., in presence of electrolytes; for alkaline industrial/household cleaners/degreasers, textile scouring, dye carriers, demulsification of petrol. oil emulsions, oilfield chems.; stabilizer for syn. latexes; glass mold release agent in silicone emulsions; FDA, EPA compliance; wh. waxy solid; sol. in water, chlorinated solvs., polar org. solvs.; m.w. 1079; sp.gr. 1.041; dens. 8.66 lb/gal (40 C); visc. 49 cP (50 C); HLB 16.4; hyd. no. 52; pour pt. 32 C; cloud pt. > 100 C (1% aq.); flash pt. (PMCC) 246 C; pH 6.6 (1% aq.); surf. tens. 35 dynes/cm (0.1% aq.); 100% act.

Tergitol® 15-S-30. [Union Carbide] C11-15 pareth-30; CAS 68131-40-8; nonionic; emulsifier, detergent, wetting agent for use at elevated temps., in presence of strong electrolytes; for alkaline industrial/household cleaners/degreasers, textile scouring, dye carriers, vinyl acetate and acrylate polymerization; stabilizer for syn. latexes; demulsifier for petrol. oil emulsions; FDA, EPA compliance; wh. waxy solid; sol. in water, chlorinated solvs., polar org. solvs.; m.w. 1558; sp.gr. 1.055; dens. 8.78 lb/gal (55 C); visc. 92 cP (50 C); HLB 17.5; hyd. no. 36; pour pt. 39 C; cloud pt. > 100 C (1% aq.); flash pt. (PMCC) 249 C; pH 6.5 (1% aq.); surf. tens. 39 dynes/cm (0.1% aq.); 100% act.

Tergitol® 15-S-40. [Union Carbide] C11-15 pareth-40; CAS 68131-40-8; nonionic; emulsifier, detergent, wetting agent for use at elevated temps. in presence of strong electrolytes; for alkaline industrial/household cleaners/degreasers, textile scouring, dye carriers, vinyl acetate and acrylate polymerization; stabilizer for syn. latexes; demulsifier for petrol. oil emulsions; FDA, EPA compliance; wh. waxy solid; sol. in water, chlorinated solvs., polar org. solvs.; m.w. 2004; sp.gr. 1.061; dens. 8.83 lb/gal (55 C); visc. 166 cP (50 C); HLB 18.0; hyd. no. 28; pour pt. 44 C; cloud pt. > 100 C (1% aq.); flash pt. (PMCC) 252 C; pH 7.0 (1% aq.); surf. tens. 42 dynes/cm (0.1% aq.); 100% act.

Tergitol® 24-L-3. [Union Carbide] Primary alcohol PEG ether; nonionic; detergent, emulsifier, surfactant, defoamer, intermediate; liq.; oil-sol.; HLB 7.7; biodeg.; 100% conc.

Tergitol® 24-L-45. [Union Carbide] C12-14 pareth-6 (6.3 EO); CAS 68439-50-9; nonionic; biodeg. surfactant, detergent, wetting/spreading agent, emulsifier, foaming agent, intermediate, dispersant for household and industrial cleaners, textile wet processing, paper processing, agric. formulations; FDA, EPA compoliance; APHA 20 color, low, mild char. odor; m.w. 479; misc. with propylene glycol; sp.gr. 0.987; visc. 28 cSt (37.8 C); HLB 11.6; hyd. no. 117; pour pt. 12 C; cloud pt. 45 C (1% aq.); flash pt. (PMCC) 182 C; pH 6.0 (1% aq.).; surf. tens. 29 dynes/cm (1% aq.).

Tergitol® 24-L-50. [Union Carbide] Primary alcohol PEG ether; nonionic; biodeg. detergent, wetting agent, emulsifier, surfactant used in household and industrial detergents; liq.; HLB 12.4; 100% conc.

Tergitol® 24-L-60. [Union Carbide] C12-14 pareth-7 (7.2 EO); nonionic; biodeg. surfactant, detergent, wetting/spreading agent, emulsifier, foaming agent, intermediate, dispersant for household and industrial cleaners, textile wet processing, paper processing, agric. formulations; FDA, EPA compliance; APHA 20 color, mild, low, char. odor; m.w. 519; misc. with propylene glycol, toluene; sp.gr. 0.99; visc. 31 cSt (37.8 C); HLB 12.2; hyd. no. 108; pour pt. 17 C; cloud pt. 60 C (1% aq.); flash pt. (PMCC) 174 C; pH 6.0 (1% aq.).; surf. tens. 29 dynes/cm (1% aq.).

Tergitol® 24-L-60N. [Union Carbide] C12-14 pareth-7; nonionic; biodeg. surfactant, detergent, wetting/spreading agent, emulsifier, foaming agent, intermediate, dispersant for household and industrial cleaners, textile wet processing, paper processing, agric. formulations; FDA, EPA compliance; APHA 20 color, mild, low, char. odor; m.w. 510; misc. with propylene glycol, toluene; sp.gr. 0.99 (30/20 C); visc. 28 cSt (37.8 C); HLB 12.1; hyd. no. 110; pour pt. 16 C; cloud pt. 60 C (1% aq.); flash pt. (PMCC) 178 C; pH 6.0 (1% aq.); surf. tens. 31 dyne/cm (1% aq.).

Tergitol® 24-L-75. [Union Carbide] C12 and C14 primary alcohol ethoxylate (8.3 EO); nonionic; biodeg. surfactant, detergent, wetting/spreading agent, emulsifier, foaming agent, intermediate, dispersant for household and industrial cleaners, textile wet processing, paper processing, agric. formulations; APHA 20 hazy liq. to wh. semisolid; low, mild, char. odor; m.w. 567; misc. with ethyl acetate, propylene glycol; sp.gr. 0.99 (30/20 C); visc. 35 cSt (37.8 C); HLB 12.9; cloud pt. 75 C (1% aq.); flash pt. (PMCC) 168 C; pour pt. 21 C; surf. tens. 31 dyne/cm (1% aq.); pH 6.0 (1% aq.).

Tergitol® 24-L-92. [Union Carbide] C12 and C14 primary alcohol ethoxylate (10.6 EO); nonionic; biodeg. surfactant, detergent, wetting/spreading agent, emulsifier, foaming agent, intermediate, dispersant for household and industrial cleaners, textile wet processing, paper processing, agric. formulations; wh. solid, low, mild, char. odor; m.w. 668; misc. with propylene glycol, toluene; sp.gr. 1.00 (30 C); visc. 42 cSt (37.8 C); HLB 14.0; cloud pt. 92 C (1% aq.); flash pt. (PMCC) 185 C; pour pt. 29 C; surf. tens. 34 dyne/cm (1% aq.); pH 6.0 (1% aq.).

Tergitol® 24-L-98N. [Union Carbide] C12 and C14 primary alcohol ethoxylate (11.3 EO); nonionic; biodeg. surfactant, detergent, wetting/spreading agent, emulsifier, foaming agent, intermediate, dispersant for household and industrial cleaners, textile wet processing, paper processing, agric. formulations; APHA 20 solid; low, mild, char. odor; m.w. 701; misc. with propylene glycol, toluene; sp.gr. 1.01 (30 C); visc. 77 cSt (37.8 C); HLB 14.2; cloud pt. 98 C (1% aq.); flash pt. (PMCC) 204 C; pour pt. 28 C; surf. tens. 36 dyne/cm (1% aq.); pH 6.0 (1% aq.).

Tergitol® 25-L-3. [Union Carbide] C12-15 pareth-3; nonionic; detergent, wetting agent, emulsifier, textile surfactant and lubricant; Pt-Co 50 max. liq.; m.w. 341; dens. 7.73 lb/gal; sp.gr. 0.929; visc. 36 cs; HLB 7.7; cloud pt. < 0 C (1% aq.); flash pt. 260 F (COC); pour pt. 8 C; pH 5–7; biodeg.; 100% conc.

Tergitol® 25-L-5. [Union Carbide] C12-15 pareth-5; nonionic; detergent, wetting agent, emulsifier, textile surfactant and lubricant; Pt.Co 50 max. liq.; m.w. 440; dens. 7.94 lb/gal (30 C); sp.gr. 0.955; visc. 95 cs; HLB 10.4; cloud pt. < 0 C (1% aq.); flash pt. 325 F (COC); pour pt. 16 C; pH 5–7 (1%); biodeg.; 100% conc.

Tergitol® 25-L-7. [Union Carbide] C12-15 pareth-7; nonionic; detergent, wetting agent, emulsifier, textile surfactant and lubricant; also in desizing; Pt.Co 50 max. liq.; m.w. 550; dens. 8.19 lb/gal (30 C); sp.gr. 0.985; visc. 51 cps (38 C); HLB 12.4; cloud pt. 50 C (1% aq.); flash pt. 340 F (COC); pour pt. 23 C; surf. tens. 28.7 dynes/cm (0.1%); pH 5–7 (1%); biodeg.; 100% conc.

Tergitol® 25-L-9. [Union Carbide] C12-15 pareth-9; nonionic; detergent, wetting agent, emulsifier, textile surfactant and lubricant; also in desizing; Pt.Co 50 max. semisolid; m.w. 585; dens. 8.18 lb/gal (40 C); sp.gr. 0.991; visc. 33 cps (40 C); HLB 12.8; cloud pt. 60 C (1% aq.); flash pt. 285 F (COC); pour pt. 24 C; surf. tens. 29.0 dynes/cm (0.1%); pH 5–7 (1%); biodeg.; 100% conc.

Tergitol® 25-L-12. [Union Carbide] C12-15 pareth-12; nonionic; detergent, wetting agent, emulsifier, textile surfactant and lubricant; also in desizing; Pt.Co 100 max. solid; m.w. 730; dens. 8.37 lb/gal (40 C); sp.gr. 1.013; visc. 51 cs (38 C); HLB 14.2; cloud pt. 90 C (1% aq.); flash pt. 410 F (COC); pour pt. 30 C; surf. tens. 32.0 dynes/cm (0.1%); pH 5–7 (1%); biodeg.; 100% conc.

Tergitol® 26-L-1.6. [Union Carbide] C12-16 pareth-1.5; CAS 68551-12-02; nonionic; surfactant, detergent, wetting/spreading agent, emulsifier, foaming agent, intermediate, dispersant, lubricant for household and hard surf. cleaners, textile processing, pulp processing, agric. formulations; APHA 10 clear to slightly hazy liq.; mild char. odor; m.w. 264; misc. with ethyl acetate, ethanol, propylene glycol, IPA, toluene, hexane; sp.gr. 0.895; visc. 14 cSt (37.8 C); HLB 5.0; flash pt. (PMCC) 143 C; pour pt. 10 C; pH 6.5 (1% in 10:6 IPA/water).

Tergitol® 26-L-3. [Union Carbide] C12-16 pareth-3; nonionic; surfactant, emulsifier, intermediate for sulfation; used for prewash spotters, coning oil, hydrocarbon-based cleaners, agric.; as sulfated prod. in cosmetics, hand dishwash, lt. duty detergents; FDA and EPA compliance; APHA 10 clear to slightly hazy liq.; mild char. odor; misc. with ethyl acetate, ethanol, propylene glycol, IPA, toluene, hexane; m.w. 328; sp.gr. 0.917; visc. 19 cSt (37.8 C); HLB 8.0; hyd. no. 168-178; flash pt. (PMCC) 157 C; pour pt. 6 C; pH 6.9 (1% in 10:6 IPA/water).

Tergitol® 26-L-5. [Union Carbide] C12-16 pareth-5; nonionic; surfactant, detergent, wetting/spreading agent, emulsifier, foaming agent, intermediate, dispersant, lubricant for household and hard surf. cleaners, textile processing, pulp processing, agric. formulations; APHA 10 hazy liq.; mild char. odor; m.w. 419; misc. with propylene glycol, butyl Cellosolve; sp.gr. 0.969; visc. 26 cSt (37.8 C); HLB 10.5; flash pt. (PMCC) 177 C; pour pt. 8 C; pH 6.5 (1% in 10:6 IPA/water).

Tergitol® D-683. [Union Carbide] Alkoxylated alkylphenol; CAS 37251-69-7; nonionic; emulsifier for fiber finishing operations; dispersant for pigments in resins, plastics and for abrasives in hard surf. cleaners; improves wetting of oil-based materials in coatings and adhesives; clear liq.; sol. in water, chlorinated solvs., most org. solvs.; m.w. 1004; sp.gr. 1.019 (30/20 C); dens. 8.48 lb/gal (50 C); visc. 320 cP; hyd. no. 56; pour pt. -1 C; cloud pt. 21 C (1% aq.); flash pt. (PMCC) 210 C; pH 6.4 (1% aq.); surf. tens. 32 dynes/cm (0.1% aq.); Ross-Miles foam 38 mm (0.1%, initial); 100% act.

Tergitol® Min-Foam 1X. [Union Carbide] C11-15 alcohols reacted with EO and PO; CAS 68551-14-4; nonionic; biodeg. surfactant, foam depressant, wetting agent, detergent for household/industrial cleaners, drycleaning, textile processing, metal cleaning, circuit board cleaners, leather, paper deinking; clear liq., mild char. odor; sol. in water, xylene, butyl Cellosolve, anhyd. IPA, perchloroethylene, oils; m.w. 645; sp.gr. 0.995; dens. 8.28 lb/gal; visc. 57 cP; HLB 11.4; hyd. no. 87; solid. pt. –42 C; pour pt. -38 C; cloud pt. 40 C (1% aq.); flash pt. (PMCC) 111 C; pH 6.7 (1% aq.); surf. tens. 29 dynes/cm (0.1% aq.); Draves wetting 6 s (0.1% aq.); Ross-Miles foam 120 mm (0.1% aq., initial); 100% act.

Tergitol® Min-Foam 2X. [Union Carbide] C11-15

alcohols reacted with EO and PO; CAS 68551-14-4; nonionic; biodeg. surfactant, antifoam, detergent for household/industrial detergents, machine dishwashing, rinse aid, acid metal cleaners, textile dyeing/processing/scouring, circuit board cleaners, paper deinking, water treatment; completely rinseable; clear liq.; sol. in xylene, butyl Cellosolve, anhyd. IPA, perchloroethylene, oils; m.w. 630; sp.gr. 0.978; dens. 8.14 lb/gal; visc. 49 cP; HLB 13.8; hyd. no. 89; solid. pt. –47 C; pour pt. -42 C; cloud pt. 20 C (1% aq.); flash pt. (PMCC) 155 C; pH 6.6 (1% aq.); surf. tens. 29 dynes/cm (0.1% aq.); Ross-Miles foam 45 mm (0.1% aq., initial); 100% act.

Tergitol® NP-4. [Union Carbide] Nonoxynol-4; CAS 9016-45-9; nonionic; detergent, wetting agent, emulsifier used in dry-cleaning detergents, household and industrial cleaners; colorless clear liq.; char. odor; oil-sol.; m.w. 396; sol. in kerosene, aliphatic hydrocarbons, butyl acetate, butyl Cellosolve, corn and wh. min. oil, diesel fuel, anhyd. IPA, Stod., toluene; sp.gr. 1.031; dens. 8.57 lb/gal; visc. 445 cs; HLB 8.9; b.p. > 300 C; flash pt. 480 F (COC); cloud pt. 0 C; 100% act.

Tergitol® NP-5. [Union Carbide] Nonoxynol-5; CAS 9016-45-9; nonionic; detergent, wetting agent, emulsifier for household and industrial applics.; liq.; HLB 10.0; 100% conc.

Tergitol® NP-6. [Union Carbide] Nonoxynol-6; CAS 9016-45-9; nonionic; detergent, wetting agent, emulsifier for household and industrial applics.; colorless clear liq.; char. odor; m.w. 484; sol. see Tergitol NP-4; sp.gr. 1.055; dens. 8.67 lb/gal; visc. 373 cs; HLB 10.9; b.p. > 300 C; flash pt. 360 F (PMCC); cloud pt. 0 C (0.5% aq.); pH 5–7.5; 100% act.

Tergitol® NP-7. [Union Carbide] Nonoxynol-7; CAS 9016-45-9; nonionic; emulsifier, wetting agent, dispersant used in agric. toxicant formulations, cleaners and santizers; cleaning booster; Pt.Co 75 max. clear liq.; mild char. odor; m.w. 528; sol. in butyl acetate, butyl Cellosolve, corn oil, anhyd. IPA, toluene, water; sp.gr. 1.055; dens. 8.71 lb/gal; visc. 338 cs; HLB 11.7; solid. pt. –6 C; b.p. > 250 C; flash pt. 525 F (COC); cloud pt. 20 C; pH 5–8 (10% aq.); 100% act.

Tergitol® NP-8. [Union Carbide] Nonoxynol-8; CAS 9016-45-9; nonionic; removal or emulsification of grease or oils; wetting agent, dispersant, emulsifier; Pt.Co 50 max. clear liq.; mild char. odor; m.w. 572; sol. in butyl acetate, butyl Cellosolve, ethylene glycol, anhyd. IPA, toluene, water; sp.gr. 1.056; visc. 325; HLB 12.3; solid. pt. –3 C; b.p. 250 C; flash pt. 400 F (PMCC); cloud pt. 43 C; pH 5.0–8.0 (10% aq.); surf. tens. 30 dynes/cm (0.1% aq.); 100% act.

Tergitol® NP-9. [Union Carbide] Nonoxynol-9; CAS 9016-45-9; nonionic; detergent, wetting agent, dispersant, emulsifier; yel. clear liq.; mild char. odor; m.w. 616; sol. see Tergitol NP-8; sp.gr. 1.057; dens. 8.80 lb/gal; visc. 318 cs; HLB 12.9; solid. pt. 0 C; b.p. > 250 C; flash pt. 540 F (COC); cloud pt. 54 C; pH 5–8 (10% aq.); surf. tens. 30 dynes/cm (0.1% aq.); 100% act.

Tergitol® NP-10. [Union Carbide] Nonoxynol-10; CAS 9016-45-9; nonionic; detergent, wetting agent, emulsifier for household and industrial applics.; yel. clear liq.; mild char. odor; m.w. 682; sol. see Tergitol NP-8; sp.gr. 1.062; dens. 8.84 lb/gal; visc. 327 cs; HLB 13.6; solid. pt. 7 C; b.p. > 250 C; flash pt. 500 F (COC); cloud pt. 63 C (0.5% aq.); 100% act.

Tergitol® NP-13. [Union Carbide] Nonoxynol-13; CAS 9016-45-9; nonionic; detergent, wetting agent, emulsifier for household and industrial applics.; yel. clear liq.; mild char. odor; m.w. 792; sol. see Tergitol NP-8; sp.gr. 1.071; dens. 8.90 lb/gal; visc. 410 cs; HLB 14.4; solid. pt. 16 C; b.p. > 250 C; cloud pt. 83 C (0.5% aq.); pH 5.0–8.0 (10% aq.); surf. tens. 34 dynes/cm (0.1% aq.); 100% act.

Tergitol® NP-14. [Union Carbide] Nonoxynol-4; CAS 9016-45-9; nonionic; detergent, wetting agent, emulsifier, intermediate; used in textile applics.; Pt.Co 100 max. liq.; m.w. 396; dens. 8.57 lb/gal; sp.gr. 1.031; visc. 448 cs; HLB 8.9; cloud pt. < 0 C (1% aq.); flash pt. 480 F (COC); solid. pt. –40 C; pH 5–8 (10%); 100% conc.

Tergitol® NP-15. [Union Carbide] Nonoxynol-15; CAS 9016-45-9; nonionic; detergent, wetting agent, emulsifier for household and industrial applics.; liq.; HLB 15.0; 100% conc.

Tergitol® NP-27. [Union Carbide] Nonoxynol-7; CAS 9016-45-9; nonionic; detergent, wetting agent, emulsifier, intermediate; used in textile applics.; Pt.Co 75 max. liq.; m.w. 528; dens. 8.71 lb/gal; sp.gr. 1.038; visc. 338 cs; HLB 11.7; cloud pt. 22 C (1% aq.); flash pt. 525 F (COC); solid. pt. –6 C; surf. tens. 30.4 dynes/cm (0.1%); pH 5–8 (10%); 100% conc.

Tergitol® NP-40 (70% Aq.). [Union Carbide] Nonoxynol-40; CAS 9016-45-9; nonionic; surfactant for emulsion polymerization; leveling agent; clear liq.; mild odor; water-sol.; sp.gr. 1.104; dens. 9.19 lb/gal; visc. 533 cs; solid. pt. –2 C; 70% act.

Tergitol® NP-40. [Union Carbide] Nonoxynol-40; CAS 9016-45-9; nonionic; detergent, wetting agent, asphalt emulsifier, emulsion polymerization, leveling agent; wh. solid; mild char. odor; m.w. 1980; water-sol.; sp.gr. 1.080; dens. 8.97 lb/gal (55 C); HLB 17.8; m.p. 48 C; b.p. > 250 C; flash pt. 525 F(COC); cloud pt. 100 (0.5%); pH 4.0–8.0 (10% aq.); surf. tens. 45 dynes/cm (0.1% aq.); 100% act.

Tergitol® NP-44. [Union Carbide] Nonoxynol-40; CAS 9016-45-9; nonionic; detergent, wetting agent, emulsifier, intermediate; used in textile applics.; Gardner 5 max. solid; m.w. 1980; water-sol.; dens. 8.97 lb/gal (55 C); sp.gr. 1.082; HLB 17.8; cloud pt. > 100 C (1% aq.); flash pt. 525 F (COC); solid. pt. 48 C; pH 4–8 (10%); 100% conc.

Tergitol® NP-55, 70% Aq. [Union Carbide] Nonoxynol-55; CAS 9016-45-9; nonionic; surfactant for emulsion polymerization; leveling agent; solid; 70% conc.

Tergitol® NP-70, 70% Aq. [Union Carbide] Nonoxynol-70; CAS 9016-45-9; nonionic; surfactant for emulsion polymerization; leveling agent; solid; 70% conc.

Tergitol® NPX. [Union Carbide] Nonoxynol-10 (10.5 EO); CAS 9016-45-9; nonionic; detergent, wetting agent, emulsifier, intermediate; used in textile applics.; Pt.Co. 50 max. liq.; m.w. 682; dens. 8.84 lb/gal; sp.gr. 1.062; visc. 318 cs; HLB 13.6; cloud pt. 63 C (0.5% aq.); flash pt. 500 F (COC); solid. pt. 7 C; pH 5–8 (10%); 100% conc.

Tergitol® TMN-3. [Union Carbide] Isolaureth-3; CAS 60828-78-6; nonionic; emulsifier, wetting agent, coupler, penetrant, leveling agent for textile processing, lubricants, water treatment, solv. cleaners/degreasers, metalworking fluids, drycleaning, oilfield chems., pulp/paper deinking; defoamer for aq. systems; intermediate for anionic surfactants used in household, industrial, and personal care prods.; FDA, EPA compliance; clear liq.; sol. in chlorinated solvs., most polar and nonpolar solvs, oils; m.w. 312; sp.gr. 0.930; dens. 7.74 lb/gal; visc. 19 cP; solid. pt. –40 C; HLB 8.1; hyd. no. 180; pour pt. -49 C; cloud pt. < 0 C (1% aq.); flash pt. (PMCC) 130 C; pH 7.2 (10% in aq. IPA); 100% act.

Tergitol® TMN-6. [Union Carbide] Isolaureth-6; CAS 60828-78-6; nonionic; wetting agent, penetrant, spreading agent, spreading agent, rewetting agent, detergent, dispersant, leveling agent for textile wet processing, fiber lubricants, hard surf. cleaners, agric., water treatment, circuit board cleaners; leather; FDA, EPA compliance; clear liq.; sol. in water, most polar and nonpolar solvs., oils; m.w. 543; dens. 8.40 lb/gal; sp.gr. 1.009; visc. 71 cP; solid. pt. –22 C; HLB 11.7; hyd. no. 103; pour pt. -29 C; cloud pt. 37 C (0.5% aq.); flash pt. (PMCC) none; pH 5.4 (10%); surf. tens. 26 dynes/cm (0.1% aq.); Draves wetting 3 s (0.1% aq.); Ross-Miles foam 110 mm (0.1% aq., initial); 90% act. in water.

Tergitol® TMN-10. [Union Carbide] Isolaureth-10; CAS 60828-78-6; nonionic; emulsifier, wetting/spreading agent, penetrant, detergent, leveling agent for high temp. applics., textile wet processing, fiber lubricants, agric., paper deinking, metal cleaners, acid cleaners, water treatment, leather; circuit board cleaners; FDA, EPA compliance; clear liq.; sol. in water, most polar solvs.; m.w. 683; sp.gr. 1.044; dens. 8.69 lb/gal; visc. 96 cP; HLB 14.1; hyd. no. 82; pour pt. -16 C; cloud pt. 77 C (1% aq.); solid. pt. –7.5 C; flash pt. (PMCC) none; pH 6.4 (10% aq.); surf. tens. 27 dynes/cm (0.1% aq.); Draves wetting 9 s (0.1% aq.); Ross-Miles foam 117 mm (0.1% aq., initial); 90% act. in water.

Tergitol® TP-9. [Union Carbide] Nonoxynol-9; CAS 9016-45-9; nonionic; detergent, wetting agent, emulsifier used in dry-cleaning detergents, household and industrial cleaners; Pt-Co 50 max. liq.; m.w. 616; dens. 8.80 lb/gal; sp.gr. 1.057; visc. 318 cs; HLB 12.9; cloud pt. 54 C (0.5% aq.); flash pt. 540 F (COC); solid. pt. 0 C; surf. tens. 31.2 dynes/cm (0.1%); pH 5–8 (10%); 100% act.

Tergitol® XD. [Union Carbide] PPG-24-buteth-27; CAS 9038-95-3; nonionic; emulsifier, dispersant, stabilizer for agric. insecticides/herbicides, latex polymerization, iodophor mfg. for germicidal cleaning, latex paints, dye pigments, leather; emulsifier for silicone oils, diacyl peroxides; FDA, EPA compliance; wh. waxy solid; sol. in chlorinated solvs., many polar org. solvs.; m.w. 3117; sp.gr. 1.041; dens. 8.66 lb/gal (40 C); visc. 251 cP (50 C); solid. pt. 33 C; hyd. no. 18; pour pt. 35 C; cloud pt. 76 C (1% aq.); flash pt. (PMCC) 208 C; pH 6.5 (20% aq.); surf. tens. 38 dynes/cm (0.1% aq.); Ross-Miles foam 76 mm (0.1%, initial); 100% act.

Tergitol® XH. [Union Carbide] EO/PO copolymer; CAS 9038-95-3; nonionic; emulsifier, dispersant for agric., latex polymerization, iodophor mfg. for germicidal cleaning, latex paints, dye pigments, leather finishes, toilet bowl cleaners; emulsifier for silicone oils, diacyl peroxides; FDA, EPA compliance; wh. waxy solid; sol. in chlorinated solvs., many polar org. solvs.; m.w. 3740; sp.gr. 1.048; dens. 8.72 lb/gal (50 C); visc. 319 cP (50 C); solid. pt. 41 C; hyd. no. 15; pour pt. 44 C; cloud pt. 99 C (1% aq.); flash pt. (PMCC) 144 C; pH 5.4 (10% aq.); surf. tens. 39 dynes/cm (0.1% aq.); Ross-Miles foam 58 mm (0.1% aq., initial); 100% act.

Tergitol® XJ. [Union Carbide] EO/PO copolymer; CAS 9038-95-3; nonionic; emulsifier, dispersant, stabilizer for latex polymerization, agric., latex paints, iodophors, dye pigments, leather finishes; emulsifier for silicone oils; FDA, EPA compliance; wh. waxy solid; sol. in chlorinated solvs., many polar org. solvs.; m.w. 2550; sp.gr. 1.023; dens. 8.51 lb/gal (45 C); visc. 149 cP (50 C); hyd. no. 22; pour pt. 26 C; cloud pt. 50 C (1% aq.); flash pt. (PMCC) 216 C; pH 6.5 (1% aq.); surf. tens. 36 dynes/cm (0.1% aq.); Ross-Miles foam 50 mm (0.1% aq., initial); 100% act.

Teric 9A2. [ICI Australia] C9-11 ethoxylate (2 EO); nonionic; surfactant for sulfation feedstock, polishes and waxes, drycleaning, solv. cleaners/degreasers; APHA 100 color; sol. in aromatics, ethyl acetate, glycol ethers, trichlorethylene, ethanol, kerosene, min. oil, olein; sp.gr. 0.90; visc. 25 cps; HLB 8.7; hyd. no. 220-232; pour pt. < 0 C; pH 6-8; 100% act.

Teric 9A5. [ICI Australia] C9-11 ethoxylate (5 EO); nonionic; emulsifier for agric.; chem. mfg., window cleaners, solv. cleaners/degreasers; APHA 100 color; sol. in water, aromatics, ethyl acetate, glycol ethers, trichlorethylene, ethanol, olein, ethylene glycol; sp.gr. 0.976; visc. 52 cps; HLB 11.6; pour pt. < 0 C; cloud pt. 33-38 C; pH 6-8; surf. tens. 20.1 dynes/cm (0.1%); 100% act.

Teric 9A6. [ICI Australia] C9-11 ethoxylate (6 EO); nonionic; emulsifier for agric.; filter cake dewatering; dust supression in coal industry; domestic, laundry, hard surf. and window cleaners; leather; paper deinking; skin care prods.; textile wetting, carbonizing, leveling, scouring; APHA 100 color; sol. in water, aromatics, ethyl acetate, glycol ethers, trichlorethylene, ethanol, ethylene glycol; sp.gr. 0.987; visc. 38 cps; HLB 12.4; pour pt. < 0 C; cloud pt. 50-55 C; pH 6-8; surf. tens. 24.0 dynes/cm (0.1%); 100% act.

Teric 9A8. [ICI Australia] C9-11 ethoxylate (8 EO); nonionic; surfactant for domestic, hard surf., laundry cleaners; paper deinking; wetting agent for textiles; APHA 100 color; sol. in water, aromatics, ethyl acetate, glycol ethers, trichlorethylene, ethanol, ethylene glycol; sp.gr. 0.988 (50 C); visc. 21 cps (50 C); HLB 13.7; pour pt. 13 C; cloud pt. 75-80 C; pH 6-8; surf. tens. 27.0 dynes/cm (0.1%); 100% act.

Teric 9A10. [ICI Australia] C9-11 ethoxylate (10 EO); nonionic; surfactant; APHA 100 color; sp.gr. 1.044 (50 C); visc. 25 cps (50 C); HLB 14.7; pour pt. 22 C; cloud pt. 94-100 C; pH 6-8; surf. tens. 32.5 dynes/cm (0.1%); 100% act.

Teric 9A12. [ICI Australia] C9-11 ethoxylate (12 EO); nonionic; surfactant for laundry prods., metal soaking, leather finishing; APHA 100 color; sol. in water, aromatics, ethyl acetate, glycol ethers, trichlorethylene, ethanol, ethylene glycol; sp.gr. 1.023 (50 C); visc. 40 cps (50 C); HLB 15.4; hyd. no. 77-86; pour pt. 24 C; pH 6-8; surf. tens. 34.5 dynes/cm (0.1%); 100% act.

Teric 12A2. [ICI Australia] C12-15 pareth-2; nonionic; intermediate for phosphate ester prod., sulfation; surfactant for solv. cleaners/degreasers; fiber lubricant/antistat for textile spinning; APHA 80 color; sol. in aromatics, ethyl acetate, glycol ethers, trichlorethylene, ethanol, kerosene, min. oil, olein; sp.gr. 0.872; visc. 25 cps; HLB 6.0; hyd. no. 192-200; pour pt. < 0 C; pH 6-8; 100% act.

Teric 12A3. [ICI Australia] C12-15 pareth-3 (2.6 EO); nonionic; wetting agent, dispersant, coemulsifier, detergent, intermediate used in oil and solv.-based systems; horticultural applics.; w/o emulsions; household/industrial cleaners; skin care prods.; mfg. of sulfate and phosphate surfactants; Hazen 80 liq.; sol. in benzene, ethyl acetate, ethyl Icinol, perchlorethylene, ethanol, kerosene, min. and veg. oil, olein; sp.gr. 0.912; visc. 28 cps; m.p. < 0 C; HLB 7.1; hyd. no. 174-178; pH 6–8 (1% aq.); biodeg.; 100% act.

Teric 12A4. [ICI Australia] C12-15 pareth-4; nonionic; emulsifier for agric.; filter cake dewatering; polishes and waxes; window cleaners; solv. cleaners/degreasers; cutting oils; APHA 100 color; sol. in aromatics, ethyl acetate, glycol ethers, trichlorethylene, ethanol, kerosene, min. oil, olein; sp.gr. 0.950; visc. 40 cps; HLB 9.8; hyd. no. 142-150; pour pt. 3 C; pH 5.5-7.5; 100% act.

Teric 12A6. [ICI Australia] C12-15 pareth-6; nonionic; wetting agent, dispersant, detergent, chemical intermediate; used in textile processing, hard surface abrasives, metal descaling, detergent for household and industrial use in laundries; metal cleaning, sanitizer; Hazen 100 liq. to paste; sol. in benzene, ethyl acetate, ethyl Icinol, perchlorethylene, ethanol, kerosene, min., veg. and paraffin oil, olein; disp. in water; sp.gr. 0.955 (50 C); visc. 24 cps (50 C); m.p. 10 ± 2 C; HLB 11.2; cloud pt. 37-42 C; surf. tens. 27.8 dynes/cm; pH 6–8 (1% aq.); biodeg.; 100% act.

Teric 12A7. [ICI Australia] C12-15 pareth-7; nonionic; surfactant for skin care gels; dairy cleaning; laundry prods.; lt. duty detergents; paper processing; fiber lubricant/antistat for textile spinning, cotton dye leveling; APHA 100 color; sol. in water, aromatics, ethyl acetate, glycol ethers, trichlorethylene, ethanol, ethylene glycol; sp.gr. 0.958 (50 C); visc. 26 cps (50 C); HLB 12.1; pour pt. 15 C; cloud pt. 48-52 C; pH 6-8; surf. tens. 30 dynes/cm (0.1%); 100% act.

Teric 12A8. [ICI Australia] C12-15 pareth-8; nonionic; surfactant for dairy cleaning, hard surf. cleaners, leather tanning, wool scouring; APHA 100 color; sol. in water, aromatics, ethyl acetate, glycol ethers, trichlorethylene, ethanol, ethylene glycol; sp.gr. 0.970 (50 C); visc. 29 cps (50 C); HLB 12.6; pour pt. 16 C; cloud pt. 54-62 C; pH 5.5-7.5; surf. tens. 27.4 dynes/cm (0.1%); 100% act.

Teric 12A9. [ICI Australia] C12-15 pareth-9; nonionic; wetting agent, dispersant, detergent, chemical intermediate; used in textile processing, hard surface abrasives, metal descaling, detergent for household and industrial use in laundries; metal cleaning, sanitizer; Hazen 100 paste; sol. in water, benzene, ethyl acetate, ethyl Icinol, perchlorethylene, ethanol; sp.gr. 0.984 (50 C); visc. 32 cps (50 C); m.p. 20 ± 2 C; HLB 13.6; cloud pt. 82 ± 2 C (1%); surf. tens. 30.2 dynes; pH 6–8 (1% aq.); biodeg.; 100% act.

Teric 12A12. [ICI Australia] C12-15 pareth-12 (11.5 EO); nonionic; wetting agent, dispersant, detergent, chemical intermediate; used in textile processing, hard surface abrasives, metal descaling, detergent for household and industrial use in laundries; metal cleaning, sanitizer; Hazen 100 solid; sol. in benzene, ethyl acetate, ethyl Icinol, perchlorethylene, ethanol; sp.gr. 0.993 (50 C); visc. 37 cps (50 C); m.p. 24 ± 2 C; HLB 14.4; cloud pt. 92 ± 2 C (1%); surf. tens. 35.6 dynes/cm; pH 6–8 (1% aq.); biodeg.; 100% act.

Teric 12A16. [ICI Australia] C12-15 pareth-16; nonionic; wetting agent, dispersant, emulsifier, detergent; used in degreasing compds., industrial and household hard surface cleaners; sanitizers; Hazen 100 solid; sol. in water, benzene, ethyl acetate, ethyl Icinol, perchlorethylene, ethanol; sp.gr. 1.009 (50 C); visc. 38 cps (50 C); m.p. 27 ± 2 C; HLB 15.3; cloud pt. 98 ± 2 C; surf. tens. 33.7 dynes/cm; pH 6–8 (1% aq.); biodeg.; 100% act.

Teric 12A23. [ICI Australia] C12-15 pareth-23; nonionic; wetting agent, dispersant, emulsifier, detergent; used in degreasing compds., industrial and household hard surface cleaners; sanitizers; Hazen 100 solid; sol. in water, benzene, ethyl acetate, ethyl Icinol, perchlorethylene, ethanol; sp.gr. 1.032 (50 C); visc. 75 cps (50 C); m.p. 35 ± 2 C; HLB 16.6; cloud pt. > 100 C; surf. tens. 35.7 dynes/cm; pH 6–8 (1% aq.); biodeg.; 100% act.

Teric 12M2. [ICI Australia] PEG-2 cocamine; CAS 61791-14-8; nonionic; dispersant, emulsifier, stabilizer; wetting agent of hydrophobic surfaces; used in metal, stone, paper and textile processing; formulation of lubricants and dye bath auxs.; emulsion stabilization; fat liquoring compds.; liq.; sol. in benzene, ethyl acetate, ethyl Icinol, perchlorethylene, ethanol, kerosene, min. and veg. oil; disp. in water; sp.gr. 0.913; visc. 166 cps; m.p. –1 ± 2 C; HLB 11.4; surf. tens. 28.2 dynes/cm; pH 8–10 (1% aq.); 100% act.

Teric 12M5. [ICI Australia] PEG-5 cocamine; CAS 61791-14-8; nonionic; see Teric 12M2; liq.; sol. in benzene, ethyl acetate, ethyl Icinol, perchlorethylene, ethanol, min., veg. and paraffin oil, olein; disp. in water; sp.gr. 0.971; visc. 311 cps; m.p. –9 ± 2 C; HLB 12.4; surf. tens. 29.2 dynes/cm; pH 8–10 (1% aq.); 100% act.

Teric 12M15. [ICI Australia] PEG-15 cocamine; CAS 61791-14-8; nonionic; see Teric 12M2; liq.; sol. in water, benzene, ethyl acetate, ethyl Icinol, perchlorethylene, ethanol, veg. oil, olein; sp.gr. 1.038; visc. 343 cps; m.p. –11 ± 2 C; HLB 15.7; cloud pt. > 100 C; surf. tens. 39.6 dynes/cm; pH 8–10 (1% aq.); 100% act.

Teric 13A5. [ICI Australia] Trideceth-5; CAS 24938-91-8; nonionic; emulsifier for agric.; filter cake dewatering; polishes/waxes; solv. cleaners/degreasers; APHA 100 color; sol. in aromatics, ethyl acetate, glycol ethers, trichlorethylene, ethanol, kerosene, olein, ethylene glycol; sp.gr. 0.980; visc. 70 cps; HLB 10.4; hyd. no. 130-138; pour pt. < 0 C; pH 5.5-7.5; surf. tens. 24 dynes/cm (0.1%); 100% act.

Teric 13A7. [ICI Australia] Trideceth-7; CAS 24938-91-8; nonionic; emulsifier for agric.; intermediate for carboxymethylate mfg., sulfation; filter cake dewatering; dust suppression in coal industry; domestic and window cleaners; leather finishing; wetting agent for paper and cellulose, textiles; APHA 100 color; sol. in water, aromatics, ethyl acetate, glycol ethers, trichlorethylene, ethanol, olein, ethylene glycol; sp.gr. 0.984; visc. 80 cps; HLB 11.8; hyd. no. 112-118; pour pt. 0 C; pH 5.5-7.5; surf. tens. 26 dynes/cm (0.1%); 100% act.

Teric 13A9. [ICI Australia] Trideceth-9; CAS 24938-91-8; nonionic; wetting agent and dispersant, emulsifier, detergent used in pigment disps. and agric. and horticultural applics.; solv. emulsion cleaners; paper; wool dye leveling; Hazen 150 liq.; sol. in water, ethyl acetate, ethyl Icinol, ethanol; sp.gr. 1.018; visc. 142 cps; m.p. < 0 C; HLB 13.3; cloud pt. 63 ± 2 C; surf. tens. 27.9 dynes/cm; pH 6–8 (1% aq.); 95% act.

Teric 13M15. [ICI Australia] PEG-15 C13-15 fatty amine; nonionic; biodeg. emulsifier, solubilizer for agric. preps.; liq.; 100% conc.

Teric 15A11. [ICI Australia] C14-15 pareth-11; nonionic; in biodeg. detergent formulation, low foam laundry prods., bottle washing, metal soaking; Hazen 100 solid; sol. in water, aromatics, ethyl acetate, glycol ethers, trichlorethylene, ethanol, ethylene glycol; sp.gr. 0.99 (50 C); visc. 40 cps (50 C); HLB 13.8; pour pt. 26 C; cloud pt. 84-89 C; pH 5.5-7.5; surf. tens. 31 dynes/cm (0.1%); 100% conc.

Teric 16A16. [ICI Australia] Ceteareth-16; CAS 68439-49-6; nonionic; wetting agent, emulsifier, detergent, solubilizer, dispersant in hydrophobic conditions; dye and pigment carriers; textile applics.; mfg. of wax emulsions and polishes for household and industrial use; antistat; Hazen 100 flakes; sol. in water, benzene, ethyl acetate, ethyl Icinol, perchlorethylene, ethanol; sp.gr. 1.01 (50 C); visc. 65 cps (50 C); m.p. 38 ± 2 C; HLB 14.9; cloud pt. > 100 C; surf. tens. 37.1 dynes/cm; pH 6–8 (1% aq.); biodeg.; 100% act.

Teric 16A22. [ICI Australia] Ceteareth-22; CAS 68439-49-6; nonionic; emulsifier for agric.; enzyme coating; laundry detergents; metal soaking; rubber latex stabilization; Hazen 100 flakes; sol. in water, benzene, ethyl acetate, ethyl Icinol, perchlorethylene, ethanol; sp.gr. 1.01 (50 C); visc. 92 cps (50 C); m.p. 40 ± 2 C; HLB 15.8; cloud pt. > 100 C; surf. tens. 41.3 dynes/cm; pH 6–8 (1% aq.); biodeg.; 100% act.

Teric 16A29. [ICI Australia] Ceteareth-29; CAS 68439-49-6; nonionic; surfactant for enzyme coating, toilet blocks; Hazen 100 flakes; sol. in water, benzene, ethyl acetate, ethyl Icinol, perchlorethylene, ethanol; sp.gr. 1.05 (50 C); visc. 105 cps (50 C); m.p. 45 ± 2 C; HLB 17.7; cloud pt. > 100 C; surf. tens. 46.0 dynes/cm; pH 6–8 (1% aq.); biodeg.; 100% act.

Teric 16A50. [ICI Australia] Ceteareth-50; CAS 68439-49-6; nonionic; surfactant for toilet blocks; APHA 150 color; sol. in water, aromatics, ethyl acetate, glycol ethers, trichlorethylene, ethanol; sp.gr. 1.04 (50 C); visc. 220 cps (50 C); HLB 18.3; hyd. no. 16-19; pour pt. 54 C; pH 6-8; surf. tens. 48 dynes/cm (0.1%); 100% act.

Teric 16M2. [ICI Australia] PEG-2 soya amine; CAS 61791-24-0; nonionic; wetting agent, dispersant, emulsifier for waxes and fats; formulation of leather dressing and metal cleaning compds., fiber lubricant applics. in the bldg. industry; liq.; sol. in benzene, ethyl acetate, ethyl Icinol, perchlorethylene, ethanol, kerosene, min., paraffin and veg. oil, olein; sp.gr. 0.906; visc. 191 cps; m.p. –15 ± 2 C; HLB 9.5; surf. tens. 26.6 dynes/cm; pH 8–10 (1% aq.); 100% act.

Teric 16M5. [ICI Australia] PEG-5 soya amine; CAS 61791-24-0; nonionic; wetting agent and dispersant; used in metal, cleaning, leather dressing and dye leveling compds., agric. sprays; emulsifier for fats and waxes; fiber lubricant; wax emulsion for the bldg. industry; liq.; sol. in benzene, ethyl acetate, ethyl Icinol, perchlorethylene, ethanol, kerosene, min., paraffin and veg. oil, olein; sp.gr. 0.954; visc. 182 cps; m.p. –16 ± 2 C; HLB 10.5; surf. tens. 29.7 dynes/cm; pH 8–10 (1% aq.); 100% act.

Teric 16M10. [ICI Australia] PEG-10 soya amine; CAS 61791-24-0; nonionic; wetting agent, dispersant, emulsifier for fats and waxes; used in leather dressing compds., fiber lubricant; wax emulsions for fiber board pkg. and particle board for bldg. industry; liq.; sol. in water, benzene, ethyl acetate, ethyl Icinol, perchlorethylene, ethanol, kerosene, min., veg. and paraffin oil, olein; sp.gr. 0.994; visc. 219 cps; m.p. –22 ± 2 C; HLB 12.1; cloud pt. 89 ± 2 C; surf. tens. 36.1 dynes/cm; pH 8–10 (1% aq.); 100% act.

Teric 16M15. [ICI Australia] PEG-15 soya amine; CAS 61791-24-0; nonionic; wetting agent and dispersant; used in metal, cleaning, leather dressing and dye leveling compds., agric. sprays; emulsifier for fats and waxes; fiber lubricant; wax emulsion for the bldg. industry; liq.; sol. in water, benzene, ethyl acetate, ethyl Icinol, perchlorethylene, ethanol; sp.gr. 1.021; visc. 266 cps; m.p. –18 ± 2 C; HLB 13.8; cloud pt. > 100 C; surf. tens. 37.4 dynes/cm; pH 8–10 (1% aq.); 100% act.

Teric 17A2. [ICI Australia] Cetoleth-2; nonionic; emulsifier, mfg. of textile lubricants, solv. and waterless hand cleaners; coemulsifier; detergent additive in petrol. oils; intermediate for anionic surfactants; Hazen 150 liq.; sol. in benzene, ethyl acetate, ethyl Icinol, perchlorethylene, ethanol, kerosene, veg., paraffin and min. oils, olein; disp. in water; sp.gr. 0.902; visc. 37 cps; m.p. 11 ± 2 C; HLB 6.0; pH 6–8 (1% aq.); biodeg.; 100% act.

Teric 17A3. [ICI Australia] Cetoleth-3; nonionic; emulsifier, mfg. of textile lubricants, solv. and waterless hand cleaners; coemulsifier; detergent additive in petrol. oils; intermediate for anionic surfactants; Hazen 150 liq.; sol. in benzene, ethyl acetate, ethyl Icinol, perchlorethylene, ethanol, kerosene, veg., paraffin and min. oils, olein; disp. in water; sp.gr. 0.918; visc. 41 cps; m.p. 13 ± 2 C; HLB 7.5; pH 6–8 (1% aq.); biodeg.; 100% act.

Teric 17A6. [ICI Australia] Cetoleth-6; nonionic; emulsifier, mfg. of textile lubricants, solv. and waterless hand cleaners; coemulsifier; detergent additive in petrol. oils; intermediate for anionic surfactants; Hazen 150 liq.; sol. in benzene, ethyl acetate, ethyl Icinol, perchlorethylene, ethanol, veg. oil, olein; disp. in water; sp.gr. 0.957; visc. 60 cps; m.p. 15 ± 2 C; HLB 10.2; pH 6–8 (1% aq.); biodeg.; 100% act.

Teric 17A8. [ICI Australia] Cetoleth-8; nonionic; wetting agent, dispersant, emulsifier; used in processing of yarns and fabrics; textile dyeing; horticultural sprays; removal of oil slicks; Hazen 150 paste; sol. in water, benzene, ethyl acetate, ethyl Icinol, perchlorethylene, ethanol, olein; sp.gr. 0.963 (50 C); visc. 35 cps (50 C); m.p. 17 ± 2 C; HLB 11.6; cloud pt. 41 ± 2 C; surf. tens. 30.8 dynes/cm; pH 6–8 (1% aq.); biodeg.; 100% act.

Teric 17A10. [ICI Australia] Cetoleth-10; nonionic; wetting agent, dispersant, emulsifier; used in processing of yarns and fabrics; textile dyeing; horticultural sprays; removal of oil slicks; solubilizer for sanitation chemicals; Hazen 150 solid; sol. in water, benzene, ethyl acetate, ethyl Icinol, perchlorethylene, ethanol, olein; sp.gr. 0.973 (50 C); visc. 43 cps (50 C); m.p. 24 ± 2 C; HLB 12.7; cloud pt. 72 ± 2 C; surf. tens. 32.9 dynes/cm; pH 6–8 (1% aq.); biodeg.; 100% act.

Teric 17A13. [ICI Australia] Cetoleth-13; nonionic; wetting agent, dispersant, emulsifier; used in processing of yarns and fabrics; textile dyeing; horticultural sprays; removal of oil slicks; solubilizer for sanitation chemicals; Hazen 150 solid; sol. in water, benzene, ethyl acetate, ethyl Icinol, perchlorethylene, ethanol; sp.gr. 0.986 (50 C); visc. 45 cps (50 C); m.p. 25 ± 2 C; HLB 13.8; cloud pt. 84 ± 2 C; surf. tens. 33.0 dynes/cm; pH 6–8 (1% aq.); biodeg.;

100% act.

Teric 17A25. [ICI Australia] Ceteoleth-25; nonionic; wetting agent, dispersant, emulsifier; used in processing of yarns and fabrics; textile dyeing; horticultural sprays; removal of oil slicks; solubilizer for sanitation chemicals; Hazen 150 solid; sol. in water, benzene, ethyl acetate, ethyl Icinol, perchlorethylene, ethanol; sp.gr. 1.030 (50 C); visc. 92 cps (50 C); m.p. 37 ± 2 C; HLB 16.2; cloud pt. > 100 C; surf. tens. 42.0 dynes/cm; biodeg.; 100% act.

Teric 17M2. [ICI Australia] PEG-2 tallow amine; nonionic; wetting agent and dispersant; dewatering agent in elec. components and road aggregates; emulsifier for fats, waxes, min. oils and metal working lubricants; corrosion inhibitor; paste; sol. in benzene, ethyl acetate, ethyl Icinol, perchlorethylene, ethanol, kerosene, min. paraffin and veg. oil, olein; sp.gr. 0.896; visc. 56 cps (40 C); m.p. 25 ± 2 C; HLB 9; surf. tens. 26.5 dynes/cm; pH 8–10 (1% aq.); 100% act.

Teric 17M5. [ICI Australia] PEG-5 tallow amine; nonionic; wetting agent and dispersant; dewatering agent in elec. components and road aggregates; emulsifier for fats, waxes, min. oils and metal working lubricants; corrosion inhibitor; semiliq.; sol. in benzene, ethyl acetate, ethyl Icinol, perchlorethylene, ethanol, kerosene, min. paraffin and veg. oil, olein; sp.gr. 0.953; visc. 177 cps; m.p. 11 ± 2 C; HLB 10; surf. tens. 31.5 dynes/cm; pH 8–10 (1% aq.); 100% act.

Teric 17M15. [ICI Australia] PEG-15 tallow amine; nonionic; wetting agent and dispersant; dewatering agent in elec. components and road aggregates; emulsifier for fats, waxes, min. oils and metal working lubricants; corrosion inhibitor; liq.; sol. in water, benzene, ethyl acetate, ethyl Icinol, perchlorethylene, ethanol, veg. oil., olein; sp.gr. 1.028; visc. 428 cps; m.p. –19 ± 2 C; HLB 13.3; cloud pt. > 100 C; surf. tens. 40.0 dynes/cm; pH 8–10 (1% aq.); 100% act.

Teric 18M2. [ICI Australia] PEG-2 stearamine; nonionic; wetting agent, dispersant; emulsion stabilizer; emulsifier used in agric. toxicants and processing of textiles, paper, leather and bldg. board; corrosion inhibitor in lubricants and greases; solid; sol. in benzene, ethyl acetate, ethyl Icinol, perchlorethylene, ethanol, kerosene, min., paraffin and veg. oil, olein; sp.gr. 0.887 (50 C); visc. 75 cps (50 C); m.p. 50 ± 2 C; HLB 8.5; pH 8–10 (1% aq.); 100% act.

Teric 18M5. [ICI Australia] PEG-5 stearamine; CAS 26635-92-7; nonionic; wetting agent, dispersant; emulsion stabilizer; emulsifier used in agric. toxicants and processing of textiles, paper, leather and bldg. board; corrosion inhibitor in lubricants and greases; softener and antistat in solv. cleaning of textiles and compd. of plastics; solid; sol. in benzene, ethyl acetate, ethyl Icinol, perchlorethylene, ethanol, kerosene, min., paraffin and veg. oil, olein; disp. in water; sp.gr. 0.935 (50 C); visc. 46 cps (50 C); m.p. 35 ± 2 C; HLB 9.5; pH 8–10 (1% aq.); 100% act.

Teric 18M10. [ICI Australia] PEG-10 stearamine; CAS 26635-92-7; nonionic; wetting agent, dispersant; emulsion stabilizer; emulsifier used in agric. toxicants, textiles, paper, leather and bldg. board; corrosion inhibitor in lubricants and greases; softener, antistat for plastics, textiles; liq.; sol. in water, benzene, ethyl acetate, ethyl Icinol, perchlorethylene, ethanol, kerosene, veg. oil, olein; sp.gr. 0.997; visc. 255 cps; m.p. 18 ± 2 C; HLB 11.2; cloud pt. > 100 C; surf. tens. 34.0 dynes/cm; pH 8–10 (1% aq.); 100% act.

Teric 18M20. [ICI Australia] PEG-20 stearamine; CAS 26635-92-7; nonionic; wetting agent, dispersant; emulsion stabilizer; emulsifier used in agric. toxicants and processing of textiles, paper, leather and bldg. board; corrosion inhibitor in lubricants and greases; paste; sol. in water, benzene, ethyl acetate, ethyl Icinol, perchlorethylene, ethanol, olein; sp.gr. 1.025 (50 C); visc. 66 cps (50 C); m.p. 28 ± 2 C; HLB 14.5; cloud pt. > 100 C; surf. tens. 40.3 dynes/cm; pH 8–10 (1% aq.); 100% act.

Teric 18M30. [ICI Australia] PEG-30 stearamine; CAS 26635-92-7; nonionic; wetting agent, dispersant; emulsion stabilizer; emulsifier used in agric. toxicants and processing of textiles, paper, leather and bldg. board; corrosion inhibitor in lubricants and greases; solid; sol. in water, benzene, ethyl acetate, ethyl Icinol, perchlorethylene, ethanol; sp.gr. 1.038 (50 C); visc. 116 cps (50 C); m.p. 28 ± 2 C; HLB 17.8; cloud pt. > 100 C; surf. tens. 43.0 dynes/cm; pH 8–10 (1% aq.); 100% act.

Teric 121. [ICI Australia] Fatty acid EO condensate; nonionic; antifoam agent, emulsifier; coemulsifier of veg. and min oils, paraffinic waxes and solvs.; lubricant for textile processing; foam control agent in paper mfg., industrial fermentation and distillation, textile dyeing; adhesives and waste water treatment; Hazen > 250 liq.; sol. in benzene, ethyl acetate, ethyl Icinol, perchlorethylene, ethanol, veg. oil, olein; sp.gr. 0.987; visc. 353 cps; m.p. < 0 C; HLB 7.4; pH 8 (1% aq.); biodeg.; 100% act.

Teric 124. [ICI Australia] Fatty acid EO condensate; nonionic; antifoam agent, emulsifier; coemulsifier of veg. and min oils, paraffinic waxes and solvs.; lubricant for textile processing; foam control agent in paper mfg., industrial fermentation and distillation, textile dyeing; Hazen > 250 liq.; sol. in benzene, ethyl acetate, ethyl Icinol, perchlorethylene, ethanol, veg. oil, olein; also sol. in min. oil; sp.gr. 0.966; visc. 118 cps; m.p. < 0 C; HLB 4.5; pH 8 (1% aq.); biodeg.; 100% act.

Teric 127. [ICI Australia] Fatty acid EO condensate; nonionic; antifoam agent, emulsifier; coemulsifier of veg. and min oils, paraffinic waxes and solvs.; lubricant for textile processing; foam control agent in paper mfg., industrial fermentation and distillation, textile dyeing; Hazen > 250 liq.; sol. in benzene, ethyl acetate, ethyl Icinol, perchlorethylene, ethanol, veg. oil, olein; sp.gr. 0.978; visc. 87 cps; m.p. < 0 C; HLB 7.7; pH 8 (1% aq.); 100% act.

Teric 128. [ICI Australia] Ethoxylated alkyl phenol; nonionic; low foam wetting agent, deinking agent for paper recycling; emulsifier; liq.; 100% conc.

Teric 129. [ICI Australia] Glycerol alkoxylate; defoamer for min. processing and industrial applics.; liq.; 100% conc.

Teric 151. [ICI Australia] Fatty acid EO condensate; nonionic; emulsifier, antistat, lubricant; textile processing aid for fibers; o/w emulsions; Hazen > 250 solid; sol. in benzene, ethyl acetate, ethyl Icinol, perchlorethylene, ethanol, kerosene, veg. oil, olein; disp. in water; sp.gr. 0.994 (50 C); visc. 45 cps (50 C); m.p. 30 ± 2 C; HLB 11.6; pH 7–9 (1% aq.); biodeg.; 100% act.

Teric 152. [ICI Australia] Blend; nonionic; emulsifier for degreasing compds., textile applics.; liq.; 100% conc.

Teric 154. [ICI Australia] Fatty acid EO condensate; nonionic; antistat and emulsifier; formulation of textile processing aids; veg. and lt. petrol. oils; Hazen > 250 liq.; sol. in water, benzene, ethyl acetate, ethyl Icinol, perchlorethylene, ethanol, kerosene, veg. oil, olein; sp.gr. 1.031; visc. 89 cps; m.p. 7 ± 2 C; HLB 13.3; surf. tens. 30.0 dynes/cm; pH 6–8 (1% aq.); biodeg.; 100% act.

Teric 157. [ICI Australia] Ethoxylated alkyl phenol; nonionic; wetting agent, detergent for use in industrial applics., powd. blends; powd.; 70% conc.

Teric 158. [ICI Australia] Alkyl phenol ethoxylate; nonionic; low foam wetting agent; deinking agent for paper recycling; emulsifier; formulation of powds. for agric., surf. coatings and ceramic applics.; Hazen > 250 fine powd.; HLB 11.6; pH 6–8 (1% aq.); 100% conc.

Teric 160. [ICI Australia] Alkylaryl polyoxyalkylene ether; nonionic; low foam surfactant, wetting agent and dispersant; mfg. of paints, alkaline cleaners, automatic dishwashers, heavy-duty detergents, metal cleaning, wettable powds., and sanitizer; Hazen 250 liq.; sol. in water, benzene, ethyl acetate, ethyl Icinol, perchlorethylene, ethanol, veg. oil, olein; sp.gr. 1.032; visc. 312 cps; m.p. < 0 C; HLB 11.3; cloud pt. 29 ± 2 C; surf. tens. 30.4 dynes/cm; pH 6–8 (1% aq.); 100% act.

Teric 161. [ICI Australia] Alkylaryl polyoxyalkylene ether; nonionic; see Teric 160; Hazen 250 liq.; sol. in water, benzene, ethyl acetate, ethyl Icinol, perchlorethylene, ethanol, veg. oil, olein; sp.gr. 1.034; visc. 340 cps; m.p. < 0 C; HLB 12.3; cloud pt. 35 ± 2 C; surf. tens. 33.4 dynes/cm; pH 6–8 (1% aq.); 100% act.

Teric 163. [ICI Australia] Alkylaryl polyoxyalkylene ether; nonionic; demulsifier, detergent for machine dishwashing, in-situ cleaning of milking plants, detergent sanitizer and iodophor mfg; Hazen 250 liq.; sol. in water, benzene, ethyl acetate, ethyl Icinol, perchlorethylene, ethanol, veg. oil, olein; sp.gr. 1.110; visc. 1240 cps; m.p. < 0 C; HLB 14.9; cloud pt. 97 ± 2 C; surf. tens. 50.4 dynes/cm; pH 4–6 (1% aq.); 100% act.

Teric 164. [ICI Australia] Alcohol ethyoxylate, modified; nonionic; biodeg. low foam surfactant for use in alkaline cleaners, automatic dishwashers, heavy-duty detergents, metal cleaning; Hazen 250 liq.; sol. in water, benzene, ethyl acetate, ethyl Icinol, perchlorethylene, ethanol, kerosene, veg. oil, olein; sp.gr. 0.937; visc. 65 cps; m.p. 2 ± 2 C; HLB 10.8; cloud pt. 40 ± 2 C; surf. tens. 30.5 dynes/cm; pH 6–8; 100% min. act.

Teric 165. [ICI Australia] Alcohol ethoxylate, modified; nonionic; biodeg. low foam wetting and dedusting agent for solid caustic and highly alkaline detergents; dk. br. liq.; sol. in water, ethyl Icinol, ethanol, olein; sp.gr. 1.035; visc. 68 cps; HLB 12.5; cloud pt. 30–35 C (1% aq.); pour pt. 2–3 C; surf. tens. 28.2 dynes/cm (0.1%); pH 5; 89% min. act.

Teric 200. [ICI Australia] Alkylene oxide derivative; nonionic; emulsifier, dispersant; mfg. of concs. of insecticides, pesticides and herbicides in solvs.; Hazen 150 paste; sol. in water, benzene, ethyl acetate, ethyl Icinol, perchlorethylene, ethanol, olein; sp.gr. 1.034 (50 C); visc. 123 cps (50 C); m.p. 30 ± 2 C; HLB 16.1; cloud pt. 90 ± 2 C; surf. tens. 36.0 dynes/cm; pH 6–8 (1% aq.); 100% act.

Teric 203. [ICI Australia] Alkylene oxide derivative; nonionic; agric. emulsifier; dispersant for conc. formulations; paste; 100% conc.

Teric 225. [ICI Australia] Alkoxylated amine; nonionic; agric. wetter, sticker; liq.; 100% conc.

Teric 226. [ICI Australia] Alkylene oxide deriv.; nonionic; emulsifier blend for agric. emulsions; paste; 100% conc.

Teric 304. [ICI Australia] Modified fatty amine ethoxylate; nonionic; demulsifier, anticaking agent; liq.; 100% conc.

Teric 305. [ICI Australia] Phosphate ester of nonionic surfactant; anionic; biodeg. detergent, hydrotrope for metal cleaning applics.; sol. and stable in strong alkali sol'ns.; liq.; 100% conc.

Teric 306. [ICI Australia] Phosphate ester; anionic; biodeg. detergent, hydrotrope for heavy-duty liq. detergents; stable in alkaline formulations; liq.; 100% conc.

Teric 313. [ICI Australia] Alkylaryl POE; nonionic; solubilizer, emulsifier; liq.; 100% conc.

Teric 350. [ICI Australia] Fatty acid condensate; nonionic; emulsifier and coemulsifier for solvs. and min. oils; formulation of bases for solv. cleaners, degreasing and decarbonizing preparations; lubricant and corrosion inhibitor in lubricant systems used in metal working industry; Hazen > 250 liq.; sol. in benzene, ethyl Icinol, perchlorethylene, ethanol, olein; disp. in water; sp.gr. 1.010; visc. 600 cps; HLB 16.6; pH 8–10 (1% aq.); 100% conc.

Teric 351B. [ICI Australia] Fatty acid condensate; nonionic; emulsifier and coemulsifier for solvs. and min. oils; formulation of bases for solv. cleaners, degreasing and decarbonizing preparations; lubricant and corrosion inhibitor in lubricant systems used in metal working industry; Hazen > 250 liq.; sol. in perchlorethylene, olein; disp. in water; sp.gr. 1.017; visc. 750 cps; HLB 8.2; pH 8–10 (1% aq.).

Teric BL8. [ICI Australia] Syn. alcohol polyalkylene oxide deriv.; nonionic; detergent; wetting agent and dispersant in printing inks, textiles, wool scouring, dust suppression, household and industrial surf. cleaning; emulsifier of solvs., greases and oils; biodeg.; Hazen 250 liq.; sol. in water, benzene, ethyl acetate, ethyl Icinol, perchlorethylene, ethanol, kerosene, veg. oil; sp.gr. 1.003; visc. 63 cps; m.p. < 0 C; HLB 13.7; cloud pt. 59 ± 2 C; surf. tens. 30.3 dynes/cm; pH 6–8 (1% aq.); 100% act.

Teric BL9. [ICI Australia] Syn. alcohol polyalkylene oxide deriv.; nonionic; see Teric BL8; Hazen 250 liq.; sol. in water, benzene, ethyl acetate, ethyl Icinol, perchlorethylene, ethanol, veg. oil; sp.gr. 1.018; visc. 100 cps; m.p. < 0 C; HLB 14.5; cloud pt. 67 ± 2 C; surf. tens. 30.0 dynes/cm; pH 6–8 (1% aq.); 98% act.

Teric C12. [ICI Australia] PEG-12 castor oil; nonionic; detergent, emulsifier and coemulsifier in formulation of sol. oils, solv. cleaners and temp. protective coatings; corrosion resistant, used in industrial and institutional cleaning preparations; die lubricant in metal forming operations; mold release agent in plastics; fiber lubricant in textile processing; Hazen > 250 liq.; sol. in water, benzene, ethyl acetate, ethyl Icinol, ethanol, veg. oil, olein; sp.gr. 1.048; visc. 5000 cps.; m.p. 3 ± 2 C; HLB 12.7; cloud pt. 67–69 C; surf. tens. 42.0 dynes/cm; pH 6–8 (1% aq.); 100% act.

Teric CDE. [ICI Australia] Cocamide DEA; nonionic; emulsifier of perchlorethylene, power kerosene; foam booster/stabilizer, thickener and detergent in domestic dishwashing formulations; biodeg.; Hazen > 250 liq.; sol. in benzene, ethyl Icinol, perchlorethylene, ethanol, olein; disp. in water; sp.gr. 1.010; visc. 600 cps; m.p. < 0 C; HLB 16.6; pH 8–10; 100% act.

Teric CME3. [ICI Australia] PEG-3 cocamide MEA; nonionic; dispersant, detergent; wetting agent for textile applics.; coemulsifier for various oils; emulsion stabilizer; foam booster in detergent blends; Hazen > 250 liq., paste; low odor; sol. in benzene, ethyl acetate, ethyl Icinol, perchlorethylene, ethanol, olein; disp. in water; sp.gr. 0.972 (50 C); visc. 56 cps (50 C); m.p. 18 ± 2 C; HLB 13.8; pH 8–10 (1% aq.); biodeg.; 100% act.

Teric CME7. [ICI Australia] PEG-7 cocamide MEA; nonionic; wetting agent for textile applics.; dispersant, detergent, foam stabilizer; cleaning compds.; biodeg.; Hazen > 250 liq.; low odor; sol. in water, benzene, ethyl acetate, ethyl Icinol, perchlorethylene, ethanol; sp.gr. 1.048; visc. 264 cps; m.p. 5 ± 2 C; HLB 15.1; surf. tens. 30.0 dynes/cm; pH 8–10; 100% act.

Teric DD5. [ICI Australia] Dodoxynol-5; nonionic; emulsifier, coemulsifier in solv. emulsion cleaners; formulation of concs. for metal lubricants, cutting, milling and grinding aids, textile processing aid and min. drilling lubricants; Hazen 250 liq.; sol. in benzene, ethyl acetate, ethyl Icinol, perchlorethylene, ethanol, kerosene, min. and veg. oil, olein; sp.gr. 1.014; visc. 550 cps; m.p. < 0 C; HLB 8.6; hyd. no. 115-122; surf. tens. 29.5 dynes/cm; pH 6–8 (1% aq.); biodeg.; 100% act.

Teric DD9. [ICI Australia] Dodoxynol-9; nonionic; see Teric DD5; Hazen 150 liq.; sol. in benzene, ethyl Icinol, ethanol, kerosene, olein; sp.gr. 1.040; visc. 453 cps; m.p. < 0 C; HLB 12.1; surf. tens. 28.2 dynes/cm; pH 6–8 (1% aq.); 100% act.

Teric G9A5. [ICI Australia] C9-11 pareth-5; nonionic; emulsifier for solvs.; formulation of hard surf. cleaners, degreasers and dispersants in industrial and domestic applics.; Hazen 250 liq.; sol. in water, veg. oil, olein; sp.gr. 0.976; visc. 52 cps; m.p. < 0 C; HLB 11.6; cloud pt. 36 ± 2 C; surf. tens. 20.1 dynes/cm; pH 5–7 (1% aq.); 100% act.

Teric G9A6. [ICI Australia] C9-11 pareth-6; see Teric G9A5; Hazen 250 liq.; sol. in water, benzene, ethyl acetate, ethyl Icinol, perchlorethylene, ethanol, veg. oil, olein; sp.gr. 0.987; visc. 38 cps; m.p. < 0 C; HLB 12.4; cloud pt. 53 ± 2 C; surf. tens. 24.0 dynes/cm; pH 5–7 (1% aq.); 100% act.

Teric G9A8. [ICI Australia] C9-11 pareth-8; nonionic; emulsifier, dispersant, wetting agent; used in detergent formulations for domestic use; degreaser in textile applics.; Hazen 250 paste; sol. in water, benzene, ethyl acetate, ethyl Icinol, perchlorethylene, ethanol, veg. oil, olein; sp.gr. 0.988 (50 C); visc. 21 cps (50 C); m.p. 13 ± 2 C; HLB 13.7; cloud pt. 78 ± 2 C; surf. tens. 27.0 dynes/cm; pH 5–7 (1% aq.); 100% act.

Teric G9A12. [ICI Australia] C9-11 pareth-12; nonionic; see Teric G9A8; Hazen 250 solid; sol. in water, benzene, ethyl acetate, ethyl Icinol, perchlorethylene, ethanol, veg. oil, olein; sp.gr. 1.023; visc. 40 cps (50 C); m.p. 24 ± 2 C; HLB 15.4; cloud pt. > 100 C; surf. tens. 34.5 dynes/cm; pH 5–7 (1% aq.); 100% act.

Teric G12A4. [ICI Australia] C12-15 pareth-4 (4.5 EO); nonionic; wetting agent and dispersant, detergent; dust suppression; aq. pigment dispersion; coemulsifier for w/o-type emulsion concs.; liq. and powd. detergents for industrial and domestic laundry, dishwashing, metal cleaning, sanitizing, textile processing aids, abrasive cleaners; intermediate in mfg. of sulfate and phosphate surfactants; Hazen 250 liq.; sol. in benzene, ethyl acetate, ethyl Icinol, perchlorethylene, ethanol, kerosene, min., paraffin and veg. oil, olein; disp. in water; sp.gr. 0.950; visc. 40 cps; m.p. 3 ± 2 C; HLB 9.8; pH 6–8 (1% aq.); 100% act.

Teric G12A6. [ICI Australia] C12-15 pareth-6; nonionic; wetting agent and dispersant, intermediate, textile processing, hard surf. abrasives, metal descaling, domestic and industrial use in laundries, intermediate for metal cleaning, sanitizing, dairy detergents; Hazen 250 liq. to paste; sp.gr. 0.955; visc. 24 cps (50 C); m.p. 10 ± 2 C; HLB 11.2; cloud pt. 40 ± 2 C; surf. tens. 27.8 dynes/cm; pH 6–8 (1% aq.); biodeg.; 100% act.

Teric G12A8. [ICI Australia] C12-15 pareth-8 (7.8 EO); nonionic; wetting agent and dispersant, intermediate, textile processing, hard surf. abrasives, metal descaling, domestic and industrial use in laundries, intermediate for metal cleaning, sanitizing, dairy detergents; Hazen 250 paste; sol. in benzene, ethyl acetate, ethyl Icinol, perchlorethylene, ethanol, kerosene, min., paraffin and veg. oil, olein; disp. in water; sp.gr. 0.971 (50 C); visc. 31 cps (50 C); m.p. 16 ± 2 C; HLB 12.6; cloud pt. 58 ± 2 C; surf. tens. 27.2 dynes/cm; pH 6–8 (1% aq.); biodeg.; 100% act.

Teric G12A12. [ICI Australia] C12-15 pareth-12 (11.5 EO); nonionic; see Teric 12A6; Hazen 250 solid; sp.gr. 0.993; visc. 37 cps; m.p. 24 ± 2 C; HLB 14.4; cloud pt. 92 ± 2 C; surf. tens. 35.6 dynes/cm; pH 6–8 (1% aq.); biodeg.; 100% act.

Teric GN Series. [ICI Australia] Nonylphenol ethoxylate, tech.; CAS 9016-45-9; nonionic; tech. grades of Teric N series; liq./solid; 100% conc.

Teric LA4. [ICI Australia] C12-15 pareth-4 (4.5 EO); nonionic; wetting agent and dispersant, detergent; dust suppression; aq. pigment dispersion; coemulsifier for w/o-type emulsion concs.; liq. and powd. detergents for industrial and domestic laundry, dishwashing, metal cleaning, sanitizing, textile processing aids, abrasive cleaners; intermediate in mfg. of sulfate and phosphate surfactants; liq.; sol. in water, ethyl acetate, ethyl Icinol, ethanol, kerosene, min. oil, olein; sp.gr. 0.955; visc. 45 cps; m.p. 0 ± 2 C; HLB 9.8; pH 6–8 (1% aq.); biodeg.; 88% act.

Teric LA8. [ICI Australia] C12-15 pareth-8 (7.8 EO); nonionic; wetting agent and dispersant, intermediate, textile processing, hard surf. abrasives, metal descaling, domestic and industrial use in laundries, intermediate for metal cleaning, sanitizing, dairy detergents; lt. amber liq.; sol. in water, benzene, ethyl acetate, ethyl Icinol, perchlorethylene, ethanol; sp.gr. 1.002; visc. 100 cps; m.p. 10 ± 2 C; HLB 12.6; cloud pt. 60 ± 2 C; surf. tens. 27.2 dynes/cm; pH 6–8 (1% aq.); biodeg.; 88% act.

Teric LAN70. [ICI Australia] PEG-70 lanolin; CAS 61790-81-6; nonionic; emulsifier, dispersant, detergent additive, sanitizer component, textile lubricant, cosmetic ingred.; yel.-amber soft waxy solid; sol. in water, benzene, ethyl Icinol, perchlorethylene, ethanol, olein; sp.gr. 1.13; visc. 246 cps (50 C); m.p. 42 ± 2 C; HLB 16.3; cloud pt. > 100 C; surf. tens. 48.1 dynes/cm; pH 4.5–7.0 (1% aq.); 100% act.

Teric N2. [ICI Australia] Nonoxynol-2; CAS 9016-45-9; nonionic; defoamer, wetting agent and dispersant in solv. and oil-based systems; coemulsifier for o/w emulsions; emulsifier for w/o emulsions; used in agric. toxicant prods., industrial solv.-cleaning systems; surf. coating preparations; intermediate for prod. of anionics; Hazen 100 liq.; sol. in benzene, ethyl acetate, ethyl Icinol, perchlorethylene, ethanol, kerosene, min., paraffin and veg. oil, olein; disp. in water; sp.gr. 1.001; visc. 620 cps; m.p. < 0 C; HLB 5.7; hyd. no. 178-185; pour pt. < 0 C; pH 6–8 (1% aq.); 100% act.

Teric N3. [ICI Australia] Nonoxynol-3 (3.5 EO); CAS 9016-45-9; nonionic; defoamer, wetting agent and dispersant in solv. and oil-based systems; coemulsifier for o/w emulsions; emulsifier for w/o emulsions; used in agric. toxicant prods., industrial solv.-cleaning systems; surf. coating preparations; intermediate for prod. of anionics; Hazen 100 liq.; sol. in benzene, ethyl acetate, ethyl Icinol, perchlorethylene, ethanol, kerosene, min., paraffin and veg. oil, olein; disp. in water; sp.gr. 1.017; visc. 395 cps; m.p. < 0 C; HLB 8.2; hyd. no. 145-153; pH 6–8 (1% aq.); 100% act.

Teric N4. [ICI Australia] Nonoxynol-4; CAS 9016-45-9; nonionic; surfactant for agric., drycleaning, metalworking lubricants, tank cleaning, textile finishes, antistat, lubricant; Hazen 100 liq.; sol. in benzene, ethyl acetate, ethyl Icinol, perchlorethylene, ethanol, kerosene, min., paraffin and veg. oil, olein; disp. in water; sp.gr. 1.023; visc. 370 cps; m.p. < 0 C; HLB 8.9; hyd. no. 138-145; pH 6–8 (1% aq.); 100% act.

Teric N5. [ICI Australia] Nonoxynol-5 (5.5 EO); CAS 9016-45-9; nonionic; surfactant for agric., drycleaning, metalworking lubricants, tank cleaning, textile finishes, antistat, lubricant; Hazen 100 liq.; sol. in benzene, ethyl acetate, ethyl Icinol, perchlorethylene, ethanol, kerosene, veg. oil, olein; disp. in water; sp.gr. 1.035; visc. 355 cps; m.p. < 0 C; HLB 10.5; hyd. no. 118-126; pH 6–8 (1% aq.); 100% act.

Teric N8. [ICI Australia] Nonoxynol-8 (8.5 EO); CAS 9016-45-9; nonionic; wetting agent, dispersant, emulsifier, solubilizer, detergent; used in concrete mfg., agric. sprays, solv. cleaners, paints; detergents; Hazen 100 liq.; sol. in water, benzene, ethyl acetate, ethyl Icinol, perchlorethylene, ethanol, veg. oil, olein; sp.gr. 1.056; visc. 350 cps; m.p. 0 ± 2 C; HLB 12.3; cloud pt. 30-34 C; surf. tens. 29.4 dynes/cm; pH 6–8 (1% aq.); 100% act.

Teric N9. [ICI Australia] Nonoxynol-9; CAS 9016-45-9; nonionic; wetting agent, dispersant, emulsifier, solubilizer, detergent; used in concrete mfg., agric. sprays, solv. cleaners, paints; detergents; Hazen 100 liq.; sol. in water, benzene, ethyl acetate, ethyl Icinol, perchorethylene, veg. oil, olein; sp.gr. 1.060; visc. 330 cps; m.p. 0 ± 2 C; HLB 12.8; cloud pt. 50-55 C; surf. tens. 30.6 dynes/cm; pH 6–8 (1% aq.); 100% act.

Teric N10. [ICI Australia] Nonoxynol-10; CAS 9016-45-9; nonionic; wetting agent, dispersant, emulsifier, solubilizer, detergent; used in concrete mfg., agric. sprays, solv. cleaners, paints; detergents; Hazen 100 liq.; sol. in water, benzene, ethyl acetate, ethyl Icinol, perchorethylene, veg. oil, olein; sp.gr. 1.063; visc. 360 cps; m.p. 5 ± 2 C; HLB 13.3; cloud pt. 67 ± 2 C; surf. tens. 30.6 dynes/cm; pH 6–8 (1% aq.); 100% act.

Teric N11. [ICI Australia] Nonoxynol-11; CAS 9016-45-9; nonionic; wetting agent, dispersant, emulsifier, solubilizer, detergent; used in concrete mfg., agric. sprays, solv. cleaners, paints; detergents; Hazen 100 liq.; sol. in water, benzene, ethyl acetate, ethyl Icinol, perchlorethylene, ethanol, olein; sp.gr. 1.069; visc. 410 cps; m.p. 7 ± 2 C; HLB 13.7; cloud pt. 74 ± 2 C; surf. tens. 34.8 dynes/cm; pH 6–8 (1% aq.); 100% act.

Teric N12. [ICI Australia] Nonoxynol-12; CAS 9016-45-9; nonionic; wetting agent, dispersant, emulsifier, solubilizer, detergent; used in concrete mfg., agric. sprays, solv. cleaners, paints; detergents; Hazen 100 liq.; sol. in water, benzene, ethyl acetate, ethyl Icinol, perchlorethylene, ethanol, olein; sp.gr. 1.045 (50 C); visc. 67 cps (50 C); m.p. 11 ± 2 C; HLB 13.9; cloud pt. 82 ± 2 C; surf. tens. 35.2 dyens/cm; pH 6–8 (1% aq.); 100% act.

Teric N13. [ICI Australia] Nonoxynol-13; CAS 9016-45-9; nonionic; wetting agent, dispersant, emulsifier, solubilizer, detergent; used in concrete mfg., agric. sprays, solv. cleaners, paints; detergents; Hazen 100 liq.; sol. in water, benzene, ethyl acetate, ethyl Icinol, perchlorethylene, ethanol, olein; sp.gr. 1.049 (50 C); visc. 75 cps (50 C); m.p. 14+2 C; HLB 14.4; cloud pt. 89 ± 2 C; surf. tens. 34.9 dynes/cm; pH 6–8 (1% aq.); 100% act.

Teric N15. [ICI Australia] Nonoxynol-15; CAS 9016-45-9; nonionic; wetting agent, coemulsifier; used in paints, perfumes, detergents and sanitizers; Hazen 100 paste; sol. in water, benzene, ethyl acetate, ethyl Icinol, perchlorethylene, ethanol, olein; sp.gr. 1.051 (50 C); visc. 82 cps (50 C); m.p. 21 ± 2 C; HLB 15.0; cloud pt. 95 ± 2 C; surf. tens. 33.4 dynes/cm; pH 6–8 (1% aq.); 100% act.

Teric N20. [ICI Australia] Nonoxynol-20; CAS 9016-45-9; nonionic; wetting agent, coemulsifier; used in paints, perfumes, detergents and sanitizers; Hazen 100 solid; sol. in water, benzene, ethyl acetate, ethyl Icinol, perchlorethylene, ethanol, olein; sp.gr. 1.061 (50 C); visc. 100 cps (50 C); m.p. 30 ± 2 C; HLB 16.0; hyd. no. 42-57; cloud pt. > 100 C; surf. tens. 41.7 dynes/cm; pH 6–8 (1% aq.); 100% act.

Teric N30. [ICI Australia] Nonoxynol-30; CAS 9016-45-9; nonionic; wetting agent, coemulsifier; used in paints, perfumes, detergents and sanitizers; Hazen 150 flakes; sol. in water, benzene, ethyl acetate, ethyl Icinol, perchlorethylene, ethanol; sp.gr. 1.066 (50 C); visc. 150 cps (50 C); m.p. 40 ± 2 C; HLB 17.2; hyd. no. 31-36; cloud pt. > 100 C; surf. tens. 42.8 dynes/cm; pH 6–8 (1% aq.); 100% act.

Teric N30L. [ICI Australia] Nonoxynol-30; CAS 9016-45-9; nonionic; APHA 100 color; sp.gr. 1.095; visc. 2000 cps; HLB 17.2; pour pt. 2 C; pH 6-8; surf. tens. 42.8 dynes/cm (0.1%); 70% act.

Teric N40. [ICI Australia] Nonoxynol-40; CAS 9016-45-9; nonionic; wetting agent, coemulsifier; used in paints, perfumes, detergents and sanitizers; Hazen 200 flakes; sol. in water, benzene, ethyl acetate, ethyl Icinol, perchlorethylene, ethanol; sp.gr. 1.080 (50 C); visc. 250 cps (50 C); m.p. 45 ± 2 C; HLB 18.0; hyd. no. 25-32; cloud pt. > 100 C; surf. tens. 41.0 dynes/cm; pH 6–8 (1% aq.); 100% act.

Teric N40L. [ICI Australia] Nonoxynol-40; CAS 9016-45-9; nonionic; APHA 100 color; sp.gr. 1.099; visc. 1975 cps; HLB 18.0; pour pt. 5 C; pH 6-8; surf. tens. 42.0 dynes/cm (0.1%); 70% act.

Teric N100. [ICI Australia] Nonoxynol-100; CAS 9016-45-9; nonionic; wetting agent, coemulsifier; used in paints, perfumes, detergents and sanitizers; Hazen 100 flakes; sol. in water, benzene, ethyl acetate, ethyl Icinol, perchlorethylene, ethanol; sp.gr. 1.113 (60 C); m.p. 52 ± 2 C; HLB 19.1; hyd. no. 11.2-13.5; cloud pt. > 100 C; surf. tens. 48 dynes/cm (0.1%); pH 6–8 (1% aq.); 100% act.

Teric OF4. [ICI Australia] PEG-4 oleate; CAS 9004-96-0; nonionic; emulsifier used in o/w emulsions of veg. and min. oils, fats, waxes and solvs.; formulation of cutting oils; metal lubricant in metal cleaning formulations and in solv. degreasers; textile and paper finishing applics.; antistat in textile processing and mfg. of syn. fibers; Hazen > 250 liq.; sol. in benzene, ethyl acetate, ethyl Icinol, perchlorethylene, ethanol, kerosene, veg. oil, olein; disp. in water; sp.gr. 0.978; visc. 87 cps; m.p. < 0 C; HLB 7.7; pH 8 (1% aq.); biodeg.; 100% act.

Teric OF6. [ICI Australia] PEG-6 oleate; CAS 9004-96-0; nonionic; see Teric OF4; Hazen > 250 liq.; sol. in benzene, ethyl acetate, ethyl Icinol, perchlorethylene, ethanol, kerosene, veg. oil, olein; disp. in water; sp.gr. 0.992; visc. 82 cps; m.p. < 0 C; HLB 9.7; surf. tens. 33.9 dynes/cm; pH 7–9 (1% aq.); biodeg.; 100% act.

Teric OF8. [ICI Australia] PEG-8 oleate; CAS 9004-96-0; nonionic; emulsifier, antistat; aux. product and antistat in textile and paper finishing applics.; Hazen > 250 liq.; sol. in benzene, ethyl acetate, ethyl Icinol, perchlorethylene, ethanol, kerosene, veg. oil, olein; disp. in water; sp.gr. 1.012; visc. 120 cps; m.p. < 0 C; HLB 11.1; pH 8–10 (1% aq.); biodeg.; 100% act.

Teric PE61. [ICI Australia] POP + 5.7 EO; nonionic; demulsifier, intermediate, emulsifier, detergent, wetting agent and dispersant; pigments in latex paints; detergent sanitizer and alkaline cleaner; dewatering aid for treatment of crude oil emulsions; defoamer in paper pulp liquors, starch sizing, glues, and boiler water systems; Hazen 100 liq.; sol. in benzene, ethyl acetate, ethyl Icinol, perchlorethylene, veg. oil, olein; sp.gr. 1.017; visc. 388 cps; m.p. < 0 C; HLB 3; cloud pt. 24 ± 2 C; surf. tens. 39.0 dynes/cm; pH 6 (1% aq.); 100% act.

Teric PE62. [ICI Australia] POP + 17 EO; nonionic; see Teric PE61; also for bloat control in cattle; Hazen 100 liq.; sol. in water, benzene, ethyl acetate, ethyl Icinol, perchlorethylene, ethanol, veg. oil, olein; sp.gr. 1.032; visc. 440 cps; m.p. < 0 C; HLB 7; cloud pt. 32 ± 2 C; surf. tens. 42.1 dynes/cm; pH 6 (1% aq.); 100% act.

Teric PE64. [ICI Australia] POP + 25.5 EO; nonionic; wetting agent, dispersant, emulsifier, detergent, intermediate, demulsifier; used for pigments in latex paints, dairy cleaners and sanitizers; bloat control in cattle; oil well applics.; Hazen 100 liq.; sol. in water, benzene, ethyl acetate, ethyl Icinol, perchlorethylene, ethanol, olein; sp.gr. 1.051; visc. 1332 cps; m.p. 8 ± 2 C; HLB 15; cloud pt. 58 ± 2 C; surf. tens. 43.2 dynes/cm; pH 6 (1% aq.); 100% act.

Teric PE68. [ICI Australia] POP + 150 EO; nonionic; wetting agent, dispersant, emulsifier, detergent, intermediate; stabilizer in mfg. of latex paints, paper coatings, dairy cleaners and santizers; plasticizer for resins; Hazen 100 flakes; sol. in water, benzene, ethyl acetate, ethyl Icinol, perchlorethylene, ethanol, olein; m.p. 50 ± 2 C; HLB 29; cloud pt. > 100 C; surf. tens. 49.0 dynes/cm; pH 6 (1% aq.); 100% act.

Teric PEG 200. [ICI Australia] PEG 200; biodeg.; binder for glazing; intermediate for PEG esters, methacrylate resins, PU foams; plasticizer/solv. for cork; toiletries; metalworking lubricants; paints/resins; paper/film; printing inks; textile emulsifier; APHA 25 color; sol. in water and most polar org. solvs.; m.w. 190-210; sp.gr. 1.127; visc. 4.2 cst (99 C); pour pt. -50 C; flash pt. (OC) > 170 C; ref. index 1.459; pH 4.5-7.5; surf. tens. 44.6 dynes/cm; 99% act.

Teric PEG 300. [ICI Australia] PEG 300; pesticide solubilizer/carrier; visc. modifier for brake fluids; intermediate for PEG esters, PU foams; plasticizer/solv.; cosmetics; metalworking lubricants; paints/resins; paper/film; pharmaceuticals; printing inks; rubber; textile aux.; biodeg.; APHA 25 color; sol. in water and most polar org. solvs.; m.w. 285-315; sp.gr. 1.128; visc. 5.8 cst (99 C); pour pt. -12 C; flash pt. (OC) > 190 C; ref. index 1.463; pH 4.5-7.5; surf. tens. 44.6 dynes/cm; 99% act.

Teric PEG 400. [ICI Australia] PEG 400; biodeg.; emulsifier, antistat for textiles; rubber; inks; cosmetics; pharmaceuticals; paper/film; pesticide solubilizer/carrier; intermediate for PEG esters, PU foams; plasticizer/solv. for cork; metalworking lubricants; paints/resins; APHA 25 color; sol. in water and most polar org. solvs.; m.w. 380-420; sp.gr. 1.130; visc. 7.3 cst (99 C); pour pt. 6 C; flash pt. (OC) > 215 C; ref. index 1.465; pH 4.5-7.5; surf. tens. 44.6 dynes/cm; 99% act.

Teric PEG 600. [ICI Australia] PEG 600; biodeg.; emulsifier for textiles; pesticide solubilizer/carrier; binder for ceramics; intermediate for PEG esters, PU foams; cosmetics; pharmaceuticals; metalworking lubricants; resins; paper/film; inks; rubber; APHA 10 color (25% aq.); sol. in water and most polar org. solvs.; m.w. 560-630; sp.gr. 1.127; visc. 10.4 cst (99 C); pour pt. 19 C; flash pt. (OC) > 230 C; ref. index 1.454; pH 4.5-7.5; surf. tens. 44.6 dynes/cm; 99% act.

Teric PEG 800. [ICI Australia] PEG 800; biodeg.; textile aux.; pharmaceutical tabletting; pesticide solubilizer/carrier; intermediate for PEG esters; toiletries; metalworking lubricants; APHA 30 color (25% aq.); sol. in water and most polar org. solvs.; m.w. 760-840; sp.gr. 1.184; visc. 13.8 cst (99 C); pour pt. 28 C; flash pt. (OC) > 235 C; ref. index 1.455; pH 4.5-7.5; 99% act.

Teric PEG 1000. [ICI Australia] PEG 1000; biodeg.; intermediate for PEG esters; cosmetics/toiletries/soaps; metalworking lubricants; wood processing; APHA 30 color; sol. in water and most polar org. solvs.; m.w. 950-1050; sp.gr. 1.198; visc. 17.8 cst (99 C); pour pt. 37 C; flash pt. (OC) > 245 C; ref. index 1.455; pH 4.5-7.5; surf. tens. 54.2 dynes/cm (50% aq.); 100% act.

Teric PEG 1500. [ICI Australia] PEG 1500; biodeg.; textile finishing and sizing; wood processing; latex prod.; printing inks; pharmaceuticals; paper/film; APHA 30 color (25% aq.); sol. in water and most polar org. solvs.; m.w. 1430-1570; sp.gr. 1.208; visc. 28.4 cst (99 C); pour pt. 46 C; flash pt. (OC) > 260 C; ref. index 1.456; pH 4.5-7.5; surf. tens. 53.1 dynes/cm (50% aq.); 100% act.

Teric PEG 3350. [ICI Australia] PEG 3350; biodeg.; intermediates for PEG esters; spinning aid for textiles; pharmaceuticals; APHA 30 color (25% aq.); sol. in water and most polar org. solvs.; m.w. 3000-3700; sp.gr. 1.215; visc. 70-120 cst (99 C); pour pt. 53 C; flash pt. (OC) > 260 C; ref. index 1.456; pH 4.5-7.5; surf. tens. 54.0 dynes/cm (50% aq.); 100% act.

Teric PEG 4000. [ICI Australia] PEG 4000; biodeg.; binder/plasticizer for ceramics; intermediate for copolymers, PEG esters; cosmetics; pharmaceuticals; metalworking lubricants and electropolishes; resins; paper/film; printing inks; rubber antistat, release, compding. aid; textile aux.; APHA 50 color (25% aq.); sol. in water and most polar org. solvs.; m.w. 3300-4000; sp.gr. 1.217; visc. 130-180 cst (99 C); pour pt. 56 C; flash pt. (OC) > 260 C; ref. index 1.456; pH 4.5-7.5; surf. tens. 54.4 dynes/cm (50% aq.); 100% act.

Teric PEG 6000. [ICI Australia] PEG 6000; biodeg.; cosmetics; pharmaceuticals; thickener for inks;

wood processing; binder/plasticizer for ceramics; intermediate for copolymers, PEG esters; metalworking lubricants; resins; APHA 50 color (25% aq.); sol. in water and most polar org. solvs.; m.w. 6000-7500; sp.gr. 1.217; visc. 330-500 cst (99 C); pour pt. 55-60 C; flash pt. (OC) > 260 C; ref. index 1.456; pH 4.5-7.5; surf. tens. 55.3 dynes/cm (50% aq.); 100% act.

Teric PEG 8000. [ICI Australia] PEG 8000; biodeg.; intermediate for copolymers, PEG esters; cosmetics; pharmaceuticals; thickener for inks; release for rubber molding; sol. in water and most polar org. solvs.; m.w. 7000-9000; sp.gr. 1.2; visc. 700-900 cst (99 C); pour pt. 55-60 C; flash pt. (OC) > 260 C; ref. index 1.456; pH 4.5-7.5; 100% act.

Teric PEG 12000. [ICI Australia] PEG 12000; biodeg.; APHA 50 color (25% aq.); sol. in water and most polar org. solvs.; m.w. 8000-9000; sp.gr. 1.2;pour pt. 55-60 C; flash pt. (OC) > 260 C; ref. index 1.456; pH 4.5-7.5; 100% act.

Teric PPG 400. [ICI Australia] PPG 400; biodeg.; intermediate for surfactants, ethers, esters; metalworking lubricants; solv. for paints/varnishes, printing inks, veg. oils; textile lubricants; APHA 100 color; sol. in polar org. solvs.; sp.gr. 1.009; visc. 82 cps; hyd. no. 220-250; pour pt. -26 C; flash pt. (OC) 204 C; ref. index 1.445; pH 5-8 (1% aq. disp.); surf. tens. 31.1 dynes/cm; 99% act.

Teric PPG 1000. [ICI Australia] PPG 1000; biodeg.; intermediate for surfactants, ethers, esters; antifoam for ceramics, rubber; hydraulic brake fluids; dyeing; metalworking lubricants; plasticizer for plastics; latex coagulant; textile lubricant; APHA 100 color; sol. in polar org. solvs.; sp.gr. 1.006; visc. 196 cps; hyd. no. 106-118; pour pt. -36 C; flash pt. (OC) 227 C; ref. index 1.448; surf. tens. 31.3 dynes/cm; 99% act.

Teric PPG 1650. [ICI Australia] PPG 1650; biodeg.; intermediate for surfactants, ethers, esters; antifoam for ceramics, rubber; lubricant/softener for leather; metalworking lubricant; demulsifier for petrol. industry; plasticizer for plastics; sol. in polar org. solvs.; sp.gr. 1.006; visc. 330 cps; hyd. no. 67-71; pour pt. -32 C; flash pt. (OC) 230 C; ref. index 1.449; pH 5-8 (1% aq. disp.); surf. tens. 31.8 dynes/cm; 99% act.

Teric PPG 2250. [ICI Australia] PPG 2250; biodeg.; intermediate for surfactants, ethers, esters; hydraulic brake fluids; cosmetics; dyeing; lubricant/softener for leather; metalworking lubricant; demulsifier for petrol.; latex coagulant; rubber release agent; solv. for veg. oils; APHA 50 color; sol. in polar org. solvs.; sp.gr. 1.005; visc. 482 cps; hyd. no. 47-54; pour pt. -29 C; flash pt. (OC) 240 C; ref. index 1.449; pH 4.0-7.5 (10% in aq. methanol); surf. tens. 32.1 dynes/cm; 99% act.

Teric PPG 4000. [ICI Australia] PPG 4000; biodeg.; intermediate for surfactants, ethers, esters; cosmetics; lubricant/softener for leather; metalworking lubricants; demulsifier for petrol. industry; mold release for rubber; APHA 250 color; sol. in polar org. solvs.; sp.gr. 1.004; visc. 1232 cps; hyd. no. 26-30; pour pt. -30 C; flash pt. (OC) 240 C; ref. index 1.450; pH 5-8 (10% in aq. methanol); surf. tens. 32.2 dynes/cm; 99% act.

Teric SF9. [ICI Australia] PEG-9 stearate; CAS 9004-99-3; nonionic; coemulsifier, preparation of self-emulsifying waxes for waterproofing of fiber and chip boards, bldg. sheets, and cardboard pkg.; component of floor wax emulsions and wax strippers; Hazen 250 solid; sol. in water, benzene, ethyl acetate, ethyl Icinol, perchlorethylene, ethanol, kerosene, veg. oil, olein; sp.gr. 0.994 (50 C); visc. 45 cps (50 C); m.p. 30 ± 2 C; HLB 10.8; pH 8 (1% aq.); biodeg.; 100% act.

Teric SF15. [ICI Australia] PEG-15 stearate; CAS 9004-99-3; nonionic; coemulsifier; textile and leather dressing oils and a dye bath additive; Hazen 250 solid; sol. in water, benzene, ethyl acetate, ethyl Icinol, perchlorethylene; sp.gr. 1.014 (50 C); visc. 55 cps (50 C); m.p. 32 ± 2 C; HLB 14.0; surf. tens. 40.7 dynes/cm; pH 8 (1% aq.); biodeg.; 100% act.

Teric T2. [ICI Australia] PEG-2 tallate; CAS 61791-00-2; nonionic; wetting agent and dispersant for oil-based systems; emulsifier, detergent, grinding aid for wet milling of pigments and resins; mfg. of lubricants for metal and textile industries, solv. cleaners, and powd. detergent formulations; Hazen > 250 liq.; sol. in benzene, ethyl acetate, ethyl Icinol, perchlorethylene, ethanol, min. and veg. oil, olein; disp. in water; sp.gr. 0.966; visc. 118 cps; m.p. < 0 C; HLB 4.5; pH 6–8 (1% aq.); biodeg.; 100% act.

Teric T5. [ICI Australia] PEG-5 tallate; CAS 61791-00-2; nonionic; wetting agent and dispersant for oil-based systems; emulsifier, detergent, grinding aid for wet milling of pigments and resins; mfg. of lubricants for metal and textile industries, solv. cleaners, and powd. detergent formulations; Hazen > 250 liq.; sol. in benzene, ethyl acetate, ethyl Icinol, perchlorethylene, ethanol, min. and veg. oil, olein; disp. in water; sp.gr. 1.018; visc. 196 cps; m.p. < 0 C; HLB 8.2; pH 8–10 (1% aq.); biodeg.; 100% act.

Teric T7. [ICI Australia] PEG-7 tallate; CAS 61791-00-2; nonionic; wetting agent and dispersant for oil-based systems; emulsifier, detergent, grinding aid for wet milling of pigments and resins; mfg. of lubricants for metal and textile industries, solv. cleaners, and powd. detergent formulations; Hazen > 250 liq.; sol. in benzene, ethyl acetate, ethyl Icinol, perchlorethylene, ethanol, veg. oil, olein; sp.gr. 1.031; visc. 151 cps; m.p. < 0 C; HLB 10.0; pH 8–10 (1% aq.); biodeg.; 100% act.

Teric T10. [ICI Australia] PEG-10 tallate; CAS 61791-00-2; nonionic; wetting agent and dispersant for oil-based systems; emulsifier, detergent, grinding aid for wet milling of pigments and resins; mfg. of lubricants for metal and textile industries, solv. cleaners, and powd. detergent formulations; Hazen > 250 liq.; sol. in benzene, ethyl acetate, ethyl Icinol, perchlorethylene, ethanol, veg. oil, olein; also disp. in water; sp.gr. 1.042; visc. 253 cps; m.p. 5 ± 2 C; HLB 11.7; pH 8–10 (1% aq.); biodeg.; 100% act.

Teric X5. [ICI Australia] Octoxynol-5; CAS 9002-93-1; nonionic; wetting agent and dispersant, detergent used in alkaline and metal cleaners, agric. powds., emulsions, pigment and wax dispersions; textile auxs.; paints; Hazen 100 liq.; sol. in benzene, ethyl acetate, ethyl Icinol, perchlorethylene, ethanol, kerosene, min., paraffin and veg. oil, olein; sp.gr. 1.046; visc. 405 cps; m.p. < 0 ± 2 C; HLB 10.4; hyd. no. 125-133; pH 6–8 (1% aq.); 100% act.

Teric X7. [ICI Australia] Octoxynol-7 (7.5 EO); CAS 9002-93-1; nonionic; see Teric X5; Hazen 100 liq.; sol. in water, benzene, ethyl acetate, ethyl Icinol, perchlorethylene, ethanol, veg. oil, olein; sp.gr. 1.059; visc. 420 cps; m.p. < 0 ± 2 C; HLB 12.3; cloud pt. 28 ± 2 C; surf. tens. 29.5 dynes/cm; pH 6–8 (1% aq.); 100% act.

Teric X8. [ICI Australia] Octoxynol-8 (8.5 EO); CAS 9002-93-1; nonionic; see Teric X5; Hazen 100 liq.;

sol. in water, benzene, ethyl acetate, ethyl Icinol, perchlorethylene, ethanol, olein; sp.gr. 1.060; visc. 402 cps; m.p. < 0 ± 2 C; HLB 12.6; cloud pt. 47 ± 2 C; surf. tens. 30.2 dynes/cm; pH 6–8 (1% aq.); 100% act.

Teric X10. [ICI Australia] Octoxynol-10; CAS 9002-93-1; nonionic; see Teric X5; Hazen 100 liq.; sol. in water, benzene, ethyl acetate, ethyl Icinol, perchlorethylene, ethanol, olein; sp.gr. 1.062; visc. 393 cps; m.p. 7 ± 2 C; HLB 13.6; cloud pt. 65 ± 2 C; surf. tens. 31.8 dynes/cm; pH 6–8 (1% aq.); 100% act.

Teric X11. [ICI Australia] Octoxynol-11; CAS 9002-93-1; nonionic; see Teric X5; Hazen 100 liq.; sol. in water, benzene, ethyl acetate, ethyl Icinol, perchlorethylene, ethanol, olein; sp.gr. 1.067; visc. 371 cps; m.p. 8 ± 2 C; HLB 14.0; cloud pt. 80 ± 2 C; surf. tens. 31.0 dynes/cm; pH 6–8 (1% aq.); 100% act.

Teric X13. [ICI Australia] Octoxynol-13; CAS 9002-93-1; nonionic; see Teric X5; Hazen 100 liq.; sol. in water, benzene, ethyl acetate, ethyl Icinol, perchlorethylene, ethanol, olein; sp.gr. 1.052; visc. 70 cps; m.p. 15 ± 2 C; HLB 14.7; cloud pt. 90 ± 2 C; surf. tens. 32.0 dynes/cm; pH 6–8 (1% aq.); 100% act.

Teric X16. [ICI Australia] Octoxynol-16; CAS 9002-93-1; nonionic; see Teric X5; Hazen 100 semisolid; sol. in benzene, ethyl acetate, ethyl Icinol, perchlorethylene, ethanol, kerosene, min., paraffin and veg. oil, olein; sp.gr. 1.059 (50 C); visc. 95 cps (50 C); m.p. 24 ± 2 C; HLB 15.2; cloud pt. > 100 C; surf. tens. 35.6 dynes/cm; pH 6–8 (1% aq.); 100% act.

Teric X40. [ICI Australia] Octoxynol-40; CAS 9002-93-1; nonionic; see Teric X5; Hazen 100 solid; sol. in water, benzene, ethyl acetate, ethyl Icinol, perchlorethylene, ethanol; sp.gr. 1.082 (50 C); visc. 245 cps (50 C); m.p. 44 ± 2 C; HLB 17.6; cloud pt. > 100 C; surf. tens. 36.4 dynes/cm; pH 6–8 (1% aq.); 100% act.

Teric X40L. [ICI Australia] Octoxynol-40; CAS 9002-93-1; nonionic; APHA 100 color; sp.gr. 1.098; visc. 975 cps; HLB 17.6; pour pt. 5 C; pH 7-9; surf. tens. 36.4 dynes/cm (0.1%); 70% act.

Tetradecene-1. [Ethyl] C14 alpha olefins; intermediate for surfactants and industrial chemicals; liq.; 100% conc.

Tetranol. [Sandoz] Sodium butyl oleate; nonionic; wetting and leveling agent for dyeing processes; assistant for vat dyeing; emulsifier and detergent in scouring processes; water-sol.; pH 5.8–6.0.

Tetraterge D-101, NFF. [Reilly-Whiteman] Detergent for textile scouring.

Tetrawet DWN. [Reilly-Whiteman] Wetting agent for textile processing of cellulosics and blends.

Tetronic® 50R1. [BASF] EO/PO ethylene diamine block copolymer; nonionic; surfactant series functioning as emulsion stabilizers, solubilizers, dispersants, wetting agents, antistats, penetrants, plasticizers, defoaming agents, demulsifiers in the petrol., paint, paper, cement, ink, cosmetic; drug, plastic, detergent, and metalworking industries; rubber activator; R series for low foaming applics.; liq.; m.w. 2640; visc. 670 cps; HLB 1–7; pour pt. –18 C; cloud pt. 29 C (1% aq.); surf. tens. 40 dynes/cm (0.1%); toxicology: minimal skin and minimal to mild eye irritation; 100% act.

Tetronic® 50R4. [BASF] EO/PO ethylene diamine block copolymer; see Tetronic 50R1; liq.; m.w. 3740; visc. 1080 cps; HLB 7–12; pour pt. –4 C; cloud pt. 59 C (1% aq.); surf. tens. 46 dynes/cm (0.1%); toxicology: minimal skin and minimal to mild eye irritation; 100% act.

Tetronic® 50R8. [BASF] EO/PO ethylene diamine block copolymer; CAS 107397-59-1; see Tetronic 50R1; solid; m.w. 10,200; visc. 650 cps (77 C); HLB 12–18; m.p. 38 C; cloud pt. 88 C (1% aq.); surf. tens. 53 dynes/cm (0.1%); toxicology: minimal skin and minimal to mild eye irritation; 100% act.

Tetronic® 70R1. [BASF] EO/PO ethylene diamine block copolymer; see Tetronic 50R1; liq.; m.w. 3400; visc. 800 cps; HLB 1–7; pour pt. –17 C; cloud pt. 25 C (1% aq.); surf. tens. 39 dynes/cm (0.1%); toxicology: minimal skin and minimal to mild eye irritation; 100% act.

Tetronic® 70R2. [BASF] EO/PO ethylene diamine block copolymer; see Tetronic 50R1; liq.; m.w. 3870; visc. 880 cps; HLB 1–7; pour pt. –22 C; cloud pt. 31 C (1% aq.); surf. tens. 40 dynes/cm (0.1%); toxicology: minimal skin and minimal to mild eye irritation; 100% act.

Tetronic® 70R4. [BASF] EO/PO ethylene diamine block copolymer; CAS 107397-59-1; see Tetronic 50R1; liq.; m.w. 5230; visc. 2160 cps; HLB 7–12; pour pt. 6 C; cloud pt. 42 C (1% aq.); surf. tens. 43 dynes/cm (0.1%); toxicology: minimal skin and minimal to mild eye irritation; 100% act.

Tetronic® 90R1. [BASF] EO/PO ethylene diamine block copolymer; see Tetronic 50R1; liq.; m.w. 4580; visc. 940 cps; HLB 1–7; pour pt. –17 C; cloud pt. 21 C (1% aq.); surf. tens. insol. (0.1%); toxicology: minimal skin and minimal to mild eye irritation; 100% act.

Tetronic® 90R4. [BASF] EO/PO ethylene diamine block copolymer; CAS 107397-59-1; nonionic; see Tetronic 50R1; liq.; m.w. 7240; visc. 3870 cps; HLB 1–7; pour pt. 12 C; cloud pt. 43 C (1% aq.); surf. tens. 43 dynes/cm (0.1%); toxicology: minimal skin and minimal to mild eye irritation; 100% act.

Tetronic® 90R8. [BASF] EO/PO ethylene diamine block copolymer; see Tetronic 50R1; solid; m.w. 18,700; visc. 4000 cps (77 C); HLB 12–18; m.p. 47 C; cloud pt. 88 C (1% aq.); surf. tens. 50 dynes/cm (0.1%); toxicology: minimal skin and minimal to mild eye irritation; 100% act.

Tetronic® 110R1. [BASF] EO/PO ethylene diamine block copolymer; see Tetronic 50R1; liq.; m.w. 5220; visc. 1000 cps; HLB 1–7; pour pt. –18 C; cloud pt. 21 C (1% aq.); surf. tens. insol. (0.1%); toxicology: minimal skin and minimal to mild eye irritation; 100% act.

Tetronic® 110R2. [BASF] EO/PO ethylene diamine block copolymer; see Tetronic 50R1; liq.; m.w. 5900; visc. 1320 cps; HLB 1–7; pour pt. –20 C; cloud pt. 27 C (1% aq.); surf. tens. 38 dynes/cm (0.1%); toxicology: minimal skin and minimal to mild eye irritation; 100% act.

Tetronic® 110R7. [BASF] EO/PO ethylene diamine block copolymer; see Tetronic 50R1; solid; m.w. 13,200; visc. 900 cps (77 C); HLB 7–12; m.p. 47 C; cloud pt. 64 C (1% aq.); surf. tens. 46 dynes/cm (0.1%); toxicology: minimal skin and minimal to mild eye irritation; 100% act.

Tetronic® 130R1. [BASF] EO/PO ethylene diamine block copolymer; see Tetronic 50R1; liq.; m.w. 6800; visc. 1240 cps; HLB 1–7; pour pt. –18 C; cloud pt. 20 C (1% aq.); surf. tens. insol. (0.1%); toxicology: minimal skin and minimal to mild eye irritation; 100% act.

Tetronic® 130R2. [BASF] EO/PO ethylene diamine block copolymer; see Tetronic 50R1; liq.; m.w. 7740; visc. 1880 cps; HLB 1–7; pour pt. –3 C; cloud pt. 25 C (1% aq.); surf. tens. 35 dynes/cm (0.1%);

toxicology: minimal skin and minimal to mild eye irritation; 100% act.

Tetronic® 150R1. [BASF] EO/PO ethylene diamine block copolymer; CAS 107397-59-1; nonionic; see Tetronic 50R1; liq.; m.w. 8000; visc. 1840 cps; HLB 1–7; pour pt. –17 C; cloud pt. 20 C (1% aq.); surf. tens. insol. (0.1%); toxicology: minimal skin and minimal to mild eye irritation; 100% act.

Tetronic® 150R4. [BASF] EO/PO ethylene diamine block copolymer; see Tetronic 50R1; paste; m.w. 11,810; visc. 850 cps (60 C); HLB 1–7; pour pt. 34 C; cloud pt. 29 C (1% aq.); surf. tens. 39 dynes/cm (0.1%); toxicology: minimal skin and minimal to mild eye irritation; 100% act.

Tetronic® 150R8. [BASF] EO/PO ethylene diamine block copolymer; see Tetronic 50R1; solid; m.w. 20,400; visc. 6300 cps (77 C); HLB 7–12; m.p. 53 C; cloud pt. 77 C (1% aq.); surf. tens. 44 dynes/cm (0.1%); toxicology: minimal skin and minimal to mild eye irritation; 100% act.

Tetronic® 304. [BASF] Poloxamine 304; nonionic; emulsifier, thickener, wetting agent, dispersant, solubilizer, stabilizer for cosmetics and pharmaceuticals; demulsifier in petrol. industry; detergent ingred.; antistat for polyethylene and resin molding powds.; metal treatment; emulsion polymerization; used in latex-based paints, aq.-based syn. cutting fluids and vulcanization of rubber; colorless liq.; m.w. 1650; ref. index 1.4649; water-sol.; sp.gr. 1.06; visc. 450 cps; HLB 16; cloud pt. 94 C (1%); pour pt. –11 C; surf. tens. 53.0 dynes/cm (0.1%); toxicology: none to mild eye and minimal to moderate skin irritation; 100% act.

Tetronic® 504. [BASF] Poloxamine 504; nonionic; see Tetronic 304; colorless liq.; m.w. 3400; ref. index 1.4612; water-sol.; sp.gr. 1.04; visc. 800 cps; HLB 15.5; cloud pt. 68 C (1%); pour pt. 7 C; surf. tens. 44.2 dynes/cm (0.1%); toxicology: none to mild eye and minimal to moderate skin irritation; 100% act.

Tetronic® 701. [BASF] Poloxamine 701; nonionic; see Tetronic 304; colorless liq.; m.w. 3400; ref. index. 1.4553; sp.gr. 1.02; visc. 575 cps; HLB 3; cloud pt. 22 C (1%); pour pt. –21 C; surf. tens. 36.1 dynes/cm; toxicology: none to mild eye and minimal to moderate skin irritation; 100% act.

Tetronic® 702. [BASF] Poloxamine 702; nonionic; see Tetronic 304; colorless liq.; m.w. 4000; ref. index 1.4572; water-sol.; sp.gr. 1.03; visc. 770 cps; HLB 7; cloud pt. 27 C (1%); pour pt. –7 C; surf. tens. 36.8 dynes/cm (0.1%); toxicology: none to mild eye and minimal to moderate skin irritation; 100% act.

Tetronic® 704. [BASF] Poloxamine 704; nonionic; see Tetronic 304; colorless liq.; m.w. 5500; ref. index 1.4613; water-sol.; sp.gr. 1.04; visc. 850 cps; HLB 15; cloud pt. 65 C (1%); pour pt. 18 C; surf. tens. 40.3 dynes/cm (0.1%); toxicology: none to mild eye and minimal to moderate skin irritation; 100% act.

Tetronic® 707. [BASF] Poloxamine 707; nonionic; see Tetronic 304; flakes; m.w. 12,000; water-sol.; m.p. 49 C; HLB 27; cloud pt. > 100 C (1%); surf. tens. 47.6 dynes/cm (0.1%); toxicology: none to mild eye and minimal to moderate skin irritation; 100% act.

Tetronic® 901. [BASF] Poloxamine 901; nonionic; see Tetronic 304; colorless liq.; m.w. 4750; ref. index 1.4545; sp.gr. 1.02; visc. 700 cps; HLB 2.5; cloud pt. 20 C (1%); pour pt. –23 C; toxicology: none to mild eye and minimal to moderate skin irritation; 100% act.

Tetronic® 904. [BASF] Poloxamine 904; nonionic; see Tetronic 304; colorless liq.; m.w. 7500; ref. index 1.4604; water-sol.; sp.gr. 1.04; visc. 6000 cps; HLB 14.5; cloud pt. 64 C (1%); pour pt. 29 C; surf. tens. 35.4 dynes; toxicology: none to mild eye and minimal to moderate skin irritation; 100% act.

Tetronic® 908. [BASF] Poloxamine 908; nonionic; see Tetronic 304; flakes; m.w. 27,000; water-sol.; m.p. 58 C; HLB 30.5; cloud pt. > 100 C (1%); surf. tens. 45.7 dynes/cm (0.1%); toxicology: none to mild eye and minimal to moderate skin irritation; 100% act.

Tetronic® 909. [BASF] EO/PO ethylene diamine block copolymer; see Tetronic 304; solid; m.w. 30,000; visc. 20,000 cps (77 C); HLB > 24; m.p. 59 C; cloud pt. > 100 C (1% aq.); surf. tens. 56 dynes/cm (0.1%); toxicology: none to mild eye and minimal to moderate skin irritation; 100% act.

Tetronic® 1101. [BASF] Poloxamine 1101; nonionic; see Tetronic 304; colorless liq.; m.w. 5600; ref. index 1.4540; sp.gr. 1.02; visc. 700 cps; HLB 2; cloud pt. 17 C (1%); pour pt. –15 C; surf. tens. 34.0 dynes/cm; toxicology: none to mild eye and minimal to moderate skin irritation; 100% act.

Tetronic® 1102. [BASF] Poloxamine 1102; nonionic; see Tetronic 304; colorless liq.; m.w. 6300; ref. index 1.4557; water-sol.; sp.gr. 1.03; visc. 820 cps; HLB 6; cloud pt. 31 C (1%); pour pt. 7 C; surf. tens. 33.7 dynes/cm (0.1%); toxicology: none to mild eye and minimal to moderate skin irritation; 100% act.

Tetronic® 1104. [BASF] Poloxamine 1104; nonionic; see Tetronic 304; paste; m.w. 8300; water-sol.; m.p. 34 C; HLB 14; cloud pt. 72 C (1%); surf. tens. 36.3 dynes/cm (0.1%); toxicology: none to mild eye and minimal to moderate skin irritation; 100% act.

Tetronic® 1107. [BASF] Poloxamine 1107; nonionic; see Tetronic 304; solid; m.w. 14,500; water-sol.; m.p. 51 C; HLB 24; cloud pt. > 100 C; surf. tens. 42.9 dynes/cm (0.1%); toxicology: none to mild eye and minimal to moderate skin irritation; 100% act.

Tetronic® 1301. [BASF] Poloxamine 1301; nonionic; see Tetronic 304; colorless liq.; m.w. 6800; ref. index 1.4545; sp.gr. 1.02; visc. 1000 cps; HLB 1.5; cloud pt. 16 C (1%); pour pt. –9 C; surf. tens. 33.4 dynes/cm; toxicology: none to mild eye and minimal to moderate skin irritation; 100% act.

Tetronic® 1302. [BASF] Poloxamine 1302; nonionic; see Tetronic 304; colorless liq.; m.w. 7800; ref. index 1.4562; water-sol.; sp.gr. 1.03; visc. 1300 cps; HLB 5.5; cloud pt. 20 C (1%) pour pt. 13 C; surf. tens. 34.1 dynes/cm (0.1%); toxicology: none to mild eye and minimal to moderate skin irritation; 100% act.

Tetronic® 1304. [BASF] Poloxamine 1304; nonionic; see Tetronic 304; paste; m.w. 10,500; water-sol.; m.p. 36 C; HLB 13.5; cloud pt. 78 C (1%); surf. tens. 35.5 dynes/cm (0.1%); toxicology: none to mild eye and minimal to moderate skin irritation; 100% act.

Tetronic® 1307. [BASF] Poloxamine 1307; nonionic; see Tetronic 304; solid; m.w. 18,600; water-sol.; m.p. 54 C; HLB 23.5; cloud pt. > 100 C (1%); surf. tens. 43.8 dynes/cm (0.1%); toxicology: none to mild eye and minimal to moderate skin irritation; 100% act.

Tetronic® 1501. [BASF] Poloxamine 1501; nonionic; see Tetronic 304; colorless liq.; m.w. 7900; ref. index 1.4537; sp.gr. 1.02; visc. 1170 cps; HLB 1.0; cloud pt. 15 C (1%); pour pt. –4 C; surf. tens. 33.9 dynes/cm; toxicology: none to mild eye and minimal to moderate skin irritation; 100% act.

Tetronic® 1502. [BASF] Poloxamine 1502; nonionic; see Tetronic 304; colorless liq.; m.w. 9000; ref. index 1.4560; water-sol.; sp.gr. 1.03; visc. 1570 cps; HLB 5.0; cloud pt. 70 C (1%); pour pt. 18 C; surf. tens. 34.1 dynes/cm (0.1%); toxicology: none to mild eye and minimal to moderate skin irritation; 100% act.

Tetronic® 1504. [BASF] Poloxamine 1504; nonionic; see Tetronic 304; paste; m.w. 12,500; water-sol.; m.p. 41 C; HLB 13.0; cloud pt. 90 C (1%); surf. tens. 37.3 dynes/cm (0.1%); toxicology: none to mild eye and minimal to moderate skin irritation; 100% act.

Tetronic® 1508. [BASF] Poloxamine 1508; nonionic; see Tetronic 304; solid; m.w. 27,000; water-sol.; m.p. 60 C; HLB 27; cloud pt. > 100 C (1%); surf. tens. 43.8 dynes/cm (0.1%); toxicology: none to mild eye and minimal to moderate skin irritation; 100% act.

Texadril 2010. [Henkel/Cospha; Henkel/Functional Prods.] EO/PO block copolymer; nonionic; low foaming wetting agent and coemulsifier; liq.; 100% act.

Texadril 8780. [Henkel/Functional Prods.] EO/PO block copolymer; nonionic; stabilizer, emulsifier esp. for vinyl acetate homo- and copolymers; solid; 100% act.

Texadril LT 1285. [Henkel/Functional Prods.] Polyglycol ether; nonionic; antifoaming agent for aq. systems; high alkali and acid stability; completely biodeg.; liq.; 85% act.

Texamin PD. [Henkel/Cospha; Henkel Canada] Fatty acid polydialkanolamide; nonionic; emulsifier for metalworking fluids; corrosion inhibitor; 100% conc.

Texamine 84(L). [Zohar Detergent Factory] Alkanolamide and ethanolamine alkylbenzene sulfonate; anionic/nonionic; raw material for mfg. of liq. detergents; paste; 84% conc.

Texapon 842. [Henkel Canada; Henkel KGaA] Sodium octyl sulfate; anionic; wetting agent for hydrophobic powds.; liq.; 40% conc.

Texapon 890. [Henkel KGaA] Sodium octyl sulfate; anionic; wetting agent for hydrophobic powds.; powd.; 100% conc.

Texapon 1030, 1090. [Henkel KGaA] Sodium decyl sulfate; anionic; wetting agent and foamer for fire fighting foams; liq. and powd. resp.; 30 and 90% conc.

Texapon EA-K. [Henkel/Cospha; Henkel Canada] Proprietary blend; anionic; ready-to-dilute conc. for clear shampoos, foam baths, cleansing preps., liq. soaps, dishwash; liq.; 60% conc.

Texapon F. [Henkel] Fatty alcohol sulfate and alkylaryl sulfate compd.; anionic; detergent, dishwashing agents; powd.; 45% conc.

Texapon F35. [Henkel] Fatty alcohol sulfate and alkylaryl sulfate compd.; anionic; detergent, dishwashing agents; powd.; 35% conc.

Texapon GYP 1. [Henkel KGaA] Surfactant blend; anionic; foaming agent for plaster boards; liq.; 43% conc.

Texapon GYP 2. [Henkel KGaA] Fatty alcohol polyglycol ether; anionic; foamer for prod. of plaster boards; liq.; 40% conc.

Texapon GYP 3. [Henkel KGaA] Anionic/nonionic surfactants with foam stabilizer; anionic; foaming agent for plaster boards; liq.; 60% conc.

Texapon GYP 4. [Henkel KGaA] Anionic/nonionic surfactants with foam stabilizer; anionic; foaming agent for plaster boards; liq.; 69% conc.

Texapon HD-L 2. [Henkel KGaA] Surfactant/fatty acid blend; anionic; optimized surfactant conc. for heavy-duty liqs.; liq.; 80% conc.

Texapon K-12. [Henkel/Cospha; Henkel/Functional Prods.] Sodium lauryl sulfate; CAS 151-21-3; anionic; emulsifier for emulsion polymerization; detergent for high-solids cleansers; also for dispersants, wettable powds.; wh. powd.; > 90% act.

Texapon K-1296. [Henkel/Cospha; Henkel/Functional Prods.] Sodium lauryl sulfate; CAS 151-21-3; anionic; wetting agent and detergent for cleaning formulations; additive for mech. latex foaming; wh. powd.; > 96% act.

Texapon K-12 Granules. [Henkel/Cospha; Henkel KGaA] Sodium lauryl sulfate; CAS 151-21-3; anionic; foamer for pharmaceuticals, detergents, dispersants, wettable powds.; wh. gran.; 85% act.

Texapon K-12 Needles. [Henkel/Cospha; Henkel KGaA] Sodium lauryl sulfate; CAS 151-21-3; anionic; foamer for toothpaste, detergents, wettable powds.; wh. needles; 85% act.

Texapon L-100. [Henkel/Cospha] Sodium lauryl sulfate; CAS 151-21-3; wetting and cleansing agent, dispersant used in dentifrice mfg., high-solids cleansers, wettable powds.; wh. powd.; pH 7.0 (1% aq.); 98% act.

Texapon LLS. [Henkel KGaA] Lithium lauryl sulfate; anionic; wetting agent for carpet shampoos, dry foam cleaners; liq.; 30% conc.

Texapon LS 35. [Henkel KGaA] Sodium lauryl sulfate (C12-14); CAS 151-21-3; anionic; raw material for detergents, hand cleaners; paste; 35% conc.

Texapon LS Highly Conc. [Henkel/Functional Prods.] Sodium lauryl sulfate (C12-14); CAS 151-21-3; anionic; foaming agent for acrylate dispersions, carpet and upholstery cleaners; powd.; 86-90% act.

Texapon LS Highly Conc. Needles. [Henkel KGaA] Sodium lauryl sulfate (C12-C14); CAS 151-21-3; anionic; detergent, foamer used in personal care prods. and detergents; needles; 90% conc.

Texapon LT-327. [Pulcra SA] Sodium lauryl ether (2.7) sulfate; CAS 9004-82-4; anionic; for shampoos and lt. duty detergents; liq.; 27% conc.

Texapon MGLS. [Henkel Canada] Magnesium lauryl sulfate; anionic; mild foamer for shampoos, bubble baths, cleansing preps., carpet shampoos; liq.; 30% conc.

Texapon N 25. [Henkel KGaA; Pulcra SA] Sodium laureth (2) sulfate; CAS 9004-82-4; anionic; base for personal care prods.; dishwashing agent; fire-fighting foam concs.; solubilizer for perfumes; liq.; m.w. 382; visc. 2500-2900 mPa•s; pH 6.4-7.5 (10%); 26.5-27.5% act.

Texapon N 40. [Henkel KGaA; Pulcra SA] Sodium laureth (2) sulfate; CAS 9004-82-4; 1335-72-4; anionic; detergent for liq. shampoos, bubble baths, lt. duty detergents; liq.; m.w. 382; visc. 4000-4500 mPa•s; pH 6.4-7.5 (10%); 26.5-27.5% act.

Texapon N 70. [Henkel/Cospha; Henkel Canada; Henkel KGaA; Pulcra SA] Sodium laureth (2) sulfate; CAS 9004-82-4; 1335-72-4; anionic; detergent in personal care prods., lt. duty detergents; solubilizer for perfumes; water-wh. paste; m.w. 388; visc. < 9000 mPa•s; pH 7-9 (3%); 68-73% act.

Texapon N 70 LS. [Henkel/Cospha; Henkel KGaA] Sodium laureth sulfate; CAS 9004-82-4; anionic; detergent for liq. shampoos, bubble baths, lt. duty detergents; paste; 74% act.

Texapon N 103. [Henkel KGaA] Sodium laureth

sulfate (3 EO); CAS 9004-82-4; anionic; liq. shampoo and bubble bath base; dishwashing agent and fire-fighting foam concs.; liq.; 27–28% conc.

Texapon NSE. [Pulcra SA] Sodium laureth sulfate; CAS 9004-82-4; anionic; surfactant for emulsion polymerization; liq.; m.w. 405; visc. < 350 mPa•s; pH 6.5-7.5 (10%); 26-28% act.

Texapon NSO. [Henkel KGaA; Pulcra SA] Sodium lauryl ether (2) sulfate; CAS 9004-82-4; anionic; detergent for liq. shampoos, bubble baths, shower gels, mfg. of household detergents; wetting agent and detergent for textile fibers, esp. wool and blends; liq.; m.w. 382; visc. < 200 mPa•s; pH 6.4-7.5 (10%); 26.5-27.5% act.

Texapon OT Highly Conc. Needles. [Henkel/Functional Prods.; Henkel KGaA] Sodium lauryl sulfate C12-C18; CAS 151-21-3; anionic; detergent, shampoo base; for bubble bath, soaps; emulsifier for emulsion polymerization, additive for mech. latex foaming, carpet and upholstery cleaners; needles; 91-92% act.

Texapon P. [Henkel/Cospha; Henkel/Functional Prods.; Henkel Canada; Henkel KGaA] Alkyl sulfate compd.; anionic; detergent, wetting agent; all-purpose detergent base for dishwashing, the canning industry; emulsifier for vinyl acetate homo and copolymers; liq.; 26-28% solids.

Texapon PLT-227. [Pulcra SA] Sodium laureth-2 sulfate; CAS 9004-82-4; 1335-72-4; anionic; for shampoos and liq. detergents; liq.; 27% conc.

Texapon PN-235. [Pulcra SA] Sodium laureth-2.35 sulfate; CAS 9004-82-4; 1335-72-4; anionic; surfactant for mfg. of shampoos, foam baths, personal foam washers, household detergents; solubilizer for perfumes; wetting agent/detergent for textile fibers, esp. wool and blends; liq.; m.w. 403; visc. < 350 mPa•s; pH 6.5-7.5 (10%); 26.5-27.5% act.

Texapon PN-254. [Pulcra SA] Sodium laureth-2.4 sulfate; CAS 9004-82-4; anionic; surfactant for mfg. of shampoos, foam baths, personal foam washers, household detergents; solubilizer for perfumes; paste; m.w. 388; visc. 5000-8000 mPa•s; pH 6.5-8.0 (1%); 53-55% act.

Texapon PNA. [Pulcra SA] Ammonium lauryl sulfate; anionic; base for liq. shampoos, foam baths, in mfg. of household liq. detergents, fire-fighting foams; visc. liq.; m.w. 296; pH 6.3-6.8 (10%); 26-28% act.

Texapon PNA-127. [Pulcra SA] Ammonium lauryl ether (1) sulfate; CAS 32612-48-9; anionic; wetting agent for acid or gel-like shampoos, lt. duty detergents, window cleaners; stable foam; low eye and skin irritation; liq./paste; m.w. 336; pH 6.5-7.0 (10%); 25-26% act.

Texapon SCO. [Henkel/Cospha; Henkel Canada] Sulfated/sulfonated castor oil; anionic; wetting agent; liq.; 75% conc.

Texapon SP 60 N. [Henkel KGaA] Surfactant blend; anionic; optimized surfactant conc. for dishwashing agents, detergents; paste; 60% conc.

Texapon SP 100. [Henkel/Cospha; Henkel Canada; Henkel KGaA] Proprietary blend; anionic/nonionic; ready-to-dilute conc. for all-purpose detergents, dishwashing agents for all temp. and hard water conditions; liq.; 100% conc.

Texapon T 35. [Henkel] TEA-lauryl sulfate; anionic; detergent in personal care prods., lt. duty detergents; solubilizer for perfumes; liq.; 35% conc.

Texapon T 42. [Henkel KGaA; Pulcra SA] TEA lauryl sulfate; CAS 139-96-8; anionic; detergent, emulsifier for hair and rug shampoos, bubble baths, shower gels, fire fighting foams; liq.; m.w. 420; visc. < 100 mPa•s; pH 7.2-7.5 (10%); 40-43% act.

Texapon VHC Needles. [Henkel/Cospha; Henkel/Functional Prods.; Henkel Canada; Henkel KGaA] Sodium lauryl sulfate; CAS 151-21-3; anionic; wetting agent, foamer, emulsifier, detergent and cosmetic base for personal care prods., scouring agents, pigment dispersions, emulsion polymerization; wh. fine needles; pH 7–8 (1% aq.); 86-90% act.

Texapon VHC Powd. [Henkel/Cospha; Henkel KGaA] Sodium lauryl sulfate; CAS 151-21-3; anionic; foamer, dispersant, wetting agent for bubble baths, cosmetic cleansers; raw material for mfg. of detergents and syndet bars; powd.; 88% conc.

Texapon Z. [Henkel KGaA] Sodium lauryl sulfate C12-C14; CAS 151-21-3; anionic; detergent, wetting agent, dispersant for personal care, pharmaceutical preparations, pesticides, paint; needles, powd.; 58–62% conc.

Texapon ZHC Needles. [Henkel/Cospha; Henkel KGaA] Sodium lauryl sulfate; CAS 151-21-3; anionic; foaming agent, dispersant, wetting agent for foaming bubble baths, cosmetic cleansing creams and emulsions; air entraining agent; wh. needles; 85% conc.

Texapon ZHC Powder. [Henkel/Cospha; Henkel/Functional Prods.] Sodium lauryl sulfate C12-C18; CAS 151-21-3; foaming dispersion and wetting agent for bubble baths, cosmetic cleansing creams and emulsions, emulsion polymerization, mech. latex foaming, carpet and upholstery cleaners; wh. fine powd.; pH 7.5–8.5 (0.25% aq.); 88% act.

Texi TD. [CNC Int'l.] Replacement for sodium hydrosulfite for reduction clearing, stripping dyes, and equip. cleaning in textile industry.

Texin 128. [Henkel Canada; Henkel KGaA] Sodium lauryl sulfosuccinate; anionic; detergent, wetting agent, emulsifier for lt. duty household and industrial cleaners, hand soaps, carpet shampoos; powd.; 90% conc.

Texin DOS 75. [Henkel Canada; Henkel KGaA] Sodium diisooctyl sulfosuccinate; anionic; wetting agent, low foam emulsifier for glass, metal cleaners, dust repellents, firefighting foams; liq.; 70% conc.

Texin SB 3. [Henkel KGaA] Disodium sulfosuccinic acid monoalkyl polyglycol; anionic; mild surfactant for hand cleaners, detergents; liq.; 40% conc.

Texnol R 5. [Nippon Nyukazai] Alkyl benzyl ammonium salt; anionic; emulsifier, fungicide, softener; liq.; 100% act.

Texo 227. [Texo] Alkylaryl sulfonate; anionic; general purpose, nonphosphated, biodeg. cleaners; liq.; HLB 11.0.

Texo 284. [Texo] Alkylaryl sulfonate; nonionic; wetting agent; liq.; HLB 10.0; biodeg.

Texo 1060. [Texo] Alkylaryl sulfonate; anionic; general purpose, nonphosphated biodeg. cleaners; liq.; HLB 11.0.

Texo LP 528A. [Texo] Ethoxylated alkylaryl; nonionic; mold release agent; liq.; 40% conc.

Texofor A1P. [Rhone-Poulenc UK] Ceteth-20; nonionic; surfactant; solid; HLB 16.2; 100% conc.

Texofor B1. [Rhone-Poulenc UK] Alcohol ethoxylate; nonionic; surfactant for aq. systems; solid; HLB 15.8; 100% conc.

Textamine 05. [Henkel] Fatty imidazoline; cationic; oil-emulsifying agent, corrosion inhibitor and ink dispersant; liq.; 100% act.

Textamine 1839. [Henkel] Fatty-amido tert. amine; cationic; foamer, thickener, ore flotation agent and starting material for quats., betaines, biocides and amine oxides; liq.; 100% act.

Textamine A-5-D. [Henkel] Fatty imidazoline; cationic; corrosion inhibitor, antistripping agent, emulsion breaker, dispersant, and raw material for quat.; liq.; 100% act.

Textamine A-3417. [Henkel] Fatty amido diamine; cationic; corrosion inhibitor, asphalt emulsifier; flotation, wetting, and antistripping agent; intermediate for spray wax emulsifier formulations for car washes; visc. liq.; dens. 7.8 lb/gal; 100% conc.

Textamine A-W-5. [Henkel] Fatty imidazoline; cationic; corrosion inhibitor, emulsion breaker, dispersant; liq.; oil-sol., water-disp.; 100% act.

Textamine Carbon Detergent K. [Henkel/Cospha; Henkel Canada] Fatty amino complex; nonionic; emulsifier for R.T. solv. degreasing formulations; aids in carbon removal from aircraft, diesels, and metal surfaces; soft paste; dens. 7.8 lb/gal; 84% conc.

Textamine O-1. [Henkel] 1-Hydroxyethyl-2-oleyl imidazoline; nonionic; emulsifier for hydrocarbons and nonpolar solvs., agric. sprays; dispersant; lubricant for metalworking compds.; fuel additive for sludge dispersions; used in fungicide, pesticide, and herbicide formulations; thickener and corrosion inhibitor for paints; used in industrial applics.; amber liq.; sol. in alcohol, chlorinated hydrocarbons, oils; disp. in water; dens. 7.66 lb/gal; pH 10.5 (10); 90% imidazoline.

Textamine O-5. [Henkel] Oleo-fatty imidazoline; cationic; corrosion inhibitor for gasoline and diesel fuels; emulsifier, emulsion breaker, dispersant, textile softener and leather working agent; liq.; oil-sol., water-disp.; 100% conc.

Textamine Oxide CA. [Henkel] Cocoyl amidoalkylamine; nonionic/cationic; wetting agent and foamer in brines and conc. acid and alkaline sol'ns.; liq.; 50% act.

Textamine Oxide CAW. [Henkel] Cocoyl amidoalkylamine oxide; nonionic/cationic; wetting agent and foamer in brines and conc. acid and alkaline sol'ns.; liq.; 30% act.

Textamine Oxide LMW. [Henkel] Lauryl dimethylamine oxide; nonionic/cationic; wetting agent and foamer in brines and conc. acid and alkaline sol'ns.; liq.; 30% act.

Textamine Oxide TA. [Henkel] Tallow amidoalkylamine; nonionic/cationic; foamer, conditioner, and thickener for use in household and industrial cleaners; liq.; 50% act.

Textamine Polymer. [Henkel] Complex resinous fatty amine; cationic; corrosion inhibitor, wetting agent, emulsion breaker, asphalt wetting agent; liq.; 100% act.

Textamine T-1. [Henkel/Cospha; Henkel Canada] 1-Hydroxyethyl-2-tall oil imidazoline; nonionic; corrosion inhibitor, emulsifier, wetting agent for metalworking fluids; dispersant; amber liq.; sol. in alcohol, chlorinated hydrocarbons, aliphatic, and aromatic hydrocarbons, oils; disp. in water; dens. 7.75 lb/gal; pH 11 (10%); 90% min. imidazoline.

Textamine T-5-D. [Henkel/Cospha; Henkel Canada] Fatty amido imidazoline; cationic; emulsifier, corrosion inhibitor, dispersant; liq.; oil-sol.; 100% act.

Textol 80 (L). [Zohar Detergent Factory] Ethanolamine alkylbenzene sulfonate; anionic; raw material for mfg. of liq. detergents; paste; 80% conc.

Tex-Wet 1001. [Intex] Sodium dioctyl sulfosuccinate; CAS 577-11-7; anionic; wetting and rewetting agent used in textile wet processing; biodeg.; clear liq.; dens. 8.8 lb/gal; pH 6 ± 0.5 (1%); 60 ± solids.

Tex-Wet 1002. [Intex] Phosphated alcohol; anionic; dispersant, wetting agent and penetrant for textile wet processing; dyeing assistant for woolen goods; lt. amber liq.; water-sol.; dens. 8.75 lb/gal; pH 7.0 ± 0.5; 55% conc.

Tex-Wet 1004. [Intex] Alkylaryl sulfonate and sulfated nonionic; anionic; detergent, wetting and rewetting agent; wet processing, preparation and/or desizing assistant; dyeing assistant; dens. 8.2 lb/gal; pH 7–8; biodeg.; 35 ± 0.5% solids.

Tex-Wet 1010. [Intex] Long chain phosphated alcohol, modified; anionic; wetting agent, dyeing assistant, detergent, penetrant, boil-off assistant for textiles; liq.; 50% conc.

Tex-Wet 1034. [Intex] Phosphate ester; anionic; wetting agent, detergent, emulsifier, antistat for textile processing; liq.; 70% conc.

Tex-Wet 1048. [Intex] POE linear alcohol; nonionic; wetting agent, penetrant in textile prods.; liq.; HLB 12.5; 20% conc.

Tex-Wet 1070. [Intex] Phosphate ester; anionic; wetting agent in kier boiling, caustic saturation, peroxide bleaching, soaping-off; liq.; 68% conc.

Tex-Wet 1103. [Intex] Phosphated alcohol; anionic; intermediate for preparing penetrants; wetting agent for textile processing; liq.; 100% conc.

Tex-Wet 1104. [Intex] Phosphate, acid form; anionic; detergent, wetting agent, emulsifier for heavy-duty all-purpose liqs., pesticides, polymerization, drycleaning; liq.; 100% conc.

Tex-Wet 1131. [Intex] Coconut alkanolamide; nonionic; detergent, emulsifier, foam stabilizer; liq.; 100% conc.

Tex-Wet 1140. [Intex] Sodium xylene sulfonate; CAS 1300-72-7; anionic; solubilizer, hydrotrope for alkylbenzene sulfonate sol'ns. and detergent formulations; liq.; 40% act.

Tex-Wet 1143. [Intex] Phosphate ester, acid form; anionic; detergent, wetting agent, emulsifier for pesticides, polymerization, drycleaning, dye carriers; liq.; 100% conc.

Tex-Wet 1155. [Intex] Ethoxylated alkyl phenol; nonionic; detergent, wetting, emulsifier for insecticides and herbicides; household and industrial cleaners, textile processing; liq.; HLB 13.5; 100% conc.

Tex-Wet 1158. [Intex] Sodium alcohol ether sulfate; anionic; wetting agent, dispersant, emulsifier, detergent, dyeing assistant; fabric scouring agent; cotton processing; stabilizer for dye bath; dyeing of syns.; pale amber liq.; dens. 8.7 lb/gal; 57% min. act.

Tex-Wet 1197. [Intex] Dodecylbenzene sulfonic acid; anionic; detergent intermediate; emulsifier, coemulsifier used in textile dyeing and scouring; dish detergent and cleaning formulations; biodeg.; redbr. liq.; dens. 8.8 lb/gal; 97% act.

Thorowet G-40 3230. [Clough] Sodium dioctyl sulfosuccinate; anionic; wetting and rewetting agent; stable to dilute acids and alkali; dewatering agent for mining ores; liq.; pH 6.5-8.0; 42–45% conc.

Thorowet G-60 3142. [Clough] Sodium dioctyl sulfosuccinate; anionic; wetting and rewetting agent; stable to dilute acids and alkali; dewatering agent for mining ores; liq.; pH 6.5-7.5; 60% conc.

Thorowet G-60 SS 0336. [Clough] Sodium dioctyl sulfosuccinate; anionic; wetting and rewetting agent; stable to dilute acids and alkali; dewatering agent for mining ores; liq.; pH 6-7; 60% min. act.

Thorowet G-75 1060. [Clough] Sodium dioctyl sulfosuccinate; anionic; wetting and rewetting agent; stable to dilute acids and alkali; dewatering agent for mining ores; liq.; pH 5-6; 73-76% conc.

Tinegal® KD. [Ciba-Geigy/Dyestuffs] EO condensate, modified; amphoteric; dyeing and printing aux. for nylon; liq.

Tinegal® MR-50. [Ciba-Geigy/Dyestuffs] Quat. ammonium salt; cationic; leveling agent for acrylic fibers; liq.

Tinegal® NA. [Ciba-Geigy/Dyestuffs] Polyoxyethylated fatty alcohol; nonionic; dyebath stabilizer, surfactant, emulsifier; liq.

Tinegal® NT. [Ciba-Geigy/Dyestuffs] Proprietary fatty ethoxylate; nonionic; polester dyeing aux.; controls dye strike rate; replaces carrier in high temp. dyeing; liq.

Tinegal® PM. [Ciba-Geigy] Alkylamine polyglycol ether; cationic; leveling agent for nylon; liq.

Tinegal® Resist N. [Ciba-Geigy/Dyestuffs] Sulfonated compd.; anionic; retarding agent for acid dyes on nylon and wool blends; liq.

Tinegal® TP. [Ciba-Geigy/Dyestuffs] Ethoxy aliphatic compd.; cationic; surfactant for washing off printed nylon and polyester fabrics; liq.

Tinegal® W. [Ciba-Geigy/Dyestuffs] Complex aliphatic; amphoteric; leveling agent for dyeing mixed grades of wool, cotton, rayon, and acetate; retarding and stripping agent for dyes; liq.

Tinofix® ECO. [Ciba-Geigy/Dyestuffs] Formaldehyde-free condensate; cationic; fixative for fiber reactive dyes on cellulosic fibers; liq.

Tinofix® EW Sol. [Ciba-Geigy/Dyestuffs] Amine condensate; cationic; fixative for direct, diazo and sulfur colors; liq.

Titadine TA. [Titan] Alkyl naphthalene sulfonic acid ester and sulfated alcohol mixt.; wetting agent; textile, leather, paper, household detergents; liq.

Titadine TCP. [Titan] Sulfonated alcohol deriv.; wetting agent, emulsifier for textiles, leather, paper, and household detergents; liq.

Titan Castor No. 75. [Titan] Sulfonated castor oil; anionic; emulsifier for textile processing; liq.; 62% conc.

Titan Decitrene. [Titan] Alkylated aromatic sulfonate; wetting agent; textile, leather, paper, household detergents.

Titanole RMA. [Titan] Alkylated aryl sodium sulfonate; wetting agent; textile, leather, paper, household detergents; liq.

Titanterge CAC. [Titan] Coconut amine condensate; nonionic; detergent for textile processing; liq.; 100% conc.

Titanterge DBS Conc. [Titan] Dodecylbenzene sulfonate; anionic; wetting agent, detergent for texile processing; liq.; 92% conc.

Titapene Conc. [Titan] Sodium sulfosuccinate, modified; anionic; wetting agent in textile processing; liq.

Titazole HTX. [Titan] Fatty acid amino sulfonate; anionic; detergent for textile processing; liq.

Titazole SA. [Titan] Sodium alkyl naphthalene sulfonate; wetting agent; textile, leather, paper, household detergents; liq.

T-Maz® 20. [PPG/Specialty Chem.] Polysorbate 20; CAS 9005-64-5; nonionic; emulsifier, solubilizer, wetting agent, antistat, stabilizer, dispersant, visc. modifier, suspending agent used in the food, cosmetic, drug, textile and metalworking industries; yel. liq.; sol. in water, ethanol, acetone, toluene, veg. oil; sp.gr. 1.1; visc. 400 cps; HLB 16.7; sapon. no. 40–50; hyd. no. 96-108; flash pt. (PMCC) > 350 F; 97% act.

T-Maz® 28. [PPG/Specialty Chem.] PEG-80 sorbitan laurate; emulsifier and solubilizer of essential oils, wetting agent, visc. modifier, antistat, stabilizer and dispersant used in food, cosmetic, drug, textile and metalworking industries; yel. liq.; sol. in water; sp.gr. 1.0; visc. 1100 cps; HLB 19.2; acid no. 2 max.; sapon. no. 5–15; hyd. no. 25-40; flash pt. (PMCC) > 350 F; 30% max. water.

T-Maz® 40. [PPG/Specialty Chem.] Polysorbate 40; CAS 9005-66-7; nonionic; emulsifier, solubilizer, wetting agent, antistat, stabilizer, dispersant, visc. modifier, suspending agent used in the food, cosmetic, drug, textile and metalworking industries; yel. liq.; sol. in water; disp. in toluene, veg. oil; sp.gr. 1.0; visc. 600 cps; HLB 15.8; sapon. no. 41-52; hyd. no. 85-105; flash pt. (PMCC) > 350 F; 97% min. act.

T-Maz® 60. [PPG/Specialty Chem.] Polysorbate 60; CAS 9005-67-8; emulsifier, solubilizer, wetting agent, antistat, stabilizer, dispersant, visc. modifier, suspending agent used in the food, cosmetic, drug, textile and metalworking industries; yel. gel; sol. in water, min. spirits, toluol; sp.gr. 1.1; visc. 550 cps; HLB 14.9; pour pt. 23–25 C; sapon. no. 45–55; hyd. no. 81-96; flash pt. (PMCC) > 350 F; 97% min. act.

T-Maz® 61. [PPG/Specialty Chem.] Polysorbate 61; CAS 9005-67-8; nonionic; emulsifier, solubilizer, wetting agent, antistat, stabilizer, dispersant, visc. modifier, suspending agent used in the food, cosmetic, drug, textile and metalworking industries; Gardner 5 paste; disp. @ 5% in water, min. oils, toluene, veg. oils; HLB 9.5; sapon. no. 98-113; hyd. no. 170-200; flash pt. (PMCC) > 350 F; 100% conc.

T-Maz® 65. [PPG/Specialty Chem.] Polysorbate 65; CAS 9005-71-4; nonionic; emulsifier, solubilizer, wetting agent, antistat, stabilizer, dispersant, visc. modifier, suspending agent used in the food, cosmetic, drug, textile and metalworking industries; sol. in ethanol, acetone, naphtha; disp. in water; sp.gr. 1.1; m.p. 30–32 C; HLB 10.5; sapon. no. 88–98; 97% min. act.

T-Maz® 80. [PPG/Specialty Chem.] Polysorbate 80; CAS 9005-65-6; nonionic; emulsifier, solubilizer, wetting agent, antistat, stabilizer, dispersant, visc. modifier, suspending agent used in the food, cosmetic, drug, textile and metalworking industries; yel. liq.; sol. in water, ethanol, veg. oil, toluol; sp.gr. 1.0; visc. 400 cps; HLB 15.0; sapon. no. 45–55; hyd. no. 65-80; flash pt. (PMCC) > 350 F; 97% min. act.

T-Maz® 80KLM. [PPG/Specialty Chem.] Polysorbate 80, kosher, low melt pt.; CAS 9005-65-6; emulsifier, solubilizer, wetting agent, antistat, stabilizer, dispersant, visc. modifier, suspending agent used in the food, cosmetic, drug, textile and metalworking industries; Gardner 5 liq.; sol. @ 5% in water, veg. oil; disp. in toluene; HLB 15.0; sapon. no. 45–55; hyd. no. 65-80; flash pt. (PMCC) > 350 F.

T-Maz® 81. [PPG/Specialty Chem.] Polysorbate 81; CAS 9005-65-6; nonionic; emulsifier, solubilizer, wetting agent, antistat, stabilizer, dispersant, visc. modifier, suspending agent used in the food, cosmetic, drug, textile and metalworking industries; Gardner 6 liq.; sol. in min. spirits; disp. in water, min. oils, toluene, veg. oils; HLB 10.0; sapon. no. 96-104; hyd. no. 134-150; flash pt. (PMCC) > 350 F; 100% conc.

T-Maz® 85. [PPG/Specialty Chem.] Polysorbate 85; CAS 9005-70-3; nonionic; emulsifier, solubilizer, wetting agent, antistat, stabilizer, dispersant, visc. modifier, suspending agent used in the food, cosmetic, drug, textile and metalworking industries; yel. liq.; sol. in ethanol, min. spirits; disp. in water; sp.gr. 1.0; visc. 300 cps; HLB 11.1; sapon. no. 83–93; hyd. no. 39-52; flash pt. (PMCC) > 350 F; 95% min. act.

T-Maz® 90. [PPG/Specialty Chem.] PEG-20 sorbitan tallate; nonionic; emulsifier, solubilizer, wetting agent, antistat, stabilizer, dispersant, visc. modifier, suspending agent used in the food, cosmetic, drug, textile and metalworking industries; amber liq.; sol. in water, ethanol, toluol, veg. oil; sp.gr. 1.0; visc. 500 cps; HLB 14.9; sapon. no. 45–55; hyd. no. 65-80; flash pt. (PMCC) > 350 F; 97% min. act.

T-Maz® 95. [PPG/Specialty Chem.] PEG-20 sorbitan tritallate; emulsifier, solubilizer, wetting agent, visc. modifier, antistat, stabilizer, dispersant for food, cosmetic, drug, textile and metalworking industries; amber liq.; sol. in min. spirits; disp. in water, min. oils, toluene, veg. oils; sp.gr. 1.0; visc. 350 cps; HLB 11.0; sapon. no. 83-93; hyd. no. 39-52; flash pt. (PMCC) > 350 F; 3.0% max. water.

T-Mulz® 63. [Harcros] Calcium alkylaryl sulfonate/POE ether blend; anionic/nonionic; emulsifier for ethyl parathion/methyl parathion pesticide formulations; FDA compliance; dk. amber liq.; sp.gr. 0.99; dens. 8.2 lb/gal; pour pt. 30 F; flash pt. (PMCC) 100 F; pH 2-4 (1% aq.); 100% conc.

T-Mulz® 66H. [Harcros] Phosphate ester, potassium salt; anionic; hydrotrope for heavy-duty liq. alkaline cleaners; clear liq.; sol. in ethanol; sp.gr. 1.25; dens. 10.4 lb/gal; pour pt. < 5 F; flash pt. (PMCC) > 200 F; pH 6.5-7.5 (1% aq.); 50% min. act.

T-Mulz® 339. [Harcros] Blend; nonionic; agric. emulsifier for Dimethoate; FDA compliance; lt. amber liq.; sp.gr. 0.99; dens. 8.2 lb/gal; pour pt. 23 F; flash pt. (PMCC) 105 F; pH 5-7 (1% aq.); 100% conc.

T-Mulz® 391. [Harcros] Blend; nonionic; emulsifier for oil-based agric. formulations; FDA compliance; lt. amber liq.; sp.gr. 1.02; dens. 8.5 lb/gal; pour pt. < 0 F; flash pt. (PMCC) > 250 F; pH 5-7 (1% in 15% IPA/water); 100% conc.

T-Mulz® 392. [Harcros] Blend; nonionic; emulsifier for oil-based agric. formulations; FDA compliance; lt. amber liq.; sp.gr. 1.04; dens. 8.7 lb/gal; pour pt. < 0 F; flash pt. (PMCC) > 250 F; pH 5-7 (1% in 15% IPA/water); 100% conc.

T-Mulz® 426. [Harcros] Phosphate ester, free acid; anionic; drycleaning detergent, industrial formulations, solv. degreasers, textile lubricant, agric. applics.; FDA compliance; amber visc. liq.; sol. in ethanol, ethylene, xyelne; disp. in min. oil, kerosene; sp.gr. 1.00; dens. 8.3 lb/gal; acid no. 97 (pH 5.5); pour pt. 25 F; flash pt. (PMCC) > 350 F; pH 2-3 (1% aq.); surf. tens. 34.8 dynes/cm (0.05% aq.); 99.5% min. act.

T-Mulz® 540. [Harcros] Calcium alkylaryl sulfonate/POE ether blend; anionic/nonionic; emulsifier for phenoxy herbicide formulations; liq.

T-Mulz® 565. [Harcros] Phosphate ester, free acid; anionic; detergent for heavy-duty liq. formulations, dry cleaning, petrol. lubricant, agric. emulsifier; FDA compliance; amber visc. liq.; sol. in ethanol, ethylene, xylene; sp.gr. 1.10; dens. 9.2 lb/gal; acid no. 62 (pH 5.5); pour pt. 36 F; flash pt. (PMCC) > 350 F; pH 2-3 (1% aq.); surf. tens. 36.6 dynes/cm (0.05% aq.); 99.5% min. act.

T-Mulz® 596. [Harcros] Phosphate ester, free acid; anionic; emulsion polymerization surfactant; FDA compliance; amber waxy solid; sol. in ethanol, ethylene, xylene; sp.gr. 1.09 (140 F); dens. 9.1 lb/gal (140 F); acid no. 103 (pH 5.5); pour pt. 110 F; flash pt. (PMCC) > 350 F; pH 2-3 (1% aq.); surf. tens. 40.5 dynes/cm (0.05% aq.); 97% min. act.

T-Mulz® 598. [Harcros] Phosphate ester, free acid; anionic; detergent, emulsifier, hydrotrope for highly alkaline sol'ns., dry cleaning, general purpose cleaners, agric. formulations, down hole scale inhibitor, emulsion polymerization; FDA compliance; amber visc. liq.; sol. in ethanol, ethylene, xylene; disp. in min. oil, kerosene; sp.gr. 1.10; dens. 9.2 lb/gal; acid no. 65 (pH 5.5); pour pt. 23 F; flash pt. (PMCC) > 350 F; pH 2-3 (1% aq.); surf. tens. 34.4 dynes/cm (0.05% aq.); 99.5% min. act.

T-Mulz® 734-2. [Harcros] Phosphate ester, free acid; anionic; emulsifier, hydrotrope, detergent for agric. formulations, solvs., detergents, emulsion polymerization, alkaline cleaners; amber clear visc. liq.; sol. in ethanol, ethylene, xylene; sp.gr. 1.18 (68 F); dens. 9.8 lb/gal (68 F); acid no. 107 (pH 5.5); pour pt. 50 F; flash pt. > 350 F (PMCC); pH 2–3 (1% aq.); surf. tens. 38.4 dynes/cm (0.05% aq.); 99.5% min. act.

T-Mulz® 800. [Harcros] Phosphate ester, free acid; anionic; detergent, emulsifier, hydrotrope for heavy-duty alkaline cleaners, textiles, conc. electrolyte sol'ns.; pale yel. liq.; sol. in ethanol, xylene; sp.gr. 1.10; dens. 9.2 lb/gal; acid no. 116 (pH 5.5); pour pt. 40 F; flash pt. (PMCC) > 350 F; pH 2-3 (1% aq.); surf. tens. 30.8 dynes/cm (0.05% aq.); 99.5% min. act.

T-Mulz® 808A. [Harcros] Blend; nonionic; emulsifier for agric. crop oil formulations; FDA compliance; straw clear liq.; dens. 8.33 lb/gal (68 F); sp.gr. 1.00 (68 F); HLB 10.8; flash pt. > 200 F (TCC); pour pt. < 10 F; pH 4–6 (1% aq. disp.); 100% conc.

T-Mulz® 844. [Harcros] Phosphate ester, free acid; anionic; detergent, wetting agent, hydrotrope for alkaline cleaners, textiles; pale yel. liq.; sol. in ethanol; sp.gr. 1.19; dens. 9.9 lb/gal; acid no. 156 (pH 5.5); pour pt. 25 F; flash pt. (PMCC) > 200 F; pH 2-3 (1% aq.); surf. tens. 33.4 dynes/cm (0.05% aq.); 80% act.

T-Mulz® 979. [Harcros] Blend; anionic/nonionic; single-blend general-purpose emulsifier for pesticides; FDA compliance; dk. amber liq.; sp.gr. 1.04; dens. 8.7 lb/gal; pour pt. 70 F; flash pt. (PMCC) 108 F; pH 5-7 (1% aq.); 100% conc.

T-Mulz® 1120. [Harcros] POE ether blend; nonionic; emulsifier, spreader sticker spray adjuvant for herbicides, fungicides, and insecticides; FDA compliance; lt. amber liq.; sp.gr. 0.99; dens. 8.3 lb/gal; pour pt. < 10 F; flash pt. (TCC) 67 F; pH 4-5 (1% aq.); 90% conc.

T-Mulz® 1158. [Harcros] Phosphate ester, free acid; anionic; emulsifier for min. oil, metal processing, solv. cleaners, textile processing, agric. formulations, corrosion inhibition; antistat; oilfield applics.; FDA compliance; amber visc. liq.; sol. @ 5% in min. oil, kerosene, ethanol, ethylene, xylene; sp.gr. 1.10; dens. 9.2 lb/gal; acid no. 85 (pH 5.5); pour pt. 50 F; flash pt. (PMCC) > 350 F; pH 2-3 (1% aq.); surf. tens. 32.4 dynes/cm (0.05% aq.); 99.5% min. act.

T-Mulz® 8015. [Harcros] POE ether blend; nonionic; emulsifier, spreader sticker for agric. herbicides, fungicides, insecticides; FDA compliance; clear liq.; sp.gr. 1.03; dens. 8.6 lb/gal; pour pt. 0 F; flash pt. (PMCC) 100 F; pH 5-7 (1% aq.).

T-Mulz® AO2. [Harcros] Blend; nonionic; emulsifier for spray oils used in conj. with pesticides; FDA compliance; lt. amber liq.; oil-sol.; sp.gr. 1.00; dens. 8.3 lb/gal; HLB 11.3; pour pt. 20 F; flash pt. (PMCC) > 500 F; pH 5-7 (1% aq.); 100% conc.

T-Mulz® AS-1151. [Harcros] Amine alkylaryl sulfonate/polyethylene ether blend; anionic/nonionic; emulsifier for phenoxy herbicide formulations; FDA compliance; dk. amber liq.; sp.gr. 1.02; dens. 8.5 lb/gal; HLB 11.5; pour pt. 25 F; flash pt. (PMCC) 104 F; pH 5-7 (1% aq.).

T-Mulz® AS-1152. [Harcros] Amine alkylaryl sulfonate/polyethylene ether blend; anionic/nonionic; emulsifier for phenoxy herbicide formulations; FDA compliance; dk. amber liq.; sp.gr. 1.04; dens. 8.7 lb/gal; HLB 12.5; pour pt. 35 F; flash pt. (PMCC) 115 F; pH 5-7 (1% aq.); 100% conc.

T-Mulz® AS-1153. [Harcros] Amine alkylaryl sulfonate/polyethylene ether blend; anionic/nonionic; emulsifier for phenoxy herbicide formulations; FDA compliance; dk. amber liq.; sp.gr. 1.02; dens. 8.5 lb/gal; HLB 12.6; pour pt. 40 F; flash pt. (PMCC) 108 F; pH 5-7 (1% aq.); 100% conc.

T-Mulz® AS-1180. [Harcros] Amine alkylaryl sulfonate/polyethylene ether blend; anionic/nonionic; emulsifier for phenoxy herbicide formulations; FDA compliance; dk. amber liq.; sp.gr. 1.02; dens. 8.5 lb/gal; HLB 12.5; pour pt. 50 F; flash pt. (PMCC) 115 F; pH 5-7 (1% aq.); 100% conc.

T-Mulz® COC. [Harcros] Blended POE ethers; nonionic; emulsifier for agric. crop oils; FDA compliance; lt. amber liq.; oil-sol.; sp.gr. 0.99; dens. 8.2 lb/gal; HLB 10.8; pour pt. < 10 F; flash pt. (PMCC) > 200 F; pH 4-6 (1% aq.).

T-Mulz® FCO. [Harcros] Blend; nonionic; emulsifier for Florida citrus oils; FDA compliance; lt. amber liq.; sp.gr. 1.03; dens. 8.6 lb/gal; pour pt. < 0 F; flash pt. (PMCC) > 400 F; pH 5-7 (1% in 15% IPA/water); 100% conc.

T-Mulz® FGO-17. [Harcros] POE ether blend; nonionic; emulsifier for conc. agric. crop oils; FDA compliance; dk. amber liq.; sp.gr. 0.98; dens. 8.1 lb/gal; HLB 10.8; pour pt. 26 F; flash pt. (PMCC) > 200 F; pH 4-6 (1% aq.).

T-Mulz® FGO-17A. [Harcros] POE ether blend; nonionic; emulsifier for conc. agric. crop oils; FDA compliance; dk. amber liq.; sp.gr. 0.95; dens. 7.9 lb/gal; HLB 10.8; pour pt. 20 f; flash pt. (PMCC) > 200 F; pH 4-6 (1% aq.).

T-Mulz® HCC. [Harcros] Blend; emulsifier and detergent for waterless hand cleaners and degreasers; liq.; 100% conc.

T-Mulz® Mal 5. [Harcros] Calcium alkylaryl sulfonate/POE ether blend; nonionic; high flash emulsifier for 5 lb/gal Malathion; FDA compliance; dk. amber liq.; sp.gr. 1.01; dens. 8.4 lb/gal; HLB 13.5; pour pt. 25 F; flash pt. (PMCC) 106 F; pH 5-7 (1% aq.); 100% conc.

T-Mulz® O. [Harcros] Calcium alkylaryl sulfonate/POE ether blend; anionic/nonionic; matched pair emulsifier with T-Mulz W for pesticide formulation; FDA compliance; dk. amber liq.; sp.gr. 1.04; dens. 8.7 lb/gal; HLB 10.7; pour pt. 55 F; flash pt. (PMCC) 108 F; pH 5-7 (1% aq.); 100% conc.

T-Mulz® PB. [Harcros] Blend; nonionic; emulsifier for Pyrethrum pesticide formulations; FDA compliance; lt. amber liq.; sp.gr. 1.04; dens. 8.7 lb/gal; HLB 11.0; pour pt. < 0 F; flash pt. (PMCC) > 350 F; pH 5-7 (1% in 15% IPA/water); 100% conc.

T-Mulz® VO. [Harcros] Blend; nonionic; emulsifier for soybean oil; FDA compliance; lt. amber liq.; oil-sol.; sp.gr. 1.01; dens. 8.4 lb/gal; HLB 10.0; pour pt. < 0 F; flash pt. (PMCC) > 350 F; pH 5-7 (1% in 15% IPA/water); 100% conc.

T-Mulz® W. [Harcros] Calcium alkylaryl sulfonate/POE ether blend; anionic/nonionic; matched pair emulsifier with T-Mulz O for pesticide formulations; FDA compliance; amber liq.; sp.gr. 1.04; dens. 8.7 lb/gal; HLB 14.4; pour pt. 64 F; flash pt. (PMCC) 109 F; pH 5-7 (1% aq.); 100% conc.

TN Cleaner S. [Tokai Seiyu Ind.] Mixt. of ether-type nonionic active agents; nonionic; for scouring and washing of syn. fibers and cotton; paste; 55% conc.

TO-33-F. [Hefti Ltd.] Sorbitan trioleate; CAS 28266-58-0; nonionic; w/o emulsifier for cosmetics, pharmaceuticals; antifoamer, solubilizer; for various tech. applics.; liq.; HLB 2.5; 100% conc.

TO-55-E. [Hefti Ltd.] PEG-18 sorbitan trioleate; CAS 9005-70-3; nonionic; o/w and w/o emulsifier for min. oils, veg. oils, train oils, waxes, etc.; for cattle feed, textiles, biocides, paints, varnishes, plastics, leather, fur, tech. applics., cosmetics, pharmaceuticals; liq.; HLB 10.5; 100% conc.

TO-55-EL. [Hefti Ltd.] PEG-17 sorbitan trioleate; CAS 9005-70-3; nonionic; o/w and w/o emulsifier for min. oils, veg. oils, train oils, waxes, etc.; for cattle feed, textiles, biocides, paints, varnishes, plastics, leather, fur, tech. applics., cosmetics, pharmaceuticals; liq.; HLB 10.0; 100% conc.

TO-55-F. [Hefti Ltd.] Polysorbate 85; CAS 9005-70-3; nonionic; o/w and w/o emulsifier cosmetics, pharmaceuticals, tech. emulsions (wood preservation, pigment stabilization, furniture polishes); liq.; HLB 11.0; 100% conc.

Tohol N-220. [Toho Chem. Industry] Cocamide DEA; nonionic; foam stabilizer and thickener for shampoo, detergent, toothpaste; liq.; 100% conc.

Tohol N-220X. [Toho Chem. Industry] Cocamide DEA; nonionic; foam stabilizer and thickener for shampoo, detergent, toothpaste; solid; 100% conc.

Tohol N-230. [Toho Chem. Industry] Lauramide DEA; CAS 120-40-1; nonionic; foam stabilizer and thickener for shampoo, detergent, toothpaste; liq.; 100% conc.

Tohol N-230X. [Toho Chem. Industry] Lauramide DEA; CAS 120-40-1; nonionic; foam stabilizer and thickener for shampoo, detergent, and toothpaste; solid; 100% conc.

Toho Me-PEG Series. [Toho Chem. Industry] Methoxy polyethylene glycol (m.w. 225, 350, 550, 705, 1000); base material for surfactant, syn. resin, plasticizer, lubricating industries; wetting, softening, penetrating, lubricating and cleaning agent for textile, paper, ink, pigments, etc.; liq./solid.

Toho PEG Series. [Toho Chem. Industry] Polyethylene glycol (m.w. 200, 300, 400, 600, 1000, 1500, 1540, 2000, 4000); base material for surfactant, syn. resin, plasticizer, lubricating industries; wetting, softening, penetrating, lubricating and cleaning agent for textile, paper, ink, pigments, etc.; liq./paste/solid.

Toho Salt A-5. [Toho Chem. Industry] Anionic complex; anionic; dyeing assistant, dispersant for disperse dyestuffs; liq.

Toho Salt A-10. [Toho Chem. Industry] Anionic complex; anionic; dyeing assistant, dispersant for disperse dyestuffs; liq.

Toho Salt CP-60. [Toho Chem. Industry] Blend; nonionic; solubilizer for dyestuffs; liq.

Toho Salt OK-70. [Toho Chem. Industry] Blend; anionic/nonionic; dyeing assistant for rapid textile dyeing; liq.

Toho Salt SM-151, TH. [Toho Chem. Industry] POE alkyl ester; dispersant and leveling agent for polyester fibers; liq.

Toho Salt TM-1. [Toho Chem. Industry] Blend; anionic/nonionic; dispersant and leveling agent for polyester/cotton fibers in high temp. and high pressure rapid dyeing; liq.

Toho Salt UF-350C. [Toho Chem. Industry] Blend; anionic/nonionic; dyeing assistant for rapid textile dyeing; liq.

Toho Salt UF-650C. [Toho Chem. Industry] Blend; anionic/nonionic; dyeing assistant, dispersant and leveling agent for disperse dyestuff in high temp. and high pressure rapid dyeing; liq.; 85% conc.

Tolcide MBT. [Albright & Wilson UK] Methylenebis thiocyanate; biocide for use in water treatment, paper, antifoulant paint, leather, timber preservation; powd.

Tomah AO-14-2. [Exxon/Tomah] Bishydroxyethylisodecyloxypropylamine oxide; cationic; foam stabilizers/boosters in liq. detergents, shampoos, hard surf. cleaners, laundry detergents; grease emulsifier, soil suspension aid; forms synergistic surfactant base for built household, instititional and industrial cleaners with quats. and nonionics; Gardner 2 clear liq.; sp.gr. 0.956 (15 C); pour pt. < 20 F; amine no. 83-89; flamm.; 50% act.

Tomah AO-728 Special. [Exxon/Tomah] Amine oxide; detergent, foam booster/stabilizer for industrial and household detergents, dishwash, personal care prods.; Gardner 3 max. liq.; sp.gr. 0.958; pour pt. < 20 F; amine no. 72-79; flamm.; 50% act.

Tomah DA-14. [Exxon/Tomah] N-Isodecyloxypropyl-1,3-diaminopropane; CAS 72162-46-0; cationic; corrosion inhibitor for metalworking fluids; antistat; flotation collector; additive for fuel, lubricant, petrol. refining; intermediate for surfactants, textile foamers, ethoxylates and agric. chem.; crosslinking agent for epoxies; bactericidal props.; lt. amber liq.; m.w. 295; sp.gr. 0.86; pour pt. –50 F; amine no. 375-395; flash pt. (COC) 108 C; 100% act.; 90% min. diamine.

Tomah DA-16. [Exxon/Tomah] N-Isododecyloxypropyl-1,3-diaminopropane; cationic; corrosion inhibitor for metalworking fluids; antistat; flotation collector; additive for fuel, lubricant, petrol. refining; intermediate for surfactants, textile foamers, ethoxylates and agric. chem.; crosslinking agent for epoxies; liq.; m.w. 328; pour pt. -30 F; amine no. 335-355; flash pt. (COC) 160 C; 100% conc.

Tomah DA-17. [Exxon/Tomah] N-Isotridecyloxypropyl-1,3-diaminopropane; cationic; emulsifier; corrosion inhibitor for metalworking fluids; antistat; flotation collector; additive for fuel, lubricant, petrol. refining; intermediate for surfactants, textile foamers, ethoxylates and agric. chem.; crosslinking agent for epoxies; amber liq.; m.w. 340; sp.gr. 0.87; pour pt. –30 F; amine no. 325-350; flash pt. (COC) 160 C; 95% min. amine.

Tomah E-14-2. [Exxon/Tomah] Bis-(2-hydroxyethyl) isodecyloxypropylamine; CAS 34360-00-4; cationic; emulsifier, corrosion inhibitor, lubricant used in min. acid inhibition, textile processing; amber liq.; sp.gr. 0.94; HLB 8.3; amine no. 170-185; alkaline corrosive liq.; toxicology: can cause burns or irritation to skin and eyes; 95–100% act.

Tomah E-14-5. [Exxon/Tomah] PEG-5 isodecyloxypropylamine; cationic; emulsifier, corrosion inhibitor, lubricant used in min. acid inhibition, textile processing; amber liq.; sp.gr. 0.99; HLB 5.0; amine no. 123-129; alkaline corrosive liq.; alkaline corrosive liq.; toxicology: can cause burns or irritation to skin and eyes; 95–100% act.

Tomah E-17-2. [Exxon/Tomah] Bis-(2-hydroxyethyl) isotridecyloxypropylamine; cationic; emulsifier, corrosion inhibitor, lubricant used in min. acid inhibition, textile processing; amber liq.; sp.gr. 0.93; HLB 5.6; amine no. 150-165; alkaline corrosive liq.; toxicology: can cause burns or irritation to skin and eyes; 95–100% act.

Tomah E-18-2. [Exxon/Tomah] Bis (2-hydroxyethyl) octadecyloxypropylamine; cationic; emulsifier, corrosion inhibitor, lubricant used in min. acid inhibition, textile processing; wax; sp.gr. 0.95; HLB 3.0; amine no. 153-163; alkaline corrosive liq.; toxicology: can cause burns or irritation to skin and eyes; 95–100% act.

Tomah E-18-5. [Exxon/Tomah] PEG-5 stearyloxypropylamine; cationic; emulsifier, corrosion inhibitor, lubricant used in min. acid inhibition, textile processing; paste; sp.gr. 0.97; HLB 11.0; amine no. 112-117; alkaline corrosive liq.; toxicology: can cause burns or irritation to skin and eyes; 95–100% act.

Tomah E-18-8. [Exxon/Tomah] PEG-8 stearyloxypropylamine; emulsifier, corrosion inhibitor, lubricant for min. acid inhibition, textile processing; sp.gr. 1.01; amine no. 89-94; alkaline corrosive liq.; toxicology: can cause burns or irritation to skin and eyes; 100% act.

Tomah E-18-10. [Exxon/Tomah] PEG-10 stearyloxypropylamine; emulsifier, corrosion inhibitor, lubricant used in min. acid inhibition, textile processing; amber liq.; sp.gr. 1.02; HLB 14.0; amine no. 77-82; alkaline corrosive liq.; 100% act.

Tomah E-18-15. [Exxon/Tomah] PEG-15 stearyloxypropylamine; cationic; emulsifier, corrosion inhibitor, lubricant used in min. acid inhibition, textile processing; amber liq.; sp.gr. 1.03; HLB 16.0; amine no. 58-63; alkaline corrosive liq.; 100% act.

Tomah E-19-2. [Exxon/Tomah] Bis (2-hydroxyethyl) linear alkyloxypropylamine; surfactant to modify emulsification, surf. tension, solubility; for acid thickeners, antistats, petrol. prod. and refining, agric. adjuvants, textile processing aids, corrosion inhibition, detergent boosters; chem. intermediate; Gardner 6 max. color; sp.gr. 0.95; amine no. 150-160; alkaline corrosive liq.; toxicology: can cause burns or irritation to skin and eyes; 95% min. tert. amine.

Tomah E-24-2. [Exxon/Tomah] PEG-2 Guerbet C20 alcohol amine; surfactant, corrosion inhibitor.

Tomah E-DT-3. [Exxon/Tomah] PEG-3 1,3-diaminopropane; surfactant for acid thickeners, antistat, cationic emulsification, petrol. prod. and refining, agric. adjuvants, textile processing aids, corrosion inhibition, detergent boosters; chem. intermediate; Gardner 17 max. color; sp.gr. 0.95; HLB 7.5; amine no. 225-240; alkaline corrosive liq.; toxicology: can cause burns or irritation to skin and eyes; 95% min. tert. amine.

Tomah E-S-2. [Exxon/Tomah] PEG-2 soyamine; CAS 61791-24-0; emulsifier, corrosion inhibitor, lubricant used in min. acid inhibition, textile processing; amber liq.; sp.gr. 0.95; HLB 4.5; amine no. 150-160; alkaline corrosive liq.; toxicology: can cause burns or irritation to skin and eyes; 95–100% act.

Tomah E-S-5. [Exxon/Tomah] PEG-5 soyamine; CAS 61791-24-0; emulsifier, corrosion inhibitor, lubricant used in min. acid inhibition, textile processing;

amber liq.; sp.gr. 0.95; HLB 5.0; alkaline corrosive liq.; toxicology: can cause burns or irritation to skin and eyes; 95–100% act.

Tomah E-S-15. [Exxon/Tomah] PEG-15 soya amine; CAS 61791-24-0; emulsifier, corrosion inhibitor, lubricant used in min. acid inhibition, textile processing; amber liq.; sp.gr. 1.03; HLB 15.0; amine no. 57-63; alkaline corrosive liq.; toxicology: can cause burns or irritation to skin and eyes; 95–100% act.

Tomah E-T-2. [Exxon/Tomah] PEG-2 tallowamine; surfactant for acid thickeners, antistat, cationic emulsification, petrol. prod. and refining, agric. adjuvants, textile processing aids, corrosion inhibition, detergent boosters; chem. intermediate; Gardner 16 max. color; sp.gr. 0.91; HLB 5.0; amine no. 153-161; alkaline corrosive liq.; toxicology: can cause burns or irritation to skin and eyes; 95% min. tert. amine.

Tomah E-T-5. [Exxon/Tomah] PEG-5 tallowamine; surfactant for acid thickeners, antistat, cationic emulsification, petrol. prod. and refining, agric. adjuvants, textile processing aids, corrosion inhibition, detergent boosters; chem. intermediate; Gardner 16 max. color; sp.gr. 0.95; HLB 14.0; amine no. 112-120; alkaline corrosive liq.; toxicology: can cause burns or irritation to skin and eyes; 100% act.

Tomah E-T-15. [Exxon/Tomah] PEG-15 tallowamine; surfactant for acid thickeners, antistat, cationic emulsification, petrol. prod. and refining, agric. adjuvants, textile processing aids, corrosion inhibition, detergent boosters; chem. intermediate; Gardner 16 max. color; sp.gr. 1.02; HLB 17.0; amine no. 59-63; alkaline corrosive liq.; toxicology: can cause burns or irritation to skin and eyes.

Tomah PA-10. [Exxon/Tomah] Hexyloxypropylamine; corrosion inhibitor for metalworking fluids; antistat; flotation collector; additive for fuel, lubricant, petrol. refining; intermediate for surfactants, textile foaming agents, ethoxylates and agric. chem.; crosslinking agent for epoxy resins; lt. amber liq.; m.w. 165; water-sol.; sp.gr. 0.84; pour pt. –30 F; amine no. 325-340; flash pt. (COC) 84 C; 95% min. amine.

Tomah PA-12EH. [Exxon/Tomah] 2-Ethylhexyloxypropylamine; CAS 5397-31-9; detergent intermediate; clear liq., ammoniacal odor; insol. in water; sp.gr. 0.85; visc. 6 cSt; f.p. -30 F; b.p. 525 F; amine no. 285; flash pt. (PMCC) 201 F; toxicology: eye and skin corrosive; 100% act.

Tomah PA-13i. [Exxon/Tomah] Isononyloxypropylamine; CAS 29317-52-0; detergent intermediate; clear liq., ammoniacal odor; insol. in water; sp.gr. 0.85; visc. 6 cSt; f.p. -50 f; b.p. 525 F; amine no. 260; flash pt. (PMCC) 201 F; toxicology: eye and skin corrosive; 100% act.

Tomah PA-14. [Exxon/Tomah] Isodecyloxypropylamine; CAS 7617-78-9; cationic; emulsifier; corrosion inhibitor for metalworking fluids; antistat; flotation collector; additive for fuel, lubricant, petrol. refining; intermediate for surfactants, textile foamers, ethoxylates and agric. chem.; crosslinking agent for epoxies; colorless to lt. amber liq.; m.w. 229; sp.gr. 0.84; pour pt. –50 F; amine no. 240-255; flash pt. (COC) 104 C; 95% min. amine.

Tomah PA-14 Acetate. [Exxon/Tomah] Isodecyl oxypropyl amine acetate; cationic; patented emulsifier, gellation/wetting agent for clays, fillers, and fibers in organic coatings and sealants, e.g., roof coatings, tile adhesives, caulks, pipe coatings, automotive undercoatings, alkyd paints, foundry coatings, polymers/elastomers; bactericidal props.; lt. amber liq.; sp.gr. 0.93; dens. 7.74 lb/gal; visc. 1540 cps; HLB 15.5; pour pt. 15 F; acid no. 185-205; amine no. 185-205; 95% act.

Tomah PA-16. [Exxon/Tomah] Isododecyloxypropylamine; corrosion inhibitor for metalworking fluids; antistat; flotation collector; additive for fuel, lubricant, petrol. refining; intermediate for surfactants, textile foamers, ethoxylates and agric. chem.; crosslinking agent for epoxies; liq.; m.w. 253; pour pt. -30 F; amine no. 215-230; flash pt. (COC) 120 C; 100% conc.

Tomah PA-17. [Exxon/Tomah] Isotridecyloxypropylamine; corrosion inhibitor, cationic emulsification and replacement for oleyl and soya amines; chemical intermediate; colorless to lt. amber liq.; m.w. 274; sp.gr. 0.85; pour pt. –30 F; amine no. 195-215; flash pt. (COC) 125 C; 95% min. amine.

Tomah PA-19. [Exxon/Tomah] Linear C12-15 alkyloxypropylamine; CAS 68610-26-4; corrosion inhibitor for metalworking fluids; antistat; flotation collector; additive for fuel, lubricant, petrol. refining; intermediate for surfactants, textile foamers, ethoxylates and agric. chem.; crosslinking agent for epoxies; clear liq., ammoniacal odor; insol. in water; m.w. 272; sp.gr. 0.85; visc. 6 cSt; f.p. 50 f; b.p. 400 F min.; pour pt. 50 F; amine no. 195-215; flash pt. (PMCC) 201 F; toxicology: eye and skin corrosive.

Tomah PA-24. [Exxon/Tomah] Isoarachidyloxypropylamine; Guerbet C20 alcohol primary amine; detergent intermediate; experimental prod. for research and development; clear liq., ammoniacal odor; insol. in water; sp.gr. 0.85; visc. 22 cSt; f.p. -30 f; b.p. 570 F; flash pt. (PMCC) 201 F; toxicology: eye and skin corrosive.

Tomah PA-1214. [Exxon/Tomah] Octyl/decyloxypropylamine; corrosion inhibitor for metalworking fluids; antistat; flotation collector; additive for fuel, lubricant, petrol. refining; intermediate for surfactants, textile foaming agents, ethoxylates and agric. chem.; crosslinking agent for epoxy resins; sp.gr. 0.85.

Tomah Q-14-2. [Exxon/Tomah] Isodecyloxypropyl dihydroxyethyl methyl ammonium chloride; CAS 125740-36-5; cationic; quat. used as acid corrosion inhibitor, plastics and textile antistat, and emulsifier; bactericidal props.; lt. amber liq.; water-sol.; sp.gr. 0.97; HLB 19.5; pH 6–9 (1% aq.); flamm.; 74% act.

Tomah Q-14-2PG. [Exxon/Tomah] Isodecyloxypropyl dihydroxyethyl methyl ammonium chloride, propylene glycol; cationic; quat. surfactant for use as emulsifiers, antistats, corrosion inhibitors, nonionic detergency booster in laundry prods., in nonbutyl cleaning systems; amber liq.; sp.gr. 0.97; 75% act.

Tomah Q-17-2. [Exxon/Tomah] Isotridecyloxypropyl dihydroxyethyl methyl ammonium chloride, IPA; cationic; emulsifier; boosts efficiency of nonionic surfactants; used in hard surf. cleaners, laundry, transportation cleaners; bactericidal props.; amber liq.; insol. in water; sp.gr. 0.97; dens. 8.5 lb/gal; visc. 472 cSt (100 F); HLB 8.9; pour pt. –10 C; flash pt. 430 F; cloud pt. < 25 C; pH 6–9 (5% aq.); flamm.; 74% in IPA-water solv. blend.

Tomah Q-17-2PG. [Exxon/Tomah] Isotridecyloxypropyl dihydroxyethyl methyl ammonium chloride, propylene glycol; cationic; quat. surfactant for use as emulsifiers, antistats, corrosion inhibitors, nonionic detergency booster in laundry prods., in nonbutyl cleaning systems; amber liq.; sp.gr. 1.02; 75% act.

Tomah Q-18-2. [Exxon/Tomah] Octadecyl dihydroxyethyl methyl ammonium chloride, IPA; cationic; quat. surfactant for use as emulsifiers, antistats, corrosion inhibitors, nonionic detergency booster in laundry prods., in nonbutyl cleaning systems; amber liq.; sp.gr. 0.91; pH 6–9 (5% aq.); flamm.; 50% in IPA-water solv. blend.

Tomah Q-18-15. [Exxon/Tomah] Stearyl PEG-15 methyl ammonium chloride; cationic; quat. used as acid corrosion inhibitor, plastics and textile antistat, and emulsifier; amber liq.; sp.gr. 1.06; HLB 16.3-18.0; pH 6–9 (5% aq.); flamm.; 100% act.

Tomah Q-2C. [Exxon/Tomah] Dicoco dimonium chloride, IPA; detergency booster with biocidal activity; amber liq.; sol. in water; sp.gr. 0.94; visc. 65 cSt; f.p. 30 F; b.p. 175 F (IPA); flash pt. (PMCC) 75 F; flamm.; toxicology: eye and skin corrosive; 75% act.

Tomah Q-24-2. [Exxon/Tomah] Methyl dihydroxyethyl isoarachidaloxypropyl ammonium chloride, IPA; Guerbet C20 alcohol dihydroxyethyl ammonium chloride; detergency booster with biocidal activity; experimental material; sp.gr. 0.91; visc. 400 cSt; f.p. 32 F; b.p. 180 F (IPA); flash pt. (TCC) 75 F; flamm.; toxicology: corrosive to skin and eyes; 75% act.

Tomah Q-311. [Exxon/Tomah] Monosoya amidoamine quat.; detergency booster with biocidal activity; 75% act.

Tomah Q-511. [Exxon/Tomah] Monococo amidoamine quat.; detergency booster with biocidal activity; 75% act.

Tomah Q-C-15. [Exxon/Tomah] PEG-15 cocomonium chloride; cationic; quat. surfactant for use as emulsifiers, antistats, corrosion inhibitors, nonionic detergency booster in laundry prods., in nonbutyl cleaning systems; amber liq.; sp.gr. 1.06; HLB 15.0; pH 6-9 (5%); 100% act.

Tomah Q-D-T. [Exxon/Tomah] Tallow dimethyl trimethyl propylene diammonium chloride, IPA; cationic; quat. used as acid corrosion inhibitor, plastics and textile antistat, and emulsifier; amber liq.; sp.gr. 0.95; HLB 15.0; pH 6–9 (5% aq.); flamm.; 50% act. in IPA-water.

Tomah Q-DT-HG. [Exxon/Tomah] Tallow diamine quat. in hexylene glycol; quat. surfactant for use as emulsifiers, antistats, corrosion inhibitors, nonionic detergency booster in laundry prods., in nonbutyl cleaning systems; amber liq.; sp.gr. 0.98; pH 5-8 (5%); 70% act.

Tomah Q-S. [Exxon/Tomah] Soya trimethyl ammonium chloride, IPA; cationic; quat. surfactant for use as emulsifiers, antistats, corrosion inhibitors, nonionic detergency booster in laundry prods., in nonbutyl cleaning systems; amber liq.; sp.gr. 0.91; HLB 15.0; pH 6–9 (5% aq.); flamm.; 50% act. in IPA-water.

Tomah Q-S-80. [Exxon/Tomah] Soytrimonium chloride; cationic; quat. surfactant for use as emulsifiers, antistats, corrosion inhibitors, nonionic detergency booster in laundry prods., in nonbutyl cleaning systems; amber liq.; sp.gr. 0.91; pH 6-9 (5%); flamm.; 80% act.

Tomah Q-ST-50. [Exxon/Tomah] Steartrimonium chloride; quat. surfactant for use as emulsifiers, antistats, corrosion inhibitors, nonionic detergency booster in laundry prods., in nonbutyl cleaning systems; amber liq.; sp.gr. 0.88; pH 5-9 (5%); flamm.; 50% act.

Tonerclean 208, 209. [Nippon Nyukazai] Blend; nonionic; deinking agent for reclaimed paper prod.; liq.; 100% conc.

Topcithin. [Lucas Meyer] Soya lecithin, highly purified; CAS 8002-43-5; nonionic; improves emulsifying and flowing properties and physical appearance; extends shelf-life; liq.; 100% conc.

Totablan. [Ceca SA] Anionic; hydrogen peroxide bleaching stabilizer and wetting agent.

Toxanon 500. [Sanyo Chem. Industries] Surfactant blend; anionic/nonionic; emulsifier for org. chlorinated insecticides; liq.

Toxanon AH. [Sanyo Chem. Industries] Surfactant; nonionic; emulsifier for org. phosphate insecticides; liq.

Toxanon AM Series. [Sanyo Chem. Industries] Surfactant blend; anionic/nonionic; emulsifier for machine oil; 2-3% use level; liq.

Toxanon GR-31A. [Sanyo Chem. Industries] Polycarboxyl-type polymeric surfactant; anionic; dispersant for agric. pesticides and herbicides in gran. form; liq.; 40% conc.

Toxanon P-8H, P-8L. [Sanyo Chem. Industries] Surfactant blend; anionic/nonionic; emulsifier pair for org. phosphate insecticides; liq.

Toxanon P-900. [Sanyo Chem. Industries] Surfactant blend; anionic/nonionic; emulsifier for org. phosphate insecticides; 5-10% use level; liq.

Toxanon SC-4. [Sanyo Chem. Industries] Surfactant blend; anionic/nonionic; emulsifier for Sumicidin; 6% use level; liq.

Toxanon V-1D. [Sanyo Chem. Industries] Surfactant blend; anionic/nonionic; emulsifier for sanitary insecticides incl. Lindane, Malathion, Diazinon, DDVP, Sumithion; liq.

Toxanon XK-30. [Sanyo Chem. Industries] Anionic; emulsifier for o-dichlorobenzene and m-cresol; liq.

Toximul® 60 Series. [Stepan Europe] Surfactant blend; anionic/nonionic; emulsifier for agric. emulsifiable concs. with 20-40% dimethoate; wh. liq. to paste.

Toximul® 360A. [Stepan; Stepan Canada; Stepan Europe] Sulfonate/nonionic blend; anionic/nonionic; matched pair emulsifier with Toximul 360B for Chlordane and Heptachlor pesticides; dk. brn. liq.; HLB 10.4.

Toximul® 360B. [Stepan Europe] Surfactant blend; anionic/nonionic; matched pair emulsifier with Toximul 360A for Chlordane and Heptachlor pesticides; dk. brn. liq.; HLB 13.5.

Toximul® 360B-HF. [Stepan; Stepan Canada] Sulfonate/nonionic blend; high flash version of Toximul 360B; liq.

Toximul® 374. [Stepan Europe] Surfactant blend; nonionic; emulsifier for paraffinic oil-based agric. formulations; pale yel. visc. liq.; HLB 8.0.

Toximul® 425, 475. [Stepan; Stepan Canada] Sulfonate/nonionic blend; matched pair emulsifiers for Malathion; liq.

Toximul® 500. [Stepan; Stepan Canada] Sulfonate nonionic blend; anionic/nonionic; emulsifier for insecticides; EPA cleared; liq.; sol. in xylene; HLB 10.5.

Toximul® 600. [Stepan; Stepan Canada] Sulfonate/nonionic blend; nonionic; general purpose emulsifier for pesticides; EPA cleared; liq.; sol. in xylene; HLB 10.5.

Toximul® 701, 702. [Stepan; Stepan Canada] Formulated prod.; anionic/nonionic; matched pair emulsifiers for pesticide formulations; liq.; 100% conc.
Toximul® 705. [Stepan Europe] Surfactant blend; nonionic; emulsifier for agric. paraffinic spray oils; water-wh. to pale yel. liq.; HLB 8.0.
Toximul® 707. [Stepan Europe] Surfactant blend; nonionic; emulsifier for agric. paraffinic spray oils; water-wh. to pale yel. liq.
Toximul® 709. [Stepan; Stepan Canada] Sulfonate nonionic blend; nonionic; matched pair emulsifier with Toximul 710 for dinitro herbicide formulations; EPA cleared; amber liq.; sol. in xylene; HLB 10.5; 100% act.
Toximul® 710. [Stepan; Stepan Canada; Stepan Europe] Sulfonate nonionic blend; nonionic; matched emulsifier pair with Toximul 709 for dinitro herbicides; EPA cleared; amber liq.; sol. in xylene, water; HLB 13.0.
Toximul® 713. [Stepan; Stepan Canada] Formulated prod.; nonionic; adjuvant for use with Toximul 701/702 to improve high poundage organo-phosphate pesticide formulations; liq.; 100% conc.
Toximul® 715. [Stepan; Stepan Canada; Stepan Europe] Sulfonate nonionic blend; nonionic; matched emulsifier pair with Toximul 716 for dinitroaniline herbicides; EPA cleared; liq.; sol. in xylene; HLB 10.5.
Toximul® 716. [Stepan; Stepan Canada; Stepan Europe] Sulfonate nonionic blend; nonionic; matched emulsifier pair with Toximul 715 for dinitroaniline herbicides; EPA cleared; liq.; sol. in xylene; HLB 12.0.
Toximul® 800. [Stepan; Stepan Canada] Formulated prod.; anionic/nonionic; emulsifier for propanol; liq.
Toximul® 804. [Stepan; Stepan Canada; Stepan Europe] Blend; anionic; emulsifier for propanil agric. formulations; EPA cleared; amber liq.; sol. in xylene; HLB 11.0.
Toximul® 811. [Stepan; Stepan Canada] Sulfonate/anionic blend; anionic; emulsifier for Betasan agric. formulations; EPA cleared; amber liq.; sol. in xylene, water; HLB 13.5.
Toximul® 812. [Stepan; Stepan Canada] Formulated prod.; anionic/nonionic; emulsifier for 5 lb/gal Dinitro; liq.
Toximul® 850. [Stepan; Stepan Canada] Formulated prod.; nonionic; adjuvant-emulsifier for paraffinic crop oils; liq.
Toximul® 852. [Stepan; Stepan Canada] Formulated prod.; nonionic; adjuvant-emulsifier for veg. crop oils; liq.
Toximul® 856N. [Stepan; Stepan Canada] Formulated prod.; nonionic; adjuvant wetting agent for pesticides; liq.; 80% conc.
Toximul® 857N. [Stepan; Stepan Canada] Formulated prod.; nonionic; adjuvant-spreader/sticker for pesticides; liq.; 90% conc.
Toximul® 858N. [Stepan; Stepan Canada] Formulated prod.; nonionic; adjuvant-sticker for pesticides; liq.
Toximul® 8170. [Stepan; Stepan Canada] Formulated prod.; anionic; adjuvant-compatibility agent for pesticides; liq.; 100% conc.
Toximul® 8240. [Stepan; Stepan Canada; Stepan Europe] PEG-36 castor oil; emulsifier for agric. formulations; EPA cleared; amber liq.; sol. in xylene, water; HLB 13.0.
Toximul® 8241. [Stepan; Stepan Canada; Stepan Europe] PEG-30 castor oil; emulsifier for agric. formulations; EPA cleared; amber liq.; sol. in xylene, water; HLB 12.0.
Toximul® 8242. [Stepan; Stepan Canada; Stepan Europe] PEG-40 castor oil; emulsifier for agric. formulations; EPA cleared; amber liq.; sol. in xylene, water; HLB 13.0.
Toximul® 8301N. [Stepan; Stepan Canada] Formulated prod.; nonionic; adjuvant-wetting agent for pesticides; liq.; 100% conc.
Toximul® 8320. [Stepan; Stepan Canada] Butyl EO/PO block copolymer; nonionic; emulsifier component, flowable surfactant, intermediate for pesticides; EPA cleared; lt. amber liq.; sol. in xylene, water; HLB 12.0; 100% act.
Toximul® 8321. [Stepan; Stepan Canada] EO/PO block copolymer; nonionic; emulsifier, wetting agent, intermediate for pesticides; EPA cleared; amber liq.; sol. in xylene; HLB 5.5; 100% act.
Toximul® 8322. [Stepan; Stepan Canada] EO/PO block copolymer; nonionic; emulsifier, intermediate for pesticides; EPA cleared; amber liq. to paste; sol. in xylene, water; HLB 14.0; 100% conc.
Toximul® 8323. [Stepan; Stepan Canada] EO/PO block copolymer; nonionic; emulsifier component, flowable surfactant, intermediate for pesticides; EPA cleared; amber liq. to paste; sol. in xylene, water; HLB 17.0; 100% conc.
Toximul® D. [Stepan; Stepan Canada; Stepan Europe] Sulfonate/nonionic blend; nonionic; matched pair emulsifier with Toximul H-HF for pesticide formulations; dispersant, stabilizer, hydrophobic agent; EPA cleared; dk. brn. liq.; sol. in xylene; HLB 10.5.
Toximul® FF. [Stepan Europe] Surfactant blends; nonionic; hydrophilic emulsifier for agric. emulsifiable concs., DDVP; used with Toximul S; liq. to paste.
Toximul® H-HF. [Stepan; Stepan Canada; Stepan Europe] Sulfonate/nonionic blend; anionic/nonionic; matched pair emulsifier with Toximul D for pesticides; dispersant, stabilizer, hydrophilic agent; EPA cleared; dk. brn. liq.; sol. in xylene; HLB 13.5.
Toximul® MP. [Stepan; Stepan Canada; Stepan Europe] Sulfonate nonionic blend; anionic/nonionic; emulsifier for Parathion and Malathion-based insecticides; dispersant, stabilizer; EPA cleared; dk. brn. liq.; sol. in xylene; HLB 11.0.
Toximul® MP-10. [Stepan Europe] Sulfonate nonionic blend; anionic/nonionic; emulsifier for insecticides; dispersant, stabilizer for solv.-free blends containing methyl parathion, ethyl parathion; used with Toximul® D and H-HF; EPA cleared; dk. brn. liq.; sol. in xylene; HLB 12.0.
Toximul® MP-26. [Stepan; Stepan Canada; Stepan Europe] Sulfonate/nonionic blend; anionic/nonionic; emulsifier, dispersant, stabilizer for phosphate insecticides; emulsifier for high poundage organophosphates; EPA cleared; dk. brn. liq.; sol. in xylene, water; HLB 12.5.
Toximul® MP-HF. [Stepan; Stepan Canada] Sulfonate/nonionic blend; high flash version of Toximul MP; liq.
Toximul® R-HF. [Stepan; Stepan Canada; Stepan Europe] Sulfonate/nonionic blend; anionic/nonionic; matched pair emulsifier with Toximul S-HF for herbicides; dispersant, stabilizer, hydrophilic agent; EPA cleared; dk. brn. liq.; sol. in xylene; HLB 10.5.
Toximul® SEE-340. [Stepan; Stepan Canada; Stepan Europe] PEG-20 sorbitan tritallate; emulsifier for agric. formulations; EPA cleared; liq.; sol. in xy-

lene; HLB 11.0.

Toximul® SF Series. [Stepan Europe] Surfactant blend; anionic/nonionic; emulsifier for solv.-free agric. emulsifiable concs. of Malathion, Parathion, Fenitrothion, Ethion; paste.

Toximul® S-HF. [Stepan; Stepan Canada] Sulfonate/nonionic blend; anionic/nonionic; matched emulsifier pair with Toximul R-HF for herbicides; dispersant, stabilizer, hydrophilic agent; EPA cleared; dk. brn. liq.; sol. in xylene; HLB 13.0.

Toximul® TA-2. [Stepan; Stepan Canada; Stepan Europe] PEG-2 tallowamine; emulsifier for agric. formulations; EPA cleared; amber liq.; sol. in xylene; HLB 5.0.

Toximul® TA-5. [Stepan; Stepan Canada; Stepan Europe] PEG-5 tallowamine; emulsifier for agric. formulations; EPA cleared; amber liq.; sol. in xylene; HLB 9.0.

Toximul® TA-15. [Stepan; Stepan Canada; Stepan Europe] PEG-15 tallowamine; emulsifier for agric. formulations; EPA cleared; amber liq.; sol. in xylene; HLB 14.5.

Trans-10, -10K. [Trans-Chemco] Silicone emulsion; -10K kosher grade; defoamer for aq. systems incl. food processing; disp. in water; 10% act.

Trans-20, -20K. [Trans-Chemco] Silicone emulsion; -20K kosher grade; defoamer for aq. systems incl. food processing; water-disp.; 20% act.

Trans-25, -25K. [Trans-Chemco] Silicone emulsion; -25K kosher grade; defoamer for aq. systems incl. food processing; water-disp.; 25% act.

Trans-30, -30K. [Trans-Chemco] Silicone emulsion; -30K kosher grade; defoamer for aq. systems incl. food processing; water-disp.; 30% act.

Trans-100. [Trans-Chemco] Silicone compd.; defoamer for detergents, petrol. refining, cutting oils, insect repellents, chemical and food processing; sol. in aliphatic, aromatic, and chlorinated hydrocarbons; 100% act.

Trans-107. [Trans-Chemco] Nonsilicone; defoamer for alkaline systems; slightly disp. in water.

Trans-109. [Trans-Chemco] Nonsilicone; defoamer for paints, coolants, lubricants, cutting oils; slightly disp. in water; 100% act.

Trans-134S. [Trans-Chemco] Emulsion with 11% silicone; defoamer for mfg. of water-based inks; water-disp.

Trans-134S K. [Trans-Chemco] Emulsion with 11% silicone, kosher; defoamer for aq. systems incl. food processing; water-disp.

Trans-135. [Trans-Chemco] Nonsilicone; defoamer for paper/paperboard mfg.; water-disp.

Trans-137. [Trans-Chemco] Nonsilicone; defoamer for alkaline systems, paper and paperboard mfg., wastewater treatment; water-disp.

Trans-138. [Trans-Chemco] Nonsilicone; defoamer for water-based inks, wastewater treatment; disp. in water; sol. in hydrocarbon solvs.; 100% act.

Trans-143. [Trans-Chemco] Nonsilicone; defoamer for mfg. of syn. latex; water-disp.

Trans-149. [Trans-Chemco] Nonsilicone; defoamer for paper/paperboard mfg.; water-disp.

Trans-154, -156. [Trans-Chemco] Nonsilicone; defoamer for starch processing; sol. in hydrocarbon solvs.; slightly disp. in water; 100% act.

Trans-166. [Trans-Chemco] Nonsilicone; defoamer for mfg. of syn. latex, sol. cutting oils; water-disp.

Trans-175, -176. [Trans-Chemco] Nonsilicone; defoamer for mfg. of soya flour and isolate, egg washing; water-disp.

Trans-177S, -177S K. [Trans-Chemco] Emulsion with 5% silicone; -177 S K kosher grade; defoamer for aq. systems incl. food processing, wastewater treatment; water-disp.

Trans-179. [Trans-Chemco] Nonsilicone, water-based; defoamer for mfg. edible phosphates; water-sol.

Trans-198. [Trans-Chemco] Nonsilicone, water-based; defoamer for mfg. of phosphoric acid; water-disp.

Trans-220. [Trans-Chemco] Nonsilicone, water-based; long-lasting antifoam for food processing; water-disp.

Trans-222. [Trans-Chemco] Emulsion with 3% silicone; defoamer/antifoam for aq. systems; water-disp.

Trans-224. [Trans-Chemco] Nonsilicone, water-based; aq. food processing defoamer; water-disp.

Trans-225S. [Trans-Chemco] Silicone emulsion; aq. food processing antifoam; water-disp.; 12.5% act.

Trans-262. [Trans-Chemco] Nonsilicone; defoamer and drainage aid for brn. stock washers; water-insol.; 100% act.

Trans-264. [Trans-Chemco] Surfactants and hydrophobed silica; defoamer and drainage aid for brn. stock washers; insol. in water; 100% act.

Trans-265. [Trans-Chemco] Nonsilicone; water-based ink defoamer; partly sol. in water; 100% act.

Trans-266. [Trans-Chemco] Surfactants and hydrophobed silica; defoamer for paper machine stock, bleach stock, effluent, and red liquor; partly sol. in water; 100% act.

Trans-1030. [Trans-Chemco] Emulsion with 3% silicone; defoamer for sauna baths, whirlpools, rug shampoos; water-disp.

Trans-1030 K. [Trans-Chemco] Emulsion with 3% silicone, kosher; defoamer for aq. systems incl. food processing; water-disp.

Trans-FG2. [Trans-Chemco] Nonsilicone, water-based; defoamer for veg. canning, food processing; water-disp.

Trans-N1. [Trans-Chemco] Nonsilicone, water based; defoamer for paper/paperboard mfg.; water-disp.

Trans-RC1. [Trans-Chemco] Nonsilicone; defoamer for mfg. of coatings for paper/paperboard, in clay coatings; sol. in hydrocarbon solvs., water-insol.; 100% act.

Transcutol. [Gattefosse SA] Ethyldiglycol; CAS 111-90-0; solv. for active ingreds. in pharmaceutical preps.; cosurfactant for microemulsions; liq.; 100% conc.

Trem-LF-40. [Henkel] Sodium alkylallyl sulfosuccinate; polymerizable surfactant, solubilizer; provides low foaming emulsions with improved water resistance; lt. yel. liq.; water-sol.; 40% act.

Triameen T. [Akzo] N-Tallowalkyl dipropylene triamine; CAS 61791-57-9; cationic; industrial surfactant; Gardner 8 max. color; sp.gr. 0.85 (50 C); m.p. 34 C; iodine no. 25; amine no. 415; flash pt. (PMCC) > 176 C.

Tricol M-20. [Takemoto Oil & Fat] Surfactant/min. oil blend; anionic/nonionic; coning oil for polyester yarn; liq.; 100% conc.

Trilon® A. [BASF] Trisodium NTA; CAS 5064-31-3; low m.w. general purpose chelate; complexes iron in acid pH range; detergent builder for non-phosphate liqs.; in soaps, detergents, water treatment, metal finishing and plating, pulp and paper mfg., synthesis of polymers, photography; textiles, chemical cleaning; liq.; m.w. 257; chelating act. 150 mg $CaCO_3$/g; 38% conc. in water.

Trinoram C. [Ceca SA] N-Coco dipropylene triamine; cationic; emulsifier for bitumen; antistripping agent for roadmaking.

Trinoram O. [Ceca SA] Oleic dipropylene triamine; cationic.

Trinoram S. [Ceca SA] N-tallow dipropylene triamine; CAS 61791-57-9; cationic; emulsifier for bitumen; antistripping agent for roadmaking; paste.

Tris Amino® (Crystals). [Angus] Tris (hydroxymethyl) aminomethane; pigment dispersant, neutralizing amine, corrosion inhibitor, acid-salt catalyst, pH buffer, chemical and pharmaceutical intermediate, solubilizer; m.w. 121.1; sol. 80 g/100 ml water; m.p. 171 C; b.p. 219 C; pH 10.4 (0.1M aq. sol'n.).

Triton® 770 Conc. [Union Carbide; Union Carbide Europe] Sodium alkylaryl polyether sulfate; anionic; detergent, emulsifier, wetting agent, penetrant used in detergents, cleaners, textile and leather processing, emulsion polymerization; degreasing agent on skins prior to tanning; amber clear liq.; sol. in water, acid and alkaline sol'ns.; sp.gr. 0.98; dens. 8.3 lb/gal; visc. 15 cps; flash pt. 78 F (PMCC); pour pt. -20 F; pH 7.5 (5% aq.); surf. tens. 28 dynes/cm (1%); Ross-Miles foam 165 mm (0.1%, initial, 120 F); 30% act., 23% IPA.

Triton® AG-98. [Union Carbide] Octylphenoxy polyethoxy ethanol; nonionic; low-foam spray adjuvant to enhance herbicidal activity; agric. adjuvant; liq.; 80% conc.

Triton® B-1956. [Union Carbide] Modified phthalic glyceryl alkyl resin; nonionic; spreader in herbicides, miticide, insecticide, and fungicide sprays for fruits and vegetables; emulsifier for insecticides; detergent; liq.; oil-sol.; 77% conc.

Triton® BG-10. [Union Carbide; Union Carbide Europe] Glucoside; nonionic; surfactant for alkali bottle washing, industrial cleaner, metal cleaning, agric. formulations; liq.; sol. in 50% sodium hydroxide sol'n.; 70% conc.

Triton® CF-10. [Union Carbide; Union Carbide Europe] Alkylaryl polyether; nonionic; low-foaming detergent for mechanical dishwashing, rinse aids, laundering, metal and dairy cleaning, textile wetting; nylon dyeing assistant; defoamer for food soils; lt. amber sl. hazy liq.; sol. in water, alcohols, glycols, acetone, aromatics, chlorinated solvs.; sp.gr. 1.07; dens. 8.9 lb/gal; visc. 250 cps; HLB 14.0; pour pt. 60 F; cloud pt. 28 C (1% aq.); flash pt. (TOC) > 500 F; surf. tens. 31 dynes/cm (1%); 100% act.

Triton® CF-21. [Union Carbide; Union Carbide Europe] Alkylaryl polyether; nonionic; low-foaming detergent, wetting agent for mechanical dishwashing, textiles, final rinse additive, laundering, metal and dairy cleaning; lt. amber clear liq.; sol. in water, alcohols, glycols, acetone, aromatics, chlorinated solvs.; sp.gr. 1.04; dens. 8.7 lb/gal; visc. 250 cps; HLB 12.9; pour pt. -35 C; cloud pt. 38 C (1% aq.); flash pt. (TOC) > 249 C; pH 6.0 (5% aq.); surf. tens. 31 dynes/cm (1%); 100% act.

Triton® CF-32. [Union Carbide; Union Carbide Europe] Amine polyglycol condensate; nonionic; wetting and antifoamer for protein soils; low-foaming detergent for mechanical dishwashing, metal and dairy cleaning; VCS 5 liq.; dens. 8.6 lb/gal; visc. 550 cps; cloud pt. 25 C (1%); flash pt. > 300 F (TOC); pour pt. 15 F; surf. tens. 32 dynes/cm (1%); 95% act.

Triton® CF-54. [Union Carbide; Union Carbide Europe] Polyethoxy adduct, modified; nonionic; low-foaming detergent, wetting agent for mech. dishwashing, rinse additive, automatic laundering, metal cleaning, dairy equip. cleaner; caustic stable; food soil defoamer; lt. straw clear liq.; sp.gr. 1.04; dens. 8.7 lb/gal; visc. 175 cP; HLB 13.6; pour pt. 35 F; cloud pt. 38 C (1% aq.); flash pt. (PMCC) > 200 F; pH 6 (5% aq.); surf. tens. 29 dynes/cm (0.01%); 100% act.

Triton® CF-76. [Union Carbide; Union Carbide Europe] Polyethoxy adduct, modified; nonionic; low-foaming detergent, wetting agent for mech. dishwashing, metal cleaning, bottle washing, food process equip. cleaners; food soil/soap defoamer; caustic-stable; lt. yel. clear liq.; sol. in many common solvs.; sp.gr. 1.04; dens. 8.7 lb/gal; visc. 295 cps; HLB 12.6; pour pt. 5 C; cloud pt. 31 C (1%); flash pt. (TOC) > 300 F; pH 9 (5% aq.); surf. tens. 29 dynes/cm (0.01% aq.); 100% act.

Triton® CF-87. [Union Carbide; Union Carbide Europe] Alkylaryl ether, modified; nonionic; low-foam surfactant; rinse aid in mechanical dishwashing; metal cleaning; textile wetting; straw liq.; dens. 9.0 lb/gal; visc. 240 cps; HLB 12.7; cloud pt. 32 C (1%); flash pt. > 215 F (TOC); pour pt. 30 F; surf. tens. 32 dynes/cm (0.01%); 90% act.

Triton® CG-110. [Union Carbide; Union Carbide Europe] Alkyl glucoside; nonionic; low irritation surfactant, foaming agent for shampoos, skin creams, lotions, bar soaps, industrial cleaners, alkali bottlewashing, food processing; biodeg.; liq.; sol. in 50% sodium hydroxide sol'n.; 60% conc.

Triton® CS-7. [Union Carbide] Alkylaryl polyethoxylate/sodium alkyl sulfonate blend; agric. spreader-binder in insecticide, fungicide concs.; penetrant and dew suppressant for ornamental turf grasses; liq.; 60% conc.

Triton® DF-12. [Union Carbide; Union Carbide Europe] Polyethoxylated alcohol, modified; nonionic; biodeg. low-foaming detergent for mech. dishwashing, rinse additive, automatic laundering, metal cleaning, floor scrubbing, dairy equip. cleaners, textile wetting; stable in acid, caustic sol'ns.; lt. straw clear liq.; sol. in water, alcohols, aromatics, chlorinated solvs.; sp.gr. 1.035; dens. 8.6 lb/gal; visc. 60 cps; HLB 10.6; pour pt. 65 F; cloud pt. 17 C (1% aq.); flash pt. (COC) 430 F; pH 6 (5% aq.); surf. tens. 30 dynes/cm (0.1%); 100% act.

Triton® DF-16. [Union Carbide; Union Carbide Europe] Ethoxylated linear alcohol; nonionic; low-foam detergent, wetting agent for paper and textile processing, rinse aids, mech. dishwashing, metal cleaners, hard surf. cleaners; biodeg.; APHA < 200 clear liq.; sol. in water < 97 F, many common solvs.; sp.gr. 0.987; dens. 8.2 lb/gal; visc. 35 cps; HLB 11.6; pour pt. 20 F; cloud pt. 36 C (1% aq.); flash pt. (PMCC) > 93 C; pH 6.0 (5% aq.); surf. tens. 28 dynes/cm (0.1%); Ross-Miles foam 2 mm (0.1%, initial, 120 F); 100% act.

Triton® DF-18. [Union Carbide; Union Carbide Europe] Polyethoxylated alcohol, modified; nonionic; biodeg. low-foam detergent for mech. dishwashing, food processing equip. cleaners, spray metal cleaner; food soil defoamer; yel. clear liq.; sp.gr. 1.02; dens. 8.5 lb/gal; visc. 140 cps; HLB 11.3; cloud pt. < 0 C (1% aq.); flash pt. > 220 F (TOC); pour pt. 65 F; pH 7.0 (5% aq.); surf. tens. 32 dynes/cm (0.01%); 90% act.

Triton® DF-20. [Union Carbide; Union Carbide Europe] Modified ethoxylate, acid form; anionic; low foaming detergent for highly alkaline built liq. conc. cleaners, metal cleaners, hard surf. cleaners, steam cleaners, dishwash, food process equip. cleaners, paint stripping; biodeg.; straw clear liq.; sp.gr. 1.02; dens. 8.5 lb/gal; visc. 630 cP; pour pt. 5 F; flash pt. (TOC) > 300 F; 100% act.

Triton® GR-5M. [Union Carbide; Union Carbide Europe] Dioctyl sodium sulfosuccinate; anionic; high speed wetting and rewetting agent, emulsifier, dispersant for paints, textiles; pale yel. clear liq.; sol. in water; sp.gr. 1.005; dens. 8.4 lb/gal; visc. 40 cps; pour pt. -60 F; flash pt. (Seta CC) 75 F; pH 6.0 (1% aq.); surf. tens. 27 dynes/cm (0.1%); Ross-Miles foam 195 mm (0.1%, initial); 60% act. in aq. IPA.

Triton® GR-7M. [Union Carbide; Union Carbide Europe] Dioctyl sodium sulfosuccinate; anionic; high speed wetting and rewetting agent, emulsifier, dispersant for paints, textiles, drycleaning detergents, agric. emulsions; lt. amber clear liq.; sp.gr. 1.04; dens. 8.7 lb/gal; visc. 110 cps; pour pt. –70 F; flash pt. (Seta CC) 133 F; pH 7.0 (1% aq.); 64% act. in lt. petrol. distillate.

Triton® H-55. [Union Carbide] Phosphate ester, potassium salt; anionic; hydrotrope/solubilizer for built liq. concs.; as surfactant for alkaline builder sol'ns.; lt. amber clear liq.; sp.gr. 1.35; dens. 11.2 lb/gal; visc. 40 cps; pour pt. -10 F; flash pt. (PMCC) 220 F; pH 8-10; 50% act.

Triton® H-66. [Union Carbide] Phosphate ester, potassium salt; anionic; hydrotrope/solubilizer for surfactants in built liq. concs.; lt. yel. clear liq.; sp.gr. 1.26; dens. 10.5 lb/gal; visc. 120 cps; pour pt. -5 F; flash pt. (PMCC) > 200 F; pH 8-10; 50% act.

Triton® N-42. [Union Carbide; Union Carbide Europe] Nonoxynol-4; CAS 9016-45-9; nonionic; detergent, emulsifier; colorless-yel. clear visc. liq.; m.w. 405; water-disp., oil-sol.; sp.gr. 1.068; dens. 8.5 lb/gal; visc. 250 cps; HLB 9.1; pour pt. -26 C; flash pt. (TOC) 149 C; pH 7.5; surf. tens. 29 dynes/cm (0.01% aq.); 100% conc.

Triton® N-57. [Union Carbide; Union Carbide Europe] Nonoxynol-5; CAS 9016-45-9; nonionic; detergent, emulsifier for solv. cleaners; intermediate; APHA 100 clear liq.; oil-sol.; misc. with most polar org. solvs., aromatic hydrocarbons; insol. in water; m.w. 440; sp.gr. 1.029; dens. 8.57 lb/gal; visc. 240 cps; HLB 10.0; cloud pt. < 0 C (1%); flash pt. > 515 F (COC); pour pt. –25 F; surf. tens. 29 dynes/cm (0.1%); 100% act.

Triton® N-60. [Union Carbide; Union Carbide Europe] Nonylphenol ethoxylate; nonionic; antifogging agent in plasticized PVC; corrosion inhibitor; dispersant for petrol. oils; liq.; 100% conc.

Triton® N-101. [Union Carbide; Union Carbide Europe] Nonoxynol-9; CAS 9016-45-9; nonionic; wetting agent, detergent, dispersant, solubilizer, emulsifier for pesticides, household and industrial cleaners, textile processing, hard surf. cleaners, laundry, metal cleaning, detergent-sanitizers; APHA 100 clear liq.; sol. in water, benzene, toluene, xylene, trichlorethylene, ethanol, IPA; m.w. 642; sp.gr. 1.046; dens. 8.7 lb/gal; visc. 240 cps; HLB 13.4; pour pt. 4 C; cloud pt. 54 C (1% aq.); flash pt. (TOC) > 300 F; pH 6 (5% aq.); surf. tens. 29 dynes/cm (1%); Ross-Miles foam 51 mm (0.1%, initial, 125 F); 100% act.

Triton® N-111. [Union Carbide; Union Carbide Europe] Nonoxynol-11; CAS 9016-45-9; nonionic; wetting agent, detergent, emulsifier for textile processing; APHA 100 liq.; m.w. 704; water-sol.; dens. 8.8 lb/gal; visc. 310 cps; HLB 13.8; cloud pt. 72 C (1%); flash pt. (TOC) > 300 F; pour pt. 55 F; 100% conc.

Triton® N-150. [Union Carbide; Union Carbide Europe] Nonoxynol-15; CAS 9016-45-9; nonionic; detergent, emulsifier; APHA 160 liq.; m.w. 880; dens. 9.0 lb/gal; visc. 4350 cps; HLB 15.0; cloud pt. 95 C (1%); pour pt. 65 F; surf. tens. 33 dynes/cm (1%); 100% conc.

Triton® QS-15. [Union Carbide; Union Carbide Europe] Sodium salt of amphoteric surfactant; amphoteric; detergent for hard surf. cleaners, highly built alkaline cleaners, bottle washing, metal cleaning; caustic-stable; amber liq.; sol. in water, alcohols, aromatics, chlorinated solvs., glycol ethers, ketones; dens. 9.3 lb/gal; visc. 8500 cP; pour pt. -4 C; flash pt. (TOC) > 300 F; pH 10.5 (5% aq.); 100% act.

Triton® QS-44. [Union Carbide; Union Carbide Europe] Phosphate surfactant, free acid; anionic; hydrotrope, detergent, wetting agent; solubilizer for nonionic surfactants in alkaline cleaning baths, metal cleaning and for nonionic and anionic surfactants in built concs.; lt. amber clear liq.; sol. in water, alcohols, esters, ketones; sp.gr. 1.18; dens. 9.8 lb/gal; visc. 8000 cps; pour pt. 35 F; flash pt. (TOC) > 200 F; pH 1.3-2.0 (5%); surf. tens. 34 dyne/cm (0.1%); Ross-Miles foam 145 mm (pH 7, 0.125% aq., 50 C); 80% act.

Triton® RW-20. [Union Carbide; Union Carbide Europe] Alkylamine ethoxylate (2 EO); nonionic/cationic; emulsifier, detergent, degreaser for sec. oil recovery, waste treatment, transport cleaners, pipeline/refinery equip./chem. plant cleaning, metal cleaning, metalworking fluids, textiles; lt. amber clear liq.; amine odor; sol. in 10% HCl, IPA, butoxyethanol, xylene, kerosene, CCl_4; insol. in water; sp.gr. 0.913; dens. 7.61 lb/gal; visc. 240 cps; HLB 6–8; b.p. 73 C; flash pt. (Seta CC) > 167 C; pH 9.5–11.0 (5% aq.); 99% conc.

Triton® RW-50. [Union Carbide; Union Carbide Europe] Alkylamine ethoxylate (5 EO); nonionic/cationic; emulsifier, detergent, degreaser for sec. oil recovery, waste treatment, transport cleaners, pipeline/refinery equip./chem. plant cleaning, metal cleaning, metalworking fluids, textiles; lt. amber clear liq.; amine odor; sol. @ 10% in water, 10% HCl, IPA, butoxyethanol, xylene, kerosene, CCl_4; sp.gr. 0.968; dens. 8.06 lb/gal; HLB 12–14; pour pt. –38 F; flash pt. 149 C (CC); cloud pt. < 0 C (1% aq.); pH 10.0–11.5 (5% aq.); surf. tens. 32 dynes/cm (0.01% aq.); Ross-Miles foam 0 mm (pH 7, initial); 99% conc.

Triton® RW-75. [Union Carbide; Union Carbide Europe] Alkylamine ethoxylate (7.5 EO); nonionic/cationic; emulsifier, detergent, degreaser for sec. oil recovery, waste treatment, transport cleaners, pipeline/refinery equip./chem. plant cleaning, metal cleaning, metalworking fluids, textiles; lt. amber clear liq.; amine odor; sol. @ 10% in water, 10% HCl, IPA, butoxyethanol, xylene, CCl_4; sp.gr. 0.981; dens. 8.17 lb/gal; HLB 14–16; flash pt. > 149 C (CC); cloud pt. 32 C (1% aq.); pH 9.5–11.0 (5% aq.); 99% conc.

Triton® RW-100. [Union Carbide; Union Carbide Europe] Alkylamine ethoxylate (10 EO); nonionic/cationic; emulsifier, detergent, degreaser for sec. oil recovery, waste treatment, transport cleaners, pipeline/refinery equip./chem. plant cleaning, metal cleaning, metalworking fluids, textiles; lt. amber sl. hazy liq.; amine odor; sol. in water, 10% HCl, IPA, butoxyethanol, xylene, CCl_4; sp.gr. 1.004; dens. 8.36 lb/gal; visc. 180 cps; HLB 16; pour pt. –25 F; flash pt. 149 C (CC); cloud pt. 67 C (1% aq.); pH 9.5–11.0 (5% aq.); surf. tens. 39 dynes/cm (0.01% aq.); Ross-Miles foam 12 mm (pH 7, initial); 99% conc.

Triton® RW-150. [Union Carbide; Union Carbide Europe] Alkylamine ethoxylate (15 EO); nonionic/cationic; emulsifier, detergent, degreaser for sec. oil recovery, waste treatment, transport cleaners, pipeline/refinery equip./chem. plant cleaning, metal cleaning, metalworking fluids, textiles; lt. amber sl. hazy liq.; amine odor; sol. in water, 10% HCl, IPA, butoxyethanol, xylene, CCl_4; sp.gr. 1.024; dens. 8.58 lb/gal; visc. 480 cps; HLB > 16; pour pt. –8 F; flash pt. > 149 C (CC); cloud pt. 96 C (1% aq.); pH 9.5–11.0 (5% aq.); surf. tens. 46 dynes/cm (0.01% aq.); Ross-Miles foam 22 mm (pH 7, initial); 99% conc.

Triton® W-30 Conc. [Union Carbide; Union Carbide Europe] Sodium alkylaryl ether sulfate; anionic; detergent, wetting agent, penetrant for sizing, dyeing, desizing, textile processing; emulsifier in emulsion polymerization; amber clear liq.; sol. in water, acid and alkaline sol'ns.; sp.gr. 0.98; dens. 8.2 lb/gal; visc. 9 cps; flash pt. 74 F (PMCC); pour pt. –15 F; pH 7.5 (5% aq.); surf. tens. 29 dynes/cm (1%); Ross-Miles foam 110 mm (0.1%, initial, 120 F); 27% act., 27% IPA.

Triton® X-15. [Union Carbide; Union Carbide Europe] Octoxynol-1; CAS 9002-93-1; nonionic; surfactant, coupling agent, emulsifier for industrial/household cleaners, emulsion polymerization, agric., latex stabilizer; FDA, EPA compliance; APHA 250 liq.; sol. in aliphatic hydrocarbons; misc. with aromatics, alcohols, glycols, ethers, ketones; m.w. 250; sp.gr. 0.985; dens. 8.2 lb/gal; visc. 790 cps; HLB 3.6; flash pt. > 300 F (TOC); pour pt. 15 F; 100% act.

Triton® X-35. [Union Carbide; Union Carbide Europe] Octoxynol-3; CAS 9002-93-1; nonionic; surfactant, coupling agent, emulsifier for industrial/household cleaners, emulsion polymerization, agric., latex stabilizer; FDA, EPA compliance; APHA 125 liq.; sol. in aliphatic hydrocarbons; misc. with aromatic hydrocarbons, alcohols, glycols, ethers, ketones; m.w. 338; sp.gr. 1.023; dens. 8.5 lb/gal; visc. 370 cps; HLB 7.8; flash pt. > 300 F (TOC); pour pt. –10 F; surf. tens. 29 dynes/cm (0.01%); Ross-Miles foam 5 mm (0.1%, initial, 120 F); 100% act.

Triton® X-45. [Union Carbide; Union Carbide Europe] Octoxynol-5; CAS 9002-93-1; nonionic; emulsifier, detergent, dispersant, wetting agent for solv. cleaners, metal cleaners, drycleaning, insecticides; FDA, EPA compliance; clear liq.; sol. in oil; misc. with alcohol, glycols, ethers, ketones, aromatic hydrocarbons; m.w. 426; sp.gr. 1.040; dens. 8.7 lb/gal; visc. 290 cps; HLB 10.4; cloud pt. < 0 C (1%); flash pt. > 300 F (TOC); pour pt. –26 C; surf. tens. 28 dynes/cm (0.1%); Ross-Miles foam 16 mm (0.1%, initial, 120 F); 100% act.

Triton® X-100. [Union Carbide; Union Carbide Europe] Octoxynol-9 (9–10 EO); CAS 9002-93-1; nonionic; wetting agent, dispersant, detergent, household and industrial cleaners; metal cleaners; sanitizers; textile processing, wool scouring; emulsifier for insecticides and herbicides; solubilizer of perfumes; FDA, EPA compliance; clear liq.; sol. in water, toluene, xylene, trichlorethylene, ethylene glycol, alcohols; m.w. 628; sp.gr. 1.065; dens. 8.9 lb/gal; visc. 240 cps; HLB 13.5; cloud pt. 65 C (1% aq.); flash pt. > 300 F (TOC); pour pt. 45 F; pH 6 (5% aq.); surf. tens. 30 dynes/cm (1%); Ross-Miles foam 110 mm (0.1%, 120 F); 100% act.

Triton® X-100 CG. [Union Carbide] Octoxynol-9 (9-10 EO); CAS 9002-93-1; nonionic; detergent, wetting agent for household/industrial cleaners, hard surf. cleaners, metal cleaners, sanitizers, textiles, pesticides; EPA compliance; clear liq.; sol. in water, toluene, xylene, trichlorethylene, ethylene glycol, alcohols; sp.gr. 1.07; dens. 8.9 lb/gal; visc. 240 cP; HLB 13.5; pour pt. 7 C; cloud pt. 65 C (1% aq.); flash pt. (TOC) > 300 F; pH 6 (5% aq.); surf. tens. 30 dynes/cm (0.1%); Ross-Miles foam 110 mm (0.1%, 120 F); 100% act.

Triton® X-102. [Union Carbide; Union Carbide Europe] Octoxynol-13 (12-13 EO); CAS 9002-93-1; nonionic; detergent, wetting agent, foam stabilizer at high temps., in presence of electrolytes; metal cleaning, industrial, household liq. detergents and cleaners, sanitizers; solubilizer of anionic detergents; FDA, EPA compliance; lt. color clear liq.; sol. in water, many common org. solvs.; m.w. 756; sp.gr. 1.07; dens. 8.9 lb/gal; visc. 330 cps; HLB 14.6; cloud pt. 88 C (1% aq.); flash pt. > 300 F (TOC); pour pt. 60 F; surf. tens. 32 dynes/cm (1%); Ross-Miles foam 130 mm (0.1%, 120 F); 100% act.

Triton® X-114. [Union Carbide; Union Carbide Europe] Octoxynol-8 (7-8 EO); CAS 9002-93-1; nonionic; controlled foam detergent, wetting agent used in household and industrial laundry, metal cleaners, industrial/institutional hard surf. cleaners; biodispersant and oil emulsifier in cooling towers; FDA, EPA compliance; clear liq.; sol. in water, many common solvs.; m.w. 536; sp.gr. 1.05-1.06; dens. 8.8 lb/gal; visc. 260 cps; HLB 12.4; cloud pt. 25 C (1% aq.); flash pt. > 300 F (TOC); pour pt. 15 F; pH 6 (5% aq.); surf. tens. 29 dynes/cm (1%); Ross-Miles foam 25 mm (0.1%, 50 C); 100% act.

Triton® X-114SBHF. [Union Carbide] Alkylaryl POE glycol; nonionic; wetting agent, penetrant, spreader-adjuvant for herbicide, fungicide, and insecticide spray; assists water-in-soil penetration; EPA compliance; liq.; 80% conc.

Triton® X-120. [Union Carbide; Union Carbide Europe] Octoxynol-9 (9-10 EO); CAS 9002-93-1; nonionic; wetting agent, dipsersant for agric. wettable powds.; wh. solid; m.w. 628; 40% conc.

Triton® X-155-90%. [Union Carbide; Union Carbide Europe] Methylenebisdiamyl phenoxy polyethoxy ethanol; nonionic; surfactant in emulsification of aromatic hydrocarbons; wetting agent, detergent; antistat, lubricant for textile fiber processing; stabilizer for polymer emulsions; yel.-amber clear liq.; mild odor; sol. in water, aromatic hydrocarbons; sp.gr. 1.02; dens. 8.50 lb/gal; visc. 250 cps; HLB 12.5; pour pt. –20 F; cloud pt. < 10 C (1% aq.); flash pt. (Seta CC) 53 C; pH 7 (5%); pH 8.50 (5%); surf. tens. 30 dynes/cm (0.1% aq.); Ross-Miles foam 17 mm (0.1%, initial); 90% act. in aq. IPA.

Triton® X-165-70%. [Union Carbide; Union Carbide Europe] Octoxynol-16; CAS 9002-93-1; nonionic; detergent, emulsifier, wetting agent for industrial/household cleaners, emulsion polymerization, latex stabilizer; FDA compliance; APHA 125 aq. sol'n.; sol. in inorg. salt sol'ns., aq. min. acids; misc. with water, alcohols, glycols, ethers, ketones; m.w. 910;

sp.gr. 1.080; dens. 9.0 lb/gal; visc. 540 cps; HLB 15.8; cloud pt. > 100 C (1%); flash pt. > 300 F (TOC); pour pt. 55 F; surf. tens. 35 dynes/cm (0.1%); Ross-Miles foam 145 mm (0.1%, initial, 120 F); 70% act.

Triton® X-180. [Union Carbide] Alkylaryl polyether alcohol/org. sulfonate blend; anionic; emulsifier for pesticides; hazy amber liq.; dens. 8.6 lb/gal; visc. 130 cps; flash pt. (Seta CC) 54 F; pour pt. 40 F; 100% conc.

Triton® X-185. [Union Carbide] Alkylaryl polyether alcohol/org. sulfonate blend; anionic; emulsifier for pesticides; clear amber liq.; dens. 8.7 lb/gal; visc. 465 cps; flash pt. (Seta CC) 34 F; pour pt. 40 F; 100% conc.

Triton® X-190. [Union Carbide] Alkylaryl polyether alcohol/org. sulfonate blend; anionic; emulsifier for pesticides; clear amber liq.; dens. 8.8 lb/gal; visc. 715 cps; flash pt. (Seta CC) 46 F; pour pt. -15 F; 100% conc.

Triton® X-193. [Union Carbide] Alkylaryl polyether alcohol/org. sulfonate blend; anionic; emulsifier for pesticides; hazy amber liq.; dens. 8.7 lb/gal; visc. 740 cps; flash pt. (Seta CC) 74 F; pour pt. 60 F; 100% conc.

Triton® X-200. [Union Carbide; Union Carbide Europe] Sodium octoxynol-2 ethane sulfonate; anionic; detergent for metal cleaning, pickling and plating baths, household cleaners; polymerization emulsifier, latex post-stabilizer; wetting agent in strong alkaline baths; dispersant for fulling in lime soap; dye leveling agent for acid dyestuffs; wh. liq.; dens. 8.9 lb/gal; visc. 7000 cps; flash pt. (TOC) > 300 F; pour pt. 25 F; surf. tens. 29 dynes/cm (1%); 28% conc.

Triton® X-207. [Union Carbide; Union Carbide Europe] Alkylaryl polyether alcohol; nonionic; emulsifier, lubricant for agric. sprays, industrial oils, syn. fiber finishes; amber clear liq.; disp. in water; misc. with aromatic solvs., kerosene, chlorinated solvs.; sp.gr. 0.98; dens. 8.2 lb/gal; visc. 600 cps; HLB 10.7; pour pt. –10 F; flash pt. > 200 F (PMCC); 100% act.

Triton® X-301. [Union Carbide; Union Carbide Europe] Sodium alkylaryl polyether sulfate; anionic; wetting agent, emulsifier, penetrant, high-foam detergent for household and industrial cleaners, emulsion polymerization; wh. paste; sol. in water and acid and alkaline sol'ns.; sp.gr. 1.05; dens. 8.8 lb/gal; visc. 4200 cps; flash pt. > 300 F (TOC); pour pt. 30 F; pH 7.5 (5% aq.); surf. tens. 28 dynes/cm (0.1%); Ross-Miles foam 145 mm (0.1%, 120 F, initial); 20% act. in water.

Triton® X-305-70%. [Union Carbide; Union Carbide Europe] Octoxynol-30; CAS 9002-93-1; nonionic; detergent, emulsifier, wetting agent for household/industrial cleaners, emulsion polymerization, agric., latex stabilizer; FDA, EPA compliance; APHA 150 liq.; sol. in inorg. salt sol'ns., aq. min. acids; misc. with water, alcohols, glycols, ethers, ketones; m.w. 1526; sp.gr. 1.095; dens. 9.1 lb/gal; visc. 470 cps; HLB 17.3; pour pt. 35 F; cloud pt. > 100 C (1%); flash pt. > 300 F (TOC); surf. tens. 37 dynes/cm (0.1%); Ross-Miles foam 150 mm (0.1%, initial, 120 F); 70% act.

Triton® X-363M. [Union Carbide; Union Carbide Europe] Alkylaryl polyether ethanol; nonionic; emulsifier for agric. spray oils, industrial oils; food contact uses; lt. straw liq.; dens. 8.2 lb/gal; visc. 100 cps; HLB 9.1; flash pt. (Seta CC) > 198 F; pour pt. -25 F; 100% conc.

Triton® X-405-70%. [Union Carbide; Union Carbide Europe] Octoxynol-40; CAS 9002-93-1; nonionic; detergent, emulsifier, wetting agent for household/industrial cleaners, emulsion polymerization, agric.; latex stabilizer; coupling agent in naphthol dyeing; FDA, EPA compliance; APHA 250 liq.; sol. in inorg. salt sol'ns., aq. min. acids; misc. with water, alcohols, glycols, ethers, ketones; m.w. 1966; sp.gr. 1.102; dens. 9.2 lb/gal; visc. 490 cps; HLB 17.9; cloud pt. > 100 C (1%); flash pt. > 212 F (TOC); pour pt. 25 F; surf. tens. 37 dynes/cm (0.1%); Ross-Miles foam 126 mm (0.1%, initial, 120 F); 70% act.

Triton® X-705-70%. [Union Carbide; Union Carbide Europe] Octoxynol-70; CAS 9002-93-1; nonionic; emulsifier for emulsion polymerization; FDA, EPA compliance; APHA 500 liq.; sol. in inorg. salt sol'ns., aq. min. acids; misc. with water, alcohols, glycols, ethers, ketones; sp.gr. 1.1; dens. 9.2 lb/gal; visc. 505 cP; HLB 18.7; pour pt. 43 F; cloud pt. > 100 C (1% aq.); flash pt. (Seta CC) > 230 F; surf. tens. 39 dynes/cm (0.1%); Ross-Miles foam 95 mm (0.1%, initial, 120 F); 70% act.

Triton® XL-80N. [Union Carbide] Alcohol alkoxylate; nonionic; biodeg. detergent, wetting agent, leveling agent, coupling agent, emulsifier, dispersant, stabilizer for industrial/institutional/household cleaners, metal cleaning, laundry, paper deinking, textile processing, fiber lubricants, paints; oilfield chems., water treatment, circuit board cleaners, leather operations; APHA 25 clear to hazy liq., low odor; m.w. 442; sp.gr. 0.98; dens. 8.13 lb/gal; visc. 25 cps; pour pt. -2 C; cloud pt. 50 C (1% aq.); flash pt. (PMCC) 127 C; pH 6.5 (1% aq.); surf. tens. 28 dynes/cm (0.1% aq.); Draves wetting 4.6 s (0.1% aq.); Ross-Miles foam 95 mm (0.1% initial); 100% act.

Triumphnetzer ZSG. [Zschimmer & Schwarz] Diisooctyl sulfosuccinate; anionic; wetting agent for textile and chem-tech prods., household/industrial cleaners, drycleaning, textile, ceramic, and varnish industries; colorless to ylsh. liq.; sol. in water, ethanol; dens. 1.1 g/cc; pH 4.5-6.5 (5%); 67% act., 72% solids.

Triumphnetzer ZSN. [Zschimmer & Schwarz] Diisooctyl sulfosuccinate; anionic; wetting agent for textile and chem-tech prods.; liq.; 75% conc.

Troykyd® Anti-Gel. [Troy] Wetting agent designed to prevent excessive bodying, gellation, and seeding of pigment pastes and pigmented solv.-based coatings; stabilizes paste visc. and permits higher pigment of the finished paint; esp. effective in systems containing reactive pigments such as zinc oxide and red lead, and with high absorptive pigments such as carbon blk.; used in the pigment disp. step of paint mfg. at 0.5–1.0% based on total wt. of paint; clear almost water-wh. liq.; sp.gr. 0.95–0.98; dens. 7.9–8.2 lb/gal; visc. A-3 to A-5; acid no. 1–3.

Troykyd® D44. [Troy] Stearate-modified hydrocarbon; defoamer for aq. systems, latex paints, pigment disps., adhesives, caulks, sealants, cement compds.; liq.; sp.gr. 0.91; usage level: 0.2-0.6%; 99% act.

Troykyd® D55. [Troy] Hydrophobic inorg. disp.; nonionic; defoamer for printing inks, latex paints, pigment disps., joint cement compds., water-reducible systems; liq.; sp.gr. 0.93; usage level: 0.2-0.6%; 100% act.

Troykyd® D126. [Troy] Glycol-treated carbonate; nonionic; defoamer for printing inks, adhesives/

caulks/sealants, spackling joint cement compds., concrete/tile/grout, drilling muds; powd.; dens. 1.50 g/cc; usage level: 4-8 lb/ton; 100% act.

Troykyd® D333. [Troy] Modified glycol/hydrocarbon blend; defoamer for metalworking fluids, printing inks, latex paints, adhesives/caulks/sealants, wood stains, floor wax cleaning compds.;; FDA compliance; liq.; sp.gr. 0.85-0.88; usage level: 0.2-0.6%; 100% act.

Troykyd® D666. [Troy] Hydrocarbon-based surfactant; defoamer for metalworking fluids, printing inks, latex paints, adhesives/caulks/sealants, wood stains, paper coatings, water-reducible systems, warp sizes; FDA compliance; liq.; sp.gr. 0.84-0.87; usage level: 0.1-0.5%; 100% act.

Troykyd® D777. [Troy] Silicone and glycol ester-modified hydrocarbon; defoamer for printing inks, latex paints, adhesives/caulks/sealants, water-reducible systems; liq.; sp.gr. 0.86-0.89; usage level: 0.2-0.6%; 100% act.

Troykyd® D999. [Troy] Fatty acid/hydrocarbon blend; nonionic; defoamer for metalworking fluids, printing inks, latex paints, adhesives/caulks/sealants, wood stains, paper coatings, water-reducible systems; liq.; sp.gr. 0.86-0.89; usage level: 0.2-0.5%; 100% act.

Troykyd® Emulso Wet. [Troy] Nonionic; surfactant; wetting agent, disp. stabilizer for latex paints when improved stability and compatibility are needed for min. side effects; emulsion stabilizer for any aq. system; improves color acceptance; esp. with red iron oxide and other inorg. pigments; useful in semigloss latex paint systems; should be added in the initial pigment disp. step of paint mfg. at 6 lb/100 gal.; lt. yel. mobile liq.; sol. in water and most solvs. except aliphatic hydrocarbons; sp.gr. 1.07–1.09; dens. 8.9–9.2 lb/gal; visc. (Gardner): K-P; HLB 15.

Troysol 307. [Troy] Silicone/nonsilicone blend; nonionic; surfactant, defoamer for solv.-based printing inks, urethane systems, lacquers, polyester gel coats, industrial baking systems; liq.; sp.gr. 0.88; usage level: 0.2-0.6%; 30% act.

Troysol 98C. [Troy] Fatty amine soap; amphoteric; wetting and dispersing agent for org. pigments, carbon blk. magnetic media, iron oxides; for nonaq. systems, paints/coatings, alkyd, acrylic, epoxy, urethane, polyester, lacquers, printing inks; liq.; usage level: 0.25-1.0%; 95% conc.

Troysol AFL. [Troy] Polymeric ester blend; nonionic; surfactant, defoamer, antifloat agent, air release agent for nonaq. coatings (alkyd, acrylic, epoxy, polyester, lacquer), printing inks; liq.; sp.gr. 0.87; usage level: 0.2-0.6%; 28% act.

Troysol CD1. [Troy] Polymeric ester; nonionic; pigment wetting agent and dispersant for solv.-based paints (alkyd, acrylic, epoxy, urethane, polyester), printing inks; liq.; usage level: 0.25-1.0%; 70% act.

Troysol CD2. [Troy] Modified veg. oil; nonionic; pigment wetting agent for solv.-based paint, printing inks; liq.; usage level: 0.25-1.0%; 100% act.

Troysol LAC. [Troy] Alkyl surfactant; anionic; substrate wetting agent, flow control additive, leveling agent for aq. systems, paints, printing inks, adhesives and coating sfor polyethylene and wax coated film and pkg.; liq.; usage level: 0.1-0.4%; 48% act.

Troysol Q148. [Troy] Modified siloxane copolymer; nonionic; surface active defoamer, flow/leveling agent, substrate wetting agent for aq., nonaq. and high solids systems, paints/coatings, alkyd, acrylic, polyester, printing inks; liq.; usage level: 0.2-0.6%; 27% act.

Troysol S366. [Troy] Modified siloxane copolymer; nonionic; surface active flow/leveling agent, substrate wetting agent for aq. and nonaq. systems, paints/coatings, alkyd, acrylic, epoxy, urethane, polyester, lacquer, printing inks; liq.; usage level: 0.2-0.6%; 60% act.

Troysol UGA. [Troy] Surfactant/solvent blend; nonionic; pigment wetting and dispersing agent for aq. and oil systems; liq.; 95% conc.

Trycol® 5874. [Henkel/Emery] Trideceth-14; CAS 24938-91-8; nonionic; hydrophilic general purpose emulsifier; for agric. applics.; EPA-exempt; Gardner 1 liq.; sol. in water; disp. in min. oil; dens. 8.7 lb/gal; visc. 78 cSt (100 F); HLB 15.1; cloud pt. 80 C (5% saline); pour pt. 10 C; 75% act. in water.

Trycol® 5877. [Henkel/Emery] Ethoxylated lauryl alcohol; nonionic; emulsifier, solubilizer; solid; water-sol.; HLB 16.7; 100% conc.

Trycol® 5878. [Henkel/Emery] Ethoxylated fatty alcohol ethers; nonionic; emulsifier, lubricant for textiles and detergent shampoo base; solubilizer; personal care prods.; liq.; HLB 9.5; 100% conc.

Trycol® 5882. [Henkel/Emery; Henkel/Textile] Laureth-4; nonionic; coemulsifier for silicone in polishes, mold release agents; emulsifier in industrial lubricants, agric., textile applics.; intermediate for shampoo bases; biodeg.; EPA-exempt; Gardner 1 liq.; sol. in butyl stearate, glycerol trioleate, xylene; disp. in water, min. oil; dens. 7.7 lb/gal; visc. 20 cSt (100 F); HLB 9.2; cloud pt. < 25 C; flash pt. 325 F; pour pt. 12 C; 100% conc.

Trycol® 5887. [Henkel/Emery] Ethoxylated octyl alcohol; nonionic; emulsifier, solubilizer; solid; water-sol.; HLB 15.3; 100% conc.

Trycol® 5888. [Henkel/Emery; Henkel/Textile] Steareth-20; CAS 9005-00-9; nonionic; emulsifier, solubilizer, solv. emulsifier in textile dye carriers, agric. formulations; stabilizer in latices; used in dyeing assistants, fruit coatings; EPA-exempt; Gardner 1 solid; sol. in water, xylene; HLB 15.3; m.p. 40 C; cloud pt. 91 C (5% saline); flash pt. 460 F.

Trycol® 5940. [Henkel/Emery; Henkel/Textile] Trideceth-6; CAS 24938-91-8; nonionic; detergent, wetting agent, emulsifier, dispersant, foam builder, solubilizer, coupling agent, rewetting agent for institutional, industrial, and household cleaners, degreasers, cutting oils, wool scouring, agric. applics.; intermediate; EPA-exempt; Gardner 1 liq.; sol. @ 5% in butyl stearate, glycerol trioleate, Stod., xylene; disp. in water, min. oil; dens. 8.2 lb/gal; visc. 32 cSt (100 F); HLB 11.4; pour pt. 12 C; cloud pt. < 25 C; flash pt. 345 F; 100% conc.

Trycol® 5941. [Henkel/Emery; Henkel/Textile] Trideceth-9; CAS 24938-91-8; nonionic; wetting, rewetting agent, detergent for lt. and heavy-duty high-foaming cleaners, paper towels, raw wool detergent, agric. formulations; coemulsifier; EPA-exempt; Gardner 1 liq.; sol. in water, glycerol trioleate, Stod., xylene; disp. in min. oil; dens. 8.3 lb/gal; visc. 43 cSt (100 F); HLB 13.0; cloud pt. 54 C; flash pt. 390 F; pour pt. 20 C; 100% conc.

Trycol® 5942. [Henkel/Emery] Trideceth-11; CAS 24938-91-8; nonionic; foam builder, detergent for lt. and heavy-duty cleaners; emulsifier for pesticides; EPA-exempt; Gardner 1 liq.; sol. in water, glycerol trioleate, Stod., xylene; disp. in min. oil; dens. 8.4 lb/gal; visc. 47 cSt (100 F); HLB 13.8; cloud pt. 73 C; flash pt. 420 F; pour pt. 17 C; 100%

conc.

Trycol® 5943. [Henkel/Emery; Henkel/Textile] Trideceth-12; CAS 24938-91-8; nonionic; foam builder, detergent for lt. and heavy-duty cleaners, textile processing; emulsifier for pesticides; EPA-exempt; Gardner 1 semisolid; sol. in water, xylene; dens. 8.4 lb/gal; HLB 14.5; m.p. 16 C; cloud pt. 70 C; flash pt. 440 F.; 100% conc.

Trycol® 5944. [Henkel/Emery] Trideceth-9; CAS 24938-91-8; nonionic; wetting, rewetting agent, detergent for lt. and heavy-duty high-foaming cleaners, paper towels, raw wool detergent, agric. applics.; coemulsifier; EPA-exempt; Gardner 1 liq.; sol. in water, glycerol trioleate, Stod., xylene; disp. in min. oil; dens. 8.4 lb/gal; visc. 61 cSt (100 F); HLB 13.0; cloud pt. 54 C; pour pt. < –10 C; 85% act. in water.

Trycol® 5946. [Henkel/Emery; Henkel/Textile] Trideceth-18; CAS 24938-91-8; nonionic; emulsifier, dispersant, detergent, wetting agent used in paper, agric. and textile industries; leveling agent, solubilizer, foam stabilizer; intermediate in mfg. anionic surfactants; EPA-exempt; Gardner 1 solid; sol. in water; dens. 8.5 lb/gal (40 C); HLB 16.0; m.p. 38 C; cloud pt. 83 C (5% saline); flash pt. 475 F; 100% conc.

Trycol® 5949. [Henkel/Emery; Henkel/Textile] Trideceth-8; CAS 24938-91-8; nonionic; foam builder, solubilizer, coemulsifier for high foaming detergent, textile and agric. formulations ; EPA-exempt; Gardner 1 liq.; sol. in water, glycerol trioleate, Stod., xylene; disp. in min. oil; dens. 8.3 lb/gal; visc. 39 cSt (100 F); HLB 12.5; cloud pt. 43 C; flash pt. 370 F; pour pt. 8 C; 100% conc.

Trycol® 5950. [Henkel/Emery; Henkel/Textile] Deceth-4; nonionic; wetting agent and penetrant for textile and agric. applics.; intermediate; EPA-exempt; Gardner 1 liq.; sol. in xylene, glycerol trioleate; disp. in water, min. oil; dens. 7.9 lb/gal; visc. 17 cSt (100 F); HLB 10.5; cloud pt. < 25 C; flash pt. 260 F; pour pt. –5 C; 100% conc.

Trycol® 5951. [Henkel/Emery] Deceth-5; nonionic; penetrant, wetting agent and emulsifier for heavy duty cleaners, agric. formulations; EPA-exempt; Gardner 1 liq.; sol. in butyl stearate, glycerol trioleate, xylene; disp. in water, min. oil, Stod.; dens. 8.2 lb/gal; visc. 23 cs; HLB 11.6; pour pt. 8 C; cloud pt. < 25 C; flash pt. 340 F; 100% act.

Trycol® 5952. [Henkel/Emery; Henkel/Textile] Deceth-6; nonionic; textile wetting agent; emulsifier for polyethylene emulsions used in water-repellent applics.; agric. formulations; EPA-exempt; Gardner 1 liq.; sol. in water, butyl stearate, glycerol trioleate, Stod., xylene; disp. in min. oil; dens. 8.3 lb/gal; visc. 26 cSt (100 F); HLB 12.4; cloud pt. 42 C; flash pt. 280 F; pour pt. 8 C; 97% conc.

Trycol® 5953. [Henkel/Emery; Henkel/Textile] Deceth-6; nonionic; textile wetting agent; emulsifier for polyethylene emulsions used in water-repellent applics.; agric. formulations; EPA-exempt; Gardner 1 liq.; sol. in water; disp. in min. oil; dens. 8.3 lb/gal; visc. 28 cSt (100 F); HLB 12.4; cloud pt. 42 C; flash pt. 315 F; pour pt. –15 C; 90% act. in water.

Trycol® 5956. [Henkel/Emery; Henkel/Textile] Deceth-9; nonionic; hydrophilic general-purpose surfactant; for agric. and textile applics.; EPA-exempt; Gardner 1 liq.; sol. @ 5% in water, xylene, methyl oleate; disp. in min. oil, Stod.; dens. 8.5 lb/gal; visc. 37 cSt (100 F); pour pt. 15 C; HLB 14.3; cloud pt. 80.5 C; flash pt. 220 F.

Trycol® 5957. [Henkel/Emery] Trideceth-10; CAS 24938-91-8; nonionic; wetting/rewetting agent, coemulsifier for aromatic and aliphatic solvs.; agric. formulations; EPA-exempt; Gardner 1 liq.; disp. @ 5% in Stod.; dens. 8.3 lb/gal; visc. 57 cSt (100 F); pour pt. 18 C; HLB 13.8; cloud pt. 63 C; flash pt. 390 F.

Trycol® 5963. [Henkel/Emery; Henkel/Textile] Laureth-8; CAS 9002-92-0; nonionic; biodeg. wetting agent, detergent, emulsifier for industrial, institutional and household cleaners, agric. applics.; EPA-exempt; Gardner 1 liq.; sol. in water, glycerol trioleate, xylene; dens. 8.2 lb/gal; visc. 35 cSt (100 F); HLB 12.6; cloud pt. 63 C; flash pt. 415 F; pour pt. 25 C; biodeg.

Trycol® 5964. [Henkel/Emery; Henkel/Textile] Laureth-23; CAS 9002-92-0; nonionic; emulsifier, solubilizer, lubricant used in textiles, agric., and detergent shampoo bases, personal care prods.; coemulsifier for silicone in polishes and mold release agents; solv. emulsifier for textile dye carriers; EPA-exempt; Gardner 1 solid; sol. in water, xylene; dens. 8.6 lb/gal (70 C); HLB 16.7; m.p. 40 C; cloud pt. 93 C (5% saline); flash pt. 440 F; 100% conc.

Trycol® 5966. [Henkel/Emery; Henkel/Textile] Ethoxylated lauryl alcohol; CAS 9002-92-0; nonionic; general purpose, low foaming emulsifier esp. useful for min. oils, textiles, agric. formulation; EPA-exempt; Gardner 1 liq.; sol. in min. oil, butyl stearate, glycerol trioleate, Stod.; disp. in water, xylene; dens. 7.8 lb/gal; visc. 23 cSt (100 F); HLB 8.7; cloud pt. < 25 C; flash pt. 340 F; pour pt. –7 C; 100% conc.

Trycol® 5967. [Henkel/Emery] Laureth-12; CAS 9002-92-0; nonionic; biodeg. detergent, emulsifier; intermediate for shampoo base; textile lubricant; agric. formulations; EPA-exempt; Gardner 1 solid; sol. 5% in water, xylene; dens. 8.4 lb/gal; HLB 14.8; m.p. 32 C; cloud pt. 90 C; flash pt. 465 F; 100% conc.

Trycol® 5968. [Henkel/Emery; Henkel/Textile] Trideceth-8; CAS 24938-91-8; nonionic; foam builder, solubilizer, coemulsifier for high foaming detergent, textile, and agric. formulations; Gardner 1 liq.; sol. in water, glycerol trioleate, Stod., xylene; disp. in min. oil; dens. 8.4 lb/gal; visc. 51 cSt (100 F); HLB 12.5; cloud pt. 43 C; pour pt. < –10 C; 90% act. in water.

Trycol® 5971. [Henkel/Emery; Henkel/Textile] Oleth-20; CAS 9004-98-2; nonionic; emulsifier, dispersant, solubilizer; textile lubricant; intermediate for shampoo base; Gardner 1 solid; sol. in water; disp. in butyl stearate, glycerol trioleate, Stod., xylene; dens. 8.5 lb/gal; HLB 15.3; m.p. 39 C; cloud pt. 87 C (5% saline); flash pt. 500 F.

Trycol® 5972. [Henkel/Emery; Henkel/Textile] Oleth-23; CAS 9004-98-2; nonionic; emulsifier, dispersant, solubilizer, detergent, stabilizer, anticoagulant, dyeing assistant, lubricant used in textiles, cosmetics and processing of animal fibers; Gardner 2 solid; sol. in water, xylene; HLB 15.8; m.p. 47 C; cloud pt. 89 C (5% saline); flash pt. 440 F; 100% conc.

Trycol® 5993. [Henkel/Emery; Henkel/Textile] Trideceth-3; CAS 24938-91-8; nonionic; emulsifier, antifoam for textiles, agric.; intermediate; EPA-exempt; Gardner 1 liq.; sol. in butyl stearate, glycerol trioleate, Stod., xylene; disp. in water, min. oil; dens. 7.8 lb/gal; visc. 19 cSt (100 F); HLB 7.9; cloud pt. < 25 C; flash pt. 300 F; pour pt. –15 C.

Trycol® 5999. [Henkel/Textile] Deceth-6; nonionic;

surfactant for textile use; Gardner 1 liq.; sol. @ 5% in water, butyl stearate, glycerol trioleate, Stod., xylene; disp. in min. oil; dens. 8.3 lb/gal; visc. 26 cSt (100 F); HLB 12.4; pour pt. 8 C; cloud pt. 42 C; flash pt. 280 F.

Trycol® 6720. [Henkel/Emery] POE/POP isodecyl alcohol; nonionic; low foaming detergent, wetting agent controlling foam and providing penetration for textile wet processing, metal processing rinses; Gardner 1 liq.; sol. in water, glycerol trioleate, Stod., xylene; dens. 8.2 lb/gal; visc. 39 cSt (100 F); cloud pt. 24 C; flash pt. 360 F.; 100% conc.

Trycol® 6802. [Henkel/Emery] Nonionic; surfactant, dyeing assistant, mild scouring agent; Gardner 1 liq.; sol. in Stod.; disp. in water, min. oil; dens. 8.8 lb/gal; HLB 15.3; cloud pt. 80–90 C (5% saline); flash pt. 485 F; pour pt. 28 C.

Trycol® 6940. [Henkel/Emery; Henkel/Textile] Nonoxynol-5; CAS 9016-45-9; nonionic; emulsifier, detergent, wetting agent; moderate to low foam; for agric. and textile formulations; EPA-exempt; Gardner 2 liq.; sol. in Stod., xylene; disp. in water, min. oil; dens. 8.5 lb/gal; visc. 102 cSt (100 F); HLB 9.9; cloud pt. < 25 C; flash pt. 500 F; pour pt. 11 C; 100% conc.

Trycol® 6941. [Henkel/Emery] Ethoxylated nonylphenol; CAS 9016-45-9; nonionic; moderately high foam surfactant, detergent, emulsifier for emulsion polymerization; liq.; HLB 18.6; 70% conc.

Trycol® 6942. [Henkel/Emery] Nonoxynol-100; CAS 9016-45-9; nonionic; wetting agent in high electrolyte sol'ns.; stabilizer in syn. latices; Gardner 1 solid; sol. in water; HLB 19.0; m.p. 56 C; cloud pt. 72 C (10% saline); flash pt. 500 F.

Trycol® 6943. [Henkel/Textile] Octoxynol-30; CAS 9002-93-1; nonionic; surfactant for textile use; Gardner 1 liq.; sol. @ 5% in water; dens. 9.0 lb/gal; visc. 260 cSt (100 F); HLB 17.1; pour pt. 5 C; cloud pt. 92 C (5% saline); 70% act.

Trycol® 6951. [Henkel/Emery] Nonoxynol-50; CAS 9016-45-9; nonionic; wetting agent, dispersant, coemulsifier for agric. formulations, high electrolyte sol'ns.; EPA-exempt; Gardner 1 liq.; sol. @ 5% in water; dens. 9.0 lb/gal; visc. 614 cSt (100 F); pour pt. 9 C; HLB 18.2; cloud pt. 76 C (10% saline); 50% act.

Trycol® 6952. [Henkel/Emery; Henkel/Textile] Nonoxynol-15; CAS 9016-45-9; nonionic; mod. high foaming surfactant, wetting agent, emulsifier for agric. formulations, textile processing; Gardner 1 liq.; sol. in water, xylene; dens. 8.9 lb/gal; visc. 139 cSt (100 F); HLB 15.0; cloud pt. 97 C; flash pt. 525 F; pour pt. 20 C; 100% conc. DISCONTINUED.

Trycol® 6953. [Henkel/Emery; Henkel/Textile] Nonoxynol-12; CAS 9016-45-9; nonionic; detergent, wetting agent, emulsifier for detergent formulations, emulsions, textiles, agric. formulations; EPA-exempt; Gardner 1 liq.; sol. in water, Stod., xylene; dens. 8.8 lb/gal; visc. 131 cSt (100 F); HLB 14.1; cloud pt. 80 C; flash pt. 430 F; pour pt. 10 C; 100% conc.

Trycol® 6954. [Henkel/Emery] Nonoxynol-150; CAS 9016-45-9; nonionic; wetting agent, dispersant; emulsifier for emulsion polymerization; Gardner 1 solid; sol. in water; HLB 19.3; m.p. 60 C; cloud pt. 69 C (10% saline); flash pt. 500 F.

Trycol® 6956. [Henkel/Textile] Octoxynol-40; CAS 9002-93-1; nonionic; surfactant for textile use; salt-free; Gardner 1 liq.; sol. @ 5% in water; dens. 9.0 lb/gal; visc. 220 cSt (100 F); HLB 17.9; pour pt. 13 C; cloud pt. 74 C (5% saline); 70% act.

Trycol® 6957. [Henkel/Emery; Henkel/Textile] Nonoxynol-40; CAS 9016-45-9; nonionic; detergent, wetting agent, dispersant, coemulsifier, stabilizer; for emulsion polymerization, agric. formulations, textile processing; EPA-exempt; Gardner 1 solid; sol. in water, xylene; HLB 17.8; cloud pt. 90 C (5% saline); flash pt. 560 F; 100% conc.

Trycol® 6958. [Henkel/Emery] Nonoxynol-13; CAS 9016-45-9; nonionic; detergent, wetting agent, emulsifier for agric. formulations; EPA-exempt; Gardner 1 liq.; sol. in water, xylene; dens. 8.8 lb/gal; visc. 126 cSt (100 F); HLB 14.4; cloud pt. 83 C; flash pt. 550 F; pour pt. 10 C.

Trycol® 6960. [Henkel/Emery; Henkel/Textile] Nonoxynol-1; CAS 9016-45-9; nonionic; surfactant, coemulsifier, defoamer; for textile processing; Gardner 1 liq.; sol. in min. oil, xylene; dens. 8.3 lb/gal; visc. 150 cSt (100 F); HLB 4.6; cloud pt. < 25 C; flash pt. 385 F; pour pt. –10 C; 100% conc.

Trycol® 6961. [Henkel/Emery; Henkel/Textile] Nonoxynol-4; CAS 9016-45-9; nonionic; detergent and emulsifier for agric. and textile applics.; chemical intermediate; corrosion inhibitor in two-cycle engine oils; EPA-exempt; Gardner 2 liq.; sol. in Stod., xylene; disp. in min. oil; dens. 8.5 lb/gal; visc. 472 cSt (100 F); HLB 8.9; cloud pt. < 25 C; flash pt. 430 F; pour pt. –10 C; 100% conc.

Trycol® 6962. [Henkel/Emery; Henkel/Textile] Nonoxynol-6; CAS 9016-45-9; nonionic; dispersant, wetting agent, coemulsifier in acid cleaning sol'ns., solv. emulsions, detergents, textiles, agric. formulations; moderate to low foam; EPA-exempt; Gardner 2 liq.; sol. in Stod., xylene; disp. in water, min. oil; dens. 8.5 lb/gal; visc. 100 cSt (100 F); HLB 10.9; cloud pt. < 25 C; flash pt. 515 F; pour pt. –10 C; 100% conc.

Trycol® 6963. [Henkel/Emery; Henkel/Textile] Nonoxynol-7; CAS 9016-45-9; nonionic; emulsifier, dispersant, wetting agent; moderate to low foam; base for spreader for agric. formulation, textile processing; EPA-exempt; Gardner 1 liq.; sol. in Stod., xylene; disp. in water, min. oil; dens. 8.6 lb/gal; visc. 100 cSt (100 F); HLB 11.7; cloud pt. < 25 C; flash pt. 495 F; pour pt. –10 C; 100% conc.

Trycol® 6964. [Henkel/Emery; Henkel/Textile] Nonoxynol-9; CAS 9016-45-9; nonionic; detergent, wetting agent, emulsifier for household and industrial cleaning, textiles, and agric. formulations; EPA-exempt; Gardner 1 liq.; sol. in water, Stod., xylene; dens. 8.8 lb/gal; visc. 112 cSt (100 F); HLB 13.0; cloud pt. 54 C; flash pt. 510 F; pour pt. 5 C; 100% conc.

Trycol® 6965. [Henkel/Emery] Nonoxynol-11; CAS 9016-45-9; nonionic; surfactant for detergent formulations, emulsions, agric. formulations, for high temps.; tolerant of builders; EPA-exempt; Gardner 2 liq.; sol. in water, Stod., xylene; dens. 8.7 lb/gal; visc. 116 cSt (100 F); HLB 13.5; cloud pt. 71 C; flash pt. 480 F; pour pt. 5 C; 100% conc.

Trycol® 6967. [Henkel/Emery; Henkel/Textile] Nonoxynol-20; CAS 9016-45-9; nonionic; mod. high foaming detergent, wetting agent, emulsfier for emulsion polymerization, textile processing; Gardner 1 solid; sol. in water, xylene; HLB 16.0; m.p. 34 C; cloud pt. 88 C (5% saline); flash pt. 515 F; 100% conc.

Trycol® 6968. [Henkel/Emery; Henkel/Textile] Nonoxynol-30; CAS 9016-45-9; nonionic; detergent, wetting agent; emulsifier in emulsion polymerization, for fats, oils, and waxes, agric. formulation, textile processing; EPA-exempt; Gardner 1 solid; sol. in water, xylene; HLB 17.1; m.p. 43 C; cloud pt. 93 C (5% saline); flash pt. 515 F; 100% conc.

Trycol® 6969. [Henkel/Emery; Henkel/Textile] Nonoxynol-30; CAS 9016-45-9; nonionic; detergent, wetting agent; emulsifier in emulsion polymerization, for fats, oils, and waxes, agric. formulations, textile processing; EPA-exempt; Gardner 1 liq.; sol. in water; dens. 9.0 lb/gal; visc. 260 cSt (100 F); HLB 17.1; cloud pt. 92 C (5% saline); pour pt. 5 C; 70% act. in water.

Trycol® 6970. [Henkel/Emery; Henkel/Textile] Nonoxynol-40; CAS 9016-45-9; nonionic; detergent, wetting agent, dispersant, coemulsifier, stabilizer; for emulsion polymerization, agric. formulations, textile processing; EPA-exempt; Gardner 1 liq.; sol. in water; dens. 9.2 lb/gal; visc. 385 cSt (100 F); HLB 17.8; cloud pt. 90 C (5% saline); pour pt. 7 C; 70% act. in water.

Trycol® 6971. [Henkel/Emery] Nonoxynol-50; CAS 9016-45-9; nonionic; moderately high foaming detergent, emulsifier; for emulsion polymerization, agric. formulations; EPA-exempt; Gardner 1 solid; sol. in water, xylene; HLB 18.2; m.p. 54 C; cloud pt. 76 C (10% saline); flash pt. 520 F; 100% conc.

Trycol® 6972. [Henkel/Emery] Nonoxynol-50; CAS 9016-45-9; nonionic; detergent, wetting agent, dispersant, coemulsifier, stabilizer; for emulsion polymerization; Gardner 1 liq.; sol. in water; dens. 9.0 lb/gal; visc. 440 cSt (100 F); HLB 18.2; cloud pt. 76 C (10% saline); pour pt. –4 C; 70% act.

Trycol® 6974. [Henkel/Emery; Henkel/Textile] Nonoxynol-10; CAS 9016-45-9; nonionic; dispersant, emulsifier, wetting agent for agric. formulations, textile processing; EPA-exempt; Gardner 1 liq.; sol. in water, Stod., xylene; dens. 8.7 lb/gal; visc. 111 cSt (100 F); HLB 13.2; cloud pt. 62 C; flash pt. 530 F; pour pt. 11 C.

Trycol® 6975. [Henkel/Emery] Octoxynol-30; CAS 9002-93-1; nonionic; emulsifier, wetting agent, detergent, dispersant, solubilizer, coupler for industrial/institutional/household cleaners, metal cleaning, oil drilling, agric. formulations; EPA-exempt; Gardner 1 liq.; sol. in water; dens. 9.0 lb/gal; visc. 260 cSt (100 F); HLB 17.1; cloud pt. 92 C (10% saline); pour pt. 5 C; 70% act. in water.

Trycol® 6979. [Henkel/Emery] Alkylaryl POE glycol; nonionic; emulsifier, detergent; surfactant; gel; oil-sol.; water-disp.; HLB 9.8; 100% conc.

Trycol® 6981. [Henkel/Emery] Nonoxynol-100; CAS 9016-45-9; nonionic; moderately high foaming detergent, emulsifier; for emulsion polymerization; Gardner 1 liq.; sol. in water; dens. 9.2 lb/gal; visc. 564 cSt (100 F); HLB 19.0; cloud pt. 72 C (10% saline); pour pt. 18 C; 70% act.

Trycol® 6984. [Henkel/Emery; Henkel/Textile] Octoxynol-40; CAS 9002-93-1; nonionic; dispersant, wetting agent; emulsifier for emulsion polymerization of acrylic and vinyl monomers; textile processing; agric. formulations; EPA-exempt; Gardner 1 liq.; sol. in water; dens. 9.0 lb/gal; visc. 220 cSt (100 F); HLB 17.9; cloud pt. 74 C (10% saline); pour pt. 13 C; 70% act. in water.

Trycol® 6985. [Henkel/Emery; Henkel/Textile] Nonyl nonoxynol-8; CAS 9014-93-1; nonionic; emulsifier in textile dye carrier applics., insecticides, wax emulsions; foam control agent; spreading agent in pigment printing; post-stabilizer in emulsion polymerization; intermediate; EPA-exempt; Gardner 2 liq.; sol. in xylene, glycerol trioleate; disp. in water, min. oil; dens. 8.4 lb/gal; visc. 173 cSt (100 F); HLB 10.4; cloud pt. < 25 C; flash pt. 525 F; pour pt. 9 C; 100% conc.

Trycol® 6989. [Henkel/Emery; Henkel/Textile] Nonyl nonoxynol-150; CAS 9014-93-1; nonionic; emulsifier, dispersant, wetting agent for built detergents, hard surf. cleaners, dairy detergents, pesticides, textiles; leveling agent; EPA-exempt; Gardner 1 liq.; sol. in water; dens. 9.0 lb/gal; visc. 367 cSt (100 F); HLB 19.0; cloud pt. 85 C (5% saline); pour pt. 23 C; 50% act. in water.

Trycol® CE. [Henkel/Emery] Ethoxylated alcohol; emulsifier for min. oils and dye carriers; Gardner 1 liq.; disp. in water; dens. 7.8 lb/gal; visc. 23 cSt (100 F); HLB 8.7; pour pt. –7 C; flash pt. 340 F; cloud pt. < 25 C.

Trycol® DA-4. [Henkel/Emery] Deceth-4; CAS 26183-52-8; nonionic; wetting agent, penetrant for yarn and fabrics in dyeing systems and resin pad-bath applics.; intermediate in the prod. of anionics; Gardner < 1 liq.; sol. in glycerol trioleate; disp. in water; dens. 7.9 lb/gal; visc. 30 cs; HLB 10.5; cloud pt. < 25 C; 100% act.

Trycol® DA-6. [Henkel/Emery] Deceth-6; CAS 26183-52-8; nonionic; wetting agent in clay soils, fire fighting; penetrant and emulsifier in repellent applics.; Gardner 1 liq.; dens. 8.3 lb/gal; visc. 15 cs; HLB 12.5; cloud pt. 40 C; 100% act.

Trycol® DA-69. [Henkel/Emery] Deceth-6; CAS 26183-52-8; nonionic; wetting agent in clay soils, fire fighting; penetrant and emulsifier in repellent applics.; Gardner 1 liq.; dens. 8.3 lb/gal; visc. 28 cSt (100 F); HLB 12.4; pour pt. –15 C; flash pt. 315 F; cloud pt. 42 C; 90% act.

Trycol® DNP-8. [Henkel/Emery] Nonyl nonoxynol-8; CAS 9014-93-1; nonionic; foam control agent; emulsifier for polar and nonpolar solvs. and textile jet dye carrier applics.; spreading agent in pigment printing; post-stabilizer in emulsion polymerization; intermediate in mfg. of anionics; surfactant component in acidic cleaners, aerosols, insecticides and wax emulsions; Gardner 2 liq.; disp. in water; dens. 8.4 lb/gal; visc. 410 cs; HLB 10.4; cloud pt. < 25 C; 100% act.

Trycol® DNP-150. [Henkel/Emery] Nonyl nonoxynol-150; CAS 9014-93-1; nonionic; dispersant and wetting agent for pesticides; emulsifier, detergent and leveling agent in textile industry; controlled-subs component in detergents and cleaners; Gardner 1 solid, flakes; m.p. 55 C; HLB 19.0; cloud pt. > 100 C; 100% act.

Trycol® DNP-150/50. [Henkel/Emery] Nonyl nonoxynol-150; CAS 9014-93-1; nonionic; dispersant and wetting agent for pesticides; emulsifier, detergent and leveling agent in textile industry; controlled-subs component in detergents and cleaners; Gardner < 1 liq.; dens. 9.0 lb/gal; HLB 19.0; cloud pt. > 100 C; 50% act.

Trycol® LAL-4. [Henkel/Emery] Laureth-4; CAS 9002-92-0; nonionic; intermediate; emulsifier for oils and fats in industrial lubricants, textile coning oils; coemulsifier for silicone in cleaner polishes and mold release agents; Gardner 1 liq.; disp. in water; dens. 7.7 lb/gal; visc. 35 cs; HLB 9.2; cloud pt. < 25 C; biodeg.; 100% act.

Trycol® LAL-8. [Henkel/Emery] Laureth-8; CAS 9002-92-0; nonionic; detergent, emulsifier, wetting agent; Gardner 1 liq.; dens. 8.2 lb/gal; HLB 12.6; cloud pt. 62 C; biodeg.; 100% act.

Trycol® LAL-12. [Henkel/Emery] Laureth-12; CAS 9002-92-0; nonionic; detergent, emulsifier, wetting agent; Gardner 1 solid; m.p. 32 C; HLB 14.0; cloud pt. 90 C; biodeg.; 100% act.

Trycol® LAL-23. [Henkel/Emery] Laureth-23; CAS 9002-92-0; nonionic; solubilizer; coemulsifier for silicone in cleaner polishes and mold release agents; solv. emulsifier for textile dye carriers; Gardner < 1 solid; m.p. 46 C; HLB 16.9; cloud pt. > 100 C; 100% act.

Trycol® NP-1. [Henkel/Emery] Nonoxynol-1; nonionic; defoamer; surfactant and coemulsifier; Gardner 1 liq.; sol. in min. oil, xylene; dens. 8.3 lb/gal; visc. 460 cs; HLB 4.6; cloud pt. < 25 C; 100% act.

Trycol® NP-4. [Henkel/Emery] Nonoxynol-4; CAS 9016-45-9; nonionic; coemulsifier often used as a corrosion inhibitor in two cycle engine oils; Gardner 1 liq.; dens. 8.5 lb/gal; visc. 300 cs; HLB 8.9; cloud pt. < 25 C; 100% act.

Trycol® NP-6. [Henkel/Emery] Nonoxynol-6; CAS 9016-45-9; nonionic; dispersant and wetting agent; coemulsifier in acid cleaning sol'ns, solv. emulsions and detergents; Gardner 2 liq.; disp. in water; dens. 8.5 lb/gal; visc. 100 cSt (100 F); HLB 10.9; pour pt. −10 C; flash pt. 515 F; cloud pt. < 25 C.

Trycol® NP-7. [Henkel/Emery] Nonoxynol-7; CAS 9016-45-9; nonionic; wetting agent in solv. emulsions, cleaners and detergents; Gardner 1 liq.; disp. in water; dens. 8.6 lb/gal; visc. 325 cs; HLB 12.3; cloud pt. 20 C; 100% act.

Trycol® NP-9. [Henkel/Emery] Nonoxynol-9; CAS 9016-45-9; nonionic; detergent, wetting agent, emulsifier; Gardner 1 liq.; dens. 8.8 lb/gal; visc. 325 cs; HLB 12.9; cloud pt. 55 C; 100% act.

Trycol® NP-11. [Henkel/Emery] Nonoxynol-11; CAS 9016-45-9; nonionic; surfactant used in formulation for detergents and emulsification; Gardner 2 liq.; dens. 8.7 lb/gal; visc. 116 cSt (100 F); HLB 13.5; pour pt. 5 C; flash pt. 480 F; cloud pt. 71 C.

Trycol® NP-20. [Henkel/Emery] Nonoxynol-20; CAS 9016-45-9; nonionic; surfactant used as detergent, wetting agent and emulsifier for emulsion polymerization; Gardner 1 solid; HLB 16.0; m.p. 34 C; flash pt. 515 F; cloud pt. 88 C.

Trycol® NP-30. [Henkel/Emery] Nonoxynol-30; CAS 9016-45-9; nonionic; detergent, wetting agent; emulsifier for fats, oils, and waxes and emulsion polymerization; Gardner 1 solid; m.p. 40 C; HLB 17.1; cloud pt. > 100 C; 100% act.

Trycol® NP-40. [Henkel/Emery] Nonoxynol-40; CAS 9016-45-9; nonionic; detergent; wetting agent in sol'ns. of electrolytes; stabilizer for syn. fabrics; coemulsifier for fats, waxes and oils; Gardner 1 solid; HLB 17.8; m.p. 40 C; flash pt. 560 F; cloud pt. 90 C.

Trycol® NP-50. [Henkel/Emery] Nonoxynol-50; CAS 9016-45-9; nonionic; detergent; wetting agent in sol'ns. of electrolytes; stabilizer for syn. fabrics; coemulsifier for fats, waxes and oils; Gardner 1 solid; HLB 18.2; m.p. 54 C; flash pt. 520 F; cloud pt. 76 C.

Trycol® NP-307. [Henkel/Emery] Nonoxynol-30; CAS 9016-45-9; nonionic; detergent and wetting agent; emulsifier in emulsion polymerization and for fats, oils and waxes; Gardner 1 liq.; dens. 9.0 lb/gal; visc. 260 cSt (100 F); HLB 17.1; pour pt. 5 C; cloud pt. 92 C.

Trycol® NP-407. [Henkel/Emery] Nonoxynol-40; CAS 9016-45-9; nonionic; detergent; wetting agent in sol'ns. of electrolytes; stabilizer for syn. fabrics; coemulsifier for fats, waxes and oils; Gardner 1 liq.; dens. 9.2 lb/gal; visc. 385 cSt (100 F); HLB 17.8; pour pt. 7 C; cloud pt. 90 C.

Trycol® NP-507. [Henkel/Emery] Nonoxynol-50; CAS 9016-45-9; nonionic; detergent; wetting agent in sol'ns. of electrolytes; stabilizer for syn. fabrics; coemulsifier for fats, waxes and oils; Gardner 1 liq.; dens. 9.0 lb/gal; visc. 440 cSt (100 F); HLB 18.2; pour pt. –4 C; cloud pt. 76 C.

Trycol® NP-1007. [Henkel/Emery] Nonoxynol-100; CAS 9016-45-9; nonionic; surfactant used as wetting agent in high electrolyte sol'ns.; stabilizer in syn. latices; Gardner 1 liq.; dens. 9.2 lb/gal; HLB 19.0; pour pt. 18 C; cloud pt. 72 C; 70% act.

Trycol® OAL-23. [Henkel/Emery] Oleth-23; CAS 9004-98-2; nonionic; dispersant, solubilizer, detergent, stabilizer and anticoagulant for latices and dye pastes; dyeing assistant for wool/acrylic blends; detergent and lubricant for fiber/fabric scouring; emulsifier for waxes used in coating citrus fruits; Gardner 2 solid; m.p. 40 C; HLB 15.9; cloud pt. > 100 C; 100% act.

Trycol® OP-407. [Henkel/Emery] Octoxynol-40; CAS 9002-93-1; nonionic; hydrophilic dispersant and wetting agent; primary emulsifier in emulsion of polymerization of acrylic and vinyl monomers; Gardner 1 liq.; dens. 9.0 lb/gal; visc. 220 cSt (100 F); HLB 17.9; pour pt. 13 C; cloud pt. 74 C.

Trycol® SAL-20. [Henkel/Emery] Steareth-20; CAS 9005-00-9; nonionic; emulsifier and solubilizer used in dyeing assistants and textile dye carriers; stabilizer in latices; wax emulsifier in coatings for citrus fruits; Gardner 1 solid; HLB 15.3; m.p. 40 C; flash pt. 560 F; cloud pt. 91 C.

Trycol® TDA-3. [Henkel/Emery] Trideceth-3; CAS 24938-91-8; nonionic; emulsifier and antifoam in textile formulations; intermediate for prod. of anionics by sulfation and phosphation; Gardner < 1 liq.; disp. in water; dens. 7.8 lb/gal; visc. 30 cs; HLB 8.0; cloud pt. < 25 C; 100% act.

Trycol® TDA-6. [Henkel/Emery] Trideceth-6; CAS 24938-91-8; nonionic; dispersant and wetting agent, emulsifier and detergent in degreasers and cutting oils; wool scouring; intermediate for prod. of surfactants by sulfation or phosphation; Gardner < 1 liq.; disp. in water; dens. 8.2 lb/gal; visc. 50 cs; HLB 11.4; cloud pt. < 25 C; 100% act.

Trycol® TDA-8. [Henkel/Emery] Trideceth-8; CAS 24938-91-8; nonionic; coemulsifier, foam builder and solubilizer for alkylaryl sulfonates, essential oils, aromatic solvs., waxes and fats; Gardner < 1 liq.; dens. 8.3 lb/gal; visc. 65 cs; HLB 12.5; cloud pt. 44 C; 100% act.

Trycol® TDA-9. [Henkel/Emery] Trideceth-9; CAS 24938-91-8; nonionic; wetting agent, surfactant used with alkylaryl sulfonates in cleaners; rewetter for paper towels; raw wool detergent; coemulsifier for aromatic and aliphatic solvs.; Gardner < 1 liq.; dens. 8.3 lb/gal; visc. 75 cs; HLB 13.2; cloud pt. 55 C; 100% act.

Trycol® TDA-11. [Henkel/Emery] Trideceth-11; CAS 24938-91-8; nonionic; detergent; lt. and heavy duty cleaning formulations; Gardner < 1 liq.; dens. 8.4 lb/gal; HLB 13.7; cloud pt. 74 C; 100% act.

Trycol® TDA-12. [Henkel/Emery] Trideceth-12; CAS 24938-91-8; nonionic; hydrophilic surfactant for lt. and heavy duty cleaning formulations; Gardner 1 semisolid; dens. 8.4 lb/gal; HLB 14.5; m.p. 16 C; flash pt. 440 F; cloud pt. 70 C.

Trycol® TDA-18. [Henkel/Emery] Trideceth-18; CAS 24938-91-8; nonionic; dispersant, solubilizer, emulsifier in cleaners; intermediate in mfg. of anionic surfactants; textile acid dye leveling agent and wet processing detergent; foam stabilizer; Gardner < 1 solid; m.p. 39 C; HLB 16.0; cloud pt. > 100 C; 100% act.

Trycol® TP-2. [Henkel/Emery] PEG-2 tridecyl phenol; nonionic; coemulsifier, foam reducing agent; Gardner 6 liq.; dens. 8.1 lb/gal; visc. 600 cs; HLB 4.6; cloud pt. < 25 C; 100% act.

Trycol® TP-6. [Henkel/Emery] PEG-6 tridecyl phenol; nonionic; dispersant, detergent, emulsifier in metal cleaning, cutting oil and drycleaning formulations, emulsifiable solv. cleaners; Gardner 6 liq.; disp. in water; dens. 8.5 lb/gal; visc. 300 cs; HLB 9.9; cloud pt. < 25 C; 100% act.

Trydet 19. [Henkel/Emery] Ethoxylated mixed rosin and fatty acids, anhyd.; nonionic; detergent and emulsifier; textile leveling agent in dyeing applics.; coemulsifier for xylene, kerosene, trichlorobenzene, o-dichlorobenzene; Gardner 12 liq.; dens. 8.9 lb/gal; visc. 200 cSt (100 F); HLB 13.4; pour pt. 15 C; flash pt. 570 F; cloud pt. 50 C.

Trydet 20. [Henkel/Emery] Ethoxylated mixed esters of rosin and fatty acids; nonionic; detergent and emulsifier; textile leveling agent in dyeing applics.; coemulsifier for xylene, kerosene, trichlorobenzene, o-dichlorobenzene; dens. 9.0 lb/gal; visc. 480 cs; HLB 13.4; cloud pt. > 100 C; 96% act.

Trydet 22. [Henkel/Emery] Ethoxylated tall oil fatty acids; nonionic; surfactant used in formulating controlled foaming detergents for commercial laundries, textile scouring, metal and hard surface cleaners and degreasers; Gardner 12 liq.; dens. 9.0 lb/gal; visc. 196 cSt (100 F); HLB 12.1; pour pt. 20 C; flash pt. 530 F; cloud pt. 35 C.

Trydet 2610. [Henkel/Emery] PEG-23 stearate; CAS 9004-99-3; nonionic; emulsifier, thickener for agric. formulations; EPA-exempt; Gardner 1 solid; sol. @ 5% in water, min. oil, xylene; m.p. 36 C; HLB 15.7; cloud pt. 86-90 C; flash pt. 430 F.

Trydet 2636. [Henkel/Emery] PEG-7 stearate; CAS 9004-99-3; nonionic; emulsifier, softener, lubricant for textiles, leathers, industrial lubricants; Gardner 1 solid; sol. in xylene; disp. in water; visc. 57 cSt (100 F); HLB 10.1; m.p. 28 C; cloud pt. < 25 C; flash pt. 500 F.

Trydet 2640. [Henkel/Emery] PEG-8 stearate; CAS 9004-99-3; nonionic; surfactant, emollient, emulsifier, wetting agent, textile softener, lubricant, defoamer, stabilizer, visc. control agent, pigment wetting agent, mold release agent; solid; HLB 12.0; 100% conc.

Trydet 2644. [Henkel/Emery; Henkel/Textile] PEG 400 isostearate; CAS 56002-14-3; nonionic; surfactant, lubricant for fiber lubricants, textile processing aids, fabric softeners; Gardner 1 liq.; sol. in xylene; disp. in water; dens. 8.5 lb/gal; visc. 70 cSt (100 F); HLB 11.3; cloud pt. < 25 C; flash pt. 450 F; pour pt. 10 C.

Trydet 2670. [Henkel/Emery; Henkel/Textile] PEG-5 stearate; CAS 9004-99-3; nonionic; emulsifier, lubricant, softener for textiles, leather, polishes, agric.; EPA-exempt; Gardner 1 liq.; sol. in xylene; disp. in water; dens. 8.6 lb/gal; visc. 44 cSt (100 F); HLB 9.2; pour pt. 25 C; cloud pt. < 25 C; flash pt. 500 F; 100% conc.

Trydet 2671. [Henkel/Emery; Henkel/Textile] PEG-8 stearate; CAS 9004-99-3; nonionic; o/w emulsifier, SE lubricant, softener for textile, agric. and other industrial applics.; stabilizer in starch sol'ns.; EPA-exempt; Gardner 1 solid; sol. in xylene; disp. in water; dens. 8.5 lb/gal; HLB 11.4; m.p. 30 C; cloud pt. < 25 C; flash pt. 500 F; 100% conc.

Trydet 2672. [Henkel/Emery] PEG-40 stearate; CAS 9004-99-3; nonionic; emulsifier, lubricant used in prod. of textile lubricants and softeners, agric. formulations; stabilizer, antigellant for starch sol'ns.; Gardner 1 solid; sol. in water, xylene; HLB 17.3; m.p. 50 C; cloud pt. 75–81 C (5% saline); flash pt. 540 F. DISCONTINUED.

Trydet 2675. [Henkel/Emery; Henkel/Textile] PEG-50 stearate; CAS 9004-99-3; nonionic; surfactant, emollient, emulsifier, wetting agent, textile softener, lubricant, defoamer, stabilizer, visc. control agent, pigment wetting agent, mold release agent; for agric. formulations; EPA-exempt; Gardner 1 liq.; sol. @ 5% in water; dens. 8.5 lb/gal; visc. 671 cST (100 F); HLB 17.8; pour pt. 0 C; cloud pt. 81 C (5% saline); flash pt. 540 F; 30% act. in water.

Trydet 2676. [Henkel/Emery; Henkel/Textile] PEG-10 oleate; CAS 9004-96-0; nonionic; emulsifier, lubricant for pesticides, metal cleaners, textile detergents and dyeing assistants, leather; rewetting agent for paper; EPA-exempt; Gardner 2 liq.; sol. in xylene; disp. in water; dens. 8.5 lb/gal; visc. 54 cSt (100 F); HLB 12.2; cloud pt. < 25 C; flash pt. 560 F; pour pt. 14 C.

Trydet 2681. [Henkel/Emery] Ethoxylated tall oil; nonionic; used in formulating controlled-foam detergents for commercial laundries, textile scouring, metal and hard surf. cleaners, degreasers; Gardner 12 liq.; sol. in water, glycerol trioleate; dens. 9.0 lb/gal; visc. 196 cSt (100 F); HLB 12.1; cloud pt. 35 C; flash pt. 530 F; pour pt. 20 C.

Trydet 2682. [Henkel/Emery; Henkel/Textile] PEG-16 tallate; CAS 61791-00-2; nonionic; detergent, emulsifier, textile leveling agent; coemulsifier; Gardner 10 liq.; sol. in water; disp. in glycerol trioleate, xylene; dens. 9.0 lb/gal; visc. 209 cSt (100 F); HLB 13.4; cloud pt. 42–53 C (5% saline); flash pt. 590 F; pour pt. 17 C.

Trydet 2685. [Henkel/Emery] Ethoxylated fatty acid, nonionic; emulsifier for min. oils and fats; lubricant; Gardner 1 solid; sol. in min. oil, butyl stearate, glycerol trioleate, Stod., xylene; disp. in water; HLB 7.5; m.p. 37 C; cloud pt. < 25 C; flash pt. 350 F.

Trydet 2691. [Henkel/Textile] PEG-4 monomer acid; nonionic; surfactant for textile use; Gardner 5 liq.; sol. @ 5% in butyl stearate, glycerol trioleate; disp. in water, min. oil; dens. 8.1 lb/gal; visc. 44 cSt (100 F); HLB 6.6; pour pt. -3 C; cloud pt. < 25 C; flash pt. 425 F.

Trydet 2692. [Henkel/Textile] PEG-8 monomer acid; nonionic; surfactant for textile use; Gardner 8 liq.; sol. @ 5% in min. oil, glycerol trioleate, xylene; disp. in water; dens. 8.6 lb/gal; visc. 59 cSt (100 F); HLB 10.1; pour pt. 10 C; cloud pt. < 25 C; flash pt. 445 F.

Trydet DO-9. [Henkel/Emery] PEG-9 dioleate; CAS 9005-07-6; nonionic; lipophilic emulsifier and solubilizer for min. oils, fats and solvs.; emulsifier for kerosene in agric. and pesticides sprays; emulsification of latex paints, metalworking fluids, solvs., specialty and industrial lubricants; Gardner 3 liq.; disp. in water; dens. 8.1 lb/gal; visc. 45 cSt (100 F); HLB 8.8; pour pt. -6 C; flash pt. 515 F; cloud pt. < 25 C.

Trydet ISA-4. [Henkel/Emery] PEG-4 isostearate;

CAS 56002-14-3; nonionic; lubricant used in textile fiber processing, conc. fabric softeners; Gardner 2 liq.; dens. 8.2 lb/gal; visc. 100 cs; HLB 7.7; cloud pt. < 25 C; 100% act.

Trydet ISA-9. [Henkel/Emery] PEG-9 isostearate; CAS 56002-14-3; nonionic; lubricant used in textile fiber processing, conc. fabric softeners; Gardner 1 liq.; dens. 8.5 lb/gal; visc. 125 cs; HLB 11.7; cloud pt. < 25 C; 100% act.

Trydet ISA-14. [Henkel/Emery] PEG-14 isostearate; CAS 56002-14-3; nonionic; surfactant used in fiber lubricants and processing aids; emulsifier in finish formulations; Gardner 5 liq.; disp. in water; dens. 8.6 lb/gal; visc. 92 cSt (100 F); HLB 13.0; pour pt. 10 C; flash pt. 490 F; cloud pt. < 25 C.

Trydet LA-5. [Henkel/Emery] PEG-5 laurate; CAS 9004-81-3; nonionic; emulsifier and coupling agent; defoamer in water base coatings; visc. depressant in vinyl plastic sols.; control additive in hair rinse formulations; paper softener; Gardner liq.; dens. 8.2 lb/gal; visc. 123 cSt (100 F); HLB 9.3; pour pt. 9 C; flash pt. 445 F; cloud pt. < 25 C.

Trydet LA-7. [Henkel/Emery] PEG-7 laurate; CAS 9004-81-3; nonionic; emulsifier; lubricant component and scrooping agent for textile fibers and yarns; visc. control agent for plastisols; Gardner 2 liq.; dens. 8.4 lb/gal; visc. 55 cs; HLB 11.8; cloud pt. < 25 C; 100% act.

Trydet MP-9. [Henkel/Emery] PEG-9 pelargonic acid; nonionic; surfactant as a base lubricant for syn. fiber finishes; coemulsifier and coupling agent; Gardner 1 liq.; dens. 8.7 lb/gal; visc. 34 cSt (100 F); HLB 14.3; pour pt. 5 C; flash pt. 440 F; cloud pt. 37 C.

Trydet OA-5. [Henkel/Emery] PEG-5 oleate; CAS 9004-96-0; nonionic; emulsifier for min., cutting, fatty oils and solvs.; solv. emulsifier in metal cleaners and degreasers; w/o emulsifier for pesticide control; textile processing; softener and lubricant leather during tanning; Gardner 3 liq.; disp. in water; dens. 8.1 lb/gal; visc. 55 cs; HLB 8.2; cloud pt. < 25 C; 100% act.

Trydet OA-7. [Henkel/Emery] PEG-7 oleate; CAS 9004-96-0; nonionic; emulsifier for min., cutting, fatty oils and solvs.; solv. emulsifier in metal cleaners and degreasers; w/o emulsifier for pesticide control; textile processing; softener and lubricant leather during tanning; Gardner 3 liq.; dens. 8.3 lb/gal; visc. 160 cs; HLB 10.2; cloud pt. < 25 C; 100% act.

Trydet OA-10. [Henkel/Emery] PEG-10 oleate; CAS 9004-96-0; nonionic; lubricant, emulsifier for solvs. in pesticide carriers, metal cleaners and neatsfoot oil in leather fat liquoring; textile specialty detergent and dyeing assistant; rewetting agent for paper; Gardner 2 liq.; dens. 8.5 lb/gal; visc. 110 cs; HLB 12.1; cloud pt < 25 C; 100% act.

Trydet SA-5. [Henkel/Emery] PEG-5 stearate; CAS 9004-99-3; nonionic; emulsifier for min. oils and fats; lubricant in leather industry; napping softener in textile industry; min. oil and lard oil emulsions for polish and metal buffing compds.; Gardner 1 solid; m.p. 35 C; HLB 8.7; cloud pt. < 25 C; 100% act.

Trydet SA-7. [Henkel/Emery] PEG-7 stearate; CAS 9004-99-3; nonionic; waxy emulsifier for oils and fats in industrial lubricants; softener and lubricant to textiles and leather; Gardner < 1 solid; m.p. 30 C; HLB 11.1; cloud pt. < 25 C; 100% act.

Trydet SA-8. [Henkel/Emery] PEG-8 stearate; CAS 9004-99-3; nonionic; lubricant, softener and scrooping agent for textile and industrial applics.; stabilizer in starch sol'ns.; Gardner < 1 solid; m.p. 30 C; HLB 11.1; cloud pt. < 25 C; 100% act.

Trydet SA-23. [Henkel/Emery] PEG-23 stearate; CAS 9004-99-3; nonionic; emulsifier for textile lubricants and softeners; emulsifier and thickener in personal care prods. and pharmaceuticals; antigellant in starch sol'ns.; Gardner 1 solid; m.p. 45 C; HLB 15.6; cloud pt. > 100 C; 100% act.

Trydet SA-40. [Henkel/Emery] PEG-40 stearate; CAS 9004-99-3; nonionic; emulsifier for glyceryl stearate and waxy esters in prod. of conc. textile lubricants and softeners; stabilizer and antigellant for starch sol'ns.; emulsifier for cosmetics and pharmaceuticals; Gardner < 1 solid, flakes; m.p. 50 C; HLB 17.3; cloud pt. < 25 C; 100% act.

Trydet SA-50/30. [Henkel/Emery] PEG-50 stearate; CAS 9004-99-3; nonionic; hydrophilic emulsifier for preparing solubilized oils; visc. modifier, softener or plasticizer in acrylic or vinyl resin emulsions; Gardner 1 liq.; dens. 8.5 lb/gal; visc. 671 cSt (100 F); HLB 17.8; pour pt. 0 C; flash pt. 540 F; cloud pt. 81 C.

Trydet SO-9. [Henkel/Emery] PEG-9 sesquioleate; nonionic; oil and fat emulsifier in industrial and textile lubricants; Gardner 3 liq.; dens. 8.2 lb/gal; visc. 50 cSt (100 F); HLB 9.4; pour pt. –6 C; flash pt. 550 F; cloud pt. < 25 C.

Tryfac® 525-K. [Henkel/Emery] Phosphate ester; detergent and wetting agent for built liq. detergents; antistat in fiber finishes; Gardner 2 liq.; dens. 8.8 lb/gal; visc. 560 cSt (100 F); pour pt. 18 C; flash pt. 435 F; cloud pt. 76 C.

Tryfac® 610-K. [Henkel/Emery] Phosphate ester potassium salt; anionic; wetting agent, dispersant, emulsifier for chlorinated hydrocarbons in cleaners, detergents and dye carriers; component in metalworking compds.; textile emulsifier and detergent in solv. scouring; scouring agent for cotton goods; antistat in processing oils for fibers; corrosion inhibitor; Gardner < 1 liq.; dens. 9.2 lb/gal; visc. 700 cs; 88% act.

Tryfac® 910-K. [Henkel/Emery] Phosphate ester potassium salt; anionic; detergent and wetting agent in textile desizing; polymer emulsification stabilizer; Gardner 1 liq.; dens. 9.4 lb/gal; visc. 480 cs; 90% act.

Tryfac® 5365. [Henkel/Emery] Phosphate ester; anionic; detergent, wetting and antistat, emulsifier; liq.

Tryfac® 5552. [Henkel/Emery; Henkel/Textile] Phosphate ester, free acid form; anionic; surfactant intermediate; salts used as emulsifiers for aliphatic and aromatic solvs., agric. formulations, textile processing; EPA-exempt; Gardner 1 liq.; sol. in water, butyl stearate, Stod., xylene; dens. 8.8 lb/gal; visc. 170 cSt (100 F); cloud pt. 76 C; flash pt. 360 F; pour pt. < –15 C.

Tryfac® 5553. [Henkel/Emery; Henkel/Textile] Phosphate ester (potassium salt of Tryfac 5552); anionic; emulsifier, wetting agent, detergent, antistat; used in heavy duty cleaners, metalworking compds., agric., textile applics.; corrosion inhibitor, dispersant; EPA-exempt; Gardner 1 liq.; sol. in water, butyl stearate, glycerol trioleate, Stod., xylene; dens. 9.2 lb/gal; visc. 345 cSt (100 F); cloud pt. > 100 C (10% saline); pour pt. < –15 C.

Tryfac® 5554. [Henkel/Emery; Henkel/Textile] Phosphate ester, potassium salt; anionic; emulsifier, wetting agent, detergent, antistat; used in heavy duty cleaners, metalworking compds., agric., textile applics.; corrosion inhibitor, dispersant; stabilizer for polymer emulsification, starch conversion in textile desizing; EPA-exempt; Gardner 1 liq.; sol. in water, Stod., xylene; dens. 9.4 lb/gal; visc. 340 cSt (100 F); cloud pt. > 100 C (10% saline); pour pt. -9 C.

Tryfac® 5555. [Henkel/Emery; Henkel/Textile] Phosphate ester, free acid form; anionic; emulsifier for flame retardants, in textile industry, agric. formulations, for high-temp. applics.; EPA-exempt; Gardner 5 liq.; sol. in min. oil, butyl stearate, glycerol trioleate, Stod., xylene; disp. in water; dens. 8.8 lb/gal; visc. 2300 cSt (100 F); cloud pt. < 25 C; flash pt. 440 F; pour pt. -3 C.

Tryfac® 5556. [Henkel/Emery; Henkel/Textile] Complex phosphate ester, free acid form; CAS 51811-79-1; anionic; wetting agent, dispersant, antistat in textile processing; solv. emulsifier for textile scours, detergents, pesticides; drycleaning detergent; also used in emulsion polymerization; EPA-exempt; Gardner 2 liq.; sol. in water, xylene; dens. 9.3 lb/gal; visc. 1700 cSt (100 F); cloud pt. 68 C (5% saline); flash pt. 450 F; pour pt. 5 C.

Tryfac® 5557. [Henkel/Emery] Phosphate ester, potassium salt; anionic; surfactant, wetting agent, dispersant; esp. for textile scours; Gardner 2 liq.; sol. in water; dens. 9.3 lb/gal; pour pt. 5 C; 55% act.

Tryfac® 5559. [Henkel/Emery; Henkel/Textile] Phosphate ester, potassium salt; anionic; detergent, wetting agent, oily soil emulsifier in built detergents, agric. formulations; antistat in fiber finishes for polyester and PP; EPA-exempt; Gardner 2 liq.; sol. @ 5% in water, butyl stearate, glycerol trioleate, Stod., xylene; dens. 8.8 lb/gal; visc. 560 cSt (100 F); cloud pt. 76 C (5% saline); flash pt. 435 F; pour pt. 18 C.

Tryfac® 5560. [Henkel/Emery; Henkel/Textile] Phosphate ester, free acid form; anionic; emulsifier, dispersant, antistat; for textile applics.; FDA compliance; Gardner 2 liq.; sol. in water, butyl stearate, glycerol trioleate, Stod., xylene; dens. 8.8 lb/gal; flash pt. 405 F; pour pt. 5 C.

Tryfac® 5561. [Henkel/Emery] Phosphate ester; emulsifier for pesticides; EPA-exempt; Gardner 2 liq.; sol. @ 5% in xylene, methyl oleate; disp. in min. oil, Stod.; dens. 9.2 lb/gal; visc. 2196 cSt (100 F); pour pt. 5 C; flash pt. 440 F.

Tryfac® 5569. [Henkel/Emery] Phosphate ester, free acid form; hydrotrope for nonionic surfactants in alkaline sol'n.; nondiscoloring in contact with solid caustic; Gardner 6 liq.; sol. in water; dens. 10.4 lb/gal; visc. 1180 cSt (100 F); cloud pt. 51 C; flash pt. 425 F; pour pt. 5 C; 100% act.

Tryfac® 5571. [Henkel/Emery] Ethoxylated alcohol phosphate ester, potassium salt; anionic; detergent, dispersant and wetting agent used in fabric preparation, hard surface cleaning, agric. formulation, and antistat protection of fibers; EPA-exempt; Gardner 1 liq.; sol. in water, butyl stearate, glycerol trioleate, Stod., xylene; dens. 9.2 lb/gal; visc. 650 cSt (100 F); pour pt. < -15 C; cloud pt. 45 C.

Tryfac® 5573. [Henkel/Emery] Phosphate ester, free acid form; CAS 12751-23-4; anionic; mold release agent, antistat, dispersant, emulsifier; Gardner 1 solid; sol. in min. oil, butyl stearate, glycerol trioleate, Sod., xylene; disp. in water; dens. 7.8 lb/gal (40 C); m.p. 35 C; cloud pt. < 25 C; flash pt. 365 F.

Tryfac® 5576. [Henkel/Emery; Henkel/Textile] Ethoxylated alcohol phosphate ester, potassium salt; anionic; hydrophilic surfactant used in electrolytes and caustic sol'ns.; antistat, anticorrosive and dispersant; for textile applics.; Gardner 1 liq.; sol. in water, butyl stearate, xylene; dens. 9.5 lb/gal; visc. 190 cSt (100 F); pour pt. < -10 C; cloud pt. 49–54 C.

Tryfac® HWD. [Henkel/Emery] Phosphate ester in free acid form; hydrotrope, solubilizer for nonionic surfactants used in built liq. detergents for industrial and institutional cleaners, degreasers, metal, and other hard surf. cleaners; Gardner 3 liq.; sol. in highly alkaline sol'ns., (5%) in water, xylene; dens. 9.2 lb/gal; visc. 306 cSt (100 F); pour pt. 12 C; flash pt. 380 F; cloud pt. 78 C; 100% act.

Trylon® 5900. [Henkel/Emery] Ethoxylated castor oil; nonionic; emulsifier, lubricant for textiles, pigment dispersants in latex paints, solubilizers for essential oils; liq./solid; HLB 4.9; 100% conc.

Trylon® 5902. [Henkel/Emery] Ethoxylated castor oil; nonionic; emulsifier, lubricant for textiles, pigment dispersants in latex paints, solubilizers for essential oils; liq./solid; HLB 9.7; 100% conc.

Trylon® 5906. [Henkel/Emery] Ethoxylated hydrog. castor oil; nonionic; emulsifier, lubricant for textiles, pigment dispersants in latex paints, solubilizers for essential oils; liq./solid; HLB 11.7; 100% conc.

Trylon® 5909. [Henkel/Emery] Ethoxylated hydrog. castor oil; nonionic; emulsifier, lubricant for textiles, pigment dispersants in latex paints, solubilizers for essential oils; liq./solid; HLB 13.1; 100% conc.

Trylon® 5921. [Henkel/Emery] Ethoxylated hydrog. castor oil; nonionic; emulsifier, lubricant for textiles, pigment dispersants in latex paints, solubilizers for essential oils; liq./solid; HLB 8.4; 100% conc.

Trylon® 5922. [Henkel/Emery] Ethoxylated hydrog. castor oil; nonionic; emulsifier, lubricant for textiles, pigment dispersants in latex paints, solubilizers for essential oils; liq./solid; HLB 10.8; 100% conc.

Trylon® 6702. [Henkel/Emery; Henkel/Textile] Anionic; solv. emulsifier, detergent for metal degreasers, engine block cleaners, solv. scouring, textile dye applics.; Gardner 7 liq.; sol. in butyl stearate, glycerol trioleate, Stod., xylene; disp. in water; dens. 8.5 lb/gal; visc. 1250 cSt (100 F); flash pt. 530 F; pour pt. -10 C.

Trylon® 6735. [Henkel/Emery; Henkel/Textile] Alkyl polyether; nonionic; low-foaming emulsifier, wetting agent for industrial, institutional, and consumer detergents, textile scouring, bleaching and jet dyeing systems; coemulsifier for solvs.; Gardner 3 clear liq.; sol. in water, butyl stearate, glycerol trioleate, Stod., xylene; dens. 8.2 lb/gal; visc. 41 cSt (100 F); cloud pt. 37 C; flash pt. 370 F; pH 6.0 (1%); pour pt. 9 C; 100% conc.

Trylon® 6751. [Henkel OPG] Fatty ester ethoxylate; coemulsifier for solvs., oils and esters; fiber-to-metal lubricant; Gardner 3 liq.; disp. in water; dens. 8.4 lb/gal; visc. 79 cSt (100 F); pour pt. 10 C; flash pt. 530 F; cloud pt. < 25 C.

Trylon® 6837. [Henkel/Emery] Ether; nonionic; emulsifier for solvs., veg. oils, waxes, tanning chems. in leather industry; Gardner 2 liq.; sol. in butyl stearate, glycerol trioleate, xylene; disp. in water, Stod.; dens. 8.6 lb/gal; visc. 126 cSt (100 F); pour pt. 5 C; cloud pt. < 25 C; flash pt. 545 F.

Trylon® 6848. [Henkel OPG] Anionic; emulsifier, dye assistant; Gardner 7 liq.; sol. in butyl stearate, Stod., xylene; disp. in water; dens. 8.4 lb/gal; visc. 98 cSt

(100 F); flash pt. 180 F; pour pt. < –8 C.

Trylon® EW. [Henkel OPG] Nonionic ether; polymer stabilizer, resin bath penetrant, solv. scour emulsifier, and enzyme bath penetrant in textile applics.; emulsifier for tanning chemicals in the leather industry; used in cold water scours for felted fabrics; Gardner 2 liq.; sol. (5%) in butyl stearate, glycerol trioleate, xylene; disp. in water, Stoddard; dens. 8.6 lb/gal; visc. 126 cSt (100 F); cloud pt. < 25 C; flash pt. 545 F; pour pt. 5 C.

Trylox® 1086. [Henkel/Emery] Ethoxylated sorbitol hexaoleate; nonionic; o/w emulsifier for petrol. and veg. oils, and solvs.; used in metal lubricants, textile processing aids; emulsifier component for aliphatic and aromatic solvs. in textile dye carriers; Gardner 6 liq.; disp. in water; dens. 8.4 lb/gal; visc. 235 cs; HLB 10.2; cloud pt. < 25 C; 100% act.

Trylox® 1186. [Henkel/Emery] Ethoxylated sorbitol hexatallate; o/w emulsifier for petrol. and veg. oils, and solvs.; used in metal lubricants, textile processing aids; emulsifier component for aliphatic and aromatic solvs. in textile dye carriers; Gardner 9 liq.; disp. in water; dens. 8.5 lb/gal; visc. 330 cs; HLB 10.9; cloud pt. < 25 C; 100% act.

Trylox® 5900. [Henkel/Emery] PEG-5 castor oil; nonionic; emulsifier, dispersant, carrier, foam control agent, lubricant for paints, paper coatings, dye carriers, agric. formulations; EPA-exempt; Gardner 4 liq.; sol. in glycerol trioleate; dens. 8.2 lb/gal; visc. 375 cSt (100 F); HLB 4.0; flash pt. 550 F; pour pt. –3 C; 100% conc.

Trylox® 5901. [Henkel/Emery] PEG-10 castor oil; nonionic; emulsifier for agric. formulations; EPA-exempt; Gardner 4 liq.; sol. @ 5% in xylene; disp. in Stod., methyl oleate; dens. 8.3 lb/gal; visc. 332 cSt (100 F); pour pt. -5 C; HLB 6.4; flash pt. 560 F. DISCONTINUED.

Trylox® 5902. [Henkel/Emery; Henkel/Textile] PEG-16 castor oil; nonionic; emulsifier, lubricant for metalworking oils, hydraulic fluids, textiles, agric.; EPA-exempt; Gardner 3 liq.; sol. in xylene; disp. in water; dens. 8.5 lb/gal; visc. 546 cSt (100 F); HLB 8.6; cloud pt. < 25 C; flash pt. 565 F; pour pt. –22 C.

Trylox® 5904. [Henkel/Emery; Henkel/Textile] PEG-25 castor oil; nonionic; emulsifier, lubricant for formulation of sol. oils, cutting fluids, fiber finishes, agric. formulations; EPA-exempt; Gardner 4 liq.; disp. in water, butyl stearate, glycerol trioleate, Stod., xylene; dens. 8.6 lb/gal; visc. 396 cSt (100 F); HLB 10.8; cloud pt. 66 C (1% saline); flash pt. 565 F; pour pt. –5 C.

Trylox® 5906. [Henkel/Emery; Henkel/Textile] PEG-30 castor oil; nonionic; emulsifier, pigment dispersant, degreaser, lubricant; visc. and emulsion stabilizer of PVAc and water-based paints; also for paper, textile, agric. and lubricant applics.; EPA-exempt; Gardner 2 liq.; sol. in water, xylene; dens. 8.8 lb/gal; visc. 309 cSt (100 F); HLB 11.8; cloud pt. 55 C (1% saline); flash pt. 555 F; pour pt. 9 C.

Trylox® 5907. [Henkel/Emery; Henkel/Textile] PEG-36 castor oil; nonionic; emulsifier for solvs. and oils, agric. formulations; lubricant and softener for textiles and leather; coemulsifier for formulating textile dye carriers; EPA-exempt; Gardner 2 liq.; sol. in water, xylene; dens. 8.8 lb/gal; visc. 363 cSt (100 F); HLB 12.6; cloud pt. 78 C (1% saline); flash pt. 575 F; pour pt. 12 C.

Trylox® 5909. [Henkel/Emery; Henkel/Textile] PEG-40 castor oil; nonionic; emulsifier for solvs. and oils, agric. formulations; lubricant and softener for textiles and leather; coemulsifier for formulating textile dye carriers; EPA-exempt; Gardner 2 liq.; sol. in water; dens. 8.8 lb/gal; visc. 313 cSt (100 F); HLB 13.0; cloud pt. 80 C (1% saline); pour pt. 18 C.

Trylox® 5918. [Henkel/Emery; Henkel/Textile] PEG-200 castor oil; nonionic; lubricant, emulsifier, antistat, humectant for textile fiber processing; Gardner 1 liq.; sol. in water; dens. 8.5 lb/gal; visc. 1015 cSt (100 F); HLB 18.1; cloud pt. 80 C (5% saline); pour pt. 7 C; 50% act.

Trylox® 5921. [Henkel/Emery; Henkel/Textile] PEG-16 hydrog. castor oil; CAS 61788-85-0; nonionic; emulsifier, lubricant, softener; used in fabric softeners and aerosol fabric sprays; Gardner 1 liq.; sol. in glycerol trioleate, xylene; disp. in water, min. oil; dens. 8.4 lb/gal; visc. 569 cSt (100 F); HLB 8.6; cloud pt. < 25 C; flash pt. 565 F; pour pt. 7 C.

Trylox® 5922. [Henkel/Emery; Henkel/Textile] PEG-25 hydrog. castor oil; CAS 61788-85-0; nonionic; emulsifier for textile applics., resin finishing, softener-lubricant systems; coemulsifier in dye carrier solv. systems; Gardner 1 liq.; sol. in glycerol trioleate, xylene; disp. in water, min. oil; dens. 8.6 lb/gal; visc. 535 cSt (100 F); HLB 10.8; cloud pt. 25 C; flash pt. 560 F; pour pt. 5 C.

Trylox® 5925. [Henkel/Textile] PEG-200 hydrog. castor oil; CAS 61788-85-0; nonionic; surfactant for textile use; Gardner 1 liq.; sol. @ 5% in water, min. oil, glycerol trioleate, xylene; dens. 8.6 lb/gal; visc. 5200 cSt (100 F); HLB 18.1; pour pt. 5 C; cloud pt. 61 C; flash pt. 510 F; 50% act.

Trylox® 6746. [Henkel/Emery; Henkel/Textile] PEG-40 sorbitol hexaoleate; nonionic; o/w emulsifier, dispersant, wetting agent, lubricant, plasticizer, solubilizer for household/industrial/institutional prods., metal lubricants, textile and cosmetic use; Gardner 6 liq.; sol. @ 5% in butyl stearate, glycerol trioleate, Stod., xylene; disp. in water, min. oil; dens. 8.4 lb/gal; visc. 120 cSt (100 F); HLB 10.4; cloud pt. < 25 C; pour pt. < –10 C; flash pt. 515 F; 100% conc.

Trylox® 6747. [Henkel/Emery; Henkel/Textile] PEG-60 sorbitol hexaoleate; CAS 57171-56-9; nonionic; emulsifier for org. solvs., textile processing; used for paraffin oils in industrial lubricants; Gardner 4 liq.; sol. in min. oil, butyl stearate, glycerol trioleate, xylene; disp. in water, Stod.; dens. 8.4 lb/gal; visc. 300 cs; HLB 11.3; cloud pt. < 25 C; 100% act.

Trylox® 6753. [Henkel/Emery] PEG-20 sorbitol; nonionic; humectant, plasticizer; intermediate; used in surfactant sol'ns.; emulsifier for textile and cosmetic use; Gardner 1 liq.; sol. in water; dens. 9.7 lb/gal; visc. 200 cSt (100 F); HLB 15.4; cloud pt. > 100 C (10% saline); pour pt. 7 C; flash pt. 435 F; 100% conc.

Trylox® CO-5. [Henkel/Emery] PEG-5 castor oil; nonionic; emulsifier, coemulsifier for chlorinated and aromatic solvs.; dispersant for pigment slurries in water-based paint and clay; carrier for paper coatings; textile foam control agent and emulsifier in dye carriers; Gardner 4 liq.; disp. in water; dens. 8.2 lb/gal; visc. 800 cs; HLB 3.8; cloud pt. < 25 C; 100% act.

Trylox® CO-10. [Henkel/Emery] PEG-10 castor oil; nonionic; emulsifier for w/o emulsions; Gardner 4 liq.; oil-sol.; dens. 8.3 lb/gal; visc. 332 cSt (100 F); HLB 6.4; pour pt. –5 C; flash pt. 560 F.

Trylox® CO-16. [Henkel/Emery] PEG-16 castor oil; nonionic; emulsifier, lubricant, coemulsifier for metalworking oils, hydraulic fluids, rayon delustrants and fiber lubricants; component in textile leveling and dispersant for vat and naphthol dyes; Gardner 3 liq.; dens. 8.5 lb/gal; visc. 600 cs; HLB 8.6; cloud pt. < 25 C; 100% act.

Trylox® CO-25. [Henkel/Emery] PEG-25 castor oil; nonionic; emulsifier and lubricant in formulation of sol. oils, cutting fluids and fiber finishes; Gardner 4 liq.; dens. 8.6 lb/gal; visc. 396 cSt (100 F); HLB 10.8; pour pt. –5 C; flash pt. 565 F; cloud pt. 66 C.

Trylox® CO-30. [Henkel/Emery] PEG-30 castor oil; nonionic; emulsifier, degreaser, dispersant, visc. stabilizer, lubricant for oils, solvs., waxes, syn. fiber lubricants, pigments, fat liquoring, paints, urethane foams and polyester resins; coemulsifier in fabric softener and dye carrier systems; dyeing assistant; Gardner 2 liq.; dens. 8.8 lb/gal; visc. 720 cs; HLB 11.7; cloud pt. > 100 C; 100% act.

Trylox® CO-36. [Henkel/Emery] PEG-36 castor oil; nonionic; emulsifier for solvs. and oils; lubricant and softener for textiles and leather; coemulsifier with anionic emulsifier in formulating textile dye carriers; Gardner 2 liq.; dens. 8.8 lb/gal; visc. 700 cs; HLB 12.6; cloud pt > 100 C; 100% act.

Trylox® CO-40. [Henkel/Emery] PEG-40 castor oil; emulsifier for solvs. and oils; lubricant and softener for textiles and leather; coemulsifier with anionic emulsifier in formulating textile dye carriers; Gardner 2 liq.; dens. 8.8 lb/gal; visc. 650 cs; HLB 12.9; cloud pt. > 100 C; 100% act.

Trylox® CO-80. [Henkel/Emery] PEG-80 castor oil; nonionic; o/w emulsifier and solubilizer; used in fiber finish and textile lubricant formulations; Gardner 2 liq.; dens. 8.9 lb/gal; visc. 322 cSt (100 F); HLB 15.9; pour pt. 20 C; cloud pt. 77 C.

Trylox® CO-200, -200/50. [Henkel/Emery] PEG-200 castor oil; nonionic; lubricant, textile fiber processing as antistat humectant and scrooping agent; emulsifier in blended lubricants for fibers and yarns; Gardner 1 solid and Gardner < 1 liq. resp.; dens. 8.6 lb/gal; m.p. 40 C; HLB 18.1; cloud pt. > 100 C; 100 and 50% act.

Trylox® HCO-5. [Henkel/Emery] PEG-5 hydrog. castor oil; CAS 61788-85-0; nonionic; emulsifier, coemulsifier for syn. esters used in the textile industry; fiber lubricant and softener; Gardner 1 liq.; disp. in water; dens. 7.9 lb/gal; visc. 1200 cs; HLB 3.8; cloud pt. < 25 C; 100% act.

Trylox® HCO-16. [Henkel/Emery] PEG-16 hydrog. castor oil; CAS 61788-85-0; nonionic; emulsifier for castor oil; lubricant and softener in fabric softeners and aerosol fabric sprays; Gardner 1 liq.; low odor; dens. 8.4 lb/gal; visc. 2100 cs; HLB 8.2; cloud pt. < 25 C; 100% act.

Trylox® HCO-25. [Henkel/Emery] PEG-25 hydrog. castor oil; CAS 61788-85-0; nonionic; emulsifier for textile resin finishing and softener lubricant systems; coemulsifier for lanolin and dye carrier systems; Gardner 1 liq.; disp. in water, dens. 8.6 lb/gal; visc. 1100 cs; HLB 10.8; cloud pt. 25 C; 100% act.

Trylox® HCO-200/50. [Henkel/Emery] PEG-200 hydrog. castor oil; CAS 61788-85-0; nonionic; emulsifier, lubricant; Gardner < 1 clear liq.; dens. 8.8 lb/gal; HLB 18.1; cloud pt. > 100 C; 50% act.

Trylox® SS-20. [Henkel/Emery] PEG-20 sorbitol; nonionic; humectant, plasticizer; intermediate in the synthesis of fatty acid esters; Gardner 1 liq.; dens. 9.7 lb/gal; visc. 400 cs; HLB 16.1; cloud pt. > 100 C; 100% act.

Trymeen® 6601. [Henkel/Emery; Henkel/Textile] PEG-10 cocamine; CAS 61791-14-8; cationic; coemulsifier, dispersant, antistat, emulsifier, softener, lubricant for textile applics., industrial lubricants, pesticides; substantive to metals, fibers, clays, etc.; Gardner 8 liq.; sol. in water, glycerol trioleate, xylene; dens. 8.4 lb/gal; visc. 68 cSt (100 F); HLB 13.6; cloud pt. 87 C; flash pt. 445 F; pour pt. –10 C; 100% conc.

Trymeen® 6602. [Henkel/Emery] PEG-15 coco amine; CAS 61791-14-8; cationic; antistat, emulsifier, lubricant, dispersant, softener for textiles, agric.; substantive to metals, fibers, and clays; liq.; sol. in water, xylene, methyl oleate; dens. 8.6 lb/gal; HLB 15.2; cloud pt. 97 C; flash pt. 510 F; 100% conc. DISCONTINUED.

Trymeen® 6603. [Henkel/Emery] PEG-8 tallow amine; cationic; low foaming emulsifier contributing wetting, lubricating, and softening properties; solv. emulsifier for dye carriers; as dye assistant; emulsifier for polyethylene textile softeners, in industrial lubricants, agric. formulation; EPA-exempt; Gardner 8 liq.; sol. in water, butyl stearate, glycerol trioleate, Stod., xylene; dens. 8.2 lb/gal; visc. 87 cSt (100 F); HLB 11.4; cloud pt. 77 C (1% saline); flash pt. 525 F; pour pt. –2 C. DISCONTINUED.

Trymeen® 6606. [Henkel/Emery; Henkel/Textile] PEG-15 tallow amine; cationic; emulsifier, antiprecipitant for dye baths, agric. formulations; leveling agent; intermediate for quats.; antistat for syn. fiber processing; EPA-exempt; Gardner 8 liq.; sol. in water, glycerol trioleate, xylene; dens. 8.5 lb/gal; visc. 96 cSt (100 F); HLB 14.3; cloud pt. 95 C (5% saline); flash pt. 445 F; pour pt. –10 C; 100% conc.

Trymeen® 6607. [Henkel/Emery; Henkel/Textile] PEG-20 tallow amine; cationic; antistat, coemulsifier, and lubricant for textile and agric. applics.; antiprecipitant, leveling, and migrating agent in dyeing processes; EPA-exempt; Gardner 6 liq.; sol. in water; dens. 8.7 lb/gal; visc. 119 cSt (100 F); HLB 15.4; cloud pt. 87 C (10% saline); flash pt. 550 F; pour pt. –2 C; 100% conc.

Trymeen® 6609. [Henkel/Emery; Henkel/Textile] PEG-25 tallow amine; cationic; antistat, coemulsifier, and lubricant for textile and agric. applics.; antiprecipitant, leveling, and migrating agent in dyeing processes; EPA-exempt; Gardner 6 liq.; sol. in water, xylene; dens. 8.8 lb/gal; visc. 128 cSt (100 F); HLB 16.0; cloud pt. 87 C (10% saline); flash pt. 540 F; pour pt. 16 C; 100% conc.

Trymeen® 6610. [Henkel/Emery; Henkel/Textile] PEG-40 tallow amine; cationic; emulsifier for agric. formulations, textile processing; EPA-exempt; Gardner 5 solid; sol. @ 5% in water, xylene; dens. 8.8 lb/gal (50 C); m.p. 37 C; HLB 17.4; cloud pt. 85 C (10% saline; flash pt. 485 F; 100% conc.

Trymeen® 6617. [Henkel/Emery; Henkel/Textile] PEG-50 stearyl amine; CAS 26635-92-7; cationic; emulsifier, antistat for metal buffing compds., latex rubber compding., agric. formulations; anticoagulant; lubricant, leveling agent for textile applics.; EPA-exempt; Gardner 4 solid; sol. in water; dens. 8.5 lb/gal (40 C); HLB 17.8; m.p. 35 C; cloud pt. 82 C (10% saline); flash pt. 540 F; 100% conc.

Trymeen® 6620. [Henkel/Textile] PEG-30 oleamine; cationic; surfactant for textile use; Gardner 4 liq.; sol. @ 5% in water; disp. in min. oil, butyl stearate, glycerol trioleate; dens. 8.9 lb/gal; visc. 445 cSt (100 F); HLB 15.3; pour pt. 5 C; cloud pt. 87 C (10% saline); 80% act.

Trymeen® 6622. [Henkel/Emery; Henkel/Textile]

PEG-30 oleamine; cationic; emulsifier for agric. formulations, dyeing assistant, antiprecipitant, stripping agent, dye leveler for textile applics.; Gardner 6 liq.; sol. in water; dens. 8.9 lb/gal; visc. 445 cSt (100 F); HLB 15.3; cloud pt. 86 C (10% saline); pour pt. 5 C; 80% aq.

Trymeen® 6623. [Henkel/Emery] PEG-30 oleamine; cationic; emulsifier for agric. formulation, dyeing assistant, antiprecipitant, stripping agent, dye leveler for textile applics.; Gardner 8 solid; sol. in water; HLB 16.6; m.p. 35 C; cloud pt. 86 C (10% saline); flash pt. 470 F. DISCONTINUED.

Trymeen® 6637. [Henkel/Emery; Henkel/Textile] PEG-40 tallow amine; cationic; emulsifier, antistat for commercial carpet maintenance, agric. formulations, textile processing; dyeing assistant, stabilizer for latices; EPA-exempt; Gardner 4 liq.; sol. in water, xylene; dens. 9.1 lb/gal; visc. 150 cSt (100 F); HLB 17.4; cloud pt. 85 C (10% saline); pour pt. 9 C; 80% act. in water.

Trymeen® CAM-10. [Henkel/Emery] PEG-10 coconut amine; CAS 61791-14-8; cationic; coemulsifier and antistat for textile processing; dispersant for inorg. salts in viscose spinning; emulsifier for fats and oils in industrial lubricants; Gardner 8 liq.; dens. 8.4 lb/gal; visc. 260 cs; HLB 13.8; cloud pt. > 100 C; 100% act.

Trymeen® CAM-15. [Henkel/Emery] PEG-15 coconut amine; CAS 61791-14-8; cationic; emulsifier, antistat, softener; substantive to clay, metals, mins., and textiles; additive in the mfg. of rayon fiber; prevents sludge build-up in spin baths; textile processing; industrial lubricant; Gardner 6 liq.; dens. 8.6 lb/gal; visc. 120 cs; HLB 15.4; cloud pt. > 100 C; 100% act.

Trymeen® OAM 30/60. [Henkel/Textile] PEG-30 oleamine; cationic; surfactant for textile use; Gardner 1 liq.; sol. @ 5% in water; disp. in min. oil, glycerol trioleate; dens. 9.0 lb/gal; visc. 38 cSt (100 F); HLB 15.3; pour pt. -12 C; cloud pt. 87 C (10% saline); 60% act.

Trymeen® OAM-30/80. [Henkel/Emery; Henkel/Textile] PEG-30 oleamine; cationic; emulsifier; textile dyeing assistant; antiprecipitant in cross dyeing; stripping agent and dye leveler for acid dyes; Gardner 6 liq.; dens. 8.9 lb/gal; HLB 15.3; cloud pt. > 100 C; 80% act. DISCONTINUED.

Trymeen®SAM-50. [Henkel/Emery; Henkel/Textile] PEG-50 stearyl amine; CAS 26635-92-7; cationic; antistat, emulsifier in metal buffing compds.; lubricant for fiberglass; leveling agent for dye applics.; Gardner 4 solid; m.p. 50 C; HLB 17.8; cloud pt. > 100 C; 100% act.

Trymeen® TAM-8. [Henkel/Emery] PEG-8 tallow amine; cationic; emulsifier for min. and veg. oils, waxes, solvs., polyethylene textile softeners, fats and oils in industrial lubricants; solv. emulsifier for dye carriers; Gardner 8 liq.; dens. 8.2 lb/gal; visc. 200 cs; HLB 11.5; cloud pt. 80 C; 100% act.

Trymeen® TAM-15. [Henkel/Emery; Henkel/Textile] PEG-15 tallow amine; cationic; emulsifier; antiprecipitant for mixed dye baths; leveling agent for acid dyes; migrating agent for dispersed dyes; intermediate for quat. ammonium compds.; antistat for processing syn. fibers; Gardner 8 liq.; dens. 8.5 lb/gal; visc. 175 cs; HLB 14.3; cloud pt. > 100 C; 100% act. DISCONTINUED.

Trymeen® TAM-20. [Henkel/Emery; Henkel/Textile] PEG-20 tallow amine; cationic; textile antistat, coemulsifier and lubricant for wool and syn. fiber processing; antiprecipitant, leveling and migrating agent in dyeing; Gardner 6 liq.; dens. 8.7 lb/gal; visc. 240 cs; HLB 15.4; cloud pt. > 100 C; 100% act.

Trymeen® TAM-25. [Henkel/Emery; Henkel/Textile] PEG-25 tallow amine; cationic; textile antistat, coemulsifier and lubricant for wool and syn. fiber processing; antiprecipitant, leveling and migrating agent in dyeing; Gardner 6 liq.; dens. 8.8 lb/gal; visc. 170 cs; HLB 16.0; cloud pt. > 100 C; 100% act.

Trymeen® TAM-40. [Henkel/Emery; Henkel/Textile] PEG-40 tallow amine; cationic; emulsifier, antistat additive for commercial carpet maintenance; dyeing assistant and stabilizer for latices; Gardner 6 solid; m.p. 40 C; HLB 17.4; cloud pt. > 100 C; 100% act.

TS-33-F. [Hefti Ltd.] Sorbitan tristearate; CAS 26658-19-5; nonionic; w/o emulsifier for cosmetics, pharmaceuticals; stabilizer for textile printing inks; for silicone antifoams, polish mfg., wax working; flakes; HLB 2.0; 100% conc.

TSA-70. [Henkel Canada] Toluene sulfonic acid; anionic; catalyst in resin prod.; salts as hydrotropes in household and industrial cleaners, metal cleaning compds.; liq.; 70% conc.

T-Soft® SA-97. [Harcros] Linear alkylbenzene sulfonic acid; anionic; high foaming detergent intermediate; liq.; 97% conc.

T-Tergamide 1CD. [Harcros] Cocamide DEA; nonionic; foam stabilizer, thickener, visc. modifier for detergents and shampoos; liq.; 100% conc.

T-Tergamide 1PD. [Harcros] Palm kernelamide DEA; nonionic; foam stabilizer, thickener, visc. modifier for detergents and shampoos; liq.; 100% conc.

Turkey Red Oil 100%. [Zschimmer & Schwarz] Sulfated castor oil; anionic; solubilizer, refatting agent; liq.; 85% conc.

Tween® 20. [ICI Spec. Chem.; ICI Surf. Belgium] Polysorbate 20 NF; CAS 9005-64-5; nonionic; solubilizer; o/w emulsifier; detergent for shampoos; antistat and fiber lubricant used in textile industry; flavor emulsifier; pale yel. liq.; sol. in water, methanol, ethanol, IPA, propylene glycol, ethylene glycol, cottonseed oil; sp.gr. 1.1; visc. 400 cps; HLB 16.7; flash pt. > 300 F; sapon. no. 40–50; 100% act.

Tween® 20 SD. [ICI Spec. Chem.] POE sorbitan monolaurate; anionic; solubilizer; o/w emulsifier; detergent for shampoos; antistat and fiber lubricant used in textile industry; flavor emulsifier; liq.; 100% act.

Tween® 21. [ICI Spec. Chem.; ICI Surf. Belgium] Polysorbate 21; CAS 9005-64-5; nonionic; emulsifier, solubilizer; antistat, fiber lubricant for textiles; yel. oily liq.; sol. in corn oil, dioxane, Cellosolve, CCl_4, methanol, ethanol, aniline; disp. in water; sp.gr. 1.1; visc. 500 cps; HLB 13.3; flash pt. > 300 F; sapon. no. 100–115; 100% act.

Tween® 40. [ICI Spec. Chem.; ICI Surf. Belgium] Polysorbate 40 NF; CAS 9005-66-7; nonionic; emulsifier, solubilizer; textile antistat, fiber lubricant; pale yel. liq.; sol. in water, methanol, ethanol, IPA, ethylene glycol, cottonseed oil; sp.gr. 1.08; HLB 15.6; flash pt. > 300 F; sapon. no. 41–52; 100% act.

Tween® 61. [ICI Spec. Chem.; ICI Surf. Belgium] Polysorbate 61; CAS 9005-67-8; nonionic; emulsifier, solubilizer for perfume, flavor, vitamin oils; tan waxy solid; sol. in methanol, ethanol; disp. in water; sp.gr. 1.06; HLB 9.6; flash pt. > 300 F; pour pt. 100 F; sapon. no. 95–115; 100% act.

Tween® 81. [ICI Spec. Chem.; ICI Surf. Belgium] Polysorbate 81; CAS 9005-65-6; nonionic; emulsifier, solubilizer for perfume, flavor, vitamin oils; amber oily liq.; sol. in min. and corn oil, dioxane, Cellosolve, methanol, ethanol, ethyl acetate, aniline; disp. in water; sp.gr. 1; visc. 450 cps; HLB 10.0; flash pt. > 300 F; sapon. no. 96–104; 100% act.

Tween® 85. [ICI Spec. Chem.; ICI Surf. Belgium] Polysorbate 85; CAS 9005-70-3; nonionic; emulsifier, solubilizer for perfume, flavor, vitamin oils; floating bath oils; amber liq.; sol. in veg. oil, Cellosolve, lower alcohols, aromatic solvs., ethyl acetate, min. oils and spirits, acetone, dioxane, CCl_4 and ethylene glycol; disp. in water; sp.gr. 1.0; visc. 300 cps; HLB 11.0; flash pt. > 300 F; sapon. no. 80–95; 100% act.

Twitchell 6805 Oil. [Henkel/Emery] Sulfonated min. oil; anionic; rewetting agent, softener and textile fiber lubricant, deduster; liq.; 100% conc.

Twitchell 6808 Oil. [Henkel/Emery] Sulfated fatty min. deriv.; anionic; lubricant, wetting agent, and dyestuff deduster; liq.; 100% conc.

Twitchell 6809 Oil. [Henkel/Emery] Min. fatty derivs.; nonionic; wetting, lubricant, dyestuff deduster; liq.; 100% conc.

Tylose C, CB Series. [Hoechst Celanese/Colorants & Surf.] Sodium CMC; anionic; binder in pencil leads; thickener in batteries, rubber industry, cosmetics, foodstuffs, pharmaceuticals, tobacco and textile industry; dispersant, emulsifier for insecticidal, fungicidal and herbicidal prods.; plasticizer in ceramics; surface sizing in paper industry; press aid and lubricant in welding electrodes; gran., powd.

Tylose CBR Grades. [Hoechst Celanese/Colorants & Surf.; Hoechst AG] Sodium CMC; binder for coatings; sedimenting aid in mining; gelling agent/binder/thickener in chemical tech. and rubber industry; sizing for paper and textile industry; plasticizer/filler in soaps and hand cleaning pastes; graying inhibitor for detergents; gran.

Tylose H Series. [Hoechst Celanese/Colorants & Surf.; Hoechst AG] Hydroxyethylcellulose; nonionic; binder, thickener, plasticizer, visc. control agent, protective colloid in ceramics, emulsion polymerization, tobacco and textile industry, agric., cosmetics, soaps, and hand cleaning pastes; gran.; water-sol.

Tylose MHB. [Hoechst Celanese/Colorants & Surf.] Methyl hydroxyethylcellulose; CAS 9032-42-2; nonionic; binder, thickener, pigment, foam, and filler stabilizer, dispersant, emulsifier, plasticizer, visc. control and sedimenting aid, and protective colloid used in coatings, paints, resins, mining, batteries, insecticides, fungicides, herbicides; rubber, textile, and leather industry; ceramics, suspension polymerization, and pharmaceuticals; gran.; water-sol.

Tylose MH Grades. [Hoechst Celanese/Colorants & Surf.; Hoechst AG] Methyl hydroxyethylcellulose; nonionic; binder, thickener, pigment, foam, and filler stabilizer, dispersant, emulsifier, plasticizer, visc. control and sedimenting aid, and protective colloid used in coatings, paints, resins, mining, batteries, insecticides, fungicides, herbicides; rubber, textile, and leather industry; ceramics, suspension polymerization, and pharmaceuticals; gran.; water-sol.

Tylose MH-K, MH-xp, MHB-y, MHB-yp. [Hoechst Celanese/Colorants & Surf.] Methyl hydroxyethylcellulose; see Tylose MH.

U

Ucar® Acrylic 503. [Union Carbide] Modified acrylic emulsion; anionic/nonionic; surfactant for high-quality exterior, interior, and semigloss house paints; emulsion; 0.4 μ particle size; sp.gr. 1.18 (polymer dens.); dens. 9.2 lb/gal; visc. 1000 cP (latex); pH 6.0; 58% solids.

Ucar® Acrylic 505. [Union Carbide] Modified acrylic emulsion; anionic/nonionic; surfactant for high-quality exterior, interior, and semigloss house paints; emulsion; 0.6 μ particle size; sp.gr. 1.18 (polymer dens.); dens. 9.1 lb/gal; visc. 800 cP (latex); pH 6.0; 55% solids.

Ucar® Acrylic 515. [Union Carbide] Modified acrylic emulsion; anionic/nonionic; surfactant for high-quality exterior, interior, and semigloss house paints; emulsion; 0.5 μ particle size; sp.gr. 1.18 (polymer dens.); dens. 9.1 lb/gal; visc. 300 cP (latex); pH 8.5; 53% solids.

Ucar® Acrylic 516. [Union Carbide] Modified acrylic emulsion; anionic/nonionic; surfactant with exc. adhesion properties for exterior paints; emulsion; 0.5 μ particle size; sp.gr. 1.18 (polymer dens.); dens. 9.1 lb/gal; visc. 250 cP (latex); pH 9.5; 53% solids.

Ucar® Acrylic 518. [Union Carbide] Modified acrylic emulsion; anionic/nonionic; surfactant for high-quality exterior, interior, and semigloss house paints; emulsion; 0.5 μ particle size; sp.gr. 1.18 (polymer dens.); dens. 9.0 lb/gal; visc. 500 cP (latex); pH 8.5; 50% solids.

Ucar® Acrylic 522. [Union Carbide] Modified acrylic; anionic/nonionic; surfactant for trade paints; 0.5 μ particle size; sp.gr. 1.18; dens. 9.0 lb/gal; visc. 900 cP; pH 7.5; 51% solids.

Ucar® Acrylic 525. [Union Carbide] Modified acrylic; anionic/nonionic; surfactant for trade paints; 0.5 μ particle size; sp.gr. 1.18; dens. 8.7 lb/gal; visc. 1000 cP; pH 9; 54% solids.

Udet® 950. [Witco/Organics] Linear alkylaryl sulfonate; anionic; used for heavy-duty detergent powds.; powd.; 95% conc.

Ufablend DC. [Unger Fabrikker AS] Surfactant blend; anionic; biodeg. high foaming detergent conc. for mfg. of liq. detergents, dishwash, hard surf. cleaners; effective in hard and soft water; lt. yel. paste; pH 7-8 (1%); 65% act.

Ufablend E 60. [Unger Fabrikker AS] Surfactant blend; anionic/nonionic; liq. detergent and hard surf. cleaner based on biodeg. raw materials; liq./paste; 60% conc.

Ufablend HDL. [Unger Fabrikker AS] Surfactant/foam controller/optical brightener blend; anionic/nonionic; biodeg. conc. liq. laundry detergent with minimum foaming; for hand and machine washing; yel. soft paste; visc. 120 cps; cloud pt. -10 C; pH 7.5-8.0; 100% act.

Ufablend HDL85N. [Unger Fabrikker AS] Surfactant blend; anionic/nonionic; conc. for liq. laundry detergents; paste; 85% conc.

Ufablend W 100. [Unger Fabrikker AS] Surfactant blend; anionic/nonionic; biodeg. high foaming detergent conc. for washing up liqs., car shampoos, wall/floor cleaners, carpet cleaners, dishwash, hard surf. cleaners, etc.; yel. paste; pH 7-8 (1%); 100% act.

Ufablend W Conc. [Unger Fabrikker AS] Surfactant blend; anionic/nonionic; biodeg. high foaming detergent conc. for dishwash, hard surface liq. detergents, car wash, carpet cleaners, wall/floor cleaners; suitable for hard and soft water; ylsh. soft paste; pH 6.5-7.5; 63.5-66.5% act.

Ufablend W Conc. 60. [Unger Fabrikker AS] Surfactant blend; anionic/nonionic; detergent conc. for hard surface liq. detergents, dishwash; pale yel. visc. liq.; pH 6.5-7.5 (1%); 60% act.

Ufacid K. [Unger Fabrikker AS] Linear dodecylbenzene sulfonic acid; anionic; intermediate for mfg. of sulfonates used in detergent powds., liqs., emulsifiers; biodeg.; brn. liq., char. odor; m.w. 322; sp.gr. 1.03 (50 C); visc. 950 cps; flash pt. > 90 C; 96-98% act.

Ufacid KA. [Unger Fabrikker AS] 2-Phenyl alkylbenzene sulfonic acid; intermediate for mfg. of sulfonates used in liq. detergents; biodeg.; m.w. 316.

Ufacid KB. [Unger Fabrikker AS] 2-Phenyl alkylbenzene sulfonic acid; intermediate for mfg. of sulfonates used in liq. and powd. detergents; biodeg.; m.w. 322.

Ufacid KF. [Unger Fabrikker AS] 2-Phenyl alkylbenzene sulfonic acid; intermediate for mfg. of sulfonates used in liq. detergents; biodeg.; m.w. 322.

Ufacid KN. [Unger Fabrikker AS] 2-Phenyl alkylbenzene sulfonic acid; intermediate for mfg. of sulfonates used in powd., paste or liq. detergents; biodeg.; m.w. 329.

Ufacid KW. [Unger Fabrikker AS] Linear dodecylbenzene sulfonic acid; intermediate for mfg. of sulfonates used in liq., paste and powd. detergents; biodeg.; brn. liq., char. odor; m.w. 322; sp.gr. 1.03; visc. 950 cps; flash pt. > 90 C; 96-98% act.

Ufacid TPB. [Unger Fabrikker AS] Branched dodecylbenzene sulfonic acid; anionic; intermediate for mfg. of sulfonates used in detergent powds., liqs., emulsifiers; exc. foaming/detergency; partially biodeg.; brn. liq., char. odor; m.w. 322; sp.gr. 1.03; visc. 6000 cps; flash pt. > 90 C; 95-97% act.

Ufanon K-80. [Unger Fabrikker AS] Cocamide DEA (2:1); nonionic; biodeg. detergent, foaming agent, wetting agent, thickener, foam stabilizer for liq.

detergents, shampoos, cosmetics, leather industry; confers some corrosion protection; brn. visc. liq., char. amine odor; pH 9-10 (1%); 48-56% conc.

Ufanon KD-S. [Unger Fabrikker AS] Cocamide DEA (1:1); nonionic; biodeg. visc. modifier, foam booster/stabilizer for liq. detergents, shampoos, cosmetics, leather industry; pale yel. liq.; sp.gr. 0.997; sapon. no. 150; cloud pt. -11 C; pH 9.5-10.0 (1%); usage level: 1-10%; 90% conc.

Ufapol. [Unger Fabrikker AS] Surfactant blend; anionic; surfactant for emulsion polymerization; liq.

Ufarol Am 30. [Unger Fabrikker AS] Ammonium lauryl sulfate; anionic; biodeg. detergent, wetting and foaming agent for shampoos, bath prods., general-purpose detergents, laundry cleaners, carpet shampoos, furniture cleaning, textiles, leather, paints; stable in hard water and alkali, moderately stable in acids; pale yel. liq.; m.w. 289-295; visc. 2000 ± 1000 cp; pH 6-7 (10%); 28-30% act.

Ufarol Am 70. [Unger Fabrikker AS] Ammonium lauryl sulfate; anionic; biodeg. detergent, wetting and foaming agent for shampoos, bath prods., general-purpose detergents, laundry cleaners, carpet shampoos, furniture cleaning, textiles, leather, paints; stable in hard water and alkali, moderately stable in acids; water-wh. paste; m.w. 289-295; pH 6-7 (10%); 68-72% act.

Ufarol Na-30. [Unger Fabrikker AS] Sodium lauryl sulfate; CAS 151-21-3; anionic; biodeg. detergent, wetting and foaming agent for shampoos, bath prods., general-purpose detergents, laundry cleaners, carpet shampoos, furniture cleaning, textiles, leather, paints; stable in hard water and alkali, moderately stable in acids; water-wh. to pale yel. liq.; m.w. 293-299; visc. < 200 cp; pH 7.5-9.0 (2%); 29-31% act.

Ufarol TA-40. [Unger Fabrikker AS] TEA lauryl sulfate; anionic; biodeg. surfactant for shampoos, bath prods., carpet shampoos, furniture cleaning, laundry, textiles; mild to hair and scalp; water-wh. to pale yel. liq.; m.w. 412-418; visc. 100 cps; cloud pt. 0 C; pH 6.5-7.5 (1%); 39-41% act.

Ufaryl DB80. [Unger Fabrikker AS] Sodium dodecylbenzene sulfonate, branched; anionic; detergent, wetting agent, foaming agent, emulsifier for lt. and heavy-duty detergents, dairy, metal, floor, vehicle and bottle cleaners; wetting agent in insecticides, metal pickling, printing inks, paper processing, textiles; partially biodeg.; pale cream free-flowing powd.; m.w. 343-345; sp.gr. 420 g/l; pH 6-9 (1%); 79-83% act.

Ufaryl DL80 CW. [Unger Fabrikker AS] Linear sodium dodecylbenzene sulfonate; anionic; high foaming detergent, wetting agent, anticaking agent for scouring, toilet blocks, dairy and brewery cleaners, agric., concrete mfg., paper prod., metal processing, textiles; stable in hard water, acids, alkalis; wh. to lt. yel. free-flowing powd.; m.w. 343-345; sp.gr. 430 ± 40 g/l; pH 10-11 (1%); 79-83% act.

Ufaryl DL85. [Unger Fabrikker AS] Sodium dodecylbenzene sulfonate linear; anionic; high foaming detergent, emulsifier, wetting agent, anticaking agent; for scouring, dairy and brewery cleaners, agric., concrete, paper prod., metal processing, textiles, toilet blocks; wh. to lt. yel. free-flowing powd.; m.w. 343-345; sp.gr. 420 ± 40 g/l; pH 10-11 (1%); 83–87% act.

Ufaryl DL90. [Unger Fabrikker AS] Linear sodium dodecylbenzene sulfonate; anionic; high foaming detergent, wetting agent, emulsifier, anticaking agent; for scouring, dairy and brewery cleaning, agric., concrete mfg., paper prod., metal processing, textiles, toilet blocks; wh. to lt. yel. free-flowing powd.; m.w. 343-345; sp.gr. 400 ± 40 g/l; pH 7-9 (1%); 88-92% act.

Ufasan 35. [Unger Fabrikker AS] Linear sodium dodecylbenzene sulfonate; anionic; biodeg. detergent, wetting agent, foaming agent for liq. detergents, dishwash, hair and car shampoos and in plastics, metal, agric., polish, textiles, mining, oil and cement industries; stable in hard water and acids; golden visc. liq., weak char. odor; m.w. 343-345; cloud pt. -7 C; pH 7.5-8.5; 34–36% act.

Ufasan 50. [Unger Fabrikker AS] Linear sodium alkylbenzene sulfonate; anionic; surfactant, emulsifier for detergents and in plastics, metal, agric., polish, textiles, mining, oil and cement industries; pumpable liq./paste; m.w. 344; 50% act.

Ufasan 60A. [Unger Fabrikker AS] Linear sodium dodecylbenzene sulfonate; anionic; biodeg., highly sol. surfactant for use in high active liq. detergents, dishwash, industrial cleaners, car shampoos, laundry washes and in plastics, metal, agric., polish, textiles, mining, oil and cement industries; stable in hard water, acid, alkalis; very pale mobile paste, weak odor; m.w. 337-339; pH 8-10 (1%); 59-61% act.

Ufasan 62B. [Unger Fabrikker AS] Sodium alkylbenzene sulfonate; partially biodeg. detergent, wetting and foaming agent for detergent powds., and in plastics, metal, agric., polish, textiles, mining, oil and cement industries; stable in hard water, acids, alkalis; wh. paste, weak odor; m.w. 343-345; pH 8-10 (1%); 61-63% act.

Ufasan 65. [Unger Fabrikker AS] Linear sodium dodecylbenzene sulfonate; anionic; biodeg. detergent, wetting and foaming agent for liq. detergents, dishwash, industrial detergents, shampoos, laundry powds. and in plastics, metal, agric., polish, textiles, mining, oil and cement industries; stable in hard water, acids, alkalis; wh. paste, weak odor; m.w. 343-345; pH 8-10 (1%); 64–66% act.

Ufasan IPA. [Unger Fabrikker AS] Isopropylamine alkylbenzene sulfonate; biodeg. raw material for detergents, dishwash, cleaning liqs., car shampoos, industrial cleaners and in plastics, metal, agric., polish, textiles, mining, oil and cement industries; emulsifier in solv.-based waterless hand and industrial cleaners; stable in hard water, diluted acids and alkalis; golden visc. liq., weak odor; sol. in water; m.w. 381; visc. 20,000-30,000 cp; cloud pt. < -10 C; pH 6.5-8.0 (1%); 94-96% act.

Ufasan TEA. [Unger Fabrikker AS] TEA dodecylbenzene sulfonate, linear; anionic; biodeg. surfactant for liq. detergents, dishwash, hair and car shampoos and in plastics, metal, agric., polish, textiles, mining, oil and cement industries; stable in hard water, diluted acids and alkalis; golden visc. liq., weak odor; sol. in water; m.w. 469-473; flash pt. (Abel-Penksy) > 63 C; 49–51% act.

Ufasoft 75. [Unger Fabrikker AS] Quat. ammonium imidazoline methosulfate; cationic; surfactant, softener, fabric conditioner, antistat, bactericide for laundry prods.; yel. soft paste, isopropanol odor; disp. in water; m.w. 730; sp.gr. 0.91; pour pt. 15 C; flash pt. 23 C; pH 5.5-6.5 (5% aq.); toxicology: irritating to skin; damaging to eyes; 73-77% act. in IPA.

Ukanil 2262. [ICI PLC] POP/POE block polymer; nonionic; industrial and domestic detergent defoamer, rinse aid; liq.; 100% conc.

Ukanil 3000. [ICI PLC] POP/POE block polymer; nonionic; wool lubricant; liq.; 100% conc.

Ukaril 190. [ICI PLC] Modified alcohol ethoxylate; CAS 68213-23-0; nonionic; low visc. textile spin finish component; liq.; 100% conc.

Ultra Blend 60, 100. [Witco] Linear alkylaryl sulfonate, alkanolamide and ethoxylated adduct; anionic; detergent and car washing conc.; liq.; biodeg.; 60 and 100% conc. resp.

Ultra Blend P-280. [Witco] Linear alkylaryl sulfonate, alcohol ether sulfate and ethoxylated adduct; anionic; detergent, high-foaming conc.; biodeg.; liq.

Ultra NCS Liquid. [Witco] Ammonium cumene sulfonate; anionic; hydrotrope, coupling agent, solubilizer; cloud pt. depressant for liq. detergents; antiblocking agent for powd. detergents; liq.; 40% conc.

Ultra NXS Liquid. [Witco] Ammonium xylene sulfonate; hydrotrope, coupling agent, solubilizer; cloud pt. depressant for liq. detergents; antiblocking agent for powd. detergents; liq.; 45% conc.

Ultraphos Series. [Witco SA] Ethoxylated fatty alcohol phosphate ester; anionic; metal working additive; detergent when neutralized; liq.; 100% conc.

Ultra SCS Liquid. [Witco] Sodium cumene sulfonate; hydrotrope, coupling agent, solubilizer; cloud pt. depressant for liq. detergents; antiblocking agent for powd. detergents; liq.; 40% conc.

Ultra Sulfate AE-3. [Witco] Ammonium alcohol ether sulfate; anionic; wetting agent, penetrant, lubricant, emulsifier, dye dispersant, scouring aid, antistat; base for personal care products; detergent for specialty household and industrial cleaners; textile surfactant; liq.; 58% conc.

Ultra Sulfate SE-5. [Witco] Sodium alcohol ether sulfate; wetting agent, penetrant, lubricant, emulsifier, dye dispersant, scouring aid, antistat; base for personal care products; detergent for specialty household and industrial cleaners; textile surfactant; liq.; 58% conc.

Ultra Sulfate SL-1. [Witco] Sodium lauryl sulfate; CAS 151-21-3; wetting agent, penetrant, lubricant, emulsifier, dye dispersant, scouring aid, antistat; base for personal care products; detergent for specialty household and industrial cleaners; textile surfactant; emulsion polymerization and latex stabilization; liq.; 30% conc.

Ultra SXS Liq., Powd. [Witco] Sodium xylene sulfonate; CAS 1300-72-7; anionic; hydrotrope, coupling agent, solubilizer; cloud pt. depressant for liq. detergents; antiblocking agent for powd. detergents; 40% liq.; 90% powd.

Ultratex® ESB, WK. [Ciba-Geigy] Polysiloxane, modified; cationic; textile finish; liq.

Ultravon® SFN. [Ciba-Geigy/Dyestuffs] Proprietary; anionic; wetter, detergent for peroxide bleaching; liq.

Unamide® C-2. [Lonza] PEG-3 cocamide; nonionic; foam booster/stabilizer, visc. builder, emulsifier; used in personal care, household and industrial prods.; wh. solid; sapon. no. 15 max.; pH 9–10 (5%); 100% act.

Unamide® C-5. [Lonza] PEG-6 cocamide; nonionic; foam stabilizer, visc. builder, emulsifier for shampoos, lt. and heavy-duty detergents; stable over broad pH range; lt. yel. liq.; sapon. no. 12 max.; pH 9–10 (5%); 100% act.

Unamide® C-72-3. [Lonza] Cocamide DEA (2:1); nonionic; visc. builder, foam stabilizer; used for lt. duty liqs., industrial, household hard surface cleaners, surfactant, emulsifier, corrosion inhibitor, lubricant and personal care prods.; amber liq.; acid no. 0–10; pH 9–10.5 (5%); 100% act.

Unamide® C-7649. [Lonza] Cocamide DEA, modified; anionic/nonionic; detergent base for floor cleaners; foam stabilizer, emulsifier; compat. with inorganic builders; liq.; 100% conc.

Unamide® C-7944. [Lonza] Cocamide DEA and DEA-dodecylbenzenesulfonate; anionic/nonionic; for industrial cleaners, floor strippers and degreasers; liq.; 100% conc.

Unamide® D-10. [Lonza] 2:1 Coco DEA, modified; nonionic; visc. builder, foam stabilizer; used for lt. duty liqs., industrial, household hard surface cleaners, surfactant, emulsifier, corrosion inhibitor, lubricant and personal care prods.; dk. amber liq.; acid no. 55–75; pH 9–10.5 (5%); 100% act.

Unamide® J-56. [Lonza] Lauramide DEA; CAS 120-40-1; nonionic; visc. builder, foam stabilizer; used for lt. duty liqs., industrial, household hard surface cleaners, surfactant, emulsifier, corrosion inhibitor, lubricant and personal care prods.; amber liq.; acid no. 18–25; pH 9–10.5 (5%); 100% act.

Unamide® L-2. [Lonza] PEG-3 lauramide; nonionic; foam booster/stabilizer, visc. builder, emulsifier; used in personal care, household and industrial prods.; wh. solid; sapon. no. 15 max.; pH 9–10 (5%); 100% act.

Unamide® L-5. [Lonza] PEG-6 lauramide; nonionic; foam booster/stabilizer, visc. builder, emulsifier; used in personal care, household and industrial prods.; wh. soft wax, sapon. no. 12 max.; pH 9–10 (5%); 100% act.

Unamide® LDL. [Lonza] Cocamide DEA (1:1); nonionic; visc. builder, foam stabilizer; used for lt. duty liqs., industrial, household hard surface cleaners, surfactant, emulsifier, corrosion inhibitor, lubricant and personal care prods.; biodeg.; lt. straw liq.; dens. 8.2 lb/gal; sp.gr. 0.98; acid no. 9–10.5; pH 9–10.5 (5%); 100% act.

Unamide® LDX. [Lonza] Lauramide DEA; CAS 120-40-1; nonionic; foam booster, stabilizer, visc. builder used in personal care prods., lt. duty detergents; wh. solid; water-sol.; dens. 8.2 lb/gal; sp.gr. 0.99; acid no. 0–3; pH 9–10.5 (5%); biodeg.; 100% act.

Unamide® N-72-3. [Lonza] Cocamide DEA; nonionic; visc. builder, foam stabilizer; used for lt. duty liqs., industrial, household hard surface cleaners, surfactant, emulsifier, corrosion inhibitor, lubricant and personal care prods.; yel. liq.; acid no. 0–10; pH 9–10.5 (5%); 100% act.

Unamide® S. [Lonza] Stearamide DEA; nonionic; thickener and opacifier in personal care prods.; visc. builder, gelling agent, lubricant, industrial and household prods.; textile lubricant and finishing agent; wh. waxy solid; acid no. 11–15; 100% act.

Unamide® SI. [Lonza] 1:1 Stearic (ethyl amino hydroxyethyl) amide; nonionic; visc. builder, gelling agent, opacifier used in household and industrial prods.; textile lubricant and finishing agent; wh. waxy solid.

Unamide® W. [Lonza] 1:1 Stearamide DEA; nonionic; visc. builder, gelling agent, opacifier used in household and industrial prods.; textile lubricant and finishing agent; amber liq.

Unamine® C. [Lonza] Coco hydroxyethyl imidazoline; nonionic; emulsifier, surfactant, fungicide; textile antistat; leather treating; base for quats.; acid detergent and wetting agent; corrosion inhibitor; pigment flushing agent; liq.; 92% conc.

Unamine® O. [Lonza] Oleyl hydroxyethyl imidazoline; nonionic; emulsifier, demulsifier; antistat for drycleaning fluids; water displacing agent; corrosion inhibitor; base for quats.; pigment flushing agent; liq.; sol. in acidic sol'ns.; 92% conc.

Unamine® S. [Lonza] Stearyl hydroxyethyl imidazoline; nonionic; softener, antistat for syn. fabrics; surfactant; filming compd.; corrosion inhibitor; solid; oil-sol.; 92% conc.

Unamine® T. [Lonza] Tall oil hydroxyethyl imidazoline; cationic; surfactant; w/o emulsifier, corrosion inhibitor, filming agent, dewatering compd.; flushing agent for pigments; liq.; acid-sol.; 92% conc.

Ungerol AM3-60. [Unger Fabrikker AS] Ammonium lauryl ether sulfate (3 EO); anionic; for liq. detergents, shampoos; paste; 58–62% conc.

Ungerol AM3-75. [Unger Fabrikker AS] Ammonium lauryl ether (3 EO) sulfate; anionic; biodeg. detergent, wetting and foaming agent, emulsifier used in liq. detergents, car shampoos, bath foams; stable in hard water, moderately stable in acids; pale straw soft paste; m.w. 440-450; sp.gr. 1.05; flash pt. (Abel Pensky) 42 C; pH 7-9 (1%); 72-75% act.

Ungerol CG27. [Unger Fabrikker AS] Sodium lauryl ether sulfate; CAS 9004-82-4; anionic; liq. detergents and shampoos; liq.; 26–28% conc.

Ungerol LES 2-28. [Unger Fabrikker AS] Sodium laureth sulfate (2 EO); CAS 9004-82-4; anionic; biodeg. detergent, wetting and foaming agent, emulsifier for liq. detergents, shampoos, bath prods., wallboard mfg., textiles, drilling aux.; exc. stability in hard water, alkalis; moderately stable in acids; colorless liq.; m.w. 379-389; visc. 300-600 cps; cloud pt. 0 C; pH 7-8 (1%); 26-28% act.

Ungerol LES 2-70. [Unger Fabrikker AS] Sodium laureth sulfate (2 EO); CAS 9004-82-4; anionic; biodeg. detergent, wetting and foaming agent, emulsifier for liq. detergents, shampoos, bath prods., wallboard mfg., textiles, drilling aux.; exc. stability in hard water, alkalis; moderately stable in acids; pale straw pumpable paste; m.w. 379-389; pH 7-9 (1%); 68-72% act.

Ungerol LES 3-28. [Unger Fabrikker AS] Sodium laureth sulfate (3 EO); CAS 9004-82-4; anionic; biodeg. detergent, wetting and foaming agent, emulsifier for liq. detergents, shampoos, bath prods., wallboard mfg., textiles, drilling aux.; exc. stability in hard water, alkalis; moderately stable in acids; colorless liq.; m.w. 445-455; sp.gr. 1.02; visc. 500-800 cps; pH 7.0-8.5 (1%); 26-28% act.

Ungerol LES 3-54. [Unger Fabrikker AS] Sodium lauryl ether sulfate; CAS 9004-82-4; anionic; liq. detergents, shampoos; paste; 53–55% conc.

Ungerol LES 3-70. [Unger Fabrikker AS] Sodium laureth sulfate (3 EO); CAS 9004-82-4; anionic; biodeg. detergent, wetting and foaming agent, emulsifier for liq. detergents, shampoos, bath prods., wallboard mfg., textiles, drilling aux.; exc. stability in hard water, alkalis; moderately stable in acids; pale straw pumpable paste; m.w. 445-455; sp.gr. 1.05; visc. 40,000 cps; pH 7-9 (1%); 68-72% act.

Ungerol LS, LSN. [Unger Fabrikker AS] Sodium lauryl sulfate; CAS 151-21-3; anionic; carpet shampoos, furniture cleaners, polymerization emulsifier; paste; 29–41% conc.

Ungerol N2-28. [Unger Fabrikker AS] Sodium laureth sulfate (2 EO); CAS 9004-82-4; anionic; biodeg. detergent, wetting and foaming agent, emulsifier for liq. detergents, shampoos, bath prods., wallboard mfg., textiles, drilling aux.; exc. stability in hard water, alkalis; moderately stable in acids; colorless to pale straw liq.; m.w. 377-387; sp.gr. 1.02; visc. 200 cps max.; cloud pt. 0 C; pH 7-8 (1%); 27-29% act.

Ungerol N2-70. [Unger Fabrikker AS] Sodium laureth sulfate (2 EO); CAS 9004-82-4; anionic; biodeg. detergent, wetting and foaming agent, emulsifier for liq. detergents, shampoos, bath prods., wallboard mfg., textiles, drilling aux.; exc. stability in hard water, alkalis; moderately stable in acids; pale straw paste; m.w. 377-387; sp.gr. 1.05; pH 7-9 (1%); 68-72% act.

Ungerol N3-28. [Unger Fabrikker AS] Sodium laureth sulfate (3 EO); CAS 9004-82-4; anionic; biodeg. detergent, wetting and foaming agent, emulsifier for liq. detergents, shampoos, bath prods., wallboard mfg., textiles, drilling aux.; exc. stability in hard water, alkalis; moderately stable in acids; colorless liq.; m.w. 420-430; sp.gr. 1.02; visc. 150 cps; pH 7-8 (1%); 27-29% act.

Ungerol N3-70. [Unger Fabrikker AS] Sodium laureth sulfate (3 EO); CAS 9004-82-4; anionic; biodeg. detergent, wetting and foaming agent, emulsifier for liq. detergents, shampoos, bath prods., wallboard mfg., textiles, drilling aux.; exc. stability in hard water, alkalis; moderately stable in acids; pale straw pumpable paste; m.w. 420-430; sp.gr. 1.05; visc. 40,000 cps; pH 7-9 (1%); 68-72% act.

Unidri M-40. [Hart Chem. Ltd.] Anionic; dewatering agent for vacuum filtration of minerals; drainage aid for drainage improvement over lime mud washers in pulp/paper industry; liq.; 40% act.

Unidri M-60. [Hart Chem. Ltd.] Anionic; dewatering agent for vacuum filtration of minerals; liq.

Unidri M-75. [Hart Chem. Ltd.] Anionic; low foaming surfactant for dewatering metallic oxides, coal and other minerals; liq.

Unifroth B. [Hart Chem. Ltd.] Proprietary blend; frother for flotation operations where foam builds up; exc. for column flotation; liq.

Unifroth G. [Hart Chem. Ltd.] Glycol blend; frother for flotation operations where selectivity is required; liq.

Unifroth H. [Hart Chem. Ltd.] Glycol blend; frother for flotation operations where strength is required; liq.

Union Carbide® L-45 Series. [Union Carbide] Dimethylpolysiloxane polymers; emollient for creams and lotions; antifoam for nonaq. petrol. processing systems, hydraulic fluids, pigment flotation on paints; heat transfer fluid for drug mfg.; lubricant; also for cosmetic, car polish, lubricant/release applics.,; aerosol pkg.; sol. in most nonpolar solvs.; visc. 7–100,000 cstk.

Union Carbide® L-720, -721. [Union Carbide] Organosilicone; nonionic; stabilizer, lubricant, antifoamer, mold release and wetting agent, penetrant; solid; water-sol.; 100% conc.

Union Carbide® L-722, -727. [Union Carbide] Organosilicone; nonionic; stabilizer, lubricant, antifoamer, mold release and wetting agent, penetrant; liq.; sol. in alcohol and hydrocarbon; 100% conc.

Union Carbide® L-7001, -7002. [Union Carbide] Organosilicone; nonionic; wetting agent, penetrant, coemulsifier, lubricant, dispersant; liq.; water sol. and disp.; 100% conc.

Union Carbide® L-7600, -7602, -7604. [Union Carbide] Organosilicone; nonionic; wetting agent, penetrant, coemulsifier, lubricant, dispersant; liq.; water sol. and disp.; 100% conc.

Union Carbide® L-7605. [Union Carbide] Organosilicone; nonionic; wetting agent, penetrant, coemulsifier, lubricant, dispersant; solid; water sol. and disp.; 100% conc.

Union Carbide® L-7607. [Union Carbide] Organosilicone; nonionic; wetting agent, penetrant, coemulsifier, lubricant, dispersant; liq.; water sol. and disp. 100% conc.

Uniperol® AC. [BASF AG] Oxyethylated fatty amine; cationic; dyeing aux., leveling agent; used in the textile industry; lt. br. clear visc. liq.; water-sol.

Uniperol® AC Highly Conc. [BASF AG] Oxyethylated fatty amine; cationic; dyeing aux., leveling agent; used in the textile industry; br. visc. oil; water-sol.

Uniperol® EL. [BASF/Fibers; BASF AG] Oxyethylated veg. oil; nonionic; surfactant, dispersant, leveling agent, emulsifier for fatty acids, fatty oils, waxes in textile dyeing and printing; stable to hard water; yel. oil which becomes pasty or solid when cooled; sol. in water, liq. or melted fatty acids, fats, waxes, and many org. solvs.; 100% active.

Uniperol® O. [BASF/Fibers; BASF AG] Fatty alcohol-polyglycol ether; nonionic; emulsifier, dyeing aux., leveling and wetting agent, dispersant, detergent used in textile processing; colorless liq.; water-sol.; 20% act.

Uniperol® O Micropearl. [BASF AG] Fatty alcohol polyglycol ether; nonionic; emulsifier, dyeing aux., leveling and wetting agent, dispersant, detergent used in textile processing; stable to hard water, acids, alkalies, metal salts; off-wh. powd.; sol. in water; pH 6-7.5 (5% aq.); toxicology: irritant to skin and mucous membranes on prolonged exposure.

Uniperol® W, W Flakes. [BASF/Fibers; BASF AG] Aliphatic ethoxylate; anionic; surfactant, dispersant, leveling agent, protective colloid; for wool and syn. fibers; sl. wetting action, no detergency; stable to acids, alkalies, water-hardening salts, heavy metal ions, electrolytes; yel.-brn. clear liq., yel. to lt. brn. flakes; sol. in water; 40% conc.

Unisol BT. [ICI Surf. UK] EO condensate; nonionic; leveling agent and penetrant for coloring of silk and wool at low dyeing temps.; liq.

Unisol LAC GN. [ICI Surf. UK] Amine ethoxylate/naphthalene sulfonate condensate; anionic/cationic; leveling agent for applic. of acid, chrome, premetallized, and reactive dyestuffs to wool, silk, nylon; liq.

Unisol MFD. [ICI Surf. UK] Ethoxylate blend; nonionic; dyeing assistant for wool dyeing with metal complex dyes; liq.

Unisol WL. [ICI Surf. UK] Ethoxylated prods./anionic surfactant blend; cationic; low foam leveling agent for dyeing wool with acid, metal complex, chrome, and reactive dyes; liq.

Unital G. [Henkel-Nopco] Depolymerized polypeptide; nonionic; protective colloid; leveling and retarding agent for vat and naphthol dyes; liq.; 50% conc.

V

Valdet 561. [Air Prods./Valchem] Nonylphenol ethoxylate; nonionic; emulsifier, detergent, wetting agent, dispersant for desizing, dyeing, scouring fabrics; resin finish penetrant; emulsifier for wool lubricants; APHA 150 clear to sl. hazy liq., aromatic odor; sol. in water; dens. 8.7 lb/gal; sp.gr. 1.055-1.065; cloud pt. 158-165 (1%); flash pt. 427-445 F; pH 7.0 ± 1.0 (10%); 99% min. solids.

Valdet 4016. [Air Prods./Valchem] Alcohol ethoxylate, straight chain; nonionic; wetting agent; nonrewetting surfactant and emulsion stabilizer used in fluorocarbon finishes; solv. emulsion scouring operations; emulsifier; biodeg.; APHA 60 max. clear to slightly hazy liq.; water-sol.; sp.gr. 1.011; cloud pt. 39–45 C (1%); acid no. 1 max.; pH 7.0 ± 0.5 (10%); 90% act.

Valdet AC-40. [Air Prods./Valchem] Amine condensate; nonionic; detergent for industrial cleaning, textiles, compounding; liq.; 40% conc.

Valdet CC. [Air Prods./Valchem] Amine condensate; anionic; detergent for industrial cleaning, textiles, compounding; dk. amber clear liq.; coconut oil odor; misc. with water; dens. 8.3 lb/gal; sp.gr. 0.990–1.010; 100% act.

Valscour SWR-2. [Air Prods./Valchem] Phosphate ester; anionic; detergent, emulsifier used in textiles; liq.; 23% conc.

Valsof FR-4. [Air Prods./Valchem] Amine condensate; nonionic; softener, lubricant; liq.

Valsof PMS. [Air Prods./Valchem] Amine condensate; nonionic; softener, lubricant; liq.

Value 401. [Marubishi Oil Chem.] Surfactant blend; nonionic; wetting agent, penetrant, emulsifier; for textile scouring; liq.; HLB 10.6; 100% conc.

Value 402. [Marubishi Oil Chem.] Surfactant blend; nonionic; wetting agent, penetrant, emulsifier; for textile scouring; liq.; HLB 11.0; 100% conc.

Value 1209C. [Marubishi Oil Chem.] POE mono cocofatty acid ester; nonionic; emulsifier for general use; solid, liq.; water-sol.; HLB 13.4; 100% conc.

Value 1407. [Marubishi Oil Chem.] POE monooleate; nonionic; emulsifier for general use; liq.; water-disp.; HLB 10.3; 100% conc.

Value 1414. [Marubishi Oil Chem.] POE monooleate; nonionic; emulsifier for general use; liq.; HLB 13.6; 100% conc.

Value 3608. [Marubishi Oil Chem.] POE octyl phenyl ether; nonionic; emulsifier for jute batching emulsion; liq.; HLB 12.6; 100% conc.

Value 3706. [Marubishi Oil Chem.] POE nonyl phenyl ether; nonionic; dispersant, emulsifier, wetting agent, detergent; liq.; water-disp.; HLB 10.9; 100% conc.

Value 3710. [Marubishi Oil Chem.] POE nonyl phenyl ether; nonionic; emulsifier, textile scouring and wetting agent; liq.; water-sol.; HLB 13.3; 100% conc.

Value MS-1. [Marubishi Oil Chem.] Syn. sperm oil sodium sulfate; anionic; softener, emulsifier for veg., animal oil and wax; paste, liq.; 75% conc.

Value W-307. [Marubishi Oil Chem.] Surfactant blend; nonionic; detergent for raw wool scouring; liq.; HLB 12.1; 100% conc.

Valwet 092. [Air Prods./Valchem] Arylakyl sulfonate blend; anionic; wetting agent, penetrant; non-rewetting; liq.; 35% conc.

Vanisol BIS sodico-2 (redesignated Geropon® BIS/SODICO-2). [Rhone-Poulenc Geronazzo]

Vannox AC 2, ACH, ACL. [Nippon Nyukazai] Anionic/nonionic; emulsifier for org. chlorinated insecticides; liq.

Vannox PW. [Nippon Nyukazai] Anionic; wetting agent for wettable powd.; powd.

Vanseal 35. [R.T. Vanderbilt] Sodium cocoyl sarcosinate; anionic; biodeg. industrial grade surfactant with outstanding mildness, lather building, and conditioning props.; compat. with cationics; used for soaps, bath gels, shampoos, shaving creams, dentifrices, textile and leather processing; 24% act.

Vanseal CS. [R.T. Vanderbilt] Cocoyl sarcosine; anionic; biodeg. surfactant, foaming and wetting agent, detergent, foam booster for soaps, bath gels, shampoos, shaving creams, dentifrices, rug shampoos, oven cleaners, dishwash, textile/leather processing; offers tolerance to hard water, mildness; pale yel. liq.; 94% conc.

Vanseal LS. [R.T. Vanderbilt] Lauroyl sarcosine; anionic; biodeg. surfactant, foaming and wetting agent, detergent, foam booster for soaps, bath gels, shampoos, shaving creams, dentifrices, rug shampoos, oven cleaners, dishwash, textile/leather processing; offers tolerance to hard water, mildness; wh. waxy solid; 94% conc.

Vanseal NACS-30. [R.T. Vanderbilt] Sodium cocoyl sarcosinate; anionic; biodeg. surfactant, foaming and wetting agent, detergent, foam booster for soaps, bath gels, shampoos, shaving creams, dentifrices, rug shampoos, oven cleaners, dishwash, textile/leather processing; offers tolerance to hard water, mildness; liq.; 30 ± 1% conc.

Vanseal NALS-30. [R.T. Vanderbilt] Sodium lauroyl sarcosinate; anionic; biodeg. surfactant, foaming and wetting agent, detergent, foam booster for soaps, bath gels, shampoos, shaving creams, dentifrices, rug shampoos, oven cleaners, dishwash, textile/leather processing; offers tolerance to hard water, mildness; colorless liq.; 30 ± 1% conc.

Vanseal NALS-95. [R.T. Vanderbilt] Sodium lauroyl

sarcosinate; anionic; biodeg. surfactant, foaming and wetting agent, detergent, foam booster for soaps, bath gels, shampoos, shaving creams, dentifrices, rug shampoos, oven cleaners, dishwash, textile/leather processing; offers tolerance to hard water, mildness; wh. powd.; 94% conc.

Vanseal OS. [R.T. Vanderbilt] Oleoyl sarcosine; anionic; biodeg. surfactant, foaming and wetting agent, detergent, foam booster for soaps, bath gels, shampoos, shaving creams, dentifrices, rug shampoos, oven cleaners, dishwash, textile/leather processing; offers tolerance to hard water, mildness; yel. liq.; 94% conc.

Varamide® A-2. [Sherex/Div. of Witco] Refined cocamide DEA (2:1); nonionic; thickener, foam stabilizer, and detergent for soaps, hand cleaners, bubble baths, textile applics. as a scouring and fulling agent, and in rug and floor cleaners; clear yel. liq.; water-disp.; dens. 8.5 lb/gal; pH 10.2 (1%); 100% act.

Varamide® A-7. [Sherex/Div. of Witco] Oleamide DEA (2:1); nonionic; rust inhibitor and base for o/w emulsifiers, detergents, anticorrosive cleaners, and thickener for waterless hand cleaners; degreasers; amber clear liq.; dens. 8.2 lb/gal; pH 10.4 (1%); 100% act.

Varamide® A-10. [Sherex/Div. of Witco] Modified cocamide DEA (2:1); nonionic; thickener, foam stabilizer, detergent, rust inhibitor for floor cleaners and general-purpose cleaners; reduces necessary rinsing and increases visc., perfume and dye solubility in anionic base cleaners; suitable for chain belt lubricants and metal-lathe working sol'ns.; clear amber liq.; water-sol.; dens. 8.3 lb/gal; pH 8.8 (1%); 100% act.

Varamide® A-12. [Sherex/Div. of Witco] Modified cocamide DEA (2:1); anionic; see Varamide A-10; A-12 is a higher salt tolerance version designed for use at phosphate levels as high as 10%; clear amber liq.; dens. 8.2 lb/gal; pH 9 (1%); 100% act.

Varamide® A-80. [Sherex/Div. of Witco] Modified coco DEA (2:1); general detergent base which solubilizes quickly and rinses easily; for household and industrial applics.; liq.; 100% solids.

Varamide® A-83. [Sherex/Div. of Witco] Modified coco DEA (2:1); anionic; lower cost version of Varamide A-10 used in industrial, institutional, and household cleaners, wh. sidewall cleaners, wax strippers, degreasers, and for textile scouring and fulling; liq.; dens. 8.3 lb/gal; pH 9 (1%); 100% conc.

Varamide® A-84. [Sherex/Div. of Witco] Modified coco DEA (2:1); general-purpose detergent base which solubilizes quickly and rinses easily; liq.; 100% solids.

Varamide® AC-28. [Sherex/Div. of Witco] Modified mixed DEA; emulsifier, anticorrosive in cutting oils, grinding oils, and metalworking liqs.; liq.; sol. in hydrocarbons, emulsified in water; 100% solids.

Varamide® C-212. [Sherex/Div. of Witco] Cocamide MEA (1:1); CAS 68140-00-1; nonionic; foam booster and stabilizer in aq. systems; degreaser; hair conditioning agent; visc. modifier; for household and industrial applics.; flake; 100% solids.

Varamide® FBR. [Sherex/Div. of Witco] Surfactant blend; anionic/nonionic; conc. for floor and hard surf. cleaners, truckwash; liq.; 100% solids.

Varamide® L-203. [Sherex/Div. of Witco] Lauramide MEA (1:1); CAS 142-78-9; nonionic; foam booster and stabilizer for aq. systems; degreaser; hair conditioner, visc. modifier; for household and industrial applics.; flake; 100% act.

Varamide® MA-1. [Sherex/Div. of Witco] Refined cocamide DEA (1:1); nonionic; foam stabilizer and booster, thickener; basic liq. superamide for shampoos, bubble bath, and dishwasher; low cost equivalent to lauric superamide; does not require melting; gives higher visc. and foam stability than conventional 2:1 alkanolamides in liq. detergent systems; Gardner 3+ max. clear liq.; readily disp. in water; dens. 8.2 lb/gal; pH 9.8 (1%).; 100% act.

Varamide® ML-1. [Sherex/Div. of Witco] Lauramide DEA (1:1); CAS 120-40-1; nonionic; thickener and foam stabilizer for shampoo, bubble bath, and hand laundry detergent; gives the highest visc., foam level, and stability of the superamides in series; wh. wax; water-disp.; dens. 8.1 lb/gal; pH 10.2 (1%); 100% conc.

Varamide® ML-4. [Sherex/Div. of Witco] Lauramide DEA (1:1); CAS 120-40-1; nonionic; detergent, foam stabilizer for industrial cleaning applics.; wh. wax; dens. 8.2 lb/gal; pH 10.2 (1%); 100% conc.

Varamide® T-55. [Sherex/Div. of Witco] Tallow MEA ethoxylate; detergent base for floor and hard-surface cleaners; tolerant to high builder levels and hard water; liq.; 100% conc.

Varine C. [Sherex/Div. of Witco] Coco hydroxyethyl imidazoline; cationic; emulsifier, anticorrosive, raw material; shampoo base, penetrating oils, antistats, corrosion inhibitors, paints, printing inks, textiles, adhesives; Gardner 9 liq.; m.w. 274; disp. in water; sol. in polar solvs. and hydrocarbons; m.p. 31 C; 100% act.

Varine O®. [Sherex/Div. of Witco] Oleyl hydroxyethyl imidazoline; cationic; emulsifier, anticorrosive, raw material; shampoo base, penetrating oils, antistats, corrosion inhibitors, paints, printing inks, textiles, adhesives; Gardner 8 liq.; m.w. 352; sol. in polar solvs. and hydrocarbons; m.p. –10 C; 100% act.

Varine O Acetate. [Sherex/Div. of Witco] Oleic imidazoline acetate; anticorrosive; liq.; 100% solids.

Varine T. [Sherex/Div. of Witco] Tall oil hydroxyethyl imidazoline; anticorrosive for automobile industry; Gardner 8; m.w. 350; sol. in polar solvs. and hydrocarbons; disp. in water; m.p. –10 C; 100% solids, 85% tert. amine.

Varion® 2C. [Sherex/Div. of Witco] Disodium cocoamphodiacetate; amphoteric; high foaming, mild surfactant for shampoos, liq. soaps and body cleansers; Gardner 5 liq.; pH 8.0; 50% solids.

Varion® 2L. [Sherex/Div. of Witco] Disodium lauroamphodiacetate; amphoteric; high foaming, nonirritating surfactant for baby care cosmetics, household and industrial applics.; Gardner 4 liq.; pH 8.0; 50% solids.

Varion® AM-B14. [Sherex/Div. of Witco] Cocamidopropyl betaine; amphoteric; mild foam booster, visc. builder, lime soap dispersant for household, industrial, and cosmetic applics.; liq.; 40% conc.

Varion® AM-KSF-40%. [Sherex/Div. of Witco] Disodium cocoamphodipropionate; amphoteric; salt-free surfactant for industrial, institutional and household cleaners, heavy-duty cleaners, steam cleaners, metal cleaners, textile auxs.; offers coupling ability with foam and detergency in high-electrolyte systems; liq.; 40% conc.

Varion® AM-V. [Sherex/Div. of Witco] Caprylic glycinate; amphoteric; coupling agent for use in low-foam wetting, acidic or basic systems; liq.; 35%

solids.

Varion® CADG-HS. [Sherex/Div. of Witco] Cocamidopropyl betaine; amphoteric; surfactant; detergent; foam booster and visc. modifier for personal care prods., textile aux., heavy-duty detergents; LS is low-salt version; yel. clear liq.; pH 6.5–8.5; 35–37% solids.

Varion® CADG-LS. [Sherex/Div. of Witco] Cocamidopropyl betaine; amphoteric; surfactant; detergent; foam booster and visc. modifier; personal care prods. and heavy-duty detergents; low-salt version; yel. clear liq.; pH 5.0-7.0; 35-37% solids.

Varion® CADG-W. [Sherex/Div. of Witco] Cocamidopropyl betaine; amphoteric; surfactant; detergent; foam booster and visc. modifier; personal care prods. and heavy-duty detergents; liq.; 40% conc.

Varion® CAS. [Sherex/Div. of Witco] Cocamidopropyl hydroxysultaine; amphoteric; emulsifier and foamer for drilling operations, personal care prods., detergents, car care formulations; lime soap dispersant; Gardner 4 max. clear liq.; sol. in water and electrolytes; flash pt. > 200 F (PMCC); pH 6–8; surf. tens. 30.0 dynes/cm (0.1%); 50% solids.

Varion® CAS-W. [Sherex/Div. of Witco] Coco amidopropyl hydroxysultaine; amphoteric; emulsifier and foamer for drilling operations, personal care prods., detergents, car care formulations; lime soap dispersant; liq.; 50% solids.

Varion® CDG. [Sherex/Div. of Witco] Lauryl betaine; amphoteric; foamer, detergent, solubilizer, conditioner used in shampoos, acid cleaners; lime soap dispersant; liq.; sol. in water and electrolytes; 31% act.

Varion® CDG-LS. [Sherex/Div. of Witco] Lauryl betaine; amphoteric; foaming agent in acid systems; low salt version of CDG; liq.; 35% solids.

Varion® HC. [Sherex/Div. of Witco] Lauryl hydroxysultaine; amphoteric; wetting agent, foamer, surfactant, coupler; stable in 40% sodium hydroxide; liq.; 50% solids.

Varion® SDG. [Sherex/Div. of Witco] Stearyl betaine; amphoteric; detergent, softener for fabric and hair applics.; liq.; 50% conc.

Varion® TEG. [Sherex/Div. of Witco] Dihydroxyethyl tallow glycinate; amphoteric; surfactant, min. acid thickener, ingred. for hair shampoos, household and industrial applics., acid cleaners; liq.; 50% conc.

Varion® TEG-40%. [Sherex/Div. of Witco] Dihydroxyethyl tallow glycinate; amphoteric; lower visc. version of Varion TEG; for thickening of hydrochloric and phosphoric acid toilet bowl systems; liq.; 40% solids.

Variquat® 50AC. [Sherex/Div. of Witco] Benzalkonium chloride, IPA; germicidal conc. for disinfection and sanitization; liq.; 50% solids.

Variquat® 50AE. [Sherex/Div. of Witco] Benzalkonium chloride, ethanol; germicidal conc. for disinfection and sanitization; liq.; 50% solids.

Variquat® 50MC. [Sherex/Div. of Witco] Benzalkonium chloride; cationic; germicide, algicide, disinfectant, sanitizer, deodorant; used in pesticides and mfg. of sanitizers; food processing, dairy, restaurant, industrial and household prods.; Gardner 2 max. liq.; m.w. 358; flash pt. (PM) 120 F; 50% act.

Variquat® 50ME. [Sherex/Div. of Witco] Dimethyl alkyl (C12–C16) benzyl ammonium chloride (50%), ethyl alcohol (7.5%) in water; specialty quat., germicide used for disinfection and sanitizing for hospitals, beautician instruments, food processing plants; Gardner 2 max. liq.; m.w. 358; sol. in water or alcohol; sp.gr. 0.96; dens. 8.0 lb/gal; flash pt. 120 F (P-M); pH 5–8; 57.5% act.

Variquat® 60LC. [Sherex/Div. of Witco] Benzalkonium chloride; cationic; swimming pool and water treatment algicide; Gardner 2 max. liq.; m.w. 351; flash pt. (PM) 130 F; 49.5–51.5% quat.

Variquat® 66. [Sherex/Div. of Witco] Ethyl bis (polyethoxy ethanol) tallow ammonium chloride; antistat and degreaser for aq. cleaning systems; liq.; 72% solids.

Variquat® 80AC. [Sherex/Div. of Witco] Benzalkonium chloride, IPA; IPA version of Variquat 80MC; liq.; 80% solids.

Variquat® 80AE. [Sherex/Div. of Witco] Benzalkonium chloride, ethanol; ethanol version of Variquat 80MC; liq.; 80% solids.

Variquat® 80LC. [Sherex/Div. of Witco] Benzyl chloride quat.; algicide conc. for swimming pools; liq.; 80% solids.

Variquat® 80MC. [Sherex/Div. of Witco] Dimethyl alkyl (C12–16) benzyl ammonium chloride; cationic; germicide, disinfectant, deodorant used in pesticides and sanitizers; corrosion inhibitor for water inj. systems; liq.; m.w. 358; flash pt. (PM) 100 F; 80% act.

Variquat® 80ME. [Sherex/Div. of Witco] Dimethyl alkyl (C12–C16) benzyl ammonium chloride; germicidal conc. for disinfection and sanitization; liq.; Gardner 2 max.; m.w. 358; flash pt. 100 F (P-M); 80% solids.

Variquat® 638. [Sherex/Div. of Witco] PEG-2 cocomonium chloride, IPA; cationic; detergent booster, antistat, emulsifier for hard surf. cleaners and other liq. detergents; plating bath foam blanket; Gardner 9 max. liq.; m.w. 353; flash pt. 56 F (PM); 75% solids.

Variquat® B200. [Sherex/Div. of Witco] Benzyltrimethyl ammonium chloride; cationic; dispersant, dye leveler and retarder, emulsifier used in textile industry; APHA 50 max. liq.; pH 6.5–8.0; 60% act.

Variquat® B343. [Sherex/Div. of Witco] Dihydrog. tallow methyl benzyl ammonium chloride; specialty quat.; Gardner 2 max. paste; m.w. 656; flash pt. 84 F (PM); 74–76% quat.

Variquat® B345. [Sherex/Div. of Witco] Dimethyl hydrog. tallow benzyl ammonium chloride; specialty quat.; liq.; Gardner 3 max.; m.w. 417; flash pt. 81 F (PM); 75–77% solids.

Variquat® E290. [Sherex/Div. of Witco] Cetrimonium chloride; cationic; quat. for personal care prods.; Gardner 1 max. clear liq.; 28.5–30.0% in water.

Variquat® K300. [Sherex/Div. of Witco] Dimethyl dicoco ammonium chloride, IPA; cationic; low cloud pt. emulsifier, dispersant; also for laundry detergent-softeners; Gardner 5 max. liq.; m.w. 460; flash pt. 50 F (PM); 75% conc.

Variquat® K375. [Sherex/Div. of Witco] Dicoco dimethyl ammonium chloride in hexylene glycol; specialty quat.; liq.; Gardner 5 max.; m.w. 439; flash pt. > 200 F (P-M); 74–77% quat.

Variquat® K1215. [Sherex/Div. of Witco] PEG-15 cocomonium chloride; specialty quat.; Gardner 11 max. liq.; m.w. 910; flash pt. 258 F (PM); 95% quat. min.

Varisoft® 110. [Sherex/Div. of Witco] Dihydrog. tallowamidoethyl hydroxyethylmonium methosulfate, IPA/water; cationic; fabric softener conc. for home and commercial laundries, textile processing;

Gardner 6 max. paste; flash pt. (PMCC) 99 F; 74–76% solids.

Varisoft® 136-1000P. [Sherex/Div. of Witco] Proprietary; quat. for fabric softeners; Gardner 4 solid; flash pt. (PMCC) > 200 F; 100% solids.

Varisoft® 137. [Sherex/Div. of Witco] Dihydrog. tallow dimonium methosulfate, IPA; cationic; quat. for home and commercial laundry fabric and tissue softeners, debonding agent and antistat; emulsifier; Gardner 5 solid paste; m.w. 656; disp. in water; dens. 7.4 lb/gal; flash pt. 78 F (PMCC); pH 5–8; 89–91% solids.

Varisoft® 222 (75%). [Sherex/Div. of Witco] Methyl bis (tallowamidoethyl) 2-hydroxyethyl ammonium methyl sulfate, IPA; cationic; fabric softener conc. for home and commercial laundries, textile processing; Gardner 6 max. liq.; flash pt. (PMCC) 72 F; 74–76% solids.

Varisoft® 222 (90%). [Sherex/Div. of Witco] Methyl bis (tallowamidoethyl) 2-hydroxyethyl ammonium methyl sulfate, IPA; specialty quat. for rinse cycle fabric softeners; Gardner 6 max. paste; flash pt. (PMCC) 85 F; 89–91% solids.

Varisoft® 222 LM (90%). [Sherex/Div. of Witco] Methylbis (tallow amidoethyl) 2-hydroxy ethyl ammonium methyl sulfate, IPA; softener with exc. handling properties for laundry rinse-cycle softeners; Gardner 6 liq.; flash pt. (PMCC) 85 F; 90% solids.

Varisoft® 222 LT (90%). [Sherex/Div. of Witco] Methyl bis (oleyl amidoethyl) 2-hydroxyethyl ammonium methylsulfate, IPA; cationic; softener conc. for formulation of liq. detergent-softeners or high solids softener prods.; Gardner 8 liq.; flash pt. (PMCC) 75 F; 90% solids.

Varisoft® 238 (90%). [Sherex/Div. of Witco] Methyl bis (tallowamidoethyl) 2-hydroxypropyl ammonium methyl sulfate; specialty quat.; Gardner 6 max. liq.; flash pt. 85 F (P-M); 89–91% solids.

Varisoft® 445. [Sherex/Div. of Witco] Methyl (1) hydrog. tallow amidoethyl (2) hydrog. tallow imidazolinium methosulfate; fabric softener conc. for home and commercial laundries, textile processing; Gardner 7 paste; flash pt. (PMCC) 72 F; 75% solids.

Varisoft® 475. [Sherex/Div. of Witco] Quaternium-27, IPA; CAS 86088-85-9; cationic; fabric softener conc. for home and commercial laundries, textile processing; Gardner 5 max. liq.; flash pt. 72 F (PM); 75–77% solids.

Varisoft® 920. [Sherex/Div. of Witco] Tallow bis hydroxyethyl methyl ammonium chloride; cationic; softener conc. for anionic/nonionic based detergent/softeners; Gardner 6 soft paste; m.w. 400; flash pt. (PMCC) 66 F; 75% act.

Varisoft® 950. [Sherex/Div. of Witco] Quat.; for nonionic-based laundry detergent-softeners; soft paste.

Varisoft® 3690. [Sherex/Div. of Witco] Methyl (1) oleylamidoethyl (2) oleyl imidazolinium methosulfate, IPA; cationic; quat. for laundry detergent-softeners; Gardner 7 max. liq.; m.w. 723; flash pt. (PMCC) 66 F; 75% solids.

Varisoft® 3690N (90%). [Sherex/Div. of Witco] Methyl-1 oleyl amidoethyl 2-oleyl-imidazolinium methyl sulfate; fabric softener; Gardner 6 liq.; m.w. 723; flash pt. (PMCC) > 200 F; 98% solids.

Varisoft® DS-100. [Sherex/Div. of Witco] Proprietary; quat. for fabric softeners; Gardner 4 solid; flash pt. (PMCC) > 200 F; 100% solids.

Varonic® 32-E20. [Sherex/Div. of Witco] Oleth-20; CAS 9004-98-2; nonionic; emulsifier, stabilizer; paste; HLB 11.3; 100% solids.

Varonic® 2271. [Sherex/Div. of Witco] PEG-8 tallowamine; dye leveler; improves wettability and reduces dye affinity at the surf., permitting more even migration onto substrates such as hair; coemulsifier and stabilizer for emulsion systems; neutralizer, plasticizer; antistat in acidic systems; liq.; HLB 10.0; 99% solids.

Varonic® DM55. [Sherex/Div. of Witco] Methyl capped glycol ether; solv. with low toxicity and exc. grease cutting properties for surf. cleaners; liq.; water-emulsifiable; 100% solids.

Varonic® K202. [Sherex/Div. of Witco] PEG-2 cocamine; CAS 61791-14-8; nonionic; emulsifier, antistat; corrosion inhibitor in metal finishing (e.g. as cutting oil additives); detergent, antifouling, antistalling, and deicing agent in gasoline; also in textile lubricants, oil field emulsification; Gardner 2 liq.; sol. in IPA, benzene, Stod., CCl_4, MEK; insol. in water; sp.gr. 0.916; HLB 6.2; 100% solids.

Varonic® K202 SF. [Sherex/Div. of Witco] PEG-2 cocamine; CAS 61791-14-8; nonionic; dye leveler; improves wettability and reduces dye affinity at the surf., permitting more even migration onto substrates such as hair; coemulsifier and stabilizer for emulsion systems; neutralizer, plasticizer; antistat in acidic systems; liq.; HLB 6.2; 99% solids.

Varonic® K205. [Sherex/Div. of Witco] PEG-5 cocamine; CAS 61791-14-8; nonionic; corrosion inhibitor, antifouling agent in gasoline; antistat, lubricant in wool spinning; emulsifier and leveling agent in textile dyeing; insecticide and herbicide systems; metal finishing; Gardner 7 liq.; sol. in IPA, benzene, Stod., CCl_4, MEK; sp.gr. 0.977; HLB 11.0; 100% solids.

Varonic® K205 SF. [Sherex/Div. of Witco] PEG-5 cocamine; CAS 61791-14-8; nonionic; dye leveler; improves wettability and reduces dye affinity at the surf., permitting more even migration onto substrates such as hair; coemulsifier and stabilizer for emulsion systems; neutralizer, plasticizer; antistat in acidic systems; liq.; HLB 11.0; 99% solids.

Varonic® K210. [Sherex/Div. of Witco] PEG-10 cocamine; CAS 61791-14-8; nonionic; dye leveling agent, dispersant in paper industry; wetting and spreading agent and emulsifier in insecticides and herbicides; tanning of leather and furs; wetting and redeposition aids in fur treating; Gardner 7 liq.; sol. in water, IPA, benzene, CCl_4; sp.gr. 0.995; HLB 13.8; 100% solids.

Varonic® K210 SF. [Sherex/Div. of Witco] PEG-10 cocamine; CAS 61791-14-8; nonionic; dye leveler; improves wettability and reduces dye affinity at the surf., permitting more even migration onto substrates such as hair; coemulsifier and stabilizer for emulsion systems; neutralizer, plasticizer; antistat in acidic systems; liq.; HLB 13.8; 99% solids.

Varonic® K215. [Sherex/Div. of Witco] PEG-15 cocamine; CAS 61791-14-8; nonionic; wetting agent and redeposition aid in fur treatment; textile additive in spinning bath to reduce jet clogging and disperse sulfur/zinc sulfide particles to prevent cratering; in petrol. industry for corrosion resistance; Gardner 7 liq.; sol. in water, IPA, benzene, CCl_4, MEK; sp.gr. 1.040; HLB 15.4; 100% solids.

Varonic® K215 LC. [Sherex/Div. of Witco] PEG-15 cocamine; CAS 61791-14-8; low-color version of Varonic K215; wetting agent during liming stage in tanning of leathers and furs; textile aid reducing jet clogging in spinning bath and dispersing sulfur/zinc sulfide particles preventing cratering; Gardner 3 liq.; sol. in water, IPA, benzene, CCl_4, MEK; sp.gr. 1.040; HLB 15.4; 100% solids.

Varonic® K215 SF. [Sherex/Div. of Witco] PEG-15 cocamine; CAS 61791-14-8; nonionic; dye leveler; improves wettability and reduces dye affinity at the surf., permitting more even migration onto substrates such as hair; coemulsifier and stabilizer for emulsion systems; neutralizer, plasticizer; antistat in acidic systems; liq.; HLB 15.4; 99% solids.

Varonic® LI-42. [Sherex/Div. of Witco] PEG-20 glyceryl stearate; nonionic; low-irritation detergent, emulsifier, lubricant, solubilizer for household and industrial applics., personal care prods.; Gardner 2 paste; HLB 13; m.p. 27 C; surf. tens. 44 dynes/cm (1%); 100% solids.

Varonic® LI-48. [Sherex/Div. of Witco] PEG-80 glyceryl tallowate; nonionic; emulsifier, solubilizer, thickener, dispersant, antiirritant surfactant used in household and industrial applics., personal care prods.; Gardner 2 hard solid; sol. in water, methanol, ethanol, IPA, acetone, and butyl Cellosolve; HLB 18.0; m.p. 41 C; pH 7.0; surf. tens. 49.5 dynes/cm; 100% conc.

Varonic® LI-63. [Sherex/Div. of Witco] PEG-30 glyceryl cocoate; nonionic; low-irritation detergent, emulsifier, solubilizer for soaps, specialized lubricants, personal care prods.; Gardner 2 paste; HLB 15.9; m.p. 27 C; surf. tens. 40 dynes/cm (1%); 100% solids.

Varonic® LI-67. [Sherex/Div. of Witco] PEG-80 glyceryl cocoate; nonionic; low-irritation detergent, emulsifier, solubilizer for personal care prods.; liq. soaps gelling agent; Gardner 2 solid; HLB 18; m.p. 42 C; surf. tens. 47 dynes/cm (1%); 100% solids.

Varonic® LI-67-75%. [Sherex/Div. of Witco] PEG-80 glyceryl cocoate; nonionic; low-irritation detergent, emulsifier, solubilizer for personal care prods., household and industrial applics.; liq.; HLB 18.0; 75% solids.

Varonic® MT 65. [Sherex/Div. of Witco] Methyl capped alkoxylated fatty alcohol; nonionic; low-foam detergent, wetting agent, textile spinning lubricant; Gardner 2 clear liq.; sol. in polar and nonpolar solvs.; pH 3.5 (10%); surf. tens. 32 dynes/cm; 100% solids.

Varonic® Q202. [Sherex/Div. of Witco] PEG-2 oleamine; anticorrosive emulsifier for metalworking, grinding oils; liq.; HLB 4.7; 100% solids.

Varonic® Q202 SF. [Sherex/Div. of Witco] PEG-2 oleamine; dye leveler; improves wettability and reduces dye affinity at the surf., permitting more even migration onto substrates such as hair; coemulsifier and stabilizer for emulsion systems; neutralizer, plasticizer; antistat in acidic systems; liq.; HLB 4.7; 99% solids.

Varonic® Q205 SF. [Sherex/Div. of Witco] PEG-5 oleamine; dye leveler; improves wettability and reduces dye affinity at the surf., permitting more even migration onto substrates such as hair; coemulsifier and stabilizer for emulsion systems; neutralizer, plasticizer; antistat in acidic systems; liq.; HLB 8.4; 99% solids.

Varonic® S202. [Sherex/Div. of Witco] PEG-2 stearamine; polymer additive and emulsifier; solid; HLB 5.0; 100% solids.

Varonic® S202 SF. [Sherex/Div. of Witco] PEG-2 stearamine; dye leveler; improves wettability and reduces dye affinity at the surf., permitting more even migration onto substrates such as hair; coemulsifier and stabilizer for emulsion systems; neutralizer, plasticizer; antistat in acidic systems; paste; HLB 5.0; 99% solids.

Varonic® T202. [Sherex/Div. of Witco] PEG-2 tallowamine; nonionic; lubricant, softener, scouring aid, dye leveler and antistat for textiles; in syn. latex paints; emulsifier for latex, dyes, and oils; dispersant; acid cleaners; process modifier in polymer industry; raw material for quat. and amphoteric surfactants; Gardner 2 semiliq.; sol. in IPA, benzene, Stod., CCl_4; insol. in water; sp.gr. 0.915; HLB 5.1; 100% solids.

Varonic® T202 SF. [Sherex/Div. of Witco] PEG-2 tallowamine; dye leveler; improves wettability and reduces dye affinity at the surf., permitting more even migration onto substrates such as hair; coemulsifier and stabilizer for emulsion systems; neutralizer, plasticizer; antistat in acidic systems; paste; HLB 5.1; 99% solids.

Varonic® T202 SR. [Sherex/Div. of Witco] PEG-2 tallowamine; surfactant for optimum visc. control in acid cleaners; paste; HLB 5.1; 100% solids.

Varonic® T205. [Sherex/Div. of Witco] PEG-5 tallowamine; nonionic; lubricant, softener, and antistat for syn. fibers; antistat/lubricant in wool spinning; emulsifier/leveling agent in water/oil dye systems; emulsifier and spreading-wetting agent in insecticides and herbicides; process modifier for polymers; Gardner 7 liq.; sol. in IPA, benzene, Stod., CCl_4, MEK; emul. in water; sp.gr. 0.966; HLB 9.2; 100% solids.

Varonic® T205 SF. [Sherex/Div. of Witco] PEG-5 tallowamine; dye leveler; improves wettability and reduces dye affinity at the surf., permitting more even migration onto substrates such as hair; coemulsifier and stabilizer for emulsion systems; neutralizer, plasticizer; antistat in acidic systems; liq.; HLB 9.2; 99% solids.

Varonic® T210. [Sherex/Div. of Witco] PEG-10 tallowamine; nonionic; lubricant, scouring aid, dye leveler for textile industry; process modifier in polymer industry; raw material for quat. and amphoteric surfactants; liq.; HLB 12.6; 100% solids.

Varonic® T210 SF. [Sherex/Div. of Witco] PEG-10 tallowamine; dye leveler; improves wettability and reduces dye affinity at the surf., permitting more even migration onto substrates such as hair; coemulsifier and stabilizer for emulsion systems; neutralizer, plasticizer; antistat in acidic systems; liq.; HLB 12.6; 99% solids.

Varonic® T215. [Sherex/Div. of Witco] PEG-15 tallowamine; nonionic; lubricant, softener, and antistat for textiles; wetting aid/emulsifier in tanning; in petrol. industry used with caustic to break o/w crude oil emulsions and to impart corrosion resistance; raw material for quat. and amphoteric surfactants; Gardner 7 liq.; sol. in water, IPA, benzene, CCl_4, MEK; sp.gr 1.029; HLB 14.4; 100% conc.

Varonic® T215LC. [Sherex/Div. of Witco] PEG-15 tallowamine; low-color version of Varonic T215; liq.; HLB 14.4; 100% solids.

Varonic® T215 SF. [Sherex/Div. of Witco] PEG-15 tallowamine; dye leveler; improves wettability and reduces dye affinity at the surf., permitting more even migration onto substrates such as hair; coemulsifier and stabilizer for emulsion systems; neutralizer, plasticizer; antistat in acidic systems; liq.; HLB 14.4; 99% solids.

Varox® 185E. [Sherex/Div. of Witco] Dihydroxyethyl C12-15 alkoxypropylamine oxide; nonionic; detergent, foam booster/stabilizer for anionic surfactants used in liq. dishwashing, shampoos and fabric detergents; Gardner 1 liq.; pH 4.5–5.5; surf. tens. 31.5 dynes/cm; 39–43% amine oxide.

Varox®365. [Sherex/Div. of Witco] Lauramine oxide; nonionic; detergent, foam booster/stabilizer for anionic surfactants for shampoo and detergent systems; hypochlorite-stable; Gardner 1 liq.; 30% solids.

Varox®375. [Sherex/Div. of Witco] Lauramine oxide; hypochlorite-stable foam booster/stabilizer; liq.; 40% solids.

Varox® 743. [Sherex/Div. of Witco] Cocodihydroxyethylamine oxide; nonionic; detergent, foam booster and stabilizer; liq.; 50% conc.

Varox® 1770. [Sherex/Div. of Witco] Cocamidopropylamine oxide; nonionic; mild detergent, foam booster for liq. soaps, shampoos; liq.; 35% act.

Varsulf® MS30. [Sherex/Div. of Witco] Surfactant blend; foaming, dry-residue rug and upholstery shampoo conc.; liq.; 30% solids.

Varsulf® NOS-25. [Sherex/Div. of Witco] Sodium alkylphenol polyglycol ether sulfate; anionic; emulsifier for emulsion polymerization; lt. yel. clear liq.; pH 7.0 (1%); 35% conc.

Varsulf® S-1333. [Sherex/Div. of Witco] Disodium ricinoleamido MEA-sulfosuccinate aq. disp.; anionic; detergent, refatting agent used in dishwash, liq. soaps, and personal care prods.; anti-irritant for other surfactants; pale yel. liq.; dens. 9.4 lb/gal; visc. 300 cps; pH 6.5–7.5; surf. tens. 28.8 dynes/cm; 40% solids.

Varsulf® SBF-12. [Sherex/Div. of Witco] Lauryl alcohol sulfosuccinate; detergent for fine fabric wash systems; paste; 40% solids.

Varsulf®SBFA-30. [Sherex/Div. of Witco] Disodium laureth sulfosuccinate aq. disp.; anionic; detergent, refatting agent used in dishwash, fine fabric wash, liq. soaps, and personal care prods.; wh.-pale yel. liq.; dens. 9.2 lb/gal; visc. 200 cps; pH 6.5–7.5; surf. tens. 21.7 dynes/cm; 40% solids.

Varsulf®SBL-203. [Sherex/Div. of Witco] Disodium lauramido MEA-sulfosuccinate aq. disp.; anionic; detergent, refatting agent used in dishwash, lt. duty detergents, personal care prods., rug and upholstery shampoos; Gardner 3 max. liq.; dens. 9.5 lb/gal; visc. 100–500 cps; pH 6.5–7.5; surf. tens. 24.3 dynes/cm; 39.0% min. solids.

Velvetex®610L. [Henkel Canada] Sodium lauriminodipropionate; amphoteric; surfactant used in personal care prods.; sec. emulsifier with nonionics; amber clear liq.; sol. in water, acid or alkali systems, polar solvs.; sp.gr. 1.038; visc. < 100 cps; pH 7.0–8.0 (1%); 29.0–31.0% solids.

Velvetex® AB-45. [Henkel/Cospha; Henkel Canada] Coco-betaine; amphoteric; surfactant, conditioner, emulsifier, solubilizer used in industrial use, liq. detergents, cleansing emulsions, personal care prods.; visc. builder, gelling agent; lime soap dispersant; frothing agent; Gardner 2 clear liq.; water-sol.; sp.gr. 1.03; visc. < 100 cps; cloud pt. < 0 C; pH 6.5–8.5 (10%); 43–45% solids.

Velvetex® BA. [Henkel Canada] Cocamidopropyl betaine; amphoteric; surfactant used in foam drilling and blanketing; air-inhibiting agent for cement gypsum board and cleaners for oily surfaces; used in personal care prods.; liq.; dens. 8.2 lb/gal; 60% conc. in alcohol.

Velvetex® BA-35. [Henkel/Cospha; Henkel Canada] Cocamidopropyl betaine; amphoteric; dispersant, foaming and wetting agent, antistat for household and industrial detergents; lime soap dispersant; biodeg.; stable in strong acid and alkaline sol'n.; clear yel. liq.; water-sol.; sp.gr. 1.05; visc. < 100 cps; cloud pt. < 0 C; 34–37% solids.

Velvetex®OLB-50. [Henkel/Cospha; Henkel Canada] Oleyl betaine; amphoteric; mild, foaming detergent, visc. builder, conditioner used in personal care prods., mfg. of fibers and cutting oils; hair conditioner;substantive to hair and skin; amber translucent gel; water-sol.; sp.gr. 0.953 (60 C); pH 6.0–8.0 (10%); 48.0–52.0% solids.

Versilan MX123. [Harcros UK] Alcohol ethoxylate, modified; nonionic; moderate foaming, good wetting surfactant used in highly built liqs., bottle washing, in-place cleaning, spray metal cleaning, automatic dishwashing, descalers, concrete and aluminum cleaners, derusting agents; straw clear visc. liq., faint odor; sol. in electrolytes, water; sp.gr. 1.183; visc. 3023 cs; pour pt. 6 C; flash pt. (COC) > 250 F; pH 2.2 (1% aq.); surf. tens. 33.8 dynes/cm (0.1%); corrosive org. acid; toxicology: extremely irritating to eyes and skin; 100% act.

Versilan MX134. [Harcros UK] EO/PO copolymer, modified; nonionic; biodeg. moderate foaming, good wetting surfactant for highly built liqs., acid cleaners, steam cleaning, vehicle cleaning, general purpose floor and wall cleaners, descalers, concrete and aluminum cleaners, derusting agents; straw clear visc. liq., faint odor; sol. in electrolytes, water; sp.gr. 1.170; visc. 1170 cs; pour pt. 3 C; flash pt. (PMCC) > 150 C; pH 2.2 (1% aq.); surf. tens. 28.1 dynes/cm (0.1%); corrosive org. acid; toxicology: extremely irritating to eyes and skin; 100% act.

Versilan MX244. [Harcros UK] Amphoteric; foaming agent, detergent for industrial cleaning, degreasers, decarbonizers, dri-foam carpet shampoos, fire fighting foams, foamed concrete, foamed latex carpet backing, plasterboard mfg.; amber clear liq., faint odor; sol. in water; sp.gr. 1.030; visc. 45 cs; pour pt. -5 C; flash pt. (COC) > 100 C; pH 7.5 (1% aq.); 35% solids in water.

Versilan MX332. [Harcros UK] Surfactant blend; anionic/nonionic; biodeg. surfactant for built liq. cleaners, highly alkaline and acid cleaning concs.; straw clear visc. liq., faint odor; sol. in water; high sol. in high electrolyte concs.; sp.gr 1.270; visc. 9920 cs; pour pt. < 0 C; flash pt. (PMCC) > 150 C; pH 2.2 (1% aq.); surf. tens. 32 dynes/cm (0.1%); 100% act.

Verv®. [Am. Ingredients/Patco] Calcium stearoyl-2-lactylate; anionic; starch and protein complexing agent, softener for use in yeast-leavened bakery prods.; conditioning agent in dehydrated potatoes; cream powd.; mild caramel odor; slightly tart taste; acid no. 50–86; ester no. 125–164; 100% conc.

Victawet® 12. [Akzo] EO reaction prod. with 2-ethyl hexanol and P2O5; nonionic; wetting agent, penetrant, dispersant, stabilizer; for pkg. dyeing of nylon, acid-type cleaners, emulsion polymerization, starch coatings; APHA 500 max. liq.; sol. in alcohol, acetone, toluene; disp. in water; dens. 9.34 lb/gal; sp.gr. 1.121; visc. 91.2 cs (100 F); b.p. 325 F; flash pt. 250 F (COC); pour pt. –48 F; surf. tens. 29.4 dynes/cm2 (0.2%); pH 7.0–7.4 (0.5%); 100% act.

Victawet®35B. [Akzo] 2-Ethyl hexanol P2O5, NaOH reaction prod.; anionic; wetting agent, penetrant, dispersant, stabilizer; surfactant; lt. tan to wh. soft paste; sol. in water, ethylene glycol; dens. 9.7 lb/gal; sp.gr. 1.17; visc. > 100,000 cps; b.p. 200 F; flash pt. 200 F (COC); surf. tens. 26.2 dynes/cm2; pH 6.9–7.4 (0.5%); 68–72% act.

Victawet® 58B. [Akzo] Capryl alcohol, P_2O_5, NaOH reaction prod.; anionic; wetting agent, penetrant, dispersant, stabilizer, surfactant; lt.tan to wh. soft paste; sol. in water, ethylene glycol; dens. 9.77 lb/gal; sp.gr. 1.121; visc. > 100,000 cps; flash pt. 180 F (COC); surf. tens. 29.0 dynes/cm2; pH 6.8–7.3 (5%); 68–72% act.

Vifcoll CCN-40, CCN-40 Powd. [Nikko Chem. Co. Ltd.] N-Cocoyl collagen peptide, sodium salt; detergent, emulsifier used in personal care prods., pharmaceuticals, food industry, and household cleaning prods.; pale yel. liq. and wh. or pale yel. powd. resp.; m.w. 600; pH 5.5–7.5 (both, 3% aq. for powd.); 30% act. in water (CCN-40).

Vikolev. [Vikon] Proprietary; dye leveler for polyamide, polyacrylamide, and polyester fibers.

Vikomul. [Vikon] Emulsifier/detergent blend; emulsifier for waxes and oils, starch and size removal in 1-4% NaOH; liq.; 40-85% conc.

Vikosperse KDS. [Vikon] Sulfonated naphthalene formaldehyde condensate; anionic; low dusting, low foaming dispersant for disperse dyes; 95% act.

Viscasil®. [GE Silicones] Dimethicone; nonionic; defoamer, release agent, emollient in cosmetics, polishes, paint additives, and mechanical devices; lubricant in rubber or plastic-to-metal applics.; suggested for auto polish, skin care, hair care, textile softeners, antifoams for petrol. refining,; rubber and plastic mold release, film modifier in coatings, damping in mechanical/elec. applics.; water-wh. clear, oily fluid; sp.gr. 0.975–0.980; visc. 5000, 10,000, 12,500, 30,000, 60,000, 100,000, 300,000, and 600,000 cstk grades; pour pt. –58 to –25 F; flash pt. 500 F; ref. index 1.4035; surf. tens. 21.1–21.3 dynes/cm; conduct. 0.090–0.092 Btu/h-°F ft2/ft; sp. heat 0.36 Btu/lb/°F; dissip. factor 0.0001; dielec. str. 35.0 kV; dielec. const. 2.75; vol. resist. 1 ¥ 1014 ohm-cm; 100% act.

Vismul ET-20. [Toho Chem. Industry] Blend; nonionic; emulsifier, thickener in textile printing; liq.; 55% conc.

Vismul ET-55. [Toho Chem. Industry] Blend; nonionic; emulsifier, thickener in textile printing; liq.; 45% conc.

Vista C-550, -560. [Vista] Sodium dodecylbenzene sulfonate, straight chain; anionic; detergent, foamer used in liq. detergent compds.; liq.; 50 and 60% conc. resp.

Vista S-697. [Vista] Linear alkylbenzene sulfonic acid; intermediate for mfg. surfactants and emulsifier systems; dens. 8.6 lb/gal; visc. 523 cSt (100 F); toxicology: corrosive to skin and eyes; 96% act.

Vista SA-597. [Vista] Straight chain dodecylbenzene sulfonic acid; anionic; detergent intermediate; liq.; 96% conc.

Vitexol® K. [BASF] Emulsified mixt. of aliphatic hydroxyl compds. and a neutral phosphoric acid ester; defoamer for dyeing of cotton/polyester knitgoods in jet dyeing machines; lt. amber liq., char. odor; toxicology: avoid breathing vapors from hot liquors; use gloves and goggles handling conc. prod.

Volpo 25 D 3. [Croda Chem. Ltd.] PEG-3 C12-15 ether; CAS 68131-39-5; nonionic; emulsifier, dispersant, wetting agent, gelling agent, scouring and solubilizing agent for industrial and cosmetic applics.; liq.; 97% conc.

Volpo 25 D 5. [Croda Chem. Ltd.] PEG-5 C12-15 ether; CAS 68131-39-5; nonionic; emulsifier, dispersant, wetting agent, gelling agent, scouring and solubilizing agent for industrial and cosmetic applics.; liq.; 97% conc.

Volpo 25 D 10. [Croda Chem. Ltd.] PEG-10 C12-15 ether; CAS 68131-39-5; nonionic; emulsifier, dispersant, wetting agent, gelling agent, scouring and solubilizing agent for industrial and cosmetic applics.; paste; 97% conc.

Volpo 25 D 15. [Croda Chem. Ltd.] PEG-15 C12-15 ether; CAS 68131-39-5; nonionic; emulsifier, dispersant, wetting agent, gelling agent, scouring and solubilizing agent for industrial and cosmetic applics.; solid; 97% conc.

Volpo 25 D 20. [Croda Chem. Ltd.] PEG-20 C12-15 ether; CAS 68131-39-5; nonionic; emulsifier, dispersant, wetting agent, gelling agent, scouring and solubilizing agent for industrial and cosmetic applics.; solid; 97% conc.

Volpo 3. [Croda Inc.] Oleth-3; CAS 9004-98-2; 52581-71-2; nonionic; emollient, lubricant, emulsifier, solubilizer for cosmetic applic.; emulsifier in astringent creams and lotions; clear gel formation; superfatting in shampoos, foaming bath preparations, and Carbopol gels; used in cold waves, depilatories, and hair straighteners; solubilizer for bromo acids in lipsticks and liq. rouge; spreading agent for bath oils; hazy liq.; sol. in alcohols, glycols, ketones, and chlorinated and aromatic solvs., min. oil, and nonpolar oils; insol. in water; HLB 6.6; acid no. 2.0 max.; iodine no. 57-62; hyd. no. 135-150; pH 5–7 (3% aq.).; usage level: 0.5-5%.

Volpo CS-3. [Croda Chem. Ltd.] Ceteareth-3; CAS 68439-49-6; nonionic; emulsifier, dispersant, wetting agent, gellant, solubilizer for industrial and cosmetic applics.; solid; sol. in ethanol, kerosene, trichloroethylene, oleic acid, oleyl alcohol; HLB 7.3; pH 6.0-7.5 (3%); 97% conc.

Volpo CS-5. [Croda Chem. Ltd.] Ceteareth-5; CAS 68439-49-6; nonionic; emulsifier, dispersant, wetting agent, gellant, solubilizer for industrial and cosmetic applics.; off wh. soft waxy solid; sol. in ethanol, trichloroethylene, oleic acid; HLB 9.5; surf. tens. 33.0 dynes/cm (0.1% aq.); pH 6.0-7.5 (3%); 97% conc.

Volpo CS-10. [Croda Chem. Ltd.] Ceteareth-10; CAS 68439-49-6; nonionic; emulsifier, dispersant, wetting agent, gellant, solubilizer for industrial and cosmetic applics.; off wh. soft waxy solid; sol. in ethanol, trichloroethylene, oleic acid; HLB 12.9; cloud pt. 68 C (1% aq.); surf. tens. 34.5 dynes/cm (0.1% aq.); pH 6.0-7.5 (3%); 97% conc.

Volpo CS-15. [Croda Chem. Ltd.] Ceteareth-15; CAS 68439-49-6; nonionic; emulsifier, dispersant, wetting agent, gellant, solubilizer for industrial and cosmetic applics.; off wh. waxy solid; sol. in ethanol, trichloroethylene, oleic acid; HLB 14.6; cloud pt. 94 C (1% aq.); surf. tens. 35.5 dynes/cm (0.1% aq.); pH 6.0-7.5 (3%); 97% conc.

Volpo CS-20. [Croda Chem. Ltd.] Ceteareth-20; CAS 68439-49-6; nonionic; emulsifier, dispersant, wetting agent, gellant, solubilizer for industrial and cosmetic applics.; off wh. hard waxy solid; sol. in water, ethanol, trichloroethylene, oleic acid; HLB 15.6; cloud pt. 78 C (1% aq.); surf. tens. 41.5 dynes/cm (0.1% aq.); pH 6.0-7.5 (3%); 97% conc.

Volpo L4. [Croda Chem. Ltd.] Laureth-4; CAS 9002-92-0; nonionic; general purpose emulsifier and dispersant; liq.; 97% conc.

Volpo L23. [Croda Chem. Ltd.] Laureth-23; CAS 9002-92-0; nonionic; o/w emulsifier and solubilizer for cosmetics, pharmaceuticals, and household prods.; solid; 97% conc.

Volpo N3. [Croda Chem. Ltd.] Oleth-3, distilled; CAS 9004-98-2; nonionic; emulsifier, dispersant, wetting agent, gelling agent, scouring and solubilizing agent for industrial and cosmetic applics.; liq.; HLB 6.6; 97% conc.

Volpo N5. [Croda Chem. Ltd.] Oleth-5, distilled; CAS 9004-98-2; nonionic; emulsifier, dispersant, wetting agent, gelling agent, scouring and solubilizing agent for industrial and cosmetic applics.; liq.; HLB 9.0; 97% conc.

Volpo N10. [Croda Chem. Ltd.] Oleth-10, distilled; CAS 9004-98-2; nonionic; emulsifier, dispersant, wetting agent, gelling agent, scouring and solubilizing agent for industrial and cosmetic applics.; paste; HLB 12.4; 97% conc.

Volpo N15. [Croda Chem. Ltd.] Oleth-15, distilled; CAS 9004-98-2; nonionic; emulsifier, dispersant, wetting agent, gelling agent, scouring and solubilizing agent for industrial and cosmetic applics.; paste; HLB 14.2; 97% conc.

Volpo N20. [Croda Chem. Ltd.] Oleth-20, distilled; CAS 9004-98-2; nonionic; emulsifier, dispersant, wetting agent, gelling agent, scouring and solubilizing agent for industrial and cosmetic applics.; solid; HLB 15.5; 97% conc.

Volpo O3. [Croda Chem. Ltd.] Oleth-3; CAS 9004-98-2; nonionic; emulsifier, dispersant, wetting agent, gelling agent, scouring and solubilizing agent for industrial and cosmetic applics.; liq.; 97% conc.

Volpo O5. [Croda Chem. Ltd.] Oleth-5; CAS 9004-98-2; nonionic; emulsifier, dispersant, wetting agent, gelling agent, scouring and solubilizing agent for industrial and cosmetic applics.; liq.; 97% conc.

Volpo O10. [Croda Chem. Ltd.] Oleth-10; CAS 9004-98-2; nonionic; emulsifier, dispersant, wetting agent, gelling agent, scouring and solubilizing agent for industrial and cosmetic applics.; paste; 97% conc.

Volpo O15. [Croda Chem. Ltd.] Oleth-15; CAS 9004-98-2; nonionic; emulsifier, dispersant, wetting agent, gelling agent, scouring and solubilizing agent for industrial and cosmetic applics.; paste; 97% conc.

Volpo O20. [Croda Chem. Ltd.] Oleth-20; CAS 9004-98-2; nonionic; emulsifier, dispersant, wetting agent, gelling agent, scouring and solubilizing agent for industrial and cosmetic applics.; solid; 97% conc.

Volpo S-2. [Croda Inc.] Steareth-2; CAS 9005-00-9; nonionic; emulsifier for o/w systems, cosmetics; stable over wide pH range; wh. soft solid; sol. in alcohol, glycols, ketones, most chlorinated and aromatic solvs., keroesene, min. oil; cloud pt. < 55 C (1% aq.); HLB 4.9; acid no. 1.0 max.; hyd. no. 150-170; pH 6.0-7.0 (3%); usage level: 0.5-5%; 100% conc.

Volpo T-3. [Croda Chem. Ltd.] Trideceth-3; CAS 24938-91-8; nonionic; emulsifier, dispersant, wetting agent, gelling agent, scouring and solubilizing agent for industrial and cosmetic applics.; liq.; 97% conc.

Volpo T-5. [Croda Chem. Ltd.] Trideceth-5; CAS 24938-91-8; nonionic; emulsifier, dispersant, wetting agent, gelling agent, scouring and solubilizing agent for industrial and cosmetic applics.; liq.; 97% conc.

Volpo T-10. [Croda Chem. Ltd.] Trideceth-10; CAS 24938-91-8; nonionic; emulsifier, dispersant, wetting agent, gelling agent, scouring and solubilizing agent for industrial and cosmetic applics.; paste; 97% conc.

Volpo T-15. [Croda Chem. Ltd.] Trideceth-15; CAS 24938-91-8; nonionic; emulsifier, dispersant, wetting agent, gelling agent, scouring and solubilizing agent for industrial and cosmetic applics.; paste; 97% conc.

Volpo T-20. [Croda Chem. Ltd.] Trideceth-20; CAS 24938-91-8; nonionic; emulsifier, dispersant, wetting agent, gelling agent, scouring and solubilizing agent for industrial and cosmetic applics.; solid; 97% conc.

Vorite 105. [CasChem] Polymerized castor oil; pigment wetting/dispersing agent; plasticizer for resins, gums, polymers; lubricant, penetrant; coupling solv.; adhesion promoter; for cellulose lacquers, inks, adhesives, industrial lubricants, polishes, caulks, leather dressing, hydraulic fluids; rubber compding., gasket cement; Gardner 4 color; sp.gr. 0.975; visc. 26 stokes; pour pt. -5 F; acid no. 2; iodine no. 85; sapon. no. 170; hyd. no. 130.

Vorite 110. [CasChem] Polymerized castor oil; see Vorite 105; Gardner 3 color; sp.gr. 0.990; visc. 115 stokes; pour pt. 15 F; acid no. 2; iodine no. 82; sapon. no. 166; hyd. no. 102.

Vorite 115. [CasChem] Polymerized castor oil; see Vorite 105; Gardner 3 color; sp.gr. 0.995; visc. 192 stokes; pour pt. 20 F; acid no. 2; iodine no. 85; sapon. no. 165; hyd. no. 93.

Vorite 120. [CasChem] Polymerized castor oil; see Vorite 105; Gardner 4 color; sp.gr. 1.001; visc. 700 stokes; pour pt. 50 F; acid no. 2; iodine no. 82; sapon. no. 160; hyd. no. 78.

Vorite 125. [CasChem] Polymerized castor oil; see Vorite 105; Gardner 4 color; sp.gr. 1.007; visc. 900 stokes; pour pt. 55 F; acid no. 2; iodine no. 84; sapon. no. 157; hyd. no. 72.

Vykamol N/E, S/E. [Croda Chem. Ltd.] Glyceryl stearate; nonionic; emulsifier, dispersant, wetting, scouring and gelling agent, solubilizer; solid.

W180. [Hart Chem. Ltd.] Ester/quaternary amine blend; cationic; softener, rewetting agent for cotton, nylon, polyester fibers; liq.

Wachsemulsion 1864. [Zschimmer & Schwarz] Carnauba/nonionic wax emulsion; mfg. of cleaning agents with glossing effect; liq.; 20% act.

Wayfos A. [Olin] Phosphate acid, acid form; anionic; antistat, emulsifier, wetting agent, detergent, coupling agent, solubilizer, lubricant, corrosion inhibitor; for alkaline built cleaners, textiles, plastics, metal, emulsion polymerization, agric. applics.; Gardner 2-5 liq.; sol. in alcohols, glycol ethers, aromatic and aliphatic hydrocarbons, chlorinated solvs.; disp. in water; sp.gr. 1.08; dens. 8.94 lb/gal; pH 1.5-2.0 (5% aq.); toxicology: LD50 (oral, rats) < 5 g/kg; corrosive to eyes; possible skin corrosivity; 100% act.

Wayfos D-10N. [Olin] Poly(oxy-1,2-ethanediyl), nonylphenyl-hydroxy phosphate; anionic; antistat, dispersant, corrosion inhibitor, emulsifier, stabilizer, solubilizer for textile processing, alkaline and acid soak cleaners, hard surf. cleaners, dishwashes, wax removers, emulsion polymerization, agric., paper processing; Gardner 1-2 visc. liq.; sol. in water, alcohols, glycol ethers, aromatic and aliphatic hydrocarbons, chlorinated solvs.; sp.gr. 1.10; dens. 9.15 lb/gal; pH 1.7-2.0 (5% aq.); Draves wetting 44 s (0.25%); Ross-Miles foam 133; toxicology: LD50 (oral, rats) > 5 g/kg; skin irritant; corrosive to eyes; 100% act.

Wayfos M-60. [Olin] Aromatic phosphate ester, free acid; anionic; corrosion inhibitor, hydrotrope; drycleaning detergent; emulsifier for pesticides, emulsion polymerization; lt. amber visc. liq.; sol. in alcohols, glycol ethers, aromatic and aliphatic hydrocarbons, chlorinated solvs.; disp. in water; sp.gr. 1.12; dens. 9.32 lb/gal; pH 2.0-2.5 (1% aq.); toxicology: LD50 (oral, rats) > 5 g/kg; irritating to skin and eyes; 100% act.

Wayfos M-100. [Olin] Aromatic phosphate ester, free acid; anionic; antistat, detergent for all-purpose heavy-duty formulations; corrosion inhibitor for nonferrous metals; pesticide emulsifier; pitch dispersant for paper pulp processing; lt. amber visc. liq.; sol. in water, alcohols, glycol ethers, aliphatic hydrocarbons, chlorinated solvs.; disp. in aromatic hydrocarbons; sp.gr. 1.13; dens. 9.40 lb/gal; pH 2.0-2.5 (1% aq.); Draves wetting 10 s (0.25%); toxicology: LD50 (oral, rats) > 5 g/kg; skin irritant; corrosive to eyes; 100% act.

Wayhib® S. [Olin] TEA phosphate ester, sodium salt; anionic; detergent builder, sequestrant, corrosion inhibitor, pipeline scale inhibitor, water circulating systems; for air conditioning, boiler treatment compds.; Gardner 1 liq.; misc. with water; sp.gr. 1.45; dens. 12.07 lb/gal; pH 4.6 (1% aq.); toxicology: irritating to skin and eyes; 70% act.

Wayplex NTP-S. [Olin] Pentasodium aminotrimethylene phosphonate; anionic; detergent builder, scale inhibitor; liq.; 40% conc.

Westvaco® 1480. [Westvaco] Tall oil acid; CAS 61790-12-3; anionic; emulsifier for emulsion polymerization and post stabilization; liq.; iodine no. 80; 100% conc.

Westvaco® 1483. [Westvaco] Tall oil acid, stabilized; CAS 61790-12-3; anionic; replacement for oleic acid in industrial applics.; liq.; iodine no. 85; 100% conc.

Westvaco Diacid® 1550. [Westvaco] Acrylinoleic acid; CAS 53980-88-4; intermediate forming high solids, low visc. soaps, esters, polyamide derivs.; as surfactant, coupling agent; hydrotrope esp. in caustic systems; biodeg.; Gardner 9 max. liq.; dens. 8.45 lb/gal; visc. 5000 cps (100 F); acid no. 265-277; flash pt. (COC) 455 F; nontoxic.

Westvaco Diacid® 1575. [Westvaco] Acrylinoleic acid; CAS 53980-88-4; anionic; low color and high purity version of Westvaco Diacid 1550; liq.; 100% conc.

Westvaco Diacid® H-240. [Westvaco] Partial salt of Westvaco Diacid; CAS 68127-33-3; anionic; coupling agent for built detergents, phenolic disinfectants, and liq. cleaners; liq.; 40% conc.

Westvaco® M-30. [Westvaco] Stabilized tall oil rosin and fatty acids; CAS 68152-92-1; anionic; emulsifier for polymerization of styrene and butadiene for syn. tire rubber; liq.; 100% conc.

Westvaco® M-40. [Westvaco] Stabilized tall oil rosin and fatty acids; CAS 68152-92-1; anionic; emulsifier for polymerization of styrene and butadiene for syn. tire rubber; liq.; 100% conc.

Westvaco® M-70. [Westvaco] Stabilized tall oil rosin and fatty acids; CAS 68152-92-1; anionic; emulsifier for polymerization of styrene and butadiene; liq.; 100% conc.

Westvaco® Resin 90. [Westvaco] Stabilized tall oil rosin; CAS 8050-09-7; anionic; emulsifier for syn. rubber emulsion polymerization; tackifier, stabilizer, plasticizer; Gardner 6 max. liq.; dens. 8.6 lb/gal; visc. 620 cps (200 F); pour pt. 130 F; acid no. 168-178; sapon. no. 175; 100% conc.

Westvaco® Resin 95. [Westvaco] Stabilized tall oil rosin; CAS 8050-09-7; anionic; emulsifier for SBR polymerization; tackifier, stabilizer, plasticizer; Gardner 11 max. liq.; dens. 8.6 lb/gal; soften. pt. (R&B) 112 F; acid no. 158-166; 100% conc.

Westvaco® Resin 790. [Westvaco] Sodium rosinate; CAS 61790-51-0; anionic; emulsifier for emulsion polymerization; liq.; 70% conc.

Westvaco® Resin 795. [Westvaco] Sodium rosinate; CAS 61790-51-0; anionic; emulsifier for emulsion polymerization; liq.; 70% conc.

Westvaco® Resin 895. [Westvaco] Potassium rosinate; CAS 61790-50-9; anionic; emulsifier for emulsion polymerization; liq.; 80% conc.

Wetting Agent 611, CG 16. [Chem-Y BV] Org. sulfonic acid; anionic; wetting and washing agent; liq. and paste resp.; 35 and 70% conc. resp.

Wetting Agent FCGB. [Henkel/Functional Prods.] Sodium laureth phosphate; anionic; wetting agent for metal surfaces; useful in cyanide copper plating baths to produce clean finishes; liq.; 10% conc.

Wettol® D 1. [BASF AG] Sodium salt of phenolsulfonic acid condensation prod.; emulsifier, wetting agent, dispersant for formulation of wettable powds. for crop protection; powd.; water-sol.; 95% conc.

Wettol® D 2. [BASF AG] Sodium salt of condensed naphthalene sulfonic acid; emulsifier, wetting agent, dispersant for the formulation of wettable powds. for crop protection; powd.; water-sol.; 95% conc.

Wettol® EM 1. [BASF AG] Calcium alkylaryl sulfonate; anionic; emulsifier, wetting and dispersing agent for formulation of pesticides; liq.; 68–70% conc.

Wettol® EM 2. [BASF AG] Alkylphenol oxalkylation prod.; nonionic; emulsifier for crop protection emulsifiable concs.; liq.; 82.5% conc.

Wettol® EM 3. [BASF AG] Oxalkylated veg. oil; anionic; emulsifier, dispersant, wetting agent for pesticides; liq.; 85% conc.

Wettol® NT 1. [BASF AG] Sodium alkyl naphthalene sulfonate; anionic; emulsifier, wetting agent, dispersant for crop protection wettable powds.; powd.; 65% conc.

White Swan. [Croda Chem. Ltd.] Lanolin BP, anhyd.; conditioner, moisturizer; w/o emulsifier, and emollient for personal care prods., pharmaceuticals; superfatting agent in soap; aids pigment dispersion; yel. unctuous mass.

Wibarco. [Wibarco GmbH] Linear dodecylbenzene sulfonic acid; CAS 27176-87-0; anionic; biodeg. surfactant intermediate; liq.; 98% conc.

Witbreak 770. [Witco/Organics] Surfactant blend; anionic; oilfield surfactant, demulsifier, slugging compd.; liq.; sol. in xylene; disp. in IPA, kerosene, water; dens. 8.4 lb/gal; visc. 200 cps; pour pt. < 10 F; pH 6.5.

Witbreak 772. [Witco/Organics] Surfactant blend; nonionic; oilfield surfactant, demulsifier, slugging compd.; liq.; sol. in IPA, xylene; disp. in water; dens. 8.0 lb/gal; visc. 22 cps; pour pt. 8 F; pH 6.5.

Witbreak 774. [Witco/Organics] Surfactant blend; nonionic; oilfield surfactant, demulsifier, slugging compd.; liq.; sol. in IPA, xylene; disp. in water, kerosene; dens. 8.2 lb/gal; visc. 300 cps; pour pt. < 0 F; pH 6.5.

Witbreak 1390. [Witco/Organics] Alcohol ether sulfonate; anionic; oilfield surfactant, foaming agent; liq.; sol. in water, IPA; dens. 8.8 lb/gal; visc. 50 cps; pour pt. < 4 F; pH 7.8.

Witbreak DGE-75. [Witco/Organics; Witco SA] Glycol ester; nonionic; demulsifier for petrol. industry and waste disposal of processing oils in metals; liq.

Witbreak DGE-128A. [Witco/Organics; Witco SA] Glycol ester; nonionic; base for crude oil demulsifier; liq.; sol. in IPA, xylene; dens. 8.4 lb/gal; visc. 590 cps; pour pt. 10 F; pH 4.

Witbreak DGE-169. [Witco/Organics] Glycol ester; nonionic; oilfield surfactant, demulsifier; liq.; sol. in IPA, xylene; dens. 8.5 lb/gal; visc. 11,000 cps; pour pt. 20 F; pH 4.

Witbreak DGE-182. [Witco/Organics; Witco SA] Glycol ester; nonionic; base for crude oil demulsifier; liq.

Witbreak DPG-15. [Witco/Organics] POE glycol; nonionic; oilfield surfactant, demulsifier; liq.; sol. in IPA, xylene; dens. 8.5 lb/gal; visc. 700 cps; pour pt. 20 F; pH 7.

Witbreak DPG-482. [Witco/Organics] POE glycol; nonionic; oilfield surfactant, demulsifier; liq.; sol. in water, IPA, xylene; dens. 8.8 lb/gal; visc. 1300 cps; pour pt. 30 F; pH 7.

Witbreak DRA-21. [Witco/Organics] Oxyalkylated phenolic resin; nonionic; base for crude oil demulsifier; liq.; sol. in IPA, xylene; disp. in water; dens. 8.5 lb/gal; visc. 600 cps; pour pt. 10 F; pH 11.

Witbreak DRA-22. [Witco/Organics] Oxyalkylated phenolic resin; nonionic; oilfield surfactant, demulsifier; liq.; sol. in IPA, xylene, water; dens. 8.5 lb/gal; visc. 800 cps; pour pt. 50 F; pH 11.

Witbreak DRA-50. [Witco/Organics] Glycol ester; nonionic; base for crude oil demulsifier; liq.

Witbreak DRB-11. [Witco/Organics; Witco SA] Oxyalkylated phenolic resin; nonionic; base for crude oil demulsifier; liq.; sol. in IPA, xylene; dens. 8.4 lb/gal; visc. 1700 cps; pour pt. 20 F; pH 7.

Witbreak DRB-127. [Witco/Organics; Witco SA] Oxyalkylated phenolic resin; nonionic; base for crude oil demulsifier; liq.; sol. in IPA, xylene; dens. 8.2 lb/gal; visc. 9000 cps; pour pt. 20 F; pH 10.

Witbreak DRB-401. [Witco/Organics] Oxyalkylated phenolic resin; nonionic; oilfield surfactant, demulsifier; liq.; sol. in IPA, xylene, kerosene; dens. 8.2 lb/gal; visc. 1380 cps; pour pt. 5 F; pH 11.

Witbreak DRC-163. [Witco/Organics] Oxyalkylated phenolic resin; nonionic; oilfield surfactant, demulsifier; liq.; sol. in IPA, xylene; dens. 8.7 lb/gal; visc. 12,000 cps; pour pt. 35 F; pH 10.

Witbreak DRC-164. [Witco/Organics] Oxyalkylated phenolic resin; nonionic; base for crude oil demulsifier; liq.; sol. in IPA, xylene; dens. 8.5 lb/gal; visc. 11,300 cps; pour pt. 45 F; pH 10.

Witbreak DRC-165. [Witco/Organics] Oxyalkylated phenolic resin; nonionic; oilfield surfactant, demulsifier; liq.; sol. in IPA, xylene; dens. 8.7 lb/gal; visc. 21,000 cps; pour pt. 40 F; pH 10.

Witbreak DRC-168. [Witco/Organics] Oxyalkylated phenolic resin; nonionic; oilfield surfactant, demulsifier; liq.; sol. in IPA, xylene; dens. 8.7 lb/gal; visc. 8000 cps; pour pt. 30 F; pH 10.

Witbreak DRC-232. [Witco/Organics] Oxyalkylated phenolic resin; nonionic; oilfield surfactant, demulsifier; liq.; sol. in IPA, xylene; dens. 8.6 lb/gal; visc. 5000 cps; pour pt. 20 F; pH 10.

Witbreak DRE-8164. [Witco/Organics] Resin ester; nonionic; oilfield surfactant, demulsifier; liq.; disp. in water, IPA, kerosene, xylene; dens. 8.3 lb/gal; visc. 450 cps; pour pt. < 0 F; pH 6.

Witbreak DRI-9020. [Witco/Organics] Polyol; nonionic; oilfield surfactant, demulsifier; liq.; sol. in IPA, kerosene, xylene; disp. in water; dens. 8.2 lb/gal; visc. 4500 cps; pour pt. 5 F; pH 7.

Witbreak DRI-9037. [Witco/Organics] Polyol; nonionic; oilfield surfactant, demulsifier; liq.; sol. in IPA, xylene; dens. 8.3 lb/gal; visc. 800 cps; pour pt. < 30 F; pH 6.

Witbreak DRI-9038. [Witco/Organics] Polyol; nonionic; oilfield surfactant, demulsifier; liq.; sol. in IPA, xylene; disp. in kerosene; dens. 8.1 lb/gal; visc. 700 cps; pour pt. < 30 F; pH 5.

Witbreak DTG-62. [Witco/Organics] Polyoxyalkylene glycol; nonionic; oilfield surfactant; demulsifier; paste; sol. in IPA, xylene; disp. in water; dens. 8.6 lb/gal; pour pt. 80 F; pH 10.

Witbreak GBG-3172. [Witco/Organics] Polyoxyalkylated modified resin; nonionic; oilfield surfactant; demulsifier; liq.; sol. in IPA, xylene; dens. 8.6 lb/gal; visc. 10,000 cps; pour pt. 25 F; pH 7.

Witbreak RTC-323, -1315, -1316. [Witco/Organics; Witco SA] Polymeric amine salt; anionic; o/w demulsifier used in petrol. industry and for waste disposal of processing oils in metals; liq.; water-sol.

Witbreak RTC-326. [Witco/Organics] Polymeric amine; cationic; oilfield surfactant, reverse demulsifier; liq.; sol. in IPA, water; dens. 9.5 lb/gal; visc. 12,000 cps; pour pt. 20 F; pH 8.5.

Witbreak RTC-330. [Witco/Organics] Polymeric amine; cationic; oilfield surfactant, reverse demulsifier; liq.; sol. in water; dens. 10.0 lb/gal; visc. 800 cps; pour pt. < 0 F; pH 5.5.

Witbreak RTC-6010, -6020, –6030, -6040. [Witco/Organics] Polymeric amine; demulsifier in petrol. industry.

Witcamide® 82. [Witco/Organics] Cocamide DEA and diethanolamine; nonionic; foam stabilizer, visc. modifier, lubricant, conditioner, emulsifier, wetting agent, penetrant, dye dispersant, scouring aid, antistat; cosmetics and toiletries, base for scrub soap; thickener; industrial and textile surfactant; metal processing; liq.; 100% conc.

Witcamide® 128T. [Witco/Organics] Cocamide DEA; nonionic; detergent, foam booster/stabilizer, visc. modifier, substantive conditioner; liq.; sol. in water, disp. in oils; sp.gr. 0.99.

Witcamide® 272. [Witco/Organics; Witco SA] Fatty alkanolamide, modified; CAS 68081-99-2; nonionic; detergent, emulsifier, lubricant, wetting agent, penetrant, dye dispersant, scouring aid, antistat for industrial, textile, drilling fluids, metal processing; stable to strong acids, alkalies; liq.; sol. in water, high concs. of electrolytes; sp.gr. 1.16; 100% conc.

Witcamide® 511. [Witco/Organics; Witco SA; Witco Isarel] Modified fatty alkanolamide; nonionic; lubricant, wetting agent, penetrant, dye dispersant, scouring aid, antistat, w/o emulsifier for aerosol and industrial applics., drilling muds, metal processing, textiles; liq.; sol. in oils, IPA, kerosene, xylene; disp. in water; sp.gr. 0.96; dens. 8.0 lb/gal; visc. 850 cps; pour pt. < 0 F; pH 8.6.

Witcamide® 512. [Witco/Organics] Fatty alkanolamide; anionic/nonionic; detergent, o/w emulsifier, coupling agent; o/w emulsifier for drilling muds; liq.; sol. in water, IPA; sp.gr. 1.15; dens. 9.6 lb/gal; visc. 4000 cps; pour pt. 25 F; pH 8.5.

Witcamide® 1017. [Witco/Organics] Cocamide DEA; nonionic; drilling mud surfactant, detergent; liq.; sol. in water, IPA, kerosene, xylene; dens. 8.9 lb/gal; visc. 1300 cps; pour pt. 32 F; pH 9.5; 100% conc.

Witcamide® 5130. [Witco/Organics; Witco SA] Modified cocamide DEA; nonionic; detergent, emulsifier, conditioner, visc. modifier, lubricant, dispersant, suspending and wetting agent, penetrant, dye dispersant, scouring aid, antistat, foam stabilizer for aerosol formulations, textiles, industrial use, metal processing; amber clear visc. liq.; sol. (@ 5%) in water, glycols, alcohols, and chlorinated and aromatic hydrocarbons; sp.gr. 1.0; acid no. 43; pH 9.2 (1% aq.); flash pt. > 200 F (PMCC); 98% act.

Witcamide® 5133. [Witco/Organics] Cocamide DEA and diethanolamine; nonionic; emulsifier, foam stabilizer, visc. modifier, lubricant, conditioner, wetting agent, penetrant, dye dispersant, scouring aid, antistat for textiles, industrial use, personal care prods., metal processing; liq.; 100% conc.

Witcamide® 5138. [Witco/Organics; Witco SA] Fatty alkanolamide condensate; nonionic; emulsifier, detergent, lubricant, wetting agent, penetrant, dye dispersant, scouring aid, antistat for industrial use, textiles, metal processing, dry cleaning, oleoresinous coatings, petrol. industry; clear to slightly hazy liq.; sol. (5%) in kerosene, xylene, IPA; disp. in water; sp.gr. 0.99; pH 9.5 (5% aq.); flash pt. > 93 C; 100% conc.

Witcamide® 5140. [Witco/Organics] Alkanolamide, modified; nonionic; detergent, wetting agent, dispersant; industrial cleaning agent; foam stabilizer in drilling mud surfactants and aq. media; air-entraining agent; textiles; liq.; 100% conc.

Witcamide® 5145M. [Witco/Organics] Alkanolamide, modified; anionic/nonionic; detergent for built liqs., dry cleaning; liq.; 100% conc.

Witcamide® 5168. [Witco/Organics] Diethanolaminooleamide DEA; cationic; surfactant for cosmetics, textiles, detergents, and industrial uses; liq.; 100% conc.

Witcamide® 5195. [Witco/Organics] Lauramide DEA; CAS 120-40-1; nonionic; visc. modifier, foam stabilizer, lubricant, conditioner, lubricant, emulsifier, wetting agent, thickener, penetrant, dye dispersant, scouring aid, antistat; cosmetics and toiletries; industrial foamer and stabilizer; metal processing, textile surfactant; aerosol formulations; paste; water-sol.; sp.gr. 0.99.

Witcamide® 6310. [Witco/Organics] Lauramide DEA, modified; CAS 120-40-1; nonionic; conditioner, foam stabilizer, gelling agent, lubricant, and visc. modifier for cosmetics and toiletries; industrial detergent; paste; water-sol.; sp.gr. 0.97; 100% conc.

Witcamide® 6445. [Witco/Organics] Cocamide DEA, modified; nonionic; industrial detergent, o/w emulsifier, lubricant, visc. modifier, foamer and stabilizer, conditioner, coupling agent; liq.; water-sol.; sp.gr. 0.99; 100% conc.

Witcamide® 6511. [Witco/Organics] Lauramide DEA; CAS 120-40-1; nonionic; detergent, foam booster/stabilizer, substantive conditioner, visc. modifier; liq.; water-sol.; sp.gr. 0.99.

Witcamide® 6514. [Witco/Organics] Cocamide DEA; nonionic; detergent, foam booster/stabilizer, visc. modifier, substantive conditioner; liq.; water-sol.; sp.gr. 0.98.

Witcamide® 6515. [Witco/Organics] Cocamide DEA; nonionic; detergent, foam booster/stabilizer, visc. modifier, substantive conditioner; liq.; sol. in water, disp. in oil; sp.gr. 0.99.

Witcamide® 6531. [Witco/Organics] Cocamide DEA; nonionic; detergent, foam booster/stabilizer, conditioner; liq.; water-sol.; sp.gr. 1.01.

Witcamide® 6533. [Witco/Organics] Cocamide DEA, diethanolamine; nonionic; detergent, dispersant, o/w emulsifier, lubricant, visc. modifier, conditioner; liq.; water-sol.; sp.gr. 0.99.

Witcamide® 6625. [Witco/Organics] Cocamide

DEA, modified; nonionic; detergent, foam booster/stabilizer, visc. modifier, substantive conditioner; liq.; water-sol.; sp.gr. 0.98.

Witcamide® AL69-58. [Witco/Organics] Alkanolamide; cationic/nonionic; lubricant, wetting agent, penetrant, dye dispersant, scouring aid, antistat; industrial w/o emulsifier; metal processing; textile surfactant; liq.; 100% conc.

Witcamide® C. [Witco/Organics; Witco SA] Cocamide MEA; CAS 68140-00-1; nonionic; detergent, foam stabilizer for cosmetics, softener, visc. control agent; wax; 100% conc.

Witcamide® CD. [Witco/Organics] Cocamide DEA; nonionic; detergent, thickener, foam booster/stabilizer, conditioner; liq.; water-sol.; sp.gr. 1.02; 100% conc.

Witcamide® CDA. [Witco/Organics] Cocamide DEA, modified; nonionic; detergent, dispersant, o/w emulsifier, lubricant, visc. modifier, conditioner; base for scrub soap formulations; liq.; water-sol.; sp.gr. 1.00; 100% conc.

Witcamide® CDS. [Witco/Organics; Witco SA] Alkylolamide; nonionic; additive for cutting oils, corrosion inhibitor; liq.; 100% conc.

Witcamide® Coco Condensate. [Witco/Organics] Cocamide DEA, diethanolamine; nonionic; detergent, foam stabilizer, lubricant, conditioner; liq.; water-sol.; sp.gr. 1.02.

Witcamide® CPA. [Witco/Organics; Witco SA] Cocamide MIPA; CAS 68333-82-4; nonionic; detergent, foam stabilizer for cosmetics, softener, visc. control agent; wax; 100% conc.

Witcamide® LDTS. [Witco/Organics; Witco SA] Cocamide DEA; CAS 61731-31-9; nonionic; detergent, foam stabilizer, softener, visc. control for cosmetics; liq.; 100% conc.

Witcamide® M-3. [Witco/Organics] Cocamide DEA; nonionic; cleansing agent, conditioner, lubricant, and visc. modifier for cosmetics and toiletries; detergent, foamer and stabilizer for industrial detergents; liq.; water-sol.; 100% conc.

Witcamide® MAS. [Witco/Organics; Witco SA] Stearamide MEA-stearate; CAS 14351-40-7; nonionic; opacifier, conditioner, lubricant, gelling agent for personal care, household, and institutional liq. soaps; used as partial or total replacement for veg. waxes in polishes; coating agent for paper and textiles; mold release agent for industrial processing; additive for raising melting pts. of petrol. waxes or glyceride waxes and fats; ingred. of insulating coatings or barriers, and of water-repellent compds.; lt. cream waxy flakes; mild char. odor; sol. in aliphatic, aromatic, and chlorinated hydrocarbons, alcohol, ketones, esters; sp.gr. 0.844; m.p. 81 C; acid no. 5.0; 100% conc.

Witcamide® N. [Witco/Organics] Alkylolamide; nonionic; metal cleaning, insecticides; liq.; 100% conc.

Witcamide® NA. [Witco/Organics; Witco SA] Alkylolamide; nonionic; foam stabilizer, softener; liq.; 100% conc.

Witcamide® S771. [Witco/Organics] Cocamide DEA, modified; nonionic; detergent, emulsifier, foamer, stabilizer, lubricant, coupling agent for shampoos, bubble baths, industrial detergents; liq.; water-sol.; sp.gr. 0.98; 100% conc.

Witcamide® S780. [Witco/Organics] Cocamide DEA, modified; nonionic; emulsifier, coupling agent, foamer, stabilizer, lubricant for shampoos, bubble baths, industrial detergents; liq.; water-sol.; sp.gr. 0.97; 100% conc.

Witcamine® 204. [Witco/Organics] Imidazoline; cationic; oilfield surfactant, corrosion inhibitor; liq.; sol. in IPA, xylene; disp. in water, kerosene; dens. 7.9 lb/gal; visc. 2300 cps; pour pt. 23 F; pH 11.

Witcamine® 209. [Witco/Organics; Witco SA] 1-(2-Aminoethyl)-2-n-alkyl-2-imidazoline; cationic; emulsifier, lubricity and wetting agent; penetrant, dye dispersant, scouring aid, antistat, corrosion inhibitor intermediate in petrol. industry, metal processing, textiles; liq.; sol. in IPA, heavy aromatic naphtha, toluene, xylene; disp. in water; dens. 7.8 lb/gal; visc. 200 cps; pour pt. 30 F; pH 11.

Witcamine® 210. [Witco/Organics; Witco SA] Alkyl amidoamine; cationic; emulsifier, lubricant; corrosion inhibitor intermediate in petrol. industry and metal processing; paste; sol. in IPA, heavy aromatic naphtha, toluene, keroxene, xylene; insol. in water; dens. 7.6 lb/gal; visc. 2500 cps; pour pt. 91 F; pH 10.

Witcamine® 211. [Witco/Organics] Mixed 1-(2-aminoethyl)-2-n-alkyl-2-imidazoline; cationic; intermediate for oil or water sol. salts used as corrosion inhibitors, bactericides, softeners; for petrol. industry; liq.; sol. in IPA, heavy aromatic naphtha, toluene, xylene; insol. in water; dens. 7.3 lb/gal; visc. 20,000 cps; pour pt. 20 F; pH 11.

Witcamine® 235. [Witco/Organics] 1-Polyaminoethyl-2-n-alkyl-2-imidazoline; emulsifier, lubricant, wetting agent, penetrant, dye dispersant, scouring aid, antistat, corrosion inhibitor intermediate in petrol. industry, metal processing, textiles.

Witcamine® 240. [Witco/Organics] Polyamido imidazoline; cationic; oilfield surfactant, corrosion inhibitor; liq.; sol. in kerosene, xylene; disp. in IPA; dens. 8.0 lb/gal; visc. 900 cps; pour pt. 32 F; pH 9.

Witcamine® 760. [Witco/Organics] Imidazoline; cationic; oilfield surfactant, corrosion inhibitor; liq.; sol. in kerosene, xylene; disp. in water, IPA; dens. 7.6 lb/gal; visc. 100 cps; pour pt. 10 F; pH 10.

Witcamine® 3164. [Witco/Organics] Alkylamidoamine; emulsifier, lubricant, corrosion inhibitor intermediate in petrol. industry and metal processing.

Witcamine® 6606. [Witco/Organics] PEG-15 tallow amine; cationic; antistat, dispersant, o/w emulsifier, lubricant, substantivity and wetting agent; lt. amber liq.; water-sol.; sp.gr. 1.02; pH 9.5; 99% solids.

Witcamine® 6622. [Witco/Organics] PEG-30 oleamine; cationic; antistat, dispersant, o/w emulsifier, lubricant, substantivity and wetting agent; lt. amber liq.; water-sol.; sp.gr. 1.07; pH 9.5; 80% solids.

Witcamine® AL42-12. [Witco/Organics] Fatty imidazoline; cationic; corrosion inhibitor for ferrous metals; salts act as antistat, emulsifier, wetting agent, dispersant, and detergent; demulsification of anionic emulsions, ore flotation, leather and textile softening, degreasing,; and pickling operations, asphalt bonding; amber clear liq.; faint amine odor; sol. in IPA, kerosene, xylene; m.w. 350; sp.gr. 0.935; dens. 7.8 lb/gal; visc. 400 cps; pour pt. < 0 F; pH 11; 87.0% tert. amine content.

Witcamine® PA-60B. [Witco/Organics] Salt of Witcamine AL42-12; cationic; industrial antistat, corrosion inhibitor, dispersant and wetting agent; for petrol. industry; liq.; sol. in IPA, kerosene, xylene; disp. in water; dens. 8.1 lb/gal; visc. 14,500 cps; pour pt. 19 F; pH 6.5.

Witcamine® RAD 0500. [Witco/Organics; Witco SA] POE rosin amine; cationic; oilfield surfactant; corrosion inhibitor intermediate, scouring agent, softener, dye assistant, emulsifier for acid-stable emulsions; inhibitor for HCl in acid pickling; liq.; sol. in IPA, xylene; disp. in kerosene; dens. 8.8 lb/gal; visc. 25,500 cps; pour pt. 61 F; pH 10; 100% conc.

Witcamine® RAD 0515. [Witco/Organics; Witco SA] POE rosin amine; cationic; oilfield surfactant; corrosion inhibitor intermediate, scouring agent, softener, dye assistant, emulsifier for acid-stable emulsions; inhibitor for HCl in acid pickling; liq.; sol. in IPA, kerosene, xylene; dens. 8.8 lb/gal; visc. 29,500 cps; pour pt. 62 F; pH 10; 100% conc.

Witcamine® RAD 1100. [Witco/Organics; Witco SA] POE rosin amine; cationic; oilfield surfactant; corrosion inhibitor intermediate, scouring agent, softener, dye assistant, emulsifier for acid-stable emulsions; inhibitor for HCl in acid pickling; liq.; sol. in IPA, water, xylene; dens. 9.0 lb/gal; visc. 21,000 cps; pour pt. < 30 F; pH 10; 100% conc.

Witcamine® RAD 1110. [Witco/Organics; Witco SA] POE rosin amine; cationic; oilfield surfactant; corrosion inhibitor intermediate, scouring agent, softener, dye assistant, emulsifier for acid-stable emulsions; inhibitor for HCl in acid pickling; liq.; sol. in IPA, water, xylene; dens. 8.9 lb/gal; visc. 2250 cps; pour pt. < 30 F; pH 10; 100% conc.

Witcamine® TI-60. [Witco/Organics] Amido imidazoline; cationic; oilfield surfactant, corrosion inhibitor; liq.; sol. in IPA, water, xylene; dens. 8.8 lb/gal; visc. 700 cps; pour pt. 23 F; pH 10.

Witco® 94H Acid. [Witco/Organics] Alkylbenzene sulfonic acid; anionic; industrial detergent, dispersant, o/w emulsifier, wetting agent; liq.; 95% conc.

Witco® 97H Acid. [Witco/Organics] Alkylbenzene sulfonic acid; anionic; industrial detergent, dispersant, o/w emulsifier, wetting agent; liq.; sol. in water and oil; 95% conc.

Witco® 912. [Witco/Organics] POE carboxylic acid esters and sulfonates; pigment dispersant for aq. systems; solubilizer for methyl cellulose; paint surfactant.

Witco® 915. [Witco/Organics] Imidazoline; cationic; pigment dispersant and suspending agent, wetting and grinding aid for oil-based paint systems; emulsifier for latex paints.

Witco® 916. [Witco/Organics] Fatty acid ester; spreading agent for aq. paint systems.

Witco® 918. [Witco/Organics] Alkylaryl sulfonate; pigment wetting and grinding aid for paints; pigment dispersant; oil-sol.

Witco® 934. [Witco/Organics] Alkanolamide; pigment dispersant and suspending agent, wetting and grinding aid for oil-based paint systems; surfactant for latex paints.

Witco® 936. [Witco/Organics] Alkylphenol/EO adduct; pigment wetting and grinding aid in oil-based paint systems.

Witco® 942. [Witco/Organics] Glycerol monooleate; CAS 111-03-5; emulsifier for latex paints.

Witco® 960. [Witco/Organics] Alkylphenol/EO adduct; pigment wetting and grinding aid for aq. paint systems; emulsifier for polymerization.

Witco® 1298. [Witco/Organics] Dodecylbenzene sulfonic acid; anionic; detergent intermediate, o/w emulsifier, solubilizer, wetting agent, and detergent for household prods., metal cleaning; emulsion polymerization surfactant for latex stabilization and pigment dispersion; liq.; sol. in oil and water; sp.gr. 1.05; 97% act.

Witco® 1298 H. [Witco/Organics] Branched dodecylbenzene sulfonic acid; anionic; detergent, dispersant, emulsifier, wetting agent; base for industrial specialty applics.; nonbiodeg.; liq.; sol. in oil and water; sp.gr. 1.04; 97% act.

Witco® 1298 HA. [Witco/Organics] Surfactant blend; anionic; surfactant for insecticides; EPA clearance; liq.

Witco® Acid B. [Witco/Organics] Dodecylbenzene sulfonic acid; anionic; detergent intermediate; detergent, dispersant, o/w emulsifier, wetting agent for cosmetics, industrial and metal cleaning; solubilizer, emulsion polymerization surfactant for latex stabilization and pigment dispersion; liq.; sol. in oil and water; 100% conc.

Witco® Acid TPB. [Witco/Organics] Alkylaryl sulfonic acid; anionic; base for detergent formulations; liq.; 98% conc.

Witco® CHB Acid. [Witco/Organics] Alkylaryl sulfonic aicd; anionic; industrial detergent and o/w emulsifier; liq.; sol. in oil.

Witco® D51-29. [Witco/Organics] Alkylaryl sulfonic acid; anionic; industrial detergent, dispersant, o/w emulsifier; liq.; water-sol.

Witco® DTA-150, -180. [Witco/Organics] Fatty acid polymers; corrosion inhibitor intermediate; petrol. industry.

Witco® DTA-350. [Witco/Organics] Dimer-trimer acid; anionic; corrosion inhibitor for petrol. industry; liq.; sol. in IPA, kerosene, xylene; dens. 8.0 lb/gal; visc. 5560 cps; pH 4.5.

Witco® TX Acid. [Witco/Organics] Modified toluene sulfonic acid; anionic; hydrotrope, catalyst; coupler and solubilizer for liq. detergents; anticaking aid in dry neutralization; catalyst in org. reactions (e.g., esterification); liq. to 10–15 C; 95% act.

Witcodet 100. [Witco/Organics] Formulated prod.; anionic/nonionic; detergent conc. base for dishwash, carwash, laundery, all-purpose, and shampoo formulations; Gardner 5 liq.; sol. in water; sp.gr. 1.02; visc. 400 cps; pH 8.0; 98% solids.

Witcodet 793. [Witco/Organics] Formulated prod.; detergent conc. base for carwash formulations; liq.; 49% conc.

Witcodet 5480. [Witco/Organics] Formulated prod.; detergent conc. base for rug and upholstery shampoos; liq.; 35% conc.

Witcodet B-1. [Witco/Organics] Blend; anionic/nonionic; detergent conc. for dishwash; Gardner 4 liq.; water-sol.; sp.gr. 1.08; visc. 200 cps; pH 5.8; 50% solids.

Witcodet CWC. [Witco/Organics] Blend; anionic; detergent conc. for car wash applics.; Gardner 4 liq.; water-sol.; sp.gr. 1.05; visc. 300 cps; pH 8.5; 52% solids.

Witcodet DC-47. [Witco/Organics] Blend; anionic/nonionic; detergent conc. for dishwash; Gardner 5 liq.; water-sol.; sp.gr. 1.07; visc. 500 cps; pH 8.0; 47% solids.

Witcodet DLC-47. [Witco/Organics] Blend; anionic/nonionic; detergent conc. for dishwash; Gardner 5 liq.; water-sol.; sp.gr. 1.07; visc. 500 cps; pH 8.0; 50% solids.

Witcodet SC. [Witco/Organics] Formulated prod.; anionic/nonionic; detergent conc. base for liq. dishwash formulations; Gardner 3 liq.; water-sol.; sp.gr. 1.02; visc. 400 cps; pH 6.5; 57% solids.

Witcolate 1050. [Witco/Organics] Sodium C12-15 pareth sulfate; anionic; detergent, detergent base, emulsifier, foamer, wetting agent for the detergent industry; liq.; water-sol.; sp.gr. 1.04; 39% act.

Witcolate 1247H. [Witco/Organics] Alcohol ether sulfate; anionic; industrial detergent, foamer; petrol.

industry intermediate; electrolyte tolerant; liq.; sol. in IPA and water; dens. 8.8 lb/gal; visc. 175 cps; pour pt. < 4 F; pH 7.8.

Witcolate 1259. [Witco/Organics] Alcohol ether sulfate; anionic; industrial coupler, detergent, foamer, solubilizer, o/w emulsifier, wetting agent for metal cleaning, dry cleaning, petrol. industry; electrolyte tolerant; liq.; sol. in water, naphthenic oils, IPA, perchloroethylene; dens. 8.8 lb/gal; visc. 500 cps; pour pt. < 4 F; pH 7.5.

Witcolate 1276. [Witco/Organics] Ammonium alcohol ether sulfate; anionic; wetting agent, penetrant, lubricant, emulsifier, dye dispersant, scouring aid, antistat; foamer for wallboard mfg. and petrol. industry; detergent for metal cleaning; textile surfactant; emulsion polymerization and latex stabilization; liq.; sol. in water, IPA; sp.gr. 1.05; dens. 8.8 lb/gal; visc. 30 cps; pour pt. < 10 F; pH 7.8; 60% solids.

Witcolate 1390. [Witco/Organics] Alcohol ether sulfate; anionic; foamer for petrol. industry; liq.; sol. in IPA and water.

Witcolate 3220. [Witco/Organics] Alkyl ether sulfate; anionic; oilfield surfactant, foaming agent; liq.; sol. in water; disp. in IPA, xylene; dens. 9.0 lb/gal; visc. 40 cps; pour pt. < 4 F; pH 8.8.

Witcolate 3570. [Witco/Organics; Witco SA] Sodium alkyl ether sulfate; anionic; detergent for shampoos, bubble baths, household and industrial cleansing compositions; paste; 70% conc.

Witcolate 6400. [Witco/Organics] Sodium lauryl sulfate; CAS 151-21-3; anionic; detergent, foaming agent, wetting agent; liq.; water-sol.; sp.gr. 1.04; 29% act.

Witcolate 6430. [Witco/Organics] Ammonium lauryl sulfate; anionic; detergent, foaming agent, wetting agent; liq.; water-sol.; sp.gr. 1.01; 28.5% act.

Witcolate 6431 (formerly DeSonol A). [Witco/Organics] Ammonium lauryl sulfate; anionic; detergent, foaming agent, wetting agent; liq.; water-sol.; sp.gr. 1.02; 28% act.

Witcolate 6434 (formerly DeSonol T). [Witco/Organics] TEA-lauryl sulfate; anionic; detergent, wetting agent, foaming agent; liq.; water-sol.; sp.gr. 1.05; 40% act.

Witcolate 6450. [Witco/Organics] Sodium laureth (1) sulfate; CAS 9004-82-4; anionic; detergent, foaming agent, wetting agent; liq.; water-sol.; sp.gr. 1.04; 25% act.

Witcolate 6453. [Witco/Organics] Sodium laureth sulfate; CAS 9004-82-4; anionic; detergent, foaming agent, wetting agent; liq.; water-sol.; sp.gr. 1.05; 28% act.

Witcolate 6455. [Witco/Organics] Sodium laureth (2) sulfate; CAS 9004-82-4; anionic; detergent, foaming agent, wetting agent; liq.; water-sol.; sp.gr. 1.05; 26% act.

Witcolate 6462. [Witco/Organics] Sodium alkyl sulfate; anionic; detergent, dispersant, wetting agent; liq.; water-sol.; sp.gr. 1.08; 38% act.

Witcolate 6465. [Witco/Organics] Sodium 2-ethylhexyl sulfate; anionic; detergent, dispersant, wetting agent; liq.; water-sol.; sp.gr. 1.12; 39% act.

Witcolate 7031. [Witco/Organics] Sodium alkyl sulfate; anionic; industrial detergent, dispersant, foamer, wetting agent, penetrant; for steam extraction rug shampoos; electrolyte tolerance; liq.; water-sol.; sp.gr. 1.09; 39% act.

Witcolate 7093. [Witco/Organics] Sodium deceth sulfate; anionic; industrial detergent, foamer and wetting agent for pressure-spray applics.; electrolyte tolerance; liq.; water-sol.; sp.gr. 1.04; 38.5% act.

Witcolate 7103. [Witco/Organics] Ammonium alcohol ether sulfate; anionic; high foaming detergent, wetting agent in high electrolyte systems such as oil well drilling; liq.; water-sol.; sp.gr. 1.105; 58% act.

Witcolate A. [Witco/Organics] Sodium lauryl sulfate; CAS 151-21-3; anionic; cleansing agent, detergent base, foamer, emulsifier, wetting agent, solubilizer for cosmetics and toiletries, industrial detergents; liq.; water-sol; sp.gr. 1.04; 29% act.

Witcolate AE (formerly DeSonol AE). [Witco/Organics] Ammonium laureth sulfate; anionic; detergent, foaming agent, wetting agent; liq.; water-sol.; sp.gr. 1.06; 59% act.

Witcolate AE-3. [Witco/Organics] Ammonium C12-15 pareth sulfate; anionic; detergent, wetting agent, emulsifier, foamer for the detergent industry; liq.; water-sol.; sp.gr. 1.04; 59% act.

Witcolate AE3S. [Witco/Organics; Witco SA] Sodium alkyl ether sulfate; anionic; base for liq. detergents; liq.; 27% conc.

Witcolate AM. [Witco/Organics] Ammonium lauryl sulfate; anionic; detergent, foaming agent, wetting agent; liq.; water-sol.; sp.gr. 1.01; 28.5% act.

Witcolate A Powder. [Witco/Organics] Sodium lauryl sulfate; CAS 151-21-3; anionic; detergent, wetting agent, foamer used in personal care prods., wool detergents; polymerization emulsifier; powd.; water-sol.; sp.gr. 0.35; 93% act.

Witcolate C. [Witco/Organics] Sodium lauryl sulfate; CAS 151-21-3; anionic; detergent, foaming agent, wetting agent; paste; water-sol.; sp.gr. 1.06; 29% act.

Witcolate D51-51. [Witco/Organics] Nonoxynol-4 sulfate; anionic; antistat for syn. fibers and polymer prods.; surfactant, wetting agent and dispersant; emulsifier for polymerization of acrylics, vinyl acetate, vinyl acrylics, styrene, SAN, styrene acrylic, vinyl chloride; FDA compliance; clear liq.; sol. in water, ethanol; sp.gr. 1.06; visc. 2500 cps; pH 8.0 (5% aq.); flash pt. (PMCC) > 93.3 C; surf. tens. 30.0 dynes/cm (1%); Ross-Miles foam 173 mm (initial, 1%, 49 C); 34% solids.

Witcolate D51-51EP. [Witco/Organics] Sodium nonoxynol-4 sulfate; CAS 9014-90-8; emulsifier for emulsion polymerization of acrylic, vinyl acetate, vinyl acrylic, vinyl chloride; FDA compliance; Gardner 4 color; visc. 400 cps; pH 8 (10% aq.); surf. tens. 30 dynes/cm (1%); 30% act.

Witcolate D51-52. [Witco/Organics] Sodium alkylaryl polyether sulfate; anionic; emulsifier for polymerization; liq.

Witcolate D51-53. [Witco/Organics] Nonoxynol-10 sulfate; emulsifier for acrylic, vinyl acetate, vinyl acrylic, styrene, SAN, styrene acrylic, and vinyl chloride polymerization; visc. 250 cps; surf. tens. 40 dynes/cm (1%); Ross-Miles foam 218 mm (initial, 1%, 49 C); 34% solids.

Witcolate D51-53HA. [Witco/Organics] Nonoxynol-10 sulfate; emulsifier for acrylic, vinyl acetate, vinyl acrylic, styrene, SAN, styrene acrylic, and vinyl chloride polymerization; visc. 50 cps; surf. tens. 41.2 dynes/cm (1%); Ross-Miles foam 224 mm (initial, 1%, 49 C); 30% solids.

Witcolate D51-60. [Witco/Organics] Nonoxynol-30 sulfate; emulsifier for acrylic, vinyl acetate, vinyl acrylic, styrene, SAN, styrene acrylic, and vinyl chloride polymerization; visc. 300 cps; surf. tens. 43.3 dynes/cm (1%); Ross-Miles foam 234 mm (initial, 1%, 49 C); 40% solids.

Witcolate D-510. [Witco/Organics] Sodium 2-ethylhexyl sulfate; anionic; detergent, wetting agent and penetrant for industrial use and polymerization reactions; dispersant for bleaching powds.; lime soap and grease dispersant; clear liq.; sol. (5%) in water, IPA; sp.gr. 1.12; pH 10.5 (10% aq.); flash pt. > 93 C; 39% act.

Witcolate ES-2. [Witco/Organics] Sodium laureth (2) sulfate; CAS 9004-82-4; anionic; detergent, foaming agent, wetting agent; liq.; water-sol.; sp.gr. 1.05; 26% act.

Witcolate ES-3. [Witco/Organics] Sodium laureth (3) sulfate; CAS 9004-82-4; anionic; detergent, foaming agent, wetting agent; liq.; water-sol.; sp.gr. 1.05; 28% act.

Witcolate LCP. [Witco/Organics] Sodium lauryl sulfate; CAS 151-21-3; anionic; detergent, foaming agent, wetting agent; liq.; water-sol.; sp.gr. 1.04; 29% act.

Witcolate LES-60A. [Witco/Organics] Ammonium laureth sulfate; anionic; deterent, foaming agent, wetting agent; liq.; water-sol.; sp.gr. 1.06; 59% act.

Witcolate LES-60C. [Witco/Organics] Sodium laureth sulfate; CAS 9004-82-4; anionic; detergent, foaming agent, wetting agent; liq.; water-sol.; sp.gr. 1.06; 58% act.

Witcolate NH. [Witco/Organics] Ammonium lauryl sulfate; anionic; detergent, foaming agent, wetting agent; liq.; water-sol.; sp.gr. 1.01; 28.5% act.

Witcolate OME. [Witco/Organics] Sodium laureth sulfate; CAS 9004-82-4; anionic; used in detergents and personal care prods.; liq.; 24% conc.

Witcolate S (formerly DeSonol S). [Witco/Organics] Sodium lauryl sulfate; CAS 151-21-3; anionic; detergent, foaming agent, wetting agent; liq.; water-sol.; sp.gr. 1.04; 29% act.

Witcolate S1285C. [Witco/Organics] Sodium laureth sulfate; CAS 9004-82-4; anionic; cleansing agent, detergent, wetting agent, detergent base, and foamer for cosmetics, toiletries, industrial detergents, textiles; liq.; water-sol.; 60% conc.

Witcolate S1300C. [Witco/Organics] Ammonium laureth sulfate; anionic; cleansing agent, detergent, wetting agent, detergent base, and foamer for cosmetics, toiletries, industrial detergents, textiles; liq.; water-sol.; 60% conc.

Witcolate SE (formerly DeSonol SE). [Witco/Organics] Sodium laureth sulfate; CAS 9004-82-4; anionic; detergent, foaming agent, wetting agent; liq.; water-sol.; sp.gr. 1.06; 59% act.

Witcolate SE-5. [Witco/Organics] Sodium C12-15 pareth sulfate; anionic; detergent, detergent base, foamer, o/w emulsifier, wetting agent for industrial and household detergents, textiles; liq.; water-sol.; sp.gr. 1.04; 59% act.

Witcolate SL-1. [Witco/Organics] Sodium lauryl sulfate; CAS 151-21-3; emulsifier for acrylics, acrylonitrile, carboxylated SBR, chloroprene, styrene, vinyl chloride, vinyl acetate; FDA compliance; visc. 50 cps; HLB 40; surf. tens. 31.8 dynes/cm (1%); Ross-Miles foam 180 mm (initial, 1%, 49 C); 29% solids.

Witcolate T. [Witco/Organics] TEA-lauryl sulfate; anionic; cleansing agent, and foamer, emulsifier, solubilizer for cosmetics and toiletries; detergent base; wetting agent for industrial detergents; liq.; water-sol.; 40% conc.

Witcolate TLS-500. [Witco/Organics] TEA-lauryl sulfate; anionic; detergent, foaming agent, wetting agent; liq.; water-sol.; sp.gr. 1.05; 40% act.

Witcolate WAC. [Witco/Organics] Sodium lauryl sulfate; CAS 151-21-3; anionic; detergent, foaming agent, wetting agent.

Witcolate WAC-GL. [Witco/Organics] Sodium lauryl sulfate; CAS 151-21-3; anionic; detergent, foaming agent, wetting agent; liq.; water-sol.; sp.gr. 1.04; 29% act.

Witcolate WAC-LA. [Witco/Organics] Sodium lauryl sulfate; CAS 151-21-3; anionic; detergent, foaming agent, wetting agent; liq.; water-sol.; sp.gr. 1.04; 29% act.

Witcomul 78. [Witco/Organics] Sorbitan tallate; w/o emulsifier, thickener, emulsion stabilizer for industrial applics.; drilling fluid additive.

Witcomul 1054. [Witco/Organics] Blend; anionic; emulsifier, dye carrier; liq.; 100% conc.

Witcomul 1317. [Witco/Organics] Blend; anionic; emulsifier, dye carrier; liq.; 100% conc.

Witcomul 1557. [Witco/Organics] POE sorbitan monotallate; drilling fluid additive in petrol. industry.

Witcomul 3107. [Witco/Organics] Surfactant blend; anionic/nonionic; oilfield surfactant, drilling mud o/w emulsifier; liq.; sol. in water, IPA; dens. 8.8 lb/gal; visc. 450 cps; pour pt. < 0 F; pH 7.0.

Witcomul 3126. [Witco/Organics] Amine ester; anionic; drilling fluid lubricity additive for petrol. industry; liq.; sol. in water, IPA; dens. 9.9 lb/gal; visc. 660 cps; pour pt. 0 F; pH 8.5.

Witcomul 3154. [Witco/Organics] Surfactant blend; anionic/nonionic; oilfield surfactant, drilling mud w/o emulsifier; liq.; sol. in kerosene, IPA, xylene; dens. 8.6 lb/gal; visc. 1700 cps; pour pt. 0 F; pH 7.0.

Witcomul 3158. [Witco/Organics] Surfactant blend; anionic/nonionic; oilfield surfactant, drilling mud w/o emulsifier; liq.; sol. in kerosene, IPA, xylene; dens. 8.2 lb/gal; visc. 2200 cps; pour pt. 28 F; pH 10.5.

Witcomul 3230. [Witco/Organics] Fatty acid amide; cationic/nonionic; oilfield surfactant, foaming agent; liq.; sol. in kerosene, IPA, xylene; dens. 8.2 lb/gal; visc. 2300 cps; pour pt. 32 F; pH 7.

Witcomul 4016. [Witco/Organics] Complex alkylate; nonionic; oilfield surfactant, oil slick dispersant; liq.; sol. in IPA, kerosene, xylene; disp. in water; dens. 8.6 lb/gal; visc. 300 cps; pour pt. 10 F; pH 7.3.

Witcomul H-50A. [Witco/Organics] Polyalkoxy carboxylic acid ester/sulfonated oil blend; anionic/nonionic; emulsifier for emulsion cleaning; sludge dispersant; liq.; aliphatic sol.; 100% conc.

Witcomul H-52. [Witco/Organics] Sulfonates/POE ether blend; anionic/nonionic; emulsifier for aliphatic hydrocarbons, kerosene, degreasing concs.; liq.

Witconate 30DS. [Witco/Organics] Sodium dodecylbenzene sulfonate; anionic; detergent, foaming agent, wetting agent; liq.; sol. in water; sp.gr. 1.07; pH 7.5; 30% act.

Witconate 45BX. [Witco/Organics] Sodium dodecylbenzene sulfonate; anionic; detergent base; liq.; 42% conc.

Witconate 45DS. [Witco/Organics] Sodium dodecylbenzene sulfonate; anionic; detergent, foaming agent, wetting agent; liq.; sol. in water; sp.gr. 1.09; pH 7.5; 43% act.

Witconate 45 Liq. [Witco/Organics] Sodium dodecylbenzenesulfonate, sodium xylenesulfonate; anionic; detergent base, emulsifier, foamer, and wetting agent for detergent industry and industrial surfactants; electrolyte tolerant; liq.; water-sol.; sp.gr. 1.08; pH 7.0; 45% act.

Witconate 45LX. [Witco/Organics] Sodium dodecylbenzenesulfonate, sodium xylene sulfonate; anionic; detergent, wetting agent, foaming agent, base for household and industrial specialties; low cloud pt.; liq.; water-sol.; sp.gr. 1.09; pH 7.5; 43% act.

Witconate 60B. [Witco/Organics] Sodium dodecylbenzene sulfonate; anionic; cleansing agent, foamer, solubilizer for cosmetics and toiletries; detergent base, foamer and wetting agent, emulsifier for industrial detergents; liq.; water-sol.; 60% conc.

Witconate 60L. [Witco/Organics; Witco SA] Linear calcium dodecylbenzene sulfonate; anionic; basic lipophilic emulsifier for pesticide emulsifiable concs.; liq.; 60% conc.

Witconate 60T. [Witco/Organics] TEA-dodecylbenzenesulfonate; anionic; detergent, wetter, emulsifier, foaming agent; base for household, industrial, and cosmetic/toiletry specialty compds.; liq.; water-sol.; sp.gr. 1.08; pH 6.5; 58% act.

Witconate 68KN. [Witco/Organics] Blend; anionic/nonionic; detergent, foaming agent, wetting agent; liq.; sol. in water; sp.gr. 1.09; pH 7.0; 67% act.

Witconate 79S. [Witco/Organics] TEA-dodecylbenzene sulfonate; anionic; detergent, dispersant, emulsifier, wetting agent; industrial surfactant; agric. applics.; EPA clearance; liq.; water-sol.; sp.gr. 1.06; pH 7.0; 52% act.

Witconate 90F. [Witco/Organics] Sodium dodecylbenzenesulfonate; anionic; detergent base, o/w emulsifier, foamer, and wetting agent for detergent industry; wetting agent for metal cleaning; textile latex frothing agent; flake; water-sol.; sp.gr. 0.45; pH 8.0; 91% act.

Witconate 90F H. [Witco/Organics] Branched sodium dodecylbenzene sulfonate; anionic; detergent, foaming agent, wetting agent; flake; sol. in water; sp.gr. 0.40; pH 8.0; 91% act.

Witconate 93S. [Witco/Organics] Amine dodecylbenzene sulfonate; anionic; detergent, detergent base, emulsifier, foamer, wetting agent, dispersant, solubilizer for the detergent industry; biodeg.; liq.; sol. in oil, IPA, kerosene, xylene; disp. in water; sp.gr. 1.05; dens. 8.5 lb/gal; visc. 800 cps; pour pt. 25 F; pH 4.5; 91% act.

Witconate 605A. [Witco/Organics] Calcium dodecylbenzene sulfonate; anionic; industrial detergent, dispersant, o/w and w/o emulsifier, lubricant, wetting agent, demulsifier; agric., oilfield applics.; EPA clearance; liq.; sol. in oil, kerosene, xylene; disp. in IPA; sp.gr. 1.02; dens. 8.5 lb/gal; visc. 2500 cps; pour pt. < 10 F; pH 5.0; 60% act.

Witconate 605AC. [Witco/Organics; Witco SA] Branched calcium dodecylbenzene sulfonate; anionic; basic lipophilic emulsifier for pesticide emulsifiable concs.; liq.; HLB 7.0-8.0; 2% conc.

Witconate 605T. [Witco/Organics] Alkylaryl sulfonate; penetrant, lubricant, scouring aid, antistat, o/w emulsifier, dispersant, oil wetting agent for industrial uses; additive for lube oils; textile surfactant; oil-sol.

Witconate 702. [Witco/Organics] Alkylaryl sulfonic acid, ammonium salt; demulsifier for the petrol. industry.

Witconate 703. [Witco/Organics] Alkylaryl sulfonate; anionic; oilfield surfactant, demulsifier, slugging compd.; liq.; sol. in IPA, kerosene, xylene; dens. 8.3 lb/gal; visc. 1400 cps; pour pt. 5 F; pH 7.

Witconate 705. [Witco/Organics] Alkylaryl sulfonate; nonionic; oilfield surfactant, demulsifier, slugging compd.; liq.; sol. in IPA, kerosene, xylene; disp. in water; dens. 8.3 lb/gal; visc. 500 cps; pour pt. 0 F; pH 6.5.

Witconate 1075X. [Witco/Organics] Amine alkylaryl sulfonate; anionic; industrial detergent, dispersant, o/w emulsifier, foamer, and wetting agent; liq.; water-sol.

Witconate 1223H. [Witco/Organics] Sodium branched dodecylbenzene sulfonate; emulsifier for ABS, SAN, SBR, styrene, vinyl acrylic, vinyl chloride polymerization; wetting agent for latexes; FDA compliance; visc. 500 cps; surf. tens. 31.7 dynes/cm (1%); Ross-Miles foam 175 mm (initial, 1%, 49 C); 23% solids.

Witconate 1240 Slurry. [Witco/Organics] Sodium dodecylbenzenesulfonate; anionic; detergent base, emulsifier, foamer, and wetting agent for the household and industrial detergents; slurry; water-sol.; sp.gr. 1.07; pH 7.5; 40% act.

Witconate 1250 Slurry. [Witco/Organics] Sodium dodecylbenzene sulfonate; anionic; detergent, wetting agent, foaming agent, base for household and industrial detergents, emulsion polymerization; slurry; water-sol.; sp.gr. 1.08; pH 7.5; 53% act.

Witconate 1260 Slurry. [Witco/Organics] Sodium dodecylbenzenesulfonate, sodium xylenesulfonate; anionic; detergent, wetting agent, foaming agent, base for household and industrial compds., textile surfactant; slurry; water-sol.; sp.gr. 1.10; pH 7.5; 60% act.

Witconate 1840X. [Witco/Organics] Sodium salt of sulfated oleic acid; anionic; detergent, coupler, wetting agent, emulsifier, penetrant, lubricant, dye dispersant, scouring aid, antistat, corrosion inhihbitor; industrial surfactant for alkaline systems; metal processing; textile surfactant; dk. liq.; sol. in water, IPA; sp.gr. 1.10; flash pt. > 93 C; pH 6–8 (5% aq.).

Witconate 1850. [Witco/Organics] Sodium dodecylbenzene sulfonate; anionic; detergent; liq.; 40% conc.

Witconate 3009-15. [Witco/Organics] Sodium alkylaryl ether sulfate; latex emulsifier and stabilizer for polymerizations.

Witconate 3203. [Witco/Organics] Sulfonate; anionic; oilfield surfactant, foaming agent; liq.; sol. in water, IPA; dens. 9.6 lb/gal; visc. 40 cps; pour pt. 5 F; pH 7.5.

Witconate AO 24. [Witco/Organics; Witco SA] C12-14 alpha olefin sulfonate; anionic; foaming agent for drilling, concrete, tufted carpet and fire fighting applics.; liq.; 40% conc.

Witconate AOK. [Witco/Organics] Sodium C14-16 olefin sulfonate; anionic; detergent, foaming agent, wetting agent; flake; sol. in water; sp.gr. 0.40; pH 8.5; 90% act.

Witconate AOS. [Witco/Organics; Witco SA] Sodium C14-16 olefin sulfonate; CAS 68439-57-6; anionic; detergent, foamer, wetting agent, solubilizer for cosmetics and toiletries, industrial detergents, textiles, petrol. industry; emulsifier for plastics/rubber polymerization; FDA compliance; liq.; sol. in water; disp. in IPA; sp.gr. 1.05; dens. 8.8 lb/gal; visc. 70 cps; HLB 11.8; pour pt. 32 F; pH 7.7; surf. tens. 35.0 dynes/cm (1%); Ross-Miles foam 168 mm (initial, 1%, 49 C); 39% act.

Witconate AOS-EP. [Witco/Organics] Sodium alpha-olefin sulfonate; emulsifier for emulsion polymerization, latex paints, adhesives, binders; lt. clear liq.; dens. 8.92 lb/gal; visc. 200 cps; flash pt. (PMCC) >

200 F; pH 7.5-8.5 (5%); 38-40% act.

Witconate C50H. [Witco/Organics] Sodium dodecylbenzene sulfonate; anionic; industrial detergent, foamer and wetting agent; electrolyte tolerant; liq.; water-sol.; 44% conc.

Witconate CHB. [Witco/Organics] Alkyaryl sulfonic acid; demulsifier used in the petrol. industry.

Witconate D51-51. [Witco/Organics] Sodium alkylaryl polyether sulfonate; anionic; emulsion polymerization surfactant; liq.; water-sol.

Witconate DS. [Witco/Organics] Sodium decylbenzenesulfonate; anionic; detergent, foaming agent, wetting agent; flake; water-sol.; sp.gr. 0.45; pH 8.0; 91% act.

Witconate DS Dense. [Witco/Organics] Sodium dodecylbenzene sulfonate; anionic; detergent, foaming agent, wetting agent; powd; sol. in water; sp.gr. 0.50; pH 8.0; 91% act.

Witconate K. [Witco/Organics] Sodium dodecylbenzenesulfonate; anionic; detergent, foaming agent, wetting agent; flake; water-sol.; sp.gr. 0.45; pH 7.5; 91% act.

Witconate K Dense. [Witco/Organics] Sodium dodecylbenzene sulfonate; anionic; detergent, foaming agent, wetting agent; powd; sol. in water; sp.gr. 0.50; pH 7.5; 91% act.

Witconate KX. [Witco/Organics] Sodium dodecylbenzene sulfonate; anionic; detergent, foaming agent, wetting agent; flake; sol. in water; sp.gr. 0.48; pH 7.6; 91% act.

Witconate LX F. [Witco/Organics] Sodium dodecylbenzenesulfonate and sodium xylenesulfonate; anionic; detergent base, emulsifier, foamer, and wetting agent for detergent industry and industrial surfactants; flake; water-sol.; sp.gr. 0.48; pH 8.0; 91% act.

Witconate LXH. [Witco/Organics] Branched TEA-dodecylbenzene sulfonate; anionic; detergent, foaming agent, wetting agent, dispersant; liq.; water-sol.; sp.gr. 1.09; pH 7.2; 53% act.

Witconate LX Powd. [Witco/Organics] Sodium dodecylbenzenesulfonate and sodium xylenesulfonate; anionic; detergent, wetting agent; base for household and industrial specialty compds.; powd.

Witconate NAS-8. [Witco/Organics] Sodium alkane sulfonate; anionic; low foaming, biodeg. detergent base with hydrotropic props.; stable to electrolyte, acid, alkaline, and chlorine bleach; liq.; 39% conc.

Witconate NCS. [Witco/Organics] Ammonium cumene sulfonate; anionic; antiblocking agent, coupler, solubilizer, cloud pt. depressant, and hydrotrope for detergent industry; liq.; water-sol.

Witconate NIS. [Witco/Organics] Sodium isethionate; anionic; detergent, foaming agent, wetting agent; liq.; water-sol.; sp.gr. 1.37; pH 8.5; 56% act.

Witconate NXS. [Witco/Organics] Ammonium xylenesulfonate; hydrotrope, solubilizer, coupler and processing aid in detergent mfg. and industrial processes; antiblocking and anticaking agent in powd. prods.; formulates shampoos, aerosols, cutting oils, glue; textile finishing; Klett 30 liq.; dens. 1.1 g/cc; pH 8.0; 40% act.

Witconate P-1020BUST. [Witco/Organics] Surfactant blend; nonionic; agric. surfactant; EPA clearance; liq.

Witconate P-1052N. [Witco/Organics; Witco SA; Witco Israel] Dodecylbenzene sulfonic acid, amine salt; anionic; emulsifier, fuel oil additive, solubilizer, degreaser for emulsions; liq.; oil-sol.

Witconate P10-59. [Witco/Organics; Witco SA; Witco Israel] Isopropylamine dodecylbenzene sulfonate; anionic; emulsifier, solubilizer, detergent, and wetting agent for oil-based systems; dispersant in oil and water-based systems; used in dry-cleaning surfactants; hydrotrope for liq. detergents; oilfield demulsifier; amber clear liq.; sol. in water, kerosene, xylene, IPA; sp.gr. 1.02; dens. 8.5 lb/gal; visc. 10,000 cps; pour pt. 5 F; flash pt. > 93 C; pH 4.8 (20% in 25% IPA); 90% act.

Witconate P-1073F. [Witco/Organics] Alkylaryl sulfonic acid, amine salt; drilling fluid additive used in the petrol. industry.

Witconate SCS 45%. [Witco/Organics] Sodium cumene sulfonate; anionic; hydrotrope, solubilizer, coupler and processing aid in detergent mfg. and industrial processes; antiblocking and anticaking agent in powd. prods.; formulates shampoos, aerosols, cutting oils, glue; textile finishing; Klett 50 liq.; sol. in water; sp.gr. 1.16; pH 8.0; 45% act.

Witconate SCS 93%. [Witco/Organics] Sodium cumene sulfonate; anionic; hydrotrope, solubilizer, coupler and processing aid in detergent mfg. and industrial processes; antiblocking and anticaking agent in powd. prods.; formulates shampoos, aerosols, cutting oils, glue; textile finishing; powd.; sol. in water; sp.gr. 0.33; pH 8.0; 93% act.

Witconate SE-5. [Witco/Organics] Sodium alcohol ether sulfate; anionic; oilfield surfactant, foaming agent; liq.; sol. in water; dens. 8.8 lb/gal; visc. 65 cps; pour pt. 22 F; pH 7.5.

Witconate SK. [Witco/Organics] Sodium dodecylbenzene sulfonate; anionic; detergent, foaming agent, wetting agent; flake; sol. in water; sp.gr. 0.55; pH 8.0; 40% act.

Witconate STS. [Witco/Organics] Sodium toluene sulfonate; hydrotrope, solubilizer, coupler and processing aid in detergent mfg. and industrial processes; antiblocking and anticaking agent in powd. prods.; formulates shampoos, aerosols, cutting oils, glue; textile finishing; Klett 150 liq., powd.; dens. 0.32–1.2 g/cc; pH 9.0–10.0; 40% act. liq., 90% act. powd.

Witconate SXS 40%. [Witco/Organics] Sodium xylene sulfonate; CAS 1300-72-7; anionic; hydrotrope, solubilizer, coupler and processing aid in detergent mfg. and industrial processes; antiblocking and anticaking agent in powd. prods.; formulates shampoos, aerosols, cutting oils, glue; textile finishing; Klett 40 liq.; sol. in water; sp.gr. 1.18; pH 8.0; 41% act.

Witconate SXS 90%. [Witco/Organics] Sodium xylene sulfonate; CAS 1300-72-7; anionic; hydrotrope, solubilizer, coupler and processing aid in detergent mfg. and industrial processes; antiblocking and anticaking agent in powd. prods.; formulates shampoos, aerosols, cutting oils, glue; textile finishing; powd.; sol. in water; sp.gr. 0.35; pH 8.0; 93% act.

Witconate TAB. [Witco/Organics] TEA-dodecylbenzene sulfonate; anionic; detergent, foamer, wetting agent, solubilizer, emulsifier in cosmetics/toiletries, industrial surfactants, paints; electrolyte tolerant; liq.; water-sol.; 60% conc.

Witconate TDB. [Witco/Organics] Sodium tridecylbenzene sulfonate; anionic; detergent base, emulsifier, foamer, and wetting agent for the detergent industry; used in industrial surfactants; liq.; water-sol.; 44% conc.

Witconate TX Acid. [Witco/Organics; Witco SA] Modified toluene sulfonic acid; anionic; wetting agent, hydrotrope, coupler and solubilizer for liq. detergents; anticaking aid in dry neutralization; catalyst in org. reactions; liq.; sol. in oil; sp.gr. 1.30; 96% act.

Witconate YLA. [Witco/Organics] Amine dodecylbenzene sulfonate; anionic; detergent base, emulsifier, solubilizer, foaming and wetting agent for detergents, drycleaning charge soaps, solv. cleaners, metal cleaning, and textile industries; liq.; oil-sol.; 95% conc.

Witconol 14. [Witco/Organics; Witco SA] Polyglyceryl-4 oleate; nonionic; w/o and o/w emulsifier, lubricant, emollient, spreader, sticker, and antifoamer for industrial use, aerosols; Gardner 9 liq; sol. in oil, disp. in water; sp.gr. 0.99; HLB 9.4; 100% conc.

Witconol 18L. [Witco/Organics] Polyglyceryl-4 isostearate; nonionic; emulsifier, spreader, sticker, antifoaming agent, solubilizer for w/o emulsions, aerosols; liq.; oil-sol.; 100% conc.

Witconol 171, 172. [Witco/Organics] Polyalkylene glycol ether; defoamer for soap and detergent foam control; surfactant.

Witconol 1206. [Witco/Organics] Alkyl POE glycol ether; nonionic; wetting agent, detergent, o/w emulsifier for industrial uses; visc. and flow control agent for polymerization reactions; Gardner 1 liq.; water-sol.; sp.gr. 1.00; pH 7.0.

Witconol 1207. [Witco/Organics] Alkyl POE glycol ether; wetting agent for industrial uses.

Witconol 2301. [Witco/Organics] Methyl oleate; nonionic; defoamer, lubricant, moisture barrier; Gardner 5 liq.; sol. in oil; sp.gr. 0.88; pour pt. -16 C.

Witconol 2326. [Witco/Organics] Butyl stearate; nonionic; lubricant; liq.; sol. in oil; sp.gr. 0.86; pour pt. 20 C.

Witconol 2380. [Witco/Organics] Propylene glycol stearate; nonionic; o/w emulsifier, lubricant, opacifier; Gardner 2 beads; sol. in oil; sp.gr. 0.90; HLB 1.8; m.p. 36 C.

Witconol 2400. [Witco/Organics] Glyceryl stearate; nonionic; o/w emulsifier, lubricant; beads; disp. in oil; sp.gr. 0.93; HLB 3.9; m.p. 58 C.

Witconol 2401. [Witco/Organics] Glyceryl stearate; nonionic; o/w emulsifier, lubricant; beads; disp. in oil; sp.gr. 0.92; HLB 3.9; m.p. 58 C.

Witconol 2407. [Witco/Organics] Glyceryl stearate SE; nonionic; o/w emulsifier, lubricant; Gardner 3 beads; disp. in water, oils; sp.gr. 0.93; HLB 5.1; m.p. 58 C.

Witconol 2421. [Witco/Organics] Glyceryl oleate; CAS 111-03-5; nonionic; defoamer, o/w emulsifier, lubricant, moisture barrier; liq.; oil-sol.; sp.gr. 0.95; HLB 3.4; pour pt. 19 C.

Witconol 2500. [Witco/Organics] Sorbitan oleate; CAS 1338-43-8; nonionic; w/o emulsifier, lubricant, coupling agent; Gardner 8 liq.; sol. in oil; sp.gr. 1.00; HLB 4.6; pour pt. < 0 C.

Witconol 2503. [Witco/Organics] Sorbitan trioleate; CAS 26266-58-0; nonionic; w/o emulsifier, lubricant, coupling agent; Gardner 7 liq.; sol. in oil; sp.gr. 0.95; HLB 2.1; pour pt. < 0 C.

Witconol 2620. [Witco/Organics] PEG-4 laurate; CAS 9004-81-3; nonionic; o/w emulsifier, lubricant; Gardner 1 liq.; disp. in water, oil; sp.gr. 0.98; HLB 9.3; pour pt. 9 C.

Witconol 2622. [Witco/Organics] PEG-4 dilaurate; CAS 9005-02-1; nonionic; o/w emulsifier, lubricant; Gardner 2 liq.; sol. in oil, disp. in water; sp.gr. 0.96; HLB 7.6; pour pt. 0 C.

Witconol 2640. [Witco/Organics] PEG-8 stearate; CAS 9004-99-3; nonionic; o/w emulsifier, lubricant; Gardner 1 solid; disp. in water, oil; sp.gr. 1.02; HLB 12.0; m.p. 32 C.

Witconol 2642. [Witco/Organics] PEG-8 distearate; CAS 9005-08-7; nonionic; o/w emulsifier, lubricant; Gardner 2 solid; disp. in water, sol. in oil; HLB 7.5; m.p. 36 C.

Witconol 2648. [Witco/Organics] PEG-8 dioleate; CAS 9005-07-6; nonionic; o/w emulsifier, lubricant, defoamer; Gardner 4 liq.; disp. in water, sol. in oil; sp.gr. 0.97; HLB 8.8; pour pt. 6 C.

Witconol 2720. [Witco/Organics] Polysorbate 20; CAS 9005-64-5; nonionic; o/w emulsifier, dispersant, visc. modifier, coupling agent; Gardner 6 liq.; water-sol.; sp.gr. 1.10; HLB 16.7; pour pt. -10 C.

Witconol 2722. [Witco/Organics] Polysorbate 80; CAS 9005-65-6; nonionic; o/w emulsifier, dispersant, visc. modifier, coupling agent; Gardner 6 liq.; sol. in water; sp.gr. 1.08; HLB 15.0; pour pt. -12 C.

Witconol 5906. [Witco/Organics] PEG-30 castor oil; nonionic; dispersant, o/w emulsifier, lubricant; Gardner 2 liq.; water-sol.; sp.gr. 1.06; HLB 11.8; clear pt. 55 C.

Witconol 5907. [Witco/Organics] PEG-36 castor oil; nonionic; dispersant, o/w emulsifier, lubricant.

Witconol 5909. [Witco/Organics] PEG-40 castor oil; nonionic; dispersant, o/w emulsifier, lubricant; Gardner 2 liq.; water-sol.; sp.gr. 1.06; HLB 13.0; clear pt. 80 C.

Witconol 6903. [Witco/Organics] Polysorbate 85; CAS 9005-70-3; nonionic; o/w emulsifier, dispersant, lubricant, coupling agent; Gardner 7 liq.; disp. in oils, water; sp.gr. 1.03; HLB 11.1; pour pt. -15 C.

Witconol APEM. [Witco/Organics] PPG-3-myreth-3; nonionic; lubricant, emulsifier, wetting agent, penetrant, dye dispersant, scouring aid, antistat, solv. coupler; syn. oils for personal care prods.; metal processing; textile surfactant; liq.; sol. in 3A ethanol, min. oil.

Witconol APM. [Witco/Organics; Witco SA] PPG-3 myristyl ether; CAS 63793-60-2; nonionic; lubricant, emulsifier, wetting agent, penetrant, dye dispersant, scouring aid, antistat, solv. coupler; metal processing; textile surfactant; emollient oil for cosmetics and toiletries, solubilizer; Gardner 1 liq.; sol. in oil; sp.gr. 0.90; pH 7.0; 100% conc.

Witconol APS. [Witco/Organics] PPG-11 stearyl ether; nonionic; lubricant, emulsifier, wetting agent, penetrant, dye dispersant, scouring aid, antistat, solv. coupler; syn. oils for personal care prods.; metal processing; textile surfactant; emollient oil for cosmetics and toiletries, solubilizer; Gardner 1 liq.; sol. in oil; sp.gr. 0.94; pH 7.0.

Witconol CA. [Witco/Organics] Glyceryl stearate SE; nonionic; lubricant, bodying agent, emulsifier, opacifier used in industrial, cosmetic and aerosol formulations; wax; 100% conc.

Witconol CD-17. [Witco/Organics] PPG-34; nonionic; antistat, emulsifier for personal care prods.; syn. lubricant oil and foam modifier for aerosol formulations.

Witconol CD-18. [Witco/Organics; Witco SA] PPG-27 glyceryl ether; CAS 25791-96-2; nonionic; emollient, visc. control agent, spreading agent for personal care prods., cosmetic emulsifier, lubricant, antistat; liq.; sol. in lower hydrocarbon solvs., lower alcohols; 100% conc.

Witconol DOS. [Witco/Organics] Diethylene glycol monooleate; CAS 9004-96-0; nonionic; o/w emulsifier for cosmetic and industrial formulations; self-emulsifying; liq.; 100% conc.

Witconol EGMS. [Witco/Organics] Glycol stearate; nonionic; opacifier, conditioner; Gardner 2 beads; sol. in oil; sp.gr. 0.88; HLB 2.2; m.p. 50 C.

Witconol F26-46. [Witco/Organics; Witco SA] PPG-36 oleate; nonionic; emollient oil, spreading agent and coupler for cosmetic oil systems; conditioner in hair grooms; surfactant with aux. lubricating, dispersing, and coupling properties; dispersant for industrial use; spreading and anticaking agent for aerosol prods.; visc. and flow control agent; APHA 150 clear liq.; sol. in common alcohols, water/alcohol sol'ns., and hydrocarbons; sp.gr. 0.987; visc. 270 cps; acid no. 1.3; sapon. no. 25.0; flash pt. > 93 C; 0.1% moisture.

Witconol GOT. [Witco/Organics] Glycerol mono- and dioleate; nonionic; surfactant and emulsifier for industrial use; liq.; sol. in oil.; 100% conc.

Witconol H31. [Witco/Organics] PEG-8 tallate; CAS 61791-00-2; nonionic; defoamer, dispersant and o/w emulsifier for cosmetic and industrial applics.; metal processing emulsifiable oil; liq.; sol. in oil; 100% conc.

Witconol H31A. [Witco/Organics; Witco SA; Witco Israel] PEG-8 oleate; CAS 9004-96-0; nonionic; lubricant and plasticizer in oils and polymers; o/w emulsifier for min. and veg. oils, and solvs., cosmetic and industrial applics.; improves flow and leveling of coatings, increases spreadability of personal care prods.; defoamer; lt. amber liq.; sol. in alcohol, xylene, kerosene, perchloroethylene, wh. min. oil; partially sol. in water; sp.gr. 0.99; HLB 12.5; acid no. 7; pH 3.7 (3% aq.); flash pt. > 93 C (PMCC); 100% conc.

Witconol H33. [Witco/Organics] PEG-8 dioleate; CAS 9005-07-6; nonionic; industrial detergent; defoamer, o/w emulsifier, lubricant; liq.; oil-sol., water-disp.; sp.gr. 0.98; HLB 8.8; pour pt. 6 C.

Witconol H35A. [Witco/Organics] PEG-8 stearate; CAS 9004-99-3; nonionic; o/w emulsifier for cosmetic and industrial uses; wax; 100% conc.

Witconol MST. [Witco/Organics; Witco SA] Glyceryl stearate; nonionic; emulsifier for cosmetic, pharmaceutical, aerosol formulations; internal lubricant, plasticizer, and emulsifier in industrial applics.; flow control agent for polymerization reactions; dispersant; flake; disp. in oil; sp.gr. 0.93; HLB 3.9; m.p. 58 C; 100% conc.

Witconol NP-40. [Witco/Organics] Nonoxynol-4; CAS 9016-45-9; nonionic; detergent, o/w emulsifier, solubilizer for oils in metal processing; liq.; sol. in naphthenic and paraffinic oil.

Witconol NP-100. [Witco/Organics] Nonoxynol-10; CAS 9016-45-9; nonionic; pigment dispersant, emulsifier, latex stabilizer, and leveling agent for polymerization reactions; visc. 350 cps; Ross-Miles foam 158 mm (initial, 1%, 49 C); 100% solids.

Witconol NP-300. [Witco/Organics] Nonoxynol-30; CAS 9016-45-9; nonionic; paint industry emulsifier for emulsion polymerization; pigment wetting and grinding agent for aq. systems; spreading agent; solid, liq.; water-sol.

Witconol NP-330. [Witco/Organics] Nonyl phenol ethoxy/propoxy (30); emulsifier for ABS, SAN, SBR, styrene, vinyl acrylic, vinyl chloride polymerization; plasticizer; visc. 600 cps; surf. tens. 41.1 dynes/cm (1%); Ross-Miles foam 237 mm (initial, 1%, 49 C); 100% solids.

Witconol NS-108LQ. [Witco/Organics; Witco SA] Polyalkoxylated alkyphenol; nonionic; wetting agent, dispersant for agric. aq. suspension concs.; EPA clearance; paste; HLB 12.7; 100% conc.

Witconol NS-500K. [Witco/Organics; Witco SA] POE POP block copolymer; nonionic; industrial antistat, coupler, detergent, dispersant, o/w emulsifier, lubricant, solubilizer, spreading agent; for agric. emulsifiable concs., aq. suspension concs.; EPA clearance; solid; water-sol.; 100% conc.

Witconol NS-500LQ. [Witco/Organics] Surfactant blend; nonionic; agric. surfactant; EPA clearance; liq.

Witconol O. [Witco/Organics] Glycerol mono- and dioleate; industrial surfactant and emulsifier; oil-sol.

Witconol RDC-D. [Witco/Organics] Diglycol coconate; nonionic; surfactant, emulsifier; cosmetic and industrial use; liq.; oil-sol.

Witconol RHP. [Witco/Organics] Propylene glycol monostearate SE; nonionic; emollient, conditioner, emulsifier, lubricant, opacifier and visc. modifier for cosmetics and toiletries, general industrial use; paste; sol. in oil; 100% conc.

Witconol RHT. [Witco/Organics] Glyceryl stearate SE; nonionic; lubricant, plasticizer, o/w emulsifier used in industrial applics.; Gardner 2 flake; disp. in water, oil; sp.gr. 0.93; HLB 5.1; m.p. 58 C; 100% conc.

Witconol SN Series. [Witco/Organics] Straight chain fatty alcohol ethoxylates; nonionic; detergents, intermediates for household and industrial prods.; liq.; 100% conc.

Witcor 3192. [Witco/Organics] Complex amine phosphate; cationic; corrosion inhibitor for petrol. industry; liq.; sol. in IPA, kerosene, xylene; disp. in water.

Witcor 3194, 3195. [Witco/Organics] Fatty amide; cationic/nonionic; corrosion inhibitor for petrol. industry; liq.; sol. in IPA, kerosene, xylene.

Witcor 3630. [Witco/Organics] Imidazoline; cationic; oilfield surfactant, corrosion inhibitor; liq.; sol. in IPA, xylene; disp. in water, kerosene; dens. 8.4 lb/gal; pour pt. 20 F; pH 5.3.

Witcor 3635. [Witco/Organics] Imidazoline; cationic; oilfield surfactant, corrosion inhibitor; liq.; sol. in IPA, kerosene, xylene; disp. in water; dens. 8.3 lb/gal; visc. 620 cps; pour pt. 35 F; pH 6.3.

Witcor CI-1. [Witco/Organics] Complex surfactant; cationic/nonionic; oilfield surfactant, corrosion inhibitor; liq.; sol. in IPA, kerosene, xylene; disp. in water; dens. 8.0 lb/gal; visc. 75 cps; pour pt. < 30 F; pH 12.

Witcor CI-6. [Witco/Organics] Complex surfactant; cationic; corrosion inhibitor for HCl; oilfield chem.; liq.; sol. in xylene, IPA; insol. in water; dens. 8.3 lb/gal; visc. 53 cps; pour pt. < 30 F; pH 11.

Witcor CI-3117. [Witco/Organics] Amide; intermediate/finished corrosion inhibitor; sol. in xylene, IPA, kerosene; insol. in water.

Witcor PC100. [Witco/Organics] Surfactant blend; anionic; oilfield surfactant for paraffin inhibition; liq.; sol. in IPA, xylene; disp. in water, kerosene; dens. 8.2 lb/gal; visc. 140 cps; pour pt. < -5 F; pH 7.5.

Witcor PC200. [Witco/Organics] Amine alkylaryl sulfonate; anionic; oilfield surfactant for paraffin inhibition; liq.; sol. in IPA, xylene, kerosene; disp. in water; dens. 8.5 lb/gal; visc. 800 cps; pour pt. 25 F; pH 4.5.

Witcor PC205. [Witco/Organics] Amine alkylaryl sulfonate; anionic; oilfield surfactant for paraffin inhibition; liq.; sol. in IPA, xylene, kerosene; disp. in water; dens. 8.5 lb/gal; visc. 2740 cps; pour pt. 10 F; pH 7.5.

Witcor PC210. [Witco/Organics] Amine alkylaryl sulfonate; anionic; oilfield surfactant for paraffin inhibition; liq.; sol. in IPA, xylene, kerosene; disp. in water; dens. 8.5 lb/gal; visc. 10,000 cps; pour pt. 5 F; pH 5.0.

Witcor SI-3065. [Witco/Organics] Complex phosphate ester; anionic; scale inhibitor for water treatment; used in petrol. industry; liq.; sol. in water; dens. 10.8 lb/gal; visc. 40 cps; pour pt. 32 F; pH 4.5.

Witcosperse 201. [Witco] Alkylaryl polyethoxyethanol; nonionic; wetting agent for pesticide formulations; powd.; 50% conc. DISCONTINUED.

Witcosperse 205. [Witco] Polyoxyalkylated glycol; wetting agent for pesticide formulations; powd.; 50% conc. DISCONTINUED.

Witcosperse 206. [Witco] Ethoxylated sorbitan ester; wetting agent for pesticide formulations; powd.; 50% conc. DISCONTINUED.

Witcosperse 207. [Witco] Alkyl polyoxyalkylene ether; wetting agent for pesticide formulations; powd.; 50% conc. DISCONTINUED.

Witflow 60. [Witco/Organics] Alkylaryl polyether alcohol; pigment wetting agent, spreading agent, flow modifier for paints/coatings; straw-colored clear liq.; sol. @ 5% in min. spirits, disp. in water; sp.gr. 1.04; flash pt. (PMCC) > 200 F; pH 7.0; usage level: 0.75-1.5%; 99% act.

Witflow 100. [Witco/Organics] Alkylaryl polyether alcohol; pigment wetting agent, spreading agent, flow modifier for paints/coatings; straw-colored clear liq.; sol. @ 5% in water; sp.gr. 1.06; flash pt. (PMCC) > 200 F; pH 7.0; usage level: 0.75-1.5%; 99% act.

Witflow 901. [Witco/Organics] Sodium diester sulfosuccinate; pigment wetting agent, dispersant, flow modifier for paints/coatings; clear lt. visc. liq.; sol. @ 5% in min. spirits, disp. in water; sp.gr. 1.06; flash pt. (PMCC) 133 F; pH 6.5; usage level: 0.1-0.7%; 70% act.

Witflow 902. [Witco/Organics] Sodium half ester sulfosuccinate; pigment wetting agent, dispersant, flow modifier for paints/coatings; lt. yel. clear liq.; sol. @ 5% in water; sp.gr. 1.10; flash pt. (PMCC) > 200 F; pH 6.5; usage level: 0.1-0.7%; 37% act.

Witflow 910. [Witco/Organics] Polyglyceryl fatty acid ester; pigment wetting agent, spreading agent, flow modifier, defoamer for paints/coatings; hazy visc. liq.; sol. @ 5% in min. spirits, disp. in water; sp.gr. 0.99; flash pt. (PMCC) > 200 F; pH 9.0; usage level: 0.75-1.5%; 99% act.

Witflow 912. [Witco/Organics] Surfactant blend; corrosion protective additive for paints/coatings; helps disperse cellulosics in cold water; dk. amber hazy liq.; partly sol. @ 5% in min. spirits, disp. in water; sp.gr. 0.99; flash pt. (PMCC) 171 F; pH 3.5; usage level: 0.5-1.5%; 100% act.

Witflow 914. [Witco/Organics] Alkanolamide; pigment dispersant, wetting agent for paints/coatings; anti-flood aid; dk. amber clear visc. liq.; sol. @ 5% in min. spirits, disp. in water; sp.gr. 1.00; flash pt. (PMCC) > 200 F; pH 9.5; usage level: 1.0-4.0%; 98% act.

Witflow 914. [Witco/Organics] Modified alkanolamide; pigment dispersant, wetting agent for paints/coatings; anti-flood aid; dk. clear visc. liq.; sol. @ 5% in water; sp.gr. 1.00; flash pt. (PMCC) > 200 F; pH 9.5; usage level: 1.0-4.0%; 98% act.

Witflow 916. [Witco/Organics] PEG-400 oleate; CAS 9004-96-0; leveling agent, flow modifier, defoamer for paints/coatings; pale amber clear liq.; sol. @ 5% in min. spirits, disp. in water; sp.gr. 0.99; flash pt. (PMCC) > 200 F; pH 3.5; usage level: 0.5-1.25%; 100% act.

Witflow 918. [Witco/Organics] Amine sulfonate; pigment dispersant for oil or water-based paints/coatings; lt. amber clear liq.; sol. @ 5% in min. spirits, disp. in water; sp.gr. 1.02; flash pt. (PMCC) > 200 F; pH 5.0; usage level: 1.0-3.5%; 95% act.

Witflow 930. [Witco/Organics] Alkanolamide; pigment dispersant, wetting agent for paints/coatings; anti-flood aid; dk. amber clear liq.; sol. @ 5% in water; sp.gr. 0.96; flash pt. (PMCC) > 200 F; pH 9.0; usage level: 1.0-4.0%; 100% act.

Witflow 950. [Witco/Organics] Polypropoxy quat. ammonium chloride; cationic; pigment wetting, grinding and suspension aid for oil-based paints/coatings; lt. amber clear oily liq.; sol. @ 5% in water, min. spirits; sp.gr. 1.01; flash pt. (PMCC) > 200 F; pH 6.5; usage level: 1.0-3.0%; 98% act.

Witflow 953. [Witco/Organics] Polypropoxy quat. ammonium chloride; cationic; pigment wetting, grinding and suspension aid for aq. paints/coatings; lt. amber clear oily liq.; disp. @ 5% in water; sp.gr. 1.03; flash pt. (PMCC) 141 F; pH 6.5; usage level: 1.0-3.0%; 98% act.

Witflow 977. [Witco/Organics] Phosphate ester; pigment dispersant, anti-flood and anti-float aid for paints/coatings; lt. amber clear liq.; sol. @ 5% in min. spirits, water; sp.gr. 1.07; flash pt. (PMCC) > 200 F; pH 4.0; usage level: 1.0-2.0%; 100% act.

Witflow 979. [Witco/Organics] Phosphate ester; pigment dispersant, anti-flood and anti-float aid for paints/coatings; lt. amber clear liq.; sol. @ 5% in min. spirits, water; sp.gr. 1.05; flash pt. (PMCC) > 200 F; pH 4.0; usage level: 1.0-2.5%; 100% act.

Witflow 990. [Witco/Organics] Polysorbate 20; CAS 9005-64-5; nonionic; o/w emulsifier, antistatic finishing agent for paints/coatings; clear liq.; sol. @ 5% in water, min. spirits; sp.gr. 1.1; flash pt. (PMCC) > 200 F; pH 6.0; usage level: 0.1-0.5%; 100% act.

Witflow 991. [Witco/Organics] Polysorbate 80; CAS 9005-65-6; pigment dispersant for paints/coatings; clear liq.; sol. @ 5% in water, min. spirits; sp.gr. 1.08; flash pt. (PMCC) > 200 F; pH 6.0; usage level: 0.1-0.5%; 100% act.

WR Base. [Clark] Fatty melamine condensate; used to make wax-melamine textile water repellents.

XYZ

X78-2. [Reilly-Whiteman] Sulfated oil; anionic; lubricant for leather; syn. sulfated sperm oil replacement; liq.; 75% conc.

Xanthates. [Hoechst AG] Sodium xanthogenate; CAS 4741-30-4; flotation collector for sulfide and sulfidized minerals; powd.

Yeoman. [Croda Chem. Ltd.] Anhyd. lanolin BP; emollient used in personal care prods., pharmaceuticals; superfatting agent for soap; aids dispersion of pigments into anhyd. systems; yel. unctuous mass; lipophilic.

Zeeclean No. 1. [DeeZee] Alkyl sulfonate/nonionic blend; anionic/nonionic; all-purpose cleaner, grease emulsifier, machine cleaning, textile washing; liq.; 35% conc.

Zee-emul No. 35. [DeeZee] Alkyl sulfonate/nonionic blend; anionic/nonionic; emulsifier for textile dye carriers, pesticides; liq.; HLB 12.2; 100% conc.

Zee-emul No. 50. [DeeZee] Alkyl sulfonate/nonionic blend; anionic/nonionic; emulsifier for textile dye carriers, pesticides; liq.; HLB 12.2; 100% conc.

Zeescour 555. [DeeZee] Alkyl sulfonate/nonionic blend; anionic/nonionic; detergent, emulsifier for high alkaline conditions; liq.; 40% conc.

Zeeterge 444. [DeeZee] Carboxylated alcohol; nonionic; multifunctional specialty cleaner; stable at pH 4-14; liq.; 70% conc.

Zetesol 856. [Zschimmer & Schwarz] MIPA-laureth sulfate; anionic; detergent for cosmetics, shampoos, bath preps., liq. syn. soap; liq.; 56% conc.

Zetesol 856 D. [Zschimmer & Schwarz] MIPA C12-15 pareth sulfate; anionic; detergent for cosmetics, shampoos, bath preps., liq. syn. soap; liq.; 58% conc.

Zetesol 856 DT. [Zschimmer & Schwarz] MIPA laureth sulfate with betaine; anionic; detergent for cosmetics, shampoos, bath preps., liq. syn. soap; liq.; 59% conc.

Zetesol 856 T. [Zschimmer & Schwarz] MIPA-laureth sulfate, cocamidopropyl betaine; anionic; detergent for cosmetics, shampoos, bath preps., liq. syn. soap; liq.; 56% conc.

Zetesol 2056. [Zschimmer & Schwarz] MIPA-laureth sulfate; anionic; detergent for personal care prods., household and industrial cleaners; ylsh. clear liq.; misc. with cold water; dens. 1.06 g/cc; visc. 3000 mPa•s; cloud pt. 10 C; pH 6.5-7.0 (10%); 56% act. in water.

Zetesol AP. [Zschimmer & Schwarz] Ammonium C12-15 pareth sulfate, propylene glycol; anionic; detergent for cosmetics, shampoos, bath preps., liq. hand cleaners, dishwash, household cleaners; straw-colored liq.; dens. 1.05 g/cc; pH 6.0-6.8 (10%); 60% act.

Zetesol NL. [Zschimmer & Schwarz] Sodium laureth sulfate; CAS 9004-82-4; 1335-72-4; anionic; detergent for personal care, household and industrial cleaners; almost colorless clear liq.; dens. 1.04 g/cc; visc. 100 mPa•s; cloud pt. 0 C; pH 6.0-7.0 (10%); 28% act.

Zetesol SE 35. [Zschimmer & Schwarz] MIPA fatty alcohol ether sulfate with pearlescent and fatty acid amido alkyl betaine; anionic; detergent, cosmetic, emulsion and personal care prods. with pearly luster; fluid; 35% conc.

Zohar 60 SD(L). [Zohar Detergent Factory] Sodium alkylbenzene sulfonate and builders; anionic; spray-dried conc. for detergent powd. mfg.; 60% conc.

Zohar 70 SD(L). [Zohar Detergent Factory] Sodium alkylbenzene sulfonate and builders; anionic; spray-dried conc. for detergent powd. mfg.; 70% conc.

Zohar AD. [Zohar Detergent Factory] Alkaline cleaner and degreaser for aircraft MIL specs., trailers, machinery and heavily soiled hard surfaces; liq.

Zohar Amido Amine 1. [Zohar Detergent Factory] Cocamidopropyl dimethylamine; nonionic; intermediate for mfg. of amine salts, amine oxides, betaines, quats.; liq. to solid; 100% conc.

Zohar Automat SD. [Zohar Detergent Factory] Sodium alkylbenzene sulfonate and builders; anionic; spray-dried built powd. for machine washing; powd.; 60% conc.

Zohar Conc. SD. [Zohar Detergent Factory] Conc. powd. for mfg. machine washing powds. by dry mixing; powd.

Zohar EGMS. [Zohar Detergent Factory] Glycol stearate; nonionic; coemulsifier for o/w emulsions; opacifier, pearlescent, emollient, superfatting agent for mfg. of shampoos, liq. toilet soaps, bath prods.; flakes; 100% conc.

Zohar Export SD. [Zohar Detergent Factory] Sodium alkylbenzene sulfonate and builders; anionic; spray-dried built powd. for hand washing; powd.

Zohar GLST. [Zohar Detergent Factory] Glyceryl stearate; nonionic; emulsifier, thickener, superfatting agent; flakes; 100% conc.

Zohar GLST SE. [Zohar Detergent Factory] Glyceryl stearate SE; nonionic; emulsifier, coemulsifier, thickener, opacifier, superfatting agent; flakes; 100% conc.

Zohar HAS 470. [Zohar Detergent Factory] High m.w. mono alkylbenzene; anionic; intermediate for mfg. of oil-sol. sulfates used as emulsifiers, rust preventives, degreasers; liq.; oil-sol.; 81-85% conc.

Zohar KAL SD. [Zohar Detergent Factory] Sodium alkylbenzene sulfonate and builders; anionic; spray-dried built powd. for machine washing; powd.

Zoharconc A.D. [Zohar Detergent Factory] Aircraft detergent conc.

Zoharconc APL. [Zohar Detergent Factory] All-purpose liq. conc.

Zoharconc Dead Sea. [Zohar Detergent Factory] Foaming bath conc. (with Dead Sea minerals).

Zoharconc DIS. [Zohar Detergent Factory] Disinfectant liq. conc.

Zoharconc FC. [Zohar Detergent Factory] Floor cleaner conc. without wax.

Zoharconc FCW. [Zohar Detergent Factory] Floor cleaner conc. with wax.

Zoharconc FS. [Zohar Detergent Factory] Domestic fabric softener conc.

Zoharconc J-Super. [Zohar Detergent Factory] Textile softner conc. for denim.

Zoharconc LLH. [Zohar Detergent Factory] Laundry liq. conc. for hand washing.

Zoharconc LLM. [Zohar Detergent Factory] Laundry liq. conc. for household machine washing.

Zoharconc LS. [Zohar Detergent Factory] Mild liq. soap conc.

Zoharconc RA. [Zohar Detergent Factory] Rinse aid conc.

Zoharconc S.D. [Zohar Detergent Factory] Laundry powd. conc. for household machine wash.

Zoharconc Tex 80. [Zohar Detergent Factory] Dishwashing liq. conc.

Zoharex A-10. [Zohar Detergent Factory] Fatty acid polyglycol ester; nonionic; detergent for laundry liqs.; liq.; 100% conc.

Zoharex B-10. [Zohar Detergent Factory] Fatty acid polyglycol ester; nonionic; detergent for laundry powds.; liq.; 100% conc.

Zoharex N-25. [Zohar Detergent Factory] Nonionic; detergent conc.; powd.; 25% conc.

Zoharex N-60. [Zohar Detergent Factory] Conc.; nonionic; raw material for mfg. of powds.; powd.; 60% conc.

Zoharfoam. [Zohar Detergent Factory] Anionic/nonionic; foaming agent for drilling use; liq.; 40% conc.

Zoharlab. [Zohar Detergent Factory] Linear sodium alkylbenzene sulfonate; anionic; raw material for mfg. of detergents; paste; 60% conc.

Zoharpon ETA 27. [Zohar Detergent Factory] Sodium lauryl ether sulfate (2 EO); CAS 9004-82-4; anionic; raw material for personal care prods., lt. duty detergents; liq.; 27% conc.

Zoharpon ETA 70. [Zohar Detergent Factory] Sodium lauryl ether sulfate (2 EO); CAS 9004-82-4; anionic; raw material for liq. shampoos and cosmetic preparations; lt.-duty detergents; paste; 70% conc.

Zoharpon ETA 270 (OXO). [Zohar Detergent Factory] Sodium lauryl ether sulfate (3 EO); CAS 9004-82-4; anionic; raw material for shampoos, cosmetics, lt. duty detergents; liq.; 27% conc.

Zoharpon ETA 271. [Zohar Detergent Factory] Sodium lauryl ether sulfate (1 EO); CAS 9004-82-4; anionic; raw material for shampoos, cosmetics, lt. duty detergents; liq.; 25% conc.

Zoharpon ETA 273. [Zohar Detergent Factory] Sodium lauryl ether sulfate (3 EO); CAS 9004-82-4; anionic; raw material for shampoos, cosmetics, lt. duty detergents; liq.; 27% conc.

Zoharpon ETA 603. [Zohar Detergent Factory] Sodium lauryl ether sulfate (3 EO); CAS 9004-82-4; anionic; raw material for shampoos, cosmetics, lt. duty detergents; liq.; 60% conc.

Zoharpon ETA 700 (OXO). [Zohar Detergent Factory] Sodium lauryl ether sulfate (3 EO); CAS 9004-82-4; anionic; raw material for shampoos, cosmetics, lt. duty detergents; paste; 70% conc.

Zoharpon ETA 703. [Zohar Detergent Factory] Sodium lauryl ether sulfate (3 EO); CAS 9004-82-4; anionic; raw material for shampoos, cosmetics, lt. duty detergents; paste; 70% conc.

Zoharpon K. [Zohar Detergent Factory] Sodium sulfosuccinate and alkyl sulfate; anionic; conc. for carpet and upholstery shampoos; liq.; 25% conc.

Zoharpon LAD. [Zohar Detergent Factory] DEA lauryl sulfate; anionic; raw material for shampoos, cosmetics, lt. duty detergents; liq.; 35% conc.

Zoharpon LAEA 253. [Zohar Detergent Factory] Ammonium lauryl ether sulfate (3 EO); anionic; raw material for shampoos, cosmetics, lt. duty detergents; liq.; 25% conc.

Zoharpon LMT42. [Zohar Detergent Factory] Sodium lauryl methyl taurate; anionic; mild detergent, foamer, dispersant used in soap, syndet toilet bars, shampoos, bubble baths; paste; 30% conc.

Zoharpon SM. [Zohar Detergent Factory] Alkanolamide sulfosuccinate; anionic; raw material for mild shampoos, carpet shampoos; liq. to paste; 40% conc.

Zoharquat 50. [Zohar Detergent Factory] Benzalkonium chloride; cationic; disinfectant, fungicide, bacteriocide and algicide; liq.; 50% conc.

Zoharquat 80 EXP. [Zohar Detergent Factory] Benzalkonium chloride; cationic; disinfectant, fungicide, algicide.

Zoharsoft. [Zohar Detergent Factory] Quat. imidazoline deriv.; cationic; fabric softener base; liq. to paste; 75% conc.

Zoharsoft 90. [Zohar Detergent Factory] Quat. imidazoline deriv.; cationic; fabric softener base for domestic and commercial laundry use; liq. to paste; disp. in cold to hot water; 90% conc.

Zoharsoft DAS. [Zohar Detergent Factory] Fatty acid-amine deriv.; cationic; softener base for textile industry for all fibers (cotton, rayon, acetates, wool, nylon); flakes; disp. in hot water; pH 4; 100% conc.

Zoharsoft DAS-N. [Zohar Detergent Factory] Fatty acid-amine deriv.; cationic; fabric softener for textile mills for all fibers (cotton, rayon, acetates, wool, nylon); flakes; disp. in acetic acid; pH 8; 100% conc.

Zoharsoft DAS-PW. [Zohar Detergent Factory] Fatty acid condensate; cationic; fabric softener base for prep. of nonfoaming and low yel. emulsions; provides soft handle on all kinds of fibers; flakes; disp. in hot water; pH 3; 100% conc.

Zoharsoft DAS-SIL. [Zohar Detergent Factory] Silicone-enriched cationic base; cationic; fabric softener base for brushed or raised fabrics, elastomeric finishes; flakes; disp. in hot water; pH 4; 100% conc.

Zoharsyl L-30. [Zohar Detergent Factory] Sodium lauroyl sarcosinate; CAS 137-16-6; raw material for mfg. of hair shampoos, conditioners, toothpastes, carpet and upholstery shampoos; anticorrosive props.; liq.; 30% conc.

Zohartaine AB. [Zohar Detergent Factory] Lauryl betaine; CAS 683-10-3; foam booster, mild ingred. for shampoos and detergents; industrial foamer; liq.; 36% conc.

Zohartaine TM. [Zohar Detergent Factory] Dihydroxyethyltallow glycinate; amphoteric; thickener and anticorrosive agent for tech. acid formulations, industrial cleaners; component of mild shampoos; visc. liq. to paste; 40% conc.

Zoharteric D. [Zohar Detergent Factory] Cocoamphodiacetate; amphoteric; component for personal care prods.; detergent for specialty cleaners; liq. to gel; 50% conc.

Zoharteric DJ. [Zohar Detergent Factory] Disodium

lauroamphodiacetate; amphoteric; component of mild, nonirritating conditioning shampoos, bubble baths; detergent for specialty cleaners; liq.; 28% conc.

Zoharteric DO. [Zohar Detergent Factory] Disodium cocoamphodiacetate, sodium lauryl sulfate, hexylene glycol; amphoteric; component of mild, nonirritating conditioning shampoos, bubble baths, baby shampoos, cleansing materials; liq.; 47% conc.

Zoharteric DOT. [Zohar Detergent Factory] Disodium cocoamphodiacetate, sodium trideceth sulfate, hexylene glycol; amphoteric; component of extra mild, nonirritating, non-eye-stinging shampoos, cleansing materials; liq.; 50% conc.

Zoharteric D-SF. [Zohar Detergent Factory] Disodium cocoamphodipropionate, salt-free; amphoteric; component of mild, nonirritating, conditioning shampoos, bubble baths; detergent for heavy-duty household and industrial cleaners with tolerance for alkalies and electrolytes; liq.; 40% conc.

Zoharteric D-SF 70%. [Zohar Detergent Factory] Disodium cocoamphodiacetate, salt-free; amphoteric; component of mild, nonirritating, conditioning shampoos, bubble baths; detergent for heavy-duty household and industrial cleaners with tolerance for alkalies and electrolytes; liq.; 70% conc.

Zoharteric LF. [Zohar Detergent Factory] Sodium caprylоamphoacetate; amphoteric; surfactant and wetting agent for low-foam heavy-duty household and industrial cleaners; tolerance for alkalies and electrolytes; liq.; 34% conc.

Zoharteric LF-SF. [Zohar Detergent Factory] Disodium capryloamphodipropionate; amphoteric; surfactant and wetting agent for low-foam heavy-duty household and industrial cleaners; high tolerance for alkalies and electrolytes; liq.; 39% conc.

Zoharteric M. [Zohar Detergent Factory] Sodium cocoamphoacetate; amphoteric; component for personal care prods.; detergent for specialty cleaners; liq.; 44% conc.

Zonyl® A. [DuPont] EO-ester condensate; nonionic; emulsifier, lubricant; textile wetting agent for fibers and fabrics during dyeing; leveling agent and antifoam in pigment coatings in paper industry; used in metal processing; clear-pale yel. liq.; mild odor; sol. in water, polar and nonpolar solvs.; sp.gr. 1.07; HLB 6.7; surf. tens. 26 dynes/cm (0.1% aq.); 100% act.

Zonyl® FSA. [DuPont] Fluorochemical surfactant; anionic; wetting agent, emulsifier, dispersant, corrosion inhibitor, leveling agent for adhesives, agric., polishes, polymerization, pigment grinding, cleaners, coatings and paints, fire fighting, paper, ink, oil, plastics, and textile industries; liq.; sol > 2% in water, methanol; sp.gr. 1.03 mg/m³; dens. 8.6 lb/gal; flash pt. (PMCC) 21 C; surf. tens. 18 dynes/cm (0.1%); flamm.; 25% solids in water/IPA (1:1).

Zonyl® FSB. [DuPont] Flurochemical surfactant; amphoteric; wetting agent, emulsifier, dispersant, corrosion inhibitor, leveling agent for adhesives, agric., polishes, polymerization, pigment grinding, cleaners, coatings and paints, fire fighting, paper, ink, oil, plastics, and textile industries; foamer, froth flotation agent for solv. cleaners, elastomers; liq.; sol. > 2% in water, IPA, methanol; dens. 8.8 lb/gal; flash pt. 67 F (PMCC); surf. tens. 17 dynes/cm (0.1%); 40% solids in IPA/water.

Zonyl® FSC. [DuPont] Fluorochemical surfactant; cationic; wetting agent, emulsifier, dispersant, corrosion inhibitor, leveling agent for adhesives, agric., polishes, polymerization, pigment grinding, cleaners, coatings and paints, fire fighting, paper, ink, oil, plastics, and textile industries; foamer, froth flotation agent for solv. cleaners, elastomers; liq.; sol. > 2% in water, IPA, methanol, acetone; sp.gr. 1.16 mg/m³; dens. 9.7 lb/gal; flash pt. (PMCC) 21 C; surf. tens. 19 dynes/cm (0.1% aq.); flamm.; 50% solids in water/IPA (25:25).

Zonyl® FSE. [DuPont] Fluorochemical surfactant; anionic; leveling agent for emulsion, pigment dispersant, wetting and foaming agent for corrosive media; liq.; sp.gr. 1.12 mg/m³; dens. 9.3 lb/gal; flash pt. (PMCC) > 93 C; surf. tens. 20 dynes/cm (0.1% aq.); 14% conc. in water/ethylene glycol (62:24).

Zonyl® FSJ. [DuPont] Fluorochemical surfactant; anionic; wetting agent for adhesives, agric., cleaners, fire fighting, petrol., resin molds, textiles; liq.; sol. > 2% in water, methanol; sp.gr. 1.12 mg/m³; dens. 9.3 lb/gal; flash pt. (PMCC) 28 C; surf. tens. 21 dynes/cm (0.1% aq.); flamm.; 40% solids in water/IPA (45:15).

Zonyl® FSK. [DuPont] Fluorochemical surfactant; amphoteric; leveling agent for emulsion, pigment dispersant, wetting and foaming agent for corrosive media; liq.; sol. in polar solvs.; sp.gr. 1.25 mg/m³; dens. 10.4 lb/gal; flash pt. (PMCC) 40 C; surf. tens. 19 dynes/cm (0.1% aq.); corrosive; 47% solids in acetic acid.

Zonyl® FSN. [DuPont] Fluorochemical surfactant; nonionic; wetting agent, emulsifier, dispersant, corrosion inhibitor, leveling agent for adhesives, agric., polishes, polymerization, pigment grinding, cleaners, coatings and paints, fire fighting, paper, ink, oil, plastics, and textile industries; liq.; sol. > 2% in water, IPA, methanol, acetone, ethyl acetate, THF; sp.gr. 1.06 mg/m³; dens. 8.8 lb/gal; flash pt. (PMCC) 22 C; surf. tens. 23 dynes/cm (0.1%); flamm.; 40% solids in water/IPA (30:30).

Zonyl® FSN-100. [DuPont] Fluorochemical surfactant; nonionic; leveling agent for emulsion, pigment dispersant, wetting and foaming agent for corrosive media; thin paste; sol. in polar and nonpolar solvs.; sp.gr. 1.35 mg/m³; dens. 11.2 lb/gal; flash pt. > 93 C; surf. tens. 23 dynes/cm (0.1% aq.); 100% conc.

Zonyl® FSO. [DuPont] Fluorochemical surfactant; nonionic; leveling agent for emulsion, pigment dispersant, wetting and foaming agent for corrosive media; turbid liq.; sol. in polar and nonpolar solvs.; sp.gr. 1.35 mg/m³; dens. 11.2 lb/gal; flash pt. (PMCC) > 93 C; surf. tens. 18 dynes/cm (0.1% aq.); 50% conc. in water/ethylene glycol (25:25).

Zonyl® FSO-100. [DuPont] Fluorochemical surfactant; nonionic; leveling agent for emulsion, pigment dispersant, wetting and foaming agent for corrosive media; turbid liq.; sol. in polar and nonpolar solvs.; sp.gr. 1.35 mg/m³; dens. 11.2 lb/gal; flash pt. (PMCC) > 93 C; surf. tens. 23 dynes/cm (0.1%); 100% conc.

Zonyl® FSP. [DuPont] Fluorochemical surfactant; anionic; wetting agent, emulsifier, dispersant, corrosion inhibitor, leveling agent, antifoam for adhesives, agric., polishes, polymerization, pigment grinding, cleaners, coatings and paints, fire fighting, paper, ink, oil, plastics, and textile industries; liq.; sol. > 2% in water, methanol; sp.gr. 1.15 mg/m³; dens. 9.6 lb/gal; flash pt. (PMCC) 24 C; surf. tens. 21 dynes/cm (0.1%); flamm.; 35% solids in water/IPA (45:20).

Zonyl® NF. [DuPont] Fluorochemical surfactant; anionic; wetting agent, oil and grease holdouts for paper and paperboard; liq.; 95% conc.

Zonyl® RP. [DuPont] Fluorochemical surfactant; anionic; wetting agent, paper fluoridizer; amber liq.; sol. in water; dens. 9.5 lb/gal; visc. 10 cps; flash pt. > 28 C; pH 7–8; 34% conc.

Zonyl® TBS. [DuPont] Fluorochemical surfactant; anionic; leveling agent for emulsions; pigment dispersant; wetting and foaming agent for corrosive media; slurry; sp.gr. 1.20 mg/m³; dens. 10.0 lb/gal; flash pt. (PMCC) > 93 C; surf. tens. 24 dynes/cm (0.1% aq.); 33% solids in water/acetic acid (64:3).

Zonyl® UR. [DuPont] Fluorochemical surfactant; anionic; wetting agent, antifoam, corrosion inhibitor; leveling agent for floor polishes; solid/paste; sol. in polar and nopolar solvs.; sp.gr. 1.84 mg/m³; dens. 15.3 lb/gal; m.p. 20 C; flash pt. (PMCC) > 93 C; surf. tens. 28 dynes/cm (0.1% aq.); corrosive; 100% conc.

Zoramide CM. [Zohar Detergent Factory] Cocamide MEA; CAS 68140-00-1; nonionic; foam booster, thickener, superfatting agent; flakes; 100% conc.

Zoramox. [Zohar Detergent Factory] Coconut amido alkyl amine oxide; nonionic; wetting agent, visc. booster/stabilizer for shampoos, bubble baths; visc. builder for low pH shampoos, other liq. detergents; liq.; 30% conc.

Zoramox LO. [Zohar Detergent Factory] Alkyl dimethyl amine oxide; amphoteric; wetting agent, foam booster/stabilizer for shampoos, bubble baths; visc. builder for low pH shampoos, other liq. detergents; liq.; 30% conc.

Zorapol LS-30. [Zohar Detergent Factory] Modified sodium lauryl sulfate; CAS 151-21-3; foaming agent for syn. latexes, emulsion polymerization aid, esp. in mfg. of polyacrylate emulsion; liq. to paste; 32% conc.

Zorapol SN-9. [Zohar Detergent Factory] Sodium nonoxynol-9 sulfate; CAS 9014-90-8; emulsifier for emulsion polymerization of polyacrylates; in prep. of household detergents; liq.; 32% conc.

Zusolat 1004. [Zschimmer & Schwarz] Fatty alcohol polyglycol ether (4 EO); nonionic; washing and cleansing agent; liq.; water-sol.; cloud pt. 57-61 C (1% aq.); 85% act.

Zusolat 1005/85. [Zschimmer & Schwarz] Fatty alcohol polyglycol ether (5 EO); nonionic; dispersant, emulsifier, wetting agent for mfg. of washing and cleaning agents, dishwash, household/industrial and metal cleaners, textile, leather, and paper aux.; colorless liq.; water-sol.; cloud pt. 65-68 C (5 g/20 ml butyl diglycol 25%); pH 5-7 (10%); 85% act. in water.

Zusolat 1008/85. [Zschimmer & Schwarz] Fatty alcohol polyglycol ether (8 EO); nonionic; washing and cleansing agent for dishwash, industrial/household and metal cleaners, textile, leather and paper aux.; colorless liq.; water-sol.; cloud pt. 66-70 C (1% aq.); pH 5-7 (10%); 85% act. in water.

Zusolat 1010/85. [Zschimmer & Schwarz] Fatty alcohol polyglycol ether (10 EO); nonionic; washing and cleansing agent for dishwash, industrial/household and metal cleaners, textile, leather and paper aux.; colorless liq.; water-sol.; cloud pt. 87-91 C (1% aq.); pH 5-7 (10%); 85% act. in water.

Zusolat 1012/85. [Zschimmer & Schwarz] Fatty alcohol polyglycol ether (12 EO); nonionic; washing and cleansing agent for dishwash, industrial/household and metal cleaners, textile, leather and paper aux.; colorless liq.; water-sol.; cloud pt. 93-96 C (1% aq.); pH 5-7 (10%); 85% act. in water.

Zusomin C 108. [Zschimmer & Schwarz] PEG-8 cocamine; CAS 61791-14-8; nonionic; basic material and component for dyeing and textile auxs.; intermediate for quaternization; metal cleaners; liq.; amine value 18-19; 100% act.

Zusomin O 102. [Zschimmer & Schwarz] PEG-2 oleamine; nonionic; base material for chemo-tech. prods.; liq.; amine value 28-29; 100% act.

Zusomin O 105. [Zschimmer & Schwarz] PEG-5 oleamine; nonionic/cationic; basic material for textile and dyeing auxs., intermediate for quaternization, metal cleaners; liq.; amine value 18-20; 100% act.

Zusomin S 110. [Zschimmer & Schwarz] PEG-10 stearamine; CAS 26635-92-7; nonionic; basic material for dyeing and textile aux.; intermediate for quaternization and metal cleaners; liq.; amine value 12-14; 100% act.

Zusomin S 125. [Zschimmer & Schwarz] PEG-25 stearamine; CAS 26635-92-7; nonionic; base material for chemo-tech. prods.; liq.; amine value 7-8; 100% act.

Zusomin TG 102. [Zohar Detergent Factory] PEG-2 tallowamine; nonionic; basic material and component for dyeing and textile aux.; intermediate for quaternization and metal cleaners; paste; 100% conc.

Part II
Tradename Application Cross Reference

Tradename Application Cross Reference

Tradename chemicals from the first part of this reference are grouped by broad application areas derived from manufacturer's specifications. The following tradename products may not be limited to the application areas represented here.

Agricultural Chemicals

Ablumul AG-306
Ablumul AG-420
Ablumul AG-900
Ablumul AG-909
Ablumul AG-910
Ablumul AG-1214
Ablumul AG-AH
Ablumul AG-GL
Ablumul AG-H
Ablumul AG-KTM
Ablumul AG-L
Ablumul AG-MBX
Ablumul AG-SB
Ablumul AG-WP
Ablumul AG-WPS
Ablunol LA 3
Ablunol LA 5
Ablunol LA 7
Ablunol LA 9
Ablunol LA 12
Ablunol LA 16
Ablunol LA 40
Ablusol C-78
Acconon 300-MO
Acconon 400-MO
Acetamin 24
Acetamin 86
Adsee® 100-80
Adsee® 775
Adsee® 799
Adsee® 801
Adsee® 2141
Aerosol® AY-65
Aerosol® AY-100
Aerosol® OS
Aerosol® OT-75%
Aerosol® OT-B
Aerosol® OT-MSO
Aerosol® OT-S
AF 10 IND
AF 30 IND
AF 93
AF 8805
AF 8810
AF 8810 FG
AF 8820
AF 8820 FG
AF 8830
AF 8830 FG
AF 9000
Agrilan® A
Agrilan® AC
Agrilan® AEC123
Agrilan® AEC145
Agrilan® AEC156
Agrilan® AEC167
Agrilan® AEC178
Agrilan® AEC189
Agrilan® AEC200
Agrilan® AEC211
Agrilan® AEC266
Agrilan® AEC299
Agrilan® AEC310
Agrilan® BA
Agrilan® BM
Agrilan® C91D
Agrilan® D54
Agrilan® DG102
Agrilan® DG113
Agrilan® EA14
Agrilan® EA25
Agrilan® EA36
Agrilan® EA47
Agrilan® EA58
Agrilan® EA69
Agrilan® EA80
Agrilan® F460
Agrilan® F491
Agrilan® F502
Agrilan® F513
Agrilan® F524
Agrilan® F535
Agrilan® F546
Agrilan® F557
Agrilan® FS101
Agrilan® FS112
Agrilan® MC-90
Agrilan® TKA103
Agrilan® TKA114
Agrilan® TKA125
Agrilan® TKA147
Agrilan® WP101
Agrilan® WP112
Agrilan® WP123
Agrilan® WP134
Agrilan® WP145
Agrilan® WP156
Agrilan® WP167
Agrilan® WP178
Agrilan® X98
Agrilan® X109
Agrimul 26-B
Agrimul 70-A
Agrimul A-300
Agrimul N-300
Agrimul S-300
Agrisol PX401
Agrisol PX413
Agriwet 1186A
Agriwet CA, C91, C92
Agriwet FOA
Agriwet T-F
Akypogene Jod F
Akyposal 100 DE
AL 2070
Alcodet® 218
Alcodet® 260
Alcodet® SK
Alkamuls® 400-DO
Alkamuls® GMO
Alkamuls® PSMO-5
Alkasurf® NP-6
Alkasurf® NP-9
Alkasurf® NP-11
Alkasurf® OP-5
All Wet
Amsul 70
Antifoam 7800 New
Aristonate H
Aristonate L
Aristonate M
Armix 176
Armix 180-C
Armix 183
Armix 185
Armix 309
Armul 03
Armul 17
Armul 21
Armul 22
Armul 33
Armul 34
Armul 44
Armul 55
Armul 66
Armul 88
Armul 100, 101, 102
Armul 214
Armul 215
Armul 930
Armul 940
Armul 950
Armul 2404
Armul 3260
Armul 5830
Arnox 930, 940
Arquad® 2C-75
Arquad® 12-50
Arquad® 16-50
Arquad® 18-50
Arquad® B-50
Arquad® B-90
Arquad® C-50
Arquad® DMMCB-50
Arquad® T-2C-50
Arquad® T-27W
Arylan® CA
Arylan® PWS
Atlas G-1256
Atlas G-1281
Atlas G-1284
Atlas G-1285
Atlas G-1288
Atlas G-1289
Atlas G-1292
Atlas G-1295
Atlas G-1300
Atlas G-1304
Atlas G-3300B
Atlas G-8916PF
Atlas G-8936CJ
Atlox 80
Atlox 775
Atlox 804
Atlox 847
Atlox 848
Atlox 849
Atlox 1045A
Atlox 3300B
Atlox 3335 B
Atlox 3386 B
Atlox 3387 BM
Atlox 3400B
Atlox 3401
Atlox 3403F
Atlox 3404F
Atlox 3406F
Atlox 3409F
Atlox 3414F
Atlox 3422F
Atlox 3450F
Atlox 3453F
Atlox 3454F
Atlox 3455F
Atlox 4851 B
Atlox 4853 B
Atlox 4855 B
Atlox 4856 B

Agricultural Chemicals (cont'd.)

Atlox 4857 B
Atlox 4858 B
Atlox 4861B
Atlox 4862
Atlox 4868B
Atlox 4875
Atlox 4880B
Atlox 4881
Atlox 4885
Atlox 4890B
Atlox 4896
Atlox 4898
Atlox 4899B
Atlox 4901
Atlox 4911
Atlox 4912
Atlox 4990B
Atlox 4991
Atlox 4995
Atlox 5320
Atlox 5325
Atlox 5330
Atlox 8916PF
Atlox 8916TF
Atplus 300F
Atplus 401
Atplus 526
Atplus 1992
Atplus 2380
Atsurf 311
Atsurf 1910
Avirol® SO 70P
Berol 28
Berol 106
Berol 302
Berol 303
Berol 307
Berol 381
Berol 386
Berol 387
Berol 389
Berol 391
Berol 392
Berol 397
Berol 455
Berol 456
Berol 457
Berol 458
Berol 822
Berol 824
Berol 930
Berol 938
Berol 946
Berol 947
Berol 948
Berol 949
Bio-Step Series
Blancol® N
Bubble Breaker® 259, 260, 613-M, 622, 730, 737, 746, 748, 900, 913, 917
CA DBS 50 SA
Calfax 10L-45
Calfax DB-45
Calimulse EM-99
Calimulse PRS
Calsoft L-40
Calsoft L-60
Calsoft T-60
Calsuds CD-6
Cardolite® NC-507
Cardolite® NC-510
Casul® 55 HF
Casul® 70 HF
Catinex KB-42
Cedephos® FA600
Cedephos® RA600
Chemal DA-4
Chemal DA-6
Chemal DA-9
Chemax DNP-150/50
Chemax DOSS/70
Chemax DOSS-75E
Chemax E-400 MO
Chemax NP-1.5
Chemax NP-4
Chemax NP-6
Chemax NP-9
Chemax OP-3
Chemax OP-5
Chemax OP-30/70
Chemax PEG 400 DO
Chemax PEG 600 DO
Chemeen DT-3
Chemeen DT-15
Chemfac NC-0910
Chemfac PA-080
Chemfac PB-082
Chemfac PB-106
Chemfac PB-135
Chemfac PB-184
Chemfac PB-264
Chemfac PC-188
Chemfac PD-600
Chemfac PF-623
Chemfac PF-636
Chemfac PN-322
Chemphos TC-310S
Chemphos TR-495
Cithrol 2MS
Cithrol 3MS
Cithrol 4MS
Cithrol 6MS
Cithrol 10MS
Cithrol 15MS
Cithrol 40MS
Crafol AP-55
Crafol AP-85
Crapol AV-10
Crapol AV-11
Crill 1
Crill 2
Crill 3
Crill 4
Crillet 1
Crillet 3
Crillet 4
Crodamine 1.HT
Crodamine 1.O, 1.OD
Crodamine 1.T
Crodamine 1.16D
Crodamine 1.18D
Crodamine 3.A16D
Crodamine 3.A18D
Crodamine 3.ABD
Crodamine 3.AED
Crodamine 3.AOD
Croduret 10
Croduret 30
Croduret 40
Croduret 60
Croduret 100
Crown Anti-Foam
Cycloryl DCA
Farmin R 24H, R 86H
Farmin R 86H
Flo-Mo® 1X, 2X
Flo-Mo® 8X
Flo-Mo® 80/20
Flo-Mo® 1002
Flo-Mo® 1031
Flo-Mo® 1082
Flo-Mo® DEH, DEL
Flo-Mo® Low Foam
Flo-Mo® Suspend
Flo-Mo® Suspend Plus
Fluowet PL
Fluowet PP
FMB 3328-5 Quat, 3328-8 Quat
FMB 4500-5 Quat
FMB 6075-5 Quat, 6075-8 Quat
Foamaster DRY
Foamaster Soap L
Fosterge BA-14 Acid
Fosterge LF
Gantrez® AN
Gantrez® AN-139
Gantrez® AN-169
Gantrez® AN-179
Genamin C Grades
Genamin CC Grades
Genamin T-020
Genamin T-050
Genamin T-150
Genamin T-200
Genamin T-308
Genamin TA Grades
Genamine C-020
Genamine C-050
Genamine C-080
Genamine C-100
Genamine C-150
Genamine C-200
Genamine C-250
Genamine O-020
Genamine O-050
Genamine O-080
Genamine O-100
Genamine O-150
Genamine O-200
Genamine O-250
Genamine S-020
Genamine S-050
Genamine S-080
Genamine S-150
Genamine S-200
Genamine S-250
Genapol® 24-L-60
Genapol® 24-L-60N
Genapol® 24-L-75
Genapol® 24-L-92
Genapol® 24-L-98N
Genapol® 26-L-1
Genapol® 26-L-1.6
Genapol® 26-L-2
Genapol® 26-L-3
Genapol® 26-L-5
Genapol® 26-L-45
Genapol® 26-L-60
Genapol® 26-L-60N
Genapol® 42-L-3
Genapol® C-050
Genapol® C-080
Genapol® C-100
Genapol® C-150
Genapol® C-200
Genapol® L Grades
Genapol® LRO Liq., Paste
Genapol® O-050
Genapol® O-080
Genapol® O-090
Genapol® O-100
Genapol® O-120
Genapol® O-150
Genapol® O-200
Genapol® O-230
Genapol® OX Grades
Genapol® PF 10
Genapol® PF 20
Genapol® PF 40
Genapol® PF 80
Genapol® T Grades
Genapol® V 2908
Genapol® V 2909
Geronol AG-100/200 Series
Geronol AG-821
Geronol AG-900
Geronol AZ82
Geronol FF/4
Geronol FF/6
Geronol MOE/2/N
Geronol MS
Geronol PRH/4-A
Geronol PRH/4-B
Geronol RE/70
Geronol SC/120
Geronol SC/121
Geronol SC/138
Geronol SC/177
Geronol SN
Geronol V/087
Geronol V/497
Geropon® 40/D
Geropon® 99
Geropon® 111
Geropon® CET/50/P
Geropon® CYA/DEP
Geropon® FMS
Geropon® IN
Geropon® K/65
Geropon® K/202
Geropon® NK
Geropon® SC/211
Geropon® SC/213
Geropon® SDS
Geropon® TA/72
Geropon® TA/72/S

Agricultural Chemicals (cont'd.)

Geropon® TA/764
Glycidol Surfactant 10G
Glycosperse® O-20 FG, O-20 KFG
GP-330-I Antifoam Emulsion
Gradonic N-95
Hetoxamine C-2
Hetoxamine C-5
Hetoxamine O-2
Hetoxamine O-5
Hetoxamine O-15
Hetoxamine S-2
Hetoxamine S-5
Hetoxamine S-15
Hetoxamine ST-2
Hetoxamine ST-5
Hetoxamine ST-15
Hetoxamine ST-50
Hetoxamine T-2
Hetoxamine T-5
Hetoxamine T-15
Hetoxamine T-20
Hetoxamine T-30
Hetoxamine T-50-70%
Hodag Nonionic E-5
Hodag Nonionic E-6
Hodag Nonionic E-7
Hodag Nonionic E-10
Hodag Nonionic E-12
Hodag Nonionic E-20
Hodag Nonionic E-30
Hoe S 1816
Hoe S 1816-1
Hoe S 1816-2
Hoe S 1984 (TP 2279)
Hoe S 1984 (TP 2283)
Hoe S 2713
Hoe S 2713 HF
Hoe S 2749
Hoe S 2895
Hoe S 2896
Hoe S 3435
Hoe S 3618
Hoe S 3680
Hostapal 2345
Hostapal N-040
Hostapal N-060
Hostapal N-100
Hostapal TP 2347
Hostaphat AR K
Hostaphat MDAR Grades
Hostaphat MDLZ
Hostapon T Powd.
Hostapur DTC
Hostapur DTC FA
Hyonic NP-60
Hyonic NP-90
Hyonic NP-110
Hyonic NP-120
Hyonic NP-407
Hyonic NP-500
Hyonic OP-40
Hyonic OP-70
Hyonic OP-100
Hyonic OP-407
Hyonic OP-705
Igepal® CA-210
Igepal® CA-520
Igepal® CA-620
Igepal® CA-630
Igepal® CA-720
Igepal® CO-430
Igepal® CO-520
Igepal® CO-530
Igepal® CO-660
Igepal® CO-710
Igepal® CO-720
Igepal® CO-730
Igepal® CO-880
Igepal® CO-890
Igepal® CO-970
Igepal® CO-990
Igepal® DM-430
Igepal® DM-710
Igepal® DM-730
Igepal® DM-880
Igepal® DM-970 FLK
Igepal® DX-430
Igepal® KA
Igepal® NP-8
Igepal® RC-520
Igepal® RC-620
Igepal® RC-630
Incrocas 30
Incrocas 40
Interwet® 33
Ionet S-20
Ionet S-60 C
Ionet S-80
Ionet T-20 C
Ionet T-60 C
Ionet T-80 C
Jet Amine DMCD
Jet Amine DMOD
Jet Amine DMSD
Jet Amine DMTD
Jet Amine M2C
Jet Amine PC
Jet Amine PCD
Jet Amine PHT
Jet Amine PHTD
Jet Amine PO
Jet Amine POD
Jet Amine PS
Jet Amine PSD
Jet Amine PT
Jet Amine PTD
Kalcohl 10H
Kalcohl 20
Kalcohl 40
Kalcohl 60
Kemester® 104
Kessco® PEG 200 DO
Kessco® PEG 200 DS
Kessco® PEG 200 ML
Kessco® PEG 200 MO
Kessco® PEG 200 MS
Kessco® PEG 300 DL
Kessco® PEG 300 DO
Kessco® PEG 300 DS
Kessco® PEG 300 ML
Kessco® PEG 300 MO
Kessco® PEG 300 MS
Kessco® PEG 400 DL
Kessco® PEG 400 DO
Kessco® PEG 400 ML
Kessco® PEG 400 MO
Kessco® PEG 600 DO
Kessco® PEG 600 MO
Kessco® PEG 600 MS
Kessco® PEG 1000 DL
Kessco® PEG 1000 DO
Kessco® PEG 1000 DS
Kessco® PEG 1000 ML
Kessco® PEG 1000 MO
Kessco® PEG 1000 MS
Kessco® PEG 1540 DL
Kessco® PEG 1540 DO
Kessco® PEG 1540 DS
Kessco® PEG 1540 ML
Kessco® PEG 1540 MO
Kessco® PEG 1540 MS
Kessco® PEG 4000 DL
Kessco® PEG 4000 DO
Kessco® PEG 4000 DS
Kessco® PEG 4000 ML
Kessco® PEG 4000 MO
Kessco® PEG 4000 MS
Kessco® PEG 6000 DL
Kessco® PEG 6000 DO
Kessco® PEG 6000 ML
Kessco® PEG 6000 MO
Kessco® PEG 6000 MS
Klearfac® AA040
Lankropol® WA
Lankropol® WN
Leonil DB Powd.
Leonil OS
Lignosite® 260
Lignosol AXD
Lignosol B
Lignosol BD
Lignosol DXD
Lignosol HCX
Lomar® LS
Lomar® PL
Lomar® PW
Lomar® PWA
Lomar® PWA Liq.
Lonzest® PEG 4-DO
Lonzest® PEG 4-L
Mackanate DOS-70MS
Mackanate DOS-75
Macol® DNP-150
Macol® NP-9.5
Madeol AG 1989 N
Madeol AG BX
Madeol AG/TR 8, AG/TR 12
Makon® 4
Makon® 6
Makon® 7
Makon® 8
Makon® 10
Makon® 11
Makon® 12
Makon® 14
Makon® 30
Makon® 50
Makon® NI 10, NI 20, NI 30
Manoxol MA
Manoxol N
Manoxol OT
Manoxol OT 60%
Manoxol OT/B
Manro SDBS 25/30
Marasperse N-22
Marlowet® 4901
Marlowet® 4902
Marlowet® EF
Marlowet® IHF
Marlowet® ISM
Marlowet® OFA
Marlowet® R 20
Marlowet® R 22
Marlowet® R 25
Marlowet® R 32
Marlowet® R 36
Marlowet® R 40
Marlowet® R 54
Mazamide® L-298
Mazeen® 173
Mazeen® 174
Mazeen® 174-75
Mazeen® C-2
Mazeen® C-5
Mazeen® C-10
Mazeen® C-15
Mazeen® S-2
Mazeen® S-5
Mazeen® S-10
Mazeen® S-15
Mazeen® T-2
Mazeen® T-5
Mazeen® T-10
Mazeen® T-15
Mazon® 1045A
Mazon® 1086
Mazon® 1096
Mazu® 208 LPD
Mazu® DF 200SX
Mazu® DF 200SXSP
Mazu® DF 210SX
Mazu® DF 210SX Mod 1
Mazu® DF 210SXSP
Mazu® DF 215SX
Mazu® DF 230SX
Mazu® DF 230SXSP
Mazu® DF 243
Merpol® 100
Micro-Step® H-301
Micro-Step® H-302
Micro-Step® H-303
Micro-Step® H-304
Micro-Step® H-305
Micro-Step® H-306
Monafax 785
Monafax 786
Monafax 794
Monafax 872
Monafax 1293
Monawet MB-100
Monawet MM-80
Monawet MO-65-150
Monawet MO-65 PEG
Monawet MO-70-150
Monawet MO-70E
Monawet MO-70 PEG

Agricultural Chemicals (cont'd.)

Monawet MO-70R
Monawet MO-70S
Monawet MO-75E
Monawet MO-84R2W
Monawet SNO-35
Monazoline O
Monolan® 1030
Monolan® 2000
Monolan® 2800
Monolan® M
Monolan® O Range
Monolan® PC
Morwet® B
Morwet® DB
Morwet® IP
Morwet® M
Nacconol® 90G
Nansa® BXS
Nansa® EVM50
Nansa® EVM62/H
Nansa® EVM70
Nansa® EVM70/B
Nansa® EVM70/E
Nansa® HS80-AU
Nansa® LSS38/A
Nansa® S40/S
Nansa® SS 30
Nansa® SS 50
Nansa® SS 55
Nansa® SS A/S
Nansa® YS94
Naxonic NI-40
Naxonic NI-60
Naxonic NI-100
Neutronyx® 656
Newcol 506
Newcol 508
Newcol 560
Newcol 561H, 562, 564, 565
Newcol 566
Newcol 804, 808, 860
Newcol 862
Newcol 864
Newcol 865
Newkalgen 2360X1
Newkalgen 2720X75
Newkalgen 3000 A & B
Newpol PE-61
Newpol PE-62
Newpol PE-64
Newpol PE-68
Newpol PE-74
Newpol PE-75
Newpol PE-78
Newpol PE-88
Niaproof® Anionic Surfactant 08
Nikkol DDP-2
Nikkol GO-430
Nikkol GO-440
Nikkol GO-460
Nikkol NP-10
Nikkol NP-5
Nikkol NP-7.5
Nikkol NP-15
Nikkol NP-18TX
Nikkol OP-3
Nikkol OP-10
Nikkol OP-30
Nikkol TDP-2
Nikkol TDP-4
Nikkol TDP-6
Nikkol TDP-8
Nikkol TDP-10
Ninate® 401
Ninate® 401-A
Ninate® 411
Ninate® DS 70
Ninate® PA
Niox KI Series
Nipol® 2782
Nipol® 4472
Nipol® 5595
Nissan Amine AB
Nissan Amine ABT
Nissan Amine BB
Nissan Amine FB
Nissan Amine MB
Nissan Amine OB
Nissan Amine PB
Nissan Amine SB
Nissan Amine VB
Nissan Cation ABT-350, 500
Nissan Cation F2-10R, -20R, -40E, -50
Nissan Cation FB, FB-500
Nissan Cation L-207
Nissan Cation M2-100
Nissan Cation PB-40, -300
Nissan Cation S2-100
Nissan Nonion E-205
Nissan Nonion E-206
Nissan Nonion E-208
Nissan Nonion E-215
Nissan Nonion E-220
Nissan Nonion E-230
Nissan Nonion HS-204.5
Nissan Nonion HS-206
Nissan Nonion HS-208
Nissan Nonion HS-210
Nissan Nonion HS-215
Nissan Nonion HS-220
Nissan Nonion HS-240
Nissan Nonion HS-270
Nissan Nonion K-202
Nissan Nonion K-203
Nissan Nonion K-204
Nissan Nonion K-207
Nissan Nonion K-211
Nissan Nonion K-215
Nissan Nonion K-220
Nissan Nonion K-230
Nissan Nonion NS-202
Nissan Nonion NS-204.5
Nissan Nonion NS-206
Nissan Nonion NS-208.5
Nissan Nonion NS-209
Nissan Nonion NS-210
Nissan Nonion NS-212
Nissan Nonion NS-215
Nissan Nonion NS-220
Nissan Nonion NS-230
Nissan Nonion NS-240
Nissan Nonion NS-250
Nissan Nonion NS-270
Nissan Nonion P-210
Nissan Nonion P-213
Nissan Nonion S-206
Nissan Nonion S-207
Nissan Nonion S-215
Nissan Nonion S-220
Nonionic E-4
Nonipol 20
Nonipol 40
Nonipol 55
Nonipol 60
Nonipol 70
Nonipol 85
Nonipol 95
Nonipol 100
Nonipol 110
Nonipol 120
Nonipol 130
Nonipol 160
Nonipol 200
Nonipol 400
Nopco® RDY
Nopcosperse AD-6
Norfox® 40
Norfox® 90
Norfox® F-221
Norfox® KO
Norfox® T-60
NP-55-60
Nutrol 100
Octowet 40
Octowet 55
Octowet 60
Octowet 60-I
Octowet 65
Octowet 70
Octowet 70A
Octowet 70BC
Octowet 70PG
Octowet 75
Octowet 75E
Orzan® LS
PGE-600-ML
PGE-600-MS
Paracol AL, AH
Paracol SV
Pentine 1185 5432
Pestilizer® B Series
Petro® WP
Phenyl Sulphonate HFCA
Phenyl Sulphonate HSR
Phospholan® PHB14
Phospholan® PNP9
Phospholan® PRP5
Phospholipid PTD
Phospholipid PTL
Phospholipid PTZ
Plurafac® T-55
Pluraflo® E4A
Pluraflo® E4B
Pluraflo® E5A
Pluraflo® E5B
Pluraflo® E5BG
Pluraflo® E5G
Pluraflo® N5G
Pluronic® 10R5
Pluronic® 10R8
Pluronic® 12R3
Pluronic® 17R1
Pluronic® 17R2
Pluronic® 17R4
Pluronic® 17R8
Pluronic® 22R4
Pluronic® 25R1
Pluronic® 25R2
Pluronic® 25R4
Pluronic® 25R5
Pluronic® 25R8
Pluronic® 31R1
Pluronic® 31R2
Pluronic® 31R4
Pluronic® F38
Pluronic® F77
Pluronic® F87
Pluronic® F88
Pluronic® F98
Pluronic® F127
Pluronic® L31
Pluronic® L35
Pluronic® L42
Pluronic® L43
Pluronic® L44
Pluronic® L61
Pluronic® L62
Pluronic® L62D
Pluronic® L62LF
Pluronic® L63
Pluronic® L64
Pluronic® L72
Pluronic® L81
Pluronic® L92
Pluronic® P65
Pluronic® P103
Pluronic® P104
Pluronic® PE 10500
Plysurf A207H
Plysurf A208B
Plysurf A208S
Plysurf A210G
Plysurf A212C
Plysurf A215C
Plysurf A216B
Plysurf A217E
Plysurf A219B
Plysurf AL
Pogol 300
Poly-G® 200
Poly-G® 300
Poly-G® 400
Poly-G® 600
Poly-G® 1000
Poly-G® 1500
Poly-G® 2000
Poly-G® B1530
Polysurf A212E
Poly-Tergent® E-17A
Poly-Tergent® E-17B
Poly-Tergent® E-25B
Pronal EX-300
Prote-pon P-2 EHA-02-K30
Prote-pon P 2 EHA-02-Z

Agricultural Chemicals (cont'd.)

Prote-pon P 2 EHA-Z
Prote-pon P-0101-02-Z
Prote-pon P-L 201-02-K30
Prote-pon P-L 201-02-Z
Prote-pon P-NP-06-K30
Prote-pon P-NP-06-Z
Prote-pon P-NP-10-K30
Prote-pon P-NP-10-MZ
Prote-pon P-NP-10-Z
Prote-pon P-OX 101-02-K75
Prote-pon P-TD-06-K13
Prote-pon P-TD-06-K30
Prote-pon P-TD 06-K60
Prote-pon P-TD-06-Z
Prote-pon P-TD-09-Z
Prote-pon P-TD-12-Z
Prote-pon TD-09-K30
Prote-sorb SML
Prote-sorb SMO
Prote-sorb SMP
Prote-sorb SMS
Prote-sorb STO
Prote-sorb STS
Protowet C
Prox-onic DT-03
Prox-onic DT-015
Prox-onic DT-030
Prox-onic L 081-05
Prox-onic L 101-05
Prox-onic L 102-02
Prox-onic L 121-09
Prox-onic L 161-05
Prox-onic L 181-05
Prox-onic L 201-02
Prox-onic MC-02
Prox-onic MC-05
Prox-onic MO-02
Prox-onic MO-015
Prox-onic MO-030
Prox-onic MO-030-80
Prox-onic MS-02
Prox-onic MS-05
Prox-onic MS-011
Prox-onic MS-050
Prox-onic MT-02
Prox-onic MT-05
Prox-onic MT-015
Prox-onic MT-020
Prox-onic PEG-2000
Prox-onic PEG-6000
Prox-onic PEG-10,000
Prox-onic PEG-20,000
Prox-onic PEG-35,000
Prox-onic PPG-900
Prox-onic PPG-1800
Prox-onic PPG-4000
Prox-onic SML-020
Prox-onic SMO-05
Prox-onic SMO-020
Prox-onic SMP-020
Prox-onic SMS-020
Prox-onic STS-020
Prox-onic TBP-08
Prox-onic TBP-030
Prox-onic TM-06
Prox-onic TM-08
Prox-onic TM-010
Quimipol ENF 140
Quimipol ENF 170
Radiamac 6149
Radiamac 6159
Radiamac 6169
Radiamac 6179
Radiasurf® 7125
Radiasurf® 7150
Radiasurf® 7156
Radiasurf® 7196
Radiasurf® 7402
Radiasurf® 7410
Radiasurf® 7411
Radiasurf® 7444
Radiasurf® 7900
Reax® 45A
Reax® 45DA
Reax® 45DTC
Reax® 45L
Reax® 45T
Rewopal® HV 5
Rewopal® HV 9
Rewopal® HV 25
Rewopal® RO 40
Rexol 25/7
Rexol 25/8
Rexol 25/30
Rexol 25/307
Rexol 25J
Rexol 45/10
Rexol 65/4
Rexol 65/6
Rexonic L125-9
Rexonic N25-9
Rexonic N25-9(85%)
Rhodacal® 70/B
Rhodacal® 330
Rhodacal® 2283
Rhodacal® BA-77
Rhodacal® BX-78
Rhodacal® CA, 70%
Rhodacal® DDB-40
Rhodacal® DS-10
Rhodacal® DSB
Rhodacal® IN
Rhodacal® N
Rhodacal® NK
Rhodafac® MD-12-116
Rhodafac® PE-510
Rhodafac® R9-47A
Rhodafac® RD-510
Rhodafac® RE-410
Rhodafac® RE-610
Rhodafac® RE-960
Rhodafac® RM-410
Rhodafac® RM-510
Rhodafac® RM-710
Rhodafac® RS-410
Rhodafac® RS-610
Rhodafac® RS-710
Rhodameen® O-12
Rhodameen® PN-430
Rhodameen® S-20
Rhodameen® T-12
Rhodameen® T-30-90%
Rhodasurf® 860/P
Rhodasurf® BC-420
Rhodasurf® BC-610
Rhodasurf® BC-840
Rhodasurf® LA-30
Rhodasurf® ON-870
Rhodopol® 23
Rhodorsil® AF 422
Rhodorsil® AF 426R
Rhodorsil® AF 454
Rhodorsil® AF 20441
Rhodorsil® EP 6703
Rhodorsil® SC 5020
Rhodorsil® SC 5021
Sandoz Sulfonate 3B2
Sandoz Sulfonate AAS 75S
Sanimal 55HX, 55LX, 250C, K51
Sanimal L, M, H
Schercomid CCD
Schercozoline B
Schercozoline C
Schercozoline O
Schercozoline S
Secoster® DO 600
Secoster® MA 300
Sellogen DFL
Sellogen HR
Sellogen HR-90
Sellogen NS-50
Sellogen W
Sellogen WL Liq.
Serdox NNP 4
Serdox NNP 5
Serdox NNP 6
Serdox NNP 7
Serdox NNP 8.5
Serdox NNP 12
Serdox NNP 13
Sermul EA 88
Sermul EN 25
Sermul EN 26
Servoxyl VPGZ 7/100
Servoxyl VPIZ 100
Servoxyl VPNZ 10/100
Servoxyl VPQZ 9/100
Servoxyl VPTZ 3/100
Servoxyl VPTZ 100
Servoxyl VPUZ
Servoxyl VPYZ 500
Silicone AF-10 FG
Silicone AF-10 IND
Silicone AF-30 IND
Silicone AF-100 IND
Silwet® L-77
Silwet® L-7602
Sinocol L
Sinocol PQ2
Siponic® SK (redesignated Alcodet® SK)
Siponic® TD-6
Soitem 5 C/70
Soitem 8 FL/N
Soitem 13 FL/N
Soitem 70
Soitem 101
Soitem 207
Soitem 251
Soitem 258
Soitem 480
Soitem 520
Soitem B
Soitem SC/70
Soitem SFL/1
Soprofor 3D33
Soprofor FL
Soprofor PA/17, PA/19
Soprofor PS/17, PS/19, PS/21
Soprophor® 4D384
Soprophor® 40/D
Soprophor® 497/P
Soprophor® 724/P
Soprophor® 796/P
Soprophor® BSU
Soprophor® CY/8
Soprophor® FLK
Soprophor® K/202
Soprophor® S/25
Soprophor® S40-P
Soprophor® SC/167
Sorbanox AST
Sorbax HO-40
Sorbax HO-50
Sorbax MO-40
Sorbax PML-20
Sorbax PMO-5
Sorbax PMO-20
Sorbax PMP-20
Sorbax PMS-20
Sorbax PTO-20
Sorbax PTS-20
Sorbax SML
Sorbax SMO
Sorbax SMP
Sorbax SMS
Sorbax STO
Sorbax STS
Sorbeth 40HO
Sorbirol O
Sorpol 230
Sorpol 320
Sorpol 355
Sorpol 560
Sorpol 900A, 1200, 2020K, 3005X
Sorpol 900H
Sorpol 900L
Sorpol 1200K
Sorpol 2495 G
Sorpol 2676S
Sorpol 2678S
Sorpol 3005X
Sorpol 3044
Sorpol 3370
Sorpol 5037
Sorpol 5039
Sorpol 5060
Sorpol 8070
Sorpol 9939
Sorpol H-770
Sorpol L-550
Sorpol W-150
Sponto® 101
Sponto® 102
Sponto® 140T
Sponto® 150T

Agricultural Chemicals (cont'd.)

Sponto® 150 TH
Sponto® 150 TL
Sponto® 168-D
Sponto® 169-T
Sponto® 200
Sponto® 203
Sponto® 207
Sponto® 217
Sponto® 221
Sponto® 232, 234
Sponto® 232T
Sponto® 234T
Sponto® 300T
Sponto® 305
Sponto® 500T
Sponto® 710T
Sponto® 712T
Sponto® 714T
Sponto® 723T
Sponto® 2174 H
Sponto® 2224T
Sponto® 4648-23A, 4648-23B
Sponto® AC60-02D
Sponto® AD4-10N
Sponto® AD6-39A
Sponto® AG3-55T
Sponto® AG-540
Sponto® AG-1040
Sponto® AG-1265
Sponto® AK30-02BT
Sponto® AK30-23
Sponto® AK31-53
Sponto® AK31-56
Sponto® AK31-64
Sponto® AK31-66
Sponto® AK31-69
Sponto® AL69-49
Sponto® AM2-07
Sponto® CA-861
Sponto® H-44-C
Sponto® N-140B
Spreading Agent ET0672
Step-Flow 21
Step-Flow 22
Step-Flow 23
Step-Flow 24
Step-Flow 25
Step-Flow 26
Step-Flow 41
Step-Flow 42
Step-Flow 61
Step-Flow 63
Stepanol® AM
Stepanol® AM-V
Stepanol® DEA
Stepanol® ME Dry
Stepanol® MG
Stepanol® WA-100
Stepanol® WAC
Stepanol® WAC-P
Stepanol® WA Paste
Stepanol® WA Special
Stepanol® WAT
Stepfac® 8170
Stepfac® 8171
Stepfac® 8172
Stepfac® 8173
Stepfac® PN 10
Stepsperse® DF-100
Stepsperse® DF-200
Stepsperse® DF-300
Stepsperse® DF-400
Stepwet® DF-60
Stepwet® DF-90
Sul-fon-ate AA-9
Sul-fon-ate AA-10
Sulframin Acide TPB
Super Wet
Super Wet Granular 15G
Supragil® GN
Supragil® MNS/90
Supragil® NS/90
Supragil® WP
Surfactol® 318
Surfactol® 365
Surfactol® 575
Surfactol® 590
Surfam P-MEPA
Surfam P5 Dist
Surfam P5 Tech
Surfax 250
Surfax 345
Surfax 585
Surfonic® HDL
Surfonic® HDL-1
Surfonic® L12-3
Surfonic® L12-6
Surfonic® L12-8
Surfonic® L24-2
Surfonic® L24-3
Surfonic® L24-7
Surfonic® L24-9
Surfonic® L24-12
Surfonic® L46-7
Surfonic® N-10
Surfonic® N-31.5
Surfonic® N-40
Surfonic® N-60
Surfonic® N-85
Surfonic® N-95
Surfonic® N-100
Surfonic® N-102
Surfonic® N-120
Surfonic® N-130
Surfonic® N-150
Surfonic® N-200
Surfonic® N-300
Surfonic® NB-5
Surfynol® 82
Surfynol® 82S
Surfynol® 104
Surfynol® 104A
Surfynol® 104S
Surfynol® 420
Surfynol® 440
Surfynol® DF-37
Surfynol® SE
Surfynol® TG-E
Swanic D-16, -24, -52, -60, -80, –120
Syn Fac® 334
Synperonic NP13
Synperonic NPE1800
Synprolam 35A
Syntofor A03, A04, AB03, AB04, AL3, AL4, B03, B04
Syntopon 8 D
Tallates, K
T-Det® BP-1
T-Det® C-18
T-Det® DD-5
T-Det® DD-7
T-Det® EPO-61
T-Det® EPO-64L
T-Det® N-4
T-Det® N-6
T-Det® N-8
T-Det® N-9.5
T-Det® N-14
T-Det® N-50
T-Det® N-507
T-Det® O-6
T-Det® O-40
T-Det® O-307
Tensiofix 20200
Tensiofix 35300
Tensiofix 35600
Tensiofix AS
Tensiofix B7416
Tensiofix B7438
Tensiofix B7453
Tensiofix B7504
Tensiofix B8426
Tensiofix B8427
Tensiofix BC222
Tensiofix BCZ
Tensiofix BS
Tensiofix CG21
Tensiofix D40 Bio-Act
Tensiofix EDS
Tensiofix EDS3
Tensiofix LX Special
Tensiofix PO120, PO132
Tensiofix XN6, XN10
Tergitol® 15-S-5
Tergitol® 15-S-7
Tergitol® 15-S-9
Tergitol® 15-S-15
Tergitol® 24-L-45
Tergitol® 24-L-60
Tergitol® 24-L-60N
Tergitol® 24-L-75
Tergitol® 24-L-92
Tergitol® 24-L-98N
Tergitol® 26-L-1.6
Tergitol® 26-L-3
Tergitol® 26-L-5
Tergitol® NP-7
Tergitol® TMN-10
Tergitol® XD
Tergitol® XJ
Teric 9A5
Teric 9A6
Teric 12A3
Teric 12A4
Teric 13A5
Teric 13A9
Teric 13M15
Teric 16M5
Teric 16M15
Teric 17A8
Teric 17A10
Teric 17A13
Teric 17A25
Teric 18M2
Teric 18M10
Teric 18M20
Teric 18M30
Teric 158
Teric 200
Teric 203
Teric 225
Teric 226
Teric N2
Teric N3
Teric N4
Teric N5
Teric N8
Teric N9
Teric N10
Teric N11
Teric N12
Teric N13
Teric PEG 300
Teric PEG 400
Teric PEG 600
Teric PEG 800
Teric X5
Teric X7
Teric X8
Teric X10
Teric X11
Teric X13
Teric X16
Teric X40
Teric X40L
Texapon Z
Textamine O-1
Tex-Wet 1104
Tex-Wet 1143
Tex-Wet 1155
T-Mulz® 63
T-Mulz® 339
T-Mulz® 391
T-Mulz® 392
T-Mulz® 426
T-Mulz® 540
T-Mulz® 565
T-Mulz® 598
T-Mulz® 734-2
T-Mulz® 808A
T-Mulz® 979
T-Mulz® 1120
T-Mulz® 1158
T-Mulz® 8015
T-Mulz® AO2
T-Mulz® AS-1151
T-Mulz® AS-1152
T-Mulz® AS-1153
T-Mulz® AS-1180
T-Mulz® COC
T-Mulz® FGO-17
T-Mulz® FGO-17A
T-Mulz® Mal 5
T-Mulz® O
T-Mulz® PB
T-Mulz® W
Tomah DA-14

Agricultural Chemicals (cont'd.)

Tomah DA-16
Tomah DA-17
Tomah E-19-2
Tomah E-DT-3
Tomah E-T-2
Tomah E-T-5
Tomah PA-10
Tomah PA-14
Tomah PA-16
Tomah PA-1214
Toxanon 500
Toxanon AH
Toxanon AM Series
Toxanon GR-31A
Toxanon P-8H, P-8L
Toxanon P-900
Toxanon SC-4
Toxanon V-1D
Toxanon XK-30
Toximul® 60 Series
Toximul® 360A
Toximul® 360B
Toximul® 360B-HF
Toximul® 374
Toximul® 425, 475
Toximul® 500
Toximul® 600
Toximul® 701, 702
Toximul® 705
Toximul® 707
Toximul® 709
Toximul® 710
Toximul® 713
Toximul® 715
Toximul® 716
Toximul® 800
Toximul® 804
Toximul® 811
Toximul® 812
Toximul® 850
Toximul® 852
Toximul® 856N
Toximul® 857N
Toximul® 858N
Toximul® 8170
Toximul® 8240
Toximul® 8242
Toximul® 8301N
Toximul® 8320
Toximul® 8321
Toximul® 8322
Toximul® 8323
Toximul® D
Toximul® FF
Toximul® H-HF
Toximul® MP
Toximul® MP-10
Toximul® MP-26
Toximul® MP-HF
Toximul® R-HF
Toximul® S-HF
Toximul® SEE-340
Toximul® SF Series
Toximul® TA-2
Toximul® TA-5
Toximul® TA-15
Trans-100
Triton® AG-98
Triton® B-1956
Triton® BG-10
Triton® CS-7
Triton® GR-7M
Triton® N-101
Triton® X-15
Triton® X-35
Triton® X-45
Triton® X-100 CG
Triton® X-114SBHF
Triton® X-120
Triton® X-180
Triton® X-185
Triton® X-190
Triton® X-193
Triton® X-207
Triton® X-305-70%
Triton® X-363M
Triton® X-405-70%
Trycol® 5874
Trycol® 5882
Trycol® 5888
Trycol® 5940
Trycol® 5941
Trycol® 5942
Trycol® 5943
Trycol® 5964
Trycol® 5966
Trycol® 5967
Trycol® 5968
Trycol® 5993
Trycol® 6940
Trycol® 6951
Trycol® 6953
Trycol® 6957
Trycol® 6958
Trycol® 6961
Trycol® 6962
Trycol® 6963
Trycol® 6964
Trycol® 6965
Trycol® 6968
Trycol® 6969
Trycol® 6970
Trycol® 6971
Trycol® 6974
Trycol® 6975
Trycol® 6984
Trycol® 6985
Trycol® 6989
Trycol® DNP-8
Trycol® DNP-150
Trycol® DNP-150/50
Trydet 2610
Trydet 2670
Trydet 2671
Trydet 2675
Trydet 2676
Trydet DO-9
Trydet OA-7
Trydet OA-10
Tryfac® 5552
Tryfac® 5553
Tryfac® 5554
Tryfac® 5555
Tryfac® 5556
Tryfac® 5559
Tryfac® 5561
Tryfac® 5571
Trylox® 5900
Trylox® 5902
Trylox® 5904
Trylox® 5906
Trylox® 5907
Trylox® 5909
Trymeen® 6601
Trymeen® 6607
Trymeen® 6609
Trymeen® 6610
Trymeen® 6617
Trymeen® 6622
Trymeen® 6637
Tylose C, CB Series
Tylose H Series
Tylose MH Grades
Tylose MH-K, MH-xp, MHB-y, MHB-yp
Tylose MHB
Ufaryl DB80
Ufaryl DL80 CW
Ufaryl DL85
Ufasan 35
Ufasan 50
Ufasan 60A
Ufasan 62B
Ufasan IPA
Ufasan TEA
Vannox AC 2, ACH, ACL
Variquat® 80MC
Varonic® K205
Varonic® K210
Varonic® T205
Wayfos A
Wayfos D-10N
Wayfos M-60
Wayfos M-100
Wettol® D 1
Wettol® D 2
Wettol® EM 1
Wettol® EM 2
Wettol® EM 3
Wettol® NT 1
Witcamide® N
Witco® 1298 HA
Witconate 605A
Witconate 605AC
Witconate 60L
Witconate 79S
Witconate P-1020BUST
Witconol NS-500K
Witconol NS-500LQ
Zee-emul No. 35
Zee-emul No. 50
Zonyl® FSA
Zonyl® FSB
Zonyl® FSC
Zonyl® FSJ
Zonyl® FSN
Zonyl® FSP

Chemical Processing

AF 70
AF 75
AF 8805
AF 8810
AF HL-40
AF HL-52
Alphoxat S 110, 120
Antifoam 7800 New
Armeen® 3-12
Armeen® 3-16
Armeen® DM10
Armeen® DM12
Gantrez® AN
GP-210 Silicone Antifoam Emulsion
GP-310-I
Lutensit® A-BO
Lutensit® A-LBA
Lutensit® AS 2230, 2270
Lutensit® AS 3330
Lutensit® AS 3334
Lutensol® A 7
Lutensol® A 8
Lutensol® AO 3
Lutensol® AO 4
Lutensol® AO 5
Lutensol® AO 7
Lutensol® AO 8
Lutensol® AO 11
Lutensol® AO 12
Lutensol® AO 109
Lutensol® AT 11
Lutensol® AT 18
Lutensol® AT 25
Lutensol® AT 50
Lutensol® ON 30
Lutensol® ON 50
Lutensol® ON 60
Lutensol® ON 70
Lutensol® ON 80
Lutensol® ON 110
Lutensol® TO 3
Lutensol® TO 5
Lutensol® TO 7
Lutensol® TO 8
Lutensol® TO 10
Lutensol® TO 12
Lutensol® TO 15
Lutensol® TO 20
Lutensol® TO 89
Lutensol® TO 109
Lutensol® TO 129
Lutensol® TO 389
Masil® 173
Merpol® A
Nekal® BX Conc. Paste
Nekal® BX Dry
Petrowet® R
Pluracol® W170

Chemical Processing (cont'd.)

Pluracol® W260
Pluracol® W660
Pluracol® W2000
Pluracol® W3520N
Pluracol® W3520N-RL
Pluracol® W5100N
Pluracol® WD1400
Radiamine 6140
Radiamine 6141
Radiamine 6160
Radiamine 6161
Radiamine 6163, 6164
Radiamine 6170
Radiamine 6171
Radiamine 6172
Radiamine 6173
Radiamine 6540
Radiamine 6560
Radiamine 6570
Radiamine 6572
Radiasurf® 7146
Radiasurf® 7150
Radiasurf® 7152
Radiasurf® 7156
Radiasurf® 7175
Radiasurf® 7444
Radiasurf® 7900
Rexfoam B, C
Rexfoam D
Rhodorsil® AF 422
Rhodorsil® AF 454
Silwet® L-7602
Silwet® L-7605
Silwet® L-7607
Teric 9A5
Trans-100

Construction, Building Materials, Asphalt Manufacture

Acra-500
Adogen® 560
Adogen® 570-S
Adogen® 572
Aerosol® A-102
Aerosol® A-103
Agesperse 71
Alkamide® 2112
Ameenex C-18
Ascote 5, 9, 12, 12L, 14
Asfier Series
Avirol® 200
Avirol® 252 S
Calsoft L-40
Calsoft L-60
Carbowax® PEG 600
Carbowax® PEG 900
Carbowax® PEG 1000
Carbowax® PEG 1450
Catamine 101
Catigene® CA 56
Cedepal FA-406
Chemeen DT-3
Chemeen DT-15
Chemphos TDAP
Chemzoline T-33
CNC Defoamer 44 Series
Crodazoline O
Crodazoline S
Fizul 201-11
Foamaster DD-72
Foamer AD
Foamer CD
Foamkill® 639AA
Forbest 850
Geropon® 99
Geropon® WT-27
Iconol NP-30
Iconol NP-30-70%
Iconol NP-40
Iconol NP-40-70%
Iconol NP-50-70%
Iconol NP-100
Iconol NP-100-70%
Iconol OP-5
Iconol OP-7
Iconol OP-10
Iconol OP-30
Iconol OP-30-70%
Iconol OP-40
Iconol OP-40-70%
Imwitor® 940
Indulin® 206
Indulin® AQS
Indulin® AQS-IM
Indulin® AS-1
Indulin® AS-Special
Indulin® AT
Indulin® MQK
Indulin® MQK-IM
Indulin® SA-L
Indulin® W-1
Indulin® W-3
Indulin® XD-70
Interwet® 33
Jet Amine TET
Jet Amine TRT
Jet Amine TT
Jet Quat 2C-75
Jet Quat 2HT-75
Jet Quat C-50
Jet Quat DT-50
Jet Quat S-50
Jet Quat T-27W
Jet Quat T-50
Kemamide® B
Kemamide® E
Kemamide® O
Kemamide® S
Kemamide® U
Kemamide® W-20
Kemamide® W-39
Kemamide® W-40
Kemamide® W-40/300
Kemamide® W-45
Kemamine® D-190
Kemamine® D-650
Kemamine® D-970
Kemamine® D-974
Kemamine® D-989
Kemamine® D-999
Lamacit AP 6
Lankrocell® D15L
Lankrocell® KLOP
Lankrocell® KLOP/CV
Lignosol AXD
Lignosol B
Lignosol BD
Lignosol DXD
Lilamin VP75
Lilamuls EM 24, EM 26, EM 33
Lomar® PL
Lomar® PW
Manro DES 32
Maramul SS
Marasperse N-22
Marlowet® 5606
Marlowet® BIK
Marlowet® BL
Marlowet® SLS
Marlowet® T
Millifoam ODE-60A
Monateric ADA
Monawet MO-70E
Monazoline T
Nalco® 8638
Nansa® 1192
Nansa® HS80S
Nansa® HS80/SF
Nansa® HS85S
Neosolve® AD-1
Nissan Amine ABT
Nissan Amine BB
Nissan Amine DT
Nissan Amine DTH
Nissan Amine MB
Nissan Amine OB
Nissan Amine PB
Nissan Amine SB
Nissan Amine VB
Nissan Anon BF
Nissan Asphasol 10
Nissan Asphasol 20
Nissan Cation ABT-350, 500
Nissan Cation F2-10R, -20R, -40E, -50
Nissan Cation FB, FB-500
Nissan Cation L-207
Nissan Cation M2-100
Nissan Cation PB-40, -300
Nissan Cation S2-100
Nopalcol 1-L
Nopalcol 2-DL
Nopalcol 4-C
Nopalcol 4-CH
Nopalcol 4-L
Nopalcol 4-S
Nopalcol 6-DO
Nopalcol 6-DTW
Nopalcol 6-L
Nopalcol 6-R
Nopalcol 6-S
Nopalcol 10-COH
Nopalcol 12-CO
Nopalcol 12-COH
Nopalcol 19-CO
Nopalcol 30-TWH
Nopalcol 200
Nopalcol 400
Nopalcol 600
Nopalcol Series
Nopco® PD#1-D
Nopcogen 22-O
Noram 2C
Noram DMC
Noram DMS
Noram M2SH
Norfox® SLES-60
Orzan® AE
Orzan® CG
Polyram S
Quaternary O
Radiamine AA 23, 27, 57
Radiamine AA 60
Redicote AD-130
Redicote AD-141
Redicote AD-142
Redicote AD-150
Redicote AD-170
Redicote AE 1A
Redicote AE 6
Redicote AE 6-A
Redicote AE 7
Redicote AE 9
Redicote AE 22
Redicote AE 26
Redicote AE 45
Redicote AE 55
Redicote AE 80
Redicote AE 91
Redicote AE 93
Redicote Series
Remanol 1300/35
Rewopon® JMBT
Rexol 25/100-70%
Rhodacal® BX-78
Rhodameen® O-12
Rhodameen® S-20
Rhodapex® CD-128
Rhodapon® BOS
Rhodapon® CAV
Schercozoline I
Schercozoline L
Schercozoline O
Schercozoline S
Steposol® CA-207
Surfonic® HDL
Surfonic® L10-3
Surfonic® L610-3
Surfonic® N-557
Tallow Amine
Tallow Diamine
Tallow Tetramine
Tallow Triamine

Construction, Building Materials, Asphalt Manufacture (cont'd.)

T-Det® N-40
T-Det® N-50
T-Det® N-70
T-Det® N-307
T-Det® N-407
T-Det® N-507
T-Det® N-705
T-Det® N-1007
Teepol PB
Tembind A 002
Teric 16M15
Teric 17A10
Teric 17M2
Teric 17M5
Teric 17M15
Teric 18M2
Teric 18M5
Teric 18M10
Teric 18M20
Teric SF9
Texapon GYP 1
Texapon GYP 2
Texapon GYP 3
Texapon GYP 4
Textamine A-3417
Textamine Polymer
Trinoram C
Trinoram S
Ungerol LES 2-28
Ungerol LES 2-70
Ungerol LES 3-28
Ungerol LES 3-70
Ungerol N2-28
Ungerol N2-70
Ungerol N3-28
Ungerol N3-70
Velvetex® BA
Versilan MX244
Vorite 105
Vorite 110
Vorite 115
Vorite 120
Vorite 125
Witcolate 1276

Dry Cleaning

Aerosol® OT-75%
Aerosol® OT-100%
Aerosol® OT-MSO
Aerosol® OT-S
Alkadet DCB-100
Alkasurf® OP-5
Aminol COR-4
Aristonate H
Aristonate L
Aristonate M
Armul 908
Arnox 908
Avirol® SO 70P
Calimulse PRS
Carsonon® N-4
Catinex KB-13
Chemax DOSS/70
Chemax DOSS-75E
Chemfac NC-0910
Chemfac PC-188
Chemfac PD-600
Chemfac PF-623
Chemfac PF-636
Chemphos TR-495
Chimin P1A
CPA-Alpha®
Crill 4
Gardilene IPA/94
Geropon® 99
Geropon® SS-O-75
Geropon® WT-27
Igepal® CA-210
Igepal® CA-420
Igepal® RC-520
Lankropol® KO
Laurelphos E-61
Mackanate DOS-70MS
Mackanate DOS-75
Manoxol MA
Manoxol N
Manoxol OT
Manoxol OT 60%
Manoxol OT/B
Maphos® 15
Maphos® 18
Maphos® 30
Maphos® 54
Maphos® 55
Maphos® 56
Maphos® 76
Maphos® 76 NA
Maphos® 77
Maphos® 151
Maphos® 236
Maphos® FDEO
Maphos® L 13
Marlon® AMX
Marlophor® DG-Acid
Marlophor® F1-Acid
Marlophor® LN-Acid
Marlophor® MO 3-Acid
Marlophor® N5-Acid
Marlophor® ND
Marlophor® ND-Acid
Marlophor® ND DEA Salt
Marlophor® ND NA-Salt
Marlophor® NP5-Acid, NP6-Acid, NP7-Acid
Marlophor® OC5-Acid
Marlophor® T6-Acid
Marlophor® T10-Acid
Marlophor® T10-DEA Salt
Marlophor® T10-Sodium Salt
Marlophor® UW12-Acid
Mazamide® L-298
Mazawet® 77
Mazawet® DOSS 70
Monafax 785
Monafax 786
Monafax 794
Monafax 872
Monafax 1293
Monamid® 7-153 CS
Monawet MO-65 PEG
Monawet MO-70S
Mulsifan RT 18
Naxel AAS-Special 3
Ninate® 411
Niox KF-22
Niox KQ-33
Nissan Chloropearl
Nissan New Roayl P
Nissan Nonion DS-60HN
Nissan Sun Flora
Nitrene C
Nitrene N
Nopalcol 1-L
Nopalcol 1-S
Nopalcol 1-TW
Nopalcol 2-DL
Nopalcol 4-C
Nopalcol 4-CH
Nopalcol 4-L
Nopalcol 4-S
Nopalcol 6-DO
Nopalcol 6-DTW
Nopalcol 6-L
Nopalcol 6-R
Nopalcol 6-S
Nopalcol 10-COH
Nopalcol 12-CO
Nopalcol 12-COH
Nopalcol 19-CO
Nopalcol 30-TWH
Nopalcol 200
Nopalcol 400
Nopalcol 600
Nopalcol Series
Norfox® DCSA
Norfox® F-342
Norfox® OP-45
Pentine 1185 5432
Petronate® K
Petronate® S
Petrosul® H-50
Petrosul® H-60
Petrosul® H-70
Petrosul® HM-62, HM-70
Petrosul® L-60
Petrosul® M-50
Petrosul® M-60
Petrosul® M-70
Phospholan® PDB3
Prote-pon P-2 EHA-02-K30
Prote-pon P 2 EHA-02-Z
Prote-pon P 2 EHA-Z
Prote-pon P-0101-02-Z
Prote-pon P-L 201-02-K30
Prote-pon P-L 201-02-Z
Prote-pon P-NP-06-K30
Prote-pon P-NP-06-Z
Prote-pon P-NP-10-K30
Prote-pon P-NP-10-MZ
Prote-pon P-NP-10-Z
Prote-pon P-OX 101-02-K75
Prote-pon P-TD-06-K13
Prote-pon P-TD-06-K30
Prote-pon P-TD 06-K60
Prote-pon P-TD-06-Z
Prote-pon P-TD-09-Z
Prote-pon P-TD-12-Z
Prote-pon TD-09-K30
Protowet 5171
Quimipol EA 2504
Quimipol ENF 65
Quimipol ENF 80
Radiasurf® 7125
Rewopol® SBDO 70
Rewoquat CR 3099
Rexol 65/4
Rexol 65/6
Rexonic N23-3
Rhodacal® 330
Rhodacal® IPAM
Rhodafac® L3-64A
Rhodafac® PE-510
Rhodafac® R9-47A
Rhodafac® RE-610
Rhodafac® RE-960
Rhodafac® RM-410
Rhodafac® RM-510
Rhodafac® RM-710
Rhodafac® RS-410
Rhodafac® RS-610
Rhodafac® RS-710
Rhodaterge® DCB-100
Sandoz Sulfonate AAS-Special 3
Schercomid 304
Schercophos NP-9
Schercopon 2WD
Schercoteric O-AA
Servoxyl VPGZ 7/100
Servoxyl VPIZ 100
Servoxyl VPNZ 10/100
Servoxyl VPQZ 9/100
Servoxyl VPTZ 3/100
Servoxyl VPTZ 100
Servoxyl VPUZ
Servoxyl VPYZ 500
Siponic® TD-6
Sovatex C Series
Stafoam
Stafoam DF-1
Stafoam DF-4
Stafoam DL
Surfonic® HDL-1
Surfonic® JL-80X
T-Det® DD-5
T-Det® N-4
T-Det® N-6
T-Det® O-4

Dry Cleaning (cont'd.)

Tergitol® 15-S-3
Tergitol® 15-S-5
Tergitol® NP-4
Tergitol® TMN-3
Tergitol® TP-9
Teric 9A2
Teric N4
Teric N5
Tex-Wet 1104
Tex-Wet 1143
T-Mulz® 426
T-Mulz® 565
T-Mulz® 598
Triton® GR-7M
Triton® X-45
Triumphnetzer ZSG
Trycol® TP-6
Tryfac® 5556
Unamine® O
Witcamide® 5138
Witcamide® 5145M
Witcolate 1259
Witconate P10-59
Witconate YLA

Electronics Industry

Fluorad FC-93
Fluorad FC-95
Fluorad FC-99
Fluorad FC-100
Foamkill® 652
Foamkill® 652B
Mazeen® 173
Mazeen® 174
Mazeen® 174-75
Mazeen® 241-3
Tergitol® 15-S-7
Tergitol® 15-S-9
Tergitol® Min-Foam 1X
Tergitol® Min-Foam 2X
Tergitol® TMN-6
Tergitol® TMN-10
Teric 17M2
Teric 17M5
Teric 17M15
Triton® XL-80N

Fire Fighting

Aerosol® OT-75%
Amphoteric N
Avirol® 200
Avirol® 270A
Avirol® 280 S
Avirol® 603 A
Chemal DA-6
Chemal DA-9
Chemal LA-12
Chemal LA-23
Geropon® WT-27
Karasurf AS-26
Manro ALES 60
Manro BES 60
Manro BES 70
Miranol® H2C-HA
Monawet MO-70
Monawet MO-70R
Montosol IL-13
Montosol PL-14
Perlankrol® EAD60
Perlankrol® ESS25
Phospholipid PTD
Phospholipid PTL
Phospholipid PTZ
Rewopol® TLS 45/B
Rhodapon® L-22, L-22/C
Schercozoline B
Sellogen HR
Stepantan® AS-40
Sulfotex DOS
Surfonic® HDL
Texapon 1030, 1090
Texapon N 25
Texapon N 103
Texapon PNA
Texapon T 42
Texin DOS 75
Trycol® DA-6
Trycol® DA-69
Versilan MX244
Witconate AO 24
Zonyl® FSA
Zonyl® FSB
Zonyl® FSC
Zonyl® FSJ
Zonyl® FSN
Zonyl® FSP

Industrial, Institutional and Consumer Cleaning, Germicides/Sanitizers

Abil® B 88184
Ablumine 12
Ablumine 1214
Ablumine 3500
Ablunol LA 3
Ablunol LA 5
Ablunol LA 7
Ablunol LA 9
Ablunol LA 12
Ablunol LA 16
Ablunol LA 40
Ablunol NP 8
Ablunol NP 15
Abluphat AP Series
Abluphat LP Series
Abluphat OP Series
Ablusol CDE
Ablusol DBD
Ablusol DBM
Ablusol DBT
Ablusol LDE
Abluter BE
Accomid C
Accomid PK
Accoquat 2C-75
Accosoft 550, 620
Accosoft 707
Accosoft 748
Accosoft A-155
Acetoquat CPB
Acetoquat CPC
Acetoquat CTAB
Acid Thickener
Actrafos SN-306
Actrafos SN-314
Actrasol 6092
Actrasol C-50, C-75, C-85
Actrasol CS-75
Actrasol SBO
Actrasol SP
Actrasol SP 175K
Actrasol SR 75
Actrasol SRK 75
Acylglutamate AS-12
Acylglutamate CS-11
Acylglutamate CS-21
Acylglutamate CT-12
Acylglutamate DL-12
Acylglutamate GS-11
Acylglutamate GS-21
Acylglutamate HS-11
Acylglutamate HS-21
Acylglutamate LS-11
Acylglutamate LT-12
Acylglutamate MS-11
AD-700
AD-709
AD-710
AD-713
AD-713C
AD-716
AD-742
AD-742C
AD-747
AD-747C
AD-748
Adinol OT16
Adol® 63
Aerosol® 18
Aerosol® 19
Aerosol® AY-65
Aerosol® AY-100
Aerosol® C-61
Aerosol® GPG
Aerosol® NPES 458
Aerosol® OS
Aerosol® OT-100%
AF 8820
AF CM Conc.
AF HL-23
Akypo LF 1
Akypo LF 2
Akypo LF 3
Akypo LF 4
Akypo LF 4N
Akypo LF 5
Akypo LF 6
Akypo MB 1585
Akypo MB 1614/1
Akypo MB 1614/2
Akypo MB 2528S
Akypo MB 2621 S
Akypo MB 2705 S
Akypo NTS
Akypo RLMQ 38
Akypo TFC-S
Akypo TFC-SN
Akypogene Jod F
Akypogene Jod MB 1918
Akypogene VSM
Akypogene VSM-N
Akypoquat 40
Akypoquat 129
Akyposal BA 28
Akyposal RLM 56 S
Akyposal TLS 42
Alcodet® 218
Alcodet® 260
Alcodet® HSC-1000
Alcodet® IL-3500
Alcodet® MC-2000
Alcodet® SK
Alcojet®
Alconox®
Alcotabs®
Alfonic® 610-50R
Alfonic® 810-40
Alfonic® 810-60

Industrial, Institutional and Consumer Cleaning (cont'd.)

Alfonic® 1012-40
Alfonic® 1012-60
Alkali Surfactant NM
Alkamide® 101 CG
Alkamide® 200 CGN
Alkamide® 206 CGN
Alkamide® 210 CGN
Alkamide® 2104
Alkamide® 2106
Alkamide® 2110
Alkamide® 2112
Alkamide® 2124
Alkamide® 2204
Alkamide® 2204A
Alkamide® C-212
Alkamide® CDM
Alkamide® CDO
Alkamide® CL63
Alkamide® CME
Alkamide® DC-212
Alkamide® DC-212/M
Alkamide® DC-212/S
Alkamide® DC-212/SE
Alkamide® DIN 100
Alkamide® DL-203
Alkamide® L7DE-PG
Alkamide® L9DE
Alkamide® L-203
Alkamide® SDO
Alkamide® SODI
Alkamuls® 400-MO
Alkamuls® TD-41
Alkanol® 189-S
Alkanol® WXN
Alkasurf® NP-4
Alkasurf® NP-6
Alkasurf® NP-9
Alkasurf® NP-10
Alkasurf® NP-11
Alkasurf® NP-12
Alkasurf® OP-8
Alkasurf® OP-10
Alkasurf® OP-12
Alkawet® N
Alkawet® NP-6
Alkenyl Succinic Anhydrides
Alkylate 215
Alkylate 225
Alkylate 227
Alkylate 230
Alpha-Step® ML-40
Alpha-Step® ML-A
Alphenate HM
Alrosol B
Amerlate® LFA
Ameroxol® OE-2
Amidex 1248
Amidex 1285
Amidex 1351
Amidex C
Amidex CA
Amidex CE
Amidex CME
Amidex CO-1
Amidex KME
Amidex L-9
Amidex LD
Amidex OE
Amidex TD
Amidex WD
Amido Betaine C
Amido Betaine C-45
Amidox® C-2
Amidox® C-5
Amidox® L-5
Amine C
Aminol CM, CM Flakes, CM-C Flakes, CM-D Flakes
Aminol COR-2
Aminol HCA
Aminol LM-30C, LM-30C Special
Aminol VR-14
Aminoxid WS 35
Amisoft CS-11
Amisoft CT-12
Amisoft GS-11
Amisoft HS-11
Amisoft LS-11
Amisoft MS-11
Ammonium Cumene Sulfonate 60
Ammonyx® CDO
Ammonyx® CO
Ammonyx® DMCD-40
Ammonyx® LO
Ammonyx® MO
Ampholak 7CX
Ampholak 7TX
Ampholak 7TY
Ampholak BCA-30
Ampholak BTH-35
Ampholak XCE
Ampholak XCO-30
Ampholak XJO
Ampholak XO7
Ampholak XO7-SD 55
Ampholak XOO-30
Ampholak YCA/P
Ampholak YCE
Ampholak YJH
Ampholan® B-171
Ampholyt JA 120
Ampholyt JA 140
Ampholyt JB 130
Ampholyte KKE
Amphoram CB A30
Amphoram CP1
Amphoram CT 30
Amphosol CA
Amphosol CG
Amphosol DM
Amphoteen 24
Amphotensid B4
Amphotensid CT
Amphotensid D1
Amphoterge® J-2
Amphoterge® K
Amphoterge® K-2
Amphoterge® KJ-2
Amphoterge® NX
Amphoterge® SB
Amphoterge® W
Amphoterge® W-2
Amphoteric 400
Amphoteric L
Amphoteric N
Amphoteric SC
Amyx CDO 3599
Amyx CO 3764
Amyx LO 3594
Amyx SO 3734
Anedco DF-6002
Anionyx® 12S
Antarox® BL-225
Antarox® BL-236
Antarox® BL-240
Antarox® BL-330
Antarox® EGE 31-1
Antarox® FM 33
Antarox® FM 53
Antarox® FM 63
Antarox® LA-EP 15
Antarox® LA-EP 16
Antarox® LA-EP 25
Antarox® LA-EP 25LF
Antarox® LA-EP 45
Antarox® LA-EP 59
Antarox® LA-EP 65
Antarox® LA-EP 73
Antarox® LF-330
Antifoam CM Conc.
AO-14-2
AO-728 Special
APG® 300 Glycoside
APG® 325 Glycoside
Arflow 168
Arkomon A Conc.
Arkopon Brands
Armid® 18
Armid® C
Armid® HT
Armid® O
Armix 146
Armul 1005
Aromox® C/12
Aromox® C/12-W
Aromox® DM14D-W
Aromox® DM16
Aromox® DMB
Aromox® DMC
Aromox® DMC-W
Aromox® DMHTD
Aromox® DMMCD-W
Aromox® T/12
Arosurf® 42-PE10
Arosurf® 66-E10
Arquad® 2HT-75
Arquad® B-100
Arquad® DM14B-90
Arquad® DMCB-80
Arquad® DMHTB-75
Arquad® DMMCB-50
Arquad® S-2C-50
Arquad® T-2C-50
Arquad® T-27W
Arstim RRC
Arylan® CA
Arylan® HAL
Arylan® LQ
Arylan® PWS
Arylan® SBC Acid
Arylan® SC Acid
Arylan® SKN Acid
Arylan® SP Acid
Arylan® SX85
Autopur WK 4121
Avanel® S-35
Avanel® S-74
Avirol® 252 S
Avirol® 270A
Avirol® 280 S
Avirol® 300
Avirol® 400 T
Avirol® 603 A
Avirol® 603 S
Avirol® A
Avirol® SL 2010
Avirol® SL 2015
Avirol® SL 2020
Avirol® SO 70P
Avirol® T 40
Aviscour HQ50
Bactistep® MH 80
Berol 048
Berol 079
Berol 087
Berol 173
Berol 185
Berol 223
Berol 225
Berol 226
Berol 259
Berol 260
Berol 272
Berol 370
Berol 452
Berol 475
Berol 521
Berol 522
Berol 525
Berol 556
Berol 563
Berol 716
Berol 733
Berol 784
Berol 797
Berol 806
Berol WASC
Beycostat 211 A
Beycostat 714 A
Beycostat B 070 A
Beycostat B 080 A
Beycostat B 089 A
Beycostat B 151
Beycostat LP 4 A
Beycostat LP 12 A
Bio-Soft® 9283
Bio-Soft® CS 50
Bio-Soft® D-233
Bio-Soft® D-35X
Bio-Soft® D-40
Bio-Soft® D-53
Bio-Soft® D-62
Bio-Soft® EA-8
Bio-Soft® EA-10
Bio-Soft® ERM
Bio-Soft® JN
Bio-Soft® LAS-40S
Bio-Soft® LD-32

Industrial, Institutional and Consumer Cleaning (cont'd.)

Bio-Soft® LD-47
Bio-Soft® LD-95
Bio-Soft® LD-145
Bio-Soft® LD-150
Bio-Soft® LD-190
Bio-Soft® LDL-4
Bio-Soft® LF 77A
Bio-Soft® N-300
Bio-Soft® N-411
Bio-Soft® Ninex 21
Bio-Soft® PG 4
Bio-Soft® S-100
Bio-Soft® S-130
Bio-Surf PBC-430
Bio-Terge® AS-40
Bio-Terge® AS-90 Beads
Bio-Terge® PAS-8S
Biopal® LF-20
Biopal® NR-20
Biopal® NR-20 W
Biozan
Blancol®
Briquest® 301-50A
Briquest® 543-45AS
Briquest® ADPA-60AW
BTC® 50 USP
BTC® 65 USP
BTC® 818
BTC® 818-80
BTC® 824
BTC® 824 P100
BTC® 835
BTC® 885
BTC® 885 P40
BTC® 888
BTC® 1010-80
BTC® 2125, 2125-80, 2125 P-40
BTC® 2125M
BTC® 2125M-80, 2125M P-40
BTC® 8248
BTC® 8249
BTC® 8358
Bubble Breaker® 1840X
Bug Remover Conc.
Burco Anionic APS
Burco Anionic APS-LF
Burco CS-LF
Burco FAE
Burco HCS-50NF
Burco NCS-80
Burco NF-225
Burco NPS-50%
Burco NPS-225
Burco TME
Burcofac 1060
Burcofac 9125
Burcoterge DG-40
Burcowet TMW
CAE
Calamide C
Calamide CW-100
Calamide S
Calcium Sulfonate C-50N
Calester
Calfax 10L-45
Calfax DB-45
Calfoam AAL
Calfoam ES-30
Calfoam LLD
Calfoam NEL-60
Calfoam NLS-30
Calfoam SEL-60
Calfoam SLS-30
Caloxylate N-9
Calsoft AOS-40
Calsoft F-90
Calsoft L-40
Calsoft L-60
Calsoft LAS-99
Calsoft T-60
Calsuds 81 Conc.
Calsuds A
Calsuds CD-6
Carbon Detergent K
Carbopol® 615, 616, 617
Carbowax® MPEG 350
Carbowax® MPEG 550
Carbowax® MPEG 2000
Carbowax® MPEG 5000
Carbowax® PEG 3350
Carbowax® PEG 4600
Carbowax® PEG 8000
Carnauba Spray 200
Carsamide® C-3
Carsamide® CA
Carsamide® SAC
Carsamide® SAL-7
Carsamide® SAL-9
Carsofoam® 211
Carsofoam® MS Conc.
Carsofoam® T-60-L
Carsonol® ALS-S
Carsonol® ANS
Carsonol® AOS
Carsonol® MLS
Carsonol® SES-A
Carsonol® SES-S
Carsonol® SLES
Carsonol® SLS
Carsonol® SLS-R
Carsonol® SLS-S
Carsonon® L-985
Carsonon® LF-5
Carsonon® LF-46
Carsonon® N-4
Carsonon® N-6
Carsonon® N-8
Carsonon® N-9
Carsonon® N-10
Carsonon® N-11
Carsonon® N-12
Carsonon® N-50
Carsonon® ND-317
Carsonon® TD-10
Carsonon® TD-11, TD-11 70%
Carsosulf SXS-Liq.
Carspray #2
Carspray 205
Carspray 300
Carspray 300HF
Carspray 375
Carspray 500
Carspray 650
Carspray 700
Carspray CW
Catigene® 824
Catigene® 8248
Catigene® CETAC 30
Catinex KB-15
Catinex KB-18
Catinex KB-19
Catinex KB-20
Catinex KB-22
Catinex KB-32
Catinex KB-40
Catinex KB-41
Catinex KB-42
Catinex KB-43
Catinex KB-48
Catinex KB-49
Catinex KB-51
Cedemide AX
Cedepal SS-203, -306, -403, -406
Cedephos® FA600
Cedephos® RA600
Cedepon LT-40
Cetalox 8, 25
Cetalox 50
Cetalox AT
Chemal 2EH-2
Chemal 2EH-5
Chemal BP 261
Chemal BP-262
Chemal BP-262LF
Chemal BP-2101
Chemal DA-4
Chemal DA-6
Chemal DA-9
Chemal LA-4
Chemal LA-9
Chemal LA-12
Chemal LA-23
Chemal LF 14B, 25B, 40B
Chemal LFL-10, -17, -19, -28, -38, -47
Chemal TDA-3
Chemal TDA-6
Chemal TDA-9
Chemal TDA-12
Chemal TDA-15
Chemal TDA-18
Chemax AR-497
Chemax DF-30
Chemax DF-100
Chemax DNP-8
Chemax DNP-15
Chemax DNP-150
Chemax DNP-150/50
Chemax DOSS/70
Chemax DOSS-75E
Chemax NP-4
Chemax NP-6
Chemax NP-9
Chemax OP-30/70
Chemax PEG 400 DO
Chemax PEG 600 DO
Chemax TO-10
Chemax TO-16
Chembetaine CAS
Chemfac NC-0910
Chemfac PA-080
Chemfac PB-082
Chemfac PB-106
Chemfac PB-135
Chemfac PB-184
Chemfac PB-264
Chemfac PC-188
Chemfac PD-600
Chemfac PF-623
Chemfac PF-636
Chemfac PN-322
Chemoxide LM-30
Chemphos TC-227
Chemphos TC-310D
Chemphos TC-310S
Chemphos TR-421
Chemphos TR-513
Chemphos TX-625D
Chemzoline T-11
Chimin P1A
Chimin P40
Chimin P45
Chimin RI
Chimipal OLD
Chimipal OS 2
Chimipal PE 300, PE 302
Chimipal PE 402, PE 403, PE 405
Chimipal PE 520
Chimipon FC
Chimipon HD
Chimipon LD
Chimipon LDP
Chimipon TSB
CHT Lavotan DS
CHT Subitol HLF Conc.
CHT Subitol LS-N
CHT Subitol SAN
Cithrol A
Cithrol DGDL N/E
Cithrol DGDL S/E
Cithrol DGDO N/E
Cithrol DGDO S/E
Cithrol DGDS N/E
Cithrol DGDS S/E
Cithrol DGML N/E
Cithrol DGML S/E
Cithrol DGMO N/E
Cithrol DGMO S/E
Cithrol DGMS N/E
Cithrol DGMS S/E
Cithrol DPGML N/E
Cithrol DPGML S/E
Cithrol DPGMO S/E
Cithrol DPGMS N/E
Cithrol DPGMS S/E
Cithrol EGDL N/E
Cithrol EGDL S/E
Cithrol EGDO N/E
Cithrol EGDO S/E
Cithrol EGDS N/E

Industrial, Institutional and Consumer Cleaning (cont'd.)

Cithrol EGDS S/E
Cithrol EGML N/E
Cithrol EGML S/E
Cithrol EGMO N/E
Cithrol EGMO S/E
Cithrol EGMR N/E
Cithrol EGMR S/E
Cithrol EGMS N/E
Cithrol EGMS S/E
Cithrol GDL N/E
Cithrol GDL S/E
Citranox®
Cleary's Waterless Hand Cleaner
Clink A-26
Clink A-70
Closyl 30 2089
Closyl LA 3584
CNC Detergent E
CNC Sol UE
CNC Wet CP
Code 8059
Comperlan 100
Comperlan KDO
Comperlan LD
Comperlan LM
Comperlan LP
Comperlan OD
Comperlan P 100
Comperlan PD
Comperlan PKDA
Comperlan PVD
Corexit CL578
Corexit CL8500
Corexit CL8569
Corexit CL8594
Corexit CL8662
Corexit CL8685
Cosmopon 35
Cosmopon BL
Cosmopon LE 50
Cosmopon MO
Cosmopon SES
Cosmopon TR
Crafol AP-60
Crafol AP-63
Cralane AT-17
Crapol AU-40
Crapol AV-10
Crapol AV-11
Crillet 1
Crillon LDE
Crodalan AWS
Crodapearl Liq.
Crodapearl NI Liquid
Crodasinic LS30
Crodasinic LS35
Crodasinic LT40
Crodasinic OS35
Croduret 10
Croduret 30
Croduret 40
Croduret 60
Croduret 100
Cromeen
Crown Anti-Foam
Cyclomox® L
Cycloryl LDC
Cycloteric CAPA
Cycloteric SLIP
Findet AD-18
Findet SB
Flexricin® 9
Flexricin® 13
Flexricin® 15
Flo-Mo® AJ-85
Flo-Mo® AJ-100
Fluilan
Fluorad FC-95
Fluorad FC-109
Fluorad FC-129
Fluorad FC-135
Fluorad FC-171
Foamer AD
Foamer CD
Foamer
Foamkill® 2947
Foamkill® 810F
Foamkill® 830F
Foamkill® 836B
Foamkill® CPD
Foamkill® MS Conc.
Foamkill® MSF Conc.
Foamole M
Fosfamide N
Fosterge BA-14 Acid
Fosterge LF
Fosterge LFD
Fosterge LFS
G-1000-S Antifoam Compd.
GP-209
GP-210 Silicone Antifoam Emulsion
GP-215
GP-217
GP-226
GP-227
GP-262 Defoamer
GP-295 Defoamer
GP-300-I Antifoam Compd.
GP-310-I
Gantrez® AN-119
Gantrez® AN-149
Gantrez® AN-169
Gantrez® AN-179
Gardinol CX
Gemtex SC-75
Gemtex WBT
Gemtex WBT-9
Genagen CA-050
Genapol® 24-L-45
Genapol® 24-L-50
Genapol® 24-L-60
Genapol® 24-L-60N
Genapol® 24-L-92
Genapol® 24-L-98N
Genapol® 26-L-1
Genapol® 26-L-1.6
Genapol® 26-L-2
Genapol® 26-L-3
Genapol® 26-L-5
Genapol® 26-L-45
Genapol® 26-L-60N
Genapol® 26-L-98N
Genapol® 42-L-3
Genapol® ARO
Genapol® PGM Conc.
Genapol® X Grades
Genapol® ZRO Liq., Paste
Genopur ASA
Geropon® 99
Geropon® DOS
Geropon® DOS FP
Geropon® SBL-203
Geropon® SS-L7DE
Geropon® T-33
Geropon® T-43
Geropon® T-51
Geropon® T-77
Geropon® TA/K
Geropon® WT-27
Glucopon 225
Glucopon 425
Glucopon 600
Glucopon 625
Good-rite® K-702
Good-rite® K-732
Good-rite® K-739
Good-rite® K-752
Good-rite® K-759
Good-rite® K-7058
Good-rite® K-7058D
Good-rite® K-7058N
Good-rite® K-7200N
Good-rite® K-7600N
Good-rite® K-7658
Gradonic FA-20
Gradonic LFA Series
Gradonic N-95
Hampfoam 35
Hamposyl® C
Hamposyl® C-30
Hamposyl® L
Hamposyl® L-30
Hamposyl® L-95
Hamposyl® M
Hamposyl® M-30
Hamposyl® O
Hamposyl® S
Hartamide LDA
Hartamide LMEA
Hartamide OD
Hartasist 46
Hartenol LAS-30
Hartenol LES 60
Hartofix 2X
Hartopol 25R2
Hartopol 31R1
Hartopol L42
Hartopol L44
Hartopol L62
Hartopol L62LF
Hartopol L64
Hartopol LF-1
Hartopol LF-2
Hartopol LF-5
Hartopol P65
Hartosolve OL
Hartotrope AXS
Hartotrope KTS 50
Hartotrope STS-40, Powd.
Hartotrope SXS 40, Powd.
Heavy Duty Cleaner HDC
Hercules® AR150
Hercules® AR160
Hetamide LA
Hetamide MMC, OC
Hetoxide C-30
Hetoxide C-40
Hetoxide C-60
Hetoxide C-200
Hetoxide DNP-5
Hetoxide DNP-10
Hetoxide HC-16
Hetoxol 15 CSA
Hetoxol CA-2
Hetoxol CA-20
Hetoxol CAWS
Hetoxol CS-15
Hetoxol CS-20
Hetoxol CS-30
Hetoxol CS-50
Hetoxol CS-50 Special
Hetoxol CSA-15
Hetoxol L-3N
Hetoxol L-4N
Hetoxol L-9
Hetoxol L-9N
Hetoxol LS-9
Hetoxol TD-3
Hetoxol TD-6
Hetoxol TD-12
Hetoxol TDEP-15
Hetoxol TDEP-63
Hetsulf 40, 40X
Hetsulf 50A
Hexaryl D 60 L
Hipochem AM-99
Hipochem AMC
Hodag Nonionic 1017-R
Hodag Nonionic 1025-R
Hodag Nonionic 2017-R
Hodag Nonionic 2025-R
Hodag Nonionic 4017-R
Hodag Nonionic 4025-R
Hodag Nonionic 5025-R
Hodag Sole-Mulse B
Hostapal CVH
Hostapon KA Powd.
Hostapon T Powd.
Hostapur OS Brands
Hostapur SAS 60
Hybase® C-300
Hybase® C-400
Hybase® C-500
Hymolon CWC
Hymolon K90
Hyonic CPG 745
Hyonic NP-40
Hyonic NP-90
Hyonic NP-110
Hyonic NP-407
Hyonic OP-40
Hyonic OP-100
Hyonic PE-40
Hyonic PE-90

Industrial, Institutional and Consumer Cleaning (cont'd.)

Hyonic PE-100
Hystrene® 3675C
Hystrene® 3680
Hystrene® 3687
Hystrene® 3695
Iconol DA-4
Iconol DA-6
Iconol DA-6-90%
Iconol DA-9
Iconol DDP-10
Iconol DNP-8
Iconol DNP-24
Iconol DNP-150
Iconol NP-7
Iconol NP-12
Iconol NP-30
Iconol NP-30-70%
Iconol NP-40
Iconol NP-40-70%
Iconol NP-50
Iconol NP-50-70%
Iconol NP-100
Iconol NP-100-70%
Iconol OP-5
Iconol OP-7
Iconol OP-10
Iconol OP-30
Iconol OP-30-70%
Iconol OP-40
Iconol OP-40-70%
Iconol TDA-3
Iconol TDA-6
Iconol TDA-8
Iconol TDA-8-90%
Iconol TDA-10
Igepal® CA-210
Igepal® CA-420
Igepal® CA-520
Igepal® CA-620
Igepal® CA-630
Igepal® CA-720
Igepal® CO-630
Igepal® CO-660
Igepal® CO-710
Igepal® CO-720
Igepal® CO-730
Igepal® CO-850
Igepal® CO-880
Igepal® CO-887
Igepal® DM-530
Igepal® DM-710
Igepal® DM-880
Igepal® DM-970 FLK
Igepal® KA
Igepal® LAVE
Igepal® NP-2
Igepal® NP-5
Igepal® NP-6
Igepal® NP-8
Igepal® NP-9
Igepal® NP-12
Igepal® NP-14
Igepal® NP-17
Igepal® NP-20
Igepal® NP-30
Igepal® O
Igepal® OD-410
Igepal® RC-520
Igepal® RC-620
Igepal® RC-630
Incrocas 30
Incrocas 40
Incromide CA
Incromide CM
Incromide L-90
Incromide LA
Incromide LM-70
Incromide LR
Incromine Oxide I
Incromine Oxide ISMO
Incromine Oxide L
Incromine Oxide L-40
Incromine Oxide MC
Incromine Oxide O
Incromine Oxide OD-50
Incronam 30
Incronam CD-30
Incropol CS-20
Incropol CS-50
Incropol L-7
Incrosoft 100
Incrosoft 100P
Incrosul LMS
Industrol® DW-5
Inhibitor 60Q
Intravon® JF
Intravon® JU
Intravon® SO
Ionet MO-200
Ionet MO-400
Ionet MS-400
Ionet MS-1000
Jet Amine DMCD
Jet Amine DMOD
Jet Amine DMSD
Jet Amine DMTD
Jet Amine M2C
Jet Quat S-2C-50
Kadif 50 Flakes
Kamar BL
Karafac 78
Karafac 78 (LF)
Karamide 121
Karamide 221
Karamide 363
Karamide 442-M
Karamide CO2A
Karamide CO9A
Karamide CO22
Karamide HTDA
Karapeg 200-MO
Karaphos HSPE
Karaphos SWPE
Karaphos XFA
Karasurf AS-26
Kemamide® B
Kemamide® E
Kemamide® O
Kemamide® S
Kemamide® U
Kemamide® W-20
Kemamide® W-39
Kemamide® W-40
Kemamide® W-40/300
Kemamide® W-45
Kerasol 1398
Klearfac® AA040
Korantin® PA
Korantin® PAT
Korantin® SMK
Korantin® TD
Lan-Aqua-Sol 50
Lan-Aqua-Sol 100
Lancare
Laneto 40
Laneto 50
Laneto 60
Lankropol® OPA
Lankropol® WA
Lankrosol SXS-30
Laural LS
Laurel PDW
Laurel R-50
Laurel SD-101
Laurel SD-120
Laurel SD-140N
Laurel SD-150
Laurel SD-180
Laurel SD-350
Laurel SD-900M
Laurel SD-1031
Laurel SDW
Laurelox 12
Laurelphos A-600
Laurelphos D 44 B
Laurelphos OL-529
Laurelphos P-71
Laureltex FMC
Lauridit® KD, KDG
Lauridit® KM
Lauridit® LM
Lauridit® OD
Lauridit® PD
Lauridit® PPD
Lauridit® SDG
Lebon 101H, 105
Lebon 2000
Lexaine® C
Lexaine® CG-30
Lexaine® CS
Lexaine® CSB-50
Lexaine® LM
Lexate BPQ
Lexemul® EGDS
Lexemul® EGMS
Lignosite® 431
Lignosite® 458
Lignosol AXD
Lignosol DXD
Lilaminox M4
Lilaminox M24
Lipocol O-5
Liqui-Nox®
Loropan CME
Loropan LD
Loropan LM
Lubrol N5
Lumo Stabil S 80
Lumo WW 75
Lumorol 4153
Lumorol 4154
Lumorol 4192
Lumorol 4290
Lumorol GG 65
Lumorol RK
Lumorol W 5058
Lumorol W 5157
Lutensit® A-EP
Lutensit® A-ES
Lutensit® A-FK
Lutensit® A-PS
Lutensit® AN 10
Lutensit® AN 30
Lutensit® AN 40
Lutensit® K-HP
Lutensit® K-LC, K-LC 80
Lutensit® K-OC
Lutensit® K-TI
Lutensol® A 7
Lutensol® AO 3
Lutensol® AO 4
Lutensol® AO 5
Lutensol® AO 7
Lutensol® AO 8
Lutensol® AO 11
Lutensol® AO 12
Lutensol® AO 109
Lutensol® AO 3109
Lutensol® AP 6
Lutensol® AP 7
Lutensol® AP 8
Lutensol® AP 9
Lutensol® AT 11
Lutensol® AT 18
Lutensol® AT 25
Lutensol® AT 50
Lutensol® AT 80
Lutensol® ED 140
Lutensol® ED 310
Lutensol® ED 370
Lutensol® ED 610
Lutensol® FA 12
Lutensol® FSA 10
Lutensol® GD 50, GD 70
Lutensol® LF 220, 221, 223, 224
Lutensol® LF 400
Lutensol® LF 401
Lutensol® LF 403
Lutensol® LF 404, 405
Lutensol® LF 431
Lutensol® LF 600
Lutensol® LF 700
Lutensol® LF 711
Lutensol® LF 1300
Lutensol® LSV
Lutensol® LT 30
Lutensol® ON 30
Lutensol® ON 50
Lutensol® ON 60
Lutensol® ON 70
Lutensol® ON 110
Lutensol® TO 3
Lutensol® TO 5
Lutensol® TO 7
Lutensol® TO 8
Lutensol® TO 10
Lutensol® TO 12
Lutensol® TO 15
Lutensol® TO 20

Industrial, Institutional and Consumer Cleaning (cont'd.)

Lutensol® TO 89
Lutensol® TO 109
Lutensol® TO 129
Lutensol® TO 389
Mackam 1L
Mackam 1L-30
Mackam 2C
Mackam 2C-75
Mackam 2C-SF
Mackam 2CY
Mackam 2CYSF
Mackam 2L
Mackam 2LSF
Mackam 2W
Mackam 35
Mackam 35 HP
Mackam CAP
Mackam CB-35
Mackam CB-LS
Mackam CET
Mackam CSF
Mackam ISA
Mackam J
Mackam LAP
Mackam LMB
Mackam LT
Mackam MLT
Mackam OB-30
Mackam RA
Mackam TM
Mackam WGB
Mackamide O
Mackamine CO
Mackamine IAO
Mackamine ISMO
Mackamine LAO
Mackamine LO
Mackamine O2
Mackamine OAO
Mackamine SAO
Mackamine SO
Mackamine WGO
Mackanate CM
Mackanate CM-100
Mackazoline C
Mackazoline CY
Mackazoline L
Mackazoline O
Mackine 101
Mackine 201
Mackine 301
Mackine 321
Mackine 401
Mackine 421
Mackine 501
Mackine 601
Mackine 701
Mackine 801
Mackine 901
Macol® 21
Macol® 24
Macol® 25
Macol® 26
Macol® 30
Macol® 45
Macol® 300
Macol® 660
Macol® 3520
Macol® 5100
Macol® CA-2
Macol® CSA-2
Macol® CSA-4
Macol® CSA-10
Macol® CSA-15
Macol® CSA-20
Macol® CSA-40
Macol® CSA-50
Macol® DNP-5
Macol® DNP-10
Macol® DNP-15
Macol® DNP-21
Macol® DNP-150
Macol® LA-4
Macol® LA-9
Macol® LA-12
Macol® LA-23
Macol® LA-790
Macol® LF-110
Macol® LF-111
Macol® LF-115
Macol® LF-120
Macol® LF-125
Macol® NP-4
Macol® NP-5
Macol® NP-6
Macol® NP-8
Macol® NP-11
Macol® NP-12
Macol® NP-15
Macol® NP-20
Macol® NP-20(70)
Macol® NP-30(70)
Macol® NP-100
Macol® OA-2
Macol® OA-4
Macol® OA-5
Macol® OP-3
Macol® OP-5
Macol® OP-8
Macol® OP-10
Macol® SA-2
Macol® SA-5
Macol® SA-10
Macol® SA-15
Macol® SA-20
Macol® SA-40
Macol® TD-3
Macol® TD-8
Macol® TD-10
Macol® TD-12
Macol® TD-15
Macol® TD-100
Macol® TD-610
Mafo® CAB
Makon® 4
Makon® 6
Makon® 7
Makon® 8
Makon® 10
Makon® 11
Makon® 12
Makon® 14
Makon® 30
Makon® 50
Makon® NF-5
Makon® NF-12
Makon® OP-6
Makon® OP-9
Manro ADS 35
Manro ALEC 27
Manro ALS 30
Manro AO 30C
Manro AT 1200
Manro BA Acid
Manro BES 27
Manro BES 60
Manro BES 70
Manro CD
Manro CD/G
Manro CDS
Manro CMEA
Manro DB 30
Manro DB 56
Manro DB 98
Manro DES 32
Manro DS 35
Manro HA Acid
Manro HCS
Manro NA Acid
Manro SDBS 25/30
Manro SDBS 60
Manro SLS 28
Manro TDBS 60
Manro TL 40
Manromid 150-ADY
Manromid 853
Manromid CD
Manromid CDG
Manromid CDS
Manromid CDX
Manromid CMEA
Manromid LMA
Manrosol ACS60
Manrosol SCS40
Manrosol SCS93
Manrosol STS40
Manrosol STS90
Manrosol SXS30
Manrosol SXS40
Manrosol SXS93
Manroteric CAB
Manroteric CDX38
Manroteric CEM38
Manroteric CyNa50
Manroteric NAB
Manroteric SAB
Manrowet MO70S
Maphos® 76
Maphos® 8135
Maphos® DT
Maphos® JA 60
Maphos® JP 70
Maphos® L 13
Maprosyl® 30
Maquat 4450-E
Maquat DLC-1214
Maquat MC-1412
Maquat MC-1416
Maquat MC-6025-50%
Maquat MQ-2525
Maquat MQ-2525M
Maquat SC-18
Maquat SC-1632
Maquat TC-76
Maranil A
Maranil ABS
Maranil CB-22
Maranil DBS
Maranil Paste A 55, A 75
Maranil Powd. A
Marasperse N-22
Marlamid® D 1218
Marlamid® D 1885
Marlamid® DF 1818
Marlamid® KL
Marlamid® M 1218
Marlamid® M 1618
Marlazin® 7102
Marlazin® 7265
Marlazin® 8567
Marlazin® KC 21/50
Marlazin® L 2
Marlazin® L 10
Marlazin® L 410
Marlazin® OL 2
Marlazin® OL 20
Marlazin® S 10
Marlazin® S 40
Marlazin® T 10
Marlazin® T 50
Marlican®
Marlinat® 24/28
Marlinat® 24/70
Marlinat® 242/70
Marlinat® 242/70 S
Marlinat® 243/28
Marlinat® 243/70
Marlinat® 5303
Marlinat® CM 20
Marlinat® CM 40
Marlinat® CM 45
Marlinat® CM 100
Marlinat® CM 105
Marlinat® CM 105/80
Marlinat® DF 8
Marlinat® SRN 30
Marlipal® 013/200
Marlipal® 013/30
Marlipal® 013/400
Marlipal® 013/50
Marlipal® 013/60
Marlipal® 013/70
Marlipal® 013/80
Marlipal® 013/90
Marlipal® 013/939
Marlipal® 24/30
Marlipal® 24/40
Marlipal® 24/50
Marlipal® 24/60
Marlipal® 24/70
Marlipal® 24/80
Marlipal® 24/90
Marlipal® 24/120
Marlipal® 24/939
Marlipal® 34/30, /50, /60, /70, /79, /99, /100, /109, /110, /119, /120, /140
Marlipal® 104
Marlipal® 1012/4
Marlipal® 1012/6

Industrial, Institutional and Consumer Cleaning (cont'd.)

Marlipal® 1618/8
Marlipal® 1618/10
Marlipal® 1618/11
Marlipal® 1618/18
Marlipal® 1618/25
Marlipal® 1618/25 P 6000
Marlipal® 1618/40
Marlipal® 1618/80
Marlipal® 1850/10
Marlipal® 1850/30
Marlipal® 1850/40
Marlipal® 1850/5
Marlipal® 1850/80
Marlipal® BS
Marlipal® FS
Marlipal® KE
Marlipal® KF
Marlipal® ML
Marlipal® NE
Marlipal® O11/30
Marlipal® O11/50
Marlipal® O11/79
Marlipal® O11/88
Marlipal® O13/20
Marlipal® O13/30
Marlipal® O13/40
Marlipal® O13/50
Marlipal® O13/60
Marlipal® O13/70
Marlipal® O13/80
Marlipal® O13/90
Marlipal® O13/100
Marlipal® O13/120
Marlipal® O13/150
Marlipal® O13/170
Marlipal® O13/200
Marlipal® O13/400
Marlipal® O13/500
Marlipal® O13/939
Marlipal® SU
Marlon® A 350
Marlon® A 360
Marlon® A 365
Marlon® A 375
Marlon® A 390
Marlon® AFM 40, 40 N, 43, 50N
Marlon® AFO 40
Marlon® AFO 50
Marlon® AFR
Marlon® AM 80
Marlon® AMX
Marlon® ARL
Marlon® AS3
Marlon® AS3-R
Marlon® PF 40
Marlon® PS 30
Marlon® PS 60
Marlon® PS 60 W
Marlon® PS 65
Marlophen® 81
Marlophen® 82
Marlophen® 83
Marlophen® 83N
Marlophen® 84
Marlophen® 84N
Marlophen® 85
Marlophen® 85N
Marlophen® 86
Marlophen® 86N
Marlophen® 86N/S
Marlophen® 87
Marlophen® 87N
Marlophen® 88
Marlophen® 88N
Marlophen® 89
Marlophen® 89N
Marlophen® 89.5N
Marlophen® 810
Marlophen® 810N
Marlophen® 812
Marlophen® 812N
Marlophen® 814
Marlophen® 814N
Marlophen® 820
Marlophen® 820N
Marlophen® 825
Marlophen® 830
Marlophen® 830N
Marlophen® 840N
Marlophen® 850
Marlophen® 850N
Marlophen® 1028
Marlophen® 1028N
Marlophor® AS-Acid
Marlophor® CS-Acid
Marlophor® DS-Acid
Marlophor® FC-Acid
Marlophor® FC-Sodium Salt
Marlophor® ID-Acid
Marlophor® MN-60
Marlophor® NP5-Acid, NP6-Acid, NP7-Acid
Marlopon® ADS 50
Marlopon® ADS 65
Marlopon® AMS 60
Marlopon® AT
Marlopon® AT 50
Marlopon® CA
Marlosol® R70
Marlowet® 4508
Marlowet® 4800
Marlowet® 4900
Marlowet® 4901
Marlowet® 4902
Marlowet® 5311
Marlowet® 5400
Marlowet® 5440
Marlowet® 5609
Marlowet® 5622
Marlowet® 5626
Marlowet® 5635
Marlowet® FOX
Marlowet® GFN
Marlowet® GFW
Marlowet® IHF
Marlowet® ISM
Marlowet® PW
Marlowet® SLS
Marlowet® T
Marlowet® TM
Marlox® B 24/50
Marlox® B 24/60
Marlox® B 24/80
Marlox® FK 14
Marlox® FK 64
Marlox® FK 69
Marlox® FK 86
Marlox® FK 1614
Marlox® LM 25/30
Marlox® LM 55/18
Marlox® LM 75/30
Marlox® LP 90/20
Marlox® M 606/1
Marlox® M 606/2
Marlox® MO 124
Marlox® MO 145
Marlox® MO 154
Marlox® MO 174
Marlox® MO 244
Marlox® NP 109
Marlox® S 58
Masil® 1066C
Masil® SF 5
Masil® SF 10
Masil® SF 20
Masil® SF 50
Masil® SF 100
Masil® SF 200
Masil® SF 350
Masil® SF 350 FG
Masil® SF 500
Masil® SF 1000
Masil® SF 5000
Masil® SF 10,000
Masil® SF 12,500
Masil® SF 30,000
Masil® SF 60,000
Masil® SF 100,000
Masil® SF 300,000
Masil® SF 500,000
Masil® SF 600,000
Masil® SF 1,000,000
Maypon 4C
Mazamide® 65
Mazamide® 66
Mazamide® 68
Mazamide® 70
Mazamide® 80
Mazamide® C-5
Mazamide® CCO
Mazamide® CMEA Extra
Mazamide® CS 148
Mazamide® J 10
Mazamide® JR 100
Mazamide® JR 300
Mazamide® JR 400
Mazamide® JT 128
Mazamide® L-298
Mazamide® LLD
Mazamide® LM
Mazamide® LM 20
Mazamide® O 20
Mazamide® PCS
Mazamide® RO
Mazamide® SS 20
Mazamide® TC
Mazamide® WC Conc.
Mazawet® 36
Mazawet® 77
Mazawet® DF
Mazawet® DOSS 70
Mazclean EP
Mazclean W
Mazclean W-10
Mazon® 21
Mazon® 23
Mazon® 27
Mazon® 29
Mazon® 40
Mazon® 40A
Mazon® 41
Mazon® 60T
Mazon® 70
Mazon® 85
Mazon® DWD-100
Mazox® CAPA
Mazox® CAPA-37
Mazox® CDA
Mazox® KCAO
Mazox® LDA
Mazox® MDA
Mazox® ODA
Mazox® SDA
Mazu® 319
Mazu® DF 205SX
Mazu® DF 210SX Mod 1
Mazu® DF 215SX
Mazu® DF 255
Merbron R
Merpol® 100
Merpol® A
Merpol® SH
Mersolat H 30
Mersolat H 40
Mersolat H 68
Mersolat H 76
Mersolat H 95
Mersolat W 40
Mersolat W 68
Mersolat W 76
Mersolat W 93
Miramine® C
Miranol® 2CIB
Miranol® C2M Anhyd. Acid
Miranol® C2M Conc. NP
Miranol® C2M Conc. OP
Miranol® C2M-SF 70%
Miranol® C2M-SF Conc.
Miranol® CM Conc. NP
Miranol® CM Conc. OP
Miranol® CM-SF Conc.
Miranol® CS Conc.
Miranol® DM
Miranol® DM Conc. 45%
Miranol® FA-NP
Miranol® FAS
Miranol® FB-NP
Miranol® FBS
Miranol® H2C-HA
Miranol® HM Conc.
Miranol® J2M Conc.
Miranol® JA

Industrial, Institutional and Consumer Cleaning (cont'd.)

Miranol® JAS-50
Miranol® JBS
Miranol® JEM Conc.
Miranol® JS Conc.
Miranol® L2M-SF Conc.
Miranol® LB
Miranol® SM Conc.
Mirataine® A2P-TS-30
Mirataine® ASC
Mirataine® BB
Mirataine® CBC
Mirataine® CB/M
Mirataine® CBS, CBS Mod.
Mirataine® FM
Mirataine® H2C-HA
Mirataine® TM
Miravon B12DF
Miravon B79R
Mona AT-1200
Mona NF-10
Mona NF-15
Mona NF-25
Monafax 060
Monafax 785
Monafax 786
Monafax 794
Monafax 831
Monafax 872
Monafax 1214
Monafax 1293
Monamate C-1142
Monamate CPA-40
Monamate CPA-100
Monamate LA-100
Monamid® 150-AD
Monamid® 150-ADD
Monamid® 150-DR
Monamid® 150-LMWC
Monamid® 150-LW
Monamid® 150-LWA
Monamid® 664-MC
Monamid® 716
Monamid® 853
Monamid® 1007
Monamid® CMA
Monamid® LIPA
Monamid® LMA
Monamid® LMIPA
Monamid® LMMA
Monamine 779
Monamine ALX-80SS
Monamine ALX-100 S
Monamine I-76
Monamine R8-26
Monamine T-100
Monamulse 653-C
Monamulse 947
Monamulse dL-1273
Monaquat AT-1074
Monaquat TG
Monastat 1195
Monaterge 85
Monaterge LF-945
Monateric 811
Monateric 1188M
Monateric CA-35
Monateric CEM-38
Monateric COAB
Monateric CyA-50
Monateric CyNa-50
Monateric LFNa-50
Monateric TDB-35
Monatrope 1250
Monatrope 1296
Monawet MO-70E
Monazoline C
Monazoline O
Monazoline T
Monolan® 2000
Monolan® 2800
Monolan® 3000 E/50
Monolan® 3000 E/60, 8000 E/80
Monolan® PB
Monolan® PM7
Monolan® PT
Montosol IL-13
Montosol PF-16, -18
Montosol PG-12
Montosol PL-14
Montosol PQ-11, -15
Montosol TQ-11
Montovol GL-13
Montovol RF-10
Montovol RF-11
Montovol RJ-13
Montovol RL-10
Mulsifan ABN
Mulsifan K 326 Spezial
Mulsifan RT 18
Mulsifan RT 63
Mulsifan RT 110
Mulsifan RT 231
Mulsifan RT 248
Mulsifan STK
Nacconol® 90G
Na Cumene Sulfonate 40, Sulfonate Powd.
Nalkylene® 575L
Nansa® 1042
Nansa® 1339
Nansa® 1340
Nansa® 1347
Nansa® 1385
Nansa® 1389
Nansa® 1390/E
Nansa® 1400 Series
Nansa® 1909
Nansa® 7052
Nansa® 7053/E
Nansa® 7069
Nansa® AS 40
Nansa® HAD
Nansa® HS40-AU
Nansa® HS40/S
Nansa® HS80-AU
Nansa® HS80P
Nansa® HS80S
Nansa® HS80SK
Nansa® HS85
Nansa® HS85S
Nansa® HSA/L
Nansa® LES 42
Nansa® LSS38/A
Nansa® MA30
Nansa® SSAL
Nansa® TS 60
Nansa® UCA/S, UCP/S
Nansa® YS94
Na Toluene Sulfonate 30, 40
Naturechem® EGHS
Naturechem® GMHS
Naturechem® OHS
Naturechem® PGHS
Naturechem® PGR
Naturechem® THS-200
Naxchem CD-6M
Naxchem Detergent CNB
Naxchem Dispersant K
Naxchem N-Foam 802
Naxel AAS-40S
Naxel AAS-45S
Naxel AAS-60S
Naxel AAS-Special 3
Naxel DDB 500
Naxide 1230
Naxolate WA-97
Naxolate WA Special
Naxolate WAG
Naxonac 510
Naxonac 600
Naxonac 610
Naxonic NI-40
Naxonic NI-100
Naxonol CO
Naxonol PN 66
Naxonol PO
Neodol® 1-3
Neodol® 1-5
Neodol® 1-7
Neodol® 1-9
Neodol® 23-1
Neodol® 23-3
Neodol® 23-5
Neodol® 23-6.5
Neodol® 23-12
Neodol® 25-3
Neodol® 25-3A
Neodol® 25-3S
Neodol® 25-7
Neodol® 25-9
Neodol® 25-12
Neodol® 45-2.25
Neodol® 45-7
Neodol® 45-13
Neodol® 91-2.5
Neodol® 91-6
Neodol® 91-8
Neopelex No. 6, No. 25, No. 6F Powder, F-25, F-65
Neoscoa MSC-80
Neoscoa TH-102
Neutronyx® 656
Newpol PE-61
Newpol PE-62
Newpol PE-64
Newpol PE-68
Newpol PE-74
Newpol PE-75
Newpol PE-78
Newpol PE-88
Niaproof® Anionic Surfactant 08
Nikkol CCK-40
Nikkol CCN-40
Nikkol DDP-2
Nikkol DDP-4
Nikkol DDP-6
Nikkol DDP-8
Nikkol DDP-10
Ninol® 201
Ninol® 1281
Ninol® 1285
Ninol® 1301
Ninol® 40-CO
Ninol® 4821 F
Ninol® 5024
Ninol® A-10MM
Ninol® B
Ninol® CMP
Ninol® LDL 2
Ninol® LMP
Ninol® SR-100
Ninox® FCA
Ninox® L
Ninox® M
Ninox® SO
Niox EO-12
Niox EO-13
Niox EO-14, 23
Niox EO-32, -35
Niox KF-12
Niox KF-13
Niox KF-17
Niox KF-18
Niox KF-26
Niox KG-14
Niox KH Series
Niox KJ-55
Niox KJ-56
Niox KJ-61
Niox KJ-66
Niox KL-16
Niox KL-19
Niox KQ-20
Niox KQ-70, LQ-13
Nissan Cation BB
Nissan Diapon K
Nissan Nonion E-205
Nissan Nonion E-206
Nissan Nonion E-208
Nissan Nonion E-215
Nissan Nonion E-220
Nissan Nonion E-230
Nissan Nonion HS-204.5
Nissan Nonion HS-206
Nissan Nonion HS-208
Nissan Nonion HS-210
Nissan Nonion HS-215
Nissan Nonion HS-220
Nissan Nonion HS-240
Nissan Nonion HS-270
Nissan Nonion K-202
Nissan Nonion K-203
Nissan Nonion K-204
Nissan Nonion K-207
Nissan Nonion K-211

Industrial, Institutional and Consumer Cleaning (cont'd.)

Nissan Nonion K-215
Nissan Nonion K-220
Nissan Nonion K-230
Nissan Nonion NS-202
Nissan Nonion NS-204.5
Nissan Nonion NS-206
Nissan Nonion NS-208.5
Nissan Nonion NS-209
Nissan Nonion NS-210
Nissan Nonion NS-212
Nissan Nonion NS-215
Nissan Nonion NS-220
Nissan Nonion NS-230
Nissan Nonion NS-240
Nissan Nonion NS-250
Nissan Nonion NS-270
Nissan Nonion P-208
Nissan Nonion P-210
Nissan Nonion P-213
Nissan Nonion S-206
Nissan Nonion S-207
Nissan Nonion S-215
Nissan Nonion S-220
Nissan Ohsen A
Nissan Persoft NK-60, -100
Nissan Soft Osen 550A
Nissan Stafoam DL
Nissan Stafoam DO, DOS
Nissan Stafoam L
Nissan Sunalpha T
Nissan Sunamide C-3, CF-3, CF-10
Nissan Sunbase, Powder
Nitrene 100 SD
Nitrene 11120
Nitrene 11230
Nitrene 13026
Nitrene A-309
Nitrene A-567
Nitrene C
Nitrene C Extra
Nitrene L-76
Nitrene L-90
Nitrene N
Nonasol 3922
Nonasol N4AS
Nonasol N4SS
Nonipol Soft SS-50
Nonipol Soft SS-70
Nonipol Soft SS-90
Nonisol 100
Nopco® 1179
Nopcogen 22-O
Nopcosulf CA-60, -70
Nopcosulf TA-45V
Nopcosurf CA
Noramac C 26
Noramac O
Noramac S
Noramac SH
Noramox C2
Noramox C5
Noramox C11
Noramox C15
Noramox O2
Noramox O5
Noramox O11
Noramox O15
Noramox O20
Noramox S1
Noramox S2
Noramox S5
Noramox S7
Noramox S11
Noramox S15
Noramox S20
Norfox® 1101
Norfox® 1115
Norfox® 40
Norfox® 90
Norfox® ALES-60
Norfox® ALKA
Norfox® ALPHA XL
Norfox® ALS
Norfox® CMA
Norfox® Coco Powder
Norfox® DC
Norfox® DC-38
Norfox® DCO
Norfox® DCS
Norfox® DCSA
Norfox® DOSA
Norfox® EGMS
Norfox® GMS
Norfox® Hercules Conc.
Norfox® IM-38
Norfox® KO
Norfox® NP-7
Norfox® NP-9
Norfox® NP-11
Norfox® OP-100
Norfox® OP-102
Norfox® OP-114
Norfox® PEA-N
Norfox® SLES-02
Norfox® SLES-30
Norfox® SLES-60
Norfox® SLS
Norfox® T-60
Norfox® TB Granules
Norfox® X
NP-55-80
NSA-17
Nutrapon AL 30
Nutrapon AL 60
Nutrapon DL 3891
Nutrapon DW 0266
Nutrapon HA 3841
Nutrapon KPC 0156
Nutrapon LD 0206
Nutrapon PP 3563
Nutrapon RS 1147
Nutrol 622
Nutrol 640
Nutrol 656
Nutrol Betaine OL 3798
Oakite Ladd
Obanol 516
Octenyl Succinic Anhydride, n-
Octosol A-18
Orzan® LS
Orzan® LS-50
Oxetal 500/85
Oxetal 800/85
Oxetal C 110
Oxetal D 104
Oxetal ID 104
Oxetal O 108
Oxetal O 112
Oxetal T 106
Oxetal T 110
Oxetal TG 111
Oxetal TG 118
Pantex
Pegosperse® 200 DL
Pegosperse® 400 DOT
Pegosperse® 600 DOT
Pegosperse® PMS CG
Pentine Acid 5431
Peramit MLN
Perlankrol® DAF25
Perlankrol® DGS
Perlankrol® EAD60
Perlankrol® ESD
Perlankrol® ESK29
Perlankrol® ESS25
Perlankrol® FF
Perlankrol® FN65
Perlankrol® O
Perlankrol® PA Conc.
Perlankrol® SN
Petro® 11
Petro® 22
Petro® BA
Petro® BAF
Petro® LBA
Petro® LBAF
Petro® P
Petro® ULF
Petro® WP
Petrosul® L-60
Petrowet® R
Phosfac 1004, 1006, 1044, 1044FA, 1066, 1066FA, 1068FA
Phosfetal 201
Phosfetal 204
Phosfetal 205
Phosfetal 600
Phosfetal 601
Phosfetal 602
Phosfetal 603
Phospholan® AD-1
Phospholan® BH14
Phospholan® KPE4
Phospholan® PNP9
Phospholan® PSP6
Phospholipid PTD
Phospholipid PTL
Phospholipid PTZ
Phosphoteric® T-C6
Plantaren APG 225
Plantaren CG 60
Pluracol® E400 NF
Pluracol® E600
Pluracol® E1450
Pluracol® E1450 NF
Pluracol® E2000
Pluracol® E4000 NF
Pluracol® E4500
Pluracol® E8000 NF
Pluradot HA-410
Pluradot HA-420
Pluradot HA-430
Pluradot HA-433
Pluradot HA-440
Pluradot HA-450
Pluradot HA-510
Pluradot HA-520
Pluradot HA-530
Pluradot HA-540
Pluradot HA-550
Plurafac® A-24
Plurafac® A-27
Plurafac® A-38
Plurafac® A-39
Plurafac® A-46
Plurafac® B-25-5
Plurafac® B-26
Plurafac® C-17
Plurafac® D-25
Plurafac® LF 120
Plurafac® LF 131
Plurafac® LF 132
Plurafac® LF 220
Plurafac® LF 221
Plurafac® LF 223
Plurafac® LF 224
Plurafac® LF 231
Plurafac® LF 400
Plurafac® LF 401, LF 403
Plurafac® LF 404, LF 405
Plurafac® LF 431
Plurafac® LF 600
Plurafac® LF 700, LF 711
Plurafac® LF 1430
Plurafac® RA-20
Plurafac® RA-30
Plurafac® RA-40
Plurafac® RA-43
Plurafac® RA-50
Plurafac® T-55
Pluriol® E 200
Pluriol® E 300
Pluriol® E 400
Pluriol® E 600
Pluriol® PE 4300
Pluriol® PE 6100
Pluriol® PE 6200
Pluriol® PE 6400
Pluriol® PE 10100
Pluriol® PE 10500
Pluronic® 10R5
Pluronic® 10R8
Pluronic® 12R3
Pluronic® 17R1
Pluronic® 17R2
Pluronic® 17R4
Pluronic® 17R8
Pluronic® 22R4
Pluronic® 25R1
Pluronic® 25R2
Pluronic® 25R4
Pluronic® 25R5
Pluronic® 25R8
Pluronic® 31R1

Industrial, Institutional and Consumer Cleaning (cont'd.)

Pluronic® 31R2
Pluronic® 31R4
Plysurf A207H
Plysurf A208B
Plysurf A208S
Plysurf A210G
Plysurf A212C
Plysurf A215C
Plysurf A216B
Plysurf A217E
Plysurf A219B
Plysurf AL
Polyaldo® DGHO
Polycarboxylate AMC 60
Poly-G® 200
Polysurf A212E
Poly-Tergent® 3B2
Poly-Tergent® 3B2 Acid
Poly-Tergent® 4C3
Poly-Tergent® CS-1
Poly-Tergent® E-17A
Poly-Tergent® E-17B
Poly-Tergent® E-25B
Poly-Tergent® P-17A
Poly-Tergent® P-17B
Poly-Tergent® P-17BLF
Poly-Tergent® P-17BX
Poly-Tergent® P-17D
Poly-Tergent® P-22A
Poly-Tergent® P-32A
Poly-Tergent® P-32D
Poly-Tergent® P-9E
Poly-Tergent® RCS-43
Poly-Tergent® S-205LF
Poly-Tergent® S-305LF
Poly-Tergent® S-405LF
Poly-Tergent® S-505LF
Poly-Tergent® SL-42
Poly-Tergent® SL-62
Poly-Tergent® SL-92
Poly-Tergent® SLF-18
Produkt GS 5001
Produkt RT 63
Profan ME-20
Progalan X-13
Progasol 230
Progasol 443, 457
Propetal 99
Propetal 103
Propetal 241
Propetal 281
Propetal 340
Propetal 341
Propomeen 2HT-11
Prote-pon P-2 EHA-02-K30
Prote-pon P 2 EHA-02-Z
Prote-pon P 2 EHA-Z
Prote-pon P-0101-02-Z
Prote-pon P-L 201-02-K30
Prote-pon P-L 201-02-Z
Prote-pon P-NP-06-K30
Prote-pon P-NP-06-Z
Prote-pon P-NP-10-K30
Prote-pon P-NP-10-MZ
Prote-pon P-NP-10-Z
Prote-pon P-OX 101-02-K75
Prote-pon P-TD-06-K13
Prote-pon P-TD-06-K30
Prote-pon P-TD 06-K60
Prote-pon P-TD-06-Z
Prote-pon P-TD-09-Z
Prote-pon P-TD-12-Z
Prote-pon TD-09-K30
Prote-sorb SML
Prote-sorb SMO
Prote-sorb SMP
Prote-sorb SMS
Prote-sorb STO
Prote-sorb STS
Protowet 5171
Prox-onic 2EHA-1/02
Prox-onic 2EHA-1/05
Prox-onic CSA-1/04
Prox-onic CSA-1/06
Prox-onic CSA-1/010
Prox-onic CSA-1/015
Prox-onic CSA-1/020
Prox-onic CSA-1/030
Prox-onic CSA-1/050
Prox-onic DA-1/04
Prox-onic DA-1/06
Prox-onic DA-1/09
Prox-onic DDP-09
Prox-onic DNP-08
Prox-onic DNP-0150
Prox-onic DNP-0150/50
Prox-onic L 081-05
Prox-onic L 101-05
Prox-onic L 102-02
Prox-onic L 121-09
Prox-onic L 161-05
Prox-onic L 181-05
Prox-onic L 201-02
Prox-onic OCA-1/06
Prox-onic OP-09
Prox-onic OP-016
Prox-onic PEG-4000
Prox-onic PEG-6000
Prox-onic PEG-10,000
Prox-onic PEG-20,000
Prox-onic PEG-35,000
Prox-onic SA-1/02
Prox-onic SA-1/010
Prox-onic SA-1/020
Prox-onic TBP-08
Prox-onic TBP-030
Prox-onic TD-1/03
Prox-onic TD-1/06
Prox-onic TD-1/09
Prox-onic TD-1/012
Prox-onic TM-06
Prox-onic TM-08
Prox-onic TM-010
Prox-onic UA-03
Prox-onic UA-06
Prox-onic UA-09
Prox-onic UA-012
Purton CFD
Purton SFD
Quadrilan® BC
Quadrilan® MY211
Quaternary O
Querton 14Br-40
Querton 16Cl-29
Querton 16Cl-50
Quimipol 9108
Quimipol EA 4505
Quimipol EA 4508
Quimipol EA 6801
Quimipol EA 6802
Quimipol EA 9108
Quimipol ENF 120
Radiaquat 6410
Radiaquat 6412
Radiaquat 6442
Radiaquat 6462
Radiaquat 6470
Radiaquat 6475
Radiaquat 6480
Radiasurf® 7000
Radiasurf® 7146
Radiasurf® 7152
Radiasurf® 7153
Radiasurf® 7175
Radiasurf® 7196
Radiasurf® 7402
Remcopal 229
Renex® 20
Renex® 22
Renex® 25
Renex® 30
Renex® 31
Renex® 35
Renex® 36
Renex® 648
Renex® 678
Renex® 688
Renex® 690
Renex® 703
Renex® 704
Renex® 707
Renex® 714
Renex® 720
Rewomat B 2003
Rewomid® C 212
Rewomid® DC 212 S
Rewomid® DL 203 S
Rewomid® DL 240
Rewomid® DO 280
Rewomid® IPL 203
Rewopal® HV 5
Rewopal® HV 9
Rewopal® HV 14
Rewopal® HV 25
Rewopal® MT 65
Rewopal® O 8
Rewopal® PO
Rewopal® TA 25
Rewophat NP 90
Rewophat TD 40
Rewophat TD 70
Rewopol® AL 3
Rewopol® BW
Rewopol® BWA
Rewopol® CL 30
Rewopol® CT 65
Rewopol® FBR
Rewopol® HD 50 L
Rewopol® MLS 30
Rewopol® MLS 35
Rewopol® NEHS 40
Rewopol® NL 2
Rewopol® NL 3
Rewopol® NL 3-28
Rewopol® NL 3-70
Rewopol® PGK 2000
Rewopol® S 1954
Rewopol® S 2311
Rewopol® SBC 212
Rewopol® SBC 212 G
Rewopol® SBDO 70
Rewopol® SBF 12
Rewopol® SBF 12 P
Rewopol® SBFA 30
Rewopol® SBL 203
Rewopol® SBL 203 G, 203 P
Rewopol® SBLC, SBLC G
Rewopol® SBV
Rewopol® SBZ
Rewopol® SCK 2040
Rewopol® TLS 40
Rewopol® TS 25
Rewopol® TS 35
Rewopol® TS 40 P
Rewopol® TS 100
Rewopol® TSK 30
Rewopol® TSSP 25
Rewopon® AM-V
Reworyl® ACS 60
Reworyl® NCS 40
Reworyl® NKS 50
Reworyl® NKS 100
Reworyl® NTS 40
Reworyl® TKS 90 F
Reworyl® TKS 90/L
Rewoteric AM B13
Rewoteric AM B14
Rewoteric AM CAS
Rewoteric AM DML
Rewoteric AM KSF 40
Rewoteric AM TEG
Rewoteric AM V
Rewoteric AM VSF
Rewoteric QAM 50
Rexobase BAT
Rexoclean 200N
Rexoclean APC
Rexoclean APXX
Rexoclean HAC
Rexoclean NFC
Rexoclean SA
Rexoclean SRT
Rexol 25/8
Rexol 25/9
Rexol 25/15
Rexol 25J
Rexol 35/11
Rexol 45/7
Rexol 45/10
Rexol 45/12
Rexol 45/16
Rexol 65/9
Rexol 65/10
Rexol 65/11
Rexol 65/14

Industrial, Institutional and Consumer Cleaning (cont'd.)

Rexonic 1012-6
Rexonic 1218-6
Rexonic N23-6.5
Rexonic N25-14
Rexonic N25-14(85%)
Rexonic N25-9
Rexonic N25-9(85%)
Rexonic N91-6
Rexonic P-1
Rexonic P-4
Rexonic P-5
Rexonic P-6
Rexonic P-9
Rexophos 25/67
Rexophos 25/97
Rexophos 4668
Rexophos JV 5015
Rhodacal® 330
Rhodacal® A-246L
Rhodacal® BA-77
Rhodacal® BX-78
Rhodacal® DDB-40
Rhodacal® DSB
Rhodacal® LA Acid
Rhodacal® N
Rhodafac® BG-510
Rhodafac® BP-769
Rhodafac® BX-660
Rhodafac® L3-15A
Rhodafac® L4-27A
Rhodafac® L6-36A
Rhodafac® LO-529
Rhodafac® MD-12-116
Rhodafac® PA-15
Rhodafac® PA-17
Rhodafac® PA-19
Rhodafac® PA-23
Rhodafac® PS-17
Rhodafac® PS-23
Rhodafac® R5-09/S
Rhodafac® R9-47A
Rhodafac® RA-600
Rhodafac® RE-410
Rhodafac® RE-610
Rhodafac® RE-960
Rhodafac® RM-510
Rhodafac® RM-710
Rhodafac® RS-410
Rhodafac® RS-610
Rhodafac® RS-710
Rhodamox® CAPO
Rhodamox® LO
Rhodapex® CO-433
Rhodapex® CO-436
Rhodapex® ES
Rhodapex® EST-30
Rhodapex® ESY
Rhodapex® MA360
Rhodapon® BOS
Rhodapon® CAV
Rhodapon® L-22, L-22/C
Rhodapon® LM
Rhodapon® LSB, LSB/CT
Rhodapon® LT-6
Rhodapon® OLS
Rhodapon® SM Special
Rhodaquat® DAET-90
Rhodaquat® T
Rhodasurf® 25-7
Rhodasurf® 91-6
Rhodasurf® 870
Rhodasurf® 1012-6
Rhodasurf® BC-610
Rhodasurf® BC-720
Rhodasurf® BC-840
Rhodasurf® DA-530
Rhodasurf® DA-630
Rhodasurf® DB 311
Rhodasurf® LA-3
Rhodasurf® LA-7
Rhodasurf® LA-9
Rhodasurf® LA-12
Rhodasurf® LA-15
Rhodasurf® ON-870
Rhodasurf® T
Rhodasurf® T50
Rhodasurf® TB-970
Rhodaterge® 206C
Rhodaterge® CAN
Rhodaterge® DCA
Rhodaterge® FL
Rhodaterge® LD-50Q
Rhodaterge® LD-60
Rhodaterge® RS-25
Rhodaterge® SMC
Ritachol® 1000
Rueterg 60-T
Rueterg 97-G
Rueterg 97-S
Sandet ALH
Sandopan® DTC
Sandopan® DTC-100
Sandopan® DTC-Acid
Sandopan® DTC Linear P
Sandopan® DTC Linear P Acid
Sandopan® JA-36
Sandopan® KST
Sandopan® LA-8
Sandopan® MA-18
Sandopan® TFL Conc.
Sandoxylate® 206
Sandoxylate® 224
Sandoxylate® 408
Sandoxylate® 412
Sandoxylate® 418
Sandoxylate® 424
Sandoxylate® AC-9
Sandoxylate® AC-24
Sandoxylate® AC-46
Sandoxylate® AD-4
Sandoxylate® AD-6
Sandoxylate® AD-9
Sandoxylate® AL-4
Sandoxylate® AO-12
Sandoxylate® AO-20
Sandoxylate® AO-60
Sandoxylate® AT-6.5
Sandoxylate® AT-12
Sandoxylate® C-10
Sandoxylate® C-15
Sandoxylate® C-32
Sandoxylate® FO-9
Sandoxylate® FO-30/70
Sandoxylate® FS-9
Sandoxylate® FS-35
Sandoxylate® NC-5
Sandoxylate® NC-15
Sandoxylate® NSO-30
Sandoxylate® NT-5
Sandoxylate® NT-15
Sandoxylate® PDN-7
Sandoxylate® PN-6
Sandoxylate® PN-9
Sandoxylate® PN-10.9
Sandoxylate® PO-5
Sandoxylate® SX-208
Sandoxylate® SX-224
Sandoxylate® SX-408
Sandoxylate® SX-412
Sandoxylate® SX-418
Sandoxylate® SX-424
Sandoxylate® SX-602
Sandoz Amide CO
Sandoz Amide NP
Sandoz Amide NT
Sandoz Amide PE
Sandoz Amide PL
Sandoz Amide PN
Sandoz Amide PO
Sandoz Amide PS
Sandoz Amine Oxide XA-C
Sandoz Amine Oxide XA-L
Sandoz Amine Oxide XA-M
Sandoz Phosphorester 510
Sandoz Phosphorester 600
Sandoz Phosphorester 690
Sandoz Sulfate 216
Sandoz Sulfate 219
Sandoz Sulfate 1030
Sandoz Sulfate A
Sandoz Sulfate K
Sandoz Sulfate W2-30
Sandoz Sulfate WA Dry
Sandoz Sulfate WAG
Sandoz Sulfate WAS
Sandoz Sulfate WA Special
Sandoz Sulfonate AAS 35S
Sandoz Sulfonate AAS 40FS
Sandoz Sulfonate AAS 45S
Sandoz Sulfonate AAS 50MS
Sandoz Sulfonate AAS 60S
Sandoz Sulfonate AAS 70S
Sandoz Sulfonate AAS 75S
Sandoz Sulfonate AAS 90
Sarkosyl® L
Sarkosyl® NL-30
Sarkosyl® O
Schercamox C-AA
Schercamox DMA
Schercamox DMM
Schercomid 1-102
Schercomid 1214
Schercomid CCD
Schercomid CDA
Schercomid CDO-Extra
Schercomid LD
Schercomid MME
Schercomid SCE
Schercomid SCO-Extra
Schercomid SL-Extra
Schercomid TO-2
Schercophos NP-9
Schercopol CMS-Na
Schercopol OMS-Na
Schercopon 2WD
Schercoteric CY-2
Schercoteric CY-SF-2
Schercoteric I-AA
Schercoteric LS-2
Schercoteric LS-EP
Schercoteric MS-2
Schercoteric MS-2 Modified
Schercoteric MS-2ES Modified
Schercoteric MS-2TE Modified
Schercoteric MS-SF
Schercoteric MS-SF (38%)
Schercoteric MS-SF (70%)
Schercoteric MS
Schercoteric O-AA
Schercoteric OS-SF
Schercozoline B
Schercozoline C
Secomix® E40
Secosol® AS
Secosol® DOS 70
Secosol® EA/40
Secoster® A
Secoster® DO 600
Secoster® MA 300
Secoster® MO 400
Secosyl
Sellig N 8 100
Sellig N 9 100
Sellig SU 30 100
Sellogen HR
Serdet DCK 3/70
Serdet DCK 30
Serdet DCN 30
Serdet DFK 40
Serdet DPK 30
Serdox NES 8/85
Serdox NNP 12
Serdox NNP 13
Servamine KAC 412
Servamine KAC 422
Servamine KEP 4527
Servoxyl VLB 1123
Setacin 103 Spezial

Industrial, Institutional and Consumer Cleaning (cont'd.)

Setacin F Spezial Paste
Setacin M
Silwet® L-77
Silwet® L-7001
Silwet® L-7200
Silwet® L-7210
Silwet® L-7230
Silwet® L-7600
Silwet® L-7604
Sinopol 707
Sinopol 714
Siponic® F-300
Siponic® F-400
Siponic® NP-9
Siponic® NP-15
Siponic® NP-40
Siponic® TD-3
Sipothix® 1941
S-Maz® 20
S-Maz® 40
S-Maz® 60
S-Maz® 60KHM
S-Maz® 65K
S-Maz® 80
S-Maz® 80K
S-Maz® 85
S-Maz® 85K
S-Maz® 90
S-Maz® 95
Sochamine OX 30
Sodium Octyl Sulfate Powd.
Soft Detergent 60
Softigen® 767
Sokalan® HP 22
Sokalan® HP 50
Sokalan® HP 53
Solar Soap Powd.
Solricin® 135
Solricin® 235
Solricin® 535
Soprofor NPF/10
Sorbit P
So/San 30M
Sovatex C Series
Span® 20
Span® 40
Span® 60, 60K
Span® 65
Span® 85
Spraywax 660-A Conc.
Stafoam
Stafoam DF-1
Stafoam F
Stamid HT 3901
Standamid® CD
Standamid® CMG
Standamid® KDM
Standamid® KDO
Standamid® KM
Standamid® LD
Standamid® LDO
Standamid® LDS
Standamid® LM
Standamid® PD
Standamid® SD
Standamid® SDO
Standamid® SM
Standamid® SOD
Standamid® SOMD
Standamox C 30
Standamox LAO-30
Standamox LMW-30
Standamul® B-1
Standamul® B-2
Standamul® B-3
Standapol® A
Standapol® AL-60
Standapol® AP-60
Standapol® AP Blend
Standapol® BW
Standapol® EA-K
Standapol® ES-50
Standapol® ES-250
Standapol® ES-350
Standapol® LF
Standapol® S
Standapol® SL-60
Standapol® SP-60
Standapol® WAQ-LC
Standapol® WAQ-LCX
Steol® 4N
Steol® CA-460
Steol® CS-460
Steol® KA-460
Steol® KS-460
Steol® OS 28
Stepan-Mild® SL3
Stepanol® AM
Stepanol® AM-V
Stepanol® DEA
Stepanol® DFS
Stepanol® GP-3 Conc.
Stepanol® HDL-50
Stepanol® LDL-3
Stepanol® ME Dry
Stepanol® MG
Stepanol® RS
Stepanol® SPT
Stepanol® WA-100
Stepanol® WAC
Stepanol® WAC-P
Stepanol® WA Paste
Stepanol® WAQ
Stepanol® WA Special
Stepanol® WAT
Stepanquat® F/T
Stepfac® 8170
Steposol® CA-60H
Steposol® CA-207
Steposol® CA-319
Straight Alkaline Cleaner SAC
Strodex® P-100
Strodex® PK-90
Sul-fon-ate AA-9
Sul-fon-ate AA-10
Sulfetal 4105
Sulfetal 4187
Sulfetal AF
Sulfetal C 38
Sulfetal C 90
Sulfetal CCO 50
Sulfetal CJOT 38
Sulfetal FA 40
Sulfetal TC 50
Sulfetal TC 50 W
Sulfochem 25-3A
Sulfochem 25-3S
Sulfochem 436
Sulfochem ES-2
Sulfochem MG
Sulfochem SLN
Sulfochem SLP
Sulfochem SLP-95
Sulfochem SLS
Sulfochem SLX
Sulfonic 800
Sulfonic Acid LS
Sulfopon 101, 101 Special
Sulfopon 101 Spez
Sulfopon K35
Sulfopon KT 115
Sulfopon KT 115-50
Sulfopon P-40
Sulfopon T 55
Sulfopon T Powd.
Sulfopon WA 1
Sulfopon WAQ LCX
Sulfopon WAQ Special
Sulfosil P-491
Sulfotex 110
Sulfotex A
Sulfotex LAS-90
Sulfotex LCX
Sulfotex LMS-E
Sulfotex OA
Sulfotex OT
Sulfotex PAI
Sulfotex PAI-S
Sulfotex RIF
Sulfotex SAL, SAT
Sulfotex SXS-40
Sulfotex UBL-100
Sulfotex WA
Sulfotex WAQ-LCX
Sulframin 40
Sulframin 40DA
Sulframin 40RA
Sulframin 40T
Sulframin 45
Sulframin 45LX
Sulframin 60T
Sulframin 85
Sulframin 90
Sulframin 1230
Sulframin 1240, 1245
Sulframin 1250, 1260
Sulframin 1255
Sulframin 1288
Sulframin 1298
Sulframin 1388
Sulframin 4010D, 4010R
Sulframin CSA
Sulframin HD, Low-Foam HD
Sulframin LX
Sulframin Phos-Free HD
Sulphonic Acid LS
SuperWash DZ
Surfactol® 13
Surfactol® 318
Surfactol® 365
Surfactol® 380
Surfactol® 575
Surfactol® 590
Surfagene FAD 106
Surfam P5 Dist
Surfam P5 Tech
Surfam P-MEPA
Surfax 150
Surfax CN
Surfax MG
Surfine AZI-A
Surfine T-A
Surfine WLG-A
Surfine WNT-A
Surfine WNT Conc.
Surfine WNT Gel
Surfine WNT LC
Surfine WNT-LS
Surfonic® HDL
Surfonic® HDL-1
Surfonic® JL-80X
Surfonic® L10-3
Surfonic® L12-3
Surfonic® L12-6
Surfonic® L12-8
Surfonic® L24-2
Surfonic® L24-3
Surfonic® L24-4
Surfonic® L24-7
Surfonic® L24-9
Surfonic® L24-12
Surfonic® L42-3
Surfonic® L46-7
Surfonic® L610-3
Surfonic® LF-17
Surfonic® N-31.5
Surfonic® N-40
Surfonic® N-60
Surfonic® N-85
Surfonic® N-95
Surfonic® N-100
Surfonic® N-102
Surfonic® N-120
Surfonic® N-130
Surfonic® N-150
Surfonic® N-200
Surfonic® N-300
Surfonic® NB-5
Surfynol® 61
Surfynol® 82
Surfynol® 104
Surmax® CS-504
Surmax® CS-515
Surmax® CS-521
Surmax® CS-522
Surmax® CS-555
Surmax® CS-586
Swanic 52L
Swascol 1PC
Swascol 3L
Swascol 4L
Swastik Detergent Powder
Sway
Synotol ME 90
Synperonic 3S60A
Synperonic 3S60S

Industrial, Institutional and Consumer Cleaning (cont'd.)

Synperonic 3S70
Synperonic 13/8
Synperonic 13/9
Synperonic 13/10
Synperonic 13/12
Synperonic 87K
Synperonic 91/2.5
Synperonic 91/4
Synperonic 91/5
Synperonic 91/6
Synperonic 91/7
Synperonic 91/8
Synperonic 91/10
Synperonic 91/12
Synperonic A2
Synperonic A3
Synperonic A4
Synperonic A6
Synperonic A7
Synperonic A9
Synperonic A11
Synperonic LF/B26
Synperonic LF/D25
Synperonic LF/RA30
Synperonic LF/RA40
Synperonic LF/RA43
Synperonic LF/RA50
Synperonic LF/RA260
Synperonic LF/RA290
Synperonic LF/RA343
Synperonic N
Synperonic NXP
Synperonic OP12.5
Synperonic OP16.5
Synperonic PE30/10
Synperonic PE30/20
Synperonic PE30/40
Synprolam 35BQC (50)
Synprolam 35BQC (80)
Synprolam 35DMA
Synprolam 35DMBQC
Synprolam 35MX1/O
Synprolam 35X2/O
Synprolam VC
Syntaryl Series
Syntens KMA 55
Syntens KMA 70-W-02
Syntens KMA 70-W-07
Syntens KMA 70-W-12
Syntens KMA 70-W-17
Syntens KMA 70-W-22
Syntens KMA 70-W-32
Syntergent DMC
Synthrapol KB
Syntopon F
Syntopon G
Syntopon S 493, 630, 1030
Tanapon NF-200
Tanaterge PE-67
Taski TR101
Tauranol M-35
Tauranol ML
Tauranol MS
Tauranol WS, WS Conc.
Tauranol WSP
T-Det® 25-3A
T-Det® 25-3S
T-Det® A-026
T-Det® D-150
T-Det® DD-5
T-Det® DD-7
T-Det® DD-9
T-Det® DD-10
T-Det® EPO-62L
T-Det® EPO-64L
T-Det® LF-416
T-Det® N-4
T-Det® N-8
T-Det® N-9.5
T-Det® N-10.5
T-Det® N-14
T-Det® N-100
T-Det® O-9
T-Det® O-12
T-Det® RQ1
T-Det® RY2
T-Det® TDA-65
Teepol CM 44
Teepol GC 56P
Teepol GD7P
Teepol GD53
Teepol HB 6
Teepol HB 7
Tegamineoxide WS-35
Tego®-Betaine C
Tego®-Betaine L-90
Tego®-Betaine S
Tego®-Betaine T
Tegotain A 4080
Tekstim 8741
Tensagex DLM 670
Tensagex DLS 670
Tensagex DLS 970
Tensagex EOC 670
Tequat RO
Terg-A-Zyme®
Tergenol 1122
Tergenol 3964
Tergitol® 15-S-3
Tergitol® 15-S-5
Tergitol® 15-S-7
Tergitol® 15-S-15
Tergitol® 15-S-20
Tergitol® 15-S-30
Tergitol® 15-S-40
Tergitol® 24-L-45
Tergitol® 24-L-50
Tergitol® 24-L-60
Tergitol® 24-L-60N
Tergitol® 24-L-75
Tergitol® 24-L-92
Tergitol® 24-L-98N
Tergitol® 26-L-1.6
Tergitol® 26-L-3
Tergitol® 26-L-5
Tergitol® D-683
Tergitol® Min-Foam 1X
Tergitol® Min-Foam 2X
Tergitol® NP-4
Tergitol® NP-5
Tergitol® NP-6
Tergitol® NP-7
Tergitol® NP-10
Tergitol® NP-13
Tergitol® NP-15
Tergitol® TMN-3
Tergitol® TMN-6
Tergitol® TMN-10
Tergitol® TP-9
Tergitol® XD
Tergitol® XH
Teric 9A2
Teric 9A5
Teric 9A6
Teric 9A8
Teric 9A12
Teric 12A2
Teric 12A3
Teric 12A4
Teric 12A6
Teric 12A7
Teric 12A8
Teric 12A9
Teric 12A12
Teric 12A16
Teric 12A23
Teric 13A5
Teric 13A7
Teric 13A9
Teric 15A11
Teric 16A16
Teric 16A22
Teric 17A2
Teric 17A3
Teric 17A6
Teric 17A10
Teric 17A13
Teric 17A25
Teric 160
Teric 161
Teric 163
Teric 164
Teric 165
Teric 306
Teric 350
Teric 351B
Teric BL8
Teric BL9
Teric C12
Teric CDE
Teric CME7
Teric DD5
Teric DD9
Teric G9A5
Teric G9A6
Teric G9A8
Teric G9A12
Teric G12A4
Teric G12A6
Teric G12A8
Teric G12A12
Teric LA4
Teric LA8
Teric LAN70
Teric N2
Teric N3
Teric N4
Teric N5
Teric N8
Teric N9
Teric N10
Teric N11
Teric N12
Teric N13
Teric N15
Teric N20
Teric N30
Teric N40
Teric N100
Teric OF4
Teric OF6
Teric PE61
Teric PE62
Teric PE64
Teric PE68
Teric T2
Teric T5
Teric T7
Teric T10
Teric X5
Teric X7
Teric X8
Teric X10
Teric X11
Teric X13
Teric X16
Teric X40
Teric X40L
Texapon EA-K
Texapon F
Texapon F35
Texapon HD-L 2
Texapon K-12
Texapon K-1296
Texapon L-100
Texapon LLS
Texapon LS Highly Conc.
Texapon LT-327
Texapon MGLS
Texapon N 25
Texapon N 40
Texapon N 70
Texapon N 70 LS
Texapon N 103
Texapon NSO
Texapon OT Highly Conc. Needles
Texapon P
Texapon PLT-227
Texapon PN-235
Texapon PN-254
Texapon PNA
Texapon PNA-127
Texapon SP 60 N
Texapon SP 100
Texapon T 35
Texapon T 42
Texapon ZHC Powder
Texin 128
Texin DOS 75
Texo 1060
Textamine A-3417
Textamine Carbon Detergent K
Textamine Oxide TA
Textol 80 (L)
Tex-Wet 1104
Tex-Wet 1155
Tex-Wet 1197
Titadine TA

Industrial, Institutional and Consumer Cleaning (cont'd.)

Titadine TCP
Titan Decitrene
Titanole RMA
Titazole SA
T-Mulz® 66H
T-Mulz® 426
T-Mulz® 598
T-Mulz® 734-2
T-Mulz® 800
T-Mulz® 844
Tomah AO-14-2
Tomah AO-728 Special
Tomah Q-14-2PG
Tomah Q-17-2
Tomah Q-17-2PG
Tomah Q-18-2
Tomah Q-C-15
Tomah Q-DT-HG
Tomah Q-S
Tomah Q-S-80
Tomah Q-ST-50
Triton® 770 Conc.
Triton® BG-10
Triton® CF-10
Triton® CF-21
Triton® CF-32
Triton® CF-54
Triton® CF-76
Triton® CF-87
Triton® CG-110
Triton® DF-12
Triton® DF-16
Triton® DF-18
Triton® DF-20
Triton® N-57
Triton® N-101
Triton® QS-15
Triton® QS-44
Triton® RW-20
Triton® RW-50
Triton® RW-150
Triton® X-15
Triton® X-35
Triton® X-100
Triton® X-100 CG
Triton® X-102
Triton® X-114
Triton® X-165-70%
Triton® X-200
Triton® X-301
Triton® X-305-70%
Triton® X-405-70%
Triton® XL-80N
Triumphnetzer ZSG
Troykyd® D333
Trycol® 5940
Trycol® 5941
Trycol® 5942
Trycol® 5943
Trycol® 5944
Trycol® 5964
Trycol® 6962
Trycol® 6964
Trycol® 6975
Trycol® 6989
Trycol® DNP-8
Trycol® DNP-150
Trycol® DNP-150/50
Trycol® LAL-23
Trycol® NP-6
Trycol® TDA-11
Trycol® TDA-12
Trycol® TDA-18
Trycol® TP-6
Trydet 22
Trydet 2681
Tryfac® 525-K
Tryfac® 610-K
Tryfac® 5553
Tryfac® 5571
Tryfac® HWD
Trylon® 6702
Trylon® 6735
Trylox® 6746
Trymeen® 6637
Trymeen® TAM-40
Udet® 950
Ufablend DC
Ufablend E 60
Ufablend HDL
Ufablend HDL85N
Ufablend W 100
Ufablend W Conc.
Ufablend W Conc. 60
Ufacid K
Ufacid KA
Ufacid KB
Ufacid KF
Ufacid KN
Ufacid KW
Ufacid TPB
Ufarol Am 30
Ufarol Am 70
Ufarol Na-30
Ufarol TA-40
Ufaryl DB80
Ufaryl DL80 CW
Ufaryl DL85
Ufaryl DL90
Ufasan 35
Ufasan 60A
Ufasan 65
Ufasan IPA
Ufasoft 75
Ukanil 2262
Ultra Blend 60, 100
Ultra NCS Liquid
Ultra NXS Liquid
Ultra SCS Liquid
Ultra SXS Liq., Powd.
Ultra Sulfate AE-3
Ultra Sulfate SE-5
Ultra Sulfate SL-1
Unamide® C-2
Unamide® C-5
Unamide® C-72-3
Unamide® C-7649
Unamide® C-7944
Unamide® D-10
Unamide® J-56
Unamide® L-2
Unamide® L-5
Unamide® LDL
Unamide® LDX
Unamide® N-72-3
Unamide® S
Unamide® SI
Unamide® W
Ungerol AM3-75
Ungerol CG27
Ungerol LES 2-28
Ungerol LES 2-70
Ungerol LES 3-28
Ungerol LES 3-70
Ungerol LS, LSN
Ungerol N2-28
Ungerol N2-70
Ungerol N3-28
Ungerol N3-70
Union Carbide® L-45 Series
Valdet AC-40
Valdet CC
Vanseal CS
Vanseal LS
Vanseal NACS-30
Vanseal NALS-30
Vanseal NALS-95
Vanseal OS
Varamide® A-2
Varamide® A-7
Varamide® A-10
Varamide® A-80
Varamide® A-83
Varamide® A-84
Varamide® C-212
Varamide® FBR
Varamide® L-203
Varamide® MA-1
Varamide® ML-1
Varamide® ML-4
Varamide® T-55
Varion® 2C
Varion® 2L
Varion® AM-B14
Varion® AM-KSF-40%
Varion® CADG-HS
Varion® CADG-LS
Varion® CADG-W
Varion® CAS
Varion® CAS-W
Varion® CDG
Varion® CDG-LS
Varion® TEG
Varion® TEG-40%
Variquat® 50AC
Variquat® 50AE
Variquat® 50MC
Variquat® 50ME
Variquat® 80ME
Variquat® 638
Variquat® K300
Varonic® DM55
Varonic® LI-42
Varonic® LI-48
Varonic® LI-67-75%
Varonic® T202 SR
Varox® 1770
Varox® 185E
Varsulf MS30
Varsulf® S-1333
Varsulf® SBF-12
Varsulf® SBFA-30
Varsulf® SBL-203
Velvetex® BA
Velvetex® BA-35
Versilan MX123
Versilan MX134
Versilan MX244
Versilan MX332
Victawet® 12
Vifcoll CCN-40, CCN-40 Powd.
Vista C-550, -560
Wachsemulsion 1864
Wayfos A
Wayfos D-10N
Wayfos M-60
Wayfos M-100
Witcamide® 5140
Witcamide® 6310
Witcamide® 6445
Witcamide® M-3
Witcamide® MAS
Witcamide® S771
Witcamide® S780
Witco® 94H Acid
Witco® 97H Acid
Witco® 1298
Witco® CHB Acid
Witco® TX Acid
Witcodet 100
Witcodet 793
Witcodet 5480
Witcodet B-1
Witcodet CWC
Witcodet DC-47
Witcodet DLC-47
Witcodet SC
Witcolate 1247H
Witcolate 3570
Witcolate 7031
Witcolate 7093
Witcolate A
Witcolate AE-3
Witcolate AE3S
Witcolate S1285C
Witcolate S1300C
Witcolate SE-5
Witcolate T
Witcomul H-50A
Witconate 45 Liq.
Witconate 45LX
Witconate 60B
Witconate 60T
Witconate 79S
Witconate 1075X
Witconate 1240 Slurry
Witconate 1250 Slurry
Witconate 1260 Slurry
Witconate 1840X
Witconate AOS
Witconate LX F
Witconate LX Powd.
Witconol H33
Witconol SN Series
Zeeclean No. 1
Zeeterge 444
Zetesol 2056
Zetesol AP
Zetesol NL
Zohar 60 SD(L)

Industrial, Institutional and Consumer Cleaning (cont'd.)

Zohar 70 SD(L)
Zohar AD
Zohar Automat SD
Zohar KAL SD
Zoharconc A.D
Zoharconc APL
Zoharconc DIS
Zoharconc FC
Zoharconc FCW
Zoharconc LLH
Zoharconc LLM
Zoharconc S.D
Zoharconc Tex 80
Zoharex A-10
Zoharex B-10
Zoharex N-25
Zoharpon ETA 27
Zoharpon ETA 70
Zoharpon ETA 270 (OXO)
Zoharpon ETA 271
Zoharpon ETA 273
Zoharpon ETA 603
Zoharpon ETA 700 (OXO)
Zoharpon ETA 703
Zoharpon K
Zoharpon LAD
Zoharpon LAEA 253
Zoharpon LMT42
Zoharpon SM
Zoharquat 50
Zoharquat 80 EXP
Zoharsyl L-30
Zohartaine AB
Zohartaine TM
Zoharteric D-SF
Zoharteric D-SF 70%
Zoharteric D
Zoharteric DJ
Zoharteric LF
Zoharteric M
Zonyl® A
Zonyl® FSA
Zonyl® FSB
Zonyl® FSC
Zonyl® FSJ
Zonyl® FSN
Zonyl® FSP
Zoramox
Zoramox LO
Zorapol SN-9
Zusolat 1004
Zusolat 1005/85
Zusolat 1008/85
Zusolat 1010/85
Zusolat 1012/85

Leather and Fur Processing

Ablunol NP 9
Ablunol S-80
Ablusol C-78
AF 72
AF 75
Alkamuls® CO-40
Alkamuls® EL-719
Alkamuls® PSTO-20
Alkamuls® SMO
Alkamuls® STO
Alkanol® XC
Amollan®
Amollan® A
Amollan® S
Amphoteric 400
Antifoam 7800 New
Aristonate H
Aristonate L
Aristonate M
Armul 906
Arnox 906
Atlas Defoamer AFC
Atlas EMJ-2
Atlas EMJ-C
Atlasol 103
Atlasol 155
Atlasol 160-S
Atlasol 6920
Atlasol 6920-VF
Atlasol KAD
Atlasol KMM
Atlas Sul. Neats L-2
Atlas Sul. Oil HC
Avitone® A
Barre® Common Degras
Berol 28
Berol 108
Berol 190
Berol 191
Berol 198
Berol 302
Berol 303
Berol 307
Berol 381
Berol 386
Berol 387
Berol 389
Berol 391
Berol 392
Berol 397
Berol 455
Berol 456
Berol 457
Berol 458
Blancol®
Chemax CO-16
Chemax CO-25
Chemax CO-28
Chemax CO-30
Chemax CO-36
Chemax CO-40
Chemax CO-200/50
Chemax HCO-5
Chemax HCO-16
Chemax HCO-25
Chemax HCO-200/50
Chemax NP-15
Chemax NP-20
Chemax NP-30
Chemax NP-30/70
Chemax PEG 400 DO
Chemax PEG 600 DO
Chimin KF3
Chimin KSP
Chimin P50
Cithrol 6ML
Colorol E
Coralon F
Coralon Grades
Crafol AP-240
Crafol AP-260
Crafol AP-261
Crill 1
Crill 2
Crillet 4
Crodamine 1.HT
Crodamine 1.O, 1.OD
Crodamine 1.T
Crodamine 1.16D
Crodamine 1.18D
Crodamine 3.A16D
Crodamine 3.A18D
Crodamine 3.ABD
Crodamine 3.AED
Crodamine 3.AOD
Fancor LFA
Feliderm CS
Forbest VP 13
Forbest VP 20
Gardinol CX
Gardinol CXM
Genapol® C-050
Genapol® C-080
Genapol® C-100
Genapol® C-150
Genapol® C-200
Genapol® O-020
Genapol® O-050
Genapol® O-080
Genapol® O-090
Genapol® O-100
Genapol® S-020
Genapol® S-050
Genapol® S-080
Genapol® S-100
Genapol® S-150
Genapol® S-200
Genapol® S-250
Genapol® X Grades
Geropon® TBS
GP-210 Silicone Antifoam Emulsion
Glycidol Surfactant 10G
Hetoxamate FA-5
Hetoxamate LA-5
Hetoxamate LA-9
Hetoxamate MO-2
Hetoxamate MO-5
Hetoxamate MO-9
Hetoxamate MO-15
Hetoxamate SA-5
Hetoxamate SA-7
Hetoxamate SA-9
Hetoxamate SA-13
Hetoxamate SA-23
Hetoxamate SA-35
Hetoxamate SA-40
Hetoxamate SA-90
Hetoxamate SA-90F
Hetoxamine C-2
Hetoxamine C-5
Hetoxamine O-2
Hetoxamine O-5
Hetoxamine O-15
Hetoxamine S-2
Hetoxamine S-5
Hetoxamine S-15
Hetoxamine ST-2
Hetoxamine ST-5
Hetoxamine ST-15
Hetoxamine ST-50
Hetoxamine T-2
Hetoxamine T-5
Hetoxamine T-15
Hetoxamine T-20
Hetoxamine T-30
Hetoxamine T-50-70%
Hostapal 3634 Highly Conc.
Hostapon T Powd.
Hydrolene
Igepal® CO-630
Igepal® CO-660
Igepal® CO-710
Igepal® CO-720
Igepal® DM-430
Igepal® DM-710
Igepal® DM-730
Igepal® DX-430
Kemester® 104
Kemester® 143
Kemester® 1000
Kemester® 4000
Lamigen ES 30
Lamigen ES 60
Lamigen ES 100
Lamigen ET 20
Lamigen ET 70
Lamigen ET 90
Lamigen ET 180
Lankropol® ADF
Lauroxal 8
Lignosol TSD
Lignosol X
Lignosol XD
Lipotin 100, 100J, SB
Lipotin A
Lipotin H
Lutensol® AP 6
Lutensol® AP 7
Lutensol® AP 8
Lutensol® AP 9
Lutensol® FSA 10
Marlipal® KF
Marlowet® 4800
Marlowet® 5600
Marlowet® FOX
Marlowet® LVS
Marlowet® LVX
Marlowet® NF
Marlowet® R 11
Mazamide® L-298
Mazu® DF 205SX
Mazu® DF 215SX

Leather and Fur Processing (cont'd.)

Mazu® DF 230SX
Mazu® DF 230SXSP
Merpol® HCS
Mersolat H 30
Mersolat H 40
Mersolat H 68
Mersolat H 76
Mersolat H 95
Mersolat W 40
Mersolat W 68
Mersolat W 76
Mersolat W 93
Minemal 320, 325, 330
Mirataine® T2C-30
Monosulf
Mulsifan RT 7
Mulsifan RT 18
NP-55-90
Nansa® BXS
Nansa® EVM50
Nansa® EVM70/B
Nansa® EVM70/E
Naxel AAS-45S
Naxonol PN 66
Neatsan D
Niaproof® Anionic Surfactant 4
Niaproof® Anionic Surfactant 7
Niox KI Series
Niox KL-16
Nissan Nonion L-2
Nissan Nonion L-4
Nissan Nonion O-2
Nissan Nonion O-3
Nissan Nonion O-4
Nissan Nonion O-6
Nopalcol 1-S
Nopalcol 1-TW
Nopalcol 2-DL
Nopalcol 4-C
Nopalcol 4-CH
Nopalcol 4-L
Nopalcol 4-O
Nopalcol 6-DO
Nopalcol 6-DTW
Nopalcol 6-L
Nopalcol 6-R
Nopalcol 6-S
Nopalcol 10-COH
Nopalcol 12-CO
Nopalcol 12-COH
Nopalcol 19-CO
Nopalcol 30-TWH
Nopalcol 200
Nopalcol 400
Nopalcol 600
Nopalcol Series
Norfox® 916
Nuva L
Oxetal 500/85
Oxetal D 104
Oxetal ID 104
Oxetal O 108
Oxetal O 112
Oxetal TG 111
Oxetal TG 118
Pale 16
Pale 170
Pale 1000
Petronate® CR
Petronate® HL
Petronate® L
Petronate® S
Petrosul® H-50
Petrosul® H-60
Petrosul® H-70
Petrosul® HM-62, HM-70
Petrosul® L-60
Petrosul® M-50
Petrosul® M-60
Petrosul® M-70
Product BCO
Propetal 241
Prote-sorb SML
Prote-sorb SMO
Prote-sorb SMP
Prote-sorb SMS
Prote-sorb STS
Prox-onic HR-05
Prox-onic HR-016
Prox-onic HR-025
Prox-onic HR-030
Prox-onic HR-036
Prox-onic HR-040
Prox-onic HR-080
Prox-onic HR-0200
Prox-onic HR-0200/50
Prox-onic HRH-05
Prox-onic HRH-016
Prox-onic HRH-025
Prox-onic HRH-0200
Prox-onic HRH-0200/50
Prox-onic OA-2/020
Prox-onic OL-1/05
Prox-onic OL-1/09
Prox-onic OL-1/014
Prox-onic SML-020
Prox-onic SMO-05
Prox-onic SMO-020
Prox-onic SMP-020
Prox-onic SMS-020
Prox-onic STS-020
Radiaquat 6444
Radiaquat 6471
Radiasurf® 7125
Radiasurf® 7146
Radiasurf® 7150
Radiasurf® 7151
Radiasurf® 7156
Radiasurf® 7175
Radiasurf® 7201
Radiasurf® 7206
Radiasurf® 7270
Radiasurf® 7402
Radiasurf® 7410
Radiasurf® 7411
Radiasurf® 7420
Radiasurf® 7444
Radiasurf® 7900
Remolgan CX, PM
Rewomine IM-CA
Rexobase HD
Rexol 25/8
Rexol 25/9
Rexol 25/10
Rexol 25J
Rexol 25JWC
Rexol 35/5
Rexonic 1012-6
Rexonic 1218-6
Rexonic N23-6.5
Rexonic N25-7
Rexonic N25-14(85%)
Rhodacal® BA-77
Rhodacal® BX-78
Rhodafac® MC-470
Rhodasurf® ON-870
Rychem® 21
Rychem® 33
Sandopan® TFL Conc.
Schercozoline B
Schercozoline C
Serdas GLN
Serdox NBS 4
Serdox NBS 5.5
Serdox NBS 6
Serdox NBS 6.6
Serdox NES 8/85
Serdox NNP 8.5
Serdox NNP 12
Serdox NNP 13
Servirox OEG 45
Servirox OEG 55
Servirox OEG 65
Servirox OEG 90/50
Silicone AF-10 FG
Sinol LDA-6
Sinol NPX
Siponic® NP-9
Siponic® NP-15
Softenol® 3900
Solocod G
Solulan® PB-2
Solulan® PB-5
Solulan® PB-10
Solulan® PB-20
Sorbax PML-20
Sorbax PMO-5
Sorbax PMP-20
Sorbax PMS-20
Sorbax PTO-20
Sorbax PTS-20
Sorbax SML
Sorbax SMO
Sorbax SMP
Sorbax SMS
Sorbax STO
Sorbax STS
Sovatex C Series
Standamul® B-2
Standamul® B-3
Stepanol® AM
Stepanol® AM-V
Stepanol® DEA
Stepanol® ME Dry
Stepanol® MG
Stepanol® WA-100
Stepanol® WAC
Stepanol® WAC-P
Stepanol® WA Paste
Stepanol® WAQ
Stepanol® WA Special
Stepanol® WAT
Strodex® MO-100
Sul-fon-ate AA-9
Sul-fon-ate AA-10
Surfactol® 13
Surfactol® 318
Surfactol® 340
Surfactol® 365
Surfactol® 380
Surfactol® 575
Surfactol® 590
Surfonic® HDL
Syntens KMA 55
Syntergent K
Syntofor A03, A04, AB03, AB04, AL3, AL4, B03, B04
Syntopon 8 A
T-Det® C-40
T-Det® N-6
T-Det® N-8
T-Det® N-9.5
T-Det® O-6
T-Det® O-8
Tek-Wet 955
Tekstim 8741
Tergitol® 15-S-7
Tergitol® 15-S-9
Tergitol® Min-Foam 1X
Tergitol® TMN-6
Tergitol® TMN-10
Tergitol® XD
Tergitol® XH
Tergitol® XJ
Teric 9A12
Teric 9A6
Teric 12A8
Teric 13A7
Teric 16M2
Teric 16M5
Teric 16M15
Teric 18M2
Teric 18M5
Teric 18M10
Teric 18M20
Teric 18M30
Teric PPG 1650
Teric PPG 2250
Teric PPG 4000
Teric SF15
Textamine O-5
Titadine TA
Titadine TCP
Titan Decitrene
Titanole RMA
Titazole SA
TO-55-E
TO-55-EL
Tolcide MBT
Triton® 770 Conc.
Triton® XL-80N
Trydet 2636
Trydet 2670
Trydet 2676
Trydet OA-7
Trydet OA-10
Trydet SA-5
Trydet SA-7

Leather and Fur Processing (cont'd.)

Trylon® 6837
Trylon® EW
Trylox® 5907
Trylox® 5909
Trylox® CO-36
Trylox® CO-40
Tylose MHB
Ufanon K-80
Ufanon KD-S
Ufarol Am 30
Ufarol Na-30
Unamine® C
Vanseal 35
Vanseal CS
Vanseal LS
Vanseal NACS-30
Vanseal NALS-30
Vanseal NALS-95
Vanseal OS
Varonic® K210
Varonic® K215
Varonic® K215 LC
Varonic® T215
Varonic® T215LC
Vorite 105
Vorite 110
Vorite 115
Vorite 120
Vorite 125
Witcamine® AL42-12
X78-2
Zusolat 1005/85
Zusolat 1008/85
Zusolat 1010/85
Zusolat 1012/85

Metalworking, Cleaning, Processing, Corrosion Inhibiting, Cutting and Drilling Oils

AB®
Ablunol LA 3
Ablunol LA 5
Ablunol LA 7
Ablunol LA 9
Ablunol LA 12
Ablunol LA 16
Ablunol LA 40
Ablunol NP 9
Acid Aid X
Acid Foamer
Acid Thickener
Acto 450
Actrabase 31-A
Actrabase 215
Actrabase 264
Actrabase PS-470
Actrabase SS-503
Actrabase SS-523
Actrafos 110, 110A
Actrafos 152A
Actrafos 161
Actrafos 186
Actrafos 208
Actrafos 216
Actrafos 306
Actrafos 314, 315
Actrafos 800
Actrafos 822
Actrafos SA-208
Actrafos SA-216
Actrafos SN-315
Actrafos T
Actralube SOS
Actralube Syn-147
Actralube Syn-153
Actramide 202
Actramide 5264
Actrasol C-50, C-75, C-85
Actrasol EO
Actrasol MY-75
Actrasol PSR
Actrasol SS
Actrol 4DP
Actrol 4MP, 628
Adogen® 461
Adogen® 462
Adogen® 464
Adogen® 471
Adogen® 560
Adogen® 570-S
Adogen® 572
Adol® 42
Adol® 61 NF
Adol® 62 NF
Adol® 80
AEPD®
Aerosol® 19
Aerosol® AY-65
Aerosol® AY-100
Aerosol® IB-45
Aerosol® MA-80
Aerosol® OS
Aerosol® OT-75%
Aerosol® OT-MSO
Aerosol® TR-70
AF 10 IND
AF 70
AF 75
AF 8810
AF 8810 FG
AF 9000
AF HL-23
AF HL-26
AF HL-40
Ageflex FM-1
Akypo MB 2621 S
Akypo MB 2705 S
Akypo OP 80
Akypo OP 190
Akypo RLM 25
Akypo RO 20
Akypo RO 50
Akypo RO 90
Akypo TBP 180
Akypopress DB
Alcodet® 218
Alcodet® MC-2000
Alcodet® SK
Alfol® 1012 CDC, 1012 HA
Alfol® 1216
Alkamide® 200 CGN
Alkamide® 2104
Alkamide® 2204
Alkamide® 2204A
Alkamide® DO-280
Alkamide® SDO
Alkamide® WRS 1-66
Alkamuls® 400-DO
Alkamuls® A
Alkamuls® AP
Alkamuls® B
Alkamuls® BR
Alkamuls® EL-620
Alkamuls® PSMO-5
Alkamuls® SML
Alkamuls® SMO
Alkamuls® TD-41
Alkanol® 189-S
Alkasurf® NP-4
Alkasurf® NP-6
Alkasurf® OP-10
Alkasurf® OP-12
Alkaterge®-C
Alkaterge®-E
Alkaterge®-T
Alkaterge®-T-IV
Alkenyl Succinic Anhydrides
Ameenex 70 WS
Ameenex 73 WS
Ameenex C-20
Ameenex Polymer
Amidex CO-1
Amidex OE
Amine BG
Amine BGD
Amine C
Amine HBG
Amine HBGD
Amine KK
Amine KKD
Amine O
Amine OL
Amine OLD
Amine T
Aminol TEC N
AMP
AMP-95
AMPD
Amphoteric 400
Amphoteric N
Anedco DF-6002
Antarox® BL-214
Antarox® BL-225
Antarox® BL-240
Antarox® BL-330
Antarox® FM 33
Antarox® FM 53
Antarox® FM 63
Antarox® LF-222
Antarox® RA 40
Aristonate H
Aristonate L
Aristonate M
Arkomon A Conc.
Arkomon SO
Arkopon Brands
Armac® 18D-40
Armac® C
Armac® T
Armeen® 2C
Armeen® 2HT
Armeen® 2T
Armeen® 12D
Armeen® 16D
Armeen® 18
Armeen® C
Armeen® CD
Armeen® DM8
Armeen® DM10
Armeen® DM12
Armeen® DM14
Armeen® DM16
Armeen® DMC
Armeen® DMHT
Armeen® DMO
Armeen® DMT
Armeen® HT
Armeen® HTD
Armeen® L8D
Armeen® O
Armeen® OL
Armeen® OLD
Armeen® S
Armeen® SD
Armeen® T
Armeen® TD
Armid® HT
Armid® O
Armotan® MO
Armul 950
Arnox 500, 515
Arnox 950
Aromox® C/12
Aromox® C/12-W
Arquad® 2C-75
Arquad® 2HT-75
Arquad® 12-37W
Arquad® 12-50
Arquad® 16-29W
Arquad® 16-50
Arquad® 18-50
Arquad® C-33-W
Arquad® C-50
Arquad® S-2C-50
Arquad® S-50
Arquad® T-2C-50
Arquad® T-27W
Arquad® T-50

Metalworking, Cleaning, Processing, Corrosion Inhibiting (cont'd.)

Arsul DDB
Arsul LAS
Artrads 6524
Artrads 6923
Artrads 7522
Atsolyn PE 36
Atsolyn TD 50
AZdry 40
AZdry 70
Barium Petronate 50-S Neutral
Barre® Common Degras
Base 75
Base 76
Base 85
Base 865
Base 7800
Base 8000
Base 8000P
Base ML
Base MO
Base MT
Berol 28
Berol 108
Berol 190
Berol 191
Berol 198
Berol 302
Berol 303
Berol 307
Berol 381
Berol 386
Berol 387
Berol 389
Berol 391
Berol 392
Berol 397
Berol 455
Berol 456
Berol 457
Berol 458
Berol 521
Berol 594
Beycostat 256A
Beycostat B 706 A
Bio-Soft® EA-8
Bio-Terge® PAS-8S
Bohrmittel Hoechst
Briquest® 301-30SH
Briquest® ADPA-60AW
Bubble Breaker® 1840X
Burco FAE
Burco HCS-50NF
Burco NPS-225
Calamide C
Calcium Petronate
Calcium Sulfonate C-50N
Calcium Sulfonate EP-163
Calfax DB-45
Calfax DBA-40
Calfax DBA-70
Calsoft F-90
Carbowax® PEG 8000
Carsofoam® 211
Carsonon® LF-5
Carsonon® LF-46
Carsoquat® 621
Carsoquat® 621 (80%)
Catinex KB-27
Catinex KB-43
Catinex KB-48
Catisol AO 100
Catisol AO C
Cedephos® FA600
Cedephos® RA600
Chemal 2EH-2
Chemal 2EH-5
Chemal BP 261
Chemal BP-262
Chemal BP-262LF
Chemal BP-2101
Chemal DA-4
Chemal DA-6
Chemal DA-9
Chemal LF 14B
Chemal LF 25B
Chemal LF 40B
Chemal LFL-10
Chemal LFL-17
Chemal LFL-19
Chemal LFL-28
Chemal LFL-38
Chemal LFL-47
Chemax CO-16
Chemax DF-30
Chemax DF-100
Chemax DNP-150/50
Chemax E-200 MO
Chemax E-200 MS
Chemax E-400 MO
Chemax E-400 MS
Chemax E-600 ML
Chemax E-600 MO
Chemax HCO-5
Chemax HCO-16
Chemax HCO-25
Chemax HCO-200/50
Chemax NP-1.5
Chemax NP-4
Chemax NP-6
Chemax NP-9
Chemax NP-10
Chemax NP-15
Chemax NP-20
Chemax NP-30
Chemax NP-30/70
Chemax OP-7
Chemax OP-9
Chemax OP-30/70
Chemax PEG 400 DO
Chemax PEG 600 DO
Chemax SBO
Chemax SCO
Chemazine 18, C, O, TO
Chembetaine CAS
Chemeen 18-2
Chemeen 18-5
Chemeen 18-50
Chemeen C-2
Chemeen C-5
Chemeen C-10
Chemeen C-15
Chemeen DT-3
Chemeen DT-15
Chemeen HT-2
Chemeen HT-5
Chemeen HT-50
Chemfac NC-0910
Chemfac PA-080
Chemfac PB-082
Chemfac PB-106
Chemfac PB-135
Chemfac PB-184
Chemfac PB-264
Chemfac PC-188
Chemfac PD-600
Chemfac PF-623
Chemfac PF-636
Chemfac PN-322
Chemphos TC-227
Chemphos TC-310
Chemphos TC-310S
Chemphos TR-513
Chemphos TX-625
Chemphos TX-625D
Chemquat 12-33
Chemquat 12-50
Chemquat 16-50
Chemquat C/33W
Chemsulf SBO/65
Chemsulf SCO/75
Chemzoline 1411
Chemzoline C-22
Chemzoline T-11
Chemzoline T-33
Chemzoline T-44
Chimin P1A
Cithrol 2ML
Cithrol 2MO
Cithrol 4ML
Cithrol 4MO
Cithrol 6ML
Cithrol 6MO
Cithrol 10ML
Cithrol 10MO
Cithrol 40MO
Cithrol 60ML
Cithrol 60MO
Cithrol A
Cithrol DGDL N/E
Cithrol DGDL S/E
Cithrol DGDO N/E
Cithrol DGDO S/E
Cithrol DGDS N/E
Cithrol DGDS S/E
Cithrol DGML N/E
Cithrol DGML S/E
Cithrol DGMO N/E
Cithrol DGMO S/E
Cithrol DGMS N/E
Cithrol DGMS S/E
Cithrol DPGML N/E
Cithrol DPGML S/E
Cithrol DPGMO S/E
Cithrol DPGMS N/E
Cithrol DPGMS S/E
Cithrol EGDL N/E
Cithrol EGDL S/E
Cithrol EGDO N/E
Cithrol EGDO S/E
Cithrol EGDS N/E
Cithrol EGDS S/E
Cithrol EGML N/E
Cithrol EGML S/E
Cithrol EGMO N/E
Cithrol EGMO S/E
Cithrol EGMR N/E
Cithrol EGMR S/E
Cithrol EGMS N/E
Cithrol EGMS S/E
Cithrol GDL N/E
Cithrol GDL S/E
Clink A-26
Clink A-70
Closyl 30 2089
Closyl LA 3584
Comperlan PD
Comperlan PVD
Compound 170
Conco X-200
Crafol AP-53
Crapol AV-10
Crapol AV-11
Crill 1
Crill 2
Crill 3
Crill 4
Crillet 1
Crillon CDY
Crillon ODE
Crodafos N3 Acid
Crodamine 1.HT
Crodamine 1.O, 1.OD
Crodamine 1.T
Crodamine 1.16D
Crodamine 1.18D
Crodamine 3.A16D
Crodamine 3.A18D
Crodamine 3.ABD
Crodamine 3.AED
Crodasinic L
Crodasinic LS30
Crodasinic LS35
Crodasinic OS35
Crodazoline S
Croduret 10
Croduret 30
Croduret 40
Croduret 60
Croduret 100
Crown Anti-Foam
Crown Foamer 20
Crown Foamer 20X
Crown Foamer 50
Fancol OA 95
Fancor LFA
Farmin 20, 60, 68, 80, 86, AB, C
Farmin DM20
Farmin DM40, 60, 80, 86
Farmin DMC
Farmin HT
Farmin M2C
Farmin M2TH-L
Farmin O
Farmin R 86H
Farmin S
Farmin T
Finazoline OA

Metalworking, Cleaning, Processing, Corrosion Inhibiting (cont'd.)

Flexricin® 100
Fluorad FC-93
Fluorad FC-95
Fluorad FC-98
Fluorad FC-99
Fluorad FC-100
Fluorad FC-430
Fluowet 40 M
Foamkill® 8J-1
Foamkill® 8R
Foamkill® 649
Foamkill® 654NS
Foamkill® 836B
Forlanit P
Fosfamide CPD-170
Fosterge LF
Fosterge R
Fosterge RD
Fosterge W
G-1000-S Antifoam Compd.
Genapol® 2317
Genapol® B
Genapol® PS
Geropon® BIS/SODICO-2
Geropon® T-33
Geropon® T-51
Geropon® T-77
Geropon® TBS
Geropon® WT-27
GP-210 Silicone Antifoam Emulsion
GP-262 Defoamer
GP-295 Defoamer
GP-300-I Antifoam Compd.
GP-330-I Antifoam Emulsion
Gradonic LFA Series
Hetoxamate FA-5
Hetoxamate LA-5
Hetoxamate LA-9
Hetoxamate MO-5
Hetoxamate MO-9
Hetoxamate MO-15
Hetoxamate SA-5
Hetoxamate SA-35
Hetoxamate SA-90
Hetoxamate SA-90F
Hetoxamine C-2
Hetoxamine C-5
Hetoxamine O-2
Hetoxamine O-5
Hetoxamine O-15
Hetoxamine S-2
Hetoxamine S-5
Hetoxamine S-15
Hetoxamine ST-2
Hetoxamine ST-5
Hetoxamine ST-15
Hetoxamine ST-50
Hetoxamine T-2
Hetoxamine T-5
Hetoxamine T-15
Hetoxamine T-20
Hetoxamine T-30
Hetoxamine T-50-70%
Hetoxide BN-13
Hetoxide BY-1.8
Hetoxide C-30
Hetsorb L-4
Hetsorb L-10
Hetsorb L-20
Hetsorb O-20
Hetsorb TS-20
Hodag Amine C-100-L
Hodag Amine C-100-O
Hodag Amine C-100-S
Hodag Nonionic 1035-L
Hodag Nonionic 1044-L
Hodag Nonionic 1061-L
Hodag Nonionic 1062-L
Hodag Nonionic 1064-L
Hodag Nonionic 1068-F
Hodag Nonionic 1088-F
Hodag PE-005
Hodag PE-104
Hodag PE-106
Hodag PE-109
Hodag PE-206
Hodag PE-209
Hodag SML
Hodag SMO
Hodag SMP
Hodag STO
Hodag STS
Hostacor 2098
Hostacor 2125
Hostacor 2270
Hostacor 2272
Hostacor 2291
Hostacor 2292
Hostacor 2732
Hostacor BBM
Hostacor BF
Hostacor BK
Hostacor BM
Hostacor BS
Hostacor DT
Hostacor E
Hostacor H Liq. N
Hostacor TP 2445
Hostaphat AW
Hostawet TDC
Hybase® C-300
Hybase® C-400
Hybase® C-500
Hybase® M-300
Hybase® M-400
Hydrolene
Hyonic CPG 745
Hyonic NP-40
Hyonic NP-110
Hyonic OP-40
Hyonic PE-40
Hystrene® 3675
Hystrene® 3675C
Hystrene® 3680
Hystrene® 3687
Hystrene® 3695
Igepal® CA-620
Igepal® CA-630
Igepal® CA-720
Igepal® CO-210
Igepal® CO-520
Igepal® CO-530
Igepal® CO-610
Igepal® CO-630
Igepal® CO-660
Igepal® CO-710
Igepal® CO-720
Igepal® CO-730
Igepal® CO-850
Igepal® DM-710
Igepal® DM-730
Igepal® LAVE
Igepal® NP-2
Igepal® NP-6
Igepal® NP-9
Igepal® NP-10
Igepal® NP-12
Igepal® NP-20
Igepal® O
Igepal® OD-410
Incrocas 30
Incrocas 40
Industrene® 105
Industrene® 143
Industrene® 205
Industrene® 206
Industrene® 225
Industrene® 226
Industrene® 325
Industrene® 328
Industrene® 365
Industrene® 4516
Industrene® 4518
Industrene® 5016
Industrene® 6018
Industrene® 7018
Industrene® 9018
Industrene® B
Industrene® D
Industrene® R
Inhibitor 60Q
Inhibitor 212
Inipol OO2
Inipol OT2
Inipol S 43
Invermul
Inversol 140
Inversol 170
Ionet AT-140
Ionet DL-200
Ionet DO-200
Ionet DO-400
Ionet DO-600
Ionet DO-1000
Ionet DS-300
Ionet DS-400
Ionet S-20
Ionet S-60 C
Ionet S-80
Ionet T-20 C
Ionet T-60 C
Ionet T-80 C
Jet Amine DC
Jet Amine DE-13
Jet Amine DE 810
Jet Amine DO
Jet Amine DT
Jet Amine PC
Jet Amine PCD
Jet Amine PE 08/10
Jet Amine PE 1214
Jet Amine PHT
Jet Amine PHTD
Jet Amine POD
Jet Amine PS
Jet Amine PSD
Jet Amine PT
Jet Amine PTD
Jet Amine TET
Jet Amine TRT
Jet Amine TT
Kalcohl 5-24, 6-24, 7-24
Karafac 78
Karafac 78 (LF)
Karaphos HSPE
Karaphos SWPE
Karaphos XFA
Kelig 100
Kemamide® B
Kemamide® E
Kemamide® O
Kemamide® S
Kemamide® U
Kemamide® W-20
Kemamide® W-39
Kemamide® W-40
Kemamide® W-40/300
Kemamide® W-45
Kemamine® BQ-2802C
Kemamine® BQ-9702C
Kemamine® BQ-9742C
Kemamine® P-150, P-150D
Kemamine® P-190, P-190D
Kemamine® P-650D
Kemamine® P-880, P-880D
Kemamine® P-970
Kemamine® P-970D
Kemamine® P-974D
Kemamine® P-989D
Kemamine® P-990, P-990D
Kemamine® P-999
Kemamine® Q-1902C
Kemamine® Q-2802C
Kemamine® Q-6502C
Kemamine® Q-9702C
Kemamine® Q-9743C
Kemamine® Q-9743CHGW
Kemamine® T-9742D
Kemamine® T-9902D
Kemamine® T-9992D
Kemester® 115
Kemester® 143
Kemester® 205
Kemester® 213
Kemester® 226
Kemester® 1000
Kemester® 2050
Kemester® 4516
Kemester® 9022
Kemester® EGDS
Kemester® EGMS
Kessco® 653

Metalworking, Cleaning, Processing, Corrosion Inhibiting (cont'd.)

Kessco® 874
Kessco® 887
Kessco® 891
Kessco® 894
Kessco® GMO
Kessco® GMS
Kessco® PEG 200 DL
Kessco® PEG 400 DS
Kessco® PEG 600 DL
Kessco® PEG 600 DS
Kessco® PEG 600 ML
Kessco® PEG 6000 DS
Klearfac® AA270
Korantin® BH Liq.
Korantin® BH Solid
Korantin® CD
Korantin® LUB
Korantin® MAT
Korantin® PA
Korantin® PAT
Korantin® SH
Korantin® SMK
Korantin® TD
Lankropol® OPA
Lanpolamide 5
Lauramide S
Laurel M-10-257
Laurel PEG 400 DT
Laurel PEG 400 MO
Laurel PEG 600 DT
Laurel PEG 600 MT
Laurel R-50
Laurel R-75
Laurel SD-101
Laurel SD-300
Laurel SD-400
Laurel SD-520T
Laurel SD-570
Laurel SD-580
Laurel SD-590
Laurel SD-750
Laurel SD-800
Laurel SD-LOA
Laurel SRO
Laurelphos 39
Laurelphos 400
Laurelphos L-50
Laurelphos P-71
Lauridit® PD
Lauropal 11
Lauropal 1150
Lauroxal 8
Lexaine® CSB-50
Lexemul® PEG-200 DL
Liposorb O
Liposorb P
Liposorb S
Lonzaine® C
Lonzaine® CO
Lonzest® PEG 4-O
Lonzest® SML
Lonzest® SMO
Lonzest® SMP
Lonzest® SMS
Lonzest® STO
Lonzest® STS
LSP 33
Lubrhophos® LB-400
Lubrhophos® LE-500
Lubrhophos® LE-600
Lubrhophos® LE-700
Lubrhophos® LF-200
Lubrhophos® LK-500
Lubrhophos® LL-550
Lubrhophos® LM-400
Lubrhophos® LM-600
Lubrhophos® LP-700
Lubrhophos® LS-500
Lubricant EHS
Lubrizol® 5375
Lutensol® LF 400
Lutensol® LF 401
Lutensol® LF 711
Mackam 2CYSF
Mackam 2LSF
Mackam CSF
Mackamide CD-6
Mackamide CD-8
Mackamide CD-25
Mackamide CDM
Mackamide CDX
Mackamide CSA
Mackazoline C
Mackazoline CY
Mackazoline L
Mackazoline O
Mackester EGDS
Mackester EGMS
Mackester IDO
Mackester IP
Mackester SP
Mackester TD-88
Mackine 101
Mackine 201
Mackine 301
Mackine 321
Mackine 401
Mackine 421
Mackine 501
Mackine 601
Mackine 701
Mackine 801
Mackine 901
Macol® 1
Macol® 2
Macol® 2D
Macol® 2LF
Macol® 4
Macol® 8
Macol® 18
Macol® 19
Macol® 22
Macol® 33
Macol® 40
Macol® 300
Macol® 660
Macol® 3520
Macol® 5100
Macol® CA-2
Macol® CSA-4
Macol® CSA-2
Macol® CSA-10
Macol® CSA-15
Macol® CSA-20
Macol® CSA-40
Macol® CSA-50
Macol® DNP-5
Macol® DNP-10
Macol® DNP-15
Macol® DNP-21
Macol® LA-4
Macol® LA-9
Macol® LA-12
Macol® LA-23
Macol® LA-790
Macol® LF-110
Macol® LF-111
Macol® LF-120
Macol® NP-4
Macol® NP-5
Macol® NP-6
Macol® NP-8
Macol® NP-11
Macol® NP-12
Macol® NP-15
Macol® NP-20
Macol® NP-20(70)
Macol® NP-30(70)
Macol® NP-100
Macol® OA-2
Macol® OA-4
Macol® OA-5
Macol® OP-3
Macol® OP-5
Macol® OP-8
Macol® OP-10
Macol® P-500
Macol® P-1200
Macol® P-1750
Macol® P-2000
Macol® P-3000
Macol® P-4000
Macol® SA-2
Macol® SA-5
Macol® SA-10
Macol® SA-15
Macol® SA-20
Macol® SA-40
Macol® TD-3
Macol® TD-8
Macol® TD-10
Macol® TD-12
Macol® TD-15
Macol® TD-100
Macol® TD-610
Mafo® 13
Mafo® 13 MOD 1
Mafo® CAB
Mafo® CAB SP
Makon® 4
Makon® 6
Makon® 8
Makon® 10
Makon® 12
Makon® 14
Makon® 30
Makon® 8240
Makon® NF-5
Makon® NF-12
Makon® OP-9
Manro AT 1200
Manro BA Acid
Manro DNNS/B
Manro DNNS/C
Manro DS 35
Manro NA Acid
Manromid 150-ADY
Manromid 853
Manromid CD
Manromid CDG
Manromid CDX
Manroteric CEM38
Manroteric CyNa50
Mapeg® 200 DL
Mapeg® 200 DO
Mapeg® 200 DOT
Mapeg® 200 DS
Mapeg® 200 ML
Mapeg® 200 MO
Mapeg® 200 MOT
Mapeg® 200 MS
Mapeg® 400 DL
Mapeg® 400 DO
Mapeg® 400 DOT
Mapeg® 400 DS
Mapeg® 400 ML
Mapeg® 400 MO
Mapeg® 400 MOT
Mapeg® 400 MS
Mapeg® 600 DL
Mapeg® 600 DO
Mapeg® 600 DOT
Mapeg® 600 DS
Mapeg® 600 ML
Mapeg® 600 MO
Mapeg® 6000 DS
Mapeg® CO-25
Mapeg® CO-25H
Mapeg® EGDS
Mapeg® EGMS
Mapeg® S-40
Mapeg® TAO-15
Maphos® 30
Maphos® 33
Maphos® 41A
Maphos® 55
Maphos® 56
Maphos® 58
Maphos® 60A
Maphos® 66H
Maphos® 79
Maphos® 8135
Maphos® JA 60
Maphos® JM 51
Maphos® JM 71
Maphos® JP 70
Maphos® L 4
Maphos® L-6
Maphos® L 13
Marlon® AMX
Marlophor® AS-Acid
Marlophor® CS-Acid
Marlophor® DS-Acid
Marlowet® 1072
Marlowet® 4530
Marlowet® 4530 LF
Marlowet® 4534
Marlowet® 4536
Marlowet® 4538
Marlowet® 4539
Marlowet® 4539 LF
Marlowet® 4541

Metalworking, Cleaning, Processing, Corrosion Inhibiting (cont'd.)

Marlowet® 4700
Marlowet® 4702
Marlowet® 4703
Marlowet® 4900
Marlowet® 4938
Marlowet® 4940
Marlowet® 5165
Marlowet® 5301
Marlowet® 5320
Marlowet® 5324
Marlowet® 5361
Marlowet® 5440
Marlowet® 5459
Marlowet® 5480
Marlowet® 5622
Marlowet® BIK
Marlowet® LVS
Marlowet® MA
Marlowet® OAM, OAM Spec.
Marlowet® OCM
Marlowet® OFA
Marlowet® OFW
Marlowet® OTS
Marlowet® PW
Marlowet® SAF
Marlowet® SDT
Marlowet® SW
Marlowet® SWN
Marlowet® T
Marlowet® TM
Masil® 1066C
Masil® 1066D
Masil® SF 5
Masil® SF 10
Masil® SF 20
Masil® SF 50
Masil® SF 100
Masil® SF 200
Masil® SF 350
Masil® SF 350 FG
Masil® SF 500
Masil® SF 1000
Masil® SF 5000
Masil® SF 10,000
Masil® SF 12,500
Masil® SF 30,000
Masil® SF 60,000
Masil® SF 100,000
Masil® SF 300,000
Masil® SF 500,000
Masil® SF 1,000,000
Mayphos 45
Mazamide® 65
Mazamide® 68
Mazamide® 70
Mazamide® C-5
Mazamide® J 10
Mazamide® JR 300
Mazamide® JR 400
Mazamide® L-5
Mazamide® L-298
Mazamide® LLD
Mazamide® LM
Mazamide® LM 20
Mazamide® O 20
Mazamide® PCS
Mazamide® RO
Mazamide® SS 10
Mazamide® SS 20
Mazamide® T 20
Mazamide® TC
Mazawet® 77
Mazawet® DF
Mazclean W
Mazclean W-10
Mazclean WRI
Mazeen® C-2
Mazeen® C-15
Mazeen® T-2
Mazeen® T-3.5
Mazeen® T-5
Mazol® GMO
Mazol® GMO #1
Mazol® GMO Ind
Mazol® GMS-D
Mazol® PETO
Mazol® PETO Mod 1
Mazon® 60T
Mazon® 86
Mazon® 114
Mazon® 114A
Mazon® 1045A
Mazon® 1086
Mazon® 1096
Mazon® RI 6
Mazon® RI 37
Maztreat 246
Mazu® 197
Mazu® DF 197
Mazu® DF 204
Mazu® DF 205SX
Mazu® DF 210SX
Mazu® DF 215SX
Mazu® DF 230SX
Mazu® DF 230SXSP
Mazu® DF 255
Merpol® 100
Merpol® HCS
Merpol® SH
Metasol HP-500
Metolat TH 75
Miramine® C
Miranate® LEC
Miranol® C2M Anhyd. Acid
Miranol® C2M Conc. NP LV
Miranol® C2M NP LA
Miranol® C2M-SF Conc.
Miranol® CM Conc. NP
Miranol® CM Conc. OP
Miranol® CS Conc.
Miranol® FA-NP
Miranol® FBS
Miranol® JA
Miranol® JEM Conc.
Miranol® JS Conc.
Miranol® L2M-SF Conc.
Mirapol® WT
Mirataine® T2C-30
Miravon B12DF
Miravon B79R
Mona AT-1200
Monafax 057
Monafax 785
Monafax 786
Monafax 794
Monafax 831
Monafax 872
Monafax 939
Monafax 1293
Monalube 780
Monamid® 150-ADY
Monamid® 150-IS
Monamine CD-100
Monamine LM-100
Monamulse CI
Monaquat ISIES
Monateric 1000
Monateric COAB
Monateric CyA-50
Monateric CyNa-50
Monazoline C
Monazoline CY
Monazoline IS
Monazoline O
Monazoline T
Monolan® 2500 E/30
Monolan® 3000 E/60, 8000 E/80
Mulsifan RT 1
Mulsifan RT 2
Mulsifan RT 7
Mulsifan RT 18
Mulsifan RT 37
Mulsifan RT 113
Nacconol® 40G
Nacconol® 90G
Nansa® HS80-AU
Nansa® HS80S
Nansa® HS85S
Nansa® S40/S
Naxchem Emulsifier 700
Naxel AAS-40S
Naxel AAS-Special 3
Naxonol CO
Naxonol PN 66
Neoscoa 363
Neutronyx® 656
Newcol 3-80, 3-85
Newcol 20
Newcol 25
Newcol 40
Newcol 45
Newcol 60
Newcol 65
Newcol 80
Newcol 85
Newcol 405, 410, 420
Niaproof® Anionic Surfactant 4
Niaproof® Anionic Surfactant 7
Niaproof® Anionic Surfactant 08
Nikkol TDP-2
Nikkol TDP-4
Nikkol TDP-6
Nikkol TDP-8
Nikkol TDP-10
Ninate® 411
Ninol® 11-CM
Ninol® 1281
Ninol® 1301
Ninol® 201
Ninol® SR-100
Niox KI Series
Nissan Amine 2-OLR
Nissan Amine DT
Nissan Cation ABT-350, 500
Nissan Cation FB, FB-500
Nissan Cation L-207
Nissan Cation M2-100
Nissan Cation PB-40, -300
Nissan Cation S2-100
Nissan Nonion E-205
Nissan Nonion E-206
Nissan Nonion E-208
Nissan Nonion E-215
Nissan Nonion E-220
Nissan Nonion E-230
Nissan Nonion HS-204.5
Nissan Nonion HS-206
Nissan Nonion HS-208
Nissan Nonion HS-210
Nissan Nonion HS-215
Nissan Nonion HS-220
Nissan Nonion HS-240
Nissan Nonion HS-270
Nissan Nonion K-202
Nissan Nonion K-203
Nissan Nonion K-204
Nissan Nonion K-207
Nissan Nonion K-211
Nissan Nonion K-215
Nissan Nonion K-230
Nissan Nonion P-208
Nissan Nonion P-210
Nissan Nonion P-213
Nissan Nonion S-207
Nissan Nonion S-215
Nissan Nonion S-220
Nissan Nymeen DT-203, -208
Nissan Nymeen L-201
Nissan Nymeen L-202, 207
Nissan Nymeen S-202, S-204
Nissan Nymeen S-210
Nissan Nymeen S-215, -220
Nissan Nymeen T2-206, -210
Nissan Nymeen T2-230, -260
Nissan Plonon 102
Nissan Plonon 104
Nissan Plonon 108
Nissan Plonon 171
Nissan Plonon 172
Nissan Plonon 201
Nissan Plonon 204
Nissan Plonon 208
Nissan Tert. Amine AB
Nissan Tert. Amine ABT

Metalworking, Cleaning, Processing, Corrosion Inhibiting (cont'd.)

Nissan Tert. Amine BB
Nissan Tert. Amine FB
Nissan Tert. Amine MB
Nissan Tert. Amine PB
Nissan Unister
Nitrene 100 SD
Nitrene C Extra
Nitrene OE
Nopco® 2272-R
Nopcogen 22-O
Noram 2C
Noram DMC
Noram DMS
Noram M2SH
Noram SH
Noramac C
Noramac C 26
Noramac SH
Noramox C2
Noramox S1
Norfox® IM-38
Norfox® KO
Norfox® PEA-N
Norfox® PE-W
Nutrol 100
Ocenol 50/55+2EO
OS-2
Paracol OP
Petro® 11
Petro® BA
Petro® LBA
Petromix®
Petronate® 25C, 25H
Petronate® CR
Petronate® HL
Petronate® HMW
Petronate® L
Petronate® S
Petrosan 102
Petrosul® H-50
Petrosul® L-60
Petrosul® M-50
Petrosul® M-60
Petrosul® M-70
Petrowet® R
Phosfac 1004, 1006, 1044, 1044FA, 1066, 1066FA, 1068FA
Phosfac 1068 C-FA
Phosfac 8000
Phosfac 8608
Phosfac 9604
Phosfac 9609
Phosfetal 601
Phosphanol Series
Phospholan® PNP9
Pluracol® E200
Pluracol® E400 NF
Pluracol® E600
Pluracol® E1450
Pluracol® E1450 NF
Pluracol® E2000
Pluracol® E4000 NF
Pluracol® E4500
Pluracol® E8000
Pluracol® E8000 NF
Pluracol® W170
Pluracol® W260
Pluracol® W660
Pluracol® W2000
Pluracol® W3520N
Pluracol® W3520N-RL
Pluracol® W5100N
Pluracol® WD90K
Pluracol® WD1400
Plurafac® A-38
Plurafac® A-39
Plurafac® B-25-5
Plurafac® B-26
Plurafac® D-25
Pluriol® E 200
Pluriol® E 1500
Pluriol® PE 6100
Pluriol® PE 6400
Pluronic® 10R5
Pluronic® 10R8
Pluronic® 12R3
Pluronic® 17R1
Pluronic® 17R2
Pluronic® 17R4
Pluronic® 17R8
Pluronic® 22R4
Pluronic® 25R1
Pluronic® F38
Pluronic® F68
Pluronic® F68LF
Pluronic® F77
Pluronic® F87
Pluronic® F88
Pluronic® F98
Pluronic® F127
Pluronic® L10
Pluronic® L31
Pluronic® L35
Pluronic® L42
Pluronic® L43
Pluronic® L44
Pluronic® L61
Pluronic® L62
Pluronic® L62D
Pluronic® L62LF
Pluronic® L63
Pluronic® L64
Pluronic® L72
Pluronic® L92
Pluronic® L101
Pluronic® L121
Pluronic® L122
Pluronic® P65
Pluronic® P75
Pluronic® P84
Pluronic® P85
Pluronic® P94
Pluronic® P103
Pluronic® P104
Pluronic® P105
Pluronic® P123
Plysurf A207H
Plysurf A208B
Plysurf A208S
Plysurf A210G
Plysurf A212C
Plysurf A215C
Plysurf A217E
Plysurf A219B
Plysurf AL
Pogol 200
Poly-G® 200
Polysurf A212E
Poly-Tergent® E-17A
Poly-Tergent® E-17B
Poly-Tergent® E-25B
Poly-Tergent® P-22A
Poly-Tergent® P-32A
Poly-Tergent® S-305LF
Poly-Tergent® S-405LF
Poly-Tergent® S-505LF
Poly-Tergent® SL-42
Poly-Tergent® SL-62
Poly-Tergent® SL-92
Priolene 6900
Priolene 6905
Priolene 6906
Priolene 6910
Priolene 6933
Priosorine 3501
Product VN-11
Propetal 241
Prote-pon P-2 EHA-02-K30
Prote-pon P 2 EHA-02-Z
Prote-pon P 2 EHA-Z
Prote-pon P-0101-02-Z
Prote-pon P-L 201-02-K30
Prote-pon P-L 201-02-Z
Prote-pon P-NP-06-K30
Prote-pon P-NP-06-Z
Prote-pon P-NP-10-K30
Prote-pon P-NP-10-MZ
Prote-pon P-NP-10-Z
Prote-pon P-OX 101-02-K75
Prote-pon P-TD-06-K13
Prote-pon P-TD-06-K30
Prote-pon P-TD 06-K60
Prote-pon P-TD-06-Z
Prote-pon P-TD-09-Z
Prote-pon P-TD-12-Z
Prote-pon TD-09-K30
Prote-sorb SML
Prote-sorb SMO
Prote-sorb SMP
Prote-sorb SMS
Protowet 5171
Prox-onic 2EHA-1/02
Prox-onic 2EHA-1/05
Prox-onic CC-05
Prox-onic CC-09
Prox-onic CC-014
Prox-onic CSA-1/04
Prox-onic CSA-1/06
Prox-onic CSA-1/010
Prox-onic CSA-1/015
Prox-onic CSA-1/020
Prox-onic CSA-1/030
Prox-onic CSA-1/050
Prox-onic DA-1/04
Prox-onic DA-1/06
Prox-onic DA-1/09
Prox-onic DNP-08
Prox-onic DNP-0150
Prox-onic DNP-0150/50
Prox-onic DT-03
Prox-onic DT-015
Prox-onic DT-030
Prox-onic EP 1090-1
Prox-onic EP 1090-2
Prox-onic EP 2080-1
Prox-onic EP 4060-1
Prox-onic MC-02
Prox-onic MC-05
Prox-onic MO-02
Prox-onic MO-015
Prox-onic MO-030
Prox-onic MO-030-80
Prox-onic MS-02
Prox-onic MS-05
Prox-onic MS-011
Prox-onic MS-050
Prox-onic MT-02
Prox-onic MT-05
Prox-onic MT-015
Prox-onic MT-020
Prox-onic NP-1.5
Prox-onic NP-010
Prox-onic NP-015
Prox-onic OA-2/020
Prox-onic OCA-1/06
Prox-onic OL-1/05
Prox-onic OL-1/09
Prox-onic OL-1/014
Prox-onic OP-09
Prox-onic OP-016
Prox-onic PEG-2000
Prox-onic PEG-4000
Prox-onic PEG-6000
Prox-onic PEG-10,000
Prox-onic PEG-20,000
Prox-onic PEG-35,000
Prox-onic SA-1/02
Prox-onic SA-1/010
Prox-onic SA-1/020
Prox-onic SML-020
Prox-onic SMO-05
Prox-onic SMO-020
Prox-onic SMP-020
Prox-onic SMS-020
Prox-onic ST-05
Prox-onic ST-09
Prox-onic ST-014
Prox-onic ST-023
Prox-onic STO-020
Prox-onic STS-020
Prox-onic TBP-08
Prox-onic TBP-030
Prox-onic TD-1/03
Prox-onic TD-1/06
Prox-onic TD-1/09
Prox-onic TD-1/012
Prox-onic TM-06
Prox-onic TM-08
Prox-onic TM-010
Prox-onic UA-03
Prox-onic UA-06
Prox-onic UA-09
Prox-onic UA-012
Quartamin DCP
Quaternary O
Quimipol EA 9106
R-60 Z-5
Radiamac 6149

Metalworking, Cleaning, Processing, Corrosion Inhibiting (cont'd.)

Radiamac 6159
Radiamac 6169
Radiamac 6179
Radiamine 6140
Radiamine 6141
Radiamine 6160
Radiamine 6161
Radiamine 6163
Radiamine 6164
Radiamine 6170
Radiamine 6171
Radiamine 6172
Radiamine 6173
Radiamine 6540
Radiamine 6560
Radiamine 6570
Radiamine 6572
Radiasurf® 7000
Radiasurf® 7125
Radiasurf® 7135
Radiasurf® 7136
Radiasurf® 7137
Radiasurf® 7140
Radiasurf® 7141
Radiasurf® 7144
Radiasurf® 7145
Radiasurf® 7147
Radiasurf® 7151
Radiasurf® 7152
Radiasurf® 7153
Radiasurf® 7155
Radiasurf® 7157
Radiasurf® 7175
Radiasurf® 7196
Radiasurf® 7201
Radiasurf® 7206
Radiasurf® 7269
Radiasurf® 7270
Radiasurf® 7345
Radiasurf® 7372
Radiasurf® 7400
Radiasurf® 7402
Radiasurf® 7403
Radiasurf® 7404
Radiasurf® 7412
Radiasurf® 7413
Radiasurf® 7414
Radiasurf® 7417
Radiasurf® 7420
Radiasurf® 7421
Radiasurf® 7422
Radiasurf® 7423
Radiasurf® 7431
Radiasurf® 7432
Radiasurf® 7443
Radiasurf® 7453
Radiasurf® 7454
Radiasurf® 7600
Renex® 22
Renex® 30
Renex® 31
Renex® 35
Renex® 648
Renex® 690
Rewocoros RAB 90
Rewolub KSM 80
Rewomat B 2003
Rewomid® AC 28
Rewomid® DO 280 S
Rewomid® RE
Rewomine IM-BT
Rewomine IM-CA
Rewomine IM-OA
Rewopal® EO 70
Rewopal® HV 5
Rewopal® HV 9
Rewopal® HV 25
Rewopal® LA 3
Rewopal® M 365
Rewopal® RO 40
Rewophat E 1027
Rewophat EAK 8190
Rewopol® BWA
Rewopol® FBR
Rewopol® NEHS 40
Rewopol® SBDO 70
Rewopol® SK 275
Rewopon® IM OA
Rewopon® IM-CA
Rewopon® JMCA
Rewopon® JMOA
Rewoteric AM DML
Rewoteric AM V
Rexfoam D
Rexol 25/14
Rexol 25/15
Rexol 35/11
Rexol 45/12
Rexol 45/16
Rexonic N23-6.5
Rexonic N25-14
Rexonic P-4
Rexonic P-9
Rexophos 25/67
Rhodacal® 330
Rhodacal® DSB
Rhodafac® BG-510
Rhodafac® GB-520
Rhodafac® L6-36A
Rhodafac® LO-529
Rhodafac® MC-470
Rhodafac® MD-12-116
Rhodafac® PA-17
Rhodafac® PA-19
Rhodafac® PE-510
Rhodafac® R5-09/S
Rhodafac® RB-400
Rhodafac® RD-510
Rhodafac® RE-610
Rhodafac® RE-960
Rhodafac® RK-500
Rhodafac® RM-410
Rhodafac® RS-410
Rhodafac® RS-610
Rhodafac® RS-710
Rhodameen® PN-430
Rhodameen® T-15
Rhodamox® CAPO
Rhodamox® LO
Rhodapon® BOS
Rhodapon® CAV
Rhodapon® OLS
Rhodasurf® 860/P
Rhodasurf® L-4
Rhodasurf® L-25
Rhodasurf® L-790
Rhodasurf® LA-7
Rhodasurf® LA-9
Rhodasurf® LA-12
Rhodasurf® LA-15
Rhodasurf® LA-30
Rhodasurf® ON-870
Rhodasurf® T
Rhodasurf® T50
Rhodaterge® FL
Rhodorsil® AF 422
Rilanit G 16
Runox Series
Sandoz Amide CO
Sandoz Amide NP
Sandoz Sulfonate 2A1
Sandoz Sulfonate AAS 50MS
Sandoz Sulfonate AAS 70S
Sandoz Sulfonate AAS-Special 3
Sarkosyl® L
Sarkosyl® O
Schercomid ODA
Schercomid TO-2
Schercophos NP-6
Schercozoline I
Schercozoline L
Schercozoline O
Schercozoline S
Secomine TA 02
Secoster® 874
Secoster® 887
Secoster® DO 600
Secoster® MA 300
Serdolamide POF 61 C
Serdox NJAD 15
Serdox NJAD 20
Serdox NJAD 30
Servoxyl VPGZ 7/100
Servoxyl VPIZ 100
Servoxyl VPNZ 10/100
Servoxyl VPQZ 9/100
Servoxyl VPTZ 3/100
Servoxyl VPTZ 100
Servoxyl VPUZ
Servoxyl VPYZ 500
Silicone AF-10 IND
Silicone AF-30 IND
Silwet® L-7001
Silwet® L-7607
Sinonate 263B
Sinonate 263M
Sinopol 60
Sinopol 80
Sinopol 405, 410
Sinopol 420
Sinopol 1830
Siponic® TD-3
Siponic® TD-6
S-Maz® 80
S-Maz® 85
S-Maz® 90
S-Maz® 95
Smithol 22LD
Smithol 50
Softenol® 3991
Soi Mul 130
Soi Mul 235
Solricin® 135
Solricin® 285
Solton Series
Soprophor® BO/318
Sorban AO 1
Sorban AST
Sorban CO
Sorbax HO-40
Sorbax HO-50
Sorbax MO-40
Sorbax PML-20
Sorbax PMO-5
Sorbax PMP-20
Sorbax PMS-20
Sorbax PTO-20
Sorbax PTS-20
Sorbax SML
Sorbax SMO
Sorbax SMP
Sorbax SMS
Sorbax STO
Sorbax STS
Sovatex IM Series
Span® 80
Stafoam
Stafoam DF-1
Stafoam DF-4
Stafoam F
Standamid® SOD
Stepan C-40
Stepan C-65
Stepan C-68
Stepanol® AM-V
Stepanol® DEA
Stepanol® ME Dry
Stepanol® MG
Stepanol® WA-100
Stepanol® WAC
Stepanol® WAC-P
Stepanol® WA Paste
Stepanol® WA Special
Stepanol® WAT
Stepantex® DO 90
Stepfac® 8170
Strodex® MOK-70
Strodex® P-100
Strodex® PK-90
Strodex® PK-95G
Sul-fon-ate AA-9
Sul-fon-ate AA-10
Sul-fon-ate OA-5
Sulfetal 4105
Sulfetal 4187
Sulfetal C 90
Sulfetal FA 40
Sulframin 40
Sulframin 40DA
Sulframin 40RA
Sulframin 40T
Sulframin 45
Sulframin 45LX
Sulframin 60T
Sulframin 85
Sulframin 90
Sulframin 1240, 1245
Sulframin 1250, 1260
Sulframin LX

Metalworking, Cleaning, Processing, Corrosion Inhibiting (cont'd.)

Surfac® P14B
Surfac® P24M
Surfactol® 13
Surfactol® 318
Surfactol® 340
Surfactol® 365
Surfactol® 380
Surfactol® 575
Surfactol® 590
Surfagene FAD 106
Surfam P5 Dist
Surfam P5 Tech
Surfam P10
Surfam P12B
Surfam P14B
Surfam P17B
Surfam P24M
Surfam P86M
Surfam P89MB
Surfax 8916/A
Surfonic® HDL
Surfonic® HDL-1
Surfonic® JL-80X
Surfonic® N-10
Surfonic® N-31.5
Surfonic® N-40
Surfonic® N-60
Surfonic® N-85
Surfonic® N-95
Surfonic® N-100
Surfonic® N-102
Surfonic® N-120
Surfonic® N-130
Surfonic® N-150
Surfonic® N-200
Surfonic® N-300
Surfonic® NB-5
Surfynol® 82
Surfynol® 82S
Surfynol® 104A
Surfynol® 104H
Surfynol® 104PA
Surfynol® 104PG
Surfynol® 440
Surfynol® 465
Surfynol® 485
Surfynol® DF-110, DF-110S
Surfynol® DF-37
Surfynol® PG-50
Surmax® CS-504
Surmax® CS-515
Surmax® CS-521
Surmax® CS-522
Surmax® CS-555
Surmax® CS-586
Sylfam 2042
Sylfam 2082
Sylfan 20
Sylfan 96
Sylfat® D-1
Sylfat® DX
Sylfat® MM
Sylfat® RD-1
Sylvacote® K
Synperonic LF/B26
Synperonic LF/D25
Synperonic LF/RA30
Synperonic LF/RA260
Synperonic LF/RA290
Synperonic N
Synprolam 35
Synprolam 35BQC (50)
Synprolam 35BQC (80)
Synprolam 35DM
Synprolam 35DMBQC
Synprolam 35M
Synprolam 35N3
Synprolam 35N3X3, 35N3X5, 35N3X10, 35N3X15, 35N3X25
Syntergent K
Syntofor A03, A04, AB03, AB04, AL3, AL4, B03, B04
Syntopon 8 B
Syntopon E
Tallow Amine
Tallow Diamine
Tallow Tetramine
Tallow Triamine
T-Det® C-40
T-Det® EPO-61
T-Det® EPO-62L
T-Det® EPO-64L
T-Det® LF-416
T-Det® N-4
T-Det® N-6
T-Det® N-12
T-Det® N-14
T-Det® RQ1
T-Det® RY2
T-Det® TDA-65
Tenax 2010
Tergitol® 15-S-3
Tergitol® 15-S-5
Tergitol® 15-S-7
Tergitol® 15-S-9
Tergitol® Min-Foam 1X
Tergitol® Min-Foam 2X
Tergitol® TMN-3
Tergitol® TMN-10
Teric 9A12
Teric 12A6
Teric 12A9
Teric 12A12
Teric 12M2
Teric 12M5
Teric 12M15
Teric 15A11
Teric 16A22
Teric 16M2
Teric 16M5
Teric 16M15
Teric 17M2
Teric 17M5
Teric 17M15
Teric 18M2
Teric 18M10
Teric 18M20
Teric 18M30
Teric 160
Teric 161
Teric 305
Teric 350
Teric 351B
Teric C12
Teric DD5
Teric DD9
Teric G12A4
Teric G12A6
Teric G12A8
Teric G12A12
Teric LA4
Teric N4
Teric N5
Teric OF4
Teric OF6
Teric PEG 200
Teric PEG 300
Teric PEG 400
Teric PEG 600
Teric PEG 800
Teric PEG 4000
Teric PEG 6000
Teric PPG 400
Teric PPG 1000
Teric PPG 1650
Teric PPG 2250
Teric PPG 4000
Teric T2
Teric T5
Teric T7
Teric T10
Teric X5
Teric X7
Teric X8
Teric X10
Teric X11
Teric X13
Teric X16
Teric X40
Teric X40L
Tetronic® 50R1
Tetronic® 50R4
Tetronic® 50R8
Tetronic® 70R1
Tetronic® 70R2
Tetronic® 70R4
Tetronic® 90R1
Tetronic® 90R4
Tetronic® 90R8
Tetronic® 110R1
Tetronic® 110R2
Tetronic® 110R7
Tetronic® 130R1
Tetronic® 130R2
Tetronic® 150R1
Tetronic® 150R4
Tetronic® 150R8
Tetronic® 304
Tetronic® 504
Tetronic® 701
Tetronic® 702
Tetronic® 704
Tetronic® 901
Tetronic® 904
Tetronic® 908
Tetronic® 909
Tetronic® 1101
Tetronic® 1102
Tetronic® 1104
Tetronic® 1107
Tetronic® 1301
Tetronic® 1302
Tetronic® 1304
Tetronic® 1307
Tetronic® 1501
Tetronic® 1502
Tetronic® 1504
Tetronic® 1508
Texamin PD
Texin DOS 75
Textamine 05
Textamine A-5-D
Textamine A-W-5
Textamine Carbon Detergent K
Textamine O-1
Textamine Polymer
Textamine T-1
Textamine T-5-D
T-Maz® 20
T-Maz® 28
T-Maz® 40
T-Maz® 60
T-Maz® 61
T-Maz® 65
T-Maz® 80
T-Maz® 80KLM
T-Maz® 81
T-Maz® 85
T-Maz® 90
T-Maz® 95
T-Mulz® 1158
Tomah DA-14
Tomah DA-16
Tomah DA-17
Tomah E-14-2
Tomah E-14-5
Tomah E-17-2
Tomah E-18-2
Tomah E-18-5
Tomah E-18-8
Tomah E-18-10
Tomah E 18-15
Tomah E-19-2
Tomah E-24-2
Tomah E-S-5
Tomah E-S-15
Tomah E-T-5
Tomah E-T-15
Tomah PA-10
Tomah PA-16
Tomah PA-17
Tomah PA-19
Tomah PA-1214
Tomah Q-14-2PG
Tomah Q-17-2PG
Tomah Q-18-2
Tomah Q-18-15
Tomah Q-C-15
Tomah Q-D-T
Tomah Q-DT-HG
Tomah Q-S
Tomah Q-S-80
Tomah Q-ST-50
Trans-100
Trans-109
Trans-166
Trilon® A
Tris Amino® (Crystals)

Metalworking, Cleaning, Processing, Corrosion Inhibiting (cont'd.)

Triton® BG-10
Triton® CF-10
Triton® CF-21
Triton® CF-32
Triton® CF-54
Triton® CF-76
Triton® CF-87
Triton® DF-12
Triton® DF-16
Triton® DF-18
Triton® DF-20
Triton® N-60
Triton® N-101
Triton® QS-15
Triton® QS-44
Triton® RW-20
Triton® RW-50
Triton® RW-75
Triton® RW-100
Triton® RW-150
Triton® X-45
Triton® X-100
Triton® X-100 CG
Triton® X-102
Triton® X-114
Triton® X-200
Triton® XL-80N
Troykyd® D666
Troykyd® D999
Trycol® 5940
Trycol® 6720
Trycol® 6961
Trycol® 6975
Trycol® NP-4
Trycol® TDA-6
Trycol® TP-6
Trydet 22
Trydet 2676
Trydet DO-9
Trydet OA-7
Trydet OA-10
Trydet SA-5
Tryfac® 610-K
Tryfac® 5553
Tryfac® 5554
Tryfac® HWD
Trylox® 1186
Trylox® 5902
Trylox® 5904
Trylox® 6746
Trylox® CO-16
Trylox® CO-25
Trymeen® 6601
Trymeen® 6617
Trymeen® CAM-15
Trymeen® SAM-50
Tylose C, CB Series
Ufanon K-80
Ufaryl DB80
Ufaryl DL80 CW
Ufaryl DL85
Ufaryl DL90
Ufasan 50
Ufasan 60A
Ufasan 62B
Ufasan 65
Ufasan IPA
Ufasan TEA
Ultraphos Series
Unamide® C-72-3
Unamide® D-10
Unamide® J-56
Unamide® LDL
Unamide® N-72-3
Unamine® O
Unamine® T
Varamide® A-7
Varamide® A-10
Varamide® AC-28
Varine C
Varine O®
Varine O Acetate
Varine T
Varion® AM-KSF-40%
Variquat® 80MC
Variquat® 638
Varonic® K202
Varonic® K205
Varonic® K215
Varonic® Q202
Velvetex® OLB-50
Versilan MX123
Versilan MX134
Wayfos A
Wayfos D-10N
Wayfos M-60
Wayfos M-100
Wayhib® S
Wetting Agent FCGB
Witbreak DGE-75
Witcamide® 272
Witcamide® 511
Witcamide® 5130
Witcamide® 5133
Witcamide® 5138
Witcamide® 5195
Witcamide® AL69-58
Witcamide® CDS
Witcamine® 204
Witcamine® 209
Witcamine® 210
Witcamine® 211
Witcamine® 235
Witcamine® 240
Witcamine® 3164
Witcamine® AL42-12
Witcamine® PA-60B
Witcamine® RAD 0500
Witcamine® RAD 0515
Witcamine® RAD 1100
Witcamine® RAD 1110
Witcamine® TI-60
Witco® 1298
Witco® Acid B
Witco® DTA-150, -180
Witco® DTA-350
Witcolate 1259
Witcolate 1276
Witconate 90F
Witconate 1840X
Witconate NXS
Witconate SCS 45%
Witconate SCS 93%
Witconate STS
Witconate SXS 40%
Witconate SXS 90%
Witconate YLA
Witconol APEM
Witconol APM
Witconol APS
Witconol H31
Witconol NP-40
Witcor 3192
Witcor 3194, 3195
Witcor 3630
Witcor 3635
Witcor CI-1
Witcor CI-6
Zohar HAS 470
Zonyl® A
Zonyl® FSA
Zonyl® FSB
Zonyl® FSC
Zonyl® FSN
Zusolat 1005/85
Zusolat 1008/85
Zusolat 1010/85
Zusolat 1012/85
Zusomin C 108
Zusomin O 105
Zusomin S 110
Zusomin TG 102

Mining and Mineral Processing

Acetamin 24
Acetamin 86
Acetamin C
Acetamin HT
Acetamin T
Adol® 66
Adol® 80
Adol® 520 NF
Adol® 620 NF
Aerosol® AY-65
Aerosol® AY-100
Aerosol® MA-80
Aerosol® OS
Agesperse 71
Agesperse 80
Akypomine® AT
Akypomine® BC 50
Ameenex C-18
Aristonate H
Aristonate L
Aristonate M
Arizona DR-22
Armac® 18D-40
Armac® C
Armac® T
Armeen® 2C
Armeen® 2HT
Armeen® 2T
Armeen® 12D
Armeen® 16D
Armeen® 18
Armeen® C
Armeen® CD
Armeen® HT
Armeen® HTD
Armeen® OL
Armeen® OLD
Armeen® S
Armeen® SD
Armeen® TD
Armid® HT
Armid® O
Armoflote Series
Beycostat LP 12 A
Calsoft F-90
Calsoft L-40
Calsoft L-60
Carbowax® MPEG 350
Carbowax® MPEG 550
Carbowax® PEG 4600
Carbowax® PEG 8000
Crodamine 1.HT
Crodamine 1.O, 1.OD
Crodamine 1.T
Crodamine 1.16D
Crodamine 1.18D
Crodamine 3.A16D
Crodamine 3.A18D
Crodamine 3.ABD
Crodamine 3.AED
Crodamine 3.AOD
Farmin 20
Farmin 60
Farmin 68
Farmin 80
Farmin 86
Farmin AB
Farmin C
Farmin HT
Farmin O
Farmin S
Farmin T
Finazoline CA
Flotanol Grades
Flotigam Grades
Flotigol CS
Flotinor FS-2
Flotinor S Grades
Flotinor SM 15
Flotol Grades
Geropon® DOS
Geropon® DOS FP
Geropon® SS-O-75
Hostaflot L Grades
Hostaflot X Grades
Jet Amine DC
Jet Amine DO
Jet Amine DT
Jet Amine PC
Jet Amine PCD
Jet Amine PE 08/10
Jet Amine PHT
Jet Amine PHTD
Jet Amine PO
Jet Amine POD
Jet Amine PS

Mining and Mineral Processing (cont'd.)

Jet Amine PSD
Jet Amine PT
Jet Amine PTD
Kemamine® P-150
Kemamine® P-150D
Kemamine® P-190
Kemamine® P-190D
Kemamine® P-880
Kemamine® P-880D
Kemamine® P-970
Kemamine® P-989D
Kemamine® P-990
Kemamine® P-990D
Kemamine® P-999
Lilamac S1
Marlophor® NP5-Acid
Marlophor® NP6-Acid
Marlophor® NP7-Acid
Marlowet® 5641
Melioran 118
Monawet MM-80
Monawet MO-65 PEG
Monawet MO-65-150
Monawet MO-70
Monawet MO-70-150
Monawet MO-70E
Monawet MO-70 PEG
Monawet MO-70R
Monawet MO-70S
Monawet MO-75E
Monawet MO-84R2W
Monawet MO-85P
Nacconol® 40G
Nacconol® 90G
Nissan Amine DT
Nissan Amine DTH
Noramac C
Octowet 40
Octowet 55
Octowet 60
Octowet 60-I
Octowet 65
Octowet 70
Octowet 70A
Octowet 70BC
Octowet 70PG
Octowet 75
Octowet 75E
Petrosul® H-60
Petrosul® H-70
Petrosul® HM-62
Petrosul® HM-70
Phosokresol Grades
Phospholan® PSP6
Phospholipid PTD
Phospholipid PTL
Phospholipid PTZ
Pluracol® P-410
Pluracol® P-710
Pluracol® P-1010
Pluracol® P-2010
Pluracol® P-3010
Pluracol® P-4010
Quaternary O
Radiamine 6140
Radiamine 6141
Radiamine 6160
Radiamine 6161
Radiamine 6163, 6164
Radiamine 6170
Radiamine 6171
Radiamine 6172
Radiamine 6173
Rhodapon® EC111
Silwet® L-77
Silwet® L-7604
Silwet® L-7607
SK Flot 1
SK Flot 2
SK Flot 3
SK Flot 4
SK Flot FA 1
SK Flot FA 2
SK Flot FA 3
SK Flot FA 4
Sole-Terge 8
Stepanflote® 24
Stepanflote® 85L
Stepanflote® 97A
Sul-fon-ate AA-10
Surfac® P14B
Surfac® P24M
Surfax 500
Synprolam 35A
Tallow Amine
Tallow Diamine
Tallow Tetramine
Tallow Triamine
Textamine 1839
Thorowet G-40 3230
Thorowet G-60 3142
Thorowet G-60 SS 0336
Thorowet G-75 1060
Tylose CBR Grades
Tylose MHB
Tylose MH Grades
Tylose MH-K, MH-xp, MHB-y, MHB-yp
Ufasan 35
Ufasan 50
Ufasan 60A
Ufasan 62B
Ufasan 65
Ufasan IPA
Ufasan TEA
Unidri M-40
Unidri M-60
Unidri M-75
Witcamine® AL42-12

Paints, Coatings, Lacquers, Inks, Adhesives

AA#2 Lime Additive
Ablunol 200ML
Ablunol 200MO
Ablunol 200MS
Ablunol 400ML
Ablunol 400MO
Ablunol 400MS
Ablunol 600ML
Ablunol 600MO
Ablunol 600MS
Ablunol 1000MO
Ablunol 1000MS
Ablunol 6000DS
Ablunol NP 30
Ablunol NP 30 70%
Ablunol NP 40
Ablunol NP 40 70%
Ablunol NP 50
Ablunol NP 50 70%
Adol® 42
Adol® 61 NF
Adol® 62 NF
Adol® 80
Adol® 520 NF
Adol® 620 NF
Aerosol® A-103
Aerosol® C-61
Aerosol® OT-75%
Aerosol® OT-100%
Aerosol® OT-S
Aerosol® TR-70
AF 10 IND
AF 60
AF 112
AF 8810
AF 8810 FG
AF 9020
AF GN-11-P
AF HL-21
AF HL-26
AF HL-27
AF HL-40
AF HL-52
Ageflex FM-1
Agesperse 71
Agesperse 80
Ajidew A-100
Ajidew N-50
Ajidew SP-100
Alcolec® 439-C
Alcolec® 440-WD
Alkamuls® EL-719
Alkamuls® L-9
Alkamuls® PSMO-5
Alkasurf® NP-6
Alkaterge®-C
Amihope LL-11
Amine C
AMP
AMP-95
AMPD
Anti-Terra®-202
Anti-Terra®-U
Anti-Terra®-U80
Armeen® C
Armeen® CD
Armeen® Z
Armid® 18
Armid® C
Armid® HT
Armid® O
Armul 1007
Armul 1009
Atlas G-4809
Atra Polymer 10
Autopoon GK Series
Autopoon NI
Avirol® SA 4106
Avirol® SO 70P
Berol 108
Berol 190
Berol 191
Berol 198
Berol 594
Berol WASC
Biozan
Bubble Breaker® 259, 260, 613-M, 622, 730, 737, 746, 748, 900, 913, 917
Bubble Breaker® 776, 3017-A
Bubble Breaker® 3056-A
BYK®-020
BYK®-022
BYK®-023
BYK®-024
BYK®-025
BYK®-031
BYK®-032
BYK®-033
BYK®-034
BYK®-035
BYK®-045
BYK®-051
BYK®-052
BYK®-053
BYK®-075
BYK®-080
BYK®-085
BYK®-141
BYK®-151
BYK®-307
BYK®-320
BYK®-321
BYK®-331
BYK®-336
BYK®-341
BYK®-344
BYK®-354
BYK®-A500
BYK®-A501
BYK®-P 104
BYK®-P 104S
BYK®-P 105
Bykumen®
Bykumen®-WS
Calcium Stearate, Regular
Calsoft F-90
Carbopol® 940
Carbowax® MPEG 350
Carbowax® MPEG 550
Carbowax® MPEG 750
Carbowax® MPEG2000

Paints, Coatings, Lacquers, Inks, Adhesives (cont'd.)

Carbowax® MPEG 5000
Carbowax® PEG 200
Carbowax® PEG 300
Carbowax® PEG 400
Carbowax® PEG 600
Carbowax® PEG 900
Carbowax® PEG 1000
Carbowax® PEG 1450
Carbowax® PEG 3350
Carbowax® PEG 4600
Carbowax® PEG 8000
Carsonon® L-985
Catinex KB-15
Catinex KB-32
Chemax CO-5
Chemax CO-16
Chemax CO-25
Chemax CO-28
Chemax CO-30
Chemax CO-36
Chemax CO-40
Chemax CO-200/50
Chemax DF-30
Chemax DF-100
Chemax DNP-150/50
Chemax DOSS/70
Chemax DOSS-75E
Chemax E-200 MO
Chemax HCO-5
Chemax HCO-16
Chemax HCO-25
Chemax HCO-200/50
Chemax NP-1.5
Chemax NP-4
Chemax NP-6
Chemax NP-9
Chemax NP-10
Chemax NP-15
Chemax NP-20
Chemax NP-30
Chemax NP-30/70
Chemax OP-3
Chemax OP-5
Chemax OP-7
Chemax OP-9
Chemax OP-30/70
Chemphos TC-310S
Chemzoline T-11
Chemzoline T-33
Clearate LV
Clearate Special Extra
Clearate WD
CNC Antifoam 30-FG
CNC Antifoam 77
CNC Antifoam 495, 495-M
CNC Defoamer 229
Colorol 20
Colorol 46
Colorol 70
Colorol Aquasorb
Colorol E
Colorol F
Colorol Standard
Crill 1
Crill 2
Crill 4
Crillet 1
Crodazoline O
Crodazoline S
Cromul 1540
Fancol OA 95
Fancor LFA
Fluorad FC-95
Fluorad FC-109
Fluorad FC-121
Fluorad FC-129
Fluorad FC-170-C
Fluorad FC-171
Fluorad FC-430
Fluorad FC-431
Foamaster 8034
Foamaster DD-72
Foamaster DF-122NS
Foamaster DF-177-F, -178
Foamaster DF-198-L
Foamaster DNH-1
Foamaster JMY
Foamaster NDW
Foamaster NS-20
Foamaster NXZ
Foamaster P
Foamaster PD-1
Foamaster TDB
Foamaster VL
Foamkill® 8R
Foamkill® 30C
Foamkill® 400A
Foamkill® 608
Foamkill® 614
Foamkill® 618 Series
Foamkill® 629
Foamkill® 639
Foamkill® 639AA
Foamkill® 639J
Foamkill® 639JOH
Foamkill® 639L
Foamkill® 639P
Foamkill® 639Q
Foamkill® 644 Series
Foamkill® 649
Foamkill® 649C
Foamkill® 652B
Foamkill® 663J
Foamkill® 679
Foamkill® 684 Series
Foamkill® 810
Foamkill® 810F
Foamkill® 830
Foamkill® 830F
Foamkill® 836A
Foamkill® 836B
Foamkill® 852
Foamkill® 1001 Series
Foamkill® CPD
Foamkill® DF#4
Foamkill® FBF
Foamkill® FPF
Foamkill® MS Conc.
Foamkill® MS-1
Foamkill® MSF Conc.
Foamkill® SEA
Forbest 13
Forbest 18
Forbest 20
Forbest 30
Forbest 33
Forbest 50
Forbest 150
Forbest 560
Forbest 610
Forbest 620
Forbest 850
Forbest 1500W
Forbest 2000C
Forbest 8209
Forbest G23
Forbest S 7
Forbest VP 13
Forbest VP 18
Forbest VP 20
Forbest VP 33
Forbest VP S7
Forbest WP
G-1000-S Antifoam Compd.
Gantrez® AN-119
Gantrez® AN-139
Gantrez® AN-149
Gantrez® AN-169
Gantrez® AN-179
Genamine C-020
Genamine C-050
Genamine C-080
Genamine C-100
Genamine C-150
Genamine C-200
Genamine C-250
Genamine O-020
Genamine O-050
Genamine O-080
Genamine O-100
Genamine O-150
Genamine O-200
Genamine O-250
Genamine S-020
Genamine S-050
Genamine S-080
Genamine S-100
Genamine S-150
Genamine S-200
Genamine S-250
Geropon® 99
Geropon® CYA/DEP
Geropon® SS-O-75
Geropon® TA/K
Geropon® TX/99
Geropon® WS-25, WS-25-I
Glycidol Surfactant 10G
GP-210 Silicone Antifoam Emulsion
GP-214
GP-217
GP-218
GP-219
GP-226
GP-262 Defoamer
GP-295 Defoamer
GP-300-I Antifoam Compd.
GP-310-I
GP-330-I Antifoam Emulsion
GP-7000
Gradonic 400-ML
Haroil SCO-65
Haroil SCO-75, -7525
Hartomer GP 2164
Hercules® 831 Defoamer
Hercules® 845 Defoamer
Hetoxide G-26
Hydrolene
Hyonic NP-60
Hyonic OP-70
Hyonic PE-90
Idet 5LP
Igepal® CO-630
Igepal® CO-660
Igepal® CO-710
Igepal® CO-720
Igepal® CTA-639W
Igepal® DM-970 FLK
Igepal® LAVE
Igepal® OD-410
Imperon Emulsifier 774
Industrene® D
Inipol OO2
Inipol OT2
Inipol S 43
Interwet® 33
Karapeg 400-ML
Karapeg 600-ML
Kemamide® B
Kemamide® E
Kemamide® O
Kemamide® S
Kemamide® U
Kemamide® W-20
Kemamide® W-39
Kemamide® W-40
Kemamide® W-40/300
Kemamide® W-45
Kilfoam
Klucel® E
Klucel® G
Klucel® H
Klucel® J
Klucel® L
Klucel® M
Lamigen ES 30
Lamigen ES 60
Lamigen ES 100
Lamigen ET 20
Lamigen ET 70
Lamigen ET 90
Lamigen ET 180
Lankropol® WA
Lankropol® WN
Laurel PEG 400 DT
Laurel PEG 400 MO
Laurel PEG 600 DT
Laurel R-50
Laurel R-75
Lecithin W.D
Lexemul® PEG-200 DL
Lipotin 100, 100J, SB
Lipotin A

Paints, Coatings, Lacquers, Inks, Adhesives (cont'd.)

Lipotin H
Lomar® PW
Lomar® PWA
Lubrizol® 2152
Lubrizol® 2153
Lubrizol® 2155
Lumiten® E
Lumiten® I
Lumiten® N
Lutensol® AP 6
Lutensol® AP 7
Lutensol® AP 8
Lutensol® AP 9
Macol® 65
Macol® 90
Macol® 90(70)
Macol® 99A
Macol® 625
Macol® 626
Macol® 627
Macol® NP-9.5
Manro DL 28
Manro PTSA/C
Marlinat® DF 8
Marlipal® 1618/8
Marlipal® 1618/10
Marlipal® 1618/11
Marlipal® 1618/18
Masil® 2132
Masil® 2133
Masil® 2134
Masil® SF 5
Masil® SF 10
Masil® SF 20
Masil® SF 50
Masil® SF 100
Masil® SF 350
Masil® SF 350 FG
Masil® SF 500
Masil® SF 1000
Masil® SF 5000
Masil® SF 10,000
Masil® SF 12,500
Masil® SF 30,000
Masil® SF 60,000
Masil® SF 100,000
Masil® SF 300,000
Masil® SF 500,000
Masil® SF 600,000
Mazawet® 77
Mazawet® DF
Mazawet® DOSS 70
Mazeen® 173
Mazeen® 174
Mazeen® 174-75
Mazeen® 241-3
Mazeen® C-2
Mazeen® C-5
Mazeen® C-10
Mazeen® C-15
Mazeen® DBA-1
Mazeen® S-2
Mazeen® S-5
Mazeen® S-10
Mazeen® S-15
Mazeen® T-2
Mazeen® T-10
Mazeen® T-15
Mazu® 68 C
Mazu® 108 L
Mazu® 151 PY
Mazu® 160 CA
Mazu® 161
Mazu® 208 L
Mazu® 251
Mazu® 252
Mazu® 252 A
Mazu® 255
Mazu® 290
Mazu® 309
Mazu® 320
Mazu® 321
Mazu® 2501
Mazu® 2502
Mazu® DF 200SX
Mazu® DF 200SXSP
Mazu® DF 210SX
Mazu® DF 210SX Mod 1
Mazu® DF 210SXSP
Mazu® DF 230SX
Mazu® DF 230SXSP
Mazu® DF 243
Merpol® A
Merpol® HCS
Metolat FC 355
Metolat FC 388
Metolat FC 514
Metolat FC 515
Metolat FC 530
Metolat LA 524
Metolat LA 571
Metolat LA 573
Metolat P 853
Miranol® HM Conc.
Monawet 1240
Monawet MM-80
Monawet MO-70
Monawet MO-70-150
Monawet MO-70E
Monawet MO-70 PEG
Monawet MO-70S
Monawet MO-75E
Monawet MO-84R2W
Monawet MO-85P
Monawet MT-70
Monawet MT-70E
Monawet MT-80H2W
Monawet SNO-35
Monazoline C
Monazoline CY
Monazoline O
Monazoline T
Monolan® PPG440, PPG1100, PPG2200
Nalco® 2301
Nalco® 2305
Nalco® 2311
Nalco® 2314
Nansa® EVM50
Nansa® EVM70/B
Nansa® EVM70/E
Nansa® S40/S
Niaproof® Anionic Surfactant 4
Niaproof® Anionic Surfactant 7
Niaproof® Anionic Surfactant 08
Nikkol GO-430
Nikkol GO-440
Nikkol GO-460
Nissan Nonion HS-204.5
Nissan Nonion HS-206
Nissan Nonion HS-208
Nissan Nonion HS-210
Nissan Nonion HS-215
Nissan Nonion HS-240
Nissan Nonion HS-270
Nissan Nonion NS-202
Nissan Nonion NS-204.5
Nissan Nonion NS-206
Nissan Nonion NS-208.5
Nissan Nonion NS-209
Nissan Nonion NS-210
Nissan Nonion NS-212
Nissan Nonion NS-215
Nissan Nonion NS-220
Nissan Nonion NS-230
Nissan Nonion NS-240
Nissan Nonion NS-250
Nissan Nonion NS-270
Noigen ES 120
Noigen ES 140
Noigen ES 160
Nopalcol 4-O
Nopco® 1419-A
Nopco® JMY
Nopco® NDW
Nopco® NXZ
Nopco® PD#1-D
Noram DMC
Noram M2SH
Noramium DA.50
Octosol 496
Octosol A-1
Octowet 60-I
Octowet 70BC
Octowet 70PG
Ombrelub FC 533
Pale 4
Pale 16
Pale 170
Pale 1000
Patcote® 305
Patcote® 306
Patcote® 307
Patcote® 309
Patcote® 310
Patcote® 311
Patcote® 315
Patcote® 460
Patcote® 512
Patcote® 513
Patcote® 519
Patcote® 520
Patcote® 525
Patcote® 531
Patcote® 550
Patcote® 555
Patcote® 555K
Patcote® 577
Patcote® 597
Patcote® 801
Patcote® 803
Patcote® 806
Patcote® 811
Patcote® 812
Patcote® 841M
Patcote® 845
Patcote® 847
Perenol EI
Petronate® Neutral (50-S)
Petrosul® H-50
Petrosul® H-60
Petrosul® H-70
Petrosul® HM-62, HM-70
Petrosul® M-50
Petrosul® M-60
Petrosul® M-70
Pluracol® E200
Pluracol® E400 NF
Pluracol® E1450
Pluracol® E1450 NF
Pluracol® E2000
Pluracol® E4000 NF
Pluracol® E4500
Pluracol® E8000
Pluriol® E 200
Pluriol® E 300
Pluriol® E 400
Pluriol® E 600
Pluriol® E 1500
Pluriol® E 4000
Pluriol® E 6000
Pluriol® E 9000
Pluronic® L10
Poly-G® 200
Poly-G® 300
Poly-G® 400
Poly-G® 600
Poly-G® 1000
Poly-G® 1500
Poly-G® 2000
Poly-G® B1530
Polystep® B-19
Polystep® B-20
Poly-Tergent® P-17A
Poly-Tergent® P-17B
Poly-Tergent® P-17BLF
Poly-Tergent® P-17BX
Poly-Tergent® P-17D
Polywet ND-1
Polywet ND-2
Prifrac 2912
Pripol 1025
Pronal 502, 502A
Pronal P-805
Prox-onic DNP-08
Prox-onic DNP-0150
Prox-onic DNP-0150/50
Prox-onic EP 1090-1
Prox-onic EP 1090-2
Prox-onic EP 2080-1
Prox-onic EP 4060-1
Prox-onic HR-05
Prox-onic HR-016
Prox-onic HR-025
Prox-onic HR-030
Prox-onic HR-036

Paints, Coatings, Lacquers, Inks, Adhesives (cont'd.)

Prox-onic HR-040
Prox-onic HR-080
Prox-onic HR-0200
Prox-onic HR-0200/50
Prox-onic HRH-05
Prox-onic HRH-016
Prox-onic HRH-025
Prox-onic HRH-0200
Prox-onic HRH-0200/50
Prox-onic TBP-08
Prox-onic TBP-030
Radiasurf® 7146
Radiasurf® 7150
Radiasurf® 7151
Radiasurf® 7156
Radiasurf® 7175
Radiasurf® 7196
Radiasurf® 7402
Radiasurf® 7410
Radiasurf® 7411
Radiasurf® 7444
Radiasurf® 7900
Ram Polymer 110
Redicote TR-1114X
Redicote TR-1130
Redicote TR-1134
Renex® 36
Rewomine IM-CA
Rewomine IM-OA
Rexfoam 150-A
Rexfoam B, C
Rexfoam D
Rexol 25/10
Rexol 25/40
Rexol 25/50
Rexol 25/100-70%
Rexol 25/507
Rexol 25J
Rexol 45/10
Rexonic 1012-6
Rexonic L125-9
Rexonic N23-6.5
Rexonic N25-7
Rexonic N25-9
Rexonic N25-9(85%)
Rexonic N25-14(85%)
Rheodol SP-O10
Rheodol SP-O30
Rheodol SP-S10
Rheotol
Rhodacal® 330
Rhodacal® BA-77
Rhodacal® RM/77-D
Rhodasurf® ON-870
Rotolan®
Schercozoline C
Scripset 520
Secosol® DOS 70
Ser-Ad FA 153
Ser-Ad FN Series
Serdas GBS, GBU
Serdas GE 4010
Serdas GE 4050
Serdas GLN
SF81-50
SF96®
Silicone AF-10 IND
Silicone AF-30 FG
Silicone AF-30 IND
Silicone AF-100 FG
Silicone AF-100 IND
Silwet® L-77
Silwet® L-7001
Silwet® L-7200
Silwet® L-7210
Silwet® L-7230
Silwet® L-7500
Silwet® L-7602
Silwet® L-7604
Silwet® L-7607
Silwet® L-7622
Siponic® NP-9
Siponic® NP-15
Sole-Onic CDS
Sopralub ACR 275
Sperse Polymer IV
Strodex® MO-100
Strodex® MOK-70
Strodex® MRK-98
Strodex® PK-80A
Strodex® PK-90
Strodex® PK-95G
Strodex® SEK-50
Sulfotex SXS-40
Supragil® WP
Surfactol® 13
Surfactol® 340
Surfactol® 365
Surfactol® 380
Surfactol® 575
Surfactol® 590
Surfam P5 Dist.
Surfam P5 Tech.
Surfonic® HDL-1
Surfonic® LF-17
Surfonic® N-10
Surfonic® N-31.5
Surfonic® N-85
Surfonic® N-102
Surfonic® N-120
Surfonic® N-130
Surfonic® N-150
Surfonic® N-200
Surfonic® N-300
Surfonic® NB-5
Surfynol® 61
Surfynol® 82
Surfynol® 82S
Surfynol® 104
Surfynol® 104A
Surfynol® 104BC
Surfynol® 104H
Surfynol® 104PA
Surfynol® 104PG
Surfynol® 420
Surfynol® 465
Surfynol® 485
Surfynol® CT-136
Surfynol® D-101
Surfynol® D-201
Surfynol® DF-08
Surfynol® DF-34
Surfynol® DF-37
Surfynol® DF-58
Surfynol® DF-60
Surfynol® DF-70
Surfynol® DF-75
Surfynol® DF-110
Surfynol® DF-110S
Surfynol® DF-110D
Surfynol® DF-110L
Surfynol® DF-210
Surfynol® DF-574
Surfynol® DF-695
Surfynol® PC
Surfynol® PG-50
Surfynol® PSA-204
Surfynol® PSA-216
Surfynol® SE
Surfynol® SE-F
Surfynol® TG
Surmax® CS-504
Swanic 6L, 7L
Sylfat® DX
Sylfat® MM
Sylvaros® 20
Syn Fac® 334
Syn Fac® 8216
Synperonic PE30/80
T-Det® C-40
T-Det® DD-10
T-Det® N 1.5
T-Det® N-4
T-Det® N-8
T-Det® N-9.5
T-Det® N-10.5
T-Det® N-407
T-Det® N-507
T-Det® O-4
T-Det® O-8
Tego® Dispers 610
Tego® Dispers 610 S
Tego® Dispers 700
Tego® Dispers 705
Tego® Flow 425
Tego® Foamex 800
Tego® Foamex 805
Tego® Foamex 1435
Tego® Foamex 1488
Tego® Foamex 3062
Tego® Foamex 7447
Tego® Foamex KS 6
Tego® Foamex KS 10
Tego® Foamex L 808
Tego® Foamex L 822
Tego® Foamex N
Tego® Glide 100
Tego® Glide 405
Tego® Glide 406
Tego® Glide 410
Tego® Glide 470
Tego® Glide A 115
Tego® Glide B 1484
Tego® Glide ZG 400
Tego® Hammer 300000
Tegotens 4100
Tegotens BL 130
Tegotens BL 150
Tego® Wet KL 245
Tego® Wet ZFS 453
Tego® Wet ZFS 454
Tergitol® 15-S-7
Tergitol® 15-S-9
Tergitol® D-683
Tergitol® XD
Tergitol® XH
Tergitol® XJ
Teric 16A29
Teric 121
Teric 158
Teric 160
Teric 161
Teric BL8
Teric BL9
Teric N2
Teric N3
Teric N8
Teric N9
Teric N10
Teric N11
Teric N12
Teric N13
Teric N15
Teric N20
Teric N30
Teric N40
Teric N100
Teric PE61
Teric PE62
Teric PE64
Teric PE68
Teric PEG 200
Teric PEG 300
Teric PEG 400
Teric PEG 1500
Teric PPG 400
Teric X5
Teric X7
Teric X8
Teric X10
Teric X11
Teric X13
Teric X16
Teric X40
Teric X40L
Tetronic® 50R1
Tetronic® 50R4
Tetronic® 50R8
Tetronic® 90R1
Tetronic® 110R1
Tetronic® 110R2
Tetronic® 110R7
Tetronic® 130R1
Tetronic® 130R2
Tetronic® 150R1
Tetronic® 150R4
Tetronic® 150R8
Tetronic® 304
Tetronic® 504
Tetronic® 701
Tetronic® 702
Tetronic® 704
Tetronic® 901
Tetronic® 904
Tetronic® 908
Tetronic® 909
Tetronic® 1101
Tetronic® 1102
Tetronic® 1104
Tetronic® 1107
Tetronic® 1301
Tetronic® 1302

Paints, Coatings, Lacquers, Inks, Adhesives (cont'd.)

Tetronic® 1304
Tetronic® 1307
Tetronic® 1502
Tetronic® 1504
Tetronic® 1508
Texapon Z
Textamine O-1
TO-55-E
TO-55-EL
Toho Me-PEG Series
Toho PEG Series
Tolcide MBT
Tomah PA-14 Acetate
Trans-109
Trans-134S
Trans-138
Trans-RC1
Triton® GR-5M
Triton® GR-7M
Triton® XL-80N
Triumphnetzer ZSG
Troykyd® Anti-Gel
Troykyd® D44
Troykyd® D55
Troykyd® D126
Troykyd® D333
Troykyd® D666
Troykyd® D777
Troykyd® D999
Troykyd® Emulso Wet
Troysol 98C
Troysol 307
Troysol AFL
Troysol CD1
Troysol CD2
Troysol LAC
Troysol Q148
Troysol S366
Trydet DO-9
Trydet LA-5
Trylon® 5900
Trylon® 5902
Trylon® 5906
Trylon® 5909
Trylon® 5921
Trylon® 5922
Trylox® 5900
Trylox® 5906
Trylox® CO-5
Trylox® CO-30
Tylose CBR Grades
Tylose MHB
Tylose MH Grades
Tylose MH-K, MH-xp, MHB-y, MHB-yp
Ucar® Acrylic 503
Ucar® Acrylic 505
Ucar® Acrylic 515
Ucar® Acrylic 516
Ucar® Acrylic 518
Ucar® Acrylic 522
Ucar® Acrylic 525
Ufarol Am 30
Ufarol Na-30
Ufaryl DB80
Union Carbide® L-45 Series
Varine C
Varine O®
Varonic® T202
Viscasil®
Vorite 105
Vorite 110
Vorite 115
Vorite 120
Vorite 125
Witcamide® 5138
Witcamide® MAS
Witco® 912
Witco® 915
Witco® 916
Witco® 918
Witco® 934
Witco® 936
Witco® 942
Witco® 960
Witconate AOS-EP
Witconate TAB
Witconol NP-300
Witflow 60
Witflow 100
Witflow 901
Witflow 902
Witflow 910
Witflow 912
Witflow 914
Witflow 914
Witflow 916
Witflow 918
Witflow 930
Witflow 950
Witflow 953
Witflow 977
Witflow 979
Witflow 990
Witflow 991
Zonyl® FSA
Zonyl® FSB
Zonyl® FSC
Zonyl® FSJ
Zonyl® FSN
Zonyl® FSP

Paper and Pulp Processing

Ablunol NP 9
Ablunol NP 30
Ablunol NP 30 70%
Ablunol NP 40
Ablunol NP 40 70%
Ablunol NP 50
Ablunol NP 50 70%
Acid Felt Scour
Actrasol 6092
Actrasol C-50, C-75, C-85
Actrasol CS-75
Actrasol OY-75
Actrasol PSR
Actrasol SBO
Actrasol SP
Actrasol SP 175K
Actrasol SR 75
Actrasol SRK 75
Adol® 66
Adol® 80
Adol® 520 NF
Adol® 620 NF
Advantage 5 Defoamer
Advantage 6 Defoamer
Advantage 7 Defoamer
Advantage 10 Defoamer
Advantage 52-B
Advantage 52EH Defoamer
Advantage 52-JS
Advantage 70DYX
Advantage 70PHE
Advantage 70WLH
Advantage 91WW
Advantage 136 Defoamer
Advantage 136Z Defoamer
Advantage 187 Defoamer
Advantage 187Z Defoamer
Advantage 344 Defoamer
Advantage 357 Defoamer
Advantage 388 Defoamer
Advantage 470A Defoamer
Advantage 491A Defoamer
Advantage 831 Defoamer
Advantage 833 Defoamer
Advantage 951 Defoamer
Advantage 1007B Defoamer
Advantage 1275PD
Advantage 1280PD
Advantage 1512 Defoamer
Advantage 5271 Production Aid
Advantage DF 110
Advantage DF 244
Advantage DF 285
Advantage Eff-101 Defoamer
Advantage M104 Defoamer
Advantage M133A Defoamer
Advantage M201 Defoamer
Advantage M1250 Defoamer
Advantage M1251 Production Aid
Aerosol® 18
Aerosol® 19
Aerosol® C-61
Aerosol® OT-75%
AF 75
AF GN-23
AF HL-40
AF HL-52
Afranil®
Ageflex FM-1
Agesperse 71
Agesperse 80
Ajidew A-100
Ajidew N-50
Ajidew SP-100
Alfol® 20+
Alfol® 22+
Alkamide® STEDA
Alkamuls® EL-719
Alkamuls® PSMS-20
Alkanol® WXN
Alkaterge®-C
Alkaterge®-E
Amergel® 100
Amergel® 200
Amergel® 500
Antarox® EGE 25-2
Antifoam 7800 New
Antifoam CM Conc.
Armid® 18
Armid® C
Armid® HT
Armid® O
Armul 910
Arnox 910
Aromox® C/12
Aromox® C/12-W
Arquad® 2C-75
Arquad® 2HT-75
Arquad® 12-50
Arquad® 16-50
Arquad® C-50
Arquad® T-27W
ASA
Atlas EMJ-2
Atlas EMJ-C
Avirol® SA 4106
Avitone® A
Blancol®
Blancol® N
Briquest® 543-45AS
Briquest® ADPA-60AW
Bubble Breaker® 259, 260, 613-M, 622, 730, 737, 746, 748, 900, 913, 917
Calcium Stearate, Regular
Carbopol® 934
Carsonon® L-985
Carsonon® LF-5
Carsonon® LF-46

Paper and Pulp Processing (cont'd.)

Carsonon® TD-10
Chemal BP 261
Chemal BP-262
Chemal BP-262LF
Chemal BP-2101
Chemax CO-5
Chemax CO-16
Chemax CO-25
Chemax CO-28
Chemax CO-30
Chemax CO-36
Chemax CO-40
Chemax CO-200/50
Chemax DF-30
Chemax DF-100
Chemax DNP-150/50
Chemax HCO-5
Chemax HCO-16
Chemax HCO-25
Chemax HCO-200/50
Chemax NP-1.5
Chemax NP-4
Chemax NP-6
Chemax NP-9
Chemax NP-10
Chemax NP-15
Chemax NP-20
Chemax NP-30
Chemax NP-30/70
Chemax OP-30/70
Chemquat 12-33
Chemquat 12-50
Chemquat 16-50
Chemquat C/33W
Chimin P40
Chimin P45
Cithrol A
Cithrol DGDL N/E
Cithrol DGDL S/E
Cithrol DGDO N/E
Cithrol DGDO S/E
Cithrol DGDS N/E
Cithrol DGDS S/E
Cithrol DGML N/E
Cithrol DGML S/E
Cithrol DGMO N/E
Cithrol DGMO S/E
Cithrol DGMS N/E
Cithrol DGMS S/E
Cithrol DPGML N/E
Cithrol DPGML S/E
Cithrol DPGMO S/E
Cithrol DPGMS N/E
Cithrol DPGMS S/E
Cithrol EGDL N/E
Cithrol EGDL S/E
Cithrol EGDO N/E
Cithrol EGDO S/E
Cithrol EGDS N/E
Cithrol EGDS S/E
Cithrol EGML N/E
Cithrol EGML S/E
Cithrol EGMO N/E
Cithrol EGMO S/E
Cithrol EGMR N/E
Cithrol EGMR S/E
Cithrol EGMS N/E
Cithrol EGMS S/E
Cithrol GDL N/E
Cithrol GDL S/E
CNC Antifoam 30-FG
CNC Antifoam 495, 495-M
CNC Defoamer 12, 34, 407
CNC Defoamer 69, 97
CNC Defoamer 229
CNC PAL 210 T
CNC PAL 1000
CNC Solv 809
CNC Wet CP
CNC Wet CP-X
Cosmopon BN
Fancol OA 95
Felton 3T
Fluorad FX-8
Foamaster DR-187
Foamkill® 30C
Foamkill® 30HP
Foamkill® 400A
Foamkill® 608
Foamkill® 614
Foamkill® 618 Series
Foamkill® 627
Foamkill® 628A
Foamkill® 639
Foamkill® 639L
Foamkill® 639P
Foamkill® 644 Series
Foamkill® 649
Foamkill® 649N, 649P
Foamkill® 652L
Foamkill® 660, 660F
Foamkill® 663J
Foamkill® 679
Foamkill® 684 Series
Foamkill® 684A
Foamkill® 684P
Foamkill® 700, 700 Conc.
Foamkill® 810F
Foamkill® 830
Foamkill® 830F
Foamkill® 836A
Foamkill® 852
Foamkill® 1001 Series
Foamkill® CMP
Foamkill® CP
Foamkill® FPF
Foamkill® MS Conc.
Foamkill® MS-1
Foamkill® NSP-1, NSP-3, NSP-4, NSP-5
Forbest VP 13
Gantrez® AN
Genapol® C-050
Genapol® C-080
Genapol® C-100
Genapol® C-150
Genapol® C-200
Genapol® O-020
Genapol® O-050
Genapol® O-080
Genapol® O-090
Genapol® O-100
Genapol® O-120
Genapol® O-150
Genapol® O-200
Genapol® O-230
Genapol® S-020
Genapol® S-050
Genapol® S-080
Genapol® S-100
Genapol® S-150
Genapol® S-200
Genapol® S-250
Genapol® X Grades
Geropon® 99
Geropon® T-43
Geropon® T-51
Geropon® TBS
Geropon® WT-27
Glycidol Surfactant 10G
GP-295 Defoamer
GP-300-I Antifoam Compd.
GP-310-I
Gradonic LFA Series
Harol RG-71L
Hartopol L64
Hartowet CW
Hercules® 4 Defoamer
Hercules® 137 Defoamer
Hercules® 187 Defoamer
Hercules® 388 Defoamer
Hercules® 491 Defoamer
Hercules® 492 Defoamer
Hercules® 831 Defoamer
Hercules® 845 Defoamer
Hercules® 1512 Defoamer
Hercules® 2051GS Defoamer
Hercules® 2470 Defoamer
Hercules® Eff-101 Defoamer
Hetoxol D
Hodag Amine C-100-L
Hodag Amine C-100-O
Hodag Amine C-100-S
Hyonic NP-90
Hyonic NP-110
Hyonic OP-100
Hyonic PE-90
Hyonic PE-100
Igepal® CA-620
Igepal® CO-530
Igepal® CO-630
Igepal® CO-660
Igepal® CO-710
Igepal® CO-720
Igepal® DM-710
Igepal® DM-730
Igepal® DM-970 FLK
Igepal® RC-520
Incroquat SDQ-25
Kelco® HV
Kelco® LV
Kelgin® F
Kelgin® HV, LV, MV
Kelgin® QL
Kelgin® XL
Kelset®
Kemamide® W-20
Kemamide® W-39
Kemamide® W-40
Kemamide® W-40/300
Kemamide® W-45
Kemester® 5500
Kemester® 6000SE
Kito 703
Klucel® E, G, H, J, L, M
Lamacit AP 6
Lanolin Alcohols LO
Lanquell 206, 217
Lipolan G
Lomar® PW
Lutensol® AP 6
Lutensol® AP 7
Lutensol® AP 8
Lutensol® AP 9
Mackester EGDS
Mackester EGMS
Mackester IDO
Mackester IP
Mackester SP
Mackester TD-88
Macol® 1
Macol® 2
Macol® 2D
Macol® 2LF
Macol® 4
Macol® 8
Macol® 15
Macol® 15-20
Macol® 18
Macol® 19
Macol® 20
Macol® 22
Macol® 27
Macol® 31
Macol® 32
Macol® 33
Macol® 34
Macol® 35
Macol® 40
Macol® 42
Macol® 44
Macol® 46
Macol® 72
Macol® 77
Macol® 85
Macol® 88
Macol® 101
Macol® 108
Macol® P-500
Macol® P-1200
Macol® P-1750
Macol® P-2000
Macol® P-3000
Macol® P-4000
Makon® 4
Makon® 6
Makon® 8

Paper and Pulp Processing (cont'd.)

Makon® 10
Makon® 12
Makon® 14
Makon® 30
Mapeg® CO-25
Mapeg® CO-25H
Marlican®
Marlinat® DF 8
Marlophen® DNP 16
Marlophen® DNP 18
Marlophen® DNP 30
Marlophen® DNP 100
Marlophen® DNP 150
Marlophor® F1-Acid
Marlophor® LN-Acid
Marlophor® N5-Acid
Marlophor® ND
Marlophor® ND-Acid
Marlophor® ND DEA Salt
Marlowet® 4857
Marlowet® 5401
Marlowet® 5600
Mazawet® 77
Mazawet® DOSS 70
Mazon® 43, 43LF
Mazu® DF 210SX
Mazu®DF210SX Mod 1
Mazu® DF 210SXSP
Mazu® DF 243
Merpol® A
Merpol® C
Merpol® HCS
Merpol® SE
Merpol® SH
Monafax 1214
Monafax 1293
Monawet MB-45
Monawet MM-80
Monosulf
Mulsifan RT 110
Mulsifan RT 258
Nacconol® 40G
Nacconol® 90G
Nansa® BXS
Nansa® HS80S
Nansa® HS85S
Nansa® S40/S
Naxide 1230
Neoscoa 203C
Neoscoa 363
Neoscoa 500C
Neoscoa 2326
Neoscoa CM-40
Neoscoa CM-57
Neoscoa ED-201C
Neoscoa FS-100
Neoscoa GF 3C
Neoscoa GF-2000
Neoscoa MSC-80
Neoscoa OT-80E
Neoscoa PRA-8C
Neoscoa SS-10
Neoscoa TH-102
Nissan Cation AB
Nissan Disfoam C Series
Nissan Plonon 102
Nissan Plonon 104
Nissan Plonon 108
Nissan Plonon 171
Nissan Plonon 172
Nissan Plonon 201
Nissan Plonon 204
Nissan Plonon 208
Nonatell 1002
Nonatell 1003
Nopalcol 1-L
Nopalcol 2-DL
Nopalcol 4-C
Nopalcol 4-CH
Nopalcol 4-L
Nopalcol 4-O
Nopalcol 4-S
Nopalcol 6-DO
Nopalcol 6-DTW
Nopalcol 6-L
Nopalcol 6-R
Nopalcol 6-S
Nopalcol 10-COH
Nopalcol 12-CO
Nopalcol 12-COH
Nopalcol 19-CO
Nopalcol 30-TWH
Nopalcol 200
Nopalcol 400
Nopalcol 600
Nopalcol Series
Nopco® 2272-R
Nopcogen 22-O
Norfox® 90
Norfox® 916
Norfox® DCS
NP-55-90
Oakite Defoamant
Oakite Defoamant RC
Octowet 40
Octowet 55
Octowet 60
Octowet 60-I
Octowet 65
Octowet 70
Octowet 70A
Octowet 70BC
Octowet 70PG
Octowet 75
Octowet 75E
Ogtac-85
Oxetal 500/85
Oxetal D 104
Oxetal ID 104
Oxetal O 108
Oxetal O 112
Oxetal TG 111
Oxetal TG 118
Pegosperse® 400 DOT
Pegosperse® 600 DOT
Pepol BS-184
Pepol BS-403
Peregal® ST
Petrowet® R
PGE-400-MS
Pluracol® E400 NF
Pluracol® E600
Pluracol® E1450
Pluracol® E1450 NF
Pluracol® E2000
Pluracol® E4000 NF
Pluracol® E4500
Pluracol® E8000 NF
Pluracol® P-410
Pluracol® P-710
Pluracol® P-1010
Pluracol® P-2010
Pluracol® P-3010
Pluracol® P-4010
Pluriol® E 200
Pluriol® E 1500
Pluriol® E 4000
Pluriol® E 6000
Pluriol® E 9000
Pluronic® 10R5
Pluronic® 10R8
Pluronic® 12R3
Pluronic® 17R1
Pluronic® 17R2
Pluronic® 17R4
Pluronic® 17R8
Pluronic® 22R4
Pluronic® 25R2
Pluronic® 25R4
Pluronic® 25R5
Pluronic® 31R2
Pluronic® 31R4
Pluronic® F127
Pluronic® F38
Pluronic® F68
Pluronic® F68LF
Pluronic® F77
Pluronic® F87
Pluronic® F88
Pluronic® F98
Pluronic® L10
Pluronic® L31
Pluronic® L35
Pluronic® L42
Pluronic® L43
Pluronic® L44
Pluronic® L61
Pluronic® L62
Pluronic® L62D
Pluronic® L62LF
Pluronic® L63
Pluronic® L64
Pluronic® L72
Pluronic® L81
Pluronic® L92
Pluronic® L101
Pluronic® L121
Pluronic® L122
Pluronic® P65
Pluronic® P75
Pluronic® P84
Pluronic® P85
Pluronic® P94
Pluronic® P103
Pluronic® P104
Pluronic® P105
Pluronic® P123
Pogol 600
Poly-G® 200
Poly-G® 300
Poly-G® 400
Poly-G® 600
Poly-G® 1000
Poly-G® 1500
Poly-G® 2000
Poly-G® B1530
Poly-Tergent® 4C3
Poly-Tergent® E-17A
Poly-Tergent® E-17B
Poly-Tergent® E-25B
Poly-Tergent® J-200
Poly-Tergent® J-300
Poly-Tergent® P-22A
Poly-Tergent® P-32A
Poly-Tergent® S-205LF
Polywet ND-1
Polywet ND-2
Product BCO
Pronal 502
Pronal 502A
Pronal 2200
Pronal 3300
Propetal 241
Prox-onic DNP-08
Prox-onic DNP-0150
Prox-onic DNP-0150/50
Prox-onic EP 1090-1
Prox-onic EP 1090-2
Prox-onic EP 2080-1
Prox-onic EP 4060-1
Prox-onic HR-05
Prox-onic HR-016
Prox-onic HR-025
Prox-onic HR-030
Prox-onic HR-036
Prox-onic HR-040
Prox-onic HR-0200/50
Prox-onic HRH-05
Prox-onic HRH-016
Prox-onic HRH-025
Prox-onic HRH-0200
Prox-onic HRH-0200/50
Prox-onic NP-1.5
Prox-onic NP-010
Prox-onic PEG-2000
Prox-onic PEG-4000
Prox-onic PEG-10,000
Prox-onic PEG-20,000
Prox-onic TBP-08
Prox-onic TBP-030
Prox-onic TD-1/03
Prox-onic TD-1/06
Prox-onic TD-1/09
Prox-onic TD-1/012
Radiasurf® 7000
Radiasurf® 7125
Radiasurf® 7146
Radiasurf® 7151
Radiasurf® 7175
Rewopal® HV 5
Rewopal® HV 9
Rewopal® HV 25
Rewopol® B 2003
Rewopol® SBDO 70
Rewopol® SMS 35
Rexfoam D
Rexol 25/8
Rexol 25/10
Rexol 25J
Rexol 25JM1
Rexol 45/10

Paper and Pulp Processing (cont'd.)

Rexonic 1012-6
Rexonic 1218-6
Rexonic L125-9
Rexonic N23-3
Rexonic N23-6.5
Rexonic N25-7
Rexonic N25-9
Rexonic N25-9(85%)
Rexonic N25-14(85%)
Rhodacal® BX-78
Rhodafac® RS-610
Rhodamox® CAPO
Rhodamox® LO
Rhodasurf® E 400
Rhodasurf® L-4
Rhodasurf® L-25
Rhodasurf® TDA-8.5
Rycofax® 618
Rycofax® O
Schercozoline I
Schercozoline O
Serdas GLN
Serdox NNP 10
Serdox NNP 11
Serdox NOP 9
Servirox OEG 45
Servirox OEG 55
Servirox OEG 65
Servirox OEG 90/50
Silicone AF-10 FG
Silwet® L-77
Silwet® L-720
Silwet® L-7001
Silwet® L-7500
Siponic® L-12
Siponic® NP-9
Siponic® NP-15
Standamul® B-3
Sul-fon-ate AA-9
Sul-fon-ate AA-10
Sulfotex OA
Superloid®
Surfonic® JL-80X
Surfonic® LF-17
Surfynol® 61
Surfynol® 104BC
Surfynol® 104H
Surfynol® 104PA
Surfynol® 104PG
Surfynol® 420
Surfynol® 440
Surfynol® 485
Surfynol® DF-75
Surfynol® DF-110D
Surfynol® DF-110L
Surfynol® PC
Surfynol® PG-50
Surfynol® SE
Surfynol® SE-F
Surfynol® TG
Surmax® CS-504
Surmax® CS-515
Surmax® CS-521
Surmax® CS-555
Surmax® CS-586
Surpasol E-436
Sylvaros® 20
Sylvaros® 80
Synperonic OP 7.5
Synperonic OP8
Synperonic OP10
Synperonic OP10.5
Synperonic OP11
Syntens KMA 55
Syntergent K
T-Det® C-40
T-Det® DD-10
T-Det® EPO-61
T-Det® EPO-62L
T-Det® EPO-64L
T-Det® N-6
T-Det® N-8
T-Det® N-9.5
T-Det® RQ1
T-Det® RY2
Tergitol® 15-S-3
Tergitol® 15-S-5
Tergitol® 15-S-7
Tergitol® 15-S-9
Tergitol® 15-S-15
Tergitol® 24-L-45
Tergitol® 24-L-60
Tergitol® 24-L-60N
Tergitol® 24-L-92
Tergitol® 24-L-98N
Tergitol® 26-L-5
Tergitol® Min-Foam 1X
Tergitol® Min-Foam 2X
Tergitol® TMN-3
Tergitol® TMN-10
Teric 12A7
Teric 12M2
Teric 12M5
Teric 12M15
Teric 13A7
Teric 13A9
Teric 18M2
Teric 18M5
Teric 18M10
Teric 18M20
Teric 18M30
Teric 121
Teric 124
Teric 127
Teric 128
Teric OF4
Teric OF6
Teric OF8
Teric PE61
Teric PE62
Teric PE68
Teric PEG 200
Teric PEG 400
Teric PEG 600
Teric PEG 1500
Teric PEG 8000
Tetronic® 50R1
Tetronic® 50R4
Tetronic® 50R8
Tetronic® 70R1
Tetronic® 70R2
Tetronic® 70R4
Tetronic® 90R1
Tetronic® 90R4
Tetronic® 90R8
Tetronic® 110R1
Tetronic® 110R2
Tetronic® 110R7
Tetronic® 130R1
Tetronic® 130R2
Titadine TA
Titadine TCP
Titan Decitrene
Titanole RMA
Titazole SA
Toho Me-PEG Series
Toho PEG Series
Tolcide MBT
Tonerclean 208, 209
Trans-135
Trans-137
Trans-149
Trans-266
Trans-N1
Trans-RC1
Trilon® A
Triton® DF-16
Triton® XL-80N
Troykyd® D666
Troykyd® D999
Trycol® 5941
Trycol® 5944
Trycol® TDA-9
Trydet 2676
Trydet LA-5
Trylox® 5900
Trylox® 5906
Trylox® CO-5
Tylose C, CB Series
Tylose CBR Grades
Ufaryl DB80
Ufaryl DL80 CW
Ufaryl DL85
Ufaryl DL90
Unidri M-40
Varonic® K210
Wayfos D-10N
Wayfos M-100
Witcamide® MAS
Zonyl® A
Zonyl® FSA
Zonyl® FSB
Zonyl® FSC
Zonyl® FSN
Zonyl® FSP
Zonyl® NF
Zonyl® RP
Zusolat 1008/85
Zusolat 1010/85
Zusolat 1012/85

Petroleum, Oil and Gas Processing

Ablunol NP 4
Actrasol MY-75
AD-749
AD-750, AD-750C
ADF-600
ADM-407
ADM-407C
ADM-408
ADM-408C
ADM-409
ADM-409C
ADM-410
ADM-410C
ADM-411
ADM-412
ADM-412C
ADM-456
ADM-456C
ADM-457
ADM-457C
ADM-458
ADM-467
ADM-477
ADM-477C
ADM-487
ADM-487C
Adma® 18
Adogen® 461
Adogen® 462
Adogen® 464
Adogen® 471
Adol® 42
Adol® 61 NF
Adol® 62 NF
Adol® 66
Adol® 80
Adol® 520 NF
Adol® 620 NF
Aerosol® OT-75%
AF 10 IND
AF 70
AF-800
AF-801
AF-802
AF 9000
AF GN-11-P
Akypo LF 4
Alfonic® 610-50R
Alfonic® 810-40
Alfonic® 810-60
Alkasurf® NP-1
Alkasurf® NP-20, NP-20 70%
Alkenyl Succinic Anhydrides
Ameenex 70 WS
Ameenex 73 WS
Ameenex C-18
Ameenex C-20
Ameenex Polymer
Amidex CO-1
Anedco DF-6002
Anedco DF-6031
Anedco DF-6130

Petroleum, Oil and Gas Processing (cont'd.)

Anedco DF-6131
Anedco DF-6231
Anedco DF-6233
Anedco DF-6300
Antarox® 17-R-2
Antarox® 25-R-2
Antarox® 31-R-1
Antarox® F88 FLK
Antarox® L-61
Antarox® L-62
Antarox® L-62 LF
Antarox® L-64
Antarox® P-84
Antarox® P-104
Aquafoam 9451
Aquafoam 9452
Aquamul Series
Ardril DMD
Ardril DME
Ardril DMS
Arfoam 2213
Arfoam 2386
Aristonate H
Aristonate L
Aristonate M
Arizona DR-22
Armeen® O
Armid® O
Armul 1007
Armul 1009
Arnox 500, 515
Arnox BP Series
Aromox® C/12
Aromox® C/12-W
Arquad® 2C-75
Arquad® 12-50
Arquad® 16-50
Arquad® 18-50
Arquad® B-100
Arquad® C-50
Arquad® DMCB-80
Arquad® S-2C-50
Arquad® T-2C-50
Arquad® T-27W
Arstim RRC
Atlosol Series
Atpet 545
Atpet 787
Atpet 900
Barium Petronate 50-S Neutral
Barium Petronate Basic
Base ML
Base MO
Base MT
Bio-Soft® EA-8
Biozan
Briquest® 301-50A
Briquest® 543-45AS
Bubble Breaker® 3073-7, D
Carbonox
Carbopol® 907
Catigene® 1011
Catigene® CETAC 30
Catinex KB-11
Catinex KB-15
Catinex KB-31
Catinex KB-32
Chembetaine BC-50
Chemfac RD-1200
Chemphonate 22
Chemphonate AMP-S
Chemphonate HEDP
Chemphonate N
Chemphonate NP
Chemphos TC-227
Chemphos TC-310S
Chemphos TR-414W
Chemzoline 1411
Chemzoline C-22
Chemzoline T-11
Chemzoline T-44
Crodasinic O
Cromul 1540
Fancol OA 95
Fluorad FC-760
Gantrez® AN-139
Gantrez® AN-149
Gantrez® AN-169
Gantrez® AN-179
Gardiquat 12H
Genapol® UD-030
Genapol® UD-050
Genapol® UD-079
Genapol® UD-080
Genapol® UD-088
Genapol® UD-110
Geropon® DOS
Geropon® DOS FP
Geropon® TK-32
Geropon® X2152
Hartasist 16
Hartasist 20, 37
Hartasist DF-28
Hartbreak Series
Hodag SML
Hodag SMO
Hodag SMP
Hodag STO
Hodag STS
Hostamer Brands
Hybase® C-300
Hybase® C-400
Hybase® C-500
Hybase® M-12
Hybase® M-300
Hybase® M-400
Hydroace Series
Hyonic NP-40
Hyonic OP-40
Hystrene® 3675
Hystrene® 3675C
Hystrene® 3680
Hystrene® 3687
Hystrene® 3695
Igepal® CO-430
Igepal® CO-520
Igepal® CO-530
Igepal® CO-850
Igepal® CO-880
Igepal® CO-887
Igepal® DM-970 FLK
Indulin® W-1
Industrene® D
Inipol S 43
Jet Amine DC
Jet Amine DO
Jet Amine DT
Jet Amine PC
Jet Amine PCD
Jet Amine PE 08/10
Jet Amine PHT
Jet Amine PHTD
Jet Amine PO
Jet Amine POD
Jet Amine PS
Jet Amine PSD
Jet Amine TET
Jet Amine TRT
Jet Quat 2C-75
Jet Quat 2HT-75
Jet Quat C-50
Jet Quat DT-50
Jet Quat S-2C-50
Jet Quat S-50
Jet Quat T-27W
Jet Quat T-50
Kemamide® B
Kemamide® E
Kemamide® O
Kemamide® S
Kemamide® U
Kemamide® W-20
Kemamide® W-39
Kemamide® W-40
Kemamide® W-40/300
Kemamine® D-190
Kemamine® D-650
Kemamine® D-970
Kemamine® D-974
Kemamine® D-989
Kemamine® D-999
Kemamine® P-150, P-150D
Kemamine® P-190, P-190D
Kemamine® P-650D
Kemamine® P-880, P-880D
Kemamine® P-970
Kemamine® P-970D
Kemamine® P-989D
Kemamine® P-990, P-990D
Kemamine® P-999
Kemamine® T-1902D
Kemamine® T-6501
Kemamine® T-6502D
Kemamine® T-9701
Kemamine® T-9702D
Kemamine® T-9742D
Kemamine® T-9902
Kemamine® T-9902D
Kemamine® T-9972D
Kemamine® T-9992D
Lankro Mud-Aids
Lankro Mud-Emuls
Lankromul OSD
Latol MOD
Latol MTO
Laurex® 810
Lonzest® PEG 4-DO
Lubran FPD Series
Makon® 4
Makon® 6
Makon® 7
Makon® 8
Makon® 10
Makon® 11
Makon® 12
Makon® 14
Makon® 30
Makon® NF-5
Makon® NF-12
Masil® SF 5
Masil® SF 10
Masil® SF 20
Masil® SF 50
Masil® SF 100
Masil® SF 200
Masil® SF 350
Masil® SF 350 FG
Masil® SF 500
Masil® SF 1000
Masil® SF 5000
Masil® SF 10,000
Masil® SF 12,500
Masil® SF 60,000
Masil® SF 100,000
Masil® SF 300,000
Masil® SF 600,000
Masil® SF 1,000,000
Mazamide® L-298
Mazu® DF 200SX
Mazu® DF 200SXSP
Mazu® DF 210SX
Mazu® DF 210SX Mod 1
Mazu® DF 210SXSP
Mazu® DF 230SX
Mazu® DF 230SXSP
Mazu® DF 243
Merpol® A
Mona AT-1200
Monamulse 748
Monaquat AT-1074
Monateric ADA
Monateric COAB
Monateric CyA-50
Monawet MO-65-150
Monazoline C
Nansa® EVM62/H
Nansa® EVM70
Nansa® LSS38/A
Naxchem Dispersant K
Naxel AAS-Special 3
Naxonic NI-40
Naxonic NI-60
Nissan Amine DT
Nissan Amine DTH
Nissan Cation AB
Nonisol 210
Noram 2C
Noram 2SH
Noram DMC
Noram DMS
Noram M2SH
Noram S
Norfox® F-221
Norfox® SLES-60
Oakite Defoamant

Petroleum, Oil and Gas Processing (cont'd.)

Ofax® Series
Onyxide® 200
Peltex
Petronate® 25C, 25H
Petronate® Basic
Petronate® CR
Petronate® HL
Petronate® L
Petronate® Neutral (50-S)
Petronate® S
Petrostep 420
Petrostep 465
Petrostep HMW
Petrostep LMC
Petrostep MMW
Petrosul® H-50
Petrosul® H-60
Petrosul® H-70
Petrosul® HM-62
Petrosul® HM-70
Petrosul® L-60
Petrosul® M-50
Petrosul® M-60
Petrosul® M-70
Petrowet® R
Phospholipid PTD
Phospholipid PTL
Phospholipid PTZ
Pluriol® PE 6200
Pluronic® 10R5
Pluronic® 10R8
Pluronic® 12R3
Pluronic® 17R1
Pluronic® 17R4
Pluronic® 17R8
Pluronic® 22R4
Pluronic® 25R1
Pluronic® 25R2
Pluronic® 25R5
Pluronic® 31R1
Pluronic® 31R4
Pluronic® L10
Poly-G® 200
Poly-G® 300
Poly-G® 400
Poly-G® 600
Poly-G® 1000
Poly-G® 1500
Poly-G® 2000
Poly-G® B1530
Pronal EX-100
Pronal EX-300
Prox-onic EP 1090-1
Prox-onic EP 1090-2
Prox-onic EP 2080-1
Prox-onic EP 4060-1
Prox-onic SML-020
Prox-onic SMO-05
Prox-onic SMO-020
Prox-onic SMP-020
Prox-onic SMS-020
Pyronate® 40
Quatrene 7670
Quatrene CE
Quatrex 152
Quatrex 162
Quatrex 172
Quatrex 182
Querton KKBCl-50
Rewopal® HV 4
Rewopal® HV 6
Rewopal® HV 8
Rewopol® CT 65
Rexfoam D
Rexol 25/6
Rexol 25/407
Rexol 45/3
Rhodafac® MD-12-116
Rhodapex® CO-433
Rhodapex® CO-436
Rhodasurf® BC-420
Rhodasurf® BC-610
Rhodasurf® BC-630
Rhodorsil® AF 454
Rhodorsil® AF 20432
Seapol-06
Secosol® DOS 70
Secoster® A
Secoster® DO 600
Secoster® MO 400
Sellig N 6 100
Sellig N 8 100
Sepabase R Grades
Separol AF 27
Separol WF 22
Separol WF 34
Separol WF 41
Separol WF 221
Sepawet®
Servo CK 492
Servo CK 494
Servo CK 601
Silwet® L-7001
Silwet® L-7500
Silwet® L-7602
Silwet® L-7607
Sorbax HO-50
Sorbax PMO-5
Sorbax PMP-20
Sorbax PMS-20
Sorbax PTO-20
Sorbax PTS-20
Sorbax SML
Sorbeth 40HO
Stepanflo 20
Stepanflo 30
Stepanflo 40
Stepanflo 50
Stepantan® H-100
Steposol® CA-207
Sul-fon-ate AA-9
Sul-fon-ate AA-10
Sulfochem RF
Sulfotex PAI
Sulfotex PAI-S
Sulfotex PAW
Sulfotex RIF
Sulfotex SAL, SAT
Sulframin 40
Surfactol® 13
Surfam P10
Surfam P12B
Surfam P14B
Surfam P17B
Surfam P24M
Surfam P86M
Surfam P89MB
Surflo® S32
Surflo® S40
Surflo® S41
Surflo® S242
Surflo® S302
Surflo® S412
Surflo® S568
Surflo® S586
Surflo® S596
Surflo® S970
Surflo® S1143
Surflo® S1259
Surflo® S2023
Surflo® S2024
Surflo® S2181
Surflo® S2221
Surflo® S7650
Surflo® S7655
Surflo® S8536
Surfonic® HDL-1
Surfonic® N-10
Surfonic® N-40
Surfonic® N-60
Surfonic® N-85
Surfonic® N-95
Surfonic® N-100
Surfonic® N-102
Surfonic® N-120
Surfonic® N-130
Surfonic® N-150
Surfonic® N-200
Surfonic® N-300
Surfonic® NB-5
Surfynol® 420
Surfynol® 485
Surpasol 53
Surpasol NT-57
Sylfam 2042
Sylfam 2082
Sylfan 20
Sylfan 96
Sylfat® DX
Sylfat® MM
Sylfat® RD-1
Sylvacote® K
Syntopon H
Tallow Amine
Tallow Diamine
Tallow Tetramine
Tallow Triamine
T-Det® EPO-61
T-Det® N 1.5
T-Det® N-14
T-Det® N-20
T-Det® O-8
Tego® Antifoam
Tergitol® 15-S-3
Tergitol® 15-S-5
Tergitol® 15-S-7
Tergitol® 15-S-9
Tergitol® 15-S-15
Tergitol® 15-S-20
Tergitol® 15-S-30
Tergitol® 15-S-40
Teric 17A2
Teric 17A3
Teric 17A6
Teric 154
Teric PE61
Teric PE62
Teric PPG 1650
Teric PPG 2250
Teric PPG 4000
Tetronic® 130R1
Tetronic® 304
Tetronic® 504
Tetronic® 701
Tetronic® 702
Tetronic® 704
Tetronic® 707
Tetronic® 901
Tetronic® 904
Tetronic® 908
Tetronic® 909
Tetronic® 1101
Tetronic® 1102
Tetronic® 1104
Tetronic® 1107
Tetronic® 1301
Tetronic® 1302
Tetronic® 1304
Tetronic® 1307
Tetronic® 1501
Tetronic® 1502
Tetronic® 1504
Tetronic® 1508
T-Mulz® 565
T-Mulz® 1158
Tomah DA-14
Tomah DA-16
Tomah E-19-2
Tomah E-DT-3
Tomah E-T-2
Tomah E-T-5
Tomah E-T-15
Tomah PA-10
Tomah PA-14
Tomah PA-16
Tomah PA-19
Tomah PA-1214
Trans-100
Triton® N-60
Triton® RW-20
Triton® RW-50
Triton® RW-75
Triton® RW-100
Triton® RW-150
Triton® XL-80N
Trycol® 6975
Trylox® 1086
Trylox® 1186
Ufasan 35
Ufasan 50
Ufasan 60A
Ufasan 62B
Ufasan TEA
Union Carbide® L-45 Series
Varonic® K202
Varonic® K215
Varonic® T215
Varonic® T215LC
Viscasil®
Witbreak 770

Petroleum, Oil and Gas Processing (cont'd.)

Witbreak 772
Witbreak 774
Witbreak 1390
Witbreak DGE-75
Witbreak DPG-15
Witbreak DPG-482
Witbreak DRA-21
Witbreak DRA-22
Witbreak DRA-50
Witbreak DRB-11
Witbreak DRB-127
Witbreak DRB-401
Witbreak DRC-163
Witbreak DRC-164
Witbreak DRC-165
Witbreak DRC-168
Witbreak DRC-232
Witbreak DRE-8164
Witbreak DRI-9020
Witbreak DRI-9037
Witbreak DRI-9038
Witbreak DTG-62
Witbreak GBG-3172
Witbreak RTC-323
Witbreak RTC-326
Witbreak RTC-330
Witbreak RTC-1315
Witbreak RTC-1316
Witbreak RTC-6010
Witbreak RTC-6020
Witbreak RTC-6030, -6040
Witcamide® 5138
Witcamine® 209
Witcamine® 210
Witcamine® 211
Witcamine® 235
Witcamine® 240
Witcamine® 3164
Witcamine® PA-60B
Witcamine® RAD 0500
Witcamine® RAD 1100
Witcamine® RAD 1110
Witco® DTA-150, -180
Witco® DTA-350
Witcolate 1259
Witcolate 1276
Witcolate 1390
Witcolate 3220
Witcolate 7103
Witcomul 1557
Witcomul 3107
Witcomul 3126
Witcomul 3154
Witcomul 3158
Witcomul 3230
Witcomul 4016
Witconate 605A
Witconate 702
Witconate 703
Witconate 705
Witconate 3203
Witconate AOS
Witconate CHB
Witconate P10-59
Witconate P-1073F
Witconate SE-5
Witcor 3192
Witcor 3194, 3195
Witcor 3635
Witcor CI-1
Witcor CI-6
Witcor PC100
Witcor PC200
Witcor PC205
Witcor SI-3065

Plastics, Rubbers and Resin Manufacture, Emulsion Polymerization, Latex Processing

AA Standard
Abex® 12S
Abex® 18S
Abex® 22S
Abex® 23S
Abex® 26S
Abex® 33S
Abex® 1404
Abex® AAE-301
Abex® EP-110
Abex® EP-115
Abex® EP-120
Abex® JKB
Abex® VA 50
Abil® B 9806, B 9808
Ablumine 08
Ablumine 10
Ablumine 1214
Ablumox T-15
Ablumox T-20
Ablunol GML
Ablunol GMO
Ablunol NP 30
Ablunol NP 30 70%
Ablunol NP 40
Ablunol NP 40 70%
Ablunol NP 50
Ablunol NP 50 70%
Ablusol C-78
Ablusol DA
Ablusol M-75
Acconon 300-MO
Acconon 400-MO
Acintol® 736
Acintol® 746
Acintol® EPG
ACtone® 1
ACtone® 2000V
ACtone® 2010, 2010P
ACtone® 2461
ACtone® N
Adma® 18
Adol® 62 NF
Adol® 63
Adol® 66
Adol® 85 NF
Adol® 630
Adol® 640
Aerosol® 18
Aerosol® 19
Aerosol® 501
Aerosol® A-102
Aerosol® A-103
Aerosol® A-196-40
Aerosol® A-196-85
Aerosol® A-268
Aerosol® C-61
Aerosol® DPOS-45
Aerosol® IB-45
Aerosol® MA-80
Aerosol® NPES 458
Aerosol® NPES 930
Aerosol® NPES 2030
Aerosol® NPES 3030
Aerosol® OT-75%
Aerosol® OT-100%
Aerosol® OT-B
Aerosol® OT-S
Aerosol® TR-70
AF 75
AF 93
AF 9000
AF 9020
AF HL-27
Ageflex FM-1
Agesperse 71
Agesperse 80
Agesperse 81
Agesperse 82
Akypo OP 80
Akypo RLMQ 38
Akyporox NP 1200V
Akyporox OP 400V
Akyporox RTO 70
Akyporox SAL SAS
Akyposal ALS 33
Akyposal BA 28
Akyposal BD
Akyposal EO 20 MW
Akyposal NPS 60
Akyposal NPS 100
Akyposal NPS 250
Alfol® 16
Alfol® 18
Alkamide® STEDA
Alkamuls® EL-719
Alkamuls® PSML-4
Alkamuls® R81
Alkamuls® SML
Alkanol® 189-S
Alkasil® NE 58-50
Alkasil® NEP 73-70
Alkasurf® NP-4
Alkasurf® NP-6
Alkasurf® OP-30, 70%
Alkasurf® OP-40, 70%
Alphenate TH 454
Alrosol Conc.
Amine C
AMP
AMP-95
Anti-Terra®-204
Anti-Terra®-207
Anti-Terra®-P
Antifoam 7800 New
Arizona DR-24
Arizona DR-25
Arizona DRS-40
Arizona DRS-42
Arizona DRS-43
Arizona DRS-44
Arizona DRS-50
Arizona DRS-51E
Arkofix Grades
Arkopon Brands
Armeen® 18
Armeen® 18D
Armeen® DM8
Armeen® DM10
Armeen® DM12
Armeen® DM14
Armeen® DM16
Armeen® DMC
Armeen® DMHT
Armeen® DMO
Armeen® DMT
Armeen® M2-10D
Armeen® Z
Armid® 18
Armid® C
Armid® E
Armid® HT
Armid® O
Armul 906
Armul 930
Armul 940
Arnox 906
Arnox 930, 940
Aromox® C/12
Aromox® C/12-W
Arquad® 2C-75
Arquad® 12-50
Arquad® 16-50
Arquad® 18-50
Arquad® C-50
Arquad® T-2C-50
Arquad® T-27W
Arylan® SBC25
Arylan® SBC Acid
Arylan® SC15
Arylan® SC30
Arylan® SC Acid
Arylan® SO60 Acid
Arylan® SP Acid
Arylan® SX85
Arylan® SY30
Arylan® SY Acid
Arylan® TE/C
Astrowet 102
Atsolyn PE 27
Avirol® 125 E
Avirol® A

Plastics, Rubbers, and Resin Manufacture (cont'd.)

Avirol® AE 3003
Avirol® AOO 1080
Avirol® FES 996
Avirol® SA 4106
Avirol® SA 4110
Avirol® SA 4113
Avirol® SE 3002
Avirol® SE 3003
Avirol® SL 2010
Avirol® SL 2015
Avirol® SL 2020
Avirol® T 40
Avitone® A
Berol 02
Berol 28
Berol 108
Berol 190
Berol 191
Berol 198
Berol 269
Berol 277
Berol 278
Berol 281
Berol 291
Berol 292
Berol 295
Berol 302
Berol 303
Berol 307
Berol 374
Berol 381
Berol 386
Berol 387
Berol 389
Berol 391
Berol 392
Berol 397
Berol 452
Berol 455
Berol 456
Berol 457
Berol 458
Berol 752
Beycostat 656 A
Beycostat LA
Beycostat LP 9 A
Beycostat LP 12 A
Beycostat NA
Beycostat NE
BYK®-024
BYK®-066
BYK®-070
Calcium Stearate, Regular
Calfoam ES-30
Calfoam NLS-30
Calfoam SLS-30
Capcure Emulsifier 37S
Capcure Emulsifier 65
Carbopol® 934
Cardolite® NC-507
Cardolite® NC-510
Carsonol® ALS-S
Carsonol® SLS-R
Carsonol® SLS-S
Carsonol® TLS
Carsonon® N-30
Catalyst NKS
Catalyst PAT
Catinex KB-44
Catinex KB-45
Catinex KB-50
Cedepal CA-890
Cedepal CO-436
Cedephos® FA600
Cedephos® RA600
Chemal DA-4
Chemal DA-6
Chemal DA-9
Chemax DOSS/70
Chemax DOSS-75E
Chemax E-200 MO
Chemax E-200 MS
Chemax E-400 MO
Chemax E-400 MS
Chemax E-600 ML
Chemax E-600 MO
Chemax HCO-5
Chemax HCO-16
Chemax HCO-25
Chemax HCO-200/50
Chemax NP-40
Chemax NP-40/70
Chemax NP-50
Chemax NP-50/70
Chemax NP-100
Chemax NP-100/70
Chemax OP-40/70
Chemeen 18-2
Chemeen 18-5
Chemeen 18-50
Chemeen HT-5
Chemeen HT-50
Chemfac 100
Chemfac NC-0910
Chemfac PC-099E
Chemfac PC-188
Chemfac PD-600
Chemfac PF-623
Chemfac PF-636
Chemfac PN-322
Chemphos TC-227
Chemphos TC-310
Chemphos TC-337
Chemquat 12-33
Chemquat 12-50
Chemquat 16-50
Chemquat C/33W
Chimin P1A
Chimin P45
Chimipal APG 400
Chimipon GT
Chimipon NA
Cirrasol® AEN-XB
Cirrasol® ALN-GM
Cithrol A
Cithrol DGDL N/E
Cithrol DGDL S/E
Cithrol DGDO N/E
Cithrol DGDO S/E
Cithrol DGDS N/E
Cithrol DGDS S/E
Cithrol DGML N/E
Cithrol DGML S/E
Cithrol DGMO S/E
Cithrol DGMS N/E
Cithrol DGMS S/E
Cithrol DPGML S/E
Cithrol DPGMS N/E
Cithrol DPGMS S/E
Cithrol EGDL N/E
Cithrol EGDL S/E
Cithrol EGDO N/E
Cithrol EGDO S/E
Cithrol EGDS N/E
Cithrol EGDS S/E
Cithrol EGML N/E
Cithrol EGML S/E
Cithrol EGMO N/E
Cithrol EGMO S/E
Cithrol EGMR N/E
Cithrol EGMR S/E
Cithrol EGMS N/E
Cithrol EGMS S/E
Cithrol GDL N/E
CNC Antifoam 77
Colorin 301
Colorin 302
Colorol 20
Colorol 46
Cosmopon BN
Crill 1
Crill 2
Crodamet Series
Crodamine 1.HT
Crodamine 1.O, 1.OD
Crodamine 1.T
Crodamine 1.16D
Crodamine 1.18D
Crodamine 3.A16D
Crodamine 3.A18D
Crodamine 3.ABD
Crodamine 3.AED
Crodamine 3.AOD
Crodasinic O
Croduret 10
Croduret 30
Croduret 40
Croduret 60
Croduret 100
Fancol OA 95
Fizul 201-11
Fizul M-440
Fizul MD-318
Flexricin® 13
Fluorad FC-118
Fluorad FC-126
Foamaster 8034
Foamaster JMY
Foamaster NDW
Foamaster TBD-1
Foamaster TDB
Foamkill® 30C
Foamkill® 614
Foamkill® 614NS
Foamkill® 639
Foamkill® 639Q
Foamkill® 652B
Gantrez® AN-119
Gantrez® AN-139
Gantrez® AN-149
Gantrez® AN-169
Gantrez® AN-179
Gardilene S25L
Gemtex 445
Gemtex 680
Gemtex 691-40
Gemtex PA-75
Gemtex PA-85P
Gemtex PAX-60
Gemtex SC-75E, SC Powd.
Geronol ACR/4
Geronol ACR/9
Geropon® 99
Geropon® AB/20
Geropon® ACR/4
Geropon® ACR/9
Geropon® BIS/SODICO-2
Geropon® CYA/45
Geropon® CYA/60
Geropon® CYA/DEP
Geropon® IN
Geropon® MLS/A
Geropon® SDS
Geropon® SS-O-75
Geropon® T/36-DF
Geropon® WT-27
Glycidol Surfactant 10G
GP-209
GP-214
GP-215
GP-218
GP-219
GP-7000
Grindtek PGE 25
Grindtek PGE 55
Grindtek PGE 55-6
Grindtek PGE-DSO
Hamposyl® AL-30
Hartomer 4900
Hartomer GP 2164
Hartomer GP 4935
Hartomer JV 4091
Hartomul PE-30
Hartopol L64
Hostapal BV Conc.
Hyonic GL 400
Hyonic NP-60
Hyonic NP-407
Hyonic NP-500
Hyonic OP-70
Hystrene® 3675
Hystrene® 3675C
Hystrene® 3680
Hystrene® 3687
Hystrene® 3695
Iconol NP-30
Iconol NP-30-70%
Iconol NP-40
Iconol NP-40-70%
Iconol NP-50-70%
Iconol NP-100
Iconol NP-100-70%
Iconol OP-5
Iconol OP-7
Iconol OP-10
Iconol OP-30
Iconol OP-30-70%
Iconol OP-40
Iconol OP-40-70%

Plastics, Rubbers, and Resin Manufacture (cont'd.)

Igepal® CA-880
Igepal® CA-887
Igepal® CA-890
Igepal® CA-897
Igepal® CO-430
Igepal® CO-530
Igepal® CO-850
Igepal® CO-880
Igepal® CO-887
Igepal® CO-890
Igepal® CO-897
Igepal® CO-970
Igepal® CO-977
Igepal® CO-980
Igepal® CO-987
Igepal® CO-990
Igepal® CO-997
Igepal® DM-430
Igepal® DM-730
Igepal® DM-880
Igepal® DX-430
Igepal® LAVE
Igepal® NP-10
Igepal® NP-20
Igepal® NP-30
Igepal® O
Igepal® OD-410
Incropol L-7
Industrene® 105
Industrene® 143
Industrene® 205
Industrene® 206
Industrene® 225
Industrene® 226
Industrene® 325
Industrene® 328
Industrene® 365
Industrene® 4516
Industrene® 4518
Industrene® 5016
Industrene® 6018
Industrene® 7018
Industrene® 9018
Industrene® B
Industrene® D
Industrene® R
Ionet DL-200
Ionet DO-200
Ionet DO-400
Ionet DO-600
Ionet DO-1000
Ionet DS-300
Karawet DOSS
Kemamide® B
Kemamide® E
Kemamide® O
Kemamide® S
Kemamide® S-221
Kemamide® U
Kemamide® W-20
Kemamide® W-39
Kemamide® W-40
Kemamide® W-40/300
Kemamide® W-45
Kemamine® P-150, P-150D
Kemamine® P-190, P-190D
Kemamine® P-970
Kemamine® P-970D
Kemamine® P-974D
Kemamine® P-989D
Kemamine® P-990
Kemamine® P-990D
Kemamine® P-999
Kemamine® T-1902D
Kemamine® T-6501
Kemamine® T-6502D
Kemamine® T-9701
Kemamine® T-9702D
Kemamine® T-9902D
Kemamine® T-9972D
Kemester® 143
Kemester® 4000
Kemester® 5221SE
Kemester® 6000
Kessco® PEG 200 DO
Kessco® PEG 200 DS
Kessco® PEG 200 ML
Kessco® PEG 200 MO
Kessco® PEG 200 MS
Kessco® PEG 300 DL
Kessco® PEG 300 DO
Kessco® PEG 300 DS
Kessco® PEG 300 ML
Kessco® PEG 300 MO
Kessco® PEG 300 MS
Kessco® PEG 400 DL
Kessco® PEG 400 DO
Kessco® PEG 400 ML
Kessco® PEG 400 MO
Kessco® PEG 600 DO
Kessco® PEG 600 MO
Kessco® PEG 600 MS
Kessco® PEG 1000 DL
Kessco® PEG 1000 DO
Kessco® PEG 1000 DS
Kessco® PEG 1000 ML
Kessco® PEG 1000 MO
Kessco® PEG 1000 MS
Kessco® PEG 1540 DL
Kessco® PEG 1540 DO
Kessco® PEG 1540 DS
Kessco® PEG 1540 ML
Kessco® PEG 1540 MO
Kessco® PEG 1540 MS
Kessco® PEG 4000 DL
Kessco® PEG 4000 ML
Kessco® PEG 4000 MO
Kessco® PEG 4000 MS
Kessco® PEG 6000 DL
Kessco® PEG 6000 DO
Kessco® PEG 6000 ML
Kessco® PEG 6000 MO
Kessco® PEG 6000 MS
Klearfac® AA040
Lamigen ES 30
Lamigen ES 60
Lamigen ES 100
Lamigen ET 20
Lamigen ET 70
Lamigen ET 90
Lamigen ET 180
Lankropol® KMA
Lankropol® KN51
Lankropol® KO2
Lankropol® OPA
Larostat® 264 A Anhyd.
Larostat® 264 A Conc.
Laural LS
Laurex® CS
Laurex® NC
Leocon 1070B
Leomin AN
Leukonöl LBA-2
Levelan® P148
Levelan® P208
Levelan® P307
Lipolan LB-440
Lipolan PJ-400
Liponox NC 2Y
Liponox NC-500
Lipophos PE9
Lipophos PL6
Lomar® HP
Lomar® LS
Lomar® LS Liq.
Lomar® PL
Lomar® PW
Lomar® PWA
Lomar® PWA Liq.
Loxiol G 52
Loxiol G 53
Loxiol P 1420
Loxiol VPG 1354
Loxiol VPG 1451
Loxiol VPG 1496
Loxiol VPG 1743
Mackester EGDS
Mackester EGMS
Mackester IDO
Mackester IP
Mackester SP
Mackester TD-88
Macol® 65
Macol® 90
Macol® 90(70)
Macol® 300
Macol® 625
Macol® 626
Macol® 627
Macol® 660
Macol® 3520
Macol® 5100
Macol® P-500
Macol® P-1200
Macol® P-1750
Macol® P-2000
Macol® P-3000
Macol® P-4000
Manro BA Acid
Manro DES 32
Manro DL 28
Manro MA 35
Manro PTSA/E
Manro PTSA/H
Manro PTSA/LS
Manro SDBS 25/30
Manro SLS 28
Manro TDBS 60
Mapeg® CO-25
Mapeg® CO-25H
Maphos® 15
Maphos® 17
Maphos® 18
Maphos® 30
Maphos® 41A
Maphos® 54
Maphos® 55
Maphos® 56
Maphos® 76
Maphos® 76 NA
Maphos® 77
Maphos® 151
Maphos® 236
Maphos® FDEO
Maphos® L-6
Maphos® L 13
Maranil CB-22
Maranil Powd. A
Masil® 1066C
Masil® 1066D
Masil® 2132
Masil® 2133
Masil® 2134
Masil® SF 5
Masil® SF 10
Masil® SF 50
Masil® SF 100
Masil® SF 200
Masil® SF 350 FG
Masil® SF 500
Masil® SF 5000
Masil® SF 10,000
Masil® SF 12,500
Masil® SF 30,000
Masil® SF 60,000
Masil® SF 100,000
Masil® SF 300,000
Masil® SF 600,000
Masil® SF 1,000,000
Mazawet® 77
Mazawet® DF
Mazeen® 173
Mazeen® 174
Mazeen® 174-75
Mazeen® 175
Mazeen® 176
Mazeen® C-2
Mazeen® C-5
Mazeen® C-10
Mazeen® S-2
Mazeen® S-5
Mazeen® S-10
Mazeen® S-15
Mazeen® T-2
Mazeen® T-5
Mazeen® T-10
Mazeen® T-15
Mazol® 159
Mazol® GMO
Mazol® GMS-90
Mazol® GMS-D
Mazon® 1045A
Mazon® 1086
Mazon® 1096
Mazu® DF 200SX
Mazu® DF 200SXSP
Mazu® DF 210SXSP
Merpol® A
Merpol® HCS
Miranol® CM-SF Conc.

Plastics, Rubbers, and Resin Manufacture (cont'd.)

Miranol® J2M Conc.
Miranol® SM Conc.
Modicol L
Monafax 785
Monafax 794
Monafax 1293
Monawet 1240
Monawet MB-45
Monawet MM-80
Monawet MO-70
Monawet MO-70E
Monawet MT-70
Monawet MT-70E
Monawet SNO-35
Monawet TD-30
Monolan® 2500 E/30
Monolan® 3000 E/60, 8000 E/80
Monolan® 8000 E/80
Monolan® O Range
Monolan® PPG440, PPG1100, PPG2200
Montosol PB-25
Montovol RF-10
Nacconol® 40G
Nacconol® 90G
Nalco® 2340
Nalco® 8669
Nansa® 1042
Nansa® 1042/P
Nansa® 1106/P
Nansa® 1169/P
Nansa® EVM50
Nansa® EVM70/B
Nansa® HS85S
Nansa® SS 30
Nansa® SS 50
Nansa® SS 55
Nansa® SSA
Nansa® SSA/P
Naxchem N-Foam 802
Naxonac 510
Naxonac 690-70
Newcol 180T
Newcol 261A, 271A
Newcol 506
Newcol 508
Newcol 560SF
Newcol 560SN
Newcol 568
Newcol 607, 610
Newcol 614, 623
Newcol 704, 707
Newcol 707SF
Newcol 710, 714, 723
Newcol 861S
Newcol 1305SN, 1310SN
Newpol PE-61
Newpol PE-62
Newpol PE-64
Newpol PE-68
Niaproof® Anionic Surfactant 4
Niaproof® Anionic Surfactant 08
Nikkol GO-430
Nikkol GO-440
Nikkol GO-460
Niox KQ-34
Nissan Nonion HS-204.5
Nissan Nonion HS-206
Nissan Nonion HS-208
Nissan Nonion HS-210
Nissan Nonion HS-215
Nissan Nonion HS-240
Nissan Nonion HS-270
Nissan Nonion NS-202
Nissan Nonion NS-204.5
Nissan Nonion NS-206
Nissan Nonion NS-208.5
Nissan Nonion NS-209
Nissan Nonion NS-210
Nissan Nonion NS-212
Nissan Nonion NS-215
Nissan Nonion NS-230
Nissan Nonion NS-240
Nissan Nonion NS-250
Nissan Nonion NS-270
Nissan Persoft EK
Nissan Plonon 102
Nissan Plonon 104
Nissan Plonon 108
Nissan Plonon 171
Nissan Plonon 172
Nissan Plonon 204
Nissan Plonon 208
Nissan Rapisol B-30, B-80, C-70
Nissan Tert. Amine AB
Nissan Tert. Amine ABT
Nissan Tert. Amine BB
Nissan Tert. Amine FB
Nissan Tert. Amine MB
Nissan Tert. Amine PB
Nissan Trax K-300
Nissan Trax N-300
Noiox AK-41
Nonipol 20
Nonipol 40
Nonipol 55
Nonipol 60
Nonipol 70
Nonipol 85
Nonipol 95
Nonipol 100
Nonipol 110
Nonipol 120
Nonipol 130
Nonipol 160
Nonipol 200
Nonipol 400
Nopalcol 1-L
Nopalcol 2-DL
Nopalcol 4-C
Nopalcol 4-CH
Nopalcol 4-L
Nopalcol 4-S
Nopalcol 6-DO
Nopalcol 6-DTW
Nopalcol 6-L
Nopalcol 6-R
Nopalcol 6-S
Nopalcol 10-COH
Nopalcol 12-CO
Nopalcol 12-COH
Nopalcol 19-CO
Nopalcol 30-TWH
Nopalcol 200
Nopalcol 400
Nopalcol 600
Nopco® 2031
Octasol IB-45
Octosol 400
Octosol 449
Octosol 571
Octosol A-1
Octosol A-18
Octosol A-18-A
Octosol HA-80
Octosol IB-45
Octosol SLS
Octosol SLS-1
Octosol TH-40
Octowet 70BC
Octowet 70PG
Ogtac 85 V
Pale 4
Pale 16
Pale 170
Pale 1000
Patcote® 520
Pationic® 901
Pationic® 902
Pationic® 907
Pationic® 909
Pationic® 919
Pationic® 925
Pationic® 930
Pationic® 940
Pationic® 1042
Pationic® 1052
Pationic® 1064
Pationic® 1074
Pationic® 1230
Pationic® 1240
Pationic® 1264
Pegosperse® 200 DL
Pegosperse® PMS CG
Pelex OT-P
Pelex SS-L
Perlankrol® ATL40
Perlankrol® DGS
Perlankrol® DSA
Perlankrol® EAD60
Perlankrol® EP12
Perlankrol® EP24
Perlankrol® EP36
Perlankrol® ESD
Perlankrol® ESD60
Perlankrol® FB25
Perlankrol® FD63
Perlankrol® FF
Perlankrol® FN65
Perlankrol® FT58
Perlankrol® FV70
Perlankrol® FX35
Perlankrol® PA Conc.
Perlankrol® RN75
Perlankrol® SN
PGE-400-DS
PGE-600-DS
PGE-600-ML
Phosphanol Series
Phospholan® PNP9
Pluracol® E200
Pluracol® E400 NF
Pluracol® E600
Pluracol® E1450
Pluracol® E1450 NF
Pluracol® E2000
Pluracol® E4000 NF
Pluracol® E4500
Pluracol® E8000
Pluracol® E8000 NF
Pluracol® P-410
Pluracol® P-710
Pluracol® P-1010
Pluracol® P-2010
Pluracol® P-3010
Pluracol® P-4010
Pluracol® W170
Pluracol® W260
Pluracol® W660
Pluracol® W2000
Pluracol® W3520N
Pluracol® W3520N-RL
Pluracol® W5100N
Pluracol® WD90K
Pluracol® WD1400
Pluriol® E 9000
Pluriol® P 600
Pluriol® P 900
Pluriol® P 2000
Pluriol® PE 6200
Pluriol® PE 6400
Pluriol® PE 6800
Pluriol® PE 10500
Plysurf A207H
Plysurf A208B
Plysurf A208S
Plysurf A210G
Plysurf A212C
Plysurf A215C
Plysurf A217E
Plysurf A219B
Plysurf AL
Pogol 1500
Polirol 1BS
Polirol 4, 6
Polirol 10
Polirol 23
Polirol 215
Polirol C5
Polirol DS
Polirol L400
Polirol LS
Polirol NF80
Polirol O55
Polirol SE 301
Polirol TR/LNA
Poly-Tergent® 2EP
Polystep® A-4
Polystep® A-7
Polystep® A-11
Polystep® A-13
Polystep® A-15
Polystep® A-15-30K
Polystep® A-16
Polystep® A-16-22
Polystep® A-17
Polystep® A-18

Plastics, Rubbers, and Resin Manufacture (cont'd.)

Polystep® B-1
Polystep® B-3
Polystep® B-5
Polystep® B-7
Polystep® B-11
Polystep® B-12
Polystep® B-19
Polystep® B-20
Polystep® B-22
Polystep® B-23
Polystep® B-24
Polystep® B-25
Polystep® B-27
Polystep® B-29
Polystep® F-1
Polystep® F-2
Polystep® F-3
Polystep® F-4
Polystep® F-5
Polystep® F-6
Polystep® F-9
Polystep® F-10, F-10 Ec
Polystep® F-95B
Polystep® PN 209
Polysurf A212E
Polywet AX-7
Polywet KX-3
Polywet KX-4
Prifac 5902, 5904, 5905
Prifrac 2920
Prifrac 2922
Priolene 6900
Priolene 6901
Priolene 6907
Priolene 6911
Priolene 6930
Priosorine 3501
Pripol 1004
Pripol 1017, 1022
Prote-pon P-2 EHA-02-K30
Prote-pon P 2 EHA-02-Z
Prote-pon P 2 EHA-Z
Prote-pon P-0101-02-Z
Prote-pon P-L 201-02-K30
Prote-pon P-L 201-02-Z
Prote-pon P-NP-06-K30
Prote-pon P-NP-06-Z
Prote-pon P-NP-10-K30
Prote-pon P-NP-10-MZ
Prote-pon P-NP-10-Z
Prote-pon P-OX 101-02-K75
Prote-pon P-TD-06-K13
Prote-pon P-TD-06-K30
Prote-pon P-TD-06-Z
Prote-pon P-TD-09-Z
Prote-pon P-TD-12-Z
Prote-pon TD-09-K30
Prox-onic CSA-1/04
Prox-onic CSA-1/06
Prox-onic CSA-1/010
Prox-onic CSA-1/015
Prox-onic CSA-1/020
Prox-onic CSA-1/030
Prox-onic CSA-1/050
Prox-onic DT-03
Prox-onic DT-015
Prox-onic DT-030
Prox-onic EP 1090-1
Prox-onic EP 1090-2
Prox-onic EP 2080-1
Prox-onic EP 4060-1
Prox-onic HR-05
Prox-onic HR-016
Prox-onic HR-025
Prox-onic HR-036
Prox-onic HR-080
Prox-onic HR-0200/50
Prox-onic HRH-05
Prox-onic HRH-016
Prox-onic HRH-025
Prox-onic HRH-0200
Prox-onic HRH-0200/50
Prox-onic LA-1/02
Prox-onic LA-1/04
Prox-onic LA-1/09
Prox-onic LA-1/012
Prox-onic MC-02
Prox-onic MC-05
Prox-onic MO-02
Prox-onic MO-015
Prox-onic MO-030
Prox-onic MO-030-80
Prox-onic MS-05
Prox-onic MS-011
Prox-onic MS-050
Prox-onic MT-015
Prox-onic MT-020
Prox-onic OA-1/04
Prox-onic OA-1/09
Prox-onic OA-2/020
Prox-onic OP-09
Prox-onic OP-030/70
Prox-onic OP-040/70
Prox-onic PEG-2000
Prox-onic PEG-20,000
Puxol CB-22
Puxol FB-11
Quadrilan® AT
Quimipol 9106B
Quimipol EA 2503
Quimipol EA 9106B
Quimipol ENF 40
Quimipol ENF 200
Quimipol ENF 230
Quimipol ENF 300
Radiamine 6140
Radiamine 6141
Radiamine 6160
Radiamine 6161
Radiamine 6163, 6164
Radiamine 6170
Radiamine 6171
Radiamine 6172
Radiamine 6173
Radiasurf® 7000
Radiasurf® 7125
Radiasurf® 7140
Radiasurf® 7146
Radiasurf® 7150
Radiasurf® 7151
Radiasurf® 7153
Radiasurf® 7196
Radiasurf® 7410
Radiasurf® 7411
Remcopal 6
Remcopal 229
Remcopal 31250
Remcopal 33820
Rewoderm® S 1333
Rewomat B 2003
Rewomat TMS
Rewopal® HV 5
Rewopal® HV 9
Rewopal® HV 25
Rewophat E 1027
Rewophat NP 90
Rewopol® 15/L
Rewopol® B 1003
Rewopol® B 2003
Rewopol® NL 2-28
Rewopol® NL 3
Rewopol® NL 3-28
Rewopol® NL 3-70
Rewopol® NLS 15 L
Rewopol® NLS 28
Rewopol® NLS 30 L
Rewopol® NOS 5
Rewopol® NOS 8
Rewopol® NOS 10
Rewopol® NOS 25
Rewopol® SBDB 45
Rewopol® SBDC 40
Rewopol® SBDD 65
Rewopol® SBDO 75
Rewopol® SBFA 50
Rewopol® SBMB 80
Rewopol® SMS 35
Rewopol® TMSF
Rewoquat CPEM
Reworyl® NKS 50
Rexfoam D
Rexol 25/4
Rexol 25/6
Rexol 25/7
Rexol 25/8
Rexol 25/9
Rexol 25/10
Rexol 25/12
Rexol 25/307
Rexol 25/407
Rexol 45/10
Rexol 45/307
Rexol 45/407
Rheodol TW-P120
Rhodacal® 301-10
Rhodacal® 330
Rhodacal® A-246L
Rhodacal® BA-77
Rhodacal® BX-78
Rhodacal® DOV
Rhodacal® DS-4
Rhodacal® DS-10
Rhodacal® DSB
Rhodacal® LDS-22
Rhodacal® RM/77-D
Rhodafac® BX-660
Rhodafac® MC-470
Rhodafac® PE-510
Rhodafac® PS-17
Rhodafac® PS-19
Rhodafac® RE-610
Rhodafac® RE-960
Rhodameen® OS-12
Rhodamox® CAPO
Rhodapex® AB-20
Rhodapex® CO-433
Rhodapex® CO-436
Rhodapex® EST-30
Rhodapon® CAV
Rhodapon® L-22, L-22/C
Rhodapon® L-22HNC
Rhodapon® LCP
Rhodapon® OLS
Rhodapon® OS
Rhodapon® SB
Rhodasurf® E 400
Rhodasurf® L-4
Rhodasurf® L-25
Rhodasurf® L-790
Rhodasurf® ON-877
Rhodasurf® PEG 8000
Rhodorsil® AF 422
Sandet 60
Sandoz Phosphorester 610
Sandoz Phosphorester 690
Sandoz Sulfonate 2A1
Santone® 3-1-SH
Schercophos NP-6
Schercowet DOS-70
Secolat
Secosol® ALL40
Secosol® DOS 70
Sellogen HR
SEM-35
Serdet DFK 40
Serdet DJK 30
Serdet DM
Serdet DNK 30
Serdet DSK 40
Serdet Perle Conc.
Serdox NNP 25
Serdox NNP 30/70
Serdox NOP 30/70
Serdox NOP 40/70
Serdox NSG 400
Sermul EA 30
Sermul EA 54
Sermul EA 88
Sermul EA 129
Sermul EA 136
Sermul EA 139
Sermul EA 146
Sermul EA 150
Sermul EA 151
Sermul EA 152
Sermul EA 176
Sermul EA 188
Sermul EA 211
Sermul EA 214
Sermul EA 221
Sermul EA 224
Sermul EA 242
Sermul EA 370
Sermul EN 20/70
Sermul EN 30/70
Sermul EN 145

Plastics, Rubbers, and Resin Manufacture (cont'd.)

Sermul EN 155
Sermul EN 237
Sermul EN 312
Servoxyl VPGZ 7/100
Servoxyl VPIZ 100
Servoxyl VPNZ 10/100
Servoxyl VPQZ 9/100
Servoxyl VPTZ 3/100
Servoxyl VPTZ 100
Servoxyl VPUZ
Servoxyl VPYZ 500
SF18
SF96®
Silicone AF-100 FG
Silicone AF-100 IND
Silwet® L-7001
Silwet® L-7500
Silwet® L-7600
Silwet® L-7602
Sinonate 960SF
Sinonate 960SN
Sinopol 610
Sinopol 623
Sinopol 707
Sinopol 714
Sinopol 806
Sinopol 808
Sinopol 908
Sinopol 910
Siponic® F-300
Siponic® F-400
Siponic® L-12
Siponic® NP-8
Siponic® NP-9
Siponic® NP-9.5
Siponic® NP-10
Siponic® NP-13
Siponic® NP-15
Siponic® NP-407
Soft Detergent 95
Softenol® 3900
Softenol® 3991
Sole-Terge 8
Solricin® 135
Solricin® 235
Solricin® 435
Solricin® 535
Soprofor NPF/10
Sorbax HO-40
Sorbax HO-50
Sorbax MO-40
Steol® 4N
Steol® CA-460
Steol® COS 433
Stepan C-40
Stepan C-65
Stepan C-68
Stepanol® WA-100
Stepfac® 8170
Steramine CD
Sul-fon-ate AA-9
Sul-fon-ate AA-10
Sulfochem 436
Sulfochem 437
Sulfochem 438
Sulfopon 101/POL
Sulfopon 101 Spez
Sulfopon 102
Sulfopon P-40
Sulfotex DOS
Sulframin 90
Sulframin 1230
Sulframin 1255
Sulframin Acide TPB
Sulframin AOS
Sunnol DOS
Sunnol DP-2630
Superol
Supragil® NK
Surfagene FAD 105
Surfagene FAZ 109
Surfax 215
Surfax 218
Surfax 220
Surfax 495
Surfax 502
Surfax 536
Surfax 539
Surfax 550
Surfax 560
Surfonic® HDL
Surfonic® HDL-1
Surfonic® N-557
Surfynol® 82S
Surfynol® 465
Swascol 1P
Sylvaros® 20
Sylvaros® 80
Synperonic OP16
Synperonic OP30
Synperonic PE30/80
Synprolam 35
Synprolam 35DM
Synprolam 35MX1
Synprolam 35MX3
Synprolam 35MX5
Synprolam 35X2
Synprolam 35X2/O
Synprolam 35X10
Synprolam 35X15
Synprolam 35X20
T-Det® BP-1
T-Det® N-20
T-Det® N-40
T-Det® N-50
T-Det® N-70
T-Det® N-100
T-Det® N-307
T-Det® N-407
T-Det® N-507
T-Det® N-705
T-Det® N-707
T-Det® N-1007
T-Det® O-40
T-Det® O-307
Tegiloxan®
Tego® Dispers 610
Tego® Dispers 610 S
Tego® Dispers 620
Tego® Dispers 620 S
Tego® Dispers 630
Tego® Dispers 700
Tensopol A 79
Tensopol A 795
Tensopol ACL 79
Tensopol S 30 LS
Tergitol® 15-S-3
Tergitol® 15-S-5
Tergitol® 15-S-30
Tergitol® 15-S-40
Tergitol® NP-40
Tergitol® NP-40, 70% Aq.
Tergitol® NP-55, 70% Aq.
Tergitol® NP-70, 70% Aq.
Tergitol® XD
Tergitol® XH
Tergitol® XJ
Teric 16A22
Teric 18M5
Teric 18M10
Teric C12
Teric PE68
Teric PEG 200
Teric PEG 300
Teric PEG 400
Teric PEG 600
Teric PEG 4000
Teric PEG 6000
Teric PEG 8000
Teric PPG 1000
Teric PPG 1650
Teric PPG 2250
Teric PPG 4000
Tetronic® 50R1
Tetronic® 50R4
Tetronic® 50R8
Tetronic® 70R1
Tetronic® 70R2
Tetronic® 70R4
Tetronic® 90R1
Tetronic® 90R4
Tetronic® 90R8
Tetronic® 110R1
Tetronic® 110R2
Tetronic® 110R7
Tetronic® 130R2
Tetronic® 150R1
Tetronic® 150R4
Tetronic® 150R8
Tetronic® 304
Tetronic® 504
Tetronic® 701
Tetronic® 702
Tetronic® 704
Tetronic® 707
Tetronic® 901
Tetronic® 904
Tetronic® 908
Tetronic® 909
Tetronic® 1101
Tetronic® 1102
Tetronic® 1104
Tetronic® 1107
Tetronic® 1301
Tetronic® 1302
Tetronic® 1304
Tetronic® 1307
Tetronic® 1501
Tetronic® 1502
Tetronic® 1504
Tetronic® 1508
Texapon K-12
Texapon K-1296
Texapon NSE
Texapon OT Highly Conc. Needles
Texapon VHC Needles
Tex-Wet 1104
Tex-Wet 1143
T-Mulz® 596
T-Mulz® 598
T-Mulz® 734-2
TO-55-E
TO-55-EL
Tomah DA-14
Tomah DA-16
Tomah DA-17
Tomah PA-10
Tomah PA-14
Tomah PA-16
Tomah PA-19
Tomah PA-1214
Tomah Q-14-2
Tomah Q-18-15
Triton® 770 Conc.
Triton® N-60
Triton® W-30 Conc.
Triton® X-15
Triton® X-35
Triton® X-165-70%
Triton® X-200
Triton® X-301
Triton® X-305-70%
Triton® X-405-70%
Triton® X-705-70%
Trycol® 6941
Trycol® 6954
Trycol® 6957
Trycol® 6967
Trycol® 6968
Trycol® 6969
Trycol® 6970
Trycol® 6971
Trycol® 6972
Trycol® 6981
Trycol® 6984
Trycol® 6985
Trycol® DNP-8
Trycol® NP-20
Trycol® NP-30
Trycol® NP-307
Trycol® NP-407
Trydet SA-50/30
Tryfac® 910-K
Tryfac® 5556
Trylox® CO-30
Trymeen® 6617
Tylose C, CB Series
Tylose H Series
Tylose MHB
Tylose MH Grades
Tylose MH-K, MH-xp, MHB-y, MHB-yp
Ufapol
Ufasan 50
Ufasan 65
Ufasan IPA
Ufasan TEA
Ultra Sulfate SL-1

Plastics, Rubbers, and Resin Manufacture (cont'd.)

Ungerol LS, LSN
Varsulf® NOS-25
Victawet® 12
Viscasil®
Vorite 105
Vorite 110
Vorite 115
Vorite 120
Vorite 125
Wayfos A
Wayfos D-10N
Wayfos M-60
Westvaco® 1480
Westvaco® M-30
Westvaco® M-40
Westvaco® M-70
Westvaco® Resin 90
Westvaco® Resin 95
Westvaco® Resin 790
Westvaco® Resin 795
Westvaco® Resin 895
Witco® 960
Witco® 1298
Witcolate 1276
Witcolate A Powder
Witcolate D-510
Witcolate D51-51
Witcolate D51-51EP
Witcolate D51-52
Witcolate D51-53
Witcolate D51-53HA
Witcolate SL-1
Witconate 1223H
Witconate 1250 Slurry
Witconate 3009-15
Witconate AOS
Witconate AOS-EP
Witconate D51-51
Witconol NP-100
Witconol NP-300
Witconol NP-330
Witco® Acid B
Zonyl® FSA
Zonyl® FSB
Zonyl® FSC
Zonyl® FSJ
Zonyl® FSN
Zonyl® FSP
Zorapol LS-30
Zorapol SN-9

Refractories, Masonry, Ceramics

Ablusol ML
Agesperse 71
Agesperse 80
Agesperse 81
Agesperse 82
Albrite MALP
Alkanol® XC
Ampholan® B-171
Armid® 18
Armid® C
Armid® HT
Armid® O
Avirol® A
Avirol® SL 2010
Avirol® SL 2015
Avirol® SL 2020
Avirol® T 40
Blancol®
Calcium Stearate, Regular
Calsoft L-40
Calsoft L-60
Carbowax® PEG 400
Carbowax® PEG 600
Carbowax® PEG 900
Carbowax® PEG 1000
Carbowax® PEG 1450
Carbowax® PEG 3350
Carbowax® PEG 4600
Carbowax® PEG 8000
Chupol C
Chupol EX
Foamaster NDW
Foamaster PD-1
Foamaster TDB
Glutrin
Goulac
Hi-Fluid
Igepal® NP-9
Indulin® W-1
Lankropol® KNB22
Lankropol® KSG72
Lignosite® 458
Lignosol B
Lignosol BD
Lignosol TS
Lignosol TSD
Lomar® PL
Lomar® PW
Lomar® PWA
Lumiten® E
Lumiten® I
Lumiten® N
Marlophen® X
Marlophor® AS-Acid
Marlophor® DS-Acid
Monateric ADA
Mulsifan RT 110
Mulsifan RT 258
Nacconol® 40G
Nacconol® 90G
Nansa® BMC
Nansa® MA30
Nissan Amine AB
Nissan Amine ABT
Nissan Amine BB
Nissan Amine FB
Nissan Amine MB
Nissan Amine OB
Nissan Amine PB
Nissan Amine SB
Nissan Amine VB
Nissan Cation ABT-350, 500
Nissan Cation F2-10R, -20R, -40E, -50
Nissan Cation FB, FB-500
Nissan Cation L-207
Nissan Cation M2-100
Nissan Cation PB-40, -300
Nissan Cation S2-100
Norfox® Agent 2A-2S
Ombrelub FC 533
Pale 4
Pluracol® P-410
Pluracol® P-710
Pluracol® P-1010
Pluracol® P-2010
Pluracol® P-3010
Pluracol® P-4010
Pluriol® E 200
Pluriol® E 300
Pluriol® E 400
Pluriol® E 600
Pluriol® E 1500
Pluriol® E 4000
Pluriol® E 6000
Pluriol® E 9000
Polefine 510N
Prox-onic PEG-4000
Prox-onic PEG-6000
Prox-onic PEG-10,000
Prox-onic PEG-35,000
Radiasurf® 7125
Rhodacal® N
Rhodacal® RM/210
Schercomid SCO-Extra
Serdox NNP 8.5
Stepanform® 1440
Stepantan® AS-40
Stepantan® DT-60
Stepantan® H-100
Steposol® CA-207
Sul-fon-ate AA-9
Sul-fon-ate AA-10
Surfonic® HDL
Surfonic® JL-80X
Surfynol® 82S
Surfynol® 104BC
Surfynol® 104H
Surfynol® 104PA
Surfynol® 104PG
Surfynol® DF-37
Surfynol® DF-58
Surfynol® DF-110
Surfynol® DF-110D
Surfynol® DF-110L
Surfynol® DF-110S
Surfynol® PG-50
Teric 158
Teric N8
Teric N9
Teric N10
Teric N11
Teric N12
Teric N13
Teric PEG 600
Teric PEG 4000
Teric PEG 6000
Teric PPG 1650
Tetronic® 50R1
Tetronic® 50R4
Tetronic® 50R8
Tetronic® 70R1
Tetronic® 70R2
Tetronic® 70R4
Tetronic® 90R1
Tetronic® 90R4
Tetronic® 90R8
Tetronic® 110R1
Tetronic® 110R2
Tetronic® 110R7
Tetronic® 130R1
Tetronic® 130R2
Tetronic® 150R1
Tetronic® 150R4
Tetronic® 150R8
Triumphnetzer ZSG
Troykyd® D44
Troykyd® D55
Troykyd® D126
Tylose C, CB Series
Tylose H Series
Tylose MHB
Tylose MH Grades
Tylose MH-K, MH-xp, MHB-y, MHB-yp
Ufaryl DL80 CW
Ufaryl DL85
Ufaryl DL90
Ufasan 35
Ufasan 50
Ufasan 60A
Ufasan 62B
Ufasan 65
Ufasan IPA
Ufasan TEA
Versilan MX123
Versilan MX134
Versilan MX244
Witconate AO 24

Textiles and Fibers

Ablunol 200ML
Ablunol 200MO
Ablunol 200MS
Ablunol 400ML
Ablunol 400MO
Ablunol 400MS
Ablunol 600ML
Ablunol 600MO
Ablunol 600MS
Ablunol 1000MO
Ablunol 1000MS
Ablunol 6000DS
Ablunol GMS
Ablunol LA 3
Ablunol LA 5
Ablunol LA 7
Ablunol LA 9
Ablunol LA 12
Ablunol LA 16
Ablunol LA 40
Ablunol NP 9
Ablunol NP 20
Ablunol NP 30
Ablunol NP 30 70%
Ablunol NP 40
Ablunol NP 40 70%
Ablunol NP 50
Ablunol NP 50 70%
Ablunol S-60
Ablunol S-85
Abluphat LP Series
Ablusoft A
Ablusoft ND
Ablusoft PE
Ablusoft SN
Ablusoft SNC
Ablusol C-78
Ablusol NL
Abluton 7000
Abluton CDL
Abluton CMN
Abluton CTP
Abluton EP
Abluton LMO
Abluton N
Abluton N
Abluton T
Abluton T
Acetamin 24
Acetamin 86
Acrylic Resin AS
Actrasol 6092
Actrasol C-50, C-75, C-85
Actrasol CS-75
Actrasol OY-75
Actrasol PSR
Actrasol SBO
Actrasol SP
Actrasol SP 175K
Actrasol SR 75
Actrasol SRK 75
Adinol OT16
Adma® 18
Adol® 42
Adol® 61 NF
Adol® 62 NF
Adol® 63
Adol® 66
Adol® 80
Adol® 520 NF
Adol® 620 NF
AE-7
AE-1214/3
AE-1214/6
Aerosol® 501
Aerosol® A-102
Aerosol® A-103
Aerosol® C-61
Aerosol® MA-80
Aerosol® NPES 458
Aerosol® OT-75%
AF 10 IND
AF 72
AF 75
AF 93
AF 8810
AF 8810 FG
AF HL-40
Afilan EHS
Afilan ICS
Afilan ODA
Afilan POD
Afilan PP
Afilan SME
Afilan TDA
Afilan TDS
Afilan TMOD
Afilan TMPP
Afilan TXE
Agesperse 80
Agesperse 81
Agesperse 82
Ahcovel Base 500
Ahcovel Base 700
Ahcovel Base N-15
Ahcovel Base N-62
Ahcovel Base OB
Ahcovel R Base
Ahcowet RS
Airrol CT-1
Ajidew A-100
Ajidew N-50
Ajidew SP-100
Akypo ITD 70 BV
Akyporox NP 200
Akyporox RO 90
Albatex® OR
Albegal® A
Albegal® B
Albegal® BMD
Albigen® A
Alcodet® 218
Alcodet® 260
Alcodet® SK
Alcodet® TX 4000
Alcolec® 439-C
Aldosperse® ML 23
Aldosperse® MS-20
Alkamide® SDO
Alkamide® SODI
Alkamuls® 400-DO
Alkamuls® 400-DS
Alkamuls® B
Alkamuls® BR
Alkamuls® CO-40
Alkamuls® EL-719
Alkamuls® EL-985
Alkamuls® GMO
Alkamuls® MS-40
Alkamuls® PSML-20
Alkamuls® PSMO-5
Alkamuls® PSMS-20
Alkamuls® PSTO-20
Alkamuls® R81
Alkamuls® S-6
Alkamuls® S-8
Alkamuls® S-65-8
Alkamuls® S-65-40
Alkamuls® SMO
Alkamuls® STO
Alkamuls® TD-41
Alkanol® 189-S
Alkanol® 6112
Alkanol® A-CN
Alkanol® ND
Alkanol® WXN
Alkanol® XC
Alkasil® NE 58-50
Alkasil® NEP 73-70
Alkasurf® NP-9
Alkasurf® NP-10
Alkasurf® NP-11
Alkasurf® NP-20, NP-20 70%
Alkasurf® OP-10
Alkaterge®-C
Alkaterge®-E
Alphenate GA 65
Alphenate PE Extra
Alphoxat S 110, 120
Alrosperse 11P Flake
Alrowet® D-65
Amerchol® CAB
Amerchol® L-101
Amgard Series
Amiet CD/14
Amiet CD/17
Amiet CD/22
Amiet CD/27
Amiet DT/15
Amiet DT/20
Amiet DT/32
Amiet OD/14
Amiet TD/14
Amiet TD/17
Amiet TD/22
Amiet TD/27
Amiet THD/14
Amiet THD/17
Amiet THD/22
Amiet THD/27
Amiladin
Amiladin C-1802
Amine C
Aminol CA-2
Amphoterge® S
Antarox® BL-214
Antarox® BL-225
Antarox® BL-240
Antarox® EGE 25-2
Antarox® FM 33
Antarox® FM 53
Antarox® FM 63
Anti-foam TP
Antifoam CM Conc.
Aphrogene 5001
Appretan Grades
Aquaperle D34
Aquasol® AR 90
Aquasol® W 90
Arkofil Brands
Arkomon A Conc.
Arlatone® T
Armid® 18
Armid® C
Armid® HT
Armid® O
Armul 910
Armul 912
Armul 1007
Armul 1009
Arnox 912
Aromox® C/12
Arquad® 12-37W
Arquad® 12-50
Arquad® 16-29W
Arquad® 16-50
Arquad® 18-50
Arquad® B-50
Arquad® B-90
Arquad® B-100
Arquad® C-33-W
Arquad® C-50
Arquad® DM14B-90
Arquad® DMHTB-75
Arquad® DMMCB-50
Arquad® S-2C-50
Arquad® S-50
Arquad® T-2C-50
Arquad® T-27W
Arquad® T-50
Arylan® SX85
Arylene M60
Atcowet C
Atlas Defoamer AFC
Atlas EM-2
Atlas EM-13
Atlas EMJ-2
Atlas EMJ-C
Atlas G-1086
Atlas G-1096
Atlas G-1256
Atlas G-1281
Atlas G-1284
Atlas G-1285
Atlas G-1288
Atlas G-1289
Atlas G-1292
Atlas G-1295
Atlas G-1300
Atlas G-1304
Atlas G-1530
Atlas G-1554
Atlas G-1556
Atlas G-1564
Atlas G-2109
Atlas G-2127
Atlas G-2203
Atlasol 103
Atlasol KAD
Atlox 1045A

Textiles and Fibers (cont'd.)

Atlox 1285
Atranonic Polymer 20
Atranonic Polymer N200
Atsolyn PE 27
Avistin® FD
Avistin® PN
Avitone® A
Avitone® F
Avitone® T
Avivan® SFC
AZdry 40
AZdry 70
Babinar 715
Babinar 801C
Barisol Super BRM
Basojet® PEL 200%
Basol® WS
Basophen® M
Basophen® NB-U
Basophen® RA
Basophen® RA
Basophen® RBD
Basophor A
Basopon® LN
Basopon® LN
Basopon® TX-110
Berol 28
Berol 108
Berol 190
Berol 191
Berol 198
Berol 302
Berol 303
Berol 307
Berol 381
Berol 386
Berol 387
Berol 389
Berol 391
Berol 392
Berol 397
Berol 455
Berol 456
Berol 457
Berol 458
Berol 475
Beycostat 319 P
Beycostat B 706 E
Bio-Soft® AS-40
Bio-Soft® CS 50
Bio-Soft® D-40
Bio-Soft® D-62
Bio-Soft® EA-8
Bio-Soft® S-100
Bio-Terge® PAS-8S
Bozemine N 60
Bozemine N 609
Bozemine NSI
Briquest® ADPA-60AW
Bubble Breaker® 259, 260, 613-M, 622, 730, 737, 746, 748, 900, 913, 917
Burco Anionic APS
Burco CS-LF
Burcofac 1060
Burcofac 9125
Cadoussant AS
Calsoft T-60
Calsuds CD-6
Caplube 8385
Caplube 8410
Caplube 8445
Caplube 8448
Caplube 8508
Caplube 8540B
Caplube 8540C
Caplube 8540D
Carbopol® 910
Carbopol® 934
Carbostat 2203
Carriant Series
Carsamide® 7644
Carsonol® SLS Paste B
Carsonon® L-985
Carsonon® TD-10
Carsosoft® S-75
Carsosoft® S-90-M
Carsosoft® T-90
Cassurit Grades
Catigene® T 50
Catigene® T 80
Catinal HTB
Catinex KB-42
Cation DS
Cation SF-10
Catisol AO 100
Catisol AO C
Cedephos® FA600
Cenegen® 7
Cenegen® B
Cenegen® CJB
Cenegen® EKD
Cenegen® NWA
Cenekol® 1141
Cenekol® FT Supra
Cenekol® Liq.
Cenekol® NCS Liq.
Cerfak 1400
Charlab Condensate-K
Charlab Leveler AT Special
Charlab Leveler DSL
Charlab LPC
Chemal 2EH-2
Chemal 2EH-5
Chemal BP 261
Chemal BP-262
Chemal BP-262LF
Chemal BP-2101
Chemal DA-4
Chemal DA-6
Chemal DA-9
Chemal TDA-3
Chemal TDA-6
Chemal TDA-9
Chemal TDA-12
Chemal TDA-15
Chemal TDA-18
Chemax CO-5
Chemax CO-16
Chemax CO-25
Chemax CO-28
Chemax CO-30
Chemax CO-36
Chemax CO-40
Chemax CO-200/50
Chemax DF-30
Chemax DF-100
Chemax DNP-8
Chemax DNP-15
Chemax DNP-150
Chemax DNP-150/50
Chemax DOSS/70
Chemax DOSS-75E
Chemax E-200 MO
Chemax E-200 MS
Chemax E-400 MO
Chemax E-400 MS
Chemax E-600 ML
Chemax E-600 MO
Chemax E-600 MS
Chemax E-1000 MS
Chemax HCO-5
Chemax HCO-16
Chemax HCO-25
Chemax NP-1.5
Chemax NP-6
Chemax NP-9
Chemax NP-10
Chemax NP-15
Chemax NP-20
Chemax NP-30
Chemax NP-30/70
Chemax OP-30/70
Chemax PEG 400 DO
Chemax PEG 600 DO
Chemax SBO
Chemax SCO
Chemax TO-8
Chemax TO-16
Chemeen 18-2
Chemeen 18-5
Chemeen 18-50
Chemeen C-2
Chemeen C-5
Chemeen C-10
Chemeen C-15
Chemeen HT-2
Chemeen HT-5
Chemeen HT-50
Chemeen O-30, O-30/80
Chemeen T-2
Chemeen T-5
Chemeen T-10
Chemeen T-15
Chemeen T-20
Chemfac NC-0910
Chemfac PC-188
Chemfac PD-600
Chemfac PF-623
Chemfac PF-636
Chemfac PN-322
Chemphos TR-495
Chemphos TR-517
Chemquat 12-33
Chemquat 12-50
Chemquat 16-50
Chemquat C/33W
Chemsulf SBO/65
Chemsulf SCO/75
Chimin KSP
Chimin P40
Chimin P50
CHPTA
Chromosol SS
CHT Heptol NWS
Cibaphasol® AS
Cirrasol® AEN-XB
Cirrasol® AEN-XF
Cirrasol® AEN-XZ
Cirrasol® ALN-FP
Cirrasol® ALN-GM
Cirrasol® ALN-TF
Cirrasol® ALN-TS
Cirrasol® ALN-TV
Cirrasol® ALN-WF
Cirrasol® ALN-WY
Cirrasol® EN-MB
Cirrasol® EN-MP
Cirrasol® GM
Cirrasol® LAN-SF
Cirrasol® LC-HK
Cirrasol® LC-PQ
Cirrasol® LN-GS
Cithrol 2DS
Cithrol 2MO
Cithrol 3DS
Cithrol 4DS
Cithrol 4ML
Cithrol 4MO
Cithrol 6DS
Cithrol 6ML
Cithrol 6MO
Cithrol 10DS
Cithrol 10ML
Cithrol 10MO
Cithrol 40MO
Cithrol 60ML
Cithrol 60MO
Cithrol A
Cithrol DGDL N/E
Cithrol DGDL S/E
Cithrol DGDO N/E
Cithrol DGDO S/E
Cithrol DGDS N/E
Cithrol DGDS S/E
Cithrol DGML N/E
Cithrol DGML S/E
Cithrol DGMO N/E
Cithrol DGMO S/E
Cithrol DGMS N/E
Cithrol DGMS S/E
Cithrol DPGML N/E
Cithrol DPGML S/E
Cithrol DPGMO S/E
Cithrol DPGMS N/E
Cithrol DPGMS S/E
Cithrol EGDL N/E
Cithrol EGDL S/E
Cithrol EGDO N/E
Cithrol EGDO S/E
Cithrol EGDS N/E
Cithrol EGDS S/E
Cithrol EGML N/E
Cithrol EGML S/E
Cithrol EGMO N/E
Cithrol EGMO S/E
Cithrol EGMR N/E
Cithrol EGMR S/E
Cithrol EGMS N/E
Cithrol EGMS S/E

Textiles and Fibers (cont'd.)

Cithrol GDL N/E
Cithrol GDL S/E
CNC Antifoam 1-A
CNC Antifoam 1-AP
CNC Antifoam 10-FG
CNC Antifoam 30-FG
CNC Antifoam 495
CNC Antifoam 495-M
CNC Defoamer 229
CNC Defoamer 544-C
CNC Detergent E
CNC Dispersion PE
CNC Foam Assist AA
CNC Foam Assist AA-100
CNC Leveler JH
CNC PAL 1000
CNC PAL AN
CNC PAL DS
CNC PAL NRW
CNC PAL V-8 Supra
CNC Product ST
CNC Sol 72-N Series
CNC Sol BD
CNC Sol UE
CNC Sol XN
CNC Sol XNN #11
CNC Wet CP Conc.
CNC Wet SS-80
Colsol
Compound S.A
Coning Oil C Special
Consamine 15
Consamine CA
Consamine DS Powd.
Consamine K-Gel
Consamine OM
Consamine P
Consamine PA
Consamine X
Consoft CP-50
Consolevel JBT
Consolevel N
Consoluble 71
Consonyl FIX
Consos Castor Oil
Consoscour 47
Consoscour M
Consoscour TEK
Coralon GP
Cordex DJ
Cordon AES-65
Cordon COT
Cordon N-400
Cordon NU 890/75
Coursemin SWG-7
Coursemin SWG-10
Crafol AP-11
Crafol AP-16
Crafol AP-20
Crafol AP-36
Crafol AP-63
Crafol AP-65
Crafol AP-67
Crafol AP-201
Crafol AP-202
Crafol AP-203
Crafol AP-241
Crafol AP-260
Crafol AP-261
Crafol AP-262
Cralane AT-17
Cralane AU-10
Crestex REM 55
Crestolan NF
Crestopene 5X
Crestosolve 630
Crill 1
Crill 2
Crill 3
Crill 4
Crillet 1
Crodamet Series
Crodamine 1.HT
Crodamine 1.O, 1.OD
Crodamine 1.T
Crodamine 1.16D
Crodamine 1.18D
Crodamine 3.A16D
Crodamine 3.A18D
Crodamine 3.ABD
Crodamine 3.AED
Crodamine 3.AOD
Crodasinic LT40
Crodazoline O
Croduret 10
Croduret 30
Croduret 40
Croduret 60
Croduret 100
Cyclomatic Dur
Cycloton® D261C/70
Cyncal® 80%
Farmin 2C
Farmin 20, 60, 68, 80, 86, AB, C
Farmin D86
Farmin HT
Farmin O
Farmin S
Farmin T
Finapal E
Finazoline OA
Findet AD-18
Findet CF-4
Findet CF-440
Findet DD
Findet NHP
Findet OJP-5
Findet OJP-25
Findet SB
Finsist C-2 Conc.
Finsist WW
Fixogene CD Liq.
Flexricin® 9
Flexricin® 15
Flexricin® 100
Fluorad FX-8
Fluowet OL
Fluowet OTN
Fluowet SB
Foamaster 206-A, 267-A, 335-A
Foamaster 340
Foamaster 371-S
Foamaster KF-99
Foamkill® 685
Foamkill® 2890 Conc.
Foamkill® CMP
Foamkill® CPD
Foamkill® D-1
Foamkill® DP
Foamkill® FBF
Foamkill® FCD
Foamkill® MS
Fosfamide N
Fuman 630
G-250
G-263
G-265
G-1045A
G-1086
G-1087
G-1089
G-1096
G-1120
G-1121
G-1292
G-1293
G-1300
G-1556
G-1564
G-2109
G-2162
G-2200
G-2207
G-3300
G-3634A
G-3780-A
G-3886
G-3890
G-7076
Gantrez® AN
Gardinol CX
Gardinol CXM
Gardinol WA Paste
Geitol RC-100
Gemtex PA-75
Gemtex PA-85P
Gemtex PAX-60
Gemtex SC-40
Gemtex SC-70
Gemtex SC-75
Gemtex SC-75E, SC Powd.
Gemtex SM-33
Gemtex WBT-9
Genagen C-100
Genagen CD
Genagen KFC-100
Genagen O-090
Genagen OD-090
Genagen P-070
Genagen PL-090
Genagen S-080
Genagen S-400
Genagen TA-080
Genagen TA-120
Genagen TA-160
Genamin T-020
Genamin T-050
Genamin T-150
Genamin T-200
Genamin T-308
Genamin XET
Genaminox KC
Genapol® 2299
Genapol® C-050
Genapol® C-080
Genapol® C-100
Genapol® C-150
Genapol® C-200
Genapol® DA-040
Genapol® DA-060
Genapol® GC-050
Genapol® GEV
Genapol® O-020
Genapol® O-050
Genapol® O-080
Genapol® O-090
Genapol® O-100
Genapol® O-120
Genapol® O-150
Genapol® O-200
Genapol® O-230
Genapol® PF 10
Genapol® PF 20
Genapol® PF 40
Genapol® PF 80
Genapol® PS
Genapol® S-020
Genapol® S-050
Genapol® S-080
Genapol® S-100
Genapol® S-150
Genapol® S-200
Genapol® S-250
Genapol® T-250
Genapol® X-040
Genapol® X-060
Genapol® X-080
Genapol® X-100
Geropon® 99
Geropon® CYA/DEP
Geropon® DOS
Geropon® DOS FP
Geropon® SS-O-75
Geropon® T-33
Geropon® T-43
Geropon® T-51
Geropon® T-77
Geropon® TBS
Geropon® WS-25, WS-25-I
Geropon® WT-27
Glazamine DP2
Glazamine M
Glycox® PETC
Glytex® 203
Glytex® 213
Glytex® 273
Glytex® 513
Glytex® 558
Glytex® 663
Glytex® 1085
Glytex® EL 176
Glytex® EL 882
Glytex® EL 905
Glytex® L 154
Glytex® L 203
GMS Base
GMS/SE Base

Textiles and Fibers (cont'd.)

GP-218
GP-219
GP-226
GP-7000
Gradonic LFA Series
Gradonic N-95
Gran UP A-600
Gran UP AX-15
Gran UP CS-500
Gran UP CS-700F
Gran UP US-800
Hartolon 1328
Hartolon 5683
Hartolon AL
Hartolon HVH Base
Hartolon NA
Hartolon PC
Hartomul PE-30
Hartonyl L531
Hartonyl L535
Hartonyl L537
Hartosoft 171
Hartosoft CN
Hartosoft GF
Hartosoft S5793
Hartowet 5917
Hartowet MSW
Hartowet SLN
Hetoxamate FA-5
Hetoxamate LA-5
Hetoxamate LA-9
Hetoxamate MO-5
Hetoxamate MO-9
Hetoxamate MO-15
Hetoxamate SA-5
Hetoxamate SA-9
Hetoxamate SA-35
Hetoxamate SA-90
Hetoxamate SA-90F
Hetoxamine C-5
Hetoxamine O-2
Hetoxamine O-5
Hetoxamine O-15
Hetoxamine S-2
Hetoxamine S-5
Hetoxamine S-15
Hetoxamine ST-2
Hetoxamine ST-5
Hetoxamine ST-15
Hetoxamine ST-50
Hetoxamine T-2
Hetoxamine T-5
Hetoxamine T-15
Hetoxamine T-20
Hetoxamine T-30
Hetoxamine T-50
Hetoxamine T-50-70%
Hetoxide BN-13
Hetoxide BP-3
Hetoxide BY-3
Hetoxide C-2
Hetoxide C-9
Hetoxide C-15
Hetoxide C-25
Hetoxide C-30
Hetoxide C-40
Hetoxide C-60
Hetoxide C-200
Hetoxide DNP-5
Hetoxide DNP-10
Hetoxide HC-16
Hetoxol 15 CSA
Hetoxol CA-2
Hetoxol CA-20
Hetoxol CAWS
Hetoxol CS-15
Hetoxol CS-20
Hetoxol CS-30
Hetoxol CS-50
Hetoxol CS-50 Special
Hetoxol CSA-15
Hetoxol D
Hetoxol L-3N
Hetoxol L-4N
Hetoxol L-9
Hetoxol L-9N
Hetoxol LS-9
Hetoxol TD-3
Hetoxol TD-6
Hetoxol TD-12
Hipochem ADN
Hipochem AR-100
Hipochem B-3-M
Hipochem Base MC
Hipochem BSM
Hipochem C-95
Hipochem CAD
Hipochem Carrier 761
Hipochem Carrier TA-3
Hipochem CDL
Hipochem Compatibilizer WMC
Hipochem D-6-H
Hipochem Dispersol GTO
Hipochem Dispersol SB
Hipochem Dispersol SCO
Hipochem Dispersol SP
Hipochem DZ
Hipochem EFK
Hipochem EK-18
Hipochem EK-6
Hipochem Finish 178
Hipochem GM
Hipochem HPEL
Hipochem Jet Dye T
Hipochem Jet Scour
Hipochem JN-6
Hipochem LCA
Hipochem LH-Soap
Hipochem M-51
Hipochem Migrator J
Hipochem MS-BW
Hipochem MS-LF
Hipochem MTD
Hipochem NAC
Hipochem No. 3
Hipochem No. 40-L
Hipochem No. 641
Hipochem NOC
Hipochem PDO
Hipochem PND-11
Hipochem Retarder CJ
Hipochem RPS
Hipochem SO
Hipochem SRC
Hipochem WSS
Hipofix 491
Hipolon New
Hiposcour® 1
Hiposcour® 3-80
Hiposcour® 6
Hiposcour® ARG-2
Hiposcour® BFS
Hiposcour® NFMS-2
Hipowet IBS
Hodag Amine C-100-L
Hodag Amine C-100-O
Hodag Amine C-100-S
Honol 405
Honol GA
Honol MGR
Honoralin PL
Hostapal 3634 Highly Conc.
Hostapal BV Conc.
Hostapal CV Brands
Hostapal CVH
Hostapal N-040
Hostapal N-060R
Hostapal N-300
Hostapal SF
Hostaphat 2122
Hostaphat 2188
Hostaphat 2204
Hostaphat F Brands
Hostaphat FL-340 N
Hostaphat FO-380
Hostaphat HI
Hostaphat MDIT
Hostaphat OD
Hostapon IDC
Hostapon T Powd.
Hostapur CX Highly Conc.
Hostapur DOS Hi Conc.
Hostapur SAS 60
Hydrolene
Hymolon CWC
Hymolon K90
Hyonic NP-60
Hyonic NP-90
Hyonic OP-7
Hyonic OP-10
Hyonic OP-55
Hyonic OP-70
Hyonic OP-100
Hyonic PE-90
Hystrene® 1835
Hystrene® 3022
Hystrene® 4516
Hystrene® 5012
Hystrene® 5016
Hystrene® 5522
Hystrene® 7018
Hystrene® 7022
Hystrene® 9014
Hystrene® 9016
Hystrene® 9514
Hystrene® 9718 NF FG
Hystrene® 9912
Iberpenetrant-114
Iberpol Gel
Iberpon W
Iberscour AC
Iberscour P
Iberscour W Conc.
Iberterge 65
Iberterge CO-40
Iberwet BO
Iberwet E
Iberwet W-100
Ice #2
Icomeen® O-30
Icomeen® O-30-80%
Icomeen® S-5
Icomeen® T-2
Icomeen® T-5
Icomeen® T-7
Icomeen® T-25
Icomeen® T-25 CWS
Icomeen® T-40
Icomeen® T-40-80%
Iconol DA-4
Iconol DA-6
Iconol DA-6-90%
Iconol DA-9
Iconol TDA-3
Iconol TDA-6
Iconol TDA-8
Iconol TDA-8-90%
Iconol TDA-10
Idet 5L SP NF
Idet 10, 20–P
Igepal® CA-620
Igepal® CA-880
Igepal® CA-887
Igepal® CA-890
Igepal® CA-897
Igepal® CO-430
Igepal® CO-630
Igepal® CO-660
Igepal® CO-710
Igepal® CO-720
Igepal® CO-880
Igepal® CO-887
Igepal® DM-530
Igepal® DM-730
Igepal® DM-970 FLK
Igepal® KA
Igepal® NP-2
Igepal® NP-5
Igepal® NP-8
Igepal® NP-10
Igepal® NP-12
Igepal® NP-20
Igepal® O
Igepal® RC-520
Imacol JN
Imacol S Liq.
Imerol SS Liq.
Imerol VLF Liq.
Incrocas 30
Incrocas 40
Incromate SDL
Incropol CS-20
Incropol CS-50
Incroquat SDQ-25
Incrosoft 100
Incrosoft 100P
Incrosoft 248

Textiles and Fibers (cont'd.)

Incrosoft CFI-75
Incrosoft S-75
Incrosoft S-90
Incrosoft S-90M
Incrosoft T-75
Incrosoft T-90
Industrene® 223
Intex Scour 707
Intracarrier® ATM
Intrafomil® AK
Intralan® Salt HA
Intralan® Salt N
Intraphasol COP
Intraphasol PC
Intrapol 1014
Intraquest® TA Sol'n
Intrassist® LA-LF
Intratex® A
Intratex® AN
Intratex® B
Intratex® BD
Intratex® C
Intratex® CA-2
Intratex® DD
Intratex® DD-LF
Intratex® JD
Intratex® JD-E
Intratex® N
Intratex® N-1
Intratex® OR
Intratex® POK
Intratex® W New
Intravon® AN
Intravon® JET
Intravon® JF
Intravon® JU
Intravon® NI
Intravon® SO
Intravon® SOL
Intravon® SOL-N
Intravon® SOL-W
Ionet 300
Ionet LD-7-200
Ionet RAP-50
Ionet RAP-80
Ionet RAP-250
Ionet S-20
Ionet S-60 C
Ionet S-80
Ionet S-85
Ionet T-20 C
Ionet T-60 C
Ionet T-80 C
Irgalev® PBF
Irgalube® 53
J Wet 19A
Jet Quat 2C-75
Jet Quat 2HT-75
Jet Quat C-50
Jet Quat DT-50
Jet Quat S-2C-50
Jet Quat S-50
Jet Quat T-27W
Jet Quat T-50
Kamar BL
Karafac 78
Karafac 78 (LF)
Karamide 121
Karamide 363
Karamide HTDA
Karamide SDA
Karapeg DEG-MO
Karapeg DEG-MS
Karaphos HSPE
Karaphos SWPE
Karaphos XFA
Karasoft YB-11
Karawet DOSS
Katapol® OA-860
Kelco® HV
Kelco® LV
Kelgin® F
Kelgin® HV, LV, MV
Kelgin® QL
Kelgin® XL
Kelset®
Kemamide® B
Kemamide® E
Kemamide® O
Kemamide® S
Kemamide® U
Kemamide® W-20
Kemamide® W-39
Kemamide® W-40
Kemamide® W-40/300
Kemamide® W-45
Kemamine® BQ-2802C
Kemamine® BQ-9702C
Kemamine® BQ-9742C
Kemamine® Q-1902C
Kemamine® Q-2802C
Kemamine® Q-6502C
Kemamine® Q-9702C
Kemamine® Q-9743C
Kemamine® Q-9743CHGW
Kemamine® T-1902D
Kemamine® T-6501
Kemamine® T-6502D
Kemamine® T-9701
Kemamine® T-9702D
Kemamine® T-9892D
Kemamine® T-9902
Kemamine® T-9972D
Kemamine® T-9992D
Kemester® 1000
Kemester® 2000
Kemester® 4000
Kemester® 5221SE
Kemester® 5500
Kemester® 6000SE
Kerasol 1014
Kerinol C 109
Kessco® 874
Kessco® 887
Kessco® 891
Kessco® BS
Kessco® GMO
Kessco® GMS
Kessco® IPM
Kessco® IPP
Kessco® OP
Kessco® PEG 200 DL
Kessco® PEG 400 DS
Kessco® PEG 600 DL
Kessco® PEG 600 DS
Kessco® PEG 600 ML
Kessco® PEG 6000 DS
Kieralon® B, B Highly Conc.
Kieralon® C
Kieralon® D
Kieralon® ED
Kieralon® JET
Kieralon® KB
Kieralon® NB-150
Kieralon® NB-ED
Kieralon® NB-OL
Kieralon® OL
Kieralon® TX-199
Kieralon® TX-410 Conc.
Kilfoam
Kingoil S-10
Klearfac® AA040
Klearfac® AA420
Klearfac® AB270
Lamigen ES 30
Lamigen ES 60
Lamigen ES 100
Lamigen ET 20
Lamigen ET 70
Lamigen ET 90
Lamigen ET 180
LAN-401
Lankropol® ADF
Lankropol® WA
Lankropol® WN
Lanolin Alcohols LO
Larosol ALM-1
Larosol DBL
Larosol DBL-3
Larosol DBL-3 Conc.
Larosol NLA-25
Larosol NRL Conc.
Larosol NRL-40
Larosol PDQ-2
Larosol PNC
Larostat® 143
Larostat® 264 A Anhyd.
Larostat® 264 A Conc.
Larostat® 300 A
Larostat® 1084
Lauramide D
Laurel M-10-257
Laurel M-10-257
Laurel PEG 400 DT
Laurel PEG 400 DT
Laurel PEG 400 MO
Laurel PEG 400 MO
Laurel PEG 400 MT
Laurel PEG 400 MT
Laurel PEG 600 DT
Laurel PEG 600 DT
Laurel PEG 600 MT
Laurel R-50
Laurel R-75
Laurel SBT
Laurel SD-101
Laurel SMR
Laurelox 12
Laurelphos E-61
Laurelterge 837, 1390
Laureltex 308, 308 LF
Laureltex 6030, 6030S
Laureltex CW
Lauridit® KM
Lauridit® LM
Lauropal 9
Lauropal 11
Lauropal 0205
Lauropal 950
Lauropal 1150
Lauroxal 6
Laventin® CW
Laventin® W
Lenetol B Conc.
Lenetol HP-LFN
Lenetol KWB
Lenetol PS
Lenetol WLF 125
Leomin AN
Leomin CN
Leomin FA
Leomin FANF
Leomin HSG
Leomin KP
Leomin LS
Leomin OR
Leonil DB Powd.
Leonil EBL
Leonil L
Leonil UN
Leophen® BN
Leophen® LG
Leophen® M
Leophen® ML
Leophen® RA
Leophen® RBD
Leophen® U
Levelan NKD
Levelan R-15, -200
Levelan WS
Levelan® P208
Levelan® P307
Levelan® PE 304
Levelan® PG 434
Levelene
Levenol A Conc.
Levenol DS-1
Levenol PW
Levenol RK
Levenol TD-326
Levenol WX, WZ
Lipolan AO
Lipolan AOL
Lipolan TE, TE(P)
Lipomin CH
Liponox NC 6E, NCG, NCI, NCT
Liponox NC-70
Lipophos PE9
Lipophos PL6
Lipotac TE
Lipotac TE-P
Lipotin 100, 100J, SB
Lipotin A
Lipotin H
Lomar® LS
Lomar® PW
Lonzaine® C
Lonzaine® CO

Textiles and Fibers (cont'd.)

Lonzest® PEG 4-DO
Lonzest® PEG 4-L
Lonzest® SML
Lonzest® SMO
Lonzest® SMP
Lonzest® SMS
Lonzest® STO
Lonzest® STS
Lufibrol® E
Lufibrol® FW
Lufibrol® KB Liq.
Lufibrol® KE
Lufibrol® NB-7
Lufibrol® NB-T
Lufibrol® O
Luprintol® PE
Lutopon SN
Lutostat 171
Lyogen AFS Liq.
Lyogen BE Liq.
Lyogen DFT Liq.
Lyogen F Liq.
Lyogen KF Liq.
Lyogen MS Liq.
Lyogen NL Liq., P Liq.
Lyogen PAA Liq.
Lyogen SF Liq.
Lyogen SMK-40 Liq., SMK Paste
Lyogen V (U) Liq.
Lyogen WD Liq.
Mackester EGDS
Mackester EGMS
Mackester IDO
Mackester IP
Mackester SP
Mackester TD-88
Macol® 1
Macol® 2
Macol® 2D
Macol® 2LF
Macol® 4
Macol® 8
Macol® 18
Macol® 19
Macol® 22
Macol® 27
Macol® 31
Macol® 32
Macol® 33
Macol® 34
Macol® 35
Macol® 40
Macol® 42
Macol® 44
Macol® 46
Macol® 72
Macol® 77
Macol® 85
Macol® 88
Macol® 101
Macol® 108
Macol® CA-2
Macol® CSA-2
Macol® CSA-4
Macol® CSA-10
Macol® CSA-15
Macol® CSA-20
Macol® CSA-50
Macol® DNP-5
Macol® DNP-10
Macol® DNP-15
Macol® DNP-21
Macol® LA-4
Macol® LA-9
Macol® LA-12
Macol® LA-23
Macol® LA-790
Macol® NP-4
Macol® NP-5
Macol® NP-6
Macol® NP-8
Macol® NP-9.5
Macol® NP-11
Macol® NP-12
Macol® NP-15
Macol® NP-20
Macol® NP-20(70)
Macol® NP-30(70)
Macol® NP-100
Macol® OA-2
Macol® OA-4
Macol® OA-5
Macol® OP-3
Macol® OP-5
Macol® OP-8
Macol® OP-10
Macol® P-500
Macol® P-1200
Macol® P-1750
Macol® P-2000
Macol® P-3000
Macol® P-4000
Macol® SA-2
Macol® SA-5
Macol® SA-10
Macol® SA-15
Macol® SA-20
Macol® SA-40
Macol® TD-3
Macol® TD-8
Macol® TD-10
Macol® TD-12
Macol® TD-15
Macol® TD-610
Makon® 4
Makon® 6
Makon® 8
Makon® 10
Makon® 11
Makon® 12
Makon® 14
Makon® 30
Makon® 40
Makon® NF-5
Makon® NF-12
Makon® OP-9
Manro SDBS 25/30
Mapeg® 200 DL
Mapeg® 200 DO
Mapeg® 200 DS
Mapeg® 200 ML
Mapeg® 200 MO
Mapeg® 200 MOT
Mapeg® 200 MS
Mapeg® 400 DL
Mapeg® 400 DO
Mapeg® 400 DOT
Mapeg® 400 DS
Mapeg® 400 ML
Mapeg® 400 MO
Mapeg® 400 MOT
Mapeg® 400 MS
Mapeg® 600 DL
Mapeg® 600 DO
Mapeg® 600 DOT
Mapeg® 600 ML
Mapeg® 600 MO
Mapeg® 600 MOT
Mapeg® 600 MS
Mapeg® 6000 DS
Mapeg® EGDS
Mapeg® EGMS
Mapeg® S-40
Mapeg® TAO-15
Maphos® 60A
Maphos® DT
Maranil DBS
Markwet NR-25
Marlamid® A 18
Marlamid® A 18 E
Marlamid® AS 18
Marlamid® D 1218
Marlamid® DF 1218
Marlamid® O 18
Marlamid® OS 18
Marlazin® L 10
Marlazin® L 410
Marlazin® OL 2
Marlazin® S 10
Marlazin® S 40
Marlazin® T 10
Marlazin® T 15/2
Marlazin® T 16/1
Marlazin® T 50
Marlinat® DF 8
Marlinat® HA 12
Marlipal® 013/30
Marlipal® 013/50
Marlipal® 013/400
Marlipal® 24/30
Marlipal® 24/40
Marlipal® 24/50
Marlipal® 24/60
Marlipal® 24/70
Marlipal® 24/80
Marlipal® 24/90
Marlipal® 24/120
Marlipal® 24/939
Marlipal® 34/30, /50, /60, /70, /79, /99, /100, /109, /110, /119, /120, /140
Marlipal® 124
Marlipal® 1850/5
Marlipal® 1850/10
Marlipal® 1850/30
Marlipal® 1850/40
Marlipal® 1850/80
Marlipal® KF
Marlipal® NE
Marlipal® O11/30
Marlipal® O11/50
Marlipal® O11/79
Marlipal® O11/88
Marlipal® O13/20
Marlipal® O13/30
Marlipal® O13/40
Marlipal® O13/50
Marlipal® O13/80
Marlipal® O13/90
Marlipal® O13/100
Marlipal® O13/120
Marlipal® O13/150
Marlipal® O13/170
Marlipal® O13/200
Marlipal® O13/400
Marlipal® O13/500
Marlon® A 350
Marlon® A 360
Marlon® A 365
Marlon® A 375
Marlon® AS3
Marlon® AS3-R
Marlon® PS 30
Marlon® PS 60
Marlon® PS 60 W
Marlon® PS 65
Marlophen® 87
Marlophen® 87N
Marlophen® 88
Marlophen® 88N
Marlophen® 89
Marlophen® 89.5N
Marlophen® 810
Marlophen® 810N
Marlophen® 812
Marlophen® 812N
Marlophen® 1028
Marlophen® 1028N
Marlophen® DNP 16
Marlophen® DNP 18
Marlophen® DNP 30
Marlophen® DNP 100
Marlophen® DNP 150
Marlophor® CS-Acid
Marlophor® DS-Acid
Marlophor® F1-Acid
Marlophor® FC-Sodium Salt
Marlophor® IH-Acid
Marlophor® LN-Acid
Marlophor® MO 3-Acid
Marlophor® N5-Acid
Marlophor® ND
Marlophor® ND-Acid
Marlophor® ND DEA Salt
Marlophor® ND NA-Salt
Marlophor® OC5-Acid
Marlophor® ON3-Acid, ON5-Acid, ON7-Acid
Marlophor® T6-Acid
Marlophor® T10-Acid
Marlophor® UW12-Acid
Marlosoft® A 18 M
Marlosoft® B 18 M
Marlosoft® IQ 75
Marlosoft® IQ 90

Textiles and Fibers (cont'd.)

Marlosol® 183
Marlosol® 186
Marlosol® 188
Marlosol® 189
Marlosol® 1820
Marlosol® 1825
Marlosol® BS
Marlosol® F08
Marlosol® FS
Marlosol® OL2
Marlosol® OL10
Marlosol® OL15
Marlosol® OL20
Marlosol® TF3
Marlosol® TF4
Marlowet® 1072
Marlowet® 4536
Marlowet® 4538
Marlowet® 4541
Marlowet® 4603
Marlowet® 4703
Marlowet® 4800
Marlowet® 4857
Marlowet® 5001
Marlowet® 5311
Marlowet® 5400
Marlowet® 5401
Marlowet® BL
Marlowet® EF
Marlowet® FOX
Marlowet® GFW
Marlowet® LVX
Marlowet® MA
Marlowet® NF
Marlowet® OFA
Marlowet® OFW
Marlowet® OTS
Marlowet® PW
Marlowet® R 20
Marlowet® R 22
Marlowet® R 25
Marlowet® R 32
Marlowet® R 36
Marlowet® R 40
Marlowet® R 54
Marlowet® SAF
Marlowet® SLM
Marlowet® SLS
Marlowet® SW
Marlowet® SWN
Marlowet® WOE
Marlowet® WSD
Marlox® 3000
Marlox® B 24/50
Marlox® B 24/60
Marlox® B 24/80
Marlox® FK 14
Marlox® FK 64
Marlox® FK 69
Marlox® FK 86
Marlox® FK 1614
Marlox® L 6
Marlox® MO 124
Marlox® MO 145
Marlox® MO 174
Marlox® MO 244
Marlox® MS 48
Marlox® ND 121
Marlox® OD 105
Marlox® Q 286
Marlox® T 50/5
Marvanfix NDF
Marvanfix® ATA
Marvanfix® C
Marvanfix® FNC
Marvanlube 1031
Marvanol® Aftertreat 2AF
Marvanol® BAN
Marvanol® BVD
Marvanol® CO
Marvanol® DC
Marvanol® Defoamer AM-2
Marvanol® Defoamer MOB
Marvanol® Defoamer MR-30A
Marvanol® Defoamer S-22
Marvanol® GC
Marvanol® KMA
Marvanol® KXL
Marvanol® Leveler DL
Marvanol® Leveltone 7.5
Marvanol® LSL
Marvanol® MRB
Marvanol® NHM
Marvanol® Penetrant 35
Marvanol® POL-41
Marvanol® Pretreat GD-P
Marvanol® Pretreat HPC
Marvanol® RD2-1852
Marvanol® RD2-2284
Marvanol® RD2-2581
Marvanol® RDF
Marvanol® RE-1274
Marvanol® RE-1281
Marvanol® RE-1824
Marvanol® REAC A-213
Marvanol® REACT 1051
Marvanol® SBO (60%)
Marvanol® Scour 2 Base
Marvanol® Scour 05
Marvanol® Scour 2582
Marvanol® Scour FRM
Marvanol® Scour LF
Marvanol® Scour PCO
Marvanol® Solvent Scour 34
Marvanol® SOR
Marvanol® SPO (60%)
Marvanscour® KW
Marvanscour® LF
Marvantex BS
Marvantex RBDS
Marvantex T-100
Marvelin W-50
Masil® 1066C
Masil® 1066D
Masil® 2132
Masil® 2133
Masil® 2134
Masil® SF 5
Masil® SF 10
Masil® SF 20
Masil® SF 50
Masil® SF 100
Masil® SF 350
Masil® SF 350 FG
Masil® SF 500
Masil® SF 1000
Masil® SF 5000
Masil® SF 10,000
Masil® SF 12,500
Masil® SF 30,000
Masil® SF 60,000
Masil® SF 100,000
Masil® SF 500,000
Masil® SF 600,000
Masil® SF 1,000,000
Matexil AA-NS
Matexil BA-PK
Matexil Binder AS
Matexil Binder BD
Matexil BN PA
Matexil CA MN
Matexil DA N
Matexil DN VL 200
Matexil FA MIV
Matexil FA N
Matexil FA SN Liq.
Matexil FC ER
Matexil FC PN
Matexil Fixer SF
Matexil LA NS
Matexil LC CWL
Matexil LC RA
Matexil LN RD
Matexil PA Liq.
Matexil PA SNX Liq.
Matexil PN DG
Matexil PN MFC
Matexil PN PR
Matexil Softener GK
Matexil Thickener CP
Matexil WA HS
Matexil WA KBN
Matexil WN PB
Maxitol No. 10
May-Tein CT
May-Tein SK
Mazamide® L-298
Mazawet® 36
Mazawet® 77
Mazawet® DOSS 70
Mazeen® 173
Mazeen® 174
Mazeen® 174-75
Mazeen® C-2
Mazeen® C-5
Mazeen® C-10
Mazeen® C-15
Mazeen® S-2
Mazeen® S-5
Mazeen® S-10
Mazeen® S-15
Mazeen® T-2
Mazeen® T-5
Mazeen® T-10
Mazeen® T-15
Mazol® 159
Mazol® GMS-90
Mazol® GMS-D
Mazon® 60T
Mazox® CDA
Mazox® SDA
Mazu® DF 210SX
Mazu® DF 210SXSP
Mazu® DF 243
Medialan Brands
Melatex AS-80
Merbron R
Merce Assist ADB
Mercerol AW-LF
Mercerol GVC-65 Liq.
Mercerol NL
Mercerol QW
Mercerol SM
Merpol® A
Merpol® C
Merpol® CH-196
Merpol® DA
Merpol® HCS
Merpol® HCW
Merpol® LF-H
Merpol® OJS
Merpol® SE
Merpol® SH
Mersitol 2434 AP
Mersolat H 30
Mersolat H 40
Mersolat H 68
Mersolat H 76
Mersolat H 95
Mersolat W 40
Mersolat W 68
Mersolat W 76
Mersolat W 93
Metachloron® A4 Liq.
Metrosol AZ
Michelene 10
Michelene 15
Migregal 2N
Migregal NC-2
Miramine® C
Miranol® DM
Miranol® DM Conc. 45%
Mirapol® A-15
Mirapol® WT
Mirataine® T2C-30
Mitin® FF High Conc.
Monafax 1293
Monamid® 15-70W
Monamid® 853
Monamine ALX-100 S
Monaquat ISIES
Monaquat TG
Monateric 1188M
Monateric CM-36S
Monateric CyNa-50
Monatrope 1250
Monawet MM-80
Monawet MO-70
Monawet MO-70E

Textiles and Fibers (cont'd.)

Monawet MO-70R
Monawet SNO-35
Monawet TD-30
Monazoline C
Monazoline CY
Monazoline O
Monazoline T
Monogen
Monolan® 2500 E/30
Monolan® 3000 E/60, 8000 E/80
Monolan® PB
Monolan® PEG 300
Monolan® PEG 400
Monolan® PEG 600
Monolan® PEG 1000
Monolan® PEG 1500
Monolan® PEG 4000
Monolan® PEG 6000
Monolan® PT
Monopole Oil
Monosulf
Montaline 9575 M
Mulsifan RT 1
Mulsifan RT 2
Mulsifan RT 7
Mulsifan RT 23
Mulsifan RT 24
Mulsifan RT 37
Mulsifan RT 110
Mulsifan RT 113
Mulsifan RT 258
Nacconol® 40G
Nansa® BXS
Nansa® EVM50
Nansa® EVM70/B
Nansa® EVM70/E
Nansa® HS80S
Nansa® HS85S
Nansa® LSS38/A
Nansa® MA30
Nansa® TS 60
Natrex D 3
Natrex DSP 213
Natrex GA 251
Natrex J 3
Natrex SO
Naxchem CD-6M
Naxchem Detergent CNB
Naxchem Emulsifier 700
Naxel AAS-40S
Naxel AAS-45S
Naxel AAS-60S
Naxide 1230
Naxolate WAG
Naxonic NI-40
Naxonic NI-60
Naxonol CO
Naxonol PN 66
Naxonol PO
Nekanil® 907
Nekanil® 910
Nekanil® LN
Neodol® 1-5
Neoscoa 203C
Neoscoa 363
Neoscoa 500C
Neoscoa 2326
Neoscoa CM-40
Neoscoa CM-57
Neoscoa ED-201C
Neoscoa GF 3C
Neoscoa GF-2000
Neoscoa OT-80E
Neoscoa PRA-8C
Neospinol 264
Neospinol 358
Neustrene® 045
Neustrene® 053
Neustrene® 060
Neustrene® 064
Newcol 405, 410, 420
Newlon K-1
Niaproof® Anionic Surfactant 08
Ninate® 411
Ninol® 11-CM
Ninol® 40-CO
Ninol® 201
Ninol® 1301
Niox KF-25
Niox KG-14
Niox KH Series
Niox KI Series
Niox KL-16
Niox KP-68
Niox KP-69
Niox KQ-32
Niox KQ-36
Niox KQ-55, -56
Niox KQ-70, LQ-13
Nissan Amine AB
Nissan Amine ABT
Nissan Amine BB
Nissan Amine FB
Nissan Amine MB
Nissan Amine OB
Nissan Amine PB
Nissan Amine SB
Nissan Amine VB
Nissan Anon BF
Nissan Cation AB
Nissan Cation ABT-350, 500
Nissan Cation BB
Nissan Cation F2-10R, -20R, -40E, -50
Nissan Cation FB, FB-500
Nissan Cation L-207
Nissan Cation M2-100
Nissan Cation PB-40, -300
Nissan Cation S2-100
Nissan Diapon TO
Nissan Dispanol LS-100
Nissan Dispanol N-100
Nissan Dispanol TOC
Nissan Nonion E-205
Nissan Nonion E-206
Nissan Nonion E-208
Nissan Nonion E-215
Nissan Nonion E-220
Nissan Nonion E-230
Nissan Nonion HS-204.5
Nissan Nonion HS-206
Nissan Nonion HS-208
Nissan Nonion HS-210
Nissan Nonion HS-215
Nissan Nonion HS-220
Nissan Nonion HS-240
Nissan Nonion HS-270
Nissan Nonion K-202
Nissan Nonion K-203
Nissan Nonion K-204
Nissan Nonion K-207
Nissan Nonion K-211
Nissan Nonion K-215
Nissan Nonion K-220
Nissan Nonion K-230
Nissan Nonion L-2
Nissan Nonion L-4
Nissan Nonion LP-20R, LP-20RS
Nissan Nonion NS-202
Nissan Nonion NS-204.5
Nissan Nonion NS-206
Nissan Nonion NS-209
Nissan Nonion NS-210
Nissan Nonion NS-212
Nissan Nonion NS-215
Nissan Nonion NS-220
Nissan Nonion NS-230
Nissan Nonion NS-240
Nissan Nonion NS-250
Nissan Nonion NS-270
Nissan Nonion O-2
Nissan Nonion O-3
Nissan Nonion O-4
Nissan Nonion O-6
Nissan Nonion P-6
Nissan Nonion P-208
Nissan Nonion P-210
Nissan Nonion P-213
Nissan Nonion S-2
Nissan Nonion S-4
Nissan Nonion S-6
Nissan Nonion S-10
Nissan Nonion S-15
Nissan Nonion S-15.4
Nissan Nonion S-40
Nissan Nonion S-206
Nissan Nonion S-207
Nissan Nonion S-215
Nissan Nonion S-220
Nissan Nonion T-15
Nissan Nymeen DT-203, -208
Nissan Nymeen L-201
Nissan Nymeen L-202, 207
Nissan Nymeen S-202, S-204
Nissan Nymeen S-210
Nissan Nymeen S-215, -220
Nissan Nymeen T2-206, -210
Nissan Nymeen T2-230, -260
Nissan Persoft EK
Nissan Persoft NK-60, -100
Nissan Softer 706
Nissan Softer 1000
Nissan Stafoam DF
Nissan Trax H-45
Nissan Trax K-40
Nofome 2510
Nonal 206
Nonal 208
Nonal 210
Nonal 310
Nonex C5E
Nonex DL-2
Nonex S3E
Nonfome IDC
Nonipol BX
Nonipol D-160
Nonipol T-20
Nonipol T-28
Nonipol T-100
Nonipol TH
Nonisol 100
Nonisol 210
Nopalcol 1-S
Nopalcol 1-TW
Nopalcol Series
Nopco® 1179
Nopco® 2272-R
Nopcogen 14-S
Nopcogen 16-L
Nopcogen 22-O
Nopcosperse 28-B
Nopcosperse WEZ
Nopcosulf TA-30
Nopcosulf TA-45V
Noram O
Noram S
Norfox® 916
Norfox® DCS
Norfox® Oleic Flakes
Norfox® PEA-N
Norfox® Vertex Flakes
NP-55-80
NP-55-85
NP-55-90
Nutrol S-60 5350
Nuva F, FH
Nylomine Assistant DN
Nysist
Oakite Defoamant RC
Obazoline 662Y
Octosol A-1
Octosol A-18
Octowet 40
Octowet 55
Octowet 60
Octowet 60-I
Octowet 70BC
Octowet 75
Ogtac-85
Olapon ND-9, SW
Osimol Grunau 109
Osimol Grunau 110
Osimol Grunau DP
Osimol Grunau EFA
Osimol Grunau MA
Osimol Grunau PHT
Osimol Grunau RAC
Osimol Grunau SF

Textiles and Fibers (cont'd.)

Ospin Salt ON
Ospin TAN
Ospol 790
Oxetal 500/85
Oxetal D 104
Oxetal ID 104
Oxetal O 108
Oxetal O 112
Oxetal TG 111
Oxetal TG 118
Pamolyn® 100
Patogen 311
Patogen 345
Patogen 353
Patogen 378
Patogen 393
Patogen AO-30
Patogen CAC
Patogen P-10 Acid
Patogen PD-3
Patogen SME
Pat-Wet LF-55
Pat-Wet Q-4
Pat-Wet SP
Pat-Wet SW
Pegnol OA-400
Pegnol PDS-60
Pegosperse® 200 DL
Pegosperse® 400 DOT
Pegosperse® PMS CG
Pelex OT-P
Pepol BS-184
Pepol BS-201
Pepol BS-403
Peregal® ST
Perlankrol® DGS
Perlankrol® FV70
Permalene A-100
Permalose TM
Peronal MTB
Peroxal 36
Petronate® CR
Petronate® HL
Petronate® K
Petronate® L
Petronate® S
Petrosul® H-50
Petrosul® H-60
Petrosul® H-70
Petrosul® HM-62, HM-70
Petrosul® L-60
Petrosul® M-50
Petrosul® M-60
Petrowet® R
PG No. 4
PGE-600-ML
PGE-600-MS
Phosfac 1068 C-FA
Phosfac 5508
Phosphanol Series
Phospholan® PNP9
Phospholan® PSP6
Phospholipid PTD
Phospholipid PTL
Phospholipid PTZ
Pinamine A
Pinamine K
Pineotrene K
Plantaren 600 CS UP
Pluracol® E200
Pluracol® E400 NF
Pluracol® E600
Pluracol® E1450
Pluracol® E1450 NF
Pluracol® E2000
Pluracol® E4000 NF
Pluracol® E4500
Pluracol® E8000
Pluracol® E8000 NF
Pluracol® W170
Pluracol® W260
Pluracol® W660
Pluracol® W2000
Pluracol® W3520N
Pluracol® W3520N-RL
Pluracol® W5100N
Pluracol® WD90K
Pluracol® WD1400
Pluriol® E 200
Pluriol® E 300
Pluriol® E 400
Pluriol® E 600
Pluriol® E 1500
Pluriol® E 4000
Pluriol® E 6000
Pluronic® 10R5
Pluronic® 10R8
Pluronic® 17R1
Pluronic® 17R8
Pluronic® 22R4
Pluronic® 25R1
Pluronic® F38
Pluronic® F68
Pluronic® F68LF
Pluronic® F77
Pluronic® L10
Pluronic® L31
Pluronic® L35
Pluronic® L42
Pluronic® L43
Pluronic® L44
Pluronic® L61
Pluronic® L62
Pluronic® L62D
Pluronic® L62LF
Pluronic® L63
Pluronic® L64
Pluronic® L72
Pluronic® L81
Pluronic® L92
Pluronic® L101
Pluronic® L121
Pluronic® L122
Pluronic® P65
Pluronic® P75
Pluronic® P84
Pluronic® P85
Pluronic® P94
Pluronic® P103
Pluronic® P104
Pluronic® P105
Pluronic® P123
Pogol 200
Polyanthrene KS Liq. New
Poly-G® 200
Poly-G® 300
Poly-G® 400
Poly-G® 600
Poly-G® 1000
Poly-G® 1500
Poly-G® 2000
Poly-G® B1530
Polyglycol B-11-50
Polyglycol B-11-100
Polyglycol B-11-150
Polyglycol B-11-260
Polyglycol B-11-660
Polyglycol P-41-300
Polylev #745
Polylube #745
Polylube 4507
Polylube 5713
Polylube GK
Polylube RE
Polylube SC
Polylube Wax
Polylube WS
Polysol J
Polyspin PA
Poly-Tergent® 2A1 Acid
Poly-Tergent® 2A1-L
Poly-Tergent® 3B2
Poly-Tergent® 3B2 Acid
Poly-Tergent® 4C3
Poly-Tergent® E-17A
Poly-Tergent® E-17B
Poly-Tergent® E-25B
Poly-Tergent® J-200
Poly-Tergent® J-300
Poly-Tergent® P-17A
Poly-Tergent® P-17B
Poly-Tergent® P-17BLF
Poly-Tergent® P-17BX
Poly-Tergent® P-17D
Poly-Tergent® P-32D
Poly-Tergent® S-305LF
Poly-Tergent® S-405LF
Poly-Tergent® S-505LF
Poly-Tergent® SL-42
Poly-Tergent® SL-92
Prechem 90
Prechem 120
Prechem 2000
Prechem NFE
Prechem NPX
Prestogen® K
Prifac 9428
Prifac 9429
Primasol® FP
Primasol® KW
Primasol® NB-NF
Primasol® SD
Printac CN-230
Pripol 1004
Product BCO
Progalan X-13
Progasol 40
Progasol 230
Progasol 443, 457
Progasol FSD
Propetal 241
Prosol 518
Prosol 525
Prosol CT30
Prosol RS
Prote-pon P-2 EHA-02-K30
Prote-pon P 2 EHA-02-Z
Prote-pon P 2 EHA-Z
Prote-pon P-0101-02-Z
Prote-pon P-L 201-02-K30
Prote-pon P-L 201-02-Z
Prote-pon P-NP-06-K30
Prote-pon P-NP-06-Z
Prote-pon P-NP-10-K30
Prote-pon P-NP-10-MZ
Prote-pon P-NP-10-Z
Prote-pon P-OX 101-02-K75
Prote-pon P-TD-06-K13
Prote-pon P-TD-06-K30
Prote-pon P-TD 06-K60
Prote-pon P-TD-06-Z
Prote-pon P-TD-09-Z
Prote-pon P-TD-12-Z
Prote-pon TD-09-K30
Prote-sorb SML
Prote-sorb SMO
Prote-sorb SMP
Prote-sorb STO
Prote-sorb STS
Protowet 100
Protowet 5171
Protowet C
Protowet D-75
Protowet E-4
Protowet MB
Prox-onic 2EHA-1/02
Prox-onic 2EHA-1/05
Prox-onic CC-05
Prox-onic CC-09
Prox-onic CC-014
Prox-onic CSA-1/04
Prox-onic CSA-1/06
Prox-onic CSA-1/010
Prox-onic CSA-1/015
Prox-onic CSA-1/020
Prox-onic CSA-1/030
Prox-onic CSA-1/050
Prox-onic DA-1/04
Prox-onic DA-1/06
Prox-onic DA-1/09
Prox-onic DNP-08
Prox-onic DNP-0150
Prox-onic DNP-0150/50
Prox-onic DT-03
Prox-onic DT-015
Prox-onic DT-030
Prox-onic EP 1090-1
Prox-onic HR-05
Prox-onic HR-016
Prox-onic HR-025
Prox-onic HR-036
Prox-onic HR-040
Prox-onic HR-080
Prox-onic HR-0200
Prox-onic HRH-05

Textiles and Fibers (cont'd.)

Prox-onic HRH-016
Prox-onic HRH-025
Prox-onic HRH-0200
Prox-onic HRH-0200/50
Prox-onic LA-1/02
Prox-onic LA-1/04
Prox-onic LA-1/09
Prox-onic LA-1/012
Prox-onic MC-02
Prox-onic MC-05
Prox-onic MO-02
Prox-onic MO-015
Prox-onic MS-02
Prox-onic MS-05
Prox-onic MS-011
Prox-onic MS-050
Prox-onic MT-02
Prox-onic MT-05
Prox-onic MT-020
Prox-onic NP-1.5
Prox-onic NP-010
Prox-onic OA-1/04
Prox-onic OA-1/09
Prox-onic OCA-1/06
Prox-onic OL-1/05
Prox-onic OL-1/09
Prox-onic OL-1/014
Prox-onic OP-09
Prox-onic SA-1/02
Prox-onic SA-1/010
Prox-onic SA-1/020
Prox-onic SML-020
Prox-onic SMO-05
Prox-onic SMO-020
Prox-onic SMP-020
Prox-onic SMS-020
Prox-onic ST-05
Prox-onic ST-09
Prox-onic ST-014
Prox-onic ST-023
Prox-onic STO 020
Prox-onic STS-020
Prox-onic TBP-08
Prox-onic TBP-030
Prox-onic TD-1/03
Prox-onic TD-1/06
Prox-onic TD-1/09
Prox-onic TD-1/012
Prox-onic UA-03
Prox-onic UA-06
Prox-onic UA-09
Prox-onic UA-012
Puxol FB-11
Quadrilan® AT
Quartamin 86P
Quimipol EA 2506
Quimipol EA 2507
Quimipol EA 6803
Quimipol EA 6804
Quimipol EA 6806
Quimipol EA 6807
Quimipol EA 6808
Quimipol EA 9106
Quimipol ENF 65
Quimipol ENF 80
Quimipol ENF 90
Quimipol ENF 95
Quimipol ENF 100
Quimipol ENF 110
Radiamine 6140
Radiamine 6141
Radiamine 6160
Radiamine 6161
Radiamine 6163, 6164
Radiamine 6170
Radiamine 6171
Radiamine 6172
Radiamine 6173
Radiaquat 6444
Radiaquat 6462
Radiaquat 6471
Radiasurf® 7000
Radiasurf® 7135
Radiasurf® 7137
Radiasurf® 7145
Radiasurf® 7146
Radiasurf® 7147
Radiasurf® 7150
Radiasurf® 7151
Radiasurf® 7152
Radiasurf® 7153
Radiasurf® 7155
Radiasurf® 7156
Radiasurf® 7157
Radiasurf® 7175
Radiasurf® 7196
Radiasurf® 7201
Radiasurf® 7206
Radiasurf® 7269
Radiasurf® 7270
Radiasurf® 7345
Radiasurf® 7402
Radiasurf® 7410
Radiasurf® 7411
Radiasurf® 7420
Radiasurf® 7900
Ram Polymer 110
Reginol 2701
Regitant S-78
Rematard Grades
Remcopal L9
Remcopal LC
Remol Brands
Renex® 20
Renex® 22
Renex® 30
Renex® 35
Renex® 36
Renex® 688
Renex® 690
Renex® 751
Reserve Salt Flake
Resogen® 35 Conc.
Resogen® DM
Retardine
Rewomid® DL 203 S
Rewopal® EO 70
Rewopal® HV 8
Rewopal® LA 3
Rewopal® MT 2455
Rewopal® MT 2540
Rewopal® MT 5722
Rewopal® MT 65
Rewopal® RO 40
Rewophat E 1027
Rewophat EAK 8190
Rewophat NP 90
Rewophat OP 80
Rewopol® CT 65
Rewopol® CTN
Rewopol® NI 56
Rewopol® NL 2-28
Rewopol® NL 3-28
Rewopol® NL 3-70
Rewopol® SBDO 70
Rewopol® TLS 45/B
Rewopol® WS 11
Rewopol® WS 12
Reworyl® NKS 50
Reworyl® NKS 100
Rexobase NC
Rexobase TR
Rexoclean 25X
Rexoclean HAC
Rexoclean
Rexoclean JA
Rexoclean PCL
Rexol 25/10
Rexol 25/11
Rexol 25/100-70%
Rexol 25/407
Rexol 25J
Rexol 25JWC
Rexol 35/5
Rexol 35/8
Rexol 45/10
Rexol 2000 HWM
Rexol CCN
Rexonic 1006
Rexonic 1012-6
Rexonic 1218-6
Rexonic L125-9
Rexonic N25-14(85%)
Rexonic N25-7
Rexonic N25-9
Rexonic N25-9(85%)
Rexonic P-1
Rexopal 3928
Rexopal EP-1
Rexophos 25/97
Rexophos 35/98
Rexophos BP-2
Rexopon E
Rexopon RS
Rexopon RSN
Rexopon SK
Rexopon V
Rexoscour
Rexoscour SF
Rexosolve BCT
Rexosolve OA, OAC
Rexowet A, A-25, A Conc.
Rexowet ASG-81
Rexowet CR
Rexowet GA-1
Rexowet GR
Rexowet GR-LF
Rexowet GRS
Rexowet LS
Rexowet MS
Rexowet NF, NFX
Rexowet RW
Rexowet RW Conc.
Rexowet RWF
Rexowet VL
Rhodacal® BA-77
Rhodacal® BX-78
Rhodacal® CA, 70%
Rhodacal® N
Rhodafac® GB-520
Rhodafac® L3-64A
Rhodafac® MC-470
Rhodafac® PA-17
Rhodafac® PA-19
Rhodafac® PE-510
Rhodafac® PEH
Rhodafac® PS-17
Rhodafac® PS-19
Rhodafac® RA-600
Rhodafac® RB-400
Rhodafac® RD-510
Rhodafac® RE-610
Rhodafac® RE-960
Rhodafac® RM-410
Rhodafac® RS-410
Rhodafac® RS-610
Rhodafac® RS-710
Rhodameen® OA-910
Rhodameen® OS-12
Rhodameen® T-5
Rhodameen® T-12
Rhodameen® T-15
Rhodameen® T-30-90%
Rhodameen® T-50
Rhodamox® LO
Rhodapex® CO-433
Rhodapex® CO-436
Rhodapex® EST-30
Rhodapon® BOS
Rhodapon® CAV
Rhodapon® EC111
Rhodapon® OLS
Rhodaquat® T
Rhodasurf® 870
Rhodasurf® DA-4
Rhodasurf® DA-6
Rhodasurf® DA-530
Rhodasurf® DA-630
Rhodasurf® DA-639
Rhodasurf® L-4
Rhodasurf® L-25
Rhodasurf® L-790
Rhodasurf® LA-3
Rhodasurf® LA-30
Rhodasurf® LA-42
Rhodasurf® LA-90
Rhodasurf® ON-870
Rhodasurf® ON-877
Rhodasurf® T
Rhodasurf® T50
Rhodasurf® TB-970
Rhodasurf® TDA-5
Rhodasurf® TDA-6
Rhodasurf® TDA-8.5
Rhodorsil® AF 422
Rhodorsil® AF 20432
Ridafoam NS-221
Romie 802
Rychem® 21
Rychem® 33
Salt 100, 200, 300M

Textiles and Fibers (cont'd.)

Sandolube NV
Sandolube NVJ
Sandolube NVN
Sandolube NVS
Sandopan® BFN
Sandopan® CBH Paste
Sandopan® CBN
Sandopan® KD
Sandopan® KD (U) Liquid
Sandopan® LF Liquid
Sandopan® LFW
Sandopan® TFL Liquid
Sandoxylate® SX-208
Sandoxylate® SX-224
Sandoxylate® SX-408
Sandoxylate® SX-412
Sandoxylate® SX-418
Sandoxylate® SX-424
Sandoxylate® SX-602
Sandoz Amide CO
Sandoz Amide NP
Sandoz Amide PS
Sandoz Amine Oxide XA-C
Sandoz Amine Oxide XA-L
Sandoz Amine Oxide XA-M
Sandoz Sulfate 216
Sandoz Sulfate 219
Sandoz Sulfate WE
Sandoz Sulfonate 2A1
Sandoz Sulfonate AAS 45S
Sandoz Sulfonate AAS 60S
Sandozin® AM
Sandozin® NA Liq.
Sandozin® NE
Sandozin® NIT
Sanfix 555
Sanleaf CL-533M
Sanmorin 11
Sanmorin AM
Sansilic 11
Sanstat 1200
Santone® 3-1-SH
Savosellig REAC 4
Sawaclean AO
Sawaclean AOL
Schercomid EAC
Schercomid EAC-S
Schercopol DS-120
Schercopol DS-140
Schercotaine IAB
Schercotaine OAB
Schercotarder
Schercoterge 140
Schercoteric LS-2
Schercoteric LS-EP
Schercowet DOS-70
Schercozoline B
Schercozoline C
Schercozoline I
Schercozoline L
Schercozoline O
Schercozoline S
Scour NFP
Scourol 700
Secosol® DOS 70
Secoster® DO 600
Sellig 10 Mode
Sellig 13
Sellig Antimousse S
Sellig Dispersant FPZ
Sellig Hydrophilisant
Sellig LA 1150
Sellig Mouillant 9083 NI/AL
Sellig Mouillant EMPT 80
Sellig N 9 100
Sellig NK 729064
Selligon SP
Selligor 860 SP
Selligor SAN 1
Sellogen 641
Sellogen HR
SEM-35
Serdas GLN
Serdox NBS 4
Serdox NBS 5.5
Serdox NBS 6
Serdox NBS 6.6
Serdox NDI 100
Serdox NES 8/85
Serdox NNP 8.5
Serdox NNP 11
Serdox NSG 400
Serica 300
Serica 830
Sermul EA 30
Servamine KAC 412
Servirox OEG 45
Servirox OEG 55
Servirox OEG 65
Servirox OEG 90/50
Sevestat ML 300
Sevestat NDE
SF96®
Silicone AF-10 FG
Silicopearl SR
Silvatol® AS Conc.
Silvatol® SO
Silwet® L-7001
Silwet® L-7200
Silwet® L-7210
Silwet® L-7230
Silwet® L-7500
Silwet® L-7600
Silwet® L-7604
Silwet® L-7607
Silwet® L-7622
Sinopol 254A
Sinopol 860
Sinopol 864
Sinopol 864H
Sinopol 865
Sinopol 964
Sinopol 964H
Sinopol 965
Siponic® F-300
Siponic® F-400
Siponic® NP-9
Siponic® TD-3
Size CB
Sizing Wax PA, PT, SM
S-Maz® 20
S-Maz® 40
S-Maz® 60
S-Maz® 60KHM
S-Maz® 65K
S-Maz® 80
S-Maz® 80K
S-Maz® 85
S-Maz® 85K
S-Maz® 90
S-Maz® 95
Smithol 22LD
So/San 30M
Sofbon C-1-S
Sofnon 105G
Sofnon B-3
Sofnon GF-2, GF-5
Sofnon HG-180
Sofnon LA-75
Sofnon SP-852R
Sofnon SP-1000
Sofnon SP-9400
Sofnon TK-11
Softal 300
Softal A-3
Softal IT-7
Softal MR-30
Softenol® 3900
Softlon AC-4
Softlon FC-28
Softlon FC-136
Softyne CLS
Softyne CSN
Softyne H
Solegal W Conc.
Sole-Terge TS-2-S
Solupret Brands
Solusoft NK, WA, WL
Solutene TER
Solvent Scour 25/27
Solvent-Scour 263
Solvent-Scour 880
Sopralub ACR 265
Sopralub ACR 275
Sorbax HO-50
Sorbax PML-20
Sorbax PMO-5
Sorbax PMP-20
Sorbax PMS-20
Sorbax PTS-20
Sorbax SML
Sorbax SMO
Sorbax SMP
Sorbax SMS
Sorbax STO
Sorbax STS
Sorbit P
Sovatex C Series
Sovatex DS/C5
Spanscour EFS
Spanscour GR
Spanscour N20
Span® 20
Span® 40
Span® 60, 60K
Span® 65
Span® 80
Span® 85
Sperse Polymer IV
Stafoam
Stafoam DF-1
Stafoam DF-4
Stafoam DL
Stafoam F
Standamul® B-2
Standamul® STC-25
Standapol® 1610
Standapol® AK-43
Standapon 4149 Conc.
Stanlev R-276
Stanol 212F
Stansperse 506
Stantex Antistat F
Stantex MOR
Stantex MOR Special
Stantex PENE 40-DF
Stantex T-14 DF
Steol® 4N
Steol® CA-460
Steol® CS-460
Steol® KA-460
Steol® KS-460
Stepan C-40
Stepan C-65
Stepan C-68
Stepanol® AM
Stepanol® AM-V
Stepanol® DEA
Stepanol® ME Dry
Stepanol® MG
Stepanol® WA-100
Stepanol® WAC
Stepanol® WAC-P
Stepanol® WA Paste
Stepanol® WAQ
Stepanol® WA Special
Stepanol® WAT
Stepantex® 130
Stepantex® B-29
Stepantex® CO-30
Stepantex® CO-36
Stepantex® CO-40
Stepantex® DA-6
Stepantex® GS 90
Stepantex® TD14
Stepantex® TM10
Stepantex® TM15
Stepantex® VS 90
Steramine 49
Steramine CD
Steramine CGL
Steramine CR 25
Steramine FPA 197
Steramine GS
Steramine PNA 75
Steramine S2
Steramine TV
Strodex® Super V-8
Sufatol LS/3
Sufatol LX/B
Sufatol LX/C
Sul-fon-ate OE-500
Sulfonated Red Oil
Sulfonic 800

Textiles and Fibers (cont'd.)

Sulfotex OA
Sulframin 40
Sulframin 40DA
Sulframin 40RA
Sulframin 40T
Sulframin 45
Sulframin 45LX
Sulframin 60T
Sulframin 85
Sulframin 90
Sulframin 1240, 1245
Sulframin 1250, 1260
Sulframin 1288
Sulframin 1298
Sulframin 1388
Sulframin LX
Sunaptol LT
Sunaptol P Extra Liq.
Sunnol NES
Sunsoflon CK
Sunsoflon K-2
Sunsoflon MT-100
Sunsolt RZ-2, -6
Superkleen C
Superloid®
Supersol ICS
Supersurf AFX
Supersurf JM
Supersurf WAF
SuperWash DZ
Supragil® WP
Surfactol® 13
Surfactol® 318
Surfactol® 340
Surfactol® 365
Surfactol® 575
Surfactol® 590
Surfam P5 Dist
Surfam P5 Tech
Surfam P10
Surfam P12B
Surfam P14B
Surfam P17B
Surfam P24M
Surfam P86M
Surfam P89MB
Surfam P-MEPA
Surfax 8916/A
Surfonic® HDL-1
Surfonic® LF-17
Surfonic® N-31.5
Surfonic® N-40
Surfonic® N-85
Surfonic® N-95
Surfonic® N-100
Surfonic® N-102
Surfonic® N-120
Surfonic® N-130
Surfonic® N-150
Surfonic® N-200
Surfonic® N-300
Surfonic® NB-5
Swanic 6L, 7L
Swanic 110
Swanic CWT
Swanic T-51
Sylfat® D-1
Sylfat® RD-1
Syn Fac® 334
Syn Fac® 905
Syn Fac® 8210
Syn Fac® 8216
Syn Fac® TDA-92
Syn Fac® TEA-97
Syn Lube 106 (60%)
Syn Lube 107
Syn Lube 728
Syn Lube 1632H
Syn Lube 6277-A
Synperonic 10/5-100%
Synperonic 10/5-90%
Synperonic 10/6-100%
Synperonic 87K
Synperonic 91/5
Synperonic 91/6
Synperonic 91/7
Synperonic 91/8
Synperonic 91/10
Synperonic A2
Synperonic A3
Synperonic A4
Synperonic A6
Synperonic A7
Synperonic A9
Synperonic A11
Synperonic BD100
Synperonic FS
Synperonic K87
Synperonic KB
Synperonic LF/B26
Synperonic LF/D25
Synperonic N
Synperonic NP8
Synperonic NP9
Synperonic NX
Synperonic OP 7.5
Synperonic OP8
Synperonic OP10
Synperonic OP10.5
Synperonic OP11
Synprolam 35BQC (50)
Synprolam 35BQC (80)
Synprolam 35DMBQC
Synprolam 35MX1
Synprolam 35MX1QC
Synprolam 35N3X3, 35N3X5, 35N3X10, 35N3X15, 35N3X25
Synprolam 35X10
Synprolam 35X15
Synprolam 35X2
Synprolam 35X2/O
Synprolam 35X20
Synprolam 35X25
Synprolam 35X2QS, 35X5QS, 35X10QS
Synprolam FS
Syntens KMA 70-W-02
Syntens KMA 70-W-07
Syntens KMA 70-W-12
Syntens KMA 70-W-22
Syntens KMA 70-W-32
Syntergent 55-A
Syntergent 647-M
Syntergent K
Syntergent SF
Syntergent TER-1
Synthrapol LA-DN
Synthrapol N
Synthrapol SP
Syntofor A03, A04, AB03, AB04, AL3, AL4, B03, B04
Syntophos O6N
Syntopon 8 A
Syntopon 8 C
Syntopon B
Syntopon B 300
Syntopon C
Tally® 100 Plus
Tanabron W
Tanalev® 221
Tanapal® LD-3, LD-3T
Tanapal® NC
Tanapure® AC
Tanaquad T-85
Tanawet AR
Tanawet FBW
Tanawet NRWG
Tauranol M-35
Tauranol T-Gel
T-Det® A-026
T-Det® C-40
T-Det® DD-10
T-Det® EPO-61
T-Det® EPO-62L
T-Det® N-8
T-Det® N-9.5
T-Det® N-10.5
T-Det® N-14
T-Det® N-30
T-Det® N-70
T-Det® N-100
T-Det® N-307
T-Det® N-705
T-Det® N-707
T-Det® O-6
T-Det® O-40
T-Det® RQ1
T-Det® RY2
T-Det® TDA-65
Tephal Grunau AB Conc.
Tergenol G
Tergenol S Liq.
Tergenol Slurry
Tergitol® 15-S-3
Tergitol® 15-S-5
Tergitol® 15-S-7
Tergitol® 15-S-9
Tergitol® 15-S-12
Tergitol® 15-S-15
Tergitol® 15-S-20
Tergitol® 15-S-30
Tergitol® 15-S-40
Tergitol® 24-L-45
Tergitol® 24-L-60
Tergitol® 24-L-60N
Tergitol® 24-L-75
Tergitol® 24-L-92
Tergitol® 24-L-98N
Tergitol® 25-L-5
Tergitol® 25-L-7
Tergitol® 25-L-9
Tergitol® 25-L-12
Tergitol® 26-L-1.6
Tergitol® 26-L-5
Tergitol® Min-Foam 1X
Tergitol® Min-Foam 2X
Tergitol® NP-14
Tergitol® NP-27
Tergitol® NP-44
Tergitol® NPX
Tergitol® TMN-3
Tergitol® TMN-6
Tergitol® TMN-10
Teric 9A6
Teric 9A8
Teric 12A2
Teric 12A6
Teric 12A7
Teric 12A8
Teric 12A9
Teric 12A12
Teric 12M2
Teric 12M5
Teric 12M15
Teric 13A7
Teric 13A9
Teric 16A16
Teric 16M2
Teric 16M5
Teric 16M15
Teric 17A2
Teric 17A6
Teric 17A8
Teric 17A10
Teric 17A13
Teric 17A25
Teric 18M2
Teric 18M5
Teric 18M10
Teric 18M20
Teric 18M30
Teric 121
Teric 124
Teric 127
Teric 151
Teric 152
Teric 154
Teric BL8
Teric BL9
Teric C12
Teric CME3
Teric CME7
Teric DD5
Teric DD9
Teric G9A8
Teric G9A12
Teric G12A4
Teric G12A6
Teric G12A8
Teric G12A12
Teric LA4
Teric LA8
Teric LAN70
Teric N4
Teric N5
Teric OF4
Teric OF6
Teric OF8
Teric PEG 200

Textiles and Fibers (cont'd.)

Teric PEG 300
Teric PEG 400
Teric PEG 600
Teric PEG 1500
Teric PEG 3350
Teric PEG 4000
Teric PPG 400
Teric PPG 1000
Teric SF9
Teric SF15
Teric T2
Teric T5
Teric T7
Teric T10
Teric X5
Teric X7
Teric X8
Teric X10
Teric X11
Teric X13
Teric X16
Teric X40
Teric X40L
Tetraterge D-101, NFF
Tetrawet DWN
Texapon NSO
Texapon PN-235
Texi TD
Textamine O-5
Tex-Wet 1001
Tex-Wet 1002
Tex-Wet 1004
Tex-Wet 1010
Tex-Wet 1034
Tex-Wet 1048
Tex-Wet 1070
Tex-Wet 1103
Tex-Wet 1155
Tex-Wet 1158
Tex-Wet 1197
Tinegal® KD
Tinegal® MR-50
Tinegal® NA
Tinegal® NT
Tinegal® PM
Tinegal® Resist N
Tinegal® TP
Tinegal® W
Tinofix® ECO
Tinofix® EW Sol
Titadine TA
Titadine TCP
Titan Castor No. 75
Titan Decitrene
Titanole RMA
Titanterge CAC
Titanterge DBS Conc.
Titapene Conc.
Titazole HTX
Titazole SA
T-Maz® 20
T-Maz® 28
T-Maz® 40
T-Maz® 60
T-Maz® 61
T-Maz® 65
T-Maz® 80
T-Maz® 80KLM
T-Maz® 81
T-Maz® 85
T-Maz® 90
T-Maz® 95
T-Mulz® 426
T-Mulz® 800
T-Mulz® 844
T-Mulz® 1158
TN Cleaner S
TO-55-E
Toho Me-PEG Series
Toho PEG Series
Toho Salt A-5
Toho Salt A-10
Toho Salt CP-60
Toho Salt OK-70
Toho Salt TM-1
Toho Salt UF-350C
Toho Salt UF-650C
Tomah DA-14
Tomah DA-16
Tomah DA-17
Tomah E-14-2
Tomah E-14-5
Tomah E-17-2
Tomah E-18-2
Tomah E-18-5
Tomah E-18-8
Tomah E-18-10
Tomah E-18-15
Tomah E-DT-3
Tomah E-S-2
Tomah E-S-5
Tomah E-S-15
Tomah E-T-2
Tomah E-T-15
Tomah PA-14
Tomah PA-16
Tomah PA-19
Tomah PA-1214
Tomah Q-14-2
Tomah Q-18-15
Tomah Q-D-T
Tricol M-20
Trilon® A
Triton® CF-10
Triton® CF-87
Triton® DF-12
Triton® DF-16
Triton® GR-5M
Triton® GR-7M
Triton® N-101
Triton® N-111
Triton® RW-20
Triton® RW-50
Triton® RW-75
Triton® RW-100
Triton® RW-150
Triton® W-30 Conc.
Triton® X-100
Triton® X-100 CG
Triton® X-155-90%
Triumphnetzer ZSG
Triumphnetzer ZSN
Trycol® 5878
Trycol® 5882
Trycol® 5888
Trycol® 5940
Trycol® 5941
Trycol® 5943
Trycol® 5964
Trycol® 5966
Trycol® 5967
Trycol® 5968
Trycol® 5971
Trycol® 5972
Trycol® 5993
Trycol® 5999
Trycol® 6720
Trycol® 6940
Trycol® 6943
Trycol® 6953
Trycol® 6956
Trycol® 6957
Trycol® 6960
Trycol® 6961
Trycol® 6962
Trycol® 6963
Trycol® 6964
Trycol® 6968
Trycol® 6969
Trycol® 6970
Trycol® 6974
Trycol® 6984
Trycol® 6985
Trycol® 6989
Trycol® DNP-8
Trycol® DNP-150
Trycol® DNP-150/50
Trycol® LAL-4
Trycol® LAL-23
Trycol® NP-40
Trycol® NP-50
Trycol® NP-407
Trycol® NP-507
Trycol® OAL-23
Trycol® SAL-20
Trycol® TDA-3
Trycol® TDA-6
Trycol® TDA-9
Trycol® TDA-18
Trydet 19
Trydet 20
Trydet 2636
Trydet 2640
Trydet 2644
Trydet 2670
Trydet 2671
Trydet 2676
Trydet 2681
Trydet 2682
Trydet 2691
Trydet 2692
Trydet ISA-4
Trydet ISA-9
Trydet ISA-14
Trydet LA-7
Trydet MP-9
Trydet OA-7
Trydet OA-10
Trydet SA-5
Trydet SA-7
Trydet SA-8
Trydet SA-23
Trydet SA-40
Trydet SO-9
Tryfac® 525-K
Tryfac® 610-K
Tryfac® 910-K
Tryfac® 5552
Tryfac® 5554
Tryfac® 5555
Tryfac® 5556
Tryfac® 5557
Tryfac® 5559
Tryfac® 5560
Tryfac® 5576
Trylon® 5900
Trylon® 5902
Trylon® 5906
Trylon® 5909
Trylon® 5921
Trylon® 5922
Trylon® 6702
Trylon® 6735
Trylon® EW
Trylox® 1086
Trylox® 1186
Trylox® 5902
Trylox® 5904
Trylox® 5907
Trylox® 5909
Trylox® 5918
Trylox® 5921
Trylox® 5922
Trylox® 5925
Trylox® 6746
Trylox® 6747
Trylox® 6753
Trylox® CO-5
Trylox® CO-16
Trylox® CO-25
Trylox® CO-30
Trylox® CO-36
Trylox® CO-40
Trylox® CO-80
Trylox® CO-200, -200/50
Trylox® HCO-5
Trylox® HCO-16
Trylox® HCO-25
Trymeen® 6601
Trymeen® 6606
Trymeen® 6607
Trymeen® 6609
Trymeen® 6610
Trymeen® 6617
Trymeen® 6620
Trymeen® 6622
Trymeen® 6637
Trymeen® CAM-10
Trymeen® CAM-15
Trymeen® OAM 30/60
Trymeen® TAM-8
Trymeen® TAM-20
Trymeen® TAM-25
Tween® 20
Tween® 21
Tween® 40
Twitchell 6805 Oil
Tylose C, CB Series
Tylose CBR Grades
Tylose H Series
Tylose MH Grades

Textiles and Fibers (cont'd.)

Tylose MH-K, MH-xp, MHB-y, MHB-yp
Ufarol Am 30
Ufarol Am 70
Ufarol Na-30
Ufarol TA-40
Ufaryl DB80
Ufaryl DL85
Ufaryl DL90
Ufasan 35
Ufasan 50
Ufasan 60A
Ufasan 62B
Ufasan 65
Ufasan IPA
Ufasan TEA
Ufasoft 75
Ukanil 3000
Ukaril 190
Ultra Sulfate AE-3
Ultra Sulfate SE-5
Ultra Sulfate SL-1
Ultratex® ESB, WK
Unamide® S
Unamide® SI
Unamide® W
Unamine® C
Ungerol LES 2-28
Ungerol LES 2-70
Ungerol LES 3-28
Ungerol LES 3-70
Ungerol N2-28
Ungerol N2-70
Ungerol N3-28
Ungerol N3-70
Uniperol® AC
Uniperol® AC Highly Conc.
Uniperol® EL
Uniperol® O
Uniperol® O Micropearl
Uniperol® W, W Flakes
Unisol BT
Unisol LAC GN
Unisol MFD
Unisol WL
Valdet 561
Valdet CC
Valscour SWR-2
Value 401
Value 402
Value 3608
Value 3710
Value W-307
Vanseal 35
Vanseal CS
Vanseal LS
Vanseal NACS-30
Vanseal NALS-30
Vanseal NALS-95
Vanseal OS
Varamide® A-2
Varamide® A-83
Varine C
Varine O®
Varion® CADG-HS
Varion® SDG
Variquat® B200
Varisoft® 110
Varisoft® 136-1000P
Varisoft® 137
Varisoft® 222 (75%)
Varisoft® 222 (90%)
Varisoft® 222 LM (90%)
Varisoft® 222 LT (90%)
Varisoft® 238 (90%)
Varisoft® 445
Varisoft® 475
Varisoft® 920
Varisoft® 950
Varisoft® 3690
Varisoft® 3690N (90%)
Varisoft® DS-100
Varonic® K202
Varonic® K205
Varonic® K215
Varonic® K215 LC
Varonic® MT 65
Varonic® T202
Varonic® T210
Varonic® T215
Varonic® T215LC
Victawet® 12
Vikolev
Viscasil®
Vismul ET-20
Vismul ET-55
Vitexol® K
W180
Wayfos A
Wayfos D-10N
Witcamide® 82
Witcamide® 272
Witcamide® 5130
Witcamide® 5133
Witcamide® 5140
Witcamide® 5168
Witcamide® 5195
Witcamide® AL69-58
Witcamide® MAS
Witcamine® 209
Witcamine® 235
Witcamine® AL42-12
Witcolate 1276
Witcolate A Powder
Witcolate S1285C
Witcolate S1300C
Witcolate SE-5
Witconate 90F
Witconate 605T
Witconate 1260 Slurry
Witconate 1840X
Witconate AOS
Witconate NXS
Witconate SCS 45%
Witconate SCS 93%
Witconate STS
Witconate SXS 40%
Witconate SXS 90%
Witconate YLA
Witconol APEM
Witconol APM
Witconol APS
WR Base
Zeeclean No. 1
Zee-emul No. 35
Zee-emul No. 50
Zoharconc FS
Zoharconc J-Super
Zoharsoft
Zoharsoft 90
Zoharsoft DAS
Zoharsoft DAS-N
Zoharsoft DAS-PW
Zoharsoft DAS-SIL
Zonyl® FSA
Zonyl® FSB
Zonyl® FSC
Zonyl® FSJ
Zonyl® FSN
Zonyl® FSP
Zusolat 1005/85
Zusolat 1008/85
Zusolat 1010/85
Zusolat 1012/85
Zusomin C 108
Zusomin O 105
Zusomin S 110
Zusomin TG 102

Water and Waste Treatment

Actrafoam A, B, C, S
Advantage 470A Defoamer
Advantage 1007B Defoamer
Advantage 1275PD
Advantage 1280PD
Advantage M1250 Defoamer
Advantage M1251 Production Aid
Advantage M133A Defoamer
Advantage M201 Defoamer
Aerosol® C-61
AF 10 IND
AF 30 IND
AF 72
AF 8810
AF 8810 FG
AF 8820
AF 9020
AF GN-11-P
AF GN-23
AF HL-23
AF HL-26
Ageflex FM-1
Agesperse 71
Agesperse 80
Agesperse 81
Agesperse 82
Amergel® 100
Amergel® 200
Amergel® 500
Anedco DF-6002
Armeen® Z
Armid® HT
Arquad® DMCB-80
Arquad® S-50
Briquest® 301-50A
Briquest® 462-23K
Briquest® ADPA-60AW
BTC® 99
BTC® 776
BTC® 824
BTC® 835
BTC® 885 P40
BTC® 1010-80
BTC® 2125, 2125-80, 2125 P-40
BTC® 2125M
BTC® 2125M-80, 2125M P-40
BTC® 2565
BTC® 2568
BTC® 8248
BTC® 8249
BTC® 8358
Calamide C
Carbopol® 907
Catigene® 50 USP
Catigene® 65 USP
Catigene® 776
Catigene® 818
Catigene® 818-80
Catigene® 824
Catigene® 885
Catigene® 888
Catigene® 1011
Catigene® 2125 M
Catigene® 2125M P40
Catigene® 2125 P40
Catigene® 2565
Catigene® 4513-50
Catigene® 4513-80
Catigene® 4513-80 M
Catigene® 8248
Catigene® B 50
Catigene® B 80
Catigene® DC 100
Chemax DF-100
Chemax DF-30
Chemoxide LM-30
Chemphonate AMP
Chemphonate AMP-S
CNC Antifoam 77
CNC Antifoam 495, 495-M
CNC Defoamer 229
CNC Dispersant WB Series
Crill 1
Crill 2
CWT AF-200 Antifoam
FMB 65-15 Quat.

Water and Waste Treatment (cont'd.)

FMB65-28 Quat
FMB 210-8 Quat, 210-15 Quat
FMB 302-8 Quat
FMB 451-8 Quat
FMB 504-5 Quat
FMB 3328-5 Quat, 3328-8 Quat
FMB 4500-5 Quat
FMB 6075-5 Quat, 6075-8 Quat
Foamkill® 30HP
Foamkill® 627
Foamkill® 660, 660F
Foamkill® 663J
Foamkill® 684 Series
Foamkill® 684A
Foamkill® 684P
Foamkill® 687D
Foamkill® 687DS
Foamkill® 700, 700 Conc.
Foamkill® 836B
Foamkill® CMP
Foamkill® DF#4
Foamkill® EFT
Foamkill® MS
G-1000-S Antifoam Compd.
Gardiquat 1450
Gardiquat 1480
Gardiquat SV 480
Geropon® X2152
GP-210 Silicone Antifoam Emulsion
GP-262 Defoamer
GP-295 Defoamer
GP-330-I Antifoam Emulsion
Hercules® 388 Defoamer
Hercules® Eff-101 Defoamer
Igepal® NP-9
JAQ Powdered Quat
Kemamine® BQ-9702C
Kemamine® Q-1902C
Kemamine® Q-6502C
Leocon 1020B
Lignosol AXD
Maquat 4450-E
Mazu® 43 C
Mazu® 68 C
Mazu® 112
Mazu® 140
Mazu® 141
Mazu® 142
Mazu® DF 205SX
Mazu® DF 215SX
Mazu® DF 230S
Mazu® DF 230SP
Mazu® DF 230SX
Mazu® DF 230SXSP
Mazu® DF 243
Monawet MB-45
Monawet MB-100
Monawet MM-80
Monawet MO-65 PEG
Monawet MO-65-150
Monawet MO-70
Monawet MO-70-150
Monawet MO-70E
Monawet MO-70 PEG
Monawet MO-70R
Monawet MO-70S
Monawet MO-75E
Monawet MO-84R2W
Monawet MO-85P
Niaproof® Anionic Surfactant 4
Nissan Amine AB
Nissan Amine ABT
Nissan Amine DT
Nissan Amine DTH
Nissan Amine FB
Nissan Amine MB
Nissan Amine OB
Nissan Amine PB
Nissan Amine SB
Nissan Amine VB
Nissan Cation AB
Nissan Cation ABT-350, 500
Nissan Cation F2-10R, -20R, -40E, -50
Nissan Cation FB, FB-500
Nissan Cation L-207
Nissan Cation PB-40, -300
Nissan Cation S2-100
Pegosperse® 200 DL
Pegosperse® 400 DOT
Pegosperse® 600 DOT
Pegosperse® PMS CG
Pluracol® WD90K
Pluriol® PE 6100
Pluronic® 10R5
Pluronic® 10R8
Pluronic® 12R3
Pluronic® 17R1
Pluronic® 17R2
Pluronic® 17R8
Pluronic® 22R4
Pluronic® 25R1
Pluronic® F127
Pluronic® F38
Pluronic® F68LF
Pluronic® F77
Pluronic® F87
Pluronic® F88
Pluronic® F98
Pluronic® L31
Pluronic® L35
Pluronic® L42
Pluronic® L43
Pluronic® L44
Pluronic® L61
Pluronic® L62
Pluronic® L62D
Pluronic® L62LF
Pluronic® L63
Pluronic® L64
Pluronic® L72
Pluronic® L81
Pluronic® L92
Pluronic® L101
Pluronic® L121
Pluronic® L122
Pluronic® P65
Pluronic® P75
Pluronic® P84
Pluronic® P85
Pluronic® P94
Pluronic® P103
Pluronic® P104
Pluronic® P105
Pluronic® P123
Poly-Tergent® P-17A
Poly-Tergent® P-17B
Poly-Tergent® P-17BLF
Poly-Tergent® P-17BX
Poly-Tergent® P-17D
Poly-Tergent® P-32D
Polywet ND-1
Polywet ND-2
Querton KKBCl-50
Rexfoam B, C
Rexfoam D
Sedipol®
Sevefilm 20
Silicone AF-30 FG
Smithol 22LD
Surfagene FPT
T-Det® EPO-61
Tergitol® 15-S-3
Tergitol® 15-S-5
Tergitol® 15-S-7
Tergitol® 15-S-9
Tergitol® Min-Foam 2X
Tergitol® TMN-3
Tergitol® TMN-6
Tergitol® TMN-10
Teric 121
Teric PE61
Teric PE62
Trans-137
Trans-138
Trans-177S, -177S K
Trilon® A
Triton® RW-50
Triton® RW-75
Triton® RW-100
Triton® X-114
Triton® XL-80N
Variquat® 60LC
Variquat® 66
Variquat® 80LC
Wayhib® S
Witcor SI-3065

Part III
Chemical Component
Cross Reference

Chemical Component Cross Reference

This section is a cross reference between major chemical components and the tradename products that contain these chemicals. Not all the chemicals in this section are surfactants, but all are chemicals contained in the tradename products that function as surfactants.

Acetamide MEA (CTFA)
CAS 142-26-7; EINECS 205-530-8
Synonyms: N-Acetyl ethanolamine; N-(2-Hydroxyethyl) acetamide
Tradenames: Hetamide MA.; Mackamide AME-75, AME-100.; Schercomid AME.

Acetylated hydrogenated lard glyceride (CTFA)
CAS 8029-91-2
Synonyms: Glycerides, lard mono-, hydrogenated, acetates
Tradenames: Tegin® E-61; Tegin® E-61 NSE

Acetylated hydrogenated tallow glyceride
CAS 68990-58-9; EINECS 273-612-0
Synonyms: Glycerides, tallow mono-, hydrogenated, acetates
Tradenames: Tegin® E-41

Acetylated hydrogenated tallow glycerides (CTFA)
Synonyms: Glycerides, tallow mono-, di- and tri-, hydrogenated, acetates
Tradenames: Tegin® E-41 NSE

Acetylated lanolin (CTFA)
CAS 61788-48-5; EINECS 262-979-2
Tradenames: Ritacetyl®

Acetylated lanolin alcohol (CTFA)
CAS 61788-49-6; EINECS 262-980-8
Synonyms: Lanolin, alcohols, acetates
Tradenames containing: Crodalan AWS

Acetylated lard glyceride (CTFA)
CAS 8029-92-3
Synonyms: Glycerides, lard mono-, acetates
Tradenames: Tegin® E-66; Tegin® E-66 NSE

Acetylenic glycol
CAS 126-86-3
Tradenames: Surfynol® PC
Tradenames containing: Surfynol® DF-110D; Surfynol® DF-110L; Surfynol® TG-E.

Acrylates copolymer (CTFA)
Synonyms: Acrylic/acrylate copolymer
Tradenames: Acusol® 810; Acusol® 820; Acusol® 830; Acusol® 840
Tradenames containing: Forbest VP S7

Acrylate/steareth-20 methacrylate copolymer (CTFA)
Tradenames: Sipothix® 1941

Acrylic resin
Synonyms: Acrylic polymer; Acrylic fiber; Nitrile rubber
Tradenames: Ucar® Acrylic 503; Ucar® Acrylic 505; Ucar® Acrylic 515; Ucar® Acrylic 516; Ucar® Acrylic 518; Ucar® Acrylic 522; Ucar® Acrylic 525

Acrylinoleic acid
CAS 53980-88-4
Synonyms: C21-dicarboxylic acid; 5 (or 6)-carboxy-4-hexyl-2-cyclohexene-1-octanoic acid
Tradenames: Westvaco Diacid® 1550; Westvaco Diacid® 1575

Acyl amido-amine
Tradenames: Tequat RO

Alcohols C9-11. *See C9-11 alcohols*

Alcohols C10. *See n-Decyl alcohol*

Alcohols C11. *See Undecyl alcohol*

Alcohols C12-13. *See C12-13 alcohols*

Alcohols C12-15. *See C12-15 alcohols*

Alcohols, lanolin. *See Lanolin alcohol*

Algin (CTFA)
CAS 9005-38-3
Synonyms: Sodium alginate; Sodium polymannuronate
Tradenames: Kelco® HV; Kelco® LV; Kelcosol®; Kelgin® F; Kelgin® HV, LV, MV; Kelgin® QL; Kelgin® XL; Kelvis®

Alginic acid (CTFA)
CAS 9005-32-7; EINECS 232-680-1
Synonyms: Norgine
Tradenames: Kelacid®

Alkenyl succinic anhydride
Tradenames: Alkenyl Succinic Anhydrides; ASA

Alkyl dimethyl benzyl ammonium chloride. *See Benzalkonium chloride*

Alkyl dimethyl dichlorobenzyl ammonium chloride
CAS 8023-53-8
Synonyms: Alkyl (C_8H_{17} to $C_{18}H_{37}$) dimethyl-3,4-dichlorobenzyl ammonium chloride
Tradenames: Maquat DLC-1214

Alkyl dimethyl ethylbenzyl ammonium chloride
Tradenames containing: Catigene® 2125 M; Catigene® 2125 P40; Catigene® 2125M P40; Catigene® 4513-50; Catigene® 4513-80; Catigene® 4513-80 M; FMB 6075-5 Quat, 6075-8 Quat; Maquat MQ-2525; Maquat MQ-2525M

Alkyl dimethyl hydroxyethyl ammonium chloride
CAS 85736-63-6
Tradenames: Synprolam 35MX1QC

Alkyl trimethyl ammonium bromide
Tradenames: Catinal HTB; Empigen® CHB40

Alkyl trimethyl ammonium chloride
CAS 68391-03-7; EINECS 269-922-0
Tradenames: Arquad® S; Quartamin 86P; Querton 280

Aluminum orthophosphate
CAS 7784-30-7
Synonyms: Aluminum phosphate; Aluminophosphoric acid; Phosphoric acid, aluminum salt (1:1)
Tradenames: Albrite MALP

Amine dodecylbenzene sulfonate
Tradenames: Amsul 70; Marlon® AMX; Marlopon® AMS 60; Ninate® 411; Rhodacal® 2283 (formerly Soprophor® 2283); Witconate 93S; Witconate P-1052N; Witconate YLA

2-Amino-1-butanol (CTFA)
CAS 96-20-8; 5856-63-3; EINECS 202-488-2
Synonyms: 2-Amino-n-butyl alcohol
Tradenames: AB®

2-Amino-2-ethyl-1,3-propanediol (CTFA)
CAS 115-70-8; EINECS 204-101-2
Synonyms: AEPD; Aminoamylene glycol
Tradenames: AEPD®

2-Amino-2-methyl-1,3-propanediol (CTFA)
CAS 115-69-5; EINECS 204-100-7
Synonyms: AMPD; Aminobutylene glycol; Butanedioleamine; 2-Amino-2-methyl-1,3-propanediol
Tradenames: AMPD

2-Amino-2-methyl-1-propanol (CTFA)
CAS 124-68-5; EINECS 204-709-8
Synonyms: AMP; Isobutanolamine; Aminomethyl propanol; Isobutanol-2 amine
Tradenames: AMP; AMP-95

Aminotrimethylene phosphonic acid (CTFA)
CAS 6419-19-8; EINECS 229-146-5
Synonyms: Amino tris (methylene phosphonic acid); Nitrilotris (methylene) triphosphonic acid
Tradenames: Chemphonate AMP

Ammonia (CTFA)
CAS 7664-41-7; EINECS 231-635-3
Synonyms: Ammonia gas; Ammonia anhydrous; Spirit of Hartshorn
Tradenames containing: Daxad® 32S

Ammonium acrylate
Tradenames: Ram Polymer 110

Ammonium alginate (CTFA)
CAS 9005-34-9
Synonyms: Ammonium polymannuronate; Alginic acid, ammonium salt
Tradenames: Superloid®
Tradenames containing: Keltose®

Ammonium C12-15 pareth sulfate (CTFA)
Synonyms: Ammonium pareth-25 sulfate; POE (1–4) C12-15 alcohol ether sulfate
Tradenames: Neodol® 25-3A; Standapol® AP-60; Witcolate AE-3
Tradenames containing: Zetesol AP

Ammonium caseinate (CTFA)
CAS 1336-21-6; 9005-42-9
Synonyms: Casein, ammonium salt
Tradenames: Atrasein 115

Ammonium chloride (CTFA)
CAS 12125-02-9; EINECS 235-186-4
Synonyms: Sal ammoniac; Salmiac; Ammonium muriate
Tradenames containing: Schercotaine CAB-A; Schercotaine SCAB-A

Ammonium cocoyl sarcosinate (CTFA)
Synonyms: Amides, coconut oil, with sarcosine, ammonium salts; Glycine, N-methyl-, N-coco amido deriv., ammonium salt; Ammonium N-cocoyl sarcosinate
Tradenames: Hamposyl® AC-30

Ammonium cumenesulfonate (CTFA)
CAS 37475-88-0; EINECS 253-519-1
Synonyms: Benzenesulfonic acid, (1-methylethyl)-, ammonium salt
Tradenames: Ammonium Cumene Sulfonate 60; Eltesol® AC60; Eltesol® ACS 60; Manrosol ACS60; Reworyl® ACS 60; Ultra NCS Liquid; Witconate NCS

Ammonium deceth sulfate
Tradenames: Nutrapon FA-50 0066

Ammonium dodecylbenzene sulfonate (CTFA)
CAS 1331-61-9; EINECS 215-559-8
Synonyms: Ammonium lauryl benzene sulfonate
Tradenames: Ablusol DBM; Hetsulf 50A; Nansa® AS 40; Newcol 210; Sandoz Sulfonate AAS 50MS

Ammonium 2-ethylhexanol sulfate
Tradenames: 2-EHS Base; Karasurf AS-26

Ammonium 2-ethylhexyl sulfate. *See Ammonium octyl sulfate*

Ammonium laureth sulfate (CTFA)
CAS 32612-48-9 (generic); 67762-19-0
Synonyms: Ammonium lauryl ether sulfate
Tradenames: Avirol® 603 A; Avirol® AE 3003; Calfoam NEL-60; Carsonol® SES-A; Laural EC; Manro ALEC 27; Manro ALES 60; Montosol PF-16, -18; Montosol PG-12; Montosol PQ-11, -15; Montosol TQ-11; Nonasol N4AS; Nutrapon AL 2; Nutrapon AL 30; Nutrapon AL 60; Polystep® B-11; Rewopol® AL 3; Rhodapex® AB-20 (formerly Geropon AB/20); Rhodapex® MA360 (formerly Sipon® MA360); Sandoz Sulfate 216; Standapol® AL-60; Steol® CA-460; Sulfotex OT; Texapon PNA-127; Ungerol AM3-60; Ungerol AM3-75; Witcolate AE (formerly DeSonol AE); Witcolate LES-60A; Witcolate S1300C; Zoharpon LAEA 253
Tradenames containing: Stepanol® AEG

Ammonium laureth-9 sulfate (CTFA)
CAS 32612-48-9 (generic)
Synonyms: PEG-9 lauryl ether sulfate, ammonium salt; Ammonium PEG (450) lauryl ether sulfate; POE (9) lauryl ether sulfate, ammonium salt
Tradenames: Geropon® AB/20

Ammonium laureth-12 sulfate (CTFA)
CAS 32612-48-9 (generic)
Tradenames: Polystep® B-22

Ammonium laureth-30 sulfate
Tradenames: Polystep® B-20

Ammonium lauroyl sarcosinate (CTFA)
CAS 68003-46-3; EINECS 268-130-2

Synonyms: N-methyl-N-(1-oxododecyl) glycine, ammonium salt
Tradenames: Hamposyl® AL-30

Ammonium lauryl benzene sulfonate. *See Ammonium dodecylbenzene sulfonate*

Ammonium lauryl ether sulfate. *See Ammonium laureth sulfate*

Ammonium lauryl sulfate (CTFA)
CAS 2235-54-3; 68081-96-9; EINECS 218-793-9
Synonyms: Sulfuric acid, monododecyl ester, ammonium salt
Tradenames: Akyposal ALS 33; Avirol® 200; Avirol® 270A; Avirol® A; Calfoam NLS-30; Carsonol® ALS-R; Carsonol® ALS-S; Colonial ALS; Empicol® AL30; Manro ALS 30; Marlinat® DFN 30; Norfox® ALS; Nutrapon HA 3841; Nutrapon PP 3563; Octosol ALS-28; Peramit MLN; Perlankrol® DAF25; Permalose TM; Polystep® B-7; Rhodapon® L-22, L-22/C (formerly Sipon® L-22); Rhodapon® L-22HNC (formerly Sipon® L-22HNC); Sandoz Sulfate A; Sermul EA 129; Standapol® A; Stepanol® AM-V; Stepanol® AM; Sulfochem ALS; Surfax 215; Texapon PNA; Ufarol Am 30; Ufarol Am 70; Witcolate 6430; Witcolate 6431 (formerly DeSonol A); Witcolate AM; Witcolate NH
Tradenames containing: Stepanol® AEG

Ammonium lignosulfonate
CAS 8061-53-8
Tradenames: Lignosite® 17; Lignosol TS; Lignosol TSD; Tembind A 002

Ammonium myreth sulfate (CTFA)
CAS 27731-61-9
Synonyms: Ammonium myristyl ether sulfate
Tradenames containing: Standapol® EA-K

Ammonium myristyl ether sulfate. *See Ammonium myreth sulfate*

Ammonium naphthalene-formaldehyde sulfonate
Tradenames: Dehscofix 929; Dehscofix 930

Ammonium naphthalene sulfonate
Tradenames: Emery® 5366 (Lomar PWA Liq.); Emery® 5367 Lomar PWA; Lomar® PWA Liq; Lomar® PWA

Ammonium nonoxynol-4 sulfate (CTFA)
CAS 31691-97-1 (generic); 63351-73-5
Tradenames: Polystep® B-1; Rhodapex® CO-436 (formerly Alipal® CO-436); Sulfochem 436

Ammonium nonoxynol-9 sulfate
Tradenames: Abex® EP-110 (formerly Alipal® EP-110)

Ammonium nonoxynol-10 sulfate
Tradenames: Sermul EA 152

Ammonium nonoxynol-20 sulfate
Tradenames: Abex® EP-115 (formerly Alipal® EP-115)

Ammonium nonoxynol-30 sulfate
Tradenames: Abex® EP-120 (formerly Alipal® EP-120)

Ammonium nonoxynol-77 sulfate
Tradenames: Abex® EP-227 (formerly Alipal® EP-227)

Ammonium octyl sulfate
Synonyms: Ammonium 2-ethylhexyl sulfate
Tradenames: Penetrol 2-EHS

Ammonium oleic sulfate
Tradenames: Actrasol SR 75

Ammonium pareth-25 sulfate. *See Ammonium C12-15 pareth sulfate*

Ammonium perfluorooctanoate
CAS 3825-26-1
Synonyms: Ammonium perfluorocaprylate; Ammonium pentadecafluorooctanoate
Tradenames: Fluorad FC-118; Fluorad FC-143

Ammonium polyacrylate (CTFA)
CAS 9003-03-6
Synonyms: Poly(acrylic acid), ammonium salt; 2-Propenoic acid, homopolymer, ammonium salt
Tradenames: Daxad® 37LA7
Tradenames containing: Akypogene KTS

Ammonium polymethacrylate
Tradenames: Daxad® 32; Daxad® 34A9

Ammonium ricinoleate
Tradenames: Solricin® 285

Ammonium stearate (CTFA)
CAS 1002-89-7; EINECS 213-695-2
Synonyms: Octadecanoic acid, ammonium salt
Tradenames: Ammonium Stearate 33% Liq

Ammonium xylenesulfonate (CTFA)
CAS 26447-10-9; EINECS 247-710-9
Synonyms: Benzenesulfonic acid, dimethyl-, ammonium salt
Tradenames: Eltesol® AX 40; Hartotrope AXS; Ultra NXS Liquid; Witconate NXS

Anhydrous lanolin. *See Lanolin*

Apricotamidopropyl betaine (CTFA)
CAS 133934-08-4
Synonyms: Quat. ammonium compd., (carboxymethyl)(3-apricotamidopropyl)dimethyl, hydroxide, inner salt; Apricotamidopropyl dimethyl glycine; Apricot amide propylbetaine
Tradenames: Schercotaine APAB

Arachidyl alcohol. *See Eicosanol*

Arachidyl behenyl amine
Tradenames: Kemamine® P-150, P-150D; Kemamine® P-190, P-190D

Arachidyl-behenyl 1,3-propylene diamine
Tradenames: Kemamine® D-190

Azelaic acid
CAS 123-99-9
Synonyms: Nonanedoic acid; 1,7-Heptanedicarboxylic acid; Anchoic acid
Tradenames: Emerox® 1110; Emerox® 1144

Babassuamide DEA (CTFA)
Synonyms: N,N-Bis(2-hydroxyethyl)babassu fatty acid amide; Diethanolamine babassu fatty acid condensate; Babassu diethanolamide
Tradenames: Incromide BAD

Barium dinonylnaphthalene sulfonate
CAS 25619-56-1
Tradenames: Manro DNNS/B

Barium petroleum sulfonate
CAS 61790-48-5
Tradenames: Barium Petronate 50-S Neutral, Basic

Beeswax (CTFA)
CAS 8006-40-4 (white), 8012-89-3 (yellow)
Tradenames: Syncrowax BB4

Behenalkonium chloride (CTFA)
CAS 16841-14-8; EINECS 240-865-3

Synonyms: Behenyl dimethyl benzyl ammonium chloride; Benzyldocosyldimethylammonium chloride
Tradenames: Kemamine® BQ-2802C
Tradenames containing: Incroquat Behenyl BDQ/P; Incroquat Behenyl TMC/P

Behenamide (CTFA)
CAS 3061-75-4; EINECS 221-304-1
Synonyms: Behenic acid amide; Docosanamide
Tradenames: Kemamide® B

Behenamidopropyl dimethylamine (CTFA)
CAS 60270-33-9; EINECS 262-134-8
Synonyms: Dimethylaminopropyl behenamide; N-[3-(Dimethylamino)propyl]docosanamide
Tradenames: Mackine 601

Behenamine
Synonyms: Behenyl amine
Tradenames: Nissan Amine VB

Behenic acid (CTFA)
CAS 112-85-6; EINECS 204-010-8
Synonyms: Docosanoic acid
Tradenames: Hystrene® 5522; Hystrene® 7022; Hystrene® 9022; Prifrac 2989

Behenic acid amide. *See Behenamide*

Behenoyl-PG-trimonium chloride (CTFA)
CAS 69537-38-8; EINECS 274-033-6
Synonyms: (3-Behenoyloxy-2-hydroxypropyl) trimethyl ammonium chloride
Tradenames containing: Akypoquat 40

Behenyl alcohol (CTFA)
CAS 661-19-8; EINECS 211-546-6
Synonyms: 1-Docosanol
Tradenames: Adol® 60; Emery® 3304; Loxiol VPG 1451

Behenyl amine. *See Behenamine*

Behenyl dimethyl amine. *See Dimethyl behenamine*

Behenyl dimethyl benzyl ammonium chloride. *See Behenalkonium chloride*

Behenyl hydroxyethyl imidazoline (CTFA)
Synonyms: Behenyl imidazoline; 1H-Imidazole-1-ethanol, 4,5-dihydro-2-docosanyl-; Behenic imidazoline
Tradenames: Schercozoline B

Behenyl imidazoline. *See Behenyl hydroxyethyl imidazoline*

Benzalkonium chloride (CTFA)
CAS 8001-54-5; 61789-71-7; 68391-01-5; 68424-85-1; 68989-00-4; 85409-22-9; EINECS 263-080-8; 269-919-4; 270-325-2; 287-089-1
Synonyms: Alkyl dimethyl benzyl ammonium chloride
Tradenames: Ablumine 08; Ablumine 10; Ablumine 12; Ablumine 18; Ablumine 1214; Ablumine 3500; Alkaquat® DMB-451-50, DMB-451-80; Arquad® DMMCB-50; Arquad® DMMCB-75; BTC® 50 USP; BTC® 65 USP; BTC® 835; BTC® 8358; Carsoquat® 621; Carsoquat® 621 (80%); Catigene® 50 USP; Catigene® 65 USP; Catigene® 824; Catigene® 2565; Catigene® 8248; Catigene® B 50; Catigene® B 80; Catigene® T 50; Catigene® T 80; Empigen® BAC50; Empigen® BAC50/BP; Empigen® BAC80; Empigen® BAC90; Empigen® BCB50; Empigen® BCF 80; Exameen 3500 3714; Exameen 3580 3719; Gardiquat 12H; Gardiquat 1450; Gardiquat 1480; Gardiquat SV 480; Lebon GM; Lutensit® K-LC, K-LC 80; Lutensit® K-OC; Maquat LC-12S; Maquat MC-1412; Maquat MC-1416; Maquat MC-6025-50%; Protectol® KLC 50, 80; Quadrilan® BC; Rhodaquat® A-50 (formerly Cequartyl A-50); Synprolam 35DMBQC; Variquat® 50MC; Variquat® 60LC; Variquat® 80MC; Variquat® 80ME; Zoharquat 50; Zoharquat 80 EXP
Tradenames containing: Arquad® B-100; BTC® 776; BTC® 885; BTC® 885 P40; BTC® 888; BTC® 2125M-80, 2125M P-40; Catigene® 776; Catigene® 885; Catigene® 888; Catigene® 2125 M; Catigene® 2125 P40; Catigene® 2125M P40; Catigene® 4513-50; Catigene® 4513-80; Catigene® 4513-80 M; FMB 6075-5 Quat, 6075-8 Quat; Maquat MQ-2525; Maquat MQ-2525M; Maquat TC-76; Variquat® 50AC; Variquat® 50AE; Variquat® 50ME; Variquat® 80AC; Variquat® 80AE;

Benztrimonium chloride. *See Benzyl trimethyl ammonium chloride*

Benzyl trimethyl ammonium chloride
CAS 56-93-9
Synonyms: TMBAC; Trimethylbenzylammonium chloride; Benzyl trimethyl ammonium chloride; Benztrimonium chloride
Tradenames: Hipochem Migrator J; Variquat® B200

Bis (acyloxyethyl) hydroxyethyl methylammonium methosulfate
Tradenames containing: Dehyquart AU-36; Dehyquart AU-56

Bis hydrogenated tallowalkyl-2-hydroxypropyl amine
CAS 71060-61-2
Tradenames: Propomeen 2HT-11

Bis (2-hydroxyethyl). *See also Dihydroxyethyl*

Bis-hydroxyethyl dihydroxypropyl stearaminium chloride (CTFA)
Tradenames containing: Monaquat TG

Bis-(2-hydroxyethyl) isodecyloxypropylamine
CAS 34360-00-4
Tradenames: Tomah E-14-2

Bishydroxyethylisodecyloxypropylamine oxide
Tradenames: Tomah AO-14-2

Bis-(2-hydroxyethyl) isotridecyloxypropylamine
Tradenames: Tomah E-17-2

Bis (2-hydroxyethyl) octadecyloxypropylamine
Tradenames: Tomah E-18-2

Boric acid DEA
Tradenames: Rewocoros RAB 90

Buteth-2 carboxylic acid (CTFA)
Synonyms: PEG-2 butyl ether carboxylic acid; PEG 100 butyl ether carboxylic acid
Tradenames: Akypo LF 5
Tradenames containing: Akypo LF 6

Butoxy diglycol (CTFA)
CAS 112-34-5; EINECS 203-961-6
Synonyms: PEG-2 butyl ether; Diethylene glycol butyl ether; 2-(2-Butoxyethoxy) ethanol
Tradenames containing: Mackanate DOS-70BC; Monawet MO-70

Butoxyethanol (CTFA)
CAS 111-76-2; EINECS 203-905-0

Synonyms: 2-Butoxyethanol; Ethylene glycol monobutyl ether; Ethylene glycol butyl ether
Tradenames containing: Surfynol® 104BC

Butoxynol-5 carboxylic acid (CTFA)
Synonyms: PEG-5 butyl phenyl ether carboxylic acid; POE (5) butyl phenyl ether carboxylic acid
Tradenames: Akypo TBP 40

Butoxynol-19 carboxylic acid
CAS 104909-82-2
Tradenames: Akypo TBP 180

Butyl alcohol (CTFA)
CAS 71-36-3; EINECS 200-751-6
Synonyms: n-Butyl alcohol; 1-Butanol; Propyl carbinol
Tradenames: Alfol® 4

Butyl benzoate (CTFA)
CAS 136-60-7; EINECS 205-252-7
Synonyms: Benzoic acid, butyl ester; n-Butyl benzoate
Tradenames: Hipochem B-3-M

Butyl ethylene. *See 1-Hexene*

Butyl naphthalene sodium sulfonate. *See Sodium butyl naphthalene sulfonate*

Butyl naphthalene sulfonate
Tradenames: Dyasulf 9268-A; Emkal BNS Acid

Butyl octadecanoate. *See Butyl stearate*

Butyl oleate (CTFA)
CAS 142-77-8; EINECS 205-559-6
Synonyms: Butyl 9-octadecenoate
Tradenames: Kemester® 4000
Tradenames containing: Iberwet BO

Butyl phosphate
Tradenames: Findet DD

Butyl stearate (CTFA)
CAS 123-95-5; EINECS 204-666-5
Synonyms: Butyl octadecanoate; n-Butyl octadecanoate; Octadecanoic acid butyl ester
Tradenames: Emerest® 2325; Kessco® BS; Witconol 2326

But-2-yne-1,4-diol
CAS 110-65-6
Synonyms: 1,4-Butynediol
Tradenames: Korantin® BH Liq; Korantin® BH Solid

C12-14 acids
CAS 68002-90-4
Tradenames: C-1214

C18-36 acid (CTFA)
CAS 68476-03-9
Tradenames: Syncrowax AW1-C

C18-36 acid glycol ester (CTFA)
Tradenames: Syncrowax ERL-C

C18-36 acid triglyceride (CTFA)
Tradenames: Syncrowax HGL-C

Calcium alginate (CTFA)
CAS 9005-35-0
Synonyms: Alginic acid, calcium salt
Tradenames containing: Keltose®

Calcium behenate (CTFA)
CAS 3578-72-1; EINECS 222-700-7
Synonyms: Calcium docosanoate; Docosanoic acid, calcium salt
Tradenames containing: Syncrowax HRS-C

Calcium dinonylnaphthalene sulfonate
Tradenames: Manro DNNS/C

Calcium dodecylbenzene sulfonate
CAS 26264-06-2; 68953-96-8
Tradenames: Ablusol DBC; Agrilan® X98; Arylan® CA; CA DBS 50 SA; Casul® 70 HF; Emcol® P 50-20 B; Emcol® P-1020B; Emulson AG/CAL; Geronol V/087; Nansa® EVM50; Phenyl Sulphonate CA, CAL; Phenyl Sulphonate HFCA; Rhodacal® 70/B (formerly Soprophor® 70/B); Rhodacal® CA, 70% (formerly Alkasurf® CA); Sandoz Sulfonate AAS 75S; Sermul EA 88; Witconate 60L; Witconate 605A; Witconate 605AC
Tradenames containing: Geronol AZ82; Geronol FF/4; Geronol FF/6; Geronol MS; Geronol RE/70; Geronol SC/120; Geronol SC/121; Geronol SC/177; Geronol SN; Geronol V/497; Nansa® EVM62/H; Nansa® EVM70; Nansa® EVM70/B; Nansa® EVM70/E

Calcium lactylate
Tradenames: Pationic® 1230; Pationic® 1240

Calcium lignosulfonate (CTFA)
CAS 8061-52-7
Synonyms: Lignosulfonic acid, calcium salt
Tradenames: Glutrin; Goulac; Lignosite®; Lignosite® 401; Lignosol B; Lignosol BD; Lignosol SFX

Calcium/sodium stearoyl lactylate
Tradenames: Pationic® 925

Calcium stearate (CTFA)
CAS 1592-23-0; EINECS 216-472-8
Synonyms: Calcium octadecanoate; Stearic acid, calcium salt
Tradenames: Calcium Stearate, Regular

Calcium stearoyl lactylate (CTFA)
CAS 5793-94-2; EINECS 227-335-7
Synonyms: Calcium stearoyl-2-lactylate; Calcium stearyl-2-lactylate; Calcium stelate
Tradenames: Grindtek FAL 2; Pationic® 930; Pationic® 940; Verv®

Calcium sulfonate
CAS 61789-86-4
Tradenames: Hybase® C-300; Hybase® C-400; Hybase® C-500; Lubrizol® 2152

C6-10 alcohol
Tradenames: Alfol® 610; Alfol® 610ADE; Alfol® 610 AFC

C8 alcohols. *See Caprylic alcohol*

C8-10 alcohols
CAS 68603-15-5
Tradenames: Alfol® 810; Alfol® 810FD; CO-810

C9-11 alcohols (CTFA)
CAS 66455-17-2; 68551-08-6
Synonyms: Alcohols, C9-11
Tradenames: Dobanol 91; Neodol® 91

C10-12 alcohols
Tradenames: Alfol® 1012 CDC, 1012 HA

C11 alcohol. *See Undecyl alcohol*

C12 alcohol. *See Lauryl alcohol*

C12-13 alcohols (CTFA)
CAS 75782-86-4
Synonyms: Alcohols, C12-13
Tradenames: Dobanol 23; Neodol® 23

C12-14 alcohols
CAS 67762-41-8; 68002-90-4
Tradenames: Alfol® 1214; Alfol® 1214 GC; Alfol® 1216 CO; Alfol® 1412; CO-1214; CO-1270

C12-15 alcohols (CTFA)
CAS 63393-82-8
Synonyms: Alcohols, C12-15
Tradenames: Dobanol 25; Neodol® 25

C12-16 alcohols (CTFA)
CAS 68855-56-1
Tradenames: Alfol® 1216; Alfol® 1216 DCBA

C12-18 alcohols
CAS 67762-25-8
Tradenames: CO-1218; Loxiol VPG 1496

C13 alcohol. *See Tridecyl alcohol*

C14 alcohol. *See Myristyl alcohol*

C14–15 alcohols (CTFA)
CAS 75782-87-5
Tradenames: Dobanol 45; Neodol® 45

C14-18 alcohols
Tradenames: Alfol® 1418 DDB

C15 alcohol. *See Pentadecyl alcohol*

C16-18 alcohols. *See Cetearyl alcohol*

C16-20 alcohols
Tradenames: Alfol® 1620

C18-32 alcohols
Tradenames containing: Epal® 20+

C13-15 alkyl bis (2-hydroxyethyl) amine oxide
Tradenames: Synprolam 35X2/O

C12-14 alkyl dimethylamine n-oxide
CAS 85408-49-7; EINECS 2870116
Tradenames: Lilaminox M24

C13-15 alkyl-1,3-propane diamine
Tradenames: Synprolam 35N3

C13-15 alkyl trimonium chloride
Tradenames: Synprolam 35TMQC

C4 alpha olefin
CAS 106-98-9
Tradenames: Ethyl® Butene-1; Ethyl® Tetradecene-1; Gulftene 4

C6 alpha olefin. *See 1-Hexene*

C8 alpha olefin. *See Octene-1*

C10 alpha olefin. *See Decene-1*

C12 alpha olefin. *See Dodecene-1*

C12-14 alpha olefin
CAS 64743-02-8
Tradenames: Neodene® 1214

C14 alpha olefin. *See Tetradecene-1*
CAS 1120-36-1

C14-16 alpha olefins
CAS 64743-02-8
Tradenames: Ethyl® C1416; Neodene® 1416

C14-18 alpha olefins
Tradenames: Neodene® 14/16/18

C16 alpha olefin. *See Hexadecene-1*
CAS 629-73-2

C16-18 alpha olefin
CAS 64743-02-8
Tradenames: Ethyl® C1618; Neodene® 1618

C18 alpha olefin. *See Octadecene-1*

C20-24 alpha olefin
Tradenames: Gulftene 20-24

C24-28 alpha olefin
Tradenames: Gulftene 24-28

C30 alpha olefin
Tradenames: Gulftene 30+

C13-15 amine
CAS 68155-27-1
Tradenames: Synprolam 35

Capramide DEA (CTFA)
CAS 136-26-5; EINECS 205-234-9
Synonyms: Capric acid diethanolamide; N,N-Bis(2-hydroxyethyl) decanamide
Tradenames: Alrosol C; Comperlan CD; Hetamide 1069; Mackamide CD-10
Tradenames containing: Emid® 6544; Standamid® CD

Capric acid (CTFA)
CAS 334-48-5; EINECS 206-376-4
Synonyms: n-Decanoic acid; n-Capric acid
Tradenames: C-1095; Emery® 659

Capric acid diethanolamide. *See Capramide DEA*

Capric/caprylic amidopropyl betaine

Capric dimethyl amine oxide. *See Decylamine oxide*

Caproamphocarboxyglycinate. *See Disodium caproamphodiacetate*

Caproamphocarboxypropionate. *See Disodium caproamphodipropionate*

Caproamphoglycinate. *See Sodium caproamphoacetate*

Capryleth-4 carboxylic acid (CTFA)
Synonyms: PEG-4 caprylyl ether carboxylic acid; PEG 200 caprylyl ether carboxylic acid; POE (4) caprylyl ether carboxylic acid
Tradenames: Akypo MB 1614/1

Capryleth-6 carboxylic acid (CTFA)
Synonyms: PEG-6 caprylyl ether carboxylic acid; PEG 300 caprylyl ether carboxylic acid; POE (6) caprylyl ether carboxylic acid
Tradenames: Akypo LF 1

Capryleth-9 carboxylic acid (CTFA)
CAS 107600-33-9
Synonyms: PEG-9 caprylyl ether carboxylic acid; PEG 450 caprylyl ether carboxylic acid; POE (9) caprylyl ether carboxylic acid
Tradenames: Akypo LF 2
Tradenames containing: Akypo LF 4; Akypo LF 6

Capryl hydroxyethyl imidazoline (CTFA)
CAS 37478-68-5; EINECS 253-521-2
Synonyms: Caprylic imidazoline; 2-Nonyl-4,5-dihydro-1H-imidazole-1-ethanol; 1H-Imidazole-1-ethanol, 4,5-dihydro-2-nonyl-
Tradenames: Mackazoline CY; Monazoline CY

Caprylic acid (CTFA)
CAS 124-07-2; EINECS 204-677-5
Synonyms: n-Octanoic acid; Octoic acid
Tradenames: C-895; C-899; Emery® 657; Prifrac 2901; Prifrac 2910

Caprylic alcohol (CTFA)
CAS 111-87-5; EINECS 203-917-6
Synonyms: n-Octyl alcohol; 1- or n-Octanol; C8 alcohols
Tradenames: Alfol® 8; Emery® 3322; Emery® 3324; Emery® 3329; Epal® 8; Lorol C8; Lorol

C8-98; Lorol C8 Chemically Pure
Tradenames containing: Epal® 108; Epal® 610; Epal® 810

Caprylic/capric acid
CAS 67762-36-1
Tradenames: C-810; C-810L; Emery® 658; Industrene® 365; Philacid 0810; Prifrac 2912

Caprylic/capric glycerides (CTFA)
Tradenames: Imwitor® 742

Caprylic/capric triglyceride (CTFA)
CAS 65381-09-1
Synonyms: Octanoic/decanoic acid triglyceride
Tradenames: Miglyol® 812

Caprylic glycinate
Tradenames: Varion® AM-V

Caprylic imidazoline. *See Capryl hydroxyethyl imidazoline*

Capryloamphoacetate. *See Sodium caprylo-amphoacetate*

Capryloamphocarboxyglycinate. *See Disodium capryloamphodiacetate*

Capryloamphocarboxypropionate. *See Disodium capryloamphodipropionate*

Capryloamphodiacetate. *See Disodium capryloamphodiacetate*

Capryloamphodipropionate. *See Disodium capryloamphodipropionate*

Capryloamphoglycinate. *See Sodium caprylo-amphoacetate*

Capryloamphopropionate. *See Sodium caprylo-amphopropionate*

Capryloamphopropylsulfonate. *See Sodium capryloamphohydroxypropyl sulfonate*

Carbomer (CTFA)
CAS 9007-16-3; 9003-01-4; 9007-17-4; 76050-42-5
Tradenames: Carbopol® 910; Carbopol® 934; Carbopol® 940

Carboxymethylcellulose sodium (CTFA)
CAS 9004-32-4
Synonyms: CMC; Cellulose gum (CTFA); Sodium carboxymethylcellulose; Sodium CMC
Tradenames: Tylose C, CB Series, CBR Grades

Castor oil (CTFA)
CAS 1323-38-2; 8001-79-4; EINECS 232-293-8
Synonyms: Ricinus oil; Oil of Palma Christi; Tangantangan oil
Tradenames: AA Standard; #1 Oil; Surfactol® 13

Castor oil, hydrogenated. *See Hydrogenated castor oil*

Castor oil, polymerized
Tradenames: #15 Oil; #30 Oil; #40 Oil; Pale 4; Pale 16; Pale 170; Pale 1000; Vorite 105; Vorite 110; Vorite 115; Vorite 120; Vorite 125

Castor oil sulfated. *See Sulfated castor oil*

C44 dimer acid
Tradenames: Pripol 1004

C12-16 dimethylamine
Tradenames: Empigen® AF; Empigen® AG

C10-14 dodecylbenzene
Tradenames: Nalkylene® 550L

Cellulose gum (CTFA). *See Carboxymethylcellulose sodium*

Cetamine oxide. *See Palmitamine oxide*

Ceteareth-2 (CTFA)
CAS 68439-49-6 (generic)
Synonyms: PEG-2 cetyl/stearyl ether; POE (2) cetyl/stearyl ether; PEG 100 cetyl/stearyl ether
Tradenames: Macol® CSA-2; Rhodasurf® A-2 (formerly Texafor A-2)

Ceteareth-3 (CTFA)
CAS 68439-49-6 (generic)
Synonyms: PEG-3 cetyl/stearyl ether; POE (3) cetyl/stearyl ether
Tradenames: Volpo CS-3

Ceteareth-4 (CTFA)
CAS 68439-49-6 (generic)
Synonyms: PEG-4 cetyl/stearyl ether; POE (4) cetyl/stearyl ether
Tradenames: Hetoxol CS-4; Lipocol SC-4; Macol® CSA-4; Prox-onic CSA-1/04; Sellig SU 4 100

Ceteareth-5 (CTFA)
CAS 68439-49-6 (generic)
Synonyms: PEG-5 cetyl/stearyl ether; POE (5) cetyl/stearyl ether
Tradenames: Hetoxol CS-5; Volpo CS-5

Ceteareth-6 (CTFA)
CAS 68439-49-6 (generic)
Synonyms: PEG-6 cetyl/stearyl ether; POE (6) cetyl/stearyl ether
Tradenames: Dehydol PCS 6; Marlipal® 1618/6; Marlowet® TA 6; Prox-onic CSA-1/06; Rhodasurf® A-6 (formerly Texafor A-6)

Ceteareth-8 (CTFA)
CAS 68439-49-6 (generic)
Synonyms: PEG-8 cetyl/stearyl ether; PEG 400 cetyl/stearyl ether; POE (8) cetyl/stearyl ether
Tradenames: Marlipal® 1618/8; Marlowet® TA 8

Ceteareth-9
CAS 68439-49-6 (generic)
Synonyms: PEG-9 cetyl/stearyl ether; POE (9) cetyl/stearyl ether
Tradenames: Hetoxol CS-9

Ceteareth-10 (CTFA)
CAS 68439-49-6 (generic)
Synonyms: PEG-10 cetyl/stearyl ether; POE (10) cetyl/stearyl ether
Tradenames: Macol® CSA-10; Marlipal® 1618/10; Marlowet® TA 10; Prox-onic CSA-1/010; Volpo CS-10

Ceteareth-11 (CTFA)
CAS 68439-49-6 (generic)
Synonyms: PEG-11 cetyl/stearyl ether; POE (11) cetyl/stearyl ether
Tradenames: Empilan® KM 11; Marlipal® 1618/11; Rolfor HT 11

Ceteareth-12 (CTFA)
CAS 68439-49-6 (generic)
Synonyms: PEG-12 cetyl/stearyl ether; POE (12) cetyl/stearyl ether
Tradenames: Eumulgin B1; Standamul® B-1

Ceteareth-14
CAS 68439-49-6 (generic)
Synonyms: PEG-14 cetyl/stearyl ether; POE (14) cetyl/stearyl ether
Tradenames: Dehydol PCS 14

Ceteareth-15 (CTFA)
CAS 68439-49-6 (generic)

Synonyms: PEG-15 cetyl/stearyl ether; POE (15) cetyl/stearyl ether
Tradenames: Hetoxol 15 CSA; Hetoxol CS-15; Hetoxol CSA-15; Lipocol SC-15; Macol® CSA-15; Prox-onic CSA-1/015; Volpo CS-15

Ceteareth-16 (CTFA)
CAS 68439-49-6 (generic)
Synonyms: PEG-16 cetyl/stearyl ether; POE (16) cetyl/stearyl ether
Tradenames: Teric 16A16

Ceteareth-18 (CTFA)
CAS 68439-49-6 (generic)
Synonyms: PEG-18 cetyl/stearyl ether; POE (18) cetyl/stearyl ether
Tradenames: Marlipal® 1618/18; Marlowet® TA 18; Sellig SU 18 100

Ceteareth-20 (CTFA)
CAS 68439-49-6 (generic)
Synonyms: PEG-20 cetyl/stearyl ether; POE (20) cetyl/stearyl ether
Tradenames: Acconon W230; Brij® 68; Empilan® KM 20; Eumulgin B2; Hetoxol CS-20; Incropol CS-20; Lipocol SC-20; Macol® CSA-20; Prox-onic CSA-1/020; Remcopal 238; Rewopal® CSF 20; Sellig SU 25 100; Standamul® B-2; Volpo CS-20
Tradenames containing: Hetoxol CS-20D; Hetoxol D; Hetoxol G; Hetoxol J; Paramul® J; Teginacid®; Teginacid® X

Ceteareth-22
CAS 68439-49-6 (generic)
Synonyms: PEG-22 cetyl/stearyl ether; POE (22) cetyl/stearyl ether
Tradenames: Teric 16A22

Ceteareth-23 (CTFA)
CAS 68439-49-6 (generic)
Synonyms: PEG-23 cetyl/stearyl ether; PEG (23) cetyl/stearyl ether; POE (23) cetyl/stearyl ether
Tradenames: Rhodasurf® A-1P (formerly Texafor A-1P)

Ceteareth-25 (CTFA)
CAS 68439-49-6 (generic)
Synonyms: PEG-25 cetyl/stearyl ether; POE (25) cetyl/stearyl ether
Tradenames: Hetoxol CS-25; Industrol® CSS-25; Marlipal® 1618/25; Marlowet® TA 25; Remcopal 229; Rolfor HT 25
Tradenames containing: Marlipal® 1618/25 P 6000

Ceteareth-27 (CTFA)
CAS 68439-49-6 (generic)
Synonyms: PEG-27 cetyl/stearyl ether; POE (27) cetyl/stearyl ether
Tradenames: Plurafac® A-38

Ceteareth-28 (CTFA)
CAS 68439-49-6 (generic)
Synonyms: PEG-28 cetyl/stearyl ether; POE (28) cetyl/stearyl ether
Tradenames: Marlowet® FOX

Ceteareth-29
CAS 68439-49-6 (generic)
Synonyms: PEG-29 cetyl/stearyl ether; POE (29) cetyl/stearyl ether
Tradenames: Teric 16A29

Ceteareth-30 (CTFA)
CAS 68439-49-6 (generic)
Synonyms: PEG-30 cetyl/stearyl ether; POE (30) cetyl/stearyl ether
Tradenames: Eumulgin B3; Hetoxol CS-30; Prox-onic CSA-1/030; Standamul® B-3
Tradenames containing: Hetoxol L

Ceteareth-32
Tradenames: Sellig SU 30 100

Ceteareth-40 (CTFA)
CAS 68439-49-6 (generic)
Synonyms: PEG-40 cetyl/stearyl ether; POE (40) cetyl/stearyl ether
Tradenames: Hetoxol CS-40W; Macol® CSA-40; Marlipal® 1618/40

Ceteareth-46
Tradenames: Sellig SU 50 100

Ceteareth-50 (CTFA)
CAS 68439-49-6 (generic)
Synonyms: PEG-50 cetyl/stearyl ether; POE (50) cetyl/stearyl ether
Tradenames: Empilan® KM 50; Hetoxol CS-50; Hetoxol CS-50 Special; Incropol CS-50; Macol® CSA-50; Prox-onic CSA-1/050; Teric 16A50

Ceteareth-55 (CTFA)
CAS 68439-49-6 (generic)
Synonyms: PEG-55 cetyl/stearyl ether; POE (55) cetyl/stearyl ether
Tradenames: Plurafac® A-39

Ceteareth-60 (CTFA)
CAS 68439-49-6 (generic)
Synonyms: PEG-60 cetyl/stearyl ether; POE (60) cetyl/stearyl ether
Tradenames: Rhodasurf® A-60 (formerly Texafor A-60)

Ceteareth-80
Tradenames: Marlipal® 1618/80

Ceteareth-7 carboxylic acid
CAS 68954-89-2
Tradenames: Akypo RCS 60

Ceteareth-2 phosphate
Tradenames: Crodafos CS2 Acid

Ceteareth-5 phosphate (CTFA)
Tradenames: Crodafos CS5 Acid

Ceteareth-10 phosphate
Tradenames: Crodafos CS10 Acid

Cetearyl alcohol (CTFA)
CAS 8005-44-5; 67762-30-5
Synonyms: Cetostearyl alcohol; Cetyl/stearyl alcohol; C16-18 alcohols
Tradenames: Adol® 63; Adol® 630; Adol® 640; Alfol® 1618; Alfol® 1618 CG; Epal® 618; Hetoxol CS; Laurex® 4550; Laurex® CS; Laurex® CS/D; Laurex® CS/W; Loxiol G 52; Loxiol G 53; Loxiol P 1420; Philcohol 1618; TA-1618
Tradenames containing: Carsoquat® 816-C; Crodex C; Crodex N; Dermalcare® POL (formerly Cyclochem® POL); Hetoxol CS-20D; Hetoxol D; Hetoxol J; Hetoxol L; Maquat SC-1632; Paramul® J; Ritachol® 1000; Ritachol® 3000; Ritapro 100

Ceteth-2 (CTFA)
CAS 9004-95-9 (generic)
Synonyms: PEG-2 cetyl ether; POE (2) cetyl ether; PEG 100 cetyl ether

Tradenames: Ethosperse® CA-2; Hetoxol CA-2; Lipocol C-2; Macol® CA-2; Simulsol 52

Ceteth-5 (CTFA)
CAS 9004-95-9 (generic)
Synonyms: PEG-5 cetyl ether; POE (5) cetyl ether
Tradenames containing: Solulan® 5

Ceteth-10 (CTFA)
CAS 9004-95-9 (generic)
Synonyms: PEG-10 cetyl ether; POE (10) cetyl ether
Tradenames: Brij® 56; Hetoxol CA-10; Lipocol C-10; Macol® CA-10; Simulsol 56

Ceteth-15 (CTFA)
CAS 9004-95-9 (generic)
Synonyms: PEG-15 cetyl ether; POE (15) cetyl ether
Tradenames: Nikkol BC-15TX, -15TX(FF)

Ceteth-16 (CTFA)
CAS 9004-95-9 (generic)
Synonyms: PEG-16 cetyl ether; POE (16) cetyl ether
Tradenames containing: Solulan® 16

Ceteth-20 (CTFA)
CAS 9004-95-9 (generic)
Synonyms: PEG-20 cetyl ether; POE (20) cetyl ether; Cetomacrogol 1000
Tradenames: Brij® 58; Hetoxol CA-20; Lipocol C-20; Texofor A1P
Tradenames containing: Crodex N; Dermalcare® POL (formerly Cyclochem® POL); Hetoxol CAWS; Teginacid® H

Ceteth-24 (CTFA)
CAS 9004-95-9 (generic)
Synonyms: PEG-24 cetyl ether; POE (24) cetyl ether
Tradenames containing: Hetoxol C-24; Solulan® C-24

Cetethyl morpholinium ethosulfate (CTFA)
CAS 78-21-7; EINECS 201-094-8
Synonyms: Cethyl ethyl morpholinium ethosulfate; Quaternium-25; Morpholinium, 4-ethyl-4-hexadecyl, ethyl sulfate
Tradenames: G-263

Cetin. *See Cetyl palmitate*

Cetoleth-2
Synonyms: PEG-2 cetyl/oleyl ether; POE (2) cetyl/oleyl ether
Tradenames: Eumulgin EP2; Teric 17A2

Cetoleth-3
Synonyms: PEG-3 cetyl/oleyl ether; POE (3) cetyl/oleyl ether
Tradenames: Ethylan® 172; Teric 17A3

Cetoleth-4
Tradenames: Remcopal 234

Cetoleth-5
Tradenames: Disponil O 5

Cetoleth-6
Synonyms: PEG-6 cetyl/oleyl ether; POE (6) cetyl/oleyl ether
Tradenames: Empilan® KL 6; Ethylan® ME; Prox-onic OCA-1/06; Teric 17A6

Cetoleth-8
Synonyms: PEG-8 cetyl/oleyl ether; POE (8) cetyl/oleyl ether
Tradenames: Oxetal O 108; Sellig SP 8 100; Teric 17A8

Cetoleth-10
Synonyms: PEG-10 cetyl/oleyl ether; POE (10) cetyl/oleyl ether
Tradenames: Disponil O 10; Empilan® KL 10; Remcopal 10; Teric 17A10

Cetoleth-11
CAS 68920-66-1
Tradenames: Rolfor CO 11

Cetoleth-12
Tradenames: Oxetal O 112

Cetoleth-13
Synonyms: PEG-13 cetyl/oleyl ether; PEG (13) cetyl/oleyl ether
Tradenames: Ethylan® OE; Teric 17A13

Cetoleth-16
Tradenames: Sellig SP 16 100

Cetoleth-18
Tradenames: Remcopal 18; Sellig SP 20 100

Cetoleth-19
Tradenames: Ethylan® R

Cetoleth-20
Synonyms: PEG-20 cetyl/oleyl ether; POE (20) cetyl/oleyl ether
Tradenames: Disponil O 20; Empilan® KL 20

Cetoleth-23
Synonyms: PEG-23 cetyl/oleyl ether; POE (23) cetyl/oleyl ether
Tradenames: Remcopal D

Cetoleth-25 (CTFA)
Synonyms: PEG-25 cetyl/oleyl ether; POE (25) cetyl/oleyl ether
Tradenames: Disponil O 250; Marlipal® SU; Remcopal 25; Remcopal 220; Teric 17A25

Cetoleth-27
Tradenames: Sellig SP 25 50; Sellig SP 25 100

Cetoleth-30
Tradenames: Mergital OC 30E; Sellig SP 3020; Sellig SP 30 100

Cetoleth-55
CAS 68920-66-1
Tradenames: Polirol O55

Cetrimide BP
Tradenames containing: Crodex C

Cetrimonium bromide (CTFA)
CAS 57-09-0; EINECS 200-311-3
Synonyms: Cetyltrimethylammonium bromide; Hexadecyltrimethylammonium bromide; N,N,N-Trimethyl-1-hexadecanaminium bromide
Tradenames: Acetoquat CTAB; Bromat; Cetrimide BP

Cetrimonium chloride (CTFA)
CAS 112-02-7; EINECS 203-928-6
Synonyms: Cetyl trimethyl ammonium chloride; Palmityl trimethyl ammonium chloride; Hexadecyl trimethyl ammonium chloride
Tradenames: Adogen® 444; Arquad® 16-29; Arquad® 16-29W; Chemquat 16-50; Dehyquart A; Nissan Cation PB-40, -300; Radiaquat 6444; Variquat® E290; Querton 16Cl-29; Querton 16Cl-50
Tradenames containing: Arquad® 16-50

Cetyl acetate (CTFA)
CAS 629-70-9; EINECS 211-103-7
Synonyms: Hexadecyl acetate

Tradenames containing: Crodalan AWS

Cetyl alcohol (CTFA)
CAS 36653-82-4; 36311-34-9; EINECS 253-149-0
Synonyms: Palmityl alcohol; C16 linear primary alcohol; 1-Hexadecanol
Tradenames: Adol® 52 NF; Adol® 520 NF; Alfol® 16; Alfol® 16 NF; CO-1695; Emery® 3336; Emery® 3337; Emery® 3338; Emery® 3339; Epal® 16NF; Kalcohl 60; Lorol C16; Lorol C16-95; Lorol C16-98; Lorol C16 Chemically Pure; Loxiol VPG 1743; Philcohol 1600
Tradenames containing: Emery® 3310; Emery® 3311; Emery® 3313; Emery® 3314; Emery® 3315; Emery® 3316; Emulgade EO-10; Epal® 1214; Epal® 1218; Epal® 1416; Epal® 1416-LD; Epal® 1418; Epal® 1618; Epal® 1618T; Epal® 1618RT; HD-Ocenol 45/50; HD-Ocenol 50/55; HD-Ocenol 50/55III; HD-Ocenol 60/65; HD-Ocenol 70/75; HD-Ocenol 80/85; Ritachol® 3000; Ritachol® 4000

Cetyl amine. *See Palmitamine*

Cetyl betaine (CTFA)
CAS 693-33-4; EINECS 211-748-4
Synonyms: N-(Carboxymethyl)-N,N-dimethyl-1-hexadecanaminium hydroxide, inner salt
Tradenames: Lonzaine® 16SP; Mackam CET; Product BCO
Tradenames containing: Darvan® NS

Cetyldiethanolaminephosphate
CAS 90388-14-0
Tradenames: Crodafos CDP

Cetyl dimethicone copolyol (CTFA)
Tradenames: Abil® B 9806, B 9808

Cetyl dimethylamine. *See Dimethyl palmitamine*

Cetyl ethyl morpholinium ethosulfate. *See Cetethyl morpholinium ethosulfate*

Cetylic acid. *See Palmitic acid*

Cetyl oleth-7 phosphate
Tradenames: Servoxyl VPGZ 7/100

Cetyl palmitate (CTFA)
CAS 540-10-3; EINECS 208-736-6
Synonyms: Hexadecanoic acid, hexadecyl ester; Palmitic acid, hexadecyl ester; Cetin
Tradenames: Kessco® 653

Cetylpyridinium bromide
CAS 140-72-7
Synonyms: Hexadecylpyridinium bromide
Tradenames: Acetoquat CPB

Cetylpyridinium chloride (CTFA)
CAS 123-03-5; 6004-24-6; EINECS 204-593-9
Synonyms: 1-Hexadecylpyridinium chloride
Tradenames: Acetoquat CPC

Cetyl/stearyl alcohol. *See Cetearyl alcohol*

Cetyltrimethylammonium bromide. *See Cetrimonium bromide*

Cetyltrimethyl ammonium chloride. *See Cetrimonium chloride*

Chloro-2-hydroxypropyl trimonium chloride
CAS 3327-22-8
Synonyms: 3-Chloro-2-hydroxypropyltrimethyl ammonium chloride
Tradenames: CHPTA 65%

Choleth-24 (CTFA)
CAS 27321-96-6 (generic)
Synonyms: PEG-24 cholesteryl ether
Tradenames containing: Solulan® C-24

C24-40 hydrocarbons
Tradenames containing: Epal® 20+

Cocamide (CTFA)
CAS 61789-19-3; EINECS 263-039-4
Synonyms: Coconut oil amides; Coconut acid amide
Tradenames: Armid® C; Schercomid 304

Cocamide DEA (CTFA)
CAS 8051-30-7; 61791-31-9; 68603-42-9; EINECS 263-163-9
Synonyms: Coconut diethanolamide; Cocoyl diethanolamide; N,N-bis (2-hydroxyethyl) coco amides
Tradenames: Ablumide CDE; Accomid C; Alkamide® 200 CGN (formerly Cyclomide 200 CGN); Alkamide® 2104; Alkamide® 2110; Alkamide® 2204; Alkamide® CDE; Alkamide® CDM; Alkamide® CDO; Alkamide® CL63; Alkamide® DC-212/M (formerly Cyclomide DC-212/M); Alkamide® DC-212/S (formerly Cyclomide DC-212/S); Alkamide® DC-212/SE (formerly Cyclomide DC-212/SE); Alkamide® KD (formerly Cyclomide KD); Alrosol B; Amidex 1285; Amidex 1351; Amidex C; Amidex CA; Amidex CE; Amidex CO-1; Aminol COR-2; Aminol COR-4; Aminol COR-4C; Aminol HCA; Aminol KDE; Calamide C; Calamide CW-100; Carsamide® 7644; Carsamide® C-3; Carsamide® C-7944; Carsamide® CA; Carsamide® SAC; Chimipal DCL/M; Comperlan KDO; Comperlan PD; Comperlan PKDA; Comperlan SD; Compound S.A; Condensate PC; Crillon CDY; Emid® 6514; Emid® 6515; Emid® 6521; Emid® 6529; Emid® 6538; Empilan® 2502; Empilan® CDE; Empilan® CDX; Empilan® FD; Empilan® FD20; Empilan® FE; Esi-Det CDA; Esi-Terge 10; Esi-Terge B-15; Esi-Terge S-10; Esi-Terge T-5; Ethox 2449; Ethox COA; Ethylan® A15; Ethylan® LD; Ethylan® LDA-37; Ethylan® LDA-48; Ethylan® LDG; Ethylan® LDS; Hartamide KL; Hartamide OD; Hetamide DSUC; Hetamide MC; Hetamide MCS; Hetamide RC; Hymolon CWC; Hymolon K90; Iconol 28; Iconol COA; Incromide CA; Karamide 121; Karamide 221; Karamide 363; Karamide 442-M; Karamide CO22; Karamide CO2A; Karamide CO9A; Lauramide 11; Lauramide D; Lauramide ME; Laurel SD-101; Laurel SD-300; Laurel SD-900M; Laurel SD-950; Lauridit® KD, KDG; Loropan KD; Mackamide 100-A; Mackamide C; Mackamide CD; Mackamide CS; Mackamide EC; Mackamide MC; Manro CD; Manro CDS; Manro CDX; Manromid 853; Manromid CD; Manromid CDG; Manromid CDS; Manromid CDX; Marlamid® D 1218; Marlamid® DF 1218; Mazamide® 68; Mazamide® 80; Mazamide® 524; Mazamide® 1281; Mazamide® CCO; Mazamide® CS 148; Mazamide® JT 128; Mazamide® WC Conc; Monamid® 7-153 CS; Monamid® 150-AD; Monamid® 150-ADD; Monamid® 150-DR; Monamid® 150-LWA; Monamine ADD-100; Monamine ADS-100; Monamine I-76; Naxonol CO; Naxonol PN 66; Naxonol PO; Ninol® 11-CM; Ninol® 40-CO; Ninol® 49-CE (formerly Onyxol SD); Ninol® 4821 F; Nissan Stafoam DF; Nitrene 11120;

Nitrene 11230; Nitrene 13026; Nitrene A-309; Nitrene A-567; Nitrene C; Nitrene C Extra; Nitrene N; Norfox® DC-38; Norfox® DCO; Norfox® DCS; Norfox® DCSA; Norfox® DESA; Norfox® DOSA; Norfox® PEA-N; Norfox® X; Purton CFD; Quimipol DEA OC; Rewomid® C 220SE; Rewomid® DC 212 S; Rewomid® DC 220 SE; Rhodaterge® DCA (formerly Cycloryl DCA); Schercomid 1-102; Schercomid EAC; Schercomid EAC-S; Schercomid SCE; Schercomid SCO-Extra; Stafoam DF-1; Stafoam DF-4; Stafoam F; Stamid HT 3901; Standamid® KD; Standamid® KDM; Standamid® KDO; Standamid® PK-KD; Standamid® PK-KDO; Standamid® PK-KDS; Standamid® PK-SD; Standamid® SD; Standamid® SDO; Synotol 119 N; Synotol CN 20; Synotol CN 60; Synotol CN 80; Synotol CN 90; Synotol Detergent E; T-Tergamide 1CD; Teric CDE; Tohol N-220; Tohol N-220X; Ufanon K-80; Ufanon KD-S; Unamide® C-72-3; Unamide® C-7649; Unamide® D-10; Unamide® LDL; Unamide® N-72-3; Varamide® A-10; Varamide® A-12; Varamide® A-2; Varamide® A-80; Varamide® A-83; Varamide® A-84; Varamide® MA-1; Witcamide® 128T; Witcamide® 1017; Witcamide® 5130; Witcamide® 6445; Witcamide® 6514; Witcamide® 6515; Witcamide® 6531; Witcamide® 6625; Witcamide® CD; Witcamide® CDA; Witcamide® LDTS; Witcamide® M-3; Witcamide® S771; Witcamide® S780

Tradenames containing: Akypogene WSW-W; Alkamide® DC-212 (formerly Cyclomide DC-212); Aminol COR-2C; Calsuds CD-6; Comperlan LS; Emid® 6531; Emid® 6533; Mackamide CD-25; Mackamide CD-6; Mackamide CD-8; Mackamide CDC; Mackamide CDM; Mackamide CDS-80; Mackamide CDT; Manro CD/G; Mazamide® 70; Monamine 779; Monamine AA-100; Monamine AC-100; Monamine AD-100; Monamine ALX-100 S; Monamine CF-100 M; Monaterge 85 HF; Ninol® SR-100; Norfox® DC; Norfox® KD; Rewomid® DL 240; Rewopol® SBV; Schercomid CCD; Schercomid CDA; Schercomid CDO-Extra; Standamid® PD; Standapol® AP Blend; Standapol® BW; Standapol® EA-K; Stepanol® AEG; Unamide® C-7944; Witcamide® 82; Witcamide® 5133; Witcamide® 6533; Witcamide® Coco Condensate;

Cocamide MEA (CTFA)

CAS 68140-00-1; EINECS 268-770-2

Synonyms: Coconut monoethanolamide; N-(2-hydroxyethyl) coco fatty acid amide; Coconut fatty acid monoethanolamide

Tradenames: Ablumide CME; Alkamide® C-212 (formerly Cyclomide C-212); Alkamide® CME; Amidex CME; Amidex KME; Aminol CM, CM Flakes, CM-C Flakes, CM-D Flakes; Chimipal MC; Comperlan 100; Comperlan KM; Comperlan P 100; Emid® 6500; Empilan® CM; Empilan® CM/F; Empilan® CME; Foamole M; Hetamide CMA; Hetamide CME; Hetamide CME-CO; Incromide CM; Laurel SD-1000; Laurel SD-1050; Lauridit® KM; Loropan CME; Mackamide CMA; Manro CMEA; Manromid CMEA; Marlamid® M 1218; Mazamide® CFAM; Mazamide® CMEA Extra; Monamide; Monamid® CMA; Ninol® CMP; Norfox® CMA; P & G Amide No. 27; Rewomid® C 212; Rewomid® CD; Schercomid CME; Standamid® 100; Standamid® CM; Standamid® CMG; Standamid® KM; Standamid® SM; Swanic 51; Synotol ME 90; Varamide® C-212; Witcamide® C; Zoramide CM

Tradenames containing: Equex STM; Genapol® PGM Conc; Marlamid® PG 20; Standapol® S

Cocamide MIPA (CTFA)

CAS 68333-82-4, 68440-05-1; 8039-67-6; EINECS 269-793-0

Synonyms: Coconut monoisopropanolamide

Tradenames: Empilan® CIS; Rewomid® IPP 240; Schercomid CMI; Witcamide® CPA

Cocamide TEA

Tradenames: Laurel SD-590

Cocamidoamine oxide

Tradenames: Standamox C 30

Cocamidopropylamine oxide (CTFA)

CAS 68155-09-9; EINECS 268-938-5

Synonyms: Cocamidopropyl dimethylamine oxide; Coco amides, N-[3-(dimethylamino)propyl], N-oxide; N-[3-(Dimethylamino)propyl]coco amides-N-oxide

Tradenames: Ablumox CAPO; Aminoxid WS 35; Ammonyx® CDO; Amyx CDO 3599; Barlox® C; Chemoxide CAW; Empigen® OS/A; Mackamine CAO; Mazox® CAPA-37; Mazox® CAPA; Ninox® FCA; Patogen AO-30; Rhodamox® CAPO (formerly Alkamox® CAPO); Schercamox C-AA; Standamox CAW; Standamox PCAW; Varox® 1770

Cocamidopropyl betaine (CTFA)

CAS 61789-40-0; 70851-07-9; 83138-08-3; 86438-79-1; EINECS 263-058-8

Synonyms: CADG; Cocamidopropyl dimethyl glycine

Tradenames: Abluter BE; Ampholak BCA-30; Ampholyt JB 130; Amphosol® CA; Amphosol® CG; Amphotensid B4; Chembetaine C; Dehyton® K; Dehyton® KE; Deriphat BAW; Emcol® 5430; Emcol® 6748; Emcol® Coco Betaine; Emcol® DG; Emcol® NA-30; Emery® 6744; Empigen® BS; Empigen® BS/H; Empigen® BS/P; Incronam 30; Lebon 2000; Lexaine® C; Lexaine® CG-30; Lexaine® CS; Lonzaine® C; Lonzaine® CO; Mackam 35; Mackam 35 HP; Mackam J; Mafo® CAB; Mafo® CAB 425; Mafo® CAB SP; Mafo® CFA 35; Manroteric CAB; Mirataine® BD-R; Mirataine® BET-C-30 (formerly Cycloteric BET-C-30); Mirataine® BET-W (formerly Cycloteric BET-W); Mirataine® CB/M; Mirataine® CBC; Mirataine® CBR; Mirataine® CCB; Monateric ADA; Monateric COAB; Nutrol Betaine OL 3798; Ralufon® 414; Rewoteric AM B13; Rewoteric AM B14; Tegotain D; Tegotain L 7; Tegotain L 5351; Tegotain S; Tego®-Betaine C; Tego®-Betaine F; Tego®-Betaine L-5351; Tego®-Betaine S; Tego®-Betaine T; Varion® AM-B14; Varion® CADG-HS; Varion® CADG-LS; Varion® CADG-W; Velvetex® BA; Velvetex® BA-35

Tradenames containing: Miracare® BC-10; Mirasheen® 202 (formerly Cyclosheen 202); Schercotaine CAB-A; Schercotaine CAB-K;

Standapol® AP Blend; Stepanol® AEG; Zetesol 856 T

Cocamidopropyl dimethylamine (CTFA)
CAS 68140-01-2; EINECS 268-771-8
Synonyms: N-[3-Dimethylamino)propyl]coco amides
Tradenames: Empigen® AS; Mackine 101; Schercodine C; Zohar Amido Amine 1

Cocamidopropyl dimethylamine betaine. *See Cocamidopropyl betaine*
Synonyms: Cocoyl amido propyl dimethylamine betaine

Cocamidopropyl dimethylamine lactate (CTFA)
CAS 68425-42-3
Synonyms: N-[3-(Dimethylamino)propyl] cocamide lactate
Tradenames: Incromate CDL

Cocamidopropyl dimethylamine oxide. *See Cocamidopropylamine oxide*

Cocamidopropyl dimethylamine propionate (CTFA)
CAS 68425-43-4
Synonyms: N-[3-(Dimethylamino)propyl]coco amides, propionates
Tradenames: Emcol® 1655; Mackam CAP

N-[3-Cocamido)-propyl]-N,N-dimethyl betaine, potassium salt
Tradenames: Schercotaine CAB-KG

Cocamidopropyl dimethyl glycine. *See Cocamidopropyl betaine*

Cocamidopropyl hydroxysultaine (CTFA)
CAS 68139-30-0; 70851-08-0; EINECS 268-761-3
Synonyms: (3-Cocamidopropyl)(2-hydroxy-3-sulfopropyl) dimethyl quaternary ammonium compounds, hydroxides, inner salt
Tradenames: Chembetaine CAS; Crosultaine C-50; Lexaine® CSB-50; Lonzaine® CS; Lonzaine® JS; Mafo® CSB 50; Mirataine® CBS, CBS Mod; Rewoteric AM CAS; Varion® CAS-W; Varion® CAS
Tradenames containing: Miracare® MS-1 (formerly Compound MS-1); Miracare® MS-2 (formerly Compound MS-2); Miracare® MS-4; Schercotaine SCAB-A; Schercotaine SCAB-KG

Cocamidopropyl lauryl ether (CTFA)
Tradenames: Marlamid® KL
Tradenames containing: Marlamid® KLP

Cocamine (CTFA)
CAS 61788-46-3; EINECS 262-977-1
Synonyms: Coconut amine
Tradenames: Amine KK; Amine KKD; Armeen® C; Armeen® CD; Genamin CC Grades; Jet Amine PC; Jet Amine PCD; Kemamine® P-650D; Nissan Amine FB; Noram C; Radiamine 6160; Radiamine 6161

Cocamine acetate
CAS 61790-57-6
Tradenames: Acetamin 24; Acetamin C; Armac® C; Noramac C 26; Noramac C; Radiamac 6169

Cocamine oxide (CTFA)
CAS 61788-90-7; EINECS 263-016-9
Synonyms: Coco dimethylamine oxide; Coconut dimethylamine oxide; Dimethyl cocoalkylamine oxide
Tradenames: Aminoxid A 4080; Aromox® DMC; Aromox® DMC-W; Barlox® 12; Empigen® 5083; Genaminox KC; Karox AO-30; Mackamine CO; Naxide 1230; Noxamine C2-30; Noxamine CA 30; Schercamox DMC; Sochamine OX 30
Tradenames containing: Aromox® DMCD

Cocaminobutyric acid (CTFA)
CAS 68649-05-8; EINECS 272-021-5
Synonyms: Butanoic acid, 3-amino-, N-coco alkyl derivatives; 3-Aminobutanoic acid, n-coco alkyl derivatives
Tradenames: Armeen® Z

Cocaminopropionic acid (CTFA)
CAS 1462-54-0; 84812-94-2
Synonyms: N-coco-2-aminopropionic acid
Tradenames: Ampholyte KKE; Amphoram CP1

Coceth-5 (CTFA)
CAS 61791-13-7 (generic)
Synonyms: PEG-5 coconut alcohol; POE (5) coconut ether
Tradenames: Genapol® C-050; Genapol® GC-050; Marlowet® CA 5

Coceth-7
Tradenames: Dehydol LT 7; Dehydol LT 7 L

Coceth-8 (CTFA)
CAS 61791-13-7 (generic)
Synonyms: PEG-8 coconut alcohol; POE (8) coconut ether; PEG 400 coconut ether
Tradenames: Dehydol PLT 8; Genapol® C-080
Tradenames containing: Standapol® S

Coceth-10 (CTFA)
CAS 61791-13-7 (generic)
Synonyms: PEG-10 coconut alcohol; POE (10) coconut ether; PEG 500 coconut ether
Tradenames: Genapol® C-100; Marlowet® CA 10; Oxetal C 110

Coceth-15
Tradenames: Genapol® C-150

Coceth-20
Tradenames: Genapol® C-200

Coceth-27
Tradenames: Dehydol LT 3

Cocoalkonium chloride (CTFA)
CAS 61789-71-7; 139-07-1; EINECS 263-080-8
Synonyms: Coco dimonium chloride; Coco dimethyl benzyl ammonium chloride; Cocoalkyl dimethyl benzyl ammonium chloride
Tradenames: Marlazin® KC 21/50; Nissan Cation F2-10R, -20R, -40E, -50; Noramium DA.50; Servamine KAC 422; Querton KKBCl-50
Tradenames containing: Arquad® DMCB-80

Cocoamphoacetate. *See Sodium cocoamphoacetate*

Cocoamphocarboxyglycinate. *See Disodium cocoamphodiacetate*

Cocoamphocarboxypropionate. *See Disodium cocoamphodipropionate*

Cocoamphocarboxypropionic acid. *See Cocoamphodipropionic acid*

Cocoamphodiacetate. *See Disodium cocoamphodiacetate*

Cocoamphodipropionate. *See Disodium cocoamphodipropionate*

Cocoamphodipropionic acid (CTFA)
CAS 68919-40-4

Synonyms: Cocoamphocarboxypropionic acid
Tradenames: Miranol® C2M Anhyd. Acid

Cocoamphoglycinate. *See Sodium cocoamphoacetate*

Cocoamphohydroxypropylsulfonate. *See Sodium cocoamphohydroxypropyl sulfonate*

Cocoamphopolycarboxyglycinate
CAS 97659-53-5; EINECS 307-458-3
Tradenames: Ampholak 7CX

Cocoamphopropionate. *See Sodium cocoamphopropionate*

Cocoamphopropylsulfonate. *See Sodium cocoamphohydroxypropyl sulfonate*

Cocobetainamido amphopropionate (CTFA)
CAS 100085-64-1; EINECS 309-206-8
Tradenames: Rewoteric QAM 50

Coco-betaine (CTFA)
CAS 68424-94-2
Synonyms: Quat. ammonium compds., carboxymethyl (coco alkyl) dimethyl hydroxides, inner salts; Coconut betaine; Coco dimethyl glycine
Tradenames: Accobetaine CL; Ampho B11-34; Ampholan® E210; Amphoram CB A30; Chembetaine BW; Emcol® CC-37-18; Ethox 2650; Incronam CD-30; Lonzaine® 12C; Mackam CB-35; Mackam CB-LS; Mafo® CB 40; Surfax ACB; Velvetex® AB-45

Coco dialkyl benzyl ammonium chloride
Tradenames: Quatrene CE

Coco diamine
CAS 61791-63-7
Tradenames: Jet Amine DC

Cocodiaminopropane diacetate
Tradenames: Duomac® C

Cocodihydroxyethylamine oxide (CTFA)
Synonyms: Coco bis-2-hydroxyethylamine oxide
Tradenames: Varox® 743

Coco dimethyl amine. *See Dimethyl cocamine*

Coco dimethylamino oxide. *See Cocamine oxide*

Coco dimethyl betaine
Tradenames: Empigen® BB-AU; Mirataine® D-40 (formerly Ambiteric D-40); Nissan Anon BF

Coco dimonium chloride. *See Cocoalkonium chloride*

Cocodimonium hydroxypropyl hydrolyzed wheat protein (CTFA)
Tradenames: Hydrotriticum QM

N-Coco dipropylene triamine
Tradenames: Trinoram C

Coco-EDTA-amide
Tradenames: Rewopol® CHT 12

Cocoiminodipropionate
CAS 91995-05-0; 97659-50-2
Tradenames: Ampholak YCA/P; Ampholak YCE; Ampholan® U 203

Coco nitrile
Tradenames: Arneel® C

Coconut acid (CTFA)
CAS 61788-47-4; 67701-05-7; EINECS 262-978-7
Synonyms: Coco fatty acids; Coconut oils acids
Tradenames: C-108; C-110; Emery®621; Emery® 622; Emery® 626; Emery® 627; Emery® 629; Industrene® 325; Industrene® 328; Philacid 0818; Philacid 1218
Tradenames containing: Emery® 515; Emery® 516; Emery® 517; Hystrene® 1835; TC-1005; TC-1010; TC-1010T

Coconut alcohol (CTFA)
CAS 68425-37-6; EINECS 270-351-4
Synonyms: Coconut fatty alcohol
Tradenames: Laurex® CH; Philcohol 1214; Philcohol 1218

Coconut diethanolamide. *See Cocamide DEA*

Coconut dimethylamine oxide. *See Cocamine oxide*

Coconut hydroxyethyl imidazoline. *See Cocoyl hydroxyethyl imidazoline*

Coconut imidazoline betaine
Tradenames: Empigen® CDR10; Empigen® CDR30

Coconut monoethanolamide. *See Cocamide MEA*

Coconut monoisopropanolamide. *See Cocamide MIPA*

Coconut oil (CTFA)
CAS 8001-31-8; EINECS 232-282-8
Synonyms: Copra oil
Tradenames: Esi-Terge 40% Coconut Oil Soap; Hipochem LH-Soap
Tradenames containing: LHS 40% Coconut Oil Soap

Coconut oil amides. *See Cocamide*

Coconut trimethyl ammonium chloride. *See Cocotrimonium chloride*

Cocopropylenediamine
CAS 61791-63-7
Synonyms: Coco diamino propane; Coconut oil propane diamine
Tradenames: Diamine KKP; Dinoram C; Duomeen® C; Duomeen® CD; Kemamine® D-650; Radiamine 6560

Coco propylene diamine acetate
Tradenames: Dinoramac C

Coco taurine
Tradenames: Amphoram CT 30

Cocotrimonium chloride (CTFA)
CAS 61789-18-2; EINECS 263-038-9
Synonyms: Coconut trimethyl ammonium chloride; Cocoyl trimethyl ammonium chloride; Quaternary ammonium compds., coco alkyl trimethyl, chlorides
Tradenames: Adogen® 461; Arquad® C-33W; Chemquat C/33W; Jet Quat C-50; Nissan Cation FB, FB-500; Noramium MC 50; Quartamin CPR; Servamine KAC 412
Tradenames containing: Arquad® C-33; Arquad® C-50

Cocoyl hydroxyethyl imidazoline (CTFA)
CAS 61791-38-6; EINECS 263-170-7
Synonyms: 1H-Imidazole-1-ethanol, 4,5-dihydro-2-norcocoyl-; Cocoyl imidazoline; Coconut hydroxyethyl imidazoline
Tradenames: Chemzoline C-22; Crapol AV-10; Finazoline CA; Mackazoline C; Miramine® C (formerly Alkazine® C); Monazoline C; Schercozoline C; Unamine® C; Varine C

Cocoyl sarcosine (CTFA)
CAS 68411-97-2; EINECS 270-156-4

Synonyms: N-Cocoyl-N-methyl glycine; N-methyl-N-(1-coconut alkyl) glycine
Tradenames: Hamposyl® C; Sarkosyl® LC; Vanseal CS

Collagen hydrolysates. *See Hydrolyzed collagen*

Copra oil. *See Coconut oil*

C6-10 pareth-3
Tradenames: Alfonic® 610-50R

C8-10 pareth-2
Tradenames: Alfonic® 810-40

C8-10 pareth-5
Tradenames: Alfonic® 810-60

C9-11 pareth-2
Synonyms: Pareth-91-2; PEG-2 C9-11 alcohol ether
Tradenames: Teric 9A2

C9-11 pareth-3 (CTFA)
CAS 68439-46-3 (generic)
Synonyms: Pareth-91-3; PEG-3 C9-11 alcohol ether
Tradenames: Dobanol 91-2.5; Neodol® 91-2.5

C9-11 pareth-4
CAS 68439-45-2
Synonyms: Pareth-91-4; PEG-4 alkyl C9-11 alcohol
Tradenames: Berol 260; Serdox NBS 4

C9-11 pareth-5
Synonyms: Pareth-91-5; PEG-5 alkyl C9-11 alcohol
Tradenames: Dobanol 91-5; Serdox NBS 5.5; Teric 9A5; Teric G9A5

C9-11 pareth-6 (CTFA)
CAS 68439-46-3 (generic)
Synonyms: Pareth-91-6; PEG-6 C9-11 alcohol ether
Tradenames: Dobanol 91-6; Neodol® 91-6; Rhodasurf® 91-6 (formerly Siponic® 91-6); Serdox NBS 6; Serdox NBS 6.6; Teric 9A6; Teric G9A6

C9-11 pareth-8 (CTFA)
CAS 68439-46-3 (generic)
Synonyms: Pareth-91-8
Tradenames: Dobanol 91-8; Neodol® 91-8; Serdox NBS 8.5; Teric 9A8; Teric G9A8

C9-11 pareth-10
Tradenames: Teric 9A10

C9-11 pareth-12
Tradenames: Teric 9A12; Teric G9A12

C10-12 pareth-3
Tradenames: Alfonic® 1012-40; Surfonic® L12-3

C10-12 pareth-5
Synonyms: PEG-5 C10-12 alcohol
Tradenames: Alfonic® 1012-60; Bio-Soft® ET-630

C10-12 pareth-6
Tradenames: Surfonic® L12-6

C10-12 pareth-8
Tradenames: Surfonic® L12-8

C11-15 pareth-3 (CTFA)
CAS 68131-40-8 (generic)
Synonyms: Pareth-15-3
Tradenames: Tergitol® 15-S-3

C11-15 pareth-5 (CTFA)
CAS 68131-40-8 (generic)
Synonyms: Pareth-15-5
Tradenames: Tergitol® 15-S-5

C11-15 pareth-7 (CTFA)
CAS 68131-40-8 (generic)
Synonyms: Pareth-15-7
Tradenames: Tergitol® 15-S-7

C11-15 pareth-9 (CTFA)
CAS 68131-40-8 (generic)
Tradenames: Tergitol® 15-S-9

C11-15 pareth-12 (CTFA)
CAS 6813-14-0 (generic)
Synonyms: Pareth-15-12
Tradenames: Tergitol® 15-S-12

C11-15 pareth-15 (CTFA)
Synonyms: Pareth-15-15
Tradenames: Tergitol® 15-S-15

C11-15 pareth-20 (CTFA)
CAS 68131-40-8 (generic)
Synonyms: Pareth-15-20
Tradenames: Tergitol® 15-S-20

C11-15 pareth-30 (CTFA)
CAS 68131-40-8 (generic)
Synonyms: Pareth-15-30
Tradenames: Tergitol® 15-S-30

C11-15 pareth-40 (CTFA)
CAS 68131-40-8 (generic)
Synonyms: Pareth-15-40
Tradenames: Tergitol® 15-S-40

C12-13 pareth-1
Tradenames: Neodol® 23-1

C12-13 pareth-3 (CTFA)
CAS 66455-14-9 (generic)
Synonyms: Pareth-23-3; PEG-3 C12-13 fatty alcohol ether
Tradenames: Neodol® 23-3

C12-13 pareth-5
Tradenames: Neodol® 23-5

C12-13 pareth-6
Tradenames: Dobanol 23-6.5

C12-13 pareth-7 (CTFA)
CAS 66455-14-9 (generic)
Synonyms: Pareth-23-7; PEG-7 C12–13 fatty alcohol ether
Tradenames: Neodol® 23-6.5; Neodol® 23-6.5T

C12-13 pareth-12
Tradenames: Neodol® 23-12

C12-14 pareth-2
Synonyms: PEG-2 C12-14 alcohol
Tradenames: Alfonic® 1214-GC-30; Surfonic® L24-2

C12-14 pareth-3
CAS 68439-50-9
Synonyms: PEG-3 C12-14 alcohol
Tradenames: Alfonic® 1214-GC-40; Alfonic® 1412-40; Genapol® 24-L-3; Genapol® 42-L-3; Surfonic® L24-3

C12-14 pareth-4
Tradenames: Surfonic® L24-4

C12-14 pareth-6
CAS 68439-50-9
Tradenames: Genapol® 24-L-45; Tergitol® 24-L-45

C12-14 pareth-7
CAS 68439-50-9
Tradenames: Alfonic® 1412-60; Genapol® 24-L-50; Genapol® 24-L-60; Genapol® 24-L-60N; Surfonic® L24-7; Tergitol® 24-L-60; Tergitol® 24-L-60N

C12-14 pareth-8
Synonyms: PEG-8 C12-14 fatty alcohol
Tradenames: Genapol® 24-L-75; Lutensol® A 8; Surfonic® L24-9

C12-14 pareth-10
Tradenames: Genapol® 24-L-92

C12-14 pareth-11
Tradenames: Genapol® 24-L-98N; Surfonic® L24-12

C12-14 pareth-12
Synonyms: PEG-12 C12-14 alkyl ether
Tradenames: Serdox NEL 12/80

C12-15 pareth-2
Tradenames: Teric 12A2

C12-15 pareth-3 (CTFA)
CAS 68131-39-5 (generic)
Synonyms: Pareth-25-3; PEG-3 C12–15 fatty alcohol ether
Tradenames: Bio-Soft® E-400; Dobanol 25-3; Neodol® 25-3; Rhodasurf® LA-3 (formerly Alkasurf® LA-3); Tergitol® 25-L-3; Teric 12A3; Volpo 25 D 3

C12-15 pareth-4 (CTFA)
CAS 79131-39-5 (generic)
Synonyms: Pareth-25-4
Tradenames: Teric 12A4; Teric G12A4; Teric LA4

C12-15 pareth-5 (CTFA)
CAS 68131-39-5 (generic)
Synonyms: PEG-5 C12-15 alkyl ether; Pareth-25-5
Tradenames: Tergitol® 25-L-5; Volpo 25 D 5

C12-15 pareth-6
Synonyms: PEG-6 C12-15 alcohol; Pareth-25-6
Tradenames: Teric 12A6; Teric G12A6

C12-15 pareth-7 (CTFA)
CAS 68131-39-5 (generic)
Synonyms: Pareth-25-7; PEG-7 C12–15 fatty alcohol ether
Tradenames: Bio-Soft® EN 600; Dobanol 25-7; Neodol® 25-7; Rhodasurf® 25-7 (formerly Siponic® 25-7); Rhodasurf® LA-7 (formerly Alkasurf® LA-7); Tergitol® 25-L-7; Teric 12A7

C12-15 pareth-8
Synonyms: PEG-8 C12-15 alkyl ether; Pareth-25-8
Tradenames: Serdox NES 8/85; Teric 12A8; Teric G12A8; Teric LA8

C12-15 pareth-9 (CTFA)
CAS 68131-39-5 (generic)
Synonyms: Pareth-25-9
Tradenames: Dobanol 25-9; Neodol® 25-9; Tergitol® 25-L-9; Teric 12A9

C12-15 pareth-10
Synonyms: Pareth-25-10
Tradenames: Volpo 25 D 10

C12-15 pareth-12 (CTFA)
CAS 68131-39-5 (generic)
Synonyms: Pareth 25-12
Tradenames: Neodol® 25-12; Rhodasurf® LA-12 (formerly Alkasurf® LA-12); Tergitol® 25-L-12; Teric 12A12; Teric G12A12

C12-15 pareth-15
Tradenames: Volpo 25 D 15

C12-15 pareth-16
Tradenames: Teric 12A16

C12-15 pareth-20
Tradenames: Volpo 25 D 20

C12-15 pareth-23
Tradenames: Teric 12A23

C12-15 pareth-7 carboxylic acid (CTFA)
CAS 88497-58-9 (generic)
Synonyms: Pareth-25-7 carboxylic acid; PEG-7 C12-15 alkyl ether carboxylic acid
Tradenames: Sandopan® DTC Linear P Acid; Surfine WNT-A

C12-15 pareth-2 phosphate (CTFA)
Synonyms: Pareth-25-2 phosphate; PEG-2-C12-15 alcohols phosphate
Tradenames: Nikkol TDP-2

C12-16 pareth-1
CAS 68551-12-2
Tradenames: Alfonic® 1216-22; Genapol® 26-L-1; Tergitol® 26-L-1.6

C12-16 pareth-2
Tradenames: Alfonic® 1216-30; Genapol® 26-L-1.6; Genapol® 26-L-2

C12-16 pareth-3
Synonyms: PEG-3 C12,C14,C16 alcohols
Tradenames: Genapol® 26-L-3; Tergitol® 26-L-3

C12-16 pareth-5
Tradenames: Genapol® 26-L-5; Tergitol® 26-L-5

C12-16 pareth-6
Tradenames: Genapol® 26-L-45

C12-16 pareth-7
Tradenames: Genapol® 26-L-50; Genapol® 26-L-60; Genapol® 26-L-60N

C12-16 pareth-8
Tradenames: Genapol® 26-L-75

C12-16 pareth-11
Tradenames: Genapol® 26-L-98N

C12-18 pareth-3
Synonyms: PEG-3 C12-18 alcohol
Tradenames: Ethal 3328

C12-18 pareth-12
Tradenames: Bio-Soft® E-710

C13-15 pareth-2
Tradenames: Renex® 702

C13-15 pareth-3
Tradenames: Lutensol® AO 3; Renex® 703

C13-15 pareth-4
Tradenames: Lutensol® AO 4

C13-15 pareth-5
Tradenames: Lutensol® AO 5

C13-15 pareth-7
Tradenames: Lutensol® AO 7; Renex® 707

C13-15 pareth-8
Tradenames: Lutensol® AO 8

C13-15 pareth-10
Tradenames: Lutensol® AO 10; Lutensol® AO 109

C13-15 pareth-11
Tradenames: Lutensol® AO 11; Renex® 711

C13-15 pareth-12
Tradenames: Lutensol® AO 12

C13-15 pareth-20
Tradenames: Renex® 720

C13-15 pareth-30
Tradenames: Lutensol® AO 30

C14-15 pareth-2
Tradenames: Neodol® 45-2.25

C14-15 pareth-7 (CTFA)
Synonyms: Pareth 45-7; PEG-7 C14-15 alcohol ether
Tradenames: Dobanol 45-7; Neodol® 45-7; Neodol® 45-7T

C14-15 pareth-11 (CTFA)
Synonyms: Pareth-45-11
Tradenames: Neodol® 45-11; Teric 15A11

C14-15 pareth-12
Tradenames: Neodol® 45-12T

C14-15 pareth-13 (CTFA)
Synonyms: Pareth-45-13; PEG-13 C14-15 alcohol ether
Tradenames: Neodol® 45-13

C14-15 pareth-8 carboxylic acid (CTFA)
Synonyms: PEG-8 C14-15 alkyl ether carboxylic acid; PEG 400 C14-15 alkyl ether carboxylic acid
Tradenames: Surfine WNG-A

C14-16 pareth-7
Tradenames: Surfonic® L46-7

C16-18 pareth-3
Tradenames: Ethal 368

C16-18 pareth-7
Synonyms: PEG-7 C16-18 fatty alcohol
Tradenames: Lutensol® A 7

C16-18 pareth-11
Synonyms: PEG-11 sat. C16-18 alcohol
Tradenames: Lutensol® AT 11

C16-18 pareth-18
Tradenames: Lutensol® AT 18

C16-18 pareth-25
Tradenames: Ethal CSA-25; Lutensol® AT 25

C16-18 pareth-50
Tradenames: Lutensol® AT 50

C16-18 pareth-80
Tradenames: Lutensol® AT 80

C8-10 phosphate
Tradenames: Hostaphat OD

Cresylic acid
EINECS 215-293-2
Tradenames: Flotigol CS
Tradenames containing: Rexowet CR

C12-14 tridecylbenzene
Tradenames: Nalkylene® 575L; Nalkylene® 600L

Cumene sulfonic acid
CAS 28631-63-2
Tradenames: Eltesol® CA 65; Eltesol® CA 96; Sulframin CSA

C10-12 undecylbenzene
Tradenames: Nalkylene® 500

DEA-acrylinoleate
Synonyms: DEA-C21-dicarboxylate
Tradenames containing: Monaterge 85 HF

DEA-caprate
Tradenames containing: Mackamide CD-6

DEA-cetyl phosphate (CTFA)
CAS 69331-39-1
Synonyms: Cetyl DEA phosphate
Tradenames: Amphisol

DEA-coconate
Tradenames containing: Mackamide CDC

DEA-dodecylbenzene sulfonate (CTFA)
CAS 26545-53-9; EINECS 247-784-2
Synonyms: Benzenesulfonic acid, dodecyl-, compd. with 2,2′-iminobis [ethanol] (1:1)
Tradenames: Ablusol DBD; Marlopon® ADS 50; Marlopon® ADS 65
Tradenames containing: Mackamide CDS-80; Monamine ALX-100 S; Monaterge 85 HF; Unamide® C-7944

DEA-laureth sulfate (CTFA)
CAS 58855-36-0; 55353-19-0; 54351-50-7; 81859-24-7
Synonyms: Diethanolamine laureth sulfate
Tradenames: Cedepal SD-409
Tradenames containing: Monamine 779

DEA lauryl sulfate (CTFA)
CAS 143-00-0; 68585-44-4
Synonyms: Sulfuric acid, monododecyl ester, compd. with 2,2′-iminodiethnaol (1:1); Diethanolamine lauryl sulfate
Tradenames: Duponol® EP; Empicol® DA; Stepanol® DEA; Zoharpon LAD

DEA oleate
Tradenames containing: Mackamide CDM; Mackamide ODM

DEA-oleth-3 phosphate (CTFA)
CAS 58855-63-3
Synonyms: Diethanolamine oleth-3 phosphate; Diethanolammonium POE (3) oleyl ether phosphate
Tradenames: Crodafos N3 Neutral

DEA-oleth-10 phosphate (CTFA)
CAS 58855-63-3
Synonyms: Diethanolamine oleth-10 phosphate; Diethanolammonium POE (10) oleyl ether phosphate
Tradenames: Crodafos N10 Neutral

Decaglyceryl. *See Polyglyceryl-10*

1-Decanol. *See n-Decyl alcohol*

Decene-1
CAS 872-05-9
Synonyms: Linear C10 alpha olefin; Decylene
Tradenames: Ethyl® Decene-1; Gulftene 10; Neodene® 10

Deceth-2
Tradenames: Bio-Soft® FF 400

Deceth-4 (CTFA)
CAS 26183-52-8 (generic)
Synonyms: PEG-4 decyl ether; PEG 200 decyl ether; POE (4) decyl ether
Tradenames: Chemal DA-4; Desonic® DA-4; Ethal DA-4; Genapol® DA-040; Iconol DA-4; Marlipal® 1012/4; Oxetal D 104; Prox-onic DA-1/04; Trycol® 5950; Trycol® DA-4

Deceth-5
CAS 26183-52-8 (generic)
Tradenames: Trycol® 5951

Deceth-6 (CTFA)
CAS 26183-52-8 (generic)
Synonyms: PEG-6 decyl ether; PEG 300 decyl

ether; POE (6) decyl ether
Tradenames: Chemal DA-6; Desonic® DA-6; Ethal DA-6; Genapol® DA-060; Iconol DA-6; Iconol DA-6-90%; Marlipal® KF; Prox-onic DA-1/06; Rhodasurf® DA-639 (formerly Emulphogene® DA-639); Trycol® 5952; Trycol® 5953; Trycol® 5999; Trycol® DA-6; Trycol® DA-69

Deceth-9
CAS 26183-52-8 (generic)
Tradenames: Chemal DA-9; Ethal DA-9; Iconol DA-9; Prox-onic DA-1/09; Trycol® 5956

Deceth-4 phosphate (CTFA)
CAS 52019-36-0 (generic); 9004-80-2 (generic)
Synonyms: PEG-4 decyl ether phosphate; PEG 200 decyl ether phosphate
Tradenames: Cedephos® FA600; Monafax 1214; Monafax 831; Rhodafac® RA-600 (formerly Gafac® RA-600)

Deceth-6 phosphate (CTFA)
CAS 52019-36-0 (generic)
Synonyms: PEG-6 decyl ether phosphate; PEG 300 decyl ether phosphate; POE (6) decyl ether phosphate
Tradenames: Findet AD-18

n-Decyl alcohol (CTFA)
CAS 112-30-1; 68526-85-2; EINECS 203-956-9
Synonyms: Alcohol C-10; Noncarbinol; 1-Decanol
Tradenames: Alfol® 10; Emery® 3323; Emery® 3330; Epal® 10; Kalcohl 10H; Lorol C10; Lorol C10-98
Tradenames containing: Epal® 108; Epal® 610; Epal® 810; Epal® 1012

Decylamine oxide (CTFA)
CAS 2605-79-0; EINECS 220-020-5
Synonyms: Capric dimethyl amine oxide; Decyl dimethyl amine oxide
Tradenames: Barlox® 10S

c-Decyl betaine (CTFA)
CAS 2644-45-3; EINECS 220-152-3
Synonyms: N-(Carboxymethyl)-N,N-dimethyl-1-decanaminium hydroxide, inner salt; Decyl dimethyl glycine
Tradenames containing: Darvan® NS

Decyl dimethyl octyl ammonium chloride. *See Quaternium-24*

Decyl diphenyl ether disulfonic acid
CAS 70191-74-1
Tradenames: Poly-Tergent® 3B2 Acid

Decyl diphenyloxide disulfonate
Tradenames: Dowfax 3B0

Dialkyl dimethyl ammonium chloride
Synonyms: Quaternium 31
Tradenames: Radiaquat 6443

Dialkyl methyl benzyl ammonium chloride
Tradenames containing: BTC® 776; Catigene® 776; Maquat TC-76

Diammonium cocoyl sulfosuccinate
Tradenames: Empimin® MSS

Diammonium lauryl sulfosuccinate (CTFA)
Synonyms: Butanedioic acid, sulfo-, 1-dodecyl ester, diammonium salt; Sulfobutanedioic acid, 1-dodecyl ester, diammonium salt
Tradenames: Incrosul LSA

Diammonium stearyl sulfosuccinamate
CAS 68128-59-6
Synonyms: Diammonium octadecyl sulfosuccinamate
Tradenames: Octosol A-18-A

Diamyl sodium sulfosuccinate (CTFA)
CAS 922-80-5; EINECS 213-085-6
Synonyms: Sodium diamyl sulfosuccinate; Sulfobutanedioic acid 1,4-dipentyl ester sodium salt; Sulfosuccinic acid dipentyl ester sodium salt
Tradenames: Aerosol® AY-65; Aerosol® AY-100; Geropon® AY
Tradenames containing: Mackanate AY-65TD

Dibehenyl/diarachidyl dimonium chloride (CTFA)
Tradenames: Kemamine® Q-1902C

Dibehenyldimonium chloride (CTFA)
CAS 26597-36-4
Tradenames: Kemamine® Q-2802C

Dibehenyl methylamine (CTFA)
CAS 61372-91-6; EINECS 262-740-2
Synonyms: N-Docosyl-N-methyl-1-docosamine; Methyl dibehenylamine
Tradenames: Kemamine® T-2801

Dibutyl maleate
CAS 105-76-0; EINECS 203-328-4
Synonyms: DBM; 2-Butenedioic acid, dibutyl ester
Tradenames: DBM

Dicapryl/dicaprylyl dimonium chloride (CTFA)
CAS 68424-95-3
Synonyms: Quaternary ammonium compds., di-C8-10-alkyldimethyl, chlorides
Tradenames: FMB 302-8 Quat
Tradenames containing: FMB 504-5 Quat

Diceteareth-10 phosphate (CTFA)
Tradenames: Marlophor® T10-Acid

Dicetyl dimonium chloride (CTFA)
CAS 1812-53-9; EINECS 217-325-0
Synonyms: Quaternium-31; N-Hexadecyl-N,N-dimethyl-1-hexadecanaminium chloride; Dicetyl dimethyl ammonium chloride
Tradenames containing: FMB 1210 5 Quat, 1210-8 Quat

o-Dichlorobenzene
CAS 95-50-1
Synonyms: 1,2-Dichlorobenzene; Orthodichlorobenzene
Tradenames: Solutene TER

Dicocamine (CTFA)
CAS 61789-76-2; EINECS 263-086-0
Synonyms: Dicoco alkyl amine; Amines, dicoco alkyl
Tradenames: Armeen® 2C; Noram 2C

Dicocodimonium chloride (CTFA)
CAS 61789-77-3; EINECS 263-087-6
Synonyms: Dicoco dimethyl ammonium chloride; Quaternium-34
Tradenames: Accoquat 2C-75; Accoquat 2C-75H; Dodigen 1490; Jet Quat 2C-75; Kemamine® Q-6502C; Quartamin DCP; Radiaquat 6462
Tradenames containing: Adogen® 462; Arquad® 2C-75; Arquad® S-2C-50; Arquad® T-2C-50; M-Quat® 2475; Tomah Q-2C; Variquat® K300; Variquat® K375

Dicoco distearyl pentaerythrityl citrate
Tradenames: Dehymuls FCE

Dicoco methylamine
CAS 61788-62-3
Synonyms: Dicocoalkyl methylamine
Tradenames: Armeen® M2C; Jet Amine M2C; Kemamine® T-6501; Noram M2C

Dicoco nitrite
CAS 71487-01-9
Tradenames: Arquad® 2C-70 Nitrite

Dicyandiamide formaldehyde resin
Tradenames: Matexil FC PN

Dicyclohexyl sodium sulfosuccinate (CTFA)
CAS 23386-52-9; EINECS 245-629-3
Synonyms: Sodium dicyclohexyl sulfosuccinate; Succinic acid, sulfo-, 1,4-dicyclohexyl ester, sodium salt
Tradenames: Aerosol® A-196-40; Aerosol® A-196-85; Gemtex 691-40; Octosol TH-40; Rewopol® SBDC 40

Didecylamine
CAS 1120-49-6
Tradenames: Armeen® 2-10

Didecyl dimethylamine oxide
CAS 100545-50-4
Tradenames: Damox® 1010

Didecyldimonium chloride (CTFA)
CAS 7173-51-5; EINECS 230-525-2
Synonyms: Didecyl dimethyl ammonium chloride; N-Decyl-N,N-dimethyl-1-decanaminium chloride; Dimethyl didecyl ammonium chloride
Tradenames: BTC® 99; BTC® 1010; Catigene® 1011; FMB 210-8 Quat, 210-15 Quat; Maquat 4450-E; Radiaquat 6410; Radiaquat 6412
Tradenames containing: Arquad® 210-50; BTC® 885; Catigene® 818; Catigene® 818-80; Catigene® 885; Catigene® 888

Didecyl dimonium methosulfate
Tradenames: Ablumine D10

Didecyl methylamine
CAS 7396-58-9
Synonyms: Methyl decyl-1-amino decane
Tradenames: Armeen® M2-10D; Dama® 1010

Diethanolamidooleamide DEA (CTFA)
Tradenames: Witcamide® 5168

Diethanolamine (CTFA)
CAS 111-42-2; EINECS 203-868-0
Synonyms: DEA; 2,2′-Iminobisethanol; Di(2-hydroxyethyl) amine; 2,2′-Iminodiethanol
Tradenames containing: Alkamide® DC-212 (formerly Cyclomide DC-212); Alkamide® DO-280 (formerly Cyclomide DO-280); Aminol COR-2C; Aminol N-1918; Emid® 6531; Emid® 6533; Emid® 6541; Emid® 6544; Empilan® LDX; Hetamide DO; Mazamide® 70; Monamine AA-100; Monamine ACO-100; Monamine AD-100; Monamine ADY-100; Monamine ALX-100 S; Monamine CD-100; Monamine CF-100 M; Monamine LM-100; Monamine R8-26; Monamine T-100; Norfox® DC; Norfox® KD; Rewomid® DL 203; Rewomid® DL 203 S; Rewomid® DL 240; Rewomid® DO 280; Schercomid 1214; Schercomid CCD; Schercomid CDA; Schercomid CDO-Extra; Schercomid LD; Schercomid ODA; Schercomid TO-2; Standamid® CD; Standamid® PD; Witcamide® 82; Witcamide® 5133; Witcamide® 6533; Witcamide® Coco Condensate

Diethanolamine laureth sulfate. *See DEA-laureth sulfate*

Diethylene glycol (CTFA) ... *See also PEG-2...*
CAS 111-46-6; EINECS 203-872-2
Synonyms: DEG; Dihydroxydiethyl ether; Diglycol; 2,2′-Oxybisethanol
Tradenames containing: Mackanate DOS-70DEG; Propoquad® T/12

Diethylene glycol butyl ether. *See Butoxy diglycol*

Diethylene glycol dodecyl ether. *See Laureth-2*

Diethylene glycol monoethyl ether. *See Ethoxydiglycol*

Diethylenetriamine pentakis(methylene phosphonic acid)
CAS 22042-96-2
Tradenames: Briquest® 543-45AS

Diglyceryl borate sesquioleate
Tradenames: Emulbon S-83

Diglycol. *See Diethylene glycol. See also PEG-2...*

Dihexyl sodium sulfosuccinate (CTFA)
CAS 3006-15-3; 2373-38-8; 6001-97-4; EINECS 221-109-1
Synonyms: Sodium dihexyl sulfosuccinate; Butanedioic acid, sulfo-, 1,4-dihexyl ester, sodium salt
Tradenames: Aerosol® MA-80; Empimin® MA; Gemtex 680; Octosol HA-80
Tradenames containing: Lankropol® KMA; Monawet MM-80

Di(hydrogenated tallowalkyl) (2-hydroxy-2-methylethyl) quaternary ammonium chloride
Tradenames containing: Propoquad® 2HT/11

Dihydrogenated tallowamidoethyl hydroxyethylmonium methosulfate (CTFA)
Tradenames containing: Varisoft® 110

Dihydrogenated tallow amine. *See Hydrogenated ditallowamine*
CAS 61789-79-5
Tradenames: Amine 2HBG; Armeen® 2HT

Dihydrogenated tallow benzylmonium chloride
CAS 61789-73-9
Synonyms: Dihydrogenated tallow methyl benzyl ammonium chloride; Di(hydrogenated tallow) benzyl methyl ammonium chloride
Tradenames: Variquat® B343
Tradenames containing: Arquad® M2HTB-80

Dihydrogenated tallow dimonium methosulfate
Synonyms: Dimethyl dihydrogenated tallow ammonium methyl sulfate
Tradenames: Ablumine DHT75; Ablumine DHT90
Tradenames containing: Varisoft® 137

Dihydrogenated tallow methylamine (CTFA)
CAS 61788-63-4; 67700-99-6; EINECS 262-991-8
Tradenames: Amine M2HBG; Armeen® M2HT; Kemamine® T-9701; Noram M2SH

Dihydroxyethyl C12-15 alkoxypropylamine oxide (CTFA)
Tradenames: Varox® 185E

Dihydroxyethyl cocamine oxide (CTFA)
CAS 61791-47-7; EINECS 263-180-1
Synonyms: N,N (bis (2-hydroxyethyl) cocamine oxide; Coco di(hdyroxyethyl) amine oxide;

Ethanol, 2,2′-iminobis, N-coco alkyl, N-oxide
Tradenames: Aromox® C/12-W
Tradenames containing: Aromox® C/12

Dihydroxyethyl tallowamine oxide (CTFA)
CAS 61791-46-6; EINECS 263-179-6
Synonyms: Bis (2-hydroxyethyl) tallow amine oxide; 2,2′-Iminobisethanol, N-tallow alkyl, N-oxide; Amines, tallow alkyl dihydroxyethyl, oxides
Tradenames containing: Aromox® T/12

Dihydroxyethyl tallow glycinate (CTFA)
CAS 61791-25-1; 70750-46-8
Synonyms: Quaternary ammonium compds., (carboxymethyl)(tallow alkyl)dihydroxyethyl, hydroxides, inner salts; Tallow dihydroxyethyl betaine; Tallow dihydroxyethyl glycine
Tradenames: Ampholak BTH-35; Mackam TM; Miranol® TM; Mirataine® TM; Rewoteric AM TEG; Tego®-Betaine N-192; Varion® TEG; Varion® TEG-40%; Zohartaine TM

Diisobutyl maleate
CAS 14234-82-3; EINECS 238-102-4
Tradenames: DIBM

Diisobutyl sodium sulfosuccinate (CTFA)
CAS 127-39-9; EINECS 204-839-5
Synonyms: Sodium diisobutyl sulfosuccinate; 1,4-Bis(2-methylpropyl) sulfobutanedioate, sodium salt; Sulfosuccinic acid diisobutyl ester sodium salt
Tradenames: Aerosol® IB-45; Gemtex 445; Geropon® CYA/45; Geropon® CYA/DEP; Manoxol IB; Monawet MB-45; Monawet MB-100; Octasol IB-45; Rewopol® SBDB 45

Diisodecyl sodium sulfosuccinate
Tradenames: Rewopol® SBDD 65

Diisohexyl sulfosuccinate
Tradenames: Rewopol® SBMB 80

Diisooctyl maleate
CAS 1330-76-3
Tradenames: DIOM

Diisooctyl sodium sulfosuccinate
Tradenames: Texin DOS 75

Dilaureth-10 phosphate (CTFA)
Tradenames: Nikkol DLP-10

Dilaurin. *See Glyceryl dilaurate*

Dilauryl acetyl dimonium chloride (CTFA)
CAS 90283-04-8
Tradenames: Schercoquat ALA

Dilinoleic acid (CTFA)
CAS 6144-28-1; 61788-89-4
Synonyms: 9,12-Octadecadienoic acid, dimer; Dimer acid
Tradenames: Empol® 1004; Empol® 1007; Empol® 1010; Empol® 1014; Empol® 1016; Empol® 1018; Empol® 1020; Empol® 1022; Empol® 1026; Empol® 1061; Hystrene® 3675; Hystrene® 3675C; Hystrene® 3680; Hystrene® 3687; Hystrene® 3695; Industrene® D; Pripol 1025

Dimer acid. *See Dilinoleic acid*

Dimethicone (CTFA)
CAS 9006-65-9; 9016-00-6; 63148-62-9; 68037-74-1 (branched)
Synonyms: PDMS; Dimethyl polysiloxane; Dimethyl silicone; Polydimethylsiloxane
Tradenames: AF 60; AF 66; AF 93; AF 9020; AF 9021; Alkasil® HNM 1223-15 (70%); Anedco DF-6130; Anedco DF-6131; Anedco DF-6231; Anedco DF-6233; Aphrogene Jet; BYK®-331; BYK®-336; BYK®-341; BYK®-344; Dabco® DC1630; Dow Corning® 200 Fluid; Dow Corning® 200 Fluid, Food Grade; Dow Corning® FS-1265 Fluid; Foamkill® 810F; Foamkill® 830F; Masil® SF 5; Masil® SF 10; Masil® SF 20; Masil® SF 50; Masil® SF 100; Masil® SF 200; Masil® SF 350; Masil® SF 350 FG; Masil® SF 500; Masil® SF 1000; Masil® SF 5000; Masil® SF 10,000; Masil® SF 12,500; Masil® SF 30,000; Masil® SF 60,000; Masil® SF 100,000; Masil® SF 300,000; Masil® SF 500,000; Masil® SF 600,000; Masil® SF 1,000,000; Rhodorsil® AF 422; Rhodorsil® AF 454; Rhodorsil® AF 461 LV; Rhodorsil® AF 20432; SEM-35; SF18; SF81-50; SF96®; Silicone AF-10 FG; Silicone AF-10 IND; Silicone AF-30 FG; Silicone AF-30 IND; Union Carbide® L-45 Series; Viscasil®
Tradenames containing: Dow Corning® 1500 Compd; Dow Corning® 1520 Silicone Antifoam

Dimethicone copolyol (CTFA)
CAS 64365-23-7; 67762-96-3
Synonyms: Dimethylsiloxane-glycol copolymer
Tradenames: Abil® B 8843, B 8847; Abil® B 88184; Dow Corning® 190 Surfactant; Dow Corning® 193 Surfactant; Masil® 1066C; Masil® 1066D; Silwet® L-7001; Silwet® L-7002; Silwet® L-7004; Silwet® L-720; Silwet® L-722; Silwet® L-7500; Silwet® L-7600; Silwet® L-7602; Silwet® L-7604; Silwet® L-7605; Silwet® L-7614; Tego® Foamex 1488; Tego® Foamex 3062; Tego® Foamex 7447; Tego® Glide 100; Tego® Glide 405; Tego® Glide 406; Tego® Glide 410; Tego® Glide 470; Tego® Glide B 1484; Tego® Glide ZG 400; Tego® Wet KL 245

Dimethylaminoethyl methacrylate (CTFA)
CAS 2867-47-2; EINECS 220-688-8
Synonyms: 2-(Dimethylamino)ethyl 2-methyl-2-propenoate
Tradenames: Ageflex FM-1

2-Dimethylamino-2-methyl-1-propanol
CAS 7005-47-2
Synonyms: 2-Dimethylaminomethyl propanol
Tradenames: DMAMP-80

Dimethylaminopropyl ricinoleamide benzyl chloride
Tradenames containing: ES-1239

Dimethyl arachidyl-behenyl amine
Tradenames: Kemamine® T-1902D

Dimethyl behenamine (CTFA)
CAS 215-42-9; 21542-96-1
Synonyms: Behenyl dimethyl amine; N,N-Dimethyl-1-docosanamine
Tradenames: Adogen® MA-112 SF; Crodamine 3.ABD; Kemamine® T-2802D

Dimethyl cocamine (CTFA)
CAS 61788-93-0; EINECS 263-020-0
Synonyms: Coco dimethyl amine; Amines, coco alkyl dimethyl; Dimethyl coconut amine
Tradenames: Amine 2MKKD; Armeen® DMCD; Jet Amine DMCD; Kemamine® T-6502;

Kemamine® T-6502D; Noram DMC; Noram DMCD

Dimethyl cocamine oxide. *See Cocamine oxide*

Dimethyl decylamine
CAS 1120-24-7
Synonyms: Decyl dimethylamine; C10 alkyl dimethylamine; N,N-Dimethyl decylamine
Tradenames: Adma® 10
Tradenames containing: Adma® WC

Dimethyl di(hydrogenated tallow)ammonium chloride. *See Quaternium-18*

Dimethyl erucylamine
Tradenames: Crodamine 3.AED

Dimethyl hexynol
CAS 107-54-0
Synonyms: 3,5-Dimethyl-1-hexyne-3-ol
Tradenames: Surfynol® 61

Dimethyl hydrogenated tallow amine (CTFA)
CAS 61788-95-2; EINECS 263-022-1
Synonyms: Hydrogenated tallow dimethylamine; Hydrogenated tallowalkyl dimethylamine
Tradenames: Amine 2MHBGD; Armeen® DMHTD; Kemamine® T-9702D; Nissan Tert. Amine ABT; Noram DMSH D

Dimethyl isosorbide (CTFA)
CAS 5306-85-4; EINECS 226-159-8
Synonyms: DMI; Isosorbide dimethyl ether; D-Glucitol, 1,4:3,6-dianhydro-2,5-di-o-methyl
Tradenames: Arlasolve® DMI

Dimethyl lauramine (CTFA)
CAS 112-18-5; 67700-98-5; EINECS 203-943-8
Synonyms: Lauryl dimethylamine; Dodecyldimethylamine
Tradenames: Adma® 12; Amine 2M12D; Armeen® DM12D; Empigen® AB; Nissan Tert. Amine BB
Tradenames containing: Adma® 246-451; Adma® 246-621; Adma® 1214; Adma® 1416; Adma® WC

Dimethyl myristamine (CTFA)
CAS 112-75-4; 68439-70-3; EINECS 204-002-4
Synonyms: Dimethyl myristylamine; Myristyl dimethylamine; Tetradecyl dimethylamine
Tradenames: Adma® 14; Amine 2M14D; Armeen® DM14D; Empigen® AH; Nissan Tert. Amine MB;
Tradenames containing: Adma® 246-451; Adma® 246-621; Adma® 1214; Adma® 1416; Adma® WC

Dimethyl octylamine
CAS 7378-99-6
Synonyms: DMOA; C8 alkyl dimethylamine; Octyl dimethylamine; N,N-dimethyloctylamine
Tradenames: Adma® 8
Tradenames containing: Adma® WC

Dimethyl octynediol (CTFA)
CAS 1321-87-5; 78-66-0
Synonyms: 3,6-Dimethyl-4-octyne-3,6-diol
Tradenames: Surfynol® 82
Tradenames containing: Surfynol® 82S

Dimethyl oleamide
Tradenames: Hallcomid® M-18-OL

Dimethyl oleamine
CAS 14727-68-5; 28061-69-0
Synonyms: Oleyl dimethylamine
Tradenames: Armeen® DMOD; Crodamine 3.AOD; Jet Amine DMOD; Kemamine® T-9892D

Dimethyl oleic-linolenic amine
CAS 68037-96-7
Tradenames: Kemamine® T-9992D

Dimethyl palmitamine (CTFA)
CAS 112-69-6; 68037-93-4; EINECS 203-997-2
Synonyms: Palmityl dimethylamine; Hexadecyl dimethylamine; Cetyl dimethylamine
Tradenames: Adma® 16; Amine 2M16D; Armeen® DM16D; Crodamine 3.A16D; Kemamine® T-8902; Nissan Tert. Amine PB
Tradenames containing: Adma® 246-451; Adma® 246-621; Adma® 1416; Adma® WC

Dimethyl polysiloxane. *See Dimethicone*

Dimethyl silicone. *See Dimethicone*

Dimethylsiloxane-glycol copolymer. *See Dimethicone copolyol*

Dimethyl soyamine (CTFA)
CAS 61788-91-8; EINECS 263-017-4
Synonyms: Soya dimethyl amine
Tradenames: Armeen® DMSD; Jet Amine DMSD; Kemamine® T-9972; Kemamine® T-9972D

Dimethyl stearamine (CTFA)
CAS 124-28-7; EINECS 204-694-8
Synonyms: Stearyl dimethyl amine; Octadecyl dimethylamine; N,N-Dimethyl-1-octadecanamine
Tradenames: Adma® 18; Amine 2M18D; Armeen® DM18D; Crodamine 3.A18D; Kemamine® T-9902; Kemamine® T-9902D; Nissan Tert. Amine AB
Tradenames containing: Adma® 1416; Adma® WC

Dimethyl tallowamine (CTFA)
CAS 68814-69-7
Synonyms: Tallow dimethylamine; Tallow alkyl dimethylamine
Tradenames: Armeen® DMTD; Jet Amine DMTD; Kemamine® T-9742D; Noram DMS; Noram DMSD

Dioctadecylamine
CAS 112-99-2
Synonyms: Distearylamine
Tradenames: Armeen® 2-18

Dioctyl ammonium sulfosuccinate
Tradenames: Geropon® X2152 (formerly Rhodiasurf X2152); Octowet 70A

Dioctyl dimonium chloride
Synonyms: Dioctyl dimethyl ammonium chloride
Tradenames containing: BTC® 885; Catigene® 818; Catigene® 818-80; Catigene® 885; Catigene® 888

Dioctyl maleate (CTFA)
CAS 142-16-5; 2915-53-9; 56235-92-8; EINECS 205-524-5
Synonyms: Bis (2-ethylhexyl) maleate; Di-N-octyl maleate; Di-(2-ethylhexyl) maleate
Tradenames: DOM

Dioctyl phthalate (CTFA)
CAS 117-81-7; EINECS 204-211-0
Synonyms: DOP; Di(2-ethylhexyl) phthalate; Di-s-octyl phthalate; 1,2-Benzenedicarboxylic acid dioctyl ester
Tradenames: Imperon Softener D

Dioctyl sodium sulfosuccinate (CTFA)
CAS 577-11-7; 1369-66-3; EINECS 209-406-4
Synonyms: DSS; Sodium dioctyl sulfosuccinate; Sodium di(2-ethylhexyl) sulfosuccinate; Docusate sodium
Tradenames: Ablusol C-78; Aerosol® GPG; Aerosol® OT-75%; Aerosol® OT-100%; Aerosol® OT-S; Alconate® SBDO (see Geropon® SS-O-75); Alrowet® D-65; Arowet SC-75; Arylene M40; Arylene M60; Arylene M75; Astrowet O-75; Atlas WA-100; Avirol® SO 70P; Chemax DOSS/70; Chemax DOSS-75E; Chimin DOS 70; Coptal WA OSN; Denwet CM; Depasol AS-27; Disponil SUS IC 8; Drewfax® 0007; Drewfax® S-700; Elfanol® 883; Emcol® 4500; Emcol® 4560; Emcol® DOSS; Empimin® OP45; Empimin® OP70; Empimin® OT; Empimin® OT75; Gemtex PA-75E; Gemtex PAX-60; Gemtex SC-40; Gemtex SC-70; Gemtex SC-75E, SC Powd; Gemtex SM-33; Geropon® CYA/60; Geropon® DOS (formerly Rhodiasurf DOS); Geropon® DOS FP (formerly Rhodiasurf DOS FP); Geropon® SDS; Geropon® SS-O-70PG (formerly Alkasurf® SS-O-70PG); Geropon® SS-O-75 (formerly Alkasurf® SS-O-75); Geropon® WT-27 (formerly Nekal® WT-27); Hipochem EK-18; Hodag DOSS-70; Hodag DOSS-75; Imbirol OT/Na 70; Karawet DOSS; Leonil OS; Mackanate DOS-40; Mackanate DOS-70; Mackanate DOS-75; Manoxol OT; Marlinat® DF 8; Mazawet® DOSS 70; Monawet MO-65-150; Monawet MO-65 PEG; Monawet MO-70-150; Monawet MO-70 PEG; Nikkol OTP-100S; Ninate® DS 70; Nissan Rapisol B-30, B-80, C-70; Octowet 40; Octowet 55; Octowet 60; Octowet 60-I; Octowet 65; Octowet 70; Octowet 70BC; Octowet 70PG; Octowet 75; Octowet 75E; Protowet 5171; Protowet D-75; Rewopol® SBDO 70; Rewopol® SBDO 75; Rexowet ASG-81; Sanmorin OT 70; Schercopol DOS-70; Schercopol DOS-PG-70; Schercopol DOS-PG-85; Schercowet DOS-70; Secosol® DOS 70; Servoxyl VLA 2170; Serwet WH 170; Supermontaline SLT65; Surfax 550; Surfax 585; Tex-Wet 1001; Thorowet G-40 3230; Thorowet G-60 3142; Thorowet G-60 SS 0336; Thorowet G-75 1060; Triton® GR-5M; Triton® GR-7M
Tradenames containing: Aerosol® OT-70 PG; Aerosol® OT-B; Aerosol® OT-MSO; Agrilan® AEC266; Agrilan® AEC299; Astrowet O-70-PG; Gemtex PA-70P; Gemtex PA-75; Gemtex PA-85P; Gemtex SC-75; Geropon® 99 (formerly Pentex® 99); Lankropol® KO; Lankropol® KO2; Lutensit® A-BO; Mackanate DOS-70BC; Mackanate DOS-70DEG; Mackanate DOS-70MS; Mackanate DOS-70N; Mackanate DOS-70PG; Manoxol OT 60%; Manoxol OT/B; Monawet MO-70; Monawet MO-70E; Monawet MO-70R; Monawet MO-70S; Monawet MO-75E; Monawet MO-84R2W; Monawet MO-85P

Dipropylene glycol (CTFA)
CAS 110-98-5; EINECS 203-821-4
Synonyms: Di-1,2-propylene glycol; 1,1′-Oxybis-2-propanol
Tradenames containing: Larostat® 377 DPG; Surfynol® DF-110D.

Disodium C12-15 pareth sulfosuccinate (CTFA)
CAS 39354-47-5
Synonyms: Disodium pareth-25 sulfosuccinate; Sulfobutanedioic acid, C12-15 pareth ester, disodium salt
Tradenames: Emcol® 4300

Disodium caproamphodiacetate (CTFA)
CAS 54849-16-0; 70750-05-9
Synonyms: Caproamphocarboxyglycinate
Tradenames containing: Mackam 2CT

Disodium capryloamphodiacetate (CTFA)
CAS 7702-01-4; 68608-64-0; EINECS 231-721-0
Synonyms: 1H-Imidazolium, 1-[2-(carboxymethoxy)ethyl]-1-(carboxymethyl)-2-heptyl-4,5-dihydro-, hydroxide, disodium salt; Capryloamphocarboxyglycinate; Capryloamphodiacetate
Tradenames: Ampholak XJO; Amphoterge® J-2; Mackam 2CY; Miranol® J2M Conc; Miranol® JB (formerly Mirapon JB); Schercoteric CY-2; Sochamine A 8955

Disodium capryloamphodipropionate (CTFA)
CAS 68815-55-4
Synonyms: Capryloamphocarboxypropionate; Capryloamphodipropionate
Tradenames: Amphoterge® KJ-2; Mackam 2CYSF; Miranol® J2M-SF Conc; Miranol® JBS (formerly Mirapon JBS); Monateric 811; Monateric 1000; Schercoteric CY-SF-2; Zoharteric LF-SF

Disodium cetearyl sulfosuccinamate
Tradenames: Empimin® MK/B; Empimin® MKK/L

Disodium cetearyl sulfosuccinate (CTFA)
Synonyms: Disodium cetyl-stearyl sulfosuccinate; Sulfobutanedioic acid, cetyl/stearyl ester, disodium salt
Tradenames: Empicol® STT; Empimin® MKK98

Disodium cetyl stearyl sulfosuccinamate. *See Disodium cetearyl sulfosuccinamate*

Disodium cocamido MEA-sulfosuccinate (CTFA)
CAS 61791-66-0; 68784-08-7; EINECS 272-219-1
Synonyms: Sulfobutanedioic acid, C-(2-cocamidoethyl) esters, disodium salts; Disodium cocoylmonoethanolamide sulfosuccinate
Tradenames: Mackanate CM; Mackanate CM-100; Rewopol® SBC 212; Rewopol® SBC 212 G; Schercopol CMS-Na

Disodium cocamido MIPA-sulfosuccinate (CTFA)
CAS 68515-65-1; EINECS 271-102-2
Synonyms: Sulfobutanedioic acid, 2-cocamido-1-methylethyl esters, disodium salts
Tradenames: Monamate C-1142; Monamate CPA-40; Monamate CPA-100

Disodium cocoamphodiacetate (CTFA)
CAS 61791-32-0; 68650-39-5; 68647-53-0; EINECS 272-043-5
Synonyms: N-Cocamidoethyl-N-2-hydroxyethyl-N-carboxyethylglycine, sodium salt; Cocoamphodiacetate; Cocoamphocarboxyglycinate
Tradenames: Amphoterge® W-2; Chemteric 2C; Mackam 2C; Mackam 2C-75; Manroteric CDX38; Miranol® 2CIB (formerly Alkateric® 2CIB); Miranol® C2M Conc. NP; Miranol® C2M Conc. OP; Miranol® C2M NP LA; Miranol® FB-NP; Monateric CDX-38; Mona-

teric CDX-38 Mod; Schercoteric MS-2; Varion® 2C; Zoharteric D; Zoharteric D-SF 70%
Tradenames containing: Miranol® C2M Conc. NP-PG; Schercoteric MS-2 Modified; Schercoteric MS-2ES Modified; Schercoteric MS-2TE Modified; Zoharteric DO; Zoharteric DOT

Disodium cocoamphodipropionate (CTFA)
CAS 86438-35-9; 83138-08-3; 86438-79-1; 68411-57-4; 68604-71-7; 68910-41-5; EINECS 270-131-8
Synonyms: Cocoamphocarboxypropionate; Cocoamphodipropionate
Tradenames: Amphoterge® K-2; Chemteric SF; Dehyton® G-SF; Mackam 2C-SF; Miranol® C2M-SF 70%; Miranol® C2M-SF Conc; Miranol® FBS; Monateric CEM-38; Varion® AM-KSF-40%; Zoharteric D-SF

Disodium cocoyl glutamate
Tradenames: Acylglutamate CS-21

Disodium cocoyl sulfosuccinamate
Tradenames: Empimin® MH

Disodium cocoyl sulfosuccinate
Tradenames: Empimin® MHH

Disodium cocoyl tallowyl glutamate
Tradenames: Acylglutamate GS-21

Disodium deceth-6 sulfosuccinate (CTFA)
CAS 68311-03-5 (generic); 39354-45-5
Synonyms: Sulfobutanedioic acid, deceth-6 ester, disodium salt
Tradenames: Aerosol® A-102; Monawet TD-30

Disodium dodecyloxy propyl sulfosuccinamate
CAS 58353-68-7
Tradenames: Octosol A-1

Disodium isodecyl sulfosuccinate (CTFA)
CAS 37294-49-8; EINECS 253-452-8
Synonyms: Sulfobutanedioic acid, 4-isodecyl ester, disodium salt
Tradenames: Aerosol® A-268

Disodium lauramido MEA-sulfosuccinate (CTFA)
CAS 25882-44-4; EINECS 247-310-4
Synonyms: Sulfobutanedioic acid, 1-ester with N-(2-hydroxyethyl) dodecanamide, disodium salt
Tradenames: Geropon® SBL-203 (formerly Alconate® SBL-203); Incrosul LMS; Rewopol® SBL 203; Rewopol® SBL 203 G, 203 P; Varsulf® SBL-203
Tradenames containing: Marlinat® SRN 30

Disodium lauraminopropionate
Tradenames containing: Miracare® MS-1 (formerly Compound MS-1); Miracare® MS-2 (formerly Compound MS-2); Miracare® MS-4

Disodium laureth sulfosuccinate (CTFA)
CAS 39354-45-5 (generic); 40754-59-4; 42016-08-0; 58450-52-5; EINECS 255-062-3
Synonyms: Sulfobutanedioic acid, 4-[2-[2-[2-(dodecyloxy) ethoxy]ethoxy]ethyl]ester, disodium salt; Disodium lauryl ether sulfosuccinate
Tradenames: Empicol® SDD; Geropon® ACR/4 (formerly Geronol ACR/4); Mackanate L-101, -102; Marlinat® SL 3/40; Rewopol® SBFA 30; Rolpon SE 138; Schercopol LPS; Setacin 103 Spezial; Stepan-Mild® SL3; Varsulf® SBFA-30
Tradenames containing: Rewopol® SBV; Stepan-Mild® LSB

Disodium lauriminodipropionate (CTFA)
CAS 3655-00-3; EINECS 222-899-0
Synonyms: Disodium N-lauryl-beta-iminodipropionate; Disodium 3,3′-(dodecylimino) dipropionate; N-(2-Carboxyethyl)-N-dodecyl-beta-alanine, disodium salt
Tradenames: Deriphat 160; Miranol® H2C-HA (formerly Alkateric® A2P-LPS); Monateric 1188M

Disodium lauroamphodiacetate (CTFA)
CAS 14350-97-1; 68608-66-2; EINECS 238-306-3
Synonyms: 1H-Imidazolium, 1-[2-(carboxymethoxy) ethyl]-1-carboxymethyl)-4,5-dihydro-2-undecyl-, hydroxide, disodium salt; Lauroamphocarboxyglycinate; Lauroamphodiacetate
Tradenames: Mackam 2L; Miranol® LB (formerly Mirapon LB); Schercoteric LS-2; Varion® 2L; Zoharteric DJ
Tradenames containing: Miracare® 2MHT (formerly Miranol® 2MHT); Mackam LT

Disodium lauroamphodipropionate (CTFA)
CAS 68610-43-5; 68929-04-4; 68920-18-3
Synonyms: 1H-Imidazolium,1-[2-(2-carboxyethoxy) ethyl-1-(2-carboxyethyl)-4,5-dihydro-2-undecyl-, hydroxide, disodium salt; Lauroamphocarboxypropionate; Lauroamphodipropionate
Tradenames: Mackam 2LSF

Disodium lauryl sulfosuccinate (CTFA)
CAS 13192-12-6; 19040-44-9; 26838-05-1; 26838-05-1; 36409-57-1; EINECS 248-030-5; 236-149-5
Synonyms: Sulfobutanedioic acid, 1-dodecyl ester, disodium salt
Tradenames: Empicol® SLL; Empicol® SLL/P; Monamate LA-100; Rewopol® SBF 12; Rewopol® SBF 12 P; Rewopol® SBR 12-Powder; Setacin F Spezial Paste
Tradenames containing: Rhodaterge® SSB (formerly Miranate® SSB)

Disodium methylenebis (naphthalene sulfonic acid)
Tradenames: Atlox 4862

Disodium myristamido MEA-sulfosuccinate (CTFA)
CAS 37767-42-3
Synonyms: Sulfobutanedioic acid, 2[(1-oxotetradecyl) amino]ethyl ester, disodium salt; Disodium monomyristamido MEA-sulfosuccinate
Tradenames: Emcol® 4100M

Disodium nonoxynol-10 sulfosuccinate (CTFA)
CAS 67999-57-9 (generic); 9040-38-4
Synonyms: Sulfobutanedioic acid, nonoxynol-10 ester, disodium salt
Tradenames: Aerosol® A-103; Fizul 301; Geronol ACR/9; Monawet 1240

Disodium oleamido MEA-sulfosuccinate (CTFA)
CAS 68479-64-1; 79702-63-0; EINECS 270-864-3
Synonyms: Sulfobutanedioic acid, mono[2-[(1-oxo-9-octadecenyl)amino]ethyl]ester, disodium salt
Tradenames: Schercopol OMS-Na

Disodium oleamido MIPA-sulfosuccinate (CTFA)
CAS 43154-85-4; EINECS 256-120-0
Synonyms: Sulfobutanedioic acid, 4-[1-methyl-2-[(1-oxo-9-octadecenyl)amino]ethyl]ester, disodium salt; Disodium oleoyl isopropanolamide

sulfosuccinate; Disodium monooleamido MIPA-sulfosuccinate
Tradenames: Emcol® 4161L; Emcol® K8300

Disodium oleamido PEG-2 sulfosuccinate (CTFA)
CAS 56388-43-3; EINECS 260-143-1
Synonyms: Sulfobutanedioic acid, C-[2-[2-[(1-oxo-9-octadecenyl)amino]ethoxy]ethyl]ester, disodium salt; Disodium oleamido diglycol sulfosuccinate
Tradenames: Anionyx® 12S; Monamate OPA-30; Monamate OPA-100

Disodium oleoamphodiacetate
CAS 97659-53-5
Synonyms: Oleoamphocarboxyglycinate
Tradenames: Ampholak XO7; Ampholak XO7-SD-55; Ampholak XOO-30

Disodium N-oleyl sulfosuccinamate
Tradenames: Empimin® MTT

Disodium oleyl sulfosuccinate (CTFA)
Synonyms: Sulfobutanedioic acid, 1-(9-octadecenyl) ester, disodium salt
Tradenames: Empimin® MTT/A

Disodium pareth-25 sulfosuccinate. *See Disodium C12-15 pareth sulfosuccinate*

Disodium PEG-4 cocamido MIPA sulfosuccinate (CTFA)
Tradenames: Rewopol® SBZ

Disodium ricinoleamido MEA-sulfosuccinate (CTFA)
CAS 40754-60-7; 65277-54-5; 67893-42-9; EINECS 267-617-7; 265-672-1
Synonyms: Sulfobutanedioic acid, 1-[2-[(12-hydroxy-1-oxo-9-octadecenyl)amino]ethyl]ester, disodium salt; Disodium ricinoleyl monoethanolamide sulfosuccinate
Tradenames: Rewoderm® S 1333; Rewoderm® S 1333 P; Schercopol RMS-Na; Varsulf® S-1333

Disodium stearoyl glutamate
Tradenames: Acylglutamate HS-21

Disodium stearyl sulfosuccinamate (CTFA)
CAS 14481-60-8; EINECS 238-479-5
Synonyms: Sulfobutanedioic acid, monooctadecyl ester, disodium salt; Disodium octadecyl sulfosuccinamate
Tradenames: Aerosol® 18; Lankropol® ODS/LS; Lankropol® ODS/PT; Octosol A-18; Rewopol® SBF 18; Surfax 500

Disodium tallamphodipropionate
Tradenames: Miranol® L2M-SF Conc. (formerly Mirapon L2M-SF Conc.)

Disodium tallow aminodipropionate
Tradenames: Mirataine® A2P-TS-30 (formerly Alkateric® A2P-TS)

Disodium taillow iminodipropionate (CTFA)
CAS 61791-56-8
Synonyms: Disodium N-tallow-β iminodipropionate; N-(2-Carboxyethyl)-N-(tallow acyl)-β-alanine
Tradenames: Deriphat 154; Deriphat 154L; Mirataine® T2C-30; Monateric TDB-35

Disodium tallow sulfosuccinamate (CTFA)
CAS 90268-48-7; EINECS 290-850-0
Synonyms: Sulfobutanedioic acid, tallow ester, disodium salt
Tradenames: Rewopol® B 1003

Disodium undecylenamido MEA-sulfosuccinate (CTFA)
CAS 26650-05-5; 37311-67-4; 40839-40-5; 65277-52-3; EINECS 247-873-6
Synonyms: Sulfobutanedioic acid, 4-[2-[(1-oxo-10-undecenyl)amino]ethyl]ester, disodium salt; Disodium undecylenoyl monoethanolamide sulfosuccinate
Tradenames: Empicol® SEE; Schercopol UMS-Na

Disodium wheatgermamphodiacetate (CTFA)
Tradenames: Mackam 2W

Distearylamine. *See Dioctadecylamine*

Distearyldimonium chloride (CTFA)
CAS 107-64-2; EINECS 203-508-2
Synonyms: Distearyl dimethyl ammonium chloride; Quaternium-5; Dioctadecyl dimethyl ammonium chloride
Tradenames: Arquad® 218-100; Cation DS; Dehyquart DAM
Tradenames containing: Arquad® 218-75

Ditallowalkonium chloride. *See Quaternium-18*

Ditallowamine
CAS 68783-24-4
Tradenames: Armeen® 2T

Ditallow dimonium chloride (CTFA)
CAS 68153-32-2; 68783-78-8; EINECS 272-207-6
Synonyms: Ditallow dimethyl ammonium chloride; Quaternium-48; Dimethyl ditallow ammonium chloride
Tradenames: Adogen® 470; Radiaquat 6470
Tradenames containing: Arquad® 2T-75

Ditallow dimonium methosulfate
Tradenames: Ablumine DT

Ditridecyl adipate (CTFA)
CAS 26401-35-4; 16958-92-2; EINECS 247-660-8
Synonyms: Ditridecyl hexanedioate; Hexanedioic acid, ditridecyl ester
Tradenames: Afilan TDA

Ditridecyl sodium sulfosuccinate (CTFA)
CAS 2673-22-5; EINECS 220-219-7
Synonyms: Sodium bistridecyl sulfosuccinate; Sodium ditridecyl sulfosuccinate; Sulfobutanedioic acid, 1,4-ditridecyl ester, sodium salt
Tradenames: Aerosol® TR-70; Emcol® 4600; Geropon® BIS/SODICO-2 (formerly Vanisol BIS/SODICO-2); Monawet MT-70E; Polirol TR/LNA
Tradenames containing: Monawet MT-70; Monawet MT-80H2W

Docosanoic acid. *See Behenic acid*

Docusate sodium. *See Dioctyl sodium sulfosuccinate*

Dodecanoic acid. *See Lauric acid*

1-Dodecanol. *See Lauryl alcohol*

Dodecene-1
CAS 112-41-4; 6842-15-5
Synonyms: C12 alpha olefin; alpha-Dodecylene; Tetrapropylene
Tradenames: Ethyl® Dodecene-1; Gulftene 12; Neodene® 12

Dodecenyl succinic anhydride
CAS 25377-73-5; 26544-38-7; DDSA
Tradenames: Dodecenyl Succinic Anhydride

Dodecyl alcohol. *See Lauryl alcohol*

Dodecylbenzene sulfonic acid (CTFA)
CAS 27176-87-0; 68411-32-5; 68584-22-5; 68608-88-8; 85536-14-7; EINECS 248-289-4
Synonyms: DDBSA
Tradenames: Alarsol AL; Arsul DDB; Arsul LAS; Arylan® S Acid; Arylan® SBC Acid; Arylan® SC Acid; Arylan® SKN Acid; Arylan® SP Acid; Bio-Soft® JN; Bio-Soft® S-100; Calsoft LAS-99; Carsosulf UL-100 Acid; DDBS Special; DeSonate SA; DeSonate SA-H; Elfan® WA Sulphonic Acid; Emka DDBSA; Manro BA Acid; Manro HA Acid; Manro NA Acid; Maranil ABS; Maranil DBS; Marlon® AS3; Mazon® 85; Nansa® 1042; Nansa® 1042/P; Nansa® 1909; Nansa® SBA; Nansa® SSA; Nansa® SSA/P; Naxel AAS-98S; Pentine Acid 5431; Polystep® A-13; Polystep® A-17; Puxol XB-10; Reworyl® K; Reworyl® Sulfonic Acid K; Rhodacal® LA Acid (formerly Alkasurf® LA Acid); Rueterg SA; Sandoz Sulfonate AAS 98S; Serdet DM; Stepantan® H-100; Sulfotex UBL-100; Sulframin Acide B; Sulframin Acide TPB; Tex-Wet 1197; Ufacid K; Ufacid KW; Ufacid TPB; Vista SA-597; Wibarco; Witco® 1298; Witco® 1298 H; Witco® Acid B

Dodecyl diphenyl ether disulfonic acid
CAS 80260-73-2
Tradenames: Poly-Tergent® 2A1 Acid

Dodecyl diphenyl oxide disulfonic acid
Tradenames: Dowfax 2A0

Dodecyl phosphate
Tradenames: Hostaphat MDL

Dodoxynol-4
CAS 9014-92-0 (generic); 26401-47-8 (generic)
Synonyms: PEG-4 dodecyl phenyl ether; POE (4) dodecyl phenyl ether
Tradenames: Rexol 65/4

Dodoxynol-5 (CTFA)
CAS 9014-92-0 (generic); 26401-47-8 (generic)
Synonyms: PEG-5 dodecyl phenyl ether; POE (5) dodecyl phenyl ether
Tradenames: Desonic® 5D; T-Det® DD-5; Teric DD5

Dodoxynol-6 (CTFA)
CAS 9014-92-0 (generic); 26401-47-8 (generic)
Synonyms: PEG-6 dodecyl phenyl ether; POE (6) dodecyl phenyl ether; PEG 300 dodecyl phenyl ether
Tradenames: Desonic® 6D; Igepal® RC-520; Rexol 65/6

Dodoxynol-7 (CTFA)
CAS 9014-92-0 (generic); 26401-47-8 (generic)
Synonyms: PEG-7 dodecyl phenyl ether; POE (7) dodecyl phenyl ether
Tradenames: T-Det® DD-7

Dodoxynol-9 (CTFA)
CAS 9014-92-0 (generic); 26401-47-8 (generic)
Synonyms: PEG-9 dodecyl phenyl ether; POE (9) dodecyl phenyl ether; PEG 450 dodecyl phenyl ether
Tradenames: Desonic® 9D; Prox-onic DDP-09; Rexol 65/9; T-Det® DD-9; Teric DD9

Dodoxynol-10
CAS 9014-92-0 (generic); 26401-47-8 (generic)
Synonyms: PEG-10 dodecyl phenyl ether; POE (10) dodecyl phenyl ether
Tradenames: Desonic® 10D; Iconol DDP-10; Igepal® RC-620; Igepal® RC-630; Rexol 65/10; T-Det® DD-10

Dodoxynol-11
CAS 9014-92-0 (generic); 26401-47-8 (generic)
Synonyms: PEG-11 dodecyl phenyl ether; POE (11) dodecyl phenyl ether
Tradenames: Rexol 65/11; T-Det® DD-11

Dodoxynol-12 (CTFA)
CAS 9014-92-0 (generic); 26401-47-8 (generic)
Synonyms: PEG-12 dodecyl phenyl ether; POE (12) dodecyl phenyl ether; PEG 600 dodecyl phenyl ether
Tradenames: Desonic® 12D; Prox-onic DDP-012

Dodoxynol-14
CAS 9014-92-0 (generic); 26401-47-8 (generic)
Synonyms: PEG-14 dodecyl phenyl ether; POE (14) dodecyl phenyl ether
Tradenames: Desonic® 14D; Rexol 65/14; T-Det® DD-14

Eicosanol
CAS 629-96-9
Synonyms: Arachidyl alcohol; 1-Eicosanol
Tradenames containing: Epal® 1618RT

Eicosene-1
CAS 3452-07-1
Synonyms: Linear C20 alpha olefin
Tradenames: Neodene® 20

Emulsifying wax NF
Tradenames: Dermalcare® NI (formerly Cyclochem® NI); Hetoxol P

Epoxidized soybean oil (CTFA)
CAS 8013-07-8; EINECS 232-391-0
Synonyms: Soybean oil, epoxidized
Tradenames: Hoe S 3680

Erucamide (CTFA)
CAS 112-84-5; EINECS 204-009-2
Synonyms: Erucic acid amide; 13-Docosenamide; cis 13-Docosenamide
Tradenames: Armid® E; Kemamide® E

Erucamide TEA
Tradenames: Laurel SD-520T

Erucamidopropyl hydroxysultaine (CTFA)
Tradenames: Crosultaine E-30

Erucyl stearamide
Tradenames: Kemamide® S-221

Ethanol. *See Ethyl alcohol*

Ethoxydiglycol (CTFA)
CAS 111-90-0; EINECS 203-919-7
Synonyms: Diethylene glycol monoethyl ether; "Carbitol"; 2-(2-Ethoxyethoxy) ethanol
Tradenames: Transcutol
Tradenames containing: Marlowet® RNP/K

Ethyl alcohol
CAS 64-17-5; EINECS 200-578-6
Synonyms: EtOH; Ethanol; Alcohol (CTFA); Absolute alcohol
Tradenames containing: Agrilan® AEC266; Arquad® 2T-75; Arquad® 210-50; Empimin® KSN60; Lankropol® KMA; Lankropol® KO2; Monawet MO-70E; Monawet MO-75E; Variquat® 50AE; Variquat® 50ME; Variquat® 80AE

Ethylene alcohol. *See Glycol*

Ethylene dioleamide (CTFA)
CAS 110-31-6; EINECS 203-756-1
Synonyms: N,N'-1,2-Ethanediylbis-9-octadecenamide; 9-Octadecenamide, N,N'-1,2-ethanediylbis-
Tradenames: Kemamide® W-20

Ethylene distearamide (CTFA)
CAS 110-30-5; EINECS 203-755-6
Synonyms: N,N'-Ethylene bisstearamide; N,N'-1,2-Ethanediylbisoctadecanamide
Tradenames: Alkamide® STEDA; Kemamide® W-39; Kemamide® W-40; Kemamide® W-40/300; Kemamide® W-45

Ethylene glycol. *See Glycol*

Ethylene glycol monobutyl ether. *See Butoxyethanol*

Ethylene glycol monophenyl ether. *See Phenoxyethanol*

Ethylene glycol nonyl phenyl ether. *See Nonoxynol-1*

Ethylene glycol octyl phenyl ether. *See Octoxynol-1*

2-Ethylhexanol
CAS 104-76-7; EINECS 203-234-3
Synonyms: 2-EH; 2-Ethylhexyl alcohol; Octyl alcohol
Tradenames containing: Surfynol® 104A

2-Ethylhexanol phosphate
CAS 78-42-2
Synonyms: Tris (2-ethylhexyl) phosphate; Trioctyl phosphate
Tradenames: Base 104

2-Ethylhexylamine
CAS 104-75-6
Synonyms: 2-Ethyl-1-hexylamine; Octylamine
Tradenames: Armeen® L8D

2-Ethylhexyl hydrogenated tallowalkyl methosulfate
Tradenames: Arquad® HTL8(W) MS-85

2-Ethylhexyl phosphate
Tradenames: Servoxyl VPTZ 100

Ethyl hydroxymethyl oleyl oxazoline (CTFA)
CAS 68140-98-7; 88543-32-2; EINECS 268-820-3
Synonyms: 4-Ethyl-2-(8-heptadecenyl)-4,5-dihydro-4-oxazolemethanol
Tradenames: Alkaterge®-E

Ferro-chromium lignosulfonate
Tradenames: Peltex

Glycereth-7 (CTFA)
CAS 31694-55-0 (generic)
Synonyms: PEG-7 glyceryl ether; POE (7) glyceryl ether
Tradenames: Hetoxide G-7

Glycereth-26 (CTFA)
CAS 31694-55-0 (generic)
Synonyms: PEG-26 glyceryl ether; POE (26) glyceryl ether
Tradenames: Acconon ETG; Ethosperse® G-26; Hetoxide G-26; Liponic EG-1

Glycereth-25 PCA isostearate (CTFA)
Synonyms: PEG-25 glyceryl ether PCA isostearate
Tradenames: Pyroter GPI-25

Glycerin (CTFA)
CAS 56-81-5; EINECS 200-289-5
Synonyms: Glycerol; 1,2,3-Propanetriol; Glycerine
Tradenames: Emery® 912; Emery® 916; Emery® 918; Superol
Tradenames containing: Manro CD/G; Mirasheen® 202 (formerly Cyclosheen 202)

Glycerol. *See Glycerin*

Glyceryl caprate (CTFA)
CAS 26402-22-2; EINECS 247-667-6
Synonyms: Glyceryl monocaprate; Decanoic acid, monoester with 1,2,3-propanetriol
Tradenames: Imwitor® 910

Glyceryl caprylate (CTFA)
CAS 26402-26-6; EINECS 247-668-1
Synonyms: Glyceryl monocaprylate; Octanoic acid, monoester with 1,2,3-propanetriol; Monooctanoin
Tradenames: Imwitor® 908

Glyceryl caprylate/caprate (CTFA)
Tradenames: Capmul® MCM

Glyceryl citrate/lactate/linoleate/oleate (CTFA)
Tradenames: Imwitor® 375

Glyceryl cocoate (CTFA)
CAS 61789-05-7; EINECS 263-027-9
Synonyms: Glycerides, coconut oil mono-; Glycerol mono coconut oil; Glyceryl coconate
Tradenames: Aldo® MC; Radiasurf® 7144

Glyceryl dilaurate (CTFA)
CAS 27638-00-2; EINECS 248-586-9
Synonyms: Dilaurin; Dodecanoic acid, diester with 1,2,3-propanetriol
Tradenames: Cithrol GDL N/E; Kessco® GDL

Glyceryl dilaurate SE
Tradenames: Cithrol GDL S/E

Glyceryl dioleate (CTFA)
CAS 25637-84-7; EINECS 247-144-2
Synonyms: 9-Octadecenoic acid, diester with 1,2,3-propanetriol
Tradenames: Cithrol GDO N/E; Emerest® 2419; Nikkol DGO-80

Glyceryl dioleate SE
Tradenames: Cithrol GDO S/E

Glyceryl distearate (CTFA)
CAS 1323-83-7; EINECS 215-359-0
Synonyms: Octadecanoic acid, diester with 1,2,3-propanetriol
Tradenames: Cithrol GDS N/E; Kessco® GDS 386F; Nikkol DGS-80

Glyceryl distearate SE
Tradenames: Cithrol GDS S/E

Glyceryl hydroxystearate (CTFA)
CAS 1323-42-8; EINECS 215-355-9
Synonyms: Glyceryl 12-hydroxystearate; Hydroxystearic acid, monoester with glycerol
Tradenames: Naturechem® GMHS

Glyceryl isostearate (CTFA)
CAS 32057-14-0; 66085-00-5
Synonyms: Glyceryl monoisostearate; Isooctadecanoic acid, monoester with 1,2,3-propanetriol
Tradenames: Emerest® 2410; Imwitor® 780; Peceol Isostearique

Glyceryl lanolate (CTFA)
CAS 97404-50-7
Synonyms: Glyceryl monolanolate; Lanolin acid, monoester with 1,2,3-propanetriol
Tradenames: Lanesta G

Glyceryl laurate (CTFA)
CAS 142-18-7; EINECS 205-526-6
Synonyms: Glyceryl monolaurate; Dodecanoic acid, monoester with 1,2,3-propanetriol; Dodecanoic acid, 2,3-dihydroxypropyl ester
Tradenames: Ablunol GML; Aldo® MLD; Cithrol GML N/E; Imwitor® 312; Kessco® GML
Tradenames containing: Acconon CON

Glyceryl laurate SE (CTFA)
Tradenames: Cithrol GML S/E

Glyceryl mono/diisostearate
CAS 68958-48-5
Tradenames: Tegin® ISO

Glyceryl mono/dilaurate
Tradenames: Aldo® ML

Glyceryl mono/dioleate
CAS 25496-72-4
Tradenames: Aldo® MOD FG; Aldo® MO FG; Caplube 8350; Tegin® O

Glyceryl mono/distearate
Tradenames: Teginacid® H-SE; Teginacid® SE; Teginacid® X-SE
Tradenames containing: Teginacid® ML-SE

Glyceryl myristate (CTFA)
CAS 589-68-4; 67701-33-1
Synonyms: Glyceryl monomyristate; Monomyristin; Tetradecanoic acid, monoester with 1,2,3-propanetriol
Tradenames: Imwitor® 914

Glyceryl oleate (CTFA)
CAS 111-03-5; 37220-82-9; EINECS 203-827-7; 253-407-2
Synonyms: Glyceryl monooleate; Monoolein; 9-Octadecenoic acid, monoester with 1,2,3-propanetriol
Tradenames: Ablunol GMO; Aldo® MO; Aldo® MO Tech; Cithrol GMO N/E; Emerest® 2421 (see Witconol 2421); Kemester® 2000; Kessco® GMO; Mazol® 300 K; Mazol® GMO; Mazol® GMO #1; Mazol® GMO Ind; Mazol® GMO K; Pationic® 1064; Pationic® 1074; Radiasurf® 7150; Radiasurf® 7152; Tegin® O NSE; Tegin® O Special; Tegin® O Special NSE; Witco® 942; Witconol 2421

Glyceryl oleate SE (CTFA)
Tradenames: Aldo® MOD; Cithrol GMO S/E; Radiasurf® 7151

Glyceryl palmitate/stearate
Synonyms: Glyceryl stearate palmitate
Tradenames: Imwitor® 940 K

Glyceryl ricinoleate (CTFA)
CAS 141-08-2; EINECS 205-455-0
Synonyms: 12-Hydroxy-9-octadecenoic acid, monoester with 1,2,3-propanetriol; Monoricinolein; Glyceryl monoricinoleate
Tradenames: Aldo® MR; Cithrol GMR N/E; Drewmulse® GMRO; Flexricin® 13; Mazol® GMR; Radiasurf® 7153; Tegin® RZ; Tegin® RZ NSE

Glyceryl ricinoleate SE
Synonyms: Glyceryl triricinoleate SE
Tradenames: Cithrol GMR S/E

Glyceryl stearate (CTFA)
CAS 123-94-4; 11099-07-3; 31566-31-1; 85666-92-8; EINECS 250-705-4; 234-325-6; 204-664-4
Synonyms: Monostearin; Glyceryl monostearate; Octadecanoic acid, monoester with 1,2,3-propanetriol
Tradenames: Ablunol GMS; Ahcovel Base N-15; Aldo® HMS; Aldo® MS; Aldo® MSA; Aldo® MS Industrial; Aldo® MS LG; Alkamuls® GMS/C (formerly Cyclochem GMS, Dermalcare® GMS); Arlacel® 129; Cithrol GMS N/E; CPH-53-N; CPH-250-SE; Drewmulse® GMS; Dur-Em® 117; Edenor GMS; Emerest® 2400; Emerest® 2401; Empilan® GMS NSE40; Geleol; GMS Base; Imwitor® 191; Imwitor® 900; Kemester® 5500; Kemester® 6000; Kessco® GMS; Lexemul® 503; Lexemul® 515; Mazol® GMS-90; Norfox® GMS; Pationic® 901; Pationic® 902; Pationic® 909; Pationic® 1042; Pationic® 1052; Radiasurf® 7140; Radiasurf® 7600; Radiasurf® 7900; Secoster® SDG; Sinochem GMS; Softenol® 3900; Softenol® 3991; Sterol GMS; Tegin® GRB; Tegin® M; Teginacid® Special SE; Vykamol N/E, S/E; Witconol 2400; Witconol 2401; Witconol MST; Zohar GLST
Tradenames containing: Aldosperse® O-20 FG; Ice #2; Kessco® GMS SE/AS; Lipomulse 165; Tally® 100 Plus; Tegacid® Regular VA; Teginacid®; Teginacid® H; Teginacid® ML; Teginacid® X

Glyceryl stearate SE (CTFA)
CAS 86418-55-5; 31566-31-1; 11099-07-3; 85666-92-8; 977053-96-5
Synonyms: Glyceryl monostearate SE
Tradenames: Aldo® MSD; Chemsperse GMS-SE; Cithrol GMS Acid Stable; Cithrol GMS S/E; Emerest® 2407; Empilan® GMS LSE40; Empilan® GMS LSE80; Empilan® GMS MSE40; Empilan® GMS NSE90; Empilan® GMS SE32; Empilan® GMS SE40; Empilan® GMS SE70; GMS/SE Base; Kemester® 6000SE; Kessco® GMS SE; Lexemul® 55SE; Lexemul® 530; Lexemul® T; Mazol® GMS-D; Radiasurf® 7141; Tegacid® Special VA; Witconol 2407; Witconol CA; Witconol RHT; Zohar GLST SE
Tradenames containing: Tegin® C-611

Glyceryl stearate citrate (CTFA)
CAS 39175-72-9
Synonyms: 2-Hydroxy-1,2,3-propanetricarboxylic acid, monoester with 1,2,3-propanetriol monooctadecanoate
Tradenames: Imwitor® 370

Glyceryl stearate lactate (CTFA)
Tradenames: Durlac® 100W

Glyceryl stearate palmitate. *See Glyceryl palmitate/stearate*

Glyceryl tallate
Tradenames: EM-40

Glyceryl triacetyl ricinoleate (CTFA)
CAS 101-34-8; EINECS 202-935-1
Synonyms: 9-Octadecenoic acid, 12-(acetyloxy)-, 1,2,3-propanetriol ester; 1,2,3-Propanetriyl 12-(acetyloxy)-9-octadecenoate
Tradenames: Naturechem® GTR

Glyceryl tribehenate (CTFA)
CAS 18641-57-1
Synonyms: Tribehenin; Docosanoic acid, 1,2,3-propanetriyl ester

Tradenames: Syncrowax HR-C
Tradenames containing: Syncrowax HRS-C

Glyceryl trioleate. *See Triolein*

Glyceryl tristearate. *See Tristearin*

Glycidyl trimethyl ammonium chloride
CAS 3033-77-0
Synonyms: (2,3-Epoxypropyl) trimethylammonium chloride
Tradenames: Ogtac-85; Ogtac 85 V

Glycol (CTFA)
CAS 107-21-1; EINECS 203-473-3
Synonyms: Ethylene glycol; 1,2-Ethanediol; Ethylene alcohol
Tradenames containing: Surfynol® 104E; Surfynol® 104H; Surfynol® TG

Glycol dilaurate (CTFA)
CAS 624-04-4
Synonyms: Ethylene glycol dilaurate; Lauric acid, 1,2-ethanediyl ester; Dodecanoic acid 1,2-ethanediyl ester
Tradenames: Cithrol EGDL N/E

Glycol dilaurate SE
Tradenames: Cithrol EGDL S/E

Glycol dioleate
CAS 928-24-5
Synonyms: Ethylene glycol dioleate
Tradenames: Cithrol EGDO N/E

Glycol dioleate SE
Tradenames: Cithrol EGDO S/E

Glycol distearate (CTFA)
CAS 627-83-8; EINECS 211-014-3
Synonyms: EGDS; Ethylene glycol distearate; Octadecanoic acid, 1,2-ethanediyl ester
Tradenames: Cithrol EGDS N/E; Drewmulse® EGDS; Emerest® 2355; Kemester® EGDS; Kessco® EGDS; Lexemul® EGDS; Lipo EGDS; Mackester EGDS; Mapeg® EGDS; Pegosperse® 50 DS; Radiasurf® 7269
Tradenames containing: Genapol® PGM Conc.

Glycol distearate SE
Tradenames: Cithrol EGDS S/E

Glycol ditallowate (CTFA)
Synonyms: Ethylene glycol ditallowate; Tallow fatty acid, 1,2-ethanediyl ester
Tradenames containing: Marlamid® PG 20

Glycol hydroxystearate (CTFA)
CAS 33907-46-9; EINECS 251-732-4
Synonyms: Ethylene glycol monohydroxystearate; Glycol monohydroxystearate; Hydroxyoctadecanoic acid, 2-hydroxyethyl ester
Tradenames: Naturechem® EGHS

Glycol laurate
CAS 4219-48-1
Synonyms: Ethylene glycol monolaurate
Tradenames: Cithrol EGML N/E; EG-ML

Glycol laurate SE
Tradenames: Cithrol EGML S/E

Glycol MIPA stearate (CTFA)
Tradenames containing: Crodapearl Liq.

Glycol oleate (CTFA)
CAS 4500-01-0; EINECS 224-806-9
Synonyms: Ethylene glycol monooleate; Glycol monooleate; 2-Hydroxyethyl 9-octadecenoate
Tradenames: Cithrol EGMO N/E; Tegin® GO

Glycol oleate SE
Tradenames: Cithrol EGMO S/E

Glycol propargyl ether
Synonyms: Ethylene glycol monopropargyl ether
Tradenames: Hetoxol PA-1

Glycol ricinoleate (CTFA)
CAS 106-17-2; EINECS 203-369-8
Synonyms: Ethylene glycol monoricinoleate; Glycol monoricinoleate; 2-Hydroxyethyl 12-hydroxy-9-octadecenoate
Tradenames: Cithrol EGMR N/E; Flexricin® 15

Glycol ricinoleate SE
Tradenames: Cithrol EGMR S/E

Glycol stearate (CTFA)
CAS 111-60-4; 97281-23-7; EINECS 203-886-9; 306-522-8
Synonyms: EGMS; Ethylene glycol monostearate; Glycol monostearate; 2-Hydoxyethyl octadecanoate
Tradenames: Ablunol EGMS; Alkamuls® EGMS/C (formerly Cyclochem EGMS/C); Alkamuls® SEG (formerly Cyclochem® SEG); Cithrol EGMS N/E; Drewmulse® EGMS; Emerest® 2350; Hodag EGMS; Kemester® EGMS; Kessco® EGMS; Lauramide EG; Lexemul® EGMS; Lipo EGMS; Mackester EGMS; Mackester IP; Mackester SP; Mapeg® EGMS; Mapeg® EGMS-K; Monthyle; Norfox® EGMS; Pegosperse® EGMS-70; Radiasurf® 7270; Ritasynt IP; Sterol ST 1; Tegin® G 6100; Witconol EGMS; Zohar EGMS
Tradenames containing: Dermalcare® POL (formerly Cyclochem® POL); Mirasheen® 202 (formerly Cyclosheen 202); Nutrapon B 1365; Standapol® S

Glycol stearate SE (CTFA)
Synonyms: Ethylene glycol monostearate SE
Tradenames: Cithrol EGMS S/E

Gum rosin. *See Rosin*

2-Heptadecenyl-4,4-bishydroxymethyl oxazoline
Tradenames: Anfomul 01T

Heptadecenyl hydroxyethyl imidazoline
Tradenames: Rewomine IM-CA; Rewomine IM-OA

Hexadecanoic acid. *See Palmitic acid*

1-Hexadecanol. *See Cetyl alcohol*

Hexadecene-1
CAS 629-73-2
Synonyms: Linear C16 alpha olefin; Cetene
Tradenames: Ethyl® Hexadecene-1; Gulftene 16; Neodene® 16

Hexadecyl. *See Cetyl, Palmityl*

Hexadecyl trimethyl ammonium chloride. *See Cetrimonium chloride*

Hexaethylene glycol. *See PEG-6*

Hexaglyceryl. *See Polyglyceryl-6*

Hexahydro-1,3,5-tris (2-hydroxyethyl)-s-triazine
Tradenames: Onyxide® 200

Hexamethylol melamine resin
CAS 531-18-0
Synonyms: Hexa (hydroxymethyl) melamine; 2,4,6-Tris (bis (hydroxymethyl) amino)-s-triazine
Tradenames: Glazamine M

1-Hexene
CAS 592-41-6
Synonyms: C6 linear alpha olefin; Hexylene; Butyl ethylene; Hexylene
Tradenames: Ethyl® Hexene-1; Gulftene 6; Neodene® 6

Hexeth-4 carboxylic acid (CTFA)
CAS 105391-15-9
Synonyms: PEG-4 hexyl ether carboxylic acid; PEG 200 hexyl ether carboxyic acid; POE (4) hexyl ether carboxylic acid
Tradenames: Akypo LF 3; Akypo MB 2528S
Tradenames containing: Akypo LF 4

Hexyl alcohol (CTFA)
CAS 111-27-3; 68526-79-4; EINECS 203-852-3
Synonyms: 1- or n-Hexanol; Pentylcarbinol; Amylcarbinol
Tradenames: Alfol® 6; Emery® 3321; Epal® 6; Lorol C6
Tradenames containing: Epal® 108; Epal® 610; Nansa® EVM70/B

Hexylene glycol (CTFA)
CAS 107-41-5; EINECS 203-489-0
Synonyms: 2-Methyl-2,4-pentanediol; 4-Methyl-2,4-pentanediol; α,α,α'-Trimethyltrimethyleneglycol
Tradenames containing: Mackam 2CT; Miracare® 2MHT (formerly Miranol® 2MHT); Monawet MT-70; Monawet MT-80H2W; Schercoteric MS-2 Modified; Variquat® K375; Zoharteric DO; Zoharteric DOT

Hexyloxypropylamine
Tradenames: Tomah PA-10

Hexyl phosphate
Tradenames: Findet NHP; Hostaphat HI

Hydrocarbon oil
Tradenames containing: Empicryl® 6059; Empicryl® 6070

Hydrogenated coconut acid (CTFA)
CAS 68938-15-8
Synonyms: Acids, coconut, hydrogenated; Coconut acid, hydrogenated; Fatty acids, coco, hydrogenated
Tradenames: Emery® 625; Hystrene® 5012; Industrene® 223

Hydrogenated ditallowamine (CTFA)
CAS 61789-79-5; EINECS 263-089-7
Synonyms: Dihydrogenated tallow amine; Bis(hydrogenated tallow alkyl) amines; Amines, bis(hydrogenated tallow alkyl)-
Tradenames: Noram 2SH

Hydrogenated lanolin (CTFA)
CAS 8031-44-5; EINECS 232-452-1
Synonyms: Lanolin, hydrogenated
Tradenames: Lipolan

Hydrogenated lard glyceride (CTFA)
CAS 8040-05-9
Synonyms: Glycerides, hydrogenated lard mono-
Tradenames: Lactomul 925

Hydrogenated menhaden acid (CTFA)
Synonyms: Acids, menhaden, hydrogenated; Menhaden acid, hydrogenated
Tradenames: Hystrene® 3022

Hydrogenated menhaden oil (CTFA)
Synonyms: Menhaden oil, hydrogenated; Oils, menhaden, hydrogenated
Tradenames: Neustrene® 045; Neustrene® 053

Hydrogenated methyl tallowate
Tradenames: Emery® 2204

Hydrogenated soybean oil (CTFA)
CAS 8016-70-4; EINECS 232-410-2
Synonyms: Oils, soybean, hydrogenated; Soybean oil, hydrogenated
Tradenames: Neustrene® 064
Tradenames containing: Tally® 100 Plus

Hydrogenated stearic acid
Tradenames: Industrene® B; Industrene® R

Hydrogenated tallow acid (CTFA)
CAS 61790-38-3; EINECS 263-130-9
Synonyms: Acids, tallow, hydrogenated; Tallow acid, hydrogenated
Tradenames: Prifac 9428; Prifac 9429

Hydrogenated tallowalkonium chloride (CTFA)
CAS 61789-72-8
Synonyms: Hydrogenated tallow dimethyl benzyl ammonium chloride
Tradenames: Empigen® BCM75, BCM75/A; Kemamine® BQ-9702C; Querton 441-BC; Variquat® B345
Tradenames containing: Arquad® DMHTB-75

Hydrogenated tallow amide (CTFA)
CAS 61790-31-6; EINECS 263-123-0
Synonyms: Amides, tallow, hydrogenated; Tallow amides, hydrogenated
Tradenames: Armid® HT

Hydrogenated tallowamine (CTFA)
CAS 61788-45-2; EINECS 262-976-6
Synonyms: Amines, hydrogenated tallow alkyl; Tallow amine, hydrogenated
Tradenames: Amine HBG; Amine HBGD; Armeen® HT; Armeen® HTD; Crodamine 1.HT; Jet Amine PHT; Jet Amine PHTD; Kemamine® P-970; Kemamine® P-970D; Nissan Amine ABT; Noram SH; Radiamine 6140; Radiamine 6141

Hydrogenated tallowamine acetate
CAS 61790-59-8
Synonyms: Hydrogenated tallowalkyl amine acetates
Tradenames: Acetamin HT; Armac® HT; Dinoramac SH; Noramac SH; Radiamac 6149

Hydrogenated tallow dimethylamine oxide
CAS 68390-99-8
Tradenames: Aromox® DMHTD

Hydrogenated tallow glyceride (CTFA)
CAS 61789-09-1; EINECS 263-031-0
Synonyms: Hydrogenated tallow monoglyceride; Glycerides, hydrogenated tallow mono-
Tradenames: Monomuls 90-25

Hydrogenated tallow glyceride citrate (CTFA)
CAS 68990-59-0; EINECS 273-613-6
Synonyms: Glycerides, tallow mono-, hydrogenated, citrates
Tradenames: Grindtek CA-P
Tradenames containing: Tegin® C-611

Hydrogenated tallow glycerides (CTFA)
CAS 68308-54-3; EINECS 269-658-6
Synonyms: Hydrogenated tallow mono-, di- and triglycerides; Glycerides, tallow mono-, di- and tri-, hydrogenated
Tradenames: Neustrene® 059; Neustrene® 060

Hydrogenated tallow 1,3-propylene diamine
CAS 68603-64-5
Tradenames: Diamin HT; Dinoram SH; Kemamine® D-970; Nissan Amine DTH; Radiamine 6540

Hydrogenated tallowtrimonium chloride (CTFA)
CAS 61788-78-1; EINECS 263-005-9
Synonyms: Hydrogenated tallow trimethyl ammonium chloride
Tradenames: Nissan Cation ABT-350, 500; Noramium MSH 50; Quartamin HTPR

Hydrogenated vegetable glycerides phosphate (CTFA)
CAS 85411-01-4; 25212-19-5
Tradenames: Emphos F27-85

Hydrolyzed animal protein. *See Hydrolyzed collagen*

Hydrolyzed collagen (CTFA)
CAS 9015-54-7
Synonyms: Collagen hydrolysates; Hydrolyzed animal protein; Proteins, collagen, hydrolysate
Tradenames: Remanol 1300/35

Hydroxyethylcellulose (CTFA)
CAS 9004-62-0
Synonyms: Cellulose, 2-hydroxyethyl ether; H.E. cellulose
Tradenames: Tylose H Series

1-Hydroxyethyl-1-diphosphonic acid
Tradenames: Chemphonate HEDP

1-Hydroxyethylidene-1,1-diphosphonic acid
CAS 2809-21-4
Synonyms: HEDPA; Acetodiphosphonic acid
Tradenames: Briquest® ADPA-60A; Briquest® ADPA-60AW

Hydroxyethyl stearamide-MIPA (CTFA)
Tradenames containing: Crodapearl NI Liquid

Hydroxylated lanolin (CTFA)
CAS 68424-66-8; EINECS 270-315-8
Synonyms: Lanolin, hydroxylated
Tradenames: Hydroxylan; OHlan®

Hydroxylated lecithin (CTFA)
CAS 8029-76-3; EINECS 232-440-6
Synonyms: Lecithin, hydroxylated
Tradenames: Centrolene® A, S; Lipotin H

Hydroxyoctacosanyl hydroxystearate (CTFA)
Synonyms: 12-Hydroxystearic acid, beta-hydroxyoctacosanyl ester
Tradenames: Elfacos® C26

Hydroxypropyl bis-isostearamidopropyldimonium chloride (CTFA)
CAS 11381-09-0; 111381-08-9
Tradenames: Schercoquat 2IAE; Schercoquat 2IAP

Hydroxypropylcellulose (CTFA)
CAS 9004-64-2
Synonyms: Cellulose, 2-hydroxypropyl ether; Oxypropylated cellulose
Tradenames: Klucel® E, G, H, J, L, M

Hydroxystearic acid (CTFA)
CAS 106-14-9; EINECS 203-366-1
Synonyms: HSA; 12-Hydroxyoctadecanoic acid; 12-Hydroxystearic acid; Octadecanoic acid, 12-hydroxy-
Tradenames: 12-HSA

2,2´-Iminobisethanol. *See Diethanolamine*

2,2´-Iminodiethanol. *See Diethanolamine*

Iodine
CAS 7553-56-2; EINECS 231-442-4
Tradenames containing: Akypogene Jod F

Isoarachidyloxypropylamine
Tradenames: Tomah PA-24

Isobutyl alcohol
CAS 78-83-1; EINECS 201-148-0
Synonyms: Isobutanol; Isopropylcarbinol; 2-Methyl-1-propanol
Tradenames containing: Nansa® EVM62/H; Nansa® EVM70; Nansa® EVM70/E

Isoceteth-20 (CTFA)
Synonyms: PEG-20 isocetyl ether; POE (20) isohexadecyl ether; PEG 1000 isocetyl ether
Tradenames: Arlasolve® 200

Isocetyl alcohol (CTFA)
CAS 36311-34-9; EINECS 252-964-9
Synonyms: Isohexadecanol; Isohexadecyl alcohol; Isopalmityl alcohol
Tradenames: Rilanit G 16

Isocetyl stearate (CTFA)
CAS 25339-09-7; EINECS 246-868-6
Tradenames: Afilan ICS

Isodeceth-3
Tradenames: Ethylan® CD103; Remcopal 273; Remcopal LO 2B

Isodeceth-4 (CTFA)
Synonyms: PEG-4 isodecyl ether; POE (4) isodecyl ether; PEG 200 isodecyl ether
Tradenames: Oxetal ID 104; Rhodasurf® DA-530 (formerly Emulphogene® DA-530)

Isodeceth-6 (CTFA)
Synonyms: PEG-6 isodecyl ether; POE (6) isodecyl ether; PEG 300 isodecyl ether
Tradenames: Rhodasurf® 860/P (formerly Soprophor® 860/P); Rhodasurf® DA-630 (formerly Emulphogene® DA-630)

Isodeceth-7
Tradenames: Ethylan® CD107

Isodeceth-9
Tradenames: Ethylan® CD109

Isodecyl oleate (CTFA)
CAS 59231-34-4; EINECS 261-673-6
Synonyms: 9-Octadecenoic acid, isodecyl ester
Tradenames: Mackester IDO

Isodecyloxypropylamine
CAS 7617-78-9
Tradenames: Tomah PA-14

Isodecyl oxypropyl amine acetate
Tradenames: Tomah PA-14 Acetate

N-Isodecyloxypropyl 1-1,3-diaminopropane
CAS 72162-46-0
Tradenames: Tomah DA-14

Isodecyloxypropyl dihydroxyethyl methyl ammonium chloride
CAS 125740-36-5
Tradenames: Tomah Q-14-2
Tradenames containing: Tomah Q-14-2PG

Isododecyloxypropylamine
Tradenames: Tomah PA-16

N-Isododecyloxypropyl-1,3-diaminopropane
Tradenames: Tomah DA-16

Isoeicosanol
Tradenames: Disponil G 200

Isolaureth-3 (CTFA)
CAS 60828-78-6
Synonyms: PEG-3 isolauryl ether; POE (3) isolauryl ether
Tradenames: Tergitol® TMN-3

Isolaureth-6 (CTFA)
CAS 60828-78-6
Synonyms: PEG-6 isolauryl ether; POE (6) isolauryl ether; PEG 300 isolauryl ether
Tradenames: Tergitol® TMN-6

Isolaureth-10 (CTFA)
CAS 60828-76-6
Synonyms: PEG-10 isolauryl ether; POE (10) isolauryl ether; PEG 500 isolauryl ether
Tradenames: Tergitol® TMN-10

Isononyloxypropylamine
CAS 29317-52-0
Tradenames: Tomah PA-13i

Isopropyl alcohol (CTFA)
CAS 67-63-0; EINECS 200-661-7
Synonyms: IPA; Isopropanol; 2-Propanol; Dimethyl carbinol
Tradenames containing: Adogen® 462; Adogen® 471; Aromox® C/12; Aromox® DM16; Aromox® DMCD; Aromox® T/12; Arquad® 2C-75; Arquad® 2HT-75; Arquad® 12-33; Arquad® 12-50; Arquad® 16-50; Arquad® 18-50; Arquad® 218-75; Arquad® B-100; Arquad® C-33; Arquad® C-50; Arquad® DMCB-80; Arquad® DMHTB-75; Arquad® M2HTB-80; Arquad® S-50; Arquad® T-50; Carsosoft® S-75; Dehyquart AU-36; Dehyquart AU-56; Duoquad® O-50; Duoquad® T-50; Ethoduoquad® T/15-50; Ethoquad® 18/12; Ethoquad® C/12; Ethoquad® C/12 Nitrate; Ethoquad® CB/12; Ethoquad® O/12; Gemtex PA-75; Gemtex SC-75; M-Quat® 257; M-Quat® 2475; Monawet MM-80; Propoquad® 2HT/11; Surfynol® 104PA; Tauranol ML; Tomah Q-2C; Tomah Q-17-2; Tomah Q-18-2; Tomah Q-24-2; Tomah Q-D-T; Tomah Q-S; Variquat® 50AC; Variquat® 80AC; Variquat® 638; Variquat® K300; Varisoft® 110; Varisoft® 137; Varisoft® 222 (75%); Varisoft® 222 (90%); Varisoft® 222 LM (90%); Varisoft® 222 LT (90%); Varisoft® 475; Varisoft® 3690

Isopropylamine dodecylbenzenesulfonate (CTFA)
CAS 26264-05-1; 68584-24-7; EINECS 247-556-2
Synonyms: Dodecylbenzenesulfonic acid, comp. with 2-propanamine (1:1)
Tradenames: Arylan® PWS; Calimulse PRS; Manro HCS; Nansa® YS94; Naxel AAS-Special 3; Pentine 1185 5432; Polystep® A-11; Rhodacal® 330 (formerly Siponate® 330); Rhodacal® IPAM (formerly Alkasurf® IPAM); Sandoz Sulfonate AAS-Special 3; Witconate P10-59

Isopropyl lanolate (CTFA)
CAS 63393-93-1; EINECS 264-119-1
Synonyms: Lanolin fatty acids, isopropyl esters
Tradenames: Amerlate® P; Emerest® 1723; Emerest® 11723

Isopropyl myristate (CTFA)
CAS 110-27-0; EINECS 203-751-4
Synonyms: IPM; 1-Methylethyl tetradecanoate; Tetradecanoic acid, 1-methylethyl ester
Tradenames: Emerest® 2314; Kessco® IPM

Isopropyl oleate sulfonate
Tradenames containing: Rexowet CR

Isopropyl palmitate (CTFA)
CAS 142-91-6; EINECS 205-571-1
Synonyms: IPP; Isopropyl n-hexadecanoate; Hexadecanoic acid, 1-methylethyl ester; 1-Methylethyl hexandecanoate
Tradenames: Emerest® 2316; Kessco® IPP

Isosorbide laurate (CTFA)
Tradenames: Arlamol® ISML

Isostearamide DEA (CTFA)
CAS 52794-79-3; EINECS 258-193-4
Synonyms: N,N-Bis(2-hydroxyethyl) isooctadecanamide
Tradenames: Mackamide ISA; Monamid® 150-IS; Standamid® ID

Isostearamido dimethylamine propionate
Tradenames: Mackam ISP

Isostearamidopropylamine oxide (CTFA)
Synonyms: N-[3-(Dimethylamino) propyl] isooctadecanamide, N-oxide; Amides, isostearic, N-[3-(dimethylamino)propyl], N-oxide
Tradenames: Incromine Oxide I; Mackamine IAO

Isostearamidopropyl betaine (CTFA)
CAS 6179-44-8; 63566-37-0; EINECS 228-227-2
Synonyms: N-(Carboxymethyl)-N,N-dimethyl-3-[(1-oxoisooctadecyl)amino]-1-propanaminium hydroxide, inner salt
Tradenames: Mackam ISA; Schercotaine IAB

Isostearamidopropyl dimethylamine (CTFA)
CAS 67799-04-6; EINECS 267-101-1
Synonyms: N-[3-(Dimethylamino)propyl]isooctadecanamide
Tradenames: Mackine 401

Isostearamidopropyl morpholine (CTFA)
Synonyms: Isooctadecanamide, N-[3-(4-morpholinyl) propyl]-; N-[3-(4-Morpholinyl) propyl]isooctadecanamide
Tradenames: Mackine 421

Isostearamidopropyl morpholine lactate (CTFA)
CAS 72300-24-4; 80145-09-1
Synonyms: Propanoic acid, 2-hydroxy-, compd. with N-[3-(4-morpholinyl) propyl] isooctadecanamide
Tradenames: Emcol® ISML

Isostearamidopropyl morpholine oxide (CTFA)
Synonyms: Amides, isostearic, N-[3-(4-morpholinyl) propyl], N-oxide; Isooctadecanamide, N-[3-(4-morpholinyl) propyl]-N-oxide; N-[3-(4-Morpholinyl) propyl] isooctadecanamide-N-oxide
Tradenames: Incromine Oxide ISMO; Mackamine ISMO

Isosteareth-2 (CTFA)
CAS 52292-17-8 (generic)
Synonyms: PEG-2 isostearyl ether; POE (2) isostearyl ether; PEG 100 isostearyl ether
Tradenames: Arosurf® 66-E2; Hetoxol IS-2

Isosteareth-10 (CTFA)
CAS 52292-17-8 (generic)
Synonyms: PEG-10 isostearyl ether; POE (10) isostearyl ether; PEG 500 isostearyl ether
Tradenames: Arosurf® 66-E10

Isosteareth-20 (CTFA)
CAS 52292-17-8 (generic)
Synonyms: PEG-20 isostearyl ether; POE (20) isostearyl ether; PEG 1000 isostearyl ether
Tradenames: Arosurf® 66-E20

Isostearic acid (CTFA)
CAS 2724-58-5; EINECS 220-336-3
Synonyms: Heptadecanoic acid, 16-methyl-; Isooctadecanoic acid; 16-Methylheptadecanoic acid
Tradenames: Emersol® 871; Emersol® 875; Priosorine 3501; Priosorine 3505
Tradenames containing: Akypo TFC-SN

Isostearoyl PG-trimonium chloride (CTFA)
Tradenames: Akypoquat 129

Isostearyl alcohol (CTFA)
CAS 27458-93-1; 70693-04-8; EINECS 248-470-8
Synonyms: 1-Heptadecanol, 16-methyl-; Isooctadecanol; 16-Methyl-1-heptadecanol
Tradenames: Adol® 66

Isostearyl benzoate (CTFA)
CAS 68411-27-8
Synonyms: Benzoic acid, isostearyl ester
Tradenames: Finsolv® SB

Isostearyl diglyceryl succinate (CTFA)
CAS 66085-00-5
Tradenames: Imwitor® 780 K

Isostearyl ethylimidonium ethosulfate (CTFA)
Synonyms: Quaternium-32
Tradenames: Monaquat ISIES

Isostearyl hydroxyethyl imidazoline (CTFA)
CAS 68966-38-1; EINECS 273-429-6
Synonyms: Isostearyl imidazoline; 4,5-Dihydro-2-isoheptadecyl-1H-imidazole-1-ethanol
Tradenames: Monazoline IS; Schercozoline I

Isotrideceth-6
Tradenames: Oxetal T 106

Isotrideceth-10
Tradenames: Oxetal T 110

Isotridecyloxypropylamine
Tradenames: Tomah PA-17

N-Isotridecyloxypropyl 1,3-diaminopropane
Tradenames: Tomah DA-17

Isotridecyloxypropyl dihydroxyethyl methyl ammonium chloride
Tradenames containing: Tomah Q-17-2; Tomah Q-17-2PG

Isoundeceth-3
Tradenames: Prox-onic UA-03

Isoundeceth-6
Tradenames: Prox-onic UA-06

Isoundeceth-9
Tradenames: Prox-onic UA-09

Isoundeceth-12
Tradenames: Prox-onic UA-012

Lactamide MEA (CTFA)
CAS 5422-34-4; EINECS 226-546-1
Synonyms: 2-Hydroxy-N-(2-hydroxyethyl) propanamide; Lactic acid monoethanolamide; Monoethanolamine lactic acid amide
Tradenames: Mackamide LME; Schercomid LME

Lactic acid (CTFA)
CAS 50-21-5; EINECS 200-018-0
Synonyms: 2-Hydroxypropanoic acid; 2-Hydroxypropionic acid; Milk acid
Tradenames: Patlac® LA

Laneth-5 (CTFA)
CAS 61791-20-6 (generic)
Synonyms: PEG-5 lanolin ether; POE (5) lanolin ether
Tradenames: Polychol 5
Tradenames containing: Solulan® 5

Laneth-10 (CTFA)
CAS 61791-20-6 (generic)
Synonyms: PEG-10 lanolin ether; POE (10) lanolin ether; PEG 500 lanolin ether
Tradenames: Polychol 10

Laneth-16 (CTFA)
CAS 61791-20-6 (generic)
Synonyms: PEG-16 lanolin ether; POE (16) lanolin ether
Tradenames containing: Ritachol® 3000; Solulan® 16

Laneth-20 (CTFA)
CAS 61791-20-6 (generic)
Synonyms: PEG-20 lanolin ether; POE (20) lanolin ether; PEG 1000 lanolin ether
Tradenames: Aqualose W20; Aqualose W20/50; Polychol 20

Laneth-40 (CTFA)
CAS 61791-20-6 (generic)
Synonyms: PEG-40 lanolin ether; POE (40) lanolin ether; PEG 2000 lanolin ether
Tradenames: Polychol 40

Lanolic acids. *See Lanolin acid*

Lanolin (CTFA)
CAS 8006-54-0 (anhyd.), 8020-84-6 (hyd.); EINECS 232-348-6
Synonyms: Anhydrous lanolin; Wool wax; Wool fat
Tradenames: Emery® 1650; Emery® 1656; Lanolin Alcohols LO; Lanolin Fatty Acids O; Paralan; White Swan; Yeoman
Tradenames containing: Amerchol® C; Aquaphil K; Nimcolan® 1747

Lanolin acid (CTFA)
CAS 68424-43-1; EINECS 270-302-7
Synonyms: Lanolic acids; Lanolin fatty acids; Acids, lanolin
Tradenames: Amerlate® LFA; Amerlate® WFA; Fancor LFA; Skliro Distilled

Lanolin alcohol (CTFA)
CAS 8027-33-6; EINECS 232-430-1
Synonyms: Alcohols, lanolin; Wool wax alcohol
Tradenames: Argowax Cosmetic Super; Argowax Dist; Argowax Standard; Ceralan®; Dusoran MD; Emery® 1780; Fancol LA; Nimco® 1780
Tradenames containing: Amerchol® C; Amerchol® CAB; Amerchol® L-101; Aquaphil K; Fancol LAO; Nimcolan® 1747

Lanolin, hydrogenated. *See Hydrogenated lanolin*

Lanolin, hydroxylated. *See Hydroxylated lanolin*

Lanolin oil (CTFA)
CAS 8038-43-5; 70321-63-0; 8006-54-0
Synonyms: Dewaxed lanolin; Oils, lanolin
Tradenames: Fluilan

Lanolin wax (CTFA)
CAS 68201-49-0; EINECS 269-220-4
Synonyms: Deoiled lanolin

Tradenames: Albalan; Lanfrax®; Lanfrax® 1776
Tradenames containing: Juniorlan 1664

Lapyrium chloride (CTFA)
CAS 6272-74-8; EINECS 228-464-1
Synonyms: 1-(2-Hydroxyethyl)carbamoyl methyl pyridinium chloride laurate; N-(Lauryl col-amino formyl methyl) pyridinium chloride
Tradenames: Emcol® E-607L

Lauralkonium bromide (CTFA)
CAS 7281-04-1; EINECS 230-698-4
Synonyms: Benzenemethanaminium, N-dodecyl-N,N-dimethyl-, bromide; N-Dodecyl-N,N-dimethyl benzenemethanaminium bromide; Lauryl dimethyl benzyl ammonium bromide
Tradenames: Amonyl BR 1244

Lauralkonium chloride (CTFA)
CAS 139-07-1; EINECS 205-351-5
Synonyms: Lauryl dimethyl benzyl ammonium chloride; N,N-Dimethyl-N-dodecylbenzene-methanaminium chloride
Tradenames: Catinal CB-50; Catinal MB-50A; Dehyquart LDB; Retarder N; Retarder N-85

Lauramide DEA (CTFA)
CAS 120-40-1; 52725-64-1; EINECS 204-393-1
Synonyms: Lauric diethanolamide; N,N-Bis(2-hydroxyethyl)dodecanamide; Diethanolamine lauric acid amide
Tradenames: Ablumide LDE; Alkamide® DL-203/S (formerly Cyclomide DL-203/S); Alkamide® DL-207/S (formerly Cyclomide DL-207/S); Alkamide® L9DE; Amidex L-9; Amidex LD; Aminol LM-30C, LM-30C Special; Carsamide® SAL-7; Carsamide® SAL-9; Cedemide AX; Chimipal LDA; Comperlan LD; Comperlan LMD; Crillon LDE; Deteryl AG; Emid® 6510; Emid® 6511; Emid® 6518; Emid® 6519; Empilan® LDE; Ethylan® MLD; Hartamide LDA; Hetamide LL; Hetamide ML; Hetamide MOC; Incromide L-90; Incromide LM-70; Incromide LR; Intermediate 512; Loropan LD; Mackamide L10; Mackamide L95; Mackamide LLM; Mackamide LMD; Manromid 1224; Mazamide® 1214; Mazamide® L-298; Mazamide® LM; Mazamide® LM 20; Monamid® 150-LMWC; Monamid® 150-LW; Monamid® 716; Ninol® 30-LL; Ninol® 55-LL; Ninol® 70-SL; Ninol® 96-SL; Ninol® AX; Nissan Stafoam DL; Nissan Stafoam L; Nitrene L-90; Rewomid® DLMS; Schercomid SL-Extra; Schercomid SL-ML, SL-ML-LC; Schercomid SLM-C; Schercomid SLM-LC; Schercomid SLM-S; Schercomid SLMC-75; Stafoam DL; Standamid® KDS; Standamid® LD; Standamid® LDO; Standamid® LDS; Synotol L 60; Synotol L 90; Synotol LM 60; Tohol N-230; Tohol N-230X; Unamide® J-56; Unamide® LDX; Varamide® ML-1; Varamide® ML-4; Witcamide® 5195; Witcamide® 6310; Witcamide® 6511
Tradenames containing: Alkamide® L7DE; Bio-Soft® LD-95; Emid® 6541; Empilan® LDX; Hetamide LML; Loropan LMD; Mirasheen® 202 (formerly Cyclosheen 202); Monamid® 1007; Monamine ACO-100; Monamine LM-100; Nitrene L-76; Rewomid® DL 203; Rewomid® DL 203 S; Schercomid 1214; Schercomid LD; Schercomid SLA; Standamid® LD 80/20; Synotol LM 90

Lauramide MEA (CTFA)
CAS 142-78-9; EINECS 205-560-1
Synonyms: Lauric monoethanolamide; N-(2-Hydroxyethyl)dodecanamide; Monoethanolamine lauric acid amide
Tradenames: Ablumide LME; Alkamide® L-203 (formerly Cyclomide L-203); Comperlan LM; Empilan® LME; Hartamide LMEA; Incromide LCL; Lauridit® LM; Loropan LM; Mackamide LMM; Manromid LMA; Monamid® LMA; Monamid® LMMA; Ninol® LMP; Rewomid® L 203; Standamid® LM; Varamide® L-203

Lauramide MIPA (CTFA)
CAS 142-54-1; EINECS 205-541-8
Synonyms: Lauric monoisopropanolamide; N-(2-Hydroxypropyl)dodecanamide; Monoisopropanolamine lauric acid amide
Tradenames: Comperlan LP; Empilan® LIS; Lauridit® LMI; Monamid® LIPA; Monamid® LMIPA; Rewomid® IPL 203; Standamid® LP
Tradenames containing: Standapol® S

Lauramidopropylamine oxide (CTFA)
CAS 61792-31-2; EINECS 263-218-7
Synonyms: Lauramidopropyl dimethylamine oxide; N-[3-(Dimethylamino) propyl]dodecanamide-N-oxide; Amides, lauric, N-[3-(dimethylamino)propyl], N-oxide
Tradenames: Mackamine LAO

Lauramidopropyl betaine (CTFA)
CAS 4292-10-8; 86438-78-0; EINECS 224-292-6
Synonyms: N-(Carboxymethyl)N,N-dimethyl-3-[(1-oxododecyl)amino]-1-propanaminium hydroxide, inner salt
Tradenames: Chemoxide L; Lexaine® LM; Mackam LMB; Mirataine® BB; Tego®-Betaine L-90
Tradenames containing: Lexate BPQ

Lauramidopropyl dimethylamine (CTFA)
CAS 3179-80-4; EINECS 221-661-3
Synonyms: N-[3-(Dimethylamino) propyl] dodecanamide; Dimethylaminopropyl lauramide
Tradenames: Mackine 801; Schercodine L

Lauramidopropyl dimethylamine oxide. *See Lauramidopropylamine oxide*

Lauramidopropyl dimethylamine propionate (CTFA)
Synonyms: Dimethylaminopropyl lauramide propionate
Tradenames: Mackam LAP

Lauramidopropyl PEG-dimonium chloride phosphate
CAS 83682-78-4
Tradenames: Phospholipid PTD

Lauramine (CTFA)
CAS 124-22-1; EINECS 204-690-6
Synonyms: Lauryl amine; 1-Dodecanamine; Dodecylamine
Tradenames: Amine 12-98D; Armeen® 12; Armeen® 12D; Nissan Amine BB; Radiamine 6163, 6164

Lauramine oxide (CTFA)
CAS 1643-20-5; 70592-80-2; EINECS 216-700-6
Synonyms: Lauryl dimethylamine oxide; N,N-Dimethyl-1-dodecanamine-N-oxide
Tradenames: Ablumox LO; Ammonyx® DMCD-40; Ammonyx® LO; Amyx LO 3594; Aro-

mox® DMMC-W; Chemoxide LM-30; Emcol® L; Emcol® LO; Empigen® OB; Incromine Oxide L; Incromine Oxide L-40; Karox LO; Laurelox 12; Mackamine LO; Mazox® LDA; Ninox® L; Oxamin LO; Rhodamox® LO (formerly Alkamox® LO, Cyclomox® L); Sandoz Amine Oxide XA-L; Schercamox DML; Textamine Oxide LMW; Varox® 365; Varox® 375

Tradenames containing: Standamox PL

Lauraminopropionic acid (CTFA)
CAS 1462-54-0; 3614-12-8; EINECS 215-968-1
Synonyms: N-Dodecyl-beta-alanine; N-Lauryl, myristyl beta-aminopropionic acid
Tradenames: Deriphat 151C; Deriphat 170C

Laurdimonium hydroxypropyl hydrolyzed wheat protein
Tradenames: Hydrotriticum QL

Laureth-1 (CTFA)
CAS 4536-30-5; EINECS 224-886-5
Synonyms: PEG-1 lauryl ether; Ethylene glycol monolauryl ether; 2-(Dodecyloxy)ethanol
Tradenames: Bio-Soft® E-200; Hetoxol L-1; Nikkol BL-1SY

Laureth-2 (CTFA)
CAS 3055-93-4 (generic); 9002-92-0; EINECS 221-279-7
Synonyms: Diethylene glycol dodecyl ether; PEG-2 lauryl ether; 2-[2-(Dodecyloxy)ethoxy] ethanol
Tradenames: Akyporox RLM 22; Bio-Soft® E-300; Dehydol LS 2; Empilan® KB 2; Hetoxol L-2; Marlipal® 24/20; Marlowet® LMA 2; Nikkol BL-2; Nikkol BL-2SY; Oxetal VD 20; Prox-onic LA-1/02; Rolfor N 24/2

Laureth-3 (CTFA)
CAS 3055-94-5; EINECS 221-280-2
Synonyms: Triethylene glycol dodecyl ether; 2-[2-[2-(Dodecyloxy)ethoxy]ethoxy]ethanol; PEG-3 lauryl ether
Tradenames: AE-1214/3; Ablunol LA-3; Dehydol LS 3; Empilan® KB 3; Empilan® KC 3; Ethal 326; Hetoxol L-3N; Marlipal® 24/30; Marlowet® LMA 3; Nikkol BL-3SY; Oxetal VD 28; Remcopal 121; Rewopal® LA 3; Rolfor N 24/3

Laureth-4 (CTFA)
CAS 5274-68-0; EINECS 226-097-1
Synonyms: PEG-4 lauryl ether; PEG 200 lauryl ether; 3,6,9,12-Tetraoxatetracosan-1-ol
Tradenames: Akyporox RLM 40; Ameroxol® LE-4; Chemal LA-4; Dehydol LS 4; Emthox® 5882; Ethal LA-4; Ethosperse® LA-4; Hetoxol L-4; Hetoxol L-4N; Lipocol L-4; Macol® LA-4; Marlipal® 24/40; Marlipal® 124; Marlowet® LA 4; Marlowet® LMA 4; Nikkol BL-4.2; Nikkol BL-4SY; Prox-onic LA-1/04; Remcopal 4; Rhodasurf® L-4 (formerly Siponic® L-4); Simulsol P4; Trycol® 5882; Trycol® LAL-4; Volpo L4
Tradenames containing: Marlowet® SAF/K

Laureth-5 (CTFA)
CAS 3055-95-6; EINECS 221-281-8
Synonyms: PEG-5 lauryl ether; POE (5) lauryl ether; 3,6,9,12,15-Pentaoxyheptacosan-1-ol
Tradenames: Ablunol LA-5; Marlipal® 24/50; Mulsifan RT 23; Nikkol BL-5SY

Laureth-6 (CTFA)
CAS 3055-96-7; EINECS 221-282-3
Synonyms: PEG-6 lauryl ether; POE (6) lauryl ether; 3,6,9,12,15,18-Hexaoxatriacontan-1-ol
Tradenames: Dehydol PID 6; Marlipal® 24/60; Nikkol BL-6SY; Rewopal® LA 6; Rewopal® LA 6-90

Laureth-7 (CTFA)
CAS 9002-92-0 (generic); 3055-97-8; EINECS 221-283-9
Synonyms: PEG-7 lauryl ether; POE (6) lauryl ether; 3,6,9,12,15,18,21-Heptaoxatritriacontan-1-ol
Tradenames: Ablunol LA-7; Bio-Soft® EC-600; Ethal LA-7; Incropol L-7; Macol® LA-790; Marlipal® 24/70; Marlowet® LA 7; Marlowet® LMA 7; Nikkol BL-7SY; Rhodasurf® B-7 (formerly Texafor B-7); Rhodasurf® L-790 (formerly Siponic® L-7-90)

Laureth-8 (CTFA)
CAS 9002-92-0 (generic); 3055-98-9
Synonyms: PEG-8 lauryl ether; POE (8) lauryl ether; 3,6,9,12,15,18,21,24-Octaoxahexatriacontan-1-ol
Tradenames: Akyporox RLM 80; Akyporox RLM 80V; Marlipal® 24/80; Nikkol BL-8SY; Trycol® 5963; Trycol® LAL-8

Laureth-9 (CTFA)
CAS 9002-92-0 (generic); 3055-99-0; 68439-50-9; EINECS 221-284-4
Synonyms: PEG-9 lauryl ether; POE (9) lauryl ether; 3,6,9,12,15,18,21,24,27-Nonaoxanonatriacontan-1-ol
Tradenames: Ablunol LA-9; Bio-Soft® E-670; Carsonon® L-985; Chemal LA-9; Ethal 926; Hetoxol L-9; Hetoxol L-9N; Macol® LA-9; Marlipal® 24/90; Marlipal® 129; Nikkol BL-9EX, -9EX(FF); Prox-onic LA-1/09; Remcopal 258; Remcopal L9; Remcopal LC; Remcopal LP; Rolfor Z 24/9; Sellig LA 9 100
Tradenames containing: Hetoxol LS-9

Laureth-10 (CTFA)
CAS 9002-92-0 (generic); 6540-99-4
Synonyms: PEG-10 lauryl ether; POE (10) lauryl ether; 3,6,9,12,15,18,21,24,27,30-Dexaoxadotetracontan-1-ol
Tradenames: Dehydol 100; Marlipal® 24/100; Marlowet® LMA 10; Remcopal L12; Rewopal® LA 10; Rewopal® LA 10-80

Laureth-11 (CTFA)
CAS 9002-92-0 (generic)
Synonyms: PEG-11 lauryl ether; POE (11) lauryl ether
Tradenames: Marlipal® 24/110; Remcopal 21411

Laureth-12 (CTFA)
CAS 9002-92-0 (generic); 3056-00-6; EINECS 221-286-5
Synonyms: PEG-12 lauryl ether; POE (12) lauryl ether; PEG 600 lauryl ether
Tradenames: Ablunol LA-12; Chemal LA-12; Ethosperse® LA-12; Lipocol L-12; Macol® LA-12; Marlipal® 24/120; Marlowet® CA 12; Prox-onic LA-1/012; Remcopal 21912 AL; Siponic® L-12 (see Rhodasurf® LA-12); Trycol® 5967; Trycol® LAL-12
Tradenames containing: Comperlan LS

Laureth-14 (CTFA)
CAS 9002-92-0 (generic)
Synonyms: PEG-14 lauryl ether; POE (14) lauryl ether
Tradenames: Marlipal® 24/140

Laureth-15 (CTFA)
CAS 9002-92-0 (generic)
Synonyms: PEG-15 lauryl ether; POE (15) lauryl ether
Tradenames: Marlipal® 24/150

Laureth-16 (CTFA)
CAS 9002-92-0 (generic)
Synonyms: PEG-16 lauryl ether; POE (16) lauryl ether
Tradenames: Ablunol LA-16; Akyporox RLM 160; Rhodasurf® B-1 (formerly Texafor B-1)

Laureth-20 (CTFA)
CAS 9002-92-0 (generic)
Synonyms: PEG-20 lauryl ether; POE (20) lauryl ether
Tradenames: Marlipal® 24/200; Marlowet® LMA 20; Remcopal 20; Sellig LA 1150; Sellig LA 11 100

Laureth-21
Synonyms: PEG-21 lauryl ether; POE (21) lauryl ether
Tradenames: Nikkol BL-21

Laureth-23 (CTFA)
CAS 9002-92-0 (generic)
Synonyms: PEG-23 lauryl ether; POE (23) lauryl ether
Tradenames: Ameroxol® LE-23; Bio-Soft® E-840; Brij® 35; Brij® 35 Liq; Brij® 35 SP; Brij® 35 SP Liq; Chemal LA-23; Ethosperse® LA-23; Hetoxol L-23; Lipocol L-23; Macol® LA-23; Prox-onic LA-1/023; Rhodasurf® L-25 (formerly Siponic® L-25); Rhodasurf® LAN-23-75% (formerly Alkasurf® LAN-23); Rolfor Z 24/23; Trycol® 5964; Trycol® LAL-23; Volpo L23

Laureth-25 (CTFA)
CAS 9002-92-0 (generic)
Synonyms: PEG-25 lauryl ether; POE (25) lauryl ether
Tradenames: Nikkol BL-25

Laureth-30 (CTFA)
CAS 9002-92-0 (generic)
Synonyms: PEG-30 lauryl ether; POE (30) lauryl ether
Tradenames: Marlipal® 24/300

Laureth-40 (CTFA)
CAS 9002-92-0 (generic)
Synonyms: PEG-40 lauryl ether; POE (40) lauryl ether
Tradenames: Ablunol LA-40

Laureth-3 carboxylic acid (CTFA)
Synonyms: PEG-3 lauryl ether carboxylic acid
Tradenames: Rewopol® CL 30

Laureth-4 carboxylic acid (CTFA)
CAS 68954-89-2
Synonyms: PEG-4 lauryl ether carboxylic acid
Tradenames: Akypo RLM 25

Laureth-5 carboxylic acid (CTFA)
CAS 68954-89-2
Synonyms: PEG-5 lauryl ether carboxylic acid; POE (5) lauryl ether carboxylic acid
Tradenames: Akypo 1690 S; Akypo RLM 38; Akypo RLMQ 38; Marlinat® CM 40; Sandopan® LA-8
Tradenames containing: Akypo TFC-S; Akypo TFC-SN

Laureth-6 carboxylic acid (CTFA)
Synonyms: PEG-6 lauryl ether carboxylic acid; POE (6) lauryl ether carboxylic acid; PEG 300 lauryl ether carboxylic acid
Tradenames: Akypo RLM 45

Laureth-11 carboxylic acid (CTFA)
Synonyms: PEG-11 lauryl ether carboxylic acid; POE (11) lauryl ether carboxylic acid
Tradenames: Akypo RLM 100; Marlinat® CM 100

Laureth-14 carboxylic acid (CTFA)
Synonyms: PEG-14 lauryl ether carboxylic acid; POE (14) lauryl ether carboxylic acid
Tradenames: Akypo RLM 130

Laureth-17 carboxylic acid (CTFA)
CAS 27306-90-7; 68954-89-2
Synonyms: PEG-17 lauryl ether carboxylic acid; POE (17) lauryl ether carboxylic acid
Tradenames: Akypo RLM 160

Laureth phosphate (CTFA)
Tradenames: Surfagene FDD 402

Laureth-3 phosphate (CTFA)
CAS 39464-66-9 (generic); 25852-45-3
Synonyms: PEG-3 lauryl ether phosphate; Poly (oxy-1,2-ethanediyl) α-phosphono-ω-(dodecyloxy)-; POE (3) lauryl ether phosphate
Tradenames: Marlophor® MO 3-Acid; Rewophat EAK 8190

Laureth-4 phosphate (CTFA)
CAS 39464-66-9 (generic)
Synonyms: PEG-4 lauryl ether phosphate; PEG 200 lauryl ether phosphate; POE (4) lauryl ether phosphate
Tradenames: Rhodafac® RD-510 (formerly Gafac® RD-510)

Laureth-7 phosphate (CTFA)
CAS 39464-66-9 (generic)
Synonyms: PEG-7 lauryl ether phosphate
Tradenames: Akypomine® MW 05

Lauric acid (CTFA)
CAS 143-07-7; EINECS 205-582-1
Synonyms: n-Dodecanoic acid; Dodecanoic acid; Dodecoic acid
Tradenames: C-1298; Emery® 650; Emery® 651; Emery® 652; Hetamide LA; Hystrene® 9512; Hystrene® 9912; Philacid 1200; Prifrac 2920; Prifrac 2922
Tradenames containing: Philacid 1214

Lauric diethanolamide. *See Lauramide DEA*

Lauric monoethanolamide. *See Lauramide MEA*

Lauric monoisopropanolamide. *See Lauramide MIPA*

Lauric/myristic dimethylethyl ammonium ethosulfate
Tradenames containing: Larostat® 377 DPG

Lauric/myristic MEA
Tradenames containing: Alpha-Step® ML-A

Lauroamphoacetate. *See Sodium lauroamphoacetate*

Lauroamphocarboxyglycinate. *See Disodium lauroamphodiacetate*

Lauroamphocarboxypropionate. *See Disodium lauroamphodipropionate*

Lauroamphodiacetate. *See Disodium lauroamphodiacetate*

Lauroamphodipropionate. *See Disodium lauroamphodipropionate*

Lauroamphoglycinate. *See Sodium lauroamphoacetate*

Lauroamphohydroxypropylsulfonate. *See Sodium lauroamphohydroxypropylsulfonate*

Lauroamphopropylsulfonate. *See Sodium lauroamphohydroxypropylsulfonate*

Lauroyl lysine (CTFA)
CAS 52315-75-0; EINECS 257-843-4
Synonyms: Lauroyl-1-lysine; Lauroyl-L-lysine
Tradenames: Amihope LL-11

Lauroyl sarcosine (CTFA)
CAS 97-78-9; EINECS 202-608-3
Synonyms: N-Methyl-N-(1-oxododecyl) glycine
Tradenames: Crodasinic L; Hamposyl® L; Sarkosyl® L; Vanseal LS

Laurtrimonium chloride (CTFA)
CAS 112-00-5; EINECS 203-927-0
Synonyms: Lauryl trimethyl ammonium chloride; Dodecyl trimethyl ammonium chloride; N,N,N-Trimethyl-1-dodecanaminium chloride
Tradenames: Arquad® 12-37W; Chemquat 12-33; Chemquat 12-50; Dehyquart LT; Empigen® 5089; Octosol 562; Octosol 571
Tradenames containing: Arquad® 12-33; Arquad® 12-50

Lauryl alcohol (CTFA)
CAS 112-53-8; 68526-86-3; EINECS 203-982-0
Synonyms: 1-Dodecanol; C12 linear primary alcohol; Dodecyl alcohol
Tradenames: Alfol® 12; Cachalot® L-50; Cachalot® L-90; Emery® 3326; Emery® 3327; Emery® 3328; Emery® 3331; Emery® 3332; Emery® 3335 Chemically Pure; Emery® 3346; Emery® 3351; Emery® 3352; Emery® 3353; Emery® 3357; Epal® 12; Kalcohl 20; Laurex® L1; Laurex® NC; Lorol C8-C10; Lorol C8-C10 Special; Lorol C8-C18; Lorol C10-C12; Lorol C12; Lorol C12-99; Lorol C12 Chemically Pure; Lorol C12-C14; Lorol C12-C16; Lorol C12-C18; Lorol Lauryl Alcohol Tech; Lorol Special; Philcohol 1200
Tradenames containing: Epal® 12/70; Epal® 12/85; Epal® 1012; Epal® 1214; Epal® 1218; Epal® 1412

Lauryl amine. *See Lauramine*

Lauryl betaine (CTFA)
CAS 683-10-3; 11140-78-6; 66455-29-6; EINECS 211-669-5
Synonyms: 1-Dodecanaminium, N-(carboxymethyl)-N,N-dimethyl-, hydroxide, inner salt; Lauryl dimethyl glycine; Lauryl dimethylamine betaine
Tradenames: Chimin BX; Empigen® BB; Rewoteric AM DML; Swanol AM-301; Varion® CDG; Varion® CDG-LS; Zohartaine AB

Lauryl dimethylamine. *See Dimethyl lauramine*

Lauryl dimethylamine oxide. *See Lauramine oxide*

Lauryl dimethyl benzyl ammonium chloride. *See Lauralkonium chloride*

Lauryl dimethyl glycine. *See Lauryl betaine*

Lauryl hydroxyethyl imidazoline (CTFA)
CAS 136-99-2; EINECS 205-271-0
Synonyms: Lauryl imidazoline; 1H-Imidazole-1-ethanol, 4,5-dihydro-2-undecyl-
Tradenames: Crapol AV-11; Mackazoline L; Schercozoline L

Laurylhydroxysulfobetaine. *See Lauryl hydroxysultaine*

Lauryl hydroxysultaine (CTFA)
CAS 13197-76-7; EINECS 236-164-7
Synonyms: Laurylhydroxysulfobetaine
Tradenames: Rewoteric AM HC; Varion® HC

Lauryl imidazoline. *See Lauryl hydroxyethyl imidazoline*

Lauryl/myristyl alcohol
Tradenames: Kalcohl 5-24, 6-24, 7-24

Lauryl/myristyl dimethylamine oxide
Tradenames: Bio-Surf PBC-460; Standamox LAO-30; Standamox LMW-30

Laurylpyridinium bisulfate
CAS 17342-21-1
Tradenames: Dehyquart D

Laurylpyridinium chloride (CTFA)
CAS 104-74-5; EINECS 203-232-2
Synonyms: 1-Dodecylpyridinium chloride
Tradenames: Charlab LPC; Dehyquart C; Dehyquart C Crystals

Lauryl trimethyl ammonium chloride. *See Laurtrimonium chloride*

Lecithin (CTFA)
CAS 8002-43-5; EINECS 232-307-2
Tradenames: Alcolec® 439-C; Alcolec® 440-WD; Alcolec® 495; Alcolec® BS; Alcolec® Extra A; Alcolec® F-100; Alcolec® FF-100; Alcolec® Granules; Alcolec® S; Asol; Blendmax Series; Centrol® 2FSB, 2FUB, 3FSB, 3FUB; Centrol® CA; Centromix® CPS; Centromix® E; Centrophase® 31; Centrophase® 152; Centrophase® C; Centrophase® HR2B, HR2U; Centrophase® HR4B, HR4U; Centrophase® HR6B; Centrophase® NV; Centrophil® K; Centrophil® M, W; Clearate B-60; Clearate LV; Clearate Special Extra; Clearate WD; Lecithin W.D; Lexin K; Lipotin 100, 100J, SB; M-C-Thin; M-C-Thin 45; Phospholipon 50G; Topcithin

Lecithin, hydroxylated. *See Hydroxylated lecithin*

Linoleamide DEA (CTFA)
CAS 56863-02-6; EINECS 260-410-2
Synonyms: Linoleic diethanolamide; N,N-Bis(2-hydroxyethyl)-9,12-octadecadienamide; Diethanolamine linoleic acid amide
Tradenames: Alkamide® DIN295 (formerly Cyclomide DIN295); Alkamide® DIN295/S (formerly Cyclomide DIN295/S); Hetamide LN; Hetamide LNO; Incromide LA; Mackamide LOL; Mazamide® LLD; Mazamide® SS 10; Mazamide® SS 20; Monamid® 15-70W; Monamid® 150-ADY; Nitrene 100 SD; Purton SFD; Schercomid SLE; Standamid® SOD; Standamid® SOMD
Tradenames containing: Hetamide LML; Mona-

mid® 1007; Monamine AC-100; Monamine ADY-100; Monamine CD-100; Monamine R8-26

Linoleic acid (CTFA)
CAS 60-33-3; EINECS 200-470-9
Synonyms: 9,12-Octadecadienoic acid; Linolic acid
Tradenames: Emersol® 315

Linoleic diethanolamide. *See Linoleamide DEA*

Linolenic acid (CTFA)
CAS 463-40-1; EINECS 207-334-8
Synonyms: 9,12,15-Octadecatrienoic acid; alpha-Linolenic acid
Tradenames: Industrene® 120; Industrene® 130

Linoleyl alcohol
Tradenames containing: Emery® 3318; HD-Ocenol 110/130

Lithium cocoyl glutamate
Tradenames: Acylglutamate DL-12

Lithium lauryl sulfate
CAS 2044-56-6
Tradenames: Texapon LLS

Magnesium dodecylbenzene sulfonate
Tradenames: Nansa® MS45

Magnesium laureth-11 carboyxlate (CTFA)
CAS 99330-44-6
Synonyms: PEG-11 lauryl ether carboxylic acid, magnesium salt; Magnesium PEG (11) lauryl ether carboxylate
Tradenames: Akypo®-Soft 100 MgV

Magnesium lauryl sulfate (CTFA)
CAS 3097-08-3; 68081-97-0; EINECS 221-450-6
Synonyms: Magnesium monododecyl sulfate; Sulfuric acid, monododecyl ester, magnesium salt
Tradenames: Carsonol® MLS; Montovol GL-13; Norfox® MLS; Rhodapon® LM (formerly Sipon® LM); Stepanol® MG; Sulfetal MG 30; Sulfochem MG; Surfax MG; Texapon MGLS

Magnesium sulfonate
CAS 71786-47-5
Tradenames: Hybase® M-300; Hybase® M-400
Tradenames containing: Hybase® M-12

Magnesium xylene sulfonate
Tradenames: Eltesol® MGX

Mannide oleate
Tradenames: Arlacel® A

MEA-dodecylbenzene sulfonate
Tradenames: Rueterg 97-G

MEA-laureth-6 carboxylate (CTFA)
Synonyms: PEG-6 lauryl ether carboxylic acid, monoethanolamine salt
Tradenames containing: Akypogene WSW-W

MEA laureth phosphate
Tradenames: Crafol AP-262

MEA-lauryl sulfate (CTFA)
CAS 4722-98-9; 68908-44-1; EINECS 225-214-3
Synonyms: Sulfuric acid, monododecyl ester, compd. with 2-aminoethanol (1:1); Monoethanolamine lauryl sulfate
Tradenames: Cosmopon MO; Elfan® 240M; Empicol® LQ33/T; Rewopol® MLS 30; Rewopol® MLS 35

MEA-PPG-6-laureth-6-carboxylate (CTFA)
Synonyms: PEG-7-PPG-6 lauryl ether carboxylic acid, monoethanolamine salt
Tradenames containing: Akypogene SO

Melamine/formaldehyde resin (CTFA)
CAS 9003-08-1]
Synonyms: Melamine resin; 1,3,5-Triazine,2,4,6-triamine, polymer with formaldehyde
Tradenames: Glazamine DP2

Melamine resin. *See Melamine/formaldehyde resin*

Menhaden oil, hydrogenated. *See Hydrogenated menhaden oil*

Meroxapol 105 (CTFA)
CAS 9003-11-6 (generic)
Tradenames: Macol® 15; Pluronic® 10R5

Meroxapol 108 (CTFA)
CAS 9003-11-6 (generic)
Tradenames: Macol® 16; Pluronic® 10R8

Meroxapol 171 (CTFA)
CAS 9003-11-6 (generic)
Tradenames: Hodag Nonionic 1017-R; Macol® 18; Pluronic® 17R1

Meroxapol 172 (CTFA)
CAS 9003-11-6 (generic)
Tradenames: Hodag Nonionic 2017-R; Macol® 19; Pluronic® 17R2

Meroxapol 174 (CTFA)
CAS 9003-11-6 (generic)
Tradenames: Pluronic® 17R4

Meroxapol 178 (CTFA)
CAS 9003-11-6 (generic)
Tradenames: Pluronic® 17R8

Meroxapol 251 (CTFA)
CAS 9003-11-6 (generic)
Tradenames: Macol® 32; Pluronic® 25R1

Meroxapol 252 (CTFA)
CAS 9003-11-6 (generic)
Tradenames: Macol® 40; Pluronic® 25R2

Meroxapol 254 (CTFA)
CAS 9003-11-6 (generic)
Tradenames: Macol® 34; Pluronic® 25R4

Meroxapol 255 (CTFA)
CAS 9003-11-6 (generic)
Tradenames: Pluronic® 25R5

Meroxapol 258 (CTFA)
CAS 9003-11-6 (generic)
Tradenames: Pluronic® 25R8

Meroxapol 311 (CTFA)
CAS 9003-11-6 (generic)
Tradenames: Macol® 33; Pluronic® 31R1

Meroxapol 312 (CTFA)
CAS 9003-11-6 (generic)
Tradenames: Pluronic® 31R2

Meroxapol 314 (CTFA)
CAS 9003-11-6 (generic)
Tradenames: Pluronic® 31R4

Methacrylate copolymer
Synonyms: Methacrylic acid copolymer
Tradenames containing: Empicryl® 6059; Empicryl® 6070

Methoxyethoxypropylamine
Tradenames: Surfam P-MEPA

Methoxy PEG 400 monolaurate
Tradenames: Ethox 2672

Methoxy PEG-400 monooleate
Tradenames: Emery® 6779

Methoxypropylamine
CAS 5332-73-0
Synonyms: 3-MPA
Tradenames: Surfam P5 Dist; Surfam P5 Tech

Methylated spirits
Tradenames containing: Manoxol OT 60%

Methyl behenate
CAS 929-77-1
Synonyms: Methyl docosanoate
Tradenames: Kemester® 9022

Methyl bis (hydrogenated tallow amido ethyl) 2-hydroxyethyl ammonium chloride
Tradenames: Incrosoft 100; Incrosoft 100P

Methyl bis (hydrogenated tallow amidoethyl) 2-hydroxyethyl ammonium methosulfate
Tradenames: Accosoft 440-75

Methyl bis (oleylamidoethyl) 2-hydroxyethyl ammonium methosulfate
Tradenames: Accosoft 750
Tradenames containing: Varisoft® 222 LT (90%)

Methyl bis (tallowamidoethyl) 2-hydroxyethyl ammonium methosulfate
Tradenames: Accosoft 540 HC; Accosoft 550-90 HHV; Accosoft 550L-90
Tradenames containing: Varisoft® 222 (75%); Varisoft® 222 (90%); Varisoft® 222 LM (90%)

Methyl bis (tallowamidoethyl) 2-hydroxypropyl ammonium methosulfate
Tradenames: Accosoft 620-90; Varisoft® 238 (90%)

Methyl caprylate/caprate (CTFA)
CAS 67762-39-4
Tradenames: Emery® 2209

Methyl cocoate (CTFA)
CAS 61788-59-8; EINECS 262-988-1
Synonyms: Fatty acids, coco, methyl esters
Tradenames: Emery® 2253; Emery® 2254

Methyl dihydroxyethyl isoarachidaloxypropyl ammonium chloride
Tradenames containing: Tomah Q-24-2

Methyl distearamine
Tradenames: Amine M218

Methyl eicosenate
CAS 1120-28-1
Synonyms: Methyl arachidate
Tradenames: Kemester® 2050

Methylenebisdiamyl phenoxy polyethoxy ethanol
Tradenames: Triton® X-155-90%

Methylenebis (thiocyanate)
CAS 6317-18-6
Synonyms: MBT
Tradenames: Tolcide MBT

Methyl gluceth-10 (CTFA)
CAS 68239-42-9
Synonyms: PEG-10 methyl glucose ether; POE (10) methyl glucose ether
Tradenames: Glucam® E-10

Methyl gluceth-20 (CTFA)
CAS 68239-43-0
Synonyms: PEG-20 methyl glucose ether; POE (20) methyl glucose ether
Tradenames: Glucam® E-20

Methyl glucose dioleate (CTFA)
CAS 82933-91-3
Tradenames: Glucate® DO

Methyl glucose sesquistearate (CTFA)
CAS 68936-95-8; EINECS 273-049-0
Synonyms: D-Glucopyranoside, methyl, octadecanoate (2:3)
Tradenames: Glucate® SS

Methyl glycol. *See Propylene glycol*

Methyl (1) hydrogenated tallow amidoethyl (2) hydrogenated tallow imidazolinium methosulfate
Tradenames: Accosoft 808HT; Varisoft® 445

Methyl hydroxyethylcellulose (CTFA)
CAS 9032-42-2
Tradenames: Tylose MH Grades; Tylose MHB; Tylose MH-K, MH-xp, MHB-y, MHB-yp

Methyl hydroxystearate (CTFA)
CAS 141-23-1; EINECS 205-471-8
Synonyms: Methyl 12-hydroxyoctadecanoate; Methyl 12-hydroxystearate; 12-Hydroxyoctadecanoic acid, methyl ester
Tradenames: Naturechem® MHS

Methyl lardate
Tradenames: Base ML; Emery® Methyl Lardate; Estrasan 4L

Methyl laurate (CTFA)
CAS 111-82-0; 67762-40-7; EINECS 203-911-3
Synonyms: Methyl dodecanoate; Dodecanoic acid, methyl ester
Tradenames: Emery® 2270; Emery® 2290; Emery® 2296; Stepan C-40
Tradenames containing: Degreez

Methyl myristate (CTFA)
CAS 124-10-7; EINECS 204-680-1
Synonyms: Methyl tetradecanoate; Tetradecanoic acid, methyl ester
Tradenames: Emery® 2214
Tradenames containing: Degreez

Methyl naphthalene
Tradenames: Abluton CMN; Matexil CA MN

Methyl oleate (CTFA)
CAS 112-62-9; EINECS 203-992-5
Synonyms: Methyl 9-octadecenoate; 9-Octadecenoic acid, methyl ester
Tradenames: Emerest® 2301; Emery® 2219; Emery® Methyl Oleate; Kemester® 104; Kemester® 115; Kemester® 205; Witconol 2301
Tradenames containing: Stepan C-68; Degreez

Methyl oleate/linoleate
Tradenames: Kemester® 213

Methyl oleylamidoethyl oleyl imidazolinium methosulfate
Tradenames: Varisoft® 3690N (90%)
Tradenames containing: Varisoft® 3690

Methyl palmitate (CTFA)
CAS 112-39-0; EINECS 203-966-3
Synonyms: Methyl hexadecanoate; Hexadecanoic acid, methyl ester
Tradenames: Emery® 2216
Tradenames containing: Degreez

Methyl palmitate oleate
Tradenames: Petrosan 102; Stepan C-65

Methyl palm kernelate
Tradenames: Emery® 2255

2-Methyl-2,4-pentanediol. *See Hexylene glycol*

Methyl polysiloxane
Tradenames: Tego® Hammer 300000; Tego® Wet ZFS 454

1-Methyl-2-pyrrolidinone
CAS 872-50-4
Tradenames: Partsprep Degreaser

Methyl ricinoleate (CTFA)
CAS 141-24-2; EINECS 205-472-3
Synonyms: 12-Hydroxy-9-octadecenoic acid, methyl ester; Methyl 12-hydroxy-9-octadecenoate; castor oil acid, methyl ester
Tradenames: Estrasan 1

Methyl soyate
Tradenames: Kemester® 226

Methyl stearate (CTFA)
CAS 112-61-8; EINECS 203-990-4
Synonyms: Methyl octadecanoate; Octadecanoic acid, methyl ester
Tradenames: Emery® 2218; Kemester® 4516
Tradenames containing: Degreez; Stepan C-68

Methyl tallowate
Tradenames: Base MT; Emery® 2203; Estrasan 3; Kemester® 143

Methyl tri (C8-10) ammonium chloride
Tradenames: Adogen® 464

Mineral oil (CTFA)
CAS 8012-95-1; 8042-47-5; EINECS 232-384-2
Synonyms: Heavy or light mineral oil; Paraffin oil; Liquid paraffin
Tradenames: LAN-401; Prosol CT30; Tego® Foamex KS 6
Tradenames containing: Aerosol® OT-MSO; Amerchol® L-101; Fancol LAO; Lankropol® KO; Tego® Foamex KS 10

Mineral spirits (CTFA)
CAS 8032-32-4; 64475-85-0; EINECS 232-453-7
Synonyms: White spirits; Ligroin; Petroleum spirits
Tradenames containing: Mackanate DOS-70MS; Monawet MO-70S

Minkamide DEA (CTFA)
Synonyms: N,N-Bis(2-hydroxyethyl)mink fatty acid amide; Diethanolamine mink fatty acid condensate; Mink fatty acid diethanolamide
Tradenames: Incromide Mink D

Minkamidopropylamine oxide (CTFA)
Synonyms: n-[3-(Dimethylamino)propyl[mink amides-N-oxide
Tradenames: Incromine Oxide Mink

MIPA C12-15 pareth sulfate (CTFA)
Tradenames: Zetesol 856 D

MIPA-dodecylbenzenesulfonate (CTFA)
CAS 42504-46-1; 54590-52-2; EINECS 255-854-9
Synonyms: Dodecylbenzenesulfonic acid, compd. with 1-amino-2-propanol (1:1); Monoisopropanolamine dodecylbenzenesulfonate
Tradenames: Hetsulf IPA

MIPA-laureth sulfate (CTFA)
CAS 83016-76-6; 9062-04-8
Synonyms: Poly(oxy-1,2-ethanediyl), α-sulfo-ω-(dodecyloxy)-, compd. with 1-amino-2-propanol; Monoisopropanolamine lauryl ether sulfate
Tradenames: Montosol IL-13; Zetesol 856; Zetesol 2056
Tradenames containing: Zetesol 856 T

MIPA-lauryl sulfate (CTFA)
CAS 21142-28-9; EINECS 244-238-5
Synonyms: Sulfuric acid, monododecyl ester, compd. with 1-amino-2-propanol (1:1); Monoisopropanolamine lauryl sulfate; Dodecyl sulfate, compd. with 1-amino-2-propanol (1:1)
Tradenames: Sulfetal CJOT 38

Mixed isopropanolamines lauryl sulfate (CTFA)
CAS 68877-25-8
Tradenames: Carsonol® ILS

Monoethanolamine lauric acid amide. *See Lauramide MEA*

Monomyristin. *See Glyceryl myristate*

Monoolein. *See Glyceryl oleate*

Monoricinolein. *See Glyceryl ricinoleate*

Monostearin. *See Glyceryl stearate*

Montan acid wax (CTFA)
CAS 68476-03-9; EINECS 270-664-6
Synonyms: Fatty acids, montan wax; Waxes, montan fatty acids
Tradenames: BASF Wax LS

Myreth-3 (CTFA)
CAS 27306-79-2 (generic)
Synonyms: PEG-3 myristyl ether; POE (3) myristyl ether
Tradenames: Hetoxol M-3

Myristalkonium chloride (CTFA)
CAS 139-08-2; EINECS 205-352-0
Synonyms: N,N-Dimethyl-N-tetradecylbenzenemethanaminium chloride; Myristyl dimethyl benzyl ammonium chloride; Tetradecyl dimethyl benzyl ammonium chloride
Tradenames: Arquad® DM14B-90; BTC® 824; BTC® 824 P100; BTC® 2565; BTC® 2568; BTC® 8248; BTC® 8249; Catigene® DC 100; Exameen 824 3724; Exameen 8248 3729; FMB 65-15 Quat, 65-28 Quat; FMB 451-8 Quat; FMB 4500-5 Quat; JAQ Powdered Quat; Nissan Cation M2-100
Tradenames containing: BTC® 2125, 2125-80, 2125 P-40; BTC® 2125M; Exameen 2125 M 80 3709; Exameen 2125 M 3704; FMB 504-5 Quat; FMB 1210-5 Quat, 1210-8 Quat; FMB 3328-5 Quat, 3328-8 Quat

Myristamide DEA (CTFA)
CAS 7545-23-5; EINECS 231-426-7
Synonyms: Myristic diethanolamide; N,N-Bis(2-hydroxyethyl)myristamide; N,N-Bis(2-hydroxyethyl) tetradecanamide
Tradenames: Hetamide M; Monamid® 150-MW
Tradenames containing: Alkamide® L7DE; Loropan LMD; Nitrene L-76; Schercomid SLA; Standamid® LD 80/20; Synotol LM 90

Myristamide MEA (CTFA)
CAS 142-58-5; EINECS 205-546-5
Synonyms: Myristic monoethanolamide; N-(2-Hydroxyethyl)tetradecanamide; Myristoyl monoethanolamide
Tradenames: Schercomid MME

Myristamine
CAS 2016-42-4
Synonyms: Myristyl amine; Tetradecylamine
Tradenames: Amine 14D; Nissan Amine MB

Myristamine oxide (CTFA)
CAS 3332-27-2; EINECS 222-059-3

Synonyms: Myristyl dimethyl amine oxide; Tetradecyl dimethyl amine oxide; N,N-Dimethyl-1-tetradecanamine-N-oxide
Tradenames: Admox® 14-85; Ammonyx® MCO; Ammonyx® MO; Aromox® DM14D-W; Barlox® 14; Emcol® M; Empigen® OH25; Incromine Oxide M; Lilaminox M4; Mazox® MDA; Ninox® M; Sandoz Amine Oxide XA-M; Schercamox DMA; Schercamox DMM
Tradenames containing: Incromine Oxide MC; Standamox PL

Myristic acid (CTFA)
CAS 544-63-8; EINECS 208-875-2
Synonyms: Tetradecanoic acid
Tradenames: C-1495; Emery® 654; Emery® 655; Hystrene® 9014; Hystrene® 9514; Philacid 1400; Prifrac 2940; Prifrac 2942
Tradenames containing: Philacid 1214

Myristic diethanolamide. *See Myristamide DEA*

Myristic monoethanolamide. *See Myristamide MEA*

Myristoyl sarcosine (CTFA)
CAS 52558-73-3; EINECS 258-007-1
Synonyms: N-Methyl-N-(1-oxotetradecyl)glycine; Myristoyl N-methylglycine
Tradenames: Hamposyl® M

Myristyl alcohol (CTFA)
CAS 112-72-1; EINECS 204-000-3
Synonyms: C14 linear primary alcohol; 1-Tetradecanol; Tetradecyl alcohol
Tradenames: Alfol® 14; Emery® 3334; Emery® 3345; Emery® 3347; Epal® 14; Kalcohl 40; Lorol C14; Lorol C14-98; Lorol C14 Chemically Pure; Philcohol 1400
Tradenames containing: Epal® 12/70; Epal® 12/85; Epal® 1214; Epal® 1218; Epal® 1412; Epal® 1416; Epal® 1416-LD; Epal® 1418

Myristyl amine. *See Myristamine*

Myristyl dimethylamine. *See Dimethyl myristamine*

Myristyl dimethyl amine oxide. *See Myristamine oxide*

Myristyl trimethyl ammonium bromide. *See Myrtrimonium bromide*

Myrtrimonium bromide (CTFA)
CAS 1119-97-7; EINECS 214-291-9
Synonyms: Myristyl trimethyl ammonium bromide; N,N,N-Trimethyl-1-tetradecanaminium bromide; Tetradonium bromide
Tradenames: Empigen® CHB; Querton 14Br-40

2-Naphthalenesulfonic acid
CAS 120-18-3
Synonyms: beta-Naphthalenesulfonic acid
Tradenames: Dehscofix 918

Neopentyl glycol
CAS 126-30-7
Synonyms: 2,2-Dimethyl-1,3-propanediol
Tradenames containing: Lutensit® A-BO

Nitrilotris (methylene phosphonic acid)
Synonyms: Amino tris (methylene phosphonic acid)
Tradenames: Briquest® 301-50A

Nonanoic acid. *See Pelargonic acid*

Nonoxynol-1 (CTFA)
CAS 26027-38-3 (generic); 37205-87-1 (generic); 27986-36-3; EINECS 248-762-5
Synonyms: Ethylene glycol nonyl phenyl ether; PEG-1 nonyl phenyl ether; 2-(Nonylphenoxy) ethanol
Tradenames: Akyporox NP 15; Alkasurf® NP-1; Cedepal CO-210; Chemax NP-1.5; Desonic® 1.5N; Emthox® 6960; Ethal NP-1.5; Ethylan® NP 1; Iconol NP-1.5; Marlophen® 81N; Norfox® NP-1; Prox-onic NP-1.5; Rexol 25/1; Surfonic® N-10; T-Det® N 1.5; Trycol® 6960; Trycol® NP-1

Nonoxynol-2 (CTFA)
CAS 26027-38-3 (generic); 37205-87-1 (generic); 27176-93-8 (generic); 9016-45-9 (generic); EINECS 248-291-5
Synonyms: PEG-2 nonyl phenyl ether; POE (2) nonyl phenyl ether; PEG 100 nonyl phenyl ether
Tradenames: Etophen 102; Igepal® CO-210; Marlophen® 82N; Remcopal 3112; Teric N2
Tradenames containing: Akypogene SO

Nonoxynol-3 (CTFA)
CAS 27176-95-0 (generic); 84562-92-5 (generic); 51437-95-7 (generic); 9016-45-9 (generic)
Synonyms: PEG-3 nonyl phenyl ether; POE (3) nonyl phenyl ether; 2-[2-[2-(Nonylphenoxy) ethoxy] ethoxy] ethanol
Tradenames: Akyporox NP 30; Etophen 103; Marlophen® 83N; Sermul EN 3; Surfonic® N-31.5; Teric N3

Nonoxynol-4 (CTFA)
CAS 7311-27-5; 9016-45-9 (generic); 26027-38-3 (generic); 37205-87-1 (generic); 27176-97-2; EINECS 230-770-5
Synonyms: PEG-4 nonyl phenyl ether; POE (4) nonyl phenyl ether; PEG 200 nonyl phenyl ether
Tradenames: Ablunol NP4; Akyporox NP 40; Alkasurf® NP-4; Arkopal N040; Carsonon® N-4; Cedepal CO-430; Chemax NP-4; Desonic® 4N; Emthox® 6961; Emulgator U4; Ethal NP-4; Ethylan® 44; Hetoxide NP-4; Hostapal N-040; Hyonic NP-40; Iconol NP-4; Igepal® CO-430; Macol® NP-4; Makon® 4; Marlophen® 84N; Naxonic NI-40; Norfox® NP-4; Nutrol 622; Polystep® F-1; Prox-onic NP-04; Remcopal 334; Renex® 647; Rewopal® HV 4; Rexol 25/4; Sellig N 4 100; Serdox NNP 4; Siponic® NP-4 (see Igepal® CO-430); Surfonic® N-40; Synperonic NP4; Syntopon A 100; T-Det® N-4; Tergitol® NP-4; Tergitol® NP-14; Teric N4; Triton® N-42; Trycol® 6961; Trycol® NP-4; Witconol NP-40

Nonoxynol-5 (CTFA)
CAS 9016-45-9 (generic); 26027-38-3 (generic); 37205-87-1 (generic); 26264-02-8; 20636-48-0; EINECS 247-555-7
Synonyms: PEG-5 nonyl phenyl ether; POE (5) nonyl phenyl ether; 14-(Nonylphenoxy)-3,6,9,12-tetraoxatetradecan-1-ol
Tradenames: Desonic® 5N; Ethylan® 55; Etophen 105; Iconol NP-5; Igepal® CO-520; Macol® NP-5; Marlophen® 85N; Nikkol NP-5; Poly-Tergent® B-150; Renex® 648; Rewopal® HV 5; Sellig N 5 100; Serdox NNP 5; Synperonic NP5; Tergitol® NP-5; Teric N5; Triton® N-57; Trycol® 6940

Nonoxynol-6 (CTFA)
CAS 9016-45-9 (generic); 26027-38-3 (generic); 37205-87-1 (generic); 27177-01-1; 27177-05-5
Synonyms: PEG-6 nonyl phenyl ether; POE (6) nonyl phenyl ether; PEG 300 nonyl phenyl ether

Tradenames: Ablunol NP6; Alkasurf® NP-6; Arkopal N060; Carsonon® N-6; Cedepal CO-530; Chemax NP-6; Dehydrophen PNP 4; Dehydrophen PNP 6; Desonic® 6N; Emthox® 6962; Emulgator U6; Ethal NP-6; Ethylan® 77; Etophen 106; Hetoxide NP-6; Hostapal N-060; Hostapal N-060R; Hyonic NP-60; Iconol NP-6; Igepal® CO-530; Macol® NP-6; Makon® 6; Marlophen® 86N; Marlophen® 86N/S; Naxonic NI-60; Nissan Nonion NS-206; Norfox® NP-6; NP-55-60; Poly-Tergent® B-200; Polystep® F-2; Prox-onic NP-06; Remcopal 666; Renex® 697; Rewopal® HV 6; Rexol 25/6; Sellig N 6 100; Serdox NNP 6; Siponic® NP-6 (see Igepal® CO-530); Surfonic® N-60; Synperonic NP6; Syntopon A; T-Det® N-6; Tergitol® NP-6; Trycol® 6962; Trycol® NP-6

Nonoxynol-7 (CTFA)

CAS 9016-45-9 (generic); 26027-38-3 (generic); 27177-05-5; 37205-87-1 (generic); EINECS 248-292-0

Synonyms: PEG-7 nonyl phenyl ether; POE (7) nonyl phenyl ether

Tradenames: Desonic® 7N; Etophen 107; Iconol NP-7; Makon® 7; Marlophen® 87N; Norfox® NP-7; Rexol 25/7; Serdox NNP 7; Synperonic NP7; Tergitol® NP-7; Tergitol® NP-27; Trycol® 6963; Trycol® NP-7

Nonoxynol-8 (CTFA)

CAS 9016-45-9 (generic); 26027-38-3 (generic); 37205-87-1 (generic); 26571-11-9; 27177-05-5; EINECS 248-293-6; 247-816-5

Synonyms: PEG-8 nonyl phenyl ether; POE (8) nonyl phenyl ether; PEG 400 nonyl phenyl ether

Tradenames: Ablunol NP8; Alkasurf® NP-8; Arkopal N080; Carsonon® N-8; Cedepal CO-610; Dehydrophen PNP 8; Ethylan® TU; Etophen 108; Hostapal N-080; Igepal® CO-610; Macol® NP-8; Makon® 8; Marlophen® 88N; Nikkol NP-7.5; NP-55-80; NP-55-85; Nutrol 611; Polystep® F-3; Quimipol ENF 80; Remcopal 349; Renex® 688; Rewopal® HV 8; Rexol 25/8; Sellig N 8 100; Serdox NNP 8.5; Siponic® NP-8 (see Alkasurf® NP-8); Surfonic® N-85; Synperonic NP8; Syntopon B; T-Det® N-8; Tergitol® NP-8; Teric N8

Tradenames containing: Surfonic® HDL

Nonoxynol-9 (CTFA)

CAS 9016-45-9 (generic); 26027-38-3 (generic); 26571-11-9; 37205-87-1 (generic); 14409-72-4

Synonyms: PEG-9 nonyl phenyl ether; POE (9) nonyl phenyl ether; PEG 450 nonyl phenyl ether

Tradenames: Ablunol NP9; Akyporox NP 90; Alkasurf® NP-9; Arkopal N090; Caloxylate N-9; Carsonon® N-9; Cedepal CO-630; Chemax NP-9; Desonic® 9N; Empilan® NP9; Emulgator U9; Ethal NP-9; Ethylan® BCP; Ethylan® KEO; Etophen 109; Gradonic N-95; Hetoxide NP-9; Hostapal N-090; Hyonic NP-90; Iconol NP-9; Igepal® CO-630; Igepal® FN-9.5 (formerly Texafor FN-9.5); Macol® NP-9.5; Marlophen® 89N; Marlophen® 1028N; Norfox® NP-9; NP-55-90; Nutrol 600; Poly-Tergent® B-300; Prox-onic NP-09; Remcopal 29; Remcopal 6110; Renex® 698; Rewopal® HV 9; Rexol 25/9; Sellig N 9 100; Siponic® NP-9 (see Igepal® CO-630); Siponic® NP-9.5 (see Igepal® CO-630); Syn Fac® 905; Synperonic NP9; Syntopon C; Tergitol® NP-9; Tergitol® TP-9; Teric N9; Triton® N-101; Trycol® 6964; Trycol® NP-9

Tradenames containing: Mackanate DOS-70N

Nonoxynol-10 (CTFA)

CAS 9016-45-9 (generic); 26027-38-3 (generic); 27177-08-8; 37205-87-1 (generic); 27942-26-3; EINECS 248-294-1

Synonyms: PEG-10 nonyl phenyl ether; POE (10) nonyl phenyl ether; PEG 500 nonyl phenyl ether

Tradenames: Ablunol NP10; Akyporox NP 95; Akyporox NP 105; Alkasurf® NP-10; Arkopal N100; Carsonon® N-10; Cedepal CO-710; Chemax NP-10; Dehydrophen PNP 10; Desonic® 10N; Etophen 110; Eumulgin 286; Hetoxide NP-10; Hostapal N-100; Hyonic NP-100; Hyonic PE-100; Iconol NP-10; Igepal® CO-660; Igepal® CO-710; Lutensol® AP 10; Makon® 10; Marlophen® 810N; Naxonic NI-100; Nikkol NP-10; Poly-Tergent® B-315; Polystep® F-4; Prox-onic NP-010; Renex® 690; Rewopal® HV 10; Rexol 25/10; Sellig N 10 100; Serdox NNP 10; Simulsol 1030 NP; Siponic® NP-10 (see Igepal® CO-660); Surfonic® N-95; Surfonic® N-100; Surfonic® N-102; Synperonic NP10; Syntopon D; T-Det® N-9.5; Tergitol® NP-10; Tergitol® NPX; Teric N10; Trycol® 6974; Witconol NP-100

Tradenames containing: Surfonic® HDL-1

Nonoxynol-11 (CTFA)

CAS 9016-45-9 (generic); 26027-38-3 (generic); 37205-87-1 (generic)

Synonyms: PEG-11 nonyl phenyl ether; POE (11) nonyl phenyl ether

Tradenames: Alkasurf® NP-11; Arkopal N110; Carsonon® N-11; Desonic® 11N; Hostapal N-110; Hyonic NP-110; Igepal® FN-11 (formerly Texafor FN-11); Macol® NP-11; Makon® 11; Neutronyx® 656; Norfox® NP-11; Nutrol 656; Poly-Tergent® B-350; Remcopal PONF; Rexol 25/11; Sellig N 11 100; Serdox NNP 11; T-Det® N-10.5; Teric N11; Triton® N-111; Trycol® 6965; Trycol® NP-11

Nonoxynol-12 (CTFA)

CAS 9016-45-9 (generic); 26027-38-3 (generic); 37205-87-1 (generic)

Synonyms: PEG-12 nonyl phenyl ether; POE (12) nonyl phenyl ether; PEG 600 nonyl phenyl ether

Tradenames: Ablunol NP12; Alkasurf® NP-12; Carsonon® N-12; Dehydrophen PNP 12; Desonic® 12N; Emulgator U12; Ethylan® DP; Etophen 112; Hetoxide NP-12; Hyonic NP-120; Iconol NP-12; Igepal® CO-720; Macol® NP-12; Makon® 12; Marlophen® 812N; Nissan Nonion NS-212; Polystep® F-5; Remcopal 3712; Renex® 682; Sellig N 12 100; Serdox NNP 12; Sermul EN 229; Surfonic® N-120; Synperonic NP12; T-Det® N-12; Teric N12; Trycol® 6953

Nonoxynol-13 (CTFA)

CAS 9016-45-9 (generic); 26027-38-3 (generic); 37205-87-1 (generic)

Synonyms: PEG-13 nonyl phenyl ether; POE (13) nonyl phenyl ether

Tradenames: Arkopal N130; Desonic® 13N; Hostapal N-130; Remcopal 3515; Renex® 679; Serdox NNP 13; Siponic® NP-13 (see Igepal® CO-720); Surfonic® N-130; Synperonic NP13; Syntopon E; Tergitol® NP-13; Teric N13;

Trycol® 6958

Nonoxynol-14 (CTFA)
CAS 9016-45-9 (generic); 26027-38-3 (generic); 37205-87-1 (generic)
Synonyms: PEG-14 nonyl phenyl ether; POE (14) nonyl phenyl ether
Tradenames: Ethylan® BV; Etophen 114; Makon® 14; Marlophen® 814N; Polystep® F-6; Rewopal® HV 14; Rexol 25/14; Serdox NNP 14; T-Det® N-14

Nonoxynol-15 (CTFA)
CAS 9016-45-9 (generic); 26027-38-3 (generic); 37205-87-1 (generic)
Synonyms: PEG-15 nonyl phenyl ether; POE (15) nonyl phenyl ether
Tradenames: Ablunol NP15; Akyporox NP 150; Alkasurf® NP-15; Alkasurf® NP-15, 80%; Arkopal N150; Chemax NP-15; Dehydrophen PNP 15; Desonic® 15N; Hetoxide NP-15-85%; Iconol NP-15; Igepal® CO-730; Macol® NP-15; Nikkol NP-15; Nutrol 640; Poly-Tergent® B-500; Prox-onic NP-015; Renex® 678; Rexol 25/15; Serdox NNP 15; Sermul EN 15; Siponic® NP-15 (see Igepal® CO-730); Surfonic® N-150; Synperonic NP15; Syntopon F; Tergitol® NP-15; Teric N15; Triton® N-150

Nonoxynol-16
Tradenames: Ablunol NP16; Sellig N 15 100

Nonoxynol-17
Tradenames: Sellig N 1780; Sellig N 17 100

Nonoxynol-18 (CTFA)
CAS 9016-45-9 (generic); 37205-87-1 (generic); 26027-38-3 (generic)
Synonyms: PEG-18 nonyl phenyl ether; POE (18) nonyl phenyl ether
Tradenames: Nikkol NP-18TX

Nonoxynol-19
Tradenames: Sellig N 20 80; Sellig N 20 100

Nonoxynol-20 (CTFA)
CAS 9016-45-9 (generic); 26027-38-3 (generic); 37205-87-1 (generic)
Synonyms: PEG-20 nonyl phenyl ether; POE (20) nonyl phenyl ether; PEG 1000 nonyl phenyl ether
Tradenames: Ablunol NP20; Akyporox NP 200; Chemax NP-20; Dehydrophen PNP 20; Desonic® 20N; Ethal NP-20; Ethylan® 20; Etophen 120; Iconol NP-20; Igepal® CO-850; Lutensol® AP 20; Macol® NP-20; Macol® NP-20(70); Marlophen® 820N; Prox-onic NP-020; Remcopal 3820; Remcopal 33820; Renex® 649; Rexol 25/20; Sermul EN 20/70; Soprofor NP/20; Surfonic® N-200; Synperonic NP20; Syntopon G; T-Det® N-20; Teric N20; Trycol® 6967; Trycol® NP-20

Nonoxynol-23 (CTFA)
CAS 9016-45-9 (generic); 26027-38-3 (generic); 37205-87-1 (generic)
Synonyms: PEG-23 nonyl phenyl ether; POE (23) nonyl phenyl ether
Tradenames: Arkopal N230; Hostapal N-230

Nonoxynol-25
Tradenames: Igepal® CA-877; Rewopal® HV 25; Sellig N 30 70; Sellig N 30 100; Serdox NNP 25

Nonoxynol-27
Tradenames: Remcopal L30; Remcopal NP 30

Nonoxynol-30 (CTFA)
CAS 9016-45-9 (generic); 26027-38-3 (generic); 37205-87-1 (generic)
Synonyms: PEG-30 nonyl phenyl ether; POE (30) nonyl phenyl ether
Tradenames: Ablunol NP30; Ablunol NP30 70%; Akyporox NP 300V; Alkasurf® NP-30, 70%; Carsonon® N-30; Chemax NP-30; Chemax NP-30/70; Dehydrophen PNP 30; Desonic® 30N; Desonic® 30N70; Ethylan® N30; Hetoxide NP-30; Hostapal N-300; Iconol NP-30; Iconol NP-30-70%; Igepal® CO-880; Igepal® CO-887; Igepal® FN-30 (formerly Texafor FN-30); Macol® NP-30(70); Makon® 30; Marlophen® 830N; Polystep® F-9; Prox-onic NP-030; Prox-onic NP-030/70; Renex® 650; Rexol 25/30; Rexol 25/307; Serdox NNP 30/70; Sermul EN 30/70; Sermul EN 145; Soprofor NP/30; Surfonic® N-300; Surfonic® NB-5; Synperonic NP30; Synperonic NP30/70; Syntopon H; T-Det® N-30; T-Det® N-307; Teric N30; Teric N30L; Trycol® 6968; Trycol® 6969; Trycol® NP-30; Trycol® NP-307; Witconol NP-300

Nonoxynol-34
Tradenames: Polystep® F-95B

Nonoxynol-35
Synonyms: PEG-35 nonyl phenyl ether; POE (35) nonyl phenyl ether
Tradenames: Ethylan® HA Flake; Synperonic NP35

Nonoxynol-40 (CTFA)
CAS 9016-45-9 (generic); 26027-38-3 (generic); 37205-87-1 (generic)
Synonyms: PEG-40 nonyl phenyl ether; POE (40) nonyl phenyl ether; PEG 2000 nonyl phenyl ether
Tradenames: Ablunol NP40; Ablunol NP40 70%; Alkasurf® NP-40, 70%; Carsonon® N-40, 70%; Chemax NP-40; Chemax NP-40/70; Dehydrophen PNP 40; Desonic® 40N; Desonic® 40N70; Emthox® 6957; Ethal NP-407; Hetoxide NP-40; Hyonic NP-407; Iconol NP 40; Iconol NP-40-70%; Igepal® CO-890; Igepal® CO-897; Marlophen® 840N; Polystep® F-10, F-10 Ec; Prox-onic NP-040; Prox-onic NP-040/70; Rexol 25/40; Rexol 25/407; Sellig N 1050; Siponic® NP-40 (see Igepal® CO-890); Siponic® NP-407 (see Igepal® CO-897); Synperonic NP40; T-Det® N-40; T-Det® N-407; Tergitol® NP-40; Tergitol® NP-40 (70% Aq.); Tergitol® NP-44; Teric N40; Teric N40L; Trycol® 6957; Trycol® 6970; Trycol® NP-40; Trycol® NP-407

Nonoxynol-50 (CTFA)
CAS 9016-45-9 (generic); 26027-38-3 (generic); 37205-87-1 (generic)
Synonyms: PEG-50 nonyl phenyl ether; POE (50) nonyl phenyl ether
Tradenames: Ablunol NP50; Ablunol NP50 70%; Alkasurf® NP-50 70%; Carsonon® N-50; Chemax NP-50; Chemax NP-50/70; Desonic® 50N; Desonic® 50N70; Hetoxide NP-50; Hyonic NP-500; Iconol NP-50; Iconol NP-50-70%; Igepal® CO-970; Igepal® CO-977; Makon® 50; Marlophen® 850N; Nopco® RDY; Prox-onic NP-050; Prox-onic NP-050/70; Remcopal 31250; Rewopal® HV 50; Rexol

25/50; Rexol 25/507; Synperonic NP50; T-Det® N-50; T-Det® N-507; Trycol® 6951; Trycol® 6971; Trycol® 6972; Trycol® NP-50; Trycol® NP-507

Nonoxynol-52
Tradenames: Sellig N 50 100

Nonoxynol-55
Tradenames: Surfonic® N-557; Tergitol® NP-55, 70% Aq

Nonoxynol-70
Synonyms: PEG-70 nonyl phenyl ether; POE (70) nonyl phenyl ether
Tradenames: Iconol NP-70; Iconol NP-70-70%; Igepal® CO-980; Igepal® CO-987; T-Det® N-70; T-Det® N-705; T-Det® N-707; Tergitol® NP-70, 70% Aq.

Nonoxynol-100 (CTFA)
CAS 9016-45-9 (generic); 26027-38-3 (generic); 37205-87-1 (generic)
Synonyms: PEG-100 nonyl phenyl ether; POE (100) nonyl phenyl ether
Tradenames: Carsonon® N-100, 70%; Chemax NP-100; Chemax NP-100/70; Desonic® 100N; Desonic® 100N70; Iconol NP-100; Iconol NP-100-70%; Igepal® CO-990; Igepal® CO-997; Macol® NP-100; Prox-onic NP-0100; Prox-onic NP-0100/70; Rexol 25/100-70%; Soprofor NP/100; T-Det® N-100; T-Det® N-1007; Teric N100; Trycol® 6942; Trycol® 6981; Trycol® NP-1007

Nonoxynol-120 (CTFA)
CAS 9016-45-9 (generic); 37205-87-1 (generic); 26027-38-3 (generic)
Synonyms: PEG-120 nonyl phenyl ether; POE (120) nonyl phenyl ether
Tradenames: Akyporox NP 1200V

Nonoxynol-150
Tradenames: Trycol® 6954

Nonoxynol-8 carboxylic acid (CTFA)
CAS 3115-49-9; 28212-44-4 (generic)
Synonyms: PEG-8 nonyl phenyl ether carboxylic acid; PEG 400 nonyl phenyl ether carboxylic acid; POE (8) nonyl phenyl ether carboxylic acid
Tradenames: Akypo NP 70

Nonoxynol-10 carboxylic acid (CTFA)
CAS 28212-44-4 (generic)
Synonyms: PEG-10 nonyl phenyl ether carboxylic acid; PEG 500 nonyl phenyl ether carboxylic acid; POE (10) nonyl phenyl ether carboxylic acid
Tradenames: Sandopan® MA-18; Surfine AZI-A

Nonoxynol-9 iodine
Tradenames: Biopal® VRO-20

Nonoxynol-12 iodine (CTFA)
Synonyms: PEG-12 nonyl phenyl ether iodine complex; POE (12) nonyl phenyl ether iodine complex; PEG 600 nonyl phenyl ether iodine complex
Tradenames: Biopal® NR-20

Nonoxynol-6 phosphate (CTFA)
CAS 51811-79-1
Synonyms: PEG-6 nonyl phenyl ether phosphate; POE (6) nonyl phenyl ether phosphate; PEG 300 nonyl phenyl ether phosphate
Tradenames: Emphos CS-136; Monafax 786; Schercophos NP-6; Sermul EA 211

Nonoxynol-9 phosphate (CTFA)
CAS 51811-79-1; 66197-78-2; EINECS 266-231-6
Synonyms: PEG-9 nonyl phenyl ether phosphate; POE (9) nonyl phenyl ether phosphate; PEG 450 nonyl phenyl ether phosphate
Tradenames: Monafax 785; Rhodafac® RE-610 (formerly Gafac® RE-610); Schercophos NP-9; Surfagene FAZ 109; Surfagene FAZ 109 NV

Nonoxynol-10 phosphate (CTFA)
Synonyms: PEG-10 nonyl phenyl ether phosphate; POE (10) nonyl phenyl ether phosphate; PEG 500 nonyl phenyl ether phosphate
Tradenames: Emphos CS-141; Sermul EA 188; Servoxyl VPNZ 10/100

Nonoxynol-15 phosphate
Tradenames: Sermul EA 136

Nonoxynol-20 phosphate
Tradenames: Chemphos TC-337

Nonoxynol-50 phosphate
Tradenames: Sermul EA 205

Nonoxynol-4 sulfate
Synonyms: Nonyl phenol ethoxy (4) sulfate
Tradenames: Witcolate D51-51

Nonoxynol-10 sulfate
Synonyms: Nonyl phenol ethoxy (10) sulfate
Tradenames: Witcolate D51-53; Witcolate D51-53HA

Nonoxynol-30 sulfate
Tradenames: Witcolate D51-60

Nonyl naphthalene sodium sulfonate
Tradenames: Emkal NNS

Nonyl naphthalene sulfonic acid
Tradenames: Emkal NNS Acid

Nonyl nonoxynol-4
CAS 9014-93-1 (generic)
Synonyms: PEG-4 dinonyl phenyl ether; POE (4) dinonyl phenyl ether
Tradenames: Hetoxide DNP-4

Nonyl nonoxynol-5 (CTFA)
CAS 9014-93-1 (generic)
Synonyms: PEG-5 dinonyl phenyl ether; POE (5) dinonyl phenyl ether
Tradenames: Hetoxide DNP-5; Macol® DNP-5

Nonyl nonoxynol-7
Tradenames: Igepal® DM-430

Nonyl nonoxynol-8
Synonyms: PEG-8 dinonyl phenyl ether; POE (8) dinonyl phenyl ether
Tradenames: Chemax DNP-8; Ethal DNP-8; Iconol DNP-8; Prox-onic DNP-08; Trycol® 6985; Trycol® DNP-8

Nonyl nonoxynol-9
Synonyms: PEG-9 dinonyl phenyl ether; POE (9) dinonyl phenyl ether
Tradenames: Igepal® DM-530

Nonyl nonoxynol-10 (CTFA)
CAS 9014-93-1 (generic)
Synonyms: PEG-10 dinonyl phenyl ether; POE (10) dinonyl phenyl ether; PEG 500 dinonyl phenyl ether
Tradenames: Hetoxide DNP-9.6; Hetoxide DNP-10; Macol® DNP-10; Sellig DN 10 100

Nonyl nonoxynol-15
Synonyms: PEG-15 dinonyl phenyl ether; POE (15) dinonyl phenyl ether
Tradenames: Chemax DNP-15; Igepal® DM-710; Macol® DNP-15

Nonyl nonoxynol-16
Tradenames: Marlophen® DNP 16

Nonyl nonoxynol-18
Synonyms: PEG-18 dinonyl phenyl ether; POE (18) dinonyl phenyl ether
Tradenames: Chemax DNP-18; Ethal DNP-18; Marlophen® DNP 18

Nonyl nonoxynol-21
Tradenames: Macol® DNP-21

Nonyl nonoxynol-22
Tradenames: Sellig DN 22 100

Nonyl nonoxynol-24
Synonyms: PEG-24 dinonyl phenyl ether; POE (24) dinonyl phenyl ether
Tradenames: Iconol DNP-24; Igepal® DM-730

Nonylnonoxynol-30
Tradenames: Marlophen® DNP 30

Nonyl nonoxynol-49 (CTFA)
CAS 9014-93-1 (generic)
Synonyms: PEG-49 dinonyl phenyl ether; POE (49) dinonyl phenyl ether
Tradenames: Igepal® DM-880

Nonyl nonoxynol-70
Synonyms: PEG-70 dinonyl phenyl ether; POE (70) dinonyl phenyl ether
Tradenames: T-Det® D-70

Nonyl nonoxynol-100 (CTFA)
CAS 9014-93-1 (generic)
Synonyms: PEG-100 dinonyl phenyl ether; POE (100) dinonyl phenyl ether
Tradenames: Serdox NDI 100

Nonyl nonoxynol-150 (CTFA)
CAS 9014-93-1 (generic)
Synonyms: PEG-150 dinonyl phenyl ether; POE (150) dinonyl phenyl ether
Tradenames: Chemax DNP-150; Chemax DNP-150/50; Ethal DNP-150/50%; Iconol DNP-150; Igepal® DM-970 FLK; Macol® DNP-150; Marlophen® DNP 150; Prox-onic DNP-0150; Prox-onic DNP-0150/50; T-Det® D-150; Trycol® 6989; Trycol® DNP-150; Trycol® DNP-150/50

Nonyl nonoxynol-7 phosphate (CTFA)
CAS 66172-78-9; 66172-83-6; EINECS 266-215-9
Synonyms: PEG-7 dinonyl phenyl ether phosphate; POE (7) dinonyl phenyl ether phosphate
Tradenames: Rhodafac® RM-410 (formerly Gafac® RM-410)

Nonyl nonoxynol-9 phosphate (CTFA)
CAS 66172-82-5; EINECS 266-218-5
Synonyms: PEG-9 dinonyl phenyl ether phosphate; POE (9) dinonyl phenyl ether phosphate; PEG 450 dinonyl phenyl ether phosphate
Tradenames: Servoxyl VPQZ 9/100

Nonyl nonoxynol-10 phosphate (CTFA)
Synonyms: PEG-10 dinonyl phenyl ether phosphate; POE (10) dinonyl phenyl ether phosphate; PEG 500 dinonyl phenyl ether phosphate
Tradenames: Rhodafac® RM-510 (formerly Gafac® RM-510)

Nonyl nonoxynol-15 phosphate (CTFA)
Synonyms: PEG-15 dinonyl phenyl ether phosphate; POE (15) dinonyl phenyl ether phosphate
Tradenames: Rhodafac® RM-710 (formerly Gafac® RM-710)

Octadecene-1
CAS 112-88-9
Synonyms: Linear C18 alpha olefin
Tradenames: Gulftene 18; Neodene® 18

Octadecene nitrile
Tradenames: Arneel® OD

9-Octadecenoic acid, diester with 1,2,3-propanetriol. *See Glyceryl dioleate*

Octadecyl. *See Stearyl*

n-Octanoic acid. *See Caprylic acid*

Octanol. *See Caprylic alcohol*

Octene-1
CAS 111-66-0
Synonyms: Linear C8 alpha olefin
Tradenames: Ethyl® Octene-1; Gulftene 8; Neodene® 8

Octenyl succinic anhydride
CAS 26680-54-6
Synonyms: OSA
Tradenames: n-Octenyl Succinic Anhydride

Octeth-4
Synonyms: PEG-4 octyl ether; POE (4) octyl ether
Tradenames: Dehydol O4; Dehydol O4 DEO; Dehydol O4 Special

Octeth-3 carboxylic acid (CTFA)
Synonyms: PEG-3 octyl ether carboxylic acid; POE (3) octyl ether carboxylic acid
Tradenames: Defoamer B 90

Octoic acid. *See Caprylic acid*

Octoxynol-1 (CTFA)
CAS 9002-93-1 (generic); 9036-19-5 (generic); 9004-87-9 (generic); 2315-67-5
Synonyms: Ethylene glycol octyl phenyl ether; PEG-1 octyl phenyl ether; 2-[p-(1,1,3,3-Tetramethylbutyl) phenoxy]ethanol
Tradenames: Alkasurf® OP-1; Cedepal CA-210; Igepal® CA-210; Rexol 45/1; Triton® X-15

Octoxynol-3 (CTFA)
CAS 9002-93-1 (generic); 9004-87-9 (generic); 9036-19-5 (generic); 2315-62-0; 27176-94-9
Synonyms: 2-[2-[2-[p-(1,1,3,3-Tetramethylbutyl)-phenoxy]ethoxy]ethoxy]ethanol; PEG-3 octyl phenyl ether; POE (3) octyl phenyl ether
Tradenames: Chemax OP-3; Igepal® CA-420; Macol® OP-3; Nikkol OP-3; Rexol 45/3; Synperonic OP3; Triton® X-35

Octoxynol-4
Tradenames: Dehydrophen POP 4; Desonic® S-45; Hyonic OP-40; Sellig O 4 100; T-Det® O-4

Octoxynol-5 (CTFA)
CAS 9002-93-1 (generic); 9036-19-5 (generic); 9004-87-9 (generic); 2315-64-2; 27176-99-4
Synonyms: PEG-5 octyl phenyl ether; POE (5) octyl phenyl ether; 14-(Octylphenoxy)-3,6,9,12-tetraoxatetradecan-1-ol
Tradenames: Alkasurf® OP-5; Cedepal CA-520; Chemax OP-5; Iconol OP-5; Igepal® CA-520; Macol® OP-5; Marlophen® 85; Remcopal 306; Rexol 45/5; Sellig O 5 100; Synperonic OP4.5; Teric X5; Triton® X-45

Octoxynol-6
Synonyms: PEG-6 octyl phenyl ether; POE (6) octyl phenyl ether
Tradenames: Makon® OP-6; Marlophen® 86; Sellig O 6 100; T-Det® O-6

Octoxynol-7 (CTFA)
CAS 9002-93-1 (generic); 9004-87-9 (generic); 9036-19-5 (generic); 27177-02-2
Synonyms: PEG-7 octyl phenyl ether; POE (7) octyl phenyl ether; 20-(Octylphenoxy)-3,6,9,12,15,18-hexaoxaeicosan-1-ol
Tradenames: Chemax OP-7; Desonic® S-114; Hyonic OP-7; Hyonic OP-70; Iconol OP-7; Igepal® CA-620; Marlophen® 87; Rexol 45/7; Teric X7

Octoxynol-8 (CTFA)
CAS 9004-87-9 (generic); 9036-19-5 (generic); 9002-93-1 (generic)
Synonyms: PEG-8 octyl phenyl ether; PEG 400 octyl phenyl ether; POE (8) octyl phenyl ether
Tradenames: Alkasurf® OP-8; Dehydrophen POP 8; Macol® OP-8; Marlophen® 88; Sellig O 8 100; Synperonic OP7.5; T-Det® O-8; Teric X8; Triton® X-114

Octoxynol-9 (CTFA)
CAS 9002-93-1 (generic); 9004-87-9 (generic); 9010-43-9; 9036-19-5 (generic); 42173-90-0
Synonyms: PEG-9 octyl phenyl ether; POE (9) octyl phenyl ether; PEG 450 octyl phenyl ether
Tradenames: Cedepal CA-630; Chemax OP-9; Desonic® S-100; Hyonic OP-100; Hyonic PE-250; Igepal® CA-630; Igepal® F-85 (formerly Texafor F-85); Igepal® KA (formerly Rhodiasurf KA); Makon® OP-9; Marlophen® 89; Marlophen® 1028; Nutrol 100; Prox-onic OP-09; Remcopal O9; Sellig O 9 100; Serdox NOP 9; Syntopon 8 C; T-Det® O-9; Triton® X-100; Triton® X-100 CG; Triton® X-120
Tradenames containing: Marlowet® RNP/K

Octoxynol-10 (CTFA)
CAS 9002-93-1 (generic); 9004-87-9 (generic); 9036-19-5 (generic); 2315-66-4; 27177-07-7
Synonyms: PEG-10 octyl phenyl ether; POE (10) octyl phenyl ether; PEG 500 octyl phenyl ether
Tradenames: Akyporox OP 100; Alkasurf® OP-10; Dehydrophen POP 10; Hyonic OP-10; Iconol OP-10; Igepal® O (formerly Rhodiasurf O); Macol® OP-10; Macol® OP-10 SP; Marlophen® 810; Nikkol OP-10; Renex® 750; Rexol 45/10; Synperonic OP10; Teric X10

Octoxynol-11 (CTFA)
CAS 9004-87-9 (generic); 9036-19-5 (generic); 9002-93-1 (generic)
Synonyms: PEG-11 octyl phenyl ether; POE (11) octyl phenyl ether
Tradenames: Remcopal O11; Sellig O 11 100; Synperonic OP11; Teric X11

Octoxynol-12 (CTFA)
CAS 9002-93-1 (generic); 9036-19-5 (generic); 9004-87-9 (generic)
Synonyms: PEG-12 octyl phenyl ether; POE (12) octyl phenyl ether; PEG 600 octyl phenyl ether
Tradenames: Akyporox OP 115 SPC; Alkasurf® OP-12; Cedepal CA-720; Desonic® S-102; Macol® OP-12; Marlophen® 812; Remcopal O12; Rexol 45/12; Sellig O 12 100; T-Det® O-12

Octoxynol-13 (CTFA)
CAS 9002-93-1 (generic); 9004-87-9 (generic); 9036-19-5 (generic)
Synonyms: PEG-13 octyl phenyl ether; POE (13) octyl phenyl ether
Tradenames: Igepal® CA-720; Synperonic OP12.5; Teric X13; Triton® X-102

Octoxynol-14
Tradenames: Marlophen® 814

Octoxynol-16 (CTFA)
CAS 9004-87-9 (generic); 9036-19-5 (generic); 9002-93-1 (generic)
Synonyms: PEG-16 octyl phenyl ether; POE (16) octyl phenyl ether
Tradenames: Macol® OP-16(75); Prox-onic OP-016; Rexol 45/16; Synperonic OP16; Teric X16; Triton® X-165-70%

Octoxynol-17
Tradenames: Dehydrophen POP 17; Dehydrophen POP 17/80

Octoxynol-20 (CTFA)
CAS 9002-93-1 (generic); 9036-19-5 (generic); 9004-87-9 (generic)
Synonyms: PEG-20 octyl phenyl ether; POE (20) octyl phenyl ether; PEG 1000 octyl phenyl ether
Tradenames: Akyporox OP 200; Marlophen® 820; Sellig O 20 100; Synperonic OP20

Octoxynol-25 (CTFA)
CAS 9002-93-1 (generic); 9036-19-5 (generic); 9004-87-9 (generic)
Synonyms: PEG-25 octyl phenyl ether; POE (25) octyl phenyl ether
Tradenames: Akyporox OP 250 V; Synperonic OP25

Octoxynol-30 (CTFA)
CAS 9004-87-9 (generic); 9036-19-5 (generic); 9002-93-1 (generic)
Synonyms: PEG-30 octyl phenyl ether; POE (30) octyl phenyl ether
Tradenames: Alkasurf® OP-30, 70%; Chemax OP-30/70; Iconol OP-30; Iconol OP-30-70%; Igepal® CA-880; Igepal® CA-887; Macol® OP-30(70); Nikkol OP-30; Prox-onic OP-030/70; Rexol 45/307; Serdox NOP 30/70; Siponic® F-300 (see Igepal® CA-887); Synperonic OP30; T-Det® O-307; Triton® X-305-70%; Trycol® 6943; Trycol® 6975

Octoxynol-33 (CTFA)
CAS 9002-93-1 (generic); 9036-19-5 (generic); 9004-87-9 (generic)
Synonyms: PEG-33 octyl phenyl ether; POE (33) octyl phenyl ether
Tradenames containing: Abex® VA 50

Octoxynol-40 (CTFA)
CAS 9002-93-1 (generic); 9004-87-9 (generic); 9036-19-5 (generic)
Synonyms: PEG-40 octyl phenyl ether; POE (40) octyl phenyl ether
Tradenames: Akyporox OP 400V; Alkasurf® OP-40, 70%; Cedepal CA-890; Cedepal CA-897; Chemax OP-40; Chemax OP-40/70; Desonic® S-405; Hyonic OP-407; Iconol OP-40; Iconol OP-40-70%; Igepal® CA-890; Igepal® CA-897; Macol® OP-40(70); Nissan Nonion HS-240; Prox-onic OP-040/70; Rexol 45/407; Serdox NOP 40/70; Siponic® F-400 (see

Igepal® CA-897); Synperonic OP40; Synperonic OP40/70; T-Det® O-40; T-Det® O-407; Teric X40; Teric X40L; Triton® X-405-70%; Trycol® 6956; Trycol® 6984; Trycol® OP-407

Octoxynol-70 (CTFA)
CAS 9004-87-9 (generic); 9036-19-5 (generic); 9002-93-1 (generic)
Synonyms: PEG-70 octyl phenyl ether; POE (70) octyl phenyl ether
Tradenames: Hyonic OP-705; Nissan Nonion HS-270; Triton® X-705-70%

Octoxynol-92
Tradenames: Igepal® F-920 (formerly Texafor F-920)

Octoxynol-200
CAS 9002-93-1
Synonyms: Poly(oxyethylene)-p-tert-octylphenyl ether
Tradenames containing: Defoamer A 50

Octoxynol-9 carboxylic acid (CTFA)
CAS 72160-13-5; 107628-08-0
Synonyms: PEG-9 octyl phenyl ether carboxylic acid; PEG 450 octyl phenyl ether carboxylic acid; POE (9) octyl phenyl ether carboxylic acid
Tradenames: Akypo OP 80; Akyposal OP 80

Octoxynol-20 carboxylic acid (CTFA)
CAS 72160-13-5; 107628-08-0
Synonyms: PEG-20 octyl phenyl ether carboxylic acid; PEG 1000 octyl phenyl ether carboxylic acid; POE (20) octyl phenyl ether carboxylic acid
Tradenames: Akypo OP 190

Octyl alcohol. *See 2-Ethylhexanol*

n-Octyl alcohol. *See Caprylic alcohol*

Octyl-decyl alcohol
Tradenames: Laurex® 810

Octyl/decyloxypropylamine
Tradenames: Tomah PA-1214

Octyldodecyl benzoate (CTFA)
Synonyms: Benzoic acid, 2-octyldodecyl ester
Tradenames: Finsolv® BOD

Octyl hydroxystearate (CTFA)
CAS 29383-26-4; 29710-25-6
Synonyms: 2-Ethylhexyl oxystearate; 12-Hydroxyoctadecanoic acid, 2-ethylhexyl ester
Tradenames: Naturechem® OHS

Octyliminodipropionate
CAS 52663-87-3
Tradenames: Ampholak YJH

Octyloxypropylamine
CAS 5397-31-9
Synonyms: 2-Ethylhexyloxy propylamine
Tradenames: Tomah PA-12EH

Octyl palmitate (CTFA)
CAS 29806-73-3; EINECS 249-862-1
Synonyms: 2-Ethylhexyl palmitate; 2-Ethylhexyl hexadecanoate
Tradenames: Kessco® OP

Octyl stearate (CTFA)
CAS 22047-49-0; EINECS 244-754-0
Synonyms: 2-Ethylhexyl stearate; 2-Ethylhexyl octadecanoate; Octadecanoic acid, 2-ethylhexyl ester
Tradenames: Afilan EHS

Oleamide (CTFA)
CAS 301-02-0; EINECS 206-103-9
Synonyms: 9-Octadecenamide; Oleyl amide
Tradenames: Armid® O; Kemamide® O; Kemamide® U

Oleamide DEA (CTFA)
CAS 93-83-4; EINECS 202-281-7
Synonyms: Oleic diethanolamide; Diethanolamine oleic acid amide; N,N-Bis(2-hydroxyethyl)9-octadecenamide
Tradenames: Alrosol O; Calamide O; Chimipal OLD; Comperlan OD; Crillon ODE; Emid® 6545; Hartamide 9137; Hetamide OC; Incromide OD; Laurel SD-400; Lauridit® OD; Mackamide MO; Mackamide NOA; Mackamide O; Marlamid® D 1885; Mazamide® O 20; Ninol® 201; Nissan Stafoam DO, DOS; Norfox® F-221; Rewomid® DO 280 SE; Schercomid SO-A; Serdolamide POF 61; Serdolamide POF 61 C; Varamide® A-7
Tradenames containing: Alkamide® DO-280 (formerly Cyclomide DO-280); Hetamide DO; Mackamide ODM; Ninol® SR-100; Rewomid® DO 280; Schercomid ODA

Oleamide MIPA (CTFA)
CAS 111-05-7; EINECS 203-828-2
Synonyms: Oleic monoisopropanolamide; N-(2-Hydroxypropyl)-9-octadecenamide; Monoisopropanolamine oleic acid amide
Tradenames: Mackamide OP; Schercomid OMI

Oleamidopropylamine oxide (CTFA)
CAS 25159-40-4; EINECS 246-684-6
Synonyms: 9-Octadecenamide, N-[3-(dimethylamino) propyl]-, N-oxide; Oleamidopropyl dimethylamine oxide; N-[3-(Dimethylamino) propyl]-9-octadecenamide-N-oxide
Tradenames: Incromine Oxide O; Mackamine OAO

Oleamidopropyl betaine (CTFA)
CAS 25054-76-6; EINECS 246-584-2
Synonyms: N-(Carboxymethyl)-N,N-dimethyl-3-[(1-oxooctadecenyl)amino]-1-propanaminium hydroxide, inner salt; Oleamidopropyl dimethyl glycine
Tradenames: Mackam HV; Mirataine® BET-O-30 (formerly Cycloteric BET-O-30); Schercotaine OAB;

Oleamidopropyl dihydroxypropyl dimonium chloride
Tradenames containing: Lexate BPQ

Oleamidopropyl dimethylamine (CTFA)
CAS 109-28-4; EINECS 203-661-5
Synonyms: Dimethylaminopropyl oleamide; N-[3-Dimethylamino)propyl]-9-octadecenamide
Tradenames: Mackine 501; Schercodine O

Oleamine (CTFA)
CAS 112-90-3; EINECS 204-015-5
Synonyms: Oleyl amine; 9-Octadcecen-1-amine
Tradenames: Amine OL; Amine OLD; Armeen® O; Armeen® OD; Armeen® OL; Armeen® OLD; Crodamine 1.O, 1.OD; Jet Amine PO; Jet Amine POD; Kemamine® P-989D; Nissan Amine OB; Noram O; Radiamine 6172; Radiamine 6173

Oleamine acetate
Tradenames: Catisol AO 100; Catisol AO C; Noramac O

Oleamine bishydroxypropyltrimonium chloride (CTFA)
Synonyms: Oleamine hydroxypropylbistrimonium chloride
Tradenames containing: Akypomine® P 191

Oleamine oxide (CTFA)
CAS 14351-50-9; EINECS 238-311-0
Synonyms: Oleyl dimethyl amine oxide; Oleylamine oxide; N,N-Dimethyl-9-octadecen-1-amine-N-oxide
Tradenames: Incromine Oxide OD-50; Mackamine O2; Mazox® ODA; Noxamine O2-30; Standamox O1

Oleic acid (CTFA)
CAS 112-80-1; EINECS 204-007-1
Synonyms: cis-9-Octadecenoic acid; Red oil; 9-Octadecenoic acid
Tradenames: Emersol® 210; Emersol® 213 NF; Emersol® 221 NF; Emersol® 233 LL; Emersol® 6313 NF; Emersol® 6321 NF; Emersol® 6333 NF; Emersol® 7021; Industrene® 104; Industrene® 105; Industrene® 106; Industrene® 205; Industrene® 206; Industrene® 206LP; Oleine D; Pamolyn® 100; Priolene 6900; Priolene 6901; Priolene 6905; Priolene 6906; Priolene 6907; Priolene 6910; Priolene 6922; Priolene 6930; Priolene 6933

Oleic diethanolamide. *See Oleamide DEA*

Oleic dipropylene triamine
Tradenames: Trinoram O

Oleic imidazoline acetate
Tradenames: Varine O Acetate

Oleic-linoleic acid
Tradenames: Industrene® 224

Oleic/linoleic amine
Tradenames: Kemamine® P-999

Oleic monoisopropanolamide. *See Oleamide MIPA*

Oleic-stearic acid
Tradenames: Industrene® M

Olein. *See Triolein*

Oleoamphocarboxyglycinate. *See Disodium oleoamphodiacetate*

Oleoamphohydroxypropylsulfonate. *See Sodium oleoamphohydroxypropylsulfonate*

Oleoamphopropionate. *See Sodium oleoamphopropionate*

Oleoamphopropylsulfonate. *See Sodium oleoamphohydroxypropylsulfonate*

Oleoyl PG-trimonium chloride (CTFA)
Tradenames containing: Akypoquat 40

Oleoyl sarcosine (CTFA)
CAS 110-25-8; EINECS 203-749-3
Synonyms: N-Methyl-N-(1-oxo-9-octadecenyl)glycine; Oleyl methylaminoethanoic acid; Oleyl sarcosine
Tradenames: Crodasinic O; Hamposyl® O; Sarkosyl® O; Vanseal OS

Oleth-2 (CTFA)
CAS 9004-98-2 (generic)
Synonyms: PEG-2 oleyl ether; POE (2) oleyl ether; PEG 100 oleyl ether
Tradenames: Ameroxol® OE-2; Eumulgin PWM2; Genapol® O-020; Hetoxol OL-2; Lipocol O-2; Macol® OA-2; Nikkol BO-2; Ritoleth 2; Serdox NOL 2; Simulsol 92

Oleth-3 (CTFA)
CAS 9004-98-2 (generic); 52581-71-2
Synonyms: PEG-3 oleyl ether; POE (3) oleyl ether
Tradenames: Hetoxol OA-3 Special; Volpo 3; Volpo N3; Volpo O3

Oleth-4 (CTFA)
CAS 9004-98-2 (generic); 5353-26-4
Synonyms: PEG-4 oleyl ether; POE (4) oleyl ether; PEG 200 oleyl ether
Tradenames: Chemal OA-4; Hetoxol OL-4; Macol® OA-4; Prox-onic OA-1/04

Oleth-5 (CTFA)
CAS 9004-98-2 (generic); 5353-27-5
Synonyms: PEG-5 oleyl ether; POE (5) oleyl ether; 3,6,9,12,15-Pentaoxatriacont-24-en-1-ol
Tradenames: Chemal OA-5; Eumulgin O5; Eumulgin PWM5; Eumulgin WM5; Genapol® O-050; Hetoxol OA-5 Special; Hetoxol OL-5; Lipocol O-5; Macol® OA-5; Marlipal® 1850/5; Marlowet® OA 5; Marlowet® WOE; Ritoleth 5; Volpo N5; Volpo O5
Tradenames containing: Eumulgin M8; Marlowet® SAF/K; Solulan® 5

Oleth-6 (CTFA)
CAS 9004-98-2 (generic)
Synonyms: PEG-6 oleyl ether; POE (6) oleyl ether; PEG 300 oleyl ether
Tradenames: Ablunol OA-6

Oleth-7 (CTFA)
CAS 9004-98-2 (generic)
Synonyms: PEG-7 oleyl ether; POE (7) oleyl ether
Tradenames: Ablunol OA-7; Akyporox RTO 70; Nikkol BO-7

Oleth-8 (CTFA)
CAS 9004-98-2 (generic)
Synonyms: PEG-8 oleyl ether; POE (8) oleyl ether; PEG 400 oleyl ether
Tradenames: Genapol® O-080; Rolfor O 8; Serdox NOL 8

Oleth-9 (CTFA)
CAS 9004-98-2 (generic)
Synonyms: PEG-9 oleyl ether; POE (9) oleyl ether; PEG 450 oleyl ether
Tradenames: Akyporox RO 90; Chemal OA-9; Genapol® O-090; Prox-onic OA-1/09

Oleth-10 (CTFA)
CAS 9004-98-2 (generic)
Synonyms: PEG-10 oleyl ether; POE (10) oleyl ether; PEG 500 oleyl ether
Tradenames: Ethal OA-10; Eumulgin O10; Eumulgin PWM10; Genapol® O-100; Hetoxol OA-10 Special; Hetoxol OL-10; Hetoxol OL-10H; Lipocol O-10; Macol® OA-10; Marlipal® 1850/10; Marlowet® OA 10; Nikkol BO-10TX; Ritoleth 10; Simulsol 96; Volpo N10; Volpo O10
Tradenames containing: Eumulgin M8

Oleth-12 (CTFA)
CAS 9004-98-2 (generic)
Synonyms: PEG-12 oleyl ether; POE (12) oleyl ether; PEG 600 oleyl ether
Tradenames: Genapol® O-120

Oleth-15 (CTFA)
CAS 9004-98-2 (generic)
Synonyms: PEG-15 oleyl ether; POE (15) oleyl ether
Tradenames: Genapol® O-150; Nikkol BO-15TX;

Serdox NOL 15; Volpo N15; Volpo O15

Oleth-16 (CTFA)
CAS 9004-98-2 (generic); 25190-05-0 (generic)
Synonyms: PEG-16 oleyl ether; POE (16) oleyl ether
Tradenames containing: Solulan® 16

Oleth-18
Synonyms: PEG-18 oleyl ether; POE (18) oleyl ether
Tradenames: Eumulgin PWM17

Oleth-20 (CTFA)
CAS 9004-98-2 (generic)
Synonyms: PEG-20 oleyl ether; POE (20) oleyl ether; PEG 1000 oleyl ether
Tradenames: Ahco 3998; Arosurf® 32-E20; Brij® 98G; Brij® 99; Chemal OA-20/70CWS; Chemal OA-20G; Genapol® O-200; Hetoxol OA-20 Special; Hetoxol OL-20; Hostacerin O-20; Industrol® LG-70; Industrol® LG-100; Industrol® OAL-20; Lipocol O-20; Macol® OA-20; Nikkol BO-20TX; Prox-onic OA-1/020; Prox-onic OA-2/020; Rhodasurf® ON-870 (formerly Emulphor® ON-870); Rhodasurf® ON-877 (formerly Emulphor® ON-877); Ritoleth 20; Simulsol 98; Trycol® 5971; Varonic® 32-E20; Volpo N20; Volpo O20

Oleth-22
Tradenames: Rolfor O 22

Oleth-23 (CTFA)
CAS 9004-98-2 (generic)
Synonyms: PEG-23 oleyl ether; POE (23) oleyl ether
Tradenames: Ethal OA-23; Genapol® O-230; Hetoxol OL-23; Trycol® 5972; Trycol® OAL-23

Oleth-24
Tradenames containing: Hetoxol C-24

Oleth-25 (CTFA)
CAS 9004-98-2 (generic)
Synonyms: PEG-25 oleyl ether; POE (25) oleyl ether
Tradenames: Eumulgin PWM25; Mulsifan RT 27
Tradenames containing: Emulgade EO-10

Oleth-30 (CTFA)
CAS 9004-98-2 (generic)
Synonyms: PEG-30 oleyl ether; POE (30) oleyl ether
Tradenames: Marlipal® 1850/30; Marlowet® OA 30

Oleth-40 (CTFA)
CAS 9004-98-2 (generic)
Synonyms: PEG-40 oleyl ether; POE (40) oleyl ether; PEG 2000 oleyl ether
Tradenames: Hetoxol OL-40; Marlipal® 1850/40

Oleth-50 (CTFA)
CAS 9004-98-2 (generic)
Synonyms: PEG-50 oleyl ether; POE (50) oleyl ether
Tradenames: Nikkol BO-50

Oleth-80
Tradenames: Marlipal® 1850/80

Oleth-3 carboxylic acid (CTFA)
Synonyms: PEG-3 oleyl ether carboxylic acid; POE (3) oleyl ether carboxylic acid
Tradenames: Akypo RO 20

Oleth-6 carboxylic acid (CTFA)
Synonyms: PEG-6 oleyl ether carboxylic acid; POE (6) oleyl ether carboxylic acid
Tradenames: Akypo RO 50

Oleth-10 carboxylic acid (CTFA)
Synonyms: PEG-10 oleyl ether carboxylic acid; POE (10) oleyl ether carboxylic acid; PEG 500 oleyl ether carboxylic acid
Tradenames: Akypo RO 90

Oleth-3 phosphate (CTFA)
CAS 39464-69-2 (generic)
Synonyms: PEG-3 oleyl ether phosphate; POE (3) oleyl ether phosphate; Oleyl triethoxy mono diphosphate
Tradenames: Crafol AP-11; Crodafos N3 Acid

Oleth-4 phosphate (CTFA)
CAS 39464-69-2 (generic)
Synonyms: PEG-4 oleyl ether phosphate; POE (4) oleyl ether phosphate; PEG 200 oleyl ether phosphate
Tradenames: Chemfac PB-184

Oleth-5 phosphate
Synonyms: PEG-5 oleyl ether phosphate; POE (5) oleyl ether phosphate
Tradenames: Crodafos N5 Acid

Oleth-10 phosphate (CTFA)
CAS 39464-69-2 (generic)
Synonyms: PEG-10 oleyl ether phosphate; POE (10) oleyl ether phosphate; PEG 200 oleyl ether phosphate
Tradenames: Crodafos N10 Acid

Oleyl alcohol (CTFA)
CAS 143-28-2; EINECS 205-597-3
Synonyms: 9-Octadecen-1-ol; cis-9-Octadecen-1-ol
Tradenames: Adol® 80; Adol® 85 NF; Adol® 90 NF; Cachalot® O-3; Cachalot® O-8; Cachalot® O-27; Emery® 3312; Emery® 3317; Fancol OA 95; HD-Ocenol 90/95; HD-Ocenol 92/96
Tradenames containing: Emery® 3310; Emery® 3311; Emery® 3313; Emery® 3314; Emery® 3315; Emery® 3316; Emery® 3318; HD-Ocenol 45/50; HD-Ocenol 50/55; HD-Ocenol 50/55III; HD-Ocenol 60/65; HD-Ocenol 70/75; HD-Ocenol 80/85; HD-Ocenol 110/130

Oleyl amidoethyl oleyl imidazoline
CAS 68310-76-9
Tradenames: Servamine KOO 330 B

Oleyl amine. *See Oleamine*

Oleyl betaine (CTFA)
CAS 871-37-4; EINECS 212-806-1
Synonyms: N-(Carboxymethyl)-N,N-dimethyl-9-octadecen-1-aminium hydroxide, inner salt; Oleyl dimethyl glycine
Tradenames: Mackam OB-30; Velvetex® OLB-50

Oleyl/cetyl sulfate
Tradenames: Montapol CST; Sulfopon O

Oleyl diamine
CAS 7173-62-8
Tradenames: Jet Amine DO

Oleyl dimethylamine. *See Dimethyl oleamine*

Oleyl dimethylamine oxide
Tradenames: Chemoxide O1

Oleyl dimethyl benzyl ammonium chloride. *See Olealkonium chloride*

Oleyl dimethylethyl ammonium ethosulfate
Tradenames: Larostat® 143

Oleyl dimethyl glycine. *See Oleyl betaine*

Oleyl hydroxyethyl imidazoline (CTFA)
CAS 95-38-5; 21652-27-7; 27136-73-8; EINECS 248-248-0; 244-501-4; 202-414-9
Synonyms: 2-(8-Heptadecenyl)-4,5-dihydro-1H-imidazole-1-ethanol; 1-Hydroxyethyl-2-oleyl imidazoline; Oleyl imidazoline
Tradenames: Amine O; Crodazoline O; Finazoline OA; Mackazoline O; Miramine® O (formerly Alkazine® O); Miramine® OC; Monazoline O; Nopcogen 22-O; Schercozoline O; Textamine O-1; Textamine O-5; Unamine® O; Varine O®

Oleyl imidazoline methosulfate
Tradenames: Empigen® FRH75S

Oleyl imidazolinium hydrochloride
CAS 62449-33-6
Tradenames: Norfox® IM-38

Oleyl propanediamine
CAS 68037-97-8
Tradenames: Duomeen® O; Radiamine 6572

Oleyl 1,3-propylene diamine
CAS 7173-62-8
Synonyms: Oleyl diamino propane
Tradenames: Diamin O; Diamine OL; Dinoram O; Duomeen® OL; Kemamine® D-989

Oleyl propylene diamine acetate
Tradenames: Dinoramac O

Oleyl propylene diamine dioleate
CAS 34140-91-5
Tradenames: Inipol OO2

Oleyl propylene diamine ditallate
Tradenames: Inipol OT2

Oleyl sarcosinic acid
Tradenames: Rewopol® SK 275

Oleyltrimonium chloride
Synonyms: Oleyl trimethyl ammonium chloride
Tradenames: Noramium MO 50
Tradenames containing: Arquad® S-2C-50

Olivamide DEA (CTFA)
Synonyms: N,N-Bis(2-hydoxyethyl)olive fatty acid amide; Diethanolamine olive fatty acid condensate; Olive oil fatty acid diethanolamide
Tradenames: Incromide OLD

Olivamidopropylamine oxide (CTFA)
Synonyms: N-[3-(Dimethylamino)propyl]olive amides-N-oxide
Tradenames: Incromine Oxide OL

Oxazolidine
CAS 497-25-6; 51200-87-4; 7747-35-5; 6542-37-6
Synonyms: 2-Oxazolidine
Tradenames: Amine CS-1135®

Palmitamide DEA (CTFA)
CAS 7545-24-6
Synonyms: N,N-Bis(2-hydroxyethyl)hexadecanamide; N,N-Bis (2-hydroxyethyl) palmitamide; Diethanolamine palmitic acid amide
Tradenames containing: Schercomid SLA

Palmitamidopropyl trimonium chloride
Tradenames: Servamine KEP 4527

Palmitamine (CTFA)
CAS 143-27-1; EINECS 205-596-8
Synonyms: Cetyl amine; Palmityl amine; 1-Hexadecanamine
Tradenames: Amine 16D; Armeen® 16; Armeen® 16D; Crodamine 1.16D; Kemamine® P-880, P-880D; Nissan Amine PB

Palmitamine oxide (CTFA)
CAS 7128-91-8; EINECS 230-429-0
Synonyms: Cetamine oxide; Cetyl dimethyl amine oxide; Hexadecyl dimethylamine oxide
Tradenames: Ammonyx® CO; Amyx CO 3764; Barlox® 16S; Mazox® CDA; Sandoz Amine Oxide XA-C; Sandoz Amine XA-Q
Tradenames containing: Aromox® DM16; Incromine Oxide MC

Palmitic acid (CTFA)
CAS 57-10-3; EINECS 200-312-9
Synonyms: Hexadecanoic acid; Cetylic acid; Hexadecylic acid
Tradenames: Emersol® 143; Hystrene® 8016; Hystrene® 9016; Industrene® 4516; Prifrac 2960

Palmitic/oleic acids
CAS 67701-08-0
Tradenames: HK-1618

Palmitic/stearic acid mono/diglycerides
Synonyms: Palmitic/stearic acid glycerol monodiester
Tradenames: Tegotens 4100; Tegotens BL 130

Palmitoyl PG-trimonium chloride (CTFA)
Tradenames containing: Akypoquat 40

Palmityl. *See also Cetyl*

Palmityl alcohol. *See Cetyl alcohol*

Palmityl dimethylamine. *See Dimethyl palmitamine*

Palm kernel alcohol (CTFA)
Synonyms: Alcohols, palm kernel
Tradenames: Laurex® PKH

Palm kernelamide DEA (CTFA)
Synonyms: Palm kernel oil acid diethanolamide; Diethanolamine palm kernel oil acid amide; N,N-Bis(2-hydroxyethyl)palm kernel oil acid amide
Tradenames: Accomid 50; Accomid PK; Lauridit® PPD; Mackamide PK; T-Tergamide 1PD

Palm kernelamide MEA (CTFA)
Synonyms: N-(2-Hydroxyethyl) palm kernel oil acid amide; Monoethanolamine palm kernel oil acid amide; Palm kernel oil acid monoethanolamide
Tradenames: Mackamide PKM

Palm oil glycerides (CTFA)
Synonyms: Glycerides, palm oil mono-, di- and tri-
Tradenames: Imwitor® 940

Paraffin oil. *See Mineral oil*

Pareth-15-3. *See C11-15 pareth-3*

Pareth-23-3. *See C12-13 pareth-3*

Pareth-25-3. *See C12-15 pareth-3*

Pareth-45-7. *See C14-15 pareth-7*

Pareth-91-2. *See C9-11 pareth-2*

PCA (CTFA)
CAS 98-79-3; EINECS 202-700-3
Synonyms: Pyrrolidonecarboxylic acid; 5-Oxo-L-proline; L-Pyroglutamic acid

Tradenames: Ajidew A-100
Tradenames containing: Ajidew SP-100

PCA ethyl N-cocoyl-L-arginate (CTFA)
Synonyms: N2 Cocoyl-L-arginine ethyl ester DL-pyrrolidone carboxylic acid salt
Tradenames: CAE

PCA Soda. *See Sodium PCA*

Peanutamide MEA (CTFA)
Synonyms: N-(2-Hydroxyethyl) peanut acid amide; Monoethanolamine peanut acid amide; Peaty fatty acid monoethanolamide
Tradenames: Rewomid® OM 101/G

Peanutamide MIPA (CTFA)
Synonyms: N-(2-Hydroxypropyl)peanut acid amide
Tradenames: Rewomid® OM 101/IG

PEG-4 (CTFA)
CAS 25322-68-3 (generic); 112-60-7; EINECS 203-989-9
Synonyms: PEG 200; POE (4); 2,2′-[Oxybis(2,1-ethanediyloxy)bisethanol
Tradenames: Carbowax® PEG 200; Droxol 200; Emery® 6773; Nopalcol 200; Pluracol® E200; Poly-G® 200; Teric PEG 200

PEG-6 (CTFA)
CAS 25322-68-3 (generic); 2615-15-8; EINECS 220-045-1
Synonyms: PEG 300; Hexaethylene glycol; Macrogol 300
Tradenames: Carbowax® PEG 300; Emery® 6687; Lutrol® E 300; Pluracol® E300; Poly-G® 300; Teric PEG 300
Tradenames containing: Pluracol® E1500

PEG-8 (CTFA)
CAS 25322-68-3 (generic); 5117-19-1; EINECS 225-856-4
Synonyms: PEG 400; POE (8); 3,6,9,12,15,18,21-Heptaoxatricosane-1,23-diol
Tradenames: Carbowax® PEG 400; Droxol 400; Lutrol® E 400; Nopalcol 400; Pluracol® E400; Pluracol® E400 NF; Poly-G® 400; Teric PEG 400

PEG-12 (CTFA)
CAS 25322-68-3 (generic); 6790-09-6; EINECS 229-859-1
Synonyms: PEG 600; POE (12); Macrogol 600
Tradenames: Carbowax® PEG 600; Droxol 600; Emery® 6686; Nopalcol 600; Pluracol® E600; Pluracol® E600 NF; Poly-G® 600; Teric PEG 600

PEG-14 (CTFA)
CAS 25322-68-3 (generic)
Synonyms: POE (14)
Tradenames: Rhodasurf® PEG 600 (formerly Alkapol PEG 600)

PEG-16 (CTFA)
CAS 25322-68-3 (generic)
Synonyms: PEG 800; POE (16)
Tradenames: Teric PEG 800

PEG-20 (CTFA)
CAS 25322-68-3 (generic)
Synonyms: PEG 1000; Macrogol 1000; POE (20)
Tradenames: Carbowax® PEG 900; Carbowax® PEG 1000; Pluracol® E1000; Poly-G® 1000; Teric PEG 1000

PEG-32 (CTFA)
CAS 25322-68-3 (generic)
Synonyms: PEG 1540; Macrogol 1540; POE (32)
Tradenames: Carbowax® PEG 1450; Lutrol® E 1500
Tradenames containing: Pluracol® E1500

PEG-40 (CTFA)
CAS 25322-68-3 (generic)
Synonyms: PEG 2000; POE (40)
Tradenames: Pluracol® E2000; Poly-G® 2000

PEG-75 (CTFA)
CAS 25322-68-3 (generic)
Synonyms: PEG 4000; POE (75)
Tradenames: Carbowax® PEG 3350; Lutrol® E 4000; Pluracol® E4000; Pluracol® E4000 NF; Teric PEG 4000

PEG-100 (CTFA)
CAS 25322-68-3 (generic)
Synonyms: PEG (100); POE (100)
Tradenames: Carbowax® PEG 4600

PEG 100. *See PEG-2*

PEG-150 (CTFA)
CAS 25322-68-3 (generic)
Synonyms: PEG 6000; Macrogol 6000; POE (150)
Tradenames: Carbowax® PEG 8000; Lutrol® E 6000; Lutrol® E 8000; Pluracol® E6000; Teric PEG 6000

PEG 200. *See PEG-4*

PEG 300. *See PEG-6*

PEG 400. *See PEG-8*

PEG 500. *See PEG-10*

PEG 600. *See PEG-12*

PEG 800. *See PEG-16*

PEG 1000. *See PEG-20*

PEG 1540. *See PEG-32*

PEG 2000. *See PEG-40*

PEG 4000. *See PEG-75*

PEG 6000. *See PEG-150*

PEG-2M (CTFA)
CAS 25322-68-3 (generic)
Synonyms: PEG-2000; POE (2000)
Tradenames: Prox-onic PEG-2000

PEG-4M
Tradenames: Prox-onic PEG-4000

PEG-6M
Tradenames: Prox-onic PEG-6000

PEG-8M
Synonyms: PEG (8000)
Tradenames: Rhodasurf® PEG 8000; Teric PEG 8000

PEG-10M
Tradenames: Prox-onic PEG-10,000

PEG-12M
Synonyms: PEG (12000)
Tradenames: Teric PEG 12000.

PEG-20M (CTFA)
CAS 25322-68-3 (generic)
Synonyms: PEG-20000; POE (20000)
Tradenames: Prox-onic PEG-20,000

PEG-35M
Synonyms: PEG (35000)
Tradenames: Prox-onic PEG-35,000

PEG-6 abietate
Synonyms: POE (6) abietate; PEG 300 abietate
Tradenames: Secoster® MA 300

PEG-60 almond glycerides (CTFA)
Synonyms: PEG 3000 almond glycerides; POE (60) almond glycerides
Tradenames: Crovol A70

PEG-6 betanaphthol
Tradenames: Igepal® 132

PEG-8 betanaphthol
Tradenames: Igepal® 131 (formerly Rhodiasurf 131)

PEG-13 betanaphthol
CAS 35545-57-4
Tradenames: Hetoxide BN-13

PEG-6 bisphenol A
Tradenames: Ethal BPA-6

PEG-2 butyl ether. *See Butoxy diglycol*

PEG-2 butyl ether carboxylic acid. *See Buteth-2 carboxylic acid*

PEG-2 butynediol
Tradenames: Hetoxide BY-1.8

PEG-3 butynediol
Tradenames: Hetoxide BY-3

PEG-2 C13-15 alkyl amine
Tradenames: Synprolam 35X2

PEG-5 C13-15 alkyl amine
Tradenames: Synprolam 35X5

PEG-10 C13-15 alkyl amine
Tradenames: Synprolam 35X10

PEG-15 C13-15 alkyl amine
Tradenames: Synprolam 35X15; Teric 13M15

PEG-20 C13-15 alkyl amine
Tradenames: Synprolam 35X20

PEG-25 C13-15 alkyl amine
Tradenames: Synprolam 35X25

PEG-35 C13-15 alkyl amine
Tradenames: Synprolam 35X35

PEG-50 C13-15 alkyl amine
Tradenames: Synprolam 35X50

PEG-1 C13-15 alkyl methyl amine;
CAS 92112-62-4
Tradenames: Synprolam 35MX1

PEG-3 C13-15 alkylmethylamine
Tradenames: Synprolam 35MX3

PEG-5 C13-15 alkylmethylamine
Tradenames: Synprolam 35MX5

PEG-6 caprylic/capric glycerides (CTFA)
CAS 52504-24-2
Synonyms: PEG 300 caprylic/capric glycerides; POE (6) caprylic/capric glycerides
Tradenames: Softigen® 767; Sterol CC 595

PEG-8 caprylic/capric glycerides (CTFA)
CAS 57307-99-0
Synonyms: PEG 400 caprylate/caprate glycerides
Tradenames: L.A.S.; Labrasol

PEG-12 caprylic ether
Synonyms: PEG-12 octyl alcohol
Tradenames: Rhodasurf® N-12 (formerly Texafor N-12)

PEG-2 castor oil (CTFA)
CAS 61791-12-6 (generic)
Synonyms: POE (2) castor oil; PEG 100 castor oil
Tradenames: Hetoxide C-2

PEG-3 castor oil (CTFA)
CAS 61791-12-6 (generic)
Synonyms: POE (3) castor oil
Tradenames: Akyporox RZO 30

PEG-5 castor oil (CTFA)
CAS 61791-12-6 (generic)
Synonyms: POE (5) castor oil
Tradenames: Ablunol CO 5; Acconon CA-5; Chemax CO-5; Emulsogen EL-050; Ethox CO-5; Prox-onic HR-05; Surfactol® 318; Trylox® 5900; Trylox® CO-5

PEG-6 castor oil
CAS 61691-12-6 (generic)
Tradenames: Desonic® 6C

PEG-8 castor oil (CTFA)
CAS 61791-12-6 (generic)
Synonyms: POE (8) castor oil; PEG 400 castor oil
Tradenames: Acconon CA-8

PEG-9 castor oil (CTFA)
CAS 61791-12-6 (generic)
Synonyms: POE (9) castor oil; PEG 450 castor oil
Tradenames: Acconon CA-9; Hetoxide C-9

PEG-10 castor oil (CTFA)
CAS 61791-12-6 (generic)
Synonyms: POE (10) castor oil; PEG 500 castor oil
Tradenames: Ablunol CO 10; Alkamuls® D-10 (formerly Texafor D-10); Etocas 10; Trylox® CO-10

PEG-11 castor oil (CTFA)
CAS 61791-12-6 (generic)
Synonyms: PEG (11) castor oil; POE (11) castor oil
Tradenames: Marlowet® R 11/K

PEG-12 castor oil
CAS 61791-12-6 (generic)
Synonyms: POE (12) castor oil
Tradenames: Teric C12

PEG-15 castor oil (CTFA)
CAS 61791-12-6 (generic)
Synonyms: POE (15) castor oil
Tradenames: Ablunol CO 15; Acconon CA-15; Alkamuls® CO-15 (formerly Alkasurf® CO-15); Hetoxide C-15

PEG-16 castor oil
CAS 61791-12-6 (generic)
Synonyms: POE (16) castor oil
Tradenames: Chemax CO-16; Ethox CO-16; Mapeg® CO-16; Prox-onic HR-016; Trylox® 5902; Trylox® CO-16

PEG-17 castor oil
CAS 61791-12-6 (generic)
Synonyms: POE (17) castor oil
Tradenames: Servirox OEG 45

PEG-18 castor oil
Tradenames: Alkamuls® R81 (formerly Rhodiasurf R81)

PEG-20 castor oil (CTFA)
CAS 61791-12-6 (generic)
Synonyms: POE (20) castor oil; PEG 1000 castor oil
Tradenames: Alkamuls® D-20 (formerly Texafor D-20); Berol 829; Etocas 20; Nopalcol 10-CO; Sellig R 20 100

PEG-21 castor oil
Tradenames: Sellig HR 18 100

PEG-22 castor oil
Tradenames: Alkamuls® RC (formerly Soprophor RC)

PEG-23 castor oil
Tradenames: Remcopal 4018

PEG-25 castor oil (CTFA)
CAS 61791-12-6 (generic)
Synonyms: POE (25) castor oil
Tradenames: Cerex EL 250; Chemax CO-25; Emulsogen EL-250; Emulson CO 25; Ethox CO-25; Hetoxide C-25; Industrol® CO-25; Mapeg® CO-25; Prox-onic HR-025; Trylox® 5904; Trylox® CO-25

PEG-26 castor oil
CAS 61791-12-6 (generic)
Synonyms: POE (26) castor oil
Tradenames: Servirox OEG 55

PEG-28 castor oil
CAS 61791-12-6 (generic)
Synonyms: POE (28) castor oil
Tradenames: Berol 106; Chemax CO-28

PEG-30 castor oil (CTFA)
CAS 61791-12-6 (generic)
Synonyms: POE (30) castor oil
Tradenames: Ablunol CO 30; Alkamuls® D-30; Alkamuls® EL-620 (formerly Emulphor® EL-620); Alkamuls® EL-620L (formerly Emulphor® EL-620L); Cerex EL 300; Chemax CO-30; Desonic® 30C; Emulsogen EL-300; Ethox CO-30; Etocas 30; Hetoxide C-30; Incrocas 30; Industrol® CO-30; Mapeg® CO-30; Prox-onic HR-030; Sellig R 3395 SP; Stepantex® CO-30; Toximul® 8241; Trylox® 5906; Trylox® CO-30; Witconol 5906

PEG-31 castor oil
Tradenames: Remcopal 40; Remcopal 4000

PEG-32 castor oil
CAS 61791-12-6 (generic)
Synonyms: POE (32) castor oil
Tradenames: Berol 195; Berol 199; Sellig R 3395-C435; Servirox OEG 65

PEG-33 castor oil (CTFA)
CAS 61791-12-6 (generic)
Synonyms: POE (33) castor oil
Tradenames: Alkamuls® B (formerly Rhodiasurf B); Alkamuls® BR (formerly Rhodiasurf BR); Sellig R 3395

PEG-35 castor oil (CTFA)
CAS 61791-12-6 (generic)
Synonyms: POE (35) castor oil
Tradenames: Emulson CO 40 N; Etocas 35

PEG-36 castor oil (CTFA)
CAS 61791-12-6 (generic)
Synonyms: POE (36) castor oil; PEG 1800 castor oil
Tradenames: Alkamuls® OR/36 (formerly Soprophor® OR/36); Cerex EL 360; Chemax CO-36; Desonic® 36C; Emulsogen EL-360; Ethox CO-36; Eumulgin PRT 36; Industrol® CO-36; Makon® 8240; Mapeg® CO-36; Prox-onic HR-036; Stepantex® CO-36; Toximul® 8240; Trylox® 5907; Trylox® CO-36; Witconol 5907

PEG-40 castor oil (CTFA)
CAS 61791-12-6 (generic)
Synonyms: POE (40) castor oil; PEG 2000 castor oil
Tradenames: Alkamuls® CO-40 (formerly Alkasurf® CO-40); Alkamuls® EL-719 (formerly Emulphor® EL-719); Alkamuls® EL-719L (formerly Emulphor® EL-719L); Alkasurf® CO-40 (redesignated Alkamuls® CO-40); Berol 108; Cerex EL 400; Chemax CO-40; Desonic® 40C; Emulpon EL 40; Emulsogen EL-400; Emulson CO-40; Ethox CO-40; Etocas 40; Eumulgin PRT 40; Eumulgin RO 40; Hetoxide C-40; Incrocas 40; Industrol® CO-40; Marlowet® R 40/K; Mulsifan RT 69; Prox-onic HR-040; Remcopal 40 S3; Remcopal 40 S3 LE; Sellig R 4095; Simulsol OL 50; Stepantex® CO-40; Surfactol® 365; T-Det® C-40; Toximul® 8242; Trylox® 5909; Trylox® CO-40; Witconol 5909
Tradenames containing: Carsoquat® 816-C; Maquat SC-1632; Marlowet® RNP/K

PEG-44 castor oil
Tradenames: Sellig R 4495

PEG-45 castor oil
CAS 61791-12-6 (generic)
Synonyms: POE (45) castor oil
Tradenames: Ablunol CO 45

PEG-50 castor oil (CTFA)
CAS 61791-12-6 (generic)
Synonyms: POE (50) castor oil
Tradenames: Etocas 50

PEG-54 castor oil (CTFA)
CAS 61791-12-6 (generic)
Synonyms: POE (54) castor oil
Tradenames: Desonic® 54C

PEG-55 castor oil (CTFA)
CAS 61791-12-6 (generic)
Synonyms: PEG (55) castor oil; POE (55) castor oil
Tradenames: Emulson EL

PEG-56 castor oil
Tradenames: Eumulgin PRT 56

PEG-60 castor oil (CTFA)
CAS 61791-12-6 (generic)
Synonyms: POE (60) castor oil
Tradenames: Alkamuls® 14/R (formerly Soprophor 14/R); Etocas 60; Hetoxide C-60; Simulsol 1285

PEG-70 castor oil
Tradenames: Marlosol® R70

PEG-75 castor oil
Tradenames: Berol 190

PEG-80 castor oil
CAS 61791-12-6 (generic)
Synonyms: POE (80) castor oil
Tradenames: Chemax CO-80; Ethox CO-81; Industrol® CO-80-80%; Prox-onic HR-080; Trylox® CO-80

PEG-81 castor oil
CAS 61791-12-6
Tradenames: Emulson AG 81C

PEG-100 castor oil (CTFA)
CAS 61791-12-6 (generic)
Synonyms: POE (100) castor oil; PEG (100) castor oil
Tradenames: Etocas 100

PEG-160 castor oil
CAS 61791-12-6 (generic)
Tradenames: Berol 198

PEG-180 castor oil
CAS 61791-12-6 (generic)
Synonyms: POE (180) castor oil
Tradenames: Servirox OEG 90/50

PEG-200 castor oil (CTFA)
CAS 61791-12-6 (generic)
Synonyms: POE (200) castor oil; PEG (200) castor oil
Tradenames: Alkamuls® EL-980 (formerly Emulphor® EL-980); Alkamuls® EL-985 (formerly Emulphor® EL-985); Berol 191; Chemax CO-200/50; Emulsogen EL-2000; Emulson EL 200; Ethox CO-200; Ethox CO-200/50%; Etocas 200; Eumulgin PRT 200; Hetoxide C-200; Hetoxide C-200-50%; Industrol® CO-200; Industrol® CO-200-50%; Mapeg® CO-200; Prox-onic HR-0200; Prox-onic HR-0200/50; Trylox® 5918; Trylox® CO-200, -200/50

PEG-18 castor oil dioleate (CTFA)
Synonyms: POE (18) castor oil dioleate
Tradenames: Marlowet® LVS/K
Tradenames containing: Marlowet® LVX/K

PEG-30 castor oil glycerides
Tradenames: Dacospin 1735-A

PEG cetyl ether. *See Ceteth Series*

PEG cetyl/oleyl ether. *See Cetoleth Series*

PEG cetyl/stearyl ether. *See Ceteareth Series*

PEG-2 cocamide
Tradenames: Eumulgin PC 2

PEG-3 cocamide (CTFA)
CAS 61791-08-0 (generic)
Synonyms: PEG (3) coconut amide; POE (3) coconut amide
Tradenames: Amidox® C-2; Unamide® C-2

PEG-4 cocamide
Tradenames: Eumulgin PC 4

PEG-5 cocamide (CTFA)
CAS 61791-08-0 (generic)
Synonyms: POE (5) coconut amide
Tradenames: Eumulgin C4; Genagen CA-050

PEG-6 cocamide (CTFA)
CAS 61791-08-0 (generic)
Synonyms: POE (6) coconut amide; PEG 300 coconut amide
Tradenames: Amidox® C-5; Empilan® MAA; Rewopal® C 6; Unamide® C-5

PEG-10 cocamide
Tradenames: Eumulgin PC 10; Eumulgin PC 10/85

PEG-3 cocamide MEA
Tradenames: Teric CME3

PEG-6 cocamide MEA
Tradenames: Mazamide® C-5

PEG-7 cocamide MEA
Tradenames: Teric CME7

PEG-2 cocamine (CTFA)
CAS 61791-14-8 (generic); 61791-31-9
Synonyms: PEG 100 coconut amine; POE (2) coconut amine; Bis (2-hydroxyethyl) coco amine
Tradenames: Accomeen C2; Berol 307; Chemeen C-2; Ethomeen® C/12; Ethox CAM-2; Hetoxamine C-2; Mazeen® C-2; Noramox C2; Prox-onic MC-02; Teric 12M2; Varonic® K202; Varonic® K202 SF

PEG-5 cocamine (CTFA)
CAS 61791-14-8 (generic)
Synonyms: POE (5) coconut amine
Tradenames: Accomeen C5; Alkaminox® C-5; Chemeen C-5; Ethomeen® C/15; Hetoxamine C-5; Mazeen® C-5; Noramox C5; Prox-onic MC-05; Teric 12M5; Varonic® K205; Varonic® K205 SF

PEG-7 cocamine
Tradenames: Ablumox C-7

PEG-8 cocamine
Tradenames: Zusomin C 108

PEG-10 cocamine (CTFA)
CAS 61791-14-8 (generic)
Synonyms: POE (10) coconut amine; PEG 500 coconut amine
Tradenames: Accomeen C10; Chemeen C-10; Ethomeen® C/20; Ethylan® TN-10; Mazeen® C-10; Trymeen® 6601; Trymeen® CAM-10; Varonic® K210; Varonic® K210 SF

PEG-11 cocamine
Tradenames: Noramox C11

PEG-12 cocamine
Tradenames: Eumulgin PA 12

PEG-14 cocamine
Tradenames: Amiet CD/14

PEG-15 cocamine (CTFA)
CAS 8051-52-3 (generic); 61791-14-8 (generic)
Synonyms: POE (15) coconut amine
Tradenames: Accomeen C15; Berol 397; Chemeen C-15; Ethomeen® C/25; Ethox CAM-15; Ethylan® TC; Hetoxamine C-15; Mazeen® C-15; Noramox C15; Prox-onic MC-015; Teric 12M15; Trymeen® CAM-15; Varonic® K215; Varonic® K215 LC; Varonic® K215 SF

PEG-17 cocamine
Tradenames: Amiet CD/17

PEG-22 cocamine
Tradenames: Amiet CD/22

PEG-27 cocamine
Tradenames: Amiet CD/27

PEG-2 cocoate
Synonyms: PEG 100 cocoate; POE (2) monococoate; Diglycol coconate
Tradenames: Witconol RDC-D

PEG-5 cocoate (CTFA)
CAS 61791-29-5 (generic)
Synonyms: POE (5) monococoate
Tradenames: Ethofat® C/15; Prox-onic CC-05

PEG-8 cocoate (CTFA)
CAS 61791-29-5 (generic)
Synonyms: POE (8) monococoate; PEG 400 monococoate
Tradenames: Emulsan K; Nopalcol 4-C; Nopalcol 4-CH; Pegosperse® 400 MC

PEG-9 cocoate
CAS 67762-35-0
Tradenames: Prox-onic CC-09; Rolfat C 9

PEG-10 cocoate
CAS 61791-29-5 (generic)

Synonyms: POE (10) monococoate; PEG 500 monococoate
Tradenames: Crodet C10; Genagen C-100; Nonex C5E

PEG-14 cocoate
Tradenames: Prox-onic CC-014

PEG-15 cocoate (CTFA)
CAS 61791-29-5 (generic)
Synonyms: POE (15) monococoate
Tradenames: Ethofat® C/25

PEG-23 cocoate
Tradenames: Eumulgin PK 23

PEG-200 cocoate
Tradenames: Polirol C5

PEG-2 coco-benzonium chloride (CTFA)
CAS 61789-68-2
Synonyms: PEG-2 cocobenzyl ammonium chloride; PEG 100 coco-benzonium chloride; POE (2) coco-benzonium chloride
Tradenames containing: Ethoquad® CB/12

PEG-2 cocomonium chloride (CTFA)
CAS 70750-47-9
Synonyms: PEG 100 cocomonium chloride; POE (2) cocomonium chloride; Methyl bis (2-hydroxyethyl) cocammonium chloride
Tradenames containing: Ethoquad® C/12; Variquat® 638

PEG-15 cocomonium chloride (CTFA)
CAS 61791-10-4
Synonyms: POE (15) cocomonium chloride; Methylpolyoxyethylene (15) coco ammonium chloride; PEG-15 coco methyl ammonium chloride
Tradenames: Ethoquad® C/25; Tomah Q-C-15; Variquat® K1215

PEG-5 cocomonium methosulfate (CTFA)
CAS 68989-03-7
Synonyms: POE (5) cocomonium methosulfate; Coconut pentaethoxy methyl ammonium methyl sulfate
Tradenames: Rewoquat CPEM

PEG-2 cocomonium nitrate
CAS 71487-00-8
Synonyms: PEG-2 cocomethyl ammonium nitrate
Tradenames containing: Ethoquad® C/12 Nitrate

PEG coconut ether. *See Coceth Series*

PEG decyl ether. *See Deceth Series*

PEG-3 1,3-diaminopropane
Tradenames: Tomah E-DT-3

PEG-4 dicocoate
Synonyms: Tetraethylene dicocoate
Tradenames: Mapeg® 200 DC

PEG-2 dilaurate (CTFA)
CAS 9005-02-1 (generic)
Synonyms: PEG 100 dilaurate; POE (2) dilaurate
Tradenames: Cithrol DGDL N/E

PEG-2 dilaurate SE
Tradenames: Cithrol DGDL S/E

PEG-4 dilaurate (CTFA)
CAS 9005-02-1 (generic)
Synonyms: PEG 200 dilaurate; POE (4) dilaurate
Tradenames: Acconon 200-DL; Cithrol 2DL; Emerest® 2622; Emerest® 2704; Ethox DL-5; Hetoxamate 200 DL; Hodag 22-L; Karapeg 200-DL; Kessco® PEG 200 DL; Lexemul® PEG-200 DL; Mapeg® 200 DL; Nonex DL-2; Nopalcol 2-DL; Pegosperse® 200 DL; Witconol 2622

PEG-6 dilaurate (CTFA)
CAS 9005-02-1 (generic)
Synonyms: POE (6) dilaurate; PEG 300 dilaurate
Tradenames: Kessco® PEG 300 DL

PEG-8 dilaurate (CTFA)
CAS 9005-02-1 (generic)
Synonyms: POE (8) dilaurate; PEG 400 dilaurate
Tradenames: CPH-79-N; Cithrol 4DL; Emerest® 2652; Ethox DL-9; Hodag 42-L; Industrol® DL-9; Karapeg 400-DL; Kessco® PEG 400 DL; Lexemul® PEG-400 DL; Mapeg® 400 DL; Pegosperse® 400 DL

PEG-12 dilaurate (CTFA)
CAS 9005-02-1 (generic)
Synonyms: POE (12) dilaurate; PEG 600 dilaurate
Tradenames: Cithrol 6DL; Karapeg 600-DL; Kessco® PEG 600 DL; Mapeg® 600 DL

PEG-20 dilaurate (CTFA)
CAS 9005-02-1 (generic)
Synonyms: POE (20) dilaurate; PEG 1000 dilaurate
Tradenames: Cithrol 10DL; Kessco® PEG 1000 DL

PEG-32 dilaurate (CTFA)
CAS 9005-02-1 (generic)
Synonyms: POE (32) dilaurate; PEG 1540 dilaurate
Tradenames: Kessco® PEG 1540 DL

PEG-75 dilaurate (CTFA)
CAS 9005-02-1 (generic)
Synonyms: POE (75) dilaurate; PEG 4000 dilaurate
Tradenames: Kessco® PEG 4000 DL

PEG-150 dilaurate (CTFA)
CAS 9005-02-1 (generic)
Synonyms: POE (150) dilaurate; PEG 6000 dilaurate
Tradenames: Kessco® PEG 6000 DL

PEG dinonyl phenyl ether. *See Nonyl nonoxynol Series*

PEG-3 dioctoate
Synonyms: Triethylene glycol dioctoate
Tradenames: Mackester TD-88

PEG-2 dioleate
CAS 9005-07-6 (generic); 52668-97-0 (generic)
Synonyms: Diethylene glycol dioleate; POE (2) dioleate; PEG 100 dioleate
Tradenames: Cithrol DGDO N/E; Karapeg DEG-DO

PEG-2 dioleate SE
Tradenames: Cithrol DGDO S/E

PEG-4 dioleate (CTFA)
CAS 9005-07-6 (generic); 52688-97-0 (generic)
Synonyms: POE (4) dioleate; PEG 200 dioleate
Tradenames: Chemax PEG 200 DO; Cithrol 2DO; Karapeg 200-DO; Kessco® PEG 200 DO; Mapeg® 200 DO

PEG-6 dioleate (CTFA)
CAS 9005-07-6 (generic); 52688-97-0 (generic)
Synonyms: POE (6) dioleate; PEG 300 dioleate
Tradenames: Kessco® PEG 300 DO

PEG-6-32 dioleate (CTFA)
CAS 9005-07-6 (generic); 52688-97-0 (generic)
Synonyms: POE (1500) dioleate; PEG 1500 dioleate
Tradenames: Pegosperse® 1500 DO

PEG-8 dioleate (CTFA)
CAS 9005-07-6 (generic); 52688-97-0 (generic)
Synonyms: POE (8) dioleate; PEG 400 dioleate
Tradenames: Acconon 400-DO; Alkamuls® 400-DO; Chemax PEG 400 DO; Cithrol 4DO; Emerest® 2648; Ethox DO-9; Hodag 42-O; Industrol® DO-9; Karapeg 400-DO; Kessco® PEG 400 DO; Lonzest® PEG 4-DO; Mapeg® 400 DO; Nonex DO-4; Nonisol 210; Pegosperse® 400 DO; Radiasurf® 7443; Surfax 8916/A; Witconol 2648; Witconol H33

PEG-9 dioleate
Synonyms: POE (9) dioleate
Tradenames: Trydet DO-9

PEG-12 dioleate (CTFA)
CAS 9005-07-6 (generic); 52688-97-0 (generic)
Synonyms: POE (12) dioleate; PEG 600 dioleate
Tradenames: Alkamuls® 600-DO; Chemax PEG 600 DO; Cithrol 6DO; Dyafac PEG 6DO; Emerest® 2665; Ethox DO-14; Hodag 62-O; Industrol® DO-13; Karapeg 600-DO; Kessco® PEG 600 DO; Mapeg® 600 DO; Marlipal® FS; Marlosol® FS; Nopalcol 6-DO; Radiasurf® 7444; Secoster® DO 600
Tradenames containing: Marlowet® LVX/K

PEG-20 dioleate (CTFA)
CAS 9005-07-6 (generic); 52688-97-0 (generic)
Synonyms: POE (20) dioleate; PEG 1000 dioleate
Tradenames: Cithrol 10DO; Kessco® PEG 1000 DO

PEG-32 dioleate (CTFA)
CAS 9005-07-6 (generic); 52688-97-0 (generic)
Synonyms: POE (32) dioleate; PEG 1540 dioleate
Tradenames: Kessco® PEG 1540 DO

PEG-75 dioleate (CTFA)
CAS 9005-07-6 (generic); 52688-97-0 (generic)
Synonyms: POE (75) dioleate; PEG 4000 dioleate
Tradenames: Kessco® PEG 4000 DO

PEG-150 dioleate (CTFA)
CAS 9005-07-6 (generic); 52688-97-0 (generic)
Synonyms: POE (150) dioleate; PEG 6000 dioleate
Tradenames: Kessco® PEG 6000 DO

PEG-2 distearate (CTFA)
CAS 109-30-8; EINECS 203-663-6
Synonyms: POE (2) distearate; PEG 100 distearate
Tradenames: Cithrol DGDS N/E

PEG-2 distearate SE
Tradenames: Cithrol DGDS S/E

PEG-4 distearate (CTFA)
CAS 9005-08-7 (generic)
Synonyms: POE (4) distearate; PEG 200 distearate
Tradenames: Cithrol 2DS; Kessco® PEG 200 DS; Mapeg® 200 DS

PEG-6 distearate (CTFA)
CAS 9005-08-7 (generic)
Synonyms: POE (6) distearate; PEG 300 distearate
Tradenames: Cithrol 3DS; Kessco® PEG 300 DS

PEG-8 distearate (CTFA)
CAS 9005-08-7 (generic)
Synonyms: POE (8) distearate; PEG 400 distearate
Tradenames: Alkamuls® 400-DS; Cithrol 4DS; Emerest® 2642; Emerest® 2712; Hetoxamate 400 DS; Hodag 42-S; Karapeg 400-DS; Kessco® PEG 400 DS; Mapeg® 400 DS; PGE-400-DS; Pegosperse® 400 DS; Radiasurf® 7453; Witconol 2642

PEG-12 distearate (CTFA)
CAS 9005-08-7 (generic)
Synonyms: POE (12) distearate; PEG 600 distearate
Tradenames: Cithrol 6DS; Kessco® PEG 600 DS; Mapeg® 600 DS; Marlosol® BS; PGE-600-DS; Radiasurf® 7454

PEG-20 distearate (CTFA)
CAS 9005-08-7 (generic)
Synonyms: POE (20) distearate; PEG 1000 distearate
Tradenames: Cithrol 10DS; Kessco® PEG 1000 DS

PEG-32 distearate (CTFA)
CAS 9005-08-7 (generic)
Synonyms: POE (32) distearate; PEG 1540 distearate
Tradenames: Kessco® PEG 1540 DS; Mapeg® 1540 DS

PEG-75 distearate (CTFA)
CAS 9005-08-7 (generic)
Synonyms: POE (75) distearate; PEG 4000 distearate
Tradenames: Kessco® PEG 4000 DS

PEG-150 distearate (CTFA)
CAS 9005-08-7 (generic)
Synonyms: POE (150) distearate; PEG 6000 distearate
Tradenames: Ablunol 6000DS; Karapeg 6000-DS; Kessco® PEG 6000 DS; Mapeg® 6000 DS; Pegosperse® 6000 DS
Tradenames containing: Miracare® BC-10; Miracare® MS-1 (formerly Compound MS-1); Miracare® MS-2 (formerly Compound MS-2); Miracare® MS-4

PEG-4 ditallate
CAS 61791-01-3 (generic)
Synonyms: POE (4) ditallate; PEG 200 ditallate
Tradenames: Mapeg® 200 DOT

PEG-8 ditallate (CTFA)
CAS 61791-01-3 (generic)
Synonyms: POE (8) ditallate; PEG 400 ditallate
Tradenames: Ethox DTO-9A; Laurel PEG 400 DT; Mapeg® 400 DOT; Pegosperse® 400 DOT

PEG-12 ditallate (CTFA)
CAS 61791-01-3 (generic)
Synonyms: POE (12) ditallate; PEG 600 ditallate
Tradenames: Industrol® DT-13; Laurel PEG 600 DT; Mapeg® 600 DOT; Pegosperse® 600 DOT; R3124 Ester

PEG-8 ditallowate
Tradenames: Nopalcol 4-DTW

PEG-12 ditallowate
Tradenames: Nopalcol 6-DTW

PEG-8 di/triricinoleate (CTFA)
Synonyms: POE (8) di-tri-ricinoleate; PEG 400 di-tri-ricinoleate
Tradenames: Pegosperse® 400 DTR

PEG-22/dodecyl glycol copolymer (CTFA)
Tradenames: Elfacos® ST 37

PEG-45/dodecyl glycol copolymer (CTFA)
Tradenames: Elfacos® ST 9

PEG dodecyl phenyl ether. *See Dodoxynol Series*

PEG-2 2-ethylhexyl ether
Synonyms: PEG-2 2-ethylhexanol; POE (2) 2-ethylhexanol
Tradenames: Chemal 2EH-2; Ethal EH-2; Prox-onic 2EHA-1/02

PEG-3 ethylhexyl ether
Tradenames: Hetoxol CD-3

PEG-4-ethylhexyl ether
Tradenames: Hetoxol CD-4

PEG-5-ethylhexyl ether
Tradenames: Chemal 2EH-5; Ethal EH-5; Prox-onic 2EHA-1/05

PEG-7 glyceryl cocoate (CTFA)
CAS 66105-29-1; 68201-46-7 (generic)
Synonyms: POE (7) glyceryl monococoate; PEG (7) glyceryl monococoate
Tradenames: Mazol® 159

PEG-30 glyceryl cocoate (CTFA)
CAS 68201-46-7 (generic)
Synonyms: POE (30) glyceryl monococoate
Tradenames: Varonic® LI-63

PEG-78 glyceryl cocoate (CTFA)
CAS 68201-46-7 (generic)
Synonyms: POE (78) glyceryl monococoate
Tradenames: Simulsol CG

PEG-80 glyceryl cocoate (CTFA)
CAS 68201-46-7 (generic)
Synonyms: POE (80) glyceryl monococoate
Tradenames: Varonic® LI-67; Varonic® LI-67-75%

PEG-12 glyceryl dioleate (CTFA)
Synonyms: PEG 600 glyceryl dioleate; POE (12) glyceryl dioleate
Tradenames: Marlowet® G 12 DO

PEG glyceryl ether. *See Glycereth Series*

PEG-20 glyceryl isostearate (CTFA)
CAS 69468-44-6
Synonyms: PEG 1000 glyceryl isostearate; POE (20) glyceryl isostearate
Tradenames: Tagat® I2

PEG-30 glyceryl isostearate (CTFA)
CAS 69468-44-6
Synonyms: POE (30) glyceryl isostearate
Tradenames: Tagat® I

PEG-8 glyceryl laurate
Synonyms: POE (8) glyceryl laurate
Tradenames: Sterol LG 491

PEG-20 glyceryl laurate (CTFA)
CAS 59070-56-3 (generic); 51248-32-9
Synonyms: POE (20) glyceryl monolaurate; PEG 1000 glyceryl monolaurate
Tradenames: Lamacit GML 20; Tagat® L2

PEG-23 glyceryl laurate (CTFA)
CAS 59070-56-3 (generic)
Synonyms: POE (23) glyceryl laurate
Tradenames: Aldosperse® ML 23

PEG-30 glyceryl laurate (CTFA)
CAS 59070-56-3 (generic); 51248-32-9
Synonyms: POE (30) glyceryl laurate
Tradenames: Tagat® L

PEG-20 glyceryl oleate (CTFA)
CAS 68889-49-6 (generic); 51192-09-7
Synonyms: POE (20) glyceryl oleate; PEG 1000 glyceryl monooleate
Tradenames: Tagat® O2

PEG-30 glyceryl oleate (CTFA)
CAS 68889-49-6 (generic); 51192-09-7
Synonyms: POE (30) glyceryl oleate
Tradenames: Tagat® O

PEG-15 glyceryl ricinoleate (CTFA)
CAS 51142-51-9 (generic); 39310-72-0
Synonyms: POE (15) glyceryl monoricinoleate
Tradenames: Tagat® RI

PEG-5 glyceryl sesquioleate (CTFA)
Synonyms: POE (5) glyceryl sesquioleate
Tradenames: Marlowet® GDO 4

PEG-20 glyceryl stearate (CTFA)
CAS 68553-11-7; 51158-08-8
Synonyms: PEG 1000 glyceryl monostearate; POE (20) glyceryl monostearate
Tradenames: Aldo® MS-20 FG; Aldosperse® MS-20; Durfax® EOM; Radiasurf® 7000; Tagat® S2; Varonic® LI-42
Tradenames containing: Tally® 100 Plus

PEG-30 glyceryl stearate (CTFA)
CAS 51158-08-8
Synonyms: POE (30) glyceryl monostearate
Tradenames: Tagat® S

PEG-200 glyceryl stearate (CTFA)
Synonyms: POE (200) glyceryl stearate
Tradenames: Simulsol 220 TM

PEG-80 glyceryl tallowate (CTFA)
Synonyms: POE (80) glyceryl monotallowate
Tradenames: Varonic® LI-48

PEG-25 glyceryl trioleate (CTFA)
CAS 68958-64-5
Synonyms: POE (25) glyceryl trioleate
Tradenames: Tagat® TO

PEG-10 glycol tallate
Tradenames: Ethofat® 142/20

PEG-5 hydrogenated castor oil (CTFA)
CAS 61788-85-0 (generic)
Synonyms: POE (5) hydrogenated castor oil; PEG (5) hydrogenated castor oil
Tradenames: Alkamuls® COH-5 (formerly Emulphor® COH-5); Chemax HCO-5; Emulsogen HEL-050; Prox-onic HRH-05; Trylox® HCO-5

PEG-7 hydrogenated castor oil (CTFA)
CAS 61788-85-0 (generic)
Synonyms: POE (7) hydrogenated castor oil
Tradenames: Dehymuls HRE 7; Remcopal HC 7

PEG-10 hydrogenated castor oil
CAS 61788-85-0 (generic)
Synonyms: POE (10) hydrogenated castor oil; PEG 500 hydrogenated castor oil
Tradenames: Croduret 10

PEG-16 hydrogenated castor oil (CTFA)
CAS 61788-85-0 (generic)
Synonyms: POE (16) hydrogenated castor oil
Tradenames: Chemax HCO-16; Emulsogen HEL-160R; Ethox HCO-16; Hetoxide HC-16; Mapeg® CO-16H; Prox-onic HRH-016; Trylox® 5921; Trylox® HCO-16

PEG-20 hydrogenated castor oil (CTFA)
CAS 61788-85-0 (generic)
Synonyms: POE (20) hydrogenated castor oil
Tradenames: Nopalcol 10-COH; Remcopal HC 20

PEG-25 hydrogenated castor oil (CTFA)
CAS 61788-85-0 (generic)
Synonyms: POE (25) hydrogenated castor oil
Tradenames: Arlatone® G; Chemax HCO-25; Dacospin POE(25)HRG; Emulsogen HEL-250; Emulson EL/H25; Ethox HCO-25; Hetoxide HC-25; Industrol® COH-25; Mapeg® CO-25H; Prox-onic HRH-025; Trylox® 5922; Trylox® HCO-25

PEG-30 hydrogenated castor oil (CTFA)
CAS 61788-85-0 (generic)
Synonyms: POE (30) hydrogenated castor oil
Tradenames: Croduret 30

PEG-33 hydrogenated castor oil
Tradenames: Remcopal HC 33

PEG-40 hydrogenated castor oil (CTFA)
CAS 61788-85-0 (generic)
Synonyms: POE (40) hydrogenated castor oil
Tradenames: Akyporox CO 400; Croduret 40; Hetoxide HC-40; Remcopal HC 40; Rolfor COH 40; Tagat® R40

PEG-60 hydrogenated castor oil (CTFA)
CAS 61788-85-0 (generic)
Synonyms: POE (60) hydrogenated castor oil
Tradenames: Akyporox CO 600; Croduret 60; Hetoxide HC-60; Remcopal HC 60; Tagat® R60
Tradenames containing: Tagat® R63

PEG-100 hydrogenated castor oil (CTFA)
CAS 61788-85-0 (generic)
Synonyms: POE (100) hydrogenated castor oil; PEG (100) hydrogenated castor oil
Tradenames: Croduret 100

PEG-200 hydrogenated castor oil (CTFA)
CAS 61788-85-0 (generic)
Synonyms: POE (200) hydrogenated castor oil; PEG (200) hydrogenated castor oil
Tradenames: Chemax HCO-200/50; Croduret 200; Ethox HCO-200/50%; Industrol® COH-200; Prox-onic HRH-0200; Prox-onic HRH-0200/50; Trylox® 5925; Trylox® HCO-200/50

PEG-40 hydrogenated castor oil PCA isostearate (CTFA)
Synonyms: PEG-40 hydrogenated castor oil pyroglutamic isostearic diester; PEG 2000 hydrogenated castor oil PCA isostearate
Tradenames: Pyroter CPI-40

PEG-9 hydrogenated coconut acid
Tradenames: Dacospin 869

PEG-20 hydrogenated lanolin (CTFA)
CAS 68648-27-1 (generic)
Synonyms: POE (20) hydrogenated lanolin; PEG 1000 hydrogenated lanolin
Tradenames: Satexlan 20; Super-Sat AWS-4

PEG-24 hydrogenated lanolin (CTFA)
CAS 68648-27-1 (generic)
Synonyms: POE (24) hydrogenated lanolin
Tradenames: Super-Sat AWS-24

PEG-13 hydrogenated tallow amide (CTFA)
CAS 68155-24-8
Synonyms: Ethoxylated (13) hydrogenated tallowamide; POE (13) hydrogenated tallow amide
Tradenames: Ethomid® HT/23

PEG-50 hydrogenated tallow amide
CAS 68155-24-8
Tradenames: Ethomid® HT/60; Schercomid HT-60

PEG-2 hydrogenated tallow amine (CTFA)
CAS 61791-26-2 (generic)
Synonyms: PEG 100 tallow amine; POE (2) tallow amine; PEG-2 tallow amine
Tradenames: Chemeen HT-2

PEG-5 hydrogenated tallow amine (CTFA)
CAS 61791-26-2 (generic)
Synonyms: POE (5) tallow amine
Tradenames: Chemeen HT-5; Prox-onic MHT-05

PEG-14 hydrogentaed tallowamine
Tradenames: Amiet THD/14

PEG-15 hydrogenated tallow amine (CTFA)
CAS 61791-26-2 (generic)
Synonyms: POE (15) tallow amine
Tradenames: Chemeen HT-15; Prox-onic MHT-015

PEG-17 hydrogenated tallowamine
Tradenames: Amiet THD/17

PEG-22 hydrogenated tallowamine
Tradenames: Amiet THD/22

PEG-27 hydrogenated tallowamine
Tradenames: Amiet THD/27

PEG-50 hydrogenated tallow amine (CTFA)
CAS 61791-26-2 (generic)
Synonyms: POE (50) tallow amine
Tradenames: Chemeen HT-50

PEG-60 hydrogenated tallowate
Tradenames: Nopalcol 30-TWH

PEG-3 hydrogenated tallow propylene diamine
Tradenames: Dinoramox SH 3

PEG isocetyl ether. *See Isoceteth Series*

PEG isodecyl ether. *See Isodeceth Series*

PEG-5 isodecyloxypropylamine
Tradenames: Tomah E-14-5

PEG isolauryl ether. *See Isolaureth Series*

PEG-6 isolauryl thioether (CTFA)
Synonyms: POE (6) isolauryl thioether; PEG 300 isolauryl thioether
Tradenames: Alcodet® 260 (formerly Siponic® 260); Prox-onic TM-06

PEG-8 isolauryl thioether (CTFA)
Synonyms: POE (8) isolauryl thioether; PEG 400 isolauryl thioether
Tradenames: Alcodet® SK (formerly Siponic® SK); Prox-onic TM-08

PEG-10 isolauryl thioether (CTFA)
Synonyms: POE (10) isolauryl thioether; PEG 500 isolauryl thioether
Tradenames: Alcodet® 218 (formerly Siponic® 218); Prox-onic TM-010

PEG-4 isostearate (CTFA)
CAS 56002-14-3 (generic)
Synonyms: POE (4) monoisostearate; PEG 200 monoisostearate
Tradenames: Emerest® 2625; Trydet ISA-4

PEG-8 isostearate (CTFA)
CAS 56002-14-3 (generic)

Synonyms: POE (8) isostearate; PEG 400 isostearate
Tradenames: Emerest® 2644; Trydet 2644

PEG-9 isostearate
CAS 56002-14-3 (generic)
Synonyms: POE (9) isostearate
Tradenames: Ethox MI-9; Industrol® MIS-9; Trydet ISA-9

PEG-12 isostearate (CTFA)
CAS 56002-14-3 (generic)
Synonyms: POE (12) isostearate; PEG 600 monoisostearate
Tradenames: Emerest® 2664

PEG-14 isostearate
CAS 56002-14-3 (generic)
Synonyms: POE (14) isostearate
Tradenames: Ethox MI-14; Trydet ISA-14

PEG isostearyl ether. *See Isosteareth Series*

PEG-5 lanolate (CTFA)
CAS 68459-50-7 (generic)
Synonyms: PEG-5 lanolin acids; POE (5) lanolate
Tradenames containing: Lanpolamide 5

PEG-27 lanolin (CTFA)
CAS 61790-81-6; 8051-81-8
Synonyms: POE (27) lanolin
Tradenames: Lanogel® 21

PEG-30 lanolin (CTFA)
CAS 61790-81-6 (generic)
Synonyms: POE (30) lanolin
Tradenames: Aqualose L30

PEG-40 lanolin (CTFA)
CAS 8051-82-9; 61790-81-6 (generic)
Synonyms: POE (40) lanolin
Tradenames: Laneto 40; Lanogel® 31

PEG-55 lanolin (CTFA)
CAS 61790-81-6 (generic)
Synonyms: POE (55) lanolin
Tradenames: Solan E

PEG-60 lanolin (CTFA)
CAS 61790-81-6 (generic)
Synonyms: POE (60) lanolin
Tradenames: Laneto 60; Solan 50
Tradenames containing: Ritachol® 3000; Ritachol® 4000

PEG-70 lanolin
CAS 61790-81-6 (generic)
Synonyms: POE (70) lanolin
Tradenames: Teric LAN70

PEG-75 lanolin (CTFA)
CAS 8039-09-6; 61790-81-6 (generic)
Synonyms: POE (75) lanolin; PEG 4000 lanolin
Tradenames: Aqualose L75; Aqualose L75/50; Ethoxylan® 1685; Ethoxylan® 1686; Lan-Aqua-Sol 50; Lan-Aqua-Sol 100; Laneto 50; Lanogel® 41; Solan; Solangel 401; Solulan® 75; Solulan® L-575; Super Solan Flaked

PEG-85 lanolin (CTFA)
CAS 61790-81-6 (generic)
Synonyms: POE (85) lanolin
Tradenames: Lanogel® 61

PEG lanolin acids. *See PEG lanolate Series*

PEG-5 lanolinamide (CTFA)
Synonyms: POE (5) lanolinamide
Tradenames containing: Lanpolamide 5

PEG lanolin ether. *See Laneth Series*

PEG-3 lauramide (CTFA)
CAS 26635-75-6 (generic)
Synonyms: POE (3) lauryl amide; PEG (3) lauryl amide
Tradenames: Amidox® L-2; Unamide® L-2

PEG-5 lauramide (CTFA)
CAS 26635-75-6 (generic)
Synonyms: POE (5) lauryl amide
Tradenames: Amidox® L-5

PEG-6 lauramide (CTFA)
CAS 26635-75-6 (generic)
Synonyms: POE (6) lauryl amide; PEG 300 lauryl amide
Tradenames: Unamide® L-5

PEG-6 lauramide DEA
Tradenames: Mazamide® L-5

PEG-10 lauramine
Tradenames: Marlazin® L 10

PEG-3 lauramine oxide (CTFA)
Synonyms: POE (3) lauryl dimethyl amine oxide; PEG (3) lauryl dimethyl amine oxide
Tradenames: Empigen® OY

PEG-2 laurate (CTFA)
CAS 141-20-8; EINECS 205-468-1
Synonyms: Diethylene glycol laurate; Diglycol laurate; PEG 100 monolaurate
Tradenames: Cithrol DGML N/E; Hodag DGL; Lipo Diglycol Laurate; Mapeg® DGLD; Nopalcol 1-L; Pegosperse® 100 L; Pegosperse® 100 ML; Radiasurf® 7420; Radiasurf® 7421; Sole-Onic CDS

PEG-2 laurate SE (CTFA)
Synonyms: Diethylene glycol monolaurate self-emulsifying; PEG 100 monolaurate self-emulsifying; POE (2) monolaurate self-emulsifying
Tradenames: Cithrol DGML S/E; Lipo DGLS

PEG-4 laurate (CTFA)
CAS 9004-81-3 (generic); 10108-24-4
Synonyms: POE (4) monolaurate; PEG 200 monolaurate
Tradenames: Ablunol 200ML; Alkamuls® PE/220 (formerly Soprophor® PE/220); CPH-27-N; Chemax E-200 ML; Cithrol 2ML; Crodet L4; Emerest® 2620; Hodag 20-L; Karapeg 200-ML; Kessco® PEG 200 ML; Mapeg® 200 ML; Nopalcol 2-L; Pegosperse® 200 ML; Radiasurf® 7422; Witconol 2620

PEG-5 laurate
CAS 9004-81-3 (generic)
Synonyms: POE (5) monolaurate
Tradenames: Ethox ML-5; Hetoxamate LA-5; Industrol® ML-5; Trydet LA-5

PEG-6 laurate (CTFA)
CAS 9004-81-3 (generic); 2370-64-1; EINECS 219-136-9
Synonyms: POE (6) monolaurate; PEG 300 monolaurate
Tradenames: Emerest® 2630; Kessco® PEG 300 ML; Secoster® ML 300

PEG-7 laurate
CAS 9004-81-3 (generic)
Synonyms: POE (7) monolaurate
Tradenames: Trydet LA-7

PEG-8 laurate (CTFA)
CAS 9004-81-3 (generic); 35179-86-3
Synonyms: POE (8) monolaurate; PEG 400

monolaurate
Tradenames: Ablunol 400ML; Acconon 400-ML; Alkamuls® PE/400 (formerly Soprophor PE/400; CPH-30-N; Chemax E-400 ML; Cithrol 4ML; Crodet L8; Emerest® 2650; Gradonic 400-ML; Hodag 40-L; Karapeg 400-ML; Kessco® PEG 400 ML; Lexemul® PEG-400ML; Lonzest® PEG 4-L; Mapeg® 400 ML; Nonisol 100; Nopalcol 4-L; Pegosperse® 400 ML; Radiasurf® 7423

PEG-9 laurate (CTFA)
CAS 106-08-1; 9004-81-3 (generic); EINECS 203-359-3
Synonyms: POE (9) monolaurate
Tradenames: Alkamuls® L-9 (formerly Alkasurf® L-9); Ethox ML-9; Hetoxamate LA-9; Industrol® ML-9

PEG-12 laurate (CTFA)
CAS 9004-81-3 (generic)
Synonyms: POE (12) monolaurate; PEG 600 monolaurate
Tradenames: Ablunol 600ML; Alkamuls® 600-GML; Cithrol 6ML; Crodet L12; Emerest® 2661; Hodag 60-L; Karapeg 600-ML; Kessco® PEG 600 ML; Mapeg® 600 ML; Nopalcol 6-L; PGE-600-ML; Pegosperse® 600 ML

PEG-14 laurate (CTFA)
CAS 9004-81-3 (generic)
Synonyms: POE (14) monolaurate
Tradenames: Chemax E-600 ML; Ethox ML-14; Industrol® ML-14

PEG-20 laurate (CTFA)
CAS 9004-81-3 (generic)
Synonyms: POE (20) monolaurate; PEG 1000 monolaurate
Tradenames: Cithrol 10ML; Kessco® PEG 1000 ML

PEG-24 laurate
CAS 9004-81-3 (generic)
Synonyms: POE (24) monolaurate
Tradenames: Crodet L24

PEG-32 laurate (CTFA)
CAS 9004-81-3 (generic)
Synonyms: POE (32) monolaurate; PEG 1540 monolaurate
Tradenames: Kessco® PEG 1540 ML

PEG-40 laurate
CAS 9004-81-3 (generic)
Synonyms: POE (40) monolaurate
Tradenames: Crodet L40

PEG-75 laurate (CTFA)
CAS 9004-81-3 (generic)
Synonyms: POE (75) monolaurate; PEG 4000 monolaurate
Tradenames: Kessco® PEG 4000 ML; Secoster® ML 4000

PEG-100 laurate
CAS 9004-81-3 (generic)
Synonyms: POE (100) monolaurate; PEG (100) monolaurate
Tradenames: Crodet L100

PEG-150 laurate (CTFA)
CAS 9004-81-3 (generic)
Synonyms: POE (150) monolaurate; PEG 6000 monolaurate
Tradenames: Cithrol 60ML; Kessco® PEG 6000 ML

PEG-400 laurate
Tradenames: DyaFac LA9

PEG lauryl ether. *See Laureth Series*

PEG-20 mannitan laurate
Tradenames: Atlas G-9046T

PEG-6 methyl ether (CTFA)
CAS 9004-74-4 (generic)
Synonyms: POE (6) methyl ether; PEG 300 methyl ether; PEG-6 monomethyl ether
Tradenames: Carbowax® MPEG 350

PEG-10 methyl ether
CAS 9004-74-4 (generic)
Synonyms: POE (10) methyl ether; PEG 500 methyl ether
Tradenames: Carbowax® MPEG 550

PEG-16 methyl ether
CAS 9004-74-4 (generic)
Synonyms: POE (16) methyl ether
Tradenames: Carbowax® MPEG 750

PEG-40 methyl ether
CAS 9004-74-4 (generic)
Synonyms: POE (40) methyl ether
Tradenames: Carbowax® MPEG 2000

PEG-100 methyl ether
CAS 9004-74-4 (generic)
Synonyms: POE (100) methyl ether
Tradenames: Carbowax® MPEG 5000

PEG methyl glucose ether. *See Methyl gluceth Series*

PEG-20 methyl glucose sesquistearate (CTFA)
CAS 68389-70-8
Synonyms: POE (20) methyl glucose sesquistearate; PEG 1000 methyl glucose sesquistearate
Tradenames: Glucamate® SSE-20

PEG-8 monomerate
Tradenames: Ethox MA-8

PEG-15 monomerate
Tradenames: Ethox MA-15

PEG myristyl ether. *See Myreth Series*

PEG-13 naphthole
Tradenames: Rewopal® BN 13

PEG octyl ether. *See Octeth Series*

PEG octyl phenyl ether. *See Octoxynol Series*

PEG-3 oleamide
CAS 26027-372; 31799-71-0
Tradenames: Dionil® OC

PEG-5 oleamide (CTFA)
CAS 31799-71-0
Synonyms: POE (5) oleyl amide; PEG (5) oleyl amide
Tradenames: Ethomid® O/15

PEG-6 oleamide (CTFA)
CAS 26027-37-2
Synonyms: PEG 300 oleyl amide; POE (6) oleyl amide
Tradenames: Dionil® SH 100

PEG-7 oleamide (CTFA)
CAS 26027-37-2
Synonyms: Ethoxylated (7) oleamide; POE (7) oleamide
Tradenames: Ethomid® O/17

PEG-9 oleamide (CTFA)
Synonyms: POE (9) oleyl amide

Tradenames: Rewopal® O 8

PEG-14 oleamide
Tradenames: Dionil® W 100

PEG-15 oleamide
Synonyms: POE (15) oleyl amide
Tradenames: Rewopal® O 15

PEG-2 oleamine (CTFA)
Synonyms: PEG-2 oleyl amine; POE (2) oleyl amine; PEG 100 oleyl amine
Tradenames: Berol 302; Ethomeen® O/12; Hetoxamine O-2; Marlazin® OL 2; Noramox O2; Prox-onic MO-02; Rhodameen® O-2 (formerly Catafor O-2); Rhodameen® O-12 (formerly Rhodiasurf O-12); Varonic® Q202; Varonic® Q202 SF; Zusomin O 102

PEG-5 oleamine (CTFA)
Synonyms: POE (5) oleyl amine
Tradenames: Ethomeen® O/15; Hetoxamine O-5; Noramox O5; Varonic® Q205 SF; Zusomin O 105

PEG-6 oleamine
Tradenames: Rhodameen® O-6

PEG-7 oleamine
CAS 26635-93-8
Tradenames: Berol 28

PEG-11 oleamine
Tradenames: Noramox O11

PEG-12 oleamine
Tradenames: Berol 303

PEG-14 oleamine
Tradenames: Amiet OD/14

PEG-15 oleamine (CTFA)
Synonyms: POE (15) oleyl amine
Tradenames: Ethomeen® O/25; Hetoxamine O-15; Noramox O15; Prox-onic MO-015; Rhodameen® RAM/7 Base (formerly Soprophor RAM/7)

PEG-20 oleamine
Tradenames: Marlazin® OL 20; Noramox O20

PEG-25 oleamine
Tradenames: Rhodameen® RAM/8 Base (formerly Soprophor RAM/8)

PEG-30 oleamine (CTFA)
Synonyms: POE (30) oleyl amine
Tradenames: Chemeen O-30, O-30/80; Ethox OAM-308; Eumulgin PA 30; Icomeen® O-30; Icomeen® O-30-80%; Prox-onic MO-030; Prox-onic MO-030-80; Rhodameen® OA-910 (formerly Katapol® OA-910); Trymeen® 6620; Trymeen® 6622; Trymeen® OAM 30/60; Witcamine® 6622

PEG-2 oleammonium chloride (CTFA)
CAS 18448-65-2
Synonyms: PEG 100 oleamonium chloride; POE (2) oleamonium chloride
Tradenames containing: Ethoquad® O/12

PEG-15 oleammonium chloride (CTFA)
CAS 28880-55-9
Synonyms: POE (15) oleamonium chloride
Tradenames: Ethoquad® O/25

PEG-2 oleate (CTFA)
CAS 106-12-7; EINECS 203-364-0
Synonyms: Diethylene glycol monooleate; Diglycol oleate; POE (2) monooleate
Tradenames: Cithrol DGMO N/E; Hetoxamate MO-2; Hodag DGO; Karapeg DEG-MO; Marlosol® OL2; Pegosperse® 100 O; Radiasurf® 7400; Secoster® MO 100; Witconol DOS

PEG-2 oleate SE (CTFA)
Synonyms: Diethylene glycol monooleate self-emulsifying; PEG 100 monooleate self-emulsifying; POE (2) monooleate self-emulsifying
Tradenames: Cithrol DGMO S/E

PEG-4 oleate (CTFA)
CAS 9004-96-0 (generic); 10108-25-5; EINECS 233-293-0
Synonyms: POE (4) monooleate; PEG 200 monooleate
Tradenames: Ablunol 200MO; CPH-39-N; Cithrol 2MO; Crodet O4; Emerest® 2624; Ethylan® A2; Hetoxamate MO-4; Karapeg 200-MO; Kessco® PEG 200 MO; Mapeg® 200 MO; Radiasurf® 7402; Remcopal 207; Soprofor M/52; Teric OF4

PEG-5 oleate (CTFA)
CAS 9004-96-0 (generic)
Synonyms: POE (5) monooleate
Tradenames: Alphoxat O 105; Chemax E-200 MO; Ethofat® O/15; Ethox MO-5; Hetoxamate MO-5; Industrol® MO-5; Prox-onic OL-1/05; Trydet OA-5

PEG-6 oleate (CTFA)
CAS 9004-96-0 (generic)
Synonyms: POE (6) monooleate; PEG 300 monooleate
Tradenames: Acconon 300-MO; Alkamuls® A (formerly Rhodiasurf A); Alkamuls® AP (formerly Rhodiasurf AP); Alkamuls® M-6 (formerly Texafor M-6); Emerest® 2632; Ethylan® A3; Industrol® MO-6; Kessco® PEG 300 MO; Radiasurf® 7431; Remcopal 6; Rolfat OL 6; Sellig AO 6 100; Teric OF6

PEG-7 oleate (CTFA)
CAS 9004-96-0 (generic)
Synonyms: POE (7) monooleate
Tradenames: Industrol® MO-7; Marlosol® OL7; Rolfat OL 7; Trydet OA-7

PEG-8 oleate (CTFA)
CAS 9004-96-0 (generic)
Synonyms: POE (8) monooleate; PEG 400 monooleate
Tradenames: Ablunol 400MO; Acconon 400-MO; Cithrol 4MO; Cithrol A; Crodet O8; Emerest® 2646; Empilan® BQ 100; Emulsan O; Ethylan® A4; Hodag 40-O; Karapeg 400-MO; Kessco® PEG 400 MO; Laurel PEG 400 MO; Lonzest® PEG 4-O; Mapeg® 400 MO; Marlosol® OL8; Noiox AK-44; Nonex O4E; Nopalcol 4-O; Pegosperse® 400 MO; Radiasurf® 7403; Secoster® MO 400; Teric OF8; Witconol H31A; Witflow 916

PEG-9 oleate (CTFA)
CAS 9004-96-0 (generic)
Synonyms: POE (9) monooleate; PEG 450 monooleate
Tradenames: Alkamuls® 400-MO; Chemax E-400 MO; Ethox MO-9; Genagen O-090; Hetoxamate MO-9; Industrol® MO-9; Prox-onic OL-1/09; Rolfat OL 9; Sellig AO 9 100

PEG-10 oleate (CTFA)
CAS 9004-96-0 (generic)

Synonyms: POE (10) monooleate; PEG 500 monooleate; POE (10) oleic acid
Tradenames: Alphoxat O 110; Atlas G-2143; Ethofat® O/20; Marlosol® OL10; Serdox NOG 440; Trydet 2676; Trydet OA-10

PEG-12 oleate (CTFA)
CAS 9004-96-0 (generic)
Synonyms: POE (12) monooleate; PEG 600 monooleate
Tradenames: Ablunol 600MO; Alkamuls® 600-MO; CPH-41-N; Cithrol 6MO; Crodet O12; Emerest® 2660; Ethylan® A6; Industrol® MO-13; Karapeg 600-MO; Kessco® PEG 600 MO; Mapeg® 600 MO; Nopalcol 6-O; Radiasurf® 7404

PEG-14 oleate (CTFA)
CAS 9004-96-0 (generic)
Synonyms: POE (14) monooleate
Tradenames: Alkamuls® O-14 (formerly Alkasurf® O-14); Chemax E-600 MO; Ethox MO-14; Pegosperse® 700 TO; Prox-onic OL-1/014

PEG-15 oleate (CTFA)
CAS 9004-96-0 (generic)
Synonyms: POE (15) monooleate
Tradenames: Alphoxat O 115; Hetoxamate MO-15; Marlosol® OL15; Sellig AO 15 100

PEG-20 oleate (CTFA)
CAS 9004-96-0 (generic)
Synonyms: POE (20) monooleate; PEG 1000 monooleate
Tradenames: Ablunol 1000MO; Chemax E-1000 MO; Cithrol 10MO; Kessco® PEG 1000 MO; Marlosol® OL20

PEG-24 oleate
Tradenames: Crodet O24

PEG-25 oleate
Tradenames: Sellig AO 25 100

PEG-32 oleate (CTFA)
CAS 9004-96-0 (generic)
Synonyms: POE (32) monooleate; PEG 1540 monooleate
Tradenames: Kessco® PEG 1540 MO

PEG-40 oleate
Tradenames: Crodet O40

PEG-60 oleate
Tradenames: Alkamuls® 783/P (formerly Geronol 783/P)

PEG-75 oleate (CTFA)
CAS 9004-96-0 (generic)
Synonyms: POE (75) monooleate; PEG 4000 monooleate
Tradenames: Cithrol 40MO; Emerest® 2618; Kessco® PEG 4000 MO

PEG-100 oleate
Tradenames: Crodet O100

PEG-150 oleate (CTFA)
CAS 9004-96-0 (generic)
Synonyms: POE (150) monooleate; PEG 6000 monooleate
Tradenames: Cithrol 60MO; Emerest® 2617; Emerest® 2619; Kessco® PEG 6000 MO

PEG-200 oleate
Synonyms: POE (200) monooleate
Tradenames: Niox AK-40

PEG-400 oleate
Synonyms: POE (400) monooleate
Tradenames: Niox AK-44

PEG oleyl ethers. *See Oleth Series*

PEG-2 oleyl/stearyl amine
Tradenames: Rhodameen® OS-12 (formerly Rhodiasurf OS-12)

PEG-7 palmitate
CAS 9004-94-8 (generic)
Synonyms: PEG (7) monopalmitate; POE (7) monopalmitate
Tradenames: Genagen P-070

PEG-20 palmitate (CTFA)
CAS 9004-94-8 (generic)
Synonyms: POE (20) monopalmitate; PEG 1000 monopalmitate
Tradenames: Atlas G-2079

PEG-6 pelargonate
Synonyms: POE (6) monopelargonate; PEG 300 monopelargonate
Tradenames: Emerest® 2634

PEG-8 pelargonate
Synonyms: POE (8) pelargonate; PEG 400 pelargonate
Tradenames: Emerest® 2654; Emerest® 2658

PEG-9 pelargonate
CAS 31621-91-7
Synonyms: POE (9) pelargonate
Tradenames: Ethox 1122; Genagen PL-090; Trydet MP-9

PEG-5 phytosterol
Tradenames: Nikkol BPS-5

PEG-10 phytosterol
Tradenames: Nikkol BPS-10

PEG-15 phytosterol
Tradenames: Nikkol BPS-15

PEG-20 phytosterol
Tradenames: Nikkol BPS-20

PEG-25 phytosterol (CTFA)
Tradenames: Nikkol BPS-25

PEG-30 phytosterol
Tradenames: Nikkol BPS-30

PEG-6 PPG-2.5 C9-C11 alcohols ether
Tradenames: Hetoxol 916P

PEG-20-PPG-10 glyceryl stearate (CTFA)
Tradenames: Acconon TGH

PEG-6 PPG-3 tridecylether
Tradenames: Hetoxol TDEP-63

PEG-10 PPG-15 tridecyl ether
Tradenames: Hetoxol TDEP-15

PEG-10 propylene glycol (CTFA)
Synonyms: POE (10) propylene glycol; PEG 500 propylene glycol
Tradenames containing: Acconon CON

PEG-55 propylene glycol oleate (CTFA)
Synonyms: POE (55) propylene glycol oleate
Tradenames containing: Antil® 141 Liq.

PEG-25 propylene glycol stearate (CTFA)
Synonyms: POE (25) propylene glycol monostearate
Tradenames: Atlas G-2162; G-2162; Simulsol PS20

PEG-4 rapeseedamide (CTFA)
CAS 85536-23-8
Synonyms: POE (4) rapeseedamide; PEG 200 rapeseedamide
Tradenames: Aminol TEC N

PEG-3 rapeseed ester
Tradenames: Marlosol® RF3

PEG-2 ricinoleate (CTFA)
CAS 5401-17-2; 9004-97-1 (generic); EINECS 226-448-9
Synonyms: Diethylene glycol monoricinoleate; PEG-2 monoricinoleate; POE (2) ricinoleate
Tradenames: Pegosperse® 100 MR

PEG-8 ricinoleate (CTFA)
CAS 9004-97-1 (generic)
Synonyms: PEG 400 ricinoleate; POE (8) ricinoleate
Tradenames: Hodag 40-R

PEG-9 ricinoleate (CTFA)
CAS 9004-97-1 (generic)
Synonyms: PEG 450 ricinoleate; POE (9) ricinoleate
Tradenames: Cerex EL 4929

PEG-12 ricinoleate
CAS 9004-97-1 (generic)
Synonyms: POE (12) monoricinoleate
Tradenames: Nopalcol 6-R

PEG-15 rosin acid
Tradenames: Hercules® AR150

PEG-16 rosin acid
Tradenames: Hercules® AR160

PEG-15 rosinate
Tradenames: Chemax AR-497

PEG-8 sesquioleate (CTFA)
Synonyms: PEG 400 sesquioleate; POE (8) sesquioleate
Tradenames: Emerest® 2647; Ethox 2966; Ethox SO-9

PEG-9 sesquioleate
Tradenames: Trydet SO-9

PEG-12 sesquioleate
Tradenames: Industrol® SO-13

PEG-40 sorbitan diisostearate (CTFA)
Synonyms: POE (40) sorbitan diisostearate
Tradenames: Emsorb® 2726

PEG-40 sorbitan hexaoleate (CTFA)
CAS 57171-56-9
Synonyms: PEG-40 sorbitol hexaoleate; POE (40) sorbitol hexaoleate; PEG 2000 sorbitol hexaoleate
Tradenames: Atlas G-1086; Sorbeth 40; Sorbeth 40HO; Trylox® 6746

PEG-50 sorbitan hexaoleate (CTFA)
CAS 57171-56-9
Synonyms: PEG-50 sorbitol hexaoleate; POE (50) sorbitol hexaoleate
Tradenames: Atlas G-1096; Ethox HO-50

PEG-55 sorbitan hexaoleate
Tradenames: Sorbeth 55HO

PEG-60 sorbitan hexaoleate
CAS 57171-56-9
Tradenames: Trylox® 6747

PEG-40 sorbitan hexatallate
Tradenames: Glycosperse® HTO-40

PEG-5 sorbitan isostearate (CTFA)
CAS 66794-58-9 (generic)
Synonyms: POE (5) sorbitan isostearate
Tradenames: Montanox 71

PEG-20 sorbitan isostearate (CTFA)
CAS 66794-58-9 (generic)
Synonyms: PEG 1000 sorbitan monoisostearate; POE (20) sorbitan monoisostearate; Polysorbate 120
Tradenames: Crillet 6; Montanox 70

PEG-5 sorbitan laurate
Synonyms: POE (5) sorbitan monolaurate
Tradenames: Montanox 21

PEG-10 sorbitan laurate (CTFA)
CAS 9005-64-5 (generic)
Synonyms: POE (10) sorbitan monolaurate; PEG 500 sorbitan monolaurate
Tradenames: Atlas G-7596-J; Glycosperse® L-10; Hetsorb L-10; Liposorb L-10

PEG-20 sorbitan laurate. *See Polysorbate 20*

PEG-80 sorbitan laurate (CTFA)
CAS 9005-64-5 (generic)
Synonyms: POE (80) sorbitan monolaurate
Tradenames: G-4280; T-Maz® 28

PEG-5 sorbitan oleate. *See Polysorbate 81*

PEG-80 sorbitan palmitate (CTFA)
CAS 9005-66-7 (generic)
Synonyms: PEG (80) sorbitan monopalmitate; POE (80) sorbitan monopalmitate
Tradenames: G-4252
Tradenames containing: Miracare® BC-10; Miracare® MS-1 (formerly Compound MS-1); Miracare® MS-2 (formerly Compound MS-2); Miracare® MS-4

PEG-40 sorbitan peroleate (CTFA)
Synonyms: POE (40) sorbitan peroleate; PEG 2000 sorbitan peroleate; POE (40) sorbitol septaoleate
Tradenames: Arlatone® T

PEG-4 sorbitan stearate. *See Polysorbate 61*

PEG-20 sorbitan stearate. *See Polysorbate 60*

PEG-16 sorbitan tallate
Tradenames: Atlox 8916P

PEG-20 sorbitan tallate
Tradenames: Desonic® SMT-20; Desotan® SMT-20; T-Maz® 90

PEG-30 sorbitan tetraoleate (CTFA)
Synonyms: POE (30) sorbitan tetraoleate
Tradenames: Nikkol GO-430

PEG-40 sorbitan tetraoleate (CTFA)
CAS 9003-11-6
Synonyms: POE (40) sorbitan tetraoleate; PEG 2000 sorbitan tetraoleate
Tradenames: Nikkol GO-440

PEG-60 sorbitan tetraoleate (CTFA)
Synonyms: POE (60) sorbitan tetraoleate
Tradenames: Nikkol GO-460

PEG-16 sorbitan trioleate
Tradenames: Emsorb® 6917

PEG-17 sorbitan trioleate
Tradenames: TO-55-EL

PEG-18 sorbitan trioleate
CAS 9005-70-3
Tradenames: TO-55-E

PEG-20 sorbitan trioleate. *See Polysorbate 85*

PEG-16 sorbitan tristearate
Tradenames: Atlas G-7166P; Emsorb® 6908

PEG-20 sorbitan tristearate. *See Polysorbate 65*

PEG-20 sorbitan tritallate
Tradenames: T-Maz® 95; Toximul® SEE-340

PEG sorbitol ethers. *See Sorbeth Series*

PEG-6 soya fatty acid
CAS 61791-07-9
Tradenames: Rolfat SO 6

PEG-2 soyamine (CTFA)
CAS 61791-24-0 (generic)
Synonyms: POE (2) soya amine; PEG 100 soya amine; Bis (2-hydroxyethyl) soya amine
Tradenames: Accomeen S2; Chemeen S-2; Ethomeen® S/12; Hetoxamine S-2; Mazeen® S-2; Teric 16M2; Tomah E-S-2

PEG-5 soyamine (CTFA)
CAS 61791-24-0 (generic)
Synonyms: POE (5) soya amine
Tradenames: Accomeen S5; Chemeen S-5; Ethomeen® S/15; Hetoxamine S-5; Icomeen® S-5; Mazeen® S-5; Teric 16M5; Tomah E-S-5

PEG-10 soyamine (CTFA)
CAS 61791-24-0 (generic)
Synonyms: POE (10) soya amine; PEG 500 soya amine
Tradenames: Accomeen S10; Ethomeen® S/20; Mazeen® S-10; Teric 16M10

PEG-15 soyamine (CTFA)
CAS 61791-24-0 (generic)
Synonyms: POE (15) soya amine
Tradenames: Accomeen S15; Ethomeen® S/25; Hetoxamine S-15; Mazeen® S-15; Teric 16M15; Tomah E-S-15

PEG-30 soyamine
Synonyms: POE (30) soya amine
Tradenames: Chemeen S-30; Chemeen S-30/80

PEG-25 soya sterol (CTFA)
Synonyms: POE (25) soya sterol
Tradenames: Generol® 122E25

PEG-2 stearamine (CTFA)
CAS 10213-78-2; EINECS 233-520-3
Synonyms: PEG-2 stearyl amine; PEG 100 stearyl amine; Bis-2-hydroxyethyl stearamine
Tradenames: Chemeen 18-2; Ethomeen® 18/12; Ethox SAM-2; Hetoxamine ST-2; Prox-onic MS-02; Teric 18M2; Varonic® S202; Varonic® S202 SF

PEG-5 stearamine (CTFA)
CAS 26635-92-7
Synonyms: POE (5) stearyl amine
Tradenames: Chemeen 18-5; Ethomeen® 18/15; Hetoxamine ST-5; Icomeen® 18-5; Prox-onic MS-05; Teric 18M5

PEG-10 stearamine (CTFA)
CAS 26635-92-7
Synonyms: POE (10) stearyl amine; PEG 500 stearyl amine
Tradenames: Ethomeen® 18/20; Ethox SAM-10; Marlazin® S 10; Teric 18M10; Zusomin S 110

PEG-11 stearamine
CAS 26635-92-7
Tradenames: Prox-onic MS-011

PEG-15 stearamine (CTFA)
CAS 26635-92-7
Synonyms: POE (15) stearyl amine
Tradenames: Ethomeen® 18/25; Hetoxamine ST-15

PEG-20 stearamine
CAS 26635-92-7
Synonyms: POE (20) stearyl amine
Tradenames: Teric 18M20

PEG-25 stearamine
Tradenames: Zusomin S 125

PEG-30 stearamine
CAS 26635-92-7
Synonyms: POE (30) stearyl amine
Tradenames: Teric 18M30

PEG-40 stearamine
Tradenames: Marlazin® S 40

PEG-50 stearamine (CTFA)
CAS 26635-92-7
Synonyms: POE (50) stearyl amine; Ethoxylated 1-aminooctadecane
Tradenames: Chemeen 18-50; Ethomeen® 18/60; Ethox SAM-50; Hetoxamine ST-50; Icomeen® 18-50; Prox-onic MS-050; Trymeen® 6617; Trymeen® SAM-50

PEG-2 stearate (CTFA)
CAS 106-11-6; 9004-99-3 (generic); 61791-00-2; 85116-97-8; EINECS 203-363-5; 285-550-1
Synonyms: Diethylene glycol stearate; Diglycol stearate; PEG 100 monostearate
Tradenames: Ablunol DEGMS; Cithrol DGMS N/E; DMS-33; Drewmulse® DGMS; Hodag DGS; Hydrine; Karapeg DEG-MS; Lipal DGMS; Nopalcol 1-S; Pegosperse® 100 S; Radiasurf® 7410; Sterol ST 2; Tegin® D 6100; Tegin® DGS

PEG-2 stearate SE (CTFA)
Synonyms: POE (2) monostearate self-emulsifying; Diethylene glycol monostearate self-emulsifying; PEG 100 monostearate self-emulsifying
Tradenames: Cithrol DGMS S/E; Kemester® 5221SE; Lipo DGS-SE; Radiasurf® 7411

PEG-3 stearate (CTFA)
CAS 10233-24-6; 9004-99-3 (generic); EINECS 233-562-2
Synonyms: POE (3) stearate; 2-[2-(2-Hydroxyethoxy) ethoxy]ethyl octadecanoate
Tradenames: Marlosol® 183

PEG-4 stearate (CTFA)
CAS 106-07-0; 9004-99-3 (generic); EINECS 203-358-8
Synonyms: POE (4) stearate; PEG 200 monostearate
Tradenames: Ablunol 200MS; Acconon 200-MS; Cithrol 2MS; Crodet S4; Ethofat® 18/14; Karapeg 200-MS; Kessco® PEG 200 MS; Mapeg® 200 MS; Radiasurf® 7412

PEG-5 stearate (CTFA)
CAS 9004-99-3 (generic)
Synonyms: POE (5) stearate
Tradenames: Chemax E-200 MS; Ethofat® 60/15; Eumulgin PST 5; Hetoxamate SA-5; Industrol® MS-5; Prox-onic ST-05; Trydet 2670; Trydet SA-5

PEG-6 stearate (CTFA)
CAS 9004-99-3 (generic); 10108-28-8

Synonyms: POE (6) stearate; PEG 300 monostearate
Tradenames: Alkamuls® S-6 (formerly Rhodiasurf S-6); Cithrol 3MS; Emerest® 2636; Kessco® PEG 300 MS; Marlosol® 186; Nonex S3E; Radiasurf® 7432; Sellig Stearo 6

PEG-6-32 stearate (CTFA)
CAS 9004-99-3 (generic)
Synonyms: PEG 1500 monostearate; POE 1500 monostearate
Tradenames: Hodag 150-S; Mapeg® 1500 MS; Pegosperse® 1500 MS

PEG-7 stearate (CTFA)
CAS 9004-99-3 (generic)
Synonyms: POE (7) stearate
Tradenames: Hetoxamate SA-7; Industrol® MS-7; Trydet 2636; Trydet SA-7

PEG-8 stearate (CTFA)
CAS 9004-99-3 (generic); 70802-40-3
Synonyms: POE (8) stearate; PEG 400 monostearate
Tradenames: Ablunol 400MS; Acconon 400-MS; Alkamuls® S-65-8 (formerly Alkasurf® S-65-8); Cithrol 4MS; Crodet S8; Emerest® 2640; Emerest® 2641; Emerest® 2711; Ethox MS-8; Eumulgin ST-8; Genagen S-080; Hodag 40-S; Industrol® MS-8; Kessco® PEG 400 MS; Mapeg® 400 MS; Marlosol® 188; Nonisol 300; Nopalcol 4-S; Pegosperse® 400 MS; PG No. 4; PGE-400-MS; Radiasurf® 7413; Simulsol M 45; Trydet 2640; Trydet 2671; Trydet SA-8; Witconol 2640; Witconol H35A

PEG-9 stearate (CTFA)
CAS 9004-99-3 (generic); 5349-52-0; EINECS 226-312-9
Synonyms: POE (9) stearate
Tradenames: Chemax E-400 MS; Hetoxamate SA-9; Industrol® MS-9; Marlosol® 189; Prox-onic ST-09; Rolfat SG; Serdox NSG 400; Teric SF9

PEG-10 stearate (CTFA)
CAS 9004-99-3 (generic)
Synonyms: POE (10) stearate; PEG 500 monostearate
Tradenames: Alphoxat S 110; Ethofat® 60/20

PEG-12 stearate (CTFA)
CAS 9004-99-3 (generic)
Synonyms: POE (12) stearate; PEG 600 monostearate
Tradenames: Ablunol 600MS; Cithrol 6MS; Crodet S12; Emerest® 2662; Hetoxamate SA-13; Hodag 60-S; Kessco® PEG 600 MS; Mapeg® 600 MS; Nopalcol 6-S; PGE-600-MS; Pegosperse® 600 MS; Radiasurf® 7414

PEG-14 stearate (CTFA)
CAS 9004-99-3 (generic); 10289-94-8
Synonyms: POE (14) stearate
Tradenames: Chemax E-600 MS; Ethox MS-14; Prox-onic ST-014

PEG-15 stearate
CAS 9004-99-3 (generic)
Synonyms: POE (15) stearate
Tradenames: Ethofat® 60/25; Teric SF15

PEG-20 stearate (CTFA)
CAS 9004-99-3 (generic)
Synonyms: POE (20) stearate; PEG 1000 monostearate
Tradenames: Ablunol 1000MS; Alphoxat S 120; Chemax E-1000 MS; Cithrol 10MS; Emerest® 2610; Hetoxamate SA-23; Hodag 100-S; Kessco® PEG 1000 MS; Mapeg® 1000 MS; Marlosol® 1820; Pegosperse® 1000 MS; Simulsol M 49

PEG-21 stearate
Tradenames: Stearate PEG 1000

PEG-23 stearate
CAS 9004-99-3 (generic)
Synonyms: POE (23) stearate
Tradenames: Ethox MS-23; Industrol® MS-23; Prox-onic ST-023; Trydet 2610; Trydet SA-23

PEG-24 stearate
CAS 9004-99-3 (generic)
Synonyms: POE (24) stearate
Tradenames: Crodet S24

PEG-30 stearate (CTFA)
CAS 9004-99-3 (generic)
Synonyms: POE (30) stearate
Tradenames: Atlas G-2151; Mergital ST 30/E; Sellig S 30 100; Simulsol M 51

PEG-32 stearate (CTFA)
CAS 9004-99-3 (generic)
Synonyms: PEG 1540 stearate
Tradenames: Kessco® PEG 1540 MS

PEG-35 stearate (CTFA)
CAS 9004-99-3 (generic)
Synonyms: POE (35) stearate
Tradenames: Hetoxamate SA-35

PEG-40 stearate (CTFA)
CAS 9004-99-3 (generic); 31791-00-2
Synonyms: POE (40) stearate; PEG 2000 monostearate
Tradenames: Alkamuls® S-65-40 (formerly Alkasurf® S-65-40); Atlas G-2198; Crodet S40; Emerest® 2715; Ethox MS-40; Genagen S-400; Hetoxamate SA-40; Industrol® MS-40; Mapeg® S-40; Mapeg® S-40K; Pegosperse® 1750 MS; Rolfat ST 40; RS-55-40; Simulsol M 52; Trydet SA-40
Tradenames containing: AF 72; AF 75; Teginacid® ML; Teginacid® ML-SE

PEG-50 stearate (CTFA)
CAS 9004-99-3 (generic)
Synonyms: POE (50) stearate
Tradenames: Emerest® 2675; Simulsol M 53; Trydet 2675; Trydet SA-50/30

PEG-60 stearate
CAS 9004-99-3 (generic)
Synonyms: POE (60) stearate
Tradenames: Nopalcol 30-S

PEG-75 stearate (CTFA)
CAS 9004-99-3 (generic)
Synonyms: POE (75) stearate; PEG 4000 monostearate
Tradenames: Cithrol 40MS; Kessco® PEG 4000 MS; Pegosperse® 4000 MS

PEG-90 stearate (CTFA)
CAS 9004-99-3 (generic)
Synonyms: POE (90) stearate
Tradenames: Hetoxamate SA-90; Hetoxamate SA-90F

PEG-100 stearate (CTFA)
CAS 9004-99-3 (generic)
Synonyms: POE (100) stearate; PEG (100) monostearate

Tradenames: Crodet S100; Industrol® MS-100; Rolfat ST 100; Simulsol M 59
Tradenames containing: Kessco® GMS SE/AS; Lipomulse 165

PEG-150 stearate (CTFA)
CAS 9004-99-3 (generic)
Synonyms: POE (150) stearate; PEG 6000 monostearate
Tradenames: Kessco® PEG 6000 MS
Tradenames containing: Ritachol® 1000; Ritachol® 3000

PEG-600 stearate
Tradenames: DyaFac 6-S

PEG-1500 stearate
Tradenames: Radiasurf® 7417

PEG-2 stearmonium chloride (CTFA)
CAS 3010-24-0
Synonyms: N,N-Bis(2-hydroxyethyl)-N-methyl-octadecanaminium chloride; PEG 100 stearmonium chloride; POE (2) stearmonium chloride
Tradenames containing: Ethoquad® 18/12

PEG-15 stearmonium chloride (CTFA)
CAS 28724-32-5
Synonyms: POE (15) stearmonium chloride
Tradenames: Ethoquad® 18/25

PEG-5 stearyloxypropylamine
Tradenames: Tomah E-18-5

PEG-8 stearyloxypropylamine
Tradenames: Tomah E-18-8

PEG-10 stearyloxypropylamine
Tradenames: Tomah E-18-10

PEG-15 stearyloxypropylamine
Tradenames: Tomah E-18-15

PEG-2 tallate
Synonyms: POE (2) monotallate; PEG 100 monotallate; PEG-2 tall oil acid ester
Tradenames: Teric T2

PEG-3 tallate
Tradenames: Marlosol® TF3; Sellig T 3 100

PEG-4 tallate (CTFA)
CAS 61791-00-2 (generic)
Synonyms: POE (4) monotallate; PEG 200 monotallate
Tradenames: Mapeg® 200 MOT; Marlosol® TF4

PEG-5 tallate (CTFA)
CAS 61791-00-2 (generic)
Synonyms: POE (5) monotallate
Tradenames: Hetoxamate FA-5; Teric T5

PEG-6 tallate
CAS 61791-00-2
Tradenames: Aconol X6; Rolfat R 106

PEG-7 tallate
Synonyms: POE (7) monotallate
Tradenames: Teric T7

PEG-8 tallate (CTFA)
CAS 61791-00-2 (generic)
Synonyms: POE (8) monotallate; PEG 400 monotallate
Tradenames: Chemax TO-8; Ethox TO-8; Genagen TA-080; Industrol® 400-MOT; Laurel PEG 400 MT; Mapeg® 400 MOT; Pegosperse® 400 MOT; Prox-onic TA-1/08; Witconol H31

PEG-9 tallate
Tradenames: Ethox TO-9A

PEG-10 tallate (CTFA)
CAS 61791-00-2 (generic)
Synonyms: POE (10) monotallate; PEG 500 monotallate
Tradenames: Aconol X10; Chemax TO-10; Industrol® TO-10; Prox-onic TA-1/010; Teric T10

PEG-12 tallate (CTFA)
CAS 61791-00-2 (generic)
Synonyms: POE (12) monotallate; PEG 600 monotallate
Tradenames: EM-600; Genagen TA-120; Laurel PEG 600 MT; Mapeg® 600 MOT

PEG-14 tallate
Tradenames: Alkamuls® TD-41 (formerly Rhodiasurf TD-41); Sellig T 14 100

PEG-15 tallate
CAS 61791-00-2 (generic); 65071-95-6
Synonyms: POE (15) monotallate; POE (15) tall oil acid
Tradenames: Ethofat® 242/25; Ethofat® 433

PEG-16 tallate (CTFA)
CAS 61791-00-2 (generic)
Synonyms: POE (16) monotallate
Tradenames: Chemax TO-16; Ethox TO-16; Genagen TA-160; Industrol® TO-16; Prox-onic TA-1/016; Renex® 20; Trydet 2682
Tradenames containing: Renex® 25

PEG-17 tallate
Tradenames: Sellig T 1790

PEG-20 tallate (CTFA)
CAS 61791-00-2 (generic)
Synonyms: POE (20) monotallate; PEG 1000 monotallate
Tradenames: Hetoxamate FA-20

PEG-660 tallate
Tradenames: Mapeg® TAO-15

PEG-1 tallowamine
Tradenames: Noramox S1

PEG-2 tallowamine
CAS 61791-44-4
Synonyms: Bis (2-hydroxyethyl) tallow amine
Tradenames: Accomeen T2; Berol 456; Chemeen T-2; Desomeen® TA-2; Ethomeen® T/12; Ethox TAM-2; Genamin T-020; Hetoxamine T-2; Icomeen® T-2; Mazeen® T-2; Noramox S2; Prox-onic MT-02; Teric 17M2; Tomah E-T-2; Toximul® TA-2; Varonic® T202; Varonic® T202 SF; Varonic® T202 SR; Zusomin TG 102

PEG-4 tallowamine
Tradenames: Mazeen® T-3.5

PEG-5 tallowamine
CAS 61791-44-4
Tradenames: Accomeen T5; Berol 391; Berol 457; Chemeen T-5; Desomeen® TA-5; Ethomeen® T/15; Ethox TAM-5; Genamin T-050; Hetoxamine T-5; Icomeen® T-5; Mazeen® T-5; Noramox S5; Prox-onic MT-05; Rhodameen® PN-430 (formerly Katapol® PN-430); Rhodameen® T-5 (formerly Alkaminox® T-5); Teric 17M5; Tomah E-T-5; Toximul® TA-5; Varonic® T205; Varonic® T205 SF

PEG-7 tallowamine
Tradenames: Icomeen® T-7; Noramox S7

PEG-8 tallowamine
Tradenames: Trymeen® TAM-8; Varonic® 2271

PEG-10 tallowamine
CAS 61791-26-2
Tradenames: Berol 389; Berol 458; Chemeen T-10; Ethox TAM-10; Eumulgin PA 10; Marlazin® T 10; Mazeen® T-10; Rhodameen® S-20 (formerly Rhodiasurf S-20); Varonic® T210; Varonic® T210 SF

PEG-11 tallowamine
Tradenames: Noramox S11

PEG-12 tallowamine
Tradenames: Rhodameen® T-12 (formerly Alkaminox® T-12)

PEG-14 tallowamine
Tradenames: Amiet TD/14

PEG-15 tallow amine
CAS 61791-26-2
Tradenames: Ablumox T-15; Accomeen T15; Alkaminox® T-15 (redesignated Rhodameen® T-15); Berol 381; Berol 392; Chemeen T-15; Desomeen® TA-15; Ethomeen® T/25; Ethox TAM-15; Ethylan® TT-15; Genamin T-150; Hetoxamine T-15; Icomeen® T-15; Marlazin® T 15/2; Mazeen® T-15; Noramox S15; Prox-onic MT-015; Rhodameen® S-25 (formerly Rhodiasurf S-25); Rhodameen® T-15 (formerly Alkaminox® T-15); Serdox NJAD 15; Teric 17M15; Tomah E-T-15; Toximul® TA-15; Trymeen® 6606; Varonic® T215; Varonic® T215LC; Varonic® T215 SF; Witcamine® 6606

PEG-16 tallowamine
Tradenames: Marlazin® T 16/1

PEG-17 tallowamine
Tradenames: Amiet TD/17

PEG-20 tallowamine
Tradenames: Ablumox T-20; Atlas G-3780A; Berol 386; Chemeen T-20; Desomeen® TA-20; Ethox TAM-20; Eumulgin PA 20; G-3780-A; Genamin T-200; Hetoxamine T-20; Icomeen® T-20; Noramox S20; Prox-onic MT-020; Serdox NJAD 20; Trymeen® 6607; Trymeen® TAM-20

PEG-22 tallowamine
Tradenames: Amiet TD/22

PEG-25 tallowamine
Tradenames: Ethox TAM-25; Icomeen® T-25; Trymeen® 6609; Trymeen® TAM-25

PEG-27 tallowamine
Tradenames: Amiet TD/27

PEG-30 tallowamine
Tradenames: Genamin T-308; Hetoxamine T-30; Rhodameen® T-30-45% (formerly Alkaminox® T-30); Rhodameen® T-30-90% (formerly Alkaminox® T-30); Serdox NJAD 30

PEG-40 tallowamine
Tradenames: Berol 387; Icomeen® T-40; Icomeen® T-40-80%; Trymeen® 6610; Trymeen® 6637; Trymeen® TAM-40

PEG-50 tallowamine
Tradenames: Ethomeen® T/60; Hetoxamine T-50; Hetoxamine T-50-70%; Marlazin® T 50; Rhodameen® T-50 (formerly Alkaminox® T-50)

PEG-20 tallow amine diethyl sulfate salt
Tradenames: Ethox TAM-20 DQ

PEG-3 tallow aminopropylamine (CTFA)
CAS 61790-85-0
Synonyms: POE (3) tallow aminopropylamine; PEG-3 N-tallow-1,3-diaminopropane
Tradenames: Ethoduomeen® T/13

PEG-10 tallow aminopropylamine (CTFA)
CAS 61790-85-0 (generic)
Synonyms: POE (10) tallow aminopropylamine; PEG 500 tallow aminopropylamine; PEG-10 N-tallow-1,3-diaminopropane
Tradenames: Ethoduomeen® T/20

PEG-15 tallow aminopropylamine (CTFA)
CAS 61790-85-0 (generic)
Synonyms: POE (15) tallow aminopropylamine; PEG-15 N-tallow-1,3-diaminopropane
Tradenames: Chemeen DT-15; Ethoduomeen® T/25

PEG-2 tallowate
CAS 68153-64-0 (generic)
Synonyms: Diethylene glycol monotallowate; POE (2) tallowate; PEG-2 tallow acid ester
Tradenames: Nopalcol 1-TW

PEG-3 tallow diamine
Tradenames: Berol 455; Chemeen DT-3; Prox-onic DT-03

PEG-15 tallow diamine
Tradenames: Ethox DT-15; Prox-onic DT-015

PEG-30 tallow diamine
Tradenames: Chemeen DT-30; Ethox DT-30; Prox-onic DT-030

PEG tallow ethers. *See Talloweth Series*

PEG-3 tallow propylene diamine
Tradenames: Dinoramox S3

PEG-7 tallow propylene diamine
Tradenames: Dinoramox S7

PEG-12 tallow propylene diamine
Tradenames: Dinoramox S12

PEG-3 tallow propylenedimonium dimethosulfate (CTFA)
CAS 93572-63-5; EINECS 297-495-0
Synonyms: N-Tallowalkyl-N,N′-dimethyl-N,N′-polyethyleneglycol-propylenebis-ammonium-bis-methosulfate; POE (3) tallow propylenedimonium dimethosulfate
Tradenames: Rewoquat DQ 35

PEG-3.5 tetramethyl decynediol
CAS 9014-85-1
Tradenames: Surfynol® 440

PEG-10 tetramethyl decynediol
CAS 9014-85-1
Tradenames: Surfynol® 465

PEG-30 tetramethyl decynediol
CAS 9014-85-1
Tradenames: Surfynol® 485

PEG tridecyl ether. *See Trideceth Series*

PEG-2 tridecyl phenol
Tradenames: Trycol® TP-2

PEG-6 tridecyl phenol
Tradenames: Trycol® TP-6

PEG-8 triethanolamine
Tradenames: Syn Fac® TDA-92

PEG-66 trihydroxystearin (CTFA)
CAS 61788-85-0
Synonyms: POE (66) trihydroxystearin
Tradenames: Surfactol® 575

PEG-200 trihydroxystearin (CTFA)
Synonyms: POE (200) trihydroxystearin
Tradenames: Naturechem® THS-200; Surfactol® 590

PEG-16 tristyrylphenol
Tradenames: Soprophor® BSU

PEG-20 tristyrylphenol
Tradenames: Soprophor® CY/8

PEG-25 tristyrylphenol
Tradenames: Soprophor® S/25

PEG-40 tristyrylphenol
Tradenames: Soprophor® S40-P

Pelargonic acid (CTFA)
CAS 112-05-0; EINECS 203-931-2
Synonyms: Nonanoic acid
Tradenames: Emery® 1202

Pentadecyl alcohol (CTFA)
CAS 629-76-5
Synonyms: 1-Pentadecanol; C15 linear primary alcohol
Tradenames: Neodol® 5

3-(n-Pentadecyl) phenol
CAS 501-24-6; 3158-56-3
Tradenames: Cardolite® NC-507; Cardolite® NC-510

Pentaerythrityl oleate
Tradenames: Radiasurf® 7156

Pentaerythrityl stearate
Tradenames: Radiasurf® 7175

Pentaerythrityl tetracaprylate/caprate
CAS 68441-68-9; 69226-96-6
Tradenames: Kessco® 874

Pentaerythrityl tetraoleate (CTFA)
CAS 19321-40-5; EINECS 242-960-5
Synonyms: Pentaerythritol tetraoleate
Tradenames: Mazol® PETO; Mazol® PETO Mod 1

Pentaerythrityl tetrapelargonate (CTFA)
CAS 14450-05-6; EINECS 238-430-8
Synonyms: Nonanoic acid, 2,2-bis[[(1-oxononyl)oxy] methyl] 1,3-propanediyl ester; Pentaerythritol tetrapelargonate
Tradenames: Afilan PP; Emerest® 2485

N,N,N′,N′,N′-Penta(2-hydroxyethyl)-N-tallowalkyl-1,3-propane diammonium diacetate
Tradenames containing: Ethoduoquad® T/15-50

N,N,N′,N′,N′-Pentamethyl-N-octadecenyl-1,3-diammonium dichlorides
CAS 68310-73-6
Tradenames containing: Duoquad® O-50

N,N,N′,N′,N′-Pentamethyl-n-tallow-1,3-propanediammonium dichlorides
CAS 68107-29-4
Tradenames containing: Duoquad® T-50

Pentasodium aminotrimethylene phosphonate (CTFA)
CAS 2235-43-0; EINECS 218-791-8
Synonyms: Aminotri(methylenephosphonic acid) pentasodium salt; Pentasodium [nitrilotris (methylene) trisphosphonate
Tradenames: Wayplex NTP-S

Perfluorooctanesulfonyl fluoride
CAS 307-35-7
Tradenames: Fluorad FX-8

Perfluorooctanoic acid
Tradenames: Fluorad FC-26

Petrolatum (CTFA)
CAS 8009-03-8 (NF); 8027-32-5 (USP); EINECS 232-373-2
Synonyms: Petroleum jelly; Petrolatum amber; Petrolatum white
Tradenames containing: Amerchol® C; Amerchol® CAB; Nimcolan® 1747

Phenol sulfonic acid
CAS 98-67-9; 1333-39-7
Synonyms: p-Phenolsulfonic acid; 4-Hydroxybenzenesulfonic acid; Sulfocarbolic acid
Tradenames: Eltesol® PSA 65

Phenoxyethanol (CTFA)
CAS 122-99-6; EINECS 204-589-7
Synonyms: 2-Phenoxyethanol; Phenoxytol; Ethylene glycol monophenyl ether
Tradenames: Igepal® Cephene Distilled (formerly Cephene Distilled)

2-Phosphono butane tricarboxylic acid-1,2,4
Tradenames: Bayhibit; Bayhibit® AM

Pine lignin
CAS 37203-80-8
Tradenames: Indulin® AT

Pine oil (CTFA)
CAS 8002-09-3
Synonyms: Oils, pine; Yarmor
Tradenames containing: Pinamine A; Pinamine K

POE. *See PEG*

Poloxamer 101 (CTFA)
CAS 9003-11-6 (generic)
Synonyms: Methyl oxirane polymers (generic); Polyethylenepolypropylene glycols, polymers (generic)
Tradenames: Macol® 46; Pluronic® L31

Poloxamer 105 (CTFA)
CAS 9003-11-6 (generic)
Tradenames: Macol® 35; Pluronic® L35; Synperonic PE/L35

Poloxamer 108 (CTFA)
CAS 9003-11-6 (generic)
Tradenames: Pluronic® F38; Pluronic® F68LF; Pluronic® L62D; Pluronic® L62LF; Synperonic PE/F38

Poloxamer 122 (CTFA)
CAS 9003-11-6 (generic)
Tradenames: Macol® 42; Pluronic® L42; Synperonic PE/L42

Poloxamer 123 (CTFA)
CAS 9003-11-6 (generic)
Tradenames: Pluronic® L43; Synperonic PE/L43

Poloxamer 124 (CTFA)
CAS 9003-11-6 (generic)
Tradenames: Macol® 44; Pluronic® L44; Synperonic PE/L44

Poloxamer 181 (CTFA)
CAS 9003-11-6 (generic)

Tradenames: Macol® 1; Pluronic® L61; Synperonic PE/L61

Poloxamer 182 (CTFA)
CAS 9003-11-6 (generic)
Tradenames: Macol® 2; Pluronic® L62; Synperonic PE/L62

Poloxamer 183 (CTFA)
CAS 9003-11-6 (generic)
Tradenames: Pluronic® L63

Poloxamer 184 (CTFA)
CAS 9003-11-6 (generic)
Tradenames: Macol® 4; Pluronic® L64; Synperonic PE/L64

Poloxamer 185 (CTFA)
CAS 9003-11-6 (generic)
Tradenames: Pluronic® P65

Poloxamer 188 (CTFA)
CAS 9003-11-6 (generic)
Tradenames: Macol® 8; Pluronic® F68; Synperonic PE/F68

Poloxamer 212 (CTFA)
CAS 9003-11-6 (generic)
Tradenames: Macol® 72; Pluronic® L72

Poloxamer 215 (CTFA)
CAS 9003-11-6 (generic)
Tradenames: Pluronic® P75; Synperonic PE/P75

Poloxamer 217 (CTFA)
CAS 9003-11-6 (generic)
Tradenames: Macol® 77; Pluronic® F77

Poloxamer 231 (CTFA)
CAS 9003-11-6 (generic)
Tradenames: Pluronic® L81; Synperonic PE/L81

Poloxamer 234 (CTFA)
CAS 9003-11-6 (generic)
Tradenames: Pluronic® P84; Synperonic PE/P84

Poloxamer 235 (CTFA)
CAS 9003-11-6 (generic)
Tradenames: Macol® 85; Pluronic® P85; Synperonic PE/P85

Poloxamer 237 (CTFA)
CAS 9003-11-6 (generic)
Tradenames: Antarox® PGP 23-7 (formerly Alkatronic PGP 23-7); Pluronic® F87; Synperonic PE/F87

Poloxamer 238 (CTFA)
CAS 9003-11-6 (generic)
Tradenames: Pluronic® F88; Synperonic PE/F88

Poloxamer 282 (CTFA)
CAS 9003-11-6 (generic)
Tradenames: Pluronic® L92; Synperonic PE/L92

Poloxamer 284 (CTFA)
CAS 9003-11-6 (generic)
Tradenames: Pluronic® P94; Synperonic PE/P94

Poloxamer 288 (CTFA)
CAS 9003-11-6 (generic)
Tradenames: Pluronic® F98; Synperonic PE/F98

Poloxamer 331 (CTFA)
CAS 9003-11-6 (generic)
Tradenames: Macol® 101; Pluronic® L101; Synperonic PE/L101

Poloxamer 333 (CTFA)
CAS 9003-11-6 (generic)
Tradenames: Pluronic® P103; Synperonic PE/P103

Poloxamer 334 (CTFA)
CAS 9003-11-6 (generic)
Tradenames: Pluronic® P104; Synperonic PE/P104

Poloxamer 335 (CTFA)
CAS 9003-11-6 (generic)
Tradenames: Pluronic® P105

Poloxamer 338 (CTFA)
CAS 9003-11-6 (generic)
Tradenames: Macol® 108; Pluronic® F108; Synperonic PE/F108

Poloxamer 401 (CTFA)
CAS 9003-11-6 (generic)
Tradenames: Antarox® E-100 (formerly Supronic E-100); Pluronic® L121; Synperonic PE/L121

Poloxamer 402 (CTFA)
CAS 9003-11-6 (generic)
Tradenames: Pluronic® L122

Poloxamer 403 (CTFA)
CAS 9003-11-6 (generic)
Tradenames: Macol® 23; Pluronic® P123

Poloxamer 407 (CTFA)
CAS 9003-11-6 (generic)
Tradenames: Macol® 27; Pluronic® F127; Synperonic PE/F127

Poloxamine 304 (CTFA)
CAS 11111-34-5 (generic)
Tradenames: Synperonic T/304; Tetronic® 304

Poloxamine 504 (CTFA)
CAS 11111-34-5 (generic)
Tradenames: Tetronic® 504

Poloxamine 701 (CTFA)
CAS 11111-34-5 (generic)
Tradenames: Synperonic T/701; Tetronic® 701

Poloxamine 702 (CTFA)
CAS 11111-34-5 (generic)
Tradenames: Tetronic® 702

Poloxamine 704 (CTFA)
CAS 11111-34-5 (generic)
Tradenames: Tetronic® 704

Poloxamine 707 (CTFA)
CAS 11111-34-5 (generic)
Tradenames: Synperonic T/707; Tetronic® 707

Poloxamine 901 (CTFA)
CAS 11111-34-5 (generic)
Tradenames: Tetronic® 901

Poloxamine 904 (CTFA)
CAS 11111-34-5 (generic)
Tradenames: Synperonic T/904; Tetronic® 904

Poloxamine 908 (CTFA)
CAS 11111-34-5 (generic)
Tradenames: Synperonic T/908; Tetronic® 908

Poloxamine 1101 (CTFA)
CAS 11111-34-5 (generic)
Tradenames: Tetronic® 1101

Poloxamine 1102 (CTFA)
CAS 11111-34-5 (generic)
Tradenames: Tetronic® 1102

Poloxamine 1104 (CTFA)
CAS 11111-34-5 (generic)
Tradenames: Tetronic® 1104

Poloxamine 1107
CAS 11111-34-5 (generic)
Tradenames: Tetronic® 1107

Poloxamine 1301 (CTFA)
CAS 11111-34-5 (generic)
Tradenames: Synperonic T/1301; Tetronic® 1301

Poloxamine 1302 (CTFA)
CAS 11111-34-5 (generic)
Tradenames: Synperonic T/1302; Tetronic® 1302

Poloxamine 1304 (CTFA)
CAS 11111-34-5 (generic)
Tradenames: Tetronic® 1304

Poloxamine 1307 (CTFA)
CAS 11111-34-5 (generic)
Tradenames: Tetronic® 1307

Poloxamine 1501 (CTFA)
CAS 11111-34-5 (generic)
Tradenames: Tetronic® 1501

Poloxamine 1502 (CTFA)
CAS 11111-34-5 (generic)
Tradenames: Tetronic® 1502

Poloxamine 1504 (CTFA)
CAS 11111-34-5 (generic)
Tradenames: Tetronic® 1504

Poloxamine 1508 (CTFA)
CAS 11111-34-5 (generic)
Tradenames: Tetronic® 1508

Polyacrylamide (CTFA)
CAS 9003-05-8
Synonyms: 2-Propenamide, homopolymer
Tradenames containing: Akypomine® P 191

Polyacrylate
Synonyms: Acrylic resin
Tradenames: Nopcosperse AD-6

Polyacrylic acid (CTFA)
CAS 9003-01-4
Synonyms: 2-Propenoic acid, homopolymer; Acrylic acid polymers
Tradenames: Acusol® 445; Burcotreat 900-A; Carbopol® 613, 614; Carbopol® 615, 616, 617; Carbopol® 907; Good-rite® K-702; Good-rite® K-732; Good-rite® K-752; Good-rite® K-7058N; Good-rite® K-7200N; Good-rite® K-7600N; Good-rite® K-7658
Tradenames containing: Good-rite® K-7058

Polydimethylsiloxane. *See Dimethicone*

Polyethylene (CTFA)
CAS 9002-88-4
Synonyms: Ethene, homopolymer
Tradenames: Hartolon 5683

Polyethylene glycol. *See PEG*

Polyglyceryl-10 decaisostearate
Synonyms: Decaglycerin decaisostearate
Tradenames: Nikkol Decaglyn 10-IS

Polyglyceryl-10 decaoleate (CTFA)
CAS 11094-60-3
Synonyms: Decaglycerol decaoleate; Decaglyceryl decaoleate
Tradenames: Caplube 8442; Nikkol Decaglyn 10-O

Polyglyceryl-10 decastearate (CTFA)
CAS 39529-26-5; EINECS 254-495-5
Synonyms: Decaglycerol decastearate; Decaglyceryl decastearate; Octadecanoic acid, decaester with decaglycerol
Tradenames: Caplube 8448; Nikkol Decaglyn 10-S

Polyglyceryl-3 diisostearate (CTFA)
CAS 85404-84-8; 31566-31-1; 11099-07-3; 85666-92-8
Synonyms: Triglyceryl diisostearate
Tradenames: Emerest® 2452

Polyglyceryl-10 diisostearate (CTFA)
Synonyms: Decaglyceryl diisostearate
Tradenames: Nikkol Decaglyn 2-IS

Polyglyceryl-6 dioleate (CTFA)
Synonyms: Hexaglycerol dioleate; Hexaglyceryl dioleate
Tradenames: Plurol Oleique WL 1173

Polyglyceryl-3 distearate (CTFA)
Synonyms: Triglyceryl distearate
Tradenames: Cremophor® GS-32

Polyglyceryl-6 distearate (CTFA)
Synonyms: Hexaglycerol distearate; Hexaglyceryl distearate
Tradenames: Plurol Stearique WL 1009

Polyglyceryl-10 heptaisostearate
Synonyms: Decaglycerin heptaisostearate; Decaglyceryl heptaisostearate
Tradenames: Nikkol Decaglyn 7-IS

Polyglyceryl-10 heptaoleate (CTFA)
Synonyms: Decaglycerin heptaoleate; Decaglyceryl heptaoleate
Tradenames: Nikkol Decaglyn 7-O

Polyglyceryl-10 heptastearate (CTFA)
Synonyms: Decaglycerin heptastearate; Decaglyceryl heptastearate
Tradenames: Nikkol Decaglyn 7-S

Polyglyceryl-10 hexaoleate
Tradenames: Polyaldo® DGHO

Polyglyceryl-4 isostearate (CTFA)
CAS 91824-88-3
Synonyms: Tetraglyceryl monoisostearate; Isooctanoic acid, monoester with tetraglycerol
Tradenames: Tegin® T 4753; Witconol 18L

Polyglyceryl-6 isostearate (CTFA)
Tradenames: Plurol Isostearique

Polyglyceryl-6 laurate
Synonyms: Hexaglycerin monolaurate; Hexaglyceryl monolaurate
Tradenames: Nikkol Hexaglyn 1-L

Polyglyceryl-3 oleate (CTFA)
CAS 9007-48-1 (generic)
Synonyms: Triglyceryl oleate
Tradenames: Caplube 8410; Grindtek PGE 25; Mazol® PGO-31 K; Santone® 3-1-SH

Polyglyceryl-4 oleate (CTFA)
CAS 9007-48-1 (generic)
Synonyms: Tetraglyceryl monooleate
Tradenames: Nikkol Tetraglyn 1-O; Witconol 14

Polyglyceryl-6 oleate (CTFA)
CAS 9007-48-1 (generic)
Synonyms: Hexaglycerin monooleate; Hexaglyceryl oleate
Tradenames: Nikkol Hexaglyn 1-O

Polyglyceryl-8 oleate (CTFA)
CAS 9007-48-1 (generic); 75719-56-1
Synonyms: 9-Octadecenoic acid, monoester with octaglycerol
Tradenames: Santone® 8-1-O

Polyglyceryl-10 pentaisostearate
Synonyms: Decaglycerin pentaiosostearate
Tradenames: Nikkol Decaglyn 5-IS

Polyglyceryl-4 pentaoleate
Synonyms: Tetraglyceryl pentaoleate
Tradenames: Nikkol Tetraglyn 5-O

Polyglyceryl-6 pentaoleate (CTFA)
Synonyms: Hexaglyceryl pentaoleate
Tradenames: Nikkol Hexaglyn 5-O

Polyglyceryl-10 pentaoleate (CTFA)
Synonyms: Decaglycerin pentaoleate
Tradenames: Nikkol Decaglyn 5-O

Polyglyceryl-4 pentastearate
Synonyms: Tetraglyceryl pentastearate
Tradenames: Nikkol Tetraglyn 5-S

Polyglyceryl-6 pentastearate (CTFA)
Synonyms: Hexaglyceryl pentastearate
Tradenames: Nikkol Hexaglyn 5-S

Polyglyceryl-10 pentastearate (CTFA)
Synonyms: Decaglycerin pentastearate; Decaglyceryl pentastearate
Tradenames: Nikkol Decaglyn 5-S

Polyglyceryl-6 polyricinoleate
Synonyms: Hexaglyceryl polyricinoleate
Tradenames: Nikkol Hexaglyn PR-15

Polyglyceryl-3 stearate (CTFA)
CAS 37349-34-1 (generic); 27321-72-8; EINECS 248-403-2
Synonyms: Triglyceryl stearate
Tradenames: Drewpol® 3-1-SK; Grindtek PGE 55

Polyglyceryl-3 stearate SE (CTFA)
Synonyms: Triglyceryl monostearate SE
Tradenames: Grindtek PGE 55-6

Polyglyceryl-4 stearate (CTFA)
CAS 37349-34-1 (generic); 68004-11-5
Synonyms: Tetraglyceryl monostearate; Octadecanoic acid, monoester with tetraglycerol
Tradenames: Nikkol Tetraglyn 1-S

Polyglyceryl-6 stearate
Synonyms: Hexaglycerol stearate; Hexaglyceryl stearate
Tradenames: Nikkol Hexaglyn 1-S

Polyglyceryl-10 tetracocate
Synonyms: Decaglycerol tetracocate
Tradenames: Caplube 8445

Polyglyceryl-10 tetraoleate (CTFA)
CAS 34424-98-1; EINECS 252-011-7
Synonyms: Decaglycerol tetraoleate; Decaglyceryl tetraoleate; 9-Octadecenoic acid, tetraester with decaglycerol
Tradenames: Caplube 8440; Drewpol® 10-4-OK; Mazol® PGO-104

Polyglyceryl-10 trioleate (CTFA)
Synonyms: Decaglycerin trioleate; Decaglyceryl trioleate
Tradenames: Nikkol Decaglyn 3-O

Polyglyceryl-4 tristearate
Synonyms: Tetraglyceryl tristearate
Tradenames: Nikkol Tetraglyn 3-S

Polyglyceryl-6 tristearate (CTFA)
Synonyms: Hexaglycerin tristearate
Tradenames: Nikkol Hexaglyn 3-S

Polyglyceryl-10 tristearate
Synonyms: Decaglycerin tristearate; Decaglyceryl tristearate
Tradenames: Nikkol Decaglyn 3-S

Polymethacrylic acid
CAS 25087-26-7
Tradenames: Daxad® 34; Daxad® 34S
Tradenames containing: Daxad® 32S

Polyoxypropylene. *See PPG*

Polyquaternium-2 (CTFA)
CAS 63451-27-4; 68555-36-2
Tradenames: Mirapol® A-15

Polysiloxane polyether copolymer
CAS 68937-54-2
Synonyms: Dimethicone copolyol
Tradenames: Tego® Flow 425; Tego® Foamex 800; Tego® Foamex 805; Tego® Foamex L 808; Tego® Foamex L 822
Tradenames containing: Tego® Foamex KS 10

Polysorbate 20 (CTFA)
CAS 9005-64-5 (generic)
Synonyms: POE (20) sorbitan monolaurate; PEG-20 sorbitan laurate; Sorbimacrogol laurate 300
Tradenames: Accosperse 20; Ahco 7596T; Alkamuls® PSML-20; Alkamuls® T-20 (formerly Soprofor T/20); Armotan® PML 20; Crillet 1; Disponil SML 120 F1; Drewmulse® POE-SML; Emasol L-120; Emsorb® 6915; Ethsorbox L-20; Ethylan® GEL2; Glycosperse® L-20; Hetsorb L-20; Hodag PSML-20; Industrol® L-20-S; Ionet T-20 C; Liposorb L-20; Lonzest® SML-20; ML-55-F; Montanox 20 DF; Nissan Nonion LT-221; Norfox® Sorbo T-20; Prox-onic SML-020; Radiasurf® 7137; Secoster® CL 10; Sorbax PML-20; Sorbon T-20; T-Maz® 20; Tween® 20; Witconol 2720; Witflow 990

Polysorbate 21 (CTFA)
CAS 9005-64-5 (generic)
Synonyms: POE (4) sorbitan monolaurate; PEG-4 sorbitan laurate
Tradenames: Ahco 7596D; Crillet 11; Disponil SML 104 F1; Emasol L-106; Emsorb® 6916; Ethylan® GLE-21; Hetsorb L-4; ML-55-F-4; Tween® 21

Polysorbate 40 (CTFA)
CAS 9005-66-7
Synonyms: POE (20) sorbitan monopalmitate; Sorbimacrogol palmitate 300; Sorbitan, monohexadecanoate, poly(oxy-1,2-ethaneidyl) derivs
Tradenames: Ahco DFP-156; Crillet 2; Disponil SMP 120 F1; Emasol P-120; Emsorb® 6910; Ethylan® GEP4; Glycosperse® P-20; Hetsorb P-20; Hodag PSMP-20; Liposorb P-20; Lonzest® SMP-20; Montanox 40 DF; Prox-onic SMP-020; Rheodol TW-P120; Secoster® CP 10; Sorbax PMP-20; T-Maz® 40; Tween® 40

Polysorbate 60 (CTFA)
CAS 9005-67-8 (generic)
Synonyms: POE (20) sorbitan monostearate; PEG-20 sorbitan stearate; Sorbimacrogol stearate 300
Tradenames: Accosperse 60; Ahco DFS-149; Alkamuls® PSMS-20; Alkamuls® T-60 (formerly Soprofor T/60); Crillet 3; Disponil SMS 120 F1; Drewmulse® POE-SMS; Durfax® 60; Emasol S-120; Emsorb® 2728; Ethsorbox S-20; Ethylan® GES6; Glycosperse® S-20; Hodag

PSMS-20; Industrol® S-20-S; Lonzest® SMS-20; MS-55-F; Montanox 60 DF; Prox-onic SMS-020; Radiasurf® 7147; Secoster® CS 10; Sorbax PMS-20; T-Maz® 60
Tradenames containing: Ritachol® 1000; Ritachol® 3000

Polysorbate 61 (CTFA)
CAS 9005-67-8 (generic)
Synonyms: POE (4) sorbitan monostearate; PEG-4 sorbitan stearate
Tradenames: Ahco DFS-96; Crillet 31; Emasol S-106; Emsorb® 6906; Emsorb® 6909; Hetsorb S-4; Montanox 61; T-Maz® 61; Tween® 61

Polysorbate 65 (CTFA)
CAS 9005-71-4
Synonyms: POE (20) sorbitan tristearate; PEG-20 sorbitan tristearate; Sorbimacrogol tristearate 300
Tradenames: Ahco 7166T; Crillet 35; Disponil STS 120 F1; Drewmulse® POE-STS; Durfax® 65; Emasol S-320; Ethsorbox TS-20; Glycosperse® TS-20; Hetsorb TS-20; Hodag PSTS-20; Industrol® STS-20-S; Lonzest® STS-20; Montanox 65; Prox-onic STS-020; Sorbax PTS-20; T-Maz® 65

Polysorbate 80 (CTFA)
CAS 9005-65-6 (generic); 37200-49-0
Synonyms: POE (20) sorbitan monooleate; PEG-20 sorbitan oleate; Sorbimacrogol oleate 300
Tradenames: Accosperse 80; Ahco DFO-150; Alkamuls® PSMO-20; Alkamuls® T-80 (formerly Soprofor T/80); Armotan® PMO 20; Atlas G-4905; Crillet 4; Desonic® SMO-20; Desotan® SMO-20; Disponil SMO 120 F1; Drewmulse® POE-SMO; Durfax® 80; Emasol O-120; Emsorb® 2722; Emsorb® 6900; Ethsorbox O-20; Ethylan® GEO8; Glycosperse® O-20; Glycosperse® O-20 FG, O-20 KFG; Glycosperse® O-20 Veg; Glycosperse® O-20X; Hetsorb O-20; Hodag PSMO-20; Industrol® O-20-S; Lonzest® SMO-20; Montanox 80 DF; Prox-onic SMO-020; Radiasurf® 7157; Sorbax PMO-20; Sorbilene O; T-Maz® 80; T-Maz® 80KLM; Witconol 2722; Witflow 991
Tradenames containing: Aldosperse® O-20 FG; Crodalan AWS; Ice #2

Polysorbate 81 (CTFA)
CAS 9005-65-5 (generic)
Synonyms: POE (5) sorbitan monooleate; PEG-5 sorbitan oleate
Tradenames: Ahco DFO-100; Alkamuls® PSMO-5; Crillet 41; Emasol O-106; Emsorb® 6901; Ethylan® GEO81; Ethylan® GOE-21; Glycosperse® O-5; Liposorb O-5; Montanox 81; Prox-onic SMO-05; Sorbax PMO-5; T-Maz® 81; Tween® 81

Polysorbate 85 (CTFA)
CAS 9005-70-3
Synonyms: POE (20) sorbitan trioleate; PEG-20 sorbitan trioleate; Sorbimacrogol trioleate 300
Tradenames: Ahco DFO-110; Alkamuls® PSTO-20; Alkamuls® T-85 (formerly Soprofor T/85) Crillet 45; Disponil STO 120 F1; Emasol O-320; Emsorb® 6903; Emsorb® 6913; Ethsorbox TO-20; Ethylan® GPS85; Glycosperse® TO-20; Hetsorb TO-20; Liposorb TO-20; Lonzest® STO-20; Montanox 85; Prox-onic STO-020; Sorbanox CO; Sorbax PTO-20; T-Maz® 85; TO-55-F; Tween® 85; Witconol 6903

Polyvinyl alcohol (CTFA)
CAS 9002-89-5 (super and fully hydrolyzed)
Synonyms: PVA; PVAL; Ethenol, homopolymer; PVOH
Tradenames: Polysol J

Polyvinyl isobutyl ether
Synonyms: PVI; Polyvinyl ether
Tradenames: Perenol EI

Polyvinylpyrrolidone. *See PVP*

POP. *See PPG*

Potash
CAS 584-08-7
Synonyms: Potassium carbonate; Pearl ash; Carbonic acid, dipotassium salt
Tradenames containing: LHS 40% Coconut Oil Soap

Potassium alginate (CTFA)
CAS 9005-36-1
Synonyms: Potassium polymannuronate; Alginic acid, potassium salt
Tradenames: Kelmar®; Kelmar® Improved

Potassium castorate (CTFA)
CAS 8013-05-6; EINECS 232-388-4
Synonyms: Castor oil, potassium salt
Tradenames: Solricin® 235

Potassium cetyl phosphate (CTFA)
Synonyms: Phosphoric acid, cetyl ester, potassium salt
Tradenames: Amphisol K

Potassium chloride (CTFA)
CAS 7447-40-7; EINECS 231-211-8
Tradenames containing: Schercotaine CAB-K; Schercotaine SCAB-KG

Potassium cocamidopropyl hydroxysultaine
Tradenames: Schercotaine SCAB-K

Potassium cocoate (CTFA)
CAS 61789-30-8; EINECS 263-049-9
Synonyms: Potassium coconate; Coconut acid, potassium salt; Fatty acids, coconut oil, potassium salts
Tradenames: Chimipon SK; Mackadet 40K; Norfox® 1101; Norfox® 1115
Tradenames containing: Akypogene ZA 97 SP

Potassium cocoyl hydrolyzed collagen (CTFA)
CAS 68920-65-0
Synonyms: Acid chlorides, coco, reaction prods. with protein hydrolyzates, potassium salts; Potassium cocoyl hydrolyzed animal protein; Potassium coco-hydrolyzed animal protein
Tradenames: May-Tein C; Maypon 4C; Nikkol CCK-40

Potassium dihydroxyethyl cocamine oxide phosphate (CTFA)
Synonyms: Potassium phosphated bis (hydroxyethyl) coco amine oxide
Tradenames: Mazox® KCAO

Potassium dodecylbenzene sulfonate (CTFA)
CAS 27177-77-1; EINECS 248-296-2
Synonyms: Dodecylbenzenesulfonic acid, potassium salt
Tradenames: Polystep® A-15-30K

Potassium hexamethylene diamine tetrakis (methylene phosphate)
Tradenames: Briquest® 462-23K

Potassium hydroxide (CTFA)
CAS 1310-58-3; EINECS 215-181-3
Synonyms: Caustic potash; Potassium hydrate; Lye
Tradenames containing: Det-O-Jet®

Potassium lauryl sulfate (CTFA)
CAS 4706-78-9; EINECS 225-190-4
Synonyms: Sulfuric acid, monododecyl ester, potassium salt
Tradenames: Sulfochem K

Potassium naphthalene-formaldehyde sulfonate
Tradenames: Daxad® 11KLS; Daxad® 19K

Potassium naphthalene sulfonate
Tradenames: Lomar® HP

Potassium oleate (CTFA)
CAS 143-18-0; EINECS 205-590-5
Synonyms: Potassium 9-octadecenoate; Oleic acid, potassium salt
Tradenames: Emkapol PO-18; Norfox® KO; Octosol 449

Potassium oleic sulfate
Tradenames: Actrasol SRK 75

Potassium phosphate (CTFA)
CAS 7778-77-0; EINECS 231-913-4
Synonyms: MKP; Potassium phosphate, monobasic; Potassium dihydrogen orthophosphate; Monopotassium orthophosphate
Tradenames: Beycostat 148 K; Beycostat 273 P; Beycostat 319 P; Beycostat 714 P; Beycostat DP

Potassium polyacrylate (CTFA)
CAS 25608-12-2
Synonyms: Polyacrylic acid, potassium salt
Tradenames: Daxad® 37LK9

Potassium ricinoleate (CTFA)
CAS 7492-30-0; EINECS 231-314-8
Synonyms: 12-Hydroxy-9-octadecenoic acid, monopotassium salt
Tradenames: Solricin® 135

Potassium ricinoleic sulfate
Tradenames: Actrasol PSR

Potassium rosinate
CAS 61790-50-9
Synonyms: Potassium soap of rosin
Tradenames: Arizona DRS-40; Arizona DRS-42; Arizona DRS-50; Arizona DRS-51E; Diprosin K-80; Dresinate® 90; Dresinate® 91; Dresinate® 95; Dresinate® 214; Dresinate® 515; Westvaco® Resin 895

Potassium tallate (CTFA)
CAS 61790-44-1
Synonyms: Tall oil acid, potassium salt
Tradenames: Tallates, K
Tradenames containing: Akypogene ZA 97 SP

Potassium talloweth sulfate
Synonyms: Potassium tallow ether sulfate
Tradenames: Elfan® NS 682 KS

Potassium toluenesulfonate (CTFA)
CAS 16106-44-8; 30526-22-8; EINECS 240-273-5
Synonyms: 4-Methylbenzenesulfonic acid, potassium salt
Tradenames: Eltesol® PT 45; Eltesol® PT 93; Hartotrope KTS 50

Potassium xylene sulfonate (CTFA)
Synonyms: Xylene sulfonic acid, potassium salt
Tradenames: Eltesol® PX 40; Eltesol® PX 93
Tradenames containing: Akypogene ZA 97 SP

Povidone. *See PVP*

PPG-5
Tradenames containing: Hetoxol CAWS

PPG-9 (CTFA)
CAS 25322-69-4 (generic)
Synonyms: Polyoxypropylene (9); Polypropylene glycol (9); PPG 400
Tradenames: Macol® P-500; Pluracol® P-410

PPG-12 (CTFA)
CAS 25322-69-4 (generic)
Synonyms: Polyoxypropylene (12); Polypropylene glycol (12)
Tradenames: Pluracol® P-710

PPG-17 (CTFA)
CAS 25322-69-4 (generic)
Synonyms: Polyoxypropylene (17); Polypropylene glycol (12)
Tradenames: Pluracol® P-1010

PPG-20 (CTFA)
CAS 25322-69-4 (generic)
Synonyms: Polyoxypropylene (20); Polypropylene glycol (20); PPG 1200
Tradenames: Macol® P-1200

PPG-26 (CTFA)
CAS 25322-69-4 (generic)
Synonyms: Polyoxypropylene (26); Polypropylene glycol (26); PPG 2000
Tradenames: Macol® P-2000; Pluracol® P-2010

PPG-30 (CTFA)
CAS 25322-69-4 (generic)
Synonyms: Polyoxypropylene (30); Polypropylene glycol (30); PPG 4000
Tradenames: Macol® P-4000; Pluracol® P-4010

PPG-34 (CTFA)
CAS 25322-69-4 (generic)
Synonyms: Polyoxypropylene (34); Polypropylene glycol (34)
Tradenames: Witconol CD-17

PPG-40
Tradenames: Eumulgin PPG 40

PPG-2 bisphenol A
Tradenames: Ethox 3113

PPG-5-buteth-7 (CTFA)
CAS 9038-95-3 (generic); 9065-63-8 (generic)
Synonyms: POE (7) POP (5) monobutyl ether; POP (5) POE (6) monobutyl ether
Tradenames: Pluracol® W170

PPG-7-buteth-10 (CTFA)
CAS 9038-95-3 (generic); 9065-63-8 (generic)
Synonyms: POE (10) POP (7) monobutyl ether; POP (7) POE (10) monobutyl ether
Tradenames: Macol® 300

PPG-12-buteth-16 (CTFA)
CAS 9038-95-3 (generic); 9065-63-8 (generic)
Synonyms: POE (16) POP (12) monobutyl ether; POP (12) POE (16) monobutyl ether
Tradenames: Macol® 660; Pluracol® W660

PPG-20-buteth-30 (CTFA)
CAS 9038-95-3 (generic); 9065-63-8 (generic)
Synonyms: POE (30) POP (20) monobutyl ether;

POP (20) POE (30) monobutyl ether
Tradenames: Pluracol® W2000

PPG-24-buteth-27 (CTFA)
CAS 9038-95-3 (generic); 9065-63-8 (generic)
Synonyms: POE (27) POP (24) monobutyl ether; POP (24) POE (27) monobutyl ether
Tradenames: Tergitol® XD

PPG-28-buteth-35 (CTFA)
CAS 9038-95-3 (generic); 9065-63-8 (generic)
Synonyms: POE (35) POP (28) monobutyl ether; POP (28) POE (35) monobutyl ether
Tradenames: Lutrol® W-3520; Macol® 3520

PPG-33-buteth-45 (CTFA)
CAS 9038-95-3 (generic); 9065-63-8 (generic)
Synonyms: POE (45) POP (33) monobutyl ether; POP (33) POE (45) monobutyl ether
Tradenames: Macol® 5100; Pluracol® W5100N

PPG-2-ceteareth-9 (CTFA)
Synonyms: POE (9) POP (2) cetyl/stearyl ether; POP (2) POE (9) cetyl/stearyl ether
Tradenames: Eumulgin L

PPG-4 ceteth-1 (CTFA)
CAS 9087-53-0 (generic); 37311-01-6 (generic)
Synonyms: POE (1) POP (4) cetyl ether; POP (4) POE (1) cetyl ether
Tradenames: Nikkol PBC-31

PPG-4-ceteth-10 (CTFA)
CAS 9087-53-0 (generic); 37311-01-6 (generic)
Synonyms: POE (10) POP (4) cetyl ether; POP (4) POE (10) cetyl ether
Tradenames: Nikkol PBC-33

PPG-4-ceteth-20 (CTFA)
CAS 9087-53-0 (generic); 37311-01-6 (generic)
Synonyms: POE (20) POP (4) cetyl ether
Tradenames: Nikkol PBC-34

PPG-5-ceteth-20 (CTFA)
CAS 9087-53-0 (generic); 37311-01-6 (generic)
Synonyms: POE (20) POP (5) cetyl ether; POP (5) POE (20) cetyl ether
Tradenames: Procetyl AWS
Tradenames containing: Crodapearl NI Liquid

PPG-8-ceteth-1 (CTFA)
CAS 9087-53-0 (generic); 37311-01-6 (generic)
Synonyms: POE (1) POP (8) cetyl ether; POP (8) POE (1) cetyl ether
Tradenames: Nikkol PBC-41

PPG-8-ceteth-20 (CTFA)
CAS 9087-53-0 (generic); 37311-01-6 (generic)
Synonyms: POE (20) POP (8) cetyl ether; POP (8) POE (20) cetyl ether
Tradenames: Nikkol PBC-44; Nikkol PBC-44(FF)

PPG-10 cetyl ether (CTFA)
CAS 9035-85-2 (generic)
Synonyms: POP (10) cetyl ether; PPG (10) cetyl ether
Tradenames: Procetyl 10

PPG-30 cetyl ether (CTFA)
CAS 9035-85-2 (generic)
Synonyms: POP (30) cetyl ether; PPG (30) cetyl ether
Tradenames: Procetyl 30

PPG-50 cetyl ether (CTFA)
CAS 9035-85-2 (generic)
Synonyms: POP (50) cetyl ether; PPG (50) cetyl ether
Tradenames: Procetyl 50

PPG-10 cetyl ether phosphate (CTFA)
CAS 111019-03-5
Synonyms: POP (10) cetyl ether phosphate
Tradenames: Crodafos CAP

PPG-2 cocamine (CTFA)
CAS 68516-06-3
Tradenames: Propomeen C/12

PPG-6 C12-18 pareth-11 (CTFA)
Tradenames: Plurafac® D-25

PPG-6-deceth-4 (CTFA)
Synonyms: POE (4) POP (6) decyl ether
Tradenames: Marlox® FK 64

PPG-6-deceth-9 (CTFA)
Synonyms: POE (9) POP (6) decyl ether
Tradenames: Marlox® FK 69

PPG-8 deceth-6 (CTFA)
Synonyms: POE (6) POP (8) decyl ether; POP (8) POE (6) decyl ether
Tradenames: Marlox® FK 86

PPG-9 diethylmonium chloride (CTFA)
CAS 9042-76-6
Synonyms: POP (9) methyl diethyl ammonium chloride; Quaternium-6
Tradenames: Emcol® CC-9

PPG-25 diethylmonium chloride (CTFA)
Synonyms: POP (25) methyl diethyl ammonium chloride; Quaternium-20
Tradenames: Emcol® CC-36

PPG-40 diethylmonium chloride (CTFA)
CAS 9076-43-1
Synonyms: POP (40) methyl diethyl ammonium chloride; Quaternium-21
Tradenames: Emcol® CC-42

PPG-20 ditallate
Tradenames: Ethox PPG 1025 DTO; Ross Chem PPG 1025DT

PPG 1000 ditallate
Tradenames: Ethox 2610; Industrol® 1025-DT

PPG-27 glyceryl ether (CTFA)
CAS 25791-96-2 (generic); EINECS 247-144-2
Tradenames: Witconol CD-18

PPG-2 hydrogenated tallowamine (CTFA)
Tradenames: Propomeen HT/12

PPG-2-isodeceth-4 (CTFA)
Synonyms: POE (4) POP (2) isodecyl ether; POP (2) POE (4) isodecyl ether
Tradenames: Sandoxylate® SX-408

PPG-2-isodeceth-9 (CTFA)
Synonyms: POE (9) POP (2) isodecyl ether; POP (2) POE (9) isodecyl ether
Tradenames: Sandoxylate® SX-418

PPG-2-isodeceth-12 (CTFA)
Synonyms: POE (12) POP (2) isodecyl ether; POP (2) POE (12) isodecyl ether
Tradenames: Sandoxylate® SX-412; Sandoxylate® SX-424

PPG-3-isodeceth-1 (CTFA)
Synonyms: POE (3) POP (1) isodecyl ether; POP (1) POE (3) isodecyl ether
Tradenames: Sandoxylate® SX-602

PPG-3-isosteareth-9 (CTFA)
Synonyms: POE (9) POP (3) isostearyl ether; POP (3) POE (9) isostearyl ether; PEG-9-PPG-3 isostearyl ether

Tradenames: Arosurf® 66-PE12

PPG-2 lanolin alcohol ether (CTFA)
CAS 68439-53-2 (generic)
Synonyms: POP (2) lanolin ether; PPG (2) lanolin ether; PPG-2 lanolin ether
Tradenames: Solulan® PB-2

PPG-5 lanolin alcohol ether (CTFA)
CAS 68439-53-2 (generic)
Synonyms: POP (5) lanolin ether; PPG (5) lanolin ether; PPG-5 lanolin ether
Tradenames: Solulan® PB-5

PPG-10 lanolin alcohol ether (CTFA)
CAS 68439-53-2 (generic)
Synonyms: POP (10) lanolin ether; PPG (10) lanolin ether; PPG-10 lanolin ether
Tradenames: Solulan® PB-10

PPG-20 lanolin alcohol ether (CTFA)
CAS 68439-53-2 (generic)
Synonyms: POP (20) lanolin ether; PPG (20) lanolin ether; PPG-20 lanolin ether
Tradenames: Solulan® PB-20

PPG-30 lanolin alcohol ether (CTFA)
CAS 68439-53-2 (generic)
Synonyms: POP (30) lanolin ether; PPG (30) lanolin ether; PPG-30 lanolin ether
Tradenames: Hetoxol PLA

PPG-2 laurate
Synonyms: PPG (2) laurate; POP (2) monolaurate
Tradenames: Cithrol DPGML N/E

PPG-2 laurate SE
Tradenames: Cithrol DPGML S/E

PPG-3-laureth-9 (CTFA)
Synonyms: POE (9) POP (3) lauryl ether; POP (3) POE (9) lauryl ether
Tradenames: Acconon 1300

PPG-4 laureth-2 (CTFA)
CAS 68439-51-0
Synonyms: POE (2) POP (4) lauryl ether; POP (4) POE (2) lauryl ether
Tradenames: Marlox® MO 124

PPG-4 laureth-5 (CTFA)
CAS 68439-51-0
Synonyms: POE (5) POP (4) lauryl ether; POP (4) POE (5) lauryl ether
Tradenames: Marlox® MO 154

PPG-4 laureth-7 (CTFA)
CAS 68439-51-0
Synonyms: POE (7) POP (4) lauryl ether; POP (4) POE (7) lauryl ether
Tradenames: Marlox® MO 174

PPG-25-laureth-25 (CTFA)
CAS 37311-00-5
Synonyms: POE (25) POP (25) lauryl ether; POP (25) POE (25) lauryl ether
Tradenames: Ethox 2423

PPG-7 lauryl ether (CTFA)
CAS 9064-14-6 (generic)
Synonyms: POP (7) lauryl ether
Tradenames: Marlox® L 6

PPG-10 methyl glucose ether (CTFA)
CAS 61849-72-7
Synonyms: POP (10) methyl glucose ether; PPG (10) methyl glucose ether
Tradenames: Glucam® P-10

PPG-20 methyl glucose ether (CTFA)
CAS 61849-72-7
Synonyms: POP (20) methyl glucose ether; PPG (20) methyl glucose ether
Tradenames: Glucam® P-20

PPG-3-myreth-3 (CTFA)
CAS 37311-04-9 (generic)
Synonyms: POE (3) POP (3) myristyl ether; POP (3) POE (3) myristyl ether
Tradenames: Witconol APEM

PPG-3 myristyl ether (CTFA)
CAS 63793-60-2 (generic)
Synonyms: Tripropylene glycol myristyl ether; POP (3) myristyl ether; PPG (3) myristyl ether
Tradenames: Hetoxol MP-3; Promyristyl PM-3; Witconol APM

PPG-2 oleate
Synonyms: Dipropylene glycol monooleate; POP (2) monooleate; PPG (2) monooleate
Tradenames: Cithrol DPGMO N/E

PPG-2 oleate SE
Tradenames: Cithrol DPGMO S/E

PPG-26 oleate (CTFA)
CAS 31394-71-5 (generic)
Synonyms: POP (26) monooleate; PPG (26) monooleate
Tradenames: Lutrol® OP-2000

PPG-36 oleate (CTFA)
CAS 31394-71-5 (generic)
Synonyms: POP (36) monooleate; PPG (36) monooleate
Tradenames: Witconol F26-46

PPG-50 oleyl ether (CTFA)
CAS 52581-71-2 (generic)
Synonyms: POP (50) oleyl ether; PPG (50) oleyl ether
Tradenames: Provol 50

PPG-12-PEG-50 lanolin (CTFA)
CAS 68458-88-8 (generic)
Synonyms: POE (50) POP (12) lanolin; POP (12) POE (50) lanolin
Tradenames: Laneto AWS; Lanexol AWS

PPG-12-PEG-65 lanolin oil (CTFA)
CAS 68458-88-8 (generic)
Synonyms: POE (65) POP (12) lanolin oil; POP (12) POE (65) lanolin oil
Tradenames: Fluilan AWS

PPG-40-PEG-60 lanolin oil (CTFA)
Synonyms: POE (60) POP (40) lanolin oil; POP (40) POE (60) lanolin oil
Tradenames: Aqualose LL100

PPG-2 stearate
Synonyms: Dipropylene glycol monostearate; POP (2) monostearate; PPG (2) monostearate
Tradenames: Cithrol DPGMS N/E

PPG-2 stearate SE
Tradenames: Cithrol DPGMS S/E

PPG-11 stearyl ether (CTFA)
CAS 25231-21-4 (generic)
Synonyms: POP (11) stearyl ether; PPG (11) stearyl ether
Tradenames: Arlamol® F; Witconol APS

PPG-15 stearyl ether (CTFA)
CAS 25231-21-4 (generic)

Synonyms: POP (15) stearyl ether; PPG (15) stearyl ether
Tradenames: Acconon E; Hetoxol SP-15; Prostearyl 15; Prox-onic SA1-015/P

PPG-15 stearyl ether benzoate (CTFA)
Synonyms: POP (15) stearyl ether benzoate
Tradenames: Finsolv® P

PPG-2 tallowamine (CTFA)
CAS 68951-72-4
Synonyms: N-Tallowalkyl-1,1′-iminobis-2-propanol; Dipropylene glycol tallowamine
Tradenames: Propomeen T/12

Propylene glycol (CTFA)
CAS 57-55-6; EINECS 200-338-0
Synonyms: 1,2-Propanediol; 1,2-Dihydroxypropane; Methyl glycol
Tradenames containing: Aerosol® OT-70 PG; Antil® 141 Liq; Astrowet O-70-PG; ES-1239; Gemtex PA-70P; Gemtex PA-85P; Geropon® 99 (formerly Pentex® 99); Incroquat Behenyl BDQ/P; Incroquat Behenyl TMC/P; Mackanate DOS-70PG; Miranol® C2M Conc. NP-PG; Monawet MO-70R; Monawet MO-84R2W; Rhodaterge® SSB (formerly Miranate® SSB); Surfynol® 104PG; Surfynol® PG-50; Surfynol® TG-E.; Tagat® R63; Tomah Q-14-2PG; Tomah Q-17-2PG; Zetesol AP

Propylene glycol alginate (CTFA)
CAS 9005-37-2
Synonyms: Hydroxypropyl alginate; Alginic acid, ester with 1,2-propanediol
Tradenames: Kelcoloid® D; Kelcoloid® DH, DO, DSF; Kelcoloid® HVF, LVF, O, S

Propylene glycol capreth-4 (CTFA)
Synonyms: Propylene glycol PEG (4) capryl ether
Tradenames: Marlox® FK 14

Propylene glycol ceteth-3 acetate (CTFA)
Synonyms: Propylene glycol PEG (3) cetyl ether acetate
Tradenames: Hetester PCA

Propylene glycol dicaprylate/dicaprate (CTFA)
CAS 58748-27-9; 9062-04-8; 68988-72-7
Synonyms: Decanoic acid, 1-methyl-1,2-ethanediyl ester mixed with 1-methyl-1,2-ethanediyl dioctanoate
Tradenames: Aldo® DC

Propylene glycol hydroxystearate (CTFA)
CAS 33907-47-0; EINECS 251-734-5
Synonyms: Octadecanoic acid, 12-hydroxy-, monoester with 1,2-propanediol
Tradenames: Naturechem® PGHS

Propylene glycol isoceteth-3 acetate (CTFA)
Synonyms: Propylene glycol PEG (3) isocetyl ether acetate
Tradenames: Hetester PHA

Propylene glycol laurate (CTFA)
CAS 142-55-2; 27194-74-7; EINECS 205-542-3
Synonyms: Dodecanoic acid, 2-hydroxypropyl ester; Dodecanoic acid, monoester with 1,2-propanediol; Propylene glycol monolaurate
Tradenames: Cithrol PGML N/E; Drewmulse® PGML; Kessco® PGML

Propylene glycol laurate SE
Tradenames: Cithrol PGML S/E

Propylene glycol myristate (CTFA)
CAS 29059-24-3; EINECS 249-395-3
Synonyms: Propylene glycol monomyristate; Tetradecanoic acid, monoester with 1,2-propanediol
Tradenames: Radiasurf® 7196

Propylene glycol oleate (CTFA)
CAS 1330-80-9; EINECS 215-549-3
Synonyms: 9-Octadecenoic acid, monoester with 1,2-propanediol
Tradenames: Cithrol PGMO N/E; Radiasurf® 7206

Propylene glycol oleate SE (CTFA)
Tradenames: Cithrol PGMO S/E

Propylene glycol oleth-5 (CTFA)
Synonyms: Propylene glycol PEG (5) oleyl ether
Tradenames: Marlowet® OA 4/1

Propylene glycol palmito/stearate
Tradenames: Monosteol

Propylene glycol ricinoleate (CTFA)
CAS 26402-31-3; EINECS 247-669-7
Synonyms: 12-Hydroxy-9-octadecenoic acid, monoester with 1,2-propanediol; Propylene glycol monoricinoleate
Tradenames: Cithrol PGMR N/E; Flexricin® 9; Naturechem® PGR

Propylene glycol ricinoleate SE
Tradenames: Cithrol PGMR S/E

Propylene glycol stearate (CTFA)
CAS 1323-39-3; EINECS 215-354-3
Synonyms: Propylene glycol monostearate; Octadecanoic acid, monoester with 1,2-propanediol
Tradenames: Aldo® PMS; Cithrol PGMS N/E; Drewmulse® PGMS; Emerest® 2380; Grindtek PGMS 90; Hodag PGS; Kessco® PGMS; Pegosperse® PMS CG; Radiasurf® 7201; Tegin® P-411; Witconol 2380

Propylene glycol stearate SE (CTFA)
Tradenames: Cithrol PGMS S/E; Tegin® P; Witconol RHP

Propyl oleate
Tradenames: Emerest® 2302

Proteins, collagen, hydrolysate. *See Hydrolyzed collagen*

PVM/MA copolymer (CTFA)
CAS 9011-16-9
Synonyms: Methyl vinyl ether/maleic anhydride copolymer; Poly(methyl vinyl ether/maleic anhydride); 2,5-Furandione, polymer with methoxyethylene
Tradenames: Gantrez® AN; Gantrez® AN-119; Gantrez® AN-139; Gantrez® AN-149; Gantrez® AN-169; Gantrez® AN-179; Gantrez® S-95

PVP (CTFA)
CAS 9003-39-8
Synonyms: Polyvinylpyrrolidone; Povidone; 1-Ethenyl-2-pyrrolidinone, homopolymer
Tradenames: Peregal® ST; Sokalan® HP 50; Sokalan® HP 53

Pyrrolidonecarboxylic acid. *See PCA*

Quaternium-5. *See Distearyldimonium chloride*

Quaternium-6. *See PPG-9 diethylmonium chloride*

Quaternium-7. *See Steapyrium chloride*

Quaternium-9. *See Soytrimonium chloride*

Quaternium-12
Tradenames: BTC® 1010-80

Quaternium-14 (CTFA)
CAS 27479-28-3; EINECS 248-486-5
Synonyms: Dodecyl dimethyl ethylbenzyl ammonium chloride; N-Dodecyl-*ar*-ethyl-N,N-dimethylbenzenemethanaminium chloride
Tradenames containing: BTC® 2125, 2125-80, 2125 P-40; BTC® 2125M; BTC® 2125M-80, 2125M P-40; Exameen 2125 M 3704; Exameen 2125 M 80 3709; FMB 3328-5 Quat, 3328-8 Quat

Quaternium-18 (CTFA)
CAS 61789-80-8; EINECS 263-090-2
Synonyms: Dimethyl di(hydrogenated tallow) ammonium chloride; Dihydrogenated tallow dimethyl ammonium chloride; Ditallowalkonium chloride
Tradenames: Jet Quat 2HT-75; Kemamine® Q-9702C; Quartamin D86P, D86PI, D86PL; Radiaquat 6442; Radiaquat 6475; Radiaquat 6480
Tradenames containing: Arquad® 2HT-75; M-Quat® 257

Quaternium-20. *See PPG-25 diethylmonium chloride*

Quaternium-21. *See PPG-40 diethylmonium chloride*

Quaternium-24 (CTFA)
CAS 32426-11-2; EINECS 251-035-5
Synonyms: Decyl dimethyl octyl ammonium chloride; Octyl decyl dimethyl ammonium chloride
Tradenames: BTC® 818; BTC® 818-80
Tradenames containing: BTC® 885 P40; BTC® 888; Catigene® 818; Catigene® 818-80; Catigene® 885; Catigene® 888

Quaternium-25. *See Cetethyl morpholinium ethosulfate*

Quaternium-27 (CTFA)
CAS 86088-85-9
Synonyms: Methyl-1-tallow amido ethyl-2-tallow imidazolinium methyl sulfate; Tallow imidazolinium methosulfate
Tradenames: Empigen® FRB75S; Empigen® FRC75S; Empigen® FRC90S; Empigen® FRG75S; Incrosoft S-75; Incrosoft S-90; Incrosoft S-90M
Tradenames containing: Carsosoft® S-75; Varisoft® 475

Quaternium-31. *See Dicetyl dimonium chloride*

Quaternium-32. *See Isostearyl ethylimidonium ethosulfate*

Quaternium-34. *See Dicocodimonium chloride*

Quaternium-48. *See Ditallow dimonium chloride*

Quaternium-52 (CTFA)
CAS 58069-11-7
Tradenames: Dehyquart SP

Quaternium-53 (CTFA)
CAS 130124-24-2
Synonyms: Ditallow diamido methosulfate
Tradenames: Carsosoft® T-90; Incrosoft T-75; Incrosoft T-90; Incrosoft T-90HV

Quaternium-72 (CTFA)
Synonyms: 1-Methyl-L-oleylamidoethyl-2-oleyl-imidazolinium methosulfate
Tradenames: Incrosoft 248; Incrosoft CFI-75

Ricinoleamide DEA (CTFA)
CAS 40716-42-5; EINECS 255-051-3
Synonyms: N,N-Bis(2-hydroxyethyl) ricinoleamide; Diethanolamine ricinoleic acid amide; 12-Hydroxy-N,N-bis(2-hydroxyethyl)-9-octadecenamide
Tradenames: Alkamide® RODEA (formerly Cyclomide RODEA); Aminol CA-2; Mackamide R

Ricinoleamide MEA (CTFA)
CAS 106-16-1; EINECS 203-368-2
Synonyms: Ricinoleoyl monoethanolamide; N-(2-Hydroxyethyl)-12-hydroxy-9-octadecenamide; Monoethanolamine ricinoleic acid amide
Tradenames: Rewomid® R 280

Ricinoleamidopropyl betaine (CTFA)
CAS 86089-12-5
Synonyms: N-(Carboxymethyl)-N,N-dimethyl-3[(1-oxoricinoleyl)amino]-1-propanaminium hydroxide, inner salt; Ricinoleamidopropyl dimethyl glycine
Tradenames: Mackam RA

Ricinoleamidopropyl dimethylamine (CTFA)
CAS 20457-75-4; 977010-66-4; EINECS 243-835-8
Synonyms: N-[3-(Dimethylamino) propyl] ricinoleamide; Ricinoleamide, N-[3-(dimethylamino)propyl]-
Tradenames: Mackine 201

Ricinoleic acid (CTFA)
CAS 141-22-0; EINECS 205-470-2
Synonyms: 12-Hydroxy-9-octadecenoic acid; 9-Octadecenoic acid, 12-hydroxy-; d-12-Hydroxyoleic acid
Tradenames: Flexricin® 100; P®-10 Acid

Ricinus oil. *See castor oil*

Rosin (CTFA)
CAS 8050-09-7; 8052-10-6; EINECS 232-475-7
Synonyms: Colophony; Gum rosin; Rosin gum
Tradenames: Acintol® R Type 3A; Acintol® R Type S; Acintol® R Type SB; Arizona DR-22; Arizona DR-24; Sylvaros® 20; Sylvaros® 80; Westvaco® Resin 90; Westvaco® Resin 95
Tradenames containing: Mackamide CD-25; Mackamide CDT

Sesamide DEA
Tradenames: Incromide SED

Sesamidopropylamine oxide (CTFA)
Synonyms: Amides, sesame, N-[3-(dimethylamino) propyl], N-oxide; N-[3-(Dimethylamino)propyl]sesame amides-N-oxide
Tradenames: Incromine Oxide SE

Silica (CTFA)
CAS 7631-86-9; 112945-52-5
Synonyms: Silicon dioxide, fumed; Silicon dioxide; Silicic anhydride
Tradenames: AF 72; AF 75; Bubble Breaker® 3017-A; Bubble Breaker® 3056A; Bubble Breaker® 776, 3017-A; Dow Corning® 1500 Compd; Dow Corning® 1520 Silicone Antifoam; Hercules® 137 Defoamer; Hercules® 2470 Defoamer; Rexfoam 150-A; Surfynol® 82S; Surfynol® 104S

Silica, amorphous. *See Silica*

Silicate of soda
Tradenames containing: Det-O-Jet®

Silicone compounds and emulsions. *See also Dimethicone, Simethicone*
Tradenames: Ablufoam SAE; Ablupol SAE; Ablusoft SF; AF 10 FG; AF 10 IND; AF 30 FG; AF 30 IND; AF 70; AF 100 FG; AF 100 IND; AF 1025; AF 6050; AF 8805; AF 8805 FG; AF 8810; AF 8810 FG; AF 8820; AF 8820 FG; AF 8830; AF 8830 FG; AF 9000; AF CM Conc; Agitan E 255; Agitan E 256; Anedco DF-6031; Antifoam E-20; Antifoam-G; Atlas Defoamer AFC; Chemax DF-10, DF-10A; Chemax DF-30; Chemax DF-100; CNC Antifoam 10-FG; CNC Antifoam 30-FG; CNC Antifoam 77; CNC Antifoam 100; CNC Antifoam 495, 495-M; CNC Antifoam A-107; CNC Antifoam SS; CNC Antifoam SSD; Crown Anti-Foam; Defoamer S; Defoamer SAS; Dehydran 150; Dow Corning® Antifoam 1410; Dow Corning® Antifoam 1430; Dow Corning® Antifoam 1510-US; Dow Corning® Antifoam 1520-US; Dow Corning® Antifoam 2210; Dow Corning® Antifoam B; Dow Corning® Antifoam FG-10; Dow Corning® Antifoam H-10; Dow Corning® Antifoam Y-30; Drewplus® L-407; Drewplus® L-813; Drewplus® L-833; Emka Defoam DP; Emka Defoam SD-100; Emka Defoam SMM; Foamaster 371-S; Foamaster DRY; Foamkill® 30C; Foamkill® 30HP; Foamkill® 400A; Foamkill® 679; Foamkill® 810; Foamkill® 830; Foamkill® 836B; Foamkill® 852; Foamkill® CPD; Foamkill® EFT; Foamkill® FBF; Foamkill® FPF; Foamkill® MS; Foamkill® MS Conc; Foamkill® MS-1; GP-210 Silicone Antifoam Emulsion; GP-310-I; GP-330-I Antifoam Emulsion; Hartasist 16; Hartosoft S5793; Hodag FD Series; Kilfoam; Marvanol® Defoamer AM-2; Marvanol® Defoamer MR-30A; Marvanol® Defoamer S-22; Mazu® DF 205SX; Mazu® DF 210SX; Mazu® DF 210SX Mod 1; Mazu® DF 210SXSP; Mazu® DF 215SX; Mazu® DF 230S; Mazu® DF 230SX; Mazu® DF 230SXSP; Mazu® DF 243; MY Silicone A-08; MY Silicone FT-80; Norfox® 243; Norfox® DF210SX; Octosperse TS-10; Octosperse TS-20; Octosperse TS-30; Octosperse TS-50; Patcote® 305; Patcote® 306; Patcote® 307; Patcote® 311; Patcote® 315; Patcote® 512; Patcote® 513; Patcote® 519; Patcote® 520; Patcote® 525; Patcote® 531; Patcote® 550; Patcote® 555K; Patcote® 577; Pronal EX-100; Pronal EX-300; Rexfoam B, C; Rexfoam D; Rhodorsil® AF 426R; Rhodorsil® SC 5020; Rhodorsil® SC 5021; Sellig Antimousse S; Silicone AF-100 FG; Silicone AF-100 IND; Sofnon SP-9400; Solusoft NK, WA, WL; Surfynol® DF-695; Trans-20, -20K; Trans-25, -25K; Trans-30, -30K; Trans-134S; Trans-134S K; Trans-177S, -177S K; Trans-222; Trans-225S; Trans-1030; Trans-1030 K

Silicone glycol copolymer
Tradenames: Dabco® DC193; Dabco® DC197; Dabco® DC198; Dabco® DC1315; Dabco® DC5043; Dabco® DC5098; Dabco® DC5103; Dabco® DC5125; Dabco® DC5160; Dabco® DC5418

Simethicone (CTFA)
CAS 8050-81-5
Tradenames: Dow Corning® Antifoam A; Dow Corning® Antifoam AF; Dow Corning® Antifoam C; Mazu® DF 230SP; Tego® Foamex N

Sodium acrylate/maleic acid copolymer
Tradenames: Acusol® 479N; Acusol® 479ND

Sodium alginate. *See Algin*

Sodium alpha olefin sulfonate
CAS 68188-45-5; 68439-57-6
Tradenames: Bio-Soft® AS-40; Bio-Terge® AS-90 Beads; Calsoft AOS-40; DeSonate AOS; Hostapur OS Brands; Sermul EA 214; Stepantan® AS-12; Stepantan® AS-12 Flake; Stepantan® AS-40; Stepantan® AS-90 Beads; Sulframin AOS

Sodium alpha sulfomethyl cocoate
Tradenames: Alpha-Step® MC-48

Sodium alpha sulfomethyl laurate
Tradenames containing: Alpha-Step® ML-A

Sodium aminoethyl sulfonate
Tradenames: Iberpol Gel

Sodium amino tris(methylene phosphonate)
Tradenames: Chemphonate AMP-S

Sodium benzoate (CTFA)
CAS 532-32-1; EINECS 208-534-8
Synonyms: Benzoic acid, sodium salt
Tradenames containing: Aerosol® OT-B; Manoxol OT/B; Monawet MO-85P

Sodium bis(naphthalene) sulfonate
Tradenames: Disrol SH

Sodium bistridecyl sulfosuccinate. *See Ditridecyl sodium sulfosuccinate*

Sodium borate (CTFA)
CAS 1303-96-4
Synonyms: Sodium tetraborate decahydrate; Boric acid, disodium salt
Tradenames: Borax

Sodium butoxyethoxy acetate (CTFA)
CAS 67990-17-4; EINECS 268-040-3
Synonyms: Acetic acid, (2-butoxyethoxy)-, sodium salt
Tradenames: Miranate® B (formerly Mirawet® B)

Sodium butyl naphthalene sulfonate
Synonyms: Butyl naphthalene sodium sulfonate
Tradenames: Emkal BNS; Emkal BNX Powd; Morwet® B

Sodium butyl oleate
Tradenames: Tetranol

Sodium butyl oleate sulfate
Tradenames: Actrasol SBO; Marvanol® SBO (60%)

Sodium C13-17 alkane sulfonate (CTFA)
Tradenames: Marlon® PS 30; Marlon® PS 60; Marlon® PS 65
Tradenames containing: Marlon® PF 40

Sodium C12-15 alkoxypropyl iminodipropionate (CTFA)
Tradenames: Amphoteric N

Sodium C12-15 alkyl ether sulfate
Tradenames: Serdet DPK 3/70

Sodium C14-17 alkyl sec sulfonate (CTFA)
CAS 68037-49-0; EINECS 268-213-3
Synonyms: Sodium C14-17 sec alcohol sulfonate; Sodium C14-17 alcohol sulfonate; Sodium C14-17 sec alkane sulfonate

Tradenames: Hostapur SAS 60

Sodium C12-14 alkyl sulfate
Tradenames containing: Dispersogen SI

Sodium C12-15 alkyl sulfate (CTFA)
Synonyms: Sodium C12-15 alcohols sulfate
Tradenames: Avirol® SL 2015

Sodium caproamphoacetate (CTFA)
CAS 14350-94-8; 68647-46-1; 25704-59-0; 68608-61-7; EINECS 271-951-9; 238-303-7
Synonyms: 1H-Imidazolium, 1-(carboxymethyl)-4,5-dihydro-1-(2-hydroxyethyl)-2-nonyl-, hydroxide, sodium salt; Caproamphoglycinate
Tradenames: Miranol® SM Conc.

Sodium capryleth-2 carboxylate (CTFA)
Synonyms: PEG (2) capryl ether carboxylic acid, sodium salt; Sodium PEG 100 capryl ether carboxylate; Sodium POE (2) capryl ether carboxylate
Tradenames containing: Akypo OCD 10 NV

Sodium capryleth-9 carboxylate (CTFA)
Synonyms: PEG-9 capryl ether carboxylic acid, sodium salt; Sodium PEG 450 capryl ether carboxylate; Sodium POE (9) capryl ether carboxylate
Tradenames containing: Akypo LF 4N

Sodium capryloamphoacetate (CTFA)
CAS 13039-35-5; EINECS 235-907-2
Synonyms: Capryloamphoglycinate; Capryloamphoacetate
Tradenames: Rewoteric AM V; Zoharteric LF

Sodium capryloamphohydroxypropyl sulfonate (CTFA)
CAS 68610-39-9
Synonyms: Capryloamphopropylsulfonate
Tradenames: Miranol® JS Conc. (formerly Mirapon JS Conc.)

Sodium capryloamphopropionate (CTFA)
Synonyms: Capryloamphopropionate
Tradenames: Manroteric CyNa50; Miranol® JAS-50 (formerly Mirapon JAS-50); Monateric CyNa-50; Rewoteric AM VSF

Sodium carbonate (CTFA)
CAS 497-19-8; EINECS 207-838-8
Synonyms: Soda ash; Carbonic acid, disodium salt
Tradenames containing: Alcojet®

Sodium carboxymethylcellulose. *See Carboxymethylcellulose sodium*

Sodium cetearyl sulfate (CTFA)
CAS 59186-41-3
Synonyms: Sodium cetostearyl sulfate; Sodium cetyl/stearyl sulfate
Tradenames: Dehydag Wax E; Lanette E; Meliorian 118; Rhodapon® EC111 (formerly Sipex® EC111)

Sodium ceteth-13 carboxylate (CTFA)
CAS 33939-65-0 (generic)
Synonyms: Ceteth-13 carboxylic acid, sodium salt; PEG-13 cetyl ether carboxylic acid, sodium salt
Tradenames: Sandopan® KST

Sodium cetyl/oleyl sulfate
Tradenames: Empicol® CHC 30; Gardinol CX; Gardinol CXM

Sodium cetyl stearyl sulfosuccinate
Tradenames: Secosol® AS

Sodium cetyl sulfate (CTFA)
CAS 1120-01-0; EINECS 214-292-4
Synonyms: 1-Hexadecanol, hydrogen sulfate, sodium salt
Tradenames: Nikkol SCS

Sodium CMC. *See Carboxymethylcellulose sodium*

Sodium cocaminopropionate (CTFA)
CAS 8033-69-0; 12676-37-8; 68608-68-4; EINECS 271-795-1
Synonyms: Sodium-N-coco-beta-aminopropionate; beta-Alanine, N-coco alkyl derivs., sodium salts
Tradenames: Deriphat 151

Sodium cocoamphoacetate (CTFA)
CAS 68334-21-4; 68608-65-1; 68390-66-9; 68647-53-0; EINECS 269-819-0
Synonyms: Cocoamphoacetate; Cocoamphoglycinate
Tradenames: Ampholak XCO-30; Amphoterge® W; Empigen® CDR40; Empigen® CDR60; Mackam 1C; Miranol® CM Conc. NP; Miranol® CM Conc. OP; Miranol® FA-NP (formerly Mirapon FA-NP); Monateric CM-36S; Schercoteric MS; Zoharteric M

Sodium cocoamphohydroxypropyl sulfonate (CTFA)
CAS 68604-73-9; EINECS 271-705-0
Synonyms: Cocoamphopropylsulfonate; Cocoamphohydroxypropylsulfonate
Tradenames: Amphoterge® SB; Miranol® CS Conc; Sandoteric CFL

Sodium cocoamphopropionate (CTFA)
CAS 68919-41-5
Synonyms: Cocoamphocarboxypropionate; Coconut fattyacid amidoethyl-N-2-hydroxyethylaminopropionate; Cocoamphopropionate
Tradenames: Amphoterge® K; Mackam CSF; Manroteric CEM38; Miranol® CM-SF Conc; Miranol® FAS (formerly Mirapon FAS); Monateric CA-35; Rewoteric AM KSF 40; Schercoteric MS-SF

Sodium cocoate (CTFA)
CAS 61789-31-9; EINECS 263-050-4
Synonyms: Coconut oil fatty acid, sodium salt; Fatty acids, coconut oil, sodium salts; Sodium coconut oil soap
Tradenames: Norfox® Coco Powder
Tradenames containing: Prisavon 1981

Sodium cocomonoglyceride sulfate (CTFA)
CAS 61789-04-6; EINECS 263-026-3
Synonyms: Sodium coconut monoglyceride sulfate
Tradenames: Nikkol SGC-80N

Sodium coco-sulfate (CTFA)
Synonyms: Sodium coconut sulfate; Sulfuric acid, monococoyl ester, sodium salt
Tradenames: Elfan® 280; Elfan® 280 Powd.

Sodium coco-tallow sulfate
Tradenames: Elfan® KT 550

Sodium cocoyl glutamate (CTFA)
CAS 68187-32-6; EINECS 269-087-2
Synonyms: Monosodium N-cocoyl-L-glutamate; L-Glutamic acid, N-coco acyl derivs., monosodium salts; Sodium N-cocoyl-L-glutamate
Tradenames: Acylglutamate CS-11; Amisoft CS-11
Tradenames containing: Acylglutamate AS-12; Acylglutamate GS-11; Amisoft GS-11

Sodium cocoyl hydrolyzed collagen (CTFA)
CAS 68188-38-5
Synonyms: Acid chlorides, coco, reaction prod. with protein hydrolyzates, sodium salts; Sodium cocoyl hydrolyzed animal protein; Sodium coco-hydrolyzed animal protein
Tradenames: May-Tein SK; Nikkol CCN-40

Sodium cocoyl isethionate (CTFA)
CAS 61789-32-0; 58969-27-0; EINECS 263-052-5
Synonyms: Fatty acids, coconut oil, sulfoethyl esters, sodium salts
Tradenames: Geropon® AC-78 (formerly Igepon® AC-78); Geropon® AS-200; Hostapon KA Powd; Jordapon® CI-60 Flake; Jordapon® CI-Powd; Tauranol I-78; Tauranol I-78-3; Tauranol I-78-6; Tauranol I-78E, I-78E Flakes; Tauranol I-78 Flakes;
Tradenames containing: Tauranol I-78/80

Sodium cocoyl sarcosinate (CTFA)
CAS 61791-59-1; EINECS 263-193-2
Synonyms: Amides, coconut oil, with sarcosine, sodium salts; Sodium N-cocoyl sarcosinate
Tradenames: Closyl 30 2089; Hampfoam 35; Hamposyl® C-30; Vanseal 35; Vanseal NACS-30
Tradenames containing: Nutrapon RS 1147

Sodium C12-14 olefin sulfonate (CTFA)
Tradenames containing: Marlinat® SRN 30

Sodium C14-16 olefin sulfonate (CTFA)
CAS 68439-57-6; EINECS 270-407-8
Tradenames: Bio-Terge® AS-40; Carsonol® AOS; DeSonate AUS; Elfan® OS 46; Nansa® LSS38/A; Norfox® ALPHA XL; Polystep® A-18; Rhodacal® 301-10 (formerly Siponate® 301-10); Rhodacal® A-246L (formerly Siponate® A-246L); Witconate AOK; Witconate AOS; Witconate AOS-EP

Sodium C12-15 pareth-6 carboxylate (CTFA)
CAS 70632-06-3 (generic)
Synonyms: PEG-6 C12-15 alkyl ether carboxylic acid, sodium salt; PEG 300 C12-15 alkyl ether carboxylic acid, sodium salt
Tradenames: Sandopan® DTC Linear P

Sodium C12-15 pareth-7 carboxylate (CTFA)
CAS 70632-06-3 (generic)
Synonyms: PEG-7 C12-15 alkyl ether carboxylic acid, sodium salt; POE (7) C12-15 alkyl ether carboxylic acid, sodium salt; Sodium pareth-25-7 carboxylate
Tradenames: Surfine WNT Gel; Surfine WNT LC; Surfine WNT-LS

Sodium C12-15 pareth sulfate (CTFA)
Synonyms: Sodium pareth-25 sulfate; POE (1–4) C12–15 fatty alcohol ether sulfated, sodium salt
Tradenames: Neodol® 25-3S; Serdet DPK 30; Standapol® SP-60; Witcolate 1050; Witcolate SE-5

Sodium C12-15 pareth-3 sulfonate (CTFA)
Tradenames: Avanel® S-30

Sodium C12-15 pareth-7 sulfonate
Tradenames: Avanel® S-70

Sodium C12-15 pareth-9 sulfonate
Tradenames: Avanel® S-90

Sodium C12-15 pareth-15 sulfonate (CTFA)
Tradenames: Avanel® S-150

Sodium C12-15 pareth-3 sulfosuccinate
Tradenames: Sermul EA 221

Sodium cumenesulfonate (CTFA)
CAS 32073-22-6; EINECS 250-913-5; 248-938-7
Synonyms: (1-Methylethyl)benzene, monosulfo deriv., sodium salt
Tradenames: Eltesol® SC 40; Eltesol® SC 93; Eltesol® SC Pellets; Manrosol SCS40; Manrosol SCS93; Na Cumene Sulfonate 40, Sulfonate Powd; Reworyl® NCS 40; Ultra SCS Liquid; Witconate SCS 45%; Witconate SCS 93%

Sodium deceth-2 carboxylate (CTFA)
Synonyms: PEG-2 decyl ether carboxylic acid, sodium salt; PEG 100 decyl ether carboxylic acid, sodium salt; Sodium PEG 100 decyl ether carboxylate
Tradenames containing: Akypo OCD 10 NV

Sodium deceth sulfate (CTFA)
Synonyms: Sodium decyl ether sulfate
Tradenames: Cedepal FS-406; Witcolate 7093

Sodium decylbenzene sulfonate (CTFA)
CAS 1322-98-1; EINECS 215-347-5
Synonyms: Decylbenzenesulfonic acid, sodium salt
Tradenames: Witconate DS

Sodium decyl diphenyl ether disulfonate
CAS 36445-71-3
Tradenames: Poly-Tergent® 3B2

Sodium decyl diphenyl ether sulfonate
CAS 25167-32-2
Tradenames: Sandoz Sulfonate 3B2

Sodium decyl diphenyloxide disulfonate
CAS 36445-71-3
Tradenames: Calfax 10L-45; Dowfax 3B2; Dowfax XU 40340.00

Sodium decyl sulfate (CTFA)
CAS 142-87-0; 84501-49-5
Tradenames: Atlasol 103; Avirol® SA 4110; Empimin® SDS; Polystep® B-25; Sandoz Sulfate 1030; Serdet DJK 30; Sulfotex 110; Texapon 1030, 1090
Tradenames containing: Sulfotex WAQ-LCX

Sodium diamyl sulfosuccinate. *See Diamyl sodium sulfosuccinate*

Sodium dibutyl naphthalene sulfonate
Tradenames: Dehscofix 917; Geropon® NK; Morwet® DB; Rhodacal® NK (formerly Geropon NK)

Sodium dicarboxyethyl cocophosphoethyl imidazoline (CTFA)
Tradenames: Phosphoteric® T-C6

Sodium diceteareth-10 phosphate (CTFA)
Tradenames: Marlophor® T10-Sodium Salt

Sodium dicyclohexyl sulfosuccinate. *See Dicyclohexyl sodium sulfosuccinate*

Sodium didodecylbenzene sulfonate
Tradenames: Petronate® S

Sodium di(2-ethylhexyl) sulfosuccinate. *See Dioctyl sodium sulfosuccinate*

Sodium dihexyl sulfosuccinate. *See Dihexyl sodium sulfosuccinate*

Sodium diisobutyl naphthalene sulfonate
Tradenames: Leonil DB Powd

Sodium diisobutyl sulfosuccinate. *See Diisobutyl sodium sulfosuccinate*

Sodium diisopropyl naphthalene sulfonate
CAS 1322-93-6
Tradenames: Aerosol® OS; Dehscofix 916; Dehscofix 916S; Geropon® IN; Morwet® IP; Supragil® WP

Sodium dimethylamyl sulfosuccinate
Tradenames: Manoxol MA

Sodium dimethyl naphthalene-formaldehyde sulfonate
Tradenames: Dehscofix 923; Dehscofix 926

Sodium dimethyl naphthalene sulfonate
Tradenames: Morwet® M

Sodium dinaphthalene methane sulfonate
Tradenames: Rhodacal® RM/77-D (formerly Geropon® RM/77-D)

Sodium dinonyl sulfosuccinate
Tradenames: Geropon® WS-25, WS-25-I (formerly Nekal® WS-25, WS-25-I); Manoxol N

Sodium dioctyl sulfosuccinate. *See Dioctyl sodium sulfosuccinate*

Sodium diphenyl ether disulfonate
CAS 70191-76-3
Tradenames: Poly-Tergent® 4C3

Sodium diphenyl methane sulfonate
Tradenames: Geropon® FMS

Sodium dodecylbenzenesulfonate (CTFA)
CAS 25155-30-0; 68081-81-2; 85117-50-06; EINECS 246-680-4
Synonyms: Sodium lauryl benzene sulfonate; Dodecylbenzenesulfonic acid, sodium salt; Dodecylbenzene sodium sulfonate
Tradenames: Akyposal NAF; Arylan® SBC25; Arylan® SC15; Arylan® SC30; Arylan® SX Flake; Arylan® SX85; Bio-Soft® D-40; Bio-Soft® D-62; Calsoft F-90; Calsoft L-40; Calsoft L-60; DeSonate 50-S; DeSonate 60-S; Detergent ADC; Elfan® WA; Elfan® WA Powder; Hartofol 40; Hetsulf 40, 40X; Hetsulf 60S; Hetsulf Acid; Hoe S 2713; Hoe S 2713 HF; Kadif 50 Flakes; Lumo Stabil S 80; Lumo WW 75; Manro DL 32; Manro SDBS 25/30; Manro SDBS 60; Maranil Paste A 55, A 75; Maranil Powd. A; Marlon® A 350; Marlon® A 360; Marlon® A 365; Marlon® A 375; Marlon® A 390; Marlon® AFO 40; Marlon® AFO 50; Marlon® AFR; Nacconol® 40G; Nacconol® 90G; Nansa® 1106/P; Nansa® 1169/P; Nansa® HS40-AU; Nansa® HS40/S; Nansa® HS80P; Nansa® HS80S; Nansa® HS80SK; Nansa® HS85; Nansa® HS85S; Nansa® SB62; Nansa® SL 30; Nansa® SS 30; Nansa® SS 60; Naxel AAS-40S; Naxel AAS-45S; Norfox® 40; Norfox® 85; Norfox® 90; Norfox® LAS-99; Phenyl Sulphonate HSR; Polystep® A-4; Polystep® A-7; Polystep® A-15; Polystep® A-16; Polystep® A-16-22; Puxol FB-11; Reworyl® NKS 50; Reworyl® NKS 100; Rhodacal® DDB-40 (formerly Siponate® DDB-40); Rhodacal® DS-4 (formerly Siponate® DS-4); Rhodacal® DS-10 (formerly Siponate® DS-10); Rhodacal® LDS-22 (formerly Siponate® LDS-22); Sandet 60; Sandoz Sulfonate AAS 35S; Sandoz Sulfonate AAS 40FS; Sandoz Sulfonate AAS 45S; Sandoz Sulfonate AAS 70S; Sandoz Sulfonate AAS 90; Soitem 5 C/70; Soitem SC/70; Stepantan® DS-40; Sul-fon-ate AA-9; Sul-fon-ate AA-10; Sul-fon-ate LA-10; Sulfotex LAS-90; Sulframin 1230; Sulframin 1255; Ufaryl DB80; Ufaryl DL80 CW; Ufaryl DL85; Ufaryl DL90; Ufasan 35; Ufasan 60A; Ufasan 65; Vista C-550, -560; Witconate 30DS; Witconate 45BX; Witconate 45DS; Witconate 60B; Witconate 90F; Witconate 90F H; Witconate 1223H; Witconate 1240 Slurry; Witconate 1250 Slurry; Witconate 1850; Witconate C50H; Witconate DS Dense; Witconate K; Witconate K Dense; Witconate KX; Witconate SK
Tradenames containing: Akypogene VSM-N; Akypogene WSW-W; Bio-Soft® LD-95; Marlon® AFM 40, 40N, 43, 50N; Marlon® AM 80; Marlon® ARL; Nansa® MA30; Sovatex DS/C5; Witconate 45 Liq; Witconate 45LX; Witconate 1260 Slurry; Witconate LX F; Witconate LX Powd

Sodium dodecyl diphenyl ether disulfonate
CAS 28519-02-0; 40795-56-0
Tradenames: Eleminol MON-7; Pelex SS-H; Poly-Tergent® 2A1-L; Poly-Tergent® 2EP

Sodium dodecyl diphenyl ether sulfonate
Tradenames: Sandoz Sulfonate 2A1

Sodium dodecyl diphenyloxide disulfonate
Tradenames: Dowfax 2A1; Dowfax 2EP; Dowfax XDS 30599; Dowfax XU 40333.00; Pelex SS-L; Rhodacal® DSB (formerly Siponate® DSB)

Sodium eicosyloxypropyl iminodipropionate
Tradenames: Amphoteric 300

Sodium ethyl-2 sulfolaurate (CTFA)
CAS 7381-01-3
Synonyms: Dodecanoic acid, 2-sulfoethyl ester, sodium salt; Sodium 2-sulfoethyldodecanoate

Sodium glyceryl oleate phosphate (CTFA)
Tradenames: Emphos D70-30C

Sodium glyceryl trioleate sulfate
Tradenames: Actrasol EO

Sodium heptadecyl sulfate
Tradenames: Niaproof® Anionic Surfactant 7; Rexowet 77

Sodium hexadecyl diphenyloxide disulfonate
Tradenames: Dowfax 8390; Dowfax XDS 8390.00; Dowfax XU 40341.00

Sodium hexeth-4 carboxylate (CTFA)
Synonyms: PEG-4 hexyl ether carboxylic acid, sodium salt; PEG 200 hexyl ether carboxylic acid, sodium salt; Sodium POE (4) hexyl ether carboxylate
Tradenames containing: Akypo LF 4N; Akypo TPR

Sodium hexyl diphenyloxide disulfonate
Tradenames: Dowfax XDS 8292.00

Sodium hydrogenated tallow glutamate (CTFA)
CAS 38517-23-6
Synonyms: Sodium hydrogenated tallowyl glutamate; Sodium N-hydrog. tallowyl-L-glutamate
Tradenames: Acylglutamate HS-11; Amisoft HS-11
Tradenames containing: Acylglutamate GS-11; Amisoft GS-11

Sodium hypochlorite
CAS 7681-52-9

Tradenames containing: Det-O-Jet®

Sodium isethionate (CTFA)
CAS 1562-00-1; EINECS 216-343-6
Synonyms: Sodium 2-hydroxyethanesulfonic acid; 2-Hydroxyethanesulfonic acid, sodium salt
Tradenames: Emery® 5440; Witconate NIS

Sodium isodecyl sulfate
Tradenames: Rhodapon® CAV (formerly Sipex® CAV)

Sodium isooctyl sulfate
Tradenames: Sulfetal 4105; Sulfetal 4187; Sulfetal FA 40

Sodium isopropyl naphthalene sulfonate
Tradenames: Rhodacal® IN (formerly Geropon IN)

Sodium isostearoamphopropionate (CTFA)
CAS 68630-96-6; EINECS 271-929-9
Synonyms: 1H-Imidazolium, 1-(2-carboxyethyl)-4,5-dihydro-3-(2-hydroxyethyl)-2-isoheptadeceyl-, hydroxide, inner salt; Isostearoamphopropionate (previously)
Tradenames: Monateric ISA-35; Schercoteric I-AA

Sodium isostearoyl lactylate (CTFA)
CAS 66988-04-3; EINECS 266-533-8
Tradenames: Pationic ISL

Sodium lactate (CTFA)
CAS 72-17-3; EINECS 200-772-0
Synonyms: 2-Hydroxypropanoic acid, monosodium salt; Lacolin
Tradenames: Patlac® NAL

Sodium lauramido DEA-sulfosuccinate
Tradenames: Geropon® SS-L7DE (formerly Alkasurf® SS-L7DE)

Sodium lauramido MEA-sulfosuccinate
Tradenames: Marlinat® HA 12

Sodium laureth-5 carboxylate (CTFA)
CAS 33939-64-9 (generic)
Synonyms: PEG-5 lauryl ether carboxylic acid, sodium salt; Sodium POE (5) lauryl ether carboxylate; Laureth-5 carboxylic acid, sodium salt
Tradenames: Marlinat® CM 45

Sodium laureth-6 carboxylate (CTFA)
CAS 33939-64-9 (generic); 53610-02-9
Synonyms: PEG-6 lauryl ether carboxylic acid, sodium salt; Sodium POE (6) lauryl ether carboxylate; Laureth-6 carboxylic acid, sodium salt
Tradenames: Akypo NTS
Tradenames containing: Akypogene KTS

Sodium laureth-11 carboxylate (CTFA)
CAS 33939-64-9 (generic); 53610-02-9; 68987-89-3
Synonyms: PEG-11 lauryl ether carboxylic acid, sodium salt; Sodium POE (11) lauryl ether carboxylate; Laureth-11 carboxylic acid, sodium salt
Tradenames: Akypo®-Soft 100 NV; Marlinat® CM 105; Marlinat® CM 105/80
Tradenames containing: Akypogene Jod F

Sodium laureth-13 carboxylate (CTFA)
CAS 33939-64-9 (generic); 70632-06-3
Synonyms: PEG-13 lauryl ether carboxylic acid, sodium salt; POE (13) lauryl ether carboxylic acid, sodium salt; Laureth-13 carboxylic acid, sodium salt
Tradenames: Miranate® LEC; Surfine WLL
Tradenames containing: Miracare® BC-10; Miracare® MS-1 (formerly Compound MS-1); Miracare® MS-2 (formerly Compound MS-2); Miracare® MS-4

Sodium laureth-14 carboxylate (CTFA)
CAS 33939-64-9 (generic)
Synonyms: PEG-14 lauryl ether carboxylic acid, sodium salt; POE (14) lauryl ether carboxylic acid, sodium salt; Laureth-14 carboxylic acid, sodium salt
Tradenames: Akypo®-Soft 130 NV

Sodium laureth-17 carboxylate (CTFA)
CAS 33939-64-9 (generic)
Synonyms: PEG-17 lauryl ether carboxylic acid, sodium salt; POE (17) lauryl ether carboxylic acid, sodium salt; Laureth-17 carboxylic acid, sodium salt
Tradenames: Akypo®-Soft 160 NV

Sodium laureth phosphate
Tradenames: Crafol AP-260; Crafol AP-261; Forlanit P; Forlanon; Wetting Agent FCGB

Sodium laureth-4 phosphate (CTFA)
CAS 42612-52-2 (generic)
Synonyms: Sodium POE (4) lauryl ether phosphate; Sodium PEG 200 lauryl ether phosphate
Tradenames: Rhodafac® MC-470 (formerly Gafac® MC-470)

Sodium laureth sulfate (CTFA)
CAS 1335-72-4; 3088-31-1; 9004-82-4 (generic); 13150-00-0; 15826-16-1; 68891-38-3; EINECS 221-416-0
Synonyms: Sodium lauryl ether sulfate (n=1–4); PEG (1-4) lauryl ether sulfate, sodium salt
Tradenames: Akyposal 9278 R; Akyposal EO 20 MW; Akyposal EO 20 PA; Akyposal EO 20 PA/TS; Akyposal EO 20 SF; Akyposal MS SPC; Avirol® 252 S; Avirol® 603 S; Avirol® FES 996; Avirol® SE 3002; Avirol® SE 3003; Calfoam ES-30; Calfoam SEL-60; Carsonol® SES-S; Carsonol® SLES; Cedepal SS-203, 306, 403, 406; Chemsalan RLM 28; Chemsalan RLM 56; Chemsalan RLM 70; Colonial SLE(2)S; Colonial SLE(3)S; Colonial SLES-70; Cosmopon LE 50; Disponil FES 32; Disponil FES 61; Disponil FES 77; Elfan® NS 242; Elfan® NS 242 Conc; Elfan® NS 243 S; Elfan® NS 243 S Conc; Elfan® NS 252 S; Elfan® NS 252 S Conc; Elfan® NS 423 SH; Elfan® NS 423 SH Conc; Empicol® ESB; Empicol® ESB3; Empimin® KSN27; Empimin® KSN70; Genapol® ARO; Genapol® LRO Liq., Paste; Genapol® ZRO Liq., Paste; Hartenol LES 60; Laural LS; Manro BES 27; Manro BES 60; Manro BES 70; Marlinat® 242/28; Marlinat® 242/70; Marlinat® 242/70 S; Marlinat® 243/28; Marlinat® 243/70; Nonasol N4SS; Norfox® SLES-02; Norfox® SLES-03; Norfox® SLES-60; Nutrapon ES-60 3568; Nutrapon KPC 0156; Polystep® B-12; Rewopol® NL 2; Rewopol® NL 2-28; Rewopol® NL 3; Rewopol® NL 3-28; Rewopol® NL 3-70; Rhodapex® ES (formerly Sipon® ES); Rhodapex® ESY (formerly Sipon® ESY); Rolpon 24/230; Rolpon 24/270; Rolpon 24/330 N; Sactol 2 OS 2; Sactol 2 OS 28; Sandoz Sulfate

219; Sandoz Sulfate ES-3; Sandoz Sulfate WE; Serdet DCK 3/70; Serdet DCK 30; Sermul EA 30; Sermul EA 370; Standapol® ES-250; Standapol® ES-350; Standapol® SL-60; Steol® 4N; Steol® CS-260; Steol® CS-460; Steol® OS 28; Sulfochem ES-2; Sulfochem ES-70; Sulfotex 6040; Sulfotex LMS-E; Surfax 218; Texapon LT-327; Texapon N 25; Texapon N 40; Texapon N 70; Texapon N 70 LS; Texapon N 103; Texapon NSE; Texapon NSO; Texapon PLT-227; Texapon PN-235; Texapon PN-254; Ungerol CG27; Ungerol LES 2-28; Ungerol LES 2-70; Ungerol LES 3-28; Ungerol LES 3-54; Ungerol LES 3-70; Ungerol N2-28; Ungerol N2-70; Ungerol N3-28; Ungerol N3-70; Witcolate 6450; Witcolate 6453; Witcolate 6455; Witcolate ES-2; Witcolate ES-3; Witcolate LES-60C; Witcolate OME; Witcolate S1285C; Witcolate SE (formerly DeSonol SE); Zetesol NL; Zoharpon ETA 27; Zoharpon ETA 70; Zoharpon ETA 270 (OXO); Zoharpon ETA 271; Zoharpon ETA 273; Zoharpon ETA 603; Zoharpon ETA 700 (OXO); Zoharpon ETA 703

Tradenames containing: Abex® VA 50; Akypogene WSW-W; Bio-Soft® LD-95; Crodapearl Liq; Empimin® KSN60; Genapol® PGM Conc; Marlamid® KLP; Marlon® PF 40; Rewopol® SBV; Schercoteric MS-2ES Modified; Standapol® AP Blend; Standapol® S

Sodium laureth-12 sulfate (CTFA)
CAS 9004-82-4 (generic); 66161-57-7
Synonyms: PEG (12) lauryl ether sulfate, sodium salt; PEG 600 lauryl ether sulfate, sodium salt; Sodium POE (12) lauryl ether sulfate
Tradenames: Disponil FES 92E; Polystep® B-23

Sodium laureth-30 sulfate
CAS 9004-82-4 (generic)
Synonyms: PEG (30) lauryl ether sulfate, sodium salt; Sodium POE (30) lauryl ether sulfate
Tradenames: Polystep® B-19

Sodium laureth sulfosuccinate
Tradenames: Secosol® ALL40

Sodium lauriminodipropionate (CTFA)
CAS 14960-06-6; 26256-79-1; EINECS 239-032-7
Synonyms: Sodium N-lauryl-beta-iminodipropionate; N-(2-Carboxyethyl)-N-dodecyl-beta-alanine, monosodium salt
Tradenames: Deriphat 160C; Mackam 160C; Mirataine® H2C-HA; Velvetex® 610L

Sodium lauroamphoacetate (CTFA)
CAS 14350-96-0; 68647-44-9; 26837-33-2; 68298-21-5; 68608-66-2; EINECS 271-949-8; 269-547-2; 238-305-8
Synonyms: 1H-Imidazolium, 1-(carboxymethyl)-4,5-dihydro-1-(2-hydroxyethyl)-2-undecyl-, hydroxide, sodium salt; Lauroamphoacetate; Lauroamphoglycinate
Tradenames: Ampholyt JA 140; Dehyton® PMG; Mackam 1L-30; Mackam 1L; Miranol® HM Conc; Schercoteric LS
Tradenames containing: Mackam MLT; Miracare® BC-10

Sodium lauroamphohydroxypropylsulfonate (CTFA)
CAS 68039-23-6; EINECS 268-242-1
Synonyms: Lauroamphohydroxypropylsulfonate; Lauroamphopropylsulfonate
Tradenames: Schercoteric LS-EP

Sodium lauroyl glutamate (CTFA)
CAS 29923-31-7 (L-form); 29923-34-0 (DL-form); 42926-22-7 (L-form); 98984-78-2; EINECS 249-958-3
Synonyms: N-Dodecyl-L-glutamic acid, monosodium salt; N-(1-Oxododecyl)glutamic acid, monosodium salt; Sodium N-lauroyl-L-glutamate
Tradenames: Acylglutamate LS-11; Amisoft LS-11

Sodium lauroyl sarcosinate (CTFA)
CAS 137-16-6; EINECS 205-281-5
Synonyms: N-Methyl-N-(1-oxododecyl)glycine, sodium salt
Tradenames: Closyl LA 3584; Crodasinic LS30; Crodasinic LS35; Hamposyl® L-30; Hamposyl® L-95; Maprosyl® 30; Sarkosyl® NL-30; Secosyl; Vanseal NALS-30; Vanseal NALS-95; Zoharsyl L-30

Sodium lauryl benzene sulfonate. *See Sodium dodecylbenzenesulfonate*

Sodium lauryl ether sulfate (n=1–4). *See Sodium laureth sulfate*

Sodium lauryl/oleyl sulfate
Tradenames: Duponol® D Paste

Sodium lauryl/propoxy sulfosuccinate
Tradenames: Emcol® 4910

Sodium lauryl sulfate (CTFA)
CAS 151-21-3; 68585-47-7; 68955-19-1; EINECS 205-788-1
Synonyms: SDS; Sulfuric acid, monododecyl ester, sodium salt; Sodium dodecyl sulfate
Tradenames: Akyporox SAL SAS; Akyposal NLS; Alscoap LN-40, LN-90; Avirol® 280 S; Avirol® SL 2010; Avirol® SL 2020; Calfoam SLS-30; Carsonol® SLS; Carsonol® SLS Paste B; Carsonol® SLS-R; Carsonol® SLS-S; Carsonol® SLS Special; Chemsalan NLS 30; Colonial LKP; Colonial SLS; Cosmopon 35; Duponol® ME Dry; Duponol® QC; Duponol® WA Dry; Duponol® WA Paste; Duponol® WAQ; Duponol® WAQE; Elfan® 200; Elfan® 240; Elfan® 260 S; Empicol® 0185; Empicol® 0303; Empicol® 0303V; Empicol® 0919; Empicol® LM45; Empicol® LM/T; Empicol® LS30; Empicol® LS30P; Empicol® LX; Empicol® LX28; Empicol® LXS95; Empicol® LXV; Empicol® LXV/D; Empicol® LY28/S; Empicol® LZ; Empicol® LZ/D; Empicol® LZ/E; Empicol® LZG 30; Empicol® LZGV; Empicol® LZGV/C; Empicol® LZP; Empicol® LZV/D; Empicol® LZV/E; Empicol® WA; Empicol® WAK; Empimin® LR28; Equex S; Equex SP; Equex SW; Gardinol WA Paste; Hartenol LAS-30; Laural P; Manro DL 28; Manro SLS 28; Marlinat® DFK 30; Montovol RF-10; Naxolate WA Special; Naxolate WA-97; Naxolate WAG; Nikkol SLS; Norfox® SLS; Nutrapon DL 3891; Nutrapon W 1367; Nutrapon WAC 3005; Nutrapon WAQ; Nutrapon WAQE 2364; Octosol SLS; Octosol SLS-1; Perlankrol® DSA; Polirol LS; Polystep® B-3; Polystep® B-5; Polystep® B-24; Rewopol® 15/L; Rewopol® NLS 15 L; Rewopol® NLS 28; Rewopol® NLS 30 L; Rewopol® NLS 90; Rhodapon® LCP (formerly Sipon® LCP); Rhodapon® LSB, LSB/CT (formerly Sipon® LSB); Rhodapon® SB (formerly

Sipon® SB); Rhodapon® SB-8208/S (formerly Sipon® 21LS); Rhodapon® SM Special (formerly Sipon® SM Spec.); Rolpon LSX; Sactol 2 S 3; Sandoz Sulfate WA Dry; Sandoz Sulfate WA Special; Sandoz Sulfate WAG; Sandoz Sulfate WAS; Serdet DFK 40; Sermul EA 150; Standapol® WAQ-LC; Standapol® WAQ-LCX; Stepanol® ME Dry; Stepanol® WA Extra; Stepanol® WA Paste; Stepanol® WA Special; Stepanol® WA-100; Stepanol® WAC; Stepanol® WAC-P; Stepanol® WAQ; Sulfetal C 38; Sulfetal C 90; Sulfochem SLN; Sulfochem SLP; Sulfochem SLP-95; Sulfochem SLS; Sulfochem SLX; Sulfopon 101, 101 Special; Sulfopon 101/POL; Sulfopon 101 Spez; Sulfopon 102; Sulfopon K35; Sulfopon P-40; Sulfopon WAQ LCX; Sulfopon WAQ Special; Sulfotex LCX; Sulfotex WA; Surfax 250; Surfax CN; Swascol 1P; Swascol 1PC; Swascol 3L; Texapon K-12; Texapon K-12 Granules; Texapon K-12 Needles; Texapon K-1296; Texapon L-100; Texapon LS 35; Texapon LS Highly Conc; Texapon LS Highly Conc. Needles; Texapon OT Highly Conc. Needles; Texapon VHC Needles; Texapon VHC Powd; Texapon Z; Texapon ZHC Needles; Texapon ZHC Powder; Ufarol Na-30; Ultra Sulfate SL-1; Ungerol LS, LSN; Witcolate 6400; Witcolate A; Witcolate A Powder; Witcolate C; Witcolate LCP; Witcolate S (formerly DeSonol S); Witcolate SL-1; Witcolate WAC; Witcolate WAC-GL; Witcolate WAC-LA; Zorapol LS-30

Tradenames containing: Equex STM; Nutrapon B 1365; Nutrapon RS 1147; Rhodaterge® SSB (formerly Miranate® SSB); Schercoteric MS-2 Modified; Standapol® BW; Standapol® S; Sulfetal TC 50; Sulfotex WAQ-LCX; Zoharteric DO

Sodium lauryl sulfoacetate (CTFA)
CAS 1847-58-1; EINECS 217-431-7
Synonyms: Acetic acid, sulfo-, 1-dodecyl ester, sodium salt; Sulfoacetic acid, 1-dodecyl ester, sodium salt
Tradenames: Lathanol® LAL; Nikkol LSA
Tradenames containing: Stepan-Mild® LSB

Sodium lauryl sulfosuccinate (CTFA)
Tradenames: Secosol® AL/959; Texin 128

Sodium lignate
Tradenames: Indulin® SA-L

Sodium lignosulfonate (CTFA)
CAS 8061-51-6
Synonyms: Sodium polignate; Lignosulfonic acid, sodium salt
Tradenames: Darvan® No. 2; Dyqex®; Kelig 100; Ke-Mul® 181; Ke-Mul® A97; Lignosite® 231; Lignosite® 431; Lignosite® 458; Lignosite® 823; Lignosol AXD; Lignosol D-10, D-30; Lignosol DXD; Lignosol FTA; Lignosol HCX; Lignosol NSX 110; Lignosol NSX 120; Lignosol WT; Lignosol X; Lignosol XD; Marasperse N-22; Orzan® CD; Orzan® LS; Orzan® LS-50; Reax® 45A; Reax® 45DA; Reax® 45DTC; Reax® 45L; Reax® 45T; Temsperse S 001

Sodium maleic acid/olefin copolymer
Tradenames: Acusol® 460ND

Sodium metasilicate (CTFA)
CAS 6834-92-0; EINECS 229-912-9
Synonyms: Silicic acid, disodium salt; Sodium metasilicate, anhydrous
Tradenames: Drymet®
Tradenames containing: Alcojet®

Sodium methallyl sulfonate
CAS 1561-92-8
Tradenames: Geropon® MLS/A

Sodium methyl cocoyl taurate (CTFA)
CAS 12765-39-8; 61791-42-2
Synonyms: Sodium cocoyl methyl taurate; Sodium N-cocoyl-N-methyl taurate; Sodium N-methyl-N-cocoyl taurate
Tradenames: Geropon® TC-42 (formerly Igepon® TC-42); Nikkol CMT-30; Nissan Diapon K; Tauranol WS, WS Conc; Tauranol WSP

Sodium methyl lauroyl taurate (CTFA)
Synonyms: Sodium lauroyl methyl taurate; Sodium N-lauroyl methyl taurate; Sodium N-methyl-N-lauroyl taurate
Tradenames: Zoharpon LMT42

Sodium methylnaphthalenesulfonate (CTFA)
CAS 26264-58-4; EINECS 247-561-6
Synonyms: Sodium methane napthalene sulfonate; Sodium polynapthalene methane sulfonate
Tradenames: Rhodacal® RM/210 (formerly Geropon® RM/210); Soitem 207; Supragil® MNS/90

Sodium methyl oleoyl taurate (CTFA)
CAS 137-20-2; EINECS 205-285-7
Synonyms: Sodium N-oleoyl-N-methyl taurate; Oleyl methyl tauride; Sodium N-methyl-N-oleoyl taurate
Tradenames: Adinol OT16; Arkopon T Grades; Consamine K-Gel; Geropon® T-33 (formerly Igepon® T-33); Geropon® T-43 (formerly Igepon® T-43); Geropon® T-51 (formerly Igepon® T-51); Geropon® T-77 (formerly Igepon® T-77); Hostapon T Powd; Nissan Diapon TO; Tauranol M-35; Tauranol MS; Tauranol T-Gel; Tergenol G; Tergenol S Liq; Tergenol Slurry
Tradenames containing: Tauranol ML

Sodium methyl stearoyl taurate (CTFA)
CAS 149-39-3; EINECS 205-738-9; 205-713-2
Synonyms: Sodium stearoyl methyl taurate; Sodium N-methyl-N-stearoyl taurate; Sodium N-stearoyl-N-methyl taurate
Tradenames: Nikkol SMT

Sodium methyl-2 sulfolaurate (CTFA)
Tradenames containing: Alpha-Step® ML-40

Sodium methyl tall oil acid taurate
Tradenames: Geropon® TK-32 (formerly Igepon® TK-32)

Sodium methyl tallow taurate
Tradenames: Nissan Diapon T

Sodium myreth sulfate (CTFA)
CAS 25446-80-4; EINECS 246-986-8
Synonyms: Sodium myristyl ether sulfate; PEG (1-4) myristyl ether sulfate, sodium salt
Tradenames: Standapol® ES-50

Sodium myristoyl glutamate (CTFA)
CAS 38517-37-2; 71368-20-2; EINECS 253-981-4
Synonyms: N-(1-Oxotetradecyl)glutamic acid, monosodium salt
Tradenames: Acylglutamate MS-11; Amisoft MS-11

Sodium myristoyl sarcosinate (CTFA)
CAS 30364-51-3; EINECS 250-151-3
Synonyms: N-Methyl-N-(1-oxotetradecyl)glycine, sodium salt
Tradenames: Hamposyl® M-30

Sodium myristyl ether sulfate. *See Sodium myreth sulfate*

Sodium myristyl sulfate (CTFA)
CAS 1191-50-0; 139-88-8; EINECS 214-737-2
Synonyms: Sodium tetradecyl sulfate; Sulfuric acid, monotetradecyl ester, sodium salt
Tradenames: Niaproof® Anionic Surfactant 4

Sodium naphthalene formaldehyde sulfonate. *See Sodium polynaphthalene sulfonate*

Sodium naphthalene sulfonate. *See Sodium polynaphthalene sulfonate*

Sodium nitrilotris(methylene phosphate)
Synonyms: Amino tris (methylene phosphonic acid), sodium salt
Tradenames: Briquest® 301-30SH

Sodium m-nitrobenzenesulfonate (CTFA)
CAS 127-68-4; EINECS 204-857-3
Synonyms: Benzenesulfonic acid, 3-nitro, sodium salt
Tradenames: Matexil PA Liq.; Reserve Salt Flake

Sodium nonoxynol-6 phosphate (CTFA)
CAS 12068-19-8; EINECS 235-093-9
Synonyms: Sodium POE (6) nonyl phenyl ether phosphate; PEG 300 nonyl phenyl ether phosphate, sodium salt; POE (6) nonyl phenylether phosphate, sodium salt
Tradenames: Surfagene FAD 106

Sodium nonoxynol-9 phosphate (CTFA)
Synonyms: PEG-9 nonyl phenyl ether phosphate, sodium salt; PEG 450 nonyl phenyl ether phosphate, sodium salt; Sodium PEG-9 nonyl phenyl ether phosphate
Tradenames: Emphos CS-1361

Sodium nonoxynol-4 sulfate (CTFA)
CAS 9014-90-8 (generic)
Synonyms: PEG-4 nonyl phenyl ether sulfate, sodium salt; PEG 200 nonyl phenyl ether sulfate, sodium salt; Sodium PEG-4 nonyl phenyl ether sulfate
Tradenames: Polystep® B-27; Rhodapex® CO-433 (formerly Alipal® CO-433); Serdet DNK 30; Sermul EA 54; Steol® COS 433; Witcolate D51-51EP

Sodium nonoxynol-6 sulfate (CTFA)
CAS 9014-90-8 (generic)
Synonyms: PEG-6 nonyl phenyl ether sulfate, sodium salt; POE (6) nonyl phenyl ether sulfate, sodium salt; PEG 300 nonyl phenyl ether sulfate, sodium salt
Tradenames: Akyposal NPS 60

Sodium nonoxynol-8 sulfate (CTFA)
CAS 9014-90-8 (generic)
Synonyms: PEG-8 nonyl phenyl ether sulfate, sodium salt; PEG 400 nonyl phenyl ether sulfate, sodium salt; POE (8) nonyl phenyl ether sulfate, sodium salt
Tradenames: Disponil AES 60 E

Sodium nonoxynol-9 sulfate
Tradenames: Zorapol SN-9

Sodium nonoxynol-10 sulfate (CTFA)
CAS 9014-90-8 (generic)
Tradenames: Akyposal NPS 100; Sermul EA 151

Sodium nonoxynol-15 sulfate
CAS 9014-90-8 (generic)
Tradenames: Sermul EA 146

Sodium nonoxynol-25 sulfate (CTFA)
CAS 9014-90-8 (generic)
Tradenames: Akyposal NPS 250; Montosol PB-25

Sodium nonoxynol-10 sulfosuccinate
Tradenames: Sermul EA 176

Sodium nonyl benzene sulfonate
Tradenames: Emkal NOBS

Sodium octane sulfonate
Tradenames: Bio-Terge® PAS-8S

Sodium octoxynol-2 ethane sulfonate (CTFA)
CAS 2917-94-4; 67923-87-9; EINECS 267-791-4; 220-851-3
Synonyms: 2-[2-[2-Octylphenoxy)ethoxy]ethoxy] ethanesulfonic acid, sodium salt; Entsufon
Tradenames: Triton® X-200

Sodium octoxynol-3 sulfate
Tradenames: Polystep® C-OP3S

Sodium octoxynol-6 sulfate (CTFA)
Synonyms: PEG-6 octyl phenyl ether sulfate, sodium salt; POE (6) octyl phenyl ether sulfate, sodium salt; PEG 300 octyl phenyl ether sulfate, sodium salt
Tradenames: Akyposal BD

Sodium octyl/decyl sulfate
Tradenames: Duponol® WN

Sodium N-octyl-beta-iminodipropionic acid
Tradenames: Deriphat 130-C

Sodium octylphenoxyethoxyethyl sulfonate
Tradenames: Newcol 861S

Sodium octyl sulfate (CTFA)
CAS 126-92-1; 142-31-4; EINECS 204-812-8
Synonyms: Sodium 2-ethylhexyl sulfate; Sulfuric acid, mono (2-ethylhexyl) ester, sodium salt
Tradenames: Avirol® SA 4106; Avirol® SA 4108; Carsonol® SHS; Chemsulf S2EH-Na; Cosmopon SES; Duponol® 80; Empicol® 0585/A; Niaproof® Anionic Surfactant 08; Norfox® Anionic 27; Norfox® SEHS; Polystep® B-29; Rewopol® NEHS 40; Rewopol® NI 56; Rhodapon® BOS (formerly Sipex® BOS); Rhodapon® OLS (formerly Sipex® OLS); Sandoz Sulfate 830; Sandoz Sulfate W2-30; Serdet DSK 40; Sodium Octyl Sulfate Powd; Sole-Terge TS-2-S; Standapol® LF; Stepantex® B-29; Sulfotex OA; Texapon 842; Texapon 890; Witcolate 6465; Witcolate D-510
Tradenames containing: Akypo TFC-S; Akypo TFC-SN

Sodium octyl sulfonate
Tradenames: Bio-Terge® PAS-8

Sodium oleate (CTFA)
CAS 143-19-1; EINECS 205-591-0
Synonyms: Sodium 9-octadecenoate; 9-Octadecenoic acid, sodium salt
Tradenames: Norfox® Oleic Flakes; Norfox® Vertex Flakes

Sodium oleic sulfate
CAS 67998-94-1
Tradenames: Dyasulf 2031; Sul-fon-ate OA-5; Sul-fon-ate OA-5R; Witconate 1840X

Sodium oleoamphohydroxypropylsulfonate (CTFA)
CAS 68610-38-8
Synonyms: Oleoamphopropylsulfonate; Oleoamphohydroxypropylsulfonate
Tradenames: Sandopan® TFL Conc.; Sandoteric TFL Conc.

Sodium oleoamphopropionate (CTFA)
CAS 67892-37-9; EINECS 267-569-7
Synonyms: 1H-Imidazolium, 1-(2-carboxyethyl)-2-(8-heptadecenyl)-4,5-dihydro-1-(2-hydroxyethyl)-, hydroxide, inner salt; Oleoamphopropionate
Tradenames: Schercoteric O-AA; Schercoteric OS-SF

Sodium oleoyl glutamate
Tradenames containing: Acylglutamate AS-12

Sodium N-oleoyl sarcosinate
Tradenames: Crodasinic OS35

Sodium oleyl/cetyl sulfate
Tradenames: Elfan® 680

Sodium oleyl sulfate (CTFA)
CAS 1847-55-8; EINECS 217-430-1
Synonyms: 9-Octadecen-1-ol, hydrogen sulfate, sodium salt
Tradenames: Duponol® LS Paste; Rhodapon® OS (formerly Sipex® OS); Sulfopon O 680

Sodium oleyl sulfosuccinamate
Tradenames: Cosmopon BN

Sodium N-palmityl N-cyclohexyl taurine
Tradenames: Hostapon IDC

Sodium pareth-25 sulfate. *See Sodium C12-15 pareth sulfate*

Sodium PCA (CTFA)
CAS 28874-51-3; 54571-67-4; EINECS 249-277-1
Synonyms: PCA-Na; PCA Soda; 5-Oxo-DL-proline, sodium salt; Sodium pyroglutamate
Tradenames: Ajidew N-50
Tradenames containing: Ajidew SP-100

Sodium PEG-6 cocamide carboxylate
Tradenames: Akypo®-Soft KA 250 BVC

Sodium petroleum sulfonate
CAS 68608-26-4; 78330-12-8
Tradenames: Aristonate H; Aristonate L; Aristonate M; Petronate® CR; Petronate® HL; Petronate® L; Petrosul® H-50; Petrosul® H-60; Petrosul® H-70; Petrosul® HM-62, HM-70; Petrosul® L-60; Petrosul® M-50; Petrosul® M-60; Petrosul® M-70

Sodium phenol sulfonate (CTFA)
CAS 1300-51-2; EINECS 215-087-2
Synonyms: Hydroxybenzenesulfonic acid, monosodium salt
Tradenames: Supragil® GN

Sodium polyacrylate
CAS 9003-04-7
Synonyms: Polyacrylic acid, sodium salt
Tradenames: Acusol® 410N; Acusol® 445N; Acusol® 445ND; Acusol® 480N; Acusol® 860N; Burcosperse AP Liq; Daxad® 37LN7; Daxad® 37LN10; Daxad® 37LN10-35; Daxad® 37NS; Good-rite® K-7058D; Good-rite® K-739; Good-rite® K-759
Tradenames containing: Good-rite® K-7058

Sodium polyglutamate (CTFA)
CAS 28829-38-1
Tradenames: Ajicoat SPG

Sodium polyisobutylene/maleic anhydride copolymer
Tradenames: Daxad® 31S

Sodium polymethacrylate (CTFA)
CAS 25086-62-8; 54193-36-1
Synonyms: 2-Propenoic acid, 2-methyl-, homopolymer, sodium salt
Tradenames: Daxad® 30; Daxad® 30-30; Daxad® 30S; Daxad® 34N10; Daxad® 35; Daxad® 41

Sodium polynaphthalene-methane sulfonate. *See Sodium methylnaphthalenesulfonate*

Sodium polynaphthalene sulfonate (CTFA)
CAS 9084-06-4
Synonyms: Sodium naphthalene-formaldehyde sulfonate
Tradenames: Ablusol ML; Ablusol NL; Aerosol® NS; Arylan® SNS; Chromasist 87H; Chromasist 1487A; Daxad® 11; Daxad® 11G; Daxad® 13; Daxad® 14B; Daxad® 15; Daxad® 16; Daxad® 17; Daxad® 19; Daxad® 19L-33; Dehscofix 912; Dehscofix 914; Dehscofix 914/AS; Dehscofix 914/ASL; Dehscofix 915; Dehscofix 915/AS; Dehscofix 920; Dispersogen A; Emery® 5370 Sellogen W; Emery® 5371 Sellogen WL; Emery® 5375 Sellogen DFL; Emery® 5380 Sellogen HR; Emery® 5381 Sellogen HR-90; Harol RG-71L; Lomar® D; Lomar® D SOL; Lomar® LS; Lomar® LS Liq; Lomar® PW; Petro® 11; Petro® 22; Petro® AA; Petro® BA; Petro® BAF; Petro® Dispersant 425; Petro® LBA; Petro® P; Petro® S; Petro® ULF; Petro® WP; Rexowet MS; Rhodacal® BA-77 (formerly Nekal® BA-77); Rhodacal® BX-78 (formerly Nekal® BX-78); Rhodacal® N (formerly Blancol® N); Sellogen DFL; Sellogen HR-90; Sellogen WL Liq; Wettol® D 2
Tradenames containing: Dispersogen SI

Sodium/potassium naphthalene-formaldehyde sulfonate
Tradenames: Daxad® 14C; Daxad® 19L-40

Sodium propyl oleate sulfate
Tradenames: Marvanol® SPO (60%)

Sodium ricinoleate (CTFA)
CAS 5323-95-5; EINECS 226-191-2
Synonyms: 12-Hydroxy-9-octadecenoic acid, sodium salt
Tradenames: Solricin® 435; Solricin® 535

Sodium rosinate
CAS 61790-51-0
Synonyms: Sodium soap of pale rosin
Tradenames: Arizona DR-25; Arizona DRS-43; Arizona DRS-44; Diprosin N-70; Dresinate® 81; Dresinate® 731; Dresinate® TX-60W; Dresinate® X; Dresinate® XX; Westvaco® Resin 790; Westvaco® Resin 795

Sodium soybean oil sulfate
Tradenames: Actrasol CS-75; Actrasol OY-75

Sodium stearoamphoacetate (CTFA)
CAS 30473-39-3; 68608-63-9; EINECS 250-215-0
Synonyms: 1H-Imidazolium, 1-(carboxymethyl)-2-heptadecyl-4,5-dihydro-1-(2-hydroxyethyl)-, hydroxide, disodium salt; Stearoamphoacetate; Stearoamphoglycinate

Tradenames: Amphoterge® S; Miranol® DM; Miranol® DM Conc. 45%

Sodium stearoyl lactylate (CTFA)
CAS 25383-99-7; EINECS 246-929-7
Synonyms: Octadecanoic acid, 2-(1-carboxyethoxy)-1-methyl-2-oxoethyl ester, sodium salt; Sodium stearyl-2-lactylate
Tradenames: Grindtek FAL 1; Pationic® 920

Sodium tallate
Synonyms: Tall oil rosin sodium salt
Tradenames: Dresinate® TX

Sodium tallowamide MEA-sulfosuccinate
Tradenames: Elfanol® 510

Sodium tallowate (CTFA)
CAS 8052-48-0; EINECS 232-491-4
Synonyms: Tallow, sodium salt
Tradenames: Emery® 2895 Foamaster Soap L; Foamaster Soap L
Tradenames containing: Prisavon 1981

Sodium tallow sulfate (CTFA)
CAS 8052-50-4; 68140-10-3; 68955-20-4; EINECS 232-494-0
Synonyms: Sodium tallow alcohol sulfate; Sulfuric acid, monotallow alkyl esters, sodium salts
Tradenames: Empicol® TAS30; Empicol® TAS80V; Empicol® TAS90; Montovol GJ-12
Tradenames containing: Sulfetal TC 50

Sodium tallow sulfosuccinamate
CAS 68988-69-2
Tradenames: Empimin® MKK/AU; Manro MA 35

Sodium/TEA laureth sulfate
Tradenames: Laural ED

Sodium/TEA-lauroyl hydrolyzed collagen amino acid
Tradenames: Lipoproteol LCO

Sodium/TEA-lauroyl hydrolyzed keratin amino acids (CTFA)
Tradenames: Lipoproteol LK

Sodium TEA phosphonate
Tradenames: Chemphonate NP

Sodium/TEA-undecenoyl collagen amino acids (CTFA)
Synonyms: Sodium/TEA-undecylenoyl animal collagen amino acids
Tradenames: Lipoproteol UCO

Sodium tetraborate decahydrate. *See Sodium borate*

Sodium tetradecyl sulfate. *See Sodium myristyl sulfate*

Sodium tetrahydronaphthalene sulfonate
Tradenames: Alkanol® S

Sodium toluenesulfonate (CTFA)
CAS 657-84-1; 12068-03-0; EINECS 235-088-1
Synonyms: Methylbenzenesulfonic acid, sodium salt
Tradenames: Eltesol® ST 34; Eltesol® ST 40; Eltesol® ST 90; Eltesol® ST Pellets; Hartotrope STS-40, Powd; Manrosol STS40; Manrosol STS90; Na Toluene Sulfonate 30, 40; Reworyl® NTS 40; Witconate STS
Tradenames containing: Marlon® ARL

Sodium trideceth-3 carboxylate (CTFA)
CAS 68891-17-8 (generic); 61757-59-3 (generic)
Synonyms: PEG-3 tridecyl ether carboxylic acid, sodium salt; Sodium POE (3) tridecyl ether carboxylate
Tradenames: Akypo ITD 30 N; Nikkol ECT-3NEX, ECTD-3NEX

Sodium trideceth-6 carboxylate (CTFA)
CAS 68891-17-8 (generic); 61757-59-3 (generic)
Synonyms: PEG-6 tridecyl ether carboxylic acid, sodium salt; Sodium POE (6) tridecyl ether carboxylate
Tradenames: Nikkol ECTD-6NEX

Sodium trideceth-7 carboxylate (CTFA)
CAS 68891-17-8 (generic); 61757-59-3 (generic)
Synonyms: PEG-7 tridecyl ether carboxylic acid, sodium salt; POE (7) tridecyl ether carboxylic acid, sodium salt
Tradenames: Rewopol® CTN; Rolpon C 200; Sandopan® DTC; Sandopan® DTC-100

Sodium trideceth-8 carboxylate
Tradenames: Akypo ITD 70 BV

Sodium trideceth sulfate (CTFA)
CAS 25446-78-0 (n=3); 66161-58-8 (n=4); 3026-63-9; EINECS 246-985-2
Synonyms: Sodium tridecyl ether sulfate; Sodium POE tridecyl sulfate
Tradenames: Akyposal BA 28; Cedepal TD-403 MF; Cedepal TDS 484; Rhodapex® EST-30 (formerly Sipex® EST-30); Tensagex DLM 627; Tensagex DLM 670; Tensagex DLM 927; Tensagex DLM 970; Tensagex DLS 670; Tensagex DLS 970
Tradenames containing: Mackam 2CT; Mackam MLT; Miracare® 2MHT (formerly Miranol® 2MHT); Miracare® BC-10; Miracare® MS-1 (formerly Compound MS-1); Miracare® MS-2 (formerly Compound MS-2); Miracare® MS-4; Schercoteric MS-2TE Modified; Zoharteric DOT

Sodium trideceth sulfonate
Tradenames containing: Mackam LT

Sodium tridecylbenzene sulfonate (CTFA)
CAS 26248-24-8; EINECS 247-536-3
Synonyms: Tridecylbenzenesulfonic acid, sodium salt
Tradenames: Witconate TDB

Sodium tridecyl sulfate (CTFA)
CAS 3026-63-9; 25446-78-0; EINECS 221-188-2
Synonyms: 1-Tridecanol, hydrogen sulfate, sodium salt
Tradenames: Avirol® SA 4113; Colonial STDS

Sodium vinyl sulfonate
Synonyms: SVS
Tradenames: Hartomer 4900

Sodium xanthogenate
CAS 4741-30-4; 140-90-9
Synonyms: Sodium ethylxanthate; Sodium xanthate; Ethylxanthic acid sodium salt
Tradenames: Xanthates

Sodium xylenesulfonate (CTFA)
CAS 1300-72-7; EINECS 215-090-9
Synonyms: Dimethylbenzene sulfonic acid, sodium salt
Tradenames: Carsosulf SXS-Liq; Eltesol® SX 30; Eltesol® SX 40; Eltesol® SX 93; Eltesol® SX Pellets; Esi-Terge SXS; Hartotrope SXS 40, Powd; Lankrosol SXS-30; Manrosol SXS30; Manrosol SXS40; Manrosol SXS93; Norfox®

SXS40, SXS96; Nutrol SXS 5418; Reworyl® NXS 40; Sulfotex SXS-40; Tex-Wet 1140; Ultra SXS Liq., Powd; Witconate SXS 40%; Witconate SXS 90%

Tradenames containing: Witconate 45 Liq; Witconate 45LX; Witconate 1260 Slurry; Witconate LX F; Witconate LX Powd

Sorbeth-20 (CTFA)
Synonyms: PEG-20 sorbitol ether; POE (20) sorbitol ether; PEG 1000 sorbitol ether
Tradenames: Ethosperse® SL-20; Trylox® 6753

Sorbeth-55
Synonyms: PEG-55 sorbitol ether; POE (55) sorbitol ether
Tradenames: Sorbeth 55

Sorbitan diisostearate (CTFA)
CAS 68238-87-9
Synonyms: Anhydrohexitol diisostearate
Tradenames: Emsorb® 2518

Sorbitan dioleate (CTFA)
CAS 29116-98-1; EINECS 249-448-0
Synonyms: Sorbide dioleate; Sorbitan, di-9-octadecenoate; Anhydrosorbitol dioleate
Tradenames: Atlas G-950; DO-33-F

Sorbitan distearate (CTFA)
CAS 36521-89-8
Synonyms: Anhydrosorbitol distearate; Sorbitan dioctadecanoate
Tradenames: Emasol S-20

Sorbitan isostearate (CTFA)
CAS 1338-39-2; 5959-89-7; 71902-01-7
Synonyms: Anhydrosorbitol monoisostearate; 1,4-Anhydro-D-glucitol, 6-isooctadecanoate; Sorbitan monoisooctadecanoate
Tradenames: Anfomul S6; Crill 6; Emsorb® 2516; Montane 70

Sorbitan laurate (CTFA)
CAS 1338-39-2; 5959-89-7; EINECS 215-663-3
Synonyms: Sorbitan monolaurate; Anhydrosorbitol monolaurate; Sorbitan monododecanoate
Tradenames: Ablunol S-20; Ahco 759; Alkamuls® S-20 (formerly Soprofor S/20); Alkamuls® SML; Armotan® ML; Crill 1; Disponil SML 100 F1; Drewmulse® SML; Durtan® 20; Emsorb® 2515; Ethylan® GL20; Glycomul® L; Hetan SL; Hodag SML; Ionet S-20; Liposorb L; Lonzest® SML; ML-33-F; Montane 20; Newcol 20; Nissan Nonion LP-20R, LP-20RS; Prote-sorb SML; Radiasurf® 7125; S-Maz® 20; Secoster® KL 10; Sorbax SML; Sorbon S-20; Span® 20

Sorbitan oleate (CTFA)
CAS 1338-43-8; 5938-38-5; EINECS 215-665-4
Synonyms: SMO; Sorbitan monooleate; Sorbitan mono-9-octadecenoate; Anhydrosorbitol monooleate
Tradenames: Ablunol S-80; Ahco 832; Ahco 944; Alkamuls® S-80 (formerly Soprofor S/80); Alkamuls® SMO; Anfomul S4; Anfomul S50; Armotan® MO; Atlas G-4884; Crill 4; Crill 50; Desonic® SMO; Desotan® SMO; Disponil SMO 100 F1; Drewmulse® SMO; Durtan® 80; Emsorb® 2500; Ethylan® GO80; Glycomul® O; Hetan SO; Hodag SMO; Ionet S-80; Liposorb O; Lonzest® SMO; MO-33-F; Montane 80; Newcol 3-80, 3-85; Newcol 80; Nikkol SO-10; Norfox® Sorbo S-80; Prote-sorb SMO; Radiasurf® 7155; Rheodol SP-O10; S-Maz® 80; S-Maz® 80K; Sorban AO; Sorbax SMO; Sorbirol O; Sorbon S-80; Span® 80; Witconol 2500

Sorbitan palmitate (CTFA)
CAS 26266-57-9; EINECS 247-568-8
Synonyms: Sorbitan monopalmitate; 1,4-Anhydro-D-glucitol, 6-hexadecanoate
Tradenames: Ablunol S-40; Ahco FP-67; Armotan® MP; Crill 2; Disponil SMP 100 F1; Emsorb® 2510; Ethylan® GP-40; Glycomul® P; Hodag SMP; Liposorb P; Lonzest® SMP; Montane 40; MP-33-F; Newcol 40; Prote-sorb SMP; Radiasurf® 7135; S-Maz® 40; Secoster® KP 10; Sorbax SMP; Sorbon S-40; Span® 40

Sorbitan sesquiisostearate (CTFA)
Synonyms: Sorbitan, monohexadecanoate
Tradenames: Montane 73

Sorbitan sesquioleate (CTFA)
CAS 8007-43-0; EINECS 232-360-1
Synonyms: Anhydrosorbitol sesquioleate; Anhydrohexitol sesquioleate; Sorbitan, 9-octadecenoate (2:3)
Tradenames: Anfomul S43; Arlacel® C; Crill 43; Dehymuls SSO; Disponil SSO 100 F1; Emasol O-15 R; Emsorb® 2502; Glycomul® SOC; Liposorb SQO; Montane 83; Nikkol SO-15; S-Maz® 83R

Sorbitan stearate (CTFA)
CAS 1338-41-6; EINECS 215-664-9
Synonyms: SMS; Sorbitan monostearate; Sorbitan monooctadecanoate; Anhydrosorbitol monostearate
Tradenames: Ablunol S-60; Ahco 909; Alkamuls® S-60 (formerly Soprofor S/60); Alkamuls® SMS; Armotan® MS; Crill 3; Disponil SMS 100 F1; Drewmulse® SMS; Durtan® 60; Emsorb® 2505; Emultex SMS; Ethylan® GS60; Glycomul® S; Glycomul® S FG; Glycomul® S KFG; Grindtek SMS; Hetan SS; Ionet S-60 C; Liposorb S; Lonzest® SMS; Montane 60; MS-33-F; Newcol 60; Nikkol SS-10; Prote-sorb SMS; Radiasurf® 7145; Rheodol SP-S10; S-Maz® 60; S-Maz® 60KHM; Secoster® KS 10; Sorban AST; Sorbax SMS; Sorbon S-60; Span® 60, 60K
Tradenames containing: AF 72; AF 75

Sorbitan tallate
Tradenames: Desonic® SMT; Desotan® SMT; S-Maz® 90; Witcomul 78

Sorbitan trioleate (CTFA)
CAS 26266-58-0; EINECS 247-569-3
Synonyms: STO; Anhydrosorbitol trioleate; Sorbitan tri-9-octadecenoate
Tradenames: Ablunol S-85; Ahco FO-18; Alkamuls® S-85 (formerly Soprofor S/85); Alkamuls® STO; Atlas G-4885; Atlox 4885; Crill 45; Disponil STO 100 F1; Emsorb® 2503; Ethylan® GT85; Glycomul® TO; Hodag STO; Ionet S-85; Liposorb TO; Lonzest® STO; Montane 85; Nikkol SO-30; Prote-sorb STO; Rheodol SP-O30; S-Maz® 85; S-Maz® 85K; Sorban CO; Sorbax STO; Span® 85; TO-33-F; Witconol 2503

Sorbitan tristearate (CTFA)
CAS 26658-19-5; EINECS 247-891-4
Synonyms: STS; Anhydrosorbitol tristearate; Sorbitan trioctadecanoate

Tradenames: Ahco FS-21; Alkamuls® S-65 (formerly Soprofor S/65); Alkamuls® STS; Crill 35; Disponil STS 100 F1; Drewmulse® STS; Emsorb® 2507; Glycomul® TS KFG; Glycomul® TS; Grindtek STS; Hodag STS; Liposorb TS; Lonzest® STS; Montane 65; Nikkol SS-30; Prote-sorb STS; Radiasurf® 7345; Rheodol SP-S30; S-Maz® 65K; Sorbax STS; Span® 65; TS-33-F

Sorbitan tritallate
Tradenames: S-Maz® 95

Soy acid (CTFA)
CAS 68308-53-2; 67701-08-0; EINECS 269-657-0
Synonyms: Acids, soy; Fatty acids, soya
Tradenames: Emery® 610; Emery® 618; Industrene® 126; Industrene® 225; Industrene® 226; Industrene® 226 FG; S-210

Soya diethanolamide. *See Soyamide DEA*

Soya dimethyl amine. *See Dimethyl soyamine*

Soyamide DEA (CTFA)
CAS 68425-47-8; EINECS 270-355-6
Synonyms: Soya diethanolamide; N,N-Bis (hydroxyethyl)soya amides
Tradenames: Alkamide® SDO; Lauridit® SDG; Mackamide S; Mackamide SD; Manromid 150-ADY; Marlamid® DF 1818

Soyamidopropyl dimethylamine (CTFA)
CAS 68188-30-7
Synonyms: N-[3-(Dimethylamino)propyl]soya amides; Dimethylaminopropyl soyamide
Tradenames: Mackine 901

Soyamine (CTFA)
CAS 61790-18-9; EINECS 263-112-0
Synonyms: Soya primary amine; Amines, soya alkyl; Soyaalkylamine
Tradenames: Armeen® S; Armeen® SD; Jet Amine PS; Jet Amine PSD; Nissan Amine SB

Soyaminopropylamine (CTFA)
Tradenames: Kemamine® D-999

Soyapropylenediamine
Tradenames: Diamin S

Soya trimethyl ammonium chloride. *See Soytrimonium chloride*

Soybean oil, epoxidized. *See Epoxidized soybean oil*

Soyethyldimonium ethosulfate (CTFA)
CAS 68308-67-8
Synonyms: Soya dimethyl ethyl ammonium ethyl sulfate; Soyaethyldimonium ethosulfate
Tradenames: Larostat® 88; Larostat® 264 A; Larostat® 264 A Anhyd; Larostat® 264 A Conc

Soytrimonium chloride (CTFA)
CAS 61790-41-8; EINECS 263-134-0
Synonyms: Soya trimethyl ammonium chloride; Quaternium-9; N-(Soya alkyl)-N,N,N-trimethyl ammonium chloride
Tradenames: Jet Quat S-50; Tomah Q-S-80
Tradenames containing: Arquad® S-50; Tomah Q-S

Steapyrium chloride (CTFA)
CAS 1341-08-8; 14492-68-3; 42566-92-7; EINECS 238-501-3
Synonyms: 1-[2-Oxo-2-[[(1-oxooctadecyl)oxy] ethyl] amino] ethyl] pyridinium chloride; Quaternium-7; N-(Stearoyl colamino formyl methyl) pyridinium chloride
Tradenames: Emcol® E-607S

Stearalkonium chloride (CTFA)
CAS 122-19-0; EINECS 204-527-9
Synonyms: Stearyl dimethyl benzyl ammonium chloride; Octadecyl dimethyl benzyl ammonium chloride; N,N-Dimethyl-N-octadecyl-benzenemethanaminium chloride
Tradenames: Amyx A-25-S 0040; Emcol® 4; Incroquat SDQ-25; Maquat SC-18; Nissan Cation S2-100; Standamul® STC-25
Tradenames containing: BTC® 885; Carsoquat® 816-C; Maquat SC-1632

Stearamide (CTFA)
CAS 124-26-5; EINECS 204-693-2
Synonyms: Octadecanamide; Stearic acid amide
Tradenames: Armid® 18; Kemamide® S

Stearamide DEA (CTFA)
CAS 93-82-3; EINECS 202-280-1
Synonyms: Stearic acid diethanolamide; Stearoyl diethanolamide; N,N-bis(2-hydroxyethyl) octadecanamide
Tradenames: Ablumide SDE; Alkamide® DS-280/S (formerly Cyclomide DS-280/S); Hetamide DS; Karamide ST-DEA; Monamid® 718; Nopcogen 14-S; Unamide® S; Unamide® W
Tradenames containing: Aminol N-1918

Stearamide DIBA-stearate (CTFA)
Tradenames: Paramul® SAS

Stearamide MEA (CTFA)
CAS 111-57-9; EINECS 203-883-2
Synonyms: Stearic acid monoethanolamide; Stearoyl monoethanolamide; N-(2-hydroxyethyl) octadecanamide
Tradenames: Alkamide® S-280 (formerly Cyclomide S-280); Mackamide SMA; Mazamide® SMEA; Monamid® S; Rewomid® S 280

Stearamide MEA-stearate (CTFA)
CAS 14351-40-7; EINECS 238-310-5
Synonyms: Octadecanoic acid, 2-[(1-oxooctadecyl) amino]ethyl ester; Stearic monoethanolamide stearate; 2-[(1-Oxooctadecyl)amino]ethyl octadecanoate
Tradenames: Witcamide® MAS

Stearamidoethyl diethylamine (CTFA)
CAS 16889-14-8; EINECS 240-924-3
Synonyms: Diethylaminoethyl stearamide; N-[2-Diethylamino)ethyl]octadecanamide
Tradenames containing: Tegacid® Regular VA

Stearamidoethyl ethanolamine (CTFA)
CAS 141-21-9; EINECS 205-469-7
Synonyms: Ethanolaminoethyl stearamide; N-[2-[(2-Hydroxyethyl) amino] ethyl] octadecanamide
Tradenames: Avistin® FD; Avistin® PN; Marlamid® A 18

Stearamidopropylamine oxide (CTFA)
CAS 25066-20-0
Synonyms: N-[3-(Dimethylamino) propyl] octadecanamide-N-oxide
Tradenames: Mackamine SAO

Stearamidopropyl dimethylamine (CTFA)
CAS 7651-02-7; EINECS 231-609-1
Synonyms: Dimethylaminopropyl stearamide; N-[3-(Dimethylamino)propyl]octadecanamide
Tradenames: Mackine 301; Miramine® SODI (formerly Alkamide® SODI)

Stearamidopropyl dimethylamine lactate (CTFA)
CAS 55819-53-9; EINECS 259-837-7
Synonyms: Propanoic acid, 2-hydroxy-, compd. with N-[3-(dimethylamino) propyl] octadecanamide
Tradenames: Emcol® 3780; Incromate SDL

Stearamidopropyl morpholine (CTFA)
CAS 55852-13-6
Synonyms: N-[3-(4-Morpholinyl) propyl] octadecanamide
Tradenames: Mackine 321

Stearamidopropyl PG-dimonium chloride phosphate (CTFA)
Tradenames: Phospholipid PTS

Stearamine (CTFA)
CAS 124-30-1; EINECS 204-695-3
Synonyms: Stearyl amine; Octadecylamine; 1-Octadecanamine
Tradenames: Amine 18D; Armeen® 18; Armeen® 18D; Crodamine 1.18D; Kemamine® P-990, P-990D; Nissan Amine AB
Tradenames containing: Armac® 18D-40

Stearamine acetate
CAS 2190-04-7
Synonyms: Stearyl amine acetate; Octadecylamine acetate
Tradenames: Acetamin 86; Armac® 18D
Tradenames containing: Armac® 18D-40

Stearamine oxide (CTFA)
CAS 2571-88-2; EINECS 219-919-5
Synonyms: Stearyl dimethylamine oxide; Octadecyl dimethylamine oxide; N,N-Dimethyl-1-octadecanamine-N-oxide
Tradenames: Admox® 18-85; Ammonyx® SO; Amyx SO 3734; Annonyx SO; Barlox® 18S; Incromine Oxide S; Mackamine SO; Mazox® SDA; Ninox® SO; Standamox PS

Steardimonium hydroxypropyl hydrolyzed wheat protein
Tradenames: Hydrotriticum QS

Steareth-2 (CTFA)
CAS 9005-00-9 (generic)
Synonyms: PEG-2 stearyl ether; POE (2) stearyl ether; PEG 100 stearyl ether
Tradenames: Brij® 72; Hetoxol STA-2; Lipocol S-2; Macol® SA-2; Prox-onic SA-1/02; Simulsol 72; Volpo S-2

Steareth-5 (CTFA)
CAS 9005-00-9 (generic)
Synonyms: PEG-5 stearyl ether; POE (5) stearyl ether
Tradenames: Macol® SA-5
Tradenames containing: Solulan® 5

Steareth-7 (CTFA)
CAS 9005-00-9 (generic); 66146-84-7
Synonyms: PEG-7 stearyl ether; POE (7) stearyl ether; 3,6,9,12,15,18,21-Heptaoxanonatriacontan-1-ol
Tradenames: Ablunol SA-7

Steareth-9
Tradenames containing: Hetoxol LS-9

Steareth-10 (CTFA)
CAS 9005-00-9 (generic)
Synonyms: PEG-10 stearyl ether; POE (10) stearyl ether; PEG 500 stearyl ether
Tradenames: Hetoxol STA-10; Lipocol S-10; Macol® SA-10; Prox-onic SA-1/010; Simulsol 76
Tradenames containing: Ritapro 100

Steareth-15 (CTFA)
CAS 9005-00-9 (generic)
Synonyms: PEG-15 stearyl ether; POE (15) stearyl ether
Tradenames: Macol® SA-15

Steareth-16 (CTFA)
CAS 9005-00-9 (generic)
Synonyms: PEG-16 stearyl ether; POE (16) stearyl ether
Tradenames containing: Solulan® 16

Steareth-20 (CTFA)
CAS 9005-00-9 (generic)
Synonyms: PEG-20 stearyl ether; POE (20) stearyl ether; PEG 1000 stearyl ether
Tradenames: Hetoxol STA-20; Lipocol S-20; Macol® SA-20; Prox-onic SA-1/020; Simulsol 78; Trycol® 5888; Trycol® SAL-20
Tradenames containing: Ritachol® 1000; Ritachol® 3000; Ritachol® 4000; Ritapro 100

Steareth-21 (CTFA)
CAS 9005-00-9 (generic)
Synonyms: PEG-21 stearyl ether; POE (21) stearyl ether
Tradenames: Brij® 721 S

Steareth-30 (CTFA)
CAS 9005-00-9 (generic)
Synonyms: PEG-30 stearyl ether; POE (30) stearyl ether
Tradenames: Hetoxol STA-30

Steareth-40 (CTFA)
CAS 9005-00-9 (generic)
Synonyms: PEG-40 stearyl ether; POE (40) stearyl ether; PEG 2000 stearyl ether
Tradenames: Macol® SA-40

Steareth-100 (CTFA)
CAS 9005-00-9 (generic)
Synonyms: PEG-100 stearyl ether; POE (100) stearyl ether
Tradenames: Brij® 700 S

Steareth-200
Tradenames: Rhodasurf® TB-970 (formerly Emulphogene® TB-970)

Steareth-7 carboxylic acid
CAS 68954-89-2; 59559-30-7
Tradenames: Akypo RS 60

Steareth-11 carboxylic acid
Tradenames: Akypo RS 100

Stearethyldimonium ethosulfate
Tradenames: Larostat® 451

Stearic acid (CTFA)
CAS 57-11-4; EINECS 200-313-4
Synonyms: n-Octadecanoic acid
Tradenames: Edenor ST-1; Emersol® 110; Emersol® 120; Emersol® 132 NF Lily®; Emersol® 150; Emersol® 152 NF, 153 NF; Emersol® 6320; Emersol® 6332 NF; Emersol® 6349; Emersol® 6351; Emery® 400; Emery® 404; Emery® 405; Emery® 410; Emery® 420; Emery® 422; Hystrene® 4516; Hystrene® 5016; Hystrene® 5016 NF; Hystrene® 7018; Hystrene® 7018 FG; Hystrene® 8018; Hystrene® 8718 FG; Hystrene® 9718; Hystrene® 9718 NF FG; Industrene® 4518; Industrene®

5016; Industrene® 5016 FG; Industrene® 7018; Industrene® 9018; Prifrac 2980; Prifrac 2981; Pristerene 4904; Pristerene 4905; Pristerene 4910; Pristerene 4911; Pristerene 4915; Pristerene 4921
Tradenames containing: Tauranol I-78/80

Stearic (ethyl amino hydroxyethyl) amide
Tradenames: Unamide® SI

Stearic-oleic DEA
Tradenames: Emid® 6543

Stearoamphoacetate. *See Sodium stearoamphoacetate*

Stearoamphoglycinate. *See Sodium stearoamphoacetate*

Stearoyl lactylic acid (CTFA)
Tradenames: Grindtek FAL 3

Stearoyl PG-trimonium chloride (CTFA)
Synonyms: Stearoyl propylene glycol trimethylammonium chloride
Tradenames containing: Akypoquat 40

Stearoyl sarcosine (CTFA)
CAS 142-48-3; EINECS 205-539-7
Synonyms: Stearoyl N-methylglycine; Stearoyl N-methylaminoacetic acid; N-Methyl-N-(1-oxooctadecyl)glycine
Tradenames: Hamposyl® S

Steartrimonium chloride (CTFA)
CAS 112-03-8; EINECS 203-929-1
Synonyms: Stearyl trimethyl ammonium chloride; Octadecyl trimethyl ammonium chloride; N,N,N-Trimethyl-1-octadecanaminium chloride
Tradenames: Nissan Cation AB; Octosol 474; Tomah Q-ST-50
Tradenames containing: Arquad® 18-50

Stearyl alcohol (CTFA)
CAS 112-92-5; EINECS 204-017-6
Synonyms: n-Octadecanol; 1-Octadecanol; C18 linear alcohol
Tradenames: Adol® 61 NF; Adol® 62 NF; Adol® 620 NF; Alfol® 18; Alfol® 18 NF; CO-1895; CO-1897; CO-1898; Emery® 3343; Emery® 3344; Emery® 3348; Epal® 18NF; Lorol C18; Lorol C18-98; Lorol C18 Chemically Pure; Loxiol VPG 1354; Philcohol 1800
Tradenames containing: Epal® 1218; Epal® 1418; Epal® 1618; Epal® 1618RT; Epal® 1618T; Hetoxol G

Stearyl betaine (CTFA)
CAS 820-66-6; EINECS 212-470-6
Synonyms: N-(Carboxymethyl)-N,N-dimethyl-1-octadecanaminium hydroxide, inner salt; Stearyl dimethyl glycine
Tradenames: Lonzaine® 18S; Varion® SDG

Stearyl diethanol methyl ammonium chloride
Tradenames: M-Quat® 32

Stearyl dihydroxyethyl methyl ammonium chloride
Synonyms: Octadecyl dihydroxyethyl methyl ammonium chloride
Tradenames containing: Tomah Q-18-2

Stearyl dimethyl amine. *See Dimethyl stearamine*

Stearyl dimethylamine oxide. *See Stearamine oxide*

Stearyl dimethyl benzyl ammonium chloride. *See Stearalkonium chloride*

Stearyl dimethyl glycine. *See Stearyl betaine*

Stearyl hydroxyethyl imidazoline (CTFA)
CAS 95-19-2; EINECS 202-397-8
Synonyms: Stearyl imidazoline; 2-Heptadecyl-4,5-dihydro-1H-imidazole
Tradenames: Amine S; Atlasol KAD; Crodazoline S; Schercozoline S; Unamine® S

Stearyl imidazoline. *See Stearyl hydroxyethyl imidazoline*

Stearyl nitrile
Tradenames: Arneel® 18 D

Stearyl PEG-15 methyl ammonium chloride
Tradenames: Tomah Q-18-15

Stearyl stearate (CTFA)
CAS 2778-96-3; EINECS 220-476-5
Synonyms: Octadecanoic acid, octadecyl ester
Tradenames: Hetester 412

Stearyl trimethyl ammonium chloride. *See Steartrimonium chloride*

Styrene/MA copolymer (CTFA)
CAS 9011-13-6
Synonyms: SMA; Styrene/maleic anhydride copolymer; 2,5-Furandione, polymer with ethenylbenzene
Tradenames: Scripset 520; Scripset 540; Scripset 550; Scripset 700; Scripset 720

Styrene/maleic anhydride copolymer, sodium salt
Tradenames: Scripset 500

Sucrose cocoate (CTFA)
CAS 25339-99-5
Tradenames: Grilloten LSE87K

Sucrose distearate (CTFA)
CAS 27195-16-0; EINECS 248-317-5
Synonyms: alpha-D-Glucopyranoside, beta-D-fructofuranosyl, dioctadecanoate
Tradenames containing: Crodesta F-110

Sucrose laurate (CTFA)
CAS 25339-99-5; EINECS 246-873-3
Synonyms: alpha-D-Glucopyranoside, beta-D-fructofuranosyl, monododecanoate
Tradenames: Grilloten LSE87

Sucrose ricinoleate (CTFA)
Tradenames: Grilloten ZT12, ZT40, ZT80

Sucrose stearate (CTFA)
CAS 25168-73-4; EINECS 246-705-9
Synonyms: alpha-D-Glucopyranoside, beta-D-fructofuranoysl, monooctadecanoate
Tradenames: Grilloten PSE141G
Tradenames containing: Crodesta F-110

Sulfated butyl oleate
Tradenames: Chemax SBO; Chemsulf SBO/65; Dyasulf BO-65; Hipochem Dispersol SB

Sulfated butyl tallate
CAS 42808-36-6
Tradenames: Laurel SBT

Sulfated castor oil (CTFA)
CAS 8002-33-3; EINECS 232-306-7
Synonyms: Castor oil sulfated; Sulfonated castor oil; Turkey-red oil
Tradenames: Actrasol C-50, C-75, C-85; Ahco AJ-110; Aquasol® AR 90; Aquasol® W 90; Chemax SCO; Chemsulf SCO/75; Consos Castor Oil; Cordon NU 890/75; Dyasulf 1761-A; Dyasulf C-70; Eureka 102; Eureka 102-WK; G-

7205; G-7274; Haroil SCO-65, -7525; Haroil SCO-75; Hipochem Dispersol SCO; Laurel R-50; Laurel R-75; Marvanol® SCO (50%); Monopole Oil; Monosulf; Nopcocastor; Nopcocastor L; Nopcosulf CA-60, -70; Nopco-surf CA; Sulfotex 130; Sykanol DKM 45, 80; Texapon SCO; Titan Castor No. 75; Turkey Red Oil 100%

Sulfated fish oil
Tradenames: Atlas Sul. Oil HC; Solocod G

Sulfated glyceryl trioleate
Tradenames: Hipochem Dispersol GTO

Sulfated methyl rapeseed ester
Tradenames: Laurel SMR

Sulfated methyl soyate, sodium salt
Tradenames: Actrasol MY-75

Sulfated neatsfoot oil
CAS 68585-05-7
Tradenames: Atlas Sul. Neats L-2

Sulfated oleic acid
Tradenames: Dyasulf OA-60; Sulfonic 800

Sulfated olive oil
Tradenames: Olitex 75

Sulfated peanut oil (CTFA)
CAS 73138-79-1
Synonyms: Oils, peanut, sulfated; Peanut oil, sulfated
Tradenames: Standapol® 1610

Sulfated propyl oleate
Tradenames: Hipochem Dispersol SP

Sulfated rapeseed oil
CAS 617788-68-9
Tradenames: Actrasol 6092; Laurel SRO

Sulfated tall oil
CAS 61790-35-0
Tradenames: Actrasol SS; Cordon PB-870; Eureka 392; Laurel M-10-257

Sulfated tall oil, potassium salt
Tradenames: Actrasol SP 175K

Sulfated tall oil, sodium salt
Tradenames: Actrasol SP

Sulfated tallow
Tradenames: Nopcosulf TA-30; Nopcosulf TA-45V

Sulfated vegetable oil
Tradenames: Hydrolene

alpha-Sulfo methyl laurate
Tradenames: Calester

Sulfonated fish oil
Tradenames: Eureka 400-R

Sulfonated oleyl alcohol
Tradenames: Hipochem LCA

Sulfonic acid
Tradenames: Lamepon 287 SF; Osimol Grunau DP; Osimol Grunau SF; Sulfonic 864
Tradenames containing: Eltesol® TA 65

Tallamide DEA (CTFA)
CAS 68155-20-4; EINECS 268-949-5
Synonyms: Diethanolamine tall oil acid amide; Tall oil acid diethanolamide; Tall oil diethanolamide
Tradenames: Laurel SD-750; Schercomid SO-T
Tradenames containing: Monamine T-100; Schercomid TO-2

Tallamide TEA
Tradenames: Laurel SD-580

Tall oil acid (CTFA)
CAS 61790-12-3; EINECS 263-107-3
Synonyms: Acids, tall oil; Fatty acids, tall oil
Tradenames: Acintol® 736; Acintol® 746; Acintol® 2122; Acintol® 7002; Acintol® D25LR; Acintol® D30E; Acintol® D30LR; Acintol® D40LR; Acintol® D40T; Acintol® D60LR; Acintol® DFA; Acintol® EPG; Acintol® FA-1; Acintol® FA-1 Special; Acintol® FA-2; Acintol® FA-3; Westvaco® 1480; Westvaco® 1483
Tradenames containing: Defoamer A 50

Tall oil amido-amine
Tradenames: Ameenex C-18

Tall oil amidopropyl dimethylethyl ammonium ethosulfate
Tradenames: Servamine KET 4542

Tall oil aminoethyl imidazoline
Synonyms: Aminoethyl tall oil imidazoline
Tradenames: Chemzoline 1411; Chemzoline T-11; Chemzoline T-33

Tall oil aminopropyl dimethylamine
Tradenames: Sermul EK 330

Tall oil diethanolamide. *See Tallamide DEA*

Tall oil hydroxyethyl imidazoline (CTFA)
CAS 61791-39-7; EINECS 263-171-2
Synonyms: 1-Hydroxyethyl-2-tall oil imidazoline; Tall oil imidazoline; 4,5-Dihydro-7-nortall oil-1H-imidazole-1-ethanol
Tradenames: Amine T; Chemzoline T-44; Miramine® TO (formerly Alkazine® TO); Monazoline T; Textamine T-1; Unamine® T; Varine T

Tall oil soap. *See Rosin*

Tallow acid (CTFA)
CAS 61790-37-2; 67701-06-8; EINECS 263-129-3
Synonyms: Fatty acids, tallow; Acids, tallow
Tradenames: Emery® 531; Industrene® 143; Industrene® 145; Prifac 7920; Prifac 7935; T-11; T-18; T-20; T-22
Tradenames containing: Emery® 515; Emery® 516; Emery® 517; Hystrene® 1835; TC-1005; TC-1010; TC-1010T

Tallow alcohol (CTFA)
Synonyms: Alcohols, tallow
Tradenames: Adol® 42; Emery® 3320; Hydrenol D

Tallowalkonium chloride (CTFA)
CAS 61789-75-1; EINECS 263-085-5
Synonyms: Tallow dimethyl benzyl ammonium chloride; Dimethyl benzyl tallow ammonium chloride
Tradenames: Kemamine® BQ-9742C; Noramium S 75

Tallowalkylmethyl-bis(2-hydroxy-2-methylethyl) quaternary ammonium methylsulfates
CAS 79770-97-1
Tradenames containing: Propoquad® T/12

Tallowamide DEA (CTFA)
CAS 68140-08-9; EINECS 268-772-3
Synonyms: Tallow diethanolamide; Diethanolamine tallow acid amide; N,N-Bis(2-hydroxyethyl)tallow amides

Tradenames: Amidex TD

Tallowamide MEA (CTFA)
CAS 68153-63-9; 68440-25-5
Synonyms: N-(2-Hydroxyethyl) tallow acid amide; Tallow acid monoethanolamide; Monoethanolamine tallow acid amide
Tradenames: Marlamid® M 1618

Tallowamidopropyl hydroxysultaine (CTFA)
Synonyms: Quaternary ammonium compds., (3-tallowamidopropyl)(2-hydroxy-3-sulfopropyl) dimethyl, hydroxide, inner salt
Tradenames: Crosultaine T-30

Tallow amine (CTFA)
CAS 61790-33-8; EINECS 263-125-1
Synonyms: Amines, tallow alkyl; Tallowamine; Tallowalkylamine
Tradenames: Amine BG; Amine BGD; Armeen® T; Armeen® TD; Crodamine 1.T; Genamin TA Grades; Jet Amine PT; Jet Amine PTD; Kemamine® P-974D; Noram S; Radiamine 6170; Radiamine 6171; Tallow Amine

Tallowamine acetate
CAS 2190-04-7
Tradenames: Acetamin T; Armac® T; Noramac S; Radiamac 6179

Tallowamine oxide (CTFA)
Synonyms: Tallow dimethylamine oxide; Amines, tallow alkyl dimethyl, oxides
Tradenames: Noxamine S2-30

Tallowaminopropylamine (CTFA)
CAS 68439-73-6; EINECS 270-416-7
Synonyms: Tallow trimethylene diamine
Tradenames: Kemamine® D-974

Tallowamphopolycarboxyglycinate
Tradenames: Ampholak 7TX

Tallow bis hydroxyethyl methyl ammonium chloride
Synonyms: Methyl bis (2-hydroxyethyl) tallow ammonium chloride
Tradenames: Varisoft® 920

Tallow diamine
CAS 61791-55-7
Tradenames: Jet Amine DT; Tallow Diamine

Tallow-1,3-diaminopropane. *See Tallowpropylene diamine*

Tallow diethanolamide. *See Tallowamide DEA*

Tallow dihydroxyethyl betaine. *See Dihydroxyethyl tallow glycinate*

Tallow dihydroxyethyl glycine. *See Dihydroxyethyl tallow glycinate*

Tallow dimethylamine. *See Dimethyl tallowamine*

Tallow dimethylamine oxide. *See Tallowamine oxide*

Tallow dimethyl benzyl ammonium chloride. *See Tallowalkonium chloride*

Tallow dimethyl trimethyl propylene diammonium chloride
Tradenames containing: Tomah Q-D-T

Tallow dipropylene triamine
CAS 61791-57-9
Tradenames: Triameen T; Trinoram S

Talloweth-5
Tradenames: Dehydol TA 5

Talloweth-7
Tradenames: Dehydol PTA 7

Talloweth-11
CAS 61791-28-4 (generic)
Synonyms: PEG-11 tallow ether; POE (11) tallow ether
Tradenames: Dehydol PTA 114; Dehydol TA 11; Oxetal TG 111; Rewopal® TA 11

Talloweth-12
Tradenames: Dehydol TA 12

Talloweth-14
Tradenames: Dehydol TA 14

Talloweth-18
Synonyms: PEG-18 tallow ether; POE (18) tallow ether
Tradenames: Oxetal TG 118

Talloweth-20
Synonyms: PEG-20 tallow ether; POE (20) tallow ether
Tradenames: Dehydol TA 20

Talloweth-23
Tradenames: Dehydol PTA 23; Dehydol PTA 23/E

Talloweth-25
Synonyms: PEG-25 tallow ether; POE (25) tallow ether
Tradenames: Rewopal® TA 25; Rewopal® TA 25/S

Talloweth-30
Synonyms: PEG-30 tallow ether; POE (30) tallow ether
Tradenames: Dehydol TA 30

Talloweth-40
Synonyms: PEG-40 tallow ether; POE (40) tallow ether
Tradenames: Dehydol PTA 40

Talloweth-50
Synonyms: PEG-50 tallow ether; POE (50) tallow ether
Tradenames: Rewopal® TA 50

Talloweth-80
Synonyms: PEG-80 tallow ether; POE (80) tallow ether
Tradenames: Dehydol PTA 80

Talloweth-7 carboxylic acid
Tradenames: Akypo RT 60

3-Tallow-1,3-hexahydropyrimidine
Tradenames: Duomeen® LT-4

Tallow imidazolinium methosulfate. *See Quaternium-27*

Tallow nitrile
Tradenames: Arneel® T

Tallow pentamethylpropane diammonium chloride
Tradenames: Adogen® 477

Tallow pentamine
Tradenames: Jet Amine TT

Tallow polypropylene polyamine
CAS 68911-79-5
Tradenames: Polyram S

N-Tallow-1,3-propanediamine
CAS 68439-73-6
Synonyms: Tallow-1,3-diaminopropane
Tradenames: Duomeen® TX; Radiamine 6570

Tallow propane diamine diacetate
Tradenames: Duomac® T

Tallow propane diamine dioleate
CAS 61791-53-5
Synonyms: N-Tallow-1,3-diaminopropane dioleate
Tradenames: Duomeen® TDO

Tallow propylene diamine
CAS 61791-55-7
Synonyms: N-Tallow-1,3-propanediamine; Tallow-1,3-diamino propane
Tradenames: Diamin T; Diamine BG; Diamine HBG; Dinoram S; Duomeen® T; Lilamuls BG; Nissan Amine DT

Tallow propylene diamine acetate
Tradenames: Dinoramac S

Tallow tetramine
Tradenames: Jet Amine TET; Tallow Tetramine

Tallow triamine
Tradenames: Jet Amine TRT; Tallow Triamine

Tallow trimethyl ammonium methosulfate. *See Tallowtrimonium methosulfate*

Tallowtrimonium chloride (CTFA)
CAS 8030-78-2; 7491-05-2; 68002-61-9; EINECS 232-447-4
Synonyms: Quaternary ammonium compds., tallow alkyl trimethyl, chlorides; Tallow trimethyl ammonium chloride
Tradenames: Arquad® T-27W; Jet Quat T-27W; Jet Quat T-50; Kemamine® Q-9743C; Kemamine® Q-9743CHGW; Noramium MS 50; Quartamin TPR; Radiaquat 6471
Tradenames containing: Adogen® 471; Arquad® T-2C-50; Arquad® T-50

Tallowtrimonium methosulfate
Synonyms: Tallow trimethyl ammonium methosulfate; C16-18 trimethyl ammonium methosulfate
Tradenames: Empigen® CM

TEA cocoate (CTFA)
CAS 61790-64-5; EINECS 263-155-5
Synonyms: Fatty acids, coconut oil, triethanolamine salts; Triethanolamine coconut acid
Tradenames: Akypogene FP 35 T

TEA-cocoyl-glutamate (CTFA)
CAS 68187-29-1; EINECS 269-084-6
Synonyms: L-Glutamic acid, N-coco acyl derivs., compds. with triethanolamine; Triethanolamine cocoyl glutamate
Tradenames: Acylglutamate CT-12; Amisoft CT-12

TEA-cocoyl hydrolyzed collagen (CTFA)
CAS 68952-16-9
Synonyms: TEA-coco-hydrolyzed animal protein; TEA-cocoyl hydrolyzed animal protein
Tradenames: Lexein® S620TA; Maypon 4CT; May-Tein CT
Tradenames containing: Lexate BPQ

TEA-dodecylbenzenesulfonate (CTFA)
CAS 27323-41-7; 68411-31-4; EINECS 248-406-9
Synonyms: Dodecylbenzenesulfonic acid, compd. with 2,2′,2′′-nitrilotris[ethanol] (1:1); Triethanolamine dodecylbenzene sulfonate
Tradenames: Ablusol DBT; Bio-Soft® N-300; Bio-Surf PBC-420; Calsoft T-60; Carsofoam® T-60-L; Elfan® WAT; Emulson AG 255; Esi-Terge T-60; Hartofol 60T; Hexaryl D 60 L; Hoe S 2749; Manro TDBS 60; Maranil CB-22; Marlopon® AT; Marlopon® AT 50; Marlopon® CA; Mazon® 60T; Nansa® TS 60; Naxel AAS-60S; Norfox® T-60; Puxol CB-22; Reworyl® TKS 90/L; Rhodacal® DOV (formerly Siponate® DOV); Rueterg 60-T; Rueterg 97-T; Sandoz Sulfonate AAS 60S; Stepantan® DT-60; Ufasan TEA; Witconate 60T; Witconate 79S; Witconate LXH; Witconate TAB

TEA-laureth sulfate (CTFA)
CAS 27028-82-6
Synonyms: Triethanolamine lauryl ether sulfate
Tradenames: Montosol PL-14; Sactol 2 OT

TEA-lauroyl glutamate (CTFA)
CAS 53576-49-1; 31955-67-6; EINECS 258-636-1
Synonyms: L-Glutamic acid, N-lauryl and 2,2′,2′′-nitrilotriethanol (1:1); Triethanolamine lauroyl glutamate
Tradenames: Acylglutamate LT-12

TEA lauroyl sarcosinate (CTFA)
CAS 2224-49-9; 16693-53-1; EINECS 240-736-1
Synonyms: Glycine, N-methyl-N-(oxododecyl)-, compd. with 2,2′,2′′-nitrilotris[ethanol] (1:1); Triethanolamine lauroyl sarcosinate
Tradenames: Crodasinic LT40

TEA-lauryl sulfate (CTFA)
CAS 139-96-8; 68908-44-1; EINECS 205-388-7
Synonyms: Sulfuric acid, monododecyl ester, compd. with 2,2′,2′′-nitrilotris[ethanol] (1:1); Triethanolammonium lauryl sulfate; Triethanolamine lauryl sulfate
Tradenames: Akyposal TLS 42; Avirol® 300; Avirol® 400 T; Avirol® T 40; Berol 480; Carsonol® TLS; Cedepon LT-40; Colonial TLS; Cosmopon TR; Elfan® 240T; Empicol® TL40; Empicol® TL40/T; Laural D; Manro TL 40; Marlinat® DFL 40; Montovol RF-11; Montovol RL-10; Norfox® TLS; Nutrapon TLS-500; Perlankrol® ATL40; Rewopol® TLS 40; Rhodapon® LT-6 (formerly Sipon® LT-6); Sactol 2 T; Stepanol® SPT; Stepanol® WAT; Sulfetal KT 400; Sulfochem TLS; Sulfotex WAT; Sunnol LST; Swascol 4L; Texapon T 35; Texapon T 42; Ufarol TA-40; Witcolate 6434 (formerly DeSonol T); Witcolate T; Witcolate TLS-500
Tradenames containing: Equex STM

TEA-PEG-3 cocamide sulfate (CTFA)
Tradenames: Genapol® AMS

TEA ricinoleate
Tradenames: Atlasol KMM

TEA-sodium dodecylbenzene sulfonate
Tradenames: Chimipon TSB

Tetradecanoic acid. *See Myristic acid*

Tetradecene-1
CAS 1120-36-1
Synonyms: Linear C14 alpha olefin
Tradenames: Gulftene 14; Neodene® 14; Tetradecene-1

Tetradecyl. *See Myristyl*

Tetradecylamine. *See Myristamine*

Tetraethylene glycol... *See PEG-4...*

Tetraglyceryl. *See Polyglyceryl-4*

Tetrahydroxypropyl ethylenediamine (CTFA)
CAS 102-60-3; EINECS 203-041-4

Tradenames: Mazeen® 173; Mazeen® 174; Mazeen® 174-75

Tetramethyl decynediol (CTFA)
CAS 126-86-3; EINECS 204-809-1
Synonyms: 2,4,7,9-Tetramethyl-5-decyn-4,7-diol
Tradenames: Surfynol® 104
Tradenames containing: Surfynol® 104A; Surfynol® 104BC; Surfynol® 104E; Surfynol® 104H; Surfynol® 104PA; Surfynol® 104PG; Surfynol® 104S; Surfynol® PG-50; Surfynol® TG

Tetrapotassium pyrophosphate (CTFA)
CAS 7320-34-5; EINECS 230-785-7
Synonyms: TKPP; Diphosphoric acid, tetrapotassium salt; Potassium pyrophosphate
Tradenames: Empiphos 4KP

Tetrasodium dicarboxyethyl stearyl sulfosuccinamate (CTFA)
CAS 3401-73-8; 37767-39-8; 38916-42-6; EINECS 222-273-7
Synonyms: Tetrasodium dicarboxyethyl octadecyl sulfosuccinamate
Tradenames: Aerosol® 22; Fizul M-440; Lankropol® ATE; Monawet SNO-35; Rewopol® B 2003

Tetrasodium EDTA (CTFA)
CAS 64-02-8; EINECS 200-573-9
Synonyms: EDTA Na_4; Edetate sodium; Tetrasodium edetate; Ethylene diamine tetraacetic acid, sodium salt
Tradenames: Intraquest® TA Sol'n

TIPA-lauryl sulfate (CTFA)
CAS 661-61-6
Synonyms: Sulfuric acid, monododecyl ester, compd. with 1,1′,1″-nitrilotris[2-propanol]; Triisopropanolamine lauryl sulfate
Tradenames: Rewopol® TLS 45/B

Toluene sulfonic acid (CTFA)
CAS 104-15-4; 70788-37-3; EINECS 203-180-0
Synonyms: 4-Methylbenzenesulfonic acid
Tradenames: Eltesol® TA 96; Eltesol® TA Series; Eltesol® TSX; Eltesol® TSX/A; Eltesol® TSX/SF; Manro PTSA/C; Manro PTSA/E; Manro PTSA/H; Manro PTSA/LS; Sulframin TX; TSA-70; Witco® TX Acid; Witconate TX Acid
Tradenames containing: Eltesol® TA 65

Tribehenin. *See Glyceryl tribehenate*

Tributoxyethyl phosphate
CAS 78-51-3
Synonyms: TBEP; Ethanol, 2-butoxy-phosphate (3:1)
Tradenames: TBEP

Tributyl phosphate
CAS 126-73-8
Synonyms: Tri-n-butyl phosphate
Tradenames: Phos-Ad 100

Trichlorobenzene
CAS 120-82-1
Synonyms: 1,2,4-Trichlorobenzene
Tradenames: Abluton CTP; Hipochem GM; Hipochem Jet Dye T

Trideceth-2 (CTFA)
Synonyms: PEG 100 tridecyl ether; POE (2) tridecyl ether
Tradenames: Synperonic A2
Tradenames containing: Akypo TPR; Akypogene VSM-N; Akypoquat 40

Trideceth-3 (CTFA)
CAS 4403-12-7; 24938-91-8 (generic)
Synonyms: PEG-3 tridecyl ether; PEG (3) tridecyl ether; POE (3) tridecyl ether
Tradenames: Bio-Soft® TD 400; Chemal TDA-3; Emthox® 5993; Ethal TDA-3; Hetoxol TD-3; Iconol TDA-3; Macol® TD-3; Prox-onic TD-1/03; Rhodasurf® BC-420 (formerly Emulphogene® BC-420); Siponic® TD-3 (see Rhodasurf® BC-420); Synperonic A3; Trycol® 5993; Trycol® TDA-3; Volpo T-3

Trideceth-4
CAS 24938-91-8 (generic)
Synonyms: PEG-4 tridecyl ether; POE (4) tridecyl ether
Tradenames: Genapol® X-040

Trideceth-5
CAS 24938-91-8 (generic)
Synonyms: PEG-5 tridecyl ether; POE (5) tridecyl ether
Tradenames: Teric 13A5; Volpo T-5

Trideceth-6 (CTFA)
CAS 24938-91-8 (generic)
Synonyms: PEG-6 tridecyl ether; PEG 300 tridecyl ether; POE (6) tridecyl ether
Tradenames: Ahcowet DQ-114; Chemal TDA-6; Desonic® 6T; Ethal TDA-6; Ethosperse® TDA-6; Genapol® X-060; Hetoxol TD-6; Iconol TDA-6; Lipocol TD-6; Macol® TD-610; Prox-onic TD-1/06; Renex® 36; Rhodasurf® BC-610 (formerly Emulphogene® BC-610); Rolfor TR 6; Siponic® TD-6 (see Rhodasurf® BC-610); Trycol® 5940; Trycol® TDA-6

Trideceth-7 (CTFA)
CAS 24938-91-8 (generic)
Synonyms: PEG-7 tridecyl ether; POE (7) tridecyl ether
Tradenames: Flo-Mo® AJ-85; Flo-Mo® AJ-100; Rhodasurf® BC-630 (formerly Emulphogene® BC-630); Synperonic A7; Teric 13A7

Trideceth-8
Synonyms: PEG-8 tridecyl ether; POE (8) tridecyl ether
Tradenames: Bio-Soft® TD 630; Genapol® X-080; Iconol TDA-8; Iconol TDA-8-90%; Macol® TD-8; Rhodasurf® ROX (formerly Soprophor ROX); Trycol® 5949; Trycol® 5968; Trycol® TDA-8

Trideceth-9 (CTFA)
CAS 24938-91-8 (generic)
Synonyms: PEG-9 tridecyl ether; PEG 450 tridecyl ether; POE (9) tridecyl ether
Tradenames: Chemal TDA-9; Desonic® 9T; Emthox® 5941; Ethal TDA-9; Hetoxol TD-9; Iconol TDA-9; Prox-onic TD-1/09; Rolfor TR 9; Synperonic A9; Teric 13A9; Trycol® 5941; Trycol® 5944; Trycol® TDA-9

Trideceth-10 (CTFA)
CAS 24938-91-8 (generic)
Synonyms: PEG-10 tridecyl ether; PEG 500 tridecyl ether; POE (10) tridecyl ether
Tradenames: Carsonon® TD-10; Desonic® 10T; Genapol® X-100; Iconol TDA-10; Macol® TD-10; Rhodasurf® BC-720 (formerly Emulphogene® BC-720); Rhodasurf® T-95 (formerly Texafor T-95); Trycol® 5957; Volpo T-10

Trideceth-11 (CTFA)
CAS 24938-91-8 (generic)
Synonyms: PEG-11 tridecyl ether; POE (11) tridecyl ether
Tradenames: Carsonon® TD-11, TD-11 70%; Emthox® 5942; Synperonic A11; Trycol® 5942; Trycol® TDA-11

Trideceth-12 (CTFA)
CAS 24938-91-8 (generic)
Synonyms: PEG-12 tridecyl ether; PEG 600 tridecyl ether; POE (12) tridecyl ether
Tradenames: Ahcowet DQ-145; Chemal TDA-12; Desonic® 12T; Ethal TDA-12; Hetoxol TD-12; Lipocol TD-12; Macol® TD-12; Prox-onic TD-1/012; Renex® 30; Rolfor TR 12; Trycol® 5943; Trycol® TDA-12
Tradenames containing: Renex® 35

Trideceth-14
CAS 24938-91-8 (generic)
Synonyms: PEG-14 tridecyl ether; POE (14) tridecyl ether
Tradenames: Rhodasurf® BC-737 (formerly Emulphogene® BC-737); Trycol® 5874

Trideceth-15 (CTFA)
CAS 24938-91-8 (generic)
Synonyms: PEG-15 tridecyl ether; POE (15) tridecyl ether
Tradenames: Chemal TDA-15; Desonic® 15T; Macol® TD-15; Renex® 31; Rhodasurf® BC-840 (formerly Emulphogene® BC-840); Volpo T-15

Trideceth-18
CAS 24938-91-8 (generic)
Synonyms: PEG-18 tridecyl ether; POE (18) tridecyl ether
Tradenames: Chemal TDA-18; Ethal TDA-18; Hetoxol TD-18; Iconol TDA-18-80%; Trycol® 5946; Trycol® TDA-18

Trideceth-20 (CTFA)
CAS 24938-91-8 (generic)
Synonyms: PEG-20 tridecyl ether; PEG 1000 tridecyl ether; POE (20) tridecyl ether
Tradenames: Synperonic A20; Volpo T-20

Trideceth-25
CAS 24938-91-8 (generic)
Synonyms: PEG-25 tridecyl ether; POE (25) tridecyl ether
Tradenames: Hetoxol TD-25

Trideceth-29
CAS 24938-91-8 (generic)
Synonyms: PEG-29 tridecyl ether; POE (29) tridecyl ether
Tradenames: Iconol TDA-29-80%

Trideceth-50 (CTFA)
CAS 24938-91-8 (generic)
Synonyms: POE (50) tridecyl ether
Tradenames: Synperonic A50

Trideceth-100
CAS 24938-91-8 (generic)
Synonyms: PEG-100 tridecyl ether; POE (100) tridecyl ether
Tradenames: Macol® TD-100

Trideceth-150
CAS 24938-91-8 (generic)
Synonyms: PEG-150 tridecyl ether; POE (150) tridecyl ether
Tradenames: T-Det® TDA-150

Trideceth-7 carboxylic acid (CTFA)
CAS 56388-96-6 (generic); 68412-55-5 (generic); 24938-91-8 (generic)
Synonyms: PEG-7 tridecyl ether carboxylic acid; POE (7) tridecyl ether carboxylic acid
Tradenames: Incrodet TD7-C; Rewopol® CT 65; Sandopan® DTC-Acid; Surfine T-A

Trideceth-19 carboxylic acid (CTFA)
CAS 24938-91-8 (generic); 68412-55-5 (generic)
Synonyms: POE (19) tridecyl ether carboxylic acid
Tradenames: Sandopan® JA-36

Trideceth-6 phosphate (CTFA)
CAS 9046-01-9 (generic); 24938-91-8 (generic); 73070-47-0 (generic)
Synonyms: PEG-6 tridecyl ether phosphate; PEG 300 tridecyl ether phosphate; POE (6) tridecyl ether phosphate
Tradenames: Rhodafac® RS-610 (formerly Gafac® RS-610)

Tridecyl alcohol (CTFA)
CAS 68526-86-3; 112-70-9; EINECS 203-998-8
Synonyms: Tridecanol; C13 linear primary alcohol
Tradenames: Hyonic PE-360; Neodol® 3; Synprol Alcohol
Tradenames containing: Mackanate AY-65TD

Tridecylbenzene sulfonic acid (CTFA)
CAS 25496-01-9; EINECS 247-036-5
Synonyms: Benzenesulfonic acid, tridecyl-
Tradenames: Arylan® SO60 Acid; LABS 100/H.V; Nansa® TDB

Tridecyl phosphate
Tradenames: Hostaphat MDIT

Tridecyl stearate (CTFA)
CAS 31556-45-3; EINECS 250-696-7
Synonyms: Octadecanoic acid, tridecyl ester
Tradenames: Afilan TDS; Cirrasol® LN-GS; Emerest® 2308

Triethanolamine (CTFA)
CAS 102-71-6; EINECS 203-049-8
Synonyms: TEA; 2,2′,2′′-Nitrilotris(ethanol); Trolamine; Trihydroxytriethylamine
Tradenames containing: Surfonic® HDL; Surfonic® HDL-1

Triglyceryl. *See Polyglyceryl-3*

Trihexadecylamine
CAS 67701-00-2
Tradenames: Armeen® 3-16

Trihexadecylmethyl ammonium chloride
CAS 71060-72-5
Tradenames: Arquad® 316(W)

Trilaurylamine (CTFA)
CAS 102-87-4; EINECS 203-063-4
Synonyms: Trilauramine; Tridodecyl amine; N,N-Didodecyl-1-dodecanamine
Tradenames: Armeen® 3-12

Trilauryl phosphate (CTFA)
Synonyms: Phosphoric acid, trilauryl ester
Tradenames: Marlophor® LN-Acid

Trilinoleic acid (CTFA)
CAS 7049-66-3; 68939-90-6
Synonyms: Trimer acid; Fatty acids, C18, unsaturated, trimers; 9,12-Octadecadienoic acid, trimer
Tradenames: Empol® 1040; Empol® 1041; Empol® 1043; Hystrene® 5460

Trimer acid. *See Trilinoleic acid*

N,N,N′-Trimethyl-N′-9-octadecenyl-1,3-diaminopropane
CAS 68715-87-7
Synonyms: N,N,N′-Trimethyl-N′-9-octadecenyl-1,3-propanediamine
Tradenames: Duomeen® OTM

Trimethylolpropane oleate
Tradenames: Radiasurf® 7372

Trimethylolpropane tricaprylate/tricaprate
CAS 68956-08-1
Synonyms: 2-Ethyl-2-[[(oxo-octyl/decyl)oxy] methyl]-1,3-propanediyl octanoate/decanoate; Trimethylolpropane tricaprylate/caprate
Tradenames: Kessco® 887

Trimethylolpropane tripelargonate
Tradenames: Afilan TMPP

N,N,N′-Trimethyl-N′-tallow-1,3-diaminopropane
CAS 68783-25-5
Tradenames: Duomeen® TTM

Trioctyl phosphate
Tradenames: Hostaphat 2122

Triolein (CTFA)
CAS 122-32-7; EINECS 204-534-7
Synonyms: Glyceryl trioleate; Olein; 9-Octadecenoic acid, 1,2,3-propanetriyl ester
Tradenames: Emerest® 2423; Kemester® 1000
Tradenames containing: Juniorlan 1664

Tripropylene glycol myristyl ether. *See PPG-3 myristyl ether*

Tris (hydroxymethyl) aminomethane
CAS 77-86-1
Synonyms: THAM; Tromethamine (CTFA); 2-Amino-2-(hydroxymethyl)-1,3-propanediol
Tradenames: Tris Amino® (Crystals)

Trisodium NTA (CTFA)
CAS 5064-31-3; EINECS 225-768-6
Synonyms: Trisodium nitrilotriacetate; Nitrilo triacetic acid sodium salt; N,N-Bis(carboxymethyl)glycine, trisodium salt
Tradenames: Trilon® A

Tristearin (CTFA)
CAS 555-43-1; EINECS 209-097-6
Synonyms: Glyceryl tristearate; 1,2,3-Propanetriol trioctadecanoate; Octadecanoic acid, 1,2,3-propanetriyl ester
Tradenames: Pationic® 919

Turkey-red oil. *See Sulfated castor oil*

Undeceth-3
Tradenames: Neodol® 1-3

Undeceth-5 (CTFA)
Synonyms: POE (5) undecyl ether
Tradenames: Neodol® 1-5

Undeceth-7
Tradenames: Neodol® 1-7

Undeceth-9
Tradenames: Neodol® 1-9

Undecyl alcohol (CTFA)
CAS 112-42-5
Synonyms: Alcohol, undecyl; C11 primary alcohol; Alcohols, C11
Tradenames: Neodol® 1

Undecylenamide MEA (CTFA)
CAS 20545-92-0; 25377-63-3; EINECS 243-870-9
Synonyms: Undecylenoyl monoethanolamide; N-(2-Hydroxyethyl)undecenamide
Tradenames: Rewocid® U 185; Rewomid® U 185

Urea (CTFA)
CAS 57-13-6; EINECS 200-315-5
Synonyms: Carbamide; Carbonyldiamide
Tradenames containing: Bio-Soft® LD-95; Renex® 25; Renex® 35

Urea-formaldehyde resin (CTFA)
Tradenames: Diamonine B

Vinylamide/vinylsulfonic acid copolymer
Tradenames: Hostadrill Brands; Hostamer Brands

Wheat germamide DEA
Tradenames: Incromide WGD

Wheat germamidopropylamine oxide (CTFA)
Synonyms: Wheat germ oil amides, N-[3-(dimethylamino)propyl]-, N-oxide; N-[3-(Dimethylamino) propyl]wheat germ oil amides, N-oxide
Tradenames: Incromine Oxide WG; Mackamine WGO

Wheat germamidopropyl betaine (CTFA)
Synonyms: N-(Carboxymethyl)-N,N-dimethyl-3-[(1-oxowheat germ alkyl)amino]-1-propanaminium hyroxides, inner salts; Wheat germ oil amido betaine
Tradenames: Mackam WGB

Wheat germamidopropyl dimethylamine (CTFA)
Synonyms: Dimethylaminopropyl wheat germamide
Tradenames: Mackine 701

White spirits. *See Mineral spirits*

Wool fat. *See Lanolin*

Wool wax. *See Lanolin*

Xanthan gum (CTFA)
CAS 11138-66-2; EINECS 234-394-2
Synonyms: Corn sugar gum; Xanthan
Tradenames: Biozan; Rhodopol® 23

Xylene sulfonic acid (CTFA)
CAS 25321-41-9; EINECS 246-839-8
Synonyms: Dimethylbenzenesulfonic acid; Benzenesulfonic acid, dimethyl-
Tradenames: Eltesol® 4200; Eltesol® XA; Eltesol® XA65; Eltesol® XA90

Zinc lactylate
Tradenames: Pationic® 1264

Part IV
Manufacturers Directory

Manufacturers Directory

Aceto Chemical Co., Inc.
1 Hollow Lane, Suite 201
Lake Success, NY 11042-1215
USA
Tel.: 516-627-6000
FAX 516-627-6093
Telex: 62662

Air Products and Chemicals, Inc.
7201 Hamilton Blvd.
Allentown, PA 18195
USA
Tel.: 215-481-4911; 800-345-3148
FAX 215-481-5900
Telex: 275425

Air Products Nederland B.V.
Herculesplein 359, PO Box 85075
NL 3508 AB Utrecht
Netherlands
Tel.: 31-30-511828

Air Products Pacific, Inc.
Sakurabashi Yachiyo Bldg., 5-6, Umeda 2-Chome, Kita-Ku
Osaka 530
Japan

Air Products, Valchem Div.
403 Carline Rd.
Langley, SC 29834
USA
Tel.: 803-593-4466

Ajinomoto Co., Inc.
Mina Mi 5-32, Koenji Suginami
Tokyo
Japan
Tel.: 33-314-3211
FAX 33-312-7207

Ajinomoto USA, Inc.
Glenpointe Centre West, 500 Frank W. Burr Blvd.
Teaneck, NJ 07666-6894
USA
Tel.: 201-488-1212
FAX 201-488-6472
Telex: 275425 (AJNJ)

Akzo Chemicals Inc./Chemicals Div.
300 S. Riverside Plaza
Chicago, IL 60606
USA
Tel.: 312-906-7500; 800-257-8292
FAX 312-906-7680
Telex: 25-3233

Akzo Chemicals Inc.
Asphalt Chemicals Dept.
Adamsfield Business Center
7101 Adams St., Unit #7
Willowbrook, IL 60521
USA
Tel.: 708-789-2494; 800-BITUMEN
FAX 708-789-2506

Akzo Chemicals BV
PO Box 975
3800 AZ Amersfoort
The Netherlands
Tel.: 033-643911
FAX 033-637448
Telex: 79322

Akzo Chemicals GmbH
Phillippstrasse 27, PO Box 100132
W-5160 Dueren
Germany
Tel.: 2421-492261
FAX 2421-492487
Telex: 833911

Akzo Chemicals Ltd.
PO Box 80
Parramatta N.S.W. 2150
Australia

Akzo Chemicals Ltd.
100 University Ave., Suite 906
Toronto, OntarioMSJ IV6
Canada

Akzo Chemicals Ltd.
1-5 Queens Rd., Hersham, Walton-on-Thames
Surrey KT12 5NL
UK
Tel.: 0932-247891
FAX 0932-231204
Telex: 21997

Akzo Chemie Italia SpA
Via Vismara, 20020 Arese
Milano
Italy
Tel.: 2-938 08 71
FAX 2-938 08 16
Telex: 332526

Akzo Chemie S.A.
13, Ave. Marnix
1050 Bruxelles
Belgium
Tel.: 02-5180411

Akzo Japan Ltd.
Godo Kaikan Bldg., 3-27 Kioi-cho, Chiyoda-Ku
Tokyo 102
Japan

Albright & Wilson Ltd., European Hdqtrs.
PO Box 3
210-222 Hagley Rd. West, Oldbury, Warley
West Midlands B68 0NN
UK
Tel.: 44-21-429-4942
FAX 44-21-420-5151
Telex: 336291

Albright & Wilson Americas
PO Box 26229
Richmond, VA 23260-6229
USA
Tel.: 804-550-4300; 800-446-3700
FAX 804-550-4385

Albright & Wilson Am. (Canada)
2 Gibbs Rd.
Islington, OntarioM9B 1R1
Canada
Tel.: 416-234-7000; 800-268-2520
FAX 416-237-1064

Albright & Wilson (Australia) Ltd.
PO Box 20
Yarraville, Victoria
Australia
Tel.: 3-688-7777
FAX 3-688-7788

Marchon Espanola SA
Carretera Montblanc Km 2, 4
Alcover (Tarragona)
Spain
Tel.: 846000
Telex: 56461 MESPA E

Marchon France SA
BP 19, F-55300
St. Mihiel
France
Tel.: 29 91 22 22
FAX 29 89 08 69

Marchon Italiana SpA
Casella Postale No. 30, 1-46043
Castiglione delle Stivier
Italy

Alco Chemical/
Div. of National Starch & Chem.
909 Mueller Dr., PO Box 5401
Chattanooga, TN 37406
USA
Tel.: 615-629-1405; 800-251-1080
FAX 615-698-8723
Telex: 755002

Alconox Inc.
215 Park Ave. So.
New York, NY 10003
USA
Tel.: 212-473-1300
FAX 212-353-1342

Allied-Signal/A-C® Performance Additives
PO Box 2332R
Morristown, NJ 07962-2332
USA
Tel.: 201-455-2145; 800-222-0094
FAX 201-455-6154
Telex: 990433

Alox Corp.
3943 Buffalo Ave., PO Box 517
Niagara Falls, NY 14302
USA
Tel.: 716-282-1295
FAX 716-282-2289

Amerchol Corp.
PO Box 4051
136 Talmadge Rd.
Edison, NJ 08818
USA
Tel.: 908-248-6000; 800-367-3534
FAX 908-287-4186
Telex: 833472

Amerchol, D.F. Anstead Ltd.
Victoria House, Radford Way, Bellericay
Essex CM12 0DE
UK

Amerchol Europe
Havenstraat 86
B-1800 Vilvoorde
Belgium
Tel.: 32-2-252-4012
FAX 32-2-252-4909
Telex: 846-69105

Amerchol, Ikeda Corp.
New Tokyo Bldg., No. 3-1
Marunouchi 3-Chome, Chiyoda-Ku
Tokyo 100
Japan

American Cyanamid
Corporate Headquarters
One Cyanamid Plaza
Wayne, NJ 07470
USA
Tel.: 201-831-4111; 800-922-0187
FAX 201-839-8847

American Cyanamid/Textiles
One Cyanamid Plaza
Wayne, NJ 07470
USA
Tel.: 201-831-4111; 800-922-0187
FAX 201-839-8847

Cyanamid BV
Postbus 1523, 3000 BM
Rotterdam
The Netherlands
Tel.: 010-411 6340
FAX 010-413 6788
Telex: 23554

Cyanamid Canada Inc./Carbide Products Div.
88 McNabb St.
Markham, OntarioL3R 6E6
Canada
Tel.: 416-470-3600
FAX 416-470-3852
Telex: 06-966602

Cyanamid GmbH
Pfaffenriederstrasse 7, W-8190
Wolfratshausen
Germany
Tel.: 8171-2201
FAX 8171-22411
Telex: 526364

Cyanamid India Ltd.
Nyloc House
254-D2 Dr. Annie Besant Rd.
Bombay 400 025
India

Cyanamid Quimica do Brasil Ltda.
Av. Imperatriz Leopoldina, 86
Sao Paulo
Brazil

Cyanamid Taiwan Corp.
8/F Union Commercial Bldg.
137, Nanking E. Rd., Sec. 2
Taipei
Taiwan, R.O.C.

American Emulsions Co.
PO Box 3787
Dalton, GA 30721
USA
Tel.: 404-226-7028
FAX 404-278-5183

American Ingredients Co.
14622 S. Lakeside Ave.
Dolton, IL 60419
USA
Tel.: 708-849-8590; 800-821-2250
FAX 816-561-0422

American Lecithin *See under Rhone-Poulenc*

Anedco, Inc.
10429 Koenig Rd.
Houston, TX 77034
USA
Tel.: 713-484-3900
FAX 713-484-3931

Angus Chemical Co.
2211 Sanders Rd.
Northbrook, IL 60062
USA
Tel.: 708-498-6700; 800-362-2580
FAX 708-498-6706
Telex: 275422 ANGUSUR

Angus Chemical Co.
101 Cecil St., #14-07 Tong Eng Bldg., 0106
Singapore

Angus Chemie GmbH
19, Moorgate St.
Rotherham, Yorkshire S60 2DA
UK
Tel.: 0709 377743
FAX 0709 370596

Angus Chemie GmbH
Le Bonaparte, Centre d'Affaires Paris-Nord
F-93153 Le Blanc Mesnil
Paris
France

Angus Chemie GmbH
Huyssenallee 5
4300 Essen 1,
Germany
Tel.: 0201-233531
FAX 0201-238661

Apex Chemical Co., Inc.
200 South First St.
Elizabethport, NJ 07206
USA
Tel.: 908-354-5420
FAX 908-354-2640
Telex: 178326 APEX UT

Aqualon Co.
PO Box 15417
2711 Centreville Rd.
Wilmington, DE 19850-5417
USA
Tel.: 302-996-2000; 800-345-8104
FAX 302-996-2049
Telex: 4761123

Aqualon Canada Inc.
5407 Eglinton Ave. West, Suite 103
Etobicoke, OntarioM9C 5K6
Canada
Tel.: 416-620-5400

Aqualon France BV
44, Ave. de Chatou
F-92508 Rueil Malmaison Cedex
France
Tel.: 1-4751-2919
FAX 47 77 06 14

Aqualon GmbH
PO Box 130125, Paul Thomas Strasse 58
D-4000 Düsseldorf 13
Germany
Tel.: 49-211-7491-0

Aqualon (UK) Ltd.
Genesis Centre, Garrett Field, Birchwood
Warrington, Cheshire WA3 7BH
UK
Tel.: 44-925-830077

Aquaness Chemicals
3920 Essex Lane
PO Box 27714
Houston, TX 77227
USA
Tel.: 713-599-7400
FAX 713-599-7460
Telex: 4620058

Aquatec Quimica SA
Rua Sampaio Viana, 425
CEP 0400 Postal 4885
Sao Paulo
Brazil
Tel.: 011-884-4466 ext.329
FAX 11-884-0747
Telex: 1121312

Arizona Chemical Co.
Div. of International Paper
1001 E. Business Hwy. 98
Panama City, FL 32401
USA
Tel.: 904-785-6700; 800-526-5294
FAX 904-785-2203
Telex: 514411

Arol Chemical Products Co.
649 Ferry St.
Newark, NJ 07105
USA
Tel.: 201-344-1510
FAX 201-344-7127

Atsaun Chemical Corp.
23, Janki Niwas
N.C. Kelkar Rd., Dadar
Bombay
400 028
India
Tel.: 430-1454-422-3145

Atlas Refinery, Inc.
142 Lockwood St.
Newark, NJ 07105
USA
Tel.: 201-589-2002
FAX 201-589-7377
Telex: 138-425 ATLASOIL

Atramax Inc.
PO Box 278
Hawthorne, NJ 07507
USA
Tel.: 212-882-2263
FAX 212-798-2546

Auschem SpA
Via Baertsch, 1
24100 Bergamo
Italy
Tel.: 35-346282
FAX 35-340386
Telex: 300275

BASF AG
ESA/WA-H 201
D-6700 Ludwigshafen
Germany
Tel.: 0621-60-99603
FAX 0621-60-41787
Telex: 469499-0 BAS D

BASF Corp.
100 Cherry Hill Rd.
Parsippany, NJ 07054
USA
Tel.: 201-316-3000; 800-669-BASF
FAX 201-402-1832

BASF Corp.
Chemicals Div., Thermoplastic PU, Elastomers
1609 Biddle Ave.
Wyandotte, MI 48192
USA
Tel.: 313-246-6323

BASF Belgium S.A.
Ave. Hamoirlaan 14
B-1180 Brussels
Belgium
Tel.: 2 375 21 11
FAX 2 375 10 42
Telex: 22016

BASF Canada Ltd.
PO Box 430
Montreal, Quebec H4L 4V8
Canada

BASF Espanola S.A.
Paseo de Gracia 99, E-08008
Barcelona
Spain
Tel.: 3-215 13 54
FAX 3-215 95 06
Telex: 52488

BASF India, Ltd.
Maybaker House, S.K. Ahire Marg.
PO Box 19108
Bombay 400 025
India

BASF Japan Ltd.
C.P.O. Box 1757
Tokyo 100-91
Japan

BASF PLC
PO Box 4, Earl Road, Cheadle Hulme
Cheshire SK8 6QG
UK
Tel.: 061-485-6222
FAX 061-486-0891
Telex: 669211

BASF S.A. Compagnie Francaise
MC-NT, 140, Rue Jules Guesde
92303 Levallois-Perret
France

Bernel Chemical Co., Inc.
174 Grand Ave.
Englewood, NJ 07631
USA
Tel.: 201-569-8934
FAX 201-569-1741

Berol Nobel AB
Box 11536
S-10061 Stockholm
Sweden
Tel.: 46-8-743-4000
FAX 46-8-644-3955
Telex: 10513 benobl s

Berol Nobel Inc.
Meritt 8 Corporate Park
99 Hawley Lane
Stratford, CT
06497

Berol Nobel SA
Rue Gachard 88, Bte 9
B-1050 Bruxelles
Belgium
Tel.: 32-02-640-5065
FAX 32-2-640-6997
Telex: 62812

Boliden Intertrade Inc.
3400 Peachtree Rd. NE, Suite 401
Atlanta, GA 30326
USA
Tel.: 404-239-6700; 800-241-1912
FAX 404-239-6701
Telex: 981036

Borregaard LignoTech
PO Box 162
N-1701 Sarpsborg
Norway
Tel.: (09)11 80 00
FAX (09) 11 87 70

LignoTech Canada Inc.
1950, Rue Léon Harmel, Quebec P.Q. GIN HK3
Canada
Tel.: 418-684-3000
FAX 418-684-3005

LignoTech (U.K.) Ltd.
Clayton Rd., Birchwood
Warrington, Cheshire WA3 6QQ
UK
Tel.: (0925)824511
FAX (0925) 812186

LignoTech USA, Inc.
100 Highway 51 South
Rothschild, WI 54474-1198
USA
Tel.: 715-359-6544
FAX 715-355-3648

Burlington Chemical Co., Inc.
PO Box 111
Burlington, NC 27216
USA
Tel.: 919-584-0111; 800-334-8550
FAX 919-584-3548
Telex: 9102502503

Calgene Chem. Inc.
602 E. Algonquin Rd.
Des Plaines, IL 60016
USA
Tel.: 708-298-4000
FAX 708-298-1519
Telex: 206168

Cardolite Corp.
500 Doremus Ave.
Newark, NJ 07105-4805
USA
Tel.: 201-344-5015; 800-322-7365
FAX 201-344-1197

Carroll Co.
2900 W. Kingsley
Garland, TX 75041
USA
Tel.: 214-278-1304; 800-527-5722
FAX 214-840-0678

CasChem Inc.
40 Ave. A
Bayonne, NJ 07002
USA
Tel.: 201-858-7900; 800-CASCHEM
FAX 201-437-2728
Telex: 710-729-4466

Catawba-Charlab, Inc.
5046 Pineville Rd., PO Box 240497
Charlotte, NC 240497
USA
Tel.: 704-523-4242
FAX 704-522-8142

Ceca SA/Div. of Atochem
22, place de l'Iris, La Défense 2
Cedex 54
92062 Paris-La Défense
France
Tel.: 33-147-96-9090
FAX 33-147-96-9234
Telex: 611444 ckd

Celanese. *See under Hoechst-Celanese*

Central Soya Co. of America
1300 Ft. Wayne Nat'l. Bank Bldg.
PO Box 1400
Fort Wayne, IN 46801
USA
Tel.: 219-425-5100; 800-348-0960
FAX 219-425-5485
Telex: 276170

Central Soya
PO Box 5063
3008 AB Rotterdam
Netherlands
Tel.: 31-10-42-39-600
FAX 31-10-42-30-897
Telex: 20041 CNSOY NL

Chemax, Inc.
PO Box 6067, Highway 25 South
Greenville, SC 29606
USA
Tel.: 803-277-7000; 800-334-6234
FAX 803-277-7807
Telex: 570412 IPM15SC

Chemische Fabrik Grünau GmbH. *See under Grünau*

Chemische Fabrik Wibarco GmbH. *See under Wibarco*

Chemron Corp.
PO Box 2299
Paso Robles, CA 93447
USA
Tel.: 805-239-1550
FAX 805-239-8551

Chemsal Chemicals & Co. KG
Chem-Y GmbH/Salim joint venture
Kupferstrasse 1, PO Box 100262
D-4240 Emmerich 1
Germany
Tel.: 02822 711-0
FAX 02822 18294
Telex: 8125124

Chem-Y Fabriek Van Chemische Producten B.V.
PO Box 50
2410 AB, Bodegraven
Netherlands
Tel.: 028-22/711-0
FAX 49282218294
Telex: 8125 124

Chem-Y GmbH
Kupferstrasse 1
D4240 Emmerich
Germany
Tel.: 49-2822/7110
FAX 49-2822/18294
Telex: 8125124

Chevron Chemical Co.
PO Box 3766
Houston, TX 77253
USA
Tel.: 713-754-4290
Telex: 762799

Chevron Chemical Co./Olefin & Derivs.
PO Box 3766
Houston, TX 77253
USA
Tel.: 800-231-3826

Ciba-Geigy AG
CH-4002 Basel
Switzerland
Tel.: 061 223 1341
FAX 061 231 7422
Telex: 668083

Ciba-Geigy Corp.
444 Saw Mill River Rd.
Ardsley, NY 10502
USA
Tel.: 914-478-3131; 800-431-1874

Ciba-Geigy Corp./Dyestuffs & Chemicals Div.
PO Box 18300
Greensboro, NC 27419
USA
Tel.: 800-334-9481
FAX 919-632-7098

Ciba-Geigy Corp./Dyestuffs & Chemicals (UK)
Ashton New Road, Clayton
Manchester M11 4AR
UK

Ciba-Geigy Marienberg GmbH
Postfach 1253
D-6140 Bensheim 1
Germany
Tel.: 06254 79 0
FAX 06254 79493
Telex: 625491

Ciba-Geigy PLC
30 Buckingham Gate
London SW1E 6LH
UK

Ciba-Geigy/Textile Products Div.
PO Box 18300
Greensboro, NC 27419-8300
Tel.: 919-632-6000
FAX 919-632-7665

Clark Chemical Inc.
25 Trammel St.
Marietta, GA 30064
USA
Tel.: 404-514-8909
FAX 404-514-8906

W.A. Cleary Chemical Corp.
Southview Industrial Park
178 Route #522 Suite A
Dayton, NJ 08810
USA
Tel.: 908-329-8399; 800-524-1662
FAX 908-274-0894

Climax Performance Materials Corp.
7666 W. 63rd St.
Summit, IL 60501
USA
Tel.: 708-458-8450; 800-323-3231
FAX 708-458-0286

Clough Chemical Co., Ltd.
178 St. Pierre
PO Box 1017
St-Jean-sur-Richelieu, Quebec J3B 7B5
Canada
Tel.: 514-346-6848; 800-363-9284
FAX 514-346-7263

CNC International, Limited Partnership
PO Box 3000
Woonsocket, RI 02895
USA
Tel.: 401-769-6100
FAX 401-769-4509

Colonial Chemical, Inc.
9431 Mountain Shadows Dr.
Chattanooga, TN 37421
USA
Tel.: 615-267-8947
FAX 615-266-0770

Consos, Inc.
PO Box 34186
Charlotte, NC 28234
USA
Tel.: 704-596-2813
FAX 704-596-4861

CPS Kemi Aps
Hejreskovv, 22
3490 Kvistagaard
Denmark
Tel.: 2 890533
FAX 42 23 80 77

CPS Chem. Co., Inc.
PO Box 162
Old Bridge, NJ 08857
USA
Tel.: 908-727-3100
FAX 908-727-2260
Telex: 844532-CPSOLDB

Croda Chemicals Ltd.
Cowick Hall, Snaith Goole
North Humberside DN14 9AA
UK
Tel.: 0405-8605551
FAX 0405-860205
Telex: 57601

Croda Canada Ltd.
78 Tisdale Ave.
Toronto, OntarioM4A 1Y7
Canada
Tel.: 416-751-3571
FAX 416-751-9611

Croda Chemicals Group Pty. Ltd.
PO Box 1012
Richmond, North Victoria 3121
Australia

Croda do Brazil Ltda.
Rua Croda 230 Distrito Industrial
CEP 13.053
Campinas/SP-C.P. 1098
Brazil

Croda Inc.
7 Century Dr.
Parsippany, NJ 07054-4698
USA
Tel.: 201-644-4900
FAX 201-644-9222

Croda Italiana Srl
Via Grocco, N917 27036
Mortara (PV)
Italy

Croda Japan KK
Aceman Bldg., 5F 3 7
Tokuicho 1-Chome Highashi-ku
Osaka 540
Japan
Tel.: 6-942-1791

Croda Surfactants Ltd.
Cowick Hall, Snaith, Goole
North Humberside DN14 9AA
UK
Tel.: 0405 860551
FAX 0405 860205
Telex: 57601

Croda Universal Ltd.
Cowick Hall, Snaith, Goole
North Humberside DN14 9AA
UK
Tel.: 0405 860551
FAX 0405 860205
Telex: 57601

Crompton & Knowles Corp.
Dyes & Chems. Div.
PO Box 33188
Charlotte, NC 28233
USA
Tel.: 704-372-5890
FAX 704-372-1522

Crompton & Knowles Corp.
Ingredient Tech. Div.
1595 MacArthur Blvd.
Mahwah, NJ 07430
USA
Tel.: 201-818-1200; 800-343-4860

Crown Technology, Inc./Chemical Div.
7513 E. 96 St.
PO Box 50426
Indianapolis, IN 46250
USA
Tel.: 317-845-0045
FAX 317-845-9086

Crucible Chemical Co.
PO Box 6786, Donaldson Center
Greenville, SC 29606
USA
Tel.: 803-277-1284; 800-845-8873

Cyanamid. *See under American Cyanamid*

Dai-ichi Kogyo Seiyaku Co., Ltd.
Miki Bldg., 3-12-1, Nihombashi, Chuo-ku
Tokyo 103
Japan
Tel.: 03-3274-6731
FAX 03-3274-4128
Telex: 222 6258

DeSoto. *See Witco/Organics*

DeeZee Chemical Inc.
14010 Orange Ave.
Paramount, CA 90723
USA
Tel.: 310-529-2556
FAX 310-529-8037

Dexter Chemical Corp.
845 Edgewater Rd.
Bronx, NY 10474
USA
Tel.: 212-542-7700
FAX 212-991-7684
Telex: 127061

Dow Chemical U.S.A.
2020 W.H. Dow Center
Midland, MI 48674
USA
Tel.: 517-636-1000; 800-441-4DOW

Dow Chemical Canada Inc.
1086 Modeland Rd., PO Box 1012
Sarnia, OntarioN7T 7K7
Canada
Tel.: 519-339-3131; 800-363-6250

Dow Chemical Co. Ltd.
Stana Place, Fairfield Ave., Staines
Middlesex TW18 4SX
UK

Dow Chemical Europe S.A.
Bachtobelstrasse 3
CH-8810 Horgen
Switzerland
Tel.: 41-1-728-2111
FAX 41-1-728-2935
Telex: 826940

Dow Chemical Pacific ltd.
39th Floor, Sun Hung Kai Centre
30 Harbour Rd., Wanchai, PO Box 711
Hong Kong

Dow Quimica S.A.
Sao Paulo
Brazil

Dow Corning Corp.
PO Box 0994
Midland, MI 48686-0994
USA
Tel.: 517-496-4000; 800-248-2481
Telex: 227450

Dow Corning France SA
Le Britannia A10, 20 Bid E Deruelle
69432 Lyon Cedex 03
France
Tel.: 78 60 51 48
FAX 78 62 78 98
Telex: 300537

Dow Corning Ltd.
Reading Bridge House, Reading
Berkshire RG1 4EX
UK
Tel.: 0734 507251
FAX 0734 575051
Telex: 848340

Drew Industrial Div./Div. of Ashland
One Drew Plaza
Boonton, NJ 07005
USA
Tel.: 201-263-7800; 800-526-1015 x7800
FAX 201-263-4483
Telex: DREWCHEMS BOON

Duphar BV. *See Solvay Duphar BV*

DuPont de Nemours & Co., Inc., E.I.
1007 Market St.
Wilmington, DE 19898
USA
Tel.: 302-774-7573; 800-441-9442
FAX 302-774-7573
Telex: 6717325

DuPont Canada Inc.
Box 2200, Streetsville
Mississauga, OntarioL5M 2H3
Canada
Tel.: 416-821-5612

DuPont de Nemours (France) S.A.
9, Rue de Vienne
75008-Paris
France

DuPont Far East Inc.
Kowa Bldg. No. 2
11-39 Akasaka 1-Chome, Minato-Ku
Tokyo 107
Japan
Tel.: 585-5511

DuPont S.A. de C.V.
Homero 206, Col. Polanco
Mexico 5
D.F. Mexico

DuPont (UK) Ltd.
Wedgewood Way, Stevenage
Herts SG1 4QN
UK

Eastern Color & Chemical Co.
35 Livingston St.
PO Box 6161
Providence, RI 02904
USA
Tel.: 401-331-9000
FAX 401-331-2155

Emkay Chemical Co.
319-325 Second St., PO Box 42
Elizabeth, NJ 07206
USA
Tel.: 908-352-7053
FAX 908-352-6398

Emulsion Systems Inc.
70 East Sunrise Hwy.
Valley Stream, NY 11581-1233
USA

Tel.: 516-825-3232; 800-ESI-CRYL
FAX 516-825-3233

Essential Industries Inc.
28391 Essential Rd.
PO Box 12
Merton, WI 53056-0012
USA
Tel.: 414-538-1122; 800-551-9679
FAX 414-538-1354

Ethox Chemicals, Inc.
PO Box 5094, Sta. B
Greenville, SC 29606
USA
Tel.: 803-277-1620
FAX 803-277-8981

Ethyl Corp.
451 Florida Blvd.
Baton Rouge, LA 70801
USA
Tel.: 504-388-7040; 800-535-3030
FAX 504-388-7686
Telex: 586441, 586431

Ethyl Asia Pacific Co.
#13-06 PUB Bldg.
Devonshire Wing, 111 Somerset Rd.
Singapore 0923
Rep. of Singapore

Ethyl Canada, Inc.
350 Burnhamthorpe Rd. West, Suite 600
Mississauga, OntarioL5B 3JI
Canada
Tel.: 416-566-9222
FAX 416-566-99962

Ethyl Corp. UK
Goldlay House
114 Parkway, Chelmsford
Essex CM2 7PP
UK
Tel.: 245-287-577

Ethyl Japan
Christy Bldg. 2/F
1-22 Moto Azabu 3-Chome, Minato-Ku
Tokyo 106
Japan

Ethyl S.A.
523 Ave. Louise, Box 19
B-1050 Brussels
Belgium
Tel.: 32-2-642-4411
FAX 32-2648-0560
Telex: 22549

Exxon Chemical Co.
PO Box 3272
Houston, TX 77253-3272
USA
Tel.: 713-870-6000; 800-526-0749
FAX 713-870-6661
Telex: 794588

Deutsche Exxon Chemical GmbH
Dompropst Ketzer-Str. 1-9
5000 Köln 1
Germany
Tel.: 221 16150
FAX 221 160 5320

Exxon Chemical Co./Tomah Products
1012 Terra Dr., PO Box 388
Milton, WI 53563
USA
Tel.: 608-868-6811; 800-441-0708
FAX 608-868-6810
Telex: 910-280-1401

Exxon Chemical Int'l. Marketing Inc.
Mechelsesteenweg 363
B-1950 Kraainem
Belgium
Tel.: 02-769-3111
Telex: 24733

Exxon Chemical Japan Ltd.
TBS Kaikan Bldg.
3-3, Akasaka 5-Chome, Minato-Ku
Tokyo 107
Japan
Tel.: 03-582-9243
Telex: 22846

Exxon Chemical Ltd.
Arundel Towers, Portland Terrace
Southampton SO9 2GW
UK
Tel.: 0703-634191
Telex: 47437

Exxon Chemical Mediterranea SpA
Via Paleocapa 7
20121 Milano
Italy
Tel.: 2 88031
FAX 2 8803231
Telex: 311561 ESSOCH I

Fanning Corp., The
1775 W. Diversity Pkwy.
Chicago, IL 60614-1009
USA
Tel.: 312-248-5700
FAX 312-248-6810
Telex: 910-221-1335

Ferro Corp./Keil Chemical Div.
3000 Sheffield Ave.
Hammond, IN 46320
USA
Tel.: 219-931-2630
FAX 219-931-6318
Telex: 725484

Fina PLC
Fina House, 1 Ashley Ave., Epsom
Surrey KT18 5AD
UK
Tel.: 44-03727-26226
FAX 44-03727-45821
Telex: 894317

Fina Chemicals
4, Rue Jacques de Lalaing
B-1040 Brussels
Belgium
Tel.: 02-288-91-11
FAX 288-32-99

Fina Oil and Chemical Co.
8350 North Central Expressway
PO Box 2159
Dallas, TX 75221
USA
Tel.: 214-750-2400; 800-344-FINA

Oleofina Far East
138 Cecil St., #17-01, Cecil Court
RS Singapore 0106

Finetex Inc.
418 Falmouth Ave.
PO Box 216
Elmwood Park, NJ 07407
USA
Tel.: 201-797-4686
FAX 201-797-6558
Telex: 710-988-2239

Represented by:
Pennine Chemical Ltd.
Kent Works, Thomas St.
Conglton
Cheshire CW12 1QZ
UK

Gattefosse SA
36 Chemin de Genas, BP 603
69800 Saint Priest
France
Tel.: 78-90-63-11
FAX 78-90-4567
Telex: 340 240

Gattefosse Corp.
189 Kinderkamack Rd.
Westwood, NJ 07675
USA
Tel.: 201-573-1700
FAX 201-573-9671

Represented by:
Alfa Chemicals Ltd.
Broadway House, 7-9 Shute End
Workingham
Berkshire RG11 1BH
UK

General Electric Co./Silicone Products Div.
260 Hudson River Rd.
Waterford, NY 12188
USA
Tel.: 518-237-3330; 800-255-8886

Genesee Polymers Corp.
G-5251 Fenton Rd., PO Box 7047
Flint, MI 48507-0047
USA
Tel.: 313-238-4966
FAX 313-767-3016

Georgia-Pacific Corp.
133 Peachtree St. N.E., PO Box 105605
Atlanta, GA 30348
USA
Tel.: 404-521-4711

Georgia-Pacific Chemical Div.
1754 Thorne Rd.
Tacoma, WA 98421
USA
Tel.: 206-572-8181

Goldschmidt AG, Th.
Goldschmidtstrasse 100, Postfach 101461
D-4300 Essen 1
Germany
Tel.: 0201-173-2947
FAX 201-173-2160
Telex: 857170

Goldschmidt Chemical Corp.
914 E. Randolph Rd., PO Box 1299
Hopewell, VA 23860
USA
Tel.: 804-541-8658; 800-446-1809
FAX 804-541-2783
Telex: 710-958-1350

Goldschmidt Japan KK, Th.
Rm. 1113, Shuwa Kioi-cho TBR Bldg. No. 7
5-Chome, Koji-machi, Chiyoda-ku
Tokyo 102
Japan

Goldschmidt Ltd., Th.
Tego House, Victoria Rd., Ruislip
Middlesex HA4 0YL
UK
Tel.: 01-4227788
FAX 01-8648159

Tego Chemie Service GmbH
Goldschmidstr. 100, Postfach 101461
D-4300 Essen 1
Germany
Tel.: 0201-1732571
FAX 0201-1732639
Telex: 85717-20tgd

Tego Chemie Service USA
PO Box 1299
914 E. Randolph Rd.
Hopewell, VA 23860
USA
Tel.: 804-541-8658; 800-446-1809
FAX 804-541-2783

BFGoodrich Co.
Specialty Polymers & Chem. Div.
9921 Brecksville Rd.
Brecksville, OH 44141
USA
Tel.: 216-447-5000; 800-331-1144
FAX 216-447-5720
Telex: 423313

BFGoodrich Canada
195 Columbia St. West
Waterloo, OntarioN2J 4N9
Canada

BFGoodrich Chemical (Deutschland) GmbH
Goerlitzer Str. 1
4040 Neuss 1
Germany

BFGoodrich Chemical (UK) Ltd.
The Lawn, 100 Lampton Road, Hounslow
Middlesex TW3 4EB
UK
Tel.: 0821-570 4700
FAX 081-570 0850
Telex: 265714 BFG UK

W.R. Grace/Organic Chemicals Div.

55 Hayden Ave.
Lexington, MA 02173
USA
Tel.: 617-861-6600; 800-232-6100
FAX 617-862-3869
Telex: 200076

Grace & Co. of Canada Ltd., W.R.
3455 Harvester Rd., Unit #7
Burlington, OntarioL7N 3P2
Canada
Tel.: 416-681-0285

W.R. Grace Ltd.
Northdale House, North Circular Rd.
London NW10 7UH
UK
Tel.: 081-965 0611
FAX 081-963 0928
Telex: 25139

Grace NV
Nijverheidsstraat 7
2260 Westerlo
Belgium
Tel.: 014 57 56 11
FAX 014 58 55 30
Telex: 31500

Graden Chemical Co., Inc.

426 Bryan St.
Havertown, PA 19083
USA
Tel.: 215-449-3808

Grindsted Products Inc.

201 Industrial Pkwy., PO Box 26
Industrial Airport, KS 66031
USA
Tel.: 913-764-8100; 800-255-6837
FAX 913-764-5407
Telex: 4-37295

Grindsted do Brazil
Ind. Ecom Ltda., Rodovia Regisé Bitten Court
KM 275, 5 Cx. Postal 172
06800 Embú S.P.
Brazil

Grinsted France S.A.R.L.
Parc D'Activités de Tissaloup
Ave. Jean D'Alembert
F-78190 Trappes
France

Grindstedvaerket GmbH
Roberts-Bosch Strabe
D-2085 Quickborn, Deutschland
Germany

Grindsted Products A/S
Edwin Rahrs Vej 38
DK-8220 Brabrand
Denmark
Tel.: 45-06-25-3366
FAX 45-06-25-1077
Telex: 64177

Grindsted Products Ltd.
Northern Way, Bury St. Edmunds
Suffolk IP32 6NP
UK
Tel.: 44284769631

Grünau, Chemische Fabrick GmbH

A Henkel Group Co.
Robert-Hansen Str. 1, Postfach 1063
W-7918 Illertissen, Bavaria
Germany
Tel.: (07303)13-0
FAX (07303)13206
Telex: 719114 gruea-d

Guelph Soap Co. Inc.

34 York St.
Elora, Ontario N0B 150
Canada
Tel.: 519-846-0934
FAX 519-846-9552

C.P. Hall Co.

7300 South Central Ave.
Chicago, IL 60638-0428
USA
Tel.: 708-594-6000; 800-321-8242
FAX 708-458-0428

Harcros Chemicals UK Ltd.

Specialty Chemicals Div.
Lankro House, PO Box 1, Eccles
Manchester M30 0BH
UK
Tel.: 44-61-789-7300
FAX 44-61-788-7886
Telex: 667725

Harcros Chemicals BV
Haagen House, PO Box 44
6040 AA Roermond
The Netherlands

Harcros Chemicals (Deutschland) GmbH
Wilhelm-Oswald-Strasse
W-5200 Siegburg
Germany

Tel.: 2241-54980
FAX 2241-549811
Telex: 8869605

Harcros Chemicals France S.a.r.l.
BP 40, 441220 St. Laurent
Nouan
France

Harcros Chemicals Inc.
5200 Speaker Rd., PO Box 2930
Kansas City, KS 66106-1095
USA
Tel.: 913-321-3131
FAX 913-621-7718
Telex: 477266

Harcros Chemicals Scandia ApS
Vesterbrogade 14A
1620 Copenhagen V
Denmark
Tel.: 31 21 42 00
FAX 31 21 42 27
Telex: 16 152 LANKRO DK

Lankro Chemicals
140 Chia Hsin Bldg.
96 Chung Shan N. Rd., Sec. 2
Taipei 10449
Taiwan, R.O.C.

A. Harrison & Co., Inc.
PO Box 494
Pawtucket, RI 02862
USA
Tel.: 401-725-7450; 800-544-2942
FAX 401-725-3570

Hart Chemicals Ltd.
256 Victoria Rd. South
Guelph, Ontario N1H 6K8
Canada
Tel.: 519-824-3280
FAX 519-824-0755
Telex: 06956537

Hart Products Corp.
173 Sussex St.
Jersey City, NJ 07302
USA
Tel.: 201-433-6632

Hefti Ltd. Chemical Products
PO Box 1623
CH-8048 Zurich
Switzerland
Tel.: 41-01-432-1340
FAX 41-01-432-2940
Telex: 822225

Henkel Corp./Organic Products Div.
300 Brookside Ave.
Ambler, PA 19022
USA
Tel.: 215-628-1000; 800-922-0605
FAX 215-628-1200

Henkel Argentina S.A.
Avda. E. Madero Piso 14
1106 Capital Federal
Argentina

Henkel Canada Ltd.
2290 Argentia Rd.
Mississauga, OntarioL5N 6H9
Canada
Tel.: 416-542-7550
FAX 416-542-7588

Henkel Chemicals Ltd.
Henkel House
292-308 Southbury Rd.
Enfield EN1 1TS
UK
Tel.: 081 804 3343
FAX 081 443 2777
Telex: 922708

Henkel Chemicals Ltd.
Organic Products Div.
Merit House, The Hyde
Edgeware Rd.
London NW9 5AB
UK

Henkel Corp./Cospha
300 Brookside Ave.
Ambler, PA 19002
USA
Tel.: 215-628-1476; 800-531-0815 (sales)
FAX 215-628-1450

Henkel Corp./Functional Products
300 Brookside Ave.
Ambler, PA 19002
USA
Tel.: 215-628-1466; 800-654-7588
FAX 215-628-1155

Henkel Corp./Textile Chemicals
11709 Fruehauf Dr.
Charlotte, NC 28273-6507
USA
Tel.: 800-634-2436
FAX 704-587-3804

Henkel KGaA/Cospha
Postfach 1100
D-4000, Dusseldorf 1
Germany
Tel.: 49-211-797-2289
FAX 49-211-798-7696
Telex: 085817-0

Henkel KGaA/Dehydag
Postfach 1100
D-4000 Düsseldorf 1
Germany
Tel.: 49-211 797-4221
FAX 49-211 798-8558
Telex: 085817-122

Henkel-Nopco SA/Process Chem. Div.
185 Ave. de Fontainebleau, 77310 St. Fargeau
Ponthierry
France
Tel.: 33-60-65-9090
FAX 33-60-65-7880
Telex: 692027

Henkel Corp./Emery Group
11501 Northlake Dr.
Cincinnati, OH 45249
USA
Tel.: 513-530-7300; 800-543-7370
FAX 513-530-7581

Henkel Corp./Emery Chemicals Ltd.
365 Evans Ave.
Toronto, OntarioM8Z 1K2
Canada
Tel.: 416-259-3751

Henkel Corp./Emery Group
1301 Jefferson St.
Hoboken, NJ 07030
USA
Tel.: 800-234-4365

Henkel Corp./Emery Group Japan
PO Box 191, World Trade Center
2-4-1 Hamamatsu-cho, Minato-Ku
Tokyo 105
Japan
Tel.: 4355611-2

Henkel Corp./Emery Group/OPG
3300 Westinghouse Blvd.
Charlotte, NC 28217
USA
Tel.: 800-634-2436

Hercules Inc.
Hercules Plaza-6205SW
Wilmington, DE 19894
USA
Tel.: 302-594-6500; 800-247-4372
FAX 302-594-5400
Telex: 835-479

Hercules BV
8 Veraartlaan, PO Box 5822
2280 HV Rijswijk
The Netherlands
Tel.: 31-070-150-000
Telex: 31172

Hercules Ltd./European Hdqts.
20 Red Lion St.
London WC1R 4PB
UK

Heterene Chemical Co., Inc.
PO Box 247
792 21st Ave.
Paterson, NJ 07543
USA
Tel.: 201-278-2000
FAX 201-278-7512
Telex: 883358

High Point Chemical Corp.
PO Box 2316, 243 Woodbine St.
High Point, NC 27261
USA
Tel.: 919-884-2214; 800-727-2214
FAX 919-884-5039

Hilton Davis Chemical Co.
2235 Langdon Farm Rd.
Cincinnati, OH 45237
USA
Tel.: 513-841-4000; 800-477-1022
FAX 800-477-4565

Represented by:
Censtead Ltd., D.F.,
Victoria House, Radford Way, Billericay
Essex CM12 0DE
UK

Hodag Corp. *See Calgene Chem. Inc.*

Hoechst Celanese/Int'l. Headqtrs.
26 Main St.
Chatham, NJ 07928
USA
Tel.: 201-635-2600; 800-235-2637
FAX 201-635-4330
Telex: 136346

Celanese France S.a.r.l.
Z.I. Le Broteau 69540
Irigny
France

Celanese GmbH
Hordenbachstrasse #40
D-5600 Wuppertal 21
Germany

Celanese Mexicana S.A.
Av. Revolucion No. 1425
Mexico 20, D.F.
Mexico

Hoechst AG
Postfach 80 03 20
D-6230 Frankfurt am Main 90
Germany
Tel.: 49-69-305-03113/7043
FAX 49-69-303665/66
Telex: 6990936

Hoechst Celanese/Colorants & Surfactants Div.
5200 77 Center Dr.
Charlotte, NC 28217
USA
Tel.: 704-527-6000; 800-255-6189
FAX 704-559-6323

Hoechst Celanese Far East Ltd.
801 Hong Kong Club Bldg.
3A Chater Rd., Central
Hong Kong

Hoechst Celanese Fine Chemicals Div.
Portsmouth Tech. Ctr.
3340 West Norfolk Rd.
Portsmouth, VA 23703
USA
Tel.: 804-483-7320
FAX 804-483-7460

Hoechst Japan
10-33, 4-Chome-Akasaka, Minato-Ku
Tokyo
Japan

Hoechst (UK) Ltd./Polymers Div.
Walton Manor, Wlaton, Milton Keynes
Bucks MK7 7A3
UK
Tel.: 0908 665050
FAX 0908 680516
Telex: 826300

E.F. Houghton & Co.
Valley Forge Tech. Ctr., PO Box 930
Valley Forge, PA 19482
USA
Tel.: 215-666-4000
FAX 215-666-7354
Telex: 510-6604518

Hüls AG
Postfach 1320
D-4370 Marl 1
Germany
Tel.: 49-02365-49-1
FAX 49-02365-49-2000
Telex: 829211-0

Hüls America Inc.
PO Box 456, 80 Centennial Ave.
Piscataway, NJ 08855-0456
USA
Tel.: 908-980-6800; 800-631-5275
FAX 908-980-6970

Hüls Canada, Inc.
235 Orenda Rd.
Brampton, OntarioL6T 1E6
Canada
Tel.: 416-451-3810

Hüls France SA
49-51 Quai de Dion Bouton
92815 Puteaux Cedex
France
Tel.: 1 49 06 55 00
FAX 1 47 73 97 65
Telex: 611868

Hüls (UK) Ltd.
Edinburgh House
43-51 Windsor Rd., Slough
Berks SL1 2HL
UK
Tel.: 0753-71851

Humphrey Chemical Co., Inc.,
A Cambrex Co.
Devine St.
North Haven, CT 06473-0325
USA
Tel.: 203-230-4945; 800-652-3456
FAX 203-287-9197
Telex: 994487

Huntington Laboratories, Inc.
970 East Tipton St.
Huntington, IN 46750
USA
Tel.: 219-356-8100; 800-537-5724
FAX 219-356-6485

ICI PLC/Chemicals & Polymers Group
PO Box 90, Wilton, Middlesborough
Cleveland TS6 8JE
UK
Tel.: 0642 454144
FAX 0642 432444
Telex: 587 461

ICI Americas, Inc.
New Murphy Rd. & Concord Pike
Wilmington, DE 19897
USA
Tel.: 302-886-3000; 800-456-3669
FAX 302-886-2972
Telex: 4945649

ICI Australia Operations Pty. Ltd.
ICI Surfactants Australia
ICI House, 1 Nicholson St.
Melbourne 300
Australia
Tel.: 61-03-665-7111
FAX 61-03-665-7009
Telex: 30192

ICI Japan Ltd.
Osaka Green Bldg.
1,3-Chome Kitahama, Higashi-Ku
Osaka 541
Japan

ICI Specialty Chemicals
Concord Pike & New Murphy Rd.
Wilmington, DE 19897
USA
Tel.: 302-886-3000; 800-822-8215
FAX 302-886-2972

ICI Surfactants (Belgium)
Everslaan 45
B-3078 Everberg
Belgium
Tel.: 32-02-758-9361
FAX 32-02-758-9686

ICI Surfactants Ltd. (UK)
Smith's Rd., Bolton
Lancashire BL3 2QJ
UK
Tel.: 44-204-21971/4
FAX 44-204-363676
Telex: 667841

Inolex Chemical Co.
Jackson & Swanson Sts.
Philadelphia, PA 19148-3497
USA
Tel.: 215-271-0800; 800-521-9891
FAX 215-271-2621
Telex: 834617

Represented by:
Black, Ltd., Stanley,
30/31 Islington Green
London
UK

Intex Chemical, Inc.
Div. of EZE Prods., Inc.
603 High Tech Court
Greenville, SC 29650
USA
Tel.: 803-877-5747; 800-845-1668
FAX 803-879-7196

ISP Technologies Inc.
PO Box 1006
Bound Brook, NJ 08805
USA
Tel.: 908-271-0111; 800-622-4423

Ivax Industries Inc./Textile Products Div.
PO Box 10027
Rock Hill, SC 29731
USA
Tel.: 803-366-9411; 800-343-7872
FAX 803-366-7256

Jetco Chemicals, Inc.
PO Box 1898
Corsicana, TX 75110
USA
Tel.: 903-872-3011; 800-477-5353
FAX 903-872-4216
Telex: 75110

Kao Corp. S.A.
Puig dels Tudons, 10
08210 Barbera Del Valles
Barcelona
Spain
Tel.: 34-3-729-0000
FAX 34-3-718-9829
Telex: 59749

Karlshamns Lipids for Care
S-374 82
Karlshamn
Sweden
Tel.: 46-454-823-00
FAX 46-454-129-11
Telex: 4500 fopart s

Karlshamns Lipids for Care
PO Box 569
Columbus, OH 43216
USA
Tel.: 614-299-3131; 800-526-4547
FAX 614-299-8279
Telex: 245494 capctyprdcol

Kelco/Div. of Merck & Co., Inc.
8355 Aero Dr.
San Diego, CA 92123
USA
Tel.: 619-292-4900; 800-535-2656
FAX 619-467-6520
Telex: WUD 695228

Kelco International Ltd.
Westminster Tower, 3 Albert Embankment
London SE1 7RZ
UK
Tel.: 01-735-0333
FAX 071-735 1363
Telex: 23815 KAILIL G

Kempen, Elektrochemische Fabrik Kempen GmbH
Postfach 100 260
D-4152 Kempen 1
W. Germany

Lankro. *See under Harcros*

Lever Industrial Co.
7450 Industry Dr.
North Charleston, SC 29418
USA
Tel.: 803-767-0540
FAX 803-552-6337

Lever Industriel
103 Rue DeParis
9300 Bobigny
France

LignoTech. *See under Borregaard*

Lion Corp.
2-22,1-Chome Yokoami, Sumida-ku
Tokyo 130
Japan
Tel.: 81-3-3621-6675
FAX 81-3-3621-6738
Telex: 262-2114

Lipo Chemicals, Inc.
207 19th Ave.
Paterson, NJ 07504
USA
Tel.: 201-345-8600
FAX 201-345-8365
Telex: 130117

Represented by:
Blagden Campbell & Campbell Ltd.
A.M.P. House, Dingwall Rd.
Croydon CR9 3QU
UK

Lonza Inc.
17-17 Route 208
Fair Lawn, NJ 07410
USA
Tel.: 201-794-2400; 800-777-1875 (tech.)
FAX 201-703-2028

Lubrizol Corp.
29400 Lakeland Blvd.
Wickliffe, OH 44092
USA
Tel.: 216-943-4200
FAX 216-943-5337
Telex: 4332033

Lubrizol France
Tour Europe
92400 Courbevoie
France

Lubrizol GmbH
Bogenallee 10
2000 Hamburg 13
Germany

Lubrizol Japan Ltd.
No. 23 Mori Bldg., 5th Floor
23-7 Toranomon, 1-Chome, Minato-ku
Tokyo 105
Japan

Lubrizol Ltd.
Waldron House, 57-63 Old Church St.
London SW3 5BS
UK
Tel.: 351-3311-20
FAX 351-3310

Lucas Meyer GmbH & Co.
PO Box 261665
D-2000 Hamburg 26
Germany
Tel.: 49-40-789-550
FAX 49-40-789-8329
Telex: 2163220 myer d

3M Co./Industrial Chem. Prods. Div.
3M Center Bldg. 223-6S-04
St. Paul, MN 55144-1000
USA
Tel.: 612-736-1394; 800-541-6752

3M Australia Pty. Ltd.
950 Pacific Hwy., PO Box 99
Pymble N.S.W. 2073
Australia

3M Belgium N.V./S.A.
Canadastraat 11
2730 Zwijndrecht
Belgium

3M Canada Inc.
PO Box 5757 Terminal A
1840 Oxford St. East
London, OntarioN6A 4T1
Canada

3M Deutschland GmbH
PO Box 100422
D-4040 Neuss 1
Germany

3M United Kingdom PLC
3M House, PO Box 1, Bracknell
Berkshire RG12 1JU
UK

Sumitomo/3M Ltd.
Central PO Box 490
33-1 Tamagawadai 2-Chome, Setagaya-ku
Tokyo 158
Japan

Manchem Inc.
77 Maple Dr.
Hudson, OH 44238
USA

Manchem Ltd.
Aston New Rd.
Manchester
M11 4AT
UK

Manro Products Ltd.
Bridge St., Stalybridge
Cheshire SK15 1PH
UK
Tel.: 44-61-338-5511
FAX 44-61-303-2991
Telex: 668442

Marchon. *See under Albright & Wilson*

Marlowe-Van Loan Corp.
PO Box 1851
High Point, NC 27261
USA
Tel.: 919-886-7126; 800-422-4MVL
FAX 919-889-6663
Telex: TWX: 510-926-1589

Marubishi Oil Chemical Co., Ltd.
3-7-12 Tomobuchi-cho, Miyakojima-ku
Osaka 534
Japan
Tel.: 81-6-928-0331
FAX 81-6-928-0310

Mason Chemical Co.
5253 West Belmont Ave.
Chicago, IL 60641
USA
Tel.: 312-282-0200; 800-362-1855
FAX 312-282-0821

Matsumoto Yushi-Seiyaku Co., Ltd.
1-3, 2-Chome, Shibukawa-cho, Yao City
Osaka
Japan

Maybrook Inc.
570 Broadway
PO Box 68
Lawrence, MA 01842
USA
Tel.: 508-682-1853
FAX 508-682-2544

Mayco Oil & Chemical Co.
775 Louis Dr., PO Box 2809
Warminster, PA 18974-0357
USA
Tel.: 215-672-6600; 800-523-3903
FAX 215-443-7094

Mazer. *See under PPG Industries*

McIntyre Chemical Co., Ltd.
1000 Governors Hwy.
University Park, IL 60466
USA
Tel.: 708-534-6200
FAX 708-534-6216

M. Michel & Co., Inc
90 Broad St.
New York, NY 10004
USA
Tel.: 212-344-3878
FAX 212-344-3880
Telex: 421468

Miles Inc./Organic Products Div.
Bldg. 14, Mobay Rd.
Pittsburgh, PA 15205-9741
USA
Tel.: 412-777-2000; 800-662-2927
FAX 412-777-7840
Telex: 1561261

Miles Inc./Polymers Div.
Mobay Rd.
Pittsburgh, PA 15205-9741
USA
Tel.: 412-777-2000; 800-526-4550

Milliken Chemicals
PO Box 1927
Spartanburg, SC 29304
USA
Tel.: 803-573-2200
FAX 803-573-2430
Telex: 810-282-2580

Milliken Chemicals
PO Box 817
Inman, SC 29349
USA
Tel.: 803-472-7208
FAX 803-472-4129

Mitsubishi Kasei Corp.
5-2, Marunouchi 2-Chome, Chiyoda-ku
Tokyo 100
Japan

Mitsubishi Kasei America, Inc.
81 Main St., Suite 401
White Plains, NY 10601
USA
Tel.: 914-761-9450
FAX 914-681-0760

Mitsubishi Kasei, Europe Office
Am Seestern, Niederkasseler LohWeg 8
4000 Duesseldorf 11
Germany

Mona Industries Inc.
PO Box 425, 76 E. 24th St.
Paterson, NJ 07544
USA
Tel.: 201-345-8220; 800-553-6662
FAX 201-345-3527
Telex: 130308

Monsanto Chemical Co.
800 N. Lindbergh Blvd.
St. Louis, MO 63167
USA
Tel.: 314-694-1000; 800-325-4330
FAX 314-694-7625
Telex: 44-7282

Monsanto Canada Inc.
2330 Argentia Rd.
Box 787, Streetsville Postal Station
Mississauga, OntarioL5M 2G4
Canada
Tel.: 416-826-9222

Monsanto/Detergents & Phosphates Div.;
800 N. Lindbergh Blvd.
St. Louis, MO 63167
USA
Tel.: 314-694-1000; 800-325-4330
FAX 314-694-7625
Telex: 650 397 7820

Monsanto Europe SA
Ave. de Tervuren 270-272
B-1150 Brussels
Belgium
Tel.: 2-761-41-11
FAX 2-761-40-40
Telex: 62927 Mesab

Monsanto Japan Ltd.
Room 520, Kokusai Bldg.
1-Marunouchi 3-Chome, Chiyoda-Ku
Tokyo 100
Japan

Monsanto PLC
Monsanto House, Chineham Court
Chineham, Basingstoke
Hants RG24 0UL
UK
Tel.: 0256-572-88
FAX 0256-54995
Telex: 858837

Münzing Chemie GmbH
Salzstrasse 174
D-7100 Heilbronn
Germany
Tel.: 07131/1586-0
FAX 07131/1586-25
Telex: 728614

Murphy-Phoenix Co.
PO Box 22930
Beachwood, OH 44122
USA

Nalco Chemical Co.
One Nalco Center
Naperville, IL 60563-1198
USA
Tel.: 708-305-1000; 800-527-7753

Nalco/Process Chems. Div.
6216 W. 66th Pl.
Chicago, IL 60638
USA
Tel.: 708-496-5041; 800-435-0861
FAX 708-496-5290

National Starch & Chemical Corp.
Box 6500, 10 Finderne Ave.
Bridgewater, NJ 08807
USA
Tel.: 908-685-5000; 800-726-0450
FAX 908-685-5005

Represented by:
National Adhesives & Resins Ltd.
Braunston Daventry
Northante NN11 7JL
UK

Niacet Corp.
PO Box 258
400 47th St.
Niagara Falls, NY 14304
USA
Tel.: 716-285-1474; 800-828-1207
FAX 716-285-1497
Telex: 6730170

Nikko Chemical Co., Ltd.
1-4-8 Nihonbashi-Bakurocho, Chuoku
Tokyo 103
Japan
Tel.: 81-3-662-0371
FAX 81-3-664-8620
Telex: 2522744 NIKKOL J

Nippon Nyukazai Co., Ltd.
9-19 Ginza 3-Chome, Chuo-ku
Tokyo 104
Japan
Tel.: 81-3-543-8571

Nippon Oils & Fats Co., Ltd.
10-1, Yaraku-Cho, 1-Chome, Chiyoda-Ku
Tokyo 100
Japan
Tel.: 81-3-283-7140
FAX 81-3-283-7134
Telex: 222-2041

Nippon Senka Chemical Industries, Ltd.
17-34 Hanaten-Higashi, 1-Chome, Tsurami-ku
Osaka
Japan

Norman, Fox & Co.
5511 S. Boyle Ave., PO Box 58727
Vernon, CA 90058
USA
Tel.: 213-583-0016; 800-632-1777
FAX 213-583-9769

Oakite Products, Inc.
50 Valley Rd.
Berkeley Hts., NJ 07922
USA
Tel.: 908-464-6900; 800-526-4473
FAX 908-464-6031

Oleofina. *See under Fina*

Olin Chemicals
120 Long Ridge Rd., PO Box 1355
Stamford, CT 06904
USA
Tel.: 203-356-2000; 800-243-9171

Olin Australia Ltd.
1-3 Atchison St., PO Box 141
St. Leonards 2065, N.S.W.
Australia

Olin Brasil Limitada
Rua Galeno de Castro, 165
Jurubatuba, Santo Amaro
04696 Sao Paulo, SP
Brazil

Olin Europe S.A.
108-110 Blvd. Haussmann
75008 Paris
France
Tel.: 33-1-293-3210

Olin Japan Inc.
Shiozaki Bldg.
7-1 Hirakawa-Cho 2-Chome, Chiyoda-ku
Tokyo 102
Japan
Tel.: 81-3-263-4615

Olin UK Ltd.
Suite 7, Kidderminster Rd., Cutnall Green
Worchestershire WR9 0NS
UK

OmniChem NV
Industrial Research Park
B-1348, Louvain-la-Neuve
Belgium
Tel.: 32-10-450031
FAX 32-10-450693

M.S. Paisner Inc.
53 Beaumont St.
PO Box 358
Canton, MA 02021
USA
Tel.: 617-828-2040
FAX 617-828-2202

Penreco, Div. of Pennzoil Prods. Co.
RD 2, Box 1
Karns City, PA 16041
USA
Tel.: 412-756-0110; 800-245-3952
FAX 412-756-1050
Telex: 866-321

Pilot Chemical Co.
11756 Burke St.
Santa Fe Springs, CA 90670
USA

Tel.: 310-723-0036
FAX 310-945-1877
Telex: 4991200 PILOT

Polymer Research Corp. of America
2186 Mill Ave.
Brooklyn, NY 11234
USA
Tel.: 718-444-4300
FAX 718-241-3930

PPG Industries/Specialty Chemicals
3938 Porett Dr.
Gurnee, IL 60031
USA
Tel.: 708-244-3410; 800-323-0856
FAX 708-244-9633
Telex: 25-3310

Mazer Chemicals (Canada)
Mississauga, OntarioL4Z 1H8
Canada

Mazer Chemicals UK Ltd.
Carrington Business Park
Carrington, Urmston
Manchester M31 4DD
UK

Mazer de Mexico S.A. de C.V.
Federico T. de La Chica No. 16-103
CD Satelite, C.P. 53100
Edo
Mexico
Tel.: 525-562-1015
FAX 525-393-5909

PPG Canada Inc./Spec. Chem.
2 Robert Speck Pkwy., Suite 750
Mississauga, OntarioL4Z 1H8
Canada
Tel.: 416-848-2500
FAX 416-848-2501
Telex: 06960351 canbiz miss

PPG Industrial do Brazil Ltda.
Edificio Grande Avenida, Paulista Ave. 1754, Suite 153
Sao Paulo
Brazil 01310
Tel.: 55-11-2840433
FAX 55-11-2892105
Telex: 391-1139104ppgbrazil

PPG Industries/Asia/Pacific Ltd.
Takanawa Court, 5th floor
12-1 Takanawa 3-Chome, Minato-Ku
Tokyo 108
Japan
Tel.: 81-03-3280-2911
FAX 82-02-3280-2920
Telex: 02-42719 PPGPACJ

PPG Industries Int'l. Inc./Taiwan
Suite 601, Worldwide House
No. 131, Min Sheng East Rd., Sec. 3
Taipei 105
Taiwan R.O.C.
Tel.: 886-2-514-8052
FAX 886-2-514-7957

PPG Industries (UK) Ltd./Specialty Chem.
Carrington Business Park, Carrington, Urmston
Manchester M31 4DD
UK
Tel.: 44-61-777-9203
FAX 44-61-777-9064
Telex: 851-94014896 mazu g

Procter & Gamble Co./Chem. Div.
120 W. Fifth St., Suite 502
Cincinnati, OH 45202
USA
Tel.: 513-562-2655; 800-543-1580
FAX 513-579-9582

Procter & Gamble Inc. Canada
4711 Yonge St.
PO Box 355, Station A
Toronto, OntarioM5W 1C5
Canada
Tel.: 416-730-4059
FAX 416-730-4122

Protex
B.P. 177, 6, rue Barbès
92305 Levallois-Paris
France
Tel.: 47-57-74-00
FAX 47-57-69-28
Telex: 620987

Pulcra SA
Sector E C/42
Barcelona
08040
Spain
Tel.: 34-3-323-5914
FAX 34-3-323-6760
Telex: 98301

Quimigal-Quimica de Portugal E.P.
Av. Infante Santo No. 2
1300 Lisboa
Portugal
Tel.: 351-1-604040
Telex: 12301

Raschig AG
Mundenheimer Strasse 100
D-6700 Ludwigshafen/Rhine
Germany
Tel.: 0621-56180
FAX 0621-532885
Telex: 464 877 ralu d

Raschig Corp.
PO Box 7656
Richmond, VA 23231
USA
Tel.: 804-222-9516
FAX 804-226-1569

Rayonier Inc.
18000 Pacific Hwy. South, Suite 900
Seattle, WA 98188
USA
Tel.: 206-246-3400
FAX 206-248-4162
Telex: 4949619

Reilly-Whiteman Inc.
801 Washington St.
Conshohocken, PA 19428
USA
Tel.: 215-828-3800; 800-533-4514
FAX 215-834-7855
Telex: 5106608845

Rewo Chemische Werke GmbH
Postfach 1160
Industriegebiet West
D-6497 Steinau
Germany
Tel.: 49-06663-540
FAX 49-06663-54-129
Telex: 493589

Rewo Chemicals Ltd.
Gorsey lane, Widnes
Cheshire
WA8 0HE
UK
Tel.: 051-495-1989
FAX 051-495-2003
Telex: 627434 SIPWID G

Rhone-Poulenc S.A.
Dept. Biochimie, 18, ave. d'Alsace-F
92400 Courbevoie, Paris
France

American Lecithin Co., Inc.
PO Box 1908
33 Turner Rd.
Danbury, CT 06813-1908
USA
Tel.: 203-790-2700
FAX 203-790-2705

Rhone-Poulenc Basic Chemical Co.
One Corporate Dr., Box 881
Shelton, CT 06484
USA
Tel.: 203-925-3300; 800-642-4200
FAX 203-925-3627

Rhone-Poulenc Chem. Ltd., Perf. Prods. Group
Woodley, Stockport
Cheshire SK6 1PQ
UK
Tel.: 44-61-430-4391
FAX 44-61-430-4364
Telex: 667835

Rhone-Poulenc Chimie (France)
Cedex 29
F-92097 Paris La Defense
France
Tel.: 33-47-68-1234
FAX 33-47-68-0900

Rhone-Poulenc Geronazzo SpA
Via Milano 78
20021 Ospiate Di Bollate
Milano
Italy
Tel.: 39-2-350-3212
FAX 39-2-350-1770
Telex: 331547 GERO I

Rhone-Poulenc, Inc./Chemicals Div.
Box 125
Monmouth Junction, NJ 08852
USA
Tel.: 201-297-0100
FAX 201-297-1597

Rhone-Poulenc, Inc./Surfactants & Specialties
CN 7500, Prospect Plains Rd.
Cranberry, NJ 08512-7500
USA
Tel.: 609-860-8300; 800-922-2189
FAX 609-860-7626

Rhone-Poulenc, Inc./Textile & Rubber Div.
PO Box 1740
Dalton, GA 30720
USA
Tel.: 404-259-4831
FAX 404-259-5979

Rhone-Poulenc Surfactants & Specialties Canada
2000 Argentia Rd.
Plaza 3, Suite 400
Mississauga, OntarioLSN 1V9
Canada
Tel.: 416-821-4450
FAX 416-821-9339

RITA Corp.
1725 Kilkenny Court
PO Box 585
Woodstock, IL 60098
USA
Tel.: 815-337-2500; 800-426-7759
FAX 815-337-2522
Telex: 72-2438

Represented by:
Maprecos
4, Rue des Passe-Loups
7770 Fontaine
Le Port
France

Rohm & Haas Co.
Independence Mall West
Philadelphia, PA 19105
USA
Tel.: 215-592-3000; 800-323-4165
FAX 215-592-2285
Telex: 845-247

Röhm GmbH Chemische Fabrik
Kirschenallee, Postfach 4242
D-6100 Darmstadt
Germany
Tel.: 6151 18 01
FAX 6151 184007

Rohm & Haas Asia Ltd.
Kaisei Bldg.
8-10 Azabudai 1-Chome, Minato-ku
Tokyo 106
Japan

Rohm & Haas (Australia) Pty. Ltd.
969 Burke Rd., PO Box 11
Camberwell, Victoria 3124
Australia

Rohm & Haas Canada Inc.
2 Manse Rd.
West Hill, OntarioM1E 3T9
Canada

Rohm & Haas Co. European Operations
Chesterfield House, Bloomsbury Way 15-19
London WC1A 2TP
UK
Tel.: 071 242 4455
FAX 071 404 4126
Telex: 24139

Rohm & Haas (UK) Ltd.
Bloomsbury Way, Chesterfield House
London
UK
Tel.: 71-242-4455
FAX 71-404-4126

Ronsheim & Moore Ltd.
Ings Lane, Castleford
Yorkshire WF10 2JT
UK
Tel.: 44-977-556565
FAX 44-977-518058
Telex: 55378

Ross Chemical, Inc.
303 Dale Dr.
Fountain Inn, SC 29644
USA
Tel.: 803-862-4474; 800-521-8246
FAX 803-862-2912

Ruetgers-Nease Chemical Co., Inc.
201 Struble Rd.
State College, PA 16801
USA
Tel.: 814-238-2424
FAX 814-238-1567

Sandoz Chemicals Corp.
4000 Monroe Rd.
Charlotte, NC 28205
USA
Tel.: 704-331-7000; 800-631-8077
FAX 704-372-5787
Telex: 704-216-922

Sandoz Chemicals Corp. Canada
Dorva, Quebec H9R 4PR
Canada

Sandoz Ltd.
Calverley Lane, Horsforth
Leeds LS18 4RP
UK
Tel.: 0532 584646
FAX 0532 390063
Telex: 557114

Sandoz Prods. Ltd./Chemicals Div.
Lichtstrasse 35
CH-4002 Basel
Switzerland
Tel.: 41-61-324-1111
FAX 41-61-324-6080
Telex: 96505049

Sanyo Chemical Industries, Ltd.
11-1 Ikkyo Nomoto-cho Higashiyama-ku
Kyoto
605
Japan
Tel.: 81-75-541-4311
FAX 81-75-551-2557
Telex: 05422110

Scher Chemicals, Inc.
Industrial West & Styertowne Rd.
PO Box 4317
Clifton, NJ 07012
USA
Tel.: 201-471-1300
FAX 201-471-3783
Telex: 642643 Scherclif

Represented by:
Chesham Chemicals Ltd.
Cunningham House
Bessborough Rd.
Harrow HA1 3DU
UK

Seppic
75 Quai d'Orsay
75321 Paris Cedex 07
France
Tel.: 33-40-62-59-01
Telex: 290665

Seppic, Inc./Subsid. of Seppic France
30 Two Bridges Rd., Suite 370
Fairfield, NJ 07004
USA
Tel.: 201-882-5597
FAX 201-882-5178

Servo Delden B.V.
Postbus 1
7490 AA Delden
The Netherlands
Tel.: 31-5407-63535
FAX 31-5407-64125
Telex: 44347

Shell Chemical Co.
PO Box 2463, 1 Shell Plaza
Houston, TX 77002
USA
Tel.: 713-241-0981
FAX 713-241-6916
Telex: 762248

For Int'l. Sales:
Pecten Chemicals, Inc.,
One Shell Plaza
Houston, TX 77252-9932
USA
Tel.: 713-241-6161

Shell Chemicals UK Ltd.
1 Northumberland Ave.
London WC2N 5LA
UK
Tel.: 71-934-1234

Shell Chimie France
27 Rue de Berri
7539 Paris Cedex 08
France

Shell Nederland Chemie B.V.
PO Box 187
2501 CD The Hague
The Netherlands

Sherex. *See under Witco*

Sherex Polymers, Inc.
2525 S. Combee Rd.
Lakeland, FL 33801
USA
Tel.: 813-665-6226

Sino-Japan Chemical Co.
3 fl. 237 Sec. 1, Chien Kuo South Rd.
Taipei, Hsien
Taiwan
Tel.: 886-2-700-1422
FAX 886-2-707-3921

Werner G. Smith, Inc.,
1730 Train Ave.
Cleveland, OH 44113
USA
Tel.: 216-861-3676; 800-535-8343
FAX 216-861-3680

Societa Italiana Emulsionanti
Via R. Cozzi 34
20125 Milano
Italy
Tel.: 39-2-642-4041
FAX 39-2-643-0820
Telex: 315006

Soitem-Societa Italiana Emulsionanti Srl. *See Societa Italiana Emulsionanti*

Soluol Chemical Co.
Green Hill & Market Sts., Box 112
W. Warwick, RI 02893
USA
Tel.: 401-821-8100
FAX 401-823-6673

Solvay & Cie SA
Rue du Prince Albert 33
1050 Brussels
Belgium
Tel.: 2/509-6111
FAX 2/509-6617
Telex: 21337

Solvay Chemicals Ltd.
Unit 1, Grovelands Business Centre
Boundary Way, Hemel Hempstead
Herts HP2 7TE
UK
Tel.: 0442-236555

Solvay Duphar BV
Postbus 900
1380 DA Weesp
The Netherlands
Tel.: 31-2940-77711
FAX 31-2940-80253
Telex: 14232

Standard Chemical Co.
Mill Lane, Cheadle
Cheshire SK8 2NX
UK
Tel.: 44-61-428-5225
FAX 44-61-428-0890
Telex: 666413

Stepan Co.
22 West Frontage Rd.
Northfield, IL 60093
USA
Tel.: 708-446-7500; 800-228-8312
FAX 708-501-2443

Stepan Canada
90 Matheson Blvd. W., Suite 201
Mississauga, OntarioL5R 3P3
Canada
Tel.: 416-507-1631
FAX 416-507-1633

Stepan Co./PVO Dept.
100 West Hunter Ave.
Maywood, NJ 07607
USA
Tel.: 201-845-3030
FAX 201-845-6754
Telex: 710-990-5170

Stepan Europe SA
BP127
38340 Voreppe
France
Tel.: 33-7650-8133
FAX 33-7656-7165
Telex: 320511 F

Stockhausen, Inc.
2408 Doyle St.
Greensboro, NC 27406
USA
Tel.: 919-333-3500
FAX 919-333-3545
Telex: 574405

Swastik Household & Industrial Products Ltd.
Shahibag House, 13 Walchand Hirachand Marg
Ballard Estate
Bombay 400 038
India

Sybron Chemicals Inc.
PO Box 125
Wellford, SC 29385
USA
Tel.: 803-439-6333; 800-677-3500
FAX 803-439-1612

Sybron Chemicals Canada Ltd.
120 Norfinch Dr., Unit 1
Downsview, OntarioM3N 1X2
Canada
Tel.: 416-663-7166

Taiwan Surfactant Corp.
No. 106, 8-1 Floor, Sec. 2
Chung An E. Rd.
Taipei
Taiwan, R.O.C.
Tel.: 886-2-507-9155
FAX 886-2-507-7011
Telex: 27568 surfact

Takemoto Oil & Fat Co., Ltd.
No. 5, Sec. 2, Minato-Machi
Gamagori
Aichi 443
Japan
Tel.: 81-533-68-2117
FAX 81-533-67-3496
Telex: 4324604

Tego. *See under Goldschmidt*

Temfibre Inc.
C.P. 3000
Temiscaming, Quebec J0Z 3R0
Canada
Tel.: 819-627-9505
FAX 819-627-3622
Telex: 067-76281

La Tessilchimica SpA. *See Auschem SpA*

Texaco Chemical Co.
PO Box 15730
Austin, TX 78761
USA
Tel.: 512-483-0053; 800-231-3107
FAX 512-483-0925
Telex: 776-408

S.A. Texaco Belgium N.V.
Int'l. Congress Center, Citadel Park
B-900 Ghent
Belgium
Tel.: 011-32-91-41-5920

Texaco Chemical Deutschland GmbH
Baumwall 5
2000 Hamburg 11
Germany
Tel.: 011-49-40-36-3737

Texaco France S.A.
5, rue Bellini, Tour Arago
F-92806 Puteaux Cedex
France
Tel.: 011-33-1-47-78-1655

Texaco Ltd.
195 Knightsbridge
London SW7 1RU
UK
Tel.: 011-411-584-5000

Texo Corp.
2801 Highland Ave.
Cincinnati, OH 45212
USA
Tel.: 513-731-3400

Textile Rubber & Chem. Co. *See Tiarco*

Tiarco Chemical Div./Textile Rubber & Chemical Co.
1300 Tiarco Dr.
Dalton, GA 30720
USA
Tel.: 706-277-1300
FAX 706-277-3738

Titan Chemical Products, Inc.
PO Box 20
Short Hills, NJ 07078
USA

Toho Chemical Industry Co., Ltd.
No. 2-5, 1-chome, Ningyo-cho
Nihonbashi, Chuo-ku
Tokyo 103
Japan
Tel.: 81-3-3668-2271
FAX 81-3-3668-2278
Telex: 252-2332 TOHO K J

Tokai Seiyu Ind. Co. Ltd.
67, 2-Chome, Yamadahigashimachi
Higashi-ku
Nagoya
Japan
Tel.: 81-52-721-2611
FAX 81-52-721-8775

Trans-Chemco, Inc.
19235 84th St.
Bristol, WI 53104
USA
Tel.: 414-857-2363

Witco Corp.
520 Madison Ave.
New York, NY 10022
USA
Tel.: 212-605-3680
FAX 212-486-4198

Sherex Chemical Co., Inc./Div. of Witco
PO Box 646
5777 Frantz Rd.
Dublin, OH 43017
USA
Tel.: 614-764-6500; 800-848-7370
FAX 614-764-6650
Telex: 245356

Witco B.V.
PO Box 5
NL-1540 LZ Koog aan de Zaan
The Netherlands
Tel.: 75-283854
FAX 75-210811
Telex: 19270

Witco Canada Ltd.
2 Lansing Sq., Suite 1200
Willowdale, OntarioM2J 4Z4
Canada
Tel.: 416-497-9991

Witco Chemical Ltd. (UK)
Union Lane, Droitwich
Worcester WR9 9BB
UK

Witco Corp./Humko Chem. Div.
PO Box 125
Memphis, TN 38101-0125
USA
Tel.: 901-684-7000; 800-238-9150
FAX 901-682-6531
Telex: 53-928

Witco Corp./Organics Div.
1000 Convery Blvd.
Perth Amboy, NJ 08862-1932
USA
Tel.: 201-826-7777; 800-231-1542

Witco Corp./Sonneborn Div.
520 Madison Ave.
New York, NY 10022-4236
USA
Tel.: 212-605-3981
FAX 212-754-5676
Telex: 62470

Witco Ltd.
PO Box 10245
26112 Haifa Bay
Israel
Tel.: 972-4-469-111
FAX 972-4-469-137
Telex: 45198

Witco SA
10 Rue Cambaceres
75008 Paris
France
Tel.: 42-65-99-03
FAX 42-65-67-61
Telex: 290233

Yorkshire Pat-Chem Inc.
11 Worley Rd., PO Box 1926
Greenville, SC 29602
USA
Tel.: 803-233-3941; 800-443-9358
FAX 803-232-3542

Yoshimura Oil Chemical Co., Ltd.
Minami 5-Chome-1-1, Honan-cho
Toyonaki-shi
Osaki
561
Japan
Tel.: 81-6-334-3331-7
FAX 81-6-331-4078

Zohar Detergent Factory
PO Box 11 300
Tel-Aviv
61 112
Israel
Tel.: 03-528-7236
FAX 03-5287239
Telex: 33557 zohar il

Zschimmer & Schwarz GmbH & Co.
4-5 Max-Schwarz-Str., Postfach 2179
D-5420 Lahnstein
Germany
Tel.: 49-02621 12 0
FAX 2621-12407
Telex: 86 9816 750

Zschimmer & Schwarz Argentina S.A.
Bdo. de Irigoyen 556-5 B
Buenos Aires
Argentina

Zschimmer & Schwarz France S.a.r.l.
10 rue Saint-Marc
F-75002 Paris
France
Tel.: 42 33 10 33
FAX 40 26 23 81
Telex: 670465 ZS F

Zschimmer & Schwarz Italiana SpA
Casella Postale N.1
I-13038 Tricerro (Vc)
Italy

Appendices

CAS Number-to-Tradename Cross Reference

CAS numbers reference specific chemicals or chemical groups. Tradename products are associated with CAS numbers through their major chemical constituent.

CAS 56-81-5	Superol
CAS 57-10-3	Emersol® 143
CAS 57-10-3	Hystrene® 8016
CAS 57-10-3	Hystrene® 9016
CAS 57-10-3	Industrene® 4516
CAS 57-10-3	Prifrac 2960
CAS 57-11-4	Edenor ST-1
CAS 57-11-4	Emersol® 110
CAS 57-11-4	Emersol® 120
CAS 57-11-4	Emersol® 132 NF Lily®
CAS 57-11-4	Emersol® 150
CAS 57-11-4	Emersol® 152 NF, 153 NF
CAS 57-11-4	Emersol® 6320
CAS 57-11-4	Emersol® 6332 NF
CAS 57-11-4	Emersol® 6349
CAS 57-11-4	Emersol® 6351
CAS 57-11-4	Emery® 400
CAS 57-11-4	Emery® 404
CAS 57-11-4	Emery® 405
CAS 57-11-4	Emery® 410
CAS 57-11-4	Emery® 420
CAS 57-11-4	Emery® 422
CAS 57-11-4	Hystrene® 4516
CAS 57-11-4	Hystrene® 5016
CAS 57-11-4	Hystrene® 5016 NF
CAS 57-11-4	Hystrene® 7018
CAS 57-11-4	Hystrene® 7018 FG
CAS 57-11-4	Hystrene® 8018
CAS 57-11-4	Hystrene® 8718 FG
CAS 57-11-4	Hystrene® 9718
CAS 57-11-4	Hystrene® 9718 NF FG
CAS 57-11-4	Industrene® 4518
CAS 57-11-4	Industrene® 5016
CAS 57-11-4	Industrene® 5016 FG
CAS 57-11-4	Industrene® 7018
CAS 57-11-4	Industrene® 9018
CAS 57-11-4	Industrene® B
CAS 57-11-4	Industrene® R
CAS 57-11-4	Prifrac 2980
CAS 57-11-4	Prifrac 2981
CAS 57-11-4	Pristerene 4904
CAS 57-11-4	Pristerene 4905
CAS 57-11-4	Pristerene 4910
CAS 57-11-4	Pristerene 4911
CAS 57-11-4	Pristerene 4915
CAS 57-11-4	Pristerene 4921
CAS 78-66-0	Surfynol® 82
CAS 93-83-4	Alrosol O
CAS 93-83-4	Chimipal OLD
CAS 93-83-4	Comperlan OD
CAS 93-83-4	Crillon ODE
CAS 93-83-4	Emid® 6545
CAS 93-83-4	Marlamid® D 1885
CAS 93-83-4	Schercomid SO-A
CAS 95-19-2	Crodazoline S
CAS 95-38-5	Crodazoline O
CAS 95-38-5	Marlowet® 5440
CAS 95-38-5	Schercozoline O
CAS 97-78-9	Crodasinic L
CAS 102-87-4	Armeen® 3-12
CAS 104-74-5	Dehyquart C
CAS 104-74-5	Dehyquart C Crystals
CAS 104-75-6	Armeen® L8D
CAS 106-11-6	Hydrine
CAS 106-11-6	Lipal DGMS
CAS 106-11-6	Sterol ST 2
CAS 106-17-2	Cithrol EGMR N/E
CAS 106-98-9	Gulftene 4
CAS 107-54-0	Surfynol® 61
CAS 107-64-2	Arquad® 218-75
CAS 107-64-2	Arquad® 218-100
CAS 109-28-4	Schercodine O
CAS 110-25-8	Crodasinic O
CAS 111-03-5	Ablunol GMO
CAS 111-03-5	Aldo® MO
CAS 111-03-5	Aldo® MO Tech
CAS 111-03-5	Cithrol GMO N/E
CAS 111-03-5	Kemester® 2000
CAS 111-03-5	Kessco® GMO
CAS 111-03-5	Mazol® 300 K
CAS 111-03-5	Mazol® GMO
CAS 111-03-5	Mazol® GMO #1
CAS 111-03-5	Mazol® GMO K
CAS 111-03-5	Pationic® 1064
CAS 111-03-5	Pationic® 1074
CAS 111-03-5	Radiasurf® 7150
CAS 111-03-5	Radiasurf® 7152
CAS 111-03-5	Tegin® O NSE
CAS 111-03-5	Tegin® O Special
CAS 111-03-5	Tegin® O Special NSE
CAS 111-03-5	Witco® 942
CAS 111-03-5	Witconol 2421
CAS 111-05-7	Schercomid OMI
CAS 111-27-3	Epal® 6
CAS 111-60-4	Cithrol EGMS N/E
CAS 111-60-4	Monthyle
CAS 111-60-4	Sterol ST 1
CAS 111-60-4	Tegin® G 6100
CAS 111-66-0	Gultfene 8
CAS 111-66-0	Neodene® 8

CAS 111-87-5	Epal® 8
CAS 111-90-0	Transcutol
CAS 112-00-5	Arquad® 12-37W
CAS 112-00-5	Arquad® 12-50
CAS 112-00-5	Dehyquart LT
CAS 112-02-7	Arquad® 16-29W
CAS 112-02-7	Arquad® 16-50
CAS 112-02-7	Dehyquart A
CAS 112-02-7	Querton 16Cl-29
CAS 112-02-7	Querton 16Cl-50
CAS 112-02-7	Radiaquat 6444
CAS 112-03-8	Arquad® 18-50
CAS 112-18-5	Adma® 12
CAS 112-18-5	Armeen® DM12D
CAS 112-18-5	Farmin DM20
CAS 112-30-1	Epal® 10
CAS 112-41-4	Gulftene 12
CAS 112-41-4	Neodene® 12
CAS 112-42-5	Neodol® 1
CAS 112-53-8	Alfol® 12
CAS 112-53-8	Cachalot® L-50
CAS 112-53-8	Cachalot® L-90
CAS 112-53-8	Emery® 3326
CAS 112-53-8	Emery® 3327
CAS 112-53-8	Emery® 3328
CAS 112-53-8	Emery® 3331
CAS 112-53-8	Emery® 3332
CAS 112-53-8	Emery® 3335 Chemically Pure
CAS 112-53-8	Emery® 3346
CAS 112-53-8	Emery® 3351
CAS 112-53-8	Emery® 3352
CAS 112-53-8	Emery® 3353
CAS 112-53-8	Emery® 3357
CAS 112-53-8	Epal® 12
CAS 112-53-8	Kalcohl 20
CAS 112-53-8	Laurex® L1
CAS 112-53-8	Laurex® NC
CAS 112-53-8	Lorol C8-C10
CAS 112-53-8	Lorol C8-C10 Special
CAS 112-53-8	Lorol C8-C18
CAS 112-53-8	Lorol C10-C12
CAS 112-53-8	Lorol C12
CAS 112-53-8	Lorol C12-99
CAS 112-53-8	Lorol C12 Chemically Pure
CAS 112-53-8	Lorol C12-C14
CAS 112-53-8	Lorol C12-C16
CAS 112-53-8	Lorol C12-C18
CAS 112-53-8	Lorol Lauryl Alcohol Tech
CAS 112-53-8	Lorol Special
CAS 112-53-8	Philcohol 1200
CAS 112-69-6	Adma® 16
CAS 112-69-6	Armeen® DM16D
CAS 112-69-6	Crodamine 3.A16D
CAS 112-70-9	Neodol® 3
CAS 112-72-1	Alfol® 14
CAS 112-72-1	Emery® 3334
CAS 112-72-1	Emery® 3345
CAS 112-72-1	Emery® 3347
CAS 112-72-1	Epal® 14
CAS 112-72-1	Kalcohl 40
CAS 112-72-1	Lorol C14
CAS 112-72-1	Lorol C14-98
CAS 112-72-1	Lorol C14 Chemically Pure
CAS 112-72-1	Philcohol 1400
CAS 112-75-4	Adma® 14
CAS 112-75-4	Armeen® DM14D
CAS 112-75-4	Empigen® AH
CAS 112-80-1	Emersol® 210
CAS 112-80-1	Emersol® 213 NF
CAS 112-80-1	Emersol® 221 NF
CAS 112-80-1	Emersol® 233 LL
CAS 112-80-1	Emersol® 6313 NF
CAS 112-80-1	Emersol® 6321 NF
CAS 112-80-1	Emersol® 6333 NF
CAS 112-80-1	Emersol® 7021
CAS 112-80-1	Industrene® 104
CAS 112-80-1	Industrene® 105
CAS 112-80-1	Industrene® 106
CAS 112-80-1	Industrene® 205
CAS 112-80-1	Industrene® 206
CAS 112-80-1	Industrene® 206LP
CAS 112-80-1	Oleine D
CAS 112-80-1	Pamolyn® 100
CAS 112-80-1	Priolene 6900
CAS 112-80-1	Priolene 6901
CAS 112-80-1	Priolene 6905
CAS 112-80-1	Priolene 6906
CAS 112-80-1	Priolene 6907
CAS 112-80-1	Priolene 6910
CAS 112-80-1	Priolene 6922
CAS 112-80-1	Priolene 6930
CAS 112-80-1	Priolene 6933
CAS 112-84-5	Armid® E
CAS 112-85-6	Hystrene® 5522
CAS 112-85-6	Hystrene® 7022
CAS 112-88-9	Gulftene 18
CAS 112-88-9	Neodene® 18
CAS 112-90-3	Amine OL
CAS 112-90-3	Amine OLD
CAS 112-90-3	Armeen® O
CAS 112-90-3	Armeen® OD
CAS 112-90-3	Armeen® OL
CAS 112-90-3	Armeen® OLD
CAS 112-90-3	Crodamine 1.O, 1.OD
CAS 112-90-3	Jet Amine PO
CAS 112-90-3	Jet Amine POD
CAS 112-90-3	Kemamine® P-989D
CAS 112-90-3	Nissan Amine OB
CAS 112-90-3	Noram O
CAS 112-90-3	Radiamine 6172
CAS 112-90-3	Radiamine 6173
CAS 112-92-5	Adol® 61 NF
CAS 112-92-5	Adol® 62 NF
CAS 112-92-5	Adol® 620 NF
CAS 112-92-5	Alfol® 18
CAS 112-92-5	Alfol® 18 NF
CAS 112-92-5	CO-1895
CAS 112-92-5	CO-1897
CAS 112-92-5	CO-1898
CAS 112-92-5	Emery® 3343
CAS 112-92-5	Emery® 3344
CAS 112-92-5	Emery® 3348
CAS 112-92-5	Epal® 18NF
CAS 112-92-5	Lorol C18
CAS 112-92-5	Lorol C18 Chemically Pure
CAS 112-92-5	Lorol C18-98
CAS 112-92-5	Loxiol VPG 1354
CAS 112-92-5	Philcohol 1800
CAS 112-99-2	Armeen® 2-18
CAS 115-70-8	AEPD®
CAS 120-40-1	Ablumide LDE
CAS 120-40-1	Alkamide® DL-203/S
CAS 120-40-1	Alkamide® DL-207/S
CAS 120-40-1	Alkamide® L9DE
CAS 120-40-1	Amidex L-9
CAS 120-40-1	Amidex LD
CAS 120-40-1	Aminol LM-30C, LM-30C Spec.
CAS 120-40-1	Carsamide® SAL-7
CAS 120-40-1	Carsamide® SAL-9
CAS 120-40-1	Cedemide AX

CAS 120-40-1 Chimipal LDA
CAS 120-40-1 Comperlan LD
CAS 120-40-1 Crillon LDE
CAS 120-40-1 Deteryl AG
CAS 120-40-1 Emid® 6510
CAS 120-40-1 Emid® 6511
CAS 120-40-1 Emid® 6518
CAS 120-40-1 Emid® 6519
CAS 120-40-1 Empilan® LDE
CAS 120-40-1 Ethylan® MLD
CAS 120-40-1 Hartamide LDA
CAS 120-40-1 Hetamide LL
CAS 120-40-1 Hetamide ML
CAS 120-40-1 Hetamide MOC
CAS 120-40-1 Incromide L-90
CAS 120-40-1 Incromide LM-70
CAS 120-40-1 Incromide LR
CAS 120-40-1 Intermediate 512
CAS 120-40-1 Loropan LD
CAS 120-40-1 Mackamide L10
CAS 120-40-1 Mackamide L95
CAS 120-40-1 Mackamide LLM
CAS 120-40-1 Mackamide LMD
CAS 120-40-1 Manromid 1224
CAS 120-40-1 Mazamide® 1214
CAS 120-40-1 Mazamide® L-298
CAS 120-40-1 Mazamide® LM
CAS 120-40-1 Mazamide® LM 20
CAS 120-40-1 Monamid® 150-LMWC
CAS 120-40-1 Monamid® 150-LW
CAS 120-40-1 Monamid® 716
CAS 120-40-1 Ninol® 30-LL
CAS 120-40-1 Ninol® 55-LL
CAS 120-40-1 Ninol® 70-SL
CAS 120-40-1 Ninol® 96-SL
CAS 120-40-1 Ninol® AX
CAS 120-40-1 Nissan Stafoam DL
CAS 120-40-1 Nissan Stafoam L
CAS 120-40-1 Nitrene L-90
CAS 120-40-1 Rewomid® DLMS
CAS 120-40-1 Schercomid SL-Extra
CAS 120-40-1 Schercomid SL-ML, SL-ML-LC
CAS 120-40-1 Schercomid SLM-C
CAS 120-40-1 Schercomid SLM-LC
CAS 120-40-1 Schercomid SLM-S
CAS 120-40-1 Schercomid SLMC-75
CAS 120-40-1 Stafoam DL
CAS 120-40-1 Standamid® KDS
CAS 120-40-1 Standamid® LD
CAS 120-40-1 Standamid® LDO
CAS 120-40-1 Standamid® LDS
CAS 120-40-1 Synotol L 60
CAS 120-40-1 Synotol L 90
CAS 120-40-1 Synotol LM 60
CAS 120-40-1 Tohol N-230
CAS 120-40-1 Tohol N-230X
CAS 120-40-1 Unamide® J-56
CAS 120-40-1 Unamide® LDX
CAS 120-40-1 Varamide® ML-1
CAS 120-40-1 Varamide® ML-4
CAS 120-40-1 Witcamide® 5195
CAS 120-40-1 Witcamide® 6310
CAS 120-40-1 Witcamide® 6511
CAS 122-19-0 Incroquat SDQ-25
CAS 122-32-7 Emerest® 2423
CAS 123-94-4 Aldo® HMS
CAS 123-94-4 Aldo® MS
CAS 123-94-4 Aldo® MSA
CAS 123-94-4 Aldo® MSD
CAS 123-94-4 Aldo® MS Industrial
CAS 123-94-4 Aldo® MS LG
CAS 124-07-2 C-895
CAS 124-07-2 C-899
CAS 124-22-1 Amine 12-98D
CAS 124-22-1 Armeen® 12
CAS 124-22-1 Armeen® 12D
CAS 124-26-5 Armid® 18
CAS 124-28-7 Adma® 18
CAS 124-28-7 Armeen® DM18D
CAS 124-28-7 Crodamine 3.A18D
CAS 124-28-7 Kemamine® T-9902D
CAS 124-30-1 Amine 18D
CAS 124-30-1 Armeen® 18
CAS 124-30-1 Armeen® 18D
CAS 124-30-1 Crodamine 1.18D
CAS 124-68-5 AMP
CAS 124-68-5 AMP-95
CAS 126-86-3 Surfynol® 104
CAS 126-86-3 Surfynol® 104A
CAS 126-86-3 Surfynol® 104BC
CAS 126-86-3 Surfynol® 104E
CAS 126-86-3 Surfynol® 104H
CAS 126-86-3 Surfynol® 104PA
CAS 126-86-3 Surfynol® 104PG
CAS 126-86-3 Surfynol® 104S
CAS 126-86-3 Surfynol® TG
CAS 126-92-1 Cosmopon SES
CAS 126-92-1 Niaproof® Anionic Surfactant 08
CAS 127-39-9 Aerosol® IB-45
CAS 136-26-5 Alrosol C
CAS 136-26-5 Standamid® CD
CAS 136-99-2 Schercozoline L
CAS 137-16-6 Crodasinic LS30
CAS 137-16-6 Crodasinic LS35
CAS 137-16-6 Zoharsyl L-30
CAS 137-20-2 Adinol OT16
CAS 137-20-2 Hostapon T Powd
CAS 139-07-1 Dehyquart LDB
CAS 139-07-1 Noramium DA.50
CAS 139-08-2 Arquad® DM14B-90
CAS 139-88-8 Niaproof® Anionic Surfactant 4
CAS 139-96-8 Avirol® 300
CAS 139-96-8 Avirol® 400 T
CAS 139-96-8 Avirol® T 40
CAS 139-96-8 Berol 480
CAS 139-96-8 Cosmopon TR
CAS 139-96-8 Empicol® TL40/T
CAS 139-96-8 Norfox® TLS
CAS 139-96-8 Perlankrol® ATL40
CAS 139-96-8 Texapon T 42
CAS 140-72-7 Acetoquat CPB
CAS 141-20-8 Cithrol DGML N/E
CAS 141-21-9 Avistin® FD
CAS 141-21-9 Avistin® PN
CAS 141-21-9 Marlamid® A 18
CAS 141-22-0 Flexricin® 100
CAS 141-22-0 P®-10 Acid
CAS 142-31-4 Bio-Terge® PAS-8
CAS 142-54-1 Comperlan LP
CAS 142-54-1 Empilan® LIS
CAS 142-54-1 Lauridit® LMI
CAS 142-54-1 Monamid® LIPA
CAS 142-54-1 Monamid® LMIPA
CAS 142-54-1 Rewomid® IPL 203
CAS 142-54-1 Standamid® LP
CAS 142-78-9 Ablumide LME
CAS 142-78-9 Alkamide® L-203
CAS 142-78-9 Comperlan LM
CAS 142-78-9 Empilan® LME
CAS 142-78-9 Hartamide LMEA

CAS 142-78-9 Incromide LCL
CAS 142-78-9 Lauridit® LM
CAS 142-78-9 Loropan LM
CAS 142-78-9 Mackamide LMM
CAS 142-78-9 Manromid LMA
CAS 142-78-9 Monamid® LMA
CAS 142-78-9 Monamid® LMMA
CAS 142-78-9 Ninol® LMP
CAS 142-78-9 Rewomid® L 203
CAS 142-78-9 Standamid® LM
CAS 142-78-9 Varamide® L-203
CAS 142-87-0 Akyporox SAL SAS
CAS 143-07-7 C-1298
CAS 143-07-7 Emery® 650
CAS 143-07-7 Emery® 651
CAS 143-07-7 Emery® 652
CAS 143-07-7 Hetamide LA
CAS 143-07-7 Hystrene® 9512
CAS 143-07-7 Hystrene® 9912
CAS 143-07-7 Philacid 1200
CAS 143-07-7 Prifrac 2920
CAS 143-07-7 Prifrac 2922
CAS 143-27-1 Amine 16D
CAS 143-27-1 Armeen® 16
CAS 143-27-1 Armeen® 16D
CAS 143-27-1 Crodamine 1.16D
CAS 143-28-2 Adol® 80
CAS 143-28-2 Adol® 85 NF
CAS 143-28-2 Adol® 90 NF
CAS 143-28-2 Cachalot® O-3
CAS 143-28-2 Cachalot® O-8
CAS 143-28-2 Cachalot® O-27
CAS 143-28-2 Emery® 3312
CAS 143-28-2 Emery® 3317
CAS 143-28-2 Fancol OA 95
CAS 143-28-2 HD-Ocenol 90/95
CAS 143-28-2 HD-Ocenol 92/96
CAS 151-21-3 Akyposal NLS
CAS 151-21-3 Alscoap LN-40, LN-90
CAS 151-21-3 Avirol® 280 S
CAS 151-21-3 Avirol® SL 2010
CAS 151-21-3 Avirol® SL 2020
CAS 151-21-3 Calfoam SLS-30
CAS 151-21-3 Carsonol® SLS
CAS 151-21-3 Carsonol® SLS Paste B
CAS 151-21-3 Carsonol® SLS-R
CAS 151-21-3 Carsonol® SLS-S
CAS 151-21-3 Carsonol® SLS Special
CAS 151-21-3 Chemsalan NLS 30
CAS 151-21-3 Colonial LKP
CAS 151-21-3 Colonial SLS
CAS 151-21-3 Cosmopon 35
CAS 151-21-3 Duponol® ME Dry
CAS 151-21-3 Duponol® QC
CAS 151-21-3 Duponol® WA Dry
CAS 151-21-3 Duponol® WA Paste
CAS 151-21-3 Duponol® WAQ
CAS 151-21-3 Duponol® WAQE
CAS 151-21-3 Elfan® 200
CAS 151-21-3 Elfan® 240
CAS 151-21-3 Elfan® 260 S
CAS 151-21-3 Empicol® 0185
CAS 151-21-3 Empicol® 0303
CAS 151-21-3 Empicol® 0303V
CAS 151-21-3 Empicol® 0919
CAS 151-21-3 Empicol® LM45
CAS 151-21-3 Empicol® LM/T
CAS 151-21-3 Empicol® LS30
CAS 151-21-3 Empicol® LX28
CAS 151-21-3 Empicol® LXS95
CAS 151-21-3 Empicol® LXV
CAS 151-21-3 Empicol® LXV/D
CAS 151-21-3 Empicol® LY28/S
CAS 151-21-3 Empicol® LZ/E
CAS 151-21-3 Empicol® LZG 30
CAS 151-21-3 Empicol® LZGV
CAS 151-21-3 Empicol® LZGV/C
CAS 151-21-3 Empicol® LZP
CAS 151-21-3 Empicol® LZV/D
CAS 151-21-3 Empicol® LZV/E
CAS 151-21-3 Empicol® WA
CAS 151-21-3 Empicol® WAK
CAS 151-21-3 Empimin® LR28
CAS 151-21-3 Equex S
CAS 151-21-3 Equex SP
CAS 151-21-3 Equex SW
CAS 151-21-3 Gardinol WA Paste
CAS 151-21-3 Hartenol LAS-30
CAS 151-21-3 Laural P
CAS 151-21-3 Manro DL 28
CAS 151-21-3 Manro SLS 28
CAS 151-21-3 Marlinat® DFK 30
CAS 151-21-3 Montovol RF-10
CAS 151-21-3 Naxolate WA-97
CAS 151-21-3 Naxolate WA Special
CAS 151-21-3 Naxolate WAG
CAS 151-21-3 Nikkol SLS
CAS 151-21-3 Norfox® SLS
CAS 151-21-3 Nutrapon DL 3891
CAS 151-21-3 Nutrapon W 1367
CAS 151-21-3 Nutrapon WAC 3005
CAS 151-21-3 Nutrapon WAQ
CAS 151-21-3 Nutrapon WAQE 2364
CAS 151-21-3 Octosol SLS
CAS 151-21-3 Octosol SLS-1
CAS 151-21-3 Perlankrol® DSA
CAS 151-21-3 Polirol LS
CAS 151-21-3 Polystep® B-3
CAS 151-21-3 Polystep® B-5
CAS 151-21-3 Polystep® B-24
CAS 151-21-3 Rewopol® 15/L
CAS 151-21-3 Rewopol® NLS 15 L
CAS 151-21-3 Rewopol® NLS 28
CAS 151-21-3 Rewopol® NLS 30 L
CAS 151-21-3 Rewopol® NLS 90
CAS 151-21-3 Rhodapon® LCP
CAS 151-21-3 Rhodapon® LSB, LSB/CT
CAS 151-21-3 Rhodapon® SB
CAS 151-21-3 Rhodapon® SB-8208/S
CAS 151-21-3 Rhodapon® SM Special
CAS 151-21-3 Rolpon LSX
CAS 151-21-3 Sactol 2 S 3
CAS 151-21-3 Sandoz Sulfate WA Dry
CAS 151-21-3 Sandoz Sulfate WAG
CAS 151-21-3 Sandoz Sulfate WAS
CAS 151-21-3 Sandoz Sulfate WA Special
CAS 151-21-3 Serdet DFK 40
CAS 151-21-3 Sermul EA 150
CAS 151-21-3 Standapol® WAQ-LC
CAS 151-21-3 Standapol® WAQ-LCX
CAS 151-21-3 Stepanol® ME Dry
CAS 151-21-3 Stepanol® WA-100
CAS 151-21-3 Stepanol® WAC
CAS 151-21-3 Stepanol® WAC-P
CAS 151-21-3 Stepanol® WA Extra
CAS 151-21-3 Stepanol® WA Paste
CAS 151-21-3 Stepanol® WA Special
CAS 151-21-3 Stepanol® WAQ
CAS 151-21-3 Sulfetal C 38
CAS 151-21-3 Sulfetal C 90

CAS 151-21-3 Sulfochem SLN
CAS 151-21-3 Sulfochem SLP
CAS 151-21-3 Sulfochem SLP-95
CAS 151-21-3 Sulfochem SLS
CAS 151-21-3 Sulfochem SLX
CAS 151-21-3 Sulfopon 101 Spez
CAS 151-21-3 Sulfopon 101, 101 Special
CAS 151-21-3 Sulfopon 101/POL
CAS 151-21-3 Sulfopon 102
CAS 151-21-3 Sulfopon K35
CAS 151-21-3 Sulfopon P-40
CAS 151-21-3 Sulfopon WAQ LCX
CAS 151-21-3 Sulfopon WAQ Special
CAS 151-21-3 Sulfotex LCX
CAS 151-21-3 Sulfotex WA
CAS 151-21-3 Surfax 250
CAS 151-21-3 Surfax CN
CAS 151-21-3 Swascol 1P
CAS 151-21-3 Swascol 1PC
CAS 151-21-3 Swascol 3L
CAS 151-21-3 Texapon K-12
CAS 151-21-3 Texapon K-1296
CAS 151-21-3 Texapon K-12 Granules
CAS 151-21-3 Texapon K-12 Needles
CAS 151-21-3 Texapon L-100
CAS 151-21-3 Texapon LS 35
CAS 151-21-3 Texapon LS Highly Conc
CAS 151-21-3 Texapon LS Highly Conc. Needles
CAS 151-21-3 Texapon OT Highly Conc. Needles
CAS 151-21-3 Texapon VHC Needles
CAS 151-21-3 Texapon VHC Powd
CAS 151-21-3 Texapon Z
CAS 151-21-3 Texapon ZHC Needles
CAS 151-21-3 Texapon ZHC Powder
CAS 151-21-3 Ufarol Na-30
CAS 151-21-3 Ultra Sulfate SL-1
CAS 151-21-3 Ungerol LS, LSN
CAS 151-21-3 Witcolate 6400
CAS 151-21-3 Witcolate A
CAS 151-21-3 Witcolate A Powder
CAS 151-21-3 Witcolate C
CAS 151-21-3 Witcolate LCP
CAS 151-21-3 Witcolate S
CAS 151-21-3 Witcolate SL-1
CAS 151-21-3 Witcolate WAC
CAS 151-21-3 Witcolate WAC-GL
CAS 151-21-3 Witcolate WAC-LA
CAS 151-21-3 Zorapol LS-30
CAS 301-02-0 Armid® O
CAS 307-35-7 Fluorad FX-8
CAS 334-48-5 C-1095
CAS 335-72-4 Avirol® 252 S
CAS 335-72-4 Avirol® SE 3002
CAS 335-72-4 Avirol® SE 3003
CAS 335-72-4 Texapon N 40
CAS 335-72-4 Texapon N 70
CAS 335-72-4 Texapon PLT-227
CAS 335-72-4 Texapon PN-235
CAS 335-72-4 Zetesol NL
CAS 544-63-8 C-1495
CAS 544-63-8 Emery® 654
CAS 544-63-8 Emery® 655
CAS 544-63-8 Hystrene® 9014
CAS 544-63-8 Hystrene® 9514
CAS 544-63-8 Philacid 1400
CAS 544-63-8 Prifrac 2940
CAS 544-63-8 Prifrac 2942
CAS 577-11-7 Astrowet O-70-PG
CAS 577-11-7 Astrowet O-75
CAS 577-11-7 Chimin DOS 70
CAS 577-11-7 Drewfax® 0007
CAS 577-11-7 Empimin® OT
CAS 577-11-7 Empimin® OT75
CAS 577-11-7 Imbirol OT/Na 70
CAS 577-11-7 Lankropol® KO2
CAS 577-11-7 Marlinat® DF 8
CAS 577-11-7 Schercopol DOS-70
CAS 577-11-7 Schercopol DOS-PG-85
CAS 577-11-7 Tex-Wet 1001
CAS 592-41-6 Gulftene 6
CAS 592-41-6 Neodene® 6
CAS 624-04-4 Cithrol EGDL N/E
CAS 627-83-8 Cithrol EGDS N/E
CAS 629-73-2 Gulftene 16
CAS 629-73-2 Neodene® 16
CAS 629-76-5 Neodol® 5
CAS 657-84-1 Eltesol® ST 34
CAS 657-84-1 Eltesol® ST 40
CAS 657-84-1 Eltesol® ST 90
CAS 657-84-1 Eltesol® ST Pellets
CAS 683-10-3 Chimin BX
CAS 683-10-3 Zohartaine AB
CAS 820-66-6 Lonzaine® 18S
CAS 872-05-9 Gulftene 10
CAS 872-05-9 Neodene® 10
CAS 928-24-5 Cithrol EGDO N/E
CAS 1119-97-7 Querton 14Br-40
CAS 1120-24-7 Adma® 10
CAS 1120-36-1 Gulftene 14
CAS 1120-36-1 Neodene® 14
CAS 1120-49-6 Armeen® 2-10
CAS 1300-72-7 Carsosulf SXS-Liq
CAS 1300-72-7 Eltesol® SX 30
CAS 1300-72-7 Eltesol® SX 40
CAS 1300-72-7 Eltesol® SX 93
CAS 1300-72-7 Eltesol® SX Pellets
CAS 1300-72-7 Esi-Terge SXS
CAS 1300-72-7 Hartotrope SXS 40, Powd
CAS 1300-72-7 Lankrosol SXS-30
CAS 1300-72-7 Manrosol SXS30
CAS 1300-72-7 Manrosol SXS40
CAS 1300-72-7 Manrosol SXS93
CAS 1300-72-7 Norfox® SXS40, SXS96
CAS 1300-72-7 Nutrol SXS 5418
CAS 1300-72-7 Reworyl® NXS 40
CAS 1300-72-7 Sulfotex SXS-40
CAS 1300-72-7 Tex-Wet 1140
CAS 1300-72-7 Ultra SXS Liq., Powd
CAS 1300-72-7 Witconate SXS 40%
CAS 1300-72-7 Witconate SXS 90%
CAS 1321-94-4 Arosolve MN-LF
CAS 1323-38-2 Surfactol® 13
CAS 1323-39-3 Cithrol PGMS N/E
CAS 1323-39-3 Hodag PGS
CAS 1323-83-7 Cithrol GDS N/E
CAS 1323-83-7 Kessco® GDS 386F
CAS 1323-83-7 Nikkol DGS-80
CAS 1330-76-3 DIOM
CAS 1330-80-9 Cithrol PGMO N/E
CAS 1333-39-7 Eltesol® PSA 65
CAS 1338-39-2 Ablunol S-20
CAS 1338-39-2 Ahco 759
CAS 1338-39-2 Alkamuls® S-20
CAS 1338-39-2 Alkamuls® SML
CAS 1338-39-2 Armotan® ML
CAS 1338-39-2 Crill 1
CAS 1338-39-2 Disponil SML 100 F1
CAS 1338-39-2 Drewmulse® SML
CAS 1338-39-2 Durtan® 20
CAS 1338-39-2 Emsorb® 2515

CAS 1338-39-2	Ethylan® GL20
CAS 1338-39-2	Glycomul® L
CAS 1338-39-2	Hetan SL
CAS 1338-39-2	Hodag SML
CAS 1338-39-2	Ionet S-20
CAS 1338-39-2	Liposorb L
CAS 1338-39-2	Lonzest® SML
CAS 1338-39-2	ML-33-F
CAS 1338-39-2	Montane 20
CAS 1338-39-2	Newcol 20
CAS 1338-39-2	Nissan Nonion LP-20R, LP-20RS
CAS 1338-39-2	Prote-sorb SML
CAS 1338-39-2	Radiasurf® 7125
CAS 1338-39-2	S-Maz® 20
CAS 1338-39-2	Secoster® KL 10
CAS 1338-39-2	Sorbax SML
CAS 1338-39-2	Sorbon S-20
CAS 1338-39-2	Span® 20
CAS 1338-41-6	Ablunol S-60
CAS 1338-41-6	Ahco 909
CAS 1338-41-6	Alkamuls® S-60
CAS 1338-41-6	Alkamuls® SMS
CAS 1338-41-6	Armotan® MS
CAS 1338-41-6	Crill 3
CAS 1338-41-6	Disponil SMS 100 F1
CAS 1338-41-6	Drewmulse® SMS
CAS 1338-41-6	Durtan® 60
CAS 1338-41-6	Emsorb® 2505
CAS 1338-41-6	Emultex SMS
CAS 1338-41-6	Ethylan® GS60
CAS 1338-41-6	Glycomul® S
CAS 1338-41-6	Glycomul® S FG
CAS 1338-41-6	Glycomul® S KFG
CAS 1338-41-6	Grindtek SMS
CAS 1338-41-6	Hetan SS
CAS 1338-41-6	Ionet S-60 C
CAS 1338-41-6	Liposorb S
CAS 1338-41-6	Lonzest® SMS
CAS 1338-41-6	Montane 60
CAS 1338-41-6	MS-33-F
CAS 1338-41-6	Newcol 60
CAS 1338-41-6	Nikkol SS-10
CAS 1338-41-6	Prote-sorb SMS
CAS 1338-41-6	Radiasurf® 7145
CAS 1338-41-6	Rheodol SP-S10
CAS 1338-41-6	Secoster® KS 10
CAS 1338-41-6	S-Maz® 60
CAS 1338-41-6	S-Maz® 60KHM
CAS 1338-41-6	Sorban AST
CAS 1338-41-6	Sorbax SMS
CAS 1338-41-6	Sorbon S-60
CAS 1338-41-6	Span® 60, 60K
CAS 1338-43-8	Ablunol S-80
CAS 1338-43-8	Ahco 832
CAS 1338-43-8	Ahco 944
CAS 1338-43-8	Alkamuls® S-80
CAS 1338-43-8	Alkamuls® SMO
CAS 1338-43-8	Anfomul S4
CAS 1338-43-8	Anfomul S50
CAS 1338-43-8	Armotan® MO
CAS 1338-43-8	Atlas G-4884
CAS 1338-43-8	Crill 4
CAS 1338-43-8	Crill 50
CAS 1338-43-8	Desonic® SMO
CAS 1338-43-8	Desotan® SMO
CAS 1338-43-8	Disponil SMO 100 F1
CAS 1338-43-8	Drewmulse® SMO
CAS 1338-43-8	Durtan® 80
CAS 1338-43-8	Emsorb® 2500
CAS 1338-43-8	Ethylan® GO80
CAS 1338-43-8	Glycomul® O
CAS 1338-43-8	Hetan SO
CAS 1338-43-8	Hodag SMO
CAS 1338-43-8	Ionet S-80
CAS 1338-43-8	Liposorb O
CAS 1338-43-8	Lonzest® SMO
CAS 1338-43-8	MO-33-F
CAS 1338-43-8	Montane 80
CAS 1338-43-8	Newcol 3-80, 3-85
CAS 1338-43-8	Newcol 80
CAS 1338-43-8	Nikkol SO-10
CAS 1338-43-8	Norfox® Sorbo S-80
CAS 1338-43-8	Prote-sorb SMO
CAS 1338-43-8	Radiasurf® 7155
CAS 1338-43-8	Rheodol SP-O10
CAS 1338-43-8	S-Maz® 80
CAS 1338-43-8	S-Maz® 80K
CAS 1338-43-8	Sorban AO
CAS 1338-43-8	Sorbax SMO
CAS 1338-43-8	Sorbirol O
CAS 1338-43-8	Sorbon S-80
CAS 1338-43-8	Span® 80
CAS 1338-43-8	Witconol 2500
CAS 1369-66-3	Empimin® OP70
CAS 1462-54-0	Amphoram CP1
CAS 1643-20-5	Empigen® OB
CAS 1838-08-0	Farmin R 86H
CAS 2016-42-4	Amine 14D
CAS 2016-56-0	Acetamin 24
CAS 2016-56-0	Nopcogen 16-L
CAS 2190-04-7	Acetamin 86
CAS 2190-04-7	Noramac S
CAS 2235-54-3	Akyposal ALS 33
CAS 2235-54-3	Avirol® 200
CAS 2235-54-3	Sermul EA 129
CAS 2235-54-3	Surfax 215
CAS 2311-27-5	Dehydrophen PNP 4
CAS 2571-88-2	Admox® 18-85
CAS 2571-88-2	Incromine Oxide S
CAS 2571-88-2	Standamox PS
CAS 2605-79-0	Barlox® 10S
CAS 2673-22-5	Polirol TR/LNA
CAS 2867-47-2	Ageflex FM-1
CAS 2915-53-9	DOM
CAS 3006-15-3	Empimin® MA
CAS 3010-24-0	Ethoquad® 18/12
CAS 3026-63-9	Avirol® SA 4113
CAS 3033-77-0	Ogtac 85 V
CAS 3055-33-4	Lauropal 2
CAS 3055-93-4	Akyporox RLM 22
CAS 3055-94-5	Dehydol PLS 3
CAS 3055-95-6	Mulsifan RT 23
CAS 3088-31-1	Akyposal RLM 56 S
CAS 3088-31-1	Surfax 218
CAS 3115-49-9	Akypo NP 70
CAS 3179-80-4	Schercodine L
CAS 3327-22-8	CHPTA 65%
CAS 3332-27-2	Admox® 14-85
CAS 3332-27-2	Aromox® DM14D-W
CAS 3332-27-2	Empigen® OH25
CAS 3332-27-2	Incromine Oxide M
CAS 3332-27-2	Lilaminox M4
CAS 3332-27-2	Schercamox DMA
CAS 3332-27-2	Schercamox DMM
CAS 3452-07-1	Neodene® 20
CAS 3567-25-7	Mitin® FF High Conc
CAS 3614-12-8	Deriphat 170C
CAS 3655-00-3	Deriphat 160
CAS 3655-00-3	Mirataine® H2C-HA
CAS 4219-48-1	Cithrol EGML N/E

CAS Number	Tradename
CAS 4292-10-8	Lexaine® LM
CAS 4500-01-0	Cithrol EGMO N/E
CAS 4722-98-9	Cosmopon MO
CAS 4722-98-9	Empicol® LQ33/T
CAS 4741-30-4	Xanthates
CAS 5064-31-3	Trilon® A
CAS 5274-68-0	Akyporox RLM 40
CAS 5274-68-0	Ameroxol® LE-4
CAS 5274-68-0	Hetoxol L-4
CAS 5397-31-9	Tomah PA-12EH
CAS 6001-97-4	Lankropol® KMA
CAS 6179-44-8	Schercotaine IAB
CAS 7128-91-8	Aromox® DM16
CAS 7173-51-5	Arquad® 210-50
CAS 7173-62-8	Dinoram O
CAS 7173-62-8	Duomeen® OL
CAS 7173-62-8	Jet Amine DO
CAS 7205-87-1	Marlophen® 82N
CAS 7205-87-1	Marlophen® 83N
CAS 7205-87-1	Marlophen® 84N
CAS 7205-87-1	Marlophen® 85N
CAS 7205-87-1	Marlophen® 86N
CAS 7205-87-1	Marlophen® 88N
CAS 7205-87-1	Marlophen® 89N
CAS 7205-87-1	Marlophen® 810N
CAS 7205-87-1	Marlophen® 812N
CAS 7205-87-1	Marlophen® 814N
CAS 7205-87-1	Marlophen® 820N
CAS 7205-87-1	Marlophen® 830N
CAS 7205-87-1	Marlophen® 840N
CAS 7205-87-1	Marlophen® 850N
CAS 7205-87-1	Marlophen® 1028N
CAS 7378-99-6	Adma® 8
CAS 7396-58-9	Armeen® M2-10D
CAS 7396-58-9	Dama® 1010
CAS 7617-78-9	Tomah PA-14
CAS 7702-01-4	Schercoteric CY-2
CAS 7702-01-4	Sochamine A 8955
CAS 8002-43-5	Alcolec® 439-C
CAS 8002-43-5	Alcolec® 440-WD
CAS 8002-43-5	Alcolec® 495
CAS 8002-43-5	Alcolec® BS
CAS 8002-43-5	Alcolec® Extra A
CAS 8002-43-5	Alcolec® F-100
CAS 8002-43-5	Alcolec® FF-100
CAS 8002-43-5	Alcolec® Granules
CAS 8002-43-5	Alcolec® S
CAS 8002-43-5	Asol
CAS 8002-43-5	Blendmax Series
CAS 8002-43-5	Centrol® 2FSB, 2FUB, 3FSB, 3FUB
CAS 8002-43-5	Centrol® CA
CAS 8002-43-5	Centromix® CPS
CAS 8002-43-5	Centromix® E
CAS 8002-43-5	Centrophase® 31
CAS 8002-43-5	Centrophase® 152
CAS 8002-43-5	Centrophase® C
CAS 8002-43-5	Centrophase® HR2B, HR2U
CAS 8002-43-5	Centrophase® HR4B, HR4U
CAS 8002-43-5	Centrophase® HR6B
CAS 8002-43-5	Centrophase® NV
CAS 8002-43-5	Centrophil® K
CAS 8002-43-5	Centrophil® M, W
CAS 8002-43-5	Clearate B-60
CAS 8002-43-5	Clearate LV
CAS 8002-43-5	Clearate Special Extra
CAS 8002-43-5	Clearate WD
CAS 8002-43-5	Lecithin W.D
CAS 8002-43-5	Lexin K
CAS 8002-43-5	Lipotin 100, 100J, SB
CAS 8002-43-5	M-C-Thin
CAS 8002-43-5	M-C-Thin 45
CAS 8002-43-5	Phospholipon 50G
CAS 8002-43-5	Topcithin
CAS 8006-54-0	Fluilan
CAS 8006-54-0	Lanolin Alcohols LO
CAS 8007-43-0	Anfomul S43
CAS 8007-43-0	Arlacel® C
CAS 8007-43-0	Crill 43
CAS 8007-43-0	Dehymuls SSO
CAS 8007-43-0	Disponil SSO 100 F1
CAS 8007-43-0	Emasol O-15 R
CAS 8007-43-0	Emsorb® 2502
CAS 8007-43-0	Glycomul® SOC
CAS 8007-43-0	Liposorb SQO
CAS 8007-43-0	Montane 83
CAS 8007-43-0	Nikkol SO-15
CAS 8007-43-0	S-Maz® 83R
CAS 8020-84-6	Lanolin Fatty Acids O
CAS 8027-33-6	Argowax Cosmetic Super
CAS 8027-33-6	Dusoran MD
CAS 8030-78-2	Arquad® T-27W
CAS 8030-78-2	Arquad® T-50
CAS 8030-78-2	Jet Quat T-27W
CAS 8030-78-2	Jet Quat T-50
CAS 8030-78-2	Noramium MS 50
CAS 8039-09-6	Lantrol® PLN
CAS 8050-09-7	Westvaco® Resin 90
CAS 8050-09-7	Westvaco® Resin 95
CAS 8051-30-7	Empilan® 2502
CAS 8051-30-7	Empilan® FD
CAS 8061-51-6	Lignosite® 231
CAS 8061-51-6	Lignosite® 431
CAS 8061-51-6	Lignosite® 458
CAS 8061-51-6	Lignosite® 823
CAS 8061-51-6	Lignosol AXD
CAS 8061-51-6	Lignosol D-10, D-30
CAS 8061-51-6	Lignosol DXD
CAS 8061-51-6	Lignosol FTA
CAS 8061-51-6	Lignosol HCX
CAS 8061-51-6	Lignosol NSX 110
CAS 8061-51-6	Lignosol NSX 120
CAS 8061-51-6	Lignosol TS
CAS 8061-51-6	Lignosol WT
CAS 8061-51-6	Lignosol X
CAS 8061-51-6	Lignosol XD
CAS 8061-52-7	Lignosol SFX
CAS 8061-53-8	Lignosite® 17
CAS 9002-32-0	Lauropal 1150
CAS 9002-92-0	Akyporox RLM 80
CAS 9002-92-0	Akyporox RLM 160
CAS 9002-92-0	Brij® 35
CAS 9002-92-0	Brij® 35 Liq
CAS 9002-92-0	Brij® 35 SP
CAS 9002-92-0	Brij® 35 SP Liq
CAS 9002-92-0	Ethosperse® LA-4
CAS 9002-92-0	Ethosperse® LA-12
CAS 9002-92-0	Ethosperse® LA-23
CAS 9002-92-0	Lauropal 9
CAS 9002-92-0	Lauropal 11
CAS 9002-92-0	Marlipal® 124
CAS 9002-92-0	Marlowet® BL
CAS 9002-92-0	Noigen ET83
CAS 9002-92-0	Noigen ET102
CAS 9002-92-0	Noigen ET143
CAS 9002-92-0	Noigen ET160
CAS 9002-92-0	Noigen ET170
CAS 9002-92-0	Noigen ET190
CAS 9002-92-0	Noigen YX400, YX500
CAS 9002-92-0	Trycol® 5963

CAS 9002-92-0 Trycol® 5964
CAS 9002-92-0 Trycol® 5966
CAS 9002-92-0 Trycol® 5967
CAS 9002-92-0 Trycol® LAL-4
CAS 9002-92-0 Trycol® LAL-8
CAS 9002-92-0 Trycol® LAL-12
CAS 9002-92-0 Trycol® LAL-23
CAS 9002-92-0 Volpo L4
CAS 9002-92-0 Volpo L23
CAS 9002-93-1 Akyporox OP 100
CAS 9002-93-1 Akyporox OP 115 SPC
CAS 9002-93-1 Akyporox OP 200
CAS 9002-93-1 Akyporox OP 250 V
CAS 9002-93-1 Akyporox OP 400V
CAS 9002-93-1 Alkasurf® OP-1
CAS 9002-93-1 Alkasurf® OP-5
CAS 9002-93-1 Alkasurf® OP-8
CAS 9002-93-1 Alkasurf® OP-10
CAS 9002-93-1 Alkasurf® OP-12
CAS 9002-93-1 Alkasurf® OP-30, 70%
CAS 9002-93-1 Alkasurf® OP-40, 70%
CAS 9002-93-1 Cedepal CA-210
CAS 9002-93-1 Cedepal CA-520
CAS 9002-93-1 Cedepal CA-630
CAS 9002-93-1 Cedepal CA-720
CAS 9002-93-1 Cedepal CA-890
CAS 9002-93-1 Cedepal CA-897
CAS 9002-93-1 Chemax OP-3
CAS 9002-93-1 Chemax OP-5
CAS 9002-93-1 Chemax OP-7
CAS 9002-93-1 Chemax OP-9
CAS 9002-93-1 Chemax OP-30/70
CAS 9002-93-1 Chemax OP-40
CAS 9002-93-1 Chemax OP-40/70
CAS 9002-93-1 Dehydrophen POP 4
CAS 9002-93-1 Dehydrophen POP 8
CAS 9002-93-1 Dehydrophen POP 10
CAS 9002-93-1 Dehydrophen POP 17
CAS 9002-93-1 Dehydrophen POP 17/80
CAS 9002-93-1 Desonic® S-45
CAS 9002-93-1 Desonic® S-100
CAS 9002-93-1 Desonic® S-102
CAS 9002-93-1 Desonic® S-114
CAS 9002-93-1 Desonic® S-405
CAS 9002-93-1 Hyonic OP-7
CAS 9002-93-1 Hyonic OP-10
CAS 9002-93-1 Hyonic OP-40
CAS 9002-93-1 Hyonic OP-70
CAS 9002-93-1 Hyonic OP-100
CAS 9002-93-1 Hyonic OP-407
CAS 9002-93-1 Hyonic OP-705
CAS 9002-93-1 Hyonic PE-250
CAS 9002-93-1 Iconol OP-5
CAS 9002-93-1 Iconol OP-7
CAS 9002-93-1 Iconol OP-10
CAS 9002-93-1 Iconol OP-30
CAS 9002-93-1 Iconol OP-30-70%
CAS 9002-93-1 Iconol OP-40
CAS 9002-93-1 Iconol OP-40-70%
CAS 9002-93-1 Igepal® CA-210
CAS 9002-93-1 Igepal® CA-420
CAS 9002-93-1 Igepal® CA-520
CAS 9002-93-1 Igepal® CA-620
CAS 9002-93-1 Igepal® CA-630
CAS 9002-93-1 Igepal® CA-720
CAS 9002-93-1 Igepal® CA-880
CAS 9002-93-1 Igepal® CA-887
CAS 9002-93-1 Igepal® CA-890
CAS 9002-93-1 Igepal® CA-897
CAS 9002-93-1 Igepal® F-85
CAS 9002-93-1 Igepal® F-920
CAS 9002-93-1 Igepal® KA
CAS 9002-93-1 Igepal® O
CAS 9002-93-1 Macol® OP-3
CAS 9002-93-1 Macol® OP-5
CAS 9002-93-1 Macol® OP-8
CAS 9002-93-1 Macol® OP-10
CAS 9002-93-1 Macol® OP-10 SP
CAS 9002-93-1 Macol® OP-12
CAS 9002-93-1 Macol® OP-16(75)
CAS 9002-93-1 Macol® OP-30(70)
CAS 9002-93-1 Macol® OP-40(70)
CAS 9002-93-1 Makon® OP-6
CAS 9002-93-1 Makon® OP-9
CAS 9002-93-1 Marlophen® 85
CAS 9002-93-1 Marlophen® 86
CAS 9002-93-1 Marlophen® 87
CAS 9002-93-1 Marlophen® 88
CAS 9002-93-1 Marlophen® 89
CAS 9002-93-1 Marlophen® 810
CAS 9002-93-1 Marlophen® 812
CAS 9002-93-1 Marlophen® 814
CAS 9002-93-1 Marlophen® 820
CAS 9002-93-1 Marlophen® 1028
CAS 9002-93-1 Nikkol OP-3
CAS 9002-93-1 Nikkol OP-10
CAS 9002-93-1 Nikkol OP-30
CAS 9002-93-1 Nissan Nonion HS-240
CAS 9002-93-1 Nissan Nonion HS-270
CAS 9002-93-1 Nutrol 100
CAS 9002-93-1 Prox-onic OP-09
CAS 9002-93-1 Prox-onic OP-016
CAS 9002-93-1 Prox-onic OP-030/70
CAS 9002-93-1 Prox-onic OP-040/70
CAS 9002-93-1 Remcopal 306
CAS 9002-93-1 Remcopal O9
CAS 9002-93-1 Remcopal O11
CAS 9002-93-1 Remcopal O12
CAS 9002-93-1 Renex® 750
CAS 9002-93-1 Rexol 45/1
CAS 9002-93-1 Rexol 45/3
CAS 9002-93-1 Rexol 45/5
CAS 9002-93-1 Rexol 45/7
CAS 9002-93-1 Rexol 45/10
CAS 9002-93-1 Rexol 45/12
CAS 9002-93-1 Rexol 45/16
CAS 9002-93-1 Rexol 45/307
CAS 9002-93-1 Rexol 45/407
CAS 9002-93-1 Sellig O 4 100
CAS 9002-93-1 Sellig O 5 100
CAS 9002-93-1 Sellig O 6 100
CAS 9002-93-1 Sellig O 8 100
CAS 9002-93-1 Sellig O 9 100
CAS 9002-93-1 Sellig O 11 100
CAS 9002-93-1 Sellig O 12 100
CAS 9002-93-1 Sellig O 20 100
CAS 9002-93-1 Serdox NOP 9
CAS 9002-93-1 Serdox NOP 30/70
CAS 9002-93-1 Serdox NOP 40/70
CAS 9002-93-1 Siponic® F-300
CAS 9002-93-1 Siponic® F-400
CAS 9002-93-1 Synperonic OP3
CAS 9002-93-1 Synperonic OP4.5
CAS 9002-93-1 Synperonic OP7.5
CAS 9002-93-1 Synperonic OP10
CAS 9002-93-1 Synperonic OP11
CAS 9002-93-1 Synperonic OP12.5
CAS 9002-93-1 Synperonic OP16
CAS 9002-93-1 Synperonic OP20
CAS 9002-93-1 Synperonic OP25

CAS 9002-93-1	Synperonic OP30
CAS 9002-93-1	Synperonic OP40
CAS 9002-93-1	Synperonic OP40/70
CAS 9002-93-1	Syntopon 8 C
CAS 9002-93-1	T-Det® O-4
CAS 9002-93-1	T-Det® O-6
CAS 9002-93-1	T-Det® O-8
CAS 9002-93-1	T-Det® O-9
CAS 9002-93-1	T-Det® O-12
CAS 9002-93-1	T-Det® O-40
CAS 9002-93-1	T-Det® O-307
CAS 9002-93-1	T-Det® O-407
CAS 9002-93-1	Teric X5
CAS 9002-93-1	Teric X7
CAS 9002-93-1	Teric X8
CAS 9002-93-1	Teric X10
CAS 9002-93-1	Teric X11
CAS 9002-93-1	Teric X13
CAS 9002-93-1	Teric X16
CAS 9002-93-1	Teric X40
CAS 9002-93-1	Teric X40L
CAS 9002-93-1	Triton® X-15
CAS 9002-93-1	Triton® X-35
CAS 9002-93-1	Triton® X-45
CAS 9002-93-1	Triton® X-100
CAS 9002-93-1	Triton® X-100 CG
CAS 9002-93-1	Triton® X-102
CAS 9002-93-1	Triton® X-114
CAS 9002-93-1	Triton® X-120
CAS 9002-93-1	Triton® X-165-70%
CAS 9002-93-1	Triton® X-305-70%
CAS 9002-93-1	Triton® X-405-70%
CAS 9002-93-1	Triton® X-705-70%
CAS 9002-93-1	Trycol® 6943
CAS 9002-93-1	Trycol® 6956
CAS 9002-93-1	Trycol® 6975
CAS 9002-93-1	Trycol® 6984
CAS 9002-93-1	Trycol® OP-407
CAS 9002-97-5	Akypopress DB
CAS 9003-03-6	Size CB
CAS 9003-11-6	Antarox® PGP 23-7
CAS 9003-11-6	Berol 370
CAS 9003-11-6	Berol 374
CAS 9003-11-6	Chimipal PE 300, PE 302
CAS 9003-11-6	Epan 710
CAS 9003-11-6	Epan 720
CAS 9003-11-6	Epan 740
CAS 9003-11-6	Epan 750
CAS 9003-11-6	Industrol® N3
CAS 9003-11-6	Monolan® 8000 E/80
CAS 9003-11-6	Monolan® 12,000 E/80
CAS 9003-11-6	Monolan® PC
CAS 9003-11-6	Newpol PE-61
CAS 9003-11-6	Newpol PE-62
CAS 9003-11-6	Newpol PE-64
CAS 9003-11-6	Newpol PE-68
CAS 9003-11-6	Newpol PE-74
CAS 9003-11-6	Newpol PE-75
CAS 9003-11-6	Newpol PE-78
CAS 9003-11-6	Newpol PE-88
CAS 9003-11-6	Separol WF 34
CAS 9003-11-6	Separol WF 41
CAS 9003-11-6	Synperonic P105
CAS 9003-11-6	Synperonic PE/F38
CAS 9003-11-6	Synperonic PE/L31
CAS 9003-11-6	Synperonic PE/L62LF
CAS 9003-11-6	Synperonic PE/P75
CAS 9003-39-8	Sokalan® HP 50
CAS 9004-64-2	Klucel® E, G, H, J, L, M
CAS 9004-78-8	Marlophen® P 7
CAS 9004-80-2	Crafol AP-60
CAS 9004-81-3	Ablunol 200ML
CAS 9004-81-3	Ablunol 400ML
CAS 9004-81-3	Ablunol 600ML
CAS 9004-81-3	Acconon 400-ML
CAS 9004-81-3	Alkamuls® 600-GML
CAS 9004-81-3	Alkamuls® L-9
CAS 9004-81-3	Alkamuls® PE/220
CAS 9004-81-3	Alkamuls® PE/400
CAS 9004-81-3	Atlas G-2127
CAS 9004-81-3	Chemax E-200 ML
CAS 9004-81-3	Chemax E-400 ML
CAS 9004-81-3	Chemax E-600 ML
CAS 9004-81-3	Chimipal APG 400
CAS 9004-81-3	Cithrol 2ML
CAS 9004-81-3	Cithrol 4ML
CAS 9004-81-3	Cithrol 6ML
CAS 9004-81-3	Cithrol 10ML
CAS 9004-81-3	Cithrol 60ML
CAS 9004-81-3	Cithrol DGML N/E
CAS 9004-81-3	Cithrol DGML S/E
CAS 9004-81-3	CPH-27-N
CAS 9004-81-3	CPH-30-N
CAS 9004-81-3	Crodet L4
CAS 9004-81-3	Crodet L8
CAS 9004-81-3	Crodet L12
CAS 9004-81-3	Crodet L24
CAS 9004-81-3	Crodet L40
CAS 9004-81-3	Crodet L100
CAS 9004-81-3	DyaFac LA9
CAS 9004-81-3	Emerest® 2620
CAS 9004-81-3	Emerest® 2630
CAS 9004-81-3	Emerest® 2650
CAS 9004-81-3	Emerest® 2661
CAS 9004-81-3	Ethox ML-5
CAS 9004-81-3	Ethox ML-9
CAS 9004-81-3	Ethox ML-14
CAS 9004-81-3	Gradonic 400-ML
CAS 9004-81-3	Hetoxamate LA-5
CAS 9004-81-3	Hetoxamate LA-9
CAS 9004-81-3	Hodag 20-L
CAS 9004-81-3	Hodag 40-L
CAS 9004-81-3	Hodag 60-L
CAS 9004-81-3	Hodag DGL
CAS 9004-81-3	Industrol® ML-5
CAS 9004-81-3	Industrol® ML-9
CAS 9004-81-3	Industrol® ML-14
CAS 9004-81-3	Karapeg 200-ML
CAS 9004-81-3	Karapeg 400-ML
CAS 9004-81-3	Karapeg 600-ML
CAS 9004-81-3	Kessco® PEG 200 ML
CAS 9004-81-3	Kessco® PEG 300 ML
CAS 9004-81-3	Kessco® PEG 400 ML
CAS 9004-81-3	Kessco® PEG 600 ML
CAS 9004-81-3	Kessco® PEG 1000 ML
CAS 9004-81-3	Kessco® PEG 1540 ML
CAS 9004-81-3	Kessco® PEG 4000 ML
CAS 9004-81-3	Kessco® PEG 6000 ML
CAS 9004-81-3	Lexemul® PEG-400ML
CAS 9004-81-3	Lipo DGLS
CAS 9004-81-3	Lipo Diglycol Laurate
CAS 9004-81-3	Lonzest® PEG 4-L
CAS 9004-81-3	Mapeg® 200 ML
CAS 9004-81-3	Mapeg® 400 ML
CAS 9004-81-3	Mapeg® 600 ML
CAS 9004-81-3	Mapeg® DGLD
CAS 9004-81-3	Nonisol 100
CAS 9004-81-3	Nopalcol 1-L
CAS 9004-81-3	Nopalcol 2-L
CAS 9004-81-3	Nopalcol 4-L

CAS 9004-81-3 Nopalcol 6-L
CAS 9004-81-3 Pegosperse® 100 L
CAS 9004-81-3 Pegosperse® 100 ML
CAS 9004-81-3 Pegosperse® 200 ML
CAS 9004-81-3 Pegosperse® 400 ML
CAS 9004-81-3 Pegosperse® 600 ML
CAS 9004-81-3 PGE-600-ML
CAS 9004-81-3 Polirol L400
CAS 9004-81-3 Radiasurf® 7420
CAS 9004-81-3 Radiasurf® 7421
CAS 9004-81-3 Radiasurf® 7422
CAS 9004-81-3 Radiasurf® 7423
CAS 9004-81-3 Secoster® ML 300
CAS 9004-81-3 Secoster® ML 4000
CAS 9004-81-3 Sole-Onic CDS
CAS 9004-81-3 Trydet LA-5
CAS 9004-81-3 Trydet LA-7
CAS 9004-81-3 Witconol 2620
CAS 9004-82-4 Akyposal 9278 R
CAS 9004-82-4 Akyposal EO 20 MW
CAS 9004-82-4 Akyposal EO 20 PA
CAS 9004-82-4 Akyposal EO 20 PA/TS
CAS 9004-82-4 Akyposal EO 20 SF
CAS 9004-82-4 Akyposal MS SPC
CAS 9004-82-4 Avirol® 252 S
CAS 9004-82-4 Avirol® 603 S
CAS 9004-82-4 Avirol® FES 996
CAS 9004-82-4 Avirol® SE 3002
CAS 9004-82-4 Avirol® SE 3003
CAS 9004-82-4 Calfoam ES-30
CAS 9004-82-4 Calfoam SEL-60
CAS 9004-82-4 Carsonol® SES-S
CAS 9004-82-4 Carsonol® SLES
CAS 9004-82-4 Cedepal SS-203, -306, -403, -406
CAS 9004-82-4 Chemsalan RLM 28
CAS 9004-82-4 Chemsalan RLM 56
CAS 9004-82-4 Chemsalan RLM 70
CAS 9004-82-4 Colonial SLE(2)S
CAS 9004-82-4 Colonial SLE(3)S
CAS 9004-82-4 Colonial SLES-70
CAS 9004-82-4 Cosmopon LE 50
CAS 9004-82-4 Disponil FES 32
CAS 9004-82-4 Disponil FES 61
CAS 9004-82-4 Disponil FES 77
CAS 9004-82-4 Elfan® NS 242
CAS 9004-82-4 Elfan® NS 242 Conc
CAS 9004-82-4 Elfan® NS 243 S
CAS 9004-82-4 Elfan® NS 243 S Conc
CAS 9004-82-4 Elfan® NS 252 S
CAS 9004-82-4 Elfan® NS 252 S Conc
CAS 9004-82-4 Elfan® NS 423 SH
CAS 9004-82-4 Elfan® NS 423 SH Conc
CAS 9004-82-4 Empicol® ESB
CAS 9004-82-4 Empicol® ESB3
CAS 9004-82-4 Empimin® KSN27
CAS 9004-82-4 Empimin® KSN70
CAS 9004-82-4 Genapol® ARO
CAS 9004-82-4 Genapol® LRO Liq., Paste
CAS 9004-82-4 Genapol® ZRO Liq., Paste
CAS 9004-82-4 Hartenol LES 60
CAS 9004-82-4 Laural LS
CAS 9004-82-4 Manro BES 27
CAS 9004-82-4 Manro BES 60
CAS 9004-82-4 Manro BES 70
CAS 9004-82-4 Marlinat® 242/28
CAS 9004-82-4 Marlinat® 242/70
CAS 9004-82-4 Marlinat® 242/70 S
CAS 9004-82-4 Marlinat® 243/28
CAS 9004-82-4 Marlinat® 243/70
CAS 9004-82-4 Nonasol N4SS
CAS 9004-82-4 Norfox® SLES-02
CAS 9004-82-4 Norfox® SLES-03
CAS 9004-82-4 Norfox® SLES-60
CAS 9004-82-4 Nutrapon ES-60 3568
CAS 9004-82-4 Nutrapon KPC 0156
CAS 9004-82-4 Polystep® B-12
CAS 9004-82-4 Rewopol® NL 2
CAS 9004-82-4 Rewopol® NL 2-28
CAS 9004-82-4 Rewopol® NL 3
CAS 9004-82-4 Rewopol® NL 3-28
CAS 9004-82-4 Rewopol® NL 3-70
CAS 9004-82-4 Rhodapex® ES
CAS 9004-82-4 Rhodapex® ESY
CAS 9004-82-4 Rolpon 24/230
CAS 9004-82-4 Rolpon 24/270
CAS 9004-82-4 Rolpon 24/330 N
CAS 9004-82-4 Sactol 2 OS 2
CAS 9004-82-4 Sactol 2 OS 28
CAS 9004-82-4 Sandoz Sulfate 219
CAS 9004-82-4 Sandoz Sulfate ES-3
CAS 9004-82-4 Sandoz Sulfate WE
CAS 9004-82-4 Serdet DCK 3/70
CAS 9004-82-4 Serdet DCK 30
CAS 9004-82-4 Sermul EA 30
CAS 9004-82-4 Sermul EA 370
CAS 9004-82-4 Standapol® ES-250
CAS 9004-82-4 Standapol® ES-350
CAS 9004-82-4 Standapol® SL-60
CAS 9004-82-4 Steol® 4N
CAS 9004-82-4 Steol® CS-260
CAS 9004-82-4 Steol® CS-460
CAS 9004-82-4 Steol® OS 28
CAS 9004-82-4 Sulfochem ES-2
CAS 9004-82-4 Sulfochem ES-70
CAS 9004-82-4 Sulfotex 6040
CAS 9004-82-4 Sulfotex LMS-E
CAS 9004-82-4 Surfax 218
CAS 9004-82-4 Texapon LT-327
CAS 9004-82-4 Texapon N 25
CAS 9004-82-4 Texapon N 40
CAS 9004-82-4 Texapon N 70
CAS 9004-82-4 Texapon N 70 LS
CAS 9004-82-4 Texapon N 103
CAS 9004-82-4 Texapon NSE
CAS 9004-82-4 Texapon NSO
CAS 9004-82-4 Texapon PLT-227
CAS 9004-82-4 Texapon PN-235
CAS 9004-82-4 Texapon PN-254
CAS 9004-82-4 Ungerol CG27
CAS 9004-82-4 Ungerol LES 2-28
CAS 9004-82-4 Ungerol LES 2-70
CAS 9004-82-4 Ungerol LES 3-28
CAS 9004-82-4 Ungerol LES 3-54
CAS 9004-82-4 Ungerol LES 3-70
CAS 9004-82-4 Ungerol N2-28
CAS 9004-82-4 Ungerol N2-70
CAS 9004-82-4 Ungerol N3-28
CAS 9004-82-4 Ungerol N3-70
CAS 9004-82-4 Witcolate 6450
CAS 9004-82-4 Witcolate 6453
CAS 9004-82-4 Witcolate 6455
CAS 9004-82-4 Witcolate ES-2
CAS 9004-82-4 Witcolate ES-3
CAS 9004-82-4 Witcolate LES-60C
CAS 9004-82-4 Witcolate OME
CAS 9004-82-4 Witcolate S1285C
CAS 9004-82-4 Witcolate SE
CAS 9004-82-4 Zetesol NL
CAS 9004-82-4 Zoharpon ETA 27
CAS 9004-82-4 Zoharpon ETA 70

CAS 9004-82-4	Zoharpon ETA 270 (OXO)
CAS 9004-82-4	Zoharpon ETA 271
CAS 9004-82-4	Zoharpon ETA 273
CAS 9004-82-4	Zoharpon ETA 603
CAS 9004-82-4	Zoharpon ETA 700 (OXO)
CAS 9004-82-4	Zoharpon ETA 703
CAS 9004-83-5	Alcodet® HSC-1000
CAS 9004-83-5	Alcodet® IL-3500
CAS 9004-83-5	Alcodet® MC-2000
CAS 9004-83-5	Alcodet® TX 4000
CAS 9004-92-2	Hetoxol OL-2
CAS 9004-95-9	Brij® 56
CAS 9004-95-9	Brij® 58
CAS 9004-95-9	Cirrasol® ALN-WF
CAS 9004-95-9	Lipocol C-20
CAS 9004-95-9	Rolfor 162
CAS 9004-96-0	Ablunol 200MO
CAS 9004-96-0	Ablunol 400MO
CAS 9004-96-0	Ablunol 600MO
CAS 9004-96-0	Ablunol 1000MO
CAS 9004-96-0	Acconon 300-MO
CAS 9004-96-0	Acconon 400-MO
CAS 9004-96-0	Alkamuls® 400-MO
CAS 9004-96-0	Alkamuls® 600-MO
CAS 9004-96-0	Alkamuls® 783/P
CAS 9004-96-0	Alkamuls® A
CAS 9004-96-0	Alkamuls® AP
CAS 9004-96-0	Alkamuls® M-6
CAS 9004-96-0	Alkamuls® O-14
CAS 9004-96-0	Alphoxat O 105
CAS 9004-96-0	Alphoxat O 110
CAS 9004-96-0	Alphoxat O 115
CAS 9004-96-0	Atlas G-2143
CAS 9004-96-0	Chemax E-200 MO
CAS 9004-96-0	Chemax E-400 MO
CAS 9004-96-0	Chemax E-600 MO
CAS 9004-96-0	Chemax E-1000 MO
CAS 9004-96-0	Cithrol 2MO
CAS 9004-96-0	Cithrol 4MO
CAS 9004-96-0	Cithrol 6MO
CAS 9004-96-0	Cithrol 10MO
CAS 9004-96-0	Cithrol 40MO
CAS 9004-96-0	Cithrol 60MO
CAS 9004-96-0	Cithrol A
CAS 9004-96-0	Cithrol DGMO N/E
CAS 9004-96-0	Cithrol DGMO S/E
CAS 9004-96-0	CPH-39-N
CAS 9004-96-0	CPH-41-N
CAS 9004-96-0	Crodet O4
CAS 9004-96-0	Crodet O8
CAS 9004-96-0	Crodet O12
CAS 9004-96-0	Crodet O24
CAS 9004-96-0	Crodet O40
CAS 9004-96-0	Crodet O100
CAS 9004-96-0	Emerest® 2617
CAS 9004-96-0	Emerest® 2618
CAS 9004-96-0	Emerest® 2619
CAS 9004-96-0	Emerest® 2624
CAS 9004-96-0	Emerest® 2632
CAS 9004-96-0	Emerest® 2646
CAS 9004-96-0	Emerest® 2660
CAS 9004-96-0	Empilan® BQ 100
CAS 9004-96-0	Emulan® A
CAS 9004-96-0	Emulsan O
CAS 9004-96-0	Ethofat® O/15
CAS 9004-96-0	Ethofat® O/20
CAS 9004-96-0	Ethox MO-5
CAS 9004-96-0	Ethox MO-9
CAS 9004-96-0	Ethox MO-14
CAS 9004-96-0	Ethylan® A2
CAS 9004-96-0	Ethylan® A3
CAS 9004-96-0	Ethylan® A4
CAS 9004-96-0	Ethylan® A6
CAS 9004-96-0	Eumulgin PLT 4
CAS 9004-96-0	Eumulgin PLT 5
CAS 9004-96-0	Eumulgin PLT 6
CAS 9004-96-0	Genagen O-090
CAS 9004-96-0	Hetoxamate MO-2
CAS 9004-96-0	Hetoxamate MO-4
CAS 9004-96-0	Hetoxamate MO-5
CAS 9004-96-0	Hetoxamate MO-9
CAS 9004-96-0	Hetoxamate MO-15
CAS 9004-96-0	Hodag 40-O
CAS 9004-96-0	Hodag DGO
CAS 9004-96-0	Industrol® MO-5
CAS 9004-96-0	Industrol® MO-6
CAS 9004-96-0	Industrol® MO-7
CAS 9004-96-0	Industrol® MO-9
CAS 9004-96-0	Industrol® MO-13
CAS 9004-96-0	Ionet MO-200
CAS 9004-96-0	Karapeg 200-MO
CAS 9004-96-0	Karapeg 400-MO
CAS 9004-96-0	Karapeg 600-MO
CAS 9004-96-0	Karapeg DEG-MO
CAS 9004-96-0	Kessco® PEG 200 MO
CAS 9004-96-0	Kessco® PEG 300 MO
CAS 9004-96-0	Kessco® PEG 400 MO
CAS 9004-96-0	Kessco® PEG 600 MO
CAS 9004-96-0	Kessco® PEG 1000 MO
CAS 9004-96-0	Kessco® PEG 1540 MO
CAS 9004-96-0	Kessco® PEG 4000 MO
CAS 9004-96-0	Kessco® PEG 6000 MO
CAS 9004-96-0	Laurel PEG 400 MO
CAS 9004-96-0	Lonzest® PEG 4-O
CAS 9004-96-0	Mapeg® 200 MO
CAS 9004-96-0	Mapeg® 400 MO
CAS 9004-96-0	Mapeg® 600 MO
CAS 9004-96-0	Marlosol® OL2
CAS 9004-96-0	Marlosol® OL7
CAS 9004-96-0	Marlosol® OL8
CAS 9004-96-0	Marlosol® OL10
CAS 9004-96-0	Marlosol® OL15
CAS 9004-96-0	Marlosol® OL20
CAS 9004-96-0	Niox AK-40
CAS 9004-96-0	Niox AK-44
CAS 9004-96-0	Noigen ES 90
CAS 9004-96-0	Noigen ES 120
CAS 9004-96-0	Noigen ES 140
CAS 9004-96-0	Noigen ES 160
CAS 9004-96-0	Noiox AK-44
CAS 9004-96-0	Nonex O4E
CAS 9004-96-0	Nopalcol 4-O
CAS 9004-96-0	Nopalcol 6-O
CAS 9004-96-0	Pegosperse® 100 O
CAS 9004-96-0	Pegosperse® 400 MO
CAS 9004-96-0	Pegosperse® 700 TO
CAS 9004-96-0	Prox-onic OL-1/05
CAS 9004-96-0	Prox-onic OL-1/09
CAS 9004-96-0	Prox-onic OL-1/014
CAS 9004-96-0	Radiasurf® 7400
CAS 9004-96-0	Radiasurf® 7402
CAS 9004-96-0	Radiasurf® 7403
CAS 9004-96-0	Radiasurf® 7404
CAS 9004-96-0	Radiasurf® 7431
CAS 9004-96-0	Remcopal 6
CAS 9004-96-0	Remcopal 207
CAS 9004-96-0	Rolfat OL 6
CAS 9004-96-0	Rolfat OL 7
CAS 9004-96-0	Rolfat OL 9
CAS 9004-96-0	Secoster® MO 100

CAS 9004-96-0	Secoster® MO 400
CAS 9004-96-0	Sellig AO 6 100
CAS 9004-96-0	Sellig AO 9 100
CAS 9004-96-0	Sellig AO 15 100
CAS 9004-96-0	Sellig AO 25 100
CAS 9004-96-0	Serdox NOG 440
CAS 9004-96-0	Soprofor M/52
CAS 9004-96-0	Teric OF4
CAS 9004-96-0	Teric OF6
CAS 9004-96-0	Teric OF8
CAS 9004-96-0	Trydet 2676
CAS 9004-96-0	Trydet OA-5
CAS 9004-96-0	Trydet OA-7
CAS 9004-96-0	Trydet OA-10
CAS 9004-96-0	Witconol DOS
CAS 9004-96-0	Witconol H31A
CAS 9004-96-0	Witflow 916
CAS 9004-97-1	Cerex EL 4929
CAS 9004-97-1	Hodag 40-R
CAS 9004-97-1	Nopalcol 6-R
CAS 9004-97-1	Pegosperse® 100 MR
CAS 9004-98-2	Ablunol OA-6
CAS 9004-98-2	Ablunol OA-7
CAS 9004-98-2	Ahco 3998
CAS 9004-98-2	Akyporox RO 90
CAS 9004-98-2	Akyporox RTO 70
CAS 9004-98-2	Ameroxol® OE-2
CAS 9004-98-2	Arosurf® 32-E20
CAS 9004-98-2	Brij® 98G
CAS 9004-98-2	Brij® 99
CAS 9004-98-2	Chemal OA-4
CAS 9004-98-2	Chemal OA-5
CAS 9004-98-2	Chemal OA-9
CAS 9004-98-2	Chemal OA-20/70CWS
CAS 9004-98-2	Chemal OA-20G
CAS 9004-98-2	Dehydol 100
CAS 9004-98-2	Ethal OA-10
CAS 9004-98-2	Ethal OA-23
CAS 9004-98-2	Eumulgin M8
CAS 9004-98-2	Eumulgin O5
CAS 9004-98-2	Eumulgin O10
CAS 9004-98-2	Eumulgin PWM2
CAS 9004-98-2	Eumulgin PWM5
CAS 9004-98-2	Eumulgin PWM10
CAS 9004-98-2	Eumulgin PWM17
CAS 9004-98-2	Eumulgin PWM25
CAS 9004-98-2	Eumulgin WM5
CAS 9004-98-2	Genapol® O-020
CAS 9004-98-2	Genapol® O-050
CAS 9004-98-2	Genapol® O-080
CAS 9004-98-2	Genapol® O-090
CAS 9004-98-2	Genapol® O-100
CAS 9004-98-2	Genapol® O-120
CAS 9004-98-2	Genapol® O-150
CAS 9004-98-2	Genapol® O-200
CAS 9004-98-2	Genapol® O-230
CAS 9004-98-2	Hetoxol C-24
CAS 9004-98-2	Hetoxol OA-3 Special
CAS 9004-98-2	Hetoxol OA-5 Special
CAS 9004-98-2	Hetoxol OA-10 Special
CAS 9004-98-2	Hetoxol OA-20 Special
CAS 9004-98-2	Hetoxol OL-4
CAS 9004-98-2	Hetoxol OL-5
CAS 9004-98-2	Hetoxol OL-10
CAS 9004-98-2	Hetoxol OL-10H
CAS 9004-98-2	Hetoxol OL-20
CAS 9004-98-2	Hetoxol OL-23
CAS 9004-98-2	Hetoxol OL-40
CAS 9004-98-2	Hostacerin O-20
CAS 9004-98-2	Industrol® LG-70
CAS 9004-98-2	Industrol® LG-100
CAS 9004-98-2	Industrol® OAL-20
CAS 9004-98-2	Lipocol O-2
CAS 9004-98-2	Lipocol O-5
CAS 9004-98-2	Lipocol O-10
CAS 9004-98-2	Lipocol O-20
CAS 9004-98-2	Macol® OA-2
CAS 9004-98-2	Macol® OA-4
CAS 9004-98-2	Macol® OA-5
CAS 9004-98-2	Macol® OA-10
CAS 9004-98-2	Macol® OA-20
CAS 9004-98-2	Marlipal® 1850/5
CAS 9004-98-2	Marlipal® 1850/10
CAS 9004-98-2	Marlipal® 1850/30
CAS 9004-98-2	Marlipal® 1850/40
CAS 9004-98-2	Marlipal® 1850/80
CAS 9004-98-2	Marlowet® OA 5
CAS 9004-98-2	Marlowet® OA 10
CAS 9004-98-2	Marlowet® OA 30
CAS 9004-98-2	Marlowet® WOE
CAS 9004-98-2	Merpol® OJ
CAS 9004-98-2	Mulsifan RT 27
CAS 9004-98-2	Nikkol BO-2
CAS 9004-98-2	Nikkol BO-7
CAS 9004-98-2	Nikkol BO-10TX
CAS 9004-98-2	Nikkol BO-15TX
CAS 9004-98-2	Nikkol BO-20TX
CAS 9004-98-2	Nikkol BO-50
CAS 9004-98-2	Noigen ET60
CAS 9004-98-2	Noigen ET77
CAS 9004-98-2	Noigen ET80
CAS 9004-98-2	Noigen ET100
CAS 9004-98-2	Noigen ET120
CAS 9004-98-2	Noigen ET140
CAS 9004-98-2	Noigen ET180
CAS 9004-98-2	Noigen ET190S
CAS 9004-98-2	Prox-onic OA-1/04
CAS 9004-98-2	Prox-onic OA-1/09
CAS 9004-98-2	Prox-onic OA-1/020
CAS 9004-98-2	Prox-onic OA-2/020
CAS 9004-98-2	Rhodasurf® ON-870
CAS 9004-98-2	Rhodasurf® ON-877
CAS 9004-98-2	Ritoleth 2
CAS 9004-98-2	Ritoleth 5
CAS 9004-98-2	Ritoleth 10
CAS 9004-98-2	Ritoleth 20
CAS 9004-98-2	Rolfor O 8
CAS 9004-98-2	Rolfor O 22
CAS 9004-98-2	Serdox NOL 2
CAS 9004-98-2	Serdox NOL 8
CAS 9004-98-2	Serdox NOL 15
CAS 9004-98-2	Simulsol 92
CAS 9004-98-2	Simulsol 96
CAS 9004-98-2	Simulsol 98
CAS 9004-98-2	Trycol® 5971
CAS 9004-98-2	Trycol® 5972
CAS 9004-98-2	Trycol® OAL-23
CAS 9004-98-2	Varonic® 32-E20
CAS 9004-98-2	Volpo 3
CAS 9004-98-2	Volpo N3
CAS 9004-98-2	Volpo N5
CAS 9004-98-2	Volpo N10
CAS 9004-98-2	Volpo N15
CAS 9004-98-2	Volpo N20
CAS 9004-98-2	Volpo O3
CAS 9004-98-2	Volpo O5
CAS 9004-98-2	Volpo O10
CAS 9004-98-2	Volpo O15
CAS 9004-98-2	Volpo O20
CAS 9004-99-3	Ablunol 200MS

CAS 9004-99-3 Ablunol 400MS
CAS 9004-99-3 Ablunol 600MS
CAS 9004-99-3 Ablunol 1000MS
CAS 9004-99-3 Ablunol DEGMS
CAS 9004-99-3 Acconon 200-MS
CAS 9004-99-3 Acconon 400-MS
CAS 9004-99-3 Alkamuls® S-6
CAS 9004-99-3 Alkamuls® S-65-8
CAS 9004-99-3 Alkamuls® S-65-40
CAS 9004-99-3 Alphoxat S 110
CAS 9004-99-3 Alphoxat S 120
CAS 9004-99-3 Atlas G-2151
CAS 9004-99-3 Atlas G-2198
CAS 9004-99-3 Chemax E-200 MS
CAS 9004-99-3 Chemax E-400 MS
CAS 9004-99-3 Chemax E-600 MS
CAS 9004-99-3 Chemax E-1000 MS
CAS 9004-99-3 Cithrol 2MS
CAS 9004-99-3 Cithrol 3MS
CAS 9004-99-3 Cithrol 4MS
CAS 9004-99-3 Cithrol 6MS
CAS 9004-99-3 Cithrol 10MS
CAS 9004-99-3 Cithrol 40MS
CAS 9004-99-3 Cithrol DGMS N/E
CAS 9004-99-3 Cithrol DGMS S/E
CAS 9004-99-3 Crodet S4
CAS 9004-99-3 Crodet S8
CAS 9004-99-3 Crodet S12
CAS 9004-99-3 Crodet S24
CAS 9004-99-3 Crodet S40
CAS 9004-99-3 Crodet S100
CAS 9004-99-3 DMS-33
CAS 9004-99-3 Drewmulse® DGMS
CAS 9004-99-3 DyaFac 6-S
CAS 9004-99-3 Emerest® 2610
CAS 9004-99-3 Emerest® 2636
CAS 9004-99-3 Emerest® 2640
CAS 9004-99-3 Emerest® 2641
CAS 9004-99-3 Emerest® 2662
CAS 9004-99-3 Emerest® 2675
CAS 9004-99-3 Emerest® 2711
CAS 9004-99-3 Emerest® 2715
CAS 9004-99-3 Ethofat® 18/14
CAS 9004-99-3 Ethofat® 60/15
CAS 9004-99-3 Ethofat® 60/20
CAS 9004-99-3 Ethofat® 60/25
CAS 9004-99-3 Ethox MS-8
CAS 9004-99-3 Ethox MS-14
CAS 9004-99-3 Ethox MS-23
CAS 9004-99-3 Ethox MS-40
CAS 9004-99-3 Eumulgin PST 5
CAS 9004-99-3 Eumulgin ST-8
CAS 9004-99-3 Genagen S-080
CAS 9004-99-3 Genagen S-400
CAS 9004-99-3 Hetoxamate SA-5
CAS 9004-99-3 Hetoxamate SA-7
CAS 9004-99-3 Hetoxamate SA-9
CAS 9004-99-3 Hetoxamate SA-13
CAS 9004-99-3 Hetoxamate SA-23
CAS 9004-99-3 Hetoxamate SA-35
CAS 9004-99-3 Hetoxamate SA-40
CAS 9004-99-3 Hetoxamate SA-90
CAS 9004-99-3 Hetoxamate SA-90F
CAS 9004-99-3 Hodag 40-S
CAS 9004-99-3 Hodag 60-S
CAS 9004-99-3 Hodag 100-S
CAS 9004-99-3 Hodag 150-S
CAS 9004-99-3 Hodag DGS
CAS 9004-99-3 Hydrine
CAS 9004-99-3 Industrol® MS-5
CAS 9004-99-3 Industrol® MS-7
CAS 9004-99-3 Industrol® MS-8
CAS 9004-99-3 Industrol® MS-9
CAS 9004-99-3 Industrol® MS-23
CAS 9004-99-3 Industrol® MS-40
CAS 9004-99-3 Industrol® MS-100
CAS 9004-99-3 Ionet MS-400
CAS 9004-99-3 Karapeg 200-MS
CAS 9004-99-3 Karapeg DEG-MS
CAS 9004-99-3 Kemester® 5221SE
CAS 9004-99-3 Kessco® PEG 200 MS
CAS 9004-99-3 Kessco® PEG 300 MS
CAS 9004-99-3 Kessco® PEG 400 MS
CAS 9004-99-3 Kessco® PEG 600 MS
CAS 9004-99-3 Kessco® PEG 1000 MS
CAS 9004-99-3 Kessco® PEG 1540 MS
CAS 9004-99-3 Kessco® PEG 4000 MS
CAS 9004-99-3 Kessco® PEG 6000 MS
CAS 9004-99-3 Lipal DGMS
CAS 9004-99-3 Lipo DGS-SE
CAS 9004-99-3 Mapeg® 200 MS
CAS 9004-99-3 Mapeg® 400 MS
CAS 9004-99-3 Mapeg® 600 MS
CAS 9004-99-3 Mapeg® 1000 MS
CAS 9004-99-3 Mapeg® 1500 MS
CAS 9004-99-3 Mapeg® S-40
CAS 9004-99-3 Mapeg® S-40K
CAS 9004-99-3 Marlosol® 183
CAS 9004-99-3 Marlosol® 186
CAS 9004-99-3 Marlosol® 188
CAS 9004-99-3 Marlosol® 189
CAS 9004-99-3 Marlosol® 1820
CAS 9004-99-3 Mergital ST 30/E
CAS 9004-99-3 Nonex S3E
CAS 9004-99-3 Nonisol 300
CAS 9004-99-3 Nopalcol 1-S
CAS 9004-99-3 Nopalcol 4-S
CAS 9004-99-3 Nopalcol 6-S
CAS 9004-99-3 Nopalcol 30-S
CAS 9004-99-3 Pegosperse® 50 MS
CAS 9004-99-3 Pegosperse® 100 S
CAS 9004-99-3 Pegosperse® 400 MS
CAS 9004-99-3 Pegosperse® 600 MS
CAS 9004-99-3 Pegosperse® 1000 MS
CAS 9004-99-3 Pegosperse® 1500 MS
CAS 9004-99-3 Pegosperse® 1750 MS
CAS 9004-99-3 Pegosperse® 4000 MS
CAS 9004-99-3 PG No. 4
CAS 9004-99-3 PGE-400-MS
CAS 9004-99-3 PGE-600-MS
CAS 9004-99-3 Prox-onic ST-05
CAS 9004-99-3 Prox-onic ST-09
CAS 9004-99-3 Prox-onic ST-014
CAS 9004-99-3 Prox-onic ST-023
CAS 9004-99-3 Radiasurf® 7410
CAS 9004-99-3 Radiasurf® 7411
CAS 9004-99-3 Radiasurf® 7412
CAS 9004-99-3 Radiasurf® 7413
CAS 9004-99-3 Radiasurf® 7414
CAS 9004-99-3 Radiasurf® 7417
CAS 9004-99-3 Radiasurf® 7432
CAS 9004-99-3 Rolfat SG
CAS 9004-99-3 Rolfat ST 40
CAS 9004-99-3 Rolfat ST 100
CAS 9004-99-3 RS-55-40
CAS 9004-99-3 Sellig S 30 100
CAS 9004-99-3 Sellig Stearo 6
CAS 9004-99-3 Serdox NSG 400
CAS 9004-99-3 Simulsol M 45
CAS 9004-99-3 Simulsol M 49

CAS 9004-99-3 Simulsol M 51
CAS 9004-99-3 Simulsol M 52
CAS 9004-99-3 Simulsol M 53
CAS 9004-99-3 Simulsol M 59
CAS 9004-99-3 Stearate PEG 1000
CAS 9004-99-3 Sterol ST 2
CAS 9004-99-3 Tegin® D 6100
CAS 9004-99-3 Tegin® DGS
CAS 9004-99-3 Teric SF15
CAS 9004-99-3 Teric SF9
CAS 9004-99-3 Trydet 2610
CAS 9004-99-3 Trydet 2636
CAS 9004-99-3 Trydet 2640
CAS 9004-99-3 Trydet 2670
CAS 9004-99-3 Trydet 2671
CAS 9004-99-3 Trydet 2675
CAS 9004-99-3 Trydet SA-5
CAS 9004-99-3 Trydet SA-7
CAS 9004-99-3 Trydet SA-8
CAS 9004-99-3 Trydet SA-23
CAS 9004-99-3 Trydet SA-40
CAS 9004-99-3 Trydet SA-50/30
CAS 9004-99-3 Witconol 2640
CAS 9004-99-3 Witconol H35A
CAS 9005-00-9 Ablunol SA-7
CAS 9005-00-9 Brij® 72
CAS 9005-00-9 Brij® 700 S
CAS 9005-00-9 Brij® 721 S
CAS 9005-00-9 Hetoxol STA-2
CAS 9005-00-9 Hetoxol STA-10
CAS 9005-00-9 Hetoxol STA-20
CAS 9005-00-9 Hetoxol STA-30
CAS 9005-00-9 Lipocol S-2
CAS 9005-00-9 Lipocol S-10
CAS 9005-00-9 Lipocol S-20
CAS 9005-00-9 Macol® SA-2
CAS 9005-00-9 Macol® SA-5
CAS 9005-00-9 Macol® SA-10
CAS 9005-00-9 Macol® SA-15
CAS 9005-00-9 Macol® SA-20
CAS 9005-00-9 Macol® SA-40
CAS 9005-00-9 Prox-onic SA-1/02
CAS 9005-00-9 Prox-onic SA-1/010
CAS 9005-00-9 Prox-onic SA-1/020
CAS 9005-00-9 Rhodasurf® TB-970
CAS 9005-00-9 Simulsol 72
CAS 9005-00-9 Simulsol 76
CAS 9005-00-9 Simulsol 78
CAS 9005-00-9 Trycol® 5888
CAS 9005-00-9 Trycol® SAL-20
CAS 9005-00-9 Volpo S-2
CAS 9005-02-1 Acconon 200-DL
CAS 9005-02-1 Cithrol 2DL
CAS 9005-02-1 Cithrol 4DL
CAS 9005-02-1 Cithrol 6DL
CAS 9005-02-1 Cithrol 10DL
CAS 9005-02-1 Cithrol DGDL N/E
CAS 9005-02-1 Cithrol DGDL S/E
CAS 9005-02-1 CPH-79-N
CAS 9005-02-1 Emerest® 2622
CAS 9005-02-1 Emerest® 2652
CAS 9005-02-1 Emerest® 2704
CAS 9005-02-1 Ethox DL-5
CAS 9005-02-1 Ethox DL-9
CAS 9005-02-1 Hetoxamate 200 DL
CAS 9005-02-1 Hodag 22-L
CAS 9005-02-1 Hodag 42-L
CAS 9005-02-1 Industrol® DL-9
CAS 9005-02-1 Ionet DL-200
CAS 9005-02-1 Karapeg 200-DL
CAS 9005-02-1 Karapeg 400-DL
CAS 9005-02-1 Karapeg 600-DL
CAS 9005-02-1 Kessco® PEG 200 DL
CAS 9005-02-1 Kessco® PEG 300 DL
CAS 9005-02-1 Kessco® PEG 400 DL
CAS 9005-02-1 Kessco® PEG 600 DL
CAS 9005-02-1 Kessco® PEG 1000 DL
CAS 9005-02-1 Kessco® PEG 1540 DL
CAS 9005-02-1 Kessco® PEG 4000 DL
CAS 9005-02-1 Kessco® PEG 6000 DL
CAS 9005-02-1 Lexemul® PEG-200 DL
CAS 9005-02-1 Lexemul® PEG-400 DL
CAS 9005-02-1 Mapeg® 200 DL
CAS 9005-02-1 Mapeg® 400 DL
CAS 9005-02-1 Mapeg® 600 DL
CAS 9005-02-1 Nonex DL-2
CAS 9005-02-1 Nopalcol 2-DL
CAS 9005-02-1 Pegosperse® 200 DL
CAS 9005-02-1 Pegosperse® 400 DL
CAS 9005-02-1 Witconol 2622
CAS 9005-07-6 Acconon 400-DO
CAS 9005-07-6 Alkamuls® 400-DO
CAS 9005-07-6 Alkamuls® 600-DO
CAS 9005-07-6 Chemax PEG 200 DO
CAS 9005-07-6 Chemax PEG 400 DO
CAS 9005-07-6 Chemax PEG 600 DO
CAS 9005-07-6 Cithrol 2DO
CAS 9005-07-6 Cithrol 4DO
CAS 9005-07-6 Cithrol 6DO
CAS 9005-07-6 Cithrol 10DO
CAS 9005-07-6 Cithrol DGDO N/E
CAS 9005-07-6 Cithrol DGDO S/E
CAS 9005-07-6 Dyafac PEG 6DO
CAS 9005-07-6 Emerest® 2648
CAS 9005-07-6 Emerest® 2665
CAS 9005-07-6 Ethox DO-9
CAS 9005-07-6 Ethox DO-14
CAS 9005-07-6 Hodag 42-O
CAS 9005-07-6 Hodag 62-O
CAS 9005-07-6 Industrol® DO-9
CAS 9005-07-6 Industrol® DO-13
CAS 9005-07-6 Ionet DO-200
CAS 9005-07-6 Ionet DO-400
CAS 9005-07-6 Ionet DO-600
CAS 9005-07-6 Ionet DO-1000
CAS 9005-07-6 Karapeg 200-DO
CAS 9005-07-6 Karapeg 400-DO
CAS 9005-07-6 Karapeg 600-DO
CAS 9005-07-6 Karapeg DEG-DO
CAS 9005-07-6 Kessco® PEG 200 DO
CAS 9005-07-6 Kessco® PEG 300 DO
CAS 9005-07-6 Kessco® PEG 400 DO
CAS 9005-07-6 Kessco® PEG 600 DO
CAS 9005-07-6 Kessco® PEG 1000 DO
CAS 9005-07-6 Kessco® PEG 1540 DO
CAS 9005-07-6 Kessco® PEG 4000 DO
CAS 9005-07-6 Kessco® PEG 6000 DO
CAS 9005-07-6 Lonzest® PEG 4-DO
CAS 9005-07-6 Mapeg® 200 DO
CAS 9005-07-6 Mapeg® 400 DO
CAS 9005-07-6 Mapeg® 600 DO
CAS 9005-07-6 Marlipal® FS
CAS 9005-07-6 Marlosol® FS
CAS 9005-07-6 Nonex DO-4
CAS 9005-07-6 Nonisol 210
CAS 9005-07-6 Nopalcol 6-DO
CAS 9005-07-6 Pegosperse® 400 DO
CAS 9005-07-6 Pegosperse® 1500 DO
CAS 9005-07-6 Radiasurf® 7443
CAS 9005-07-6 Radiasurf® 7444

CAS 9005-07-6 Secoster® DO 600
CAS 9005-07-6 Surfax 8916/A
CAS 9005-07-6 Trydet DO-9
CAS 9005-07-6 Witconol 2648
CAS 9005-07-6 Witconol H33
CAS 9005-08-7 Ablunol 6000DS
CAS 9005-08-7 Alkamuls® 400-DS
CAS 9005-08-7 Cithrol 2DS
CAS 9005-08-7 Cithrol 3DS
CAS 9005-08-7 Cithrol 4DS
CAS 9005-08-7 Cithrol 6DS
CAS 9005-08-7 Cithrol 10DS
CAS 9005-08-7 Cithrol DGDS N/E
CAS 9005-08-7 Cithrol DGDS S/E
CAS 9005-08-7 Emerest® 2642
CAS 9005-08-7 Emerest® 2712
CAS 9005-08-7 Emulmin 862
CAS 9005-08-7 Hetoxamate 400 DS
CAS 9005-08-7 Hodag 42-S
CAS 9005-08-7 Ionet DS-300
CAS 9005-08-7 Ionet DS-400
CAS 9005-08-7 Karapeg 400-DS
CAS 9005-08-7 Karapeg 6000-DS
CAS 9005-08-7 Kessco® PEG 200 DS
CAS 9005-08-7 Kessco® PEG 300 DS
CAS 9005-08-7 Kessco® PEG 400 DS
CAS 9005-08-7 Kessco® PEG 600 DS
CAS 9005-08-7 Kessco® PEG 1000 DS
CAS 9005-08-7 Kessco® PEG 1540 DS
CAS 9005-08-7 Kessco® PEG 4000 DS
CAS 9005-08-7 Kessco® PEG 6000 DS
CAS 9005-08-7 Mapeg® 200 DS
CAS 9005-08-7 Mapeg® 400 DS
CAS 9005-08-7 Mapeg® 600 DS
CAS 9005-08-7 Mapeg® 1540 DS
CAS 9005-08-7 Mapeg® 6000 DS
CAS 9005-08-7 Marlosol® BS
CAS 9005-08-7 Pegosperse® 400 DS
CAS 9005-08-7 Pegosperse® 6000 DS
CAS 9005-08-7 PGE-400-DS
CAS 9005-08-7 PGE-600-DS
CAS 9005-08-7 Radiasurf® 7453
CAS 9005-08-7 Radiasurf® 7454
CAS 9005-08-7 Witconol 2642
CAS 9005-64-5 Accosperse 20
CAS 9005-64-5 Ahco 7596D
CAS 9005-64-5 Ahco 7596T
CAS 9005-64-5 Alkamuls® PSML-20
CAS 9005-64-5 Alkamuls® T-20
CAS 9005-64-5 Armotan® PML 20
CAS 9005-64-5 Crillet 1
CAS 9005-64-5 Crillet 11
CAS 9005-64-5 Disponil SML 104 F1
CAS 9005-64-5 Disponil SML 120 F1
CAS 9005-64-5 Drewmulse® POE-SML
CAS 9005-64-5 Emasol L-106
CAS 9005-64-5 Emasol L-120
CAS 9005-64-5 Emsorb® 6915
CAS 9005-64-5 Emsorb® 6916
CAS 9005-64-5 Ethsorbox L-20
CAS 9005-64-5 Ethylan® GEL2
CAS 9005-64-5 Ethylan® GLE-21
CAS 9005-64-5 Glycosperse® L-20
CAS 9005-64-5 Hetsorb L-4
CAS 9005-64-5 Hetsorb L-20
CAS 9005-64-5 Hodag PSML-20
CAS 9005-64-5 Industrol® L-20-S
CAS 9005-64-5 Ionet T-20 C
CAS 9005-64-5 Liposorb L-20
CAS 9005-64-5 Lonzest® SML-20
CAS 9005-64-5 ML-55-F
CAS 9005-64-5 ML-55-F-4
CAS 9005-64-5 Montanox 20 DF
CAS 9005-64-5 Nissan Nonion LT-221
CAS 9005-64-5 Norfox® Sorbo T-20
CAS 9005-64-5 Prox-onic SML-020
CAS 9005-64-5 Radiasurf® 7137
CAS 9005-64-5 Secoster® CL 10
CAS 9005-64-5 Sorbax PML-20
CAS 9005-64-5 Sorbon T-20
CAS 9005-64-5 T-Maz® 20
CAS 9005-64-5 Tween® 20
CAS 9005-64-5 Tween® 21
CAS 9005-64-5 Witconol 2720
CAS 9005-64-5 Witflow 990
CAS 9005-65-6 Accosperse 80
CAS 9005-65-6 Ahco DFO-100
CAS 9005-65-6 Ahco DFO-150
CAS 9005-65-6 Alkamuls® PSMO-5
CAS 9005-65-6 Alkamuls® PSMO-20
CAS 9005-65-6 Alkamuls® T-80
CAS 9005-65-6 Armotan® PMO 20
CAS 9005-65-6 Atlas G-4905
CAS 9005-65-6 Atlox 8916TF
CAS 9005-65-6 Crillet 4
CAS 9005-65-6 Crillet 41
CAS 9005-65-6 Desonic® SMO-20
CAS 9005-65-6 Desotan® SMO-20
CAS 9005-65-6 Disponil SMO 120 F1
CAS 9005-65-6 Drewmulse® POE-SMO
CAS 9005-65-6 Durfax® 80
CAS 9005-65-6 Emasol O-106
CAS 9005-65-6 Emasol O-120
CAS 9005-65-6 Emsorb® 2722
CAS 9005-65-6 Emsorb® 6900
CAS 9005-65-6 Emsorb® 6901
CAS 9005-65-6 Ethsorbox O-20
CAS 9005-65-6 Ethylan® GEO8
CAS 9005-65-6 Ethylan® GEO81
CAS 9005-65-6 Ethylan® GOE-21
CAS 9005-65-6 Glycosperse® O-5
CAS 9005-65-6 Glycosperse® O-20
CAS 9005-65-6 Glycosperse® O-20 FG, O-20 KFG
CAS 9005-65-6 Glycosperse® O-20 Veg
CAS 9005-65-6 Glycosperse® O-20X
CAS 9005-65-6 Hetsorb O-20
CAS 9005-65-6 Hodag PSMO-20
CAS 9005-65-6 Industrol® O-20-S
CAS 9005-65-6 Ionet T-80 C
CAS 9005-65-6 Liposorb O-5
CAS 9005-65-6 Lonzest® SMO-20
CAS 9005-65-6 Montanox 80 DF
CAS 9005-65-6 Montanox 81
CAS 9005-65-6 Prox-onic SMO-05
CAS 9005-65-6 Prox-onic SMO-020
CAS 9005-65-6 Radiasurf® 7157
CAS 9005-65-6 Sorbanox AOM
CAS 9005-65-6 Sorbax PMO-5
CAS 9005-65-6 Sorbax PMO-20
CAS 9005-65-6 Sorbilene O
CAS 9005-65-6 T-Maz® 80
CAS 9005-65-6 T-Maz® 80KLM
CAS 9005-65-6 T-Maz® 81
CAS 9005-65-6 Tween® 81
CAS 9005-65-6 Witconol 2722
CAS 9005-65-6 Witflow 991
CAS 9005-66-7 Ahco DFP-156
CAS 9005-66-7 Crillet 2
CAS 9005-66-7 Disponil SMP 120 F1
CAS 9005-66-7 Emasol P-120

CAS 9005-66-7	Emsorb® 6910
CAS 9005-66-7	Ethylan® GEP4
CAS 9005-66-7	Glycosperse® P-20
CAS 9005-66-7	Hetsorb P-20
CAS 9005-66-7	Hodag PSMP-20
CAS 9005-66-7	Liposorb P-20
CAS 9005-66-7	Lonzest® SMP-20
CAS 9005-66-7	Montanox 40 DF
CAS 9005-66-7	Prox-onic SMP-020
CAS 9005-66-7	Rheodol TW-P120
CAS 9005-66-7	Secoster® CP 10
CAS 9005-66-7	Sorbax PMP-20
CAS 9005-66-7	T-Maz® 40
CAS 9005-66-7	Tween® 40
CAS 9005-67-8	Accosperse 60
CAS 9005-67-8	Ahco DFS-96
CAS 9005-67-8	Ahco DFS-149
CAS 9005-67-8	Alkamuls® PSMS-20
CAS 9005-67-8	Alkamuls® T-60
CAS 9005-67-8	Crillet 3
CAS 9005-67-8	Crillet 31
CAS 9005-67-8	Disponil SMS 120 F1
CAS 9005-67-8	Drewmulse® POE-SMS
CAS 9005-67-8	Durfax® 60
CAS 9005-67-8	Emasol S-106
CAS 9005-67-8	Emasol S-120
CAS 9005-67-8	Emsorb® 2728
CAS 9005-67-8	Emsorb® 6906
CAS 9005-67-8	Emsorb® 6909
CAS 9005-67-8	Ethsorbox S-20
CAS 9005-67-8	Ethylan® GES6
CAS 9005-67-8	Glycosperse® S-20
CAS 9005-67-8	Hetsorb S-4
CAS 9005-67-8	Hodag PSMS-20
CAS 9005-67-8	Industrol® S-20-S
CAS 9005-67-8	Ionet T-60 C
CAS 9005-67-8	Lonzest® SMS-20
CAS 9005-67-8	Montanox 60 DF
CAS 9005-67-8	Montanox 61
CAS 9005-67-8	MS-55-F
CAS 9005-67-8	Prox-onic SMS-020
CAS 9005-67-8	Radiasurf® 7147
CAS 9005-67-8	Secoster® CS 10
CAS 9005-67-8	Sorbanox AST
CAS 9005-67-8	Sorbax PMS-20
CAS 9005-67-8	T-Maz® 60
CAS 9005-67-8	T-Maz® 61
CAS 9005-67-8	Tween® 61
CAS 9005-70-3	Ahco DFO-110
CAS 9005-70-3	Alkamuls® PSTO-20
CAS 9005-70-3	Alkamuls® T-85
CAS 9005-70-3	Crillet 45
CAS 9005-70-3	Disponil STO 120 F1
CAS 9005-70-3	Emasol O-320
CAS 9005-70-3	Emsorb® 6903
CAS 9005-70-3	Emsorb® 6913
CAS 9005-70-3	Ethsorbox TO-20
CAS 9005-70-3	Ethylan® GPS85
CAS 9005-70-3	Glycosperse® TO-20
CAS 9005-70-3	Hetsorb TO-20
CAS 9005-70-3	Liposorb TO-20
CAS 9005-70-3	Lonzest® STO-20
CAS 9005-70-3	Montanox 85
CAS 9005-70-3	Prox-onic STO-020
CAS 9005-70-3	Sorbanox CO
CAS 9005-70-3	Sorbax PTO-20
CAS 9005-70-3	T-Maz® 85
CAS 9005-70-3	TO-55-E
CAS 9005-70-3	TO-55-EL
CAS 9005-70-3	TO-55-F
CAS 9005-70-3	Tween® 85
CAS 9005-70-3	Witconol 6903
CAS 9005-71-4	Ahco 7166T
CAS 9005-71-4	Crillet 35
CAS 9005-71-4	Disponil STS 120 F1
CAS 9005-71-4	Drewmulse® POE-STS
CAS 9005-71-4	Durfax® 65
CAS 9005-71-4	Emasol S-320
CAS 9005-71-4	Ethsorbox TS-20
CAS 9005-71-4	Glycosperse® TS-20
CAS 9005-71-4	Hetsorb TS-20
CAS 9005-71-4	Hodag PSTS-20
CAS 9005-71-4	Industrol® STS-20-S
CAS 9005-71-4	Lonzest® STS-20
CAS 9005-71-4	Montanox 65
CAS 9005-71-4	Prox-onic STS-020
CAS 9005-71-4	Sorbax PTS-20
CAS 9005-71-4	T-Maz® 65
CAS 9010-43-9	Syntopon 8 C
CAS 9011-16-9	Gantrez® AN-179
CAS 9014-85-1	Surfynol® 440
CAS 9014-85-1	Surfynol® 465
CAS 9014-85-1	Surfynol® 485
CAS 9014-90-8	Akyposal NPS 100
CAS 9014-90-8	Akyposal NPS 250
CAS 9014-90-8	Gardisperse AC
CAS 9014-90-8	Sermul EA 54
CAS 9014-90-8	Sermul EA 146
CAS 9014-90-8	Sermul EA 151
CAS 9014-90-8	Witcolate D51-51EP
CAS 9014-90-8	Zorapol SN-9
CAS 9014-92-0	Noigen EA 33
CAS 9014-92-0	Noigen EA 73
CAS 9014-92-0	Noigen EA 80
CAS 9014-92-0	Noigen EA 143
CAS 9014-92-0	T-Det® DD-5
CAS 9014-92-0	T-Det® DD-7
CAS 9014-92-0	T-Det® DD-9
CAS 9014-92-0	T-Det® DD-10
CAS 9014-92-0	T-Det® DD-11
CAS 9014-92-0	T-Det® DD-14
CAS 9014-93-1	Chemax DNP-8
CAS 9014-93-1	Chemax DNP-15
CAS 9014-93-1	Chemax DNP-18
CAS 9014-93-1	Chemax DNP-150
CAS 9014-93-1	Chemax DNP-150/50
CAS 9014-93-1	Ethal DNP-8
CAS 9014-93-1	Ethal DNP-18
CAS 9014-93-1	Ethal DNP-150/50%
CAS 9014-93-1	Hetoxide DNP-4
CAS 9014-93-1	Hetoxide DNP-5
CAS 9014-93-1	Hetoxide DNP-9.6
CAS 9014-93-1	Hetoxide DNP-10
CAS 9014-93-1	Iconol DNP-8
CAS 9014-93-1	Iconol DNP-24
CAS 9014-93-1	Iconol DNP-150
CAS 9014-93-1	Igepal® DM-430
CAS 9014-93-1	Igepal® DM-530
CAS 9014-93-1	Igepal® DM-710
CAS 9014-93-1	Igepal® DM-730
CAS 9014-93-1	Igepal® DM-880
CAS 9014-93-1	Igepal® DM-970 FLK
CAS 9014-93-1	Macol® DNP-5
CAS 9014-93-1	Macol® DNP-10
CAS 9014-93-1	Macol® DNP-15
CAS 9014-93-1	Macol® DNP-21
CAS 9014-93-1	Macol® DNP-150
CAS 9014-93-1	Marlophen® DNP 16
CAS 9014-93-1	Marlophen® DNP 18
CAS 9014-93-1	Marlophen® DNP 30

CAS Number	Tradename
CAS 9014-93-1	Marlophen® DNP 150
CAS 9014-93-1	Nonipol D-160
CAS 9014-93-1	Prox-onic DNP-08
CAS 9014-93-1	Prox-onic DNP-0150
CAS 9014-93-1	Prox-onic DNP-0150/50
CAS 9014-93-1	Sellig DN 10 100
CAS 9014-93-1	Sellig DN 22 100
CAS 9014-93-1	Serdox NDI 100
CAS 9014-93-1	T-Det® D-70
CAS 9014-93-1	T-Det® D-150
CAS 9014-93-1	Trycol® 6985
CAS 9014-93-1	Trycol® 6989
CAS 9014-93-1	Trycol® DNP-8
CAS 9014-93-1	Trycol® DNP-150
CAS 9014-93-1	Trycol® DNP-150/50
CAS 9016-45-9	Ablunol NP4
CAS 9016-45-9	Ablunol NP6
CAS 9016-45-9	Ablunol NP8
CAS 9016-45-9	Ablunol NP9
CAS 9016-45-9	Ablunol NP10
CAS 9016-45-9	Ablunol NP12
CAS 9016-45-9	Ablunol NP15
CAS 9016-45-9	Ablunol NP16
CAS 9016-45-9	Ablunol NP20
CAS 9016-45-9	Ablunol NP30
CAS 9016-45-9	Ablunol NP30 70%
CAS 9016-45-9	Ablunol NP40
CAS 9016-45-9	Ablunol NP40 70%
CAS 9016-45-9	Ablunol NP50
CAS 9016-45-9	Ablunol NP50 70%
CAS 9016-45-9	Akyporox NP 30
CAS 9016-45-9	Akyporox NP 40
CAS 9016-45-9	Akyporox NP 90
CAS 9016-45-9	Akyporox NP 95
CAS 9016-45-9	Akyporox NP 105
CAS 9016-45-9	Akyporox NP 150
CAS 9016-45-9	Akyporox NP 200
CAS 9016-45-9	Akyporox NP 300V
CAS 9016-45-9	Akyporox NP 1200V
CAS 9016-45-9	Alkasurf® NP-4
CAS 9016-45-9	Alkasurf® NP-6
CAS 9016-45-9	Alkasurf® NP-8
CAS 9016-45-9	Alkasurf® NP-9
CAS 9016-45-9	Alkasurf® NP-10
CAS 9016-45-9	Alkasurf® NP-11
CAS 9016-45-9	Alkasurf® NP-12
CAS 9016-45-9	Alkasurf® NP-15
CAS 9016-45-9	Alkasurf® NP-15, 80%
CAS 9016-45-9	Alkasurf® NP-30, 70%
CAS 9016-45-9	Alkasurf® NP-40, 70%
CAS 9016-45-9	Alkasurf® NP-50 70%
CAS 9016-45-9	Arkopal N040
CAS 9016-45-9	Arkopal N060
CAS 9016-45-9	Arkopal N080
CAS 9016-45-9	Arkopal N090
CAS 9016-45-9	Arkopal N100
CAS 9016-45-9	Arkopal N110
CAS 9016-45-9	Arkopal N130
CAS 9016-45-9	Arkopal N150
CAS 9016-45-9	Arkopal N230
CAS 9016-45-9	Caloxylate N-9
CAS 9016-45-9	Carsonon® N-4
CAS 9016-45-9	Carsonon® N-6
CAS 9016-45-9	Carsonon® N-8
CAS 9016-45-9	Carsonon® N-9
CAS 9016-45-9	Carsonon® N-10
CAS 9016-45-9	Carsonon® N-11
CAS 9016-45-9	Carsonon® N-12
CAS 9016-45-9	Carsonon® N-30
CAS 9016-45-9	Carsonon® N-40, 70%
CAS 9016-45-9	Carsonon® N-50
CAS 9016-45-9	Carsonon® N-100, 70%
CAS 9016-45-9	Cedepal CO-430
CAS 9016-45-9	Cedepal CO-530
CAS 9016-45-9	Cedepal CO-610
CAS 9016-45-9	Cedepal CO-630
CAS 9016-45-9	Cedepal CO-710
CAS 9016-45-9	Chemax NP-1.5
CAS 9016-45-9	Chemax NP-4
CAS 9016-45-9	Chemax NP-6
CAS 9016-45-9	Chemax NP-9
CAS 9016-45-9	Chemax NP-10
CAS 9016-45-9	Chemax NP-15
CAS 9016-45-9	Chemax NP-20
CAS 9016-45-9	Chemax NP-30
CAS 9016-45-9	Chemax NP-30/70
CAS 9016-45-9	Chemax NP-40
CAS 9016-45-9	Chemax NP-40/70
CAS 9016-45-9	Chemax NP-50
CAS 9016-45-9	Chemax NP-50/70
CAS 9016-45-9	Chemax NP-100
CAS 9016-45-9	Chemax NP-100/70
CAS 9016-45-9	Dehydrophen PNP 4
CAS 9016-45-9	Dehydrophen PNP 6
CAS 9016-45-9	Dehydrophen PNP 8
CAS 9016-45-9	Dehydrophen PNP 10
CAS 9016-45-9	Dehydrophen PNP 12
CAS 9016-45-9	Dehydrophen PNP 15
CAS 9016-45-9	Dehydrophen PNP 20
CAS 9016-45-9	Dehydrophen PNP 30
CAS 9016-45-9	Dehydrophen PNP 40
CAS 9016-45-9	Desonic® 4N
CAS 9016-45-9	Desonic® 5N
CAS 9016-45-9	Desonic® 6N
CAS 9016-45-9	Desonic® 7N
CAS 9016-45-9	Desonic® 9N
CAS 9016-45-9	Desonic® 10N
CAS 9016-45-9	Desonic® 11N
CAS 9016-45-9	Desonic® 12N
CAS 9016-45-9	Desonic® 13N
CAS 9016-45-9	Desonic® 15N
CAS 9016-45-9	Desonic® 20N
CAS 9016-45-9	Desonic® 30N
CAS 9016-45-9	Desonic® 30N70
CAS 9016-45-9	Desonic® 40N
CAS 9016-45-9	Desonic® 40N70
CAS 9016-45-9	Desonic® 50N
CAS 9016-45-9	Desonic® 50N70
CAS 9016-45-9	Desonic® 100N
CAS 9016-45-9	Desonic® 100N70
CAS 9016-45-9	Elfapur® N 50
CAS 9016-45-9	Elfapur® N 70
CAS 9016-45-9	Elfapur® N 90
CAS 9016-45-9	Elfapur® N 120
CAS 9016-45-9	Elfapur® N 150
CAS 9016-45-9	Empilan® NP9
CAS 9016-45-9	Emthox® 6957
CAS 9016-45-9	Emthox® 6961
CAS 9016-45-9	Emthox® 6962
CAS 9016-45-9	Emulgator U4
CAS 9016-45-9	Emulgator U6
CAS 9016-45-9	Emulgator U9
CAS 9016-45-9	Emulgator U12
CAS 9016-45-9	Emulmin 140
CAS 9016-45-9	Emulmin 240
CAS 9016-45-9	Emulson AG 2A
CAS 9016-45-9	Emulson AG 4B
CAS 9016-45-9	Emulson AG 7B
CAS 9016-45-9	Emulson AG 9B
CAS 9016-45-9	Emulson AG 10B

CAS 9016-45-9	Emulson AG 12B
CAS 9016-45-9	Emulson AG 20B
CAS 9016-45-9	Ethal NP-4
CAS 9016-45-9	Ethal NP-6
CAS 9016-45-9	Ethal NP-9
CAS 9016-45-9	Ethal NP-20
CAS 9016-45-9	Ethal NP-407
CAS 9016-45-9	Ethylan® 20
CAS 9016-45-9	Ethylan® 44
CAS 9016-45-9	Ethylan® 55
CAS 9016-45-9	Ethylan® 77
CAS 9016-45-9	Ethylan® BCP
CAS 9016-45-9	Ethylan® BV
CAS 9016-45-9	Ethylan® DP
CAS 9016-45-9	Ethylan® HA Flake
CAS 9016-45-9	Ethylan® KEO
CAS 9016-45-9	Ethylan® N30
CAS 9016-45-9	Ethylan® TU
CAS 9016-45-9	Etophen 102
CAS 9016-45-9	Etophen 103
CAS 9016-45-9	Etophen 105
CAS 9016-45-9	Etophen 106
CAS 9016-45-9	Etophen 107
CAS 9016-45-9	Etophen 108
CAS 9016-45-9	Etophen 109
CAS 9016-45-9	Etophen 110
CAS 9016-45-9	Etophen 112
CAS 9016-45-9	Etophen 114
CAS 9016-45-9	Etophen 120
CAS 9016-45-9	Eumulgin 286
CAS 9016-45-9	Gradonic N-95
CAS 9016-45-9	Hetoxide NP-4
CAS 9016-45-9	Hetoxide NP-6
CAS 9016-45-9	Hetoxide NP-9
CAS 9016-45-9	Hetoxide NP-10
CAS 9016-45-9	Hetoxide NP-12
CAS 9016-45-9	Hetoxide NP-15-85%
CAS 9016-45-9	Hetoxide NP-30
CAS 9016-45-9	Hetoxide NP-40
CAS 9016-45-9	Hetoxide NP-50
CAS 9016-45-9	Hostapal N-040
CAS 9016-45-9	Hostapal N-060
CAS 9016-45-9	Hostapal N-060R
CAS 9016-45-9	Hostapal N-080
CAS 9016-45-9	Hostapal N-090
CAS 9016-45-9	Hostapal N-100
CAS 9016-45-9	Hostapal N-110
CAS 9016-45-9	Hostapal N-130
CAS 9016-45-9	Hostapal N-230
CAS 9016-45-9	Hostapal N-300
CAS 9016-45-9	Hyonic NP-40
CAS 9016-45-9	Hyonic NP-60
CAS 9016-45-9	Hyonic NP-90
CAS 9016-45-9	Hyonic NP-100
CAS 9016-45-9	Hyonic NP-110
CAS 9016-45-9	Hyonic NP-120
CAS 9016-45-9	Hyonic NP-407
CAS 9016-45-9	Hyonic NP-500
CAS 9016-45-9	Hyonic PE-100
CAS 9016-45-9	Hyonic PE-120
CAS 9016-45-9	Iconol NP-4
CAS 9016-45-9	Iconol NP-5
CAS 9016-45-9	Iconol NP-6
CAS 9016-45-9	Iconol NP-7
CAS 9016-45-9	Iconol NP-9
CAS 9016-45-9	Iconol NP-10
CAS 9016-45-9	Iconol NP-12
CAS 9016-45-9	Iconol NP-15
CAS 9016-45-9	Iconol NP-20
CAS 9016-45-9	Iconol NP-30
CAS 9016-45-9	Iconol NP-30-70%
CAS 9016-45-9	Iconol NP-40
CAS 9016-45-9	Iconol NP-40-70%
CAS 9016-45-9	Iconol NP-50
CAS 9016-45-9	Iconol NP-50-70%
CAS 9016-45-9	Iconol NP-70
CAS 9016-45-9	Iconol NP-70-70%
CAS 9016-45-9	Iconol NP-100
CAS 9016-45-9	Iconol NP-100-70%
CAS 9016-45-9	Igepal® CA-877
CAS 9016-45-9	Igepal® CO-210
CAS 9016-45-9	Igepal® CO-430
CAS 9016-45-9	Igepal® CO-520
CAS 9016-45-9	Igepal® CO-530
CAS 9016-45-9	Igepal® CO-610
CAS 9016-45-9	Igepal® CO-630
CAS 9016-45-9	Igepal® CO-660
CAS 9016-45-9	Igepal® CO-710
CAS 9016-45-9	Igepal® CO-720
CAS 9016-45-9	Igepal® CO-730
CAS 9016-45-9	Igepal® CO-850
CAS 9016-45-9	Igepal® CO-880
CAS 9016-45-9	Igepal® CO-887
CAS 9016-45-9	Igepal® CO-890
CAS 9016-45-9	Igepal® CO-897
CAS 9016-45-9	Igepal® CO-970
CAS 9016-45-9	Igepal® CO-977
CAS 9016-45-9	Igepal® CO-980
CAS 9016-45-9	Igepal® CO-987
CAS 9016-45-9	Igepal® CO-990
CAS 9016-45-9	Igepal® CO-997
CAS 9016-45-9	Igepal® FN-9.5
CAS 9016-45-9	Igepal® FN-11
CAS 9016-45-9	Igepal® FN-30
CAS 9016-45-9	Indulin® XD-70
CAS 9016-45-9	Lutensol® AP 10
CAS 9016-45-9	Lutensol® AP 20
CAS 9016-45-9	Macol® NP-4
CAS 9016-45-9	Macol® NP-5
CAS 9016-45-9	Macol® NP-6
CAS 9016-45-9	Macol® NP-8
CAS 9016-45-9	Macol® NP-9.5
CAS 9016-45-9	Macol® NP-11
CAS 9016-45-9	Macol® NP-12
CAS 9016-45-9	Macol® NP-15
CAS 9016-45-9	Macol® NP-20
CAS 9016-45-9	Macol® NP-20(70)
CAS 9016-45-9	Macol® NP-30(70)
CAS 9016-45-9	Macol® NP-100
CAS 9016-45-9	Makon® 4
CAS 9016-45-9	Makon® 6
CAS 9016-45-9	Makon® 7
CAS 9016-45-9	Makon® 8
CAS 9016-45-9	Makon® 10
CAS 9016-45-9	Makon® 11
CAS 9016-45-9	Makon® 12
CAS 9016-45-9	Makon® 14
CAS 9016-45-9	Makon® 30
CAS 9016-45-9	Makon® 50
CAS 9016-45-9	Marlophen® 82N
CAS 9016-45-9	Marlophen® 83N
CAS 9016-45-9	Marlophen® 84N
CAS 9016-45-9	Marlophen® 85N
CAS 9016-45-9	Marlophen® 86N
CAS 9016-45-9	Marlophen® 86N/S
CAS 9016-45-9	Marlophen® 87N
CAS 9016-45-9	Marlophen® 88N
CAS 9016-45-9	Marlophen® 89N
CAS 9016-45-9	Marlophen® 810N
CAS 9016-45-9	Marlophen® 812N

CAS 9016-45-9 Marlophen® 814N
CAS 9016-45-9 Marlophen® 820N
CAS 9016-45-9 Marlophen® 830N
CAS 9016-45-9 Marlophen® 840N
CAS 9016-45-9 Marlophen® 850N
CAS 9016-45-9 Marlophen® 1028N
CAS 9016-45-9 Naxonic NI-40
CAS 9016-45-9 Naxonic NI-60
CAS 9016-45-9 Naxonic NI-100
CAS 9016-45-9 Neutronyx® 656
CAS 9016-45-9 Nikkol NP-5
CAS 9016-45-9 Nikkol NP-7.5
CAS 9016-45-9 Nikkol NP-10
CAS 9016-45-9 Nikkol NP-15
CAS 9016-45-9 Nikkol NP-18TX
CAS 9016-45-9 Nissan Nonion NS-206
CAS 9016-45-9 Nissan Nonion NS-212
CAS 9016-45-9 Noigen EA 50
CAS 9016-45-9 Noigen EA 70
CAS 9016-45-9 Noigen EA 130T
CAS 9016-45-9 Noigen EA 150
CAS 9016-45-9 Nonipol 20
CAS 9016-45-9 Nopco® RDY
CAS 9016-45-9 Norfox® NP-4
CAS 9016-45-9 Norfox® NP-6
CAS 9016-45-9 Norfox® NP-7
CAS 9016-45-9 Norfox® NP-9
CAS 9016-45-9 Norfox® NP-11
CAS 9016-45-9 NP-55-60
CAS 9016-45-9 NP-55-80
CAS 9016-45-9 NP-55-85
CAS 9016-45-9 NP-55-90
CAS 9016-45-9 Nutrol 600
CAS 9016-45-9 Nutrol 611
CAS 9016-45-9 Nutrol 622
CAS 9016-45-9 Nutrol 640
CAS 9016-45-9 Nutrol 656
CAS 9016-45-9 Polirol NF80
CAS 9016-45-9 Polystep® F-1
CAS 9016-45-9 Polystep® F-2
CAS 9016-45-9 Polystep® F-3
CAS 9016-45-9 Polystep® F-4
CAS 9016-45-9 Polystep® F-5
CAS 9016-45-9 Polystep® F-6
CAS 9016-45-9 Polystep® F-9
CAS 9016-45-9 Polystep® F-10, F-10 Ec
CAS 9016-45-9 Polystep® F-95B
CAS 9016-45-9 Poly-Tergent® B-150
CAS 9016-45-9 Poly-Tergent® B-200
CAS 9016-45-9 Poly-Tergent® B-300
CAS 9016-45-9 Poly-Tergent® B-315
CAS 9016-45-9 Poly-Tergent® B-350
CAS 9016-45-9 Poly-Tergent® B-500
CAS 9016-45-9 Prox-onic NP-04
CAS 9016-45-9 Prox-onic NP-06
CAS 9016-45-9 Prox-onic NP-09
CAS 9016-45-9 Prox-onic NP-010
CAS 9016-45-9 Prox-onic NP-015
CAS 9016-45-9 Prox-onic NP-020
CAS 9016-45-9 Prox-onic NP-030
CAS 9016-45-9 Prox-onic NP-030/70
CAS 9016-45-9 Prox-onic NP-040
CAS 9016-45-9 Prox-onic NP-040/70
CAS 9016-45-9 Prox-onic NP-050
CAS 9016-45-9 Prox-onic NP-050/70
CAS 9016-45-9 Prox-onic NP-0100
CAS 9016-45-9 Prox-onic NP-0100/70
CAS 9016-45-9 Quimipol ENF 80
CAS 9016-45-9 Remcopal 29
CAS 9016-45-9 Remcopal 334
CAS 9016-45-9 Remcopal 349
CAS 9016-45-9 Remcopal 666
CAS 9016-45-9 Remcopal 3112
CAS 9016-45-9 Remcopal 3515
CAS 9016-45-9 Remcopal 3712
CAS 9016-45-9 Remcopal 3820
CAS 9016-45-9 Remcopal 6110
CAS 9016-45-9 Remcopal 31250
CAS 9016-45-9 Remcopal 33820
CAS 9016-45-9 Remcopal L30
CAS 9016-45-9 Remcopal NP 30
CAS 9016-45-9 Remcopal PONF
CAS 9016-45-9 Renex® 647
CAS 9016-45-9 Renex® 648
CAS 9016-45-9 Renex® 649
CAS 9016-45-9 Renex® 650
CAS 9016-45-9 Renex® 678
CAS 9016-45-9 Renex® 679
CAS 9016-45-9 Renex® 682
CAS 9016-45-9 Renex® 688
CAS 9016-45-9 Renex® 690
CAS 9016-45-9 Renex® 697
CAS 9016-45-9 Renex® 698
CAS 9016-45-9 Rewopal® HV 4
CAS 9016-45-9 Rewopal® HV 5
CAS 9016-45-9 Rewopal® HV 6
CAS 9016-45-9 Rewopal® HV 8
CAS 9016-45-9 Rewopal® HV 9
CAS 9016-45-9 Rewopal® HV 10
CAS 9016-45-9 Rewopal® HV 14
CAS 9016-45-9 Rewopal® HV 25
CAS 9016-45-9 Rewopal® HV 50
CAS 9016-45-9 Rexol 25/4
CAS 9016-45-9 Rexol 25/6
CAS 9016-45-9 Rexol 25/7
CAS 9016-45-9 Rexol 25/8
CAS 9016-45-9 Rexol 25/9
CAS 9016-45-9 Rexol 25/10
CAS 9016-45-9 Rexol 25/11
CAS 9016-45-9 Rexol 25/14
CAS 9016-45-9 Rexol 25/15
CAS 9016-45-9 Rexol 25/20
CAS 9016-45-9 Rexol 25/30
CAS 9016-45-9 Rexol 25/40
CAS 9016-45-9 Rexol 25/50
CAS 9016-45-9 Rexol 25/100-70%
CAS 9016-45-9 Rexol 25/307
CAS 9016-45-9 Rexol 25/407
CAS 9016-45-9 Rexol 25/507
CAS 9016-45-9 Rexol 25J
CAS 9016-45-9 Rioklen NF 2
CAS 9016-45-9 Rioklen NF 4
CAS 9016-45-9 Rioklen NF 6
CAS 9016-45-9 Rioklen NF 7
CAS 9016-45-9 Rioklen NF 8
CAS 9016-45-9 Rioklen NF 9
CAS 9016-45-9 Rioklen NF 10
CAS 9016-45-9 Rioklen NF 12
CAS 9016-45-9 Rioklen NF 15
CAS 9016-45-9 Rioklen NF 20
CAS 9016-45-9 Rioklen NF 30
CAS 9016-45-9 Rioklen NF 40
CAS 9016-45-9 Sellig N 4 100
CAS 9016-45-9 Sellig N 5 100
CAS 9016-45-9 Sellig N 6 100
CAS 9016-45-9 Sellig N 8 100
CAS 9016-45-9 Sellig N 9 100
CAS 9016-45-9 Sellig N 10 100
CAS 9016-45-9 Sellig N 11 100
CAS 9016-45-9 Sellig N 12 100

CAS 9016-45-9	Sellig N 15 100
CAS 9016-45-9	Sellig N 17 100
CAS 9016-45-9	Sellig N 20 80
CAS 9016-45-9	Sellig N 20 100
CAS 9016-45-9	Sellig N 30 70
CAS 9016-45-9	Sellig N 30 100
CAS 9016-45-9	Sellig N 50 100
CAS 9016-45-9	Sellig N 1050
CAS 9016-45-9	Sellig N 1780
CAS 9016-45-9	Serdox NNP 4
CAS 9016-45-9	Serdox NNP 5
CAS 9016-45-9	Serdox NNP 6
CAS 9016-45-9	Serdox NNP 7
CAS 9016-45-9	Serdox NNP 8.5
CAS 9016-45-9	Serdox NNP 10
CAS 9016-45-9	Serdox NNP 11
CAS 9016-45-9	Serdox NNP 12
CAS 9016-45-9	Serdox NNP 13
CAS 9016-45-9	Serdox NNP 14
CAS 9016-45-9	Serdox NNP 15
CAS 9016-45-9	Serdox NNP 25
CAS 9016-45-9	Serdox NNP 30/70
CAS 9016-45-9	Sermul EN 3
CAS 9016-45-9	Sermul EN 15
CAS 9016-45-9	Sermul EN 20/70
CAS 9016-45-9	Sermul EN 30/70
CAS 9016-45-9	Sermul EN 145
CAS 9016-45-9	Sermul EN 229
CAS 9016-45-9	Simulsol 1030 NP
CAS 9016-45-9	Siponic® NP-4
CAS 9016-45-9	Siponic® NP-6
CAS 9016-45-9	Siponic® NP-8
CAS 9016-45-9	Siponic® NP-9
CAS 9016-45-9	Siponic® NP-9.5
CAS 9016-45-9	Siponic® NP-10
CAS 9016-45-9	Siponic® NP-13
CAS 9016-45-9	Siponic® NP-15
CAS 9016-45-9	Siponic® NP-40
CAS 9016-45-9	Siponic® NP-407
CAS 9016-45-9	Soprofor NP/20
CAS 9016-45-9	Soprofor NP/30
CAS 9016-45-9	Soprofor NP/100
CAS 9016-45-9	Surfonic® HDL
CAS 9016-45-9	Surfonic® HDL-1
CAS 9016-45-9	Surfonic® N-31.5
CAS 9016-45-9	Surfonic® N-40
CAS 9016-45-9	Surfonic® N-60
CAS 9016-45-9	Surfonic® N-85
CAS 9016-45-9	Surfonic® N-95
CAS 9016-45-9	Surfonic® N-100
CAS 9016-45-9	Surfonic® N-102
CAS 9016-45-9	Surfonic® N-120
CAS 9016-45-9	Surfonic® N-130
CAS 9016-45-9	Surfonic® N-150
CAS 9016-45-9	Surfonic® N-200
CAS 9016-45-9	Surfonic® N-300
CAS 9016-45-9	Surfonic® N-557
CAS 9016-45-9	Surfonic® NB-5
CAS 9016-45-9	Syn Fac® 905
CAS 9016-45-9	Synperonic NP4
CAS 9016-45-9	Synperonic NP5
CAS 9016-45-9	Synperonic NP6
CAS 9016-45-9	Synperonic NP7
CAS 9016-45-9	Synperonic NP8
CAS 9016-45-9	Synperonic NP9
CAS 9016-45-9	Synperonic NP10
CAS 9016-45-9	Synperonic NP12
CAS 9016-45-9	Synperonic NP13
CAS 9016-45-9	Synperonic NP15
CAS 9016-45-9	Synperonic NP20
CAS 9016-45-9	Synperonic NP30
CAS 9016-45-9	Synperonic NP30/70
CAS 9016-45-9	Synperonic NP35
CAS 9016-45-9	Synperonic NP40
CAS 9016-45-9	Synperonic NP50
CAS 9016-45-9	Synthrapol N
CAS 9016-45-9	Syntopon A
CAS 9016-45-9	Syntopon A 100
CAS 9016-45-9	Syntopon B
CAS 9016-45-9	Syntopon C
CAS 9016-45-9	Syntopon D
CAS 9016-45-9	Syntopon E
CAS 9016-45-9	Syntopon F
CAS 9016-45-9	Syntopon G
CAS 9016-45-9	Syntopon H
CAS 9016-45-9	T-Det® N-4
CAS 9016-45-9	T-Det® N-6
CAS 9016-45-9	T-Det® N-8
CAS 9016-45-9	T-Det® N-9.5
CAS 9016-45-9	T-Det® N-10.5
CAS 9016-45-9	T-Det® N-12
CAS 9016-45-9	T-Det® N-14
CAS 9016-45-9	T-Det® N-20
CAS 9016-45-9	T-Det® N-30
CAS 9016-45-9	T-Det® N-40
CAS 9016-45-9	T-Det® N-50
CAS 9016-45-9	T-Det® N-70
CAS 9016-45-9	T-Det® N-307
CAS 9016-45-9	T-Det® N-407
CAS 9016-45-9	T-Det® N-507
CAS 9016-45-9	T-Det® N-705
CAS 9016-45-9	T-Det® N-707
CAS 9016-45-9	T-Det® N-100
CAS 9016-45-9	T-Det® N-1007
CAS 9016-45-9	Tergitol® NP-4
CAS 9016-45-9	Tergitol® NP-5
CAS 9016-45-9	Tergitol® NP-6
CAS 9016-45-9	Tergitol® NP-7
CAS 9016-45-9	Tergitol® NP-8
CAS 9016-45-9	Tergitol® NP-9
CAS 9016-45-9	Tergitol® NP-10
CAS 9016-45-9	Tergitol® NP-13
CAS 9016-45-9	Tergitol® NP-14
CAS 9016-45-9	Tergitol® NP-15
CAS 9016-45-9	Tergitol® NP-27
CAS 9016-45-9	Tergitol® NP-40
CAS 9016-45-9	Tergitol® NP-40, 70% Aq.
CAS 9016-45-9	Tergitol® NP-44
CAS 9016-45-9	Tergitol® NP-55, 70% Aq.
CAS 9016-45-9	Tergitol® NP-70, 70% Aq.
CAS 9016-45-9	Tergitol® NPX
CAS 9016-45-9	Tergitol® TP-9
CAS 9016-45-9	Teric GN Series
CAS 9016-45-9	Teric N2
CAS 9016-45-9	Teric N3
CAS 9016-45-9	Teric N4
CAS 9016-45-9	Teric N5
CAS 9016-45-9	Teric N8
CAS 9016-45-9	Teric N9
CAS 9016-45-9	Teric N10
CAS 9016-45-9	Teric N11
CAS 9016-45-9	Teric N12
CAS 9016-45-9	Teric N13
CAS 9016-45-9	Teric N15
CAS 9016-45-9	Teric N20
CAS 9016-45-9	Teric N30
CAS 9016-45-9	Teric N30L
CAS 9016-45-9	Teric N40
CAS 9016-45-9	Teric N40L
CAS 9016-45-9	Teric N100

CAS 9016-45-9 Triton® N-42
CAS 9016-45-9 Triton® N-57
CAS 9016-45-9 Triton® N-101
CAS 9016-45-9 Triton® N-111
CAS 9016-45-9 Triton® N-150
CAS 9016-45-9 Trycol® 6940
CAS 9016-45-9 Trycol® 6941
CAS 9016-45-9 Trycol® 6942
CAS 9016-45-9 Trycol® 6951
CAS 9016-45-9 Trycol® 6953
CAS 9016-45-9 Trycol® 6954
CAS 9016-45-9 Trycol® 6957
CAS 9016-45-9 Trycol® 6958
CAS 9016-45-9 Trycol® 6960
CAS 9016-45-9 Trycol® 6961
CAS 9016-45-9 Trycol® 6962
CAS 9016-45-9 Trycol® 6963
CAS 9016-45-9 Trycol® 6964
CAS 9016-45-9 Trycol® 6965
CAS 9016-45-9 Trycol® 6967
CAS 9016-45-9 Trycol® 6968
CAS 9016-45-9 Trycol® 6969
CAS 9016-45-9 Trycol® 6970
CAS 9016-45-9 Trycol® 6971
CAS 9016-45-9 Trycol® 6972
CAS 9016-45-9 Trycol® 6974
CAS 9016-45-9 Trycol® 6981
CAS 9016-45-9 Trycol® NP-4
CAS 9016-45-9 Trycol® NP-6
CAS 9016-45-9 Trycol® NP-7
CAS 9016-45-9 Trycol® NP-9
CAS 9016-45-9 Trycol® NP-11
CAS 9016-45-9 Trycol® NP-20
CAS 9016-45-9 Trycol® NP-30
CAS 9016-45-9 Trycol® NP-40
CAS 9016-45-9 Trycol® NP-50
CAS 9016-45-9 Trycol® NP-307
CAS 9016-45-9 Trycol® NP-407
CAS 9016-45-9 Trycol® NP-507
CAS 9016-45-9 Trycol® NP-1007
CAS 9016-45-9 Witconol NP-40
CAS 9016-45-9 Witconol NP-100
CAS 9016-45-9 Witconol NP-300
CAS 9017-71-4 Erional® PA
CAS 9032-42-2 Tylose MHB
CAS 9035-85-2 Procetyl 10
CAS 9035-85-2 Procetyl 30
CAS 9035-85-2 Procetyl 50
CAS 9036-19-5 Hyonic PE-240, -260
CAS 9036-19-5 Nekanil® LN
CAS 9036-19-5 Noigen EA 92
CAS 9036-19-5 Noigen EA 102
CAS 9036-19-5 Noigen EA 110
CAS 9036-19-5 Noigen EA 112
CAS 9036-19-5 Noigen EA 120
CAS 9036-19-5 Noigen EA 120B
CAS 9036-19-5 Noigen EA 140
CAS 9036-19-5 Noigen EA 142
CAS 9036-19-5 Noigen EA 160
CAS 9036-19-5 Noigen EA 170
CAS 9036-19-5 Noigen EA 190D
CAS 9038-95-3 Pluracol® W3520N
CAS 9038-95-3 Tergitol® XD
CAS 9038-95-3 Tergitol® XH
CAS 9038-95-3 Tergitol® XJ
CAS 9040-38-4 Aerosol® A-103
CAS 9043-30-5 Berol 048
CAS 9043-30-5 Marlipal® 013/100
CAS 9046-01-9 Dextrol OC-40
CAS 9046-01-9 Plysurf A212C
CAS 9051-57-4 Aerosol® NPES 458
CAS 9051-57-4 Aerosol® NPES 930
CAS 9051-57-4 Aerosol® NPES 2030
CAS 9051-57-4 Aerosol® NPES 3030
CAS 9062-73-1 Glycosperse® L-20
CAS 9063-89-2 Octapol 60
CAS 9063-89-2 Octapol 100
CAS 9063-89-2 Octapol 300
CAS 9063-89-2 Octapol 400
CAS 9064-14-6 Marlox® L 6
CAS 9084-06-4 Lomar® LS
CAS 9084-06-4 Lomar® PW
CAS 9087-53-0 Procetyl AWS
CAS 10213-78-2 Ethomeen® 18/12
CAS 10277-04-0 Emulan® FM
CAS 11111-34-5 Synperonic T/304
CAS 11140-78-6 Rewoteric AM DML
CAS 12645-31-7 Marlophor® IH-Acid
CAS 12751-23-4 Tryfac® 5573
CAS 13127-82-7 Berol 302
CAS 13197-76-7 Rewoteric AM HC
CAS 14234-82-3 DIBM
CAS 14350-96-0 Dehyton® PMG
CAS 14351-40-7 Witcamide® MAS
CAS 14351-50-9 Incromine Oxide OD-50
CAS 14481-60-8 Aerosol® 18
CAS 14481-60-8 Lankropol® ODS/LS
CAS 14727-68-5 Jet Amine DMOD
CAS 15268-40-3 Diamine B11
CAS 15826-16-1 Cosmopon LE 50
CAS 15826-16-1 Rolpon 24/230
CAS 15826-16-1 Rolpon 24/270
CAS 15826-16-1 Rolpon 24/330 N
CAS 16693-53-1 Crodasinic LT40
CAS 17342-21-1 Dehyquart D
CAS 18448-65-2 Ethoquad® O/12
CAS 18641-57-1 Syncrowax HR-C
CAS 21142-28-9 Sulfetal CJOT 38
CAS 21542-96-1 Crodamine 3.ABD
CAS 21652-27-7 Amine O
CAS 22023-23-0 Jet Amine DE-13
CAS 22042-96-2 Briquest® 543-45AS
CAS 24938-91-8 Ahcowet DQ-114
CAS 24938-91-8 Ahcowet DQ-145
CAS 24938-91-8 Akypogene VSM
CAS 24938-91-8 Bio-Soft® TD 400
CAS 24938-91-8 Bio-Soft® TD 630
CAS 24938-91-8 Carsonon® TD-10
CAS 24938-91-8 Carsonon® TD-11, TD-11 70%
CAS 24938-91-8 Chemal TDA-3
CAS 24938-91-8 Chemal TDA-6
CAS 24938-91-8 Chemal TDA-9
CAS 24938-91-8 Chemal TDA-12
CAS 24938-91-8 Chemal TDA-15
CAS 24938-91-8 Chemal TDA-18
CAS 24938-91-8 Dehydol PLT 6
CAS 24938-91-8 Desonic® 6T
CAS 24938-91-8 Desonic® 9T
CAS 24938-91-8 Desonic® 10T
CAS 24938-91-8 Desonic® 12T
CAS 24938-91-8 Desonic® 15T
CAS 24938-91-8 Emthox® 5941
CAS 24938-91-8 Emthox® 5942
CAS 24938-91-8 Emthox® 5993
CAS 24938-91-8 Ethal TDA-3
CAS 24938-91-8 Ethal TDA-6
CAS 24938-91-8 Ethal TDA-9
CAS 24938-91-8 Ethal TDA-12
CAS 24938-91-8 Ethal TDA-18
CAS 24938-91-8 Ethosperse® TDA-6

CAS 24938-91-8	Flo-Mo® AJ-85
CAS 24938-91-8	Flo-Mo® AJ-100
CAS 24938-91-8	Genapol® X-040
CAS 24938-91-8	Genapol® X-060
CAS 24938-91-8	Genapol® X-080
CAS 24938-91-8	Genapol® X-100
CAS 24938-91-8	Hetoxol TD-3
CAS 24938-91-8	Hetoxol TD-6
CAS 24938-91-8	Hetoxol TD-9
CAS 24938-91-8	Hetoxol TD-12
CAS 24938-91-8	Hetoxol TD-18
CAS 24938-91-8	Hetoxol TD-25
CAS 24938-91-8	Hostapur CX Highly Conc
CAS 24938-91-8	Iconol TDA-3
CAS 24938-91-8	Iconol TDA-6
CAS 24938-91-8	Iconol TDA-8
CAS 24938-91-8	Iconol TDA-8-90%
CAS 24938-91-8	Iconol TDA-9
CAS 24938-91-8	Iconol TDA-10
CAS 24938-91-8	Iconol TDA-18-80%
CAS 24938-91-8	Iconol TDA-29-80%
CAS 24938-91-8	Lipocol TD-6
CAS 24938-91-8	Lipocol TD-12
CAS 24938-91-8	Macol® TD-3
CAS 24938-91-8	Macol® TD-8
CAS 24938-91-8	Macol® TD-10
CAS 24938-91-8	Macol® TD-12
CAS 24938-91-8	Macol® TD-15
CAS 24938-91-8	Macol® TD-100
CAS 24938-91-8	Macol® TD-610
CAS 24938-91-8	Merpol® SH
CAS 24938-91-8	Prox-onic TD-1/03
CAS 24938-91-8	Prox-onic TD-1/06
CAS 24938-91-8	Prox-onic TD-1/09
CAS 24938-91-8	Prox-onic TD-1/012
CAS 24938-91-8	Renex® 30
CAS 24938-91-8	Renex® 31
CAS 24938-91-8	Renex® 36
CAS 24938-91-8	Rhodasurf® BC-420
CAS 24938-91-8	Rhodasurf® BC-610
CAS 24938-91-8	Rhodasurf® BC-630
CAS 24938-91-8	Rhodasurf® BC-720
CAS 24938-91-8	Rhodasurf® BC-737
CAS 24938-91-8	Rhodasurf® BC-840
CAS 24938-91-8	Rhodasurf® ROX
CAS 24938-91-8	Rhodasurf® T-95
CAS 24938-91-8	Rolfor TR 6
CAS 24938-91-8	Rolfor TR 9
CAS 24938-91-8	Rolfor TR 12
CAS 24938-91-8	Siponic® TD-3
CAS 24938-91-8	Siponic® TD-6
CAS 24938-91-8	Synperonic 13/3
CAS 24938-91-8	Synperonic 13/5
CAS 24938-91-8	Synperonic 13/6.5
CAS 24938-91-8	Synperonic 13/8
CAS 24938-91-8	Synperonic 13/9
CAS 24938-91-8	Synperonic 13/10
CAS 24938-91-8	Synperonic 13/12
CAS 24938-91-8	Synperonic 13/15
CAS 24938-91-8	Synperonic 13/18
CAS 24938-91-8	Synperonic 13/20
CAS 24938-91-8	Synperonic A3
CAS 24938-91-8	Synperonic A7
CAS 24938-91-8	Synperonic A9
CAS 24938-91-8	Synperonic A11
CAS 24938-91-8	Synperonic A20
CAS 24938-91-8	Synperonic A50
CAS 24938-91-8	T-Det® TDA-150
CAS 24938-91-8	Teric 13A5
CAS 24938-91-8	Teric 13A7
CAS 24938-91-8	Teric 13A9
CAS 24938-91-8	Trycol® 5874
CAS 24938-91-8	Trycol® 5940
CAS 24938-91-8	Trycol® 5941
CAS 24938-91-8	Trycol® 5942
CAS 24938-91-8	Trycol® 5943
CAS 24938-91-8	Trycol® 5944
CAS 24938-91-8	Trycol® 5946
CAS 24938-91-8	Trycol® 5949
CAS 24938-91-8	Trycol® 5957
CAS 24938-91-8	Trycol® 5968
CAS 24938-91-8	Trycol® 5993
CAS 24938-91-8	Trycol® TDA-3
CAS 24938-91-8	Trycol® TDA-6
CAS 24938-91-8	Trycol® TDA-8
CAS 24938-91-8	Trycol® TDA-9
CAS 24938-91-8	Trycol® TDA-11
CAS 24938-91-8	Trycol® TDA-12
CAS 24938-91-8	Trycol® TDA-18
CAS 24938-91-8	Volpo T-3
CAS 24938-91-8	Volpo T-5
CAS 24938-91-8	Volpo T-10
CAS 24938-91-8	Volpo T-15
CAS 24938-91-8	Volpo T-20
CAS 25155-30-0	Nansa® HS80S
CAS 25155-30-0	Nansa® HS85S
CAS 25155-30-0	Phenyl Sulphonate HSR
CAS 25155-30-0	Sandet 60
CAS 25155-30-0	Sulframin 1255
CAS 25159-40-4	Incromine Oxide O
CAS 25167-32-2	Aerosol® DPOS-45
CAS 25321-41-9	Eltesol® XA65
CAS 25322-68-3	Lutrol® E 400
CAS 25322-68-3	Lutrol® E 4000
CAS 25322-68-3	Pluracol® E400
CAS 25322-68-3	Pluracol® E8000
CAS 25496-72-4	Tegin® O
CAS 25637-84-7	Cithrol GDO N/E
CAS 25637-84-7	Emerest® 2419
CAS 25637-84-7	Nikkol DGO-80
CAS 25791-96-2	Witconol CD-18
CAS 26027-37-2	Dionil® OC
CAS 26027-37-2	Dionil® SH 100
CAS 26027-37-2	Dionil® W 100
CAS 26027-37-2	Ethomid® O/17
CAS 26027-38-3	Akyporox NP 150
CAS 26027-38-3	Iconol NP-100
CAS 26027-38-3	Rexol 25/6
CAS 26183-52-8	Chemal DA-4
CAS 26183-52-8	Marlipal® 104
CAS 26183-52-8	Trycol® DA-4
CAS 26183-52-8	Trycol® DA-6
CAS 26183-52-8	Trycol® DA-69
CAS 26256-79-1	Deriphat 160C
CAS 26264-05-1	Arylan® PWS
CAS 26264-06-2	Emulson AG/CAL
CAS 26264-06-2	Sermul EA 88
CAS 26266-57-9	Ablunol S-40
CAS 26266-57-9	Ahco FP-67
CAS 26266-57-9	Armotan® MP
CAS 26266-57-9	Crill 2
CAS 26266-57-9	Disponil SMP 100 F1
CAS 26266-57-9	Emsorb® 2510
CAS 26266-57-9	Ethylan® GP-40
CAS 26266-57-9	Glycomul® P
CAS 26266-57-9	Hodag SMP
CAS 26266-57-9	Liposorb P
CAS 26266-57-9	Lonzest® SMP
CAS 26266-57-9	Montane 40
CAS 26266-57-9	MP-33-F

CAS 26266-57-9 Newcol 40
CAS 26266-57-9 Prote-sorb SMP
CAS 26266-57-9 Radiasurf® 7135
CAS 26266-57-9 S-Maz® 40
CAS 26266-57-9 Secoster® KP 10
CAS 26266-57-9 Sorbax SMP
CAS 26266-57-9 Sorbon S-40
CAS 26266-57-9 Span® 40
CAS 26266-58-0 Ablunol S-85
CAS 26266-58-0 Ahco FO-18
CAS 26266-58-0 Alkamuls® S-85
CAS 26266-58-0 Alkamuls® STO
CAS 26266-58-0 Atlas G-4885
CAS 26266-58-0 Atlox 4885
CAS 26266-58-0 Crill 45
CAS 26266-58-0 Disponil STO 100 F1
CAS 26266-58-0 Emsorb® 2503
CAS 26266-58-0 Ethylan® GT85
CAS 26266-58-0 Glycomul® TO
CAS 26266-58-0 Hodag STO
CAS 26266-58-0 Ionet S-85
CAS 26266-58-0 Liposorb TO
CAS 26266-58-0 Lonzest® STO
CAS 26266-58-0 Montane 85
CAS 26266-58-0 Nikkol SO-30
CAS 26266-58-0 Prote-sorb STO
CAS 26266-58-0 Rheodol SP-O30
CAS 26266-58-0 S-Maz® 85
CAS 26266-58-0 S-Maz® 85K
CAS 26266-58-0 Sorban CO
CAS 26266-58-0 Sorbax STO
CAS 26266-58-0 Span® 85
CAS 26266-58-0 Witconol 2503
CAS 26392-63-2 Cirrasol® 185A
CAS 26402-26-6 Imwitor® 742
CAS 26544-38-7 Dodecenyl Succinic Anhydride
CAS 26545-58-4 Irgasol® DA Liq., Powd
CAS 26591-12-8 Matexil FC PN
CAS 26635-92-7 Amiladin C-1802
CAS 26635-92-7 Chemeen 18-50
CAS 26635-92-7 Ethomeen® 18/15
CAS 26635-92-7 Ethomeen® 18/20
CAS 26635-92-7 Ethomeen® 18/25
CAS 26635-92-7 Ethomeen® 18/60
CAS 26635-92-7 Ethox SAM-10
CAS 26635-92-7 Ethox SAM-50
CAS 26635-92-7 Hetoxamine ST-5
CAS 26635-92-7 Hetoxamine ST-15
CAS 26635-92-7 Hetoxamine ST-50
CAS 26635-92-7 Icomeen® 18-5
CAS 26635-92-7 Icomeen® 18-50
CAS 26635-92-7 Marlazin® S 10
CAS 26635-92-7 Marlazin® S 40
CAS 26635-92-7 Prox-onic MS-05
CAS 26635-92-7 Prox-onic MS-011
CAS 26635-92-7 Prox-onic MS-050
CAS 26635-92-7 Teric 18M5
CAS 26635-92-7 Teric 18M10
CAS 26635-92-7 Teric 18M20
CAS 26635-92-7 Teric 18M30
CAS 26635-92-7 Trymeen® 6617
CAS 26635-92-7 Trymeen® SAM-50
CAS 26635-92-7 Zusomin S 110
CAS 26635-92-7 Zusomin S 125
CAS 26635-93-8 Berol 28
CAS 26635-93-8 Berol 303
CAS 26635-93-8 Chimipal NH 2
CAS 26635-93-8 Chimipal NH 6
CAS 26635-93-8 Chimipal NH 10
CAS 26635-93-8 Chimipal NH 20
CAS 26635-93-8 Chimipal NH 25
CAS 26635-93-8 Chimipal NH 30
CAS 26635-93-8 Marlazin® OL 2
CAS 26635-93-8 Marlowet® 5400
CAS 26658-19-5 Ahco FS-21
CAS 26658-19-5 Alkamuls® S-65
CAS 26658-19-5 Alkamuls® STS
CAS 26658-19-5 Crill 35
CAS 26658-19-5 Disponil STS 100 F1
CAS 26658-19-5 Drewmulse® STS
CAS 26658-19-5 Emsorb® 2507
CAS 26658-19-5 Glycomul® TS
CAS 26658-19-5 Glycomul® TS KFG
CAS 26658-19-5 Grindtek STS
CAS 26658-19-5 Hodag STS
CAS 26658-19-5 Liposorb TS
CAS 26658-19-5 Lonzest® STS
CAS 26658-19-5 Montane 65
CAS 26658-19-5 Nikkol SS-30
CAS 26658-19-5 Prote-sorb STS
CAS 26658-19-5 Radiasurf® 7345
CAS 26658-19-5 Rheodol SP-S30
CAS 26658-19-5 S-Maz® 65K
CAS 26658-19-5 Sorbax STS
CAS 26658-19-5 Span® 65
CAS 26658-19-5 TS-33-F
CAS 26680-54-6 Octenyl Succinic Anhydride, n-
CAS 27172-87-0 Alarsol AL
CAS 27176-87-0 Nansa® 1042/P
CAS 27176-87-0 Pentine Acid 5431
CAS 27176-87-0 Polirol DS
CAS 27176-87-0 Sulframin Acide B
CAS 27176-87-0 Wibarco
CAS 27177-05-5 NP-55-60
CAS 27177-05-5 NP-55-80
CAS 27177-05-5 NP-55-85
CAS 27177-05-5 NP-55-90
CAS 27306-79-2 Hetoxol M-3
CAS 27306-90-7 Akypo RLM 160
CAS 27323-41-7 Emulson AG 255
CAS 27323-41-7 Hexaryl D 60 L
CAS 27638-00-2 Cithrol GDL N/E
CAS 27986-36-3 Akyporox NP 15
CAS 28061-69-0 Armeen® DMOD
CAS 28061-69-0 Crodamine 3.AOD
CAS 28266-58-0 TO-33-F
CAS 28519-02-0 Calfax DB-45
CAS 28519-02-0 Eleminol MON-7
CAS 28631-63-2 Eltesol® CA 65
CAS 28724-32-5 Ethoquad® 18/12
CAS 28724-32-5 Ethoquad® 18/25
CAS 28984-69-2 Alkaterge®-T
CAS 29116-98-1 Atlas G-950
CAS 29116-98-1 DO-33-F
CAS 29317-52-0 Tomah PA-13i
CAS 29381-93-9 Marlopon® AT
CAS 29923-31-7 Amisoft LS-11
CAS 30260-72-1 Calfax DBA-40
CAS 30260-72-1 Calfax DBA-70
CAS 30342-62-2 Crodateric S
CAS 30416-77-4 Perlankrol® PA Conc
CAS 31556-45-3 Cirrasol® LN-GS
CAS 31566-31-1 Cithrol GMS Acid Stable
CAS 31566-31-1 Cithrol GMS N/E
CAS 31566-31-1 Cithrol GMS S/E
CAS 31566-31-1 Empilan® GMS NSE40
CAS 31566-31-1 Empilan® GMS SE40
CAS 31566-31-1 Geleol
CAS 31566-31-1 Imwitor® 191
CAS 31566-31-1 Sterol GMS

CAS 31621-91-7	Ethox 1122
CAS 31694-55-0	Liponic EG-1
CAS 31791-00-2	RS-55-40
CAS 31799-71-0	Dionil® OC
CAS 31799-71-0	Ethomid® O/15
CAS 31799-71-0	Lutensol® FSA 10
CAS 32426-11-2	BTC® 818
CAS 32426-11-2	BTC® 818-80
CAS 32456-28-6	Crodateric O, O.100
CAS 32612-48-9	Texapon PNA-127
CAS 34140-91-5	Inipol OO2
CAS 34360-00-4	Tomah E-14-2
CAS 34938-91-8	Dehydol PID 6
CAS 35545-57-4	Hetoxide BN-13
CAS 36409-57-1	Empicol® SLL
CAS 36409-57-1	Empicol® SLL/P
CAS 36521-89-8	Emasol S-20
CAS 36653-82-4	Adol® 52 NF
CAS 36653-82-4	Adol® 520 NF
CAS 36653-82-4	Alfol® 16
CAS 36653-82-4	Alfol® 16 NF
CAS 36653-82-4	CO-1695
CAS 36653-82-4	Emery® 3336
CAS 36653-82-4	Emery® 3337
CAS 36653-82-4	Emery® 3338
CAS 36653-82-4	Emery® 3339
CAS 36653-82-4	Epal® 16NF
CAS 36653-82-4	Kalcohl 60
CAS 36653-82-4	Lorol C16
CAS 36653-82-4	Lorol C16-95
CAS 36653-82-4	Lorol C16-98
CAS 36653-82-4	Lorol C16 Chemically Pure
CAS 36653-82-4	Loxiol VPG 1743
CAS 36653-82-4	Philcohol 1600
CAS 36863-52-2	Phospholan® KPE4
CAS 37200-49-0	Armotan® PMO 20
CAS 37205-87-1	Marlophen® 87N
CAS 37205-87-1	Marlophen® 89.5N
CAS 37205-87-1	Marlophen® 890N
CAS 37205-87-1	Marlowet® ISM
CAS 37220-82-9	Emerest® 2421
CAS 37251-69-7	Berol 948
CAS 37251-69-7	Tergitol® D-683
CAS 37354-45-5	Empicol® SDD
CAS 37971-36-1	Bayhibit® AM
CAS 38411-30-3	Marlon® A 365
CAS 38517-23-6	Amisoft HS-11
CAS 38618-12-1	Emulamid FO-5DF
CAS 38916-42-6	Aerosol® 22
CAS 39310-72-0	Tagat® RI
CAS 39354-45-5	Aerosol® A-102
CAS 39354-45-5	Rolpon SE 138
CAS 39354-45-5	Schercopol LPS
CAS 39407-03-9	Marlophor® HS-Acid
CAS 39464-64-7	Plysurf A207H
CAS 39464-66-9	Crafol AP-63
CAS 39464-69-2	Crafol AP-11
CAS 39464-69-2	Crodafos N3 Acid
CAS 39464-69-2	Crodafos N5 Acid
CAS 39464-69-2	Crodafos N10 Acid
CAS 39464-69-2	Laurelphos 400
CAS 39464-70-5	Chemfac PC-006
CAS 39464-70-5	Marlowet® 5324
CAS 40716-42-5	Alkamide® RODEA
CAS 40716-42-5	Aminol CA-2
CAS 40716-42-5	Mackamide R
CAS 51158-08-8	Aldosperse® MS-20
CAS 51158-08-8	Rolfat 5347
CAS 51158-08-8	Tagat® S
CAS 51158-08-8	Tagat® S2
CAS 51192-09-7	Tagat® O
CAS 51192-09-7	Tagat® O2
CAS 51200-87-4	Amine CS-1135®
CAS 51248-32-9	Lamacit GML 20
CAS 51248-32-9	Tagat® L
CAS 51248-32-9	Tagat® L2
CAS 51811-79-1	Dextrol OC-20
CAS 51811-79-1	Emulson AG 7000
CAS 51811-79-1	Norfox® PE-LF
CAS 51811-79-1	Norfox® PEW
CAS 51811-79-1	Phospholan® PNP9
CAS 51811-79-1	Polirol 10
CAS 51811-79-1	Tryfac® 5556
CAS 52019-36-0	Chemfac PD-600
CAS 52229-50-2	Gantrez® AN-119
CAS 52229-50-2	Gantrez® AN-139
CAS 52229-50-2	Gantrez® AN-149
CAS 52229-50-2	Gantrez® AN-169
CAS 52292-17-8	Hetoxol IS-2
CAS 52315-75-0	Amihope LL-11
CAS 52503-47-6	Lutensol® ED 310
CAS 52503-47-6	Lutensol® ED 370
CAS 52504-24-2	Softigen® 767
CAS 52581-71-2	Volpo 3
CAS 52623-95-7	Laurelphos RH-44
CAS 52624-57-4	Pluradot HA-433
CAS 52624-57-4	Pluradot HA-510
CAS 52624-57-4	Pluradot HA-530
CAS 52663-87-3	Ampholak YJH
CAS 52668-97-0	Marlosol® FS
CAS 52668-97-0	Marlowet® 4702
CAS 52668-97-0	Marlowet® 4703
CAS 52725-64-1	Comperlan LMD
CAS 53610-02-9	Surfine AZI-A
CAS 53980-88-4	Westvaco Diacid® 1550
CAS 53980-88-4	Westvaco Diacid® 1575
CAS 54045-08-8	Marlipal® SU
CAS 56002-14-3	Emerest® 2625
CAS 56002-14-3	Emerest® 2644
CAS 56002-14-3	Emerest® 2664
CAS 56002-14-3	Ethox MI-9
CAS 56002-14-3	Ethox MI-14
CAS 56002-14-3	Industrol® MIS-9
CAS 56002-14-3	Trydet 2644
CAS 56002-14-3	Trydet ISA-4
CAS 56002-14-3	Trydet ISA-9
CAS 56002-14-3	Trydet ISA-14
CAS 56863-02-6	Schercomid SLE
CAS 57171-56-9	Atlas G-1086
CAS 57171-56-9	Atlas G-1096
CAS 57171-56-9	Trylox® 6747
CAS 57307-99-0	Labrasol
CAS 57635-48-0	Akypo RO 50
CAS 57635-48-0	Akypo RO 90
CAS 58069-11-7	Dehyquart SP
CAS 58353-68-7	Octosol A-1
CAS 58855-63-3	Crodafos N3 Neutral
CAS 58855-63-3	Crodafos N10 Neutral
CAS 59355-61-2	Empigen® OY
CAS 59559-30-7	Akypo RS 60
CAS 60828-78-6	Tergitol® TMN-3
CAS 60828-78-6	Tergitol® TMN-6
CAS 60828-78-6	Tergitol® TMN-10
CAS 61731-12-6	Emulpon EL 20
CAS 61731-12-6	Emulpon EL 40
CAS 61731-31-9	Witcamide® LDTS
CAS 61788-45-2	Amine HBG
CAS 61788-45-2	Amine HBGD
CAS 61788-45-2	Armeen® HT
CAS 61788-45-2	Armeen® HTD

CAS 61788-45-2 Crodamine 1.HT
CAS 61788-45-2 Jet Amine PHT
CAS 61788-45-2 Jet Amine PHTD
CAS 61788-45-2 Kemamine® P-970
CAS 61788-45-2 Kemamine® P-970D
CAS 61788-45-2 Noram SH
CAS 61788-45-2 Radiamine 6140
CAS 61788-45-2 Radiamine 6141
CAS 61788-46-3 Amine KK
CAS 61788-46-3 Amine KKD
CAS 61788-46-3 Armeen® C
CAS 61788-46-3 Armeen® CD
CAS 61788-46-3 Jet Amine PC
CAS 61788-46-3 Jet Amine PCD
CAS 61788-46-3 Radiamine 6160
CAS 61788-46-3 Radiamine 6161
CAS 61788-47-4 Industrene® 325
CAS 61788-47-4 Industrene® 328
CAS 61788-62-3 Armeen® M2C
CAS 61788-62-3 Jet Amine M2C
CAS 61788-62-3 Noram M2C
CAS 61788-63-4 Armeen® M2HT
CAS 61788-63-4 Kemamine® T-9701
CAS 61788-63-6 Noram M2SH
CAS 61788-66-3 Noram C
CAS 61788-85-0 Akyporox CO 400
CAS 61788-85-0 Akyporox CO 600
CAS 61788-85-0 Alkamuls® COH-5
CAS 61788-85-0 Arlatone® G
CAS 61788-85-0 Chemax HCO-5
CAS 61788-85-0 Chemax HCO-16
CAS 61788-85-0 Chemax HCO-25
CAS 61788-85-0 Chemax HCO-200/50
CAS 61788-85-0 Croduret 10
CAS 61788-85-0 Croduret 30
CAS 61788-85-0 Croduret 40
CAS 61788-85-0 Croduret 60
CAS 61788-85-0 Croduret 100
CAS 61788-85-0 Croduret 200
CAS 61788-85-0 Dacospin POE(25)HRG
CAS 61788-85-0 Dehymuls HRE 7
CAS 61788-85-0 Emulsogen HEL-050
CAS 61788-85-0 Emulsogen HEL-160R
CAS 61788-85-0 Emulsogen HEL-250
CAS 61788-85-0 Emulson EL/H25
CAS 61788-85-0 Ethox HCO-16
CAS 61788-85-0 Ethox HCO-25
CAS 61788-85-0 Ethox HCO-200/50%
CAS 61788-85-0 Hetoxide HC-16
CAS 61788-85-0 Hetoxide HC-25
CAS 61788-85-0 Hetoxide HC-40
CAS 61788-85-0 Hetoxide HC-60
CAS 61788-85-0 Industrol® COH-25
CAS 61788-85-0 Industrol® COH-200
CAS 61788-85-0 Mapeg® CO-16H
CAS 61788-85-0 Mapeg® CO-25H
CAS 61788-85-0 Nopalcol 10-COH
CAS 61788-85-0 Nopalcol 12-COH
CAS 61788-85-0 Prox-onic HRH-05
CAS 61788-85-0 Prox-onic HRH-025
CAS 61788-85-0 Prox-onic HRH-016
CAS 61788-85-0 Prox-onic HRH-0200
CAS 61788-85-0 Prox-onic HRH-0200/50
CAS 61788-85-0 Remcopal HC 7
CAS 61788-85-0 Remcopal HC 20
CAS 61788-85-0 Remcopal HC 33
CAS 61788-85-0 Remcopal HC 40
CAS 61788-85-0 Remcopal HC 60
CAS 61788-85-0 Rolfor COH 40
CAS 61788-85-0 Surfactol® 575
CAS 61788-85-0 Surfactol® 590
CAS 61788-85-0 Tagat® R40
CAS 61788-85-0 Tagat® R60
CAS 61788-85-0 Tagat® R63
CAS 61788-85-0 Trylox® 5921
CAS 61788-85-0 Trylox® 5922
CAS 61788-85-0 Trylox® 5925
CAS 61788-85-0 Trylox® HCO-5
CAS 61788-85-0 Trylox® HCO-16
CAS 61788-85-0 Trylox® HCO-25
CAS 61788-85-0 Trylox® HCO-200/50
CAS 61788-90-7 Aminoxid A 4080
CAS 61788-90-7 Aromox® DMC
CAS 61788-90-7 Aromox® DMC-W
CAS 61788-90-7 Naxide 1230
CAS 61788-90-7 Noxamine CA 30
CAS 61788-90-7 Schercamox DMC
CAS 61788-90-7 Sochamine OX 30
CAS 61788-91-8 Armeen® DMSD
CAS 61788-93-0 Amine 2MKKD
CAS 61788-93-0 Armeen® DMCD
CAS 61788-93-0 Noram DMCD
CAS 61788-95-2 Amine 2MHBGD
CAS 61788-95-2 Armeen® DMHTD
CAS 61788-95-2 Noram DMSH D
CAS 61789-18-2 Arquad® C-33W
CAS 61789-18-2 Arquad® C-50
CAS 61789-18-2 Jet Quat C-50
CAS 61789-18-2 Noramium MC 50
CAS 61789-19-3 Armid® C
CAS 61789-30-8 Chimipon SK
CAS 61789-30-8 Norfox® 1101
CAS 61789-30-8 Norfox® 1115
CAS 61789-40-0 Empigen® BS/P
CAS 61789-40-0 Incronam 30
CAS 61789-40-0 Rewoteric AM B13
CAS 61789-40-0 Rewoteric AM B14
CAS 61789-40-0 Tego®-Betaine F
CAS 61789-40-0 Tego® Betaine L-5351
CAS 61789-68-2 Ethoquad® CB/12
CAS 61789-71-7 Arquad® DMCB-80
CAS 61789-71-7 Querton KKBCl-50
CAS 61789-72-8 Arquad® DMHTB-75
CAS 61789-73-9 Arquad® M2HTB-80
CAS 61789-75-2 Noram 2C
CAS 61789-76-2 Armeen® 2C
CAS 61789-77-3 Arquad® 2C-75
CAS 61789-77-3 Jet Quat 2C-75
CAS 61789-77-3 Radiaquat 6462
CAS 61789-79-5 Armeen® 2HT
CAS 61789-79-5 Noram 2SH
CAS 61789-80-8 Arquad® 2HT-75
CAS 61789-80-8 Jet Quat 2HT-75
CAS 61789-80-8 Kemamine® Q-9702C
CAS 61789-80-8 M-Quat® 257
CAS 61789-80-8 Quartamin D86P, D86PL, D86PI
CAS 61789-80-8 Radiaquat 6442
CAS 61789-80-8 Radiaquat 6475
CAS 61789-80-8 Radiaquat 6480
CAS 61790-12-3 Acintol® 736
CAS 61790-12-3 Acintol® 746
CAS 61790-12-3 Acintol® 2122
CAS 61790-12-3 Acintol® 7002
CAS 61790-12-3 Acintol® D25LR
CAS 61790-12-3 Acintol® D30E
CAS 61790-12-3 Acintol® D30LR
CAS 61790-12-3 Acintol® D40LR
CAS 61790-12-3 Acintol® D40T
CAS 61790-12-3 Acintol® D60LR
CAS 61790-12-3 Acintol® DFA

CAS 61790-12-3	Acintol® EPG
CAS 61790-12-3	Acintol® FA-1
CAS 61790-12-3	Acintol® FA-1 Special
CAS 61790-12-3	Acintol® FA-2
CAS 61790-12-3	Acintol® FA-3
CAS 61790-12-3	Westvaco® 1480
CAS 61790-12-3	Westvaco® 1483
CAS 61790-18-9	Armeen® S
CAS 61790-18-9	Jet Amine PS
CAS 61790-18-9	Jet Amine PSD
CAS 61790-18-9	Nissan Amine SB
CAS 61790-31-6	Armid® HT
CAS 61790-33-8	Amine BG
CAS 61790-33-8	Amine BGD
CAS 61790-33-8	Armeen® T
CAS 61790-33-8	Armeen® TD
CAS 61790-33-8	Crodamine 1.T
CAS 61790-33-8	Genamin TA Grades
CAS 61790-33-8	Jet Amine PT
CAS 61790-33-8	Jet Amine PTD
CAS 61790-33-8	Kemamine® P-974D
CAS 61790-33-8	Noram S
CAS 61790-33-8	Radiamine 6170
CAS 61790-33-8	Radiamine 6171
CAS 61790-33-8	Tallow Amine
CAS 61790-37-2	Industrene® 143
CAS 61790-41-8	Arquad® S-50
CAS 61790-41-8	Jet Quat S-50
CAS 61790-50-9	Westvaco® Resin 895
CAS 61790-51-0	Westvaco® Resin 790
CAS 61790-51-0	Westvaco® Resin 795
CAS 61790-57-6	Noramac C 26
CAS 61790-59-8	Armac® HT
CAS 61790-59-8	Noramac SH
CAS 61790-63-4	Marlamid® DF 1218
CAS 61790-81-6	Aqualose L30
CAS 61790-81-6	Aqualose L75
CAS 61790-81-6	Aqualose L75/50
CAS 61790-81-6	Ethoxylan® 1685
CAS 61790-81-6	Ethoxylan® 1686
CAS 61790-81-6	Lan-Aqua-Sol 50
CAS 61790-81-6	Lan-Aqua-Sol 100
CAS 61790-81-6	Laneto 40
CAS 61790-81-6	Laneto 50
CAS 61790-81-6	Laneto 60
CAS 61790-81-6	Lanogel® 21
CAS 61790-81-6	Lanogel® 31
CAS 61790-81-6	Lanogel® 41
CAS 61790-81-6	Lanogel® 61
CAS 61790-81-6	Solan 50
CAS 61790-81-6	Solan E
CAS 61790-81-6	Solan
CAS 61790-81-6	Solangel 401
CAS 61790-81-6	Solulan® 75
CAS 61790-81-6	Solulan® L-575
CAS 61790-81-6	Super Solan Flaked
CAS 61790-81-6	Teric LAN70
CAS 61790-85-0	Chemeen DT-15
CAS 61790-85-0	Dinoramox S3
CAS 61790-85-0	Ethoduomeen® T/13
CAS 61790-85-0	Ethoduomeen® T/20
CAS 61790-85-0	Ethoduomeen® T/25
CAS 61791-00-2	Aconol X6
CAS 61791-00-2	Aconol X10
CAS 61791-00-2	Alkamuls® TD-41
CAS 61791-00-2	Chemax TO-8
CAS 61791-00-2	Chemax TO-10
CAS 61791-00-2	Chemax TO-16
CAS 61791-00-2	EM-600
CAS 61791-00-2	Ethofat® 242/25
CAS 61791-00-2	Ethofat® 433
CAS 61791-00-2	Ethox TO-8
CAS 61791-00-2	Ethox TO-9A
CAS 61791-00-2	Ethox TO-16
CAS 61791-00-2	Genagen TA-080
CAS 61791-00-2	Genagen TA-120
CAS 61791-00-2	Genagen TA-160
CAS 61791-00-2	Hetoxamate FA-5
CAS 61791-00-2	Hetoxamate FA-20
CAS 61791-00-2	Industrol® 400-MOT
CAS 61791-00-2	Industrol® TO-10
CAS 61791-00-2	Industrol® TO-16
CAS 61791-00-2	Laurel PEG 400 MT
CAS 61791-00-2	Laurel PEG 600 MT
CAS 61791-00-2	Mapeg® 200 MOT
CAS 61791-00-2	Mapeg® 400 MOT
CAS 61791-00-2	Mapeg® 600 MOT
CAS 61791-00-2	Mapeg® TAO-15
CAS 61791-00-2	Marlosol® TF3
CAS 61791-00-2	Marlosol® TF4
CAS 61791-00-2	Pegosperse® 400 MOT
CAS 61791-00-2	Prox-onic TA-1/08
CAS 61791-00-2	Prox-onic TA-1/010
CAS 61791-00-2	Prox-onic TA-1/016
CAS 61791-00-2	Renex® 20
CAS 61791-00-2	Rolfat R 106
CAS 61791-00-2	Sellig T 1790
CAS 61791-00-2	Sellig T 3 100
CAS 61791-00-2	Sellig T 14 100
CAS 61791-00-2	Teric T2
CAS 61791-00-2	Teric T5
CAS 61791-00-2	Teric T7
CAS 61791-00-2	Teric T10
CAS 61791-00-2	Trydet 2682
CAS 61791-00-2	Witconol H31
CAS 61791-01-3	Ethox DTO-9A
CAS 61791-01-3	Industrol® DT-13
CAS 61791-01-3	Laurel PEG 400 DT
CAS 61791-01-3	Laurel PEG 600 DT
CAS 61791-01-3	Mapeg® 200 DOT
CAS 61791-01-3	Mapeg® 400 DOT
CAS 61791-01-3	Mapeg® 600 DOT
CAS 61791-01-3	Pegosperse® 400 DOT
CAS 61791-01-3	Pegosperse® 600 DOT
CAS 61791-01-3	R3124 Ester
CAS 61791-07-9	Rolfat SO 6
CAS 61791-08-0	Empilan® LP10
CAS 61791-08-0	Empilan® MAA
CAS 61791-08-0	Eumulgin PC 2
CAS 61791-08-0	Eumulgin PC 4
CAS 61791-08-0	Eumulgin PC 10
CAS 61791-08-0	Eumulgin PC 10/85
CAS 61791-10-4	Ethoquad® C/25
CAS 61791-12-6	Berol 106
CAS 61791-12-6	Berol 108
CAS 61791-12-6	Berol 190
CAS 61791-12-6	Berol 191
CAS 61791-12-6	Berol 195
CAS 61791-12-6	Berol 198
CAS 61791-12-6	Berol 199
CAS 61791-12-6	Berol 829
CAS 61791-12-6	Cerex EL 250
CAS 61791-12-6	Cerex EL 300
CAS 61791-12-6	Cerex EL 360
CAS 61791-12-6	Cerex EL 400
CAS 61791-12-6	Emulson AG 81C
CAS 61791-12-6	Emulson AG/CO 25
CAS 61791-12-6	Emulson AG/COH
CAS 61791-12-6	Emulson AG/EL
CAS 61791-12-6	Emulson CO 25

CAS 61791-12-6	Emulson CO-40
CAS 61791-12-6	Emulson CO 40 N
CAS 61791-12-6	Emulson EL
CAS 61791-12-6	Emulson EL 200
CAS 61791-12-6	Etocas 10
CAS 61791-12-6	Marlosol® R70
CAS 61791-12-6	Marlowet® R 11
CAS 61791-12-6	Mulsifan RT 69
CAS 61791-12-6	Surfactol® 318
CAS 61791-12-6	Surfactol® 340
CAS 61791-12-6	Surfactol® 365
CAS 61791-12-6	Surfactol® 380
CAS 61791-13-7	Dehydol LT 3
CAS 61791-13-7	Dehydol LT 7
CAS 61791-13-7	Dehydol LT 7 L
CAS 61791-13-7	Dehydol PLT 8
CAS 61791-13-7	Genapol® C-050
CAS 61791-13-7	Genapol® C-080
CAS 61791-13-7	Genapol® C-100
CAS 61791-13-7	Genapol® C-150
CAS 61791-13-7	Genapol® C-200
CAS 61791-13-7	Genapol® GC-050
CAS 61791-13-7	Marlowet® CA 5
CAS 61791-13-7	Marlowet® CA 10
CAS 61791-13-7	Oxetal C 110
CAS 61791-14-8	Ablumox C-7
CAS 61791-14-8	Accomeen C2
CAS 61791-14-8	Accomeen C5
CAS 61791-14-8	Accomeen C10
CAS 61791-14-8	Accomeen C15
CAS 61791-14-8	Alkaminox® C-5
CAS 61791-14-8	Amiet CD/14
CAS 61791-14-8	Amiet CD/17
CAS 61791-14-8	Amiet CD/22
CAS 61791-14-8	Amiet CD/27
CAS 61791-14-8	Berol 307
CAS 61791-14-8	Berol 397
CAS 61791-14-8	Chemeen C-2
CAS 61791-14-8	Chemeen C-5
CAS 61791-14-8	Chemeen C-10
CAS 61791-14-8	Chemeen C-15
CAS 61791-14-8	Ethomeen® C/12
CAS 61791-14-8	Ethomeen® C/15
CAS 61791-14-8	Ethomeen® C/20
CAS 61791-14-8	Ethomeen® C/25
CAS 61791-14-8	Ethox CAM-2
CAS 61791-14-8	Ethox CAM-15
CAS 61791-14-8	Ethylan® TC
CAS 61791-14-8	Ethylan® TN-10
CAS 61791-14-8	Eumulgin PA 12
CAS 61791-14-8	Hetoxamine C-2
CAS 61791-14-8	Hetoxamine C-5
CAS 61791-14-8	Hetoxamine C-15
CAS 61791-14-8	Mazeen® C-2
CAS 61791-14-8	Mazeen® C-5
CAS 61791-14-8	Mazeen® C-10
CAS 61791-14-8	Mazeen® C-15
CAS 61791-14-8	Noramox C2
CAS 61791-14-8	Noramox C5
CAS 61791-14-8	Noramox C11
CAS 61791-14-8	Noramox C15
CAS 61791-14-8	Prox-onic MC-02
CAS 61791-14-8	Prox-onic MC-05
CAS 61791-14-8	Prox-onic MC-015
CAS 61791-14-8	Rolamet C 11
CAS 61791-14-8	Teric 12M2
CAS 61791-14-8	Teric 12M5
CAS 61791-14-8	Teric 12M15
CAS 61791-14-8	Trymeen® 6601
CAS 61791-14-8	Trymeen® CAM-10
CAS 61791-14-8	Trymeen® CAM-15
CAS 61791-14-8	Varonic® K202
CAS 61791-14-8	Varonic® K202 SF
CAS 61791-14-8	Varonic® K205
CAS 61791-14-8	Varonic® K205 SF
CAS 61791-14-8	Varonic® K210
CAS 61791-14-8	Varonic® K210 SF
CAS 61791-14-8	Varonic® K215
CAS 61791-14-8	Varonic® K215 LC
CAS 61791-14-8	Varonic® K215 SF
CAS 61791-14-8	Zusomin C 108
CAS 61791-20-6	Polychol 5
CAS 61791-20-6	Polychol 10
CAS 61791-20-6	Polychol 20
CAS 61791-20-6	Polychol 40
CAS 61791-24-0	Accomeen S2
CAS 61791-24-0	Accomeen S5
CAS 61791-24-0	Accomeen S10
CAS 61791-24-0	Accomeen S15
CAS 61791-24-0	Chemeen S-2
CAS 61791-24-0	Chemeen S-5
CAS 61791-24-0	Chemeen S-30
CAS 61791-24-0	Chemeen S-30/80
CAS 61791-24-0	Ethomeen® S/12
CAS 61791-24-0	Ethomeen® S/15
CAS 61791-24-0	Ethomeen® S/20
CAS 61791-24-0	Ethomeen® S/25
CAS 61791-24-0	Hetoxamine S-2
CAS 61791-24-0	Hetoxamine S-5
CAS 61791-24-0	Hetoxamine S-15
CAS 61791-24-0	Icomeen® S-5
CAS 61791-24-0	Mazeen® S-2
CAS 61791-24-0	Mazeen® S-5
CAS 61791-24-0	Mazeen® S-10
CAS 61791-24-0	Mazeen® S-15
CAS 61791-24-0	Teric 16M2
CAS 61791-24-0	Teric 16M5
CAS 61791-24-0	Teric 16M10
CAS 61791-24-0	Teric 16M15
CAS 61791-24-0	Tomah E-S-2
CAS 61791-24-0	Tomah E-S-5
CAS 61791-24-0	Tomah E-S-15
CAS 61791-25-1	Rewoteric AM TEG
CAS 61791-25-1	Tego®-Betaine N-192
CAS 61791-26-2	Berol 381
CAS 61791-26-2	Berol 386
CAS 61791-26-2	Berol 387
CAS 61791-26-2	Berol 389
CAS 61791-26-2	Berol 391
CAS 61791-26-2	Berol 392
CAS 61791-26-2	Berol 457
CAS 61791-26-2	Berol 458
CAS 61791-26-2	Emulgin PA Series
CAS 61791-26-2	Ethomeen® T/15
CAS 61791-26-2	Ethomeen® T/25
CAS 61791-28-4	Dehydol PTA 7
CAS 61791-28-4	Dehydol PTA 23
CAS 61791-28-4	Dehydol PTA 23/E
CAS 61791-28-4	Dehydol PTA 40
CAS 61791-28-4	Dehydol PTA 114
CAS 61791-28-4	Dehydol TA 5
CAS 61791-28-4	Dehydol TA 11
CAS 61791-28-4	Dehydol TA 12
CAS 61791-28-4	Dehydol TA 14
CAS 61791-28-4	Dehydol TA 20
CAS 61791-28-4	Dehydol TA 30
CAS 61791-28-4	Emulmin 40
CAS 61791-28-4	Emulmin 50
CAS 61791-28-4	Emulmin 70
CAS 61791-28-4	Oxetal TG 111

CAS 61791-28-4	Oxetal TG 118
CAS 61791-28-4	Rewopal® TA 11
CAS 61791-28-4	Rewopal® TA 25
CAS 61791-28-4	Rewopal® TA 25/S
CAS 61791-28-4	Rewopal® TA 50
CAS 61791-29-5	Crodet C10
CAS 61791-31-9	Berol 307
CAS 61791-31-9	Ethomeen® C/12
CAS 61791-31-9	Norfox® DC
CAS 61791-31-9	Norfox® DCSA
CAS 61791-31-9	Norfox® KD
CAS 61791-38-6	Schercozoline C
CAS 61791-44-4	Berol 456
CAS 61791-44-4	Ethomeen® T/12
CAS 61791-53-5	Duomeen® TDO
CAS 61791-55-7	Dinoram S
CAS 61791-55-7	Duomeen® T
CAS 61791-55-7	Jet Amine DT
CAS 61791-55-7	Lilamuls BG
CAS 61791-56-8	Monateric TDB-35
CAS 61791-57-9	Jet Amine TRT
CAS 61791-57-9	LSP 33
CAS 61791-57-9	Triameen T
CAS 61791-57-9	Trinoram S
CAS 61791-63-7	Dinoram C
CAS 61791-63-7	Duomeen® C
CAS 61791-63-7	Duomeen® CD
CAS 61791-63-7	Jet Amine DC
CAS 61791-63-7	Radiamine 6560
CAS 61791-66-6	Noxamine O2-30
CAS 61827-42-7	Synperonic 10/3-90%
CAS 61827-42-7	Synperonic 10/3-100%
CAS 61827-42-7	Synperonic 10/5-90%
CAS 61827-42-7	Synperonic 10/5-100%
CAS 61827-42-7	Synperonic 10/6-90%
CAS 61827-42-7	Synperonic 10/6-100%
CAS 61827-42-7	Synperonic 10/7-90%
CAS 61827-42-7	Synperonic 10/7-100%
CAS 61827-42-7	Synperonic 10/8-90%
CAS 61827-42-7	Synperonic 10/8-100%
CAS 61827-42-7	Synperonic 10/11-90%
CAS 61827-42-7	Synperonic 10/11-100%
CAS 61827-42-7	Synperonic KB
CAS 61970-18-9	Armeen® SD
CAS 62705-16-2	Sanac C
CAS 63011-36-5	Marlipal® 013/20
CAS 63393-82-8	Neodol® 25
CAS 63451-23-0	Crodateric Cy
CAS 63793-60-2	Witconol APM
CAS 64743-02-8	Neodene® 14/16/18
CAS 64743-02-8	Neodene® 1214
CAS 64743-02-8	Neodene® 1416
CAS 64743-02-8	Neodene® 1618
CAS 65071-95-6	Ethofat® 242/25
CAS 65071-95-6	Ethofat® 433
CAS 65143-89-7	Calfax 16L-35
CAS 65381-09-1	Miglyol® 812
CAS 66085-00-5	Imwitor® 780 K
CAS 66085-00-5	Peceol Isostearique
CAS 66197-78-2	Crafol AP-53
CAS 66455-15-0	Marlipal® KF
CAS 66455-17-2	Neodol® 91
CAS 66455-29-6	Amphoteen 24
CAS 66455-29-6	Empigen® BB
CAS 66455-29-6	Empigen® BB-AU
CAS 66794-58-9	Crillet 6
CAS 67700-98-5	Amine 2M12D
CAS 67700-98-5	Empigen® AB
CAS 67700-99-6	Amine M2HBG
CAS 67701-00-2	Armeen® 3-16
CAS 67701-05-7	C-108
CAS 67701-05-7	C-110
CAS 67701-05-7	Hystrene® 1835
CAS 67701-06-8	T-11
CAS 67701-06-8	T-18
CAS 67701-06-8	T-20
CAS 67701-06-8	T-22
CAS 67701-08-0	HK-1618
CAS 67701-08-0	Industrene® 225
CAS 67701-08-0	Industrene® 226
CAS 67701-08-0	S-210
CAS 67701-27-3	Neustrene® 059
CAS 67701-27-3	Neustrene® 060
CAS 67762-25-8	CO-1218
CAS 67762-30-5	TA-1618
CAS 67762-35-0	Polirol C5
CAS 67762-35-0	Rolfat C 9
CAS 67762-36-1	C-810
CAS 67762-36-1	C-810L
CAS 67762-96-3	Tego® Foamex 3062
CAS 67774-74-7	Marlican®
CAS 67892-37-9	Schercoteric O-AA
CAS 67990-17-4	Miranate® B
CAS 68002-61-9	Radiaquat 6471
CAS 68002-71-1	Neustrene® 064
CAS 68002-72-2	Neustrene® 045
CAS 68002-72-2	Neustrene® 053
CAS 68002-90-4	C-1214
CAS 68002-96-0	Polirol 215
CAS 68002-97-1	Empilan® KB 2
CAS 68002-97-1	Empilan® KB 3
CAS 68037-93-4	Amine 2M16D
CAS 68037-96-7	Kemamine® T-9992D
CAS 68037-97-8	Radiamine 6572
CAS 68071-35-2	Chimin P40
CAS 68071-35-2	Chimin P45
CAS 68071-48-7	Berol 305
CAS 68081-81-2	Gardilene S25L
CAS 68081-81-2	Nansa® HS80-AU
CAS 68081-81-2	Nansa® HS80P
CAS 68081-81-2	Nansa® SS 60
CAS 68081-96-9	Empicol® AL30
CAS 68081-99-2	Witcamide® 272
CAS 68092-28-4	Emulamid TO-21
CAS 68127-33-3	Westvaco Diacid® H-240
CAS 68130-43-8	Akypomine® BC 50
CAS 68131-39-5	Volpo 25 D 3
CAS 68131-39-5	Volpo 25 D 5
CAS 68131-39-5	Volpo 25 D 10
CAS 68131-39-5	Volpo 25 D 15
CAS 68131-39-5	Volpo 25 D 20
CAS 68131-40-8	Nonipol Soft SS-50
CAS 68131-40-8	Nonipol Soft SS-70
CAS 68131-40-8	Nonipol Soft SS-90
CAS 68131-40-8	Tergitol® 15-S-3
CAS 68131-40-8	Tergitol® 15-S-5
CAS 68131-40-8	Tergitol® 15-S-7
CAS 68131-40-8	Tergitol® 15-S-9
CAS 68131-40-8	Tergitol® 15-S-12
CAS 68131-40-8	Tergitol® 15-S-15
CAS 68131-40-8	Tergitol® 15-S-20
CAS 68131-40-8	Tergitol® 15-S-30
CAS 68131-40-8	Tergitol® 15-S-40
CAS 68139-30-0	Rewoteric AM CAS
CAS 68139-89-9	Tenax 2010
CAS 68140-00-1	Ablumide CME
CAS 68140-00-1	Alkamide® C-212
CAS 68140-00-1	Alkamide® CME
CAS 68140-00-1	Amidex CME
CAS 68140-00-1	Amidex KME

CAS 68140-00-1 Aminol CM, CM Flakes, CM-C Flakes, CM-D Flakes
CAS 68140-00-1 Chimipal MC
CAS 68140-00-1 Comperlan 100
CAS 68140-00-1 Comperlan KM
CAS 68140-00-1 Comperlan P 100
CAS 68140-00-1 Emid® 6500
CAS 68140-00-1 Empilan® CM
CAS 68140-00-1 Empilan® CME
CAS 68140-00-1 Empilan® CM/F
CAS 68140-00-1 Foamole M
CAS 68140-00-1 Hetamide CMA
CAS 68140-00-1 Hetamide CME
CAS 68140-00-1 Hetamide CME-CO
CAS 68140-00-1 Incromide CM
CAS 68140-00-1 Laurel SD-1050
CAS 68140-00-1 Lauridit® KM
CAS 68140-00-1 Loropan CME
CAS 68140-00-1 Mackamide CMA
CAS 68140-00-1 Manro CMEA
CAS 68140-00-1 Manromid CMEA
CAS 68140-00-1 Marlamid® M 1218
CAS 68140-00-1 Mazamide® CFAM
CAS 68140-00-1 Mazamide® CMEA Extra
CAS 68140-00-1 Monamide
CAS 68140-00-1 Monamid® CMA
CAS 68140-00-1 Ninol® CMP
CAS 68140-00-1 Norfox® CMA
CAS 68140-00-1 P & G Amide No. 27
CAS 68140-00-1 Rewomid® C 212
CAS 68140-00-1 Rewomid® CD
CAS 68140-00-1 Schercomid CME
CAS 68140-00-1 Standamid® 100
CAS 68140-00-1 Standamid® CM
CAS 68140-00-1 Standamid® CMG
CAS 68140-00-1 Standamid® KM
CAS 68140-00-1 Standamid® SM
CAS 68140-00-1 Swanic 51
CAS 68140-00-1 Synotol ME 90
CAS 68140-00-1 Varamide® C-212
CAS 68140-00-1 Witcamide® C
CAS 68140-00-1 Zoramide CM
CAS 68140-01-2 Schercodine C
CAS 68152-92-1 Westvaco® M-30
CAS 68152-92-1 Westvaco® M-40
CAS 68152-92-1 Westvaco® M-70
CAS 68153-32-2 Radiaquat 6470
CAS 68153-58-2 Estersulf 1HH
CAS 68153-63-9 Marlamid® M 1618
CAS 68154-97-2 Marlox® FK 64
CAS 68154-97-2 Marlox® FK 69
CAS 68154-97-2 Marlox® FK 86
CAS 68155-09-9 Aminoxid WS 35
CAS 68155-09-9 Empigen® OS/A
CAS 68155-09-9 Standamox PCAW
CAS 68155-20-4 Schercomid SO-T
CAS 68155-24-8 Ethomid® HT/23
CAS 68155-24-8 Ethomid® HT/60
CAS 68155-27-1 Synprolam 35
CAS 68213-23-0 Synperonic A3
CAS 68213-23-0 Synperonic A4
CAS 68213-23-0 Synperonic A7
CAS 68213-23-0 Synperonic A9
CAS 68213-23-0 Synperonic A11
CAS 68213-23-0 Synperonic M2
CAS 68213-23-0 Synperonic M3
CAS 68213-23-0 Ukaril 190
CAS 68308-54-3 Imwitor® 191
CAS 68310-73-6 Duoquad® O-50
CAS 68310-76-9 Servamine KOO 330 B
CAS 68333-82-4 Witcamide® CPA
CAS 68334-21-4 Empigen® CDR40
CAS 68334-21-4 Empigen® CDR60
CAS 68334-21-4 Schercoteric MS
CAS 68389-70-8 Glucamate® SSE-20
CAS 68391-01-5 Arquad® B-100
CAS 68391-01-5 Synprolam 35DMBQC
CAS 68391-03-7 Querton 280
CAS 68391-04-8 Synprolam 35DM
CAS 68391-07-1 Kemamine® T-9742D
CAS 68411-00-7 Neodene® 1112 IO
CAS 68411-00-7 Neodene® 1314 IO
CAS 68411-00-7 Neodene® 1518 IO
CAS 68411-31-4 Marlopon® AT 50
CAS 68411-32-5 Sulframin Acide TPB
CAS 68412-54-4 Berol 02
CAS 68412-54-4 Berol 09
CAS 68412-54-4 Berol 26
CAS 68412-54-4 Berol 259
CAS 68412-54-4 Berol 267
CAS 68412-54-4 Berol 277
CAS 68412-54-4 Berol 278
CAS 68412-54-4 Berol 281
CAS 68412-54-4 Berol 291
CAS 68412-54-4 Berol 292
CAS 68412-54-4 Berol 295
CAS 68412-54-4 Berol WASC
CAS 68412-54-4 Synperonic N
CAS 68412-54-4 Synperonic NP4
CAS 68412-54-4 Synperonic NX
CAS 68412-54-4 Synperonic NXP
CAS 68424-94-2 Incronam CD-30
CAS 68424-94-2 Surfax ACB
CAS 68425-47-8 Marlamid® DF 1818
CAS 68439-39-4 Marlophor® DS-Acid
CAS 68439-39-4 Marlophor® FC-Acid
CAS 68439-45-2 Berol 260
CAS 68439-45-2 Empilan® KA5/90
CAS 68439-45-2 Empilan® KA8/80
CAS 68439-45-2 Empilan® KA10/80
CAS 68439-45-2 Syntens KMA 55
CAS 68439-46-3 Dehydol PO 5
CAS 68439-46-3 Synperonic 91/2.5
CAS 68439-46-3 Synperonic 91/4
CAS 68439-46-3 Synperonic 91/5
CAS 68439-46-3 Synperonic 91/7
CAS 68439-46-3 Synperonic 91/12
CAS 68439-46-3 Synperonic 91/20
CAS 68439-49-6 Acconon W230
CAS 68439-49-6 Brij® 68
CAS 68439-49-6 Dehydol PCS 6
CAS 68439-49-6 Dehydol PCS 14
CAS 68439-49-6 Dehydol PTA 23/E
CAS 68439-49-6 Dehydol PTA 80
CAS 68439-49-6 Empilan® KM 11
CAS 68439-49-6 Empilan® KM 20
CAS 68439-49-6 Empilan® KM 50
CAS 68439-49-6 Eumulgin B1
CAS 68439-49-6 Eumulgin B2
CAS 68439-49-6 Eumulgin B3
CAS 68439-49-6 Hetoxol 15 CSA
CAS 68439-49-6 Hetoxol CS-4
CAS 68439-49-6 Hetoxol CS-5
CAS 68439-49-6 Hetoxol CS-9
CAS 68439-49-6 Hetoxol CS-15
CAS 68439-49-6 Hetoxol CS-20
CAS 68439-49-6 Hetoxol CS-25
CAS 68439-49-6 Hetoxol CS-30
CAS 68439-49-6 Hetoxol CS-40W
CAS 68439-49-6 Hetoxol CS-50

CAS No.	Trade name
CAS 68439-49-6	Hetoxol CS-50 Special
CAS 68439-49-6	Hetoxol CSA-15
CAS 68439-49-6	Incropol CS-20
CAS 68439-49-6	Incropol CS-50
CAS 68439-49-6	Industrol® CSS-25
CAS 68439-49-6	Lipocol SC-4
CAS 68439-49-6	Lipocol SC-15
CAS 68439-49-6	Lipocol SC-20
CAS 68439-49-6	Macol® CSA-2
CAS 68439-49-6	Macol® CSA-4
CAS 68439-49-6	Macol® CSA-10
CAS 68439-49-6	Macol® CSA-15
CAS 68439-49-6	Macol® CSA-20
CAS 68439-49-6	Macol® CSA-40
CAS 68439-49-6	Macol® CSA-50
CAS 68439-49-6	Marlipal® 1618/6
CAS 68439-49-6	Marlipal® 1618/8
CAS 68439-49-6	Marlipal® 1618/10
CAS 68439-49-6	Marlipal® 1618/11
CAS 68439-49-6	Marlipal® 1618/18
CAS 68439-49-6	Marlipal® 1618/25
CAS 68439-49-6	Marlipal® 1618/40
CAS 68439-49-6	Marlipal® 1618/80
CAS 68439-49-6	Marlowet® 4800
CAS 68439-49-6	Marlowet® 4857
CAS 68439-49-6	Marlowet® FOX
CAS 68439-49-6	Marlowet® PW
CAS 68439-49-6	Marlowet® TA 6
CAS 68439-49-6	Marlowet® TA 8
CAS 68439-49-6	Marlowet® TA 10
CAS 68439-49-6	Marlowet® TA 18
CAS 68439-49-6	Marlowet® TA 25
CAS 68439-49-6	Plurafac® A-38
CAS 68439-49-6	Plurafac® A-39
CAS 68439-49-6	Prox-onic CSA-1/04
CAS 68439-49-6	Prox-onic CSA-1/06
CAS 68439-49-6	Prox-onic CSA-1/010
CAS 68439-49-6	Prox-onic CSA-1/015
CAS 68439-49-6	Prox-onic CSA-1/020
CAS 68439-49-6	Prox-onic CSA-1/030
CAS 68439-49-6	Prox-onic CSA-1/050
CAS 68439-49-6	Remcopal 229
CAS 68439-49-6	Remcopal 238
CAS 68439-49-6	Rewopal® CSF 20
CAS 68439-49-6	Rhodasurf® A-1P
CAS 68439-49-6	Rhodasurf® A-2
CAS 68439-49-6	Rhodasurf® A-6
CAS 68439-49-6	Rhodasurf® A-60
CAS 68439-49-6	Rolfor HT 11
CAS 68439-49-6	Rolfor HT 25
CAS 68439-49-6	Sellig SU 4 100
CAS 68439-49-6	Sellig SU 18 100
CAS 68439-49-6	Sellig SU 25 100
CAS 68439-49-6	Sellig SU 30 100
CAS 68439-49-6	Sellig SU 50 100
CAS 68439-49-6	Standamul® B-1
CAS 68439-49-6	Standamul® B-2
CAS 68439-49-6	Standamul® B-3
CAS 68439-49-6	Teric 16A16
CAS 68439-49-6	Teric 16A22
CAS 68439-49-6	Teric 16A29
CAS 68439-49-6	Teric 16A50
CAS 68439-49-6	Volpo CS-3
CAS 68439-49-6	Volpo CS-5
CAS 68439-49-6	Volpo CS-10
CAS 68439-49-6	Volpo CS-15
CAS 68439-49-6	Volpo CS-20
CAS 68439-50-9	Berol 050
CAS 68439-50-9	Berol 173
CAS 68439-50-9	Marlipal® 24/20
CAS 68439-50-9	Rolfor N 24/2
CAS 68439-50-9	Rolfor N 24/3
CAS 68439-50-9	Rolfor Z 24/23
CAS 68439-50-9	Rolfor Z 24/9
CAS 68439-50-9	Tergitol® 24-L-45
CAS 68439-51-0	Berol 087
CAS 68439-51-0	Berol 185
CAS 68439-51-0	Marlox® MO 124
CAS 68439-51-0	Marlox® MO 154
CAS 68439-51-0	Marlox® MO 174
CAS 68439-53-2	Solulan® PB-2
CAS 68439-54-3	Rolfor E 270, E 527
CAS 68439-57-6	Sermul EA 214
CAS 68439-57-6	Sulframin AOS
CAS 68439-57-6	Witconate AOS
CAS 68439-70-3	Amine 2M14D
CAS 68439-73-6	Radiamine 6570
CAS 68440-05-1	Empilan® CIS
CAS 68479-64-1	Schercopol OMS-Na
CAS 68511-41-1	Jet Amine PE 1214
CAS 68515-73-1	Burco NPS-225
CAS 68515-73-1	Glucopon 225
CAS 68515-73-1	Glucopon 425
CAS 68516-06-3	Propomeen C/12
CAS 68551-12-02	Tergitol® 26-L-1.6
CAS 68551-13-3	Empilan® KCMP 0703/F
CAS 68551-13-3	Empilan® KCMP 0705/F
CAS 68551-13-3	Rolfor EP 237, EP 526
CAS 68551-14-4	Tergitol® Min-Foam 1X
CAS 68551-14-4	Tergitol® Min-Foam 2X
CAS 68554-09-6	Propoquad® 2HT/11
CAS 68584-22-4	Nansa® SS A/S
CAS 68584-22-5	Nansa® 1042
CAS 68584-22-5	Nansa® SSA
CAS 68584-24-7	Gardilene IPA/94
CAS 68584-24-7	Nansa® YS94
CAS 68584-25-8	Nansa® TS 50
CAS 68585-34-2	Empicol® ESB
CAS 68585-34-2	Empicol® ESB3
CAS 68585-34-2	Empimin® SQ25
CAS 68585-34-2	Empimin® SQ70
CAS 68585-34-2	Tensagex EOC 628
CAS 68585-34-2	Tensagex EOC 670
CAS 68585-44-4	Empicol® DA
CAS 68585-47-4	Empicol® LX
CAS 68585-47-7	Empicol® LS30P
CAS 68603-15-5	CO-810
CAS 68603-42-9	Aminol KDE
CAS 68603-42-9	Chimipal DCL/M
CAS 68603-42-9	Empilan® CDX
CAS 68603-42-9	Schercomid SCE
CAS 68603-42-9	Schercomid SCO-Extra
CAS 68603-42-9	Synotol CN 60
CAS 68603-64-5	Dinoram SH
CAS 68604-71-7	Miranol® C2M-SF 70%
CAS 68604-71-7	Miranol® C2M-SF Conc
CAS 68604-71-7	Miranol® FBS
CAS 68604-73-9	Miranol® CS Conc
CAS 68606-27-3	Laurel SD-1000
CAS 68607-29-4	Duoquad® T-50
CAS 68607-29-4	Jet Quat DT-50
CAS 68608-61-7	Miranol® SM Conc
CAS 68608-64-0	Ampholak XJO
CAS 68608-64-0	Miranol® J2M Conc
CAS 68608-64-0	Miranol® JB
CAS 68608-65-1	Ampholak XCO-30
CAS 68608-65-1	Miranol® CM Conc. NP
CAS 68608-65-1	Miranol® CM Conc. OP
CAS 68608-65-1	Miranol® FA-NP
CAS 68608-66-2	Miranol® HM Conc

CAS 68608-88-8 Nansa® SBA
CAS 68610-26-4 Tomah PA-19
CAS 68610-39-9 Miranol® JS Conc.
CAS 68630-96-6 Schercoteric I-AA
CAS 68648-87-3 Alkylate 215
CAS 68648-87-3 Alkylate 225
CAS 68649-55-8 Emulsifiant 33 AD
CAS 68650-39-5 Miranol® C2M Conc. NP
CAS 68650-39-5 Miranol® C2M NP LA
CAS 68650-39-5 Miranol® FB-NP
CAS 68650-39-5 Schercoteric MS-2
CAS 68715-87-7 Duomeen® OTM
CAS 68783-22-2 Schercomid HT-60
CAS 68783-24-4 Armeen® 2T
CAS 68783-25-5 Duomeen® TTM
CAS 68783-78-8 Arquad® 2T-75
CAS 68784-08-7 Cosmopon BL
CAS 68784-08-7 Schercopol CMS-Na
CAS 68814-69-7 Armeen® DMTD
CAS 68814-69-7 Noram DMSD
CAS 68891-11-2 Synperonic NPE1800
CAS 68891-17-8 Akypo ITD 30 N
CAS 68891-21-4 Berol 269
CAS 68891-21-4 Berol 272
CAS 68891-21-4 Berol 716
CAS 68891-38-3 Berol 452
CAS 68891-38-3 Sermul EA 30
CAS 68908-44-1 Empicol® LQ33/T
CAS 68908-44-1 Empicol® TL40/T
CAS 68909-63-7 Mirataine® FM
CAS 68909-65-9 Norfox® PE-600
CAS 68911-79-5 Jet Amine TET
CAS 68911-79-5 Polyram S
CAS 68919-40-4 Miranol® C2M Anhyd. Acid
CAS 68919-41-5 Miranol® CM-SF Conc
CAS 68920-66-1 Marlowet® FOX
CAS 68920-66-1 Polirol O55
CAS 68920-66-1 Rolfor CO 11
CAS 68936-95-8 Glucate® SS
CAS 68937-55-3 Abil® B 8843, B 8847
CAS 68937-55-3 Abil® B 88184
CAS 68937-66-6 Syntens KMA 70-W-02
CAS 68937-66-6 Syntens KMA 70-W-07
CAS 68937-66-6 Syntens KMA 70-W-12
CAS 68937-66-6 Syntens KMA 70-W-17
CAS 68937-66-6 Syntens KMA 70-W-22
CAS 68937-66-6 Syntens KMA 70-W-32
CAS 68937-90-6 Hystrene® 5460
CAS 68951-72-4 Propomeen T/12
CAS 68953-96-8 Nansa® EVM50
CAS 68954-89-2 Akypo RCS 60
CAS 68954-89-2 Akypo RLM 25
CAS 68954-89-2 Akypo RLM 45
CAS 68954-89-2 Akypo RLM 130
CAS 68954-89-2 Akypo RLMQ 38
CAS 68954-89-2 Akypo RS 60
CAS 68954-89-2 Akypo RS 100
CAS 68954-89-2 Akypo RT 60
CAS 68955-19-1 Empicol® LZ
CAS 68955-19-1 Empicol® LZ/D
CAS 68955-19-1 Tensopol ACL 79
CAS 68955-20-4 Empicol® TAS30
CAS 68958-48-5 Tegin® ISO
CAS 68958-64-5 Tagat® TO
CAS 68966-38-1 Schercozoline I
CAS 68987-89-3 Akypo NTS
CAS 68987-89-3 Akypo®-Soft 100 NV
CAS 68987-89-3 Akypo®-Soft 130 NV
CAS 68987-89-3 Akypo®-Soft 160 NV
CAS 68987-90-6 Synperonic OP3
CAS 68988-69-2 Empimin® MKK/AU
CAS 68989-00-4 Empigen® BCB50
CAS 68990-52-3 Methyl Ester B
CAS 69011-84-3 Akyposal BD
CAS 69013-18-9 Surfonic® LF-17
CAS 69227-20-9 Berol 07
CAS 69227-20-9 Berol 08 Powd
CAS 69227-21-0 Marlox® MS 48
CAS 69227-21-0 Synperonic 87K
CAS 69227-21-0 Synperonic BD100
CAS 69227-21-0 Synperonic LF/B26
CAS 69227-21-0 Synperonic LF/D25
CAS 69227-21-0 Synperonic LF/RA30
CAS 69227-21-0 Synperonic LF/RA40
CAS 69227-21-0 Synperonic LF/RA43
CAS 69227-21-0 Synperonic LF/RA50
CAS 69227-21-0 Synperonic LF/RA260
CAS 69227-21-0 Synperonic LF/RA290
CAS 69227-21-0 Synperonic LF/RA343
CAS 69468-44-6 Tagat® I
CAS 69468-44-6 Tagat® I2
CAS 70592-80-2 Empigen® OB
CAS 70750-46-8 Ampholak BTH-35
CAS 70750-47-9 Ethoquad® C/12
CAS 70788-37-3 Eltesol® TSX
CAS 70851-07-9 Ampholak BCA-30
CAS 70851-08-0 Mirataine® CBS, CBS Mod
CAS 71060-61-2 Propomeen 2HT-11
CAS 71060-72-5 Arquad® 316(W)
CAS 71487-00-8 Ethoquad® C/12 Nitrate
CAS 71487-01-9 Arquad® 2C-70 Nitrite
CAS 71902-01-7 Crill 6
CAS 71902-01-7 Montane 70
CAS 72160-13-5 Akypo OP 80
CAS 72160-13-5 Akypo OP 190
CAS 72162-46-0 Tomah DA-14
CAS 73296-89-6 Tensopol A 79
CAS 73296-89-6 Tensopol A 795
CAS 73296-89-6 Tensopol PCL 94
CAS 73296-89-6 Tensopol S 30 LS
CAS 74623-31-7 Pluracol® W170
CAS 74623-31-7 Pluracol® W660
CAS 74623-31-7 Pluracol® W2000
CAS 74623-31-7 Pluracol® W5100N
CAS 75422-21-0 Empimin® LAM30/AU
CAS 75499-49-9 Alkaterge®-T
CAS 75782-86-4 Neodol® 23
CAS 75782-87-5 Neodol® 45
CAS 76483-21-1 Marlophor® AS-Acid
CAS 76483-21-1 Marlophor® CS-Acid
CAS 79770-97-1 Propoquad® T/12
CAS 83933-91-3 Glucate® DO
CAS 84501-49-5 Empimin® SDS
CAS 84501-87-1 T-9
CAS 84812-94-2 Ampholyte KKE
CAS 85005-55-6 Marlophen® 81N
CAS 85116-97-8 Radiasurf® 7410
CAS 85117-50-6 Nansa® HS85
CAS 85408-49-7 Lilaminox M24
CAS 85536-14-7 Marlon® AS3
CAS 85536-23-8 Aminol TEC N
CAS 85736-63-6 Synprolam 35MX1QC
CAS 86088-85-9 Carsosoft® S-75
CAS 86088-85-9 Incrosoft S-75
CAS 86088-85-9 Incrosoft S-90
CAS 86088-85-9 Incrosoft S-90M
CAS 86088-85-9 Varisoft® 475
CAS 88543-32-2 Alkaterge®-E
CAS 90388-14-0 Crodafos CDP
CAS 91648-56-5 Rolpon 24/330

CAS 91648-56-5	Rolpon 24/370
CAS 91744-38-6	Imwitor® 370
CAS 91995-05-0	Ampholak YCA/P
CAS 92112-62-4	Synprolam 35MX1
CAS 93455-78-8	Radiasurf® 7400
CAS 93820-52-1	Rewoteric AM KSF 40
CAS 95706-86-8	Alkaterge®-T-IV
CAS 97281-23-7	Radiasurf® 7270
CAS 97404-50-7	Lanesta G
CAS 97488-62-5	Ampholak 7TY
CAS 97659-50-2	Ampholak YCE
CAS 97659-51-3	Ampholak XCE
CAS 97659-53-5	Ampholak 7CX
CAS 97659-53-5	Ampholak 7TX
CAS 97659-53-5	Ampholak XO7
CAS 97659-53-5	Ampholak XO7-SD-55
CAS 97659-53-5	Ampholak XOO-30
CAS 97999-44-5	Marlowet® 5311
CAS 99330-44-6	Akypo®-Soft 100 MgV
CAS 99821-01-9	Atlas G-5000
CAS 104042-16-2	Newcol 707SF
CAS 104042-16-2	Newcol 710, 714, 723
CAS 104909-82-2	Akypo TBP 180
CAS 105391-15-9	Akypo LF 3
CAS 105391-15-9	Akypo LF 5
CAS 105391-15-9	Akypo MB 1585
CAS 106392-12-5	Dowfax 30C05, 30C10, 50C15
CAS 106392-12-5	Epan 785
CAS 106392-12-5	Lutrol® F 127
CAS 106392-12-5	Monolan® 2800
CAS 107397-59-1	Tetronic® 50R8
CAS 107397-59-1	Tetronic® 70R4
CAS 107397-59-1	Tetronic® 90R4
CAS 107397-59-1	Tetronic® 150R1
CAS 107498-00-0	Pluradot HA-430
CAS 107628-08-0	Akypo OP 80
CAS 107628-08-0	Akypo OP 115
CAS 107628-08-0	Akypo OP 190
CAS 109464-53-1	Sokalan® HP 22
CAS 110152-58-4	Indulin® W-1
CAS 110615-47-9	APG® 600 Glycoside
CAS 110615-47-9	APG® 625 Glycoside
CAS 110615-47-9	Glucopon 425
CAS 110615-47-9	Glucopon 600
CAS 110615-47-9	Glucopon 625
CAS 111019-03-5	Crodafos CAP
CAS 111798-26-6	Marlophor® FC-Sodium Salt
CAS 113976-90-2	APG® 325 Glycoside
CAS 125740-36-5	Tomah Q-14-2
CAS 130124-24-2	Carsosoft® T-90
CAS 130124-24-2	Incrosoft T-75
CAS 130124-24-2	Incrosoft T-90
CAS 130124-24-2	Incrosoft T-90HV
CAS 133934-08-4	Schercotaine APAB
CAS 678213-23-0	Synperonic A2
CAS 977057-32-1	Syntopon A
CAS 977057-34-3	Syntopon B
CAS 977057-35-4	Syntopon D

CAS Number-to-Chemical Compound Cross Reference

CAS 50-21-5	Lactic acid
CAS 56-81-5	Glycerin
CAS 56-93-9	Benzyl trimethyl ammonium chloride
CAS 57-09-0	Cetrimonium bromide
CAS 57-10-3	Palmitic acid
CAS 57-11-4	Stearic acid
CAS 57-13-6	Urea
CAS 57-55-6	Propylene glycol
CAS 57-88-5	Cholesterol
CAS 60-33-3	Linoleic acid
CAS 64-02-8	Tetrasodium EDTA
CAS 64-17-5	Ethyl alcohol
CAS 67-63-0	Isopropyl alcohol
CAS 71-36-3	Butyl alcohol
CAS 72-17-3	Sodium lactate
CAS 77-86-1	Tris (hydroxymethyl) aminomethane
CAS 78-21-7	Cetethyl morpholinium ethosulfate
CAS 78-51-3	Tributoxyethyl phosphate
CAS 78-66-0	Dimethyl octynediol
CAS 78-83-1	Isobutyl alcohol
CAS 93-82-3	Stearamide DEA
CAS 93-83-4	Oleamide DEA
CAS 95-19-2	Stearyl hydroxyethyl imidazoline
CAS 95-38-5	Oleyl hydroxyethyl imidazoline
CAS 95-50-1	o-Dichlorobenzene
CAS 96-20-8	2-Amino-1-butanol
CAS 97-78-9	Lauroyl sarcosine
CAS 98-67-9	Phenol sulfonic acid
CAS 98-79-3	PCA
CAS 101-34-8	Glyceryl triacetyl ricinoleate
CAS 102-60-3	Tetrahydroxypropyl ethylenediamine
CAS 102-71-6	Triethanolamine
CAS 102-87-4	Trilaurylamine
CAS 104-15-4	Toluene sulfonic acid
CAS 104-74-5	Laurylpyridinium chloride
CAS 104-75-6	2-Ethylhexylamine
CAS 104-76-7	2-Ethylhexanol
CAS 105-76-0	Dibutyl maleate
CAS 106-07-0	PEG-4 stearate
CAS 106-08-1	PEG-9 laurate
CAS 106-11-6	PEG-2 stearate
CAS 106-12-7	PEG-2 oleate
CAS 106-14-9	Hydroxystearic acid
CAS 106-16-1	Ricinoleamide MEA
CAS 106-17-2	Glycol ricinoleate
CAS 106-98-9	C4 alpha olefin
CAS 107-21-1	Glycol
CAS 107-41-5	Hexylene glycol
CAS 107-54-0	Dimethyl hexynol
CAS 107-64-2	Distearyldimonium chloride
CAS 109-28-4	Oleamidopropyl dimethylamine
CAS 109-30-8	PEG-2 distearate
CAS 110-25-8	Oleoyl sarcosine
CAS 110-27-0	Isopropyl myristate
CAS 110-30-5	Ethylene distearamide
CAS 110-31-6	Ethylene dioleamide
CAS 110-65-6	But-2-yne-1,4-diol
CAS 110-98-5	Dipropylene glycol
CAS 111-03-5	Glyceryl oleate
CAS 111-05-7	Oleamide MIPA
CAS 111-27-3	Hexyl alcohol
CAS 111-42-2	Diethanolamine
CAS 111-46-6	Diethylene glycol
CAS 111-57-9	Stearamide MEA
CAS 111-60-4	Glycol stearate
CAS 111-66-0	Octene-1
CAS 111-76-2	Butoxyethanol
CAS 111-82-0	Methyl laurate
CAS 111-87-5	Caprylic alcohol
CAS 111-90-0	Ethoxydiglycol
CAS 112-00-5	Laurtrimonium chloride
CAS 112-02-7	Cetrimonium chloride
CAS 112-03-8	Steartrimonium chloride
CAS 112-05-0	Pelargonic acid
CAS 112-18-5	Dimethyl lauramine
CAS 112-30-1	n-Decyl alcohol
CAS 112-34-5	Butoxy diglycol
CAS 112-39-0	Methyl palmitate
CAS 112-41-4	Dodecene-1
CAS 112-42-5	Undecyl alcohol
CAS 112-53-8	Lauryl alcohol
CAS 112-60-7	PEG-4
CAS 112-61-8	Methyl stearate
CAS 112-62-9	Methyl oleate
CAS 112-69-6	Dimethyl palmitamine
CAS 112-70-9	Tridecyl alcohol
CAS 112-72-1	Myristyl alcohol
CAS 112-75-4	Dimethyl myristamine
CAS 112-80-1	Oleic acid
CAS 112-84-5	Erucamide
CAS 112-85-6	Behenic acid
CAS 112-88-9	Octadecene-1
CAS 112-90-3	Oleamine
CAS 112-92-5	Stearyl alcohol
CAS 112-99-2	Dioctadecylamine
CAS 115-69-5	2-Amino-2-methyl-1,3-propanediol
CAS 115-70-8	2-Amino-2-ethyl-1,3-propanediol
CAS 117-81-7	Dioctyl phthalate
CAS 120-18-3	2-Naphthalenesulfonic acid
CAS 120-40-1	Lauramide DEA
CAS 122-19-0	Stearalkonium chloride
CAS 122-32-7	Triolein
CAS 122-99-6	Phenoxyethanol
CAS 123-03-5	Cetylpyridinium chloride
CAS 123-94-4	Glyceryl stearate

CAS 123-95-5	Butyl stearate
CAS 124-07-2	Caprylic acid
CAS 124-10-7	Methyl myristate
CAS 124-22-1	Lauramine
CAS 124-26-5	Stearamide
CAS 124-28-7	Dimethyl stearamine
CAS 124-30-1	Stearamine
CAS 124-68-5	2-Amino-2-methyl-1-propanol
CAS 126-30-7	Neopentyl glycol
CAS 126-73-8	Tributyl phosphate
CAS 126-86-3	Tetramethyl decynediol
CAS 126-92-1	Sodium octyl sulfate
CAS 127-39-9	Diisobutyl sodium sulfosuccinate
CAS 127-68-4	Sodium m-nitrobenzenesulfonate
CAS 136-26-5	Capramide DEA
CAS 136-60-7	Butyl benzoate
CAS 136-99-2	Lauryl hydroxyethyl imidazoline
CAS 137-16-6	Sodium lauroyl sarcosinate
CAS 137-20-2	Sodium methyl oleoyl taurate
CAS 139-07-1	Lauralkonium chloride
CAS 139-08-2	Myristalkonium chloride
CAS 139-88-8	Sodium myristyl sulfate
CAS 139-96-8	TEA-lauryl sulfate
CAS 140-72-7	Cetylpyridinium bromide
CAS 141-08-2	Glyceryl ricinoleate
CAS 141-20-8	PEG-2 laurate
CAS 141-21-9	Stearamidoethyl ethanolamine
CAS 141-22-0	Ricinoleic acid
CAS 141-23-1	Methyl hydroxystearate
CAS 141-24-2	Methyl ricinoleate
CAS 142-16-5	Dioctyl maleate
CAS 142-18-7	Glyceryl laurate
CAS 142-26-7	Acetamide MEA
CAS 142-31-4	Sodium octyl sulfate
CAS 142-48-3	Stearoyl sarcosine
CAS 142-54-1	Lauramide MIPA
CAS 142-55-2	Propylene glycol laurate
CAS 142-58-5	Myristamide MEA
CAS 142-77-8	Butyl oleate
CAS 142-78-9	Lauramide MEA
CAS 142-87-0	Sodium decyl sulfate
CAS 142-91-6	Isopropyl palmitate
CAS 143-00-0	DEA-lauryl sulfate
CAS 143-07-7	Lauric acid
CAS 143-18-0	Potassium oleate
CAS 143-19-1	Sodium oleate
CAS 143-27-1	Palmitamine
CAS 143-28-2	Oleyl alcohol
CAS 149-39-3	Sodium methyl stearoyl taurate
CAS 151-21-3	Sodium lauryl sulfate
CAS 215-42-9	Dimethyl behenamine
CAS 301-02-0	Oleamide
CAS 307-35-7	Perfluorooctanesulfonyl fluoride
CAS 334-48-5	Capric acid
CAS 463-40-1	Linolenic acid
CAS 497-19-8	Sodium carbonate
CAS 497-25-6	Oxazolidine
CAS 501-24-6	3-(n-Pentadecyl) phenol
CAS 532-32-1	Sodium benzoate
CAS 540-10-3	Cetyl palmitate
CAS 544-63-8	Myristic acid
CAS 555-43-1	Tristearin
CAS 577-11-7	Dioctyl sodium sulfosuccinate
CAS 589-68-4	Glyceryl myristate
CAS 592-41-6	1-Hexene
CAS 624-04-4	Glycol dilaurate
CAS 627-83-8	Glycol distearate
CAS 629-70-9	Cetyl acetate
CAS 629-73-2	Hexadecene-1
CAS 629-76-5	Pentadecyl alcohol
CAS 629-96-9	Eicosanol
CAS 657-84-1	Sodium toluenesulfonate
CAS 661-19-8	Behenyl alcohol
CAS 661-61-6	TIPA-lauryl sulfate
CAS 683-10-3	Lauryl betaine
CAS 693-33-4	Cetyl betaine
CAS 820-66-6	Stearyl betaine
CAS 871-37-4	Oleyl betaine
CAS 872-05-9	Decene-1
CAS 872-50-4	1-Methyl-2-pyrrolidinone
CAS 922-80-5	Diamyl sodium sulfosuccinate
CAS 928-24-5	Glycol dioleate
CAS 929-77-1	Methyl behenate
CAS 1002-89-7	Ammonium stearate
CAS 1119-97-7	Myrtrimonium bromide
CAS 1120-01-0	Sodium cetyl sulfate
CAS 1120-24-7	Dimethyl decylamine
CAS 1120-28-1	Methyl eicosenate
CAS 1120-36-1	Tetradecene-1
CAS 1120-49-6	Didecylamine
CAS 1191-50-0	Sodium myristyl sulfate
CAS 1300-51-2	Sodium phenol sulfonate
CAS 1300-72-7	Sodium xylenesulfonate
CAS 1310-58-3	Potassium hydroxide
CAS 1321-87-5	Dimethyl octynediol
CAS 1322-93-6	Sodium diisopropyl naphthalene sulfonate
CAS 1322-98-1	Sodium decylbenzene sulfonate
CAS 1323-38-2	Castor oil
CAS 1323-39-3	Propylene glycol stearate
CAS 1323-42-8	Glyceryl hydroxystearate
CAS 1323-83-7	Glyceryl distearate
CAS 1330-76-3	Diisooctyl maleate
CAS 1330-80-9	Propylene glycol oleate
CAS 1331-61-9	Ammonium dodecylbenzene sulfonate
CAS 1333-39-7	Phenol sulfonic acid
CAS 1335-72-4	Sodium laureth sulfate
CAS 1336-21-6	Ammonium caseinate
CAS 1338-39-2	Sorbitan laurate
CAS 1338-41-6	Sorbitan stearate
CAS 1338-43-8	Sorbitan oleate
CAS 1341-08-8	Steapyrium chloride
CAS 1369-66-3	Dioctyl sodium sulfosuccinate
CAS 1462-54-0	Lauraminopropionic acid
CAS 1561-92-8	Sodium methallyl sulfonate
CAS 1562-00-1	Sodium isethionate
CAS 1592-23-0	Calcium stearate
CAS 1643-20-5	Lauramine oxide
CAS 1812-53-9	Dicetyl dimonium chloride
CAS 1847-55-8	Sodium oleyl sulfate
CAS 1847-58-1	Sodium lauryl sulfoacetate
CAS 2016-42-4	Myristamine
CAS 2044-56-6	Lithium lauryl sulfate
CAS 2190-04-7	Stearamine acetate
CAS 2224-49-9	TEA lauroyl sarcosinate
CAS 2235-43-0	Pentasodium aminotrimethylene phosphonate
CAS 2235-54-3	Ammonium lauryl sulfate
CAS 2315-62-0	Octoxynol-3
CAS 2315-64-2	Octoxynol-5
CAS 2315-66-4	Octoxynol-10
CAS 2315-67-5	Octoxynol-1
CAS 2370-64-1	PEG-6 laurate
CAS 2373-38-8	Dihexyl sodium sulfosuccinate
CAS 2571-88-2	Stearamine oxide
CAS 2605-79-0	Decylamine oxide
CAS 2615-15-8	PEG-6
CAS 2644-45-3	c-Decyl betaine
CAS 2673-22-5	Ditridecyl sodium sulfosuccinate

CAS Number	Chemical Compound
CAS 2724-58-5	Isostearic acid
CAS 2778-96-3	Stearyl stearate
CAS 2809-21-4	1-Hydroxyethylidene-1,1-diphosphonic acid
CAS 2867-47-2	Dimethylaminoethyl methacrylate
CAS 2915-53-9	Dioctyl maleate
CAS 2917-94-4	Sodium octoxynol-2 ethane sulfonate
CAS 3006-15-3	Dihexyl sodium sulfosuccinate
CAS 3010-24-0	PEG-2 stearmonium chloride
CAS 3026-63-9	Sodium tridecyl sulfate
CAS 3033-77-0	Glycidyl trimethyl ammonium chloride
CAS 3055-93-4	Laureth-2
CAS 3055-94-5	Laureth-3
CAS 3055-95-6	Laureth-5
CAS 3055-96-7	Laureth-6
CAS 3055-97-8	Laureth-7
CAS 3055-98-9	Laureth-8
CAS 3055-99-0	Laureth-9
CAS 3056-00-6	Laureth-12
CAS 3061-75-4	Behenamide
CAS 3088-31-1	Sodium laureth sulfate
CAS 3097-08-3	Magnesium lauryl sulfate
CAS 3115-49-9	Nonoxynol-8 carboxylic acid
CAS 3158-56-3	3-(n-Pentadecyl) phenol
CAS 3179-80-4	Lauramidopropyl dimethylamine
CAS 3327-22-8	Chloro-2-hydroxypropyl trimonium chloride
CAS 3332-27-2	Myristamine oxide
CAS 3401-73-8	Tetrasodium dicarboxyethyl stearyl sulfosuccinamate
CAS 3452-07-1	Eicosene-1
CAS 3578-72-1	Calcium behenate
CAS 3614-12-8	Lauraminopropionic acid
CAS 3655-00-3	Disodium lauriminodipropionate
CAS 4219-48-1	Glycol laurate
CAS 4292-10-8	Lauramidopropyl betaine
CAS 4403-12-7	Trideceth-3
CAS 4500-01-0	Glycol oleate
CAS 4536-30-5	Laureth-1
CAS 4706-78-9	Potassium lauryl sulfate
CAS 4722-98-9	MEA-lauryl sulfate
CAS 4741-30-4	Sodium xanthogenate
CAS 5064-31-3	Trisodium NTA
CAS 5117-19-1	PEG-8
CAS 5274-68-0	Laureth-4
CAS 5306-85-4	Dimethyl isosorbide
CAS 5323-95-5	Sodium ricinoleate
CAS 5332-73-0	Methoxypropylamine
CAS 5349-52-0	PEG-9 stearate
CAS 5353-26-4	Oleth-4
CAS 5353-27-5	Oleth-5
CAS 5397-31-9	Octyloxypropylamine
CAS 5401-17-2	PEG-2 ricinoleate
CAS 5422-34-4	Lactamide MEA
CAS 5793-94-2	Calcium stearoyl lactylate
CAS 5856-63-3	2-Amino-1-butanol
CAS 5938-38-5	Sorbitan oleate
CAS 5959-89-7	Sorbitan laurate
CAS 6001-97-4	Dihexyl sodium sulfosuccinate
CAS 6004-24-6	Cetylpyridinium chloride
CAS 6144-28-1	Dilinoleic acid
CAS 6179-44-8	Isostearamidopropyl betaine
CAS 6272-74-8	Lapyrium chloride
CAS 6317-18-6	Methylenebis (thiocyanate)
CAS 6419-19-8	Aminotrimethylene phosphonic acid
CAS 6540-99-4	Laureth-10
CAS 6542-37-6	Oxazolidine
CAS 6790-09-6	PEG-12
CAS 6813-14-0	C11-15 pareth-12
CAS 6834-92-0	Sodium metasilicate
CAS 6842-15-5	Dodecene-1
CAS 7005-47-2	2-Dimethylamino-2-methyl-1-propanol
CAS 7049-66-3	Trilinoleic acid
CAS 7128-91-8	Palmitamine oxide
CAS 7173-51-5	Didecyldimonium chloride
CAS 7173-62-8	Oleyl 1,3-propylene diamine
CAS 7281-04-1	Lauralkonium bromide
CAS 7311-27-5	Nonoxynol-4
CAS 7320-34-5	Tetrapotassium pyrophosphate
CAS 7378-99-6	Dimethyl octylamine
CAS 7381-01-3	Sodium ethyl-2 sulfolaurate
CAS 7396-58-9	Didecyl methylamine
CAS 7447-40-7	Potassium chloride
CAS 7491-05-2	Tallowtrimonium chloride
CAS 7492-30-0	Potassium ricinoleate
CAS 7545-23-5	Myristamide DEA
CAS 7617-78-9	Isodecyloxypropylamine
CAS 7631-86-9	Silica
CAS 7651-02-7	Stearamidopropyl dimethylamine
CAS 7664-41-7	Ammonia
CAS 7681-52-9	Sodium hypochlorite
CAS 7702-01-4	Disodium capryloamphodiacetate
CAS 7747-35-5	Oxazolidine
CAS 7778-77-0	Potassium phosphate
CAS 8001-31-8	Coconut oil
CAS 8001-54-5	Benzalkonium chloride
CAS 8001-79-4	Castor oil
CAS 8002-09-3	Pine oil
CAS 8002-33-3	Sulfated castor oil
CAS 8002-43-5	Lecithin
CAS 8005-44-5	Cetearyl alcohol
CAS 8006-54-0	Lanolin (anhyd.)
CAS 8006-54-0	Lanolin oil
CAS 8007-43-0	Sorbitan sesquioleate
CAS 8009-03-8	Petrolatum (NF)
CAS 8012-95-1	Mineral oil
CAS 8013-05-6	Potassium castorate
CAS 8013-07-8	Epoxidized soybean oil
CAS 8016-70-4	Hydrogenated soybean oil
CAS 8020-84-6	Lanolin, hydrous
CAS 8023-53-8	Alkyl dimethyl dichlorobenzyl ammonium chloride
CAS 8027-32-5	Petrolatum (USP)
CAS 8027-33-6	Lanolin alcohol
CAS 8029-76-3	Hydroxylated lecithin
CAS 8029-91-2	Acetylated hydrogenated lard glyceride
CAS 8029-92-3	Acetylated lard glyceride
CAS 8030-78-2	Tallowtrimonium chloride
CAS 8031-44-5	Hydrogenated lanolin
CAS 8032-32-4	Mineral spirits
CAS 8033-69-0	Sodium cocaminopropionate
CAS 8038-43-5	Lanolin oil
CAS 8039-09-6	PEG-75 lanolin
CAS 8039-67-6	Cocamide MIPA
CAS 8040-05-9	Hydrogenated lard glyceride
CAS 8042-47-5	Mineral oil
CAS 8050-09-7	Rosin
CAS 8050-81-5	Simethicone
CAS 8051-30-7	Cocamide DEA
CAS 8051-52-3	PEG-15 cocamine
CAS 8051-81-8	PEG-27 lanolin
CAS 8051-82-9	PEG-40 lanolin
CAS 8052-10-6	Rosin
CAS 8052-48-0	Sodium tallowate

CAS 8052-50-4	Sodium tallow sulfate
CAS 8061-51-6	Sodium lignosulfonate
CAS 8061-52-7	Calcium lignosulfonate
CAS 8061-53-8	Ammonium lignosulfonate
CAS 9002-88-4	Polyethylene
CAS 9002-89-5	Polyvinyl alcohol (super and fully hydrolyzed)
CAS 9002-92-0	Laureth series
CAS 9002-93-1	Octoxynol series
CAS 9003-01-4	Carbomer
CAS 9003-01-4	Polyacrylic acid
CAS 9003-03-6	Ammonium polyacrylate
CAS 9003-04-7	Sodium polyacrylate
CAS 9003-05-8	Polyacrylamide
CAS 9003-08-1	Melamine/formaldehyde resin
CAS 9003-11-6	Meroxapol series
CAS 9003-11-6	Poloxamer series
CAS 9003-39-8	PVP
CAS 9004-32-4	Carboxymethylcellulose sodium
CAS 9004-62-0	Hydroxyethylcellulose
CAS 9004-64-2	Hydroxypropylcellulose
CAS 9004-74-4	PEG methyl ether series
CAS 9004-80-2	Deceth-4 phosphate
CAS 9004-81-3	PEG laurate series
CAS 9004-82-4	Sodium laureth sulfate
CAS 9004-82-4	Sodium laureth-12 sulfate
CAS 9004-82-4	Sodium laureth-30 sulfate
CAS 9004-87-9	Octoxynol series
CAS 9004-94-8	PEG palmitate series
CAS 9004-95-9	Ceteth series
CAS 9004-96-0	PEG oleate series
CAS 9004-97-1	PEG ricinoleate series
CAS 9004-98-2	Oleth series
CAS 9004-99-3	PEG stearate series
CAS 9005-00-9	Steareth series
CAS 9005-02-1	PEG dilaurate series
CAS 9005-07-6	PEG dioleate series
CAS 9005-08-7	PEG distearate series
CAS 9005-32-7	Alginic acid
CAS 9005-34-9	Ammonium alginate
CAS 9005-35-0	Calcium alginate
CAS 9005-36-1	Potassium alginate
CAS 9005-37-2	Propylene glycol alginate
CAS 9005-38-3	Algin
CAS 9005-42-9	Ammonium caseinate
CAS 9005-64-5	PEG-10 sorbitan laurate
CAS 9005-64-5	PEG-80 sorbitan laurate
CAS 9005-64-5	Polysorbate 20
CAS 9005-64-5	Polysorbate 21
CAS 9005-65-5	Polysorbate 81
CAS 9005-65-6	Polysorbate 80
CAS 9005-66-7	PEG-80 sorbitan palmitate
CAS 9005-66-7	Polysorbate 40
CAS 9005-67-8	Polysorbate 60
CAS 9005-67-8	Polysorbate 61
CAS 9005-70-3	PEG-18 sorbitan trioleate
CAS 9005-70-3	Polysorbate 85
CAS 9005-71-4	Polysorbate 65
CAS 9006-65-9	Dimethicone
CAS 9007-16-3	Carbomer
CAS 9007-17-4	Carbomer
CAS 9007-48-1	Polyglyceryl oleate series
CAS 9010-43-9	Octoxynol-9
CAS 9011-13-6	Styrene/MA copolymer
CAS 9011-16-9	PVM/MA copolymer
CAS 9014-85-1	PEG tetramethyl decynediol series
CAS 9014-90-8	Sodium nonoxynol sulfate series
CAS 9014-92-0	Dodoxynol series
CAS 9014-93-1	Nonyl nonoxynol series
CAS 9015-54-7	Hydrolyzed collagen
CAS 9016-00-6	Dimethicone
CAS 9016-45-9	Nonoxynol series
CAS 9032-42-2	Methyl hydroxyethylcellulose
CAS 9035-85-2	PPG cetyl ether series
CAS 9036-19-5	Octoxynol series
CAS 9038-95-3	PPG buteth series
CAS 9040-38-4	Disodium nonoxynol-10 sulfosuccinate
CAS 9042-76-6	PPG-9 diethylmonium chloride
CAS 9046-01-9	Trideceth-6 phosphate
CAS 9062-04-8	MIPA-laureth sulfate
CAS 9062-04-8	Propylene glycol dicaprylate/ dicaprate
CAS 9064-14-6	PPG-7 lauryl ether
CAS 9065-63-8	PPG buteth series
CAS 9076-43-1	PPG-40 diethylmonium chloride
CAS 9084-06-4	Sodium polynaphthalene sulfonate
CAS 9087-53-0	PPG ceteth series
CAS 10108-24-4	PEG-4 laurate
CAS 10108-25-5	PEG-4 oleate
CAS 10108-28-8	PEG-6 stearate
CAS 10213-78-2	PEG-2 stearamine
CAS 10233-24-6	PEG-3 stearate
CAS 10289-94-8	PEG-14 stearate
CAS 11094-60-3	Polyglyceryl-10 decaoleate
CAS 11099-07-3	Glyceryl stearate
CAS 11099-07-3	Glyceryl stearate SE
CAS 11099-07-3	Polyglyceryl-3 diisostearate
CAS 11111-34-5	Poloxamine series
CAS 11138-66-2	Xanthan gum
CAS 11140-78-6	Lauryl betaine
CAS 11381-09-0	Hydroxypropyl bis-isostear-amidopropyldimonium chloride
CAS 12068-03-0	Sodium toluenesulfonate
CAS 12068-19-8	Sodium nonoxynol-6 phosphate
CAS 12125-02-9	Ammonium chloride
CAS 12676-37-8	Sodium cocaminopropionate
CAS 12765-39-8	Sodium methyl cocoyl taurate
CAS 13039-35-5	Sodium capryloamphoacetate
CAS 13150-00-0	Sodium laureth sulfate
CAS 13192-12-6	Disodium lauryl sulfosuccinate
CAS 13197-76-7	Lauryl hydroxysultaine
CAS 14234-82-3	Diisobutyl maleate
CAS 14350-94-8	Sodium caproamphoacetate
CAS 14350-96-0	Sodium lauroamphoacetate
CAS 14350-97-1	Disodium lauroamphodiacetate
CAS 14351-40-7	Stearamide MEA-stearate
CAS 14351-50-9	Oleamine oxide
CAS 14409-72-4	Nonoxynol-9
CAS 14450-05-6	Pentaerythrityl tetrapelargonate
CAS 14481-60-8	Disodium stearyl sulfosuccinamate
CAS 14492-68-3	Steapyrium chloride
CAS 14727-68-5	Dimethyl oleamine
CAS 14960-06-6	Sodium lauriminodipropionate
CAS 15826-16-1	Sodium laureth sulfate
CAS 16106-44-8	Potassium toluenesulfonate
CAS 16693-53-1	TEA lauroyl sarcosinate
CAS 16841-14-8	Behenalkonium chloride
CAS 16889-14-8	Stearamidoethyl diethylamine
CAS 16958-92-2	Ditridecyl adipate
CAS 17342-21-1	Laurylpyridinium bisulfate
CAS 18448-65-2	PEG-2 oleammonium chloride
CAS 18641-57-1	Glyceryl tribehenate
CAS 19040-44-9	Disodium lauryl sulfosuccinate
CAS 19321-40-5	Pentaerythrityl tetraoleate
CAS 20457-75-4	Ricinoleamidopropyl dimethylamine

CAS 20545-92-0	Undecylenamide MEA
CAS 20636-48-0	Nonoxynol-5
CAS 21142-28-9	MIPA-lauryl sulfate
CAS 21542-96-1	Dimethyl behenamine
CAS 21652-27-7	Oleyl hydroxyethyl imidazoline
CAS 22042-96-2	Diethylenetriamine pentakis (methylene phosphonic acid)
CAS 22047-49-0	Octyl stearate
CAS 23386-52-9	Dicyclohexyl sodium sulfosuccinate
CAS 24938-91-8	Trideceth series
CAS 24938-91-8	Trideceth carboxylic acid series
CAS 24938-91-8	Trideceth phosphate series
CAS 25054-76-6	Oleamidopropyl betaine
CAS 25066-20-0	Stearamidopropylamine oxide
CAS 25086-62-8	Sodium polymethacrylate
CAS 25087-26-7	Polymethacrylic acid
CAS 25155-30-0	Sodium dodecylbenzenesulfonate
CAS 25159-40-4	Oleamidopropylamine oxide
CAS 25167-32-2	Sodium decyl diphenyl ether sulfonate
CAS 25168-73-4	Sucrose stearate
CAS 25190-05-0	Oleth-16
CAS 25212-19-5	Hydrogenated vegetable glycerides phosphate
CAS 25231-21-4	PPG stearyl ether series
CAS 25321-41-9	Xylene sulfonic acid
CAS 25322-68-3	PEG series
CAS 25322-69-4	PPG series
CAS 25339-09-7	Isocetyl stearate
CAS 25339-99-5	Sucrose laurate
CAS 25377-63-3	Undecylenamide MEA
CAS 25377-73-5	Dodecenyl succinic anhydride
CAS 25383-99-7	Sodium stearoyl lactylate
CAS 25446-78-0	Sodium trideceth sulfate
CAS 25446-80-4	Sodium myreth sulfate
CAS 25496-01-9	Tridecylbenzene sulfonic acid
CAS 25496-72-4	Glyceryl mono/dioleate
CAS 25608-12-2	Potassium polyacrylate
CAS 25619-56-1	Barium dinonylnaphthalene sulfonate
CAS 25637-84-7	Glyceryl dioleate
CAS 25704-59-0	Sodium caproamphoacetate
CAS 25791-96-2	PPG-27 glyceryl ether
CAS 25852-45-3	Laureth-3 phosphate
CAS 25882-44-4	Disodium lauramido MEA-sulfosuccinate
CAS 26027-37-2	PEG oleamide series
CAS 26027-38-3	Nonoxynol series
CAS 26183-52-8	Deceth series
CAS 26248-24-8	Sodium tridecylbenzene sulfonate
CAS 26256-79-1	Sodium lauriminodipropionate
CAS 26264-02-8	Nonoxynol-5
CAS 26264-05-1	Isopropylamine dodecylbenzenesulfonate
CAS 26264-06-2	Calcium dodecylbenzene sulfonate
CAS 26264-58-4	Sodium methylnaphthalenesulfonate
CAS 26266-57-9	Sorbitan palmitate
CAS 26266-58-0	Sorbitan trioleate
CAS 26401-35-4	Ditridecyl adipate
CAS 26401-47-8	Dodoxynol series
CAS 26402-22-2	Glyceryl caprate
CAS 26402-26-6	Glyceryl caprylate
CAS 26402-31-3	Propylene glycol ricinoleate
CAS 26447-10-9	Ammonium xylenesulfonate
CAS 26544-38-7	Dodecenyl succinic anhydride
CAS 26545-53-9	DEA-dodecylbenzene sulfonate
CAS 26571-11-9	Nonoxynol-8
CAS 26571-11-9	Nonoxynol-9
CAS 26597-36-4	Dibehenyldimonium chloride
CAS 26635-75-6	PEG lauramide series
CAS 26635-92-7	PEG stearamine series
CAS 26635-93-8	PEG-7 oleamine
CAS 26650-05-5	Disodium undecylenamido MEA-sulfosuccinate
CAS 26658-19-5	Sorbitan tristearate
CAS 26680-54-6	Octenyl succinic anhydride
CAS 26837-33-2	Sodium lauroamphoacetate
CAS 26838-05-1	Disodium lauryl sulfosuccinate
CAS 27028-82-6	TEA-laureth sulfate
CAS 27136-73-8	Oleyl hydroxyethyl imidazoline
CAS 27176-87-0	Dodecylbenzene sulfonic acid
CAS 27176-93-8	Nonoxynol-2
CAS 27176-94-9	Octoxynol-3
CAS 27176-95-0	Nonoxynol-3
CAS 27176-97-2	Nonoxynol-4
CAS 27176-99-4	Octoxynol-5
CAS 27177-01-1	Nonoxynol-6
CAS 27177-02-2	Octoxynol-7
CAS 27177-05-5	Nonoxynol-6, -7, -8
CAS 27177-07-7	Octoxynol-10
CAS 27177-08-8	Nonoxynol-10
CAS 27177-77-1	Potassium dodecylbenzene sulfonate
CAS 27194-74-7	Propylene glycol laurate
CAS 27195-16-0	Sucrose distearate
CAS 27306-79-2	Myreth-3
CAS 27306-90-7	Laureth-17 carboxylic acid
CAS 27321-72-8	Polyglyceryl-3 stearate
CAS 27321-96-6	Choleth-24
CAS 27323-41-7	TEA-dodecylbenzenesulfonate
CAS 27458-93-1	Isostearyl alcohol
CAS 27479-28-3	Quaternium-14
CAS 27638-00-2	Glyceryl dilaurate
CAS 27731-61-9	Ammonium myreth sulfate
CAS 27942-26-3	Nonoxynol-10
CAS 27986-36-3	Nonoxynol-1
CAS 28061-69-0	Dimethyl oleamine
CAS 28212-44-4	Nonoxynol carboxylic acid series
CAS 28519-02-0	Sodium dodecyl diphenyl ether disulfonate
CAS 28631-63-2	Cumene sulfonic acid
CAS 28724-32-5	PEG-15 stearmonium chloride
CAS 28829-38-1	Sodium polyglutamate
CAS 28874-51-3	Sodium PCA
CAS 28880-55-9	PEG-15 oleammonium chloride
CAS 29059-24-3	Propylene glycol myristate
CAS 29116-98-1	Sorbitan dioleate
CAS 29317-52-0	Isononyloxypropylamine
CAS 29383-26-4	Octyl hydroxystearate
CAS 29710-25-6	Octyl hydroxystearate
CAS 29806-73-3	Octyl palmitate
CAS 29923-31-7	Sodium lauroyl glutamate (L)
CAS 29923-34-0	Sodium lauroyl glutamate (DL)
CAS 30364-51-3	Sodium myristoyl sarcosinate
CAS 30473-39-3	Sodium stearoamphoacetate
CAS 30526-22-8	Potassium toluenesulfonate
CAS 31394-71-5	PPG oleate series
CAS 31556-45-3	Tridecyl stearate
CAS 31566-31-1	Glyceryl stearate
CAS 31566-31-1	Glyceryl stearate SE
CAS 31566-31-1	Polyglyceryl-3 diisostearate
CAS 31621-91-7	PEG-9 pelargonate
CAS 31691-97-1	Ammonium nonoxynol-4 sulfate
CAS 31694-55-0	Glycereth series
CAS 31791-00-2	PEG-40 stearate
CAS 31799-71-0	PEG oleamide series

CAS 31955-67-6 TEA-lauroyl glutamate
CAS 32057-14-0 Glyceryl isostearate
CAS 32073-22-6 Sodium cumenesulfonate
CAS 32426-11-2 Quaternium-24
CAS 32612-48-9 Ammonium laureth sulfate
CAS 32612-48-9 Ammonium laureth-9 sulfate
CAS 32612-48-9 Ammonium laureth-12 sulfate
CAS 33907-46-9 Glycol hydroxystearate
CAS 33907-47-0 Propylene glycol hydroxystearate
CAS 33939-64-9 Sodium laureth carboxylate series
CAS 33939-65-0 Sodium ceteth-13 carboxylate
CAS 34140-91-5 Oleyl propylene diamine dioleate
CAS 34360-00-4 Bis-(2-hydroxyethyl) isodecyloxypropylamine
CAS 34424-98-1 Polyglyceryl-10 tetraoleate
CAS 35179-86-3 PEG-8 laurate
CAS 35545-57-4 PEG-13 betanaphthol
CAS 36311-34-9 Isocetyl alcohol
CAS 36409-57-1 Disodium lauryl sulfosuccinate
CAS 36445-71-3 Sodium decyl diphenyl ether disulfonate
CAS 36445-71-3 Sodium decyl diphenyloxide disulfonate
CAS 36521-89-8 Sorbitan distearate
CAS 36653-82-4 Cetyl alcohol
CAS 37200-49-0 Polysorbate 80
CAS 37203-80-8 Pine lignin
CAS 37205-87-1 Nonoxynol series
CAS 37220-82-9 Glyceryl oleate
CAS 37294-49-8 Disodium isodecyl sulfosuccinate
CAS 37311-00-5 PPG-25-laureth-25
CAS 37311-01-6 PPG-4-ceteth-10
CAS 37311-04-9 PPG-3-myreth-3
CAS 37311-67-4 Disodium undecylenamido MEA-sulfosuccinate
CAS 37349-34-1 Polyglyceryl stearate series
CAS 37475-88-0 Ammonium cumenesulfonate
CAS 37478-68-5 Capryl hydroxyethyl imidazoline
CAS 37767-39-8 Tetrasodium dicarboxyethyl stearyl sulfosuccinamate
CAS 37767-42-3 Disodium myristamido MEA-sulfosuccinate
CAS 38517-23-6 Sodium hydrogenated tallow glutamate
CAS 38517-37-2 Sodium myristoyl glutamate
CAS 38916-42-6 Tetrasodium dicarboxyethyl stearyl sulfosuccinamate
CAS 39175-72-9 Glyceryl stearate citrate
CAS 39310-72-0 PEG-15 glyceryl ricinoleate
CAS 39354-45-5 Disodium laureth sulfosuccinate
CAS 39354-47-5 Disodium C12-15 pareth sulfosuccinate
CAS 39464-66-9 Laureth phosphate series
CAS 39464-69-2 Oleth phosphate series
CAS 39529-26-5 Polyglyceryl-10 decastearate
CAS 40716-42-5 Ricinoleamide DEA
CAS 40754-59-4 Disodium laureth sulfosuccinate
CAS 40754-60-7 Disodium ricinoleamido MEA-sulfosuccinate
CAS 40795-56-0 Sodium dodecyl diphenyl ether disulfonate
CAS 40839-40-5 Disodium undecylenamido MEA-sulfosuccinate
CAS 42016-08-0 Disodium laureth sulfosuccinate
CAS 42173-90-0 Octoxynol-9
CAS 42504-46-1 MIPA-dodecylbenzenesulfonate
CAS 42566-92-7 Steapyrium chloride
CAS 42612-52-2 Sodium laureth-4 phosphate
CAS 42808-36-6 Sulfated butyl tallate
CAS 42926-22-7 Sodium lauroyl glutamate (L)
CAS 43154-85-4 Disodium oleamido MIPA-sulfosuccinate
CAS 51142-51-9 PEG-15 glyceryl ricinoleate
CAS 51158-08-8 PEG glyceryl stearate series
CAS 51192-09-7 PEG glyceryl oleate series
CAS 51200-87-4 Oxazolidine
CAS 51248-32-9 PEG glyceryl laurate series
CAS 51437-95-7 Nonoxynol-3
CAS 51811-79-1 Nonoxynol phosphate series
CAS 52019-36-0 Deceth phosphate series
CAS 52292-17-8 Isosteareth series
CAS 52315-75-0 Lauroyl lysine
CAS 52504-24-2 PEG-6 caprylic/capric glycerides
CAS 52558-73-3 Myristoyl sarcosine
CAS 52581-71-2 PPG oleyl ether series
CAS 52663-87-3 Octyliminodipropionate
CAS 52688-97-0 PEG dioleate series
CAS 52725-64-1 Lauramide DEA
CAS 52794-79-3 Isostearamide DEA
CAS 53576-49-1 TEA-lauroyl glutamate
CAS 53610-02-9 Sodium laureth carboxylate series
CAS 53980-88-4 Acrylinoleic acid
CAS 54193-36-1 Sodium polymethacrylate
CAS 54351-50-7 DEA-laureth sulfate
CAS 54571-67-4 Sodium PCA
CAS 54590-52-2 MIPA-dodecylbenzenesulfonate
CAS 54849-16-0 Disodium caproamphodiacetate
CAS 55353-19-0 DEA-laureth sulfate
CAS 55819-53-9 Stearamidopropyl dimethylamine lactate
CAS 55852-13-6 Stearamidopropyl morpholine
CAS 56002-14-3 PEG isostearate series
CAS 56235-92-8 Dioctyl maleate
CAS 56388-43-3 Disodium oleamido PEG-2 sulfosuccinate
CAS 56388-96-6 Trideceth-7 carboxylic acid
CAS 56863-02-6 Linoleamide DEA
CAS 57171-56-9 PEG sorbitan hexaoleate series
CAS 57307-99-0 PEG-8 caprylic/capric glycerides
CAS 58069-11-7 Quaternium-52
CAS 58353-68-7 Disodium dodecyloxy propyl sulfosuccinamate
CAS 58450-52-5 Disodium laureth sulfosuccinate
CAS 58748-27-9 Propylene glycol dicaprylate/dicaprate
CAS 58855-36-0 DEA-laureth sulfate
CAS 58855-63-3 DEA-oleth phosphate series
CAS 58969-27-0 Sodium cocoyl isethionate
CAS 59070-56-3 PEG glyceryl laurate series
CAS 59186-41-3 Sodium cetearyl sulfate
CAS 59231-34-4 Isodecyl oleate
CAS 59559-30-7 Steareth-7 carboxylic acid
CAS 60270-33-9 Behenamidopropyl dimethylamine
CAS 60828-78-6 Isolaureth series
CAS 61372-91-6 Dibehenyl methylamine
CAS 61691-12-6 PEG-6 castor oil
CAS 61757-59-3 Sodium trideceth carboxylate series
CAS 61788-45-2 Hydrogenated tallowamine
CAS 61788-46-3 Cocamine
CAS 61788-47-4 Coconut acid
CAS 61788-48-5 Acetylated lanolin
CAS 61788-49-6 Acetylated lanolin alcohol
CAS 61788-59-8 Methyl cocoate
CAS 61788-62-3 Dicoco methylamine
CAS 61788-63-4 Dihydrogenated tallow methylamine
CAS 61788-78-1 Hydrogenated tallowtrimonium chloride

CAS 61788-85-0 PEG hydrogenated castor oil series
CAS 61788-85-0 PEG-66 trihydroxystearin
CAS 61788-89-4 Dilinoleic acid
CAS 61788-90-7 Cocamine oxide
CAS 61788-91-8 Dimethyl soyamine
CAS 61788-93-0 Dimethyl cocamine
CAS 61788-95-2 Dimethyl hydrogenated tallow amine
CAS 61789-04-6 Sodium cocomonoglyceride sulfate
CAS 61789-05-7 Glyceryl cocoate
CAS 61789-09-1 Hydrogenated tallow glyceride
CAS 61789-18-2 Cocotrimonium chloride
CAS 61789-19-3 Cocamide
CAS 61789-30-8 Potassium cocoate
CAS 61789-31-9 Sodium cocoate
CAS 61789-32-0 Sodium cocoyl isethionate
CAS 61789-40-0 Cocamidopropyl betaine
CAS 61789-68-2 PEG-2 coco-benzonium chloride
CAS 61789-71-7 Benzalkonium chloride
CAS 61789-72-8 Hydrogenated tallowalkonium chloride
CAS 61789-73-9 Dihydrogenated tallow benzylmonium chloride
CAS 61789-75-1 Tallowalkonium chloride
CAS 61789-76-2 Dicocamine
CAS 61789-77-3 Dicocodimonium chloride
CAS 61789-79-5 Hydrogenated ditallowamine
CAS 61789-80-8 Quaternium-18
CAS 61789-86-4 Calcium sulfonate
CAS 61790-12-3 Tall oil acid
CAS 61790-18-9 Soyamine
CAS 61790-31-6 Hydrogenated tallow amide
CAS 61790-33-8 Tallow amine
CAS 61790-35-0 Sulfated tall oil
CAS 61790-37-2 Tallow acid
CAS 61790-38-3 Hydrogenated tallow acid
CAS 61790-41-8 Soytrimonium chloride
CAS 61790-44-1 Potassium tallate
CAS 61790-48-5 Barium petroleum sulfonate
CAS 61790-50-9 Potassium rosinate
CAS 61790-51-0 Sodium rosinate
CAS 61790-57-6 Cocamine acetate
CAS 61790-59-8 Hydrogenated tallowamine acetate
CAS 61790-64-5 TEA cocoate
CAS 61790-81-6 PEG lanolin series
CAS 61790-85-0 PEG tallow aminopropylamine series
CAS 61791-00-2 PEG tallate series
CAS 61791-01-3 PEG ditallate series
CAS 61791-07-9 PEG-6 soya fatty acid
CAS 61791-08-0 PEG cocamide series
CAS 61791-10-4 PEG-15 cocomonium chloride
CAS 61791-12-6 PEG castor oil series
CAS 61791-13-7 Coceth series
CAS 61791-14-8 PEG cocamine series
CAS 61791-20-6 Laneth series
CAS 61791-24-0 PEG soyamine series
CAS 61791-25-1 Dihydroxyethyl tallow glycinate
CAS 61791-26-2 PEG hydrogenated tallow amine series
CAS 61791-28-4 Talloweth-11
CAS 61791-29-5 PEG cocoate series
CAS 61791-31-9 Cocamide DEA
CAS 61791-32-0 Disodium cocoamphodiacetate
CAS 61791-38-6 Cocoyl hydroxyethyl imidazoline
CAS 61791-39-7 Tall oil hydroxyethyl imidazoline
CAS 61791-42-2 Sodium methyl cocoyl taurate
CAS 61791-44-4 Bis (2-hydroxyethyl) tallow amine
CAS 61791-44-4 PEG tallowamine series
CAS 61791-46-6 Dihydroxyethyl tallowamine oxide
CAS 61791-47-7 Dihydroxyethyl cocamine oxide
CAS 61791-53-5 Tallow propane diamine dioleate
CAS 61791-55-7 Tallow propylene diamine
CAS 61791-56-8 Disodium tallowiminodipropionate
CAS 61791-57-9 Tallow dipropylene triamine
CAS 61791-59-1 Sodium cocoyl sarcosinate
CAS 61791-63-7 Coco diamine
CAS 61791-63-7 Cocopropylenediamine
CAS 61791-66-0 Disodium cocamido MEA-sulfosuccinate
CAS 61792-31-2 Lauramidopropylamine oxide
CAS 61849-72-7 PPG methyl glucose ether series
CAS 62449-33-6 Oleyl imidazolinium hydrochloride
CAS 63148-62-9 Dimethicone
CAS 63351-73-5 Ammonium nonoxynol-4 sulfate
CAS 63393-82-8 C12-15 alcohols
CAS 63393-93-1 Isopropyl lanolate
CAS 63394-02-5 Silicone elastomer
CAS 63451-27-4 Polyquaternium-2
CAS 63566-37-0 Isostearamidopropyl betaine
CAS 63793-60-2 PPG-3 myristyl ether
CAS 64365-23-7 Dimethicone copolyol
CAS 64475-85-0 Mineral spirits
CAS 64743-02-8 C12-14 alpha olefins
CAS 64743-02-8 C14-16 alpha olefins
CAS 64743-02-8 C14-18 alpha olefins
CAS 64743-02-8 C16-18 alpha olefins
CAS 65071-95-6 PEG-15 tallate
CAS 65277-52-3 Disodium undecylenamido MEA-sulfosuccinate
CAS 65277-54-5 Disodium ricinoleamido MEA-sulfosuccinate
CAS 65381-09-1 Caprylic/capric triglyceride
CAS 66085-00-5 Isostearyl diglyceryl succinate
CAS 66085-00-5 Glyceryl isostearate
CAS 66105-29-1 PEG-7 glyceryl cocoate
CAS 66146-84-7 Steareth-7
CAS 66161-57-7 Sodium laureth-12 sulfate
CAS 66161-58-8 Sodium trideceth sulfate
CAS 66172-78-9 Nonyl nonoxynol-7 phosphate
CAS 66172-82-5 Nonyl nonoxynol-9 phosphate
CAS 66172-83-6 Nonyl nonoxynol-7 phosphate
CAS 66197-78-2 Nonoxynol-9 phosphate
CAS 66455-14-9 C12-13 pareth series
CAS 66455-17-2 C9-11 alcohols
CAS 66455-29-6 Lauryl betaine
CAS 66794-58-9 PEG sorbitan isostearate series
CAS 66988-04-3 Sodium isostearoyl lactylate
CAS 67700-98-5 Dimethyl lauramine
CAS 67700-99-6 Dihydrogenated tallow methylamine
CAS 67701-00-2 Trihexadecylamine
CAS 67701-05-7 Coconut acid
CAS 67701-06-8 Tallow acid
CAS 67701-08-0 Palmitic/oleic acids
CAS 67701-08-0 Soy acid
CAS 67701-33-1 Glyceryl myristate
CAS 67762-19-0 Ammonium laureth sulfate
CAS 67762-25-8 C12-18 alcohols
CAS 67762-30-5 Cetearyl alcohol
CAS 67762-35-0 PEG-9 cocoate
CAS 67762-36-1 Caprylic/capric acid
CAS 67762-39-4 Methyl caprylate/caprate

CAS 67762-40-7 Methyl laurate
CAS 67762-96-3 Dimethicone copolyol
CAS 67799-04-6 Isostearamidopropyl dimethylamine
CAS 67892-37-9 Sodium oleoamphopropionate
CAS 67893-42-9 Disodium ricinoleamido MEA-sulfosuccinate
CAS 67923-87-9 Sodium octoxynol-2 ethane sulfonate
CAS 67990-17-4 Sodium butoxyethoxy acetate
CAS 67998-94-1 Sodium oleic sulfate
CAS 67999-57-9 Disodium nonoxynol-10 sulfosuccinate
CAS 68002-61-9 Tallowtrimonium chloride
CAS 68002-90-4 C12-14 acids
CAS 68003-46-3 Ammonium lauroyl sarcosinate
CAS 68004-11-5 Polyglyceryl-4 stearate
CAS 68037-49-0 Sodium C14-17 alkyl sec sulfonate
CAS 68037-74-1 Dimethicone
CAS 68037-93-4 Dimethyl palmitamine
CAS 68037-96-7 Dimethyl oleic-linolenic amine
CAS 68037-97-8 Oleyl propanediamine
CAS 68039-23-6 Sodium lauroamphohydroxypropylsulfonate
CAS 68081-81-2 Sodium dodecylbenzenesulfonate
CAS 68081-96-9 Ammonium lauryl sulfate
CAS 68081-97-0 Magnesium lauryl sulfate
CAS 68107-29-4 N,N,N′,N′,N′-Pentamethyl-n-tallow-1,3-propanediammonium dichlorides
CAS 68128-59-6 Diammonium stearyl sulfosuccinamate
CAS 68131-39-5 C12-15 pareth series
CAS 68131-40-8 C11-15 pareth series
CAS 68139-30-0 Cocamidopropyl hydroxysultaine
CAS 68140-00-1 Cocamide MEA
CAS 68140-01-2 Cocamidopropyl dimethylamine
CAS 68140-08-9 Tallowamide DEA
CAS 68140-10-3 Sodium tallow sulfate
CAS 68140-98-7 Ethyl hydroxymethyl oleyl oxazoline
CAS 68153-32-2 Ditallow dimonium chloride
CAS 68153-63-9 Tallowamide MEA
CAS 68153-64-0 PEG-2 tallowate
CAS 68155-09-9 Cocamidopropylamine oxide
CAS 68155-20-4 Tallamide DEA
CAS 68155-24-8 PEG hydrogenated tallow amide series
CAS 68155-27-1 C13-15 amine
CAS 68187-29-1 TEA-cocoyl-glutamate
CAS 68187-32-6 Sodium cocoyl glutamate
CAS 68188-30-7 Soyamidopropyl dimethylamine
CAS 68188-38-5 Sodium cocoyl hydrolyzed collagen
CAS 68188-45-5 Sodium alpha olefin sulfonate
CAS 68201-46-7 PEG glyceryl cocoate series
CAS 68201-49-0 Lanolin wax
CAS 68238-87-9 Sorbitan diisostearate
CAS 68239-42-9 Methyl gluceth-10
CAS 68239-43-0 Methyl gluceth-20
CAS 68298-21-5 Sodium lauroamphoacetate
CAS 68308-53-2 Soy acid
CAS 68308-54-3 Hydrogenated tallow glycerides
CAS 68308-67-8 Soyethyldimonium ethosulfate
CAS 68310-73-6 N,N,N′,N′,N′-Pentamethyl-N-octadecenyl-1,3-diammonium dichlorides
CAS 68310-76-9 Oleyl amidoethyl oleyl imidazoline
CAS 68311-03-5 Disodium deceth-6 sulfosuccinate
CAS 68333-82-4 Cocamide MIPA
CAS 68334-21-4 Sodium cocoamphoacetate
CAS 68389-70-8 PEG-20 methyl glucose sesquistearate
CAS 68390-66-9 Sodium cocoamphoacetate
CAS 68390-99-8 Hydrogenated tallow dimethylamine oxide
CAS 68391-01-5 Benzalkonium chloride
CAS 68391-03-7 Alkyl trimethyl ammonium chloride
CAS 68411-27-8 Isostearyl benzoate
CAS 68411-31-4 TEA-dodecylbenzenesulfonate
CAS 68411-32-5 Dodecylbenzene sulfonic acid
CAS 68411-57-4 Disodium cocoamphodipropionate
CAS 68411-97-2 Cocoyl sarcosine
CAS 68412-55-5 Trideceth carboxylic acid series
CAS 68424-43-1 Lanolin acid
CAS 68424-66-8 Hydroxylated lanolin
CAS 68424-85-1 Benzalkonium chloride
CAS 68424-94-2 Coco-betaine
CAS 68424-95-3 Dicapryl/dicaprylyl dimonium chloride
CAS 68425-37-6 Coconut alcohol
CAS 68425-42-3 Cocamidopropyl dimethylamine lactate
CAS 68425-43-4 Cocamidopropyl dimethylamine propionate
CAS 68425-47-8 Soyamide DEA
CAS 68439-46-3 C9-11 pareth series
CAS 68439-49-6 Ceteareth series
CAS 68439-50-9 C12-14 pareth series
CAS 68439-50-9 Laureth-9
CAS 68439-51-0 PPG laureth series
CAS 68439-53-2 PPG lanolin alcohol ether series
CAS 68439-57-6 Sodium alpha olefin sulfonate
CAS 68439-57-6 Sodium C14-16 olefin sulfonate
CAS 68439-70-3 Dimethyl myristamine
CAS 68439-73-6 N-Tallow-1,3-propanediamine
CAS 68440-05-1 Cocamide MIPA
CAS 68440-25-5 Tallowamide MEA
CAS 68441-68-9 Pentaerythrityl tetracaprylate/caprate
CAS 68458-88-8 PPG/PEG lanolin series
CAS 68459-50-7 PEG-5 lanolate
CAS 68476-03-9 Montan acid wax
CAS 68479-64-1 Disodium oleamido MEA-sulfosuccinate
CAS 68515-65-1 Disodium cocamido MIPA-sulfosuccinate
CAS 68516-06-3 PPG-2 cocamine
CAS 68526-79-4 Hexyl alcohol
CAS 68526-85-2 n-Decyl alcohol
CAS 68526-86-3 Lauryl alcohol
CAS 68526-86-3 Tridecyl alcohol
CAS 68551-08-6 C9-11 alcohols
CAS 68551-12-2 C12-16 pareth-1
CAS 68553-11-7 PEG-20 glyceryl stearate
CAS 68555-36-2 Polyquaternium-2
CAS 68584-22-5 Dodecylbenzene sulfonic acid
CAS 68584-24-7 Isopropylamine dodecylbenzenesulfonate
CAS 68585-05-7 Sulfated neatsfoot oil
CAS 68585-44-4 DEA-lauryl sulfate
CAS 68585-47-7 Sodium lauryl sulfate
CAS 68603-15-5 C8-10 alcohols
CAS 68603-42-9 Cocamide DEA
CAS 68603-64-5 Hydrogenated tallow 1,3-

	propylene diamine
CAS 68604-71-7	Disodium cocoamphodipropionate
CAS 68604-73-9	Sodium cocoamphohydroxypropyl sulfonate
CAS 68608-26-4	Sodium petroleum sulfonate
CAS 68608-61-7	Sodium caproamphoacetate
CAS 68608-63-9	Sodium stearoamphoacetate
CAS 68608-64-0	Disodium capryloamphodiacetate
CAS 68608-65-1	Sodium cocoamphoacetate
CAS 68608-66-2	Disodium lauroamphodiacetate
CAS 68608-66-2	Sodium lauroamphoacetate
CAS 68608-68-4	Sodium cocaminopropionate
CAS 68608-88-8	Dodecylbenzene sulfonic acid
CAS 68610-38-8	Sodium oleoamphohydroxypropylsulfonate
CAS 68610-39-9	Sodium capryloamphohydroxypropyl sulfonate
CAS 68610-43-5	Disodium lauroamphodipropionate
CAS 68630-96-6	Sodium isostearoamphopropionate
CAS 68647-44-9	Sodium lauroamphoacetate
CAS 68647-46-1	Sodium caproamphoacetate
CAS 68647-53-0	Sodium cocoamphoacetate
CAS 68648-27-1	PEG hydrogenated lanolin series
CAS 68649-05-8	Cocaminobutyric acid
CAS 68650-39-5	Disodium cocoamphodiacetate
CAS 68715-87-7	N,N,N′-Trimethyl-N′-9-octadecenyl-1,3-diaminopropane
CAS 68783-24-4	Ditallowamine
CAS 68783-25-5	N,N,N′-Trimethyl-N′-tallow-1,3-diaminopropane
CAS 68783-78-8	Ditallow dimonium chloride
CAS 68784-08-7	Disodium cocamido MEA-sulfosuccinate
CAS 68814-69-7	Dimethyl tallowamine
CAS 68815-55-4	Disodium capryloamphodipropionate
CAS 68855-56-1	C12-16 alcohols
CAS 68877-25-8	Mixed isopropanolamines lauryl sulfate
CAS 68889-49-6	PEG glyceryl oleate series
CAS 68891-17-8	Sodium trideceth carboxylate series
CAS 68891-38-3	Sodium laureth sulfate
CAS 68908-44-1	MEA-lauryl sulfate
CAS 68908-44-1	TEA-lauryl sulfate
CAS 68910-41-5	Disodium cocoamphodipropionate
CAS 68911-79-5	Tallow polypropylene polyamine
CAS 68919-40-4	Cocoamphodipropionic acid
CAS 68919-41-5	Sodium cocoamphopropionate
CAS 68920-18-3	Disodium lauroamphodipropionate
CAS 68920-65-0	Potassium cocoyl hydrolyzed collagen
CAS 68920-66-1	Cetoleth series
CAS 68929-04-4	Disodium lauroamphodipropionate
CAS 68936-95-8	Methyl glucose sesquistearate
CAS 68937-54-2	Polysiloxane polyether copolymer
CAS 68938-15-8	Hydrogenated coconut acid
CAS 68939-90-6	Trilinoleic acid
CAS 68951-72-4	PPG-2 tallowamine
CAS 68952-16-9	TEA-cocoyl hydrolyzed collagen
CAS 68953-96-8	Calcium dodecylbenzene sulfonate
CAS 68954-89-2	Ceteareth-7 carboxylic acid
CAS 68954-89-2	Laureth-17 carboxylic acid
CAS 68954-89-2	Laureth-4 carboxylic acid
CAS 68954-89-2	Laureth-5 carboxylic acid
CAS 68954-89-2	Steareth-7 carboxylic acid
CAS 68955-19-1	Sodium lauryl sulfate
CAS 68955-20-4	Sodium tallow sulfate
CAS 68956-08-1	Trimethylolpropane tricaprylate/tricaprate
CAS 68958-48-5	Glyceryl mono/diisostearate
CAS 68958-64-5	PEG-25 glyceryl trioleate
CAS 68966-38-1	Isostearyl hydroxyethyl imidazoline
CAS 68987-89-3	Sodium laureth-11 carboxylate
CAS 68988-69-2	Sodium tallow sulfosuccinamate
CAS 68988-72-7	Propylene glycol dicaprylate/dicaprate
CAS 68989-00-4	Benzalkonium chloride
CAS 68989-03-7	PEG-5 cocomonium methosulfate
CAS 68990-58-9	Acetylated hydrogenated tallow glyceride
CAS 68990-59-0	Hydrogenated tallow glyceride citrate
CAS 69226-96-6	Pentaerythrityl tetracaprylate/caprate
CAS 69331-39-1	DEA-cetyl phosphate
CAS 69468-44-6	PEG glyceryl isostearate series
CAS 69537-38-8	Behenoyl-PG-trimonium chloride
CAS 70191-74-1	Decyl diphenyl ether disulfonic acid
CAS 70191-76-3	Sodium diphenyl ether disulfonate
CAS 70321-63-0	Lanolin oil
CAS 70592-80-2	Lauramine oxide
CAS 70632-06-3	Sodium C12-15 pareth carboxylate series
CAS 70632-06-3	Sodium laureth-13 carboxylate
CAS 70693-04-8	Isostearyl alcohol
CAS 70750-05-9	Disodium caproamphodiacetate
CAS 70750-46-8	Dihydroxyethyl tallow glycinate
CAS 70750-47-9	PEG-2 cocomonium chloride
CAS 70788-37-3	Toluene sulfonic acid
CAS 70802-40-3	PEG-8 stearate
CAS 70851-07-9	Cocamidopropyl betaine
CAS 70851-08-0	Cocamidopropyl hydroxysultaine
CAS 71060-61-2	Bis hydrogenated tallowalkyl-2-hydroxypropyl amine
CAS 71060-72-5	Trihexadecylmethyl ammonium chloride
CAS 71368-20-2	Sodium myristoyl glutamate
CAS 71487-00-8	PEG-2 cocomonium nitrate
CAS 71487-01-9	Dicoco nitrite
CAS 71786-47-5	Magnesium sulfonate
CAS 71902-01-7	Sorbitan isostearate
CAS 72160-13-5	Octoxynol carboxylic acid series
CAS 72162-46-0	N-Isodecyloxypropyl 1-1,3-diaminopropane
CAS 72300-24-4	Isostearamidopropyl morpholine lactate
CAS 73070-47-0	Trideceth-6 phosphate
CAS 73138-79-1	Sulfated peanut oil
CAS 75719-56-1	Polyglyceryl-8 oleate
CAS 75782-86-4	C12-13 alcohols
CAS 75782-87-5	C14–15 alcohols
CAS 76050-42-5	Carbomer
CAS 78330-12-8	Sodium petroleum sulfonate
CAS 79131-39-5	C12-15 pareth-4
CAS 79702-63-0	Disodium oleamido MEA-

CAS No.	Name
	sulfosuccinate
CAS 79770-97-1	Tallowalkylmethyl-bis(2-hydroxy-2-methylethyl)quaternary ammonium methylsulfates
CAS 80145-09-1	Isostearamidopropyl morpholine lactate
CAS 80260-73-2	Dodecyl diphenyl ether disulfonic acid
CAS 81859-24-7	DEA-laureth sulfate
CAS 82933-91-3	Methyl glucose dioleate
CAS 83016-76-6	MIPA-laureth sulfate
CAS 83138-08-3	Cocamidopropyl betaine
CAS 83138-08-3	Disodium cocoamphodipropionate
CAS 83682-78-4	Lauramidopropyl PEG-dimonium chloride phosphate
CAS 84501-49-5	Sodium decyl sulfate
CAS 84562-92-5	Nonoxynol-3
CAS 84812-94-2	Cocaminopropionic acid
CAS 85116-97-8	PEG-2 stearate
CAS 85117-50-6	Sodium dodecylbenzenesulfonate
CAS 85404-84-8	Polyglyceryl-3 diisostearate
CAS 85408-49-7	C12-14 alkyl dimethylamine n-oxide
CAS 85409-22-9	Benzalkonium chloride
CAS 85411-01-4	Hydrogenated vegetable glycerides phosphate
CAS 85536-14-7	Dodecylbenzene sulfonic acid
CAS 85536-23-8	PEG-4 rapeseedamide
CAS 85666-92-8	Glyceryl stearate
CAS 85666-92-8	Glyceryl stearate SE
CAS 85666-92-8	Polyglyceryl-3 diisostearate
CAS 85736-63-6	Alkyl dimethyl hydroxyethyl ammonium chloride
CAS 86088-85-9	Quaternium-27
CAS 86089-12-5	Ricinoleamidopropyl betaine
CAS 86418-55-5	Glyceryl stearate SE
CAS 86438-35-9	Disodium cocoamphodipropionate
CAS 86438-78-0	Lauramidopropyl betaine
CAS 86438-79-1	Cocamidopropyl betaine
CAS 86438-79-1	Disodium cocoamphodipropionate
CAS 88497-58-9	C12-15 pareth-7 carboxylic acid
CAS 88543-32-2	Ethyl hydroxymethyl oleyl oxazoline
CAS 90268-48-7	Disodium tallow sulfosuccinamate
CAS 90283-04-8	Dilauryl acetyl dimonium chloride
CAS 90388-14-0	Cetyldiethanolaminephosphate
CAS 91824-88-3	Polyglyceryl-4 isostearate
CAS 91995-05-0	Cocoiminodipropionate
CAS 92112-62-4	PEG-1 C13-15 alkyl methyl amine
CAS 93572-63-5	PEG-3 tallow propylenedimonium dimethosulfate
CAS 94349-40-3	Nonoxynol-9 iodine
CAS 97281-23-7	Glycol stearate
CAS 97404-50-7	Glyceryl lanolate
CAS 97659-50-2	Cocoiminodipropionate
CAS 97659-53-5	Cocoamphopolycarboxyglycinate
CAS 97659-53-5	Disodium oleoamphodiacetate
CAS 98984-78-2	Sodium lauroyl glutamate
CAS 99330-44-6	Magnesium laureth-11 carboyxlate
CAS 100085-64-1	Cocobetainamido amphopropionate
CAS 100545-50-4	Didecyl dimethylamine oxide
CAS 104909-82-2	Butoxynol-19 carboxylic acid
CAS 105391-15-9	Hexeth-4 carboxylic acid
CAS 107600-33-9	Capryleth-9 carboxylic acid
CAS 107628-08-0	Octoxynol carboxylic acid series
CAS 110332-91-7	PPG-2 isoceteth-20 acetate
CAS 111019-03-5	PPG-10 cetyl ether phosphate
CAS 111381-08-9	Hydroxypropyl bis-isostearamidopropyldimonium chloride
CAS 112945-52-5	Silica
CAS 125740-36-5	Isodecyloxypropyl dihydroxyethyl methyl ammonium chloride
CAS 130124-24-2	Quaternium-53
CAS 133934-08-4	Apricotamidopropyl betaine
CAS 617788-68-9	Sulfated rapeseed oil
CAS 977010-66-4	Ricinoleamidopropyl dimethylamine
CAS 977053-96-5	Glyceryl stearate SE

EINECS-to-CAS Number Cross Reference

EINECS	CAS	Chemical
200-018-0	50-21-5	Lactic acid
200-289-5	56-81-5	Glycerin
200-311-3	57-09-0	Cetrimonium bromide
200-312-9	57-10-3	Palmitic acid
200-313-4	57-11-4	Stearic acid
200-315-5	57-13-6	Urea
200-338-0	57-55-6	Propylene glycol
200-353-2	57-88-5	Cholesterol
200-470-9	60-33-3	Linoleic acid
200-573-9	64-02-8	Tetrasodium EDTA
200-578-6	64-17-5	Ethyl alcohol
200-661-7	67-63-0	Isopropyl alcohol
200-751-6	71-36-3	Butyl alcohol
200-772-0	72-17-3	Sodium lactate
201-094-8	78-21-7	Cetethyl morpholinium ethosulfate
201-148-0	78-83-1	Isobutyl alcohol
202-280-1	93-82-3	Stearamide DEA
202-281-7	93-83-4	Oleamide DEA
202-397-8	95-19-2	Stearyl hydroxyethyl imidazoline
202-414-9	95-38-5	Oleyl hydroxyethyl imidazoline
202-488-2	96-20-8	2-Amino-1-butanol
202-608-3	97-78-9	Lauroyl sarcosine
202-700-3	98-79-3	PCA
202-935-1	101-34-8	Glyceryl triacetyl ricinoleate
203-041-4	102-60-3	Tetrahydroxypropyl ethylenediamine
203-049-8	102-71-6	Triethanolamine
203-063-4	102-87-4	Trilaurylamine
203-180-0	104-15-4	Toluene sulfonic acid
203-232-2	104-74-5	Laurylpyridinium chloride
203-234-3	104-76-7	2-Ethylhexanol
203-328-4	105-76-0	Dibutyl maleate
203-358-8	106-07-0	PEG-4 stearate
203-359-3	106-08-1	PEG-9 laurate
203-363-5	106-11-6	PEG-2 stearate
203-364-0	106-12-7	PEG-2 oleate
203-366-1	106-14-9	Hydroxystearic acid
203-368-2	106-16-1	Ricinoleamide MEA
203-369-8	106-17-2	Glycol ricinoleate
203-473-3	107-21-1	Glycol
203-489-0	107-41-5	Hexylene glycol
203-508-2	107-64-2	Distearyldimonium chloride
203-661-5	109-28-4	Oleamidopropyl dimethylamine
203-663-6	109-30-8	PEG-2 distearate
203-749-3	110-25-8	Oleoyl sarcosine
203-751-4	110-27-0	Isopropyl myristate
203-755-6	110-30-5	Ethylene distearamide
203-756-1	110-31-6	Ethylene dioleamide
203-821-4	110-98-5	Dipropylene glycol
203-827-7	111-03-5	Glyceryl oleate
203-828-2	111-05-7	Oleamide MIPA
203-852-3	111-27-3	Hexyl alcohol
203-868-0	111-42-2	Diethanolamine
203-872-2	111-46-6	Diethylene glycol
203-883-2	111-57-9	Stearamide MEA
203-886-9	111-60-4	Glycol stearate
203-905-0	111-76-2	Butoxyethanol
203-911-3	111-82-0	Methyl laurate
203-917-6	111-87-5	C8 alcohol. See Caprylic alcohol
203-917-6	111-87-5	Caprylic alcohol
203-919-7	111-90-0	Ethoxydiglycol
203-927-0	112-00-5	Laurtrimonium chloride
203-928-6	112-02-7	Cetrimonium chloride
203-929-1	112-03-8	Steartrimonium chloride
203-931-2	112-05-0	Pelargonic acid
203-943-8	112-18-5	Dimethyl lauramine
203-956-9	112-30-1	n-Decyl alcohol
203-961-6	112-34-5	Butoxy diglycol
203-966-3	112-39-0	Methyl palmitate
203-982-0	112-53-8	Lauryl alcohol
203-989-9	112-60-7	PEG-4
203-990-4	112-61-8	Methyl stearate
203-992-5	112-62-9	Methyl oleate
203-997-2	112-69-6	Dimethyl palmitamine
203-998-8	112-70-9	Tridecyl alcohol
204-000-3	112-72-1	Myristyl alcohol
204-002-4	112-75-4	Dimethyl myristamine
204-007-1	112-80-1	Oleic acid
204-009-2	112-84-5	Erucamide
204-010-8	112-85-6	Behenic acid
204-015-5	112-90-3	Oleamine
204-017-6	112-92-5	Stearyl alcohol
204-100-7	115-69-5	2-Amino-2-methyl-1,3-propanediol
204-101-2	115-70-8	2-Amino-2-ethyl-1,3-propanediol
204-211-0	117-81-7	Dioctyl phthalate
204-393-1	120-40-1	Lauramide DEA
204-527-9	122-19-0	Stearalkonium chloride
204-534-7	122-32-7	Triolein
204-589-7	122-99-6	Phenoxyethanol
204-593-9	123-03-5	Cetylpyridinium chloride
204-664-4	123-94-4	Glyceryl stearate
204-666-5	123-95-5	Butyl stearate
204-677-5	124-07-2	Caprylic acid
204-680-1	124-10-7	Methyl myristate
204-690-6	124-22-1	Lauramine
204-693-2	124-26-5	Stearamide
204-694-8	124-28-7	Dimethyl stearamine
204-695-3	124-30-1	Stearamine
204-709-8	124-68-5	2-Amino-2-methyl-1-propanol
204-809-1	126-86-3	Tetramethyl decynediol

EINECS	CAS	Chemical
204-812-8	126-92-1	Sodium octyl sulfate
204-839-5	127-39-9	Diisobutyl sodium sulfosuccinate
204-857-3	127-68-4	Sodium m-nitrobenzenesulfonate
205-234-9	136-26-5	Capramide DEA
205-252-7	136-60-7	Butyl benzoate
205-271-0	136-99-2	Lauryl hydroxyethyl imidazoline
205-281-5	137-16-6	Sodium lauroyl sarcosinate
205-285-7	137-20-2	Sodium methyl oleoyl taurate
205-351-5	139-07-1	Lauralkonium chloride
205-352-0	139-08-2	Myristalkonium chloride
205-388-7	139-96-8	TEA-lauryl sulfate
205-455-0	141-08-2	Glyceryl ricinoleate
205-468-1	141-20-8	PEG-2 laurate
205-469-7	141-21-9	Stearamidoethyl ethanolamine
205-470-2	141-22-0	Ricinoleic acid
205-471-8	141-23-1	Methyl hydroxystearate
205-472-3	141-24-2	Methyl ricinoleate
205-524-5	142-16-5	Dioctyl maleate
205-526-6	142-18-7	Glyceryl laurate
205-530-8	142-26-7	Acetamide MEA
205-539-7	142-48-3	Stearoyl sarcosine
205-541-8	142-54-1	Lauramide MIPA
205-542-3	142-55-2	Propylene glycol laurate
205-546-5	142-58-5	Myristamide MEA
205-559-6	142-77-8	Butyl oleate
205-560-1	142-78-9	Lauramide MEA
205-571-1	142-91-6	Isopropyl palmitate
205-582-1	143-07-7	Lauric acid
205-590-5	143-18-0	Potassium oleate
205-591-0	143-19-1	Sodium oleate
205-596-8	143-27-1	Palmitamine
205-597-3	143-28-2	Oleyl alcohol
205-713-2	149-39-3	Sodium methyl stearoyl taurate
205-738-9	149-39-3	Sodium methyl stearoyl taurate
205-788-1	151-21-3	Sodium lauryl sulfate
206-103-9	301-02-0	Oleamide
206-376-4	334-48-5	Capric acid
207-334-8	463-40-1	Linolenic acid
207-838-8	497-19-8	Sodium carbonate
208-534-8	532-32-1	Sodium benzoate
208-736-6	540-10-3	Cetyl palmitate
208-875-2	544-63-8	Myristic acid
209-097-6	555-43-1	Tristearin
209-406-4	577-11-7	Dioctyl sodium sulfosuccinate
211-014-3	627-83-8	Glycol distearate
211-103-7	629-70-9	Cetyl acetate
211-546-6	661-19-8	Behenyl alcohol
211-669-5	683-10-3	Lauryl betaine
211-748-4	693-33-4	Cetyl betaine
212-470-6	820-66-6	Stearyl betaine
212-806-1	871-37-4	Oleyl betaine
213-085-6	922-80-5	Diamyl sodium sulfosuccinate
213-695-2	1002-89-7	Ammonium stearate
214-291-9	1119-97-7	Myrtrimonium bromide
214-292-4	1120-01-0	Sodium cetyl sulfate
214-737-2	1191-50-0	Sodium myristyl sulfate
215-087-2	1300-51-2	Sodium phenol sulfonate
215-090-9	1300-72-7	Sodium xylenesulfonate
215-181-3	1310-58-3	Potassium hydroxide
215-293-2		Cresylic acid

EINECS	CAS	Chemical
215-347-5	1322-98-1	Sodium decylbenzene sulfonate
215-354-3	1323-39-3	Propylene glycol stearate
215-355-9	1323-42-8	Glyceryl hydroxystearate
215-359-0	1323-83-7	Glyceryl distearate
215-549-3	1330-80-9	Propylene glycol oleate
215-559-8	1331-61-9	Ammonium dodecylbenzene sulfonate
215-663-3	1338-39-2	Sorbitan laurate
215-664-9	1338-41-6	Sorbitan stearate
215-665-4	1338-43-8	Sorbitan oleate
215-968-1	1462-54-0	Lauraminopropionic acid
216-343-6	1562-00-1	Sodium isethionate
216-472-8	1592-23-0	Calcium stearate
216-700-6	1643-20-5	Lauramine oxide
217-325-0	1812-53-9	Dicetyl dimonium chloride
217-430-1	1847-55-8	Sodium oleyl sulfate
217-431-7	1847-58-1	Sodium lauryl sulfoacetate
218-791-8	2235-43-0	Pentasodium aminotrimethylene phosphonate
218-793-9	2235-54-3	Ammonium lauryl sulfate
219-136-9	2370-64-1	PEG-6 laurate
219-919-5	2571-88-2	Stearamine oxide
220-020-5	2605-79-0	Decylamine oxide
220-045-1	2615-15-8	PEG-6
220-152-3	2644-45-3	c-Decyl betaine
220-219-7	2673-22-5	Ditridecyl sodium sulfosuccinate
220-336-3	2724-58-5	Isostearic acid
220-476-5	2778-96-3	Stearyl stearate
220-688-8	2867-47-2	Dimethylaminoethyl methacrylate
220-851-3	2917-94-4	Sodium octoxynol-2 ethane sulfonate
221-109-1	3006-15-3	Dihexyl sodium sulfosuccinate
221-188-2	3026-63-9	Sodium tridecyl sulfate
221-279-7	9002-92-0	Laureth-2
221-280-2	3055-94-5	Laureth-3
221-281-8	3055-95-6	Laureth-5
221-282-3	3055-96-7	Laureth-6
221-283-9	3055-97-8	Laureth-7
221-284-4	3055-99-0	Laureth-9
221-286-5	3056-00-6	Laureth-12
221-304-1	3061-75-4	Behenamide
221-416-0	1335-72-4	Sodium laureth sulfate
221-450-6	3097-08-3	Magnesium lauryl sulfate
221-661-3	3179-80-4	Lauramidopropyl dimethylamine
222-059-3	3332-27-2	Myristamine oxide
222-273-7	3401-73-8	Tetrasodium dicarboxyethyl stearyl sulfosuccinamate
222-700-7	3578-72-1	Calcium behenate
222-899-0	3655-00-3	Disodium lauriminodipropionate
224-292-6	4292-10-8	Lauramidopropyl betaine
224-806-9	4500-01-0	Glycol oleate
224-886-5	4536-30-5	Laureth-1
225-190-4	4706-78-9	Potassium lauryl sulfate
225-214-3	4722-98-9	MEA-lauryl sulfate
225-768-6	5064-31-3	Trisodium NTA
225-856-4	5117-19-1	PEG-8
226-097-1	5274-68-0	Laureth-4
226-159-8	5306-85-4	Dimethyl isosorbide
226-191-2	5323-95-5	Sodium ricinoleate
226-312-9	5349-52-0	PEG-9 stearate
226-448-9	5401-17-2	PEG-2 ricinoleate

EINECS	CAS	Chemical
226-546-1	5422-34-4	Lactamide MEA
227-335-7	5793-94-2	Calcium stearoyl lactylate
228-227-2	6179-44-8	Isostearamidopropyl betaine
228-464-1	6272-74-8	Lapyrium chloride
229-146-5	6419-19-8	Aminotrimethylene phosphonic acid
229-859-1	6790-09-6	PEG-12
229-912-9	6834-92-0	Sodium metasilicate
230-429-0	7128-91-8	Palmitamine oxide
230-525-2	7173-51-5	Didecyldimonium chloride
230-698-4	7281-04-1	Lauralkonium bromide
230-770-5	7311-27-5	Nonoxynol-4
230-785-7	7320-34-5	Tetrapotassium pyrophosphate
231-211-8	7447-40-7	Potassium chloride
231-314-8	7492-30-0	Potassium ricinoleate
231-426-7	7545-23-5	Myristamide DEA
231-442-4		Iodine
231-609-1	7651-02-7	Stearamidopropyl dimethylamine
231-635-3	7664-41-7	Ammonia
231-721-0	7702-01-4	Disodium capryloamphodiacetate
231-913-4	7778-77-0	Potassium phosphate
232-282-8	8001-31-8	Coconut oil
232-293-8	8001-79-4	Castor oil
232-306-7	8002-33-3	Sulfated castor oil
232-307-2	8002-43-5	Lecithin
232-348-6	8006-54-0	Lanolin
232-360-1	8007-43-0	Sorbitan sesquioleate
232-373-2	8009-03-8	Petrolatum
232-384-2	8012-95-1	Mineral oil
232-388-4	8013-05-6	Potassium castorate
232-391-0	8013-07-8	Epoxidized soybean oil
232-410-2	8016-70-4	Hydrogenated soybean oil
232-430-1	8027-33-6	Lanolin alcohol
232-440-6	8029-76-3	Hydroxylated lecithin
232-447-4	8030-78-2	Tallowtrimonium chloride
232-452-1	8031-44-5	Hydrogenated lanolin
232-453-7	8032-32-4	Mineral spirits
232-475-7	8050-09-7	Rosin
232-491-4	8052-48-0	Sodium tallowate
232-494-0	8052-50-4	Sodium tallow sulfate
232-680-1	9005-32-7	Alginic acid
233-293-0	10108-25-5	PEG-4 oleate
233-520-3	10213-78-2	PEG-2 stearamine
233-562-2	9004-99-3	PEG-3 stearate
234-325-6	123-94-4	Glyceryl stearate
234-394-2	11138-66-2	Xanthan gum
235-088-1	12068-03-0	Sodium toluenesulfonate
235-093-9	12068-19-8	Sodium nonoxynol-6 phosphate
235-186-4	12125-02-9	Ammonium chloride
235-907-2	13039-35-5	Sodium capryloamphoacetate
236-149-5	13192-12-6	Disodium lauryl sulfosuccinate
236-164-7	13197-76-7	Lauryl hydroxysultaine
238-102-4	14234-82-3	Diisobutyl maleate
238-303-7	14350-94-8	Sodium caproamphoacetate
238-305-8	14350-96-0	Sodium lauroamphoacetate
238-306-3	14350-97-1	Disodium lauroamphodiacetate
238-310-5	14351-40-7	Stearamide MEA-stearate
238-311-0	14351-50-9	Oleamine oxide
238-430-8	14450-05-6	Pentaerythrityl tetrapelargonate
238-479-5	14481-60-8	Disodium stearyl sulfosuccinamate
238-501-3	1341-08-8	Steapyrium chloride
239-032-7	14960-06-6	Sodium lauriminodipropionate
240-273-5	16106-44-8	Potassium toluenesulfonate
240-736-1	16693-53-1	TEA lauroyl sarcosinate
240-865-3	16841-14-8	Behenalkonium chloride
240-924-3	16889-14-8	Stearamidoethyl diethylamine
242-960-5	19321-40-5	Pentaerythrityl tetraoleate
243-835-8	20457-75-4	Ricinoleamidopropyl dimethylamine
243-870-9	20545-92-0	Undecylenamide MEA
244-238-5	21142-28-9	MIPA-lauryl sulfate
244-501-4	95-38-5	Oleyl hydroxyethyl imidazoline
244-754-0	22047-49-0	Octyl stearate
245-629-3	23386-52-9	Dicyclohexyl sodium sulfosuccinate
246-584-2	25054-76-6	Oleamidopropyl betaine
246-680-4	25155-30-0	Sodium dodecylbenzenesulfonate
246-684-6	25159-40-4	Oleamidopropylamine oxide
246-705-9	25168-73-4	Sucrose stearate
246-839-8	25321-41-9	Xylene sulfonic acid
246-868-6	25339-09-7	Isocetyl stearate
246-873-3	25339-99-5	Sucrose laurate
246-929-7	25383-99-7	Sodium stearoyl lactylate
246-985-2	25446-78-0	Sodium trideceth sulfate
246-986-8	25446-80-4	Sodium myreth sulfate
247-036-5	25496-01-9	Tridecylbenzene sulfonic acid
247-144-2	25637-84-7	Glyceryl dioleate
247-144-2	25791-96-2	PPG-27 glyceryl ether
247-310-4	25882-44-4	Disodium lauramido MEA-sulfosuccinate
247-536-3	26248-24-8	Sodium tridecylbenzene sulfonate
247-555-7	26264-02-8	Nonoxynol-5
247-556-2	26264-05-1	Isopropylamine dodecylbenzenesulfonate
247-561-6	26264-58-4	Sodium methylnaphthalenesulfonate
247-568-8	26266-57-9	Sorbitan palmitate
247-569-3	26266-58-0	Sorbitan trioleate
247-660-8	26401-35-4	Ditridecyl adipate
247-667-6	26402-22-2	Glyceryl caprate
247-668-1	26402-26-6	Glyceryl caprylate
247-669-7	26402-31-3	Propylene glycol ricinoleate
247-710-9	26447-10-9	Ammonium xylenesulfonate
247-784-2	26545-53-9	DEA-dodecylbenzene sulfonate
247-816-5	26571-11-9	Nonoxynol-8
247-873-6	26650-05-5	Disodium undecylenamido MEA-sulfosuccinate
247-891-4	26658-19-5	Sorbitan tristearate
248-030-5	13192-12-6	Disodium lauryl sulfosuccinate
248-248-0	95-38-5	Oleyl hydroxyethyl imidazoline
248-289-4	27176-87-0	Dodecylbenzene sulfonic acid
248-291-5	9016-45-9	Nonoxynol-2
248-292-0	27177-05-5	Nonoxynol-7
248-293-6	26571-11-9	Nonoxynol-8
248-294-1	27177-08-8	Nonoxynol-10
248-296-2	27177-77-1	Potassium dodecylbenzene sulfonate
248-317-5	27195-16-0	Sucrose distearate
248-403-2	27321-72-8	Polyglyceryl-3 stearate

EINECS	CAS	Chemical
248-406-9	27323-41-7	TEA-dodecylbenzenesulfonate
248-470-8	27458-93-1	Isostearyl alcohol
248-486-5	27479-28-3	Quaternium-14
248-586-9	27638-00-2	Glyceryl dilaurate
248-762-5	26027-38-3	Nonoxynol-1
248-938-7	32073-22-6	Sodium cumenesulfonate
249-277-1	28874-51-3	Sodium PCA
249-395-3	29059-24-3	Propylene glycol myristate
249-448-0	29116-98-1	Sorbitan dioleate
249-862-1	29806-73-3	Octyl palmitate
249-958-3	29923-31-7	Sodium lauroyl glutamate (L-form)
249-958-3	29923-34-0	Sodium lauroyl glutamate (DL-form)
249-958-3	42926-22-7	Sodium lauroyl glutamate (L-form)
249-958-3	98984-78-2	Sodium lauroyl glutamate
250-151-3	30364-51-3	Sodium myristoyl sarcosinate
250-215-0	30473-39-3	Sodium stearoamphoacetate
250-696-7	31556-45-3	Tridecyl stearate
250-705-4	123-94-4;	Glyceryl stearate
250-913-5	32073-22-6	Sodium cumenesulfonate
251-035-5	32426-11-2	Quaternium-24
251-732-4	33907-46-9	Glycol hydroxystearate
251-734-5	33907-47-0	Propylene glycol hydroxystearate
252-011-7	34424-98-1	Polyglyceryl-10 tetraoleate
252-964-9	36311-34-9	Isocetyl alcohol
253-149-0	36653-82-4	Cetyl alcohol
253-407-2	111-03-5	Glyceryl oleate
253-452-8	37294-49-8	Disodium isodecyl sulfosuccinate
253-519-1	37475-88-0	Ammonium cumenesulfonate
253-521-2	37478-68-5	Capryl hydroxyethyl imidazoline
253-981-4	38517-37-2	Sodium myristoyl glutamate
254-495-5	39529-26-5	Polyglyceryl-10 decastearate
255-051-3	40716-42-5	Ricinoleamide DEA
255-062-3	39354-45-5	Disodium laureth sulfosuccinate
255-854-9	42504-46-1	MIPA-dodecylbenzene-sulfonate
256-120-0	43154-85-4	Disodium oleamido MIPA-sulfosuccinate
257-843-4	52315-75-0	Lauroyl lysine
258-007-1	52558-73-3	Myristoyl sarcosine
258-193-4	52794-79-3	Isostearamide DEA
258-636-1	53576-49-1	TEA-lauroyl glutamate
259-837-7	55819-53-9	Stearamidopropyl dimethylamine lactate
260-143-1	56388-43-3	Disodium oleamido PEG-2 sulfosuccinate
260-410-2	56863-02-6	Linoleamide DEA
261-673-6	59231-34-4	Isodecyl oleate
262-134-8	60270-33-9	Behenamidopropyl dimethylamine
262-740-2	61372-91-6	Dibehenyl methylamine
262-976-6	61788-45-2	Hydrogenated tallowamine
262-977-1	61788-46-3	Cocamine
262-978-7	61788-47-4	Coconut acid
262-979-2	61788-48-5	Acetylated lanolin
262-980-8	61788-49-6	Acetylated lanolin alcohol
262-988-1	61788-59-8	Methyl cocoate
262-991-8	61788-63-4	Dihydrogenated tallow methylamine
263-005-9	61788-78-1	Hydrogenated tallowtrimonium chloride
263-016-9	61788-90-7	Cocamine oxide
263-017-4	61788-91-8	Dimethyl soyamine
263-020-0	61788-93-0	Dimethyl cocamine
263-022-1	61788-95-2	Dimethyl hydrogenated tallow amine
263-026-3	61789-04-6	Sodium cocomonoglyceride sulfate
263-027-9	61789-05-7	Glyceryl cocoate
263-031-0	61789-09-1	Hydrogenated tallow glyceride
263-038-9	61789-18-2	Cocotrimonium chloride
263-039-4	61789-19-3	Cocamide
263-049-9	61789-30-8	Potassium cocoate
263-050-4	61789-31-9	Sodium cocoate
263-052-5	61789-32-0	Sodium cocoyl isethionate
263-058-8	61789-40-0	Cocamidopropyl betaine
263-080-8	8001-54-5	Benzalkonium chloride
263-080-8	61789-71-7	Cocoalkonium chloride
263-085-5	61789-75-1	Tallowalkonium chloride
263-086-0	61789-76-2	Dicocamine
263-087-6	61789-77-3	Dicocodimonium chloride
263-089-7	61789-79-5	Hydrogenated ditallowamine
263-090-2	61789-80-8	Quaternium-18
263-107-3	61790-12-3	Tall oil acid
263-112-0	61790-18-9	Soyamine
263-123-0	61790-31-6	Hydrogenated tallow amide
263-125-1	61790-33-8	Tallow amine
263-129-3	61790-37-2	Tallow acid
263-130-9	61790-38-3	Hydrogenated tallow acid
263-134-0	61790-41-8	Soytrimonium chloride
263-155-5	61790-64-5	TEA cocoate
263-163-9	61791-31-9	Cocamide DEA
263-163-9	68603-42-9	Cocamide DEA
263-170-7	61791-38-6	Cocoyl hydroxyethyl imidazoline
263-171-2	61791-39-7	Tall oil hydroxyethyl imidazoline
263-179-6	61791-46-6	Dihydroxyethyl tallowamine oxide
263-180-1	61791-47-7	Dihydroxyethyl cocamine oxide
263-193-2	61791-59-1	Sodium cocoyl sarcosinate
263-218-7	61792-31-2	Lauramidopropylamine oxide
264-119-1	63393-93-1	Isopropyl lanolate
265-672-1	40754-60-7	Disodium ricinoleamido MEA-sulfosuccinate
266-215-9	66172-78-9	Nonyl nonoxynol-7 phosphate
266-218-5	66172-82-5	Nonyl nonoxynol-9 phosphate
266-231-6	66197-78-2	Nonoxynol-9 phosphate
266-533-8	66988-04-3	Sodium isostearoyl lactylate
267-101-1	67799-04-6	Isostearamidopropyl dimethylamine
267-569-7	67892-37-9	Sodium oleoamphopropionate
267-617-7	40754-60-7	Disodium ricinoleamido MEA-sulfosuccinate
267-791-4	2917-94-4	Sodium octoxynol-2 ethane sulfonate
268-040-3	67990-17-4	Sodium butoxyethoxy acetate
268-130-2	68003-46-3	Ammonium lauroyl sarcosinate
268-213-3	68037-49-0	Sodium C14-17 alkyl sec sulfonate
268-242-1	68039-23-6	Sodium lauroamphohydroxy-propylsulfonate
268-761-3	68139-30-0	Cocamidopropyl hydroxysultaine
268-770-2	68140-00-1	Cocamide MEA

EINECS	CAS	Chemical
268-771-8	68140-01-2	Cocamidopropyl dimethylamine
268-772-3	68140-08-9	Tallowamide DEA
268-820-3	68140-98-7	Ethyl hydroxymethyl oleyl oxazoline
268-938-5	68155-09-9	Cocamidopropylamine oxide
268-949-5	68155-20-4	Tallamide DEA
269-084-6	68187-29-1	TEA-cocoyl-glutamate
269-087-2	68187-32-6	Sodium cocoyl glutamate
269-220-4	68201-49-0	Lanolin wax
269-547-2	14350-96-0	Sodium lauroamphoacetate
269-657-0	68308-53-2	Soy acid
269-658-6	68308-54-3	Hydrogenated tallow glycerides
269-793-0	68333-82-4	Cocamide MIPA
269-819-0	68334-21-4	Sodium cocoamphoacetate
269-919-4	8001-54-5	Benzalkonium chloride
269-922-0	68391-03-7	Alkyl trimethyl ammonium chloride
270-131-8	86438-35-9	Disodium cocoamphodipropionate
270-156-4	68411-97-2	Cocoyl sarcosine
270-302-7	68424-43-1	Lanolin acid
270-315-8	68424-66-8	Hydroxylated lanolin
270-325-2	8001-54-5	Benzalkonium chloride
270-351-4	68425-37-6	Coconut alcohol
270-355-6	68425-47-8	Soyamide DEA
270-407-8	68439-57-6	Sodium C14-16 olefin sulfonate
270-416-7	68439-73-6	Tallowaminopropylamine
270-664-6	68476-03-9	Montan acid wax
270-864-3	68479-64-1	Disodium oleamido MEA-sulfosuccinate
271-102-2	68515-65-1	Disodium cocamido MIPA-sulfosuccinate
271-705-0	68604-73-9	Sodium cocoamphohydroxypropyl sulfonate
271-795-1	8033-69-0	Sodium cocaminopropionate
271-929-9	68630-96-6	Sodium isostearoamphopropionate
271-949-8	14350-96-0	Sodium lauroamphoacetate
271-951-9	14350-94-8	Sodium caproamphoacetate
272-021-5	68649-05-8	Cocaminobutyric acid
272-043-5	68650-39-5	Disodium cocoamphodiacetate
272-207-6	68783-78-8	Ditallow dimonium chloride
272-219-1	68784-08-7	Disodium cocamido MEA-sulfosuccinate
273-049-0	68936-95-8	Methyl glucose sesquistearate
273-429-6	68966-38-1	Isostearyl hydroxyethyl imidazoline
273-612-0	68990-58-9	Acetylated hydrogenated tallow glyceride
273-613-6	68990-59-0	Hydrogenated tallow glyceride citrate
274-033-6	69537-38-8	Behenoyl-PG-trimonium chloride
285-550-1	106-11-6	PEG-2 stearate
287-011-6	85408-49-7	C12-14 alkyl dimethylamine n-oxide
287-089-1	8001-54-5	Benzalkonium chloride
290-850-0	90268-48-7	Disodium tallow sulfosuccinamate
297-495-0	93572-63-5	PEG-3 tallow propylenedimonium dimethosulfate
306-522-8	111-60-4	Glycol stearate
307-458-3	97659-53-5	Cocoamphopolycarboxyglycinate
309-206-8	100085-64-1	Cocobetainamido amphopropionate

Ionic Classification

AMPHOTERIC:

Abluter BE
Abluter GL Series
Agrilan® TKA114
Agrilan® WP167
Albegal® A
Albegal® B
Alkali Surfactant NM
Alkanol® CNR
Alkaterge®-E
Alkaterge®-T
Alkawet® CF
Alkawet® LF
Ambiteric D40
Amido Betaine C
Amido Betaine C-45
Amido Betaine C Conc.
Amihope LL-11
Amphionic 25B
Amphionic SFB
Amphionic XL
Ampho B11-34
Ampho T-35
Ampholak 7CX
Ampholak 7TX
Ampholak 7TY
Ampholak BCA-30
Ampholak BTH-35
Ampholak XCE
Ampholak XCO-30
Ampholak XJO
Ampholak XO7
Ampholak XO7-SD-55
Ampholak XOO-30
Ampholak YCA/P
Ampholak YCE
Ampholak YJH
Ampholan® B-171
Ampholan® E210
Ampholan® U 203
Ampholyt JA 120
Ampholyt JA 140
Ampholyt JB 130
Ampholyte KKE
Amphoram CB A30
Amphoram CP1
Amphoram CT 30
Amphosol® CA
Amphosol® CB3
Amphosol® CG
Amphosol® DM
Amphoteen 24
Amphotensid B4
Amphotensid CT
Amphotensid D1
Amphoterge® J-2
Amphoterge® K
Amphoterge® K-2
Amphoterge® KJ-2
Amphoterge® NX
Amphoterge® S
Amphoterge® SB
Amphoterge® W
Amphoterge® W-2
Amphoteric 300
Amphoteric 400
Amphoteric L
Amphoteric N
Amphoteric SC
Antaron FC-34
Antaron MC-44
Armeen® SZ
Armeen® Z
Avivan® SFC
Blendmax Series
Cenegen® NWA
Centrol® 2FSB, 2FUB, 3FSB, 3FUB
Centrol® CA
Centrolene® A, S
Centromix® CPS
Centromix® E
Centrophase® 31
Centrophase® 152
Centrophase® C
Centrophase® HR2B, HR2U
Centrophase® HR4B, HR4U
Centrophase® HR6B
Centrophase® NV
Centrophil® K
Centrophil® M, W
Chembetaine BC-50
Chembetaine C
Chembetaine CAS
Chemoxide L
Chemteric 2C
Chemteric SF
Chimin BX
Clink A-70
Colorol 20
Colorol 70
Colorol E
Colorol F
Colorol Standard
Crodateric C
Crodateric Cy
Crodateric L
Crodateric O, O.100
Crodateric S
Crosultaine C-50
Crosultaine E-30
Crosultaine T-30
Dehyton® G-SF
Dehyton® K
Dehyton® KE
Dehyton® PMG
Deriphat 130-C
Deriphat 151
Deriphat 151C
Deriphat 154
Deriphat 154L
Deriphat 160
Deriphat 160C
Deriphat 170C
Deriphat BAW
Deriphat BC, BCW
Elec AC
Eleton 1100
Emcol® 5430
Emcol® 6748
Emcol® CC-37-18
Emcol® Coco Betaine
Emcol® DG
Emcol® NA-30
Emery® 6744
Emery® 6752
Emkapon AMP
Empigen® 5107
Empigen® BB
Empigen® BB-AU
Empigen® BS/H
Empigen® BS/P
Empigen® CDR10
Empigen® CDR30
Empigen® CDR40
Empigen® CDR60
Examide-CS
Fluorad FC-100
Fluorad FC-751
Foramousse WO 107
Forbest VP S7
Hartaine CB-40
Hipochem CAD
Hypan®
Hypan® SA100H
Incronam 30
Incronam CD-30
Indulin® AQS
Intratex® A
Intratex® B
Intratex® C
Intratex® DD
Laurelox 12
Lebon 101H, 105
Lebon 2000
Levelan NKD
Levelan® NKS
Levelan WS
Lexaine® C
Lexaine® CG-30
Lexaine® CS
Lexaine® CSB-50
Lexaine® LM
Lilaminox M24
Lilaminox M4
Lipomin CH
Lipotin A
Lipotin H
Lonzaine® 12C
Lonzaine® C
Lonzaine® CO
Lonzaine® CS
Lonzaine® JS
Lutostat 171
Mackam 1C
Mackam 1C-SF
Mackam 1L
Mackam 1L-30
Mackam 2C
Mackam 2C-75
Mackam 2C-SF
Mackam 2CT
Mackam 2CY
Mackam 2CYSF
Mackam 2L
Mackam 2LSF
Mackam 2W
Mackam 35
Mackam 35 HP
Mackam 160C
Mackam CB-35
Mackam CB-LS
Mackam CSF
Mackam HV
Mackam ISA
Mackam LMB
Mackam LT
Mackam MEJ
Mackam MLT
Mackam OB-30
Mackam TM
Mackamine CO
Mafo® 13
Mafo® 13 MOD 1
Mafo® CAB
Mafo® CAB 425

AMPHOTERIC (CONT'D.)

Mafo® CAB SP
Mafo® CB 40
Mafo® CFA 35
Mafo® CSB 50
Mafo® SBAO 110
Manro AO 3OC
Manro AT 1200
Manro DPM 2169
Manroteric CAB
Manroteric CDX38
Manroteric CEM38
Manroteric CyNa50
Manroteric NAB
Manroteric SAB
Markwet 851
Mazon® 114
M-C-Thin
Miracare® 2MHT
Miranol® 2CIB
Miranol® C2M Anhyd. Acid
Miranol® C2M Conc. NP
Miranol® C2M Conc. NP LV
Miranol® C2M Conc. OP
Miranol® C2M NP LA
Miranol® C2M-SF 70%
Miranol® C2M-SF Conc.
Miranol® CM Conc. NP
Miranol® CM Conc. OP
Miranol® CM-SF Conc.
Miranol® CS Conc.
Miranol® DM
Miranol® DM Conc. 45%
Miranol® FA-NP
Miranol® FAS
Miranol® FB-NP
Miranol® FBS
Miranol® H2C-HA
Miranol® HMA
Miranol® HM Conc.
Miranol® J2M Conc.
Miranol® J2M-SF Conc.
Miranol® JA
Miranol® JAS-50
Miranol® JB
Miranol® JBS
Miranol® JEM Conc.
Miranol® JEM-SF
Miranol® JS Conc.
Miranol® L2M-SF Conc.
Miranol® LB
Miranol® SM Conc.
Mirataine® A2P-TS-30
Mirataine® ASC
Mirataine® BB
Mirataine® BET-C-30
Mirataine® BET-O-30
Mirataine® CB/M
Mirataine® CBC
Mirataine® CBR
Mirataine® CBS, CBS Mod
Mirataine® D-40
Mirataine® FM
Mirataine® H2C-HA
Mirataine® JC-HA
Mirataine® T2C-30
Mirataine® TM
Mona AT-1200
Monaterge LF-945
Monateric 811
Monateric 1000
Monateric 1188M
Monateric ADA
Monateric CA-35
Monateric CDX-38
Monateric CDX-38 Mod
Monateric CEM-38
Monateric CM-36S
Monateric CNa-40
Monateric COAB
Monateric CyA-50
Monateric CyNa-50
Monateric ISA-35
Monateric LFNa-50
Monateric TDB-35
Natrex EFB 171
Nissan Anon BF
Norfox® ALKA
Nutrol Betaine OL 3798
Obazoline 662Y
Patogen AO-30
Phosphoteric® T-C6
Product BCO
Ralufon® 414
Ralufon® CA
Ralufon® DCH
Ralufon® DL
Ralufon® DM
Ralufon® DP
Ralufon® DS
Ralufon® DT
Ralufon® MDS
Retarder N
Retarder N-85
Rewopon® AM-V
Rewoteric AM B13
Rewoteric AM B14
Rewoteric AM CAS
Rewoteric AM DML
Rewoteric AM HC
Rewoteric AM KSF 40
Rewoteric AM V
Rewoteric AM VSF
Rhodaterge® FL
Sanac C
Sanac S
Sanac T
Sandolube NV
Sandolube TRA
Sandopan® TFL Conc.
Sandopan® TFL Liquid
Sandopur DK
Sandoz Amine XA-Q
Schercotaine APAB
Schercotaine CAB-A
Schercotaine CAB-K
Schercotaine CAB-KG
Schercotaine IAB
Schercotaine OAB
Schercotaine SCAB-A
Schercotaine SCAB-K
Schercoteric CY-2
Schercoteric CY-SF-2
Schercoteric I-AA
Schercoteric LS
Schercoteric LS-2
Schercoteric LS-EP
Schercoteric MS
Schercoteric MS-2
Schercoteric MS-2 Modified
Schercoteric MS-2ES Modified
Schercoteric MS-2TE Modified
Schercoteric MS-SF
Schercoteric MS-SF (38%)
Schercoteric MS-SF (70%)
Schercoteric O-AA
Schercoteric OS-SF
Sochamine A 8955
Sovatex MP/1
Stepanon CG
Surfax ACB
Swanol AM-301
Tego®-Betaine C
Tego®-Betaine F
Tego®-Betaine L-5351
Tego®-Betaine L-90
Tego®-Betaine N-192
Tego®-Betaine S
Tego®-Betaine T
Tegotain D
Tegotain L 7
Tegotain L 5351
Tegotain N 192
Tegotain S
Tegotain WS 35
Tinegal® KD
Tinegal® W
Triton® QS-15
Troysol 98C
Varion® 2C
Varion® 2L
Varion® AM-B14
Varion® AM-KSF-40%
Varion® AM-V
Varion® CADG-HS
Varion® CADG-LS
Varion® CADG-W
Varion® CAS
Varion® CAS-W
Varion® CDG
Varion® CDG-LS
Varion® HC
Varion® SDG
Varion® TEG
Varion® TEG-40%
Velvetex® 610L
Velvetex® AB-45
Velvetex® BA
Velvetex® BA-35
Velvetex® OLB-50
Versilan MX244
Zohartaine TM
Zoharteric D
Zoharteric D-SF
Zoharteric D-SF 70%
Zoharteric DJ
Zoharteric DO
Zoharteric DOT
Zoharteric LF
Zoharteric LF-SF
Zoharteric M
Zonyl® FSB
Zonyl® FSK
Zoramox LO

AMPHOTERIC/ ANIONIC:

Carsonol® BD
Intratex® DD-LF
Lumorol RK
Produkt GM 4300

AMPHOTERIC/ NONIONIC:

Arolev IDD
Sofnon LA-75

ANIONIC:

Abex® 12S
Abex® 18S
Abex® 22S
Abex® 23S
Abex® 26S
Abex® 33S
Abex® AAE-301
Abex® EP-110
Abex® EP-115
Abex® EP-120
Abex® EP-227
Abex® JKB
Abex® LIV/2330
Abex® LIV/30
Abex® VA 50
Ablumul OCE
Abluphat AP Series
Abluphat OP Series
Ablusol BX Series
Ablusol C-70
Ablusol C-78
Ablusol CDE
Ablusol DA
Ablusol DBC
Ablusol DBD
Ablusol DBM
Ablusol DBT
Ablusol LDE
Ablusol LME
Ablusol M-75
Ablusol ML
Ablusol N
Ablusol NL
Ablusol OA
Ablusol OK
Ablusol PM
Ablusol P Series
Ablusol SF Series
Ablusol SN Series
Ablusol TA
Abluton CDL
Acrilev AM, AM-Special
Acrilev OJP-25N
Acto 450
Acto 500
Acto 630
Acto 632
Acto 636
Acto 639
Actrabase 31-A

ANIONIC (CONT'D.)
Actrabase 163
Actrabase PS-470
Actrafos 110, 110A
Actrafos 152A
Actrafos 161, 315
Actrafos 186
Actrafos 822
Actrafos SA-208
Actrafos SA-216
Actrafos SN-306
Actrafos SN-314
Actrafos SN-315
Actrafos T
Actralube Syn-153
Actrasol C-50, C-75, C-85
Actrasol EO
Actrasol MY-75
Actrasol PSR
Actrasol SP
Actrasol SP 175K
Actrasol SR 75
Actrasol SRK 75
Actrasol SS
Acylglutamate CS-11
Acylglutamate CT-12
Acylglutamate MS-11
ADF-630
Adinol OT16
ADM-407
ADM-407C
Advantage M1250 Defoamer
Aerosol® 18
Aerosol® 22
Aerosol® 501
Aerosol® A-102
Aerosol® A-103
Aerosol® A-196-40
Aerosol® A-196-85
Aerosol® A-268
Aerosol® AY-65
Aerosol® AY-100
Aerosol® DPOS-45
Aerosol® GPG
Aerosol® IB-45
Aerosol® MA-80
Aerosol® NPES 458
Aerosol® NPES 930
Aerosol® NPES 2030
Aerosol® NPES 3030
Aerosol® NS
Aerosol® OS
Aerosol® OT-100%
Aerosol® OT-70 PG
Aerosol® OT-75%
Aerosol® OT-B
Aerosol® OT-MSO
Aerosol® OT-S
Aerosol® TR-70
Agitan 285
Agitan 288
Agrilan® AEC178
Agrilan® AEC266
Agrilan® AEC299
Agrilan® C91D
Agrilan® DG102
Agrilan® F491
Agrilan® F513
Agrilan® F524
Agrilan® F535
Agrilan® F546
Agrilan® F557
Agrilan® TKA125
Agrilan® TKA147
Agrilan® WP101
Agrilan® WP145
Agrilan® WP156
Agrilan® WP178
Agrilan® X98
Agrilan® X109
Agrimul N-300
Agrimul VUQ
Agriwet CA, C91, C92
Agriwet FOA
Agriwet T-F
Ahco A-117
Ahco AB 100
Ahco AB 118
Ahco AB 120
Ahco AB 135
Ahco AB 146
Ahco AB 160
Ahco AB 228
Ahco AJ-110
Ahco AY 2200
Ahcovel R Base
Ahcowet RS
Airrol CT-1
Ajicoat SPG
Akypo ITD 30 N
Akypo ITD 70 BV
Akypo LF 1
Akypo LF 2
Akypo LF 3
Akypo LF 4
Akypo LF 4N
Akypo LF 5
Akypo LF 6
Akypo MB 1585
Akypo MB 1614/1
Akypo MB 1614/2
Akypo MB 2528S
Akypo MB 2621 S
Akypo MB 2705 S
Akypo MB 2717 S
Akypo NP 70
Akypo NTS
Akypo RLM 38 NV
Akypo RLM 100 MGV
Akypo RLM 160 NV
Akypo RO 50
Akypo RO 90
Akypogene FP 35 T
Akypogene Jod MB 1918
Akypogene KTS
Akypogene ZA 97 SP
Akypomine® AT
Akypomine® BC 50
Akypopress DB
Akyporox SAL SAS
Akyposal 100 DE
Akyposal 100 DEG
Akyposal ALS 33
Akyposal BA 28
Akyposal BD
Akyposal EO 20 MW
Akyposal EO 20 PA
Akyposal EO 20 PA/TS
Akyposal EO 20 SF
Akyposal MS SPC
Akyposal NLS
Akyposal NPS 60
Akyposal NPS 100
Akyposal NPS 250
Akyposal RLM 56 S
Akyposal TLS 42
Akypo®-Soft 100 MgV
Akypo®-Soft 100 NV
Akypo®-Soft 130 NV
Akypo®-Soft 160 NV
Alarsol AL
Albatex® FFC
Alconate® SBDO
Alconox®
Alcotabs®
Alfonic® 1412-A Ether Sulfate
Alfonic® 1412-S Ether Sulfate
Alkamide® 206 CGN
Alkanol® 189-S
Alkanol® DW
Alkanol® ND
Alkanol® S
Alkanol® WXN
Alkanol® XC
Alkawet® NP-6
Alkylate 215
Alkylate 225
Alkylate 227
Alkylate 230
Alkylate H230H, H230L
Alkylene Series
Alpha-Step® ML-40
Alpha-Step® ML-A
Alphenate TFC-76
Alphenate TH 454
Alrowet® D-65
Alscoap LN-40, LN-90
Amerlate® LFA
Amerlate® WFA
Amisoft CS-11
Amisoft CT-12
Amisoft GS-11
Amisoft HS-11
Amisoft LS-11
Amisoft MS-11
Amlev MDV
Amlev MOD
Amlev OMF
Ammonium Stearate 33% Liq.
Amphisol
Amphisol K
Amsul 70
Anedco ADM-407
Anedco ADM-407C
Anedco ADM-408, ADM-410
Anedco ADM-456, ADM-457
Anedco AW-398
Anedco DF-6031
Anedco DF-6130
Anedco DF-6131
Anedco DF-6231
Anedco DF-6233
Anionyx® 12S
Anti-Terra®-207
Aquafoam 9451
Aquafoam 9452
Aquasol® AR 90
Aquasol® W 90
Arbyl ASN
Arfoam 2213
Arfoam 2386
Aristonate H
Aristonate L
Aristonate M
Arizona DRS-40
Arizona DRS-42
Arizona DRS-43
Arizona DRS-44
Arizona DRS-50
Arizona DRS-51E
Arkomon A Conc.
Arkomon SO
Arkopon Brands
Arkopon T Grades
Armix 146
Armix 185
Arodet TA-8
Arolev CDD
Arowet 70 E
Arowet S
Arowet SC-75
Arsul DDB
Arsul LAS
Arylan® CA
Arylan® HE Acid
Arylan® LQ
Arylan® PWS
Arylan® S Acid
Arylan® S90 Flake
Arylan® SBC Acid
Arylan® SBC25
Arylan® SC15
Arylan® SC30
Arylan® SC Acid
Arylan® SP Acid
Arylan® SX Flake
Arylan® SX85
Arylan® SY30
Arylan® SY Acid
Arylan® TE/C
Arylene M40
Arylene M60
Arylene M75
Astrowet O-70-PG
Astrowet O-75
Atcowet C
Atlas G-711
Atlas G-2203
Atlas G-3300
Atlas Sul. Neats L-2
Atlas Sul. Oil HC
Atlas WA-100
Atlasol 103
Atlasol 160-S
Atlasol 6920
Atlasol 6920-VF
Atlox 2081 B
Atlox 3300 B
Atlox 3335 B
Atlox 3386 B
Atlox 4851 B
Atlox 4855 B
Atlox 4856 B
Atlox 4857 B

ANIONIC (CONT'D.)
Atlox 4858 B
Atlox 4861B
Atlox 4862
Atlox 5320
Atlox 5325
Atolene RW
Atplus 400
Atplus 401
Atplus 535
Atplus 1403
Atra Polymer 10
Atrasein 115
Atsolyn PE 36
Atsowet P 50
Atsurf 1910
Avanel® S-30
Avanel® S-35
Avanel® S-70
Avanel® S-74
Avanel® S-150
Avirol® 125 E
Avirol® 252 S
Avirol® 270A
Avirol® 280 S
Avirol® 400 T
Avirol® 603 A
Avirol® 603 S
Avirol® A
Avirol® AE 3003
Avirol® AOO 1080
Avirol® FES 996
Avirol® SA 4106
Avirol® SA 4108
Avirol® SA 4110
Avirol® SA 4113
Avirol® SE 3002
Avirol® SE 3003
Avirol® SL 2010
Avirol® SL 2015
Avirol® SL 2020
Avirol® SO 70P
Avirol® T 40
Avitone® F
Avitone® T
AW-398
Barisol Super BRM
Barium Petronate 50-S Neutral
Barium Petronate Basic
Base 75
Base 7800
Basojet® PEL 200%
Basol® WS
Basophen® RA
Basophen® RBD
Berol 475
Berol 480
Berol 521
Berol 522
Berol 733
Berol 752
Berol 822
Berol 824
Beycopon 345
Beycopon EC
Beycopon S 3A
Beycopon S 50
Beycopon TB
Beycopon TL
Beycostat 148 K
Beycostat 211 A
Beycostat 231
Beycostat 256A
Beycostat 273 P
Beycostat 319 P
Beycostat 714 P
Beycostat B 070 A
Beycostat B 080 A
Beycostat B 089 A
Beycostat B 151
Beycostat B 231
Beycostat B 327
Beycostat B 337
Beycostat B 706
Beycostat B 706 A
Beycostat B 706 E
Beycostat C 103
Beycostat C 213
Beycostat DP
Beycostat LA
Beycostat LP 12 A
Beycostat NE
Beycostat NED
Biodet B/D
Biodet B/TA
Biodet D
Biodet TA
Biopal® NR-20
Biopal® NR-20 W
Bio-Soft® AS-40
Bio-Soft® D-35X
Bio-Soft® D-40
Bio-Soft® D-53
Bio-Soft® D-62
Bio-Soft® JN
Bio-Soft® LAS-40S
Bio-Soft® LD-145
Bio-Soft® LD-150
Bio-Soft® LD-190
Bio-Soft® LD-32
Bio-Soft® LD-47
Bio-Soft® N-300
Bio-Soft® N-411
Bio-Soft® Ninex 21
Bio-Soft® S-100
Bio-Soft® S-130
Bio-Surf PBC-420
Bio-Surf PBC-430
Bio-Terge® AS-40
Bio-Terge® AS-90 Beads
Bio-Terge® PAS-8
Bio-Terge® PAS-8S
Bohrmittel Hoechst
Burco TM-HF
Burco Anionic APS
Burco Anionic APS-LF
Burcofac 9125
BYK®-151
BYK®-P104
BYK®-P104S
BYK®-P105
Bykumen®
Bykumen®-WS
CA DBS 50 SA
Calcium Petronate
Calcium Sulfonate C-50N
Calcium Sulfonate EP-163
Calfax 10L-45
Calfax 16L-35
Calfax DB-45
Calfax DBA-40
Calfax DBA-70
Calfoam AAL
Calfoam ES-30
Calfoam LLD
Calfoam NEL-60
Calfoam NLS-30
Calfoam SEL-60
Calfoam SLS-30
Calimulse PRS
Calsoft F-90
Calsoft L-40
Calsoft L-60
Calsoft LAS-99
Calsoft T-60
Calsolene Oil HSA
Calsolene Oil HSAD
Calsuds A
Calsuds CD-6
Carbonox
Carbopol® 615, 616, 617
Carbopol® 907
Carbopol® 910
Carbopol® 934
Carbopol® 940
Carsofoam® T-60-L
Carsonol® ALS-R
Carsonol® ALS-S
Carsonol® ANS
Carsonol® ILS
Carsonol® MLS
Carsonol® SES-A
Carsonol® SES-S
Carsonol® SHS
Carsonol® SLES
Carsonol® SLS
Carsonol® SLS Paste B
Carsonol® SLS-R
Carsonol® SLS-S
Carsonol® SLS Special
Carsonol® TLS
Carsosulf SXS-Liq.
Casul® 55 HF
Casul® 70 HF
Cedepal CO-436
Cedepal FS-406
Cedepal SS-203, -306, -403, -406
Cedepal TD-400
Cedephos® CP610
Cedephos® FA600
Cedephos® FE610
Cedepon A 100
Cedepon AM
Cedepon LT-40
Cedepon SX-55
Cedepon T
Cenegen® 7
Cenekol® 1141
Cenekol® FT Supra
Cenekol® Liq., NCS Liq.
Charlab Leveler AT Special
Charlab Leveler DSL
Chemax DOSS-75E
Chemfac PA-080
Chemfac PB-082
Chemfac PB-106
Chemfac PB-135
Chemfac PB-184
Chemfac PB-264
Chemfac PC-006
Chemfac PC-099E
Chemfac PC-188
Chemfac PD-600
Chemfac PD-990
Chemfac PF-623
Chemfac PF-636
Chemfac PN-322
Chemfac PX-322
Chemphos TC-310
Chemphos TC-337
Chemphos TDAP
Chemsalan NLS 30
Chemsalan RLM 28
Chemsalan RLM 56
Chemsalan RLM 70
Chemsulf 094
Chemsulf S2EH-Na
Chimin DOS 70
Chimin KF3
Chimin P10
Chimin P1A
Chimin P40
Chimin P45
Chimin RI
Chimipon GT
Chimipon NA
Chimipon SK
Chimipon TSB
Chromasist 6B
Chupol EX
Cibaphasol® AS
Cithrol DGDL S/E
Cithrol DGDO S/E
Cithrol DGDS S/E
Cithrol DGML S/E
Cithrol DGMO S/E
Cithrol DGMS S/E
Cithrol DPGML S/E
Cithrol DPGMO S/E
Cithrol DPGMS S/E
Cithrol EGDL S/E
Cithrol EGDO S/E
Cithrol EGDS S/E
Cithrol EGML S/E
Cithrol EGMO S/E
Cithrol EGMR S/E
Cithrol EGMS S/E
Cithrol GDL S/E
Cithrol GDO S/E
Cithrol GDS S/E
Cithrol GML S/E
Cithrol GMO S/E
Cithrol GMR S/E
Cithrol GMS S/E
Cithrol PGML S/E
Cithrol PGMO S/E
Cithrol PGMR S/E
Cithrol PGMS S/E
Closyl 30 2089
Closyl LA 3584
CNC Detergent E
CNC Dispersant WB Series
CNC Gel Series
CNC Product ST
CNC Sol 72-N Series
CNC Wet SS-80
Colloid 111-D
Collone HV

ANIONIC (CONT'D.)
Compound 170
Compound 535
Compound S.A
Conco X-200
Consamine DS Powd
Consamine K-Gel
Consamine P
Consamine X
Consolevel JBT
Consoluble 71
Consonyl FIX
Consos Castor Oil
Consoscour 47
Coptal WA OSN
Cordon AES-65
Cordon COT
Cordon DA-B
Cordon N-400
Cordon NU 890/75
Cordon PB-870
Cosmopon 35
Cosmopon BL
Cosmopon BN
Cosmopon LE 50
Cosmopon MO
Cosmopon SES
Cosmopon TR
Coursemin SWG-10
Coursemin SWG-7
CPH-250-SE
Crafol AP-11
Crafol AP-16
Crafol AP-20
Crafol AP-21
Crafol AP-35
Crafol AP-36
Crafol AP-53
Crafol AP-55
Crafol AP-63
Crafol AP-64
Crafol AP-65
Crafol AP-67
Crafol AP-68
Crafol AP-69
Crafol AP-70
Crafol AP-85
Crafol AP-201
Crafol AP-202
Crafol AP-203
Crafol AP-240
Crafol AP-241
Crafol AP-260
Crafol AP-261
Crafol AP-262
Crodafos CAP
Crodafos CDP
Crodafos CS2 Acid
Crodafos CS5 Acid
Crodafos CS10 Acid
Crodafos N3 Acid
Crodafos N3 Neutral
Crodafos N5 Acid
Crodafos N10 Acid
Crodafos N10 Neutral
Crodasinic LS30
Crodasinic LS35
Crodasinic LT40
Crodasinic OS35
Cromeen
Cropol
Cyclomatic Dur
Dacospin PE-47
Dacospin PE-146
Dapral® AS
Darvan® ME
Darvan® WAQ
Daxad® 37LA7
DDBS 100
DDBS-100 SP
DDBS Special
Deceresol Surfactant OT 75%
Deceresol Surfactant OT Special
Defoamer SF
Dehscofix 904
Dehscofix 905
Dehscofix 907
Dehscofix 911
Dehscofix 912
Dehscofix 914
Dehscofix 914/AS
Dehscofix 914/ASL
Dehscofix 915
Dehscofix 915/AS
Dehscofix 916
Dehscofix 916S
Dehscofix 917
Dehscofix 918
Dehscofix 920
Dehscofix 923
Dehscofix 926
Dehscofix 929
Dehscofix 930
Dehscotex DT 809
Dehydag Wax E
Dehypon Conc.
Delion 624
Delion 662
Delion 6067
Delion A-016
Delion A-160
Demelan CB-60
Denphos 623
Denphos P-610
Densol 284
Densol 6920
Densol P-82
Denwet CM
Denwet RG-7
Depasol AS-27
Depasol CM-41
Depsodye AR
Depsolube ACA
DeSonate 50-S
DeSonate 60-S
DeSonate AOS
DeSonate AUS
Desophos® 5 AP
Desophos® 5 BMP
Desophos® 6 NP
Desophos® 14 DNP
Detergent ADC
Deterpal 832
Deterpal LC
Detersol WR
Deteryl 955 FS
Dextrol OC-15
Dextrol OC-20
Dextrol OC-40
Dextrol OC-50
Dextrol OC-70
Dikssol 201, 202
Dikssol 301, 302
Disperbyk®
Dispersing Agent SS Dry
Dispersogen A
Dispersogen SI
Disponil AEP 5300
Disponil AEP 8100
Disponil AEP 9525
Disponil AES 13
Disponil AES 21
Disponil AES 42
Disponil AES 48
Disponil AES 60
Disponil AES 60 E
Disponil AES 72
Disponil FES 32
Disponil FES 61
Disponil FES 77
Disponil FES 92E
Disponil MGS 65
Disponil MGS 935
Disponil SUS 29 L
Disponil SUS 65
Disponil SUS 87 Special
Disponil SUS 90
Dodecenyl Succinic Anhydride
Dodiflood Brands
Doittol 14
Doittol 891
Doittol APS Conc.
Dowfax 2A0
Dowfax 2A1
Dowfax 2EP
Dowfax 3B0
Dowfax 3B2
Dowfax 8390
Dowfax XDS 8292.00
Dowfax XDS 8390.00
Dowfax XDS 30599
Dowfax XU 40333.00
Dowfax XU 40340.00
Dowfax XU 40341.00
Dresinate® 81
Dresinate® 90
Dresinate® 91
Dresinate® 95
Dresinate® 214
Dresinate® 515
Dresinate® TX
Dresinate® X
Dresinate® XX
Drewfax® S-700
Duofol T
Duponol® 80
Duponol® D Paste
Duponol® EP
Duponol® FAS
Duponol® G
Duponol® LS Paste
Duponol® ME Dry
Duponol® QC
Duponol® RA
Duponol® SP
Duponol® WA Dry
Duponol® WA Paste
Duponol® WAQ
Duponol® WAQE
Duponol® WN
Dyasulf 9268-A
Dyasulf BO-65
Dyasulf C-70
Dyasulf OA-60
Dymsol® 31-P
Dymsol® 38-C
Dymsol® 2031
Dymsol® LP
Dymsol® PA
Dymsol® S
Dyqex®
Eccoclean C-90
Eccoful DL Conc.
Eccoful FC Conc.
Eccolene OW
Ecco Leveler 700
Ecconol 61
Ecconol 606
Ecconol 2818
Ecconol 2833
Eccoscour D-7
Eccoscour SNP
Eccoterge 35-S
Eccoterge ASB
Eccoterge Conc.
Eccoterge MV Conc.
Eccoterge S-35
Eccotex P. Conc.
Eccowet® LF Conc.
Eccowet® W-50
Edenor ST-1
Edunine PA
Eganal PS
Eganal SZ
Elec RC
Elecut S-507
Eleminol ES-70
Eleminol MON-7
Elfan® 200
Elfan® 240
Elfan® 240M
Elfan® 240T
Elfan® 280
Elfan® 280 Powd
Elfan® 680
Elfan® KM 550
Elfan® KM 730
Elfan® KT 550
Elfan® NS 242
Elfan® NS 242 Conc.
Elfan® NS 243 S
Elfan® NS 243 S Conc.
Elfan® NS 252 S
Elfan® NS 252 S Conc.
Elfan® NS 423 SH
Elfan® NS 423 SH Conc.
Elfan® NS 682 KS
Elfan® OS 46
Elfan® SP 325
Elfan® SP 400
Elfan® SP 500
Elfan® WA
Elfan® WA Powder
Elfan® WA Sulphonic Acid
Elfan® WAT
Elfanol® 510

ANIONIC (CONT'D.)
Elfanol® 616
Elfanol® 850
Elfanol® 883
Elimina 254
Elimina 505
Eltesol® 4402
Eltesol® 4443M
Eltesol® 5400 Series
Eltesol® 7200 Series
Eltesol® AC60
Eltesol® AX 40
Eltesol® CA 65
Eltesol® CA 96
Eltesol® PSA 65
Eltesol® PT 93
Eltesol® PX 40
Eltesol® PX 93
Eltesol® SC 40
Eltesol® SC 93
Eltesol® SC Pellets
Eltesol® ST 34
Eltesol® ST 40
Eltesol® ST 90
Eltesol® ST Pellets
Eltesol® SX 30
Eltesol® SX 40
Eltesol® SX 93
Eltesol® SX Pellets
Eltesol® TA 65
Eltesol® TA 96
Eltesol® TA Series
Eltesol® TSX
Eltesol® TSX/A
Eltesol® TSX/SF
Eltesol® XA
Eltesol® XA65
Eltesol® XA90
Emal 20C
Emalox C 102
Emalox OEP-1
Emcol® 4100M
Emcol® 4161L
Emcol® 4300
Emcol® 4500
Emcol® 4580PG
Emcol® 4600
Emcol® 4930, 4940
Emcol® CN-6
Emcol® CS-143, -151, -165
Emcol® K8300
Emcol® P 50-20 B
Emcol® P-1020 BU
Emcol® P-1020B
Emcol® P-1045
Emcol® P-1059B
Emery® 5440
Emkabase
Emkabase ODC-2
Emkafol D
Emkafol OT
Emkagen 49
Emkal BNS
Emkal BNS Acid
Emkal BNX Powd
Emkal NNS
Emkal NNS Acid
Emkal NOBS
Emkalar Base E-55
Emkalar Base NC
Emkalite BAC
Emkane Acid
Emkane HAL
Emkane HAX
Emkanol NC, NCD 25, 35, 45, 55
Emkapol PO-18
Emkapon 4S, DS, SS, TS
Emkapon CT
Emkapon Jel BS
Emkapon K
Emkapon KW
Emkapon L
Emkasol DE
Emkasol WAS, WAS-T
Emkatex AA
Emkatex AA-80
Emkatex AES
Emkatex CF 32
Emkatex DEM
Emkatex DX, DXP
Emkatex LE
Emkatex PX29
Emkatex RA
Emkatex WNX
Emphos AM2-10C
Emphos CS-121
Emphos CS-136
Emphos CS-141
Emphos CS-147
Emphos CS-151
Emphos CS-330
Emphos CS-733
Emphos CS-735
Emphos CS-1361
Emphos D70-30C
Emphos D70-31
Emphos F27-85
Emphos PS-21A
Emphos PS-121
Emphos PS-220
Emphos PS-222
Emphos PS-236
Emphos PS-331
Emphos PS-400
Emphos PS-410
Emphos PS-413
Emphos PS-415M
Emphos PS-440
Emphos PS-810
Emphos PS-900
Emphos TS-230
Empicol® 0185
Empicol® 0216
Empicol® 0303
Empicol® 0303V
Empicol® 0585/A
Empicol® AL30
Empicol® CHC 30
Empicol® DA
Empicol® ESB
Empicol® ESB3
Empicol® LQ33/T
Empicol® LS30P
Empicol® LX
Empicol® LX28
Empicol® LXS95
Empicol® LXV
Empicol® LXV/D
Empicol® LZ
Empicol® LZ/D
Empicol® LZ/E
Empicol® LZG 30
Empicol® LZGV
Empicol® LZV/D
Empicol® LZV/E
Empicol® SDD
Empicol® SEE
Empicol® SFF
Empicol® SGG
Empicol® SHH
Empicol® SLL
Empicol® SLL/P
Empicol® STT
Empicol® TAS30
Empicol® TL40/T
Empicol® WAK
Empicol® XM 17
Empimin® 3631
Empimin® AQ60
Empimin® BMA
Empimin® BMB
Empimin® KSN27
Empimin® KSN60
Empimin® KSN70
Empimin® LAM30/AU
Empimin® LR28
Empimin® LSM30
Empimin® LSM30-AU
Empimin® MA
Empimin® MH
Empimin® MK/B
Empimin® MKK/AU
Empimin® MKK/L
Empimin® MKK98
Empimin® MTT/A
Empimin® OP45
Empimin® OP70
Empimin® OT
Empimin® OT75
Empimin® SDS
Empimin® SQ25
Empimin® SQ70
Emulgit 60
Emulgit 70
Emulsifiant 33 AD
Emulsifier K 30 40%
Emulsifier K 30 68%
Emulsifier K 30 76%
Emulsifier K 30 95%
Emulsifier Q
Emulsifier WHC
Emulsogen B2M
Emulsogen BB
Emulsogen CP 136
Emulsogen H
Emulsogen HFIT
Emulsogen IC
Emulsogen ICL
Emulsogen IP 400
Emulsogen ITL
Emulsogen STH
Emulson AG 255
Emulson AG 7000
Emulson AG/CAL
Ensidom O
Ensital AP 115
Equex S
Equex SP
Equex STM
Equex SW
Erional® NW
Erional® PA
Erional® RN
Esi-Det EP-20
Esi-Terge 40% Coconut Oil Soap
Esi-Terge 320
Esi-Terge 330
Esi-Terge RT-61
Esi-Terge SXS
Esi-Terge T-60
Estersulf 1HH
Ethfac 102
Ethfac 104
Ethfac 106
Ethfac 133
Ethfac 136
Ethfac 140
Ethfac 142W
Ethfac 161
Ethfac 324
Ethfac 353
Ethfac 361
Ethfac 363
Ethfac 391
Ethfac 1018
Ethfac NP-16
Ethfac NP-110
Ethfac PB-1
Ethfac PB-2
Ethfac PD-6
Ethfac PD-990
Ethfac PP-16
Ethfac PP-36/50%
Ethox 1358
Ethox 2195
Ethox 2684
Ethox 2928
Ethyl Foaming Agent
Eureka 102
Eureka 102-WK
Eureka 392
Eureka 400-R
Eureka 800-R
Findet A-100-UN
Findet CF-4
Findet CF-440
Findet DD
Findet NHP
Findet OJP-5
Findet OJP-25
Findet SB
Fizul M-440
Fizul MD-318
Flo-Mo® 50H
Flo-Mo® Suspend
Flo-Mo® Suspend Plus
Fluorad FC-93
Fluorad FC-95
Fluorad FC-98
Fluorad FC-99
Fluorad FC-109
Fluorad FC-118
Fluorad FC-120
Fluorad FC-121
Fluorad FC-126
Fluorad FC-129
Fluorad FC-143

ANIONIC (CONT'D.)
Fluowet 40 M
Fluowet PL
Fluowet PP
Foamer
Foamer
Foamer AD
Foamer CD
Forbest 18
Forbest VP 18
Forlanit P
Forlanon
Fosfamide CPD-170
Fosfamide N
Fosterge BA-14 Acid
Fosterge LF
Fosterge LFD
Fosterge LFS
Fosterge R
Fosterge RD
Fosterge W
G-2200
G-2207
G-3300
G-7205
G-7274
Gantrez® AN
Gantrez® AN-119
Gantrez® AN-139
Gantrez® AN-149
Gantrez® AN-169
Gantrez® AN-179
Gantrez® S-95
Gardilene IPA/94
Gardilene S25L
Gardinol CX
Gardinol CXM
Gardinol WA Paste
Gardisperse AC
Gemtex 445
Gemtex 680
Gemtex 691-40
Gemtex PA-70P
Gemtex PA-75
Gemtex PA-75E
Gemtex PA-85P
Gemtex PAX-60
Gemtex SC-40
Gemtex SC-70
Gemtex SC-75
Gemtex SC-75E, SC Powd
Gemtex SM-33
Gemtex WBT
Gemtex WBT-9
Gemtex WNT-Conc
Genapol® AMS
Genapol® ARO
Genapol® CRT 40
Genapol® LRO Liq., Paste
Genapol® PGM Conc.
Genapol® ZRO Liq., Paste
Geronol ACR/4
Geronol ACR/9
Geronol PRH/4-A, -B
Geropon® 99
Geropon® 111
Geropon® AB/20
Geropon® AC-78
Geropon® ACR/9
Geropon® AY
Geropon® BIS/SODICO-2
Geropon® CYA/45
Geropon® CYA/60
Geropon® CYA/DEP
Geropon® DOS
Geropon® DOS FP
Geropon® FA-82
Geropon® FMS
Geropon® IN
Geropon® MLS/A
Geropon® NK
Geropon® S-1585
Geropon® SC/211
Geropon® SC/213
Geropon® SDS
Geropon® SS-L7DE
Geropon® SS-O-70PG
Geropon® SS-O-75
Geropon® T-33
Geropon® T/36-DF
Geropon® T-43
Geropon® T-51
Geropon® T-77
Geropon® TA/72
Geropon® TA/72/S
Geropon® TA/764
Geropon® TA/K
Geropon® TBS
Geropon® TC-42
Geropon® TK-32
Geropon® TX/99
Geropon® WS-25, WS-25-I
Geropon® WT-27
Geropon® X2152
Glutrin
Good-rite® K-732
Good-rite® K-7058
Good-rite® K-7200N
Good-rite® K-7600N
Goulac
Hampfoam 35
Hamposyl® AC-30
Hamposyl® AL-30
Hamposyl® C
Hamposyl® C-30
Hamposyl® L
Hamposyl® L-30
Hamposyl® L-95
Hamposyl® M
Hamposyl® M-30
Hamposyl® O
Hamposyl® S
Haroil SCO-65, -7525
Haroil SCO-75
Harol RG-71L
Hartamide 9137
Hartamide KL
Hartamide LMEA
Hartasist DF-28
Hartenol LAS-30
Hartenol LES 60
Hartofol 40
Hartofol 60T
Hartomer 4900
Hartosolve OL
Hartotrope AXS
Hartotrope KTS 50
Hartotrope STS-40, Powd
Hartotrope SXS 40, Powd
Hartowet MSW
Hetsulf 40, 40X
Hetsulf 50A
Hetsulf 60S
Hetsulf Acid
Hetsulf IPA
Hexaryl D 60 L
Hi-Fluid
Hipochem Dispersol GTO
Hipochem Dispersol SB
Hipochem Dispersol SCO
Hipochem Dispersol SP
Hipochem EK-18
Hipochem EK-6
Hipochem FNL
Hipochem GM
Hipochem LCA
Hipochem LH-Soap
Hipochem MT-1
Hipochem NAC
Hipochem No. 3
Hipochem No. 40-L
Hipochem No. 641
Hipochem PND-11
Hipochem SO
Hipochem SRC
Hiposcour® BFS
Hiposcour® NFMS-2
Hipowet IBS
Hodag DOSS-70
Hodag DOSS-75
Hodag DTSS-70
Hodag PE-005
Hodag PE-104
Hodag PE-106
Hodag PE-109
Hodag PE-206
Hodag PE-209
Hodag Sole-Mulse B
Hoe S 1984 (TP 2279)
Hoe S 1984 (TP 2283)
Hoe S 2713
Hoe S 2713 HF
Hoe S 2749
Hoe S 2895
Hoe S 2896
Hoe S 3618
Honoralin PL
Hostacor 2098
Hostacor 2125
Hostacor 2270
Hostacor 2272
Hostacor 2291
Hostacor 2292
Hostacor 2732
Hostacor BBM
Hostacor BF
Hostacor BK
Hostacor BM
Hostacor BS
Hostacor E
Hostacor H Liq. N
Hostacor TP 2445
Hostapal 2345
Hostapal BV Conc.
Hostapal CVS
Hostapal TP 2347
Hostaphat AR K
Hostaphat F Brands
Hostaphat MDAR Grades
Hostaphat MDLZ
Hostapon KA Powd
Hostapon T Powd
Hostapur DTC
Hostapur DTC FA
Hostapur OS Brands
Hostawet TDC
Humectol C, C Highly Conc.
Hybase® C-300
Hybase® C-400
Hybase® C-500
Hybase® M-12
Hybase® M-300, -400
Hybase® M-400
Hymolon CWC
Iberpenetrant-114
Iberpol Gel
Iberpon W
Iberscour AC
Iberterge CO-40
Iberwet B.I.G
Idet 5L
Idet 5LP
Idet 5L SP NF
Idet 10, 20–P
Imbirol OT/Na 70
Incrodet TD7-C
Incrosul LMS
Incrosul LSA
Indulin® AQS-IM
Indulin® AT
Indulin® SA-L
Intex Scour 707
Intralan® Salt HA
Intraphasol COP
Intraphasol PC
Intratex® BD
Intratex® JD
Intratex® JD-E
Intratex® N
Intratex® N-1
Intratex® POK
Ionet 300
Ionet LD-7-200
Ionet RAP-50
Ionet RAP-80
Ionet RAP-250
Irgalev® PBF
Irgasol® DA Liq., Powd
Jordapon® CI-60 Flake
Jordapon® CI-Powd
J Wet 19A
Kamar BL
Karafac 78
Karafac 78 (LF)
Kara Lube AL
Kara Lube AL-Conc.
Karawet CS
Karawet DOSS
Kelig 100
Kemester® 5221SE
Kemester® 6000SE
Ke-Mul® 181
Ke-Mul® A97
Kieralon® C
Kieralon® NB-OL
Kingoil S-10
Klearfac® AA040
Klearfac® AA270

ANIONIC (CONT'D.)
Klearfac® AA420
Klearfac® AB270
Kleen-Paste
Korantin® SH
LABS-100
LABS 100/H.V
LABS-100 SP
Lactimon®
Lactimon®-WS
Lamacit AP 6
Lamepon 287 SF
Lamepon A
Lanapex HTS
Lanapex R
Lan-Aqua-Sol 50
Lan-Aqua-Sol 100
Lanette E
Lankropol® ADF
Lankropol® ATE
Lankropol® KMA
Lankropol® KN51
Lankropol® KNB22
Lankropol® KO
Lankropol® KO2
Lankropol® KSG72
Lankropol® ODS/PT
Lankropol® OPA
Lankropol® WA
Lankropol® WN
Lankrosol HS101
Lankrosol HS112
Lankrosol SXS-30
Lanolin Fatty Acids O
Larosol DBL
Larosol DBL-3
Larosol DBL-3 Conc.
Larosol NRL Conc.
Larosol NRL-40
Lathanol® LAL
Laural D
Laural EC
Laural ED
Laural LS
Laural P
Laurel M-10-257
Laurel R-50
Laurel R-75
Laurel SBT
Laurel SD-150
Laurel SRO
Laurelphos 60G
Laurelphos 400
Laurelphos A-600
Laurelphos C-30LF
Laurelphos D 44 B
Laurelphos E-61
Laurelphos OL-529
Laurelphos P-71
Laurelphos RH-44
Lebon A-5000
Lenetol PS
Lenetol WLF 125
Leomin AN
Leonil DB Powd
Leonil EBL
Leonil KS
Leonil OS
Leophen® BN
Leophen® LG
Leophen® M
Leophen® ML
Leophen® RA
Leophen® RBD
Leukonöl LBA-2
Lexein® S620TA
Lexemul® 55SE
Lexemul® 530
Lexemul® T
Lignosite®
Lignosite® 17
Lignosite® 231
Lignosite® 260
Lignosite® 401
Lignosite® 431
Lignosite® 458
Lignosite® 823
Lignosite® AC
Lignosite® L
Lignosol AXD
Lignosol B
Lignosol BD
Lignosol D-10, D-30
Lignosol DXD
Lignosol FTA
Lignosol HCX
Lignosol NSX 110
Lignosol SFX
Lignosol TS
Lignosol TSD
Lignosol WT
Lignosol X
Lignosol XD
Lipolan 327 F
Lipolan 1400
Lipolan AO
Lipolan AOL
Lipolan G
Lipolan LB-440
Lipolan LB-840
Lipolan PB-800
Lipolan PJ-400
Lipolan TE, TE (P)
Lipomix G
Lipon PS-206
Lipophos PE9
Lipophos PL6
Lipoproteol LCOK
Lipoproteol LK
Lipoproteol UCO
Lipotac TE
Lipotac TE-P
Lomar® D
Lomar® D SOL
Lomar® HP
Lomar® LS
Lomar® LS Liq.
Lomar® PW
Lomar® PWA
Lomar® PWA Liq.
LSF-54
Lubrhophos® LB-400
Lubrhophos® LE-500
Lubrhophos® LE-600
Lubrhophos® LE-700
Lubrhophos® LF-200
Lubrhophos® LK-500
Lubrhophos® LL-550
Lubrhophos® LM-400
Lubrhophos® LM-600
Lubrhophos® LP-700
Lubrhophos® LS-500
Lubricant EHS
Lubrizol® 2152
Lubrizol® 5369, 5372
Lubrizol® 5375
Lufibrol® KB Liq.
Lufibrol® KE
Lumiten® I
Lumorol 4290
Lumo Stabil S 80
Lumo WW 75
Lutensit® A-BO
Lutensit® A-EP
Lutensit® A-ES
Lutensit® A-FK
Lutensit® A-LBA
Lutensit® A-PS
Lutensit® AN 40
Lutensit® AS 2230, 2270
Lutensit® AS 3330
Lutensit® AS 3334
Lutopon SN
LZ 5362 A
LZ 5364
Mackadet 40K
Mackam CAP
Mackam ISP
Mackam LAP
Mackanate CM
Mackanate CM-100
Mackanate DOS-70MS
Mackanate DOS-75
Mackanate L-101, -102
Madeol AG 1989 N
Madeol AG BX
Manoxol IB
Manoxol MA
Manoxol N
Manoxol OT
Manoxol OT 60%
Manoxol OT/B
Manro ALES 60
Manro ALS 30
Manro BA Acid
Manro BES 27
Manro BES 60
Manro BES 70
Manro CD
Manro CD/G
Manro CDS
Manro CDX
Manro CMEA
Manro DB 56
Manro DES 32
Manro DL 28
Manro DL 32
Manro DS 35
Manro HA Acid
Manro HCS
Manro MA 35
Manro NA Acid
Manro PTSA/E
Manro PTSA/H
Manro PTSA/LS
Manro SDBS 25/30
Manro SDBS 60
Manro SLS 28
Manro TDBS 60
Manro TL 40
Manrosol SXS30
Manrosol SXS40
Manrosol SXS93
Maphos® 17
Maphos® 76 NA
Maphos® 78
Maphos® JP 70
Maphos® L-6
Maprosyl® 30
Maramul SS
Maranil A
Maranil ABS
Maranil CB-22
Maranil DBS
Maranil Paste A 55, A 75
Maranil Powd. A
Marasperse 52 CP
Marasperse N-22
Markwet 3003
Markwet WL-12
Marlican®
Marlinat® 24/28
Marlinat® 24/70
Marlinat® 242/28
Marlinat® 242/70
Marlinat® 242/70 S
Marlinat® 243/28
Marlinat® 243/70
Marlinat® CM 20
Marlinat® CM 40
Marlinat® CM 45
Marlinat® CM 100
Marlinat® CM 105
Marlinat® CM 105/80
Marlinat® DF 8
Marlinat® DFK 30
Marlinat® DFL 40
Marlinat® DFN 30
Marlinat® HA 12
Marlinat® SL 3/40
Marlinat® SRN 30
Marlon® A 350
Marlon® A 360
Marlon® A 365
Marlon® A 375
Marlon® A 390
Marlon® AFR
Marlon® AMX
Marlon® ARL
Marlon® AS3
Marlon® AS3-R
Marlon® PS 30
Marlon® PS 60
Marlon® PS 60 W
Marlon® PS 65
Marlophor® AS-Acid
Marlophor® CS-Acid
Marlophor® DG-Acid
Marlophor® DS-Acid
Marlophor® F1-Acid
Marlophor® FC-Acid
Marlophor® FC-Sodium Salt
Marlophor® HS-Acid
Marlophor® ID-Acid
Marlophor® IH-Acid
Marlophor® LN-Acid
Marlophor® MN-60
Marlophor® MO 3-Acid
Marlophor® N5-Acid

ANIONIC (CONT'D.)
Marlophor® ND
Marlophor® ND-Acid
Marlophor® NP5-Acid, NP6-Acid, NP7-Acid
Marlophor® OC5-Acid
Marlophor® ON3-Acid, ON5-Acid, ON7-Acid
Marlophor® T10-Acid
Marlophor® T10-DEA Salt
Marlophor® T10-Sodium Salt
Marlophor® T6-Acid
Marlophor® UW12-Acid
Marlopon® ADS 50
Marlopon® ADS 65
Marlopon® AT 50
Marlopon® CA
Marlowet® 1072
Marlowet® 4508
Marlowet® 4530
Marlowet® 4534
Marlowet® 4538
Marlowet® 4539
Marlowet® 5301
Marlowet® 5311
Marlowet® 5320
Marlowet® 5324
Marlowet® 5600
Marlowet® 5606
Marlowet® 5609
Marlowet® 5622
Marlowet® 5626
Marlowet® 5635
Marlowet® T
Marvanol® GC
Marvanol® Leveltone 7.5
Marvanol® Penetrant 35
Marvanol® REAC A-213
Marvanol® SBO (60%)
Marvanol® SCO (50%)
Marvanol® Scour FRM
Marvanol® Solvent Scour 34
Marvanol® SPO (60%)
Marvantex BS
Marvantex GS
Matexil AA-NS
Matexil BA-PK
Matexil Binder BD
Matexil CA DPL
Matexil CA MN
Matexil DA N
Matexil FA MIV
Matexil FA N
Matexil FA SN Liq.
Matexil LA NS
Matexil PA Liq.
Matexil PA SNX Liq.
Matexil WA HS
Matexil WA KBN
Maxitol No. 10
Mayphos 45
Maypon 4C
May-Tein C
May-Tein CT
May-Tein SK
Mazawet® DOSS 70
Mazon® 21
Mazon® 60T
Mazon® 70
Mazon® 85
Medialan Brands
Melatex AS-80
Melioran 118
Meliorian 118
Merce Assist ADB
Mercerol NL
Mercerol QW
Mercerol SM
Merpol® C
Mersitol 2434 AP
Mersolat H 30
Mersolat H 40
Mersolat H 68
Mersolat H 76
Mersolat H 95
Mersolat W 40
Mersolat W 68
Mersolat W 76
Mersolat W 93
Metolat FC 514
Metolat FC 515
Metolat FC 530
Metolat LA 571
Metolat LA 573
Metolat TH 75
Migregal 2N
Miranate® B
Miranate® LEC
Mitin® FF High Conc.
Mona NF-10
Mona NF-15
Mona NF-25
Monafax 057
Monafax 060
Monafax 785
Monafax 786
Monafax 794
Monafax 831
Monafax 872
Monafax 939
Monafax 1214
Monafax 1293
Monamate CPA-40
Monamate CPA-100
Monamate LA-100
Monamate OPA-30
Monamate OPA-100
Monamulse 947
Monatrope 1250
Monatrope 1296
Monawet 1240
Monawet MB-45
Monawet MB-100
Monawet MM-80
Monawet MO-65-150
Monawet MO-70
Monawet MO-70-150
Monawet MO-70E
Monawet MO-70 PEG
Monawet MO-70R
Monawet MO-70S
Monawet MO-75E
Monawet MO-84R2W
Monawet MO-85P
Monawet MT-70
Monawet MT-70E
Monawet MT-80H2W
Monawet SNO-35
Monawet TD-30
Monogen
Monopole Oil
Monosulf
Montaline 1054
Montaline 9575 M
Montaline RH
Montapol CST
Montopol CST
Montosol IL-13
Montosol IQ-15
Montosol PB-25
Montosol PF-16, -18
Montosol PG-12
Montosol PL-14
Montosol PQ-11, -15
Montosol TQ-11
Montovol GJ-12
Montovol GL-13
Montovol RF-10
Montovol RF-11
Montovol RJ-13
Montovol RL-10
Morwet® B
Morwet® DB
Morwet® EFW
Morwet® IP
Morwet® M
Multinol C
Nacconol® 35SL
Nacconol® 40G
Nacconol® 90G
Na Cumene Sulfonate 40, Sulfonate Powd.
Nansa® 1042
Nansa® 1042/P
Nansa® AS 40
Nansa® BMC
Nansa® BXS
Nansa® EVM50
Nansa® EVM62/H
Nansa® EVM70
Nansa® EVM70/B
Nansa® EVM70/E
Nansa® HS40/S
Nansa® HS80-AU
Nansa® HS80P
Nansa® HS80S
Nansa® HS85
Nansa® HS85S
Nansa® HSA/L
Nansa® LSS38/A
Nansa® MS45
Nansa® S40/S
Nansa® SB 30
Nansa® SB62
Nansa® SBA
Nansa® SL 30
Nansa® SS 30
Nansa® SS 50
Nansa® SS 55
Nansa® SS 60
Nansa® SSA
Nansa® SSAL
Nansa® SS A/S
Nansa® TDB
Nansa® TS 50
Nansa® TS 60
Nansa® UCA/S, UCP/S
Nansa® YS94
Natrex J 3
Natrex SO
Naxel AAS-40S
Naxel AAS-45S
Naxel AAS-60S
Naxel AAS-98S
Naxel AAS-Special 3
Naxolate WA-97
Naxolate WAG
Naxolate WA Special
Naxonac 510
Naxonac 600
Naxonac 610
Naxonac 690-70
Neatsan D
Nekal® BX Conc. Paste
Nekal® BX Dry
Neodol® 25-3A
Neodol® 25-3S
Neopelex FS
Neopelex No. 6, No. 25, No. 6F Powder, F-25, F-65
Neoscoa CM-40
Neoscoa CM-57
Neoscoa ED-201C
Neoscoa OT-80E
Neospinol 264
Newcol 210
Newcol 261A, 271A
Newcol 290K, 290M, 291PG
Newcol 560SF
Newcol 560SN
Newcol 707SF
Newcol 710, 714, 723
Newcol 861S
Newcol 1305SN, 1310SN
Niaproof® Anionic Surfactant 4
Niaproof® Anionic Surfactant 7
Niaproof® Anionic Surfactant 08
Nikkol CCK-40
Nikkol CCN-40
Nikkol CMT-30
Nikkol DDP-10
Nikkol DDP-2
Nikkol DDP-4
Nikkol DDP-6
Nikkol DDP-8
Nikkol DLP-10
Nikkol ECT-3NEX, ECTD-3NEX
Nikkol ECTD-6NEX
Nikkol LSA
Nikkol OTP-100S
Nikkol SCS
Nikkol SGC-80N
Nikkol SLS
Nikkol SMT
Nikkol TDP-2
Nikkol TDP-4
Nikkol TDP-6
Nikkol TDP-8
Nikkol TDP-10
Ninate® 401
Ninate® 411
Ninate® DS 70

ANIONIC (CONT'D.)
Ninate® PA
Nissan Diapon K
Nissan Diapon T
Nissan Diapon TO
Nissan Newrex
Nissan Ohsen A
Nissan Persoft EK
Nissan Persoft SK
Nissan Rapisol B-30, B-80, C-70
Nissan Soft Osen 550A
Nissan Sunalpha T
Nissan Sunamide C-3
Nissan Sunamide CF-3, CF-10
Nissan Sunbase, Powder
Nissan Trax H-45
Nissan Trax K-40
Nissan Trax K-300
Nissan Trax N-300
Nonasol 3922
Nonasol LD-50
Nonasol LD-51
Nonasol MBS
Nonasol N4AS
Nonasol N4SS
Nonasol SNS-30
Nopco® 1186A
Nopco® 1471
Nopco® 2031
Nopco® 2272-R
Nopcocastor
Nopcocastor L
Nopcosperse AD-6
Nopcosulf CA-60, -70
Nopcosulf TA-30
Nopcosulf TA-45V
Nopcosurf CA
Norfox® 40
Norfox® 85
Norfox® 90
Norfox® 1101
Norfox® 1115
Norfox® ALES-60
Norfox® ALPHA XL
Norfox® ALS
Norfox® Anionic 27
Norfox® Coco Powder
Norfox® KO
Norfox® LAS-99
Norfox® MLS
Norfox® Oleic Flakes
Norfox® PE-600
Norfox® PE-LF
Norfox® PEA-N
Norfox® PEW
Norfox® SEHS
Norfox® SLES-02
Norfox® SLES-03
Norfox® SLES-30
Norfox® SLES-60
Norfox® SLS
Norfox® SXS40, SXS96
Norfox® T-60
Norfox® TB Granules
Norfox® Vertex Flakes
Nutrapon AL 2
Nutrapon AL 30
Nutrapon AL 60
Nutrapon ES-60 3568
Nutrapon HA 3841
Nutrapon KPC 0156
Nutrapon RS 1147
Nutrapon TD 3792
Nutrapon TLS-500
Nutrapon W 1367
Nutrapon WAC 3005
Nutrapon WAQ
Nutrol S-60 5350
Nutrol SXS 5418
Nuva F, FH
Nylomine Assistant DN
Nysist
Nysist LSO
Octaron PS 80
Octosol 449
Octosol 496
Octosol A-1
Octosol A-18
Octosol A-18-A
Octosol ALS-28
Octosol HA-80
Octosol IB-45
Octosol SLS
Octosol SLS-1
Octosol TH-40
Octowet 40
Octowet 55
Octowet 60
Octowet 60-I
Octowet 65
Octowet 70
Octowet 70A
Octowet 70BC
Octowet 70PG
Octowet 75
Octowet 75E
Oleine D
Olitex 75
Orapret WTNB 25
Orgozon CC 1118
Orgozon Conc. 0680
Osimol Grunau 109
Osimol Grunau 110
Osimol Grunau DP
Osimol Grunau SF
Amlev LD
Lankropol® ODS/LS
Paracol OP
Pationic ISL
Patogen 393
Patogen P-10 Acid
Patogen SME
Pat-Wet SP
Pat-Wet SW
Pelex NBL, NB Paste
Pelex OT-P
Pelex SS-H
Pelex SS-L
Pelex TA
Peltex
Penetral NA 20
Penetron OT-30
Pentine 1185 5432
Pentine Acid 5431
Pentrone S127
Peramit MLN
Perlankrol® ACM2
Perlankrol® ATL40
Perlankrol® DAF25
Perlankrol® DGS
Perlankrol® DSA
Perlankrol® EAD60
Perlankrol® EP12
Perlankrol® EP24
Perlankrol® EP36
Perlankrol® ESD
Perlankrol® ESD60
Perlankrol® ESK29
Perlankrol® ESS25
Perlankrol® FB25
Perlankrol® FD63
Perlankrol® FF
Perlankrol® FN65
Perlankrol® FT58
Perlankrol® FV70
Perlankrol® FX35
Perlankrol® O
Perlankrol® PA Conc.
Perlankrol® RN75
Perlankrol® SN
Peronal MTB
Petro® 11
Petro® 22
Petro® P
Petro® S
Petro® ULF
Petro® WP
Petromix®
Petronate® 25C, 25H
Petronate® Basic
Petronate® CR
Petronate® HL
Petronate® K
Petronate® L
Petronate® Neutral (50-S)
Petronate® S
Petrostep 420
Petrostep 465
Petrostep HMW
Petrostep LMC
Petrostep MMW
Petrosul® H-50
Petrosul® H-60
Petrosul® H-70
Petrosul® HM-62, HM-70
Petrosul® L-60
Petrosul® M-50
Petrosul® M-60
Petrosul® M-70
Petrowet® R
Phenyl Sulphonate CA, CAL
Phenyl Sulphonate HFCA
Phenyl Sulphonate HSR
Phosfac 1004, 1006, 1044, 1044FA, 1066, 1066FA, 1068FA
Phosfac 5500
Phosfac 5508
Phosfac 5513
Phosfac 5520
Phosfac 8000
Phosfac 8608
Phosfac 9604
Phosfac 9609
Phosfetal 601
Phospholan® AD-1
Phospholan® ALF5
Phospholan® ALF16
Phospholan® BH14
Phospholan® KPE4
Phospholan® PDB3
Phospholan® PHB14
Phospholan® PNP9
Phospholan® PRP5
Phospholan® PSP6
Pinamine A
Plysurf A207H
Plysurf A208B
Plysurf A208S
Plysurf A210G
Plysurf A212C
Plysurf A215C
Plysurf A216B
Plysurf A217E
Plysurf A219B
Plysurf AL
Polefine 510N
Polirol 4, 6
Polirol 10
Polirol DS
Polirol LS
Polirol SE 301
Polirol TR/LNA
Polystep® A-4
Polystep® A-7
Polystep® A-11
Polystep® A-15
Polystep® A-15-30K
Polystep® A-16
Polystep® A-16-22
Polystep® A-17
Polystep® A-18
Polystep® B-1
Polystep® B-3
Polystep® B-5
Polystep® B-7
Polystep® B-11
Polystep® B-12
Polystep® B-19
Polystep® B-20
Polystep® B-22
Polystep® B-23
Polystep® B-24
Polystep® B-25
Polystep® B-27
Polystep® B-29
Polystep® B-LCP
Polystep® C-OP3S
Polystep® CM 4 S
Polystep® PN 209
Polysurf A212E
Poly-Tergent® 2A1 Acid
Poly-Tergent® 2A1-L
Poly-Tergent® 2EP
Poly-Tergent® 3B2
Poly-Tergent® 3B2 Acid
Poly-Tergent® 4C3
Poly-Tergent® CS-1
Polywet AX-7
Polywet KX-3
Polywet ND-1
Polywet ND-2
Polywet Z1766
Prechem 90
Prechem 120
Primasol® FP
Primasol® NB-NF

ANIONIC (CONT'D.)
Primasol® NF
Progalan X-13
Progasol 40
Protowet 5171
Protowet C
Protowet D-75
Protowet MB
Protowet NBF
Protowet XL
Puxol CB-22
Puxol FB-11
Puxol XB-10
Pyronate® 40
Ralufon® EA 15-90
Ralufon® F
Ralufon® N
Ralufon® NAPE 14-90
Ram Polymer 110
Reax® 45A
Reax® 45DA
Reax® 45DTC
Reax® 45L
Reax® 45T
Redicote AE 80
Regitant S-78
Reserve Salt Flake
Rewocoros RAB 90
Rewoderm® S 1333
Rewoderm® S 1333 P
Rewomat B 2003
Rewomat TMS
Rewophat E 1027
Rewophat EAK 8190
Rewophat NP 90
Rewophat OP 80
Rewophat TD 40
Rewophat TD 70
Rewopol® 15/L
Rewopol® AL 3
Rewopol® B 1003
Rewopol® B 2003
Rewopol® CHT 12
Rewopol® CL 30
Rewopol® CT 65
Rewopol® CTN
Rewopol® HD 50 L
Rewopol® MLS 30
Rewopol® MLS 35
Rewopol® NEHS 40
Rewopol® NI 56
Rewopol® NL 2-28
Rewopol® NL 3
Rewopol® NL 3-28
Rewopol® NL 3-70
Rewopol® NLS 15 L
Rewopol® NLS 28
Rewopol® NLS 30 L
Rewopol® NLS 90
Rewopol® NOS 5
Rewopol® NOS 8
Rewopol® NOS 10
Rewopol® NOS 25
Rewopol® PGK 2000
Rewopol® S 1954
Rewopol® S 2311
Rewopol® SBC 212
Rewopol® SBC 212 G
Rewopol® SBDB 45
Rewopol® SBDC 40
Rewopol® SBDD 65
Rewopol® SBDO 70
Rewopol® SBDO 75
Rewopol® SBF 12
Rewopol® SBF 12 P
Rewopol® SBF 18
Rewopol® SBFA 30
Rewopol® SBFA 50
Rewopol® SBL 203
Rewopol® SBL 203 G, 203 P
Rewopol® SBLC, SBLC G
Rewopol® SBMB 80
Rewopol® SBR 12-Powder
Rewopol® SBV
Rewopol® SBZ
Rewopol® SCK 2040
Rewopol® SK 275
Rewopol® SMS 35
Rewopol® TLS 40
Rewopol® TLS 45/B
Rewopol® TMSF
Rewopol® TS 25
Rewopol® TS 35
Rewopol® TS 40 P
Rewopol® TS 100
Rewopol® TSK 30
Rewopol® TSSP 25
Rewopol® WS 11
Rewopol® WS 12
Reworyl® ACS 60
Reworyl® K
Reworyl® NCS 40
Reworyl® NKS 50
Reworyl® NKS 100
Reworyl® NTS 40
Reworyl® NXS 40
Reworyl® Sulfonic Acid K
Reworyl® TKS 90 F
Reworyl® TKS 90/L
Rexobase PCL
Rexobase XX
Rexoclean JA
Rexoclean PCL
Rexopene
Rexophos 25/67
Rexophos 25/97
Rexophos 35/98
Rexophos 4668
Rexophos BP-2
Rexophos JV 5015
Rexophos N25-38
Rexopon V
Rexoscour
Rexosolve 150
Rexowet 77
Rexowet CR
Rexowet GR
Rexowet MS
Rexowet NF, NFX
Rexowet RW
Rhodacal® 70/B
Rhodacal® 301-10
Rhodacal® 330
Rhodacal® 2283
Rhodacal® A-246L
Rhodacal® BA-77
Rhodacal® BX-78
Rhodacal® CA, 70%
Rhodacal® DDB-40
Rhodacal® DOV
Rhodacal® DS-4
Rhodacal® DS-10
Rhodacal® DSB
Rhodacal® IN
Rhodacal® IPAM
Rhodacal® LA Acid
Rhodacal® LDS-22
Rhodacal® N
Rhodacal® NK
Rhodacal® RM/77-D
Rhodacal® RM/210
Rhodafac® BG-510
Rhodafac® BP-769
Rhodafac® BX-660
Rhodafac® CB
Rhodafac® GB-520
Rhodafac® HA-70
Rhodafac® L3-15A
Rhodafac® L3-64A
Rhodafac® L4-27A
Rhodafac® L6-36A
Rhodafac® LO-529
Rhodafac® MB
Rhodafac® MC-470
Rhodafac® MD-12-116
Rhodafac® PA-15
Rhodafac® PA-17
Rhodafac® PA-19
Rhodafac® PA-23
Rhodafac® PA-35
Rhodafac® PE-510
Rhodafac® PS-17
Rhodafac® PS-19
Rhodafac® PS-23
Rhodafac® PV-27
Rhodafac® R5-09/S
Rhodafac® R9-47A
Rhodafac® RA-600
Rhodafac® RB-400
Rhodafac® RD-510
Rhodafac® RE-410
Rhodafac® RE-610
Rhodafac® RE-960
Rhodafac® RK-500
Rhodafac® RM-410
Rhodafac® RM-510
Rhodafac® RM-710
Rhodafac® RP-710
Rhodafac® RS-410
Rhodafac® RS-610
Rhodafac® RS-710
Rhodapex® AB-20
Rhodapex® CD-128
Rhodapex® CO-433
Rhodapex® CO-436
Rhodapex® ES
Rhodapex® EST-30
Rhodapex® ESY
Rhodapex® F-85/SD
Rhodapex® F-775/SD
Rhodapex® MA360
Rhodapon® BOS
Rhodapon® CAV
Rhodapon® EC111
Rhodapon® L-22, L-22/C
Rhodapon® LCP
Rhodapon® LM
Rhodapon® LSB, LSB/CT
Rhodapon® LT-6
Rhodapon® OLS
Rhodapon® OS
Rhodapon® SB
Rhodapon® SM Special
Rolpon 24/230
Rolpon 24/270
Rolpon 24/330
Rolpon 24/330 N
Rolpon 24/370
Rolpon C 200
Rolpon LSX
Rolpon SE 138
Rueterg 60-T
Rueterg 97-G
Rueterg 97-S
Rueterg 97-T
Rueterg IPA-HP
Rueterg SA
Runox 1000C
Sactol 2 OS 2
Sactol 2 OS 28
Sactol 2 OT
Sactol 2 S 3
Sactol 2 T
Sandet 60
Sandet ALH
Sandopan® B
Sandopan® CBH Paste
Sandopan® CBN
Sandopan® DTC
Sandopan® DTC-100
Sandopan® DTC-Acid
Sandopan® DTC Linear P
Sandopan® DTC Linear P Acid
Sandopan® JA-36
Sandopan® KD
Sandopan® KD (U) Liquid
Sandopan® KST
Sandopan® LA-8
Sandopan® MA-18
Sandopan® RS-8
Sandozin® AM
Sandozin® AMP
Sandozin® NE
Sandozol KB
Sandoz Phosphorester 510
Sandoz Sulfate 216
Sandoz Sulfate 219
Sandoz Sulfate 830
Sandoz Sulfate 1030
Sandoz Sulfate W2-30
Sandoz Sulfonate 2A1
Sandoz Sulfonate 3B2
Sandoz Sulfonate AAS 35S
Sandoz Sulfonate AAS 40FS
Sandoz Sulfonate AAS 45S
Sandoz Sulfonate AAS 50MS
Sandoz Sulfonate AAS 60S
Sandoz Sulfonate AAS 70S
Sandoz Sulfonate AAS 75S
Sandoz Sulfonate AAS 90
Sandoz Sulfonate AAS 98S
Sandoz Sulfonate AAS-Special 3
Sanmorin AM
Sanmorin OT 70
Sansilic 11

ANIONIC (CONT'D.)
Sanstat 230
Sarkosyl® L
Savosellig REAC 4
Sawaclean AO
Sawaclean AOL
Schercomul K
Schercophos L
Schercophos NP-6
Schercophos NP-9
Schercopol CMS-Na
Schercopol DOS-70
Schercopol DOS-PG-85
Schercopol OMS-Na
Schercopol RMS-Na
Schercopol UMS-Na
Schercowet DOS-70
Scripset 500
Scripset 700
Scripset 720
Secolat
Secosol® AL/959
Secosol® ALL40
Secosol® AS
Secosol® DOS 70
Secosol® EA/40
Secosyl
Sellig 10 Mode
Sellig 13
Sellig Dispersant FPZ
Sellig JN 25
Sellig MRC
Selligon SP
Sellogen 641
Sellogen DFL
Sellogen HR
Sellogen HR-90
Sellogen NS-50
Sellogen W
Sellogen WL Acid
Sellogen WL Liq.
Ser-Ad FA 153
Serdet DCK 3/70
Serdet DCK 30
Serdet DCN 30
Serdet DFK 40
Serdet DJK 30
Serdet DM
Serdet DNK 30
Serdet DPK 3/70
Serdet DPK 30
Serdet DSK 40
Serdet Perle Conc.
Sermul EA 30
Sermul EA 54
Sermul EA 88
Sermul EA 129
Sermul EA 136
Sermul EA 139
Sermul EA 146
Sermul EA 150
Sermul EA 151
Sermul EA 152
Sermul EA 176
Sermul EA 188
Sermul EA 205
Sermul EA 211
Sermul EA 214
Sermul EA 221
Sermul EA 224
Sermul EA 242
Sermul EA 370
Sermul EN 7
Servoxyl VLA 2170
Servoxyl VLB 1123
Servoxyl VPGZ 7/100
Servoxyl VPNZ 10/100
Servoxyl VPQZ 9/100
Servoxyl VPTZ 3/100
Servoxyl VPTZ 100
Servoxyl VPYZ 500
Serwet WH 170
Setacin 103 Spezial
Setacin F Spezial Paste
Setacin M
Silvatol® AS Conc.
Silvatol® PBS
Silvatol® SO
Simulsol G 101
Sinonate 263B
Sinonate 263M
Sinonate 290M
Sinonate 290MH
Sinonate 960SF
Sinonate 960SN
Sodium Octyl Sulfate Powd
Sofnon GF-2, GF-5
Sofnon SP-1000
Soft Detergent 60
Soft Detergent 95
Soitem 5 C/70
Soitem 8 FL/N
Soitem 13 FL/N
Soitem B
Soitem SC/70
Sole-Onic CDS
Sole-Terge TS-2-S
Solocod G
Solricin® 135
Solumin F Range
Solutene TER
Soprofor 3D33
Soprofor FL
Soprofor NPF/10
Soprofor PA/17, PA/19
Soprofor PS/17, PS/19, PS/21
Soprophor® 4D384
Soprophor® FLK
Sorbit P
Sorpol 2495 G
Sorpol 5037
Sorpol 5039
Sorpol 5060
Sorpol 8070
Sovatex C Series
Spanscour EFS
Sperse Polymer IV
Sponto® 168-D
Sponto® 169-T
Sponto® 207
Sponto® AK30-23
Stabilisal S Liq.
Stabilizer CB
Stabilizer CS
Stabilizer SIFA
Standapol® 1610
Standapol® A
Standapol® ES-50
Standapol® ES-250
Standapol® ES-350
Standapol® LF
Standapol® WAQ-LCX
Standapon 4149 Conc.
Stanol 212F
Stantex 322
Stantex MOR
Stantex MOR Special
Stantex PENE 20
Stantex PENE 40-DF
Stantex T-14 DF
Steol® 4N
Steol® CA-460
Steol® COS 433
Steol® CS-460
Steol® KA-460
Steol® KS-460
Steol® OS 28
Stepanflo 20
Stepanflo 30
Stepanflo 40
Stepanflo 50
Stepanflote® 24
Stepanflote® 85L
Stepanflote® 97A
Stepanform® 1050
Stepanform® 1440
Stepanform® 1750
Stepanform® 1850
Stepanform® 2160
Stepanform® 3040
Stepanform® 5012
Stepanform® 5040
Stepanol® AEG
Stepanol® AM
Stepanol® AM-V
Stepanol® DEA
Stepanol® DFS
Stepanol® ME Dry
Stepanol® MG
Stepanol® RS
Stepanol® SPT
Stepanol® WA-100
Stepanol® WAC
Stepanol® WA Extra
Stepanol® WA Paste
Stepanol® WAQ
Stepanol® WA Special
Stepanol® WAT
Stepantan® AS-40
Stepantan® CG
Stepantan® DS-40
Stepantan® DT-60
Stepantan® H-100
Stepantan® HP-90
Stepfac® 8170
Stepfac® 8171
Stepfac® 8172
Stepfac® 8173
Stepfac® PN 10
Step-Flow 41
Step-Flow 42
Step-Flow 61
Steposol® CA-207
Stepsperse® DF-200
Stepsperse® DF-300
Stepwet® DF-60
Stepwet® DF-90
Strodex® MO-100
Strodex® MOK-70
Strodex® MR-100
Strodex® MRK-98
Strodex® P-100
Strodex® PK-90
Strodex® PSK-28
Strodex® SE-100
Strodex® SEK-50
Strodex® Super V-8
Sufatol LX/B
Sulfetal 4069
Sulfetal 4105
Sulfetal 4187
Sulfetal C 38
Sulfetal C 90
Sulfetal CJOT 38
Sulfetal FA 40
Sulfetal KT 400
Sulfetal MG 30
Sulfetal TC 50
Sulfochem 436
Sulfochem 437
Sulfochem 438
Sulfochem ALS
Sulfochem B-221
Sulfochem B-221OP
Sulfochem ES-2
Sulfochem ES-70
Sulfochem K
Sulfochem MG
Sulfochem SLS
Sulfochem TLS
Sul-fon-ate AA-9
Sul-fon-ate AA-10
Sul-fon-ate OA-5
Sul-fon-ate OE-500
Sulfonated Castor Oil 50%
Sulfonated Castor Oil 75%
Sulfonated Castor Oil GTO
Sulfonated Red Oil
Sulfonic 800
Sulfonic 864
Sulfonic Acid LS
Sulfopon 101 Spez
Sulfopon 101, 101 Special
Sulfopon 101/POL
Sulfopon 102
Sulfopon K35
Sulfopon KT 115
Sulfopon KT 115-50
Sulfopon O
Sulfopon O 680
Sulfopon P-40
Sulfopon T 55
Sulfopon T Powd
Sulfopon WA 1
Sulfopon WAQ LCX
Sulfopon WAQ Special
Sulfotex 110
Sulfotex 130
Sulfotex 6040
Sulfotex A
Sulfotex DOS
Sulfotex LCX
Sulfotex LMS-E
Sulfotex OA
Sulfotex OT
Sulfotex PAI
Sulfotex PAI-S
Sulfotex PAW
Sulfotex RAW

ANIONIC (CONT'D.)
Sulfotex RIF
Sulfotex SAL, SAT
Sulfotex SXS-40
Sulfotex UBL-100
Sulfotex WA
Sulfotex WAQ-LCX
Sulfotex WAT
Sulframin 40
Sulframin 40DA
Sulframin 40RA
Sulframin 40T
Sulframin 45
Sulframin 45LX
Sulframin 85
Sulframin 90
Sulframin 1230
Sulframin 1240, 1245
Sulframin 1250, 1260
Sulframin 1255
Sulframin 1288
Sulframin 1298
Sulframin 1388
Sulframin Acide B
Sulframin Acide TPB
Sulframin AOS
Sulframin LX
Sulphonic Acid LS
Sunnol 710 H
Sunnol CM-1470
Sunnol DL-1430
Sunnol DOS
Sunnol DP-2630
Sunnol LM-1130
Sunnol LST
Sunnol NES
Sunnol NP-2030
Sunsolt RZ-2, -6
Supermontaline SLT65
Supragil® GN
Supragil® MNS/90
Supragil® NK
Supragil® NS/90
Supragil® WP
Surfagene FAD 105
Surfagene FAZ 109
Surfagene FAZ 109 NV
Surfagene FDD 402
Surfax 150
Surfax 215
Surfax 218
Surfax 220
Surfax 250
Surfax 345
Surfax 495
Surfax 500
Surfax 502
Surfax 536
Surfax 539
Surfax 550
Surfax 560
Surfax 585
Surfax CN
Surfax MG
Surfax WO
Surfine WLL
Surfine WNT Conc.
Surfine WNT Gel
Surfine WNT LC
Surfine WNT-LS
Surflo® 390
Surflo® CW3
Surflo® S30
Surflo® S32
Surflo® S40
Surflo® S302
Surflo® S362
Surflo® S375
Surflo® S398
Surflo® S412
Surflo® S568
Surflo® S586
Surflo® S970
Surflo® S1143
Surflo® S1720
Surflo® S1748
Surflo® S1815
Surflo® S2023
Surflo® S3902
Surmax® CS-504
Surmax® CS-515
Surmax® CS-521
Surmax® CS-522
Surmax® CS-555
Surmax® CS-586
Swanic 52L
Swanic CWT
Swanic T-51
Swascol 1P
Swascol 1PC
Swascol 3L
Swascol 4L
Sykanol DKM 45, 80
Syncrowax AW1-C
Synperonic 3S60A
Synperonic 3S70
Syntaryl Series
Syntergent TER-1
Synthrapol LA-DN
Synthrapol SP
Syntophos O6N
Syntopon 8 A
Syntopon 8 B
Syntopon 8 C
Syntopon 8 D
Syntopon S 493, 630, 1030
Tallates, K
Tanalev® 221
Tanapal® ME Acid
Tanapon NF-200
Tanapure® AC
Tanaterge FTD
Tanaterge PE-67
Tanawet® RCN
Tauranol I-78
Tauranol I-78-3
Tauranol I-78-6
Tauranol I-78/80
Tauranol I-78E, I-78E Flakes
Tauranol I-78 Flakes
Tauranol M-35
Tauranol ML
Tauranol MS
Tauranol T-Gel
Tauranol WS, WS Conc.
Tauranol WSP
T-Det® 25-3A
T-Det® 25-3S
Teepol CM 44
Teepol HB 7
Tegacid® Special VA
Tegin® C-63
Tegin® C-611
Tegin® DGS
Tegin® GO
Tegin® ISO
Tegin® M
Tegin® O NSE
Tegin® O Special
Tegin® O Special NSE
Tegin® P
Teginacid® Special SE
Tembind A 002
Temsperse S 001
Temsperse S 003
Tenax 2010
Tensagex DLM 627
Tensagex DLM 670
Tensagex DLM 927
Tensagex DLM 970
Tensagex DLS 670
Tensagex DLS 970
Tensagex EOC 628
Tensagex EOC 670
Tensianol 399 ISL
Tensianol 399 KS1
Tensianol 399 N1
Tensianol 399 SCI-L
Tensiofix LX Special
Tensopol A 79
Tensopol A 795
Tensopol ACL 79
Tensopol PCL 94
Tensopol S 30 LS
Tephal Grunau AB Conc.
Tephal Grunau FL Conc.
Terg-A-Zyme®
Tergenol 1122
Tergenol S Liq.
Tergenol Slurry
Teric 305
Teric 306
Texapon 842
Texapon 890
Texapon 1030, 1090
Texapon EA-K
Texapon F
Texapon F35
Texapon GYP 1
Texapon GYP 2
Texapon GYP 3
Texapon GYP 4
Texapon HD-L 2
Texapon K-12
Texapon K-1296
Texapon K-12 Granules
Texapon K-12 Needles
Texapon LLS
Texapon LS 35
Texapon LS Highly Conc.
Texapon LS Highly Conc. Needles
Texapon LT-327
Texapon MGLS
Texapon N 25
Texapon N 40
Texapon N 70
Texapon N 70 LS
Texapon N 103
Texapon NSE
Texapon NSO
Texapon OT Highly Conc. Needles
Texapon P
Texapon PLT-227
Texapon PN-235
Texapon PN-254
Texapon PNA
Texapon PNA-127
Texapon SCO
Texapon SP 60 N
Texapon T 35
Texapon T 42
Texapon VHC Needles
Texapon VHC Powd
Texapon Z
Texapon ZHC Needles
Texin 128
Texin DOS 75
Texin SB 3
Texnol R 5
Texo 227
Texo 1060
Textol 80 (L)
Tex-Wet 1001
Tex-Wet 1002
Tex-Wet 1004
Tex-Wet 1010
Tex-Wet 1034
Tex-Wet 1070
Tex-Wet 1103
Tex-Wet 1104
Tex-Wet 1140
Tex-Wet 1143
Tex-Wet 1158
Tex-Wet 1197
Thorowet G-40 3230
Thorowet G-60 3142
Thorowet G-60 SS 0336
Thorowet G-75 1060
Tinegal® Resist N
Titan Castor No. 75
Titanterge DBS Conc.
Titapene Conc.
Titazole HTX
T-Mulz® 66H
T-Mulz® 426
T-Mulz® 565
T-Mulz® 596
T-Mulz® 598
T-Mulz® 734-2
T-Mulz® 800
T-Mulz® 844
T-Mulz® 1158
Toho Salt A-5
Toho Salt A-10
Totablan
Toxanon GR-31A
Toxanon XK-30
Toximul® 804
Toximul® 811
Toximul® 8170
Triton® 770 Conc.
Triton® DF-20
Triton® GR-5M
Triton® GR-7M
Triton® H-55
Triton® H-66
Triton® QS-44

ANIONIC (CONT'D.)
Triton® W-30 Conc.
Triton® X-180
Triton® X-185
Triton® X-190
Triton® X-193
Triton® X-200
Triton® X-301
Triumphnetzer ZSG
Triumphnetzer ZSN
Troysol LAC
Tryfac® 610-K
Tryfac® 910-K
Tryfac® 5365
Tryfac® 5552
Tryfac® 5553
Tryfac® 5554
Tryfac® 5555
Tryfac® 5556
Tryfac® 5557
Tryfac® 5559
Tryfac® 5560
Tryfac® 5571
Tryfac® 5573
Tryfac® 5576
Trylon® 6702
Trylon® 6848
TSA-70
T-Soft® SA-97
Turkey Red Oil 100%
Tween® 20 SD
Twitchell 6805 Oil
Twitchell 6808 Oil
Tylose C, CB Series
Udet® 950
Ufablend DC
Ufacid K
Ufacid TPB
Ufapol
Ufarol Am 30
Ufarol Am 70
Ufarol Na-30
Ufarol TA-40
Ufaryl DB80
Ufaryl DL80 CW
Ufaryl DL85
Ufaryl DL90
Ufasan 35
Ufasan 50
Ufasan 60A
Ufasan 65
Ufasan TEA
Ultra Blend 60, 100
Ultra Blend P-280
Ultra NCS Liquid
Ultraphos Series
Ultra Sulfate AE-3
Ultra SXS Liq., Powd
Ultravon® SFN
Ungerol AM3-60
Ungerol AM3-75
Ungerol CG27
Ungerol LES 2-28
Ungerol LES 2-70
Ungerol LES 3-28
Ungerol LES 3-54
Ungerol LES 3-70
Ungerol LS, LSN
Ungerol N2-28
Ungerol N2-70
Ungerol N3-28
Ungerol N3-70
Unidri M-40
Unidri M-60
Unidri M-75
Uniperol® W, W Flakes
Valdet CC
Valscour SWR-2
Value MS-1
Valwet 092
Vannox PW
Vanseal 35
Vanseal CS
Vanseal LS
Vanseal NACS-30
Vanseal NALS-30
Vanseal NALS-95
Vanseal OS
Varamide® A-12
Varamide® A-83
Varsulf® NOS-25
Varsulf® S-1333
Varsulf® SBFA-30
Varsulf® SBL-203
Verv®
Victawet® 35B
Victawet® 58B
Vikosperse KDS
Vista C-550, -560
Vista SA-597
Wayfos A
Wayfos D-10N
Wayfos M-60
Wayfos M-100
Wayhib® S
Wayplex NTP-S
Westvaco® 1480
Westvaco® 1483
Westvaco Diacid® 1575
Westvaco Diacid® H-240
Westvaco® M-30
Westvaco® M-40
Westvaco® M-70
Westvaco® Resin 90
Westvaco® Resin 95
Westvaco® Resin 790
Westvaco® Resin 795
Westvaco® Resin 895
Wetting Agent 611, CG 16
Wetting Agent FCGB
Wettol® EM 1
Wettol® EM 3
Wettol® NT 1
Wibarco
Witbreak 770
Witbreak 1390
Witbreak RTC-323, -1315, -1316
Witco® 94H Acid
Witco® 97H Acid
Witco® 1298
Witco® 1298 H
Witco® 1298 HA
Witco® Acid B
Witco® Acid TPB
Witco® CHB Acid
Witco® D51-29
Witco® DTA-350
Witco® TX Acid
Witcodet CWC
Witcolate 1050
Witcolate 1247H
Witcolate 1259
Witcolate 1276
Witcolate 1390
Witcolate 3220
Witcolate 3570
Witcolate 6400
Witcolate 6430
Witcolate 6431
Witcolate 6434
Witcolate 6450
Witcolate 6453
Witcolate 6455
Witcolate 6462
Witcolate 6465
Witcolate 7031
Witcolate 7093
Witcolate 7103
Witcolate A
Witcolate A Powder
Witcolate AE
Witcolate AE-3
Witcolate AE3S
Witcolate AM
Witcolate C
Witcolate D-510
Witcolate D51-51
Witcolate D51-52
Witcolate ES-2
Witcolate ES-3
Witcolate LCP
Witcolate LES-60A
Witcolate LES-60C
Witcolate NH
Witcolate OME
Witcolate S
Witcolate S1285C
Witcolate S1300C
Witcolate SE
Witcolate SE-5
Witcolate T
Witcolate TLS-500
Witcolate WAC
Witcolate WAC-GL
Witcolate WAC-LA
Witcomul 1054
Witcomul 1317
Witcomul 3126
Witconate 30DS
Witconate 45 Liq.
Witconate 45BX
Witconate 45DS
Witconate 45LX
Witconate 60B
Witconate 60L
Witconate 60T
Witconate 79S
Witconate 90F
Witconate 90F H
Witconate 93S
Witconate 605A
Witconate 605AC
Witconate 703
Witconate 1075X
Witconate 1240 Slurry
Witconate 1250 Slurry
Witconate 1260 Slurry
Witconate 1840X
Witconate 1850
Witconate 3203
Witconate AO 24
Witconate AOK
Witconate AOS
Witconate C50H
Witconate D51-51
Witconate DS
Witconate DS Dense
Witconate K
Witconate K Dense
Witconate KX
Witconate LX F
Witconate LX Powd
Witconate LXH
Witconate NAS-8
Witconate NCS
Witconate NIS
Witconate P-1052N
Witconate P10-59
Witconate SCS 45%
Witconate SCS 93%
Witconate SE-5
Witconate SK
Witconate SXS 40%
Witconate SXS 90%
Witconate TAB
Witconate TDB
Witconate TX Acid
Witconate YLA
Witcor PC100
Witcor PC200
Witcor PC205
Witcor PC210
Witcor SI-3065
X78-2
Zetesol 856
Zetesol 856 D
Zetesol 856 DT
Zetesol 856 T
Zetesol 2056
Zetesol AP
Zetesol NL
Zetesol SE 35
Zohar 60 SD (L)
Zohar 70 SD (L)
Zohar Automat SD
Zohar Export SD
Zohar HAS 470
Zohar KAL SD
Zoharlab
Zoharpon ETA 27
Zoharpon ETA 70
Zoharpon ETA 270 (OXO)
Zoharpon ETA 271
Zoharpon ETA 273
Zoharpon ETA 603
Zoharpon ETA 700 (OXO)
Zoharpon ETA 703
Zoharpon K
Zoharpon LAD
Zoharpon LAEA 253
Zoharpon LMT42
Zoharpon SM
Zonyl® FSA
Zonyl® FSE
Zonyl® FSJ
Zonyl® FSP
Zonyl® NF
Zonyl® RP
Zonyl® TBS

Zonyl® UR

ANIONIC/ CATIONIC:

Alkamide® 210 CGN
Aquamul Series
Cenegen® B
Chembetaine BW
Depsodye HN
Unisol LAC GN

ANIONIC/ NONIONIC:

Abluhide DS
Abluhide I
Ablumul AG-306
Ablumul AG-420
Ablumul AG-910
Ablumul AG-GL
Ablumul AG-H
Ablumul AG-KP3
Ablumul AG-KTM
Ablumul AG-L
Ablumul AG-MBX
Ablumul AG-SB
Ablumul AG-WP
Ablumul AG-WPS
Ablumul M
Ablumul MI
Ablumul OP
Ablumul S2S
Ablumul S Series
Ablumul T
Ablumul T2
Abluton 700
Abluton A
Abluton BT
Abluton CMN
Abluton CTP
Abluton RT430
Acrylic Resin AS
AD-716
Adsee® 2141
Agent 2A-2S
Agrilan® A
Agrilan® AC
Agrilan® AEC189
Agrilan® AEC200
Agrilan® AEC211
Agrilan® AEC310
Agrilan® BA
Agrilan® BM
Agrilan® F460
Agrilan® MC-90
Akypo OP 80
Akypo OP 115
Akypo OP 190
Akypo RLM 45
Akypo RLM 100
Akypo RLM 160
Akypogene VSM
Akypogene VSM-N
Alkamide® 2104
Alkamide® 2112
Alkamide® 2204
Alkamide® 2204A
Alkamide® WRS 1-66
Alkawet® N
Alphenate PE Extra
Arbreak Series
Ardril DMD
Ardril DME
Armix 309
Armul 22
Armul 33
Armul 34
Armul 44
Armul 55
Armul 66
Armul 88
Armul 100, 101, 102
Armul 214
Armul 215
Armul 3260
Armul 5830
Arodet MKD
Arodet N-100 Special
Arodet WIN
Arolube MIT-1
Arosolve 570-HF
Arosolve 570 Special
Arosolve B-950
Arosolve CON
Arosolve DBM
Arosolve IWS
Arosolve MRC-A
Arosolve MRC-HF
Arosolve RCB
Arylan® DA 36
Arylan® HAL
Atlas L-801-LF
Atlasol 155
Atlox 3387
Atlox 3387 BM
Atlox 3400B
Atlox 3401
Atlox 3403F
Atlox 3404F
Atlox 3406F
Atlox 3409F
Atlox 3414F
Atlox 3422F
Atlox 3450F
Atlox 3453F
Atlox 3454F
Atlox 3455F
Atlox 4868B
Atlox 4880B
Atlox 4890B
Atlox 4899B
Atlox 4990B
Atlox 5330
Atolex LS/3
Atsurf 311
Aviscour 98/70
Aviscour 125
Aviscour HP50
Aviscour HQ50
Basophen® M
Basophen® NB-U
Berol 525
Berol 784
Berol 797
Berol 930
Berol 947
Berol 949
Beycostat 273 A
Beycostat 319 A
Beycostat 656 A
Beycostat 714 A
Beycostat DA
Beycostat LP 4 A
Beycostat LP 9 A
Beycostat NA
Beycostat QA
Bio-Soft® LD-95
Bio-Step Series
Burco CS-LF
Burcoterge DG-40
Calsuds 81 Conc.
Carding Oil A-4 Super
Carriant Series
Carsamide® 7644
Chimipon FC
Chimipon HD
Chimipon LD
Chimipon LDP
Citranox®
CNC Antifoam 495, 495-M
CNC PAL AN
Consolevel N
Cromul 1540
Delion 342
Delion 964
Delion F-200 Series
Delion F-500 Series
Demelan CB-10
Demelan CB-70
Demelan CM-95
Demulfer Series
Depsolube CP
Detergent Concentrate 840
Deteryl AG
Disperbyk®-181
Dodifoam Brands
Drynol E/20, E/30, E/40
Duoteric MB1, MB2
Eccoscour CB
Eganal LFI
Emcol® 150 T
Emcol® AC 62-36
Emcol® AD 27-21
Emcol® AG 4-48N
Emcol® AK 16-97N
Emcol® AK 18-72A
Emcol® AL 26-43H
Emcol® AL 26-43L
Emcol® H 400X
Emcol® N Series
Emid® 6521
Emid® 6529
Emid® 6533
Emkane HAD
Empimin® 3060
Empimin® II56
Emulsogen IT
Emulsogen T
Emulson AG 3020
Emulson AG 3080
Emulson AG 3490
Emulson AG 4020
Emulson AG 4080
Emulson AG 5190
Emulson AG 5590
Emulson AG 5790
Emulson AG 7760
Emulson AG/FLS
Emulson AG/FN
Esi-Terge HA-20
Esi-Terge L-75
Ethox 1345K
Finapal E
Flo-Mo® 1X, 2X
Flo-Mo® 8X
Flo-Mo® DEH, DEL
Foramousse D
Geronol AG-100/200 Series
Geronol AZ82
Geronol FF/4
Geronol FF/6
Geronol MS
Geronol RE/70
Geronol SC/120
Geronol SC/121
Geronol SN
Geronol V/087
Geronol V/497
Geropon® ACR/4
Geropon® K/65
Gran UP A-600
Gran UP AX-15
Gran UP CS-500
Gran UP CS-700F
Gran UP US-800
Hartomer JV 4091
Hartowet 5917
Hipochem B-3-M
Hipochem RPS
Hiposcour® ARG-2
Honol 405
Honol MGR
Hydroace Series
Imerol XN
Intravon® JET
Intravon® SO
Intravon® SOL
Intravon® SOL-N
Intravon® SOL-W
Kerasol 1398
Kieralon® B, B Highly Conc.
Kieralon® ED
Kieralon® KB
Kieralon® NB-ED
Kieralon® OL
Kilfoam
Lanapex JS Conc.
Lancare
Laurel PDW
Laurel SDW
Lenetol 416
Lenetol 527
Lenetol KWB
Leonil L
Leonil UN
Leophen® U
Levelox TYF-10
Liqui-Nox®
Lumorol 4153
Lumorol 4154
Lumorol 4192
Lumorol GG 65
Lumorol W 5058
Lumorol W 5157
Lutensit® AN 10
Lutensit® AN 30

ANIONIC/NONIONIC (CONT'D.)
Manro DB 30
Manro DB 98
Maphos® 15
Maphos® 18
Maphos® 30
Maphos® 33
Maphos® 41A
Maphos® 54
Maphos® 55
Maphos® 56
Maphos® 58
Maphos® 60A
Maphos® 66H
Maphos® 76
Maphos® 77
Maphos® 79
Maphos® 91
Maphos® 151
Maphos® 236
Maphos® 8135
Maphos® DT
Maphos® FDEO
Maphos® JA 60
Maphos® JM 51
Maphos® JM 71
Maphos® L 4
Maphos® L 13
Marlamid® KLP
Marlon® AFM 40, 40 N, 43, 50N
Marlon® AFO 40
Marlon® AFO 50
Marlon® AM 80
Marlon® PF 40
Marlophor® ND DEA Salt
Marlophor® ND NA-Salt
Marlopon® AMS 60
Marlowet® BIK
Marlowet® OFA
Marlowet® RA
Marlox® M 606/1
Marlox® M 606/2
Marvanscour® LF
Matexil Binder AS
Mazamide® 65
Mazamide® L-298
Mazamide® LM 20
Mazamide® SS 20
Mazamide® T 20
Mazon® DWD-100
Merinol LH-200
Migregal NC-2
Mindust Series
Minemal 350
Miracare® MS-1
Miracare® MS-2
Miracare® MS-4
Monalube 780
Monamid® 7-153 CS
Monamine 779
Monamine AA-100
Monamine AC-100
Monamine ACO-100
Monamine ADS-100
Monamine ALX-80SS
Monamine ALX-100 S
Monamine CF-100 M
Monamine I-76
Monamine LM-100
Monamine R8-26
Monamine T-100
Monamulse 653-C
Monamulse dL-1273
Monaterge 85
Monaterge 85 HF
Mulsifan ABN
Mulsifan K 326 Spezial
Mulsifan STK
Nansa® HAD
Nansa® MA30
Natrex D 3
Natrex DSP 213
Neoscoa 500C
Neoscoa GF 3C
Neoscoa PRA-8C
Neoscoa TH-102
Newkalgen 135R
Newkalgen 2360X1
Newkalgen 2720X75
Newkalgen 3000 A & B
Nitrene 11230
Nonipol T-20
Nonipol T-100
Nonipol TH
Norfox® DCO
Norfox® DC3
Norfox® Hercules Conc.
Norfox® Unimulse OW
Nutrapon B 1365
Paracol AL, AH
Paracol SV
Peroxal 36
Prechem 2000
Printac CN-230
Protowet 5218
Ralufon® CM 15-13
Rhodaterge® DCA
Rhodaterge® DCB-100
Rhodaterge® LD-50Q
Rhodaterge® LD-60
Rhodaterge® WHC-347
Runox Series
Sandopan® 2N
Sandopan® BFN
Sanimal 55HX, 55LX, 250C, K51
Sanimal L, M, H
Schercomid CDA
Schercomid CDO-Extra
Schercomid ODA
Schercomid SO-A
Schercomid SO-T
Schercomid TO-2
Schercopon 2WD
Sinocol L
Sinocol MBX
Sinol LDA-6
Sinol NPX
Sinol TWS
Sinopol 254A
Sofnon B-3
Sofnon TK-11
Soi Mul 130
Soi Mul 235
Soitem 101
Soitem 251
Soitem 258
Soitem 480
Soitem 520
Solvent-Scour 263
Sorpol 320
Sorpol 355
Sorpol 560
Sorpol 900A, 1200, 2020K, 3005X
Sorpol 3044
Sorpol H-770
Sorpol L-550
Sovatex DS/C5
Sponto® 101
Sponto® 102
Sponto® 140T
Sponto® 150 TH
Sponto® 150 TL
Sponto® 150T
Sponto® 203
Sponto® 221
Sponto® 232, 234
Sponto® 232T
Sponto® 234T
Sponto® 300T
Sponto® 305
Sponto® 500T
Sponto® 710T
Sponto® 712T
Sponto® 714T
Sponto® 723T
Sponto® 2174 H
Sponto® 2224T
Sponto® 4648-23A, 4648-23B
Sponto® AD4-10N
Sponto® AD6-39A
Sponto® AM2-07
Stabilizer AWN
Standapol® BW
Standapol® S
Stepanform® HP-95
Stepanform® HP-116
Stepantex® 130
Stepsperse® DF-100
Stepsperse® DF-400
Sufatol LS/3
Supersol ICS
Supersurf JM
Surfynol® CT-136
Tanabron W
Teepol GD7P
Teepol GD53
Tego® Foamex 1488
Tensiofix AS
Tensiofix B7416
Tensiofix B7438
Tensiofix B7453
Tensiofix B7504
Tensiofix B8426
Tensiofix B8427
Tensiofix BC222
Tensiofix BS
Tensiofix CG11
Tensiofix CG21
Tensiofix CS
Tensiofix EDS
Tensiofix EDS3
Texamine 84 (L)
Texapon SP 100
T-Mulz® 63
T-Mulz® 540
T-Mulz® 979
T-Mulz® AS-1151
T-Mulz® AS-1152
T-Mulz® AS-1153
T-Mulz® AS-1180
T-Mulz® O
T-Mulz® W
Toho Salt OK-70
Toho Salt TM-1
Toho Salt UF-350C
Toho Salt UF-650C
Toxanon 500
Toxanon AM Series
Toxanon P-8H, P-8L
Toxanon P-900
Toxanon SC-4
Toxanon V-1D
Toximul® 60 Series
Toximul® 360A
Toximul® 360B
Toximul® 500
Toximul® 701, 702
Toximul® 800
Toximul® 812
Toximul® H-HF
Toximul® MP
Toximul® MP-10
Toximul® MP-26
Toximul® R-HF
Toximul® S-HF
Toximul® SF Series
Tricol M-20
Ucar® Acrylic 503
Ucar® Acrylic 505
Ucar® Acrylic 515
Ucar® Acrylic 516
Ucar® Acrylic 518
Ucar® Acrylic 522
Ucar® Acrylic 525
Ufablend E 60
Ufablend HDL
Ufablend HDL85N
Ufablend W 100
Ufablend W Conc.
Ufablend W Conc. 60
Unamide® C-7649
Unamide® C-7944
Vannox AC 2, ACH, ACL
Varamide® FBR
Versilan MX332
Witcamide® 512
Witcamide® 5145M
Witcodet 100
Witcodet B-1
Witcodet DC-47
Witcodet DLC-47
Witcodet SC
Witcomul 3107
Witcomul 3154
Witcomul 3158
Witcomul H-50A
Witcomul H-52
Witconate 68KN
Zeeclean No. 1
Zee-emul No. 35
Zee-emul No. 50
Zeescour 555
Zoharfoam

ANIONIC/ NONIONIC/ CATIONIC:

Berol 806

CATIONIC:

Ablumine 08
Ablumine 10
Ablumine 12
Ablumine 18
Ablumine 230
Ablumine 1214
Ablumine 3500
Ablumine AN
Ablumine D10
Ablumine DHT75
Ablumine DHT90
Ablumine DT
Ablumine PN
Ablusoft A
Ablusoft ND
Ablusoft SN
Ablusoft SNC
Ablusol PAC
Accoquat 2C-75
Accoquat 2C-75H
Accosoft 440-75
Accosoft 540
Accosoft 550-75
Accosoft 550-90 HHV
Accosoft 550 HC
Accosoft 550L-90
Accosoft 580
Accosoft 580 HC
Accosoft 620-90
Accosoft 750
Accosoft 808
Accosoft 808HT
Accosoft 870
Acetamin 24
Acetamin C
Acetamin HT
Acetamin T
Acetoquat CPB
Acetoquat CPC
Acetoquat CTAB
Acid Foamer
Acid Thickener
Acra-500
Adma® 8
Adma® 10
Adma® 12
Adma® 14
Adma® 16
Adma® 246-451
Adma® 246-621
Adma® 1214
Adma® 1416
Adma® WC
Adogen® 444
Adogen® 461
Adogen® 462
Adogen® 464
Adogen® 471
Adogen® 477
Adogen® 560
Adogen® 570-S
Adogen® 572
Aerosol® C-61
Agrilan® TKA103
Ahco C330
Ahcovel Base 500
Ahcovel Base 700
Ahcovel Base OB
Akypomine® P 191
Akypoquat 40
Akypoquat 129
Akypoquat 1295
Algepon AK
Alkaminox® T-15
Alkaquat® DMB-451-50, DMB-451-80
Alkaterge®-C
Ameenex C-20
Amiet DT/15
Amiet DT/20
Amiet DT/32
Amine 2HBG
Amine 2M12D
Amine 2M14D
Amine 2M16D
Amine 2M18D
Amine 2MHBGD
Amine 2MKKD
Amine 12-98D
Amine 14D
Amine 16D
Amine 18D
Amine BG
Amine BGD
Amine C
Amine HBG
Amine HBGD
Amine KK
Amine KKD
Amine M218
Amine M2HBG
Amine O
Amine OL
Amine OLD
Amine S
Amyx A-25-S 0040
Amyx CO 3764
Amyx LO 3594
Amyx SO 3734
Anedco AW-395
Anedco AW-396
Anti-Terra®-P
Aquaperle D34
Arkophob NCS
Armac® 18D
Armac® 18D-40
Armac® C
Armac® HT
Armac® T
Armeen® 2-10
Armeen® 2-18
Armeen® 2C
Armeen® 2HT
Armeen® 2T
Armeen® 3-12
Armeen® 3-16
Armeen® 12D
Armeen® 16
Armeen® 16D
Armeen® 18
Armeen® 18D
Armeen® C
Armeen® CD
Armeen® DM8
Armeen® DM10
Armeen® DM12
Armeen® DM12D
Armeen® DM14
Armeen® DM14D
Armeen® DM16
Armeen® DM16D
Armeen® DM18D
Armeen® DMC
Armeen® DMCD
Armeen® DMHT
Armeen® DMHTD
Armeen® DMO
Armeen® DMOD
Armeen® DMSD
Armeen® DMT
Armeen® DMTD
Armeen® HT
Armeen® HTD
Armeen® L8D
Armeen® M2-10D
Armeen® M2C
Armeen® M2HT
Armeen® O
Armeen® OD
Armeen® OL
Armeen® OLD
Armeen® S
Armeen® SD
Armeen® T
Armeen® TD
Armoblen® S
Armoflote Series
Arolev ADL-30
Arolev ADL-86
Aromox® C/12
Aromox® C/12-W
Aromox® DM16
Aromox® DMC
Aromox® DMC-W
Aromox® DMHTD
Aromox® DMMC-W
Arosoft LC-15
Arquad® 2C-75
Arquad® 2HT-75
Arquad® 12-33
Arquad® 12-37W
Arquad® 12-50
Arquad® 16-29
Arquad® 16-29W
Arquad® 16-50
Arquad® 18-50
Arquad® 210-50
Arquad® B-50
Arquad® B-90
Arquad® B-100
Arquad® C-33
Arquad® C-33W
Arquad® C-50
Arquad® DM14B-90
Arquad® DMMCB-50
Arquad® HTL8 (W) MS-85
Arquad® L-11
Arquad® L-15
Arquad® S-2C-50
Arquad® S-50
Arquad® T-2C-50
Arquad® T-50
Ascote 5, 9, 12, 12L, 14
Asfier Series
Atlasol KAD
Atlasol KMM
Autopoon GK 4003
Autopoon GK 4004
Autopoon NI
Avistin® FD
Avistin® PN
AW-395
Babinar 715
Babinar 801C
Bactistep® MH 80
Berol 302
Berol 556
Berol 563
Berol 594
Bio-Surf DC-730
Bromat
BTC® 50 USP
BTC® 65 USP
BTC® 776
BTC® 818
BTC® 818-80
BTC® 824
BTC® 824 P100
BTC® 835
BTC® 885
BTC® 885 P40
BTC® 888
BTC® 1010
BTC® 1010-80
BTC® 2125, 2125-80, 2125 P-40
BTC® 2125M
BTC® 2125M-80, 2125M P-40
BTC® 2565
BTC® 2568
BTC® 8248
BTC® 8249
BTC® 8358
Burco FAE
Cadoussant AS
CAE
Carbostat 2203
Carsoquat® 621
Carsoquat® 621 (80%)
Carsoquat® 816-C
Carsosoft® S-75
Carsosoft® S-90-M
Carsosoft® T-90
Catalyst PAT
Catamine 101
Catigene® 50 USP
Catigene® 65 USP
Catigene® 776
Catigene® 818
Catigene® 818-80
Catigene® 824
Catigene® 885
Catigene® 888
Catigene® 1011
Catigene® 2125 M
Catigene® 2125M P40
Catigene® 2125 P40
Catigene® 2565

CATIONIC (CONT'D.)
Catigene® 4513-50
Catigene® 4513-80
Catigene® 4513-80 M
Catigene® 8248
Catigene® B 50
Catigene® B 80
Catigene® CA 56
Catigene® CETAC 30
Catigene® DC 100
Catigene® T 50
Catigene® T 80
Catinal CB-50
Catinal HTB
Catinal MB-50A
Cation DS
Cation SF-10
Catisol AO C
Charlab LPC
Chemazine 18, C, O, TO
Chemeen 18-2
Chemeen 18-50
Chemeen C-2
Chemeen C-5
Chemeen C-15
Chemeen DT-3
Chemeen DT-15
Chemeen HT-5
Chemeen HT-15
Chemeen O-30, O-30/80
Chemeen T-2
Chemeen T-5
Chemeen T-15
Chemeen T-20
Chemquat 12-33
Chemquat 12-50
Chemquat 16-50
Chemquat C/33W
Chimipal NH 2
Chimipal NH 6
Chimipal NH 10
Chimipal NH 20
Chimipal NH 25
Chimipal NH 30
CHPTA 65%
Cirrasol® LC-HK
Cirrasol® LC-PQ
Clearbreak TEB
CM-88®
Code 8059
Colsol
Consoft CP-50
Cordex DJ
Cralane AU-10
Crapol AU-40
Crapol AV-10
Crapol AV-11
Crodamet Series
Crodamine 1.HT
Crodamine 1.16D
Crodamine 1.18D
Crodamine 1.O, 1.OD
Crodamine 1.T
Crodamine 3.A16D
Crodamine 3.A18D
Crodamine 3.ABD
Crodamine 3.AED
Crodamine 3.AOD
Crodazoline O
Crodazoline S
Crodex C
Dama® 810
Dama® 1010
Defoamer A 50
Dehyquart A
Dehyquart AU-36
Dehyquart AU-56
Dehyquart C
Dehyquart C Crystals
Dehyquart D
Dehyquart DAM
Dehyquart LDB
Dehyquart LT
Dehyquart SP
Delestat P35
Demelan AU-40
Demulsifier 3837
Desomeen® TA-2
Desomeen® TA-5
Desomeen® TA-15
Desonic® TA-2, -15
Desonic® TA-25CWS
Diable AS
Diamiet 503, 508, 520, AB, C
Diamin DO, DT
Diamin HT
Diamin O
Diamin S
Diamin T
Diamine B11
Diamine BG
Diamine HBG
Diamine KKP
Diamine OL
Dinoram C
Dinoram O
Dinoram S
Dinoram SH
Dinoramac C
Dinoramac O
Dinoramac S
Dinoramac SH
Dinoramox S7
Dinoramox S12
Dinoramox SH 3
Dismulgan Brands
Disperbyk®-130
Disperbyk®-160
Disperbyk®-161
Disperbyk®-162
Disperbyk®-163
Disperbyk®-182
Dodicor 2565
Dodigen 1490
Duomac® C
Duomac® T
Duomeen® C
Duomeen® CD
Duomeen® HT
Duomeen® LT-4
Duomeen® O
Duomeen® OL
Duomeen® OTM
Duomeen® T
Duomeen® TDO
Duomeen® TTM
Duoquad® T-50
Edunine SC-L
Edunine SE 1010
Edunine SE 1060
Edunine V Fluid
Eganal GES
Elec QN
EM-900
Emcol® 4
Emcol® 1655
Emcol® 3780
Emcol® CC-9
Emcol® CC-36
Emcol® CC-42
Emcol® CC-55
Emcol® CC-57
Emcol® CC-422
Emcol® E-607L
Emcol® E-607S
Emcol® ISML
Emcol® L
Emcol® M
Emka Defoam PWC
Emkabase CA
Emkafix RXC
Emkalon AVR
Emkalon Base C-100
Emkalon KLA
Emkasan QA-50
Emkastat PC
Emkatex DLW
Empigen® 5089
Empigen® AB
Empigen® AD
Empigen® AF
Empigen® AG
Empigen® AH
Empigen® AM
Empigen® AS
Empigen® AT
Empigen® AY
Empigen® BAC50
Empigen® BAC50/BP
Empigen® BAC90
Empigen® BCB50
Empigen® BCF 80
Empigen® BCM75, BCM75/A
Empigen® CHB40
Empigen® CM
Empigen® FKC75K
Empigen® FKC75L
Empigen® FKH75L
Empigen® FRB75S
Empigen® FRC75S
Empigen® FRC90S
Empigen® FRG75S
Empigen® FRH75S
Emulsifier 4
ES-1239
Ethoduomeen® T/13
Ethoduomeen® T/20
Ethoduomeen® T/25
Ethoduoquad® T/20
Ethomeen® 18/12
Ethomeen® 18/15
Ethomeen® 18/20
Ethomeen® 18/25
Ethomeen® 18/60
Ethomeen® C/12
Ethomeen® C/15
Ethomeen® C/20
Ethomeen® C/25
Ethomeen® O/12
Ethomeen® O/15
Ethomeen® S/12
Ethomeen® S/15
Ethomeen® S/20
Ethomeen® S/25
Ethomeen® T/12
Ethomeen® T/15
Ethomeen® T/25
Ethoquad® 18/12
Ethoquad® 18/25
Ethoquad® C/12
Ethoquad® C/12 Nitrate
Ethoquad® C/25
Ethoquad® CB/12
Ethoquad® O/12
Ethoquad® O/25
Ethox CAM-2
Ethox CAM-15
Ethox OAM-308
Ethox SAM-2
Ethox SAM-10
Ethox SAM-50
Ethox TAM-2
Ethox TAM-5
Ethox TAM-10
Ethox TAM-15
Ethox TAM-20
Ethox TAM-25
Ethoxamine SF11
Eureka E-2
Farmin R 24H
Farmin R 86H
Fastgene PNG-708
Finazoline CA
Finsist C-2 Conc.
Finsist WW
Fixogene CD Liq.
Flocculant T-9
Fluorad FC-135
Fluorad FC-750
Fluorad FC-754
Forbest 610
Forbest 850
G-250
G-263
G-265
G-3634A
Genamin C Grades
Genamin CC Grades
Genamin O, S, Brands
Genamin T-020
Genamin T-050
Genamin T-150
Genamin T-200
Genamin T-308
Genamin TA Grades
Genamine C-020
Genamine C-050
Genamine C-080
Genamine C-100
Genamine C-150
Genamine C-200
Genamine C-250
Genamine O-020
Genamine O-050
Genamine O-080
Genamine O-100
Genamine O-150
Genamine O-200

CATIONIC (CONT'D.)
Genamine O-250
Genamine S-020
Genamine S-050
Genamine S-080
Genamine S-100
Genamine S-150
Genamine S-200
Genamine S-250
Geronol Aminox/3
Glokill Series
Hartofix 2X
Hartolon 1328
Hartonyl L531
Hartonyl L535
Hartonyl L537
Hartosoft 171
Hartosoft CN
Hartosoft GF
Hartosoft S5793
Hetoxamine C-2
Hetoxamine C-5
Hetoxamine C-15
Hetoxamine S-2
Hetoxamine S-5
Hetoxamine S-15
Hetoxamine T-2
Hetoxamine T-5
Hipochem BSM
Hipochem C-95
Hipochem DZ
Hipochem Finish 178
Hipochem M-51
Hipochem Migrator J
Hipochem Retarder CJ
Hipofix 491
Hipolon New
Hodag Amine C-100-L
Hodag Amine C-100-O
Hodag Amine C-100-S
Hydrotriticum QL
Hydrotriticum QM
Hydrotriticum QS
Icomeen® 18-50
Icomeen® T-15
Imacol S Liq.
Incromate CDL
Incromate SDL
Incroquat Behenyl BDQ/P
Incroquat Behenyl TMC/P
Incroquat SDQ-25
Incrosoft 100P
Incrosoft CFI-75
Indulin® AS-1
Indulin® AS-Special
Indulin® MQK
Indulin® MQK-IM
Indulin® W-1
Indulin® W-3
Inhibitor 60Q
Inset XL-300
Intrassist® LA-LF
Intratex® CA-2
Jet Amine DC
Jet Amine DE 810
Jet Amine DMCD
Jet Amine DMOD
Jet Amine DMSD
Jet Amine DMTD
Jet Amine DO
Jet Amine DT
Jet Amine M2C
Jet Amine PC
Jet Amine PCD
Jet Amine PE 08/10
Jet Amine PHT
Jet Amine PHTD
Jet Amine PO
Jet Amine POD
Jet Amine PS
Jet Amine PSD
Jet Amine PT
Jet Amine PTD
Jet Quat 2C-75
Jet Quat 2HT-75
Jet Quat C-50
Jet Quat DT-50
Jet Quat S-2C-50
Jet Quat S-50
Jet Quat T-2C-50
Jet Quat T-27W
Jet Quat T-50
Kemamine® BQ-2802C
Kemamine® BQ-9702C
Kemamine® BQ-9742C
Kemamine® D-190
Kemamine® D-650
Kemamine® D-970
Kemamine® D-974
Kemamine® D-989
Kemamine® P-150, P-150D
Kemamine® P-190, P-190D
Kemamine® P-650D
Kemamine® P-880, P-880D
Kemamine® P-970
Kemamine® P-970D
Kemamine® P-974D
Kemamine® P-989D
Kemamine® P-990, P-990D
Kemamine® Q-1902C
Kemamine® Q-6502C
Kemamine® Q-9702C
Kemamine® Q-9743C
Kemamine® Q-9743 CHGW
Kemamine® T-1902D
Kemamine® T-6501
Kemamine® T-6502D
Kemamine® T-9701
Kemamine® T-9702D
Kemamine® T-9742D
Kemamine® T-9902D
Kemamine® T-9972D
Kieralon® TX-410 Conc.
Larosol ALM-1
Larosol PDQ-2
Larostat® 88
Larostat® 143
Larostat® 264 A
Larostat® 264 A Conc.
Larostat® 451
Leomin FA
Leomin FANF
Leomin KP
Leomin TR
Levelan R-15, -200
Levenol RK
Lilamac S1
Lilamin VP75
Lilamuls BG
Lilamuls EM 24, EM 26, EM 33
LSP 33
Lutensit® K-HP
Lutensit® K-LC, K-LC 80
Lutensit® K-OC
Mackamine CAO
Mackine 101
Mackine 201
Mackine 301
Mackine 321
Mackine 401
Mackine 421
Mackine 501
Marlamid® A 18
Marlamid® A 18 E
Marlamid® AS 18
Marlazin® 7102
Marlazin® 7265
Marlazin® 8567
Marlazin® KC 21/50
Marlazin® L 10
Marlazin® OK 1
Marlazin® OL 2
Marlazin® OL 20
Marlazin® S 10
Marlazin® S 40
Marlazin® T 10
Marlazin® T 15/2
Marlazin® T 16/1
Marlosoft® IQ 75
Marlosoft® IQ 90
Marlowet® 5165
Marlowet® 5401
Marlowet® 5440
Marvanol® Leveler DL
Marvanol® RE-1274
Marvanol® RE-1281
Matexil FC ER
Matexil FC PN
Matexil LC CWL
Matexil LC RA
Mazeen® 174
Mazeen® 174-75
Mazeen® 175
Mazeen® 176
Mazeen® 241-3
Mazeen® C-2
Mazeen® C-5
Mazeen® C-10
Mazeen® C-15
Mazeen® DBA-1
Mazeen® S-2
Mazeen® S-5
Mazeen® S-10
Mazeen® S-15
Mazeen® T-2
Mazeen® T-3.5
Mazeen® T-5
Mazeen® T-10
Mazeen® T-15
Michelene 10
Michelene 15
Migrassist® D
Migrassist® NYL
Miramine® C
Miramine® O
Miramine® OC
Miramine® TO
Mirapol® A-15
Mirapol® WT
Monaquat AT-1074
Monaquat ISIES
Monaquat TG
Monastat 1195
Monazoline C
Monazoline CY
Monazoline IS
Monazoline O
Monazoline T
M-Quat® 32
M-Quat® 2475
M-Soft-1
M-Soft-10
M-Soft-J
M-Soft-JHV
M-Soft-J-Super
MY Silicone A-08
Neocation G
Nissan Amine 2-OLR
Nissan Asphasol 10
Nissan Asphasol 20
Nissan Cation AB
Nissan Cation ABT-350, 500
Nissan Cation AR-4
Nissan Cation BB
Nissan Cation F2-10R, -20R, -40E, -50
Nissan Cation FB, FB-500
Nissan Cation L-207
Nissan Cation M2-100
Nissan Cation PB-40, -300
Nissan Cation S2-100
Nissan Softer 706
Nissan Softer 1000
Nopcogen 22-O
Noram 2C
Noram 2SH
Noram C
Noram DMC
Noram DMCD
Noram DMSD
Noram DMSH D
Noram M2C
Noram M2SH
Noram O
Noram S
Noram SH
Noramac C
Noramac C 26
Noramac O
Noramac S
Noramac SH
Noramium DA.50
Noramium MC 50
Noramium MO 50
Noramium MS 50
Noramium MSH 50
Noramium S 75
Noramox C2
Noramox C5
Noramox C11
Noramox C15
Noramox O2
Noramox O5

CATIONIC (CONT'D.)
Noramox O11
Noramox O15
Noramox O20
Noramox S1
Noramox S2
Noramox S5
Noramox S7
Noramox S11
Noramox S15
Noramox S20
Norfox® IM-38
Noxamium C2-15
Noxamium S2-11
Noxamium S2-50
Noxamium Series
Octosol 400
Octosol 474
Octosol 562
Octosol 571
Ogtac-85
Ogtac 85 V
Osimol Grunau EFA
Osimol Grunau MA
Osimol Grunau PHT
Osimol Grunau RAC
Ospin Salt ON
Ospin TAN
Peregal® ST
Phospholipid PTD
Phospholipid PTL
Phospholipid PTS
Phospholipid PTZ
Plasfalt HM, SS
Polyanthrene KS Liq. New
Polyram C
Polyram O
Polyram S
Propomeen C/12
Propomeen T/12
Propoquad® 2HT/11
Protamine-45
Protectol® KLC 50, 80
Prox-onic DT-03
Prox-onic DT-015
Prox-onic DT-030
Prox-onic MC-02
Prox-onic MC-05
Prox-onic MC-015
Prox-onic MHT-05
Prox-onic MHT-015
Prox-onic MO-02
Prox-onic MO-015
Prox-onic MO-030
Prox-onic MO-030-80
Prox-onic MS-02
Prox-onic MS-05
Prox-onic MS-011
Prox-onic MS-050
Prox-onic MT-02
Prox-onic MT-05
Prox-onic MT-015
Prox-onic MT-020
Quadrilan® AT
Quadrilan® BC
Quadrilan® MY211
Quadrilan® SK
Quartamin 86P
Quartamin CPR
Quartamin D86P, D86PL,
D86PI
Quartamin DCP
Quartamin HTPR
Quartamin TPR
Quaternary O
Quatrene 7670
Quatrene CE
Quatrex 162
Quatrex 172
Quatrex 182
Querton 14Br-40
Querton 16Cl-29
Querton 16Cl-50
Querton 280
Querton 441-BC
Querton KKBCl-50
RAC-100
Radiamac 6149
Radiamac 6159
Radiamac 6169
Radiamac 6179
Radiamine 6140
Radiamine 6141
Radiamine 6160
Radiamine 6161
Radiamine 6163, 6164
Radiamine 6170
Radiamine 6171
Radiamine 6172
Radiamine 6173
Radiamine 6540
Radiamine 6560
Radiamine 6570
Radiamine 6572
Radiamine AA 23, 27, 57
Radiamine AA 60
Radiaquat 6410
Radiaquat 6412
Radiaquat 6442
Radiaquat 6443
Radiaquat 6444
Radiaquat 6462
Radiaquat 6470
Radiaquat 6471
Radiaquat 6475
Radiaquat 6480
Redicote AD-130
Redicote AD-141
Redicote AD-142
Redicote AD-150
Redicote AD-164
Redicote AD-170
Redicote AE 1A
Redicote AE 6
Redicote AE 6-A
Redicote AE 7
Redicote AE 9
Redicote AE 22
Redicote AE 26
Redicote AE 45
Redicote AE 55
Redicote AE 91
Redicote AE 93
Redicote Series
Redicote TR-1114X
Redicote TR-1130
Redicote TR-1134
Rematard Grades
Resogen® 35 Conc.
Resogen® DM
Retentol RM
Rewomine IM-BT
Rewomine IM-CA
Rewomine IM-OA
Rewopon® IM OA
Rewopon® IM-CA
Rewopon® JMBT
Rewopon® JMCA
Rewopon® JMOA
Rewoquat CPEM
Rewoquat CR 3099
Rewoquat DQ 35
Rewoteric AM TEG
Rexobase PW
Rhodameen® 220
Rhodameen® O-2
Rhodameen® O-6
Rhodameen® O-12
Rhodameen® OA-910
Rhodameen® OS-12
Rhodameen® PN-430
Rhodameen® RAM/7 Base
Rhodameen® RAM/8 Base
Rhodameen® S-20
Rhodameen® S-25
Rhodameen® T-5
Rhodameen® T-12
Rhodameen® T-15
Rhodameen® T-30-45%
Rhodameen® T-30-90%
Rhodameen® T-50
Rhodamox® LO
Rhodaquat® A-50
Rhodaquat® DAET-90
Rhodaquat® T
Rhodaquat® VV-328
Romie 802
Rycofax® 618
Sandolube NVJ
Sanfix 555
Sanfix 555-200
Sanfix 555C
Sanfix 555C-200
Sanleaf CL-533M
Sanstat 1200
Schercodine C
Schercodine L
Schercodine O
Schercoquat 2IAE
Schercoquat 2IAP
Schercoquat ALA
Schercotarder
Schercozoline B
Schercozoline C
Schercozoline I
Schercozoline L
Schercozoline O
Schercozoline S
Sermul EK 330
Sermul EN 134
Servamine KEP 4527
Servamine KET 4542
Servamine KOO 330 B
Sevefilm 20
SK Flot 1
SK Flot 2
SK Flot 3
SK Flot 4
SK Flot FA 1
SK Flot FA 2
SK Flot FA 3
SK Flot FA 4
Sochamine OX 30
Sofbon C-1-S
Sofnon 105G
Sofnon HG-180
Softal 300
Softal A-3
Softlon AC-4
Softlon FC-28
Softlon FC-136
Softyne CLS
Softyne CSN
Softyne H
Solidegal SR
Standamox LAO-30
Standamox LMW–30
Standamul® STC-25
Stepanquat® F/T
Stepantex® DO 90
Stepantex® GS 90
Stepantex® VS 90
Steramine CGL
Steramine CR 25
Steramine FPA 197
Steramine GS
Sunsoflon CK
Sunsoflon K-2
Sunsoflon MT-100
Surfac® P14B
Surfac® P24M
Surfam P10
Surfam P12B
Surfam P14B
Surfam P17B
Surfam P24M
Surfam P86M
Surfam P89MB
Surflo® S2024
Surflo® S7655
Synperonic FS
Synprolam 35
Synprolam 35A
Synprolam 35BQC (50)
Synprolam 35BQC (80)
Synprolam 35DM
Synprolam 35DMA
Synprolam 35DMBQC
Synprolam 35M
Synprolam 35MX1
Synprolam 35MX3
Synprolam 35MX5
Synprolam 35N3
Synprolam 35N3X3,
35N3X5, 35N3X10,
35N3X15, 35N3X25
Synprolam 35TMQC
Synprolam 35X2
Synprolam 35X2QS,
35X5QS, 35X10QS
Synprolam 35X5
Synprolam 35X10
Synprolam 35X15
Synprolam 35X20
Synprolam 35X25
Synprolam 35X35
Synprolam 35X50
Synprolam FS
Synprolam VC
Tallow Amine

CATIONIC (CONT'D.)
Tallow Diamine
Tallow Triamine
Tanaquad T-85
Tanassist® JCR
Tegacid® Regular VA
Teginacid® ML-SE
Tequat RO
Textamine 05
Textamine 1839
Textamine A-5-D
Textamine A-3417
Textamine A-W-5
Textamine O-5
Textamine Polymer
Textamine T-5-D
Tinegal® MR-50
Tinegal® PM
Tinegal® TP
Tinofix® ECO
Tinofix® EW Sol
Tomah AO-14-2
Tomah DA-14
Tomah DA-16
Tomah DA-17
Tomah E-14-2
Tomah E-14-5
Tomah E-17-2
Tomah E-18-2
Tomah E-18-5
Tomah E-18-15
Tomah PA-14
Tomah PA-14 Acetate
Tomah Q-14-2
Tomah Q-14-2PG
Tomah Q-17-2
Tomah Q-17-2PG
Tomah Q-18-2
Tomah Q-18-15
Tomah Q-C-15
Tomah Q-D-T
Tomah Q-S
Tomah Q-S-80
Triameen T
Trinoram C
Trinoram O
Trinoram S
Trymeen® 6601
Trymeen® 6606
Trymeen® 6607
Trymeen® 6609
Trymeen® 6610
Trymeen® 6617
Trymeen® 6620
Trymeen® 6622
Trymeen® 6637
Trymeen® CAM-10
Trymeen® CAM-15
Trymeen® OAM 30/60
Trymeen® SAM-50
Trymeen® TAM-8
Trymeen® TAM-20
Trymeen® TAM-25
Trymeen® TAM-40
Ufasoft 75
Ultratex® ESB, WK
Unamine® T
Uniperol® AC
Uniperol® AC Highly Conc.
Unisol WL
Varine C
Varine O®
Variquat® 50MC
Variquat® 60LC
Variquat® 80MC
Variquat® 638
Variquat® B200
Variquat® E290
Variquat® K300
Varisoft® 110
Varisoft® 137
Varisoft® 222 (75%)
Varisoft® 222 LT (90%)
Varisoft® 475
Varisoft® 920
Varisoft® 3690
W180
Witbreak RTC-326
Witbreak RTC-330
Witcamide® 5168
Witcamine® 204
Witcamine® 209
Witcamine® 210
Witcamine® 211
Witcamine® 240
Witcamine® 760
Witcamine® 6606
Witcamine® 6622
Witcamine® AL42-12
Witcamine® PA-60B
Witcamine® RAD 0500
Witcamine® RAD 0515
Witcamine® RAD 1100
Witcamine® RAD 1110
Witcamine® TI-60
Witco® 915
Witcor 3192
Witcor 3630
Witcor 3635
Witcor CI-6
Witflow 950
Witflow 953
Zoharquat 50
Zoharquat 80 EXP
Zoharsoft
Zoharsoft 90
Zoharsoft DAS
Zoharsoft DAS-N
Zoharsoft DAS-PW
Zoharsoft DAS-SIL
Zonyl® FSC

CATIONIC/ AMPHOTERIC:

Rewoteric QAM 50

CATIONIC/ NONIONIC:

Ablumul AG-900
Ablumul AG-1214
Abluton BN
Alrosperse® 100
Amiet 102, 105, 110, 115, 202, 205, 210, 215, 302, 305, 310, 315, 402, 405, 410,
Amiet CD/14
Amiet CD/17
Amiet CD/22
Amiet CD/27
Amiet OD/14
Amiet TD/14
Amiet TD/17
Amiet TD/22
Amiet TD/27
Amiet THD/14
Amiet THD/17
Amiet THD/22
Amiet THD/27
Amlev RDC
Atlas G-1530
Autopur WK 4121
AW-396
Barlox® 12
Barlox® 14
Barlox® C
Berol 225
Berol 226
Bio-Surf PBC-460
Burco NCS-80
Cenegen® CJB
Cenegen® EKD
Dehscotex SN Series
Delion 966
Dinoramox S3
Edunine RWT
Edunine S82
Edunine SE
Hartox DMCD
Icomeen® 18-5
Icomeen® O-30
Icomeen® O-30-80%
Icomeen® S-5
Icomeen® T-2
Icomeen® T-5
Icomeen® T-7
Icomeen® T-20
Icomeen® T-25
Icomeen® T-25 CWS
Icomeen® T-40
Icomeen® T-40-80%
Intratex® W New
Mazox® CAPA
Mazox® CAPA-37
Mazox® CDA
Mazox® KCAO
Mazox® LDA
Mazox® MDA
Mazox® ODA
Mazox® SDA
Merbron R
Nissan Chloropearl
Nissan New Royal P
Nissan Sun Flora
Norfox® X
Produkt GM 6115
Rhodameen® VP-532/SPB
Rhodamox® CAPO
Schercomul QW
Schercopol DS-120
Schercopol DS-140
Sinocol PQ2
Softal MR-30
Tensiofix D40 Bio-Act
Textamine Oxide CA
Textamine Oxide CAW
Textamine Oxide LMW
Textamine Oxide TA
Triton® RW-20
Triton® RW-50
Triton® RW-75
Triton® RW-100
Triton® RW-150
Witcamide® AL69-58
Witcomul 3230
Witcor 3194, 3195
Witcor CI-1
Zusomin O 105

NONIONIC:

Abil® B 8843, B 8847
Abil® B 88184
Ablumide CDE
Ablumide CME
Ablumide LDE
Ablumide LME
Ablumide SDE
Ablumox C-7
Ablumox CAPO
Ablumox LO
Ablumox T-15
Ablumox T-20
Ablumul AG-909
Ablumul AG-AH
Ablumul EP
Ablumul TN
Ablunol 200ML
Ablunol 200MO
Ablunol 200MS
Ablunol 400ML
Ablunol 400MO
Ablunol 400MS
Ablunol 600ML
Ablunol 600MO
Ablunol 600MS
Ablunol 1000MO
Ablunol 1000MS
Ablunol 6000DS
Ablunol CO 5
Ablunol CO 10
Ablunol CO 15
Ablunol CO 30
Ablunol CO 45
Ablunol DEGMS
Ablunol EGMS
Ablunol GML
Ablunol GMO
Ablunol GMS
Ablunol LA-3
Ablunol LA-5
Ablunol LA-7
Ablunol LA-9
Ablunol LA-12
Ablunol LA-16
Ablunol LA-40
Ablunol LMO
Ablunol LN
Ablunol NP4
Ablunol NP6
Ablunol NP8
Ablunol NP9
Ablunol NP10

NONIONIC (CONT'D.)
Ablunol NP12
Ablunol NP15
Ablunol NP16
Ablunol NP20
Ablunol NP30
Ablunol NP30 70%
Ablunol NP40
Ablunol NP40 70%
Ablunol NP50
Ablunol NP50 70%
Ablunol OA-6
Ablunol OA-7
Ablunol S-20
Ablunol S-40
Ablunol S-60
Ablunol S-80
Ablunol S-85
Ablunol SA-7
Ablusoft PE
Abluton EP
Abluton LMO
Abluton LN
Abluton N
Abluton T
Accomeen C2
Accomeen C5
Accomeen C10
Accomeen C15
Accomeen S2
Accomeen S5
Accomeen S10
Accomeen T2
Accomeen T5
Accomeen T15
Accomid C
Accomid PK
Acconon 200-DL
Acconon 200-MS
Acconon 300-MO
Acconon 400-DO
Acconon 400-ML
Acconon 400-MO
Acconon 400-MS
Acconon 1300
Acconon CA-5
Acconon CA-8
Acconon CA-9
Acconon CA-15
Acconon CON
Acconon E
Acconon ETG
Acconon TGH
Acconon W230
Accosperse 20
Accosperse 60
Accosperse 80
Accosperse TGH
Acid Felt Scour
Aconol X6
Aconol X10
Actralube SOS
Actralube Syn-147
Actramide 202
Actramide 410
Actrol 4DP
Actrol 4MP, 628
Admox® 1214
Adol® 42
Adol® 52 NF
Adol® 61 NF
Adol® 62 NF
Adol® 63
Adol® 66
Adol® 80
Adol® 85 NF
Adol® 90 NF
Adol® 320, 330, 340
Adol® 520 NF
Adol® 620 NF
Adol® 630
Adol® 640
Adsee® 775
Adsee® 799
Adsee® 801
Adsee® 1080
Adsee® AK31-73
AE-3
AE-7
AE-1214/3
AE-1214/6
AEPD®
AF 10 FG
AF 10 IND
AF 30 FG
AF 30 IND
AF 60
AF 72
AF 75
AF 8805 FG
AF 8810
AF 8810 FG
AF 8820
AF 8820 FG
AF 8830
AF 8830 FG
AF 9020
AF CM Conc.
AF GN-11-P
AF HL-21
AF HL-26
AF HL-40
AF HL-52
Agitan 290
Agitan 633
Agrilan® AEC123
Agrilan® AEC145
Agrilan® AEC156
Agrilan® AEC167
Agrilan® D54
Agrilan® DG113
Agrilan® EA47
Agrilan® F502
Agrilan® FS101
Agrilan® FS112
Agrilan® WP112
Agrimul 70-A
Agrisol PX401
Agrisol PX413
Ahco 759
Ahco 832
Ahco 909
Ahco 944
Ahco 3998
Ahco 7166T
Ahco 7596D
Ahco 7596T
Ahco DFO-100
Ahco DFO-110
Ahco DFO-150
Ahco DFP-156
Ahco DFS-96
Ahco DFS-149
Ahco DHS-111
Ahco EO-102
Ahco EO-114
Ahco FO-18
Ahco FP-67
Ahco FS-21
Ahcovel Base N-15
Ahcovel Base N-62
Ahcovel Base N-64
Ahcowet DQ-114
Ahcowet DQ-145
Akypo RLM 25
Akypo RLM 130
Akypo RLMQ 38
Akypo TBP 180
Akyporox CO 400
Akyporox CO 600
Akyporox NP 15
Akyporox NP 30
Akyporox NP 40
Akyporox NP 90
Akyporox NP 95
Akyporox NP 105
Akyporox NP 150
Akyporox NP 200
Akyporox NP 300
Akyporox NP 300V
Akyporox NP 1200V
Akyporox OP 40
Akyporox OP 100
Akyporox OP 115 SPC
Akyporox OP 200
Akyporox OP 250 V
Akyporox OP 400V
Akyporox RLM 22
Akyporox RLM 40
Akyporox RLM 80
Akyporox RLM 80V
Akyporox RLM 160
Akyporox RO 90
Akyporox RTO 70
AL 2070
Albalan
Albatex® OR
Albegal® BMD
Albigen® A
Alcodet® 218
Alcodet® 260
Alcodet® HSC-1000
Alcodet® IL-3500
Alcodet® MC-2000
Alcodet® SK
Alcodet® TX 4000
Alcojet®
Alcolec® 439-C
Alcolec® 440-WD
Alcolec® 495
Alcolec® BS
Alcolec® Extra A
Alcolec® Granules
Alcolec® S
Aldo® DC
Aldo® HMS
Aldo® MC
Aldo® ML
Aldo® MLD
Aldo® MO
Aldo® MO FG
Aldo® MO Tech
Aldo® MOD
Aldo® MOD FG
Aldo® MR
Aldo® MS
Aldo® MS-20 FG
Aldo® MSA
Aldo® MSD
Aldo® MS Industrial
Aldo® MS LG
Aldo® PMS
Aldosperse® ML 23
Aldosperse® MS-20
Aldosperse® O-20 FG
Alfonic® 610-50R
Alfonic® 810-40
Alfonic® 810-60
Alfonic® 1012-40
Alfonic® 1012-60
Alfonic® 1216-22
Alfonic® 1216-30
Alfonic® 1412-40
Alfonic® 1412-60
Alfonic® 1618-65
Alkadet 15
Alkamide® 200 CGN
Alkamide® 2110
Alkamide® C-2
Alkamide® C-212
Alkamide® CDE
Alkamide® CDM
Alkamide® CDO
Alkamide® CL63
Alkamide® CME
Alkamide® CP-1255
Alkamide® CP-6565
Alkamide® DC-212
Alkamide® DC-212/M
Alkamide® DC-212/MP
Alkamide® DC-212/S
Alkamide® DC-212/SE
Alkamide® DIN295/S
Alkamide® DL-203/S
Alkamide® DL-207/S
Alkamide® DO-280
Alkamide® DS-280/S
Alkamide® KD
Alkamide® L7DE
Alkamide® L9DE
Alkamide® L-203
Alkamide® LIPA/C
Alkamide® S-280
Alkamide® SDO
Alkamide® STEDA
Alkamuls® 14/R
Alkamuls® 400-DO
Alkamuls® 400-DS
Alkamuls® 400-MO
Alkamuls® 600-DO
Alkamuls® 600-MO
Alkamuls® 783/P
Alkamuls® A
Alkamuls® AG-821
Alkamuls® AG-900
Alkamuls® AP
Alkamuls® B
Alkamuls® BR
Alkamuls® CO-15
Alkamuls® CO-40

NONIONIC (CONT'D.)
Alkamuls® COH-5
Alkamuls® D-10
Alkamuls® D-20
Alkamuls® D-30
Alkamuls® EGMS/C
Alkamuls® EL-620
Alkamuls® EL-620L
Alkamuls® EL-719
Alkamuls® EL-719L
Alkamuls® EL-980
Alkamuls® EL-985
Alkamuls® GMR-55LG
Alkamuls® GMS/C
Alkamuls® L-9
Alkamuls® M-6
Alkamuls® O-14
Alkamuls® OR/36
Alkamuls® PE/220
Alkamuls® PE/400
Alkamuls® PSML-20
Alkamuls® PSMO-5
Alkamuls® PSMO-20
Alkamuls® PSMS-20
Alkamuls® PSTO-20
Alkamuls® R81
Alkamuls® RC
Alkamuls® S-6
Alkamuls® S-8
Alkamuls® S-20
Alkamuls® S-60
Alkamuls® S-65
Alkamuls® S-65-8
Alkamuls® S-65-40
Alkamuls® S-80
Alkamuls® S-85
Alkamuls® SEG
Alkamuls® SML
Alkamuls® SMO
Alkamuls® SMS
Alkamuls® STO
Alkamuls® STS
Alkamuls® T-20
Alkamuls® T-60
Alkamuls® T-65
Alkamuls® T-80
Alkamuls® T-85
Alkamuls® TD-41
Alkanol® 6112
Alkanol® A-CN
Alkasil® NE 58-50
Alkasil® NEP 73-70
Alkasurf® CO-40
Alkasurf® NP-1
Alkasurf® NP-4
Alkasurf® NP-6
Alkasurf® NP-8
Alkasurf® NP-9
Alkasurf® NP-10
Alkasurf® NP-11
Alkasurf® NP-12
Alkasurf® NP-15
Alkasurf® NP-15, 80%
Alkasurf® NP-30, 70%
Alkasurf® NP-40, 70%
Alkasurf® NP-50 70%
Alkasurf® OP-1
Alkasurf® OP-5
Alkasurf® OP-8
Alkasurf® OP-10
Alkasurf® OP-12
Alkasurf® OP-30, 70%
Alkasurf® OP-40, 70%
Alkaterge®-T-IV
Alkylox P1904
All Wet
Alox® 1689
Alphenate GA 65
Alphenate HM
Alphoxat O 105
Alphoxat O 110
Alphoxat O 115
Alphoxat S 110
Alphoxat S 120
Alrodyne 6104
Alrosperse 11P Flake
Amerchol® C
Amerchol® CAB
Amerchol® L-101
Amerlate® P
Ameroxol® LE-4
Ameroxol® LE-23
Ameroxol® OE-2
Amidex CE
Amidex CME
Amidex CO-1
Amidex KME
Amidex L-9
Amidex LD
Amidex OE
Amidox® C-2
Amidox® C-5
Amidox® L-2
Amidox® L-5
Amiladin
Amiladin C-1802
Aminol CA-2
Aminol CM, CM Flakes, CM-C Flakes, CM-D Flakes
Aminol COR-2
Aminol COR-2C
Aminol COR-4
Aminol COR-4C
Aminol HCA
Aminol KDE
Aminol LM-30C, LM-30C Special
Aminol N-1918
Aminol TEC N
Aminol VR-14
Aminoxid A 4080
Aminoxid WS 35
Amlev KC-2
Amlev KOF-7
Amlev PNL
Ammonyx® CDO
Ammonyx® CO
Ammonyx® DMCD-40
Ammonyx® LO
Ammonyx® MCO
Ammonyx® MO
Ammonyx® SO
AMP
AMP-95
Amsol GMS
Amyx CDO 3599
Anedco AW-397
Anfomul 01T
Anfomul PL
Anfomul PLR
Anfomul PO
Anfomul S4
Anfomul S6
Anfomul S43
Anfomul S50
Antarox® 17-R-2
Antarox® 25-R-2
Antarox® 31-R-1
Antarox® 461/P
Antarox® 487/P
Antarox® AA-60
Antarox® B-10
Antarox® B-25
Antarox® BA-PE 70
Antarox® BA-PE 80
Antarox® BL-214
Antarox® BL-225
Antarox® BL-236
Antarox® BL-240
Antarox® BL-330
Antarox® BL-344
Antarox® BO/327
Antarox® EGE 25-2
Antarox® EGE 31-1
Antarox® F88
Antarox® FM 33
Antarox® FM 53
Antarox® FM 63
Antarox® L-61
Antarox® L-62
Antarox® L-62 LF
Antarox® L-64
Antarox® LA-EP 15
Antarox® LA-EP 16
Antarox® LA-EP 25
Antarox® LA-EP 25LF
Antarox® LA-EP 45
Antarox® LA-EP 59
Antarox® LA-EP 73
Antarox® LA-EPB-17
Antarox® LF-222
Antarox® LF-224
Antarox® LF-330
Antarox® LF-344
Antarox® P-84
Antarox® P-104
Antarox® PGP 18-1
Antarox® PGP 18-2
Antarox® PGP 18-2D
Antarox® PGP 18-2LF
Antarox® PGP 18-4
Antarox® PGP 18-8
Antarox® PGP 23-7
Antarox® PL/Series
Antarox® RA 40
Antil® 141 Liq.
Antispumin ZU
APG® 325 Glycoside
APG® 600 Glycoside
APG® 625 Glycoside
Aphrogene 5001
Aphrogene Jet
Aquabase
Aqualose L30
Aqualose L75
Aqualose L75/50
Aqualose LL100
Aqualose W20
Aqualose W20/50
Aquaphil K
Aramide CDM-4
Arbyl 18/50
Arbyl N
Arbyl R Conc.
Arbylen Conc.
Ardril DMS
Argobase EU
Argobase LI
Argobase MS-5
Argobase SI
Argowax Cosmetic Super
Argowax Dist
Argowax Standard
Arkopal N040
Arkopal N060
Arkopal N080
Arkopal N090
Arkopal N100
Arkopal N110
Arkopal N130
Arkopal N150
Arkopal N230
Arlacel® 129
Arlacel® 780
Arlacel® A
Arlacel® C
Arlamol® F
Arlamol® ISML
Arlasolve® 200
Arlasolve® DMI
Arlatone® 285
Arlatone® 289
Arlatone® 983
Arlatone® 985
Arlatone® G
Arlatone® T
Armeen® DMMCD
Armix 176
Armix 180-C
Armix 183
Armotan® ML
Armotan® MO
Armotan® MP
Armotan® MS
Armotan® PML 20
Armotan® PMO 20
Armul 03
Armul 16
Armul 17
Armul 21
Armul 906
Armul 908
Armul 910
Armul 912
Armul 930
Armul 940
Armul 950
Armul 1003
Armul 1005
Armul 1007
Armul 1009
Armul 2404
Arnox BP Series
Arodet AN-100
Arodet BLN Special
Arodet E-15
Arodet HCS
Arodet MER-3
Arodet N-100

NONIONIC (CONT'D.)
Arodet UG
Arofoam SNI
Arol Biodet
Arol Defoamer NA-3 Special
Arol Defoamer NA2X
Arolev HPM
Arolev MTR-7
Aromox® DM14D-W
Aromox® DMB
Aromox® DMMCD-W
Aromox® T/12
Arosolve XNF-1
Arosurf® 32-E20
Arosurf® 42-E6
Arosurf® 42-PE10
Arosurf® 66-E2
Arosurf® 66-E10
Arosurf® 66-E20
Arosurf® 66-PE12
Arowet ODA
Arsurf Series
Asol
Atlas EM-2
Atlas EM-13
Atlas EMJ-2
Atlas EMJ-C
Atlas G-950
Atlas G-1045
Atlas G-1045A
Atlas G-1052
Atlas G-1069
Atlas G-1086
Atlas G-1087
Atlas G-1096
Atlas G-1186
Atlas G-1223
Atlas G-1256
Atlas G-1281
Atlas G-1284
Atlas G-1285
Atlas G-1288
Atlas G-1289
Atlas G-1292
Atlas G-1295
Atlas G-1300
Atlas G-1304
Atlas G-1554
Atlas G-1556
Atlas G-1564
Atlas G-1649
Atlas G-2079
Atlas G-2109
Atlas G-2127
Atlas G-2143
Atlas G-2151
Atlas G-2162
Atlas G-2198
Atlas G-2684
Atlas G-3300B
Atlas G-3780A
Atlas G-3801
Atlas G-3811
Atlas G-4809
Atlas G-4822
Atlas G-4884
Atlas G-4885
Atlas G-4905
Atlas G-4909
Atlas G-5000
Atlas G-5002
Atlas G-7166P
Atlas G-7596-J
Atlas G-7596H
Atlas G-8916PF
Atlas G-8936CJ
Atlas G-9046T
Atlasol Base Oil S
Atlox 80
Atlox 775
Atlox 804
Atlox 847
Atlox 848
Atlox 849
Atlox 1045A
Atlox 1087
Atlox 1096
Atlox 1196
Atlox 1256
Atlox 1285
Atlox 4873
Atlox 4881
Atlox 4885
Atlox 4896
Atlox 4898
Atlox 4901
Atlox 4911
Atlox 4912
Atlox 4991
Atlox 4995
Atlox 4996 B
Atlox 8916P
Atlox 8916PF
Atlox 8916TF
Atolex LDA/40
Atolex NI/100B
Atpet 545
Atpet 787
Atpet 900
Atplus 526
Atplus 555
Atplus 989
Atplus 1364
Atplus 1399, 1400, 1401, 1402
Atplus 1992
Atplus 2380
Atranonic Polymer 20
Atranonic Polymer N200
Atsolyn BH 100
Atsolyn GMO
Atsolyn P
Atsolyn PE 27
Atsolyn SUC 100
Atsolyn T
Atsolyn TD 50
Atsurf 2801
Atsurf 2802
Atsurf 2803
Atsurf 2821
Atsurf 2822
Atsurf 2823
Atsurf S-80
Atsurf S-85
Atsurf T-80
Atsurf T-85
AW-397
Barlox® 10S
Barre® Common Degras
Base 76
Base 85
Basophor A
Basopon® LN
Basopon® TX-110
Berol 02
Berol 07
Berol 08 Powd
Berol 09
Berol 26
Berol 048
Berol 050
Berol 79
Berol 087
Berol 106
Berol 108
Berol 173
Berol 185
Berol 190
Berol 191
Berol 195
Berol 198
Berol 199
Berol 223
Berol 259
Berol 260
Berol 267
Berol 269
Berol 272
Berol 277
Berol 278
Berol 281
Berol 291
Berol 292
Berol 295
Berol 370
Berol 374
Berol 716
Berol 829
Berol 938
Berol 946
Berol 948
Berol WASC
Bio-Soft® 9283
Bio-Soft® CS 50
Bio-Soft® D-233
Bio-Soft® E-200
Bio-Soft® E-300
Bio-Soft® E-400
Bio-Soft® E-670
Bio-Soft® E-710
Bio-Soft® E-840
Bio-Soft® EA-4
Bio-Soft® EA-8
Bio-Soft® EA-10
Bio-Soft® EC-600
Bio-Soft® EL-719
Bio-Soft® EN 600
Bio-Soft® ERM
Bio-Soft® ET-630
Bio-Soft® FF 400
Bio-Soft® LF 77A
Bio-Soft® PG 4
Bio-Soft® TD 400
Bio-Soft® TD 630
Biodet T
Brij® 35
Brij® 35 Liq.
Brij® 35 SP
Brij® 35 SP Liq.
Brij® 56
Brij® 58
Brij® 68
Brij® 72
Brij® 98G
Brij® 99
Brij® 700 S
Brij® 721 S
Bubble Breaker® DMD-1
Bubble Breaker® DMD-2
Burco LAF-6
Burco LAF-125
Burco NPS-50%
Burco NPS-225
Burco TME
Calamide C
Calamide CW-100
Calamide O
Calamide S
Caloxylate N-9
Capcure Emulsifier 37S
Capcure Emulsifier 65
Caplube 8350
Caplube 8385
Caplube 8410
Caplube 8440
Caplube 8442
Caplube 8445
Caplube 8448
Caplube 8508
Caplube 8540B
Caplube 8540C
Caplube 8540D
Carbon Detergent K
Carsamide® CA
Carsamide® SAC
Carsamide® SAL-7
Carsamide® SAL-9
Carsonon® L-985
Carsonon® LF-5
Carsonon® LF-46
Carsonon® N-4
Carsonon® N-6
Carsonon® N-8
Carsonon® N-9
Carsonon® N-10
Carsonon® N-11
Carsonon® N-12
Carsonon® N-30
Carsonon® N-40, 70%
Carsonon® N-50
Carsonon® N-100, 70%
Carsonon® ND-317
Carsonon® TD-10
Carsonon® TD-11, TD-11 70%
Carspray 205
Cassapret SRH
Catinex KB-10
Catinex KB-11
Catinex KB-13
Catinex KB-15
Catinex KB-16
Catinex KB-19
Catinex KB-20
Catinex KB-22
Catinex KB-23
Catinex KB-24
Catinex KB-25
Catinex KB-26

NONIONIC (CONT'D.)
Catinex KB-27
Catinex KB-31
Catinex KB-32
Catinex KB-40
Catinex KB-41
Catinex KB-42
Catinex KB-43
Catinex KB-44
Catinex KB-45
Catinex KB-48
Catinex KB-49
Catinex KB-50
Catinex KB-51
Cedemide AX
Cedepal CA-210
Cedepal CA-520
Cedepal CA-630
Cedepal CA-720
Cedepal CA-890
Cedepal CA-897
Cedepal CO-210
Cedepal CO-430
Cedepal CO-500
Cedepal CO-530
Cedepal CO-610
Cedepal CO-630
Cedepal CO-710
Cedepal CO-730
Cedepal CO-880
Cedepal CO-887
Cedepal CO-890
Cedepal CO-897
Cedepal CO-970
Cedepal CO-977
Cedepal CO-990
Cedepal CO-997
Cedepal FA-406
Cedepal TD-630
Ceralan®
Cerex EL 250
Cerex EL 300
Cerex EL 360
Cerex EL 400
Cerex EL 4929
Cerfak 1400
Cerfax N-100
Cetalox 8, 25
Cetalox 50
Cetalox AT
Charlab Condensate-K
Chemal BP 261
Chemal BP-2101
Chemal BP-262
Chemal BP-262LF
Chemal DA-4
Chemal DA-6
Chemal DA-9
Chemal LA-4
Chemal LA-9
Chemal LA-12
Chemal LA-23
Chemal LF 14B, 25B, 40B
Chemal LFL-10, -17, -19, -28, -38, -47
Chemal OA-4
Chemal OA-5
Chemal OA-9
Chemal OA-20/70CWS
Chemal OA-20G
Chemal TDA-3
Chemal TDA-6
Chemal TDA-9
Chemal TDA-12
Chemal TDA-15
Chemal TDA-18
Chemax AR-497
Chemax CO-5
Chemax CO-16
Chemax CO-25
Chemax CO-28
Chemax CO-30
Chemax CO-36
Chemax CO-40
Chemax CO-80
Chemax CO-200/50
Chemax DNP-8
Chemax DNP-15
Chemax DNP-18
Chemax DNP-150
Chemax DNP-150/50
Chemax E-200 ML
Chemax E-200 MO
Chemax E-200 MS
Chemax E-400 ML
Chemax E-400 MO
Chemax E-400 MS
Chemax E-600 ML
Chemax E-600 MO
Chemax E-600 MS
Chemax E-1000 MO
Chemax E-1000 MS
Chemax HCO-5
Chemax HCO-16
Chemax HCO-25
Chemax HCO-200/50
Chemax NP-1.5
Chemax NP-4
Chemax NP-6
Chemax NP-9
Chemax NP-10
Chemax NP-15
Chemax NP-20
Chemax NP-30
Chemax NP-30/70
Chemax NP-40
Chemax NP-40/70
Chemax NP-50
Chemax NP-50/70
Chemax NP-100
Chemax NP-100/70
Chemax OP-3
Chemax OP-5
Chemax OP-7
Chemax OP-9
Chemax OP-30/70
Chemax OP-40
Chemax OP-40/70
Chemax TO-8
Chemax TO-10
Chemax TO-16
Chemfac 100
Chemfac RD-1200
Chemoxide CAW
Chemoxide LM-30
Chemoxide O1
Chemsperse GMS-SE
Chimin KSP
Chimin P50
Chimipal APG 400
Chimipal DCL/M
Chimipal DE 7
Chimipal LDA
Chimipal MC
Chimipal OLD
Chimipal OS 2
Chimipal PE 300, PE 302
Chimipal PE 402, PE 403, PE 405
Chimipal PE 520
Chupol C
Cirrasol® AEN-XB
Cirrasol® AEN-XF
Cirrasol® AEN-XZ
Cirrasol® ALN-FP
Cirrasol® ALN-GM
Cirrasol® ALN-TF
Cirrasol® ALN-TS
Cirrasol® ALN-TV
Cirrasol® ALN-WF
Cirrasol® ALN-WY
Cirrasol® EN-MB
Cirrasol® EN-MP
Cirrasol® GM
Cirrasol® LAN-SF
Cithrol 2DL
Cithrol 2DO
Cithrol 2DS
Cithrol 2ML
Cithrol 2MO
Cithrol 2MS
Cithrol 3DS
Cithrol 4DL
Cithrol 4DO
Cithrol 4DS
Cithrol 4ML
Cithrol 4MO
Cithrol 4MS
Cithrol 6DL
Cithrol 6DO
Cithrol 6DS
Cithrol 6ML
Cithrol 6MO
Cithrol 6MS
Cithrol 10DL
Cithrol 10DO
Cithrol 10DS
Cithrol 10ML
Cithrol 10MO
Cithrol 10MS
Cithrol 15MS
Cithrol 40MS
Cithrol A
Cithrol DGDL N/E
Cithrol DGDO N/E
Cithrol DGDS N/E
Cithrol DGML N/E
Cithrol DGMO N/E
Cithrol DGMS N/E
Cithrol DPGML N/E
Cithrol DPGMO N/E
Cithrol DPGMS N/E
Cithrol EGDL N/E
Cithrol EGDO N/E
Cithrol EGDS N/E
Cithrol EGML N/E
Cithrol EGMO N/E
Cithrol EGMR N/E
Cithrol EGMS N/E
Cithrol GDL N/E
Cithrol GDO N/E
Cithrol GDS N/E
Cithrol GML N/E
Cithrol GMO N/E
Cithrol GMR N/E
Cithrol GMS Acid Stable
Cithrol GMS N/E
Cithrol PGML N/E
Cithrol PGMO N/E
Cithrol PGMR N/E
Cithrol PGMS N/E
Clearate B-60
Clearate LV
Clearate Special Extra
Clearate WD
Clink A-26
CNC Antifoam 30-FG
CNC PAL 210 T
CNC PAL NRW
Collone AC
Collone NI
Comperlan 100
Comperlan CD
Comperlan KDO
Comperlan LD
Comperlan LM
Comperlan LMD
Comperlan LP
Comperlan OD
Comperlan P 100
Comperlan PD
Comperlan PKDA
Comperlan PVD
Comperlan SD
Coning Oil C Special
Consamine 15
Consamine CA
Consamine OM
Consamine PA
Consoscour M
Consoscour TEK
Corexit CL578
Corexit CL8500
Corexit CL8569
Corexit CL8594
Corexit CL8662
Corexit CL8685
CPH-27-N
CPH-30-N
CPH-39-N
CPH-41-N
CPH-53-N
CPH-79-N
CPH-376-N
Cralane AT-17
Cralane KR-13, -14
Cralane LR-11
Crill 1
Crill 2
Crill 3
Crill 4
Crill 6
Crill 35
Crill 43
Crill 45
Crill 50
Crillet 1
Crillet 2
Crillet 3
Crillet 4

NONIONIC (CONT'D.)
Crillet 6
Crillet 11
Crillet 31
Crillet 35
Crillet 41
Crillet 45
Crillon CDY
Crillon LDE
Crillon ODE
Crodalan AWS
Crodapearl NI Liquid
Crodasinic L
Crodasinic O
Crodesta F-110
Crodet C10
Crodet L4
Crodet L8
Crodet L12
Crodet L24
Crodet L40
Crodet L100
Crodet O4
Crodet O8
Crodet O12
Crodet O24
Crodet O40
Crodet O100
Crodet S4
Crodet S8
Crodet S12
Crodet S24
Crodet S40
Crodet S100
Crodex N
Croduret 10
Croduret 40
Croduret 60
Croduret 100
Croduret 200
Crothix
Crovol A70
Crown Anti-Foam
Crystal Inhibitor #5
Dacospin 869
Dacospin 1735-A
Dacospin LA-704
Dacospin LD-605
Dacospin POE (25)HRG
Daisurf C
Damox® 1010
Darvan® NS
DB-19 Antifoam Compd
Deceresol Surfactant NI Conc.
Defomax
Dehscofix 906
Dehscofix 908
Dehscofix 909
Dehscoxid 700 Series
Dehscoxid 730/740 Series
Dehydol 25
Dehydol 100
Dehydol 737
Dehydol 980
Dehydol D 3
Dehydol G 202
Dehydol G 205
Dehydol HD-FC 1
Dehydol HD-FC 2
Dehydol HD-FC 4
Dehydol HD-FC 6
Dehydol HD-L 1
Dehydol LS 2
Dehydol LS 3
Dehydol LS 4
Dehydol LT 2
Dehydol LT 3
Dehydol LT 5
Dehydol LT 6
Dehydol LT 7
Dehydol LT 7 L
Dehydol O4
Dehydol O4 DEO
Dehydol O4 Special
Dehydol PCS 6
Dehydol PCS 14
Dehydol PD 253
Dehydol PEH 2
Dehydol PID 6
Dehydol PIT 6
Dehydol PLS 1
Dehydol PLS 3
Dehydol PLS 4
Dehydol PLS 6
Dehydol PLS 8
Dehydol PLS 11/80
Dehydol PLS 12
Dehydol PLS 15
Dehydol PLS 21
Dehydol PLS 21E
Dehydol PLS 235
Dehydol PLT 5
Dehydol PLT 6
Dehydol PLT 8
Dehydol PO 5
Dehydol PTA 7
Dehydol PTA 23
Dehydol PTA 23/E
Dehydol PTA 40
Dehydol PTA 80
Dehydol PTA 114
Dehydol TA 5
Dehydol TA 6
Dehydol TA 11
Dehydol TA 12
Dehydol TA 14
Dehydol TA 20
Dehydol TA 30
Dehydol WM
Dehydol WM 90
Dehydrophen 65
Dehydrophen 100
Dehydrophen 150
Dehydrophen PNP 4
Dehydrophen PNP 6
Dehydrophen PNP 8
Dehydrophen PNP 10
Dehydrophen PNP 12
Dehydrophen PNP 15
Dehydrophen PNP 20
Dehydrophen PNP 30
Dehydrophen PNP 40
Dehydrophen POP 4
Dehydrophen POP 8
Dehydrophen POP 10
Dehydrophen POP 17
Dehydrophen POP 17/80
Dehymuls FCE
Dehymuls HRE 7
Dehymuls SSO
Dehypon G 2084
Dehypon LS-24
Dehypon LS-36
Dehypon LS-45
Dehypon LS-54
Dehypon LT 24
Dehypon LT 054
Dehypon LT 104
Dehypon OD 044
Delion F-400 Series
Delion F-4000 Series
Delvet 68, 70
Demelan CM-33
Densol BP-61, -62
Deplastol
Depsodye TCA
Depsolube LNP
Depuma®
Depuma® C-306
Depuma® OB New
Dermalcare® NI
Dermalcare® POL
Desmuldo DS 3
Desonic® 1.5N
Desonic® 3K
Desonic® 4N
Desonic® 5D
Desonic® 5K
Desonic® 5N
Desonic® 6C
Desonic® 6D
Desonic® 6N
Desonic® 6T
Desonic® 7K
Desonic® 7N
Desonic® 9D
Desonic® 9K
Desonic® 9N
Desonic® 9T
Desonic® 10D
Desonic® 10N
Desonic® 10T
Desonic® 11N
Desonic® 12D
Desonic® 12K
Desonic® 12N
Desonic® 12T
Desonic® 13N
Desonic® 14D
Desonic® 15N
Desonic® 15T
Desonic® 20N
Desonic® 30C
Desonic® 30N
Desonic® 30N70
Desonic® 36C
Desonic® 40C
Desonic® 40N
Desonic® 40N70
Desonic® 50N
Desonic® 50N70
Desonic® 54C
Desonic® 100N
Desonic® 100N70
Desonic® AJ-85
Desonic® AJ-100
Desonic® DA-4
Desonic® DA-6
Desonic® LFA 144
Desonic® LFA 198
Desonic® LFB 65
Desonic® LFC 50
Desonic® LFD 97
Desonic® LFE 89
Desonic® LFO 97
Desonic® S-45
Desonic® S-100
Desonic® S-102
Desonic® S-114
Desonic® S-405
Desonic® SMO
Desonic® SMO-20
Desonic® SMT
Desonic® SMT-20
Desonic® TDA-9
Desophos® 4 NP
Desophos® 6 MPNa
Desophos® 6 NPNa
Desophos® 9 NP
Desophos® 10TP
Desophos® 30 NP
Desotan® SMO
Desotan® SMO-20
Desotan® SMT
Desotan® SMT-20
Deterflo A 210
Deterflo A 215
Deterflo A 233
Detergent 8®
Deterpal 843
Detersol HF
Detersol MC-10
Detersol X-66
Det-O-Jet®
Dexopal 555
Diamonine B
Dianol
Dianol 300
Diazital O Extra Conc.
Dionil® OC
Dionil® RS
Dionil® S 37
Dionil® SD
Dionil® SH 100
Dionil® W 100
Dispersogen ASN
Disponil AAP 307
Disponil AAP 436
Disponil APG 110
Disponil LS 4
Disponil LS 12
Disponil LS 30
Disponil O 5
Disponil O 10
Disponil O 20
Disponil O 250
Disponil RO 40
Disponil SML 100 F1
Disponil SML 104 F1
Disponil SML 120 F1
Disponil SMO 100 F1
Disponil SMO 120 F1
Disponil SMP 100 F1
Disponil SMP 120 F1
Disponil SMS 100 F1
Disponil SMS 120 F1
Disponil SSO 100 F1
Disponil STO 100 F1
Disponil STO 120 F1

NONIONIC (CONT'D.)
Empigen® OH25
Empigen® OS/A
Empigen® OY
Empilan® 0004
Empilan® 2502
Empilan® 7132
Empilan® BQ 100
Empilan® CDE
Empilan® CDX
Empilan® CIS
Empilan® CM
Empilan® CM/F
Empilan® CME
Empilan® DL 40
Empilan® FD
Empilan® FD20
Empilan® FE
Empilan® GMS LSE40
Empilan® GMS LSE80
Empilan® GMS MSE40
Empilan® GMS NSE40
Empilan® GMS NSE90
Empilan® GMS SE32
Empilan® GMS SE40
Empilan® GMS SE70
Empilan® KA3
Empilan® KA5
Empilan® KA5/90
Empilan® KA8/80
Empilan® KA10/80
Empilan® KB 2
Empilan® KB 3
Empilan® KC 3
Empilan® KCA Series
Empilan® KCB Series
Empilan® KCL Series
Empilan® KCMP 0703/F
Empilan® KCMP 0705/F
Empilan® KCP Series
Empilan® KCX Series
Empilan® KI Series
Empilan® KL 6
Empilan® KL 10
Empilan® KL 20
Empilan® KM 11
Empilan® KM 20
Empilan® KM 50
Empilan® KS Series
Empilan® LDE
Empilan® LDX
Empilan® LIS
Empilan® LME
Empilan® LP2
Empilan® LP10
Empilan® MAA
Empilan® NP9
Empilan® OPE9.5
Empilan® P7061
Empilan® P7062
Empilan® P7087
Empilan® PF7158
Empilan® PF7159
Emsorb® 2500
Emsorb® 2502
Emsorb® 2503
Emsorb® 2505
Emsorb® 2507
Emsorb® 2510
Emsorb® 2515
Emsorb® 2516
Emsorb® 2518
Emsorb® 2722
Emsorb® 2728
Emsorb® 6900
Emsorb® 6901
Emsorb® 6903
Emsorb® 6906
Emsorb® 6908
Emsorb® 6909
Emsorb® 6910
Emsorb® 6913
Emsorb® 6915
Emsorb® 6916
Emsorb® 6917
Emthox® 5882
Emthox® 5941
Emthox® 5942
Emthox® 5993
Emthox® 6957
Emthox® 6960
Emthox® 6961
Emthox® 6962
Emulamid FO-5DF
Emulamid TO-21
Emulan® A
Emulan® AF
Emulan® AT 9
Emulan® EL
Emulan® FJ
Emulan® FM
Emulan® NP 2080
Emulan® NP 3070
Emulan® OC
Emulan® OG
Emulan® OK 5
Emulan® OP 25
Emulan® OSN
Emulan® OU
Emulan® P
Emulan® PO
Emulbon GB-90
Emulbon S-83
Emulbon T-20
Emulbon T-80
Emulgane E
Emulgane FL 7
Emulgane O
Emulgane OP
Emulgateur SO
Emulgator 64
Emulgator E-2155 SE
Emulgator F-8
Emulgator U4
Emulgator U6
Emulgator U9
Emulgator U12
Emulgeant 710
Emulgeant 900
Emulgin PA Series
Emulmin 40
Emulmin 50
Emulmin 70
Emulmin 140
Emulmin 240
Emulmin 862
Emulmin L-380
Emulpon A
Emulpon EL 18
Emulpon EL 20
Emulpon EL 33
Emulpon EL 40
Emulpon HT
Emulsan A-67
Emulsan BRW
Emulsan K
Emulsan O
Emulsifier 7X
Emulsifier 632/90%
Emulsifier DMR
Emulsit 16
Emulsit 25
Emulsit 49
Emulsit 100
Emulsogen 2144
Emulsogen A
Emulsogen AN
Emulsogen EL-050
Emulsogen EL-300
Emulsogen EL-400
Emulsogen M
Emulsogen N-060
Emulsogen NP 11-300
Emulsol CO 45
Emulson AG 2A
Emulson AG 4B
Emulson AG 7B
Emulson AG 8A
Emulson AG 9B
Emulson AG 10B
Emulson AG 12B
Emulson AG 13A
Emulson AG 15A
Emulson AG 20B
Emulson AG 24A
Emulson AG 81C
Emulson AG/CO 25
Emulson AG/COH
Emulson AG/EL
Emulson AG/PE
Emulson CO 25
Emulson CO-40
Emulson CO 40 N
Emulson EL
Emulson EL 200
Emulson EL/H25
Emultex 1302
Emultex 1502, 1515
Emultex 1602
Emultex SMS
Emultex WS
Emulvin S
Emulvin W
Ensital CAR
Ensital CF 2708C
Ensital FMH
Ensital FPA 205
Epan 710
Epan 720
Epan 740
Epan 750
Epan 785
Epan U 102
Epan U 103
Epan U 105
Epan U 108
Esi-Det CDA
Esi-Terge 10
Esi-Terge B-15
Esi-Terge N-100
Esi-Terge S-10
Esi-Terge T-5
Estrasan 1
Estrasan 3
Estrasan 4G
Estrasan 4L
Ethal 326
Ethal 368
Ethal 926
Ethal 3328
Ethal CSA-25
Ethal DA-4
Ethal DA-6
Ethal DA-9
Ethal DNP-8
Ethal DNP-18
Ethal DNP-150/50%
Ethal EH-2
Ethal EH-5
Ethal LA-4
Ethal LA-7
Ethal NP-1.5
Ethal NP-4
Ethal NP-6
Ethal NP-9
Ethal NP-20
Ethal NP-407
Ethal OA-10
Ethal OA-23
Ethal TDA-3
Ethal TDA-6
Ethal TDA-9
Ethal TDA-12
Ethal TDA-18
Ethofat® 18/14
Ethofat® 60/15
Ethofat® 60/20
Ethofat® 60/25
Ethofat® 242/25
Ethofat® 433
Ethofat® C/15
Ethofat® C/25
Ethofat® O/15
Ethofat® O/20
Ethomid® HT/15
Ethomid® HT/23
Ethomid® HT/60
Ethomid® O/15
Ethomid® O/17
Ethosperse® CA-2
Ethosperse® G-26
Ethosperse® LA-4
Ethosperse® LA-12
Ethosperse® LA-23
Ethosperse® SL-20
Ethosperse® TDA-6
Ethotal CH 5
Ethox 2423
Ethox 2610
Ethox 2672
Ethox 2966
Ethox CO-5
Ethox CO-16
Ethox CO-25
Ethox CO-30
Ethox CO-36
Ethox CO-40
Ethox CO-200
Ethox CO-200/50%
Ethox DO-9

NONIONIC (CONT'D.)
Ethox DO-14
Ethox DTO-9A
Ethox HCO-16
Ethox HCO-25
Ethox HCO-200/50%
Ethox HO-50
Ethox MA-8
Ethox MA-15
Ethox ML-5
Ethox ML-9
Ethox ML-14
Ethox MO-9
Ethox MO-14
Ethox MS-8
Ethox MS-14
Ethox MS-23
Ethox MS-40
Ethox PPG 1025 DTO
Ethox SO-9
Ethox TO-8
Ethox TO-9A
Ethox TO-16
Ethsorbox L-20
Ethsorbox O-20
Ethsorbox S-20
Ethsorbox TO-20
Ethsorbox TS-20
Ethylan® 20
Ethylan® 44
Ethylan® 55
Ethylan® 77
Ethylan® 172
Ethylan® A2
Ethylan® A3
Ethylan® A4
Ethylan® A6
Ethylan® A15
Ethylan® ABB10
Ethylan® ABC20
Ethylan® BAB20
Ethylan® BBC31
Ethylan® BCD42
Ethylan® BCP
Ethylan® BKL130
Ethylan® BNE15
Ethylan® BV
Ethylan® BZA
Ethylan® C12AH
Ethylan® C30
Ethylan® C35
Ethylan® C40AH
Ethylan® C75AH
Ethylan® C160
Ethylan® C404
Ethylan® CD103
Ethylan® CD107
Ethylan® CD109
Ethylan® CD122
Ethylan® CD123
Ethylan® CD124
Ethylan® CD127
Ethylan® CD128
Ethylan® CD129
Ethylan® CD175
Ethylan® CD802
Ethylan® CD913
Ethylan® CD916
Ethylan® CD919
Ethylan® CD964
Ethylan® CD1210
Ethylan® CD1230
Ethylan® CD1260
Ethylan® CD4511
Ethylan® CD9112
Ethylan® CD9130
Ethylan® CDP2
Ethylan® CDP3
Ethylan® CDP16
Ethylan® CF71
Ethylan® CH
Ethylan® CPG 630
Ethylan® CPG 660
Ethylan® CPG 745
Ethylan® CPG 945
Ethylan® CPG 7545
Ethylan® CRS
Ethylan® CS20
Ethylan® CX138
Ethylan® CX308
Ethylan® D252
Ethylan® D253
Ethylan® D254
Ethylan® D256
Ethylan® D257
Ethylan® D259
Ethylan® D2512
Ethylan® D2560 Flake
Ethylan® DNP16
Ethylan® DP
Ethylan® ENTX
Ethylan® FO30
Ethylan® FO60
Ethylan® GD
Ethylan® GEL2
Ethylan® GEO8
Ethylan® GEO81
Ethylan® GEP4
Ethylan® GES6
Ethylan® GL20
Ethylan® GLE-21
Ethylan® GMF
Ethylan® GO80
Ethylan® GOE-21
Ethylan® GP-40
Ethylan® GPS85
Ethylan® GS60
Ethylan® GT85
Ethylan® HA Flake
Ethylan® HB Series
Ethylan® HB1
Ethylan® HB4
Ethylan® HB15
Ethylan® HB30
Ethylan® HP
Ethylan® KELD
Ethylan® KEO
Ethylan® LD
Ethylan® LDA-37
Ethylan® LDA-48
Ethylan® LDG
Ethylan® LDS
Ethylan® LM
Ethylan® LM2
Ethylan® ME
Ethylan® MLD
Ethylan® MPA
Ethylan® N30
Ethylan® N50
Ethylan® N92
Ethylan® NK4
Ethylan® NP 1
Ethylan® OE
Ethylan® PQ
Ethylan® R
Ethylan® TB345
Ethylan® TC
Ethylan® TCO
Ethylan® TD3
Ethylan® TD10
Ethylan® TD15
Ethylan® TF
Ethylan® TH-2
Ethylan® TH-30
Ethylan® TLM
Ethylan® TN-10
Ethylan® TT-05
Ethylan® TT-07
Ethylan® TT-15
Ethylan® TT-30
Ethylan® TT-40
Ethylan® TT-203
Ethylan® TU
Ethylan® VPK
Etocas 10
Etocas 20
Etocas 35
Etocas 40
Etocas 50
Etocas 60
Etocas 100
Etocas 200
Etophen 102
Etophen 103
Etophen 105
Etophen 106
Etophen 107
Etophen 108
Etophen 109
Etophen 110
Etophen 112
Etophen 114
Etophen 120
Eumulgin 286
Eumulgin 535
Eumulgin 2142
Eumulgin 2312
Eumulgin B1
Eumulgin B2
Eumulgin B3
Eumulgin C4
Eumulgin C8
Eumulgin EP 2
Eumulgin EP 2L
Eumulgin EP 5L
Eumulgin ET 2
Eumulgin ET 5
Eumulgin ET 5L
Eumulgin ET 10
Eumulgin KP92
Eumulgin L
Eumulgin M8
Eumulgin O5
Eumulgin O10
Eumulgin PA 10
Eumulgin PA 12
Eumulgin PA 20
Eumulgin PA 30
Eumulgin PAEH 4
Eumulgin PC 2
Eumulgin PC 4
Eumulgin PC 10
Eumulgin PC 10/85
Eumulgin PK 23
Eumulgin PLT 4
Eumulgin PLT 5
Eumulgin PLT 6
Eumulgin PPG 40
Eumulgin PRT 36
Eumulgin PRT 40
Eumulgin PRT 56
Eumulgin PRT 200
Eumulgin PST 5
Eumulgin PTL 4
Eumulgin PTL 5
Eumulgin PTL 6
Eumulgin PWM2
Eumulgin PWM5
Eumulgin PWM10
Eumulgin PWM17
Eumulgin PWM25
Eumulgin RO 40
Eumulgin RT 5
Eumulgin RT 11
Eumulgin RT 20
Eumulgin RT 40
Eumulgin ST-8
Eumulgin TI 60
Eumulgin TL 30
Eumulgin TL 55
Eumulgin WM5
Eumulgin WM 7
Eumulgin WM 10
Eumulgin WO 7
Examide N-LS
Examide-DA
Fancol LA
Fancol LAO
Fancol OA 95
Fancor LFA
Farmin 2C
Farmin 20, 60, 68, 80, 86, AB, C
Farmin D86
Farmin DM20
Farmin DM40, 60, 80, 86
Farmin DMC
Farmin HT
Farmin O
Farmin S
Farmin T
Felton 3T
Flexricin® 9
Flexricin® 13
Flexricin® 15
Flo-Mo® 80/20
Flo-Mo® 1002
Flo-Mo® 1031
Flo-Mo® AJ-100
Flo-Mo® AJ-85
Flo-Mo® Low Foam
Fluilan
Fluorad FC-170-C
Fluorad FC-171
Fluorad FC-430
Fluorad FC-431
Fluorad FC-740
Fluorad FC-742
Fluorad FC-760
Foamaster 206A,267A

NONIONIC (CONT'D.)
Foamaster 1407-50
Foamaster 1719-A
Foamaster 8034
Foamaster B
Foamaster DD-72
Foamaster DF-122NS
Foamaster DF-177-F, -178
Foamaster DNH-1
Foamaster DR-187
Foamaster JMY
Foamaster NDW
Foamaster NS-20
Foamaster P
Foamaster PD-1
Foamaster TDB
Foamaster V
Foamaster VL
Foamkill® 652
Foamkill® 652B
Foamkill® CMP
Forbest 560
Forbest 620
Forbest 680
Forbest 1000B
Forbest 1500W
Forbest 2000C
Forbest 8209
G-1045A
G-1086
G-1087
G-1089
G-1096
G-1120
G-1121
G-1292
G-1293
G-1300
G-1556
G-1564
G-2109
G-2162
G-3780-A
G-3886
G-3890
G-4252
G-4280
G-5000
G-7076
Geitol RC-100
Geleol
Genagen CA-050
Genagen CD
Genaminox KC
Genapol® 24-L-3
Genapol® 24-L-45
Genapol® 24-L-50
Genapol® 24-L-60
Genapol® 24-L-60N
Genapol® 24-L-75
Genapol® 24-L-92
Genapol® 24-L-98N
Genapol® 26-L-1
Genapol® 26-L-1.6
Genapol® 26-L-2
Genapol® 26-L-3
Genapol® 26-L-5
Genapol® 26-L-45
Genapol® 26-L-50
Genapol® 26-L-60
Genapol® 26-L-60N
Genapol® 26-L-75
Genapol® 26-L-98N
Genapol® 42-L-3
Genapol® 2317
Genapol® 2822, 2908, 2909
Genapol® 3504
Genapol® AMG
Genapol® B
Genapol® C-050
Genapol® C-080
Genapol® C-100
Genapol® C-150
Genapol® C-200
Genapol® GEV
Genapol® L Grades
Genapol® O-020
Genapol® O-050
Genapol® O-080
Genapol® O-090
Genapol® O-100
Genapol® O-120
Genapol® O-150
Genapol® O-200
Genapol® O-230
Genapol® OX Grades
Genapol® PAF, PL, PN Brands
Genapol® PF 10
Genapol® PF 20
Genapol® PF 40
Genapol® PF 80
Genapol® PL 120
Genapol® PN-30
Genapol® PS
Genapol® S-020
Genapol® S-050
Genapol® S-080
Genapol® S-100
Genapol® S-150
Genapol® S-200
Genapol® S-250
Genapol® T Grades
Genapol® V 2908
Genapol® V 2909
Genapol® X Grades
Genapol® X-040
Genapol® X-060
Genapol® X-080
Genapol® X-100
Generol® 122E25
Geronol AG-821
Geronol AG-900
Geronol MOE/2/N
Geronol SC/138
Geronol SC/177
Geropon® 40/D
Geropon® CET/50/P
Geropon® K/202
Glazamine DP2
Glazamine M
Glucam® E-10
Glucam® P-10
Glucam® P-20
Glucamate® SSE-20
Glucate® DO
Glucate® SS
Glucopon 225
Glucopon 425
Glucopon 600
Glucopon 625
Glycidol Surfactant 10G
Glycomul® L
Glycomul® MA
Glycomul® O
Glycomul® P
Glycomul® S
Glycomul® S FG
Glycomul® S KFG
Glycomul® SOC
Glycomul® TAO
Glycomul® TO
Glycomul® TS
Glycomul® TS KFG
Glycosperse® L-20
Glycosperse® O-20
Glycosperse® O-20 FG, O-20 KFG
Glycosperse® O-20 Veg
Glycosperse® O-20X
Glycosperse® O-5
Glycosperse® P-20
Glycosperse® S-20
Glycosperse® TO-20
Glycosperse® TS-20
Glycox® PETC
GP-209
GP-210 Silicone Antifoam Emulsion
GP-214
GP-310-I
GP-330-I Antifoam Emulsion
Gradonic 400-ML
Gradonic FA-20
Gradonic LFA Series
Gradonic N-95
Grilloten LSE87
Grilloten LSE87K
Grilloten PSE141G
Grindtek SMS
Grindtek STS
Griton Series
Hallcomid® M-18-OL
Hartamide AD
Hartamide LDA
Hartamide OD
Hartolon 5683
Hartolon AL
Hartolon HVH Base
Hartolon NA
Hartolon PC
Hartomul PE-30
Hartopol 25R2
Hartopol 31R1
Hartopol L42
Hartopol L44
Hartopol L62
Hartopol L62LF
Hartopol L64
Hartopol L81
Hartopol LF-1
Hartopol LF-2
Hartopol LF-5
Hartopol P65
Hartopol P85
Hartowet CW
Hartowet SLN
Hercules® AR150
Hercules® AR160
Hercules® Surfactant AR 150
Hercules® Surfactant AR 160
Hetamide LA
Hetamide MC
Hetamide ML
Hetamide MMC, OC
Hetamide RC
Hetan SL
Hetan SO
Hetan SS
Hetester PCA
Hetester PHA
Hetoxamate 200 DL
Hetoxamate FA-5
Hetoxamate FA-20
Hetoxamate LA-5
Hetoxamate LA-9
Hetoxamate MO-5
Hetoxamate MO-9
Hetoxamate MO-15
Hetoxamate SA-5
Hetoxamate SA-7
Hetoxamate SA-9
Hetoxamate SA-35
Hetoxamate SA-40
Hetoxamate SA-90
Hetoxamate SA-90F
Hetoxamine O-2
Hetoxamine O-5
Hetoxamine O-15
Hetoxamine ST-2
Hetoxamine ST-5
Hetoxamine ST-15
Hetoxamine ST-50
Hetoxamine T-15
Hetoxamine T-20
Hetoxamine T-30
Hetoxide BN-13
Hetoxide BP-3
Hetoxide BY-3
Hetoxide C-2
Hetoxide C-9
Hetoxide C-15
Hetoxide C-25
Hetoxide C-30
Hetoxide C-40
Hetoxide C-60
Hetoxide C-200
Hetoxide DNP-4
Hetoxide DNP-5
Hetoxide DNP-9.6
Hetoxide DNP-10
Hetoxide G-7
Hetoxide G-26
Hetoxide HC-16
Hetoxide HC-25
Hetoxide HC-40
Hetoxide NP-4
Hetoxide NP-6
Hetoxide NP-9
Hetoxide NP-10
Hetoxide NP-12
Hetoxide NP-15-85%
Hetoxide NP-30
Hetoxide NP-40
Hetoxide NP-50
Hetoxol C-24

NONIONIC (CONT'D.)
Igepal® DX-430
Igepal® ES
Igepal® F-85
Igepal® F-920
Igepal® FN-11
Igepal® FN-30
Igepal® FN-9.5
Igepal® KA
Igepal® LAVE
Igepal® NP-2
Igepal® NP-4
Igepal® NP-5
Igepal® NP-6
Igepal® NP-8
Igepal® NP-9
Igepal® NP-10
Igepal® NP-12
Igepal® NP-14
Igepal® NP-17
Igepal® NP-20
Igepal® NP-30
Igepal® O
Igepal® OD-410
Igepal® RC-520
Igepal® RC-620
Igepal® RC-630
Igepal® SS-837
Imacol JN
Imperon Emulsifier 774
Imperon Thickener A
Imwitor® 191
Imwitor® 312
Imwitor® 370
Imwitor® 375
Imwitor® 595
Imwitor® 742
Imwitor® 780
Imwitor® 780 K
Imwitor® 900
Imwitor® 908
Imwitor® 910
Imwitor® 914
Imwitor® 940
Imwitor® 940 K
Incrocas 30
Incrocas 40
Incromide BAD
Incromide CA
Incromide CM
Incromide L-90
Incromide LA
Incromide LM-70
Incromide LR
Incromide Mink D
Incromide OD
Incromide OLD
Incromide SED
Incromide WGD
Incromine Oxide I
Incromine Oxide ISMO
Incromine Oxide L
Incromine Oxide L-40
Incromine Oxide M
Incromine Oxide MC
Incromine Oxide Mink
Incromine Oxide O
Incromine Oxide OD-50
Incromine Oxide OL
Incromine Oxide S
Incromine Oxide SE
Incromine Oxide WG
Incropol CS-20
Incropol CS-50
Incropol L-7
Indulin® XD-70
Industrol® 400-MOT
Industrol® 1025-DT
Industrol® CO-25
Industrol® CO-30
Industrol® CO-36
Industrol® CO-40
Industrol® CO-80-80%
Industrol® CO-200
Industrol® CO-200-50%
Industrol® COH-25
Industrol® COH-200
Industrol® CSS-25
Industrol® DL-9
Industrol® DO-9
Industrol® DO-13
Industrol® DT-13
Industrol® DW-5
Industrol® L-20-S
Industrol® LG-70
Industrol® LG-100
Industrol® MIS-9
Industrol® ML-5
Industrol® ML-9
Industrol® ML-14
Industrol® MO-5
Industrol® MO-6
Industrol® MO-13
Industrol® MS-5
Industrol® MS-7
Industrol® MS-8
Industrol® MS-9
Industrol® MS-23
Industrol® MS-40
Industrol® MS-100
Industrol® O-20-S
Industrol® OAL-20
Industrol® S-20-S
Industrol® SO-13
Industrol® STS-20-S
Industrol® TFA-8
Industrol® TFA-15, TFA-15-80%
Industrol® TO-10
Industrol® TO-16
Infrasan C
Intermediate 300
Intermediate 325
Intermediate 512
Interwet® 33
Intex Scour Base 706
Intralan® Salt N
Intrapol 1014
Intratex® AN
Intratex® OR
Intratex® SCS
Intravon® JF
Intravon® JU
Intravon® NI
Inversol 140
Inversol 170
Ionet AT-140
Ionet DL-200
Ionet DO-200
Ionet DO-400
Ionet DO-600
Ionet DO-1000
Ionet DS-300
Ionet DS-400
Ionet MO-200
Ionet MO-400
Ionet MS-400
Ionet MS-1000
Ionet S-20
Ionet S-60 C
Ionet S-80
Ionet S-85
Ionet T-20 C
Ionet T-60 C
Ionet T-80 C
Irgalube® 53
Juniorlan 1664
Kamar L-15
Kamar LF-7
Karamide CO2A
Karamide CO9A
Karamide CO22
Karamide ST-DEA
Kemester® 104
Kemester® 1000
Kemester® 2000
Kemester® 4000
Kemester® 5500
Kemester® 6000
Kerapol 791
Kerasol 1014
Kerinol 2012 F
Kerinol C 109
Kessco® 653
Kessco GMS
Kessco PGMS
Kieralon® D
Kieralon® JET
Kieralon® NB-150
Kieralon® TX-199
Kito 703
Klucel® E, G, H, J, L, M
Labrafil ISO
Labrafil WL 1958 CS
Labrasol
Lactomul 461
Lactomul 463
Lactomul 466
Lactomul 468
Lactomul 843
Lactomul 925
Lactomul CN-28
Lamacit GML 20
Lamigen ES 30
Lamigen ES 60
Lamigen ES 100
Lamigen ET 20
Lamigen ET 70
Lamigen ET 90
Lamigen ET 180
LAN-401
Lanapex TM
Landemul DCC Series
Lanesta G
Laneto 40
Laneto 50
Laneto AWS
Lanexol AWS
Lanfrax®
Lanfrax® 1776
Lanogel® 21
Lanogel® 31
Lanogel® 41
Lanogel® 61
Lanolin Alcohols LO
Lanoxyl 30
Lantrol® PLN
Lantrol® PLN/50
Larosol NLA-25
Larosol PNC
L.A.S
Lauramide 11
Lauramide D
Lauramide EG
Lauramide ME
Lauramide R
Lauramide S
Lauramide Special
Laurel PEG 400 DT
Laurel PEG 400 MT
Laurel PEG 600 DT
Laurel PEG 600 MT
Laurel SD-101
Laurel SD-120
Laurel SD-140N
Laurel SD-180
Laurel SD-300
Laurel SD-400
Laurel SD-520T
Laurel SD-750
Laurel SD-800
Laurel SD-900M
Laurel SD-950
Laurel SD-1000
Laurel SD-1031
Laurel SD-1050
Laurel SD-LOA
Laurex® CS
Laurex® CS/D
Laurex® CS/W
Lauridit® KD, KDG
Lauridit® KM
Lauridit® LM
Lauridit® LMI
Lauridit® OD
Lauridit® PPD
Lauridit® SDG
Lauropal 2
Lauropal 9
Lauropal 11
Lauropal 0205
Lauropal 0207L
Lauropal 0227
Lauropal 950
Lauropal 1150
Lauropal X 1003
Lauropal X 1005
Lauropal X 1007
Lauropal X 1103
Lauropal X 1105
Lauropal X 1107
Lauropal X 1203
Lauropal X 1207
Lauroxal 3
Lauroxal 6
Lauroxal 8
Laventin® CW
Laventin® W
Lenetol B Conc.
Lenetol HP-LFN

NONIONIC (CONT'D.)
Leocon 1020B
Leocon 1070B
Leocon PL-71L
Leomin HSG
Leomin OR
Levelan® A0192
Levelan® P148
Levelan® P208
Levelan® P357
Levelan® PG 434
Levelene
Levenol A Conc.
Levenol DS-1
Levenol PW
Levenol TD-326
Levenol WX, WZ
Lexemul® 503
Lexemul® 515
Lexemul® EGDS
Lexemul® EGMS
Lexemul® PEG-200 DL
Lexemul® PEG-400 DL
Lexemul® PEG-400ML
Lexin K
Lipal DGMS
Lipo DGLS
Lipo EGMS
Lipocol C-2
Lipocol C-10
Lipocol C-20
Lipocol L-4
Lipocol L-12
Lipocol L-23
Lipocol O-2
Lipocol O-5
Lipocol O-10
Lipocol O-20
Lipocol S-2
Lipocol S-10
Lipocol S-20
Lipocol SC-4
Lipocol SC-15
Lipocol SC-20
Lipocol TD-6
Lipocol TD-12
Lipolan
Lipomulse 165
Liponic EG-1
Liponox LCF
Liponox NC 2Y
Liponox NC 6E, NCG, NCI, NCT
Liponox NC-70
Liponox NC-95
Liponox NC-200
Liponox NC-300
Liponox NC-500
Liponox OCS
Lipoproteol LCO
Liposorb L
Liposorb L-10
Liposorb L-20
Liposorb O
Liposorb O-5
Liposorb P
Liposorb P-20
Liposorb S
Liposorb SQO
Liposorb TO
Liposorb TO-20
Liposorb TS
Liptol 40C
Liptol R-4000
Liptol S-2800
Lonzest® PEG 4-DO
Lonzest® PEG 4-L
Lonzest® PEG 4-O
Lonzest® SML
Lonzest® SML-20
Lonzest® SMO
Lonzest® SMO-20
Lonzest® SMP
Lonzest® SMP-20
Lonzest® SMS
Lonzest® SMS-20
Lonzest® STO
Lonzest® STO-20
Lonzest® STS
Lonzest® STS-20
Loropan CME
Loropan KD
Loropan LD
Loropan LM
Loropan LMD
Loxiol VPG 1354
Loxiol VPG 1743
Lubrizol® 2153
Lubrizol® 2155
Lubrol N5
Lumiten® N
Luprintol® PE
Lutensol® A 8
Lutensol® A 80
Lutensol® AO 3
Lutensol® AO 4
Lutensol® AO 5
Lutensol® AO 7
Lutensol® AO 8
Lutensol® AO 10
Lutensol® AO 11
Lutensol® AO 12
Lutensol® AO 30
Lutensol® AO 109
Lutensol® AO 3109
Lutensol® AP 6
Lutensol® AP 7
Lutensol® AP 8
Lutensol® AP 9
Lutensol® AP 10
Lutensol® AP 14
Lutensol® AP 20
Lutensol® AP 30
Lutensol® AT 11
Lutensol® AT 18
Lutensol® AT 25
Lutensol® AT 50
Lutensol® AT 80
Lutensol® ED 140
Lutensol® ED 310
Lutensol® ED 370
Lutensol® ED 610
Lutensol® FA 12
Lutensol® FSA 10
Lutensol® GD 50, GD 70
Lutensol® LF 220, 221, 223, 224
Lutensol® LF 400
Lutensol® LF 401
Lutensol® LF 403
Lutensol® LF 404, 405
Lutensol® LF 431
Lutensol® LF 600
Lutensol® LF 700
Lutensol® LF 711
Lutensol® LF 1300
Lutensol® LSV
Lutensol® LT 30
Lutensol® ON 30
Lutensol® ON 50
Lutensol® ON 60
Lutensol® ON 70
Lutensol® ON 80
Lutensol® ON 110
Lutensol® TO 3
Lutensol® TO 5
Lutensol® TO 7
Lutensol® TO 8
Lutensol® TO 10
Lutensol® TO 12
Lutensol® TO 15
Lutensol® TO 20
Lutensol® TO 89
Lutensol® TO 109
Lutensol® TO 129
Lutensol® TO 389
Lutrol® E 1500
Lutrol® E 4000
Lutrol® E 6000
Lutrol® F 127
Mackamide 100-A
Mackamide AME-75, AME-100
Mackamide AN55
Mackamide C
Mackamide CD
Mackamide CD-6
Mackamide CD-8
Mackamide CD-10
Mackamide CD-25
Mackamide CDC
Mackamide CDM
Mackamide CDS-80
Mackamide CDT
Mackamide CDX
Mackamide CMA
Mackamide CS
Mackamide CSA
Mackamide EC
Mackamide ISA
Mackamide L10
Mackamide L95
Mackamide LLM
Mackamide LMD
Mackamide LME
Mackamide LMM
Mackamide LOL
Mackamide MC
Mackamide MO
Mackamide NOA
Mackamide O
Mackamide ODM
Mackamide OP
Mackamide PK
Mackamide PKM
Mackamide R
Mackamide S
Mackamide SD
Mackamide SMA
Mackamine IAO
Mackamine LAO
Mackamine LO
Mackamine O2
Mackamine OAO
Mackazoline C
Mackazoline L
Mackazoline O
Mackine 601
Macol® 1
Macol® 2
Macol® 2D
Macol® 2LF
Macol® 4
Macol® 8
Macol® 10
Macol® 15
Macol® 15-20
Macol® 19
Macol® 20
Macol® 21
Macol® 22
Macol® 23
Macol® 27
Macol® 30
Macol® 31
Macol® 32
Macol® 33
Macol® 34
Macol® 35
Macol® 40
Macol® 42
Macol® 44
Macol® 46
Macol® 72
Macol® 77
Macol® 85
Macol® 88
Macol® 97
Macol® 99A
Macol® 101
Macol® 108
Macol® 300
Macol® 660
Macol® 3520
Macol® 5100
Macol® CA-2
Macol® CA-10
Macol® CSA-2
Macol® CSA-4
Macol® CSA-10
Macol® CSA-15
Macol® CSA-20
Macol® CSA-40
Macol® CSA-50
Macol® DNP-5
Macol® DNP-10
Macol® DNP-15
Macol® DNP-21
Macol® DNP-150
Macol® LA-4
Macol® LA-9
Macol® LA-12
Macol® LA-23
Macol® LA-790
Macol® LF-110
Macol® LF-111
Macol® LF-120
Macol® NP-4
Macol® NP-5
Macol® NP-6

NONIONIC (CONT'D.)
Macol® NP-8
Macol® NP-9.5
Macol® NP-11
Macol® NP-12
Macol® NP-15
Macol® NP-20
Macol® NP-20 (70)
Macol® NP-30 (70)
Macol® NP-100
Macol® OA-2
Macol® OA-4
Macol® OA-5
Macol® OA-10
Macol® OA-20
Macol® OP-3
Macol® OP-5
Macol® OP-8
Macol® OP-10
Macol® OP-10 SP
Macol® OP-12
Macol® OP-16 (75)
Macol® OP-30 (70)
Macol® OP-40 (70)
Macol® P-500
Macol® P-1200
Macol® P-2000
Macol® P-4000
Macol® SA-2
Macol® SA-5
Macol® SA-10
Macol® SA-15
Macol® SA-20
Macol® SA-40
Macol® TD-3
Macol® TD-8
Macol® TD-10
Macol® TD-12
Macol® TD 15
Macol® TD-100
Macol® TD-610
Madeol AG/TR 8
Madeol AG/TR 12
Makon® 4
Makon® 6
Makon® 7
Makon® 8
Makon® 10
Makon® 11
Makon® 12
Makon® 14
Makon® 30
Makon® 40
Makon® 50
Makon® NF-5
Makon® NF-12
Makon® NI 10, NI 20, NI 30
Makon® OP-6
Makon® OP-9
Manromid 150-ADY
Manromid 1224
Manromid LMA
Manromine 853
Mapeg® 200 DC
Mapeg® 200 DL
Mapeg® 200 DO
Mapeg® 200 DOT
Mapeg® 200 DS
Mapeg® 200 ML
Mapeg® 200 MO
Mapeg® 200 MOT
Mapeg® 200 MS
Mapeg® 400 DL
Mapeg® 400 DO
Mapeg® 400 DOT
Mapeg® 400 DS
Mapeg® 400 ML
Mapeg® 400 MO
Mapeg® 400 MOT
Mapeg® 400 MS
Mapeg® 600 DL
Mapeg® 600 DO
Mapeg® 600 DOT
Mapeg® 600 DS
Mapeg® 600 ML
Mapeg® 600 MO
Mapeg® 600 MOT
Mapeg® 600 MS
Mapeg® 1000 MS
Mapeg® 1500 MS
Mapeg® 1540 DS
Mapeg® 6000 DS
Mapeg® CO-16
Mapeg® CO-16H
Mapeg® CO-25
Mapeg® CO-25H
Mapeg® CO-30
Mapeg® CO-36
Mapeg® CO-200
Mapeg® DGLD
Mapeg® EGDS
Mapeg® EGMS
Mapeg® S-40
Mapeg® S-40K
Mapeg® TAO-15
Marchon® DC 1102
Markwet NR-25
Marlamid® D 1218
Marlamid® D 1885
Marlamid® DF 1218
Marlamid® DF 1818
Marlamid® KL
Marlamid® M 1218
Marlamid® M 1618
Marlamid® PG 20
Marlazin® L 2
Marlazin® L 410
Marlazin® T 50
Marlazin® T 410
Marlipal® 1/12
Marlipal® 011/30
Marlipal® 011/50
Marlipal® 011/79
Marlipal® 011/88
Marlipal® 011/110
Marlipal® 013/20
Marlipal® 013/30
Marlipal® 013/50
Marlipal® 013/60
Marlipal® 013/70
Marlipal® 013/80
Marlipal® 013/90
Marlipal® 013/100
Marlipal® 013/120
Marlipal® 013/150
Marlipal® 013/170
Marlipal® 013/200
Marlipal® 013/400
Marlipal® 013/939
Marlipal® 24/20
Marlipal® 24/30
Marlipal® 24/40
Marlipal® 24/50
Marlipal® 24/60
Marlipal® 24/70
Marlipal® 24/80
Marlipal® 24/90
Marlipal® 24/100
Marlipal® 24/110
Marlipal® 24/120
Marlipal® 24/140
Marlipal® 24/150
Marlipal® 24/200
Marlipal® 24/300
Marlipal® 24/939
Marlipal® 34/30, /50, /60, /70, /79, /99, /100, /109, /110, /119, /120, /140
Marlipal® 104
Marlipal® 124
Marlipal® 129
Marlipal® 1012/4
Marlipal® 1012/6
Marlipal® 1218/5
Marlipal® 1618/6
Marlipal® 1618/8
Marlipal® 1618/10
Marlipal® 1618/11
Marlipal® 1618/18
Marlipal® 1618/25
Marlipal® 1618/25 P 6000
Marlipal® 1618/40
Marlipal® 1618/80
Marlipal® 1850/5
Marlipal® 1850/10
Marlipal® 1850/30
Marlipal® 1850/40
Marlipal® 1850/80
Marlipal® BS
Marlipal® FS
Marlipal® KE
Marlipal® KF
Marlipal® ML
Marlipal® NE
Marlipal® O11/30
Marlipal® O11/50
Marlipal® O11/79
Marlipal® O11/88
Marlipal® O11/110
Marlipal® O13/20
Marlipal® O13/30
Marlipal® O13/40
Marlipal® O13/50
Marlipal® O13/60
Marlipal® O13/70
Marlipal® O13/80
Marlipal® O13/90
Marlipal® O13/100
Marlipal® O13/120
Marlipal® O13/150
Marlipal® O13/170
Marlipal® O13/200
Marlipal® O13/400
Marlipal® O13/500
Marlipal® O13/939
Marlipal® SU
Marlophen® 81
Marlophen® 81N
Marlophen® 82
Marlophen® 82N
Marlophen® 83
Marlophen® 83N
Marlophen® 84
Marlophen® 84N
Marlophen® 85
Marlophen® 85N
Marlophen® 86
Marlophen® 86N
Marlophen® 86N/S
Marlophen® 87
Marlophen® 87N
Marlophen® 88
Marlophen® 88N
Marlophen® 89
Marlophen® 89N
Marlophen® 89.5N
Marlophen® 810
Marlophen® 810N
Marlophen® 812
Marlophen® 812N
Marlophen® 814
Marlophen® 814N
Marlophen® 820
Marlophen® 820N
Marlophen® 825
Marlophen® 830
Marlophen® 830N
Marlophen® 840N
Marlophen® 850
Marlophen® 850N
Marlophen® 890
Marlophen® 890N
Marlophen® 1028
Marlophen® 1028N
Marlophen® DNP 16
Marlophen® DNP 18
Marlophen® DNP 30
Marlophen® DNP 150
Marlophen® P 1
Marlophen® P 4
Marlophen® P 7
Marlophen® X
Marlosol® 183
Marlosol® 186
Marlosol® 188
Marlosol® 189
Marlosol® 1820
Marlosol® 1825
Marlosol® BS
Marlosol® F08
Marlosol® FS
Marlosol® OL2
Marlosol® OL7
Marlosol® OL8
Marlosol® OL10
Marlosol® OL15
Marlosol® OL20
Marlosol® R70
Marlosol® RF3
Marlosol® TF3
Marlosol® TF4
Marlowet® 4603
Marlowet® 4700
Marlowet® 4702
Marlowet® 4703
Marlowet® 4800
Marlowet® 4857
Marlowet® 4862
Marlowet® 4900

NONIONIC (CONT'D.)
Marlowet® 4901
Marlowet® 4902
Marlowet® 4930
Marlowet® 4938
Marlowet® 4939
Marlowet® 4940
Marlowet® 4941
Marlowet® 5001
Marlowet® 5400
Marlowet® 5459
Marlowet® 5480
Marlowet® 5641
Marlowet® BL
Marlowet® EF
Marlowet® FOX
Marlowet® GFN
Marlowet® GFW
Marlowet® IHF
Marlowet® ISM
Marlowet® LVS
Marlowet® LVX
Marlowet® MA
Marlowet® NF
Marlowet® OA 5
Marlowet® OA 10
Marlowet® OA 30
Marlowet® OFW
Marlowet® OTS
Marlowet® PW
Marlowet® R 11
Marlowet® R 40
Marlowet® RNP
Marlowet® SAF
Marlowet® SAF/K
Marlowet® SLM
Marlowet® SLS
Marlowet® SW
Marlowet® SWN
Marlowet® TM
Marlowet® WOE
Marlowet® WSD
Marlox® 3000
Marlox® B 24/50
Marlox® B 24/60
Marlox® B 24/80
Marlox® FK 14
Marlox® FK 64
Marlox® FK 69
Marlox® FK 86
Marlox® FK 1614
Marlox® L 6
Marlox® LM 25/30
Marlox® LM 55/18
Marlox® LM 75/30
Marlox® LP 90/20
Marlox® MO 124
Marlox® MO 145
Marlox® MO 154
Marlox® MO 174
Marlox® MO 244
Marlox® MS 48
Marlox® ND 121
Marlox® NP 109
Marlox® OD 105
Marlox® Q 286
Marlox® S 58
Marlox® T 50/5
Marvanlube 1031
Marvanol® DC
Marvanol® Defoamer AM-2
Marvanol® Defoamer MOB
Marvanol® KXL
Marvanol® LSL
Marvanol® MRB
Marvanol® POL-41
Marvanol® RD2-2581
Marvanol® RE-1824
Marvanol® Scour 2 Base
Marvanol® Scour 05
Marvanol® Scour 2582
Marvanol® Scour LF
Marvanscour® KW
Marvelin W-50
Matexil BN PA
Matexil DN VL 200
Matexil Fixer SF
Matexil LN RD
Matexil PN
Matexil PN DG
Matexil PN MFC
Matexil PN PR
Matexil Softener GK
Matexil Thickener CP
Matexil WN PB
Mayco Base BFO
Mazamide® 66
Mazamide® 68
Mazamide® 70
Mazamide® 80
Mazamide® 1281
Mazamide® C-5
Mazamide® CCO
Mazamide® CFAM
Mazamide® CMEA Extra
Mazamide® CS 148
Mazamide® JR 100
Mazamide® JT 128
Mazamide® L-5
Mazamide® O 20
Mazamide® SS 10
Mazamide® WC Conc.
Mazawet® 36
Mazawet® 77
Mazawet® DF
Mazclean W
Mazclean WRI
Mazol® 300 K
Mazol® GMO
Mazol® GMO #1
Mazol® PGO-31 K
Mazon® 18A
Mazon® 40
Mazon® 40A
Mazon® 86
Mazon® 1045A
Mazon® 1086
Mazon® 1096
Mazu® DF 210SX
Mazu® DF 210SX Mod 1
Mazu® DF 230S
Mazu® DF 230SX
Mazu® DF 243
M-C-Thin 45
Mercerol AW-LF
Mergital OC 30E
Mergital ST 30/E
Merpol® 100
Merpol® A
Merpol® CH-196
Merpol® DA
Merpol® HCS
Merpol® HCW
Merpol® LF-H
Merpol® OJ
Merpol® OJS
Merpol® SE
Merpol® SH
Metasol HP-500
Methyl Ester B
Metolat FC 355
Metolat FC 388
Metolat P 853
Micro-Step® H-301
Micro-Step® H-302
Micro-Step® H-303
Micro-Step® H-304
Micro-Step® H-305
Micro-Step® H-306
Micro-Step® H-307
Miglyol® 812
Minemal 320, 325, 330
Miravon B12DF
Miravon B79R
ML-33-F
ML-55-F
ML-55-F-4
MO-33-F
Modicol L
Modicol N
Monamide
Monamid® 15-70W
Monamid® 150-AD
Monamid® 150-ADD
Monamid® 150-ADY
Monamid® 150-DR
Monamid® 150-IS
Monamid® 150-LMWC
Monamid® 150-LW
Monamid® 150-LWA
Monamid® 150-MW
Monamid® 664-MC
Monamid® 716
Monamid® 718
Monamid® 853
Monamid® CMA
Monamid® LIPA
Monamid® LMA
Monamid® LMIPA
Monamid® LMMA
Monamid® S
Monamine AD-100
Monamine ADD-100
Monamine ADY-100
Monamine CD-100
Monamulse 748
Monolan® 1030
Monolan® 1206/2
Monolan® 2000
Monolan® 2000 E/12
Monolan® 2500
Monolan® 2500 E/30
Monolan® 2800
Monolan® 3000 E/50
Monolan® 3000 E/60, 8000 E/80
Monolan® 8000 E/80
Monolan® 12,000 E/80
Monolan® M
Monolan® O Range
Monolan® OM 48
Monolan® OM 59
Monolan® OM 81
Monolan® P222
Monolan® PB
Monolan® PC
Monolan® PEG 300
Monolan® PEG 400
Monolan® PEG 600
Monolan® PEG 1000
Monolan® PEG 1500
Monolan® PEG 4000
Monolan® PEG 6000
Monolan® PK
Monolan® PL
Monolan® PM7
Monolan® PPG440, PPG1100, PPG2200
Monolan® PT
Monomuls 90-25
Monosteol
Montaline SPCV
Montane 20
Montane 40
Montane 60
Montane 65
Montane 70
Montane 80
Montane 83
Montane 85
Montanox 20 DF
Montanox 21
Montanox 40 DF
Montanox 60 DF
Montanox 61
Montanox 65
Montanox 70
Montanox 80 DF
Montanox 81
Montanox 85
Montegal 150 RG
Montegal AP 80
Montegal OL 50
Montegal SH 25
Monthyle
Montoxyl NM Conc.
MP-33-F
MS-33-F
MS-55-F
Mulsifan RT 1
Mulsifan RT 2
Mulsifan RT 7
Mulsifan RT 11
Mulsifan RT 18
Mulsifan RT 19
Mulsifan RT 23
Mulsifan RT 24
Mulsifan RT 27
Mulsifan RT 37
Mulsifan RT 63
Mulsifan RT 69
Mulsifan RT 110
Mulsifan RT 113
Mulsifan RT 163
Mulsifan RT 231
Mulsifan RT 248
Mulsifan RT 258
Mulsifan RT 282

NONIONIC (CONT'D.)
Mulsifan RT 324
Nalco® 2343
Natrex GA 251
Naturechem® EGHS
Naturechem® GMHS
Naturechem® PGHS
Naturechem® THS-200
Naxchem Emulsifier 700
Naxonic NI-40
Naxonic NI-60
Naxonic NI-100
Naxonol CO
Naxonol PN 66
Naxonol PO
Nekanil® 907
Nekanil® 910
Nekanil® LN
Neodol® 1
Neodol® 1-3
Neodol® 1-5
Neodol® 1-7
Neodol® 1-9
Neodol® 23-1
Neodol® 23-3
Neodol® 23-5
Neodol® 23-6.5
Neodol® 23-6.5T
Neodol® 23-12
Neodol® 25-3
Neodol® 25-7
Neodol® 25-9
Neodol® 25-12
Neodol® 45-2.25
Neodol® 45-7
Neodol® 45-7T
Neodol® 45-11
Neodol® 45-12T
Neodol® 45-13
Neodol® 91
Neodol® 91-2.5
Neodol® 91-6
Neodol® 91-8
Neolisal HCN
Neoscoa 203C
Neoscoa 363
Neoscoa 2326
Neoscoa FS-100
Neoscoa GF-2000
Neoscoa MSC-80
Neoscoa SS-10
Neospinol 358
Neutronyx® 656
Newcol 3-80, 3-85
Newcol 20
Newcol 25
Newcol 40
Newcol 45
Newcol 60
Newcol 65
Newcol 80
Newcol 85
Newcol 150
Newcol 170
Newcol 180
Newcol 180T
Newcol 405, 410, 420
Newcol 506
Newcol 508
Newcol 560
Newcol 561H, 562, 564, 565
Newcol 565FH
Newcol 566
Newcol 568
Newcol 569E
Newcol 607, 610, 614, 623
Newcol 704, 707
Newcol 804, 808, 860
Newcol 862
Newcol 864
Newcol 865
Newcol 1010, 1020
Newcol 1100
Newcol 1105
Newcol 1110
Newcol 1120
Newcol 1200
Newcol 1203
Newcol 1204, 1208, 1210
Newcol 1305
Newcol 1310
Newcol 1515, 1525, 1545
Newcol 1610, 1620
Newcol 1807, 1820
Newcol B4, B10, B18
Newlon K-1
Newpol PE-61
Newpol PE-62
Newpol PE-64
Newpol PE-68
Newpol PE-74
Newpol PE-75
Newpol PE-78
Newpol PE-88
Nikkol BC-15TX, -15TX (FF)
Nikkol BC-1SY thru BC-8SY
Nikkol BD-1SY thru BD-8SY
Nikkol BL-1SY
Nikkol BL-2
Nikkol BL-2SY
Nikkol BL-3SY
Nikkol BL-4SY
Nikkol BL-4.2
Nikkol BL-5SY
Nikkol BL-6SY
Nikkol BL-7SY
Nikkol BL-8SY
Nikkol BL-9EX, -9EX (FF)
Nikkol BL-21
Nikkol BL-25
Nikkol BM-1SY thru BM-8SY
Nikkol BO-2
Nikkol BO-7
Nikkol BO-10TX
Nikkol BO-15TX
Nikkol BO-20TX
Nikkol BO-50
Nikkol BPS-5
Nikkol BPS-10
Nikkol BPS-15
Nikkol BPS-20
Nikkol BPS-25
Nikkol BPS-30
Nikkol Decaglyn 2-IS
Nikkol Decaglyn 3-O
Nikkol Decaglyn 3-S
Nikkol Decaglyn 5-IS
Nikkol Decaglyn 5-O
Nikkol Decaglyn 5-S
Nikkol Decaglyn 7-IS
Nikkol Decaglyn 7-O
Nikkol Decaglyn 7-S
Nikkol Decaglyn 10-IS
Nikkol Decaglyn 10-O
Nikkol Decaglyn 10-S
Nikkol DGO-80
Nikkol DGS-80
Nikkol GO-430
Nikkol GO-440
Nikkol GO-460
Nikkol Hexaglyn 1-L
Nikkol Hexaglyn 1-O
Nikkol Hexaglyn 1-S
Nikkol Hexaglyn 3-S
Nikkol Hexaglyn 5-O
Nikkol Hexaglyn 5-S
Nikkol Hexaglyn PR-15
Nikkol NP-5
Nikkol NP-7.5
Nikkol NP-10
Nikkol NP-15
Nikkol NP-18TX
Nikkol OP-3
Nikkol OP-10
Nikkol OP-30
Nikkol PBC-31
Nikkol PBC-33
Nikkol PBC-34
Nikkol PBC-41
Nikkol PBC-44
Nikkol PBC-44 (FF)
Nikkol SO-10
Nikkol SO-15
Nikkol SO-30
Nikkol SS-10
Nikkol SS-30
Nikkol Tetraglyn 1-O
Nikkol Tetraglyn 1-S
Nikkol Tetraglyn 3-S
Nikkol Tetraglyn 5-O
Nikkol Tetraglyn 5-S
Nilfom 2X
Nilo VON
Nimco® 1780
Nimcolan® 1740
Nimcolan® 1747
Ninol® 11-CM
Ninol® 30-LL
Ninol® 40-CO
Ninol® 49-CE
Ninol® 55-LL
Ninol® 70-SL
Ninol® 96-SL
Ninol® 201
Ninol® 1281
Ninol® 1301
Ninol® 4821 F
Ninol® 5024
Ninol® A-10MM
Ninol® B
Ninol® CX
Ninol® LDL 2
Ninol® LMP
Ninol® SR-100
Ninox® FCA
Ninox® L
Ninox® M
Ninox® SO
Niox EO-12
Niox EO-13
Niox EO-14, 23
Niox EO-32, -35
Niox KF-12
Niox KF-13
Niox KF-17
Niox KF-18
Niox KF-22
Niox KF-25
Niox KF-26
Niox KG-11
Niox KG-14
Niox KH Series
Niox KI Series
Niox KJ-55
Niox KJ-56
Niox KJ-61
Niox KJ-66
Niox KL-16
Niox KL-19
Niox KP-62, -63, -67
Niox KP-68
Niox KP-69
Niox KQ-20
Niox KQ-30
Niox KQ-32
Niox KQ-33
Niox KQ-34
Niox KQ-36
Niox KQ-51
Niox KQ-55, -56
Niox KQ-70, LQ-13
Nissan Dispanol 16A
Nissan Dispanol LS-100
Nissan Dispanol N-100
Nissan Dispanol TOC
Nissan Nonion DS-60HN
Nissan Nonion E-205
Nissan Nonion E-206
Nissan Nonion E-208
Nissan Nonion E-215
Nissan Nonion E-220
Nissan Nonion E-230
Nissan Nonion HS-204.5
Nissan Nonion HS-206
Nissan Nonion HS-208
Nissan Nonion HS-210
Nissan Nonion HS-215
Nissan Nonion HS-220
Nissan Nonion HS-240
Nissan Nonion HS-270
Nissan Nonion K-202
Nissan Nonion K-203
Nissan Nonion K-204
Nissan Nonion K-207
Nissan Nonion K-211
Nissan Nonion K-215
Nissan Nonion K-220
Nissan Nonion K-230
Nissan Nonion L-2
Nissan Nonion L-4
Nissan Nonion LP-20R, LP-20RS
Nissan Nonion LT-221
Nissan Nonion NS-202
Nissan Nonion NS-204.5

NONIONIC (CONT'D.)
Nissan Nonion NS-206
Nissan Nonion NS-208.5
Nissan Nonion NS-209
Nissan Nonion NS-210
Nissan Nonion NS-212
Nissan Nonion NS-215
Nissan Nonion NS-220
Nissan Nonion NS-230
Nissan Nonion NS-240
Nissan Nonion NS-250
Nissan Nonion NS-270
Nissan Nonion O-2
Nissan Nonion O-3
Nissan Nonion O-4
Nissan Nonion O-6
Nissan Nonion P-6
Nissan Nonion P-208
Nissan Nonion P-210
Nissan Nonion P-213
Nissan Nonion S-2
Nissan Nonion S-4
Nissan Nonion S-6
Nissan Nonion S-10
Nissan Nonion S-15
Nissan Nonion S-15.4
Nissan Nonion S-40
Nissan Nonion S-206
Nissan Nonion S-207
Nissan Nonion S-215
Nissan Nonion S-220
Nissan Nonion T-15
Nissan Nymeen DT-203, -208
Nissan Nymeen L-201
Nissan Nymeen L-202, 207
Nissan Nymeen S-202, S-204
Nissan Nymeen S-210
Nissan Nymeen S-215, -220
Nissan Nymeen T2-206, -210
Nissan Nymeen T2-230, -260
Nissan Nymide MT-215
Nissan Panacete 810
Nissan Persoft NK-60, -100
Nissan Plonon 102
Nissan Plonon 104
Nissan Plonon 108
Nissan Plonon 171
Nissan Plonon 172
Nissan Plonon 201
Nissan Plonon 204
Nissan Plonon 208
Nissan Stafoam DF
Nissan Stafoam DL
Nissan Stafoam DO, DOS
Nissan Stafoam L
Nissan Unisafe A-LE
Nissan Unisafe A-LM
Nissan Unister
Nitrene 100 SD
Nitrene C
Nitrene C Extra
Nitrene L-76
Nitrene L-90
Nitrene N
Nitrene OE
Noigen 140L
Noigen EA 33
Noigen EA 50
Noigen EA 70
Noigen EA 73
Noigen EA 80
Noigen EA 80E
Noigen EA 83
Noigen EA 92
Noigen EA 102
Noigen EA 110
Noigen EA 112
Noigen EA 120
Noigen EA 120B
Noigen EA 130T
Noigen EA 140
Noigen EA 142
Noigen EA 143
Noigen EA 150
Noigen EA 152
Noigen EA 160
Noigen EA 160P
Noigen EA 170
Noigen EA 190D
Noigen ES 90
Noigen ES 120
Noigen ES 140
Noigen ES 160
Noigen ET60
Noigen ET65
Noigen ET77
Noigen ET80
Noigen ET83
Noigen ET95
Noigen ET97
Noigen ET100
Noigen ET102
Noigen ET107
Noigen ET115
Noigen ET120
Noigen ET127
Noigen ET135
Noigen ET140
Noigen ET143
Noigen ET147
Noigen ET150
Noigen ET157
Noigen ET160
Noigen ET165
Noigen ET167
Noigen ET170
Noigen ET180
Noigen ET187
Noigen ET190
Noigen ET190S
Noigen ET207
Noigen O100
Noigen YX400, YX500
Noiox AK-40
Noiox AK-41
Noiox AK-44
Nonal 206
Nonal 208
Nonal 210
Nonal 310
Nonarox 575, 730
Nonarox 1030, 1230
Nonex DL-2
Nonex DO-4
Nonex O4E
Nonex S3E
Nonicol 100
Nonicol 190
Nonionic E-4
Non Ionic Emulsifier T-9
Nonipol 20
Nonipol 40
Nonipol 55
Nonipol 60
Nonipol 70
Nonipol 85
Nonipol 100
Nonipol 110
Nonipol 120
Nonipol 130
Nonipol 160
Nonipol 200
Nonipol 400
Nonipol BX
Nonipol D-160
Nonipol Soft SS-50
Nonipol Soft SS-70
Nonipol Soft SS-90
Nonipol T-28
Nonisol 100
Nonisol 210
Nonisol 300
Nopalcol 1-L
Nopalcol 1-S
Nopalcol 1-TW
Nopalcol 2-DL
Nopalcol 2-L
Nopalcol 4-C
Nopalcol 4-CH
Nopalcol 4-DTW
Nopalcol 4-L
Nopalcol 4-O
Nopalcol 4-S
Nopalcol 6-DO
Nopalcol 6-DTW
Nopalcol 6-L
Nopalcol 6-O
Nopalcol 6-R
Nopalcol 6-S
Nopalcol 10-CO
Nopalcol 10-COH
Nopalcol 12-CO
Nopalcol 12-COH
Nopalcol 12-R
Nopalcol 19-CO
Nopalcol 30-S
Nopalcol 30-TWH
Nopalcol 200
Nopalcol 400
Nopalcol 600
Nopco® 1179
Nopco® Colorsperse 188-A
Nopco® RDY
Nopcogen 14-LT
Nopcosperse 28-B
Nopcosperse WEZ
Norfox® 1 Polyol
Norfox® 2 Polyol
Norfox® 2LF
Norfox® 4 Polyol
Norfox® 916
Norfox® CMA
Norfox® DC
Norfox® DCSA
Norfox® DESA
Norfox® DOSA
Norfox® EGMS
Norfox® F-221
Norfox® F-342
Norfox® GMS
Norfox® KD
Norfox® NP-1
Norfox® NP-4
Norfox® NP-6
Norfox® NP-7
Norfox® NP-9
Norfox® NP-11
Norfox® OP-45
Norfox® OP-100
Norfox® OP-102
Norfox® OP-114
Norfox® Sorbo S-80
Norfox® Sorbo T-20
Novel® 1412-70 Ethoxylate
Noxamine CA 30
Noxamine O2-30
NP-55-60
NP-55-80
NP-55-85
NP-55-90
NSA-17
Nutrol 100
Nutrol 600
Nutrol 611
Nutrol 622
Nutrol 640
Nutrol 656
Oakite Defoamant RC
Ocenol 50/55+2EO
Ocetox 5525
Octapol 60
Octapol 100
Octapol 300
Octapol 400
OHlan®
Ombrelub FC 533
Orapol HC
Ospin L-800
Ospol 790
Oxamin LO
Oxetal 500/85
Oxetal 800/85
Oxetal D 104
Oxetal ID 104
Oxetal O 108
Oxetal O 112
Oxetal TG 111
Oxetal TG 118
Oxetal VD 20
Oxetal VD 28
Paraffin Wax Emulsifier CB0674
Paraffin Wax Emulsifier CB0680
Paralan
Paramul® J
Paramul® SAS
Pat-Wet LF-55
Patogen 311
Patogen 345
Peceol Isostearique
Pegnol C-14
Pegnol C-18
Pegnol C-20
Pegnol HA-120

NONIONIC (CONT'D.)
Pegnol L-6
Pegnol L-8
Pegnol L-10
Pegnol L-12
Pegnol L-15
Pegnol L-20
Pegnol O-6
Pegnol O-16
Pegnol OA-400
Pegnol PDS-60
Pegnol PDS-60A
Pegol® L-62D
Pegosperse® 50 DS
Pegosperse® 50 MS
Pegosperse® 100 L
Pegosperse® 100 ML
Pegosperse® 100 MR
Pegosperse® 100 O
Pegosperse® 100 S
Pegosperse® 200 DL
Pegosperse® 200 ML
Pegosperse® 400 DL
Pegosperse® 400 DO
Pegosperse® 400 DOT
Pegosperse® 400 DS
Pegosperse® 400 DTR
Pegosperse® 400 MC
Pegosperse® 400 ML
Pegosperse® 400 MO
Pegosperse® 400 MOT
Pegosperse® 400 MS
Pegosperse® 600 DOT
Pegosperse® 600 ML
Pegosperse® 600 MS
Pegosperse® 700 TO
Pegosperse® 1000 MS
Pegosperse® 1500 DO
Pegosperse® 1500 MS
Pegosperse® 1750 MS
Pegosperse® 4000 MS
Pegosperse® 6000 DS
Pegosperse® 9000 CO
Pegosperse® MFE
Pegosperse® PMS CG
Penestrol N-160
Pepol A-0638
Pepol A-0858
Pepol AX-1
Pepol B
Pepol B-182
Pepol B-184
Pepol BS-184
Pepol BS-201
Pepol BS-403
Pepol D-301
Pepol D-304
Permalene A-100
Permalose TM
Petrosan 102
P & G Amide No. 27
PG No. 4
PGE-400-DS
PGE-400-MS
PGE-600-DS
PGE-600-ML
PGE-600-MS
Phospholipon 50G
Pinamine K
Plantaren 600 CS UP
Plantaren APG 225
Plantaren CG 60
Pluradot HA-410
Plurafac® A-24
Plurafac® A-27
Plurafac® A-38
Plurafac® A-39
Plurafac® B-25-5
Plurafac® B-26
Plurafac® C-17
Plurafac® D-25
Plurafac® LF 120
Plurafac® LF 131
Plurafac® LF 132
Plurafac® LF 220
Plurafac® LF 221
Plurafac® LF 223
Plurafac® LF 224
Plurafac® LF 231
Plurafac® LF 400
Plurafac® LF 401, LF 403
Plurafac® LF 404, LF 405
Plurafac® LF 431
Plurafac® LF 600
Plurafac® LF 700, LF 711
Plurafac® LF 1300
Plurafac® LF 1430
Plurafac® RA-20
Plurafac® RA-30
Plurafac® RA-40
Plurafac® RA-43
Plurafac® RA-50
Plurafac® T-55
Pluraflo® E4A
Pluraflo® E4B
Pluraflo® E5A
Pluraflo® E5B
Pluraflo® E5BG
Pluraflo® E5G
Pluraflo® N5G
Pluriol® E 200
Pluriol® E 300
Pluriol® E 400
Pluriol® E 600
Pluriol® E 1500
Pluriol® E 4000
Pluriol® E 6000
Pluriol® E 9000
Pluriol® P 600
Pluriol® P 900
Pluriol® P 2000
Pluriol® PE 3100
Pluriol® PE 4300
Pluriol® PE 6100
Pluriol® PE 6101
Pluriol® PE 6200
Pluriol® PE 6400
Pluriol® PE 6800
Pluriol® PE 8100
Pluriol® PE 9200
Pluriol® PE 9400
Pluriol® PE 10100
Pluriol® PE 10500
Pluriol® RPE 2540
Pluriol® RPE 3110
Pluroic® PE 6800
Plurol Isostearique
Plurol Oleique WL 1173
Plurol Stearique WL 1009
Pluronic® 10R5
Pluronic® 10R8
Pluronic® 12R3
Pluronic® 17R1
Pluronic® 17R2
Pluronic® 17R4
Pluronic® 25R1
Pluronic® 25R2
Pluronic® 25R4
Pluronic® 25R5
Pluronic® 25R8
Pluronic® 31R1
Pluronic® 31R2
Pluronic® 31R4
Pluronic® F38
Pluronic® F68
Pluronic® F68LF
Pluronic® F77
Pluronic® F87
Pluronic® F88
Pluronic® F98
Pluronic® F108
Pluronic® F127
Pluronic® L10
Pluronic® L31
Pluronic® L35
Pluronic® L42
Pluronic® L43
Pluronic® L44
Pluronic® L61
Pluronic® L62
Pluronic® L62D
Pluronic® L62LF
Pluronic® L63
Pluronic® L64
Pluronic® L72
Pluronic® L81
Pluronic® L92
Pluronic® L101
Pluronic® L121
Pluronic® L122
Pluronic® P65
Pluronic® P75
Pluronic® P84
Pluronic® P85
Pluronic® P94
Pluronic® P103
Pluronic® P104
Pluronic® P105
Pluronic® P123
Pluronic® PE 3100
Pluronic® PE 4300
Pluronic® PE 6100
Pluronic® PE 6200
Pluronic® PE 6400
Pluronic® PE 8100
Pluronic® PE 9200
Pluronic® PE 9400
Pluronic® PE 10100
Pluronic® PE 10500
Pluronic® RPE 2520
Pluronic® RPE 3110
Pogol 200
Pogol 300
Pogol 400 NF
Pogol 600
Pogol 1000
Pogol 1500
Polirol 1BS
Polirol 23
Polirol 215
Polirol C5
Polirol L400
Polirol NF80
Polirol O55
Polyaldo® DGHO
Polychol 5
Polychol 10
Polychol 20
Polychol 40
Polyester 1606
Polyester N-95
Polylube 4507
Polylube 5713
Polylube DDL
Polylube GK
Polylube RE
Polylube SC
Polylube WS
Polysol J
Polyspin MP-7-29
Polyspin PA
Polyspin PF
Polystep® F-1
Polystep® F-2
Polystep® F-3
Polystep® F-4
Polystep® F-5
Polystep® F-6
Polystep® F-9
Polystep® F-10, F-10 Ec
Polystep® F-95B
Poly-Tergent® B-150
Poly-Tergent® B-200
Poly-Tergent® B-300
Poly-Tergent® B-315
Poly-Tergent® B-350
Poly-Tergent® B-500
Poly-Tergent® E-17A
Poly-Tergent® E-17B
Poly-Tergent® E-25B
Poly-Tergent® J-200
Poly-Tergent® J-300
Poly-Tergent® P-17A
Poly-Tergent® P-17B
Poly-Tergent® P-17BLF
Poly-Tergent® P-17BX
Poly-Tergent® P-17D
Poly-Tergent® P-22A
Poly-Tergent® P-32A
Poly-Tergent® P-32D
Poly-Tergent® P-9E
Poly-Tergent® RCS-43
Poly-Tergent® RCS-48
Poly-Tergent® S-205LF
Poly-Tergent® S-305LF
Poly-Tergent® S-405LF
Poly-Tergent® S-505LF
Poly-Tergent® SL-42
Poly-Tergent® SL-62
Poly-Tergent® SL-92
Poly-Tergent® SLF-18
Prechem NFE
Prechem NPX
Primasol® SD
Procetyl 10
Procetyl AWS
Procoal 20
Product VN-11
Produkt GM 4111
Produkt RT 63

NONIONIC (CONT'D.)
Profan ME-20
Progasol 230
Progasol 443, 457
Progasol COG
Pronal 502, 502A
Pronal 2200
Pronal 3300
Pronal P-805
Propetal 99
Propetal 103
Propetal 241
Propetal 254
Propetal 281
Propetal 340
Propetal 341
Prosol 518
Prote-sorb SML
Prote-sorb SMO
Prote-sorb SMP
Prote-sorb SMS
Prote-sorb STO
Protowet 100
Protowet E-4
Prox-onic DNP-08
Prox-onic DNP-0150
Prox-onic DNP-0150/50
Prox-onic EP 1090-1
Prox-onic EP 1090-2
Prox-onic EP 2080-1
Prox-onic EP 4060-1
Prox-onic HR-05
Prox-onic HR-016
Prox-onic HR-025
Prox-onic HR-030
Prox-onic HR-036
Prox-onic HR-040
Prox-onic HR-080
Prox-onic HR-0200
Prox-onic HR-0200/50
Prox-onic HRH-05
Prox-onic HRH-016
Prox-onic HRH-025
Prox-onic HRH-0200
Prox-onic HRH-0200/50
Prox-onic NP-1.5
Prox-onic NP-04
Prox-onic NP-06
Prox-onic NP-09
Prox-onic NP-010
Prox-onic NP-015
Prox-onic NP-020
Prox-onic NP-030
Prox-onic NP-030/70
Prox-onic NP-040
Prox-onic NP-040/70
Prox-onic NP-050
Prox-onic NP-050/70
Prox-onic NP-0100
Prox-onic NP-0100/70
Prox-onic OA-1/04
Prox-onic OA-1/09
Prox-onic OA-1/020
Prox-onic OA-2/020
Prox-onic OP-09
Prox-onic OP-016
Prox-onic OP-030/70
Prox-onic OP-040/70
Prox-onic SA-1/02
Prox-onic SA-1/010
Prox-onic SA-1/020
Prox-onic SML-020
Prox-onic SMP-020
Prox-onic STO-020
Prox-onic STS-020
Prox-onic TD-1/03
Prox-onic TD-1/06
Prox-onic TD-1/09
Prox-onic TD-1/012
Purton CFD
Purton SFD
Pyroter CPI-40
Pyroter GPI-25
Quimipol 9106B
Quimipol 9108
Quimipol DEA OC
Quimipol EA 2503
Quimipol EA 2504
Quimipol EA 2506
Quimipol EA 2507
Quimipol EA 2508
Quimipol EA 2509
Quimipol EA 2512
Quimipol EA 4505
Quimipol EA 4508
Quimipol EA 6801
Quimipol EA 6802
Quimipol EA 6803
Quimipol EA 6804
Quimipol EA 6806
Quimipol EA 6807
Quimipol EA 6808
Quimipol EA 6810
Quimipol EA 6812
Quimipol EA 6814
Quimipol EA 6818
Quimipol EA 6820
Quimipol EA 6823
Quimipol EA 6825
Quimipol EA 6850
Quimipol EA 9105
Quimipol EA 9106
Quimipol EA 9106B
Quimipol EA 9108
Quimipol ED 2021
Quimipol ED 2022
Quimipol ENF 15
Quimipol ENF 20
Quimipol ENF 30
Quimipol ENF 40
Quimipol ENF 55
Quimipol ENF 65
Quimipol ENF 80
Quimipol ENF 90
Quimipol ENF 95
Quimipol ENF 100
Quimipol ENF 110
Quimipol ENF 120
Quimipol ENF 140
Quimipol ENF 170
Quimipol ENF 200
Quimipol ENF 230
Quimipol ENF 300
R3124 Ester
Radiasurf® 7000
Radiasurf® 7125
Radiasurf® 7135
Radiasurf® 7136
Radiasurf® 7137
Radiasurf® 7140
Radiasurf® 7141
Radiasurf® 7144
Radiasurf® 7145
Radiasurf® 7146
Radiasurf® 7147
Radiasurf® 7151
Radiasurf® 7152
Radiasurf® 7153
Radiasurf® 7155
Radiasurf® 7156
Radiasurf® 7157
Radiasurf® 7175
Radiasurf® 7196
Radiasurf® 7201
Radiasurf® 7206
Radiasurf® 7269
Radiasurf® 7270
Radiasurf® 7345
Radiasurf® 7372
Radiasurf® 7400
Radiasurf® 7402
Radiasurf® 7403
Radiasurf® 7404
Radiasurf® 7410
Radiasurf® 7411
Radiasurf® 7412
Radiasurf® 7413
Radiasurf® 7414
Radiasurf® 7417
Radiasurf® 7420
Radiasurf® 7421
Radiasurf® 7422
Radiasurf® 7423
Radiasurf® 7431
Radiasurf® 7432
Radiasurf® 7443
Radiasurf® 7444
Radiasurf® 7453
Radiasurf® 7454
Radiasurf® 7600
Radiasurf® 7900
Ralufon® EN 16-80
Reginol 2701
Remcopal 4
Remcopal 6
Remcopal 10
Remcopal 18
Remcopal 20
Remcopal 25
Remcopal 29
Remcopal 40
Remcopal 40 S3
Remcopal 40 S3 LE
Remcopal 121
Remcopal 207
Remcopal 220
Remcopal 229
Remcopal 234
Remcopal 238
Remcopal 258
Remcopal 273
Remcopal 306
Remcopal 334
Remcopal 349
Remcopal 666
Remcopal 3112
Remcopal 3515
Remcopal 3712
Remcopal 3820
Remcopal 4000
Remcopal 4018
Remcopal 6110
Remcopal 21411
Remcopal 21912 AL
Remcopal 31250
Remcopal 33820
Remcopal D
Remcopal HC 7
Remcopal HC 20
Remcopal HC 33
Remcopal HC 40
Remcopal HC 60
Remcopal L9
Remcopal L12
Remcopal L30
Remcopal LC
Remcopal LO 2B
Remcopal LP
Remcopal NP 30
Remcopal O11
Remcopal O12
Remcopal O9
Remcopal PONF
Renex® 20
Renex® 25
Renex® 26
Renex® 30
Renex® 31
Renex® 36
Renex® 647
Renex® 648
Renex® 649
Renex® 650
Renex® 678
Renex® 679
Renex® 682
Renex® 688
Renex® 690
Renex® 697
Renex® 698
Renex® 702
Renex® 703
Renex® 704
Renex® 707
Renex® 711
Renex® 714
Renex® 720
Renex® 750
Renex® 751
Resolvyl 610
Resolvyl BC
Rewocid® U 185
Rewolub KSM 14
Rewolub KSM 80
Rewomid® AC 28
Rewomid® C 212
Rewomid® C 220SE
Rewomid® CD
Rewomid® DC 212 S
Rewomid® DC 220 SE
Rewomid® DL 203
Rewomid® DL 203 S
Rewomid® DL 240
Rewomid® DLMS
Rewomid® DO 280
Rewomid® DO 280 S
Rewomid® DO 280 SE
Rewomid® IPL 203
Rewomid® IPP 240
Rewomid® L 203

NONIONIC (CONT'D.)
Rewomid® OM 101/G
Rewomid® OM 101/IG
Rewomid® R 280
Rewomid® S 280
Rewomid® U 185
Rewopal® BN 13
Rewopal® C 6
Rewopal® CSF 20
Rewopal® EO 70
Rewopal® HV 4
Rewopal® HV 5
Rewopal® HV 6
Rewopal® HV 8
Rewopal® HV 9
Rewopal® HV 10
Rewopal® HV 14
Rewopal® HV 25
Rewopal® HV 50
Rewopal® LA 3
Rewopal® LA 6
Rewopal® LA 6-90
Rewopal® LA 10
Rewopal® LA 10-80
Rewopal® M 365
Rewopal® MT 65
Rewopal® MT 2455
Rewopal® MT 2540
Rewopal® MT 5722
Rewopal® O 8
Rewopal® O 15
Rewopal® PO
Rewopal® RO 40
Rewopal® TA 11
Rewopal® TA 25
Rewopal® TA 25/S
Rewopal® TA 50
Rewopal® TPD 30
Rewopol® BW
Rewopol® BWA
Rewopol® DBC
Rewopol® FBR
Rexfoam D
Rexobase HD
Rexobase LN
Rexobase PBX
Rexol 25/1
Rexol 25/4
Rexol 25/6
Rexol 25/7
Rexol 25/8
Rexol 25/9
Rexol 25/10
Rexol 25/11
Rexol 25/12
Rexol 25/14
Rexol 25/15
Rexol 25/20
Rexol 25/30
Rexol 25/40
Rexol 25/50
Rexol 25/100-70%
Rexol 25/307
Rexol 25/407
Rexol 25/507
Rexol 25J
Rexol 25JM1
Rexol 25JWC
Rexol 35/3
Rexol 35/5
Rexol 35/8
Rexol 35/11
Rexol 35/100
Rexol 45/1
Rexol 45/3
Rexol 45/5
Rexol 45/7
Rexol 45/10
Rexol 45/12
Rexol 45/16
Rexol 45/307
Rexol 45/407
Rexol 65/4
Rexol 65/6
Rexol 65/9
Rexol 65/10
Rexol 65/11
Rexol 65/14
Rexol 130
Rexol 2000 HWM
Rexol 4736
Rexol AE-1
Rexol AE-2
Rexol AE-3
Rexol CCN
Rexonic 1006
Rexonic 1012-6
Rexonic 1218-6
Rexonic N23-3
Rexonic N23-6.5
Rexonic N25-3
Rexonic N25-7
Rexonic N25-9
Rexonic N25-9 (85%)
Rexonic N25-12
Rexonic N25-14
Rexonic N25-14 (85%)
Rexonic N91-1.6
Rexonic N91-2.5
Rexonic N91-6
Rexonic N91-8
Rexonic P-1
Rexonic P-3
Rexonic P-4
Rexonic P-5
Rexonic P-6
Rexonic P-9
Rexonic RL
Rexopal 3928
Rexopal EP-1
Rexopal SM-5
Rexopon E
Rexopon RSN
Rheodol SP-O10
Rheodol SP-O30
Rheodol SP-S10
Rheodol SP-S30
Rheodol TW-P120
Rhodafac® PEH
Rhodasurf® 110
Rhodasurf® 840
Rhodasurf® 860/P
Rhodasurf® 870
Rhodasurf® 1012-6
Rhodasurf® A-1P
Rhodasurf® A-2
Rhodasurf® A-6
Rhodasurf® A 24
Rhodasurf® A-60
Rhodasurf® B-1
Rhodasurf® B-7
Rhodasurf® BC-420
Rhodasurf® BC-610
Rhodasurf® BC-630
Rhodasurf® BC-720
Rhodasurf® BC-737
Rhodasurf® BC-840
Rhodasurf® DA-4
Rhodasurf® DA-6
Rhodasurf® DA-530
Rhodasurf® DA-630
Rhodasurf® DA-639
Rhodasurf® DB 311
Rhodasurf® E 400
Rhodasurf® E 600
Rhodasurf® L-4
Rhodasurf® L-25
Rhodasurf® L-790
Rhodasurf® LA-3
Rhodasurf® LA-7
Rhodasurf® LA-9
Rhodasurf® LA-12
Rhodasurf® LA-15
Rhodasurf® LA-30
Rhodasurf® LA-42
Rhodasurf® LA-90
Rhodasurf® LAN-23-75%
Rhodasurf® N-12
Rhodasurf® ON-870
Rhodasurf® ON-877
Rhodasurf® RHS
Rhodasurf® ROX
Rhodasurf® T
Rhodasurf® T50
Rhodasurf® T-95
Rhodasurf® TB-970
Rhodasurf® TDA-5
Rhodasurf® TDA-6
Rhodasurf® TDA-8.5
Rhodasurf® TR Series
Rhodaterge® 206C
Rhodaterge® CAN
Rhodaterge® EPS
Rhodaterge® RS 25
Rhodaterge® SMC
Rhodorsil® AF 422
Rhodorsil® AF 20432
Ridafoam NS-221
Rilanit G 16
Rioklen NF 2
Rioklen NF 4
Rioklen NF 6
Rioklen NF 7
Rioklen NF 8
Rioklen NF 9
Rioklen NF 10
Rioklen NF 12
Rioklen NF 15
Rioklen NF 20
Rioklen NF 30
Rioklen NF 40
Ritacetyl®
Ritachol® 1000
Ritachol® 3000
Ritachol® 4000
Ritapeg 100 DS
Ritapro 100
Ritasynt IP
Ritoleth 2
Ritoleth 5
Ritoleth 10
Ritoleth 20
Rolamet C 11
Rolfat 5347
Rolfat C 9
Rolfat OL 6
Rolfat OL 7
Rolfat OL 9
Rolfat R 106
Rolfat SG
Rolfat SO 6
Rolfat ST 40
Rolfat ST 100
Rolfor 162
Rolfor C
Rolfor CO 11
Rolfor COH 40
Rolfor E 270, E 527
Rolfor EP 237, EP 526
Rolfor HT 11
Rolfor HT 25
Rolfor N 24/2
Rolfor N 24/3
Rolfor O 8
Rolfor O 22
Rolfor TR 6
Rolfor TR 9
Rolfor TR 12
Rolfor Z 24/9
Rolfor Z 24/23
Ross Chem 743
Ross Chem PEG 600 DT
Ross Chem PPG 1025DT
Ross Emulsifier DF
Ross Emulsifier LA-4
RS-55-40
Rychem® 17
Rychem® 21
Rychem® 33
Rycofax® O
Rycomid 2120
Salt 100, 200, 300M
Sandolube NVN
Sandolube NVS
Sandopan® LF Liquid
Sandopan® LFW
Sandoxylate® 206
Sandoxylate® 224
Sandoxylate® 408
Sandoxylate® 412
Sandoxylate® 418
Sandoxylate® 424
Sandoxylate® AC-9
Sandoxylate® AC-24
Sandoxylate® AC-46
Sandoxylate® AD-4
Sandoxylate® AD-6
Sandoxylate® AD-9
Sandoxylate® AL-4
Sandoxylate® AO-12
Sandoxylate® AO-20
Sandoxylate® AO-60
Sandoxylate® AT-6.5
Sandoxylate® AT-12
Sandoxylate® C-10
Sandoxylate® C-15
Sandoxylate® C-32
Sandoxylate® FO-9
Sandoxylate® FO-30/70
Sandoxylate® FS-9

NONIONIC (CONT'D.)
Sandoxylate® FS-35
Sandoxylate® NC-5
Sandoxylate® NC-15
Sandoxylate® NSO-30
Sandoxylate® NT-5
Sandoxylate® NT-15
Sandoxylate® PDN-7
Sandoxylate® PN-6
Sandoxylate® PN-9
Sandoxylate® PN-10.9
Sandoxylate® PO-5
Sandoxylate® SX-208
Sandoxylate® SX-224
Sandoxylate® SX-408
Sandoxylate® SX-412
Sandoxylate® SX-418
Sandoxylate® SX-424
Sandoxylate® SX-602
Sandoz Amide NT
Sandoz Amine Oxide XA-C
Sandoz Amine Oxide XA-L
Sandoz Amine Oxide XA-M
Sandozin® NA Liq.
Sandozin® NIT
Sanmorin 11
Santone® 3-1-SH
Santone® 8-1-O
Sapogenat T Brands
Satexlan 20
Schercamox C-AA
Schercamox DMA
Schercamox DML
Schercamox DMM
Schercomid 1-102
Schercomid 304
Schercomid 1214
Schercomid AME
Schercomid CCD
Schercomid CME
Schercomid CMI
Schercomid EAC
Schercomid EAC-S
Schercomid HT-60
Schercomid LME
Schercomid MME
Schercomid OMI
Schercomid SCE
Schercomid SCO-Extra
Schercomid SL-Extra
Schercomid SL-ML, SL-ML-LC
Schercomid SLA
Schercomid SLE
Schercomid SLM-C
Schercomid SLM-LC
Schercomid SLM-S
Schercomul G
Schercomul H
Schercoterge 140
Scour 1161
Scourol 700
Scripset 520
Sebase
Secomine TA 02
Secomix® E40
Secoster® A
Secoster® CL 10
Secoster® CP 10
Secoster® CS 10
Secoster® DO 600
Secoster® KL 10
Secoster® KP 10
Secoster® KS 10
Secoster® MA 300
Secoster® ML 300
Secoster® ML 4000
Secoster® MO 100
Secoster® MO 400
Secoster® SDG
Sedoran FF-180, FF-200, FF-210, FF-220
Sellig Antimousse S
Sellig AO 6 100
Sellig AO 9 100
Sellig AO 15 100
Sellig AO 25 100
Sellig DN 10 100
Sellig DN 22 100
Sellig HR 18 100
Sellig Hydrophilisant
Sellig LA 1150
Sellig LA 9 100
Sellig LA 11 100
Sellig Mouillant 9083 NI/AL
Sellig Mouillant EMPT 80
Sellig N 1050
Sellig N 1780
Sellig N 20 80
Sellig N 30 70
Sellig N 4 100
Sellig N 5 100
Sellig N 6 100
Sellig N 8 100
Sellig N 9 100
Sellig N 10 100
Sellig N 11 100
Sellig N 12 100
Sellig N 15 100
Sellig N 17 100
Sellig N 20 100
Sellig N 30 100
Sellig N 50 100
Sellig NK 59
Sellig NK 83
Sellig NK 729064
Sellig O 4 100
Sellig O 5 100
Sellig O 6 100
Sellig O 8 100
Sellig O 9 100
Sellig O 11 100
Sellig O 12 100
Sellig O 20 100
Sellig R 20 100
Sellig R 3395
Sellig R 3395 SP
Sellig R 3395-C435
Sellig R 4095
Sellig R 4495
Sellig S 30 100
Sellig SP 25 50
Sellig SP 3020
Sellig SP 8 100
Sellig SP 16 100
Sellig SP 20 100
Sellig SP 25 100
Sellig SP 30 100
Sellig Stearo 6
Sellig SU 4 100
Sellig SU 18 100
Sellig SU 25 100
Sellig SU 30 100
Sellig SU 50 100
Sellig T 3 100
Sellig T 14 100
Sellig T 1790
Selligor 860 SP
Selligor SAN 1
SEM-35
Sepabase A Grades
Sepabase B Grades
Sepabase R Grades
Separol AF 27
Separol WF 22
Separol WF 34
Separol WF 41
Separol WF 221
Separol WK 25
Ser-Ad FN Series
Serdolamide POF 61
Serdolamide POF 61 C
Serdox NBS 4
Serdox NBS 5.5
Serdox NBS 6
Serdox NBS 6.6
Serdox NBS 8.5
Serdox NBSQ 5/5
Serdox NDI 100
Serdox NEL 12/80
Serdox NES 8/85
Serdox NJAD 15
Serdox NJAD 20
Serdox NJAD 30
Serdox NNP 4
Serdox NNP 5
Serdox NNP 6
Serdox NNP 7
Serdox NNP 8.5
Serdox NNP 10
Serdox NNP 11
Serdox NNP 12
Serdox NNP 13
Serdox NNP 14
Serdox NNP 15
Serdox NNP 25
Serdox NNP 30/70
Serdox NNPQ 7/11
Serdox NOG 200, 400
Serdox NOG 440
Serdox NOL 2
Serdox NOL 8
Serdox NOL 15
Serdox NOP 9
Serdox NOP 30/70
Serdox NOP 40/70
Serdox NSG 400
Serica 300
Serica 830
Sermul EN 3
Sermul EN 15
Sermul EN 20/70
Sermul EN 25
Sermul EN 26
Sermul EN 30/70
Sermul EN 145
Sermul EN 155
Sermul EN 229
Sermul EN 237
Sermul EN 259
Sermul EN 312
Sermulen 56
Servirox OEG 45
Servirox OEG 55
Servirox OEG 65
Servirox OEG 90/50
Sevestat ML 300
Sevestat NDE
SF96®
Silicone AF-10 FG
Silicone AF-10 IND
Silicone AF-30 FG
Silicone AF-30 IND
Silicopearl SR
Silwet® L-77
Silwet® L-720
Silwet® L-722
Silwet® L-7001
Silwet® L-7002
Silwet® L-7004
Silwet® L-7200
Silwet® L-7210
Silwet® L-7230
Silwet® L-7500
Silwet® L-7600
Silwet® L-7602
Silwet® L-7604
Silwet® L-7605
Silwet® L-7607
Silwet® L-7614
Silwet® L-7622
Simulsol 52
Simulsol 56
Simulsol 72
Simulsol 76
Simulsol 78
Simulsol 92
Simulsol 96
Simulsol 98
Simulsol 1030 NP
Simulsol A
Simulsol A 686
Simulsol M 45
Simulsol M 49
Simulsol M 51
Simulsol M 52
Simulsol M 53
Simulsol M 59
Simulsol NP 575
Simulsol O
Simulsol P4
Sinochem GMS
Sinol T4-1
Sinopol 20
Sinopol 25
Sinopol 60
Sinopol 65
Sinopol 80
Sinopol 85
Sinopol 150
Sinopol 150T
Sinopol 170
Sinopol 170F
Sinopol 170I
Sinopol 170I2
Sinopol 170N

NONIONIC (CONT'D.)
Sinopol 170N2
Sinopol 180I
Sinopol 180I2
Sinopol 180N
Sinopol 180N2
Sinopol 405, 410
Sinopol 410ST, 415ST
Sinopol 420
Sinopol 430ST
Sinopol 610
Sinopol 623
Sinopol 707
Sinopol 714
Sinopol 806
Sinopol 808
Sinopol 860
Sinopol 864
Sinopol 864H
Sinopol 865
Sinopol 908
Sinopol 910
Sinopol 960
Sinopol 964
Sinopol 964H
Sinopol 965
Sinopol 965FH
Sinopol 966FH
Sinopol 1100H
Sinopol 1102
Sinopol 1103
Sinopol 1109
Sinopol 1112
Sinopol 1120
Sinopol 1203
Sinopol 1207
Sinopol 1210
Sinopol 1225
Sinopol 1305
Sinopol 1310
Sinopol 1510
Sinopol 1525
Sinopol 1545
Sinopol 1610
Sinopol 1620
Sinopol 1807
Sinopol 1830
Sinopol NN-15
Sinoponic PE 61
Sinoponic PE 62
Sinoponic PE 64
Sinoponic PE 68
Siponic® F-300
Siponic® F-400
Siponic® L-12
Siponic® NP-4
Siponic® NP-6
Siponic® NP-8
Siponic® NP-9
Siponic® NP-9.5
Siponic® NP-10
Siponic® NP-13
Siponic® NP-15
Siponic® NP-40
Siponic® NP-407
Siponic® TD-3
Siponic® TD-6
S-Maz® 20
S-Maz® 40
S-Maz® 60
S-Maz® 60KHM
S-Maz® 65K
S-Maz® 80
S-Maz® 80K
S-Maz® 83R
S-Maz® 85
S-Maz® 85K
S-Maz® 90
S-Maz® 95
Smoothar DT-130
Smoothar F 828
Sofnon SP-852R
Sofnon SP-9400
Softal IT-7
Softenol® 3900
Softenol® 3991
Softigen® 767
Soitem SFL/1
Sokalan® HP 22
Solan
Solan 50
Solan E
Solangel 401
Solegal W Conc.
Sole-Terge 8
Solidokoll Brands
Solulan® 5
Solulan® 16
Solulan® 75
Solulan® C-24
Solulan® L-575
Solulan® PB-2
Solulan® PB-5
Solulan® PB-10
Solulan® PB-20
Solvent Scour 25/27
Solvent-Scour 880
Sopralub ACR 265
Sopralub ACR 275
Soprofor BC/Series
Soprofor DO/64
Soprofor M/52
Soprofor NP/20
Soprofor NP/30
Soprofor NP/100
Soprophor® 37
Soprophor® 40/D
Soprophor® 497/P
Soprophor® 724/P
Soprophor® 796/P
Soprophor® BO/318
Soprophor® BSU
Soprophor® CY/8
Soprophor® K/202
Soprophor® S/25
Soprophor® S40-P
Soprophor® SC/167
Sorban AL
Sorban AO
Sorban AO 1
Sorban AST
Sorban CO
Sorbanox AL
Sorbanox AOM
Sorbanox AP
Sorbanox AST
Sorbanox CO
Sorbax PML-20
Sorbax PMO-5
Sorbax PMO-20
Sorbax PMP-20
Sorbax PMS-20
Sorbax PTO-20
Sorbax PTS-20
Sorbax SML
Sorbax SMO
Sorbax SMP
Sorbax SMS
Sorbax STO
Sorbax STS
Sorbeth 40
Sorbeth 40HO
Sorbeth 55
Sorbeth 55HO
Sorbilene O
Sorbirol O
Sorbon S-20
Sorbon S-40
Sorbon S-60
Sorbon S-80
Sorbon T-20
Sorbon T-40
Sorbon T-60
Sorbon T-80
Sorbon TR 814
Sorbon TR 843
Sorpol 230
Sorpol 900H
Sorpol 900L
Sorpol 1200K
Sorpol 2676S
Sorpol 2678S
Sorpol 3005X
Sorpol 3370
Sorpol 9939
Sorpol W-150
Span® 20
Span® 40
Span® 60, 60K
Span® 65
Span® 80
Span® 85
Sponto® 200
Sponto® 217
Sponto® AC60-02D
Sponto® AG3-55T
Sponto® AG-540
Sponto® AG-1040
Sponto® AG-1265
Sponto® AK30-02BT
Sponto® AK31-53
Sponto® AK31-56
Sponto® AK31-64
Sponto® AK31-66
Sponto® AK31-69
Sponto® AL69-49
Sponto® CA-861
Sponto® H-44-C
Sponto® N-140B
Spreading Agent ET0672
Stamid HT 3901
Standamid® CD
Standamid® CM
Standamid® ID
Standamid® KD
Standamid® KDM
Standamid® KDO
Standamid® KDS
Standamid® LD
Standamid® LD 80/20
Standamid® LDO
Standamid® LDS
Standamid® LP
Standamid® PD
Standamid® PK-KD
Standamid® PK-KDO
Standamid® PK-KDS
Standamid® PK-SD
Standamid® SD
Standamid® SDO
Standamid® SOD
Standamid® SOMD
Standamox CAW
Standamox O1
Standamox PCAW
Standamox PL
Standamox PS
Standamul® B-1
Standamul® B-2
Standamul® B-3
Standapol® AK-43
Stansperse 506
Stantex Antistat F
Statik-Blok® FDA-3
Stearate PEG 1000
Stepantex® CO-30
Stepantex® CO-36
Stepantex® CO-40
Step-Flow 21
Step-Flow 22
Step-Flow 23
Step-Flow 24
Step-Flow 25
Step-Flow 26
Step-Flow 63
Steramine 49
Steramine CD
Steramine S2
Steramine TV
Sterol CC 595
Sterol GMS
Sterol LG 491
Sterol ST 1
Sterol ST 2
Sunaptol DL Conc.
Sunaptol LT
Sunaptol P Extra Liq.
Super-Sat AWS-4
Super-Sat AWS-24
Supersurf AFX
Supersurf WAF
Super Wet
Super Wet Granular 15G
Surfactol® 13
Surfactol® 318
Surfactol® 365
Surfactol® 590
Surfax 8916/A
Surfine AZI-A
Surfine T-A
Surfine WNG-A
Surfine WNT-A
Surflo® OW-1
Surflo® S24
Surflo® S242
Surflo® S596
Surflo® S1259
Surflo® S2181
Surflo® S2221
Surflo® S7650

NONIONIC (CONT'D.)
Surflo® S8536
Surfonic® HDL
Surfonic® HDL-1
Surfonic® JL-80X
Surfonic® L10-3
Surfonic® L12-3
Surfonic® L12-6
Surfonic® L12-8
Surfonic® L24-1
Surfonic® L24-2
Surfonic® L24-3
Surfonic® L24-4
Surfonic® L24-7
Surfonic® L24-9
Surfonic® L24-12
Surfonic® L42-3
Surfonic® L46-7
Surfonic® L46-8X
Surfonic® L610-3
Surfonic® LF-17
Surfonic® LF-27
Surfonic® LF-37
Surfonic® N-10
Surfonic® N-31.5
Surfonic® N-40
Surfonic® N-60
Surfonic® N-85
Surfonic® N-95
Surfonic® N-100
Surfonic® N-102
Surfonic® N-120
Surfonic® N-130
Surfonic® N-150
Surfonic® N-200
Surfonic® N-300
Surfonic® N-557
Surfonic® NB-5
Surfynol® 61
Surfynol® 82
Surfynol® 82S
Surfynol® 104
Surfynol® 104A
Surfynol® 104BC
Surfynol® 104E
Surfynol® 104H
Surfynol® 104PA
Surfynol® 104PG
Surfynol® 104S
Surfynol® 420
Surfynol® 440
Surfynol® 465
Surfynol® 485
Surfynol® DF-34
Surfynol® DF-110D
Surfynol® DF-110L
Surfynol® GA
Surfynol® PC
Surfynol® PG-50
Surfynol® SE
Surfynol® SE-F
Surfynol® TG
Surfynol® TG-E
Swanic 6L, 7L
Swanic 110
Swanic C-30
Swanic CD-48
Swanic CD-72
Swanic D-16, -24, -52, -60, -80, –120
Swanic SD-36
Swanic X-100
Syn Fac® 334
Syn Fac® 905
Syn Fac® 8210
Syn Fac® 8216
Syn Fac® TDA-92
Syn Fac® TEA-97
Syn Lube 106 (60%)
Syn Lube 107
Syn Lube 728
Syn Lube 1632H
Syn Lube 6277-A
Synotol 119 N
Synotol CN 20
Synotol CN 60
Synotol CN 80
Synotol CN 90
Synotol ME 90
Synperonic 10/3-90%
Synperonic 10/3-100%
Synperonic 10/5-90%
Synperonic 10/5-100%
Synperonic 10/6-90%
Synperonic 10/6-100%
Synperonic 10/7-90%
Synperonic 10/7-100%
Synperonic 10/8-90%
Synperonic 10/8-100%
Synperonic 10/11-90%
Synperonic 10/11-100%
Synperonic 13/3
Synperonic 13/5
Synperonic 13/6.5
Synperonic 13/8
Synperonic 13/9
Synperonic 13/10
Synperonic 13/12
Synperonic 13/15
Synperonic 13/18
Synperonic 13/20
Synperonic 87K
Synperonic 91/2.5
Synperonic 91/4
Synperonic 91/5
Synperonic 91/6
Synperonic 91/7
Synperonic 91/8
Synperonic 91/10
Synperonic 91/12
Synperonic 91/20
Synperonic A2
Synperonic A3
Synperonic A4
Synperonic A7
Synperonic A9
Synperonic A11
Synperonic A14
Synperonic A20
Synperonic A50
Synperonic BD100
Synperonic K87
Synperonic KB
Synperonic LF/B26
Synperonic LF/D25
Synperonic LF/RA30
Synperonic LF/RA40
Synperonic LF/RA43
Synperonic LF/RA50
Synperonic LF/RA260
Synperonic LF/RA290
Synperonic LF/RA343
Synperonic M2
Synperonic M3
Synperonic N
Synperonic NP4
Synperonic NP5
Synperonic NP6
Synperonic NP7
Synperonic NP8
Synperonic NP8.75
Synperonic NP9
Synperonic NP10
Synperonic NP12
Synperonic NP13
Synperonic NP15
Synperonic NP20
Synperonic NP30
Synperonic NP30/70
Synperonic NP35
Synperonic NP40
Synperonic NP50
Synperonic NPE1800
Synperonic NX
Synperonic NXP
Synperonic OP3
Synperonic OP4.5
Synperonic OP7.5
Synperonic OP8
Synperonic OP10
Synperonic OP10.5
Synperonic OP11
Synperonic OP12.5
Synperonic OP16
Synperonic OP16.5
Synperonic OP20
Synperonic OP25
Synperonic OP30
Synperonic OP40
Synperonic OP40/70
Synperonic P105
Synperonic PE/F38
Synperonic PE/F68
Synperonic PE/F87
Synperonic PE/F88
Synperonic PE/F108
Synperonic PE/F127
Synperonic PE/L31
Synperonic PE/L35
Synperonic PE/L42
Synperonic PE/L43
Synperonic PE/L44
Synperonic PE/L61
Synperonic PE/L62
Synperonic PE/L62LF
Synperonic PE/L64
Synperonic PE/L81
Synperonic PE/L92
Synperonic PE/L101
Synperonic PE/L121
Synperonic PE/P75
Synperonic PE/P84
Synperonic PE/P85
Synperonic PE/P94
Synperonic PE/P103
Synperonic PE30/10
Synperonic PE30/20
Synperonic PE30/40
Synperonic PE30/80
Synperonic PE39/70
Synperonic T/304
Synperonic T/701
Synperonic T/707
Synperonic T/904
Synperonic T/908
Synperonic T/1301
Synperonic T/1302
Synperonic TO5
Synprolam 35MX1/O
Synprolam 35X2/O
Syntens KMA 55
Syntens KMA 70-W-02
Syntens KMA 70-W-07
Syntens KMA 70-W-12
Syntens KMA 70-W-17
Syntens KMA 70-W-22
Syntens KMA 70-W-32
Syntergent K
Synthionic Series
Synthrapol KB
Synthrapol LN-VLA
Synthrapol N
Synthrapol WN-K
Syntofor A03, A04, AB03, AB04, AL3, AL4, B03, B04
Syntopon A
Syntopon A 100
Syntopon B
Syntopon B 300
Syntopon C
Syntopon D
Syntopon E
Syntopon F
Syntopon G
Syntopon H
T-9
Tagat® I
Tagat® I2
Tagat® L
Tagat® L2
Tagat® O
Tagat® O2
Tagat® R40
Tagat® R60
Tagat® R63
Tagat® RI
Tagat® S
Tagat® S2
Tagat® TO
Tally® 100 Plus
Tanapal® LD-3, LD-3T
Tanapal® NC
Tanapal® TTD
Tanawet® AR
Tanawet® FBW
Tanawet® NRWG
T-Det® A-026
T-Det® BP-1
T-Det® C-18
T-Det® C-40
T-Det® D-70
T-Det® D-150
T-Det® DD-5
T-Det® DD-7
T-Det® DD-9
T-Det® DD-10
T-Det® DD-11
T-Det® DD-14
T-Det® EPO-35

NONIONIC (CONT'D.)
T-Det® EPO-61
T-Det® EPO-62L
T-Det® EPO-64L
T-Det® LF-416
T-Det® N 1.5
T-Det® N-4
T-Det® N-6
T-Det® N-8
T-Det® N-9.5
T-Det® N-10.5
T-Det® N-12
T-Det® N-14
T-Det® N-20
T-Det® N-30
T-Det® N-40
T-Det® N-50
T-Det® N-70
T-Det® N-100
T-Det® N-307
T-Det® N-407
T-Det® N-507
T-Det® N-705
T-Det® N-707
T-Det® N-1007
T-Det® O-4
T-Det® O-6
T-Det® O-8
T-Det® O-9
T-Det® O-12
T-Det® O-40
T-Det® O-307
T-Det® O-407
T-Det® RQ1
T-Det® RY2
T-Det® TDA-60
T-Det® TDA-65
T-Det® TDA-70
T-Det® TDA-150
Tegamineoxide WS-35
Tegin® D 6100
Tegin® E-41
Tegin® E-41 NSE
Tegin® E-61
Tegin® E-61 NSE
Tegin® E-66
Tegin® E-66 NSE
Tegin® G 6100
Tegin® GRB
Tegin® O
Tegin® P-411
Tegin® RZ
Tegin® RZ NSE
Tegin® T 4753
Teginacid®
Teginacid® H
Teginacid® H-SE
Teginacid® SE
Teginacid® X
Teginacid® X-SE
Tego® Foamex 800
Tego® Foamex 805
Tego® Foamex 1435
Tego® Foamex 7447
Tego® Foamex L 808
Tegotain A 4080
Tegotens 4100
Tegotens BL 130
Tegotens BL 150
Tegotens I
Tegotens I 2
Tegotens L
Tegotens L 2
Tegotens O
Tegotens O 2
Tegotens R 1
Tegotens R 40
Tegotens R 60
Tegotens S
Tegotens S 2
Tegotens TO
Tego® Wet ZFS 453
Tego® Wet ZFS 454
Tekstim 8741
Tek-Wet 951
Tek-Wet 955
Tensiofix 20200
Tensiofix 35300
Tensiofix 35600
Tensiofix PO120, PO132
Tergenol 3964
Tergitol® 15-S-3
Tergitol® 15-S-5
Tergitol® 15-S-7
Tergitol® 15-S-9
Tergitol® 15-S-12
Tergitol® 15-S-15
Tergitol® 15-S-20
Tergitol® 15-S-30
Tergitol® 15-S-40
Tergitol® 24-L-3
Tergitol® 24-L-45
Tergitol® 24-L-50
Tergitol® 24-L-60
Tergitol® 24-L-60N
Tergitol® 24-L-75
Tergitol® 24-L-92
Tergitol® 24-L-98N
Tergitol® 25-L-3
Tergitol® 25-L-5
Tergitol® 25-L-7
Tergitol® 25-L-9
Tergitol® 25-L-12
Tergitol® 26-L-1.6
Tergitol® 26-L-3
Tergitol® 26-L-5
Tergitol® D-683
Tergitol® Min-Foam 1X
Tergitol® Min-Foam 2X
Tergitol® NP-4
Tergitol® NP-5
Tergitol® NP-6
Tergitol® NP-7
Tergitol® NP-8
Tergitol® NP-9
Tergitol® NP-10
Tergitol® NP-13
Tergitol® NP-14
Tergitol® NP-15
Tergitol® NP-27
Tergitol® NP-40
Tergitol® NP-40 (70% Aq.)
Tergitol® NP-44
Tergitol® NP-55, 70% Aq
Tergitol® NP-70, 70% Aq
Tergitol® NPX
Tergitol® TMN-3
Tergitol® TMN-6
Tergitol® TMN-10
Tergitol® TP-9
Tergitol® XD
Tergitol® XH
Tergitol® XJ
Teric 9A2
Teric 9A5
Teric 9A6
Teric 9A8
Teric 9A10
Teric 9A12
Teric 12A2
Teric 12A3
Teric 12A4
Teric 12A6
Teric 12A7
Teric 12A8
Teric 12A9
Teric 12A12
Teric 12A16
Teric 12A23
Teric 12M2
Teric 12M5
Teric 12M15
Teric 13A5
Teric 13A7
Teric 13A9
Teric 13M15
Teric 15A11
Teric 16A16
Teric 16A22
Teric 16A29
Teric 16A50
Teric 16M2
Teric 16M5
Teric 16M10
Teric 16M15
Teric 17A2
Teric 17A3
Teric 17A6
Teric 17A8
Teric 17A10
Teric 17A13
Teric 17A25
Teric 17M2
Teric 17M5
Teric 17M15
Teric 18M2
Teric 18M5
Teric 18M10
Teric 18M20
Teric 18M30
Teric 121
Teric 124
Teric 127
Teric 128
Teric 151
Teric 152
Teric 154
Teric 157
Teric 158
Teric 160
Teric 161
Teric 163
Teric 164
Teric 165
Teric 200
Teric 203
Teric 225
Teric 226
Teric 304
Teric 313
Teric 350
Teric 351B
Teric BL8
Teric BL9
Teric C12
Teric CDE
Teric CME3
Teric CME7
Teric DD5
Teric DD9
Teric G9A5
Teric G9A8
Teric G9A12
Teric G12A4
Teric G12A6
Teric G12A8
Teric G12A12
Teric GN Series
Teric LA4
Teric LA8
Teric LAN70
Teric N2
Teric N3
Teric N4
Teric N5
Teric N8
Teric N9
Teric N10
Teric N11
Teric N12
Teric N13
Teric N15
Teric N20
Teric N30
Teric N30L
Teric N40
Teric N40L
Teric N100
Teric OF4
Teric OF6
Teric OF8
Teric PE61
Teric PE62
Teric PE64
Teric PE68
Teric SF9
Teric SF15
Teric T2
Teric T5
Teric T7
Teric T10
Teric X5
Teric X7
Teric X8
Teric X10
Teric X11
Teric X13
Teric X16
Teric X40
Teric X40L
Tetranol
Tetronic® 50R1
Tetronic® 90R4
Tetronic® 150R1
Tetronic® 304
Tetronic® 504
Tetronic® 701
Tetronic® 702
Tetronic® 704
Tetronic® 707

NONIONIC (CONT'D.)
Tetronic® 901
Tetronic® 904
Tetronic® 908
Tetronic® 1101
Tetronic® 1102
Tetronic® 1104
Tetronic® 1107
Tetronic® 1301
Tetronic® 1302
Tetronic® 1304
Tetronic® 1307
Tetronic® 1501
Tetronic® 1502
Tetronic® 1504
Tetronic® 1508
Texadril 2010
Texadril 8780
Texadril LT 1285
Texamin PD
Texo 284
Texo LP 528A
Texofor A1P
Texofor B1
Textamine Carbon Detergent K
Textamine O-1
Textamine T-1
Tex-Wet 1048
Tex-Wet 1131
Tex-Wet 1155
Tinegal® NA
Tinegal® NT
Titanterge CAC
T-Maz® 20
T-Maz® 40
T-Maz® 61
T-Maz® 65
T-Maz® 80
T-Maz® 81
T-Maz® 85
T-Maz® 90
T-Mulz® 339
T-Mulz® 391
T-Mulz® 392
T-Mulz® 808A
T-Mulz® 1120
T-Mulz® 8015
T-Mulz® AO2
T-Mulz® COC
T-Mulz® FCO
T-Mulz® FGO-17
T-Mulz® FGO-17A
T-Mulz® Mal 5
T-Mulz® PB
T-Mulz® VO
TN Cleaner S
TO-33-F
TO-55-E
TO-55-EL
TO-55-F
Tohol N-220
Tohol N-220X
Tohol N-230
Tohol N-230X
Toho Salt CP-60
Tonerclean 208, 209
Topcithin
Toxanon AH
Toximul® 374
Toximul® 600
Toximul® 705
Toximul® 707
Toximul® 709
Toximul® 710
Toximul® 713
Toximul® 715
Toximul® 716
Toximul® 850
Toximul® 852
Toximul® 856N
Toximul® 857N
Toximul® 858N
Toximul® 8301N
Toximul® 8320
Toximul® 8321
Toximul® 8322
Toximul® 8323
Toximul® D
Toximul® FF
Triton® AG-98
Triton® B-1956
Triton® BG-10
Triton® CF-10
Triton® CF-21
Triton® CF-32
Triton® CF-54
Triton® CF-76
Triton® CF-87
Triton® CG-110
Triton® DF-12
Triton® DF-16
Triton® DF-18
Triton® N-42
Triton® N-57
Triton® N-60
Triton® N-101
Triton® N-111
Triton® N-150
Triton® X-15
Triton® X-35
Triton® X-45
Triton® X-100
Triton® X-100 CG
Triton® X-102
Triton® X-114
Triton® X-114SBHF
Triton® X-120
Triton® X-155-90%
Triton® X-165-70%
Triton® X-207
Triton® X-305-70%
Triton® X-363M
Triton® X-405-70%
Triton® X-705-70%
Triton® XL-80N
Troykyd® D55
Troykyd® D126
Troykyd® D999
Troykyd® Emulso Wet
Troysol 307
Troysol AFL
Troysol CD1
Troysol CD2
Troysol Q148
Troysol S366
Troysol UGA
Trycol® 5874
Trycol® 5877
Trycol® 5878
Trycol® 5882
Trycol® 5887
Trycol® 5888
Trycol® 5940
Trycol® 5941
Trycol® 5942
Trycol® 5943
Trycol® 5944
Trycol® 5946
Trycol® 5949
Trycol® 5950
Trycol® 5951
Trycol® 5952
Trycol® 5953
Trycol® 5956
Trycol® 5957
Trycol® 5963
Trycol® 5964
Trycol® 5966
Trycol® 5967
Trycol® 5968
Trycol® 5971
Trycol® 5972
Trycol® 5993
Trycol® 5999
Trycol® 6720
Trycol® 6802
Trycol® 6940
Trycol® 6941
Trycol® 6942
Trycol® 6943
Trycol® 6951
Trycol® 6953
Trycol® 6954
Trycol® 6956
Trycol® 6957
Trycol® 6958
Trycol® 6960
Trycol® 6961
Trycol® 6962
Trycol® 6963
Trycol® 6964
Trycol® 6965
Trycol® 6967
Trycol® 6968
Trycol® 6969
Trycol® 6970
Trycol® 6971
Trycol® 6972
Trycol® 6974
Trycol® 6975
Trycol® 6979
Trycol® 6981
Trycol® 6984
Trycol® 6985
Trycol® 6989
Trycol® DA-4
Trycol® DA-6
Trycol® DA-69
Trycol® DNP-8
Trycol® DNP-150
Trycol® DNP-150/50
Trycol® LAL-4
Trycol® LAL-8
Trycol® LAL-12
Trycol® LAL-23
Trycol® NP-1
Trycol® NP-4
Trycol® NP-6
Trycol® NP-7
Trycol® NP-9
Trycol® NP-11
Trycol® NP-20
Trycol® NP-30
Trycol® NP-40
Trycol® NP-50
Trycol® NP-307
Trycol® NP-407
Trycol® NP-507
Trycol® NP-1007
Trycol® OAL-23
Trycol® OP-407
Trycol® SAL-20
Trycol® TDA-3
Trycol® TDA-6
Trycol® TDA-8
Trycol® TDA-9
Trycol® TDA-11
Trycol® TDA-12
Trycol® TDA-18
Trycol® TP-2
Trycol® TP-6
Trydet 19
Trydet 20
Trydet 22
Trydet 2610
Trydet 2636
Trydet 2640
Trydet 2644
Trydet 2670
Trydet 2671
Trydet 2675
Trydet 2676
Trydet 2681
Trydet 2682
Trydet 2685
Trydet 2691
Trydet 2692
Trydet DO-9
Trydet ISA-4
Trydet ISA-9
Trydet ISA-14
Trydet LA-5
Trydet LA-7
Trydet MP-9
Trydet OA-5
Trydet OA-7
Trydet OA-10
Trydet SA-5
Trydet SA-7
Trydet SA-8
Trydet SA-23
Trydet SA-40
Trydet SA-50/30
Trydet SO-9
Trylon® 5900
Trylon® 5902
Trylon® 5906
Trylon® 5909
Trylon® 5921
Trylon® 5922
Trylon® 6735
Trylon® 6837
Trylox® 1086
Trylox® 5900
Trylox® 5902
Trylox® 5904
Trylox® 5906
Trylox® 5907
Trylox® 5909

NONIONIC (CONT'D.)
Trylox® 5918
Trylox® 5921
Trylox® 5922
Trylox® 5925
Trylox® 6746
Trylox® 6747
Trylox® 6753
Trylox® CO-5
Trylox® CO-10
Trylox® CO-16
Trylox® CO-25
Trylox® CO-30
Trylox® CO-36
Trylox® CO-80
Trylox® CO-200, -200/50
Trylox® HCO-5
Trylox® HCO-16
Trylox® HCO-25
Trylox® HCO-200/50
Trylox® SS-20
TS-33-F
T-Tergamide 1CD
T-Tergamide 1PD
Tween® 20
Tween® 21
Tween® 40
Tween® 61
Tween® 81
Tween® 85
Twitchell 6809 Oil
Tylose H Series
Tylose MH Grades
Tylose MHB
Ufanon K-80
Ufanon KD-S
Ukanil 2262
Ukanil 3000
Ukaril 190
Unamide® C-2
Unamide® C-5
Unamide® C-72-3
Unamide® D-10
Unamide® J 56
Unamide® L-2
Unamide® L-5
Unamide® LDL
Unamide® LDX
Unamide® N-72-3
Unamide® S
Unamide® SI
Unamide® W
Unamine® C
Unamine® O
Unamine® S
Union Carbide® L-720, -721
Union Carbide® L-722, -727
Union Carbide® L-7001, -7002
Union Carbide® L-7600, -7602, -7604
Union Carbide® L-7605
Union Carbide® L-7607
Uniperol® EL
Uniperol® O
Uniperol® O Micropearl
Unisol BT
Unisol MFD
Unital G
Valdet 561
Valdet 4016
Valdet AC-40
Valsof FR-4
Valsof PMS
Value 401
Value 402
Value 1209C
Value 1407
Value 1414
Value 3608
Value 3706
Value 3710
Value W-307
Varamide® A-2
Varamide® A-7
Varamide® A-10
Varamide® C-212
Varamide® L-203
Varamide® MA-1
Varamide® ML-1
Varamide® ML-4
Varonic® 32-E20
Varonic® K202
Varonic® K202 SF
Varonic® K205
Varonic® K205 SF
Varonic® K210
Varonic® K210 SF
Varonic® K215
Varonic® K215 SF
Varonic® LI-42
Varonic® LI-48
Varonic® LI-63
Varonic® LI-67
Varonic® LI-67-75%
Varonic® MT 65
Varonic® T202
Varonic® T205
Varonic® T210
Varonic® T215
Varox® 185E
Varox® 365
Varox® 743
Varox® 1770
Versilan MX123
Versilan MX134
Victawet® 12
Viscasil®
Vismul ET-20
Vismul ET-55
Volpo 3
Volpo 25 D 3
Volpo 25 D 5
Volpo 25 D 10
Volpo 25 D 15
Volpo 25 D 20
Volpo CS-3
Volpo CS-5
Volpo CS-10
Volpo CS-15
Volpo CS-20
Volpo L4
Volpo L23
Volpo N3
Volpo N5
Volpo N10
Volpo N15
Volpo N20
Volpo O3
Volpo O5
Volpo O10
Volpo O15
Volpo O20
Volpo S-2
Volpo T-3
Volpo T-5
Volpo T-10
Volpo T-15
Volpo T-20
Vykamol N/E, S/E
Wettol® EM 2
Witbreak 772
Witbreak 774
Witbreak DGE-75
Witbreak DGE-128A
Witbreak DGE-169
Witbreak DGE-182
Witbreak DPG-15
Witbreak DPG-482
Witbreak DRA-21
Witbreak DRA-22
Witbreak DRA-50
Witbreak DRB-11
Witbreak DRB-127
Witbreak DRB-401
Witbreak DRC-163
Witbreak DRC-164
Witbreak DRC-165
Witbreak DRC-168
Witbreak DRC-232
Witbreak DRE-8164
Witbreak DRI-9020
Witbreak DRI-9037
Witbreak DRI-9038
Witbreak DTG-62
Witbreak GBG-3172
Witcamide® 82
Witcamide® 128T
Witcamide® 272
Witcamide® 511
Witcamide® 1017
Witcamide® 5130
Witcamide® 5133
Witcamide® 5138
Witcamide® 5140
Witcamide® 5195
Witcamide® 6310
Witcamide® 6445
Witcamide® 6511
Witcamide® 6514
Witcamide® 6515
Witcamide® 6531
Witcamide® 6533
Witcamide® 6625
Witcamide® C
Witcamide® CD
Witcamide® CDA
Witcamide® CDS
Witcamide® Coco Condensate
Witcamide® CPA
Witcamide® LDTS
Witcamide® M-3
Witcamide® MAS
Witcamide® N
Witcamide® NA
Witcamide® S771
Witcamide® S780
Witcomul 4016
Witconate 705
Witconate P-1020BUST
Witconol 14
Witconol 18L
Witconol 1206
Witconol 2301
Witconol 2326
Witconol 2380
Witconol 2400
Witconol 2401
Witconol 2407
Witconol 2421
Witconol 2500
Witconol 2503
Witconol 2620
Witconol 2622
Witconol 2640
Witconol 2642
Witconol 2648
Witconol 2720
Witconol 2722
Witconol 5906
Witconol 5907
Witconol 5909
Witconol 6903
Witconol APEM
Witconol APM
Witconol APS
Witconol CA
Witconol CD-17
Witconol CD-18
Witconol DOS
Witconol EGMS
Witconol F26-46
Witconol GOT
Witconol H31
Witconol H31A
Witconol H33
Witconol H35A
Witconol MST
Witconol NP-40
Witconol NP-100
Witconol NP-300
Witconol NS-108LQ
Witconol NS-500K
Witconol NS-500LQ
Witconol RDC-D
Witconol RHP
Witconol RHT
Witconol SN Series
Witflow 990
Zeeterge 444
Zohar Amido Amine 1
Zohar EGMS
Zohar GLST
Zohar GLST SE
Zoharex A-10
Zoharex B-10
Zoharex N-25
Zoharex N-60
Zonyl® A
Zonyl® FSN
Zonyl® FSN-100
Zonyl® FSO
Zonyl® FSO-100
Zoramide CM
Zoramox
Zusolat 1004
Zusolat 1005/85

NONIONIC (CONT'D.)

Zusolat 1008/85
Zusolat 1010/85
Zusolat 1012/85
Zusomin C 108
Zusomin O 102
Zusomin S 110
Zusomin S 125
Zusomin TG 102

HLB Classification

HLB	Product
0.6	Emerest® 2423
1.0	Chemal BP-2101
1.0	Cithrol EGDL N/E
1.0	Cithrol EGDO N/E
1.0	Macol® 101
1.0	Pegosperse® 50 DS
1.0	Pluronic® L101
1.0	Prox-onic EP 4060-1
1.0-7.0	Pluronic® 22R4
1.0-7.0	Tetronic® 90R4
1.0-7.0	Tetronic® 150R1
1.1	Cithrol EGMR N/E
1.3	Emerest® 2355
1.3	Lexemul® EGDS
1.3	Nonicol 100
1.4	Mapeg® EGDS
1.5	Cithrol DGDL N/E
1.5	Cithrol EGDS N/E
1.5	Dow Corning® 1248 Fluid
1.5	Ethylan® GT85
1.5	Radiasurf® 7269
1.6	Emerest® 2419
1.7	Hetoxide C-2
1.7	Pluronic® 31R1
1.8	Ablunol S-85
1.8	Alkamuls® STO
1.8	Atlas G-4885
1.8	Atlox 4885
1.8	Atsurf S-85
1.8	Cithrol DGDO N/E
1.8	Cithrol EGMO N/E
1.8	Disponil STO 100 F1
1.8	Emerest® 2380
1.8	Glycomul® TO
1.8	Hodag STO
1.8	Ionet S-85
1.8	Liposorb TO
1.8	Montane 85
1.8	Pegosperse® 400 DTR
1.8	Prote-sorb STO
1.8	Remcopal HC 7
1.8	Sorbax STO
1.8	Span® 85
1.8	Witconol 2380
1.9	Radiasurf® 7146
1.9	Radiasurf® 7201
1.9	S-Maz® 95
2.0	Aldo® DC
2.0	Cithrol EGDL S/E
2.0	Cithrol EGDO S/E
2.0	Cithrol EGMR S/E
2.0	Cithrol EGMS N/E
2.0	Cithrol GDL N/E
2.0	Cithrol GDO N/E
2.0	Glycomul® TS KFG
2.0	Hartopol L81
2.0	Lipo EGMS
2.0	Naturechem® EGHS
2.0	Pegosperse® 50 MS
2.0	Pegosperse® EGMS-70
2.0	Pluronic® L81
2.0	Sorban CO
2.0	Syncrowax AW1-C
2.0	TS-33-F
2.1	Crill 35
2.1	Disponil STS 100 F1
2.1	Drewmulse® SMS
2.1	Drewmulse® STS
2.1	Emsorb® 2503
2.1	Glycomul® TS
2.1	Hodag STS
2.1	Liposorb TS
2.1	Nikkol SS-30
2.1	Prote-sorb STS
2.1	Radiasurf® 7156
2.1	Radiasurf® 7270
2.1	S-Maz® 85
2.1	S-Maz® 85K
2.1	Sorbax STS
2.1	Span® 65
2.1	Witconol 2503
2.2	Cithrol DGDL S/E
2.2	Emerest® 2350
2.2	Emsorb® 2507
2.2	Lexemul® EGMS
2.2	Macol® 27
2.2	S-Maz® 65K
2.2	Witconol EGMS
2.3	Radiasurf® 7175
2.3	Radiasurf® 7206
2.4	Cithrol EGDS S/E
2.4	Cithrol PGMS N/E
2.4	Durlac® 100W
2.5	Nikkol Hexaglyn 3-S
2.5	Radiasurf® 7153
2.5	TO-33-F
2.5	Tetronic® 901
2.6	Argowax Cosmetic Super
2.6	Argowax Dist
2.6	Argowax Standard
2.6	Naturechem® PGHS
2.7	Caplube 8350
2.7	Cithrol DGDO S/E
2.7	Cithrol EGML N/E
2.7	Cithrol EGMO S/E
2.7	Cithrol GMR N/E
2.7	Cithrol PGML N/E
2.7	Cithrol PGMR N/E
2.7	Emulpon EL 20
2.7	Montane 65
2.7	Radiasurf® 7150
2.7	Sterol ST 1
2.8	Aldo® HMS
2.8	Dur-Em® 117
2.8	Quimipol EA 6801
2.8	Radiasurf® 7152
2.8	Tegin® D 6100
2.9	Ablunol EGMS
2.9	Alkamuls® EGMS/C
2.9	Cithrol EGMS S/E
2.9	Cithrol GDO S/E
2.9	Emerest® 2410
2.9	Mapeg® EGMS
2.9	Radiasurf® 7140
3.0	Aldo® MO FG
3.0	Aldo® PMS
3.0	Alkamuls® GMR-55LG
3.0	Antarox® L-61
3.0	Antarox® PGP 18-1
3.0	Chemal BP 261
3.0	Cithrol DGDS N/E
3.0	EM-40
3.0	Emery® 6220
3.0	Emsorb® 2518
3.0	Geleol
3.0	Hartopol LF-1
3.0	Imwitor® 900
3.0	Imwitor® 908
3.0	Imwitor® 910
3.0	Imwitor® 940
3.0	Macol® 1
3.0	Macol® 20
3.0	Macol® 44
3.0	Monolan® 2000
3.0	Monthyle
3.0	Nissan Plonon 171
3.0	Nissan Plonon 201
3.0	Norfox® 1 Polyol
3.0	Peceol Isostearique
3.0	Pegosperse® PMS CG
3.0	Pepol B
3.0	Pluronic® L61
3.0	Prox-onic EP 1090-1
3.0	Quimipol ED 2021
3.0	Radiasurf® 7144
3.0	Sinoponic PE 61
3.0	Surfynol® DF-110D
3.0	Surfynol® DF-110L
3.0	T-Det® EPO-61
3.0	T-Det® RQ1

3.0	Teric PE61
3.0	Tetronic® 701
3.0	Tomah E-18-2
3.0-5.0	Emulan® FJ
3.0-5.0	Emulan® FM
3.0-24.0	Arnox BP Series
3.1	Cithrol PGMO N/E
3.2	Caplube 8442
3.2	Cithrol PGMS S/E
3.2	Drewpol® HL-13788
3.2	Lexemul® 515
3.2	Sterol GMS
3.2	Tegin® G 6100
3.3	Aldo® MS LG
3.3	Cithrol GMO N/E
3.3	Marlophen® 81
3.3	Marlophen® 81N
3.3	Radiasurf® 7151
3.3	Radiasurf® 7600
3.3	Tegin® O
3.3	Tegin® O NSE
3.4	Aldo® MO
3.4	Aldo® MO Tech
3.4	Alkamuls® GMS/C
3.4	Cithrol GDS N/E
3.4	Cithrol GMS N/E
3.4	Emerest® 2421
3.4	Naturechem® GMHS
3.4	Surfonic® N-10
3.4	Tegin® ISO
3.4	Witconol 2421
3.5	Arlacel® 129
3.5	Caplube 8448
3.5	Cedepal CA-210
3.5	DO-33-F
3.5	Hodag EGMS
3.5	Nikkol Decaglyn 5-IS
3.5	Nikkol Decaglyn 5-S
3.5	Pegosperse® 100 O
3.5	Radiasurf® 7410
3.6	Aldo® MS Industrial
3.6	Cithrol EGML S/E
3.6	Cithrol GMR S/E
3.6	Cithrol PGML S/E
3.6	Cithrol PGMR S/E
3.6	Hetoxol L-1
3.6	Rexol 45/1
3.6	Triton® X-15
3.7	Ablunol DEGMS
3.7	Arlacel® C
3.7	Crill 43
3.7	Dehydol PLS 1
3.7	Dehydol PLS 15
3.7	Disponil SSO 100 F1
3.7	Emasol O-15 R
3.7	Genapol® 26-L-1
3.7	Imwitor® 780 K
3.7	Liposorb SQO
3.7	Montane 83
3.7	Neodol® 23-1
3.8	Chemax CO-5
3.8	Chemax HCO-5
3.8	Kessco® GMO
3.8	Macol® OA-2
3.8	Mazol® 300 K
3.8	Mazol® GMO
3.8	Mazol® GMO #1
3.8	Mazol® GMO K
3.8	Monomuls 90-25
3.8	Nopalcol 1-S
3.8	Pegosperse® 100 S
3.8	Prox-onic HR-05
3.8	Prox-onic HRH-05
3.8	Radiasurf® 7400
3.8	Surfactol® 318
3.8	Tegin® M
3.8	Trylox® CO-5
3.8	Trylox® HCO-5
3.9	Alkamuls® COH-5
3.9	Cithrol PGMO S/E
3.9	Emerest® 2400
3.9	Emerest® 2401
3.9	Lexemul® 503
3.9	Mazol® GMS-90
3.9	Radiasurf® 7196
3.9	Witconol 2400
3.9	Witconol 2401
3.9	Witconol MST
4.0	Ablunol CO 5
4.0	Aldo® MOD FG
4.0	Aldo® MS
4.0	Alkaterge®-T
4.0	Antarox® 31-R-1
4.0	Argobase EU
4.0	Argobase LI
4.0	Argobase SI
4.0	Bio-Soft® E-200
4.0	Emultex WS
4.0	Ethox CO-5
4.0	Glycomul® SOC
4.0	Imwitor® 191
4.0	Imwitor® 312
4.0	Imwitor® 742
4.0	Labrafil ISO
4.0	Macol® 33
4.0	Monosteol
4.0	Nikkol Decaglyn 5-O
4.0	Nikkol SO-30
4.0	OHlan®
4.0	Pluronic® L122
4.0	Surfynol® 104
4.0	Surfynol® 104A
4.0	Surfynol® 104BC
4.0	Surfynol® 104E
4.0	Surfynol® 104H
4.0	Surfynol® 104PA
4.0	Surfynol® 104PG
4.0	Surfynol® 104S
4.0	Surfynol® 420
4.0	T-Det® RY2
4.0	Trylox® 5900
4.0-5.0	Alkaterge®-E
4.0-5.0	Surfynol® SE
4.0-5.0	Surfynol® SE-F
4.1	Cithrol GMO S/E
4.1	Desotan® SMT
4.1	Noigen EA 33
4.1	Nopalcol 1-TW
4.1	Prox-onic MO-02
4.1	Radiasurf® 7411
4.2	Cithrol GDS S/E
4.2	Ethox TAM-2
4.2	Rexol AE-1
4.2	Sinochem GMS
4.2	Tegacid® Regular VA
4.3	Ablunol S-80
4.3	Alkamuls® SMO
4.3	Arlacel® A
4.3	Atlas G-4884
4.3	Atplus 300F
4.3	Atplus 401
4.3	Atsolyn GMO
4.3	Atsurf S-80
4.3	Crill 4
4.3	Crill 45
4.3	Desotan® SMO
4.3	Disponil SMO 100 F1
4.3	Ethylan® GO80
4.3	Glycomul® O
4.3	Glycomul® TAO
4.3	Hodag SMO
4.3	Ionet S-80
4.3	Liposorb O
4.3	Montane 80
4.3	Norfox® Sorbo S-80
4.3	Prote-sorb SMO
4.3	S-Maz® 90
4.3	Sorbax SMO
4.3	Sorbirol O
4.3	Span® 80
4.4	Alfonic® 1216-22
4.4	Cithrol DGMS N/E
4.4	Cithrol GMS S/E
4.4	Emasol S-20
4.4	Imwitor® 595
4.4	MO-33-F
4.4	Radiasurf® 7372
4.4	Tegin® P
4.5	Aquaphil K
4.5	Emsorb® 2502
4.5	Emultex SMS
4.5	Ethomeen® T/12
4.5	Ethylan® NP 1
4.5	Lanesta G
4.5	Macol® 10
4.5	Nikkol SO-15
4.5	Norfox® NP-1
4.5	Pluronic® L31
4.5	Quimipol ENF 15
4.5	Sorban AO
4.5	Surfonic® L24-1
4.5	Teric T2
4.5	Tomah E-S-2
4.6	Alkasurf® NP-1
4.6	Arosurf® 66-E2
4.6	Catinex KB-31
4.6	Cedepal CO-210
4.6	Chemax NP-1.5
4.6	Crill 6
4.6	Emsorb® 2500
4.6	Emsorb® 2516
4.6	Emthox® 6960
4.6	Ethal NP-1.5
4.6	Iconol NP-1.5
4.6	Igepal® CO-210
4.6	Prox-onic NP-1.5
4.6	Rexol 25/1
4.6	S-Maz® 80
4.6	S-Maz® 80K
4.6	S-Maz® 83R
4.6	T-Det® N 1.5
4.6	Trycol® 6960
4.6	Trycol® NP-1
4.6	Trycol® TP-2
4.6	Witconol 2500
4.7	Ablunol S-60
4.7	Akyporox NP 15
4.7	Alkamuls® SMS
4.7	Arlacel® 780
4.7	Crill 3

HLB	Product
4.7	Disponil SMS 100 F1
4.7	Drewmulse® SMO
4.7	Durtan® 60
4.7	Emulson AG 2A
4.7	Ethylan® GS60
4.7	Hodag DGO
4.7	Hodag DGS
4.7	Ionet S-60 C
4.7	Liposorb S
4.7	Macol® SA-2
4.7	Mapeg® 200 DS
4.7	Montane 60
4.7	Nikkol SS-10
4.7	Prote-sorb SMS
4.7	Radiasurf® 7155
4.7	S-Maz® 60
4.7	Sorbax SMS
4.7	Span® 60, 60K
4.7	Sterol ST 2
4.7	Varonic® Q202
4.8	Chemeen 18-2
4.8	Igepal® CA-210
4.8	Pegosperse® 100 MR
4.8	Radiasurf® 7900
4.9	Berol 302
4.9	Berol 456
4.9	Brij® 72
4.9	Chemeen DT-3
4.9	Cithrol GML N/E
4.9	Ethox SAM-2
4.9	Hetoxol OL-2
4.9	Hetoxol STA-2
4.9	Lipocol O-2
4.9	Lipocol S-2
4.9	Macol® CA-2
4.9	Prox-onic SA-1/02
4.9	Ritoleth 2
4.9	Simulsol 72
4.9	Simulsol 92
4.9	Tegacid® Special VA
4.9	Trylon® 5900
4.9	Volpo S-2
5.0	Albalan
5.0	Aldo® MOD
5.0	Aldosperse® O-20 FG
5.0	Ameroxol® OE-2
5.0	Argobase MS-5
5.0	Chemeen T-2
5.0	Cithrol DGMO N/E
5.0	Cithrol DGMS S/E
5.0	DMS-33
5.0	Ethylan® TT-203
5.0	Eumulgin EP2
5.0	Eumulgin PWM2
5.0	Genapol® 26-L-1.6
5.0	Glucate® DO
5.0	Glycomul® S
5.0	Glycomul® S FG
5.0	Glycomul® S KFG
5.0	Hydrine
5.0	Icomeen® T-2
5.0	MS-33-F
5.0	Mulsifan STK
5.0	Nikkol SO-10
5.0	Norfox® F-221
5.0	Pluronic® L121
5.0	Polyaldo® DGHO
5.0	Prox-onic MT-02
5.0	Quimipol EA 6802
5.0	Radiasurf® 7145
5.0	Serdox NOL 2
5.0	Soitem 520
5.0	Tergitol® 26-L-1.6
5.0	Tomah E-14-5
5.0	Tomah E-T-2
5.0	Toximul® TA-2
5.0	Varonic® S202
5.0-6.0	Alcolec® 495
5.0-8.0	Silwet® L-77
5.0-8.0	Silwet® L-722
5.0-8.0	Silwet® L-7002
5.0-8.0	Silwet® L-7500
5.0-8.0	Silwet® L-7602
5.0-8.0	Silwet® L-7614
5.1	Emerest® 2407
5.1	Hetoxol CA-2
5.1	Macol® CSA-2
5.1	Varonic® T202
5.1	Varonic® T202 SR
5.1	Witconol 2407
5.1	Witconol RHT
5.2	Aldo® ML
5.2	Cithrol 2DS
5.2	Emsorb® 2505
5.2	Ice #2
5.2	Igepal® NP-2
5.2	Lexemul® 530
5.3	Hetoxamate MO-2
5.3	Ionet DO-200
5.3	Lexemul® T
5.3	Lipocol C-2
5.3	Remcopal 3112
5.3	Rolfor 162
5.3	Simulsol 52
5.3	Surfonic® L24-2
5.4	Lexemul® 55SE
5.5	Atlox 4912
5.5	Cithrol DGDS S/E
5.5	Pluronic® L92
5.5	Prox-onic DT-03
5.5	Rioklen NF 2
5.5	Toximul® 8321
5.6	Cithrol GML S/E
5.6	Ethylan® D252
5.6	Eumulgin ET 2
5.6	Plysurf AL
5.6	Tomah E-17-2
5.7	Berol 259
5.7	Cithrol DGMO S/E
5.7	Marlophen® 82
5.7	Marlophen® 82N
5.7	Nissan Nonion NS-202
5.7	Nonipol 20
5.7	Radiasurf® 7420
5.7	Teric N2
5.8	Eumulgin PC 2
5.8	Newpol PE-61
5.8	Quimipol ENF 20
5.9	Berol 307
5.9	Kessco® PEG 200 DL
5.9	Lexemul® PEG-200 DL
5.9	Nopalcol 2-DL
5.9	Pationic ISL
5.9	Renex® 702
5.9	Synperonic A2
6.0	Aldo® MR
6.0	Aldo® MSD
6.0	Alfonic® 1216-30
6.0	Antarox® 25-R-2
6.0	Bio-Soft® E-300
6.0	Centrol® CA
6.0	Cirrasol® EN-MB
6.0	Cirrasol® EN-MP
6.0	Cithrol 2DL
6.0	Drewpol® 10-4-OK
6.0	Ethosperse® CA-2
6.0	Ethox CAM-2
6.0	Ethylan® FO30
6.0	Genapol® 26-L-2
6.0	Glucate® SS
6.0	Hetoxide C-9
6.0	Industrol® DW-5
6.0	Macol® 18
6.0	Mapeg® 200 DO
6.0	Mapeg® 200 DOT
6.0	Mazol® GMR
6.0	Mazol® GMS-D
6.0	Monolan® 2500 E/30
6.0	Mulsifan RT 163
6.0	Nikkol Hexaglyn 5-O
6.0	Nikkol OP-3
6.0	Nikkol Tetraglyn 1-O
6.0	Nikkol Tetraglyn 1-S
6.0	Nissan Nonion K-202
6.0	Noigen EA 50
6.0	Noigen ET60
6.0	Noigen ET65
6.0	Nopalcol 1-L
6.0	Pegosperse® 100 ML
6.0	Plurafac® A-24
6.0	Sandopan® B
6.0	Sellig T 3 100
6.0	Syn Lube 1632H
6.0	Teric 12A2
6.0	Teric 17A2
6.0-7.0	Surfynol® 61
6.0-8.0	Triton® RW-20
6.1	Chemeen C-2
6.1	Cithrol DGML N/E
6.1	Ethox DL-5
6.1	Hetoxol L-2
6.1	Marlipal® 013/20
6.1	Prox-onic MC-02
6.1	Rexol AE-2
6.1	Rexonic N91-1.6
6.1	Synperonic M2
6.1	Tally® 100 Plus
6.2	Cithrol 2MO
6.2	Dehydol LS 2
6.2	Ethylan® CDP2
6.2	Lauropal 2
6.2	Marlipal® 24/20
6.2	Mazol® PGO-31 K
6.2	Mazol® PGO-104
6.2	Sinopol 1102
6.2	Varonic® K202
6.2	Varonic® K202 SF
6.3	Akyporox RLM 22
6.3	Cithrol 2MS
6.3	Croduret 10
6.3	Etocas 10
6.3	Eumulgin TL 30
6.3	Macol® 31
6.3	Neodol® 45-2.25
6.3	Newpol PE-62
6.3	Pluronic® 25R2
6.3	Radiasurf® 7135
6.3	Radiasurf® 7421
6.4	Ethomeen® C/12
6.4	Hetoxol OA-3 Special

HLB	Trade name
6.4	Prox-onic LA-1/02
6.4	Rolfor N 24/2
6.4	Sinopol 60
6.4	Sinopol 80
6.4	Trylox® CO-10
6.5	Caplube 8540B
6.5	Chemal BP-262LF
6.5	Emsorb® 2510
6.5	Hodag DGL
6.5	MP-33-F
6.5	Nikkol DDP-2
6.5	Nikkol Decaglyn 3-O
6.5	Nikkol Decaglyn 3-S
6.5	Pluronic® L72
6.5	Prox-onic EP 1090-2
6.5	S-Maz® 40
6.6	Ablunol CO 10
6.6	Catinex KB-10
6.6	Epan U 103
6.6	Hartopol L62LF
6.6	Hodag 22-L
6.6	Ionet DL-200
6.6	Macol® 2LF
6.6	Plysurf A208B
6.6	Sinopol 1203
6.6	Sinopol 1510
6.6	Trydet 2691
6.6	Volpo 3
6.6	Volpo N3
6.7	Ablunol S-40
6.7	Caplube 8440
6.7	Cithrol DGML S/E
6.7	Crill 2
6.7	Disponil SMP 100 F1
6.7	Emerest® 2452
6.7	Glycomul® P
6.7	Hetoxide DNP-4
6.7	Hodag SMP
6.7	Liposorb P
6.7	Merpol® A
6.7	Montane 40
6.7	Prote-sorb SMP
6.7	Quimipol EA 6803
6.7	Sorbax SMP
6.7	Span® 40
6.8	Aldo® MC
6.8	Aldo® MLD
6.8	Hartopol LF-5
6.9	Dehydol PLS 235
6.9	Ethal 368
6.9	Prox-onic MS-02
7.0	Antarox® L-62
7.0	Antarox® L-62 LF
7.0	Antarox® LA-EP 15
7.0	Antarox® LA-EP 25
7.0	Antarox® LA-EP 25LF
7.0	Antarox® PGP 18-2
7.0	Atplus 1992
7.0	Chemal BP-262
7.0	Dow Corning® FF-400
7.0	Drewpol® 3-1-SK
7.0	Elfacos® ST 9
7.0	Emery® 6221 Monolan 2500
7.0	Ethylan® 172
7.0	Ethylan® A2
7.0	Ethylan® C12AH
7.0	Hartopol L62
7.0	Hartopol LF-2
7.0	Macol® 2
7.0	Macol® 21
7.0	Mapeg® 200 DC
7.0	Monolan® 2500
7.0	Nikkol TDP-2
7.0	Nikkol TDP-4
7.0	Nissan Plonon 102
7.0	Nissan Plonon 172
7.0	Noigen ET77
7.0	Norfox® 2LF
7.0	Norfox® 2 Polyol
7.0	Pegol® L-62D
7.0	Pegosperse® 200 DL
7.0	Pepol B-182
7.0	Pepol D-301
7.0	Plurafac® RA-40
7.0	Plurafac® RA-43
7.0	Pluronic® L62
7.0	Plysurf A208S
7.0	Prox-onic EP 2080-1
7.0	Quimipol ED 2022
7.0	Rexonic P-1
7.0	Sinoponic PE 62
7.0	T-Det® EPO-62L
7.0	Teric PE62
7.0	Tetronic® 702
7.0-8.0	Emcol® P-1020B
7.0-8.0	Witconate 605AC
7.1	Plysurf A207H
7.1	Surfax 8916/A
7.1	Synperonic NP30
7.1	Synperonic OP3
7.1	Teric 12A3
7.2	Alkamuls® 400-DO
7.2	Nonex DO-4
7.2	Santone® 3-1-SH
7.3	Armul 88
7.3	Cithrol 3DS
7.3	Dehydol LT 3
7.3	Ionet DS-300
7.3	Polychol 5
7.3	Rhodasurf® A 24
7.3	Volpo CS-3
7.4	Durtan® 20
7.4	Nonex DL-2
7.4	Pegosperse® 100 L
7.4	Radiasurf® 7402
7.4	Radiasurf® 7443
7.5	Arlatone® 985
7.5	Cirrasol® LAN-SF
7.5	Emerest® 2642
7.5	Marlophen® 83
7.5	Marlophen® 83N
7.5	Marlowet® 4939
7.5	Nikkol BO-2
7.5	Radiasurf® 7412
7.5	Spreading Agent ET0672
7.5	Teric 17A3
7.5	Tomah E-DT-3
7.5	Trydet 2685
7.5	Witconol 2642
7.6	Emerest® 2622
7.6	Emerest® 2704
7.6	Hetoxol M-3
7.6	Macol® 2D
7.6	Mapeg® 200 DL
7.6	Quimipol ENF 30
7.6	Radiasurf® 7125
7.6	Surfonic® L42-3
7.6	Witconol 2622
7.7	Crodet S4
7.7	Dacospin LD-605
7.7	Emulgeant 710
7.7	Genapol® 42-L-3
7.7	Sinopol 170
7.7	Surfonic® N-31.5
7.7	Teric OF4
7.7	Trydet ISA-4
7.8	Alkamuls® 400-DS
7.8	Dehydol PD 253
7.8	Deterpal 843
7.8	Emery® 6885
7.8	Ethylan® D253
7.8	Macol® OP-3
7.8	Pegosperse® 400 DS
7.8	Pegosperse® 1500 DO
7.8	Quimipol EA 2503
7.8	Radiasurf® 7453
7.8	Renex® 703
7.8	Rexol 45/3
7.8	Rexonic N25-3
7.8	Rhodasurf® LA-3
7.8	Sellig SU 4 100
7.8	Synperonic A3
7.8	Triton® X-35
7.9	Ablunol 200MO
7.9	Ablunol LA-3
7.9	Akyporox NP 40
7.9	Chemal OA-4
7.9	Chemal TDA-3
7.9	Ethal TDA-3
7.9	Hetoxol L-3N
7.9	Hetoxol OL-4
7.9	Hetoxol TD-3
7.9	Neodol® 23-3
7.9	Neodol® 25-3
7.9	Nissan Nonion O-2
7.9	Prox-onic OA-1/04
7.9	Prox-onic TD-1/03
7.9	Siponic® TD-3
7.9	Synperonic M3
7.9	Trycol® 5993
8.0	Ablunol 200MS
8.0	Acconon 200-MS
8.0	Acconon CA-5
8.0	Alfonic® 810-40
8.0	Alfonic® 1012-40
8.0	Alfonic® 1412-40
8.0	Amerchol® L-101
8.0	Antarox® 17-R-2
8.0	Bio-Soft® E-400
8.0	Bio-Soft® TD 400
8.0	Blendmax Series
8.0	Chemal 2EH-2
8.0	Dehydol PEH 2
8.0	Dehydol PLS 3
8.0	Emsorb® 2515
8.0	Emulmin 40
8.0	Ethal 326
8.0	Ethylan® CD802
8.0	Ethylan® CDP3
8.0	Ethylan® GL20
8.0	Genapol® 24-L-3
8.0	Genapol® 26-L-3
8.0	Genapol® UD-030
8.0	Glycosperse® L-10
8.0	Hartopol L42
8.0	Hyonic OP-40
8.0	Iconol TDA-3
8.0	Igepal® CA-420

HLB	Product
8.0	Lauroxal 3
8.0	Lipocol SC-4
8.0	ML-33-F
8.0	Macol® 19
8.0	Macol® 35
8.0	Macol® OA-4
8.0	Macol® TD-3
8.0	Mapeg® 200 MS
8.0	Marlipal® 013/30
8.0	Mulsifan RT 324
8.0	Nikkol TDP-6
8.0	Nissan Nonion K-203
8.0	Nissan Nonion S-2
8.0	Noigen EA 70
8.0	Noigen ET80
8.0	Noigen ET83
8.0	PGE-400-DS
8.0	Pegosperse® 400 DO
8.0	Pegosperse® 400 DOT
8.0	Pluronic® 17R2
8.0	Pluronic® P123
8.0	Prox-onic 2EHA-1/02
8.0	Prox-onic CSA-1/04
8.0	Quimipol EA 6804
8.0	Remcopal 207
8.0	Remcopal 234
8.0	Remcopal LO 2B
8.0	Rexol 35/3
8.0	Rexol 65/4
8.0	Rhodasurf® LA-30
8.0	S-Maz® 20
8.0	Sebase
8.0	Solulan® 5
8.0	Solulan® PB-2
8.0	Surfonic® L24-3
8.0	Surfynol® 440
8.0	Tergitol® 26-L-3
8.0	Toximul® 374
8.0	Toximul® 705
8.0	Trycol® TDA-3
8.1	Armul 1003
8.1	Emerest® 2712
8.1	Empilan® KB 3
8.1	Ethal 3328
8.1	Ethal EH-2
8.1	Ethox MO-5
8.1	Mapeg® 400 DS
8.1	Marlipal® 24/30
8.1	Neodol® 91-2.5
8.1	Rexonic N23-3
8.1	Rexonic N91-2.5
8.1	Sinopol 20
8.1	Siponic® NP-4
8.1	Tergitol® TMN-3
8.2	Eumulgin PLT 4
8.2	Eumulgin PTL 4
8.2	Hetoxol CS-4
8.2	Hodag 42-S
8.2	Macol® DNP-5
8.2	Macol® OA-5
8.2	Rexol AE-3
8.2	Synperonic 91/2.5
8.2	Teric N3
8.2	Teric T5
8.2	Trydet OA-5
8.2	Trylox® HCO-16
8.3	Chemax E-200 MO
8.3	Cithrol 4DO
8.3	Dacospin LA-704
8.3	Emerest® 2624
8.3	Emerest® 2625
8.3	Hetoxide C-15
8.3	Industrol® MO-5
8.3	Lauropal X 1203
8.3	Mapeg® 200 MO
8.3	Mapeg® 200 MOT
8.3	Mapeg® DGLD
8.3	Remcopal 121
8.3	Rhodasurf® 840
8.3	Rolfor N 24/3
8.3	Sinopol 1103
8.3	Tergitol® 15-S-3
8.3	Tomah E-14-2
8.4	Berol 050
8.4	Ethylan® CD123
8.4	GP-219
8.4	GP-226
8.4	Hetsorb L-10
8.4	Hodag 42-O
8.4	Ionet DO-400
8.4	Ionet MO-200
8.4	Trylon® 5921
8.5	Ablunol CO 15
8.5	Alkaterge®-T-IV
8.5	Chemax E-200 MS
8.5	Eumulgin C4
8.5	Industrol® DO-9
8.5	Ionet DS-400
8.5	Kessco® PEG 400 DS
8.5	Remcopal 334
8.5	Sellig N 4 100
8.5	T-Det® O-4
8.5	Teric 18M2
8.6	Ablunol S-20
8.6	Alkamuls® SML
8.6	Atlox 4861B
8.6	Chemax CO-16
8.6	Chemax HCO-16
8.6	Crill 1
8.6	Disponil SML 100 F1
8.6	Drewmulse® SML
8.6	Ethofat® O/15
8.6	Ethox CO-16
8.6	Ethox HCO-16
8.6	Glycomul® L
8.6	Hetoxide HC-16
8.6	Hodag SML
8.6	Ionet S-20
8.6	Laurel PEG 400 DT
8.6	Liposorb L
8.6	Mapeg® CO-16H
8.6	Marlipal® 011/30
8.6	Montane 20
8.6	Nissan Nonion LP-20R, LP-20RS
8.6	Nonionic E-4
8.6	Pegosperse® 200 ML
8.6	Prote-sorb SML
8.6	Prox-onic HR-016
8.6	Prox-onic HRH-016
8.6	Rexol 25/4
8.6	Sorbax SML
8.6	Span® 20
8.6	Synperonic 13/3
8.6	Teric DD5
8.6	Trylox® 5902
8.6	Trylox® 5921
8.6	Trylox® CO-16
8.7	Armul 5830
8.7	Atlox 4856 B
8.7	Hetoxide NP-4
8.7	Lauropal X 1103
8.7	Neodol® 1-3
8.7	Remcopal 273
8.7	Teric 9A2
8.7	Trycol® 5966
8.7	Trycol® CE
8.7	Trydet SA-5
8.8	Caplube 8410
8.8	Caplube 8445
8.8	Cedepal CO-430
8.8	Chemax PEG 400 DO
8.8	Cithrol 2ML
8.8	Desonic® 4N
8.8	Emerest® 2648
8.8	Emulgator U4
8.8	Ethox DO-9
8.8	Ethox DTO-9A
8.8	Ethox TAM-5
8.8	Ethylan® CD913
8.8	Ethylan® TT-05
8.8	Eumulgin PC 4
8.8	Hetoxamate FA-5
8.8	Hetoxamate MO-5
8.8	Hetoxamate SA-5
8.8	Igepal® CO-430
8.8	Igepal® NP-4
8.8	Kessco® 894
8.8	Lipocol O-5
8.8	Mapeg® 400 DO
8.8	Mapeg® 400 DOT
8.8	Nutrol 622
8.8	Trydet DO-9
8.8	Witconol 2648
8.8	Witconol H33
8.9	Ablunol NP4
8.9	Berol 26
8.9	Catinex KB-11
8.9	Chemax NP-4
8.9	Dehydrophen PNP 4
8.9	Empilan® KA3
8.9	Emthox® 6961
8.9	Emulson AG 4B
8.9	Ethal NP-4
8.9	Ethylan® A3
8.9	Eumulgin PLT 5
8.9	Eumulgin PTL 5
8.9	Eumulgin TL 55
8.9	Hyonic NP-40
8.9	Icomeen® 18-5
8.9	Iconol NP-4
8.9	Macol® NP-4
8.9	Marlophen® 84
8.9	Marlophen® 84N
8.9	Marlowet® 4938
8.9	Nonipol 40
8.9	Prox-onic NP-04
8.9	Renex® 647
8.9	Surfonic® N-40
8.9	Synperonic NP4
8.9	T-Det® N-4
8.9	Tergitol® NP-4
8.9	Teric N4
8.9	Tomah Q-17-2
8.9	Trycol® 6961
8.9	Trycol® NP-4
9.0	Agrilan® AEC156
9.0	Alkasurf® NP-4
9.0	Amerchol® CAB
9.0	Amerlate® P

9.0	Arlatone® T
9.0	Atlox 775
9.0	Atlox 849
9.0	Berol 391
9.0	Berol 457
9.0	Carsonon® N-4
9.0	Chemeen HT-5
9.0	Chemeen T-5
9.0	Dehydol D 3
9.0	Emsorb® 6906
9.0	Emulmin 50
9.0	Ethylan® 44
9.0	Ethylan® ENTX
9.0	Hetoxol OA-5 Special
9.0	Hetoxol OL-5
9.0	Icomeen® S-5
9.0	Macol® 30
9.0	Macol® SA-5
9.0	Mulsifan RT 24
9.0	Mulsifan RT 113
9.0	Nikkol DDP-4
9.0	Nikkol DDP-6
9.0	Nissan Nonion E-205
9.0	Noigen EA 73
9.0	Noigen EA 80E
9.0	Noigen EA 92
9.0	Noigen ES 90
9.0	Noigen ET95
9.0	Noigen ET97
9.0	Norfox® NP-4
9.0	Pegosperse® 600 DOT
9.0	Plurafac® RA-30
9.0	Plurol Stearique WL 1009
9.0	Pluronic® P103
9.0	Prox-onic MHT-05
9.0	Prox-onic MS-05
9.0	Prox-onic MT-05
9.0	Quimipol ENF 40
9.0	Rexonic P-3
9.0	Rhodameen® T-5
9.0	Serdox NNP 4
9.0	Surfonic® L12-3
9.0	Syn Lube 728
9.0	Syntopon A 100
9.0	Teric 17M2
9.0	Toximul® TA-5
9.0	Volpo N5
9.0-10.0	Mulsifan ABN
9.0-10.0	Surfynol® TG
9.0-10.0	Surfynol® TG-E
9.0-12.0	Silwet® L-720
9.0-12.0	Silwet® L-7001
9.0-15.0	Geronol AG-10.00/200 Series
9.1	Eumulgin PST 5
9.1	Eumulgin PWM5
9.1	Industrol® MS-5
9.1	Lauropal X 1003
9.1	Nopalcol 4-DTW
9.1	Rioklen NF 4
9.1	Surfonic® L10-3
9.1	Synperonic A4
9.1	Triton® N-42
9.1	Triton® X-363M
9.2	Atlox 4855 B
9.2	Chemal LA-4
9.2	Dehydrophen POP 4
9.2	Deplastol
9.2	Disponil TA 5
9.2	Emulson AG 5590
9.2	Ethal LA-4
9.2	Eumulgin ET 5
9.2	Eumulgin WM5
9.2	G-1087
9.2	Hetoxol CS-5
9.2	Icomeen® T-5
9.2	Nissan Nonion K-204
9.2	Prox-onic LA-1/04
9.2	Rolfat OL 6
9.2	Ross Chem PEG 600 DT
9.2	Trycol® 5882
9.2	Trycol® LAL-4
9.2	Trydet 2670
9.2	Varonic® T205
9.3	Armul 03
9.3	Armul 214
9.3	Chemax E-200 ML
9.3	Cithrol 4DS
9.3	Emerest® 2620
9.3	Ethox ML-5
9.3	Mapeg® 200 ML
9.3	Sellig O 4 100
9.3	Trydet LA-5
9.3	Witconol 2620
9.4	Akyporox RLM 40
9.4	Atlox 3386 B
9.4	Dow Corning® Q4-3667 Fluid
9.4	Emerest® 2647
9.4	Emthox® 5882
9.4	Ethylan® CD103
9.4	Hetoxol L-4
9.4	Igepal® DM-430
9.4	Igepal® DX-430
9.4	Nikkol PBC-31
9.4	Plysurf A212C
9.4	Pyroter CPI-40
9.4	Sinopol 150
9.4	Surfonic® L24-4
9.4	Surfonic® L610-3
9.4	Synperonic OP4.5
9.4	Trydet SO-9
9.4	Witconol 14
9.5	Ablunol 200ML
9.5	Amerchol® C
9.5	Dehydol PLS 4
9.5	Disponil LS 4
9.5	Disponil O 5
9.5	Ethosperse® LA-4
9.5	Macol® LA-4
9.5	Marlipal® 24/40
9.5	Nikkol BL-2
9.5	Nikkol BPS-5
9.5	Nikkol Hexaglyn 1-O
9.5	Nikkol Hexaglyn 1-S
9.5	Nikkol PBC-41
9.5	Nissan Nonion L-2
9.5	Nissan Nonion NS-204.5
9.5	Poly-Tergent® B-150
9.5	Radiasurf® 7431
9.5	Remcopal HC 20
9.5	Ross Emulsifier DF
9.5	Sellig R 20 100
9.5	T-Maz® 61
9.5	Teric 16M2
9.5	Teric 18M5
9.5	Trycol® 5878
9.5	Volpo CS-5
9.6	Dehydol PLT 5
9.6	Emasol S-106
9.6	Emsorb® 6909
9.6	Etocas 20
9.6	Eumulgin PLT 6
9.6	Eumulgin PTL 6
9.6	Eumulgin RT 20
9.6	Montanox 61
9.6	Nonex S3E
9.6	Pegnol O-6
9.6	Plysurf A210G
9.6	Radiasurf® 7422
9.6	Rhodasurf® LA-42
9.6	Ritoleth 5
9.6	Sellig AO 6 100
9.6	T-Det® DD-5
9.6	Tween® 61
9.7	Alkamuls® A
9.7	Ameroxol® LE-4
9.7	Armul 66
9.7	Arquad® 2HT-75
9.7	Atlas G-1256
9.7	Atlas G-1281
9.7	Dow Corning® 5103 Surfactant
9.7	Ethylan® TD3
9.7	Hetoxol L-4N
9.7	Industrol® MO-6
9.7	Lipocol L-4
9.7	Marlipal® 124
9.7	Radiasurf® 7432
9.7	Remcopal 4
9.7	Rhodasurf® L-4
9.7	Ross Emulsifier LA-4
9.7	Simulsol P4
9.7	Sinopol 170F
9.7	Teric OF6
9.7	Trylon® 5902
9.8	Crodet L4
9.8	Ethomeen® S/25
9.8	Ethox 2966
9.8	Ethox SO-9
9.8	Ethox TO-8
9.8	Ethylan® D254
9.8	Industrol® ML-5
9.8	Nissan Nonion HS-204.5
9.8	Quimipol EA 2504
9.8	Teric 12A4
9.8	Teric G12A4
9.8	Teric LA4
9.8	Trycol® 6979
9.9	Epan U 102
9.9	Eumulgin TI 60
9.9	Lonzest® PEG 4-DO
9.9	Nissan Nonion E-206
9.9	Nissan Nonion S-206
9.9	Radiasurf® 7444
9.9	Sandoxylate® PDN-7
9.9	Trycol® 6940
9.9	Trycol® TP-6
10.0	Ablunol OA-6
10.0	Acconon CON
10.0	Ahco AB 100
10.0	Akypo LF 5
10.0	Alfonic® 610-50R
10.0	Alkamuls® 600-DO
10.0	Alkamuls® PSMO-5
10.0	Antarox® LA-EP 45

HLB	Name
10.0	Antarox® LA-EP 59
10.0	Aquabase
10.0	Atlox 848
10.0	Atlox 3401
10.0	Atlox 3404F
10.0	Caplube 8540C
10.0	Catinex KB-13
10.0	Cedepal CA-520
10.0	Crillet 41
10.0	Desonic® 6D
10.0	Dyafac PEG 6DO
10.0	Emasol O-105 R
10.0	Emasol O-106
10.0	Emery® 1650
10.0	Emsorb® 6901
10.0	Emsorb® 6908
10.0	Emsorb® 6917
10.0	Emulan® PO
10.0	Emulgator 64
10.0	Emulgator U6
10.0	Emulsifier 4
10.0	Emulsifier 7X
10.0	Ethylan® GEO81
10.0	Ethylan® ME
10.0	Glycosperse® HTO-40
10.0	Glycosperse® O-5
10.0	Hamposyl® C
10.0	Hamposyl® O
10.0	Hodag 20-L
10.0	Hodag 42-L
10.0	Hodag 62-O
10.0	Iconol DNP-8
10.0	Iconol NP-5
10.0	Iconol OP-5
10.0	Igepal® CA-520
10.0	Igepal® CO-520
10.0	Igepal® NP-5
10.0	Igepal® RC-520
10.0	Industrol® TFA-8
10.0	Liposorb O-5
10.0	Macol® NP-5
10.0	Marlophen® 85
10.0	Marlophen® 85N
10.0	Marlowet® TM
10.0	Merpol® LF-H
10.0	Montanox 81
10.0	Mulsifan K 326 Spezial
10.0	Mulsifan RT 1
10.0	Mulsifan RT 2
10.0	Mulsifan RT 7
10.0	Mulsifan RT 110
10.0	Mulsifan RT 258
10.0	Noigen EA 80
10.0	Noigen EA 83
10.0	Noigen ET100
10.0	Noigen ET102
10.0	Noigen ET107
10.0	Pegosperse® 400 DL
10.0	Plurafac® D-25
10.0	Plurafac® RA-20
10.0	Plurol Isostearique
10.0	Prox-onic SMO-05
10.0	Quimipol EA 4505
10.0	Quimipol EA 6806
10.0	Renex® 648
10.0	Rexol 65/6
10.0	Rexonic P-5
10.0	Rhodameen® PN-430
10.0	Sandopan® DTC Linear P Acid
10.0	Sellig DN 10 100
10.0	Serdox NNP 5
10.0	Solulan® PB-5
10.0	Sorbax PMO-5
10.0	Synperonic 10/3-100%
10.0	T-Maz® 81
10.0	T-Mulz® VO
10.0	Tergitol® NP-5
10.0	Teric 17M5
10.0	Teric T7
10.0	TO-55-EL
10.0	Triton® N-57
10.0	Tween® 81
10.0-12.0	Sandoxylate® 408
10.0-14.0	Geronol AG-900
10.1	Berol 455
10.1	Emerest® 2636
10.1	Ethal DNP-8
10.1	Industrol® MS-7
10.1	Newpol PE-64
10.1	Newpol PE-74
10.1	Prox-onic CSA-1/06
10.1	Trydet 2636
10.1	Trydet 2692
10.2	Akypo LF 4N
10.2	Akypo MB 2621 S
10.2	Akypo MB 2705 S
10.2	Akypo RLMQ 38
10.2	Atlas G-1086
10.2	Berol 829
10.2	Cirrasol® AEN-XB
10.2	Cithrol 4DL
10.2	Dehydol PCS 6
10.2	Emgard® 2033
10.2	G-1086
10.2	Rioklen NF 6
10.2	Sellig N 5 100
10.2	Sorbax MO-40
10.2	Teric 17A6
10.2	Trydet OA-7
10.2	Trylox® 1086
10.3	Chemax PEG 600 DO
10.3	Emerest® 2665
10.3	Ethox DO-14
10.3	Ethylan® A4
10.3	Mapeg® 600 DO
10.3	Mapeg® 600 DOT
10.3	Nissan Nonion O-3
10.3	Nopalcol 10-CO
10.3	Nopalcol 10-COH
10.3	Polysurf A212E
10.3	Remcopal 666
10.3	Value 1407
10.4	Alkasurf® OP-5
10.4	Chemax DNP-8
10.4	Chemeen C-5
10.4	Cithrol 6DO
10.4	Collone AC
10.4	Emerest® 2632
10.4	Ethylan® CD124
10.4	G-1089
10.4	Industrol® MO-7
10.4	Ionet DO-600
10.4	Macol® OP-5
10.4	Mazon® 1086
10.4	Norfox® OP-45
10.4	Prox-onic DNP-08
10.4	Prox-onic MC-05
10.4	Remcopal 4018
10.4	Rexol 45/5
10.4	Sandoxylate® PO-5
10.4	Sellig HR 18 100
10.4	Sellig O 5 100
10.4	Sorbax HO-40
10.4	Teric 13A5
10.4	Teric X5
10.4	Toximul® 360A
10.4	Triton® X-45
10.4	Trycol® 6985
10.4	Trycol® DNP-8
10.4	Trylox® 6746
10.5	Ablunol LA-5
10.5	Aconol X6
10.5	Agrimul 70-A
10.5	Berol 260
10.5	Berol 269
10.5	Casul® 70 HF
10.5	Chemal DA-4
10.5	Chemax TO-8
10.5	Crillet 35
10.5	Desonic® DA-4
10.5	Disponil STS 120 F1
10.5	Drewmulse® POE-STS
10.5	Durfax® 65
10.5	Emasol S-320
10.5	Emery® 6927 Agrimul 70-A
10.5	Emulgator E-2155 SE
10.5	Ethal DA-4
10.5	Ethylan® 55
10.5	G-3886
10.5	Hetoxamate LA-5
10.5	Hetoxamate SA-7
10.5	Hetsorb TS-20
10.5	Hodag PSTS-20
10.5	Industrol® DT-13
10.5	Industrol® STS-20-S
10.5	Marlipal® 013/50
10.5	Marlipal® 104
10.5	Merpol® SE
10.5	Montanox 65
10.5	Nikkol BO-7
10.5	Nikkol PBC-33
10.5	Nonipol 55
10.5	Nonipol Soft SS-50
10.5	Nopalcol 2-L
10.5	PGE-600-DS
10.5	Paraffin Wax Emulsifier CB0680
10.5	Prox-onic DA-1/04
10.5	Prox-onic STS-020
10.5	Prox-onic TA-1/08
10.5	Quimipol ENF 55
10.5	Remcopal 306
10.5	Rhodacal® CA, 70%
10.5	Rhodasurf® DA-4
10.5	Rhodasurf® DA-530
10.5	Rolfat OL 7
10.5	Serdox NBS 4
10.5	Sinopol 1305
10.5	Sorbax PTS-20
10.5	Sorbeth 40HO
10.5	Synperonic NP5
10.5	T-Maz® 65
10.5	TO-55-E
10.5	Tergitol® 26-L-5
10.5	Teric 16M5
10.5	Teric N5
10.5	Toximul® 500
10.5	Toximul® 600

10.5	Toximul® 709
10.5	Toximul® 715
10.5	Toximul® D
10.5	Toximul® R-HF
10.5	Trycol® 5950
10.5	Trycol® DA-4
10.6	Berol 28
10.6	Ethofat® C/15
10.6	Ethylan® TT-07
10.6	G-7076
10.6	Genapol® 26-L-5
10.6	Hodag Nonionic E-6
10.6	Igepal® DM-530
10.6	Industrol® DO-13
10.6	Laurel PEG 600 DT
10.6	Mapeg® 600 DS
10.6	Marlipal® 24/50
10.6	Paraffin Wax Emulsifier CB0674
10.6	Radiasurf® 7454
10.6	Rexol 35/5
10.6	Rhodasurf® TDA-5
10.6	Rhodasurf® TDA-6
10.6	Tergitol® 15-S-5
10.6	Triton® DF-12
10.6	Value 401
10.7	Ablunol OA-7
10.7	Ablunol SA-7
10.7	Cithrol 3MS
10.7	Duomac® T
10.7	Emulgeant 900
10.7	Ethal DDP-7
10.7	Ethylan® FO60
10.7	Hetoxide NP-6
10.7	Kessco® PEG 600 DS
10.7	Neodol® 23-5
10.7	Newpol PE-75
10.7	Nissan Dispanol 16A
10.7	Nissan Nonion S-207
10.7	Polychol 10
10.7	Prox-onic OCA-1/06
10.7	Remcopal 6
10.7	Sinol T4-1
10.7	Sinopol 1207
10.7	Sinopol 1807
10.7	Siponic® NP-6
10.7	T-Mulz® O
10.7	Triton® X-207
10.8	Arlatone® G
10.8	Armac® T
10.8	Atlas G-1292
10.8	Atlox 4990B
10.8	Cedepal CO-530
10.8	Cerex EL 250
10.8	Chemax CO-25
10.8	Chemax HCO-25
10.8	Cithrol 6DS
10.8	Crodet O8
10.8	Crodet S8
10.8	Dacospin POE(25)HRG
10.8	Dehydol PTA 7
10.8	Desonic® 6N
10.8	Disponil RO 40
10.8	Emerest® 2652
10.8	Emulmin 70
10.8	Emulson AG/CO 25
10.8	Emulson AG/COH
10.8	Ethox CO-25
10.8	Ethox DL-9
10.8	Ethox HCO-25
10.8	G-1292
10.8	Hetoxide C-25
10.8	Hetoxide HC-25
10.8	Icomeen® T-7
10.8	Iconol NP-6
10.8	Igepal® CO-530
10.8	Industrol® CO-25
10.8	Industrol® DL-9
10.8	Macol® LA-790
10.8	Mapeg® 400 DL
10.8	Mapeg® CO-25
10.8	Mapeg® CO-25H
10.8	Prox-onic HR-025
10.8	Prox-onic HRH-025
10.8	Rexol 25/6
10.8	Sellig O 6 100
10.8	Synperonic 91/4
10.8	T-Mulz® 808A
10.8	T-Mulz® COC
10.8	T-Mulz® FGO-17
10.8	T-Mulz® FGO-17A
10.8	Teric 164
10.8	Teric SF9
10.8	Trylon® 5922
10.8	Trylox® 5904
10.8	Trylox® 5922
10.8	Trylox® CO-25
10.8	Trylox® HCO-25
10.9	Ablunol NP6
10.9	Armul 906
10.9	Atlox 3400B
10.9	Berol 02
10.9	Catinex KB-15
10.9	Chemax NP-6
10.9	Cithrol GMS Acid Stable
10.9	Dehydrophen PNP 6
10.9	Emthox® 6962
10.9	Emulson AG 3490
10.9	Ethal NP-6
10.9	Ethylan® 77
10.9	Hyonic NP-60
10.9	Igepal® NP-6
10.9	Macol® NP-6
10.9	Marlophen® 86
10.9	Marlophen® 86N
10.9	Marlowet® 4900
10.9	Nissan Nonion NS-206
10.9	Nonal 206
10.9	Nonipol 60
10.9	Poly-Tergent® B-200
10.9	Prox-onic NP-06
10.9	Quimipol ENF 65
10.9	Renex® 697
10.9	Sinopol 960
10.9	Surfonic® N-60
10.9	Synperonic NP6
10.9	Syntopon A
10.9	T-Det® N-6
10.9	Tergitol® NP-6
10.9	Trycol® 6962
10.9	Trycol® NP-6
10.9	Trylox® 1186
10.9	Value 3706
11.0	Acconon 1300
11.0	Akypo RLM 45
11.0	Alcodet® 260
11.0	Aldo® MSA
11.0	Alkamuls® 400-MO
11.0	Alkamuls® PSTO-20
11.0	Alkasurf® NP-6
11.0	Atlox 3387 BM
11.0	Atlox 3454F
11.0	Atsolyn P
11.0	Atsurf T-85
11.0	Carsonon® N-6
11.0	Cirrasol® ALN-WY
11.0	Cithrol 4MS
11.0	Crillet 45
11.0	Disponil STO 120 F1
11.0	Emasol O-320
11.0	Emulan® A
11.0	Emulan® AF
11.0	Emulan® OK 5
11.0	Emulan® P
11.0	Emulson CO 25
11.0	Emulson EL/H25
11.0	Ethosperse® TDA-6
11.0	Ethox MI-9
11.0	Ethylan® GPS85
11.0	Eumulgin M8
11.0	Eumulgin PAEH 4
11.0	G-7205
11.0	G-7274
11.0	Genapol® UD-050
11.0	Glycosperse® TO-20
11.0	Glycosperse® TS-20
11.0	Hetoxide DNP-9.6
11.0	Hetsorb TO-20
11.0	Iconol DA-4
11.0	Iconol PD-8-90%
11.0	Imwitor® 375
11.0	Industrol® 400-MOT
11.0	Industrol® COH-25
11.0	Lamigen ES 30
11.0	Lauropal X 1105
11.0	Lipomulse 165
11.0	Lonzest® STO-20
11.0	Lonzest® STS-20
11.0	Montanox 85
11.0	Mulsifan RT 23
11.0	Mulsifan RT 37
11.0	NP-55-60
11.0	Noigen EA 102
11.0	Noigen EA 110
11.0	Noigen ET115
11.0	Nonex O4E
11.0	Norfox® 4 Polyol
11.0	Norfox® NP-6
11.0	Pegosperse® 400 MO
11.0	Pegosperse® 400 MOT
11.0	Pluronic® L63
11.0	Polirol C5
11.0	Prox-onic STO-020
11.0	Prox-onic TM-06
11.0	Quimipol EA 6807
11.0	Radiamac 6149
11.0	Radiamac 6169
11.0	Radiamac 6179
11.0	Sandopan® KST
11.0	Sandoxylate® SX-208
11.0	Sandoxylate® SX-408
11.0	Serdox NNP 6
11.0	Sinopol 1525
11.0	Soitem 251
11.0	Sorbanox CO
11.0	Sorbax PTO-20
11.0	Step-Flow 63
11.0	Sterol LG 491
11.0	T-Maz® 95

11.0	T-Mulz® PB
11.0	TO-55-F
11.0	Texo 227
11.0	Texo 1060
11.0	Tomah E-18-5
11.0	Toximul® 804
11.0	Toximul® MP
11.0	Toximul® SEE-340
11.0	Tween® 85
11.0	Value 402
11.0	Varonic® K205
11.0	Varonic® K205 SF
11.0-12.0	Emulsifier K 30 40%
11.0-12.0	Emulsifier K 30 68%
11.0-12.0	Emulsifier K 30 76%
11.0-12.0	Emulsifier K 30 95%
11.1	Akypo MB 1614/1
11.1	Chemax CO-28
11.1	Emsorb® 6903
11.1	Emsorb® 6913
11.1	Ethsorbox TO-20
11.1	Ethsorbox TS-20
11.1	Hodag 40-S
11.1	Simulsol M 45
11.1	T-Det® DD-7
11.1	T-Maz® 85
11.1	Teric OF8
11.1	Trydet SA-7
11.1	Trydet SA-8
11.1	Witconol 6903
11.2	Alkamuls® S-8
11.2	Alkamuls® S-65-8
11.2	Armac® C
11.2	Armul 22
11.2	Armul 34
11.2	Atlas G-1556
11.2	Atlox 847
11.2	Catinex KB-40
11.2	Duomac® C
11.2	Emulson AG 4080
11.2	Ethylan® NK4
11.2	G-1556
11.2	Industrol® TO-10
11.2	Macol® TD-610
11.2	Marlipal® 011/50
11.2	Mazon® 1096
11.2	Neodol® 1-5
11.2	Nissan Nonion HS-206
11.2	Nopalcol 6-DTW
11.2	Pegosperse® 400 MS
11.2	Sellig N 6 100
11.2	Sellig R 3395-C435
11.2	Sinopol 860
11.2	Sorbon TR 843
11.2	Synperonic 13/5
11.2	Syntopon 8 A
11.2	Teginacid® H
11.2	Teginacid® H-SE
11.2	Teric 12A6
11.2	Teric 18M10
11.2	Teric G12A6
11.3	Arosurf® 32-E20
11.3	Emerest® 2644
11.3	Emgard® 2063
11.3	Hetoxol TD-6
11.3	Macol® DNP-10
11.3	Nopalcol 4-S
11.3	Nopalcol 12-CO
11.3	Nopalcol 12-COH
11.3	Octapol 60
11.3	Rolfat OL 9
11.3	Rolfat R 106
11.3	T-Mulz® AO2
11.3	Tagat® TO
11.3	Tegotens TO
11.3	Teric 160
11.3	Triton® DF-18
11.3	Trydet 2644
11.3	Trylox® 6747
11.4	Atlas EM-2
11.4	Atlas G-1096
11.4	Atlas G-3300B
11.4	Atlox 1045A
11.4	Atlox 3300B
11.4	Cerex EL 4929
11.4	Chemal TDA-6
11.4	Cithrol 4MO
11.4	Cithrol A
11.4	Desonic® 6T
11.4	Emerest® 2711
11.4	Empilan® KA5/90
11.4	Ethal TDA-6
11.4	Ethox HO-50
11.4	Ethox MS-8
11.4	Ethylan® D256
11.4	Eumulgin C8
11.4	G-1045A
11.4	G-1096
11.4	Hodag 40-O
11.4	Iconol TDA-6
11.4	Industrol® MS-8
11.4	Lauropal 0205
11.4	Lonzest® PEG 4-O
11.4	Marlipal® 013/60
11.4	Prox-onic TD-1/06
11.4	Quimipol EA 2506
11.4	Renex® 36
11.4	Rhodasurf® 860/P
11.4	Rhodasurf® BC-610
11.4	Rolfor TR 6
11.4	Sellig AO 9 100
11.4	Tergitol® Min-Foam 1X
11.4	Teric 12M2
11.4	Trycol® 5940
11.4	Trycol® TDA-6
11.4	Trydet 2671
11.5	Ablunol 400MO
11.5	Alkamuls® B
11.5	APG® 600 Glycoside
11.5	Armul 1005
11.5	Armul 3260
11.5	Armul 44
11.5	Berol 087
11.5	Chemax TO-10
11.5	Dehydol O4
11.5	Dehydol PLS 6
11.5	Empilan® KA5
11.5	Emulson AG 3080
11.5	G-1293
11.5	Hetoxamate SA-9
11.5	Hetoxol CD-3
11.5	Hetoxol CD-4
11.5	Lauroxal 6
11.5	Mapeg® 400 MS
11.5	Marlipal® 24/60
11.5	Nikkol BL-4.2
11.5	Nikkol DDP-8
11.5	Nikkol GO-430
11.5	Nikkol OP-10
11.5	Nikkol TDP-8
11.5	Nissan Nonion O-4
11.5	PGE-400-MS
11.5	Plysurf A215C
11.5	Prox-onic TA-1/010
11.5	Quimipol EA 6808
11.5	Radiasurf® 7403
11.5	Sandoxylate® AT-6.5
11.5	Serdox NOL 8
11.5	Soitem 101
11.5	Sorbeth 55HO
11.5	T-Mulz® AS-1151
11.5	Trymeen® TAM-8
11.6	Ablunol 400MS
11.6	Armul 215
11.6	Atlox 4858 B
11.6	Atlox 4901
11.6	Catinex KB-32
11.6	Cirrasol® ALN-FP
11.6	Dehydol PO 5
11.6	Ethylan® C30
11.6	Ethylan® TD10
11.6	Genapol® 26-L-45
11.6	Glucopon 600
11.6	Hodag 40-R
11.6	Lauropal X 1005
11.6	Marlipal® 1618/8
11.6	Nissan Nonion S-4
11.6	Quimipol EA 9105
11.6	Rexol 25/7
11.6	Rolfat SG
11.6	Rolfat SO 6
11.6	Rolfor O 8
11.6	Sellig R 3395 SP
11.6	Sinopol 180I
11.6	Surfonic® L46-7
11.6	Syn Fac® 334
11.6	T-Det® O-6
11.6	Tergitol® 24-L-45
11.6	Teric 151
11.6	Teric 158
11.6	Teric 17A8
11.6	Teric 9A5
11.6	Teric G9A5
11.6	Triton® DF 16
11.6	Trycol® 5951
11.7	Armul 33
11.7	Chemax CO-30
11.7	Dehydol PLT 6
11.7	Desonic® 30C
11.7	Desonic® 7N
11.7	Emulson AG 7B
11.7	Ethox MA-8
11.7	Etocas 30
11.7	G-3300
11.7	Hetoxamate MO-9
11.7	Incrocas 30
11.7	Industrol® CO-30
11.7	Industrol® MO-9
11.7	Kessco® PEG 600 DL
11.7	Marlophen® 87
11.7	Marlophen® 87N
11.7	Marlowet® 4901
11.7	Nonipol 70
11.7	Nopalcol 4-O
11.7	Pegnol L-6
11.7	Prox-onic HR-030
11.7	Rhodacal® 330
11.7	Sinopol 170I
11.7	Sinopol 180N
11.7	Synperonic NP7

11.7	Synperonic OP7.5
11.7	Tergitol® NP-7
11.7	Tergitol® TMN-6
11.7	Teric T10
11.7	Trycol® 6963
11.7	Trydet ISA-9
11.7	Trylon® 5906
11.7	Trylox® CO-30
11.8	Ablunol CO 30
11.8	Ahco AB 118
11.8	Armul 16
11.8	Chemax E-400 MO
11.8	Dacospin 1735-A
11.8	Dehydol PID 6
11.8	Dehydol PIT 6
11.8	Emerest® 2646
11.8	Ethox CO-30
11.8	Genapol® 24-L-45
11.8	Hetoxol L-9N
11.8	Incropol L-7
11.8	Industrol® MIS-9
11.8	Ionet MO-400
11.8	Mapeg® 400 MO
11.8	Mapeg® 400 MOT
11.8	Mapeg® CO-30
11.8	Neodol® 45-7
11.8	Ospol 790
11.8	Rioklen NF 7
11.8	Synperonic 91/5
11.8	Teric 13A7
11.8	Trydet LA-7
11.8	Trylox® 5906
11.8	Witconate AOS
11.8	Witconol 5906
11.9	Atlox 4853 B
11.9	Atlox 4868B
11.9	Chemal OA-9
11.9	Dehydol LT 7
11.9	Iconol NP-7
11.9	Ionet MS-400
11.9	Liponox NC-70
11.9	Nissan Nonion P-208
11.9	Prox-onic OA-1/09
11.9	Radiasurf® 7413
11.9	Remcopal 40
11.9	Sellig S 30 100
11.9	Sellig Stearo 6
11.9	Surfonic® L24-7
12.0	Acconon 400-MS
12.0	Acconon CA-9
12.0	Ahco AB 120
12.0	Akypo LF 3
12.0	Alcodet® HSC-1000
12.0	Alcodet® IL-3500
12.0	Alcodet® MC-2000
12.0	Alfonic® 1012-60
12.0	Alfonic® 1412-60
12.0	Alfonic® 810-60
12.0	Alkamuls® EL-620
12.0	Alkasurf® NP-8
12.0	APG® 625 Glycoside
12.0	Armul 17
12.0	Arosurf® 66-E10
12.0	Atlox 3406F
12.0	Atlox 3409F
12.0	Atlox 3455F
12.0	Berol 195
12.0	Burco FAE
12.0	Chemax E-400 MS
12.0	Cithrol 6DL
12.0	Crodesta F-110
12.0	Emerest® 2640
12.0	Emerest® 2641
12.0	Emery® 6222 Monolan 1030
12.0	Empilan® 0004
12.0	Ethox MO-9
12.0	Ethox TO-9A
12.0	Eumulgin B1
12.0	Eumulgin O5
12.0	Eumulgin O10
12.0	G-3890
12.0	Hyonic OP-7
12.0	Hyonic OP-70
12.0	Igepal® CA-620
12.0	Industrol® MS-9
12.0	Lamigen ET 20
12.0	Laurel PEG 400 MT
12.0	ML-55-F-4
12.0	Macol® 42
12.0	Micro-Step® H-301
12.0	Micro-Step® H-302
12.0	Micro-Step® H-304
12.0	Micro-Step® H-305
12.0	Micro-Step® H-306
12.0	Micro-Step® H-307
12.0	Monolan® 1030
12.0	Mulsifan RT 231
12.0	Mulsifan RT 248
12.0	NP-55-80
12.0	Neodol® 23-6.5
12.0	Nissan Dispanol TOC
12.0	Noigen EA 112
12.0	Noigen EA 120
12.0	Noigen EA 120B
12.0	Noigen ES 120
12.0	Noigen ET120
12.0	Noigen ET127
12.0	Non Ionic Emulsifier T-9
12.0	OS-2
12.0	Plurafac® B-25-5
12.0	Pluronic® L43
12.0	Remcopal HC 33
12.0	Rewopal® MT 65
12.0	Rexol 65/9
12.0	Rexonic 1012-6
12.0	Rexonic N23-6.5
12.0	Rhodasurf® 1012-6
12.0	Rhodasurf® LA-7
12.0	Sellig R 3395
12.0	Sellig SP 8 100
12.0	Serdox NBS 5.5
12.0	Serdox NNP 7
12.0	Sinopol 707
12.0	Solulan® PB-10
12.0	Step-Flow 23
12.0	Syn Lube 6277-A
12.0	Synperonic 10/5-100%
12.0	Synperonic 87K
12.0	Synperonic BD100
12.0	T-9
12.0	T-Det® DD-9
12.0	T-Det® TDA-60
12.0	Teginacid®
12.0	Teginacid® SE
12.0	Teginacid® X
12.0	Teginacid® X-SE
12.0	Toximul® 716
12.0	Toximul® 8241
12.0	Toximul® 8320
12.0	Toximul® MP-10
12.0	Trydet 2640
12.0	Witconol 2640
12.0-13.0	Norfox® NP-7
12.0-14.0	Sandoxylate® 412
12.0-14.0	Triton® RW-50
12.1	Ablunol LA-7
12.1	Akyporox RO 90
12.1	Alcodet® TX 4000
12.1	Emerest® 2630
12.1	Emsorb® 6916
12.1	G-1120
12.1	Genapol® 24-L-50
12.1	Genapol® 24-L-60N
12.1	Glucopon 625
12.1	Industrol® SO-13
12.1	Marlipal® 013/70
12.1	Marlipal® 1012/6
12.1	Nissan Nonion K-207
12.1	Rexonic N25-7
12.1	Rhodasurf® L-790
12.1	Rioklen NF 8
12.1	Sinol NPX
12.1	Tergitol® 24-L-60N
12.1	Teric 12A7
12.1	Teric 16M10
12.1	Teric DD9
12.1	Trydet 22
12.1	Trydet 2681
12.1	Trydet OA-10
12.1	Value W-307
12.2	Akypo MB 2528S
12.2	Armul 55
12.2	Aromox® DM14D-W
12.2	Arosurf® 66-PE12
12.2	Bio-Soft® EN 600
12.2	Cedepal CO-610
12.2	Emulpon EL 33
12.2	Ethal LA-7
12.2	Ethylan® TU
12.2	Genapol® 24-L-60
12.2	Genapol® 26-L-50
12.2	Genapol® 26-L-60N
12.2	Hetoxol CS-9
12.2	Igepal® CO-610
12.2	Igepal® NP-8
12.2	Mapeg® 600 DL
12.2	Neodol® 25-7
12.2	Nutrol 611
12.2	Quimipol EA 2507
12.2	Quimipol ENF 80
12.2	Remcopal 10
12.2	Renex® 707
12.2	Surfonic® LF-17
12.2	Synperonic A7
12.2	Tergitol® 24-L-60
12.2	Trydet 2676
12.2	Zee-emul No. 35
12.2	Zee-emul No. 50
12.3	Ablunol NP8
12.3	Armul 908
12.3	Atlox 4857 B
12.3	Berol 199
12.3	Berol 267
12.3	Catinex KB-16
12.3	Cirrasol® AEN-XZ
12.3	Dehydrophen PNP 8
12.3	Emulson AG 5790
12.3	Ethox SAM-10

HLB	Product
12.3	Ethylan® A6
12.3	Macol® CSA-10
12.3	Macol® NP-8
12.3	Macol® OP-8
12.3	Macol® SA-10
12.3	Marlipal® 24/70
12.3	Marlophen® 88
12.3	Marlophen® 88N
12.3	Neodol® 45-7T
12.3	Nonal 208
12.3	Quimipol EA 4508
12.3	Remcopal 349
12.3	Renex® 688
12.3	Sellig N 8 100
12.3	Siponic® NP-8
12.3	Synperonic KB
12.3	Synperonic NP8
12.3	Synperonic NX
12.3	Syntopon 8 B
12.3	Tergitol® NP-8
12.3	Teric 161
12.3	Teric N8
12.3	Teric X7
12.3	Trycol® NP-7
12.4	Atlox 4890B
12.4	Atlox 4899B
12.4	Berol 389
12.4	Berol 458
12.4	Chemal DA-6
12.4	Emulsifier 632/90%
12.4	Ethal DA-6
12.4	Ethylan® BCD42
12.4	Ethylan® C35
12.4	Ethylan® CD127
12.4	Ethylan® VPK
12.4	Genapol® 26-L-60
12.4	Hetoxol OA-10 Special
12.4	Hetoxol OL-10
12.4	Hetoxol STA-10
12.4	Iconol OP-7
12.4	Lipocol O-10
12.4	Lipocol S-10
12.4	Macol® TD-8
12.4	Norfox® OP-114
12.4	Prox-onic CSA-1/010
12.4	Prox-onic DA-1/06
12.4	Prox-onic SA-1/010
12.4	Rexol 25/8
12.4	Rexol 45/7
12.4	Ritoleth 10
12.4	Simulsol 76
12.4	Simulsol 96
12.4	Sinopol 1210
12.4	Surfonic® L12-6
12.4	T-Det® N-8
12.4	T-Det® O-8
12.4	Tergitol® 15-S-7
12.4	Tergitol® 24-L-50
12.4	Teric 9A6
12.4	Teric 12M5
12.4	Teric G9A6
12.4	Triton® X-114
12.4	Trycol® 5952
12.4	Trycol® 5953
12.4	Trycol® 5999
12.4	Trycol® DA-69
12.4	Volpo N10
12.5	Akypo MB 1614/2
12.5	Alkasurf® OP-8
12.5	Atlas G-8936CJ
12.5	Berol 173
12.5	Bio-Soft® TD 630
12.5	Cerex EL 300
12.5	Desonic® DA-6
12.5	Disponil O 10
12.5	Emulson AG/FN
12.5	Ethox TAM-10
12.5	Etocas 35
12.5	Eumulgin PA 10
12.5	Grilloten LSE87
12.5	Grilloten LSE87K
12.5	Iconol DA-6-90%
12.5	Industrol® CO-36
12.5	Lauropal X 1207
12.5	Macol® OA-10
12.5	Merpol® OJ
12.5	NP-55-85
12.5	Neodol® 91-6
12.5	Nikkol BPS-10
12.5	Nikkol GO-440
12.5	Nikkol PBC-44
12.5	Nipol® 4472
12.5	Nissan Dispanol N-100
12.5	Norfox® 916
12.5	Quimipol EA 6810
12.5	Quimipol EA 9106
12.5	Rexonic N91-6
12.5	Rhodasurf® 91-6
12.5	Rhodasurf® DA-6
12.5	Rhodasurf® DA-630
12.5	Rhodasurf® TDA-8.5
12.5	Serdox NBS 6
12.5	Serdox NES 8/85
12.5	Serdox NNP 8.5
12.5	Sinopol 964
12.5	Sinopol 965FH
12.5	Syn Fac® TDA-92
12.5	Synperonic 10/6-100%
12.5	Synperonic 13/6.5
12.5	Synperonic 91/6
12.5	Synperonic A9
12.5	Syntens KMA 55
12.5	T-Mulz® AS-1152
12.5	T-Mulz® AS-1180
12.5	Teric 165
12.5	Tex-Wet 1048
12.5	Toximul® MP-26
12.5	Triton® X-155-90%
12.5	Trycol® 5949
12.5	Trycol® 5968
12.5	Trycol® DA-6
12.5	Trycol® TDA-8
12.5	Witconol H31A
12.6	Atlox 3335 B
12.6	Catinex KB-41
12.6	Chemal 2EH-5
12.6	Chemax CO-36
12.6	Dehydol PLT 8
12.6	Dehydrophen POP 8
12.6	Desonic® 36C
12.6	Ethal EH-5
12.6	Ethal OA-10
12.6	Ethox 1212
12.6	Ethox CO-36
12.6	Eumulgin ET 10
12.6	Eumulgin PWM10
12.6	Hetoxol LS-9
12.6	Hodag Nonionic E-10
12.6	Iconol DDP-10
12.6	Igepal® RC-620
12.6	Mapeg® CO-36
12.6	Marlipal® 1618/10
12.6	Marlophen® 1028
12.6	Marlophen® 1028N
12.6	Neodol® 23-6.5T
12.6	Nissan Nonion HS-208
12.6	Nissan Nonion NS-208.5
12.6	Nonipol 85
12.6	Prox-onic 2EHA-1/05
12.6	Prox-onic HR-036
12.6	Quimipol EA 2508
12.6	Quimipol EA 9106B
12.6	Remcopal 29
12.6	Rexonic 1218-6
12.6	Sellig O 8 100
12.6	Sinol LDA-6
12.6	Sinopol 864
12.6	Surfonic® N-85
12.6	Synperonic OP8
12.6	Syntopon B
12.6	T-Mulz® AS-1153
12.6	Teric 12A8
12.6	Teric G12A8
12.6	Teric LA8
12.6	Teric X8
12.6	Triton® CF-76
12.6	Trycol® 5963
12.6	Trycol® LAL-8
12.6	Trylox® 5907
12.6	Trylox® CO-36
12.6	Value 3608
12.6	Varonic® T210
12.7	Alcodet® SK
12.7	Atranonic Polymer 20
12.7	Berol 947
12.7	Crodet L8
12.7	Ethox 2672
12.7	Eumulgin PRT 36
12.7	G-1121
12.7	Hetoxide NP-9
12.7	Hetoxol CA-10
12.7	Iconol TDA-8
12.7	Iconol TDA-8 90%
12.7	Igepal® RC-630
12.7	Nopalcol 6-DO
12.7	Prox-onic TM-08
12.7	Sellig N 9 100
12.7	Siponic® NP-9
12.7	Synperonic NP8.75
12.7	Syntopon C
12.7	Teric 17A10
12.7	Teric C12
12.7	Triton® CF-87
12.7	Witconol NS-108LQ
12.8	Akyporox RLM 80
12.8	Alkamuls® L-9
12.8	Atlas G-2127
12.8	Atlox 4880B
12.8	Catinex KB-18
12.8	Cedepal FA-406
12.8	Dehydol PLS 8
12.8	Desonic® 9N
12.8	Emerest® 2634
12.8	Emulgator U9
12.8	Emulson AG 9B
12.8	Ethoxamine SF11
12.8	Ethylan® CD916
12.8	Hetoxol OL-10H
12.8	Hodag 40-L

12.8 Kessco® 891
12.8 Lauropal X 1107
12.8 Marlipal® 011/79
12.8 Marlipal® 013/80
12.8 Marlophen® 89
12.8 Marlophen® 89N
12.8 Marlowet® 4902
12.8 Merpol® SH
12.8 Nissan Nonion S-15
12.8 Nissan Nonion T-15
12.8 Nonipol Soft SS-70
12.8 Radiasurf® 7423
12.8 Remcopal 6110
12.8 Remcopal LC
12.8 Remcopal LP
12.8 Rexol 35/8
12.8 Synperonic NP9
12.8 Teric N9
12.9 Ablunol NP9
12.9 Akyporox NP 90
12.9 Armul 910
12.9 Brij® 56
12.9 Cerex EL 360
12.9 Chemax CO-40
12.9 Croduret 40
12.9 Ethylan® BCP
12.9 Eumulgin KP92
12.9 Genapol® 24-L-75
12.9 Industrol® CO-40
12.9 Ionet DO-1000
12.9 Lipocol C-10
12.9 Macol® NP-9.5
12.9 Marlipal® 24/80
12.9 Neodol® 1-7
12.9 Nissan Nonion NS-209
12.9 Nissan Nonion P-210
12.9 Poly-Tergent® B-300
12.9 Prox-onic HR-040
12.9 Prox-onic MS-011
12.9 Quimipol ENF 90
12.9 Remcopal 4000
12.9 Remcopal HC 40
12.9 Rolfat C 9
12.9 Rolfor CO 11
12.9 Sandoxylate® PN-9
12.9 Sellig R 4095
12.9 Simulsol 56
12.9 Sinopol 1100H
12.9 Sinopol 1610
12.9 Sinopol 964H
12.9 Surfonic® N-95
12.9 Tergitol® NP-9
12.9 Triton® CF-21
12.9 Trycol® NP-9
12.9 Trylox® CO-40
12.9 Volpo CS-10
13.0 Aconol X10
13.0 Agrilan® DG113
13.0 Agrilan® WP112
13.0 Akypo LF 1
13.0 Aldosperse® MS-20
13.0 Aldo® MS-20 FG
13.0 Alkamuls® 600-MO
13.0 Alkamuls® CO-40
13.0 Alkasurf® CO-40
13.0 Antarox® P-104
13.0 APG® 325 Glycoside
13.0 Aqualose LL100
13.0 Arquad® T-2C-50
13.0 Atsolyn T
13.0 Berol 716
13.0 Berol 938
13.0 Carsonon® N-9
13.0 Cedepal CA-630
13.0 Chemal TDA-9
13.0 Chemax AR-497
13.0 Chemax NP-9
13.0 Chemeen DT-15
13.0 Chimipal APG 400
13.0 Dehydol 25
13.0 Dehydol 100
13.0 Dehydol TA 11
13.0 EM-600
13.0 Emerest® 2664
13.0 Emthox® 5941
13.0 Emulan® AT 9
13.0 Emulson CO 40 N
13.0 Emultex 1302
13.0 Ethal NP-9
13.0 Ethal TDA-9
13.0 Ethox 2423
13.0 Ethox CO-40
13.0 Ethox MI-14
13.0 Ethylan® BKL130
13.0 Ethylan® CX308
13.0 Ethylan® KEO
13.0 Etocas 40
13.0 Eumulgin RT 40
13.0 G-2109
13.0 Genapol® 26-L-75
13.0 Genapol® UD-079
13.0 Hamposyl® L
13.0 Hyonic NP-90
13.0 Hyonic OP-100
13.0 Iconol DA-6
13.0 Iconol NP-9
13.0 Iconol WA-1
13.0 Iconol WA-4
13.0 Igepal® CA-630
13.0 Igepal® CO-630
13.0 Igepal® DM-710
13.0 Igepal® NP-9
13.0 Igepal® NP-10
13.0 Imwitor® 370
13.0 Incrocas 40
13.0 Lamigen ES 60
13.0 Lauropal 0207L
13.0 Lauroxal 8
13.0 Liponox NC-95
13.0 Macol® DNP-15
13.0 Makon® 8240
13.0 Marlophen® 89.5N
13.0 Mazol® 159
13.0 Mazon® 1045A
13.0 Merpol® CH-196
13.0 Micro-Step® H-303
13.0 Mulsifan RT 18
13.0 Mulsifan RT 19
13.0 Mulsifan RT 359
13.0 NP-55-90
13.0 Nikkol Hexaglyn 1-L
13.0 Noigen EA 130T
13.0 Noigen ET135
13.0 Nonex C5E
13.0 Norfox® NP-9
13.0 Nutrol 100
13.0 Nutrol 600
13.0 Pegosperse® MFE
13.0 Plurafac® A-46
13.0 Pluronic® P104
13.0 Poly-Tergent® SL-42
13.0 Prox-onic DT-015
13.0 Prox-onic NP-09
13.0 Prox-onic TD-1/09
13.0 Quimipol ENF 95
13.0 Remcopal 258
13.0 Remcopal 40 S3
13.0 Remcopal 40 S3 LE
13.0 Remcopal O9
13.0 Renex® 698
13.0 Rexol 25/9
13.0 Rexol 25J
13.0 Rexol 35/100
13.0 Rioklen NF 9
13.0 Sandopan® DTC-Acid
13.0 Sandopan® RS-8
13.0 Sandoxylate® SX-412
13.0 Santone® 8-1-O
13.0 Serdox NBS 6.6
13.0 Serdox NOP 9
13.0 Sinopol 966FH
13.0 Step-Flow 24
13.0 Step-Flow 26
13.0 Surfonic® L24-9
13.0 Surfynol® 465
13.0 Surfynol® CT-136
13.0 Surfynol® GA
13.0 Syntopon 8 C
13.0 T-Det® C-18
13.0 T-Det® TDA-65
13.0 Tagat® R40
13.0 Tegotens R 40
13.0 Toximul® 710
13.0 Toximul® 8240
13.0 Toximul® 8242
13.0 Toximul® S-HF
13.0 Trycol® 5941
13.0 Trycol® 5944
13.0 Trycol® 6964
13.0 Trydet ISA-14
13.0 Trylox® 5909
13.0 Varonic® LI-42
13.0 Witconol 5909
13.0-14.0 Burco NPS-225
13.0-14.0 Norfox® NP-11
13.0-17.0 Silwet® L-7600
13.0-17.0 Silwet® L-7604
13.0-17.0 Silwet® L-7605
13.0-17.0 Silwet® L-7607
13.1 Ablunol 400ML
13.1 Antarox® LA-EP 16
13.1 Armul 1007
13.1 Atlas G-1284
13.1 Catinex KB-19
13.1 Cithrol 4ML
13.1 Cithrol 6MO
13.1 Desonic® 40C
13.1 Emulson AG 13A
13.1 Ethox DT-15
13.1 Ethox ML-9
13.1 Ethylan® CD107
13.1 Ethylan® CD128
13.1 Eumulgin PRT 40
13.1 Glucopon 425
13.1 Hetoxide C-40
13.1 Hetoxide HC-40
13.1 Igepal® KA
13.1 Industrol® ML-9
13.1 Lexemul® PEG-400ML
13.1 Lonzest® PEG 4-L

13.1	Marlipal® 1618/11	13.3	Marlowet® 4940	13.5	Ethomid® HT/15
13.1	Nissan Nonion L-4	13.3	Montanox 21	13.5	Ethox MO-14
13.1	Nonipol 95	13.3	Neodol® 25-9	13.5	Ethylan® C40AH
13.1	Nonipol 100	13.3	Nissan Nonion HS-210	13.5	Ethylan® C404
13.1	Nopalcol 4-L	13.3	Nissan Nonion NS-210	13.5	Eumulgin PC 10
13.1	Pegnol L-8	13.3	Nonal 210	13.5	Hetoxamate MO-15
13.1	Poly-Tergent® B-315	13.3	Nonipol D-160	13.5	Iconol NP-10
13.1	Rexol 65/10	13.3	Prox-onic LA-1/09	13.5	Industrol® TO-16
13.1	Rexonic L125-9	13.3	Renex® 690	13.5	Makon® OP-9
13.1	Rexonic N25-9(85%)	13.3	Sellig R 4495	13.5	Merpol® 100
13.1	Rexonic N25-9	13.3	Sinopol 965	13.5	Nikkol DDP-10
13.1	Rexonic P-9	13.3	Surfactol® 365	13.5	Nissan Nonion O-6
13.1	Rhodasurf® LA-9	13.3	Synperonic 13/8	13.5	Nissan Plonon 204
13.1	Rolfor HT 11	13.3	Synperonic NP10	13.5	Norfox® OP-100
13.1	Sellig N 10 100	13.3	Synperonic NXP	13.5	Pegosperse® 700 TO
13.1	Sellig O 9 100	13.3	Tergitol® 15-S-9	13.5	PGE-600-MS
13.1	Surfonic® JL-80X	13.3	Teric 13A9	13.5	Pluronic® P94
13.1	T-Det® DD-10	13.3	Teric 154	13.5	Poly-Tergent® B-350
13.1	T-Det® N-9.5	13.3	Teric 17M15	13.5	Prox-onic NP-010
13.1	Trylon® 5909	13.3	Teric N10	13.5	Prox-onic OP-09
13.2	Atlox 4851 B	13.3	Tween® 21	13.5	Quimipol EA 6812
13.2	Berol 303	13.3	Value 3710	13.5	Radiasurf® 7414
13.2	Caloxylate N-9	13.4	Alkasurf® NP-9	13.5	Remcopal 21411
13.2	Carsonon® N-10	13.4	Atlas EMJ-2	13.5	Remcopal PONF
13.2	Chemax E-400 ML	13.4	Berol 272	13.5	Rexol 45/10
13.2	Cithrol 10DO	13.4	Chemax TO-16	13.5	Rexol 65/11
13.2	Dacospin 869	13.4	Crodet O12	13.5	Rhodameen® T-12
13.2	Dehydrophen POP 10	13.4	Crodet S12	13.5	Rhodasurf® LA-90
13.2	Desonic® 10D	13.4	Dehydol TA 12	13.5	Sellig LA 9 100
13.2	Emerest® 2650	13.4	Ethox TO-16	13.5	Sellig N 11 100
13.2	Hetoxol TD-9	13.4	Ethylan® DNP16	13.5	Serdox NNP 10
13.2	Hyonic NP-100	13.4	Ethylan® TD15	13.5	Synperonic 10/7-100%
13.2	Iconol TDA-9	13.4	Hetoxamate SA-13	13.5	Synperonic OP10.5
13.2	Igepal® CO-660	13.4	Macol® OP-10	13.5	T-Det® O-9
13.2	Lauropal 9	13.4	Macol® OP-10 SP	13.5	T-Mulz® Mal 5
13.2	Lauropal X 1007	13.4	Marlipal® 011/88	13.5	Tex-Wet 1155
13.2	Mapeg® 400 ML	13.4	Marlipal® 24/90	13.5	Toximul® 360B
13.2	Nonipol Soft SS-90	13.4	Marlophen® DNP 16	13.5	Toximul® 811
13.2	Nopalcol 6-S	13.4	Prox-onic TA-1/016	13.5	Toximul® H-HF
13.2	Pegosperse® 600 MS	13.4	Pyroter GPI-25	13.5	Triton® X-100
13.2	Quimipol EA 2509	13.4	Quimipol ENF 100	13.5	Triton® X-100 CG
13.2	Radiasurf® 7404	13.4	Rexol 25/10	13.5	Trycol® 6965
13.2	Sinopol 864H	13.4	Rexol 25JWC	13.5	Trycol® NP-11
13.2	Siponic® NP-9.5	13.4	Rioklen NF 10	13.6	Ablunol 600MS
13.2	Surfonic® N-100	13.4	Rolfor Z 24/9	13.6	Akyporox OP 100
13.2	Syntopon D	13.4	Sellig T 14 100	13.6	Alkamuls® EL-719
13.2	Trycol® 6974	13.4	Surfonic® N-102	13.6	Armeen® SZ
13.2	Trycol® TDA-9	13.4	Synperonic 91/7	13.6	Atlox 4991
13.3	Ablunol LA-9	13.4	T-Det® N-10.5	13.6	Atlox 4995
13.3	Ablunol NP10	13.4	Triton® N-101	13.6	Cedepal CO-710
13.3	Atlas G-2109	13.4	Trydet 19	13.6	Chemax E-600 MO
13.3	Berol 09	13.4	Trydet 20	13.6	Cirrasol® ALN-TF
13.3	Berol 108	13.4	Trydet 2682	13.6	Cirrasol® ALN-TS
13.3	Catinex KB-20	13.4	Value 1209C	13.6	Desonic® 11N
13.3	Catinex KB-42	13.5	Ablumul AG-AH	13.6	Dow Corning® 193 Surfactant
13.3	Chemal LA-9	13.5	Ablumul AG-KTM		
13.3	Dehydrophen PNP 10	13.5	Ablunol 600MO	13.6	Emerest® 2660
13.3	Desonic® 9T	13.5	Ahco AB 135	13.6	Empilan® KA8/80
13.3	Disponil SML 104 F1	13.5	Akyporox NP 105	13.6	Glucopon 225
13.3	Emasol L-106	13.5	Alkamuls® O-14	13.6	Hodag 60-S
13.3	Emulson AG 10B	13.5	Alkasurf® NP-10	13.6	Hyonic OP-10
13.3	Emulson AG 5190	13.5	Alkasurf® OP-10	13.6	Igepal® CO-710
13.3	Ethal 926	13.5	Arquad® S-2C-50	13.6	Igepal® O
13.3	Hetoxamate LA-9	13.5	Berol 048	13.6	Industrol® MO-13
13.3	Hetoxol L-9	13.5	Burco TME	13.6	Laurel PEG 600 MT
13.3	Hetsorb L-4	13.5	Caplube 8540D	13.6	Macol® TD-10
13.3	Macol® LA-9	13.5	Chemax NP-10	13.6	Mapeg® 600 MO
13.3	Marlipal® 013/90	13.5	Durfax® EOM	13.6	Mapeg® 600 MOT
13.3	Marlophen® 810	13.5	Emulsol CO 45	13.6	Mapeg® 600 MS
13.3	Marlophen® 810N	13.5	Emulson AG 4020	13.6	Nissan Nonion S-6

HLB	Product
13.6	Nopalcol 6-O
13.6	Nutrol 656
13.6	Octapol 100
13.6	Remcopal L9
13.6	Renex® 750
13.6	Sinopol 610
13.6	Sinopol 865
13.6	Sinopol 1109
13.6	Synperonic 13/9
13.6	Synperonic OP10
13.6	Tergitol® NP-10
13.6	Teric 12A9
13.6	Teric X10
13.6	Triton® CF-54
13.6	Trymeen® 6601
13.6	Value 1414
13.7	Ablumul AG-L
13.7	Armul 1009
13.7	Cithrol 10DS
13.7	Emulgator F-8
13.7	Emulson AG 15A
13.7	Ethox MA-15
13.7	Ethylan® CD4511
13.7	Iconol TDA-10
13.7	Macol® NP-11
13.7	Marlipal® 013/100
13.7	Nonal 310
13.7	Rhodasurf® 870
13.7	Sinocol L
13.7	Sinopol 150T
13.7	Sinopol 170N
13.7	Surfonic® L12-8
13.7	Teric 9A8
13.7	Teric BL8
13.7	Teric G9A8
13.7	Teric N11
13.7	Trycol® TDA-11
13.8	Ablunol CO 45
13.8	Alkasurf® NP-11
13.8	Carsonon® N-11
13.8	Chemax E-600 MS
13.8	Cirrasol® AEN-XF
13.8	Emerest® 2662
13.8	Emthox® 5942
13.8	Emulson CO-40
13.8	Ethox MS-14
13.8	Hyonic NP-110
13.8	Mapeg® TAO-15
13.8	Neodol® 45-11
13.8	Nissan Nonion P-6
13.8	Nonipol 110
13.8	Pegosperse® 1500 MS
13.8	Quimipol EA 9108
13.8	Quimipol ENF 110
13.8	Renex® 20
13.8	Rexol 25/11
13.8	Rhodasurf® BC-720
13.8	Rolfor TR 9
13.8	Sellig DN 22 100
13.8	Sinopol 1310
13.8	Sinopol 1545
13.8	Tergitol® Min-Foam 2X
13.8	Teric 15A11
13.8	Teric 16M15
13.8	Teric 17A13
13.8	Teric CME3
13.8	Triton® N-111
13.8	Trycol® 5942
13.8	Trycol® 5957
13.8	Trymeen® CAM-10
13.8	Varonic® K210
13.8	Varonic® K210 SF
13.9	Ablumul AG-H
13.9	Ablumul AG-KP3
13.9	Alcodet® 218
13.9	Alkasurf® NP-12
13.9	Emulgator U12
13.9	Ethal DNP-18
13.9	Ethomeen® C/15
13.9	Marlipal® 24/100
13.9	Marlophen® DNP 18
13.9	Neodol® 1-9
13.9	Nopalcol 4-C
13.9	Nopalcol 4-CH
13.9	Nopalcol 6-R
13.9	Pegosperse® 400 MC
13.9	Pegosperse® 400 ML
13.9	Prox-onic TM-010
13.9	Remcopal 3712
13.9	Renex® 682
13.9	Renex® 711
13.9	Synperonic 91/8
13.9	Synperonic A11
13.9	Synperonic NP12
13.9	Teric N12
14.0	Ablumul AG-MBX
14.0	Akypo LF 6
14.0	Alkamuls® TD-41
14.0	Antarox® LA-EP 73
14.0	Antarox® P-84
14.0	Aqualose L30
14.0	Atlox 804
14.0	Atlox 3403F
14.0	Atlox 3453F
14.0	Atlox 4911
14.0	Cithrol 6MS
14.0	Cralane AT-17
14.0	DRA-1500
14.0	Durfax® 60
14.0	Emulan® EL
14.0	Ethomid® O/15
14.0	Ethylan® CF71
14.0	Ethylan® DP
14.0	Ethylan® OE
14.0	Ethylan® TN-10
14.0	Ethylan® TT-15
14.0	Genapol® 24-L-92
14.0	Genapol® UD-080
14.0	Genapol® UD-088
14.0	Hodag Nonionic E-12
14.0	Iconol DA-9
14.0	Iconol NP-12
14.0	Iconol OP-10
14.0	Industrol® TFA-15, TFA-15-80%
14.0	L.A.S
14.0	Labrasol
14.0	Lamigen ET 70
14.0	Macol® NP-12
14.0	Neodol® 91-8
14.0	Nikkol GO-460
14.0	Nikkol NP-7.5
14.0	Nikkol TDP-10
14.0	Nipol® 2782
14.0	Noigen 140L
14.0	Noigen EA 140
14.0	Noigen EA 142
14.0	Noigen EA 143
14.0	Noigen ES 140
14.0	Noigen ET140
14.0	Noigen ET143
14.0	Noigen ET147
14.0	Nopco® 1179
14.0	Phospholipid PTS
14.0	Plurafac® B-26
14.0	Pluronic® L10
14.0	Pluronic® P84
14.0	Polirol L400
14.0	Poly-Tergent® SL-62
14.0	Polychol 20
14.0	Quimipol EA 6814
14.0	Quimipol ENF 120
14.0	Remcopal L12
14.0	Remcopal O11
14.0	Rewopol® SBFA 50
14.0	Rioklen NF 12
14.0	Rolfor COH 40
14.0	Sellig AO 15 100
14.0	Sellig O 11 100
14.0	Serdox NBS 8.5
14.0	Serdox NNP 11
14.0	Serdox NNP 12
14.0	Sinocol MBX
14.0	Solulan® C-24
14.0	Solulan® PB-20
14.0	Step-Flow 25
14.0	Synperonic OP11
14.0	Syntopon E
14.0	T-Det® TDA-70
14.0	Tagat® RI
14.0	Tegotens R 1
14.0	Teric SF15
14.0	Teric X11
14.0	Tomah E-18-10
14.0	Tomah E-T-5
14.0	Toximul® 8322
14.0	Triton® CF-10
14.0	Trycol® LAL-12
14.0-16.0	Triton® RW-75
14.1	Ablunol NP12
14.1	Armul 912
14.1	Berol WASC
14.1	Catinex KB-22
14.1	Dehydrophen PNP 12
14.1	Disponil APG 110
14.1	Emulson AG 12B
14.1	Ethylan® CD1210
14.1	Hyonic NP-120
14.1	Icomeen® T-15
14.1	Macol® TD-12
14.1	Marlophen® 812
14.1	Marlophen® 812N
14.1	Nissan Nonion K-211
14.1	Nissan Nonion NS-212
14.1	Nissan Nonion P-213
14.1	Nonipol 120
14.1	Pegnol L-10
14.1	Rexonic N91-8
14.1	Sinocol PQ2
14.1	Surfonic® N-120
14.1	Synperonic 13/10
14.1	T-Det® N-12
14.1	Tergitol® TMN-10
14.1	Trycol® 6953
14.2	Akypo MB 2717 S
14.2	Arquad® T-27W
14.2	Arquad® T-50
14.2	Berol 381
14.2	Berol 452
14.2	Carsonon® N-12

14.2 Dehydol PCS 14
14.2 Emulmin 140
14.2 Ethylan® D2512
14.2 Genapol® 24-L-98N
14.2 Hetoxol CS-15
14.2 Hetoxol CSA-15
14.2 Igepal® CO-720
14.2 Macol® CSA-15
14.2 Nissan Nonion E-215
14.2 Nissan Nonion S-215
14.2 Prox-onic MO-015
14.2 Quimipol EA 2512
14.2 Rexol 35/11
14.2 Rhodameen® T-15
14.2 Sellig N 12 100
14.2 Sellig SP 16 100
14.2 Sellig T 1790
14.2 Synperonic 10/8-100%
14.2 T-Det® C-40
14.2 Tagat® I2
14.2 Tegotens I 2
14.2 Volpo N15
14.2 Berol 392
14.3 Chemal DA-9
14.3 Chemeen HT-15
14.3 Chemeen T-15
14.3 Emerest® 2654
14.3 Emerest® 2658
14.3 Ethal DA-9
14.3 Ethox TAM-15
14.3 Ethylan® CD109
14.3 Eumulgin PA 12
14.3 Lipocol SC-15
14.3 Macol® SA-15
14.3 Marlipal® 24/110
14.3 Neodol® 45-12T
14.3 Pluronic® 25R4
14.3 Prox-onic CSA-1/015
14.3 Prox-onic DA-1/09
14.3 Prox-onic MHT-015
14.3 Prox-onic MT-015
14.3 Remcopal 3515
14.3 Synperonic OP12.5
14.3 Trycol® 5956
14.3 Trydet MP-9
14.3 Trymeen® 6606
14.4 Alkanol® A-CN
14.4 Arlatone® 285
14.4 Arlatone® 289
14.4 Atlas G-1285
14.4 Atlas G-1289
14.4 Atlox 1285
14.4 Cerex EL 400
14.4 Desonic® 54C
14.4 Dowfax XDS 8390.00
14.4 Duoquad® T-50
14.4 Empilan® KCMP 0705/F
14.4 Emulson AG/EL
14.4 Ethox 1122
14.4 Ethylan® CD919
14.4 Genapol® 26-L-98N
14.4 Hetoxol CAWS
14.4 Lauropal 11
14.4 Lauropal 1150
14.4 Neodol® 25-12
14.4 Neodol® 45-13
14.4 Pegnol C-14
14.4 Plysurf A216B
14.4 Renex® 679
14.4 Rexonic N25-12
14.4 Rhodasurf® LA-12
14.4 Surfonic® L24-12
14.4 Surfonic® N-130
14.4 Synperonic NP13
14.4 T-Mulz® W
14.4 Tergitol® NP-13
14.4 Teric 12A12
14.4 Teric G12A12
14.4 Teric N13
14.4 Trycol® 6958
14.4 Varonic® T215
14.4 Varonic® T215LC
14.5 Ablunol LA-12
14.5 Alkasurf® OP-12
14.5 Caplube 8385
14.5 Chemal TDA-12
14.5 Crodet L12
14.5 Dehydol PLS 12
14.5 Emulson AG 3020
14.5 Emulson EL
14.5 Ethal TDA-12
14.5 Ethylan® BV
14.5 Eumulgin PRT 56
14.5 Hetoxol TD-12
14.5 Lamigen ES 100
14.5 Lipocol L-12
14.5 Marlipal® 013/120
14.5 Nikkol BL-9EX, 9EX(FF)
14.5 Nikkol BO-10TX
14.5 Ninol® 11-CM
14.5 Nonipol 130
14.5 Pegnol O-16
14.5 Plurafac® A-27
14.5 Prox-onic LA-1/012
14.5 Prox-onic TD-1/012
14.5 Quimipol ENF 140
14.5 Remcopal O12
14.5 Renex® 30
14.5 Rexol 45/12
14.5 Rexol 65/14
14.5 Sellig LA 1150
14.5 Serdox NEL 12/80
14.5 Serdox NNP 13
14.5 Serdox NNP 14
14.5 Serdox NOL 15
14.5 Teric 18M20
14.5 Teric BL9
14.5 Tetronic® 904
14.5 Toximul® TA-15
14.5 Trycol® 5943
14.5 Trycol® TDA-12
14.6 Ahco AB 146
14.6 Atlas G-8916PF
14.6 Atlox 8916PF
14.6 Carsonon® TD-11, TD-11 70%
14.6 Cedepal CA-720
14.6 Empilan® KA10/80
14.6 Ethomeen® S/12
14.6 Igepal® CA-720
14.6 Kessco® PEG 600 ML
14.6 Lipocol TD-12
14.6 Macol® LA-12
14.6 Macol® OP-12
14.6 Marlipal® 24/120
14.6 Neodol® 23-12
14.6 Newpol PE-88
14.6 Norfox® OP-102
14.6 Pegosperse® 600 ML
14.6 Remcopal HC 60
14.6 Rexol 25/14
14.6 Rolfor TR 12
14.6 Sellig O 12 100
14.6 Sinopol 1112
14.6 Sinopol 1225
14.6 Siponic® L-12
14.6 Triton® X-102
14.6 Volpo CS-15
14.7 Berol 930
14.7 Catinex KB-23
14.7 Catinex KB-43
14.7 Cithrol 10DL
14.7 Croduret 60
14.7 Disponil LS 12
14.7 Hetoxide P-6
14.7 Macol® OA-20
14.7 Marlophen® 814
14.7 Marlophen® 814N
14.7 Remcopal 18
14.7 Synperonic 91/10
14.7 T-Det® N-14
14.7 Tergitol® 15-S-12
14.7 Teric 9A10
14.7 Teric X13
14.8 Akypo RLM 100
14.8 Chemax E-600 ML
14.8 Emerest® 2661
14.8 Ethox DT-30
14.8 Eumulgin PWM17
14.8 Hetoxide C-60
14.8 Hetoxide HC-60
14.8 Hodag 60-L
14.8 Hodag PSMS-20
14.8 Macol® DNP-21
14.8 Mapeg® 600 ML
14.8 Mapeg® 1540 DS
14.8 Marlipal® 011/110
14.8 Newpol PE-78
14.8 Pegnol L-12
14.8 Remcopal 21912 AL
14.8 Rexol 25/15
14.8 Synperonic 13/12
14.8 Trycol® 5967
14.9 Alkamuls® PSMS-20
14.9 Cirrasol® ALN-WF
14.9 Crillet 3
14.9 Crillet 6
14.9 Disponil SMS 120 F1
14.9 Drewmulse® POE-SMS
14.9 Emasol S-120
14.9 Ethox ML-14
14.9 Hetoxamate FA-20
14.9 Igepal® NP-14
14.9 Industrol® ML-14
14.9 Industrol® S-20-S
14.9 Ionet T-60 C
14.9 Liposorb L-10
14.9 Montanox 60 DF
14.9 Newpol PE-68
14.9 Nopalcol 19-CO
14.9 Plysurf A217E
14.9 Prox-onic SMS-020
14.9 Radiasurf® 7157
14.9 Sellig SU 18 100
14.9 Sorbax PMS-20
14.9 T-Maz® 60
14.9 T-Maz® 90
14.9 Teric 163

14.9	Teric 16A16
15.0	Ablunol 600ML
15.0	Ablunol NP15
15.0	Acconon ETG
15.0	Acconon W230
15.0	Alkamuls® PSMO-20
15.0	Alkasurf® NP-15
15.0	Alkasurf® NP-15, 80%
15.0	Antarox® L-64
15.0	Antarox® PGP 18-4
15.0	Atlas G-4905
15.0	Atlox 4898
15.0	Atsurf T-80
15.0	Berol 07
15.0	Berol 106
15.0	Burco NPS-50%
15.0	Catinex KB-27
15.0	Cedepal CO-730
15.0	Chemax NP-15
15.0	Chemeen C-15
15.0	Crillet 4
15.0	Crovol A70
15.0	Dehydrophen PNP 15
15.0	Desotan® SMO-20
15.0	Desotan® SMT-20
15.0	Disponil SMO 120 F1
15.0	Drewmulse® POE-SMO
15.0	Durfax® 80
15.0	Emasol O-120
15.0	Emery® 5874
15.0	Emery® 6223 Monolan 2800
15.0	Emsorb® 2722
15.0	Emsorb® 6900
15.0	Emultex 1502, 1515
15.0	Ethosperse® LA-12
15.0	Ethsorbox O-20
15.0	Ethylan® GEO8
15.0	Ethylan® GES6
15.0	Ethylan® TC
15.0	Genapol® UD-110
15.0	Glucamate® SSE-20
15.0	Glycosperse® O-20
15.0	Glycosperse® O-20 FG, O-20 KFG
15.0	Glycosperse® S-20
15.0	Grilloten PSE141G
15.0	Hartopol L64
15.0	Hetsorb O-20
15.0	Hodag PSMO-20
15.0	Iconol DNP-24
15.0	Iconol NP-15
15.0	Igepal® CO-730
15.0	Industrol® O-20-S
15.0	Ionet T-80 C
15.0	Lamigen ET 90
15.0	Lanogel® 21
15.0	Lanogel® 41
15.0	Lonzest® SMO-20
15.0	Lonzest® SMS-20
15.0	MS-55-F
15.0	Macol® 15-20
15.0	Macol® NP-15
15.0	Monolan® 2800
15.0	Montanox 80 DF
15.0	Mulsifan RT 157
15.0	Nikkol BPS-15
15.0	Nipol® 5595
15.0	Nissan Nonion HS-215
15.0	Nissan Nonion NS-215
15.0	Nissan Plonon 104
15.0	Noigen EA 150
15.0	Noigen EA 152
15.0	Noigen ET150
15.0	Noigen ET157
15.0	Nopalcol 6-L
15.0	Nutrol 640
15.0	PGE-600-ML
15.0	Pepol B-184
15.0	Pepol D-304
15.0	Pluronic® L64
15.0	Pluronic® P105
15.0	Poly-Tergent® B-500
15.0	Poly-Tergent® SL-92
15.0	Prox-onic MC-015
15.0	Prox-onic NP-015
15.0	Prox-onic SMO-020
15.0	Quimipol EA 6818
15.0	Renex® 678
15.0	Rhodameen® VP-532/SPB
15.0	Sandopan® DTC-100
15.0	Sandopan® DTC Linear P
15.0	Sandoxylate® C-32
15.0	Sandoxylate® SX-224
15.0	Serdox NNP 15
15.0	Simulsol M 49
15.0	Sinopol 714
15.0	Sinoponic PE 64
15.0	Solulan® 16
15.0	Solulan® 75
15.0	Solulan® L-575
15.0	Sorbanox AOM
15.0	Sorbax PMO-20
15.0	Sorbilene O
15.0	Step-Flow 21
15.0	Surfactol® 575
15.0	Surfonic® N-150
15.0	Syn Fac® 8216
15.0	Synperonic NP15
15.0	Syntopon F
15.0	Tagat® O2
15.0	Tagat® R60
15.0	Tagat® S2
15.0	T-Det® EPO-64L
15.0	Tegotens R 60
15.0	Tegotens S 2
15.0	Tergitol® NP-15
15.0	Teric N15
15.0	Teric PE64
15.0	Tetronic® 704
15.0	T-Maz® 80
15.0	Tomah E-S-15
15.0	Tomah Q-C-15
15.0	Tomah Q-D-T
15.0	Tomah Q-S
15.0	Triton® N-150
15.0	Witconol 2722
15.1	Drewmulse® POE-SML
15.1	Igepal® DM-730
15.1	Radiasurf® 7147
15.1	Remcopal 238
15.1	Ritoleth 20
15.1	Sellig N 15 100
15.1	Teric CME7
15.1	Trycol® 5874
15.2	Ablumul AG-420
15.2	Ablunol 1000MO
15.2	Ablunol 1000MS
15.2	Ablunol LA-16
15.2	Ablunol NP16
15.2	Berol 278
15.2	Emsorb® 2728
15.2	Ethox CAM-15
15.2	Ethsorbox S-20
15.2	Macol® CSA-20
15.2	Marlipal® 24/140
15.2	Nissan Nonion K-215
15.2	Nissan Nonion S-10
15.2	Nonipol 160
15.2	Pegosperse® 1000 MS
15.2	Remcopal 229
15.2	Rexonic N25-14
15.2	Rexonic N25-14(85%)
15.2	Sellig SP 20 100
15.2	Teric X16
15.3	Berol 397
15.3	Brij® 98G
15.3	Brij® 99
15.3	Chemal OA-20/70CWS
15.3	Chemal OA-20G
15.3	Dow Corning® 1315 Surfactant
15.3	Hetoxol OA-20 Special
15.3	Hetoxol OL-20
15.3	Hetoxol STA-20
15.3	Lipocol O-20
15.3	Lipocol S-20
15.3	Marlipal® 013/150
15.3	Merpol® HCS
15.3	Nissan Nonion E-220
15.3	Nissan Nonion S-220
15.3	Pegnol C-18
15.3	Prox-onic OA-1/020
15.3	Prox-onic OA-2/020
15.3	Prox-onic SA-1/020
15.3	Rhodasurf® LA-15
15.3	Rioklen NF 15
15.3	Sellig N 1780
15.3	Sellig N 17 100
15.3	Simulsol 78
15.3	Simulsol 98
15.3	Sorbon T-60
15.3	Stearate PEG 1000
15.3	Synperonic OP16.5
15.3	Teric 12A16
15.3	Trycol® 5887
15.3	Trycol® 5888
15.3	Trycol® 5971
15.3	Trycol® 6802
15.3	Trycol® SAL-20
15.3	Trymeen® 6620
15.3	Trymeen® 6622
15.3	Trymeen® OAM 30/60
15.3	Berol 386
15.4	Armul 21
15.4	Atlox 80
15.4	Atlox 8916TF
15.4	Dehydol TA 20
15.4	Ethox TAM-20
15.4	Eumulgin PA 20
15.4	Hetoxol CS-20
15.4	Icomeen® T-20
15.4	Lipocol SC-20
15.4	Macol® SA-20
15.4	Quimipol ENF 170
15.4	Rhodasurf® ON-870
15.4	Teric 9A12

15.4	Teric G9A12
15.4	Trylox® 6753
15.4	Trymeen® 6607
15.4	Trymeen® CAM-15
15.4	Trymeen® TAM-20
15.4	Varonic® K215
15.4	Varonic® K215 LC
15.4	Varonic® K215 SF
15.5	Brij® 721 S
15.5	Emulson AG 24A
15.5	Ethylan® CD9112
15.5	Ethylan® CS20
15.5	Eumulgin B2
15.5	G-3780-A
15.5	Hetoxol CA-20
15.5	Igepal® NP-17
15.5	Incropol CS-20
15.5	Industrol® CO-80-80%
15.5	Nikkol BC-15TX, 15TX(FF)
15.5	Nikkol BPS-20
15.5	Quimipol EA 6820
15.5	Radiasurf® 7136
15.5	Radiasurf® 7417
15.5	Remcopal 33820
15.5	Rolfor HT 25
15.5	Rolfor O 22
15.5	Syn Lube 107
15.5	Synperonic 10/11-100%
15.5	Synperonic 13/15
15.5	Tetronic® 504
15.5	Tomah PA-14 Acetate
15.5	Volpo N20
15.6	Akypo MB 1585
15.6	Akyporox RLM 160
15.6	Arquad® S-50
15.6	Crillet 2
15.6	Disponil SMP 120 F1
15.6	Emasol P-120
15.6	Emulson AG 81C
15.6	Ethox MS-23
15.6	Ethylan® C75AH
15.6	Ethylan® TLM
15.6	Glycosperse® P-20
15.6	Hetoxamate SA-23
15.6	Hetsorb P-20
15.6	Hodag 100-S
15.6	Hodag PSMP-20
15.6	Liposorb P-20
15.6	Montanox 40 DF
15.6	Pegnol L-15
15.6	Prox-onic SMP-020
15.6	Sorbax PMP-20
15.6	Tagat® I
15.6	Tergitol® 15-S-15
15.6	Trydet SA-23
15.6	Tween® 40
15.6	Volpo CS-20
15.7	Arlasolve® 200
15.7	Arquad® 18-50
15.7	Berol 190
15.7	Brij® 58
15.7	Chemax E-1000 MS
15.7	Chemeen T-20
15.7	Cirrasol® ALN-GM
15.7	Dehydrophen POP 17
15.7	Emerest® 2610
15.7	Hetoxide NP-12
15.7	Hetoxol TD-18
15.7	Industrol® MS-23
15.7	Ionet MS-1000
15.7	Lipocol C-20
15.7	Mapeg® 1000 MS
15.7	Pegnol C-20
15.7	Prox-onic MT-020
15.7	Sinopol 1620
15.7	Sorbon T-40
15.7	Tagat® L2
15.7	T-Det® BP-1
15.7	Tegotens L 2
15.7	Teric 12M15
15.7	Trydet 2610
15.8	Arquad® 16-50
15.8	Chemax CO-80
15.8	Crodet O24
15.8	Crodet S24
15.8	Emsorb® 6910
15.8	Ethal OA-23
15.8	Hetoxol OL-23
15.8	Industrol® OAL-20
15.8	Liponox NC-200
15.8	Macol® OP-16(75)
15.8	Marlipal® 013/170
15.8	Marlophen® DNP 30
15.8	Prox-onic HR-080
15.8	Prox-onic OP-016
15.8	Quimipol EA 6823
15.8	Rexol 45/16
15.8	Sellig N 20 80
15.8	Sellig N 20 100
15.8	Surfonic® N-200
15.8	T-Maz® 40
15.8	Teric 16A22
15.8	Texofor B1
15.8	Triton® X-165-70%
15.8	Trycol® 5972
15.9	Akypo RLM 130
15.9	Atlox 4881
15.9	Atlox 4896
15.9	Chemeen DT-30
15.9	Ethox CO-81
15.9	Prox-onic DT-030
15.9	Trycol® OAL-23
15.9	Trylox® CO-80
15.9	Varonic® LI-63
16.0	Ablunol NP20
16.0	Acconon CA-15
16.0	Acconon E
16.0	Acconon TGH
16.0	Ahco AB 160
16.0	Akypo LF 2
16.0	Akyporox NP 200
16.0	Aqualose L75
16.0	Atlas G-1288
16.0	Berol 281
16.0	Berol 292
16.0	Berol 948
16.0	Catinex KB-24
16.0	Chemax NP-20
16.0	Cithrol 10MS
16.0	Dehydol PTA 23
16.0	Dehydol PTA 23/E
16.0	Dehydrophen PNP 20
16.0	Emulan® NP 2080
16.0	Emulson AG 20B
16.0	Emultex 1602
16.0	Ethal NP-20
16.0	Ethal TDA-18
16.0	Ethox TAM-25
16.0	Ethylan® 20
16.0	G-2162
16.0	Hartopol L44
16.0	Hartopol P85
16.0	Hetoxide BN-13
16.0	Icomeen® T-25
16.0	Icomeen® T-25 CWS
16.0	Iconol NP-20
16.0	Igepal® CO-850
16.0	Igepal® NP-12
16.0	Igepal® NP-20
16.0	Industrol® LG-70
16.0	Industrol® LG-100
16.0	Lamigen ET 180
16.0	Laneto AWS
16.0	Levelan® P208
16.0	Lonzest® SMP-20
16.0	Macol® 85
16.0	Macol® NP-20
16.0	Macol® NP-20(70)
16.0	Marlophen® 820
16.0	Marlophen® 820N
16.0	Marlowet® 4941
16.0	Monolan® 8000 E/80
16.0	Monolan® 12,000 E/80
16.0	Mulsifan RT 11
16.0	Mulsifan RT 125
16.0	Nikkol BO-15TX
16.0	Nissan Nonion NS-220
16.0	Noigen EA 160
16.0	Noigen EA 160P
16.0	Noigen ES 160
16.0	Noigen ET160
16.0	Noigen ET165
16.0	Noigen ET167
16.0	Nonicol 190
16.0	Nonipol 200
16.0	Nopalcol 12-R
16.0	Plurafac® C-17
16.0	Pluronic® 17R4
16.0	Pluronic® L44
16.0	Pluronic® P85
16.0	Procetyl AWS
16.0	Prox-onic NP-020
16.0	Quimipol ENF 200
16.0	Remcopal 25
16.0	Remcopal 220
16.0	Remcopal 3820
16.0	Remcopal D
16.0	Renex® 649
16.0	Rexol 25/20
16.0	Sandopan® DTC
16.0	Sandopan® JA-36
16.0	Sandoxylate® 418
16.0	Sandoxylate® 424
16.0	Sandoxylate® SX-418
16.0	Sandoxylate® SX-424
16.0	Sandoxylate® SX-602
16.0	Sellig AO 25 100
16.0	Sellig LA 11 100
16.0	Simulsol M 51
16.0	Sinopol NN-15
16.0	Synperonic NP20
16.0	Syntopon G
16.0	T-Det® N-20
16.0	Teric N20
16.0	Tetronic® 304
16.0	Tomah E-18-15
16.0	Triton® RW-100
16.0	Trycol® 5946
16.0	Trycol® 6967

HLB	Trade name
16.0	Trycol® NP-20
16.0	Trycol® TDA-18
16.0	Trymeen® 6609
16.0	Trymeen® TAM-25
16.0+	Triton® RW-150
16.1	Antarox® BA-PE 70
16.1	Emulmin 240
16.1	Industrol® CSS-25
16.1	Mapeg® 1500 MS
16.1	Teric 200
16.1	Trylox® SS-20
16.2	Cithrol 10MO
16.2	Disponil TA 25
16.2	Icomeen® O-30
16.2	Icomeen® O-30-80%
16.2	Marlipal® 1618/25
16.2	Nissan Nonion HS-220
16.2	Nissan Nonion K-220
16.2	Plysurf A219B
16.2	Quimipol EA 6825
16.2	Remcopal 20
16.2	Renex® 720
16.2	Rioklen NF 20
16.2	Sellig O 20 100
16.2	Solan E
16.2	Synperonic 13/18
16.2	Synperonic A20
16.2	Teric 17A25
16.2	Texofor A1P
16.3	Ethal CSA-25
16.3	Marlipal® 013/200
16.3	Teric LAN70
16.3-18.0	Tomah Q-18-15
16.4	Eumulgin PWM25
16.4	Hetoxide NP-15-85%
16.4	Macol® LA-23
16.4	Marlipal® 24/200
16.4	Polychol 40
16.4	Quimipol ENF 230
16.4	Rhodameen® OA-910
16.4	Sellig SP 25 50
16.4	Sellig SP 25 100
16.4	Synperonic 13/20
16.4	Tagat® O
16.4	Tagat® S
16.4	Tegotens O
16.4	Tegotens S
16.4	Tergitol® 15-S-20
16.5	Arquad® C-50
16.5	Chemeen O-30, O-30/80
16.5	Cirrasol® GM
16.5	Croduret 100
16.5	Dehydol PLS 21
16.5	Dehydol PLS 21E
16.5	Dehydol TA 30
16.5	Disponil O 20
16.5	Emsorb® 6915
16.5	Emulan® OP 25
16.5	Ethylan® GEL2
16.5	Etocas 100
16.5	Eumulgin PA 30
16.5	ML-55-F
16.5	Nikkol NP-10
16.5	Nikkol PBC-34
16.5	Pegnol L-20
16.5	Prox-onic MO-030
16.5	Radiasurf® 7137
16.5	Serdox NNP 25
16.5	Sinopol 1120
16.6	Ethosperse® SL-20
16.6	Ethox OAM-308
16.6	Ethylan® TT-30
16.6	Eumulgin PK 23
16.6	Hetoxol STA-30
16.6	Nissan Nonion E-230
16.6	Rhodameen® T-30-90%
16.6	Sellig N 30 70
16.6	Sellig N 30 100
16.6	Sellig SU 30 100
16.6	Sinopol 623
16.6	Sinopol 1830
16.6	Teric 12A23
16.6	Teric 350
16.6	Teric CDE
16.7	Alkamuls® PSML-20
16.7	Berol 946
16.7	Catinex KB-26
16.7	Chemal LA-23
16.7	Crillet 1
16.7	Crodet S40
16.7	Disponil SML 120 F1
16.7	Dowfax 2A1
16.7	Dowfax 2EP
16.7	Dowfax XU 40333.00
16.7	Emasol L-120
16.7	Ethsorbox L-20
16.7	Eumulgin B3
16.7	Glycosperse® L-20
16.7	Hetoxol CS-30
16.7	Hetoxol L-23
16.7	Hetsorb L-20
16.7	Hodag PSML-20
16.7	Industrol® L-20-S
16.7	Ionet T-20 C
16.7	Liposorb L-20
16.7	Marlophen® 825
16.7	Mergital OC 30E
16.7	Montanox 20 DF
16.7	Nissan Nonion LT-221
16.7	Nissan Nonion S-15.4
16.7	Norfox® Sorbo T-20
16.7	Prox-onic LA-1/023
16.7	Prox-onic SML-020
16.7	Quimipol EA 6850
16.7	Sellig SP 3020
16.7	Sellig SP 30 100
16.7	Sorbax PML-20
16.7	Sorbon T-20
16.7	T-Maz® 20
16.7	Trycol® 5877
16.7	Trycol® 5964
16.7	Tween® 20
16.7	Witconol 2720
16.8	Cithrol 6ML
16.8	Crodet L24
16.8	Hodag 150-S
16.8	Macol® CSA-40
16.8	Mergital ST 30/E
16.8	Remcopal L30
16.8	Remcopal NP 30
16.8	Rolfor Z 24/23
16.9	Atlas G-5000
16.9	Brij® 35
16.9	Brij® 35 SP
16.9	Brij® 35 SP Liq.
16.9	Epan U 105
16.9	Ethosperse® LA-23
16.9	G-5000
16.9	Hetoxamate SA-35
16.9	Hetoxol TD-25
16.9	Industrol® MS-40
16.9	Lipocol L-23
16.9	Mapeg® S-40K
16.9	Prox-onic CSA-1/050
16.9	Rhodasurf® L-25
16.9	Simulsol M 52
16.9	Synperonic 91/20
16.9	Trycol® LAL-23
17.0	Alkamuls® S-65-40
17.0	Ameroxol® LE-23
17.0	Berol 277
17.0	Cithrol 15MS
17.0	Emulan® NP 3070
17.0	Emulan® OC
17.0	Emulan® OG
17.0	Emulan® OSN
17.0	Emulan® OU
17.0	Ethal NP-407
17.0	Ethylan® TH-30
17.0	Generol® 122E25
17.0	Hartopol P65
17.0	Hetoxide NP-30
17.0	Hodag Nonionic E-20
17.0	Iconol NP-30
17.0	Iconol NP-30-70%
17.0	Iconol OP-30
17.0	Iconol OP-30-70%
17.0	Lamacit GML 20
17.0	Lantrol® PLN
17.0	Lonzest® SML-20
17.0	Nikkol BO-20TX
17.0	Nikkol OP-30
17.0	Noigen EA 170
17.0	Noigen ET170
17.0	Pluronic® P65
17.0	RS-55-40
17.0	Rhodasurf® LAN-23-75%
17.0	Serdox NNP 30/70
17.0	Step-Flow 22
17.0	Surfynol® 485
17.0	Syntopon H
17.0	T-Det® N-30
17.0	T-Det® N-307
17.0	Tagat® L
17.0	Tegotens L
17.0	Tomah E-T-15
17.0	Toximul® 8323
17.1	Ablunol NP30
17.1	Ablunol NP30 70%
17.1	Alkasurf® NP-30, 70%
17.1	Armul 930
17.1	Arquad® 12-50
17.1	Catinex KB-25
17.1	Cetalox AT
17.1	Chemax NP-30
17.1	Chemax NP-30/70
17.1	Dehydrophen PNP 30
17.1	Desonic® 30N
17.1	Desonic® 30N70
17.1	Igepal® NP-30
17.1	Liponox NC-300
17.1	Marlophen® 830
17.1	Marlophen® 830N
17.1	Marlowet® 4700
17.1	Marlowet® 4930
17.1	Nissan Nonion NS-230
17.1	Prox-onic NP-030
17.1	Prox-onic NP-030/70
17.1	Quimipol ENF 300

HLB	Product
17.1	Renex® 650
17.1	Surfonic® N-300
17.1	Trycol® 6943
17.1	Trycol® 6968
17.1	Trycol® 6969
17.1	Trycol® 6975
17.1	Trycol® NP-30
17.1	Trycol® NP-307
17.2	Carsonon® N-30
17.2	Cedepal CO-880
17.2	Cedepal CO-887
17.2	Ethox MS-40
17.2	Hetoxamate SA-40
17.2	Icomeen® T-40
17.2	Icomeen® T-40-80%
17.2	Igepal® CO-880
17.2	Igepal® CO-887
17.2	Igepal® DM-880
17.2	Macol® NP-30(70)
17.2	Mapeg® S-40
17.2	Rexol 25/30
17.2	Synperonic OP30
17.2	Teric N30
17.2	Teric N30L
17.3	Akyporox OP 250 V
17.3	Alkasurf® OP-30, 70%
17.3	Berol 387
17.3	Catinex KB-44
17.3	Chemax OP-30/70
17.3	Disponil AAP 307
17.3	Emerest® 2715
17.3	Macol® OP-30(70)
17.3	Nissan Nonion K-230
17.3	Octapol 300
17.3	Prox-onic OP-030/70
17.3	Rioklen NF 30
17.3	Rolfat ST 40
17.3	Sinopol 806
17.3	Siponic® F-300
17.3	T-Det® O-307
17.3	Triton® X-305-70%
17.3	Trydet SA-40
17.4	Dehydol PTA 40
17.4	Disponil TA 430
17.4	Ethylan® HA Flake
17.4	Ethylan® TT-40
17.4	Hetoxol OL-40
17.4	Igepal® CA-880
17.4	Igepal® CA-887
17.4	Macol® SA-40
17.4	Marlipal® 24/300
17.4	Pegnol OA-400
17.4	Rexol 45/307
17.4	Synperonic OP40
17.4	Synperonic OP40/70
17.4	Trymeen® 6610
17.4	Trymeen® 6637
17.4	Trymeen® TAM-40
17.5	Atlas G-1295
17.5	Disponil LS 30
17.5	Ethylan® R
17.5	Hodag Nonionic E-30
17.5	Macol® 108
17.5	Marlipal® 1618/40
17.5	Rhodameen® T-50
17.5	Serdox NOP 30/70
17.5	Sorbon TR 814
17.5	Tergitol® 15-S-30
17.6	Alkasurf® NP-40, 70%
17.6	Ethylan® C160
17.6	Ethylan® HB15
17.6	Hyonic NP-407
17.6	Sellig SU 50 100
17.6	Teric X40
17.6	Teric X40L
17.7	Atranonic Polymer N200
17.7	Berol 198
17.7	Emulmin L-380
17.7	Hetoxide NP-40
17.7	Icomeen® 18-50
17.7	Siponic® NP-40
17.7	T-Det® N-40
17.7	T-Det® N-407
17.7	Teric 16A29
17.8	Ablunol NP40 70%
17.8	Ablunol NP40
17.8	Armul 940
17.8	Berol 295
17.8	Cedepal CO-890
17.8	Cedepal CO-897
17.8	Chemax NP-40
17.8	Chemax NP-40/70
17.8	Chemeen 18-50
17.8	Dehydrophen PNP 40
17.8	Dowfax 3B2
17.8	Dowfax XU 40340.00
17.8	Emerest® 2675
17.8	Emthox® 6957
17.8	Igepal® CO-890
17.8	Igepal® CO-897
17.8	Marlophen® 840N
17.8	Nonipol 400
17.8	Prox-onic MS-050
17.8	Prox-onic NP-040
17.8	Prox-onic NP-040/70
17.8	Rexol 25/40
17.8	Rexol 25/407
17.8	Sellig N 1050
17.8	Sinopol 908
17.8	Siponic® NP-407
17.8	Synperonic NP40
17.8	Tergitol® NP-40
17.8	Teric 18M30
17.8	Trycol® 6957
17.8	Trycol® 6970
17.8	Trycol® NP-40
17.8	Trycol® NP-407
17.8	Trydet 2675
17.8	Trydet SA-50/30
17.8	Trymeen® 6617
17.8	Trymeen® SAM-50
17.9	Akyporox OP 400V
17.9	Catinex KB-45
17.9	Chemax OP-40/70
17.9	Crodet L40
17.9	Disponil AAP 436
17.9	Incropol CS-50
17.9	Macol® OP-40(70)
17.9	Merpol® DA
17.9	Nissan Nonion HS-240
17.9	Octapol 400
17.9	Prox-onic OP-040/70
17.9	Simulsol M 53
17.9	Sinopol 808
17.9	Siponic® F-400
17.9	T-Det® O-40
17.9	T-Det® O-407
17.9	Triton® X-405-70%
17.9	Trycol® 6956
17.9	Trycol® 6984
17.9	Trycol® OP-407
18.0	Ablunol LA-40
18.0	Alkasurf® OP-40, 70%
18.0	Arosurf® 66-E20
18.0	Capcure Emulsifier 37S
18.0	Capcure Emulsifier 65
18.0	Carsonon® N-50
18.0	Cedepal CA-890
18.0	Cedepal CA-897
18.0	Cithrol 10ML
18.0	Croduret 200
18.0	Emulson EL 200
18.0	Ethosperse® G-26
18.0	Ethox SAM-50
18.0	Etocas 200
18.0	Hyonic GL 400
18.0	Hyonic NP-500
18.0	Hyonic OP-407
18.0	Iconol NP-40
18.0	Iconol NP-40-70%
18.0	Iconol OP-40
18.0	Iconol OP-40-70%
18.0	Igepal® CA-897
18.0	Marlipal® 013/400
18.0	Marlipal® 013/939
18.0	Naturechem® THS-200
18.0	Nikkol BO-50
18.0	Nikkol BPS-25
18.0	Nikkol BPS-30
18.0	Nikkol NP-15
18.0	Noigen ET180
18.0	Noigen ET187
18.0	Nopalcol 30-TWH
18.0	Pegosperse® 1750 MS
18.0	Pegosperse® 4000 MS
18.0	Pegosperse® 6000 DS
18.0	Rewopol® NL 3-70
18.0	Rexol 45/407
18.0	Rhodasurf® TB-970
18.0	Serdox NOP 40/70
18.0	Surfactol® 590
18.0	T-Det® D-70
18.0	T-Det® N-50
18.0	T-Det® N-507
18.0	Tergitol® 15-S-40
18.0	Teric N40
18.0	Teric N40L
18.0	Varonic® LI-48
18.0	Varonic® LI-67
18.0	Varonic® LI-67-75%
18.1	Armul 950
18.1	Atlas G-1300
18.1	Berol 191
18.1	Chemax CO-200/50
18.1	Chemax HCO-200/50
18.1	Ethox CO-200
18.1	Ethox CO-200/50%
18.1	Ethox HCO-200/50%
18.1	Eumulgin PRT 200
18.1	G-1300
18.1	Industrol® CO-200
18.1	Industrol® CO-200-50%
18.1	Industrol® COH-200
18.1	Mapeg® CO-200
18.1	Pegosperse® 9000 CO
18.1	Prox-onic HR-0200
18.1	Prox-onic HR-0200/50
18.1	Prox-onic HRH-0200
18.1	Prox-onic HRH-0200/50

18.1	Trylox® 5918
18.1	Trylox® 5925
18.1	Trylox® CO-200, -200/50
18.1	Trylox® HCO-200/50
18.2	Ablunol NP50
18.2	Ablunol NP50 70%
18.2	Atlox 4875
18.2	Berol 291
18.2	Cedepal CO-970
18.2	Cedepal CO-977
18.2	Chemax NP-50
18.2	Chemax NP-50/70
18.2	Eleminol HA-100
18.2	Igepal® CO-970
18.2	Igepal® CO-977
18.2	Liponox NC-500
18.2	Marlophen® 850
18.2	Marlophen® 850N
18.2	Nissan Nonion NS-250
18.2	Nissan Nonion S-40
18.2	Prox-onic NP-050
18.2	Prox-onic NP-050/70
18.2	Remcopal 31250
18.2	Rexol 25/50
18.2	Rexol 25/507
18.2	Sellig N 50 100
18.2	Sinopol 910
18.2	Synperonic NP50
18.2	Trycol® 6951
18.2	Trycol® 6971
18.2	Trycol® 6972
18.2	Trycol® NP-50
18.2	Trycol® NP-507
18.3	Surfonic® N-557
18.3	Synperonic A50
18.3	Teric 16A50
18.4	Aromox® C/12-W
18.4	Disponil O 250
18.4	Kessco® PEG 6000 DS
18.4	Mapeg® 6000 DS
18.5	Alkamuls® EL-980
18.5	G-3634A
18.5	Hyonic OP-705
18.5	Macol® 46
18.5	Nopalcol 30-S
18.5	Pluronic® L35
18.5	Serdox NDI 100
18.6	Ethylan® CD1260
18.6	Ethylan® D2560 Flake
18.6	Hetoxamate SA-90
18.6	Iconol NP-70
18.6	Iconol NP-70-70%
18.6	Igepal® CO-987
18.6	Trycol® 6941
18.7	Aromox® DMMCD-W
18.7	Atlas G-1304
18.7	Berol 08 Powd
18.7	Dehydol PTA 80
18.7	Igepal® CO-980
18.7	Nissan Nonion HS-270
18.7	Nissan Nonion NS-270
18.7	T-Det® N-70
18.7	T-Det® N-707
18.7	Triton® X-705-70%
18.8	Brij® 700 S
18.8	Cithrol 40MS
18.8	Crodet O100
18.8	Crodet S100
18.8	Eleminol HA-161
18.8	Emerest® 2618
18.8	Industrol® MS-100
18.8	Rolfat ST 100
18.8	Simulsol M 59
18.9	Macol® TD-100
18.9	Marlophen® 890
18.9	Marlophen® 890N
19.0	Ablunol 6000DS
19.0	Cedepal CO-990
19.0	Cedepal CO-997
19.0	Chemax DNP-150
19.0	Chemax DNP-150/50
19.0	Chemax NP-100
19.0	Chemax NP-100/70
19.0	Dowfax XDS 8292.00
19.0	Ethal DNP-150/50%
19.0	Ethomeen® S/15
19.0	Ethomid® HT/60
19.0	G-4252
19.0	Iconol DNP-150
19.0	Iconol NP-100
19.0	Iconol NP-100-70%
19.0	Iconol NP-50
19.0	Iconol NP-50-70%
19.0	Igepal® CO-990
19.0	Igepal® CO-997
19.0	Igepal® DM-970 FLK
19.0	Macol® DNP-150
19.0	Macol® NP-100
19.0	Nikkol BL-21
19.0	Nikkol NP-18TX
19.0	Nissan Nonion DS-60HN
19.0	Noigen EA 190D
19.0	Noigen ET190
19.0	Noigen ET190S
19.0	Plurafac® A-38
19.0	Prox-onic DNP-0150
19.0	Prox-onic DNP-0150/50
19.0	Prox-onic NP-0100
19.0	Rexol 25/100-70%
19.0	Softigen® 767
19.0	T-Det® D-150
19.0	T-Det® N-100
19.0	T-Det® N-1007
19.0	Trycol® 6942
19.0	Trycol® 6981
19.0	Trycol® 6989
19.0	Trycol® DNP-150
19.0	Trycol® DNP-150/50
19.0	Trycol® NP-1007
19.1	Crodet L100
19.1	G-4280
19.1	Teric N100
19.2	Akyporox NP 1200V
19.2	Emerest® 2617
19.2	Emerest® 2619
19.2	T-Maz® 28
19.3	Ethomeen® C/25
19.3	Ethomeen® T/25
19.3	Trycol® 6954
19.4	T-Det® TDA-150
19.5	Nikkol BL-25
19.5	Tomah Q-14-2
20.0	Noigen ET207
20.0	Noigen O100
20.0	Noigen YX400, YX500
20.0	Norfox® 90
20.0	Norfox® 1101
20.0	Norfox® 1115
20.0	Norfox® KO
20.0	Norfox® LAS-99
20.0	Norfox® Oleic Flakes
20.0	Norfox® PE-LF
20.0	Norfox® SLS
20.0	Norfox® T-60
20.0	Norfox® TLS
21.0	Pluronic® 10R5
22.0	Pluronic® F127
22.8	Ahco AB 228
24.0	Antarox® PGP 23-7
24.0	Plurafac® A-39
24.0	Pluronic® F87
24.0	Tetronic® 1107
24.5	Pluronic® F77
26.0	Epan U 108
26.1	Antarox® BA-PE 80
27.0	Hampfoam 35
27.0	Hamposyl® C-30
27.0	Pluronic® F108
27.0	Tetronic® 707
27.5	Pluronic® F98
28.0	Antarox® F88
28.0	Nissan Plonon 208
28.0	Pluronic® F88
29.0	Antarox® PGP 18-8
29.0	Hamposyl® AL-30
29.0	Macol® 8
29.0	Pluronic® F68
29.0	Sinoponic PE 68
29.0	Teric PE68
30.0	G-263
30.0	Hamposyl® L-30
30.0	Hamposyl® L-95
30.3	Pluronic® 25R8
30.5	Nissan Plonon 108
30.5	Pluronic® F38
30.5	Tetronic® 908
31.0	Rhodapon® L-22, L-22/C
33.0	G-265
33.0	Pluronic® 10R8
33.6	Akypo RLM 160 NV
40.0	Rhodapon® LSB, LSB/CT
40.0	Rhodapon® SB
40.0	Witcolate SL-1

Manufacturer Successors

Alcolac .... Rhone-Poulenc Surf.
Alkaril .... Rhone-Poulenc Surf.
Allied Corp. .... Allied-Signal
American Hoechst .... Hoechst Celanese
Armak .... Akzo
Bofors Lakeway .... Rhone-Poulenc
Clintwood .... Rhone-Poulenc Surf.
Colloids Inc. .... Rhone-Poulenc
Conoco Chem. .... Vista
Continental Oil Co. .... Vista
Croda Surfactants .... Croda Inc.
Cyclo .... Rhone-Poulenc Surf.
Daishowa .... Borregaard Lignotech
Darling & Co. .... Unichema Chemicals
DeSoto .... Witco/Organics
Diamond Shamrock/Process Chem. .... Henkel/Process
Diamond Shamrock Process Chem. Europe .... Henkel-Nopco SA
Duphar .... Solvay Duphar BV
Dynamit Nobel .... Hüls
Dynamit Nobel AG .... Hüls AG
Emery .... Henkel/Emery
GAF Surfactants .... Rhone-Poulenc Surf.
Geronazzo SpA .... Rhone-Poulenc Geronazzo SpA
Glyco Chem. .... Lonza
Hamblet & Hayes .... Van Waters & Rogers
Hodag .... Calgene Chem. Inc.
Jefferson Chemical Co., Inc. .... Texaco
Jordan Chem. Co. .... PPG Specialty Chem.
Kenobel SA .... Berol Nobel
Lankro Chem. Ltd. .... Harcros Chem. UK Ltd.
Lyndal .... Rhone-Poulenc/Textile & Rubber Div.
Mars .... Rhone-Poulenc Surf.
Mazer .... PPG Specialty Chem.
Miranol .... Rhone-Poulenc Surf.
Oleofina .... Fina PLC
Pat Chem. .... Yorkshire Pat-Chem
Patco .... Am. Ingredients
PPG Mazer .... PPG Specialty Chem.
PVO Int'l. .... Stepan Co./PVO Dept.
Quantum Chem. Corp./Emery Div. .... Henkel/Emery
Reed Lignin .... Borregaard Lignotech
Rhone-Poulenc SpA .... Rhone-Poulenc Geronazzo SpA
Sherex .... Witco/Organics
Sinor Kao SA .... KAO Corp. S.A.
Soitem .... Societa Ital. Emulsionanti
Tessilchimica .... Auschem SpA
Thompson-Hayward Chem. Co. .... Harcros Chem.
Van Schuppen .... Solvay Duphar BV